FAR/AIM 2014

Federal Aviation Regulations/ Aeronautical Information Manual

Federal Aviation Administration

Skyhorse Publishing

This edition includes Federal Aviation Regulation updates through July 15, 2013 and Aeronautical Information Manual updates through September 1, 2013. Regulations are updated daily on www.faa.gov.

Skyhorse Publishing books may be purchased in bulk at special discounts for sales promotion, corporate gifts, fund-raising, or educational purposes. Special editions can also be created to specifications. For details, contact the Special Sales Department, Skyhorse Publishing, 307 West 36th Street, 11th Floor, New York, NY 10018 or info@skyhorsepublishing.com.

Visit our website at www.skyhorsepublishing.com.

10 9 8 7 6 5 4 3 2 1

ISBN: 978-1-62636-015-0

Printed in Canada

CONTENTS

FEDERAL AVIATION REGULATIONS (FAR)

AERONAUTICAL INFORMATION MANUAL (AIM)

PART 1

DEFINITIONS AND ABBREVIATIONS

Contents

AUTHORITY: 49 U.S.C. 106(g), 40113, 44701.

§ 1.1 General definitions.

As used in *Subchapters* A through K of this chapter, unless the context requires otherwise:

Administrator means the Federal Aviation Administrator or any person to whom he has delegated his authority in the matter concerned.

Aerodynamic coefficients means non-dimensional coefficients for aerodynamic forces and moments.

Air carrier means a person who undertakes directly by lease, or other arrangement, to engage in air transportation.

Air commerce means interstate, overseas, or foreign air commerce or the transportation of mail by aircraft or any operation or navigation of aircraft within the limits of any Federal airway or any operation or navigation of aircraft which directly affects, or which may endanger safety in, interstate, overseas, or foreign air commerce.

Aircraft means a device that is used or intended to be used for flight in the air.

Aircraft engine means an engine that is used or intended to be used for propelling aircraft. It includes turbosuperchargers, appurtenances, and accessories necessary for its functioning, but does not include propellers.

Airframe means the fuselage, booms, nacelles, cowlings, fairings, airfoil surfaces (including rotors but excluding propellers and rotating airfoils of engines), and landing gear of an aircraft and their accessories and controls.

Airplane means an engine-driven fixed-wing aircraft heavier than air, that is supported in flight by the dynamic reaction of the air against its wings.

Airport means an area of land or water that is used or intended to be used for the landing and takeoff of aircraft, and includes its buildings and facilities, if any.

Airship means an engine-driven lighter-than-air aircraft that can be steered.

Air traffic means aircraft operating in the air or on an airport surface, exclusive of loading ramps and parking areas.

Air traffic clearance means an authorization by air traffic control, for the purpose of preventing collision between known aircraft, for an aircraft to proceed under specified traffic conditions within controlled airspace.

Air traffic control means a service operated by appropriate authority to promote the safe, orderly, and expeditious flow of air traffic.

Air Traffic Service (ATS) route is a specified route designated for channeling the flow of traffic as necessary for the provision of air traffic services. The term "ATS route" refers to a variety of airways, including jet routes, area navigation (RNAV) routes, and arrival and departure routes. An ATS route is defined by route specifications, which may include:

(1) An ATS route designator;

(2) The path to or from significant points;

(3) Distance between significant points;

(4) Reporting requirements; and

(5) The lowest safe altitude determined by the appropriate authority.

Air transportation means interstate, overseas, or foreign air transportation or the transportation of mail by aircraft.

Alert Area. An alert area is established to inform pilots of a specific area wherein a high volume of pilot training or an unusual type of aeronautical activity is conducted.

Alternate airport means an airport at which an aircraft may land if a landing at the intended airport becomes inadvisable.

Altitude engine means a reciprocating aircraft engine having a rated takeoff power that is producible from sea level to an established higher altitude.

Amateur rocket means an unmanned rocket that:

(1) Is propelled by a motor or motors having a combined total impulse of 889,600 Newton-seconds (200,000 pound-seconds) or less; and

(2) Cannot reach an altitude greater than 150 kilometers (93.2 statute miles) above the earth's surface.

Appliance means any instrument, mechanism, equipment, part, apparatus, appurtenance, or accessory, including communications equipment, that is used or intended to be used in operating or controlling an aircraft in flight, is installed in or attached to the aircraft, and is not part of an airframe, engine, or propeller.

Approved , unless used with reference to another person, means approved by the FAA or any person to whom the FAA has delegated its authority in the matter concerned, or approved under the provisions of a bilateral agreement between the United States and a foreign country or jurisdiction.

Area navigation (RNAV) is a method of navigation that permits aircraft operations on any desired flight path.

Area navigation (RNAV) route is an ATS route based on RNAV that can be used by suitably equipped aircraft.

Armed Forces means the Army, Navy, Air Force, Marine Corps, and Coast Guard, including their regular and reserve components and members serving without component status.

Autorotation means a rotorcraft flight condition in which the lifting rotor is driven entirely by action of the air when the rotorcraft is in motion.

Auxiliary rotor means a rotor that serves either to counteract the effect of the main rotor torque on a rotorcraft or to maneuver the rotorcraft about one or more of its three principal axes.

Balloon means a lighter-than-air aircraft that is not engine driven, and that sustains flight through the use of either gas buoyancy or an airborne heater.

Brake horsepower means the power delivered at the propeller shaft (main drive or main output) of an aircraft engine.

Calibrated airspeed means the indicated airspeed of an aircraft, corrected for position and instrument error. Calibrated airspeed is equal to true airspeed in standard atmosphere at sea level.

Canard means the forward wing of a canard configuration and may be a fixed, movable, or variable geometry surface, with or without control surfaces.

Canard configuration means a configuration in which the span of the forward wing is substantially less than that of the main wing.

Category:

(1) As used with respect to the certification, ratings, privileges, and limitations of airmen, means a broad classification of aircraft. Examples include: airplane; rotorcraft; glider; and lighter-than-air; and

(2) As used with respect to the certification of aircraft, means a grouping of aircraft based upon intended use or operating limitations. Examples include: transport, normal, utility, acrobatic, limited, restricted, and provisional.

Category A, with respect to transport category rotorcraft, means multiengine rotorcraft designed with engine and system isolation features specified in Part 29 and utilizing scheduled takeoff and landing operations under a critical engine failure concept which assures adequate designated surface area and adequate performance capability for continued safe flight in the event of engine failure.

Category B, with respect to transport category rotorcraft, means single-engine or multiengine rotorcraft which do not fully meet all Category A standards. Category B rotorcraft have no guaranteed stay-up ability in the event of engine failure and unscheduled landing is assumed.

Category II operations, with respect to the operation of aircraft, means a straight-in ILS approach to the runway of an airport under a Category II ILS instrument approach procedure issued by the Administrator or other appropriate authority.

Category III operations, with respect to the operation of aircraft, means an ILS approach to, and landing on, the runway of an airport using a Category III ILS instrument approach procedure issued by the Administrator or other appropriate authority.

Ceiling means the height above the earth's surface of the lowest layer of clouds or obscuring phenomena that is reported as "broken", "overcast", or "obscuration", and not classified as "thin" or "partial".

Civil aircraft means aircraft other than public aircraft.

Class:

(1) As used with respect to the certification, ratings, privileges, and limitations of airmen, means a classification of aircraft within a category having similar operating characteristics. Examples include: single engine; multiengine; land; water; gyroplane; helicopter; airship; and free balloon; and

(2) As used with respect to the certification of aircraft, means a broad grouping of aircraft having similar characteristics of propulsion, flight, or landing. Examples include: airplane; rotorcraft; glider; balloon; landplane; and seaplane.

Clearway means:

(1) For turbine engine powered airplanes certificated after August 29, 1959, an area beyond the runway, not less than 500 feet wide, centrally located about the extended centerline of the runway, and under the control of the airport authorities. The clearway is expressed in terms of a clearway plane, extending from the end of the runway with an upward slope not exceeding 1.25 percent, above which no object nor any terrain protrudes. However, threshold lights may protrude above the plane if their height above the end of the runway is 26 inches or less and if they are located to each side of the runway.

(2) For turbine engine powered airplanes certificated after September 30, 1958, but before August 30, 1959, an area beyond the takeoff runway extending no less than 300 feet on either side of the extended centerline of the runway, at an elevation no higher than the elevation of the end of the runway, clear of all fixed obstacles, and under the control of the airport authorities.

Climbout speed, with respect to rotorcraft, means a referenced airspeed which results in a flight path clear of the height-velocity envelope during initial climbout.

Commercial operator means a person who, for compensation or hire, engages in the carriage by aircraft in air commerce of persons or property, other than as an air carrier or foreign air carrier or under the authority of Part 375 of this title. Where it is doubtful that an operation is for "compensation or hire", the test applied is whether the carriage by air is merely incidental to the person's other business or is, in itself, a major enterprise for profit.

Configuration, Maintenance, and Procedures (CMP) document means a document approved by the FAA that contains minimum configuration, operating, and maintenance requirements, hardware life-limits, and Master Minimum Equipment List (MMEL) constraints necessary for an airplane-engine combination to meet ETOPS type design approval requirements.

Consensus standard means, for the purpose of certificating light-sport aircraft, an industry-developed consensus standard that applies to aircraft design, production, and airworthiness. It includes, but is not limited to, standards for aircraft design and performance, required equipment, manufacturer quality assurance systems, production acceptance test procedures, operating instructions, maintenance and inspection procedures, identification and recording of major repairs and major alterations, and continued airworthiness.

Controlled airspace means an airspace of defined dimensions within which air traffic control service is provided to IFR flights and to VFR flights in accordance with the airspace classification.

NOTE: Controlled airspace is a generic term that covers Class A, Class B, Class C, Class D, and Class E airspace.

Controlled Firing Area. A controlled firing area is established to contain activities, which if not conducted in a controlled environment, would be hazardous to nonparticipating aircraft.

Crewmember means a person assigned to perform duty in an aircraft during flight time.

Critical altitude means the maximum altitude at which, in standard atmosphere, it is possible to maintain, at a specified rotational speed, a specified power or a specified manifold pressure. Unless otherwise stated, the critical altitude is the maximum altitude at which it is possible to maintain, at the maximum continuous rotational speed, one of the following:

(1) The maximum continuous power, in the case of engines for which this power rating is the same at sea level and at the rated altitude.

(2) The maximum continuous rated manifold pressure, in the case of engines, the maximum continuous power of which is governed by a constant manifold pressure.

Critical engine means the engine whose failure would most adversely affect the performance or handling qualities of an aircraft.

Decision altitude (DA) is a specified altitude in an instrument approach procedure at which the pilot must decide whether to initiate an immediate missed approach if the pilot does not see the required visual reference, or to continue the approach. Decision altitude is expressed in feet above mean sea level.

Decision height (DH) is a specified height above the ground in an instrument approach procedure at which the pilot must decide whether to initiate an immediate missed approach if the pilot does not see the required visual reference, or to continue the approach. Decision height is expressed in feet above ground level.

Early ETOPS means ETOPS type design approval obtained without gaining non-ETOPS service experience on the candidate airplane-engine combination certified for ETOPS.

Enhanced flight visibility (EFV) means the average forward horizontal distance, from the cockpit of an aircraft in flight, at which prominent topographical objects may be clearly distinguished and identified by day or night by a pilot using an enhanced flight vision system.

Enhanced flight vision system (EFVS) means an electronic means to provide a display of the forward external scene topography (the natural or manmade features of a place or region especially in a way to show their relative positions and elevation) through the use of imaging sensors, such as a forward looking infrared, millimeter wave radiometry, millimeter wave radar, low light level image intensifying.

Equivalent airspeed means the calibrated airspeed of an aircraft corrected for adiabatic compressible flow for the particular altitude. Equivalent airspeed is equal to calibrated airspeed in standard atmosphere at sea level.

ETOPS Significant System means an airplane system, including the propulsion system, the failure or malfunctioning of which could adversely affect the safety of an ETOPS flight, or the continued safe flight and landing of an airplane during an ETOPS diversion. Each ETOPS significant system is either an ETOPS group 1 significant system or an ETOPS group 2 significant system.

(1) An ETOPS group 1 Significant System—

(i) Has fail-safe characteristics directly linked to the degree of redundancy provided by the number of engines on the airplane.

(ii) Is a system, the failure or malfunction of which could result in an IFSD, loss of thrust control, or other power loss.

(iii) Contributes significantly to the safety of an ETOPS diversion by providing additional redundancy for any system power source lost as a result of an inoperative engine.

(iv) Is essential for prolonged operation of an airplane at engine inoperative altitudes.

(2) An ETOPS group 2 significant system is an ETOPS significant system that is not an ETOPS group 1 significant system.

Extended Operations (ETOPS) means an airplane flight operation, other than an all-cargo operation in an airplane with more than two engines, during which a portion of the flight is conducted beyond a time threshold identified in part 121 or part 135 of this chapter that is determined using an approved one-engine-inoperative cruise speed under standard atmospheric conditions in still air.

Extended over-water operation means—

(1) With respect to aircraft other than helicopters, an operation over water at a horizontal distance of more than 50 nautical miles from the nearest shoreline; and

(2) With respect to helicopters, an operation over water at a horizontal distance of more than 50 nautical miles from the nearest shoreline and more than 50 nautical miles from an off-shore heliport structure.

External load means a load that is carried, or extends, outside of the aircraft fuselage.

External-load attaching means means the structural components used to attach an external load to an aircraft, including external-load containers, the backup structure at the attachment points, and any quick-release device used to jettison the external load.

Final approach fix (FAF) defines the beginning of the final approach segment and the point where final segment descent may begin.

Final takeoff speed means the speed of the airplane that exists at the end of the takeoff path in the en route configuration with one engine inoperative.

Fireproof—

(1) With respect to materials and parts used to confine fire in a designated fire zone, means the capacity to withstand at least as well as steel in dimensions appropriate for the purpose for which they are used, the heat produced when there is a severe fire of extended duration in that zone; and

(2) With respect to other materials and parts, means the capacity to withstand the heat associated with fire at least as well as steel in dimensions appropriate for the purpose for which they are used.

Fire resistant—

(1) With respect to sheet or structural members means the capacity to withstand the heat associated with fire at least as well as aluminum alloy in dimensions appropriate for the purpose for which they are used; and

(2) With respect to fluid-carrying lines, fluid system parts, wiring, air ducts, fittings, and powerplant controls, means the capacity to perform the intended functions under the heat and other conditions likely to occur when there is a fire at the place concerned.

Flame resistant means not susceptible to combustion to the point of propagating a flame, beyond safe limits, after the ignition source is removed.

Flammable, with respect to a fluid or gas, means susceptible to igniting readily or to exploding.

Flap extended speed means the highest speed permissible with wing flaps in a prescribed extended position.

Flash resistant means not susceptible to burning violently when ignited.

Flightcrew member means a pilot, flight engineer, or flight navigator assigned to duty in an aircraft during flight time.

Flight level means a level of constant atmospheric pressure related to a reference datum of 29.92 inches of mercury. Each is stated in three digits that represent hundreds of feet. For example, flight level 250 represents a barometric altimeter indication of 25,000 feet; flight level 255, an indication of 25,500 feet.

Flight plan means specified information, relating to the intended flight of an aircraft, that is filed orally or in writing with air traffic control.

Flight simulation training device (FSTD) means a flight simulator or a flight training device.

Flight time means:

(1) Pilot time that commences when an aircraft moves under its own power for the purpose of flight and ends when the aircraft comes to rest after landing; or

(2) For a glider without self-launch capability, pilot time that commences when the glider is towed for the purpose of flight and ends when the glider comes to rest after landing.

Flight training device (FTD) means a replica of aircraft instruments, equipment, panels, and controls in an open flight deck area or an enclosed aircraft cockpit replica. It includes the equipment and computer programs necessary to represent aircraft (or set of aircraft) operations in ground and flight conditions having the full range of capabilities of the systems installed in the device as described in part 60 of this chapter and the qualification performance standard (QPS) for a specific FTD qualification level.

Flight visibility means the average forward horizontal distance, from the cockpit of an aircraft in flight, at which prominent unlighted objects may be seen and identified by day and prominent lighted objects may be seen and identified by night.

Foreign air carrier means any person other than a citizen of the United States, who undertakes directly, by lease or other arrangement, to engage in air transportation.

Foreign air commerce means the carriage by aircraft of persons or property for compensation or hire, or the carriage of mail by aircraft, or the operation or navigation of aircraft in the conduct or furtherance of a business or vocation, in commerce between a place in the United States and any place outside thereof; whether such commerce moves wholly by aircraft or partly by aircraft and partly by other forms of transportation.

Foreign air transportation means the carriage by aircraft of persons or property as a common carrier for compensation or hire, or the carriage of mail by aircraft, in commerce between a place in the United States and any place outside of the United States, whether that commerce moves wholly by aircraft or partly by aircraft and partly by other forms of transportation.

Forward wing means a forward lifting surface of a canard configuration or tandem-wing configuration airplane. The surface may be a fixed, movable, or variable geometry surface, with or without control surfaces.

Full flight simulator (FFS) means a replica of a specific type; or make, model, and series aircraft cockpit. It includes the assemblage of equipment and computer programs necessary to represent aircraft operations in ground and flight conditions, a visual system providing an out-of-the-cockpit view, a system that provides cues at least equivalent to those of a three-degree-of-freedom motion system, and has the full range of capabilities of the systems installed in the device as described in part 60 of this chapter and the qualification performance standards (QPS) for a specific FFS qualification level.

Glider means a heavier-than-air aircraft, that is supported in flight by the dynamic reaction of the air against its lifting surfaces and whose free flight does not depend principally on an engine.

Ground visibility means prevailing horizontal visibility near the earth's surface as reported by the United States National Weather Service or an accredited observer.

Go-around power or thrust setting means the maximum allowable in-flight power or thrust setting identified in the performance data.

Gyrodyne means a rotorcraft whose rotors are normally engine-driven for takeoff, hovering, and landing, and for forward flight through part of its speed range, and whose means of propulsion, consisting usually of conventional propellers, is independent of the rotor system.

Gyroplane means a rotorcraft whose rotors are not engine-driven, except for initial starting, but are made to rotate by action of the air when the rotorcraft is moving; and whose means of propulsion, consisting usually of conventional propellers, is independent of the rotor system.

Helicopter means a rotorcraft that, for its horizontal motion, depends principally on its engine-driven rotors.

Heliport means an area of land, water, or structure used or intended to be used for the landing and takeoff of helicopters.

Idle thrust means the jet thrust obtained with the engine power control level set at the stop for the least thrust position at which it can be placed.

IFR conditions means weather conditions below the minimum for flight under visual flight rules.

IFR over-the-top, with respect to the operation of aircraft, means the operation of an aircraft over-the-top on an IFR flight plan when cleared by air traffic control to maintain "VFR conditions" or "VFR conditions on top".

Indicated airspeed means the speed of an aircraft as shown on its pitot static airspeed indicator calibrated to reflect standard atmosphere adiabatic compressible flow at sea level uncorrected for airspeed system errors.

In-flight shutdown (IFSD) means, for ETOPS only, when an engine ceases to function (when the airplane is airborne) and is shutdown, whether self induced, flightcrew initiated or caused by an external influence. The FAA considers IFSD for all causes: for example, flameout, internal failure, flightcrew initiated shutdown, foreign object ingestion, icing, inability to obtain or control desired thrust or power, and cycling of the start control, however briefly, even if the engine operates normally for the remainder of the flight. This definition excludes the airborne cessation of the functioning of an engine when immediately followed by an automatic engine relight and when an engine does not achieve desired thrust or power but is not shutdown.

Instrument means a device using an internal mechanism to show visually or aurally the attitude, altitude, or operation of an aircraft or aircraft part. It includes electronic devices for automatically controlling an aircraft in flight.

Instrument approach procedure (IAP) is a series of predetermined maneuvers by reference to flight instruments with specified protection from obstacles and assurance of navigation signal reception capability. It begins from the initial approach fix, or where applicable, from the beginning of a defined arrival route to a point:

(1) From which a landing can be completed; or

(2) If a landing is not completed, to a position at which holding or en route obstacle clearance criteria apply.

Interstate air commerce means the carriage by aircraft of persons or property for compensation or hire, or the carriage of mail by aircraft, or the operation or navigation of aircraft in the conduct or furtherance of a business or vocation, in commerce between a place in any State of the United States, or the District of Columbia, and a place in any other State of the United States, or the District of Columbia; or between places in the same State of the United States through the airspace over any place outside thereof; or between places in the same territory or possession of the United States, or the District of Columbia.

Interstate air transportation means the carriage by aircraft of persons or property as a common carrier for compensation or hire, or the carriage of mail by aircraft in commerce:

(1) Between a place in a State or the District of Columbia and another place in another State or the District of Columbia;

(2) Between places in the same State through the airspace over any place outside that State; or

(3) Between places in the same possession of the United States;

Whether that commerce moves wholly by aircraft of partly by aircraft and partly by other forms of transportation.

Intrastate air transportation means the carriage of persons or property as a common carrier for compensation or hire, by turbojet-powered aircraft capable of carrying thirty or more persons, wholly within the same State of the United States.

Kite means a framework, covered with paper, cloth, metal, or other material, intended to be flown at the end of a rope or cable, and having as its only support the force of the wind moving past its surfaces.

Landing gear extended speed means the maximum speed at which an aircraft can be safely flown with the landing gear extended.

Landing gear operating speed means the maximum speed at which the landing gear can be safely extended or retracted.

Large aircraft means aircraft of more than 12,500 pounds, maximum certificated takeoff weight.

Light-sport aircraft means an aircraft, other than a helicopter or powered-lift that, since its original certification, has continued to meet the following:

(1) A maximum takeoff weight of not more than—

(i) 1,320 pounds (600 kilograms) for aircraft not intended for operation on water; or

(ii) 1,430 pounds (650 kilograms) for an aircraft intended for operation on water.

(2) A maximum airspeed in level flight with maximum continuous power (V_H) of not more than 120 knots CAS under standard atmospheric conditions at sea level.

(3) A maximum never-exceed speed (V_{NE}) of not more than 120 knots CAS for a glider.

(4) A maximum stalling speed or minimum steady flight speed without the use of lift-enhancing devices (V_{S1}) of not more than 45 knots CAS at the aircraft's maximum certificated takeoff weight and most critical center of gravity.

(5) A maximum seating capacity of no more than two persons, including the pilot.

(6) A single, reciprocating engine, if powered.

(7) A fixed or ground-adjustable propeller if a powered aircraft other than a powered glider.

(8) A fixed or feathering propeller system if a powered glider.

(9) A fixed-pitch, semi-rigid, teetering, two-blade rotor system, if a gyroplane.

(10) A nonpressurized cabin, if equipped with a cabin.

(11) Fixed landing gear, except for an aircraft intended for operation on water or a glider.

(12) Fixed or retractable landing gear, or a hull, for an aircraft intended for operation on water.

(13) Fixed or retractable landing gear for a glider.

Lighter-than-air aircraft means aircraft that can rise and remain suspended by using contained gas weighing less than the air that is displaced by the gas.

Load factor means the ratio of a specified load to the total weight of the aircraft. The specified load is expressed in terms of any of the following: aerodynamic forces, inertia forces, or ground or water reactions.

Long-range communication system (LRCS). A system that uses satellite relay, data link, high frequency, or another approved communication system which extends beyond line of sight.

Long-range navigation system (LRNS). An electronic navigation unit that is approved for use under instrument flight rules as a primary means of navigation, and has at least one source of navigational input, such as inertial navigation system, global positioning system, Omega/very low frequency, or Loran C.

Mach number means the ratio of true airspeed to the speed of sound.

Main rotor means the rotor that supplies the principal lift to a rotorcraft.

Maintenance means inspection, overhaul, repair, preservation, and the replacement of parts, but excludes preventive maintenance.

Major alteration means an alteration not listed in the aircraft, aircraft engine, or propeller specifications—

(1) That might appreciably affect weight, balance, structural strength, performance, powerplant operation, flight characteristics, or other qualities affecting airworthiness; or

(2) That is not done according to accepted practices or cannot be done by elementary operations.

Major repair means a repair:

(1) That, if improperly done, might appreciably affect weight, balance, structural strength, performance, powerplant operation, flight characteristics, or other qualities affecting airworthiness; or

(2) That is not done according to accepted practices or cannot be done by elementary operations.

Manifold pressure means absolute pressure as measured at the appropriate point in the induction system and usually expressed in inches of mercury.

Maximum engine overtorque , as it applies to turbopropeller and turboshaft engines incorporating free power turbines for all ratings except one engine inoperative (OEI) ratings of two minutes or less, means the maximum torque of the free power turbine rotor assembly, the inadvertent occurrence of which, for periods of up to 20 seconds, will not require rejection of the engine from service, or any maintenance action other than to correct the cause.

Maximum speed for stability characteristics, V_{FC}/M_{FC} means a speed that may not be less than a speed midway between maximum operating limit speed (V_{MO}/M_{MO}) and demonstrated flight diving speed (V_{DF}/M_{DF}), except that, for altitudes where the Mach number is the limiting factor, M_{FC} need not exceed the Mach number at which effective speed warning occurs.

Medical certificate means acceptable evidence of physical fitness on a form prescribed by the Administrator.

Military operations area. A military operations area (MOA) is airspace established outside Class A airspace to separate or segregate certain nonhazardous military activities from IFR Traffic and to identify for V_{FR} traffic where theses activities are conducted.

V_A means design maneuvering speed.

V_B means design speed for maximum gust intensity.

V_C means design cruising speed.

V_D means design diving speed.

V_{DF}/M_{DF} means demonstrated flight diving speed.

V_{EF} means the speed at which the critical engine is assumed to fail during takeoff.

V_F means design flap speed.

V_{FC}/M_{FC} means maximum speed for stability characteristics.

V_{FE} means maximum flap extended speed.

V_H means maximum speed in level flight with maximum continuous power.

V_{LE} means maximum landing gear extended speed.

V_{LO} means maximum landing gear operating speed.

V_{LOF} means lift-off speed.

V_{MC} means minimum control speed with the critical engine inoperative.

V_{MO}/M_{MO} means maximum operating limit speed.

V_{MU} means minimum unstick speed.

V_{NE} means never-exceed speed.

V_{NO} means maximum structural cruising speed.

V_R means rotation speed.

V_S means the stalling speed or the minimum steady flight speed at which the airplane is controllable.

Minimum descent altitude (MDA) is the lowest altitude specified in an instrument approach procedure, expressed in feet above mean sea level, to which descent is authorized on final approach or during circle-to-land maneuvering until the pilot sees the required visual references for the heliport or runway of intended landing.

Minor alteration means an alteration other than a major alteration.

Minor repair means a repair other than a major repair.

National defense airspace means airspace established by a regulation prescribed, or an order issued under, 49 U.S.C. 40103(b)(3).

Navigable airspace means airspace at and above the minimum flight altitudes prescribed by or under this chapter, including airspace needed for safe takeoff and landing.

Night means the time between the end of evening civil twilight and the beginning of morning civil twilight, as published in the Air Almanac, converted to local time.

Nonprecision approach procedure means a standard instrument approach procedure in which no electronic glide slope is provided.

Operate, with respect to aircraft, means use, cause to use or authorize to use aircraft, for the purpose (except as provided in § 91.13 of this chapter) of air navigation including the piloting of aircraft, with or without the right of legal control (as owner, lessee, or otherwise).

Operational control, with respect to a flight, means the exercise of authority over initiating, conducting or terminating a flight.

Overseas air commerce means the carriage by aircraft of persons or property for compensation or hire, or the carriage of mail by aircraft, or the operation or navigation of aircraft in the conduct or furtherance of a business or vocation, in commerce between a place in any State of the United States, or the District of Columbia, and any place in a territory or possession of the United States; or between a place in a territory or possession of the United States, and a place in any other territory or possession of the United States.

Overseas air transportation means the carriage by aircraft of persons or property as a common carrier for compensation or hire, or the carriage of mail by aircraft, in commerce:

(1) Between a place in a State or the District of Columbia and a place in a possession of the United States; or

(2) Between a place in a possession of the United States and a place in another possession of the United States; whether that commerce moves wholly by aircraft or partly by aircraft and partly by other forms of transportation.

Over-the-top means above the layer of clouds or other obscuring phenomena forming the ceiling.

Parachute means a device used or intended to be used to retard the fall of a body or object through the air.

Person means an individual, firm, partnership, corporation, company, association, joint-stock association, or governmental entity. It includes a trustee, receiver, assignee, or similar representative of any of them.

Pilotage means navigation by visual reference to landmarks.

Pilot in command means the person who:

(1) Has final authority and responsibility for the operation and safety of the flight;

(2) Has been designated as pilot in command before or during the flight; and

(3) Holds the appropriate category, class, and type rating, if appropriate, for the conduct of the flight.

Pitch setting means the propeller blade setting as determined by the blade angle measured in a manner, and at a radius, specified by the instruction manual for the propeller.

Positive control means control of all air traffic, within designated airspace, by air traffic control.

Powered parachute means a powered aircraft comprised of a flexible or semi-rigid wing connected to a fuselage so that the wing is not in position for flight until the aircraft is in motion. The fuselage of a powered parachute contains the aircraft engine, a seat for each occupant and is attached to the aircraft's landing gear.

Powered-lift means a heavier-than-air aircraft capable of vertical takeoff, vertical landing, and low speed flight that depends principally on engine-driven lift devices or engine thrust for lift during these flight regimes and on nonrotating airfoil(s) for lift during horizontal flight.

Precision approach procedure means a standard instrument approach procedure in which an electronic glide slope is provided, such as ILS and PAR.

Preventive maintenance means simple or minor preservation operations and the replacement of small standard parts not involving complex assembly operations.

Prohibited area. A prohibited area is airspace designated under part 73 within which no person may operate an aircraft without the permission of the using agency.

Propeller means a device for propelling an aircraft that has blades on an engine-driven shaft and that, when rotated, produces by its action on the air, a thrust approximately perpendicular to its plane of rotation. It includes control components normally supplied by its manufacturer, but does not include main and auxiliary rotors or rotating airfoils of engines.

Public aircraft means any of the following aircraft when not being used for a commercial purpose or to carry an individual other than a crewmember or qualified non-crewmember:

(1) An aircraft used only for the United States Government; an aircraft owned by the Government and operated by any person for purposes related to crew training, equipment development, or demonstration; an aircraft owned and operated by the government of a State, the District of Columbia, or a territory or possession of the United States or a political subdivision of one of these governments; or an aircraft exclusively leased for at least 90 continuous days by the government of a State, the District of Columbia, or a territory or possession of the United States or a political subdivision of one of these governments.

(i) For the sole purpose of determining public aircraft status, *commercial purposes* means the transportation of persons or property for compensation or hire, but does not include the operation of an aircraft by the armed forces for reimbursement when that reimbursement is required by any Federal statute, regulation, or directive, in effect on November 1, 1999, or by one government on behalf of another government under a cost reimbursement agreement if the government on whose behalf the operation is conducted certifies to the Administrator of the Federal Aviation Administration that the operation is necessary to respond to a significant and imminent threat to life or property (including natural resources) and that no service by a private operator is reasonably available to meet the threat.

(ii) For the sole purpose of determining public aircraft status, *governmental function* means an activity undertaken by a government, such as national defense, intelligence missions, firefighting, search and rescue, law enforcement (including transport of prisoners, detainees, and illegal aliens), aeronautical research, or biological or geological resource management.

(iii) For the sole purpose of determining public aircraft status, *qualified non-crewmember* means an individual, other than a member of the crew, aboard an aircraft operated by the armed forces or an intelligence agency of the United States Government, or whose presence is required to perform, or is associated with the performance of, a governmental function.

(2) An aircraft owned or operated by the armed forces or chartered to provide transportation to the armed forces if—

(i) The aircraft is operated in accordance with title 10 of the United States Code;

(ii) The aircraft is operated in the performance of a governmental function under title 14, 31, 32, or 50 of the United States Code and the aircraft is not used for commercial purposes; or

(iii) The aircraft is chartered to provide transportation to the armed forces and the Secretary of Defense (or the Secretary of the department in which the Coast Guard is operating) designates the operation of the aircraft as being required in the national interest.

(3) An aircraft owned or operated by the National Guard of a State, the District of Columbia, or any territory or possession of the United States, and that meets the criteria of paragraph (2) of this definition, qualifies as a public aircraft only to the extent that it is operated under the direct control of the Department of Defense.

Rated 30-second OEI Power, with respect to rotorcraft turbine engines, means the approved brake horsepower developed under static conditions at specified altitudes and temperatures within the operating limitations established for the engine under part 33 of this chapter, for continuation of one flight operation after the failure or shutdown of one engine in multiengine rotorcraft, for up to three periods of use no longer than 30 seconds each in any one flight, and followed by mandatory inspection and prescribed maintenance action.

Rated 2-minute OEI Power, with respect to rotorcraft turbine engines, means the approved brake horsepower developed under static conditions at specified altitudes and temperatures within the operating limitations established for the engine under part 33 of this chapter, for continuation of one flight operation after the failure or shutdown of one engine in multiengine rotorcraft, for up to three periods of use no longer than 2 minutes each in any one flight, and followed by mandatory inspection and prescribed maintenance action.

Rated continuous OEI power, with respect to rotorcraft turbine engines, means the approved brake horsepower developed under static conditions at specified altitudes and temperatures within the operating limitations established for the engine under part 33 of this chapter, and limited in use to the time required to complete the flight after the failure or shutdown of one engine of a multiengine rotorcraft.

Rated maximum continuous augmented thrust, with respect to turbojet engine type certification, means the approved jet thrust that is developed statically or in flight, in standard atmosphere at a specified altitude, with fluid injection or with the burning of fuel in a separate combustion chamber, within the engine operating limitations established under Part 33 of this chapter, and approved for unrestricted periods of use.

Rated maximum continuous power, with respect to reciprocating, turbopropeller, and turboshaft engines, means the approved brake horsepower that is developed statically or in flight, in standard atmosphere at a specified altitude, within the engine operating limitations established under part 33, and approved for unrestricted periods of use.

Rated maximum continuous thrust, with respect to turbojet engine type certification, means the approved jet thrust that is developed statically or in flight, in standard atmosphere at a specified altitude, without fluid injection and without the burning of fuel in a separate combustion chamber, within the engine operating limitations established under part 33 of this chapter, and approved for unrestricted periods of use.

Rated takeoff augmented thrust, with respect to turbojet engine type certification, means the approved jet thrust that is developed statically under standard sea level conditions, with fluid injection or with the burning of fuel in a separate combustion chamber, within the engine operating limitations established under part 33 of this chapter, and limited in use to periods of not over 5 minutes for takeoff operation.

Rated takeoff power, with respect to reciprocating, turbopropeller, and turboshaft engine type certification, means the approved brake horsepower that is developed statically under standard sea level conditions, within the engine operating limitations established under part 33, and limited in use to periods of not over 5 minutes for takeoff operation.

Rated takeoff thrust, with respect to turbojet engine type certification, means the approved jet thrust that is developed statically under standard sea level conditions, without fluid injection and without the burning of fuel in a separate combustion chamber, within the engine operating limitations established under part 33 of this chapter, and limited in use to periods of not over 5 minutes for takeoff operation.

Rated 30-minute OEI power, with respect to rotorcraft turbine engines, means the approved brake horsepower developed under static conditions at specified altitudes and temperatures within the operating limitations established for the engine under part 33 of this chapter, and limited in use to one period of use no longer than 30 minutes after the failure or shutdown of one engine of a multiengine rotorcraft.

Rated 2$\frac{1}{2}$-minute OEI power, with respect to rotorcraft turbine engines, means the approved brake horsepower developed under static conditions at specified altitudes and temperatures within the operating limitations established for the engine under part 33 of this chapter for periods of use no longer than 2$\frac{1}{2}$ minutes each after the failure or shutdown of one engine of a multiengine rotorcraft.

Rating means a statement that, as a part of a certificate, sets forth special conditions, privileges, or limitations.

Reference landing speed means the speed of the airplane, in a specified landing configuration, at the point where it descends through the 50 foot height in the determination of the landing distance.

Reporting point means a geographical location in relation to which the position of an aircraft is reported.

Restricted area. A restricted area is airspace designated under Part 73 within which the flight of aircraft, while not wholly prohibited, is subject to restriction.

Rocket means an aircraft propelled by ejected expanding gases generated in the engine from self-contained propellants and not dependent on the intake of outside substances. It includes any part which becomes separated during the operation.

Rotorcraft means a heavier-than-air aircraft that depends principally for its support in flight on the lift generated by one or more rotors.

Rotorcraft-load combination means the combination of a rotorcraft and an external-load, including the external-load attaching means. Rotorcraft-load combinations are designated as Class A, Class B, Class C, and Class D, as follows:

(1) *Class A rotorcraft-load combination* means one in which the external load cannot move freely, cannot be jettisoned, and does not extend below the landing gear.

(2) *Class B rotorcraft-load combination* means one in which the external load is jettisonable and is lifted free of land or water during the rotorcraft operation.

(3) *Class C rotorcraft-load combination* means one in which the external load is jettisonable and remains in contact with land or water during the rotorcraft operation.

(4) *Class D rotorcraft-load combination* means one in which the external-load is other than a Class A, B, or C and has been specifically approved by the Administrator for that operation.

Route segment is a portion of a route bounded on each end by a fix or navigation aid (NAVAID).

Sea level engine means a reciprocating aircraft engine having a rated takeoff power that is producible only at sea level.

Second in command means a pilot who is designated to be second in command of an aircraft during flight time.

Show, unless the context otherwise requires, means to show to the satisfaction of the Administrator.

Small aircraft means aircraft of 12,500 pounds or less, maximum certificated takeoff weight.

Special VFR conditions mean meteorological conditions that are less than those required for basic VFR flight in controlled airspace and in which some aircraft are permitted flight under visual flight rules.

Special VFR operations means aircraft operating in accordance with clearances within controlled airspace in meteorological conditions less than the basic VFR weather minima. Such operations must be requested by the pilot and approved by ATC.

Standard atmosphere means the atmosphere defined in U.S. Standard Atmosphere, 1962 (Geopotential altitude tables).

Stopway means an area beyond the takeoff runway, no less wide than the runway and centered upon the extended centerline of the runway, able to support the airplane during an aborted takeoff, without causing structural damage to the airplane, and designated by the airport authorities for use in decelerating the airplane during an aborted takeoff.

Suitable RNAV system is an RNAV system that meets the required performance established for a type of operation, e.g. IFR; and is suitable for operation over the route to be flown in terms of any performance criteria (including accuracy) established by the air navigation service provider for certain routes (e.g. oceanic, ATS routes, and IAPs). An RNAV system's suitability is dependent upon the availability of ground and/or satellite navigation aids that are needed to meet any route performance criteria that may be prescribed in route specifications to navigate the aircraft along the route to be flown. Information on suitable RNAV systems is published in FAA guidance material.

Synthetic vision means a computer-generated image of the external scene topography from the perspective of the flight deck that is derived from aircraft attitude, high-precision navigation solution, and database of terrain, obstacles and relevant cultural features.

Synthetic vision system means an electronic means to display a synthetic vision image of the external scene topography to the flight crew.

Takeoff power:

(1) With respect to reciprocating engines, means the brake horsepower that is developed under standard sea level conditions, and under the maximum conditions of crankshaft rotational speed and engine manifold pressure approved for the normal takeoff, and limited in continuous use to the period of time shown in the approved engine specification; and

(2) With respect to turbine engines, means the brake horsepower that is developed under static conditions at a specified altitude and atmospheric temperature, and under the maximum conditions of rotor shaft rotational speed and gas temperature approved for the normal takeoff, and limited in continuous use to the period of time shown in the approved engine specification.

Takeoff safety speed means a referenced airspeed obtained after lift-off at which the required one-engine-inoperative climb performance can be achieved.

Takeoff thrust, with respect to turbine engines, means the jet thrust that is developed under static conditions at a specific altitude and atmospheric temperature under the maximum conditions of rotorshaft rotational speed and gas temperature approved for the normal takeoff, and limited in continuous use to the period of time shown in the approved engine specification.

Tandem wing configuration means a configuration having two wings of similar span, mounted in tandem.

TCAS I means a TCAS that utilizes interrogations of, and replies from, airborne radar beacon transponders and provides traffic advisories to the pilot.

TCAS II means a TCAS that utilizes interrogations of, and replies from airborne radar beacon transponders and provides traffic advisories and resolution advisories in the vertical plane.

TCAS III means a TCAS that utilizes interrogation of, and replies from, airborne radar beacon transponders and provides traffic advisories and resolution advisories in the vertical and horizontal planes to the pilot.

Time in service, with respect to maintenance time records, means the time from the moment an aircraft leaves the surface of the earth until it touches it at the next point of landing.

Traffic pattern means the traffic flow that is prescribed for aircraft landing at, taxiing on, or taking off from, an airport.

True airspeed means the airspeed of an aircraft relative to undisturbed air. True airspeed is equal to equivalent airspeed multiplied by $(\rho 0/\rho)\ \frac{1}{2}$.

Type:

(1) As used with respect to the certification, ratings, privileges, and limitations of airmen, means a specific make and basic model of aircraft, including modifications thereto that do not change its handling or flight characteristics. Examples include: DC-7, 1049, and F-27; and

(2) As used with respect to the certification of aircraft, means those aircraft which are similar in design. Examples include: DC-7 and DC-7C; 1049G and 1049H; and F-27 and F-27F.

(3) As used with respect to the certification of aircraft engines means those engines which are similar in design. For example, JT8D and JT8D-7 are engines of the same type, and JT9D-3A and JT9D-7 are engines of the same type.

United States, in a geographical sense, means (1) the States, the District of Columbia, Puerto Rico, and the possessions, including the territorial waters, and (2) the airspace of those areas.

United States air carrier means a citizen of the United States who undertakes directly by lease, or other arrangement, to engage in air transportation.

V_{FR} *over-the-top,* with respect to the operation of aircraft, means the operation of an aircraft over-the-top under V_{FR} when it is not being operated on an IFR flight plan.

Warning area. A warning area is airspace of defined dimensions, extending from 3 nautical miles outward from the coast of the United States, that contains activity that may be hazardous to nonparticipating aircraft. The purpose of such warning areas is to warn nonparticipating pilots of the potential danger. A warning area may be located over domestic or international waters or both.

Weight-shift-control aircraft means a powered aircraft with a framed pivoting wing and a fuselage controllable only in pitch and roll by the pilot's ability to change the aircraft's center of gravity with respect to the wing. Flight control of the aircraft depends on the wing's ability to flexibly deform rather than the use of control surfaces.

Winglet or tip fin means an out-of-plane surface extending from a lifting surface. The surface may or may not have control surfaces.

[Doc. No. 1150, 27 FR 4588, May 15, 1962]

Editorial Note: For Federal Register citations affecting § 1.1, see the List of CFR Sections Affected, which appears in the Finding Aids section of the printed volume and at *www.fdsys.gov.*

§ 1.2 Abbreviations and symbols.

In Subchapters A through K of this chapter:

AFM means airplane flight manual.

AGL means above ground level.

ALS means approach light system.

APU means auxiliary power unit.

ASR means airport surveillance radar.

ATC means air traffic control.

ATS means Air Traffic Service.

CAMP means continuous airworthiness maintenance program.

CAS means calibrated airspeed.

CAT II means Category II.

CHDO means an FAA Flight Standards certificate holding district office.

CMP means configuration, maintenance, and procedures.

CONSOL or *CONSOLAN* means a kind of low or medium frequency long range navigational aid.

DH means decision height.

DME means distance measuring equipment compatible with TACAN.

EAS means equivalent airspeed.

EFVS means enhanced flight vision system.

Equi-Time Point means a point on the route of flight where the flight time, considering wind, to each of two selected airports is equal.

ETOPS means extended operations.

EWIS , as defined by § 25.1701 of this chapter, means electrical wiring interconnection system.

FAA means Federal Aviation Administration.

FFS means full flight simulator.

FM means fan marker.

FSTD means flight simulation training device.

FTD means flight training device.

GS means glide slope.

HIRL means high-intensity runway light system.

IAS means indicated airspeed.

ICAO means International Civil Aviation Organization.

IFR means instrument flight rules.

IFSD means in-flight shutdown.

ILS means instrument landing system.

IM means ILS inner marker.

INT means intersection.

LDA means localizer-type directional aid.

LFR means low-frequency radio range.

LMM means compass locator at middle marker.

LOC means ILS localizer.

LOM means compass locator at outer marker.

M means mach number.

MAA means maximum authorized IFR altitude.

MALS means medium intensity approach light system.

MALSR means medium intensity approach light system with runway alignment indicator lights.

MCA means minimum crossing altitude.

MDA means minimum descent altitude.

MEA means minimum en route IFR altitude.

MEL means minimum equipment list.

MM means ILS middle marker.

MOCA means minimum obstruction clearance altitude.

MRA means minimum reception altitude.

MSL means mean sea level.

NDB (ADF) means nondirectional beacon (automatic direction finder).

NM means nautical mile.

NOPAC means North Pacific area of operation.

NOPT means no procedure turn required.

OEI means one engine inoperative.

OM means ILS outer marker.

OPSPECS means operations specifications.

PACOTS means Pacific Organized Track System.

PAR means precision approach radar.

PMA means parts manufacturer approval.

PTRS means Performance Tracking and Reporting System.

RAIL means runway alignment indicator light system.

RBN means radio beacon.

RCLM means runway centerline marking.

RCLS means runway centerline light system.

REIL means runway end identification lights.

RFFS means rescue and firefighting services.

RNAV means area navigation.

RR means low or medium frequency radio range station.

RVR means runway visual range as measured in the touchdown zone area.

SALS means short approach light system.

SATCOM means satellite communications.

SSALS means simplified short approach light system.

SSALSR means simplified short approach light system with runway alignment indicator lights.

TACAN means ultra-high frequency tactical air navigational aid.

TAS means true airspeed.

TCAS means a traffic alert and collision avoidance system.

TDZL means touchdown zone lights.

TSO means technical standard order.

TVOR means very high frequency terminal omnirange station.

V_A means design maneuvering speed.

V_B means design speed for maximum gust intensity.

V_C means design cruising speed.

V_D means design diving speed.

V_{DF}/M_{DF} means demonstrated flight diving speed.

V_{EF} means the speed at which the critical engine is assumed to fail during takeoff.

V_F means design flap speed.

V_{FC}/M_{FC} means maximum speed for stability characteristics.

V_{FE} means maximum flap extended speed.

V_{FTO} means final takeoff speed.

V_H means maximum speed in level flight with maximum continuous power.

V_{LE} means maximum landing gear extended speed.

V_{LO} means maximum landing gear operating speed.

V_{LOF} means lift-off speed.

V_{MC} means minimum control speed with the critical engine inoperative.

V_{MO}/M_{MO} means maximum operating limit speed.

V_{MU} means minimum unstick speed.

V_{NE} means never-exceed speed.

V_{NO} means maximum structural cruising speed.

V_R means rotation speed.

V_{REF} means reference landing speed.

V_S means the stalling speed or the minimum steady flight speed at which the airplane is controllable.

V_{SO} means the stalling speed or the minimum steady flight speed in the landing configuration.

V_{S1} means the stalling speed or the minimum steady flight speed obtained in a specific configuration.

V_{SR} means reference stall speed.

V_{SRO} means reference stall speed in the landing configuration.

V_{SR1} means reference stall speed in a specific configuration.

V_{SW} means speed at which onset of natural or artificial stall warning occurs.

V_{TOSS} means takeoff safety speed for Category A rotorcraft.

V_X means speed for best angle of climb.

V_Y means speed for best rate of climb.

V_1 means the maximum speed in the takeoff at which the pilot must take the first action (e.g., apply brakes, reduce thrust, deploy speed brakes) to stop the airplane within the accelerate-stop distance. V_1 also means the minimum speed in the takeoff, following a failure of the critical engine at V_{EF} , at which the pilot can continue the takeoff and achieve the required height above the takeoff surface within the takeoff distance.

V_2 means takeoff safety speed.

V_{2min} means minimum takeoff safety speed.

V_{FR} means visual flight rules.

V_{HF} means very high frequency.

V_{OR} means very high frequency omnirange station.

V_{ORTAC} means collocated V_{OR} and T_{ACAN}.

[Doc. No. 1150, 27 FR 4590, May 15, 1962]

Editorial Note: For Federal Register citations affecting § 1.2, see the List of CFR Sections Affected, which appears in the Finding Aids section of the printed volume and at *www.fdsys.gov.*

§ 1.3 Rules of construction.

(a) In Subchapters A through K of this chapter, unless the context requires otherwise:

(1) Words importing the singular include the plural;

(2) Words importing the plural include the singular; and

(3) Words importing the masculine gender include the feminine.

(b) In Subchapters A through K of this chapter, the word:

(1) *Shall* is used in an imperative sense;

(2) *May* is used in a permissive sense to state authority or permission to do the act prescribed, and the words "no person may * * *" or "a person may not * * *" mean that no person is required, authorized, or permitted to do the act prescribed; and

(3) *Includes* means "includes but is not limited to".

[Doc. No. 1150, 27 FR 4590, May 15, 1962, as amended by Amdt. 1-10, 31 FR 5055, Mar. 29, 1966]

PART 23

AIRWORTHINESS STANDARDS: NORMAL, UTILITY, ACROBATIC, AND COMMUTER CATEGORY AIRPLANES

Contents

Airplane Flight Manual and Approved Manual Material

Authority: 49 U.S.C. 106(g), 40113, 44701-44702, 44704.

Source: Docket No. 4080, 29 FR 17955, Dec. 18. 1964; 30 FR 258, Jan. 9, 1965, unless otherwise noted.

Special Federal Aviation Regulation No. 23

1. *Applicability.* An applicant is entitled to a type certificate in the normal category for a reciprocating or turbopropeller multiengine powered small airplane that is to be certificated to carry more than 10 occupants and that is intended for use in operations under Part 135 of the Federal Aviation Regulations if he shows compliance with the applicable requirements of Part 23 of the Federal Aviation Regulations, as supplemented or modified by the additional airworthiness requirements of this regulation.

2. *References.* Unless otherwise provided, all references in this regulation to specific sections of Part 23 of the Federal Aviation Regulations are those sections of Part 23 in effect on March 30, 1967.

Flight Requirements

3. *General.* Compliance must be shown with the applicable requirements of Subpart B of Part 23 of the Federal Aviation Regulations in effect on March 30, 1967, as supplemented or modified in sections 4 through 10 of this regulation.

Performance

4. *General.* (a) Unless otherwise prescribed in this regulation, compliance with each applicable performance requirement in sections 4 through 7 of this regulation must be shown for ambient atmospheric conditions and still air.

(b) The performance must correspond to the propulsive thrust available under the particular ambient atmospheric conditions and the particular flight condition. The available propulsive thrust must correspond to engine power or thrust, not exceeding the approved power or thrust less—

(1) Installation losses; and

(2) The power or equivalent thrust absorbed by the accessories and services appropriate to the particular ambient atmospheric conditions and the particular flight condition.

(c) Unless otherwise prescribed in this regulation, the applicant must select the take-off, en route, and landing configurations for the airplane.

(d) The airplane configuration may vary with weight, altitude, and temperature, to the extent they are compatible with the operating procedures required by paragraph (e) of this section.

(e) Unless otherwise prescribed in this regulation, in determining the critical engine inoperative takeoff performance, the accelerate-stop distance, takeoff distance, changes in the airplane's configuration, speed, power, and thrust, must be made in accordance with procedures established by the applicant for operation in service.

(f) Procedures for the execution of balked landings must be established by the applicant and included in the Airplane Flight Manual.

(g) The procedures established under paragraphs (e) and (f) of this section must—

(1) Be able to be consistently executed in service by a crew of average skill;

(2) Use methods or devices that are safe and reliable; and

(3) Include allowance for any time delays, in the execution of the procedures, that may reasonably be expected in service.

5. *Takeoff* —(a) *General.* The takeoff speeds described in paragraph (b), the accelerate-stop distance described in paragraph (c), and the takeoff distance described in paragraph (d), must be determined for—

(1) Each weight, altitude, and ambient temperature within the operational limits selected by the applicant;

(2) The selected configuration for takeoff;

(3) The center of gravity in the most unfavorable position;

(4) The operating engine within approved operating limitation; and

(5) Takeoff data based on smooth, dry, hard-surface runway.

(b) *Takeoff speeds.* (1) The decision speed V_1 is the calibrated airspeed on the ground at which, as a result of engine failure or other reasons, the pilot is assumed to have made a decision to continue or discontinue the takeoff. The speed V_1 must be selected by the applicant but may not be less than—

(i) 1.10 V_{s1};

(ii) 1.10 V_{MC}

(iii) A speed that permits acceleration to V_1 and stop in accordance with paragraph (c) allowing credit for an overrun distance equal to that required to stop the airplane from a ground speed of 35 knots utilizing maximum braking; or

(iv) A speed at which the airplane can be rotated for takeoff and shown to be adequate to safely continue the takeoff, using normal piloting skill, when the critical engine is suddenly made inoperative.

(2) Other essential takeoff speeds necessary for safe operation of the airplane must be determined and shown in the Airplane Flight Manual.

(c) *Accelerate-stop distance.* (1) The accelerate-stop distance is the sum of the distances necessary to—

(i) Accelerate the airplane from a standing start to V_1; and

(ii) Decelerate the airplane from V_1 to a speed not greater than 35 knots, assuming that in the case of engine failure, failure of the critical engine is recognized by the pilot at the speed V_1. The landing gear must remain in the extended position and maximum braking may be utilized during deceleration.

(2) Means other than wheel brakes may be used to determine the accelerate-stop distance if that means is available with the critical engine inoperative and—

(i) Is safe and reliable;

(ii) Is used so that consistent results can be expected under normal operating conditions; and

(iii) Is such that exceptional skill is not required to control the airplane.

(d) *All engines operating takeoff distance.* The all engine operating takeoff distance is the horizontal distance required to takeoff and climb to a height of 50 feet above the takeoff surface according to procedures in FAR 23.51(a).

(e) *One-engine-inoperative takeoff.* The maximum weight must be determined for each altitude and temperature within the operational limits established for the airplane, at which the airplane has takeoff capability after failure of the critical engine at or above V_1 determined in accordance with paragraph (b) of this section. This capability may be established—

(1) By demonstrating a measurably positive rate of climb with the airplane in the takeoff configuration, landing gear extended; or

(2) By demonstrating the capability of maintaining flight after engine failure utilizing procedures prescribed by the applicant.

6. *Climb* —(a) *Landing climb: All-engines-operating.* The maximum weight must be determined with the airplane in the landing configuration, for each altitude, and ambient temperature within the operational limits established for the airplane and with the most unfavorable center of gravity and out-of-ground effect in free air, at which the steady gradient of climb will not be less than 3.3 percent, with:

(1) The engines at the power that is available 8 seconds after initiation of movement of the power or thrust controls from the mimimum flight idle to the takeoff position.

(2) A climb speed not greater than the approach speed established under section 7 of this regulation and not less than the greater of 1.05$_{MC}$ or 1.10V_{S1}.

(b) *En route climb, one-engine-inoperative.* (1) the maximum weight must be determined with the airplane in the en route configuration, the critical engine inoperative, the remaining engine at not more than maximum continuous power or thrust, and the most unfavorable center of gravity, at which the gradient at climb will be not less than—

(i) 1.2 percent (or a gradient equivalent to 0.20 V_{SO} 2, if greater) at 5,000 feet and an ambient temperature of 41°F. or

(ii) 0.6 percent (or a gradient equivalent to 0.01 V_{SO} 2, if greater) at 5,000 feet and ambient temperature of 81°F.

(2) The minimum climb gradient specified in subdivisions (i) and (ii) of subparagraph (1) of this paragraph must vary linearly between 41°F. and 81°F. and must change at the same rate up to the maximum operational temperature approved for the airplane.

7. *Landing.* The landing distance must be determined for standard atmosphere at each weight and altitude in accordance with FAR 23.75(a), except that instead of the gliding approach specified in FAR 23.75(a)(1), the landing may be preceded by a steady approach down to the 50-foot height at a gradient of descent not greater than 5.2 percent (3°) at a calibrated airspeed not less than 1.3$_{s1}$.

Trim

8. *Trim* —(a) *Lateral and directional trim.* The airplane must maintain lateral and directional trim in level flight at a speed of V_h or V_{MO}/ M_{MO}, whichever is lower, with landing gear and wing flaps retracted.

(b) *Longitudinal trim.* The airplane must maintain longitudinal trim during the following conditions, except that it need not maintain trim at a speed greater than V_{MO}/ M_{MO}.

(1) In the approach conditions specified in FAR 23.161(c)(3) through (5), except that instead of the speeds specified therein, trim must be maintained with a stick force of not more than 10 pounds down to a speed used in showing compliance with section 7 of this regulation or 1.4 V_{s1} whichever is lower.

(2) In level flight at any speed from V_H or V_{MO}/ M_{MO}, whichever is lower, to either V_x or 1.4 V_{s1}, with the landing gear and wing flaps retracted.

Stability

9. *Static longitudinal stability.* (a) In showing compliance with the provisions of FAR 23.175(b) and with paragraph (b) of this section, the airspeed must return to within ±7½ percent of the trim speed.

(b) *Cruise stability.* The stick force curve must have a stable slope for a speed range of ±50 knots from the trim speed except that the speeds need not exceed V_{FC}/ M_{FC} or be less than 1.4 V_{s1}. This speed range will be considered to begin at the outer extremes of the friction band and the stick force may not exceed 50 pounds with—

(i) Landing gear retracted;

(ii) Wing flaps retracted;

(iii) The maximum cruising power as selected by the applicant as an operating limitation for turbine engines or 75 percent of maximum continuous power for reciprocating engines except that the power need not exceed that required at V_{MO}/ M_{MO}.

(iv) Maximum takeoff weight; and

(v) The airplane trimmed for level flight with the power specified in subparagraph (iii) of this paragraph.

V_{FC}/ M_{FC} may not be less than a speed midway between V_{MO}/ M_{MO} and V_{DF}/ M_{DF}, except that, for altitudes where Mach number is the limiting factor, M_{FC} need not exceed the Mach number at which effective speed warning occurs.

(c) *Climb stability. For turbopropeller powered airplanes only.* In showing compliance with FAR 23.175(a), an applicant must in lieu of the power specified in FAR 23.175(a)(4), use the maximum power or thrust selected by the applicant as an operating limitation for use during climb at the best rate of climb speed except that the speed need not be less than 1.4 V_{s1}.

Stalls

10. *Stall warning.* If artificial stall warning is required to comply with the requirements of FAR 23.207, the warning device must give clearly distinguishable indications under expected conditions of flight. The use of a visual warning

device that requires the attention of the crew within the cockpit is not acceptable by itself.

Control Systems

11. *Electric trim tabs.* The airplane must meet the requirements of FAR 23.677 and in addition it must be shown that the airplane is safely controllable and that a pilot can perform all the maneuvers and operations necessary to effect a safe landing following any probable electric trim tab runaway which might be reasonably expected in service allowing for appropriate time delay after pilot recognition of the runaway. This demonstration must be conducted at the critical airplane weights and center of gravity positions.

Instruments: Installation

12. *Arrangement and visibility.* Each instrument must meet the requirements of FAR 23.1321 and in addition—

(a) Each flight, navigation, and powerplant instrument for use by any pilot must be plainly visible to him from his station with the minimum practicable deviation from his normal position and line of vision when he is looking forward along the flight path.

(b) The flight instruments required by FAR 23.1303 and by the applicable operating rules must be grouped on the instrument panel and centered as nearly as practicable about the vertical plane of each pilot's forward vision. In addition—

(1) The instrument that most effectively indicates the attitude must be on the panel in the top center position;

(2) The instrument that most effectively indicates airspeed must be adjacent to and directly to the left of the instrument in the top center position;

(3) The instrument that most effectively indicates altitude must be adjacent to and directly to the right of the instrument in the top center position; and

(4) The instrument that most effectively indicates direction of flight must be adjacent to and directly below the instrument in the top center position.

13. *Airspeed indicating system.* Each airspeed indicating system must meet the requirements of FAR 23.1323 and in addition—

(a) Airspeed indicating instruments must be of an approved type and must be calibrated to indicate true airspeed at sea level in the standard atmosphere with a mimimum practicable instrument calibration error when the corresponding pilot and static pressures are supplied to the instruments.

(b) The airspeed indicating system must be calibrated to determine the system error, *i.e.*, the relation between IAS and CAS, in flight and during the accelerate takeoff ground run. The ground run calibration must be obtained between 0.8 of the mimimum value of V_1 and 1.2 times the maximum value of V_1, considering the approved ranges of altitude and weight. The ground run calibration will be determined assuming an engine failure at the mimimum value of V_1.

(c) The airspeed error of the installation excluding the instrument calibration error, must not exceed 3 percent or 5 knots whichever is greater, throughout the speed range from V_{MO} to 1.3 S_1 with flaps retracted and from 1.3 V_{SO} to V_{FE} with flaps in the landing position.

(d) Information showing the relationship between IAS and CAS must be shown in the Airplane Flight Manual.

14. *Static air vent system.* The static air vent system must meet the requirements of FAR 23.1325. The altimeter system calibration must be determined and shown in the Airplane Flight Manual.

Operating Limitations and Information

15. *Maximum operating limit speed V_{MO}/M_{MO}.* Instead of establishing operating limitations based on V_{ME} and V_{NO}, the applicant must establish a maximum operating limit speed V_{MO}/M_{MO} in accordance with the following:

(a) The maximum operating limit speed must not exceed the design cruising speed V_C and must be sufficiently below V_D/M_D or V_{DF}/M_{DF} to make it highly improbable that the latter speeds will be inadvertently exceeded in flight.

(b) The speed *Vmo* must not exceed 0.8 V_D/M_D or 0.8 V_{DF}/M_{DF} unless flight demonstrations involving upsets as specified by the Administrator indicates a lower speed margin will not result in speeds exceeding V_D/M_D or V_{DF}. Atmospheric variations, horizontal gusts, and equipment errors, and airframe production variations will be taken into account.

16. *Minimum flight crew.* In addition to meeting the requirements of FAR 23.1523, the applicant must establish the minimum number and type of qualified flight crew personnel sufficient for safe operation of the airplane considering—

(a) Each kind of operation for which the applicant desires approval;

(b) The workload on each crewmember considering the following:

(1) Flight path control.
(2) Collision avoidance.
(3) Navigation.
(4) Communications.
(5) Operation and monitoring of all essential aircraft systems.
(6) Command decisions; and

(c) The accessibility and ease of operation of necessary controls by the appropriate crewmember during all normal and emergency operations when at his flight station.

17. *Airspeed indicator.* The airspeed indicator must meet the requirements of FAR 23.1545 except that, the airspeed notations and markings in terms of V_{NO} and V_{NE} must be replaced by the V_{MO}/M_{MO} notations. The airspeed indicator markings must be easily read and understood by the pilot. A placard adjacent to the airspeed indicator is an acceptable means of showing compliance with the requirements of FAR 23.1545(c).

Airplane Flight Manual

18. *General.* The Airplane Flight Manual must be prepared in accordance with the requirements of FARs 23.1583 and 23.1587, and in addition the operating limitations and performance information set forth in sections 19 and 20 must be included.

19. *Operating limitations.* The Airplane Flight Manual must include the following limitations—

(a) *Airspeed limitations.* (1) The maximum operating limit speed V_{MO}/M_{MO} and a statement that this speed limit may not be deliberately exceeded in any regime of flight (climb, cruise, or descent) unless a higher speed is authorized for flight test or pilot training;

(2) If an airspeed limitation is based upon compressibility effects, a statement to this effect and information as to any symptoms, the probable behavior of the airplane, and the recommended recovery procedures; and

(3) The airspeed limits, shown in terms of V_{MO}/M_{MO} instead of V_{NO} and V_{NE}.

(b) *Takeoff weight limitations.* The maximum takeoff weight for each airport elevation, ambient temperature, and available takeoff runway length within the range selected by the applicant. This weight may not exceed the weight at which:

(1) The all-engine operating takeoff distance determined in accordance with section 5(d) or the accelerate-stop distance determined in accordance with section 5(c), which ever is greater, is equal to the available runway length;

(2) The airplane complies with the one-engine-inoperative takeoff requirements specified in section 5(e); and

(3) The airplane complies with the one-engine-inoperative en route climb requirements specified in section 6(b), assuming that a standard temperature lapse rate exists from the airport elevation to the altitude of 5,000 feet, except that the weight may not exceed that corresponding to a temperature of 41°F at 5,000 feet.

20. *Performance information.* The Airplane Flight Manual must contain the performance information determined in accordance with the provisions of the performance requirements of this regulation. The information must include the following:

(a) Sufficient information so that the take-off weight limits specified in section 19(b) can be determined for all temperatures and altitudes within the operation limitations selected by the applicant.

(b) The conditions under which the performance information was obtained, including the airspeed at the 50-foot height used to determine landing distances.

(c) The performance information (determined by extrapolation and computed for the range of weights between the maximum landing and takeoff weights) for—

(1) Climb in the landing configuration; and
(2) Landing distance.

(d) Procedure established under section 4 of this regulation related to the limitations and information required by this section in the form of guidance material including any relevant limitations or information.

(e) An explanation of significant or unusual flight or ground handling characteristics of the airplane.

(f) Airspeeds, as indicated airspeeds, corresponding to those determined for takeoff in accordance with section 5(b).

21. *Maximum operating altitudes.* The maximum operating altitude to which operation is permitted, as limited by flight, structural, powerplant, functional, or equipment characteristics, must be specified in the Airplane Flight Manual.

22. *Stowage provision for Airplane Flight Manual.* Provision must be made for stowing the Airplane Flight Manual in a suitable fixed container which is readily accessible to the pilot.

23. *Operating procedures.* Procedures for restarting turbine engines in flight (including the effects of altitude) must be set forth in the Airplane Flight Manual.

Airframe Requirements Flight Loads

24. *Engine torque.* (a) Each turbopropeller engine mount and its supporting structure must be designed for the torque effects of—

(1) The conditions set forth in FAR 23.361(a).

(2) The limit engine torque corresponding to takeoff power and propeller speed, multiplied by a factor accounting for propeller control system malfunction, including quick feathering action, simultaneously with 1 *g* level flight loads. In the absence of a rational analysis, a factor of 1.6 must be used.

(b) The limit torque is obtained by multiplying the mean torque by a factor of 1.25.

25. *Turbine engine gyroscopic loads.* Each turbopropeller engine mount and its supporting structure must be designed for the gyroscopic loads that result, with the engines at maximum continuous r.p.m., under either—

(a) The conditions prescribed in FARs 23.351 and 23.423; or

(b) All possible combinations of the following:

(1) A yaw velocity of 2.5 radius per second.
(2) A pitch velocity of 1.0 radians per second.
(3) A normal load factor of 2.5.
(4) Maximum continuous thrust.

26. *Unsymmetrical loads due to engine failure.* (a) Turbopropeller powered airplanes must be designed for the unsymmetrical loads resulting from the failure of the critical engine including the following conditions in combination with a single malfunction of the propeller drag limiting system, considering the probable pilot corrective action on the flight controls.

(1) At speeds between V_{MC} and V_D, the loads resulting from power failure because of fuel flow interruption are considered to be limit loads.

(2) At speeds between V_{MC} and V_C, the loads resulting from the disconnection of the engine compressor from the turbine or from loss of the turbine blades are considered to be ultimate loads.

(3) The time history of the thrust decay and drag buildup occurring as a result of the prescribed engine failures must be substantiated by test or other data applicable to the particular engine-propeller combination.

(4) The timing and magnitude of the probable pilot corrective action must be conservatively estimated, considering the characteristics of the particular engine-propeller-airplane combination.

(b) Pilot corrective action may be assumed to be initiated at the time maximum yawing velocity is reached, but not earlier than two seconds after the engine failure. The magnitude of the corrective action may be based on the control forces specified in FAR 23.397 except that lower forces may be assumed where it is shown by analysis or test that these forces can control the yaw and roll resulting from the prescribed engine failure conditions.

Ground Loads

27. *Dual wheel landing gear units.* Each dual wheel landing gear unit and its supporting structure must be shown to comply with the following:

(a) *Pivoting.* The airplane must be assumed to pivot about one side of the main gear with the brakes on that side locked. The limit vertical load factor must be 1.0 and the coefficient of friction 0.8. This condition need apply only to the main gear and its supporting structure.

(b) *Unequal tire inflation.* A 60-40 percent distribution of the loads established in accordance with FAR 23.471 through FAR 23.483 must be applied to the dual wheels.

(c) *Flat tire.* (1) Sixty percent of the loads specified in FAR 23.471 through FAR 23.483 must be applied to either wheel in a unit.

(2) Sixty percent of the limit drag and side loads and 100 percent of the limit vertical load established in accordance with FARs 23.493 and 23.485 must be applied to either wheel in a unit except that the vertical load need not

exceed the maximum vertical load in paragraph (c)(1) of this section.

Fatigue Evaluation

28. *Fatigue evaluation of wing and associated structure.* Unless it is shown that the structure, operating stress levels, materials, and expected use are comparable from a fatigue standpoint to a similar design which has had substantial satisfactory service experience, the strength, detail design, and the fabrication of those parts of the wing, wing carrythrough, and attaching structure whose failure would be catastrophic must be evaluated under either—

(a) A fatigue strength investigation in which the structure is shown by analysis, tests, or both to be able to withstand the repeated loads of variable magnitude expected in service; or

(b) A fail-safe strength investigation in which it is shown by analysis, tests, or both that catastrophic failure of the structure is not probable after fatigue, or obvious partial failure, of a principal structural element, and that the remaining structure is able to withstand a static ultimate load factor of 75 percent of the critical limit load factor at V_c. These loads must be multiplied by a factor of 1.15 unless the dynamic effects of failure under static load are otherwise considered.

Design and Construction

29. *Flutter.* For Multiengine turbopropeller powered airplanes, a dynamic evaluation must be made and must include—

(a) The significant elastic, inertia, and aerodynamic forces associated with the rotations and displacements of the plane of the propeller; and

(b) Engine-propeller-nacelle stiffness and damping variations appropriate to the particular configuration.

Landing Gear

30. *Flap operated landing gear warning device.* Airplanes having retractable landing gear and wing flaps must be equipped with a warning device that functions continuously when the wing flaps are extended to a flap position that activates the warning device to give adequate warning before landing, using normal landing procedures, if the landing gear is not fully extended and locked. There may not be a manual shut off for this warning device. The flap position sensing unit may be installed at any suitable location. The system for this device may use any part of the system (including the aural warning device) provided for other landing gear warning devices.

Personnel and Cargo Accommodations

31. *Cargo and baggage compartments.* Cargo and baggage compartments must be designed to meet the requirements of FAR 23.787 (a) and (b), and in addition means must be provided to protect passengers from injury by the contents of any cargo or baggage compartment when the ultimate forward inertia force is 9 g.

32. *Doors and exits.* The airplane must meet the requirements of FAR 23.783 and FAR 23.807 (a)(3), (b), and (c), and in addition:

(a) There must be a means to lock and safeguard each external door and exit against opening in flight either inadvertently by persons, or as a result of mechanical failure. Each external door must be operable from both the inside and the outside.

(b) There must be means for direct visual inspection of the locking mechanism by crewmembers to determine whether external doors and exits, for which the initial opening movement is outward, are fully locked. In addition, there must be a visual means to signal to crewmembers when normally used external doors are closed and fully locked.

(c) The passenger entrance door must qualify as a floor level emergency exit. Each additional required emergency exit except floor level exits must be located over the wing or must be provided with acceptable means to assist the occupants in descending to the ground. In addition to the passenger entrance door:

(1) For a total seating capacity of 15 or less, an emergency exit as defined in FAR 23.807(b) is required on each side of the cabin.

(2) For a total seating capacity of 16 through 23, three emergency exits as defined in 23.807(b) are required with one on the same side as the door and two on the side opposite the door.

(d) An evacuation demonstration must be conducted utilizing the maximum number of occupants for which certification is desired. It must be conducted under simulated night conditions utilizing only the emergency exits on the most critical side of the aircraft. The participants must be representative of average airline passengers with no prior practice or rehearsal for the demonstration. Evacuation must be completed within 90 seconds.

(e) Each emergency exit must be marked with the word "Exit" by a sign which has white letters 1 inch high on a red background 2 inches high, be self-illuminated or independently internally electrically illuminated, and have a minimum luminescence (brightness) of at least 160 microlamberts. The colors may be reversed if the passenger compartment illumination is essentially the same.

(f) Access to window type emergency exits must not be obstructed by seats or seat backs.

(g) The width of the main passenger aisle at any point between seats must equal or exceed the values in the following table.

Total seating capacity	Minimum main passenger aisle width	
	Less than 25 inches from floor	25 inches and more from floor
10 through 23	9 inches	15 inches.

Miscellaneous

33. *Lightning strike protection.* Parts that are electrically insulated from the basic airframe must be connected to it through lightning arrestors unless a lightning strike on the insulated part—

(a) Is improbable because of shielding by other parts; or

(b) Is not hazardous.

34. *Ice protection.* If certification with ice protection provisions is desired, compliance with the following requirements must be shown:

(a) The recommended procedures for the use of the ice protection equipment must be set forth in the Airplane Flight Manual.

(b) An analysis must be performed to establish, on the basis of the airplane's operational needs, the adequacy of the ice protection system for the various components of the airplane. In addition, tests of the ice protection system must be conducted to demonstrate that the airplane is capable of operating safely in continuous maximum and intermittent maximum icing conditions as described in FAR 25, appendix C.

(c) Compliance with all or portions of this section may be accomplished by reference, where applicable because of similarity of the designs, to analysis and tests performed by the applicant for a type certificated model.

35. *Maintenance information.* The applicant must make available to the owner at the time of delivery of the airplane the information he considers essential for the proper maintenance of the airplane. That information must include the following:

(a) Description of systems, including electrical, hydraulic, and fuel controls.

(b) Lubrication instructions setting forth the frequency and the lubricants and fluids which are to be used in the various systems.

(c) Pressures and electrical loads applicable to the various systems.

(d) Tolerances and adjustments necessary for proper functioning.

(e) Methods of leveling, raising, and towing.

(f) Methods of balancing control surfaces.

(g) Identification of primary and secondary structures.

(h) Frequency and extent of inspections necessary to the proper operation of the airplane.

(i) Special repair methods applicable to the airplane.

(j) Special inspection techniques, including those that require X-ray, ultrasonic, and magnetic particle inspection.

(k) List of special tools.

Propulsion

General

36. *Vibration characteristics.* For turbopropeller powered airplanes, the engine installation must not result in vibration characteristics of the engine exceeding those established during the type certification of the engine.

37. *In-flight restarting of engine.* If the engine on turbopropeller powered airplanes cannot be restarted at the maximum cruise altitude, a determination must be made of the altitude below which restarts can be consistently accomplished. Restart information must be provided in the Airplane Flight Manual.

38. *Engines*—(a) *For turbopropeller powered airplanes.* The engine installation must comply with the following requirements:

(1) *Engine isolation.* The powerplants must be arranged and isolated from each other to allow operation, in at least one configuration, so that the failure or malfunction of any engine, or of any system that can affect the engine, will not—

(i) Prevent the continued safe operation of the remaining engines; or

(ii) Require immediate action by any crewmember for continued safe operation.

(2) *Control of engine rotation.* There must be a means to individually stop and restart the rotation of any engine in flight except that engine rotation need not be stopped if continued rotation could not jeopardize the safety of the airplane. Each component of the stopping and restarting system on the engine side of the firewall, and that might be exposed to fire, must be at least fire resistant. If hydraulic propeller feathering systems are used for this purpose, the feathering lines must be at least fire resistant under the operating conditions that may be expected to exist during feathering.

(3) *Engine speed and gas temperature control devices.* The powerplant systems associated with engine control devices, systems, and instrumentation must provide reasonable assurance that those engine operating limitations that adversely affect turbine rotor structural integrity will not be exceeded in service.

(b) *For reciprocating-engine powered airplanes.* To provide engine isolation, the powerplants must be arranged and isolated from each other to allow operation, in at least one configuration, so that the failure or malfunction of any engine, or of any system that can affect that engine, will not—

(1) Prevent the continued safe operation of the remaining engines; or

(2) Require immediate action by any crewmember for continued safe operation.

39. *Turbopropeller reversing systems.* (a) Turbopropeller reversing systems intended for ground operation must be designed so that no single failure or malfunction of the system will result in unwanted reverse thrust under any expected operating condition. Failure of structural elements need not be considered if the probability of this kind of failure is extremely remote.

(b) Turbopropeller reversing systems intended for in-flight use must be designed so that no unsafe condition will result during normal operation of the system, or from any failure (or reasonably likely combination of failures) of the reversing system, under any anticipated condition of operation of the airplane. Failure of structural elements need not be considered if the probability of this kind of failure is extremely remote.

(c) Compliance with this section may be shown by failure analysis, testing, or both for propeller systems that allow propeller blades to move from the flight low-pitch position to a position that is substantially less than that at the normal flight low-pitch stop position. The analysis may include or be supported by the analysis made to show compliance with the type certification of the propeller and associated installation components. Credit will be given for pertinent analysis and testing completed by the engine and propeller manufacturers.

40. *Turbopropeller drag-limiting systems.* Turbopropeller drag-limiting systems must be designed so that no single failure or malfunction of any of the systems during normal or emergency operation results in propeller drag in excess of that for which the airplane was designed. Failure of structural elements of the drag-limiting systems need not be considered if the probability of this kind of failure is extremely remote.

41. *Turbine engine powerplant operating characteristics.* For turbopropeller powered airplanes, the turbine engine powerplant operating characteristics must be investigated in flight to determine that no adverse characteristics (such as stall, surge, or flameout) are present to a hazardous degree, during normal and emergency operation within the range of operating limitations of the airplane and of the engine.

42. *Fuel flow.* (a) For turbopropeller powered airplanes—

(1) The fuel system must provide for continuous supply of fuel to the engines for normal operation without interruption due to depletion of fuel in any tank other than the main tank; and

(2) The fuel flow rate for turbopropeller engine fuel pump systems must not be less than 125 percent of the fuel flow required to develop the standard sea level atmospheric conditions takeoff power selected and included as an operating limitation in the Airplane Flight Manual.

(b) For reciprocating engine powered airplanes, it is acceptable for the fuel flow rate for each pump system (main and reserve supply) to be 125 percent of the takeoff fuel consumption of the engine.

Fuel System Components

43. *Fuel pumps.* For turbopropeller powered airplanes, a reliable and independent power source must be provided for each pump used with turbine engines which do not have provisions for mechanically driving the main pumps. It must be demonstrated that the pump installations provide a reliability and durability equivalent to that provided by FAR 23.991(a).

44. *Fuel strainer or filter.* For turbopropeller powered airplanes, the following apply:

(a) There must be a fuel strainer or filter between the tank outlet and the fuel metering device of the engine. In addition, the fuel strainer or filter must be—

(1) Between the tank outlet and the engine-driven positive displacement pump inlet, if there is an engine-driven positive displacement pump;

(2) Accessible for drainage and cleaning and, for the strainer screen, easily removable; and

(3) Mounted so that its weight is not supported by the connecting lines or by the inlet or outlet connections of the strainer or filter itself.

(b) Unless there are means in the fuel system to prevent the accumulation of ice on the filter, there must be means to automatically maintain the fuel flow if ice-clogging of the filter occurs; and

(c) The fuel strainer or filter must be of adequate capacity (with respect to operating limitations established to insure proper service) and of appropriate mesh to insure proper engine operation, with the fuel contaminated to a degree (with respect to particle size and density) that can be reasonably expected in service. The degree of fuel filtering may not be less than that established for the engine type certification.

45. *Lightning strike protection.* Protection must be provided against the ignition of flammable vapors in the fuel vent system due to lightning strikes.

Cooling

46. *Cooling test procedures for turbopropeller powered airplanes.* (a) Turbopropeller powered airplanes must be shown to comply with the requirements of FAR 23.1041 during takeoff, climb en route, and landing stages of flight that correspond to the applicable performance requirements. The cooling test must be conducted with the airplane in the configuration and operating under the conditions that are critical relative to cooling during each stage of flight. For the cooling tests a temperature is "stabilized" when its rate of change is less than 2°F. per minute.

(b) Temperatures must be stabilized under the conditions from which entry is made into each stage of flight being investigated unless the entry condition is not one during which component and engine fluid temperatures would stabilize, in which case, operation through the full entry condition must be conducted before entry into the stage of flight being investigated in order to allow temperatures to reach their natural levels at the time of entry. The takeoff cooling test must be preceded by a period during which the powerplant component and engine fluid temperatures are stabilized with the engines at ground idle.

(c) Cooling tests for each stage of flight must be continued until—

(1) The component and engine fluid temperatures stabilize;

(2) The stage of flight is completed; or

(3) An operating limitation is reached.

Induction System

47. *Air induction.* For turbopropeller powered airplanes—

(a) There must be means to prevent hazardous quantities of fuel leakage or overflow from drains, vents, or other components of flammable fluid systems from entering the engine intake system; and

(b) The air inlet ducts must be located or protected so as to minimize the ingestion of foreign matter during takeoff, landing, and taxiing.

48. *Induction system icing protection.* For turbopropeller powered airplanes, each turbine engine must be able to operate throughout its flight power range without adverse effect on engine operation or serious loss of power or thrust, under the icing conditions specified in appendix C of FAR 25. In addition, there must be means to indicate to appropriate flight crewmembers the functioning of the powerplant ice protection system.

49. *Turbine engine bleed air systems.* Turbine engine bleed air systems of turbopropeller powered airplanes must be investigated to determine—

(a) That no hazard to the airplane will result if a duct rupture occurs. This condition must consider that a failure of the duct can occur anywhere between the engine port and the airplane bleed service; and

(b) That if the bleed air system is used for direct cabin pressurization, it is not possible for hazardous contamination of the cabin air system to occur in event of lubrication system failure.

Exhaust System

50. *Exhaust system drains.* Turbopropeller engine exhaust systems having low spots or pockets must incorporate drains at such locations. These drains must discharge clear of the airplane in normal and ground attitudes to prevent the accumulation of fuel after the failure of an attempted engine start.

Powerplant Controls and Accessories

51. *Engine controls.* If throttles or power levers for turbopropeller powered airplanes are such that any position of these controls will reduce the fuel flow to the engine(s) below that necessary for satisfactory and safe idle operation of the engine while the airplane is in flight, a means must be provided to prevent inadvertent movement of the control into this position. The means provided must incorporate a positive lock or stop at this idle position and must require a separate and distinct operation by the crew to displace the control from the normal engine operating range.

52. *Reverse thrust controls.* For turbopropeller powered airplanes, the propeller reverse thrust controls must have a means to prevent their inadvertent operation. The means must have a positive lock or stop at the idle position and must require a separate and distinct operation by the crew to displace the control from the flight regime.

53. *Engine ignition systems.* Each turbopropeller airplane ignition system must be considered an essential electrical load.

54. *Powerplant accessories.* The powerplant accessories must meet the requirements of FAR 23.1163, and if the continued rotation of any accessory remotely driven by the engine is hazardous when malfunctioning occurs, there must be means to prevent rotation without interfering with the continued operation of the engine.

Powerplant Fire Protection

55. *Fire detector system.* For turbopropeller powered airplanes, the following apply:

(a) There must be a means that ensures prompt detection of fire in the engine compartment. An overtemperature switch in each engine cooling air exit is an acceptable method of meeting this requirement.

(b) Each fire detector must be constructed and installed to withstand the vibration, inertia, and other loads to which it may be subjected in operation.

(c) No fire detector may be affected by any oil, water, other fluids, or fumes that might be present.

(d) There must be means to allow the flight crew to check, in flight, the functioning of each fire detector electric circuit.

(e) Wiring and other components of each fire detector system in a fire zone must be at least fire resistant.

56. *Fire protection, cowling and nacelle skin.* For reciprocating engine powered airplanes, the engine cowling must be designed and constructed so that no fire originating in the engine compartment can enter, either through openings or by burn through, any other region where it would create additional hazards.

57. *Flammable fluid fire protection.* If flammable fluids or vapors might be liberated by the leakage of fluid systems in areas other than engine compartments, there must be means to—

(a) Prevent the ignition of those fluids or vapors by any other equipment; or

(b) Control any fire resulting from that ignition.

Equipment

58. *Powerplant instruments.* (a) The following are required for turbopropeller airplanes:

(1) The instruments required by FAR 23.1305 (a)(1) through (4), (b)(2) and (4).

(2) A gas temperature indicator for each engine.

(3) Free air temperature indicator.

(4) A fuel flowmeter indicator for each engine.

(5) Oil pressure warning means for each engine.

(6) A torque indicator or adequate means for indicating power output for each engine.

(7) Fire warning indicator for each engine.

(8) A means to indicate when the propeller blade angle is below the low-pitch position corresponding to idle operation in flight.

(9) A means to indicate the functioning of the ice protection system for each engine.

(b) For turbopropeller powered airplanes, the turbopropeller blade position indicator must begin indicating when the blade has moved below the flight low-pitch position.

(c) The following instruments are required for reciprocating-engine powered airplanes:

(1) The instruments required by FAR 23.1305.

(2) A cylinder head temperature indicator for each engine.

(3) A manifold pressure indicator for each engine.

Systems and Equipments General

59. *Function and installation.* The systems and equipment of the airplane must meet the requirements of FAR 23.1301, and the following:

(a) Each item of additional installed equipment must—

(1) Be of a kind and design appropriate to its intended function;

(2) Be labeled as to its identification, function, or operating limitations, or any applicable combination of these factors, unless misuse or inadvertent actuation cannot create a hazard;

(3) Be installed according to limitations specified for that equipment; and

(4) Function properly when installed.

(b) Systems and installations must be designed to safeguard against hazards to the aircraft in the event of their malfunction or failure.

(c) Where an installation, the functioning of which is necessary in showing compliance with the applicable requirements, requires a power supply, such installation must be considered an essential load on the power supply, and the power sources and the distribution system must be capable of supplying the following power loads in probable operation combinations and for probable durations:

(1) All essential loads after failure of any prime mover, power converter, or energy storage device.

(2) All essential loads after failure of any one engine on two-engine airplanes.

(3) In determining the probable operating combinations and durations of essential loads for the power failure conditions described in subparagraphs (1) and (2) of this paragraph, it is permissible to assume that the power loads are reduced in accordance with a monitoring procedure which is consistent with safety in the types of operations authorized.

60. *Ventilation.* The ventilation system of the airplane must meet the requirements of FAR 23.831, and in addition, for pressurized aircraft the ventilating air in flight crew and passenger compartments must be free of harmful or hazardous concentrations of gases and vapors in normal operation and in the event of reasonably probable failures or malfunctioning of the ventilating, heating, pressurization, or other systems, and equipment. If accumulation of hazardous quantities of smoke in the cockpit area is reasonably probable, smoke evacuation must be readily accomplished.

Electrical Systems and Equipment

61. *General.* The electrical systems and equipment of the airplane must meet the requirements of FAR 23.1351, and the following:

(a) *Electrical system capacity.* The required generating capacity, and number and kinds of power sources must—

(1) Be determined by an electrical load analysis, and

(2) Meet the requirements of FAR 23.1301.

(b) *Generating system.* The generating system includes electrical power sources, main power busses, transmission cables, and associated control, regulation, and protective devices. It must be designed so that—

(1) The system voltage and frequency (as applicable) at the terminals of all essential load equipment can be maintained within the limits for which the equipment is designed, during any probable operating conditions;

(2) System transients due to switching, fault clearing, or other causes do not make essential loads inoperative, and do not cause a smoke or fire hazard;

(3) There are means, accessible in flight to appropriate crewmembers, for the individual and collective disconnection of the electrical power sources from the system; and

(4) There are means to indicate to appropriate crewmembers the generating system quantities essential for the safe operation of the system, including the voltage and current supplied by each generator.

62. *Electrical equipment and installation.* Electrical equipment controls, and wiring must be installed so that operation of any one unit or system of units will not adversely affect the simultaneous operation of to the safe operation.

63. *Distribution system.* (a) For the purpose of complying with this section, the distribution system includes the distribution busses, their associated feeders and each control and protective device.

(b) Each system must be designed so that essential load circuits can be supplied in the event of reasonably probable faults or open circuits, including faults in heavy current carrying cables.

(c) If two independent sources of electrical power for particular equipment or systems are required by this regulation, their electrical energy supply must be insured by means such as duplicate electrical equipment, throwover switching, or multichannel or loop circuits separately routed.

64. *Circuit protective devices.* The circuit protective devices for the electrical circuits of the airplane must meet the requirements of FAR 23.1357, and in addition circuits for loads which are essential to safe operation must have individual and exclusive circuit protection.

[Doc. No. 8070, 34 FR 189, Jan. 7, 1969, as amended by SFAR 23-1, 34 FR 20176, Dec. 24, 1969; 35 FR 1102, Jan. 28, 1970]

Subpart A—General

§ 23.1 Applicability.

(a) This part prescribes airworthiness standards for the issue of type certificates, and changes to those certificates, for airplanes in the normal, utility, acrobatic, and commuter categories.

(b) Each person who applies under Part 21 for such a certificate or change must show compliance with the applicable requirements of this part.

[Doc. No. 4080, 29 FR 17955, Dec. 18, 1964, as amended by Amdt. 23-34, 52 FR 1825, Jan. 15, 1987]

§ 23.2 Special retroactive requirements.

(a) Notwithstanding §§ 21.17 and 21.101 of this chapter and irrespective of the type certification basis, each normal, utility, and acrobatic category airplane having a passenger seating configuration, excluding pilot seats, of nine or less, manufactured after December 12, 1986, or any such foreign airplane for entry into the United States must provide a safety belt and shoulder harness for each forward- or aft-facing seat which will protect the occupant from serious head injury when subjected to the inertia loads resulting from the ultimate static load factors prescribed in § 23.561(b)(2) of this part, or which will provide the occupant protection specified in § 23.562 of this part when that section is applicable to the airplane. For other seat orientations, the seat/restraint system must be designed to provide a level of occupant protection equivalent to that provided for forward- or aft-facing seats with a safety belt and shoulder harness installed.

(b) Each shoulder harness installed at a flight crewmember station, as required by this section, must allow the crewmember, when seated with the safety belt and shoulder harness fastened, to perform all functions necessary for flight operations.

(c) For the purpose of this section, the date of manufacture is:

(1) The date the inspection acceptance records, or equivalent, reflect that the airplane is complete and meets the FAA approved type design data; or

(2) In the case of a foreign manufactured airplane, the date the foreign civil airworthiness authority certifies the airplane is complete and issues an original standard airworthiness certificate, or the equivalent in that country.

[Amdt. 23-36, 53 FR 30812, Aug. 15, 1988]

§ 23.3 Airplane categories.

(a) The normal category is limited to airplanes that have a seating configuration, excluding pilot seats, of nine or less, a maximum certificated takeoff weight of 12,500 pounds or less, and intended for nonacrobatic operation. Nonacrobatic operation includes:

(1) Any maneuver incident to normal flying;

(2) Stalls (except whip stalls); and

(3) Lazy eights, chandelles, and steep turns, in which the angle of bank is not more than 60 degrees.

(b) The utility category is limited to airplanes that have a seating configuration, excluding pilot seats, of nine or less, a maximum certificated takeoff weight of 12,500 pounds or less, and intended for limited acrobatic operation. Airplanes certificated in the utility category may be used in any of the operations covered under paragraph (a) of this section and in limited acrobatic operations. Limited acrobatic operation includes:

(1) Spins (if approved for the particular type of airplane); and

(2) Lazy eights, chandelles, and steep turns, or similar maneuvers, in which the angle of bank is more than 60 degrees but not more than 90 degrees.

(c) The acrobatic category is limited to airplanes that have a seating configuration, excluding pilot seats, of nine or less, a maximum certificated takeoff weight of 12,500 pounds or less, and intended for use without restrictions, other than those shown to be necessary as a result of required flight tests.

(d) The commuter category is limited to multiengine airplanes that have a seating configuration, excluding pilot seats, of 19 or less, and a maximum certificated takeoff weight of 19,000 pounds or less. The commuter category operation is limited to any maneuver incident to normal flying, stalls (except whip stalls), and steep turns, in which the angle of bank is not more than 60 degrees.

(e) Except for commuter category, airplanes may be type certificated in more than one category if the requirements of each requested category are met.

[Doc. No. 4080, 29 FR 17955, Dec. 18, 1964, as amended by Amdt. 23-4, 32 FR 5934, Apr. 14, 1967; Amdt. 23-34, 52 FR 1825, Jan. 15, 1987; 52 FR 34745, Sept. 14, 1987; Amdt. 23-50, 61 FR 5183, Feb. 9, 1996; Amdt. 23-62, 76 FR 75753, Dec. 2, 2011]

Subpart B—Flight

General

§ 23.21 Proof of compliance.

(a) Each requirement of this subpart must be met at each appropriate combination of weight and center of gravity within the range of loading conditions for which certification is requested. This must be shown—

(1) By tests upon an airplane of the type for which certification is requested, or by calculations based on, and equal in accuracy to, the results of testing; and

(2) By systematic investigation of each probable combination of weight and center of gravity, if compliance cannot be reasonably inferred from combinations investigated.

(b) The following general tolerances are allowed during flight testing. However, greater tolerances may be allowed in particular tests:

Item	Tolerance
Weight	+5%, −10%.
Critical items affected by weight	+5%, −1%.
C.G	±7% total travel.

§ 23.23 Load distribution limits.

(a) Ranges of weights and centers of gravity within which the airplane may be safely operated must be established. If a weight and center of gravity combination is allowable only within certain lateral load distribution limits that could be inadvertently exceeded, these limits must be established for the corresponding weight and center of gravity combinations.

(b) The load distribution limits may not exceed any of the following:

(1) The selected limits;

(2) The limits at which the structure is proven; or

(3) The limits at which compliance with each applicable flight requirement of this subpart is shown.

[Doc. No. 26269, 58 FR 42156, Aug. 6, 1993]

§ 23.25 Weight limits.

(a) *Maximum weight.* The maximum weight is the highest weight at which compliance with each applicable requirement of this part (other than those complied with at the design landing weight) is shown. The maximum weight must be established so that it is—

(1) Not more than the least of—

(i) The highest weight selected by the applicant; or

(ii) The design maximum weight, which is the highest weight at which compliance with each applicable structural loading condition of this part (other than those complied with at the design landing weight) is shown; or

(iii) The highest weight at which compliance with each applicable flight requirement is shown, and

(2) Not less than the weight with—

(i) Each seat occupied, assuming a weight of 170 pounds for each occupant for normal and commuter category airplanes, and 190 pounds for utility and acrobatic category airplanes, except that seats other than pilot seats may be placarded for a lesser weight; and

(A) Oil at full capacity, and

(B) At least enough fuel for maximum continuous power operation of at least 30 minutes for day-VFR approved airplanes and at least 45 minutes for night-VFR and IFR approved airplanes; or

(ii) The required minimum crew, and fuel and oil to full tank capacity.

(b) *Minimum weight.* The minimum weight (the lowest weight at which compliance with each applicable requirement of this part is shown) must be established so that it is not more than the sum of—

(1) The empty weight determined under § 23.29;

(2) The weight of the required minimum crew (assuming a weight of 170 pounds for each crewmember); and

(3) The weight of—

(i) For turbojet powered airplanes, 5 percent of the total fuel capacity of that particular fuel tank arrangement under investigation, and

(ii) For other airplanes, the fuel necessary for one-half hour of operation at maximum continuous power.

[Doc. No. 4080, 29 FR 17955, Dec. 18, 1964, as amended by Amdt. 23-7, 34 FR 13086, Aug. 13, 1969; Amdt. 23-21, 43 FR 2317, Jan. 16, 1978; Amdt. 23-34, 52 FR 1825, Jan. 15, 1987; Amdt. 23-45, 58 FR 42156, Aug. 6, 1993; Amdt. 23-50, 61 FR 5183, Feb. 9, 1996]

§ 23.29 Empty weight and corresponding center of gravity.

(a) The empty weight and corresponding center of gravity must be determined by weighing the airplane with—

(1) Fixed ballast;

(2) Unusable fuel determined under § 23.959; and

(3) Full operating fluids, including—

(i) Oil;

(ii) Hydraulic fluid; and

(iii) Other fluids required for normal operation of airplane systems, except potable water, lavatory precharge water, and water intended for injection in the engines.

(b) The condition of the airplane at the time of determining empty weight must be one that is well defined and can be easily repeated.

[Doc. No. 4080, 29 FR 17955, Dec. 18, 1964; 30 FR 258, Jan. 9, 1965, as amended by Amdt. 23-21, 43 FR 2317, Jan. 16, 1978]

§ 23.31 Removable ballast.

Removable ballast may be used in showing compliance with the flight requirements of this subpart, if—

(a) The place for carrying ballast is properly designed and installed, and is marked under § 23.1557; and

(b) Instructions are included in the airplane flight manual, approved manual material, or markings and placards, for the proper placement of the removable ballast under each loading condition for which removable ballast is necessary.

[Doc. No. 4080, 29 FR 17955, Dec. 18, 1964; 30 FR 258, Jan. 9, 1965, as amended by Amdt. 23-13, 37 FR 20023, Sept. 23, 1972]

§ 23.33 Propeller speed and pitch limits.

(a) *General.* The propeller speed and pitch must be limited to values that will assure safe operation under normal operating conditions.

(b) *Propellers not controllable in flight.* For each propeller whose pitch cannot be controlled in flight—

(1) During takeoff and initial climb at the all engine(s) operating climb speed specified in § 23.65, the propeller must limit the engine r.p.m., at full throttle or at maximum

allowable takeoff manifold pressure, to a speed not greater than the maximum allowable takeoff r.p.m.; and

(2) During a closed throttle glide, at V_{NE}, the propeller may not cause an engine speed above 110 percent of maximum continuous speed.

(c) *Controllable pitch propellers without constant speed controls.* Each propeller that can be controlled in flight, but that does not have constant speed controls, must have a means to limit the pitch range so that—

(1) The lowest possible pitch allows compliance with paragraph (b)(1) of this section; and

(2) The highest possible pitch allows compliance with paragraph (b)(2) of this section.

(d) *Controllable pitch propellers with constant speed controls.* Each controllable pitch propeller with constant speed controls must have—

(1) With the governor in operation, a means at the governor to limit the maximum engine speed to the maximum allowable takeoff r.p.m.; and

(2) With the governor inoperative, the propeller blades at the lowest possible pitch, with takeoff power, the airplane stationary, and no wind, either—

(i) A means to limit the maximum engine speed to 103 percent of the maximum allowable takeoff r.p.m., or

(ii) For an engine with an approved overspeed, a means to limit the maximum engine and propeller speed to not more than the maximum approved overspeed.

[Doc. No. 4080, 29 FR 17955, Dec. 18, 1964, as amended by Amdt. 23-45, 58 FR 42156, Aug. 6, 1993; Amdt. 23-50, 61 FR 5183, Feb. 9, 1996]

PERFORMANCE

§ 23.45 General.

(a) Unless otherwise prescribed, the performance requirements of this part must be met for—

(1) Still air and standard atmosphere; and

(2) Ambient atmospheric conditions, for commuter category airplanes, for reciprocating engine-powered airplanes of more than 6,000 pounds maximum weight, and for turbine engine-powered airplanes.

(b) Performance data must be determined over not less than the following ranges of conditions—

(1) Airport altitudes from sea level to 10,000 feet; and

(2) For reciprocating engine-powered airplanes of 6,000 pounds, or less, maximum weight, temperature from standard to 30°C above standard; or

(3) For reciprocating engine-powered airplanes of more than 6,000 pounds maximum weight and turbine engine-powered airplanes, temperature from standard to 30°C above standard, or the maximum ambient atmospheric temperature at which compliance with the cooling provisions of § 23.1041 to § 23.1047 is shown, if lower.

(c) Performance data must be determined with the cowl flaps or other means for controlling the engine cooling air supply in the position used in the cooling tests required by §§ 23.1041 to 23.1047.

(d) The available propulsive thrust must correspond to engine power, not exceeding the approved power, less—

(1) Installation losses; and

(2) The power absorbed by the accessories and services appropriate to the particular ambient atmospheric conditions and the particular flight condition.

(e) The performance, as affected by engine power or thrust, must be based on a relative humidity:

(1) Of 80 percent at and below standard temperature; and

(2) From 80 percent, at the standard temperature, varying linearly down to 34 percent at the standard temperature plus 50°F.

(f) Unless otherwise prescribed, in determining the takeoff and landing distances, changes in the airplane's configuration, speed, and power must be made in accordance with procedures established by the applicant for operation in service. These procedures must be able to be executed consistently by pilots of average skill in atmospheric conditions reasonably expected to be encountered in service.

(g) The following, as applicable, must be determined on a smooth, dry, hard-surfaced runway—

(1) Takeoff distance of § 23.53(b);

(2) Accelerate-stop distance of § 23.55;

(3) Takeoff distance and takeoff run of § 23.59; and

(4) Landing distance of § 23.75.

Note: The effect on these distances of operation on other types of surfaces (for example, grass, gravel) when dry, may be determined or derived and these surfaces listed in the Airplane Flight Manual in accordance with § 23.1583(p).

(h) For multiengine jets weighing over 6,000 pounds in the normal, utility, and acrobatic category and commuter category airplanes, the following also apply:

(1) Unless otherwise prescribed, the applicant must select the takeoff, enroute, approach, and landing configurations for the airplane.

(2) The airplane configuration may vary with weight, altitude, and temperature, to the extent that they are compatible with the operating procedures required by paragraph (h)(3) of this section.

(3) Unless otherwise prescribed, in determining the critical-engine-inoperative takeoff performance, takeoff flight path, and accelerate-stop distance, changes in the airplane's configuration, speed, and power must be made in accordance with procedures established by the applicant for operation in service.

(4) Procedures for the execution of discontinued approaches and balked landings associated with the conditions prescribed in § 23.67(c)(4) and § 23.77(c) must be established.

(5) The procedures established under paragraphs (h)(3) and (h)(4) of this section must—

(i) Be able to be consistently executed by a crew of average skill in atmospheric conditions reasonably expected to be encountered in service;

(ii) Use methods or devices that are safe and reliable; and

(iii) Include allowance for any reasonably expected time delays in the execution of the procedures.

[Doc. No. 27807, 61 FR 5184, Feb. 9, 1996, as amended by Amdt. 23-62, 76 FR 75753, Dec. 2, 2011]

§ 23.49 Stalling speed.

(a) V_{SO} (maximum landing flap configuration) and V_{S1} are the stalling speeds or the minimum steady flight speeds, in knots (CAS), at which the airplane is controllable with—

(1) For reciprocating engine-powered airplanes, the engine(s) idling, the throttle(s) closed or at not more than the power necessary for zero thrust at a speed not more than 110 percent of the stalling speed;

(2) For turbine engine-powered airplanes, the propulsive thrust not greater than zero at the stalling speed, or, if the resultant thrust has no appreciable effect on the stalling speed, with engine(s) idling and throttle(s) closed;

(3) The propeller(s) in the takeoff position;

(4) The airplane in the condition existing in the test, in which V_{SO} and V_{S1} are being used;

(5) The center of gravity in the position that results in the highest value of V_{SO} and V_{S1}; and

(6) The weight used when V_{SO} and V_{S1} are being used as a factor to determine compliance with a required performance standard.

(b) V_{SO} and V_{S1} must be determined by flight tests, using the procedure and meeting the flight characteristics specified in § 23.201.

(c) Except as provided in paragraph (d) of this section, V_{SO} at maximum weight may not exceed 61 knots for—

(1) Single-engine airplanes; and

(2) Multiengine airplanes of 6,000 pounds or less maximum weight that cannot meet the minimum rate of climb specified in § 23.67(a) (1) with the critical engine inoperative.

(d) All single-engine airplanes, and those multiengine airplanes of 6,000 pounds or less maximum weight with a V_{SO} of more than 61 knots that do not meet the requirements of § 23.67(a)(1), must comply with § 23.562(d).

[Doc. No. 27807, 61 FR 5184, Feb. 9, 1996, as amended by Amdt. 23-62, 76 FR 75753, Dec. 2, 2011]

§ 23.51 Takeoff speeds.

(a) For normal, utility, and acrobatic category airplanes, rotation speed, V_R, is the speed at which the pilot makes a control input, with the intention of lifting the airplane out of contact with the runway or water surface.

(1) For multiengine landplanes, V_R, must not be less than the greater of 1.05 V_{MC}; or 1.10 V_{S1};

(2) For single-engine landplanes, V_R, must not be less than V_{S1}; and

(3) For seaplanes and amphibians taking off from water, V_R, may be any speed that is shown to be safe under all reasonably expected conditions, including turbulence and complete failure of the critical engine.

(b) For normal, utility, and acrobatic category airplanes, the speed at 50 feet above the takeoff surface level must not be less than:

(1) For multiengine airplanes, the highest of—

(i) A speed that is shown to be safe for continued flight (or emergency landing, if applicable) under all reasonably expected conditions, including turbulence and complete failure of the critical engine;

(ii) 1.10 V_{MC}; or

(iii) 1.20 V_{S1}.

(2) For single-engine airplanes, the higher of—

(i) A speed that is shown to be safe under all reasonably expected conditions, including turbulence and complete engine failure; or

(ii) 1.20 V_{S1}.

(c) For normal, utility, and acrobatic category multiengine jets of more than 6,000 pounds maximum weight and commuter category airplanes, the following apply:

(1) V_1 must be established in relation to V_{EF} as follows:

(i) V_{EF} is the calibrated airspeed at which the critical engine is assumed to fail. V_{EF} must be selected by the applicant but must not be less than 1.05 V_{MC} determined under § 23.149(b) or, at the option of the applicant, not less than V_{MCG} determined under § 23.149(f).

(ii) The takeoff decision speed, V_1, is the calibrated airspeed on the ground at which, as a result of engine failure or other reasons, the pilot is assumed to have made a decision to continue or discontinue the takeoff. The takeoff decision speed, V_1, must be selected by the applicant but must not be less than V_{EF} plus the speed gained with the critical engine inoperative during the time interval between the instant at which the critical engine is failed and the instant at which the pilot recognizes and reacts to the engine failure, as indicated by the pilot's application of the first retarding means during the accelerate-stop determination of § 23.55.

(2) The rotation speed, V_R, in terms of calibrated airspeed, must be selected by the applicant and must not be less than the greatest of the following:

(i) V_1;

(ii) 1.05 V_{MC} determined under § 23.149(b);

(iii) 1.10 V_{S1}; or

(iv) The speed that allows attaining the initial climb-out speed, V_2, before reaching a height of 35 feet above the takeoff surface in accordance with § 23.57(c)(2).

(3) For any given set of conditions, such as weight, altitude, temperature, and configuration, a single value of V_R must be used to show compliance with both the one-engine-inoperative takeoff and all-engines-operating takeoff requirements.

(4) The takeoff safety speed, V_2, in terms of calibrated airspeed, must be selected by the applicant so as to allow the gradient of climb required in § 23.67 (c)(1) and (c)(2) but mut not be less than 1.10 V_{MC} or less than 1.20 V_{S1}.

(5) The one-engine-inoperative takeoff distance, using a normal rotation rate at a speed 5 knots less than V_R, established in accordance with paragraph (c)(2) of this section, must be shown not to exceed the corresponding one-engine-inoperative takeoff distance, determined in accordance with § 23.57 and § 23.59(a)(1), using the established V. The takeoff, otherwise performed in accordance with § 23.57, must be continued safely from the point at which the airplane is 35 feet above the takeoff surface and at a speed not less than the established V minus 5 knots.

(6) The applicant must show, with all engines operating, that marked increases in the scheduled takeoff distances, determined in accordance with § 23.59(a)(2), do not result from over-rotation of the airplane or out-of-trim conditions.

[Doc. No. 27807, 61 FR 5184, Feb. 9, 1996, as amended by Amdt. 23-62, 76 FR 75753, Dec. 2, 2011]

§ 23.53 Takeoff performance.

(a) For normal, utility, and acrobatic category airplanes, the takeoff distance must be determined in accordance with paragraph (b) of this section, using speeds determined in accordance with § 23.51 (a) and (b).

(b) For normal, utility, and acrobatic category airplanes, the distance required to takeoff and climb to a height of 50 feet above the takeoff surface must be determined for each weight, altitude, and temperature within the operational limits established for takeoff with—

(1) Takeoff power on each engine;

(2) Wing flaps in the takeoff position(s); and

(3) Landing gear extended.

(c) For normal, utility, and acrobatic category multiengine jets of more than 6,000 pounds maximum weight and commuter category airplanes, takeoff performance, as required by §§ 23.55 through 23.59, must be determined with the operating engine(s) within approved operating limitations.

[Doc. No. 27807, 61 FR 5185, Feb. 9, 1996, as amended by Amdt. 23-62, 76 FR 75753, Dec. 2, 2011]

§ 23.55 Accelerate-stop distance.

For normal, utility, and acrobatic category multiengine jets of more than 6,000 pounds maximum weight and commuter category airplanes, the accelerate-stop distance must be determined as follows:

(a) The accelerate-stop distance is the sum of the distances necessary to—

(1) Accelerate the airplane from a standing start to V_{EF} with all engines operating;

(2) Accelerate the airplane from V_{EF} to V_1, assuming the critical engine fails at V_{EF}; and

(3) Come to a full stop from the point at which V_1 is reached.

(b) Means other than wheel brakes may be used to determine the accelerate-stop distances if that means—

(1) Is safe and reliable;

(2) Is used so that consistent results can be expected under normal operating conditions; and

(3) Is such that exceptional skill is not required to control the airplane.

[Amdt. 23-34, 52 FR 1826, Jan. 15, 1987, as amended by Amdt. 23-50, 61 FR 5185, Feb. 9, 1996, as amended by Amdt. 23-62, 76 FR 75753, Dec. 2, 2011]

§ 23.57 Takeoff path.

For normal, utility, and acrobatic category multiengine jets of more than 6,000 pounds maximum weight and commuter category airplanes, the takeoff path is as follows:

(a) The takeoff path extends from a standing start to a point in the takeoff at which the airplane is 1500 feet above the takeoff surface at or below which height the transition from the takeoff to the enroute configuration must be completed; and

(1) The takeoff path must be based on the procedures prescribed in § 23.45;

(2) The airplane must be accelerated on the ground to V_{EF} at which point the critical engine must be made inoperative and remain inoperative for the rest of the takeoff; and

(3) After reaching V_{EF}, the airplane must be accelerated to V_2.

(b) During the acceleration to speed V_2, the nose gear may be raised off the ground at a speed not less than V_R. However, landing gear retraction must not be initiated until the airplane is airborne.

(c) During the takeoff path determination, in accordance with paragraphs (a) and (b) of this section—

(1) The slope of the airborne part of the takeoff path must not be negative at any point;

(2) The airplane must reach V_2 before it is 35 feet above the takeoff surface, and must continue at a speed as close as practical to, but not less than V_2, until it is 400 feet above the takeoff surface;

(3) At each point along the takeoff path, starting at the point at which the airplane reaches 400 feet above the takeoff surface, the available gradient of climb must not be less than—

(i) 1.2 percent for two-engine airplanes;

(ii) 1.5 percent for three-engine airplanes;

(iii) 1.7 percent for four-engine airplanes; and

(4) Except for gear retraction and automatic propeller feathering, the airplane configuration must not be changed, and no change in power that requires action by the pilot may be made, until the airplane is 400 feet above the takeoff surface.

(d) The takeoff path to 35 feet above the takeoff surface must be determined by a continuous demonstrated takeoff.

(e) The takeoff path to 35 feet above the takeoff surface must be determined by synthesis from segments; and

(1) The segments must be clearly defined and must be related to distinct changes in configuration, power, and speed;

(2) The weight of the airplane, the configuration, and the power must be assumed constant throughout each segment and must correspond to the most critical condition prevailing in the segment; and

(3) The takeoff flight path must be based on the airplane's performance without utilizing ground effect.

[Amdt. 23-34, 52 FR 1827, Jan. 15, 1987, as amended by Amdt. 23-50, 61 FR 5185, Feb. 9, 1996, as amended by Amdt. 23-62, 76 FR 75753, Dec. 2, 2011]

§ 23.59 Takeoff distance and takeoff run.

For normal, utility, and acrobatic category multiengine jets of more than 6,000 pounds maximum weight and commuter category airplanes, the takeoff distance and, at the option of the applicant, the takeoff run, must be determined.

(a) Takeoff distance is the greater of—

(1) The horizontal distance along the takeoff path from the start of the takeoff to the point at which the airplane is 35 feet above the takeoff surface as determined under § 23.57; or

(2) With all engines operating, 115 percent of the horizontal distance from the start of the takeoff to the point at which the airplane is 35 feet above the takeoff surface, determined by a procedure consistent with § 23.57.

(b) If the takeoff distance includes a clearway, the takeoff run is the greater of—

(1) The horizontal distance along the takeoff path from the start of the takeoff to a point equidistant between the liftoff point and the point at which the airplane is 35 feet above the takeoff surface as determined under § 23.57; or

(2) With all engines operating, 115 percent of the horizontal distance from the start of the takeoff to a point equidistant between the liftoff point and the point at which the airplane is 35 feet above the takeoff surface, determined by a procedure consistent with § 23.57.

[Amdt. 23-34, 52 FR 1827, Jan. 15, 1987, as amended by Amdt. 23-50, 61 FR 5185, Feb. 9, 1996, as amended by Amdt. 23-62, 76 FR 75753, Dec. 2, 2011]

§ 23.61 Takeoff flight path.

For normal, utility, and acrobatic category multiengine jets of more than 6,000 pounds maximum weight and commuter category airplanes, the takeoff flight path must be determined as follows:

(a) The takeoff flight path begins 35 feet above the takeoff surface at the end of the takeoff distance determined in accordance with § 23.59.

(b) The net takeoff flight path data must be determined so that they represent the actual takeoff flight paths, as determined in accordance with § 23.57 and with paragraph (a) of this section, reduced at each point by a gradient of climb equal to—

(1) 0.8 percent for two-engine airplanes;

(2) 0.9 percent for three-engine airplanes; and

(3) 1.0 percent for four-engine airplanes.

(c) The prescribed reduction in climb gradient may be applied as an equivalent reduction in acceleration along that part of the takeoff flight path at which the airplane is accelerated in level flight.

[Amdt. 23-34, 52 FR 1827, Jan. 15, 1987, as amended by Amdt. 23-62, 76 FR 75753, Dec. 2, 2011]

§ 23.63 Climb: General.

(a) Compliance with the requirements of §§ 23.65, 23.66, 23.67, 23.69, and 23.77 must be shown—

(1) Out of ground effect; and

(2) At speeds that are not less than those at which compliance with the powerplant cooling requirements of §§ 23.1041 to 23.1047 has been demonstrated; and

(3) Unless otherwise specified, with one engine inoperative, at a bank angle not exceeding 5 degrees.

(b) For normal, utility, and acrobatic category reciprocating engine-powered airplanes of 6,000 pounds or less maximum weight, compliance must be shown with § 23.65(a), § 23.67(a), where appropriate, and § 23.77(a) at maximum takeoff or landing weight, as appropriate, in a standard atmosphere.

(c) For reciprocating engine-powered airplanes of more than 6,000 pounds maximum weight, single-engine turbines, and multiengine turbine airplanes of 6,000 pounds or less maximum weight in the normal, utility, and acrobatic category, compliance must be shown at weights as a function of airport altitude and ambient temperature, within the operational limits established for takeoff and landing, respectively, with—

(1) Sections 23.65(b) and 23.67(b) (1) and (2), where appropriate, for takeoff, and

(2) Section 23.67(b)(2), where appropriate, and § 23.77(b), for landing.

(d) For multiengine turbine airplanes over 6,000 pounds maximum weight in the normal, utility, and acrobatic category and commuter category airplanes, compliance must be shown at weights as a function of airport altitude and ambient temperature within the operational limits established for takeoff and landing, respectively, with—

(1) Sections 23.67(c)(1), 23.67(c)(2), and 23.67(c)(3) for takeoff; and

(2) Sections 23.67(c)(3), 23.67(c)(4), and 23.77(c) for landing.

[Doc. No. 27807, 61 FR 5186, Feb. 9, 1996, as amended by Amdt. 23-62, 76 FR 75753, Dec. 2, 2011]

§ 23.65 Climb: All engines operating.

(a) Each normal, utility, and acrobatic category reciprocating engine-powered airplane of 6,000 pounds or less maximum weight must have a steady climb gradient at sea level of at least 8.3 percent for landplanes or 6.7 percent for seaplanes and amphibians with—

(1) Not more than maximum continuous power on each engine;

(2) The landing gear retracted;

(3) The wing flaps in the takeoff position(s); and

(4) A climb speed not less than the greater of 1.1 V_{MC} and 1.2 V_{S1} for multiengine airplanes and not less than 1.2 V_{S1} for single—engine airplanes.

(b) Each normal, utility, and acrobatic category reciprocating engine-powered airplane of more than 6,000 pounds maximum weight, single-engine turbine, and multiengine turbine airplanes of 6,000 pounds or less maximum weight in the normal, utility, and acrobatic category must have a steady gradient of climb after takeoff of at least 4 percent with

(1) Take off power on each engine;

(2) The landing gear extended, except that if the landing gear can be retracted in not more than seven seconds, the test may be conducted with the gear retracted;

(3) The wing flaps in the takeoff position(s); and

(4) A climb speed as specified in § 23.65(a)(4).

[Doc. No. 27807, 61 FR 5186, Feb. 9, 1996, as amended by Amdt. 23-62, 76 FR 75753, Dec. 2, 2011]

§ 23.66 Takeoff climb: One-engine inoperative.

For normal, utility, and acrobatic category reciprocating engine-powered airplanes of more than 6,000 pounds maximum weight, and turbine engine-powered airplanes in the normal, utility, and acrobatic category, the steady gradient of climb or descent must be determined at each weight, altitude, and ambient temperature within the operational limits established by the applicant with—

(a) The critical engine inoperative and its propeller in the position it rapidly and automatically assumes;

(b) The remaining engine(s) at takeoff power;

(c) The landing gear extended, except that if the landing gear can be retracted in not more than seven seconds, the test may be conducted with the gear retracted;

(d) The wing flaps in the takeoff position(s):

(e) The wings level; and

(f) A climb speed equal to that achieved at 50 feet in the demonstration of § 23.53.

[Doc. No. 27807, 61 FR 5186, Feb. 9, 1996]

§ 23.67 Climb: One engine inoperative.

(a) For normal, utility, and acrobatic category reciprocating engine-powered airplanes of 6,000 pounds or less maximum weight, the following apply:

(1) Except for those airplanes that meet the requirements prescribed in § 23.562(d), each airplane with a V_{SO} of more than 61 knots must be able to maintain a steady climb gradient of at least 1.5 percent at a pressure altitude of 5,000 feet with the—

(i) Critical engine inoperative and its propeller in the minimum drag position;

(ii) Remaining engine(s) at not more than maximum continuous power;

(iii) Landing gear retracted;

(iv) Wing flaps retracted; and

(v) Climb speed not less than 1.2 V_{S1}.

(2) For each airplane that meets the requirements prescribed in § 23.562(d), or that has a V_{SO} of 61 knots or less, the steady gradient of climb or descent at a pressure altitude of 5,000 feet must be determined with the—

(i) Critical engine inoperative and its propeller in the minimum drag position;

(ii) Remaining engine(s) at not more than maximum continuous power;

(iii) Landing gear retracted;

(iv) Wing flaps retracted; and

(v) Climb speed not less than 1.2 V_{S1}.

(b) For normal, utility, and acrobatic category reciprocating engine-powered airplanes of more than 6,000 pounds maximum weight, and turbopropeller-powered airplanes in the normal, utility, and acrobatic category—

(1) The steady gradient of climb at an altitude of 400 feet above the takeoff must be no less than 1 percent with the—

(i) Critical engine inoperative and its propeller in the minimum drag position;

(ii) Remaining engine(s) at takeoff power;

(iii) Landing gear retracted;

(iv) Wing flaps in the takeoff position(s); and

(v) Climb speed equal to that achieved at 50 feet in the demonstration of § 23.53.

(2) The steady gradient of climb must not be less than 0.75 percent at an altitude of 1,500 feet above the takeoff surface, or landing surface, as appropriate, with the—

(i) Critical engine inoperative and its propeller in the minimum drag position;

(ii) Remaining engine(s) at not more than maximum continuous power;

(iii) Landing gear retracted;

(iv) Wing flaps retracted; and

(v) Climb speed not less than 1.2 V_{S1}.

(c) For normal, utility, and acrobatic category jets of 6,000 pounds or less maximum weight—

(1) The steady gradient of climb at an altitude of 400 feet above the takeoff must be no less than 1.2 percent with the—

(i) Critical engine inoperative;

(ii) Remaining engine(s) at takeoff power;

(iii) Landing gear retracted;

(iv) Wing flaps in the takeoff position(s); and

(v) Climb speed equal to that achieved at 50 feet in the demonstration of § 23.53.

(2) The steady gradient of climb may not be less than 0.75 percent at an altitude of 1,500 feet above the takeoff surface, or landing surface, as appropriate, with the—

(i) Critical engine inoperative;

(ii) Remaining engine(s) at not more than maximum continuous power;

(iii) Landing gear retracted;

(iv) Wing flaps retracted; and

(v) Climb speed not less than 1.2 V_{S1}.

(d) For jets over 6,000 pounds maximum weight in the normal, utility and acrobatic category and commuter category airplanes, the following apply:

(1) *Takeoff; landing gear extended.* The steady gradient of climb at the altitude of the takeoff surface must be measurably positive for two-engine airplanes, not less than 0.3 percent for three-engine airplanes, or 0.5 percent for four-engine airplanes with—

(i) The critical engine inoperative and its propeller in the position it rapidly and automatically assumes;

(ii) The remaining engine(s) at takeoff power;

(iii) The landing gear extended, and all landing gear doors open;

(iv) The wing flaps in the takeoff position(s);

(v) The wings level; and

(vi) A climb speed equal to V_2.

(2) *Takeoff; landing gear retracted.* The steady gradient of climb at an altitude of 400 feet above the takeoff surface must be not less than 2.0 percent of two-engine airplanes, 2.3 percent for three-engine airplanes, and 2.6 percent for four-engine airplanes with—

(i) The critical engine inoperative and its propeller in the position it rapidly and automatically assumes;

(ii) The remaining engine(s) at takeoff power;

(iii) The landing gear retracted;

(iv) The wing flaps in the takeoff position(s);

(v) A climb speed equal to V_2.

(3) *Enroute.* The steady gradient of climb at an altitude of 1,500 feet above the takeoff or landing surface, as appropriate, must be not less than 1.2 percent for two-engine airplanes, 1.5 percent for three-engine airplanes, and 1.7 percent for four-engine airplanes with—

(i) The critical engine inoperative and its propeller in the minimum drag position;

(ii) The remaining engine(s) at not more than maximum continuous power;

(iii) The landing gear retracted;

(iv) The wing flaps retracted; and

(v) A climb speed not less than 1.2 V_{S1}.

(4) *Discontinued approach.* The steady gradient of climb at an altitude of 400 feet above the landing surface must be not less than 2.1 percent for two-engine airplanes, 2.4 percent for three-engine airplanes, and 2.7 percent for four-engine airplanes, with—

(i) The critical engine inoperative and its propeller in the minimum drag position;

(ii) The remaining engine(s) at takeoff power;

(iii) Landing gear retracted;

(iv) Wing flaps in the approach position(s) in which V_{S1} for these position(s) does not exceed 110 percent of the V_{S1} for the related all-engines-operated landing position(s); and

(v) A climb speed established in connection with normal landing procedures but not exceeding 1.5 V_{S1}.

[Doc. No. 27807, 61 FR 5186, Feb. 9, 1996, as amended by Amdt. 23-62, 76 FR 75754, Dec. 2, 2011]

§ 23.69 Enroute climb/descent.

(a) *All engines operating.* The steady gradient and rate of climb must be determined at each weight, altitude, and ambient temperature within the operational limits established by the applicant with—

(1) Not more than maximum continuous power on each engine;

(2) The landing gear retracted;

(3) The wing flaps retracted; and

(4) A climb speed not less than 1.3 V_{S1}.

(b) *One engine inoperative.* The steady gradient and rate of climb/descent must be determined at each weight, altitude, and ambient temperature within the operational limits established by the applicant with—

(1) The critical engine inoperative and its propeller in the minimum drag position;

(2) The remaining engine(s) at not more than maximum continuous power;

(3) The landing gear retracted;

(4) The wing flaps retracted; and

(5) A climb speed not less than 1.2 V_{S1}.

[Doc. No. 27807, 61 FR 5187, Feb. 9, 1996]

§ 23.71 Glide: Single-engine airplanes.

The maximum horizontal distance traveled in still air, in nautical miles, per 1,000 feet of altitude lost in a glide, and the speed necessary to achieve this must be determined with the engine inoperative, its propeller in the minimum drag position, and landing gear and wing flaps in the most favorable available position.

[Doc. No. 27807, 61 FR 5187, Feb. 9, 1996]

§ 23.73 Reference landing approach speed.

(a) For normal, utility, and acrobatic category reciprocating engine-powered airplanes of 6,000 pounds or less maximum weight, the reference landing approach speed, V_{REF}, may not be less than the greater of V_{MC}, determined in § 23.149(b) with the wing flaps in the most extended takeoff position, and 1.3 V_{S1}.

(b) For normal, utility, and acrobatic category turbine powered airplanes of 6,000 pounds or less maximum weight, turboprops of more than 6,000 pounds maximum weight, and reciprocating engine-powered airplanes of more than 6,000 pounds maximum weight, the reference landing approach speed, V_{REF}, may not be less than the greater of V_{MC}, determined in § 23.149(c), and 1.3 V_{S1}.

(c) For normal, utility, and acrobatic category jets of more than 6,000 pounds maximum weight and commuter category airplanes, the reference landing approach speed, V_{REF}, may not be less than the greater of 1.05 V_{MC}, determined in § 23.149(c), and 1.3 V_{S1}.

[Amdt. 23-62, 76 FR 75754, Dec. 2, 2011]

§ 23.75 Landing distance.

The horizontal distance necessary to land and come to a complete stop from a point 50 feet above the landing surface must be determined, for standard temperatures at each weight and altitude within the operational limits established for landing, as follows:

(a) A steady approach at not less than V_{REF}, determined in accordance with § 23.73 (a), (b), or (c), as appropriate, must be maintained down to the 50 foot height and—

(1) The steady approach must be at a gradient of descent not greater than 5.2 percent (3 degrees) down to the 50-foot height.

(2) In addition, an applicant may demonstrate by tests that a maximum steady approach gradient steeper than 5.2 percent, down to the 50-foot height, is safe. The gradient must be established as an operating limitation and the information necessary to display the gradient must be available to the pilot by an appropriate instrument.

(b) A constant configuration must be maintained throughout the maneuver.

(c) The landing must be made without excessive vertical acceleration or tendency to bounce, nose over, ground loop, porpoise, or water loop.

(d) It must be shown that a safe transition to the balked landing conditions of § 23.77 can be made from the conditions that exist at the 50 foot height, at maximum landing weight, or at the maximum landing weight for altitude and temperature of § 23.63 (c)(2) or (d)(2), as appropriate.

(e) The brakes must be used so as to not cause excessive wear of brakes or tires.

(f) Retardation means other than wheel brakes may be used if that means—

(1) Is safe and reliable; and

(2) Is used so that consistent results can be expected in service.

(g) If any device is used that depends on the operation of any engine, and the landing distance would be increased when a landing is made with that engine inoperative, the landing distance must be determined with that engine inoperative unless the use of other compensating means will result in a landing distance not more than that with each engine operating.

[Amdt. 23-21, 43 FR 2318, Jan. 16, 1978, as amended by Amdt. 23-34, 52 FR 1828, Jan. 15, 1987; Amdt. 23-42, 56 FR 351, Jan. 3, 1991; Amdt. 23-50, 61 FR 5187, Feb. 9, 1996]

§ 23.77 Balked landing.

(a) Each normal, utility, and acrobatic category reciprocating engine-powered airplane at 6,000 pounds or less maximum weight must be able to maintain a steady gradient of climb at sea level of at least 3.3 percent with—

(1) Takeoff power on each engine;

(2) The landing gear extended;

(3) The wing flaps in the landing position, except that if the flaps may safely be retracted in two seconds or less without loss of altitude and without sudden changes of angle of attack, they may be retracted; and

(4) A climb speed equal to V_{REF}, as defined in § 23.73(a).

(b) Each normal, utility, and acrobatic category reciprocating engine-powered and single engine turbine powered airplane of more than 6,000 pounds maximum weight, and multiengine turbine engine-powered airplane of 6,000 pounds or less maximum weight in the normal, utility, and acrobatic category must be able to maintain a steady gradient of climb of at least 2.5 percent with—

(1) Not more than the power that is available on each engine eight seconds after initiation of movement of the power controls from minimum flight-idle position;

(2) The landing gear extended;

(3) The wing flaps in the landing position; and

(4) A climb speed equal to V_{REF}, as defined in § 23.73(b).

(c) Each normal, utility, and acrobatic multiengine turbine powered airplane over 6,000 pounds maximum weight and each commuter category airplane must be able to maintain a steady gradient of climb of at least 3.2 percent with—

(1) Not more than the power that is available on each engine eight seconds after initiation of movement of the power controls from the minimum flight idle position;

(2) Landing gear extended;

(3) Wing flaps in the landing position; and

(4) A climb speed equal to V_{REF}, as defined in § 23.73(c).

[Doc. No. 27807, 61 FR 5187, Feb. 9, 1996, as amended by Amdt. 23-62, 76 FR 75754, Dec. 2, 2011]

FLIGHT CHARACTERISTICS

§ 23.141 General.

The airplane must meet the requirements of §§ 23.143 through 23.253 at all practical loading conditions and operating altitudes for which certification has been requested, not exceeding the maximum operating altitude established under § 23.1527, and without requiring exceptional piloting skill, alertness, or strength.

[Doc. No. 26269, 58 FR 42156, Aug. 6, 1993]

CONTROLLABILITY AND MANEUVERABILITY

§ 23.143 General.

(a) The airplane must be safely controllable and maneuverable during all flight phases including—

(1) Takeoff;

(2) Climb;

(3) Level flight;

(4) Descent;

(5) Go-around; and

(6) Landing (power on and power off) with the wing flaps extended and retracted.

(b) It must be possible to make a smooth transition from one flight condition to another (including turns and slips) without danger of exceeding the limit load factor, under any probable operating condition (including, for

Values in pounds force applied to the relevant control	Pitch	Roll	Yaw
(a) For temporary application:			
Stick	60	30	
Wheel (Two hands on rim)	75	50	
Wheel (One hand on rim)	50	25	
Rudder Pedal			150
(b) For prolonged application	10	5	20

multiengine airplanes, those conditions normally encountered in the sudden failure of any engine).

(c) If marginal conditions exist with regard to required pilot strength, the control forces necessary must be determined by quantitative tests. In no case may the control forces under the conditions specified in paragraphs (a) and (b) of this section exceed those prescribed in the following table:

[Doc. No. 4080, 29 FR 17955, Dec. 18, 1964, as amended by Amdt. 23-14, 38 FR 31819, Nov. 19, 1973; Amdt. 23-17, 41 FR 55464, Dec. 20, 1976; Amdt. 23-45, 58 FR 42156, Aug. 6, 1993; Amdt. 23-50, 61 FR 5188, Feb. 9, 1996]

§ 23.145 Longitudinal control.

(a) With the airplane as nearly as possible in trim at 1.3 V_{S1}, it must be possible, at speeds below the trim speed, to pitch the nose downward so that the rate of increase in airspeed allows prompt acceleration to the trim speed with—

(1) Maximum continuous power on each engine;

(2) Power off; and

(3) Wing flap and landing gear—

(i) retracted, and

(ii) extended.

(b) Unless otherwise required, it must be possible to carry out the following maneuvers without requiring the application of single-handed control forces exceeding those specified in § 23.143(c). The trimming controls must not be adjusted during the maneuvers:

(1) With the landing gear extended, the flaps retracted, and the airplanes as nearly as possible in trim at 1.4 V_{S1}, extend the flaps as rapidly as possible and allow the airspeed to transition from 1.4V_{S1} to 1.4 V_{S0}:

(i) With power off; and

(ii) With the power necessary to maintain level flight in the initial condition.

(2) With landing gear and flaps extended, power off, and the airplane as nearly as possible in trim at 1.3 V_{S0}: quickly apply takeoff power and retract the flaps as rapidly as possible to the recommended go around setting and allow the airspeed to transition from 1.3 V_{S0} to 1.3 V_{S1}. Retract the gear when a positive rate of climb is established.

(3) With landing gear and flaps extended, in level flight, power necessary to attain level flight at 1.1 V_{S0}, and the airplane as nearly as possible in trim, it must be possible to maintain approximately level flight while retracting the flaps as rapidly as possible with simultaneous application of not more than maximum continuous power. If gated flat positions are provided, the flap retraction may be demonstrated in stages with power and trim reset for level flight at 1.1 V_{S1}, in the initial configuration for each stage—

(i) From the fully extended position to the most extended gated position;

(ii) Between intermediate gated positions, if applicable; and

(iii) From the least extended gated position to the fully retracted position.

(4) With power off, flaps and landing gear retracted and the airplane as nearly as possible in trim at 1.4 V_{S1}, apply takeoff power rapidly while maintaining the same airspeed.

(5) With power off, landing gear and flaps extended, and the airplane as nearly as possible in trim at V_{REF}, obtain and maintain airspeeds between 1.1 V_{S0}, and either 1.7 V_{S0} or V_{FE}, whichever is lower without requiring the application of two-handed control forces exceeding those specified in § 23.143(c).

(6) With maximum takeoff power, landing gear retracted, flaps in the takeoff position, and the airplane as nearly as possible in trim at V_{FE} appropriate to the takeoff flap position, retract the flaps as rapidly as possible while maintaining constant speed.

(c) At speeds above V_{MO}/M_{MO}, and up to the maximum speed shown under § 23.251, a maneuvering capability of 1.5 g must be demonstrated to provide a margin to recover from upset or inadvertent speed increase.

(d) It must be possible, with a pilot control force of not more than 10 pounds, to maintain a speed of not more than V_{REF} during a power-off glide with landing gear and wing flaps extended, for any weight of the airplane, up to and including the maximum weight.

(e) By using normal flight and power controls, except as otherwise noted in paragraphs (e)(1) and (e)(2) of this section, it must be possible to establish a zero rate of descent at an attitude suitable for a controlled landing without exceeding the operational and structural limitations of the airplane, as follows:

(1) For single-engine and multiengine airplanes, without the use of the primary longitudinal control system.

(2) For multiengine airplanes—

(i) Without the use of the primary directional control; and

(ii) If a single failure of any one connecting or transmitting link would affect both the longitudinal and directional primary control system, without the primary longitudinal and directional control system.

[Doc. No. 26269, 58 FR 42157, Aug. 6, 1993; Amdt. 23-45, 58 FR 51970, Oct. 5, 1993, as amended by Amdt. 23-50, 61 FR 5188, Feb. 9, 1996]

§ 23.147 Directional and lateral control.

(a) For each multiengine airplane, it must be possible, while holding the wings level within five degrees, to make sudden changes in heading safely in both directions. This ability must be shown at 1.4 V_{S1} with heading changes up to 15 degrees, except that the heading change at which the rudder force corresponds to the limits specified in § 23.143 need not be exceeded, with the—

(1) Critical engine inoperative and its propeller in the minimum drag position;

(2) Remaining engines at maximum continuous power;

(3) Landing gear—

(i) Retracted; and

(ii) Extended; and

(4) Flaps retracted.

(b) For each multiengine airplane, it must be possible to regain full control of the airplane without exceeding a bank angle of 45 degrees, reaching a dangerous attitude or encountering dangerous characteristics, in the event of a sudden and complete failure of the critical engine, making allowance for a delay of two seconds in the initiation of recovery action appropriate to the situation, with the airplane initially in trim, in the following condition:

(1) Maximum continuous power on each engine;

(2) The wing flaps retracted;

(3) The landing gear retracted;

(4) A speed equal to that at which compliance with § 23.69(a) has been shown; and

(5) All propeller controls in the position at which compliance with § 23.69(a) has been shown.

(c) For all airplanes, it must be shown that the airplane is safely controllable without the use of the primary lateral control system in any all-engine configuration(s) and at any speed or altitude within the approved operating envelope. It must also be shown that the airplane's flight characteristics are not impaired below a level needed to permit continued safe flight and the ability to maintain attitudes suitable for a controlled landing without exceeding the operational and structural limitations of the airplane. If a single failure of any one connecting or transmitting link in the lateral control system would also cause the loss of additional control system(s), compliance with the above requirement must be shown with those additional systems also assumed to be inoperative.

[Doc. No. 27807, 61 FR 5188, Feb. 9, 1996]

§ 23.149 Minimum control speed.

(a) V_{MC} is the calibrated airspeed at which, when the critical engine is suddenly made inoperative, it is possible to maintain control of the airplane with that engine still inoperative, and thereafter maintain straight flight at the same speed with an angle of bank of not more than 5 degrees. The method used to simulate critical engine failure must represent the most critical mode of powerplant failure expected in service with respect to controllability.

(b) V_{MC} for takeoff must not exceed 1.2 V_{S1}, where V_{S1} is determined at the maximum takeoff weight. V_{MC} must be determined with the most unfavorable weight and center of gravity position and with the airplane airborne and the ground effect negligible, for the takeoff configuration(s) with—

(1) Maximum available takeoff power initially on each engine;

(2) The airplane trimmed for takeoff;

(3) Flaps in the takeoff position(s);

(4) Landing gear retracted; and

(5) All propeller controls in the recommended takeoff position throughout.

(c) For all airplanes except reciprocating engine-powered airplanes of 6,000 pounds or less maximum weight, the conditions of paragraph (a) of this section must also be met for the landing configuration with—

(1) Maximum available takeoff power initially on each engine;

(2) The airplane trimmed for an approach, with all engines operating, at V_{REF}, at an approach gradient equal to the steepest used in the landing distance demonstration of § 23.75;

(3) Flaps in the landing position;

(4) Landing gear extended; and

(5) All propeller controls in the position recommended for approach with all engines operating.

(d) A minimum speed to intentionally render the critical engine inoperative must be established and designated as the safe, intentional, one-engine-inoperative speed, V_{SSE}.

(e) At V_{MC}, the rudder pedal force required to maintain control must not exceed 150 pounds and it must not be necessary to reduce power of the operative engine(s). During the maneuver, the airplane must not assume any dangerous attitude and it must be possible to prevent a heading change of more than 20 degrees.

(f) At the option of the applicant, to comply with the requirements of § 23.51(c)(1), V_{MCG} may be determined. V_{MCG} is the minimum control speed on the ground, and is the calibrated airspeed during the takeoff run at which, when the critical engine is suddenly made inoperative, it is possible to maintain control of the airplane using the rudder control alone (without the use of nosewheel steering), as limited by 150 pounds of force, and using the lateral control to the extent of keeping the wings level to enable the takeoff to be safely continued. In the determination of V_{MCG}, assuming that the path of the airplane accelerating with all engines operating is along the centerline of the runway, its path from the point at which the critical engine is made inoperative to the point at which recovery to a direction parallel to the centerline is completed may not deviate more than 30 feet laterally from the centerline at any point. V_{MCG} must be established with—

(1) The airplane in each takeoff configuration or, at the option of the applicant, in the most critical takeoff configuration;

(2) Maximum available takeoff power on the operating engines;

(3) The most unfavorable center of gravity;

(4) The airplane trimmed for takeoff; and

(5) The most unfavorable weight in the range of takeoff weights.

[Doc. No. 27807, 61 FR 5189, Feb. 9, 1996]

§ 23.151 Acrobatic maneuvers.

Each acrobatic and utility category airplane must be able to perform safely the acrobatic maneuvers for which certification is requested. Safe entry speeds for these maneuvers must be determined.

§ 23.153 Control during landings.

It must be possible, while in the landing configuration, to safely complete a landing without exceeding the one-hand control force limits specified in § 23.143(c) following an approach to land—

(a) At a speed of V_{REF} minus 5 knots;

(b) With the airplane in trim, or as nearly as possible in trim and without the trimming control being moved throughout the maneuver;

(c) At an approach gradient equal to the steepest used in the landing distance demonstration of § 23.75; and

(d) With only those power changes, if any, that would be made when landing normally from an approach at V_{REF}.

[Doc. No. 27807, 61 FR 5189, Feb. 9, 1996]

§ 23.155 Elevator control force in maneuvers.

(a) The elevator control force needed to achieve the positive limit maneuvering load factor may not be less than:

(1) For wheel controls, W/100 (where W is the maximum weight) or 20 pounds, whichever is greater, except that it need not be greater than 50 pounds; or

(2) For stick controls, W/140 (where W is the maximum weight) or 15 pounds, whichever is greater, except that it need not be greater than 35 pounds.

(b) The requirement of paragraph (a) of this section must be met at 75 percent of maximum continuous power for reciprocating engines, or the maximum continuous power for turbine engines, and with the wing flaps and landing gear retracted—

(1) In a turn, with the trim setting used for wings level flight at V_O; and

(2) In a turn with the trim setting used for the maximum wings level flight speed, except that the speed may not exceed V_{NE} or V_{MO}/M_{MO}, whichever is appropriate.

(c) There must be no excessive decrease in the gradient of the curve of stick force versus maneuvering load factor with increasing load factor.

[Amdt. 23-14, 38 FR 31819, Nov. 19, 1973; 38 FR 32784, Nov. 28, 1973, as amended by Amdt. 23-45, 58 FR 42158, Aug. 6, 1993; Amdt. 23-50, 61 FR 5189 Feb. 9, 1996]

§ 23.157 Rate of roll.

(a) *Takeoff.* It must be possible, using a favorable combination of controls, to roll the airplane from a steady 30-degree banked turn through an angle of 60 degrees, so as to reverse the direction of the turn within:

(1) For an airplane of 6,000 pounds or less maximum weight, 5 seconds from initiation of roll; and

(2) For an airplane of over 6,000 pounds maximum weight,

(W+500)/1,300

seconds, but not more than 10 seconds, where W is the weight in pounds.

(b) The requirement of paragraph (a) of this section must be met when rolling the airplane in each direction with—

(1) Flaps in the takeoff position;

(2) Landing gear retracted;

(3) For a single-engine airplane, at maximum takeoff power; and for a multiengine airplane with the critical engine inoperative and the propeller in the minimum drag position, and the other engines at maximum takeoff power; and

(4) The airplane trimmed at a speed equal to the greater of 1.2 V_{S1} or 1.1 V_{MC}, or as nearly as possible in trim for straight flight.

(c) *Approach.* It must be possible, using a favorable combination of controls, to roll the airplane from a steady 30-degree banked turn through an angle of 60 degrees, so as to reverse the direction of the turn within:

(1) For an airplane of 6,000 pounds or less maximum weight, 4 seconds from initiation of roll; and

(2) For an airplane of over 6,000 pounds maximum weight,

(W+2,800)/2,200

seconds, but not more than 7 seconds, where W is the weight in pounds.

(d) The requirement of paragraph (c) of this section must be met when rolling the airplane in each direction in the following conditions—

(1) Flaps in the landing position(s);

(2) Landing gear extended;

(3) All engines operating at the power for a 3 degree approach; and

(4) The airplane trimmed at V_{REF}.

[Amdt. 23-14, 38 FR 31819, Nov. 19, 1973, as amended by Amdt. 23-45, 58 FR 42158, Aug. 6, 1993; Amdt. 23-50, 61 FR 5189, Feb. 9, 1996]

Trim

§ 23.161 Trim.

(a) *General.* Each airplane must meet the trim requirements of this section after being trimmed and without further pressure upon, or movement of, the primary controls or their corresponding trim controls by the pilot or the automatic pilot. In addition, it must be possible, in other conditions of loading, configuration, speed and power to ensure that the pilot will not be unduly fatigued or distracted by the need to apply residual control forces exceeding those for prolonged application of § 23.143(c). This applies in normal operation of the airplane and, if applicable, to those conditions associated with the failure of one engine for which performance characteristics are established.

(b) *Lateral and directional trim.* The airplane must maintain lateral and directional trim in level flight with the landing gear and wing flaps retracted as follows:

(1) For normal, utility, and acrobatic category airplanes, at a speed of 0.9 V_H, V_C, or V_{MO}/M_O, whichever is lowest; and

(2) For commuter category airplanes, at all speeds from 1.4 V_{S1} to the lesser of V_H or V_{MO}/M_{MO}.

(c) *Longitudinal trim.* The airplane must maintain longitudinal trim under each of the following conditions:

(1) A climb with—

(i) Takeoff power, landing gear retracted, wing flaps in the takeoff position(s), at the speeds used in determining the climb performance required by § 23.65; and

(ii) Maximum continuous power at the speeds and in the configuration used in determining the climb performance required by § 23.69(a).

(2) Level flight at all speeds from the lesser of V_H and either V_{NO} or V_{MO}/M_{MO} (as appropriate), to 1.4 V_{S1}, with the landing gear and flaps retracted.

(3) A descent at V_{NO} or V_{MO}/M_{MO}, whichever is applicable, with power off and with the landing gear and flaps retracted.

(4) Approach with landing gear extended and with—

(i) A 3 degree angle of descent, with flaps retracted and at a speed of 1.4 V_{S1};

(ii) A 3 degree angle of descent, flaps in the landing position(s) at V_{REF}; and

(iii) An approach gradient equal to the steepest used in the landing distance demonstrations of § 23.75, flaps in the landing position(s) at V_{REF}.

(d) In addition, each multiple airplane must maintain longitudinal and directional trim, and the lateral control force must not exceed 5 pounds at the speed used in complying with § 23.67(a), (b)(2), or (c)(3), as appropriate, with—

(1) The critical engine inoperative, and if applicable, its propeller in the minimum drag position;

(2) The remaining engines at maximum continuous power;

(3) The landing gear retracted;

(4) Wing flaps retracted; and

(5) An angle of bank of not more than five degrees.

(e) In addition, each commuter category airplane for which, in the determination of the takeoff path in accordance with § 23.57, the climb in the takeoff configuration at V_2 extends beyond 400 feet above the takeoff surface, it must be possible to reduce the longitudinal and lateral control forces to 10 pounds and 5 pounds, respectively, and the directional control force must not exceed 50 pounds at V_2 with—

(1) The critical engine inoperative and its propeller in the minimum drag position;

(2) The remaining engine(s) at takeoff power;

(3) Landing gear retracted;

(4) Wing flaps in the takeoff position(s); and

(5) An angle of bank not exceeding 5 degrees.

[Doc. No. 4080, 29 FR 17955, Dec. 18, 1964, as amended by Amdt. 23-21, 43 FR 2318, Jan. 16, 1978; Amdt. 23-34, 52 FR 1828, Jan. 15, 1987; Amdt. 23-42, 56 FR 351, Jan. 3, 1991; 56 FR 5455, Feb. 11, 1991; Amdt. 23-50, 61 FR 5189, Feb. 9, 1996]

Stability

§ 23.171 General.

The airplane must be longitudinally, directionally, and laterally stable under §§ 23.173 through 23.181. In addition, the airplane must show suitable stability and control "feel" (static stability) in any condition normally encountered in service, if flight tests show it is necessary for safe operation.

§ 23.173 Static longitudinal stability.

Under the conditions specified in § 23.175 and with the airplane trimmed as indicated, the characteristics of the elevator control forces and the friction within the control system must be as follows:

(a) A pull must be required to obtain and maintain speeds below the specified trim speed and a push required to obtain and maintain speeds above the specified trim speed. This must be shown at any speed that can be obtained, except that speeds requiring a control force in excess of 40 pounds or speeds above the maximum allowable speed or below the minimum speed for steady unstalled flight, need not be considered.

(b) The airspeed must return to within the tolerances specified for applicable categories of airplanes when the control force is slowly released at any speed within the speed range specified in paragraph (a) of this section. The applicable tolerances are—

(1) The airspeed must return to within plus or minus 10 percent of the original trim airspeed; and

(2) For commuter category airplanes, the airspeed must return to within plus or minus 7.5 percent of the original trim airspeed for the cruising condition specified in § 23.175(b).

(c) The stick force must vary with speed so that any substantial speed change results in a stick force clearly perceptible to the pilot.

[Doc. No. 4080, 29 FR 17955, Dec. 18, 1964, as amended by Amdt. 23-14, 38 FR 31820 Nov. 19, 1973; Amdt. 23-34, 52 FR 1828, Jan. 15, 1987]

§ 23.175 Demonstration of static longitudinal stability.

Static longitudinal stability must be shown as follows:

(a) *Climb.* The stick force curve must have a stable slope at speeds between 85 and 115 percent of the trim speed, with—

(1) Flaps retracted;

(2) Landing gear retracted;

(3) Maximum continuous power; and

(4) The airplane trimmed at the speed used in determining the climb performance required by § 23.69(a).

(b) *Cruise.* With flaps and landing gear retracted and the airplane in trim with power for level flight at representative cruising speeds at high and low altitudes, including speeds up to V_{NO} or V_{MO}/M_{MO}, as appropriate, except that the speed need not exceed V_H—

(1) For normal, utility, and acrobatic category airplanes, the stick force curve must have a stable slope at all speeds within a range that is the greater of 15 percent of the trim speed plus the resulting free return speed range, or 40 knots plus the resulting free return speed range, above and below the trim speed, except that the slope need not be stable—

(i) At speeds less than 1.3 V_{S1}; or

(ii) For airplanes with V_{NE} established under § 23.1505(a), at speeds greater than V_{NE}; or

(iii) For airplanes with V_{MO}/M_{MO} established under § 23.1505(c), at speeds greater than V_{FC}/M_{FC}.

(2) For commuter category airplanes, the stick force curve must have a stable slope at all speeds within a range of 50 knots plus the resulting free return speed range, above and below the trim speed, except that the slope need not be stable—

(i) At speeds less than 1.4 V_{S1}; or

(ii) At speeds greater than V_{FC}/M_{FC}; or

(iii) At speeds that require a stick force greater than 50 pounds.

(c) *Landing.* The stick force curve must have a stable slope at speeds between 1.1 V_{S1} and 1.8 V_{S1} with—

(1) Flaps in the landing position;

(2) Landing gear extended; and

(3) The airplane trimmed at—

(i) V_{REF}, or the minimum trim speed if higher, with power off; and

(ii) V_{REF} with enough power to maintain a 3 degree angle of descent.

[Doc. No. 27807, 61 FR 5190, Feb. 9, 1996]

§ 23.177 Static directional and lateral stability.

(a)(1) The static directional stability, as shown by the tendency to recover from a wings level sideslip with the rudder free, must be positive for any landing gear and flap position appropriate to the takeoff, climb, cruise, approach, and landing configurations. This must be shown with symmetrical power up to maximum continuous power, and at speeds from 1.2 V_{S1} up to V_{FE}, V_{LE}, V_{NO}, V_{FC}/M_{FC}, whichever is appropriate.

(2) The angle of sideslip for these tests must be appropriate to the type of airplane. The rudder pedal force must not reverse at larger angles of sideslip, up to that at which full rudder is used or a control force limit in § 23.143 is reached, whichever occurs first, and at speeds from 1.2 V_{S1} to V_O.

(b)(1) The static lateral stability, as shown by the tendency to raise the low wing in a sideslip with the aileron controls free, may not be negative for any landing gear and flap position appropriate to the takeoff, climb, cruise, approach, and landing configurations. This must be shown with symmetrical power from idle up to 75 percent of maximum continuous power at speeds from 1.2 V_{S1} in the takeoff configuration(s) and at speeds from 1.3 V_{S1} in other configurations, up to the maximum allowable airspeed for the configuration being investigated (V_{FE}, V_{LE}, V_{NO}, V_{FC}/M_{FC}, whichever is appropriate) in the takeoff, climb, cruise, descent, and approach configurations. For the landing configuration, the power must be that necessary to maintain a 3-degree angle of descent in coordinated flight.

(2) The static lateral stability may not be negative at 1.2 V_{S1} in the takeoff configuration, or at 1.3 V_{S1} in other configurations.

(3) The angel of sideslip for these tests must be appropriate to the type of airplane, but in no case may the constant heading sideslip angle be less than that obtainable with a 10 degree bank or, if less, the maximum bank angle obtainable with full rudder deflection or 150 pound rudder force.

(c) Paragraph (b) of this section does not apply to acrobatic category airplanes certificated for inverted flight.

(d)(1) In straight, steady slips at 1.2 V_{S1} for any landing gear and flap position appropriate to the takeoff, climb, cruise, approach, and landing configurations, and for any symmetrical power conditions up to 50 percent of maximum continuous power, the aileron and rudder control movements and forces must increase steadily, but not necessarily in constant proportion, as the angle of sideslip is increased up to the maximum appropriate to the type of airplane.

(2) At larger slip angles, up to the angle at which the full rudder or aileron control is used or a control force limit contained in § 23.143 is reached, the aileron and rudder control movements and forces may not reverse as the angle of sideslip is increased.

(3) Rapid entry into, and recovery from, a maximum sideslip considered appropriate for the airplane may not result in uncontrollable flight characteristics.

[Doc. No. 27807, 61 FR 5190, Feb. 9, 1996, as amended by Amdt. 23-62, 76 FR 75754, Dec. 2, 2011]

§ 23.181 Dynamic stability.

(a) Any short period oscillation not including combined lateral-directional oscillations occurring between the stalling speed and the maximum allowable speed appropriate to the configuration of the airplane must be heavily damped with the primary controls—

(1) Free; and

(2) In a fixed position.

(b) Any combined lateral-directional oscillations (Dutch roll) occurring between the stalling speed and the maximum allowable speed (V_{FE}, V_{LE}, V_{NO}, V_{FC}/M_{FC}) appropriate to the configuration of the airplane with the primary controls in both free and fixed position, must be damped to $1/_{10}$ amplitude in:

(1) Seven (7) cycles below 18,000 feet and

(2) Thirteen (13) cycles from 18,000 feet to the certified maximum altitude.

(c) If it is determined that the function of a stability augmentation system, reference § 23.672, is needed to meet the flight characteristic requirements of this part, the primary control requirements of paragraphs (a)(2) and (b)(2) of this section are not applicable to the tests needed to verify the acceptability of that system.

(d) During the conditions as specified in § 23.175, when the longitudinal control force required to maintain speeds differing from the trim speed by at least plus and minus 15 percent is suddenly released, the response of the airplane must not exhibit any dangerous characteristics nor be excessive in relation to the magnitude of the control force released. Any long-period oscillation of flight path, phugoid oscillation, that results must not be so unstable as to increase the pilot's workload or otherwise endanger the airplane.

[Amdt. 23-21, 43 FR 2318, Jan. 16, 1978, as amended by Amdt. 23-45, 58 FR 42158, Aug. 6, 1993; Amdt. 23-62, 76 FR 75755, Dec. 2, 2011]

Stalls

§ 23.201 Wings level stall.

(a) It must be possible to produce and to correct roll by unreversed use of the rolling control and to produce and to correct yaw by unreversed use of the directional control, up to the time the airplane stalls.

(b) The wings level stall characteristics must be demonstrated in flight as follows. Starting from a speed at least 10 knots above the stall speed, the elevator control must be pulled back so that the rate of speed reduction will not exceed one knot per second until a stall is produced, as shown by either:

(1) An uncontrollable downward pitching motion of the airplane;

(2) A downward pitching motion of the airplane that results from the activation of a stall avoidance device (for example, stick pusher); or

(3) The control reaching the stop.

(c) Normal use of elevator control for recovery is allowed after the downward pitching motion of paragraphs (b)(1) or (b)(2) of this section has unmistakably been produced, or after the control has been held against the stop for not less than the longer of two seconds or the time employed in the minimum steady slight speed determination of § 23.49.

(d) During the entry into and the recovery from the maneuver, it must be possible to prevent more than 15 degrees of roll or yaw by the normal use of controls except as provided for in paragraph (e) of this section.

(e) For airplanes approved with a maximum operating altitude at or above 25,000 feet during the entry into and the recovery from stalls performed at or above 25,000 feet, it must be possible to prevent more than 25 degrees of roll or yaw by the normal use of controls.

(f) Compliance with the requirements of this section must be shown under the following conditions:

(1) *Wing flaps:* Retracted, fully extended, and each intermediate normal operating position, as appropriate for the phase of flight.

(2) *Landing gear:* Retracted and extended as appropriate for the altitude.

(3) *Cowl flaps:* Appropriate to configuration.

(4) *Spoilers/speedbrakes:* Retracted and extended unless they have no measureable effect at low speeds.

(5) *Power:*

(i) Power/Thrust off; and

(ii) For reciprocating engine powered airplanes: 75 percent of maximum continuous power. However, if the power-to-weight ratio at 75 percent of maximum continuous power results in nose-high attitudes exceeding 30 degrees, the test may be carried out with the power required for level flight in the landing configuration at maximum landing weight and a speed of 1.4 V_{SO}, except that the power may not be less than 50 percent of maximum continuous power; or

(iii) For turbine engine powered airplanes: The maximum engine thrust, except that it need not exceed the thrust necessary to maintain level flight at 1.5 V_{S1} (where V_{S1} corresponds to the stalling speed with flaps in the approach position, the landing gear retracted, and maximum landing weight).

(6) *Trim:* At 1.5 V_{S1} or the minimum trim speed, whichever is higher.

(7) *Propeller:* Full increase r.p.m. position for the power off condition.

[Doc. No. 27807, 61 FR 5191, Feb. 9, 1996, as amended by Amdt. 23-62, 76 FR 75755, Dec. 2, 2011]

§ 23.203 Turning flight and accelerated turning stalls.

Turning flight and accelerated turning stalls must be demonstrated in tests as follows:

(a) Establish and maintain a coordinated turn in a 30 degree bank. Reduce speed by steadily and progressively tightening the turn with the elevator until the airplane is stalled, as defined in § 23.201(b). The rate of speed reduction must be constant, and—

(1) For a turning flight stall, may not exceed one knot per second; and

(2) For an accelerated turning stall, be 3 to 5 knots per second with steadily increasing normal acceleration.

(b) After the airplane has stalled, as defined in § 23.201(b), it must be possible to regain wings level flight by normal use of the flight controls, but without increasing power and without—

(1) Excessive loss of altitude;

(2) Undue pitchup;

(3) Uncontrollable tendency to spin;

(4) Exceeding a bank angle of 60 degrees in the original direction of the turn or 30 degrees in the opposite direction in the case of turning flight stalls;

(5) Exceeding a bank angle of 90 degrees in the original direction of the turn or 60 degrees in the opposite direction in the case of accelerated turning stalls; and

(6) Exceeding the maximum permissible speed or allowable limit load factor.

(c) Compliance with the requirements of this section must be shown under the following conditions:

(1) *Wings flaps:* Retracted, fully extended, and each intermediate normal operating position as appropriate for the phase of flight.

(2) *Landing gear:* Retracted and extended as appropriate for the altitude.

(3) *Cowl flaps:* Appropriate to configuration.

(4) *Spoilers/speedbrakes:* Retracted and extended unless they have no measureable effect at low speeds.

(5) *Power:*

(i) Power/Thrust off; and

(ii) For reciprocating engine powered airplanes: 75 percent of maximum continuous power. However, if the power-to-weight ratio at 75 percent of maximum continuous power results in nose-high attitudes exceeding 30 degrees, the test may be carried out with the power required for level flight in the landing configuration at maximum landing weight and a speed of 1.4 V_{SO}, except that the power may not be less than 50 percent of maximum continuous power; or

(iii) For turbine engine powered airplanes: The maximum engine thrust, except that it need not exceed the thrust necessary to maintain level flight at 1.5 V_{S1} (where V_{S1} corresponds to the stalling speed with flaps in the approach position, the landing gear retracted, and maximum landing weight).

(6) *Trim:* The airplane trimmed at 1.5 V_{S1}.

(7) *Propeller:* Full increase rpm position for the power off condition.

[Amdt. 23-14, 38 FR 31820, Nov. 19, 1973, as amended by Amdt. 23-45, 58 FR 42159, Aug. 6, 1993; Amdt. 23-50, 61 FR 5191, Feb. 9, 1996; Amdt. 23-62, 76 FR 75755, Dec. 2, 2011]

§ 23.207 Stall warning.

(a) There must be a clear and distinctive stall warning, with the flaps and landing gear in any normal position, in straight and turning flight.

(b) The stall warning may be furnished either through the inherent aerodynamic qualities of the airplane or by a device that will give clearly distinguishable indications under expected conditions of flight. However, a visual stall warning device that requires the attention of the crew within the cockpit is not acceptable by itself.

(c) During the stall tests required by § 23.201(b) and § 23.203(a)(1), the stall warning must begin at a speed exceeding the stalling speed by a margin of not less than 5 knots and must continue until the stall occurs.

(d) When following procedures furnished in accordance with § 23.1585, the stall warning must not occur during a takeoff with all engines operating, a takeoff continued with one engine inoperative, or during an approach to landing.

(e) During the stall tests required by § 23.203(a)(2), the stall warning must begin sufficiently in advance of the stall for the stall to be averted by pilot action taken after the stall warning first occurs.

(f) For acrobatic category airplanes, an artificial stall warning may be mutable, provided that it is armed automatically during takeoff and rearmed automatically in the approach configuration.

[Amdt. 23-7, 34 FR 13087, Aug. 13, 1969, as amended by Amdt. 23-45, 58 FR 42159, Aug. 6, 1993; Amdt. 23-50, 61 FR 5191, Feb. 9, 1996]

Spinning

§ 23.221 Spinning.

(a) *Normal category airplanes.* A single-engine, normal category airplane must be able to recover from a one-turn spin or a three-second spin, whichever takes longer, in not more than one additional turn after initiation of the first control action for recovery, or demonstrate compliance with the optional spin resistant requirements of this section.

(1) The following apply to one turn or three second spins:

(i) For both the flaps-retracted and flaps-extended conditions, the applicable airspeed limit and positive limit maneuvering load factor must not be exceeded;

(ii) No control forces or characteristic encountered during the spin or recovery may adversely affect prompt recovery;

(iii) It must be impossible to obtain unrecoverable spins with any use of the flight or engine power controls either at the entry into or during the spin; and

(iv) For the flaps-extended condition, the flaps may be retracted during the recovery but not before rotation has ceased.

(2) At the applicant's option, the airplane may be demonstrated to be spin resistant by the following:

(i) During the stall maneuver contained in § 23.201, the pitch control must be pulled back and held against the stop. Then, using ailerons and rudders in the proper direction, it must be possible to maintain wings-level flight within 15 degrees of bank and to roll the airplane from a 30 degree bank in one direction to a 30 degree bank in the other direction;

(ii) Reduce the airplane speed using pitch control at a rate of approximately one knot per second until the pitch control reaches the stop; then, with the pitch control pulled back and held against the stop, apply full rudder control in a manner to promote spin entry for a period of seven seconds or through a 360 degree heading change, whichever occurs first. If the 360 degree heading change is reached first, it must have taken no fewer than four seconds. This maneuver must be performed first with the ailerons in the neutral position, and then with the ailerons deflected opposite the direction of turn in the most adverse manner. Power and airplane configuration must be set in accordance with § 23.201(e) without change during the maneuver. At the end of seven seconds or a 360 degree heading change, the airplane must respond immediately and normally to primary flight controls applied to regain coordinated, unstalled flight without reversal of control effect and without exceeding the temporary control forces specified by § 23.143(c); and

(iii) Compliance with §§ 23.201 and 23.203 must be demonstrated with the airplane in uncoordinated flight, corresponding to one ball width displacement on a slip-skid indicator, unless one ball width displacement cannot be obtained with full rudder, in which case the demonstration must be with full rudder applied.

(b) *Utility category airplanes.* A utility category airplane must meet the requirements of paragraph (a) of this section. In addition, the requirements of paragraph (c) of this section and § 23.807(b)(7) must be met if approval for spinning is requested.

(c) *Acrobatic category airplanes.* An acrobatic category airplane must meet the spin requirements of paragraph (a) of this section and § 23.807(b)(6). In addition, the following requirements must be met in each configuration for which approval for spinning is requested:

(1) The airplane must recover from any point in a spin up to and including six turns, or any greater number of turns for which certification is requested, in not more than one and one-half additional turns after initiation of the first control action for recovery. However, beyond three turns, the spin may be discontinued if spiral characteristics appear.

(2) The applicable airspeed limits and limit maneuvering load factors must not be exceeded. For flaps-extended configurations for which approval is requested, the flaps must not be retracted during the recovery.

(3) It must be impossible to obtain unrecoverable spins with any use of the flight or engine power controls either at the entry into or during the spin.

(4) There must be no characteristics during the spin (such as excessive rates of rotation or extreme oscillatory motion) that might prevent a successful recovery due to disorientation or incapacitation of the pilot.

[Doc. No. 27807, 61 FR 5191, Feb. 9, 1996]

Ground and Water Handling Characteristics

§ 23.231 Longitudinal stability and control.

(a) A landplane may have no uncontrollable tendency to nose over in any reasonably expected operating condition, including rebound during landing or takeoff. Wheel brakes must operate smoothly and may not induce any undue tendency to nose over.

(b) A seaplane or amphibian may not have dangerous or uncontrollable porpoising characteristics at any normal operating speed on the water.

§ 23.233 Directional stability and control.

(a) A 90 degree cross-component of wind velocity, demonstrated to be safe for taxiing, takeoff, and landing must be established and must be not less than 0.2 V_{SO}.

(b) The airplane must be satisfactorily controllable in power-off landings at normal landing speed, without using brakes or engine power to maintain a straight path until the speed has decreased to at least 50 percent of the speed at touchdown.

(c) The airplane must have adequate directional control during taxiing.

(d) Seaplanes must demonstrate satisfactory directional stability and control for water operations up to the maximum wind velocity specified in paragraph (a) of this section.

[Doc. No. 4080, 29 FR 17955, Dec. 18, 1964, as amended by Amdt. 23-45, 58 FR 42159, Aug. 6, 1993; Amdt. 23-50, 61 FR 5192, Feb. 9, 1996]

§ 23.235 Operation on unpaved surfaces.

The airplane must be demonstrated to have satisfactory characteristics and the shock-absorbing mechanism must not damage the structure of the airplane when the airplane is taxied on the roughest ground that may reasonably be expected in normal operation and when takeoffs and landings are performed on unpaved runways having the roughest surface that may reasonably be expected in normal operation.

[Doc. No. 27807, 61 FR 5192, Feb. 9, 1996]

§ 23.237 Operation on water.

A wave height, demonstrated to be safe for operation, and any necessary water handling procedures for seaplanes and amphibians must be established.

[Doc. No. 27807, 61 FR 5192, Feb. 9, 1996]

§ 23.239 Spray characteristics.

Spray may not dangerously obscure the vision of the pilots or damage the propellers or other parts of a seaplane or amphibian at any time during taxiing, takeoff, and landing.

Miscellaneous Flight Requirements

§ 23.251 Vibration and buffeting.

(a) There must be no vibration or buffeting severe enough to result in structural damage, and each part of the airplane must be free from excessive vibration, under any appropriate speed and power conditions up to V_D/M_D, or V_{DF}/M_{DF} for turbojets. In addition, there must be no buffeting in any normal flight condition, including configuration changes during cruise, severe enough to interfere with the satisfactory control of the airplane or cause excessive fatigue to the flight crew. Stall warning buffeting within these limits is allowable.

(b) There must be no perceptible buffeting condition in the cruise configuration in straight flight at any speed up to V_{MO}/M_{MO}, except stall buffeting, which is allowable.

(c) For airplanes with M_D greater than M 0.6 or a maximum operating altitude greater than 25,000 feet, the positive maneuvering load factors at which the onset of perceptible buffeting occurs must be determined with the airplane in the cruise configuration for the ranges of airspeed or Mach number, weight, and altitude for which the airplane is to be certificated. The envelopes of load factor, speed, altitude, and weight must provide a sufficient range of speeds and load factors for normal operations. Probable inadvertent excursions beyond the boundaries of the buffet onset envelopes may not result in unsafe conditions.

[Amdt. 23-62, 76 FR 75755, Dec. 2, 2011]

§ 23.253 High speed characteristics.

If a maximum operating speed V_{MO}/M_{MO} is established under § 23.1505(c), the following speed increase and recovery characteristics must be met:

(a) Operating conditions and characteristics likely to cause inadvertent speed increases (including upsets in pitch and roll) must be simulated with the airplane trimmed at any likely speed up to V_{MO}/M_{MO}. These conditions and characteristics include gust upsets, inadvertent control movements, low stick force gradients in relation to control friction, passenger movement, leveling off from climb, and descent from Mach to airspeed limit altitude.

(b) Allowing for pilot reaction time after occurrence of the effective inherent or artificial speed warning specified in § 23.1303, it must be shown that the airplane can be recovered to a normal attitude and its speed reduced to V_{MO}/M_{MO}, without—

(1) Exceptional piloting strength or skill;

(2) Exceeding V_D/M_D, or V_{DF}/M_{DF} for turbojets, the maximum speed shown under § 23.251, or the structural limitations; and

(3) Buffeting that would impair the pilot's ability to read the instruments or to control the airplane for recovery.

(c) There may be no control reversal about any axis at any speed up to the maximum speed shown under § 23.251. Any reversal of elevator control force or tendency of the airplane to pitch, roll, or yaw must be mild and readily controllable, using normal piloting techniques.

(d) *Maximum speed for stability characteristics, V_{FC}/M_{FC}.* V_{FC}/M_{FC} may not be less than a speed midway between V_{MO}/M_{MO} and V_{DF}/M_{DF} except that, for altitudes where Mach number is the limiting factor, M_{FC} need not exceed the Mach number at which effective speed warning occurs.

[Amdt. 23-7, 34 FR 13087, Aug. 13, 1969; as amended by Amdt. 23-26, 45 FR 60170, Sept. 11, 1980; Amdt. 23-45, 58 FR 42160, Aug. 6, 1993; Amdt. 23-50, 61 FR 5192, Feb. 9, 1996; Amdt. 23-62, 76 FR 75755, Dec. 2, 2011]

§ 23.255 Out of trim characteristics.

For airplanes with an M_D greater than M 0.6 and that incorporate a trimmable horizontal stabilizer, the following requirements for out-of-trim characteristics apply:

(a) From an initial condition with the airplane trimmed at cruise speeds up to V_{MO}/M_{MO}, the airplane must have satisfactory maneuvering stability and controllability with the degree of out-of-trim in both the airplane nose-up and nose-down directions, which results from the greater of the following:

(1) A three-second movement of the longitudinal trim system at its normal rate for the particular flight condition with no aerodynamic load (or an equivalent degree of trim for airplanes that do not have a power-operated trim system), except as limited by stops in the trim system, including those required by § 23.655(b) for adjustable stabilizers; or

(2) The maximum mistrim that can be sustained by the autopilot while maintaining level flight in the high speed cruising condition.

(b) In the out-of-trim condition specified in paragraph (a) of this section, when the normal acceleration is varied from +l g to the positive and negative values specified in paragraph (c) of this section, the following apply:

(1) The stick force versus g curve must have a positive slope at any speed up to and including V_{FC}/M_{FC}; and

(2) At speeds between V_{FC}/M_{FC} and V_{DF}/M_{DF}, the direction of the primary longitudinal control force may not reverse.

(c) Except as provided in paragraphs (d) and (e) of this section, compliance with the provisions of paragraph (a) of this section must be demonstrated in flight over the acceleration range as follows:

(1) –1 g to+2.5 g; or

(2) 0 g to 2.0 g, and extrapolating by an acceptable method to –1 g and +2.5 g.

(d) If the procedure set forth in paragraph (c)(2) of this section is used to demonstrate compliance and marginal conditions exist during flight test with regard to reversal of primary longitudinal control force, flight tests must be accomplished from the normal acceleration at which a marginal condition is found to exist to the applicable limit specified in paragraph (b)(1) of this section.

(e) During flight tests required by paragraph (a) of this section, the limit maneuvering load factors, prescribed in §§ 23.333(b) and 23.337, need not be exceeded. In addition, the entry speeds for flight test demonstrations at normal acceleration values less than 1 g must be limited to the extent necessary to accomplish a recovery without exceeding V_{DF}/M_{DF}.

(f) In the out-of-trim condition specified in paragraph (a) of this section, it must be possible from an overspeed condition at V_{DF}/M_{DF} to produce at least 1.5 g for recovery by applying not more than 125 pounds of longitudinal control force using either the primary longitudinal control alone or the primary longitudinal control and the longitudinal trim system. If the longitudinal trim is used to assist in producing the required load factor, it must be shown at V_{DF}/M_{DF} that the longitudinal trim can be actuated in the airplane nose-up direction with the primary surface loaded to correspond to the least of the following airplane nose-up control forces:

(1) The maximum control forces expected in service, as specified in §§ 23.301 and 23.397.

(2) The control force required to produce 1.5 g.

(3) The control force corresponding to buffeting or other phenomena of such intensity that it is a strong deterrent to further application of primary longitudinal control force.

[Doc. No. FAA-2009-0738, 76 FR 75755, Dec. 2, 2011]

Subpart C—Structure

General

§ 23.301 Loads.

(a) Strength requirements are specified in terms of limit loads (the maximum loads to be expected in service) and ultimate loads (limit loads multiplied by prescribed factors of safety). Unless otherwise provided, prescribed loads are limit loads.

(b) Unless otherwise provided, the air, ground, and water loads must be placed in equilibrium with inertia forces, considering each item of mass in the airplane. These loads must be distributed to conservatively approximate or closely represent actual conditions. Methods used to determine load intensities and distribution on canard and tandem wing configurations must be validated by flight test measurement unless the methods used for determining those loading conditions are shown to be reliable or conservative on the configuration under consideration.

(c) If deflections under load would significantly change the distribution of external or internal loads, this redistribution must be taken into account.

(d) Simplified structural design criteria may be used if they result in design loads not less than those prescribed in §§ 23.331 through 23.521. For airplane configurations described in appendix A, § 23.1, the design criteria of appendix A of this part are an approved equivalent of §§ 23.321 through 23.459. If appendix A of this part is used, the entire appendix must be substituted for the corresponding sections of this part.

[Doc. No. 4080, 29 FR 17955, Dec. 18, 1964; 30 FR 258, Jan. 9, 1965, as amended by Amdt. 23-28, 47 FR 13315, Mar. 29, 1982; Amdt. 23-42, 56 FR 352, Jan. 3, 1991; Amdt. 23-48, 61 FR 5143, Feb. 9, 1996]

§ 23.302 Canard or tandem wing configurations.

The forward structure of a canard or tandem wing configuration must:

(a) Meet all requirements of subpart C and subpart D of this part applicable to a wing; and

(b) Meet all requirements applicable to the function performed by these surfaces.

[Amdt. 23-42, 56 FR 352, Jan. 3, 1991]

§ 23.303 Factor of safety.

Unless otherwise provided, a factor of safety of 1.5 must be used.

§ 23.305 Strength and deformation.

(a) The structure must be able to support limit loads without detrimental, permanent deformation. At any load up to limit loads, the deformation may not interfere with safe operation.

(b) The structure must be able to support ultimate loads without failure for at least three seconds, except local failures or structural instabilities between limit and ultimate load are acceptable only if the structure can sustain the required ultimate load for at least three seconds. However when proof of strength is shown by dynamic tests simulating actual load conditions, the three second limit does not apply.

[Doc. No. 4080, 29 FR 17955, Dec. 18, 1964, as amended by Amdt. 23-45, 58 FR 42160, Aug. 6, 1993]

§ 23.307 Proof of structure.

(a) Compliance with the strength and deformation requirements of § 23.305 must be shown for each critical load condition. Structural analysis may be used only if the structure conforms to those for which experience has shown this method to be reliable. In other cases, substantiating load tests must be made. Dynamic tests, including structural flight tests, are acceptable if the design load conditions have been simulated.

(b) Certain parts of the structure must be tested as specified in Subpart D of this part.

Flight Loads

§ 23.321 General.

(a) Flight load factors represent the ratio of the aerodynamic force component (acting normal to the assumed longitudinal axis of the airplane) to the weight of the airplane. A positive flight load factor is one in which the aerodynamic force acts upward, with respect to the airplane.

(b) Compliance with the flight load requirements of this subpart must be shown—

(1) At each critical altitude within the range in which the airplane may be expected to operate;

(2) At each weight from the design minimum weight to the design maximum weight; and

(3) For each required altitude and weight, for any practicable distribution of disposable load within the operating limitations specified in §§ 23.1583 through 23.1589.

(c) When significant, the effects of compressibility must be taken into account.

[Doc. No. 4080, 29 FR 17955, Dec. 18, 1964, as amended by Amdt. 23-45, 58 FR 42160, Aug. 6, 1993]

§ 23.331 Symmetrical flight conditions.

(a) The appropriate balancing horizontal tail load must be accounted for in a rational or conservative manner when determining the wing loads and linear inertia loads corresponding to any of the symmetrical flight conditions specified in §§ 23.333 through 23.341.

(b) The incremental horizontal tail loads due to maneuvering and gusts must be reacted by the angular inertia of the airplane in a rational or conservative manner.

(c) Mutual influence of the aerodynamic surfaces must be taken into account when determining flight loads.

[Doc. No. 4080, 29 FR 17955, Dec. 18, 1964; 30 FR 258, Jan. 9, 1965, as amended by Amdt. 23-42, 56 FR 352, Jan. 3, 1991]

§ 23.333 Flight envelope.

(a) *General.* Compliance with the strength requirements of this subpart must be shown at any combination of airspeed and load factor on and within the boundaries of a flight envelope (similar to the one in paragraph (d) of this section) that represents the envelope of the flight loading conditions specified by the maneuvering and gust criteria of paragraphs (b) and (c) of this section respectively.

(b) *Maneuvering envelope.* Except where limited by maximum (static) lift coefficients, the airplane is assumed to be subjected to symmetrical maneuvers resulting in the following limit load factors:

(1) The positive maneuvering load factor specified in § 23.337 at speeds up to V_D;

(2) The negative maneuvering load factor specified in § 23.337 at V_C; and

(3) Factors varying linearly with speed from the specified value at V_C to 0.0 at V_D for the normal and commuter category, and −1.0 at V_D for the acrobatic and utility categories.

(c) *Gust envelope.* (1) The airplane is assumed to be subjected to symmetrical vertical gusts in level flight. The resulting limit load factors must correspond to the conditions determined as follows:

(i) Positive (up) and negative (down) gusts of 50 f.p.s. at V_C must be considered at altitudes between sea level and 20,000 feet. The gust velocity may be reduced linearly from 50 f.p.s. at 20,000 feet to 25 f.p.s. at 50,000 feet.

(ii) Positive and negative gusts of 25 f.p.s. at V_D must be considered at altitudes between sea level and 20,000 feet. The gust velocity may be reduced linearly from 25 f.p.s. at 20,000 feet to 12.5 f.p.s. at 50,000 feet.

(iii) In addition, for commuter category airplanes, positive (up) and negative (down) rough air gusts of 66 f.p.s. at VB must be considered at altitudes between sea level and 20,000 feet. The gust velocity may be reduced linearly from 66 f.p.s. at 20,000 feet to 38 f.p.s. at 50,000 feet.

(2) The following assumptions must be made:

(i) The shape of the gust is—

$$U = \frac{U_{de}}{2}\left(1 - cos\frac{2\pi s}{25C}\right)$$

Where—

s=Distance penetrated into gust (ft.);

C=Mean geometric chord of wing (ft.); and

Ude=Derived gust velocity referred to in subparagraph (1) of this section.

(ii) Gust load factors vary linearly with speed between V_C and V_D.

(d) *Flight envelope.*

[Doc. No. 4080, 29 FR 17955, Dec. 18, 1964, as amended by Amdt. 23-7, 34 FR 13087, Aug. 13, 1969; Amdt. 23-34, 52 FR 1829, Jan. 15, 1987]

§ 23.335 Design airspeeds.

Except as provided in paragraph (a)(4) of this section, the selected design airspeeds are equivalent airspeeds (EAS).

(a) *Design cruising speed, V_C.* For V_C the following apply:

(1) Where W/S'=wing loading at the design maximum takeoff weight, V_C (in knots) may not be less than—

(i) 33 √(W/S) (for normal, utility, and commuter category airplanes);

(ii) 36 √(W/S) (for acrobatic category airplanes).

(2) For values of *W/S* more than 20, the multiplying factors may be decreased linearly with *W/S* to a value of 28.6 where *W/S*=100.

(3) V_C need not be more than 0.9 V_H at sea level.

(4) At altitudes where an M_D is established, a cruising speed M_C limited by compressibility may be selected.

(b) *Design dive speed V_D.* For V_D, the following apply:

(1) V_D/M_D may not be less than 1.25 V_C/M_C; and

(2) With V_{Cmin}, the required minimum design cruising speed, V_D (in knots) may not be less than—

(i) 1.40 V_{Cmin} (for normal and commuter category airplanes);

(ii) 1.50 V_{Cmin} (for utility category airplanes); and

(iii) 1.55 V_{Cmin} (for acrobatic category airplanes).

(3) For values of *W/S* more than 20, the multiplying factors in paragraph (b)(2) of this section may be decreased linearly with *W/S* to a value of 1.35 where *W/S*=100.

(4) Compliance with paragraphs (b)(1) and (2) of this section need not be shown if V_D/M_D is selected so that the minimum speed margin between V_C/M_C and V_D/M_D is the greater of the following:

(i) The speed increase resulting when, from the initial condition of stabilized flight at V_C/M_C, the airplane is assumed to be upset, flown for 20 seconds along a flight path 7.5° below the initial path, and then pulled up with a load factor of 1.5 (0.5 g. acceleration increment). At least 75 percent maximum continuous power for reciprocating engines, and maximum cruising power for turbines, or, if less, the power required for V_C/M_C for both kinds of engines, must be assumed until the pullup is initiated, at which point power reduction and pilot-controlled drag devices may be used; and either—

(ii) Mach 0.05 for normal, utility, and acrobatic category airplanes (at altitudes where M_D is established); or

(iii) Mach 0.07 for commuter category airplanes (at altitudes where M_D is established) unless a rational analysis, including the effects of automatic systems, is used to determine a lower margin. If a rational analysis is used, the minimum speed margin must be enough to provide for atmospheric variations (such as horizontal gusts), and the penetration of jet streams or cold fronts), instrument errors, airframe production variations, and must not be less than Mach 0.05.

(c) *Design maneuvering speed V_A.* For V_A, the following applies:

(1) V_A may not be less than $V_S \sqrt{n}$ where—

(i) V_S is a computed stalling speed with flaps retracted at the design weight, normally based on the maximum airplane normal force coefficients, C_{NA}; and

(ii) *n* is the limit maneuvering load factor used in design

(2) The value of V_A need not exceed the value of V_C used in design.

(d) *Design speed for maximum gust intensity, V_B.* For V_B, the following apply:

(1) V_B may not be less than the speed determined by the intersection of the line representing the maximum positive lift, C_{NMAX}, and the line representing the rough air gust velocity on the gust V-n diagram, or $V_{S1} \sqrt{n_g}$, whichever is less, where:

(i) n_g the positive airplane gust load factor due to gust, at speed V_C (in accordance with § 23.341), and at the particular weight under consideration; and

(ii) V_{S1} is the stalling speed with the flaps retracted at the particular weight under consideration.

(2) V_B need not be greater than V_C.

[Doc. No. 4080, 29 FR 17955, Dec. 18, 1964, as amended by Amdt. 23-7, 34 FR 13088, Aug. 13, 1969; Amdt. 23-16, 40 FR 2577, Jan. 14, 1975; Amdt. 23-34, 52 FR 1829, Jan. 15, 1987; Amdt. 23-24, 52 FR 34745, Sept. 14, 1987; Amdt. 23-48, 61 FR 5143, Feb. 9, 1996]

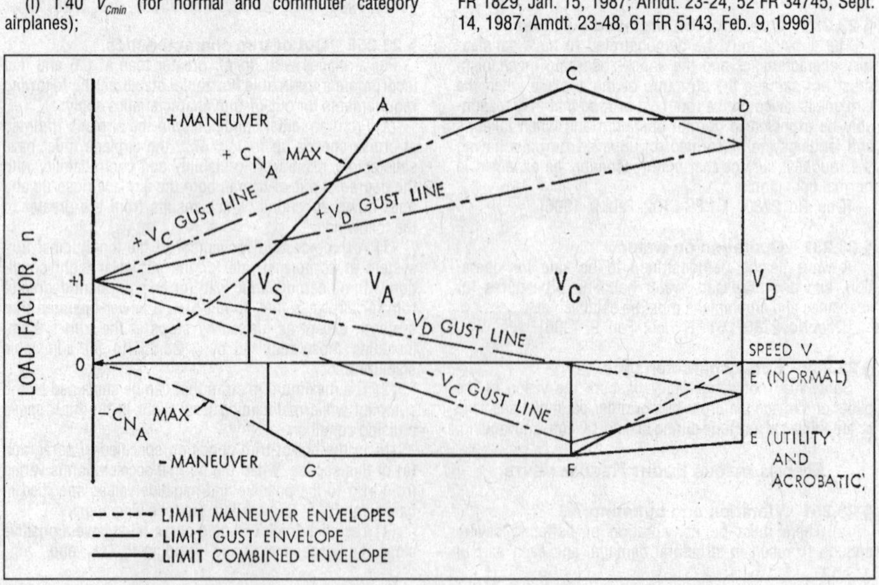

§ 23.337 Limit maneuvering load factors.

(a) The positive limit maneuvering load factor *n* may not be less than—

(1) 2.1+(24,000÷(W+10,000)) for normal and commuter category airplanes, where W=design maximum takeoff weight, except that n need not be more than 3.8;

(2) 4.4 for utility category airplanes; or

(3) 6.0 for acrobatic category airplanes.

(b) The negative limit maneuvering load factor may not be less than—

(1) 0.4 times the positive load factor for the normal utility and commuter categories; or

(2) 0.5 times the positive load factor for the acrobatic category.

(c) Maneuvering load factors lower than those specified in this section may be used if the airplane has design features that make it impossible to exceed these values in flight.

[Doc. No. 4080, 29 FR 17955, Dec. 18, 1964, as amended by Amdt. 23-7, 34 FR 13088, Aug. 13, 1969; Amdt. 23-34, 52 FR 1829, Jan. 15, 1987; Amdt. 23-48, 61 FR 5144, Feb. 9, 1996]

§ 23.341 Gust loads factors.

(a) Each airplane must be designed to withstand loads on each lifting surface resulting from gusts specified in § 23.333(c).

(b) The gust load for a canard or tandem wing configuration must be computed using a rational analysis, or may be computed in accordance with paragraph (c) of this section, provided that the resulting net loads are shown to be conservative with respect to the gust criteria of § 23.333(c).

(c) In the absence of a more rational analysis, the gust load factors must be computed as follows—

$$n = 1 + \frac{K_g U_{de} V a}{498(W / S)}$$

Where—

K_g=0.88μ_g/5.3+μ_g=gust alleviation factor;

μ_g=2(W/S)/ρ Cag=airplane mass ratio;

U_{de}=Derived gust velocities referred to in § 23.333(c) (f.p.s.);

ρ=Density of air (slugs/cu.ft.);

W/S=Wing loading (p.s.f.) due to the applicable weight of the airplane in the particular load case.

W/S=Wing loading (p.s.f.);

C=Mean geometric chord (ft.);

g=Acceleration due to gravity (ft./sec.^{2})

V=Airplane equivalent speed (knots); and

a=Slope of the airplane normal force coefficient curve C_{NA} per radian if the gust loads are applied to the wings and horizontal tail surfaces simultaneously by a rational method. The wing lift curve slope C_L per radian may be used when the gust load is applied to the wings only and the horizontal tail gust loads are treated as a separate condition.

[Amdt. 23-7, 34 FR 13088, Aug. 13, 1969, as amended by Amdt. 23-42, 56 FR 352, Jan. 3, 1991; Amdt. 23-48, 61 FR 5144, Feb. 9, 1996]

§ 23.343 Design fuel loads.

(a) The disposable load combinations must include each fuel load in the range from zero fuel to the selected maximum fuel load.

(b) If fuel is carried in the wings, the maximum allowable weight of the airplane without any fuel in the wing tank(s) must be established as "maximum zero wing fuel weight," if it is less than the maximum weight.

(c) For commuter category airplanes, a structural reserve fuel condition, not exceeding fuel necessary for 45 minutes of operation at maximum continuous power, may be selected. If a structural reserve fuel condition is selected, it must be used as the minimum fuel weight condition for showing compliance with the flight load requirements prescribed in this part and—

(1) The structure must be designed to withstand a condition of zero fuel in the wing at limit loads corresponding to:

(i) Ninety percent of the maneuvering load factors defined in § 23.337, and

(ii) Gust velocities equal to 85 percent of the values prescribed in § 23.333(c).

(2) The fatigue evaluation of the structure must account for any increase in operating stresses resulting from the design condition of paragraph (c)(1) of this section.

(3) The flutter, deformation, and vibration requirements must also be met with zero fuel in the wings.

[Doc. No. 27805, 61 FR 5144, Feb. 9, 1996]

§ 23.345 High lift devices.

(a) If flaps or similar high lift devices are to be used for takeoff, approach or landing, the airplane, with the flaps fully extended at V_F, is assumed to be subjected to symmetrical maneuvers and gusts within the range determined by—

(1) Maneuvering, to a positive limit load factor of 2.0; and

(2) Positive and negative gust of 25 feet per second acting normal to the flight path in level flight.

(b) V_F must be assumed to be not less than 1.4 V_S or 1.8 V_{SF}, whichever is greater, where—

(1) V_S is the computed stalling speed with flaps retracted at the design weight; and

(2) V_{SF} is the computed stalling speed with flaps fully extended at the design weight.

(3) If an automatic flap load limiting device is used, the airplane may be designed for the critical combinations of airspeed and flap position allowed by that device.

(c) In determining external loads on the airplane as a whole, thrust, slipstream, and pitching acceleration may be assumed to be zero.

(d) The flaps, their operating mechanism, and their supporting structures, must be designed to withstand the conditions prescribed in paragraph (a) of this section. In addition, with the flaps fully extended at V_F, the following conditions, taken separately, must be accounted for:

(1) A head-on gust having a velocity of 25 feet per second (EAS), combined with propeller slipstream corresponding to 75 percent of maximum continuous power; and

(2) The effects of propeller slipstream corresponding to maximum takeoff power.

[Doc. No. 27805, 61 FR 5144, Feb. 9, 1996]

§ 23.347 Unsymmetrical flight conditions.

(a) The airplane is assumed to be subjected to the unsymmetrical flight conditions of §§ 23.349 and 23.351. Unbalanced aerodynamic moments about the center of gravity must be reacted in a rational or conservative manner, considering the principal masses furnishing the reacting inertia forces.

(b) Acrobatic category airplanes certified for flick maneuvers (snap roll) must be designed for additional asymmetric loads acting on the wing and the horizontal tail.

[Doc. No. 4080, 29 FR 17955, Dec. 18, 1964, as amended by Amdt. 23-48, 61 FR 5144, Feb. 9, 1996]

§ 23.349 Rolling conditions.

The wing and wing bracing must be designed for the following loading conditions:

(a) Unsymmetrical wing loads appropriate to the category. Unless the following values result in unrealistic loads, the rolling accelerations may be obtained by modifying the symmetrical flight conditions in § 23.333(d) as follows:

(1) For the acrobatic category, in conditions A and F, assume that 100 percent of the semispan wing airload acts on one side of the plane of symmetry and 60 percent of this load acts on the other side.

(2) For normal, utility, and commuter categories, in Condition A, assume that 100 percent of the semispan wing airload acts on one side of the airplane and 75 percent of this load acts on the other side.

(b) The loads resulting from the aileron deflections and speeds specified in § 23.455, in combination with an airplane load factor of at least two thirds of the positive maneuvering load factor used for design. Unless the following values result in unrealistic loads, the effect of aileron displacement on wing torsion may be accounted for by adding the following increment to the basic airfoil moment coefficient over the aileron portion of the span in the critical condition determined in § 23.333(d):

$\Delta c_m = -0.01\delta$

where—

Δc_m is the moment coefficient increment; and

δ is the down aileron deflection in degrees in the critical condition.

[Doc. No. 4080, 29 FR 17955, Dec. 18, 1964, as amended by Amdt. 23-7, 34 FR 13088, Aug. 13, 1969; Amdt. 23-34, 52 FR 1829, Jan. 15, 1987; Amdt. 23-48, 61 FR 5144, Feb. 9, 1996]

§ 23.351 Yawing conditions.

The airplane must be designed for yawing loads on the vertical surfaces resulting from the loads specified in §§ 23.441 through 23.445.

[Doc. No. 4080, 29 FR 17955, Dec. 18, 1964; 30 FR 258, Jan. 9, 1965, as amended by Amdt. 23-42, 56 FR 352, Jan. 3, 1991]

§ 23.361 Engine torque.

(a) Each engine mount and its supporting structure must be designed for the effects of—

(1) A limit engine torque corresponding to takeoff power and propeller speed acting simultaneously with 75 percent of the limit loads from flight condition A of § 23.333(d);

(2) A limit engine torque corresponding to maximum continuous power and propeller speed acting simultaneously with the limit loads from flight condition A of § 23.333(d); and

(3) For turbopropeller installations, in addition to the conditions specified in paragraphs (a)(1) and (a)(2) of this section, a limit engine torque corresponding to takeoff power and propeller speed, multiplied by a factor accounting for propeller control system malfunction, including quick feathering, acting simultaneously with lg level flight loads. In the absence of a rational analysis, a factor of 1.6 must be used.

(b) For turbine engine installations, the engine mounts and supporting structure must be designed to withstand each of the following:

(1) A limit engine torque load imposed by sudden engine stoppage due to malfunction or structural failure (such as compressor jamming).

(2) A limit engine torque load imposed by the maximum acceleration of the engine.

(c) The limit engine torque to be considered under paragraph (a) of this section must be obtained by multiplying the mean torque by a factor of—

(1) 1.25 for turbopropeller installations;

(2) 1.33 for engines with five or more cylinders; and

(3) Two, three, or four, for engines with four, three, or two cylinders, respectively.

[Amdt. 23-26, 45 FR 60171, Sept. 11, 1980, as amended by Amdt. 23-45, 58 FR 42160, Aug. 6, 1993]

§ 23.363 Side load on engine mount.

(a) Each engine mount and its supporting structure must be designed for a limit load factor in a lateral direction, for the side load on the engine mount, of not less than—

(1) 1.33, or

(2) One-third of the limit load factor for flight condition A.

(b) The side load prescribed in paragraph (a) of this section may be assumed to be independent of other flight conditions.

§ 23.365 Pressurized cabin loads.

For each pressurized compartment, the following apply:

(a) The airplane structure must be strong enough to withstand the flight loads combined with pressure differential loads from zero up to the maximum relief valve setting.

(b) The external pressure distribution in flight, and any stress concentrations, must be accounted for.

(c) If landings may be made with the cabin pressurized, landing loads must be combined with pressure differential loads from zero up to the maximum allowed during landing.

(d) The airplane structure must be strong enough to withstand the pressure differential loads corresponding to the maximum relief valve setting multiplied by a factor of 1.33, omitting other loads.

(e) If a pressurized cabin has two or more compartments separated by bulkheads or a floor, the primary structure must be designed for the effects of sudden release of pressure in any compartment with external doors or windows. This condition must be investigated for the effects of failure of the largest opening in the compartment. The effects of intercompartmental venting may be considered.

§ 23.367 Unsymmetrical loads due to engine failure.

(a) Turbopropeller airplanes must be designed for the unsymmetrical loads resulting from the failure of the critical engine including the following conditions in combination with a single malfunction of the propeller drag limiting system, considering the probable pilot corrective action on the flight controls:

(1) At speeds between V_{MC} and V_D, the loads resulting from power failure because of fuel flow interruption are considered to be limit loads.

(2) At speeds between V_{MC} and V_C, the loads resulting from the disconnection of the engine compressor from the turbine or from loss of the turbine blades are considered to be ultimate loads.

(3) The time history of the thrust decay and drag buildup occurring as a result of the prescribed engine failures must be substantiated by test or other data applicable to the particular engine-propeller combination.

(4) The timing and magnitude of the probable pilot corrective action must be conservatively estimated, considering the characteristics of the particular engine-propeller-airplane combination.

(b) Pilot corrective action may be assumed to be initiated at the time maximum yawing velocity is reached, but not earlier than 2 seconds after the engine failure. The magnitude of the corrective action may be based on the limit pilot forces specified in § 23.397 except that lower forces may be assumed where it is shown by analysis or test that these forces can control the yaw and roll resulting from the prescribed engine failure conditions.

[Amdt. 23-7, 34 FR 13089, Aug. 13, 1969]

§ 23.369 Rear lift truss.

(a) If a rear lift truss is used, it must be designed to withstand conditions of reversed airflow at a design speed of—

V=8.7 √(W/S) + 8.7 (knots), where W/S=wing loading at design maximum takeoff weight.

(b) Either aerodynamic data for the particular wing section used, or a value of C_L equalling −0.8 with a chordwise distribution that is triangular between a peak at the trailing edge and zero at the leading edge, must be used.

[Doc. No. 4080, 29 FR 17955, Dec. 18, 1964, as amended by Amdt. 23-7, 34 FR 13089, Aug. 13, 1969; 34 FR 17509, Oct. 30, 1969; Amdt. 23-45, 58 FR 42160, Aug. 6, 1993; Amdt. 23-48, 61 FR 5145, Feb. 9, 1996]

§ 23.371 Gyroscopic and aerodynamic loads.

(a) Each engine mount and its supporting structure must be designed for the gyroscopic, inertial, and aerodynamic loads that result, with the engine(s) and propeller(s), if applicable, at maximum continuous r.p.m., under either:

(1) The conditions prescribed in § 23.351 and § 23.423; or

(2) All possible combinations of the following—

(i) A yaw velocity of 2.5 radians per second;

(ii) A pitch velocity of 1.0 radian per second;

(iii) A normal load factor of 2.5; and

(iv) Maximum continuous thrust.

(b) For airplanes approved for aerobatic maneuvers, each engine mount and its supporting structure must meet the requirements of paragraph (a) of this section and be designed to withstand the load factors expected during combined maximum yaw and pitch velocities.

(c) For airplanes certificated in the commuter category, each engine mount and its supporting structure must meet the requirements of paragraph (a) of this section and the gust conditions specified in § 23.341 of this part.

[Doc. No. 27805, 61 FR 5145, Feb. 9, 1996]

§ 23.373 Speed control devices.

If speed control devices (such as spoilers and drag flaps) are incorporated for use in enroute conditions—

(a) The airplane must be designed for the symmetrical maneuvers and gusts prescribed in §§ 23.333, 23.337, and 23.341, and the yawing maneuvers and lateral gusts in §§ 23.441 and 23.443, with the device extended at speeds up to the placard device extended speed; and

(b) If the device has automatic operating or load limiting features, the airplane must be designed for the maneuver and gust conditions prescribed in paragraph (a) of this section at the speeds and corresponding device positions that the mechanism allows.

[Amdt. 23-7, 34 FR 13089, Aug. 13, 1969]

Control Surface and System Loads

§ 23.391 Control surface loads.

The control surface loads specified in §§ 23.397 through 23.459 are assumed to occur in the conditions described in §§ 23.331 through 23.351.

[Doc. No. 4080, 29 FR 17955, Dec. 18, 1964, as amended by Amdt. 23-48, 61 FR 5145, Feb. 9, 1996]

§ 23.393 Loads parallel to hinge line.

(a) Control surfaces and supporting hinge brackets must be designed to withstand inertial loads acting parallel to the hinge line.

(b) In the absence of more rational data, the inertial loads may be assumed to be equal to KW, where—

(1) K=24 for vertical surfaces;

(2) K=12 for horizontal surfaces; and

(3) W=weight of the movable surfaces.

[Doc. No. 27805, 61 FR 5145, Feb. 9, 1996]

§ 23.395 Control system loads.

(a) Each flight control system and its supporting structure must be designed for loads corresponding to at least 125 percent of the computed hinge moments of the movable control surface in the conditions prescribed in §§ 23.391 through 23.459. In addition, the following apply:

(1) The system limit loads need not exceed the higher of the loads that can be produced by the pilot and automatic devices operating the controls. However, autopilot forces need not be added to pilot forces. The system must be designed for the maximum effort of the pilot or autopilot, whichever is higher. In addition, if the pilot and the autopilot act in opposition, the part of the system between them may be designed for the maximum effort of the one that imposes the lesser load. Pilot forces used for design need not exceed the maximum forces prescribed in § 23.397(b).

(2) The design must, in any case, provide a rugged system for service use, considering jamming, ground gusts, taxiing downwind, control inertia, and friction. Compliance with this subparagraph may be shown by designing for loads resulting from application of the minimum forces prescribed in § 23.397(b).

(b) A 125 percent factor on computed hinge moments must be used to design elevator, aileron, and rudder systems. However, a factor as low as 1.0 may be used if hinge moments are based on accurate flight test data, the exact reduction depending upon the accuracy and reliability of the data.

(c) Pilot forces used for design are assumed to act at the appropriate control grips or pads as they would in flight, and to react at the attachments of the control system to the control surface horns.

[Doc. No. 4080, 29 FR 17955, Dec. 18, 1964, as amended by Amdt. 23-7, 34 FR 13089, Aug. 13, 1969]

§ 23.397 Limit control forces and torques.

(a) In the control surface flight loading condition, the airloads on movable surfaces and the corresponding deflections need not exceed those that would result in flight from the application of any pilot force within the ranges specified in paragraph (b) of this section. In applying this criterion, the effects of control system boost and servo-mechanisms, and the effects of tabs must be considered. The automatic pilot effort must be used for design if it alone can produce higher control surface loads than the human pilot.

(b) The limit pilot forces and torques are as follows:

Control	Maximum forces or torques for design weight, weight equal to or less than 5,000 pounds1	Minimum forces or torques2
Aileron:		
Stick	67 lbs	40 lbs.
Wheel[3]	50 D in.-lbs[4]	40 D in.-lbs.[4]
Elevator:		
Stick	167 lbs	100 lbs.
Wheel (symmetrical)	200 lbs	100 lbs.
Wheel (unsymmetrical)[5]		100 lbs.
Rudder	200 lbs	150 lbs.

[1] For design weight (*W*) more than 5,000 pounds, the specified maximum values must be increased linearly with weight to 1.18 times the specified values at a design weight of 12,500 pounds and for commuter category airplanes, the specified values must be increased linearly with weight to 1.35 times the specified values at a design weight of 19,000 pounds.

[2] If the design of any individual set of control systems or surfaces makes these specified minimum forces or torques inapplicable, values corresponding to the present hinge moments obtained under § 23.415, but not less than 0.6 of the specified minimum forces or torques, may be used.

[3] The critical parts of the aileron control system must also be designed for a single tangential force with a limit value of 1.25 times the couple force determined from the above criteria.

[4] D=wheel diameter (inches).

[5] The unsymmetrical force must be applied at one of the normal handgrip points on the control wheel.

[Doc. No. 4080, 29 FR 17955, Dec. 18, 1964, as amended by Amdt. 23-7, 34 FR 13089, Aug. 13, 1969; Amdt. 23-17, 41 FR 55464, Dec. 20, 1976; Amdt. 23-34, 52 FR 1829, Jan. 15, 1987; Amdt. 23-45, 58 FR 42160, Aug. 6, 1993]

§ 23.399 Dual control system.

(a) Each dual control system must be designed to withstand the force of the pilots operating in opposition, using individual pilot forces not less than the greater of—

(1) 0.75 times those obtained under § 23.395; or

(2) The minimum forces specified in § 23.397(b).

(b) Each dual control system must be designed to withstand the force of the pilots applied together, in the same direction, using individual pilot forces not less than 0.75 times those obtained under § 23.395.

[Doc. No. 27805, 61 FR 5145, Feb. 9, 1996]

§ 23.405 Secondary control system.

Secondary controls, such as wheel brakes, spoilers, and tab controls, must be designed for the maximum forces that a pilot is likely to apply to those controls.

§ 23.407 Trim tab effects.

The effects of trim tabs on the control surface design conditions must be accounted for only where the surface loads are limited by maximum pilot effort. In these cases, the tabs are considered to be deflected in the direction that would assist the pilot. These deflections must correspond to the maximum degree of "out of trim" expected at the speed for the condition under consideration.

§ 23.409 Tabs.

Control surface tabs must be designed for the most severe combination of airspeed and tab deflection likely to be obtained within the flight envelope for any usable loading condition.

§ 23.415 Ground gust conditions.

(a) The control system must be investigated as follows for control surface loads due to ground gusts and taxiing downwind:

(1) If an investigation of the control system for ground gust loads is not required by paragraph (a)(2) of this section, but the applicant elects to design a part of the control system of these loads, these loads need only be carried from control surface horns through the nearest stops or gust locks and their supporting structures.

(2) If pilot forces less than the minimums specified in § 23.397(b) are used for design, the effects of surface loads due to ground gusts and taxiing downwind must be investigated for the entire control system according to the formula:

$H = K\,c\,S\,q$

where—

H=limit hinge moment (ft.-lbs.);

c=mean chord of the control surface aft of the hinge line (ft.);

S=area of control surface aft of the hinge line (sq. ft.);

q=dynamic pressure (p.s.f.) based on a design speed not less than 14.6 √(W/S) + 14.6 (f.p.s.) where W/S=wing loading at design maximum weight, except that the design speed need not exceed 88 (f.p.s.);

K=limit hinge moment factor for ground gusts derived in paragraph (b) of this section. (For ailerons and elevators, a positive value of K indicates a moment tending to depress the surface and a negative value of K indicates a moment tending to raise the surface).

(b) The limit hinge moment factor K for ground gusts must be derived as follows:

(c) At all weights between the empty weight and the maximum weight declared for tie-down stated in the appropriate manual, any declared tie-down points and surrounding structure, control system, surfaces and associated gust locks, must be designed to withstand the limit load conditions that exist when the airplane is tied down and that result from wind speeds of up to 65 knots horizontally from any direction.

[Doc. No. 4080, 29 FR 17955, Dec. 18, 1964, as amended by Amdt. 23-7, 34 FR 13089, Aug. 13, 1969; Amdt. 23-45, 58 FR 42160, Aug. 6, 1993; Amdt. 23-48, 61 FR 5145, Feb. 9, 1996]

Surface	K	Position of controls
(a) Aileron	0.75	Control column locked lashed in mid-position.
(b) Aileron	±0.50	Ailerons at full throw; + moment on one aileron, – moment on the other.
(c) Elevator	±0.75	(c) Elevator full up (–).
(d) Elevator		(d) Elevator full down (+).
(e) Rudder	±0.75	(e) Rudder in neutral.
(f) Rudder		(f) Rudder at full throw.

Horizontal Stabilizing and Balancing Surfaces

§ 23.421 Balancing loads.

(a) A horizontal surface balancing load is a load necessary to maintain equilibrium in any specified flight condition with no pitching acceleration.

(b) Horizontal balancing surfaces must be designed for the balancing loads occurring at any point on the limit maneuvering envelope and in the flap conditions specified in § 23.345.

[Doc. No. 4080, 29 FR 17955, Dec. 18, 1964, as amended by Amdt. 23-7, 34 FR 13089, Aug. 13, 1969; Amdt. 23-42, 56 FR 352, Jan. 3, 1991]

§ 23.423 Maneuvering loads.

Each horizontal surface and its supporting structure, and the main wing of a canard or tandem wing configuration, if that surface has pitch control, must be designed for the maneuvering loads imposed by the following conditions:

(a) A sudden movement of the pitching control, at the speed V_A, to the maximum aft movement, and the maximum forward movement, as limited by the control stops, or pilot effort, whichever is critical.

(b) A sudden aft movement of the pitching control at speeds above V_A, followed by a forward movement of the pitching control resulting in the following combinations of normal and angular acceleration:

where—

(1) n_m=positive limit maneuvering load factor used in the design of the airplane; and

(2) V=initial speed in knots.

The conditions in this paragraph involve loads corresponding to the loads that may occur in a "checked maneuver" (a maneuver in which the pitching control is suddenly displaced in one direction and then suddenly moved in the opposite direction). The deflections and timing of the "checked maneuver" must avoid exceeding the limit maneuvering load factor. The total horizontal surface load for both nose-up and nose-down pitching conditions is the sum of the balancing loads at V and the specified value of the normal load factor n, plus the maneuvering load increment due to the specified value of the angular acceleration.

Condition	Normal acceleration (n)	Angular acceleration (radian/sec2)
Nose-up pitching	1.0	$+39n_m \div V \times (n_m - 1.5)$
Nose-down pitching	n_m	$-39n_m \div V \times (n_m - 1.5)$

[Amdt. 23-42, 56 FR 353, Jan. 3, 1991; 56 FR 5455, Feb. 11, 1991]

§ 23.425 Gust loads.

(a) Each horizontal surface, other than a main wing, must be designed for loads resulting from—

(1) Gust velocities specified in § 23.333(c) with flaps retracted; and

(2) Positive and negative gusts of 25 f.p.s. nominal intensity at V_F corresponding to the flight conditions specified in § 23.345(a)(2).

(b) [Reserved]

(c) When determining the total load on the horizontal surfaces for the conditions specified in paragraph (a) of this section, the initial balancing loads for steady unaccelerated flight at the pertinent design speeds V_F, V_C, and V_D must first be determined. The incremental load resulting from the gusts must be added to the initial balancing load to obtain the total load.

(d) In the absence of a more rational analysis, the incremental load due to the gust must be computed as follows only on airplane configurations with aft-mounted, horizontal surfaces, unless its use elsewhere is shown to be conservative:

$$\Delta L_{ht} = \frac{K_g U_{de} V a_{ht} S_{ht}}{498}\left(1 - \frac{dE}{d\alpha}\right)$$

where—

ΔL_{ht}=Incremental horizontal tailload (lbs.);

K_g=Gust alleviation factor defined in § 23.341;

U_{de}=Derived gust velocity (f.p.s.);

V=Airplane equivalent speed (knots);

a_{ht}=Slope of aft horizontal lift curve (per radian)

S_{ht}=Area of aft horizontal lift surface (ft²); and

$$\left(1 - \frac{dE}{d\alpha}\right) = \text{Downwash factor}$$

[Doc. No. 4080, 29 FR 17955, Dec. 18, 1964, as amended by Amdt. 23-7, 34 FR 13089 Aug. 13, 1969; Amdt. 23-42, 56 FR 353, Jan. 3, 1991]

§ 23.427 Unsymmetrical loads.

(a) Horizontal surfaces other than main wing and their supporting structure must be designed for unsymmetrical loads arising from yawing and slipstream effects, in combination with the loads prescribed for the flight conditions set forth in §§ 23.421 through 23.425.

(b) In the absence of more rational data for airplanes that are conventional in regard to location of engines, wings, horizontal surfaces other than main wing, and fuselage shape:

(1) 100 percent of the maximum loading from the symmetrical flight conditions may be assumed on the surface on one side of the plane of symmetry; and

(2) The following percentage of that loading must be applied to the opposite side:

Percent=100–10 (n–1), where n is the specified positive maneuvering load factor, but this value may not be more than 80 percent.

(c) For airplanes that are not conventional (such as airplanes with horizontal surfaces other than main wing having appreciable dihedral or supported by the vertical tail surfaces) the surfaces and supporting structures must be designed for combined vertical and horizontal surface loads resulting from each prescribed flight condition taken separately.

[Amdt. 23-14, 38 FR 31820, Nov. 19, 1973, as amended by Amdt. 23-42, 56 FR 353, Jan. 3, 1991]

Vertical Surfaces

§ 23.441 Maneuvering loads.

(a) At speeds up to V_A, the vertical surfaces must be designed to withstand the following conditions. In computing the loads, the yawing velocity may be assumed to be zero:

(1) With the airplane in unaccelerated flight at zero yaw, it is assumed that the rudder control is suddenly displaced to the maximum deflection, as limited by the control stops or by limit pilot forces.

(2) With the rudder deflected as specified in paragraph (a)(1) of this section, it is assumed that the airplane yaws to the overswing sideslip angle. In lieu of a rational analysis, an overswing angle equal to 1.5 times the static sideslip angle of paragraph (a)(3) of this section may be assumed.

(3) A yaw angle of 15 degrees with the rudder control maintained in the neutral position (except as limited by pilot strength).

(b) For commuter category airplanes, the loads imposed by the following additional maneuver must be substantiated at speeds from V_A to V_D/M_D. When computing the tail loads—

(1) The airplane must be yawed to the largest attainable steady state sideslip angle, with the rudder at maximum deflection caused by any one of the following:

(i) Control surface stops;

(ii) Maximum available booster effort;

(iii) Maximum pilot rudder force as shown below:

(2) The rudder must be suddenly displaced from the maximum deflection to the neutral position.

(c) The yaw angles specified in paragraph (a)(3) of this section may be reduced if the yaw angle chosen for a particular speed cannot be exceeded in—

(1) Steady slip conditions;

(2) Uncoordinated rolls from steep banks; or

(3) Sudden failure of the critical engine with delayed corrective action.

[Doc. No. 4080, 29 FR 17955, Dec. 18, 1964, as amended by Amdt. 23-7, 34 FR 13090, Aug. 13, 1969; Amdt. 23-14, 38 FR 31821, Nov. 19, 1973; Amdt. 23-28, 47 FR 13315, Mar. 29, 1982; Amdt. 23-42, 56 FR 353, Jan. 3, 1991; Amdt. 23-48, 61 FR 5145, Feb. 9, 1996]

§ 23.443 Gust loads.

(a) Vertical surfaces must be designed to withstand, in unaccelerated flight at speed V_C, lateral gusts of the values prescribed for V_C in § 23.333(c).

(b) In addition, for commuter category airplanes, the airplane is assumed to encounter derived gusts normal to the plane of symmetry while in unaccelerated flight at V_B, V_C, V_D, and V_F. The derived gusts and airplane speeds corresponding to these conditions, as determined by §§ 23.341 and 23.345, must be investigated. The shape of the gust must be as specified in § 23.333(c)(2)(i).

(c) In the absence of a more rational analysis, the gust load must be computed as follows:

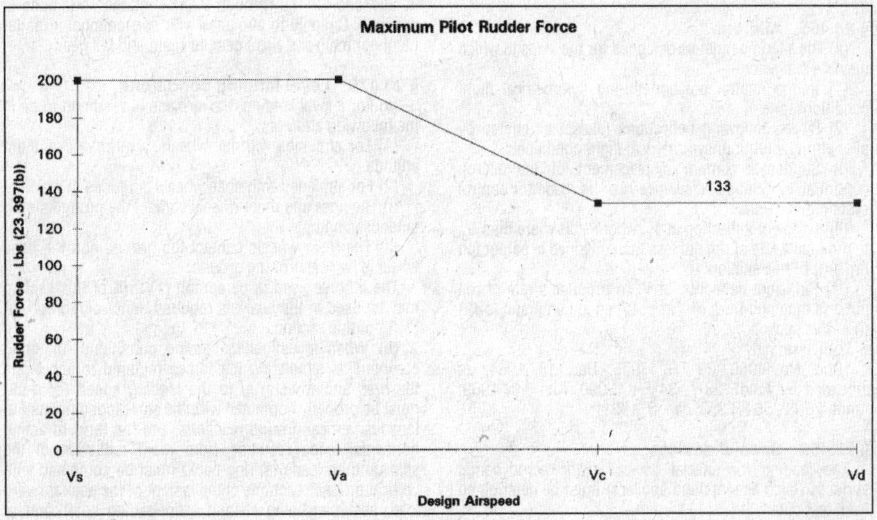

$$L_{vt} = \frac{K_{gt} U_{de} V a_{vt} S_{vt}}{498}$$

Where—

L_{vt}=Vertical surface loads (lbs.);

$$K_{gt} = \frac{0.88\mu_{gt}}{5.3 + \mu_{gt}} = \text{gust alleviation factor;}$$

$$\mu_{gt} = \frac{2W}{\rho c_t g a_{vt} S_{vt}} \frac{K^2}{l_{vt}} = \text{lateralmassratio;}$$

U_{de}=Derived gust velocity (f.p.s.);

ρ=Air density (slugs/cu.ft.);

W=the applicable weight of the airplane in the particular load case (lbs.);

S_{vt}=Area of vertical surface (ft.²);

$\bar{c}_t$=Mean geometric chord of vertical surface (ft.);

a_{vt}=Lift curve slope of vertical surface (per radian);

K=Radius of gyration in yaw (ft.);

l_{vt}=Distance from airplane c.g. to lift center of vertical surface (ft.);

g=Acceleration due to gravity (ft./sec.²); and

V=Equivalent airspeed (knots).

[Amdt. 23-7, 34 FR 13090, Aug. 13, 1969, as amended by Amdt. 23-34, 52 FR 1830, Jan. 15, 1987; 52 FR 7262, Mar. 9, 1987; Amdt. 23-24, 52 FR 34745, Sept. 14, 1987; Amdt. 23-42, 56 FR 353, Jan. 3, 1991; Amdt. 23-48, 61 FR 5147, Feb. 9, 1996]

§ 23.445 Outboard fins or winglets.

(a) If outboard fins or winglets are included on the horizontal surfaces or wings, the horizontal surfaces or wings must be designed for their maximum load in combination with loads induced by the fins or winglets and moments or forces exerted on the horizontal surfaces or wings by the fins or winglets.

(b) If outboard fins or winglets extend above and below the horizontal surface, the critical vertical surface loading (the load per unit area as determined under §§ 23.441 and 23.443) must be applied to—

(1) The part of the vertical surfaces above the horizontal surface with 80 percent of that loading applied to the part below the horizontal surface; and

(2) The part of the vertical surfaces below the horizontal surface with 80 percent of that loading applied to the part above the horizontal surface.

(c) The end plate effects of outboard fins or winglets must be taken into account in applying the yawing conditions of §§ 23.441 and 23.443 to the vertical surfaces in paragraph (b) of this section.

(d) When rational methods are used for computing loads, the maneuvering loads of § 23.441 on the vertical surfaces and the one-g horizontal surface load, including induced loads on the horizontal surface and moments or forces exerted on the horizontal surfaces by the vertical surfaces, must be applied simultaneously for the structural loading condition.

[Doc. No. 4080, 29 FR 17955, Dec. 18, 1964, as amended by Amdt. 23-14, 38 FR 31821, Nov. 19, 1973; Amdt. 23-42, 56 FR 353, Jan. 3, 1991]

Ailerons and Special Devices

§ 23.455 Ailerons.

(a) The ailerons must be designed for the loads to which they are subjected—

(1) In the neutral position during symmetrical flight conditions; and

(2) By the following deflections (except as limited by pilot effort), during unsymmetrical flight conditions:

(i) Sudden maximum displacement of the aileron control at V_A. Suitable allowance may be made for control system deflections.

(ii) Sufficient deflection at V_C, where V_C is more than V_A, to produce a rate of roll not less than obtained in paragraph (a)(2)(i) of this section.

(iii) Sufficient deflection at V_D to produce a rate of roll not less than one-third of that obtained in paragraph (a)(2)(i) of this section.

(b) [Reserved]

[Doc. No. 4080, 29 FR 17955, Dec. 18, 1964, as amended by Amdt. 23-7, 34 FR 13090, Aug. 13, 1969; Amdt. 23-42, 56 FR 353, Jan. 3, 1991]

§ 23.459 Special devices.

The loading for special devices using aerodynamic surfaces (such as slots and spoilers) must be determined from test data.

Ground Loads

§ 23.471 General.

The limit ground loads specified in this subpart are considered to be external loads and inertia forces that act upon an airplane structure. In each specified ground load condition, the external reactions must be placed in equilibrium with the linear and angular inertia forces in a rational or conservative manner.

§ 23.473 Ground load conditions and assumptions.

(a) The ground load requirements of this subpart must be complied with at the design maximum weight except that §§ 23.479, 23.481, and 23.483 may be complied with at a design landing weight (the highest weight for landing conditions at the maximum descent velocity) allowed under paragraphs (b) and (c) of this section.

(b) The design landing weight may be as low as—

(1) 95 percent of the maximum weight if the minimum fuel capacity is enough for at least one-half hour of operation at maximum continuous power plus a capacity equal to a fuel weight which is the difference between the design maximum weight and the design landing weight; or

(2) The design maximum weight less the weight of 25 percent of the total fuel capacity.

(c) The design landing weight of a multiengine airplane may be less than that allowed under paragraph (b) of this section if—

(1) The airplane meets the one-engine-inoperative climb requirements of § 23.67(b)(1) or (c); and

(2) Compliance is shown with the fuel jettisoning system requirements of § 23.1001.

(d) The selected limit vertical inertia load factor at the center of gravity of the airplane for the ground load conditions prescribed in this subpart may not be less than that which would be obtained when landing with a descent velocity (V), in feet per second, equal to 4.4 $(W/S)^{1/4}$, except that this velocity need not be more than 10 feet per second and may not be less than seven feet per second.

(e) Wing lift not exceeding two-thirds of the weight of the airplane may be assumed to exist throughout the landing impact and to act through the center of gravity. The ground reaction load factor may be equal to the inertia load factor minus the ratio of the above assumed wing lift to the airplane weight.

(f) If energy absorption tests are made to determine the limit load factor corresponding to the required limit descent velocities, these tests must be made under § 23.723(a).

(g) No inertia load factor used for design purposes may be less than 2.67, nor may the limit ground reaction load factor be less than 2.0 at design maximum weight, unless these lower values will not be exceeded in taxiing at speeds up to takeoff speed over terrain as rough as that expected in service.

[Doc. No. 4080, 29 FR 17955, Dec. 18, 1964, as amended by Amdt. 23-7, 34 FR 13090, Aug. 13, 1969; Amdt. 23-28, 47 FR 13315, Mar. 29, 1982; Amdt. 23-45, 58 FR 42160, Aug. 6, 1993; Amdt. 23-48, 61 FR 5147, Feb. 9, 1996]

§ 23.477 Landing gear arrangement.

Sections 23.479 through 23.483, or the conditions in appendix C, apply to airplanes with conventional arrangements of main and nose gear, or main and tail gear.

§ 23.479 Level landing conditions.

(a) For a level landing, the airplane is assumed to be in the following attitudes:

(1) For airplanes with tail wheels, a normal level flight attitude.

(2) For airplanes with nose wheels, attitudes in which—

(i) The nose and main wheels contact the ground simultaneously; and

(ii) The main wheels contact the ground and the nose wheel is just clear of the ground.

The attitude used in paragraph (a)(2)(i) of this section may be used in the analysis required under paragraph (a)(2)(ii) of this section.

(b) When investigating landing conditions, the drag components simulating the forces required to accelerate the tires and wheels up to the landing speed (spin-up) must be properly combined with the corresponding instantaneous vertical ground reactions, and the forward-acting horizontal loads resulting from rapid reduction of the spin-up drag loads (spring-back) must be combined with vertical ground reactions at the instant of the peak forward load, assuming wing lift and a tire-sliding coefficient of friction of 0.8. However, the drag loads may not be less than 25 percent of the maximum vertical ground reactions (neglecting wing lift).

(c) In the absence of specific tests or a more rational analysis for determining the wheel spin-up and spring-back loads for landing conditions, the method set forth in appendix D of this part must be used. If appendix D of this part is used, the drag components used for design must not be less than those given by appendix C of this part.

(d) For airplanes with tip tanks or large overhung masses (such as turbo-propeller or jet engines) supported by the wing, the tip tanks and the structure supporting the tanks or overhung masses must be designed for the effects of dynamic responses under the level landing conditions of either paragraph (a)(1) or (a)(2)(ii) of this section. In evaluating the effects of dynamic response, an airplane lift equal to the weight of the airplane may be assumed.

[Doc. No. 4080, 29 FR 17955, Dec. 18, 1964, as amended by Amdt. 23-17, 41 FR 55464, Dec. 20, 1976; Amdt. 23-45, 58 FR 42160, Aug. 6, 1993]

§ 23.481 Tail down landing conditions.

(a) For a tail down landing, the airplane is assumed to be in the following attitudes:

(1) For airplanes with tail wheels, an attitude in which the main and tail wheels contact the ground simultaneously.

(2) For airplanes with nose wheels, a stalling attitude, or the maximum angle allowing ground clearance by each part of the airplane, whichever is less.

(b) For airplanes with either tail or nose wheels, ground reactions are assumed to be vertical, with the wheels up to speed before the maximum vertical load is attained.

§ 23.483 One-wheel landing conditions.

For the one-wheel landing condition, the airplane is assumed to be in the level attitude and to contact the ground on one side of the main landing gear. In this attitude, the ground reactions must be the same as those obtained on that side under § 23.479.

§ 23.485 Side load conditions.

(a) For the side load condition, the airplane is assumed to be in a level attitude with only the main wheels contacting the ground and with the shock absorbers and tires in their static positions.

(b) The limit vertical load factor must be 1.33, with the vertical ground reaction divided equally between the main wheels.

(c) The limit side inertia factor must be 0.83, with the side ground reaction divided between the main wheels so that—

(1) 0.5 (W) is acting inboard on one side; and

(2) 0.33 (W) is acting outboard on the other side.

(d) The side loads prescribed in paragraph (c) of this section are assumed to be applied at the ground contact point and the drag loads may be assumed to be zero.

[Doc. No. 4080, 29 FR 17955, Dec. 18, 1964, as amended by Amdt. 23-45, 58 FR 42160, Aug. 6, 1993]

§ 23.493 Braked roll conditions.

Under braked roll conditions, with the shock absorbers and tires in their static positions, the following apply:

(a) The limit vertical load factor must be 1.33.

(b) The attitudes and ground contacts must be those described in § 23.479 for level landings.

(c) A drag reaction equal to the vertical reaction at the wheel multiplied by a coefficient of friction of 0.8 must be applied at the ground contact point of each wheel with brakes, except that the drag reaction need not exceed the maximum value based on limiting brake torque.

§ 23.497 Supplementary conditions for tail wheels.

In determining the ground loads on the tail wheel and affected supporting structures, the following apply:

(a) For the obstruction load, the limit ground reaction obtained in the tail down landing condition is assumed to act up and aft through the axle at 45 degrees. The shock absorber and tire may be assumed to be in their static positions.

(b) For the side load, a limit vertical ground reaction equal to the static load on the tail wheel, in combination with a side component of equal magnitude, is assumed. In addition—

(1) If a swivel is used, the tail wheel is assumed to be swiveled 90 degrees to the airplane longitudinal axis with the resultant ground load passing through the axle;

(2) If a lock, steering device, or shimmy damper is used, the tail wheel is also assumed to be in the trailing position with the side load acting at the ground contact point; and

(3) The shock absorber and tire are assumed to be in their static positions.

(c) If a tail wheel, bumper, or an energy absorption device is provided to show compliance with § 23.925(b), the following apply:

(1) Suitable design loads must be established for the tail wheel, bumper, or energy absorption device; and

(2) The supporting structure of the tail wheel, bumper, or energy absorption device must be designed to withstand the loads established in paragraph (c)(1) of this section.

[Doc. No. 4080, 29 FR 17955, Dec. 18, 1964, as amended by Amdt. 23-48, 61 FR 5147, Feb. 9, 1996]

§ 23.499 Supplementary conditions for nose wheels.

In determining the ground loads on nose wheels and affected supporting structures, and assuming that the shock absorbers and tires are in their static positions, the following conditions must be met:

(a) For aft loads, the limit force components at the axle must be—

(1) A vertical component of 2.25 times the static load on the wheel; and

(2) A drag component of 0.8 times the vertical load.

(b) For forward loads, the limit force components at the axle must be—

(1) A vertical component of 2.25 times the static load on the wheel; and

(2) A forward component of 0.4 times the vertical load.

(c) For side loads, the limit force components at ground contact must be—

(1) A vertical component of 2.25 times the static load on the wheel; and

(2) A side component of 0.7 times the vertical load.

(d) For airplanes with a steerable nose wheel that is controlled by hydraulic or other power, at design takeoff weight with the nose wheel in any steerable position, the application of 1.33 times the full steering torque combined with a vertical reaction equal to 1.33 times the maximum static reaction on the nose gear must be assumed. However, if a torque limiting device is installed, the steering torque can be reduced to the maximum value allowed by that device.

(e) For airplanes with a steerable nose wheel that has a direct mechanical connection to the rudder pedals, the mechanism must be designed to withstand the steering torque for the maximum pilot forces specified in § 23.397(b).

[Doc. No. 4080, 29 FR 17955, Dec. 18, 1964, as amended by Amdt. 23-48, 61 FR 5147, Feb. 9, 1996]

§ 23.505 Supplementary conditions for skiplanes.

In determining ground loads for skiplanes, and assuming that the airplane is resting on the ground with one main ski frozen at rest and the other skis free to slide, a limit side force equal to 0.036 times the design maximum weight must be applied near the tail assembly, with a factor of safety of 1.

[Amdt. 23-7, 34 FR 13090, Aug. 13, 1969]

§ 23.507 Jacking loads.

(a) The airplane must be designed for the loads developed when the aircraft is supported on jacks at the design maximum weight assuming the following load factors for landing gear jacking points at a three-point attitude and for primary flight structure jacking points in the level attitude:

(1) Vertical-load factor of 1.35 times the static reactions.

(2) Fore, aft, and lateral load factors of 0.4 times the vertical static reactions.

(b) The horizontal loads at the jack points must be reacted by inertia forces so as to result in no change in the direction of the resultant loads at the jack points.

(c) The horizontal loads must be considered in all combinations with the vertical load.

[Amdt. 23-14, 38 FR 31821, Nov. 19, 1973]

§ 23.509 Towing loads.

The towing loads of this section must be applied to the design of tow fittings and their immediate attaching structure.

(a) The towing loads specified in paragraph (d) of this section must be considered separately. These loads must be applied at the towing fittings and must act parallel to the ground. In addition:

(1) A vertical load factor equal to 1.0 must be considered acting at the center of gravity; and

(2) The shock struts and tires must be in there static positions.

(b) For towing points not on the landing gear but near the plane of symmetry of the airplane, the drag and side tow load components specified for the auxiliary gear apply. For towing points located outboard of the main gear, the drag and side tow load components specified for the main gear apply. Where the specified angle of swivel cannot be reached, the maximum obtainable angle must be used.

(c) The towing loads specified in paragraph (d) of this section must be reacted as follows:

(1) The side component of the towing load at the main gear must be reacted by a side force at the static ground line of the wheel to which the load is applied.

(2) The towing loads at the auxiliary gear and the drag components of the towing loads at the main gear must be reacted as follows:

(i) A reaction with a maximum value equal to the vertical reaction must be applied at the axle of the wheel to which the load is applied. Enough airplane inertia to achieve equilibrium must be applied.

(ii) The loads must be reacted by airplane inertia.

(d) The prescribed towing loads are as follows, where W is the design maximum weight:

Tow point	Position	Load		
		Magnitude	No.	Direction
Main gear		0.225W	1 2 3 4	Forward, parallel to drag axis. Forward, at 30° to drag axis. Aft, parallel to drag axis. Aft, at 30° to drag axis.
Auxiliary gear	Swiveled forward	0.3W	5 6	Forward. Aft.
	Swiveled aft	0.3W	7 8	Forward. Aft.
	Swiveled 45° from forward	0.15W	9 10	Forward, in plane of wheel. Aft, in plane of wheel.
	Swiveled 45° from aft	0.15W	11 12	Forward, in plane of wheel. Aft, in plane of wheel.

[Amdt. 23-14, 38 FR 31821, Nov. 19, 1973]

§ 23.511 Ground load; unsymmetrical loads on multiple-wheel units.

(a) *Pivoting loads.* The airplane is assumed to pivot about on side of the main gear with—

(1) The brakes on the pivoting unit locked; and

(2) Loads corresponding to a limit vertical load factor of 1, and coefficient of friction of 0.8 applied to the main gear and its supporting structure.

(b) *Unequal tire loads.* The loads established under §§ 23.471 through 23.483 must be applied in turn, in a 60/40 percent distribution, to the dual wheels and tires in each dual wheel landing gear unit.

(c) *Deflated tire loads.* For the deflated tire condition—

(1) 60 percent of the loads established under §§ 23.471 through 23.483 must be applied in turn to each wheel in a landing gear unit; and

(2) 60 percent of the limit drag and side loads, and 100 percent of the limit vertical load established under §§ 23.485 and 23.493 or lesser vertical load obtained under paragraph (c)(1) of this section, must be applied in turn to each wheel in the dual wheel landing gear unit.

[Amdt. 23-7, 34 FR 13090, Aug. 13, 1969]

WATER LOADS

§ 23.521 Water load conditions.

(a) The structure of seaplanes and amphibians must be designed for water loads developed during takeoff and landing with the seaplane in any attitude likely to occur in normal operation at appropriate forward and sinking velocities under the most severe sea conditions likely to be encountered.

(b) Unless the applicant makes a rational analysis of the water loads, §§ 23.523 through 23.537 apply.

[Doc. No. 4080, 29 FR 17955, Dec. 18, 1964, as amended by Amdt. 23-45, 58 FR 42160, Aug. 6, 1993; Amdt. 23-48, 61 FR 5147, Feb. 9, 1996]

§ 23.523 Design weights and center of gravity positions.

(a) *Design weights.* The water load requirements must be met at each operating weight up to the design landing weight except that, for the takeoff condition prescribed in § 23.531, the design water takeoff weight (the maximum weight for water taxi and takeoff run) must be used.

(b) *Center of gravity positions.* The critical centers of gravity within the limits for which certification is requested must be considered to reach maximum design loads for each part of the seaplane structure.

[Doc. No. 26269, 58 FR 42160, Aug. 6, 1993]

§ 23.525 Application of loads.

(a) Unless otherwise prescribed, the seaplane as a whole is assumed to be subjected to the loads corresponding to the load factors specified in § 23.527.

(b) In applying the loads resulting from the load factors prescribed in § 23.527, the loads may be distributed over the hull or main float bottom (in order to avoid excessive local shear loads and bending moments at the location of water load application) using pressures not less than those prescribed in § 23.533(c).

(c) For twin float seaplanes, each float must be treated as an equivalent hull on a fictitious seaplane with a weight equal to one-half the weight of the twin float seaplane.

(d) Except in the takeoff condition of § 23.531, the aerodynamic lift on the seaplane during the impact is assumed to be ⅔ of the weight of the seaplane.

[Doc. No. 26269, 58 FR 42161, Aug. 6, 1993; 58 FR 51970, Oct. 5, 1993]

§ 23.527 Hull and main float load factors.

(a) Water reaction load factors n_w must be computed in the following manner:

(1) For the step landing case

$$n_w = \frac{C_1 V_{so}^2}{\left(Tan^{\frac{2}{3}}\beta\right) W^{\frac{1}{3}}}$$

(2) For the bow and stern landing cases

$$n_w = \frac{C_1 V_{so}^2}{\left(Tan^{\frac{2}{3}}\beta\right) W^{\frac{1}{3}}} \times \frac{K_1}{\left(1+r_x^2\right)^{\frac{2}{3}}}$$

(b) The following values are used:

(1) n_w=water reaction load factor (that is, the water reaction divided by seaplane weight).

(2) C_1=empirical seaplane operations factor equal to 0.012 (except that this factor may not be less than that necessary to obtain the minimum value of step load factor of 2.33).

(3) V_{SO}=seaplane stalling speed in knots with flaps extended in the appropriate landing position and with no slipstream effect.

(4) β=Angle of dead rise at the longitudinal station at which the load factor is being determined in accordance with figure 1 of appendix I of this part.

(5) W=seaplane landing weight in pounds.

(6) K_1=empirical hull station weighing factor, in accordance with figure 2 of appendix I of this part.

(7) r_x=ratio of distance, measured parallel to hull reference axis, from the center of gravity of the seaplane to the hull longitudinal station at which the load factor is being computed to the radius of gyration in pitch of the seaplane, the hull reference axis being a straight line, in the plane of symmetry, tangential to the keel at the main step.

(c) For a twin float seaplane, because of the effect of flexibility of the attachment of the floats to the seaplane, the factor K_1 may be reduced at the bow and stern to 0.8 of the value shown in figure 2 of appendix I of this part.

This reduction applies only to the design of the carrythrough and seaplane structure.

[Doc. No. 26269, 58 FR 42161, Aug. 6, 1993; 58 FR 51970, Oct. 5, 1993]

§ 23.529 Hull and main float landing conditions.

(a) *Symmetrical step, bow, and stern landing.* For symmetrical step, bow, and stern landings, the limit water reaction load factors are those computed under § 23.527. In addition—

(1) For symmetrical step landings, the resultant water load must be applied at the keel, through the center of gravity, and must be directed perpendicularly to the keel line;

(2) For symmetrical bow landings, the resultant water load must be applied at the keel, one-fifth of the longitudinal distance from the bow to the step, and must be directed perpendicularly to the keel line; and

(3) For symmetrical stern landings, the resultant water load must be applied at the keel, at a point 85 percent of the longitudinal distance from the step to the stern post, and must be directed perpendicularly to the keel line.

(b) *Unsymmetrical landing for hull and single float seaplanes.* Unsymmetrical step, bow, and stern landing conditions must be investigated. In addition—

(1) The loading for each condition consists of an upward component and a side component equal, respectively, to 0.75 and 0.25 tan β times the resultant load in the corresponding symmetrical landing condition; and

(2) The point of application and direction of the upward component of the load is the same as that in the symmetrical condition, and the point of application of the side component is at the same longitudinal station as the upward component but is directed inward perpendicularly to the plane of symmetry at a point midway between the keel and chine lines.

(c) *Unsymmetrical landing; twin float seaplanes.* The unsymmetrical loading consists of an upward load at the step of each float of 0.75 and a side load of 0.25 tan β at one float times the step landing load reached under § 23.527. The side load is directed inboard, perpendicularly to the plane of symmetry midway between the keel and chine lines of the float, at the same longitudinal station as the upward load.

[Doc. No. 26269, 58 FR 42161, Aug. 6, 1993]

§ 23.531 Hull and main float takeoff condition.

For the wing and its attachment to the hull or main float—

(a) The aerodynamic wing lift is assumed to be zero; and

(b) A downward inertia load, corresponding to a load factor computed from the following formula, must be applied:

$$n = \frac{C_{TO} V_{S1}{}^2}{\left(Tan^{\frac{2}{3}}\beta\right) W^{\frac{1}{3}}}$$

Where—

n=inertia load factor;

C_{TO}=empirical seaplane operations factor equal to 0.004;

V_{S1}=seaplane stalling speed (knots) at the design takeoff weight with the flaps extended in the appropriate takeoff position;

β=angle of dead rise at the main step (degrees); and

W=design water takeoff weight in pounds.

[Doc. No. 26269, 58 FR 42161, Aug. 6, 1993]

§ 23.533 Hull and main float bottom pressures.

(a) *General.* The hull and main float structure, including frames and bulkheads, stringers, and bottom plating, must be designed under this section.

(b) *Local pressures.* For the design of the bottom plating and stringers and their attachments to the supporting structure, the following pressure distributions must be applied:

(1) For an unflared bottom, the pressure at the chine is 0.75 times the pressure at the keel, and the pressures between the keel and chine vary linearly, in accordance with figure 3 of appendix I of this part. The pressure at the keel (p.s.i.) is computed as follows:

$$P_K = \frac{C_2 K_2 V_{S1}{}^2}{Tan\beta_x}$$

where—

P_K=pressure (p.s.i.) at the keel;

C_2=0.00213;

K_2=hull station weighing factor, in accordance with figure 2 of appendix I of this part;

V_{S1}=seaplane stalling speed (knots) at the design water takeoff weight with flaps extended in the appropriate takeoff position; and

β_K=angle of dead rise at keel, in accordance with figure 1 of appendix I of this part.

(2) For a flared bottom, the pressure at the beginning of the flare is the same as that for an unflared bottom, and the pressure between the chine and the beginning of the flare varies linearly, in accordance with figure 3 of appendix I of this part. The pressure distribution is the same as that prescribed in paragraph (b)(1) of this section for an unflared bottom except that the pressure at the chine is computed as follows:

$$P_{ch} = \frac{C_3 K_2 V_{S1}{}^2}{Tan\beta}$$

where—

P_{ch}=pressure (p.s.i.) at the chine;

C_3=0.0016;

K_2=hull station weighing factor, in accordance with figure 2 of appendix I of this part;

V_{S1}=seaplane stalling speed (knots) at the design water takeoff weight with flaps extended in the appropriate takeoff position; and

β=angle of dead rise at appropriate station.

The area over which these pressures are applied must simulate pressures occurring during high localized impacts on the hull or float, but need not extend over an area that would induce critical stresses in the frames or in the overall structure.

(c) *Distributed pressures.* For the design of the frames, keel, and chine structure, the following pressure distributions apply:

(1) Symmetrical pressures are computed as follows:

$$P = \frac{C_4 K_2 V_{S0}{}^2}{Tan\beta}$$

where—

P=pressure (p.s.i.);

C_4=0.078 C_1 (with C_1 computed under § 23.527);

K_2=hull station weighing factor, determined in accordance with figure 2 of appendix I of this part;

V_{S0}=seaplane stalling speed (knots) with landing flaps extended in the appropriate position and with no slipstream effect; and

β=angle of dead rise at appropriate station.

(2) The unsymmetrical pressure distribution consists of the pressures prescribed in paragraph (c)(1) of this section on one side of the hull or main float centerline and one-half of that pressure on the other side of the hull or main float centerline, in accordance with figure 3 of appendix I of this part.

(3) These pressures are uniform and must be applied simultaneously over the entire hull or main float bottom. The loads obtained must be carried into the sidewall structure of the hull proper, but need not be transmitted in a fore and aft direction as shear and bending loads.

[Doc. No. 26269, 58 FR 42161, Aug. 6, 1993; 58 FR 51970, Oct. 5, 1993]

§ 23.535 Auxiliary float loads.

(a) *General.* Auxiliary floats and their attachments and supporting structures must be designed for the conditions prescribed in this section. In the cases specified in paragraphs (b) through (e) of this section, the prescribed water loads may be distributed over the float bottom to avoid excessive local loads, using bottom pressures not less than those prescribed in paragraph (g) of this section.

(b) *Step loading.* The resultant water load must be applied in the plane of symmetry of the float at a point three-fourths of the distance from the bow to the step and must be perpendicular to the keel. The resultant limit load is computed as follows, except that the value of L need not exceed three times the weight of the displaced water when the float is completely submerged:

$$L = \frac{C_5 V_{S0}{}^2 W^{\frac{2}{3}}}{Tan\beta\left(1 + ry^2\right)^{\frac{2}{3}}}$$

where—

L=limit load (lbs.);

C_5=0.0053;

V_{S0}=seaplane stalling speed (knots) with landing flaps extended in the appropriate position and with no slipstream effect;

W=seaplane design landing weight in pounds;

β_s=angle of dead rise at a station 3/4 of the distance from the bow to the step, but need not be less than 15 degrees; and

r_y=ratio of the lateral distance between the center of gravity and the plane of symmetry of the float to the radius of gyration in roll.

(c) *Bow loading.* The resultant limit load must be applied in the plane of symmetry of the float at a point one-fourth of the distance from the bow to the step and must be perpendicular to the tangent to the keel line at that point. The magnitude of the resultant load is that specified in paragraph (b) of this section.

(d) *Unsymmetrical step loading.* The resultant water load consists of a component equal to 0.75 times the load specified in paragraph (a) of this section and a side component equal to 0.025 tan β times the load specified in paragraph (b) of this section. The side load must be applied perpendicularly to the plane of symmetry of the float at a point midway between the keel and the chine.

(e) *Unsymmetrical bow loading.* The resultant water load consists of a component equal to 0.75 times the load specified in paragraph (b) of this section and a side component equal to 0.25 tan β times the load specified in paragraph (c) of this section. The side load must be applied perpendicularly to the plane of symmetry at a point midway between the keel and the chine.

(f) *Immersed float condition.* The resultant load must be applied at the centroid of the cross section of the float at a point one-third of the distance from the bow to the step. The limit load components are as follows:

$$\text{Vertical} = PgV$$

$$aft = \frac{C_X P V^{\frac{2}{3}} \left(K V_{S0}\right)^2}{2}$$

$$side = \frac{C_Y P V^{\frac{2}{3}} \left(K V_{S0}\right)^2}{2}$$

where—

P=mass density of water (slugs/ft.^{3})

V=volume of float (ft.^{3});

C_x=coefficient of drag force, equal to 0.133;

C_y=coefficient of side force, equal to 0.106;

K=0.8, except that lower values may be used if it is shown that the floats are incapable of submerging at a speed of 0.8 V_{S0} in normal operations;

V_{S0}=seaplane stalling speed (knots) with landing flaps extended in the appropriate position and with no slipstream effect; and

g=acceleration due to gravity (ft/sec^2).

(g) *Float bottom pressures.* The float bottom pressures must be established under § 23.533, except that the value of K_2 in the formulae may be taken as 1.0. The angle of dead rise to be used in determining the float bottom pressures is set forth in paragraph (b) of this section.

[Doc. No. 26269, 58 FR 42162, Aug. 6, 1993; 58 FR 51970, Oct. 5, 1993]

§ 23.537 Seawing loads.

Seawing design loads must be based on applicable test data.

[Doc. No. 26269, 58 FR 42163, Aug. 6, 1993]

Emergency Landing Conditions

§ 23.561 General.

(a) The airplane, although it may be damaged in emergency landing conditions, must be designed as prescribed in this section to protect each occupant under those conditions.

(b) The structure must be designed to give each occupant every reasonable chance of escaping serious injury when—

(1) Proper use is made of the seats, safety belts, and shoulder harnesses provided for in the design;

(2) The occupant experiences the static inertia loads corresponding to the following ultimate load factors—

(i) Upward, 3.0g for normal, utility, and commuter category airplanes, or 4.5g for acrobatic category airplanes;

(ii) Forward, 9.0g;

(iii) Sideward, 1.5g; and

(iv) Downward, 6.0g when certification to the emergency exit provisions of § 23.807(d)(4) is requested; and

(3) The items of mass within the cabin, that could injure an occupant, experience the static inertia loads corresponding to the following ultimate load factors—

(i) Upward, 3.0g;

(ii) Forward, 18.0g; and

(iii) Sideward, 4.5g.

(c) Each airplane with retractable landing gear must be designed to protect each occupant in a landing—

(1) With the wheels retracted;

(2) With moderate descent velocity; and

(3) Assuming, in the absence of a more rational analysis—

(i) A downward ultimate inertia force of 3 *g;* and

(ii) A coefficient of friction of 0.5 at the ground.

(d) If it is not established that a turnover is unlikely during an emergency landing, the structure must be designed to protect the occupants in a complete turnover as follows:

(1) The likelihood of a turnover may be shown by an analysis assuming the following conditions—

(i) The most adverse combination of weight and center of gravity position;

(ii) Longitudinal load factor of 9.0g;

(iii) Vertical load factor of 1.0g; and

(iv) For airplanes with tricycle landing gear, the nose wheel strut failed with the nose contacting the ground.

(2) For determining the loads to be applied to the inverted airplane after a turnover, an upward ultimate inertia load factor of 3.0g and a coefficient of friction with the ground of 0.5 must be used.

(e) Except as provided in § 23.787(c), the supporting structure must be designed to restrain, under loads up to those specified in paragraph (b)(3) of this section, each item of mass that could injure an occupant if it came loose in a minor crash landing.

(1) For engines mounted inside the fuselage, aft of the cabin, it must be shown by test or analysis that the engine and attached accessories, and the engine mounting structure—

(i) Can withstand a forward acting static ultimate inertia load factor of 18.0 g plus the maximum takeoff engine thrust; or

(ii) The airplane structure is designed to preclude the engine and its attached accessories from entering or protruding into the cabin should the engine mounts fail.

(2) [Reserved]

[Doc. No. 4080, 29 FR 17955, Dec. 18, 1964, as amended by Amdt. 23-7, 34 FR 13090, Aug. 13, 1969; Amdt. 23-24, 52 FR 34745, Sept. 14, 1987; Amdt. 23-36, 53 FR 30812, Aug. 15, 1988; Amdt. 23-46, 59 FR 25772, May 17, 1994; Amdt. 23-48, 61 FR 5147, Feb. 9, 1996; Amdt. 23-62, 76 FR 75756, Dec. 2, 2011]

§ 23.562 Emergency landing dynamic conditions.

(a) Each seat/restraint system for use in a normal, utility, or acrobatic category airplane, or in a commuter category jet airplane, must be designed to protect each occupant during an emergency landing when—

(1) Proper use is made of seats, safety belts, and shoulder harnesses provided for in the design; and

(2) The occupant is exposed to the loads resulting from the conditions prescribed in this section.

(b) Except for those seat/restraint systems that are required to meet paragraph (d) of this section, each seat/restraint system for crew or passenger occupancy in a normal, utility, or acrobatic category airplane, or in a commuter category jet airplane, must successfully complete dynamic tests or be demonstrated by rational analysis supported by dynamic tests, in accordance with each of the following conditions. These tests must be conducted with an occupant simulated by an anthropomorphic test dummy (ATD) defined by 49 CFR part 572, subpart B, or an FAA-approved equivalent, with a nominal weight of 170 pounds and seated in the normal upright position.

(1) For the first test, the change in velocity may not be less than 31 feet per second. The seat/restraint system must be oriented in its nominal position with respect to the airplane and with the horizontal plane of the airplane pitched up 60 degrees, with no yaw, relative to the impact vector. For seat/restraint systems to be installed in the first row of the airplane, peak deceleration must occur in not more than 0.05 seconds after impact and must reach a minimum of 19g. For all other seat/restraint systems, peak deceleration must occur in not more than 0.06 seconds after impact and must reach a minimum of 15g.

(2) For the second test, the change in velocity may not be less than 42 feet per second. The seat/restraint system must be oriented in its nominal position with respect to the airplane and with the vertical plane of the airplane yawed 10 degrees, with no pitch, relative to the impact vector in a direction that results in the greatest load on the shoulder harness. For seat/restraint systems to be installed in the first row of the airplane, peak deceleration must occur in not more than 0.05 seconds after impact and must reach a minimum of 26g. For all other seat/restraint systems, peak deceleration must occur in not more than 0.06 seconds after impact and must reach a minimum of 21g.

(3) To account for floor warpage, the floor rails or attachment devices used to attach the seat/restraint system to the airframe structure must be preloaded to misalign with respect to each other by at least 10 degrees vertically (i.e., pitch out of parallel) and one of the rails or attachment devices must be preloaded to misalign by 10 degrees in roll prior to conducting the test defined by paragraph (b)(2) of this section.

(c) Compliance with the following requirements must be shown during the dynamic tests conducted in accordance with paragraph (b) of this section:

(1) The seat/restraint system must restrain the ATD although seat/restraint system components may experience deformation, elongation, displacement, or crushing intended as part of the design.

(2) The attachment between the seat/restraint system and the test fixture must remain intact, although the seat structure may have deformed.

(3) Each shoulder harness strap must remain on the ATD's shoulder during the impact.

(4) The safety belt must remain on the ATD's pelvis during the impact.

(5) The results of the dynamic tests must show that the occupant is protected from serious head injury.

(i) When contact with adjacent seats, structure, or other items in the cabin can occur, protection must be provided so that the head impact does not exceed a head injury criteria (HIC) of 1,000.

(ii) The value of HIC is defined as—

$$HIC = \left\{ (t_2 - t_1) \left[\frac{1}{(t_2 - t_1)} \int_{t_1}^{t_2} a(t)\,dt \right]^{2.5} \right\}_{Max}$$

Where—

t_1 is the initial integration time, expressed in seconds, t_2 is the final integration time, expressed in seconds, and a(t) is the total acceleration vs. time curve for the head strike expressed as a multiple of g (units of gravity).

(iii) Compliance with the HIC limit must be demonstrated by measuring the head impact during dynamic testing as prescribed in paragraphs (b)(1) and (b)(2) of this section or by a separate showing of compliance with the head injury criteria using test or analysis procedures.

(6) Loads in individual shoulder harness straps may not exceed 1,750 pounds. If dual straps are used for retaining the upper torso, the total strap loads may not exceed 2,000 pounds.

(7) The compression load measured between the pelvis and the lumbar spine of the ATD may not exceed 1,500 pounds.

(d) For all single-engine airplanes with a V_{SO} of more than 61 knots at maximum weight, and those multiengine airplanes of 6,000 pounds or less maximum weight with a V_{SO} of more than 61 knots at maximum weight that do not comply with § 23.67(a)(1);

(1) The ultimate load factors of § 23.561(b) must be increased by multiplying the load factors by the square of the ratio of the increased stall speed to 61 knots. The increased ultimate load factors need not exceed the values reached at a V_{SO} of 79 knots. The upward ultimate load factor for acrobatic category airplanes need not exceed 5.0g.

(2) The seat/restraint system test required by paragraph (b)(1) of this section must be conducted in accordance with the following criteria:

(i) The change in velocity may not be less than 31 feet per second.

(ii)(A) The peak deceleration (g_p) of 19g and 15g must be increased and multiplied by the square of the ratio of the increased stall speed to 61 knots:

$g_p = 19.0\,(V_{SO}/61)^2$ or $g_p = 15.0\,(V_{SO}/61)^2$

(B) The peak deceleration need not exceed the value reached at a V_{SO} of 79 knots.

(iii) The peak deceleration must occur in not more than time (t_r), which must be computed as follows:

$$t_r = \frac{31}{32.2(g_p)} = \frac{.96}{g_p}$$

where—

g_p=The peak deceleration calculated in accordance with paragraph (d)(2)(ii) of this section

t_r=The rise time (in seconds) to the peak deceleration.

(e) An alternate approach that achieves an equivalent, or greater, level of occupant protection to that required by this section may be used if substantiated on a rational basis.

[Amdt. 23-36, 53 FR 30812, Aug. 15, 1988, as amended by Amdt. 23-44, 58 FR 38639, July 19, 1993; Amdt. 23-50, 61 FR 5192, Feb. 9, 1996; Amdt. 23-62, 76 FR 75756, Dec. 2, 2011]

Fatigue Evaluation

§ 23.571 Metallic pressurized cabin structures.

For normal, utility, and acrobatic category airplanes, the strength, detail design, and fabrication of the metallic structure of the pressure cabin must be evaluated under one of the following:

(a) A fatigue strength investigation in which the structure is shown by tests, or by analysis supported by test evidence, to be able to withstand the repeated loads of variable magnitude expected in service; or

(b) A fail safe strength investigation, in which it is shown by analysis, tests, or both that catastrophic failure of the structure is not probable after fatigue failure, or obvious partial failure, of a principal structural element, and that the remaining structures are able to withstand a static ultimate load factor of 75 percent of the limit load factor at V_C, considering the combined effects of normal operating pressures, expected external aerodynamic pressures, and flight loads. These loads must be multiplied by a factor of 1.15 unless the dynamic effects of failure under static load are otherwise considered.

(c) The damage tolerance evaluation of § 23.573(b).

(d) If certification for operation above 41,000 feet is requested, a damage tolerance evaluation of the fuselage pressure boundary per § 23.573(b) must be conducted.

[Doc. No. 4080, 29 FR 17955, Dec. 18, 1964, as amended by Amdt. 23-14, 38 FR 31821, Nov. 19, 1973; Amdt. 23-45, 58 FR 42163, Aug. 6, 1993; Amdt. 23-48, 61 FR 5147, Feb. 9, 1996; Amdt. 23-62, 76 FR 75756, Dec. 2, 2011]

§ 23.572 Metallic wing, empennage, and associated structures.

(a) For normal, utility, and acrobatic category airplanes, the strength, detail design, and fabrication of those parts of the airframe structure whose failure would be catastrophic must be evaluated under one of the following unless it is shown that the structure, operating stress level, materials and expected uses are comparable, from a fatigue standpoint, to a similar design that has had extensive satisfactory service experience:

(1) A fatigue strength investigation in which the structure is shown by tests, or by analysis supported by test evidence, to be able to withstand the repeated loads of variable magnitude expected in service; or

(2) A fail-safe strength investigation in which it is shown by analysis, tests, or both, that catastrophic failure of the structure is not probable after fatigue failure, or obvious partial failure, of a principal structural element, and that the remaining structure is able to withstand a static ultimate load factor of 75 percent of the critical limit load factor at V_C. These loads must be multiplied by a factor of 1.15 unless the dynamic effects of failure under static load are otherwise considered.

(3) The damage tolerance evaluation of § 23.573(b).

(b) Each evaluation required by this section must—

(1) Include typical loading spectra (e.g. taxi, ground-air-ground cycles, maneuver, gust);

(2) Account for any significant effects due to the mutual influence of aerodynamic surfaces; and

(3) Consider any significant effects from propeller slipstream loading, and buffet from vortex impingements.

[Amdt. 23-7, 34 FR 13090, Aug. 13, 1969, as amended by Amdt. 23-14, 38 FR 31821, Nov. 19, 1973; Amdt. 23-34, 52 FR 1830, Jan. 15, 1987; Amdt. 23-38, 54 FR 39511, Sept. 26, 1989; Amdt. 23-45, 58 FR 42163, Aug. 6, 1993; Amdt. 23-48, 61 FR 5147, Feb. 9, 1996]

§ 23.573 Damage tolerance and fatigue evaluation of structure.

(a) *Composite airframe structure.* Composite airframe structure must be evaluated under this paragraph instead of §§ 23.571 and 23.572. The applicant must evaluate the composite airframe structure, the failure of which would result in catastrophic loss of the airplane, in each wing (including canards, tandem wings, and winglets), empennage, their carrythrough and attaching structure, moveable control surfaces and their attaching structure fuselage, and pressure cabin using the damage-tolerance criteria prescribed in paragraphs (a)(1) through (a)(4) of this section unless shown to be impractical. If the applicant establishes that damage-tolerance criteria is impractical for a particular structure, the structure must be evaluated in accordance with paragraphs (a)(1) and (a)(6) of this section. Where bonded joints are used, the structure must also be evaluated in accordance with paragraph (a)(5) of this section. The effects of material variability and environmental conditions on the strength and durability properties of the composite materials must be accounted for in the evaluations required by this section.

(1) It must be demonstrated by tests, or by analysis supported by tests, that the structure is capable of carrying ultimate load with damage up to the threshold of detectability considering the inspection procedures employed.

(2) The growth rate or no-growth of damage that may occur from fatigue, corrosion, manufacturing flaws or impact damage, under repeated loads expected in service, must be established by tests or analysis supported by tests.

(3) The structure must be shown by residual strength tests, or analysis supported by residual strength tests, to be able to withstand critical limit flight loads, considered as ultimate loads, with the extent of detectable damage consistent with the results of the damage tolerance evaluations. For pressurized cabins, the following loads must be withstood:

(i) Critical limit flight loads with the combined effects of normal operating pressure and expected external aerodynamic pressures.

(ii) The expected external aerodynamic pressures in 1g flight combined with a cabin differential pressure equal to 1.1 times the normal operating differential pressure without any other load.

(4) The damage growth, between initial detectability and the value selected for residual strength demonstrations, factored to obtain inspection intervals, must allow development of an inspection program suitable for application by operation and maintenance personnel.

(5) For any bonded joint, the failure of which would result in catastrophic loss of the airplane, the limit load capacity must be substantiated by one of the following methods—

(i) The maximum disbonds of each bonded joint consistent with the capability to withstand the loads in paragraph (a)(3) of this section must be determined by analysis, tests, or both. Disbonds of each bonded joint greater than this must be prevented by design features; or

(ii) Proof testing must be conducted on each production article that will apply the critical limit design load to each critical bonded joint; or

(iii) Repeatable and reliable non-destructive inspection techniques must be established that ensure the strength of each joint.

(6) Structural components for which the damage tolerance method is shown to be impractical must be shown by component fatigue tests, or analysis supported by tests, to be able to withstand the repeated loads of variable magnitude expected in service. Sufficient component, subcomponent, element, or coupon tests must be done to establish the fatigue scatter factor and the environmental effects. Damage up to the threshold of detectability and ultimate load residual strength capability must be considered in the demonstration.

(b) *Metallic airframe structure.* If the applicant elects to use § 23.571(c) or § 23.572(a)(3), then the damage tolerance evaluation must include a determination of the probable locations and modes of damage due to fatigue, corrosion, or accidental damage. Damage at multiple sites due to fatigue must be included where the design is such that this type of damage can be expected to occur. The evaluation must incorporate repeated load and static analyses supported by test evidence. The extent of damage for residual strength evaluation at any time within the operational life of the airplane must be consistent with the initial detectability and subsequent growth under repeated loads. The residual strength evaluation must show that the remaining structure is able to withstand critical limit flight loads, considered as ultimate, with the extent of detectable damage consistent with the results of the damage tolerance evaluations. For pressurized cabins, the following load must be withstood:

(1) The normal operating differential pressure combined with the expected external aerodynamic pressures applied simultaneously with the flight loading conditions specified in this part, and

(2) The expected external aerodynamic pressures in 1g flight combined with a cabin differential pressure equal to 1.1 times the normal operating differential pressure without any other load.

[Doc. No. 26269, 58 FR 42163, Aug. 6, 1993; 58 FR 51970, Oct. 5, 1993, as amended by Amdt. 23-48, 61 FR 5147, Feb. 9, 1996; 73 FR 19746, Apr. 11, 2008]

§ 23.574 Metallic damage tolerance and fatigue evaluation of commuter category airplanes.

For commuter category airplanes—

(a) *Metallic damage tolerance.* An evaluation of the strength, detail design, and fabrication must show that catastrophic failure due to fatigue, corrosion, defects, or damage will be avoided throughout the operational life of the airplane. This evaluation must be conducted in accordance with the provisions of § 23.573, except as specified in paragraph (b) of this section, for each part of the structure that could contribute to a catastrophic failure.

(b) *Fatigue (safe-life) evaluation.* Compliance with the damage tolerance requirements of paragraph (a) of this section is not required if the applicant establishes that the application of those requirements is impractical for a particular structure. This structure must be shown, by analysis supported by test evidence, to be able to withstand the repeated loads of variable magnitude expected during its service life without detectable cracks. Appropriate safe-life scatter factors must be applied.

[Doc. No. 27805, 61 FR 5148, Feb. 9, 1996]

§ 23.575 Inspections and other procedures.

Each inspection or other procedure, based on an evaluation required by §§ 23.571, 23.572, 23.573 or 23.574, must be established to prevent catastrophic failure and must be included in the Limitations Section of the Instructions for Continued Airworthiness required by § 23.1529.

[Doc. No. 27805, 61 FR 5148, Feb. 9, 1996]

Subpart D—Design and Construction

§ 23.601 General.

The suitability of each questionable design detail and part having an important bearing on safety in operations, must be established by tests.

§ 23.603 Materials and workmanship.

(a) The suitability and durability of materials used for parts, the failure of which could adversely affect safety, must—

(1) Be established by experience or tests;

(2) Meet approved specifications that ensure their having the strength and other properties assumed in the design data; and

(3) Take into account the effects of environmental conditions, such as temperature and humidity, expected in service.

(b) Workmanship must be of a high standard.

[Doc. No. 4080, 29 FR 17955, Dec. 18, 1964, as amended by Amdt. 23-17, 41 FR 55464, Dec. 20, 1976; Amdt. 23-23, 43 FR 50592, Oct. 10, 1978]

§ 23.605 Fabrication methods.

(a) The methods of fabrication used must produce consistently sound structures. If a fabrication process (such as gluing, spot welding, or heat-treating) requires close control to reach this objective, the process must be performed under an approved process specification.

(b) Each new aircraft fabrication method must be substantiated by a test program.

[Doc. No. 4080, 29 FR 17955, Dec. 18, 1964; 30 FR 258, Jan. 9, 1965, as amended by Amdt. 23-23, 43 FR 50592, Oct. 10, 1978]

§ 23.607 Fasteners.

(a) Each removable fastener must incorporate two retaining devices if the loss of such fastener would preclude continued safe flight and landing.

(b) Fasteners and their locking devices must not be adversely affected by the environmental conditions associated with the particular installation.

(c) No self-locking nut may be used on any bolt subject to rotation in operation unless a non-friction locking device is used in addition to the self-locking device.

[Doc. No. 27805, 61 FR 5148, Feb. 9, 1996]

§ 23.609 Protection of structure.

Each part of the structure must—

(a) Be suitably protected against deterioration or loss of strength in service due to any cause, including—

(1) Weathering;

(2) Corrosion; and

(3) Abrasion; and

(b) Have adequate provisions for ventilation and drainage.

§ 23.611 Accessibility provisions.

For each part that requires maintenance, inspection, or other servicing, appropriate means must be incorporated into the aircraft design to allow such servicing to be accomplished.

[Doc. No. 27805, 61 FR 5148, Feb. 9, 1996]

§ 23.613 Material strength properties and design values.

(a) Material strength properties must be based on enough tests of material meeting specifications to establish design values on a statistical basis.

(b) Design values must be chosen to minimize the probability of structural failure due to material variability. Except as provided in paragraph (e) of this section, compliance with this paragraph must be shown by selecting design values that ensure material strength with the following probability:

(1) Where applied loads are eventually distributed through a single member within an assembly, the failure of which would result in loss of structural integrity of the component; 99 percent probability with 95 percent confidence.

(2) For redundant structure, in which the failure of individual elements would result in applied loads being safely distributed to other load carrying members; 90 percent probability with 95 percent confidence.

(c) The effects of temperature on allowable stresses used for design in an essential component or structure must be considered where thermal effects are significant under normal operating conditions.

(d) The design of the structure must minimize the probability of catastrophic fatigue failure, particularly at points of stress concentration.

(e) Design values greater than the guaranteed minimums required by this section may be used where only guaranteed minimum values are normally allowed if a "premium selection" of the material is made in which a specimen of each individual item is tested before use to determine that the actual strength properties of that particular item will equal or exceed those used in design.

[Doc. No. 4080, 29 FR 17955, Dec. 18, 1964; 30 FR 258, Jan. 9, 1965, as amended by Amdt. 23-23, 43 FR 50592, Oct. 30, 1978; Amdt. 23-45, 58 FR 42163, Aug. 6, 1993]

§ 23.619 Special factors.

The factor of safety prescribed in § 23.303 must be multiplied by the highest pertinent special factors of safety prescribed in §§ 23.621 through 23.625 for each part of the structure whose strength is—

(a) Uncertain;

(b) Likely to deteriorate in service before normal replacement; or

(c) Subject to appreciable variability because of uncertainties in manufacturing processes or inspection methods.

[Amdt. 23-7, 34 FR 13091, Aug. 13, 1969]

§ 23.621 Casting factors.

(a) *General.* The factors, tests, and inspections specified in paragraphs (b) through (d) of this section must be applied in addition to those necessary to establish foundry quality control. The inspections must meet approved specifications. Paragraphs (c) and (d) of this section apply to any structural castings except castings that are pressure tested as parts of hydraulic or other fluid systems and do not support structural loads.

(b) *Bearing stresses and surfaces.* The casting factors specified in paragraphs (c) and (d) of this section—

(1) Need not exceed 1.25 with respect to bearing stresses regardless of the method of inspection used; and

(2) Need not be used with respect to the bearing surfaces of a part whose bearing factor is larger than the applicable casting factor.

(c) *Critical castings.* For each casting whose failure would preclude continued safe flight and landing of the airplane or result in serious injury to occupants, the following apply:

(1) Each critical casting must either—

(i) Have a casting factor of not less than 1.25 and receive 100 percent inspection by visual, radiographic, and either magnetic particle, penetrant or other approved equivalent non-destructive inspection method; or

Casting factor	Inspection
2.0 or more	100 percent visual.
Less than 2.0 but more than 1.5	100 percent visual, and magnetic particle or penetrant or equivalent nondestructive inspection methods.
1.25 through 1.50	100 percent visual, magnetic particle or penetrant, and radiographic, or approved equivalent nondestructive inspection methods.

(ii) Have a casting factor of not less than 2.0 and receive 100 percent visual inspection and 100 percent approved non-destructive inspection. When an approved quality control procedure is established and an acceptable statistical analysis supports reduction, non-destructive inspection may be reduced from 100 percent, and applied on a sampling basis.

(2) For each critical casting with a casting factor less than 1.50, three sample castings must be static tested and shown to meet—

(i) The strength requirements of § 23.305 at an ultimate load corresponding to a casting factor of 1.25; and

(ii) The deformation requirements of § 23.305 at a load of 1.15 times the limit load.

(3) Examples of these castings are structural attachment fittings, parts of flight control systems, control surface hinges and balance weight attachments, seat, berth, safety belt, and fuel and oil tank supports and attachments, and cabin pressure valves.

(d) *Non-critical castings.* For each casting other than those specified in paragraph (c) or (e) of this section, the following apply:

(1) Except as provided in paragraphs (d)(2) and (3) of this section, the casting factors and corresponding inspections must meet the following table:

(2) The percentage of castings inspected by nonvisual methods may be reduced below that specified in subparagraph (d)(1) of this section when an approved quality control procedure is established.

(3) For castings procured to a specification that guarantees the mechanical properties of the material in the casting and provides for demonstration of these properties by test of coupons cut from the castings on a sampling basis—

(i) A casting factor of 1.0 may be used; and

(ii) The castings must be inspected as provided in paragraph (d)(1) of this section for casting factors of "1.25 through 1.50" and tested under paragraph (c)(2) of this section.

(e) *Non-structural castings.* Castings used for non-structural purposes do not require evaluation, testing or close inspection.

[Doc. No. 4080, 29 FR 17955, Dec. 18, 1964, as amended by Amdt. 23-45, 58 FR 42164, Aug. 6, 1993]

§ 23.623 Bearing factors.

(a) Each part that has clearance (free fit), and that is subject to pounding or vibration, must have a bearing factor large enough to provide for the effects of normal relative motion.

(b) For control surface hinges and control system joints, compliance with the factors prescribed in §§ 23.657 and 23.693, respectively, meets paragraph (a) of this section.

[Amdt. 23-7, 34 FR 13091, Aug. 13, 1969]

§ 23.625 Fitting factors.

For each fitting (a part or terminal used to join one structural member to another), the following apply:

(a) For each fitting whose strength is not proven by limit and ultimate load tests in which actual stress conditions are simulated in the fitting and surrounding structures, a fitting factor of at least 1.15 must be applied to each part of—

(1) The fitting;

(2) The means of attachment; and

(3) The bearing on the joined members.

(b) No fitting factor need be used for joint designs based on comprehensive test data (such as continuous joints in metal plating, welded joints, and scarf joints in wood).

(c) For each integral fitting, the part must be treated as a fitting up to the point at which the section properties become typical of the member.

(d) For each seat, berth, safety belt, and harness, its attachment to the structure must be shown, by analysis, tests, or both, to be able to withstand the inertia forces prescribed in § 23.561 multiplied by a fitting factor of 1.33.

[Doc. No. 4080, 29 FR 17955, Dec. 18, 1964, as amended by Amdt. 23-7, 34 FR 13091, Aug. 13, 1969]

§ 23.627 Fatigue strength.

The structure must be designed, as far as practicable, to avoid points of stress concentration where variable stresses above the fatigue limit are likely to occur in normal service.

§ 23.629 Flutter.

(a) It must be shown by the methods of paragraph (b) and either paragraph (c) or (d) of this section, that the airplane is free from flutter, control reversal, and divergence for any condition of operation within the limit V-n envelope and at all speeds up to the speed specified for the selected method. In addition—

(1) Adequate tolerances must be established for quantities which affect flutter, including speed, damping, mass balance, and control system stiffness; and

(2) The natural frequencies of main structural components must be determined by vibration tests or other approved methods.

(b) Flight flutter tests must be made to show that the airplane is free from flutter, control reversal and divergence and to show that—

(1) Proper and adequate attempts to induce flutter have been made within the speed range up to V_D/M_D, or V_{DF}/M_{DF} for jets;

(2) The vibratory response of the structure during the test indicates freedom from flutter;

(3) A proper margin of damping exists at V_D/M_D, or V_{DF}/M_{DF} for jets; and

(4) As V_D/M_D (or V_{DF}/M_{DF} for jets) is approached, there is no large or rapid reduction in damping.

(c) Any rational analysis used to predict freedom from flutter, control reversal and divergence must cover all speeds up to 1.2 V_D/1.2 M_D, limited to Mach 1.0 for subsonic airplanes.

(d) Compliance with the rigidity and mass balance criteria (pages 4-12), in Airframe and Equipment Engineering Report No. 45 (as corrected) "Simplified Flutter Prevention Criteria" (published by the Federal Aviation Administration) may be accomplished to show that the airplane is free from flutter, control reversal, or divergence if—

(1) V_D/M_D for the airplane is less than 260 knots (EAS) and less than Mach 0.5,

(2) The wing and aileron flutter prevention criteria, as represented by the wing torsional stiffness and aileron balance criteria, are limited in use to airplanes without large mass concentrations (such as engines, floats, or fuel tanks in outer wing panels) along the wing span, and

(3) The airplane—

(i) Does not have a T-tail or other unconventional tail configurations;

(ii) Does not have unusual mass distributions or other unconventional design features that affect the applicability of the criteria, and

(iii) Has fixed-fin and fixed-stabilizer surfaces.

(e) For turbopropeller-powered airplanes, the dynamic evaluation must include—

(1) Whirl mode degree of freedom which takes into account the stability of the plane of rotation of the propeller and significant elastic, inertial, and aerodynamic forces, and

(2) Propeller, engine, engine mount, and airplane structure stiffness and damping variations appropriate to the particular configuration.

(f) Freedom from flutter, control reversal, and divergence up to V_D/M_D must be shown as follows:

(1) For airplanes that meet the criteria of paragraphs (d)(1) through (d)(3) of this section, after the failure, malfunction, or disconnection of any single element in any tab control system.

(2) For airplanes other than those described in paragraph (f)(1) of this section, after the failure, malfunction, or disconnection of any single element in the primary flight control system, any tab control system, or any flutter damper.

(g) For airplanes showing compliance with the fail-safe criteria of §§ 23.571 and 23.572, the airplane must be shown by analysis to be free from flutter up to V_D/M_D after fatigue failure, or obvious partial failure, of a principal structural element.

(h) For airplanes showing compliance with the damage tolerance criteria of § 23.573, the airplane must be shown by analysis to be free from flutter up to V_D/M_D with the extent of damage for which residual strength is demonstrated.

(i) For modifications to the type design that could affect the flutter characteristics, compliance with paragraph (a) of this section must be shown, except that analysis based on previously approved data may be used alone to show freedom from flutter, control reversal and divergence, for all speeds up to the speed specified for the selected method.

[Amdt. 23-23, 43 FR 50592, Oct. 30, 1978, as amended by Amdt. 23-31, 49 FR 46867, Nov. 28, 1984; Amdt. 23-45, 58 FR 42164, Aug. 6, 1993; 58 FR 51970, Oct. 5, 1993; Amdt. 23-48, 61 FR 5148, Feb. 9, 1996; Amdt. 23-62, 76 FR 75756, Dec. 2, 2011]

PART 23

Wings

§ 23.641 Proof of strength.

The strength of stressed-skin wings must be proven by load tests or by combined structural analysis and load tests.

Control Surfaces

§ 23.651 Proof of strength.

(a) Limit load tests of control surfaces are required. These tests must include the horn or fitting to which the control system is attached.

(b) In structural analyses, rigging loads due to wire bracing must be accounted for in a rational or conservative manner.

§ 23.655 Installation.

(a) Movable surfaces must be installed so that there is no interference between any surfaces, their bracing, or adjacent fixed structure, when one surface is held in its most critical clearance positions and the others are operated through their full movement.

(b) If an adjustable stabilizer is used, it must have stops that will limit its range of travel to that allowing safe flight and landing.

[Doc. No. 4080, 29 FR 17955, Dec. 18, 1964, as amended by Amdt. 23-45, 58 FR 42164, Aug. 6, 1993]

§ 23.657 Hinges.

(a) Control surface hinges, except ball and roller bearing hinges, must have a factor of safety of not less than 6.67 with respect to the ultimate bearing strength of the softest material used as a bearing.

(b) For ball or roller bearing hinges, the approved rating of the bearing may not be exceeded.

[Doc. No. 4080, 29 FR 17955, Dec. 18, 1964, as amended by Amdt. 23-48, 61 FR 5148, Feb. 9, 1996]

§ 23.659 Mass balance.

The supporting structure and the attachment of concentrated mass balance weights used on control surfaces must be designed for—

(a) 24 g normal to the plane of the control surface;

(b) 12 g fore and aft; and

(c) 12 g parallel to the hinge line.

Control Systems

§ 23.671 General.

(a) Each control must operate easily, smoothly, and positively enough to allow proper performance of its functions.

(b) Controls must be arranged and identified to provide for convenience in operation and to prevent the possibility of confusion and subsequent inadvertent operation.

§ 23.672 Stability augmentation and automatic and power-operated systems.

If the functioning of stability augmentation or other automatic or power-operated systems is necessary to show compliance with the flight characteristics requirements of this part, such systems must comply with § 23.671 and the following:

(a) A warning, which is clearly distinguishable to the pilot under expected flight conditions without requiring the pilot's attention, must be provided for any failure in the stability augmentation system or in any other automatic or power-operated system that could result in an unsafe condition if the pilot was not aware of the failure. Warning systems must not activate the control system.

(b) The design of the stability augmentation system or of any other automatic or power-operated system must

FAR

permit initial counteraction of failures without requiring exceptional pilot skill or strength, by either the deactivation of the system or a failed portion thereof, or by overriding the failure by movement of the flight controls in the normal sense.

(c) It must be shown that, after any single failure of the stability augmentation system or any other automatic or power-operated system—

(1) The airplane is safely controllable when the failure or malfunction occurs at any speed or altitude within the approved operating limitations that is critical for the type of failure being considered;

(2) The controllability and maneuverability requirements of this part are met within a practical operational flight envelope (for example, speed, altitude, normal acceleration, and airplane configuration) that is described in the Airplane Flight Manual (AFM); and

(3) The trim, stability, and stall characteristics are not impaired below a level needed to permit continued safe flight and landing.

[Doc. No. 26269, 58 FR 42164, Aug. 6, 1993]

§ 23.673 Primary flight controls.

Primary flight controls are those used by the pilot for the immediate control of pitch, roll, and yaw.

[Doc. No. 4080, 29 FR 17955, Dec. 18, 1964, as amended by Amdt. 23-48, 61 FR 5148, Feb. 9, 1996]

§ 23.675 Stops.

(a) Each control system must have stops that positively limit the range of motion of each movable aerodynamic surface controlled by the system.

(b) Each stop must be located so that wear, slackness, or takeup adjustments will not adversely affect the control characteristics of the airplane because of a change in the range of surface travel.

(c) Each stop must be able to withstand any loads corresponding to the design conditions for the control system.

[Amdt. 23-17, 41 FR 55464, Dec. 20, 1976]

§ 23.677 Trim systems.

(a) Proper precautions must be taken to prevent inadvertent, improper, or abrupt trim tab operation. There must be means near the trim control to indicate to the pilot the direction of trim control movement relative to airplane motion. In addition, there must be means to indicate to the pilot the position of the trim device with respect to both the range of adjustment and, in the case of lateral and directional trim, the neutral position. This means must be visible to the pilot and must be located and designed to prevent confusion. The pitch trim indicator must be clearly marked with a position or range within which it has been demonstrated that take-off is safe for all center of gravity positions and each flap position approved for takeoff.

(b) Trimming devices must be designed so that, when any one connecting or transmitting element in the primary flight control system fails, adequate control for safe flight and landing is available with—

(1) For single-engine airplanes, the longitudinal trimming devices; or

(2) For multiengine airplanes, the longitudinal and directional trimming devices.

(c) Tab controls must be irreversible unless the tab is properly balanced and has no unsafe flutter characteristics. Irreversible tab systems must have adequate rigidity and reliability in the portion of the system from the tab to the attachment of the irreversible unit to the airplane structure.

(d) It must be demonstrated that the airplane is safely controllable and that the pilot can perform all maneuvers and operations necessary to effect a safe landing following any probable powered trim system runaway that reasonably might be expected in service, allowing for appropriate time delay after pilot recognition of the trim system runaway. The demonstration must be conducted at critical airplane weights and center of gravity positions.

[Doc. No. 4080, 29 FR 17955, Dec. 18, 1964, as amended by Amdt. 23-7, 34 FR 13091, Aug. 13, 1969; Amdt. 23-34, 52 FR 1830, Jan. 15, 1987; Amdt. 23-42, 56 FR 353, Jan. 3, 1991; Amdt. 23-49, 61 FR 5165, Feb. 9, 1996]

§ 23.679 Control system locks.

If there is a device to lock the control system on the ground or water:

(a) There must be a means to—

(1) Give unmistakable warning to the pilot when lock is engaged; or

(2) Automatically disengage the device when the pilot operates the primary flight controls in a normal manner.

(b) The device must be installed to limit the operation of the airplane so that, when the device is engaged, the pilot receives unmistakable warning at the start of the takeoff.

(c) The device must have a means to preclude the possibility of it becoming inadvertently engaged in flight.

[Doc. No. 26269, 58 FR 42164, Aug. 6, 1993]

§ 23.681 Limit load static tests.

(a) Compliance with the limit load requirements of this part must be shown by tests in which—

(1) The direction of the test loads produces the most severe loading in the control system; and

(2) Each fitting, pulley, and bracket used in attaching the system to the main structure is included.

(b) Compliance must be shown (by analyses or individual load tests) with the special factor requirements for control system joints subject to angular motion.

§ 23.683 Operation tests.

(a) It must be shown by operation tests that, when the controls are operated from the pilot compartment with the system loaded as prescribed in paragraph (b) of this section, the system is free from—

(1) Jamming;

(2) Excessive friction; and

(3) Excessive deflection.

(b) The prescribed test loads are—

(1) For the entire system, loads corresponding to the limit airloads on the appropriate surface, or the limit pilot forces in § 23.397(b), whichever are less; and

(2) For secondary controls, loads not less than those corresponding to the maximum pilot effort established under § 23.405.

[Doc. No. 4080, 29 FR 17955, Dec. 18, 1964, as amended by Amdt. 23-7, 34 FR 13091, Aug. 13, 1969]

§ 23.685 Control system details.

(a) Each detail of each control system must be designed and installed to prevent jamming, chafing, and interference from cargo, passengers, loose objects, or the freezing of moisture.

(b) There must be means in the cockpit to prevent the entry of foreign objects into places where they would jam the system.

(c) There must be means to prevent the slapping of cables or tubes against other parts.

(d) Each element of the flight control system must have design features, or must be distinctively and permanently marked, to minimize the possibility of incorrect assembly that could result in malfunctioning of the control system.

[Doc. No. 4080, 29 FR 17955, Dec. 18, 1964, as amended by Amdt. 23-17, 41 FR 55464, Dec. 20, 1976]

§ 23.687 Spring devices.

The reliability of any spring device used in the control system must be established by tests simulating service conditions unless failure of the spring will not cause flutter or unsafe flight characteristics.

§ 23.689 Cable systems.

(a) Each cable, cable fitting, turnbuckle, splice, and pulley used must meet approved specifications. In addition—

(1) No cable smaller than $\frac{1}{8}$ inch diameter may be used in primary control systems;

(2) Each cable system must be designed so that there will be no hazardous change in cable tension throughout the range of travel under operating conditions and temperature variations; and

(3) There must be means for visual inspection at each fairlead, pulley, terminal, and turnbuckle.

(b) Each kind and size of pulley must correspond to the cable with which it is used. Each pulley must have closely fitted guards to prevent the cables from being misplaced or fouled, even when slack. Each pulley must lie in the plane passing through the cable so that the cable does not rub against the pulley flange.

(c) Fairleads must be installed so that they do not cause a change in cable direction of more than three degrees.

(d) Clevis pins subject to load or motion and retained only by cotter pins may not be used in the control system.

(e) Turnbuckles must be attached to parts having angular motion in a manner that will positively prevent binding throughout the range of travel.

(f) Tab control cables are not part of the primary control system and may be less than $\frac{1}{8}$ inch diameter in airplanes that are safely controllable with the tabs in the most adverse positions.

[Doc. No. 4080, 29 FR 17955, Dec. 18, 1964, as amended by Amdt. 23-7, 34 FR 13091, Aug. 13, 1969]

§ 23.691 Artificial stall barrier system.

If the function of an artificial stall barrier, for example, stick pusher, is used to show compliance with § 23.201(c), the system must comply with the following:

(a) With the system adjusted for operation, the plus and minus airspeeds at which downward pitching control will be provided must be established.

(b) Considering the plus and minus airspeed tolerances established by paragraph (a) of this section, an airspeed must be selected for the activation of the downward pitching control that provides a safe margin above any airspeed at which any unsatisfactory stall characteristics occur.

(c) In addition to the stall warning required § 23.07, a warning that is clearly distinguishable to the pilot under all expected flight conditions without requiring the pilot's attention, must be provided for faults that would prevent the system from providing the required pitching motion.

(d) Each system must be designed so that the artificial stall barrier can be quickly and positively disengaged by the pilots to prevent unwanted downward pitching of the airplane by a quick release (emergency) control that meets the requirements of § 23.1329(b).

(e) A preflight check of the complete system must be established and the procedure for this check made available in the Airplane Flight Manual (AFM). Preflight checks that are critical to the safety of the airplane must be included in the limitations section of the AFM.

(f) For those airplanes whose design includes an autopilot system:

(1) A quick release (emergency) control installed in accordance with § 23.1329(b) may be used to meet the requirements of paragraph (d), of this section, and

(2) The pitch servo for that system may be used to provide the stall downward pitching motion.

(g) In showing compliance with § 23.1309, the system must be evaluated to determine the effect that any announced or unannounced failure may have on the continued safe flight and landing of the airplane or the ability of the crew to cope with any adverse conditions that may result from such failures. This evaluation must consider the hazards that would result from the airplane's flight characteristics if the system was not provided, and the hazard that may result from unwanted downward pitching motion, which could result from a failure at airspeeds above the selected stall speed.

[Doc. No. 27806, 61 FR 5165, Feb. 9, 1996]

§ 23.693 Joints.

Control system joints (in push-pull systems) that are subject to angular motion, except those in ball and roller bearing systems, must have a special factor of safety of not less than 3.33 with respect to the ultimate bearing strength of the softest material used as a bearing. This factor may be reduced to 2.0 for joints in cable control systems. For ball or roller bearings, the approved ratings may not be exceeded.

§ 23.697 Wing flap controls.

(a) Each wing flap control must be designed so that, when the flap has been placed in any position upon which compliance with the performance requirements of this part is based, the flap will not move from that position unless the control is adjusted or is moved by the automatic operation of a flap load limiting device.

(b) The rate of movement of the flaps in response to the operation of the pilot's control or automatic device must give satisfactory flight and performance characteristics under steady or changing conditions of airspeed, engine power, and attitude.

(c) If compliance with § 23.145(b)(3) necessitates wing flap retraction to positions that are not fully retracted, the wing flap control lever settings corresponding to those positions must be positively located such that a definite change of direction of movement of the lever is necessary to select settings beyond those settings.

[Doc. No. 4080, 29 FR 17955, Dec. 18, 1964, as amended by Amdt. 23-49, 61 FR 5165, Feb. 9, 1996]

§ 23.699 Wing flap position indicator.

There must be a wing flap position indicator for—

(a) Flap installations with only the retracted and fully extended position, unless—

(1) A direct operating mechanism provides a sense of "feel" and position (such as when a mechanical linkage is employed); or

(2) The flap position is readily determined without seriously detracting from other piloting duties under any flight condition, day or night; and

(b) Flap installation with intermediate flap positions if—

(1) Any flap position other than retracted or fully extended is used to show compliance with the performance requirements of this part; and

(2) The flap installation does not meet the requirements of paragraph (a)(1) of this section.

§ 23.701 Flap interconnection.

(a) The main wing flaps and related movable surfaces as a system must—

(1) Be synchronized by a mechanical interconnection between the movable flap surfaces that is independent of the flap drive system; or by an approved equivalent means; or

(2) Be designed so that the occurrence of any failure of the flap system that would result in an unsafe flight characteristic of the airplane is extremely improbable; or

(b) The airplane must be shown to have safe flight characteristics with any combination of extreme positions of individual movable surfaces (mechanically interconnected surfaces are to be considered as a single surface).

(c) If an interconnection is used in multiengine airplanes, it must be designed to account for the unsummetrical loads resulting from flight with the engines on one side of the plane of symmetry inoperative and the remaining engines at takeoff power. For single-engine airplanes, and multiengine airplanes with no slipstream effects on the flaps, it may be assumed that 100 percent of the critical air load acts on one side and 70 percent on the other.

[Doc. No. 4080, 29 FR 17955, Dec. 18, 1964, as amended by Amdt. 23-14, 38 FR 31821, Nov. 19, 1973; Amdt. 23-42, 56 FR 353, Jan. 3, 1991; 56 FR 5455, Feb. 11, 1991; Amdt. 23-49, 61 FR 5165, Feb. 9, 1996]

§ 23.703 Takeoff warning system.

For all airplanes with a maximum weight more than 6,000 pounds and all jets, unless it can be shown that a lift or longitudinal trim device that affects the takeoff performance of the airplane would not give an unsafe takeoff configuration when selected out of an approved takeoff position, a takeoff warning system must be installed and meet the following requirements:

(a) The system must provide to the pilots an aural warning that is automatically activated during the initial portion of the takeoff role if the airplane is in a configuration that would not allow a safe takeoff. The warning must continue until—

(1) The configuration is changed to allow safe takeoff, or

(2) Action is taken by the pilot to abandon the takeoff roll.

(b) The means used to activate the system must function properly for all authorized takeoff power settings and procedures and throughout the ranges of takeoff weights, altitudes, and temperatures for which certification is requested.

(c) For the purpose of this section, an unsafe takeoff configuration is the inability to rotate or the inability to prevent an immediate stall after rotation.

[Doc. No. 27806, 61 FR 5166, Feb. 9, 1996, as amended by Amdt. 23-62, 76 FR 75757, Dec. 2, 2011]

Landing Gear

§ 23.721 General.

For commuter category airplanes that have a passenger seating configuration, excluding pilot seats, of 10 or more, the following general requirements for the landing gear apply:

(a) The main landing-gear system must be designed so that if it fails due to overloads during takeoff and landing (assuming the overloads to act in the upward and aft directions), the failure mode is not likely to cause the spillage of enough fuel from any part of the fuel system to consitute a fire hazard.

(b) Each airplane must be designed so that, with the airplane under control, it can be landed on a paved runway with any one or more landing-gear legs not extended without sustaining a structural component failure that is likely to cause the spillage of enough fuel to consitute a fire hazard.

(c) Compliance with the provisions of this section may be shown by analysis or tests, or both.

[Amdt. 23-34, 52 FR 1830, Jan. 15, 1987]

§ 23.723 Shock absorption tests.

(a) It must be shown that the limit load factors selected for design in accordance with § 23.473 for takeoff and landing weights, respectively, will not be exceeded. This must be shown by energy absorption tests except that analysis based on tests conducted on a landing gear system with identical energy absorption characteristics may be used for increases in previously approved takeoff and landing weights.

(b) The landing gear may not fail, but may yield, in a test showing its reserve energy absorption capacity, simulating a descent velocity of 1.2 times the limit descent velocity, assuming wing lift equal to the weight of the airplane.

[Doc. No. 4080, 29 FR 17955, Dec. 18, 1964; 30 FR 258, Jan. 9, 1965, as amended by Amdt. 23-23, 43 FR 50593, Oct. 30, 1978; Amdt. 23-49, 61 FR 5166, Feb. 9, 1996]

§ 23.725 Limit drop tests.

(a) If compliance with § 23.723(a) is shown by free drop tests, these tests must be made on the complete airplane, or on units consisting of wheel, tire, and shock absorber, in their proper relation, from free drop heights not less than those determined by the following formula:

h (inches)=3.6 $(W/S)^{1/2}$

However, the free drop height may not be less than 9.2 inches and need not be more than 18.7 inches.

(b) If the effect of wing lift is provided for in free drop tests, the landing gear must be dropped with an effective weight equal to

$$W_e = W\frac{[h+(1-L)d]}{(h+d)}$$

where—

W_e=the effective weight to be used in the drop test (lbs.);

h=specified free drop height (inches);

d=deflection under impact of the tire (at the approved inflation pressure) plus the vertical component of the axle travel relative to the drop mass (inches);

$W=W_M$ for main gear units (lbs), equal to the static weight on that unit with the airplane in the level attitude (with the nose wheel clear in the case of nose wheel type airplanes);

$W=W_T$ for tail gear units (lbs.), equal to the static weight on the tail unit with the airplane in the tail-down attitude;

$W=W_N$ for nose wheel units lbs.), equal to the vertical component of the static reaction that would exist at the nose wheel, assuming that the mass of the airplane acts at the center of gravity and exerts a force of 1.0 *g* downward and 0.33 *g* forward; and

L=the ratio of the assumed wing lift to the airplane weight, but not more than 0.667.

(c) The limit inertia load factor must be determined in a rational or conservative manner, during the drop test, using a landing gear unit attitude, and applied drag loads, that represent the landing conditions.

(d) The value of d used in the computation of W_e in paragraph (b) of this section may not exceed the value actually obtained in the drop test.

(e) The limit inertia load factor must be determined from the drop test in paragraph (b) of this section according to the following formula:

$$n = n_j\frac{W_e}{W} + L$$

where—

n_j=the load factor developed in the drop test (that is, the acceleration (dv/dt) in *g* s recorded in the drop test) plus 1.0; and

W_e, W, and L are the same as in the drop test computation.

(f) The value of n determined in accordance with paragraph (e) may not be more than the limit inertia load factor used in the landing conditions in § 23.473.

[Doc. No. 4080, 29 FR 17955, Dec. 18, 1964, as amended by Amdt. 23-7, 34 FR 13091, Aug. 13, 1969; Amdt. 23-48, 61 FR 5148, Feb. 9, 1996]

§ 23.726 Ground load dynamic tests.

(a) If compliance with the ground load requirements of §§ 23.479 through 23.483 is shown dynamically by drop test, one drop test must be conducted that meets § 23.725 except that the drop height must be—

(1) 2.25 times the drop height prescribed in § 23.725(a); or

(2) Sufficient to develop 1.5 times the limit load factor.

(b) The critical landing condition for each of the design conditions specified in §§ 23.479 through 23.483 must be used for proof of strength.

[Amdt. 23-7, 34 FR 13091, Aug. 13, 1969]

§ 23.727 Reserve energy absorption drop test.

(a) If compliance with the reserve energy absorption requirement in § 23.723(b) is shown by free drop tests, the drop height may not be less than 1.44 times that specified in § 23.725.

(b) If the effect of wing lift is provided for, the units must be dropped with an effective mass equal to $W_e=Wh/(h+d)$, when the symbols and other details are the same as in § 23.725.

[Doc. No. 4080, 29 FR 17955, Dec. 18, 1964, as amended by Amdt. 23-7, 34 FR 13091, Aug. 13, 1969]

§ 23.729 Landing gear extension and retraction system.

(a) *General.* For airplanes with retractable landing gear, the following apply:

(1) Each landing gear retracting mechanism and its supporting structure must be designed for maximum flight load factors with the gear retracted and must be designed for the combination of friction, inertia, brake torque, and air loads, occurring during retraction at any airspeed up to 1.6 V_{S1} with flaps retracted, and for any load factor up to those specified in § 23.345 for the flaps-extended condition.

(2) The landing gear and retracting mechanism, including the wheel well doors, must withstand flight loads, including loads resulting from all yawing conditions specified in § 23.351, with the landing gear extended at any speed up to at least 1.6 V_{S1} with the flaps retracted.

(b) *Landing gear lock.* There must be positive means (other than the use of hydraulic pressure) to keep the landing gear extended.

(c) *Emergency operation.* For a landplane having retractable landing gear that cannot be extended manually, there must be means to extend the landing gear in the event of either—

(1) Any reasonably probable failure in the normal landing gear operation system; or

(2) Any reasonably probable failure in a power source that would prevent the operation of the normal landing gear operation system.

(d) *Operation test.* The proper functioning of the retracting mechanism must be shown by operation tests.

(e) *Position indicator.* If a retractable landing gear is used, there must be a landing gear position indicator (as well as necessary switches to actuate the indicator) or other means to inform the pilot that each gear is secured in the extended (or retracted) position. If switches are used, they must be located and coupled to the landing gear mechanical system in a manner that prevents an erroneous indication of either "down and locked" if each gear is not in the fully extended position, or "up and locked" if each landing gear is not in the fully retracted position.

(f) *Landing gear warning.* For landplanes, the following aural or equally effective landing gear warning devices must be provided:

(1) A device that functions continuously when one or more throttles are closed beyond the power settings normally used for landing approach if the landing gear is not fully extended and locked. A throttle stop may not be used in place of an aural device. If there is a manual shutoff for the warning device prescribed in this paragraph, the warning system must be designed so that when the warning has been suspended after one or more throttles are closed, subsequent retardation of any throttle to, or beyond, the position for normal landing approach will activate the warning device.

(2) A device that functions continuously when the wing flaps are extended beyond the maximum approach flap position, using a normal landing procedure, if the landing gear is not fully extended and locked. There may not be a manual shutoff for this warning device. The flap position sensing unit may be installed at any suitable location. The system for this device may use any part of the system (including the aural warning device) for the device required in paragraph (f)(1) of this section.

(g) *Equipment located in the landing gear bay.* If the landing gear bay is used as the location for equipment other than the landing gear, that equipment must be designed and installed to minimize damage from items such as a tire burst, or rocks, water, and slush that may enter the landing gear bay.

[Doc. No. 4080, 29 FR 17955, Dec. 18, 1964, as amended by Amdt. 23-7, 34 FR 13091, Aug. 13, 1969; Amdt. 23-21, 43 FR 2318, Jan. 1978; Amdt. 23-26, 45 FR

60171, Sept. 11, 1980; Amdt. 23-45, 58 FR 42164, Aug. 6, 1993; Amdt. 23-49, 61 FR 5166, Feb. 9, 1996]

§ 23.731 Wheels.

(a) The maximum static load rating of each wheel may not be less than the corresponding static ground reaction with—

(1) Design maximum weight; and

(2) Critical center of gravity.

(b) The maximum limit load rating of each wheel must equal or exceed the maximum radial limit load determined under the applicable ground load requirements of this part.

[Doc. No. 4080, 29 FR 17955, Dec. 18, 1964, as amended by Amdt. 23-45, 58 FR 42165, Aug. 6, 1993]

§ 23.733 Tires.

(a) Each landing gear wheel must have a tire whose approved tire ratings (static and dynamic) are not exceeded—

(1) By a load on each main wheel tire) to be compared to the static rating approved for such tires) equal to the corresponding static ground reaction under the design maximum weight and critical center of gravity; and

(2) By a load on nose wheel tires (to be compared with the dynamic rating approved for such tires) equal to the reaction obtained at the nose wheel, assuming the mass of the airplane to be concentrated at the most critical center of gravity and exerting a force of 1.0 W downward and 0.31 W forward (where W is the design maximum weight), with the reactions distributed to the nose and main wheels by the principles of statics and with the drag reaction at the ground applied only at wheels with brakes.

(b) If specially constructed tires are used, the wheels must be plainly and conspicuously marked to that effect. The markings must include the make, size, number of plies, and identification marking of the proper tire.

(c) Each tire installed on a retractable landing gear system must, at the maximum size of the tire type expected in service, have a clearance to surrounding structure and systems that is adequate to prevent contact between the tire and any part of the structure of systems.

[Doc. No. 4080, 29 FR 17955, Dec. 18, 1964, as amended by Amdt. 23-7, 34 FR 13092, Aug. 13, 1969; Amdt. 23-17, 41 FR 55464, Dec. 20, 1976; Amdt. 23-45, 58 FR 42165, Aug. 6, 1993]

§ 23.735 Brakes.

(a) Brakes must be provided. The landing brake kinetic energy capacity rating of each main wheel brake assembly must not be less than the kinetic energy absorption requirements determined under either of the following methods:

(1) The brake kinetic energy absorption requirements must be based on a conservative rational analysis of the sequence of events expected during landing at the design landing weight.

(2) Instead of a rational analysis, the kinetic energy absorption requirements for each main wheel brake assembly may be derived from the following formula:

KE=0.0443 WV^2/N

where—

KE=Kinetic energy per wheel (ft.-lb.);

W=Design landing weight (lb.);

V=Airplane speed in knots. V must be not less than $V_S\sqrt{}$, the poweroff stalling speed of the airplane at sea level, at the design landing weight, and in the landing configuration; and

N=Number of main wheels with brakes.

(b) Brakes must be able to prevent the wheels from rolling on a paved runway with takeoff power on the critical engine, but need not prevent movement of the airplane with wheels locked.

(c) During the landing distance determination required by § 23.75, the pressure on the wheel braking system must not exceed the pressure specified by the brake manufacturer.

(d) If antiskid devices are installed, the devices and associated systems must be designed so that no single probable malfunction or failure will result in a hazardous loss of braking ability or directional control of the airplane.

(e) For airplanes required to meet § 23.55, the rejected takeoff brake kinetic energy capacity rating of each main wheel brake assembly may not be less than the kinetic energy absorption requirements determined under either of the following methods—

(1) The brake kinetic energy absorption requirements must be based on a conservative rational analysis of the sequence of events expected during a rejected takeoff at the design takeoff weight.

(2) Instead of a rational analysis, the kinetic energy absorption requirements for each main wheel brake assembly may be derived from the following formula—

KE=0.0443 WV^2/N where;

KE=Kinetic energy per wheel (ft.-lbs.);

W=Design takeoff weight (lbs.);

V=Ground speed, in knots, associated with the maximum value of V_1 selected in accordance with § 23.51(c)(1);

N=Number of main wheels with brakes.

[Amdt. 23-7, 34 FR 13092, Aug. 13, 1969, as amended by Amdt. 23-24, 44 FR 68742, Nov. 29, 1979; Amdt. 23-42, 56 FR 354, Jan. 3, 1991; Amdt. 23-49, 61 FR 5166, Feb. 9, 1996; Amdt. 23-62, 76 FR 75757, Dec. 2, 2011]

§ 23.737 Skis.

The maximum limit load rating for each ski must equal or exceed the maximum limit load determined under the applicable ground load requirements of this part.

[Doc. No. 26269, 58 FR 42165, Aug. 6, 1993]

§ 23.745 Nose/tail wheel steering.

(a) If nose/tail wheel steering is installed, it must be demonstrated that its use does not require exceptional pilot skill during takeoff and landing, in crosswinds, or in the event of an engine failure; or its use must be limited to low speed maneuvering.

(b) Movement of the pilot's steering control must not interfere with the retraction or extension of the landing gear.

[Doc. No. 27806, 61 FR 5166, Feb. 9, 1996]

Floats and Hulls

§ 23.751 Main float buoyancy.

(a) Each main float must have—

(1) A buoyancy of 80 percent in excess of the buoyancy required by that float to support its portion of the maximum weight of the seaplane or amphibian in fresh water; and

(2) Enough watertight compartments to provide reasonable assurance that the seaplane or amphibian will stay afloat without capsizing if any two compartments of any main float are flooded.

(b) Each main float must contain at least four watertight compartments approximately equal in volume.

[Doc. No. 4080, 29 FR 17955, Dec. 18, 1964, as amended by Amdt. 23-45, 58 FR 42165, Aug. 6, 1993]

§ 23.753 Main float design.

Each seaplane main float must meet the requirements of § 23.521.

[Doc. No. 26269, 58 FR 42165, Aug. 6, 1993]

§ 23.755 Hulls.

(a) The hull of a hull seaplane or amphibian of 1,500 pounds or more maximum weight must have watertight compartments designed and arranged so that the hull auxiliary floats, and tires (if used), will keep the airplane afloat without capsizing in fresh water when—

(1) For airplanes of 5,000 pounds or more maximum weight, any two adjacent compartments are flooded; and

(2) For airplanes of 1,500 pounds up to, but not including, 5,000 pounds maximum weight, any single compartment is flooded.

(b) Watertight doors in bulkheads may be used for communication between compartments.

[Doc. No. 4080, 29 FR 17955, Dec. 18, 1964, as amended by Amdt. 23-45, 58 FR 42165, Aug. 6, 1993; Amdt. 23-48, 61 FR 5148, Feb. 9, 1996]

§ 23.757 Auxiliary floats.

Auxiliary floats must be arranged so that, when completely submerged in fresh water, they provide a righting moment of at least 1.5 times the upsetting moment caused by the seaplane or amphibian being tilted.

Personnel and Cargo Accommodations

§ 23.771 Pilot compartment.

For each pilot compartment—

(a) The compartment and its equipment must allow each pilot to perform his duties without unreasonable concentration or fatigue;

(b) Where the flight crew are separated from the passengers by a partition, an opening or openable window or door must be provided to facilitate communication between flight crew and the passengers; and

(c) The aerodynamic controls listed in § 23.779, excluding cables and control rods, must be located with respect to the propellers so that no part of the pilot or the controls lies in the region between the plane of rotation of any inboard propeller and the surface generated by a line passing through the center of the propeller hub making an angle of 5 degrees forward or aft of the plane of rotation of the propeller.

[Doc. No. 4080, 29 FR 17955, Dec. 18, 1964, as amended by Amdt. 23-14, 38 FR 31821, Nov. 19, 1973]

§ 23.773 Pilot compartment view.

(a) Each pilot compartment must be—

(1) Arranged with sufficiently extensive, clear and undistorted view to enable the pilot to safely taxi, takeoff, approach, land, and perform any maneuvers within the operating limitations of the airplane.

(2) Free from glare and reflections that could interfere with the pilot's vision. Compliance must be shown in all operations for which certification is requested; and

(3) Designed so that each pilot is protected from the elements so that moderate rain conditions do not unduly impair the pilot's view of the flight path in normal flight and while landing.

(b) Each pilot compartment must have a means to either remove or prevent the formation of fog or frost on an area of the internal portion of the windshield and side windows sufficiently large to provide the view specified in paragraph (a)(1) of this section. Compliance must be shown under all expected external and internal ambient operating conditions, unless it can be shown that the windshield and side windows can be easily cleared by the pilot without interruption of normal pilot duties.

[Doc. No. 26269, 58 FR 42165, Aug. 6, 1993; 71 FR 537, Jan. 5, 2006]

§ 23.775 Windshields and windows.

(a) The internal panels of windshields and windows must be constructed of a nonsplintering material, such as nonsplintering safety glass.

(b) The design of windshields, windows, and canopies in pressurized airplanes must be based on factors peculiar to high altitude operation, including—

(1) The effects of continuous and cyclic pressurization loadings;

(2) The inherent characteristics of the material used; and

(3) The effects of temperatures and temperature gradients.

(c) On pressurized airplanes, if certification for operation up to and including 25,000 feet is requested, an enclosure canopy including a representative part of the installation must be subjected to special tests to account for the combined effects of continuous and cyclic pressurization loadings and flight loads, or compliance with the fail-safe requirements of paragraph (d) of this section must be shown.

(d) If certification for operation above 25,000 feet is requested the windshields, window panels, and canopies must be strong enough to withstand the maximum cabin pressure differential loads combined with critical aerodynamic pressure and temperature effects, after failure of any load-carrying element of the windshield, window panel, or canopy.

(e) The windshield and side windows forward of the pilot's back when the pilot is seated in the normal flight position must have a luminous transmittance value of not less than 70 percent.

(f) Unless operation in known or forecast icing conditions is prohibited by operating limitations, a means must be provided to prevent or to clear accumulations of ice from the windshield so that the pilot has adequate view for taxi, takeoff, approach, landing, and to perform any maneuvers within the operating limitations of the airplane.

(g) In the event of any probable single failure, a transparency heating system must be incapable of raising the temperature of any windshield or window to a point where there would be—

(1) Structural failure that adversely affects the integrity of the cabin; or

(2) There would be a danger of fire.

(h) In addition, for commuter category airplanes, the following applies:

(1) Windshield panes directly in front of the pilots in the normal conduct of their duties, and the supporting structures for these panes, must withstand, without penetration, the impact of a two-pound bird when the velocity of the airplane (relative to the bird along the airplane's flight path) is equal to the airplane's maximum approach flap speed.

(2) The windshield panels in front of the pilots must be arranged so that, assuming the loss of vision through any

one panel, one or more panels remain available for use by a pilot seated at a pilot station to permit continued safe flight and landing.

[Doc. No. 4080, 29 FR 17955, Dec. 18, 1964, as amended by Amdt. 23-7, 34 FR 13092, Aug. 13, 1969; Amdt. 23-45, 58 FR 42165, Aug. 6, 1993; 58 FR 51970, Oct. 5, 1993; Amdt. 23-49, 61 FR 5166, Feb. 9, 1996]

§ 23.777 Cockpit controls.

(a) Each cockpit control must be located and (except where its function is obvious) identified to provide convenient operation and to prevent confusion and inadvertent operation.

(b) The controls must be located and arranged so that the pilot, when seated, has full and unrestricted movement of each control without interference from either his clothing or the cockpit structure.

(c) Powerplant controls must be located—

(1) For multiengine airplanes, on the pedestal or overhead at or near the center of the cockpit;

(2) For single and tandem seated single-engine airplanes, on the left side console or instrument panel;

(3) For other single-engine airplanes at or near the center of the cockpit, on the pedestal, instrument panel, or overhead; and

(4) For airplanes, with side-by-side pilot seats and with two sets of powerplant controls, on left and right consoles.

(d) When separate and distinct control levers are co-located (such as located together on the pedestal), the control location order from left to right must be power (thrust) lever, propeller (rpm control), and mixture control (condition lever and fuel cut-off for turbine-powered airplanes). Power (thrust) levers must be easily distinguishable from other controls, and provide for accurate, consistent operation. Carburetor heat or alternate air control must be to the left of the throttle or at least eight inches from the mixture control when located other than on a pedestal. Carburetor heat or alternate air control, when located on a pedestal, must be aft or below the power (thrust) lever. Supercharger controls must be located below or aft of the propeller controls. Airplanes with tandem seating or single-place airplanes may utilize control locations on the left side of the cabin compartment; however, location order from left to right must be power (thrust) lever, propeller (rpm control), and mixture control.

(e) Identical powerplant controls for each engine must be located to prevent confusion as to the engines they control.

(1) Conventional multiengine powerplant controls must be located so that the left control(s) operates the left engines(s) and the right control(s) operates the right engine(s).

(2) On twin-engine airplanes with front and rear engine locations (tandem), the left powerplant controls must operate the front engine and the right powerplant controls must operate the rear engine.

(f) Wing flap and auxiliary lift device controls must be located—

(1) Centrally, or to the right of the pedestal or powerplant throttle control centerline; and

(2) Far enough away from the landing gear control to avoid confusion.

(g) The landing gear control must be located to the left of the throttle centerline or pedestal centerline.

(h) Each fuel feed selector control must comply with § 23.995 and be located and arranged so that the pilot can see and reach it without moving any seat or primary flight control when his seat is at any position in which it can be placed.

(1) For a mechanical fuel selector:

(i) The indication of the selected fuel valve position must be by means of a pointer and must provide positive identification and feel (detent, etc.) of the selected position.

(ii) The position indicator pointer must be located at the part of the handle that is the maximum dimension of the handle measured from the center of rotation.

(2) For electrical or electronic fuel selector:

(i) Digital controls or electrical switches must be properly labelled.

(ii) Means must be provided to indicate to the flight crew the tank or function selected. Selector switch position is not acceptable as a means of indication. The "off" or "closed" position must be indicated in red.

(3) If the fuel valve selector handle or electrical or digital selection is also a fuel shut-off selector, the off position marking must be colored red. If a separate emergency shut-off means is provided, it also must be colored red.

[Doc. No. 4080, 29 FR 17955, Dec. 18, 1964, as amended by Amdt. 23-7, 34 FR 13092, Aug. 13, 1969; Amdt. 23-33, 51 FR 26656, July 24, 1986; Amdt. 23-51, 61 FR 5136, Feb. 9, 1996; Amdt. 23-62, 76 FR 75757, Dec. 2, 2011]

§ 23.779 Motion and effect of cockpit controls.

Cockpit controls must be designed so that they operate in accordance with the following movement and actuation:

(a) Aerodynamic controls:

	Motion and effect
(1) *Primary controls:*	
Aileron	Right (clockwise) for right wing down.
Elevator	Rearward for nose up.
Rudder	Right pedal forward for nose right.
(2) *Secondary controls:*	
Flaps (or auxiliary lift devices)	Forward or up for flaps up or auxiliary device stowed; rearward or down for flaps down or auxiliary device deployed.
Trim tabs (or equivalent)	Switch motion or mechanical rotation of control to produce similar rotation of the airplane about an axis parallel to the axis control. Axis of roll trim control may be displaced to accommodate comfortable actuation by the pilot. For single-engine airplanes, direction of pilot's hand movement must be in the same sense as airplane response for rudder trim if only a portion of a rotational element is accessible.

(b) Powerplant and auxiliary controls:

	Motion and effect
(1) *Powerplant controls:*	
Power (thrust) lever	Forward to increase forward thrust and rearward to increase rearward thrust.
Propellers	Forward to increase rpm.
Mixture	Forward or upward for rich.
Fuel	Forward for open.
Carburetor, air heat or alternate air	Forward or upward for cold.
Supercharger	Forward or upward for low blower.
Turbosuper-chargers	Forward, upward, or clockwise to increase pressure.
Rotary controls	Clockwise from off to full on.
(2) *Auxiliary controls:*	
Fuel tank selector	Right for right tanks, left for left tanks.
Landing gear	Down to extend.
Speed brakes	Aft to extend.

[Amdt. 23-33, 51 FR 26656, July 24, 1986, as amended by Amdt. 23-51, 61 FR 5136, Feb. 9, 1996]

§ 23.781 Cockpit control knob shape.

(a) Flap and landing gear control knobs must conform to the general shapes (but not necessarily the exact sizes or specific proportions) in the following figure:

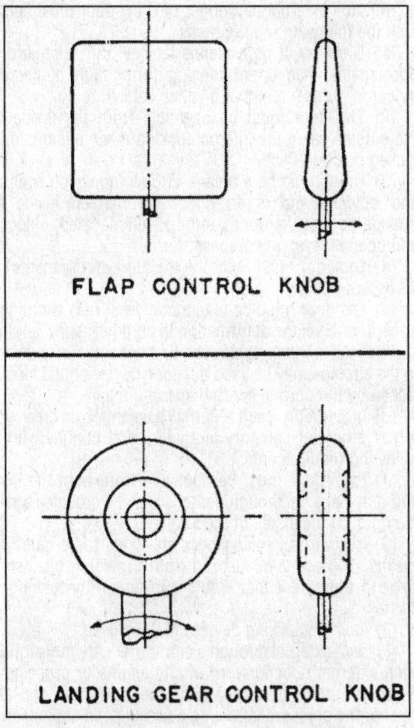

(b) Powerplant control knobs must conform to the general shapes (but not necessarily the exact sizes or specific proportions) in the following figure:

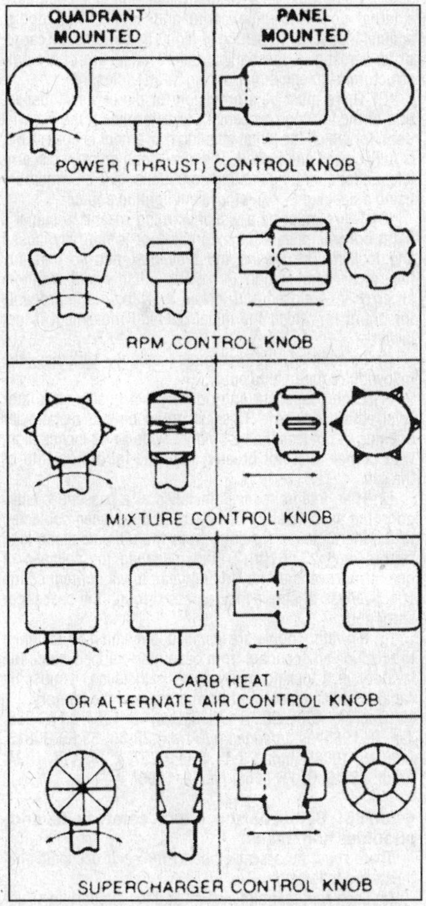

[Doc. No. 4080, 29 FR 17955, Dec. 18, 1964; 30 FR 258, Jan. 9, 1965, as amended by Amdt. 23-33, 51 FR 26657, July 24, 1986]

§ 23.783 Doors.

(a) Each closed cabin with passenger accommodations must have at least one adequate and easily accessible external door.

(b) Passenger doors must not be located with respect to any propeller disk or any other potential hazard so as to endanger persons using the door.

(c) Each external passenger or crew door must comply with the following requirements:

(1) There must be a means to lock and safeguard the door against inadvertent opening during flight by persons, by cargo, or as a result of mechanical failure.

(2) The door must be openable from the inside and the outside when the internal locking mechanism is in the locked position.

(3) There must be a means of opening which is simple and obvious and is arranged and marked inside and outside so that the door can be readily located, unlocked, and opened, even in darkness.

(4) The door must meet the marking requirements of § 23.811 of this part.

(5) The door must be reasonably free from jamming as a result of fuselage deformation in an emergency landing.

(6) Auxiliary locking devices that are actuated externally to the airplane may be used but such devices must be overridden by the normal internal opening means.

(d) In addition, each external passenger or crew door, for a commuter category airplane, must comply with the following requirements:

(1) Each door must be openable from both the inside and outside, even though persons may be crowded against the door on the inside of the airplane.

(2) If inward opening doors are used, there must be a means to prevent occupants from crowding against the door to the extent that would interfere with opening the door.

(3) Auxiliary locking devices may be used.

(e) Each external door on a commuter category airplane, each external door forward of any engine or propeller on a normal, utility, or acrobatic category airplane, and each door of the pressure vessel on a pressurized airplane must comply with the following requirements:

(1) There must be a means to lock and safeguard each external door, including cargo and service type doors, against inadvertent opening in flight, by persons, by cargo, or as a result of mechanical failure or failure of a single structural element, either during or after closure.

(2) There must be a provision for direct visual inspection of the locking mechanism to determine if the external door, for which the initial opening movement is not inward, is fully closed and locked. The provisions must be discernible, under operating lighting conditions, by a crewmember using a flashlight or an equivalent lighting source.

(3) There must be a visual warning means to signal a flight crewmember if the external door is not fully closed and locked. The means must be designed so that any failure, or combination of failures, that would result in an erroneous closed and locked indication is improbable for doors for which the initial opening movement is not inward.

(f) In addition, for commuter category airplanes, the following requirements apply:

(1) Each passenger entry door must qualify as a floor level emergency exit. This exit must have a rectangular opening of not less than 24 inches wide by 48 inches high, with corner radii not greater than one-third the width of the exit.

(2) If an integral stair is installed at a passenger entry door, the stair must be designed so that, when subjected to the inertia loads resulting from the ultimate static load factors in § 23.561(b)(2) and following the collapse of one or more legs of the landing gear, it will not reduce the effectiveness of emergency egress through the passenger entry door.

(g) If lavatory doors are installed, they must be designed to preclude an occupant from becoming trapped inside the lavatory. If a locking mechanism is installed, it must be capable of being unlocked from outside of the lavatory.

[Doc. No. 4080, 29 FR 17955, Dec. 18, 1964; 30 FR 258, Jan. 9, 1965, as amended by Amdt. 23-36, 53 FR 30813, Aug. 15, 1988; Amdt. 23-46, 59 FR 25772, May 17, 1994; Amdt. 23-49, 61 FR 5166, Feb. 9, 1996]

§ 23.785 Seats, berths, litters, safety belts, and shoulder harnesses.

There must be a seat or berth for each occupant that meets the following:

(a) Each seat/restraint system and the supporting structure must be designed to support occupants weighing at least 215 pounds when subjected to the maximum load factors corresponding to the specified flight and ground load conditions, as defined in the approved operating envelope of the airplane. In addition, these loads must be multiplied by a factor of 1.33 in determining the strength of all fittings and the attachment of—

(1) Each seat to the structure; and

(2) Each safety belt and shoulder harness to the seat or structure.

(b) Each forward-facing or aft-facing seat/restraint system in normal, utility, or acrobatic category airplanes must consist of a seat, a safety belt, and a shoulder harness, with a metal-to-metal latching device, that are designed to provide the occupant protection provisions required in § 23.562. Other seat orientations must provide the same level of occupant protection as a forward-facing or aft-facing seat with a safety belt and a shoulder harness, and must provide the protection provisions of § 23.562.

(c) For commuter category airplanes, each seat and the supporting structure must be designed for occupants weighing at least 170 pounds when subjected to the inertia loads resulting from the ultimate static load factors prescribed in § 23.561(b)(2) of this part. Each occupant must be protected from serious head injury when subjected to the inertia loads resulting from these load factors by a safety belt and shoulder harness, with a metal-to-metal latching device, for the front seats and a safety belt, or a safety belt and shoulder harness, with a metal-to-metal latching device, for each seat other than the front seats.

(d) Each restraint system must have a single-point release for occupant evacuation.

(e) The restraint system for each crewmember must allow the crewmember, when seated with the safety belt and shoulder harness fastened, to perform all functions necessary for flight operations.

(f) Each pilot seat must be designed for the reactions resulting from the application of pilot forces to the primary flight controls as prescribed in § 23.395 of this part.

(g) There must be a means to secure each safety belt and shoulder harness, when not in use, to prevent interference with the operation of the airplane and with rapid occupant egress in an emergency.

(h) Unless otherwise placarded, each seat in a utility or acrobatic category airplane must be designed to accommodate an occupant wearing a parachute.

(i) The cabin area surrounding each seat, including the structure, interior walls, instrument panel, control wheel, pedals, and seats within striking distance of the occupant's head or torso (with the restraint system fastened) must be free of potentially injurious objects, sharp edges, protuberances, and hard surfaces. If energy absorbing designs or devices are used to meet this requirement, they must protect the occupant from serious injury when the occupant is subjected to the inertia loads resulting from the ultimate static load factors prescribed in § 23.561(b)(2) of this part, or they must comply with the occupant protection provisions of § 23.562 of this part, as required in paragraphs (b) and (c) of this section.

(j) Each seat track must be fitted with stops to prevent the seat from sliding off the track.

(k) Each seat/restraint system may use design features, such as crushing or separation of certain components, to reduce occupant loads when showing compliance with the requirements of § 23.562 of this part; otherwise, the system must remain intact.

(l) For the purposes of this section, a front seat is a seat located at a flight crewmember station or any seat located alongside such a seat.

(m) Each berth, or provisions for a litter, installed parallel to the longitudinal axis of the airplane, must be designed so that the forward part has a padded end-board, canvas diaphragm, or equivalent means that can withstand the load reactions from a 215-pound occupant when subjected to the inertia loads resulting from the ultimate static load factors of § 23.561(b)(2) of this part. In addition—

(1) Each berth or litter must have an occupant restraint system and may not have corners or other parts likely to cause serious injury to a person occupying it during emergency landing conditions; and

(2) Occupant restraint system attachments for the berth or litter must withstand the inertia loads resulting from the ultimate static load factors of § 23.561(b)(2) of this part.

(n) Proof of compliance with the static strength requirements of this section for seats and berths approved as part of the type design and for seat and berth installations may be shown by—

(1) Structural analysis, if the structure conforms to conventional airplane types for which existing methods of analysis are known to be reliable;

(2) A combination of structural analysis and static load tests to limit load; or

(3) Static load tests to ultimate loads.

[Amdt. 23-36, 53 FR 30813, Aug. 15, 1988; Amdt. 23-36, 54 FR 50737, Dec. 11, 1989; Amdt. 23-49, 61 FR 5167, Feb. 9, 1996]

§ 23.787 Baggage and cargo compartments.

(a) Each baggage and cargo compartment must:

(1) Be designed for its placarded maximum weight of contents and for the critical load distributions at the appropriate maximum load factors corresponding to the flight and ground load conditions of this part.

(2) Have means to prevent the contents of any compartment from becoming a hazard by shifting, and to protect any controls, wiring, lines, equipment or accessories whose damage or failure would affect safe operations.

(3) Have a means to protect occupants from injury by the contents of any compartment, located aft of the occupants and separated by structure, when the ultimate forward inertial load factor is 9g and assuming the maximum allowed baggage or cargo weight for the compartment.

(b) Designs that provide for baggage or cargo to be carried in the same compartment as passengers must have a means to protect the occupants from injury when the baggage or cargo is subjected to the inertial loads resulting from the ultimate static load factors of § 23.561(b)(3), assuming the maximum allowed baggage or cargo weight for the compartment.

(c) For airplanes that are used only for the carriage of cargo, the flightcrew emergency exits must meet the requirements of § 23.807 under any cargo loading conditions.

[Doc. No. 27806, 61 FR 5167, Feb. 9, 1996]

§ 23.791 Passenger information signs.

For those airplanes in which the flightcrew members cannot observe the other occupants' seats or where the flightcrew members' compartment is separated from the passenger compartment, there must be at least one illuminated sign (using either letters or symbols) notifying all passengers when seat belts should be fastened. Signs that notify when seat belts should be fastened must:

(a) When illuminated, be legible to each person seated in the passenger compartment under all probable lighting conditions; and

(b) Be installed so that a flightcrew member can, when seated at the flightcrew member's station, turn the illumination on and off.

[Doc. No. 27806, 61 FR 5167, Feb. 9, 1996]

§ 23.803 Emergency evacuation.

(a) For commuter category airplanes, an evacuation demonstration must be conducted utilizing the maximum number of occupants for which certification is desired. The demonstration must be conducted under simulated night conditions using only the emergency exits on the most critical side of the airplane. The participants must be representative of average airline passengers with no prior practice or rehearsal for the demonstration. Evacuation must be completed within 90 seconds.

(b) In addition, when certification to the emergency exit provisions of § 23.807(d)(4) is requested, only the emergency lighting system required by § 23.812 may be used to provide cabin interior illumination during the evacuation demonstration required in paragraph (a) of this section.

[Amdt. 23-34, 52 FR 1831, Jan. 15, 1987, as amended by Amdt. 23-46, 59 FR 25773, May 17, 1994]

§ 23.805 Flightcrew emergency exits.

For airplanes where the proximity of the passenger emergency exits to the flightcrew area does not offer a convenient and readily accessible means of evacuation for the flightcrew, the following apply:

(a) There must be either one emergency exit on each side of the airplane, or a top hatch emergency exit, in the flightcrew area;

(b) Each emergency exit must be located to allow rapid evacuation of the crew and have a size and shape of at least a 19- by 20-inch unobstructed rectangular opening; and

(c) For each emergency exit that is not less than six feet from the ground, an assisting means must be provided. The assisting means may be a rope or any other means demonstrated to be suitable for the purpose. If the assisting means is a rope, or an approved device equivalent to a rope, it must be—

(1) Attached to the fuselage structure at or above the top of the emergency exit opening or, for a device at a pilot's emergency exit window, at another approved location if the stowed device, or its attachment, would reduce the pilot's view; and

(2) Able (with its attachment) to withstand a 400-pound static load.

[Doc. No. 26324, 59 FR 25773, May 17, 1994]

§ 23.807 Emergency exits.

(a) *Number and location.* Emergency exits must be located to allow escape without crowding in any probable crash attitude. The airplane must have at least the following emergency exits:

(1) For all airplanes with a seating capacity of two or more, excluding airplanes with canopies, at least one emergency exit on the opposite side of the cabin from the main door specified in § 23.783 of this part.

(2) [Reserved]

(3) If the pilot compartment is separated from the cabin by a door that is likely to block the pilot's escape in a minor crash, there must be an exit in the pilot's compartment. The number of exits required by paragraph (a)(1) of this section must then be separately determined for the passenger compartment, using the seating capacity of that compartment.

(4) Emergency exits must not be located with respect to any propeller disk or any other potential hazard so as to endanger persons using that exit.

(b) *Type and operation.* Emergency exits must be movable windows, panels, canopies, or external doors, openable from both inside and outside the airplane, that provide a clear and unobstructed opening large enough to admit a 19-by-26-inch ellipse. Auxiliary locking devices used to secure the airplane must be designed to be overridden by the normal internal opening means. The inside handles of emergency exits that open outward must be adequately protected against inadvertent operation. In addition, each emergency exit must—

(1) Be readily accessible, requiring no exceptional agility to be used in emergencies;

(2) Have a method of opening that is simple and obvious;

(3) Be arranged and marked for easy location and operation, even in darkness;

(4) Have reasonable provisions against jamming by fuselage deformation; and

(5) In the case of acrobatic category airplanes, allow each occupant to abandon the airplane at any speed between V_{SO} and V_D; and

(6) In the case of utility category airplanes certificated for spinning, allow each occupant to abandon the airplane at the highest speed likely to be achieved in the maneuver for which the airplane is certificated.

(c) *Tests.* The proper functioning of each emergency exit must be shown by tests.

(d) *Doors and exits.* In addition, for commuter category airplanes, the following requirements apply:

(1) In addition to the passenger entry door—

(i) For an airplane with a total passenger seating capacity of 15 or fewer, an emergency exit, as defined in paragraph (b) of this section, is required on each side of the cabin; and

(ii) For an airplane with a total passenger seating capacity of 16 through 19, three emergency exits, as defined in paragraph (b) of this section, are required with one on the same side as the passenger entry door and two on the side opposite the door.

(2) A means must be provided to lock each emergency exit and to safeguard against its opening in flight, either inadvertently by persons or as a result of mechanical failure. In addition, a means for direct visual inspection of the locking mechanism must be provided to determine that each emergency exit for which the initial opening movement is outward is fully locked.

(3) Each required emergency exit, except floor level exits, must be located over the wing or, if not less than six feet from the ground, must be provided with an acceptable means to assist the occupants to descend to the ground. Emergency exits must be distributed as uniformly as practical, taking into account passenger seating configuration.

(4) Unless the applicant has complied with paragraph (d)(1) of this section, there must be an emergency exit on the side of the cabin opposite the passenger entry door, provided that—

(i) For an airplane having a passenger seating configuration of nine or fewer, the emergency exit has a rectangular opening measuring not less than 19 inches by 26 inches high with corner radii not greater than one-third the width of the exit, located over the wing, with a step up inside the airplane of not more than 29 inches and a step down outside the airplane of not more than 36 inches;

(ii) For an airplane having a passenger seating configuration of 10 to 19 passengers, the emergency exit has a rectangular opening measuring not less than 20 inches wide by 36 inches high, with corner radii not greater than one-third the width of the exit, and with a step up inside the airplane of not more than 20 inches. If the exit is located over the wing, the step down outside the airplane may not exceed 27 inches; and

(iii) The airplane complies with the additional requirements of §§ 23.561(b)(2)(iv), 23.803(b), 23.811(c), 23.812, 23.813(b), and 23.815.

(e) For multiengine airplanes, ditching emergency exits must be provided in accordance with the following requirements, unless the emergency exits required by paragraph (a) or (d) of this section already comply with them:

(1) One exit above the waterline on each side of the airplane having the dimensions specified in paragraph (b) or (d) of this section, as applicable; and

(2) If side exits cannot be above the waterline, there must be a readily accessible overhead hatch emergency exit that has a rectangular opening measuring not less than 20 inches wide by 36 inches long, with corner radii not greater than one-third the width of the exit.

(3) In lieu of paragraph (e)(2) of this section, if any side exit(s) cannot be above the waterline, a device may be placed at each of such exit(s) prior to ditching. This device must slow the inflow of water when such exit(s) is opened with the airplane ditched. For commuter category airplanes, the clear opening of such exit(s) must meet the requirements defined in paragraph (d) of this section.

[Doc. No. 4080, 29 FR 17955, Dec. 18, 1964, as amended by Amdt. 23-7, 34 FR 13092, Aug. 13, 1969; Amdt. 23-10, 36 FR 2864, Feb. 11, 1971; Amdt. 23-34, 52 FR 1831, Jan. 15, 1987; Amdt. 23-36, 53 FR 30814, Aug. 15, 1988; 53 FR 34194, Sept. 2, 1988; Amdt. 23-46, 59 FR 25773, May 17, 1994; Amdt. 23-49, 61 FR 5167, Feb. 9, 1996; Amdt. 23-62, 76 FR 75757, Dec. 2, 2011]

§ 23.811 Emergency exit marking.

(a) Each emergency exit and external door in the passenger compartment must be externally marked and readily identifiable from outside the airplane by—

(1) A conspicuous visual identification scheme; and

(2) A permanent decal or placard on or adjacent to the emergency exit which shows the means of opening the emergency exit, including any special instructions, if applicable.

(b) In addition, for commuter category airplanes, these exits and doors must be internally marked with the word "exit" by a sign which has white letters 1 inch high on a red background 2 inches high, be self-illuminated or independently, internally electrically illuminated, and have a minimum brightness of at least 160 microlamberts. The color may be reversed if the passenger compartment illumination is essentially the same.

(c) In addition, when certification to the emergency exit provisions of § 23.807(d)(4) is requested, the following apply:

(1) Each emergency exit, its means of access, and its means of opening, must be conspicuously marked;

(2) The identity and location of each emergency exit must be recognizable from a distance equal to the width of the cabin;

(3) Means must be provided to assist occupants in locating the emergency exits in conditions of dense smoke;

(4) The location of the operating handle and instructions for opening each emergency exit from inside the airplane must be shown by marking that is readable from a distance of 30 inches;

(5) Each passenger entry door operating handle must—

(i) Be self-illuminated with an initial brightness of at least 160 microlamberts; or

(ii) Be conspicuously located and well illuminated by the emergency lighting even in conditions of occupant crowding at the door;

(6) Each passenger entry door with a locking mechanism that is released by rotary motion of the handle must be marked—

(i) With a red arrow, with a shaft of at least three-fourths of an inch wide and a head twice the width of the shaft, extending along at least 70 degrees of arc at a radius approximately equal to three-fourths of the handle length;

(ii) So that the center line of the exit handle is within ± one inch of the projected point of the arrow when the handle has reached full travel and has released the locking mechanism;

(iii) With the word "open" in red letters, one inch high, placed horizontally near the head of the arrow; and

(7) In addition to the requirements of paragraph (a) of this section, the external marking of each emergency exit must—

(i) Include a 2-inch colorband outlining the exit; and

(ii) Have a color contrast that is readily distinguishable from the surrounding fuselage surface. The contrast must be such that if the reflectance of the darker color is 15 percent or less, the reflectance of the lighter color must be at least 45 percent. "Reflectance" is the ratio of the luminous flux reflected by a body to the luminous flux it receives. When the reflectance of the darker color is greater than 15 percent, at least a 30 percent difference between its reflectance and the reflectance of the lighter color must be provided.

[Amdt. 23-36, 53 FR 30814, Aug. 15, 1988; 53 FR 34194, Sept. 2, 1988, as amended by Amdt. 23-46, 59 FR 25773, May 17, 1994]

§ 23.812 Emergency lighting.

When certification to the emergency exit provisions of § 23.807(d)(4) is requested, the following apply:

(a) An emergency lighting system, independent of the main cabin lighting system, must be installed. However, the source of general cabin illumination may be common to both the emergency and main lighting systems if the power supply to the emergency lighting system is independent of the power supply to the main lighting system.

(b) There must be a crew warning light that illuminates in the cockpit when power is on in the airplane and the emergency lighting control device is not armed.

(c) The emergency lights must be operable manually from the flightcrew station and be provided with automatic activation. The cockpit control device must have "on," "off," and "armed" positions so that, when armed in the cockpit, the lights will operate by automatic activation.

(d) There must be a means to safeguard against inadvertent operation of the cockpit control device from the "armed" or "on" positions.

(e) The cockpit control device must have provisions to allow the emergency lighting system to be armed or activated at any time that it may be needed.

(f) When armed, the emergency lighting system must activate and remain lighted when—

(1) The normal electrical power of the airplane is lost; or

(2) The airplane is subjected to an impact that results in a deceleration in excess of 2g and a velocity change in excess of 3.5 feet-per-second, acting along the longitudinal axis of the airplane; or

(3) Any other emergency condition exists where automatic activation of the emergency lighting is necessary to aid with occupant evacuation.

(g) The emergency lighting system must be capable of being turned off and reset by the flightcrew after automatic activation.

(h) The emergency lighting system must provide internal lighting, including—

(1) Illuminated emergency exit marking and locating signs, including those required in § 23.811(b);

(2) Sources of general illumination in the cabin that provide an average illumination of not less than 0.05 foot-candle and an illumination at any point of not less than 0.01 foot-candle when measured along the center line of the main passenger aisle(s) and at the seat armrest height; and

(3) Floor proximity emergency escape path marking that provides emergency evacuation guidance for the airplane occupants when all sources of illumination more than 4 feet above the cabin aisle floor are totally obscured.

(i) The energy supply to each emergency lighting unit must provide the required level of illumination for at least 10 minutes at the critical ambient conditions after activation of the emergency lighting system.

(j) If rechargeable batteries are used as the energy supply for the emergency lighting system, they may be recharged from the main electrical power system of the airplane provided the charging circuit is designed to preclude inadvertent battery discharge into the charging circuit faults. If the emergency lighting system does not include a charging circuit, battery condition monitors are required.

(k) Components of the emergency lighting system, including batteries, wiring, relays, lamps, and switches, must be capable of normal operation after being subjected to the inertia forces resulting from the ultimate load factors prescribed in § 23.561(b)(2).

(l) The emergency lighting system must be designed so that after any single transverse vertical separation of the fuselage during a crash landing:

(1) At least 75 percent of all electrically illuminated emergency lights required by this section remain operative; and

(2) Each electrically illuminated exit sign required by § 23.811 (b) and (c) remains operative, except those that are directly damaged by the fuselage separation.

[Doc. No. 26324, 59 FR 25774, May 17, 1994]

§ 23.813 Emergency exit access.

(a) For commuter category airplanes, access to window-type emergency exits may not be obstructed by seats or seat backs.

(b) In addition, when certification to the emergency exit provisions of § 23.807(d)(4) is requested, the following emergency exit access must be provided:

(1) The passageway leading from the aisle to the passenger entry door must be unobstructed and at least 20 inches wide.

(2) There must be enough space next to the passenger entry door to allow assistance in evacuation of passengers without reducing the unobstructed width of the passageway below 20 inches.

(3) If it is necessary to pass through a passageway between passenger compartments to reach a required emergency exit from any seat in the passenger cabin, the passageway must be unobstructed; however, curtains may be used if they allow free entry through the passageway.

(4) No door may be installed in any partition between passenger compartments unless that door has a means to latch it in the open position. The latching means must be able to withstand the loads imposed upon it by the door when the door is subjected to the inertia loads resulting from the ultimate static load factors prescribed in § 23.561(b)(2).

(5) If it is necessary to pass through a doorway separating the passenger cabin from other areas to reach a required emergency exit from any passenger seat, the door must have a means to latch it in the open position. The latching means must be able to withstand the loads imposed upon it by the door when the door is subjected to the inertia loads resulting from the ultimate static load factors prescribed in § 23.561(b)(2).

[Amdt. 23-36, 53 FR 30815, Aug. 15, 1988, as amended by Amdt. 23-46, 59 FR 25774, May 17, 1994]

§ 23.815 Width of aisle.

(a) Except as provided in paragraph (b) of this section, for commuter category airplanes, the width of the main passenger aisle at any point between seats must equal or exceed the values in the following table:

Number of passenger seats	Minimum main passenger aisle width	
	Less than 25 inches from floor	25 inches and more from floor
10 through 19	9 inches	15 inches.

(b) When certification to the emergency exist provisions of § 23.807(d)(4) is requested, the main passenger aisle width at any point between the seats must equal or exceed the following values:

Number of passenger seats	Minimum main passenger aisle width (inches)	
	Less than 25 inches from floor	25 inches and more from floor
10 or fewer	[1]12	15
11 through 19	12	20

[1] A narrower width not less than 9 inches may be approved when substantiated by tests found necessary by the Administrator.

[Amdt. 23-34, 52 FR 1831, Jan. 15, 1987, as amended by Amdt. 23-46, 59 FR 25774, May 17, 1994]

§ 23.831 Ventilation.

(a) Each passenger and crew compartment must be suitably ventilated. Carbon monoxide concentration may not exceed one part in 20,000 parts of air.

(b) For pressurized airplanes, the ventilating air in the flightcrew and passenger compartments must be free of harmful or hazardous concentrations of gases and vapors in normal operations and in the event of reasonably probable failures or malfunctioning of the ventilating, heating, pressurization, or other systems and equipment. If accumulation of hazardous quantities of smoke in the cockpit area is reasonably probable, smoke evacuation must be readily accomplished starting with full pressurization and without depressurizing beyond safe limits.

(c) For jet pressurized airplanes that operate at altitudes above 41,000 feet, under normal operating conditions and in the event of any probable failure conditions of any system which would adversely affect the ventilating air, the ventilation system must provide reasonable passenger comfort. The ventilation system must also provide a sufficient amount of uncontaminated air to enable the flight crew members to perform their duties without undue discomfort or fatigue. For normal operating conditions, the ventilation system must be designed to provide each occupant with at least 0.55 pounds of fresh air per minute. In the event of the loss of one source of fresh air, the supply of fresh airflow may not be less than 0.4 pounds per minute for any period exceeding five minutes.

(d) For jet pressurized airplanes that operate at altitudes above 41,000 feet, other probable and improbable Environmental Control System failure conditions that adversely affect the passenger and flight crew compartment environmental conditions may not affect flight crew performance so as to result in a hazardous condition, and no occupant shall sustain permanent physiological harm.

[Doc. No. 4080, 29 FR 17955, Dec. 18, 1964; 30 FR 258, Jan. 9, 1965, as amended by Amdt. 23-34, 52 FR 1831, Jan. 15, 1987; Amdt. 23-42, 56 FR 354, Jan. 3, 1991; Amdt. 23-62, 76 FR 75757, Dec. 2, 2011]

PRESSURIZATION

§ 23.841 Pressurized cabins.

(a) If certification for operation above 25,000 feet is requested, the airplane must be able to maintain a cabin pressure altitude of not more than 15,000 feet, in the event of any probable failure condition in the pressurization system. During decompression, the cabin altitude may not exceed 15,000 feet for more than 10 seconds and 25,000 feet for any duration.

(b) Pressurized cabins must have at least the following valves, controls, and indicators, for controlling cabin pressure:

(1) Two pressure relief valves to automatically limit the positive pressure differential to a predetermined value at the maximum rate of flow delivered by the pressure source. The combined capacity of the relief valves must be large enough so that the failure of any one valve would not cause an appreciable rise in the pressure differential. The pressure differential is positive when the internal pressure is greater than the external.

(2) Two reverse pressure differential relief valves (or their equivalent) to automatically prevent a negative pressure differential that would damage the structure. However, one valve is enough if it is of a design that reasonably precludes its malfunctioning.

(3) A means by which the pressure differential can be rapidly equalized.

(4) An automatic or manual regulator for controlling the intake or exhaust airflow, or both, for maintaining the required internal pressures and airflow rates.

(5) Instruments to indicate to the pilot the pressure differential, the cabin pressure altitude, and the rate of change of cabin pressure altitude.

(6) Warning indication at the pilot station to indicate when the safe or preset pressure differential is exceeded and when a cabin pressure altitude of 10,000 feet is exceeded. The 10,000 foot cabin altitude warning may be increased up to 15,000 feet for operations from high altitude airfields (10,000 to 15,000 feet) provided:

(i) The landing or the take off modes (normal or high altitude) are clearly indicated to the flight crew.

(ii) Selection of normal or high altitude airfield mode requires no more than one flight crew action and goes to normal airfield mode at engine stop.

(iii) The pressurization system is designed to ensure cabin altitude does not exceed 10,000 feet when in flight above flight level (FL) 250.

(iv) The pressurization system and cabin altitude warning system is designed to ensure cabin altitude warning at 10,000 feet when in flight above FL250.

(7) A warning placard for the pilot if the structure is not designed for pressure differentials up to the maximum relief valve setting in combination with landing loads.

(8) A means to stop rotation of the compressor or to divert airflow from the cabin if continued rotation of an engine-driven cabin compressor or continued flow of any compressor bleed air will create a hazard if a malfunction occurs.

(c) If certification for operation above 41,000 feet and not more than 45,000 feet is requested—

(1) The airplane must prevent cabin pressure altitude from exceeding the following after decompression from any probable pressurization system failure in conjunction with any undetected, latent pressurization system failure condition:

(i) If depressurization analysis shows that the cabin altitude does not exceed 25,000 feet, the pressurization system must prevent the cabin altitude from exceeding the cabin altitude-time history shown in Figure 1 of this section.

(ii) Maximum cabin altitude is limited to 30,000 feet. If cabin altitude exceeds 25,000 feet, the maximum time the cabin altitude may exceed 25,000 feet is 2 minutes; time starting when the cabin altitude exceeds 25,000 feet and ending when it returns to 25,000 feet.

(2) The airplane must prevent cabin pressure altitude from exceeding the following after decompression from any single pressurization system failure in conjunction with any probable fuselage damage:

(i) If depressurization analysis shows that the cabin altitude does not exceed 37,000 feet, the pressurization system must prevent the cabin altitude from exceeding the cabin altitude-time history shown in Figure 2 of this section.

(ii) Maximum cabin altitude is limited to 40,000 feet. If cabin altitude exceeds 37,000 feet, the maximum time the cabin altitude may exceed 25,000 feet is 2 minutes; time starting when the cabin altitude exceeds 25,000 feet and ending when it returns to 25,000 feet.

(3) In showing compliance with paragraphs (c)(1) and (c)(2) of this section, it may be assumed that an emergency descent is made by an approved emergency procedure. A 17-second flight crew recognition and reaction time must be applied between cabin altitude warning and the initiation of an emergency descent. Fuselage structure, engine and system failures are to be considered in evaluating the cabin decompression.

(d) If certification for operation above 45,000 feet and not more than 51,000 feet is requested—

(1) Pressurized cabins must be equipped to provide a cabin pressure altitude of not more than 8,000 feet at the

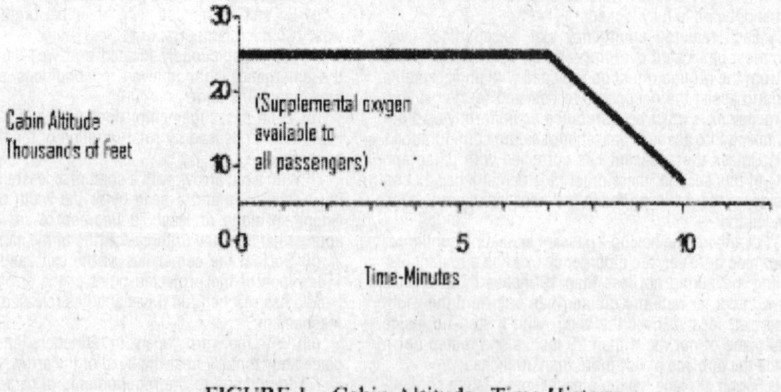

FIGURE 1—Cabin Altitude--Time History

Note: For Figure 1, time starts at the moment cabin altitude exceeds 10,000 feet during decompression.

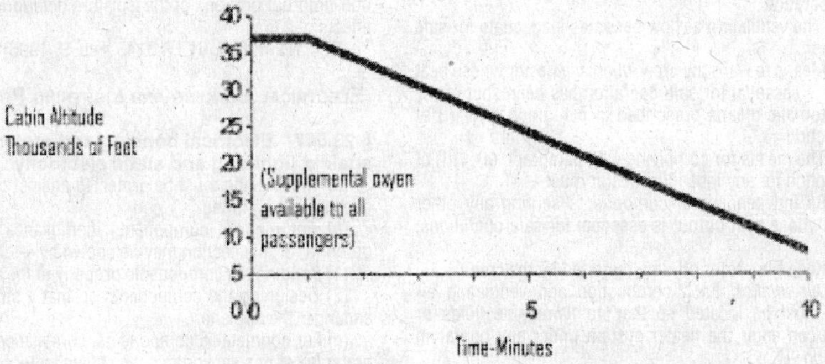

FIGURE 2—Cabin Altitude—Time History

Note: For Figure 2, time starts at the moment cabin altitude exceeds 10,000 feet during decompression.

maximum operating altitude of the airplane under normal operating conditions.

(2) The airplane must prevent cabin pressure altitude from exceeding the following after decompression from any failure condition not shown to be extremely improbable:

(i) Twenty-five thousand (25,000) feet for more than 2 minutes; or

(ii) Forty thousand (40,000) feet for any duration.

(3) Fuselage structure, engine and system failures are to be considered in evaluating the cabin decompression.

(4) In addition to the cabin altitude indicating means in (b)(6) of this section, an aural or visual signal must be provided to warn the flight crew when the cabin pressure altitude exceeds 10,000 feet.

(5) The sensing system and pressure sensors necessary to meet the requirements of (b)(5), (b)(6), and (d)(4) of this section and § 23.1447(e), must, in the event of low cabin pressure, actuate the required warning and automatic presentation devices without any delay that would significantly increase the hazards resulting from decompression.

[Amdt. 23-14, 38 FR 31822, Nov. 19, 1973, as amended by Amdt. 23-17, 41 FR 55464, Dec. 20, 1976; Amdt. 23-49, 61 FR 5167, Feb. 9, 1996; Amdt. 23-62, 76 FR 75757, Dec. 2, 2011]

§ 23.843 Pressurization tests.

(a) *Strength test.* The complete pressurized cabin, including doors, windows, canopy, and valves, must be tested as a pressure vessel for the pressure differential specified in § 23.365(d).

(b) *Functional tests.* The following functional tests must be performed:

(1) Tests of the functioning and capacity of the positive and negative pressure differential valves, and of the emergency release valve, to simulate the effects of closed regulator valves.

(2) Tests of the pressurization system to show proper functioning under each possible condition of pressure, temperature, and moisture, up to the maximum altitude for which certification is requested.

(3) Flight tests, to show the performance of the pressure supply, pressure and flow regulators, indicators, and warning signals, in steady and stepped climbs and descents at rates corresponding to the maximum attainable within the operating limitations of the airplane, up to the maximum altitude for which certification is requested.

(4) Tests of each door and emergency exit, to show that they operate properly after being subjected to the flight tests prescribed in paragraph (b)(3) of this section.

Fire Protection

§ 23.851 Fire extinguishers.

(a) There must be at least one hand fire extinguisher for use in the pilot compartment that is located within easy access of the pilot while seated.

(b) There must be at least one hand fire extinguisher located conveniently in the passenger compartment—

(1) Of each airplane accommodating more than 6 passengers; and

(2) Of each commuter category airplane.

(c) For hand fire extinguishers, the following apply:

(1) The type and quantity of each extinguishing agent used must be appropriate to the kinds of fire likely to occur where that agent is to be used.

(2) Each extinguisher for use in a personnel compartment must be designed to minimize the hazard of toxic gas concentrations.

[Doc. No. 26269, 58 FR 42165, Aug. 6, 1993]

§ 23.853 Passenger and crew compartment interiors.

For each compartment to be used by the crew or passengers:

(a) The materials must be at least flame-resistant;

(b) [Reserved]

(c) If smoking is to be prohibited, there must be a placard so stating, and if smoking is to be allowed—

(1) There must be an adequate number of self-contained, removable ashtrays; and

(2) Where the crew compartment is separated from the passenger compartment, there must be at least one illuminated sign (using either letters or symbols) notifying all passengers when smoking is prohibited. Signs which notify when smoking is prohibited must—

(i) When illuminated, be legible to each passenger seated in the passenger cabin under all probable lighting conditions; and

(ii) Be so constructed that the crew can turn the illumination on and off; and

(d) In addition, for commuter category airplanes the following requirements apply:

(1) Each disposal receptacle for towels, paper, or waste must be fully enclosed and constructed of at least fire resistant materials and must contain fires likely to occur in it under normal use. The ability of the disposal receptacle to contain those fires under all probable conditions of wear, misalignment, and ventilation expected in service must be demonstrated by test. A placard containing the legible words "No Cigarette Disposal" must be located on or near each disposal receptacle door.

(2) Lavatories must have "No Smoking" or "No Smoking in Lavatory" placards located conspicuously on each side of the entry door.

(3) Materials (including finishes or decorative surfaces applied to the materials) used in each compartment occupied by the crew or passengers must meet the following test criteria as applicable:

(i) Interior ceiling panels, interior wall panels, partitions, galley structure, large cabinet walls, structural flooring, and materials used in the construction of stowage compartments (other than underseat stowage compartments and compartments for stowing small items such as magazines and maps) must be self-extinguishing when tested vertically in accordance with the applicable portions of appendix F of this part or by other equivalent methods. The average burn length may not exceed 6 inches and the average flame time after removal of the flame source may not exceed 15 seconds. Drippings from the test specimen may not continue to flame for more than an average of 3 seconds after falling.

(ii) Floor covering, textiles (including draperies and upholstery), seat cushions, padding, decorative and nondecorative coated fabrics, leather, trays and galley furnishings, electrical conduit, thermal and acoustical insulation and insulation covering, air ducting, joint and edge covering, cargo compartment liners, insulation blankets, cargo covers and transparencies, molded and thermoformed parts, air ducting joints, and trim strips (decorative and chafing), that are constructed of materials not covered in paragraph (d)(3)(iv) of this section must be self extinguishing when tested vertically in accordance with the applicable portions of appendix F of this part or other approved equivalent methods. The average burn length may not exceed 8 inches and the average flame time after removal of the flame source may not exceed 15 seconds. Drippings from the test specimen may not continue to flame for more than an average of 5 seconds after falling.

(iii) Motion picture film must be safety film meeting the Standard Specifications for Safety Photographic Film PH1.25 (available from the American National Standards Institute, 1430 Broadway, New York, N.Y. 10018) or an FAA approved equivalent. If the film travels through ducts, the ducts must meet the requirements of paragraph (d)(3)(ii) of this section.

(iv) Acrylic windows and signs, parts constructed in whole or in part of elastomeric materials, edge-lighted instrument assemblies consisting of two or more instruments in a common housing, seatbelts, shoulder harnesses, and cargo and baggage tiedown equipment, including containers, bins, pallets, etc., used in passenger or crew compartments, may not have an average burn rate greater than 2.5 inches per minute when tested horizontally in accordance with the applicable portions of appendix F of this part or by other approved equivalent methods.

(v) Except for electrical wire cable insulation, and for small parts (such as knobs, handles, rollers, fasteners, clips, grommets, rub strips, pulleys, and small electrical parts) that the Administrator finds would not contribute significantly to the propagation of a fire, materials in items not specified in paragraphs (d)(3)(i), (ii), (iii), or (iv) of this section may not have a burn rate greater than 4.0 inches per minute when tested horizontally in accordance with the applicable portions of appendix F of this part or by other approved equivalent methods.

(e) Lines, tanks, or equipment containing fuel, oil, or other flammable fluids may not be installed in such compartments unless adequately shielded, isolated, or otherwise protected so that any breakage or failure of such an item would not create a hazard.

(f) Airplane materials located on the cabin side of the firewall must be self-extinguishing or be located at such a distance from the firewall, or otherwise protected, so that ignition will not occur if the firewall is subjected to a flame temperature of not less than 2,000 degrees F for 15 minutes. For self-extinguishing materials (except electrical wire and cable insulation and small parts that the Administrator finds would not contribute significantly to the propagation of a fire), a vertifical self-extinguishing test must be conducted in accordance with appendix F of this part or an equivalent method approved by the Administrator. The average burn length of the material may not exceed 6 inches and the average flame time after removal of the flame source may not exceed 15 seconds. Drippings from the material test specimen may not continue to flame for more than an average of 3 seconds after falling.

[Amdt. 23-14, 23 FR 31822, Nov. 19, 1973, as amended by Amdt. 23-23, 43 FR 50593, Oct. 30, 1978; Amdt. 23-25, 45 FR 7755, Feb. 4, 1980; Amdt. 23-34, 52 FR 1831, Jan. 15, 1987; Amdt. 23-62, 76 FR 75759, Dec. 2, 2011]

§ 23.855 Cargo and baggage compartment fire protection.

(a) Sources of heat within each cargo and baggage compartment that are capable of igniting the compartment contents must be shielded and insulated to prevent such ignition.

(b) Each cargo and baggage compartment must be constructed of materials that meet the appropriate provisions of § 23.853(d)(3).

(c) In addition, for commuter category airplanes, each cargo and baggage compartment must:

(1) Be located where the presence of a fire would be easily discovered by the pilots when seated at their duty station, or it must be equipped with a smoke or fire detector system to give a warning at the pilots' station, and provide sufficient access to enable a pilot to effectively reach any part of the compartment with the contents of a hand held fire extinguisher, or

(2) Be equipped with a smoke or fire detector system to give a warning at the pilots' station and have ceiling and sidewall liners and floor panels constructed of materials that have been subjected to and meet the 45 degree angle test of appendix F of this part. The flame may not penetrate (pass through) the material during application of the flame or subsequent to its removal. The average flame time after removal of the flame source may not exceed 15 seconds, and the average glow time may not exceed 10 seconds. The compartment must be constructed to provide fire protection that is not less than that required of its individual panels; or

(3) Be constructed and sealed to contain any fire within the compartment.

[Doc. No. 27806, 61 FR 5167, Feb. 9, 1996]

§ 23.856 Thermal/acoustic insulation materials.

Thermal/acoustic insulation material installed in the fuselage must meet the flame propagation test requirements of part II of Appendix F to this part, or other approved equivalent test requirements. This requirement does not apply to "small parts," as defined in § 23.853(d)(3)(v).

[Amdt. 23-62, 76 FR 75759, Dec. 2, 2011]

§ 23.859 Combustion heater fire protection.

(a) *Combustion heater fire regions.* The following combustion heater fire regions must be protected from fire in accordance with the applicable provisions of §§ 23.1182 through 23.1191 and 23.1203:

(1) The region surrounding the heater, if this region contains any flammable fluid system components (excluding the heater fuel system) that could—

(i) Be damaged by heater malfunctioning; or

(ii) Allow flammable fluids or vapors to reach the heater in case of leakage.

(2) The region surrounding the heater, if the heater fuel system has fittings that, if they leaked, would allow fuel vapor to enter this region.

(3) The part of the ventilating air passage that surrounds the combustion chamber.

(b) *Ventilating air ducts.* Each ventilating air duct passing through any fire region must be fireproof. In addition—

(1) Unless isolation is provided by fireproof valves or by equally effective means, the ventilating air duct downstream of each heater must be fireproof for a distance great enough to ensure that any fire originating in the heater can be contained in the duct; and

(2) Each part of any ventilating duct passing through any region having a flammable fluid system must be constructed or isolated from that system so that the malfunctioning of any component of that system cannot introduce flammable fluids or vapors into the ventilating airstream.

(c) *Combustion air ducts.* Each combustion air duct must be fireproof for a distance great enough to prevent damage from backfiring or reverse flame propagation. In addition—

(1) No combustion air duct may have a common opening with the ventilating airstream unless flames from backfires or reverse burning cannot enter the ventilating airstream under any operating condition, including reverse flow or malfunctioning of the heater or its associated components; and

(2) No combustion air duct may restrict the prompt relief of any backfire that, if so restricted, could cause heater failure.

(d) *Heater controls: general.* Provision must be made to prevent the hazardous accumulation of water or ice on or in any heater control component, control system tubing, or safety control.

(e) *Heater safety controls.* (1) Each combustion heater must have the following safety controls:

(i) Means independent of the components for the normal continuous control of air temperature, airflow, and fuel flow must be provided to automatically shut off the ignition and fuel supply to that heater at a point remote from that heater when any of the following occurs:

(A) The heater exchanger temperature exceeds safe limits.

(B) The ventilating air temperature exceeds safe limits.

(C) The combustion airflow becomes inadequate for safe operation.

(D) The ventilating airflow becomes inadequate for safe operation.

(ii) Means to warn the crew when any heater whose heat output is essential for safe operation has been shut off by the automatic means prescribed in paragraph (e)(1)(i) of this section.

(2) The means for complying with paragraph (e)(1)(i) of this section for any individual heater must—

(i) Be independent of components serving any other heater whose heat output is essential for safe operations; and

(ii) Keep the heater off until restarted by the crew.

(f) *Air intakes.* Each combustion and ventilating air intake must be located so that no flammable fluids or vapors can enter the heater system under any operating condition—

(1) During normal operation; or

(2) As a result of the malfunctioning of any other component.

(g) *Heater exhaust.* Heater exhaust systems must meet the provisions of §§ 23.1121 and 23.1123. In addition, there must be provisions in the design of the heater exhaust system to safely expel the products of combustion to prevent the occurrence of—

(1) Fuel leakage from the exhaust to surrounding compartments;

(2) Exhaust gas impingement on surrounding equipment or structure;

(3) Ignition of flammable fluids by the exhaust, if the exhaust is in a compartment containing flammable fluid lines; and

(4) Restrictions in the exhaust system to relieve backfires that, if so restricted, could cause heater failure.

(h) *Heater fuel systems.* Each heater fuel system must meet each powerplant fuel system requirement affecting safe heater operation. Each heater fuel system component within the ventilating airstream must be protected by shrouds so that no leakage from those components can enter the ventilating airstream.

(i) *Drains.* There must be means to safely drain fuel that might accumulate within the combustion chamber or the heater exchanger. In addition—

(1) Each part of any drain that operates at high temperatures must be protected in the same manner as heater exhausts; and

(2) Each drain must be protected from hazardous ice accumulation under any operating condition.

[Amdt. 23-27, 45 FR 70387, Oct. 23, 1980]

§ 23.863 Flammable fluid fire protection.

(a) In each area where flammable fluids or vapors might escape by leakage of a fluid system, there must be means to minimize the probability of ignition of the fluids and vapors, and the resultant hazard if ignition does occur.

(b) Compliance with paragraph (a) of this section must be shown by analysis or tests, and the following factors must be considered:

(1) Possible sources and paths of fluid leakage, and means of detecting leakage.

(2) Flammability characteristics of fluids, including effects of any combustible or absorbing materials.

(3) Possible ignition sources, including electrical faults, overheating of equipment, and malfunctioning of protective devices.

(4) Means available for controlling or extinguishing a fire, such as stopping flow of fluids, shutting down equipment, fireproof containment, or use of extinguishing agents.

(5) Ability of airplane components that are critical to safety of flight to withstand fire and heat.

(c) If action by the flight crew is required to prevent or counteract a fluid fire (e.g. equipment shutdown or actuation of a fire extinguisher), quick acting means must be provided to alert the crew.

(d) Each area where flammable fluids or vapors might escape by leakage of a fluid system must be identified and defined.

[Amdt. 23-23, 43 FR 50593, Oct. 30, 1978]

§ 23.865 Fire protection of flight controls, engine mounts, and other flight structure.

Flight controls, engine mounts, and other flight structure located in designated fire zones, or in adjacent areas that would be subjected to the effects of fire in the designated fire zones, must be constructed of fireproof material or be shielded so that they are capable of withstanding the effects of a fire. Engine vibration isolators must incorporate suitable features to ensure that the engine is retained if the non-fireproof portions of the isolators deteriorate from the effects of a fire.

[Doc. No. 27805, 61 FR 5148, Feb. 9, 1996]

Electrical Bonding and Lightning Protection

§ 23.867 Electrical bonding and protection against lightning and static electricity.

(a) The airplane must be protected against catastrophic effects from lightning.

(b) For metallic components, compliance with paragraph (a) of this section may be shown by—

(1) Bonding the components properly to the airframe; or

(2) Designing the components so that a strike will not endanger the airplane.

(c) For nonmetallic components, compliance with paragraph (a) of this section may be shown by—

(1) Designing the components to minimize the effect of a strike; or

(2) Incorporating acceptable means of diverting the resulting electrical current so as not to endanger the airplane.

[Amdt. 23-7, 34 FR 13092, Aug. 13, 1969]

Miscellaneous

§ 23.871 Leveling means.

There must be means for determining when the airplane is in a level position on the ground.

[Amdt. 23-7, 34 FR 13092, Aug. 13, 1969]

Subpart E—Powerplant

General

§ 23.901 Installation.

(a) For the purpose of this part, the airplane powerplant installation includes each component that—

(1) Is necessary for propulsion; and

(2) Affects the safety of the major propulsive units.

(b) Each powerplant installation must be constructed and arranged to—

(1) Ensure safe operation to the maximum altitude for which approval is requested.

(2) Be accessible for necessary inspections and maintenance.

(c) Engine cowls and nacelles must be easily removable or openable by the pilot to provide adequate access to and exposure of the engine compartment for preflight checks.

(d) Each turbine engine installation must be constructed and arranged to—

(1) Result in carcass vibration characteristics that do not exceed those established during the type certification of the engine.

(2) Ensure that the capability of the installed engine to withstand the ingestion of rain, hail, ice, and birds into the engine inlet is not less than the capability established for the engine itself under § 23.903(a)(2).

(e) The installation must comply with—

(1) The instructions provided under the engine type certificate and the propeller type certificate.

(2) The applicable provisions of this subpart.

(f) Each auxiliary power unit installation must meet the applicable portions of this part.

[Doc. No. 4080, 29 FR 17955, Dec. 18, 1964, as amended by Amdt. 23-7, 34 FR 13092, Aug. 13, 1969; Amdt. 23-18, 42 FR 15041, Mar. 17, 1977; Amdt. 23-29, 49 FR 6846, Feb. 23, 1984; Amdt. 23-34, 52 FR 1832, Jan. 15, 1987; Amdt. 23-34, 52 FR 34745, Sept. 14, 1987; Amdt. 23-43, 58 FR 18970, Apr. 9, 1993; Amdt. 23-51, 61 FR 5136, Feb. 9, 1996; Amdt. 23-53, 63 FR 14797, Mar. 26, 1998]

§ 23.903 Engines.

(a) *Engine type certificate.* (1) Each engine must have a type certificate and must meet the applicable requirements of part 34 of this chapter.

(2) Each turbine engine and its installation must comply with one of the following:

(i) Sections 33.76, 33.77 and 33.78 of this chapter in effect on December 13, 2000, or as subsequently amended; or

(ii) Sections 33.77 and 33.78 of this chapter in effect on April 30, 1998, or as subsequently amended before December 13, 2000; or

(iii) Section 33.77 of this chapter in effect on October 31, 1974, or as subsequently amended before April 30,

1998, unless that engine's foreign object ingestion service history has resulted in an unsafe condition; or

(iv) Be shown to have a foreign object ingestion service history in similar installation locations which has not resulted in any unsafe condition.

Note: § 33.77 of this chapter in effect on October 31, 1974, was published in 14 CFR parts 1 to 59, Revised as of January 1, 1975. See 39 FR 35467, October 1, 1974.

(b) *Turbine engine installations.* For turbine engine installations—

(1) Design precautions must be taken to minimize the hazards to the airplane in the event of an engine rotor failure or of a fire originating inside the engine which burns through the engine case.

(2) The powerplant systems associated with engine control devices, systems, and instrumentation must be designed to give reasonable assurance that those operating limitations that adversely affect turbine rotor structural integrity will not be exceeded in service.

(3) For engines embedded in the fuselage behind the cabin, the effects of a fan exiting forward of the inlet case (fan disconnect) must be addressed, the passengers must be protected, and the airplane must be controllable to allow for continued safe flight and landing.

(c) *Engine isolation.* The powerplants must be arranged and isolated from each other to allow operation, in at least one configuration, so that the failure or malfunction of any engine, or the failure or malfunction (including destruction by fire in the engine compartment) of any system that can affect an engine (other than a fuel tank if only one fuel tank is installed), will not:

(1) Prevent the continued safe operation of the remaining engines; or

(2) Require immediate action by any crewmember for continued safe operation of the remaining engines.

(d) *Starting and stopping (piston engine).* (1) The design of the installation must be such that risk of fire or mechanical damage to the engine or airplane, as a result of starting the engine in any conditions in which starting is to be permitted, is reduced to a minimum. Any techniques and associated limitations for engine starting must be established and included in the Airplane Flight Manual, approved manual material, or applicable operating placards. Means must be provided for—

(i) Restarting any engine of a multiengine airplane in flight, and

(ii) Stopping any engine in flight, after engine failure, if continued engine rotation would cause a hazard to the airplane.

(2) In addition, for commuter category airplanes, the following apply:

(i) Each component of the stopping system on the engine side of the firewall that might be exposed to fire must be at least fire resistant.

(ii) If hydraulic propeller feathering systems are used for this purpose, the feathering lines must be at least fire resistant under the operating conditions that may be expected to exist during feathering.

(e) *Starting and stopping (turbine engine).* Turbine engine installations must comply with the following:

(1) The design of the installation must be such that risk of fire or mechanical damage to the engine or the airplane, as a result of starting the engine in any conditions in which starting is to be permitted, is reduced to a minimum. Any techniques and associated limitations must be established and included in the Airplane Flight Manual, approved manual material, or applicable operating placards.

(2) There must be means for stopping combustion within any engine and for stopping the rotation of any engine if continued rotation would cause a hazard to the airplane. Each component of the engine stopping system located in any fire zone must be fire resistant. If hydraulic propeller feathering systems are used for stopping the engine, the hydraulic feathering lines or hoses must be fire resistant.

(3) It must be possible to restart an engine in flight. Any techniques and associated limitations must be established and included in the Airplane Flight Manual, approved manual material, or applicable operating placards.

(4) It must be demonstrated in flight that when restarting engines following a false start, all fuel or vapor is discharged in such a way that it does not constitute a fire hazard.

(f) *Restart envelope.* An altitude and airspeed envelope must be established for the airplane for in-flight engine restarting and each installed engine must have a restart capability within that envelope.

(g) *Restart capability.* For turbine engine powered airplanes, if the minimum windmilling speed of the engines, following the in-flight shutdown of all engines, is insufficient to provide the necessary electrical power for engine ignition, a power source independent of the engine-driven electrical power generating system must be provided to permit in-flight engine ignition for restarting.

[Amdt. 23-14, 38 FR 31822, Nov. 19, 1973]

Editorial Note: For Federal Register citations affecting § 23.903, see the List of CFR Sections Affected, which appears in the Finding Aids section of the printed volume and at *www.fdsys.gov.*

§ 23.904 Automatic power reserve system.

If installed, an automatic power reserve (APR) system that automatically advances the power or thrust on the operating engine(s), when any engine fails during takeoff, must comply with appendix H of this part.

[Doc. No. 26344, 58 FR 18970, Apr. 9, 1993]

§ 23.905 Propellers.

(a) Each propeller must have a type certificate.

(b) Engine power and propeller shaft rotational speed may not exceed the limits for which the propeller is certificated.

(c) Each featherable propeller must have a means to unfeather it in flight.

(d) The propeller blade pitch control system must meet the requirements of §§ 35.21, 35.23, 35.42 and 35.43 of this chapter.

(e) All areas of the airplane forward of the pusher propeller that are likely to accumulate and shed ice into the propeller disc during any operating condition must be suitably protected to prevent ice formation, or it must be shown that any ice shed into the propeller disc will not create a hazardous condition.

(f) Each pusher propeller must be marked so that the disc is conspicuous under normal daylight ground conditions.

(g) If the engine exhaust gases are discharged into the pusher propeller disc, it must be shown by tests, or analysis supported by tests, that the propeller is capable of continuous safe operation.

(h) All engine cowling, access doors, and other removable items must be designed to ensure that they will not separate from the airplane and contact the pusher propeller.

[Doc. No. 4080, 29 FR 17955, Dec. 18, 1964, as amended by Amdt. 23-26, 45 FR 60171, Sept. 11, 1980; Amdt. 23-29, 49 FR 6847, Feb. 23, 1984; Amdt. 23-43, 58 FR 18970, Apr. 9, 1993; Amdt. 23-59, 73 FR 63345, Oct. 24, 2008]

§ 23.907 Propeller vibration and fatigue.

This section does not apply to fixed-pitch wood propellers of conventional design.

(a) The applicant must determine the magnitude of the propeller vibration stresses or loads, including any stress peaks and resonant conditions, throughout the operational envelope of the airplane by either:

(1) Measurement of stresses or loads through direct testing or analysis based on direct testing of the propeller on the airplane and engine installation for which approval is sought; or

(2) Comparison of the propeller to similar propellers installed on similar airplane installations for which these measurements have been made.

(b) The applicant must demonstrate by tests, analysis based on tests, or previous experience on similar designs that the propeller does not experience harmful effects of flutter throughout the operational envelope of the airplane.

(c) The applicant must perform an evaluation of the propeller to show that failure due to fatigue will be avoided throughout the operational life of the propeller using the fatigue and structural data obtained in accordance with part 35 of this chapter and the vibration data obtained from compliance with paragraph (a) of this section. For the purpose of this paragraph, the propeller includes the hub, blades, blade retention component and any other propeller component whose failure due to fatigue could be catastrophic to the airplane. This evaluation must include:

(1) The intended loading spectra including all reasonably foreseeable propeller vibration and cyclic load patterns, identified emergency conditions, allowable overspeeds and overtorques, and the effects of temperatures and humidity expected in service.

(2) The effects of airplane and propeller operating and airworthiness limitations.

[Amdt. 23-59, 73 FR 63345, Oct. 24, 2008]

§ 23.909 Turbocharger systems.

(a) Each turbocharger must be approved under the engine type certificate or it must be shown that the turbocharger system, while in its normal engine installation and operating in the engine environment—

(1) Can withstand, without defect, an endurance test of 150 hours that meets the applicable requirements of § 33.49 of this subchapter; and

(2) Will have no adverse effect upon the engine.

(b) Control system malfunctions, vibrations, and abnormal speeds and temperatures expected in service may not damage the turbocharger compressor or turbine.

(c) Each turbocharger case must be able to contain fragments of a compressor or turbine that fails at the highest speed that is obtainable with normal speed control devices inoperative.

(d) Each intercooler installation, where provided, must comply with the following—

(1) The mounting provisions of the intercooler must be designed to withstand the loads imposed on the system;

(2) It must be shown that, under the installed vibration environment, the intercooler will not fail in a manner allowing portions of the intercooler to be ingested by the engine; and

(3) Airflow through the intercooler must not discharge directly on any airplane component (e.g., windshield) unless such discharge is shown to cause no hazard to the airplane under all operating conditions.

(e) Engine power, cooling characteristics, operating limits, and procedures affected by the turbocharger system installations must be evaluated. Turbocharger operating procedures and limitations must be included in the Airplane Flight Manual in accordance with § 23.1581.

[Amdt. 23-7, 34 FR 13092, Aug. 13, 1969, as amended by Amdt. 23-43, 58 FR 18970, Apr. 9, 1993]

§ 23.925 Propeller clearance.

Unless smaller clearances are substantiated, propeller clearances, with the airplane at the most adverse combination of weight and center of gravity, and with the propeller in the most adverse pitch position, may not be less than the following:

(a) *Ground clearance.* There must be a clearance of at least seven inches (for each airplane with nose wheel landing gear) or nine inches (for each airplane with tail wheel landing gear) between each propeller and the ground with the landing gear statically deflected and in the level, normal takeoff, or taxing attitude, whichever is most critical. In addition, for each airplane with conventional landing gear struts using fluid or mechanical means for absorbing landing shocks, there must be positive clearance between the propeller and the ground in the level takeoff attitude with the critical tire completely deflated and the corresponding landing gear strut bottomed. Positive clearance for airplanes using leaf spring struts is shown with a deflection corresponding to 1.5 g.

(b) *Aft-mounted propellers.* In addition to the clearances specified in paragraph (a) of this section, an airplane with an aft mounted propeller must be designed such that the propeller will not contact the runway surface when the airplane is in the maximum pitch attitude attainable during normal takeoffs and landings.

(c) *Water clearance.* There must be a clearance of at least 18 inches between each propeller and the water, unless compliance with § 23.239 can be shown with a lesser clearance.

(d) *Structural clearance.* There must be—

(1) At least one inch radial clearance between the blade tips and the airplane structure, plus any additional radial clearance necessary to prevent harmful vibration;

(2) At least one-half inch longitudinal clearance between the propeller blades or cuffs and stationary parts of the airplane; and

(3) Positive clearance between other rotating parts of the propeller or spinner and stationary parts of the airplane.

[Doc. No. 4080, 29 FR 17955, Dec. 18, 1964, as amended by Amdt. 23-43, 58 FR 18971, Apr. 9, 1993; Amdt. 23-51, 61 FR 5136, Feb. 9, 1996; Amdt. 23-48, 61 FR 5148, Feb. 9, 1996]

§ 23.929 Engine installation ice protection.

Propellers (except wooden propellers) and other components of complete engine installations must be protected against the accumulation of ice as necessary to enable satisfactory functioning without appreciable loss of thrust when operated in the icing conditions for which certification is requested.

[Amdt. 23-14, 33 FR 31822, Nov. 19, 1973, as amended by Amdt. 23-51, 61 FR 5136, Feb. 9, 1996]

§ 23.933 Reversing systems.

(a) *For turbojet and turbofan reversing systems.* (1) Each system intended for ground operation only must be designed so that, during any reversal in flight, the engine will produce no more than flight idle thrust. In addition, it must be shown by analysis or test, or both, that—

(i) Each operable reverser can be restored to the forward thrust position; or

(ii) The airplane is capable of continued safe flight and landing under any possible position of the thrust reverser.

(2) Each system intended for in-flight use must be designed so that no unsafe condition will result during normal operation of the system, or from any failure, or likely combination of failures, of the reversing system under any operating condition including ground operation. Failure of structural elements need not be considered if the probability of this type of failure is extremely remote.

(3) Each system must have a means to prevent the engine from producing more than idle thrust when the reversing system malfunctions; except that it may produce any greater thrust that is shown to allow directional control to be maintained, with aerodynamic means alone, under the most critical reversing condition expected in operation.

(b) *For propeller reversing systems.* (1) Each system must be designed so that no single failure, likely combination of failures or malfunction of the system will result in unwanted reverse thrust under any operating condition. Failure of structural elements need not be considered if the probability of this type of failure is extremely remote.

(2) Compliance with paragraph (b)(1) of this section must be shown by failure analysis, or testing, or both, for propeller systems that allow the propeller blades to move from the flight low-pitch position to a position that is substantially less than the normal flight, low-pitch position. The analysis may include or be supported by the analysis made to show compliance with § 35.21 for the type certification of the propeller and associated installation components. Credit will be given for pertinent analysis and testing completed by the engine and propeller manufacturers.

[Doc. No. 26344, 58 FR 18971, Apr. 9, 1993, as amended by Amdt. 23-51, 61 FR 5136, Feb. 9, 1996]

§ 23.934 Turbojet and turbofan engine thrust reverser systems tests.

Thrust reverser systems of turbojet or turbofan engines must meet the requirements of § 33.97 of this chapter or it must be demonstrated by tests that engine operation and vibratory levels are not affected.

[Doc. No. 26344, 58 FR 18971, Apr. 9, 1993]

§ 23.937 Turbopropeller-drag limiting systems.

(a) Turbopropeller-powered airplane propeller-drag limiting systems must be designed so that no single failure or malfunction of any of the systems during normal or emergency operation results in propeller drag in excess of that for which the airplane was designed under the structural requirements of this part. Failure of structural elements of the drag limiting systems need not be considered if the probability of this kind of failure is extremely remote.

(b) As used in this section, drag limiting systems include manual or automatic devices that, when actuated after engine power loss, can move the propeller blades toward the feather position to reduce windmilling drag to a safe level.

[Amdt. 23-7, 34 FR 13093, Aug. 13, 1969, as amended by Amdt. 23-43, 58 FR 18971, Apr. 9, 1993]

§ 23.939 Powerplant operating characteristics.

(a) Turbine engine powerplant operating characteristics must be investigated in flight to determine that no adverse characteristics (such as stall, surge, or flameout) are present, to a hazardous degree, during normal and emergency operation within the range of operating limitations of the airplane and of the engine.

(b) Turbocharged reciprocating engine operating characteristics must be investigated in flight to assure that no adverse characteristics, as a result of an inadvertent overboost, surge, flooding, or vapor lock, are present during normal or emergency operation of the engine(s) throughout the range of operating limitations of both airplane and engine.

(c) For turbine engines, the air inlet system must not, as a result of airflow distortion during normal operation, cause vibration harmful to the engine.

[Amdt. 23-7, 34 FR 13093 Aug. 13, 1969, as amended by Amdt. 23-14, 38 FR 31823, Nov. 19, 1973; Amdt. 23-18, 42 FR 15041, Mar. 17, 1977; Amdt. 23-42, 56 FR 354, Jan. 3, 1991]

§ 23.943 Negative acceleration.

No hazardous malfunction of an engine, an auxiliary power unit approved for use in flight, or any component or system associated with the powerplant or auxiliary power unit may occur when the airplane is operated at the negative accelerations within the flight envelopes prescribed in § 23.333. This must be shown for the greatest value and duration of the acceleration expected in service.

[Amdt. 23-18, 42 FR 15041, Mar. 17, 1977, as amended by Amdt. 23-43, 58 FR 18971, Apr. 9, 1993]

Fuel System

§ 23.951 General.

(a) Each fuel system must be constructed and arranged to ensure fuel flow at a rate and pressure established for proper engine and auxiliary power unit functioning under each likely operating condition, including any maneuver for which certification is requested and during which the engine or auxiliary power unit is permitted to be in operation.

(b) Each fuel system must be arranged so that—

(1) No fuel pump can draw fuel from more than one tank at a time; or

(2) There are means to prevent introducing air into the system.

(c) Each fuel system for a turbine engine must be capable of sustained operation throughout its flow and pressure range with fuel initially saturated with water at 80°F and having 0.75cc of free water per gallon added and cooled to the most critical condition for icing likely to be encountered in operation.

(d) Each fuel system for a turbine engine powered airplane must meet the applicable fuel venting requirements of part 34 of this chapter.

[Amdt. 23-15, 39 FR 35459, Oct. 1, 1974, as amended by Amdt. 23-40, 55 FR 32861, Aug. 10, 1990; Amdt. 23-43, 58 FR 18971, Apr. 9, 1993]

§ 23.953 Fuel system independence.

(a) Each fuel system for a multiengine airplane must be arranged so that, in at least one system configuration, the failure of any one component (other than a fuel tank) will not result in the loss of power of more than one engine or require immediate action by the pilot to prevent the loss of power of more than one engine.

(b) If a single fuel tank (or series of fuel tanks interconnected to function as a single fuel tank) is used on a multiengine airplane, the following must be provided:

(1) Independent tank outlets for each engine, each incorporating a shut-off valve at the tank. This shutoff valve may also serve as the fire wall shutoff valve required if the line between the valve and the engine compartment does not contain more than one quart of fuel (or any greater amount shown to be safe) that can escape into the engine compartment.

(2) At least two vents arranged to minimize the probability of both vents becoming obstructed simultaneously.

(3) Filler caps designed to minimize the probability of incorrect installation or inflight loss.

(4) A fuel system in which those parts of the system from each tank outlet to any engine are independent of each part of the system supplying fuel to any other engine.

[Doc. No. 4080, 29 FR 17955, Dec. 18, 1964, as amended by Amdt. 23-7, 34 FR 13093 Aug. 13, 1969; Amdt. 23-43, 58 FR 18971, Apr. 9, 1993]

§ 23.954 Fuel system lightning protection.

The fuel system must be designed and arranged to prevent the ignition of fuel vapor within the system by—

(a) Direct lightning strikes to areas having a high probability of stroke attachment;

(b) Swept lightning strokes on areas where swept strokes are highly probable; and

(c) Corona or streamering at fuel vent outlets.

[Amdt. 23-7, 34 FR 13093, Aug. 13, 1969]

§ 23.955 Fuel flow.

(a) *General.* The ability of the fuel system to provide fuel at the rates specified in this section and at a pressure sufficient for proper engine operation must be shown in the attitude that is most critical with respect to fuel feed and quantity of unusable fuel. These conditions may be simulated in a suitable mockup. In addition—

(1) The quantity of fuel in the tank may not exceed the amount established as the unusable fuel supply for that tank under § 23.959(a) plus that quantity necessary to show compliance with this section.

(2) If there is a fuel flowmeter, it must be blocked during the flow test and the fuel must flow through the meter or its bypass.

(3) If there is a flowmeter without a bypass, it must not have any probable failure mode that would restrict fuel flow below the level required for this fuel demonstration.

(4) The fuel flow must include that flow necessary for vapor return flow, jet pump drive flow, and for all other purposes for which fuel is used.

(b) *Gravity systems.* The fuel flow rate for gravity systems (main and reserve supply) must be 150 percent of the takeoff fuel consumption of the engine.

(c) *Pump systems.* The fuel flow rate for each pump system (main and reserve supply) for each reciprocating engine must be 125 percent of the fuel flow required by the engine at the maximum takeoff power approved under this part.

(1) This flow rate is required for each main pump and each emergency pump, and must be available when the pump is operating as it would during takeoff;

(2) For each hand-operated pump, this rate must occur at not more than 60 complete cycles (120 single strokes) per minute.

(3) The fuel pressure, with main and emergency pumps operating simultaneously, must not exceed the fuel inlet pressure limits of the engine unless it can be shown that no adverse effect occurs.

(d) *Auxiliary fuel systems and fuel transfer systems.* Paragraphs (b), (c), and (f) of this section apply to each auxiliary and transfer system, except that—

(1) The required fuel flow rate must be established upon the basis of maximum continuous power and engine rotational speed, instead of takeoff power and fuel consumption; and

(2) If there is a placard providing operating instructions, a lesser flow rate may be used for transferring fuel from any auxiliary tank into a larger main tank. This lesser flow rate must be adequate to maintain engine maximum continuous power but the flow rate must not overfill the main tank at lower engine powers.

(e) *Multiple fuel tanks.* For reciprocating engines that are supplied with fuel from more than one tank, if engine power loss becomes apparent due to fuel depletion from the tank selected, it must be possible after switching to any full tank, in level flight, to obtain 75 percent maximum continuous power on that engine in not more than—

(1) 10 seconds for naturally aspirated single-engine airplanes;

(2) 20 seconds for turbocharged single-engine airplanes, provided that 75 percent maximum continuous naturally aspirated power is regained within 10 seconds; or

(3) 20 seconds for multiengine airplanes.

(f) *Turbine engine fuel systems.* Each turbine engine fuel system must provide at least 100 percent of the fuel flow required by the engine under each intended operation condition and maneuver. The conditions may be simulated in a suitable mockup. This flow must—

(1) Be shown with the airplane in the most adverse fuel feed condition (with respect to altitudes, attitudes, and other conditions) that is expected in operation; and

(2) For multiengine airplanes, notwithstanding the lower flow rate allowed by paragraph (d) of this section, be automatically uninterrupted with respect to any engine until all the fuel scheduled for use by that engine has been consumed. In addition—

(i) For the purposes of this section, "fuel scheduled for use by that engine" means all fuel in any tank intended for use by a specific engine.

(ii) The fuel system design must clearly indicate the engine for which fuel in any tank is scheduled.

(iii) Compliance with this paragraph must require no pilot action after completion of the engine starting phase of operations.

(3) For single-engine airplanes, require no pilot action after completion of the engine starting phase of operations unless means are provided that unmistakenly alert the pilot to take any needed action at least five minutes prior to the needed action; such pilot action must not cause any change in engine operation; and such pilot action must not distract pilot attention from essential flight duties during any phase of operations for which the airplane is approved.

[Doc. No. 4080, 29 FR 17955, Dec. 18, 1964, as amended by Amdt. 23-7, 34 FR 13093, Aug. 13, 1969; Amdt. 23-43, 58 FR 18971, Apr. 9, 1993; Amdt. 23-51, 61 FR 5136, Feb. 9, 1996]

§ 23.957 Flow between interconnected tanks.

(a) It must be impossible, in a gravity feed system with interconnected tank outlets, for enough fuel to flow between the tanks to cause an overflow of fuel from any tank vent under the conditions in § 23.959, except that full tanks must be used.

(b) If fuel can be pumped from one tank to another in flight, the fuel tank vents and the fuel transfer system must be designed so that no structural damage to any airplane component can occur because of overfilling of any tank.

[Doc. No. 4080, 29 FR 17955, Dec. 18, 1964, as amended by Amdt. 23-43, 58 FR 18972, Apr. 9, 1993]

§ 23.959 Unusable fuel supply.

(a) The unusable fuel supply for each tank must be established as not less than that quantity at which the first evidence of malfunctioning occurs under the most adverse fuel feed condition occurring under each intended operation and flight maneuver involving that tank. Fuel system component failures need not be considered.

(b) The effect on the usable fuel quantity as a result of a failure of any pump shall be determined.

[Amdt. 23-7, 34 FR 13093, Aug. 13, 1969, as amended by Amdt. 23-18, 42 FR 15041, Mar. 17, 1977; Amdt. 23-51, 61 FR 5136, Feb. 9, 1996]

§ 23.961 Fuel system hot weather operation.

Each fuel system must be free from vapor lock when using fuel at its critical temperature, with respect to vapor formation, when operating the airplane in all critical operating and environmental conditions for which approval is requested. For turbine fuel, the initial temperature must be 110°F, -0°, +5°F or the maximum outside air temperature for which approval is requested, whichever is more critical.

[Doc. No. 26344, 58 FR 18972, Apr. 9, 1993; 58 FR 27060, May 6, 1993]

§ 23.963 Fuel tanks: General.

(a) Each fuel tank must be able to withstand, without failure, the vibration, inertia, fluid, and structural loads that it may be subjected to in operation.

(b) Each flexible fuel tank liner must be shown to be suitable for the particular application.

(c) Each integral fuel tank must have adequate facilities for interior inspection and repair.

(d) The total usable capacity of the fuel tanks must be enough for at least one-half hour of operation at maximum continuous power.

(e) Each fuel quantity indicator must be adjusted, as specified in § 23.1337(b), to account for the unusable fuel supply determined under § 23.959(a).

[Doc. No. 4080, 29 FR 17955, Dec. 18, 1964; 30 FR 258, Jan. 9, 1965, as amended by Amdt 23-34, 52 FR 1832, Jan. 15, 1987; Amdt. 23-43, 58 FR 18972, Apr. 9, 1993; Amdt. 23-51, 61 FR 5136, Feb. 9, 1996]

§ 23.965 Fuel tank tests.

(a) Each fuel tank must be able to withstand the following pressures without failure or leakage:

(1) For each conventional metal tank and nonmetallic tank with walls not supported by the airplane structure, a pressure of 3.5 p.s.i., or that pressure developed during maximum ultimate acceleration with a full tank, whichever is greater.

(2) For each integral tank, the pressure developed during the maximum limit acceleration of the airplane with a full tank, with simultaneous application of the critical limit structural loads.

(3) For each nonmetallic tank with walls supported by the airplane structure and constructed in an acceptable manner using acceptable basic tank material, and with actual or simulated support conditions, a pressure of 2 p.s.i. for the first tank of a specific design. The supporting structure must be designed for the critical loads occurring in the flight or landing strength conditions combined with the fuel pressure loads resulting from the corresponding accelerations.

(b) Each fuel tank with large, unsupported, or unstiffened flat surfaces,whose failure or deformation could cause fuel leakage, must be able to withstand the following test without leakage, failure, or excessive deformation of the tank walls:

(1) Each complete tank assembly and its support must be vibration tested while mounted to simulate the actual installation.

(2) Except as specified in paragraph (b)(4) of this section, the tank assembly must be vibrated for 25 hours at a total displacement of not less than $1/_{32}$ of an inch (unless another displacement is substantiated) while $2/_{3}$ filled with water or other suitable test fluid.

(3) The test frequency of vibration must be as follows:

(i) If no frequency of vibration resulting from any rpm within the normal operating range of engine or propeller speeds is critical, the test frequency of vibration is:

(A) The number of cycles per minute obtained by multiplying the maximum continuous propeller speed in rpm by 0.9 for propeller-driven airplanes, and

(B) For non-propeller driven airplanes the test frequency of vibration is 2,000 cycles per minute.

(ii) If only one frequency of vibration resulting from any rpm within the normal operating range of engine or propeller speeds is critical, that frequency of vibration must be the test frequency.

(iii) If more than one frequency of vibration resulting from any rpm within the normal operating range of engine or propeller speeds is critical, the most critical of these frequencies must be the test frequency.

(4) Under paragraph (b)(3) (ii) and (iii) of this section, the time of test must be adjusted to accomplish the same number of vibration cycles that would be accomplished in 25 hours at the frequency specified in paragraph (b)(3)(i) of this section.

(5) During the test, the tank assembly must be rocked at a rate of 16 to 20 complete cycles per minute, through an angle of 15° on either side of the horizontal (30° total), about an axis parallel to the axis of the fuselage, for 25 hours.

(c) Each integral tank using methods of construction and sealing not previously proven to be adequate by test data or service experience must be able to withstand the vibration test specified in paragraphs (b)(1) through (4) of this section.

(d) Each tank with a nonmetallic liner must be subjected to the sloshing test outlined in paragraph (b)(5) of this section, with the fuel at room temperature. In addition, a specimen liner of the same basic construction as that to be used in the airplane must, when installed in a suitable test tank, withstand the sloshing test with fuel at a temperature of 110°F.

[Doc. No. 4080, 29 FR 17955, Dec. 18, 1964, as amended by Amdt. 23-43, 58 FR 18972, Apr. 9, 1993; Amdt. 23-43, 61 FR 253, Jan. 4, 1996; Amdt. 23-51, 61 FR 5136, Feb. 9, 1996]

§ 23.967 Fuel tank installation.

(a) Each fuel tank must be supported so that tank loads are not concentrated. In addition—

(1) There must be pads, if necessary, to prevent chafing between each tank and its supports;

(2) Padding must be nonabsorbent or treated to prevent the absorption of fuel;

(3) If a flexible tank liner is used, it must be supported so that it is not required to withstand fluid loads;

(4) Interior surfaces adjacent to the liner must be smooth and free from projections that could cause wear, unless—

(i) Provisions are made for protection of the liner at those points; or

(ii) The construction of the liner itself provides such protection; and

(5) A positive pressure must be maintained within the vapor space of each bladder cell under any condition of operation, except for a particular condition for which it is shown that a zero or negative pressure will not cause the bladder cell to collapse; and

(6) Syphoning of fuel (other than minor spillage) or collapse of bladder fuel cells may not result from improper securing or loss of the fuel filler cap.

(b) Each tank compartment must be ventilated and drained to prevent the accumulation of flammable fluids or vapors. Each compartment adjacent to a tank that is an integral part of the airplane structure must also be ventilated and drained.

(c) No fuel tank may be on the engine side of the firewall. There must be at least one-half inch of clearance between the fuel tank and the firewall. No part of the engine nacelle skin that lies immediately behind a major air opening from the engine compartment may act as the wall of an integral tank.

(d) Each fuel tank must be isolated from personnel compartments by a fume-proof and fuel-proof enclosure that is vented and drained to the exterior of the airplane. The required enclosure must sustain any personnel compartment pressurization loads without permanent deformation or failure under the conditions of §§ 23.365 and 23.843 of this part. A bladder-type fuel cell, if used, must have a retaining shell at least equivalent to a metal fuel tank in structural integrity.

(e) Fuel tanks must be designed, located, and installed so as to retain fuel:

(1) When subjected to the inertia loads resulting from the ultimate static load factors prescribed in § 23.561(b)(2) of this part; and

(2) Under conditions likely to occur when the airplane lands on a paved runway at a normal landing speed under each of the following conditions:

(i) The airplane in a normal landing attitude and its landing gear retracted.

(ii) The most critical landing gear leg collapsed and the other landing gear legs extended.

In showing compliance with paragraph (e)(2) of this section, the tearing away of an engine mount must be considered unless all the engines are installed above the wing or on the tail or fuselage of the airplane.

[Doc. No. 4080, 29 FR 17955, Dec. 18, 1964, as amended by Amdt. 23-7, 34 FR 13903, Aug. 13, 1969; Amdt. 23-14, 38 FR 31823, Nov. 19, 1973; Amdt. 23-18, 42 FR 15041, Mar. 17, 1977; Amdt. 23-26, 45 FR 60171, Sept. 11, 1980; Amdt. 23-36, 53 FR 30815, Aug. 15, 1988; Amdt. 23-43, 58 FR 18972, Apr. 9, 1993]

§ 23.969 Fuel tank expansion space.

Each fuel tank must have an expansion space of not less than two percent of the tank capacity, unless the tank vent discharges clear of the airplane (in which case no expansion space is required). It must be impossible to fill the expansion space inadvertently with the airplane in the normal ground attitude.

§ 23.971 Fuel tank sump.

(a) Each fuel tank must have a drainable sump with an effective capacity, in the normal ground and flight attitudes, of 0.25 percent of the tank capacity, or $1/_{16}$ gallon, whichever is greater.

(b) Each fuel tank must allow drainage of any hazardous quantity of water from any part of the tank to its sump with the airplane in the normal ground attitude.

(c) Each reciprocating engine fuel system must have a sediment bowl or chamber that is accessible for drainage; has a capacity of 1 ounce for every 20 gallons of fuel tank capacity; and each fuel tank outlet is located so that, in the normal flight attitude, water will drain from all parts of the tank except the sump to the sediment bowl or chamber.

(d) Each sump, sediment bowl, and sediment chamber drain required by paragraphs (a), (b), and (c) of this section must comply with the drain provisions of § 23.999(b)(1) and (b)(2).

[Doc. No. 26344, 58 FR 18972, Apr. 9, 1993; 58 FR 27060, May 6, 1993]

§ 23.973 Fuel tank filler connection.

(a) Each fuel tank filler connection must be marked as prescribed in § 23.1557(c).

(b) Spilled fuel must be prevented from entering the fuel tank compartment or any part of the airplane other than the tank itself.

(c) Each filler cap must provide a fuel-tight seal for the main filler opening. However, there may be small openings in the fuel tank cap for venting purposes or for the purpose of allowing passage of a fuel gauge through the cap provided such openings comply with the requirements of § 23.975(a).

(d) Each fuel filling point, except pressure fueling connection points, must have a provision for electrically bonding the airplane to ground fueling equipment.

(e) For airplanes with engines requiring gasoline as the only permissible fuel, the inside diameter of the fuel filler opening must be no larger than 2.36 inches.

(f) For airplanes with turbine engines, the inside diameter of the fuel filler opening must be no smaller than 2.95 inches.

[Doc. No. 4080, 29 FR 17955, Dec. 18, 1964; 30 FR 258, Jan. 9, 1965, as amended by Amdt. 23-18, 42 FR 15041, Mar. 17, 1977; Amdt. 23-43, 58 FR 18972, Apr. 9, 1993; Amdt. 23-51, 61 FR 5136, Feb. 9, 1996]

§ 23.975 Fuel tank vents and carburetor vapor vents.

(a) Each fuel tank must be vented from the top part of the expansion space. In addition—

(1) Each vent outlet must be located and constructed in a manner that minimizes the possibility of its being obstructed by ice or other foreign matter;

(2) Each vent must be constructed to prevent siphoning of fuel during normal operation;

(3) The venting capacity must allow the rapid relief of excessive differences of pressure between the interior and exterior of the tank;

(4) Airspaces of tanks with interconnected outlets must be interconnected;

(5) There may be no point in any vent line where moisture can accumulate with the airplane in either the ground

or level flight attitudes, unless drainage is provided. Any drain valve installed must be accessible for drainage;

(6) No vent may terminate at a point where the discharge of fuel from the vent outlet will constitute a fire hazard or from which fumes may enter personnel compartments; and

(7) Vents must be arranged to prevent the loss of fuel, except fuel discharged because of thermal expansion, when the airplane is parked in any direction on a ramp having a one-percent slope.

(b) Each carburetor with vapor elimination connections and each fuel injection engine employing vapor return provisions must have a separate vent line to lead vapors back to the top of one of the fuel tanks. If there is more than one tank and it is necessary to use these tanks in a definite sequence for any reason, the vapor vent line must lead back to the fuel tank to be used first, unless the relative capacities of the tanks are such that return to another tank is preferable.

(c) For acrobatic category airplanes, excessive loss of fuel during acrobatic maneuvers, including short periods of inverted flight, must be prevented. It must be impossible for fuel to siphon from the vent when normal flight has been resumed after any acrobatic maneuver for which certification is requested.

[Doc. No. 4080, 29 FR 17955, Dec. 18, 1964; 30 FR 258, Jan. 9, 1965, as amended by Amdt. 23-18, 42 FR 15041, Mar. 17, 1977; Amdt. 23-29, 49 FR 6847, Feb. 23, 1984; Amdt. 23-43, 58 FR 18973, Apr. 9, 1993; Amdt. 23-51, 61 FR 5136, Feb. 9, 1996]

§ 23.977 Fuel tank outlet.

(a) There must be a fuel strainer for the fuel tank outlet or for the booster pump. This strainer must—

(1) For reciprocating engine powered airplanes, have 8 to 16 meshes per inch; and

(2) For turbine engine powered airplanes, prevent the passage of any object that could restrict fuel flow or damage any fuel system component.

(b) The clear area of each fuel tank outlet strainer must be at least five times the area of the outlet line.

(c) The diameter of each strainer must be at least that of the fuel tank outlet.

(d) Each strainer must be accessible for inspection and cleaning.

[Amdt. 23-17, 41 FR 55465, Dec. 20, 1976, as amended by Amdt. 23-43, 58 FR 18973, Apr. 9, 1993]

§ 23.979 Pressure fueling systems.

For pressure fueling systems, the following apply:

(a) Each pressure fueling system fuel manifold connection must have means to prevent the escape of hazardous quantities of fuel from the system if the fuel entry valve fails.

(b) An automatic shutoff means must be provided to prevent the quantity of fuel in each tank from exceeding the maximum quantity approved for that tank. This means must—

(1) Allow checking for proper shutoff operation before each fueling of the tank; and

(2) For commuter category airplanes, indicate at each fueling station, a failure of the shutoff means to stop the fuel flow at the maximum quantity approved for that tank.

(c) A means must be provided to prevent damage to the fuel system in the event of failure of the automatic shutoff means prescribed in paragraph (b) of this section.

(d) All parts of the fuel system up to the tank which are subjected to fueling pressures must have a proof pressure of 1.33 times, and an ultimate pressure of at least 2.0 times, the surge pressure likely to occur during fueling.

[Amdt. 23-14, 38 FR 31823, Nov. 19, 1973, as amended by Amdt. 23-51, 61 FR 5137, Feb. 9, 1996]

Fuel System Components

§ 23.991 Fuel pumps.

(a) *Main pumps.* For main pumps, the following apply:

(1) For reciprocating engine installations having fuel pumps to supply fuel to the engine, at least one pump for each engine must be directly driven by the engine and must meet § 23.955. This pump is a main pump.

(2) For turbine engine installations, each fuel pump required for proper engine operation, or required to meet the fuel system requirements of this subpart (other than those in paragraph (b) of this section), is a main pump. In addition—

(i) There must be at least one main pump for each turbine engine;

(ii) The power supply for the main pump for each engine must be independent of the power supply for each main pump for any other engine; and

(iii) For each main pump, provision must be made to allow the bypass of each positive displacement fuel pump other than a fuel injection pump approved as part of the engine.

(b) *Emergency pumps.* There must be an emergency pump immediately available to supply fuel to the engine if any main pump (other than a fuel injection pump approved as part of an engine) fails. The power supply for each emergency pump must be independent of the power supply for each corresponding main pump.

(c) *Warning means.* If both the main pump and emergency pump operate continuously, there must be a means to indicate to the appropriate flight crewmembers a malfunction of either pump.

(d) Operation of any fuel pump may not affect engine operation so as to create a hazard, regardless of the engine power or thrust setting or the functional status of any other fuel pump.

[Doc. No. 4080, 29 FR 17955, Dec. 18, 1964, as amended by Amdt. 23-7, 34 FR 13093, Aug. 13, 1969; Amdt. 23-26, 45 FR 60171, Sept. 11, 1980; Amdt. 23-43, 58 FR 18973, Apr. 9, 1993]

§ 23.993 Fuel system lines and fittings.

(a) Each fuel line must be installed and supported to prevent excessive vibration and to withstand loads due to fuel pressure and accelerated flight conditions.

(b) Each fuel line connected to components of the airplane between which relative motion could exist must have provisions for flexibility.

(c) Each flexible connection in fuel lines that may be under pressure and subjected to axial loading must use flexible hose assemblies.

(d) Each flexible hose must be shown to be suitable for the particular application.

(e) No flexible hose that might be adversely affected by exposure to high temperatures may be used where excessive temperatures will exist during operation or after engine shutdown.

[Doc. No. 4080, 29 FR 17955, Dec. 18, 1964, as amended by Amdt. 23-43, 58 FR 18973, Apr. 9, 1993]

§ 23.994 Fuel system components.

Fuel system components in an engine nacelle or in the fuselage must be protected from damage which could result in spillage of enough fuel to constitute a fire hazard as a result of a wheels-up landing on a paved runway.

[Amdt. 23-29, 49 FR 6847, Feb. 23, 1984]

§ 23.995 Fuel valves and controls.

(a) There must be a means to allow appropriate flight crew members to rapidly shut off, in flight, the fuel to each engine individually.

(b) No shutoff valve may be on the engine side of any firewall. In addition, there must be means to—

(1) Guard against inadvertent operation of each shutoff valve; and

(2) Allow appropriate flight crew members to reopen each valve rapidly after it has been closed.

(c) Each valve and fuel system control must be supported so that loads resulting from its operation or from accelerated flight conditions are not transmitted to the lines connected to the valve.

(d) Each valve and fuel system control must be installed so that gravity and vibration will not affect the selected position.

(e) Each fuel valve handle and its connections to the valve mechanism must have design features that minimize the possibility of incorrect installation.

(f) Each check valve must be constructed, or otherwise incorporate provisions, to preclude incorrect assembly or connection of the valve.

(g) Fuel tank selector valves must—

(1) Require a separate and distinct action to place the selector in the "OFF" position; and

(2) Have the tank selector positions located in such a manner that it is impossible for the selector to pass through the "OFF" position when changing from one tank to another.

[Doc. No. 4080, 29 FR 17955, Dec. 18, 1964, as amended by Amdt. 23-14, 38 FR 31823, Nov. 19, 1973; Amdt. 23-17, 41 FR 55465, Dec. 20, 1976; Amdt. 23-18, 42 FR 15041, Mar. 17, 1977; Amdt. 23-29, 49 FR 6847, Feb. 23, 1984]

§ 23.997 Fuel strainer or filter.

There must be a fuel strainer or filter between the fuel tank outlet and the inlet of either the fuel metering device or an engine driven positive displacement pump, whichever is nearer the fuel tank outlet. This fuel strainer or filter must—

(a) Be accessible for draining and cleaning and must incorporate a screen or element which is easily removable;

(b) Have a sediment trap and drain except that it need not have a drain if the strainer or filter is easily removable for drain purposes;

(c) Be mounted so that its weight is not supported by the connecting lines or by the inlet or outlet connections of the strainer or filter itself, unless adequate strength margins under all loading conditions are provided in the lines and connections; and

(d) Have the capacity (with respect to operating limitations established for the engine) to ensure that engine fuel system functioning is not impaired, with the fuel contaminated to a degree (with respect to particle size and density) that is greater than that established for the engine during its type certification.

(e) In addition, for commuter category airplanes, unless means are provided in the fuel system to prevent the accumulation of ice on the filter, a means must be provided to automatically maintain the fuel flow if ice clogging of the filter occurs.

[Amdt. 23-15, 39 FR 35459, Oct. 1, 1974, as amended by Amdt. 23-29, 49 FR 6847, Feb. 23, 1984; Amdt. 23-34, 52 FR 1832, Jan. 15, 1987; Amdt. 23-43, 58 FR 18973, Apr. 9, 1993]

§ 23.999 Fuel system drains.

(a) There must be at least one drain to allow safe drainage of the entire fuel system with the airplane in its normal ground attitude.

(b) Each drain required by paragraph (a) of this section and § 23.971 must—

(1) Discharge clear of all parts of the airplane;

(2) Have a drain valve—

(i) That has manual or automatic means for positive locking in the closed position;

(ii) That is readily accessible;

(iii) That can be easily opened and closed;

(iv) That allows the fuel to be caught for examination;

(v) That can be observed for proper closing; and

(vi) That is either located or protected to prevent fuel spillage in the event of a landing with landing gear retracted.

[Doc. No. 4080, 29 FR 17955, Dec. 18, 1964, as amended by Amdt. 23-17, 41 FR 55465, Dec. 20, 1976; Amdt. 23-43, 58 FR 18973, Apr. 9, 1993]

§ 23.1001 Fuel jettisoning system.

(a) If the design landing weight is less than that permitted under the requirements of § 23.473(b), the airplane must have a fuel jettisoning system installed that is able to jettison enough fuel to bring the maximum weight down to the design landing weight. The average rate of fuel jettisoning must be at least 1 percent of the maximum weight per minute, except that the time required to jettison the fuel need not be less than 10 minutes.

(b) Fuel jettisoning must be demonstrated at maximum weight with flaps and landing gear up and in—

(1) A power-off glide at 1.4 V_{S1};

(2) A climb, at the speed at which the one-engine-inoperative enroute climb data have been established in accordance with § 23.69(b), with the critical engine inoperative and the remaining engines at maximum continuous power; and

(3) Level flight at 1.4 V_{S1}, if the results of the tests in the conditions specified in paragraphs (b)(1) and (2) of this section show that this condition could be critical.

(c) During the flight tests prescribed in paragraph (b) of this section, it must be shown that—

(1) The fuel jettisoning system and its operation are free from fire hazard;

(2) The fuel discharges clear of any part of the airplane;

(3) Fuel or fumes do not enter any parts of the airplane; and

(4) The jettisoning operation does not adversely affect the controllability of the airplane.

(d) For reciprocating engine powered airplanes, the jettisoning system must be designed so that it is not possible to jettison the fuel in the tanks used for takeoff and landing below the level allowing 45 minutes flight at 75 percent maximum continuous power. However, if there is an auxiliary control independent of the main jettisoning control, the system may be designed to jettison all the fuel.

(e) For turbine engine powered airplanes, the jettisoning system must be designed so that it is not possible to jettison fuel in the tanks used for takeoff and landing below the level allowing climb from sea level to 10,000 feet and thereafter allowing 45 minutes cruise at a speed for maximum range.

(f) The fuel jettisoning valve must be designed to allow flight crewmembers to close the valve during any part of the jettisoning operation.

(g) Unless it is shown that using any means (including flaps, slots, and slats) for changing the airflow across or around the wings does not adversely affect fuel jettisoning, there must be a placard, adjacent to the jettisoning control, to warn flight crewmembers against jettisoning fuel while the means that change the airflow are being used.

(h) The fuel jettisoning system must be designed so that any reasonably probable single malfunction in the system will not result in a hazardous condition due to unsymmetrical jettisoning of, or inability to jettison, fuel.

[Amdt. 23-7, 34 FR 13094, Aug. 13, 1969, as amended by Amdt. 23-43, 58 FR 18973, Apr. 9, 1993; Amdt. 23-51, 61 FR 5137, Feb. 9, 1996]

Oil System

§ 23.1011 General.

(a) For oil systems and components that have been approved under the engine airworthiness requirements and where those requirements are equal to or more severe than the corresponding requirements of subpart E of this part, that approval need not be duplicated. Where the requirements of subpart E of this part are more severe, substantiation must be shown to the requirements of subpart E of this part.

(b) Each engine must have an independent oil system that can supply it with an appropriate quantity of oil at a temperature not above that safe for continuous operation.

(c) The usable oil tank capacity may not be less than the product of the endurance of the airplane under critical operating conditions and the maximum oil consumption of the engine under the same conditions, plus a suitable margin to ensure adequate circulation and cooling.

(d) For an oil system without an oil transfer system, only the usable oil tank capacity may be considered. The amount of oil in the engine oil lines, the oil radiator, and the feathering reserve, may not be considered.

(e) If an oil transfer system is used, and the transfer pump can pump some of the oil in the transfer lines into the main engine oil tanks, the amount of oil in these lines that can be pumped by the transfer pump may be included in the oil capacity.

[Doc. No. 4080, 29 FR 17955, Dec. 18, 1964, as amended by Amdt. 23-43, 58 FR 18973, Apr. 9, 1993]

§ 23.1013 Oil tanks.

(a) *Installation.* Each oil tank must be installed to—

(1) Meet the requirements of § 23.967 (a) and (b); and

(2) Withstand any vibration, inertia, and fluid loads expected in operation.

(b) *Expansion space.* Oil tank expansion space must be provided so that—

(1) Each oil tank used with a reciprocating engine has an expansion space of not less than the greater of 10 percent of the tank capacity or 0.5 gallon, and each oil tank used with a turbine engine has an expansion space of not less than 10 percent of the tank capacity; and

(2) It is impossible to fill the expansion space inadvertently with the airplane in the normal ground attitude.

(c) *Filler connection.* Each oil tank filler connection must be marked as specified in § 23.1557(c). Each recessed oil tank filler connection of an oil tank used with a turbine engine, that can retain any appreciable quantity of oil, must have provisions for fitting a drain.

(d) *Vent.* Oil tanks must be vented as follows:

(1) Each oil tank must be vented to the engine from the top part of the expansion space so that the vent connection is not covered by oil under any normal flight condition.

(2) Oil tank vents must be arranged so that condensed water vapor that might freeze and obstruct the line cannot accumulate at any point.

(3) For acrobatic category airplanes, there must be means to prevent hazardous loss of oil during acrobatic maneuvers, including short periods of inverted flight.

(e) *Outlet.* No oil tank outlet may be enclosed by any screen or guard that would reduce the flow of oil below a safe value at any operating temperature. No oil tank outlet diameter may be less than the diameter of the engine oil pump inlet. Each oil tank used with a turbine engine must have means to prevent entrance into the tank itself, or into the tank outlet, of any object that might obstruct the flow of oil through the system. There must be a shutoff valve at the outlet of each oil tank used with a turbine engine, unless the external portion of the oil system (including oil tank supports) is fireproof.

(f) *Flexible liners.* Each flexible oil tank liner must be of an acceptable kind.

(g) Each oil tank filler cap of an oil tank that is used with an engine must provide an oiltight seal.

[Doc. No. 4080, 29 FR 17955, Dec. 18, 1964, as amended by Amdt. 23-15, 39 FR 35459 Oct. 1, 1974; Amdt. 23-43, 58 FR 18973, Apr. 9, 1993; Amdt. 23-51, 61 FR 5137, Feb. 9, 1996]

§ 23.1015 Oil tank tests.

Each oil tank must be tested under § 23.965, except that—

(a) The applied pressure must be five p.s.i. for the tank construction instead of the pressures specified in § 23.965(a);

(b) For a tank with a nonmetallic liner the test fluid must be oil rather than fuel as specified in § 23.965(d), and the slosh test on a specimen liner must be conducted with the oil at 250°F.; and

(c) For pressurized tanks used with a turbine engine, the test pressure may not be less than 5 p.s.i. plus the maximum operating pressure of the tank.

[Doc. No. 4080, 29 FR 17955, Dec. 18, 1964, as amended by Amdt. 23-15, 39 FR 35460, Oct. 1, 1974]

§ 23.1017 Oil lines and fittings.

(a) *Oil lines.* Oil lines must meet § 23.993 and must accommodate a flow of oil at a rate and pressure adequate for proper engine functioning under any normal operating condition.

(b) *Breather lines.* Breather lines must be arranged so that—

(1) Condensed water vapor or oil that might freeze and obstruct the line cannot accumulate at any point;

(2) The breather discharge will not constitute a fire hazard if foaming occurs, or cause emitted oil to strike the pilot's windshield;

(3) The breather does not discharge into the engine air induction system; and

(4) For acrobatic category airplanes, there is no excessive loss of oil from the breather during acrobatic maneuvers, including short periods of inverted flight.

(5) The breather outlet is protected against blockage by ice or foreign matter.

[Doc. No. 4080, 29 FR 17955, Dec. 18, 1964, as amended by Amdt. 23-7, 34 FR 13094, Aug. 13, 1969; Amdt. 23-14, 38 FR 31823, Nov. 19, 1973]

§ 23.1019 Oil strainer or filter.

(a) Each turbine engine installation must incorporate an oil strainer or filter through which all of the engine oil flows and which meets the following requirements:

(1) Each oil strainer or filter that has a bypass, must be constructed and installed so that oil will flow at the normal rate through the rest of the system with the strainer or filter completely blocked.

(2) The oil strainer or filter must have the capacity (with respect to operating limitations established for the engine) to ensure that engine oil system functioning is not impaired when the oil is contaminated to a degree (with respect to particle size and density) that is greater than that established for the engine for its type certification.

(3) The oil strainer or filter, unless it is installed at an oil tank outlet, must incorporate a means to indicate contamination before it reaches the capacity established in accordance with paragraph (a)(2) of this section.

(4) The bypass of a strainer or filter must be constructed and installed so that the release of collected contaminants is minimized by appropriate location of the bypass to ensure that collected contaminants are not in the bypass flow path.

(5) An oil strainer or filter that has no bypass, except one that is installed at an oil tank outlet, must have a means to connect it to the warning system required in § 23.1305(c)(9).

(b) Each oil strainer or filter in a powerplant installation using reciprocating engines must be constructed and installed so that oil will flow at the normal rate through the rest of the system with the strainer or filter element completely blocked.

[Amdt. 23-15, 39 FR 35460, Oct. 1, 1974, as amended by Amdt. 23-29, 49 FR 6847, Feb. 23, 1984; Amdt. 23-43, 58 FR 18973, Apr. 9, 1993]

§ 23.1021 Oil system drains.

A drain (or drains) must be provided to allow safe drainage of the oil system. Each drain must—

(a) Be accessible;

(b) Have drain valves, or other closures, employing manual or automatic shut-off means for positive locking in the closed position; and

(c) Be located or protected to prevent inadvertent operation.

[Amdt. 23-29, 49 FR 6847, Feb. 23, 1984, as amended by Amdt. 23-43, 58 FR 18973, Apr. 9, 1993]

§ 23.1023 Oil radiators.

Each oil radiator and its supporting structures must be able to withstand the vibration, inertia, and oil pressure loads to which it would be subjected in operation.

§ 23.1027 Propeller feathering system.

(a) If the propeller feathering system uses engine oil and that oil supply can become depleted due to failure of any part of the oil system, a means must be incorporated to reserve enough oil to operate the feathering system.

(b) The amount of reserved oil must be enough to accomplish feathering and must be available only to the feathering pump.

(c) The ability of the system to accomplish feathering with the reserved oil must be shown.

(d) Provision must be made to prevent sludge or other foreign matter from affecting the safe operation of the propeller feathering system.

[Doc. No. 4080, 29 FR 17955, Dec. 18, 1964, as amended by Amdt. 23-14, 38 FR 31823, Nov. 19, 1973; Amdt. 23-43, 58 FR 18973, Apr. 9, 1993]

Cooling

§ 23.1041 General.

The powerplant and auxiliary power unit cooling provisions must maintain the temperatures of powerplant components and engine fluids, and auxiliary power unit components and fluids within the limits established for those components and fluids under the most adverse ground, water, and flight operations to the maximum altitude and maximum ambient atmospheric temperature conditions for which approval is requested, and after normal engine and auxiliary power unit shutdown.

[Doc. No. 26344, 58 FR 18973, Apr. 9, 1993, as amended by Amdt. 23-51, 61 FR 5137, Feb. 9, 1996]

§ 23.1043 Cooling tests.

(a) *General.* Compliance with § 23.1041 must be shown on the basis of tests, for which the following apply:

(1) If the tests are conducted under ambient atmospheric temperature conditions deviating from the maximum for which approval is requested, the recorded powerplant temperatures must be corrected under paragraphs (c) and (d) of this section, unless a more rational correction method is applicable.

(2) No corrected temperature determined under paragraph (a)(1) of this section may exceed established limits.

(3) The fuel used during the cooling tests must be of the minimum grade approved for the engine.

(4) For turbocharged engines, each turbocharger must be operated through that part of the climb profile for which operation with the turbocharger is requested.

(5) For a reciprocating engine, the mixture settings must be the leanest recommended for climb.

(b) *Maximum ambient atmospheric temperature.* A maximum ambient atmospheric temperature corresponding to sea level conditions of at least 100 degrees F must be established. The assumed temperature lapse rate is 3.6 degrees F per thousand feet of altitude above sea level until a temperature of –69.7 degrees F is reached, above which altitude the temperature is considered constant at –69.7 degrees F. However, for winterization installations, the applicant may select a maximum ambient atmospheric temperature corresponding to sea level conditions of less than 100 degrees F.

(c) *Correction factor (except cylinder barrels).* Temperatures of engine fluids and powerplant components (except cylinder barrels) for which temperature limits are established, must be corrected by adding to them the difference between the maximum ambient atmospheric temperature for the relevant altitude for which approval has been requested and the temperature of the ambient air at the time of the first occurrence of the maximum fluid or component temperature recorded during the cooling test.

(d) *Correction factor for cylinder barrel temperatures.* Cylinder barrel temperatures must be corrected by adding to them 0.7 times the difference between the maximum ambient atmospheric temperature for the relevant altitude for which approval has been requested and the temperature of the ambient air at the time of the first occurrence of the maximum cylinder barrel temperature recorded during the cooling test.

[Doc. No. 4080, 29 FR 17955, Dec. 18, 1964, as amended by Amdt. 23-7, 34 FR 13094, Aug. 13, 1969; Amdt. 23-21, 43 FR 2319, Jan. 16, 1978; Amdt. 23-51, 61 FR 5137, Feb. 9, 1996]

§ 23.1045 Cooling test procedures for turbine engine powered airplanes.

(a) Compliance with § 23.1041 must be shown for all phases of operation. The airplane must be flown in the configurations, at the speeds, and following the procedures recommended in the Airplane Flight Manual for the relevant stage of flight, that correspond to the applicable performance requirements that are critical to cooling.

(b) Temperatures must be stabilized under the conditions from which entry is made into each stage of flight being investigated, unless the entry condition normally is not one during which component and engine fluid temperatures would stabilize (in which case, operation through the full entry condition must be conducted before entry into the stage of flight being investigated in order to allow temperatures to reach their natural levels at the time of entry). The takeoff cooling test must be preceded by a period during which the powerplant component and engine fluid temperatures are stabilized with the engines at ground idle.

(c) Cooling tests for each stage of flight must be continued until—

(1) The component and engine fluid temperatures stabilize;

(2) The stage of flight is completed; or

(3) An operating limitation is reached.

[Amdt. 23-7, 34 FR 13094, Aug. 13, 1969, as amended by Amdt. 23-51, 61 FR 5137, Feb. 9, 1996]

§ 23.1047 Cooling test procedures for reciprocating engine powered airplanes.

Compliance with § 23.1041 must be shown for the climb (or, for multiengine airplanes with negative one-engine-inoperative rates of climb, the descent) stage of flight. The airplane must be flown in the configurations, at the speeds and following the procedures recommended in the Airplane Flight Manual, that correspond to the applicable performance requirements that are critical to cooling.

[Amdt. 23-51, 61 FR 5137, Feb. 9, 1996]

Liquid Cooling

§ 23.1061 Installation.

(a) *General.* Each liquid-cooled engine must have an independent cooling system (including coolant tank) installed so that—

(1) Each coolant tank is supported so that tank loads are distributed over a large part of the tank surface;

(2) There are pads or other isolation means between the tank and its supports to prevent chafing.

(3) Pads or any other isolation means that is used must be nonabsorbent or must be treated to prevent absorption of flammable fluids; and

(4) No air or vapor can be trapped in any part of the system, except the coolant tank expansion space, during filling or during operation.

(b) *Coolant tank.* The tank capacity must be at least one gallon, plus 10 percent of the cooling system capacity. In addition—

(1) Each coolant tank must be able to withstand the vibration, inertia, and fluid loads to which it may be subjected in operation;

(2) Each coolant tank must have an expansion space of at least 10 percent of the total cooling system capacity; and

(3) It must be impossible to fill the expansion space inadvertently with the airplane in the normal ground attitude.

(c) *Filler connection.* Each coolant tank filler connection must be marked as specified in § 23.1557(c). In addition—

(1) Spilled coolant must be prevented from entering the coolant tank compartment or any part of the airplane other than the tank itself; and

(2) Each recessed coolant filler connection must have a drain that discharges clear of the entire airplane.

(d) *Lines and fittings.* Each coolant system line and fitting must meet the requirements of § 23.993, except that the inside diameter of the engine coolant inlet and outlet lines may not be less than the diameter of the corresponding engine inlet and outlet connections.

(e) *Radiators.* Each coolant radiator must be able to withstand any vibration, inertia, and coolant pressure load to which it may normally be subjected. In addition—

(1) Each radiator must be supported to allow expansion due to operating temperatures and prevent the transmittal of harmful vibration to the radiator; and

(2) If flammable coolant is used, the air intake duct to the coolant radiator must be located so that (in case of fire) flames from the nacelle cannot strike the radiator.

(f) *Drains.* There must be an accessible drain that—

(1) Drains the entire cooling system (including the coolant tank, radiator, and the engine) when the airplane is in the normal ground altitude;

(2) Discharges clear of the entire airplane; and

(3) Has means to positively lock it closed.

[Doc. No. 4080, 29 FR 17955, Dec. 18, 1964, as amended by Amdt. 23-43, 58 FR 18973, Apr. 9, 1993]

§ 23.1063 Coolant tank tests.

Each coolant tank must be tested under § 23.965, except that—

(a) The test required by § 23.965(a)(1) must be replaced with a similar test using the sum of the pressure developed during the maximum ultimate acceleration with a full tank or a pressure of 3.5 pounds per square inch, whichever is greater, plus the maximum working pressure of the system; and

(b) For a tank with a nonmetallic liner the test fluid must be coolant rather than fuel as specified in § 23.965(d), and the slosh test on a specimen liner must be conducted with the coolant at operating temperature.

Induction System

§ 23.1091 Air induction system.

(a) The air induction system for each engine and auxiliary power unit and their accessories must supply the air required by that engine and auxiliary power unit and their accessories under the operating conditions for which certification is requested.

(b) Each reciprocating engine installation must have at least two separate air intake sources and must meet the following:

(1) Primary air intakes may open within the cowling if that part of the cowling is isolated from the engine accessory section by a fire-resistant diaphragm or if there are means to prevent the emergence of backfire flames.

(2) Each alternate air intake must be located in a sheltered position and may not open within the cowling if the emergence of backfire flames will result in a hazard.

(3) The supplying of air to the engine through the alternate air intake system may not result in a loss of excessive power in addition to the power loss due to the rise in air temperature.

(4) Each automatic alternate air door must have an override means accessible to the flight crew.

(5) Each automatic alternate air door must have a means to indicate to the flight crew when it is not closed.

(c) For turbine engine powered airplanes—

(1) There must be means to prevent hazardous quantities of fuel leakage or overflow from drains, vents, or other components of flammable fluid systems from entering the engine intake system; and

(2) The airplane must be designed to prevent water or slush on the runway, taxiway, or other airport operating surfaces from being directed into the engine or auxiliary power unit air intake ducts in hazardous quantities. The air intake ducts must be located or protected so as to minimize the hazard of ingestion of foreign matter during takeoff, landing, and taxiing.

[Doc. No. 4080, 29 FR 17955, Dec. 18, 1964, as amended by Amdt. 23-7, 34 FR 13095, Aug. 13, 1969; Amdt. 23-43, 58 FR 18973, Apr. 9, 1993; 58 FR 27060, May 6, 1993; Amdt. 23-51, 61 FR 5137, Feb. 9, 1996]

§ 23.1093 Induction system icing protection.

(a) *Reciprocating engines.* Each reciprocating engine air induction system must have means to prevent and eliminate icing. Unless this is done by other means, it must be shown that, in air free of visible moisture at a temperature of 30°F—

(1) Each airplane with sea level engines using conventional venturi carburetors has a preheater that can provide a heat rise of 90°F. with the engines at 75 percent of maximum continuous power;

(2) Each airplane with altitude engines using conventional venturi carburetors has a preheater that can provide a heat rise of 120°F. with the engines at 75 percent of maximum continuous power;

(3) Each airplane with altitude engines using fuel metering device tending to prevent icing has a preheater that, with the engines at 60 percent of maximum continuous power, can provide a heat rise of—

(i) 100°F.; or

(ii) 40°F., if a fluid deicing system meeting the requirements of §§ 23.1095 through 23.1099 is installed;

(4) Each airplane with sea level engine(s) using fuel metering device tending to prevent icing has a sheltered alternate source of air with a preheat of not less than 60°F with the engines at 75 percent of maximum continuous power;

(5) Each airplane with sea level or altitude engine(s) using fuel injection systems having metering components on which impact ice may accumulate has a preheater capable of providing a heat rise of 75°F when the engine is operating at 75 percent of its maximum continuous power; and

(6) Each airplane with sea level or altitude engine(s) using fuel injection systems not having fuel metering components projecting into the airstream on which ice may form, and introducing fuel into the air induction system downstream of any components or other obstruction on which ice produced by fuel evaporation may form, has a sheltered alternate source of air with a preheat of not less than 60°F with the engines at 75 percent of its maximum continuous power.

(b) *Turbine engines.* (1) Each turbine engine and its air inlet system must operate throughout the flight power range of the engine (including idling), without the accumulation of ice on engine or inlet system components that would adversely affect engine operation or cause a serious loss of power or thrust—

(i) Under the icing conditions specified in appendix C of part 25 of this chapter; and

(ii) In snow, both falling and blowing, within the limitations established for the airplane for such operation.

(2) Each turbine engine must idle for 30 minutes on the ground, with the air bleed available for engine icing protection at its critical condition, without adverse effect, in an atmosphere that is at a temperature between 15° and 30°F (between -9° and -1°C) and has a liquid water content not less than 0.3 grams per cubic meter in the form of drops having a mean effective diameter not less than 20 microns, followed by momentary operation at takeoff power or thrust. During the 30 minutes of idle operation, the engine may be run up periodically to a moderate power or thrust setting in a manner acceptable to the Administrator.

(c) *Reciprocating engines with Superchargers.* For airplanes with reciprocating engines having superchargers to pressurize the air before it enters the fuel metering device, the heat rise in the air caused by that supercharging at any altitude may be utilized in determining compliance with paragraph (a) of this section if the heat rise utilized is that which will be available, automatically, for the applicable altitudes and operating condition because of supercharging.

[Amdt. 23-7, 34 FR 13095, Aug. 13, 1969, as amended by Amdt. 23-15, 39 FR 35460, Oct. 1, 1974; Amdt. 23-17, 41 FR 55465, Dec. 20, 1976; Amdt. 23-18, 42 FR 15041, Mar. 17, 1977; Amdt. 23-29, 49 FR 6847, Feb. 23, 1984; Amdt. 23-43, 58 FR 18973, Apr. 9, 1993; Amdt. 23-51, 61 FR 5137, Feb. 9, 1996]

§ 23.1095 Carburetor deicing fluid flow rate.

(a) If a carburetor deicing fluid system is used, it must be able to simultaneously supply each engine with a rate of fluid flow, expressed in pounds per hour, of not less than 2.5 times the square root of the maximum continuous power of the engine.

(b) The fluid must be introduced into the air induction system—

(1) Close to, and upstream of, the carburetor; and

(2) So that it is equally distributed over the entire cross section of the induction system air passages.

§ 23.1097 Carburetor deicing fluid system capacity.

(a) The capacity of each carburetor deicing fluid system—

(1) May not be less than the greater of—

(i) That required to provide fluid at the rate specified in § 23.1095 for a time equal to three percent of the maximum endurance of the airplane; or

(ii) 20 minutes at that flow rate; and

(2) Need not exceed that required for two hours of operation.

(b) If the available preheat exceeds 50°F. but is less than 100°F., the capacity of the system may be decreased in proportion to the heat rise available in excess of 50°F.

§ 23.1099 Carburetor deicing fluid system detail design.

Each carburetor deicing fluid system must meet the applicable requirements for the design of a fuel system, except as specified in §§ 23.1095 and 23.1097.

§ 23.1101 Induction air preheater design.

Each exhaust-heated, induction air preheater must be designed and constructed to—

(a) Ensure ventilation of the preheater when the induction air preheater is not being used during engine operation;

(b) Allow inspection of the exhaust manifold parts that it surrounds; and

(c) Allow inspection of critical parts of the preheater itself.

[Doc. No. 4080, 29 FR 17955, Dec. 18, 1964, as amended by Amdt. 23-43, 58 FR 18974, Apr. 9, 1993]

§ 23.1103 Induction system ducts.

(a) Each induction system duct must have a drain to prevent the accumulation of fuel or moisture in the normal ground and flight attitudes. No drain may discharge where it will cause a fire hazard.

(b) Each duct connected to components between which relative motion could exist must have means for flexibility.

(c) Each flexible induction system duct must be capable of withstanding the effects of temperature extremes, fuel, oil, water, and solvents to which it is expected to be exposed in service and maintenance without hazardous deterioration or delamination.

(d) For reciprocating engine installations, each induction system duct must be—

(1) Strong enough to prevent induction system failures resulting from normal backfire conditions; and

(2) Fire resistant in any compartment for which a fire extinguishing system is required.

(e) Each inlet system duct for an auxiliary power unit must be—

(1) Fireproof within the auxiliary power unit compartment;

(2) Fireproof for a sufficient distance upstream of the auxiliary power unit compartment to prevent hot gas reverse flow from burning through the duct and entering any other compartment of the airplane in which a hazard would be created by the entry of the hot gases;

(3) Constructed of materials suitable to the environmental conditions expected in service, except in those areas requiring fireproof or fire resistant materials; and

(4) Constructed of materials that will not absorb or trap hazardous quantities of flammable fluids that could be ignited by a surge or reverse-flow condition.

(f) Induction system ducts that supply air to a cabin pressurization system must be suitably constructed of material that will not produce hazardous quantities of toxic gases or isolated to prevent hazardous quantities of toxic gases from entering the cabin during a powerplant fire.

[Doc. No. 4080, 29 FR 17955, Dec. 18, 1964, as amended by Amdt. 23-7, 34 FR 13095, Aug. 13, 1969; Amdt. 23-43, 58 FR 18974, Apr. 9, 1993]

§ 23.1105 Induction system screens.

If induction system screens are used—

(a) Each screen must be upstream of the carburetor or fuel injection system.

(b) No screen may be in any part of the induction system that is the only passage through which air can reach the engine, unless—

(1) The available preheat is at least 100°F.; and

(2) The screen can be deiced by heated air;

(c) No screen may be deiced by alcohol alone; and

(d) It must be impossible for fuel to strike any screen.

[Doc. No. 4080, 29 FR 17955, Dec. 18, 1964; 30 FR 258, Jan. 9, 1996, as amended by Amdt. 23-51, 61 FR 5137, Feb. 9, 1996]

§ 23.1107 Induction system filters.

If an air filter is used to protect the engine against foreign material particles in the induction air supply—

(a) Each air filter must be capable of withstanding the effects of temperature extremes, rain, fuel, oil, and solvents to which it is expected to be exposed in service and maintenance; and

(b) Each air filter shall have a design feature to prevent material separated from the filter media from interfering with proper fuel metering operation.

[Doc. No. 26344, 58 FR 18974, Apr. 9, 1993, as amended by Amdt. 23-51, 61 FR 5137, Feb. 9, 1996]

§ 23.1109 Turbocharger bleed air system.

The following applies to turbocharged bleed air systems used for cabin pressurization:

(a) The cabin air system may not be subject to hazardous contamination following any probable failure of the turbocharger or its lubrication system.

(b) The turbocharger supply air must be taken from a source where it cannot be contaminated by harmful or hazardous gases or vapors following any probable failure or malfunction of the engine exhaust, hydraulic, fuel, or oil system.

[Amdt. 23-42, 56 FR 354, Jan. 3, 1991]

§ 23.1111 Turbine engine bleed air system.

For turbine engine bleed air systems, the following apply:

(a) No hazard may result if duct rupture or failure occurs anywhere between the engine port and the airplane unit served by the bleed air.

(b) The effect on airplane and engine performance of using maximum bleed air must be established.

(c) Hazardous contamination of cabin air systems may not result from failures of the engine lubricating system.

[Amdt. 23-7, 34 FR 13095, Aug. 13, 1969, as amended by Amdt. 23-17, 41 FR 55465, Dec. 20, 1976]

Exhaust System

§ 23.1121 General.

For powerplant and auxiliary power unit installations, the following apply—

(a) Each exhaust system must ensure safe disposal of exhaust gases without fire hazard or carbon monoxide contamination in any personnel compartment.

(b) Each exhaust system part with a surface hot enough to ignite flammable fluids or vapors must be located or shielded so that leakage from any system carrying flammable fluids or vapors will not result in a fire caused by impingement of the fluids or vapors on any part of the exhaust system including shields for the exhaust system.

(c) Each exhaust system must be separated by fireproof shields from adjacent flammable parts of the airplane that are outside of the engine and auxiliary power unit compartments.

(d) No exhaust gases may discharge dangerously near any fuel or oil system drain.

(e) No exhaust gases may be discharged where they will cause a glare seriously affecting pilot vision at night.

(f) Each exhaust system component must be ventilated to prevent points of excessively high temperature.

(g) If significant traps exist, each turbine engine and auxiliary power unit exhaust system must have drains discharging clear of the airplane, in any normal ground and flight attitude, to prevent fuel accumulation after the failure of an attempted engine or auxiliary power unit start.

(h) Each exhaust heat exchanger must incorporate means to prevent blockage of the exhaust port after any internal heat exchanger failure.

(i) For the purpose of compliance with § 23.603, the failure of any part of the exhaust system will be considered to adversely affect safety.

[Doc. No. 4080, 29 FR 17955, Dec. 18, 1964, as amended by Amdt. 23-7, 34 FR 13095, Aug. 13, 1969; Amdt. 23-18, 42 FR 15042, Mar. 17, 1977; Amdt. 23-43, 58 FR 18974, Apr. 9, 1993; Amdt. 23-51, 61 FR 5137, Feb. 9, 1996]

§ 23.1123 Exhaust system.

(a) Each exhaust system must be fireproof and corrosion-resistant, and must have means to prevent failure due to expansion by operating temperatures.

(b) Each exhaust system must be supported to withstand the vibration and inertia loads to which it may be subjected in operation.

(c) Parts of the system connected to components between which relative motion could exist must have means for flexibility.

[Doc. No. 4080, 29 FR 17955, Dec. 18, 1964, as amended by Amdt. 23-43, 58 FR 18974, Apr. 9, 1993]

§ 23.1125 Exhaust heat exchangers.

For reciprocating engine powered airplanes the following apply:

(a) Each exhaust heat exchanger must be constructed and installed to withstand the vibration, inertia, and other loads that it may be subjected to in normal operation. In addition—

(1) Each exchanger must be suitable for continued operation at high temperatures and resistant to corrosion from exhaust gases;

(2) There must be means for inspection of critical parts of each exchanger; and

(3) Each exchanger must have cooling provisions wherever it is subject to contact with exhaust gases.

(b) Each heat exchanger used for heating ventilating air must be constructed so that exhaust gases may not enter the ventilating air.

[Doc. No. 4080, 29 FR 17955, Dec. 18, 1964, as amended by Amdt. 23-17, 41 FR 55465, Dec. 20, 1976]

Powerplant Controls and Accessories

§ 23.1141 Powerplant controls: General.

(a) Powerplant controls must be located and arranged under § 23.777 and marked under § 23.1555(a).

(b) Each flexible control must be shown to be suitable for the particular application.

(c) Each control must be able to maintain any necessary position without—

(1) Constant attention by flight crew members; or

(2) Tendency to creep due to control loads or vibration.

(d) Each control must be able to withstand operating loads without failure or excessive deflection.

(e) For turbine engine powered airplanes, no single failure or malfunction, or probable combination thereof, in any powerplant control system may cause the failure of any powerplant function necessary for safety.

(f) The portion of each powerplant control located in the engine compartment that is required to be operated in the event of fire must be at least fire resistant.

(g) Powerplant valve controls located in the cockpit must have—

(1) For manual valves, positive stops or in the case of fuel valves suitable index provisions, in the open and closed position; and

(2) For power-assisted valves, a means to indicate to the flight crew when the valve—

(i) Is in the fully open or fully closed position; or

(ii) Is moving between the fully open and fully closed position.

[Doc. No. 4080, 29 FR 17955, Dec. 18, 1964, as amended by Amdt. 23-7, 34 FR 13095, Aug. 13, 1969; Amdt. 23-14, 38 FR 31823, Nov. 19, 1973; Amdt. 23-18, 42 FR 15042, Mar. 17, 1977; Amdt. 23-51, 61 FR 5137, Feb. 9, 1996]

§ 23.1142 Auxiliary power unit controls.

Means must be provided on the flight deck for the starting, stopping, monitoring, and emergency shutdown of each installed auxiliary power unit.

[Doc. No. 26344, 58 FR 18974, Apr. 9, 1993]

§ 23.1143 Engine controls.

(a) There must be a separate power or thrust control for each engine and a separate control for each supercharger that requires a control.

(b) Power, thrust, and supercharger controls must be arranged to allow—

(1) Separate control of each engine and each supercharger; and

(2) Simultaneous control of all engines and all superchargers.

(c) Each power, thrust, or supercharger control must give a positive and immediate responsive means of controlling its engine or supercharger.

(d) The power, thrust, or supercharger controls for each engine or supercharger must be independent of those for every other engine or supercharger.

(e) For each fluid injection (other than fuel) system and its controls not provided and approved as part of the engine, the applicant must show that the flow of the injection fluid is adequately controlled.

(f) If a power, thrust, or a fuel control (other than a mixture control) incorporates a fuel shutoff feature, the control must have a means to prevent the inadvertent movement of the control into the off position. The means must—

(1) Have a positive lock or stop at the idle position; and

(2) Require a separate and distinct operation to place the control in the shutoff position.

(g) For reciprocating single-engine airplanes, each power or thrust control must be designed so that if the control separates at the engine fuel metering device, the airplane is capable of continued safe flight and landing.

[Amdt. 23-7, 34 FR 13095, Aug. 13, 1969, as amended by Amdt. 23-17, 41 FR 55465, Dec. 20, 1976; Amdt. 23-29, 49 FR 6847, Feb. 23, 1984; Amdt. 23-43, 58 FR 18974, Apr. 9, 1993; Amdt. 23-51, 61 FR 5137, Feb. 9, 1996]

§ 23.1145 Ignition switches.

(a) Ignition switches must control and shut off each ignition circuit on each engine.

(b) There must be means to quickly shut off all ignition on multiengine airplanes by the grouping of switches or by a master ignition control.

(c) Each group of ignition switches, except ignition switches for turbine engines for which continuous ignition is not required, and each master ignition control must have a means to prevent its inadvertent operation.

[Doc. No. 4080, 29 FR 17955, Dec. 18, 1964; 30 FR 258, Jan. 9, 1965, as amended by Amdt. 23-18, 42 FR 15042, Mar. 17, 1977; Amdt. 23-43, 58 FR 18974, Apr. 9, 1993]

§ 23.1147 Mixture controls.

(a) If there are mixture controls, each engine must have a separate control, and each mixture control must have guards or must be shaped or arranged to prevent confusion by feel with other controls.

(1) The controls must be grouped and arranged to allow—

(i) Separate control of each engine; and

(ii) Simultaneous control of all engines.

(2) The controls must require a separate and distinct operation to move the control toward lean or shut-off position.

(b) For reciprocating single-engine airplanes, each manual engine mixture control must be designed so that, if the control separates at the engine fuel metering device, the airplane is capable of continued safe flight and landing.

[Doc. No. 4080, 29 FR 17955, Dec. 18, 1964, as amended by Amdt. 23-7, 34 FR 13096, Aug. 13, 1969; Amdt. 23-33, 51 FR 26657, July 24, 1986; Amdt. 23-43, 58 FR 18974, Apr. 9, 1993]

§ 23.1149 Propeller speed and pitch controls.

(a) If there are propeller speed or pitch controls, they must be grouped and arranged to allow—

(1) Separate control of each propeller; and

(2) Simultaneous control of all propellers.

(b) The controls must allow ready synchronization of all propellers on multiengine airplanes.

§ 23.1153 Propeller feathering controls.

If there are propeller feathering controls installed, it must be possible to feather each propeller separately. Each control must have a means to prevent inadvertent operation.

[Doc. No. 27804, 61 FR 5138, Feb. 9, 1996]

§ 23.1155 Turbine engine reverse thrust and propeller pitch settings below the flight regime.

For turbine engine installations, each control for reverse thrust and for propeller pitch settings below the flight regime must have means to prevent its inadvertent operation. The means must have a positive lock or stop at the flight idle position and must require a separate and distinct operation by the crew to displace the control from the flight regime (forward thrust regime for turbojet powered airplanes).

[Amdt. 23-7, 34 FR 13096, Aug. 13, 1969]

§ 23.1157 Carburetor air temperature controls.

There must be a separate carburetor air temperature control for each engine.

§ 23.1163 Powerplant accessories.

(a) Each engine mounted accessory must—

(1) Be approved for mounting on the engine involved and use the provisions on the engines for mounting; or

(2) Have torque limiting means on all accessory drives in order to prevent the torque limits established for those drives from being exceeded; and

(3) In addition to paragraphs (a)(1) or (a)(2) of this section, be sealed to prevent contamination of the engine oil system and the accessory system.

(b) Electrical equipment subject to arcing or sparking must be installed to minimize the probability of contact with any flammable fluids or vapors that might be present in a free state.

(c) Each generator rated at or more than 6 kilowatts must be designed and installed to minimize the probability of a fire hazard in the event it malfunctions.

(d) If the continued rotation of any accessory remotely driven by the engine is hazardous when malfunctioning occurs, a means to prevent rotation without interfering with the continued operation of the engine must be provided.

(e) Each accessory driven by a gearbox that is not approved as part of the powerplant driving the gearbox must—

(1) Have torque limiting means to prevent the torque limits established for the affected drive from being exceeded;

(2) Use the provisions on the gearbox for mounting; and

(3) Be sealed to prevent contamination of the gearbox oil system and the accessory system.

[Doc. No. 4080, 29 FR 17955, Dec. 18, 1964, as amended by Amdt. 23-14, 38 FR 31823, Nov. 19, 1973; Amdt. 23-29, 49 FR 6847, Feb. 23, 1984; Amdt. 23-34, 52 FR 1832, Jan. 15, 1987; Amdt. 23-42, 56 FR 354, Jan. 3, 1991]

§ 23.1165 Engine ignition systems.

(a) Each battery ignition system must be supplemented by a generator that is automatically available as an alternate source of electrical energy to allow continued engine operation if any battery becomes depleted.

(b) The capacity of batteries and generators must be large enough to meet the simultaneous demands of the engine ignition system and the greatest demands of any electrical system components that draw from the same source.

(c) The design of the engine ignition system must account for—

(1) The condition of an inoperative generator;

(2) The condition of a completely depleted battery with the generator running at its normal operating speed; and

(3) The condition of a completely depleted battery with the generator operating at idling speed, if there is only one battery.

(d) There must be means to warn appropriate crewmembers if malfunctioning of any part of the electrical system is causing the continuous discharge of any battery used for engine ignition.

(e) Each turbine engine ignition system must be independent of any electrical circuit that is not used for assisting, controlling, or analyzing the operation of that system.

(f) In addition, for commuter category airplanes, each turbine engine ignition system must be an essential electrical load.

[Doc. No. 4080, 29 FR 17955, Dec. 18, 1964, as amended by Amdt. 23-17, 41 FR 55465 Dec. 20, 1976; Amdt. 23-34, 52 FR 1833, Jan. 15, 1987; Amdt. 23-62, 76 FR 75759, Dec. 2, 2011]

Powerplant Fire Protection

§ 23.1181 Designated fire zones; regions included.

Designated fire zones are—

(a) For reciprocating engines—

(1) The power section;

(2) The accessory section;

(3) Any complete powerplant compartment in which there is no isolation between the power section and the accessory section.

(b) For turbine engines—

(1) The compressor and accessory sections;

(2) The combustor, turbine and tailpipe sections that contain lines or components carrying flammable fluids or gases.

(3) Any complete powerplant compartment in which there is no isolation between compressor, accessory, combustor, turbine, and tailpipe sections.

(c) Any auxiliary power unit compartment; and

(d) Any fuel-burning heater, and other combustion equipment installation described in § 23.859.

[Doc. No. 26344, 58 FR 18975, Apr. 9, 1993, as amended by Amdt. 23-51, 61 FR 5138, Feb. 9, 1996]

§ 23.1182 Nacelle areas behind firewalls.

Components, lines, and fittings, except those subject to the provisions of § 23.1351(e), located behind the engine-compartment firewall must be constructed of such materials and located at such distances from the firewall that they will not suffer damage sufficient to endanger the airplane if a portion of the engine side of the firewall is subjected to a flame temperature of not less than 2000°F for 15 minutes.

[Amdt. 23-14, 38 FR 31816, Nov. 19, 1973]

§ 23.1183 Lines, fittings, and components.

(a) Except as provided in paragraph (b) of this section, each component, line, and fitting carrying flammable fluids, gas, or air in any area subject to engine fire conditions must be at least fire resistant, except that flammable fluid tanks and supports which are part of and attached to the engine must be fireproof or be enclosed by a fireproof shield unless damage by fire to any non-fireproof part will not cause leakage or spillage of flammable fluid. Components must be shielded or located so as to safeguard against the ignition of leaking flammable fluid. Flexible hose assemblies (hose and end fittings) must be shown to be suitable for the particular application. An integral oil sump of less than 25-quart capacity on a reciprocating engine need not be fireproof nor be enclosed by a fireproof shield.

(b) Paragraph (a) of this section does not apply to—

(1) Lines, fittings, and components which are already approved as part of a type certificated engine; and

(2) Vent and drain lines, and their fittings, whose failure will not result in, or add to, a fire hazard.

[Doc. No. 4080, 29 FR 17955, Dec. 18, 1964, as amended by Amdt. 23-5, 32 FR 6912, May 5, 1967; Amdt. 23-15, 39 FR 35460, Oct. 1, 1974; Amdt. 23-29, 49 FR 6847, Feb. 23, 1984; Amdt. 23-51, 61 FR 5138, Feb. 9, 1996]

§ 23.1189 Shutoff means.

(a) For each multiengine airplane the following apply:

(1) Each engine installation must have means to shut off or otherwise prevent hazardous quantities of fuel, oil, deicing fluid, and other flammable liquids from flowing into, within, or through any engine compartment, except in lines, fittings, and components forming an integral part of an engine.

(2) The closing of the fuel shutoff valve for any engine may not make any fuel unavailable to the remaining engines that would be available to those engines with that valve open.

(3) Operation of any shutoff means may not interfere with the later emergency operation of other equipment such as propeller feathering devices.

(4) Each shutoff must be outside of the engine compartment unless an equal degree of safety is provided with the shutoff inside the compartment.

(5) Not more than one quart of flammable fluid may escape into the engine compartment after engine shutoff. For those installations where the flammable fluid that escapes after shutdown cannot be limited to one quart, it must be demonstrated that this greater amount can be safely contained or drained overboard.

(6) There must be means to guard against inadvertent operation of each shutoff means, and to make it possible for the crew to reopen the shutoff means in flight after it has been closed.

(b) Turbine engine installations need not have an engine oil system shutoff if—

(1) The oil tank is integral with, or mounted on, the engine; and

(2) All oil system components external to the engine are fireproof or located in areas not subject to engine fire conditions.

(c) Power operated valves must have means to indicate to the flight crew when the valve has reached the selected position and must be designed so that the valve will not move from the selected position under vibration conditions likely to exist at the valve location.

[Doc. No. 4080, 29 FR 17955, Dec. 18, 1964, as amended by Amdt. 23-7, 34 FR 13096, Aug. 13, 1969; Amdt. 23-14, 38 FR 31823, Nov. 19, 1973; Amdt. 23-29, 49 FR 6847, Feb. 23, 1984; Amdt. 23-43, 58 FR 18975, Apr. 9, 1993]

§ 23.1191 Firewalls.

(a) Each engine, auxiliary power unit, fuel burning heater, and other combustion equipment, must be isolated from the rest of the airplane by firewalls, shrouds, or equivalent means.

(b) Each firewall or shroud must be constructed so that no hazardous quantity of liquid, gas, or flame can pass from the compartment created by the firewall or shroud to other parts of the airplane.

(c) Each opening in the firewall or shroud must be sealed with close fitting, fireproof grommets, bushings, or firewall fittings.

(d) [Reserved]

(e) Each firewall and shroud must be fireproof and protected against corrosion.

(f) Compliance with the criteria for fireproof materials or components must be shown as follows:

(1) The flame to which the materials or components are subjected must be 2,000 ±150°F.

(2) Sheet materials approximately 10 inches square must be subjected to the flame from a suitable burner.

(3) The flame must be large enough to maintain the required test temperature over an area approximately five inches square.

(g) Firewall materials and fittings must resist flame penetration for at least 15 minutes.

(h) The following materials may be used in firewalls or shrouds without being tested as required by this section:

(1) Stainless steel sheet, 0.015 inch thick.

(2) Mild steel sheet (coated with aluminum or otherwise protected against corrosion) 0.018 inch thick.

(3) Terne plate, 0.018 inch thick.

(4) Monel metal, 0.018 inch thick.

(5) Steel or copper base alloy firewall fittings.

(6) Titanium sheet, 0.016 inch thick.

[Doc. No. 4080, 29 FR 17955, Dec. 18, 1964, as amended by Amdt. 23-43, 58 FR 18975, Apr. 9, 1993; 58 FR 27060, May 6, 1993; Amdt. 23-51, 61 FR 5138, Feb. 9, 1996]

§ 23.1192 Engine accessory compartment diaphragm.

For aircooled radial engines, the engine power section and all portions of the exhaust sytem must be isolated from the engine accessory compartment by a diaphragm that meets the firewall requirements of § 23.1191.

[Amdt. 23-14, 38 FR 31823, Nov. 19, 1973]

§ 23.1193 Cowling and nacelle.

(a) Each cowling must be constructed and supported so that it can resist any vibration, inertia, and air loads to which it may be subjected in operation.

(b) There must be means for rapid and complete drainage of each part of the cowling in the normal ground and flight attitudes. Drain operation may be shown by test, analysis, or both, to ensure that under normal aerodynamic pressure distribution expected in service each drain will operate as designed. No drain may discharge where it will cause a fire hazard.

(c) Cowling must be at least fire resistant.

(d) Each part behind an opening in the engine compartment cowling must be at least fire resistant for a distance of at least 24 inches aft of the opening.

(e) Each part of the cowling subjected to high temperatures due to its nearness to exhaust sytem ports or exhaust gas impingement, must be fire proof.

(f) Each nacelle of a multiengine airplane with supercharged engines must be designed and constructed so that with the landing gear retracted, a fire in the engine compartment will not burn through a cowling or nacelle and enter a nacelle area other than the engine compartment.

(g) In addition, for all airplanes with engine(s) embedded in the fuselage or in pylons on the aft fuselage, the airplane must be designed so that no fire originating in any engine compartment can enter, either through openings or by burn-through, any other region where it would create additional hazards.

[Doc. No. 4080, 29 FR 17955, Dec. 18, 1964; 30 FR 258, Jan. 9, 1965, as amended by Amdt. 23-18, 42 FR 15042, Mar. 17, 1977; Amdt. 23-34, 52 FR 1833, Jan. 15, 1987; 58 FR 18975, Apr. 9, 1993; Amdt. 23-62, 76 FR 75759, Dec. 2, 2011]

§ 23.1195 Fire extinguishing systems.

(a) For all airplanes with engine(s) embedded in the fuselage or in pylons on the aft fuselage, fire extinguishing systems must be installed and compliance shown with the following:

(1) Except for combustor, turbine, and tailpipe sections of turbine-engine installations that contain lines or components carrying flammable fluids or gases for which a fire originating in these sections is shown to be controllable, a fire extinguisher system must serve each engine compartment;

(2) The fire extinguishing system, the quantity of the extinguishing agent, the rate of discharge, and the discharge distribution must be adequate to extinguish fires. An individual "one shot" system may be used, except for engine(s) embedded in the fuselage, where a "two shot" system is required.

(3) The fire extinguishing system for a nacelle must be able to simultaneously protect each compartment of the nacelle for which protection is provided.

(b) If an auxiliary power unit is installed in any airplane certificated to this part, that auxiliary power unit compartment must be served by a fire extinguishing system meeting the requirements of paragraph (a)(2) of this section.

[Amdt. 23-34, 52 FR 1833, Jan. 15, 1987, as amended by Amdt. 23-43, 58 FR 18975, Apr. 9, 1993; Amdt. 23-62, 76 FR 75759, Dec. 2, 2011]

§ 23.1197 Fire extinguishing agents.

For all airplanes with engine(s) embedded in the fuselage or in pylons on the aft fuselage the following applies:

(1) Be capable of extinguishing flames emanating from any burning of fluids or other combustible materials in the area protected by the fire extinguishing system; and

(2) Have thermal stability over the temperature range likely to be experienced in the compartment in which they are stored.

(b) If any toxic extinguishing agent is used, provisions must be made to prevent harmful concentrations of fluid or fluid vapors (from leakage during normal operation of the airplane or as a result of discharging the fire extinguisher on the ground or in flight) from entering any personnel compartment, even though a defect may exist in the extinguishing system. This must be shown by test except for built-in carbon dioxide fuselage compartment fire extinguishing systems for which—

(1) Five pounds or less of carbon dioxide will be discharged, under established fire control procedures, into any fuselage compartment; or

(2) Protective breathing equipment is available for each flight crewmember on flight deck duty.

[Amdt. 23-34, 52 FR 1833, Jan. 15, 1987, as amended by Amdt. 23-62, 76 FR 75760, Dec. 2, 2011]

§ 23.1199 Extinguishing agent containers.

For all airplanes with engine(s) embedded in the fuselage or in pylons on the aft fuselage the following applies:

(a) Each extinguishing agent container must have a pressure relief to prevent bursting of the container by excessive internal pressures.

(b) The discharge end of each discharge line from a pressure relief connection must be located so that discharge of the fire extinguishing agent would not damage the airplane. The line must also be located or protected to prevent clogging caused by ice or other foreign matter.

(c) A means must be provided for each fire extinguishing agent container to indicate that the container has discharged or that the charging pressure is below the established minimum necessary for proper functioning.

(d) The temperature of each container must be maintained, under intended operating conditions, to prevent the pressure in the container from—

(1) Falling below that necessary to provide an adequate rate of discharge; or

(2) Rising high enough to cause premature discharge.

(e) If a pyrotechnic capsule is used to discharge the extinguishing agent, each container must be installed so that temperature conditions will not cause hazardous deterioration of the pyrotechnic capsule.

[Amdt. 23-34, 52 FR 1833, Jan. 15, 1987; 52 FR 34745, Sept. 14, 1987; Amdt. 23-62, 76 FR 75760, Dec. 2, 2011]

§ 23.1201 Fire extinguishing systems materials.

For all airplanes with engine(s) embedded in the fuselage or in pylons on the aft fuselage the following applies:

(a) No material in any fire extinguishing system may react chemically with any extinguishing agent so as to create a hazard.

(b) Each system component in an engine compartment must be fireproof.

[Amdt. 23-34, 52 FR 1833, Jan. 15, 1987; 52 FR 7262, Mar. 9, 1987; Amdt. 23-62, 76 FR 75760, Dec. 2, 2011]

§ 23.1203 Fire detector system.

(a) There must be means that ensure the prompt detection of a fire in—

(1) An engine compartment of—

(i) Multiengine turbine powered airplanes;

(ii) Multiengine reciprocating engine powered airplanes incorporating turbochargers;

(iii) Airplanes with engine(s) located where they are not readily visible from the cockpit; and

(iv) All commuter category airplanes.

(2) The auxiliary power unit compartment of any airplane incorporating an auxiliary power unit.

(b) Each fire detector must be constructed and installed to withstand the vibration, inertia, and other loads to which it may be subjected in operation.

(c) No fire detector may be affected by any oil, water, other fluids, or fumes that might be present.

(d) There must be means to allow the crew to check, in flight, the functioning of each fire detector electric circuit.

(e) Wiring and other components of each fire detector system in a designated fire zone must be at least fire resistant.

[Amdt. 23-18, 42 FR 15042, Mar. 17, 1977, as amended by Amdt. 23-34, 52 FR 1833, Jan. 15, 1987; Amdt. 23-43, 58 FR 18975, Apr. 9, 1993; Amdt. 23-51, 61 FR 5138, Feb. 9, 1996]

Subpart F—Equipment

General

§ 23.1301 Function and installation.

Each item of installed equipment must—

(a) Be of a kind and design appropriate to its intended function.

(b) Be labeled as to its identification, function, or operating limitations, or any applicable combination of these factors; and

(c) Be installed according to limitations specified for that equipment.

[Amdt. 23-20, 42 FR 36968, July 18, 1977, as amended by Amdt. 23-62, 76 FR 75760, Dec. 2, 2011]

§ 23.1303 Flight and navigation instruments.

The following are the minimum required flight and navigation instruments:

(a) An airspeed indicator.

(b) An altimeter.

(c) A magnetic direction indicator.

(d) For reciprocating engine-powered airplanes of more than 6,000 pounds maximum weight and turbine engine powered airplanes, a free air temperature indicator or an air-temperature indicator which provides indications that are convertible to free-air.

(e) A speed warning device for—

(1) Turbine engine powered airplanes; and

(2) Other airplanes for which *Vmo/Mmo* and V_D/M_D are established under §§ 23.335(b)(4) and 23.1505(c) if *V*mo/Mmo is greater than 0.8 V_D/M_D.

The speed warning device must give effective aural warning (differing distinctively from aural warnings used for other purposes) to the pilots whenever the speed exceeds *V*mo plus 6 knots or Mmo +0.01. The upper limit of the production tolerance for the warning device may not exceed the prescribed warning speed. The lower limit of the warning device must be set to minimize nuisance warning;

(f) When an attitude display is installed, the instrument design must not provide any means, accessible to the flightcrew, of adjusting the relative positions of the attitude reference symbol and the horizon line beyond that necessary for parallax correction.

(g) In addition, for commuter category airplanes:

(1) If airspeed limitations vary with altitude, the airspeed indicator must have a maximum allowable airspeed indicator showing the variation of V_{MO} with altitude.

(2) The altimeter must be a sensitive type.

(3) Having a passenger seating configuration of 10 or more, excluding the pilot's seats and that are approved for IFR operations, a third attitude instrument must be provided that:

(i) Is powered from a source independent of the electrical generating system;

(ii) Continues reliable operation for a minimum of 30 minutes after total failure of the electrical generating system;

(iii) Operates independently of any other attitude indicating system;

(iv) Is operative without selection after total failure of the electrical generating system;

(v) Is located on the instrument panel in a position acceptable to the Administrator that will make it plainly visible to and usable by any pilot at the pilot's station; and

(vi) Is appropriately lighted during all phases of operation.

[Doc. No. 4080, 29 FR 17955, Dec. 18, 1964, as amended by Amdt. 23-17, 41 FR 55465, Dec. 20, 1976; Amdt. 23-43, 58 FR 18975, Apr. 9, 1993; Amdt. 23-49, 61 FR 5168, Feb. 9, 1996; Amdt. 23-62, 76 FR 75760, Dec. 2, 2011]

§ 23.1305 Powerplant instruments.

The following are required powerplant instruments:

(a) *For all airplanes.* (1) A fuel quantity indicator for each fuel tank, installed in accordance with § 23.1337(b).

(2) An oil pressure indicator for each engine.

(3) An oil temperature indicator for each engine.

(4) An oil quantity measuring device for each oil tank which meets the requirements of § 23.1337(d).

(5) A fire warning means for those airplanes required to comply with § 23.1203.

(b) *For reciprocating engine-powered airplanes.* In addition to the powerplant instruments required by paragraph (a) of this section, the following powerplant instruments are required:

(1) An induction system air temperature indicator for each engine equipped with a preheater and having induction air temperature limitations that can be exceeded with preheat.

(2) A tachometer indicator for each engine.

(3) A cylinder head temperature indicator for—

(i) Each air-cooled engine with cowl flaps;

(ii) [Reserved]

(iii) Each commuter category airplane.

(4) For each pump-fed engine, a means:

(i) That continuously indicates, to the pilot, the fuel pressure or fuel flow; or

(ii) That continuously monitors the fuel system and warns the pilot of any fuel flow trend that could lead to engine failure.

(5) A manifold pressure indicator for each altitude engine and for each engine with a controllable propeller.

(6) For each turbocharger installation:

(i) If limitations are established for either carburetor (or manifold) air inlet temperature or exhaust gas or turbocharger turbine inlet temperature, indicators must be furnished for each temperature for which the limitation is established unless it is shown that the limitation will not be exceeded in all intended operations.

(ii) If its oil system is separate from the engine oil system, oil pressure and oil temperature indicators must be provided.

(7) A coolant temperature indicator for each liquid-cooled engine.

(c) *For turbine engine-powered airplanes.* In addition to the powerplant instruments required by paragraph (a) of this section, the following powerplant instruments are required:

(1) A gas temperature indicator for each engine.

(2) A fuel flowmeter indicator for each engine.

(3) A fuel low pressure warning means for each engine.

(4) A fuel low level warning means for any fuel tank that should not be depleted of fuel in normal operations.

(5) A tachometer indicator (to indicate the speed of the rotors with established limiting speeds) for each engine.

(6) An oil low pressure warning means for each engine.

(7) An indicating means to indicate the functioning of the powerplant ice protection system for each engine.

(8) For each engine, an indicating means for the fuel strainer or filter required by § 23.997 to indicate the occurrence of contamination of the strainer or filter before it reaches the capacity established in accordance with § 23.997(d).

(9) For each engine, a warning means for the oil strainer or filter required by § 23.1019, if it has no bypass, to warn the pilot of the occurrence of contamination of the strainer or filter screen before it reaches the capacity established in accordance with § 23.1019(a)(5).

(10) An indicating means to indicate the functioning of any heater used to prevent ice clogging of fuel system components.

(d) *For turbojet/turbofan engine-powered airplanes.* In addition to the powerplant instruments required by paragraphs (a) and (c) of this section, the following powerplant instruments are required:

(1) For each engine, an indicator to indicate thrust or to indicate a parameter that can be related to thrust, including a free air temperature indicator if needed for this purpose.

(2) For each engine, a position indicating means to indicate to the flight crew when the thrust reverser, if installed, is in the reverse thrust position.

(e) *For turbopropeller-powered airplanes.* In addition to the powerplant instruments required by paragraphs (a) and (c) of this section, the following powerplant instruments are required:

(1) A torque indicator for each engine.

(2) A position indicating means to indicate to the flight crew when the propeller blade angle is below the flight low pitch position, for each propeller, unless it can be shown that such occurrence is highly improbable.

[Doc. No. 26344, 58 FR 18975, Apr. 9, 1993; 58 FR 27060, May 6, 1993; Amdt. 23-51, 61 FR 5138, Feb. 9, 1996; Amdt. 23-52, 61 FR 13644, Mar. 27, 1996]

§ 23.1306 Electrical and electronic system lightning protection.

(a) Each electrical and electronic system that performs a function, for which failure would prevent the continued safe flight and landing of the airplane, must be designed and installed so that—

(1) The function is not adversely affected during and after the time the airplane is exposed to lightning; and

(2) The system automatically recovers normal operation of that function in a timely manner after the airplane is exposed to lightning.

(b) For airplanes approved for instrument flight rules operation, each electrical and electronic system that performs a function, for which failure would reduce the capability of the airplane or the ability of the flightcrew to respond to an adverse operating condition, must be designed and installed so that the function recovers normal operation in a timely manner after the airplane is exposed to lightning.

[Doc. No. FAA-2010-0224; 76 FR 33135, June 8, 2011]

§ 23.1307 Miscellaneous equipment.

The equipment necessary for an airplane to operate at the maximum operating altitude and in the kinds of operation and meteorological conditions for which certification is requested and is approved in accordance with § 23.1559 must be included in the type design.

[Doc. No. 4080, 29 FR 17955, Dec. 18, 1964; 30 FR 258, Jan. 9, 1965, as amended by Amdt. 23-23, 43 FR 50593, Oct. 30, 1978; Amdt. 23-43, 58 FR 18976, Apr. 9, 1993; Amdt. 23-49, 61 FR 5168, Feb. 9, 1996]

§ 23.1308 High-intensity Radiated Fields (HIRF) Protection.

(a) Except as provided in paragraph (d) of this section, each electrical and electronic system that performs a function whose failure would prevent the continued safe flight and landing of the airplane must be designed and installed so that—

(1) The function is not adversely affected during and after the time the airplane is exposed to HIRF environment I, as described in appendix J to this part;

(2) The system automatically recovers normal operation of that function, in a timely manner, after the airplane is exposed to HIRF environment I, as described in appendix J to this part, unless the system's recovery conflicts with other operational or functional requirements of the system; and

(3) The system is not adversely affected during and after the time the airplane is exposed to HIRF environment II, as described in appendix J to this part.

(b) Each electrical and electronic system that performs a function whose failure would significantly reduce the capability of the airplane or the ability of the flightcrew to respond to an adverse operating condition must be designed and installed so the system is not adversely affected when the equipment providing the function is exposed to equipment HIRF test level 1 or 2, as described in appendix J to this part.

(c) Each electrical and electronic system that performs a function whose failure would reduce the capability of the airplane or the ability of the flightcrew to respond to an adverse operating condition must be designed and installed so the system is not adversely affected when the equipment providing the function is exposed to equipment HIRF test level 3, as described in appendix J to this part.

(d) Before December 1, 2012, an electrical or electronic system that performs a function whose failure would prevent the continued safe flight and landing of an airplane may be designed and installed without meeting the provisions of paragraph (a) provided—

(1) The system has previously been shown to comply with special conditions for HIRF, prescribed under § 21.16, issued before December 1, 2007;

(2) The HIRF immunity characteristics of the system have not changed since compliance with the special conditions was demonstrated; and

(3) The data used to demonstrate compliance with the special conditions is provided.

[Doc. No. FAA-2006-23657, 72 FR 44024, Aug. 6, 2007]

§ 23.1309 Equipment, systems, and installations.

The requirements of this section, except as identified in paragraphs (a) through (d), are applicable, in addition to specific design requirements of part 23, to any equipment or system as installed in the airplane. This section is a regulation of general requirements and does not supersede any requirements contained in another section of part 23.

(a) The airplane equipment and systems must be designed and installed so that:

(1) Those required for type certification or by operating rules perform as intended under the airplane operating and environmental conditions, including the indirect effects of lightning strikes.

(2) Any equipment and system does not adversely affect the safety of the airplane or its occupants, or the proper functioning of those covered by paragraph (a)(1) of this section.

(b) Minor, major, hazardous, or catastrophic failure condition(s), which occur during Type Inspection Authorization or FAA flight-certification testing, must have root cause analysis and corrective action.

(c) The airplane systems and associated components considered separately and in relation to other systems, must be designed and installed so that:

(1) Each catastrophic failure condition is extremely improbable and does not result from a single failure;

(2) Each hazardous failure condition is extremely remote; and

(3) Each major failure condition is remote.

(d) Information concerning an unsafe system operating condition must be provided in a timely manner to the crew to enable them to take appropriate corrective action. An appropriate alert must be provided if immediate pilot awareness and immediate or subsequent corrective action is required. Systems and controls, including indications and annunciations, must be designed to minimize crew errors which could create additional hazards.

[Doc. No. FAA-2009-0738, 76 FR 75760, Dec. 2, 2011]

§ 23.1310 Power source capacity and distribution.

(a) Each installation whose functioning is required for type certification or under operating rules and that requires a power supply is an "essential load" on the power supply. The power sources and the system must be able to supply the following power loads in probable operating combinations and for probable durations:

(1) Loads connected to the system with the system functioning normally.

(2) Essential loads, after failure of any one prime mover, power converter, or energy storage device.

(3) Essential loads after failure of—

(i) Any one engine on two-engine airplanes; and

(ii) Any two engines on airplanes with three or more engines.

(4) Essential loads for which an alternate source of power is required, after any failure or malfunction in any one power supply system, distribution system, or other utilization system.

(b) In determining compliance with paragraphs (a)(2) and (3) of this section, the power loads may be assumed to be reduced under a monitoring procedure consistent with safety in the kinds of operation authorized. Loads not required in controlled flight need not be considered for the two-engine-inoperative condition on airplanes with three or more engines.

[Doc. No. FAA-2009-0738, 76 FR 75760, Dec. 2, 2011]

Instruments: Installation

§ 23.1311 Electronic display instrument systems.

(a) Electronic display indicators, including those with features that make isolation and independence between powerplant instrument systems impractical, must:

(1) Meet the arrangement and visibility requirements of § 23.1321.

(2) Be easily legible under all lighting conditions encountered in the cockpit, including direct sunlight, considering the expected electronic display brightness level at the end of an electronic display indictor's useful life. Specific limitations on display system useful life must be contained in the Instructions for Continued Airworthiness required by § 23.1529.

(3) Not inhibit the primary display of attitude, airspeed, altitude, or powerplant parameters needed by any pilot to set power within established limitations, in any normal mode of operation.

(4) Not inhibit the primary display of engine parameters needed by any pilot to properly set or monitor powerplant limitations during the engine starting mode of operation.

(5) For certification for Instrument Flight Rules (IFR) operations, have an independent magnetic direction indicator and either an independent secondary mechanical altimeter, airspeed indicator, and attitude instrument or an

electronic display parameters for the altitude, airspeed, and attitude that are independent from the airplane's primary electrical power system. These secondary instruments may be installed in panel positions that are displaced from the primary positions specified by § 23.1321(d), but must be located where they meet the pilot's visibility requirements of § 23.1321(a).

(6) Incorporate sensory cues that provide a quick glance sense of rate and, where appropriate, trend information to the parameter being displayed to the pilot.

(7) Incorporate equivalent visual displays of the instrument markings required by §§ 23.1541 through 23.1553, or visual displays that alert the pilot to abnormal operational values or approaches to established limitation values, for each parameter required to be displayed by this part.

(b) The electronic display indicators, including their systems and installations, and considering other airplane systems, must be designed so that one display of information essential for continued safe flight and landing will be available within one second to the crew by a single pilot action or by automatic means for continued safe operation, after any single failure or probable combination of failures.

(c) As used in this section, "instrument" includes devices that are physically contained in one unit, and devices that are composed of two or more physically separate units or components connected together (such as a remote indicating gyroscopic direction indicator that includes a magnetic sensing element, a gyroscopic unit, an amplifier, and an indicator connected together). As used in this section, "primary" display refers to the display of a parameter that is located in the instrument panel such that the pilot looks at it first when wanting to view that parameter.

[Doc. No. 27806, 61 FR 5168, Feb. 9, 1996, as amended by Amdt. 23-62, 76 FR 75760, Dec. 2, 2011]

§ 23.1321 Arrangement and visibility.

(a) Each flight, navigation, and powerplant instrument for use by any required pilot during takeoff, initial climb, final approach, and landing must be located so that any pilot seated at the controls can monitor the airplane's flight path and these instruments with minimum head and eye movement. The powerplant instruments for these flight conditions are those needed to set power within powerplant limitations.

(b) For each multiengine airplane, identical powerplant instruments must be located so as to prevent confusion as to which engine each instrument relates.

(c) Instrument panel vibration may not damage, or impair the accuracy of, any instrument.

(d) For each airplane, the flight instruments required by § 23.1303, and, as applicable, by the operating rules of this chapter, must be grouped on the instrument panel and centered as nearly as practicable about the vertical plane of each required pilot's forward vision. In addition:

(1) The instrument that most effectively indicates the attitude must be on the panel in the top center position;

(2) The instrument that most effectively indicates airspeed must be adjacent to and directly to the left of the instrument in the top center position;

(3) The instrument that most effectively indicates altitude must be adjacent to and directly to the right of the instrument in the top center position;

(4) The instrument that most effectively indicates direction of flight, other than the magnetic direction indicator required by § 23.1303(c), must be adjacent to and directly below the instrument in the top center position; and

(5) Electronic display indicators may be used for compliance with paragraphs (d)(1) through (d)(4) of this section when such displays comply with requirements in § 23.1311.

(e) If a visual indicator is provided to indicate malfunction of an instrument, it must be effective under all probable cockpit lighting conditions.

[Doc. No. 4080, 29 FR 17955, Dec. 18, 1964, as amended by Amdt. 23-14, 38 FR 31824, Nov. 19, 1973; Amdt. 23-20, 42 FR 36968, July 18, 1977; Amdt. 23-41, 55 FR 43310, Oct. 26, 1990; 55 FR 46888, Nov. 7, 1990; Amdt. 23-49, 61 FR 5168, Feb. 9, 1996]

§ 23.1322 Warning, caution, and advisory lights.

If warning, caution, or advisory lights are installed in the cockpit, they must, unless otherwise approved by the Administrator, be—

(a) Red, for warning lights (lights indicating a hazard which may require immediate corrective action);

(b) Amber, for caution lights (lights indicating the possible need for future corrective action);

(c) Green, for safe operation lights; and

(d) Any other color, including white, for lights not described in paragraphs (a) through (c) of this section, provided the color differs sufficiently from the colors prescribed in paragraphs (a) through (c) of this section to avoid possible confusion.

(e) Effective under all probable cockpit lighting conditions.

[Amdt. 23-17, 41 FR 55465, Dec. 20, 1976, as amended by Amdt. 23-43, 58 FR 18976, Apr. 9, 1993]

§ 23.1323 Airspeed indicating system.

(a) Each airspeed indicating instrument must be calibrated to indicate true airspeed (at sea level with a standard atmosphere) with a minimum practicable instrument calibration error when the corresponding pitot and static pressures are applied.

(b) Each airspeed system must be calibrated in flight to determine the system error. The system error, including position error, but excluding the airspeed indicator instrument calibration error, may not exceed three percent of the calibrated airspeed or five knots, whichever is greater, throughout the following speed ranges:

(1) 1.3 V_{S1} to V_{MO}/M_{MO} or V_{NE}, whichever is appropriate with flaps retracted.

(2) 1.3 V_{S1} to V_{FE} with flaps extended.

(c) The design and installation of each airspeed indicating system must provide positive drainage of moisture from the pitot static plumbing.

(d) If certification for instrument flight rules or flight in icing conditions is requested, each airspeed system must have a heated pitot tube or an equivalent means of preventing malfunction due to icing.

(e) In addition, for normal, utility, and acrobatic category multiengine jets of more than 6,000 pounds maximum weight and commuter category airplanes, each system must be calibrated to determine the system error during the accelerate-takeoff ground run. The ground run calibration must be determined—

(1) From 0.8 of the minimum value of V_1 to the maximum value of V_2, considering the approved ranges of altitude and weight; and

(2) The ground run calibration must be determined assuming an engine failure at the minimum value of V_1.

(f) For commuter category airplanes, where duplicate airspeed indicators are required, their respective pitot tubes must be far enough apart to avoid damage to both tubes in a collision with a bird.

[Amdt. 23-20, 42 FR 36968, July 18, 1977, as amended by Amdt. 23-34, 52 FR 1834, Jan. 15, 1987; 52 FR 34745, Sept. 14, 1987; Amdt. 23-42, 56 FR 354, Jan. 3, 1991; Amdt. 23-49, 61 FR 5168, Feb. 9, 1996; Amdt. 23-62, 76 FR 75761, Dec. 2, 2011]

§ 23.1325 Static pressure system.

(a) Each instrument provided with static pressure case connections must be so vented that the influence of airplane speed, the opening and closing of windows, airflow variations, moisture, or other foreign matter will least affect the accuracy of the instruments except as noted in paragraph (b)(3) of this section.

(b) If a static pressure system is necessary for the functioning of instruments, systems, or devices, it must comply with the provisions of paragraphs (b) (1) through (3) of this section.

(1) The design and installation of a static pressure system must be such that—

(i) Positive drainage of moisture is provided;

(ii) Chafing of the tubing, and excessive distortion or restriction at bends in the tubing, is avoided; and

(iii) The materials used are durable, suitable for the purpose intended, and protected against corrosion.

(2) A proof test must be conducted to demonstrate the integrity of the static pressure system in the following manner:

(i) *Unpressurized airplanes.* Evacuate the static pressure system to a pressure differential of approximately 1 inch of mercury or to a reading on the altimeter, 1,000 feet above the aircraft elevation at the time of the test. Without additional pumping for a period of 1 minute, the loss of indicated altitude must not exceed 100 feet on the altimeter.

(ii) *Pressurized airplanes.* Evacuate the static pressure system until a pressure differential equivalent to the maximum cabin pressure differential for which the airplane is type certificated is achieved. Without additional pumping for a period of 1 minute, the loss of indicated altitude must not exceed 2 percent of the equivalent altitude of the maximum cabin differential pressure or 100 feet, whichever is greater.

(3) If a static pressure system is provided for any instrument, device, or system required by the operating rules of this chapter, each static pressure port must be designed or located in such a manner that the correlation between air pressure in the static pressure system and true ambient atmospheric static pressure is not altered when the airplane encounters icing conditions. An antiicing means or an alternate source of static pressure may be used in showing compliance with this requirement. If the reading of the altimeter, when on the alternate static pressure system differs from the reading of the altimeter when on the primary static system by more than 50 feet, a correction card must be provided for the alternate static system.

(c) Except as provided in paragraph (d) of this section, if the static pressure system incorporates both a primary and an alternate static pressure source, the means for selecting one or the other source must be designed so that—

(1) When either source is selected, the other is blocked off; and

(2) Both sources cannot be blocked off simultaneously.

(d) For unpressurized airplanes, paragraph (c)(1) of this section does not apply if it can be demonstrated that the static pressure system calibration, when either static pressure source is selected, is not changed by the other static pressure source being open or blocked.

(e) Each static pressure system must be calibrated in flight to determine the system error. The system error, in indicated pressure altitude, at sea-level, with a standard atmosphere, excluding instrument calibration error, may not exceed ±30 feet per 100 knot speed for the appropriate configuration in the speed range between 1.3 V_{S0} with flaps extended, and 1.8 V_{S1} with flaps retracted. However, the error need not be less than 30 feet.

(f) [Reserved]

(g) For airplanes prohibited from flight in instrument meteorological or icing conditions, in accordance with § 23.1559(b) of this part, paragraph (b)(3) of this section does not apply.

[Amdt. 23-1, 30 FR 8261, June 29, 1965, as amended by Amdt. 23-6, 32 FR 7586, May 24, 1967; 32 FR 13505, Sept. 27, 1967; 32 FR 13714, Sept. 30, 1967; Amdt. 23-20, 42 FR 36968, July 18, 1977; Amdt. 23-34, 52 FR 1834, Jan. 15, 1987; Amdt. 23-42, 56 FR 354, Jan. 3, 1991; Amdt. 23-49, 61 FR 5169, Feb. 9, 1996; Amdt. 23-50, 61 FR 5192, Feb. 9, 1996]

§ 23.1326 Pitot heat indication systems.

If a flight instrument pitot heating system is installed to meet the requirements specified in § 23.1323(d), an indication system must be provided to indicate to the flight crew when that pitot heating system is not operating. The indication system must comply with the following requirements:

(a) The indication provided must incorporate an amber light that is in clear view of a flightcrew member.

(b) The indication provided must be designed to alert the flight crew if either of the following conditions exist:

(1) The pitot heating system is switched "off."

(2) The pitot heating system is switched "on" and any pitot tube heating element is inoperative.

[Doc. No. 27806, 61 FR 5169, Feb. 9, 1996]

§ 23.1327 Magnetic direction indicator.

(a) Except as provided in paragraph (b) of this section—

(1) Each magnetic direction indicator must be installed so that its accuracy is not excessively affected by the airplane's vibration or magnetic fields; and

(2) The compensated installation may not have a deviation in level flight, greater than ten degrees on any heading.

(b) A magnetic nonstabilized direction indicator may deviate more than ten degrees due to the operation of electrically powered systems such as electrically heated windshields if either a magnetic stabilized direction indicator, which does not have a deviation in level flight greater than ten degrees on any heading, or a gyroscopic direction indicator, is installed. Deviations of a magnetic nonstabilized direction indicator of more than 10 degrees must be placarded in accordance with § 23.1547(e).

[Amdt. 23-20, 42 FR 36969, July 18, 1977]

§ 23.1329 Automatic pilot system.

If an automatic pilot system is installed, it must meet the following:

(a) Each system must be designed so that the automatic pilot can—

(1) Be quickly and positively disengaged by the pilots to prevent it from interfering with their control of the airplane; or

(2) Be sufficiently overpowered by one pilot to let him control the airplane.

(b) If the provisions of paragraph (a)(1) of this section are applied, the quick release (emergency) control must be located on the control wheel (both control wheels if the airplane can be operated from either pilot seat) on the side opposite the throttles, or on the stick control, (both stick controls, if the airplane can be operated from either pilot seat) such that it can be operated without moving the hand from its normal position on the control.

(c) Unless there is automatic synchronization, each system must have a means to readily indicate to the pilot the alignment of the actuating device in relation to the control system it operates.

(d) Each manually operated control for the system operation must be readily accessible to the pilot. Each control must operate in the same plane and sense of motion as specified in § 23.779 for cockpit controls. The direction of motion must be plainly indicated on or near each control.

(e) Each system must be designed and adjusted so that, within the range of adjustment available to the pilot, it cannot produce hazardous loads on the airplane or create hazardous deviations in the flight path, under any flight condition appropriate to its use, either during normal operation or in the event of a malfunction, assuming that corrective action begins within a reasonable period of time.

(f) Each system must be designed so that a single malfunction will not produce a hardover signal in more than one control axis. If the automatic pilot integrates signals from auxiliary controls or furnishes signals for operation of other equipment, positive interlocks and sequencing of engagement to prevent improper operation are required.

(g) There must be protection against adverse interaction of integrated components, resulting from a malfunction.

(h) If the automatic pilot system can be coupled to airborne navigation equipment, means must be provided to indicate to the flight crew the current mode of operation. Selector switch position is not acceptable as a means of indication.

[Doc. No. 4080, 29 FR 17955, Dec. 18, 1964; 30 FR 258, Jan. 9, 1965, as amended by Amdt. 23-23, 43 FR 50593, Oct. 30, 1978; Amdt. 23-43, 58 FR 18976, Apr. 9, 1993; Amdt. 23-49, 61 FR 5169, Feb. 9, 1996]

§ 23.1331 Instruments using a power source.

For each instrument that uses a power source, the following apply:

(a) Each instrument must have an integral visual power annunciator or separate power indicator to indicate when power is not adequate to sustain proper instrument performance. If a separate indicator is used, it must be located so that the pilot using the instruments can monitor the indicator with minimum head and eye movement. The power must be sensed at or near the point where it enters the instrument. For electric and vacuum/pressure instruments, the power is considered to be adequate when the voltage or the vacuum/pressure, respectively, is within approved limits.

(b) The installation and power supply systems must be designed so that—

(1) The failure of one instrument will not interfere with the proper supply of energy to the remaining instrument; and

(2) The failure of the energy supply from one source will not interfere with the proper supply of energy from any other source.

(c) For certification for Instrument Flight Rules (IFR) operations and for the heading, altitude, airspeed, and attitude, there must be at least:

(1) Two independent sources of power (not driven by the same engine on multiengine airplanes), and a manual or an automatic means to select each power source; or

(2) A separate display of parameters for heading, altitude, airspeed, and attitude that has a power source independent from the airplane's primary electrical power system.

[Doc. No. 26344, 58 FR 18976, Apr. 9, 1993, as amended by Amdt. 23-62, 76 FR 75761, Dec. 2, 2011]

§ 23.1335 Flight director systems.

If a flight director system is installed, means must be provided to indicate to the flight crew its current mode of operation. Selector switch position is not acceptable as a means of indication.

[Amdt. 23-20, 42 FR 36969, July 18, 1977]

§ 23.1337 Powerplant instruments installation.

(a) *Instruments and instrument lines.* (1) Each powerplant and auxiliary power unit instrument line must meet the requirements of § 23.993.

(2) Each line carrying flammable fluids under pressure must—

(i) Have restricting orifices or other safety devices at the source of pressure to prevent the escape of excessive fluid if the line fails; and

(ii) Be installed and located so that the escape of fluids would not create a hazard.

(3) Each powerplant and auxiliary power unit instrument that utilizes flammable fluids must be installed and located so that the escape of fluid would not create a hazard.

(b) *Fuel quantity indication.* There must be a means to indicate to the flightcrew members the quantity of usable fuel in each tank during flight. An indicator calibrated in appropriate units and clearly marked to indicate those units must be used. In addition:

(1) Each fuel quantity indicator must be calibrated to read "zero" during level flight when the quantity of fuel remaining in the tank is equal to the unusable fuel supply determined under § 23.959(a);

(2) Each exposed sight gauge used as a fuel quantity indicator must be protected against damage;

(3) Each sight gauge that forms a trap in which water can collect and freeze must have means to allow drainage on the ground;

(4) There must be a means to indicate the amount of usable fuel in each tank when the airplane is on the ground (such as by a stick gauge);

(5) Tanks with interconnected outlets and airspaces may be considered as one tank and need not have separate indicators; and

(6) No fuel quantity indicator is required for an auxiliary tank that is used only to transfer fuel to other tanks if the relative size of the tank, the rate of fuel transfer, and operating instructions are adequate to—

(i) Guard against overflow; and

(ii) Give the flight crewmembers prompt warning if transfer is not proceeding as planned.

(c) *Fuel flowmeter system.* If a fuel flowmeter system is installed, each metering component must have a means to by-pass the fuel supply if malfunctioning of that component severely restricts fuel flow.

(d) *Oil quantity indicator.* There must be a means to indicate the quantity of oil in each tank—

(1) On the ground (such as by a stick gauge); and

(2) In flight, to the flight crew members, if there is an oil transfer system or a reserve oil supply system.

[Doc. No. 4080, 29 FR 17955, Dec. 18, 1964, as amended by Amdt. 23-7, 34 FR 13096, Aug. 13, 1969; Amdt. 23-18, 42 FR 15042, Mar. 17, 1977; Amdt. 23-43, 58 FR 18976, Apr. 9, 1993; Amdt. 23-51, 61 FR 5138, Feb. 9, 1996; Amdt. 23-49, 61 FR 5169, Feb. 9, 1996]

Electrical Systems and Equipment

§ 23.1351 General.

(a) *Electrical system capacity.* Each electrical system must be adequate for the intended use. In addition—

(1) Electric power sources, their transmission cables, and their associated control and protective devices, must be able to furnish the required power at the proper voltage to each load circuit essential for safe operation; and

(2) Compliance with paragraph (a)(1) of this section must be shown as follows—

(i) For normal, utility, and acrobatic category airplanes, by an electrical load analysis or by electrical measurements that account for the electrical loads applied to the electrical system in probable combinations and for probable durations; and

(ii) For commuter category airplanes, by an electrical load analysis that accounts for the electrical loads applied to the electrical system in probable combinations and for probable durations.

(b) *Function.* For each electrical system, the following apply:

(1) Each system, when installed, must be—

(i) Free from hazards in itself, in its method of operation, and in its effects on other parts of the airplane;

(ii) Protected from fuel, oil, water, other detrimental substances, and mechanical damage; and

(iii) So designed that the risk of electrical shock to crew, passengers, and ground personnel is reduced to a minimum.

(2) Electric power sources must function properly when connected in combination or independently.

(3) No failure or malfunction of any electric power source may impair the ability of any remaining source to supply load circuits essential for safe operation.

(4) In addition, for commuter category airplanes, the following apply:

(i) Each system must be designed so that essential load circuits can be supplied in the event of reasonably probable faults or open circuits including faults in heavy current carrying cables;

(ii) A means must be accessible in flight to the flight crewmembers for the individual and collective disconnection of the electrical power sources from the system;

(iii) The system must be designed so that voltage and frequency, if applicable, at the terminals of all essential load equipment can be maintained within the limits for which the equipment is designed during any probable operating conditions;

(iv) If two independent sources of electrical power for particular equipment or systems are required, their electrical energy supply must be ensured by means such as duplicate electrical equipment, throwover switching, or multichannel or loop circuits separately routed; and

(v) For the purpose of complying with paragraph (b)(5) of this section, the distribution system includes the distribution busses, their associated feeders, and each control and protective device.

(c) *Generating system.* There must be at least one generator/alternator if the electrical system supplies power to load circuits essential for safe operation. In addition—

(1) Each generator/alternator must be able to deliver its continuous rated power, or such power as is limited by its regulation system.

(2) Generator/alternator voltage control equipment must be able to dependably regulate the generator/alternator output within rated limits.

(3) Automatic means must be provided to prevent damage to any generator/alternator and adverse effects on the airplane electrical system due to reverse current. A means must also be provided to disconnect each generator/alternator from the battery and other generators/alternators.

(4) There must be a means to give immediate warning to the flight crew of a failure of any generator/alternator.

(5) Each generator/alternator must have an overvoltage control designed and installed to prevent damage to the electrical system, or to equipment supplied by the electrical system that could result if that generator/alternator were to develop an overvoltage condition.

(d) *Instruments.* A means must exist to indicate to appropriate flight crewmembers the electric power system quantities essential for safe operation.

(1) For normal, utility, and acrobatic category airplanes with direct current systems, an ammeter that can be switched into each generator feeder may be used and, if only one generator exists, the ammeter may be in the battery feeder.

(2) For commuter category airplanes, the essential electric power system quantities include the voltage and current supplied by each generator.

(e) *Fire resistance.* Electrical equipment must be so designed and installed that in the event of a fire in the engine compartment, during which the surface of the firewall adjacent to the fire is heated to 2,000°F for 5 minutes or to a lesser temperature substantiated by the applicant, the equipment essential to continued safe operation and located behind the firewall will function satisfactorily and will not create an additional fire hazard.

(f) *External power.* If provisions are made for connecting external power to the airplane, and that external power can be electrically connected to equipment other than that used for engine starting, means must be provided to ensure that no external power supply having a reverse polarity, or a reverse phase sequence, can supply power to the airplane's electrical system. The external power connection must be located so that its use will not result in a hazard to the airplane or ground personnel.

(g) It must be shown by analysis, tests, or both, that the airplane can be operated safely in VFR conditions, for a period of not less than five minutes, with the normal electrical power (electrical power sources excluding the battery and any other standby electrical sources) inoperative, with critical type fuel (from the standpoint of flameout and restart capability), and with the airplane initially at the maximum certificated altitude. Parts of the electrical system may remain on if—

(1) A single malfunction, including a wire bundle or junction box fire, cannot result in loss of the part turned off and the part turned on; and

(2) The parts turned on are electrically and mechanically isolated from the parts turned off.

[Doc. No. 4080, 29 FR 17955, Dec. 18, 1964, as amended by Amdt. 23-7, 34 FR 13096, Aug. 13, 1969; Amdt. 23-14, 38 FR 31824, Nov. 19, 1973; Amdt. 23-17, 41 FR 55465, Dec. 20, 1976; Amdt. 23-20, 42 FR 36969,

July 18, 1977; Amdt. 23-34, 52 FR 1834, Jan. 15, 1987; 52 FR 34745, Sept. 14, 1987; Amdt. 23-43, 58 FR 18976, Apr. 9, 1993; Amdt. 23-49, 61 FR 5169, Feb. 9, 1996]

§ 23.1353 Storage battery design and installation.

(a) Each storage battery must be designed and installed as prescribed in this section.

(b) Safe cell temperatures and pressures must be maintained during any probable charging and discharging condition. No uncontrolled increase in cell temperature may result when the battery is recharged (after previous complete discharge)—

(1) At maximum regulated voltage or power;

(2) During a flight of maximum duration; and

(3) Under the most adverse cooling condition likely to occur in service.

(c) Compliance with paragraph (b) of this section must be shown by tests unless experience with similar batteries and installations has shown that maintaining safe cell temperatures and pressures presents no problem.

(d) No explosive or toxic gases emitted by any battery in normal operation, or as the result of any probable malfunction in the charging system or battery installation, may accumulate in hazardous quantities within the airplane.

(e) No corrosive fluids or gases that may escape from the battery may damage surrounding structures or adjacent essential equipment.

(f) Each nickel cadmium battery installation capable of being used to start an engine or auxiliary power unit must have provisions to prevent any hazardous effect on structure or essential systems that may be caused by the maximum amount of heat the battery can generate during a short circuit of the battery or of its individual cells.

(g) Nickel cadmium battery installations capable of being used to start an engine or auxiliary power unit must have—

(1) A system to control the charging rate of the battery automatically so as to prevent battery overheating;

(2) A battery temperature sensing and over-temperature warning system with a means for disconnecting the battery from its charging source in the event of an over-temperature condition; or

(3) A battery failure sensing and warning system with a means for disconnecting the battery from its charging source in the event of battery failure.

(h)(1) In the event of a complete loss of the primary electrical power generating system, the battery must be capable of providing electrical power to those loads that are essential to continued safe flight and landing for:

(i) At least 30 minutes for airplanes that are certificated with a maximum altitude of 25,000 feet or less; and

(ii) At least 60 minutes for airplanes that are certificated with a maximum altitude over 25,000 feet.

(2) The time period includes the time to recognize the loss of generated power and to take appropriate load shedding action.

[Doc. No. 4080, 29 FR 17955, Dec. 18, 1964; 30 FR 258, Jan. 9, 1965, as amended by Amdt. 23-20, 42 FR 36969, July 18, 1977; Amdt. 23-21, 43 FR 2319, Jan. 16, 1978; Amdt. 23-49, 61 FR 5169, Feb. 9, 1996; Amdt. 23-62, 76 FR 75761, Dec. 2, 2011]

§ 23.1357 Circuit protective devices.

(a) Protective devices, such as fuses or circuit breakers, must be installed in all electrical circuits other than—

(1) Main circuits of starter motors used during starting only; and

(2) Circuits in which no hazard is presented by their omission.

(b) A protective device for a circuit essential to flight safety may not be used to protect any other circuit.

(c) Each resettable circuit protective device ("trip free" device in which the tripping mechanism cannot be overridden by the operating control) must be designed so that—

(1) A manual operation is required to restore service after tripping; and

(2) If an overload or circuit fault exists, the device will open the circuit regardless of the position of the operating control.

(d) If the ability to reset a circuit breaker or replace a fuse is essential to safety in flight, that circuit breaker or fuse must be so located and identified that it can be readily reset or replaced in flight.

(e) For fuses identified as replaceable in flight—

(1) There must be one spare of each rating or 50 percent spare fuses of each rating, whichever is greater; and

(2) The spare fuse(s) must be readily accessible to any required pilot.

[Doc. No. 4080, 29 FR 17955, Dec. 18, 1964; 30 FR 258, Jan. 9, 1965, as amended by Amdt. 23-20, 42 FR 36969, July 18, 1977]; Amdt. 23-43, 58 FR 18976, Apr. 9, 1993

§ 23.1359 Electrical system fire protection.

(a) Each component of the electrical system must meet the applicable fire protection requirements of §§ 23.863 and 23.1182.

(b) Electrical cables, terminals, and equipment in designated fire zones that are used during emergency procedures must be fire-resistant.

(c) Insulation on electrical wire and electrical cable must be self-extinguishing when tested at an angle of 60 degrees in accordance with the applicable portions of appendix F of this part, or other approved equivalent methods. The average burn length must not exceed 3 inches (76 mm) and the average flame time after removal of the flame source must not exceed 30 seconds. Drippings from the test specimen must not continue to flame for more than an average of 3 seconds after falling.

[Doc. No. 27806, 61 FR 5169, Feb. 9, 1996]

§ 23.1361 Master switch arrangement.

(a) There must be a master switch arrangement to allow ready disconnection of each electric power source from power distribution systems, except as provided in paragraph (b) of this section. The point of disconnection must be adjacent to the sources controlled by the switch arrangement. If separate switches are incorporated into the master switch arrangement, a means must be provided for the switch arrangement to be operated by one hand with a single movement.

(b) Load circuits may be connected so that they remain energized when the master switch is open, if the circuits are isolated, or physically shielded, to prevent their igniting flammable fluids or vapors that might be liberated by the leakage or rupture of any flammable fluid system; and

(1) The circuits are required for continued operation of the engine; or

(2) The circuits are protected by circuit protective devices with a rating of five amperes or less adjacent to the electric power source.

(3) In addition, two or more circuits installed in accordance with the requirements of paragraph (b)(2) of this section must not be used to supply a load of more than five amperes.

(c) The master switch or its controls must be so installed that the switch is easily discernible and accessible to a crewmember.

[Doc. No. 4080, 29 FR 17955, Dec. 18, 1964; 30 FR 258, Jan. 9, 1965, as amended by Amdt. 23-20, 42 FR 36969, July 18, 1977; Amdt. 23-43, 58 FR 18977, Apr. 9, 1993; Amdt. 23-49, 61 FR 5169, Feb. 9, 1996]

§ 23.1365 Electric cables and equipment.

(a) Each electric connecting cable must be of adequate capacity.

(b) Any equipment that is associated with any electrical cable installation and that would overheat in the event of circuit overload or fault must be flame resistant. That equipment and the electrical cables must not emit dangerous quantities of toxic fumes.

(c) Main power cables (including generator cables) in the fuselage must be designed to allow a reasonable degree of deformation and stretching without failure and must—

(1) Be separated from flammable fluid lines; or

(2) Be shrouded by means of electrically insulated flexible conduit, or equivalent, which is in addition to the normal cable insulation.

(d) Means of identification must be provided for electrical cables, terminals, and connectors.

(e) Electrical cables must be installed such that the risk of mechanical damage and/or damage cased by fluids vapors, or sources of heat, is minimized.

(f) Where a cable cannot be protected by a circuit protection device or other overload protection, it must not cause a fire hazard under fault conditions.

[Doc. No. 4080, 29 FR 17955, Dec. 18, 1964, as amended by Amdt. 23-14, 38 FR 31824, Nov. 19, 1973; Amdt. 23-43, 58 FR 18977, Apr. 9, 1993; Amdt. 23-49, 61 FR 5169, Feb. 9, 1996]

§ 23.1367 Switches.

Each switch must be—

(a) Able to carry its rated current;

(b) Constructed with enough distance or insulating material between current carrying parts and the housing so that vibration in flight will not cause shorting;

(c) Accessible to appropriate flight crewmembers; and

(d) Labeled as to operation and the circuit controlled.

Lights

§ 23.1381 Instrument lights.

The instrument lights must—

(a) Make each instrument and control easily readable and discernible;

(b) Be installed so that their direct rays, and rays reflected from the windshield or other surface, are shielded from the pilot's eyes; and

(c) Have enough distance or insulating material between current carrying parts and the housing so that vibration in flight will not cause shorting.

A cabin dome light is not an instrument light.

§ 23.1383 Taxi and landing lights.

Each taxi and landing light must be designed and installed so that:

(a) No dangerous glare is visible to the pilots.

(b) The pilot is not seriously affected by halation.

(c) It provides enough light for night operations.

(d) It does not cause a fire hazard in any configuration.

[Doc. No. 27806, 61 FR 5169, Feb. 9, 1996]

§ 23.1385 Position light system installation.

(a) *General.* Each part of each position light system must meet the applicable requirements of this section and each system as a whole must meet the requirements of §§ 23.1387 through 23.1397.

(b) *Left and right position lights.* Left and right position lights must consist of a red and a green light spaced laterally as far apart as practicable and installed on the airplane such that, with the airplane in the normal flying position, the red light is on the left side and the green light is on the right side.

(c) *Rear position light.* The rear position light must be a white light mounted as far aft as practicable on the tail or on each wing tip.

(d) *Light covers and color filters.* Each light cover or color filter must be at least flame resistant and may not change color or shape or lose any appreciable light transmission during normal use.

[Doc. No. 4080, 29 FR 17955, Dec. 18, 1964, as amended by Amdt. 23-17, 41 FR 55465, Dec. 20, 1976; Amdt. 23-43, 58 FR 18977, Apr. 9, 1993]

§ 23.1387 Position light system dihedral angles.

(a) Except as provided in paragraph (e) of this section, each position light must, as installed, show unbroken light within the dihedral angles described in this section.

(b) Dihedral angle *L* (left) is formed by two intersecting vertical planes, the first parallel to the longitudinal axis of the airplane, and the other at 110 degrees to the left of the first, as viewed when looking forward along the longitudinal axis.

(c) Dihedral angle *R* (right) is formed by two intersecting vertical planes, the first parallel to the longitudinal axis of the airplane, and the other at 110 degrees to the right of the first, as viewed when looking forward along the longitudinal axis.

(d) Dihedral angle *A* (aft) is formed by two intersecting vertical planes making angles of 70 degrees to the right and to the left, respectively, to a vertical plane passing through the longitudinal axis, as viewed when looking aft along the longitudinal axis.

(e) If the rear position light, when mounted as far aft as practicable in accordance with § 23.1385(c), cannot show unbroken light within dihedral angle A (as defined in paragraph (d) of this section), a solid angle or angles of obstructed visibility totaling not more than 0.04 steradians is allowable within that dihedral angle, if such solid angle is within a cone whose apex is at the rear position light and whose elements make an angle of 30° with a vertical line passing through the rear position light.

[Doc. No. 4080, 29 FR 17955, Dec. 18, 1964; 30 FR 258, Jan. 9, 1965, as amended by Amdt. 23-12, 36 FR 21278, Nov. 5, 1971; Amdt. 23-43, 58 FR 18977, Apr. 9, 1993]

§ 23.1389 Position light distribution and intensities.

(a) *General.* The intensities prescribed in this section must be provided by new equipment with each light cover and color filter in place. Intensities must be determined with the light source operating at a steady value equal to the average luminous output of the source at the normal operating voltage of the airplane. The light distribution and intensity of each position light must meet the requirements of paragraph (b) of this section.

(b) *Position lights.* The light distribution and intensities of position lights must be expressed in terms of minimum intensities in the horizontal plane, minimum intensities in any vertical plane, and maximum intensities in overlapping beams, within dihedral angles *L*, *R*, and *A*, and must meet the following requirements:

(1) *Intensities in the horizontal plane.* Each intensity in the horizontal plane (the plane containing the longitudinal axis of the airplane and perpendicular to the plane of symmetry of the airplane) must equal or exceed the values in § 23.1391.

(2) *Intensities in any vertical plane.* Each intensity in any vertical plane (the plane perpendicular to the horizontal plane) must equal or exceed the appropriate value in § 23.1393, where *I* is the minimum intensity prescribed in § 23.1391 for the corresponding angles in the horizontal plane.

(3) *Intensities in overlaps between adjacent signals.* No intensity in any overlap between adjacent signals may exceed the values in § 23.1395, except that higher intensities in overlaps may be used with main beam intensities substantially greater than the minima specified in §§ 23.1391 and 23.1393, if the overlap intensities in relation to the main beam intensities do not adversely affect signal clarity. When the peak intensity of the left and right position lights is more than 100 candles, the maximum overlap intensities between them may exceed the values in § 23.1395 if the overlap intensity in Area A is not more than 10 percent of peak position light intensity and the overlap intensity in Area B is not more than 2.5 percent of peak position light intensity.

(c) *Rear position light installation.* A single rear position light may be installed in a position displaced laterally from the plane of symmetry of an airplane if—

(1) The axis of the maximum cone of illumination is parallel to the flight path in level flight; and

(2) There is no obstruction aft of the light and between planes 70 degrees to the right and left of the axis of maximum illumination.

[Doc. No. 4080, 29 FR 17955, Dec. 18, 1964, as amended by Amdt. 23-43, 58 FR 18977, Apr. 9, 1993]

§ 23.1391 Minimum intensities in the horizontal plane of position lights.

Each position light intensity must equal or exceed the applicable values in the following table:

Dihedral angle (light included)	Angle from right or left of longitudinal axis, measured from dead ahead	Intensity (candles)
L and *R* (red and green)	0° to 10° 10° to 20° 20° to 110°	40 30 5
A (rear white)	110° to 180°	20

[Doc. No. 4080, 29 FR 17955, Dec. 18, 1964, as amended by Amdt. 23-43, 58 FR 18977, Apr. 9, 1993]

§ 23.1393 Minimum intensities in any vertical plane of position lights.

Each position light intensity must equal or exceed the applicable values in the following table:

Angle above or below the horizontal plane	Intensity, I
0°	1.00
0° to 5°	0.90
5° to 10°	0.80
10° to 15°	0.70
15° to 20°	0.50
20° to 30°	0.30
30° to 40°	0.10
40° to 90°	0.05

[Doc. No. 4080, 29 FR 17955, Dec. 18, 1964, as amended by Amdt. 23-43, 58 FR 18977, Apr. 9, 1993]

§ 23.1395 Maximum intensities in overlapping beams of position lights.

No position light intensity may exceed the applicable values in the following equal or exceed the applicable values in § 23.1389(b)(3):

Overlaps	Maximum intensity	
	Area A (candles)	Area B (candles)
Green in dihedral angle *L*	10	1
Red in dihedral angle *R*	10	1
Green in dihedral angle *A*	5	1
Red in dihedral angle *A*	5	1
Rear white in dihedral angle *L*	5	1
Rear white in dihedral angle *R*	5	1

Where—

(a) Area A includes all directions in the adjacent dihedral angle that pass through the light source and intersect the common boundary plane at more than 10 degrees but less than 20 degrees; and

(b) Area B includes all directions in the adjacent dihedral angle that pass through the light source and intersect the common boundary plane at more than 20 degrees.

[Doc. No. 4080, 29 FR 17955, Dec. 18, 1964, as amended by Amdt. 23-43, 58 FR 18977, Apr. 9, 1993]

§ 23.1397 Color specifications.

Each position light color must have the applicable International Commission on Illumination chromaticity coordinates as follows:

(a) *Aviation red*—

y is not greater than 0.335; and

z is not greater than 0.002.

(b) *Aviation green*—

x is not greater than 0.440–0.320 *y*;

x is not greater than *y*–0.170; and

y is not less than 0.390–0.170 *x*.

(c) *Aviation white*—

x is not less than 0.300 and not greater than 0.540;

y is not less than *x*–0.040 or y_0–0.010, whichever is the smaller; and

y is not greater than *x*+0.020 nor 0.636–0.400 *x*;

Where y_0 is the *y* coordinate of the Planckian radiator for the value of *x* considered.

[Doc. No. 4080, 29 FR 17955, Dec. 18, 1964, amended by Amdt. 23-11, 36 FR 12971, July 10, 1971]

§ 23.1399 Riding light.

(a) Each riding (anchor) light required for a seaplane or amphibian, must be installed so that it can—

(1) Show a white light for at least two miles at night under clear atmospheric conditions; and

(2) Show the maximum unbroken light practicable when the airplane is moored or drifting on the water.

(b) Externally hung lights may be used.

§ 23.1401 Anticollision light system.

(a) *General.* The airplane must have an anticollision light system that:

(1) Consists of one or more approved anticollision lights located so that their light will not impair the flight crewmembers' vision or detract from the conspicuity of the position lights; and

(2) Meets the requirements of paragraphs (b) through (f) of this section.

(b) *Field of coverage.* The system must consist of enough lights to illuminate the vital areas around the airplane, considering the physical configuration and flight characteristics of the airplane. The field of coverage must extend in each direction within at least 75 degrees above and 75 degrees below the horizontal plane of the airplane, except that there may be solid angles of obstructed visibility totaling not more than 0.5 steradians.

(c) *Flashing characteristics.* The arrangement of the system, that is, the number of light sources, beam width, speed of rotation, and other characteristics, must give an effective flash frequency of not less than 40, nor more than 100, cycles per minute. The effective flash frequency is the frequency at which the airplane's complete anticollision light system is observed from a distance, and applies to each sector of light including any overlaps that exist when the system consists of more than one light source. In overlaps, flash frequencies may exceed 100, but not 180, cycles per minute.

(d) *Color.* Each anticollision light must be either aviation red or aviation white and must meet the applicable requirements of § 23.1397.

(e) *Light intensity.* The minimum light intensities in any vertical plane, measured with the red filter (if used) and expressed in terms of "effective" intensities, must meet the requirements of paragraph (f) of this section. The following relation must be assumed:

$$I_e = \frac{\int_{t_1}^{t_2} I(t)dt}{0.2 + (t_2 - t_1)}$$

where:

I_e=effective intensity (candles).

$I(t)$=instantaneous intensity as a function of time.

$t_2 - t_1$=flash time interval (seconds).

Normally, the maximum value of effective intensity is obtained when t_2 and t_1 are chosen so that the effective intensity is equal to the instantaneous intensity at t_2 and t_1.

(f) *Minimum effective intensities for anticollision lights.* Each anticollision light effective intensity must equal or exceed the applicable values in the following table.

Angle above or below the horizontal plane	Effective intensity (candles)
0° to 5°	400
5° to 10°	240
10° to 20°	80
20° to 30°	40
30° to 75°	20

[Doc. No. 4080, 29 FR 17955, Dec. 18, 1964, as amended by Amdt. 23-11, 36 FR 12972, July 10, 1971; Amdt. 23-20, 42 FR 36969, July 18, 1977; Amdt. 23-49, 61 FR 5169, Feb. 9, 1996]

Safety Equipment

§ 23.1411 General.

(a) Required safety equipment to be used by the flight crew in an emergency, such as automatic liferaft releases, must be readily accessible.

(b) Stowage provisions for required safety equipment must be furnished and must—

(1) Be arranged so that the equipment is directly accessible and its location is obvious; and

(2) Protect the safety equipment from damage caused by being subjected to the inertia loads resulting from the ultimate static load factors specified in § 23.561(b)(3) of this part.

[Amdt. 23-17, 41 FR 55465, Dec. 20, 1976, as amended by Amdt. 23-36, 53 FR 30815, Aug. 15, 1988]

§ 23.1415 Ditching equipment.

(a) Emergency flotation and signaling equipment required by any operating rule in this chapter must be installed so that it is readily available to the crew and passengers.

(b) Each raft and each life preserver must be approved.

(c) Each raft released automatically or by the pilot must be attached to the airplane by a line to keep it alongside the airplane. This line must be weak enough to break before submerging the empty raft to which it is attached.

(d) Each signaling device required by any operating rule in this chapter, must be accessible, function satisfactorily, and must be free of any hazard in its operation.

§ 23.1416 Pneumatic de-icer boot system.

If certification with ice protection provisions is desired and a pneumatic de-icer boot system is installed—

(a) The system must meet the requirements specified in § 23.1419.

(b) The system and its components must be designed to perform their intended function under any normal system operating temperature or pressure, and

(c) Means to indicate to the flight crew that the pneumatic de-icer boot system is receiving adequate pressure and is functioning normally must be provided.

[Amdt. 23-23, 43 FR 50593, Oct. 30, 1978]

§ 23.1419 Ice protection.

If certification with ice protection provisions is desired, compliance with the requirements of this section and other applicable sections of this part must be shown:

(a) An analysis must be performed to establish, on the basis of the airplane's operational needs, the adequacy of the ice protection system for the various components of the airplane. In addition, tests of the ice protection system must be conducted to demonstrate that the airplane is capable of operating safely in continuous maximum and intermittent maximum icing conditions, as described in appendix C of part 25 of this chapter. As used in this section, "Capable of operating safely," means that airplane performance, controllability, maneuverability, and stability must not be less than that required in part 23, subpart B.

(b) Except as provided by paragraph (c) of this section, in addition to the analysis and physical evaluation prescribed in paragraph (a) of this section, the effectiveness of the ice protection system and its components must be shown by flight tests of the airplane or its components in measured natural atmospheric icing conditions and by one or more of the following tests, as found necessary to determine the adequacy of the ice protection system—

(1) Laboratory dry air or simulated icing tests, or a combination of both, of the components or models of the components.

(2) Flight dry air tests of the ice protection system as a whole, or its individual components.

(3) Flight test of the airplane or its components in measured simulated icing conditions.

(c) If certification with ice protection has been accomplished on prior type certificated airplanes whose designs include components that are thermodynamically and aerodynamically equivalent to those used on a new airplane design, certification of these equivalent components may be accomplished by reference to previously accomplished tests, required in § 23.1419 (a) and (b), provided that the applicant accounts for any differences in installation of these components.

(d) A means must be identified or provided for determining the formation of ice on the critical parts of the airplane. Adequate lighting must be provided for the use of this means during night operation. Also, when monitoring of the external surfaces of the airplane by the flight crew is required for operation of the ice protection equipment, external lighting must be provided that is adequate to enable the monitoring to be done at night. Any illumination that is used must be of a type that will not cause glare or reflection that would handicap crewmembers in the performance of their duties. The Airplane Flight Manual or other approved manual material must describe the means of determining ice formation and must contain information for the safe operation of the airplane in icing conditions.

[Doc. No. 26344, 58 FR 18977, Apr. 9, 1993]

Miscellaneous Equipment

§ 23.1431 Electronic equipment.

(a) In showing compliance with § 23.1309(a), (b), and (c) with respect to radio and electronic equipment and their installations, critical environmental conditions must be considered.

(b) Radio and electronic equipment, controls, and wiring must be installed so that operation of any unit or system of units will not adversely affect the simultaneous operation of any other radio or electronic unit, or system of units, required by this chapter.

(c) For those airplanes required to have more than one flightcrew member, or whose operation will require more than one flightcrew member, the cockpit must be evaluated to determine if the flightcrew members, when seated at their duty station, can converse without difficulty under the actual cockpit noise conditions when the airplane is being operated. If the airplane design includes provision for the use of communication headsets, the evaluation must also consider conditions where headsets are being used. If the evaluation shows conditions under which it will be difficult to converse, an intercommunication system must be provided.

(d) If installed communication equipment includes transmitter "off-on" switching, that switching means must be designed to return from the "transmit" to the "off" position when it is released and ensure that the transmitter will return to the off (non transmitting) state.

(e) If provisions for the use of communication headsets are provided, it must be demonstrated that the flightcrew members will receive all aural warnings under the actual cockpit noise conditions when the airplane is being operated when any headset is being used.

[Doc. No. 26344, 58 FR 18977, Apr. 9, 1993, as amended by Amdt. 23-49, 61 FR 5169, Feb. 9, 1996; Amdt. 23-62, 76 FR 75761, Dec. 2, 2011]

§ 23.1435 Hydraulic systems.

(a) *Design.* Each hydraulic system must be designed as follows:

(1) Each hydraulic system and its elements must withstand, without yielding, the structural loads expected in addition to hydraulic loads.

(2) A means to indicate the pressure in each hydraulic system which supplies two or more primary functions must be provided to the flight crew.

(3) There must be means to ensure that the pressure, including transient (surge) pressure, in any part of the system will not exceed the safe limit above design operating pressure and to prevent excessive pressure resulting from fluid volumetric changes in all lines which are likely to remain closed long enough for such changes to occur.

(4) The minimum design burst pressure must be 2.5 times the operating pressure.

(b) *Tests.* Each system must be substantiated by proof pressure tests. When proof tested, no part of any system may fail, malfunction, or experience a permanent set. The proof load of each system must be at least 1.5 times the maximum operating pressure of that system.

(c) *Accumulators.* A hydraulic accumulator or reservoir may be installed on the engine side of any firewall if—

(1) It is an integral part of an engine or propeller system, or

(2) The reservoir is nonpressurized and the total capacity of all such nonpressurized reservoirs is one quart or less.

[Doc. No. 4080, 29 FR 17955, Dec. 18, 1964, as amended by Amdt. 23-7, 34 FR 13096, Aug. 13, 1969; Amdt. 23-14, 38 FR 31824, Nov. 19, 1973; Amdt. 23-43, 58 FR 18977, Apr. 9, 1993; Amdt. 23-49, 61 FR 5170, Feb. 9, 1996]

§ 23.1437 Accessories for multiengine airplanes.

For multiengine airplanes, engine-driven accessories essential to safe operation must be distributed among two or more engines so that the failure of any one engine will not impair safe operation through the malfunctioning of these accessories.

§ 23.1438 Pressurization and pneumatic systems.

(a) Pressurization system elements must be burst pressure tested to 2.0 times, and proof pressure tested to 1.5 times, the maximum normal operating pressure.

(b) Pneumatic system elements must be burst pressure tested to 3.0 times, and proof pressure tested to 1.5 times, the maximum normal operating pressure.

(c) An analysis, or a combination of analysis and test, may be substituted for any test required by paragraph (a) or (b) of this section if the Administrator finds it equivalent to the required test.

[Amdt. 23-20, 42 FR 36969, July 18, 1977]

§ 23.1441 Oxygen equipment and supply.

(a) If certification with supplemental oxygen equipment is requested, or the airplane is approved for operations at or above altitudes where oxygen is required to be used by the operating rules, oxygen equipment must be provided that meets the requirements of this section and §§ 23.1443 through 23.1449. Portable oxygen equipment may be used to meet the requirements of this part if the portable equipment is shown to comply with the applicable requirements, is identified in the airplane type design, and its stowage provisions are found to be in compliance with the requirements of § 23.561.

(b) The oxygen system must be free from hazards in itself, in its method of operation, and its effect upon other components.

(c) There must be a means to allow the crew to readily determine, during the flight, the quantity of oxygen available in each source of supply.

(d) Each required flight crewmember must be provided with—

(1) Demand oxygen equipment if the airplane is to be certificated for operation above 25,000 feet.

(2) Pressure demand oxygen equipment if the airplane is to be certificated for operation above 40,000 feet.

(e) There must be a means, readily available to the crew in flight, to turn on and to shut off the oxygen supply at the high pressure source. This shutoff requirement does not apply to chemical oxygen generators.

[Amdt. 23-9, 35 FR 6386, Apr. 21, 1970, as amended by Amdt. 23-43, 58 FR 18978, Apr. 9, 1993]

§ 23.1443 Minimum mass flow of supplemental oxygen.

(a) If the airplane is to be certified above 41,000 feet, a continuous flow oxygen system must be provided for each passenger.

(b) If continuous flow oxygen equipment is installed, an applicant must show compliance with the requirements of either paragraphs (b)(1) and (b)(2) or paragraph (b)(3) of this section:

(1) For each passenger, the minimum mass flow of supplemental oxygen required at various cabin pressure altitudes may not be less than the flow required to maintain, during inspiration and while using the oxygen equipment (including masks) provided, the following mean tracheal oxygen partial pressures:

(i) At cabin pressure altitudes above 10,000 feet up to and including 18,500 feet, a mean tracheal oxygen partial pressure of 100mm Hg when breathing 15 liters per minute, Body Temperature, Pressure, Saturated (BTPS) and with a tidal volume of 700cc with a constant time interval between respirations.

(ii) At cabin pressure altitudes above 18,500 feet up to and including 40,000 feet, a mean tracheal oxygen partial pressure of 83.8mm Hg when breathing 30 liters per minute, BTPS, and with a tidal volume of 1,100cc with a constant time interval between respirations.

(2) For each flight crewmember, the minimum mass flow may not be less than the flow required to maintain, during inspiration, a mean tracheal oxygen partial pressure of 149mm Hg when breathing 15 liters per minute, BTPS, and with a maximum tidal volume of 700cc with a constant time interval between respirations.

(3) The minimum mass flow of supplemental oxygen supplied for each user must be at a rate not less than that shown in the following figure for each altitude up to and including the maximum operating altitude of the airplane.

(c) If demand equipment is installed for use by flight crewmembers, the minimum mass flow of supplemental oxygen required for each flight crewmember may not be less than the flow required to maintain, during inspiration, a mean tracheal oxygen partial pressure of 122mm Hg up to and including a cabin pressure altitude of 35,000 feet, and 95 percent oxygen between cabin pressure altitudes of 35,000 and 40,000 feet, when breathing 20 liters per minutes BTPS. In addition, there must be means to allow the flight crew to use undiluted oxygen at their discretion.

(d) If first-aid oxygen equipment is installed, the minimum mass flow of oxygen to each user may not be less than 4 liters per minute, STPD. However, there may be a means to decrease this flow to not less than 2 liters per minute, STPD, at any cabin altitude. The quantity of oxygen required is based upon an average flow rate of 3 liters per minute per person for whom first-aid oxygen is required.

(e) As used in this section:

(1) BTPS means Body Temperature, and Pressure, Saturated (which is 37°C, and the ambient pressure to which the body is exposed, minus 47mm Hg, which is the tracheal pressure displaced by water vapor pressure when the breathed air becomes saturated with water vapor at 37°C).

(2) STPD means Standard, Temperature, and Pressure, Dry (which is 0°C at 760mm Hg with no water vapor).

[Doc. No. FAA-2009-0738, 76 FR 75761, Dec. 2, 2011]

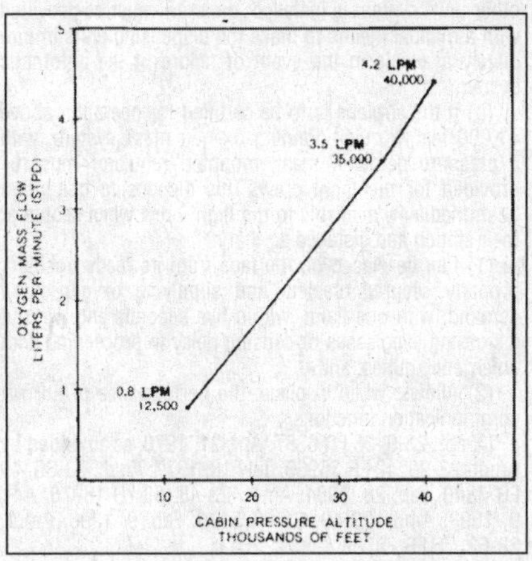

§ 23.1445 Oxygen distribution system.

(a) Except for flexible lines from oxygen outlets to the dispensing units, or where shown to be otherwise suitable to the installation, nonmetallic tubing must not be used for any oxygen line that is normally pressurized during flight.

(b) Nonmetallic oxygen distribution lines must not be routed where they may be subjected to elevated temperatures, electrical arcing, and released flammable fluids that might result from any probable failure.

(c) If the flight crew and passengers share a common source of oxygen, a means to separately reserve the minimum supply required by the flight crew must be provided.

[Doc. No. 26344, 58 FR 18978, Apr. 9, 1993, as amended by Amdt. 23-62, 76 FR 75762, Dec. 2, 2011]

§ 23.1447 Equipment standards for oxygen dispensing units.

If oxygen dispensing units are installed, the following apply:

(a) There must be an individual dispensing unit for each occupant for whom supplemental oxygen is to be supplied. Each dispensing unit must:

(1) Provide for effective utilization of the oxygen being delivered to the unit.

(2) Be capable of being readily placed into position on the face of the user.

(3) Be equipped with a suitable means to retain the unit in position on the face.

(4) If radio equipment is installed, the flightcrew oxygen dispensing units must be designed to allow the use of that equipment and to allow communication with any other required crew member while at their assigned duty station.

(b) If certification for operation up to and including 18,000 feet (MSL) is requested, each oxygen dispensing unit must:

(1) Cover the nose and mouth of the user; or

(2) Be a nasal cannula, in which case one oxygen dispensing unit covering both the nose and mouth of the user must be available. In addition, each nasal cannula or its connecting tubing must have permanently affixed—

(i) A visible warning against smoking while in use;

(ii) An illustration of the correct method of donning; and

(iii) A visible warning against use with nasal obstructions or head colds with resultant nasal congestion.

(c) If certification for operation above 18,000 feet (MSL) is requested, each oxygen dispensing unit must cover the nose and mouth of the user.

(d) For a pressurized airplane designed to operate at flight altitudes above 25,000 feet (MSL), the dispensing units must meet the following:

(1) The dispensing units for passengers must be connected to an oxygen supply terminal and be immediately available to each occupant wherever seated.

(2) The dispensing units for crewmembers must be automatically presented to each crewmember before the cabin pressure altitude exceeds 15,000 feet, or the units must be of the quick-donning type, connected to an oxygen supply terminal that is immediately available to crewmembers at their station.

(e) If certification for operation above 30,000 feet is requested, the dispensing units for passengers must be automatically presented to each occupant before the cabin pressure altitude exceeds 15,000 feet.

(f) If an automatic dispensing unit (hose and mask, or other unit) system is installed, the crew must be provided with a manual means to make the dispensing units immediately available in the event of failure of the automatic system.

(g) If the airplane is to be certified for operation above 41,000 feet, a quick-donning oxygen mask system, with a pressure demand, mask mounted regulator must be provided for the flight crew. This dispensing unit must be immediately available to the flight crew when seated at their station and installed so that it:

(1) Can be placed on the face from its ready position, properly secured, sealed, and supplying oxygen upon demand, with one hand, within five seconds and without disturbing eyeglasses or causing delay in proceeding with emergency duties; and

(2) Allows, while in place, the performance of normal communication functions.

[Amdt. 23-9, 35 FR 6387, Apr. 21, 1970, as amended by Amdt. 23-20, 42 FR 36969, July 18, 1977; Amdt. 23-30, 49 FR 7340, Feb. 28, 1984; Amdt. 23-43, 58 FR 18978, Apr. 9, 1993; Amdt. 23-49, 61 FR 5170, Feb. 9, 1996; Amdt. 23-62, 76 FR 75762, Dec. 2, 2011]

§ 23.1449 Means for determining use of oxygen.

There must be a means to allow the crew to determine whether oxygen is being delivered to the dispensing equipment.

[Amdt. 23-9, 35 FR 6387, Apr. 21, 1970]

§ 23.1450 Chemical oxygen generators.

(a) For the purpose of this section, a chemical oxygen generator is defined as a device which produces oxygen by chemical reaction.

(b) Each chemical oxygen generator must be designed and installed in accordance with the following requirements:

(1) Surface temperature developed by the generator during operation may not create a hazard to the airplane or to its occupants.

(2) Means must be provided to relieve any internal pressure that may be hazardous.

(c) In addition to meeting the requirements in paragraph (b) of this section, each portable chemical oxygen generator that is capable of sustained operation by successive replacement of a generator element must be placarded to show—

(1) The rate of oxygen flow, in liters per minute;

(2) The duration of oxygen flow, in minutes, for the replaceable generator element; and

(3) A warning that the replaceable generator element may be hot, unless the element construction is such that the surface temperature cannot exceed 100°F.

[Amdt. 23-20, 42 FR 36969, July 18, 1977]

§ 23.1451 Fire protection for oxygen equipment.

Oxygen equipment and lines must:

(a) Not be installed in any designed fire zones.

(b) Be protected from heat that may be generated in, or escape from, any designated fire zone.

(c) Be installed so that escaping oxygen cannot come in contact with and cause ignition of grease, fluid, or vapor accumulations that are present in normal operation or that may result from the failure or malfunction of any other system.

[Doc. No. 27806, 61 FR 5170, Feb. 9, 1996]

§ 23.1453 Protection of oxygen equipment from rupture.

(a) Each element of the oxygen system must have sufficient strength to withstand the maximum pressure and temperature, in combination with any externally applied loads arising from consideration of limit structural loads, that may be acting on that part of the system.

(b) Oxygen pressure sources and the lines between the source and the shutoff means must be:

(1) Protected from unsafe temperatures; and

(2) Located where the probability and hazard of rupture in a crash landing are minimized.

[Doc. No. 27806, 61 FR 5170, Feb. 9, 1996]

§ 23.1457 Cockpit voice recorders.

(a) Each cockpit voice recorder required by the operating rules of this chapter must be approved and must be installed so that it will record the following:

(1) Voice communications transmitted from or received in the airplane by radio.

(2) Voice communications of flight crewmembers on the flight deck.

(3) Voice communications of flight crewmembers on the flight deck, using the airplane's interphone system.

(4) Voice or audio signals identifying navigation or approach aids introduced into a headset or speaker.

(5) Voice communications of flight crewmembers using the passenger loudspeaker system, if there is such a system and if the fourth channel is available in accordance with the requirements of paragraph (c)(4)(ii) of this section.

(6) If datalink communication equipment is installed, all datalink communications, using an approved data message set. Datalink messages must be recorded as the output signal from the communications unit that translates the signal into usable data.

(b) The recording requirements of paragraph (a)(2) of this section must be met by installing a cockpit-mounted area microphone, located in the best position for recording voice communications originating at the first and second pilot stations and voice communications of other crewmembers on the flight deck when directed to those stations. The microphone must be so located and, if necessary, the preamplifiers and filters of the recorder must be so adjusted or supplemented, so that the intelligibility of the recorded communications is as high as practicable when recorded under flight cockpit noise conditions and played back. Repeated aural or visual playback of the record may be used in evaluating intelligibility.

(c) Each cockpit voice recorder must be installed so that the part of the communication or audio signals specified in paragraph (a) of this section obtained from each of the following sources is recorded on a separate channel:

(1) For the first channel, from each boom, mask, or handheld microphone, headset, or speaker used at the first pilot station.

(2) For the second channel from each boom, mask, or handheld microphone, headset, or speaker used at the second pilot station.

(3) For the third channel—from the cockpit-mounted area microphone.

(4) For the fourth channel from:

(i) Each boom, mask, or handheld microphone, headset, or speaker used at the station for the third and fourth crewmembers.

(ii) If the stations specified in paragraph (c)(4)(i) of this section are not required or if the signal at such a station is picked up by another channel, each microphone on the flight deck that is used with the passenger loudspeaker system, if its signals are not picked up by another channel.

(5) And that as far as is practicable all sounds received by the microphone listed in paragraphs (c)(1), (2), and (4) of this section must be recorded without interruption irrespective of the position of the interphone-transmitter key switch. The design shall ensure that sidetone for the flight crew is produced only when the interphone, public address system, or radio transmitters are in use.

(d) Each cockpit voice recorder must be installed so that:

(1)(i) It receives its electrical power from the bus that provides the maximum reliability for operation of the cockpit voice recorder without jeopardizing service to essential or emergency loads.

(ii) It remains powered for as long as possible without jeopardizing emergency operation of the airplane.

(2) There is an automatic means to simultaneously stop the recorder and prevent each erasure feature from functioning, within 10 minutes after crash impact; and

(3) There is an aural or visual means for preflight checking of the recorder for proper operation;

(4) Any single electrical failure external to the recorder does not disable both the cockpit voice recorder and the flight data recorder;

(5) It has an independent power source—

(i) That provides 10 ± 1 minutes of electrical power to operate both the cockpit voice recorder and cockpit-mounted area microphone;

(ii) That is located as close as practicable to the cockpit voice recorder; and

(iii) To which the cockpit voice recorder and cockpit-mounted area microphone are switched automatically in the event that all other power to the cockpit voice recorder is interrupted either by normal shutdown or by any other loss of power to the electrical power bus; and

(6) It is in a separate container from the flight data recorder when both are required. If used to comply with only the cockpit voice recorder requirements, a combination unit may be installed.

(e) The recorder container must be located and mounted to minimize the probability of rupture of the container as a result of crash impact and consequent heat damage to the recorder from fire.

(1) Except as provided in paragraph (e)(2) of this section, the recorder container must be located as far aft as practicable, but need not be outside of the pressurized compartment, and may not be located where aft-mounted engines may crush the container during impact.

(2) If two separate combination digital flight data recorder and cockpit voice recorder units are installed instead of one cockpit voice recorder and one digital flight data recorder, the combination unit that is installed to comply with the cockpit voice recorder requirements may be located near the cockpit.

(f) If the cockpit voice recorder has a bulk erasure device, the installation must be designed to minimize the probability of inadvertent operation and actuation of the device during crash impact.

(g) Each recorder container must:

(1) Be either bright orange or bright yellow;

(2) Have reflective tape affixed to its external surface to facilitate its location under water; and

(3) Have an underwater locating device, when required by the operating rules of this chapter, on or adjacent to the

container which is secured in such manner that they are not likely to be separated during crash impact.

[Amdt. 23-35, 53 FR 26142, July 11, 1988, as amended by Amdt. 23-58, 73 FR 12562, Mar. 7, 2008; 74 FR 32799, July 9, 2009]

§ 23.1459 Flight data recorders.

(a) Each flight recorder required by the operating rules of this chapter must be installed so that:

(1) It is supplied with airspeed, altitude, and directional data obtained from sources that meet the accuracy requirements of §§ 23.1323, 23.1325, and 23.1327, as appropriate;

(2) The vertical acceleration sensor is rigidly attached, and located longitudinally either within the approved center of gravity limits of the airplane, or at a distance forward or aft of these limits that does not exceed 25 percent of the airplane's mean aerodynamic chord;

(3)(i) It receives its electrical power from the bus that provides the maximum reliability for operation of the flight data recorder without jeopardizing service to essential or emergency loads.

(ii) It remains powered for as long as possible without jeopardizing emergency operation of the airplane.

(4) There is an aural or visual means for preflight checking of the recorder for proper recording of data in the storage medium;

(5) Except for recorders powered solely by the engine-driven electrical generator system, there is an automatic means to simultaneously stop a recorder that has a data erasure feature and prevent each erasure feature from functioning, within 10 minutes after crash impact;

(6) Any single electrical failure external to the recorder does not disable both the cockpit voice recorder and the flight data recorder; and

(7) It is in a separate container from the cockpit voice recorder when both are required. If used to comply with only the flight data recorder requirements, a combination unit may be installed. If a combination unit is installed as a cockpit voice recorder to comply with § 23.1457(e)(2), a combination unit must be used to comply with this flight data recorder requirement.

(b) Each nonejectable record container must be located and mounted so as to minimize the probability of container rupture resulting from crash impact and subsequent damage to the record from fire. In meeting this requirement the record container must be located as far aft as practicable, but need not be aft of the pressurized compartment, and may not be where aft-mounted engines may crush the container upon impact.

(c) A correlation must be established between the flight recorder readings of airspeed, altitude, and heading and the corresponding readings (taking into account correction factors) of the first pilot's instruments. The correlation must cover the airspeed range over which the airplane is to be operated, the range of altitude to which the airplane is limited, and 360 degrees of heading. Correlation may be established on the ground as appropriate.

(d) Each recorder container must:

(1) Be either bright orange or bright yellow;

(2) Have reflective tape affixed to its external surface to facilitate its location under water; and

(3) Have an underwater locating device, when required by the operating rules of this chapter, on or adjacent to the container which is secured in such a manner that they are not likely to be separated during crash impact.

(e) Any novel or unique design or operational characteristics of the aircraft shall be evaluated to determine if any dedicated parameters must be recorded on flight recorders in addition to or in place of existing requirements.

[Amdt. 23-35, 53 FR 26143, July 11, 1988, as amended by Amdt. 23-58, 73 FR 12562, Mar. 7, 2008; 74 FR 32800, July 9, 2009]

§ 23.1461 Equipment containing high energy rotors.

(a) Equipment, such as Auxiliary Power Units (APU) and constant speed drive units, containing high energy rotors must meet paragraphs (b), (c), or (d) of this section.

(b) High energy rotors contained in equipment must be able to withstand damage caused by malfunctions, vibration, abnormal speeds, and abnormal temperatures. In addition—

(1) Auxiliary rotor cases must be able to contain damage caused by the failure of high energy rotor blades; and

(2) Equipment control devices, systems, and instrumentation must reasonably ensure that no operating limitations affecting the integrity of high energy rotors will be exceeded in service.

(c) It must be shown by test that equipment containing high energy rotors can contain any failure of a high energy rotor that occurs at the highest speed obtainable with the normal speed control devices inoperative.

(d) Equipment containing high energy rotors must be located where rotor failure will neither endanger the occupants nor adversely affect continued safe flight.

[Amdt. 23-20, 42 FR 36969, July 18, 1977, as amended by Amdt. 23-49, 61 FR 5170, Feb. 9, 1996]

Subpart G—Operating Limitations and Information

§ 23.1501 General.

(a) Each operating limitation specified in §§ 23.1505 through 23.1527 and other limitations and information necessary for safe operation must be established.

(b) The operating limitations and other information necessary for safe operation must be made available to the crewmembers as prescribed in §§ 23.1541 through 23.1589.

[Amdt. 23-21, 43 FR 2319, Jan. 16, 1978]

§ 23.1505 Airspeed limitations.

(a) The never-exceed speed V_{NE} must be established so that it is—

(1) Not less than 0.9 times the minimum value of V_D allowed under § 23.335; and

(2) Not more than the lesser of—

(i) 0.9 V_D established under § 23.335; or

(ii) 0.9 times the maximum speed shown under § 23.251.

(b) The maximum structural cruising speed V_{NO} must be established so that it is—

(1) Not less than the minimum value of V_C allowed under § 23.335; and

(2) Not more than the lesser of—

(i) V_C established under § 23.335; or

(ii) 0.89 V_{NE} established under paragraph (a) of this section.

(c)(1) Paragraphs (a) and (b) of this section do not apply to turbine airplanes or to airplanes for which a design diving speed V_D/M_D is established under § 23.335(b)(4). For those airplanes, a maximum operating limit speed (V_{MO}/M_{MO} airspeed or Mach number, whichever is critical at a particular altitude) must be established as a speed that may not be deliberately exceeded in any regime of flight (climb, cruise, or descent) unless a higher speed is authorized for flight test or pilot training operations.

(2) V_{MO}/M_{MO} must be established so that it is not greater than the design cruising speed V_C/M_C and so that it is sufficiently below V_D/M_D, or V_{DF}/M_{DF} for jets, and the maximum speed shown under § 23.251 to make it highly improbable that the latter speeds will be inadvertently exceeded in operations.

(3) The speed margin between V_{MO}/M_{MO} and V_D/M_D, or V_{DF}/M_{DF} for jets, may not be less than that determined under § 23.335(b), or the speed margin found necessary in the flight tests conducted under § 23.253.

[Doc. No. 4080, 29 FR 17955, Dec. 18, 1964, as amended by Amdt. 23-7, 34 FR 13096, Aug. 13, 1969; Amdt. 23-62, 76 FR 75762, Dec. 2, 2011]

§ 23.1507 Operating maneuvering speed.

The maximum operating maneuvering speed, V_O, must be established as an operating limitation. V_O is a selected speed that is not greater than $V_S\sqrt{n}$ established in § 23.335(c).

[Doc. No. 26269, 58 FR 42165, Aug. 6, 1993]

§ 23.1511 Flap extended speed.

(a) The flap extended speed V_{FE} must be established so that it is—

(1) Not less than the minimum value of V_F allowed in § 23.345(b); and

(2) Not more than V_F established under § 23.345(a), (c), and (d).

(b) Additional combinations of flap setting, airspeed, and engine power may be established if the structure has been proven for the corresponding design conditions.

[Doc. No. 4080, 29 FR 17955, Dec. 18, 1964; 30 FR 258, Jan. 9, 1965, as amended by Amdt. 23-50, 61 FR 5192, Feb. 9, 1996]

§ 23.1513 Minimum control speed.

The minimum control speed V_{MC}, determined under § 23.149, must be established as an operating limitation.

§ 23.1519 Weight and center of gravity.

The weight and center of gravity limitations determined under § 23.23 must be established as operating limitations.

§ 23.1521 Powerplant limitations.

(a) *General.* The powerplant limitations prescribed in this section must be established so that they do not exceed the corresponding limits for which the engines or propellers are type certificated. In addition, other powerplant limitations used in determining compliance with this part must be established.

(b) *Takeoff operation.* The powerplant takeoff operation must be limited by—

(1) The maximum rotational speed (rpm);

(2) The maximum allowable manifold pressure (for reciprocating engines);

(3) The maximum allowable gas temperature (for turbine engines);

(4) The time limit for the use of the power or thrust corresponding to the limitations established in paragraphs (b)(1) through (3) of this section; and

(5) The maximum allowable cylinder head (as applicable), liquid coolant and oil temperatures.

(c) *Continuous operation.* The continuous operation must be limited by—

(1) The maximum rotational speed;

(2) The maximum allowable manifold pressure (for reciprocating engines);

(3) The maximum allowable gas temperature (for turbine engines); and

(4) The maximum allowable cylinder head, oil, and liquid coolant temperatures.

(d) *Fuel grade or designation.* The minimum fuel grade (for reciprocating engines), or fuel designation (for turbine engines), must be established so that it is not less than that required for the operation of the engines within the limitations in paragraphs (b) and (c) of this section.

(e) *Ambient temperature.* For all airplanes except reciprocating engine-powered airplanes of 6,000 pounds or less maximum weight, ambient temperature limitations (including limitations for winterization installations if applicable) must be established as the maximum ambient atmospheric temperature at which compliance with the cooling provisions of §§ 23.1041 through 23.1047 is shown.

[Doc. No. 4080, 29 FR 17955, Dec. 18, 1964; 30 FR 258, Jan. 9, 1965, as amended by Amdt. 23-21, 43 FR 2319, Jan. 16, 1978; Amdt. 23-45, 58 FR 42165, Aug. 6, 1993; Amdt. 23-50, 61 FR 5192, Feb. 9, 1996]

§ 23.1522 Auxiliary power unit limitations.

If an auxiliary power unit is installed, the limitations established for the auxiliary power must be specified in the operating limitations for the airplane.

[Doc. No. 26269, 58 FR 42166, Aug. 6, 1993]

§ 23.1523 Minimum flight crew.

The minimum flight crew must be established so that it is sufficient for safe operation considering—

(a) The workload on individual crewmembers and, in addition for commuter category airplanes, each crewmember workload determination must consider the following:

(1) Flight path control,

(2) Collision avoidance,

(3) Navigation,

(4) Communications,

(5) Operation and monitoring of all essential airplane systems,

(6) Command decisions, and

(7) The accessibility and ease of operation of necessary controls by the appropriate crewmember during all normal and emergency operations when at the crewmember flight station;

(b) The accessibility and ease of operation of necessary controls by the appropriate crewmember; and

(c) The kinds of operation authorized under § 23.1525.

[Amdt. 23-21, 43 FR 2319, Jan. 16, 1978, as amended by Amdt. 23-34, 52 FR 1834, Jan. 15, 1987]

§ 23.1524 Maximum passenger seating configuration.

The maximum passenger seating configuration must be established.

[Amdt. 23-10, 36 FR 2864, Feb. 11, 1971]

§ 23.1525 Kinds of operation.

The kinds of operation authorized (e.g. VFR, IFR, day or night) and the meteorological conditions (e.g. icing) to which the operation of the airplane is limited or from which it is prohibited, must be established appropriate to the installed equipment.

[Doc. No. 26269, 58 FR 42166, Aug. 6, 1993]

§ 23.1527 Maximum operating altitude.

(a) The maximum altitude up to which operation is allowed, as limited by flight, structural, powerplant, functional or equipment characteristics, must be established.

(b) A maximum operating altitude limitation of not more than 25,000 feet must be established for pressurized airplanes unless compliance with § 23.775(e) is shown.

[Doc. No. 26269, 58 FR 42166, Aug. 6, 1993]

§ 23.1529 Instructions for Continued Airworthiness.

The applicant must prepare Instructions for Continued Airworthiness in accordance with appendix G to this part that are acceptable to the Administrator. The instructions may be incomplete at type certification if a program exists to ensure their completion prior to delivery of the first airplane or issuance of a standard certificate of airworthiness, whichever occurs later.

[Amdt. 23-26, 45 FR 60171, Sept. 11, 1980]

Markings And Placards

§ 23.1541 General.

(a) The airplane must contain—

(1) The markings and placards specified in §§ 23.1545 through 23.1567; and

(2) Any additional information, instrument markings, and placards required for the safe operation if it has unusual design, operating, or handling characteristics.

(b) Each marking and placard prescribed in paragraph (a) of this section—

(1) Must be displayed in a conspicuous place; and

(2) May not be easily erased, disfigured, or obscured.

(c) For airplanes which are to be certificated in more than one category—

(1) The applicant must select one category upon which the placards and markings are to be based; and

(2) The placards and marking information for all categories in which the airplane is to be certificated must be furnished in the Airplane Flight Manual.

[Doc. No. 4080, 29 FR 17955, Dec. 18, 1964; 30 FR 258, Jan. 9, 1965, as amended by Amdt. 23-21, 43 FR 2319, Jan. 16, 1978]

§ 23.1543 Instrument markings: General.

For each instrument—

(a) When markings are on the cover glass of the instrument, there must be means to maintain the correct alignment of the glass cover with the face of the dial; and

(b) Each arc and line must be wide enough and located to be clearly visible to the pilot.

(c) All related instruments must be calibrated in compatible units.

[Doc. No. 4080, 29 FR 17955, Dec. 18, 1964; 30 FR 258, Jan. 9, 1965, as amended by Amdt. 23-50, 61 FR 5192, Feb. 9, 1996]

§ 23.1545 Airspeed indicator.

(a) Each airspeed indicator must be marked as specified in paragraph (b) of this section, with the marks located at the corresponding indicated airspeeds.

(b) The following markings must be made:

(1) For the never-exceed speed V_{NE}, a radial red line.

(2) For the caution range, a yellow arc extending from the red line specified in paragraph (b)(1) of this section to the upper limit of the green arc specified in paragraph (b)(3) of this section.

(3) For the normal operating range, a green arc with the lower limit at V_{S1} with maximum weight and with landing gear and wing flaps retracted, and the upper limit at the maximum structural cruising speed V_{NO} established under § 23.1505(b).

(4) For the flap operating range, a white arc with the lower limit at V_{S0} at the maximum weight, and the upper limit at the flaps-extended speed V_{FE} established under § 23.1511.

(5) For reciprocating multiengine-powered airplanes of 6,000 pounds or less maximum weight, for the speed at which compliance has been shown with § 23.69(b) relating to rate of climb at maximum weight and at sea level, a blue radial line.

(6) For reciprocating multiengine-powered airplanes of 6,000 pounds or less maximum weight, for the maximum value of minimum control speed, V_{MC}, (one-engine-inoperative) determined under § 23.149(b), a red radial line.

(c) If V_{NE} or V_{NO} vary with altitude, there must be means to indicate to the pilot the appropriate limitations throughout the operating altitude range.

(d) Paragraphs (b)(1) through (b)(4) and paragraph (c) of this section do not apply to airplanes for which a maximum operating speed V_{MO}/M_{MO} is established under § 23.1505(c). For those airplanes, there must either be a maximum allowable airspeed indication showing the variation of V_{MO}/M_{MO} with altitude or compressibility limitations (as appropriate), or a radial red line marking for V_{MO}/M_{MO} must be made at lowest value of V_{MO}/M_{MO} established for any altitude up to the maximum operating altitude for the airplane.

[Doc. No. 4080, 29 FR 17955, Dec. 18, 1964, as amended by Amdt. 23-3, 30 FR 14240, Nov. 13, 1965; Amdt. 23-7, 34 FR 13097, Aug. 13, 1969; Amdt. 23-23, 43 FR 50593, Oct. 30, 1978; Amdt. 23-50, 61 FR 5193, Feb. 9, 1996; Amdt. 23-62, 76 FR 75762, Dec. 2, 2011]

§ 23.1547 Magnetic direction indicator.

(a) A placard meeting the requirements of this section must be installed on or near the magnetic direction indicator.

(b) The placard must show the calibration of the instrument in level flight with the engines operating.

(c) The placard must state whether the calibration was made with radio receivers on or off.

(d) Each calibration reading must be in terms of magnetic headings in not more than 30 degree increments.

(e) If a magnetic nonstabilized direction indicator can have a deviation of more than 10 degrees caused by the operation of electrical equipment, the placard must state which electrical loads, or combination of loads, would cause a deviation of more than 10 degrees when turned on.

[Doc. No. 4080, 29 FR 17955, Dec. 18, 1964; 30 FR 258, Jan. 9, 1965, as amended by Amdt. 23-20, 42 FR 36969, July 18, 1977]

§ 23.1549 Powerplant and auxiliary power unit instruments.

For each required powerplant and auxiliary power unit instrument, as appropriate to the type of instruments—

(a) Each maximum and, if applicable, minimum safe operating limit must be marked with a red radial or a red line;

(b) Each normal operating range must be marked with a green arc or green line, not extending beyond the maximum and minimum safe limits;

(c) Each takeoff and precautionary range must be marked with a yellow arc or a yellow line; and

(d) Each engine, auxiliary power unit, or propeller range that is restricted because of excessive vibration stresses must be marked with red arcs or red lines.

[Amdt. 23-12, 41 FR 55466, Dec. 20, 1976, as amended by Amdt. 23-28, 47 FR 13315, Mar. 29, 1982; Amdt. 23-45, 58 FR 42166, Aug. 6, 1993]

§ 23.1551 Oil quantity indicator.

Each oil quantity indicator must be marked in sufficient increments to indicate readily and accurately the quantity of oil.

§ 23.1553 Fuel quantity indicator.

A red radial line must be marked on each indicator at the calibrated zero reading, as specified in § 23.1337(b)(1).

[Doc. No. 27807, 61 FR 5193, Feb. 9, 1996]

§ 23.1555 Control markings.

(a) Each cockpit control, other than primary flight controls and simple push button type starter switches, must be plainly marked as to its function and method of operation.

(b) Each secondary control must be suitably marked.

(c) For powerplant fuel controls—

(1) Each fuel tank selector control must be marked to indicate the position corresponding to each tank and to each existing cross feed position;

(2) If safe operation requires the use of any tanks in a specific sequence, that sequence must be marked on or near the selector for those tanks;

(3) The conditions under which the full amount of usable fuel in any restricted usage fuel tank can safely be used must be stated on a placard adjacent to the selector valve for that tank; and

(4) Each valve control for any engine of a multiengine airplane must be marked to indicate the position corresponding to each engine controlled.

(d) Usable fuel capacity must be marked as follows:

(1) For fuel systems having no selector controls, the usable fuel capacity of the system must be indicated at the fuel quantity indicator.

(2) For fuel systems having selector controls, the usable fuel capacity available at each selector control position must be indicated near the selector control.

(3) For fuel systems having a calibrated fuel quantity indication system complying with § 23.1337(b)(1) and accurately displaying the actual quantity of usable fuel in each selectable tank, no fuel capacity placards outside of the fuel quantity indicator are required.

(e) For accessory, auxiliary, and emergency controls—

(1) If retractable landing gear is used, the indicator required by § 23.729 must be marked so that the pilot can, at any time, ascertain that the wheels are secured in the extreme positions; and

(2) Each emergency control must be red and must be marked as to method of operation. No control other than an emergency control, or a control that serves an emergency function in addition to its other functions, shall be this color.

[Doc. No. 4080, 29 FR 17955, Dec. 18, 1964; 30 FR 258, Jan. 9, 1965, as amended by Amdt. 23-21, 43 FR 2319, Jan. 16, 1978; Amdt. 23-50, 61 FR 5193, Feb. 9, 1996; Amdt. 23-62, 76 FR 75763, Dec. 2, 2011]

§ 23.1557 Miscellaneous markings and placards.

(a) *Baggage and cargo compartments, and ballast location.* Each baggage and cargo compartment, and each ballast location, must have a placard stating any limitations on contents, including weight, that are necessary under the loading requirements.

(b) *Seats.* If the maximum allowable weight to be carried in a seat is less than 170 pounds, a placard stating the lesser weight must be permanently attached to the seat structure.

(c) *Fuel, oil, and coolant filler openings.* The following apply:

(1) Fuel filler openings must be marked at or near the filler cover with—

(i) For reciprocating engine-powered airplanes—

(A) The word "Avgas"; and

(B) The minimum fuel grade.

(ii) For turbine engine-powered airplanes—

(A) The words "Jet Fuel"; and

(B) The permissible fuel designations, or references to the Airplane Flight Manual (AFM) for permissible fuel designations.

(iii) For pressure fueling systems, the maximum permissible fueling supply pressure and the maximum permissible defueling pressure.

(2) Oil filler openings must be marked at or near the filler cover with the word "Oil" and the permissible oil designations, or references to the Airplane Flight Manual (AFM) for permissible oil designations.

(3) Coolant filler openings must be marked at or near the filler cover with the word "Coolant".

(d) *Emergency exit placards.* Each placard and operating control for each emergency exit must be red. A placard must be near each emergency exit control and must clearly indicate the location of that exit and its method of operation.

(e) The system voltage of each direct current installation must be clearly marked adjacent to its exernal power connection.

[Doc. No. 4080, 29 FR 17955, Dec. 18, 1964; as amended by Amdt. 23-21, 42 FR 15042, Mar. 17, 1977; Amdt. 23-23, 43 FR 50594, Oct. 30, 1978; Amdt. 23-45, 58 FR 42166, Aug. 6, 1993; 73 FR 35063, June 20, 2008]

§ 23.1559 Operating limitations placard.

(a) There must be a placard in clear view of the pilot stating—

(1) That the airplane must be operated in accordance with the Airplane Flight Manual; and

(2) The certification category of the airplane to which the placards apply.

(b) For airplanes certificated in more than one category, there must be a placard in clear view of the pilot stating that other limitations are contained in the Airplane Flight Manual.

(c) There must be a placard in clear view of the pilot that specifies the kind of operations to which the operation of the airplane is limited or from which it is prohibited under § 23.1525.

(d) The placard(s) required by this section need not be lighted.

[Doc. No. 27807, 61 FR 5193, Feb. 9, 1996, as amended by Amdt. 23-62, 76 FR 75763, Dec. 2, 2011]

§ 23.1561 Safety equipment.

(a) Safety equipment must be plainly marked as to method of operation.

(b) Stowage provisions for required safety equipment must be marked for the benefit of occupants.

§ 23.1563 Airspeed placards.

There must be an airspeed placard in clear view of the pilot and as close as practicable to the airspeed indicator. This placard must list—

(a) The operating maneuvering speed, V_O; and

(b) The maximum landing gear operating speed V_{LO}.

(c) For reciprocating multiengine-powered airplanes of more than 6,000 pounds maximum weight, and turbine engine-powered airplanes, the maximum value of the minimum control speed, V_{MC} (one-engine-inoperative) determined under § 23.149(b).

(d) The airspeed placard(s) required by this section need not be lighted if the landing gear operating speed is indicated on the airspeed indicator or other lighted area such as the landing gear control and the airspeed indicator has features such as low speed awareness that provide ample warning prior to V_{MC}.

[Amdt. 23-7, 34 FR 13097, Aug. 13, 1969, as amended by Amdt. 23-45, 58 FR 42166, Aug. 6, 1993; Amdt. 23-50, 61 FR 5193, Feb. 9, 1996; Amdt. 23-62, 76 FR 75763, Dec. 2, 2011]

§ 23.1567 Flight maneuver placard.

(a) For normal category airplanes, there must be a placard in front of and in clear view of the pilot stating: "No acrobatic maneuvers, including spins, approved."

(b) For utility category airplanes, there must be—

(1) A placard in clear view of the pilot stating: "Acrobatic maneuvers are limited to the following ___________;" (list approved maneuvers and the recommended entry speed for each); and

(2) For those airplanes that do not meet the spin requirements for acrobatic category airplanes, an additional placard in clear view of the pilot stating: "Spins Prohibited."

(c) For acrobatic category airplanes, there must be a placard in clear view of the pilot listing the approved acrobatic maneuvers and the recommended entry airspeed for each. If inverted flight maneuvers are not approved, the placard must bear a notation to this effect.

(d) For acrobatic category airplanes and utility category airplanes approved for spinning, there must be a placard in clear view of the pilot—

(1) Listing the control actions for recovery from spinning maneuvers; and

(2) Stating that recovery must be initiated when spiral characteristics appear, or after not more than six turns or not more than any greater number of turns for which the airplane has been certificated.

(e) The placard(s) required by this section need not be lighted.

[Doc. No. 4080, 29 FR 17955, Dec. 18, 1964; 30 FR 258, Jan. 9, 1965, as amended by Amdt. 23-13, 37 FR 20023, Sept. 23, 1972; Amdt. 23-21, 43 FR 2319, Jan. 16, 1978; Amdt. 23-50, 61 FR 5193, Feb. 9, 1996; Amdt. 23-62, 76 FR 75763, Dec. 2, 2011]

Airplane Flight Manual and Approved Manual Material

§ 23.1581 General.

(a) *Furnishing information.* An Airplane Flight Manual must be furnished with each airplane, and it must contain the following:

(1) Information required by §§ 23.1583 through 23.1589.

(2) Other information that is necessary for safe operation because of design, operating, or handling characteristics.

(3) Further information necessary to comply with the relevant operating rules.

(b) *Approved information.* (1) Except as provided in paragraph (b)(2) of this section, each part of the Airplane Flight Manual containing information prescribed in §§ 23.1583 through 23.1589 must be approved, segregated, identified and clearly distinguished from each unapproved part of that Airplane Flight Manual.

(2) The requirements of paragraph (b)(1) of this section do not apply to reciprocating engine-powered airplanes of 6,000 pounds or less maximum weight, if the following is met:

(i) Each part of the Airplane Flight Manual containing information prescribed in § 23.1583 must be limited to such information, and must be approved, identified, and clearly distinguished from each other part of the Airplane Flight Manual.

(ii) The information prescribed in §§ 23.1585 through 23.1589 must be determined in accordance with the applicable requirements of this part and presented in its entirety in a manner acceptable to the Administrator.

(3) Each page of the Airplane Flight Manual containing information prescribed in this section must be of a type that is not easily erased, disfigured, or misplaced, and is capable of being inserted in a manual provided by the applicant, or in a folder, or in any other permanent binder.

(c) The units used in the Airplane Flight Manual must be the same as those marked on the appropriate instruments and placards.

(d) All Airplane Flight Manual operational airspeeds, unless otherwise specified, must be presented as indicated airspeeds.

(e) Provision must be made for stowing the Airplane Flight Manual in a suitable fixed container which is readily accessible to the pilot.

(f) *Revisions and amendments.* Each Airplane Flight Manual (AFM) must contain a means for recording the incorporation of revisions and amendments.

[Amdt. 23-21, 43 FR 2319, Jan. 16, 1978, as amended by Amdt. 23-34, 52 FR 1834, Jan. 15, 1987; Amdt. 23-45, 58 FR 42166, Aug. 6, 1993; Amdt. 23-50, 61 FR 5193, Feb. 9, 1996]

§ 23.1583 Operating limitations.

The Airplane Flight Manual must contain operating limitations determined under this part 23, including the following—

(a) *Airspeed limitations.* The following information must be furnished:

(1) Information necessary for the marking of the airspeed limits on the indicator as required in § 23.1545, and the significance of each of those limits and of the color coding used on the indicator.

(2) The speeds V_{MC}, V_O, V_{LE}, and V_{LO}, if established, and their significance.

(3) In addition, for turbine powered commuter category airplanes—

(i) The maximum operating limit speed, V_{MO}/M_{MO} and a statement that this speed must not be deliberately exceeded in any regime of flight (climb, cruise or descent) unless a higher speed is authorized for flight test or pilot training;

(ii) If an airspeed limitation is based upon compressibility effects, a statement to this effect and information as to any symptoms, the probable behavior of the airplane, and the recommended recovery procedures; and

(iii) The airspeed limits must be shown in terms of V_{MO}/M_{MO} instead of V_{NO} and V_{NE}.

(b) *Powerplant limitations.* The following information must be furnished:

(1) Limitations required by § 23.1521.

(2) Explanation of the limitations, when appropriate.

(3) Information necessary for marking the instruments required by § 23.1549 through § 23.1553.

(c) *Weight.* The airplane flight manual must include—

(1) The maximum weight; and

(2) The maximum landing weight, if the design landing weight selected by the applicant is less than the maximum weight.

(3) For reciprocating engine-powered airplanes of more than 6,000 pounds maximum weight, single-engine turbines, and multiengine jets 6,000 pounds or less maximum weight in the normal, utility, and acrobatic category, performance operating limitations as follows—

(i) The maximum takeoff weight for each airport altitude and ambient temperature within the range selected by the applicant at which the airplane complies with the climb requirements of § 23.63(c)(1).

(ii) The maximum landing weight for each airport altitude and ambient temperature within the range selected by the applicant at which the airplane complies with the climb requirements of § 23.63(c)(2).

(4) For normal, utility, and acrobatic category multiengine jets over 6,000 pounds and commuter category airplanes, the maximum takeoff weight for each airport altitude and ambient temperature within the range selected by the applicant at which—

(i) The airplane complies with the climb requirements of § 23.63(d)(1); and

(ii) The accelerate-stop distance determined under § 23.55 is equal to the available runway length plus the length of any stopway, if utilized; and either:

(A) The takeoff distance determined under § 23.59(a) is equal to the available runway length; or

(B) At the option of the applicant, the takeoff distance determined under § 23.59(a) is equal to the available runway length plus the length of any clearway and the takeoff run determined under § 23.59(b) is equal to the available runway length.

(5) For normal, utility, and acrobatic category multiengine jets over 6,000 pounds and commuter category airplanes, the maximum landing weight for each airport altitude within the range selected by the applicant at which—

(i) The airplane complies with the climb requirements of § 23.63(d)(2) for ambient temperatures within the range selected by the applicant; and

(ii) The landing distance determined under § 23.75 for standard temperatures is equal to the available runway length.

(6) The maximum zero wing fuel weight, where relevant, as established in accordance with § 23.343.

(d) *Center of gravity.* The established center of gravity limits.

(e) *Maneuvers.* The following authorized maneuvers, appropriate airspeed limitations, and unauthorized maneuvers, as prescribed in this section.

(1) *Normal category airplanes.* No acrobatic maneuvers, including spins, are authorized.

(2) *Utility category airplanes.* A list of authorized maneuvers demonstrated in the type flight tests, together with recommended entry speeds and any other associated limitations. No other maneuver is authorized.

(3) *Acrobatic category airplanes.* A list of approved flight maneuvers demonstrated in the type flight tests, together with recommended entry speeds and any other associated limitations.

(4) *Acrobatic category airplanes and utility category airplanes approved for spinning.* Spin recovery procedure established to show compliance with § 23.221(c).

(5) *Commuter category airplanes.* Maneuvers are limited to any maneuver incident to normal flying, stalls, (except whip stalls) and steep turns in which the angle of bank is not more than 60 degrees.

(f) *Maneuver load factor.* The positive limit load factors in g's, and, in addition, the negative limit load factor for acrobatic category airplanes.

(g) *Minimum flight crew.* The number and functions of the minimum flight crew determined under § 23.1523.

(h) *Kinds of operation.* A list of the kinds of operation to which the airplane is limited or from which it is prohibited under § 23.1525, and also a list of installed equipment that affects any operating limitation and identification as to the equipment's required operational status for the kinds of operation for which approval has been given.

(i) *Maximum operating altitude.* The maximum altitude established under § 23.1527.

(j) *Maximum passenger seating configuration.* The maximum passenger seating configuration.

(k) *Allowable lateral fuel loading.* The maximum allowable lateral fuel loading differential, if less than the maximum possible.

(l) *Baggage and cargo loading.* The following information for each baggage and cargo compartment or zone—

(1) The maximum allowable load; and

(2) The maximum intensity of loading.

(m) *Systems.* Any limitations on the use of airplane systems and equipment.

(n) *Ambient temperatures.* Where appropriate, maximum and minimum ambient air temperatures for operation.

(o) *Smoking.* Any restrictions on smoking in the airplane.

(p) *Types of surface.* A statement of the types of surface on which operations may be conducted. (See § 23.45(g) and § 23.1587 (a)(4), (c)(2), and (d)(4)).

[Doc. No. 4080, 29 FR 17955, Dec. 18, 1964, as amended by Amdt. 23-7, 34 FR 13097, Aug. 13, 1969; Amdt. 23-10, 36 FR 2864, Feb. 11, 1971; Amdt. 23-21, 43 FR 2320, Jan. 16, 1978; Amdt. 23-23, 43 FR 50594, Oct. 30, 1978; Amdt. 23-34, 52 FR 1834, Jan. 15, 1987; Amdt. 23-45, 58 FR 42166, Aug. 6, 1993; Amdt. 23-50, 61 FR 5193, Feb. 9, 1996; Amdt. 23-62, 76 FR 75763, Dec. 2, 2011]

§ 23.1585 Operating procedures.

(a) For all airplanes, information concerning normal, abnormal (if applicable), and emergency procedures and other pertinent information necessary for safe operation and the achievement of the scheduled performance must be furnished, including—

(1) An explanation of significant or unusual flight or ground handling characteristics;

(2) The maximum demonstrated values of crosswind for takeoff and landing, and procedures and information pertinent to operations in crosswinds;

(3) A recommended speed for flight in rough air. This speed must be chosen to protect against the occurrence, as a result of gusts, of structural damage to the airplane and loss of control (for example, stalling);

(4) Procedures for restarting any turbine engine in flight, including the effects of altitude; and

(5) Procedures, speeds, and configuration(s) for making a normal approach and landing, in accordance with §§ 23.73 and 23.75, and a transition to the balked landing condition.

(6) For seaplanes and amphibians, water handling procedures and the demonstrated wave height.

(b) In addition to paragraph (a) of this section, for all single-engine airplanes, the procedures, speeds, and configuration(s) for a glide following engine failure, in accordance with § 23.71 and the subsequent forced landing, must be furnished.

(c) In addition to paragraph (a) of this section, for all multiengine airplanes, the following information must be furnished:

(1) Procedures, speeds, and configuration(s) for making an approach and landing with one engine inoperative;

(2) Procedures, speeds, and configuration(s) for making a balked landing with one engine inoperative and the conditions under which a balked landing can be performed safely, or a warning against attempting a balked landing;

(3) The V_{SSE} determined in § 23.149; and

(4) Procedures for restarting any engine in flight including the effects of altitude.

(d) In addition to paragraphs (a) and either (b) or (c) of this section, as appropriate, for all normal, utility, and acrobatic category airplanes, the following information must be furnished:

(1) Procedures, speeds, and configuration(s) for making a normal takeoff, in accordance with § 23.51 (a) and (b), and § 23.53 (a) and (b), and the subsequent climb, in accordance with § 23.65 and § 23.69(a).

(2) Procedures for abandoning a takeoff due to engine failure or other cause.

(e) In addition to paragraphs (a), (c), and (d) of this section, for all normal, utility, and acrobatic category multiengine airplanes, the information must include the following:

(1) Procedures and speeds for continuing a takeoff following engine failure and the conditions under which takeoff can safely be continued, or a warning against attempting to continue the takeoff.

(2) Procedures, speeds, and configurations for continuing a climb following engine failure, after takeoff, in accordance with § 23.67, or enroute, in accordance with § 23.69(b).

(f) In addition to paragraphs (a) and (c) of this section, for normal, utility, and acrobatic category multiengine jets weighing over 6,000 pounds, and commuter category airplanes, the information must include the following:

(1) Procedures, speeds, and configuration(s) for making a normal takeoff.

(2) Procedures and speeds for carrying out an accelerate-stop in accordance with § 23.55.

(3) Procedures and speeds for continuing a takeoff following engine failure in accordance with § 23.59(a)(1) and for following the flight path determined under § 23.57 and § 23.61(a).

(g) For multiengine airplanes, information identifying each operating condition in which the fuel system independence prescribed in § 23.953 is necessary for safety must be furnished, together with instructions for placing the fuel system in a configuration used to show compliance with that section.

(h) For each airplane showing compliance with § 23.1353 (g)(2) or (g)(3), the operating procedures for disconnecting the battery from its charging source must be furnished.

(i) Information on the total quantity of usable fuel for each fuel tank, and the effect on the usable fuel quantity, as a result of a failure of any pump, must be furnished.

(j) Procedures for the safe operation of the airplane's systems and equipment, both in normal use and in the event of malfunction, must be furnished.

[Doc. No. 27807, 61 FR 5194, Feb. 9, 1996, as amended by Amdt. 23-62, 76 FR 75763, Dec. 2, 2011]

§ 23.1587 Performance information.

Unless otherwise prescribed, performance information must be provided over the altitude and temperature ranges required by § 23.45(b).

(a) For all airplanes, the following information must be furnished—

(1) The stalling speeds V_{SO} and V_{S1} with the landing gear and wing flaps retracted, determined at maximum weight under § 23.49, and the effect on these stalling speeds of angles of bank up to 60 degrees;

(2) The steady rate and gradient of climb with all engines operating, determined under § 23.69(a);

(3) The landing distance, determined under § 23.75 for each airport altitude and standard temperature, and the type of surface for which it is valid;

(4) The effect on landing distances of operation on other than smooth hard surfaces, when dry, determined under § 23.45(g); and

(5) The effect on landing distances of runway slope and 50 percent of the headwind component and 150 percent of the tailwind component.

(b) In addition to paragraph (a) of this section, for all normal, utility, and acrobatic category reciprocating engine-powered airplanes of 6,000 pounds or less maximum weight, the steady angle of climb/descent, determined under § 23.77(a), must be furnished.

(c) In addition to paragraphs (a) and (b) of this section, if appropriate, for normal, utility, and acrobatic category airplanes, the following information must be furnished—

(1) The takeoff distance, determined under § 23.53 and the type of surface for which it is valid.

(2) The effect on takeoff distance of operation on other than smooth hard surfaces, when dry, determined under § 23.45(g);

(3) The effect on takeoff distance of runway slope and 50 percent of the headwind component and 150 percent of the tailwind component;

(4) For multiengine reciprocating engine-powered airplanes of more than 6,000 pounds maximum weight and multiengine turbine powered airplanes, the one-engine-inoperative takeoff climb/descent gradient, determined under § 23.66;

(5) For multiengine airplanes, the enroute rate and gradient of climb/descent with one engine inoperative, determined under § 23.69(b); and

(6) For single-engine airplanes, the glide performance determined under § 23.71.

(d) In addition to paragraph (a) of this section, for normal, utility, and acrobatic category multiengine jets weighing over 6,000 pounds, and commuter category airplanes, the following information must be furnished—

(1) The accelerate-stop distance determined under § 23.55;

(2) The takeoff distance determined under § 23.59(a);

(3) At the option of the applicant, the takeoff run determined under § 23.59(b);

(4) The effect on accelerate-stop distance, takeoff distance and, if determined, takeoff run, of operation on other than smooth hard surfaces, when dry, determined under § 23.45(g);

(5) The effect on accelerate-stop distance, takeoff distance, and if determined, takeoff run, of runway slope and 50 percent of the headwind component and 150 percent of the tailwind component;

(6) The net takeoff flight path determined under § 23.61(b);

(7) The enroute gradient of climb/descent with one engine inoperative, determined under § 23.69(b);

(8) The effect, on the net takeoff flight path and on the enroute gradient of climb/descent with one engine inoperative, of 50 percent of the headwind component and 150 percent of the tailwind component;

(9) Overweight landing performance information (determined by extrapolation and computed for the range of weights between the maximum landing and maximum takeoff weights) as follows—

(i) The maximum weight for each airport altitude and ambient temperature at which the airplane complies with the climb requirements of § 23.63(d)(2); and

(ii) The landing distance determined under § 23.75 for each airport altitude and standard temperature.

(10) The relationship between IAS and CAS determined in accordance with § 23.1323 (b) and (c).

(11) The altimeter system calibration required by § 23.1325(e).

[Doc. No. 27807, 61 FR 5194, Feb. 9, 1996, as amended by Amdt. 23-62, 76 FR 75763, Dec. 2, 2011]

§ 23.1589 Loading information.

The following loading information must be furnished:

(a) The weight and location of each item of equipment that can be easily removed, relocated, or replaced and that is installed when the airplane was weighed under the requirement of § 23.25.

(b) Appropriate loading instructions for each possible loading condition between the maximum and minimum weights established under § 23.25, to facilitate the center of gravity remaining within the limits established under § 23.23.

[Doc. No. 4080, 29 FR 17955, Dec. 18, 1964, as amended by Amdt. 23-45, 58 FR 42167, Aug. 6, 1993; Amdt. 23-50, 61 FR 5195, Feb. 9, 1996]

Appendix A to Part 23—Simplified Design Load Criteria

A23.1 *General.*

(a) The design load criteria in this appendix are an approved equivalent of those in §§ 23.321 through 23.459 of this subchapter for an airplane having a maximum weight of 6,000 pounds or less and the following configuration:

(1) A single engine excluding turbine powerplants;

(2) A main wing located closer to the airplane's center of gravity than to the aft, fuselage-mounted, empennage;

(3) A main wing that contains a quarter-chord sweep angle of not more than 15 degrees fore or aft;

(4) A main wing that is equipped with trailing-edge controls (ailerons or flaps, or both);

(5) A main wing aspect ratio not greater than 7;

(6) A horizontal tail aspect ratio not greater than 4;

(7) A horizontal tail volume coefficient not less than 0.34;

(8) A vertical tail aspect ratio not greater than 2;

(9) A vertical tail platform area not greater than 10 percent of the wing platform area; and

(10) Symmetrical airfoils must be used in both the horizontal and vertical tail designs.

(b) Appendix A criteria may not be used on any airplane configuration that contains any of the following design features:

(1) Canard, tandem-wing, close-coupled, or tailless arrangements of the lifting surfaces;

(2) Biplane or multiplane wing arrangements;

(3) T-tail, V-tail, or cruciform-tail (+) arrangements;

(4) Highly-swept wing platform (more than 15-degrees of sweep at the quarter-chord), delta planforms, or slatted lifting surfaces; or

(5) Winglets or other wing tip devices, or outboard fins.

A23.3 *Special symbols.*

n_1=Airplane Positive Maneuvering Limit Load Factor.

n_2=Airplane Negative Maneuvering Limit Load Factor.

n_3=Airplane Positive Gust Limit Load Factor at V_C.

n_4=Airplane Negative Gust Limit Load Factor at V_C.

n_{flap}=Airplane Positive Limit Load Factor With Flaps Fully Extended at V_F.

[1] $V_{F\min}$ = Minimum Design Flap Speed = $11.0\sqrt{n_1 W / S}$ [kts]

[2] $V_{A\min}$ = Minimum Design Maneuvering Speed = $15.0\sqrt{n_1 W / S}$ [kts]

[3] $V_{C\min}$ = Minimum Design Cruising Speed = $17.0\sqrt{n_1 W / S}$ [kts]

[4] $V_{D\min}$ = Minimum Design Dive Speed = $24.0\sqrt{n_1 W / S}$ [kts]

A23.5 *Certification in more than one category.*

The criteria in this appendix may be used for certification in the normal, utility, and acrobatic categories, or in any combination of these categories. If certification in more than one category is desired, the design category weights must be selected to make the term $n_1 W$ constant for all categories or greater for one desired category than for others. The wings and control surfaces (including wing flaps and tabs) need only be investigated for the maximum value of n_1W, or for the category corresponding to the maximum design weight, where n_1W is constant. If the acrobatic category is selected, a special unsymmetrical flight load investigation in accordance with paragraphs A23.9(c)(2) and A23.11(c)(2) of this appendix must be completed. The wing, wing carrythrough, and the horizontal tail structures must be checked for this condition. The basic fuselage structure need only be investigated for the highest load factor design category selected. The local supporting structure for dead weight items need only be designed for the highest load factor imposed when the particular items are installed in the airplane. The engine mount, however, must be designed for a higher side load factor, if certification in the acrobatic category is desired, than that required for certification in the normal and utility categories. When designing for landing loads, the landing gear and the airplane as a whole need only be investigated for the category corresponding to the maximum design weight. These simplifications apply to single-engine aircraft of conventional types for which experience is available, and the Administrator may require additional investigations for aircraft with unusual design features.

A23.7 *Flight loads.*

(a) Each flight load may be considered independent of altitude and, except for the local supporting structure for dead weight items, only the maximum design weight conditions must be investigated.

(b) Table 1 and figures 3 and 4 of this appendix must be used to determine values of n_1, n_2, n_3, and n_4, corresponding to the maximum design weights in the desired categories.

(c) Figures 1 and 2 of this appendix must be used to determine values of n_3 and n_4 corresponding to the minimum flying weights in the desired categories, and, if these load factors are greater than the load factors at the design weight, the supporting structure for dead weight items must be substantiated for the resulting higher load factors.

(d) Each specified wing and tail loading is independent of the center of gravity range. The applicant, however, must select a c.g. range, and the basic fuselage structure must be investigated for the most adverse dead weight loading conditions for the c.g. range selected.

(e) The following loads and loading conditions are the minimums for which strength must be provided in the structure:

(1) *Airplane equilibrium.* The aerodynamic wing loads may be considered to act normal to the relative wind, and to have a magnitude of 1.05 times the airplane normal loads (as determined from paragraphs A23.9 (b) and (c) of this appendix) for the positive flight conditions and a magnitude equal to the airplane normal loads for the negative conditions. Each chordwise and normal component of this wing load must be considered.

(2) *Minimum design airspeeds.* The minimum design airspeeds may be chosen by the applicant except that they may not be less than the minimum speeds found by using figure 3 of this appendix. In addition, V_{Cmin} need not exceed values of 0.9 V_H actually obtained at sea level for the lowest design weight category for which certification is desired. In computing these minimum design airspeeds, n_1 may not be less than 3.8.

(3) *Flight load factor.* The limit flight load factors specified in Table 1 of this appendix represent the ratio of the aerodynamic force component (acting normal to the assumed longitudinal axis of the airplane) to the weight of the airplane. A positive flight load factor is an aerodynamic force acting upward, with respect to the airplane.

A23.9 *Flight conditions.*

(a) *General.* Each design condition in paragraphs (b) and (c) of this section must be used to assure sufficient strength for each condition of speed and load factor on or within the boundary of a *V–n* diagram for the airplane similar to the diagram in figure 4 of this appendix. This diagram must also be used to determine the airplane structural operating limitations as specified in §§ 23.1501(c) through 23.1513 and § 23.1519.

(b) *Symmetrical flight conditions.* The airplane must be designed for symmetrical flight conditions as follows:

(1) The airplane must be designed for at least the four basic flight conditions, "A", "D", "E", and "G" as noted on the flight envelope of figure 4 of this appendix. In addition, the following requirements apply:

(i) The design limit flight load factors corresponding to conditions "D" and "E" of figure 4 must be at least as great as those specified in Table 1 and figure 4 of this appendix, and the design speed for these conditions must be at least equal to the value of V_D found from figure 3 of this appendix.

(ii) For conditions "A" and "G" of figure 4, the load factors must correspond to those specified in Table 1 of this appendix, and the design speeds must be computed using these load factors with the maximum static lift coefficient C_{NA} determined by the applicant. However, in the absence of more precise computations, these latter conditions may be based on a value of C_{NA}=±1.35 and the design speed for condition "A" may be less than V_{Amin}.

(iii) Conditions "C" and "F" of figure 4 need only be investigated when n_3 W/S or n_4 W/S are greater than n_1 W/S or n_2 W/S of this appendix, respectively.

(2) If flaps or other high lift devices intended for use at the relatively low airspeed of approach, landing, and takeoff, are installed, the airplane must be designed for the two flight conditions corresponding to the values of limit flap-down factors specified in Table 1 of this appendix with the flaps fully extended at not less than the design flap speed V_{Fmin} from figure 3 of this appendix.

(c) *Unsymmetrical flight conditions.* Each affected structure must be designed for unsymmetrical loadings as follows:

(1) The aft fuselage-to-wing attachment must be designed for the critical vertical surface load determined in accordance with paragraph SA23.11(c)(1) and (2) of this appendix.

(2) The wing and wing carry-through structures must be designed for 100 percent of condition "A" loading on one side of the plane of symmetry and 70 percent on the opposite side for certification in the normal and utility categories, or 60 percent on the opposite side for certification in the acrobatic category.

(3) The wing and wing carry-through structures must be designed for the loads resulting from a combination of 75 percent of the positive maneuvering wing loading on both sides of the plane of symmetry and the maximum wing torsion resulting from aileron displacement. The effect of aileron displacement on wing torsion at V_C or V_A using the basic airfoil moment coefficient modified over the aileron portion of the span, must be computed as follows:

(i) *Cm=Cm* +0.01$^{\delta\mu}$ (up aileron side) wing basic airfoil.

(ii) *Cm=Cm* –0.01$^{\delta\mu}$ (down aileron side) wing basic airfoil, where$^{\delta\mu}$ is the up aileron deflection and$^{\delta}$ d is the down aileron deflection.

(4) Δ critical, which is the sum of$^{\delta\mu}$ +$^{\delta}$d must be computed as follows:

(i) Compute Δα and Δ B from the formulas:

$$\Delta_\alpha = \frac{V_A}{V_C} \times \Delta_\rho \text{ and}$$

$$\Delta_\delta = 0.5\frac{V_A}{V_D} \times \Delta_\rho$$

Where Δ p=the maximum total deflection (sum of both aileron deflections) at V_A with V_A, V_C, and V_D described in subparagraph (2) of § 23.7(e) of this appendix.

(ii) Compute *K* from the formula:

$$K = \frac{(C_m - 0.01\delta_b)V_{D^2}}{(C_m - 0.01\delta_a)V_{C^2}}$$

where$^{\delta\alpha}$ is the down aileron deflection corresponding to$^{\Delta\alpha}$, and$^{\delta}$ b is the down aileron deflection corresponding to$^{\Delta}$ b as computed in step (i).

(iii) If *K* is less than 1.0,$^{\Delta\alpha}$ is Δ critical and must be used to determine$^{\delta}$u and$^{\delta}$*d*. In this case, V_C is the critical speed which must be used in computing the wing torsion loads over the aileron span.

(iv) If *K* is equal to or greater than 1.0,$^{\Delta}$b is Δ critical and must be used to determine$^{\delta}$u and$^{\delta}$d. In this case, V_d is the critical speed which must be used in computing the wing torsion loads over the aileron span.

(d) *Supplementary conditions; rear lift truss; engine torque; side load on engine mount.* Each of the following supplementary conditions must be investigated:

(1) In designing the rear lift truss, the special condition specified in § 23.369 may be investigated instead of condition "G" of figure 4 of this appendix. If this is done, and if certification in more than one category is desired, the value of *W/S* used in the formula appearing in § 23.369 must be that for the category corresponding to the maximum gross weight.

(2) Each engine mount and its supporting structures must be designed for the maximum limit torque corresponding to METO power and propeller speed acting simultaneously with the limit loads resulting from the maximum positive maneuvering flight load factor n_1. The limit torque must be obtained by multiplying the mean torque by a factor of 1.33 for engines with five or more cylinders. For 4, 3, and 2 cylinder engines, the factor must be 2, 3, and 4, respectively.

(3) Each engine mount and its supporting structure must be designed for the loads resulting from a lateral limit load factor of not less than 1.47 for the normal and utility categories, or 2.0 for the acrobatic category.

A23.11 *Control surface loads.*

(a) *General.* Each control surface load must be determined using the criteria of paragraph (b) of this section and must lie within the simplified loadings of paragraph (c) of this section.

(b) *Limit pilot forces.* In each control surface loading condition described in paragraphs (c) through (e) of this section, the airloads on the movable surfaces and the corresponding deflections need not exceed those which could be obtained in flight by employing the maximum limit pilot forces specified in the table in § 23.397(b). If the surface loads are limited by these maximum limit pilot forces, the tabs must either be considered to be deflected to their maximum travel in the direction which would assist the pilot or the deflection must correspond to the maximum degree of "out of trim" expected at the speed for the condition under consideration. The tab load, however, need not exceed the value specified in Table 2 of this appendix.

(c) *Surface loading conditions.* Each surface loading condition must be investigated as follows:

(1) Simplified limit surface loadings for the horizontal tail, vertical tail, aileron, wing flaps, and trim tabs are specified in figures 5 and 6 of this appendix.

(i) The distribution of load along the span of the surface, irrespective of the chordwise load distribution, must be assumed proportional to the total chord, except on horn balance surfaces.

(ii) The load on the stabilizer and elevator, and the load on fin and rudder, must be distributed chordwise as shown in figure 7 of this appendix.

(iii) In order to ensure adequate torsional strength and to account for maneuvers and gusts, the most severe loads must be considered in association with every center of pressure position between the leading edge and the half chord of the mean chord of the surface (stabilizer and elevator, or fin and rudder).

(iv) To ensure adequate strength under high leading edge loads, the most severe stabilizer and fin loads must be further considered as being increased by 50 percent over the leading 10 percent of the chord with the loads aft of this appropriately decreased to retain the same total load.

(v) The most severe elevator and rudder loads should be further considered as being distributed parabolically from three times the mean loading of the surface (stabilizer and elevator, or fin and rudder) at the leading edge of the elevator and rudder, respectively, to zero at the trailing edge according to the equation:

$$P(x) = 3(\overline{w})\frac{(c-x)^2}{cf^2}$$

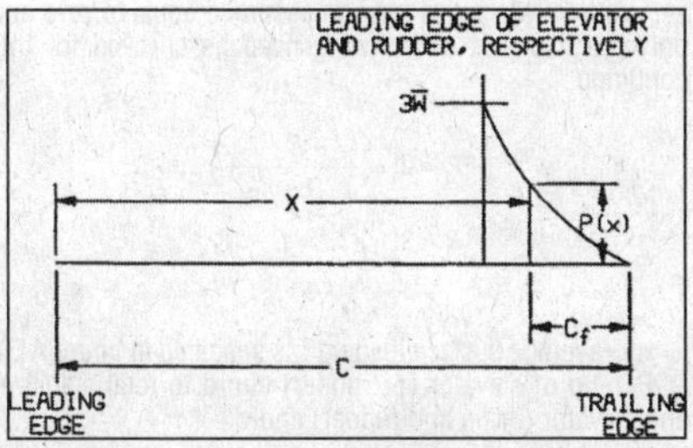

Where—

$P(x)$=local pressure at the chordwise stations x,

c=chord length of the tail surface,

c_f=chord length of the elevator and rudder respectively, and

$\bar{w}$=average surface loading as specified in Figure A5.

(vi) The chordwise loading distribution for ailerons, wing flaps, and trim tabs are specified in Table 2 of this appendix.

(2) If certification in the acrobatic category is desired, the horizontal tail must be investigated for an unsymmetrical load of 100 percent w on one side of the airplane centerline and 50 percent on the other side of the airplane centerline.

(d) *Outboard fins.* Outboard fins must meet the requirements of § 23.445.

(e) *Special devices.* Special devices must meet the requirements of § 23.459.

A23.13 *Control system loads.*

(a) *Primary flight controls and systems.* Each primary flight control and system must be designed as follows:

(1) The flight control system and its supporting structure must be designed for loads corresponding to 125 percent of the computed hinge moments of the movable control surface in the conditions prescribed in A23.11 of this appendix. In addition—

(i) The system limit loads need not exceed those that could be produced by the pilot and automatic devices operating the controls; and

(ii) The design must provide a rugged system for service use, including jamming, ground gusts, taxiing downwind, control inertia, and friction.

(2) Acceptable maximum and minimum limit pilot forces for elevator, aileron, and rudder controls are shown in the table in § 23.397(b). These pilots loads must be assumed to act at the appropriate control grips or pads as they would under flight conditions, and to be reacted at the attachments of the control system to the control surface horn.

(b) *Dual controls.* If there are dual controls, the systems must be designed for pilots operating in opposition, using individual pilot loads equal to 75 percent of those obtained in accordance with paragraph (a) of this section, except that individual pilot loads may not be less than the minimum limit pilot forces shown in the table in § 23.397(b).

(c) *Ground gust conditions.* Ground gust conditions must meet the requirements of § 23.415.

(d) *Secondary controls and systems.* Secondary controls and systems must meet the requirements of § 23.405.

Table 1
Limit Flight Load Factors

[Limit flight load factors]

Flight load factors	Normal category	Utility category	Acrobatic category
Flaps up:			
n_1	3.8	4.4	6.0
n_2	$-0.5\ n_1$		
n_3	(1)		
n_4	(2)		
Flaps down:			
n flap	$0.5\ n_1$		
n flap	[3]Zero		

[1] Find n_3 from Fig. 1

[2] Find n_4 from Fig. 2

[3] Vertical wing load may be assumed equal to zero and only the flap part of the wing need be checked for this condition.

$$P_1 = 2(\bar{w})\frac{(2 - E - 3d')}{(1 - E)}$$

$$P_2 = 2(\bar{w})(3d' + E - 1)$$

where:

$\bar{w}$=average surface loading (as specified in figure A.5)

E=ratio of elevator (or rudder) chord to total stabilizer and elevator (or fin and rudder) chord.

Table 2 - Average limit control surface loading

AVERAGE LIMIT CONTROL SURFACE LOADING			
SURFACE	DIRECTION OF LOADING	MAGNITUDE OF LOADING	CHORDWISE DISTRIBUTION
Horizontal Tail I	a) Up and Down	Figure A5 Curve (2)	See Figure A7
	b) Unsymmetrical Loading (Up and Down)	100% $\bar{w}$ on one side of airplane ℄ 65% $\bar{w}$ on other side of airplane ℄ for normal and utility categories. For acrobatic category see A23.11(c)	
Vertical Tail II	Right and Left	Figure A5 Curve (1)	Same as above
Aileron III	a) Up and Down	Figure A6 Curve (5)	(C) $\bar{W}$ ℄ Hinge
Wing Flap IV	a) Up	Figure A6 Curve (4)	(D) $2\bar{W}$ $\bar{W}$
	b) Down	.25 x Up Load (a)	
Trim Tab V	a) Up and Down	Figure A6 Curve (3)	Same as (D) above

NOTE: The surface loading I, II, III, and V above are based on speeds V_A min and V_C min.

The loading of IV is based on V_F min.

If values of speed greater than these minimums are selected for design, the appropriate surface loadings must be multiplied by the ratio $\left(\frac{V_{selected}}{V_{minimum}}\right)^2$.

For conditions I, II, III, and V the multiplying factor used must be the higher of $\left(\frac{V_A\ sel.}{V_A\ min.}\right)^2$ or $\left(\frac{V_C\ sel.}{V_C\ min.}\right)^2$

FIGURE A1—CHART FOR FINDING n_3 FACTOR AT SPEED V_C

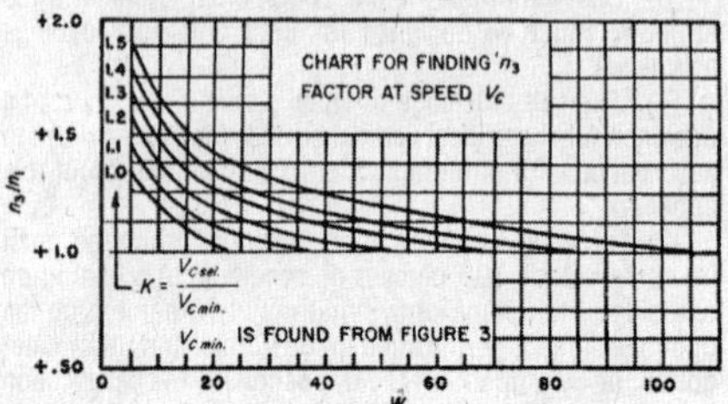

FIGURE A2 - CHART FOR FINDING n_4 FACTOR AT SPEED V_C

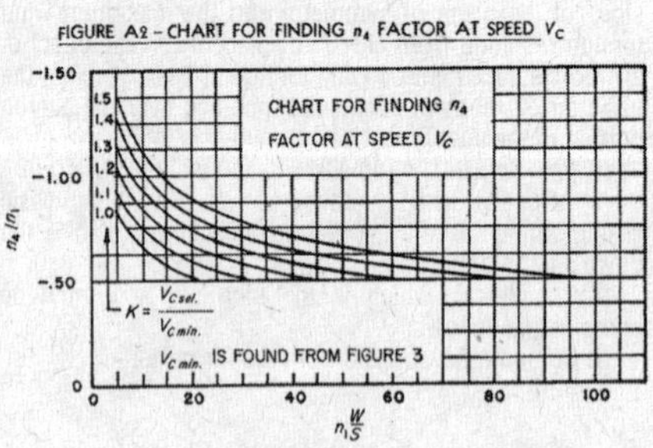

FIGURE A3—DETERMINATIONS OF MINIMUM DESIGN SPEEDS—EQUATIONS SPEEDS ARE IN KNOTS

$V_{D\,min} = 24.0\sqrt{n_1 \frac{W}{S}}$ but need not exceed $1.4\sqrt{\frac{n_1}{3.8}}V_{C\,min}$;

$V_{C\,min} = 17.0\sqrt{n_1 \frac{W}{S}}$ but need not exceed 0.9 V_H: $V_{A\,min} = 15.0\sqrt{n_1 \frac{W}{S}}$ but need not exceed V_C used in design. $V_{F\,min} = 11.0\sqrt{n_1 \frac{W}{S}}$

FIGURE A5 - AVERAGE LIMIT CONTROL SURFACE LOADING

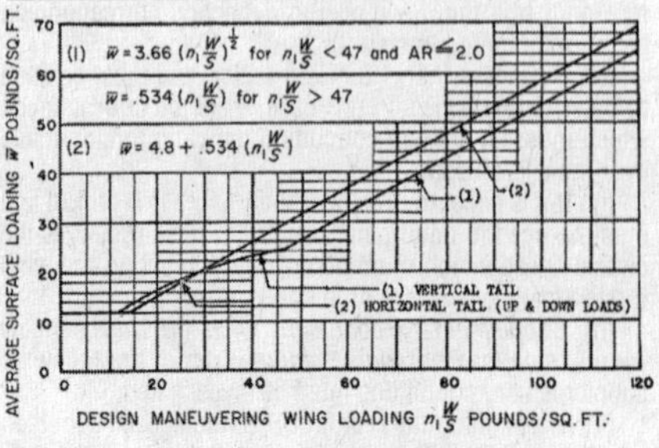

FIGURE A6 - AVERAGE LIMIT CONTROL SURFACE LOADING

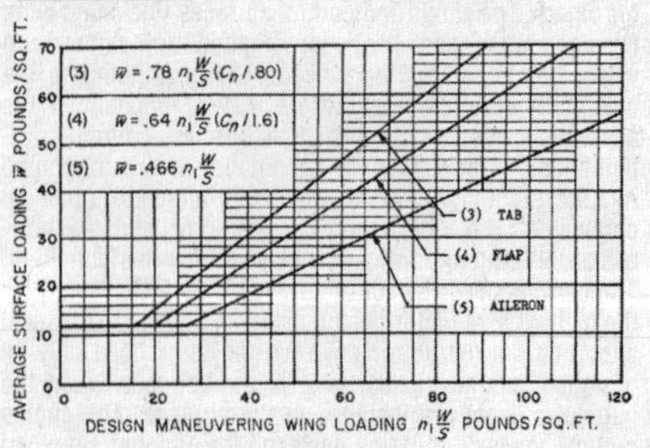

FIGURE A-4—FLIGHT ENVELOPE

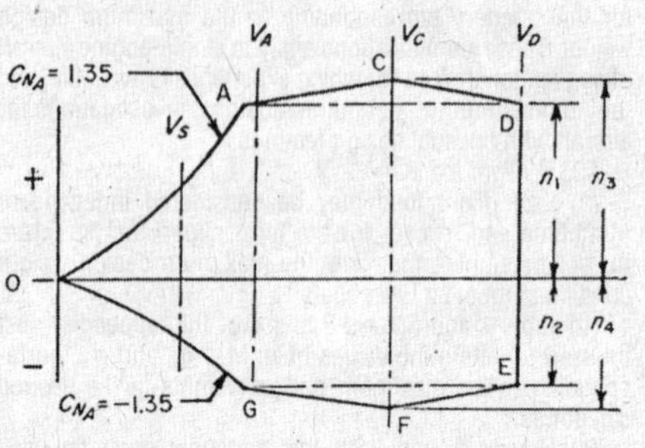

1. Conditions "C" or "F" need only be investigated when $n_3 \frac{W}{S}$ or $n_4 \frac{W}{S}$ is greater than $n_1 \frac{W}{S}$ W/w, respectively.

2. Condition "G" need not be investigated when the supplementary condition specified in § 23.369 is investigated

Figure A7
Chordwise Load Distribution for Stabilizer and Elevator or Fin and Rudder

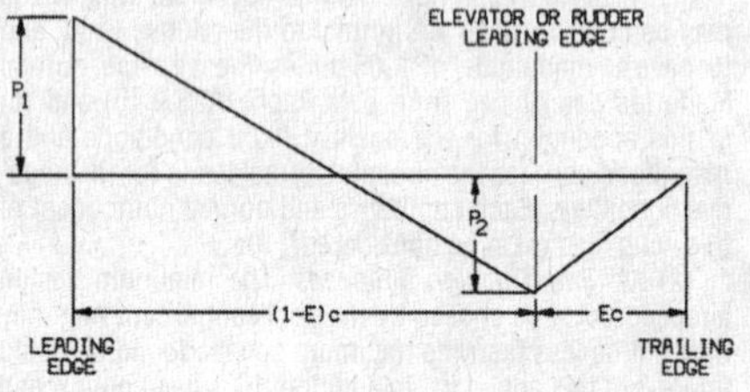

d'=ratio of distance of center of pressure of a unit spanwise length of combined stabilizer and elevator (or fin and rudder) measured from stabilizer (or fin) leading edge to the local chord. Sign convention is positive when center of pressure is behind leading edge.

c=local chord.

Note: Positive values of $\bar{w}$, P_1 and P_2 are all measured in the same direction.

[Doc. No. 4080, 29 FR 17955, Dec. 18, 1964, as amended by Amdt. 23-7, 34 FR 13097, Aug. 13, 1969; 34 FR 14727, Sept. 24, 1969; Amdt. 23-16, 40 FR 2577, Jan. 14, 1975; Amdt. 23-28, 47 FR 13315, Mar. 29, 1982; Amdt. 23-48, 61 FR 5149, Feb. 9, 1996]

Appendix B to Part 23 [Reserved]

Appendix C to Part 23—Basic Landing Conditions

[C23.1 *Basic landing conditions*]

Condition	Tail wheel type		Nose wheel type		
	Level landing	Tail-down landing	Level landing with inclined reactions	Level landing with nose wheel just clear of ground	Tail-down landing
Reference section	23.479(a)(1)	23.481(a)(1)	23.479(a)(2)(i)	23.479(a) (2)(ii)	23.481(a)(2) and (b).
Vertical component at c. g	*nW*	*nW*	*nW*	*nW*	*nW*.
Fore and aft component at c. g	*KnW*	0	*KnW*	*KnW*	0.
Lateral component in either direction at c. g	0	0	0	0	0.
Shock absorber extension (hydraulic shock absorber)	Note (2)	Note (2)	Note (2)	Note (2)	Note (2).
Shock absorber deflection (rubber or spring shock absorber), percent	100	100	100	100	100.
Tire deflection	Static	Static	Static	Static	Static.
Main wheel loads (both wheels) (*Vr*)	(*n-L*) *W*	(*n-L*) *W*b/d	(*n-L*) *W* a′/d′	(*n-L*) *W*	(*n-L*) *W*.
Main wheel loads (both wheels) (*Dr*)	*KnW*	0	*KnW* a′/d′	*KnW*	0.
Tail (nose) wheel loads (*Vf*)	0	(*n-L*) *W*a/d	(*n-L*) *W* b′/d′	0	0.
Tail (nose) wheel loads (*Df*)	0	0	*KnW* b′/d′	0	0.
Notes	(1), (3), and (4)	(4)	(1)	(1), (3), and (4)	(3) and (4).

Note (1). *K* may be determined as follows: K=0.25 for W=3,000 pounds or less; K=0.33 for W=6,000 pounds or greater, with linear variation of *K* between these weights.

Note (2). For the purpose of design, the maximum load factor is assumed to occur throughout the shock absorber stroke from 25 percent deflection to 100 percent deflection unless otherwise shown and the load factor must be used with whatever shock absorber extension is most critical for each element of the landing gear.

Note (3). Unbalanced moments must be balanced by a rational or conservative method.

Note (4). *L* is defined in §23.725(b).

Note (5). *n* is the limit inertia load factor, at the c.g. of the airplane, selected under §23.473 (d), (f), and (g).

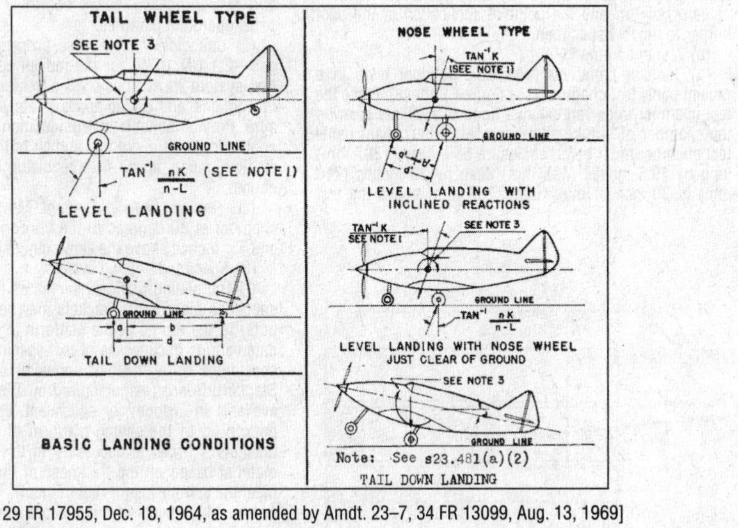

[Doc. No. 4080, 29 FR 17955, Dec. 18, 1964, as amended by Amdt. 23–7, 34 FR 13099, Aug. 13, 1969]

Appendix D to Part 23—Wheel Spin-Up and Spring-Back Loads

D23.1 *Wheel spin-up loads.*

(a) The following method for determining wheel spin-up loads for landing conditions is based on NACA T.N. 863. However, the drag component used for design may not be less than the drag load prescribed in § 23.479(b).

$F_{Hmax}=1/ r_e\sqrt{2I_w} (V_H — V_c) nF_{Vmax}/ t_s$

where—

F_{Hmax}=maximum rearward horizontal force acting on the wheel (in pounds);

r_e=effective rolling radius of wheel under impact based on recommended operating tire pressure (which may be assumed to be equal to the rolling radius under a static load of n_jW_e) in feet;

I_w=rotational mass moment of inertia of rolling assembly (in slug feet);

V_H=linear velocity of airplane parallel to ground at instant of contact (assumed to be 1.2 V_{S0}, in feet per second);

V_c=peripheral speed of tire, if prerotation is used (in feet per second) (there must be a positive means of pre-rotation before pre-rotation may be considered);

n=equals effective coefficient of friction (0.80 may be used);

F_{Vmax}=maximum vertical force on wheel (pounds)=n_jW_e, where W_e and n_j are defined in § 23.725;

t_s=time interval between ground contact and attainment of maximum vertical force on wheel (seconds). (However, if the value of F_{Vmax}, from the above equation exceeds 0.8 F_{Vmax}, the latter value must be used for F_{Hmax}.)

(b) The equation assumes a linear variation of load factor with time until the peak load is reached and under this assumption, the equation determines the drag force at the time that the wheel peripheral velocity at radius r_e equals the airplane velocity. Most shock absorbers do not exactly follow a linear variation of load factor with time. Therefore, rational or conservative allowances must be made to compensate for these variations. On most landing gears, the time for wheel spin-up will be less than the time required to develop maximum vertical load factor for the specified rate of descent and forward velocity. For exceptionally large wheels, a wheel peripheral velocity equal to the ground speed may not have been attained at the time of maximum vertical gear load. However, as stated above, the drag spin-up load need not exceed 0.8 of the maximum vertical loads.

(c) Dynamic spring-back of the landing gear and adjacent structure at the instant just after the wheels come up to speed may result in dynamic forward acting loads of considerable magnitude. This effect must be determined, in the level landing condition, by assuming that the wheel spin-up loads calculated by the methods of this appendix are reversed. Dynamic spring-back is likely to become critical for landing gear units having wheels of large mass or high landing speeds.

[Doc. No. 4080, 29 FR 17955, Dec. 18, 1964, as amended by Amdt. 23-45, 58 FR 42167, Aug. 6, 1993]

Appendix E to Part 23 [Reserved]

Appendix F to Part 23—Test Procedure

Part I—Acceptable Test Procedure for Self-Extinguishing Materials for Showing Compliance With §§ 23.853, 23.855, and 23.1359

Acceptable test procedure for self-extinguishing materials for showing compliance with §§ 23.853, 23.855 and 23.1359.

(a) *Conditioning.* Specimens must be conditioned to 70 degrees F, plus or minus 5 degrees, and at 50 percent plus or minus 5 percent relative humidity until moisture equilibrium is reached or for 24 hours. Only one specimen at a time may be removed from the conditioning environment immediately before subjecting it to the flame.

(b) *Specimen configuration.* Except as provided for materials used in electrical wire and cable insulation and in small parts, materials must be tested either as a section cut from a fabricated part as installed in the airplane or as a specimen simulating a cut section, such as: a specimen cut from a flat sheet of the material or a model of the fabricated part. The specimen may be cut from any location in a fabricated part; however, fabricated units, such as sandwich panels, may not be separated for a test. The specimen thickness must be no thicker than the minimum thickness to be qualified for use in the airplane, except that: (1) Thick foam parts, such as seat cushions, must be tested in ½ inch thickness; (2) when showing compliance with § 23.853(d)(3)(v) for materials used in small parts that must be tested, the materials must be tested in no more than ⅛ inch thickness; (3) when showing compliance with § 23.1359(c) for materials used in electrical wire and cable insulation, the wire and cable specimens must be the same size as used in the airplane. In the case of fabrics, both the warp and fill direction of the weave must be tested to determine the most critical flammability conditions. When performing the tests prescribed in paragraphs (d) and (e) of this appendix, the specimen must be mounted in a metal frame so that (1) in the vertical tests of paragraph (d) of this appendix, the two long edges and the upper edge are held securely; (2) in the horizontal test of paragraph (e) of this appendix, the two long edges and the edge away from the flame are held securely; (3) the exposed area of the specimen is at least 2 inches wide and 12 inches long, unless the actual size used in the airplane is smaller; and (4) the edge to which the burner flame is applied must not consist of the finished or protected edge of the specimen but must be representative of the actual cross section of the material or part installed in the airplane. When performing the test prescribed in paragraph (f) of this appendix, the specimen must be mounted in metal frame so that all four edges are held securely and the exposed area of the specimen is at least 8 inches by 8 inches.

(c) *Apparatus.* Except as provided in paragraph (g) of this appendix, tests must be conducted in a draft-free cabinet in accordance with Federal Test Method Standard 191 Method 5903 (revised Method 5902) which is available from the General Services Administration, Business Service Center, Region 3, Seventh and D Streets SW., Washington, D.C. 20407, or with some other approved equivalent method. Specimens which are too large for the cabinet must be tested in similar draft-free conditions.

(d) *Vertical test.* A minimum of three specimens must be tested and the results averaged. For fabrics, the direction of weave corresponding to the most critical flammability conditions must be parallel to the longest dimension. Each specimen must be supported vertically. The specimen must be exposed to a Bunsen or Tirrill burner with a nominal $^3/_8$-inch I.D. tube adjusted to give a flame of $1\frac{1}{2}$ inches in height. The minimum flame temperature measured by a calibrated thermocouple pryometer in the center of the flame must be 1550°F. The lower edge of the specimen must be three-fourths inch above the top edge of the burner. The flame must be applied to the center line of the lower edge of the specimen. For materials covered by §§ 23.853(d)(3)(i) and 23.853(f), the flame must be applied for 60 seconds and then removed. For materials covered by § 23.853(d)(3)(ii), the flame must be applied for 12 seconds and then removed. Flame time, burn length, and flaming time of drippings, if any, must be recorded. The burn length determined in accordance with paragraph (h) of this appendix must be measured to the nearest one-tenth inch.

(e) *Horizontal test.* A minimum of three specimens must be tested and the results averaged. Each specimen must be supported horizontally. The exposed surface when installed in the airplane must be face down for the test. The specimen must be exposed to a Bunsen burner or Tirrill burner with a nominal $^3/_8$-inch I.D. tube adjusted to give a flame of $1\frac{1}{2}$ inches in height. The minimum flame temperature measured by a calibrated thermocouple pyrometer in the center of the flame must be 1550°F. The specimen must be positioned so that the edge being tested is three-fourths of an inch above the top of, and on the center line of, the burner. The flame must be applied for 15 seconds and then removed. A minimum of 10 inches of the specimen must be used for timing purposes, approximately $1\frac{1}{2}$ inches must burn before the burning front reaches the timing zone, and the average burn rate must be recorded.

(f) *Forty-five degree test.* A minimum of three specimens must be tested and the results averaged. The specimens must be supported at an angle of 45 degrees to a horizontal surface. The exposed surface when installed in the aircraft must be face down for the test. The specimens must be exposed to a Bunsen or Tirrill burner with a nominal $^3/_8$ inch I.D. tube adjusted to give a flame of $1\frac{1}{2}$ inches in height. The minimum flame temperature measured by a calibrated thermocouple pyrometer in the center of the flame must be 1550°F. Suitable precautions must be taken to avoid drafts. The flame must be applied for 30 seconds with one-third contacting the material at the center of the specimen and then removed. Flame time, glow time, and whether the flame penetrates (passes through) the specimen must be recorded.

(g) *Sixty-degree test.* A minimum of three specimens of each wire specification (make and size) must be tested. The specimen of wire or cable (including insulation) must be placed at an angle of 60 degrees with the horizontal in the cabinet specified in paragraph (c) of this appendix, with the cabinet door open during the test or placed within a chamber approximately 2 feet high × 1 foot × 1 foot, open at the top and at one vertical side (front), that allows sufficient flow of air for complete combustion but is free from drafts. The specimen must be parallel to and approximately 6 inches from the front of the chamber. The lower end of the specimen must be held rigidly clamped. The upper end of the specimen must pass over a pulley or rod and must have an appropriate weight attached to it so that the specimen is held tautly throughout the flammability test. The test specimen span between lower clamp and upper pulley or rod must be 24 inches and must be marked 8 inches from the lower end to indicate the central point for flame application. A flame from a Bunsen or Tirrill burner must be applied for 30 seconds at the test mark. The burner must be mounted underneath the test mark on the specimen, perpendicular to the specimen and at an angle of 30 degrees to the vertical plane of the specimen. The burner must have a nominal bore of three-eighths inch, and must be adjusted to provide a three-inch-high flame with an inner cone approximately one-third of the flame height. The minimum temperature of the hottest portion of the flame, as measured with a calibrated thermocouple pyrometer, may not be less than 1,750°F. The burner must be positioned so that the hottest portion of the flame is applied to the test mark on the wire. Flame time, burn length, and flaming time drippings, if any, must be recorded. The burn length determined in accordance with paragraph (h) of this appendix must be measured to the nearest one-tenth inch. Breaking of the wire specimen is not considered a failure.

(h) *Burn length.* Burn length is the distance from the original edge to the farthest evidence of damage to the test specimen due to flame impingement, including areas of partial or complete consumption, charring, or embrittlement, but not including areas sooted, stained, warped, or discolored, nor areas where material has shrunk or melted away from the heat source.

Part II—Test Method To Determine the Flammability and Flame Propagation Characteristics of Thermal/Acoustic Insulation Materials

Use this test method to evaluate the flammability and flame propagation characteristics of thermal/acoustic insulation when exposed to both a radiant heat source and a flame.

(a) *Definitions.*

Flame propagation means the furthest distance of the propagation of visible flame towards the far end of the test specimen, measured from the midpoint of the ignition source flame. Measure this distance after initially applying the ignition source and before all flame on the test specimen is extinguished. The measurement is not a determination of burn length made after the test.

Radiant heat source means an electric or air propane panel.

Thermal/acoustic insulation means a material or system of materials used to provide thermal and/or acoustic protection. Examples include fiberglass or other batting material encapsulated by a film covering and foams.

Zero point means the point of application of the pilot burner to the test specimen.

(b) *Test apparatus.*

(1) *Radiant panel test chamber.* Conduct tests in a radiant panel test chamber (see figure F1 above). Place the test chamber under an exhaust hood to facilitate clearing the chamber of smoke after each test. The radiant panel test chamber must be an enclosure 55 inches (1397 mm) long by 19.5 inches (495 mm) deep by 28 inches (710 mm) to 30 inches (maximum) (762 mm) above the test

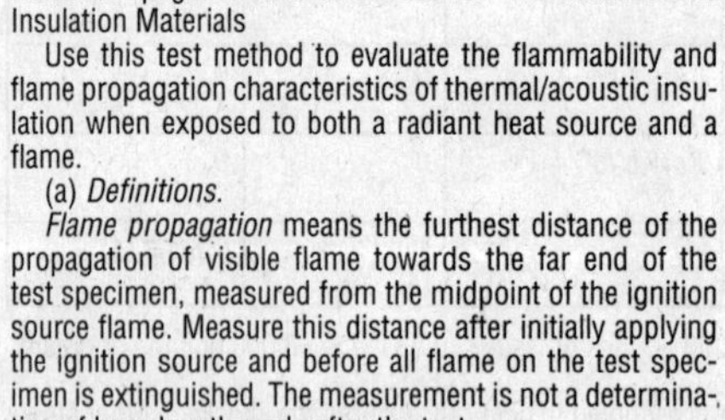

Figure 1 - Radiant Panel Test Chamber

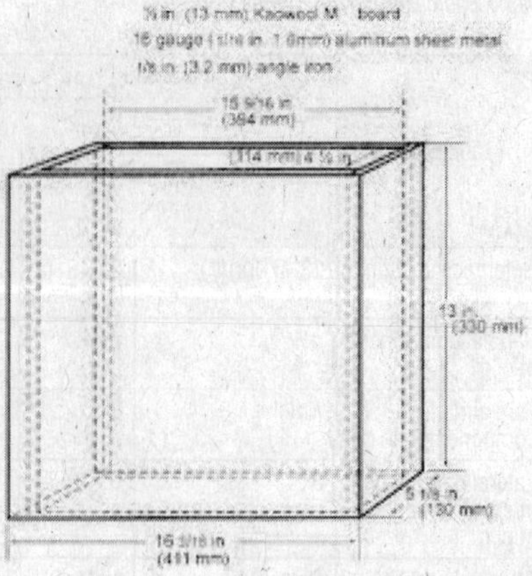

Figure 2 - Internal Chimney

specimen. Insulate the sides, ends, and top with a fibrous ceramic insulation, such as Kaowool MTM board. On the front side, provide a 52 by 12-inch (1321 by 305 mm) draft-free, high-temperature, glass window for viewing the sample during testing. Place a door below the window to provide access to the movable specimen platform holder. The bottom of the test chamber must be a sliding steel platform that has provision for securing the test specimen holder in a fixed and level position. The chamber must have an internal chimney with exterior dimensions of 5.1 inches (129 mm) wide, by 16.2 inches (411 mm) deep by 13 inches (330 mm) high at the opposite end of the chamber from the radiant energy source. The interior dimensions must be 4.5 inches (114 mm) wide by 15.6 inches (395 mm) deep. The chimney must extend to the top of the chamber (see figure F2).

(2) *Radiant heat source.* Mount the radiant heat energy source in a cast iron frame or equivalent. An electric panel must have six, 3-inch wide emitter strips. The emitter strips must be perpendicular to the length of the panel. The panel must have a radiation surface of $12\frac{7}{8}$ by $18\frac{1}{2}$ inches (327 by 470 mm). The panel must be capable of operating at temperatures up to 1300°F (704°C). An air propane panel must be made of a porous refractory material and have a radiation surface of 12 by 18 inches (305 by 457 mm). The panel must be capable of operating at temperatures up to 1,500°F (816°C). See figures F3a and F3b.

(i) *Electric radiant panel.* The radiant panel must be 3-phase and operate at 208 volts. A single-phase, 240 volt panel is also acceptable. Use a solid-state power controller and microprocessor-based controller to set the electric panel operating parameters.

(ii) *Gas radiant panel.* Use propane (liquid petroleum gas—2.1 UN 1075) for the radiant panel fuel. The panel fuel system must consist of a venturi-type aspirator for mixing gas and air at approximately atmospheric pressure. Provide suitable instrumentation for monitoring and controlling the flow of fuel and air to the panel. Include an air flow gauge, an air flow regulator, and a gas pressure gauge.

(iii) *Radiant panel placement.* Mount the panel in the chamber at 30 degrees to the horizontal specimen plane, and $7\frac{1}{2}$ inches above the zero point of the specimen.

(3) *Specimen holding system.*

(i) The sliding platform serves as the housing for test specimen placement. Brackets may be attached (via wing nuts) to the top lip of the platform in order to accommodate various thicknesses of test specimens. Place the test specimens on a sheet of Kaowool MTM board or 1260 Standard Board (manufactured by Thermal Ceramics and available in Europe), or equivalent, either resting on the bottom lip of the sliding platform or on the base of the brackets. It may be necessary to use multiple sheets of material based on the thickness of the test specimen (to meet the sample height requirement). Typically, these non-combustible sheets of material are available in $\frac{1}{4}$-inch (6 mm) thicknesses. See figure F4. A sliding platform that is deeper than the 2-inch (50.8mm) platform shown in figure F4 is also acceptable as long as the sample height requirement is met.

(ii) Attach a $\frac{1}{2}$-inch (13 mm) piece of Kaowool MTM board or other high temperature material measuring $41\frac{1}{2}$

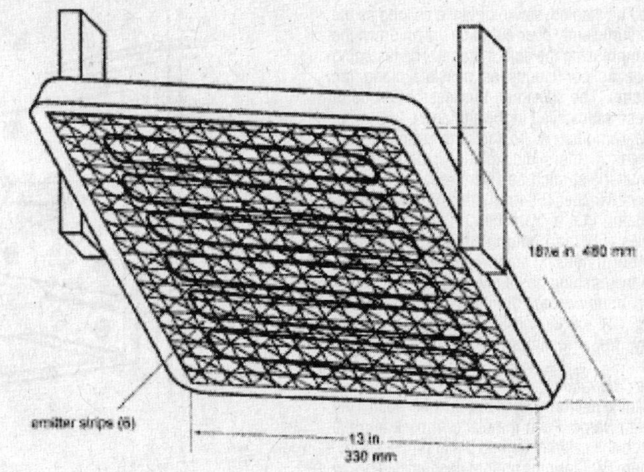

FIGURE F3a—Electric Panel

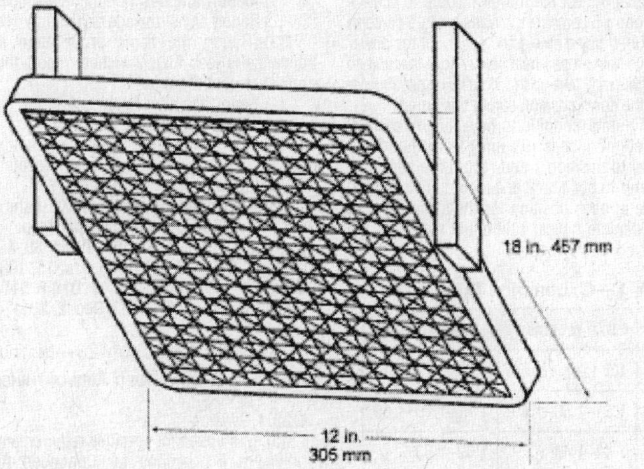

FIGURE F3b—Air Propane Radiant Panel

FIGURE F4—Sliding Platform

by $8\frac{1}{4}$ inches (1054 by 210 mm) to the back of the platform. This board serves as a heat retainer and protects the test specimen from excessive preheating. The height of this board may not impede the sliding platform movement (in and out of the test chamber). If the platform has been fabricated such that the back side of the platform is high enough to prevent excess preheating of the specimen when the sliding platform is out, a retainer board is not necessary.

(iii) Place the test specimen horizontally on the non-combustible board(s). Place a steel retaining/securing frame fabricated of mild steel, having a thickness of $\frac{1}{8}$-inch (3.2 mm) and overall dimensions of 23 by $13\frac{1}{8}$ inches (584 by 333 mm) with a specimen opening of 19

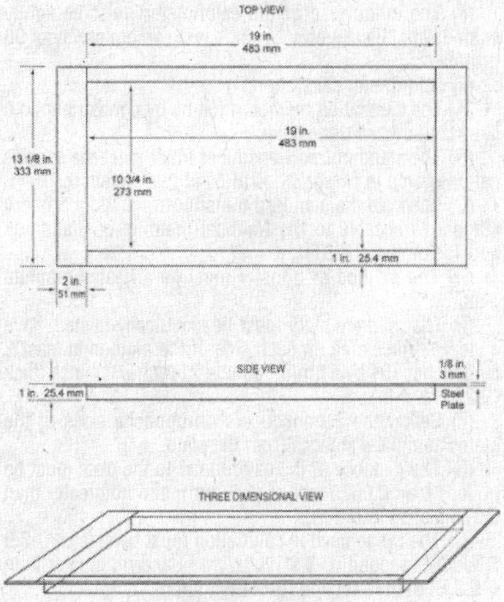

FIGURE F5: 3 Views

by $10\frac{3}{4}$ inches (483 by 273 mm) over the test specimen. The front, back, and right portions of the top flange of the frame must rest on the top of the sliding platform, and the bottom flanges must pinch all 4 sides of the test specimen. The right bottom flange must be flush with the sliding platform. See figure F5.

(4) *Pilot Burner.* The pilot burner used to ignite the specimen must be a BernzomaticTM commercial propane venturi torch with an axially symmetric burner tip and a propane supply tube with an orifice diameter of 0.006 inches (0.15 mm). The length of the burner tube must be $2\frac{7}{8}$ inches (71 mm). The propane flow must be adjusted via gas pressure through an in-line regulator to produce a blue inner cone length of $\frac{3}{4}$-inch (19 mm). A $\frac{3}{4}$-inch (19 mm) guide (such as a thin strip of metal) may be soldered to the top of the burner to aid in setting the flame height. The overall flame length must be approximately 5 inches long (127 mm). Provide a way to move the burner out of the ignition position so that the flame is horizontal and at least 2 inches (50 mm) above the specimen plane. See figure F6.

(5) *Thermocouples.* Install a 24 American Wire Gauge (AWG) Type K (Chromel- Alumel) thermocouple in the test chamber for temperature monitoring. Insert it into the chamber through a small hole drilled through the back of the chamber. Place the thermocouple so that it extends 11 inches (279 mm) out from the back of the chamber wall, $11\frac{1}{2}$ inches (292 mm) from the right side of the chamber wall, and is 2 inches (51 mm) below the radiant panel. The use of other thermocouples is optional.

(6) *Calorimeter.* The calorimeter must be a one-inch cylindrical water-cooled, total heat flux density, foil type Gardon Gage that has a range of 0 to 5 BTU/ft ^{2}-second (0 to 5.7 Watts/cm ^{2}).

(7) *Calorimeter calibration specification and procedure.*

(i) Calorimeter specification.

(A) Foil diameter must be 0.25 ±0.005 inches (6.35 ±0.13 mm).

(B) Foil thickness must be 0.0005 ±0.0001 inches (0.013 ± 0.0025 mm).

(C) Foil material must be thermocouple grade Constantan.

(D) Temperature measurement must be a Copper Constantan thermocouple.

(E) The copper center wire diameter must be 0.0005 inches (0.013 mm).

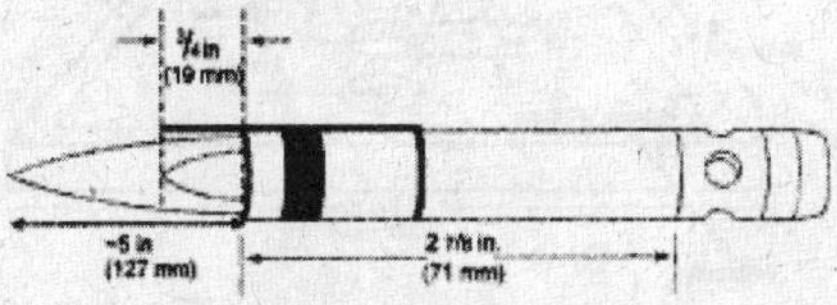

FIGURE F6—Propane Pilot Burner

(F) The entire face of the calorimeter must be lightly coated with "Black Velvet" paint having an emissivity of 96 or greater.

(ii) Calorimeter calibration.

(A) The calibration method must be by comparison to a like standardized transducer.

(B) The standardized transducer must meet the specifications given in paragraph II(b)(6) of this appendix.

(C) Calibrate the standard transducer against a primary standard traceable to the National Institute of Standards and Technology (NIST).

(D) The method of transfer must be a heated graphite plate.

(E) The graphite plate must be electrically heated, have a clear surface area on each side of the plate of at least 2 by 2 inches (51 by 51 mm), and be $\frac{1}{8}$ -inch ± $\frac{1}{16}$ -inch thick (3.2 ± 1.6 mm).

(F) Center the 2 transducers on opposite sides of the plates at equal distances from the plate.

(G) The distance of the calorimeter to the plate must be no less than 0.0625 inches (1.6 mm), and no greater than 0.375 inches (9.5 mm).

(H) The range used in calibration must be at least 0-3.5 BTUs/ft ^{2} -second (0-3.9 Watts/cm ^{2}) and no greater than 0-5.7 BTUs/ft ^{2} -second (0-6.4 Watts/cm ^{2}).

(I) The recording device used must record the 2 transducers simultaneously or at least within $\frac{1}{10}$ of each other.

(8) *Calorimeter fixture.* With the sliding platform pulled out of the chamber, install the calorimeter holding frame and place a sheet of non-combustible material in the bottom of the sliding platform adjacent to the holding frame. This will prevent heat losses during calibration. The frame must be 13$\frac{1}{8}$ inches (333 mm) deep (front to back) by 8 inches (203 mm) wide and must rest on the top of the sliding platform. It must be fabricated of $\frac{1}{8}$ -inch (3.2 mm) flat stock steel and have an opening that accommodates a $\frac{1}{2}$ -inch (12.7 mm) thick piece of refractory board, which is level with the top of the sliding platform. The board must have three 1-inch (25.4 mm) diameter holes drilled through the board for calorimeter insertion. The distance to the radiant panel surface from the centerline of the first hole ("zero" position) must be 7$\frac{1}{2}$ ± $\frac{1}{8}$ -inches (191 ± 3 mm). The distance between the centerline of the first hole to the centerline of the second hole must be 2 inches (51 mm). It must also be the same distance from the centerline of the second hole to the centerline of the third hole. See figure F7. A calorimeter holding frame that differs in construction is acceptable as long as the height from the centerline of the first hole to the radiant panel and the distance between holes is the same as described in this paragraph.

(9) *Instrumentation.* Provide a calibrated recording device with an appropriate range or a computerized data acquisition system to measure and record the outputs of the calorimeter and the thermocouple. The data acquisition system must be capable of recording the calorimeter output every second during calibration.

(10) *Timing device.* Provide a stopwatch or other device, accurate to ± 1 second/hour, to measure the time of application of the pilot burner flame.

(c) *Test specimens.*

(1) *Specimen preparation.* Prepare and test a minimum of three test specimens. If an oriented film cover material is used, prepare and test both the warp and fill directions.

(2) *Construction.* Test specimens must include all materials used in construction of the insulation (including batting, film, scrim, tape, *etc.*). Cut a piece of core material such as foam or fiberglass, and cut a piece of film cover material (if used) large enough to cover the core material. Heat sealing is the preferred method of preparing fiberglass samples, since they can be made without compressing the fiberglass ("box sample"). Cover materials that are not heat sealable may be stapled, sewn, or taped as long as the cover material is sufficiently over-cut to be drawn down the sides without compressing the core material. The fastening means should be as continuous as possible along the length of the seams. The specimen thickness must be of the same thickness as installed in the airplane.

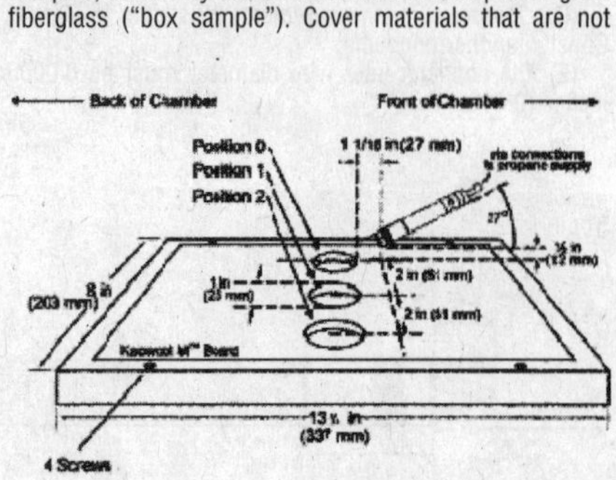

FIGURE F7—Calorimeter Holding Frame

(3) *Specimen Dimensions.* To facilitate proper placement of specimens in the sliding platform housing, cut non-rigid core materials, such as fiberglass, 12$\frac{1}{2}$ inches (318mm) wide by 23 inches (584mm) long. Cut rigid materials, such as foam, 11$\frac{1}{2}$ ± $\frac{1}{4}$ inches (292 mm ± 6mm) wide by 23 inches (584mm) long in order to fit properly in the sliding platform housing and provide a flat, exposed surface equal to the opening in the housing.

(d) *Specimen conditioning.* Condition the test specimens at 70 ± 5°F (21 ± 2°C) and 55 percent ± 10 percent relative humidity, for a minimum of 24 hours prior to testing.

(e) *Apparatus Calibration.*

(1) With the sliding platform out of the chamber, install the calorimeter holding frame. Push the platform back into the chamber and insert the calorimeter into the first hole ("zero" position). See figure F7. Close the bottom door located below the sliding platform. The distance from the centerline of the calorimeter to the radiant panel surface at this point must be 7$\frac{1}{2}$ inches ± $\frac{1}{8}$ (191 mm ± 3). Before igniting the radiant panel, ensure that the calorimeter face is clean and that there is water running through the calorimeter.

(2) Ignite the panel. Adjust the fuel/air mixture to achieve 1.5 BTUs/feet^2 -second ± 5 percent (1.7 Watts/cm^2 ± 5 percent) at the "zero" position. If using an electric panel, set the power controller to achieve the proper heat flux. Allow the unit to reach steady state (this may take up to 1 hour). The pilot burner must be off and in the down position during this time.

(3) After steady-state conditions have been reached, move the calorimeter 2 inches (51 mm) from the "zero" position (first hole) to position 1 and record the heat flux. Move the calorimeter to position 2 and record the heat flux. Allow enough time at each position for the calorimeter to stabilize. Table 1 depicts typical calibration values at the three positions.

Table 1—Calibration Table

Position	BTU/feet2sec	Watts/cm2
"Zero" Position	1.5	1.7
Position 1	1.51–1.50–1.49	1.71–1.70–1.69
Position 2	1.43–1.44	1.62–1.63

(4) Open the bottom door, remove the calorimeter and holder fixture. Use caution as the fixture is very hot.

(f) *Test Procedure.*

(1) Ignite the pilot burner. Ensure that it is at least 2 inches (51 mm) above the top of the platform. The burner may not contact the specimen until the test begins.

(2) Place the test specimen in the sliding platform holder. Ensure that the test sample surface is level with the top of the platform. At "zero" point, the specimen surface must be 7$\frac{1}{2}$ inches ± $\frac{1}{8}$ inch (191 mm ± 3) below the radiant panel.

(3) Place the retaining/securing frame over the test specimen. It may be necessary (due to compression) to adjust the sample (up or down) in order to maintain the distance from the sample to the radiant panel (7$\frac{1}{2}$ inches ± $\frac{1}{8}$ inch (191 mm ± 3) at "zero" position). With film/fiberglass assemblies, it is critical to make a slit in the film cover to purge any air inside. This allows the operator to maintain the proper test specimen position (level with the top of the platform) and to allow ventilation of gases during testing. A longitudinal slit, approximately 2 inches (51mm) in length, must be centered 3 inches ± $\frac{1}{2}$ inch (76mm ± 13mm) from the left flange of the securing frame. A utility knife is acceptable for slitting the film cover.

(4) Immediately push the sliding platform into the chamber and close the bottom door.

(5) Bring the pilot burner flame into contact with the center of the specimen at the "zero" point and simultaneously start the timer. The pilot burner must be at a 27 degree angle with the sample and be approximately $\frac{1}{2}$ inch (12 mm) above the sample. See figure F7. A stop, as shown in figure F8, allows the operator to position the burner correctly each time.

(6) Leave the burner in position for 15 seconds and then remove to a position at least 2 inches (51 mm) above the specimen.

(g) *Report.*

FIGURE F8—Propane Burner Stop

(1) Identify and describe the test specimen.

(2) Report any shrinkage or melting of the test specimen.

(3) Report the flame propagation distance. If this distance is less than 2 inches, report this as a pass (no measurement required).

(4) Report the after-flame time.

(h) *Requirements.*

(1) There must be no flame propagation beyond 2 inches (51 mm) to the left of the centerline of the pilot flame application.

(2) The flame time after removal of the pilot burner may not exceed 3 seconds on any specimen.

[Amdt. 23-23, 43 FR 50594, Oct. 30, 1978, as amended by Amdt. 23-34, 52 FR 1835, Jan. 15, 1987; 52 FR 34745, Sept. 14, 1987; Amdt. 23-49, 61 FR 5170, Feb. 9, 1996; Amdt. 23-62, 76 FR 75763, Dec. 2, 2011]

Appendix G to Part 23—Instructions for Continued Airworthiness

G23.1 *General.*

(a) This appendix specifies requirements for the preparation of Instructions for Continued Airworthiness as required by § 23.1529.

(b) The Instructions for Continued Airworthiness for each airplane must include the Instructions for Continued Airworthiness for each engine and propeller (hereinafter designated 'products'), for each appliance required by this chapter, and any required information relating to the interface of those appliances and products with the airplane. If Instructions for Continued Airworthiness are not supplied by the manufacturer of an appliance or product installed in the airplane, the Instructions for Continued Airworthiness for the airplane must include the information essential to the continued airworthiness of the airplane.

(c) The applicant must submit to the FAA a program to show how changes to the Instructions for Continued Airworthiness made by the applicant or by the manufacturers of products and appliances installed in the airplane will be distributed.

G23.2 *Format.*

(a) The Instructions for Continued Airworthiness must be in the form of a manual or manuals as appropriate for the quantity of data to be provided.

(b) The format of the manual or manuals must provide for a practical arrangement.

G23.3 *Content.*

The contents of the manual or manuals must be prepared in the English language. The Instructions for Continued Airworthiness must contain the following manuals or sections, as appropriate, and information:

(a) *Airplane maintenance manual or section.* (1) Introduction information that includes an explanation of the airplane's features and data to the extent necessary for maintenance or preventive maintenance.

(2) A description of the airplane and its systems and installations including its engines, propellers, and appliances.

(3) Basic control and operation information describing how the airplane components and systems are controlled and how they operate, including any special procedures and limitations that apply.

(4) Servicing information that covers details regarding servicing points, capacities of tanks, reservoirs, types of fluids to be used, pressures applicable to the various systems, location of access panels for inspection and servicing, locations of lubrication points, lubricants to be used, equipment required for servicing, tow instructions and limitations, mooring, jacking, and leveling information.

(b) *Maintenance instructions.* (1) Scheduling information for each part of the airplane and its engines, auxiliary power units, propellers, accessories, instruments, and equipment that provides the recommended periods at which they should be cleaned, inspected, adjusted, tested, and lubricated, and the degree of inspection, the applicable wear tolerances, and work recommended at these periods. However, the applicant may refer to an accessory, instrument, or equipment manufacturer as the source of this information if the applicant shows that the item has an exceptionally high degree of complexity requiring specialized maintenance techniques, test equipment, or expertise. The recommended overhaul periods and necessary cross reference to the Airworthiness Limitations section of the manual must also be included. In addition, the applicant must include an inspection program that includes the frequency and extent of the inspections necessary to provide for the continued airworthiness of the airplane.

(2) Troubleshooting information describing probable malfunctions, how to recognize those malfunctions, and the remedial action for those malfunctions.

(3) Information describing the order and method of removing and replacing products and parts with any necessary precautions to be taken.

(4) Other general procedural instructions including procedures for system testing during ground running, symmetry checks, weighing and determining the center of gravity, lifting and shoring, and storage limitations.

(c) Diagrams of structural access plates and information needed to gain access for inspections when access plates are not provided.

(d) Details for the application of special inspection techniques including radiographic and ultrasonic testing where such processes are specified.

(e) Information needed to apply protective treatments to the structure after inspection.

(f) All data relative to structural fasteners such as identification, discard recommendations, and torque values.

(g) A list of special tools needed.

(h) In addition, for commuter category airplanes, the following information must be furnished:

(1) Electrical loads applicable to the various systems;

(2) Methods of balancing control surfaces;

(3) Identification of primary and secondary structures; and

(4) Special repair methods applicable to the airplane.

G23.4 *Airworthiness Limitations section.*

The Instructions for Continued Airworthiness must contain a section titled Airworthiness Limitations that is segregated and clearly distinguishable from the rest of the document. This section must set forth each mandatory replacement time, structural inspection interval, and related structural inspection procedure required for type certification. If the Instructions for Continued Airworthiness consist of multiple documents, the section required by this paragraph must be included in the principal manual. This section must contain a legible statement in a prominent location that reads: "The Airworthiness Limitations section is FAA approved and specifies maintenance required under §§ 43.16 and 91.403 of the Federal Aviation Regulations unless an alternative program has been FAA approved."

[Amdt. 23-26, 45 FR 60171, Sept. 11, 1980, as amended by Amdt. 23-34, 52 FR 1835, Jan. 15, 1987; 52 FR 34745, Sept. 14, 1987; Amdt. 23-37, 54 FR 34329, Aug. 18, 1989]

Appendix H to Part 23—Installation of An Automatic Power Reserve (APR) System

H23.1, *General.*

(a) This appendix specifies requirements for installation of an APR engine power control system that automatically advances power or thrust on the operating engine(s) in the event any engine fails during takeoff.

(b) With the APR system and associated systems functioning normally, all applicable requirements (except as provided in this appendix) must be met without requiring any action by the crew to increase power or thrust.

FIGURE H1 - CRITICAL TIME INTERVAL ILLUSTRATION.

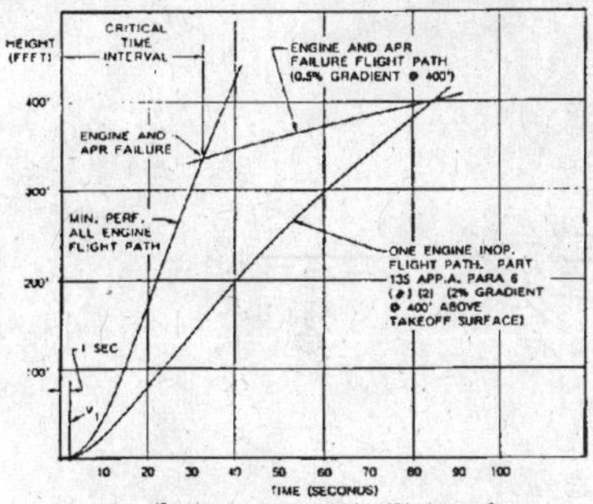

(For illustration only - typical for 120K climb speed)

H23.2, *Definitions.*

(a) *Automatic power reserve system* means the entire automatic system used only during takeoff, including all devices both mechanical and electrical that sense engine failure, transmit signals, actuate fuel controls or power levers on operating engines, including power sources, to achieve the scheduled power increase and furnish cockpit information on system operation.

(b) *Selected takeoff power,* notwithstanding the definition of "Takeoff Power" in part 1 of the Federal Aviation Regulations, means the power obtained from each initial power setting approved for takeoff.

(c) *Critical Time Interval,* as illustrated in figure H1, means that period starting at V_1 minus one second and ending at the intersection of the engine and APR failure flight path line with the minimum performance all engine flight path line. The engine and APR failure flight path line intersects the one-engine-inoperative flight path line at 400 feet above the takeoff surface. The engine and APR failure flight path is based on the airplane's performance and must have a positive gradient of at least 0.5 percent at 400 feet above the takeoff surface.

H23.3, *Reliability and performance requirements.*

(a) It must be shown that, during the critical time interval, an APR failure that increases or does not affect power on either engine will not create a hazard to the airplane, or it must be shown that such failures are improbable.

(b) It must be shown that, during the critical time interval, there are no failure modes of the APR system that would result in a failure that will decrease the power on either engine or it must be shown that such failures are extremely improbable.

(c) It must be shown that, during the critical time interval, there will be no failure of the APR system in combination with an engine failure or it must be shown that such failures are extremely improbable.

(d) All applicable performance requirements must be met with an engine failure occurring at the most critical point during takeoff with the APR system functioning normally.

H23.4, *Power setting.*

The selected takeoff power set on each engine at the beginning of the takeoff roll may not be less than—

(a) The power necessary to attain, at V_1, 90 percent of the maximum takeoff power approved for the airplane for the existing conditions;

(b) That required to permit normal operation of all safety-related systems and equipment that are dependent upon engine power or power lever position; and

(c) That shown to be free of hazardous engine response characteristics when power is advanced from the selected takeoff power level to the maximum approved takeoff power.

H23.5, *Powerplant controls—general.*

(a) In addition to the requirements of § 23.1141, no single failure or malfunction (or probable combination thereof) of the APR, including associated systems, may cause the failure of any powerplant function necessary for safety.

(b) The APR must be designed to—

(1) Provide a means to verify to the flight crew before takeoff that the APR is in an operating condition to perform its intended function;

(2) Automatically advance power on the operating engines following an engine failure during takeoff to achieve the maximum attainable takeoff power without exceeding engine operating limits;

(3) Prevent deactivation of the APR by manual adjustment of the power levers following an engine failure;

(4) Provide a means for the flight crew to deactivate the automatic function. This means must be designed to prevent inadvertent deactivation; and

(5) Allow normal manual decrease or increase in power up to the maximum takeoff power approved for the airplane under the existing conditions through the use of power levers, as stated in § 23.1141(c), except as provided under paragraph (c) of H23.5 of this appendix.

(c) For airplanes equipped with limiters that automatically prevent engine operating limits from being exceeded, other means may be used to increase the maximum level of power controlled by the power levers in the event of an APR failure. The means must be located on or forward of the power levers, must be easily identified and operated under all operating conditions by a single action of any pilot with the hand that is normally used to actuate the power levers, and must meet the requirements of § 23.777 (a), (b), and (c).

H23.6, *Powerplant instruments.*

In addition to the requirements of § 23.1305:

(a) A means must be provided to indicate when the APR is in the armed or ready condition.

(b) If the inherent flight characteristics of the airplane do not provide warning that an engine has failed, a warning system independent of the APR must be provided to give the pilot a clear warning of any engine failure during takeoff.

(c) Following an engine failure at V_1 or above, there must be means for the crew to readily and quickly verify that the APR has operated satisfactorily.

[Doc. No. 26344, 58 FR 18979, Apr. 9, 1993]

Appendix I to Part 23—Seaplane Loads

[Amdt. 23-45, 58 FR 42167, Aug. 6, 1993; 58 FR 51970, Oct. 5, 1993]

Appendix J to Part 23—HIRF Environments and Equipment HIRF Test Levels

This appendix specifies the HIRF environments and equipment HIRF test levels for electrical and electronic systems under § 23.1308. The field strength values for the HIRF environments and equipment HIRF test levels are expressed in root-mean-square units measured during the peak of the modulation cycle.

(a) HIRF environment I is specified in the following table:

Table I.—HIRF Environment I

Frequency	Field strength (volts/meter)	
	Peak	Average
10 kHz–2 MHz	50	50
2 MHz–30 MHz	100	100
30 MHz–100 MHz	50	50
100 MHz–400 MHz	100	100
400 MHz–700 MHz	700	50
700 MHz–1 GHz	700	100
GHz–2 GHz	2,000	200
2 GHz–6 GHz	3,000	200
6 GHz–8 GHz	1,000	200
8 GHz–12 GHz	3,000	300
12 GHz–18 GHz	2,000	200
18 GHz–40 GHz	600	200

PART 23

FAR

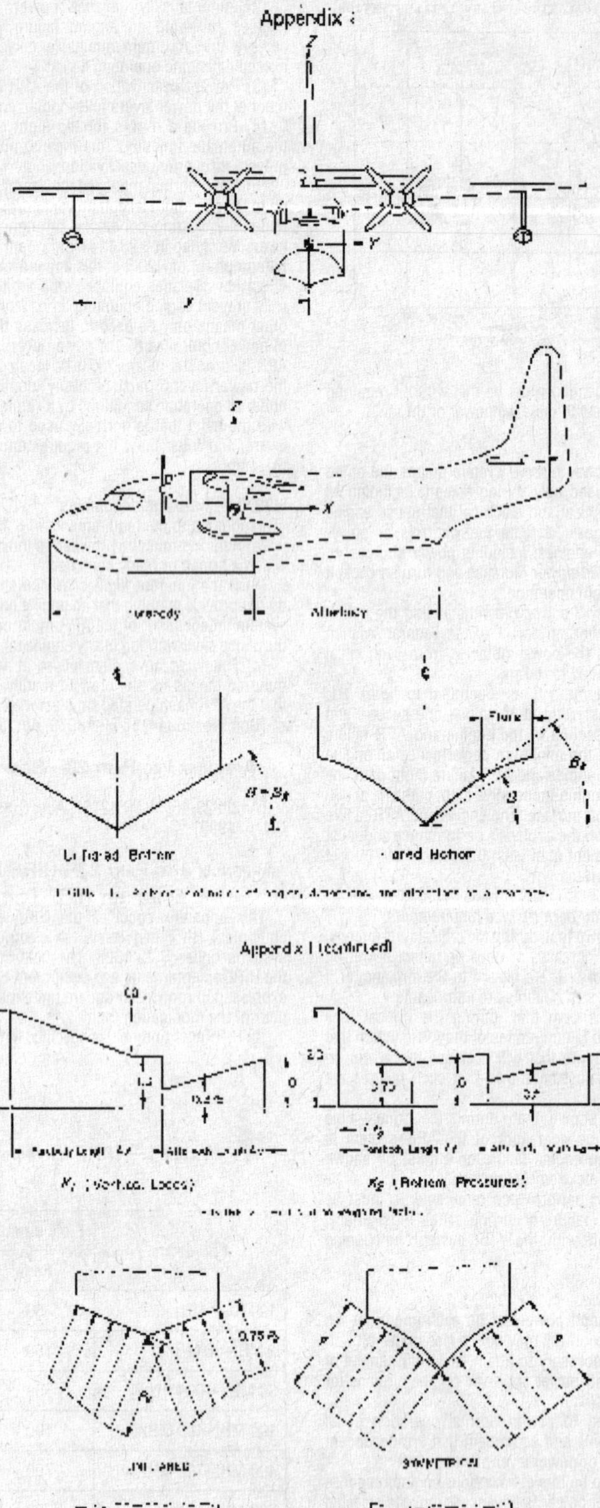

In this table, the higher field strength applies at the frequency band edges.

(b) HIRF environment II is specified in the following table:

Table II.-HIRF Environment II

Frequency	Field strength (volts/meter)	
	Peak	Average
10 kHz–500 kHz	20	20
500 kHz–2 MHz	30	30
2 MHz–30 MHz	100	100
30 MHz–100 MHz	10	10
100 MHz–200 MHz	30	10
200 MHz–400 MHz	10	10
400 MHz–1 GHz	700	40
1 GHz–2 GHz	1,300	160
2 GHz–4 GHz	3,000	120
4 GHz–6 GHz	3,000	160
6 GHz–8 GHz	400	170
8 GHz–12 GHz	1,230	230
12 GHz–18 GHz	730	190
18 GHz–40 GHz	600	150

In this table, the higher field strength applies at the frequency band edges.

(c) *Equipment HIRF Test Level 1.* (1) From 10 kilohertz (kHz) to 400 megahertz (MHz), use conducted susceptibility tests with continuous wave (CW) and 1 kHz square wave modulation with 90 percent depth or greater. The conducted susceptibility current must start at a minimum of 0.6 milliamperes (mA) at 10 kHz, increasing 20 decibels (dB) per frequency decade to a minimum of 30 mA at 500 kHz.

(2) From 500 kHz to 40 MHz, the conducted susceptibility current must be at least 30 mA.

(3) From 40 MHz to 400 MHz, use conducted susceptibility tests, starting at a minimum of 30 mA at 40 MHz, decreasing 20 dB per frequency decade to a minimum of 3 mA at 400 MHz.

(4) From 100 MHz to 400 MHz, use radiated susceptibility tests at a minimum of 20 volts per meter (*V*/m) peak with CW and 1 kHz square wave modulation with 90 percent depth or greater.

(5) From 400 MHz to 8 gigahertz (GHz), use radiated susceptibility tests at a minimum of 150 *V*/m peak with pulse modulation of 4 percent duty cycle with a 1 kHz pulse repetition frequency. This signal must be switched on and off at a rate of 1 Hz with a duty cycle of 50 percent.

(d) *Equipment HIRF Test Level 2.* Equipment HIRF test level 2 is HIRF environment II in table II of this appendix reduced by acceptable aircraft transfer function and attenuation curves. Testing must cover the frequency band of 10 kHz to 8 GHz.

(e) *Equipment HIRF Test Level 3.* (1) From 10 kHz to 400 MHz, use conducted susceptibility tests, starting at a minimum of 0.15 mA at 10 kHz, increasing 20 dB per frequency decade to a minimum of 7.5 mA at 500 kHz.

(2) From 500 kHz to 40 MHz, use conducted susceptibility tests at a minimum of 7.5 mA.

(3) From 40 MHz to 400 MHz, use conducted susceptibility tests, starting at a minimum of 7.5 mA at 40 MHz, decreasing 20 dB per frequency decade to a minimum of 0.75 mA at 400 MHz.

(4) From 100 MHz to 8 GHz, use radiated susceptibility tests at a minimum of 5 *V*/m.

[Doc. No. FAA-2006-23657, 72 FR 44025, Aug. 6, 2007]

FEDERAL AVIATION REGULATIONS (FAR)

PART 25

AIRWORTHINESS STANDARDS: TRANSPORT CATEGORY AIRPLANES

Contents

Special Federal Aviation Regulation No. 13
Special Federal Aviation Regulation No. 109

Subpart A—General

§ 25.1 Applicability.
§ 25.2 Special retroactive requirements.
§ 25.3 Special provisions for ETOPS type design approvals.
§ 25.5 Incorporations by reference.

Subpart B—Flight

General

§ 25.21 Proof of compliance.
§ 25.23 Load distribution limits.
§ 25.25 Weight limits.
§ 25.27 Center of gravity limits.
§ 25.29 Empty weight and corresponding center of gravity.
§ 25.31 Removable ballast.
§ 25.33 Propeller speed and pitch limits.

Performance

§ 25.101 General.
§ 25.103 Stall speed.
§ 25.105 Takeoff.
§ 25.107 Takeoff speeds.
§ 25.109 Accelerate-stop distance.
§ 25.111 Takeoff path.
§ 25.113 Takeoff distance and takeoff run.
§ 25.115 Takeoff flight path.
§ 25.117 Climb: general.
§ 25.119 Landing climb: All-engines-operating.
§ 25.121 Climb: One-engine-inoperative.
§ 25.123 En route flight paths.
§ 25.125 Landing.

Controllability and Maneuverability

§ 25.143 General.
§ 25.145 Longitudinal control.
§ 25.147 Directional and lateral control.
§ 25.149 Minimum control speed.

Trim

§ 25.161 Trim.

Stability

§ 25.171 General.
§ 25.173 Static longitudinal stability.
§ 25.175 Demonstration of static longitudinal stability.
§ 25.177 Static lateral-directional stability.
§ 25.181 Dynamic stability.

Stalls

§ 25.201 Stall demonstration.
§ 25.203 Stall characteristics.
§ 25.207 Stall warning.

Ground and Water Handling Characteristics

§ 25.231 Longitudinal stability and control.
§ 25.233 Directional stability and control.
§ 25.235 Taxiing condition.
§ 25.237 Wind velocities.
§ 25.239 Spray characteristics, control, and stability on water.

Miscellaneous Flight Requirements

§ 25.251 Vibration and buffeting.
§ 25.253 High-speed characteristics.
§ 25.255 Out-of-trim characteristics.

Subpart C—Structure

General

§ 25.301 Loads.
§ 25.303 Factor of safety.
§ 25.305 Strength and deformation.
§ 25.307 Proof of structure.

Flight Loads

§ 25.321 General.

Flight Maneuver and Gust Conditions

§ 25.331 Symmetric maneuvering conditions.
§ 25.333 Flight maneuvering envelope.
§ 25.335 Design airspeeds.
§ 25.337 Limit maneuvering load factors.
§ 25.341 Gust and turbulence loads.
§ 25.343 Design fuel and oil loads.
§ 25.345 High lift devices.
§ 25.349 Rolling conditions.
§ 25.351 Yaw maneuver conditions.

Supplementary Conditions

§ 25.361 Engine torque.
§ 25.363 Side load on engine and auxiliary power unit mounts.
§ 25.365 Pressurized compartment loads.
§ 25.367 Unsymmetrical loads due to engine failure.
§ 25.371 Gyroscopic loads.
§ 25.373 Speed control devices.

Control Surface and System Loads

§ 25.391 Control surface loads: General.
§ 25.393 Loads parallel to hinge line.
§ 25.395 Control system.
§ 25.397 Control system loads.
§ 25.399 Dual control system.
§ 25.405 Secondary control system.
§ 25.407 Trim tab effects.
§ 25.409 Tabs.
§ 25.415 Ground gust conditions.
§ 25.427 Unsymmetrical loads.
§ 25.445 Auxiliary aerodynamic surfaces.
§ 25.457 Wing flaps.
§ 25.459 Special devices.

Ground Loads

§ 25.471 General.
§ 25.473 Landing load conditions and assumptions.
§ 25.477 Landing gear arrangement.
§ 25.479 Level landing conditions.
§ 25.481 Tail-down landing conditions.
§ 25.483 One-gear landing conditions.
§ 25.485 Side load conditions.
§ 25.487 Rebound landing condition.
§ 25.489 Ground handling conditions.
§ 25.491 Taxi, takeoff and landing roll.
§ 25.493 Braked roll conditions.
§ 25.495 Turning.
§ 25.497 Tail-wheel yawing.
§ 25.499 Nose-wheel yaw and steering.
§ 25.503 Pivoting.
§ 25.507 Reversed braking.
§ 25.509 Towing loads.
§ 25.511 Ground load: unsymmetrical loads on multiple-wheel units.
§ 25.519 Jacking and tie-down provisions.

Water Loads

§ 25.521 General.
§ 25.523 Design weights and center of gravity positions.
§ 25.525 Application of loads.
§ 25.527 Hull and main float load factors.
§ 25.529 Hull and main float landing conditions.
§ 25.531 Hull and main float takeoff condition.
§ 25.533 Hull and main float bottom pressures.
§ 25.535 Auxiliary float loads.
§ 25.537 Seawing loads.

Emergency Landing Conditions

§ 25.561 General.
§ 25.562 Emergency landing dynamic conditions.
§ 25.563 Structural ditching provisions.

Fatigue Evaluation

§ 25.571 Damage—tolerance and fatigue evaluation of structure.

Lightning Protection

§ 25.581 Lightning protection.

Subpart D—Design and Construction

General

§ 25.601 General.
§ 25.603 Materials.
§ 25.605 Fabrication methods.
§ 25.607 Fasteners.
§ 25.609 Protection of structure.
§ 25.611 Accessibility provisions.
§ 25.613 Material strength properties and material design values.
§ 25.619 Special factors.
§ 25.621 Casting factors.
§ 25.623 Bearing factors.
§ 25.625 Fitting factors.
§ 25.629 Aeroelastic stability requirements.
§ 25.631 Bird strike damage.

Control Surfaces

§ 25.651 Proof of strength.
§ 25.655 Installation.
§ 25.657 Hinges.

Control Systems

§ 25.671 General.
§ 25.672 Stability augmentation and automatic and power-operated systems.
§ 25.675 Stops.
§ 25.677 Trim systems.
§ 25.679 Control system gust locks.
§ 25.681 Limit load static tests.
§ 25.683 Operation tests.
§ 25.685 Control system details.
§ 25.689 Cable systems.
§ 25.693 Joints.
§ 25.697 Lift and drag devices, controls.
§ 25.699 Lift and drag device indicator.
§ 25.701 Flap and slat interconnection.
§ 25.703 Takeoff warning system.

Landing Gear

§ 25.721 General.
§ 25.723 Shock absorption tests.
§§ 25.725-25.727 [Reserved]
§ 25.729 Retracting mechanism.
§ 25.731 Wheels.
§ 25.733 Tires.
§ 25.735 Brakes and braking systems.
§ 25.737 Skis.

Floats and Hulls

§ 25.751 Main float buoyancy.
§ 25.753 Main float design.
§ 25.755 Hulls.

Personnel and Cargo Accommodations

§ 25.771 Pilot compartment.
§ 25.772 Pilot compartment doors.
§ 25.773 Pilot compartment view.
§ 25.775 Windshields and windows.
§ 25.777 Cockpit controls.
§ 25.779 Motion and effect of cockpit controls.
§ 25.781 Cockpit control knob shape.
§ 25.783 Fuselage doors.
§ 25.785 Seats, berths, safety belts, and harnesses.
§ 25.787 Stowage compartments.
§ 25.789 Retention of items of mass in passenger and crew compartments and galleys.
§ 25.791 Passenger information signs and placards.
§ 25.793 Floor surfaces.
§ 25.795 Security considerations.

Emergency Provisions

§ 25.801 Ditching.
§ 25.803 Emergency evacuation.
§ 25.807 Emergency exits.
§ 25.809 Emergency exit arrangement.
§ 25.810 Emergency egress assist means and escape routes.
§ 25.811 Emergency exit marking.
§ 25.812 Emergency lighting.
§ 25.813 Emergency exit access.
§ 25.815 Width of aisle.

§ 25.817 Maximum number of seats abreast.
§ 25.819 Lower deck service compartments (including galleys).
§ 25.820 Lavatory doors.

Ventilation and Heating

§ 25.831 Ventilation.
§ 25.832 Cabin ozone concentration.
§ 25.833 Combustion heating systems.

Pressurization

§ 25.841 Pressurized cabins.
§ 25.843 Tests for pressurized cabins.

Fire Protection

§ 25.851 Fire extinguishers.
§ 25.853 Compartment interiors.
§ 25.854 Lavatory fire protection.
§ 25.855 Cargo or baggage compartments.
§ 25.856 Thermal/Acoustic insulation materials.
§ 25.857 Cargo compartment classification.
§ 25.858 Cargo or baggage compartment smoke or fire detection systems.
§ 25.859 Combustion heater fire protection.
§ 25.863 Flammable fluid fire protection.
§ 25.865 Fire protection of flight controls, engine mounts, and other flight structure.
§ 25.867 Fire protection: other components.
§ 25.869 Fire protection: systems.

Miscellaneous

§ 25.871 Leveling means.
§ 25.875 Reinforcement near propellers.
§ 25.899 Electrical bonding and protection against static electricity.

Subpart E—Powerplant

General

§ 25.901 Installation.
§ 25.903 Engines.
§ 25.904 Automatic takeoff thrust control system (ATTCS).
§ 25.905 Propellers.
§ 25.907 Propeller vibration and fatigue.
§ 25.925 Propeller clearance.
§ 25.929 Propeller deicing.
§ 25.933 Reversing systems.
§ 25.934 Turbojet engine thrust reverser system tests.
§ 25.937 Turbopropeller-drag limiting systems.
§ 25.939 Turbine engine operating characteristics.
§ 25.941 Inlet, engine, and exhaust compatibility.
§ 25.943 Negative acceleration.
§ 25.945 Thrust or power augmentation system.

Fuel System

§ 25.951 General.
§ 25.952 Fuel system analysis and test.
§ 25.953 Fuel system independence.
§ 25.954 Fuel system lightning protection.
§ 25.955 Fuel flow.
§ 25.957 Flow between interconnected tanks.
§ 25.959 Unusable fuel supply.
§ 25.961 Fuel system hot weather operation.
§ 25.963 Fuel tanks: general.
§ 25.965 Fuel tank tests.
§ 25.967 Fuel tank installations.
§ 25.969 Fuel tank expansion space.
§ 25.971 Fuel tank sump.
§ 25.973 Fuel tank filler connection.
§ 25.975 Fuel tank vents and carburetor vapor vents.
§ 25.977 Fuel tank outlet.
§ 25.979 Pressure fueling system.
§ 25.981 Fuel tank ignition prevention.

Fuel System Components

§ 25.991 Fuel pumps.
§ 25.993 Fuel system lines and fittings.
§ 25.994 Fuel system components.
§ 25.995 Fuel valves.
§ 25.997 Fuel strainer or filter.
§ 25.999 Fuel system drains.
§ 25.1001 Fuel jettisoning system.

Oil System

§ 25.1011 General.
§ 25.1013 Oil tanks.
§ 25.1015 Oil tank tests.
§ 25.1017 Oil lines and fittings.
§ 25.1019 Oil strainer or filter.
§ 25.1021 Oil system drains.
§ 25.1023 Oil radiators.
§ 25.1025 Oil valves.
§ 25.1027 Propeller feathering system.

Cooling

§ 25.1041 General.
§ 25.1043 Cooling tests.
§ 25.1045 Cooling test procedures.

Induction System

§ 25.1091 Air induction.
§ 25.1093 Induction system icing protection.
§ 25.1101 Carburetor air preheater design.
§ 25.1103 Induction system ducts and air duct systems.
§ 25.1105 Induction system screens.
§ 25.1107 Inter-coolers and after-coolers.

Exhaust System

§ 25.1121 General.
§ 25.1123 Exhaust piping.
§ 25.1125 Exhaust heat exchangers.
§ 25.1127 Exhaust driven turbo-superchargers.

Powerplant Controls and Accessories

§ 25.1141 Powerplant controls: general.
§ 25.1142 Auxiliary power unit controls.
§ 25.1143 Engine controls.
§ 25.1145 Ignition switches.
§ 25.1147 Mixture controls.
§ 25.1149 Propeller speed and pitch controls.
§ 25.1153 Propeller feathering controls.
§ 25.1155 Reverse thrust and propeller pitch settings below the flight regime.
§ 25.1157 Carburetor air temperature controls.
§ 25.1159 Supercharger controls.
§ 25.1161 Fuel jettisoning system controls.
§ 25.1163 Powerplant accessories.
§ 25.1165 Engine ignition systems.
§ 25.1167 Accessory gearboxes.

Powerplant Fire Protection

§ 25.1181 Designated fire zones; regions included.
§ 25.1182 Nacelle areas behind firewalls, and engine pod attaching structures containing flammable fluid lines.
§ 25.1183 Flammable fluid-carrying components.
§ 25.1185 Flammable fluids.
§ 25.1187 Drainage and ventilation of fire zones.
§ 25.1189 Shutoff means.
§ 25.1191 Firewalls.
§ 25.1192 Engine accessory section diaphragm.
§ 25.1193 Cowling and nacelle skin.
§ 25.1195 Fire extinguishing systems.
§ 25.1197 Fire extinguishing agents.
§ 25.1199 Extinguishing agent containers.
§ 25.1201 Fire extinguishing system materials.
§ 25.1203 Fire detector system.
§ 25.1207 Compliance.

Subpart F—Equipment

General

§ 25.1301 Function and installation.
§ 25.1302 Installed systems and equipment for use by the flightcrew.
§ 25.1303 Flight and navigation instruments.
§ 25.1305 Powerplant instruments.
§ 25.1307 Miscellaneous equipment.
§ 25.1309 Equipment, systems, and installations.
§ 25.1310 Power source capacity and distribution.
§ 25.1316 Electrical and electronic system lightning protection.
§ 25.1317 High-intensity Radiated Fields (HIRF) Protection.

Instruments: Installation

§ 25.1321 Arrangement and visibility.
§ 25.1322 Flightcrew alerting.
§ 25.1323 Airspeed indicating system.
§ 25.1325 Static pressure systems.
§ 25.1326 Pitot heat indication systems.
§ 25.1327 Magnetic direction indicator.
§ 25.1329 Flight guidance system.
§ 25.1331 Instruments using a power supply.
§ 25.1333 Instrument systems.
§ 25.1337 Powerplant instruments.

Electrical Systems and Equipment

§ 25.1351 General.
§ 25.1353 Electrical equipment and installations.
§ 25.1355 Distribution system.
§ 25.1357 Circuit protective devices.
§ 25.1360 Precautions against injury.
§ 25.1362 Electrical supplies for emergency conditions.
§ 25.1363 Electrical system tests.
§ 25.1365 Electrical appliances, motors, and transformers.

Lights

§ 25.1381 Instrument lights.
§ 25.1383 Landing lights.
§ 25.1385 Position light system installation.
§ 25.1387 Position light system dihedral angles.
§ 25.1389 Position light distribution and intensities.
§ 25.1391 Minimum intensities in the horizontal plane of forward and rear position lights.
§ 25.1393 Minimum intensities in any vertical plane of forward and rear position lights.
§ 25.1395 Maximum intensities in overlapping beams of forward and rear position lights.
§ 25.1397 Color specifications.
§ 25.1399 Riding light.
§ 25.1401 Anticollision light system.
§ 25.1403 Wing icing detection lights.

Safety Equipment

§ 25.1411 General.
§ 25.1415 Ditching equipment.
§ 25.1419 Ice protection.
§ 25.1421 Megaphones.
§ 25.1423 Public address system.

Miscellaneous Equipment

§ 25.1431 Electronic equipment.
§ 25.1433 Vacuum systems.
§ 25.1435 Hydraulic systems.
§ 25.1438 Pressurization and pneumatic systems.
§ 25.1439 Protective breathing equipment.
§ 25.1441 Oxygen equipment and supply.
§ 25.1443 Minimum mass flow of supplemental oxygen.
§ 25.1445 Equipment standards for the oxygen distributing system.
§ 25.1447 Equipment standards for oxygen dispensing units.
§ 25.1449 Means for determining use of oxygen.
§ 25.1450 Chemical oxygen generators.
§ 25.1453 Protection of oxygen equipment from rupture.
§ 25.1455 Draining of fluids subject to freezing.
§ 25.1457 Cockpit voice recorders.
§ 25.1459 Flight data recorders.
§ 25.1461 Equipment containing high energy rotors.

Subpart G—Operating Limitations and Information

§ 25.1501 General.

Operating Limitations

§ 25.1503 Airspeed limitations: general.
§ 25.1505 Maximum operating limit speed.
§ 25.1507 Maneuvering speed.
§ 25.1511 Flap extended speed.
§ 25.1513 Minimum control speed.
§ 25.1515 Landing gear speeds.
§ 25.1516 Other speed limitations.
§ 25.1517 Rough air speed, VRA.
§ 25.1519 Weight, center of gravity, and weight distribution.
§ 25.1521 Powerplant limitations.
§ 25.1522 Auxiliary power unit limitations.
§ 25.1523 Minimum flight crew.
§ 25.1525 Kinds of operation.
§ 25.1527 Ambient air temperature and operating altitude.
§ 25.1529 Instructions for Continued Airworthiness.
§ 25.1531 Maneuvering flight load factors.
§ 25.1533 Additional operating limitations.
§ 25.1535 ETOPS approval.

Markings and Placards

§ 25.1541 General.
§ 25.1543 Instrument markings: general.
§ 25.1545 Airspeed limitation information.
§ 25.1547 Magnetic direction indicator.

AIRPLANE FLIGHT MANUAL

Subpart H—Electrical Wiring Interconnection Systems (EWIS)

Subpart I—Special Federal Aviation Regulations

AUTHORITY: 49 U.S.C. 106(g), 40113, 44701, 44702 and 44704.

SOURCE: Docket No. 5066, 29 FR 18291, Dec. 24, 1964, unless otherwise noted.

Special Federal Aviation Regulation No. 13

1. *Applicability.* Contrary provisions of the Civil Air Regulations regarding certification notwithstanding,[1] this regulation shall provide the basis for approval by the Administrator of modifications of individual Douglas DC-3 and Lockheed L-18 airplanes subsequent to the effective date of this regulation.

[1] It is not intended to waive compliance with such airworthiness requirements as are included in the operating parts of the Civil Air Regulations for specific types of operation.

2. *General modifications.* Except as modified in sections 3 and 4 of this regulation, an applicant for approval of modifications to a DC-3 or L-18 airplane which result in changes in design or in changes to approved limitations shall show that the modifications were accomplished in accordance with the rules of either Part 4a or Part 4b in effect on September 1, 1953, which are applicable to the modification being made: *Provided,* That an applicant may elect to accomplish a modification in accordance with the rules of Part 4b in effect on the date of application for the modification in lieu of Part 4a or Part 4b as in effect on September 1, 1953: *And provided further,* That each specific modification must be accomplished in accordance with all of the provisions contained in the elected rules relating to the particular modification.

3. *Specific conditions for approval.* An applicant for any approval of the following specific changes shall comply with section 2 of this regulation as modified by the applicable provisions of this section.

(a) *Increase in take-off power limitation—1,200 to 1,350 horsepower.* The engine take-off power limitation for the airplane may be increased to more than 1,200 horsepower but not to more than 1,350 horsepower per engine if the increase in power does not adversely affect the flight characteristics of the airplane.

(b) *Increase in take-off power limitation to more than 1,350 horsepower.* The engine take-off power limitation for the airplane may be increased to more than 1,350 horsepower per engine if compliance is shown with the flight characteristics and ground handling requirements of Part 4b.

(c) *Installation of engines of not more than 1,830 cubic inches displacement and not having a certificated take-off rating of more than 1,350 horsepower.* Engines of not more than 1,830 cubic inches displacement and not having a certificated take-off rating of more than 1,350 horsepower which necessitate a major modification of redesign of the engine installation may be installed, if the engine fire prevention and fire protection are equivalent to that on the prior engine installation.

(d) *Installation of engines of more than 1,830 cubic inches displacement or having certificated take-off rating of more than 1,350 horsepower.* Engines of more than 1,830 cubic inches displacement or having certificated take-off rating of more than 1,350 horsepower may be installed if compliance is shown with the engine installation requirements of Part 4b: *Provided,* That where literal compliance with the engine installation requirements of Part 4b is extremely difficult to accomplish and would not contribute materially to the objective sought, and the Administrator finds that the experience with the DC-3 or L-18 airplanes justifies it, he is authorized to accept such measures of compliance as he finds will effectively accomplish the basic objective.

4. *Establishment of new maximum certificated weights.* An applicant for approval of new maximum certificated weights shall apply for an amendment of the airworthiness certificate of the airplane and shall show that the weights sought have been established, and the appropriate manual material obtained, as provided in this section.

NOTE: Transport category performance requirements result in the establishment of maximum certificated weights for various altitudes.

(a) *Weights-25,200 to 26,900 for the DC-3 and 18,500 to 19,500 for the L-18.* New maximum certificated weights of more than 25,200 but not more than 26,900 pounds for DC-3 and more than 18,500 but not more than 19,500 pounds for L-18 airplanes may be established in accordance with the transport category performance requirements of either Part 4a or Part 4b, if the airplane at the new maximum weights can meet the structural requirements of the elected part.

(b) *Weights of more than 26,900 for the DC-3 and 19,500 for the L-18.* New maximum certificated weights of more than 26,900 pounds for DC-3 and 19,500 pounds for L-18 airplanes shall be established in accordance with the structural performance, flight characteristics, and ground handling requirements of Part 4b: *Provided,* That where literal compliance with the structural requirements of Part 4b is extremely difficult to accomplish and would not contribute materially to the objective sought, and the Administrator finds that the experience with the DC-3 or L-18 airplanes justifies it, he is authorized to accept such measures of compliance as he finds will effectively accomplish the basic objective.

(c) *Airplane flight manual-performance operating information.* An approved airplane flight manual shall be provided for each DC-3 and L-18 airplane which has had new maximum certificated weights established under this section. The airplane flight manual shall contain the applicable performance information prescribed in that part of the regulations under which the new certificated weights were established and such additional information as may be necessary to enable the application of the take-off, en route, and landing limitations prescribed for transport category airplanes in the operating parts of the Civil Air Regulations.

(d) *Performance operating limitations.* Each airplane for which new maximum certificated weights are established in accordance with paragraphs (a) or (b) of this section shall be considered a transport category airplane for the purpose of complying with the performance operating limitations applicable to the operations in which it is utilized.

5. *Reference.* Unless otherwise provided, all references in this regulation to Part 4a and Part 4b are those parts of the Civil Air Regulations in effect on September 1, 1953.

This regulation supersedes Special Civil Air Regulation SR-398 and shall remain effective until superseded or rescinded by the Board.

[19 FR 5039, Aug. 11, 1954. Redesignated at 29 FR 19099, Dec. 30, 1964]

SPECIAL FEDERAL AVIATION REGULATION NO. 109

1. *Applicability.* Contrary provisions of 14 CFR parts 21, 25, and 119 of this chapter notwithstanding, an applicant is entitled to an amended type certificate or supplemental type certificate in the transport category, if the applicant complies with all applicable provisions of this SFAR.

OPERATIONS

2. *General.*

(a) The passenger capacity may not exceed 60. If more than 60 passenger seats are installed, then:

(1) If the extra seats are not suitable for occupancy during taxi, takeoff and landing, each extra seat must be clearly marked (e.g., a placard on the top of an armrest, or a placard sewn into the top of the back cushion) that the seat is not to be occupied during taxi, takeoff and landing.

(2) If the extra seats are suitable for occupancy during taxi, takeoff and landing (*i.e.,* meet all the strength and passenger injury criteria in part 25), then a note must be included in the Limitations Section of the Airplane Flight Manual that there are extra seats installed but that the number of passengers on the airplane must not exceed 60. Additionally, there must be a placard installed adjacent to each door that can be used as a passenger boarding door that states that the maximum passenger capacity is 60. The placard must be clearly legible to passengers entering the airplane.

(b) For airplanes outfitted with interior doors under paragraph 10 of this SFAR, the airplane flight manual (AFM) must include an appropriate limitation that the airplane must be staffed with at least the following number of flight attendants who meet the requirements of 14 CFR 91.533(b):

(1) The number of flight attendants required by § 91.533(a)(1) and (2) of this chapter, and

(2) At least one flight attendant if the airplane model was originally certified for 75 passengers or more.

(c) The AFM must include appropriate limitation(s) to require a preflight passenger briefing describing the appropriate functions to be performed by the passengers and the relevant features of the airplane to ensure the safety of the passengers and crew.

(d) The airplane may not be offered for common carriage or operated for hire. The operating limitations section of the AFM must be revised to prohibit any operations involving the carriage of persons or property for compensation or hire. The operators may receive remuneration to the extent consistent with parts 125 and 91, subpart F, of this chapter.

(e) A placard stating that "Operations involving the carriage of persons or property for compensation or hire are prohibited," must be located in the area of the Airworthiness Certificate holder at the entrance to the flightdeck.

(f) For passenger capacities of 45 to 60 passengers, analysis must be submitted that demonstrates that the airplane can be evacuated in less than 90 seconds under the conditions specified in § 25.803 and appendix J to part 25.

(g) In order for any airplane certified under this SFAR to be placed in part 135 or part 121 operations, the airplane must be brought back into full compliance with the applicable operational part.

EQUIPMENT AND DESIGN

3. *General.* Unless otherwise noted, compliance is required with the applicable certification basis for the airplane. Some provisions of this SFAR impose alternative requirements to certain airworthiness standards that do not apply to airplanes certificated to earlier standards. Those airplanes with an earlier certification basis are not required to comply with those alternative requirements.

4. *Occupant Protection.*

(a) Firm Handhold. In lieu of the requirements of § 25.785(j), there must be means provided to enable persons to steady themselves in moderately rough air while occupying aisles that are along the cabin sidewall, or where practicable, bordered by seats (seat backs providing a

25-pound minimum breakaway force are an acceptable means of compliance).

(b) Injury criteria for multiple occupancy side-facing seats. The following requirements are only applicable to airplanes that are subject to § 25.562.

(1) *Existing Criteria.* All injury protection criteria of § 25.562(c)(1) through (c)(6) apply to the occupants of side-facing seating. The Head Injury Criterion (HIC) assessments are only required for head contact with the seat and/or adjacent structures.

(2) *Body-to-Body Contact.* Contact between the head, pelvis, torso or shoulder area of one Anthropomorphic Test Dummy (ATD) with the head, pelvis, torso or shoulder area of the ATD in the adjacent seat is not allowed during the tests conducted in accordance with § 25.562(b)(1) and (b)(2). Contact during rebound is allowed.

(3) *Thoracic Trauma.* If the torso of an ATD at the forward-most seat place impacts the seat and/or adjacent structure during testing, compliance with the Thoracic Trauma Index (TTI) injury criterion must be substantiated by dynamic test or by rational analysis based on previous test(s) of a similar seat installation. TTI data must be acquired with a Side Impact Dummy (SID), as defined by 49 CFR part 572, subpart F, or an equivalent ATD or a more appropriate ATD and must be processed as defined in Federal Motor Vehicle Safety Standards (FMVSS) part 571.214, section S6.13.5 (49 CFR 571.214). The TTI must be less than 85, as defined in 49 CFR part 572, subpart F. Torso contact during rebound is acceptable and need not be measured.

(4) *Pelvis.* If the pelvis of an ATD at any seat place impacts seat and/or adjacent structure during testing, pelvic lateral acceleration injury criteria must be substantiated by dynamic test or by rational analysis based on previous test(s) of a similar seat installation. Pelvic lateral acceleration may not exceed 130g. Pelvic acceleration data must be processed as defined in FMVSS part 571.214, section S6.13.5 (49 CFR 571.214).

(5) *Body-to-Wall/Furnishing Contact.* If the seat is installed aft of a structure—such as an interior wall or furnishing that may contact the pelvis, upper arm, chest, or head of an occupant seated next to the structure—the structure or a conservative representation of the structure and its stiffness must be included in the tests. It is recommended, but not required, that the contact surface of the actual structure be covered with at least two inches of energy absorbing protective padding (foam or equivalent) such as Ensolite.

(6) *Shoulder Strap Loads.* Where upper torso straps (shoulder straps) are used for sofa occupants, the tension loads in individual straps may not exceed 1,750 pounds. If dual straps are used for restraining the upper torso, the total strap tension loads may not exceed 2,000 pounds.

(7) *Occupant Retention.* All side-facing seats require end closures or other means to prevent the ATD's pelvis from translating beyond the end of the seat at any time during testing.

(8) *Test Parameters.*

(i) All seat positions need to be occupied by ATDs for the longitudinal tests.

(ii) A minimum of one longitudinal test, conducted in accordance with the conditions specified in § 25.562(b)(2), is required to assess the injury criteria as follows. Note that if a seat is installed aft of structure (such as an interior wall or furnishing) that does not have a homogeneous surface, an additional test or tests may be required to demonstrate that the injury criteria are met for the area which an occupant could contact. For example, different yaw angles could result in different injury considerations and may require separate tests to evaluate.

(A) For configurations without structure (such as a wall or bulkhead) installed directly forward of the forward seat place, Hybrid II ATDs or equivalent must be in all seat places.

(B) For configurations with structure (such as a wall or bulkhead) installed directly forward of the forward seat place, a side impact dummy or equivalent ATD or more appropriate ATD must be in the forward seat place and a Hybrid II ATD or equivalent must be in all other seat places.

(C) The test may be conducted with or without deformed floor.

(D) The test must be conducted with either no yaw or 10 degrees yaw for evaluating occupant injury. Deviating from the no yaw condition may not result in the critical area of contact not being evaluated. The upper torso restraint straps, where installed, must remain on the occupant's shoulder during the impact condition of § 25.562(b)(2).

(c) For the vertical test, conducted in accordance with the conditions specified in § 25.562(b)(1), Hybrid II ATDs or equivalent must be used in all seat positions.

5. *Direct View.* In lieu of the requirements of § 25.785(h)(2), to the extent practical without compromising proximity to a required floor level emergency exit, the majority of installed flight attendant seats must be located to face the cabin area for which the flight attendant is responsible.

6. *Passenger Information Signs.* Compliance with § 25.791 is required except that for § 25.791(a), when smoking is to be prohibited, notification to the passengers may be provided by a single placard so stating, to be conspicuously located inside the passenger compartment, easily visible to all persons entering the cabin in the immediate vicinity of each passenger entry door.

7. *Distance Between Exits.* For an airplane that is required to comply with § 25.807(f)(4), in effect as of July 24, 1989, which has more than one passenger emergency exit on each side of the fuselage, no passenger emergency exit may be more than 60 feet from any adjacent passenger emergency exit on the same side of the same deck of the fuselage, as measured parallel to the airplane's longitudinal axis between the nearest exit edges, unless the following conditions are met:

(a) Each passenger seat must be located within 30 feet from the nearest exit on each side of the fuselage, as measured parallel to the airplane's longitudinal axis, between the nearest exit edge and the front of the seat bottom cushion.

(b) The number of passenger seats located between two adjacent pairs of emergency exits (commonly referred to as a passenger zone) or between a pair of exits and a bulkhead or a compartment door (commonly referred to as a "dead-end zone"), may not exceed the following:

(1) For zones between two pairs of exits, 50 percent of the combined rated capacity of the two pairs of emergency exits.

(2) For zones between one pair of exits and a bulkhead, 40 percent of the rated capacity of the pair of emergency exits.

(c) The total number of passenger seats in the airplane may not exceed 33 percent of the maximum seating capacity for the airplane model using the exit ratings listed in § 25.807(g) for the original certified exits or the maximum allowable after modification when exits are deactivated, whichever is less.

(d) A distance of more than 60 feet between adjacent passenger emergency exits on the same side of the same deck of the fuselage, as measured parallel to the airplane's longitudinal axis between the nearest exit edges, is allowed only once on each side of the fuselage.

8. *Emergency Exit Signs.* In lieu of the requirements of § 25.811(d)(1) and (2) a single sign at each exit may be installed provided:

(a) The sign can be read from the aisle while directly facing the exit, and

(b) The sign can be read from the aisle adjacent to the passenger seat that is farthest from the exit and that does not have an intervening bulkhead/divider or exit.

9. *Emergency Lighting.*

(a) *Exit Signs.* In lieu of the requirements of § 25.812(b)(1), for airplanes that have a passenger seating configuration, excluding pilot seats, of 19 seats or less, the emergency exit signs required by § 25.811(d)(1), (2), and (3) must have red letters at least 1-inch high on a white background at least 2 inches high. These signs may be internally electrically illuminated, or self illuminated by other than electrical means, with an initial brightness of at least 160 microlamberts. The color may be reversed in the case of a sign that is self-illuminated by other than electrical means.

(b) *Floor Proximity Escape Path Marking.* In lieu of the requirements of § 25.812(e)(1), for cabin seating compartments that do not have the main cabin aisle entering and exiting the compartment, the following are applicable:

(1) After a passenger leaves any passenger seat in the compartment, he/she must be able to exit the compartment to the main cabin aisle using only markings and visual features not more that 4 feet above the cabin floor, and

(2) Proceed to the exits using the marking system necessary to accomplish the actions in § 25.812(e)(1) and (e)(2).

(c) *Transverse Separation of the Fuselage.* In the event of a transverse separation of the fuselage, compliance must be shown with § 25.812(l) except as follows:

(1) For each airplane type originally type certificated with a maximum passenger seating capacity of 9 or less, not more than 50 percent of all electrically illuminated emergency lights required by § 25.812 may be rendered inoperative in addition to the lights that are directly damaged by the separation.

(2) For each airplane type originally type certificated with a maximum passenger seating capacity of 10 to 19, not more than 33 percent of all electrically illuminated emergency lights required by § 25.812 may be rendered inoperative in addition to the lights that are directly damaged by the separation.

10. *Interior doors.* In lieu of the requirements of § 25.813(e), interior doors may be installed between passenger seats and exits, provided the following requirements are met.

(a) Each door between any passenger seat, occupiable for taxi, takeoff, and landing, and any emergency exit must have a means to signal to the flightcrew, at the flightdeck, that the door is in the open position for taxi, takeoff and landing.

(b) Appropriate procedures/limitations must be established to ensure that any such door is in the open configuration for takeoff and landing.

(c) Each door between any passenger seat and any exit must have dual means to retain it in the open position, each of which is capable of reacting the inertia loads specified in § 25.561.

(d) Doors installed across a longitudinal aisle must translate laterally to open and close, e.g., pocket doors.

(e) Each door between any passenger seat and any exit must be frangible in either direction.

(f) Each door between any passenger seat and any exit must be operable from either side, and if a locking mechanism is installed, it must be capable of being unlocked from either side without the use of special tools.

11. *Width of Aisle.* Compliance is required with § 25.815, except that aisle width may be reduced to 0 inches between passenger seats during in-flight operations only, provided that the applicant demonstrates that all areas of the cabin are easily accessible by a crew member in the event of an emergency (e.g., in-flight fire, decompression). Additionally, instructions must be provided at each passenger seat for restoring the aisle width required by § 25.815. Procedures must be established and documented in the AFM to ensure that the required aisle widths are provided during taxi, takeoff, and landing.

12. *Materials for Compartment Interiors.* Compliance is required with the applicable provisions of § 25.853, except that compliance with appendix F, parts IV and V, to part 25, need not be demonstrated if it can be shown by test or a combination of test and analysis that the maximum time for evacuation of all occupants does not exceed 45 seconds under the conditions specified in appendix J to part 25.

13. *Fire Detection.* For airplanes with a type certificated passenger capacity of 20 or more, there must be means that meet the requirements of § 25.858(a) through (d) to signal the flightcrew in the event of a fire in any isolated room not occupiable for taxi, takeoff and landing, which can be closed off from the rest of the cabin by a door. The indication must identify the compartment where the fire is located. This does not apply to lavatories, which continue to be governed by § 25.854.

14. *Cooktops.* Each cooktop must be designed and installed to minimize any potential threat to the airplane, passengers, and crew. Compliance with this requirement must be found in accordance with the following criteria:

(a) Means, such as conspicuous burner-on indicators, physical barriers, or handholds, must be installed to minimize the potential for inadvertent personnel contact with hot surfaces of both the cooktop and cookware. Conditions of turbulence must be considered.

(b) Sufficient design means must be included to restrain cookware while in place on the cooktop, as well as representative contents, e.g., soup, sauces, etc., from the effects of flight loads and turbulence. Restraints must be provided to preclude hazardous movement of cookware and contents. These restraints must accommodate any cookware that is identified for use with the cooktop. Restraints must be designed to be easily utilized and effective in service. The cookware restraint system should also be designed so that it will not be easily disabled, thus rendering it unusable. Placarding must be installed which prohibits the use of cookware that cannot be accommodated by the restraint system.

(c) Placarding must be installed which prohibits the use of cooktops (i.e., power on any burner) during taxi, takeoff, and landing.

(d) Means must be provided to address the possibility of a fire occurring on or in the immediate vicinity of the cooktop. Two acceptable means of complying with this requirement are as follows:

(1) Placarding must be installed that prohibits any burner from being powered when the cooktop is unattended. (Note: This would prohibit a single person from cooking on the cooktop and intermittently serving food to passengers while any burner is powered.) A fire detector must be installed in the vicinity of the cooktop which provides an audible warning in the passenger cabin, and a fire extinguisher of appropriate size and extinguishing agent must be installed in the immediate vicinity of the cooktop. Access to the extinguisher may not be blocked by a fire on or around the cooktop.

(2) An automatic, thermally activated fire suppression system must be installed to extinguish a fire at the cooktop and immediately adjacent surfaces. The agent used in the system must be an approved total flooding agent suitable for use in an occupied area. The fire suppression system must have a manual override. The automatic activation of the fire suppression system must also automatically shut off power to the cooktop.

(e) The surfaces of the galley surrounding the cooktop which would be exposed to a fire on the cooktop surface or in cookware on the cooktop must be constructed of materials that comply with the flammability requirements of part III of appendix F to part 25. This requirement is in addition to the flammability requirements typically required of the materials in these galley surfaces. During the selection of these materials, consideration must also be given to ensure that the flammability characteristics of the materials will not be adversely affected by the use of cleaning agents and utensils used to remove cooking stains.

(f) The cooktop must be ventilated with a system independent of the airplane cabin and cargo ventilation system. Procedures and time intervals must be established to inspect and clean or replace the ventilation system to prevent a fire hazard from the accumulation of flammable oils and be included in the instructions for continued airworthiness. The ventilation system ducting must be protected by a flame arrestor. [Note:The applicant may find additional useful information in Society of Automotive Engineers, Aerospace Recommended Practice 85, Rev. E, entitled "Air Conditioning Systems for Subsonic Airplanes," dated August 1, 1991.]

(g) Means must be provided to contain spilled foods or fluids in a manner that will prevent the creation of a slipping hazard to occupants and will not lead to the loss of structural strength due to airplane corrosion.

(h) Cooktop installations must provide adequate space for the user to immediately escape a hazardous cooktop condition.

(i) A means to shut off power to the cooktop must be provided at the galley containing the cooktop and in the cockpit. If additional switches are introduced in the cockpit, revisions to smoke or fire emergency procedures of the AFM will be required.

(j) If the cooktop is required to have a lid to enclose the cooktop there must be a means to automatically shut off power to the cooktop when the lid is closed.

15. *Hand-Held Fire Extinguishers.*

(a) For airplanes that were originally type certificated with more than 60 passengers, the number of hand-held fire extinguishers must be the greater of—

(1) That provided in accordance with the requirements of § 25.851, or

(2) A number equal to the number of originally type certificated exit pairs, regardless of whether the exits are deactivated for the proposed configuration.

(b) Extinguishers must be evenly distributed throughout the cabin. These extinguishers are in addition to those required by paragraph 14 of this SFAR, unless it can be shown that the cooktop was installed in the immediate vicinity of the original exits.

16. *Security.* The requirements of § 25.795 are not applicable to airplanes approved in accordance with this SFAR.

[Doc. No. FAA-2007-28250, 74 FR 21541, May 8, 2009]

Subpart A—General

§ 25.1 Applicability.

(a) This part prescribes airworthiness standards for the issue of type certificates, and changes to those certificates, for transport category airplanes.

(b) Each person who applies under Part 21 for such a certificate or change must show compliance with the applicable requirements in this part.

§ 25.2 Special retroactive requirements.

The following special retroactive requirements are applicable to an airplane for which the regulations referenced in the type certificate predate the sections specified below—

(a) Irrespective of the date of application, each applicant for a supplemental type certificate (or an amendment to a type certificate) involving an increase in passenger seating capacity to a total greater than that for which the airplane has been type certificated must show that the airplane concerned meets the requirements of:

(1) Sections 25.721(d), 25.783(g), 25.785(c), 25.803(c) (2) through (9), 25.803 (d) and (e), 25.807 (a), (c), and (d), 25.809 (f) and (h), 25.811, 25.812, 25.813 (a), (b), and (c), 25.815, 25.817, 25.853 (a) and (b), 25.855(a), 25.993(f), and 25.1359(c) in effect on October 24, 1967, and

(2) Sections 25.803(b) and 25.803(c)(1) in effect on April 23, 1969.

(b) Irrespective of the date of application, each applicant for a supplemental type certificate (or an amendment to a type certificate) for an airplane manufactured after October 16, 1987, must show that the airplane meets the requirements of § 25.807(c)(7) in effect on July 24, 1989.

(c) Compliance with subsequent revisions to the sections specified in paragraph (a) or (b) of this section may be elected or may be required in accordance with § 21.101(a) of this chapter.

[Amdt. 25-72, 55 FR 29773, July 20, 1990, as amended by Amdt. 25-99, 65 FR 36266, June 7, 2000]

§ 25.3 Special provisions for ETOPS type design approvals.

(a) *Applicability.* This section applies to an applicant for ETOPS type design approval of an airplane:

(1) That has an existing type certificate on February 15, 2007; or

(2) For which an application for an original type certificate was submitted before February 15, 2007.

(b) *Airplanes with two engines.* (1) For ETOPS type design approval of an airplane up to and including 180 minutes, an applicant must comply with § 25.1535, except that it need not comply with the following provisions of Appendix K, K25.1.4, of this part:

(i) K25.1.4(a), fuel system pressure and flow requirements;

(ii) K25.1.4(a)(3), low fuel alerting; and

(iii) K25.1.4(c), engine oil tank design.

(2) For ETOPS type design approval of an airplane beyond 180 minutes an applicant must comply with § 25.1535.

(c) *Airplanes with more than two engines.* An applicant for ETOPS type design approval must comply with § 25.1535 for an airplane manufactured on or after February 17, 2015, except that, for an airplane configured for a three person flight crew, the applicant need not comply with Appendix K, K25.1.4(a)(3), of this part, low fuel alerting.

[Doc. No. FAA-2002-6717, 72 FR 1873, Jan. 16, 2007]

§ 25.5 Incorporations by reference.

(a) The materials listed in this section are incorporated by reference in the corresponding sections noted. These incorporations by reference were approved by the Director of the Federal Register in accordance with 5 U.S.C. 552(a) and 1 CFR part 51. These materials are incorporated as they exist on the date of the approval, and notice of any change in these materials will be published in the Federal Register. The materials are available for purchase at the corresponding addresses noted below, and all are available for inspection at the National Archives and Records Administration (NARA), and at FAA, Transport Airplane Directorate, Aircraft Certification Service, 1601 Lind Avenue, SW., Renton, Washington 98057-3356. For information on the availability of this material at NARA, call 202-741-6030, or go to: *http://www.archives.gov/federal_register/code_of_federal_regulations/ibr_locations.html.*

(b) The following materials are available for purchase from the following address: The National Technical Information Services (NTIS), Springfield, Virginia 22166.

(1) Fuel Tank Flammability Assessment Method User's Manual, dated May 2008, document number DOT/FAA/AR-05/8, IBR approved for § 25.981 and Appendix N. It can also be obtained at the following Web site: *http://www.fire.tc.faa.gov/systems/fueltank/FTFAM.stm.*

(2) [Reserved]

[73 FR 42494, July 21, 2008]

Subpart B—Flight

General

§ 25.21 Proof of compliance.

(a) Each requirement of this subpart must be met at each appropriate combination of weight and center of gravity within the range of loading conditions for which certification is requested. This must be shown—

(1) By tests upon an airplane of the type for which certification is requested, or by calculations based on, and equal in accuracy to, the results of testing; and

(2) By systematic investigation of each probable combination of weight and center of gravity, if compliance cannot be reasonably inferred from combinations investigated.

(b) [Reserved]

(c) The controllability, stability, trim, and stalling characteristics of the airplane must be shown for each altitude up to the maximum expected in operation.

(d) Parameters critical for the test being conducted, such as weight, loading (center of gravity and inertia), airspeed, power, and wind, must be maintained within acceptable tolerances of the critical values during flight testing.

(e) If compliance with the flight characteristics requirements is dependent upon a stability augmentation system or upon any other automatic or power-operated system, compliance must be shown with §§ 25.671 and 25.672.

(f) In meeting the requirements of §§ 25.105(d), 25.125, 25.233, and 25.237, the wind velocity must be measured at a height of 10 meters above the surface, or corrected for the difference between the height at which the wind velocity is measured and the 10-meter height.

(g) The requirements of this subpart associated with icing conditions apply only if the applicant is seeking certification for flight in icing conditions.

(1) Each requirement of this subpart, except §§ 25.121(a), 25.123(c), 25.143(b)(1) and (2), 25.149, 25.201(c)(2), 25.239, and 25.251(b) through (e), must be met in icing conditions. Section 25.207(c) and (d) must be met in the landing configuration in icing conditions, but need not be met for other configurations. Compliance must be shown using the ice accretions defined in appendix C, assuming normal operation of the airplane and its ice protection system in accordance with the operating limitations and operating procedures established by the applicant and provided in the Airplane Flight Manual.

(2) No changes in the load distribution limits of § 25.23, the weight limits of § 25.25 (except where limited by performance requirements of this subpart), and the center of gravity limits of § 25.27, from those for non-icing conditions, are allowed for flight in icing conditions or with ice accretion.

[Doc. No. 5066, 29 FR 18291, Dec. 24, 1964, as amended by Amdt. 25-23, 35 FR 5671, Apr. 8, 1970; Amdt. 25-42, 43 FR 2320, Jan. 16, 1978; Amdt. 25-72, 55 FR 29774, July 20, 1990; Amdt. 25-121, 72 FR 44665, Aug. 8, 2007 Amdt. 25-135, 76 FR 74654, Dec. 1, 2011]

§ 25.23 Load distribution limits.

(a) Ranges of weights and centers of gravity within which the airplane may be safely operated must be established. If a weight and center of gravity combination is allowable only within certain load distribution limits (such as spanwise) that could be inadvertently exceeded, these limits and the corresponding weight and center of gravity combinations must be established.

(b) The load distribution limits may not exceed—

(1) The selected limits;

(2) The limits at which the structure is proven; or

(3) The limits at which compliance with each applicable flight requirement of this subpart is shown.

§ 25.25 Weight limits.

(a) *Maximum weights.* Maximum weights corresponding to the airplane operating conditions (such as ramp, ground or water taxi, takeoff, en route, and landing), environmental conditions (such as altitude and temperature), and loading conditions (such as zero fuel weight, center of gravity position and weight distribution) must be established so that they are not more than—

(1) The highest weight selected by the applicant for the particular conditions; or

(2) The highest weight at which compliance with each applicable structural loading and flight requirement is shown, except that for airplanes equipped with standby power rocket engines the maximum weight must not be more than the highest weight established in accordance with appendix E of this part; or

(3) The highest weight at which compliance is shown with the certification requirements of Part 36 of this chapter.

(b) *Minimum weight.* The minimum weight (the lowest weight at which compliance with each applicable requirement of this part is shown) must be established so that it is not less than—

(1) The lowest weight selected by the applicant;

(2) The design minimum weight (the lowest weight at which compliance with each structural loading condition of this part is shown); or

(3) The lowest weight at which compliance with each applicable flight requirement is shown.

[Doc. No. 5066, 29 FR 18291, Dec. 24, 1964, as amended by Amdt. 25-23, 35 FR 5671, Apr. 8, 1970; Amdt. 25-63, 53 FR 16365, May 6, 1988]

§ 25.27 Center of gravity limits.

The extreme forward and the extreme aft center of gravity limitations must be established for each practicably separable operating condition. No such limit may lie beyond—

(a) The extremes selected by the applicant;

(b) The extremes within which the structure is proven; or

(c) The extremes within which compliance with each applicable flight requirement is shown.

§ 25.29 Empty weight and corresponding center of gravity.

(a) The empty weight and corresponding center of gravity must be determined by weighing the airplane with—

(1) Fixed ballast;

(2) Unusable fuel determined under § 25.959; and

(3) Full operating fluids, including—

(i) Oil;

(ii) Hydraulic fluid; and

(iii) Other fluids required for normal operation of airplane systems, except potable water, lavatory precharge water, and fluids intended for injection in the engine.

(b) The condition of the airplane at the time of determining empty weight must be one that is well defined and can be easily repeated.

[Doc. No. 5066, 29 FR 18291, Dec. 24, 1964, as amended by Amdt. 25-42, 43 FR 2320, Jan. 16, 1978; Amdt. 25-72, 55 FR 29774, July 20, 1990]

§ 25.31 Removable ballast.

Removable ballast may be used on showing compliance with the flight requirements of this subpart.

§ 25.33 Propeller speed and pitch limits.

(a) The propeller speed and pitch must be limited to values that will ensure—

(1) Safe operation under normal operating conditions; and

(2) Compliance with the performance requirements of §§ 25.101 through 25.125.

(b) There must be a propeller speed limiting means at the governor. It must limit the maximum possible governed engine speed to a value not exceeding the maximum allowable r.p.m.

(c) The means used to limit the low pitch position of the propeller blades must be set so that the engine does not exceed 103 percent of the maximum allowable engine rpm or 99 percent of an approved maximum overspeed, whichever is greater, with—

(1) The propeller blades at the low pitch limit and governor inoperative;

(2) The airplane stationary under standard atmospheric conditions with no wind; and

(3) The engines operating at the takeoff manifold pressure limit for reciprocating engine powered airplanes or the maximum takeoff torque limit for turbopropeller engine-powered airplanes.

[Doc. No. 5066, 29 FR 18291, Dec. 24, 1964, as amended by Amdt. 25-57, 49 FR 6848, Feb. 23, 1984; Amdt. 25-72, 55 FR 29774, July 20, 1990]

Performance

§ 25.101 General.

(a) Unless otherwise prescribed, airplanes must meet the applicable performance requirements of this subpart for ambient atmospheric conditions and still air.

(b) The performance, as affected by engine power or thrust, must be based on the following relative humidities;

(1) For turbine engine powered airplanes, a relative humidity of—

(i) 80 percent, at and below standard temperatures; and

(ii) 34 percent, at and above standard temperatures plus 50 °F.

Between these two temperatures, the relative humidity must vary linearly.

(2) For reciprocating engine powered airplanes, a relative humidity of 80 percent in a standard atmosphere. Engine power corrections for vapor pressure must be made in accordance with the following table:

Altitude H (ft.)	Vapor pressure e (In. Hg.)	Specific humidity w (Lb. moisture per lb. dry air)	Density ratio ρ / σ=0.0023769
0	0.403	0.00849	0.99508
1,000	.354	.00773	.96672
2,000	.311	.00703	.93895
3,000	.272	.00638	.91178
4,000	.238	.00578	.88514
5,000	.207	.00523	.85910
6,000	.1805	.00472	.83361
7,000	.1566	.00425	.80870
8,000	.1356	.00382	.78434
9,000	.1172	.00343	.76053
10,000	.1010	.00307	.73722
15,000	.0463	.001710	.62868
20,000	.01978	.000896	.53263
25,000	.00778	.000436	.44806

(c) The performance must correspond to the propulsive thrust available under the particular ambient atmospheric conditions, the particular flight condition, and the relative humidity specified in paragraph (b) of this section. The available propulsive thrust must correspond to engine power or thrust, not exceeding the approved power or thrust less—

(1) Installation losses; and

(2) The power or equivalent thrust absorbed by the accessories and services appropriate to the particular ambient atmospheric conditions and the particular flight condition.

(d) Unless otherwise prescribed, the applicant must select the takeoff, en route, approach, and landing configurations for the airplane.

(e) The airplane configurations may vary with weight, altitude, and temperature, to the extent they are compatible with the operating procedures required by paragraph (f) of this section.

(f) Unless otherwise prescribed, in determining the accelerate-stop distances, takeoff flight paths, takeoff distances, and landing distances, changes in the airplane's configuration, speed, power, and thrust, must be made in accordance with procedures established by the applicant for operation in service.

(g) Procedures for the execution of balked landings and missed approaches associated with the conditions prescribed in §§ 25.119 and 25.121(d) must be established.

(h) The procedures established under paragraphs (f) and (g) of this section must—

(1) Be able to be consistently executed in service by crews of average skill;

(2) Use methods or devices that are safe and reliable; and

(3) Include allowance for any time delays, in the execution of the procedures, that may reasonably be expected in service.

(i) The accelerate-stop and landing distances prescribed in §§ 25.109 and 25.125, respectively, must be determined with all the airplane wheel brake assemblies at the fully worn limit of their allowable wear range.

[Doc. No. 5066, 29 FR 18291, Dec. 24, 1964, as amended by Amdt. 25-38, 41 FR 55466, Dec. 20, 1976; Amdt. 25-92, 63 FR 8318, Feb. 18, 1998]

§ 25.103 Stall speed.

(a) The reference stall speed, V_{SR}, is a calibrated airspeed defined by the applicant. V_{SR} may not be less than a 1-g stall speed. V_{SR} is expressed as:

$$V_{SR} \geq \frac{V_{CL_{MAX}}}{\sqrt{n_{ZW}}}$$

where:

$$\left(\frac{n_{ZW}W}{qS}\right)$$

V_{CLMAX}=Calibrated airspeed obtained when the load factor-corrected lift coefficient

is first a maximum during the maneuver prescribed in paragraph (c) of this section. In addition, when the maneuver is limited by a device that abruptly pushes the nose down at a selected angle of attack (e.g., a stick pusher), V_{CLMAX} may not be less than the speed existing at the instant the device operates;

n_{ZW}=Load factor normal to the flight path at V_{CLMAX}

W=Airplane gross weight;

S=Aerodynamic reference wing area; and

q=Dynamic pressure.

(b) V_{CLMAX} is determined with:

(1) Engines idling, or, if that resultant thrust causes an appreciable decrease in stall speed, not more than zero thrust at the stall speed;

(2) Propeller pitch controls (if applicable) in the takeoff position;

(3) The airplane in other respects (such as flaps, landing gear, and ice accretions) in the condition existing in the test or performance standard in which V_{SR} is being used;

(4) The weight used when V_{SR} is being used as a factor to determine compliance with a required performance standard;

(5) The center of gravity position that results in the highest value of reference stall speed; and

(6) The airplane trimmed for straight flight at a speed selected by the applicant, but not less than 1.13V_{SR} and not greater than 1.3V_{SR}.

(c) Starting from the stabilized trim condition, apply the longitudinal control to decelerate the airplane so that the speed reduction does not exceed one knot per second.

(d) In addition to the requirements of paragraph (a) of this section, when a device that abruptly pushes the nose down at a selected angle of attack (e.g., a stick pusher) is installed, the reference stall speed, V_{SR}, may not be less than 2 knots or 2 percent, whichever is greater, above the speed at which the device operates.

[Doc. No. 28404, 67 FR 70825, Nov. 26, 2002, as amended by Amdt. 25-121, 72 FR 44665, Aug. 8, 2007]

§ 25.105 Takeoff.

(a) The takeoff speeds prescribed by § 25.107, the accelerate-stop distance prescribed by § 25.109, the takeoff path prescribed by § 25.111, the takeoff distance and takeoff run prescribed by § 25.113, and the net takeoff flight path prescribed by § 25.115, must be determined in the selected configuration for takeoff at each weight, altitude, and ambient temperature within the operational limits selected by the applicant—

(1) In non-icing conditions; and

(2) In icing conditions, if in the configuration of § 25.121(b) with the takeoff ice accretion defined in appendix C:

(i) The stall speed at maximum takeoff weight exceeds that in non-icing conditions by more than the greater of 3 knots CAS or 3 percent of V_{SR}; or

(ii) The degradation of the gradient of climb determined in accordance with § 25.121(b) is greater than one-half of the applicable actual-to-net takeoff flight path gradient reduction defined in § 25.115(b).

(b) No takeoff made to determine the data required by this section may require exceptional piloting skill or alertness.

(c) The takeoff data must be based on—

(1) In the case of land planes and amphibians:

(i) Smooth, dry and wet, hard-surfaced runways; and

(ii) At the option of the applicant, grooved or porous friction course wet, hard-surfaced runways.

(2) Smooth water, in the case of seaplanes and amphibians; and

(3) Smooth, dry snow, in the case of skiplanes.

(d) The takeoff data must include, within the established operational limits of the airplane, the following operational correction factors:

(1) Not more than 50 percent of nominal wind components along the takeoff path opposite to the direction of takeoff, and not less than 150 percent of nominal wind components along the takeoff path in the direction of takeoff.

(2) Effective runway gradients.

[Doc. No. 5066, 29 FR 18291, Dec. 24, 1964, as amended by Amdt. 25-92, 63 FR 8318, Feb. 18, 1998; Amdt. 25-121, 72 FR 44665, Aug. 8, 2007]

§ 25.107 Takeoff speeds.

(a) V_1 must be established in relation to V_{EF} as follows:

(1) V_{EF} is the calibrated airspeed at which the critical engine is assumed to fail. V_{EF} must be selected by the applicant, but may not be less than V_{MCG} determined under § 25.149(e).

(2) V_1, in terms of calibrated airspeed, is selected by the applicant; however, V_1 may not be less than V_{EF} plus the speed gained with critical engine inoperative during the time interval between the instant at which the critical engine is failed, and the instant at which the pilot recognizes and reacts to the engine failure, as indicated by the pilot's initiation of the first action (e.g., applying brakes, reducing thrust, deploying speed brakes) to stop the airplane during accelerate-stop tests.

(b) V_{2MIN}, in terms of calibrated airspeed, may not be less than—

(1) 1.13 V_{SR} for—

(i) Two-engine and three-engine turbopropeller and reciprocating engine powered airplanes; and

(ii) Turbojet powered airplanes without provisions for obtaining a significant reduction in the one-engine-inoperative power-on stall speed;

(2) 1.08 V_{SR} for—

(i) Turbopropeller and reciprocating engine powered airplanes with more than three engines; and

(ii) Turbojet powered airplanes with provisions for obtaining a significant reduction in the one-engine-inoperative power-on stall speed; and

(3) 1.10 times V_{MC} established under § 25.149.

(c) V_2, in terms of calibrated airspeed, must be selected by the applicant to provide at least the gradient of climb required by § 25.121(b) but may not be less than—

(1) V_{2MIN};

(2) V_R plus the speed increment attained (in accordance with § 25.111(c)(2)) before reaching a height of 35 feet above the takeoff surface; and

(3) A speed that provides the maneuvering capability specified in § 25.143(h).

(d) V_{MU} is the calibrated airspeed at and above which the airplane can safely lift off the ground, and con- tinue the takeoff. V_{MU} speeds must be selected by the applicant throughout the range of thrust-to-weight ratios to be certificated. These speeds may be established from free air data if these data are verified by ground takeoff tests.

(e) V_R, in terms of calibrated airspeed, must be selected in accordance with the conditions of paragraphs (e)(1) through (4) of this section:

(1) V_R may not be less than—

(i) V_1;

(ii) 105 percent of V_{MC};

(iii) The speed (determined in accordance with § 25.111(c)(2)) that allows reaching V_2 before reaching a height of 35 feet above the takeoff surface; or

(iv) A speed that, if the airplane is rotated at its maximum practicable rate, will result in a V_{LOF} of not less than —

(A) 110 percent of V_{MU} in the all-engines-operating condition, and 105 percent of V_{MU} determined at the thrust-to-weight ratio corresponding to the one-engine-inoperative condition; or

(B) If the V_{MU} attitude is limited by the geometry of the airplane (*i.e.*, tail contact with the runway), 108 percent of V_{MU} in the all-engines-operating condition, and 104 percent of V_{MU} determined at the thrust-to-weight ratio corresponding to the one-engine-inoperative condition.

(2) For any given set of conditions (such as weight, configuration, and temperature), a single value of V_R, obtained in accordance with this paragraph, must be used to show compliance with both the one-engine-inoperative and the all-engines-operating takeoff provisions.

(3) It must be shown that the one-engine-inoperative takeoff distance, using a rotation speed of 5 knots less than V_R established in accordance with paragraphs (e)(1) and (2) of this section, does not exceed the corresponding one-engine-inoperative takeoff distance using the established V_R. The takeoff distances must be determined in accordance with § 25.113(a)(1).

(4) Reasonably expected variations in service from the established takeoff procedures for the operation of the airplane (such as over-rotation of the airplane and out-of-trim conditions) may not result in unsafe flight characteristics or in marked increases in the scheduled takeoff distances established in accordance with § 25.113(a).

(f) V_{LOF} is the calibrated airspeed at which the airplane first becomes airborne.

(g) V_{FTO}, in terms of calibrated airspeed, must be selected by the applicant to provide at least the gradient of climb required by § 25.121(c), but may not be less than—

(1) 1.18 V_{SR}; and

(2) A speed that provides the maneuvering capability specified in § 25.143(h).

(h) In determining the takeoff speeds V_1, V_R, and V_2 for flight in icing conditions, the values of V_{MCG}, V_{MC}, and V_{MU} determined for non-icing conditions may be used.

[Doc. No. 5066, 29 FR 18291, Dec. 24, 1964, as amended by Amdt. 25-38, 41 FR 55466, Dec. 20, 1976; Amdt. 25-42, 43 FR 2320, Jan. 16, 1978; Amdt. 25-92, 63 FR 8318, Feb. 18, 1998; Amdt. 25-94, 63 FR 8848, Feb. 23, 1998; Amdt. 25-108, 67 FR 70826, Nov. 26, 2002; Amdt. 25-121, 72 FR 44665, Aug. 8, 2007; Amdt. 25-135, 76 FR 74654, Dec. 1, 2011]

§ 25.109 Accelerate-stop distance.

(a) The accelerate-stop distance on a dry runway is the greater of the following distances:

(1) The sum of the distances necessary to—

(i) Accelerate the airplane from a standing start with all engines operating to V_{EF} for takeoff from a dry runway;

(ii) Allow the airplane to accelerate from V_{EF} to the highest speed reached during the rejected takeoff, assuming the critical engine fails at V_{EF} and the pilot takes the first action to reject the takeoff at the V_1 for takeoff from a dry runway; and

(iii) Come to a full stop on a dry runway from the speed reached as prescribed in paragraph (a)(1)(ii) of this section; plus

(iv) A distance equivalent to 2 seconds at the V_1 for takeoff from a dry runway.

(2) The sum of the distances necessary to—

(i) Accelerate the airplane from a standing start with all engines operating to the highest speed reached during the rejected takeoff, assuming the pilot takes the first action to reject the takeoff at the V_1 for takeoff from a dry runway; and

(ii) With all engines still operating, come to a full stop on dry runway from the speed reached as prescribed in paragraph (a)(2)(i) of this section; plus

(iii) A distance equivalent to 2 seconds at the V_1 for takeoff from a dry runway.

(b) The accelerate-stop distance on a wet runway is the greater of the following distances:

(1) The accelerate-stop distance on a dry runway determined in accordance with paragraph (a) of this section; or

(2) The accelerate-stop distance determined in accordance with paragraph (a) of this section, except that the runway is wet and the corresponding wet runway values of V_{EF} and V_1 are used. In determining the wet runway accelerate-stop distance, the stopping force from the wheel brakes may never exceed:

(i) The wheel brakes stopping force determined in meeting the requirements of § 25.101(i) and paragraph (a) of this section; and

(ii) The force resulting from the wet runway braking coefficient of friction determined in accordance with paragraphs (c) or (d) of this section, as applicable, taking into account the distribution of the normal load between braked and unbraked wheels at the most adverse center-of-gravity position approved for takeoff.

Tire Pressure (psi)	Maximum Braking Coefficient (tire-to-ground)
50	$\mu_{t/g_{MAX}} = -0.0350\left(\frac{V}{100}\right)^3 + 0.306\left(\frac{V}{100}\right)^2 - 0.851\left(\frac{V}{100}\right) + 0.883$
100	$\mu_{t/g_{MAX}} = -0.0437\left(\frac{V}{100}\right)^3 + 0.320\left(\frac{V}{100}\right)^2 - 0.805\left(\frac{V}{100}\right) + 0.804$
200	$\mu_{t/g_{MAX}} = -0.0331\left(\frac{V}{100}\right)^3 + 0.252\left(\frac{V}{100}\right)^2 - 0.658\left(\frac{V}{100}\right) + 0.692$
300	$\mu_{t/g_{MAX}} = -0.0401\left(\frac{V}{100}\right)^3 + 0.263\left(\frac{V}{100}\right)^2 - 0.611\left(\frac{V}{100}\right) + 0.614$

(c) The wet runway braking coefficient of friction for a smooth wet runway is defined as a curve of friction coefficient versus ground speed and must be computed as follows:

(1) The maximum tire-to-ground wet runway braking coefficient of friction is defined as:

Where—

Tire Pressure=maximum airplane operating tire pressure (psi);

$\mu_{t/gMAX}$=maximum tire-to-ground braking coefficient;

V=airplane true ground speed (knots); and

Linear interpolation may be used for tire pressures other than those listed.

(2) The maximum tire-to-ground wet runway braking coefficient of friction must be adjusted to take into account the efficiency of the anti-skid system on a wet runway. Anti-skid system operation must be demonstrated by flight testing on a smooth wet runway, and its efficiency must be determined. Unless a specific anti-skid system efficiency is determined from a quantitative analysis of the flight testing on a smooth wet runway, the maximum tire-to-ground wet runway braking coefficient of friction determined in paragraph (c)(1) of this section must be multiplied by the efficiency value associated with the type of anti-skid system installed on the airplane:

Type of anti-skid system	Efficiency value
On-Off	0.30
Quasi-Modulating	0.50
Fully Modulating	0.80

(d) At the option of the applicant, a higher wet runway braking coefficient of friction may be used for runway surfaces that have been grooved or treated with a porous friction course material. For grooved and porous friction course runways, the wet runway braking coefficent of friction is defined as either:

(1) 70 percent of the dry runway braking coefficient of friction used to determine the dry runway accelerate-stop distance; or

(2) The wet runway braking coefficient defined in paragraph (c) of this section, except that a specific anti-skid system efficiency, if determined, is appropriate for a grooved or porous friction course wet runway, and the maximum tire-to-ground wet runway braking coefficient of friction is defined as:

Where—

Tire Pressure=maximum airplane operating tire pressure (psi);

$\mu_{t/gMAX}$=maximum tire-to-ground braking coefficient;

V=airplane true ground speed (knots); and

Linear interpolation may be used for tire pressures other than those listed.

(e) Except as provided in paragraph (f)(1) of this section, means other than wheel brakes may be used to determine the accelerate-stop distance if that means—

(1) Is safe and reliable;

(2) Is used so that consistent results can be expected under normal operating conditions; and

(3) Is such that exceptional skill is not required to control the airplane.

(f) The effects of available reverse thrust—

(1) Shall not be included as an additional means of deceleration when determining the accelerate-stop distance on a dry runway; and

(2) May be included as an additional means of deceleration using recommended reverse thrust procedures when

Tire Pressure (psi)	Maximum Braking Coefficient (tire-to-ground)
50	$\mu_{t/g_{MAX}} = 0.1470\left(\frac{V}{100}\right)^5 - 1.050\left(\frac{V}{100}\right)^4 + 2.673\left(\frac{V}{100}\right)^3 - 2.683\left(\frac{V}{100}\right)^2 + 0.403\left(\frac{V}{100}\right) + 0.859$
100	$\mu_{t/g_{MAX}} = 0.1106\left(\frac{V}{100}\right)^5 - 0.813\left(\frac{V}{100}\right)^4 + 2.130\left(\frac{V}{100}\right)^3 - 2.200\left(\frac{V}{100}\right)^2 + 0.317\left(\frac{V}{100}\right) + 0.807$
200	$\mu_{t/g_{MAX}} = 0.0498\left(\frac{V}{100}\right)^5 - 0.398\left(\frac{V}{100}\right)^4 + 1.140\left(\frac{V}{100}\right)^3 - 1.285\left(\frac{V}{100}\right)^2 + 0.140\left(\frac{V}{100}\right) + 0.701$
300	$\mu_{t/g_{MAX}} = 0.0314\left(\frac{V}{100}\right)^5 - 0.247\left(\frac{V}{100}\right)^4 + 0.703\left(\frac{V}{100}\right)^3 - 0.779\left(\frac{V}{100}\right)^2 - 0.00954\left(\frac{V}{100}\right) + 0.614$

determining the accelerate-stop distance on a wet runway, provided the requirements of paragraph (e) of this section are met.

(g) The landing gear must remain extended throughout the accelerate-stop distance.

(h) If the accelerate-stop distance includes a stopway with surface characteristics substantially different from those of the runway, the takeoff data must include operational correction factors for the accelerate-stop distance. The correction factors must account for the particular surface characteristics of the stopway and the variations in these characteristics with seasonal weather conditions (such as temperature, rain, snow, and ice) within the established operational limits.

(i) A flight test demonstration of the maximum brake kinetic energy accelerate-stop distance must be conducted with not more than 10 percent of the allowable brake wear range remaining on each of the airplane wheel brakes.

[Doc. No. 5066, 29 FR 18291, Dec. 24, 1964, as amended by Amdt. 25-42, 43 FR 2321, Jan. 16, 1978; Amdt. 25-92, 63 FR 8318, Feb. 18, 1998]

§ 25.111 Takeoff path.

(a) The takeoff path extends from a standing start to a point in the takeoff at which the airplane is 1,500 feet above the takeoff surface, or at which the transition from the takeoff to the en route configuration is completed and V_{FTO} is reached, whichever point is higher. In addition—

(1) The takeoff path must be based on the procedures prescribed in § 25.101(f);

(2) The airplane must be accelerated on the ground to V_{EF} at which point the critical engine must be made inoperative and remain inoperative for the rest of the takeoff; and

(3) After reaching V_{EF} the airplane must be accelerated to V_2.

(b) During the acceleration to speed V_2, the nose gear may be raised off the ground at a speed not less than V_R. However, landing gear retraction may not be begun until the airplane is airborne.

(c) During the takeoff path determination in accordance with paragraphs (a) and (b) of this section—

(1) The slope of the airborne part of the takeoff path must be positive at each point;

(2) The airplane must reach V_2 before it is 35 feet above the takeoff surface and must continue at a speed as close as practical to, but not less than V_2, until it is 400 feet above the takeoff surface;

(3) At each point along the takeoff path, starting at the point at which the airplane reaches 400 feet above the takeoff surface, the available gradient of climb may not be less than—

(i) 1.2 percent for two-engine airplanes;

(ii) 1.5 percent for three-engine airplanes; and

(iii) 1.7 percent for four-engine airplanes.

(4) The airplane configuration may not be changed, except for gear retraction and automatic propeller feathering, and no change in power or thrust that requires action by the pilot may be made until the airplane is 400 feet above the takeoff surface; and

(5) If § 25.105(a)(2) requires the takeoff path to be determined for flight in icing conditions, the airborne part of the takeoff must be based on the airplane drag:

(i) With the takeoff ice accretion defined in appendix C, from a height of 35 feet above the takeoff surface up to the point where the airplane is 400 feet above the takeoff surface; and

(ii) With the final takeoff ice accretion defined in appendix C, from the point where the airplane is 400 feet above the takeoff surface to the end of the takeoff path.

(d) The takeoff path must be determined by a continuous demonstrated takeoff or by synthesis from segments. If the takeoff path is determined by the segmental method—

(1) The segments must be clearly defined and must be related to the distinct changes in the configuration, power or thrust, and speed;

(2) The weight of the airplane, the configuration, and the power or thrust must be constant throughout each segment and must correspond to the most critical condition prevailing in the segment;

(3) The flight path must be based on the airplane's performance without ground effect; and

(4) The takeoff path data must be checked by continuous demonstrated takeoffs up to the point at which the airplane is out of ground effect and its speed is stabilized, to ensure that the path is conservative relative to the continous path.

The airplane is considered to be out of the ground effect when it reaches a height equal to its wing span.

(e) For airplanes equipped with standby power rocket engines, the takeoff path may be determined in accordance with section II of appendix E.

[Doc. No. 5066, 29 FR 18291, Dec. 24, 1964, as amended by Amdt. 25-6, 30 FR 8468, July 2, 1965; Amdt. 25-42, 43 FR 2321, Jan. 16, 1978; Amdt. 25-54, 45 FR 60172, Sept. 11, 1980; Amdt. 25-72, 55 FR 29774, July 20, 1990; Amdt. 25-94, 63 FR 8848, Feb. 23, 1998; Amdt. 25-108, 67 FR 70826, Nov. 26, 2002; Amdt. 25-115, 69 FR 40527, July 2, 2004; Amdt. 25-121, 72 FR 44666; Aug. 8, 2007]

§ 25.113 Takeoff distance and takeoff run.

(a) Takeoff distance on a dry runway is the greater of—

(1) The horizontal distance along the takeoff path from the start of the takeoff to the point at which the airplane is 35 feet above the takeoff surface, determined under § 25.111 for a dry runway; or

(2) 115 percent of the horizontal distance along the takeoff path, with all engines operating, from the start of the takeoff to the point at which the airplane is 35 feet above the takeoff surface, as determined by a procedure consistent with § 25.111.

(b) Takeoff distance on a wet runway is the greater of—

(1) The takeoff distance on a dry runway determined in accordance with paragraph (a) of this section; or

(2) The horizontal distance along the takeoff path from the start of the takeoff to the point at which the airplane is 15 feet above the takeoff surface, achieved in a manner consistent with the achievement of V_2 before reaching 35 feet above the takeoff surface, determined under § 25.111 for a wet runway.

(c) If the takeoff distance does not include a clearway, the takeoff run is equal to the takeoff distance. If the takeoff distance includes a clearway—

(1) The takeoff run on a dry runway is the greater of—

(i) The horizontal distance along the takeoff path from the start of the takeoff to a point equidistant between the point at which V_{LOF} is reached and the point at which the airplane is 35 feet above the takeoff surface, as determined under § 25.111 for a dry runway; or

(ii) 115 percent of the horizontal distance along the takeoff path, with all engines operating, from the start of the takeoff to a point equidistant between the point at which V_{LOF} is reached and the point at which the airplane is 35 feet above the takeoff surface, determined by a procedure consistent with § 25.111.

(2) The takeoff run on a wet runway is the greater of—

(i) The horizontal distance along the takeoff path from the start of the takeoff to the point at which the airplane is 15 feet above the takeoff surface, achieved in a manner consistent with the achievement of V_2 before reaching 35 feet above the takeoff surface, as determined under § 25.111 for a wet runway; or

(ii) 115 percent of the horizontal distance along the takeoff path, with all engines operating, from the start of the takeoff to a point equidistant between the point at which V_{LOF} is reached and the point at which the airplane is 35 feet above the takeoff surface, determined by a procedure consistent with § 25.111.

[Doc. No. 5066, 29 FR 18291, Dec. 24, 1964, as amended by Amdt. 25-23, 35 FR 5671, Apr. 8, 1970; Amdt. 25-92, 63 FR 8320, Feb. 18, 1998]

§ 25.115 Takeoff flight path.

(a) The takeoff flight path shall be considered to begin 35 feet above the takeoff surface at the end of the takeoff distance determined in accordance with § 25.113(a) or (b), as appropriate for the runway surface condition.

(b) The net takeoff flight path data must be determined so that they represent the actual takeoff flight paths (determined in accordance with § 25.111 and with paragraph (a) of this section) reduced at each point by a gradient of climb equal to—

(1) 0.8 percent for two-engine airplanes;

(2) 0.9 percent for three-engine airplanes; and

(3) 1.0 percent for four-engine airplanes.

(c) The prescribed reduction in climb gradient may be applied as an equivalent reduction in acceleration along that part of the takeoff flight path at which the airplane is accelerated in level flight.

[Doc. No. 5066, 29 FR 18291, Dec. 24, 1964, as amended by Amdt. 25-92, 63 FR 8320, Feb. 18, 1998]

§ 25.117 Climb: general.

Compliance with the requirements of §§ 25.119 and 25.121 must be shown at each weight, altitude, and ambient temperature within the operational limits established for the airplane and with the most unfavorable center of gravity for each configuration.

§ 25.119 Landing climb: All-engines-operating.

In the landing configuration, the steady gradient of climb may not be less than 3.2 percent, with the engines at the power or thrust that is available 8 seconds after initiation of movement of the power or thrust controls from the minimum flight idle to the go-around power or thrust setting—

(a) In non-icing conditions, with a climb speed of V_{REF} determined in accordance with § 25.125(b)(2)(i); and

(b) In icing conditions with the landing ice accretion defined in appendix C, and with a climb speed of V_{REF} determined in accordance with § 25.125(b)(2)(ii).

[Amdt. 25-121, 72 FR 44666; Aug. 8, 2007]

§ 25.121 Climb: One-engine-inoperative.

(a) *Takeoff; landing gear extended.* In the critical takeoff configuration existing along the flight path (between the points at which the airplane reaches V_{LOF} and at which the landing gear is fully retracted) and in the configuration used in § 25.111 but without ground effect, the steady gradient of climb must be positive for two-engine airplanes, and not less than 0.3 percent for three-engine airplanes or 0.5 percent for four-engine airplanes, at V_{LOF} and with—

(1) The critical engine inoperative and the remaining engines at the power or thrust available when retraction of the landing gear is begun in accordance with § 25.111 unless there is a more critical power operating condition existing later along the flight path but before the point at which the landing gear is fully retracted; and

(2) The weight equal to the weight existing when retraction of the landing gear is begun, determined under § 25.111.

(b) *Takeoff; landing gear retracted.* In the takeoff configuration existing at the point of the flight path at which the landing gear is fully retracted, and in the configuration used in § 25.111 but without ground effect:

(1) The steady gradient of climb may not be less than 2.4 percent for two-engine airplanes, 2.7 percent for three-engine airplanes, and 3.0 percent for four-engine airplanes, at V_2 with:

(i) The critical engine inoperative, the remaining engines at the takeoff power or thrust available at the time the landing gear is fully retracted, determined under § 25.111, unless there is a more critical power operating condition existing later along the flight path but before the point where the airplane reaches a height of 400 feet above the takeoff surface; and

(ii) The weight equal to the weight existing when the airplane's landing gear is fully retracted, determined under § 25.111.

(2) The requirements of paragraph (b)(1) of this section must be met:

(i) In non-icing conditions; and

(ii) In icing conditions with the takeoff ice accretion defined in appendix C, if in the configuration of § 25.121(b) with the takeoff ice accretion:

(A) The stall speed at maximum takeoff weight exceeds that in non-icing conditions by more than the greater of 3 knots CAS or 3 percent of V_{SR}; or

(B) The degradation of the gradient of climb determined in accordance with § 25.121(b) is greater than one-half of the applicable actual-to-net takeoff flight path gradient reduction defined in § 25.115(b).

(c) *Final takeoff.* In the en route configuration at the end of the takeoff path determined in accordance with § 25.111:

(1) The steady gradient of climb may not be less than 1.2 percent for two-engine airplanes, 1.5 percent for three-engine airplanes, and 1.7 percent for four-engine airplanes, at V_{FTO} with—

(i) The critical engine inoperative and the remaining engines at the available maximum continuous power or thrust; and

(ii) The weight equal to the weight existing at the end of the takeoff path, determined under § 25.111.

(2) The requirements of paragraph (c)(1) of this section must be met:

(i) In non-icing conditions; and

(ii) In icing conditions with the final takeoff ice accretion defined in appendix C, if in the configuration of § 25.121(b) with the takeoff ice accretion:

(A) The stall speed at maximum takeoff weight exceeds that in non-icing conditions by more than the greater of 3 knots CAS or 3 percent of V_{SR}; or

(B) The degradation of the gradient of climb determined in accordance with § 25.121(b) is greater than one-half of the applicable actual-to-net takeoff flight path gradient reduction defined in § 25.115(b).

(d) *Approach.* In a configuration corresponding to the normal all-engines-operating procedure in which V_{SR} for this configuration does not exceed 110 percent of the V_{SR} for the related all-engines-operating landing configuration:

(1) The steady gradient of climb may not be less than 2.1 percent for two-engine airplanes, 2.4 percent for three-engine airplanes, and 2.7 percent for four-engine airplanes, with—

(i) The critical engine inoperative, the remaining engines at the go-around power or thrust setting;

(ii) The maximum landing weight;

(iii) A climb speed established in connection with normal landing procedures, but not exceeding 1.4 V_{SR}; and

(iv) Landing gear retracted.

(2) The requirements of paragraph (d)(1) of this section must be met:

(i) In non-icing conditions; and

(ii) In icing conditions with the approach ice accretion defined in appendix C. The climb speed selected for non-icing conditions may be used if the climb speed for icing conditions, computed in accordance with paragraph (d)(1)(iii) of this section, does not exceed that for non-icing conditions by more than the greater of 3 knots CAS or 3 percent.

[Doc. No. 5066, 29 FR 18291, Dec. 24, 1964, as amended by Amdt. 25-84, 60 FR 30749, June 9, 1995; Amdt. 25-108, 67 FR 70826, Nov. 26, 2002; Amdt. 25-121, 72 FR 44666; Aug. 8, 2007]

§ 25.123 En route flight paths.

(a) For the en route configuration, the flight paths prescribed in paragraph (b) and (c) of this section must be determined at each weight, altitude, and ambient temperature, within the operating limits established for the airplane. The variation of weight along the flight path, accounting for the progressive consumption of fuel and oil by the operating engines, may be included in the computation. The flight paths must be determined at a speed not less than V_{FTO}, with—

(1) The most unfavorable center of gravity;

(2) The critical engines inoperative;

(3) The remaining engines at the available maximum continuous power or thrust; and

(4) The means for controlling the engine-cooling air supply in the position that provides adequate cooling in the hot-day condition.

(b) The one-engine-inoperative net flight path data must represent the actual climb performance diminished by a gradient of climb of 1.1 percent for two-engine airplanes, 1.4 percent for three-engine airplanes, and 1.6 percent for four-engine airplanes—

(1) In non-icing conditions; and

(2) In icing conditions with the en route ice accretion defined in appendix C, if:

(i) A speed of 1.18 "V_{SR0} with the en route ice accretion exceeds the en route speed selected for non-icing conditions by more than the greater of 3 knots CAS or 3 percent of V_{SR}; or

(ii) The degradation of the gradient of climb is greater than one-half of the applicable actual-to-net flight path reduction defined in paragraph (b) of this section.

(c) For three- or four-engine airplanes, the two-engine-inoperative net flight path data must represent the actual climb performance diminished by a gradient of climb of 0.3 percent for three-engine airplanes and 0.5 percent for four-engine airplanes.

[Docket No. 5066, 29 FR 18291, Dec. 24, 1964, as amended by Amdt. 25-121, 72 FR 44666; Aug. 8, 2007]

§ 25.125 Landing.

(a) The horizontal distance necessary to land and to come to a complete stop (or to a speed of approximately 3 knots for water landings) from a point 50 feet above the landing surface must be determined (for standard temperatures, at each weight, altitude, and wind within the operational limits established by the applicant for the airplane):

(1) In non-icing conditions; and

(2) In icing conditions with the landing ice accretion defined in appendix C if V_{REF} for icing conditions exceeds V_{REF} for non-icing conditions by more than 5 knots CAS at the maximum landing weight.

(b) In determining the distance in paragraph (a) of this section:

(1) The airplane must be in the landing configuration.

(2) A stabilized approach, with a calibrated airspeed of not less than V_{REF}, must be maintained down to the 50-foot height.

(i) In non-icing conditions, V_{REF} may not be less than:

(A) 1.23 V_{SR0};

(B) V_{MCL} established under § 25.149(f); and

(C) A speed that provides the maneuvering capability specified in § 25.143(h).

(ii) In icing conditions, V_{REF} may not be less than:

(A) The speed determined in paragraph (b)(2)(i) of this section;

(B) 1.23 V_{SR0} with the landing ice accretion defined in appendix C if that speed exceeds V_{REF} for non-icing conditions by more than 5 knots CAS; and

(C) A speed that provides the maneuvering capability specified in § 25.143(h) with the landing ice accretion defined in appendix C.

(3) Changes in configuration, power or thrust, and speed, must be made in accordance with the established procedures for service operation.

(4) The landing must be made without excessive vertical acceleration, tendency to bounce, nose over, ground loop, porpoise, or water loop.

(5) The landings may not require exceptional piloting skill or alertness.

(c) For landplanes and amphibians, the landing distance on land must be determined on a level, smooth, dry, hard-surfaced runway. In addition—

(1) The pressures on the wheel braking systems may not exceed those specified by the brake manufacturer;

(2) The brakes may not be used so as to cause excessive wear of brakes or tires; and

(3) Means other than wheel brakes may be used if that means—

(i) Is safe and reliable;

(ii) Is used so that consistent results can be expected in service; and

(iii) Is such that exceptional skill is not required to control the airplane.

(d) For seaplanes and amphibians, the landing distance on water must be determined on smooth water.

(e) For skiplanes, the landing distance on snow must be determined on smooth, dry, snow.

(f) The landing distance data must include correction factors for not more than 50 percent of the nominal wind components along the landing path opposite to the direction of landing, and not less than 150 percent of the nominal wind components along the landing path in the direction of landing.

(g) If any device is used that depends on the operation of any engine, and if the landing distance would be noticeably increased when a landing is made with that engine inoperative, the landing distance must be determined with that engine inoperative unless the use of compensating means will result in a landing distance not more than that with each engine operating.

[Amdt. 25-121, 72 FR 44666; Aug. 8, 2007; 72 FR 50467, Aug. 31, 2007]

CONTROLLABILITY AND MANEUVERABILITY

§ 25.143 General.

(a) The airplane must be safely controllable and maneuverable during—

(1) Takeoff;

(2) Climb;

(3) Level flight;

(4) Descent; and

(5) Landing.

(b) It must be possible to make a smooth transition from one flight condition to any other flight condition without exceptional piloting skill, alertness, or strength, and without danger of exceeding the airplane limit-load factor under any probable operating conditions, including—

(1) The sudden failure of the critical engine;

Force, in pounds, applied to the control wheel or rudder pedals	Pitch	Roll	Yaw
For short term application for pitch and roll control—two hands available for control	75	50	
For short term application for pitch and roll control—one hand available for control	50	25	
For short term application for yaw control			150
For long term application	10	5	20

(2) For airplanes with three or more engines, the sudden failure of the second critical engine when the airplane is in the en route, approach, or landing configuration and is trimmed with the critical engine inoperative; and

(3) Configuration changes, including deployment or retraction of deceleration devices.

(c) The airplane must be shown to be safely controllable and maneuverable with the critical ice accretion appropriate to the phase of flight defined in appendix C, and with the critical engine inoperative and its propeller (if applicable) in the minimum drag position:

(1) At the minimum V_2 for takeoff;

(2) During an approach and go-around; and

(3) During an approach and landing.

(d) The following table prescribes, for conventional wheel type controls, the maximum control forces permitted during the testing required by paragraph (a) through (c) of this section:

(e) Approved operating procedures or conventional operating practices must be followed when demonstrating compliance with the control force limitations for short term application that are prescribed in paragraph (d) of this section. The airplane must be in trim, or as near to being in trim as practical, in the preceding steady flight condition. For the takeoff condition, the airplane must be trimmed according to the approved operating procedures.

(f) When demonstrating compliance with the control force limitations for long term application that are prescribed in paragraph (d) of this section, the airplane must be in trim, or as near to being in trim as practical.

(g) When maneuvering at a constant airspeed or Mach number (up to V_{FC}/M_{FC}), the stick forces and the gradient of the stick force versus maneuvering load factor must lie within satisfactory limits. The stick forces must not be so great as to make excessive demands on the pilot's strength when maneuvering the airplane, and must not be so low that the airplane can easily be overstressed inadvertently. Changes of gradient that occur with changes of load factor must not cause undue difficulty in maintaining control of the airplane, and local gradients must not be so low as to result in a danger of overcontrolling.

(h) The maneuvering capabilities in a constant speed coordinated turn at forward center of gravity, as specified in the following table, must be free of stall warning or other characteristics that might interfere with normal maneuvering:

(i) When demonstrating compliance with § 25.143 in icing conditions—

(1) Controllability must be demonstrated with the ice accretion defined in appendix C that is most critical for the particular flight phase;

(2) It must be shown that a push force is required throughout a pushover maneuver down to a zero g load

Configuration	Speed	Maneuvering bank angle in a coordinated turn	Thrust/power setting
Takeoff	V2	30°	Asymmetric WAT-Limited.[1]
Takeoff	[2]V2 +XX	40°	All-engines-operating climb.[3]
En route	VFTO	40°	Asymmetric WAT-Limited.[1]
Landing	VREF	40°	Symmetric for –3° flight path angle.

[1]A combination of weight, altitude, and temperature (WAT) such that the thrust or power setting produces the minimum climb gradient specified in §25.121 for the flight condition.

[2]Airspeed approved for all-engines-operating initial climb.

[3]That thrust or power setting which, in the event of failure of the critical engine and without any crew action to adjust the thrust or power of the remaining engines, would result in the thrust or power specified for the takeoff condition at V_2, or any lesser thrust or power setting that is used for all-engines-operating initial climb procedures.

factor, or the lowest load factor obtainable if limited by elevator power or other design characteristic of the flight control system. It must be possible to promptly recover from the maneuver without exceeding a pull control force of 50 pounds; and

(3) Any changes in force that the pilot must apply to the pitch control to maintain speed with increasing sideslip angle must be steadily increasing with no force reversals, unless the change in control force is gradual and easily controllable by the pilot without using exceptional piloting skill, alertness, or strength.

(j) For flight in icing conditions before the ice protection system has been activated and is performing its intended function, it must be demonstrated in flight with the ice accretion defined in appendix C, part II(e) of this part that:

(1) The airplane is controllable in a pull-up maneuver up to 1.5 g load factor; and

(2) There is no pitch control force reversal during a pushover maneuver down to 0.5 g load factor.

[Doc. No. 5066, 29 FR 18291, Dec. 24, 1964, as amended by Amdt. 25-42, 43 FR 2321, Jan. 16, 1978; Amdt. 25-84, 60 FR 30749, June 9, 1995; Amdt. 25-108, 67 FR 70826, Nov. 26, 2002; Amdt. 25-121, 72 FR 44667, Aug. 8, 2007; Amdt. 25-129, 74 FR 38339, Aug. 3, 2009]

§ 25.145 Longitudinal control.

(a) It must be possible, at any point between the trim speed prescribed in § 25.103(b)(6) and stall identification (as defined in § 25.201(d)), to pitch the nose downward so that the acceleration to this selected trim speed is prompt with

(1) The airplane trimmed at the trim speed prescribed in § 25.103(b)(6);

(2) The landing gear extended;

(3) The wing flaps (i) retracted and (ii) extended; and

(4) Power (i) off and (ii) at maximum continuous power on the engines.

(b) With the landing gear extended, no change in trim control, or exertion of more than 50 pounds control force (representative of the maximum short term force that can be applied readily by one hand) may be required for the following maneuvers:

(1) With power off, flaps retracted, and the airplane trimmed at 1.3 V_{SR1}, extend the flaps as rapidly as possible while maintaining the airspeed at approximately 30 percent above the reference stall speed existing at each instant throughout the maneuver.

(2) Repeat paragraph (b)(1) except initially extend the flaps and then retract them as rapidly as possible.

(3) Repeat paragraph (b)(2), except at the go-around power or thrust setting.

(4) With power off, flaps retracted, and the airplane trimmed at 1.3 V_{SR1}, rapidly set go-around power or thrust while maintaining the same airspeed.

(5) Repeat paragraph (b)(4) except with flaps extended.

(6) With power off, flaps extended, and the airplane trimmed at 1.3 V_{SR1}, obtain and maintain airspeeds between V_{SW} and either 1.6 V_{SR1} or V_{FE}, whichever is lower.

(c) It must be possible, without exceptional piloting skill, to prevent loss of altitude when complete retraction of the high lift devices from any position is begun during steady, straight, level flight at 1.08 V_{SR1} for propeller powered airplanes, or 1.13 V_{SR1} for turbojet powered airplanes, with—

(1) Simultaneous movement of the power or thrust controls to the go-around power or thrust setting;

(2) The landing gear extended; and

(3) The critical combinations of landing weights and altitudes.

(d) If gated high-lift device control positions are provided, paragraph (c) of this section applies to retractions of the high-lift devices from any position from the maximum landing position to the first gated position, between gated positions, and from the last gated position to the fully retracted position. The requirements of paragraph (c) of this section also apply to retractions from each approved landing position to the control position(s) associated with the high-lift device configuration(s) used to establish the go-around procedure(s) from that landing position. In addition, the first gated control position from the maximum landing position must correspond with a configuration of the high-lift devices used to establish a go-around procedure from a landing configuration. Each gated control position must require a separate and distinct motion of the control to pass through the gated position and must have features to prevent inadvertent movement of the control through the gated position. It must only be possible to make this separate and distinct motion once the control has reached the gated position.

[Doc. No. 5066, 29 FR 18291, Dec. 24, 1964, as amended by Amdt. 25-23, 35 FR 5671, Apr. 8, 1970; Amdt. 25-72, 55 FR 29774, July 20, 1990; Amdt. 25-84, 60 FR 30749, June 9, 1995; Amdt. 25-98, 64 FR 6164, Feb. 8, 1999; 64 FR 10740, Mar. 5, 1999; Amdt. 25-108, 67 FR 70827, Nov. 26, 2002]

§ 25.147 Directional and lateral control.

(a) *Directional control; general.* It must be possible, with the wings level, to yaw into the operative engine and to safely make a reasonably sudden change in heading of up to 15 degrees in the direction of the critical inoperative engine. This must be shown at 1.3 V_S R1 for heading changes up to 15 degrees (except that the heading change at which the rudder pedal force is 150 pounds need not be exceeded), and with—

(1) The critical engine inoperative and its propeller in the minimum drag position;

(2) The power required for level flight at 1.3 V_S R1, but not more than maximum continuous power;

(3) The most unfavorable center of gravity;

(4) Landing gear retracted;

(5) Flaps in the approach position; and

(6) Maximum landing weight.

(b) *Directional control; airplanes with four or more engines.* Airplanes with four or more engines must meet the requirements of paragraph (a) of this section except that—

(1) The two critical engines must be inoperative with their propellers (if applicable) in the minimum drag position;

(2) [Reserved]

(3) The flaps must be in the most favorable climb position.

(c) *Lateral control; general.* It must be possible to make 20° banked turns, with and against the inoperative engine, from steady flight at a speed equal to 1.3 V_S R1, with—

(1) The critical engine inoperative and its propeller (if applicable) in the minimum drag position;

(2) The remaining engines at maximum continuous power;

(3) The most unfavorable center of gravity;

(4) Landing gear (i) retracted and (ii) extended;

(5) Flaps in the most favorable climb position; and

(6) Maximum takeoff weight.

(d) *Lateral control; roll capability.* With the critical engine inoperative, roll response must allow normal maneuvers. Lateral control must be sufficient, at the speeds likely to be used with one engine inoperative, to provide a roll rate necessary for safety without excessive control forces or travel.

(e) *Lateral control; airplanes with four or more engines.* Airplanes with four or more engines must be able to make 20° banked turns, with and against the inoperative engines, from steady flight at a speed equal to 1.3 V_S R1, with maximum continuous power, and with the airplane in the configuration prescribed by paragraph (b) of this section.

(f) *Lateral control; all engines operating.* With the engines operating, roll response must allow normal maneuvers (such as recovery from upsets produced by gusts and the initiation of evasive maneuvers). There must be enough excess lateral control in sideslips (up to sideslip angles that might be required in normal operation), to allow a limited amount of maneuvering and to correct for gusts. Lateral control must be enough at any speed up to V_{FC}/M_{FC} to provide a peak roll rate necessary for safety, without excessive control forces or travel.

[Doc. No. 5066, 29 FR 18291, Dec. 24, 1964, as amended by Amdt. 25-42, 43 FR 2321, Jan. 16, 1978; Amdt. 25-72, 55 FR 29774, July 20, 1990; Amdt. 25-108, 67 FR 70827, Nov. 26, 2002; Amdt. 25-115, 69 FR 40527, July 2, 2004]

§ 25.149 Minimum control speed.

(a) In establishing the minimum control speeds required by this section, the method used to simulate critical engine failure must represent the most critical mode of powerplant failure with respect to controllability expected in service.

(b) V_{MC} is the calibrated airspeed at which, when the critical engine is suddenly made inoperative, it is possible to maintain control of the airplane with that engine still inoperative and maintain straight flight with an angle of bank of not more than 5 degrees.

(c) V_{MC} may not exceed 1.13 V_{SR} with—

(1) Maximum available takeoff power or thrust on the engines;

(2) The most unfavorable center of gravity;

(3) The airplane trimmed for takeoff;

(4) The maximum sea level takeoff weight (or any lesser weight necessary to show V_{MC});

(5) The airplane in the most critical takeoff configuration existing along the flight path after the airplane becomes airborne, except with the landing gear retracted;

(6) The airplane airborne and the ground effect negligible; and

(7) If applicable, the propeller of the inoperative engine—

(i) Windmilling;

(ii) In the most probable position for the specific design of the propeller control; or

(iii) Feathered, if the airplane has an automatic feathering device acceptable for showing compliance with the climb requirements of § 25.121.

(d) The rudder forces required to maintain control at V_{MC} may not exceed 150 pounds nor may it be necessary to reduce power or thrust of the operative engines. During recovery, the airplane may not assume any dangerous attitude or require exceptional piloting skill, alertness, or strength to prevent a heading change of more than 20 degrees.

(e) V_{MCG}, the minimum control speed on the ground, is the calibrated airspeed during the takeoff run at which, when the critical engine is suddenly made inoperative, it is possible to maintain control of the airplane using the rudder control alone (without the use of nosewheel steering), as limited by 150 pounds of force, and the lateral control to the extent of keeping the wings level to enable the takeoff to be safely continued using normal piloting skill. In the determination of V_{MCG}, assuming that the path of the airplane accelerating with all engines operating is along the centerline of the runway, its path from the point at which the critical engine is made inoperative to the point at which recovery to a direction parallel to the centerline is completed may not deviate more than 30 feet laterally from the centerline at any point. V_{MCG} must be established with—

(1) The airplane in each takeoff configuration or, at the option of the applicant, in the most critical takeoff configuration;

(2) Maximum available takeoff power or thrust on the operating engines;

(3) The most unfavorable center of gravity;

(4) The airplane trimmed for takeoff; and

(5) The most unfavorable weight in the range of takeoff weights.

(f) V_{MCL}, the minimum control speed during approach and landing with all engines operating, is the calibrated airspeed at which, when the critical engine is suddenly made inoperative, it is possible to maintain control of the airplane with that engine still inoperative, and maintain straight flight with an angle of bank of not more than 5 degrees. V_{MCL} must be established with—

(1) The airplane in the most critical configuration (or, at the option of the applicant, each configuration) for approach and landing with all engines operating;

(2) The most unfavorable center of gravity;

(3) The airplane trimmed for approach with all engines operating;

(4) The most favorable weight, or, at the option of the applicant, as a function of weight;

(5) For propeller airplanes, the propeller of the inoperative engine in the position it achieves without pilot action, assuming the engine fails while at the power or thrust necessary to maintain a three degree approach path angle; and

(6) Go-around power or thrust setting on the operating engine(s).

(g) For airplanes with three or more engines, V_{MCL-2}, the minimum control speed during approach and landing with one critical engine inoperative, is the calibrated airspeed at which, when a second critical engine is suddenly made inoperative, it is possible to maintain control of the airplane with both engines still inoperative, and maintain straight flight with an angle of bank of not more than 5 degrees. V_{MCL-2} must be established with—

(1) The airplane in the most critical configuration (or, at the option of the applicant, each configuration) for approach and landing with one critical engine inoperative;

(2) The most unfavorable center of gravity;

(3) The airplane trimmed for approach with one critical engine inoperative;

(4) The most unfavorable weight, or, at the option of the applicant, as a function of weight;

(5) For propeller airplanes, the propeller of the more critical inoperative engine in the position it achieves without pilot action, assuming the engine fails while at the power or thrust necessary to maintain a three degree approach path angle, and the propeller of the other inoperative engine feathered;

(6) The power or thrust on the operating engine(s) necessary to maintain an approach path angle of three degrees when one critical engine is inoperative; and

(7) The power or thrust on the operating engine(s) rapidly changed, immediately after the second critical engine is made inoperative, from the power or thrust prescribed in paragraph (g)(6) of this section to—

(i) Minimum power or thrust; and

(ii) Go-around power or thrust setting.

(h) In demonstrations of V_{MCL} and V_{MCL-2} —

(1) The rudder force may not exceed 150 pounds;

(2) The airplane may not exhibit hazardous flight characteristics or require exceptional piloting skill, alertness, or strength;

(3) Lateral control must be sufficient to roll the airplane, from an initial condition of steady flight, through an angle of 20 degrees in the direction necessary to initiate a turn away from the inoperative engine(s), in not more than 5 seconds; and

(4) For propeller airplanes, hazardous flight characteristics must not be exhibited due to any propeller position achieved when the engine fails or during any likely subsequent movements of the engine or propeller controls.

[Doc. No. 5066, 29 FR 18291, Dec. 24, 1964, as amended by Amdt. 25-42, 43 FR 2321, Jan. 16, 1978; Amdt. 25-72, 55 FR 29774, July 20, 1990; 55 FR 37607, Sept. 12, 1990; Amdt. 25-84, 60 FR 30749, June 9, 1995; Amdt. 25-108, 67 FR 70827, Nov. 26, 2002]

Trim

§ 25.161 Trim.

(a) *General.* Each airplane must meet the trim requirements of this section after being trimmed, and without further pressure upon, or movement of, either the primary controls or their corresponding trim controls by the pilot or the automatic pilot.

(b) *Lateral and directional trim.* The airplane must maintain lateral and directional trim with the most adverse lateral displacement of the center of gravity within the relevant operating limitations, during normally expected conditions of operation (including operation at any speed from 1.3 V_{SR1} to V_{MO}/M_{MO}).

(c) *Longitudinal trim.* The airplane must maintain longitudinal trim during—

(1) A climb with maximum continuous power at a speed not more than 1.3 V_{SR1}, with the landing gear retracted, and the flaps (i) retracted and (ii) in the takeoff position;

(2) Either a glide with power off at a speed not more than 1.3 V_{SR1}, or an approach within the normal range of approach speeds appropriate to the weight and configuration with power settings corresponding to a 3 degree glidepath, whichever is the most severe, with the landing gear extended, the wing flaps (i) retracted and (ii) extended, and with the most unfavorable combination of center of gravity position and weight approved for landing; and

(3) Level flight at any speed from 1.3 V_{SR1}, to V_{MO}/M_{MO}, with the landing gear and flaps retracted, and from 1.3 V_{SR1} to V_{LE} with the landing gear extended.

(d) *Longitudinal, directional, and lateral trim.* The airplane must maintain longitudinal, directional, and lateral trim (and for the lateral trim, the angle of bank may not exceed five degrees) at 1.3 V_{SR1} during climbing flight with—

(1) The critical engine inoperative;

(2) The remaining engines at maximum continuous power; and

(3) The landing gear and flaps retracted.

(e) Airplanes with four or more engines. Each airplane with four or more engines must also maintain trim in rectilinear flight with the most unfavorable center of gravity and at the climb speed, configuration, and power required by § 25.123(a) for the purpose of establishing the en route flight paths with two engines inoperative.

[Doc. No. 5066, 29 FR 18291, Dec. 24, 1964, as amended by Amdt. 25-23, 35 FR 5671, Apr. 8, 1970; Amdt. 25-38, 41 FR 55466, Dec. 20, 1976; Amdt. 25-108, 67 FR 70827, Nov. 26, 2002; Amdt. 25-115, 69 FR 40527, July 2, 2004]

Stability

§ 25.171 General.

The airplane must be longitudinally, directionally, and laterally stable in accordance with the provisions of §§ 25.173 through 25.177. In addition, suitable stability and control feel (static stability) is required in any condition normally encountered in service, if flight tests show it is necessary for safe operation.

[Doc. No. 5066, 29 FR 18291, Dec. 24, 1964, as amended by Amdt. 25-7, 30 FR 13117, Oct. 15, 1965]

§ 25.173 Static longitudinal stability.

Under the conditions specified in § 25.175, the characteristics of the elevator control forces (including friction) must be as follows:

(a) A pull must be required to obtain and maintain speeds below the specified trim speed, and a push must be required to obtain and maintain speeds above the specified trim speed. This must be shown at any speed that can be obtained except speeds higher than the landing gear or wing flap operating limit speeds or V_{FC}/M_{FC}, whichever is appropriate, or lower than the minimum speed for steady unstalled flight.

(b) The airspeed must return to within 10 percent of the original trim speed for the climb, approach, and landing conditions specified in § 25.175 (a), (c), and (d), and must return to within 7.5 percent of the original trim speed for the cruising condition specified in § 25.175(b), when the control force is slowly released from any speed within the range specified in paragraph (a) of this section.

(c) The average gradient of the stable slope of the stick force versus speed curve may not be less than 1 pound for each 6 knots.

(d) Within the free return speed range specified in paragraph (b) of this section, it is permissible for the airplane, without control forces, to stabilize on speeds above or below the desired trim speeds if exceptional attention on the part of the pilot is not required to return to and maintain the desired trim speed and altitude.

[Amdt. 25-7, 30 FR 13117, Oct. 15, 1965]

§ 25.175 Demonstration of static longitudinal stability.

Static longitudinal stability must be shown as follows:

(a) *Climb.* The stick force curve must have a stable slope at speeds between 85 and 115 percent of the speed at which the airplane—

(1) Is trimmed, with—

(i) Wing flaps retracted;

(ii) Landing gear retracted;

(iii) Maximum takeoff weight; and

(iv) 75 percent of maximum continuous power for reciprocating engines or the maximum power or thrust selected by the applicant as an operating limitation for use during climb for turbine engines; and

(2) Is trimmed at the speed for best rate-of-climb except that the speed need not be less than 1.3 V_{SR1}.

(b) *Cruise.* Static longitudinal stability must be shown in the cruise condition as follows:

(1) With the landing gear retracted at high speed, the stick force curve must have a stable slope at all speeds within a range which is the greater of 15 percent of the trim speed plus the resulting free return speed range, or 50 knots plus the resulting free return speed range, above and below the trim speed (except that the speed range need not include speeds less than 1.3 V_{SR1}, nor speeds greater than V_{FC}/M_{FC}, nor speeds that require a stick force of more than 50 pounds), with—

(i) The wing flaps retracted;

(ii) The center of gravity in the most adverse position (see § 25.27);

(iii) The most critical weight between the maximum takeoff and maximum landing weights;

(iv) 75 percent of maximum continuous power for reciprocating engines or for turbine engines, the maximum cruising power selected by the applicant as an operating limitation (see § 25.1521), except that the power need not exceed that required at V_{MO}/M_{MO}; and

(v) The airplane trimmed for level flight with the power required in paragraph (b)(1)(iv) of this section.

(2) With the landing gear retracted at low speed, the stick force curve must have a stable slope at all speeds within a range which is the greater of 15 percent of the trim speed plus the resulting free return speed range, or 50 knots plus the resulting free return speed range, above and below the trim speed (except that the speed range need not include speeds less than 1.3 V_{SR1}, nor speeds greater than the minimum speed of the applicable speed range prescribed in paragraph (b)(1), nor speeds that require a stick force of more than 50 pounds), with—

(i) Wing flaps, center of gravity position, and weight as specified in paragraph (b)(1) of this section;

(ii) Power required for level flight at a speed equal to $(V_{MO}+1.3\ V_{SR1})/2$; and

(iii) The airplane trimmed for level flight with the power required in paragraph (b)(2)(ii) of this section.

(3) With the landing gear extended, the stick force curve must have a stable slope at all speeds within a range which is the greater of 15 percent of the trim speed plus the resulting free return speed range, or 50 knots plus the resulting free return speed range, above and below the trim speed (except that the speed range need not include speeds less than 1.3 V_{SR1}, nor speeds greater than V_{LE}, nor speeds that require a stick force of more than 50 pounds), with—

(i) Wing flap, center of gravity position, and weight as specified in paragraph (b)(1) of this section;

(ii) 75 percent of maximum continuous power for reciprocating engines or, for turbine engines, the maximum cruising power selected by the applicant as an operating limitation, except that the power need not exceed that required for level flight at V_{LE}; and

(iii) The aircraft trimmed for level flight with the power required in paragraph (b)(3)(ii) of this section.

(c) *Approach.* The stick force curve must have a stable slope at speeds between V_{SW} and 1.7 V_{SR1}, with—

(1) Wing flaps in the approach position;

(2) Landing gear retracted;

(3) Maximum landing weight; and

(4) The airplane trimmed at 1.3 V_{SR1} with enough power to maintain level flight at this speed.

(d) *Landing.* The stick force curve must have a stable slope, and the stick force may not exceed 80 pounds, at speeds between V_{SW} and 1.7 V_{SR0} with—

(1) Wing flaps in the landing position;

(2) Landing gear extended;

(3) Maximum landing weight;

(4) The airplane trimmed at 1.3 V_{SR0} with—

(i) Power or thrust off, and

(ii) Power or thrust for level flight.

(5) The airplane trimmed at 1.3 V_{SR0} with power or thrust off.

[Doc. No. 5066, 29 FR 18291, Dec. 24, 1964, as amended by Amdt. 25-7, 30 FR 13117, Oct. 15, 1965; Amdt. 25-108, 67 FR 70827, Nov. 26, 2002; Amdt. 25-115, 69 FR 40527, July 2, 2004]

§ 25.177 Static lateral-directional stability.

(a) The static directional stability (as shown by the tendency to recover from a skid with the rudder free) must be positive for any landing gear and flap position and symmetric power condition, at speeds from 1.13 V_{SR1}, up to V_{FE}, V_{LE}, or V_{FC}/M_{FC} (as appropriate for the airplane configuration).

(b) The static lateral stability (as shown by the tendency to raise the low wing in a sideslip with the aileron controls free) for any landing gear and flap position and symmetric power condition, may not be negative at any airspeed (except that speeds higher than V_{FE} need not be considered for flaps extended configurations nor speeds higher than V_{LE} for landing gear extended configurations) in the following airspeed ranges:

(1) From 1.13 V_{SR1} to V_{MO}/M_{MO}.

(2) From V_{MO}/M_{MO} to V_{FC}/M_{FC}, unless the divergence is—

(i) Gradual;

(ii) Easily recognizable by the pilot; and

(iii) Easily controllable by the pilot.

(c) The following requirement must be met for the configurations and speed specified in paragraph (a) of this section. In straight, steady sideslips over the range of sideslip angles appropriate to the operation of the airplane, the aileron and rudder control movements and forces must be substantially proportional to the angle of sideslip in a stable sense. This factor of proportionality must lie between limits found necessary for safe operation. The range of sideslip angles evaluated must include those sideslip angles resulting from the lesser of:

(1) One-half of the available rudder control input; and

(2) A rudder control force of 180 pounds.

(d) For sideslip angles greater than those prescribed by paragraph (c) of this section, up to the angle at which full rudder control is used or a rudder control force of 180 pounds is obtained, the rudder control forces may not reverse, and increased rudder deflection must be needed for increased angles of sideslip. Compliance with this requirement must be shown using straight, steady sideslips, unless full lateral control input is achieved before reaching either full rudder control input or a rudder control force of 180 pounds; a straight, steady sideslip need not be maintained after achieving full lateral control input. This requirement must be met at all approved landing gear and flap positions for the range of operating speeds and power conditions appropriate to each landing gear and flap position with all engines operating.

[Amdt. 25-135, 76 FR 74654, Dec. 1, 2011]

§ 25.181 Dynamic stability.

(a) Any short period oscillation, not including combined lateral-directional oscillations, occurring between 1.13 V_{SR} and maximum allowable speed appropriate to the configuration of the airplane must be heavily damped with the primary controls—

(1) Free; and

(2) In a fixed position.

(b) Any combined lateral-directional oscillations ("Dutch roll") occurring between 1.13 V_{SR} and maximum allowable speed appropriate to the configuration of the airplane must be positively damped with controls free, and must be controllable with normal use of the primary controls without requiring exceptional pilot skill.

[Amdt. 25-42, 43 FR 2322, Jan. 16, 1978, as amended by Amdt. 25-72, 55 FR 29775, July 20, 1990; 55 FR 37607, Sept. 12, 1990; Amdt. 25-108, 67 FR 70827, Nov. 26, 2002]

Stalls

§ 25.201 Stall demonstration.

(a) Stalls must be shown in straight flight and in 30 degree banked turns with—

(1) Power off; and

(2) The power necessary to maintain level flight at 1.5 V_{SR1} (where V_{SR1} corresponds to the reference stall speed at maximum landing weight with flaps in the approach position and the landing gear retracted).

(b) In each condition required by paragraph (a) of this section, it must be possible to meet the applicable requirements of § 25.203 with—

(1) Flaps, landing gear, and deceleration devices in any likely combination of positions approved for operation;

(2) Representative weights within the range for which certification is requested;

(3) The most adverse center of gravity for recovery; and

(4) The airplane trimmed for straight flight at the speed prescribed in § 25.103(b)(6).

(c) The following procedures must be used to show compliance with § 25.203;

(1) Starting at a speed sufficiently above the stalling speed to ensure that a steady rate of speed reduction can be established, apply the longitudinal control so that the speed reduction does not exceed one knot per second until the airplane is stalled.

(2) In addition, for turning flight stalls, apply the longitudinal control to achieve airspeed deceleration rates up to 3 knots per second.

(3) As soon as the airplane is stalled, recover by normal recovery techniques.

(d) The airplane is considered stalled when the behavior of the airplane gives the pilot a clear and distinctive indication of an acceptable nature that the airplane is stalled. Acceptable indications of a stall, occurring either individually or in combination, are—

(1) A nose-down pitch that cannot be readily arrested;

(2) Buffeting, of a magnitude and severity that is a strong and effective deterrent to further speed reduction; or

(3) The pitch control reaches the aft stop and no further increase in pitch attitude occurs when the control is held full aft for a short time before recovery is initiated.

[Doc. No. 5066, 29 FR 18291, Dec. 24, 1964, as amended by Amdt. 25-84, 60 FR 30750, June 9, 1995; Amdt. 25-108, 67 FR 70827, Nov. 26, 2002]

§ 25.203 Stall characteristics.

(a) It must be possible to produce and to correct roll and yaw by unreversed use of the aileron and rudder controls, up to the time the airplane is stalled. No abnormal nose-up pitching may occur. The longitudinal control force must be positive up to and throughout the stall. In addition, it must be possible to promptly prevent stalling and to recover from a stall by normal use of the controls.

(b) For level wing stalls, the roll occurring between the stall and the completion of the recovery may not exceed approximately 20 degrees.

(c) For turning flight stalls, the action of the airplane after the stall may not be so violent or extreme as to make it difficult, with normal piloting skill, to effect a prompt recovery and to regain control of the airplane. The maximum bank angle that occurs during the recovery may not exceed—

(1) Approximately 60 degrees in the original direction of the turn, or 30 degrees in the opposite direction, for deceleration rates up to 1 knot per second; and

(2) Approximately 90 degrees in the original direction of the turn, or 60 degrees in the opposite direction, for deceleration rates in excess of 1 knot per second.

[Doc. No. 5066, 29 FR 18291, Dec. 24, 1964, as amended by Amdt. 25-84, 60 FR 30750, June 9, 1995]

§ 25.207 Stall warning.

(a) Stall warning with sufficient margin to prevent inadvertent stalling with the flaps and landing gear in any normal position must be clear and distinctive to the pilot in straight and turning flight.

(b) The warning must be furnished either through the inherent aerodynamic qualities of the airplane or by a device that will give clearly distinguishable indications under expected conditions of flight. However, a visual stall warning device that requires the attention of the crew within the cockpit is not acceptable by itself. If a warning device is used, it must provide a warning in each of the airplane configurations prescribed in paragraph (a) of this section at the speed prescribed in paragraphs (c) and (d) of this section. Except for showing compliance with the stall warning margin prescribed in paragraph (h)(3)(ii) of this section, stall warning for flight in icing conditions must be provided by the same means as stall warning for flight in non-icing conditions.

(c) When the speed is reduced at rates not exceeding one knot per second, stall warning must begin, in each normal configuration, at a speed, V_{SW}, exceeding the speed at which the stall is identified in accordance with § 25.201(d) by not less than five knots or five percent CAS, whichever is greater. Once initiated, stall warning must continue until the angle of attack is reduced to approximately that at which stall warning began.

(d) In addition to the requirement of paragraph (c) of this section, when the speed is reduced at rates not exceeding one knot per second, in straight flight with engines idling and at the center-of-gravity position specified in § 25.103(b)(5), V_{SW}, in each normal configuration, must exceed V_{SR} by not less than three knots or three percent CAS, whichever is greater.

(e) In icing conditions, the stall warning margin in straight and turning flight must be sufficient to allow the pilot to prevent stalling (as defined in § 25.201(d)) when the pilot starts a recovery maneuver not less than three seconds after the onset of stall warning. When demonstrating compliance with this paragraph, the pilot must perform the recovery maneuver in the same way as for the airplane in non-icing conditions. Compliance with this requirement must be demonstrated in flight with the speed reduced at rates not exceeding one knot per second, with—

(1) The more critical of the takeoff ice and final takeoff ice accretions defined in appendix C for each configuration used in the takeoff phase of flight;

(2) The en route ice accretion defined in appendix C for the en route configuration;

(3) The holding ice accretion defined in appendix C for the holding configuration(s);

(4) The approach ice accretion defined in appendix C for the approach configuration(s); and

(5) The landing ice accretion defined in appendix C for the landing and go-around configuration(s).

(f) The stall warning margin must be sufficient in both non-icing and icing conditions to allow the pilot to prevent stalling when the pilot starts a recovery maneuver not less than one second after the onset of stall warning in slow-down turns with at least 1.5 g load factor normal to the flight path and airspeed deceleration rates of at least 2 knots per second. When demonstrating compliance with this paragraph for icing conditions, the pilot must perform the recovery maneuver in the same way as for the airplane in non-icing conditions. Compliance with this requirement must be demonstrated in flight with—

(1) The flaps and landing gear in any normal position;

(2) The airplane trimmed for straight flight at a speed of 1.3 V_{SR}; and

(3) The power or thrust necessary to maintain level flight at 1.3 V_{SR}.

(g) Stall warning must also be provided in each abnormal configuration of the high lift devices that is likely to be used in flight following system failures (including all configurations covered by Airplane Flight Manual procedures).

(h) For flight in icing conditions before the ice protection system has been activated and is performing its intended function, with the ice accretion defined in appendix C, part II(e) of this part, the stall warning margin in straight and turning flight must be sufficient to allow the pilot to prevent stalling without encountering any adverse flight characteristics when:

(1) The speed is reduced at rates not exceeding one knot per second;

(2) The pilot performs the recovery maneuver in the same way as for flight in non-icing conditions; and

(3) The recovery maneuver is started no earlier than:

(i) One second after the onset of stall warning if stall warning is provided by the same means as for flight in non-icing conditions; or

(ii) Three seconds after the onset of stall warning if stall warning is provided by a different means than for flight in non-icing conditions.

(i) In showing compliance with paragraph (h) of this section, if stall warning is provided by a different means in icing conditions than for non-icing conditions, compliance with § 25.203 must be shown using the accretion defined in appendix C, part II(e) of this part. Compliance with this requirement must be shown using the demonstration prescribed by § 25.201, except that the deceleration rates of § 25.201(c)(2) need not be demonstrated.

[Doc. No. 5066, 29 FR 18291, Dec. 24, 1964, as amended by Amdt. 25-7, 30 FR 13118, Oct. 15, 1965; Amdt. 25-42, 43 FR 2322, Jan. 16, 1978; Amdt. 25-108, 67 FR 70827, Nov. 26, 2002; Amdt. 25-121, 72 FR 44668, Aug. 8, 2007; Amdt. 25-129, 74 FR 38339, Aug. 3, 2009]

Ground and Water Handling Characteristics

§ 25.231 Longitudinal stability and control.

(a) Landplanes may have no uncontrollable tendency to nose over in any reasonably expected operating condition or when rebound occurs during landing or takeoff. In addition—

(1) Wheel brakes must operate smoothly and may not cause any undue tendency to nose over; and

(2) If a tail-wheel landing gear is used, it must be possible, during the takeoff ground run on concrete, to maintain any attitude up to thrust line level, at 75 percent of V_{SR1}.

(b) For seaplanes and amphibians, the most adverse water conditions safe for takeoff, taxiing, and landing, must be established.

[Docket No. 5066, 29 FR 18291, Dec. 24, 1964, as amended by Amdt. 25-108, 67 FR 70828, Nov. 26, 2002]

§ 25.233 Directional stability and control.

(a) There may be no uncontrollable ground-looping tendency in 90° cross winds, up to a wind velocity of 20 knots or 0.2 V_{SR0}, whichever is greater, except that the wind velocity need not exceed 25 knots at any speed at which the airplane may be expected to be operated on the ground. This may be shown while establishing the 90° cross component of wind velocity required by § 25.237.

(b) Landplanes must be satisfactorily controllable, without exceptional piloting skill or alertness, in power-off landings at normal landing speed, without using brakes or engine power to maintain a straight path. This may be shown during power-off landings made in conjunction with other tests.

(c) The airplane must have adequate directional control during taxiing. This may be shown during taxiing prior to takeoffs made in conjunction with other tests.

[Doc. No. 5066, 29 FR 18291, Dec. 24, 1964, as amended by Amdt. 25-23, 35 FR 5671, Apr. 8, 1970; Amdt. 25-42, 43 FR 2322, Jan. 16, 1978; Amdt. 25-94, 63 FR 8848, Feb. 23, 1998; Amdt. 25-108, 67 FR 70828, Nov. 26, 2002]

§ 25.235 Taxiing condition.

The shock absorbing mechanism may not damage the structure of the airplane when the airplane is taxied on the roughest ground that may reasonably be expected in normal operation.

§ 25.237 Wind velocities.

(a) For land planes and amphibians, the following applies:

(1) A 90-degree cross component of wind velocity, demonstrated to be safe for takeoff and landing, must be established for dry runways and must be at least 20 knots or 0.2 V_{SR0}, whichever is greater, except that it need not exceed 25 knots.

(2) The crosswind component for takeoff established without ice accretions is valid in icing conditions.

(3) The landing crosswind component must be established for:

(i) Non-icing conditions, and

(ii) Icing conditions with the landing ice accretion defined in appendix C.

(b) For seaplanes and amphibians, the following applies:

(1) A 90-degree cross component of wind velocity, up to which takeoff and landing is safe under all water conditions that may reasonably be expected in normal operation, must be established and must be at least 20 knots or 0.2 V_{SR0}, whichever is greater, except that it need not exceed 25 knots.

(2) A wind velocity, for which taxiing is safe in any direction under all water conditions that may reasonably be expected in normal operation, must be established and must be at least 20 knots or 0.2 V_{SR0}, whichever is greater, except that it need not exceed 25 knots.

[Amdt. 25-42, 43 FR 2322, Jan. 16, 1978, as amended by Amdt. 25-108, 67 FR 70827, Nov. 26, 2002; Amdt. 25-121, 72 FR 44668, Aug. 8, 2007]

§ 25.239 Spray characteristics, control, and stability on water.

(a) For seaplanes and amphibians, during takeoff, taxiing, and landing, and in the conditions set forth in paragraph (b) of this section, there may be no—

(1) Spray characteristics that would impair the pilot's view, cause damage, or result in the taking in of an undue quantity of water;

(2) Dangerously uncontrollable porpoising, bounding, or swinging tendency; or

(3) Immersion of auxiliary floats or sponsons, wing tips, propeller blades, or other parts not designed to withstand the resulting water loads.

(b) Compliance with the requirements of paragraph (a) of this section must be shown—

(1) In water conditions, from smooth to the most adverse condition established in accordance with § 25.231;

(2) In wind and cross-wind velocities, water currents, and associated waves and swells that may reasonably be expected in operation on water;

(3) At speeds that may reasonably be expected in operation on water;

(4) With sudden failure of the critical engine at any time while on water; and

(5) At each weight and center of gravity position, relevant to each operating condition, within the range of loading conditions for which certification is requested.

(c) In the water conditions of paragraph (b) of this section, and in the corresponding wind conditions, the seaplane or amphibian must be able to drift for five minutes with engines inoperative, aided, if necessary, by a sea anchor.

Miscellaneous Flight Requirements

§ 25.251 Vibration and buffeting.

(a) The airplane must be demonstrated in flight to be free from any vibration and buffeting that would prevent continued safe flight in any likely operating condition.

(b) Each part of the airplane must be demonstrated in flight to be free from excessive vibration under any appropriate speed and power conditions up to V_{DF}/M_{DF}. The maximum speeds shown must be used in establishing the operating limitations of the airplane in accordance with § 25.1505.

(c) Except as provided in paragraph (d) of this section, there may be no buffeting condition, in normal flight, including configuration changes during cruise, severe enough to interfere with the control of the airplane, to cause excessive fatigue to the crew, or to cause structural damage. Stall warning buffeting within these limits is allowable.

(d) There may be no perceptible buffeting condition in the cruise configuration in straight flight at any speed up to V_{MO}/M_{MO}, except that stall warning buffeting is allowable.

(e) For an airplane with M_D greater than .6 or with a maximum operating altitude greater than 25,000 feet, the positive maneuvering load factors at which the onset of perceptible buffeting occurs must be determined with the airplane in the cruise configuration for the ranges of airspeed or Mach number, weight, and altitude for which the airplane is to be certificated. The envelopes of load factor, speed, altitude, and weight must provide a sufficient range of speeds and load factors for normal operations. Probable inadvertent excursions beyond the boundaries of the buffet onset envelopes may not result in unsafe conditions.

[Doc. No. 5066, 29 FR 18291, Dec. 24, 1964, as amended by Amdt. 25-23, 35 FR 5671, Apr. 8, 1970; Amdt. 25-72, 55 FR 29775, July 20, 1990; Amdt. 25-77, 57 FR 28949, June 29, 1992]

§ 25.253 High-speed characteristics.

(a) *Speed increase and recovery characteristics.* The following speed increase and recovery characteristics must be met:

(1) Operating conditions and characteristics likely to cause inadvertent speed increases (including upsets in pitch and roll) must be simulated with the airplane trimmed at any likely cruise speed up to V_{MO}/M_{MO}. These conditions and characteristics include gust upsets, inadvertent control movements, low stick force gradient in relation to control friction, passenger movement, leveling off from climb, and descent from Mach to airspeed limit altitudes.

(2) Allowing for pilot reaction time after effective inherent or artificial speed warning occurs, it must be shown that the airplane can be recovered to a normal attitude and its speed reduced to V_{MO}/M_{MO}, without—

(i) Exceptional piloting strength or skill;

(ii) Exceeding V_D/M_D, V_{DF}/M_{DF}, or the structural limitations; and

(iii) Buffeting that would impair the pilot's ability to read the instruments or control the airplane for recovery.

(3) With the airplane trimmed at any speed up to V_{MO}/M_{MO}, there must be no reversal of the response to control input about any axis at any speed up to V_{DF}/M_{DF}. Any tendency to pitch, roll, or yaw must be mild and readily controllable, using normal piloting techniques. When the airplane is trimmed at V_{MO}/M_{MO}, the slope of the elevator control force versus speed curve need not be stable at speeds greater than V_{FC}/M_{FC}, but there must be a push force at all speeds up to V_{DF}/M_{DF} and there must be no sudden or excessive reduction of elevator control force as V_{DF}/M_{DF} is reached.

(4) Adequate roll capability to assure a prompt recovery from a lateral upset condition must be available at any speed up to V_{DF}/M_{DF}.

(5) With the airplane trimmed at V_{MO}/M_{MO}, extension of the speedbrakes over the available range of movements of the pilot's control, at all speeds above V_{MO}/M_{MO}, but not so high that V_{DF}/M_{DF} would be exceeded during the maneuver, must not result in:

(i) An excessive positive load factor when the pilot does not take action to counteract the effects of extension;

(ii) Buffeting that would impair the pilot's ability to read the instruments or control the airplane for recovery; or

(iii) A nose down pitching moment, unless it is small.

(b) *Maximum speed for stability characteristics,* V_{FC}/M_{FC}. V_{FC}/M_{FC} is the maximum speed at which the requirements of §§ 25.143(g), 25.147(f), 25.175(b)(1), 25.177(a) through (c), and 25.181 must be met with flaps and landing gear retracted. Except as noted in § 25.253(c), V_{FC}/M_{FC} may not be less than a speed midway between V_{MO}/M_{MO} and V_{DF}/M_{DF}, except that, for altitudes where Mach number is the limiting factor, M_{FC} need not exceed the Mach number at which effective speed warning occurs.

(c) *Maximum speed for stability characteristics in icing conditions.* The maximum speed for stability characteristics with the ice accretions defined in appendix C, at which the requirements of §§ 25.143(g), 25.147(f), 25.175(b)(1), 25.177(a) through (c), and 25.181 must be met, is the lower of:

(1) 300 knots CAS;

(2) V_{FC}; or

(3) A speed at which it is demonstrated that the airframe will be free of ice accretion due to the effects of increased dynamic pressure.

[Doc. No. 5066, 29 FR 18291, Dec. 24, 1964, as amended by Amdt. 25-23, 35 FR 5671, Apr. 8, 1970; Amdt. 25-54, 45 FR 60172, Sept. 11, 1980; Amdt. 25-72, 55 FR 29775, July 20, 1990; Amdt. 25-84, 60 FR 30750, June 9, 1995; Amdt. 25-121, 72 FR 44668, Aug. 8, 2007; Amdt. 25-135, 76 FR 74654, Dec. 1, 2011]

§ 25.255 Out-of-trim characteristics.

(a) From an initial condition with the airplane trimmed at cruise speeds up to V_{MO}/M_{MO}, the airplane must have satisfactory maneuvering stability and controllability with the degree of out-of-trim in both the airplane nose-up and nose-down directions, which results from the greater of—

(1) A three-second movement of the longitudinal trim system at its normal rate for the particular flight condition with no aerodynamic load (or an equivalent degree of trim for airplanes that do not have a power-operated trim system), except as limited by stops in the trim system, including those required by § 25.655(b) for adjustable stabilizers; or

(2) The maximum mistrim that can be sustained by the autopilot while maintaining level flight in the high speed cruising condition.

(b) In the out-of-trim condition specified in paragraph (a) of this section, when the normal acceleration is varied from +1 g to the positive and negative values specified in paragraph (c) of this section—

(1) The stick force vs. g curve must have a positive slope at any speed up to and including V_{FC}/M_{FC}; and

(2) At speeds between V_{FC}/M_{FC} and V_{DF}/M_{DF} the direction of the primary longitudinal control force may not reverse.

(c) Except as provided in paragraphs (d) and (e) of this section, compliance with the provisions of paragraph (a) of this section must be demonstrated in flight over the acceleration range—

(1) −1 g to +2.5 g; or

(2) 0 g to 2.0 g, and extrapolating by an acceptable method to −1 g and +2.5 g.

(d) If the procedure set forth in paragraph (c)(2) of this section is used to demonstrate compliance and marginal conditions exist during flight test with regard to reversal of primary longitudinal control force, flight tests must be accomplished from the normal acceleration at which a marginal condition is found to exist to the applicable limit specified in paragraph (b)(1) of this section.

(e) During flight tests required by paragraph (a) of this section, the limit maneuvering load factors prescribed in §§ 25.333(b) and 25.337, and the maneuvering load factors associated with probable inadvertent excursions beyond the boundaries of the buffet onset envelopes determined under § 25.251(e), need not be exceeded. In addition, the entry speeds for flight test demonstrations at normal acceleration values less than 1 g must be limited to the extent necessary to accomplish a recovery without exceeding V_{DF}/M_{DF}.

(f) In the out-of-trim condition specified in paragraph (a) of this section, it must be possible from an overspeed condition at V_{DF}/M_{DF} to produce at least 1.5 g for recovery by applying not more than 125 pounds of longitudinal control force using either the primary longitudinal control alone or the primary longitudinal control and the longitudinal trim system. If the longitudinal trim is used to assist in producing the required load factor, it must be shown at V_{DF}/M_{DF} that the longitudinal trim can be actuated in the airplane nose-up direction with the primary surface loaded to correspond to the least of the following airplane nose-up control forces:

(1) The maximum control forces expected in service as specified in §§ 25.301 and 25.397.

(2) The control force required to produce 1.5 g.

(3) The control force corresponding to buffeting or other phenomena of such intensity that it is a strong deterrent to further application of primary longitudinal control force.

[Amdt. 25-42, 43 FR 2322, Jan. 16, 1978]

Subpart C—Structure

General

§ 25.301 Loads.

(a) Strength requirements are specified in terms of limit loads (the maximum loads to be expected in service) and ultimate loads (limit loads multiplied by prescribed factors of safety). Unless otherwise provided, prescribed loads are limit loads.

(b) Unless otherwise provided, the specified air, ground, and water loads must be placed in equilibrium with inertia forces, considering each item of mass in the airplane. These loads must be distributed to conservatively approximate or closely represent actual conditions. Methods used to determine load intensities and distribution must be validated by flight load measurement unless the methods used for determining those loading conditions are shown to be reliable.

(c) If deflections under load would significantly change the distribution of external or internal loads, this redistribution must be taken into account.

[Doc. No. 5066, 29 FR 18291, Dec. 24, 1964, as amended by Amdt. 25-23, 35 FR 5672, Apr. 8, 1970]

§ 25.303 Factor of safety.

Unless otherwise specified, a factor of safety of 1.5 must be applied to the prescribed limit load which are considered external loads on the structure. When a loading condition is prescribed in terms of ultimate loads, a factor of safety need not be applied unless otherwise specified.

[Amdt. 25-23, 35 FR 5672, Apr. 8, 1970]

§ 25.305 Strength and deformation.

(a) The structure must be able to support limit loads without detrimental permanent deformation. At any load up to limit loads, the deformation may not interfere with safe operation.

(b) The structure must be able to support ultimate loads without failure for at least 3 seconds. However, when proof of strength is shown by dynamic tests simulating actual load conditions, the 3-second limit does not apply. Static tests conducted to ultimate load must include the ultimate deflections and ultimate deformation induced by the loading. When analytical methods are used to show compliance with the ultimate load strength requirements, it must be shown that—

(1) The effects of deformation are not significant;

(2) The deformations involved are fully accounted for in the analysis; or

(3) The methods and assumptions used are sufficient to cover the effects of these deformations.

(c) Where structural flexibility is such that any rate of load application likely to occur in the operating conditions might produce transient stresses appreciably higher than those corresponding to static loads, the effects of this rate of application must be considered.

(d) [Reserved]

(e) The airplane must be designed to withstand any vibration and buffeting that might occur in any likely operating condition up to V_D/M_D, including stall and probable inadvertent excursions beyond the boundaries of the buffet onset envelope. This must be shown by analysis, flight tests, or other tests found necessary by the Administrator.

(f) Unless shown to be extremely improbable, the airplane must be designed to withstand any forced structural vibration resulting from any failure, malfunction or adverse condition in the flight control system. These must be considered limit loads and must be investigated at airspeeds up to V_C/M_C.

[Doc. No. 5066, 29 FR 18291, Dec. 24, 1964, as amended by Amdt. 25-23, 35 FR 5672, Apr. 8, 1970; Amdt. 25-54, 45 FR 60172, Sept. 11, 1980; Amdt. 25-77, 57 FR 28949, June 29, 1992; Amdt. 25-86, 61 FR 5220, Feb. 9, 1996]

§ 25.307 Proof of structure.

(a) Compliance with the strength and deformation requirements of this subpart must be shown for each critical loading condition. Structural analysis may be used only if the structure conforms to that for which experience has shown this method to be reliable. The Administrator may require ultimate load tests in cases where limit load tests may be inadequate.

(b)-(c) [Reserved]

(d) When static or dynamic tests are used to show compliance with the requirements of § 25.305(b) for flight structures, appropriate material correction factors must be applied to the test results, unless the structure, or part thereof, being tested has features such that a number of elements contribute to the total strength of the structure and the failure of one element results in the redistribution of the load through alternate load paths.

[Doc. No. 5066, 29 FR 18291, Dec. 24, 1964, as amended by Amdt. 25-23, 35 FR 5672, Apr. 8, 1970; Amdt. 25-54, 45 FR 60172, Sept. 11, 1980; Amdt. 25-72, 55 FR 29775, July 20, 1990]

Flight Loads

§ 25.321 General.

(a) Flight load factors represent the ratio of the aerodynamic force component (acting normal to the assumed longitudinal axis of the airplane) to the weight of the airplane. A positive load factor is one in which the aerodynamic force acts upward with respect to the airplane.

(b) Considering compressibility effects at each speed, compliance with the flight load requirements of this subpart must be shown—

(1) At each critical altitude within the range of altitudes selected by the applicant;

(2) At each weight from the design minimum weight to the design maximum weight appropriate to each particular flight load condition; and

(3) For each required altitude and weight, for any practicable distribution of disposable load within the operating limitations recorded in the Airplane Flight Manual.

(c) Enough points on and within the boundaries of the design envelope must be investigated to ensure that the maximum load for each part of the airplane structure is obtained.

(d) The significant forces acting on the airplane must be placed in equilibrium in a rational or conservative manner. The linear inertia forces must be considered in equilibrium with the thrust and all aerodynamic loads, while the angular (pitching) inertia forces must be considered in equilibrium with thrust and all aerodynamic moments, including moments due to loads on components such as tail surfaces and nacelles. Critical thrust values in the range from zero to maximum continuous thrust must be considered.

[Doc. No. 5066, 29 FR 18291, Dec. 24, 1964, as amended by Amdt. 25-23, 35 FR 5672, Apr. 8, 1970; Amdt. 25-86, 61 FR 5220, Feb. 9, 1996]

Flight Maneuver and Gust Conditions

§ 25.331 Symmetric maneuvering conditions.

(a) *Procedure.* For the analysis of the maneuvering flight conditions specified in paragraphs (b) and (c) of this section, the following provisions apply:

(1) Where sudden displacement of a control is specified, the assumed rate of control surface displacement may not be less than the rate that could be applied by the pilot through the control system.

(2) In determining elevator angles and chordwise load distribution in the maneuvering conditions of paragraphs (b) and (c) of this section, the effect of corresponding pitching velocities must be taken into account. The in-trim and out-of-trim flight conditions specified in § 25.255 must be considered.

(b) *Maneuvering balanced conditions.* Assuming the airplane to be in equilibrium with zero pitching acceleration, the maneuvering conditions A through I on the maneuvering envelope in § 25.333(b) must be investigated.

(c) *Pitch maneuver conditions.* The conditions specified in paragraphs (c)(1) and (2) of this section must be investigated. The movement of the pitch control surfaces may be adjusted to take into account limitations imposed by the maximum pilot effort specified by § 25.397(b), control system stops and any indirect effect imposed by limitations in the output side of the control system (for example, stalling torque or maximum rate obtainable by a power control system.)

(1) *Maximum pitch control displacement at V_A.* The airplane is assumed to be flying in steady level flight (point A_1, § 25.333(b)) and the cockpit pitch control is suddenly moved to obtain extreme nose up pitching acceleration. In defining the tail load, the response of the airplane must be taken into account. Airplane loads that occur subsequent to the time when normal acceleration at the c.g. exceeds the positive limit maneuvering load factor (at point A_2 in § 25.333(b)), or the resulting tailplane normal load reaches its maximum, whichever occurs first, need not be considered.

(2) *Specified control displacement.* A checked maneuver, based on a rational pitching control motion vs. time profile, must be established in which the design limit load factor specified in § 25.337 will not be exceeded. Unless lesser values cannot be exceeded, the airplane response must result in pitching accelerations not less than the following:

(i) A positive pitching acceleration (nose up) is assumed to be reached concurrently with the airplane load factor of 1.0 (Points A_1 to D_1, § 25.333(b)). The positive acceleration must be equal to at least

$$\frac{39n}{v}(n-1,5),(Radians / sec.^2)$$

where—

n is the positive load factor at the speed under consideration, and Vis the airplane equivalent speed in knots.

(ii) A negative pitching acceleration (nose down) is assumed to be reached concurrently with the positive maneuvering load factor (points A_2 to D_2, § 25.333(b)). This negative pitching acceleration must be equal to at least

$$\frac{-26n}{v}(n-1,5),(Radians / sec.^2)$$

where—

n is the positive load factor at the speed under consideration; and V is the airplane equivalent speed in knots.

[Doc. No. 5066, 29 FR 18291, Dec. 24, 1964, as amended by Amdt. 25-23, 35 FR 5672, Apr. 8, 1970; Amdt. 25-46, 43 FR 50594, Oct. 30, 1978; 43 FR 52495, Nov. 13, 1978; 43 FR 54082, Nov. 20, 1978; Amdt. 25-72, 55 FR 29775, July 20, 1990; 55 FR 37607, Sept. 12, 1990; Amdt. 25-86, 61 FR 5220, Feb. 9, 1996; Amdt. 25-91, 62 FR 40704, July 29, 1997]

§ 25.333 Flight maneuvering envelope.

(a) *General.* The strength requirements must be met at each combination of airspeed and load factor on and within the boundaries of the representative maneuvering envelope (*V-n* diagram) of paragraph (b) of this section. This envelope must also be used in determining the airplane structural operating limitations as specified in § 25.1501.

(b) *Maneuvering envelope.*

[Doc. No. 5066, 29 FR 18291, Dec. 24, 1964, as amended by Amdt. 25-86, 61 FR 5220, Feb. 9, 1996]

§ 25.335 Design airspeeds.

The selected design airspeeds are equivalent airspeeds (EAS). Estimated values of V_{S0} and V_{S1} must be conservative.

(a) *Design cruising speed, V_C.* For V_C, the following apply:

(1) The minimum value of V_C must be sufficiently greater than V_B to provide for inadvertent speed increases likely to occur as a result of severe atmospheric turbulence.

(2) Except as provided in § 25.335(d)(2), V_C may not be less than V_B+1.32 U_{REF} (with U_{REF} as specified in § 25.341(a)(5)(i)). However V_C need not exceed the maximum speed

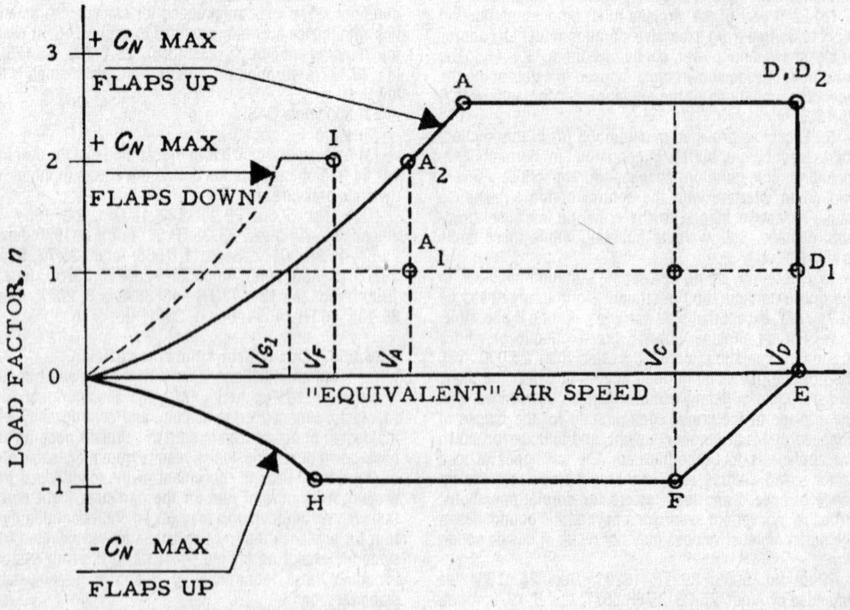

in level flight at maximum continuous power for the corresponding altitude.

(3) At altitudes where V_D is limited by Mach number, V_C may be limited to a selected Mach number.

(b) *Design dive speed,* V_D. V_D must be selected so that V_C/M_C is not greater than 0.8 V_D/M_D, or so that the minimum speed margin between V_C/M_C and V_D/M_D is the greater of the following values:

(1) From an initial condition of stabilized flight at V_C/M_C, the airplane is upset, flown for 20 seconds along a flight path 7.5° below the initial path, and then pulled up at a load factor of 1.5 *g* (0.5 *g* acceleration increment). The speed increase occurring in this maneuver may be calculated if reliable or conservative aerodynamic data is used. Power as specified in § 25.175(b)(1)(iv) is assumed until the pullup is initiated, at which time power reduction and the use of pilot controlled drag devices may be assumed;

(2) The minimum speed margin must be enough to provide for atmospheric variations (such as horizontal gusts, and penetration of jet streams and cold fronts) and for instrument errors and airframe production variations. These factors may be considered on a probability basis. The margin at altitude where M_C is limited by compressibility effects must not less than 0.07M unless a lower margin is determined using a rational analysis that includes the effects of any automatic systems. In any case, the margin may not be reduced to less than 0.05M.

(c) *Design maneuvering speed* V_A. For V_A, the following apply:

(1) V_A may not be less than $V_{S1}\sqrt{n}$ where—

(i) *n* is the limit positive maneuvering load factor at V_C; and

(ii) V_{S1} is the stalling speed with flaps retracted.

(2) V_A and V_S must be evaluated at the design weight and altitude under consideration.

(3) V_A need not be more than V_C or the speed at which the positive C_{Nmax} curve intersects the positive maneuver load factor line, whichever is less.

(d) *Design speed for maximum gust intensity,* V_B.

(1) V_B may not be less than

$$V_{S1}\left[1+\frac{K_g U_{ref} V_c a}{498w}\right]^{\frac{1}{2}}$$

where—

V_{S1}=the 1-g stalling speed based on C_{NAmax} with the flaps retracted at the particular weight under consideration;

V_c=design cruise speed (knots equivalent airspeed);

U_{ref}=the reference gust velocity (feet per second equivalent airspeed) from § 25.341(a)(5)(i);

w=average wing loading (pounds per square foot) at the particular weight under consideration.

$$K_g = \frac{.88\mu}{5.3+\mu}$$

$$\mu = \frac{2w}{pcag}$$

ρ=density of air (slugs/ft³);

c=mean geometric chord of the wing (feet);

g=acceleration due to gravity (ft/sec²);

a=slope of the airplane normal force coefficient curve, C_{NA} per radian;

(2) At altitudes where V_C is limited by Mach number—

(i) V_B may be chosen to provide an optimum margin between low and high speed buffet boundaries; and,

(ii) V_B need not be greater than V_C.

(e) *Design flap speeds,* V_F. For V_F, the following apply:

(1) The design flap speed for each flap position (established in accordance with § 25.697(a)) must be sufficiently greater than the operating speed recommended for the corresponding stage of flight (including balked landings) to allow for probable variations in control of airspeed and for transition from one flap position to another.

(2) If an automatic flap positioning or load limiting device is used, the speeds and corresponding flap positions programmed or allowed by the device may be used.

(3) V_F may not be less than—

(i) 1.6 V_{S1} with the flaps in takeoff position at maximum takeoff weight;

(ii) 1.8 V_{S1} with the flaps in approach position at maximum landing weight, and

(iii) 1.8 V_{S0} with the flaps in landing position at maximum landing weight.

(f) *Design drag device speeds,* V_{DD}. The selected design speed for each drag device must be sufficiently greater than the speed recommended for the operation of the device to allow for probable variations in speed control. For drag devices intended for use in high speed descents, V_{DD} may not be less than V_D. When an automatic drag device positioning or load limiting means is used, the speeds and corresponding drag device positions programmed or allowed by the automatic means must be used for design.

[Doc. No. 5066, 29 FR 18291, Dec. 24, 1964, as amended by Amdt. 25-23, 35 FR 5672, Apr. 8, 1970; Amdt. 25-86, 61 FR 5220, Feb. 9, 1996; Amdt. 25-91, 62 FR 40704, July 29, 1997]

§ 25.337 Limit maneuvering load factors.

(a) Except where limited by maximum (static) lift coefficients, the airplane is assumed to be subjected to symmetrical maneuvers resulting in the limit maneuvering load factors prescribed in this section. Pitching velocities appropriate to the corresponding pull-up and steady turn maneuvers must be taken into account.

(b) The positive limit maneuvering load factor *n* for any speed up to V_n may not be less than 2.1+24,000/(*W* +10,000) except that *n* may not be less than 2.5 and need not be greater than 3.8—where *W* is the design maximum takeoff weight.

(c) The negative limit maneuvering load factor—

(1) May not be less than −1.0 at speeds up to V_C; and

(2) Must vary linearly with speed from the value at V_C to zero at V_D.

(d) Maneuvering load factors lower than those specified in this section may be used if the airplane has design features that make it impossible to exceed these values in flight.

[Doc. No. 5066, 29 FR 18291, Dec. 24, 1964, as amended by Amdt. 25-23, 35 FR 5672, Apr. 8, 1970]

§ 25.341 Gust and turbulence loads.

(a) *Discrete Gust Design Criteria.* The airplane is assumed to be subjected to symmetrical vertical and lateral gusts in level flight. Limit gust loads must be determined in accordance with the provisions:

(1) Loads on each part of the structure must be determined by dynamic analysis. The analysis must take into account unsteady aerodynamic characteristics and all significant structural degrees of freedom including rigid body motions.

(2) The shape of the gust must be:

$$U = \frac{U_{ds}}{2}\left[1 - Cos\left(\frac{\pi s}{H}\right)\right]$$

for 0 ≤ s ≤ 2H

where—

s=distance penetrated into the gust (feet);

U_{ds}=the design gust velocity in equivalent airspeed specified in paragraph (a)(4) of this section; and

H=the gust gradient which is the distance (feet) parallel to the airplane's flight path for the gust to reach its peak velocity.

(3) A sufficient number of gust gradient distances in the range 30 feet to 350 feet must be investigated to find the critical response for each load quantity.

(4) The design gust velocity must be:

$$U_{ds} = U_{ref} F_g \left(H/350\right)^{1/6}$$

where—

U_{ref}=the reference gust velocity in equivalent airspeed defined in paragraph (a)(5) of this section.

F_g=the flight profile alleviation factor defined in paragraph (a)(6) of this section.

(5) The following reference gust velocities apply:

(i) At the airplane design speed V_C: Positive and negative gusts with reference gust velocities of 56.0 ft/sec EAS must be considered at sea level. The reference gust velocity may be reduced linearly from 56.0 ft/sec EAS at sea level to 44.0 ft/sec EAS at 15000 feet. The reference gust velocity may be further reduced linearly from 44.0 ft/sec EAS at 15000 feet to 26.0 ft/sec EAS at 50000 feet.

(ii) At the airplane design speed V_D: The reference gust velocity must be 0.5 times the value obtained under § 25.341(a)(5)(i).

(6) The flight profile alleviation factor, F_g, must be increased linearly from the sea level value to a value of 1.0 at the maximum operating altitude defined in § 25.1527. At sea level, the flight profile alleviation factor is determined by the following equation:

$$F_g = 0.5\left(F_{gx} + F_{gm}\right)$$

Where:

$$F_{gx} = 1 - \frac{Z_{mo}}{250000};$$

$$F_{gm} = \sqrt{R_2 Tan\left(\pi R_1/4\right)}$$

$$R_1 = \frac{\text{Maximum Landing Weight}}{\text{Maximum Take-off Weight}};$$

$$R_2 = \frac{\text{Maximum Zero Fuel Weight}}{\text{Maximum Take-off Weight}};$$

Z_{mo}=Maximum operating altitude defined in § 25.1527.

(7) When a stability augmentation system is included in the analysis, the effect of any significant system nonlinearities should be accounted for when deriving limit loads from limit gust conditions.

(b) *Continuous Gust Design Criteria.* The dynamic response of the airplane to vertical and lateral continuous turbulence must be taken into account. The continuous gust design criteria of appendix G of this part must be used to establish the dynamic response unless more rational criteria are shown.

[Doc. No. 27902, 61 FR 5221, Feb. 9, 1996; 61 FR 9533, Mar. 8, 1996]

§ 25.343 Design fuel and oil loads.

(a) The disposable load combinations must include each fuel and oil load in the range from zero fuel and oil to the selected maximum fuel and oil load. A structural reserve fuel condition, not exceeding 45 minutes of fuel under the operating conditions in § 25.1001(e) and (f), as applicable, may be selected.

(b) If a structural reserve fuel condition is selected, it must be used as the minimum fuel weight condition for showing compliance with the flight load requirements as prescribed in this subpart. In addition—

(1) The structure must be designed for a condition of zero fuel and oil in the wing at limit loads corresponding to—

(i) A maneuvering load factor of +2.25; and

(ii) The gust conditions of § 25.341(a) but assuming 85% of the design velocities prescribed in § 25.341(a)(4).

(2) Fatigue evaluation of the structure must account for any increase in operating stresses resulting from the design condition of paragraph (b)(1) of this section; and

(3) The flutter, deformation, and vibration requirements must also be met with zero fuel.

[Doc. No. 5066, 29 FR 18291, Dec. 24, 1964, as amended by Amdt. 25-18, 33 FR 12226, Aug. 30, 1968; Amdt. 25-72, 55 FR 37607, Sept. 12, 1990; Amdt. 25-86, 61 FR 5221, Feb. 9, 1996]

§ 25.345 High lift devices.

(a) If wing flaps are to be used during takeoff, approach, or landing, at the design flap speeds established for these stages of flight under § 25.335(e) and with the wing flaps in the corresponding positions, the airplane is assumed to be subjected to symmetrical maneuvers and gusts. The resulting limit loads must correspond to the conditions determined as follows:

(1) Maneuvering to a positive limit load factor of 2.0; and

(2) Positive and negative gusts of 25 ft/sec EAS acting normal to the flight path in level flight. Gust loads resulting on each part of the structure must be determined by rational analysis. The analysis must take into account the unsteady aerodynamic characteristics and rigid body motions of the aircraft. The shape of the gust must be as described in § 25.341(a)(2) except that—

U_{ds}=25 ft/sec EAS;

H=12.5 c; and

c=mean geometric chord of the wing (feet).

(b) The airplane must be designed for the conditions prescribed in paragraph (a) of this section, except that the airplane load factor need not exceed 1.0, taking into account, as separate conditions, the effects of—

(1) Propeller slipstream corresponding to maximum continuous power at the design flap speeds V_F, and with takeoff power at not less than 1.4 times the stalling speed for the particular flap position and associated maximum weight; and

(2) A head-on gust of 25 feet per second velocity (EAS).

(c) If flaps or other high lift devices are to be used in en route conditions, and with flaps in the appropriate position at speeds up to the flap design speed chosen for these conditions, the airplane is assumed to be subjected to symmetrical maneuvers and gusts within the range determined by—

(1) Maneuvering to a positive limit load factor as prescribed in § 25.337(b); and

(2) The discrete vertical gust criteria in § 25.341(a).

(d) The airplane must be designed for a maneuvering load factor of 1.5 g at the maximum take-off weight with the wing-flaps and similar high lift devices in the landing configurations.

[Doc. No. 5066, 29 FR 18291, Dec. 24, 1964, as amended by Amdt. 25-46, 43 FR 50595, Oct. 30, 1978; Amdt. 25-72, 55 FR 37607, Sept. 17, 1990; Amdt. 25-86, 61 FR 5221, Feb. 9, 1996; Amdt. 25-91, 62 FR 40704, July 29, 1997]

§ 25.349 Rolling conditions.

The airplane must be designed for loads resulting from the rolling conditions specified in paragraphs (a) and (b) of this section. Unbalanced aerodynamic moments about the center of gravity must be reacted in a rational or conservative manner, considering the principal masses furnishing the reacting inertia forces.

(a) *Maneuvering.* The following conditions, speeds, and aileron deflections (except as the deflections may be limited by pilot effort) must be considered in combination with an airplane load factor of zero and of two-thirds of the positive maneuvering factor used in design. In determining the required aileron deflections, the torsional flexibility of the wing must be considered in accordance with § 25.301(b):

(1) Conditions corresponding to steady rolling velocities must be investigated. In addition, conditions corresponding to maximum angular acceleration must be investigated for airplanes with engines or other weight concentrations outboard of the fuselage. For the angular acceleration conditions, zero rolling velocity may be assumed in the absence of a rational time history investigation of the maneuver.

(2) At V_A, a sudden deflection of the aileron to the stop is assumed.

(3) At V_C, the aileron deflection must be that required to produce a rate of roll not less than that obtained in paragraph (a)(2) of this section.

(4) At V_D, the aileron deflection must be that required to produce a rate of roll not less than one-third of that in paragraph (a)(2) of this section.

(b) *Unsymmetrical gusts.* The airplane is assumed to be subjected to unsymmetrical vertical gusts in level flight. The resulting limit loads must be determined from either the wing maximum airload derived directly from § 25.341(a), or the wing maximum airload derived indirectly from the vertical load factor calculated from § 25.341(a). It must be assumed that 100 percent of the wing air load acts on one side of the airplane and 80 percent of the wing air load acts on the other side.

[Doc. No. 5066, 29 FR 18291, Dec. 24, 1964, as amended by Amdt. 25-23, 35 FR 5672, Apr. 8, 1970; Amdt. 25-86, 61 FR 5222, Feb. 9, 1996; Amdt. 25-94, 63 FR 8848, Feb. 23, 1998]

§ 25.351 Yaw maneuver conditions.

The airplane must be designed for loads resulting from the yaw maneuver conditions specified in paragraphs (a) through (d) of this section at speeds from V_{MC} to V_D. Unbalanced aerodynamic moments about the center of gravity must be reacted in a rational or conservative manner considering the airplane inertia forces. In computing the tail loads the yawing velocity may be assumed to be zero.

(a) With the airplane in unaccelerated flight at zero yaw, it is assumed that the cockpit rudder control is suddenly displaced to achieve the resulting rudder deflection, as limited by:

(1) The control system on control surface stops; or

(2) A limit pilot force of 300 pounds from V_{MC} to V_A and 200 pounds from V_C/M_C to V_D/M_D, with a linear variation between V_A and V_C/M_C.

(b) With the cockpit rudder control deflected so as always to maintain the maximum rudder deflection available within the limitations specified in paragraph (a) of this section, it is assumed that the airplane yaws to the overswing sideslip angle.

(c) With the airplane yawed to the static equilibrium sideslip angle, it is assumed that the cockpit rudder control is held so as to achieve the maximum rudder deflection available within the limitations specified in paragraph (a) of this section.

(d) With the airplane yawed to the static equilibrium sideslip angle of paragraph (c) of this section, it is assumed that the cockpit rudder control is suddenly returned to neutral.

[Amdt. 25-91, 62 FR 40704, July 29, 1997]

Supplementary Conditions

§ 25.361 Engine torque.

(a) Each engine mount and its supporting structure must be designed for the effects of—

(1) A limit engine torque corresponding to takeoff power and propeller speed acting simultaneously with 75 percent of the limit loads from flight condition A of § 25.333(b);

(2) A limit torque corresponding to the maximum continuous power and propeller speed, acting simultaneously with the limit loads from flight condition A of § 25.333(b); and

(3) For turbopropeller installations, in addition to the conditions specified in paragraphs (a)(1) and (2) of this section, a limit engine torque corresponding to takeoff power and propeller speed, multiplied by a factor accounting for propeller control system malfunction, including quick feathering, acting simultaneously with 1g level flight loads. In the absence of a rational analysis, a factor of 1.6 must be used.

(b) For turbine engine installations, the engine mounts and supporting structure must be designed to withstand each of the following:

(1) A limit engine torque load imposed by sudden engine stoppage due to malfunction or structural failure (such as compressor jamming).

(2) A limit engine torque load imposed by the maximum acceleration of the engine.

(c) The limit engine torque to be considered under paragraph (a) of this section must be obtained by multiplying mean torque for the specified power and speed by a factor of—

(1) 1.25 for turbopropeller installations;

(2) 1.33 for reciprocating engines with five or more cylinders; or

(3) Two, three, or four, for engines with four, three, or two cylinders, respectively.

[Doc. No. 5066, 29 FR 18291, Dec. 24, 1964, as amended by Amdt. 25-23, 35 FR 5672, Apr. 8, 1970; Amdt. 25-46, 43 FR 50595, Oct. 30, 1978; Amdt. 25-72, 55 FR 29776, July 20, 1990]

§ 25.363 Side load on engine and auxiliary power unit mounts.

(a) Each engine and auxiliary power unit mount and its supporting structure must be designed for a limit load factor in lateral direction, for the side load on the engine and auxiliary power unit mount, at least equal to the maximum load factor obtained in the yawing conditions but not less than—

(1) 1.33; or

(2) One-third of the limit load factor for flight condition A as prescribed in § 25.333(b).

(b) The side load prescribed in paragraph (a) of this section may be assumed to be independent of other flight conditions.

[Doc. No. 5066, 29 FR 18291, Dec. 24, 1964, as amended by Amdt. 25-23, 35 FR 5672, Apr. 8, 1970; Amdt. 25-91, 62 FR 40704, July 29, 1997]

§ 25.365 Pressurized compartment loads.

For airplanes with one or more pressurized compartments the following apply:

(a) The airplane structure must be strong enough to withstand the flight loads combined with pressure differential loads from zero up to the maximum relief valve setting.

(b) The external pressure distribution in flight, and stress concentrations and fatigue effects must be accounted for.

(c) If landings may be made with the compartment pressurized, landing loads must be combined with pressure differential loads from zero up to the maximum allowed during landing.

(d) The airplane structure must be designed to be able to withstand the pressure differential loads corresponding to the maximum relief valve setting multiplied by a factor of 1.33 for airplanes to be approved for operation to 45,000 feet or by a factor of 1.67 for airplanes to be approved for operation above 45,000 feet, omitting other loads.

(e) Any structure, component or part, inside or outside a pressurized compartment, the failure of which could interfere with continued safe flight and landing, must be designed to withstand the effects of a sudden release of pressure through an opening in any compartment at any operating altitude resulting from each of the following conditions:

(1) The penetration of the compartment by a portion of an engine following an engine disintegration;

(2) Any opening in any pressurized compartment up to the size H_o in square feet; however, small compartments may be combined with an adjacent pressurized compartment and both considered as a single compartment for openings that cannot reasonably be expected to be confined to the small compartment. The size H_o must be computed by the following formula:

$H_o=PA_s$

where,

H_o=Maximum opening in square feet, need not exceed 20 square feet.

$P=(A_s/6240)+.024$

A_s=Maximum cross-sectional area of the pressurized shell normal to the longitudinal axis, in square feet; and

(3) The maximum opening caused by airplane or equipment failures not shown to be extremely improbable.

(f) In complying with paragraph (e) of this section, the fail-safe features of the design may be considered in determining the probability of failure or penetration and probable size of openings, provided that possible improper operation of closure devices and inadvertent door openings are also considered. Furthermore, the resulting differential pressure loads must be combined in a rational and conservative manner with 1-g level flight loads and any loads arising from emergency depressurization conditions. These loads may be considered as ultimate conditions; however, any deformations associated with these conditions must not interfere with continued safe flight and landing. The pressure relief provided by intercompartment venting may also be considered.

(g) Bulkheads, floors, and partitions in pressurized compartments for occupants must be designed to withstand the conditions specified in paragraph (e) of this section. In addition, reasonable design precautions must be taken to minimize the probability of parts becoming detached and injuring occupants while in their seats.

[Doc. No. 5066, 29 FR 18291, Dec. 24, 1964, as amended by Amdt. 25-54, 45 FR 60172, Sept. 11, 1980; Amdt. 25-71, 55 FR 13477, Apr. 10, 1990; Amdt. 25-72, 55 FR 29776, July 20, 1990; Amdt. 25-87, 61 FR 28695, June 5, 1996]

§ 25.367 Unsymmetrical loads due to engine failure.

(a) The airplane must be designed for the unsymmetrical loads resulting from the failure of the critical engine. Turbopropeller airplanes must be designed for the following conditions in combination with a single malfunction of the propeller drag limiting system, considering the probable pilot corrective action on the flight controls:

(1) At speeds between V_{MC} and V_D, the loads resulting from power failure because of fuel flow interruption are considered to be limit loads.

(2) At speeds between V_{MC} and V_C, the loads resulting from the disconnection of the engine compressor from the turbine or from loss of the turbine blades are considered to be ultimate loads.

(3) The time history of the thrust decay and drag buildup occurring as a result of the prescribed engine failures must be substantiated by test or other data applicable to the particular engine-propeller combination.

(4) The timing and magnitude of the probable pilot corrective action must be conservatively estimated, considering the characteristics of the particular engine-propeller-airplane combination.

(b) Pilot corrective action may be assumed to be initiated at the time maximum yawing velocity is reached, but not earlier than two seconds after the engine failure. The magnitude of the corrective action may be based on the control forces specified in § 25.397(b) except that lower forces may be assumed where it is shown by anaylsis or test that these forces can control the yaw and roll resulting from the prescribed engine failure conditions.

§ 25.371 Gyroscopic loads.

The structure supporting any engine or auxiliary power unit must be designed for the loads including the gyroscopic loads arising from the conditions specified in §§ 25.331, 25.341(a), 25.349, 25.351, 25.473, 25.479, and 25.481, with the engine or auxiliary power unit at the maximum rpm appropriate to the condition. For the purposes of compliance with this section, the pitch maneuver in § 25.331(c)(1) must be carried out until the positive limit maneuvering load factor (point A_2 in § 25.333(b)) is reached.

[Amdt. 25-91, 62 FR 40704, July 29, 1997]

§ 25.373 Speed control devices.

If speed control devices (such as spoilers and drag flaps) are installed for use in en route conditions—

(a) The airplane must be designed for the symmetrical maneuvers prescribed in § 25.333 and § 25.337, the yawing maneuvers prescribed in § 25.351, and the vertical and later gust conditions prescribed in § 25.341(a), at each setting and the maximum speed associated with that setting; and

(b) If the device has automatic operating or load limiting features, the airplane must be designed for the maneuver and gust conditions prescribed in paragraph (a) of this section, at the speeds and corresponding device positions that the mechanism allows.

[Doc. No. 5066, 29 FR 18291, Dec. 24, 1964, as amended by Amdt. 25-72, 55 FR 29776, July 20, 1990; Amdt. 25-86, 61 FR 5222, Feb. 9, 1996]

Control Surface and System Loads

§ 25.391 Control surface loads: General.

The control surfaces must be designed for the limit loads resulting from the flight conditions in §§ 25.331, 25.341(a), 25.349 and 25.351 and the ground gust conditions in § 25.415, considering the requirements for—

(a) Loads parallel to hinge line, in § 25.393;
(b) Pilot effort effects, in § 25.397;
(c) Trim tab effects, in § 25.407;
(d) Unsymmetrical loads, in § 25.427; and
(e) Auxiliary aerodynamic surfaces, in § 25.445.

[Doc. No. 5066, 29 FR 18291, Dec. 24, 1964, as amended by Amdt. 25-86, 61 FR 5222, Feb. 9, 1996]

§ 25.393 Loads parallel to hinge line.

(a) Control surfaces and supporting hinge brackets must be designed for inertia loads acting parallel to the hinge line.

(b) In the absence of more rational data, the inertia loads may be assumed to be equal to *KW*, where—

(1) *K*=24 for vertical surfaces;
(2) *K*=12 for horizontal surfaces; and
(3) *W*=weight of the movable surfaces.

§ 25.395 Control system.

(a) Longitudinal, lateral, directional, and drag control system and their supporting structures must be designed for loads corresponding to 125 percent of the computed hinge moments of the movable control surface in the conditions prescribed in § 25.391.

(b) The system limit loads, except the loads resulting from ground gusts, need not exceed the loads that can be produced by the pilot (or pilots) and by automatic or power devices operating the controls.

(c) The loads must not be less than those resulting from application of the minimum forces prescribed in § 25.397(c).

[Doc. No. 5066, 29 FR 18291, Dec. 24, 1964, as amended by Amdt. 25-23, 35 FR 5672, Apr. 8, 1970; Amdt. 25-72, 55 FR 29776, July 20, 1990]

§ 25.397 Control system loads.

(a) *General.* The maximum and minimum pilot forces, specified in paragraph (c) of this section, are assumed to act at the appropriate control grips or pads (in a manner simulating flight conditions) and to be reacted at the attachment of the control system to the control surface horn.

(b) *Pilot effort effects.* In the control surface flight loading condition, the air loads on movable surfaces and the corresponding deflections need not exceed those that would result in flight from the application of any pilot force within the ranges specified in paragraph (c) of this section. Two-thirds of the maximum values specified for the aileron and elevator may be used if control surface hinge moments are based on reliable data. In applying this criterion, the effects of servo mechanisms, tabs, and automatic pilot systems, must be considered.

(c) *Limit pilot forces and torques.* The limit pilot forces and torques are as follows:

Control	Maximum forces or torques	Minimum forces or torques
Aileron:		
Stick	100 lbs	40 lbs.
Wheel[1]	80 D in.-lbs[2]	40 D in.-lbs.
Elevator:		
Stick	250 lbs	100 lbs.
Wheel (symmetrical)	300 lbs	100 lbs.
Wheel (unsymmetrical)[3]		100 lbs.
Rudder	300 lbs	130 lbs.

[1] The critical parts of the aileron control system must be designed for a single tangential force with a limit value equal to 1.25 times the couple force determined from these criteria.

[2] *D*=wheel diameter (inches).

[3] The unsymmetrical forces must be applied at one of the normal handgrip points on the periphery of the control wheel.

[Doc. No. 5066, 29 FR 18291, Dec. 24, 1964, as amended by Amdt. 25-38, 41 FR 55466, Dec. 20, 1976; Amdt. 25-72, 55 FR 29776, July 20, 1990]

§ 25.399 Dual control system.

(a) Each dual control system must be designed for the pilots operating in opposition, using individual pilot forces not less than—

(1) 0.75 times those obtained under § 25.395; or
(2) The minimum forces specified in § 25.397(c).

(b) The control system must be designed for pilot forces applied in the same direction, using individual pilot forces not less than 0.75 times those obtained under § 25.395.

§ 25.405 Secondary control system.

Secondary controls, such as wheel brake, spoiler, and tab controls, must be designed for the maximum forces that a pilot is likely to apply to those controls. The following values may be used:

Pilot Control Force Limits (Secondary Controls)

Control	Limit pilot forces
Miscellaneous:	
*Crank, wheel, or lever	((1 +R) / 3)×50 lbs., but not less than 50 lbs. nor more than 150 lbs. (R=radius). (Applicable to any angle within 20° of plane of control).
Twist	133 in.–lbs.
Push-pull	To be chosen by applicant.

*Limited to flap, tab, stabilizer, spoiler, and landing gear operation controls.

§ 25.407 Trim tab effects.

The effects of trim tabs on the control surface design conditions must be accounted for only where the surface loads are limited by maximum pilot effort. In these cases, the tabs are considered to be deflected in the direction that would assist the pilot, and the deflections are—

(a) For elevator trim tabs, those required to trim the airplane at any point within the positive portion of the pertinent flight envelope in § 25.333(b), except as limited by the stops; and

(b) For aileron and rudder trim tabs, those required to trim the airplane in the critical unsymmetrical power and loading conditions, with appropriate allowance for rigging tolerances.

§ 25.409 Tabs.

(a) *Trim tabs.* Trim tabs must be designed to withstand loads arising from all likely combinations of tab setting, primary control position, and airplane speed (obtainable without exceeding the flight load conditions prescribed for the airplane as a whole), when the effect of the tab is opposed by pilot effort forces up to those specified in § 25.397(b).

(b) *Balancing tabs.* Balancing tabs must be designed for deflections consistent with the primary control surface loading conditions.

(c) *Servo tabs.* Servo tabs must be designed for deflections consistent with the primary control surface loading conditions obtainable within the pilot maneuvering effort, considering possible opposition from the trim tabs.

§ 25.415 Ground gust conditions.

(a) The control system must be designed as follows for control surface loads due to ground gusts and taxiing downwind:

(1) The control system between the stops nearest the surfaces and the cockpit controls must be designed for loads corresponding to the limit hinge moments H of paragraph (a)(2) of this section. These loads need not exceed—

(i) The loads corresponding to the maximum pilot loads in § 25.397(c) for each pilot alone; or

(ii) 0.75 times these maximum loads for each pilot when the pilot forces are applied in the same direction.

(2) The control system stops nearest the surfaces, the control system locks, and the parts of the systems (if any) between these stops and locks and the control surface horns, must be designed for limit hinge moments H, in foot pounds, obtained from the formula, $H=.0034KV^2 cS$, where—

V=65 (wind speed in knots)

K=limit hinge moment factor for ground gusts derived in paragraph (b) of this section.

c=mean chord of the control surface aft of the hinge line (ft);

S=area of the control surface aft of the hinge line (sq ft);

(b) The limit hinge moment factor K for ground gusts must be derived as follows:

Surface	K	Position of controls
(a) Aileron	0.75	Control column locked or lashed in mid-position.
(b) . . . . . . do	[1]1±0.50	Ailerons at full throw.
(c) Elevator	[1]1±0.75	(c) Elevator full down.
(d) . . . . . . do	[1]1±0.75	(d) Elevator full up.
(e) Rudder	0.75	(e) Rudder in neutral.
(f) . . . . . . do	0.75	(f) Rudder at full throw.

[1] A positive value of K indicates a moment tending to depress the surface, while a negative value of K indicates a moment tending to raise the surface.

[Doc. No. 5066, 29 FR 18291, Dec. 24, 1964, as amended by Amdt. 25-72, 55 FR 29776, July 20, 1990; Amdt. 25-91, 62 FR 40705, July 29, 1997]

§ 25.427 Unsymmetrical loads.

(a) In designing the airplane for lateral gust, yaw maneuver and roll maneuver conditions, account must be taken of unsymmetrical loads on the empennage arising from effects such as slipstream and aerodynamic interference with the wing, vertical fin and other aerodynamic surfaces.

(b) The horizontal tail must be assumed to be subjected to unsymmetrical loading conditions determined as follows:

(1) 100 percent of the maximum loading from the symmetrical maneuver conditions of § 25.331 and the vertical gust conditions of § 25.341(a) acting separately on the surface on one side of the plane of symmetry; and

(2) 80 percent of these loadings acting on the other side.

(c) For empennage arrangements where the horizontal tail surfaces have dihedral angles greater than plus or minus 10 degrees, or are supported by the vertical tail surfaces, the surfaces and the supporting structure must be designed for gust velocities specified in § 25.341(a) acting in any orientation at right angles to the flight path.

(d) Unsymmetrical loading on the empennage arising from buffet conditions of § 25.305(e) must be taken into account.

[Doc. No. 27902, 61 FR 5222, Feb. 9, 1996]

§ 25.445 Auxiliary aerodynamic surfaces.

(a) When significant, the aerodynamic influence between auxiliary aerodynamic surfaces, such as outboard fins and winglets, and their supporting aerodynamic surfaces, must be taken into account for all loading conditions including pitch, roll, and yaw maneuvers, and gusts as specified in

§ 25.341(a) acting at any orientation at right angles to the flight path.

(b) To provide for unsymmetrical loading when outboard fins extend above and below the horizontal surface, the critical vertical surface loading (load per unit area) determined under § 25.391 must also be applied as follows:

(1) 100 percent to the area of the vertical surfaces above (or below) the horizontal surface.

(2) 80 percent to the area below (or above) the horizontal surface.

[Doc. No. 5066, 29 FR 18291, Dec. 24, 1964, as amended by Amdt. 25-86, 61 FR 5222, Feb. 9, 1996]

§ 25.457 Wing flaps.

Wing flaps, their operating mechanisms, and their supporting structures must be designed for critical loads occurring in the conditions prescribed in § 25.345, accounting for the loads occurring during transition from one flap position and airspeed to another.

§ 25.459 Special devices.

The loading for special devices using aerodynamic surfaces (such as slots, slats and spoilers) must be determined from test data.

[Doc. No. 5066, 29 FR 18291, Dec. 24, 1964, as amended by Amdt. 25-72, 55 FR 29776, July 20, 1990]

Ground Loads

§ 25.471 General.

(a) *Loads and equilibrium.* For limit ground loads—

(1) Limit ground loads obtained under this subpart are considered to be external forces applied to the airplane structure; and

(2) In each specified ground load condition, the external loads must be placed in equilibrium with the linear and angular inertia loads in a rational or conservative manner.

(b) *Critical centers of gravity.* The critical centers of gravity within the range for which certification is requested must be selected so that the maximum design loads are obtained in each landing gear element. Fore and aft, vertical, and lateral airplane centers of gravity must be considered. Lateral displacements of the c.g. from the airplane centerline which would result in main gear loads not greater than 103 percent of the critical design load for symmetrical loading conditions may be selected without considering the effects of these lateral c.g. displacements on the loading of the main gear elements, or on the airplane structure provided—

(1) The lateral displacement of the c.g. results from random passenger or cargo disposition within the fuselage or from random unsymmetrical fuel loading or fuel usage; and

(2) Appropriate loading instructions for random disposable loads are included under the provisions of § 25.1583(c)(1) to ensure that the lateral displacement of the center of gravity is maintained within these limits.

(c) *Landing gear dimension data.* Figure 1 of appendix A contains the basic landing gear dimension data.

[Amdt. 25-23, 35 FR 5673, Apr. 8, 1970]

§ 25.473 Landing load conditions and assumptions.

(a) For the landing conditions specified in § 25.479 to § 25.485 the airplane is assumed to contact the ground—

(1) In the attitudes defined in § 25.479 and § 25.481;

(2) With a limit descent velocity of 10 fps at the design landing weight (the maximum weight for landing conditions at maximum descent velocity); and

(3) With a limit descent velocity of 6 fps at the design take-off weight (the maximum weight for landing conditions at a reduced descent velocity).

(4) The prescribed descent velocities may be modified if it is shown that the airplane has design features that make it impossible to develop these velocities.

(b) Airplane lift, not exceeding airplane weight, may be assumed unless the presence of systems or procedures significantly affects the lift.

(c) The method of analysis of airplane and landing gear loads must take into account at least the following elements:

(1) Landing gear dynamic characteristics.

(2) Spin-up and springback.

(3) Rigid body response.

(4) Structural dynamic response of the airframe, if significant.

(d) The landing gear dynamic characteristics must be validated by tests as defined in § 25.723(a).

(e) The coefficient of friction between the tires and the ground may be established by considering the effects of skidding velocity and tire pressure. However, this coefficient of friction need not be more than 0.8.

[Amdt. 25-91, 62 FR 40705, July 29, 1997; Amdt. 25-91, 62 FR 45481, Aug. 27, 1997; Amdt. 25-103, 66 FR 27394, May 16, 2001]

§ 25.477 Landing gear arrangement.

Sections 25.479 through 25.485 apply to airplanes with conventional arrangements of main and nose gears, or main and tail gears, when normal operating techniques are used.

§ 25.479 Level landing conditions.

(a) In the level attitude, the airplane is assumed to contact the ground at forward velocity components, ranging from V_{L1} to 1.25 V_{L2} parallel to the ground under the conditions prescribed in § 25.473 with—

(1) V_{L1} equal to V_{S0} (TAS) at the appropriate landing weight and in standard sea level conditions; and

(2) V_{L2} equal to V_{S0} (TAS) at the appropriate landing weight and altitudes in a hot day temperature of 41 degrees F. above standard.

(3) The effects of increased contact speed must be investigated if approval of downwind landings exceeding 10 knots is requested.

(b) For the level landing attitude for airplanes with tail wheels, the conditions specified in this section must be investigated with the airplane horizontal reference line horizontal in accordance with Figure 2 of Appendix A of this part.

(c) For the level landing attitude for airplanes with nose wheels, shown in Figure 2 of Appendix A of this part, the conditions specified in this section must be investigated assuming the following attitudes:

(1) An attitude in which the main wheels are assumed to contact the ground with the nose wheel just clear of the ground; and

(2) If reasonably attainable at the specified descent and forward velocities, an attitude in which the nose and main wheels are assumed to contact the ground simultaneously.

(d) In addition to the loading conditions prescribed in paragraph (a) of this section, but with maximum vertical ground reactions calculated from paragraph (a), the following apply:

(1) The landing gear and directly affected attaching structure must be designed for the maximum vertical ground reaction combined with an aft acting drag component of not less than 25% of this maximum vertical ground reaction.

(2) The most severe combination of loads that are likely to arise during a lateral drift landing must be taken into account. In absence of a more rational analysis of this condition, the following must be investigated:

(i) A vertical load equal to 75% of the maximum ground reaction of § 25.473 must be considered in combination with a drag and side load of 40% and 25% respectively of that vertical load.

(ii) The shock absorber and tire deflections must be assumed to be 75% of the deflection corresponding to the maximum ground reaction of § 25.473(a)(2). This load case need not be considered in combination with flat tires.

(3) The combination of vertical and drag components is considered to be acting at the wheel axle centerline.

[Amdt. 25-91, 62 FR 40705, July 29, 1997; Amdt. 25-91, 62 FR 45481, Aug. 27, 1997]

§ 25.481 Tail-down landing conditions.

(a) In the tail-down attitude, the airplane is assumed to contact the ground at forward velocity components, ranging from V_{L1} to V_{L2} parallel to the ground under the conditions prescribed in § 25.473 with—

(1) V_{L1} equal to V_{S0} (TAS) at the appropriate landing weight and in standard sea level conditions; and

(2) V_{L2} equal to V_{S0} (TAS) at the appropriate landing weight and altitudes in a hot day temperature of 41 degrees F. above standard.

(3) The combination of vertical and drag components considered to be acting at the main wheel axle centerline.

(b) For the tail-down landing condition for airplanes with tail wheels, the main and tail wheels are assumed to contact the ground simultaneously, in accordance with figure 3 of appendix A. Ground reaction conditions on the tail wheel are assumed to act—

(1) Vertically; and

(2) Up and aft through the axle at 45 degrees to the ground line.

(c) For the tail-down landing condition for airplanes with nose wheels, the airplane is assumed to be at an attitude corresponding to either the stalling angle or the maximum angle allowing clearance with the ground by each part of the airplane other than the main wheels, in accordance with figure 3 of appendix A, whichever is less.

[Docket No. 5066, 29 FR 18291, Dec. 24, 1964, as amended by Amdt. 25-91, 62 FR 40705, July 29, 1997; Amdt. 25-94, 63 FR 8848, Feb. 23, 1998]

§ 25.483 One-gear landing conditions.

For the one-gear landing conditions, the airplane is assumed to be in the level attitude and to contact the ground on one main landing gear, in accordance with Figure 4 of Appendix A of this part. In this attitude—

(a) The ground reactions must be the same as those obtained on that side under § 25.479(d)(1), and

(b) Each unbalanced external load must be reacted by airplane inertia in a rational or conservative manner.

[Docket No. 5066, 29 FR 18291, Dec. 24, 1964, as amended by Amdt. 25-91, 62 FR 40705, July 29, 1997]

§ 25.485 Side load conditions.

In addition to § 25.479(d)(2) the following conditions must be considered:

(a) For the side load condition, the airplane is assumed to be in the level attitude with only the main wheels contacting the ground, in accordance with figure 5 of appendix A.

(b) Side loads of 0.8 of the vertical reaction (on one side) acting inward and 0.6 of the vertical reaction (on the other side) acting outward must be combined with one-half of the maximum vertical ground reactions obtained in the level landing conditions. These loads are assumed to be applied at the ground contact point and to be resisted by the inertia of the airplane. The drag loads may be assumed to be zero.

[Docket No. 5066, 29 FR 18291, Dec. 24, 1964, as amended by Amdt. 25-91, 62 FR 40705, July 29, 1997]

§ 25.487 Rebound landing condition.

(a) The landing gear and its supporting structure must be investigated for the loads occurring during rebound of the airplane from the landing surface.

(b) With the landing gear fully extended and not in contact with the ground, a load factor of 20.0 must act on the unsprung weights of the landing gear. This load factor must act in the direction of motion of the unsprung weights as they reach their limiting positions in extending with relation to the sprung parts of the landing gear.

§ 25.489 Ground handling conditions.

Unless otherwise prescribed, the landing gear and airplane structure must be investigated for the conditions in §§ 25.491 through 25.509 with the airplane at the design ramp weight (the maximum weight for ground handling conditions). No wing lift may be considered. The shock absorbers and tires may be assumed to be in their static position.

[Doc. No. 5066, 29 FR 18291, Dec. 24, 1964, as amended by Amdt. 25-23, 35 FR 5673, Apr. 8, 1970]

§ 25.491 Taxi, takeoff and landing roll.

Within the range of appropriate ground speeds and approved weights, the airplane structure and landing gear are assumed to be subjected to loads not less than those obtained when the aircraft is operating over the roughest ground that may reasonably be expected in normal operation.

[Amdt. 25-91, 62 FR 40705, July 29, 1997]

§ 25.493 Braked roll conditions.

(a) An airplane with a tail wheel is assumed to be in the level attitude with the load on the main wheels, in accordance with figure 6 of appendix A. The limit vertical load factor is 1.2 at the design landing weight and 1.0 at the design ramp weight. A drag reaction equal to the vertical reaction multiplied by a coefficient of friction of 0.8, must be combined with the vertical ground reaction and applied at the ground contact point.

(b) For an airplane with a nose wheel the limit vertical load factor is 1.2 at the design landing weight, and 1.0 at the design ramp weight. A drag reaction equal to the vertical reaction, multiplied by a coefficient of friction of 0.8, must be combined with the vertical reaction and applied at the ground contact point of each wheel with brakes. The following two attitudes, in accordance with figure 6 of appendix A, must be considered:

(1) The level attitude with the wheels contacting the ground and the loads distributed between the main and nose gear. Zero pitching acceleration is assumed.

(2) The level attitude with only the main gear contacting the ground and with the pitching moment resisted by angular acceleration.

(c) A drag reaction lower than that prescribed in this section may be used if it is substantiated that an effective drag force of 0.8 times the vertical reaction cannot be attained under any likely loading condition.

(d) An airplane equipped with a nose gear must be designed to withstand the loads arising from the dynamic pitching motion of the airplane due to sudden application of maximum braking force. The airplane is considered to be at design takeoff weight with the nose and main gears in contact with the ground, and with a steady-state vertical load factor of 1.0. The steady-state nose gear reaction must be combined with the maximum incremental nose gear vertical reaction caused by the sudden application of maximum braking force as described in paragraphs (b) and (c) of this section.

(e) In the absence of a more rational analysis, the nose gear vertical reaction prescribed in paragraph (d) of this section must be calculated according to the following formula:

$$V_N = \frac{W_T}{A+B}\left[B + \frac{f\mu AE}{A+B+\mu E}\right]$$

Where:

V_N=Nose gear vertical reaction.

W_T=Design takeoff weight.

A=Horizontal distance between the c.g. of the airplane and the nose wheel.

B=Horizontal distance between the c.g. of the airplane and the line joining the centers of the main wheels.

E=Vertical height of the c.g. of the airplane above the ground in the 1.0 g static condition.

μ=Coefficient of friction of 0.80.

f=Dynamic response factor; 2.0 is to be used unless a lower factor is substantiated. In the absence of other information, the dynamic response factor f may be defined by the equation:

$$f = 1 + \exp\left(\frac{-\pi\xi}{\sqrt{1-\xi^2}}\right)$$

Where:

ξ is the effective critical damping ratio of the rigid body pitching mode about the main landing gear effective ground contact point.

[Doc. No. 5066, 29 FR 18291, Dec. 24, 1964, as amended by Amdt. 25-23, 35 FR 5673, Apr. 8, 1970; Amdt. 25-97, 63 FR 29072, May 27, 1998]

§ 25.495 Turning.

In the static position, in accordance with figure 7 of appendix A, the airplane is assumed to execute a steady turn by nose gear steering, or by application of sufficient differential power, so that the limit load factors applied at the center of gravity are 1.0 vertically and 0.5 laterally. The side ground reaction of each wheel must be 0.5 of the vertical reaction.

§ 25.497 Tail-wheel yawing.

(a) A vertical ground reaction equal to the static load on the tail wheel, in combination with a side component of equal magnitude, is assumed.

(b) If there is a swivel, the tail wheel is assumed to be swiveled 90° to the airplane longitudinal axis with the resultant load passing through the axle.

(c) If there is a lock, steering device, or shimmy damper the tail wheel is also assumed to be in the trailing position with the side load acting at the ground contact point.

§ 25.499 Nose-wheel yaw and steering.

(a) A vertical load factor of 1.0 at the airplane center of gravity, and a side component at the nose wheel ground contact equal to 0.8 of the vertical ground reaction at that point are assumed.

(b) With the airplane assumed to be in static equilibrium with the loads resulting from the use of brakes on one side of the main landing gear, the nose gear, its attaching structure, and the fuselage structure forward of the center of gravity must be designed for the following loads:

(1) A vertical load factor at the center of gravity of 1.0.

(2) A forward acting load at the airplane center of gravity of 0.8 times the vertical load on one main gear.

(3) Side and vertical loads at the ground contact point on the nose gear that are required for static equilibrium.

(4) A side load factor at the airplane center of gravity of zero.

(c) If the loads prescribed in paragraph (b) of this section result in a nose gear side load higher than 0.8 times the vertical nose gear load, the design nose gear side load may be limited to 0.8 times the vertical load, with unbalanced yawing moments assumed to be resisted by airplane inertia forces.

(d) For other than the nose gear, its attaching structure, and the forward fuselage structure, the loading conditions are those prescribed in paragraph (b) of this section, except that—

(1) A lower drag reaction may be used if an effective drag force of 0.8 times the vertical reaction cannot be reached under any likely loading condition; and

(2) The forward acting load at the center of gravity need not exceed the maximum drag reaction on one main gear, determined in accordance with § 25.493(b).

(e) With the airplane at design ramp weight, and the nose gear in any steerable position, the combined application of full normal steering torque and vertical force equal to 1.33 times the maximum static reaction on the nose gear must be considered in designing the nose gear, its attaching structure, and the forward fuselage structure.

[Doc. No. 5066, 29 FR 18291, Dec. 24, 1964, as amended by Amdt. 25-23, 35 FR 5673, Apr. 8, 1970; Amdt. 25-46, 43 FR 50595, Oct. 30, 1978; Amdt. 25-91, 62 FR 40705, July 29, 1997]

§ 25.503 Pivoting.

(a) The airplane is assumed to pivot about one side of the main gear with the brakes on that side locked. The limit vertical load factor must be 1.0 and the coefficient of friction 0.8.

(b) The airplane is assumed to be in static equilibrium, with the loads being applied at the ground contact points, in accordance with figure 8 of appendix A.

§ 25.507 Reversed braking.

(a) The airplane must be in a three point static ground attitude. Horizontal reactions parallel to the ground and directed forward must be applied at the ground contact point of each wheel with brakes. The limit loads must be equal to 0.55 times the vertical load at each wheel or to the load developed by 1.2 times the nominal maximum static brake torque, whichever is less.

(b) For airplanes with nose wheels, the pitching moment must be balanced by rotational inertia.

(c) For airplanes with tail wheels, the resultant of the ground reactions must pass through the center of gravity of the airplane.

§ 25.509 Towing loads.

(a) The towing loads specified in paragraph (d) of this section must be considered separately. These loads must be applied at the towing fittings and must act parallel to the ground. In addition—

(1) A vertical load factor equal to 1.0 must be considered acting at the center of gravity;

(2) The shock struts and tires must be in their static positions; and

(3) With W_T as the design ramp weight, the towing load, F_{TOW}, is—

(i) 0.3 W_T for W_T less than 30,000 pounds;

(ii) $(6W_T + 450{,}000)/70$ for W_T between 30,000 and 100,000 pounds; and

(iii) 0.15 W_T for W_T over 100,000 pounds.

(b) For towing points not on the landing gear but near the plane of symmetry of the airplane, the drag and side tow load components specified for the auxiliary gear apply. For towing points located outboard of the main gear, the drag and side tow load components specified for the main gear apply. Where the specified angle of swivel cannot be reached, the maximum obtainable angle must be used.

(c) The towing loads specified in paragraph (d) of this section must be reacted as follows:

(1) The side component of the towing load at the main gear must be reacted by a side force at the static ground line of the wheel to which the load is applied.

(2) The towing loads at the auxiliary gear and the drag components of the towing loads at the main gear must be reacted as follows:

(i) A reaction with a maximum value equal to the vertical reaction must be applied at the axle of the wheel to which the load is applied. Enough airplane inertia to achieve equilibrium must be applied.

(ii) The loads must be reacted by airplane inertia.

(d) The prescribed towing loads are as follows:

Tow point	Position	Load		
		Magnitude	No.	Direction
Main gear		0.75 *F*TOW per main gear unit	1	Forward, parallel to drag axis.
			2	Forward, at 30° to drag axis.
			3	Aft, parallel to drag axis.
			4	Aft, at 30° to drag axis.
Auxiliary gear	Swiveled forward	1.0 *F*TOW	5	Forward.
			6	Aft.
	Swiveled aft	......do	7	Forward.
			8	Aft.
	Swiveled 45° from forward	0.5 *F*TOW	9	Forward, in plane of wheel.
			10	Aft, in plane of wheel.
	Swiveled 45° from aft	......do	11	Forward, in plane of wheel.
			12	Aft, in plane of wheel.

[Doc. No. 5066, 29 FR 18291, Dec. 24, 1964, as amended by Amdt. 25-23, 35 FR 5673, Apr. 8, 1970]

§ 25.511 Ground load: unsymmetrical loads on multiple-wheel units.

(a) *General.* Multiple-wheel landing gear units are assumed to be subjected to the limit ground loads prescribed in this subpart under paragraphs (b) through (f) of this section. In addition—

(1) A tandem strut gear arrangement is a multiple-wheel unit; and

(2) In determining the total load on a gear unit with respect to the provisions of paragraphs (b) through (f) of this section, the transverse shift in the load centroid, due to unsymmetrical load distribution on the wheels, may be neglected.

(b) *Distribution of limit loads to wheels; tires inflated.* The distribution of the limit loads among the wheels of the landing gear must be established for each landing, taxiing, and ground handling condition, taking into account the effects of the following factors:

(1) The number of wheels and their physical arrangements. For truck type landing gear units, the effects of any seesaw motion of the truck during the landing impact must be considered in determining the maximum design loads for the fore and aft wheel pairs.

(2) Any differentials in tire diameters resulting from a combination of manufacturing tolerances, tire growth, and tire wear. A maximum tire-diameter differential equal to $\frac{2}{3}$ of the most unfavorable combination of diameter variations that is obtained when taking into account manufacturing tolerances, tire growth, and tire wear, may be assumed.

(3) Any unequal tire inflation pressure, assuming the maximum variation to be±5 percent of the nominal tire inflation pressure.

(4) A runway crown of zero and a runway crown having a convex upward shape that may be approximated by a slope of $1\frac{1}{2}$ percent with the horizontal. Runway crown effects must be considered with the nose gear unit on either slope of the crown.

(5) The airplane attitude.

(6) Any structural deflections.

(c) *Deflated tires.* The effect of deflated tires on the structure must be considered with respect to the loading conditions specified in paragraphs (d) through (f) of this section, taking into account the physical arrangement of the gear components. In addition—

(1) The deflation of any one tire for each multiple wheel landing gear unit, and the deflation of any two critical tires for each landing gear unit using four or more wheels per unit, must be considered; and

(2) The ground reactions must be applied to the wheels with inflated tires except that, for multiple-wheel gear units with more than one shock strut, a rational distribution of the ground reactions between the deflated and inflated tires, accounting for the differences in shock strut extensions resulting from a deflated tire, may be used.

(d) *Landing conditions.* For one and for two deflated tires, the applied load to each gear unit is assumed to be 60 percent and 50 percent, respectively, of the limit load applied to each gear for each of the prescribed landing conditions. However, for the drift landing condition of § 25.485, 100 percent of the vertical load must be applied.

(e) *Taxiing and ground handling conditions.* For one and for two deflated tires—

(1) The applied side or drag load factor, or both factors, at the center of gravity must be the most critical value up to 50 percent and 40 percent, respectively, of the limit side or drag load factors, or both factors, corresponding to the most severe condition resulting from consideration of the prescribed taxiing and ground handling conditions;

(2) For the braked roll conditions of § 25.493 (a) and (b)(2), the drag loads on each inflated tire may not be less than those at each tire for the symmetrical load distribution with no deflated tires;

(3) The vertical load factor at the center of gravity must be 60 percent and 50 percent, respectively, of the factor with no deflated tires, except that it may not be less than 1g; and

(4) Pivoting need not be considered.

(f) *Towing conditions.* For one and for two deflated tires, the towing load, F_{TOW}, must be 60 percent and 50 percent, respectively, of the load prescribed.

§ 25.519 Jacking and tie-down provisions.

(a) General. The airplane must be designed to withstand the limit load conditions resulting from the static ground load conditions of paragraph (b) of this section and, if applicable, paragraph (c) of this section at the most critical combinations of airplane weight and center of gravity. The maximum allowable load at each jack pad must be specified.

(b) Jacking. The airplane must have provisions for jacking and must withstand the following limit loads when the airplane is supported on jacks—

(1) For jacking by the landing gear at the maximum ramp weight of the airplane, the airplane structure must be designed for a vertical load of 1.33 times the vertical static reaction at each jacking point acting singly and in combination with a horizontal load of 0.33 times the vertical static reaction applied in any direction.

(2) For jacking by other airplane structure at maximum approved jacking weight:

(i) The airplane structure must be designed for a vertical load of 1.33 times the vertical reaction at each jacking point acting singly and in combination with a horizontal load of 0.33 times the vertical static reaction applied in any direction.

(ii) The jacking pads and local structure must be designed for a vertical load of 2.0 times the vertical static reaction at each jacking point, acting singly and in combination with a horizontal load of 0.33 times the vertical static reaction applied in any direction.

(c) Tie-down. If tie-down points are provided, the main tie-down points and local structure must withstand the limit loads resulting from a 65-knot horizontal wind from any direction.

[Doc. No. 26129, 59 FR 22102, Apr. 28, 1994]

WATER LOADS

§ 25.521 General.

(a) Seaplanes must be designed for the water loads developed during takeoff and landing, with the seaplane in any attitude likely to occur in normal operation, and at the appropriate forward and sinking velocities under the most severe sea conditions likely to be encountered.

(b) Unless a more rational analysis of the water loads is made, or the standards in ANC-3 are used, §§ 25.523 through 25.537 apply.

(c) The requirements of this section and §§ 25.523 through 25.537 apply also to amphibians.

§ 25.523 Design weights and center of gravity positions.

(a) *Design weights.* The water load requirements must be met at each operating weight up to the design landing weight except that, for the takeoff condition prescribed in § 25.531, the design water takeoff weight (the maximum weight for water taxi and takeoff run) must be used.

(b) *Center of gravity positions.* The critical centers of gravity within the limits for which certification is requested must be considered to reach maximum design loads for each part of the seaplane structure.

[Doc. No. 5066, 29 FR 18291, Dec. 24, 1964, as amended by Amdt. 25-23, 35 FR 5673, Apr. 8, 1970]

§ 25.525 Application of loads.

(a) Unless otherwise prescribed, the seaplane as a whole is assumed to be subjected to the loads corresponding to the load factors specified in § 25.527.

(b) In applying the loads resulting from the load factors prescribed in § 25.527, the loads may be distributed over the hull or main float bottom (in order to avoid excessive local shear loads and bending moments at the location of water load application) using pressures not less than those prescribed in § 25.533(b).

(c) For twin float seaplanes, each float must be treated as an equivalent hull on a fictitious seaplane with a weight equal to one-half the weight of the twin float seaplane.

(d) Except in the takeoff condition of § 25.531, the aerodynamic lift on the seaplane during the impact is assumed to be ⅔ of the weight of the seaplane.

§ 25.527 Hull and main float load factors.

(a) Water reaction load factors n_w must be computed in the following manner:

(1) For the step landing case

$$n_w = \frac{C_1 V_{S0}^2}{\left(Tan^{\frac{2}{3}}\beta\right)W^{\frac{1}{3}}}$$

(2) For the bow and stern landing cases

$$n_w = \frac{C_1 V_{S0}^2}{\left(Tan^{\frac{2}{3}}\beta\right)W^{\frac{1}{3}}} \times \frac{K_1}{\left(1+r_x^2\right)^{\frac{2}{3}}}$$

(b) The following values are used:

(1) n_w=water reaction load factor (that is, the water reaction divided by seaplane weight).

(2) C_1=empirical seaplane operations factor equal to 0.012 (except that this factor may not be less than that necessary to obtain the minimum value of step load factor of 2.33).

(3) V_{S0}=seaplane stalling speed in knots with flaps extended in the appropriate landing position and with no slipstream effect.

(4) β=angle of dead rise at the longitudinal station at which the load factor is being determined in accordance with figure 1 of appendix B.

(5) W=seaplane design landing weight in pounds.

(6) K_1=empirical hull station weighing factor, in accordance with figure 2 of appendix B.

(7) r_x=ratio of distance, measured parallel to hull reference axis, from the center of gravity of the seaplane to the hull longitudinal station at which the load factor is being computed to the radius of gyration in pitch of the seaplane, the hull reference axis being a straight line, in the plane of symmetry, tangential to the keel at the main step.

(c) For a twin float seaplane, because of the effect of flexibility of the attachment of the floats to the seaplane, the factor K_1 may be reduced at the bow and stern to 0.8 of the value shown in figure 2 of appendix B. This reduction applies only to the design of the carrythrough and seaplane structure.

[Doc. No. 5066, 29 FR 18291, Dec. 24, 1964, as amended by Amdt. 25-23, 35 FR 5673, Apr. 8, 1970]

§ 25.529 Hull and main float landing conditions.

(a) *Symmetrical step, bow, and stern landing.* For symmetrical step, bow, and stern landings, the limit water reaction load factors are those computed under § 25.527. In addition—

(1) For symmetrical step landings, the resultant water load must be applied at the keel, through the center of gravity, and must be directed perpendicularly to the keel line;

(2) For symmetrical bow landings, the resultant water load must be applied at the keel, one-fifth of the longitudinal distance from the bow to the step, and must be directed perpendicularly to the keel line; and

(3) For symmetrical stern landings, the resultant water load must be applied at the keel, at a point 85 percent of the longitudinal distance from the step to the stern post, and must be directed perpendicularly to the keel line.

(b) *Unsymmetrical landing for hull and single float seaplanes.* Unsymmetrical step, bow, and stern landing conditions must be investigated. In addition—

(1) The loading for each condition consists of an upward component and a side component equal, respectively, to 0.75 and 0.25 tan β times the resultant load in the corresponding symmetrical landing condition; and

(2) The point of application and direction of the upward component of the load is the same as that in the symmetrical condition, and the point of application of the side component is at the same longitudinal station as the upward component but is directed inward perpendicularly to the plane of symmetry at a point midway between the keel and chine lines.

(c) *Unsymmetrical landing; twin float seaplanes.* The unsymmetrical loading consists of an upward load at the step of each float of 0.75 and a side load of 0.25 tan β at one float times the step landing load reached under § 25.527. The side load is directed inboard, perpendicularly to the plane of symmetry midway between the keel and chine lines of the float, at the same longitudinal station as the upward load.

§ 25.531 Hull and main float takeoff condition.

For the wing and its attachment to the hull or main float—

(a) The aerodynamic wing lift is assumed to be zero; and

(b) A downward inertia load, corresponding to a load factor computed from the following formula, must be applied:

$$n = \frac{C_{TO} V_{S1}^2}{\left(Tan^{\frac{2}{3}}\beta\right)W^{\frac{1}{3}}}$$

where—

n=inertia load factor;

C_{TO}=empirical seaplane operations factor equal to 0.004;

V_{S1}=seaplane stalling speed (knots) at the design takeoff weight with the flaps extended in the appropriate takeoff position;

β=angle of dead rise at the main step (degrees); and

W=design water takeoff weight in pounds.

[Doc. No. 5066, 29 FR 18291, Dec. 24, 1964, as amended by Amdt. 25-23, 35 FR 5673, Apr. 8, 1970]

§ 25.533 Hull and main float bottom pressures.

(a) *General.* The hull and main float structure, including frames and bulkheads, stringers, and bottom plating, must be designed under this section.

(b) *Local pressures.* For the design of the bottom plating and stringers and their attachments to the supporting structure, the following pressure distributions must be applied:

(1) For an unflared bottom, the pressure at the chine is 0.75 times the pressure at the keel, and the pressures between the keel and chine vary linearly, in accordance with figure 3 of appendix B. The pressure at the keel (psi) is computed as follows:

$$P_k = C_2 \times \frac{K_2 V_{S1}^2}{\tan \beta_k}$$

where—

P_k=pressure (p.s.i.) at the keel;

C_2=0.00213;

K_2=hull station weighing factor, in accordance with figure 2 of appendix B;

V_{S1}=seaplane stalling speed (Knots) at the design water takeoff weight with flaps extended in the appropriate takeoff position; and

βk=angle of dead rise at keel, in accordance with figure 1 of appendix B.

(2) For a flared bottom, the pressure at the beginning of the flare is the same as that for an unflared bottom, and the pressure between the chine and the beginning of the flare varies linearly, in accordance with figure 3 of appendix B. The pressure distribution is the same as that prescribed in paragraph (b)(1) of this section for an unflared bottom except that the pressure at the chine is computed as follows:

$$P_{ch} = C_3 \times \frac{K_2 V_{S1}^2}{\tan \beta}$$

where—

P_{ch}=pressure (p.s.i.) at the chine;

C_3=0.0016;

K_2=hull station weighing factor, in accordance with figure 2 of appendix B;

V_{S1}=seaplane stalling speed at the design water takeoff weight with flaps extended in the appropriate takeoff position; and

β=angle of dead rise at appropriate station.

The area over which these pressures are applied must simulate pressures occurring during high localized impacts on the hull or float, but need not extend over an area that would induce critical stresses in the frames or in the overall structure.

(c) *Distributed pressures.* For the design of the frames, keel, and chine structure, the following pressure distributions apply:

(1) Symmetrical pressures are computed as follows:

$$P = C_4 \times \frac{K_2 V_{S0}{}^2}{\tan \beta}$$

where—

P=pressure (p.s.i.);

C_4=0.078 C_1 (with C_1 computed under § 25.527);

K_2=hull station weighing factor, determined in accordance with figure 2 of appendix B;

V_{S0}=seaplane stalling speed (Knots) with landing flaps extended in the appropriate position and with no slipstream effect; and

V_{S0}=seaplane stalling speed with landing flaps extended in the appropriate position and with no slipstream effect; and β=angle of dead rise at appropriate station.

(2) The unsymmetrical pressure distribution consists of the pressures prescribed in paragraph (c)(1) of this section on one side of the hull or main float centerline and one-half of that pressure on the other side of the hull or main float centerline, in accordance with figure 3 of appendix B.

These pressures are uniform and must be applied simultaneously over the entire hull or main float bottom. The loads obtained must be carried into the sidewall structure of the hull proper, but need not be transmitted in a fore and aft direction as shear and bending loads.

[Doc. No. 5066, 29 FR 18291, Dec. 24, 1964, as amended by Amdt. 25-23, 35 FR 5673, Apr. 8, 1970]

§ 25.535 Auxiliary float loads.

(a) *General.* Auxiliary floats and their attachments and supporting structures must be designed for the conditions prescribed in this section. In the cases specified in paragraphs (b) through (e) of this section, the prescribed water loads may be distributed over the float bottom to avoid excessive local loads, using bottom pressures not less than those prescribed in paragraph (g) of this section.

(b) *Step loading.* The resultant water load must be applied in the plane of symmetry of the float at a point three-fourths of the distance from the bow to the step and must be perpendicular to the keel. The resultant limit load is computed as follows, except that the value of L need not exceed three times the weight of the displaced water when the float is completely submerged:

$$L = \frac{C_5 V_{S_0}{}^2 W^{\frac{2}{3}}}{\tan^{\frac{2}{3}} \beta_S \left(1 + r_y{}^2\right)^{\frac{2}{3}}}$$

where—

L=limit load (lbs.);

C_5=0.0053;

V_{S0}=seaplane stalling speed (knots) with landing flaps extended in the appropriate position and with no slipstream effect;

W=seaplane design landing weight in pounds;

β_S=angle of dead rise at a station ¾ of the distance from the bow to the step, but need not be less than 15 degrees; and

r_y=ratio of the lateral distance between the center of gravity and the plane of symmetry of the float to the radius of gyration in roll.

(c) *Bow loading.* The resultant limit load must be applied in the plane of symmetry of the float at a point one-fourth of the distance from the bow to the step and must be perpendicular to the tangent to the keel line at that point. The magnitude of the resultant load is that specified in paragraph (b) of this section.

(d) *Unsymmetrical step loading.* The resultant water load consists of a component equal to 0.75 times the load specified in paragraph (a) of this section and a side component equal to 3.25 tan β times the load specified in paragraph (b) of this section. The side load must be applied perpendicularly to the plane of symmetry of the float at a point midway between the keel and the chine.

(e) *Unsymmetrical bow loading.* The resultant water load consists of a component equal to 0.75 times the load specified in paragraph (b) of this section and a side component equal to 0.25 tan β times the load specified in paragraph (c) of this section. The side load must be applied perpendicularly to the plane of symmetry at a point midway between the keel and the chine.

(f) *Immersed float condition.* The resultant load must be applied at the centroid of the cross section of the float at a point one-third of the distance from the bow to the step. The limit load components are as follows:

$$Vertical = {}_{\rho g}V$$

$$aft = C_{x2}{}^{\rho}V^{\frac{2}{3}}\left(KV_{s_0}\right)^2$$

$$side = C_{y2}{}^{\rho}V^{\frac{2}{3}}\left(KV_{s_0}\right)^2$$

where—

ρ=mass density of water (slugs/ft.²);

V=volume of float (ft.²);

C_x=coefficient of drag force, equal to 0.133;

C_y=coefficient of side force, equal to 0.106;

K=0.8, except that lower values may be used if it is shown that the floats are incapable of submerging at a speed of 0.8 V_{S0} in normal operations;

V_{S0}=seaplane stalling speed (knots) with landing flaps extended in the appropriate position and with no slipstream effect; and

g=acceleration due to gravity (ft./sec.²).

(g) *Float bottom pressures.* The float bottom pressures must be established under § 25.533, except that the value of K_2 in the formulae may be taken as 1.0. The angle of dead rise to be used in determining the float bottom pressures is set forth in paragraph (b) of this section.

[Doc. No. 5066, 29 FR 18291, Dec. 24, 1964, as amended by Amdt. 25-23, 35 FR 5673, Apr. 8, 1970]

§ 25.537 Seawing loads.

Seawing design loads must be based on applicable test data.

EMERGENCY LANDING CONDITIONS

§ 25.561 General.

(a) The airplane, although it may be damaged in emergency landing conditions on land or water, must be designed as prescribed in this section to protect each occupant under those conditions.

(b) The structure must be designed to give each occupant every reasonable chance of escaping serious injury in a minor crash landing when—

(1) Proper use is made of seats, belts, and all other safety design provisions;

(2) The wheels are retracted (where applicable); and

(3) The occupant experiences the following ultimate inertia forces acting separately relative to the surrounding structure:

(i) Upward, 3.0g

(ii) Forward, 9.0g

(iii) Sideward, 3.0g on the airframe; and 4.0g on the seats and their attachments.

(iv) Downward, 6.0g

(v) Rearward, 1.5g

(c) For equipment, cargo in the passenger compartments and any other large masses, the following apply:

(1) Except as provided in paragraph (c)(2) of this section, these items must be positioned so that if they break loose they will be unlikely to:

(i) Cause direct injury to occupants;

(ii) Penetrate fuel tanks or lines or cause fire or explosion hazard by damage to adjacent systems; or

(iii) Nullify any of the escape facilities provided for use after an emergency landing.

(2) When such positioning is not practical (e.g. fuselage mounted engines or auxiliary power units) each such item of mass shall be restrained under all loads up to those specified in paragraph (b)(3) of this section. The local attachments for these items should be designed to withstand 1.33 times the specified loads if these items are subject to severe wear and tear through frequent removal (e.g. quick change interior items).

(d) Seats and items of mass (and their supporting structure) must not deform under any loads up to those specified in paragraph (b)(3) of this section in any manner that would impede subsequent rapid evacuation of occupants.

[Doc. No. 5066, 29 FR 18291, Dec. 24, 1964, as amended by Amdt. 25-23, 35 FR 5673, Apr. 8, 1970; Amdt. 25-64, 53 FR 17646, May 17, 1988; Amdt. 25-91, 62 FR 40706, July 29, 1997]

§ 25.562 Emergency landing dynamic conditions.

(a) The seat and restraint system in the airplane must be designed as prescribed in this section to protect each occupant during an emergency landing condition when—

(1) Proper use is made of seats, safety belts, and shoulder harnesses provided for in the design; and

(2) The occupant is exposed to loads resulting from the conditions prescribed in this section.

(b) Each seat type design approved for crew or passenger occupancy during takeoff and landing must successfully complete dynamic tests or be demonstrated by rational analysis based on dynamic tests of a similar type seat, in accordance with each of the following emergency landing conditions. The tests must be conducted with an occupant simulated by a 170-pound anthropomorphic test dummy, as defined by 49 CFR Part 572, Subpart B, or its equivalent, sitting in the normal upright position.

(1) A change in downward vertical velocity (Δv) of not less than 35 feet per second, with the airplane's longitudinal axis canted downward 30 degrees with respect to the horizontal plane and with the wings level. Peak floor deceleration must occur in not more than 0.08 seconds after impact and must reach a minimum of 14g.

(2) A change in forward longitudinal velocity (Δv) of not less than 44 feet per second, with the airplane's longitudinal axis horizontal and yawed 10 degrees either right or left, whichever would cause the greatest likelihood of the upper torso restraint system (where installed) moving off the occupant's shoulder, and with the wings level. Peak floor deceleration must occur in not more than 0.09 seconds after impact and must reach a minimum of 16g. Where floor rails or floor fittings are used to attach the seating devices to the test fixture, the rails or fittings must be misaligned with respect to the adjacent set of rails or fittings by at least 10 degrees vertically (*i.e.*, out of Parallel) with one rolled 10 degrees.

(c) The following performance measures must not be exceeded during the dynamic tests conducted in accordance with paragraph (b) of this section:

(1) Where upper torso straps are used for crewmembers, tension loads in individual straps must not exceed 1,750 pounds. If dual straps are used for restraining the upper torso, the total strap tension loads must not exceed 2,000 pounds.

(2) The maximum compressive load measured between the pelvis and the lumbar column of the anthropomorphic dummy must not exceed 1,500 pounds.

(3) The upper torso restraint straps (where installed) must remain on the occupant's shoulder during the impact.

(4) The lap safety belt must remain on the occupant's pelvis during the impact.

(5) Each occupant must be protected from serious head injury under the conditions prescribed in paragraph (b) of this section. Where head contact with seats or other structure can occur, protection must be provided so that the head impact does not exceed a Head Injury Criterion (HIC) of 1,000 units. The level of HIC is defined by the equation:

$$HIC = \left\{ \left(t_2 - t_1\right) \left[\frac{1}{\left(t_2 - t_1\right)} \int_{t_1}^{t_2} a(t)\,dt \right]^{2.5} \right\}_{\max}$$

Where:

t_1 is the initial integration time,

t_2 is the final integration time, and

a(t) is the total acceleration vs. time curve for the head strike, and where

(t) is in seconds, and (a) is in units of gravity (g).

(6) Where leg injuries may result from contact with seats or other structure, protection must be provided to prevent axially compressive loads exceeding 2,250 pounds in each femur.

(7) The seat must remain attached at all points of attachment, although the structure may have yielded.

(8) Seats must not yield under the tests specified in paragraphs (b)(1) and (b)(2) of this section to the extent they would impede rapid evacuation of the airplane occupants.

[Amdt. 25-64, 53 FR 17646, May 17, 1988]

§ 25.563 Structural ditching provisions.

Structural strength considerations of ditching provisions must be in accordance with § 25.801(e).

FATIGUE EVALUATION

§ 25.571 Damage—tolerance and fatigue evaluation of structure.

(a) *General.* An evaluation of the strength, detail design, and fabrication must show that catastrophic failure due to fatigue, corrosion, manufacturing defects, or accidental damage, will be avoided throughout the operational life of the airplane. This evaluation must be conducted in accordance with the provisions of paragraphs (b) and (e) of this section, except as specified in paragraph (c) of this section, for each part of the structure that could contribute to a catastrophic failure (such as wing, empennage, control surfaces and their systems, the fuselage, engine mounting, landing gear, and their related primary attachments). For turbojet powered airplanes, those parts that could contribute to a catastrophic failure must also be evaluated under paragraph (d) of this section. In addition, the following apply:

(1) Each evaluation required by this section must include—

(i) The typical loading spectra, temperatures, and humidities expected in service;

(ii) The identification of principal structural elements and detail design points, the failure of which could cause catastrophic failure of the airplane; and

(iii) An analysis, supported by test evidence, of the principal structural elements and detail design points identified in paragraph (a)(1)(ii) of this section.

(2) The service history of airplanes of similar structural design, taking due account of differences in operating conditions and procedures, may be used in the evaluations required by this section.

(3) Based on the evaluations required by this section, inspections or other procedures must be established, as necessary, to prevent catastrophic failure, and must be included in the Airworthiness Limitations section of the Instructions for Continued Airworthiness required by § 25.1529. The limit of validity of the engineering data that supports the structural maintenance program (hereafter referred to as LOV), stated as a number of total accumulated flight cycles or flight hours or both, established by this section must also be included in the Airworthiness Limitations section of the Instructions for Continued Airworthiness required by § 25.1529. Inspection thresholds for the following types of structure must be established based on crack growth analyses and/or tests, assuming the structure contains an initial flaw of the maximum probable size that could exist as a result of manufacturing or service-induced damage:

(i) Single load path structure, and

(ii) Multiple load path "fail-safe" structure and crack arrest "fail-safe" structure, where it cannot be demonstrated that load path failure, partial failure, or crack arrest will be detected and repaired during normal maintenance, inspection, or operation of an airplane prior to failure of the remaining structure.

(b) *Damage-tolerance evaluation.* The evaluation must include a determination of the probable locations and modes of damage due to fatigue, corrosion, or accidental damage. Repeated load and static analyses supported by test evidence and (if available) service experience must also be incorporated in the evaluation. Special consideration for widespread fatigue damage must be included where the design is such that this type of damage could occur. An LOV must be established that corresponds to the period of time, stated as a number of total accumulated flight cycles or flight hours or both, during which it is demonstrated that widespread fatigue damage will not occur in the airplane structure. This demonstration must be by full-scale fatigue test evidence. The type certificate may be issued prior to completion of full-scale fatigue testing, provided the Administrator has approved a plan for completing the required tests. In that case, the Airworthiness Limitations section of the Instructions for Continued Airworthiness required by § 25.1529 must specify that no airplane may be operated beyond a number of cycles equal to ½ the number of cycles accumulated on the fatigue test article, until such testing is completed. The extent of damage for residual strength evaluation at any time within the operational life of the airplane must be consistent with the initial detectability and subsequent growth under repeated loads. The residual strength evaluation must show that the remaining structure is able to withstand loads (considered as static ultimate loads) corresponding to the following conditions:

(1) The limit symmetrical maneuvering conditions specified in § 25.337 at all speeds up to V_c and in § 25.345.

(2) The limit gust conditions specified in § 25.341 at the specified speeds up to V_c and in § 25.345.

(3) The limit rolling conditions specified in § 25.349 and the limit unsymmetrical conditions specified in §§ 25.367 and 25.427 (a) through (c), at speeds up to V_C.

(4) The limit yaw maneuvering conditions specified in § 25.351(a) at the specified speeds up to V_C.

(5) For pressurized cabins, the following conditions:

(i) The normal operating differential pressure combined with the expected external aerodynamic pressures applied simultaneously with the flight loading conditions specified in paragraphs (b)(1) through (4) of this section, if they have a significant effect.

(ii) The maximum value of normal operating differential pressure (including the expected external aerodynamic pressures during 1g level flight) multiplied by a factor of 1.15, omitting other loads.

(6) For landing gear and directly-affected airframe structure, the limit ground loading conditions specified in §§ 25.473, 25.491, and 25.493.

If significant changes in structural stiffness or geometry, or both, follow from a structural failure, or partial failure, the effect on damage tolerance must be further investigated.

(c) *Fatigue (safe-life) evaluation.* Compliance with the damage-tolerance requirements of paragraph (b) of this section is not required if the applicant establishes that their application for particular structure is impractical. This structure must be shown by analysis, supported by test evidence, to be able to withstand the repeated loads of variable magnitude expected during its service life without detectable cracks. Appropriate safe-life scatter factors must be applied.

(d) *Sonic fatigue strength.* It must be shown by analysis, supported by test evidence, or by the service history of airplanes of similar structural design and sonic excitation environment, that—

(1) Sonic fatigue cracks are not probable in any part of the flight structure subject to sonic excitation; or

(2) Catastrophic failure caused by sonic cracks is not probable assuming that the loads prescribed in paragraph (b) of this section are applied to all areas affected by those cracks.

(e) *Damage-tolerance (discrete source) evaluation.* The airplane must be capable of successfully completing a flight during which likely structural damage occurs as a result of—

(1) Impact with a 4-pound bird when the velocity of the airplane relative to the bird along the airplane's flight path is equal to V_c at sea level or $0.85V_c$ at 8,000 feet, whichever is more critical;

(2) Uncontained fan blade impact;

(3) Uncontained engine failure; or

(4) Uncontained high energy rotating machinery failure.

The damaged structure must be able to withstand the static loads (considered as ultimate loads) which are reasonably expected to occur on the flight. Dynamic effects on these static loads need not be considered. Corrective action to be taken by the pilot following the incident, such as limiting maneuvers, avoiding turbulence, and reducing speed, must be considered. If significant changes in structural stiffness or geometry, or both, follow from a structural failure or partial failure, the effect on damage tolerance must be further investigated.

[Amdt. 25-45, 43 FR 46242, Oct. 5, 1978, as amended by Amdt. 25-54, 45 FR 60173, Sept. 11, 1980; Amdt. 25-72, 55 FR 29776, July 20, 1990; Amdt. 25-86, 61 FR 5222, Feb. 9, 1996; Amdt. 25-96, 63 FR 15714, Mar. 31, 1998; 63 FR 23338, Apr. 28, 1998; Amdt. 25-132, 75 FR 69781, Nov. 15, 2010]

Lightning Protection

§ 25.581 Lightning protection.

(a) The airplane must be protected against catastrophic effects from lightning.

(b) For metallic components, compliance with paragraph (a) of this section may be shown by—

(1) Bonding the components properly to the airframe; or

(2) Designing the components so that a strike will not endanger the airplane.

(c) For nonmetallic components, compliance with paragraph (a) of this section may be shown by—

(1) Designing the components to minimize the effect of a strike; or

(2) Incorporating acceptable means of diverting the resulting electrical current so as not to endanger the airplane.

[Amdt. 25-23, 35 FR 5674, Apr. 8, 1970]

Subpart D—Design and Construction

General

§ 25.601 General.

The airplane may not have design features or details that experience has shown to be hazardous or unreliable. The suitability of each questionable design detail and part must be established by tests.

§ 25.603 Materials.

The suitability and durability of materials used for parts, the failure of which could adversely affect safety, must—

(a) Be established on the basis of experience or tests;

(b) Conform to approved specifications (such as industry or military specifications, or Technical Standard Orders) that ensure their having the strength and other properties assumed in the design data; and

(c) Take into account the effects of environmental conditions, such as temperature and humidity, expected in service.

[Doc. No. 5066, 29 FR 18291, Dec. 24, 1964, as amended by Amdt. 25-38, 41 FR 55466, Dec. 20 1976; Amdt. 25-46, 43 FR 50595, Oct. 30, 1978]

§ 25.605 Fabrication methods.

(a) The methods of fabrication used must produce a consistently sound structure. If a fabrication process (such as gluing, spot welding, or heat treating) requires close control to reach this objective, the process must be performed under an approved process specification.

(b) Each new aircraft fabrication method must be substantiated by a test program.

[Doc. No. 5066, 29 FR 18291, Dec. 24, 1964, as amended by Amdt. 25-46, 43 FR 50595, Oct. 30, 1978]

§ 25.607 Fasteners.

(a) Each removable bolt, screw, nut, pin, or other removable fastener must incorporate two separate locking devices if—

(1) Its loss could preclude continued flight and landing within the design limitations of the airplane using normal pilot skill and strength; or

(2) Its loss could result in reduction in pitch, yaw, or roll control capability or response below that required by Subpart B of this chapter.

(b) The fasteners specified in paragraph (a) of this section and their locking devices may not be adversely affected by the environmental conditions associated with the particular installation.

(c) No self-locking nut may be used on any bolt subject to rotation in operation unless a nonfriction locking device is used in addition to the self-locking device.

[Amdt. 25-23, 35 FR 5674, Apr. 8, 1970]

§ 25.609 Protection of structure.

Each part of the structure must—

(a) Be suitably protected against deterioration or loss of strength in service due to any cause, including—

(1) Weathering;

(2) Corrosion; and

(3) Abrasion; and

(b) Have provisions for ventilation and drainage where necessary for protection.

§ 25.611 Accessibility provisions.

(a)Means must be provided to allow inspection (including inspection of principal structural elements and control systems), replacement of parts normally requiring replacement, adjustment, and lubrication as necessary for continued airworthiness. The inspection means for each item must be practicable for the inspection interval for the item. Nondestructive inspection aids may be used to inspect structural elements where it is impracticable to provide means for direct visual inspection if it is shown that the inspection is effective and the inspection procedures are specified in the maintenance manual required by § 25.1529.

(b) EWIS must meet the accessibility requirements of § 25.1719.

[Amdt. 25-23, 35 FR 5674, Apr. 8, 1970, as amended by Amdt. 25-123, 72 FR 63404, Nov. 8, 2007]

§ 25.613 Material strength properties and material design values.

(a) Material strength properties must be based on enough tests of material meeting approved specifications to establish design values on a statistical basis.

(b) Material design values must be chosen to minimize the probability of structural failures due to material variability. Except as provided in paragraphs (e) and (f) of this section, compliance must be shown by selecting material design values which assure material strength with the following probability:

(1) Where applied loads are eventually distributed through a single member within an assembly, the failure of which would result in loss of structural integrity of the component, 99 percent probability with 95 percent confidence.

(2) For redundant structure, in which the failure of individual elements would result in applied loads being safely distributed to other load carrying members, 90 percent probability with 95 percent confidence.

(c) The effects of environmental conditions, such as temperature and moisture, on material design values used in an essential component or structure must be considered where these effects are significant within the airplane operating envelope.

(d) [Reserved]

(e) Greater material design values may be used if a "premium selection" of the material is made in which a specimen of each individual item is tested before use to determine that the actual strength properties of that particular item will equal or exceed those used in design.

(f) Other material design values may be used if approved by the Administrator.

[Doc. No. 5066, 29 FR 18291, Dec. 24, 1964, as amended by Amdt. 25-46, 43 FR 50595, Oct. 30, 1978; Amdt. 25-72, 55 FR 29776, July 20, 1990; Amdt. 25-112, 68 FR 46431, Aug. 5, 2003]

§ 25.619 Special factors.

The factor of safety prescribed in § 25.303 must be multiplied by the highest pertinent special factor of safety prescribed in §§ 25.621 through 25.625 for each part of the structure whose strength is—

(a) Uncertain;

(b) Likely to deteriorate in service before normal replacement; or

(c) Subject to appreciable variability because of uncertainties in manufacturing processes or inspection methods.

[Doc. No. 5066, 29 FR 18291, Dec. 24, 1964, as amended by Amdt. 25-23, 35 FR 5674, Apr. 8, 1970]

§ 25.621 Casting factors.

(a) *General.* The factors, tests, and inspections specified in paragraphs (b) through (d) of this section must be applied in addition to those necessary to establish foundry quality control. The inspections must meet approved specifications. Paragraphs (c) and (d) of this section apply to any structural castings except castings that are pressure tested as parts of hydraulic or other fluid systems and do not support structural loads.

(b) *Bearing stresses and surfaces.* The casting factors specified in paragraphs (c) and (d) of this section—

(1) Need not exceed 1.25 with respect to bearing stresses regardless of the method of inspection used; and

(2) Need not be used with respect to the bearing surfaces of a part whose bearing factor is larger than the applicable casting factor.

(c) *Critical castings.* For each casting whose failure would preclude continued safe flight and landing of the airplane or result in serious injury to occupants, the following apply:

(1) Each critical casting must—

(i) Have a casting factor of not less than 1.25; and

(ii) Receive 100 percent inspection by visual, radiographic, and magnetic particle or penetrant inspection methods or approved equivalent nondestructive inspection methods.

(2) For each critical casting with a casting factor less than 1.50, three sample castings must be static tested and shown to meet—

(i) The strength requirements of § 25.305 at an ultimate load corresponding to a casting factor of 1.25; and

(ii) The deformation requirements of § 25.305 at a load of 1.15 times the limit load.

(3) Examples of these castings are structural attachment fittings, parts of flight control systems, control surface hinges and balance weight attachments, seat, berth, safety belt, and fuel and oil tank supports and attachments, and cabin pressure valves.

(d) *Noncritical castings.* For each casting other than those specified in paragraph (c) of this section, the following apply:

Casting factor	Inspection
2.0 or more	100 percent visual.
Less than 2.0 but more than 1.5	100 percent visual, and magnetic particle or penetrant or equivalent nondestructive inspection methods.
1.25 through 1.50	100 percent visual, magnetic particle or penetrant, and radiographic, or approved equivalent nondestructive inspection methods.

(1) Except as provided in paragraphs (d)(2) and (3) of this section, the casting factors and corresponding inspections must meet the following table:

(2) The percentage of castings inspected by nonvisual methods may be reduced below that specified in paragraph (d)(1) of this section when an approved quality control procedure is established.

(3) For castings procured to a specification that guarantees the mechanical properties of the material in the casting and provides for demonstration of these properties by test of coupons cut from the castings on a sampling basis—

(i) A casting factor of 1.0 may be used; and

(ii) The castings must be inspected as provided in paragraph (d)(1) of this section for casting factors of "1.25 through 1.50" and tested under paragraph (c)(2) of this section.

§ 25.623 Bearing factors.

(a) Except as provided in paragraph (b) of this section, each part that has clearance (free fit), and that is subject to pounding or vibration, must have a bearing factor large enough to provide for the effects of normal relative motion.

(b) No bearing factor need be used for a part for which any larger special factor is prescribed.

§ 25.625 Fitting factors.

For each fitting (a part or terminal used to join one structural member to another), the following apply:

(a) For each fitting whose strength is not proven by limit and ultimate load tests in which actual stress conditions are simulated in the fitting and surrounding structures, a fitting factor of at least 1.15 must be applied to each part of—

(1) The fitting;

(2) The means of attachment; and

(3) The bearing on the joined members.

(b) No fitting factor need be used—

(1) For joints made under approved practices and based on comprehensive test data (such as continuous joints in metal plating, welded joints, and scarf joints in wood); or

(2) With respect to any bearing surface for which a larger special factor is used.

(c) For each integral fitting, the part must be treated as a fitting up to the point at which the section properties become typical of the member.

(d) For each seat, berth, safety belt, and harness, the fitting factor specified in § 25.785(f)(3) applies.

[Doc. No. 5066, 29 FR 18291, Dec. 24, 1964, as amended by Amdt. 25-23, 35 FR 5674, Apr. 8, 1970; Amdt. 25-72, 55 FR 29776, July 20, 1990]

§ 25.629 Aeroelastic stability requirements.

(a) *General.* The aeroelastic stability evaluations required under this section include flutter, divergence, control reversal and any undue loss of stability and control as a result of structural deformation. The aeroelastic evaluation must include whirl modes associated with any propeller or rotating device that contributes significant dynamic forces. Compliance with this section must be shown by analyses, wind tunnel tests, ground vibration tests, flight tests, or other means found necessary by the Administrator.

(b) *Aeroelastic stability envelopes.* The airplane must be designed to be free from aeroelastic instability for all configurations and design conditions within the aeroelastic stability envelopes as follows:

(1) For normal conditions without failures, malfunctions, or adverse conditions, all combinations of altitudes and speeds encompassed by the V_D/M_D versus altitude envelope enlarged at all points by an increase of 15 percent in equivalent airspeed at both constant Mach number and constant altitude. In addition, a proper margin of stability must exist at all speeds up to V_D/M_D and, there must be no large and rapid reduction in stability as V_D/M_D is approached. The enlarged envelope may be limited to Mach 1.0 when M_D is less than 1.0 at all design altitudes, and

(2) For the conditions described in § 25.629(d) below, for all approved altitudes, any airspeed up to the greater airspeed defined by;

(i) The V_D/M_D envelope determined by § 25.335(b); or,

(ii) An altitude-airspeed envelope defined by a 15 percent increase in equivalent airspeed above V_C at constant altitude, from sea level to the altitude of the intersection of 1.15 V_C with the extension of the constant cruise Mach number line, M_C, then a linear variation in equivalent airspeed to M_C +.05 at the altitude of the lowest V_C/M_C intersection; then, at higher altitudes, up to the maximum flight altitude, the boundary defined by a.05 Mach increase in M_C at constant altitude.

(c) *Balance weights.* If concentrated balance weights are used, their effectiveness and strength, including supporting structure, must be substantiated.

(d) *Failures, malfunctions, and adverse conditions.* The failures, malfunctions, and adverse conditions which must be considered in showing compliance with this section are:

(1) Any critical fuel loading conditions, not shown to be extremely improbable, which may result from mismanagement of fuel.

(2) Any single failure in any flutter damper system.

(3) For airplanes not approved for operation in icing conditions, the maximum likely ice accumulation expected as a result of an inadvertent encounter.

(4) Failure of any single element of the structure supporting any engine, independently mounted propeller shaft, large auxiliary power unit, or large externally mounted aerodynamic body (such as an external fuel tank).

(5) For airplanes with engines that have propellers or large rotating devices capable of significant dynamic forces, any single failure of the engine structure that would reduce the rigidity of the rotational axis.

(6) The absence of aerodynamic or gyroscopic forces resulting from the most adverse combination of feathered propellers or other rotating devices capable of significant dynamic forces. In addition, the effect of a single feathered propeller or rotating device must be coupled with the failures of paragraphs (d)(4) and (d)(5) of this section.

(7) Any single propeller or rotating device capable of significant dynamic forces rotating at the highest likely overspeed.

(8) Any damage or failure condition, required or selected for investigation by § 25.571. The single structural failures described in paragraphs (d)(4) and (d)(5) of this section need not be considered in showing compliance with this section if;

(i) The structural element could not fail due to discrete source damage resulting from the conditions described in § 25.571(e), and

(ii) A damage tolerance investigation in accordance with § 25.571(b) shows that the maximum extent of damage assumed for the purpose of residual strength evaluation does not involve complete failure of the structural element.

(9) Any damage, failure, or malfunction considered under §§ 25.631, 25.671, 25.672, and 25.1309.

(10) Any other combination of failures, malfunctions, or adverse conditions not shown to be extremely improbable.

(e) *Flight flutter testing.* Full scale flight flutter tests at speeds up to V_{DF}/M_{DF} must be conducted for new type designs and for modifications to a type design unless the modifications have been shown to have an insignificant effect on the aeroelastic stability. These tests must demonstrate that the airplane has a proper margin of damping at all speeds up to V_{DF}/M_{DF}, and that there is no large and rapid reduction in damping as V_{DF}/M_{DF} is approached. If a failure, malfunction, or adverse condition is simulated during flight test in showing compliance with paragraph (d) of this section, the maximum speed investigated need not exceed V_{FC}/M_{FC} if it is shown, by correlation of the flight test data with other test data or analyses, that the airplane is free from any aeroelastic instability at all speeds within the altitude-airspeed envelope described in paragraph (b)(2) of this section.

[Doc. No. 26007, 57 FR 28949, June 29, 1992]

§ 25.631 Bird strike damage.

The empennage structure must be designed to assure capability of continued safe flight and landing of the airplane after impact with an 8-pound bird when the velocity of the airplane (relative to the bird along the airplane's flight path) is equal to V_C at sea level, selected under § 25.335(a). Compliance with this section by provision of redundant structure and protected location of control system elements or protective devices such as splitter plates or energy absorbing material is acceptable. Where compliance is shown by analysis, tests, or both, use of data on airplanes having similar structural design is acceptable.

[Amdt. 25-23, 35 FR 5674, Apr. 8, 1970]

CONTROL SURFACES

§ 25.651 Proof of strength.

(a) Limit load tests of control surfaces are required. These tests must include the horn or fitting to which the control system is attached.

(b) Compliance with the special factors requirements of §§ 25.619 through 25.625 and 25.657 for control surface hinges must be shown by analysis or individual load tests.

§ 25.655 Installation.

(a) Movable tail surfaces must be installed so that there is no interference between any surfaces when one is held in its extreme position and the others are operated through their full angular movement.

(b) If an adjustable stabilizer is used, it must have stops that will limit its range of travel to the maximum for which the airplane is shown to meet the trim requirements of § 25.161.

§ 25.657 Hinges.

(a) For control surface hinges, including ball, roller, and self-lubricated bearing hinges, the approved rating of the bearing may not be exceeded. For nonstandard bearing hinge configurations, the rating must be established on the basis of experience or tests and, in the absence of a rational investigation, a factor of safety of not less than 6.67 must be used with respect to the ultimate bearing strength of the softest material used as a bearing.

(b) Hinges must have enough strength and rigidity for loads parallel to the hinge line.

[Amdt. 25-23, 35 FR 5674, Apr. 8, 1970]

Control Systems

§ 25.671 General.

(a) Each control and control system must operate with the ease, smoothness, and positiveness appropriate to its function.

(b) Each element of each flight control system must be designed, or distinctively and permanently marked, to minimize the probability of incorrect assembly that could result in the malfunctioning of the system.

(c) The airplane must be shown by analysis, tests, or both, to be capable of continued safe flight and landing after any of the following failures or jamming in the flight control system and surfaces (including trim, lift, drag, and feel systems), within the normal flight envelope, without requiring exceptional piloting skill or strength. Probable malfunctions must have only minor effects on control system operation and must be capable of being readily counteracted by the pilot.

(1) Any single failure, excluding jamming (for example, disconnection or failure of mechanical elements, or structural failure of hydraulic components, such as actuators, control spool housing, and valves).

(2) Any combination of failures not shown to be extremely improbable, excluding jamming (for example, dual electrical or hydraulic system failures, or any single failure in combination with any probable hydraulic or electrical failure).

(3) Any jam in a control position normally encountered during takeoff, climb, cruise, normal turns, descent, and landing unless the jam is shown to be extremely improbable, or can be alleviated. A runaway of a flight control to an adverse position and jam must be accounted for if such runaway and subsequent jamming is not extremely improbable.

(d) The airplane must be designed so that it is controllable if all engines fail. Compliance with this requirement may be shown by analysis where that method has been shown to be reliable.

[Doc. No. 5066, 29 FR 18291, Dec. 24, 1964, as amended by Amdt. 25-23, 35 FR 5674, Apr. 8, 1970]

§ 25.672 Stability augmentation and automatic and power-operated systems.

If the functioning of stability augmentation or other automatic or power-operated systems is necessary to show compliance with the flight characteristics requirements of this part, such systems must comply with § 25.671 and the following:

(a) A warning which is clearly distinguishable to the pilot under expected flight conditions without requiring his attention must be provided for any failure in the stability augmentation system or in any other automatic or power-operated system which could result in an unsafe condition if the pilot were not aware of the failure. Warning systems must not activate the control systems.

(b) The design of the stability augmentation system or of any other automatic or power-operated system must permit initial counteraction of failures of the type specified in § 25.671(c) without requiring exceptional pilot skill or strength, by either the deactivation of the system, or a failed portion thereof, or by overriding the failure by movement of the flight controls in the normal sense.

(c) It must be shown that after any single failure of the stability augmentation system or any other automatic or power-operated system—

(1) The airplane is safely controllable when the failure or malfunction occurs at any speed or altitude within the approved operating limitations that is critical for the type of failure being considered;

(2) The controllability and maneuverability requirements of this part are met within a practical operational flight envelope (for example, speed, altitude, normal acceleration, and airplane configurations) which is described in the Airplane Flight Manual; and

(3) The trim, stability, and stall characteristics are not impaired below a level needed to permit continued safe flight and landing.

[Amdt. 25-23, 35 FR 5675 Apr. 8, 1970]

§ 25.675 Stops.

(a) Each control system must have stops that positively limit the range of motion of each movable aerodynamic surface controlled by the system.

(b) Each stop must be located so that wear, slackness, or take-up adjustments will not adversely affect the control characteristics of the airplane because of a change in the range of surface travel.

(c) Each stop must be able to withstand any loads corresponding to the design conditions for the control system.

[Doc. No. 5066, 29 FR 18291, Dec. 24, 1964, as amended by Amdt. 25-38, 41 FR 55466, Dec. 20, 1976]

§ 25.677 Trim systems.

(a) Trim controls must be designed to prevent inadvertent or abrupt operation and to operate in the plane, and with the sense of motion, of the airplane.

(b) There must be means adjacent to the trim control to indicate the direction of the control movement relative to the airplane motion. In addition, there must be clearly visible means to indicate the position of the trim device with respect to the range of adjustment. The indicator must be clearly marked with the range within which it has been demonstrated that takeoff is safe for all center of gravity positions approved for takeoff.

(c) Trim control systems must be designed to prevent creeping in flight. Trim tab controls must be irreversible unless the tab is appropriately balanced and shown to be free from flutter.

(d) If an irreversible tab control system is used, the part from the tab to the attachment of the irreversible unit to the airplane structure must consist of a rigid connection.

[Doc. No. 5066, 29 FR 18291, Dec. 24, 1964, as amended by Amdt. 25-23, 35 FR 5675, Apr. 8, 1970; Amdt. 25-115, 69 FR 40527, July 2, 2004]

§ 25.679 Control system gust locks.

(a) There must be a device to prevent damage to the control surfaces (including tabs), and to the control system, from gusts striking the airplane while it is on the ground or water. If the device, when engaged, prevents normal operation of the control surfaces by the pilot, it must—

(1) Automatically disengage when the pilot operates the primary flight controls in a normal manner; or

(2) Limit the operation of the airplane so that the pilot receives unmistakable warning at the start of takeoff.

(b) The device must have means to preclude the possibility of it becoming inadvertently engaged in flight.

§ 25.681 Limit load static tests.

(a) Compliance with the limit load requirements of this Part must be shown by tests in which—

(1) The direction of the test loads produces the most severe loading in the control system; and

(2) Each fitting, pulley, and bracket used in attaching the system to the main structure is included.

(b) Compliance must be shown (by analyses or individual load tests) with the special factor requirements for control system joints subject to angular motion.

§ 25.683 Operation tests.

It must be shown by operation tests that when portions of the control system subject to pilot effort loads are loaded to 80 percent of the limit load specified for the system and the powered portions of the control system are loaded to the maximum load expected in normal operation, the system is free from—

(a) Jamming;

(b) Excessive friction; and

(c) Excessive deflection.

[Doc. No. 5066, 29 FR 18291, Dec. 24, 1964, as amended by Amdt. 25-23, 35 FR 5675, Apr. 8, 1970]

§ 25.685 Control system details.

(a) Each detail of each control system must be designed and installed to prevent jamming, chafing, and interference from cargo, passengers, loose objects, or the freezing of moisture.

(b) There must be means in the cockpit to prevent the entry of foreign objects into places where they would jam the system.

(c) There must be means to prevent the slapping of cables or tubes against other parts.

(d) Sections 25.689 and 25.693 apply to cable systems and joints.

[Doc. No. 5066, 29 FR 18291, Dec. 24, 1964, as amended by Amdt. 25-38, 41 FR 55466, Dec. 20, 1976]

§ 25.689 Cable systems.

(a) Each cable, cable fitting, turnbuckle, splice, and pulley must be approved. In addition—

(1) No cable smaller than $\frac{1}{8}$ inch in diameter may be used in the aileron, elevator, or rudder systems; and

(2) Each cable system must be designed so that there will be no hazardous change in cable tension throughout the range of travel under operating conditions and temperature variations.

(b) Each kind and size of pulley must correspond to the cable with which it is used. Pulleys and sprockets must have closely fitted guards to prevent the cables and chains from being displaced or fouled. Each pulley must lie in the plane passing through the cable so that the cable does not rub against the pulley flange.

(c) Fairleads must be installed so that they do not cause a change in cable direction of more than three degrees.

(d) Clevis pins subject to load or motion and retained only by cotter pins may not be used in the control system.

(e) Turnbuckles must be attached to parts having angular motion in a manner that will positively prevent binding throughout the range of travel.

(f) There must be provisions for visual inspection of fairleads, pulleys, terminals, and turnbuckles.

§ 25.693 Joints.

Control system joints (in push-pull systems) that are subject to angular motion, except those in ball and roller bearing systems, must have a special factor of safety of not less than 3.33 with respect to the ultimate bearing strength of the softest material used as a bearing. This factor may be reduced to 2.0 for joints in cable control systems. For ball or roller bearings, the approved ratings may not be exceeded.

[Amdt. 25-72, 55 FR 29777, July 20, 1990]

§ 25.697 Lift and drag devices, controls.

(a) Each lift device control must be designed so that the pilots can place the device in any takeoff, en route, approach, or landing position established under § 25.101(d). Lift and drag devices must maintain the selected positions, except for movement produced by an automatic positioning or load limiting device, without further attention by the pilots.

(b) Each lift and drag device control must be designed and located to make inadvertent operation improbable. Lift and drag devices intended for ground operation only must have means to prevent the inadvertant operation of their controls in flight if that operation could be hazardous.

(c) The rate of motion of the surfaces in response to the operation of the control and the characteristics of the automatic positioning or load limiting device must give satisfactory flight and performance characteristics under steady or changing conditions of airspeed, engine power, and airplane attitude.

(d) The lift device control must be designed to retract the surfaces from the fully extended position, during steady flight at maximum continuous engine power at any speed below V_F+9.0 (knots).

[Amdt. 25-23, 35 FR 5675, Apr. 8, 1970, as amended by Amdt. 25-46, 43 FR 50595, Oct. 30, 1978; Amdt. 25-57, 49 FR 6848, Feb. 23, 1984]

§ 25.699 Lift and drag device indicator.

(a) There must be means to indicate to the pilots the position of each lift or drag device having a separate control in the cockpit to adjust its position. In addition, an indication of unsymmetrical operation or other malfunction in the lift or drag device systems must be provided when such indication is necessary to enable the pilots to prevent or counteract an unsafe flight or ground condition, considering the effects on flight characteristics and performance.

(b) There must be means to indicate to the pilots the takeoff, en route, approach, and landing lift device positions.

(c) If any extension of the lift and drag devices beyond the landing position is possible, the controls must be clearly marked to identify this range of extension.

[Amdt. 25-23, 35 FR 5675, Apr. 8, 1970]

§ 25.701 Flap and slat interconnection.

(a) Unless the airplane has safe flight characteristics with the flaps or slats retracted on one side and extended on the other, the motion of flaps or slats on opposite sides of the plane of symmetry must be synchronized by a mechanical interconnection or approved equivalent means.

(b) If a wing flap or slat interconnection or equivalent means is used, it must be designed to account for the applicable unsymmetrical loads, including those resulting from flight with the engines on one side of the plane of symmetry inoperative and the remaining engines at takeoff power.

(c) For airplanes with flaps or slats that are not subjected to slipstream conditions, the structure must be designed for the loads imposed when the wing flaps or slats on one side are carrying the most severe load occurring in the prescribed symmetrical conditions and those on the other side are carrying not more than 80 percent of that load.

(d) The interconnection must be designed for the loads resulting when interconnected flap or slat surfaces on one side of the plane of symmetry are jammed and immovable while the surfaces on the other side are free to move and the full power of the surface actuating system is applied.

[Amdt. 25-72, 55 FR 29777, July 20, 1990]

§ 25.703 Takeoff warning system.

A takeoff warning system must be installed and must meet the following requirements:

(a) The system must provide to the pilots an aural warning that is automatically activated during the initial portion of the takeoff roll if the airplane is in a configuration, including any of the following, that would not allow a safe takeoff:

(1) The wing flaps or leading edge devices are not within the approved range of takeoff positions.

(2) Wing spoilers (except lateral control spoilers meeting the requirements of § 25.671), speed brakes, or longitudinal trim devices are in a position that would not allow a safe takeoff.

(b) The warning required by paragraph (a) of this section must continue until—

(1) The configuration is changed to allow a safe takeoff;

(2) Action is taken by the pilot to terminate the takeoff roll;

(3) The airplane is rotated for takeoff; or

(4) The warning is manually deactivated by the pilot.

(c) The means used to activate the system must function properly throughout the ranges of takeoff weights, altitudes, and temperatures for which certification is requested.

[Amdt. 25-42, 43 FR 2323, Jan. 16, 1978]

Landing Gear

§ 25.721 General.

(a) The main landing gear system must be designed so that if it fails due to overloads during takeoff and landing (assuming the overloads to act in the upward and aft directions), the failure mode is not likely to cause—

(1) For airplanes that have passenger seating configuration, excluding pilots seats, of nine seats or less, the spillage of enough fuel from any fuel system in the fuselage to constitute a fire hazard; and

(2) For airplanes that have a passenger seating configuration, excluding pilots seats, of 10 seats or more, the spillage of enough fuel from any part of the fuel system to constitute a fire hazard.

(b) Each airplane that has a passenger seating configuration excluding pilots seats, of 10 seats or more must be designed so that with the airplane under control it can be landed on a paved runway with any one or more landing gear legs not extended without sustaining a structural component failure that is likely to cause the spillage of enough fuel to constitute a fire hazard.

(c) Compliance with the provisions of this section may be shown by analysis or tests, or both.

[Amdt. 25-32, 37 FR 3969, Feb. 24, 1972]

§ 25.723 Shock absorption tests.

(a) The analytical representation of the landing gear dynamic characteristics that is used in determining the landing loads must be validated by energy absorption tests. A range of tests must be conducted to ensure that the analytical representation is valid for the design conditions specified in § 25.473.

(1) The configurations subjected to energy absorption tests at limit design conditions must include at least the design landing weight or the design takeoff weight, whichever produces the greater value of landing impact energy.

(2) The test attitude of the landing gear unit and the application of appropriate drag loads during the test must simulate the airplane landing conditions in a manner consistent with the development of rational or conservative limit loads.

(b) The landing gear may not fail in a test, demonstrating its reserve energy absorption capacity, simulating a descent velocity of 12 f.p.s. at design landing weight, assuming airplane lift not greater than airplane weight acting during the landing impact.

(c) In lieu of the tests prescribed in this section, changes in previously approved design weights and minor changes in design may be substantiated by analyses based on previous tests conducted on the same basic landing gear system that has similar energy absorption characteristics.

[Doc. No. 1999-5835, 66 FR 27394, May 16, 2001]

§§ 25.725-25.727 [Reserved]

§ 25.729 Retracting mechanism.

(a) *General.* For airplanes with retractable landing gear, the following apply:

(1) The landing gear retracting mechanism, wheel well doors, and supporting structure, must be designed for—

(i) The loads occurring in the flight conditions when the gear is in the retracted position,

(ii) The combination of friction loads, inertia loads, brake torque loads, air loads, and gyroscopic loads resulting from the wheels rotating at a peripheral speed equal to 1.23V_{SR} (with the wing-flaps in take-off position at design take-off weight), occurring during retraction and extension at any airspeed up to 1.5 V_{SR1} (with the wing-flaps in the approach position at design landing weight), and

(iii) Any load factor up to those specified in § 25.345(a) for the wing-flaps extended condition.

(2) Unless there are other means to decelerate the airplane in flight at this speed, the landing gear, the retracting mechanism, and the airplane structure (including wheel well doors) must be designed to withstand the flight loads occurring with the landing gear in the extended position at any speed up to 0.67 V_C.

(3) Landing gear doors, their operating mechanism, and their supporting structures must be designed for the yawing maneuvers prescribed for the airplane in addition to the conditions of airspeed and load factor prescribed in paragraphs (a)(1) and (2) of this section.

(b) *Landing gear lock.* There must be positive means to keep the landing gear extended in flight and on the ground. There must be positive means to keep the landing gear and doors in the correct retracted position in flight, unless it can be shown that lowering of the landing gear or doors, or flight with the landing gear or doors extended, at any speed, is not hazardous.

(c) *Emergency operation.* There must be an emergency means for extending the landing gear in the event of—

(1) Any reasonably probable failure in the normal retraction system; or

(2) The failure of any single source of hydraulic, electric, or equivalent energy supply.

(d) *Operation test.* The proper functioning of the retracting mechanism must be shown by operation tests.

(e) *Position indicator and warning device.* If a retractable landing gear is used, there must be a landing gear position indicator easily visible to the pilot or to the appropriate crew members (as well as necessary devices to actuate the indicator) to indicate without ambiguity that the retractable units and their associated doors are secured in the extended (or retracted) position. The means must be designed as follows:

(1) If switches are used, they must be located and coupled to the landing gear mechanical systems in a manner that prevents an erroneous indication of "down and locked" if the landing gear is not in a fully extended position, or of "up and locked" if the landing gear is not in the fully retracted position. The switches may be located where they are operated by the actual landing gear locking latch or device.

(2) The flightcrew must be given an aural warning that functions continuously, or is periodically repeated, if a landing is attempted when the landing gear is not locked down.

(3) The warning must be given in sufficient time to allow the landing gear to be locked down or a go-around to be made.

(4) There must not be a manual shut-off means readily available to the flightcrew for the warning required by paragraph (e)(2) of this section such that it could be operated instinctively, inadvertently, or by habitual reflexive action.

(5) The system used to generate the aural warning must be designed to minimize false or inappropriate alerts.

(6) Failures of systems used to inhibit the landing gear aural warning, that would prevent the warning system from operating, must be improbable.

(7) A flightcrew alert must be provided whenever the landing gear position is not consistent with the landing gear selector lever position.

(f) *Protection of equipment on landing gear and in wheel wells.* Equipment that is essential to the safe operation of the airplane and that is located on the landing gear and in wheel wells must be protected from the damaging effects of—

(1) A bursting tire;

(2) A loose tire tread, unless it is shown that a loose tire tread cannot cause damage.

(3) Possible wheel brake temperatures.

[Doc. No. 5066, 29 FR 18291, Dec. 24, 1964, as amended by Amdt. 25-23, 35 FR 5676, Apr. 8, 1970; Amdt. 25-42, 43 FR 2323, Jan. 16, 1978; Amdt. 25-72, 55 FR 29777, July 20, 1990; Amdt. 25-75, 56 FR 63762, Dec. 5, 1991; Amdt. 25-136, 77 FR 1617, Jan. 11, 2012]

§ 25.731 Wheels.

(a) Each main and nose wheel must be approved.

(b) The maximum static load rating of each wheel may not be less than the corresponding static ground reaction with—

(1) Design maximum weight; and

(2) Critical center of gravity.

(c) The maximum limit load rating of each wheel must equal or exceed the maximum radial limit load determined under the applicable ground load requirements of this part.

(d) *Overpressure burst prevention.* Means must be provided in each wheel to prevent wheel failure and tire burst that may result from excessive pressurization of the wheel and tire assembly.

(e) *Braked wheels.* Each braked wheel must meet the applicable requirements of § 25.735.

[Doc. No. 5066, 29 FR 18291, Dec. 24, 1964, as amended by Amdt. 25-72, 55 FR 29777, July 20, 1990; Amdt. 25-107, 67 FR 20420, Apr. 24, 2002]

§ 25.733 Tires.

(a) When a landing gear axle is fitted with a single wheel and tire assembly, the wheel must be fitted with a suitable tire of proper fit with a speed rating approved by the Administrator that is not exceeded under critical conditions and with a load rating approved by the Administrator that is not exceeded under—

(1) The loads on the main wheel tire, corresponding to the most critical combination of airplane weight (up to maximum weight) and center of gravity position, and

(2) The loads corresponding to the ground reactions in paragraph (b) of this section, on the nose wheel tire, except as provided in paragraphs (b)(2) and (b)(3) of this section.

(b) The applicable ground reactions for nose wheel tires are as follows:

(1) The static ground reaction for the tire corresponding to the most critical combination of airplane weight (up to maximum ramp weight) and center of gravity position with a force of 1.0g acting downward at the center of gravity. This load may not exceed the load rating of the tire.

(2) The ground reaction of the tire corresponding to the most critical combination of airplane weight (up to maximum landing weight) and center of gravity position combined with forces of 1.0g downward and 0.31g forward acting at the center of gravity. The reactions in this case must be distributed to the nose and main wheels by the principles of statics with a drag reaction equal to 0.31 times the vertical load at each wheel with brakes capable of producing this ground reaction. This nose tire load may not exceed 1.5 times the load rating of the tire.

(3) The ground reaction of the tire corresponding to the most critical combination of airplane weight (up to maximum ramp weight) and center of gravity position combined with forces of 1.0g downward and 0.20g forward acting at the center of gravity. The reactions in this case must be distributed to the nose and main wheels by the principles of statics with a drag reaction equal to 0.20 times the vertical load at each wheel with brakes capable of

producing this ground reaction. This nose tire load may not exceed 1.5 times the load rating of the tire.

(c) When a landing gear axle is fitted with more than one wheel and tire assembly, such as dual or dual-tandem, each wheel must be fitted with a suitable tire of proper fit with a speed rating approved by the Administrator that is not exceeded under critical conditions, and with a load rating approved by the Administrator that is not exceeded by—

(1) The loads on each main wheel tire, corresponding to the most critical combination of airplane weight (up to maximum weight) and center of gravity position, when multiplied by a factor of 1.07; and

(2) Loads specified in paragraphs (a)(2), (b)(1), (b)(2), and (b)(3) of this section on each nose wheel tire.

(d) Each tire installed on a retractable landing gear system must, at the maximum size of the tire type expected in service, have a clearance to surrounding structure and systems that is adequate to prevent unintended contact between the tire and any part of the structure or systems.

(e) For an airplane with a maximum certificated takeoff weight of more than 75,000 pounds, tires mounted on braked wheels must be inflated with dry nitrogen or other gases shown to be inert so that the gas mixture in the tire does not contain oxygen in excess of 5 percent by volume, unless it can be shown that the tire liner material will not produce a volatile gas when heated or that means are provided to prevent tire temperatures from reaching unsafe levels.

[Amdt. 25-48, 44 FR 68752, Nov. 29, 1979; Amdt. 25-72, 55 FR 29777, July 20, 1990, as amended by Amdt. 25-78, 58 FR 11781, Feb. 26, 1993]

§ 25.735 Brakes and braking systems.

(a) *Approval.* Each assembly consisting of a wheel(s) and brake(s) must be approved.

(b) *Brake system capability.* The brake system, associated systems and components must be designed and constructed so that:

(1) If any electrical, pneumatic, hydraulic, or mechanical connecting or transmitting element fails, or if any single source of hydraulic or other brake operating energy supply is lost, it is possible to bring the airplane to rest with a braked roll stopping distance of not more than two times that obtained in determining the landing distance as prescribed in § 25.125.

(2) Fluid lost from a brake hydraulic system following a failure in, or in the vicinity of, the brakes is insufficient to cause or support a hazardous fire on the ground or in flight.

(c) *Brake controls.* The brake controls must be designed and constructed so that:

(1) Excessive control force is not required for their operation.

(2) If an automatic braking system is installed, means are provided to:

(i) Arm and disarm the system, and

(ii) Allow the pilot(s) to override the system by use of manual braking.

(d) *Parking brake.* The airplane must have a parking brake control that, when selected on, will, without further attention, prevent the airplane from rolling on a dry and level paved runway when the most adverse combination of maximum thrust on one engine and up to maximum ground idle thrust on any, or all, other engine(s) is applied. The control must be suitably located or be adequately protected to prevent inadvertent operation. There must be indication in the cockpit when the parking brake is not fully released.

(e) *Antiskid system.* If an antiskid system is installed:

(1) It must operate satisfactorily over the range of expected runway conditions, without external adjustment.

(2) It must, at all times, have priority over the automatic braking system, if installed.

(f) *Kinetic energy capacity* —(1) *Design landing stop.* The design landing stop is an operational landing stop at maximum landing weight. The design landing stop brake kinetic energy absorption requirement of each wheel, brake, and tire assembly must be determined. It must be substantiated by dynamometer testing that the wheel, brake and tire assembly is capable of absorbing not less than this level of kinetic energy throughout the defined wear range of the brake. The energy absorption rate derived from the airplane manufacturer's braking requirements must be achieved. The mean deceleration must not be less than 10 fps ^{2}.

(2) *Maximum kinetic energy accelerate-stop.* The maximum kinetic energy accelerate-stop is a rejected takeoff for the most critical combination of airplane takeoff weight and speed. The accelerate-stop brake kinetic energy absorption requirement of each wheel, brake, and tire assembly must be determined. It must be substantiated by dynamometer testing that the wheel, brake, and tire assembly is capable of absorbing not less than this level of kinetic energy throughout the defined wear range of the brake. The energy absorption rate derived from the airplane manufacturer's braking requirements must be achieved. The mean deceleration must not be less than 6fps ^{2}.

(3) *Most severe landing stop.* The most severe landing stop is a stop at the most critical combination of airplane landing weight and speed. The most severe landing stop brake kinetic energy absorption requirement of each wheel, brake, and tire assembly must be determined. It must be substantiated by dynamometer testing that, at the declared fully worn limit(s) of the brake heat sink, the wheel, brake and tire assembly is capable of absorbing not less than this level of kinetic energy. The most severe landing stop need not be considered for extremely improbable failure conditions or if the maximum kinetic energy accelerate-stop energy is more severe.

(g) *Brake condition after high kinetic energy dynamometer stop(s).* Following the high kinetic energy stop demonstration(s) required by paragraph (f) of this section, with the parking brake promptly and fully applied for at least 3 minutes, it must be demonstrated that for at least 5 minutes from application of the parking brake, no condition occurs (or has occurred during the stop), including fire associated with the tire or wheel and brake assembly, that could prejudice the safe and complete evacuation of the airplane.

(h) *Stored energy systems.* An indication to the flightcrew of the usable stored energy must be provided if a stored energy system is used to show compliance with paragraph (b)(1) of this section. The available stored energy must be sufficient for:

(1) At least 6 full applications of the brakes when an antiskid system is not operating; and

(2) Bringing the airplane to a complete stop when an antiskid system is operating, under all runway surface conditions for which the airplane is certificated.

(i) *Brake wear indicators.* Means must be provided for each brake assembly to indicate when the heat sink is worn to the permissible limit. The means must be reliable and readily visible.

(j) *Overtemperature burst prevention.* Means must be provided in each braked wheel to prevent a wheel failure, a tire burst, or both, that may result from elevated brake temperatures. Additionally, all wheels must meet the requirements of § 25.731(d).

(k) *Compatibility.* Compatibility of the wheel and brake assemblies with the airplane and its systems must be substantiated.

[Doc. No. FAA-1999-6063, 67 FR 20420, Apr. 24, 2002, as amended by Amdt. 25-108, 67 FR 70827, Nov. 26, 2002; 68 FR 1955, Jan. 15, 2003]

§ 25.737 Skis.

Each ski must be approved. The maximum limit load rating of each ski must equal or exceed the maximum limit load determined under the applicable ground load requirements of this part.

Floats and Hulls

§ 25.751 Main float buoyancy.

Each main float must have—

(a) A buoyancy of 80 percent in excess of that required to support the maximum weight of the seaplane or amphibian in fresh water; and

(b) Not less than five watertight compartments approximately equal in volume.

§ 25.753 Main float design.

Each main float must be approved and must meet the requirements of § 25.521.

§ 25.755 Hulls.

(a) Each hull must have enough watertight compartments so that, with any two adjacent compartments flooded, the buoyancy of the hull and auxiliary floats (and wheel tires, if used) provides a margin of positive stability great enough to minimize the probability of capsizing in rough, fresh water.

(b) Bulkheads with watertight doors may be used for communication between compartments.

Personnel and Cargo Accommodations

§ 25.771 Pilot compartment.

(a) Each pilot compartment and its equipment must allow the minimum flight crew (established under § 25.1523) to perform their duties without unreasonable concentration or fatigue.

(b) The primary controls listed in § 25.779(a), excluding cables and control rods, must be located with respect to the propellers so that no member of the minimum flight crew (established under § 25.1523), or part of the controls, lies in the region between the plane of rotation of any inboard propeller and the surface generated by a line passing through the center of the propeller hub making an angle of five degrees forward or aft of the plane of rotation of the propeller.

(c) If provision is made for a second pilot, the airplane must be controllable with equal safety from either pilot seat.

(d) The pilot compartment must be constructed so that, when flying in rain or snow, it will not leak in a manner that will distract the crew or harm the structure.

(e) Vibration and noise characteristics of cockpit equipment may not interfere with safe operation of the airplane.

[Doc. No. 5066, 29 FR 18291, Dec. 24, 1964, as amended by Amdt. 25-4, 30 FR 6113, Apr. 30, 1965]

§ 25.772 Pilot compartment doors.

For an airplane that has a lockable door installed between the pilot compartment and the passenger compartment:

(a) For airplanes with a maximum passenger seating configuration of more than 20 seats, the emergency exit configuration must be designed so that neither crewmembers nor passengers require use of the flightdeck door in order to reach the emergency exits provided for them; and

(b) Means must be provided to enable flight crewmembers to directly enter the passenger compartment from the pilot compartment if the cockpit door becomes jammed.

(c) There must be an emergency means to enable a flight attendant to enter the pilot compartment in the event that the flightcrew becomes incapacitated.

[Doc. No. 24344, 55 FR 29777, July 20, 1990, as amended by Amdt. 25-106, 67 FR 2127, Jan. 15, 2002]

§ 25.773 Pilot compartment view.

(a) *Nonprecipitation conditions.* For nonprecipitation conditions, the following apply:

(1) Each pilot compartment must be arranged to give the pilots a sufficiently extensive, clear, and undistorted view, to enable them to safely perform any maneuvers within the operating limitations of the airplane, including taxiing takeoff, approach, and landing.

(2) Each pilot compartment must be free of glare and reflection that could interfere with the normal duties of the minimum flight crew (established under § 25.1523). This must be shown in day and night flight tests under nonprecipitation conditions.

(b) *Precipitation conditions.* For precipitation conditions, the following apply:

(1) The airplane must have a means to maintain a clear portion of the windshield, during precipitation conditions, sufficient for both pilots to have a sufficiently extensive view along the flight path in normal flight attitudes of the airplane. This means must be designed to function, without continuous attention on the part of the crew, in—

(i) Heavy rain at speeds up to 1.5 V_{SR1} with lift and drag devices retracted; and

(ii) The icing conditions specified in § 25.1419 if certification for flight in icing conditions is requested.

(2) No single failure of the systems used to provide the view required by paragraph (b)(1) of this section must cause the loss of that view by both pilots in the specified precipitation conditions.

(3) The first pilot must have a window that—

(i) Is openable under the conditions prescribed in paragraph (b)(1) of this section when the cabin is not pressurized;

(ii) Provides the view specified in paragraph (b)(1) of this section; and

(iii) Provides sufficient protection from the elements against impairment of the pilot's vision.

(4) The openable window specified in paragraph (b)(3) of this section need not be provided if it is shown that an area of the transparent surface will remain clear sufficient for at least one pilot to land the airplane safely in the event of—

(i) Any system failure or combination of failures which is not extremely improbable, in accordance with § 25.1309, under the precipitation conditions specified in paragraph (b)(1) of this section.

(ii) An encounter with severe hail, birds, or insects.

(c) *Internal windshield and window fogging.* The airplane must have a means to prevent fogging of the internal portions of the windshield and window panels over an area which would provide the visibility specified in paragraph (a) of this section under all internal and external ambient conditions, including precipitation conditions, in which the airplane is intended to be operated.

(d) Fixed markers or other guides must be installed at each pilot station to enable the pilots to position themselves in their seats for an optimum combination of outside visibility and instrument scan. If lighted markers or guides are used they must comply with the requirements specified in § 25.1381.

[Doc. No. 5066, 29 FR 18291, Dec. 24, 1964, as amended by Amdt. 25-23, 35 FR 5676, Apr. 8, 1970; Amdt. 25-46, 43 FR 50595, Oct. 30, 1978; Amdt. 25-72, 55 FR 29778, July 20, 1990; Amdt. 25-108, 67 FR 70827, Nov. 26, 2002; Amdt. 25-121, 72 FR 44669, Aug. 8, 2007; Amdt. 25-136, 77 FR 1618, Jan. 11, 2012]

§ 25.775 Windshields and windows.

(a) Internal panes must be made of nonsplintering material.

(b) Windshield panes directly in front of the pilots in the normal conduct of their duties, and the supporting structures for these panes, must withstand, without penetration, the impact of a four-pound bird when the velocity of the airplane (relative to the bird along the airplane's flight path) is equal to the value of V_C at sea level, selected under § 25.335(a).

(c) Unless it can be shown by analysis or tests that the probability of occurrence of a critical windshield fragmentation condition is of a low order, the airplane must have a means to minimize the danger to the pilots from flying windshield fragments due to bird impact. This must be shown for each transparent pane in the cockpit that—

(1) Appears in the front view of the airplane;

(2) Is inclined 15 degrees or more to the longitudinal axis of the airplane; and

(3) Has any part of the pane located where its fragmentation will constitute a hazard to the pilots.

(d) The design of windshields and windows in pressurized airplanes must be based on factors peculiar to high altitude operation, including the effects of continuous and cyclic pressurization loadings, the inherent characteristics of the material used, and the effects of temperatures and temperature differentials. The windshield and window panels must be capable of withstanding the maximum cabin pressure differential loads combined with critical aerodynamic pressure and temperature effects after any single failure in the installation or associated systems. It may be assumed that, after a single failure that is obvious to the flight crew (established under § 25.1523), the cabin pressure differential is reduced from the maximum, in accordance with appropriate operating limitations, to allow continued safe flight of the airplane with a cabin pressure altitude of not more than 15,000 feet.

(e) The windshield panels in front of the pilots must be arranged so that, assuming the loss of vision through any one panel, one or more panels remain available for use by a pilot seated at a pilot station to permit continued safe flight and landing.

[Doc. No. 5066, 29 FR 18291, Dec. 24, 1964, as amended by Amdt. 25-23, 35 FR 5676, Apr. 8, 1970; Amdt. 25-38, 41 FR 55466, Dec. 20, 1976]

§ 25.777 Cockpit controls.

(a) Each cockpit control must be located to provide convenient operation and to prevent confusion and inadvertent operation.

(b) The direction of movement of cockpit controls must meet the requirements of § 25.779. Wherever practicable, the sense of motion involved in the operation of other controls must correspond to the sense of the effect of the operation upon the airplane or upon the part operated. Controls of a variable nature using a rotary motion must move clockwise from the off position, through an increasing range, to the full on position.

(c) The controls must be located and arranged, with respect to the pilots' seats, so that there is full and unrestricted movement of each control without interference from the cockpit structure or the clothing of the minimum flight crew (established under § 25.1523) when any member of this flight crew, from 5'2" to 6'3" in height, is seated with the seat belt and shoulder harness (if provided) fastened.

(d) Identical powerplant controls for each engine must be located to prevent confusion as to the engines they control.

(e) Wing flap controls and other auxiliary lift device controls must be located on top of the pedestal, aft of the throttles, centrally or to the right of the pedestal centerline, and not less than 10 inches aft of the landing gear control.

(f) The landing gear control must be located forward of the throttles and must be operable by each pilot when seated with seat belt and shoulder harness (if provided) fastened.

(g) Control knobs must be shaped in accordance with § 25.781. In addition, the knobs must be of the same color, and this color must contrast with the color of control knobs for other purposes and the surrounding cockpit.

(h) If a flight engineer is required as part of the minimum flight crew (established under § 25.1523), the airplane must have a flight engineer station located and arranged so that the flight crewmembers can perform their functions efficiently and without interfering with each other.

[Doc. No. 5066, 29 FR 18291, Dec. 24, 1964, as amended by Amdt. 25-46, 43 FR 50596, Oct. 30, 1978]

§ 25.779 Motion and effect of cockpit controls.

Cockpit controls must be designed so that they operate in accordance with the following movement and actuation:

(a) Aerodynamic controls:

(1) *Primary.*

Controls	Motion and effect
Aileron	Right (clockwise) for right wing down.
Elevator	Rearward for nose up.
Rudder	Right pedal forward for nose right.

(2) *Secondary.*

Controls	Motion and effect
Flaps (or auxiliary lift devices)	Forward for flaps up; rearward for flaps down.
Trim tabs (or equivalent)	Rotate to produce similar rotation of the airplane about an axis parallel to the axis of the control.

(b) Powerplant and auxiliary controls:

(1) *Powerplant.*

Controls	Motion and effect
Power or thrust	Forward to increase forward thrust and rearward to increase rearward thrust.
Propellers	Forward to increase rpm.
Mixture	Forward or upward for rich.
Carburetor air heat	Forward or upward for cold.
Supercharger	Forward or upward for low blower. For turbosuperchargers, forward, upward, or clockwise, to increase pressure.

(2) *Auxiliary.*

Controls	Motion and effect
Landing gear	Down to extend.

[Doc. No. 5066, 29 FR 18291, Dec. 24, 1964, as amended by Amdt. 25-72, 55 FR 29778, July 20, 1990]

§ 25.781 Cockpit control knob shape.

Cockpit control knobs must conform to the general shapes (but not necessarily the exact sizes or specific proportions) in the following figure:

[Doc. No. 5066, 29 FR 18291, Dec. 24, 1964, as amended by Amdt. 25-72, 55 FR 29779, July 20, 1990]

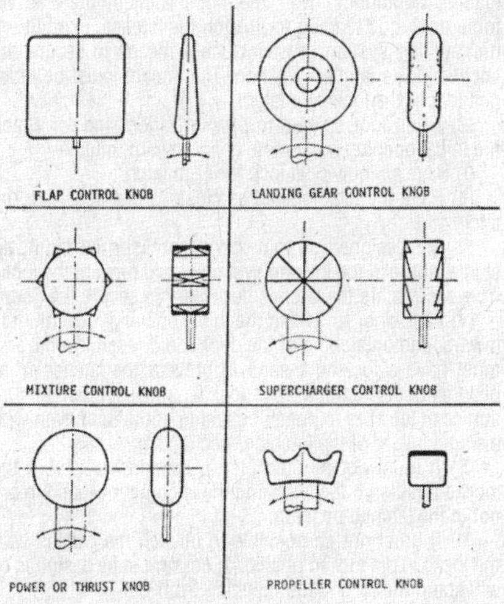

§ 25.783 Fuselage doors.

(a) *General.* This section applies to fuselage doors, which includes all doors, hatches, openable windows, access panels, covers, etc., on the exterior of the fuselage that do not require the use of tools to open or close. This also applies to each door or hatch through a pressure bulkhead, including any bulkhead that is specifically designed to function as a secondary bulkhead under the prescribed failure conditions of part 25. These doors must meet the requirements of this section, taking into account both pressurized and unpressurized flight, and must be designed as follows:

(1) Each door must have means to safeguard against opening in flight as a result of mechanical failure, or failure of any single structural element.

(2) Each door that could be a hazard if it unlatches must be designed so that unlatching during pressurized and unpressurized flight from the fully closed, latched, and locked condition is extremely improbable. This must be shown by safety analysis.

(3) Each element of each door operating system must be designed or, where impracticable, distinctively and permanently marked, to minimize the probability of incorrect assembly and adjustment that could result in a malfunction.

(4) All sources of power that could initiate unlocking or unlatching of any door must be automatically isolated from the latching and locking systems prior to flight and it must not be possible to restore power to the door during flight.

(5) Each removable bolt, screw, nut, pin, or other removable fastener must meet the locking requirements of § 25.607.

(6) Certain doors, as specified by § 25.807(h), must also meet the applicable requirements of §§ 25.809 through 25.812 for emergency exits.

(b) *Opening by persons.* There must be a means to safeguard each door against opening during flight due to inadvertent action by persons. In addition, design precautions must be taken to minimize the possibility for a person to open a door intentionally during flight. If these precautions include the use of auxiliary devices, those devices and their controlling systems must be designed so that—

(1) No single failure will prevent more than one exit from being opened; and

(2) Failures that would prevent opening of the exit after landing are improbable.

(c) *Pressurization prevention means.* There must be a provision to prevent pressurization of the airplane to an unsafe level if any door subject to pressurization is not fully closed, latched, and locked.

(1) The provision must be designed to function after any single failure, or after any combination of failures not shown to be extremely improbable.

(2) Doors that meet the conditions described in paragraph (h) of this section are not required to have a dedicated pressurization prevention means if, from every possible position of the door, it will remain open to the extent that it prevents pressurization or safely close and latch as pressurization takes place. This must also be shown with any single failure and malfunction, except that—

(i) With failures or malfunctions in the latching mechanism, it need not latch after closing; and

(ii) With jamming as a result of mechanical failure or blocking debris, the door need not close and latch if it can be shown that the pressurization loads on the jammed door or mechanism would not result in an unsafe condition.

(d) *Latching and locking.* The latching and locking mechanisms must be designed as follows:

(1) There must be a provision to latch each door.

(2) The latches and their operating mechanism must be designed so that, under all airplane flight and ground

loading conditions, with the door latched, there is no force or torque tending to unlatch the latches. In addition, the latching system must include a means to secure the latches in the latched position. This means must be independent of the locking system.

(3) Each door subject to pressurization, and for which the initial opening movement is not inward, must—

(i) Have an individual lock for each latch;

(ii) Have the lock located as close as practicable to the latch; and

(iii) Be designed so that, during pressurized flight, no single failure in the locking system would prevent the locks from restraining the latches necessary to secure the door.

(4) Each door for which the initial opening movement is inward, and unlatching of the door could result in a hazard, must have a locking means to prevent the latches from becoming disengaged. The locking means must ensure sufficient latching to prevent opening of the door even with a single failure of the latching mechanism.

(5) It must not be possible to position the lock in the locked position if the latch and the latching mechanism are not in the latched position.

(6) It must not be possible to unlatch the latches with the locks in the locked position. Locks must be designed to withstand the limit loads resulting from—

(i) The maximum operator effort when the latches are operated manually;

(ii) The powered latch actuators, if installed; and

(iii) The relative motion between the latch and the structural counterpart.

(7) Each door for which unlatching would not result in a hazard is not required to have a locking mechanism meeting the requirements of paragraphs (d)(3) through (d)(6) of this section.

(e) *Warning, caution, and advisory indications.* Doors must be provided with the following indications:

(1) There must be a positive means to indicate at each door operator's station that all required operations to close, latch, and lock the door(s) have been completed.

(2) There must be a positive means clearly visible from each operator station for any door that could be a hazard if unlatched to indicate if the door is not fully closed, latched, and locked.

(3) There must be a visual means on the flight deck to signal the pilots if any door is not fully closed, latched, and locked. The means must be designed such that any failure or combination of failures that would result in an erroneous closed, latched, and locked indication is improbable for—

(i) Each door that is subject to pressurization and for which the initial opening movement is not inward; or

(ii) Each door that could be a hazard if unlatched.

(4) There must be an aural warning to the pilots prior to or during the initial portion of takeoff roll if any door is not fully closed, latched, and locked, and its opening would prevent a safe takeoff and return to landing.

(f) *Visual inspection provision.* Each door for which unlatching of the door could be a hazard must have a provision for direct visual inspection to determine, without ambiguity, if the door is fully closed, latched, and locked. The provision must be permanent and discernible under operational lighting conditions, or by means of a flashlight or equivalent light source.

(g) *Certain maintenance doors, removable emergency exits, and access panels.* Some doors not normally opened except for maintenance purposes or emergency evacuation and some access panels need not comply with certain paragraphs of this section as follows:

(1) Access panels that are not subject to cabin pressurization and would not be a hazard if open during flight need not comply with paragraphs (a) through (f) of this section, but must have a means to prevent inadvertent opening during flight.

(2) Inward-opening removable emergency exits that are not normally removed, except for maintenance purposes or emergency evacuation, and flight deck-openable windows need not comply with paragraphs (c) and (f) of this section.

(3) Maintenance doors that meet the conditions of paragraph (h) of this section, and for which a placard is provided limiting use to maintenance access, need not comply with paragraphs (c) and (f) of this section.

(h) *Doors that are not a hazard.* For the purposes of this section, a door is considered not to be a hazard in the unlatched condition during flight, provided it can be shown to meet all of the following conditions:

(1) Doors in pressurized compartments would remain in the fully closed position if not restrained by the latches when subject to a pressure greater than ½ psi. Opening by persons, either inadvertently or intentionally, need not be considered in making this determination.

(2) The door would remain inside the airplane or remain attached to the airplane if it opens either in pressurized or unpressurized portions of the flight. This determination must include the consideration of inadvertent and intentional opening by persons during either pressurized or unpressurized portions of the flight.

(3) The disengagement of the latches during flight would not allow depressurization of the cabin to an unsafe level. This safety assessment must include the physiological effects on the occupants.

(4) The open door during flight would not create aerodynamic interference that could preclude safe flight and landing.

(5) The airplane would meet the structural design requirements with the door open. This assessment must include the aeroelastic stability requirements of § 25.629, as well as the strength requirements of subpart C of this part.

(6) The unlatching or opening of the door must not preclude safe flight and landing as a result of interaction with other systems or structures.

[Doc. No. 2003-14193, 69 FR 24501, May 3, 2004]

§ 25.785 Seats, berths, safety belts, and harnesses.

(a) A seat (or berth for a nonambulant person) must be provided for each occupant who has reached his or her second birthday.

(b) Each seat, berth, safety belt, harness, and adjacent part of the airplane at each station designated as occupiable during takeoff and landing must be designed so that a person making proper use of these facilities will not suffer serious injury in an emergency landing as a result of the inertia forces specified in §§ 25.561 and 25.562.

(c) Each seat or berth must be approved.

(d) Each occupant of a seat that makes more than an 18-degree angle with the vertical plane containing the airplane centerline must be protected from head injury by a safety belt and an energy absorbing rest that will support the arms, shoulders, head, and spine, or by a safety belt and shoulder harness that will prevent the head from contacting any injurious object. Each occupant of any other seat must be protected from head injury by a safety belt and, as appropriate to the type, location, and angle of facing of each seat, by one or more of the following:

(1) A shoulder harness that will prevent the head from contacting any injurious object.

(2) The elimination of any injurious object within striking radius of the head.

(3) An energy absorbing rest that will support the arms, shoulders, head, and spine.

(e) Each berth must be designed so that the forward part has a padded end board, canvas diaphragm, or equivalent means, that can withstand the static load reaction of the occupant when subjected to the forward inertia force specified in § 25.561. Berths must be free from corners and protuberances likely to cause injury to a person occupying the berth during emergency conditions.

(f) Each seat or berth, and its supporting structure, and each safety belt or harness and its anchorage must be designed for an occupant weight of 170 pounds, considering the maximum load factors, inertia forces, and reactions among the occupant, seat, safety belt, and harness for each relevant flight and ground load condition (including the emergency landing conditions prescribed in § 25.561). In addition—

(1) The structural analysis and testing of the seats, berths, and their supporting structures may be determined by assuming that the critical load in the forward, sideward, downward, upward, and rearward directions (as determined from the prescribed flight, ground, and emergency landing conditions) acts separately or using selected combinations of loads if the required strength in each specified direction is substantiated. The forward load factor need not be applied to safety belts for berths.

(2) Each pilot seat must be designed for the reactions resulting from the application of the pilot forces prescribed in § 25.395.

(3) The inertia forces specified in § 25.561 must be multiplied by a factor of 1.33 (instead of the fitting factor prescribed in § 25.625) in determining the strength of the attachment of each seat to the structure and each belt or harness to the seat or structure.

(g) Each seat at a flight deck station must have a restraint system consisting of a combined safety belt and shoulder harness with a single-point release that permits the flight deck occupant, when seated with the restraint system fastened, to perform all of the occupant's necessary flight deck functions. There must be a means to secure each combined restraint system when not in use to prevent interference with the operation of the airplane and with rapid egress in an emergency.

(h) Each seat located in the passenger compartment and designated for use during takeoff and landing by a flight attendant required by the operating rules of this chapter must be:

(1) Near a required floor level emergency exit, except that another location is acceptable if the emergency egress of passengers would be enhanced with that location. A flight attendant seat must be located adjacent to each Type A or B emergency exit. Other flight attendant seats must be evenly distributed among the required floor- level emergency exits to the extent feasible.

(2) To the extent possible, without compromising proximity to a required floor level emergency exit, located to provide a direct view of the cabin area for which the flight attendant is responsible.

(3) Positioned so that the seat will not interfere with the use of a passageway or exit when the seat is not in use.

(4) Located to minimize the probability that occupants would suffer injury by being struck by items dislodged from service areas, stowage compartments, or service equipment.

(5) Either forward or rearward facing with an energy absorbing rest that is designed to support the arms, shoulders, head, and spine.

(6) Equipped with a restraint system consisting of a combined safety belt and shoulder harness unit with a single point release. There must be means to secure each restraint system when not in use to prevent interference with rapid egress in an emergency.

(i) Each safety belt must be equipped with a metal to metal latching device.

(j) If the seat backs do not provide a firm handhold, there must be a handgrip or rail along each aisle to enable persons to steady themselves while using the aisles in moderately rough air.

(k) Each projecting object that would injure persons seated or moving about the airplane in normal flight must be padded.

(l) Each forward observer's seat required by the operating rules must be shown to be suitable for use in conducting the necessary enroute inspection.

[Amdt. 25-72, 55 FR 29780, July 20, 1990, as amended by Amdt. 25-88, 61 FR 57956, Nov. 8, 1996]

§ 25.787 Stowage compartments.

(a) Each compartment for the stowage of cargo, baggage, carry-on articles, and equipment (such as life rafts), and any other stowage compartment must be designed for its placarded maximum weight of contents and for the critical load distribution at the appropriate maximum load factors corresponding to the specified flight and ground load conditions, and to the emergency landing conditions of § 25.561(b), except that the forces specified in the emergency landing conditions need not be applied to compartments located below, or forward, of all occupants in the airplane. If the airplane has a passenger seating configuration, excluding pilots seats, of 10 seats or more, each stowage compartment in the passenger cabin, except for underseat and overhead compartments for passenger convenience, must be completely enclosed.

(b) There must be a means to prevent the contents in the compartments from becoming a hazard by shifting, under the loads specified in paragraph (a) of this section. For stowage compartments in the passenger and crew cabin, if the means used is a latched door, the design must take into consideration the wear and deterioration expected in service.

(c) If cargo compartment lamps are installed, each lamp must be installed so as to prevent contact between lamp bulb and cargo.

[Doc. No. 5066, 29 FR 18291, Dec. 24, 1964, as amended by Amdt. 25-32, 37 FR 3969, Feb. 24, 1972; Amdt. 25-38, 41 FR 55466, Dec. 20, 1976; Amdt. 25-51, 45 FR 7755, Feb. 4, 1980]

§ 25.789 Retention of items of mass in passenger and crew compartments and galleys.

(a) Means must be provided to prevent each item of mass (that is part of the airplane type design) in a passenger or crew compartment or galley from becoming a hazard by shifting under the appropriate maximum load factors corresponding to the specified flight and ground load conditions, and to the emergency landing conditions of § 25.561(b).

(b) Each interphone restraint system must be designed so that when subjected to the load factors specified in

§ 25.561(b)(3), the interphone will remain in its stowed position.

[Amdt. 25-32, 37 FR 3969, Feb. 24, 1972, as amended by Amdt. 25-46, 43 FR 50596, Oct. 30, 1978]

§ 25.791 Passenger information signs and placards.

(a) If smoking is to be prohibited, there must be at least one placard so stating that is legible to each person seated in the cabin. If smoking is to be allowed, and if the crew compartment is separated from the passenger compartment, there must be at least one sign notifying when smoking is prohibited. Signs which notify when smoking is prohibited must be operable by a member of the flightcrew and, when illuminated, must be legible under all probable conditions of cabin illumination to each person seated in the cabin.

(b) Signs that notify when seat belts should be fastened and that are installed to comply with the operating rules of this chapter must be operable by a member of the flightcrew and, when illuminated, must be legible under all probable conditions of cabin illumination to each person seated in the cabin.

(c) A placard must be located on or adjacent to the door of each receptacle used for the disposal of flammable waste materials to indicate that use of the receptacle for disposal of cigarettes, etc., is prohibited.

(d) Lavatories must have "No Smoking" or "No Smoking in Lavatory" placards conspicuously located on or adjacent to each side of the entry door.

(e) Symbols that clearly express the intent of the sign or placard may be used in lieu of letters.

[Amdt. 25-72, 55 FR 29780, July 20, 1990]

§ 25.793 Floor surfaces.

The floor surface of all areas which are likely to become wet in service must have slip resistant properties.

[Amdt. 25-51, 45 FR 7755, Feb. 4, 1980]

§ 25.795 Security considerations.

(a) *Protection of flightcrew compartment.* If a flightdeck door is required by operating rules:

(1) The bulkhead, door, and any other accessible boundary separating the flightcrew compartment from occupied areas must be designed to resist forcible intrusion by unauthorized persons and be capable of withstanding impacts of 300 joules (221.3 foot pounds).

(2) The bulkhead, door, and any other accessible boundary separating the flightcrew compartment from occupied areas must be designed to resist a constant 250 pound (1,113 Newtons) tensile load on accessible handholds, including the doorknob or handle.

(3) The bulkhead, door, and any other boundary separating the flightcrew compartment from any occupied areas must be designed to resist penetration by small arms fire and fragmentation devices to a level equivalent to level IIIa of the National Institute of Justice (NIJ) Standard 0101.04.

(b) Airplanes with a maximum certificated passenger seating capacity of more than 60 persons or a maximum certificated takeoff gross weight of over 100,000 pounds (45,359 Kilograms) must be designed to limit the effects of an explosive or incendiary device as follows:

(1) *Flightdeck smoke protection.* Means must be provided to limit entry of smoke, fumes, and noxious gases into the flightdeck.

(2) *Passenger cabin smoke protection.* Means must be provided to prevent passenger incapacitation in the cabin resulting from smoke, fumes, and noxious gases as represented by the initial combined volumetric concentrations of 0.59% carbon monoxide and 1.23% carbon dioxide.

(3) *Cargo compartment fire suppression.* An extinguishing agent must be capable of suppressing a fire. All cargo-compartment fire suppression systems must be designed to withstand the following effects, including support structure displacements or adjacent materials displacing against the distribution system:

(i) Impact or damage from a 0.5-inch diameter aluminum sphere traveling at 430 feet per second (131.1 meters per second);

(ii) A 15-pound per square-inch (103.4 kPa) pressure load if the projected surface area of the component is greater than 4 square feet. Any single dimension greater than 4 feet (1.22 meters) may be assumed to be 4 feet (1.22 meters) in length; and

(iii) A 6-inch (0.152 meters) displacement, except where limited by the fuselage contour, from a single point force applied anywhere along the distribution system where relative movement between the system and its attachment can occur.

(iv) Paragraphs (b)(3)(i) through (iii) of this section do not apply to components that are redundant and separated in accordance with paragraph (c)(2) of this section or are installed remotely from the cargo compartment.

(c) An airplane with a maximum certificated passenger seating capacity of more than 60 persons or a maximum certificated takeoff gross weight of over 100,000 pounds (45,359 Kilograms) must comply with the following:

(1) *Least risk bomb location.* An airplane must be designed with a designated location where a bomb or other explosive device could be placed to best protect flight-critical structures and systems from damage in the case of detonation.

(2) *Survivability of systems.* (i) Except where impracticable, redundant airplane systems necessary for continued safe flight and landing must be physically separated, at a minimum, by an amount equal to a sphere of diameter

$$D = 2\sqrt{(H_0/\pi)}$$

(where H_0 is defined under § 25.365(e)(2) of this part and D need not exceed 5.05 feet (1.54 meters)). The sphere is applied everywhere within the fuselage—limited by the forward bulkhead and the aft bulkhead of the passenger cabin and cargo compartment beyond which only one-half the sphere is applied.

(ii) Where compliance with paragraph (c)(2)(i) of this section is impracticable, other design precautions must be taken to maximize the survivability of those systems.

(3) *Interior design to facilitate searches.* Design features must be incorporated that will deter concealment or promote discovery of weapons, explosives, or other objects from a simple inspection in the following areas of the airplane cabin:

(i) Areas above the overhead bins must be designed to prevent objects from being hidden from view in a simple search from the aisle. Designs that prevent concealment of objects with volumes 20 cubic inches and greater satisfy this requirement.

(ii) Toilets must be designed to prevent the passage of solid objects greater than 2.0 inches in diameter.

(iii) Life preservers or their storage locations must be designed so that tampering is evident.

(d) *Exceptions.* Airplanes used solely to transport cargo only need to meet the requirements of paragraphs (b)(1), (b)(3), and (c)(2) of this section.

(e) *Material Incorporated by Reference.* You must use National Institute of Justice (NIJ) Standard 0101.04, Ballistic Resistance of Personal Body Armor, June 2001, Revision A, to establish ballistic resistance as required by paragraph (a)(3) of this section.

(1) The Director of the Federal Register approved the incorporation by reference of this document under 5 U.S.C. 552(a) and 1 CFR part 51.

(2) You may review copies of NIJ Standard 0101.04 at the:

(i) FAA Transport Airplane Directorate, 1601 Lind Avenue, SW., Renton, Washington 98055;

(ii) National Institute of Justice (NIJ), *http://www.ojp.usdoj.gov/nij*, telephone (202) 307-2942; or

(iii) National Archives and Records Administration (NARA). For information on the availability of this material at NARA go to *http://www.archives.gov/federal_register/code_of_federal_regulations/ibr_locations.html* or call (202) 741-6030.

(3) You may obtain copies of NIJ Standard 0101.04 from the National Criminal Justice Reference Service, P.O. Box 6000, Rockville, MD 20849-6000, telephone (800) 851-3420.

[Amdt. Nos. 25-127; 121-341, 73 FR 63879, Oct. 28, 2008, as amended at 74 FR 22819, May 15, 2009]

Emergency Provisions

§ 25.801 Ditching.

(a) If certification with ditching provisions is requested, the airplane must meet the requirements of this section and §§ 25.807(e), 25.1411, and 25.1415(a).

(b) Each practicable design measure, compatible with the general characteristics of the airplane, must be taken to minimize the probability that in an emergency landing on water, the behavior of the airplane would cause immediate injury to the occupants or would make it impossible for them to escape.

(c) The probable behavior of the airplane in a water landing must be investigated by model tests or by comparison with airplanes of similar configuration for which the ditching characteristics are known. Scoops, flaps, projections, and any other factor likely to affect the hydrodynamic characteristics of the airplane, must be considered.

(d) It must be shown that, under reasonably probable water conditions, the flotation time and trim of the airplane will allow the occupants to leave the airplane and enter the liferafts required by § 25.1415. If compliance with this provision is shown by buoyancy and trim computations, appropriate allowances must be made for probable structural damage and leakage. If the airplane has fuel tanks (with fuel jettisoning provisions) that can reasonably be expected to withstand a ditching without leakage, the jettisonable volume of fuel may be considered as buoyancy volume.

(e) Unless the effects of the collapse of external doors and windows are accounted for in the investigation of the probable behavior of the airplane in a water landing (as prescribed in paragraphs (c) and (d) of this section), the external doors and windows must be designed to withstand the probable maximum local pressures.

[Doc. No. 5066, 29 FR 18291, Dec. 24, 1964, as amended by Amdt. 25-72, 55 FR 29781, July 20, 1990]

§ 25.803 Emergency evacuation.

(a) Each crew and passenger area must have emergency means to allow rapid evacuation in crash landings, with the landing gear extended as well as with the landing gear retracted, considering the possibility of the airplane being on fire.

(b) [Reserved]

(c) For airplanes having a seating capacity of more than 44 passengers, it must be shown that the maximum seating capacity, including the number of crewmembers required by the operating rules for which certification is requested, can be evacuated from the airplane to the ground under simulated emergency conditions within 90 seconds. Compliance with this requirement must be shown by actual demonstration using the test criteria outlined in appendix J of this part unless the Administrator finds that a combination of analysis and testing will provide data equivalent to that which would be obtained by actual demonstration.

(d)-(e) [Reserved]

[Doc. No. 24344, 55 FR 29781, July 20, 1990]

§ 25.807 Emergency exits.

(a) *Type.* For the purpose of this part, the types of exits are defined as follows:

(1) *Type I.* This type is a floor-level exit with a rectangular opening of not less than 24 inches wide by 48 inches high, with corner radii not greater than eight inches.

(2) *Type II.* This type is a rectangular opening of not less than 20 inches wide by 44 inches high, with corner radii not greater than seven inches. Type II exits must be floor-level exits unless located over the wing, in which case they must not have a step-up inside the airplane of more than 10 inches nor a step-down outside the airplane of more than 17 inches.

(3) *Type III.* This type is a rectangular opening of not less than 20 inches wide by 36 inches high with corner radii not greater than seven inches, and with a step-up inside the airplane of not more than 20 inches. If the exit is located over the wing, the step-down outside the airplane may not exceed 27 inches.

(4) *Type IV.* This type is a rectangular opening of not less than 19 inches wide by 26 inches high, with corner radii not greater than 6.3 inches, located over the wing, with a step-up inside the airplane of not more than 29 inches and a step-down outside the airplane of not more than 36 inches.

(5) *Ventral.* This type is an exit from the passenger compartment through the pressure shell and the bottom fuselage skin. The dimensions and physical configuration of this type of exit must allow at least the same rate of egress as a Type I exit with the airplane in the normal ground attitude, with landing gear extended.

(6) *Tailcone.* This type is an aft exit from the passenger compartment through the pressure shell and through an openable cone of the fuselage aft of the pressure shell. The means of opening the tailcone must be simple and obvious and must employ a single operation.

(7) *Type A.* This type is a floor-level exit with a rectangular opening of not less than 42 inches wide by 72 inches high, with corner radii not greater than seven inches.

(8) *Type B.* This type is a floor-level exit with a rectangular opening of not less than 32 inches wide by 72 inches high, with corner radii not greater than six inches.

(9) *Type C.* This type is a floor-level exit with a rectangular opening of not less than 30 inches wide by 48 inches high, with corner radii not greater than 10 inches.

(b) *Step down distance.* Step down distance, as used in this section, means the actual distance between the bottom of the required opening and a usable foot hold, extending out from the fuselage, that is large enough to be effective without searching by sight or feel.

(c) *Over-sized exits.* Openings larger than those specified in this section, whether or not of rectangular shape, may be used if the specified rectangular opening can be inscribed within the opening and the base of the inscribed rectangular opening meets the specified step-up and step-down heights.

(d) *Asymmetry.* Exits of an exit pair need not be diametrically opposite each other nor of the same size; however, the number of passenger seats permitted under paragraph (g) of this section is based on the smaller of the two exits.

(e) *Uniformity.* Exits must be distributed as uniformly as practical, taking into account passenger seat distribution.

(f) *Location.* (1) Each required passenger emergency exit must be accessible to the passengers and located where it will afford the most effective means of passenger evacuation.

(2) If only one floor-level exit per side is prescribed, and the airplane does not have a tailcone or ventral emergency exit, the floor-level exits must be in the rearward part of the passenger compartment unless another location affords a more effective means of passenger evacuation.

(3) If more than one floor-level exit per side is prescribed, and the airplane does not have a combination cargo and passenger configuration, at least one floor-level exit must be located in each side near each end of the cabin.

(4) For an airplane that is required to have more than one passenger emergency exit for each side of the fuselage, no passenger emergency exit shall be more than 60 feet from any adjacent passenger emergency exit on the same side of the same deck of the fuselage, as measured parallel to the airplane's longitudinal axis between the nearest exit edges.

(g) *Type and number required.* The maximum number of passenger seats permitted depends on the type and number of exits installed in each side of the fuselage. Except as further restricted in paragraphs (g)(1) through (g)(9) of this section, the maximum number of passenger seats permitted for each exit of a specific type installed in each side of the fuselage is as follows:

Type A	110
Type B	75
Type C	55
Type I	45
Type II	40
Type III	35
Type IV	9

(1) For a passenger seating configuration of 1 to 9 seats, there must be at least one Type IV or larger overwing exit in each side of the fuselage or, if overwing exits are not provided, at least one exit in each side that meets the minimum dimensions of a Type III exit.

(2) For a passenger seating configuration of more than 9 seats, each exit must be a Type III or larger exit.

(3) For a passenger seating configuration of 10 to 19 seats, there must be at least one Type III or larger exit in each side of the fuselage.

(4) For a passenger seating configuration of 20 to 40 seats, there must be at least two exits, one of which must be a Type II or larger exit, in each side of the fuselage.

(5) For a passenger seating configuration of 41 to 110 seats, there must be at least two exits, one of which must be a Type I or larger exit, in each side of the fuselage.

(6) For a passenger seating configuration of more than 110 seats, the emergency exits in each side of the fuselage must include at least two Type I or larger exits.

(7) The combined maximum number of passenger seats permitted for all Type III exits is 70, and the combined maximum number of passenger seats permitted for two Type III exits in each side of the fuselage that are separated by fewer than three passenger seat rows is 65.

(8) If a Type A, Type B, or Type C exit is installed, there must be at least two Type C or larger exits in each side of the fuselage.

(9) If a passenger ventral or tailcone exit is installed and that exit provides at least the same rate of egress as a Type III exit with the airplane in the most adverse exit opening condition that would result from the collapse of one or more legs of the landing gear, an increase in the passenger seating configuration is permitted as follows:

(i) For a ventral exit, 12 additional passenger seats.

(ii) For a tailcone exit incorporating a floor level opening of not less than 20 inches wide by 60 inches high, with corner radii not greater than seven inches, in the pressure shell and incorporating an approved assist means in accordance with § 25.810(a), 25 additional passenger seats.

(iii) For a tailcone exit incorporating an opening in the pressure shell which is at least equivalent to a Type III emergency exit with respect to dimensions, step-up and step-down distance, and with the top of the opening not less than 56 inches from the passenger compartment floor, 15 additional passenger seats.

(h) *Other exits.* The following exits also must meet the applicable emergency exit requirements of §§ 25.809 through 25.812, and must be readily accessible:

(1) Each emergency exit in the passenger compartment in excess of the minimum number of required emergency exits.

(2) Any other floor-level door or exit that is accessible from the passenger compartment and is as large or larger than a Type II exit, but less than 46 inches wide.

(3) Any other ventral or tail cone passenger exit.

(i) *Ditching emergency exits for passengers.* Whether or not ditching certification is requested, ditching emergency exits must be provided in accordance with the following requirements, unless the emergency exits required by paragraph (g) of this section already meet them:

(1) For airplanes that have a passenger seating configuration of nine or fewer seats, excluding pilot seats, one exit above the waterline in each side of the airplane, meeting at least the dimensions of a Type IV exit.

(2) For airplanes that have a passenger seating configuration of 10 of more seats, excluding pilot seats, one exit above the waterline in a side of the airplane, meeting at least the dimensions of a Type III exit for each unit (or part of a unit) of 35 passenger seats, but no less than two such exits in the passenger cabin, with one on each side of the airplane. The passenger seat/exit ratio may be increased through the use of larger exits, or other means, provided it is shown that the evacuation capability during ditching has been improved accordingly.

(3) If it is impractical to locate side exits above the waterline, the side exits must be replaced by an equal number of readily accessible overhead hatches of not less than the dimensions of a Type III exit, except that for airplanes with a passenger configuration of 35 or fewer seats, excluding pilot seats, the two required Type III side exits need be replaced by only one overhead hatch.

(j) *Flightcrew emergency exits.* For airplanes in which the proximity of passenger emergency exits to the flightcrew area does not offer a convenient and readily accessible means of evacuation of the flightcrew, and for all airplanes having a passenger seating capacity greater than 20, flightcrew exits shall be located in the flightcrew area. Such exits shall be of sufficient size and so located as to permit rapid evacuation by the crew. One exit shall be provided on each side of the airplane; or, alternatively, a top hatch shall be provided. Each exit must encompass an unobstructed rectangular opening of at least 19 by 20 inches unless satisfactory exit utility can be demonstrated by a typical crewmember.

[Amdt. 25-72, 55 FR 29781, July 20, 1990, as amended by Amdt. 25-88, 61 FR 57956, Nov. 8, 1996; 62 FR 1817, Jan. 13, 1997; Amdt. 25-94, 63 FR 8848, Feb. 23, 1998; 63 FR 12862, Mar. 16, 1998; Amdt. 25-114, 69 FR 24502, May 3, 2004]

§ 25.809 Emergency exit arrangement.

(a) Each emergency exit, including each flightcrew emergency exit, must be a moveable door or hatch in the external walls of the fuselage, allowing an unobstructed opening to the outside. In addition, each emergency exit must have means to permit viewing of the conditions outside the exit when the exit is closed. The viewing means may be on or adjacent to the exit provided no obstructions exist between the exit and the viewing means. Means must also be provided to permit viewing of the likely areas of evacuee ground contact. The likely areas of evacuee ground contact must be viewable during all lighting conditions with the landing gear extended as well as in all conditions of landing gear collapse.

(b) Each emergency exit must be openable from the inside and the outside except that sliding window emergency exits in the flight crew area need not be openable from the outside if other approved exits are convenient and readily accessible to the flight crew area. Each emergency exit must be capable of being opened, when there is no fuselage deformation—

(1) With the airplane in the normal ground attitude and in each of the attitudes corresponding to collapse of one or more legs of the landing gear; and

(2) Within 10 seconds measured from the time when the opening means is actuated to the time when the exit is fully opened.

(3) Even though persons may be crowded against the door on the inside of the airplane.

(c) The means of opening emergency exits must be simple and obvious; may not require exceptional effort; and must be arranged and marked so that it can be readily located and operated, even in darkness. Internal exit-opening means involving sequence operations (such as operation of two handles or latches, or the release of safety catches) may be used for flightcrew emergency exits if it can be reasonably established that these means are simple and obvious to crewmembers trained in their use.

(d) If a single power-boost or single power-operated system is the primary system for operating more than one exit in an emergency, each exit must be capable of meeting the requirements of paragraph (b) of this section in the event of failure of the primary system. Manual operation of the exit (after failure of the primary system) is acceptable.

(e) Each emergency exit must be shown by tests, or by a combination of analysis and tests, to meet the requirements of paragraphs (b) and (c) of this section.

(f) Each door must be located where persons using them will not be endangered by the propellers when appropriate operating procedures are used.

(g) There must be provisions to minimize the probability of jamming of the emergency exits resulting from fuselage deformation in a minor crash landing.

(h) When required by the operating rules for any large passenger-carrying turbojet-powered airplane, each ventral exit and tailcone exit must be—

(1) Designed and constructed so that it cannot be opened during flight; and

(2) Marked with a placard readable from a distance of 30 inches and installed at a conspicuous location near the means of opening the exit, stating that the exit has been designed and constructed so that it cannot be opened during flight.

(i) Each emergency exit must have a means to retain the exit in the open position, once the exit is opened in an emergency. The means must not require separate action to engage when the exit is opened, and must require positive action to disengage.

[Doc. No. 5066, 29 FR 18291, Dec. 24, 1964, as amended by Amdt. 25-15, 32 FR 13264, Sept. 20, 1967; Amdt. 25-32, 37 FR 3970, Feb. 24, 1972; Amdt. 25-34, 37 FR 25355, Nov. 30, 1972; Amdt. 25-46, 43 FR 50597, Oct. 30, 1978; Amdt. 25-47, 44 FR 61325, Oct. 25, 1979; Amdt. 25-72, 55 FR 29782, July 20, 1990; Amdt. 25-114, 69 FR 24502, May 3, 2004; Amdt. 25-116, 69 FR 62788, Oct. 27, 2004]

§ 25.810 Emergency egress assist means and escape routes.

(a) Each non over-wing Type A, Type B or Type C exit, and any other non over-wing landplane emergency exit more than 6 feet from the ground with the airplane on the ground and the landing gear extended, must have an approved means to assist the occupants in descending to the ground.

(1) The assisting means for each passenger emergency exit must be a self-supporting slide or equivalent; and, in the case of Type A or Type B exits, it must be capable of carrying simultaneously two parallel lines of evacuees. In addition, the assisting means must be designed to meet the following requirements—

(i) It must be automatically deployed and deployment must begin during the interval between the time the exit opening means is actuated from inside the airplane and the time the exit is fully opened. However, each passenger emergency exit which is also a passenger entrance door or a service door must be provided with means to prevent deployment of the assisting means when it is opened from either the inside or the outside under nonemergency conditions for normal use.

(ii) Except for assisting means installed at Type C exits, it must be automatically erected within 6 seconds after deployment is begun. Assisting means installed at Type C exits must be automatically erected within 10 seconds from the time the opening means of the exit is actuated.

(iii) It must be of such length after full deployment that the lower end is self-supporting on the ground and provides safe evacuation of occupants to the ground after collapse of one or more legs of the landing gear.

(iv) It must have the capability, in 25-knot winds directed from the most critical angle, to deploy and, with the assistance of only one person, to remain usable after full deployment to evacuate occupants safely to the ground.

(v) For each system installation (mockup or airplane installed), five consecutive deployment and inflation tests must be conducted (per exit) without failure, and at least three tests of each such five-test series must be conducted using a single representative sample of the device. The sample devices must be deployed and inflated by the system's primary means after being subjected to the inertia forces specified in § 25.561(b). If any part of the system fails or does not function properly during the required tests, the cause of the failure or malfunction must be corrected by positive means and after that, the full series of five consecutive deployment and inflation tests must be conducted without failure.

(2) The assisting means for flightcrew emergency exits may be a rope or any other means demonstrated to be suitable for the purpose. If the assisting means is a rope, or an approved device equivalent to a rope, it must be—

(i) Attached to the fuselage structure at or above the top of the emergency exit opening, or, for a device at a pilot's emergency exit window, at another approved location if the stowed device, or its attachment, would reduce the pilot's view in flight;

(ii) Able (with its attachment) to withstand a 400-pound static load.

(b) Assist means from the cabin to the wing are required for each type A or Type B exit located above the wing and having a stepdown unless the exit without an assist-means can be shown to have a rate of passenger egress at least equal to that of the same type of non over-wing exit. If an assist means is required, it must be automatically deployed and automatically erected concurrent with the opening of the exit. In the case of assist means installed at Type C exits, it must be self-supporting within 10 seconds from the time the opening means of the exits is actuated. For all other exit types, it must be self-supporting 6 seconds after deployment is begun.

(c) An escape route must be established from each overwing emergency exit, and (except for flap surfaces suitable as slides) covered with a slip resistant surface. Except where a means for channeling the flow of evacuees is provided—

(1) The escape route from each Type A or Type B passenger emergency exit, or any common escape route from two Type III passenger emergency exits, must be at least 42 inches wide; that from any other passenger emergency exit must be at least 24 inches wide; and

(2) The escape route surface must have a reflectance of at least 80 percent, and must be defined by markings with a surface-to-marking contrast ratio of at least 5:1.

(d) Means must be provided to assist evacuees to reach the ground for all Type C exits located over the wing and, if the place on the airplane structure at which the escape route required in paragraph (c) of this section terminates is more than 6 feet from the ground with the airplane on the ground and the landing gear extended, for all other exit types.

(1) If the escape route is over the flap, the height of the terminal edge must be measured with the flap in the takeoff or landing position, whichever is higher from the ground.

(2) The assisting means must be usable and self-supporting with one or more landing gear legs collapsed and under a 25-knot wind directed from the most critical angle.

(3) The assisting means provided for each escape route leading from a Type A or B emergency exit must be capable of carrying simultaneously two parallel lines of evacuees; and, the assisting means leading from any other exit type must be capable of carrying as many parallel lines of evacuees as there are required escape routes.

(4) The assisting means provided for each escape route leading from a Type C exit must be automatically erected within 10 seconds from the time the opening means of the exit is actuated, and that provided for the escape route leading from any other exit type must be automatically erected within 10 seconds after actuation of the erection system.

(e) If an integral stair is installed in a passenger entry door that is qualified as a passenger emergency exit, the stair must be designed so that, under the following conditions, the effectiveness of passenger emergency egress will not be impaired:

(1) The door, integral stair, and operating mechanism have been subjected to the inertia forces specified in § 25.561(b)(3), acting separately relative to the surrounding structure.

(2) The airplane is in the normal ground attitude and in each of the attitudes corresponding to collapse of one or more legs of the landing gear.

[Amdt. 25-72, 55 FR 29782, July 20, 1990, as amended by Amdt. 25-88, 61 FR 57958, Nov. 8, 1996; 62 FR 1817, Jan. 13, 1997; Amdt. 25-114, 69 FR 24502, May 3, 2004]

§ 25.811 Emergency exit marking.

(a) Each passenger emergency exit, its means of access, and its means of opening must be conspicuously marked.

(b) The identity and location of each passenger emergency exit must be recognizable from a distance equal to the width of the cabin.

(c) Means must be provided to assist the occupants in locating the exits in conditions of dense smoke.

(d) The location of each passenger emergency exit must be indicated by a sign visible to occupants approaching along the main passenger aisle (or aisles). There must be—

(1) A passenger emergency exit locator sign above the aisle (or aisles) near each passenger emergency exit, or at another overhead location if it is more practical because of low headroom, except that one sign may serve more than one exit if each exit can be seen readily from the sign;

(2) A passenger emergency exit marking sign next to each passenger emergency exit, except that one sign may serve two such exits if they both can be seen readily from the sign; and

(3) A sign on each bulkhead or divider that prevents fore and aft vision along the passenger cabin to indicate emergency exits beyond and obscured by the bulkhead or divider, except that if this is not possible the sign may be placed at another appropriate location.

(e) The location of the operating handle and instructions for opening exits from the inside of the airplane must be shown in the following manner:

(1) Each passenger emergency exit must have, on or near the exit, a marking that is readable from a distance of 30 inches.

(2) Each Type A, Type B, Type C or Type I passenger emergency exit operating handle must—

(i) Be self-illuminated with an initial brightness of at least 160 microlamberts; or

(ii) Be conspicuously located and well illuminated by the emergency lighting even in conditions of occupant crowding at the exit.

(3) [Reserved]

(4) Each Type A, Type B, Type C, Type I, or Type II passenger emergency exit with a locking mechanism released by rotary motion of the handle must be marked—

(i) With a red arrow, with a shaft at least three-fourths of an inch wide and a head twice the width of the shaft, extending along at least 70 degrees of arc at a radius approximately equal to three-fourths of the handle length.

(ii) So that the centerline of the exit handle is within±1 inch of the projected point of the arrow when the handle has reached full travel and has released the locking mechanism, and

(iii) With the word "open" in red letters 1 inch high, placed horizontally near the head of the arrow.

(f) Each emergency exit that is required to be openable from the outside, and its means of opening, must be marked on the outside of the airplane. In addition, the following apply:

(1) The outside marking for each passenger emergency exit in the side of the fuselage must include a 2-inch colored band outlining the exit.

(2) Each outside marking including the band, must have color contrast to be readily distinguishable from the surrounding fuselage surface. The contrast must be such that if the reflectance of the darker color is 15 percent or less, the reflectance of the lighter color must be at least 45 percent. "Reflectance" is the ratio of the luminous flux reflected by a body to the luminous flux it receives. When the reflectance of the darker color is greater than 15 percent, at least a 30-percent difference between its reflectance and the reflectance of the lighter color must be provided.

(3) In the case of exists other than those in the side of the fuselage, such as ventral or tailcone exists, the external means of opening, including instructions if applicable, must be conspicuously marked in red, or bright chrome yellow if the background color is such that red is inconspicuous. When the opening means is located on only one side of the fuselage, a conspicuous marking to that effect must be provided on the other side.

(g) Each sign required by paragraph (d) of this section may use the word "exit" in its legend in place of the term "emergency exit".

[Amdt. 25-15, 32 FR 13264, Sept. 20, 1967, as amended by Amdt. 25-32, 37 FR 3970, Feb. 24, 1972; Amdt. 25-46, 43 FR 50597, Oct. 30, 1978; 43 FR 52495, Nov. 13, 1978; Amdt. 25-79, 58 FR 45229, Aug. 26, 1993; Amdt. 25-88, 61 FR 57958, Nov. 8, 1996]

§ 25.812 Emergency lighting.

(a) An emergency lighting system, independent of the main lighting system, must be installed. However, the sources of general cabin illumination may be common to both the emergency and the main lighting systems if the power supply to the emergency lighting system is independent of the power supply to the main lighting system. The emergency lighting system must include:

(1) Illuminated emergency exit marking and locating signs, sources of general cabin illumination, interior lighting in emergency exit areas, and floor proximity escape path marking.

(2) Exterior emergency lighting.

(b) Emergency exit signs—

(1) For airplanes that have a passenger seating configuration, excluding pilot seats, of 10 seats or more must meet the following requirements:

(i) Each passenger emergency exit locator sign required by § 25.811(d)(1) and each passenger emergency exit marking sign required by § 25.811(d)(2) must have red letters at least 1½ inches high on an illuminated white background, and must have an area of at least 21 square inches excluding the letters. The lighted background-to-letter contrast must be at least 10:1. The letter height to stroke-width ratio may not be more than 7:1 nor less than 6:1. These signs must be internally electrically illuminated with a background brightness of at least 25 foot-lamberts and a high-to-low background contrast no greater than 3:1.

(ii) Each passenger emergency exit sign required by § 25.811(d)(3) must have red letters at least 1½ inches high on a white background having an area of at least 21 square inches excluding the letters. These signs must be internally electrically illuminated or self-illuminated by other than electrical means and must have an initial brightness of at least 400 microlamberts. The colors may be reversed in the case of a sign that is self-illuminated by other than electrical means.

(2) For airplanes that have a passenger seating configuration, excluding pilot seats, of nine seats or less, that are required by § 25.811(d)(1), (2), and (3) must have red letters at least 1 inch high on a white background at least 2 inches high. These signs may be internally electrically illuminated, or self-illuminated by other than electrical means, with an initial brightness of at least 160 microlamberts. The colors may be reversed in the case of a sign that is self-illuminated by other than electrical means.

(c) General illumination in the passenger cabin must be provided so that when measured along the centerline of main passenger aisle(s), and cross aisle(s) between main aisles, at seat arm-rest height and at 40-inch intervals, the average illumination is not less than 0.05 foot-candle and the illumination at each 40-inch interval is not less than 0.01 foot-candle. A main passenger aisle(s) is considered to extend along the fuselage from the most forward passenger emergency exit or cabin occupant seat, whichever is farther forward, to the most rearward passenger emergency exit or cabin occupant seat, whichever is farther aft.

(d) The floor of the passageway leading to each floor-level passenger emergency exit, between the main aisles and the exit openings, must be provided with illumination that is not less than 0.02 foot-candle measured along a line that is within 6 inches of and parallel to the floor and is centered on the passenger evacuation path.

(e) Floor proximity emergency escape path marking must provide emergency evacuation guidance for passengers when all sources of illumination more than 4 feet above the cabin aisle floor are totally obscured. In the dark of the night, the floor proximity emergency escape path marking must enable each passenger to—

(1) After leaving the passenger seat, visually identify the emergency escape path along the cabin aisle floor to the first exits or pair of exits forward and aft of the seat; and

(2) Readily identify each exit from the emergency escape path by reference only to markings and visual features not more than 4 feet above the cabin floor.

(f) Except for subsystems provided in accordance with paragraph (h) of this section that serve no more than one assist means, are independent of the airplane's main emergency lighting system, and are automatically activated when the assist means is erected, the emergency lighting system must be designed as follows.

(1) The lights must be operable manually from the flight crew station and from a point in the passenger compartment that is readily accessible to a normal flight attendant seat.

(2) There must be a flight crew warning light which illuminates when power is on in the airplane and the emergency lighting control device is not armed.

(3) The cockpit control device must have an "on," "off," and "armed" position so that when armed in the cockpit or turned on at either the cockpit or flight attendant station the lights will either light or remain lighted upon interruption (except an interruption caused by a transverse vertical separation of the fuselage during crash landing) of the airplane's normal electric power. There must be a means to safeguard against inadvertent operation of the control device from the "armed" or "on" positions.

(g) Exterior emergency lighting must be provided as follows:

(1) At each overwing emergency exit the illumination must be—

(i) Not less than 0.03 foot-candle (measured normal to the direction of the incident light) on a 2-square-foot area where an evacuee is likely to make his first step outside the cabin;

(ii) Not less than 0.05 foot-candle (measured normal to the direction of the incident light) for a minimum width of 42 inches for a Type A overwing emergency exit and two feet for all other overwing emergency exits along the 30 percent of the slip-resistant portion of the escape route required in § 25.810(c) that is farthest from the exit; and

(iii) Not less than 0.03 foot-candle on the ground surface with the landing gear extended (measured normal to the direction of the incident light) where an evacuee using the established escape route would normally make first contact with the ground.

(2) At each non-overwing emergency exit not required by § 25.810(a) to have descent assist means the illumination must be not less than 0.03 foot-candle (measured normal to the direction of the incident light) on the ground surface with the landing gear extended where an evacuee is likely to make first contact with the ground outside the cabin.

(h) The means required in §§ 25.810(a)(1) and (d) to assist the occupants in descending to the ground must be illuminated so that the erected assist means is visible from the airplane.

(1) If the assist means is illuminated by exterior emergency lighting, it must provide illumination of not less than 0.03 foot-candle (measured normal to the direction of the incident light) at the ground end of the erected assist means where an evacuee using the established escape route would normally make first contact with the ground, with the airplane in each of the attitudes corresponding to the collapse of one or more legs of the landing gear.

(2) If the emergency lighting subsystem illuminating the assist means serves no other assist means, is independent of the airplane's main emergency lighting system, and is automatically activated when the assist means is erected, the lighting provisions—

(i) May not be adversely affected by stowage; and

(ii) Must provide illumination of not less than 0.03 foot-candle (measured normal to the direction of incident light) at the ground and of the erected assist means where an evacuee would normally make first contact with the ground, with the airplane in each of the attitudes corresponding to the collapse of one or more legs of the landing gear.

(i) The energy supply to each emergency lighting unit must provide the required level of illumination for at least 10 minutes at the critical ambient conditions after emergency landing.

(j) If storage batteries are used as the energy supply for the emergency lighting system, they may be recharged from the airplane's main electric power system: *Provided*, That, the charging circuit is designed to preclude inadvertent battery discharge into charging circuit faults.

(k) Components of the emergency lighting system, including batteries, wiring relays, lamps, and switches must be capable of normal operation after having been subjected to the inertia forces listed in § 25.561(b).

(l) The emergency lighting system must be designed so that after any single transverse vertical separation of the fuselage during crash landing—

(1) Not more than 25 percent of all electrically illuminated emergency lights required by this section are rendered inoperative, in addition to the lights that are directly damaged by the separation;

(2) Each electrically illuminated exit sign required under § 25.811(d)(2) remains operative exclusive of those that are directly damaged by the separation; and

(3) At least one required exterior emergency light for each side of the airplane remains operative exclusive of those that are directly damaged by the separation.

[Amdt. 25-15, 32 FR 13265, Sept. 20, 1967, as amended by Amdt. 25-28, 36 FR 16899, Aug. 26, 1971; Amdt. 25-32, 37 FR 3971, Feb. 24, 1972; Amdt. 25-46, 43 FR 50597, Oct. 30, 1978; Amdt. 25-58, 49 FR 43186, Oct. 26, 1984; Amdt. 25-88, 61 FR 57958, Nov. 8, 1996; Amdt. 25-116, 69 FR 62788, Oct. 27, 2004; Amdt. 25-128, 74 FR 25645, May 29, 2009]

§ 25.813 Emergency exit access.

Each required emergency exit must be accessible to the passengers and located where it will afford an effective means of evacuation. Emergency exit distribution must be as uniform as practical, taking passenger distribution into account; however, the size and location of exits on both sides of the cabin need not be symmetrical. If only one floor level exit per side is prescribed, and the airplane does not have a tailcone or ventral emergency exit, the floor level exit must be in the rearward part of the passenger compartment, unless another location affords a more effective means of passenger evacuation. Where more than one floor level exit per side is prescribed, at least one floor level exit per side must be located near each end of the cabin, except that this provision does not apply to combination cargo/passenger configurations. In addition—

(a) There must be a passageway leading from the nearest main aisle to each Type A, Type B, Type C, Type I, or Type II emergency exit and between individual passenger areas. Each passageway leading to a Type A or Type B exit must be unobstructed and at least 36 inches wide. Passageways between individual passenger areas and those leading to Type I, Type II, or Type C emergency exits must be unobstructed and at least 20 inches wide. Unless there are two or more main aisles, each Type A or B exit must be located so that there is passenger flow along the main aisle to that exit from both the forward and aft directions. If two or more main aisles are provided, there must be unobstructed cross-aisles at least 20 inches wide between main aisles. There must be—

(1) A cross-aisle which leads directly to each passageway between the nearest main aisle and a Type A or B exit; and

(2) A cross-aisle which leads to the immediate vicinity of each passageway between the nearest main aisle and a Type 1, Type II, or Type III exit; except that when two Type III exits are located within three passenger rows of each other, a single cross-aisle may be used if it leads to the vicinity between the passageways from the nearest main aisle to each exit.

(b) Adequate space to allow crewmember(s) to assist in the evacuation of passengers must be provided as follows:

(1) Each assist space must be a rectangle on the floor, of sufficient size to enable a crewmember, standing erect, to effectively assist evacuees. The assist space must not reduce the unobstructed width of the passageway below that required for the exit.

(2) For each Type A or B exit, assist space must be provided at each side of the exit regardless of whether an assist means is required by § 25.810(a).

(3) For each Type C, I or II exit installed in an airplane with seating for more than 80 passengers, an assist space must be provided at one side of the passageway regardless of whether an assist means is required by § 25.810(a).

(4) For each Type C, I or II exit, an assist space must be provided at one side of the passageway if an assist means is required by § 25.810(a).

(5) For any tailcone exit that qualifies for 25 additional passenger seats under the provisions of § 25.807(g)(9)(ii), an assist space must be provided, if an assist means is required by § 25.810(a).

(6) There must be a handle, or handles, at each assist space, located to enable the crewmember to steady himself or herself:

(i) While manually activating the assist means (where applicable) and,

(ii) While assisting passengers during an evacuation.

(c) The following must be provided for each Type III or Type IV exit—(1) There must be access from the nearest aisle to each exit. In addition, for each Type III exit in an airplane that has a passenger seating configuration of 60 or more—

(i) Except as provided in paragraph (c)(1)(ii), the access must be provided by an unobstructed passageway that is at least 10 inches in width for interior arrangements in which the adjacent seat rows on the exit side of the aisle contain no more than two seats, or 20 inches in width for interior arrangements in which those rows contain three seats. The width of the passageway must be measured with adjacent seats adjusted to their most adverse position. The centerline of the required passageway width must not be displaced more than 5 inches horizontally from that of the exit.

(ii) In lieu of one 10- or 20-inch passageway, there may be two passageways, between seat rows only, that must be at least 6 inches in width and lead to an unobstructed space adjacent to each exit. (Adjacent exits must not share a common passageway.) The width of the passageways must be measured with adjacent seats adjusted to their most adverse position. The unobstructed space adjacent to the exit must extend vertically from the floor to the ceiling (or bottom of sidewall stowage bins), inboard from the exit for a distance not less than the width of the narrowest passenger seat installed on the airplane, and from the forward edge of the forward passageway to the aft edge of the aft passageway. The exit opening must be totally within the fore and aft bounds of the unobstructed space.

(2) In addition to the access—

(i) For airplanes that have a passenger seating configuration of 20 or more, the projected opening of the exit provided must not be obstructed and there must be no interference in opening the exit by seats, berths, or other protrusions (including any seatback in the most adverse position) for a distance from that exit not less than the width of the narrowest passenger seat installed on the airplane.

(ii) For airplanes that have a passenger seating configuration of 19 or fewer, there may be minor obstructions in this region, if there are compensating factors to maintain the effectiveness of the exit.

(3) For each Type III exit, regardless of the passenger capacity of the airplane in which it is installed, there must be placards that—

(i) Are readable by all persons seated adjacent to and facing a passageway to the exit;

(ii) Accurately state or illustrate the proper method of opening the exit, including the use of handholds; and

(iii) If the exit is a removable hatch, state the weight of the hatch and indicate an appropriate location to place the hatch after removal.

(d) If it is necessary to pass through a passageway between passenger compartments to reach any required emergency exit from any seat in the passenger cabin, the passageway must be unobstructed. However, curtains may be used if they allow free entry through the passageway.

(e) No door may be installed between any passenger seat that is occupiable for takeoff and landing and any passenger emergency exit, such that the door crosses any egress path (including aisles, crossaisles and passageways).

(f) If it is necessary to pass through a doorway separating any crewmember seat (except those seats on the flightdeck), occupiable for takeoff and landing, from any emergency exit, the door must have a means to latch it in the open position. The latching means must be able to withstand the loads imposed upon it when the door is subjected to the ultimate inertia forces, relative to the surrounding structure, listed in § 25.561(b).

[Amdt. 25-1, 30 FR 3204, Mar. 9, 1965, as amended by Amdt. 25-15, 32 FR 13265, Sept. 20, 1967; Amdt. 25-32, 37 FR 3971, Feb. 24, 1972; Amdt. 25-46, 43 FR 50597, Oct. 30, 1978; Amdt. 25-72, 55 FR 29783, July 20, 1990; Amdt. 25-76, 57 FR 19244, May 4, 1992; Amdt. 25-76, 57 FR 29120, June 30, 1992; Amdt. 25-88, 61 FR 57958, Nov. 8, 1996; Amdt. 25-116, 69 FR 62788, Oct. 27, 2004; Amdt. 25-128, 74 FR 25645, May 29, 2009]

§ 25.815 Width of aisle.

The passenger aisle width at any point between seats must equal or exceed the values in the following table:

Passenger seating capacity	Minimum passenger aisle width (inches)	
	Less than 25 in. from floor	25 in. and more from floor
10 or less	[1]12	15
11 through 19	12	20
20 or more	15	20

[1] A narrower width not less than 9 inches may be approved when substantiated by tests found necessary by the Administrator.

[Amdt. 25-15, 32 FR 13265, Sept. 20, 1967, as amended by Amdt. 25-38, 41 FR 55466, Dec. 20, 1976]

§ 25.817 Maximum number of seats abreast.

On airplanes having only one passenger aisle, no more than three seats abreast may be placed on each side of the aisle in any one row.

[Amdt. 25-15, 32 FR 13265, Sept. 20, 1967]

§ 25.819 Lower deck service compartments (including galleys).

For airplanes with a service compartment located below the main deck, which may be occupied during taxi or flight but not during takeoff or landing, the following apply:

(a) There must be at least two emergency evacuation routes, one at each end of each lower deck service compartment or two having sufficient separation within each compartment, which could be used by each occupant of the lower deck service compartment to rapidly evacuate to the main deck under normal and emergency lighting conditions. The routes must provide for the evacuation of incapacitated persons, with assistance. The use of the evacuation routes may not be dependent on any powered device. The routes must be designed to minimize the possibility of blockage which might result from fire, mechanical or structural failure, or persons standing on top of or against the escape routes. In the event the airplane's main power system or compartment main lighting system should fail, emergency illumination for each lower deck service compartment must be automatically provided.

(b) There must be a means for two-way voice communication between the flight deck and each lower deck service compartment, which remains available following loss of normal electrical power generating system.

(c) There must be an aural emergency alarm system, audible during normal and emergency conditions, to enable crewmembers on the flight deck and at each required floor level emergency exit to alert occupants of each lower deck service compartment of an emergency situation.

(d) There must be a means, readily detectable by occupants of each lower deck service compartment, that indicates when seat belts should be fastened.

(e) If a public address system is installed in the airplane, speakers must be provided in each lower deck service compartment.

(f) For each occupant permitted in a lower deck service compartment, there must be a forward or aft facing seat which meets the requirements of § 25.785(d), and must be able to withstand maximum flight loads when occupied.

(g) For each powered lift system installed between a lower deck service compartment and the main deck for the carriage of persons or equipment, or both, the system must meet the following requirements:

(1) Each lift control switch outside the lift, except emergency stop buttons, must be designed to prevent the activation of the life if the lift door, or the hatch required by paragraph (g)(3) of this section, or both are open.

(2) An emergency stop button, that when activated will immediately stop the lift, must be installed within the lift and at each entrance to the lift.

(3) There must be a hatch capable of being used for evacuating persons from the lift that is openable from inside and outside the lift without tools, with the lift in any position.

[Amdt. 25-53, 45 FR 41593, June 19, 1980; 45 FR 43154, June 26, 1980; Amdt. 25-110; 68 FR 36883, June 19, 2003]

§ 25.820 Lavatory doors.

All lavatory doors must be designed to preclude anyone from becoming trapped inside the lavatory. If a locking mechanism is installed, it must be capable of being unlocked from the outside without the aid of special tools.

[Doc. No. 2003-14193, 69 FR 24502, May 3, 2004]

Ventilation and Heating

§ 25.831 Ventilation.

(a) Under normal operating conditions and in the event of any probable failure conditions of any system which would adversely affect the ventilating air, the ventilation system must be designed to provide a sufficient amount of uncontaminated air to enable the crewmembers to perform their duties without undue discomfort or fatigue and to provide reasonable passenger comfort. For normal operating conditions, the ventilation system must be designed to provide each occupant with an airflow containing at least 0.55 pounds of fresh air per minute.

(b) Crew and passenger compartment air must be free from harmful or hazardous concentrations of gases or vapors. In meeting this requirement, the following apply:

(1) Carbon monoxide concentrations in excess of 1 part in 20,000 parts of air are considered hazardous. For test purposes, any acceptable carbon monoxide detection method may be used.

(2) Carbon dioxide concentration during flight must be shown not to exceed 0.5 percent by volume (sea level equivalent) in compartments normally occupied by passengers or crewmembers.

(c) There must be provisions made to ensure that the conditions prescribed in paragraph (b) of this section are met after reasonably probable failures or malfunctioning of the ventilating, heating, pressurization, or other systems and equipment.

(d) If accumulation of hazardous quantities of smoke in the cockpit area is reasonably probable, smoke evacuation must be readily accomplished, starting with full pressurization and without depressurizing beyond safe limits.

(e) Except as provided in paragraph (f) of this section, means must be provided to enable the occupants of the following compartments and areas to control the temperature and quantity of ventilating air supplied to their compartment or area independently of the temperature and quantity of air supplied to other compartments and areas:

(1) The flight crew compartment.

(2) Crewmember compartments and areas other than the flight crew compartment unless the crewmember compartment or area is ventilated by air interchange with other compartments or areas under all operating conditions.

(f) Means to enable the flight crew to control the temperature and quantity of ventilating air supplied to the flight crew compartment independently of the temperature and quantity of ventilating air supplied to other compartments are not required if all of the following conditions are met:

(1) The total volume of the flight crew and passenger compartments is 800 cubic feet or less.

(2) The air inlets and passages for air to flow between flight crew and passenger compartments are arranged to provide compartment temperatures within 5 degrees F. of each other and adequate ventilation to occupants in both compartments.

(3) The temperature and ventilation controls are accessible to the flight crew.

(g) The exposure time at any given temperature must not exceed the values shown in the following graph after any improbable failure condition.

[Doc. No. 5066, 29 FR 18291, Dec. 24, 1964, as amended by Amdt. 25-41, 42 FR 36970, July 18, 1977; Amdt. 25-87, 61 FR 28695, June 5, 1996; Amdt. 25-89, 61 FR 63956, Dec. 2, 1996]

§ 25.832 Cabin ozone concentration.

(a) The airplane cabin ozone concentration during flight must be shown not to exceed—

(1) 0.25 parts per million by volume, sea level equivalent, at any time above flight level 320; and

(2) 0.1 parts per million by volume, sea level equivalent, time-weighted average during any 3-hour interval above flight level 270.

(b) For the purpose of this section, "sea level equivalent" refers to conditions of 25°C and 760 millimeters of mercury pressure.

(c) Compliance with this section must be shown by analysis or tests based on airplane operational procedures and performance limitations, that demonstrate that either—

(1) The airplane cannot be operated at an altitude which would result in cabin ozone concentrations exceeding the limits prescribed by paragraph (a) of this section; or

(2) The airplane ventilation system, including any ozone control equipment, will maintain cabin ozone concentrations at or below the limits prescribed by paragraph (a) of this section.

[Amdt. 25-50, 45 FR 3883, Jan. 1, 1980, as amended by Amdt. 25-56, 47 FR 58489, Dec. 30, 1982; Amdt. 25-94, 63 FR 8848, Feb. 23, 1998]

§ 25.833 Combustion heating systems.

Combustion heaters must be approved.

[Amdt. 25-72, 55 FR 29783, July 20, 1990]

Pressurization

§ 25.841 Pressurized cabins.

(a) Pressurized cabins and compartments to be occupied must be equipped to provide a cabin pressure altitude of not more than 8,000 feet at the maximum operating altitude of the airplane under normal operating conditions.

(1) If certification for operation above 25,000 feet is requested, the airplane must be designed so that occupants will not be exposed to cabin pressure altitudes in excess of 15,000 feet after any probable failure condition in the pressurization system.

(2) The airplane must be designed so that occupants will not be exposed to a cabin pressure altitude that exceeds the following after decompression from any failure condition not shown to be extremely improbable:

(i) Twenty-five thousand (25,000) feet for more than 2 minutes; or

(ii) Forty thousand (40,000) feet for any duration.

(3) Fuselage structure, engine and system failures are to be considered in evaluating the cabin decompression.

(b) Pressurized cabins must have at least the following valves, controls, and indicators for controlling cabin pressure:

(1) Two pressure relief valves to automatically limit the positive pressure differential to a predetermined value at the maximum rate of flow delivered by the pressure source. The combined capacity of the relief valves must be large enough so that the failure of any one valve would not cause an appreciable rise in the pressure differential. The pressure differential is positive when the internal pressure is greater than the external.

(2) Two reverse pressure differential relief valves (or their equivalents) to automatically prevent a negative pressure differential that would damage the structure. One valve is enough, however, if it is of a design that reasonably precludes its malfunctioning.

(3) A means by which the pressure differential can be rapidly equalized.

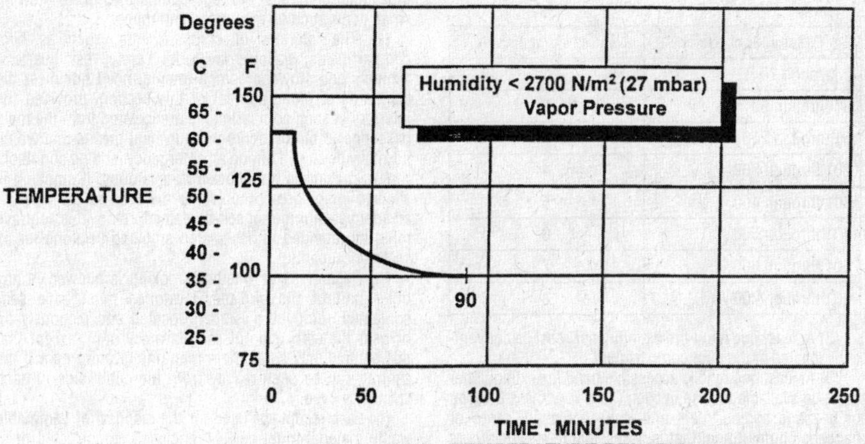

TIME - TEMPERATURE RELATIONSHIP

(4) An automatic or manual regulator for controlling the intake or exhaust airflow, or both, for maintaining the required internal pressures and airflow rates.

(5) Instruments at the pilot or flight engineer station to show the pressure differential, the cabin pressure altitude, and the rate of change of the cabin pressure altitude.

(6) Warning indication at the pilot or flight engineer station to indicate when the safe or preset pressure differential and cabin pressure altitude limits are exceeded. Appropriate warning markings on the cabin pressure differential indicator meet the warning requirement for pressure differential limits and an aural or visual signal (in addition to cabin altitude indicating means) meets the warning requirement for cabin pressure altitude limits if it warns the flight crew when the cabin pressure altitude exceeds 10,000 feet.

(7) A warning placard at the pilot or flight engineer station if the structure is not designed for pressure differentials up to the maximum relief valve setting in combination with landing loads.

(8) The pressure sensors necessary to meet the requirements of paragraphs (b)(5) and (b)(6) of this section and § 25.1447(c), must be located and the sensing system designed so that, in the event of loss of cabin pressure in any passenger or crew compartment (including upper and lower lobe galleys), the warning and automatic presentation devices, required by those provisions, will be actuated without any delay that would significantly increase the hazards resulting from decompression.

[Doc. No. 5066, 29 FR 18291, Dec. 24, 1964, as amended by Amdt. 25-38, 41 FR 55466, Dec. 20, 1976; Amdt. 25-87, 61 FR 28696, June 5, 1996]

§ 25.843 Tests for pressurized cabins.

(a) *Strength test.* The complete pressurized cabin, including doors, windows, and valves, must be tested as a pressure vessel for the pressure differential specified in § 25.365(d).

(b) *Functional tests.* The following functional tests must be performed:

(1) Tests of the functioning and capacity of the positive and negative pressure differential valves, and of the emergency release valve, to stimulate the effects of closed regulator valves.

(2) Tests of the pressurization system to show proper functioning under each possible condition of pressure, temperature, and moisture, up to the maximum altitude for which certification is requested.

(3) Flight tests, to show the performance of the pressure supply, pressure and flow regulators, indicators, and warning signals, in steady and stepped climbs and descents at rates corresponding to the maximum attainable within the operating limitations of the airplane, up to the maximum altitude for which certification is requested.

(4) Tests of each door and emergency exit, to show that they operate properly after being subjected to the flight tests prescribed in paragraph (b)(3) of this section.

Fire Protection

§ 25.851 Fire extinguishers.

(a) *Hand fire extinguishers.* (1) The following minimum number of hand fire extinguishers must be conveniently located and evenly distributed in passenger compartments:

Passenger capacity	No. of extinguishers
7 through 30	1
31 through 60	2
61 through 200	3
201 through 300	4
301 through 400	5
401 through 500	6
501 through 600	7
601 through 700	8

(2) At least one hand fire extinguisher must be conveniently located in the pilot compartment.

(3) At least one readily accessible hand fire extinguisher must be available for use in each Class A or Class B cargo or baggage compartment and in each Class E cargo or baggage compartment that is accessible to crewmembers in flight.

(4) At least one hand fire extinguisher must be located in, or readily accessible for use in, each galley located above or below the passenger compartment.

(5) Each hand fire extinguisher must be approved.

(6) At least one of the required fire extinguishers located in the passenger compartment of an airplane with a passenger capacity of at least 31 and not more than 60, and at least two of the fire extinguishers located in the passenger compartment of an airplane with a passenger capacity of 61 or more must contain Halon 1211 (bromochlorodifluoromethane $CBrC_1F_2$), or equivalent, as the extinguishing agent. The type of extinguishing agent used in any other extinguisher required by this section must be appropriate for the kinds of fires likely to occur where used.

(7) The quantity of extinguishing agent used in each extinguisher required by this section must be appropriate for the kinds of fires likely to occur where used.

(8) Each extinguisher intended for use in a personnel compartment must be designed to minimize the hazard of toxic gas concentration.

(b) Built-in fire extinguishers. If a built-in fire extinguisher is provided—

(1) Each built-in fire extinguishing system must be installed so that—

(i) No extinguishing agent likely to enter personnel compartments will be hazardous to the occupants; and

(ii) No discharge of the extinguisher can cause structural damage.

(2) The capacity of each required built-in fire extinguishing system must be adequate for any fire likely to occur in the compartment where used, considering the volume of the compartment and the ventilation rate.

[Amdt. 25-74, 56 FR 15456, Apr. 16, 1991]

§ 25.853 Compartment interiors.

For each compartment occupied by the crew or passengers, the following apply:

(a) Materials (including finishes or decorative surfaces applied to the materials) must meet the applicable test criteria prescribed in part I of appendix F of this part, or other approved equivalent methods, regardless of the passenger capacity of the airplane.

(b) [Reserved]

(c) In addition to meeting the requirements of paragraph (a) of this section, seat cushions, except those on flight crewmember seats, must meet the test requirements of part II of appendix F of this part, or other equivalent methods, regardless of the passenger capacity of the airplane.

(d) Except as provided in paragraph (e) of this section, the following interior components of airplanes with passenger capacities of 20 or more must also meet the test requirements of parts IV and V of appendix F of this part, or other approved equivalent method, in addition to the flammability requirements prescribed in paragraph (a) of this section:

(1) Interior ceiling and wall panels, other than lighting lenses and windows;

(2) Partitions, other than transparent panels needed to enhance cabin safety;

(3) Galley structure, including exposed surfaces of stowed carts and standard containers and the cavity walls that are exposed when a full complement of such carts or containers is not carried; and

(4) Large cabinets and cabin stowage compartments, other than underseat stowage compartments for stowing small items such as magazines and maps.

(e) The interiors of compartments, such as pilot compartments, galleys, lavatories, crew rest quarters, cabinets and stowage compartments, need not meet the standards of paragraph (d) of this section, provided the interiors of such compartments are isolated from the main passenger cabin by doors or equivalent means that would normally be closed during an emergency landing condition.

(f) Smoking is not allowed in lavatories. If smoking is allowed in any area occupied by the crew or passengers, an adequate number of self-contained, removable ashtrays must be provided in designated smoking sections for all seated occupants.

(g) Regardless of whether smoking is allowed in any other part of the airplane, lavatories must have self-contained, removable ashtrays located conspicuously on or near the entry side of each lavatory door, except that one ashtray may serve more than one lavatory door if the ashtray can be seen readily from the cabin side of each lavatory served.

(h) Each receptacle used for the disposal of flammable waste material must be fully enclosed, constructed of at least fire resistant materials, and must contain fires likely to occur in it under normal use. The capability of the receptacle to contain those fires under all probable conditions of wear, misalignment, and ventilation expected in service must be demonstrated by test.

[Amdt. 25-83, 60 FR 6623, Feb. 2, 1995, as amended by Amdt. 25-116, 69 FR 62788, Oct. 27, 2004]

§ 25.854 Lavatory fire protection.

For airplanes with a passenger capacity of 20 or more:

(a) Each lavatory must be equipped with a smoke detector system or equivalent that provides a warning light in the cockpit, or provides a warning light or audible warning in the passenger cabin that would be readily detected by a flight attendant; and

(b) Each lavatory must be equipped with a built-in fire extinguisher for each disposal receptacle for towels, paper, or waste, located within the lavatory. The extinguisher must be designed to discharge automatically into each disposal receptacle upon occurrence of a fire in that receptacle.

[Amdt. 25-74, 56 FR 15456, Apr. 16, 1991]

§ 25.855 Cargo or baggage compartments.

For each cargo or baggage compartment, the following apply:

(a) The compartment must meet one of the class requirements of § 25.857.

(b) Class B through Class E cargo or baggage compartments, as defined in § 25.857, must have a liner, and the liner must be separate from (but may be attached to) the airplane structure.

(c) Ceiling and sidewall liner panels of Class C compartments must meet the test requirements of part III of appendix F of this part or other approved equivalent methods.

(d) All other materials used in the construction of the cargo or baggage compartment must meet the applicable test criteria prescribed in part I of appendix F of this part or other approved equivalent methods.

(e) No compartment may contain any controls, lines, equipment, or accessories whose damage or failure would affect safe operation, unless those items are protected so that—

(1) They cannot be damaged by the movement of cargo in the compartment, and

(2) Their breakage or failure will not create a fire hazard.

(f) There must be means to prevent cargo or baggage from interfering with the functioning of the fire protective features of the compartment.

(g) Sources of heat within the compartment must be shielded and insulated to prevent igniting the cargo or baggage.

(h) Flight tests must be conducted to show compliance with the provisions of § 25.857 concerning—

(1) Compartment accessibility,

(2) The entries of hazardous quantities of smoke or extinguishing agent into compartments occupied by the crew or passengers, and

(3) The dissipation of the extinguishing agent in Class C compartments.

(i) During the above tests, it must be shown that no inadvertent operation of smoke or fire detectors in any compartment would occur as a result of fire contained in any other compartment, either during or after extinguishment, unless the extinguishing system floods each such compartment simultaneously.

(j) Cargo or baggage compartment electrical wiring interconnection system components must meet the requirements of § 25.1721.

[Amdt. 25-72, 55 FR 29784, July 20, 1990, as amended by Amdt. 25-93, 63 FR 8048, Feb. 17, 1998; Amdt. 25-116, 69 FR 62788, Oct. 27, 2004; Amdt. 25-123, 72 FR 63405, Nov. 8, 2007]

§ 25.856 Thermal/Acoustic insulation materials.

(a) Thermal/acoustic insulation material installed in the fuselage must meet the flame propagation test requirements of part VI of Appendix F to this part, or other approved equivalent test requirements. This requirement does not apply to "small parts," as defined in part I of Appendix F of this part.

(b) For airplanes with a passenger capacity of 20 or greater, thermal/acoustic insulation materials (including the means of fastening the materials to the fuselage) installed in the lower half of the airplane fuselage must meet the flame penetration resistance test requirements of part VII of Appendix F to this part, or other approved equivalent test requirements. This requirement does not apply to thermal/acoustic insulation installations that the FAA finds would not contribute to fire penetration resistance.

[Amdt. 25-111, 68 FR 45059, July 31, 2003]

§ 25.857 Cargo compartment classification.

(a) *Class A;* A Class A cargo or baggage compartment is one in which—

(1) The presence of a fire would be easily discovered by a crewmember while at his station; and

(2) Each part of the compartment is easily accessible in flight.

(b) *Class B.* A Class B cargo or baggage compartment is one in which—

(1) There is sufficient access in flight to enable a crewmember to effectively reach any part of the compartment with the contents of a hand fire extinguisher;

(2) When the access provisions are being used, no hazardous quantity of smoke, flames, or extinguishing agent, will enter any compartment occupied by the crew or passengers;

(3) There is a separate approved smoke detector or fire detector system to give warning at the pilot or flight engineer station.

(c) *Class C.* A Class C cargo or baggage compartment is one not meeting the requirements for either a Class A or B compartment but in which—

(1) There is a separate approved smoke detector or fire detector system to give warning at the pilot or flight engineer station;

(2) There is an approved built-in fire extinguishing or suppression system controllable from the cockpit.

(3) There are means to exclude hazardous quantities of smoke, flames, or extinguishing agent, from any compartment occupied by the crew or passengers;

(4) There are means to control ventilation and drafts within the compartment so that the extinguishing agent used can control any fire that may start within the compartment.

(d) [Reserved]

(e) *Class E.* A Class E cargo compartment is one on airplanes used only for the carriage of cargo and in which—

(1) [Reserved]

(2) There is a separate approved smoke or fire detector system to give warning at the pilot or flight engineer station;

(3) There are means to shut off the ventilating airflow to, or within, the compartment, and the controls for these means are accessible to the flight crew in the crew compartment;

(4) There are means to exclude hazardous quantities of smoke, flames, or noxious gases, from the flight crew compartment; and

(5) The required crew emergency exits are accessible under any cargo loading condition.

[Doc. No. 5066, 29 FR 18291, Dec. 24, 1964, as amended by Amdt. 25-32, 37 FR 3972, Feb. 24, 1972; Amdt. 25-60, 51 FR 18243, May 16, 1986; Amdt. 25-93, 63 FR 8048, Feb. 17, 1998]

§ 25.858 Cargo or baggage compartment smoke or fire detection systems.

If certification with cargo or baggage compartment smoke or fire detection provisions is requested, the following must be met for each cargo or baggage compartment with those provisions:

(a) The detection system must provide a visual indication to the flight crew within one minute after the start of a fire.

(b) The system must be capable of detecting a fire at a temperature significantly below that at which the structural integrity of the airplane is substantially decreased.

(c) There must be means to allow the crew to check in flight, the functioning of each fire detector circuit.

(d) The effectiveness of the detection system must be shown for all approved operating configurations and conditions.

[Amdt. 25-54, 45 FR 60173, Sept. 11, 1980, as amended by Amdt. 25-93, 63 FR 8048, Feb. 17, 1998]

§ 25.859 Combustion heater fire protection.

(a) *Combustion heater fire zones.* The following combustion heater fire zones must be protected from fire in accordance with the applicable provisions of §§ 25.1181 through 25.1191 and §§ 25.1195 through 25.1203;

(1) The region surrounding the heater, if this region contains any flammable fluid system components (excluding the heater fuel system), that could—

(i) Be damaged by heater malfunctioning; or

(ii) Allow flammable fluids or vapors to reach the heater in case of leakage.

(2) The region surrounding the heater, if the heater fuel system has fittings that, if they leaked, would allow fuel or vapors to enter this region.

(3) The part of the ventilating air passage that surrounds the combustion chamber. However, no fire extinguishment is required in cabin ventilating air passages.

(b) *Ventilating air ducts.* Each ventilating air duct passing through any fire zone must be fireproof. In addition—

(1) Unless isolation is provided by fireproof valves or by equally effective means, the ventilating air duct downstream of each heater must be fireproof for a distance great enough to ensure that any fire originating in the heater can be contained in the duct; and

(2) Each part of any ventilating duct passing through any region having a flammable fluid system must be constructed or isolated from that system so that the malfunctioning of any component of that system cannot introduce flammable fluids or vapors into the ventilating airstream.

(c) *Combustion air ducts.* Each combustion air duct must be fireproof for a distance great enough to prevent damage from backfiring or reverse flame propagation. In addition—

(1) No combustion air duct may have a common opening with the ventilating airstream unless flames from backfires or reverse burning cannot enter the ventilating airstream under any operating condition, including reverse flow or malfunctioning of the heater or its associated components; and

(2) No combustion air duct may restrict the prompt relief of any backfire that, if so restricted, could cause heater failure.

(d) *Heater controls; general.* Provision must be made to prevent the hazardous accumulation of water or ice on or in any heater control component, control system tubing, or safety control.

(e) *Heater safety controls.* For each combustion heater there must be the following safety control means:

(1) Means independent of the components provided for the normal continuous control of air temperature, airflow, and fuel flow must be provided, for each heater, to automatically shut off the ignition and fuel supply to that heater at a point remote from that heater when any of the following occurs:

(i) The heat exchanger temperature exceeds safe limits.

(ii) The ventilating air temperature exceeds safe limits.

(iii) The combustion airflow becomes inadequate for safe operation.

(iv) The ventilating airflow becomes inadequate for safe operation.

(2) The means of complying with paragraph (e)(1) of this section for any individual heater must—

(i) Be independent of components serving any other heater whose heat output is essential for safe operation; and

(ii) Keep the heater off until restarted by the crew.

(3) There must be means to warn the crew when any heater whose heat output is essential for safe operation has been shut off by the automatic means prescribed in paragraph (e)(1) of this section.

(f) *Air intakes.* Each combustion and ventilating air intake must be located so that no flammable fluids or vapors can enter the heater system under any operating condition—

(1) During normal operation; or

(2) As a result of the malfunctioning of any other component.

(g) *Heater exhaust.* Heater exhaust systems must meet the provisions of §§ 25.1121 and 25.1123. In addition, there must be provisions in the design of the heater exhaust system to safely expel the products of combustion to prevent the occurrence of—

(1) Fuel leakage from the exhaust to surrounding compartments;

(2) Exhaust gas impingement on surrounding equipment or structure;

(3) Ignition of flammable fluids by the exhaust, if the exhaust is in a compartment containing flammable fluid lines; and

(4) Restriction by the exhaust of the prompt relief of backfires that, if so restricted, could cause heater failure.

(h) *Heater fuel systems.* Each heater fuel system must meet each powerplant fuel system requirement affecting safe heater operation. Each heater fuel system component within the ventilating airstream must be protected by shrouds so that no leakage from those components can enter the ventilating airstream.

(i) *Drains.* There must be means to safely drain fuel that might accumulate within the combustion chamber or the heat exchanger. In addition—

(1) Each part of any drain that operates at high temperatures must be protected in the same manner as heater exhausts; and

(2) Each drain must be protected from hazardous ice accumulation under any operating condition.

[Doc. No. 5066, 29 FR 18291, Dec. 24 1964, as amended by Amdt. 25-11, 32 FR 6912, May 5, 1967; Amdt. 25-23, 35 FR 5676, Apr. 8, 1970]

§ 25.863 Flammable fluid fire protection.

(a) In each area where flammable fluids or vapors might escape by leakage of a fluid system, there must be means to minimize the probability of ignition of the fluids and vapors, and the resultant hazards if ignition does occur.

(b) Compliance with paragraph (a) of this section must be shown by analysis or tests, and the following factors must be considered:

(1) Possible sources and paths of fluid leakage, and means of detecting leakage.

(2) Flammability characteristics of fluids, including effects of any combustible or absorbing materials.

(3) Possible ignition sources, including electrical faults, overheating of equipment, and malfunctioning of protective devices.

(4) Means available for controlling or extinguishing a fire, such as stopping flow of fluids, shutting down equipment, fireproof containment, or use of extinguishing agents.

(5) Ability of airplane components that are critical to safety of flight to withstand fire and heat.

(c) If action by the flight crew is required to prevent or counteract a fluid fire (e.g., equipment shutdown or actuation of a fire extinguisher) quick acting means must be provided to alert the crew.

(d) Each area where flammable fluids or vapors might escape by leakage of a fluid system must be identified and defined.

[Amdt. 25-23, 35 FR 5676, Apr. 8, 1970, as amended by Amdt. 25-46, 43 FR 50597, Oct. 30, 1978]

§ 25.865 Fire protection of flight controls, engine mounts, and other flight structure.

Essential flight controls, engine mounts, and other flight structures located in designated fire zones or in adjacent areas which would be subjected to the effects of fire in the fire zone must be constructed of fireproof material or shielded so that they are capable of withstanding the effects of fire.

[Amdt. 25-23, 35 FR 5676, Apr. 8, 1970]

§ 25.867 Fire protection: other components.

(a) Surfaces to the rear of the nacelles, within one nacelle diameter of the nacelle centerline, must be at least fire-resistant.

(b) Paragraph (a) of this section does not apply to tail surfaces to the rear of the nacelles that could not be readily affected by heat, flames, or sparks coming from a designated fire zone or engine compartment of any nacelle.

[Amdt. 25-23, 35 FR 5676, Apr. 8, 1970]

§ 25.869 Fire protection: systems.

(a) Electrical system components:

(1) Components of the electrical system must meet the applicable fire and smoke protection requirements of §§ 25.831(c) and 25.863.

(2) Equipment that is located in designated fire zones and is used during emergency procedures must be at least fire resistant.

(3) EWIS components must meet the requirements of § 25.1713.

(b) Each vacuum air system line and fitting on the discharge side of the pump that might contain flammable vapors or fluids must meet the requirements of § 25.1183 if the line or fitting is in a designated fire zone. Other vacuum air systems components in designated fire zones must be at least fire resistant.

(c) Oxygen equipment and lines must—

(1) Not be located in any designated fire zone,

(2) Be protected from heat that may be generated in, or escape from, any designated fire zone, and

(3) Be installed so that escaping oxygen cannot cause ignition of grease, fluid, or vapor accumulations that are present in normal operation or as a result of failure or malfunction of any system.

[Amdt. 25-72, 55 FR 29784, July 20, 1990, as amended by Amdt. 25-113, 69 FR 12530, Mar. 16, 2004; Amdt. 25-123, 72 FR 63405, Nov. 8, 2007]

MISCELLANEOUS

§ 25.871 Leveling means.

There must be means for determining when the airplane is in a level position on the ground.

[Amdt. 25-23, 35 FR 5676, Apr. 8, 1970]

§ 25.875 Reinforcement near propellers.

(a) Each part of the airplane near the propeller tips must be strong and stiff enough to withstand the effects of the induced vibration and of ice thrown from the propeller.

(b) No window may be near the propeller tips unless it can withstand the most severe ice impact likely to occur.

§ 25.899 Electrical bonding and protection against static electricity.

(a) Electrical bonding and protection against static electricity must be designed to minimize accumulation of electrostatic charge that would cause—

(1) Human injury from electrical shock,

(2) Ignition of flammable vapors, or

(3) Interference with installed electrical/electronic equipment.

(b) Compliance with paragraph (a) of this section may be shown by—

(1) Bonding the components properly to the airframe; or

(2) Incorporating other acceptable means to dissipate the static charge so as not to endanger the airplane, personnel, or operation of the installed electrical/electronic systems.

[Amdt. 25-123, 72 FR 63405, Nov. 8, 2007]

Subpart E—Powerplant

GENERAL

§ 25.901 Installation.

(a) For the purpose of this part, the airplane powerplant installation includes each component that—

(1) Is necessary for propulsion;

(2) Affects the control of the major propulsive units; or

(3) Affects the safety of the major propulsive units between normal inspections or overhauls.

(b) For each powerplant—

(1) The installation must comply with—

(i) The installation instructions provided under §§ 33.5 and 35.3 of this chapter; and

(ii) The applicable provisions of this subpart;

(2) The components of the installation must be constructed, arranged, and installed so as to ensure their continued safe operation between normal inspections or overhauls;

(3) The installation must be accessible for necessary inspections and maintenance; and

(4) The major components of the installation must be electrically bonded to the other parts of the airplane.

(c) For each powerplant and auxiliary power unit installation, it must be established that no single failure or malfunction or probable combination of failures will jeopardize the safe operation of the airplane except that the failure of structural elements need not be considered if the probability of such failure is extremely remote.

(d) Each auxiliary power unit installation must meet the applicable provisions of this subpart.

[Doc. No. 5066, 29 FR 18291, Dec. 24, 1964, as amended by Amdt. 25-23, 35 FR 5676, Apr. 8, 1970; Amdt. 25-40, 42 FR 15042, Mar. 17, 1977; Amdt. 25-46, 43 FR 50597, Oct. 30, 1978; Amdt. 25-126, 73 FR 63345, Oct. 24, 2008]

§ 25.903 Engines.

(a) *Engine type certificate.* (1) Each engine must have a type certificate and must meet the applicable requirements of part 34 of this chapter.

(2) Each turbine engine must comply with one of the following:

(i) Sections 33.76, 33.77 and 33.78 of this chapter in effect on December 13, 2000, or as subsequently amended; or

(ii) Sections 33.77 and 33.78 of this chapter in effect on April 30, 1998, or as subsequently amended before December 13, 2000; or

(iii) Comply with § 33.77 of this chapter in effect on October 31, 1974, or as subsequently amended prior to April 30, 1998, unless that engine's foreign object ingestion service history has resulted in an unsafe condition; or

(iv) Be shown to have a foreign object ingestion service history in similar installation locations which has not resulted in any unsafe condition.

NOTE: § 33.77 of this chapter in effect on October 31, 1974, was published in 14 CFR parts 1 to 59, Revised as of January 1, 1975. See 39 FR 35467, October 1, 1974.

(b) *Engine isolation.* The powerplants must be arranged and isolated from each other to allow operation, in at least one configuration, so that the failure or malfunction of any engine, or of any system that can affect the engine, will not—

(1) Prevent the continued safe operation of the remaining engines; or

(2) Require immediate action by any crewmember for continued safe operation.

(c) *Control of engine rotation.* There must be means for stopping the rotation of any engine individually in flight, except that, for turbine engine installations, the means for stopping the rotation of any engine need be provided only where continued rotation could jeopardize the safety of the airplane. Each component of the stopping system on the engine side of the firewall that might be exposed to fire must be at least fire-resistant. If hydraulic propeller feathering systems are used for this purpose, the feathering lines must be at least fire resistant under the operating conditions that may be expected to exist during feathering.

(d) *Turbine engine installations.* For turbine engine installations—

(1) Design precautions must be taken to minimize the hazards to the airplane in the event of an engine rotor failure or of a fire originating within the engine which burns through the engine case.

(2) The powerplant systems associated with engine control devices, systems, and instrumentation, must be designed to give reasonable assurance that those engine operating limitations that adversely affect turbine rotor structural integrity will not be exceeded in service.

(e) *Restart capability.* (1) Means to restart any engine in flight must be provided.

(2) An altitude and airspeed envelope must be established for in-flight engine restarting, and each engine must have a restart capability within that envelope.

(3) For turbine engine powered airplanes, if the minimum windmilling speed of the engines, following the inflight shutdown of all engines, is insufficient to provide the necessary electrical power for engine ignition, a power source independent of the engine-driven electrical power generating system must be provided to permit in-flight engine ignition for restarting.

(f) *Auxiliary Power Unit.* Each auxiliary power unit must be approved or meet the requirements of the category for its intended use.

[Doc. No. 5066, 29 FR 18291, Dec. 24, 1964, as amended by Amdt. 25-23, 35 FR 5676, Apr. 8, 1970; Amdt. 25-40, 42 FR 15042, Mar. 17, 1977; Amdt. 25-57, 49 FR 6848, Feb. 23, 1984; Amdt. 25-72, 55 FR 29784, July 20, 1990; Amdt. 25-73, 55 FR 32861, Aug. 10, 1990; Amdt. 25-94, 63 FR 8848, Feb. 23, 1998; Amdt. 25-95, 63 FR 14798, Mar. 26, 1998; Amdt. 25-100, 65 FR 55854, Sept. 14, 2000]

§ 25.904 Automatic takeoff thrust control system (ATTCS).

Each applicant seeking approval for installation of an engine power control system that automatically resets the power or thrust on the operating engine(s) when any engine fails during the takeoff must comply with the requirements of appendix I of this part.

[Amdt. 25-62, 52 FR 43156, Nov. 9, 1987]

§ 25.905 Propellers.

(a) Each propeller must have a type certificate.

(b) Engine power and propeller shaft rotational speed may not exceed the limits for which the propeller is certificated.

(c) The propeller blade pitch control system must meet the requirements of §§ 35.21, 35.23, 35.42 and 35.43 of this chapter.

(d) Design precautions must be taken to minimize the hazards to the airplane in the event a propeller blade fails or is released by a hub failure. The hazards which must be considered include damage to structure and vital systems due to impact of a failed or released blade and the unbalance created by such failure or release.

[Doc. No. 5066, 29 FR 18291, Dec. 24, 1964, as amended by Amdt. 25-54, 45 FR 60173, Sept. 11, 1980; Amdt. 25-57, 49 FR 6848, Feb. 23, 1984; Amdt. 25-72, 55 FR 29784, July 20, 1990; Amdt. 25-126, 73 FR 63345, Oct. 24, 2008]

§ 25.907 Propeller vibration and fatigue.

This section does not apply to fixed-pitch wood propellers of conventional design.

(a) The applicant must determine the magnitude of the propeller vibration stresses or loads, including any stress peaks and resonant conditions, throughout the operational envelope of the airplane by either:

(1) Measurement of stresses or loads through direct testing or analysis based on direct testing of the propeller on the airplane and engine installation for which approval is sought; or

(2) Comparison of the propeller to similar propellers installed on similar airplane installations for which these measurements have been made.

(b) The applicant must demonstrate by tests, analysis based on tests, or previous experience on similar designs that the propeller does not experience harmful effects of flutter throughout the operational envelope of the airplane.

(c) The applicant must perform an evaluation of the propeller to show that failure due to fatigue will be avoided throughout the operational life of the propeller using the fatigue and structural data obtained in accordance with part 35 of this chapter and the vibration data obtained from compliance with paragraph (a) of this section. For the purpose of this paragraph, the propeller includes the hub, blades, blade retention component and any other propeller component whose failure due to fatigue could be catastrophic to the airplane. This evaluation must include:

(1) The intended loading spectra including all reasonably foreseeable propeller vibration and cyclic load patterns, identified emergency conditions, allowable overspeeds and overtorques, and the effects of temperatures and humidity expected in service.

(2) The effects of airplane and propeller operating and airworthiness limitations.

[Amdt. 25-126, 73 FR 63345, Oct. 24, 2008]

§ 25.925 Propeller clearance.

Unless smaller clearances are substantiated, propeller clearances with the airplane at maximum weight, with the most adverse center of gravity, and with the propeller in the most adverse pitch position, may not be less than the following:

(a) *Ground clearance.* There must be a clearance of at least seven inches (for each airplane with nose wheel landing gear) or nine inches (for each airplane with tail wheel landing gear) between each propeller and the ground with the landing gear statically deflected and in the level takeoff, or taxiing attitude, whichever is most critical. In addition, there must be positive clearance between the propeller and the ground when in the level takeoff attitude with the critical tire(s) completely deflated and the corresponding landing gear strut bottomed.

(b) *Water clearance.* There must be a clearance of at least 18 inches between each propeller and the water, unless compliance with § 25.239(a) can be shown with a lesser clearance.

(c) *Structural clearance.* There must be—

(1) At least one inch radial clearance between the blade tips and the airplane structure, plus any additional radial clearance necessary to prevent harmful vibration;

(2) At least one-half inch longitudinal clearance between the propeller blades or cuffs and stationary parts of the airplane; and

(3) Positive clearance between other rotating parts of the propeller or spinner and stationary parts of the airplane.

[Doc. No. 5066, 29 FR 18291, Dec. 24, 1964, as amended by Amdt. 25-72, 55 FR 29784, July 20, 1990]

§ 25.929 Propeller deicing.

(a) For airplanes intended for use where icing may be expected, there must be a means to prevent or remove hazardous ice accumulation on propellers or on accessories where ice accumulation would jeopardize engine performance.

(b) If combustible fluid is used for propeller deicing, §§ 25.1181 through 25.1185 and 25.1189 apply.

§ 25.933 Reversing systems.

(a) For turbojet reversing systems—

(1) Each system intended for ground operation only must be designed so that during any reversal in flight the engine will produce no more than flight idle thrust. In addition, it must be shown by analysis or test, or both, that—

(i) Each operable reverser can be restored to the forward thrust position; and

(ii) The airplane is capable of continued safe flight and landing under any possible position of the thrust reverser.

(2) Each system intended for inflight use must be designed so that no unsafe condition will result during normal operation of the system, or from any failure (or reasonably likely combination of failures) of the reversing system, under any anticipated condition of operation of the airplane including ground operation. Failure of structural elements need not be considered if the probability of this kind of failure is extremely remote.

(3) Each system must have means to prevent the engine from producing more than idle thrust when the reversing system malfunctions, except that it may produce any greater forward thrust that is shown to allow directional control to be maintained, with aerodynamic means alone, under the most critical reversing condition expected in operation.

(b) For propeller reversing systems—

(1) Each system intended for ground operation only must be designed so that no single failure (or reasonably likely combination of failures) or malfunction of the system will result in unwanted reverse thrust under any expected operating condition. Failure of structural elements need not be considered if this kind of failure is extremely remote.

(2) Compliance with this section may be shown by failure analysis or testing, or both, for propeller systems that allow propeller blades to move from the flight low-pitch position to a position that is substantially less than that at the normal flight low-pitch position. The analysis may include or be supported by the analysis made to show compliance with the requirements of § 35.21 of this chapter for the propeller and associated installation components.

[Amdt. 25-72, 55 FR 29784, July 20, 1990]

§ 25.934 Turbojet engine thrust reverser system tests.

Thrust reversers installed on turbojet engines must meet the requirements of § 33.97 of this chapter.

[Amdt. 25-23, 35 FR 5677, Apr. 8, 1970]

§ 25.937 Turbopropeller-drag limiting systems.

Turbopropeller power airplane propeller-drag limiting systems must be designed so that no single failure or malfunction of any of the systems during normal or emergency operation results in propeller drag in excess of that for which the airplane was designed under § 25.367. Failure of structural elements of the drag limiting systems need not be considered if the probability of this kind of failure is extremely remote.

§ 25.939 Turbine engine operating characteristics.

(a) Turbine engine operating characteristics must be investigated in flight to determine that no adverse characteristics (such as stall, surge, or flameout) are present, to a hazardous degree, during normal and emergency operation within the range of operating limitations of the airplane and of the engine.

(b) [Reserved]

(c) The turbine engine air inlet system may not, as a result of air flow distortion during normal operation, cause vibration harmful to the engine.

[Amdt. 25-11, 32 FR 6912, May 5, 1967, as amended by Amdt. 25-40, 42 FR 15043, Mar. 17, 1977]

§ 25.941 Inlet, engine, and exhaust compatibility.

For airplanes using variable inlet or exhaust system geometry, or both—

(a) The system comprised of the inlet, engine (including thrust augmentation systems, if incorporated), and exhaust must be shown to function properly under all operating conditions for which approval is sought, including all engine rotating speeds and power settings, and engine inlet and exhaust configurations;

(b) The dynamic effects of the operation of these (including consideration of probable malfunctions) upon the aerodynamic control of the airplane may not result in any condition that would require exceptional skill, alertness, or strength on the part of the pilot to avoid exceeding an operational or structural limitation of the airplane; and

(c) In showing compliance with paragraph (b) of this section, the pilot strength required may not exceed the limits set forth in § 25.143(d), subject to the conditions set forth in paragraphs (e) and (f) of § 25.143.

[Amdt. 25-38, 41 FR 55467, Dec. 20, 1976, as amended by Amdt. 25-121, 72 FR 44669, Aug. 8, 2007]

§ 25.943 Negative acceleration.

No hazardous malfunction of an engine, an auxiliary power unit approved for use in flight, or any component or system associated with the powerplant or auxiliary power unit may occur when the airplane is operated at the negative accelerations within the flight envelopes prescribed in § 25.333. This must be shown for the greatest duration expected for the acceleration.

[Amdt. 25-40, 42 FR 15043, Mar. 17, 1977]

§ 25.945 Thrust or power augmentation system.

(a) *General.* Each fluid injection system must provide a flow of fluid at the rate and pressure established for proper engine functioning under each intended operating condition. If the fluid can freeze, fluid freezing may not damage the airplane or adversely affect airplane performance.

(b) *Fluid tanks.* Each augmentation system fluid tank must meet the following requirements:

(1) Each tank must be able to withstand without failure the vibration, inertia, fluid, and structural loads that it may be subject to in operation.

(2) The tanks as mounted in the airplane must be able to withstand without failure or leakage an internal pressure 1.5 times the maximum operating pressure.

(3) If a vent is provided, the venting must be effective under all normal flight conditions.

(4) [Reserved]

(5) Each tank must have an expansion space of not less than 2 percent of the tank capacity. It must be impossible to fill the expansion space inadvertently with the airplane in the normal ground attitude.

(c) Augmentation system drains must be designed and located in accordance with § 25.1455 if—

(1) The augmentation system fluid is subject to freezing; and

(2) The fluid may be drained in flight or during ground operation.

(d) The augmentation liquid tank capacity available for the use of each engine must be large enough to allow operation of the airplane under the approved procedures for the use of liquid-augmented power. The computation of liquid consumption must be based on the maximum approved rate appropriate for the desired engine output and must include the effect of temperature on engine performance as well as any other factors that might vary the amount of liquid required.

(e) This section does not apply to fuel injection systems.

[Amdt. 25-40, 42 FR 15043, Mar. 17, 1977, as amended by Amdt. 25-72, 55 FR 29785, July 20, 1990; Amdt. 25-115, 69 FR 40527, July 2, 2004]

FUEL SYSTEM

§ 25.951 General.

(a) Each fuel system must be constructed and arranged to ensure a flow of fuel at a rate and pressure established for proper engine and auxiliary power unit functioning under each likely operating condition, including any maneuver for which certification is requested and during which the engine or auxiliary power unit is permitted to be in operation.

(b) Each fuel system must be arranged so that any air which is introduced into the system will not result in—

(1) Power interruption for more than 20 seconds for reciprocating engines; or

(2) Flameout for turbine engines.

(c) Each fuel system for a turbine engine must be capable of sustained operation throughout its flow and pressure range with fuel initially saturated with water at 80°F and having 0.75cc of free water per gallon added and cooled to the most critical condition for icing likely to be encountered in operation.

(d) Each fuel system for a turbine engine powered airplane must meet the applicable fuel venting requirements of part 34 of this chapter.

[Doc. No. 5066, 29 FR 18291, Dec. 24, 1964, as amended by Amdt. 25-23, 35 FR 5677, Apr. 8, 1970; Amdt. 25-36, 39 FR 35460, Oct. 1, 1974; Amdt. 25-38, 41 FR 55467, Dec. 20, 1976; Amdt. 25-73, 55 FR 32861, Aug. 10, 1990]

§ 25.952 Fuel system analysis and test.

(a) Proper fuel system functioning under all probable operating conditions must be shown by analysis and those tests found necessary by the Administrator. Tests, if required, must be made using the airplane fuel system or a test article that reproduces the operating characteristics of the portion of the fuel system to be tested.

(b) The likely failure of any heat exchanger using fuel as one of its fluids may not result in a hazardous condition.

[Amdt. 25-40, 42 FR 15043, Mar. 17, 1977]

§ 25.953 Fuel system independence.

Each fuel system must meet the requirements of § 25.903(b) by—

(a) Allowing the supply of fuel to each engine through a system independent of each part of the system supplying fuel to any other engine; or

(b) Any other acceptable method.

§ 25.954 Fuel system lightning protection.

The fuel system must be designed and arranged to prevent the ignition of fuel vapor within the system by—

(a) Direct lightning strikes to areas having a high probability of stroke attachment;

(b) Swept lightning strokes to areas where swept strokes are highly probable; and

(c) Corona and streamering at fuel vent outlets.

[Amdt. 25-14, 32 FR 11629, Aug. 11, 1967]

§ 25.955 Fuel flow.

(a) Each fuel system must provide at least 100 percent of the fuel flow required under each intended operating condition and maneuver. Compliance must be shown as follows:

(1) Fuel must be delivered to each engine at a pressure within the limits specified in the engine type certificate.

(2) The quantity of fuel in the tank may not exceed the amount established as the unusable fuel supply for that tank under the requirements of § 25.959 plus that necessary to show compliance with this section.

(3) Each main pump must be used that is necessary for each operating condition and attitude for which compliance with this section is shown, and the appropriate emergency pump must be substituted for each main pump so used.

(4) If there is a fuel flowmeter, it must be blocked and the fuel must flow through the meter or its bypass.

(b) If an engine can be supplied with fuel from more than one tank, the fuel system must—

(1) For each reciprocating engine, supply the full fuel pressure to that engine in not more than 20 seconds after switching to any other fuel tank containing usable fuel when engine malfunctioning becomes apparent due to the depletion of the fuel supply in any tank from which the engine can be fed; and

(2) For each turbine engine, in addition to having appropriate manual switching capability, be designed to prevent interruption of fuel flow to that engine, without attention by the flight crew, when any tank supplying fuel to that engine is depleted of usable fuel during normal operation, and any other tank, that normally supplies fuel to that engine alone, contains usable fuel.

[Doc. No. 5066, 29 FR 18291, Dec. 24, 1964, as amended by Amdt. 25-11, 32 FR 6912, May 5, 1967]

§ 25.957 Flow between interconnected tanks.

If fuel can be pumped from one tank to another in flight, the fuel tank vents and the fuel transfer system must be designed so that no structural damage to the tanks can occur because of overfilling.

§ 25.959 Unusable fuel supply.

The unusable fuel quantity for each fuel tank and its fuel system components must be established at not less than the quantity at which the first evidence of engine malfunction occurs under the most adverse fuel feed condition for all intended operations and flight maneuvers involving fuel feeding from that tank. Fuel system component failures need not be considered.

[Amdt. 25-23, 35 FR 5677, Apr. 8, 1970, as amended by Amdt. 25-40, 42 FR 15043, Mar. 17, 1977]

§ 25.961 Fuel system hot weather operation.

(a) The fuel system must perform satisfactorily in hot weather operation. This must be shown by showing that the fuel system from the tank outlets to each engine is pressurized, under all intended operations, so as to prevent vapor formation, or must be shown by climbing from the altitude of the airport elected by the applicant to the maximum altitude established as an operating limitation under § 25.1527. If a climb test is elected, there may be no evidence of vapor lock or other malfunctioning during the climb test conducted under the following conditions:

(1) For reciprocating engine powered airplanes, the engines must operate at maximum continuous power, except that takeoff power must be used for the altitudes from 1,000 feet below the critical altitude through the

critical altitude. The time interval during which takeoff power is used may not be less than the takeoff time limitation.

(2) For turbine engine powered airplanes, the engines must operate at takeoff power for the time interval selected for showing the takeoff flight path, and at maximum continuous power for the rest of the climb.

(3) The weight of the airplane must be the weight with full fuel tanks, minimum crew, and the ballast necessary to maintain the center of gravity within allowable limits.

(4) The climb airspeed may not exceed—

(i) For reciprocating engine powered airplanes, the maximum airspeed established for climbing from takeoff to the maximum operating altitude with the airplane in the following configuration:

(A) Landing gear retracted.

(B) Wing flaps in the most favorable position.

(C) Cowl flaps (or other means of controlling the engine cooling supply) in the position that provides adequate cooling in the hot-day condition.

(D) Engine operating within the maximum continuous power limitations.

(E) Maximum takeoff weight; and

(ii) For turbine engine powered airplanes, the maximum airspeed established for climbing from takeoff to the maximum operating altitude.

(5) The fuel temperature must be at least 110 °F.

(b) The test prescribed in paragraph (a) of this section may be performed in flight or on the ground under closely simulated flight conditions. If a flight test is performed in weather cold enough to interfere with the proper conduct of the test, the fuel tank surfaces, fuel lines, and other fuel system parts subject to cold air must be insulated to simulate, insofar as practicable, flight in hot weather.

[Amdt. 25-11, 32 FR 6912, May 5, 1967, as amended by Amdt. 25-57, 49 FR 6848, Feb. 23, 1984]

§ 25.963 Fuel tanks: general.

(a) Each fuel tank must be able to withstand, without failure, the vibration, inertia, fluid, and structural loads that it may be subjected to in operation.

(b) Flexible fuel tank liners must be approved or must be shown to be suitable for the particular application.

(c) Integral fuel tanks must have facilities for interior inspection and repair.

(d) Fuel tanks within the fuselage contour must be able to resist rupture and to retain fuel, under the inertia forces prescribed for the emergency landing conditions in § 25.561. In addition, these tanks must be in a protected position so that exposure of the tanks to scraping action with the ground is unlikely.

(e) Fuel tank access covers must comply with the following criteria in order to avoid loss of hazardous quantities of fuel:

(1) All covers located in an area where experience or analysis indicates a strike is likely must be shown by analysis or tests to minimize penetration and deformation by tire fragments, low energy engine debris, or other likely debris.

(2) All covers must be fire resistant as defined in part 1 of this chapter.

(f) For pressurized fuel tanks, a means with fail-safe features must be provided to prevent the buildup of an excessive pressure difference between the inside and the outside of the tank.

[Doc. No. 5066, 29 FR 18291, Dec. 24, 1964, as amended by Amdt. 25-40, 42 FR 15043, Mar. 17, 1977; Amdt. 25-69, 54 FR 40354, Sept. 29, 1989]

§ 25.965 Fuel tank tests.

(a) It must be shown by tests that the fuel tanks, as mounted in the airplane, can withstand, without failure or leakage, the more critical of the pressures resulting from the conditions specified in paragraphs (a)(1) and (2) of this section. In addition, it must be shown by either analysis or tests, that tank surfaces subjected to more critical pressures resulting from the condition of paragraphs (a)(3) and (4) of this section, are able to withstand the following pressures:

(1) An internal pressure of 3.5 psi.

(2) 125 percent of the maximum air pressure developed in the tank from ram effect.

(3) Fluid pressures developed during maximum limit accelerations, and deflections, of the airplane with a full tank.

(4) Fluid pressures developed during the most adverse combination of airplane roll and fuel load.

(b) Each metallic tank with large unsupported or unstiffened flat surfaces, whose failure or deformation could cause fuel leakage, must be able to withstand the following test, or its equivalent, without leakage or excessive deformation of the tank walls:

(1) Each complete tank assembly and its supports must be vibration tested while mounted to simulate the actual installation.

(2) Except as specified in paragraph (b)(4) of this section, the tank assembly must be vibrated for 25 hours at an amplitude of not less than $\frac{1}{32}$ of an inch (unless another amplitude is substantiated) while $\frac{2}{3}$ filled with water or other suitable test fluid.

(3) The test frequency of vibration must be as follows:

(i) If no frequency of vibration resulting from any r.p.m. within the normal operating range of engine speeds is critical, the test frequency of vibration must be 2,000 cycles per minute.

(ii) If only one frequency of vibration resulting from any r.p.m. within the normal operating range of engine speeds is critical, that frequency of vibration must be the test frequency.

(iii) If more than one frequency of vibration resulting from any r.p.m. within the normal operating range of engine speeds is critical, the most critical of these frequencies must be the test frequency.

(4) Under paragraphs (b)(3)(ii) and (iii) of this section, the time of test must be adjusted to accomplish the same number of vibration cycles that would be accomplished in 25 hours at the frequency specified in paragraph (b)(3)(i) of this section.

(5) During the test, the tank assembly must be rocked at the rate of 16 to 20 complete cycles per minute, through an angle of 15° on both sides of the horizontal (30° total), about the most critical axis, for 25 hours. If motion about more than one axis is likely to be critical, the tank must be rocked about each critical axis for $12\frac{1}{2}$ hours.

(c) Except where satisfactory operating experience with a similar tank in a similar installation is shown, nonmetallic tanks must withstand the test specified in paragraph (b)(5) of this section, with fuel at a temperature of 110 °F. During this test, a representative specimen of the tank must be installed in a supporting structure simulating the installation in the airplane.

(d) For pressurized fuel tanks, it must be shown by analysis or tests that the fuel tanks can withstand the maximum pressure likely to occur on the ground or in flight.

[Doc. No. 5066, 29 FR 18291, Dec. 24, 1964, as amended by Amdt. 25-11, 32 FR 6913, May 5, 1967; Amdt. 25-40, 42 FR 15043, Mar. 17, 1977]

§ 25.967 Fuel tank installations.

(a) Each fuel tank must be supported so that tank loads (resulting from the weight of the fuel in the tanks) are not concentrated on unsupported tank surfaces. In addition—

(1) There must be pads, if necessary, to prevent chafing between the tank and its supports;

(2) Padding must be nonabsorbent or treated to prevent the absorption of fluids;

(3) If a flexible tank liner is used, it must be supported so that it is not required to withstand fluid loads; and

(4) Each interior surface of the tank compartment must be smooth and free of projections that could cause wear of the liner unless—

(i) Provisions are made for protection of the liner at these points; or

(ii) The construction of the liner itself provides that protection.

(b) Spaces adjacent to tank surfaces must be ventilated to avoid fume accumulation due to minor leakage. If the tank is in a sealed compartment, ventilation may be limited to drain holes large enough to prevent excessive pressure resulting from altitude changes.

(c) The location of each tank must meet the requirements of § 25.1185(a).

(d) No engine nacelle skin immediately behind a major air outlet from the engine compartment may act as the wall of an integral tank.

(e) Each fuel tank must be isolated from personnel compartments by a fumeproof and fuelproof enclosure.

§ 25.969 Fuel tank expansion space.

Each fuel tank must have an expansion space of not less than 2 percent of the tank capacity. It must be impossible to fill the expansion space inadvertently with the airplane in the normal ground attitude. For pressure fueling systems, compliance with this section may be shown with the means provided to comply with § 25.979(b).

[Amdt. 25-11, 32 FR 6913, May 5, 1967]

§ 25.971 Fuel tank sump.

(a) Each fuel tank must have a sump with an effective capacity, in the normal ground attitude, of not less than the greater of 0.10 percent of the tank capacity or one-sixteenth of a gallon unless operating limitations are established to ensure that the accumulation of water in service will not exceed the sump capacity.

(b) Each fuel tank must allow drainage of any hazardous quantity of water from any part of the tank to its sump with the airplane in the ground attitude.

(c) Each fuel tank sump must have an accessible drain that—

(1) Allows complete drainage of the sump on the ground;

(2) Discharges clear of each part of the airplane; and

(3) Has manual or automatic means for positive locking in the closed position.

§ 25.973 Fuel tank filler connection.

Each fuel tank filler connection must prevent the entrance of fuel into any part of the airplane other than the tank itself. In addition—

(a) [Reserved]

(b) Each recessed filler connection that can retain any appreciable quantity of fuel must have a drain that discharges clear of each part of the airplane;

(c) Each filler cap must provide a fuel-tight seal; and

(d) Each fuel filling point must have a provision for electrically bonding the airplane to ground fueling equipment.

[Doc. No. 5066, 29 FR 18291, Dec. 24, 1964, as amended by Amdt. 25-40, 42 FR 15043, Mar. 17, 1977; Amdt. 25-72, 55 FR 29785, July 20, 1990; Amdt. 25-115, 69 FR 40527, July 2, 2004]

§ 25.975 Fuel tank vents and carburetor vapor vents.

(a) *Fuel tank vents.* Each fuel tank must be vented from the top part of the expansion space so that venting is effective under any normal flight condition. In addition—

(1) Each vent must be arranged to avoid stoppage by dirt or ice formation;

(2) The vent arrangement must prevent siphoning of fuel during normal operation;

(3) The venting capacity and vent pressure levels must maintain acceptable differences of pressure between the interior and exterior of the tank, during—

(i) Normal flight operation;

(ii) Maximum rate of ascent and descent; and

(iii) Refueling and defueling (where applicable);

(4) Airspaces of tanks with interconnected outlets must be interconnected;

(5) There may be no point in any vent line where moisture can accumulate with the airplane in the ground attitude or the level flight attitude, unless drainage is provided; and

(6) No vent or drainage provision may end at any point—

(i) Where the discharge of fuel from the vent outlet would constitute a fire hazard; or

(ii) From which fumes could enter personnel compartments.

(b) *Carburetor vapor vents.* Each carburetor with vapor elimination connections must have a vent line to lead vapors back to one of the fuel tanks. In addition—

(1) Each vent system must have means to avoid stoppage by ice; and

(2) If there is more than one fuel tank, and it is necessary to use the tanks in a definite sequence, each vapor vent return line must lead back to the fuel tank used for takeoff and landing.

§ 25.977 Fuel tank outlet.

(a) There must be a fuel strainer for the fuel tank outlet or for the booster pump. This strainer must—

(1) For reciprocating engine powered airplanes, have 8 to 16 meshes per inch; and

(2) For turbine engine powered airplanes, prevent the passage of any object that could restrict fuel flow or damage any fuel system component.

(b) [Reserved]

(c) The clear area of each fuel tank outlet strainer must be at least five times the area of the outlet line.

(d) The diameter of each strainer must be at least that of the fuel tank outlet.

(e) Each finger strainer must be accessible for inspection and cleaning.

[Amdt. 25-11, 32 FR 6913, May 5, 1967, as amended by Amdt. 25-36, 39 FR 35460, Oct. 1, 1974]

§ 25.979 Pressure fueling system.

For pressure fueling systems, the following apply:

(a) Each pressure fueling system fuel manifold connection must have means to prevent the escape of hazardous quantities of fuel from the system if the fuel entry valve fails.

(b) An automatic shutoff means must be provided to prevent the quantity of fuel in each tank from exceeding the maximum quantity approved for that tank. This means must—

(1) Allow checking for proper shutoff operation before each fueling of the tank; and

(2) Provide indication at each fueling station of failure of the shutoff means to stop the fuel flow at the maximum quantity approved for that tank.

(c) A means must be provided to prevent damage to the fuel system in the event of failure of the automatic shutoff means prescribed in paragraph (b) of this section.

(d) The airplane pressure fueling system (not including fuel tanks and fuel tank vents) must withstand an ultimate load that is 2.0 times the load arising from the maximum pressures, including surge, that is likely to occur during fueling. The maximum surge pressure must be established with any combination of tank valves being either intentionally or inadvertently closed.

(e) The airplane defueling system (not including fuel tanks and fuel tank vents) must withstand an ultimate load that is 2.0 times the load arising from the maximum permissible defueling pressure (positive or negative) at the airplane fueling connection.

[Amdt. 25-11, 32 FR 6913, May 5, 1967, as amended by Amdt. 25-38, 41 FR 55467, Dec. 20, 1976; Amdt. 25-72, 55 FR 29785, July 20, 1990]

§ 25.981 Fuel tank ignition prevention.

(a) No ignition source may be present at each point in the fuel tank or fuel tank system where catastrophic failure could occur due to ignition of fuel or vapors. This must be shown by:

(1) Determining the highest temperature allowing a safe margin below the lowest expected autoignition temperature of the fuel in the fuel tanks.

(2) Demonstrating that no temperature at each place inside each fuel tank where fuel ignition is possible will exceed the temperature determined under paragraph (a)(1) of this section. This must be verified under all probable operating, failure, and malfunction conditions of each component whose operation, failure, or malfunction could increase the temperature inside the tank.

(3) Demonstrating that an ignition source could not result from each single failure, from each single failure in combination with each latent failure condition not shown to be extremely remote, and from all combinations of failures not shown to be extremely improbable. The effects of manufacturing variability, aging, wear, corrosion, and likely damage must be considered.

(b) Except as provided in paragraphs (b)(2) and (c) of this section, no fuel tank Fleet Average Flammability Exposure on an airplane may exceed three percent of the Flammability Exposure Evaluation Time (FEET) as defined in Appendix N of this part, or that of a fuel tank within the wing of the airplane model being evaluated, whichever is greater. If the wing is not a conventional unheated aluminum wing, the analysis must be based on an assumed Equivalent Conventional Unheated Aluminum Wing Tank.

(1) Fleet Average Flammability Exposure is determined in accordance with Appendix N of this part. The assessment must be done in accordance with the methods and procedures set forth in the Fuel Tank Flammability Assessment Method User's Manual, dated May 2008, document number DOT/FAA/AR-05/8 (incorporated by reference, see § 25.5).

(2) Any fuel tank other than a main fuel tank on an airplane must meet the flammability exposure criteria of Appendix M to this part if any portion of the tank is located within the fuselage contour.

(3) As used in this paragraph,

(i) *Equivalent Conventional Unheated Aluminum Wing Tank* is an integral tank in an unheated semi-monocoque aluminum wing of a subsonic airplane that is equivalent in aerodynamic performance, structural capability, fuel tank capacity and tank configuration to the designed wing.

(ii) *Fleet Average Flammability Exposure* is defined in Appendix N to this part and means the percentage of time each fuel tank ullage is flammable for a fleet of an airplane type operating over the range of flight lengths.

(iii) *Main Fuel Tank* means a fuel tank that feeds fuel directly into one or more engines and holds required fuel reserves continually throughout each flight.

(c) Paragraph (b) of this section does not apply to a fuel tank if means are provided to mitigate the effects of an ignition of fuel vapors within that fuel tank such that no damage caused by an ignition will prevent continued safe flight and landing.

(d) Critical design configuration control limitations (CDCCL), inspections, or other procedures must be established, as necessary, to prevent development of ignition sources within the fuel tank system pursuant to paragraph (a) of this section, to prevent increasing the flammability exposure of the tanks above that permitted under paragraph (b) of this section, and to prevent degradation of the performance and reliability of any means provided according to paragraphs (a) or (c) of this section. These CDCCL, inspections, and procedures must be included in the Airworthiness Limitations section of the instructions for continued airworthiness required by § 25.1529. Visible means of identifying critical features of the design must be placed in areas of the airplane where foreseeable maintenance actions, repairs, or alterations may compromise the critical design configuration control limitations (e.g., color-coding of wire to identify separation limitation). These visible means must also be identified as CDCCL.

[Doc. No. 1999-6411, 66 FR 23129, May 7, 2001, as amended at Doc. No. FAA-2005-22997, 73 FR 42494, July 21, 2008]

Fuel System Components

§ 25.991 Fuel pumps.

(a) *Main pumps.* Each fuel pump required for proper engine operation, or required to meet the fuel system requirements of this subpart (other than those in paragraph (b) of this section, is a main pump. For each main pump, provision must be made to allow the bypass of each positive displacement fuel pump other than a fuel injection pump (a pump that supplies the proper flow and pressure for fuel injection when the injection is not accomplished in a carburetor) approved as part of the engine.

(b) *Emergency pumps.* There must be emergency pumps or another main pump to feed each engine immediately after failure of any main pump (other than a fuel injection pump approved as part of the engine).

§ 25.993 Fuel system lines and fittings.

(a) Each fuel line must be installed and supported to prevent excessive vibration and to withstand loads due to fuel pressure and accelerated flight conditions.

(b) Each fuel line connected to components of the airplane between which relative motion could exist must have provisions for flexibility.

(c) Each flexible connection in fuel lines that may be under pressure and subjected to axial loading must use flexible hose assemblies.

(d) Flexible hose must be approved or must be shown to be suitable for the particular application.

(e) No flexible hose that might be adversely affected by exposure to high temperatures may be used where excessive temperatures will exist during operation or after engine shut-down.

(f) Each fuel line within the fuselage must be designed and installed to allow a reasonable degree of deformation and stretching without leakage.

[Doc. No. 5066, 29 FR 18291, Dec. 24, 1964, as amended by Amdt. 25-15, 32 FR 13266, Sept. 20, 1967]

§ 25.994 Fuel system components.

Fuel system components in an engine nacelle or in the fuselage must be protected from damage which could result in spillage of enough fuel to constitute a fire hazard as a result of a wheels-up landing on a paved runway.

[Amdt. 25-57, 49 FR 6848, Feb. 23, 1984]

§ 25.995 Fuel valves.

In addition to the requirements of § 25.1189 for shutoff means, each fuel valve must—

(a) [Reserved]

(b) Be supported so that no loads resulting from their operation or from accelerated flight conditions are transmitted to the lines attached to the valve.

[Doc. No. 5066, 29 FR 18291, Dec. 24, 1964, as amended by Amdt. 25-40, 42 FR 15043, Mar. 17, 1977]

§ 25.997 Fuel strainer or filter.

There must be a fuel strainer or filter between the fuel tank outlet and the inlet of either the fuel metering device or an engine driven positive displacement pump, whichever is nearer the fuel tank outlet. This fuel strainer or filter must—

(a) Be accessible for draining and cleaning and must incorporate a screen or element which is easily removable;

(b) Have a sediment trap and drain except that it need not have a drain if the strainer or filter is easily removable for drain purposes;

(c) Be mounted so that its weight is not supported by the connecting lines or by the inlet or outlet connections of the strainer or filter itself, unless adequate strength margins under all loading conditions are provided in the lines and connections; and

(d) Have the capacity (with respect to operating limitations established for the engine) to ensure that engine fuel system functioning is not impaired, with the fuel contaminated to a degree (with respect to particle size and density) that is greater than that established for the engine in Part 33 of this chapter.

[Amdt. 25-36, 39 FR 35460, Oct. 1, 1974, as amended by Amdt. 25-57, 49 FR 6848, Feb. 23, 1984]

§ 25.999 Fuel system drains.

(a) Drainage of the fuel system must be accomplished by the use of fuel strainer and fuel tank sump drains.

(b) Each drain required by paragraph (a) of this section must—

(1) Discharge clear of all parts of the airplane;

(2) Have manual or automatic means for positive locking in the closed position; and

(3) Have a drain valve—

(i) That is readily accessible and which can be easily opened and closed; and

(ii) That is either located or protected to prevent fuel spillage in the event of a landing with landing gear retracted.

[Doc. No. 5066, 29 FR 18291, Dec. 24, 1964, as amended by Amdt. 25-38, 41 FR 55467, Dec. 20, 1976]

§ 25.1001 Fuel jettisoning system.

(a) A fuel jettisoning system must be installed on each airplane unless it is shown that the airplane meets the climb requirements of §§ 25.119 and 25.121(d) at maximum takeoff weight, less the actual or computed weight of fuel necessary for a 15-minute flight comprised of a takeoff, go-around, and landing at the airport of departure with the airplane configuration, speed, power, and thrust the same as that used in meeting the applicable takeoff, approach, and landing climb performance requirements of this part.

(b) If a fuel jettisoning system is required it must be capable of jettisoning enough fuel within 15 minutes, starting with the weight given in paragraph (a) of this section, to enable the airplane to meet the climb requirements of §§ 25.119 and 25.121(d), assuming that the fuel is jettisoned under the conditions, except weight, found least favorable during the flight tests prescribed in paragraph (c) of this section.

(c) Fuel jettisoning must be demonstrated beginning at maximum takeoff weight with flaps and landing gear up and in—

(1) A power-off glide at 1.3 V_{SR1};

(2) A climb at the one-engine inoperative best rate-of-climb speed, with the critical engine inoperative and the remaining engines at maximum continuous power; and

(3) Level flight at 1.3 V_{SR1}; if the results of the tests in the conditions specified in paragraphs (c)(1) and (2) of this section show that this condition could be critical.

(d) During the flight tests prescribed in paragraph (c) of this section, it must be shown that—

(1) The fuel jettisoning system and its operation are free from fire hazard;

(2) The fuel discharges clear of any part of the airplane;

(3) Fuel or fumes do not enter any parts of the airplane; and

(4) The jettisoning operation does not adversely affect the controllability of the airplane.

(e) For reciprocating engine powered airplanes, means must be provided to prevent jettisoning the fuel in the tanks used for takeoff and landing below the level allowing 45 minutes flight at 75 percent maximum continuous power. However, if there is an auxiliary control independent of the main jettisoning control, the system may be designed to jettison the remaining fuel by means of the auxiliary jettisoning control.

(f) For turbine engine powered airplanes, means must be provided to prevent jettisoning the fuel in the tanks used for takeoff and landing below the level allowing climb from sea level to 10,000 feet and thereafter allowing 45 minutes cruise at a speed for maximum range. However, if there is an auxiliary control independent of the main jettisoning control, the system may be designed to jettison the remaining fuel by means of the auxiliary jettisoning control.

(g) The fuel jettisoning valve must be designed to allow flight personnel to close the valve during any part of the jettisoning operation.

(h) Unless it is shown that using any means (including flaps, slots, and slats) for changing the airflow across or around the wings does not adversely affect fuel jettisoning, there must be a placard, adjacent to the jettisoning control, to warn flight crewmembers against jettisoning fuel while the means that change the airflow are being used.

(i) The fuel jettisoning system must be designed so that any reasonably probable single malfunction in the system will not result in a hazardous condition due to unsymmetrical jettisoning of, or inability to jettison, fuel.

[Doc. No. 5066, 29 FR 18291, Dec. 24, 1964, as amended by Amdt. 25-18, 33 FR 12226, Aug. 30, 1968; Amdt. 25-57, 49 FR 6848, Feb. 23, 1984; Amdt. 25-108, 67 FR 70827, Nov. 26, 2002]

Oil System

§ 25.1011 General.

(a) Each engine must have an independent oil system that can supply it with an appropriate quantity of oil at a temperature not above that safe for continuous operation.

(b) The usable oil capacity may not be less than the product of the endurance of the airplane under critical operating conditions and the approved maximum allowable oil consumption of the engine under the same conditions, plus a suitable margin to ensure system circulation. Instead of a rational analysis of airplane range for the purpose of computing oil requirements for reciprocating engine powered airplanes, the following fuel/oil ratios may be used:

(1) For airplanes without a reserve oil or oil transfer system, a fuel/oil ratio of 30:1 by volume.

(2) For airplanes with either a reserve oil or oil transfer system, a fuel/oil ratio of 40:1 by volume.

(c) Fuel/oil ratios higher than those prescribed in paragraphs (b)(1) and (2) of this section may be used if substantiated by data on actual engine oil consumption.

§ 25.1013 Oil tanks.

(a) *Installation.* Each oil tank installation must meet the requirements of § 25.967.

(b) *Expansion space.* Oil tank expansion space must be provided as follows:

(1) Each oil tank used with a reciprocating engine must have an expansion space of not less than the greater of 10 percent of the tank capacity or 0.5 gallon, and each oil tank used with a turbine engine must have an expansion space of not less than 10 percent of the tank capacity.

(2) Each reserve oil tank not directly connected to any engine may have an expansion space of not less than two percent of the tank capacity.

(3) It must be impossible to fill the expansion space inadvertently with the airplane in the normal ground attitude.

(c) *Filler connection.* Each recessed oil tank filler connection that can retain any appreciable quantity of oil must have a drain that discharges clear of each part of the airplane. In addition, each oil tank filler cap must provide an oil-tight seal.

(d) *Vent.* Oil tanks must be vented as follows:

(1) Each oil tank must be vented from the top part of the expansion space so that venting is effective under any normal flight condition.

(2) Oil tank vents must be arranged so that condensed water vapor that might freeze and obstruct the line cannot accumulate at any point.

(e) *Outlet.* There must be means to prevent entrance into the tank itself, or into the tank outlet, of any object that might obstruct the flow of oil through the system. No oil tank outlet may be enclosed by any screen or guard that would reduce the flow of oil below a safe value at any operating temperature. There must be a shutoff valve at the outlet of each oil tank used with a turbine engine, unless the external portion of the oil system (including the oil tank supports) is fireproof.

(f) *Flexible oil tank liners.* Each flexible oil tank liner must be approved or must be shown to be suitable for the particular application.

[Doc. No. 5066, 29 FR 18291, Dec. 24, as amended by Amdt. 25-19, 33 FR 15410, Oct. 17, 1968; Amdt. 25-23, 35 FR 5677, Apr. 8, 1970; Amdt. 25-36, 39 FR 35460, Oct. 1, 1974; Amdt. 25-57, 49 FR 6848, Feb. 23, 1984; Amdt. 25-72, 55 FR 29785, July 20, 1990]

§ 25.1015 Oil tank tests.

Each oil tank must be designed and installed so that—

(a) It can withstand, without failure, each vibration, inertia, and fluid load that it may be subjected to in operation; and

(b) It meets the provisions of § 25.965, except—

(1) The test pressure—

(i) For pressurized tanks used with a turbine engine, may not be less than 5 p.s.i. plus the maximum operating pressure of the tank instead of the pressure specified in § 25.965(a); and

(ii) For all other tanks may not be less than 5 p.s.i. instead of the pressure specified in § 25.965(a); and

(2) The test fluid must be oil at 250 °F. instead of the fluid specified in § 25.965(c).

[Doc. No. 5066, 29 FR 18291, Dec. 24, 1964, as amended by Amdt. 25-36, 39 FR 35461, Oct. 1, 1974]

§ 25.1017 Oil lines and fittings.

(a) Each oil line must meet the requirements of § 25.993 and each oil line and fitting in any designated fire zone must meet the requirements of § 25.1183.

(b) Breather lines must be arranged so that—

(1) Condensed water vapor that might freeze and obstruct the line cannot accumulate at any point;

(2) The breather discharge does not constitute a fire hazard if foaming occurs or causes emitted oil to strike the pilot's windshield; and

(3) The breather does not discharge into the engine air induction system.

§ 25.1019 Oil strainer or filter.

(a) Each turbine engine installation must incorporate an oil strainer or filter through which all of the engine oil flows and which meets the following requirements:

(1) Each oil strainer or filter that has a bypass must be constructed and installed so that oil will flow at the normal rate through the rest of the system with the strainer or filter completely blocked.

(2) The oil strainer or filter must have the capacity (with respect to operating limitations established for the engine) to ensure that engine oil system functioning is not impaired when the oil is contaminated to a degree (with respect to particle size and density) that is greater than that established for the engine under Part 33 of this chapter.

(3) The oil strainer or filter, unless it is installed at an oil tank outlet, must incorporate an indicator that will indicate contamination before it reaches the capacity established in accordance with paragraph (a)(2) of this section.

(4) The bypass of a strainer or filter must be constructed and installed so that the release of collected contaminants is minimized by appropriate location of the bypass to ensure that collected contaminants are not in the bypass flow path.

(5) An oil strainer or filter that has no bypass, except one that is installed at an oil tank outlet, must have a means to connect it to the warning system required in § 25.1305(c)(7).

(b) Each oil strainer or filter in a powerplant installation using reciprocating engines must be constructed and installed so that oil will flow at the normal rate through the rest of the system with the strainer or filter element completely blocked.

[Amdt. 25-36, 39 FR 35461, Oct. 1, 1974, as amended by Amdt. 25-57, 49 FR 6848, Feb. 23, 1984]

§ 25.1021 Oil system drains.

A drain (or drains) must be provided to allow safe drainage of the oil system. Each drain must—

(a) Be accessible; and

(b) Have manual or automatic means for positive locking in the closed position.

[Amdt. 25-57, 49 FR 6848, Feb. 23, 1984]

§ 25.1023 Oil radiators.

(a) Each oil radiator must be able to withstand, without failure, any vibration, inertia, and oil pressure load to which it would be subjected in operation.

(b) Each oil radiator air duct must be located so that, in case of fire, flames coming from normal openings of the engine nacelle cannot impinge directly upon the radiator.

§ 25.1025 Oil valves.

(a) Each oil shutoff must meet the requirements of § 25.1189.

(b) The closing of oil shutoff means may not prevent propeller feathering.

(c) Each oil valve must have positive stops or suitable index provisions in the "on" and "off" positions and must be supported so that no loads resulting from its operation or from accelerated flight conditions are transmitted to the lines attached to the valve.

§ 25.1027 Propeller feathering system.

(a) If the propeller feathering system depends on engine oil, there must be means to trap an amount of oil in the tank if the supply becomes depleted due to failure of any part of the lubricating system other than the tank itself.

(b) The amount of trapped oil must be enough to accomplish the feathering operation and must be available only to the feathering pump.

(c) The ability of the system to accomplish feathering with the trapped oil must be shown. This may be done on the ground using an auxiliary source of oil for lubricating the engine during operation.

(d) Provision must be made to prevent sludge or other foreign matter from affecting the safe operation of the propeller feathering system.

[Doc. No. 5066, 29 FR 18291, Dec. 24, 1964, as amended by Amdt. 25-38, 41 FR 55467, Dec. 20, 1976]

Cooling

§ 25.1041 General.

The powerplant and auxiliary power unit cooling provisions must be able to maintain the temperatures of powerplant components, engine fluids, and auxiliary power unit components and fluids within the temperature limits established for these components and fluids, under ground, water, and flight operating conditions, and after normal engine or auxiliary power unit shutdown, or both.

[Amdt. 25-38, 41 FR 55467, Dec. 20, 1976]

§ 25.1043 Cooling tests.

(a) *General.* Compliance with § 25.1041 must be shown by tests, under critical ground, water, and flight operating conditions. For these tests, the following apply:

(1) If the tests are conducted under conditions deviating from the maximum ambient atmospheric temperature, the recorded powerplant temperatures must be corrected under paragraphs (c) and (d) of this section.

(2) No corrected temperatures determined under paragraph (a)(1) of this section may exceed established limits.

(3) For reciprocating engines, the fuel used during the cooling tests must be the minimum grade approved for the engines, and the mixture settings must be those normally used in the flight stages for which the cooling tests are conducted. The test procedures must be as prescribed in § 25.1045.

(b) *Maximum ambient atmospheric temperature.* A maximum ambient atmospheric temperature corresponding to sea level conditions of at least 100 degrees F must be established. The assumed temperature lapse rate is 3.6 degrees F per thousand feet of altitude above sea level until a temperature of −69.7 degrees F is reached, above which altitude the temperature is considered constant at −69.7 degrees F. However, for winterization installations, the applicant may select a maximum ambient atmospheric temperature corresponding to sea level conditions of less than 100 degrees F.

(c) *Correction factor (except cylinder barrels).* Unless a more rational correction applies, temperatures of engine fluids and powerplant components (except cylinder barrels) for which temperature limits are established, must be corrected by adding to them the difference between the maximum ambient atmospheric temperature and the temperature of the ambient air at the time of the first occurrence of the maximum component or fluid temperature recorded during the cooling test.

(d) *Correction factor for cylinder barrel temperatures.* Unless a more rational correction applies, cylinder barrel temperatures must be corrected by adding to them 0.7 times the difference between the maximum ambient atmospheric temperature and the temperature of the ambient air at the time of the first occurrence of the maximum cylinder barrel temperature recorded during the cooling test.

[Doc. No. 5066, 29 FR 18291, Dec. 24, 1964, as amended by Amdt. 25-42, 43 FR 2323, Jan. 16, 1978]

§ 25.1045 Cooling test procedures.

(a) Compliance with § 25.1041 must be shown for the takeoff, climb, en route, and landing stages of flight that correspond to the applicable performance requirements. The cooling tests must be conducted with the airplane in the configuration, and operating under the conditions, that are critical relative to cooling during each stage of flight. For the cooling tests, a temperature is "stabilized" when its rate of change is less than two degrees F. per minute.

(b) Temperatures must be stabilized under the conditions from which entry is made into each stage of flight being investigated, unless the entry condition normally is not one during which component and the engine fluid temperatures would stabilize (in which case, operation through the full entry condition must be conducted before entry into the stage of flight being investigated in order to allow temperatures to reach their natural levels at the time of entry). The takeoff cooling test must be preceded by a period during which the powerplant component and engine fluid temperatures are stabilized with the engines at ground idle.

(c) Cooling tests for each stage of flight must be continued until—

(1) The component and engine fluid temperatures stabilize;

(2) The stage of flight is completed; or

(3) An operating limitation is reached.

(d) For reciprocating engine powered airplanes, it may be assumed, for cooling test purposes, that the takeoff stage of flight is complete when the airplane reaches an altitude of 1,500 feet above the takeoff surface or reaches a point in the takeoff where the transition from the takeoff to the en route configuration is completed and a speed is reached at which compliance with § 25.121(c) is shown, whichever point is at a higher altitude. The airplane must be in the following configuration:

(1) Landing gear retracted.

(2) Wing flaps in the most favorable position.

(3) Cowl flaps (or other means of controlling the engine cooling supply) in the position that provides adequate cooling in the hot-day condition.

(4) Critical engine inoperative and its propeller stopped.

(5) Remaining engines at the maximum continuous power available for the altitude.

(e) For hull seaplanes and amphibians, cooling must be shown during taxiing downwind for 10 minutes, at five knots above step speed.

[Doc. No. 5066, 29 FR 18291, Dec. 24, 1964, as amended by Amdt. 25-57, 49 FR 6848, Feb. 23, 1984]

Induction System

§ 25.1091 Air induction.

(a) The air induction system for each engine and auxiliary power unit must supply—

(1) The air required by that engine and auxiliary power unit under each operating condition for which certification is requested; and

(2) The air for proper fuel metering and mixture distribution with the induction system valves in any position.

(b) Each reciprocating engine must have an alternate air source that prevents the entry of rain, ice, or any other foreign matter.

(c) Air intakes may not open within the cowling, unless—

(1) That part of the cowling is isolated from the engine accessory section by means of a fireproof diaphragm; or

(2) For reciprocating engines, there are means to prevent the emergence of backfire flames.

(d) For turbine engine powered airplanes and airplanes incorporating auxiliary power units—

(1) There must be means to prevent hazardous quantities of fuel leakage or overflow from drains, vents, or other components of flammable fluid systems from entering the engine or auxiliary power unit intake system; and

(2) The airplane must be designed to prevent water or slush on the runway, taxiway, or other airport operating surfaces from being directed into the engine or auxiliary power unit air inlet ducts in hazardous quantities, and the air inlet ducts must be located or protected so as to minimize the ingestion of foreign matter during takeoff, landing, and taxiing.

(e) If the engine induction system contains parts or components that could be damaged by foreign objects entering the air inlet, it must be shown by tests or, if appropriate, by analysis that the induction system design can withstand the foreign object ingestion test conditions of §§ 33.76, 33.77 and 33.78(a)(1) of this chapter without failure of parts or components that could create a hazard.

[Doc. No. 5066, 29 FR 18291, Dec. 24, 1964, as amended by Amdt. 25-38, 41 FR 55467, Dec. 20, 1976; Amdt. 25-40, 42 FR 15043, Mar. 17, 1977; Amdt. 25-57, 49 FR 6849, Feb. 23, 1984; Amdt. 25-100, 65 FR 55854, Sept. 14, 2000]

§ 25.1093 Induction system icing protection.

(a) *Reciprocating engines.* Each reciprocating engine air induction system must have means to prevent and eliminate icing. Unless this is done by other means, it must be shown that, in air free of visible moisture at a temperature of 30 F., each airplane with altitude engines using—

(1) Conventional venturi carburetors have a preheater that can provide a heat rise of 120 F. with the engine at 60 percent of maximum continuous power; or

(2) Carburetors tending to reduce the probability of ice formation has a preheater that can provide a heat rise of 100°F. with the engine at 60 percent of maximum continuous power.

(b) *Turbine engines.* (1) Each turbine engine must operate throughout the flight power range of the engine (including idling), without the accumulation of ice on the engine, inlet system components, or airframe components that would adversely affect engine operation or cause a serious loss of power or thrust—

(i) Under the icing conditions specified in appendix C, and

(ii) In falling and blowing snow within the limitations established for the airplane for such operation.

(2) Each turbine engine must idle for 30 minutes on the ground, with the air bleed available for engine icing protection at its critical condition, without adverse effect, in an atmosphere that is at a temperature between 15° and 30°F (between −9° and −1°C) and has a liquid water content not less than 0.3 grams per cubic meter in the form of drops having a mean effective diameter not less than 20 microns, followed by momentary operation at takeoff power or thrust. During the 30 minutes of idle operation, the engine may be run up periodically to a moderate power or thrust setting in a manner acceptable to the Administrator.

(c) *Supercharged reciprocating engines.* For each engine having a supercharger to pressurize the air before it enters the carburetor, the heat rise in the air caused by that supercharging at any altitude may be utilized in determining compliance with paragraph (a) of this section if the heat rise utilized is that which will be available, automatically, for the applicable altitude and operating condition because of supercharging.

[Doc. No. 5066, 29 FR 18291, Dec. 24, 1964, as amended by Amdt. 25-38, 41 FR 55467, Dec. 20, 1976; Amdt. 25-40, 42 FR 15043, Mar. 17, 1977; Amdt. 25-57, 49 FR 6849, Feb. 23, 1984; Amdt. 25-72, 55 FR 29785, July 20, 1990]

§ 25.1101 Carburetor air preheater design.

Each carburetor air preheater must be designed and constructed to—

(a) Ensure ventilation of the preheater when the engine is operated in cold air;

(b) Allow inspection of the exhaust manifold parts that it surrounds; and

(c) Allow inspection of critical parts of the preheater itself.

§ 25.1103 Induction system ducts and air duct systems.

(a) Each induction system duct upstream of the first stage of the engine supercharger and of the auxiliary power unit compressor must have a drain to prevent the hazardous accumulation of fuel and moisture in the ground attitude. No drain may discharge where it might cause a fire hazard.

(b) Each induction system duct must be—

(1) Strong enough to prevent induction system failures resulting from normal backfire conditions; and

(2) Fire-resistant if it is in any fire zone for which a fire-extinguishing system is required, except that ducts for auxiliary power units must be fireproof within the auxiliary power unit fire zone.

(c) Each duct connected to components between which relative motion could exist must have means for flexibility.

(d) For turbine engine and auxiliary power unit bleed air duct systems, no hazard may result if a duct failure occurs at any point between the air duct source and the airplane unit served by the air.

(e) Each auxiliary power unit induction system duct must be fireproof for a sufficient distance upstream of the auxiliary power unit compartment to prevent hot gas reverse flow from burning through auxiliary power unit ducts and entering any other compartment or area of the airplane in which a hazard would be created resulting from the entry of hot gases. The materials used to form the remainder of the induction system duct and plenum chamber of the auxiliary power unit must be capable of resisting the maximum heat conditions likely to occur.

(f) Each auxiliary power unit induction system duct must be constructed of materials that will not absorb or trap hazardous quantities of flammable fluids that could be ignited in the event of a surge or reverse flow condition.

[Doc. No. 5066, 29 FR 18291, Dec. 24, 1964, as amended by Amdt. 25-46, 43 FR 50597, Oct. 30, 1978]

§ 25.1105 Induction system screens.

If induction system screens are used—

(a) Each screen must be upstream of the carburetor;

(b) No screen may be in any part of the induction system that is the only passage through which air can reach the engine, unless it can be deiced by heated air;

(c) No screen may be deiced by alcohol alone; and

(d) It must be impossible for fuel to strike any screen.

§ 25.1107 Inter-coolers and after-coolers.

Each inter-cooler and after-cooler must be able to withstand any vibration, inertia, and air pressure load to which it would be subjected in operation.

Exhaust System

§ 25.1121 General.

For powerplant and auxiliary power unit installations the following apply:

(a) Each exhaust system must ensure safe disposal of exhaust gases without fire hazard or carbon monoxide contamination in any personnel compartment. For test purposes, any acceptable carbon monoxide detection method may be used to show the absence of carbon monoxide.

(b) Each exhaust system part with a surface hot enough to ignite flammable fluids or vapors must be located or shielded so that leakage from any system carrying flammable fluids or vapors will not result in a fire caused by impingement of the fluids or vapors on any part of the exhaust system including shields for the exhaust system.

(c) Each component that hot exhaust gases could strike, or that could be subjected to high temperatures from exhaust system parts, must be fireproof. All exhaust system components must be separated by fireproof shields from adjacent parts of the airplane that are outside the engine and auxiliary power unit compartments.

(d) No exhaust gases may discharge so as to cause a fire hazard with respect to any flammable fluid vent or drain.

(e) No exhaust gases may discharge where they will cause a glare seriously affecting pilot vision at night.

(f) Each exhaust system component must be ventilated to prevent points of excessively high temperature.

(g) Each exhaust shroud must be ventilated or insulated to avoid, during normal operation, a temperature high enough to ignite any flammable fluids or vapors external to the shroud.

[Doc. No. 5066, 29 FR 18291, Dec. 24, 1964, as amended by Amdt. 25-40, 42 FR 15043, Mar. 17, 1977]

§ 25.1123 Exhaust piping.

For powerplant and auxiliary power unit installations, the following apply:

(a) Exhaust piping must be heat and corrosion resistant, and must have provisions to prevent failure due to expansion by operating temperatures.

(b) Piping must be supported to withstand any vibration and inertia loads to which it would be subjected in operation; and

(c) Piping connected to components between which relative motion could exist must have means for flexibility.

[Doc. No. 5066, 29 FR 18291, Dec. 24, 1964, as amended by Amdt. 25-40, 42 FR 15044, Mar. 17, 1977]

§ 25.1125 Exhaust heat exchangers.

For reciprocating engine powered airplanes, the following apply:

(a) Each exhaust heat exchanger must be constructed and installed to withstand each vibration, inertia, and other load to which it would be subjected in operation. In addition—

(1) Each exchanger must be suitable for continued operation at high temperatures and resistant to corrosion from exhaust gases;

(2) There must be means for the inspection of the critical parts of each exchanger;

(3) Each exchanger must have cooling provisions wherever it is subject to contact with exhaust gases; and

(4) No exhaust heat exchanger or muff may have any stagnant areas or liquid traps that would increase the probability of ignition of flammable fluids or vapors that might be present in case of the failure or malfunction of components carrying flammable fluids.

(b) If an exhaust heat exchanger is used for heating ventilating air—

(1) There must be a secondary heat exchanger between the primary exhaust gas heat exchanger and the ventilating air system; or

(2) Other means must be used to preclude the harmful contamination of the ventilating air.

[Doc. No. 5066, 29 FR 18291, Dec. 24, 1964, as amended by Amdt. 25-38, 41 FR 55467, Dec. 20, 1976]

§ 25.1127 Exhaust driven turbo-superchargers.

(a) Each exhaust driven turbo-supercharger must be approved or shown to be suitable for the particular application. It must be installed and supported to ensure safe operation between normal inspections and overhauls. In addition, there must be provisions for expansion and flexibility between exhaust conduits and the turbine.

(b) There must be provisions for lubricating the turbine and for cooling turbine parts where temperatures are critical.

(c) If the normal turbo-supercharger control system malfunctions, the turbine speed may not exceed its maximum allowable value. Except for the waste gate operating components, the components provided for meeting this requirement must be independent of the normal turbo-supercharger controls.

Powerplant Controls and Accessories

§ 25.1141 Powerplant controls: general.

Each powerplant control must be located, arranged, and designed under §§ 25.777 through 25.781 and marked under § 25.1555. In addition, it must meet the following requirements:

(a) Each control must be located so that it cannot be inadvertently operated by persons entering, leaving, or moving normally in, the cockpit.

(b) Each flexible control must be approved or must be shown to be suitable for the particular application.

(c) Each control must have sufficient strength and rigidity to withstand operating loads without failure and without excessive deflection.

(d) Each control must be able to maintain any set position without constant attention by flight crewmembers and without creep due to control loads or vibration.

(e) The portion of each powerplant control located in a designated fire zone that is required to be operated in the event of fire must be at least fire resistant.

(f) For powerplant valve controls located in the flight deck there must be a means:

(1) For the flightcrew to select each intended position or function of the valve; and

(2) To indicate to the flightcrew:

(i) The selected position or function of the valve; and

(ii) When the valve has not responded as intended to the selected position or function.

[Doc. No. 5066, 29 FR 18291, Dec. 24, 1964, as amended by Amdt. 25-40, 42 FR 15044, Mar. 17, 1977; Amdt. 25-72, 55 FR 29785, July 20, 1990; Amdt. 25-115, 69 FR 40527, July 2, 2004]

§ 25.1142 Auxiliary power unit controls.

Means must be provided on the flight deck for starting, stopping, and emergency shutdown of each installed auxiliary power unit.

[Amdt. 25-46, 43 FR 50598, Oct. 30, 1978]

§ 25.1143 Engine controls.

(a) There must be a separate power or thrust control for each engine.

(b) Power and thrust controls must be arranged to allow—

(1) Separate control of each engine; and

(2) Simultaneous control of all engines.

(c) Each power and thrust control must provide a positive and immediately responsive means of controlling its engine.

(d) For each fluid injection (other than fuel) system and its controls not provided and approved as part of the engine, the applicant must show that the flow of the injection fluid is adequately controlled.

(e) If a power or thrust control incorporates a fuel shutoff feature, the control must have a means to prevent the inadvertent movement of the control into the shutoff position. The means must—

(1) Have a positive lock or stop at the idle position; and

(2) Require a separate and distinct operation to place the control in the shutoff position.

[Amdt. 25-23, 35 FR 5677, Apr. 8, 1970, as amended by Amdt. 25-38, 41 FR 55467, Dec. 20, 1976; Amdt. 25-57, 49 FR 6849, Feb. 23, 1984]

§ 25.1145 Ignition switches.

(a) Ignition switches must control each engine ignition circuit on each engine.

(b) There must be means to quickly shut off all ignition by the grouping of switches or by a master ignition control.

(c) Each group of ignition switches, except ignition switches for turbine engines for which continuous ignition is not required, and each master ignition control must have a means to prevent its inadvertent operation.

[Doc. No. 5066, 29 FR 18291, Dec. 24, 1964, as amended by Amdt. 25-40, 42 FR 15044 Mar. 17, 1977]

§ 25.1147 Mixture controls.

(a) If there are mixture controls, each engine must have a separate control. The controls must be grouped and arranged to allow—

(1) Separate control of each engine; and

(2) Simultaneous control of all engines.

(b) Each intermediate position of the mixture controls that corresponds to a normal operating setting must be identifiable by feel and sight.

(c) The mixture controls must be accessible to both pilots. However, if there is a separate flight engineer station with a control panel, the controls need be accessible only to the flight engineer.

§ 25.1149 Propeller speed and pitch controls.

(a) There must be a separate propeller speed and pitch control for each propeller.

(b) The controls must be grouped and arranged to allow—

(1) Separate control of each propeller; and

(2) Simultaneous control of all propellers.

(c) The controls must allow synchronization of all propellers.

(d) The propeller speed and pitch controls must be to the right of, and at least one inch below, the pilot's throttle controls.

§ 25.1153 Propeller feathering controls.

(a) There must be a separate propeller feathering control for each propeller. The control must have means to prevent its inadvertent operation.

(b) If feathering is accomplished by movement of the propeller pitch or speed control lever, there must be means to prevent the inadvertent movement of this lever to the feathering position during normal operation.

[Doc. No. 5066, 29 FR 18291, Dec. 24, 1964, as amended by Amdt. 25-11, 32 FR 6913, May 5, 1967]

§ 25.1155 Reverse thrust and propeller pitch settings below the flight regime.

Each control for reverse thrust and for propeller pitch settings below the flight regime must have means to prevent its inadvertent operation. The means must have a positive lock or stop at the flight idle position and must require a separate and distinct operation by the crew to displace the control from the flight regime (forward thrust regime for turbojet powered airplanes).

[Amdt. 25-11, 32 FR 6913, May 5, 1967]

§ 25.1157 Carburetor air temperature controls.

There must be a separate carburetor air temperature control for each engine.

§ 25.1159 Supercharger controls.

Each supercharger control must be accessible to the pilots or, if there is a separate flight engineer station with a control panel, to the flight engineer.

§ 25.1161 Fuel jettisoning system controls.

Each fuel jettisoning system control must have guards to prevent inadvertent operation. No control may be near any fire extinguisher control or other control used to combat fire.

§ 25.1163 Powerplant accessories.

(a) Each engine mounted accessory must—

(1) Be approved for mounting on the engine involved;

(2) Use the provisions on the engine for mounting; and

(3) Be sealed to prevent contamination of the engine oil system and the accessory system.

(b) Electrical equipment subject to arcing or sparking must be installed to minimize the probability of contact with any flammable fluids or vapors that might be present in a free state.

(c) If continued rotation of an engine-driven cabin supercharger or of any remote accessory driven by the engine is hazardous if malfunctioning occurs, there must be means to prevent rotation without interfering with the continued operation of the engine.

[Doc. No. 5066, 29 FR 18291, Dec. 24, 1964, as amended by Amdt. 25-57, 49 FR 6849, Feb. 23, 1984]

§ 25.1165 Engine ignition systems.

(a) Each battery ignition system must be supplemented by a generator that is automatically available as an alternate source of electrical energy to allow continued engine operation if any battery becomes depleted.

(b) The capacity of batteries and generators must be large enough to meet the simultaneous demands of the engine ignition system and the greatest demands of any electrical system components that draw electrical energy from the same source.

(c) The design of the engine ignition system must account for—

(1) The condition of an inoperative generator;

(2) The condition of a completely depleted battery with the generator running at its normal operating speed; and

(3) The condition of a completely depleted battery with the generator operating at idling speed, if there is only one battery.

(d) Magneto ground wiring (for separate ignition circuits) that lies on the engine side of the fire wall, must be installed, located, or protected, to minimize the probability of simultaneous failure of two or more wires as a result of mechanical damage, electrical faults, or other cause.

(e) No ground wire for any engine may be routed through a fire zone of another engine unless each part of that wire within that zone is fireproof.

(f) Each ignition system must be independent of any electrical circuit, not used for assisting, controlling, or analyzing the operation of that system.

(g) There must be means to warn appropriate flight crewmembers if the malfunctioning of any part of the electrical system is causing the continuous discharge of any battery necessary for engine ignition.

(h) Each engine ignition system of a turbine powered airplane must be considered an essential electrical load.

[Doc. No. 5066, 29 FR 18291, Dec. 24, 1964, as amended by Amdt. 25-23, 35 FR 5677, Apr. 8, 1970; Amdt. 25-72, 55 FR 29785, July 20, 1990]

§ 25.1167 Accessory gearboxes.

For airplanes equipped with an accessory gearbox that is not certificated as part of an engine—

(a) The engine with gearbox and connecting transmissions and shafts attached must be subjected to the tests specified in § 33.49 or § 33.87 of this chapter, as applicable;

(b) The accessory gearbox must meet the requirements of §§ 33.25 and 33.53 or 33.91 of this chapter, as applicable; and

(c) Possible misalignments and torsional loadings of the gearbox, transmission, and shaft system, expected to result under normal operating conditions must be evaluated.

[Amdt. 25-38, 41 FR 55467, Dec. 20, 1976]

Powerplant Fire Protection

§ 25.1181 Designated fire zones; regions included.

(a) Designated fire zones are—

(1) The engine power section;

(2) The engine accessory section;

(3) Except for reciprocating engines, any complete powerplant compartment in which no isolation is provided between the engine power section and the engine accessory section;

(4) Any auxiliary power unit compartment;

(5) Any fuel-burning heater and other combustion equipment installation described in § 25.859;

(6) The compressor and accessory sections of turbine engines; and

(7) Combustor, turbine, and tailpipe sections of turbine engine installations that contain lines or components carrying flammable fluids or gases.

(b) Each designated fire zone must meet the requirements of §§ 25.863, 25.865, 25.867, 25.869, and 25.1185 through 25.1203.

[Doc. No. 5066, 29 FR 18291, Dec. 24, 1964, as amended by Amdt. 25-11, 32 FR 6913, May 5, 1967; Amdt. 25-23, 35 FR 5677, Apr. 8, 1970; Amdt. 25-72, 55 FR 29785, July 20, 1990; Amdt. 25-115, 69 FR 40527, July 2, 2004]

§ 25.1182 Nacelle areas behind firewalls, and engine pod attaching structures containing flammable fluid lines.

(a) Each nacelle area immediately behind the firewall, and each portion of any engine pod attaching structure containing flammable fluid lines, must meet each requirement of §§ 25.1103(b), 25.1165 (d) and (e), 25.1183, 25.1185(c), 25.1187, 25.1189, and 25.1195 through 25.1203, including those concerning designated fire zones. However, engine pod attaching structures need not contain fire detection or extinguishing means.

(b) For each area covered by paragraph (a) of this section that contains a retractable landing gear, compliance with that paragraph need only be shown with the landing gear retracted.

[Amdt. 25-11, 32 FR 6913, May 5, 1967]

§ 25.1183 Flammable fluid-carrying components.

(a) Except as provided in paragraph (b) of this section, each line, fitting, and other component carrying flammable fluid in any area subject to engine fire conditions, and each component which conveys or contains flammable fluid in a designated fire zone must be fire resistant, except that flammable fluid tanks and supports in a designated fire zone must be fireproof or be enclosed by a fireproof shield unless damage by fire to any non-fireproof part will not cause leakage or spillage of flammable fluid. Components must be shielded or located to safeguard against the ignition of leaking flammable fluid. An integral oil sump of less than 25-quart capacity on a reciprocating engine need not be fireproof nor be enclosed by a fireproof shield.

(b) Paragraph (a) of this section does not apply to—

(1) Lines, fittings, and components which are already approved as part of a type certificated engine; and

(2) Vent and drain lines, and their fittings, whose failure will not result in, or add to, a fire hazard.

(c) All components, including ducts, within a designated fire zone must be fireproof if, when exposed to or damaged by fire, they could—

(1) Result in fire spreading to other regions of the airplane; or

(2) Cause unintentional operation of, or inability to operate, essential services or equipment.

[Doc. No. 5066, 29 FR 18291, Dec. 24, 1964, as amended by Amdt. 25-11, 32 FR 6913, May 5, 1967; Amdt. 25-36, 39 FR 35461, Oct. 1, 1974; Amdt. 25-57, 49 FR 6849, Feb. 23, 1984; Amdt. 25-101, 65 FR 79710, Dec. 19, 2000]

§ 25.1185 Flammable fluids.

(a) Except for the integral oil sumps specified in § 25.1183(a), no tank or reservoir that is a part of a system containing flammable fluids or gases may be in a designated fire zone unless the fluid contained, the design of the system, the materials used in the tank, the shut-off means, and all connections, lines, and control provide a degree of safety equal to that which would exist if the tank or reservoir were outside such a zone.

(b) There must be at least one-half inch of clear airspace between each tank or reservoir and each firewall or shroud isolating a designated fire zone.

(c) Absorbent materials close to flammable fluid system components that might leak must be covered or treated to prevent the absorption of hazardous quantities of fluids.

[Doc. No. 5066, 29 FR 18291, Dec. 24, 1964 as amended by Amdt. 25-19, 33 FR 15410, Oct. 17, 1968; Amdt. 25-94, 63 FR 8848, Feb. 23, 1998]

§ 25.1187 Drainage and ventilation of fire zones.

(a) There must be complete drainage of each part of each designated fire zone to minimize the hazards resulting from failure or malfunctioning of any component containing flammable fluids. The drainage means must be—

(1) Effective under conditions expected to prevail when drainage is needed; and

(2) Arranged so that no discharged fluid will cause an additional fire hazard.

(b) Each designated fire zone must be ventilated to prevent the accumulation of flammable vapors.

(c) No ventilation opening may be where it would allow the entry of flammable fluids, vapors, or flame from other zones.

(d) Each ventilation means must be arranged so that no discharged vapors will cause an additional fire hazard.

(e) Unless the extinguishing agent capacity and rate of discharge are based on maximum air flow through a zone, there must be means to allow the crew to shut off sources of forced ventilation to any fire zone except the engine power section of the nacelle and the combustion heater ventilating air ducts.

§ 25.1189 Shutoff means.

(a) Each engine installation and each fire zone specified in § 25.1181(a)(4) and (5) must have a means to shut off or otherwise prevent hazardous quantities of fuel, oil, deicer, and other flammable fluids, from flowing into, within, or through any designated fire zone, except that shutoff means are not required for—

(1) Lines, fittings, and components forming an integral part of an engine; and

(2) Oil systems for turbine engine installations in which all components of the system in a designated fire zone, including oil tanks, are fireproof or located in areas not subject to engine fire conditions.

(b) The closing of any fuel shutoff valve for any engine may not make fuel unavailable to the remaining engines.

(c) Operation of any shutoff may not interfere with the later emergency operation of other equipment, such as the means for feathering the propeller.

(d) Each flammable fluid shutoff means and control must be fireproof or must be located and protected so that any fire in a fire zone will not affect its operation.

(e) No hazardous quantity of flammable fluid may drain into any designated fire zone after shutoff.

(f) There must be means to guard against inadvertent operation of the shutoff means and to make it possible for the crew to reopen the shutoff means in flight after it has been closed.

(g) Each tank-to-engine shutoff valve must be located so that the operation of the valve will not be affected by powerplant or engine mount structural failure.

(h) Each shutoff valve must have a means to relieve excessive pressure accumulation unless a means for pressure relief is otherwise provided in the system.

[Doc. No. 5066, 29 FR 18291, Dec. 24, 1964, as amended by Amdt. 25-23, 35 FR 5677, Apr. 8, 1970; Amdt. 25-57, 49 FR 6849, Feb. 23, 1984]

§ 25.1191 Firewalls.

(a) Each engine, auxiliary power unit, fuel-burning heater, other combustion equipment intended for operation in flight, and the combustion, turbine, and tailpipe sections of turbine engines, must be isolated from the rest of the airplane by firewalls, shrouds, or equivalent means.

(b) Each firewall and shroud must be—

(1) Fireproof;

(2) Constructed so that no hazardous quantity of air, fluid, or flame can pass from the compartment to other parts of the airplane;

(3) Constructed so that each opening is sealed with close fitting fireproof grommets, bushings, or firewall fittings; and

(4) Protected against corrosion.

§ 25.1192 Engine accessory section diaphragm.

For reciprocating engines, the engine power section and all portions of the exhaust system must be isolated from the engine accessory compartment by a diaphragm that complies with the firewall requirements of § 25.1191.

[Amdt. 25-23, 35 FR 5678, Apr. 8, 1970]

§ 25.1193 Cowling and nacelle skin.

(a) Each cowling must be constructed and supported so that it can resist any vibration, inertia, and air load to which it may be subjected in operation.

(b) Cowling must meet the drainage and ventilation requirements of § 25.1187.

(c) On airplanes with a diaphragm isolating the engine power section from the engine accessory section, each part of the accessory section cowling subject to flame in case of fire in the engine power section of the powerplant must—

(1) Be fireproof; and

(2) Meet the requirements of § 25.1191.

(d) Each part of the cowling subject to high temperatures due to its nearness to exhaust system parts or exhaust gas impingement must be fireproof.

(e) Each airplane must—

(1) Be designed and constructed so that no fire originating in any fire zone can enter, either through openings or by burning through external skin, any other zone or region where it would create additional hazards;

(2) Meet paragraph (e)(1) of this section with the landing gear retracted (if applicable); and

(3) Have fireproof skin in areas subject to flame if a fire starts in the engine power or accessory sections.

§ 25.1195 Fire extinguishing systems.

(a) Except for combustor, turbine, and tail pipe sections of turbine engine installations that contain lines or components carrying flammable fluids or gases for which it is shown that a fire originating in these sections can be controlled, there must be a fire extinguisher system serving each designated fire zone.

(b) The fire extinguishing system, the quantity of the extinguishing agent, the rate of discharge, and the discharge distribution must be adequate to extinguish fires. It must be shown by either actual or simulated flights tests that under critical airflow conditions in flight the discharge of the extinguishing agent in each designated fire zone specified in paragraph (a) of this section will provide an agent concentration capable of extinguishing fires in that zone and of minimizing the probability of reignition. An individual "one-shot" system may be used for auxiliary power units, fuel burning heaters, and other combustion equipment. For each other designated fire zone, two discharges must be provided each of which produces adequate agent concentration.

(c) The fire extinguishing system for a nacelle must be able to simultaneously protect each zone of the nacelle for which protection is provided.

[Doc. No. 5066, 29 FR 18291, Dec. 24, 1964, as amended by Amdt. 25-46, 43 FR 50598, Oct. 30, 1978]

§ 25.1197 Fire extinguishing agents.

(a) Fire extinguishing agents must—

(1) Be capable of extinguishing flames emanating from any burning of fluids or other combustible materials in the area protected by the fire extinguishing system; and

(2) Have thermal stability over the temperature range likely to be experienced in the compartment in which they are stored.

(b) If any toxic extinguishing agent is used, provisions must be made to prevent harmful concentrations of fluid or fluid vapors (from leakage during normal operation of the airplane or as a result of discharging the fire extinguisher on the ground or in flight) from entering any personnel compartment, even though a defect may exist in the extinguishing system. This must be shown by test except for built-in carbon dioxide fuselage compartment fire extinguishing systems for which—

(1) Five pounds or less of carbon dioxide will be discharged, under established fire control procedures, into any fuselage compartment; or

(2) There is protective breathing equipment for each flight crewmember on flight deck duty.

[Doc. No. 5066, 29 FR 18291, Dec. 24, 1964, as amended by Amdt. 25-38, 41 FR 55467, Dec. 20, 1976; Amdt. 25-40, 42 FR 15044, Mar. 17, 1977]

§ 25.1199 Extinguishing agent containers.

(a) Each extinguishing agent container must have a pressure relief to prevent bursting of the container by excessive internal pressures.

(b) The discharge end of each discharge line from a pressure relief connection must be located so that discharge of the fire extinguishing agent would not damage the airplane. The line must also be located or protected to prevent clogging caused by ice or other foreign matter.

(c) There must be a means for each fire extinguishing agent container to indicate that the container has discharged or that the charging pressure is below the established minimum necessary for proper functioning.

(d) The temperature of each container must be maintained, under intended operating conditions, to prevent the pressure in the container from—

(1) Falling below that necessary to provide an adequate rate of discharge; or

(2) Rising high enough to cause premature discharge.

(e) If a pyrotechnic capsule is used to discharge the extinguishing agent, each container must be installed so that temperature conditions will not cause hazardous deterioration of the pyrotechnic capsule.

[Doc. No. 5066, 29 FR 18291, Dec. 24, 1964, as amended by Amdt. 25-23, 35 FR 5678, Apr. 8, 1970; Amdt. 25-40, 42 FR 15044, Mar. 17, 1977]

§ 25.1201 Fire extinguishing system materials.

(a) No material in any fire extinguishing system may react chemically with any extinguishing agent so as to create a hazard.

(b) Each system component in an engine compartment must be fireproof.

§ 25.1203 Fire detector system.

(a) There must be approved, quick acting fire or overheat detectors in each designated fire zone, and in the combustion, turbine, and tailpipe sections of turbine engine installations, in numbers and locations ensuring prompt detection of fire in those zones.

(b) Each fire detector system must be constructed and installed so that—

(1) It will withstand the vibration, inertia, and other loads to which it may be subjected in operation;

(2) There is a means to warn the crew in the event that the sensor or associated wiring within a designated fire zone is severed at one point, unless the system continues to function as a satisfactory detection system after the severing; and

(3) There is a means to warn the crew in the event of a short circuit in the sensor or associated wiring within a designated fire zone, unless the system continues to function as a satisfactory detection system after the short circuit.

(c) No fire or overheat detector may be affected by any oil, water, other fluids or fumes that might be present.

(d) There must be means to allow the crew to check, in flight, the functioning of each fire or overheat detector electric circuit.

(e) Components of each fire or overheat detector system in a fire zone must be fire-resistant.

(f) No fire or overheat detector system component for any fire zone may pass through another fire zone, unless—

(1) It is protected against the possibility of false warnings resulting from fires in zones through which it passes; or

(2) Each zone involved is simultaneously protected by the same detector and extinguishing system.

(g) Each fire detector system must be constructed so that when it is in the configuration for installation it will not exceed the alarm activation time approved for the detectors using the response time criteria specified in the appropriate Technical Standard Order for the detector.

(h) EWIS for each fire or overheat detector system in a fire zone must meet the requirements of § 25.1731.

[Doc. No. 5066, 29 FR 18291, Dec. 24, 1964, as amended by Amdt. 25-23, 35 FR 5678, Apr. 8, 1970; Amdt. 25-26, 36 FR 5493, Mar. 24, 1971; Amdt. 25-123, 72 FR 63405, Nov. 8, 2007]

§ 25.1207 Compliance.

Unless otherwise specified, compliance with the requirements of §§ 25.1181 through 25.1203 must be shown by a full scale fire test or by one or more of the following methods:

(a) Tests of similar powerplant configurations;

(b) Tests of components;

(c) Service experience of aircraft with similar powerplant configurations;

(d) Analysis.

[Amdt. 25-46, 43 FR 50598, Oct. 30, 1978]

Subpart F—Equipment

General

§ 25.1301 Function and installation.

(a) Each item of installed equipment must—

(1) Be of a kind and design appropriate to its intended function;

(2) Be labeled as to its identification, function, or operating limitations, or any applicable combination of these factors;

(3) Be installed according to limitations specified for that equipment; and

(4) Function properly when installed.

(b) EWIS must meet the requirements of subpart H of this part.

[Doc.. No. 5066, Amdt. 1-6, 29 FR 18333, Dec. 24, 1964, as amended by Amdt. 25-123, 72 FR 63405, Nov. 8, 2007]

§ 25.1302 Installed systems and equipment for use by the flightcrew.

This section applies to installed systems and equipment intended for flightcrew members' use in operating the airplane from their normally seated positions on the flight deck. The applicant must show that these systems and installed equipment, individually and in combination with other such systems and equipment, are designed so that qualified flightcrew members trained in their use can safely perform all of the tasks associated with the systems' and equipment's intended functions. Such installed equipment and systems must meet the following requirements:

(a) Flight deck controls must be installed to allow accomplishment of all the tasks required to safely perform the equipment's intended function, and information must be provided to the flightcrew that is necessary to accomplish the defined tasks.

(b) Flight deck controls and information intended for the flightcrew's use must:

(1) Be provided in a clear and unambiguous manner at a resolution and precision appropriate to the task;

(2) Be accessible and usable by the flightcrew in a manner consistent with the urgency, frequency, and duration of their tasks; and

(3) Enable flightcrew awareness, if awareness is required for safe operation, of the effects on the airplane or systems resulting from flightcrew actions.

(c) Operationally-relevant behavior of the installed equipment must be:

(1) Predictable and unambiguous; and

(2) Designed to enable the flightcrew to intervene in a manner appropriate to the task.

(d) To the extent practicable, installed equipment must incorporate means to enable the flightcrew to manage errors resulting from the kinds of flightcrew interactions with the equipment that can be reasonably expected in service. This paragraph does not apply to any of the following:

(1) Skill-related errors associated with manual control of the airplane;

(2) Errors that result from decisions, actions, or omissions committed with malicious intent;

(3) Errors arising from a crewmember's reckless decisions, actions, or omissions reflecting a substantial disregard for safety; and

(4) Errors resulting from acts or threats of violence, including actions taken under duress.

[Doc. No. FAA-2010-1175, 78 FR 25846, May 3, 2013]

§ 25.1303 Flight and navigation instruments.

(a) The following flight and navigation instruments must be installed so that the instrument is visible from each pilot station:

(1) A free air temperature indicator or an air-temperature indicator which provides indications that are convertible to free-air temperature.

(2) A clock displaying hours, minutes, and seconds with a sweep-second pointer or digital presentation.

(3) A direction indicator (nonstabilized magnetic compass).

(b) The following flight and navigation instruments must be installed at each pilot station:

(1) An airspeed indicator. If airspeed limitations vary with altitude, the indicator must have a maximum allowable airspeed indicator showing the variation of V_{MO} with altitude.

(2) An altimeter (sensitive).

(3) A rate-of-climb indicator (vertical speed).

(4) A gyroscopic rate-of-turn indicator combined with an integral slip-skid indicator (turn-and-bank indicator) except that only a slip-skid indicator is required on large airplanes with a third attitude instrument system useable through flight attitudes of 360° of pitch and roll and installed in accordance with § 121.305(k) of this title.

(5) A bank and pitch indicator (gyroscopically stabilized).

(6) A direction indicator (gyroscopically stabilized, magnetic or nonmagnetic).

(c) The following flight and navigation instruments are required as prescribed in this paragraph:

(1) A speed warning device is required for turbine engine powered airplanes and for airplanes with V_{MO}/M_{MO} greater than 0.8 V_{DF}/M_{DF} or 0.8 V_D/M_D. The speed warning device must give effective aural warning (differing distinctively from aural warnings used for other purposes) to the pilots, whenever the speed exceeds V_{MO} plus 6 knots ormm_o +0.01. The upper limit of the production tolerance for the warning device may not exceed the prescribed warning speed.

(2) A machmeter is required at each pilot station for airplanes with compressibility limitations not otherwise indicated to the pilot by the airspeed indicating system required under paragraph (b)(1) of this section.

[Amdt. 25-23, 35 FR 5678, Apr. 8, 1970, as amended by Amdt. 25-24, 35 FR 7108, May 6, 1970; Amdt. 25-38, 41 FR 55467, Dec. 20, 1976; Amdt. 25-90, 62 FR 13253, Mar. 19, 1997]

§ 25.1305 Powerplant instruments.

The following are required powerplant instruments:

(a) *For all airplanes.* (1) A fuel pressure warning means for each engine, or a master warning means for all engines with provision for isolating the individual warning means from the master warning means.

(2) A fuel quantity indicator for each fuel tank.

(3) An oil quantity indicator for each oil tank.

(4) An oil pressure indicator for each independent pressure oil system of each engine.

(5) An oil pressure warning means for each engine, or a master warning means for all engines with provision for isolating the individual warning means from the master warning means.

(6) An oil temperature indicator for each engine.

(7) Fire-warning devices that provide visual and audible warning.

(8) An augmentation liquid quantity indicator (appropriate for the manner in which the liquid is to be used in operation) for each tank.

(b) *For reciprocating engine-powered airplanes.* In addition to the powerplant instruments required by paragraph (a) of this section, the following powerplant instruments are required:

(1) A carburetor air temperature indicator for each engine.

(2) A cylinder head temperature indicator for each air-cooled engine.

(3) A manifold pressure indicator for each engine.

(4) A fuel pressure indicator (to indicate the pressure at which the fuel is supplied) for each engine.

(5) A fuel flowmeter, or fuel mixture indicator, for each engine without an automatic altitude mixture control.

(6) A tachometer for each engine.

(7) A device that indicates, to the flight crew (during flight), any change in the power output, for each engine with—

(i) An automatic propeller feathering system, whose operation is initiated by a power output measuring system; or

(ii) A total engine piston displacement of 2,000 cubic inches or more.

(8) A means to indicate to the pilot when the propeller is in reverse pitch, for each reversing propeller.

(c) *For turbine engine-powered airplanes.* In addition to the powerplant instruments required by paragraph (a) of this section, the following powerplant instruments are required:

(1) A gas temperature indicator for each engine.

(2) A fuel flowmeter indicator for each engine.

(3) A tachometer (to indicate the speed of the rotors with established limiting speeds) for each engine.

(4) A means to indicate, to the flight crew, the operation of each engine starter that can be operated continuously but that is neither designed for continuous operation nor designed to prevent hazard if it failed.

(5) An indicator to indicate the functioning of the powerplant ice protection system for each engine.

(6) An indicator for the fuel strainer or filter required by § 25.997 to indicate the occurrence of contamination of the strainer or filter before it reaches the capacity established in accordance with § 25.997(d).

(7) A warning means for the oil strainer or filter required by § 25.1019, if it has no bypass, to warn the pilot of the occurrence of contamination of the strainer or filter screen before it reaches the capacity established in accordance with § 25.1019(a)(2).

(8) An indicator to indicate the proper functioning of any heater used to prevent ice clogging of fuel system components.

(d) *For turbojet engine powered airplanes.* In addition to the powerplant instruments required by paragraphs (a) and (c) of this section, the following powerplant instruments are required:

(1) An indicator to indicate thrust, or a parameter that is directly related to thrust, to the pilot. The indication must be based on the direct measurement of thrust or of parameters that are directly related to thrust. The indicator must indicate a change in thrust resulting from any engine malfunction, damage, or deterioration.

(2) A position indicating means to indicate to the flightcrew when the thrust reversing device—

(i) Is not in the selected position, and

(ii) Is in the reverse thrust position, for each engine using a thrust reversing device.

(3) An indicator to indicate rotor system unbalance.

(e) *For turbopropeller-powered airplanes.* In addition to the powerplant instruments required by paragraphs (a) and (c) of this section, the following powerplant instruments are required:

(1) A torque indicator for each engine.

(2) Position indicating means to indicate to the flight crew when the propeller blade angle is below the flight low pitch position, for each propeller.

(f) For airplanes equipped with fluid systems (other than fuel) for thrust or power augmentation, an approved means must be provided to indicate the proper functioning of that system to the flight crew.

[Amdt. 25-23, 35 FR 5678, Apr. 8, 1970, as amended by Amdt. 25-35, 39 FR 1831, Jan. 15, 1974; Amdt. 25-36, 39 FR 35461, Oct. 1, 1974; Amdt. 25-38, 41 FR 55467, Dec. 20, 1976; Amdt. 25-54, 45 FR 60173, Sept. 11, 1980; Amdt. 25-72, 55 FR 29785, July 20, 1990; Amdt. 25-115, 69 FR 40527, July 2, 2004]

§ 25.1307 Miscellaneous equipment.

The following is required miscellaneous equipment:

(a) [Reserved]

(b) Two or more independent sources of electrical energy.

(c) Electrical protective devices, as prescribed in this part.

(d) Two systems for two-way radio communications, with controls for each accessible from each pilot station, designed and installed so that failure of one system will not preclude operation of the other system. The use of a common antenna system is acceptable if adequate reliability is shown.

(e) Two systems for radio navigation, with controls for each accessible from each pilot station, designed and installed so that failure of one system will not preclude operation of the other system. The use of a common antenna system is acceptable if adequate reliability is shown.

[Amdt. 25-23, 35 FR 5678, Apr. 8, 1970, as amended by Amdt. 25-46, 43 FR 50598, Oct. 30, 1978; Amdt. 25-54, 45 FR 60173, Sept. 11, 1980; Amdt. 25-72, 55 FR 29785, July 20, 1990]

§ 25.1309 Equipment, systems, and installations.

(a) The equipment, systems, and installations whose functioning is required by this subchapter, must be designed to ensure that they perform their intended functions under any foreseeable operating condition.

(b) The airplane systems and associated components, considered separately and in relation to other systems, must be designed so that—

(1) The occurrence of any failure condition which would prevent the continued safe flight and landing of the airplane is extremely improbable, and

(2) The occurrence of any other failure conditions which would reduce the capability of the airplane or the ability of the crew to cope with adverse operating conditions is improbable.

(c) Warning information must be provided to alert the crew to unsafe system operating conditions, and to enable them to take appropriate corrective action. Systems, controls, and associated monitoring and warning means must be designed to minimize crew errors which could create additional hazards.

(d) Compliance with the requirements of paragraph (b) of this section must be shown by analysis, and where necessary, by appropriate ground, flight, or simulator tests. The analysis must consider—

(1) Possible modes of failure, including malfunctions and damage from external sources.

(2) The probability of multiple failures and undetected failures.

(3) The resulting effects on the airplane and occupants, considering the stage of flight and operating conditions, and

(4) The crew warning cues, corrective action required, and the capability of detecting faults.

(e) In showing compliance with paragraphs (a) and (b) of this section with regard to the electrical system and equipment design and installation, critical environmental conditions must be considered. For electrical generation, distribution, and utilization equipment required by or used in complying with this chapter, except equipment covered by Technical Standard Orders containing environmental test procedures, the ability to provide continuous, safe service under foreseeable environmental conditions may be shown by environmental tests, design analysis, or reference to previous comparable service experience on other aircraft.

(f) EWIS must be assessed in accordance with the requirements of § 25.1709.

[Amdt. 25-23, 35 FR 5679, Apr. 8, 1970, as amended by Amdt. 25-38, 41 FR 55467, Dec. 20, 1976; Amdt. 25-41, 42 FR 36970, July 18, 1977; Amdt. 25-123, 72 FR 63405, Nov. 8, 2007]

§ 25.1310 Power source capacity and distribution.

(a) Each installation whose functioning is required for type certification or under operating rules and that requires a power supply is an "essential load" on the power supply. The power sources and the system must be able to supply the following power loads in probable operating combinations and for probable durations:

(1) Loads connected to the system with the system functioning normally.

(2) Essential loads, after failure of any one prime mover, power converter, or energy storage device.

(3) Essential loads after failure of—

(i) Any one engine on two-engine airplanes; and

(ii) Any two engines on airplanes with three or more engines.

(4) Essential loads for which an alternate source of power is required, after any failure or malfunction in any one power supply system, distribution system, or other utilization system.

(b) In determining compliance with paragraphs (a)(2) and (3) of this section, the power loads may be assumed to be reduced under a monitoring procedure consistent with safety in the kinds of operation authorized. Loads not required in controlled flight need not be considered for the two-engine-inoperative condition on airplanes with three or more engines.

[Amdt. 25-123, 72 FR 63405, Nov. 8, 2007]

§ 25.1316 Electrical and electronic system lightning protection.

(a) Each electrical and electronic system that performs a function, for which failure would prevent the continued safe flight and landing of the airplane, must be designed and installed so that—

(1) The function is not adversely affected during and after the time the airplane is exposed to lightning; and

(2) The system automatically recovers normal operation of that function in a timely manner after the airplane is exposed to lightning.

(b) Each electrical and electronic system that performs a function, for which failure would reduce the capability of the airplane or the ability of the flightcrew to respond to an adverse operating condition, must be designed and installed so that the function recovers normal operation in a timely manner after the airplane is exposed to lightning.

[Doc. No. FAA-2010-0224, Amdt. 25-134, 76 FR 33135, June 8, 2011]

§ 25.1317 High-intensity Radiated Fields (HIRF) Protection.

(a) Except as provided in paragraph (d) of this section, each electrical and electronic system that performs a function whose failure would prevent the continued safe flight and landing of the airplane must be designed and installed so that—

(1) The function is not adversely affected during and after the time the airplane is exposed to HIRF environment I, as described in appendix L to this part;

(2) The system automatically recovers normal operation of that function, in a timely manner, after the airplane is exposed to HIRF environment I, as described in appendix L to this part, unless the system's recovery conflicts with other operational or functional requirements of the system; and

(3) The system is not adversely affected during and after the time the airplane is exposed to HIRF environment II, as described in appendix L to this part.

(b) Each electrical and electronic system that performs a function whose failure would significantly reduce the capability of the airplane or the ability of the flightcrew to respond to an adverse operating condition must be designed and installed so the system is not adversely affected when the equipment providing these functions is exposed to equipment HIRF test level 1 or 2, as described in appendix L to this part.

(c) Each electrical and electronic system that performs a function whose failure would reduce the capability of the airplane or the ability of the flightcrew to respond to an adverse operating condition must be designed and installed so the system is not adversely affected when the equipment providing the function is exposed to equipment HIRF test level 3, as described in appendix L to this part.

(d) Before December 1, 2012, an electrical or electronic system that performs a function whose failure would prevent the continued safe flight and landing of an airplane may be designed and installed without meeting the provisions of paragraph (a) provided—

(1) The system has previously been shown to comply with special conditions for HIRF, prescribed under § 21.16, issued before December 1, 2007;

(2) The HIRF immunity characteristics of the system have not changed since compliance with the special conditions was demonstrated; and

(3) The data used to demonstrate compliance with the special conditions is provided.

[Doc. No. FAA-2006-23657, 72 FR 44025, Aug. 6, 2007]

Instruments: Installation

§ 25.1321 Arrangement and visibility.

(a) Each flight, navigation, and powerplant instrument for use by any pilot must be plainly visible to him from his station with the minimum practicable deviation from his normal position and line of vision when he is looking forward along the flight path.

(b) The flight instruments required by § 25.1303 must be grouped on the instrument panel and centered as nearly as practicable about the vertical plane of the pilot's forward vision. In addition—

(1) The instrument that most effectively indicates attitude must be on the panel in the top center position;

(2) The instrument that most effectively indicates airspeed must be adjacent to and directly to the left of the instrument in the top center position;

(3) The instrument that most effectively indicates altitude must be adjacent to and directly to the right of the instrument in the top center position; and

(4) The instrument that most effectively indicates direction of flight must be adjacent to and directly below the instrument in the top center position.

(c) Required powerplant instruments must be closely grouped on the instrument panel. In addition—

(1) The location of identical powerplant instruments for the engines must prevent confusion as to which engine each instrument relates; and

(2) Powerplant instruments vital to the safe operation of the airplane must be plainly visible to the appropriate crewmembers.

(d) Instrument panel vibration may not damage or impair the accuracy of any instrument.

(e) If a visual indicator is provided to indicate malfunction of an instrument, it must be effective under all probable cockpit lighting conditions.

[Amdt. 25-23, 35 FR 5679, Apr. 8, 1970, as amended by Amdt. 25-41, 42 FR 36970, July 18, 1977]

§ 25.1322 Flightcrew alerting.

(a) Flightcrew alerts must:

(1) Provide the flightcrew with the information needed to:

(i) Identify non-normal operation or airplane system conditions, and

(ii) Determine the appropriate actions, if any.

(2) Be readily and easily detectable and intelligible by the flightcrew under all foreseeable operating conditions, including conditions where multiple alerts are provided.

(3) Be removed when the alerting condition no longer exists.

(b) Alerts must conform to the following prioritization hierarchy based on the urgency of flightcrew awareness and response.

(1) Warning: For conditions that require immediate flightcrew awareness and immediate flightcrew response.

(2) Caution: For conditions that require immediate flightcrew awareness and subsequent flightcrew response.

(3) Advisory: For conditions that require flightcrew awareness and may require subsequent flightcrew response.

(c) Warning and caution alerts must:

(1) Be prioritized within each category, when necessary.

(2) Provide timely attention-getting cues through at least two different senses by a combination of aural, visual, or tactile indications.

(3) Permit each occurrence of the attention-getting cues required by paragraph (c)(2) of this section to be acknowledged and suppressed, unless they are required to be continuous.

(d) The alert function must be designed to minimize the effects of false and nuisance alerts. In particular, it must be designed to:

(1) Prevent the presentation of an alert that is inappropriate or unnecessary.

(2) Provide a means to suppress an attention-getting component of an alert caused by a failure of the alerting function that interferes with the flightcrew's ability to safely operate the airplane. This means must not be readily available to the flightcrew so that it could be operated inadvertently or by habitual reflexive action. When an alert is suppressed, there must be a clear and unmistakable annunciation to the flightcrew that the alert has been suppressed.

(e) Visual alert indications must:

(1) Conform to the following color convention:

(i) Red for warning alert indications.

(ii) Amber or yellow for caution alert indications.

(iii) Any color except red or green for advisory alert indications.

(2) Use visual coding techniques, together with other alerting function elements on the flight deck, to distinguish between warning, caution, and advisory alert indications, if they are presented on monochromatic displays that are not capable of conforming to the color convention in paragraph (e)(1) of this section.

(f) Use of the colors red, amber, and yellow on the flight deck for functions other than flightcrew alerting must be limited and must not adversely affect flightcrew alerting.

[Amdt. 25-131, 75 FR 67209, Nov. 2, 2010]

§ 25.1323 Airspeed indicating system.

For each airspeed indicating system, the following apply:

(a) Each airspeed indicating instrument must be approved and must be calibrated to indicate true airspeed (at sea level with a standard atmosphere) with a minimum practicable instrument calibration error when the corresponding pitot and static pressures are applied.

(b) Each system must be calibrated to determine the system error (that is, the relation between IAS and CAS) in flight and during the accelerated takeoff ground run. The ground run calibration must be determined—

(1) From 0.8 of the minimum value of V_1 to the maximum value of V_2, considering the approved ranges of altitude and weight; and

(2) With the flaps and power settings corresponding to the values determined in the establishment of the takeoff path under § 25.111 assuming that the critical engine fails at the minimum value of V_1.

(c) The airspeed error of the installation, excluding the airspeed indicator instrument calibration error, may not exceed three percent or five knots, whichever is greater, throughout the speed range, from—

(1) V_{MO} to 1.23 V_{SR1}, with flaps retracted; and

(2) 1.23 V_{SR0} to V_{FE} with flaps in the landing position.

(d) From 1.23 V_{SR} to the speed at which stall warning begins, the IAS must change perceptibly with CAS and in the same sense, and at speeds below stall warning speed the IAS must not change in an incorrect sense.

(e) From V_{MO} to $V_{MO} + \frac{2}{3}(V_{DF} - V_{MO})$, the IAS must change perceptibly with CAS and in the same sense, and at higher speeds up to V_{DF} the IAS must not change in an incorrect sense.

(f) There must be no indication of airspeed that would cause undue difficulty to the pilot during the takeoff between the initiation of rotation and the achievement of a steady climbing condition.

(g) The effects of airspeed indicating system lag may not introduce significant takeoff indicated airspeed bias, or significant errors in takeoff or accelerate-stop distances.

(h) Each system must be arranged, so far as practicable, to prevent malfunction or serious error due to the entry of moisture, dirt, or other substances.

(i) Each system must have a heated pitot tube or an equivalent means of preventing malfunction due to icing.

(j) Where duplicate airspeed indicators are required, their respective pitot tubes must be far enough apart to avoid damage to both tubes in a collision with a bird.

[Doc. No. 5066, 29 FR 18291, Dec. 24, 1964, as amended by Amdt. 25-57, 49 FR 6849, Feb. 23, 1984; Amdt. 25-108, 67 FR 70828, Nov. 26, 2002; Amdt. 25-109, 67 FR 76656, Dec. 12, 2002]

§ 25.1325 Static pressure systems.

(a) Each instrument with static air case connections must be vented to the outside atmosphere through an appropriate piping system.

(b) Each static port must be designed and located in such manner that the static pressure system performance is least affected by airflow variation, or by moisture or other foreign matter, and that the correlation between air pressure in the static pressure system and true ambient atmospheric static pressure is not changed when the airplane is exposed to the continuous and intermittent maximum icing conditions defined in appendix C of this part.

(c) The design and installation of the static pressure system must be such that—

(1) Positive drainage of moisture is provided; chafing of the tubing and excessive distortion or restriction at bends in the tubing is avoided; and the materials used are durable, suitable for the purpose intended, and protected against corrosion; and

(2) It is airtight except for the port into the atmosphere. A proof test must be conducted to demonstrate the integrity of the static pressure system in the following manner:

(i) *Unpressurized airplanes.* Evacuate the static pressure system to a pressure differential of approximately 1 inch of mercury or to a reading on the altimeter, 1,000 feet above the airplane elevation at the time of the test. Without additional pumping for a period of 1 minute, the loss of indicated altitude must not exceed 100 feet on the altimeter.

(ii) *Pressurized airplanes.* Evacuate the static pressure system until a pressure differential equivalent to the maximum cabin pressure differential for which the airplane is type certificated is achieved. Without additional pumping for a period of 1 minute, the loss of indicated altitude must not exceed 2 percent of the equivalent altitude of the maximum cabin differential pressure or 100 feet, whichever is greater.

(d) Each pressure altimeter must be approved and must be calibrated to indicate pressure altitude in a standard atmosphere, with a minimum practicable calibration error when the corresponding static pressures are applied.

(e) Each system must be designed and installed so that the error in indicated pressure altitude, at sea level, with a standard atmosphere, excluding instrument calibration error, does not result in an error of more than±30 feet per 100 knots speed for the appropriate configuration in the speed range between 1.23 V_{SR0} with flaps extended and 1.7 V_{SR1} with flaps retracted. However, the error need not be less than±30 feet.

(f) If an altimeter system is fitted with a device that provides corrections to the altimeter indication, the device must be designed and installed in such manner that it can be bypassed when it malfunctions, unless an alternate altimeter system is provided. Each correction device must be fitted with a means for indicating the occurrence of reasonably probable malfunctions, including power failure, to the flight crew. The indicating means must be effective for any cockpit lighting condition likely to occur.

(g) Except as provided in paragraph (h) of this section, if the static pressure system incorporates both a primary and an alternate static pressure source, the means for selecting one or the other source must be designed so that—

(1) When either source is selected, the other is blocked off; and

(2) Both sources cannot be blocked off simultaneously.

(h) For unpressurized airplanes, paragraph (g)(1) of this section does not apply if it can be demonstrated that the static pressure system calibration, when either static pressure source is selected, is not changed by the other static pressure source being open or blocked.

[Doc. No. 5066, 29 FR 18291, Dec. 24, 1964, as amended by Amdt. 25-5, 30 FR 8261, June 29, 1965; Amdt. 25-12, 32 FR 7587, May 24, 1967; Amdt. 25-41, 42 FR 36970, July 18, 1977; Amdt. 25-108, 67 FR 70828, Nov. 26, 2002]

§ 25.1326 Pitot heat indication systems.

If a flight instrument pitot heating system is installed, an indication system must be provided to indicate to the flight crew when that pitot heating system is not operating. The indication system must comply with the following requirements:

(a) The indication provided must incorporate an amber light that is in clear view of a flight crewmember.

(b) The indication provided must be designed to alert the flight crew if either of the following conditions exist:

(1) The pitot heating system is switched "off".

(2) The pitot heating system is switched "on" and any pitot tube heating element is inoperative.

[Amdt. 25-43, 43 FR 10339, Mar. 13, 1978]

§ 25.1327 Magnetic direction indicator.

(a) Each magnetic direction indicator must be installed so that its accuracy is not excessively affected by the airplane's vibration or magnetic fields.

(b) The compensated installation may not have a deviation, in level flight, greater than 10 degrees on any heading.

§ 25.1329 Flight guidance system.

(a) Quick disengagement controls for the autopilot and autothrust functions must be provided for each pilot. The autopilot quick disengagement controls must be located on both control wheels (or equivalent). The autothrust quick disengagement controls must be located on the thrust control levers. Quick disengagement controls must be readily accessible to each pilot while operating the control wheel (or equivalent) and thrust control levers.

(b) The effects of a failure of the system to disengage the autopilot or autothrust functions when manually commanded by the pilot must be assessed in accordance with the requirements of § 25.1309.

(c) Engagement or switching of the flight guidance system, a mode, or a sensor may not cause a transient response of the airplane's flight path any greater than a minor transient, as defined in paragraph (n)(1) of this section.

(d) Under normal conditions, the disengagement of any automatic control function of a flight guidance system may not cause a transient response of the airplane's flight path any greater than a minor transient.

(e) Under rare normal and non-normal conditions, disengagement of any automatic control function of a flight guidance system may not result in a transient any greater than a significant transient, as defined in paragraph (n)(2) of this section.

(f) The function and direction of motion of each command reference control, such as heading select or vertical speed, must be plainly indicated on, or adjacent to, each control if necessary to prevent inappropriate use or confusion.

(g) Under any condition of flight appropriate to its use, the flight guidance system may not produce hazardous loads on the airplane, nor create hazardous deviations in the flight path. This applies to both fault-free operation and in the event of a malfunction, and assumes that the pilot begins corrective action within a reasonable period of time.

(h) When the flight guidance system is in use, a means must be provided to avoid excursions beyond an acceptable margin from the speed range of the normal flight envelope. If the airplane experiences an excursion outside this range, a means must be provided to prevent the flight guidance system from providing guidance or control to an unsafe speed.

(i) The flight guidance system functions, controls, indications, and alerts must be designed to minimize flightcrew errors and confusion concerning the behavior and operation of the flight guidance system. Means must be provided to indicate the current mode of operation, including any armed modes, transitions, and reversions. Selector switch position is not an acceptable means of indication. The controls and indications must be grouped and presented in a logical and consistent manner. The indications must be visible to each pilot under all expected lighting conditions.

(j) Following disengagement of the autopilot, a warning (visual and auditory) must be provided to each pilot and be timely and distinct from all other cockpit warnings.

(k) Following disengagement of the autothrust function, a caution must be provided to each pilot.

(l) The autopilot may not create a potential hazard when the flightcrew applies an override force to the flight controls.

(m) During autothrust operation, it must be possible for the flightcrew to move the thrust levers without requiring excessive force. The autothrust may not create a potential hazard when the flightcrew applies an override force to the thrust levers.

(n) For purposes of this section, a transient is a disturbance in the control or flight path of the airplane that is not consistent with response to flightcrew inputs or environmental conditions.

(1) A minor transient would not significantly reduce safety margins and would involve flightcrew actions that are well within their capabilities. A minor transient may involve a slight increase in flightcrew workload or some physical discomfort to passengers or cabin crew.

(2) A significant transient may lead to a significant reduction in safety margins, an increase in flightcrew workload, discomfort to the flightcrew, or physical distress to the passengers or cabin crew, possibly including non-fatal injuries. Significant transients do not require, in order to remain within or recover to the normal flight envelope, any of the following:

(i) Exceptional piloting skill, alertness, or strength.

(ii) Forces applied by the pilot which are greater than those specified in § 25.143(c).

(iii) Accelerations or attitudes in the airplane that might result in further hazard to secured or non-secured occupants.

[Doc. No. FAA-2004-18775, 71 FR 18191, Apr. 11, 2006]

§ 25.1331 Instruments using a power supply.

(a) For each instrument required by § 25.1303(b) that uses a power supply, the following apply:

(1) Each instrument must have a visual means integral with, the instrument, to indicate when power adequate to sustain proper instrument performance is not being supplied. The power must be measured at or near the point

where it enters the instruments. For electric instruments, the power is considered to be adequate when the voltage is within approved limits.

(2) Each instrument must, in the event of the failure of one power source, be supplied by another power source. This may be accomplished automatically or by manual means.

(3) If an instrument presenting navigation data receives information from sources external to that instrument and loss of that information would render the presented data unreliable, the instrument must incorporate a visual means to warn the crew, when such loss of information occurs, that the presented data should not be relied upon.

(b) As used in this section, "instrument" includes devices that are physically contained in one unit, and devices that are composed of two or more physically separate units or components connected together (such as a remote indicating gyroscopic direction indicator that includes a magnetic sensing element, a gyroscopic unit, an amplifier and an indicator connected together).

[Doc. No. 5066, 29 FR 18291, Dec. 24, 1964, as amended by Amdt. 25-41, 42 FR 36970, July 18, 1977]

§ 25.1333 Instrument systems.

For systems that operate the instruments required by § 25.1303(b) which are located at each pilot's station—

(a) Means must be provided to connect the required instruments at the first pilot's station to operating systems which are independent of the operating systems at other flight crew stations, or other equipment;

(b) The equipment, systems, and installations must be designed so that one display of the information essential to the safety of flight which is provided by the instruments, including attitude, direction, airspeed, and altitude will remain available to the pilots, without additional crewmember action, after any single failure or combination of failures that is not shown to be extremely improbable; and

(c) Additional instruments, systems, or equipment may not be connected to the operating systems for the required instruments, unless provisions are made to ensure the continued normal functioning of the required instruments in the event of any malfunction of the additional instruments, systems, or equipment which is not shown to be extremely improbable.

[Amdt. 25-23, 35 FR 5679, Apr. 8, 1970, as amended by Amdt. 25-41, 42 FR 36970, July 18, 1977]

§ 25.1337 Powerplant instruments.

(a) *Instruments and instrument lines.* (1) Each powerplant and auxiliary power unit instrument line must meet the requirements of §§ 25.993 and 25.1183.

(2) Each line carrying flammable fluids under pressure must—

(i) Have restricting orifices or other safety devices at the source of pressure to prevent the escape of excessive fluid if the line fails; and

(ii) Be installed and located so that the escape of fluids would not create a hazard.

(3) Each powerplant and auxiliary power unit instrument that utilizes flammable fluids must be installed and located so that the escape of fluid would not create a hazard.

(b) *Fuel quantity indicator.* There must be means to indicate to the flight crewmembers, the quantity, in gallons or equivalent units, of usable fuel in each tank during flight. In addition—

(1) Each fuel quantity indicator must be calibrated to read "zero" during level flight when the quantity of fuel remaining in the tank is equal to the unusable fuel supply determined under § 25.959;

(2) Tanks with interconnected outlets and airspaces may be treated as one tank and need not have separate indicators; and

(3) Each exposed sight gauge, used as a fuel quantity indicator, must be protected against damage.

(c) *Fuel flowmeter system.* If a fuel flowmeter system is installed, each metering component must have a means for bypassing the fuel supply if malfunction of that component severely restricts fuel flow.

(d) *Oil quantity indicator.* There must be a stick gauge or equivalent means to indicate the quantity of oil in each tank. If an oil transfer or reserve oil supply system is installed, there must be a means to indicate to the flight crew, in flight, the quantity of oil in each tank.

(e) *Turbopropeller blade position indicator.* Required turbopropeller blade position indicators must begin indicating before the blade moves more than eight degrees below the flight low pitch stop. The source of indication must directly sense the blade position.

(f) *Fuel pressure indicator.* There must be means to measure fuel pressure, in each system supplying reciprocating engines, at a point downstream of any fuel pump except fuel injection pumps. In addition—

(1) If necessary for the maintenance of proper fuel delivery pressure, there must be a connection to transmit the carburetor air intake static pressure to the proper pump relief valve connection; and

(2) If a connection is required under paragraph (f)(1) of this section, the gauge balance lines must be independently connected to the carburetor inlet pressure to avoid erroneous readings.

[Doc. No. 5066, 29 FR 18291, Dec. 24, 1964, as amended by Amdt. 25-40, 42 FR 15044, Mar. 17, 1977]

Electrical Systems and Equipment

§ 25.1351 General.

(a) *Electrical system capacity.* The required generating capacity, and number and kinds of power sources must—

(1) Be determined by an electrical load analysis; and

(2) Meet the requirements of § 25.1309.

(b) *Generating system.* The generating system includes electrical power sources, main power busses, transmission cables, and associated control, regulation, and protective devices. It must be designed so that—

(1) Power sources function properly when independent and when connected in combination;

(2) No failure or malfunction of any power source can create a hazard or impair the ability of remaining sources to supply essential loads;

(3) The system voltage and frequency (as applicable) at the terminals of all essential load equipment can be maintained within the limits for which the equipment is designed, during any probable operating condition; and

(4) System transients due to switching, fault clearing, or other causes do not make essential loads inoperative, and do not cause a smoke or fire hazard.

(5) There are means accessible, in flight, to appropriate crewmembers for the individual and collective disconnection of the electrical power sources from the system.

(6) There are means to indicate to appropriate crewmembers the generating system quantities essential for the safe operation of the system, such as the voltage and current supplied by each generator.

(c) *External power.* If provisions are made for connecting external power to the airplane, and that external power can be electrically connected to equipment other than that used for engine starting, means must be provided to ensure that no external power supply having a reverse polarity, or a reverse phase sequence, can supply power to the airplane's electrical system.

(d) *Operation without normal electrical power.* It must be shown by analysis, tests, or both, that the airplane can be operated safely in VFR conditions, for a period of not less than five minutes, with the normal electrical power (electrical power sources excluding the battery) inoperative, with critical type fuel (from the standpoint of flameout and restart capability), and with the airplane initially at the maximum certificated altitude. Parts of the electrical system may remain on if—

(1) A single malfunction, including a wire bundle or junction box fire, cannot result in loss of both the part turned off and the part turned on; and

(2) The parts turned on are electrically and mechanically isolated from the parts turned off.

[Doc. No. 5066, 29 FR 18291, Dec. 24, 1964, as amended by Amdt. 25-41, 42 FR 36970, July 18, 1977; Amdt. 25-72, 55 FR 29785, July 20, 1990]

§ 25.1353 Electrical equipment and installations.

(a) Electrical equipment and controls must be installed so that operation of any one unit or system of units will not adversely affect the simultaneous operation of any other electrical unit or system essential to safe operation. Any electrical interference likely to be present in the airplane must not result in hazardous effects on the airplane or its systems.

(b) Storage batteries must be designed and installed as follows:

(1) Safe cell temperatures and pressures must be maintained during any probable charging or discharging condition. No uncontrolled increase in cell temperature may result when the battery is recharged (after previous complete discharge)—

(i) At maximum regulated voltage or power;

(ii) During a flight of maximum duration; and

(iii) Under the most adverse cooling condition likely to occur in service.

(2) Compliance with paragraph (b)(1) of this section must be shown by test unless experience with similar batteries and installations has shown that maintaining safe cell temperatures and pressures presents no problem.

(3) No explosive or toxic gases emitted by any battery in normal operation, or as the result of any probable malfunction in the charging system or battery installation, may accumulate in hazardous quantities within the airplane.

(4) No corrosive fluids or gases that may escape from the battery may damage surrounding airplane structures or adjacent essential equipment.

(5) Each nickel cadmium battery installation must have provisions to prevent any hazardous effect on structure or essential systems that may be caused by the maximum amount of heat the battery can generate during a short circuit of the battery or of individual cells.

(6) Nickel cadmium battery installations must have—

(i) A system to control the charging rate of the battery automatically so as to prevent battery overheating;

(ii) A battery temperature sensing and over-temperature warning system with a means for disconnecting the battery from its charging source in the event of an over-temperature condition; or

(iii) A battery failure sensing and warning system with a means for disconnecting the battery from its charging source in the event of battery failure.

(c) Electrical bonding must provide an adequate electrical return path under both normal and fault conditions, on airplanes having grounded electrical systems.

[Amdt. 25-123, 72 FR 63405, Nov. 8, 2007]

§ 25.1355 Distribution system.

(a) The distribution system includes the distribution busses, their associated feeders, and each control and protective device.

(b) [Reserved]

(c) If two independent sources of electrical power for particular equipment or systems are required by this chapter, in the event of the failure of one power source for such equipment or system, another power source (including its separate feeder) must be automatically provided or be manually selectable to maintain equipment or system operation.

[Doc. No. 5066, 29 FR 18291, Dec. 24, 1964, as amended by Amdt. 25-23, 35 FR 5679, Apr. 8, 1970; Amdt. 25-38, 41 FR 55468, Dec. 20, 1976]

§ 25.1357 Circuit protective devices.

(a) Automatic protective devices must be used to minimize distress to the electrical system and hazard to the airplane in the event of wiring faults or serious malfunction of the system or connected equipment.

(b) The protective and control devices in the generating system must be designed to de-energize and disconnect faulty power sources and power transmission equipment from their associated busses with sufficient rapidity to provide protection from hazardous over-voltage and other malfunctioning.

(c) Each resettable circuit protective device must be designed so that, when an overload or circuit fault exists, it will open the circuit irrespective of the position of the operating control.

(d) If the ability to reset a circuit breaker or replace a fuse is essential to safety in flight, that circuit breaker or fuse must be located and identified so that it can be readily reset or replaced in flight. Where fuses are used, there must be spare fuses for use in flight equal to at least 50% of the number of fuses of each rating required for complete circuit protection.

(e) Each circuit for essential loads must have individual circuit protection. However, individual protection for each circuit in an essential load system (such as each position light circuit in a system) is not required.

(f) For airplane systems for which the ability to remove or reset power during normal operations is necessary, the system must be designed so that circuit breakers are not the primary means to remove or reset system power unless specifically designed for use as a switch.

(g) Automatic reset circuit breakers may be used as integral protectors for electrical equipment (such as thermal cut-outs) if there is circuit protection to protect the cable to the equipment.

[Doc. No. 5066, 29 FR 18291, Dec. 24, 1964, as amended by Amdt. 25-123, 72 FR 63405, Nov. 8, 2007]

§ 25.1360 Precautions against injury.

(a) Shock. The electrical system must be designed to minimize risk of electric shock to crew, passengers, and servicing personnel and to maintenance personnel using normal precautions.

(b) Burns. The temperature of any part that may be handled by a crewmember during normal operations must not cause dangerous inadvertent movement by the crewmember or injury to the crewmember.

[Amdt. 25-123, 72 FR 63406, Nov. 8, 2007]

§ 25.1362 Electrical supplies for emergency conditions.

A suitable electrical supply must be provided to those services required for emergency procedures after an emergency landing or ditching. The circuits for these services must be designed, protected, and installed so that the risk of the services being rendered ineffective under these emergency conditions is minimized.

[Amdt. 25-123, 72 FR 63406, Nov. 8, 2007]

§ 25.1363 Electrical system tests.

(a) When laboratory tests of the electrical system are conducted—

(1) The tests must be performed on a mock-up using the same generating equipment used in the airplane;

(2) The equipment must simulate the electrical characteristics of the distribution wiring and connected loads to the extent necessary for valid test results; and

(3) Laboratory generator drives must simulate the actual prime movers on the airplane with respect to their reaction to generator loading, including loading due to faults.

(b) For each flight condition that cannot be simulated adequately in the laboratory or by ground tests on the airplane, flight tests must be made.

§ 25.1365 Electrical appliances, motors, and transformers.

(a) Domestic appliances must be designed and installed so that in the event of failures of the electrical supply or control system, the requirements of § 25.1309(b), (c), and (d) will be satisfied. Domestic appliances are items such as cooktops, ovens, coffee makers, water heaters, refrigerators, and toilet flush systems that are placed on the airplane to provide service amenities to passengers.

(b) Galleys and cooking appliances must be installed in a way that minimizes risk of overheat or fire.

(c) Domestic appliances, particularly those in galley areas, must be installed or protected so as to prevent damage or contamination of other equipment or systems from fluids or vapors which may be present during normal operation or as a result of spillage, if such damage or contamination could create a hazardous condition.

(d) Unless compliance with § 25.1309(b) is provided by the circuit protective device required by § 25.1357(a), electric motors and transformers, including those installed in domestic systems, must have a suitable thermal protection device to prevent overheating under normal operation and failure conditions, if overheating could create a smoke or fire hazard.

[Amdt. 25-123, 72 FR 63406, Nov. 8, 2007]

Lights

§ 25.1381 Instrument lights.

(a) The instrument lights must—

(1) Provide sufficient illumination to make each instrument, switch and other device necessary for safe operation easily readable unless sufficient illumination is available from another source; and

(2) Be installed so that—

(i) Their direct rays are shielded from the pilot's eyes; and

(ii) No objectionable reflections are visible to the pilot.

(b) Unless undimmed instrument lights are satisfactory under each expected flight condition, there must be a means to control the intensity of illumination.

[Doc. No. 5066, 29 FR 18291, Dec. 24, 1964, as amended by Amdt. 25-72, 55 FR 29785, July 20, 1990]

§ 25.1383 Landing lights.

(a) Each landing light must be approved, and must be installed so that—

(1) No objectionable glare is visible to the pilot;

(2) The pilot is not adversely affected by halation; and

(3) It provides enough light for night landing.

(b) Except when one switch is used for the lights of a multiple light installation at one location, there must be a separate switch for each light.

(c) There must be a means to indicate to the pilots when the landing lights are extended.

§ 25.1385 Position light system installation.

(a) *General.* Each part of each position light system must meet the applicable requirements of this section and each system as a whole must meet the requirements of §§ 25.1387 through 25.1397.

(b) *Forward position lights.* Forward position lights must consist of a red and a green light spaced laterally as far apart as practicable and installed forward on the airplane so that, with the airplane in the normal flying position, the red light is on the left side and the green light is on the right side. Each light must be approved.

(c) *Rear position light.* The rear position light must be a white light mounted as far aft as practicable on the tail or on each wing tip, and must be approved.

(d) *Light covers and color filters.* Each light cover or color filter must be at least flame resistant and may not change color or shape or lose any appreciable light transmission during normal use.

[Doc. No. 5066, 29 FR 18291, Dec. 24, 1964, as amended by Amdt. 25-38, 41 FR 55468, Dec. 20, 1976]

§ 25.1387 Position light system dihedral angles.

(a) Except as provided in paragraph (e) of this section, each forward and rear position light must, as installed, show unbroken light within the dihedral angles described in this section.

(b) Dihedral angle *L* (left) is formed by two intersecting vertical planes, the first parallel to the longitudinal axis of the airplane, and the other at 110 degrees to the left of the first, as viewed when looking forward along the longitudinal axis.

(c) Dihedral angle *R* (right) is formed by two intersecting vertical planes, the first parallel to the longitudinal axis of the airplane, and the other at 110 degrees to the right of the first, as viewed when looking forward along the longitudinal axis.

(d) Dihedral angle *A* (aft) is formed by two intersecting vertical planes making angles of 70 degrees to the right and to the left, respectively, to a vertical plane passing through the longitudinal axis, as viewed when looking aft along the longitudinal axis.

(e) If the rear position light, when mounted as far aft as practicable in accordance with § 25.1385(c), cannot show unbroken light within dihedral angle *A* (as defined in paragraph (d) of this section), a solid angle or angles of obstructed visibility totaling not more than 0.04 steradians is allowable within that dihedral angle, if such solid angle is within a cone whose apex is at the rear position light and whose elements make an angle of 30° with a vertical line passing through the rear position light.

[Doc. No. 5066, 29 FR 18291, Dec. 24, 1964, as amended by Amdt. 25-30, 36 FR 21278, Nov. 5, 1971]

§ 25.1389 Position light distribution and intensities.

(a) *General.* The intensities prescribed in this section must be provided by new equipment with light covers and color filters in place. Intensities must be determined with the light source operating at a steady value equal to the average luminous output of the source at the normal operating voltage of the airplane. The light distribution and intensity of each position light must meet the requirements of paragraph (b) of this section.

(b) *Forward and rear position lights.* The light distribution and intensities of forward and rear position lights must be expressed in terms of minimum intensities in the horizontal plane, minimum intensities in any vertical plane, and maximum intensities in overlapping beams, within dihedral angles *L*, *R*, and *A*, and must meet the following requirements:

(1) *Intensities in the horizontal plane.* Each intensity in the horizontal plane (the plane containing the longitudinal axis of the airplane and perpendicular to the plane of symmetry of the airplane) must equal or exceed the values in § 25.1391.

(2) *Intensities in any vertical plane.* Each intensity in any vertical plane (the plane perpendicular to the horizontal plane) must equal or exceed the appropriate value in § 25.1393, where *I* is the minimum intensity prescribed in § 25.1391 for the corresponding angles in the horizontal plane.

(3) *Intensities in overlaps between adjacent signals.* No intensity in any overlap between adjacent signals may exceed the values given in § 25.1395, except that higher intensities in overlaps may be used with main beam intensities substantially greater than the minima specified in §§ 25.1391 and 25.1393 if the overlap intensities in relation to the main beam intensities do not adversely affect signal clarity. When the peak intensity of the forward position lights is more than 100 candles, the maximum overlap intensities between them may exceed the values given in § 25.1395 if the overlap intensity in Area A is not more than 10 percent of peak position light intensity and the overlap intensity in Area B is not greater than 2.5 percent of peak position light intensity.

§ 25.1391 Minimum intensities in the horizontal plane of forward and rear position lights.

Each position light intensity must equal or exceed the applicable values in the following table:

Dihedral angle (light included)	Angle from right or left of longitudinal axis, measured from dead ahead	Intensity (candles)
L and *R* (forward red and green)	0° to 10°	40
	10° to 20°	30
	20° to 110°	5
A (rear white)	110° to 180°	20

§ 25.1393 Minimum intensities in any vertical plane of forward and rear position lights.

Each position light intensity must equal or exceed the applicable values in the following table:

Angle above or below the horizontal plane	Intensity, *I*
0°	1.00
0° to 5°	0.90
5° to 10°	0.80
10° to 15°	0.70
15° to 20°	0.50
20° to 30°	0.30
30° to 40°	0.10
40° to 90°	0.05

§ 25.1395 Maximum intensities in overlapping beams of forward and rear position lights.

No position light intensity may exceed the applicable values in the following table, except as provided in § 25.1389(b)(3).

Overlaps	Maximum intensity	
	Area A (candles)	Area B (candles)
Green in dihedral angle *L*	10	1
Red in dihedral angle *R*	10	1
Green in dihedral angle *A*	5	1
Red in dihedral angle *A*	5	1
Rear white in dihedral angle *L*	5	1
Rear white in dihedral angle *R*	5	1

Where—

(a) Area A includes all directions in the adjacent dihedral angle that pass through the light source and intersect the common boundary plane at more than 10 degrees but less than 20 degrees; and

(b) Area B includes all directions in the adjacent dihedral angle that pass through the light source and intersect the common boundary plane at more than 20 degrees.

§ 25.1397 Color specifications.

Each position light color must have the applicable International Commission on Illumination chromaticity coordinates as follows:

(a) *Aviation red—*

y is not greater than 0.335; and

z is not greater than 0.002.

(b) *Aviation green—*

x is not greater than $0.440 - 0.320\,y$;

x is not greater than $y - 0.170$; and

y is not less than 0.390–0.170 x.

(c) *Aviation white*—

x is not less than 0.300 and not greater than 0.540;

y is not less than x -0.040; or y_0 -0.010, whichever is the smaller; and

y is not greater than x +0.020 nor 0.636–0.400 x;

Where y_0 is the y coordinate of the Planckian radiator for the value of x considered.

[Doc. No. 5066, 29 FR 18291, Dec. 24, 1964, as amended by Amdt. 25-27, 36 FR 12972, July 10, 1971]

§ 25.1399 Riding light.

(a) Each riding (anchor) light required for a seaplane or amphibian must be installed so that it can—

(1) Show a white light for at least 2 nautical miles at night under clear atmospheric conditions; and

(2) Show the maximum unbroken light practicable when the airplane is moored or drifting on the water.

(b) Externally hung lights may be used.

§ 25.1401 Anticollision light system.

(a) *General.* The airplane must have an anticollision light system that—

(1) Consists of one or more approved anticollision lights located so that their light will not impair the crew's vision or detract from the conspicuity of the position lights; and

(2) Meets the requirements of paragraphs (b) through (f) of this section.

(b) *Field of coverage.* The system must consist of enough lights to illuminate the vital areas around the airplane considering the physical configuration and flight characteristics of the airplane. The field of coverage must extend in each direction within at least 75 degrees above and 75 degrees below the horizontal plane of the airplane, except that a solid angle or angles of obstructed visibility totaling not more than 0.03 steradians is allowable within a solid angle equal to 0.15 steradians centered about the longitudinal axis in the rearward direction.

(c) *Flashing characteristics.* The arrangement of the system, that is, the number of light sources, beam width, speed of rotation, and other characteristics, must give an effective flash frequency of not less than 40, nor more than 100 cycles per minute. The effective flash frequency is the frequency at which the airplane's complete anticollision light system is observed from a distance, and applies to each sector of light including any overlaps that exist when the system consists of more than one light source. In overlaps, flash frequencies may exceed 100, but not 180 cycles per minute.

(d) *Color.* Each anticollision light must be either aviation red or aviation white and must meet the applicable requirements of § 25.1397.

(e) *Light intensity.* The minimum light intensities in all vertical planes, measured with the red filter (if used) and expressed in terms of "effective" intensities, must meet the requirements of paragraph (f) of this section. The following relation must be assumed:

$$I_e = \frac{\int_{t_1}^{t_2} I(t)\,dt}{0.2 + (t_2 - t_1)}$$

where:

I_e=effective intensity (candles).

$I(t)$=instantaneous intensity as a function of time.

t_2—t_1=flash time interval (seconds).

Normally, the maximum value of effective intensity is obtained when t_2 and t_1 are chosen so that the effective intensity is equal to the instantaneous intensity at t_2 and t_1.

(f) *Minimum effective intensities for anticollision lights.* Each anticollision light effective intensity must equal or exceed the applicable values in the following table.

Angle above or below the horizontal plane	Effective intensity (candles)
0° to 5°	400
5° to 10°	240
10° to 20°	80
20° to 30°	40
30° to 75°	20

[Doc. No. 5066, 29 FR 18291, Dec. 24, 1964, as amended by Amdt. 25-27, 36 FR 12972, July 10, 1971; Amdt. 25-41, 42 FR 36970, July 18, 1977]

§ 25.1403 Wing icing detection lights.

Unless operations at night in known or forecast icing conditions are prohibited by an operating limitation, a means must be provided for illuminating or otherwise determining the formation of ice on the parts of the wings that are critical from the standpoint of ice accumulation. Any illumination that is used must be of a type that will not cause glare or reflection that would handicap crewmembers in the performance of their duties.

[Amdt. 25-38, 41 FR 55468, Dec. 20, 1976]

Safety Equipment

§ 25.1411 General.

(a) *Accessibility.* Required safety equipment to be used by the crew in an emergency must be readily accessible.

(b) *Stowage provisions.* Stowage provisions for required emergency equipment must be furnished and must—

(1) Be arranged so that the equipment is directly accessible and its location is obvious; and

(2) Protect the safety equipment from inadvertent damage.

(c) *Emergency exit descent device.* The stowage provisions for the emergency exit descent devices required by § 25.810(a) must be at each exit for which they are intended.

(d) *Liferafts.* (1) The stowage provisions for the liferafts described in § 25.1415 must accommodate enough rafts for the maximum number of occupants for which certification for ditching is requested.

(2) Liferafts must be stowed near exits through which the rafts can be launched during an unplanned ditching.

(3) Rafts automatically or remotely released outside the airplane must be attached to the airplane by means of the static line prescribed in § 25.1415.

(4) The stowage provisions for each portable liferaft must allow rapid detachment and removal of the raft for use at other than the intended exits.

(e) *Long-range signaling device.* The stowage provisions for the long-range signaling device required by § 25.1415 must be near an exit available during an unplanned ditching.

(f) *Life preserver stowage provisions.* The stowage provisions for life preservers described in § 25.1415 must accommodate one life preserver for each occupant for which certification for ditching is requested. Each life preserver must be within easy reach of each seated occupant.

(g) *Life line stowage provisions.* If certification for ditching under § 25.801 is requested, there must be provisions to store life lines. These provisions must—

(1) Allow one life line to be attached to each side of the fuselage; and

(2) Be arranged to allow the life lines to be used to enable the occupants to stay on the wing after ditching.

[Doc. No. 5066, 29 FR 18291, Dec. 24, 1964, as amended by Amdt. 25-32, 37 FR 3972, Feb. 24, 1972; Amdt. 25-46, 43 FR 50598, Oct. 30, 1978; Amdt. 25-53, 45 FR 41593, June 19, 1980; Amdt. 25-70, 54 FR 43925, Oct. 27, 1989; Amdt. 25-79, 58 FR 45229, Aug. 26, 1993; Amdt. 25-116, 69 FR 62789, Oct. 27, 2004]

§ 25.1415 Ditching equipment.

(a) Ditching equipment used in airplanes to be certificated for ditching under § 25.801, and required by the operating rules of this chapter, must meet the requirements of this section.

(b) Each liferaft and each life preserver must be approved. In addition—

(1) Unless excess rafts of enough capacity are provided, the buoyancy and seating capacity beyond the rated capacity of the rafts must accommodate all occupants of the airplane in the event of a loss of one raft of the largest rated capacity; and

(2) Each raft must have a trailing line, and must have a static line designed to hold the raft near the airplane but to release it if the airplane becomes totally submerged.

(c) Approved survival equipment must be attached to each liferaft.

(d) There must be an approved survival type emergency locator transmitter for use in one life raft.

(e) For airplanes not certificated for ditching under § 25.801 and not having approved life preservers, there must be an approved flotation means for each occupant. This means must be within easy reach of each seated occupant and must be readily removable from the airplane.

[Doc. No. 5066, 29 FR 18291, Dec. 24, 1964, as amended by Amdt. 25-29, 36 FR 18722, Sept. 21, 1971; Amdt. 25-50, 45 FR 38348, June 9, 1980; Amdt. 25-72, 55 FR 29785, July 20, 1990; Amdt. 25-82, 59 FR 32057, June 21, 1994]

§ 25.1419 Ice protection.

If the applicant seeks certification for flight in icing conditions, the airplane must be able to safely operate in the continuous maximum and intermittent maximum icing conditions of appendix C. To establish this—

(a) An analysis must be performed to establish that the ice protection for the various components of the airplane is adequate, taking into account the various airplane operational configurations; and

(b) To verify the ice protection analysis, to check for icing anomalies, and to demonstrate that the ice protection system and its components are effective, the airplane or its components must be flight tested in the various operational configurations, in measured natural atmospheric icing conditions and, as found necessary, by one or more of the following means:

(1) Laboratory dry air or simulated icing tests, or a combination of both, of the components or models of the components.

(2) Flight dry air tests of the ice protection system as a whole, or of its individual components.

(3) Flight tests of the airplane or its components in measured simulated icing conditions.

(c) Caution information, such as an amber caution light or equivalent, must be provided to alert the flightcrew when the anti-ice or de-ice system is not functioning normally.

(d) For turbine engine powered airplanes, the ice protection provisions of this section are considered to be applicable primarily to the airframe. For the powerplant installation, certain additional provisions of subpart E of this part may be found applicable. (e) One of the following methods of icing detection and activation of the airframe ice protection system must be provided:

(1) A primary ice detection system that automatically activates or alerts the flightcrew to activate the airframe ice protection system;

(2) A definition of visual cues for recognition of the first sign of ice accretion on a specified surface combined with an advisory ice detection system that alerts the flightcrew to activate the airframe ice protection system; or

(3) Identification of conditions conducive to airframe icing as defined by an appropriate static or total air temperature and visible moisture for use by the flightcrew to activate the airframe ice protection system.

(f) Unless the applicant shows that the airframe ice protection system need not be operated during specific phases of flight, the requirements of paragraph (e) of this section are applicable to all phases of flight.

(g) After the initial activation of the airframe ice protection system—

(1) The ice protection system must be designed to operate continuously;

(2) The airplane must be equipped with a system that automatically cycles the ice protection system; or

(3) An ice detection system must be provided to alert the flightcrew each time the ice protection system must be cycled.

(h) Procedures for operation of the ice protection system, including activation and deactivation, must be established and documented in the Airplane Flight Manual.

[Amdt. 25-72, 55 FR 29785, July 20, 1990, as amended by Amdt. 25-121, 72 FR 44669, Aug. 8, 2007; Amdt. 25-129, 74 FR 38339, Aug. 3, 2009]

§ 25.1421 Megaphones.

If a megaphone is installed, a restraining means must be provided that is capable of restraining the megaphone when it is subjected to the ultimate inertia forces specified in § 25.561(b)(3).

[Amdt. 25-41, 42 FR 36970, July 18, 1977]

§ 25.1423 Public address system.

A public address system required by this chapter must—

(a) Be powerable when the aircraft is in flight or stopped on the ground, after the shutdown or failure of all engines and auxiliary power units, or the disconnection or failure of all power sources dependent on their continued operation, for—

(1) A time duration of at least 10 minutes, including an aggregate time duration of at least 5 minutes of announcements made by flight and cabin crewmembers, considering all other loads which may remain powered by the same source when all other power sources are inoperative; and

(2) An additional time duration in its standby state appropriate or required for any other loads that are

powered by the same source and that are essential to safety of flight or required during emergency conditions.

(b) Be capable of operation within 3 seconds from the time a microphone is removed from its stowage.

(c) Be intelligible at all passenger seats, lavatories, and flight attendant seats and work stations.

(d) Be designed so that no unused, unstowed microphone will render the system inoperative.

(e) Be capable of functioning independently of any required crewmember interphone system.

(f) Be accessible for immediate use from each of two flight crewmember stations in the pilot compartment.

(g) For each required floor-level passenger emergency exit which has an adjacent flight attendant seat, have a microphone which is readily accessible to the seated flight attendant, except that one microphone may serve more than one exit, provided the proximity of the exits allows unassisted verbal communication between seated flight attendants.

[Doc. No. 26003, 58 FR 45229, Aug. 26, 1993, as amended by Amdt. 25-115, 69 FR 40527, July 2, 2004]

Miscellaneous Equipment

§ 25.1431 Electronic equipment.

(a) In showing compliance with § 25.1309 (a) and (b) with respect to radio and electronic equipment and their installations, critical environmental conditions must be considered.

(b) Radio and electronic equipment must be supplied with power under the requirements of § 25.1355(c).

(c) Radio and electronic equipment, controls, and wiring must be installed so that operation of any one unit or system of units will not adversely affect the simultaneous operation of any other radio or electronic unit, or system of units, required by this chapter.

(d) Electronic equipment must be designed and installed such that it does not cause essential loads to become inoperative as a result of electrical power supply transients or transients from other causes.

[Docket No. 5066, 29 FR 18291, Dec. 24, 1964, as amended by Amdt. 25-113, 69 FR 12530, Mar. 16, 2004]

§ 25.1433 Vacuum systems.

There must be means, in addition to the normal pressure relief, to automatically relieve the pressure in the discharge lines from the vacuum air pump when the delivery temperature of the air becomes unsafe.

[Doc. No. 5066, 29 FR 18291, Dec. 24, 1964, as amended by Amdt. 25-72, 55 FR 29785, July 20, 1990]

§ 25.1435 Hydraulic systems.

(a) *Element design.* Each element of the hydraulic system must be designed to:

(1) Withstand the proof pressure without permanent deformation that would prevent it from performing its intended functions, and the ultimate pressure without rupture. The proof and ultimate pressures are defined in terms of the design operating pressure (DOP) as follows:

Element	Proof (xDOP)	Ultimate (xDOP)
1. Tubes and fittings.	1.5	3.0
2. Pressure vessels containing gas:		
High pressure (e.g., accumulators)	3.0	4.0
Low pressure (e.g., reservoirs)	1.5	3.0
3. Hoses	2.0	4.0
4. All other elements	1.5	2.0

(2) Withstand, without deformation that would prevent it from performing its intended function, the design operating pressure in combination with limit structural loads that may be imposed;

(3) Withstand, without rupture, the design operating pressure multiplied by a factor of 1.5 in combination with ultimate structural load that can reasonably occur simultaneously;

(4) Withstand the fatigue effects of all cyclic pressures, including transients, and associated externally induced loads, taking into account the consequences of element failure; and

(5) Perform as intended under all environmental conditions for which the airplane is certificated.

(b) *System design.* Each hydraulic system must:

(1) Have means located at a flightcrew station to indicate appropriate system parameters, if

(i) It performs a function necessary for continued safe flight and landing; or

(ii) In the event of hydraulic system malfunction, corrective action by the crew to ensure continued safe flight and landing is necessary;

(2) Have means to ensure that system pressures, including transient pressures and pressures from fluid volumetric changes in elements that are likely to remain closed long enough for such changes to occur, are within the design capabilities of each element, such that they meet the requirements defined in § 25.1435(a)(1) through (a)(5);

(3) Have means to minimize the release of harmful or hazardous concentrations of hydraulic fluid or vapors into the crew and passenger compartments during flight;

(4) Meet the applicable requirements of §§ 25.863, 25.1183, 25.1185, and 25.1189 if a flammable hydraulic fluid is used; and

(5) Be designed to use any suitable hydraulic fluid specified by the airplane manufacturer, which must be identified by appropriate markings as required by § 25.1541.

(c) *Tests.* Tests must be conducted on the hydraulic system(s), and/or subsystem(s) and elements, except that analysis may be used in place of or to supplement testing, where the analysis is shown to be reliable and appropriate. All internal and external influences must be taken into account to an extent necessary to evaluate their effects, and to assure reliable system and element functioning and integration. Failure or unacceptable deficiency of an element or system must be corrected and be sufficiently retested, where necessary.

(1) The system(s), subsystem(s), or element(s) must be subjected to performance, fatigue, and endurance tests representative of airplane ground and flight operations.

(2) The complete system must be tested to determine proper functional performance and relation to the other systems, including simulation of relevant failure conditions, and to support or validate element design.

(3) The complete hydraulic system(s) must be functionally tested on the airplane in normal operation over the range of motion of all associated user systems. The test must be conducted at the system relief pressure or 1.25 times the DOP if a system pressure relief device is not part of the system design. Clearances between hydraulic system elements and other systems or structural elements must remain adequate and there must be no detrimental effects.

[Doc. No. 28617, 66 FR 27402, May 16, 2001]

§ 25.1438 Pressurization and pneumatic systems.

(a) Pressurization system elements must be burst pressure tested to 2.0 times, and proof pressure tested to 1.5 times, the maximum normal operating pressure.

(b) Pneumatic system elements must be burst pressure tested to 3.0 times, and proof pressure tested to 1.5 times, the maximum normal operating pressure.

(c) An analysis, or a combination of analysis and test, may be substituted for any test required by paragraph (a) or (b) of this section if the Administrator finds it equivalent to the required test.

[Amdt. 25-41, 42 FR 36971, July 18, 1977]

§ 25.1439 Protective breathing equipment.

(a) Fixed (stationary, or built in) protective breathing equipment must be installed for the use of the flightcrew, and at least one portable protective breathing equipment shall be located at or near the flight deck for use by a flight crewmember. In addition, portable protective breathing equipment must be installed for the use of appropriate crewmembers for fighting fires in compartments accessible in flight other than the flight deck. This includes isolated compartments and upper and lower lobe galleys, in which crewmember occupancy is permitted during flight. Equipment must be installed for the maximum number of crewmembers expected to be in the area during any operation.

(b) For protective breathing equipment required by paragraph (a) of this section or by the applicable Operating Regulations:

(1) The equipment must be designed to protect the appropriate crewmember from smoke, carbon dioxide, and other harmful gases while on flight deck duty or while combating fires.

(2) The equipment must include—

(i) Masks covering the eyes, nose and mouth, or

(ii) Masks covering the nose and mouth, plus accessory equipment to cover the eyes.

(3) Equipment, including portable equipment, must allow communication with other crewmembers while in use. Equipment available at flightcrew assigned duty stations must also enable the flightcrew to use radio equipment.

(4) The part of the equipment protecting the eyes shall not cause any appreciable adverse effect on vision and must allow corrective glasses to be worn.

(5) The equipment must supply protective oxygen of 15 minutes duration per crewmember at a pressure altitude of 8,000 feet with a respiratory minute volume of 30 liters per minute BTPD. The equipment and system must be designed to prevent any inward leakage to the inside of the device and prevent any outward leakage causing significant increase in the oxygen content of the local ambient atmosphere. If a demand oxygen system is used, a supply of 300 liters of free oxygen at 70 °F. and 760mm. Hg. pressure is considered to be of 15-minute duration at the prescribed altitude and minute volume. If a continuous flow open circuit protective breathing system is used, a flow rate of 60 liters per minute at 8,000 feet (45 liters per minute at sea level) and a supply of 600 liters of free oxygen at 70 °F. and 760mm. Hg. pressure is considered to be of 15-minute duration at the prescribed altitude and minute volume. Continuous flow systems must not increase the ambient oxygen content of the local atmosphere above that of demand systems. BTPD refers to body temperature conditions (that is, 37°C., at ambient pressure, dry).

(6) The equipment must meet the requirements of § 25.1441.

[Doc. No. FAA-2002-13859, 69 FR 40528, July 2, 2004]

§ 25.1441 Oxygen equipment and supply.

(a) If certification with supplemental oxygen equipment is requested, the equipment must meet the requirements of this section and §§ 25.1443 through 25.1453.

(b) The oxygen system must be free from hazards in itself, in its method of operation, and in its effect upon other components.

(c) There must be a means to allow the crew to readily determine, during flight, the quantity of oxygen available in each source of supply.

(d) The oxygen flow rate and the oxygen equipment for airplanes for which certification for operation above 40,000 feet is requested must be approved.

§ 25.1443 Minimum mass flow of supplemental oxygen.

(a) If continuous flow equipment is installed for use by flight crewmembers, the minimum mass flow of supplemental oxygen required for each crewmember may not be less than the flow required to maintain, during inspiration, a mean tracheal oxygen partial pressure of 149mm. Hg. when breathing 15 liters per minute, BTPS, and with a maximum tidal volume of 700 cc. with a constant time interval between respirations.

(b) If demand equipment is installed for use by flight crewmembers, the minimum mass flow of supplemental oxygen required for each crewmember may not be less than the flow required to maintain, during inspiration, a mean tracheal oxygen partial pressure of 122mm. Hg., up to and including a cabin pressure altitude of 35,000 feet, and 95 percent oxygen between cabin pressure altitudes of 35,000 and 40,000 feet, when breathing 20 liters per minute BTPS. In addition, there must be means to allow the crew to use undiluted oxygen at their discretion.

(c) For passengers and cabin attendants, the minimum mass flow of supplemental oxygen required for each person at various cabin pressure altitudes may not be less than the flow required to maintain, during inspiration and while using the oxygen equipment (including masks) provided, the following mean tracheal oxygen partial pressures:

(1) At cabin pressure altitudes above 10,000 feet up to and including 18,500 feet, a mean tracheal oxygen partial pressure of 100mm. Hg. when breathing 15 liters per minute, BTPS, and with a tidal volume of 700 cc. with a constant time interval between respirations.

(2) At cabin pressure altitudes above 18,500 feet up to and including 40,000 feet, a mean tracheal oxygen partial pressure of 83.8mm. Hg. when breathing 30 liters per minute, BTPS, and with a tidal volume of 1,100 cc. with a constant time interval between respirations.

(d) If first-aid oxygen equipment is installed, the minimum mass flow of oxygen to each user may not be less than four liters per minute, STPD. However, there may be a means to decrease this flow to not less than two liters per minute, STPD, at any cabin altitude. The quantity of oxygen required is based upon an average flow rate of three liters per minute per person for whom first-aid oxygen is required.

(e) If portable oxygen equipment is installed for use by crewmembers, the minimum mass flow of supplemental oxygen is the same as specified in paragraph (a) or (b) of this section, whichever is applicable.

§ 25.1445 Equipment standards for the oxygen distributing system.

(a) When oxygen is supplied to both crew and passengers, the distribution system must be designed for either—

(1) A source of supply for the flight crew on duty and a separate source for the passengers and other crewmembers; or

(2) A common source of supply with means to separately reserve the minimum supply required by the flight crew on duty.

(b) Portable walk-around oxygen units of the continuous flow, diluter-demand, and straight demand kinds may be used to meet the crew or passenger breathing requirements.

§ 25.1447 Equipment standards for oxygen dispensing units.

If oxygen dispensing units are installed, the following apply:

(a) There must be an individual dispensing unit for each occupant for whom supplemental oxygen is to be supplied. Units must be designed to cover the nose and mouth and must be equipped with a suitable means to retain the unit in position on the face. Flight crew masks for supplemental oxygen must have provisions for the use of communication equipment.

(b) If certification for operation up to and including 25,000 feet is requested, an oxygen supply terminal and unit of oxygen dispensing equipment for the immediate use of oxygen by each crewmember must be within easy reach of that crewmember. For any other occupants, the supply terminals and dispensing equipment must be located to allow the use of oxygen as required by the operating rules in this chapter.

(c) If certification for operation above 25,000 feet is requested, there must be oxygen dispensing equipment meeting the following requirements:

(1) There must be an oxygen dispensing unit connected to oxygen supply terminals immediately available to each occupant, wherever seated, and at least two oxygen dispensing units connected to oxygen terminals in each lavatory. The total number of dispensing units and outlets in the cabin must exceed the number of seats by at least 10 percent. The extra units must be as uniformly distributed throughout the cabin as practicable. If certification for operation above 30,000 feet is requested, the dispensing units providing the required oxygen flow must be automatically presented to the occupants before the cabin pressure altitude exceeds 15,000 feet. The crew must be provided with a manual means of making the dispensing units immediately available in the event of failure of the automatic system.

(2) Each flight crewmember on flight deck duty must be provided with a quick-donning type oxygen dispensing unit connected to an oxygen supply terminal. This dispensing unit must be immediately available to the flight crewmember when seated at his station, and installed so that it:

(i) Can be placed on the face from its ready position, properly secured, sealed, and supplying oxygen upon demand, with one hand, within five seconds and without disturbing eyeglasses or causing delay in proceeding with emergency duties; and

(ii) Allows, while in place, the performance of normal communication functions.

(3) The oxygen dispensing equipment for the flight crewmembers must be:

(i) The diluter demand or pressure demand (pressure demand mask with a diluter demand pressure breathing regulator) type, or other approved oxygen equipment shown to provide the same degree of protection, for airplanes to be operated above 25,000 feet.

(ii) The pressure demand (pressure demand mask with a diluter demand pressure breathing regulator) type with mask-mounted regulator, or other approved oxygen equipment shown to provide the same degree of protection, for airplanes operated at altitudes where decompressions that are not extremely improbable may expose the flightcrew to cabin pressure altitudes in excess of 34,000 feet.

(4) Portable oxygen equipment must be immediately available for each cabin attendant. The portable oxygen equipment must have the oxygen dispensing unit connected to the portable oxygen supply.

[Doc. No. 5066, 29 FR 18291, Dec. 24, 1964, as amended by Amdt. 25-41, 42 FR 36971, July 18, 1977; Amdt. 25-87, 61 FR 28696, June 5, 1996; Amdt. 25-116, 69 FR 62789, Oct. 27, 2004]

§ 25.1449 Means for determining use of oxygen.

There must be a means to allow the crew to determine whether oxygen is being delivered to the dispensing equipment.

§ 25.1450 Chemical oxygen generators.

(a) For the purpose of this section, a chemical oxygen generator is defined as a device which produces oxygen by chemical reaction.

(b) Each chemical oxygen generator must be designed and installed in accordance with the following requirements:

(1) Surface temperature developed by the generator during operation may not create a hazard to the airplane or to its occupants.

(2) Means must be provided to relieve any internal pressure that may be hazardous.

(c) In addition to meeting the requirements in paragraph (b) of this section, each portable chemical oxygen generator that is capable of sustained operation by successive replacement of a generator element must be placarded to show—

(1) The rate of oxygen flow, in liters per minute;

(2) The duration of oxygen flow, in minutes, for the replaceable generator element; and

(3) A warning that the replaceable generator element may be hot, unless the element construction is such that the surface temperature cannot exceed 100 degrees F.

[Amdt. 25-41, 42 FR 36971, July 18, 1977]

§ 25.1453 Protection of oxygen equipment from rupture.

Oxygen pressure tanks, and lines between tanks and the shutoff means, must be—

(a) Protected from unsafe temperatures; and

(b) Located where the probability and hazards of rupture in a crash landing are minimized.

§ 25.1455 Draining of fluids subject to freezing.

If fluids subject to freezing may be drained overboard in flight or during ground operation, the drains must be designed and located to prevent the formation of hazardous quantities of ice on the airplane as a result of the drainage.

[Amdt. 25-23, 35 FR 5680, Apr. 8, 1970]

§ 25.1457 Cockpit voice recorders.

(a) Each cockpit voice recorder required by the operating rules of this chapter must be approved and must be installed so that it will record the following:

(1) Voice communications transmitted from or received in the airplane by radio.

(2) Voice communications of flight crewmembers on the flight deck.

(3) Voice communications of flight crewmembers on the flight deck, using the airplane's interphone system.

(4) Voice or audio signals identifying navigation or approach aids introduced into a headset or speaker.

(5) Voice communications of flight crewmembers using the passenger loudspeaker system, if there is such a system and if the fourth channel is available in accordance with the requirements of paragraph (c)(4)(ii) of this section.

(6) If datalink communication equipment is installed, all datalink communications, using an approved data message set. Datalink messages must be recorded as the output signal from the communications unit that translates the signal into usable data.

(b) The recording requirements of paragraph (a)(2) of this section must be met by installing a cockpit-mounted area microphone, located in the best position for recording voice communications originating at the first and second pilot stations and voice communications of other crewmembers on the flight deck when directed to those stations. The microphone must be so located and, if necessary, the preamplifiers and filters of the recorder must be so adjusted or supplemented, that the intelligibility of the recorded communications is as high as practicable when recorded under flight cockpit noise conditions and played back. Repeated aural or visual playback of the record may be used in evaluating intelligibility.

(c) Each cockpit voice recorder must be installed so that the part of the communication or audio signals specified in paragraph (a) of this section obtained from each of the following sources is recorded on a separate channel:

(1) For the first channel, from each boom, mask, or hand-held microphone, headset, or speaker used at the first pilot station.

(2) For the second channel from each boom, mask, or hand-held microphone, headset, or speaker used at the second pilot station.

(3) For the third channel—from the cockpit-mounted area microphone.

(4) For the fourth channel, from—

(i) Each boom, mask, or hand-held microphone, headset, or speaker used at the station for the third and fourth crew members; or

(ii) If the stations specified in paragraph (c)(4)(i) of this section are not required or if the signal at such a station is picked up by another channel, each microphone on the flight deck that is used with the passenger loudspeaker system, if its signals are not picked up by another channel.

(5) As far as is practicable all sounds received by the microphone listed in paragraphs (c)(1), (2), and (4) of this section must be recorded without interruption irrespective of the position of the interphone-transmitter key switch. The design shall ensure that sidetone for the flight crew is produced only when the interphone, public address system, or radio transmitters are in use.

(d) Each cockpit voice recorder must be installed so that—

(1)(i) It receives its electrical power from the bus that provides the maximum reliability for operation of the cockpit voice recorder without jeopardizing service to essential or emergency loads.

(ii) It remains powered for as long as possible without jeopardizing emergency operation of the airplane.

(2) There is an automatic means to simultaneously stop the recorder and prevent each erasure feature from functioning, within 10 minutes after crash impact;

(3) There is an aural or visual means for preflight checking of the recorder for proper operation;

(4) Any single electrical failure external to the recorder does not disable both the cockpit voice recorder and the flight data recorder;

(5) It has an independent power source—

(i) That provides 10±1 minutes of electrical power to operate both the cockpit voice recorder and cockpit-mounted area microphone;

(ii) That is located as close as practicable to the cockpit voice recorder; and

(iii) To which the cockpit voice recorder and cockpit-mounted area microphone are switched automatically in the event that all other power to the cockpit voice recorder is interrupted either by normal shutdown or by any other loss of power to the electrical power bus; and

(6) It is in a separate container from the flight data recorder when both are required. If used to comply with only the cockpit voice recorder requirements, a combination unit may be installed.

(e) The recorder container must be located and mounted to minimize the probability of rupture of the container as a result of crash impact and consequent heat damage to the recorder from fire.

(1) Except as provided in paragraph (e)(2) of this section, the recorder container must be located as far aft as practicable, but need not be outside of the pressurized compartment, and may not be located where aft-mounted engines may crush the container during impact.

(2) If two separate combination digital flight data recorder and cockpit voice recorder units are installed instead of one cockpit voice recorder and one digital flight data recorder, the combination unit that is installed to comply with the cockpit voice recorder requirements may be located near the cockpit.

(f) If the cockpit voice recorder has a bulk erasure device, the installation must be designed to minimize the probability of inadvertent operation and actuation of the device during crash impact.

(g) Each recorder container must—

(1) Be either bright orange or bright yellow;

(2) Have reflective tape affixed to its external surface to facilitate its location under water; and

(3) Have an underwater locating device, when required by the operating rules of this chapter, on or adjacent to the container which is secured in such manner that they are not likely to be separated during crash impact.

[Doc. No. 5066, 29 FR 18291, Dec. 24, 1964, as amended by Amdt. 25-2, 30 FR 3932, Mar. 26, 1965; Amdt. 25-16, 32 FR 13914, Oct. 6, 1967; Amdt. 25-41, 42

FR 36971, July 18, 1977; Amdt. 25-65, 53 FR 26143, July 11, 1988; Amdt. 25-124, 73 FR 12563, Mar. 7, 2008; 74 FR 32800, July 9, 2009]

§ 25.1459 Flight data recorders.

(a) Each flight recorder required by the operating rules of this chapter must be installed so that—

(1) It is supplied with airspeed, altitude, and directional data obtained from sources that meet the accuracy requirements of §§ 25.1323, 25.1325, and 25.1327, as appropriate;

(2) The vertical acceleration sensor is rigidly attached, and located longitudinally either within the approved center of gravity limits of the airplane, or at a distance forward or aft of these limits that does not exceed 25 percent of the airplane's mean aerodynamic chord;

(3)(i) It receives its electrical power from the bus that provides the maximum reliability for operation of the flight data recorder without jeopardizing service to essential or emergency loads.

(ii) It remains powered for as long as possible without jeopardizing emergency operation of the airplane.

(4) There is an aural or visual means for preflight checking of the recorder for proper recording of data in the storage medium;

(5) Except for recorders powered solely by the engine-driven electrical generator system, there is an automatic means to simultaneously stop a recorder that has a data erasure feature and prevent each erasure feature from functioning, within 10 minutes after crash impact;

(6) There is a means to record data from which the time of each radio transmission either to or from ATC can be determined;

(7) Any single electrical failure external to the recorder does not disable both the cockpit voice recorder and the flight data recorder; and

(8) It is in a separate container from the cockpit voice recorder when both are required. If used to comply with only the flight data recorder requirements, a combination unit may be installed. If a combination unit is installed as a cockpit voice recorder to comply with § 25.1457(e)(2), a combination unit must be used to comply with this flight data recorder requirement.

(b) Each nonejectable record container must be located and mounted so as to minimize the probability of container rupture resulting from crash impact and subsequent damage to the record from fire. In meeting this requirement the record container must be located as far aft as practicable, but need not be aft of the pressurized compartment, and may not be where aft-mounted engines may crush the container upon impact.

(c) A correlation must be established between the flight recorder readings of airspeed, altitude, and heading and the corresponding readings (taking into account correction factors) of the first pilot's instruments. The correlation must cover the airspeed range over which the airplane is to be operated, the range of altitude to which the airplane is limited, and 360 degrees of heading. Correlation may be established on the ground as appropriate.

(d) Each recorder container must—

(1) Be either bright orange or bright yellow;

(2) Have reflective tape affixed to its external surface to facilitate its location under water; and

(3) Have an underwater locating device, when required by the operating rules of this chapter, on or adjacent to the container which is secured in such a manner that they are not likely to be separated during crash impact.

(e) Any novel or unique design or operational characteristics of the aircraft shall be evaluated to determine if any dedicated parameters must be recorded on flight recorders in addition to or in place of existing requirements.

[Amdt. 25-8, 31 FR 127, Jan. 6, 1966, as amended by Amdt. 25-25, 35 FR 13192, Aug. 19, 1970; Amdt. 25-37, 40 FR 2577, Jan. 14, 1975; Amdt. 25-41, 42 FR 36971, July 18, 1977; Amdt. 25-65, 53 FR 26144, July 11, 1988; Amdt. 25-124, 73 FR 12563, Mar. 7, 2008; 74 FR 32800, July 9, 2009]

§ 25.1461 Equipment containing high energy rotors.

(a) Equipment containing high energy rotors must meet paragraph (b), (c), or (d) of this section.

(b) High energy rotors contained in equipment must be able to withstand damage caused by malfunctions, vibration, abnormal speeds, and abnormal temperatures. In addition—

(1) Auxiliary rotor cases must be able to contain damage caused by the failure of high energy rotor blades; and

(2) Equipment control devices, systems, and instrumentation must reasonably ensure that no operating limitations affecting the integrity of high energy rotors will be exceeded in service.

(c) It must be shown by test that equipment containing high energy rotors can contain any failure of a high energy rotor that occurs at the highest speed obtainable with the normal speed control devices inoperative.

(d) Equipment containing high energy rotors must be located where rotor failure will neither endanger the occupants nor adversely affect continued safe flight.

[Amdt. 25-41, 42 FR 36971, July 18, 1977]

Subpart G—Operating Limitations and Information

§ 25.1501 General.

(a) Each operating limitation specified in §§ 25.1503 through 25.1533 and other limitations and information necessary for safe operation must be established.

(b) The operating limitations and other information necessary for safe operation must be made available to the crewmembers as prescribed in §§ 25.1541 through 25.1587.

[Amdt. 25-42, 43 FR 2323, Jan. 16, 1978]

Operating Limitations

§ 25.1503 Airspeed limitations: general.

When airspeed limitations are a function of weight, weight distribution, altitude, or Mach number, limitations corresponding to each critical combination of these factors must be established.

§ 25.1505 Maximum operating limit speed.

The maximum operating limit speed (V_{MO}/M_{MO} airspeed or Mach Number, whichever is critical at a particular altitude) is a speed that may not be deliberately exceeded in any regime of flight (climb, cruise, or descent), unless a higher speed is authorized for flight test or pilot training operations. V_{MO}/M_{MO} must be established so that it is not greater than the design cruising speed V_C and so that it is sufficiently below V_D/M_D or V_{DF}/M_{DF} to make it highly improbable that the latter speeds will be inadvertently exceeded in operations. The speed margin between V_{MO}/M_{MO} and V_D/M_D or V_{DF}/M_{DF} may not be less than that determined under § 25.335(b) or found necessary during the flight tests conducted under § 25.253.

[Amdt. 25-23, 35 FR 5680, Apr. 8, 1970]

§ 25.1507 Maneuvering speed.

The maneuvering speed must be established so that it does not exceed the design maneuvering speed V_A determined under § 25.335(c).

§ 25.1511 Flap extended speed.

The established flap extended speed V_{FE} must be established so that it does not exceed the design flap speed V_F chosen under §§ 25.335(e) and 25.345, for the corresponding flap positions and engine powers.

§ 25.1513 Minimum control speed.

The minimum control speed V_{MC} determined under § 25.149 must be established as an operating limitation.

§ 25.1515 Landing gear speeds.

(a) The established landing gear operating speed or speeds, V_{LO}, may not exceed the speed at which it is safe both to extend and to retract the landing gear, as determined under § 25.729 or by flight characteristics. If the extension speed is not the same as the retraction speed, the two speeds must be designated as $V_{LO(EXT)}$ and $V_{LO(RET)}$, respectively.

(b) The established landing gear extended speed V_{LE} may not exceed the speed at which it is safe to fly with the landing gear secured in the fully extended position, and that determined under § 25.729.

[Doc. No. 5066, 29 FR 18291, Dec. 24, 1964, as amended by Amdt. 25-38, 41 FR 55468, Dec. 20, 1976]

§ 25.1516 Other speed limitations.

Any other limitation associated with speed must be established.

[Doc. No. 2000-8511, 66 FR 34024, June 26, 2001]

§ 25.1517 Rough air speed, VRA.

A rough air speed, V_{RA}, for use as the recommended turbulence penetration airspeed in § 25.1585(a)(8), must be established, which—

(1) Is not greater than the design airspeed for maximum gust intensity, selected for V_B; and

(2) Is not less than the minimum value of V_B specified in § 25.335(d); and

(3) Is sufficiently less than V_{MO} to ensure that likely speed variation during rough air encounters will not cause the overspeed warning to operate too frequently. In the absence of a rational investigation substantiating the use of other values, V_{RA} must be less than V_{MO}—35 knots (TAS).

[Doc. No. 27902, 61 FR 5222, Feb. 9, 1996]

§ 25.1519 Weight, center of gravity, and weight distribution.

The airplane weight, center of gravity, and weight distribution limitations determined under §§ 25.23 through 25.27 must be established as operating limitations.

§ 25.1521 Powerplant limitations.

(a) *General.* The powerplant limitations prescribed in this section must be established so that they do not exceed the corresponding limits for which the engines or propellers are type certificated and do not exceed the values on which compliance with any other requirement of this part is based.

(b) *Reciprocating engine installations.* Operating limitations relating to the following must be established for reciprocating engine installations:

(1) Horsepower or torque, r.p.m., manifold pressure, and time at critical pressure altitude and sea level pressure altitude for—

(i) Maximum continuous power (relating to unsupercharged operation or to operation in each supercharger mode as applicable); and

(ii) Takeoff power (relating to unsupercharged operation or to operation in each supercharger mode as applicable).

(2) Fuel grade or specification.

(3) Cylinder head and oil temperatures.

(4) Any other parameter for which a limitation has been established as part of the engine type certificate except that a limitation need not be established for a parameter that cannot be exceeded during normal operation due to the design of the installation or to another established limitation.

(c) *Turbine engine installations.* Operating limitations relating to the following must be established for turbine engine installations:

(1) Horsepower, torque or thrust, r.p.m., gas temperature, and time for—

(i) Maximum continuous power or thrust (relating to augmented or unaugmented operation as applicable).

(ii) Takeoff power or thrust (relating to augmented or unaugmented operation as applicable).

(2) Fuel designation or specification.

(3) Any other parameter for which a limitation has been established as part of the engine type certificate except that a limitation need not be established for a parameter that cannot be exceeded during normal operation due to the design of the installation or to another established limitation.

(d) *Ambient temperature.* An ambient temperature limitation (including limitations for winterization installations, if applicable) must be established as the maximum ambient atmospheric temperature established in accordance with § 25.1043(b).

[Amdt. 25-72, 55 FR 29786, July 20, 1990]

§ 25.1522 Auxiliary power unit limitations.

If an auxiliary power unit is installed in the airplane, limitations established for the auxiliary power unit, including categories of operation, must be specified as operating limitations for the airplane.

[Amdt. 25-72, 55 FR 29786, July 20, 1990]

§ 25.1523 Minimum flight crew.

The minimum flight crew must be established so that it is sufficient for safe operation, considering—

(a) The workload on individual crewmembers;

(b) The accessibility and ease of operation of necessary controls by the appropriate crewmember; and

(c) The kind of operation authorized under § 25.1525.

The criteria used in making the determinations required by this section are set forth in appendix D.

[Doc. No. 5066, 29 FR 18291, Dec. 24, 1964, as amended by Amdt. 25-3, 30 FR 6067, Apr. 29, 1965]

§ 25.1525 Kinds of operation.

The kinds of operation to which the airplane is limited are established by the category in which it is eligible for certification and by the installed equipment.

§ 25.1527 Ambient air temperature and operating altitude.

The extremes of the ambient air temperature and operating altitude for which operation is allowed, as limited by flight, structural, powerplant, functional, or equipment characteristics, must be established.

[Doc. No. 2000-8511, 66 FR 34024, June 26, 2001]

§ 25.1529 Instructions for Continued Airworthiness.

The applicant must prepare Instructions for Continued Airworthiness in accordance with appendix H to this part that are acceptable to the Administrator. The instructions may be incomplete at type certification if a program exists to ensure their completion prior to delivery of the first airplane or issuance of a standard certificate of airworthiness, whichever occurs later.

[Amdt. 25-54, 45 FR 60173, Sept. 11, 1980]

§ 25.1531 Maneuvering flight load factors.

Load factor limitations, not exceeding the positive limit load factors determined from the maneuvering diagram in § 25.333(b), must be established.

§ 25.1533 Additional operating limitations.

(a) Additional operating limitations must be established as follows:

(1) The maximum takeoff weights must be established as the weights at which compliance is shown with the applicable provisions of this part (including the takeoff climb provisions of § 25.121(a) through (c), for altitudes and ambient temperatures).

(2) The maximum landing weights must be established as the weights at which compliance is shown with the applicable provisions of this part (including the landing and approach climb provisions of §§ 25.119 and 25.121(d) for altitudes and ambient temperatures).

(3) The minimum takeoff distances must be established as the distances at which compliance is shown with the applicable provisions of this part (including the provisions of §§ 25.109 and 25.113, for weights, altitudes, temperatures, wind components, runway surface conditions (dry and wet), and runway gradients) for smooth, hard-surfaced runways. Additionally, at the option of the applicant, wet runway takeoff distances may be established for runway surfaces that have been grooved or treated with a porous friction course, and may be approved for use on runways where such surfaces have been designed constructed, and maintained in a manner acceptable to the Administrator.

(b) The extremes for variable factors (such as altitude, temperature, wind, and runway gradients) are those at which compliance with the applicable provisions of this part is shown.

[Doc. No. 5066, 29 FR 18291, Dec. 24, 1964, as amended by Amdt. 25-38, 41 FR 55468, Dec. 20, 1976; Amdt. 25-72, 55 FR 29786, July 20, 1990; Amdt. 25-92, 63 FR 8321, Feb. 18, 1998]

§ 25.1535 ETOPS approval.

Except as provided in § 25.3, each applicant seeking ETOPS type design approval must comply with the provisions of Appendix K of this part.

[Doc. No. FAA-2002-6717, 72 FR 1873, Jan. 16, 2007]

Markings and Placards

§ 25.1541 General.

(a) The airplane must contain—

(1) The specified markings and placards; and

(2) Any additional information, instrument markings, and placards required for the safe operation if there are unusual design, operating, or handling characteristics.

(b) Each marking and placard prescribed in paragraph (a) of this section—

(1) Must be displayed in a conspicuous place; and

(2) May not be easily erased, disfigured, or obscured.

§ 25.1543 Instrument markings: general.

For each instrument—

(a) When markings are on the cover glass of the instrument, there must be means to maintain the correct alignment of the glass cover with the face of the dial; and

(b) Each instrument marking must be clearly visible to the appropriate crewmember.

[Doc. No. 5066, 29 FR 18291, Dec. 24, 1964, as amended by Amdt. 25-72, 55 FR 29786, July 20, 1990]

§ 25.1545 Airspeed limitation information.

The airspeed limitations required by § 25.1583 (a) must be easily read and understood by the flight crew.

§ 25.1547 Magnetic direction indicator.

(a) A placard meeting the requirements of this section must be installed on, or near, the magnetic direction indicator.

(b) The placard must show the calibration of the instrument in level flight with the engines operating.

(c) The placard must state whether the calibration was made with radio receivers on or off.

(d) Each calibration reading must be in terms of magnetic heading in not more than 45 degree increments.

§ 25.1549 Powerplant and auxiliary power unit instruments.

For each required powerplant and auxiliary power unit instrument, as appropriate to the type of instrument—

(a) Each maximum and, if applicable, minimum safe operating limit must be marked with a red radial or a red line;

(b) Each normal operating range must be marked with a green arc or green line, not extending beyond the maximum and minimum safe limits;

(c) Each takeoff and precautionary range must be marked with a yellow arc or a yellow line; and

(d) Each engine, auxiliary power unit, or propeller speed range that is restricted because of excessive vibration stresses must be marked with red arcs or red lines.

[Amdt. 25-40, 42 FR 15044, Mar. 17, 1977]

§ 25.1551 Oil quantity indication.

Each oil quantity indicating means must be marked to indicate the quantity of oil readily and accurately.

[Amdt. 25-72, 55 FR 29786, July 20, 1990]

§ 25.1553 Fuel quantity indicator.

If the unusable fuel supply for any tank exceeds one gallon, or five percent of the tank capacity, whichever is greater, a red arc must be marked on its indicator extending from the calibrated zero reading to the lowest reading obtainable in level flight.

§ 25.1555 Control markings.

(a) Each cockpit control, other than primary flight controls and controls whose function is obvious, must be plainly marked as to its function and method of operation.

(b) Each aerodynamic control must be marked under the requirements of §§ 25.677 and 25.699.

(c) For powerplant fuel controls—

(1) Each fuel tank selector control must be marked to indicate the position corresponding to each tank and to each existing cross feed position;

(2) If safe operation requires the use of any tanks in a specific sequence, that sequence must be marked on, or adjacent to, the selector for those tanks; and

(3) Each valve control for each engine must be marked to indicate the position corresponding to each engine controlled.

(d) For accessory, auxiliary, and emergency controls—

(1) Each emergency control (including each fuel jettisoning and fluid shutoff must be colored red; and

(2) Each visual indicator required by § 25.729(e) must be marked so that the pilot can determine at any time when the wheels are locked in either extreme position, if retractable landing gear is used.

§ 25.1557 Miscellaneous markings and placards.

(a) *Baggage and cargo compartments and ballast location.* Each baggage and cargo compartment, and each ballast location must have a placard stating any limitations on contents, including weight, that are necessary under the loading requirements. However, underseat compartments designed for the storage of carry-on articles weighing not more than 20 pounds need not have a loading limitation placard.

(b) *Powerplant fluid filler openings.* The following apply:

(1) Fuel filler openings must be marked at or near the filler cover with—

(i) The word "fuel";

(ii) For reciprocating engine powered airplanes, the minimum fuel grade;

(iii) For turbine engine powered airplanes, the permissible fuel designations; and

(iv) For pressure fueling systems, the maximum permissible fueling supply pressure and the maximum permissible defueling pressure.

(2) Oil filler openings must be marked at or near the filler cover with the word "oil".

(3) Augmentation fluid filler openings must be marked at or near the filler cover to identify the required fluid.

(c) *Emergency exit placards.* Each emergency exit placard must meet the requirements of § 25.811.

(d) *Doors.* Each door that must be used in order to reach any required emergency exit must have a suitable placard stating that the door is to be latched in the open position during takeoff and landing.

[Doc. No. 5066, 29 FR 18291, Dec. 24, 1964, as amended by Amdt. 25-32, 37 FR 3972, Feb. 24, 1972; Amdt. 25-38, 41 FR 55468, Dec. 20, 1976; Amdt. 25-72, 55 FR 29786, July 20, 1990]

§ 25.1561 Safety equipment.

(a) Each safety equipment control to be operated by the crew in emergency, such as controls for automatic liferaft releases, must be plainly marked as to its method of operation.

(b) Each location, such as a locker or compartment, that carries any fire extinguishing, signaling, or other life saving equipment must be marked accordingly.

(c) Stowage provisions for required emergency equipment must be conspicuously marked to identify the contents and facilitate the easy removal of the equipment.

(d) Each liferaft must have obviously marked operating instructions.

(e) Approved survival equipment must be marked for identification and method of operation.

[Doc. No. 5066, 29 FR 18291, Dec. 24, 1964, as amended by Amdt. 25-46, 43 FR 50598, Oct. 30, 1978]

§ 25.1563 Airspeed placard.

A placard showing the maximum airspeeds for flap extension for the takeoff, approach, and landing positions must be installed in clear view of each pilot.

Airplane Flight Manual

§ 25.1581 General.

(a) *Furnishing information.* An Airplane Flight Manual must be furnished with each airplane, and it must contain the following:

(1) Information required by §§ 25.1583 through 25.1587.

(2) Other information that is necessary for safe operation because of design, operating, or handling characteristics.

(3) Any limitation, procedure, or other information established as a condition of compliance with the applicable noise standards of part 36 of this chapter.

(b) *Approved information.* Each part of the manual listed in §§ 25.1583 through 25.1587, that is appropriate to the airplane, must be furnished, verified, and approved, and must be segregated, identified, and clearly distinguished from each unapproved part of that manual.

(c) [Reserved]

(d) Each Airplane Flight Manual must include a table of contents if the complexity of the manual indicates a need for it.

[Amdt. 25-42, 43 FR 2323, Jan. 16, 1978, as amended by Amdt. 25-72, 55 FR 29786, July 20, 1990]

§ 25.1583 Operating limitations.

(a) *Airspeed limitations.* The following airspeed limitations and any other airspeed limitations necessary for safe operation must be furnished:

(1) The maximum operating limit speed V_{MO}/M_{MO} and a statement that this speed limit may not be deliberately exceeded in any regime of flight (climb, cruise, or descent) unless a higher speed is authorized for flight test or pilot training.

(2) If an airspeed limitation is based upon compressibility effects, a statement to this effect and information as to any symptoms, the probable behavior of the airplane, and the recommended recovery procedures.

(3) The maneuvering speed established under § 25.1507 and statements, as applicable to the particular design, explaining that:

(i) Full application of pitch, roll, or yaw controls should be confined to speeds below the maneuvering speed; and

(ii) Rapid and large alternating control inputs, especially in combination with large changes in pitch, roll, or yaw, and full control inputs in more than one axis at the same time, should be avoided as they may result in structural failures at any speed, including below the maneuvering speed.

(4) The flap extended speed V_{FE} and the pertinent flap positions and engine powers.

(5) The landing gear operating speed or speeds, and a statement explaining the speeds as defined in § 25.1515(a).

(6) The landing gear extended speed V_{LE} if greater than V_{LO} and a statement that this is the maximum speed at which the airplane can be safely flown with the landing gear extended.

(b) *Powerplant limitations.* The following information must be furnished:

(1) Limitations required by § 25.1521 and § 25.1522.

(2) Explanation of the limitations, when appropriate.

(3) Information necessary for marking the instruments required by §§ 25.1549 through 25.1553.

(c) *Weight and loading distribution.* The weight and center of gravity limitations established under § 25.1519 must be furnished in the Airplane Flight Manual. All of the following information, including the weight distribution limitations established under § 25.1519, must be presented either in the Airplane Flight Manual or in a separate weight and balance control and loading document that is incorporated by reference in the Airplane Flight Manual:

(1) The condition of the airplane and the items included in the empty weight as defined in accordance with § 25.29.

(2) Loading instructions necessary to ensure loading of the airplane within the weight and center of gravity limits, and to maintain the loading within these limits in flight.

(3) If certification for more than one center of gravity range is requested, the appropriate limitations, with regard to weight and loading procedures, for each separate center of gravity range.

(d) *Flight crew.* The number and functions of the minimum flight crew determined under § 25.1523 must be furnished.

(e) *Kinds of operation.* The kinds of operation approved under § 25.1525 must be furnished.

(f) *Ambient air temperatures and operating altitudes.* The extremes of the ambient air temperatures and operating altitudes established under § 25.1527 must be furnished.

(g) [Reserved]

(h) *Additional operating limitations.* The operating limitations established under § 25.1533 must be furnished.

(i) *Maneuvering flight load factors.* The positive maneuvering limit load factors for which the structure is proven, described in terms of accelerations, must be furnished.

[Doc. No. 5066, 29 FR 1891, Dec. 24, 1964, as amended by Amdt. 25-38, 41 FR 55468, Dec. 20, 1976; Amdt. 25-42, 43 FR 2323, Jan. 16, 1978; Amdt. 25-46, 43 FR 50598, Oct. 30, 1978; Amdt. 25-72, 55 FR 29787, July 20, 1990; Amdt. 25-105, 66 FR 34024, June 26, 2001; 75 FR 49818, Aug. 16, 2010]

§ 25.1585 Operating procedures.

(a) Operating procedures must be furnished for—

(1) Normal procedures peculiar to the particular type or model encountered in connection with routine operations;

(2) Non-normal procedures for malfunction cases and failure conditions involving the use of special systems or the alternative use of regular systems; and

(3) Emergency procedures for foreseeable but unusual situations in which immediate and precise action by the crew may be expected to substantially reduce the risk of catastrophe.

(b) Information or procedures not directly related to airworthiness or not under the control of the crew, must not be included, nor must any procedure that is accepted as basic airmanship.

(c) Information identifying each operating condition in which the fuel system independence prescribed in § 25.953 is necessary for safety must be furnished, together with instructions for placing the fuel system in a configuration used to show compliance with that section.

(d) The buffet onset envelopes, determined under § 25.251 must be furnished. The buffet onset envelopes presented may reflect the center of gravity at which the airplane is normally loaded during cruise if corrections for the effect of different center of gravity locations are furnished.

(e) Information must be furnished that indicates that when the fuel quantity indicator reads "zero" in level flight, any fuel remaining in the fuel tank cannot be used safely in flight.

(f) Information on the total quantity of usable fuel for each fuel tank must be furnished.

[Doc. No. 2000-8511, 66 FR 34024, June 26, 2001]

§ 25.1587 Performance information.

(a) Each Airplane Flight Manual must contain information to permit conversion of the indicated temperature to free air temperature if other than a free air temperature indicator is used to comply with the requirements of § 25.1303(a)(1).

(b) Each Airplane Flight Manual must contain the performance information computed under the applicable provisions of this part (including §§ 25.115, 25.123, and 25.125 for the weights, altitudes, temperatures, wind components, and runway gradients, as applicable) within the operational limits of the airplane, and must contain the following:

(1) In each case, the conditions of power, configuration, and speeds, and the procedures for handling the airplane and any system having a significant effect on the performance information.

(2) V_{SR} determined in accordance with § 25.103.

(3) The following performance information (determined by extrapolation and computed for the range of weights between the maximum landing weight and the maximum takeoff weight):

(i) Climb in the landing configuration.

(ii) Climb in the approach configuration.

(iii) Landing distance.

(4) Procedures established under § 25.101(f) and (g) that are related to the limitations and information required by § 25.1533 and by this paragraph (b) in the form of guidance material, including any relevant limitations or information.

(5) An explanation of significant or unusual flight or ground handling characteristics of the airplane.

(6) Corrections to indicated values of airspeed, altitude, and outside air temperature.

(7) An explanation of operational landing runway length factors included in the presentation of the landing distance, if appropriate.

[Doc. No. 2000-8511, 66 FR 34024, June 26, 2001, as amended by Amdt. 25-108, 67 FR 70828, Nov. 26, 2002]

Subpart H—Electrical Wiring Interconnection Systems (EWIS)

SOURCE: Docket No. FAA-2004-18379, 72 FR 63406, Nov. 8, 2007, unless otherwise noted.

§ 25.1701 Definition.

(a) As used in this chapter, electrical wiring interconnection system (EWIS) means any wire, wiring device, or combination of these, including termination devices, installed in any area of the airplane for the purpose of transmitting electrical energy, including data and signals, between two or more intended termination points. This includes:

(1) Wires and cables.

(2) Bus bars.

(3) The termination point on electrical devices, including those on relays, interrupters, switches, contactors, terminal blocks and circuit breakers, and other circuit protection devices.

(4) Connectors, including feed-through connectors.

(5) Connector accessories.

(6) Electrical grounding and bonding devices and their associated connections.

(7) Electrical splices.

(8) Materials used to provide additional protection for wires, including wire insulation, wire sleeving, and conduits that have electrical termination for the purpose of bonding.

(9) Shields or braids.

(10) Clamps and other devices used to route and support the wire bundle.

(11) Cable tie devices.

(12) Labels or other means of identification.

(13) Pressure seals.

(14) EWIS components inside shelves, panels, racks, junction boxes, distribution panels, and back-planes of equipment racks, including, but not limited to, circuit board back-planes, wire integration units, and external wiring of equipment.

(b) Except for the equipment indicated in paragraph (a)(14) of this section, EWIS components inside the following equipment, and the external connectors that are part of that equipment, are excluded from the definition in paragraph (a) of this section:

(1) Electrical equipment or avionics that are qualified to environmental conditions and testing procedures when those conditions and procedures are—

(i) Appropriate for the intended function and operating environment, and

(ii) Acceptable to the FAA.

(2) Portable electrical devices that are not part of the type design of the airplane. This includes personal entertainment devices and laptop computers.

(3) Fiber optics.

§ 25.1703 Function and installation: EWIS.

(a) Each EWIS component installed in any area of the aircraft must:

(1) Be of a kind and design appropriate to its intended function.

(2) Be installed according to limitations specified for the EWIS components.

(3) Perform the function for which it was intended without degrading the airworthiness of the airplane.

(4) Be designed and installed in a way that will minimize mechanical strain.

(b) Selection of wires must take into account known characteristics of the wire in relation to each installation and application to minimize the risk of wire damage, including any arc tracking phenomena.

(c) The design and installation of the main power cables (including generator cables) in the fuselage must allow for a reasonable degree of deformation and stretching without failure.

(d) EWIS components located in areas of known moisture accumulation must be protected to minimize any hazardous effects due to moisture.

§ 25.1705 Systems and functions: EWIS.

(a) EWIS associated with any system required for type certification or by operating rules must be considered an integral part of that system and must be considered in showing compliance with the applicable requirements for that system.

(b) For systems to which the following rules apply, the components of EWIS associated with those systems must be considered an integral part of that system or systems and must be considered in showing compliance with the applicable requirements for that system.

(1) § 25.773(b)(2) Pilot compartment view.

(2) § 25.981 Fuel tank ignition prevention.

(3) § 25.1165 Engine ignition systems.

(4) § 25.1310 Power source capacity and distribution.

(5) § 25.1316 System lightning protection.

(6) § 25.1331(a)(2) Instruments using a power supply.

(7) § 25.1351 General.

(8) § 25.1355 Distribution system.

(9) § 25.1360 Precautions against injury.

(10) § 25.1362 Electrical supplies for emergency conditions.

(11) § 25.1365 Electrical appliances, motors, and transformers.

(12) § 25.1431(c) and (d) Electronic equipment.

§ 25.1707 System separation: EWIS.

(a) Each EWIS must be designed and installed with adequate physical separation from other EWIS and airplane systems so that an EWIS component failure will not create a hazardous condition. Unless otherwise stated, for the purposes of this section, adequate physical separation must be achieved by separation distance or by a barrier that provides protection equivalent to that separation distance.

(b) Each EWIS must be designed and installed so that any electrical interference likely to be present in the airplane will not result in hazardous effects upon the airplane or its systems.

(c) Wires and cables carrying heavy current, and their associated EWIS components, must be designed and installed to ensure adequate physical separation and electrical isolation so that damage to circuits associated with essential functions will be minimized under fault conditions.

(d) Each EWIS associated with independent airplane power sources or power sources connected in combination must be designed and installed to ensure adequate physical separation and electrical isolation so that a fault in any one airplane power source EWIS will not adversely affect any other independent power sources. In addition:

(1) Airplane independent electrical power sources must not share a common ground terminating location.

(2) Airplane system static grounds must not share a common ground terminating location with any of the airplane's independent electrical power sources.

(e) Except to the extent necessary to provide electrical connection to the fuel systems components, the EWIS must be designed and installed with adequate physical separation from fuel lines and other fuel system components, so that:

(1) An EWIS component failure will not create a hazardous condition.

(2) Any fuel leakage onto EWIS components will not create a hazardous condition.

(f) Except to the extent necessary to provide electrical connection to the hydraulic systems components, EWIS must be designed and installed with adequate physical separation from hydraulic lines and other hydraulic system components, so that:

(1) An EWIS component failure will not create a hazardous condition.

(2) Any hydraulic fluid leakage onto EWIS components will not create a hazardous condition.

(g) Except to the extent necessary to provide electrical connection to the oxygen systems components, EWIS must be designed and installed with adequate physical separation from oxygen lines and other oxygen system components, so that an EWIS component failure will not create a hazardous condition.

(h) Except to the extent necessary to provide electrical connection to the water/waste systems components, EWIS must be designed and installed with adequate physical separation from water/waste lines and other water/waste system components, so that:

(1) An EWIS component failure will not create a hazardous condition.

(2) Any water/waste leakage onto EWIS components will not create a hazardous condition.

(i) EWIS must be designed and installed with adequate physical separation between the EWIS and flight or other mechanical control systems cables and associated system components, so that:

(1) Chafing, jamming, or other interference are prevented.

(2) An EWIS component failure will not create a hazardous condition.

(3) Failure of any flight or other mechanical control systems cables or systems components will not damage the EWIS and create a hazardous condition.

(j) EWIS must be designed and installed with adequate physical separation between the EWIS components and heated equipment, hot air ducts, and lines, so that:

(1) An EWIS component failure will not create a hazardous condition.

(2) Any hot air leakage or heat generated onto EWIS components will not create a hazardous condition.

(k) For systems for which redundancy is required, by certification rules, by operating rules, or as a result of the assessment required by § 25.1709, EWIS components associated with those systems must be designed and installed with adequate physical separation.

(l) Each EWIS must be designed and installed so there is adequate physical separation between it and other aircraft components and aircraft structure, and so that the EWIS is protected from sharp edges and corners, to minimize potential for abrasion/chafing, vibration damage, and other types of mechanical damage.

§ 25.1709 System safety: EWIS.

Each EWIS must be designed and installed so that:

(a) Each catastrophic failure condition—

(1) Is extremely improbable; and

(2) Does not result from a single failure.

(b) Each hazardous failure condition is extremely remote.

§ 25.1711 Component identification: EWIS.

(a) EWIS components must be labeled or otherwise identified using a consistent method that facilitates identification of the EWIS component, its function, and its design limitations, if any.

(b) For systems for which redundancy is required, by certification rules, by operating rules, or as a result of the assessment required by § 25.1709, EWIS components associated with those systems must be specifically identified with component part number, function, and separation requirement for bundles.

(1) The identification must be placed along the wire, cable, or wire bundle at appropriate intervals and in areas of the airplane where it is readily visible to maintenance, repair, or alteration personnel.

(2) If an EWIS component cannot be marked physically, then other means of identification must be provided.

(c) The identifying markings required by paragraphs (a) and (b) of this section must remain legible throughout the expected service life of the EWIS component.

(d) The means used for identifying each EWIS component as required by this section must not have an adverse effect on the performance of that component throughout its expected service life.

(e) Identification for EWIS modifications to the type design must be consistent with the identification scheme of the original type design.

§ 25.1713 Fire protection: EWIS.

(a) All EWIS components must meet the applicable fire and smoke protection requirements of § 25.831(c) of this part.

(b) EWIS components that are located in designated fire zones and are used during emergency procedures must be fire resistant.

(c) Insulation on electrical wire and electrical cable, and materials used to provide additional protection for the wire and cable, installed in any area of the airplane, must be self-extinguishing when tested in accordance with the applicable portions of Appendix F, part I, of 14 CFR part 25.

§ 25.1715 Electrical bonding and protection against static electricity: EWIS.

(a) EWIS components used for electrical bonding and protection against static electricity must meet the requirements of § 25.899.

(b) On airplanes having grounded electrical systems, electrical bonding provided by EWIS components must provide an electrical return path capable of carrying both normal and fault currents without creating a shock hazard or damage to the EWIS components, other airplane system components, or airplane structure.

§ 25.1717 Circuit protective devices: EWIS.

Electrical wires and cables must be designed and installed so they are compatible with the circuit protection devices required by § 25.1357, so that a fire or smoke hazard cannot be created under temporary or continuous fault conditions.

§ 25.1719 Accessibility provisions: EWIS.

Access must be provided to allow inspection and replacement of any EWIS component as necessary for continued airworthiness.

§ 25.1721 Protection of EWIS.

(a) No cargo or baggage compartment may contain any EWIS whose damage or failure may affect safe operation, unless the EWIS is protected so that:

(1) It cannot be damaged by movement of cargo or baggage in the compartment.

(2) Its breakage or failure will not create a fire hazard.

(b) EWIS must be designed and installed to minimize damage and risk of damage to EWIS by movement of people in the airplane during all phases of flight, maintenance, and servicing.

(c) EWIS must be designed and installed to minimize damage and risk of damage to EWIS by items carried onto the aircraft by passengers or cabin crew.

§ 25.1723 Flammable fluid fire protection: EWIS.

EWIS components located in each area where flammable fluid or vapors might escape by leakage of a fluid system must be considered a potential ignition source and must meet the requirements of § 25.863.

§ 25.1725 Powerplants: EWIS.

(a) EWIS associated with any powerplant must be designed and installed so that the failure of an EWIS component will not prevent the continued safe operation of the remaining powerplants or require immediate action by any crewmember for continued safe operation, in accordance with the requirements of § 25.903(b).

(b) Design precautions must be taken to minimize hazards to the airplane due to EWIS damage in the event of a powerplant rotor failure or a fire originating within the powerplant that burns through the powerplant case, in accordance with the requirements of § 25.903(d)(1).

§ 25.1727 Flammable fluid shutoff means: EWIS.

EWIS associated with each flammable fluid shutoff means and control must be fireproof or must be located and protected so that any fire in a fire zone will not affect operation of the flammable fluid shutoff means, in accordance with the requirements of § 25.1189.

§ 25.1729 Instructions for Continued Airworthiness: EWIS.

The applicant must prepare Instructions for Continued Airworthiness applicable to EWIS in accordance with Appendix H sections H25.4 and H25.5 to this part that are approved by the FAA.

§ 25.1731 Powerplant and APU fire detector system: EWIS.

(a) EWIS that are part of each fire or overheat detector system in a fire zone must be fire-resistant.

(b) No EWIS component of any fire or overheat detector system for any fire zone may pass through another fire zone, unless:

(1) It is protected against the possibility of false warnings resulting from fires in zones through which it passes; or

(2) Each zone involved is simultaneously protected by the same detector and extinguishing system.

(c) EWIS that are part of each fire or overheat detector system in a fire zone must meet the requirements of § 25.1203.

§ 25.1733 Fire detector systems, general: EWIS.

EWIS associated with any installed fire protection system, including those required by §§ 25.854 and 25.858, must be considered an integral part of the system in showing compliance with the applicable requirements for that system.

Subpart I—Special Federal Aviation Regulations

SOURCE: Docket No. FAA-2011-0186, Amdt. 25-133, 76 FR 12555, Mar. 8, 2011, unless otherwise noted.

§ 25.1801 SFAR No. 111—Lavatory Oxygen Systems.

The requirements of § 121.1500 of this chapter also apply to this part.

APPENDIX A TO PART 25

Figure 1—Basic landing gear dimension ata.

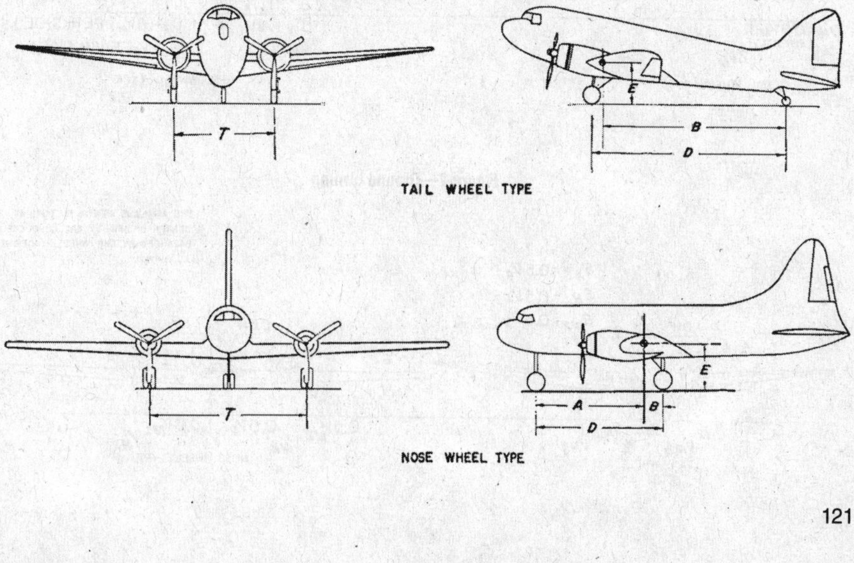

Figure 2—Level landing

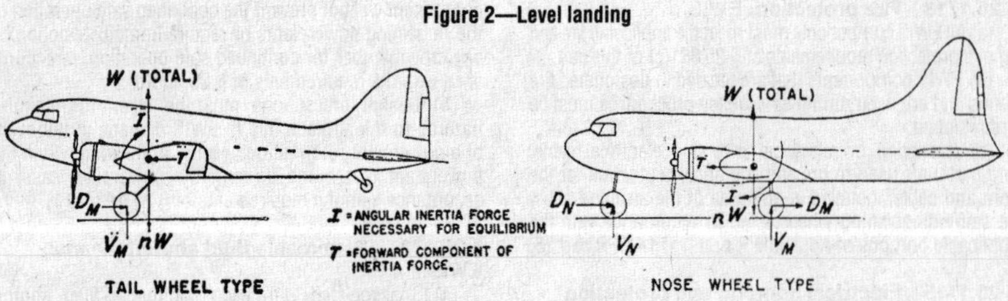

Figure 3—Tail-down landing

Figure 4

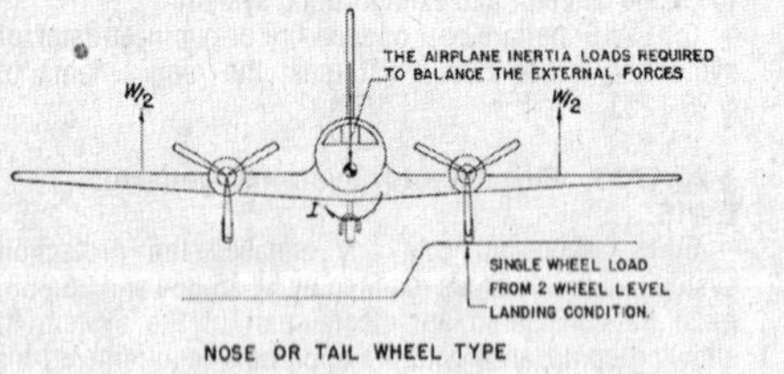

Figure 5—Lateral drift landing

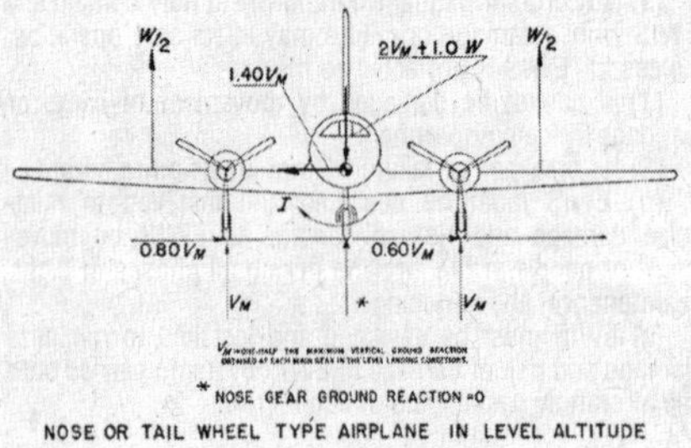

Figure 6—Braked roll

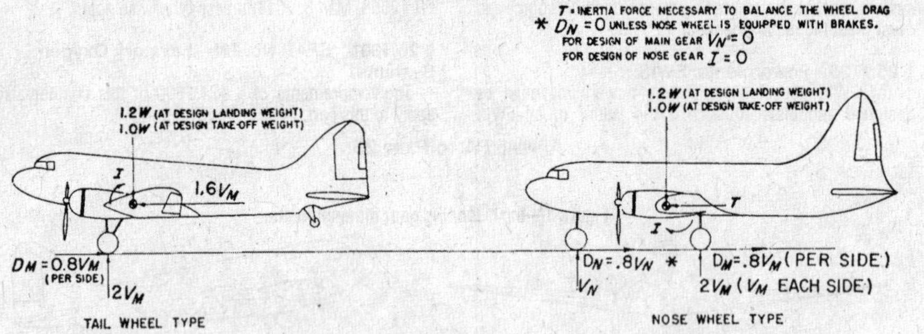

Figure 7—Ground turning

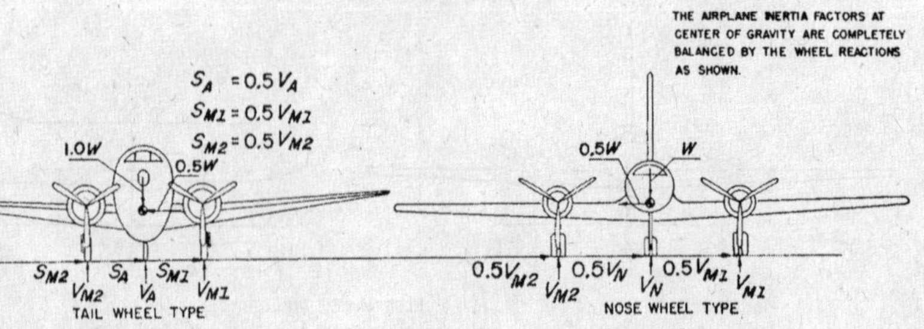

Figure 8—Pivoting, nose or tail wheel type

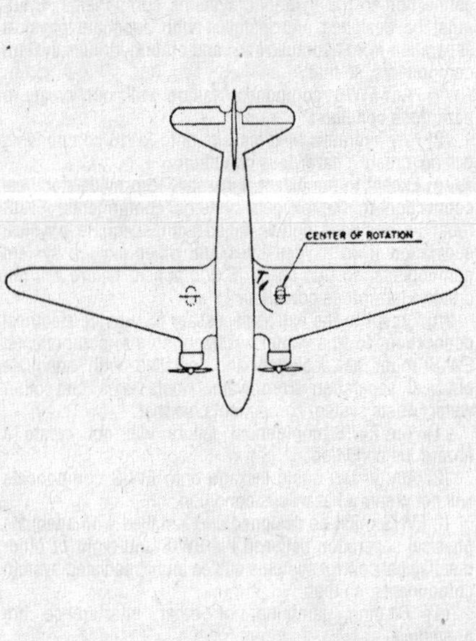

V_M and V_N are static ground reactions. For tail wheel type the airplane is in the three point attitude. Pivoting is assumed to take place about one main landing gear unit.

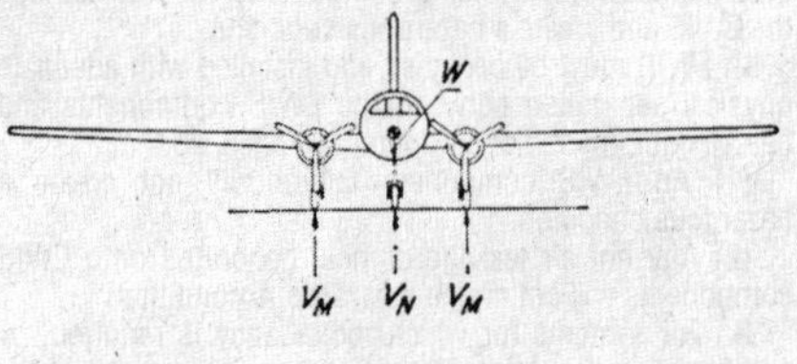

Appendix B to Part 25

Figure 1—Pictorial definition of angles, dimensions, and directions on a seaplane

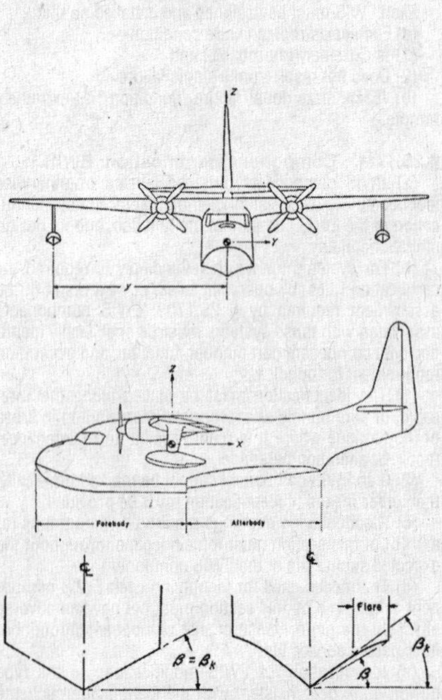

Figure 2—Hull station weighing factor.

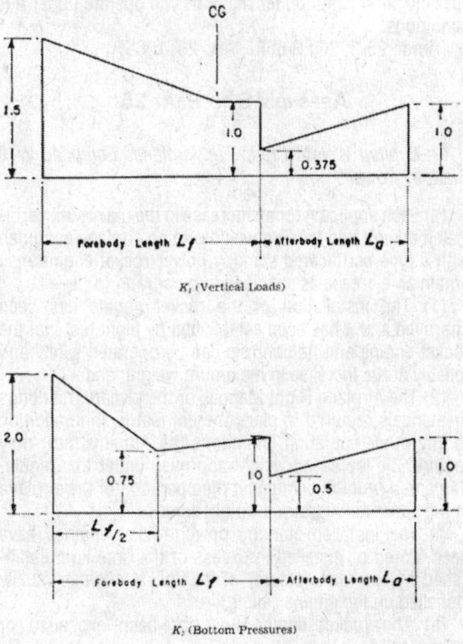

Appendix C to Part 25

Part I—Atmospheric Icing Conditions

(a) *Continuous maximum icing.* The maximum continuous intensity of atmospheric icing conditions (continuous maximum icing) is defined by the variables of the cloud liquid water content, the mean effective diameter of the cloud droplets, the ambient air temperature, and the interrelationship of these three variables as shown in figure 1 of this appendix. The limiting icing envelope in terms of altitude and temperature is given in figure 2 of this appendix. The inter-relationship of cloud liquid water content with drop diameter and altitude is determined from figures 1 and 2. The cloud liquid water content for continuous maximum icing conditions of a horizontal extent, other than 17.4 nautical miles, is determined by the value of

Figure 3—Transverse pressure distributions.

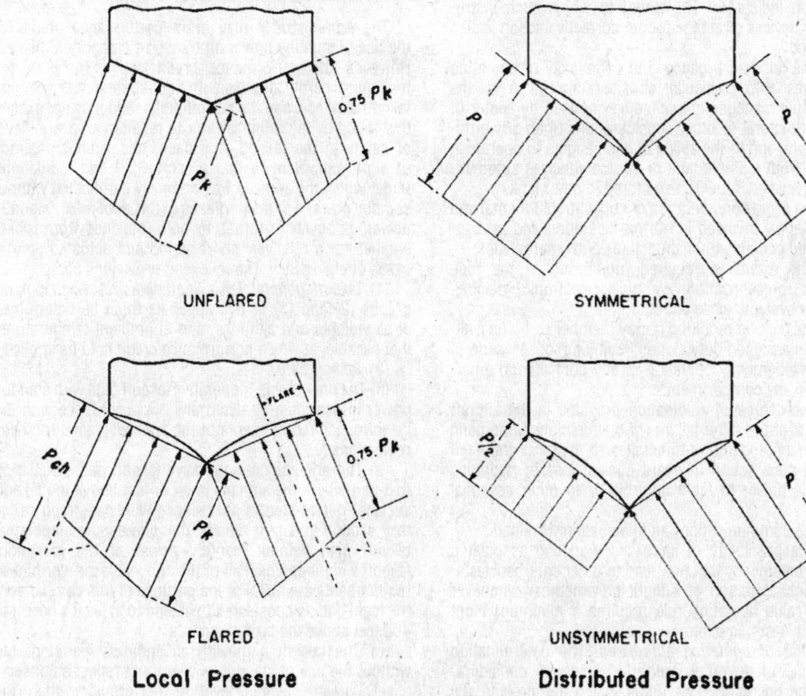

Figure 1

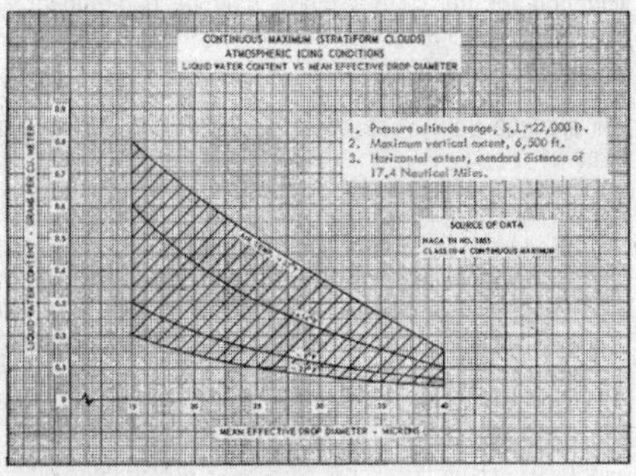

Figure 2

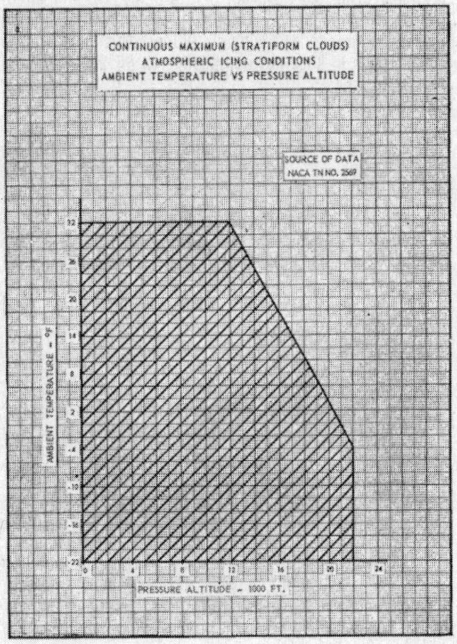

Figure 3

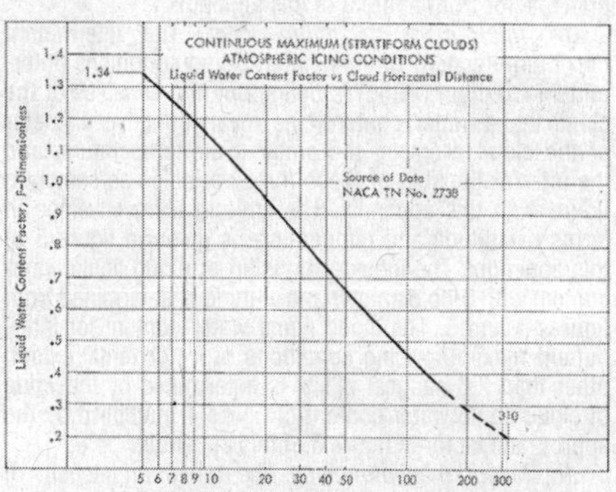

Figure 4

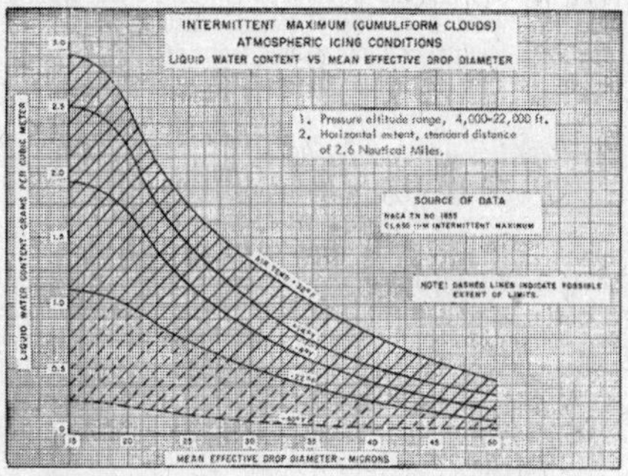

Figure 5

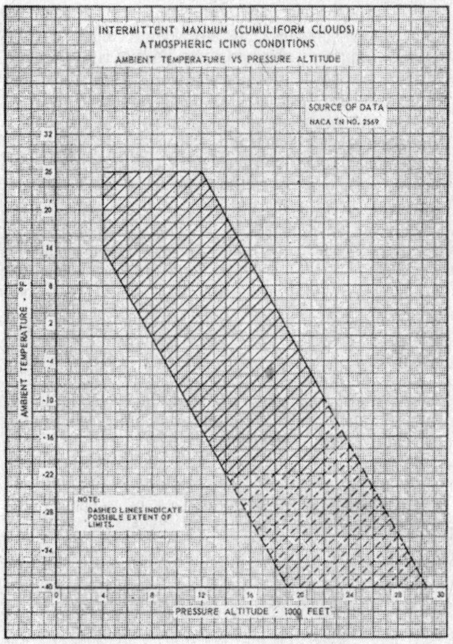

Figure 6

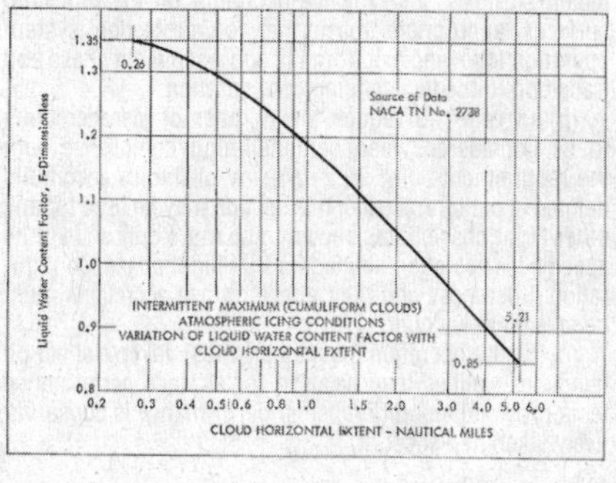

liquid water content of figure 1, multiplied by the appropriate factor from figure 3 of this appendix.

(b) *Intermittent maximum icing.* The intermittent maximum intensity of atmospheric icing conditions (intermittent maximum icing) is defined by the variables of the cloud liquid water content, the mean effective diameter of the cloud droplets, the ambient air temperature, and the interrelationship of these three variables as shown in figure 4 of this appendix. The limiting icing envelope in terms of altitude and temperature is given in figure 5 of this appendix. The inter-relationship of cloud liquid water content with drop diameter and altitude is determined from figures 4 and 5. The cloud liquid water content for intermittent maximum icing conditions of a horizontal extent, other than 2.6 nautical miles, is determined by the value of cloud liquid water content of figure 4 multiplied by the appropriate factor in figure 6 of this appendix.

.(c) *Takeoff maximum icing.* The maximum intensity of atmospheric icing conditions for takeoff (takeoff maximum icing) is defined by the cloud liquid water content of 0.35 g/m3, the mean effective diameter of the cloud droplets of 20 microns, and the ambient air temperature at ground level of minus 9 degrees Celsius (-9°C). The takeoff maximum icing conditions extend from ground level to a height of 1,500 feet above the level of the takeoff surface.

Part II—Airframe Ice Accretions for Showing Compliance With Subpart B.

(a) *Ice accretions—General.* The most critical ice accretion in terms of airplane performance and handling qualities for each flight phase must be used to show compliance with the applicable airplane performance and handling requirements in icing conditions of subpart B of this part. Applicants must demonstrate that the full range of atmospheric icing conditions specified in part I of this appendix have been considered, including the mean effective drop diameter, liquid water content, and temperature appropriate to the flight conditions (for example, configuration, speed, angle-of-attack, and altitude). The ice accretions for each flight phase are defined as follows:

(1) *Takeoff ice* is the most critical ice accretion on unprotected surfaces and any ice accretion on the protected surfaces appropriate to normal ice protection system operation, occurring between liftoff and 400 feet above the takeoff surface, assuming accretion starts at liftoff in the takeoff maximum icing conditions of part I, paragraph (c) of this appendix.

(2) *Final takeoff ice* is the most critical ice accretion on unprotected surfaces, and any ice accretion on the protected surfaces appropriate to normal ice protection system operation, between 400 feet and either 1,500 feet above the takeoff surface, or the height at which the transition from the takeoff to the en route configuration is completed and V_{FTO} is reached, whichever is higher. Ice accretion is assumed to start at liftoff in the takeoff maximum icing conditions of part I, paragraph (c) of this appendix.

(3) *En route ice* is the critical ice accretion on the unprotected surfaces, and any ice accretion on the protected surfaces appropriate to normal ice protection system operation, during the en route phase.

(4) *Holding ice* is the critical ice accretion on the unprotected surfaces, and any ice accretion on the protected surfaces appropriate to normal ice protection system operation, during the holding flight phase.

(5) *Approach ice* is the critical ice accretion on the unprotected surfaces, and any ice accretion on the protected surfaces appropriate to normal ice protection system operation following exit from the holding flight phase and transition to the most critical approach configuration.

(6) *Landing ice* is the critical ice accretion on the unprotected surfaces, and any ice accretion on the protected surfaces appropriate to normal ice protection system operation following exit from the approach flight phase and transition to the final landing configuration.

(b) In order to reduce the number of ice accretions to be considered when demonstrating compliance with the requirements of § 25.21(g), any of the ice accretions defined in paragraph (a) of this section may be used for any other flight phase if it is shown to be more critical than the specific ice accretion defined for that flight phase. Configuration differences and their effects on ice accretions must be taken into account.

(c) The ice accretion that has the most adverse effect on handling qualities may be used for airplane performance tests provided any difference in performance is conservatively taken into account.

(d) For both unprotected and protected parts, the ice accretion for the takeoff phase may be determined by calculation, assuming the takeoff maximum icing conditions defined in appendix C, and assuming that:

(1) Airfoils, control surfaces and, if applicable, propellers are free from frost, snow, or ice at the start of the takeoff;

(2) The ice accretion starts at liftoff;

(3) The critical ratio of thrust/power-to-weight;

(4) Failure of the critical engine occurs at V_{EF}; and

(5) Crew activation of the ice protection system is in accordance with a normal operating procedure provided in the Airplane Flight Manual, except that after beginning the takeoff roll, it must be assumed that the crew takes no action to activate the ice protection system until the airplane is at least 400 feet above the takeoff surface.

(e) The ice accretion before the ice protection system has been activated and is performing its intended function is the critical ice accretion formed on the unprotected and normally protected surfaces before activation and effective operation of the ice protection system in continuous maximum atmospheric icing conditions. This ice accretion only applies in showing compliance to §§ 25.143(j) and 25.207(h), and 25.207(i).

[Doc. No. 4080, 29 FR 17955, Dec. 18, 1964, as amended by Amdt. 25-121, 72 FR 44669, Aug. 8, 2007; 72 FR 50467, Aug. 31, 2007; Amdt. 25-129, 74 FR 38340, Aug. 3, 2009]

Appendix D to Part 25

Criteria for determining minimum flight crew. The following are considered by the Agency in determining the minimum flight crew under § 25.1523:

(a) *Basic workload functions.* The following basic workload functions are considered:

(1) Flight path control.

(2) Collision avoidance.

(3) Navigation.

(4) Communications.

(5) Operation and monitoring of aircraft engines and systems.

(6) Command decisions.

(b) *Workload factors.* The following workload factors are considered significant when analyzing and demonstrating workload for minimum flight crew determination:

(1) The accessibility, ease, and simplicity of operation of all necessary flight, power, and equipment controls, including emergency fuel shutoff valves, electrical controls, electronic controls, pressurization system controls, and engine controls.

(2) The accessibility and conspicuity of all necessary instruments and failure warning devices such as fire warning, electrical system malfunction, and other failure or caution indicators. The extent to which such instruments or devices direct the proper corrective action is also considered.

(3) The number, urgency, and complexity of operating procedures with particular consideration given to the specific fuel management schedule imposed by center of gravity, structural or other considerations of an airworthiness nature, and to the ability of each engine to operate at all times from a single tank or source which is automatically replenished if fuel is also stored in other tanks.

(4) The degree and duration of concentrated mental and physical effort involved in normal operation and in diagnosing and coping with malfunctions and emergencies.

(5) The extent of required monitoring of the fuel, hydraulic, pressurization, electrical, electronic, deicing, and other systems while en route.

(6) The actions requiring a crewmember to be unavailable at his assigned duty station, including: observation of systems, emergency operation of any control, and emergencies in any compartment.

(7) The degree of automation provided in the aircraft systems to afford (after failures or malfunctions) automatic crossover or isolation of difficulties to minimize the need for flight crew action to guard against loss of hydraulic or electric power to flight controls or to other essential systems.

(8) The communications and navigation workload.

(9) The possibility of increased workload associated with any emergency that may lead to other emergencies.

(10) Incapacitation of a flight crewmember whenever the applicable operating rule requires a minimum flight crew of at least two pilots.

(c) *Kind of operation authorized.* The determination of the kind of operation authorized requires consideration of the operating rules under which the airplane will be operated. Unless an applicant desires approval for a more limited kind of operation. It is assumed that each airplane certificated under this Part will operate under IFR conditions.

[Amdt. 25-3, 30 FR 6067, Apr. 29, 1965]

Appendix E to Part 25

I—Limited Weight Credit For Airplanes Equipped With Standby Power

(a) Each applicant for an increase in the maximum certificated takeoff and landing weights of an airplane equipped with a type-certificated standby power rocket engine may obtain an increase as specified in paragraph (b) if—

(1) The installation of the rocket engine has been approved and it has been established by flight test that the rocket engine and its controls can be operated safely and reliably at the increase in maximum weight; and

(2) The Airplane Flight Manual, or the placard, markings or manuals required in place thereof, set forth in addition to any other operating limitations the Administrator may require, the increased weight approved under this regulation and a prohibition against the operation of the airplane at the approved increased weight when—

(i) The installed standby power rocket engines have been stored or installed in excess of the time limit established by the manufacturer of the rocket engine (usually stenciled on the engine casing); or

(ii) The rocket engine fuel has been expended or discharged.

(b) The currently approved maximum takeoff and landing weights at which an airplane is certificated without a standby power rocket engine installation may be increased by an amount that does not exceed any of the following:

(1) An amount equal in pounds to 0.014 IN, where I is the maximum usable impulse in pounds-seconds available from each standby power rocket engine and N is the number of rocket engines installed.

(2) An amount equal to 5 percent of the maximum certificated weight approved in accordance with the applicable airworthiness regulations without standby power rocket engines installed.

(3) An amount equal to the weight of the rocket engine installation.

(4) An amount that, together with the currently approved maximum weight, would equal the maximum structural weight established for the airplane without standby rocket engines installed.

II—Performance Credit for Transport Category Airplanes Equipped With Standby Power

The Administrator may grant performance credit for the use of standby power on transport category airplanes. However, the performance credit applies only to the maximum certificated takeoff and landing weights, the takeoff distance, and the takeoff paths, and may not exceed that found by the Administrator to result in an overall level of safety in the takeoff, approach, and landing regimes of flight equivalent to that prescribed in the regulations under which the airplane was originally certificated without standby power. For the purposes of this appendix, "standby power" is power or thrust, or both, obtained from rocket engines for a relatively short period and actuated only in cases of emergency. The following provisions apply:

(1) *Takeoff; general.* The takeoff data prescribed in paragraphs (2) and (3) of this appendix must be determined at all weights and altitudes, and at ambient temperatures if applicable, at which performance credit is to be applied.

(2) *Takeoff path.*

(a) The one-engine-inoperative takeoff path with standby power in use must be determined in accordance with the performance requirements of the applicable airworthiness regulations.

(b) The one-engine-inoperative takeoff path (excluding that part where the airplane is on or just above the takeoff surface) determined in accordance with paragraph (a) of this section must lie above the one-engine-inoperative takeoff path without standby power at the maximum takeoff weight at which all of the applicable air-worthiness requirements are met. For the purpose of this comparison, the flight path is considered to extend to at least a height of 400 feet above the takeoff surface.

(c) The takeoff path with all engines operating, but without the use of standby power, must reflect a conservatively greater overall level of performance than the one-engine-inoperative takeoff path established in accordance with paragraph (a) of this section. The margin must be established by the Administrator to insure safe day-to-day operations, but in no case may it be less than 15 percent. The all-engines-operating takeoff path must be determined by a procedure consistent with that established in complying with paragraph (a) of this section.

(d) For reciprocating-engine-powered airplanes, the takeoff path to be scheduled in the Airplane Flight Manual must represent the one-engine-operative takeoff path determined in accordance with paragraph (a) of this section and modified to reflect the procedure (see paragraph (6)) established by tbe applicant for flap retraction and attainment of the en route speed. The scheduled takeoff path must have a positive slope at all points of the airborne portion and at no point must it lie above the takeoff path specified in paragraph (a) of this section.

(3) *Takeoff distance.* The takeoff distance must be the horizontal distance along the one-engine-inoperative take off path determined in accordance with paragraph (2)(a) from the start of the takeoff to the point where the airplane attains a height of 50 feet above the takeoff surface for reciprocating-engine-powered airplanes and a height of 35 feet above the takeoff surface for turbine-powered airplanes.

(4) *Maximum certificated takeoff weights.* The maximum certificated takeoff weights must be determined at all altitudes, and at ambient temperatures, if applicable, at which performance credit is to be applied and may not exceed the weights established in compliance with paragraphs (a) and (b) of this section.

(a) The conditions of paragraphs (2)(b) through (d) must be met at the maximum certificated takeoff weight.

(b) Without the use of standby power, the airplane must meet all of the en route requirements of the applicable airworthiness regulations under which the airplane was originally certificated. In addition, turbine-powered airplanes without the use of standby power must meet the final takeoff climb requirements prescribed in the applicable airworthiness regulations.

(5) *Maximum certificated landing weights.*

(a) The maximum certificated landing weights (one-engine-inoperative approach and all-engine-operating landing climb) must be determined at all altitudes, and at ambient temperatures if applicable, at which performance credit is to be applied and must not exceed that established in compliance with paragraph (b) of this section.

(b) The flight path, with the engines operating at the power or thrust, or both, appropriate to the airplane configuration and with standby power in use, must lie above the flight path without standby power in use at the maximum weight at which all of the applicable airworthiness requirements are met. In addition, the flight paths must comply with subparagraphs (i) and (ii) of this paragraph.

(i) The flight paths must be established without changing the appropriate airplane configuration.

(ii) The flight paths must be carried out for a minimum height of 400 feet above the point where standby power is actuated.

(6) *Airplane configuration, speed, and power and thrust; general.* Any change in the airplane's configuration, speed, and power or thrust, or both, must be made in accordance with the procedures established by the applicant for the operation of the airplane in service and must comply with paragraphs (a) through (c) of this section. In addition, procedures must be established for the execution of balked landings and missed approaches.

(a) The Administrator must find that the procedure can be consistently executed in service by crews of average skill.

(b) The procedure may not involve methods or the use of devices which have not been proven to be safe and reliable.

(c) Allowances must be made for such time delays in the execution of the procedures as may be reasonably expected to occur during service.

(7) *Installation and operation; standby power.* The standby power unit and its installation must comply with paragraphs (a) and (b) of this section.

(a) The standby power unit and its installation must not adversely affect the safety of the airplane.

(b) The operation of the standby power unit and its control must have proven to be safe and reliable.

[Amdt. 25-6, 30 FR 8468, July 2, 1965]

Appendix F to Part 25

Part I—Test Criteria and Procedures for Showing Compliance with § 25.853, or § 25.855.

(a) *Material test criteria* —(1) *Interior compartments occupied by crew or passengers.* (i) Interior ceiling panels, interior wall panels, partitions, galley structure, large cabinet walls, structural flooring, and materials used in the construction of stowage compartments (other than underseat stowage compartments and compartments for stowing small items such as magazines and maps) must be self-extinguishing when tested vertically in accordance with the applicable portions of part I of this appendix. The average burn length may not exceed 6 inches and the average flame time after removal of the flame source may not exceed 15 seconds. Drippings from the test specimen may not continue to flame for more than an average of 3 seconds after falling.

(ii) Floor covering, textiles (including draperies and upholstery), seat cushions, padding, decorative and nondecorative coated fabrics, leather, trays and galley furnishings, electrical conduit, air ducting, joint and edge covering, liners of Class B and E cargo or baggage compartments, floor panels of Class B, C, D, or E cargo or baggage compartments, cargo covers and transparencies, molded and thermoformed parts, air ducting joints, and trim strips (decorative and chafing), that are constructed of materials not covered in subparagraph (iv) below, must be self-extinguishing when tested vertically in accordance with the applicable portions of part I of this appendix or other approved equivalent means. The average burn length may not exceed 8 inches, and the average flame time after removal of the flame source may not exceed 15 seconds. Drippings from the test specimen may not continue to flame for more than an average of 5 seconds after falling.

(iii) Motion picture film must be safety film meeting the Standard Specifications for Safety Photographic Film PHI.25 (available from the American National Standards Institute, 1430 Broadway, New York, NY 10018). If the film travels through ducts, the ducts must meet the requirements of subparagraph (ii) of this paragraph.

(iv) Clear plastic windows and signs, parts constructed in whole or in part of elastomeric materials, edge lighted instrument assemblies consisting of two or more instruments in a common housing, seat belts, shoulder harnesses, and cargo and baggage tiedown equipment, including containers, bins, pallets, etc., used in passenger or crew compartments, may not have an average burn rate greater than 2.5 inches per minute when tested horizontally in accordance with the applicable portions of this appendix.

(v) Except for small parts (such as knobs, handles, rollers, fasteners, clips, grommets, rub strips, pulleys, and small electrical parts) that would not contribute significantly to the propagation of a fire and for electrical wire and cable insulation, materials in items not specified in paragraphs (a)(1)(i), (ii), (iii), or (iv) of part I of this appendix may not have a burn rate greater than 4.0 inches per minute when tested horizontally in accordance with the applicable portions of this appendix.

(2) *Cargo and baggage compartments not occupied by crew or passengers.*

(i) [Reserved]

(ii) A cargo or baggage compartment defined in § 25.857 as Class B or E must have a liner constructed of materials that meet the requirements of paragraph (a)(1)(ii) of part I of this appendix and separated from the airplane structure (except for attachments). In addition, such liners must be subjected to the 45 degree angle test. The flame may not penetrate (pass through) the material during application of the flame or subsequent to its removal. The average flame time after removal of the flame source may not exceed 15 seconds, and the average glow time may not exceed 10 seconds.

(iii) A cargo or baggage compartment defined in § 25.857 as Class B, C, D, or E must have floor panels constructed of materials which meet the requirements of paragraph (a)(1)(ii) of part I of this appendix and which are separated from the airplane structure (except for attachments). Such panels must be subjected to the 45 degree angle test. The flame may not penetrate (pass through) the material during application of the flame or subsequent to its removal. The average flame time after removal of the flame source may not exceed 15 seconds, and the average glow time may not exceed 10 seconds.

(iv) Insulation blankets and covers used to protect cargo must be constructed of materials that meet the requirements of paragraph (a)(1)(ii) of part I of this appendix. Tiedown equipment (including containers, bins, and pallets) used in each cargo and baggage compartment must be constructed of materials that meet the requirements of paragraph (a)(1)(v) of part I of this appendix.

(3) *Electrical system components.* Insulation on electrical wire or cable installed in any area of the fuselage must be self-extinguishing when subjected to the 60 degree test specified in part I of this appendix. The average burn length may not exceed 3 inches, and the average flame time after removal of the flame source may not exceed 30 seconds. Drippings from the test specimen may not continue to flame for more than an average of 3 seconds after falling.

(b) *Test Procedures*—(1) *Conditioning.* Specimens must be conditioned to 70±5 F., and at 50 percent±5 percent relative humidity until moisture equilibrium is reached or for 24 hours. Each specimen must remain in the conditioning environment until it is subjected to the flame.

(2) *Specimen configuration.* Except for small parts and electrical wire and cable insulation, materials must be tested either as section cut from a fabricated part as installed in the airplane or as a specimen simulating a cut section, such as a specimen cut from a flat sheet of the material or a model of the fabricated part. The specimen may be cut from any location in a fabricated part; however, fabricated units, such as sandwich panels, may not be separated for test. Except as noted below, the specimen thickness must be no thicker than the minimum thickness to be qualified for use in the airplane. Test specimens of thick foam parts, such as seat cushions, must be ½-inch in thickness. Test specimens of materials that must meet the requirements of paragraph (a)(1)(v) of part I of this appendix must be no more than ⅛-inch in thickness. Electrical wire and cable specimens must be the same size as used in the airplane. In the case of fabrics, both the warp and fill direction of the weave must be tested to determine the most critical flammability condition. Specimens must be mounted in a metal frame so that the two long edges and the upper edge are held securely during the vertical test prescribed in subparagraph (4) of this paragraph and the two long edges and the edge away from the flame are held securely during the horizontal test prescribed in subparagraph (5) of this paragraph. The exposed area of the specimen must be at least 2 inches wide and 12 inches long, unless the actual size used in the airplane is smaller. The edge to which the burner flame is applied must not consist of the finished or protected edge of the specimen but must be representative of the actual cross-section of the material or part as installed in the airplane. The specimen must be mounted in a metal frame so that all four edges are held securely and the exposed area of the specimen is at least 8 inches by 8 inches during the 45° test prescribed in subparagraph (6) of this paragraph.

(3) *Apparatus.* Except as provided in subparagraph (7) of this paragraph, tests must be conducted in a draft-free cabinet in accordance with Federal Test Method Standard 191 Model 5903 (revised Method 5902) for the vertical test, or Method 5906 for horizontal test (available from the General Services Administration, Business Service Center, Region 3, Seventh & D Streets SW., Washington, DC 20407). Specimens which are too large for the cabinet must be tested in similar draft-free conditions.

(4) *Vertical test.* A minimum of three specimens must be tested and results averaged. For fabrics, the direction of weave corresponding to the most critical flammability conditions must be parallel to the longest dimension. Each specimen must be supported vertically. The specimen must be exposed to a Bunsen or Tirrill burner with a nominal ⅜-inch I.D. tube adjusted to give a flame of 1½ inches in height. The minimum flame temperature measured by a calibrated thermocouple pyrometer in the center of the flame must be 1550 °F. The lower edge of the specimen must be ¾-inch above the top edge of the burner. The flame must be applied to the center line of the lower edge of the specimen. For materials covered by paragraph (a)(1)(i) of part I of this appendix, the flame must be applied for 60 seconds and then removed. For materials covered by paragraph (a)(1)(ii) of part I of this appendix, the flame must be applied for 12 seconds and then removed. Flame time, burn length, and flaming time of drippings, if any, may be recorded. The burn length determined in accordance with subparagraph (7) of this paragraph must be measured to the nearest tenth of an inch.

(5) *Horizontal test.* A minimum of three specimens must be tested and the results averaged. Each specimen must be supported horizontally. The exposed surface, when installed in the aircraft, must be face down for the test. The specimen must be exposed to a Bunsen or Tirrill burner with a nominal ⅜-inch I.D. tube adjusted to give a flame of 1½ inches in height. The minimum flame temperature measured by a calibrated thermocouple pyrometer in the center of the flame must be 1550 °F. The specimen must be positioned so that the edge being tested is centered ¾-inch above the top of the burner. The flame must be applied for 15 seconds and then removed. A minimum of 10 inches of specimen must be used for timing purposes, approximately 1½ inches must burn before the burning front reaches the timing zone, and the average burn rate must be recorded.

PART 25

FAR

(6) *Forty-five degree test.* A minimum of three specimens must be tested and the results averaged. The specimens must be supported at an angle of 45° to a horizontal surface. The exposed surface when installed in the aircraft must be face down for the test. The specimens must be exposed to a Bunsen or Tirrill burner with a nominal $^3/_8$-inch I.D. tube adjusted to give a flame of $1^1/_2$ inches in height. The minimum flame temperature measured by a calibrated thermocouple pyrometer in the center of the flame must be 1550 °F. Suitable precautions must be taken to avoid drafts. The flame must be applied for 30 seconds with one-third contacting the material at the center of the specimen and then removed. Flame time, glow time, and whether the flame penetrates (passes through) the specimen must be recorded.

(7) *Sixty degree test.* A minimum of three specimens of each wire specification (make and size) must be tested. The specimen of wire or cable (including insulation) must be placed at an angle of 60° with the horizontal in the cabinet specified in subparagraph (3) of this paragraph with the cabinet door open during the test, or must be placed within a chamber approximately 2 feet high by 1 foot by 1 foot, open at the top and at one vertical side (front), and which allows sufficient flow of air for complete combustion, but which is free from drafts. The specimen must be parallel to and approximately 6 inches from the front of the chamber. The lower end of the specimen must be held rigidly clamped. The upper end of the specimen must pass over a pulley or rod and must have an appropriate weight attached to it so that the specimen is held tautly throughout the flammability test. The test specimen span between lower clamp and upper pulley or rod must be 24 inches and must be marked 8 inches from the lower end to indicate the central point for flame application. A flame from a Bunsen or Tirrill burner must be applied for 30 seconds at the test mark. The burner must be mounted underneath the test mark on the specimen, perpendicular to the specimen and at an angle of 30° to the vertical plane of the specimen. The burner must have a nominal bore of $^3/_8$-inch and be adjusted to provide a 3-inch high flame with an inner cone approximately one-third of the flame height. The minimum temperature of the hottest portion of the flame, as measured with a calibrated thermocouple pyrometer, may not be less than 1750 °F. The burner must be positioned so that the hottest portion of the flame is applied to the test mark on the wire. Flame time, burn length, and flaming time of drippings, if any, must be recorded. The burn length determined in accordance with paragraph (8) of this paragraph must be measured to the nearest tenth of an inch. Breaking of the wire specimens is not considered a failure.

(8) *Burn length.* Burn length is the distance from the original edge to the farthest evidence of damage to the test specimen due to flame impingement, including areas of partial or complete consumption, charring, or embrittlement, but not including areas sooted, stained, warped, or discolored, nor areas where material has shrunk or melted away from the heat source.

PART II—FLAMMABILITY OF SEAT CUSHIONS

(a) *Criteria for Acceptance.* Each seat cushion must meet the following criteria:

(1) At least three sets of seat bottom and seat back cushion specimens must be tested.

(2) If the cushion is constructed with a fire blocking material, the fire blocking material must completely enclose the cushion foam core material.

(3) Each specimen tested must be fabricated using the principal components (i.e., foam core, flotation material, fire blocking material, if used, and dress covering) and assembly processes (representative seams and closures) intended for use in the production articles. If a different material combination is used for the back cushion than for the bottom cushion, both material combinations must be tested as complete specimen sets, each set consisting of a back cushion specimen and a bottom cushion specimen. If a cushion, including outer dress covering, is demonstrated to meet the requirements of this appendix using the oil burner test, the dress covering of that cushion may be replaced with a similar dress covering provided the burn length of the replacement covering, as determined by the test specified in § 25.853(c), does not exceed the corresponding burn length of the dress covering used on the cushion subjected to the oil burner test.

(4) For at least two-thirds of the total number of specimen sets tested, the burn length from the burner must not reach the side of the cushion opposite the burner. The burn length must not exceed 17 inches. Burn length is the perpendicular distance from the inside edge of the seat frame closest to the burner to the farthest evidence of damage to the test specimen due to flame impingement, including areas of partial or complete consumption, charring, or embrittlement, but not including areas sooted, stained, warped, or discolored, or areas where material has shrunk or melted away from the heat source.

(5) The average percentage weight loss must not exceed 10 percent. Also, at least two-thirds of the total number of specimen sets tested must not exceed 10 percent weight loss. All droppings falling from the cushions and mounting stand are to be discarded before the after-test weight is determined. The percentage weight loss for a specimen set is the weight of the specimen set before testing less the weight of the specimen set after testing expressed as the percentage of the weight before testing.

(b) *Test Conditions.* Vertical air velocity should average 25 fpm±10 fpm at the top of the back seat cushion. Horizontal air velocity should be below 10 fpm just above the bottom seat cushion. Air velocities should be measured with the ventilation hood operating and the burner motor off.

(c) *Test Specimens.* (1) For each test, one set of cushion specimens representing a seat bottom and seat back cushion must be used.

(2) The seat bottom cushion specimen must be $18\pm^1/_8$ inches (457±3mm) wide by $20\pm^1/_8$ inches (508±3mm) deep by $4\pm^1/_8$ inches (102±3mm) thick, exclusive of fabric closures and seam overlap.

(3) The seat back cushion specimen must be $18\pm^1/_8$ inches (432±3mm) wide by $25\pm^1/_8$ inches (635±3mm) high by $2\pm^1/_8$ inches (51±3mm) thick, exclusive of fabric closures and seam overlap.

(4) The specimens must be conditioned at 70±5°F (21±2°C) 55%±10% relative humidity for at least 24 hours before testing.

(d) *Test Apparatus.* The arrangement of the test apparatus is shown in Figures 1 through 5 and must include the components described in this section. Minor details of the apparatus may vary, depending on the model burner used.

(1) *Specimen Mounting Stand.* The mounting stand for the test specimens consists of steel angles, as shown in Figure 1. The length of the mounting stand legs is 12 $^1/_8$ inches (305±3mm). The mounting stand must be used for mounting the test specimen seat bottom and seat back, as shown in Figure 2. The mounting stand should also include a suitable drip pan lined with aluminum foil, dull side up.

(2) *Test Burner.* The burner to be used in testing must—

(i) Be a modified gun type;

(ii) Have an 80-degree spray angle nozzle nominally rated for 2.25 gallons/hour at 100 psi;

(iii) Have a 12-inch (305mm) burner cone installed at the end of the draft tube, with an opening 6 inches (152mm) high and 11 inches (280mm) wide, as shown in Figure 3; and

(iv) Have a burner fuel pressure regulator that is adjusted to deliver a nominal 2.0 gallon/hour of # 2 Grade kerosene or equivalent required for the test.

Burner models which have been used successfully in testing are the Lennox Model OB-32, Carlin Model 200 CRD, and Park Model DPL 3400. FAA published reports pertinent to this type of burner are: (1) Powerplant Engineering Report No. 3A, Standard Fire Test Apparatus and Procedure for Flexible Hose Assemblies, dated March 1978; and (2) Report No. DOT/FAA/RD/76/213, Reevaluation of Burner Characteristics for Fire Resistance Tests, dated January 1977.

(3) *Calorimeter.*

(i) The calorimeter to be used in testing must be a (0-15.0 BTU/ft^2 -sec. 0-17.0 W/cm^2) calorimeter, accurate±3%, mounted in a 6-inch by 12-inch (152 by 305mm) by $^3/_4$-inch (19mm) thick calcium silicate insulating board which is attached to a steel angle bracket for placement in the test stand during burner calibration, as shown in Figure 4.

(ii) Because crumbling of the insulating board with service can result in misalignment of the calorimeter, the calorimeter must be monitored and the mounting shimmed, as necessary, to ensure that the calorimeter face is flush with the exposed plane of the insulating board in a plane parallel to the exit of the test burner cone.

(4) *Thermocouples.* The seven thermocouples to be used for testing must be $^1/_{16}$ to $^1/_8$-inch metal sheathed, ceramic packed, type K, grounded thermocouples with a nominal 22 to 30 American wire gage (AWG)-size conductor. The seven thermocouples must be attached to a steel angle bracket to form a thermocouple rake for placement in the test stand during burner calibration, as shown in Figure 5.

(5) *Apparatus Arrangement.* The test burner must be mounted on a suitable stand to position the exit of the burner cone a distance of $4\pm^1/_8$ inches (102±3mm) from one side of the specimen mounting stand. The burner stand should have the capability of allowing the burner to be swung away from the specimen mounting stand during warmup periods.

(6) *Data Recording.* A recording potentiometer or other suitable calibrated instrument with an appropriate range must be used to measure and record the outputs of the calorimeter and the thermocouples.

(7) *Weight Scale.* Weighing Device—A device must be used that with proper procedures may determine the before and after test weights of each set of seat cushion specimens within 0.02 pound (9 grams). A continuous weighing system is preferred.

(8) *Timing Device.* A stopwatch or other device (calibrated to±1 second) must be used to measure the time of application of the burner flame and self-extinguishing time or test duration.

(e) *Preparation of Apparatus.* Before calibration, all equipment must be turned on and the burner fuel must be adjusted as specified in paragraph (d)(2).

(f) *Calibration.* To ensure the proper thermal output of the burner, the following test must be made:

(1) Place the calorimeter on the test stand as shown in Figure 4 at a distance of $4\pm^1/_8$ inches (102±3mm) from the exit of the burner cone.

(2) Turn on the burner, allow it to run for 2 minutes for warmup, and adjust the burner air intake damper to produce a reading of 10.5±0.5 BTU/ft^2 -sec. (11.9±0.6 w/cm^2) on the calorimeter to ensure steady state conditions have been achieved. Turn off the burner.

(3) Replace the calorimeter with the thermocouple rake (Figure 5).

(4) Turn on the burner and ensure that the thermocouples are reading 1900±100°F (1038±38°C) to ensure steady state conditions have been achieved.

(5) If the calorimeter and thermocouples do not read within range, repeat steps in paragraphs 1 through 4 and adjust the burner air intake damper until the proper readings are obtained. The thermocouple rake and the calorimeter should be used frequently to maintain and record calibrated test parameters. Until the specific apparatus has demonstrated consistency, each test should be calibrated. After consistency has been confirmed, several tests may be conducted with the pre-test calibration before and a calibration check after the series.

(g) *Test Procedure.* The flammability of each set of specimens must be tested as follows:

(1) Record the weight of each set of seat bottom and seat back cushion specimens to be tested to the nearest 0.02 pound (9 grams).

(2) Mount the seat bottom and seat back cushion test specimens on the test stand as shown in Figure 2, securing the seat back cushion specimen to the test stand at the top.

(3) Swing the burner into position and ensure that the distance from the exit of the burner cone to the side of the seat bottom cushion specimen is $4\pm^1/_8$ inches (102±3mm).

(4) Swing the burner away from the test position. Turn on the burner and allow it to run for 2 minutes to provide adequate warmup of the burner cone and flame stabilization.

(5) To begin the test, swing the burner into the test position and simultaneously start the timing device.

(6) Expose the seat bottom cushion specimen to the burner flame for 2 minutes and then turn off the burner. Immediately swing the burner away from the test position. Terminate test 7 minutes after initiating cushion exposure to the flame by use of a gaseous extinguishing agent (i.e., Halon or CO_2).

(7) Determine the weight of the remains of the seat cushion specimen set left on the mounting stand to the nearest 0.02 pound (9 grams) excluding all droppings.

(h) *Test Report.* With respect to all specimen sets tested for a particular seat cushion for which testing of compliance is performed, the following information must be recorded:

(1) An identification and description of the specimens being tested.

(2) The number of specimen sets tested.

(3) The initial weight and residual weight of each set, the calculated percentage weight loss of each set, and the calculated average percentage weight loss for the total number of sets tested.

(4) The burn length for each set tested.

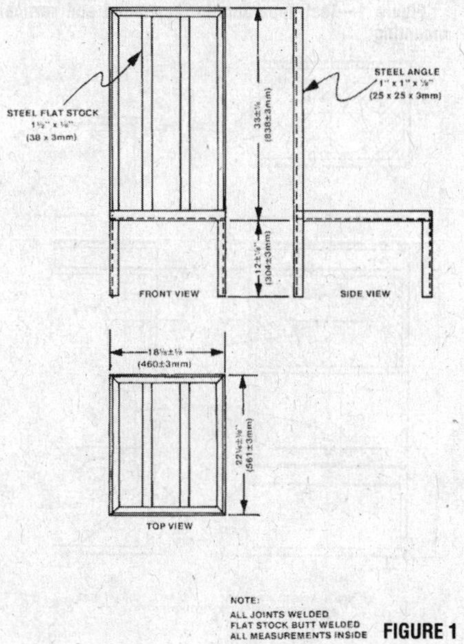

FIGURE 1

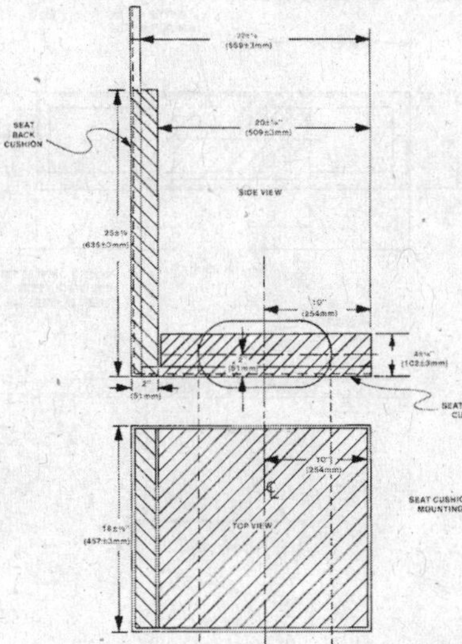

FIGURE 2

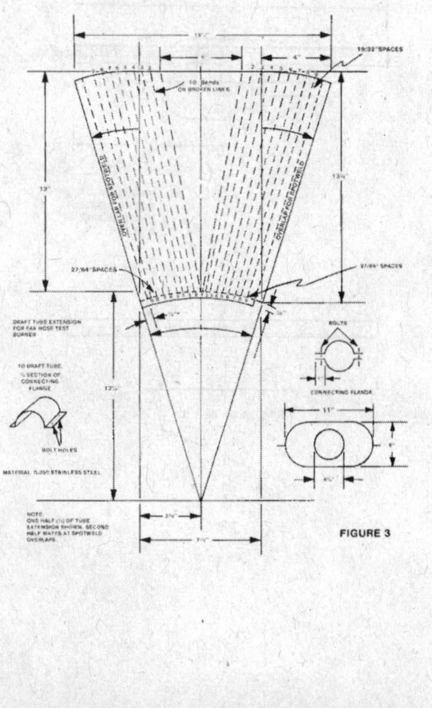

FIGURE 3

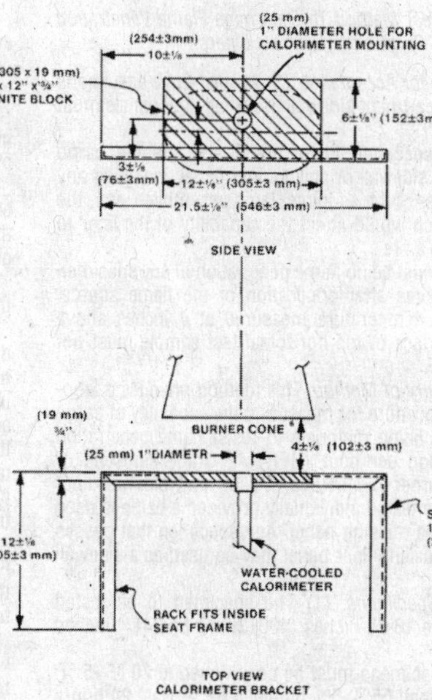

FIGURE 4

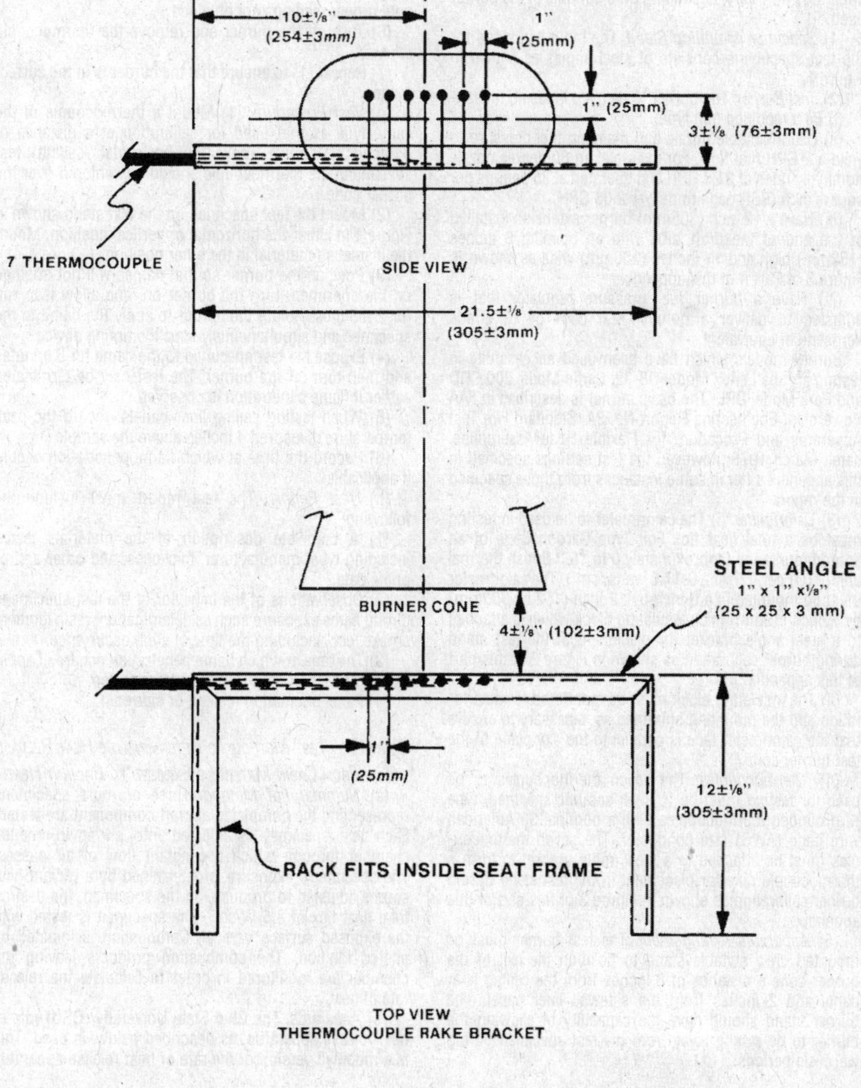

PART 25

FAR

Part III—Test Method To Determine Flame Penetration Resistance of Cargo Compartment Liners.

(a) *Criteria for Acceptance.* (1) At least three specimens of cargo compartment sidewall or ceiling liner panels must be tested.

(2) Each specimen tested must simulate the cargo compartment sidewall or ceiling liner panel, including any design features, such as joints, lamp assemblies, etc., the failure of which would affect the capability of the liner to safely contain a fire.

(3) There must be no flame penetration of any specimen within 5 minutes after application of the flame source, and the peak temperature measured at 4 inches above the upper surface of the horizontal test sample must not exceed 400 °F.

(b) *Summary of Method.* This method provides a laboratory test procedure for measuring the capability of cargo compartment lining materials to resist flame penetration with a 2 gallon per hour (GPH) #2 Grade kerosene or equivalent burner fire source. Ceiling and sidewall liner panels may be tested individually provided a baffle is used to simulate the missing panel. Any specimen that passes the test as a ceiling liner panel may be used as a sidewall liner panel.

(c) *Test Specimens.* (1) The specimen to be tested must measure 16±$^1/_8$ inches (406±3mm) by 24+$^1/_8$ inches (610±3mm).

(2) The specimens must be conditioned at 70 °F.±5 °F. (21°C.±2°C.) and 55%±5% humidity for at least 24 hours before testing.

(d) *Test Apparatus.* The arrangement of the test apparatus, which is shown in Figure 3 of Part II and Figures 1 through 3 of this part of appendix F, must include the components described in this section. Minor details of the apparatus may vary, depending on the model of the burner used.

(1) *Specimen Mounting Stand.* The mounting stand for the test specimens consists of steel angles as shown in Figure 1.

(2) *Test Burner.* The burner to be used in tesing must—

(i) Be a modified gun type.

(ii) Use a suitable nozzle and maintain fuel pressure to yield a 2 GPH fuel flow. For example: an 80 degree nozzle nominally rated at 2.25 GPH and operated at 85 pounds per square inch (PSI) gage to deliver 2.03 GPH.

(iii) Have a 12 inch (305mm) burner extension installed at the end of the draft tube with an opening 6 inches (152mm) high and 11 inches (280mm) wide as shown in Figure 3 of Part II of this appendix.

(iv) Have a burner fuel pressure regulator that is adjusted to deliver a nominal 2.0 GPH of #2 Grade kerosene or equivalent.

Burner models which have been used successfully in testing are the Lenox Model OB-32, Carlin Model 200 CRD and Park Model DPL. The basic burner is described in FAA Powerplant Engineering Report No. 3A, Standard Fire Test Apparatus and Procedure for Flexible Hose Assemblies, dated March 1978; however, the test settings specified in this appendix differ in some instances from those specified in the report.

(3) *Calorimeter.* (i) The calorimeter to be used in testing must be a total heat flux Foil Type Gardon Gage of an appropriate range (approximately 0 to 15.0 British thermal unit (BTU) per ft.^{2} sec., 0-17.0 watts/cm^2). The calorimeter must be mounted in a 6 inch by 12 inch (152 by 305mm) by $^3/_4$ inch (19mm) thick insulating block which is attached to a steel angle bracket for placement in the test stand during burner calibration as shown in Figure 2 of this part of this appendix.

(ii) The insulating block must be monitored for deterioration and the mounting shimmed as necessary to ensure that the calorimeter face is parallel to the exit plane of the test burner cone.

(4) *Thermocouples.* The seven thermocouples to be used for testing must be $^1/_{16}$ inch ceramic sheathed, type K, grounded thermocouples with a nominal 30 American wire gage (AWG) size conductor. The seven thermocouples must be attached to a steel angle bracket to form a thermocouple rake for placement in the test stand during burner calibration as shown in Figure 3 of this part of this appendix.

(5) *Apparatus Arrangement.* The test burner must be mounted on a suitable stand to position the exit of the burner cone a distance of 8 inches from the ceiling liner panel and 2 inches from the sidewall liner panel. The burner stand should have the capability of allowing the burner to be swung away from the test specimen during warm-up periods.

(6) *Instrumentation.* A recording potentiometer or other suitable instrument with an appropriate range must be used to measure and record the outputs of the calorimeter and the thermocouples.

(7) *Timing Device.* A stopwatch or other device must be used to measure the time of flame application and the time of flame penetration, if it occurs.

(e) *Preparation of Apparatus.* Before calibration, all equipment must be turned on and allowed to stabilize, and the burner fuel flow must be adjusted as specified in paragraph (d)(2).

(f) *Calibration.* To ensure the proper thermal output of the burner the following test must be made:

(1) Remove the burner extension from the end of the draft tube. Turn on the blower portion of the burner without turning the fuel or igniters on. Measure the air velocity using a hot wire anemometer in the center of the draft tube across the face of the opening. Adjust the damper such that the air velocity is in the range of 1550 to 1800 ft./min. If tabs are being used at the exit of the draft tube, they must be removed prior to this measurement. Reinstall the draft tube extension cone.

(2) Place the calorimeter on the test stand as shown in Figure 2 at a distance of 8 inches (203mm) from the exit of the burner cone to simulate the position of the horizontal test specimen.

(3) Turn on the burner, allow it to run for 2 minutes for warm-up, and adjust the damper to produce a calorimeter reading of 8.0±0.5 BTU per ft.^{2} sec. (9.1±0.6 Watts/cm^2).

(4) Replace the calorimeter with the thermocouple rake (see Figure 3).

(5) Turn on the burner and ensure that each of the seven thermocouples reads 1700 °F.±100 °F. (927°C.±38°C.) to ensure steady state conditions have been achieved. If the temperature is out of this range, repeat steps 2 through 5 until proper readings are obtained.

(6) Turn off the burner and remove the thermocouple rake.

(7) Repeat (1) to ensure that the burner is in the correct range.

(g) *Test Procedure.* (1) Mount a thermocouple of the same type as that used for calibration at a distance of 4 inches (102mm) above the horizontal (ceiling) test specimen. The thermocouple should be centered over the burner cone.

(2) Mount the test specimen on the test stand shown in Figure 1 in either the horizontal or vertical position. Mount the insulating material in the other position.

(3) Position the burner so that flames will not impinge on the specimen, turn the burner on, and allow it to run for 2 minutes. Rotate the burner to apply the flame to the specimen and simultaneously start the timing device.

(4) Expose the test specimen to the flame for 5 minutes and then turn off the burner. The test may be terminated earlier if flame penetration is observed.

(5) When testing ceiling liner panels, record the peak temperature measured 4 inches above the sample.

(6) Record the time at which flame penetration occurs if applicable.

(h) *Test Report.* The test report must include the following:

(1) A complete description of the materials tested including type, manufacturer, thickness, and other appropriate data.

(2) Observations of the behavior of the test specimens during flame exposure such as delamination, resin ignition, smoke, ect., including the time of such occurrence.

(3) The time at which flame penetration occurs, if applicable, for each of the three specimens tested.

(4) Panel orientation (ceiling or sidewall).

Part IV—Test Method To Determine the Heat Release Rate From Cabin Materials Exposed to Radiant Heat.

(a) *Summary of Method.* Three or more specimens representing the completed aircraft component are tested. Each test specimen is injected into an environmental chamber through which a constant flow of air passes. The specimen's exposure is determined by a radiant heat source adjusted to produce, on the specimen, the desired total heat flux of 3.5 W/cm^2. The specimen is tested with the exposed surface vertical. Combustion is initiated by piloted ignition. The combustion products leaving the chamber are monitored in order to calculate the release rate of heat.

(b) *Apparatus.* The Ohio State University (OSU) rate of heat release apparatus, as described below, is used. This is a modified version of the rate of heat release apparatus

Figure 1—Test apparatus for horizontal and vertical mounting

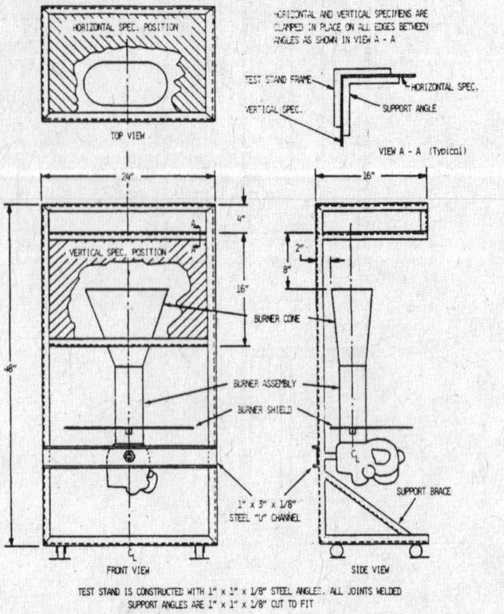

Figure 2—Calorimeter Bracket

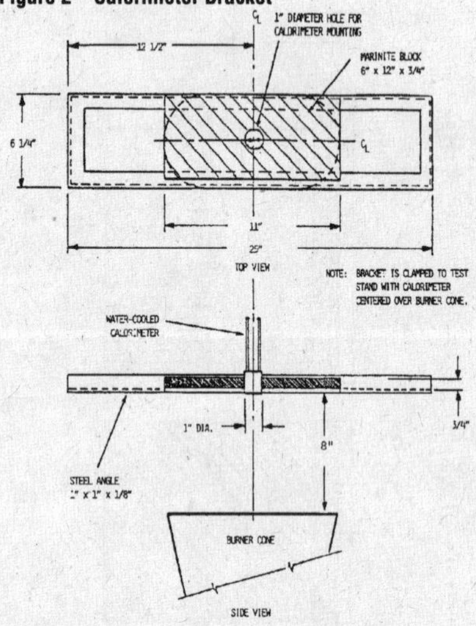

Figure 3—Thermocouple Rake Bracket

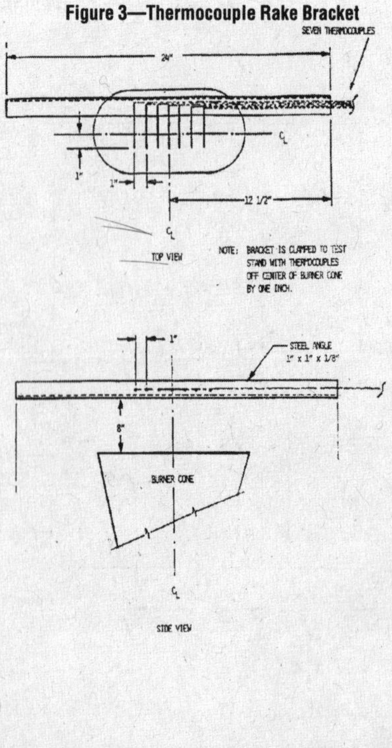

standardized by the American Society of Testing and Materials (ASTM), ASTM E-906.

(1) This apparatus is shown in Figures 1A and 1B of this part IV. All exterior surfaces of the apparatus, except the holding chamber, must be insulated with 1 inch (25mm) thick, low density, high temperature, fiberglass board insulation. A gasketed door, through which the sample injection rod slides, must be used to form an airtight closure on the specimen hold chamber.

(2) *Thermopile.* The temperature difference between the air entering the environmental chamber and that leaving must be monitored by a thermopile having five hot, and five cold, 24-guage Chromel-Alumel junctions. The hot junctions must be spaced across the top of the exhaust stack,.38 inches (10mm) below the top of the chimney. The thermocouples must have a.050±.010 inch (1.3±.3mm) diameter, ball-type, welded tip. One thermocouple must be located in the geometric center, with the other four located 1.18 inch (30mm) from the center along the diagonal toward each of the corners (Figure 5 of this part IV). The cold junctions must be located in the pan below the lower air distribution plate (see paragraph (b)(4) of this part IV). Thermopile hot junctions must be cleared of soot deposits as needed to maintain the calibrated sensitivity.

(3) *Radiation Source.* A radiant heat source incorporating four Type LL silicon carbide elements, 20 inches (508mm) long by.63 inch (16mm) O.D., must be used, as shown in Figures 2A and 2B of this part IV. The heat source must have a nominal resistance of 1.4 ohms and be capable of generating a flux up to 100 kW/m^2. The silicone carbide elements must be mounted in the stainless steel panel box by inserting them through .63 inch (16mm) holes in .03 inch (1mm) thick ceramic fiber or calcium-silicate millboard. Locations of the holes in the pads and stainless steel cover plates are shown in Figure 2B of this part IV. The truncated diamond-shaped mask of .042±.002 inch (1.07±.05mm) stainless steel must be added to provide uniform heat flux density over the area occupied by the vertical sample.

(4) *Air Distribution System.* The air entering the environmental chamber must be distributed by a.25 inch (6.3mm) thick aluminum plate having eight No. 4 drill-holes, located 2 inches (51mm) from sides on 4 inch (102mm) centers, mounted at the base of the environmental chamber. A second plate of 18 guage stainless steel having 120, evenly spaced, No. 28 drill holes must be mounted 6 inches (152mm) above the aluminum plate. A well-regulated air supply is required. The air-supply manifold at the base of the pyramidal section must have 48, evenly spaced, No. 26 drill holes located. 38 inch (10mm) from the inner edge of the manifold, resulting in an airflow split of approximately three to one within the apparatus.

(5) *Exhaust Stack.* An exhaust stack, 5.25×2.75 inches (133×70mm) in cross section, and 10 inches (254mm) long, fabricated from 28 guage stainless steel must be mounted on the outlet of the pyramidal section. A. 1.0×3.0 inch (25×76mm) baffle plate of.018±.002 inch (.50±.05mm) stainless steel must be centered inside the stack, perpendicular to the air flow, 3 inches (76mm) above the base of the stack.

(6) *Specimen Holders.* (i) The specimen must be tested in a vertical orientation. The specimen holder (Figure 3 of this part IV) must incorporate a frame that touches the specimen (which is wrapped with aluminum foil as required by paragraph (d)(3) of this Part) along only the.25 inch (6mm) perimeter. A "V" shaped spring is used to hold the assembly together. A detachable.50×.50×5.91 inch (12×12×150mm) drip pan and two.020 inch (.5mm) stainless steel wires (as shown in Figure 3 of this part IV) must be used for testing materials prone to melting and dripping. The positioning of the spring and frame may be changed to accommodate different specimen thicknesses by inserting the retaining rod in different holes on the specimen holder.

(ii) Since the radiation shield described in ASTM E-906 is not used, a guide pin must be added to the injection mechanism. This fits into a slotted metal plate on the injection mechanism outside of the holding chamber. It can be used to provide accurate positioning of the specimen face after injection. The front surface of the specimen must be 3.9 inches (100mm) from the closed radiation doors after injection.

(iii) The specimen holder clips onto the mounted bracket (Figure 3 of this part IV). The mounting bracket must be attached to the injection rod by three screws that pass through a wide-area washer welded onto a ½-inch (13mm) nut. The end of the injection rod must be threaded to screw into the nut, and a.020 inch (5.1mm) thick wide area washer must be held between two ½-inch (13mm) nuts that are adjusted to tightly cover the hole in the radiation doors through which the injection rod or calibration calorimeter pass.

(7) *Calorimeter.* A total-flux type calorimeter must be mounted in the center of a ½-inch Kaowool "M" board inserted in the sample holder to measure the total heat flux. The calorimeter must have a view angle of 180 degrees and be calibrated for incident flux. The calorimeter calibration must be acceptable to the Administrator.

(8) *Pilot-Flame Positions.* Pilot ignition of the specimen must be accomplished by simultaneously exposing the specimen to a lower pilot burner and an upper pilot burner, as described in paragraph (b)(8)(i) and (b)(8)(ii) or (b)(8)(iii) of this part IV, respectively. Since intermittent pilot flame extinguishment for more than 3 seconds would invalidate the test results, a spark ignitor may be installed to ensure that the lower pilot burner remains lighted.

(i) *Lower Pilot Burner.* The pilot-flame tubing must be.25 inch (6.3mm) O.D., .03 inch (0.8mm) wall, stainless steel tubing. A mixture of 120 cm^3/min. of methane and 850 cm^3/min. of air must be fed to the lower pilot flame burner. The normal position of the end of the pilot burner tubing is.40 inch (10mm) from and perpendicular to the exposed vertical surface of the specimen. The centerline at the outlet of the burner tubing must intersect the vertical centerline of the sample at a point .20 inch (5mm) above the lower exposed edge of the specimen.

(ii) *Standard Three-Hole Upper Pilot Burner.* The pilot burner must be a straight length of.25 inch (6.3mm) O.D., .03 inch (0.8mm) wall, stainless steel tubing that is 14 inches (360mm) long. One end of the tubing must be closed, and three No. 40 drill holes must be drilled into the tubing, 2.38 inch (60mm) apart, for gas ports, all radiating in the same direction. The first hole must be.19 inch (5mm) from the closed end of the tubing. The tube must be positioned.75 inch (19mm) above and.75 inch (19mm) behind the exposed upper edge of the specimen. The middle hole must be in the vertical plane perpendicular to the exposed surface of the specimen which passes through its vertical centerline and must be pointed toward the radiation source. The gas supplied to the burner must be methane and must be adjusted to produce flame lengths of 1 inch (25mm).

(iii) *Optional Fourteen-Hole Upper Pilot Burner.* This burner may be used in lieu of the standard three-hole burner described in paragraph (b)(8)(ii) of this part IV. The pilot burner must be a straight length of.25 inch (6.3mm) O.D., .03 inch (0.8mm) wall, stainless steel tubing that is 15.75 inches (400mm) long. One end of the tubing must be closed, and 14 No. 59 drill holes must be drilled into the tubing, .50 inch (13mm) apart, for gas ports, all radiating in the same direction. The first hole must be .50 inch (13mm) from the closed end of the tubing. The tube must be positioned above the specimen holder so that the holes are placed above the specimen as shown in Figure 1B of this part IV. The fuel supplied to the burner must be methane mixed with air in a ratio of approximately 50/50 by volume. The total gas flow must be adjusted to produce flame lengths of 1 inch (25mm). When the gas/air ratio and the flow rate are properly adjusted, approximately .25 inch (6mm) of the flame length appears yellow in color.

(c) *Calibration of Equipment*—(1) *Heat Release Rate.* A calibration burner, as shown in Figure 4, must be placed over the end of the lower pilot flame tubing using a gas tight connection. The flow of gas to the pilot flame must be at least 99 percent methane and must be accurately metered. Prior to usage, the wet test meter must be properly leveled and filled with distilled water to the tip of the internal pointer while no gas is flowing. Ambient temperature and pressure of the water are based on the internal wet test meter temperature. A baseline flow rate of approximately 1 liter/min. must be set and increased to higher preset flows of 4, 6, 8, 6 and 4 liters/min. Immediately prior to recording methane flow rates, a flow rate of 8 liters/min. must be used for 2 minutes to precondition the chamber. This is not recorded as part of calibration. The rate must be determined by using a stopwatch to time a complete revolution of the wet test meter for both the baseline and higher flow, with the flow returned to baseline before changing to the next higher flow. The thermopile baseline voltage must be measured. The gas flow to the burner must be increased to the higher preset flow and allowed to burn for 2.0 minutes, and the thermopile voltage must be measured. The sequence must be repeated until all five values have been determined. The average of the five values must be used as the calibration factor. The procedure must be repeated if the percent relative standard deviation is greater than 5 percent. Calculations are shown in paragraph (f) of this part IV.

(2) *Flux Uniformity.* Uniformity of flux over the specimen must be checked periodically and after each heating element change to determine if it is within acceptable limits of plus or minus 5 percent.

(3) As noted in paragraph (b)(2) of this part IV, thermopile hot junctions must be cleared of soot deposits as needed to maintain the calibrated sensitivity.

(d) *Preparation of Test Specimens.* (1) The test specimens must be representative of the aircraft component in regard to materials and construction methods. The standard size for the test specimens is 5.91±.03×5.91±.03 inches (149±1×149±1mm). The thickness of the specimen must be the same as that of the aircraft component it represents up to a maximum thickness of 1.75 inches (45mm). Test specimens representing thicker components must be 1.75 inches (45mm).

(2) *Conditioning.* Specimens must be conditioned as described in Part 1 of this appendix.

(3) *Mounting.* Each test specimen must be wrapped tightly on all sides of the specimen, except for the one surface that is exposed with a single layer of.001 inch (.025mm) aluminum foil.

(e) *Procedure.* (1) The power supply to the radiant panel must be set to produce a radiant flux of 3.5±.05 W/cm^2, as measured at the point the center of the specimen surface will occupy when positioned for the test. The radiant flux must be measured after the air flow through the equipment is adjusted to the desired rate.

(2) After the pilot flames are lighted, their position must be checked as described in paragraph (b)(8) of this part IV.

(3) Air flow through the apparatus must be controlled by a circular plate orifice located in a 1.5 inch (38.1mm) I.D. pipe with two pressure measuring points, located 1.5 inches (38mm) upstream and.75 inches (19mm) downstream of the orifice plate. The pipe must be connected to a manometer set at a pressure differential of 7.87 inches (200mm) of Hg. (See Figure 1B of this part IV.) The total air flow to the equipment is approximately.04 m^3/seconds. The stop on the vertical specimen holder rod must be adjusted so that the exposed surface of the specimen is positioned 3.9 inches (100mm) from the entrance when injected into the environmental chamber.

(4) The specimen must be placed in the hold chamber with the radiation doors closed. The airtight outer door must be secured, and the recording devices must be started. The specimen must be retained in the hold chamber for 60 seconds, plus or minus 10 seconds, before injection. The thermopile "zero" value must be determined during the last 20 seconds of the hold period. The sample must not be injected before completion of the "zero" value determination.

(5) When the specimen is to be injected, the radiation doors must be opened. After the specimen is injected into the environmental chamber, the radiation doors must be closed behind the specimen.

(6) [Reserved]

(7) Injection of the specimen and closure of the inner door marks time zero. A record of the thermopile output with at least one data point per second must be made during the time the specimen is in the environmental chamber.

(8) The test duration is five minutes. The lower pilot burner and the upper pilot burner must remain lighted for the entire duration of the test, except that there may be intermittent flame extinguishment for periods that do not exceed 3 seconds. Furthermore, if the optional three-hole upper burner is used, at least two flamelets must remain lighted for the entire duration of the test, except that there may be intermittent flame extinguishment of all three flamelets for periods that do not exceed 3 seconds.

(9) A minimum of three specimens must be tested.

(f) *Calculations.* (1) The calibration factor is calculated as follows:

$$K_h = \frac{(F_1 - F_0)}{(V_1 - V_0)} \times \frac{(210.8-22)k_{cal}}{mole} \times \frac{273}{T_a} \times \frac{P-P_v}{760} \times \frac{mole\ CH4STP}{22.41} \times \frac{WATT\ \min}{.01433kcal} \times \frac{kw}{1000w}$$

F_0=flow of methane at baseline (1pm)
F_1=higher preset flow of methane (1pm)
V_0=thermopile voltage at baseline (mv)
V_1=thermopile voltage at higher flow (mv)
T_a=Ambient temperature (K)
P=Ambient pressure (mm Hg)
P_v=Water vapor pressure (mm Hg)

(2) Heat release rates may be calculated from the reading of the thermopile output voltage at any instant of time as:

$$HRR = \frac{(V_m - V_b)K_n}{.2323m^2}$$

PART 25

FAR

HRR=heat release rate (kw/m²)
V_b=baseline voltage (mv)
V_m=measured thermopile voltage (mv)
K_h=calibration factor (kw/mv)

(3) The integral of the heat release rate is the total heat release as a function of time and is calculated by multiplying the rate by the data sampling frequency in minutes and summing the time from zero to two minutes.

(g) *Criteria.* The total positive heat release over the first two minutes of exposure for each of the three or more samples tested must be averaged, and the peak heat release rate for each of the samples must be averaged. The average total heat release must not exceed 65 kilowatt-minutes per square meter, and the average peak heat release rate must not exceed 65 kilowatts per square meter.

(h) *Report.* The test report must include the following for each specimen tested:

(1) Description of the specimen.

(2) Radiant heat flux to the specimen, expressed in W/cm².

(3) Data giving release rates of heat (in kW/m²) as a function of time, either graphically or tabulated at intervals no greater than 10 seconds. The calibration factor (k_h) must be recorded.

(4) If melting, sagging, delaminating, or other behavior that affects the exposed surface area or the mode of burning occurs, these behaviors must be reported, together with the time at which such behaviors were observed.

(5) The peak heat release and the 2-minute integrated heat release rate must be reported.

Part V. Test Method To Determine the Smoke Emission Characteristics of Cabin Materials

(a) *Summary of Method.* The specimens must be constructed, conditioned, and tested in the flaming mode in accordance with American Society of Testing and Materials (ASTM) Standard Test Method ASTM F814-83.

(b) *Acceptance Criteria.* The specific optical smoke density (D_s), which is obtained by averaging the reading obtained after 4 minutes with each of the three specimens, shall not exceed 200.

Part VI—Test Method To Determine the Flammability and Flame Propagation Characteristics of Thermal/Acoustic Insulation Materials

Figures to Part IV of Appendix F

Figure 1A—Rate of Heat Release Apparatus

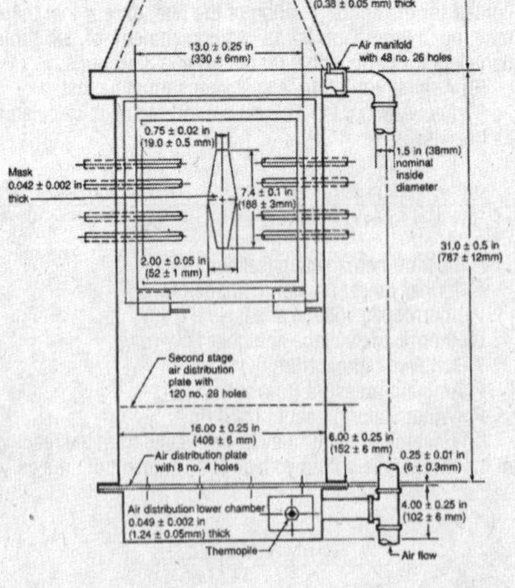

Figure 1B—Rate of Heat Release Apparatus

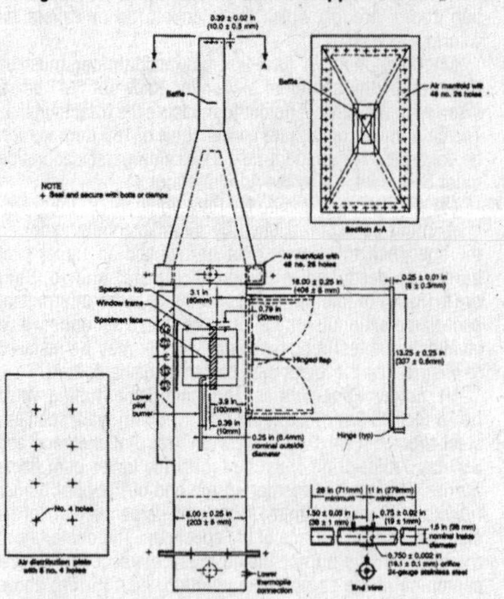

(Unless denoted otherwise all dimensions are in millimeters.)

Figure 2B—"Globar" Radiant Panel

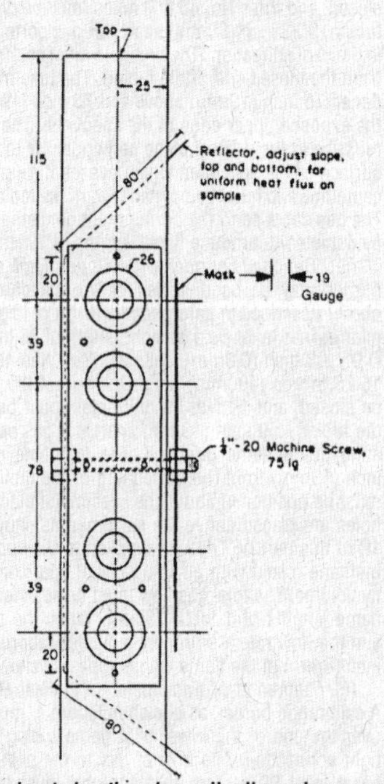

(Unless denoted otherwise all dimensions are in millimeters.)

Figure 3

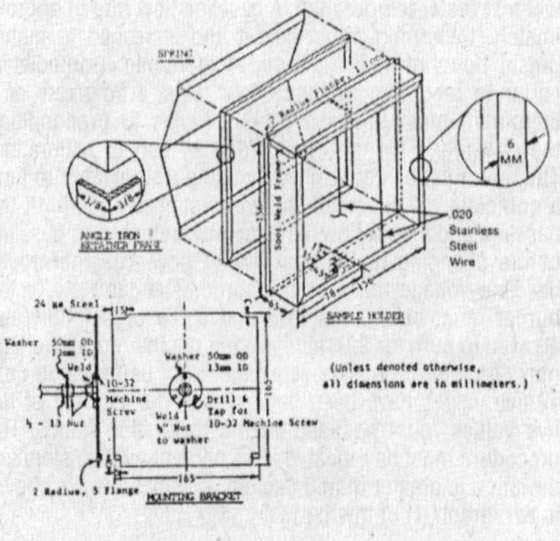

Figure 4

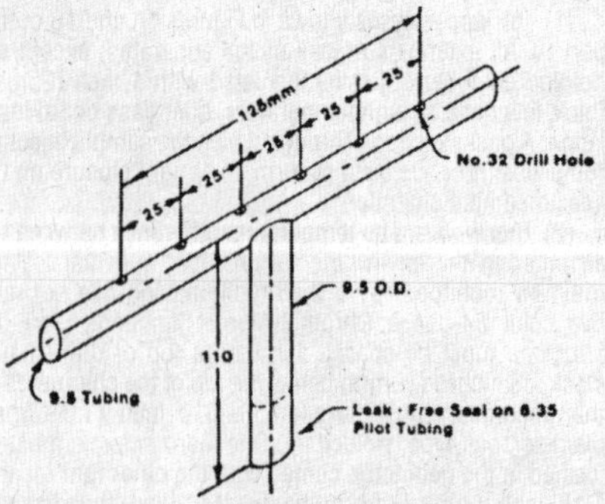

(Unless denoted otherwise all dimensions are in millimeters.)

Figure 5—Thermocouple Position

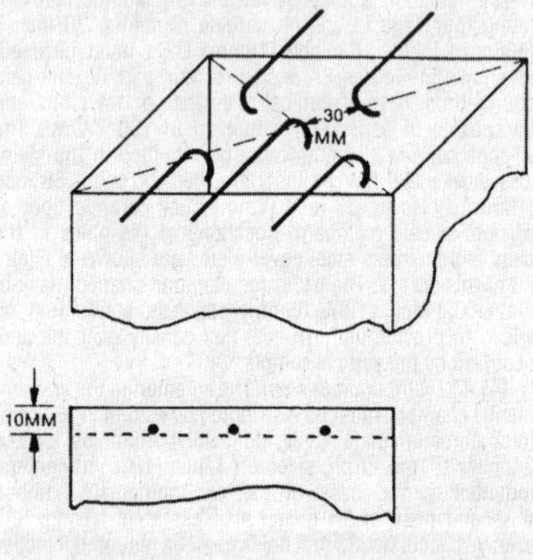

Use this test method to evaluate the flammability and flame propagation characteristics of thermal/acoustic insulation when exposed to both a radiant heat source and a flame.

(a) *Definitions.*

"Flame propagation" means the furthest distance of the propagation of visible flame towards the far end of the test specimen, measured from the midpoint of the ignition source flame. Measure this distance after initially applying the ignition source and before all flame on the test specimen is extinguished. The measurement is not a determination of burn length made after the test.

"Radiant heat source" means an electric or air propane panel.

"Thermal/acoustic insulation" means a material or system of materials used to provide thermal and/or acoustic protection. Examples include fiberglass or other batting material encapsulated by a film covering and foams.

"Zero point" means the point of application of the pilot burner to the test specimen.

(b) *Test apparatus.*

(1) *Radiant panel test chamber.* Conduct tests in a radiant panel test chamber (see figure 1 above). Place the test chamber under an exhaust hood to facilitate clearing

Figure 1—Radiant Panel Test Chamber

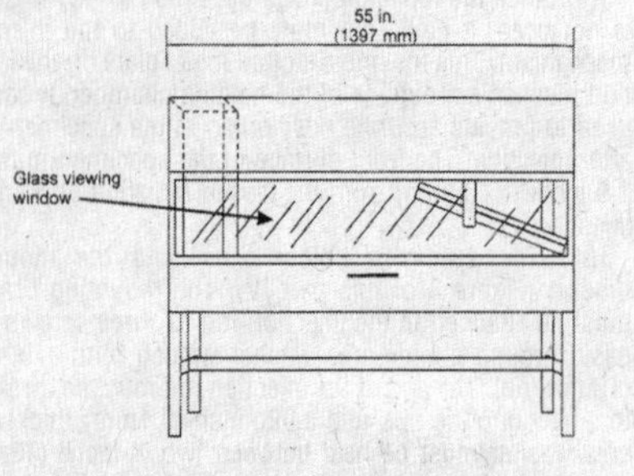

Figure 2—Internal Chimney

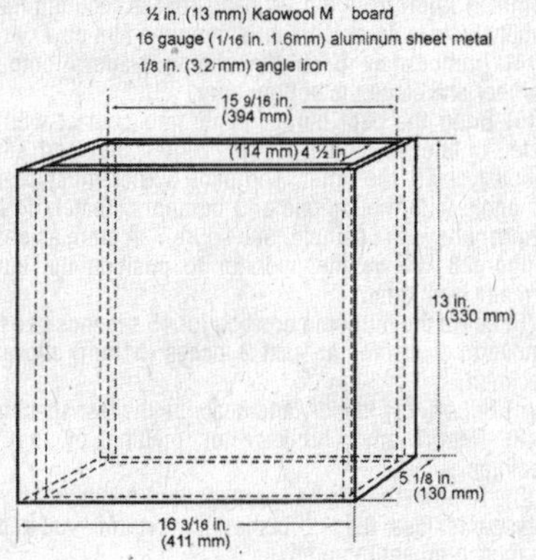

the chamber of smoke after each test. The radiant panel test chamber must be an enclosure 55 inches (1397mm) long by 19.5 (495mm) deep by 28 (710mm) to 30 inches (maximum) (762mm) above the test specimen. Insulate the sides, ends, and top with a fibrous ceramic insulation, such as Kaowool M™ board. On the front side, provide a 52 by 12-inch (1321 by 305mm) draft-free, high-temperature, glass window for viewing the sample during testing. Place a door below the window to provide access to the movable specimen platform holder. The bottom of the test chamber must be a sliding steel platform that has provision for securing the test specimen holder in a fixed and level position. The chamber must have an internal chimney with exterior dimensions of 5.1 inches (129mm) wide, by 16.2 inches (411mm) deep by 13 inches (330mm) high at the opposite end of the chamber from the radiant energy source. The interior dimensions must be 4.5 inches (114mm) wide by 15.6 inches (395mm) deep. The chimney must extend to the top of the chamber (see figure 2).

(2) *Radiant heat source.* Mount the radiant heat energy source in a cast iron frame or equivalent. An electric panel must have six, 3-inch wide emitter strips. The emitter strips must be perpendicular to the length of the panel. The panel must have a radiation surface of $12\frac{7}{8}$ by $18\frac{1}{2}$ inches (327 by 470mm). The panel must be capable of operating at temperatures up to 1300°F (704°C). An air propane panel must be made of a porous refractory material and have a radiation surface of 12 by 18 inches (305 by 457mm). The panel must be capable of operating at temperatures up to 1,500°F (816°C). *See* figures 3a and 3b.

(i) *Electric radiant panel.* The radiant panel must be 3-phase and operate at 208 volts. A single-phase, 240 volt panel is also acceptable. Use a solid-state power controller and microprocessor-based controller to set the electric panel operating parameters.

(ii) *Gas radiant panel.* Use propane (liquid petroleum gas—2.1 UN 1075) for the radiant panel fuel. The panel fuel system must consist of a venturi-type aspirator for mixing gas and air at approximately atmospheric pressure. Provide suitable instrumentation for monitoring and controlling the flow of fuel and air to the panel. Include an air flow gauge, an air flow regulator, and a gas pressure gauge.

(iii) *Radiant panel placement.* Mount the panel in the chamber at 30° to the horizontal specimen plane, and $7\frac{1}{2}$ inches above the zero point of the specimen.

(3) *Specimen holding system.*

(i) The sliding platform serves as the housing for test specimen placement. Brackets may be attached (via wing nuts) to the top lip of the platform in order to accommodate various thicknesses of test specimens. Place the test specimens on a sheet of Kaowool M™ board or 1260 Standard Board (manufactured by Thermal Ceramics and available in Europe), or equivalent, either resting on the bottom lip of the sliding platform or on the base of the brackets. It may be necessary to use multiple sheets of material based on the thickness of the test specimen (to meet the sample height requirement). Typically, these non-combustible sheets of material are available in $\frac{1}{4}$ inch (6mm) thicknesses. See figure 4. A sliding platform that is deeper than the 2-inch (50.8mm) platform shown in figure 4 is also acceptable as long as the sample height requirement is met.

(ii) Attach a $\frac{1}{2}$ inch (13mm) piece of Kaowool M™ board or other high temperature material measuring $41\frac{1}{2}$

Figure 3a—Electric Panel

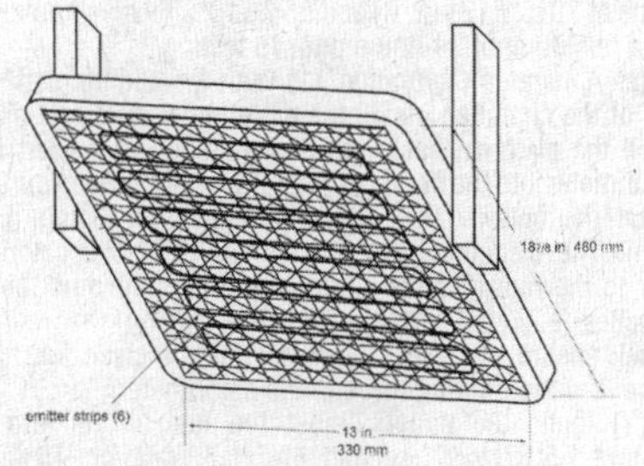

Figure 3b—Air Propane Radiant Panel

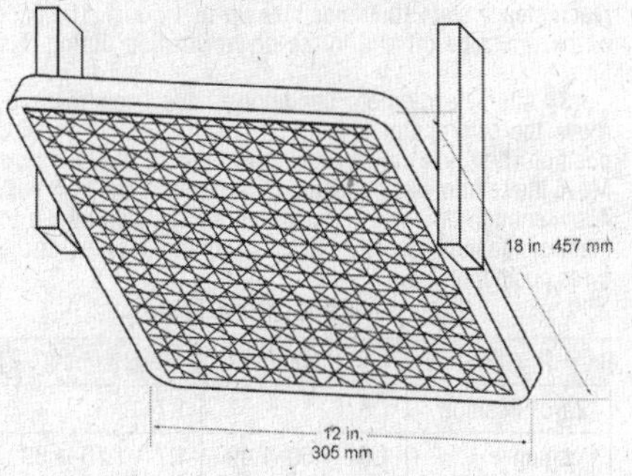

by $8\frac{1}{4}$ inches (1054 by 210mm) to the back of the platform. This board serves as a heat retainer and protects the test specimen from excessive preheating. The height of this board must not impede the sliding platform movement (in and out of the test chamber). If the platform has been fabricated such that the back side of the platform is high enough to prevent excess preheating of the specimen when the sliding platform is out, a retainer board is not necessary.

(iii) Place the test specimen horizontally on the non-combustible board(s). Place a steel retaining/securing frame fabricated of mild steel, having a thickness of $\frac{1}{8}$ inch (3.2mm) and overall dimensions of 23 by $13\frac{1}{8}$ inches (584 by 333mm) with a specimen opening of 19 by $10\frac{3}{4}$ inches (483 by 273mm) over the test specimen. The front, back, and right portions of the top flange of the frame must rest on the top of the sliding platform, and the bottom flanges must pinch all 4 sides of the test specimen. The right bottom flange must be flush with the sliding platform. See figure 5.

(4) *Pilot Burner.* The pilot burner used to ignite the specimen must be a Bernzomatic™ commercial propane venturi torch with an axially symmetric burner tip and a propane supply tube with an orifice diameter of 0.006

Figure 5—3 views

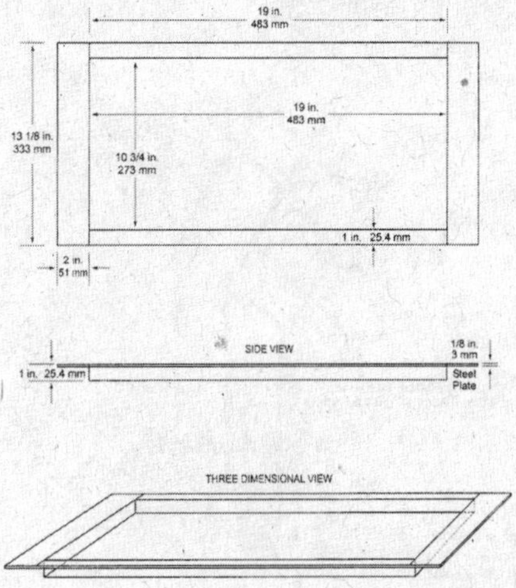

Figure 6—Propane Pilot Burner

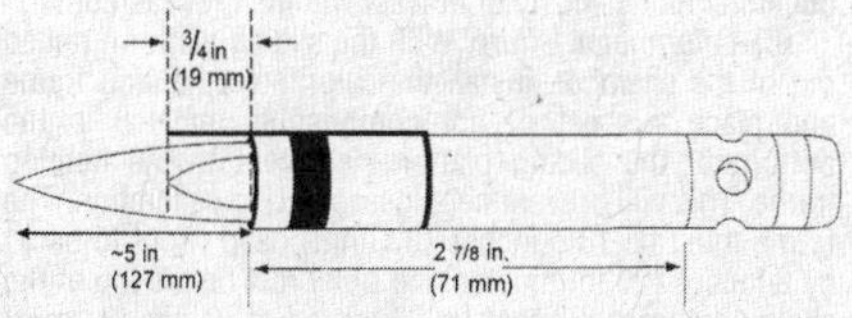

Figure 6 – Propane Pilot Burner

inches (0.15mm). The length of the burner tube must be $2\frac{7}{8}$ inches (71mm). The propane flow must be adjusted via gas pressure through an in-line regulator to produce a blue inner cone length of $\frac{3}{4}$ inch (19mm). A $\frac{3}{4}$ inch (19mm) guide (such as a thin strip of metal) may be soldered to the top of the burner to aid in setting the flame height. The overall flame length must be approximately 5 inches long (127mm). Provide a way to move the burner out of the ignition position so that the flame is horizontal and at least 2 inches (50mm) above the specimen plane. See figure 6.

(5) *Thermocouples.* Install a 24 American Wire Gauge (AWG) Type K (Chromel-Alumel) thermocouple in the test chamber for temperature monitoring. Insert it into the chamber through a small hole drilled through the back of the chamber. Place the thermocouple so that it extends 11 inches (279mm) out from the back of the chamber wall, $11\frac{1}{2}$ inches (292mm) from the right side of the chamber wall, and is 2 inches (51mm) below the radiant panel. The use of other thermocouples is optional.

(6) *Calorimeter.* The calorimeter must be a one-inch cylindrical water-cooled, total heat flux density, foil type Gardon Gage that has a range of 0 to 5 BTU/ft^2-second (0 to 5.7 Watts/cm^2).

(7) *Calorimeter calibration specification and procedure.*

(i) *Calorimeter specification.*

(A) Foil diameter must be 0.25±0.005 inches (6.35±0.13mm).

(B) Foil thickness must be 0.0005±0.0001 inches (0.013±0.0025mm).

(C) Foil material must be thermocouple grade Constantan.

(D) Temperature measurement must be a Copper Constantan thermocouple.

(E) The copper center wire diameter must be 0.0005 inches (0.013mm).

(F) The entire face of the calorimeter must be lightly coated with "Black Velvet" paint having an emissivity of 96 or greater.

(ii) *Calorimeter calibration.* (A) The calibration method must be by comparison to a like standardized transducer.

(B) The standardized transducer must meet the specifications given in paragraph VI(b)(6) of this appendix.

(C) Calibrate the standard transducer against a primary standard traceable to the National Institute of Standards and Technology (NIST).

(D) The method of transfer must be a heated graphite plate.

(E) The graphite plate must be electrically heated, have a clear surface area on each side of the plate of at least 2 by 2 inches (51 by 51mm), and be $\frac{1}{8}$ inch±$\frac{1}{16}$ inch thick (3.2±1.6mm).

(F) Center the 2 transducers on opposite sides of the plates at equal distances from the plate.

(G) The distance of the calorimeter to the plate must be no less than 0.0625 inches (1.6mm), nor greater than 0.375 inches (9.5mm).

(H) The range used in calibration must be at least 0-3.5 BTUs/ft^2 second (0-3.9 Watts/cm^2) and no greater than 0-5.7 BTUs/ft^2 second (0-6.4 Watts/cm^2).

Figure 7—Calorimeter Holding Frame

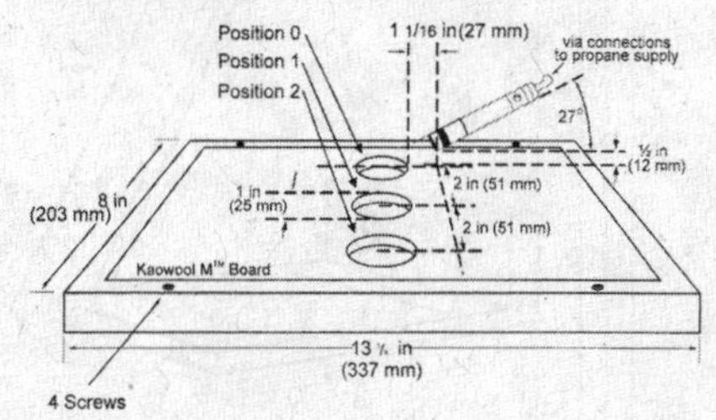

(l) The recording device used must record the 2 transducers simultaneously or at least within $\frac{1}{10}$ of each other.

(8) *Calorimeter fixture.* With the sliding platform pulled out of the chamber, install the calorimeter holding frame and place a sheet of non-combustible material in the bottom of the sliding platform adjacent to the holding frame. This will prevent heat losses during calibration. The frame must be $13\frac{1}{8}$ inches (333mm) deep (front to back) by 8 inches (203mm) wide and must rest on the top of the sliding platform. It must be fabricated of $\frac{1}{8}$ inch (3.2mm) flat stock steel and have an opening that accommodates a $\frac{1}{2}$ inch (12.7mm) thick piece of refractory board, which is level with the top of the sliding platform. The board must have three 1-inch (25.4mm) diameter holes drilled through the board for calorimeter insertion. The distance to the radiant panel surface from the centerline of the first hole ("zero" position) must be $7\frac{1}{2}\pm\frac{1}{8}$ inches (191±3mm). The distance between the centerline of the first hole to the centerline of the second hole must be 2 inches (51mm). It must also be the same distance from the centerline of the second hole to the centerline of the third hole. *See* figure 7. A calorimeter holding frame that differs in construction is acceptable as long as the height from the centerline of the first hole to the radiant panel and the distance between holes is the same as described in this paragraph.

(9) *Instrumentation.* Provide a calibrated recording device with an appropriate range or a computerized data acquisition system to measure and record the outputs of the calorimeter and the thermocouple. The data acquisition system must be capable of recording the calorimeter output every second during calibration.

(10) *Timing device.* Provide a stopwatch or other device, accurate to±1 second/hour, to measure the time of application of the pilot burner flame.

(c) *Test specimens.* (1) *Specimen preparation.* Prepare and test a minimum of three test specimens. If an oriented film cover material is used, prepare and test both the warp and fill directions.

(2) *Construction.* Test specimens must include all materials used in construction of the insulation (including batting, film, scrim, tape etc.). Cut a piece of core material such as foam or fiberglass, and cut a piece of film cover material (if used) large enough to cover the core material. Heat sealing is the preferred method of preparing fiberglass samples, since they can be made without compressing the fiberglass ("box sample"). Cover materials that are not heat sealable may be stapled, sewn, or taped as long as the cover material is over-cut enough to be drawn down the sides without compressing the core material. The fastening means should be as continuous as possible along the length of the seams. The specimen thickness must be of the same thickness as installed in the airplane.

(3) *Specimen Dimensions.* To facilitate proper placement of specimens in the sliding platform housing, cut non-rigid core materials, such as fiberglass, $12\frac{1}{2}$ inches (318mm) wide by 23 inches (584mm) long. Cut rigid materials, such as foam, $11\frac{1}{2}\pm\frac{1}{4}$ inches (292mm±6mm) wide by 23 inches (584mm) long in order to fit properly in the sliding platform housing and provide a flat, exposed surface equal to the opening in the housing.

Figure 8—Propane Burner Stop

(d) *Specimen conditioning.* Condition the test specimens at 70±5°F (21±2°C) and 55%±10% relative humidity, for a minimum of 24 hours prior to testing.

(e) *Apparatus Calibration.* (1) With the sliding platform out of the chamber, install the calorimeter holding frame. Push the platform back into the chamber and insert the calorimeter into the first hole ("zero" position). See figure 7. Close the bottom door located below the sliding platform. The distance from the centerline of the calorimeter to the radiant panel surface at this point must be $7\frac{1}{2}$inches ± $\frac{1}{8}$(191mm±3). Prior to igniting the radiant panel, ensure that the calorimeter face is clean and that there is water running through the calorimeter.

(2) Ignite the panel. Adjust the fuel/air mixture to achieve 1.5 BTUs/ft^2 -second±5% (1.7 Watts/cm^2±5%) at the "zero" position. If using an electric panel, set the power controller to achieve the proper heat flux. Allow the unit to reach steady state (this may take up to 1 hour). The pilot burner must be off and in the down position during this time.

(3) After steady-state conditions have been reached, move the calorimeter 2 inches (51mm) from the "zero" position (first hole) to position 1 and record the heat flux. Move the calorimeter to position 2 and record the heat flux. Allow enough time at each position for the calorimeter to stabilize. Table 1 depicts typical calibration values at the three positions.

Table 1—Calibration Table

Position	BTU's/ft2sec	Watts/cm2
"Zero" Position	1.5	1.7
Position 1	1.51–1.50–1.49	1.71–1.70–1.69
Position 2	1.43–1.44	1.62–1.63

(4) Open the bottom door, remove the calorimeter and holder fixture. Use caution as the fixture is very hot.

(f) *Test Procedure.* (1) Ignite the pilot burner. Ensure that it is at least 2 inches (51mm) above the top of the platform. The burner must not contact the specimen until the test begins.

(2) Place the test specimen in the sliding platform holder. Ensure that the test sample surface is level with the top of the platform. At "zero" point, the specimen surface must be $7\frac{1}{2}$ inches±$\frac{1}{8}$ inch (191mm±3) below the radiant panel.

(3) Place the retaining/securing frame over the test specimen. It may be necessary (due to compression) to adjust the sample (up or down) in order to maintain the distance from the sample to the radiant panel ($7\frac{1}{2}$ inches±$\frac{1}{8}$ inch (191mm±3) at "zero" position). With film/fiberglass assemblies, it is critical to make a slit in the film cover to purge any air inside. This allows the operator to maintain the proper test specimen position (level with the top of the platform) and to allow ventilation of gases during testing. A longitudinal slit, approximately 2 inches (51mm) in length, must be centered 3 inches±$\frac{1}{2}$ inch (76mm±13mm) from the left flange of the securing frame. A utility knife is acceptable for slitting the film cover.

(4) Immediately push the sliding platform into the chamber and close the bottom door.

(5) Bring the pilot burner flame into contact with the center of the specimen at the "zero" point and simultaneously start the timer. The pilot burner must be at a 27° angle with the sample and be approximately $\frac{1}{2}$ inch (12mm) above the sample. *See* figure 7. A stop, as shown in figure 8, allows the operator to position the burner correctly each time.

(6) Leave the burner in position for 15 seconds and then remove to a position at least 2 inches (51mm) above the specimen.

(g) Report. (1) Identify and describe the test specimen.

(2) Report any shrinkage or melting of the test specimen.

(3) Report the flame propagation distance. If this distance is less than 2 inches, report this as a pass (no measurement required).

(4) Report the after-flame time.

(h) *Requirements.* (1) There must be no flame propagation beyond 2 inches (51mm) to the left of the centerline of the pilot flame application.

(2) The flame time after removal of the pilot burner may not exceed 3 seconds on any specimen.

Figure 1—Burnthrough Test Apparatus Specimen Holder

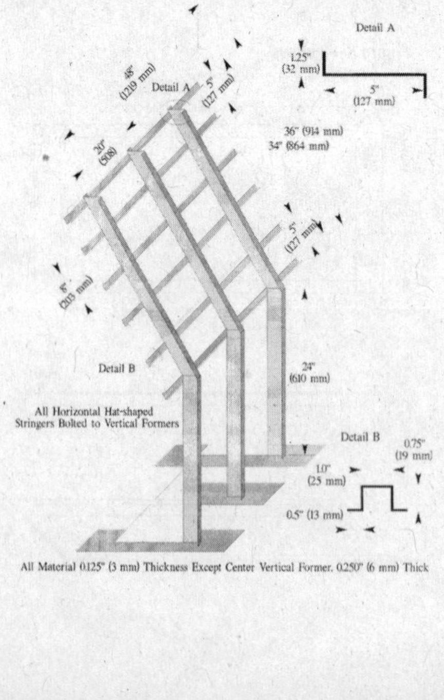

Part VII—Test Method To Determine the Burnthrough Resistance of Thermal/Acoustic Insulation Materials

Use the following test method to evaluate the burnthrough resistance characteristics of aircraft thermal/acoustic insulation materials when exposed to a high intensity open flame.

(a) *Definitions.*

Burnthrough time means the time, in seconds, for the burner flame to penetrate the test specimen, and/or the time required for the heat flux to reach 2.0 Btu/ft^2 sec (2.27 W/cm^2) on the inboard side, at a distance of 12 inches (30.5 cm) from the front surface of the insulation blanket test frame, whichever is sooner. The burnthrough time is measured at the inboard side of each of the insulation blanket specimens.

Insulation blanket specimen means one of two specimens positioned in either side of the test rig, at an angle of 30° with respect to vertical.

Specimen set means two insulation blanket specimens. Both specimens must represent the same production insulation blanket construction and materials, proportioned to correspond to the specimen size.

(b) *Apparatus.* (1) The arrangement of the test apparatus is shown in figures 1 and 2 and must include the capability of swinging the burner away from the test specimen during warm-up.

(2) *Test burner.* The test burner must be a modified gun-type such as the Park Model DPL 3400. Flame characteristics are highly dependent on actual burner setup. Parameters such as fuel pressure, nozzle depth, stator position, and intake airflow must be properly adjusted to achieve the correct flame output.

(i) *Nozzle.* A nozzle must maintain the fuel pressure to yield a nominal 6.0 gal/hr (0.378 L/min) fuel flow. A Monarch-manufactured 80° PL (hollow cone) nozzle nominally rated at 6.0 gal/hr at 100 lb/in^2 (0.71 MPa) delivers a proper spray pattern.

(ii) *Fuel Rail.* The fuel rail must be adjusted to position the fuel nozzle at a depth of 0.3125 inch (8mm) from the end plane of the exit stator, which must be mounted in the end of the draft tube.

(iii) *Internal Stator.* The internal stator, located in the middle of the draft tube, must be positioned at a depth of 3.75 inches (95mm) from the tip of the fuel nozzle. The stator must also be positioned such that the integral igniters are located at an angle midway between the 10 and 11 o'clock position, when viewed looking into the draft tube. Minor deviations to the igniter angle are acceptable if the temperature and heat flux requirements conform to the requirements of paragraph VII(e) of this appendix.

(iv) *Blower Fan.* The cylindrical blower fan used to pump air through the burner must measure 5.25 inches (133mm) in diameter by 3.5 inches (89mm) in width.

(v) *Burner cone.* Install a 12 +0.125-inch (305±3mm) burner extension cone at the end of the draft tube. The cone must have an opening 6±0.125-inch (152±3mm) high and 11±0.125-inch (280±3mm) wide (see figure 3).

(vi) *Fuel.* Use JP-8, Jet A, or their international equivalent, at a flow rate of 6.0±0.2 gal/hr (0.378±0.0126 L/min).

Figure 2—Burnthrough Test Apparatus

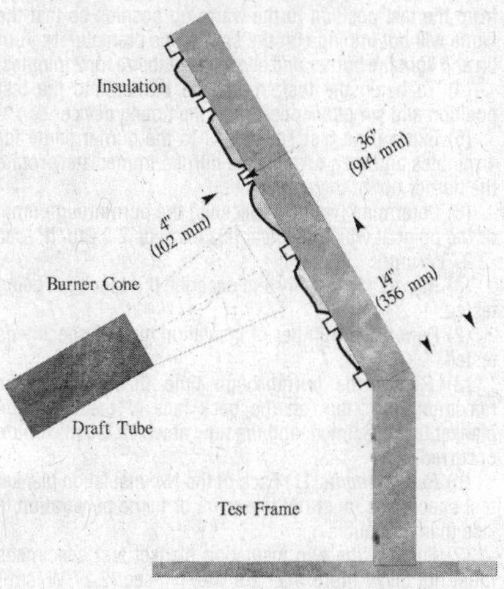

If this fuel is unavailable, ASTM K2 fuel (Number 2 grade kerosene) or ASTM D2 fuel (Number 2 grade fuel oil or Number 2 diesel fuel) are acceptable if the nominal fuel flow rate, temperature, and heat flux measurements conform to the requirements of paragraph VII(e) of this appendix.

(vii) *Fuel pressure regulator.* Provide a fuel pressure regulator, adjusted to deliver a nominal 6.0 gal/hr (0.378 L/min) flow rate. An operating fuel pressure of 100 lb/in² (0.71 MPa) for a nominally rated 6.0 gal/hr 80° spray angle nozzle (such as a PL type) delivers 6.0±0.2 gal/hr (0.378±0.0126 L/min).

(3) *Calibration rig and equipment.* (i) Construct individual calibration rigs to incorporate a calorimeter and thermocouple rake for the measurement of heat flux and temperature. Position the calibration rigs to allow movement of the burner from the test rig position to either the heat flux or temperature position with minimal difficulty.

(ii) *Calorimeter.* The calorimeter must be a total heat flux, foil type Gardon Gage of an appropriate range such as 0-20 Btu/ft² -sec (0-22.7 W/cm²), accurate to±3% of the indicated reading. The heat flux calibration method must be in accordance with paragraph VI(b)(7) of this appendix.

(iii) *Calorimeter mounting.* Mount the calorimeter in a 6- by 12-±0.125 inch (152- by 305-±3mm) by 0.75±0.125 inch (19mm±3mm) thick insulating block which is attached to the heat flux calibration rig during calibration (figure 4). Monitor the insulating block for deterioration and replace it when necessary. Adjust the mounting as necessary to ensure that the calorimeter face is parallel to the exit plane of the test burner cone.

Figure 3—Burner Draft Tube Extension Cone Diagram

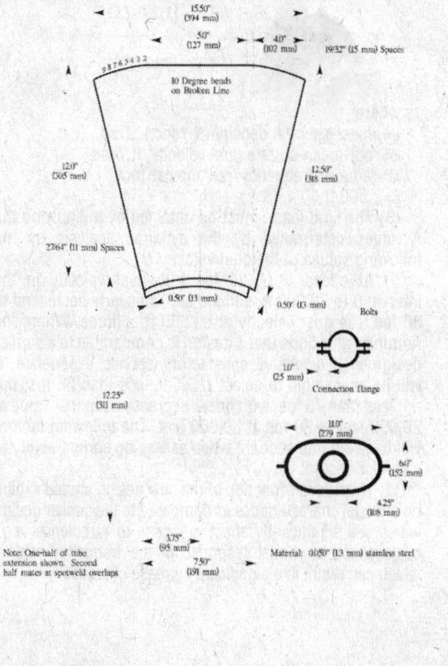

(iv) *Thermocouples.* Provide seven ⅛-inch (3.2mm) ceramic packed, metal sheathed, type K (Chromel-alumel), grounded junction thermocouples with a nominal 24 American Wire Gauge (AWG) size conductor for calibration. Attach the thermocouples to a steel angle bracket to form a thermocouple rake for placement in the calibration rig during burner calibration (figure 5).

(v) *Air velocity meter.* Use a vane-type air velocity meter to calibrate the velocity of air entering the burner. An Omega Engineering Model HH30A is satisfactory. Use a suitable adapter to attach the measuring device to the inlet side of the burner to prevent air from entering the burner other than through the measuring device, which would produce erroneously low readings. Use a flexible duct, measuring 4 inches wide (102mm) by 20 feet long (6.1 meters), to supply fresh air to the burner intake to prevent damage to the air velocity meter from ingested soot. An optional airbox permanently mounted to the burner intake area can effectively house the air velocity meter and provide a mounting port for the flexible intake duct.

Figure 4—Calorimeter Position Relative to Burner Cone

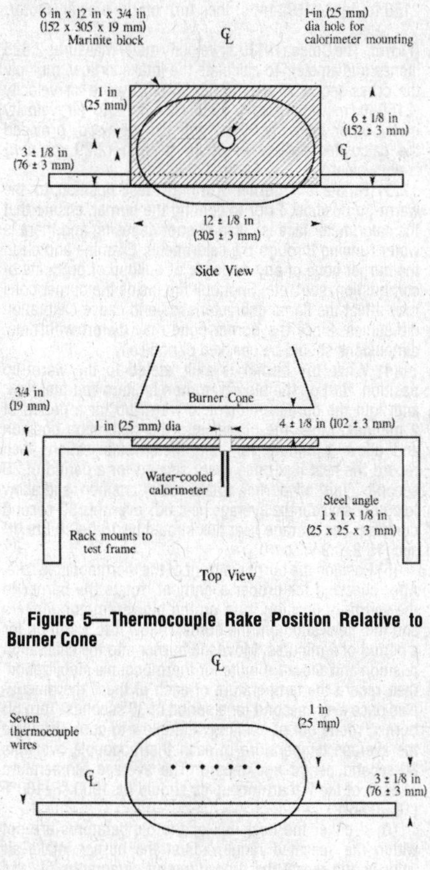

Figure 5—Thermocouple Rake Position Relative to Burner Cone

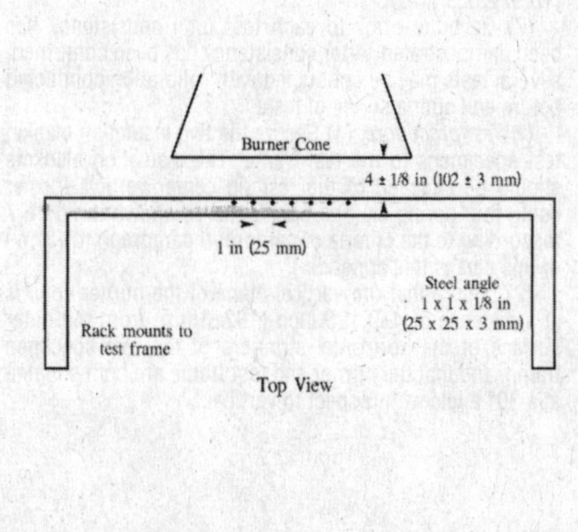

(4) *Test specimen mounting frame.* Make the mounting frame for the test specimens of ⅛-inch (3.2mm) thick steel as shown in figure 1, except for the center vertical former, which should be ¼-inch (6.4mm) thick to minimize warpage. The specimen mounting frame stringers (horizontal) should be bolted to the test frame formers (vertical) such that the expansion of the stringers will not cause the entire structure to warp. Use the mounting frame for mounting the two insulation blanket test specimens as shown in figure 2.

(5) *Backface calorimeters.* Mount two total heat flux Gardon type calorimeters behind the insulation test specimens on the back side (cold) area of the test specimen mounting frame as shown in figure 6. Position the calorimeters along the same plane as the burner cone centerline, at a distance of 4 inches (102mm) from the vertical centerline of the test frame.

(i) The calorimeters must be a total heat flux, foil type Gardon Gage of an appropriate range such as 0-5 Btu/ft² -sec (0-5.7 W/cm²), accurate to±3% of the indicated reading. The heat flux calibration method must comply with paragraph VI(b)(7) of this appendix.

(6) *Instrumentation.* Provide a recording potentiometer or other suitable calibrated instrument with an appropriate range to measure and record the outputs of the calorimeter and the thermocouples.

(7) *Timing device.* Provide a stopwatch or other device, accurate to±1%, to measure the time of application of the burner flame and burnthrough time.

(8) *Test chamber.* Perform tests in a suitable chamber to reduce or eliminate the possibility of test fluctuation due to air movement. The chamber must have a minimum floor area of 10 by 10 feet (305 by 305 cm).

(i) *Ventilation hood.* Provide the test chamber with an exhaust system capable of removing the products of combustion expelled during tests.

(c) *Test Specimens.* (1) *Specimen preparation.* Prepare a minimum of three specimen sets of the same construction and configuration for testing.

(2) *Insulation blanket test specimen.*

(i) For batt-type materials such as fiberglass, the constructed, finished blanket specimen assemblies must be 32 inches wide by 36 inches long (81.3 by 91.4 cm), exclusive of heat sealed film edges.

(ii) For rigid and other non-conforming types of insulation materials, the finished test specimens must fit into the test rig in such a manner as to replicate the actual in-service installation.

(3) *Construction.* Make each of the specimens tested using the principal components (i.e., insulation, fire barrier material if used, and moisture barrier film) and assembly processes (representative seams and closures).

(i) *Fire barrier material.* If the insulation blanket is constructed with a fire barrier material, place the fire barrier material in a manner reflective of the installed arrangement For example, if the material will be placed on the outboard side of the insulation material, inside the moisture film, place it the same way in the test specimen.

Figure 6—Position of Backface Calorimeters Relative to Test Specimen Frame

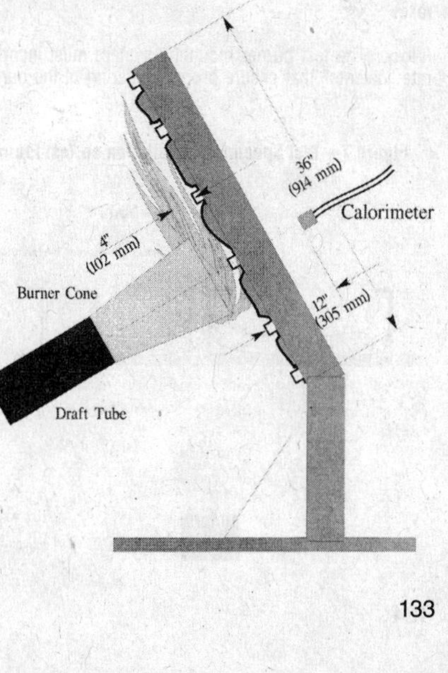

PART 25

FAR

(ii) *Insulation material.* Blankets that utilize more than one variety of insulation (composition, density, etc.) must have specimen sets constructed that reflect the insulation combination used. If, however, several blanket types use similar insulation combinations, it is not necessary to test each combination if it is possible to bracket the various combinations.

(iii) *Moisture barrier film.* If a production blanket construction utilizes more than one type of moisture barrier film, perform separate tests on each combination. For example, if a polyimide film is used in conjunction with an insulation in order to enhance the burnthrough capabilities, also test the same insulation when used with a polyvinyl fluoride film.

(iv) *Installation on test frame.* Attach the blanket test specimens to the test frame using 12 steel spring type clamps as shown in figure 7. Use the clamps to hold the blankets in place in both of the outer vertical formers, as well as the center vertical former (4 clamps per former). The clamp surfaces should measure 1 inch by 2 inches (25 by 51mm). Place the top and bottom clamps 6 inches (15.2 cm) from the top and bottom of the test frame, respectively. Place the middle clamps 8 inches (20.3 cm) from the top and bottom clamps.

(NOTE: For blanket materials that cannot be installed in accordance with figure 7 above, the blankets must be installed in a manner approved by the FAA.)

(v) *Conditioning.* Condition the specimens at 70°±5°F (21°±2°C) and 55%±10% relative humidity for a minimum of 24 hours prior to testing.

(d) *Preparation of apparatus.* (1) Level and center the frame assembly to ensure alignment of the calorimeter and/or thermocouple rake with the burner cone.

(2) Turn on the ventilation hood for the test chamber. Do not turn on the burner blower. Measure the airflow of the test chamber using a vane anemometer or equivalent measuring device. The vertical air velocity just behind the top of the upper insulation blanket test specimen must be 100±50 ft/min (0.51±0.25 m/s). The horizontal air velocity at this point must be less than 50 ft/min (0.25 m/s).

(3) If a calibrated flow meter is not available, measure the fuel flow rate using a graduated cylinder of appropriate size. Turn on the burner motor/fuel pump, after insuring that the igniter system is turned off. Collect the fuel via a plastic or rubber tube into the graduated cylinder for a 2-minute period. Determine the flow rate in gallons per hour. The fuel flow rate must be 6.0±0.2 gallons per hour (0.378±0.0126 L/min).

(e) *Calibration.* (1) Position the burner in front of the calorimeter so that it is centered and the vertical plane of the burner cone exit is 4±0.125 inches (102±3mm) from the calorimeter face. Ensure that the horizontal centerline of the burner cone is offset 1 inch below the horizontal centerline of the calorimeter (figure 8). Without disturbing the calorimeter position, rotate the burner in front of the thermocouple rake, such that the middle thermocouple (number 4 of 7) is centered on the burner cone.
Ensure that the horizontal centerline of the burner cone is also offset 1 inch below the horizontal centerline of the thermocouple tips. Re-check measurements by rotating the burner to each position to ensure proper alignment between the cone and the calorimeter and thermocouple rake.

(NOTE: The test burner mounting system must incorporate "detents" that ensure proper centering of the burner

Figure 7—Test Specimen Installation on Test Frame

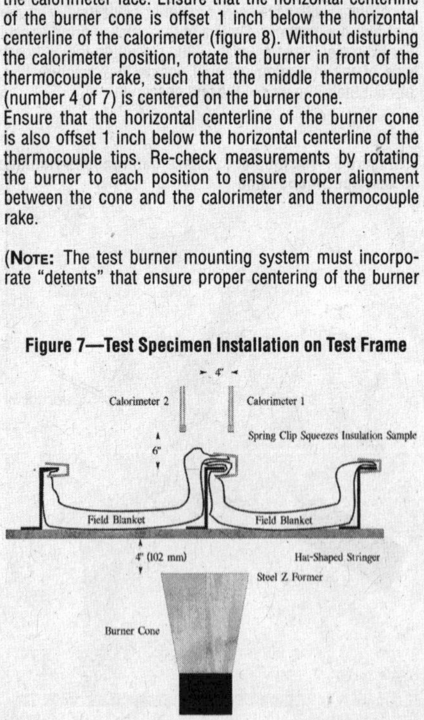

Figure 8—Burner Information and Calibration Settings

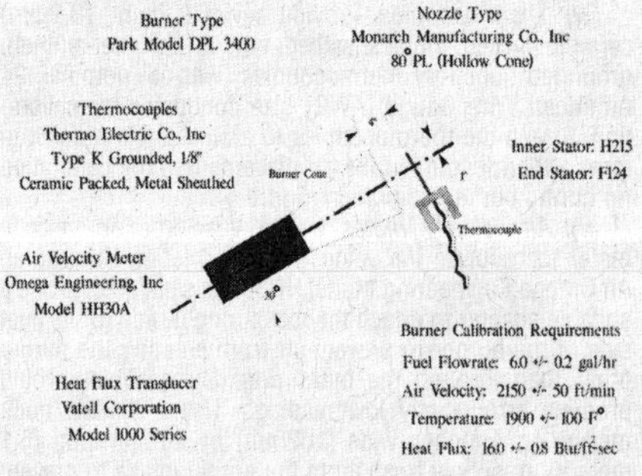

cone with respect to both the calorimeter and the thermocouple rakes, so that rapid positioning of the burner can be achieved during the calibration procedure.)
(2) Position the air velocity meter in the adapter or airbox, making certain that no gaps exist where air could leak around the air velocity measuring device. Turn on the blower/motor while ensuring that the fuel solenoid and igniters are off. Adjust the air intake velocity to a level of 2150 ft/min, (10.92 m/s) then turn off the blower/motor.

(NOTE: The Omega HH30 air velocity meter measures 2.625 inches in diameter. To calculate the intake airflow, multiply the cross-sectional area (0.03758 ft^2) by the air velocity (2150 ft/min) to obtain 80.80 ft^3/min. An air velocity meter other than the HH30 unit can be used, provided the calculated airflow of 80.80 ft^3/min (2.29 m^3/min) is equivalent.)

(3) Rotate the burner from the test position to the warm-up position. Prior to lighting the burner, ensure that the calorimeter face is clean of soot deposits, and there is water running through the calorimeter. Examine and clean the burner cone of any evidence of buildup of products of combustion, soot, etc. Soot buildup inside the burner cone may affect the flame characteristics and cause calibration difficulties. Since the burner cone may distort with time, dimensions should be checked periodically.

(4) While the burner is still rotated to the warm-up position, turn on the blower/motor, igniters and fuel flow, and light the burner. Allow it to warm up for a period of 2 minutes. Move the burner into the calibration position and allow 1 minute for calorimeter stabilization, then record the heat flux once every second for a period of 30 seconds. Turn off burner, rotate out of position, and allow to cool. Calculate the average heat flux over this 30-second duration. The average heat flux should be 16.0±0.8 Btu/ft^2 sec (18.2±0.9 W/cm^2).

(5) Position the burner in front of the thermocouple rake. After checking for proper alignment, rotate the burner to the warm-up position, turn on the blower/motor, igniters and fuel flow, and light the burner. Allow it to warm up for a period of 2 minutes. Move the burner into the calibration position and allow 1 minute for thermocouple stabilization, then record the temperature of each of the 7 thermocouples once every second for a period of 30 seconds. Turn off burner, rotate out of position, and allow to cool. Calculate the average temperature of each thermocouple over this 30-second period and record. The average temperature of each of the 7 thermocouples should be 1900 °F±100°F (1038±56°C).

(6) If either the heat flux or the temperatures are not within the specified range, adjust the burner intake air velocity and repeat the procedures of paragraphs (4) and (5) above to obtain the proper values. Ensure that the inlet air velocity is within the range of 2150 ft/min±50 ft/min (10.92±0.25 m/s).

(7) Calibrate prior to each test until consistency has been demonstrated. After consistency has been confirmed, several tests may be conducted with calibration conducted before and after a series of tests.

(f) *Test procedure.* (1) Secure the two insulation blanket test specimens to the test frame. The insulation blankets should be attached to the test rig center vertical former using four spring clamps positioned as shown in figure 7 (according to the criteria of paragraph paragraph (c)(3)(iv) of this part of this appendix).

(2) Ensure that the vertical plane of the burner cone is at a distance of 4±0.125 inch (102±3mm) from the outer surface of the horizontal stringers of the test specimen frame, and that the burner and test frame are both situated at a 30° angle with respect to vertical.

(3) When ready to begin the test, direct the burner away from the test position to the warm-up position so that the flame will not impinge on the specimens prematurely. Turn on and light the burner and allow it to stabilize for 2 minutes.

(4) To begin the test, rotate the burner into the test position and simultaneously start the timing device.

(5) Expose the test specimens to the burner flame for 4 minutes and then turn off the burner. Immediately rotate the burner out of the test position.

(6) Determine (where applicable) the burnthrough time, or the point at which the heat flux exceeds 2.0 Btu/ft^2 -sec (2.27 W/cm^2).

(g) *Report.* (1) Identify and describe the specimen being tested.

(2) Report the number of insulation blanket specimens tested.

(3) Report the burnthrough time (if any), and the maximum heat flux on the back face of the insulation blanket test specimen, and the time at which the maximum occurred.

(h) *Requirements.* (1) Each of the two insulation blanket test specimens must not allow fire or flame penetration in less than 4 minutes.

(2) Each of the two insulation blanket test specimens must not allow more than 2.0 Btu/ft^2 -sec (2.27 W/cm^2) on the cold side of the insulation specimens at a point 12 inches (30.5 cm) from the face of the test rig.

[Amdt. 25-32, 37 FR 3972, Feb. 24, 1972]

EDITORIAL NOTE: For Federal Register citations affecting appendix F to Part 25, see the List of CFR Sections Affected, which appears in the Finding Aids section of the printed volume and at *www.fdsys.gov.*

APPENDIX G TO PART 25—CONTINUOUS GUST DESIGN CRITERIA

The continuous gust design criteria in this appendix must be used in establishing the dynamic response of the airplane to vertical and lateral continuous turbulence unless a more rational criteria is used. The following gust load requirements apply to mission analysis and design envelope analysis:

(a) The limit gust loads utilizing the continuous turbulence concept must be determined in accordance with the provisions of either paragraph (b) or paragraphs (c) and (d) of this appendix.

(b) *Design envelope analysis.* The limit loads must be determined in accordance with the following:

(1) All critical altitudes, weights, and weight distributions, as specified in § 25.321(b), and all critical speeds within the ranges indicated in paragraph (b)(3) of this appendix must be considered.

(2) Values of Ā (ratio of root-mean-square incremental load root-mean-square gust velocity) must be determined by dynamic analysis. The power spectral density of the atmospheric turbulence must be as given by the equation—

$$\phi(\Omega) = \sigma^2 L/\pi \frac{1+\frac{8}{3}(1.339L\Omega)^2}{\left[1+(1.339L\Omega)^2\right]^{\frac{\Pi}{o}}}$$

where:
ϕ=power-spectral density (ft./sec.) ^{2}/rad./ft.
σ=root-mean-square gust velocity, ft./sec.
Ω=reduced frequency, radians per foot.
L=2,500 ft.

(3) The limit loads must be obtained by multiplying the Ā values determined by the dynamic analysis by the following values of the gust velocity U_σ.

(i) At speed V_c:U_σ=85 fps true gust velocity in the interval 0 to 30,000 ft. altitude and is linearly decreased to 30 fps true gust velocity at 80,000 ft. altitude. Where the Administrator finds that a design is comparable to a similar design with extensive satisfactory service experience, it will be acceptable to select U_σ at V_c less than 85 fps, but not less than 75 fps, with linear decrease from that value at 20,000 feet to 30 fps at 80,000 feet. The following factors will be taken into account when assessing comparability to a similar design:

(1) The transfer function of the new design should exhibit no unusual characteristics as compared to the similar design which will significantly affect response to turbulence; e.g., coalescence of modal response in the frequency regime which can result in a significant increase of loads.

Figure 1

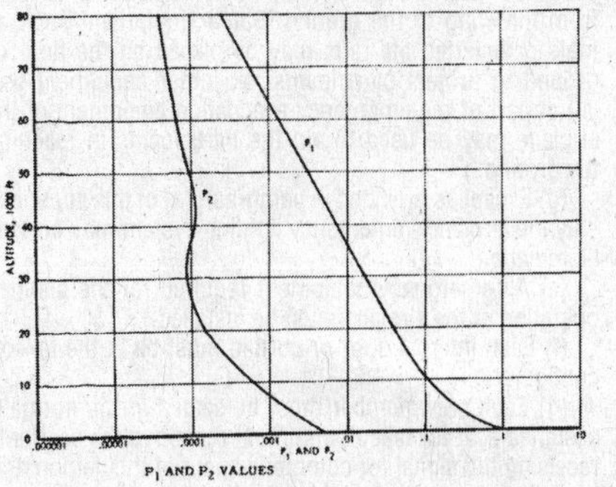

Figure 2

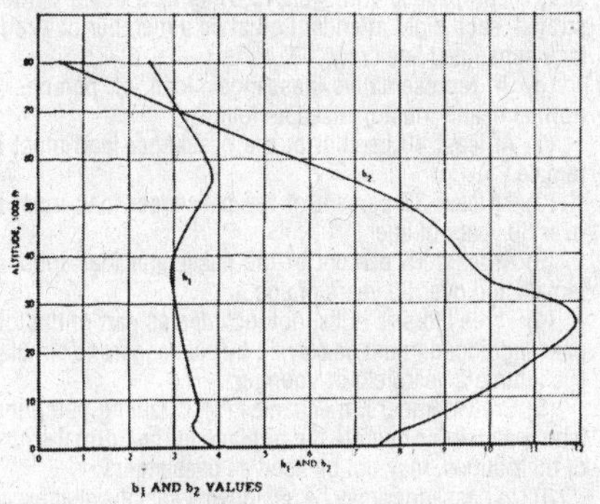

(2) The typical mission of the new airplane is substantially equivalent to that of the similar design.

(3) The similar design should demonstrate the adequacy of the U_σ selected.

(ii) At speed V_B:U_σ is equal to 1.32 times the values obtained under paragraph (b)(3)(i) of this appendix.

(iii) At speed V_D:U_σ is equal to ½ the values obtained under paragraph (b)(3)(i) of this appendix.

(iv) At speeds between V_B and V_c and between V_c and V_D:U_σ is equal to a value obtained by linear interpolation.

(4) When a stability augmentation system is included in the analysis, the effect of system nonlinearities on loads at the limit load level must be realistically or conservatively accounted for.

(c) *Mission analysis.* Limit loads must be determined in accordance with the following:

(1) The expected utilization of the airplane must be represented by one or more flight profiles in which the load distribution and the variation with time of speed, altitude, gross weight, and center of gravity position are defined. These profiles must be divided into mission segments or blocks, for analysis, and average or effective values of the pertinent parameters defined for each segment.

(2) For each of the mission segments defined under paragraph (c)(1) of this appendix, values of $\bar{A}$ and N_o must be determined by analysis. $\bar{A}$ is defined as the ratio of root-mean-square incremental load to root-mean-square gust velocity and N_o is the radius of gyration of the load power spectral density function about zero frequency. The power spectral density of the atmospheric turbulence must be given by the equation set forth in paragraph (b)(2) of this appendix.

(3) For each of the load and stress quantities selected, the frequency of exceedance must be determined as a function of load level by means of the equation—

$$N_{(r)} = \sum tN_o \left[P_1 \exp\left(-\frac{|y - y_{one-g}|}{b_1 \bar{A}}\right) + P_2 \exp\left(-\frac{|y - y_{one-g}|}{b_2 \bar{A}}\right)\right]$$

where—

t=selected time interval.

y=net value of the load or stress.

Y_{one-g}=value of the load or stress in one-g level flight.

$N(y)$=average number of exceedances of the indicated value of the load or stress in unit time.

Σ=symbol denoting summation over all mission segments.

N_o, $\bar{A}$=parameters determined by dynamic analysis as defined in paragraph (c)(2) of this appendix.

P_1, P_2, b_1, b_2=parameters defining the probability distributions of root-mean-square gust velocity, to be read from Figures 1 and 2 of this appendix.

The limit gust loads must be read from the frequency of exceedance curves at a frequency of exceedance of 2×10–5 exceedances per hour. Both positive and negative load directions must be considered in determining the limit loads.

(4) If a stability augmentation system is utilized to reduce the gust loads, consideration must be given to the fraction of flight time that the system may be inoperative. The flight profiles of paragraph (c)(1) of this appendix must include flight with the system inoperative for this fraction of the flight time. When a stability augmentation system is included in the analysis, the effect of system nonlinearities on loads at the limit load level must be conservatively accounted for.

(d) *Supplementary design envelope analysis.* In addition to the limit loads defined by paragraph (c) of this appendix, limit loads must also be determined in accordance with paragraph (b) of this appendix, except that—

(1) In paragraph (b)(3)(i) of this appendix, the value of U_σ=85 fps true gust velocity is replaced by U_σ=60 fps true gust velocity on the interval 0 to 30,000 ft. altitude, and is linearly decreased to 25 fps true gust velocity at 80,000 ft. altitude; and

(2) In paragraph (b) of this appendix, the reference to paragraphs (b)(3)(i) through (b)(3)(iii) of this appendix is to be understood as referring to the paragraph as modified by paragraph (d)(1).

[Amdt. 25-54, 45 FR 60173, Sept. 11, 1980]

Appendix H to Part 25—Instructions for Continued Airworthiness

H25.1 *General.*

(a) This appendix specifies requirements for preparation of Instructions for Continued Airworthiness as required by §§ 25.1529, 25.1729, and applicable provisions of parts 21 and 26 of this chapter.

(b) The Instructions for Continued Airworthiness for each airplane must include the Instructions for Continued Airworthiness for each engine and propeller (hereinafter designated "products"), for each appliance required by this chapter, and any required information relating to the interface of those appliances and products with the airplane. If Instructions for Continued Airworthiness are not supplied by the manufacturer of an appliance or product installed in the airplane, the Instructions for Continued Airworthiness for the airplane must include the information essential to the continued airworthiness of the airplane.

(c) The applicant must submit to the FAA a program to show how changes to the Instructions for Continued Airworthiness made by the applicant or by the manufacturers or products and appliances installed in the airplane will be distributed.

H25.2 *Format.*

(a) The Instructions for Continued Airworthiness must be in the form of a manual or manuals as appropriate for the quantity of data to be provided.

(b) The format of the manual or manuals must provide for a practical arrangement.

H25.3 *Content.*

The contents of the manual or manuals must be prepared in the English language. The Instructions for Continued Airworthiness must contain the following manuals or sections, as appropriate, and information:

(a) *Airplane maintenance manual or section.* (1) Introduction information that includes an explanation of the airplane's features and data to the extent necessary for maintenance or preventive maintenance.

(2) A description of the airplane and its systems and installations including its engines, propellers, and appliances.

(3) Basic control and operation information describing how the airplane components and systems are controlled and how they operate, including any special procedures and limitations that apply.

(4) Servicing information that covers details regarding servicing points, capacities of tanks, reservoirs, types of fluids to be used, pressures applicable to the various systems, location of access panels for inspection and servicing, locations of lubrication points, lubricants to be used, equipment required for servicing, tow instructions and limitations, mooring, jacking, and leveling information.

(b) *Maintenance instructions.* (1) Scheduling information for each part of the airplane and its engines, auxiliary power units, propellers, accessories, instruments, and equipment that provides the recommended periods at which they should be cleaned, inspected, adjusted, tested, and lubricated, and the degree of inspection, the applicable wear tolerances, and work recommended at these periods. However, the applicant may refer to an accessory, instrument, or equipment manufacturer as the source of this information if the applicant shows that the item has an exceptionally high degree of complexity requiring specialized maintenance techniques, test equipment, or expertise. The recommended overhaul periods and necessary cross references to the Airworthiness Limitations section of the manual must also be included. In addition, the applicant must include an inspection program that includes the frequency and extent of the inspections necessary to provide for the continued airworthiness of the airplane.

(2) Troubleshooting information describing probable malfunctions, how to recognize those malfunctions, and the remedial action for those malfunctions.

(3) Information describing the order and method of removing and replacing products and parts with any necessary precautions to be taken.

(4) Other general procedural instructions including procedures for system testing during ground running, symmetry checks, weighing and determining the center of gravity, lifting and shoring, and storage limitations.

(c) Diagrams of structural access plates and information needed to gain access for inspections when access plates are not provided.

(d) Details for the application of special inspection techniques including radiographic and ultrasonic testing where such processes are specified.

(e) Information needed to apply protective treatments to the structure after inspection.

(f) All data relative to structural fasteners such as identification, discard recommendations, and torque values.

(g) A list of special tools needed.

H25.4 *Airworthiness Limitations section.*

(a) The Instructions for Continued Airworthiness must contain a section titled Airworthiness Limitations that is segregated and clearly distinguishable from the rest of the document. This section must set forth—

(1) Each mandatory modification time, replacement time, structural inspection interval, and related structural inspection procedure approved under § 25.571.

(2) Each mandatory replacement time, inspection interval, related inspection procedure, and all critical design configuration control limitations approved under § 25.981 for the fuel tank system.

(3) Any mandatory replacement time of EWIS components as defined in section 25.1701.

(4) A limit of validity of the engineering data that supports the structural maintenance program (LOV), stated as a total number of accumulated flight cycles or flight hours or both, approved under § 25.571. Until the full-scale fatigue testing is completed and the FAA has approved the LOV, the number of cycles accumulated by the airplane cannot be greater than ½ the number of cycles accumulated on the fatigue test article.

(b) If the Instructions for Continued Airworthiness consist of multiple documents, the section required by this paragraph must be included in the principal manual. This section must contain a legible statement in a prominent location that reads: "The Airworthiness Limitations section is FAA-approved and specifies maintenance required under §§ 43.16 and 91.403 of the Federal Aviation Regulations, unless an alternative program has been FAA approved."

H25.5 *Electrical Wiring Interconnection System (EWIS) Instructions for Continued Airworthiness.*

(a) The applicant must prepare Instructions for Continued Airworthiness (ICA) applicable to EWIS as defined by § 25.1701 that are approved by the FAA and include the following:

(1) Maintenance and inspection requirements for the EWIS developed with the use of an enhanced zonal analysis procedure that includes:

(i) Identification of each zone of the airplane.

(ii) Identification of each zone that contains EWIS.

(iii) Identification of each zone containing EWIS that also contains combustible materials.

(iv) Identification of each zone in which EWIS is in close proximity to both primary and back-up hydraulic, mechanical, or electrical flight controls and lines.

(v) Identification of—

(A) Tasks, and the intervals for performing those tasks, that will reduce the likelihood of ignition sources and accumulation of combustible material, and

(B) Procedures, and the intervals for performing those procedures, that will effectively clean the EWIS components of combustible material if there is not an effective task to reduce the likelihood of combustible material accumulation.

(vi) Instructions for protections and caution information that will minimize contamination and accidental damage to EWIS, as applicable, during performance of maintenance, alteration, or repairs.

(2) Acceptable EWIS maintenance practices in a standard format.

(3) Wire separation requirements as determined under § 25.1707.

(4) Information explaining the EWIS identification method and requirements for identifying any changes to EWIS under § 25.1711.

(5) Electrical load data and instructions for updating that data.

(b) The EWIS ICA developed in accordance with the requirements of H25.5(a)(1) must be in the form of a document appropriate for the information to be provided, and they must be easily recognizable as EWIS ICA. This document must either contain the required EWIS ICA or specifically reference other portions of the ICA that contain this information.

[Amdt. 25-54, 45 FR 60177, Sept. 11, 1980, as amended by Amdt. 25-68, 54 FR 34329, Aug. 18, 1989; Amdt. 25-102, 66 FR 23130, May 7, 2001; Amdt. 25-123, 72 FR 63408, Nov. 8, 2007; Amdt. 25-132, 75 FR 69782, Nov. 15, 2010]

Appendix I to Part 25—Installation of an Automatic Takeoff Thrust Control System (ATTCS)

I25.1 *General.*

(a) This appendix specifies additional requirements for installation of an engine power control system that automatically resets thrust or power on operating engine(s) in the event of any one engine failure during takeoff.

(b) With the ATTCS and associated systems functioning normally as designed, all applicable requirements of Part 25, except as provided in this appendix, must be met without requiring any action by the crew to increase thrust or power.

I25.2 *Definitions.*

(a) *Automatic Takeoff Thrust Control System (ATTCS).* An ATTCS is defined as the entire automatic system used on takeoff, including all devices, both mechanical and electrical, that sense engine failure, transmit signals, actuate fuel controls or power levers or increase engine power by other means on operating engines to achieve scheduled thrust or power increases, and furnish cockpit information on system operation.

(b) *Critical Time Interval.* When conducting an ATTCS takeoff, the critical time interval is between V_1 minus 1 second and a point on the minimum performance, all-engine flight path where, assuming a simultaneous occurrence of an engine and ATTCS failure, the resulting minimum flight path thereafter intersects the Part 25 required actual flight path at no less than 400 feet above the takeoff surface. This time interval is shown in the following illustration:

I25.3 *Performance and System Reliability Requirements.*

The applicant must comply with the performance and ATTCS reliability requirements as follows:

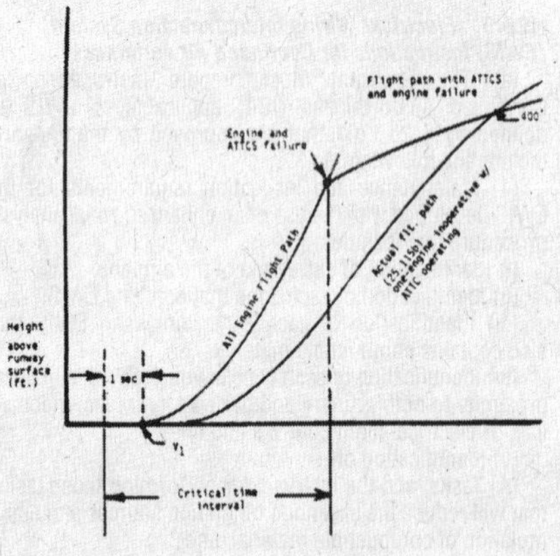

(a) An ATTCS failure or a combination of failures in the ATTCS during the critical time interval:

(1) Shall not prevent the insertion of the *maximum approved takeoff* thrust or power, or must be shown to be an improbable event.

(2) Shall not result in a significant loss or reduction in thrust or power, or must be shown to be an extremely improbable event.

(b) The concurrent existence of an ATTCS failure and an engine failure during the critical time interval must be shown to be extremely improbable.

(c) All applicable performance requirements of Part 25 must be met with an engine failure occurring at the most critical point during takeoff with the ATTCS system functioning.

I25.4 *Thrust Setting.*

The initial takeoff thrust or power setting on each engine at the beginning of the takeoff roll may not be less than any of the following:

(a) Ninety (90) percent of the thrust or power set by the ATTCS (the maximum takeoff thrust or power approved for the airplane under existing ambient conditions);

(b) That required to permit normal operation of all safety-related systems and equipment dependent upon engine thrust or power lever position; or

(c) That shown to be free of hazardous engine response characteristics when thrust or power is advanced from the initial takeoff thrust or power to the maximum approved takeoff thrust or power.

I25.5 *Powerplant Controls.*

(a) In addition to the requirements of § 25.1141, no single failure or malfunction, or probable combination thereof, of the ATTCS, including associated systems, may cause the failure of any powerplant function necessary for safety.

(b) The ATTCS must be designed to:

(1) Apply thrust or power on the operating engine(s), following any one engine failure during takeoff, to achieve the maximum approved takeoff thrust or power without exceeding engine operating limits;

(2) Permit manual decrease or increase in thrust or power up to the maximum takeoff thrust or power approved for the airplane under existing conditions through the use of the power lever. For airplanes equipped with limiters that automatically prevent engine operating limits from being exceeded under existing ambient conditions, other means may be used to increase the thrust or power in the event of an ATTCS failure provided the means is located on or forward of the power levers; is easily identified and operated under all operating conditions by a single action of either pilot with the hand that is normally used to actuate the power levers; and meets the requirements of § 25.777 (a), (b), and (c);

(3) Provide a means to verify to the flightcrew before takeoff that the ATTCS is in a condition to operate; and

(4) Provide a means for the flightcrew to deactivate the automatic function. This means must be designed to prevent inadvertent deactivation.

I25.6 *Powerplant Instruments.*

In addition to the requirements of § 25.1305:

(a) A means must be provided to indicate when the ATTCS is in the armed or ready condition; and

(b) If the inherent flight characteristics of the airplane do not provide adequate warning that an engine has failed, a warning system that is independent of the ATTCS must be provided to give the pilot a clear warning of any engine failure during takeoff.

[Amdt. 25-62, 52 FR 43156, Nov. 9, 1987]

Appendix J to Part 25—Emergency Evacuation

The following test criteria and procedures must be used for showing compliance with § 25.803:

(a) The emergency evacuation must be conducted with exterior ambient light levels of no greater than 0.3 foot-candles prior to the activation of the airplane emergency lighting system. The source(s) of the initial exterior ambient light level may remain active or illuminated during the actual demonstration. There must, however, be no increase in the exterior ambient light level except for that due to activation of the airplane emergency lighting system.

(b) The airplane must be in a normal attitude with landing gear extended.

(c) Unless the airplane is equipped with an off-wing descent means, stands or ramps may be used for descent from the wing to the ground. Safety equipment such as mats or inverted life rafts may be placed on the floor or ground to protect participants. No other equipment that is not part of the emergency evacuation equipment of the airplane may be used to aid the participants in reaching the ground.

(d) Except as provided in paragraph (a) of this appendix, only the airplane's emergency lighting system may provide illumination.

(e) All emergency equipment required for the planned operation of the airplane must be installed.

(f) Each internal door or curtain must be in the takeoff configuration.

(g) Each crewmember must be seated in the normally assigned seat for takeoff and must remain in the seat until receiving the signal for commencement of the demonstration. Each crewmember must be a person having knowledge of the operation of exits and emergency equipment and, if compliance with § 121.291 is also being demonstrated, each flight attendant must be a member of a regularly scheduled line crew.

(h) A representative passenger load of persons in normal health must be used as follows:

(1) At least 40 percent of the passenger load must be female.

(2) At least 35 percent of the passenger load must be over 50 years of age.

(3) At least 15 percent of the passenger load must be female and over 50 years of age.

(4) Three life-size dolls, not included as part of the total passenger load, must be carried by passengers to simulate live infants 2 years old or younger.

(5) Crewmembers, mechanics, and training personnel, who maintain or operate the airplane in the normal course of their duties, may not be used as passengers.

(i) No passenger may be assigned a specific seat except as the Administrator may require. Except as required by subparagraph (g) of this paragraph, no employee of the applicant may be seated next to an emergency exit.

(j) Seat belts and shoulder harnesses (as required) must be fastened.

(k) Before the start of the demonstration, approximately one-half of the total average amount of carry-on baggage, blankets, pillows, and other similar articles must be distributed at several locations in aisles and emergency exit access ways to create minor obstructions.

(l) No prior indication may be given to any crewmember or passenger of the particular exits to be used in the demonstration.

(m) The applicant may not practice, rehearse, or describe the demonstration for the participants nor may any participant have taken part in this type of demonstration within the preceding 6 months.

(n) Prior to entering the demonstration aircraft, the passengers may also be advised to follow directions of crewmembers but may not be instructed on the procedures to be followed in the demonstration, except with respect to safety procedures in place for the demonstration or which have to do with the demonstration site. Prior to the start of the demonstration, the pre-takeoff passenger briefing required by § 121.571 may be given. Flight attendants may assign demonstration subjects to assist persons from the bottom of a slide, consistent with their approved training program.

(o) The airplane must be configured to prevent disclosure of the active emergency exits to demonstration participants in the airplane until the start of the demonstration.

(p) Exits used in the demonstration must consist of one exit from each exit pair. The demonstration may be conducted with the escape slides, if provided, inflated and the exits open at the beginning of the demonstration. In this case, all exits must be configured such that the active exits are not disclosed to the occupants. If this method is used, the exit preparation time for each exit utilized must be accounted for, and exits that are not to be used in the demonstration must not be indicated before the demonstration has started. The exits to be used must be representative of all of the emergency exits on the airplane and must be designated by the applicant, subject to approval by the Administrator. At least one floor level exit must be used.

(q) Except as provided in paragraph (c) of this section, all evacuees must leave the airplane by a means provided as part of the airplane's equipment.

(r) The applicant's approved procedures must be fully utilized, except the flightcrew must take no active role in assisting others inside the cabin during the demonstration.

(s) The evacuation time period is completed when the last occupant has evacuated the airplane and is on the ground. Provided that the acceptance rate of the stand or ramp is no greater than the acceptance rate of the means available on the airplane for descent from the wing during an actual crash situation, evacuees using stands or ramps allowed by paragraph (c) of this appendix are considered to be on the ground when they are on the stand or ramp.

[Amdt. 25-72, 55 FR 29788, July 20, 1990, as amended by Amdt. 25-79, Aug. 26, 1993; Amdt. 25-117, 69 FR 67499, Nov. 17, 2004]

Appendix K to Part 25—Extended Operations (ETOPS)

This appendix specifies airworthiness requirements for the approval of an airplane-engine combination for extended operations (ETOPS). For two-engine airplanes, the applicant must comply with sections K25.1 and K25.2 of this appendix. For airplanes with more than two engines, the applicant must comply with sections K25.1 and K25.3 of this appendix.

K25.1 *Design requirements.*

K25.1.1 *Part 25 compliance.*

The airplane-engine combination must comply with the requirements of part 25 considering the maximum flight time and the longest diversion time for which the applicant seeks approval.

K25.1.2 *Human factors.*

An applicant must consider crew workload, operational implications, and the crew's and passengers' physiological needs during continued operation with failure effects for the longest diversion time for which it seeks approval.

K25.1.3 *Airplane systems.*

(a) *Operation in icing conditions.*

(1) The airplane must be certificated for operation in icing conditions in accordance with § 25.1419.

(2) The airplane must be able to safely conduct an ETOPS diversion with the most critical ice accretion resulting from:

(i) Icing conditions encountered at an altitude that the airplane would have to fly following an engine failure or cabin decompression.

(ii) A 15-minute hold in the continuous maximum icing conditions specified in Appendix C of this part with a liquid water content factor of 1.0.

(iii) Ice accumulated during approach and landing in the icing conditions specified in Appendix C of this part.

(b) *Electrical power supply.* The airplane must be equipped with at least three independent sources of electrical power.

(c) *Time limited systems.* The applicant must define the system time capability of each ETOPS significant system that is time-limited.

K25.1.4 *Propulsion systems.*

(a) *Fuel system design.* Fuel necessary to complete an ETOPS flight (including a diversion for the longest time for which the applicant seeks approval) must be available to the operating engines at the pressure and fuel-flow required by § 25.955 under any airplane failure condition not shown to be extremely improbable. Types of failures that must be considered include, but are not limited to: crossfeed valve failures, automatic fuel management system failures, and normal electrical power generation failures.

(1) If the engine has been certified for limited operation with negative engine-fuel-pump-inlet pressures, the following requirements apply:

(i) Airplane demonstration-testing must cover worst case cruise and diversion conditions involving:

(A) Fuel grade and temperature.

(B) Thrust or power variations.

(C) Turbulence and negative G.

(D) Fuel system components degraded within their approved maintenance limits.

(ii) Unusable-fuel quantity in the suction feed configuration must be determined in accordance with § 25.959.

(2) For two-engine airplanes to be certificated for ETOPS beyond 180 minutes, one fuel boost pump in each main tank and at least one crossfeed valve, or other means for transferring fuel, must be powered by an independent electrical power source other than the three power sources required to comply with section K25.1.3(b) of this appendix. This requirement does not apply if the normal fuel boost pressure, crossfeed valve actuation, or fuel transfer capability is not provided by electrical power.

(3) An alert must be displayed to the flightcrew when the quantity of fuel available to the engines falls below the level required to fly to the destination. The alert must be given when there is enough fuel remaining to safely complete a diversion. This alert must account for abnormal fuel management or transfer between tanks, and possible loss of fuel. This paragraph does not apply to airplanes with a required flight engineer.

(b) *APU design.* If an APU is needed to comply with this appendix, the applicant must demonstrate that:

(1) The reliability of the APU is adequate to meet those requirements; and

(2) If it is necessary that the APU be able to start in flight, it is able to start at any altitude up to the maximum operating altitude of the airplane, or 45,000 feet, whichever is lower, and run for the remainder of any flight.

(c) *Engine oil tank design.* The engine oil tank filler cap must comply with § 33.71(c)(4) of this chapter.

K25.1.5 *Engine-condition monitoring.*

Procedures for engine-condition monitoring must be specified and validated in accordance with Part 33, Appendix A, paragraph A33.3(c) of this chapter.

K25.1.6 *Configuration, maintenance, and procedures.*

The applicant must list any configuration, operating and maintenance requirements, hardware life limits, mmEL constraints, and ETOPS approval in a CMP document.

K25.1.7 *Airplane flight manual.*

The airplane flight manual must contain the following information applicable to the ETOPS type design approval:

(a) Special limitations, including any limitation associated with operation of the airplane up to the maximum diversion time being approved.

(b) Required markings or placards.

(c) The airborne equipment required for extended operations and flightcrew operating procedures for this equipment.

(d) The system time capability for the following:

(1) The most limiting fire suppression system for Class C cargo or baggage compartments.

(2) The most limiting ETOPS significant system other than fire suppression systems for Class C cargo or baggage compartments.

(e) This statement: "The type-design reliability and performance of this airplane-engine combination has been evaluated under 14 CFR 25.1535 and found suitable for (identify maximum approved diversion time) extended operations (ETOPS) when the configuration, maintenance, and procedures standard contained in (identify the CMP document) are met. The actual maximum approved diversion time for this airplane may be less based on its most limiting system time capability. This finding does not constitute operational approval to conduct ETOPS."

K25.2. *Two-engine airplanes.*

An applicant for ETOPS type design approval of a two-engine airplane must use one of the methods described in section K25.2.1, K25.2.2, or K25.2.3 of this appendix.

K25.2.1 *Service experience method.*

An applicant for ETOPS type design approval using the service experience method must comply with sections K25.2.1(a) and K25.2.1(b) of this appendix before conducting the assessments specified in sections K25.2.1(c) and K25.2.1(d) of this appendix, and the flight test specified in section K25.2.1(e) of this appendix.

(a) *Service experience.* The world fleet for the airplane-engine combination must accumulate a minimum of 250,000 engine-hours. The FAA may reduce this number of hours if the applicant identifies compensating factors that are acceptable to the FAA. The compensating factors may include experience on another airplane, but experience on the candidate airplane must make up a significant portion of the total service experience.

(b) *In-flight shutdown (IFSD) rates.* The demonstrated 12-month rolling average IFSD rate for the world fleet of the airplane-engine combination must be commensurate with the level of ETOPS approval being sought.

(1) For type design approval up to and including 120 minutes: An IFSD rate of 0.05 or less per 1,000 world-fleet engine-hours, unless otherwise approved by the FAA. Unless the IFSD rate is 0.02 or less per 1,000 world-fleet engine-hours, the applicant must provide a list of corrective actions in the CMP document specified in section K25.1.6 of this appendix, that, when taken, would result in an IFSD rate of 0.02 or less per 1,000 fleet engine-hours.

(2) For type design approval up to and including 180 minutes: An IFSD rate of 0.02 or less per 1,000 world-fleet engine-hours, unless otherwise approved by the FAA. If the airplane-engine combination does not meet this rate by compliance with an existing 120-minute CMP document, then new or additional CMP requirements that the applicant has demonstrated would achieve this IFSD rate must be added to the CMP document.

(3) For type design approval beyond 180 minutes: An IFSD rate of 0.01 or less per 1,000 fleet engine-hours unless otherwise approved by the FAA. If the airplane-engine combination does not meet this rate by compliance with an existing 120-minute or 180-minute CMP document, then new or additional CMP requirements that the applicant has demonstrated would achieve this IFSD rate must be added to the CMP document.

(c) *Propulsion system assessment.* (1) The applicant must conduct a propulsion system assessment based on the following data collected from the world-fleet of the airplane-engine combination:

(i) A list of all IFSD's, unplanned ground engine shutdowns, and occurrences (both ground and in-flight) when an engine was not shut down, but engine control or the desired thrust or power level was not achieved, including engine flameouts. Planned IFSD's performed during flight training need not be included. For each item, the applicant must provide—

(A) Each airplane and engine make, model, and serial number;

(B) Engine configuration, and major alteration history;

(C) Engine position;

(D) Circumstances leading up to the engine shutdown or occurrence;

(E) Phase of flight or ground operation;

(F) Weather and other environmental conditions; and

(G) Cause of engine shutdown or occurrence.

(ii) A history of unscheduled engine removal rates since introduction into service (using 6- and 12-month rolling averages), with a summary of the major causes for the removals.

(iii) A list of all propulsion system events (whether or not caused by maintenance or flightcrew error), including dispatch delays, cancellations, aborted takeoffs, turnbacks, diversions, and flights that continue to destination after the event.

(iv) The total number of engine hours and cycles, the number of hours for the engine with the highest number of hours, the number of cycles for the engine with the highest number of cycles, and the distribution of hours and cycles.

(v) The mean time between failures (MTBF) of propulsion system components that affect reliability.

(vi) A history of the IFSD rates since introduction into service using a 12-month rolling average.

(2) The cause or potential cause of each item listed in K25.2.1(c)(1)(i) must have a corrective action or actions that are shown to be effective in preventing future occurrences. Each corrective action must be identified in the CMP document specified in section K25.1.6. A corrective action is not required:

(i) For an item where the manufacturer is unable to determine a cause or potential cause.

(ii) For an event where it is technically unfeasible to develop a corrective action.

(iii) If the world-fleet IFSD rate—

(A) Is at or below 0.02 per 1,000 world-fleet engine-hours for approval up to and including 180-minute ETOPS; or

(B) Is at or below 0.01 per 1,000 world-fleet engine-hours for approval greater than 180-minute ETOPS.

(d) *Airplane systems assessment.* The applicant must conduct an airplane systems assessment. The applicant must show that the airplane systems comply with § 25.1309(b) using available in-service reliability data for ETOPS significant systems on the candidate airplane-engine combination. Each cause or potential cause of a relevant design, manufacturing, operational, and maintenance problem occurring in service must have a corrective action or actions that are shown to be effective in preventing future occurrences. Each corrective action must be identified in the CMP document specified in section K25.1.6 of this appendix. A corrective action is not required if the problem would not significantly impact the safety or reliability of the airplane system involved. A relevant problem is a problem with an ETOPS group 1 significant system that has or could result in, an IFSD or diversion. The applicant must include in this assessment relevant problems with similar or identical equipment installed on other types of airplanes to the extent such information is reasonably available.

(e) *Airplane flight test.* The applicant must conduct a flight test to validate the flightcrew's ability to safely conduct an ETOPS diversion with an inoperative engine and worst-case ETOPS Significant System failures and malfunctions that could occur in service. The flight test must validate the airplane's flying qualities and performance with the demonstrated failures and malfunctions.

K25.2.2 *Early ETOPS method.*

An applicant for ETOPS type design approval using the Early ETOPS method must comply with the following requirements:

(a) *Assessment of relevant experience with airplanes previously certificated under part 25.* The applicant must identify specific corrective actions taken on the candidate airplane to prevent relevant design, manufacturing, operational, and maintenance problems experienced on airplanes previously certificated under part 25 manufactured by the applicant. Specific corrective actions are not required if the nature of a problem is such that the problem would not significantly impact the safety or reliability of the airplane system involved. A relevant problem is a problem with an ETOPS group 1 significant system that has or could result in an IFSD or diversion. The applicant must include in this assessment relevant problems of supplier-provided ETOPS group 1 significant systems and similar or identical equipment used on airplanes built by other manufacturers to the extent such information is reasonably available.

(b) *Propulsion system design.* (1) The engine used in the applicant's airplane design must be approved as eligible for Early ETOPS in accordance with § 33.201 of this chapter.

(2) The applicant must design the propulsion system to preclude failures or malfunctions that could result in an IFSD. The applicant must show compliance with this requirement by analysis, test, in-service experience on other airplanes, or other means acceptable to the FAA. If analysis is used, the applicant must show that the propulsion system design will minimize failures and malfunctions with the objective of achieving the following IFSD rates:

(i) An IFSD rate of 0.02 or less per 1,000 world-fleet engine-hours for type design approval up to and including 180 minutes.

(ii) An IFSD rate of 0.01 or less per 1,000 world-fleet engine-hours for type design approval beyond 180 minutes.

(c) *Maintenance and operational procedures.* The applicant must validate all maintenance and operational procedures for ETOPS significant systems. The applicant must identify, track, and resolve any problems found during the validation in accordance with the problem tracking and resolution system specified in section K25.2.2(h) of this appendix.

(d) *Propulsion system validation test.* (1) The installed engine configuration for which approval is being sought must comply with § 33.201(c) of this chapter. The test engine must be configured with a complete airplane nacelle package, including engine-mounted equipment, except for any configuration differences necessary to accommodate test stand interfaces with the engine nacelle package. At the conclusion of the test, the propulsion system must be—

(i) Visually inspected according to the applicant's on-wing inspection recommendations and limits; and

(ii) Completely disassembled and the propulsion system hardware inspected to determine whether it meets the service limits specified in the Instructions for Continued Airworthiness submitted in compliance with § 25.1529.

(2) The applicant must identify, track, and resolve each cause or potential cause of IFSD, loss of thrust control, or other power loss encountered during this inspection in accordance with the problem tracking and resolution system specified in section K25.2.2 (h) of this appendix.

(e) *New technology testing.* Technology new to the applicant, including substantially new manufacturing techniques, must be tested to substantiate its suitability for the airplane design.

(f) *APU validation test.* If an APU is needed to comply with this appendix, one APU of the type to be certified with the airplane must be tested for 3,000 equivalent airplane operational cycles. Following completion of the test, the APU must be disassembled and inspected. The applicant must identify, track, and resolve each cause or potential cause of an inability to start or operate the APU in flight as intended in accordance with the problem tracking and resolution system specified in section K25.2.2(h) of this appendix.

(g) *Airplane demonstration.* For each airplane-engine combination to be approved for ETOPS, the applicant must flight test at least one airplane to demonstrate that the airplane, and its components and equipment are capable of functioning properly during ETOPS flights and diversions of the longest duration for which the applicant seeks approval. This flight testing may be performed in conjunction with, but may not substitute for the flight testing required by § 21.35(b)(2) of this chapter.

(1) The airplane demonstration flight test program must include:

(i) Flights simulating actual ETOPS, including flight at normal cruise altitude, step climbs, and, if applicable, APU operation.

(ii) Maximum duration flights with maximum duration diversions.

(iii) Maximum duration engine-inoperative diversions distributed among the engines installed on the airplanes used for the airplane demonstration flight test program. At least two one-engine-inoperative diversions must be conducted at maximum continuous thrust or power using the same engine.

(iv) Flights under non-normal conditions to demonstrate the flightcrew's ability to safely conduct an ETOPS diversion with worst-case ETOPS significant system failures or malfunctions that could occur in service.

(v) Diversions to airports that represent airports of the types used for ETOPS diversions.

(vi) Repeated exposure to humid and inclement weather on the ground followed by a long-duration flight at normal cruise altitude.

(2) The airplane demonstration flight test program must validate the adequacy of the airplane's flying qualities and performance, and the flightcrew's ability to safely conduct an ETOPS diversion under the conditions specified in section K25.2.2(g)(1) of this appendix.

(3) During the airplane demonstration flight test program, each test airplane must be operated and maintained using the applicant's recommended operating and maintenance procedures.

(4) At the completion of the airplane demonstration flight test program, each ETOPS significant system must undergo an on-wing inspection or test in accordance with the tasks defined in the proposed Instructions for Continued Airworthiness to establish its condition for continued safe operation. Each engine must also undergo a gas path inspection. These inspections must be conducted in a manner to identify abnormal conditions that could result in an IFSD or diversion. The applicant must identify, track and resolve any abnormal conditions in accordance with the problem tracking and resolution system specified in section K25.2.2(h) of this appendix.

(h) *Problem tracking and resolution system.* (1) The applicant must establish and maintain a problem tracking and resolution system. The system must:

(i) Contain a process for prompt reporting to the responsible FAA aircraft certification office of each occurrence reportable under § 21.4(a)(6) encountered during the phases of airplane and engine development used to assess Early ETOPS eligibility.

(ii) Contain a process for notifying the responsible FAA aircraft certification office of each proposed corrective action that the applicant determines necessary for each problem identified from the occurrences reported under section K25.2.2. (h)(1)(i) of this appendix. The timing of the notification must permit appropriate FAA review before taking the proposed corrective action.

(2) If the applicant is seeking ETOPS type design approval of a change to an airplane-engine combination previously approved for ETOPS, the problem tracking and resolution system need only address those problems specified in the following table, provided the applicant obtains prior authorization from the FAA:

If the change does not require a new airplane type certificate and . . .	Then the Problem Tracking and Resolution System must address . . .
(i) Requires a new engine type certificate	All problems applicable to the new engine installation, and for the remainder of the airplane, problems in changed systems only.
(ii) Does not require a new engine type certificate	Problems in changed systems only.

(i) *Acceptance criteria.* The type and frequency of failures and malfunctions on ETOPS significant systems that occur during the airplane flight test program and the airplane demonstration flight test program specified in section K25.2.2(g) of this appendix must be consistent with the type and frequency of failures and malfunctions that would be expected to occur on currently certificated airplanes approved for ETOPS.

K25.2.3. *Combined service experience and Early ETOPS method.*

An applicant for ETOPS type design approval using the combined service experience and Early ETOPS method must comply with the following requirements.

(a) A service experience requirement of not less than 15,000 engine-hours for the world fleet of the candidate airplane-engine combination.

(b) The Early ETOPS requirements of K25.2.2, except for the airplane demonstration specified in section K25.2.2(g) of this appendix; and

(c) The flight test requirement of section K25.2.1(e) of this appendix.

K25.3 ***Airplanes with more than two engines.***

An applicant for ETOPS type design approval of an airplane with more than two engines must use one of the methods described in section K25.3.1, K25.3.2, or K25.3.3 of this appendix.

K25.3.1 *Service experience method.*

An applicant for ETOPS type design approval using the service experience method must comply with section K25.3.1(a) of this appendix before conducting the airplane systems assessment specified in K25.3.1(b), and the flight test specified in section K25.3.1(c) of this appendix.

(a) *Service experience.* The world fleet for the airplane-engine combination must accumulate a minimum of 250,000 engine-hours. The FAA may reduce this number of hours if the applicant identifies compensating factors that are acceptable to the FAA. The compensating factors may include experience on another airplane, but experience on the candidate airplane must make up a significant portion of the total required service experience.

(b) *Airplane systems assessment.* The applicant must conduct an airplane systems assessment. The applicant must show that the airplane systems comply with the § 25.1309(b) using available in-service reliability data for ETOPS significant systems on the candidate airplane-engine combination. Each cause or potential cause of a relevant design, manufacturing, operational or maintenance problem occurring in service must have a corrective action or actions that are shown to be effective in preventing future occurrences. Each corrective action must be identified in the CMP document specified in section K25.1.6 of this appendix. A corrective action is not required if the problem would not significantly impact the safety or reliability of the airplane system involved. A relevant problem is a problem with an ETOPS group 1 significant system that has or could result in an IFSD or diversion. The applicant must include in this assessment relevant problems with similar or identical equipment installed on other types of airplanes to the extent such information is reasonably available.

(c) *Airplane flight test.* The applicant must conduct a flight test to validate the flightcrew's ability to safely conduct an ETOPS diversion with an inoperative engine and worst-case ETOPS significant system failures and malfunctions that could occur in service. The flight test must validate the airplane's flying qualities and performance with the demonstrated failures and malfunctions.

K25.3.2 *Early ETOPS method.*

An applicant for ETOPS type design approval using the Early ETOPS method must comply with the following requirements:

(a) *Maintenance and operational procedures.* The applicant must validate all maintenance and operational procedures for ETOPS significant systems. The applicant must identify, track and resolve any problems found during the validation in accordance with the problem tracking and resolution system specified in section K25.3.2(e) of this appendix.

(b) *New technology testing.* Technology new to the applicant, including substantially new manufacturing techniques, must be tested to substantiate its suitability for the airplane design.

(c) *APU validation test.* If an APU is needed to comply with this appendix, one APU of the type to be certified with the airplane must be tested for 3,000 equivalent airplane operational cycles. Following completion of the test, the APU must be disassembled and inspected. The applicant must identify, track, and resolve each cause or potential cause of an inability to start or operate the APU in flight as intended in accordance with the problem tracking and resolution system specified in section K25.3.2(e) of this appendix.

(d) *Airplane demonstration.* For each airplane-engine combination to be approved for ETOPS, the applicant must flight test at least one airplane to demonstrate that the airplane, and its components and equipment are capable of functioning properly during ETOPS flights and diversions of the longest duration for which the applicant seeks approval. This flight testing may be performed in conjunction with, but may not substitute for the flight testing required by § 21.35(b)(2).

(1) The airplane demonstration flight test program must include:

(i) Flights simulating actual ETOPS including flight at normal cruise altitude, step climbs, and, if applicable, APU operation.

(ii) Maximum duration flights with maximum duration diversions.

(iii) Maximum duration engine-inoperative diversions distributed among the engines installed on the airplanes used for the airplane demonstration flight test program. At least two one engine-inoperative diversions must be conducted at maximum continuous thrust or power using the same engine.

(iv) Flights under non-normal conditions to validate the flightcrew's ability to safely conduct an ETOPS diversion with worst-case ETOPS significant system failures or malfunctions that could occur in service.

(v) Diversions to airports that represent airports of the types used for ETOPS diversions.

(vi) Repeated exposure to humid and inclement weather on the ground followed by a long duration flight at normal cruise altitude.

(2) The airplane demonstration flight test program must validate the adequacy of the airplane's flying qualities and performance, and the flightcrew's ability to safely conduct an ETOPS diversion under the conditions specified in section K25.3.2(d)(1) of this appendix.

(3) During the airplane demonstration flight test program, each test airplane must be operated and maintained using the applicant's recommended operating and maintenance procedures.

(4) At the completion of the airplane demonstration, each ETOPS significant system must undergo an on-wing inspection or test in accordance with the tasks defined in the proposed Instructions for Continued Airworthiness to establish its condition for continued safe operation. Each engine must also undergo a gas path inspection. These inspections must be conducted in a manner to identify abnormal conditions that could result in an IFSD or diversion. The applicant must identify, track and resolve any abnormal conditions in accordance with the problem tracking and resolution system specified in section K25.3.2(e) of this appendix.

(e) *Problem tracking and resolution system.* (1) The applicant must establish and maintain a problem tracking and resolution system. The system must:

(i) Contain a process for prompt reporting to the responsible FAA aircraft certification office of each occurrence reportable under § 21.4(a)(6) encountered during the phases of airplane and engine development used to assess Early ETOPS eligibility.

(ii) Contain a process for notifying the responsible FAA aircraft certification office of each proposed corrective action that the applicant determines necessary for each problem identified from the occurrences reported under section K25.3.2(h)(1)(i) of this appendix. The timing of the notification must permit appropriate FAA review before taking the proposed corrective action.

(2) If the applicant is seeking ETOPS type design approval of a change to an airplane-engine combination previously approved for ETOPS, the problem tracking and resolution system need only address those problems specified in the following table, provided the applicant obtains prior authorization from the FAA:

If the change does not require a new airplane type certificate and . . .	Then the Problem Tracking and Resolution System must address . . .
(i) Requires a new engine type certificate	All problems applicable to the new engine installation, and for the remainder of the airplane, problems in changed systems only.
(ii) Does not require a new engine type certificate	Problems in changed systems only.

(f) *Acceptance criteria.* The type and frequency of failures and malfunctions on ETOPS significant systems that occur during the airplane flight test program and the airplane demonstration flight test program specified in section K25.3.2(d) of this appendix must be consistent with the type and frequency of failures and malfunctions that would be expected to occur on currently certificated airplanes approved for ETOPS.

K25.3.3 *Combined service experience and Early ETOPS method.*

An applicant for ETOPS type design approval using the Early ETOPS method must comply with the following requirements:

(a) A service experience requirement of less than 15,000 engine-hours for the world fleet of the candidate airplane-engine combination;

(b) The Early ETOPS requirements of section K25.3.2 of this appendix, except for the airplane demonstration specified in section K25.3.2(d) of this appendix; and

(c) The flight test requirement of section K25.3.1(c) of this appendix.

[Doc. No. FAA-2002-6717, 72 FR 1873, Jan. 16, 2007]

Appendix L to Part 25—HIRF Environments and Equipment HIRF Test Levels

This appendix specifies the HIRF environments and equipment HIRF test levels for electrical and electronic systems under § 25.1317. The field strength values for the HIRF environments and equipment HIRF test levels are expressed in root-mean-square units measured during the peak of the modulation cycle.

(a) HIRF environment I is specified in the following table:

Table I
HIRF Environment I

Frequency	Field strength (volts/meter)	
	Peak	Average
10 kHz–2 MHz	50	50
2 MHz–30 MHz	100	100
30 MHz–100 MHz	50	50
100 MHz–400 MHz	100	100
400 MHz–700 MHz	700	50
700 MHz–1 GHz	700	100
1 GHz–2 GHz	2,000	200
2 GHz–6 GHz	3,000	200
6 GHz–8 GHz	1,000	200
8 GHz–12 GHz	3,000	300
12 GHz–18 GHz	2,000	200
18 GHz–40 GHz	600	200

In this table, the higher field strength applies at the frequency band edges.

(b) HIRF environment II is specified in the following table:

Table II
HIRF Environment II

Frequency	Field strength (volts/meter)	
	Peak	Average
10 kHz–500 kHz	20	20
500 kHz–2 MHz	30	30
2 MHz–30 MHz	100	100
30 MHz–100 MHz	10	10
100 MHz–200 MHz	30	10
200 MHz–400 MHz	10	10
400 MHz–1 GHz	700	40
1 GHz–2 GHz	1,300	160
2 GHz–4 GHz	3,000	120
4 GHz–6 GHz	3,000	160
6 GHz–8 GHz	400	170
8 GHz–12 GHz	1,230	230
12 GHz–18 GHz	730	190
18 GHz–40 GHz	600	150

In this table, the higher field strength applies at the frequency band edges.

(c) *Equipment HIRF Test Level 1.*

(1) From 10 kilohertz (kHz) to 400 megahertz (MHz), use conducted susceptibility tests with continuous wave (CW) and 1 kHz square wave modulation with 90 percent depth or greater. The conducted susceptibility current must start at a minimum of 0.6 milliamperes (mA) at 10 kHz, increasing 20 decibels (dB) per frequency decade to a minimum of 30 mA at 500 kHz.

(2) From 500 kHz to 40 MHz, the conducted susceptibility current must be at least 30 mA.

(3) From 40 MHz to 400 MHz, use conducted susceptibility tests, starting at a minimum of 30 mA at 40 MHz, decreasing 20 dB per frequency decade to a minimum of 3 mA at 400 MHz.

(4) From 100 MHz to 400 MHz, use radiated susceptibility tests at a minimum of 20 volts per meter (V/m) peak with CW and 1 kHz square wave modulation with 90 percent depth or greater.

(5) From 400 MHz to 8 gigahertz (GHz), use radiated susceptibility tests at a minimum of 150 V/m peak with pulse modulation of 4 percent duty cycle with a 1 kHz pulse repetition frequency. This signal must be switched on and off at a rate of 1 Hz with a duty cycle of 50 percent.

(d) *Equipment HIRF Test Level 2.* Equipment HIRF test level 2 is HIRF environment II in table II of this appendix reduced by acceptable aircraft transfer function and attenuation curves. Testing must cover the frequency band of 10 kHz to 8 GHz.

(e) *Equipment HIRF Test Level 3.*

(1) From 10 kHz to 400 MHz, use conducted susceptibility tests, starting at a minimum of 0.15 mA at 10 kHz, increasing 20 dB per frequency decade to a minimum of 7.5 mA at 500 kHz.

(2) From 500 kHz to 40 MHz, use conducted susceptibility tests at a minimum of 7.5 mA.

(3) From 40 MHz to 400 MHz, use conducted susceptibility tests, starting at a minimum of 7.5 mA at 40 MHz, decreasing 20 dB per frequency decade to a minimum of 0.75 mA at 400 MHz.

(4) From 100 MHz to 8 GHz, use radiated susceptibility tests at a minimum of 5 V/m.

[Doc. No. FAA-2006-23657, 72 FR 44026, Aug. 6, 2007]

Appendix M to Part 25—Fuel Tank System Flammability Reduction Means

M25.1 *Fuel tank flammability exposure requirements.*

(a) The Fleet Average Flammability Exposure of each fuel tank, as determined in accordance with Appendix N of this part, may not exceed 3 percent of the Flammability Exposure Evaluation Time (FEET), as defined in Appendix N of this part. As a portion of this 3 percent, if flammability reduction means (FRM) are used, each of the following time periods may not exceed 1.8 percent of the FEET:

(1) When any FRM is operational but the fuel tank is not inert and the tank is flammable; and

(2) When any FRM is inoperative and the tank is flammable.

(b) The Fleet Average Flammability Exposure, as defined in Appendix N of this part, of each fuel tank may not exceed 3 percent of the portion of the FEET occurring during either ground or takeoff/climb phases of flight during warm days. The analysis must consider the following conditions.

(1) The analysis must use the subset of those flights that begin with a sea level ground ambient temperature of 80°F (standard day plus 21°F atmosphere) or above, from the flammability exposure analysis done for overall performance.

(2) For the ground and takeoff/climb phases of flight, the average flammability exposure must be calculated by dividing the time during the specific flight phase the fuel tank is flammable by the total time of the specific flight phase.

(3) Compliance with this paragraph may be shown using only those flights for which the airplane is dispatched with the flammability reduction means operational.

M25.2 ***Showing compliance.***

(a) The applicant must provide data from analysis, ground testing, and flight testing, or any combination of these, that:

(1) Validate the parameters used in the analysis required by paragraph M25.1 of this appendix;

(2) Substantiate that the FRM is effective at limiting flammability exposure in all compartments of each tank for which the FRM is used to show compliance with paragraph M25.1 of this appendix; and

(3) Describe the circumstances under which the FRM would not be operated during each phase of flight.

(b) The applicant must validate that the FRM meets the requirements of paragraph M25.1 of this appendix with any airplane or engine configuration affecting the performance of the FRM for which approval is sought.

M25.3 ***Reliability indications and maintenance access.***

(a) Reliability indications must be provided to identify failures of the FRM that would otherwise be latent and whose identification is necessary to ensure the fuel tank with an FRM meets the fleet average flammability exposure requirements listed in paragraph M25.1 of this appendix, including when the FRM is inoperative.

(b) Sufficient accessibility to FRM reliability indications must be provided for maintenance personnel or the flightcrew.

(c) The access doors and panels to the fuel tanks with FRMs (including any tanks that communicate with a tank via a vent system), and to any other confined spaces or enclosed areas that could contain hazardous atmosphere under normal conditions or failure conditions, must be permanently stenciled, marked, or placarded to warn maintenance personnel of the possible presence of a potentially hazardous atmosphere.

M25.4 ***Airworthiness limitations and procedures.***

(a) If FRM is used to comply with paragraph M25.1 of this appendix, Airworthiness Limitations must be identified for all maintenance or inspection tasks required to identify failures of components within the FRM that are needed to meet paragraph M25.1 of this appendix.

(b) Maintenance procedures must be developed to identify any hazards to be considered during maintenance of the FRM. These procedures must be included in the instructions for continued airworthiness (ICA).

M25.5 ***Reliability reporting.***

The effects of airplane component failures on FRM reliability must be assessed on an on-going basis. The applicant/holder must do the following:

(a) Demonstrate effective means to ensure collection of FRM reliability data. The means must provide data affecting FRM reliability, such as component failures.

(b) Unless alternative reporting procedures are approved by the FAA Oversight Office, as defined in part 26 of this subchapter, provide a report to the FAA every six months for the first five years after service introduction. After that period, continued reporting every six months may be replaced with other reliability tracking methods found acceptable to the FAA or eliminated if it is established that the reliability of the FRM meets, and will continue to meet, the exposure requirements of paragraph M25.1 of this appendix.

(c) Develop service instructions or revise the applicable airplane manual, according to a schedule approved by the FAA Oversight Office, as defined in part 26 of this subchapter, to correct any failures of the FRM that occur in service that could increase any fuel tank's Fleet Average Flammability Exposure to more than that required by paragraph M25.1 of this appendix.

[Doc. No. FAA-2005-22997, 73 FR 42494, July 21, 2008]

Appendix N to Part 25—Fuel Tank Flammability Exposure and Reliability Analysis

N25.1 ***General.***

(a) This appendix specifies the requirements for conducting fuel tank fleet average flammability exposure analyses required to meet § 25.981(b) and Appendix M of this part. For fuel tanks installed in aluminum wings, a qualitative assessment is sufficient if it substantiates that the tank is a conventional unheated wing tank.

(b) This appendix defines parameters affecting fuel tank flammability that must be used in performing the analysis. These include parameters that affect all airplanes within the fleet, such as a statistical distribution of ambient temperature, fuel flash point, flight lengths, and airplane descent rate. Demonstration of compliance also requires application of factors specific to the airplane model being evaluated. Factors that need to be included are maximum range, cruise mach number, typical altitude where the airplane begins initial cruise phase of flight, fuel temperature during both ground and flight times, and the performance of a flammability reduction means (FRM) if installed.

(c) The following definitions, input variables, and data tables must be used in the program to determine fleet average flammability exposure for a specific airplane model.

N25.2 ***Definitions.***

(a) *Bulk Average Fuel Temperature* means the average fuel temperature within the fuel tank or different sections of the tank if the tank is subdivided by baffles or compartments.

(b) *Flammability Exposure Evaluation Time (FEET).* The time from the start of preparing the airplane for flight, through the flight and landing, until all payload is unloaded, and all passengers and crew have disembarked. In the Monte Carlo program, the flight time is randomly selected from the Flight Length Distribution (Table 2), the pre-flight times are provided as a function of the flight time, and the post-flight time is a constant 30 minutes.

(c) *Flammable.* With respect to a fluid or gas, flammable means susceptible to igniting readily or to exploding (14 CFR Part 1, Definitions). A non-flammable ullage is one where the fuel-air vapor is too lean or too rich to burn or is inert as defined below. For the purposes of this appendix, a fuel tank that is not inert is considered flammable when the bulk average fuel temperature within the tank is within the flammable range for the fuel type being used. For any fuel tank that is subdivided into sections by baffles or compartments, the tank is considered flammable when the bulk average fuel temperature within any section of the tank, that is not inert, is within the flammable range for the fuel type being used.

(d) *Flash Point.* The flash point of a flammable fluid means the lowest temperature at which the application of a flame to a heated sample causes the vapor to ignite momentarily, or "flash." Table 1 of this appendix provides the flash point for the standard fuel to be used in the analysis.

(e) *Fleet average flammability exposure* is the percentage of the flammability exposure evaluation time (FEET) each fuel tank ullage is flammable for a fleet of an airplane type operating over the range of flight lengths in a world-wide range of environmental conditions and fuel properties as defined in this appendix.

(f) *Gaussian Distribution* is another name for the normal distribution, a symmetrical frequency distribution having a precise mathematical formula relating the mean and standard deviation of the samples. Gaussian distributions yield bell-shaped frequency curves having a preponderance of values around the mean with progressively fewer observations as the curve extends outward.

(g) *Hazardous atmosphere.* An atmosphere that may expose maintenance personnel, passengers or flight crew to the risk of death, incapacitation, impairment of ability to self-rescue (that is, escape unaided from a confined space), injury, or acute illness.

(h) *Inert.* For the purpose of this appendix, the tank is considered inert when the bulk average oxygen concentration within each compartment of the tank is 12 percent or less from sea level up to 10,000 feet altitude, then linearly increasing from 12 percent at 10,000 feet to 14.5 percent at 40,000 feet altitude, and extrapolated linearly above that altitude.

(i) *Inerting.* A process where a noncombustible gas is introduced into the ullage of a fuel tank so that the ullage becomes non-flammable.

(j) *Monte Carlo Analysis.* The analytical method that is specified in this appendix as the compliance means for assessing the fleet average flammability exposure time for a fuel tank.

(k) *Oxygen evolution* occurs when oxygen dissolved in the fuel is released into the ullage as the pressure and temperature in the fuel tank are reduced.

(l) *Standard deviation* is a statistical measure of the dispersion or variation in a distribution, equal to the square root of the arithmetic mean of the squares of the deviations from the arithmetic means.

(m) *Transport Effects.* For purposes of this appendix, transport effects are the change in fuel vapor concentration in a fuel tank caused by low fuel conditions and fuel condensation and vaporization.

(n) *Ullage.* The volume within the fuel tank not occupied by liquid fuel.

N25.3 ***Fuel tank flammability exposure analysis.***

(a) A flammability exposure analysis must be conducted for the fuel tank under evaluation to determine fleet average flammability exposure for the airplane and fuel types under evaluation. For fuel tanks that are subdivided by baffles or compartments, an analysis must be performed either for each section of the tank, or for the section of the tank having the highest flammability exposure. Consideration of transport effects is not allowed in the analysis. The analysis must be done in accordance with the methods and procedures set forth in the Fuel Tank Flammability Assessment Method User's Manual, dated May 2008, document number DOT/FAA/AR-05/8 (incorporated by reference, see § 25.5). The parameters specified in sections N25.3(b) and (c) of this appendix must be used in the fuel tank flammability exposure "Monte Carlo" analysis.

(b) The following parameters are defined in the Monte Carlo analysis and provided in paragraph N25.4 of this appendix:

(1) Cruise Ambient Temperature, as defined in this appendix.

(2) Ground Ambient Temperature, as defined in this appendix.

(3) Fuel Flash Point, as defined in this appendix.

(4) Flight Length Distribution, as defined in Table 2 of this appendix.

(5) Airplane Climb and Descent Profiles, as defined in the Fuel Tank Flammability Assessment Method User's Manual, dated May 2008, document number DOT/FAA/AR-05/8 (incorporated by reference in § 25.5).

(c) Parameters that are specific to the particular airplane model under evaluation that must be provided as inputs to the Monte Carlo analysis are:

(1) Airplane cruise altitude.

(2) Fuel tank quantities. If fuel quantity affects fuel tank flammability, inputs to the Monte Carlo analysis must be provided that represent the actual fuel quantity within the fuel tank or compartment of the fuel tank throughout each of the flights being evaluated. Input values for this data must be obtained from ground and flight test data or the approved FAA fuel management procedures.

(3) Airplane cruise mach number.

(4) Airplane maximum range.

(5) Fuel tank thermal characteristics. If fuel temperature affects fuel tank flammability, inputs to the Monte Carlo analysis must be provided that represent the actual bulk average fuel temperature within the fuel tank at each point in time throughout each of the flights being evaluated. For fuel tanks that are subdivided by baffles or compartments, bulk average fuel temperature inputs must be provided for each section of the tank. Input values for these data must be obtained from ground and flight test data or a thermal model of the tank that has been validated by ground and flight test data.

(6) Maximum airplane operating temperature limit, as defined by any limitations in the airplane flight manual.

(7) Airplane Utilization. The applicant must provide data supporting the number of flights per day and the number of hours per flight for the specific airplane model under evaluation. If there is no existing airplane fleet data to support the airplane being evaluated, the applicant must provide substantiation that the number of flights per day and the number of hours per flight for that airplane model is consistent with the existing fleet data they propose to use.

(d) *Fuel Tank FRM Model.* If FRM is used, an FAA approved Monte Carlo program must be used to show compliance with the flammability requirements of § 25.981 and Appendix M of this part. The program must determine the time periods during each flight phase when the fuel tank or compartment with the FRM would be flammable. The following factors must be considered in establishing these time periods:

(1) Any time periods throughout the flammability exposure evaluation time and under the full range of expected operating conditions, when the FRM is operating properly but fails to maintain a non-flammable fuel tank because of the effects of the fuel tank vent system or other causes,

(2) If dispatch with the system inoperative under the Master Minimum Equipment List (MMEL) is requested, the time period assumed in the reliability analysis (60 flight hours must be used for a 10-daymmEL dispatch

limit unless an alternative period has been approved by the Administrator),

(3) Frequency and duration of time periods of FRM inoperability, substantiated by test or analysis acceptable to the FAA, caused by latent or known failures, including airplane system shut-downs and failures that could cause the FRM to shut down or become inoperative.

(4) Effects of failures of the FRM that could increase the flammability exposure of the fuel tank.

(5) If an FRM is used that is affected by oxygen concentrations in the fuel tank, the time periods when oxygen evolution from the fuel results in the fuel tank or compartment exceeding the inert level. The applicant must include any times when oxygen evolution from the fuel in the tank or compartment under evaluation would result in a flammable fuel tank. The oxygen evolution rate that must be used is defined in the Fuel Tank Flammability Assessment Method User's Manual, dated May 2008, document number DOT/FAA/AR-05/8 (incorporated by reference in § 25.5).

(6) If an inerting system FRM is used, the effects of any air that may enter the fuel tank following the last flight of the day due to changes in ambient temperature, as defined in Table 4, during a 12-hour overnight period.

(e) The applicant must submit to the FAA Oversight Office for approval the fuel tank flammability analysis, including the airplane-specific parameters identified under paragraph N25.3(c) of this appendix and any deviations from the parameters identified in paragraph N25.3(b) of this appendix that affect flammability exposure, substantiating data, and any airworthiness limitations and other conditions assumed in the analysis.

N25.4 ***Variables and data tables.***

The following data must be used when conducting a flammability exposure analysis to determine the fleet average flammability exposure. Variables used to calculate fleet flammability exposure must include atmospheric ambient temperatures, flight length, flammability exposure evaluation time, fuel flash point, thermal characteristics of the fuel tank, overnight temperature drop, and oxygen evolution from the fuel into the ullage.

(a) Atmospheric Ambient Temperatures and Fuel Properties.

(1) In order to predict flammability exposure during a given flight, the variation of ground ambient temperatures, cruise ambient temperatures, and a method to compute the transition from ground to cruise and back again must be used. The variation of the ground and cruise ambient temperatures and the flash point of the fuel is defined by a Gaussian curve, given by the 50 percent value and a±1-standard deviation value.

(2) Ambient Temperature: Under the program, the ground and cruise ambient temperatures are linked by a set of assumptions on the atmosphere. The temperature varies with altitude following the International Standard Atmosphere (ISA) rate of change from the ground ambient temperature until the cruise temperature for the flight is reached. Above this altitude, the ambient temperature is fixed at the cruise ambient temperature. This results in a variation in the upper atmospheric temperature. For cold days, an inversion is applied up to 10,000 feet, and then the ISA rate of change is used.

(3) Fuel properties:

(i) For Jet A fuel, the variation of flash point of the fuel is defined by a Gaussian curve, given by the 50 percent value and a±1-standard deviation, as shown in Table 1 of this appendix.

(ii) The flammability envelope of the fuel that must be used for the flammability exposure analysis is a function of the flash point of the fuel selected by the Monte Carlo for a given flight. The flammability envelope for the fuel is defined by the upper flammability limit (UFL) and lower flammability limit (LFL) as follows:

(A) LFL at sea level=flash point temperature of the fuel at sea level minus 10 °F. LFL decreases from sea level value with increasing altitude at a rate of 1°F per 808 feet.

Table 1
Gaussian Distribution for Ground Ambient Temperature, Cruise Ambient Temperature, and Fuel Flash Point

Parameter	Temperature in deg F		
	Ground ambient temperature	Cruise ambient temperature	Fuel flash point (FP)
Mean Temp	59.95	–70	120
Neg 1 std dev	20.14	8	8
Pos 1 std dev	17.28	8	8

(B) UFL at sea level=flash point temperature of the fuel at sea level plus 63.5 °F. UFL decreases from the sea level value with increasing altitude at a rate of 1°F per 512 feet.

(4) For each flight analyzed, a separate random number must be generated for each of the three parameters (ground ambient temperature, cruise ambient temperature, and fuel flash point) using the Gaussian distribution defined in Table 1 of this appendix.

(b) The Flight Length Distribution defined in Table 2 must be used in the Monte Carlo analysis.

(c) Overnight Temperature Drop. For airplanes on which FRM is installed, the overnight temperature drop for this appendix is defined using:

(1) A temperature at the beginning of the overnight period that equals the landing temperature of the previous

Table 2
Flight Length Distribution

Flight length (NM)		Airplane maximum range—nautical miles (NM)									
From	To	1000	2000	3000	4000	5000	6000	7000	8000	9000	10000
		Distribution of flight lengths (percentage of total)									
0	200	11.7	7.5	6.2	5.5	4.7	4.0	3.4	3.0	2.6	2.3
200	400	27.3	19.9	17.0	15.2	13.2	11.4	9.7	8.5	7.5	6.7
400	600	46.3	40.0	35.7	32.6	28.5	24.9	21.2	18.7	16.4	14.8
600	800	10.3	11.6	11.0	10.2	9.1	8.0	6.9	6.1	5.4	4.8
800	1000	4.4	8.5	8.6	8.2	7.4	6.6	5.7	5.0	4.5	4.0
1000	1200	0.0	4.8	5.3	5.3	4.8	4.3	3.8	3.3	3.0	2.7
1200	1400	0.0	3.6	4.4	4.5	4.2	3.8	3.3	3.0	2.7	2.4
1400	1600	0.0	2.2	3.3	3.5	3.3	3.1	2.7	2.4	2.2	2.0
1600	1800	0.0	1.2	2.3	2.6	2.5	2.4	2.1	1.9	1.7	1.6
1800	2000	0.0	0.7	2.2	2.6	2.6	2.5	2.2	2.0	1.8	1.7
2000	2200	0.0	0.0	1.6	2.1	2.2	2.1	1.9	1.7	1.6	1.4
2200	2400	0.0	0.0	1.1	1.6	1.7	1.7	1.6	1.4	1.3	1.2
2400	2600	0.0	0.0	0.7	1.2	1.4	1.4	1.3	1.2	1.1	1.0
2600	2800	0.0	0.0	0.4	0.9	1.0	1.1	1.0	0.9	0.9	0.8
2800	3000	0.0	0.0	0.2	0.6	0.7	0.8	0.7	0.7	0.6	0.6
3000	3200	0.0	0.0	0.0	0.6	0.8	0.8	0.8	0.8	0.7	0.7
3200	3400	0.0	0.0	0.0	0.7	1.1	1.2	1.2	1.1	1.1	1.0
3400	3600	0.0	0.0	0.0	0.7	1.3	1.6	1.6	1.5	1.5	1.4
3600	3800	0.0	0.0	0.0	0.9	2.2	2.7	2.8	2.7	2.6	2.5

Flight length (NM)		Airplane maximum range—nautical miles (NM)									
From	To	1000	2000	3000	4000	5000	6000	7000	8000	9000	10000
3800	4000	0.0	0.0	0.0	0.5	2.0	2.6	2.8	2.8	2.7	2.6
4000	4200	0.0	0.0	0.0	0.0	2.1	3.0	3.2	3.3	3.2	3.1
4200	4400	0.0	0.0	0.0	0.0	1.4	2.2	2.5	2.6	2.6	2.5
4400	4600	0.0	0.0	0.0	0.0	1.0	2.0	2.3	2.5	2.5	2.4
4600	4800	0.0	0.0	0.0	0.0	0.6	1.5	1.8	2.0	2.0	2.0
4800	5000	0.0	0.0	0.0	0.0	0.2	1.0	1.4	1.5	1.6	1.5
5000	5200	0.0	0.0	0.0	0.0	0.0	0.8	1.1	1.3	1.3	1.3
5200	5400	0.0	0.0	0.0	0.0	0.0	0.8	1.2	1.5	1.6	1.6
5400	5600	0.0	0.0	0.0	0.0	0.0	0.9	1.7	2.1	2.2	2.3
5600	5800	0.0	0.0	0.0	0.0	0.0	0.6	1.6	2.2	2.4	2.5
5800	6000	0.0	0.0	0.0	0.0	0.0	0.2	1.8	2.4	2.8	2.9
6000	6200	0.0	0.0	0.0	0.0	0.0	0.0	1.7	2.6	3.1	3.3
6200	6400	0.0	0.0	0.0	0.0	0.0	0.0	1.4	2.4	2.9	3.1
6400	6600	0.0	0.0	0.0	0.0	0.0	0.0	0.9	1.8	2.2	2.5
6600	6800	0.0	0.0	0.0	0.0	0.0	0.0	0.5	1.2	1.6	1.9
6800	7000	0.0	0.0	0.0	0.0	0.0	0.0	0.2	0.8	1.1	1.3
7000	7200	0.0	0.0	0.0	0.0	0.0	0.0	0.0	0.4	0.7	0.8
7200	7400	0.0	0.0	0.0	0.0	0.0	0.0	0.0	0.3	0.5	0.7
7400	7600	0.0	0.0	0.0	0.0	0.0	0.0	0.0	0.2	0.5	0.6
7600	7800	0.0	0.0	0.0	0.0	0.0	0.0	0.0	0.1	0.5	0.7
7800	8000	0.0	0.0	0.0	0.0	0.0	0.0	0.0	0.1	0.6	0.8
8000	8200	0.0	0.0	0.0	0.0	0.0	0.0	0.0	0.0	0.5	0.8
8200	8400	0.0	0.0	0.0	0.0	0.0	0.0	0.0	0.0	0.5	1.0
8400	8600	0.0	0.0	0.0	0.0	0.0	0.0	0.0	0.0	0.6	1.3
8600	8800	0.0	0.0	0.0	0.0	0.0	0.0	0.0	0.0	0.4	1.1
8800	9000	0.0	0.0	0.0	0.0	0.0	0.0	0.0	0.0	0.2	0.8
9000	9200	0.0	0.0	0.0	0.0	0.0	0.0	0.0	0.0	0.0	0.5
9200	9400	0.0	0.0	0.0	0.0	0.0	0.0	0.0	0.0	0.0	0.2
9400	9600	0.0	0.0	0.0	0.0	0.0	0.0	0.0	0.0	0.0	0.1
9600	9800	0.0	0.0	0.0	0.0	0.0	0.0	0.0	0.0	0.0	0.1
9800	10000	0.0	0.0	0.0	0.0	0.0	0.0	0.0	0.0	0.0	0.1

Table 3
Landing Outside Air Temperature

Parameter	Landing outside air temperature °F
Mean Temperature	58.68
negative 1 std dev	20.55
positive 1 std dev	13.21

Table 4
Outside Air Temperature (OAT) Drop

Parameter	OAT drop temperature °F
Mean Temp	12.0
1 std dev	6.0

flight that is a random value based on a Gaussian distribution; and

(2) An overnight temperature drop that is a random value based on a Gaussian distribution.

(3) For any flight that will end with an overnight ground period (one flight per day out of an average number of flights per day, depending on utilization of the particular airplane model being evaluated), the landing outside air temperature (OAT) is to be chosen as a random value from the following Gaussian curve:

(4) The outside ambient air temperature (OAT) overnight temperature drop is to be chosen as a random value from the following Gaussian curve:

(d) Number of Simulated Flights Required in Analysis. In order for the Monte Carlo analysis to be valid for showing compliance with the fleet average and warm day flammability exposure requirements, the applicant must run the analysis for a minimum number of flights to ensure that the fleet average and warm day flammability exposure for the fuel tank under evaluation meets the applicable flammability limits defined in Table 5 of this appendix.

Table 5
Flammability Exposure Limit

Minimum number of flights in Monte Carlo analysis	Maximum acceptable Monte Carlo average fuel tank flammability exposure (percent) to meet 3 percent requirements	Maximum acceptable Monte Carlo average fuel tank flammability exposure (percent) to meet 7 percent part 26 requirements
10,000	2.91	6.79
100,000	2.98	6.96
1,000,000	3.00	7.00

[Doc. No. FAA-2005-22997, 73 FR 42495, July 21, 2008]

PART 26

CONTINUED AIRWORTHINESS AND SAFETY IMPROVEMENTS FOR TRANSPORT CATEGORY AIRPLANES

Contents

Authority: 49 U.S.C. 106(g), 40113, 44701, 44702 and 44704.

Source: Docket No. FAA-2004-18379, 72 FR 63409, Nov. 8, 2007, unless otherwise noted.

Subpart A—General

§ 26.1 Purpose and scope.

(a) This part establishes requirements for support of the continued airworthiness of and safety improvements for transport category airplanes. These requirements may include performing assessments, developing design changes, developing revisions to Instructions for Continued Airworthiness (ICA), and making necessary documentation available to affected persons. Requirements of this part that establish standards for design changes and revisions to the ICA are considered airworthiness requirements.

(b) Except as provided in paragraph (c) of this section, this part applies to the following persons, as specified in each subpart of this part:

(1) Holders of type certificates and supplemental type certificates.

(2) Applicants for type certificates and supplemental type certificates and changes to those certificates (including service bulletins describing design changes).

(3) Persons seeking design approval for airplane repairs, alterations, or modifications that may affect airworthiness.

(4) Holders of type certificates and their licensees producing new airplanes.

(c) An applicant for approval of a design change is not required to comply with any applicable airworthiness requirement of this part if the applicant elects or is required to comply with a corresponding amendment to part 25 of this chapter that is adopted concurrently or after that airworthiness requirement.

(d) For the purposes of this part, the word "type certificate" does not include supplemental type certificates.

§ 26.3 Definitions.

For the purposes of this part:

FAA Oversight Office is the aircraft certification office or office of the Transport Airplane Directorate with oversight responsibility for the relevant type certificate, supplemental type certificate, or manufacturer, as determined by the Administrator.

§ 26.5 Applicability table.

Table 1 of this section provides an overview of the applicability of this part. It provides guidance in identifying what sections apply to various types of entities. The specific applicability of each subpart and section is specified in the regulatory text.

[Doc. No. FAA-2006-24281, 75 FR 69782, Nov. 15, 2010]

Subpart B—Enhanced Airworthiness Program for Airplane Systems

§ 26.11 Electrical wiring interconnection systems (EWIS) maintenance program.

(a) Except as provided in paragraph (g) of this section, this section applies to transport category, turbine-powered airplanes with a type certificate issued after January 1, 1958, that, as a result of the original certification, or later increase in capacity, have—

(1) A maximum type-certificated passenger capacity of 30 or more or

(2) A maximum payload capacity of 7,500 pounds or more.

(b) Holders of, and applicants for, type certificates, as identified in paragraph (d) of this section must develop Instructions for Continued Airworthiness (ICA) for the representative airplane's EWIS in accordance with part 25, Appendix H paragraphs H25.5(a)(1) and (b) of this subchapter in effect on December 10, 2007 for each affected type design, and submit those ICA for review and approval by the FAA Oversight Office. For purposes of this section, the "representative airplane" is the configuration of each model series airplane that incorporates all variations of EWIS used in production on that series airplane, and all TC-holder-designed modifications mandated by airworthiness directive as of the effective date of this rule. Each person specified in paragraph (d) of this section must also review any fuel tank system ICA developed by that person to comply with SFAR 88 to ensure compatibility with the EWIS ICA, including minimizing redundant requirements.

(c) Applicants for amendments to type certificates and supplemental type certificates, as identified in paragraph (d) of this section, must:

(1) Evaluate whether the design change for which approval is sought necessitates a revision to the ICA required by paragraph (b) of this section to comply with the requirements of Appendix H, paragraphs H25.5(a)(1) and (b). If so, the applicant must develop and submit the necessary revisions for review and approval by the FAA Oversight Office.

(2) Ensure that any revised EWIS ICA remain compatible with any fuel tank system ICA previously developed to comply with SFAR 88 and any redundant requirements between them are minimized.

(d) The following persons must comply with the requirements of paragraph (b) or (c) of this section, as applicable, before the dates specified.

(1) Holders of type certificates (TC): December 10, 2009.

(2) Applicants for TCs, and amendments to TCs (including service bulletins describing design changes), if the date of application was before December 10, 2007 and the certificate was issued on or after December 10, 2007: December 10, 2009 or the date the certificate is issued, whichever occurs later.

(3) Unless compliance with § 25.1729 of this subchapter is required or elected, applicants for amendments to TCs, if the application was filed on or after December 10, 2007: December 10, 2009, or the date of approval of the certificate, whichever occurs later.

(4) Applicants for supplemental type certificates (STC), including changes to existing STCs, if the date of application was before December 10, 2007 and the certificate was issued on or after December 10, 2007: June 7, 2010, or the date of approval of the certificate, whichever occurs later.

(5) Unless compliance with § 25.1729 of this subchapter is required or elected, applicants for STCs, including changes to existing STCs, if the application was filed on or after December 10, 2007, June 7, 2010, or the date of approval of the certificate, whichever occurs later.

(e) Each person identified in paragraphs (d)(1), (d)(2), and (d)(4) of this section must submit to the FAA Oversight Office for approval a compliance plan by March 10, 2008. The compliance plan must include the following information:

(1) A proposed project schedule, identifying all major milestones, for meeting the compliance dates specified in paragraph (d) of this section.

(2) A proposed means of compliance with this section, identifying all required submissions, including all compliance items as mandated in part 25, Appendix H paragraphs H25.5(a)(1) and (b) of this subchapter in effect on December 10, 2007, and all data to be developed to substantiate compliance.

(3) A proposal for submitting a draft of all compliance items required by paragraph (e)(2) of this section for review by the FAA Oversight Office not less than 60 days before the compliance time specified in paragraph (d) of this section.

(4) A proposal for how the approved ICA will be made available to affected persons.

(f) Each person specified in paragraph (e) must implement the compliance plan, or later approved revisions, as approved in compliance with paragraph (e) of this section.

(g) This section does not apply to the following airplane models:

(1) Lockheed L-188
(2) Bombardier CL-44
(3) Mitsubishi YS-11
(4) British Aerospace BAC 1-11
(5) Concorde
(6) deHavilland D.H. 106 Comet 4C
(7) VFW—Vereinigte Flugtechnische Werk VFW-614
(8) Illyushin Aviation IL 96T
(9) Bristol Aircraft Britannia 305
(10) Handley Page Herald Type 300

Table 1—Applicability of Part 26 Rules

	Applicable sections			
	Subpart B EAPAS/FTS	Subpart C widespread fatigue damage	Subpart D fuel tank flammability	Subpart E damage tolerance data
Effective date of rule	December 10, 2007	January 14, 2011	December 26, 2008	January 11, 2008
Existing[1] TC Holders	26.11	26.21	26.33	26.43, 26.45, 26.49
Pending[1] TC Applicants	26.11	26.21	26.37	26.43, 26.45
Future[2] TC applicants	N/A	N/A	N/A	26.43
Existing[1] STC Holders	N/A	26.21	26.35	26.47, 26.49
Pending[1] STC/ATC applicants	26.11	26.21	26.35	26.45, 26.47, 26.49
Future[2] STC/ATC applicants	26.11	26.21	26.35	26.45, 26.47, 26.49
Manufacturers	N/A	N/A	26.39	N/A

[1] As of the effective date of the identified rule.
[2] Application made after the effective date of the identified rule.

PART 26

FAR

(11) Avions Marcel Dassault—Breguet Aviation Mercure 100C

(12) Airbus Caravelle

(13) Lockheed L-300

[Amdt. 26-0, 72 FR 63409, Nov. 8, 2007; 72 FR 68618, Dec. 5, 2007]

Subpart C—Aging Airplane Safety—Widespread Fatigue Damage

Source: Doc. No. FAA-2006-24281, 75 FR 69782, Nov. 15, 2010, unless otherwise noted.

§ 26.21 Limit of validity.

(a) *Applicability.* Except as provided in paragraph (g) of this section, this section applies to transport category, turbine-powered airplanes with a maximum takeoff gross weight greater than 75,000 pounds and a type certificate issued after January 1, 1958, regardless of whether the maximum takeoff gross weight is a result of an original type certificate or a later design change. This section also applies to transport category, turbine-powered airplanes with a type certificate issued after January 1, 1958, if a design change approval for which application is made after January 14, 2011 has the effect of reducing the maximum takeoff gross weight from greater than 75,000 pounds to 75,000 pounds or less.

(b) *Limit of validity.* Each person identified in paragraph (c) of this section must comply with the following requirements:

(1) Establish a limit of validity of the engineering data that supports the structural maintenance program (hereafter referred to as LOV) that corresponds to the period of time, stated as a number of total accumulated flight cycles or flight hours or both, during which it is demonstrated that widespread fatigue damage will not occur in the airplane. This demonstration must include an evaluation of airplane structural configurations and be supported by test evidence and analysis at a minimum and, if available, service experience, or service experience and teardown inspection results, of high-time airplanes of similar structural design, accounting for differences in operating conditions and procedures. The airplane structural configurations to be evaluated include—

(i) All model variations and derivatives approved under the type certificate; and

(ii) All structural modifications to and replacements for the airplane structural configurations specified in paragraph (b)(1)(i) of this section, mandated by airworthiness directives as of January 14, 2011.

(2) If the LOV depends on performance of maintenance actions for which service information has not been mandated by airworthiness directive as of January 14, 2011, submit the following to the FAA Oversight Office:

(i) For those maintenance actions for which service information has been issued as of the applicable compliance date specified in paragraph (c) of this section, a list identifying each of those actions.

(ii) For those maintenance actions for which service information has not been issued as of the applicable compliance date specified in paragraph (c) of this section, a list identifying each of those actions and a binding schedule for providing in a timely manner the necessary service information for those actions. Once the FAA Oversight Office approves this schedule, each person identified in paragraph (c) of this section must comply with that schedule.

(3) Unless previously accomplished, establish an Airworthiness Limitations section (ALS) for each airplane structural configuration evaluated under paragraph (b)(1) of this section.

(4) Incorporate the applicable LOV established under paragraph (b)(1) of this section into the ALS for each airplane structural configuration evaluated under paragraph (b)(1) and submit it to the FAA Oversight Office for approval.

(c) *Persons who must comply and compliance dates.* The following persons must comply with the requirements of paragraph (b) of this section by the specified date.

(1) Holders of type certificates (TC) of airplane models identified in Table 1 of this section: No later than the applicable date identified in Table 1 of this section.

(2) Applicants for TCs, if the date of application was before January 14, 2011: No later than the latest of the following dates:

(i) January 14, 2016;

(ii) The date the certificate is issued; or

(iii) The date specified in the plan approved under § 25.571(b) for completion of the full-scale fatigue testing and demonstrating that widespread fatigue damage will not occur in the airplane structure.

(3) Applicants for amendments to TCs, with the exception of amendments to TCs specified in paragraphs (c)(6) or (c)(7) of this section, if the original TC was issued before January 14, 2011: No later than the latest of the following dates:

(i) January 14, 2016;

(ii) The date the amended certificate is issued; or

(iii) The date specified in the plan approved under § 25.571(b) for completion of the full-scale fatigue testing and demonstrating that widespread fatigue damage will not occur in the airplane structure.

(4) Applicants for amendments to TCs, with the exception of amendments to TCs specified in paragraphs (c)(6) or (c)(7) of this section, if the application for the original TC was made before January 14, 2011 but the TC was not issued before January 14, 2011: No later than the latest of the following dates:

(i) January 14, 2016;

(ii) The date the amended certificate is issued; or

(iii) The date specified in the plan approved under § 25.571(b) for completion of the full-scale fatigue testing and demonstrating that widespread fatigue damage will not occur in the airplane structure.

(5) Holders of either supplemental type certificates (STCs) or amendments to TCs that increase maximum takeoff gross weights from 75,000 pounds or less to greater than 75,000 pounds: No later than July 14, 2012.

(6) Applicants for either STCs or amendments to TCs that increase maximum takeoff gross weights from 75,000 pounds or less to greater than 75,000 pounds: No later than the latest of the following dates:

(i) July 14, 2012;

(ii) The date the certificate is issued; or

(iii) The date specified in the plan approved under § 25.571(b) for completion of the full-scale fatigue testing and demonstrating that widespread fatigue damage will not occur in the airplane structure.

(7) Applicants for either STCs or amendments to TCs that decrease maximum takeoff gross weights from greater than 75,000 pounds to 75,000 pounds or less, if the date of application was after January 14, 2011: No later than the latest of the following dates:

(i) July 14, 2012;

(ii) The date the certificate is issued; or

(iii) The date specified in the plan approved under § 25.571(b) for completion of the full-scale fatigue testing and demonstrating that widespread fatigue damage will not occur in the airplane structure.

(d) *Compliance plan.* Each person identified in paragraph (e) of this section must submit a compliance plan consisting of the following:

(1) A proposed project schedule, identifying all major milestones, for meeting the compliance dates specified in paragraph (c) of this section.

(2) A proposed means of compliance with paragraphs (b)(1) through (b)(4) of this section.

(3) A proposal for submitting a draft of all compliance items required by paragraph (b) of this section for review by the FAA Oversight Office not less than 60 days before the compliance date specified in paragraph (c) of this section, as applicable.

(4) A proposal for how the LOV will be distributed.

(e) *Compliance dates for compliance plans.* The following persons must submit the compliance plan described in paragraph (d) of this section to the FAA Oversight Office by the specified date.

(1) Holders of type certificates: No later than April 14, 2011.

(2) Applicants for TCs and amendments to TCs, with the exception of amendments to TCs specified in paragraphs (e)(4), (e)(5), or (e)(6) of this section, if the date of application was before January 14, 2011 but the TC or TC amendment was not issued before January 14, 2011: No later than April 14, 2011.

(3) Holders of either supplemental type certificates or amendments to TCs that increase maximum takeoff gross weights from 75,000 pounds or less to greater than 75,000 pounds: No later than April 14, 2011.

(4) Applicants for either STCs or amendments to TCs that increase maximum takeoff gross weights from 75,000 pounds or less to greater than 75,000 pounds, if the date of application was before January 14, 2011: No later than April 14, 2011.

(5) Applicants for either STCs or amendments to TCs that increase maximum takeoff gross weights from 75,000 pounds or less to greater than 75,000 pounds, if the date of application is on or after January 14, 2011: Within 90 days after the date of application.

(6) Applicants for either STCs or amendments to TCs that decrease maximum takeoff gross weights from greater than 75,000 pounds to 75,000 pounds or less, if the date of application is on or after January 14, 2011: Within 90 days after the date of application.

(f) *Compliance plan implementation.* Each affected person must implement the compliance plan as approved in compliance with paragraph (d) of this section.

(g) *Exceptions.* This section does not apply to the following airplane models:

(1) Bombardier BD-700.

(2) Bombardier CL-44.

(3) Gulfstream GV.

(4) Gulfstream GV-SP.

(5) British Aerospace, Aircraft Group, and Societe Nationale Industrielle Aerospatiale Concorde Type 1.

(6) British Aerospace (Commercial Aircraft) Ltd., Armstrong Whitworth Argosy A.W. 650 Series 101.

(7) British Aerospace Airbus, Ltd., BAC 1-11.

(8) BAE Systems (Operations) Ltd., BAe 146.

(9) BAE Systems (Operations) Ltd., Avro 146.

(10) Lockheed 300-50A01 (USAF C141A).

(11) Boeing 707.

(12) Boeing 720.

(13) deHavilland D.H. 106 Comet 4C.

(14) Ilyushin Aviation IL-96T.

(15) Bristol Aircraft Britannia 305.

(16) Avions Marcel Dassault-Breguet Aviation Mercure 100C.

(17) Airbus Caravelle.

(18) D & R Nevada, LLC, Convair Model 22.

(19) D & R Nevada, LLC, Convair Model 23M.

[Doc. No. FAA-2006-24281, 75 FR 69782, Nov. 15, 2010, as amended at 77 FR 30878, May 24, 2012]

Table 1—Compliance Dates for Affected Airplanes

Airplane model(all existing1models)	Compliance date—(monthsafter January 14, 2011)
Airbus:	
A300 Series, A310 Series, A300–600 Series	18
A318 Series	48
A319 Series	48
A320 Series	48
A321 Series	48
A330–200, –200 Freighter, –300 Series	48
A340–200, –300, –500, –600 Series	48
A380–800 Series	60
Boeing:	
717	48
727 (all series)	18
737 (Classics): 737–100, –200, –200C, –300, –400, –500	18
737 (NG): 737–600, –700, –700C, –800, –900, –900ER	48
747 (Classics): 747–100, –100B, –100B SUD, –200B, -200C, –200F, –300, 747SP, 747SR	18
747–400: 747–400, –400D, –400F	48
757	48
767	48
777–200, –300	48
777–200LR, 777–300ER, 777F	60
Bombardier:	
CL–600: 2D15 (Regional Jet Series 705), 2D24 (Regional Jet Series 900)	60
Embraer:	
ERJ 170	60

Airplane model (all existing[1] models)	Compliance date—(months after January 14, 2011)
ERJ 190	60
Fokker:	
F.28 Mark 0070, Mark 0100	18
Lockheed:	
L-1011	18
188	18
382 (all series)	18
McDonnell Douglas:	
DC-8, -8F	18
DC-9	18
MD-80 (DC-9-81, -82, -83, -87, MD-88)	18
MD-90	48
DC-10	18
MD-10	48
MD-11, -11F	48
All Other Airplane Models Listed on a Type Certificate as of January 14, 2011	60

[1]Type certificated as of January 14, 2011.

§ 26.23 Extended limit of validity.

(a) *Applicability.* Any person may apply to extend a limit of validity of the engineering data that supports the structural maintenance program (hereafter referred to as LOV) approved under § 25.571 of this subchapter, § 26.21, or this section. Extending an LOV is a major design change. The applicant must comply with the relevant provisions of subparts D or E of part 21 of this subchapter and paragraph (b) of this section.

(b) *Extended limit of validity.* Each person applying for an extended LOV must comply with the following requirements:

(1) Establish an extended LOV that corresponds to the period of time, stated as a number of total accumulated flight cycles or flight hours or both, during which it is demonstrated that widespread fatigue damage will not occur in the airplane. This demonstration must include an evaluation of airplane structural configurations and be supported by test evidence and analysis at a minimum and, if available, service experience, or service experience and teardown inspection results, of high-time airplanes of similar structural design, accounting for differences in operating conditions and procedures. The airplane structural configurations to be evaluated include—

(i) All model variations and derivatives approved under the type certificate for which approval for an extension is sought; and

(ii) All structural modifications to and replacements for the airplane structural configurations specified in paragraph (b)(1)(i) of this section, mandated by airworthiness directive, up to the date of approval of the extended LOV.

(2) Establish a revision or supplement, as applicable, to the Airworthiness Limitations section (ALS) of the Instructions for Continued Airworthiness required by § 25.1529 of this subchapter, and submit it to the FAA Oversight Office for approval. The revised ALS or supplement to the ALS must include the applicable extended LOV established under paragraph (b)(1) of this section.

(3) Develop the maintenance actions determined by the WFD evaluation performed in paragraph (b)(1) of this section to be necessary to preclude WFD from occurring before the airplane reaches the proposed extended LOV. These maintenance actions must be documented as airworthiness limitation items in the ALS and submitted to the FAA Oversight Office for approval.

Subpart D—Fuel Tank Flammability

Source: Docket No. FAA-2005-22997, 73 FR 42499, July 21, 2008, unless otherwise noted.

§ 26.31 Definitions.

For purposes of this subpart—

(a) *Fleet Average Flammability Exposure* has the meaning defined in Appendix N of part 25 of this chapter.

(b) *Normally Emptied* means a fuel tank other than a Main Fuel Tank. Main Fuel Tank is defined in 14 CFR 25.981(b).

§ 26.33 Holders of type certificates: Fuel tank flammability.

(a) *Applicability.* This section applies to U.S. type certificated transport category, turbine-powered airplanes, other than those designed solely for all-cargo operations, for which the State of Manufacture issued the original certificate of airworthiness or export airworthiness approval on or after January 1, 1992, that, as a result of original type certification or later increase in capacity have:

(1) A maximum type-certificated passenger capacity of 30 or more, or

(2) A maximum payload capacity of 7,500 pounds or more.

(b) *Flammability Exposure Analysis.* (1) *General.* Within 150 days after December 26, 2008, holders of type certificates must submit for approval to the FAA Oversight Office a flammability exposure analysis of all fuel tanks defined in the type design, as well as all design variations approved under the type certificate that affect flammability exposure. This analysis must be conducted in accordance with Appendix N of part 25 of this chapter.

(2) *Exception.* This paragraph (b) does not apply to—

(i) Fuel tanks for which the type certificate holder has notified the FAA under paragraph (g) of this section that it will provide design changes and service instructions for Flammability Reduction Means or an Ignition Mitigation Means (IMM) meeting the requirements of paragraph (c) of this section.

(ii) Fuel tanks substantiated to be conventional unheated aluminum wing tanks.

(c) *Design Changes.* For fuel tanks with a Fleet Average Flammability Exposure exceeding 7 percent, one of the following design changes must be made.

(1) *Flammability Reduction Means (FRM).* A means must be provided to reduce the fuel tank flammability.

(i) Fuel tanks that are designed to be Normally Emptied must meet the flammability exposure criteria of Appendix M of part 25 of this chapter if any portion of the tank is located within the fuselage contour.

(ii) For all other fuel tanks, the FRM must meet all of the requirements of Appendix M of part 25 of this chapter, except, instead of complying with paragraph M25.1 of this appendix, the Fleet Average Flammability Exposure may not exceed 7 percent.

(2) *Ignition Mitigation Means (IMM).* A means must be provided to mitigate the effects of an ignition of fuel vapors within the fuel tank such that no damage caused by an ignition will prevent continued safe flight and landing.

(d) *Service Instructions.* No later than December 27, 2010, holders of type certificates required by paragraph (c) of this section to make design changes must meet the requirements specified in either paragraph (d)(1) or (d)(2) of this section. The required service instructions must identify each airplane subject to the applicability provisions of paragraph (a) of this section.

(1) *FRM.* The type certificate holder must submit for approval by the FAA Oversight Office design changes and service instructions for installation of fuel tank flammability reduction means (FRM) meeting the criteria of paragraph (c) of this section.

(2) *IMM.* The type certificate holder must submit for approval by the FAA Oversight Office design changes and service instructions for installation of fuel tank IMM that comply with 14 CFR 25.981(c) in effect on December 26, 2008.

(e) *Instructions for Continued Airworthiness (ICA).* No later than December 27, 2010, holders of type certificates required by paragraph (c) of this section to make design changes must submit for approval by the FAA Oversight Office, critical design configuration control limitations (CDCCL), inspections, or other procedures to prevent increasing the flammability exposure of any tanks equipped with FRM above that permitted under paragraph (c)(1) of this section and to prevent degradation of the performance of any IMM provided under paragraph (c)(2) of this section. These CDCCL, inspections, and procedures must be included in the Airworthiness Limitations Section (ALS) of the ICA required by 14 CFR 25.1529 or paragraph (f) of this section. Unless shown to be impracticable, visible means to identify critical features of the design must be placed in areas of the airplane where foreseeable maintenance actions, repairs, or alterations may compromise the critical design configuration limitations. These visible means must also be identified as a CDCCL.

(f) *Airworthiness Limitations.* Unless previously accomplished, no later than December 27, 2010, holders of type certificates affected by this section must establish an ALS of the maintenance manual or ICA for each airplane configuration evaluated under paragraph (b)(1) of this section and submit it to the FAA Oversight Office for approval. The ALS must include a section that contains the CDCCL, inspections, or other procedures developed under paragraph (e) of this section.

(g) *Compliance Plan for Flammability Exposure Analysis.* Within 90 days after December 26, 2008, each holder of a type certificate required to comply with paragraph (b) of this section must submit to the FAA Oversight Office a compliance plan consisting of the following:

(1) A proposed project schedule for submitting the required analysis, or a determination that compliance with paragraph (b) of this section is not required because design changes and service instructions for FRM or IMM will be developed and made available as required by this section.

(2) A proposed means of compliance with paragraph (b) of this section, if applicable.

(h) *Compliance Plan for Design Changes and Service Instructions.* Within 210 days after December 26, 2008, each holder of a type certificate required to comply with paragraph (d) of this section must submit to the FAA Oversight Office a compliance plan consisting of the following:

(1) A proposed project schedule, identifying all major milestones, for meeting the compliance dates specified in paragraphs (d), (e) and (f) of this section.

(2) A proposed means of compliance with paragraphs (d), (e) and (f) of this section.

(3) A proposal for submitting a draft of all compliance items required by paragraphs (d), (e) and (f) of this section for review by the FAA Oversight Office not less than 60 days before the compliance times specified in those paragraphs.

(4) A proposal for how the approved service information and any necessary modification parts will be made available to affected persons.

(i) Each affected type certificate holder must implement the compliance plans, or later revisions, as approved under paragraph (g) and (h) of this section.

[Doc. No. FAA-2005-22997, 73 FR 42499, July 21, 2008, as amended by Amdt. 26-3, 74 FR 31619, July 2, 2009]

§ 26.35 Changes to type certificates affecting fuel tank flammability.

(a) *Applicability.* This section applies to holders and applicants for approvals of the following design changes to any airplane subject to 14 CFR 26.33(a):

(1) Any fuel tank designed to be Normally Emptied if the fuel tank installation was approved pursuant to a supplemental type certificate or a field approval before December 26, 2008;

(2) Any fuel tank designed to be Normally Emptied if an application for a supplemental type certificate or an amendment to a type certificate was made before December 26, 2008 and if the approval was not issued before December 26, 2008; and

(3) If an application for a supplemental type certificate or an amendment to a type certificate is made on or after December 26, 2008, any of the following design changes:

(i) Installation of a fuel tank designed to be Normally Emptied,

(ii) Changes to existing fuel tank capacity, or

(iii) Changes that may increase the flammability exposure of an existing fuel tank for which FRM or IMM is required by § 26.33(c).

(b) *Flammability Exposure Analysis* —(1) *General.* By the times specified in paragraphs (b)(1)(i) and (b)(1)(ii) of this section, each person subject to this section must submit for approval a flammability exposure analysis of the auxiliary fuel tanks or other affected fuel tanks, as defined in the type design, to the FAA Oversight Office. This analysis must be conducted in accordance with Appendix N of part 25 of this chapter.

(i) Holders of supplemental type certificates and field approvals: Within 12 months of December 26, 2008,

(ii) Applicants for supplemental type certificates and for amendments to type certificates: Within 12 months after December 26, 2008, or before the certificate is issued, whichever occurs later.

(2) *Exception.* This paragraph does not apply to—

(i) Fuel tanks for which the type certificate holder, supplemental type certificate holder, or field approval holder has notified the FAA under paragraph (f) of this section that it will provide design changes and service instructions for an IMM meeting the requirements of § 25.981(c) in effect December 26, 2008; and

(ii) Fuel tanks substantiated to be conventional unheated aluminum wing tanks.

(c) *Impact Assessment.* By the times specified in paragraphs (c)(1) and (c)(2) of this section, each person subject to paragraph (a)(1) of this section holding an approval for installation of a Normally Emptied fuel tank on an airplane model listed in Table 1 of this section, and each person subject to paragraph (a)(3)(iii) of this section, must submit for approval to the FAA Oversight Office an assessment of the fuel tank system, as modified by their design change. The assessment must identify any features of the design change that compromise any critical design configuration control limitation (CDCCL) applicable to any airplane on which the design change is eligible for installation.

(1) Holders of supplemental type certificates and field approvals: Before June 26, 2011.

(2) Applicants for supplemental type certificates and for amendments to type certificates: Before June 26, 2011 or before the certificate is issued, whichever occurs later.

Table 1

Model—Boeing
747 Series
737 Series
777 Series
767 Series
757 Series
Model—Airbus
A318, A319, A320, A321 Series
A300, A310 Series
A330, A340 Series

(d) *Design Changes and Service Instructions.* By the times specified in paragraph (e) of this section, each person subject to this section must meet the requirements of paragraphs (d)(1) or (d)(2) of this section, as applicable.

(1) For holders and applicants subject to paragraph (a)(1) or (a)(3)(iii) of this section, if the assessment required by paragraph (c) of this section identifies any features of the design change that compromise any CDCCL applicable to any airplane on which the design change is eligible for installation, the holder or applicant must submit for approval by the FAA Oversight Office design changes and service instructions for Flammability Impact Mitigation Means (FIMM) that would bring the design change into compliance with the CDCCL. Any fuel tank modified as required by this paragraph must also be evaluated as required by paragraph (b) of this section.

(2) Applicants subject to paragraph (a)(2), or (a)(3)(i) of this section must comply with the requirements of 14 CFR 25.981, in effect on December 26, 2008.

(3) Applicants subject to paragraph (a)(3)(ii) of this section must comply with the requirements of 14 CFR 26.33.

(e) *Compliance Times for Design Changes and Service Instructions.* The following persons subject to this section must comply with the requirements of paragraph (d) of this section at the specified times.

(1) Holders of supplemental type certificates and field approvals: Before December 26, 2012.

(2) Applicants for supplemental type certificates and for amendments to type certificates; Before December 26, 2012, or before the certificate is issued, whichever occurs later.

(f) *Compliance Planning.* By the applicable date specified in Table 2 of this section, each person subject to paragraph (a)(1) of this section must submit for approval by the FAA Oversight Office compliance plans for the flammability exposure analysis required by paragraph (b) of this section, the impact assessment required by paragraph (c) of this section, and the design changes and service instructions required by paragraph (d) of this section. Each person's compliance plans must include the following:

(1) A proposed project schedule for submitting the required analysis or impact assessment.

(2) A proposed means of compliance with paragraph (d) of this section.

(3) For the requirements of paragraph (d) of this section, a proposal for submitting a draft of all design changes, if any are required, and Airworthiness Limitations (including CDCCLs) for review by the FAA Oversight Office not less than 60 days before the compliance time specified in paragraph (e) of this section.

(4) For the requirements of paragraph (d) of this section, a proposal for how the approved service information and any necessary modification parts will be made available to affected persons.

Table 2—Compliance Planning Dates

	Flammability exposure analysis plan	Impact assessment plan	Design changes and service instructions plan
STC and Field Approval Holders	March 26, 2009	February 26, 2011	August 26, 2011.

(g) Each person subject to this section must implement the compliance plans, or later revisions, as approved under paragraph (f) of this section.

[Doc. No. FAA-2005-22997, 73 FR 42499, July 21, 2008, as amended by Amdt. 26-3, 74 FR 31619, July 2, 2009]

§ 26.37 Pending type certification projects: Fuel tank flammability.

(a) *Applicability.* This section applies to any new type certificate for a transport category airplane, if the application was made before December 26, 2008, and if the certificate was not issued before December 26, 2008. This section applies only if the airplane would have—

(1) A maximum type-certificated passenger capacity of 30 or more, or

(2) A maximum payload capacity of 7,500 pounds or more.

Table 1

Model—Boeing
747 Series
737 Series
777 Series
767 Series

(b) If the application was made on or after June 6, 2001, the requirements of 14 CFR 25.981 in effect on December 26, 2008, apply.

[Doc. No. FAA-2005-22997, 73 FR 42499, July 21, 2008, as amended by Amdt. 26-3, 74 FR 31619, July 2, 2009]

§ 26.39 Newly produced airplanes: Fuel tank flammability.

(a) *Applicability:* This section applies to Boeing model airplanes specified in Table 1 of this section, including passenger and cargo versions of each model, when application is made for original certificates of airworthiness or export airworthiness approvals after December 27, 2010.

(b) Any fuel tank meeting all of the criteria stated in paragraphs (b)(1), (b)(2) and (b)(3) of this section must have flammability reduction means (FRM) or ignition mitigation means (IMM) that meet the requirements of 14 CFR 25.981 in effect on December 26, 2008.

(1) The fuel tank is Normally Emptied.

(2) Any portion of the fuel tank is located within the fuselage contour.

(3) The fuel tank exceeds a Fleet Average Flammability Exposure of 7 percent.

(c) All other fuel tanks that exceed an Fleet Average Flammability Exposure of 7 percent must have an IMM that meets 14 CFR 25.981(d) in effect on December 26, 2008, or an FRM that meets all of the requirements of Appendix M to this part, except instead of complying with paragraph M25.1 of that appendix, the Fleet Average Flammability Exposure may not exceed 7 percent.

[Doc. No. FAA-2005-22997, 73 FR 42499, July 21, 2008, as amended by Amdt. 26-3, 74 FR 31619, July 2, 2009]

Subpart E—Aging Airplane Safety—Damage Tolerance Data for Repairs and Alterations

SOURCE: Docket No. FAA-2005-21693, 72 FR 70505, Dec. 12, 2007, unless otherwise noted.

§ 26.41 Definitions.

Affects (or Affected) means structure has been physically repaired, altered, or modified, or the structural loads acting on the structure have been increased or redistributed.

Baseline structure means structure that is designed under the original type certificate or amended type certificate for that airplane model.

Damage Tolerance Evaluation (DTE) means a process that leads to a determination of maintenance actions necessary to detect or preclude fatigue cracking that could contribute to a catastrophic failure. As applied to repairs and alterations, a DTE includes the evaluation both of the repair or alteration and of the fatigue critical structure affected by the repair or alteration.

Damage Tolerance Inspection (DTI) means the inspection developed as a result of a DTE. A DTI includes the areas to be inspected, the inspection method, the inspection procedures, including acceptance and rejection criteria, the threshold, and any repeat intervals associated with those inspections. The DTI may specify a time limit when a repair or alteration needs to be replaced or modified. If the DTE concludes that DT-based supplemental structural inspections are not necessary, the DTI contains a statement to that effect.

DT data mean DTE documentation and the DTI.

DTE documentation means data that identify the evaluated fatigue critical structure, the basic assumptions applied in a DTE, and the results of a DTE.

Fatigue critical structure means airplane structure that is susceptible to fatigue cracking that could contribute to a catastrophic failure, as determined in accordance with § 25.571 of this chapter. Fatigue critical structure includes structure, which, if repaired or altered, could be susceptible to fatigue cracking and contribute to a catastrophic failure. Such structure may be part of the baseline structure or part of an alteration.

Implementation schedule consists of documentation that establishes the timing for accomplishing the necessary actions for developing DT data for repairs and alterations, and for incorporating those data into an operator's continuing airworthiness maintenance program. The documentation must identify times when actions must be taken as specific numbers of airplane flight hours, flight cycles, or both.

Published repair data mean instructions for accomplishing repairs, which are published for general use in structural repair manuals and service bulletins (or equivalent types of documents).

§ 26.43 Holders of and applicants for type certificates—Repairs.

(a) *Applicability.* Except as specified in paragraph (g) of this section, this section applies to transport category, turbine powered airplane models with a type certificate issued after January 1, 1958, that as a result of original type certification or later increase in capacity have—

(1) A maximum type certificated passenger seating capacity of 30 or more; or

(2) A maximum payload capacity of 7,500 pounds or more.

(b) *List of fatigue critical baseline structure.* For airplanes specified in paragraph (a) of this section, the holder of or applicant for a type certificate must—

(1) Identify fatigue critical baseline structure for all airplane model variations and derivatives approved under the type certificate; and

(2) Develop and submit to the FAA Oversight Office for review and approval, a list of the structure identified under paragraph (b)(1) of this section and, upon approval, make the list available to persons required to comply with § 26.47 and §§ 121.1109 and 129.109 of this chapter.

(c) *Existing and future published repair data.* For repair data published by a holder of a type certificate that is current as of January 11, 2008 and for all later published repair data, the holder of a type certificate must—

(1) Review the repair data and identify each repair specified in the data that affects fatigue critical baseline structure identified under paragraph (b)(1) of this section;

(2) Perform a DTE and develop the DTI for each repair identified under paragraph (c)(1) of this section, unless previously accomplished;

(3) Submit the DT data to the FAA Oversight Office or its properly authorized designees for review and approval; and

(4) Upon approval, make the DTI available to persons required to comply with §§ 121.1109 and 129.109 of this chapter.

(d) *Future repair data not published.* For repair data developed by a holder of a type certificate that are approved after January 11, 2008 and are not published, the type certificate holder must accomplish the following for repairs specified in the repair data that affect fatigue critical baseline structure:

(1) Perform a DTE and develop the DTI.

(2) Submit the DT data required in paragraph (d)(1) of this section for review and approval by the FAA Oversight Office or its properly authorized designees.

(3) Upon approval, make the approved DTI available to persons required to comply with §§ 121.1109 and 129.109 of this chapter.

(e) *Repair evaluation guidelines.* Except for airplane models whose type certificate is issued after January 11, 2008, holders of a type certificate for each airplane model subject to this section must—

(1) Develop repair evaluation guidelines for operators' use that include—

(i) A process for conducting surveys of affected airplanes that will enable identification and documentation of all existing repairs that affect fatigue critical baseline structure identified under paragraph (b)(1) of this section and § 26.45(b)(2);

(ii) A process that will enable operators to obtain the DTI for repairs identified under paragraph (e)(1)(i) of this section; and

(iii) An implementation schedule for repairs covered by the repair evaluation guidelines. The implementation schedule must identify times when actions must be taken as specific numbers of airplane flight hours, flight cycles, or both.

(2) Submit the repair evaluation guidelines to the FAA Oversight Office for review and approval.

(3) Upon approval, make the guidelines available to persons required to comply with §§ 121.1109 and 129.109 of this chapter.

(4) If the guidelines direct the operator to obtain assistance from the holder of a type certificate, make such assistance available in accordance with the implementation schedule.

(f) *Compliance times.* Holders of type certificates must submit the following to the FAA Oversight Office or its properly authorized designees for review and approval by the specified compliance time:

(1) The identified list of fatigue critical baseline structure required by paragraph (b)(2) of this section must be submitted no later than 180 days after January 11, 2008 or before issuance of the type certificate, whichever occurs later.

(2) For published repair data that are current as of January 11, 2008, the DT data required by paragraph (c)(3) of this section must be submitted by June 30, 2009.

(3) For repair data published after January 11, 2008, the DT data required by paragraph (c)(3) of this section must be submitted before FAA approval of the repair data.

(4) For unpublished repair data developed after January 11, 2008, the DT data required by paragraph (d)(1) of this section must be submitted within 12 months of the airplane's return to service or in accordance with a schedule approved by the FAA Oversight Office.

(5) The repair evaluation guidelines required by paragraph (e)(1) of this section must be submitted by December 30, 2009.

(g) *Exceptions.* The requirements of this section do not apply to the following transport category airplane models:

(1) Convair CV-240, 340, 440, if modified to include turbine engines.

(2) Vickers Armstrong Viscount, TCDS No. A-814.

(3) Douglas DC-3, if modified to include turbine engines, TCDS No. A-618.

(4) Bombardier CL-44, TCDS No. 1A20.

(5) Mitsubishi YS-11, TCDS No. A1PC.

(6) British Aerospace BAC 1-11, TCDS No. A5EU.

(7) Concorde, TCDS No. A45EU.

(8) deHavilland D.H. 106 Comet 4C, TCDS No. 7A10.

(9) deHavilland DHC-7, TCDS No. A20EA.

(10) VFW-Vereinigte Flugtechnische Werk VFW-614, TCDS No. A39EU.

(11) Illyushin Aviation IL 96T, TCDS No. A54NM.

(12) Bristol Aircraft Britannia 305, TCDS No. 7A2.

(13) Handley Page Herald Type 300, TCDS No. A21N.

(14) Avions Marcel Dassault—Breguet Aviation Mercure 100C, TCDS No. A40EU.

(15) Airbus Caravelle, TCDS No. 7A6.

(16) Lockheed L-300, TCDS No. A2SO.

(17) Boeing 707-100/-200, TCDS No. 4A21.

(18) Boeing 707-300/-400, TCDS No. 4A26.

(19) Boeing 720, TCDS No. 4A28.

[Doc. No. FAA-2005-21693, 72 FR 70505, Dec. 12, 2007, as amended by Amdt. 26-4, 75 FR 11734, Mar. 12, 2010]

§ 26.45 Holders of type certificates—Alterations and repairs to alterations.

(a) *Applicability.* This section applies to transport category airplanes subject to § 26.43.

(b) *Fatigue critical alteration structure.* For existing and future alteration data developed by the holder of a type certificate, the holder must—

(1) Review alteration data and identify all alterations that affect fatigue critical baseline structure identified under § 26.43(b)(1);

(2) For each alteration identified under paragraph (b)(1) of this section, identify any fatigue critical alteration structure;

(3) Develop and submit to the FAA Oversight Office for review and approval a list of the structure identified under paragraph (b)(2) of this section; and

(4) Upon approval, make the list required in paragraph (b)(3) of this section available to persons required to comply with §§ 121.1109 and 129.109 of this chapter.

(c) *DT Data.* For existing and future alteration data developed by the holder of a type certificate that affect fatigue critical baseline structure identified under § 26.43(b)(1), unless previously accomplished, the holder must—

(1) Perform a DTE and develop the DTI for the alteration and fatigue critical baseline structure that is affected by the alteration;

(2) Submit the DT data developed in accordance with paragraphs (c)(1) of this section to the FAA Oversight Office or its properly authorized designees for review and approval; and

(3) Upon approval, make the DTI available to persons required to comply with §§ 121.1109 and 129.109 of this chapter.

(d) *DT Data for Repairs Made to Alterations.* For existing and future repair data developed by a holder of a type certificate, the type certificate holder must—

(1) Review the repair data, and identify each repair that affects any fatigue critical alteration structure identified under paragraph (b)(2) of this section;

(2) For each repair identified under paragraph (d)(1) of this section, unless previously accomplished, perform a DTE and develop DTI;

(3) Submit the DT data developed in accordance with paragraph (d)(2) of this section to the FAA Oversight Office or its properly authorized designees for review and approval; and

(4) Upon approval, make the DTI available to persons required to comply with §§ 121.1109 and 129.109 of this chapter.

(e) *Compliance times.* Holders of type certificates must submit the following to the FAA Oversight Office or its properly authorized designees for review and approval by the specified compliance time:

(1) The list of fatigue critical alteration structure identified under paragraph (b)(3) of this section must be submitted—

(i) No later than 360 days after January 11, 2008, for alteration data approved before January 11, 2008.

(ii) No later than 30 days after March 12, 2010 or before initial approval of the alteration data, whichever occurs later, for alteration data approved on or after January 11, 2008.

(2) For alteration data developed and approved before January 11, 2008, the DT data required by paragraph (c)(2) of this section must be submitted by June 30, 2009.

(3) For alteration data approved on or after January 11, 2008, DT data required by paragraph (c)(2) of this section must be submitted before initial approval of the alteration data.

(4) For repair data developed and approved before January 11, 2008, the DT data required by paragraph (d)(2) of this section must be submitted by June 30, 2009.

(5) For repair data developed and approved after January 11, 2008, the DT data required by paragraph (d)(2) of this section must be submitted within 12 months after initial approval of the repair data and before making the DT data available to persons required to comply with §§ 121.1109 and 129.109 of this chapter.

[Doc. No. FAA-2005-21693, 72 FR 70505, Dec. 12, 2007, as amended by Amdt. 26-4, 75 FR 11734, Mar. 12, 2010]

§ 26.47 Holders of and applicants for a supplemental type certificate—Alterations and repairs to alterations.

(a) *Applicability.* This section applies to transport category airplanes subject to § 26.43.

(b) *Fatigue critical alteration structure.* For existing structural alteration data approved under a supplemental certificate, the holder of the supplemental certificate must—

(1) Review the alteration data and identify all alterations that affect fatigue critical baseline structure identified under § 26.43(b)(1);

(2) For each alteration identified under paragraph (b)(1) of this section, identify any fatigue critical alteration structure;

(3) Develop and submit to the FAA Oversight Office for review and approval a list of the structure identified under paragraph (b)(2) of this section; and

(4) Upon approval, make the list required in paragraph (b)(3) of this section available to persons required to comply with §§ 121.1109 and 129.109 of this chapter.

(c) *DT Data.* For existing and future alteration data developed by the holder of a supplemental type certificate that affect fatigue critical baseline structure identified under § 26.43(b)(1), unless previously accomplished, the holder of a supplemental type certificate must—

(1) Perform a DTE and develop the DTI for the alteration and fatigue critical baseline structure that is affected by the alteration;

(2) Submit the DT data developed in accordance with paragraphs (c)(1) of this section to the FAA Oversight Office or its properly authorized designees for review and approval; and

(3) Upon approval, make the DTI available to persons required to comply with §§ 121.1109 and 129.109 of this chapter.

(d) *DT Data for Repairs Made to Alterations.* For existing and future repair data developed by the holder of a supplemental holder of a supplemental type certificate, the holder of a supplemental type certificate must—

(1) Review the repair data, and identify each repair that affects any fatigue critical alteration structure identified under paragraph (b)(2) of this section;

(2) For each repair identified under paragraph (d)(1) of this section, unless previously accomplished, perform a DTE and develop DTI;

(3) Submit the DT data developed in accordance with paragraph (d)(2) of this section to the FAA Oversight Office or its properly authorized designees for review and approval; and

(4) Upon approval, make the DTI available to persons required to comply with §§ 121.1109 and 129.109 of this chapter.

(e) *Compliance times.* Holders of supplemental type certificates must submit the following to the FAA Oversight Office or its properly authorized designees for review and approval by the specified compliance time:

(1) The list of fatigue critical alteration structure required by paragraph (b)(3) of this section must be submitted no later than 360 days after January 11, 2008.

(2) For alteration data developed and approved before January 11, 2008, the DT data required by paragraph (c)(2) of this section must be submitted by June 30, 2009.

(3) For alteration data developed after January 11, 2008, the DT data required by paragraph (c)(2) of this section must be submitted before approval of the alteration data and making it available to persons required to comply with §§ 121.1109 and 129.109 of this chapter.

(4) For repair data developed and approved before January 11, 2008, the DT data required by paragraph (d)(2) of this section must be submitted by June 30, 2009.

(5) For repair data developed and approved after January 11, 2008, the DT data required by paragraph (d)(2) of this section, must be submitted within 12 months after initial approval of the repair data and before making the DT data available to persons required to comply with §§ 121.1109 and 129.109 of this chapter.

§ 26.49 Compliance plan.

(a) *Compliance plan.* Except for applicants for type certificates and supplemental type certificates whose applications are submitted after January 11, 2008, each person identified in §§ 26.43, 26.45, and 26.47, must submit a compliance plan consisting of the following:

(1) A project schedule identifying all major milestones for meeting the compliance times specified in §§ 26.43(f), 26.45(e), and 26.47(e), as applicable.

(2) A proposed means of compliance with §§ 26.43, 26.45, and 26.47, as applicable.

(3) A plan for submitting a draft of all compliance items required by this subpart for review by the FAA Oversight Office not less than 60 days before the applicable compliance date.

(b) *Compliance dates for compliance plans.* The following persons must submit the compliance plan described in paragraph (a) of this section to the FAA Oversight Office for approval on the following schedule:

(1) For holders of type certificates, no later than 90 days after January 11, 2008.

(2) For holders of supplemental type certificates no later than 180 days after January 11, 2008.

(3) For applicants for changes to type certificates whose application are submitted before January 11, 2008, no later than 180 days after January 11, 2008.

(c) *Compliance Plan Implementation.* Each affected person must implement the compliance plan as approved in compliance with paragraph (a) of this section.

PART 27

AIRWORTHINESS STANDARDS: NORMAL CATEGORY ROTORCRAFT

Contents

Subpart A—General

Subpart B—Flight

General

Performance

Flight Characteristics

Ground and Water Handling Characteristics

Miscellaneous Flight Requirements

Subpart C—Strength Requirements

General

Flight Loads

Control Surface and System Loads

Ground Loads

Water Loads

Main Component Requirements

Emergency Landing Conditions

Fatigue Evaluation

Subpart D—Design and Construction

General

Rotors

Control Systems

Landing Gear

Floats and Hulls

Personnel and Cargo Accommodations

Fire Protection

External Loads

Miscellaneous

Subpart E—Powerplant

General

Rotor Drive System

Fuel System

Fuel System Components

Oil System

Cooling

Induction System

Exhaust System

Powerplant Controls and Accessories

AUTHORITY: 49 U.S.C. 106(g), 40113, 44701-44702, 44704.

SOURCE: Docket No. 5074, 29 FR 15695, Nov. 24, 1964, unless otherwise noted.

Subpart A—General

§ 27.1 Applicability.

(a) This part prescribes airworthiness standards for the issue of type certificates, and changes to those certificates, for normal category rotorcraft with maximum weights of 7,000 pounds or less and nine or less passenger seats.

(b) Each person who applies under Part 21 for such a certificate or change must show compliance with the applicable requirements of this part.

(c) Multiengine rotorcraft may be type certified as Category A provided the requirements referenced in appendix C of this part are met.

[Doc. No. 5074, 29 FR 15695, Nov. 24, 1964, as amended by Amdt. 27-33, 61 FR 21906, May 10, 1996; Amdt. 27-37, 64 FR 45094, Aug. 18, 1999]

§ 27.2 Special retroactive requirements.

(a) For each rotorcraft manufactured after September 16, 1992, each applicant must show that each occupant's seat is equipped with a safety belt and shoulder harness that meets the requirements of paragraphs (a), (b), and (c) of this section.

(1) Each occupant's seat must have a combined safety belt and shoulder harness with a single-point release. Each pilot's combined safety belt and shoulder harness must allow each pilot, when seated with safety belt and shoulder harness fastened, to perform all functions necessary for flight operations. There must be a means to secure belts and harnesses, when not in use, to prevent interference with the operation of the rotorcraft and with rapid egress in an emergency.

(2) Each occupant must be protected from serious head injury by a safety belt plus a shoulder harness that will prevent the head from contacting any injurious object.

(3) The safety belt and shoulder harness must meet the static and dynamic strength requirements, if applicable, specified by the rotorcraft type certification basis.

(4) For purposes of this section, the date of manufacture is either—

(i) The date the inspection acceptance records, or equivalent, reflect that the rotorcraft is complete and meets the FAA-Approved Type Design Data; or

(ii) The date the foreign civil airworthiness authority certifies that the rotorcraft is complete and issues an original standard airworthiness certificate, or equivalent, in that country.

(b) For rotorcraft with a certification basis established prior to October 18, 1999—

(1) The maximum passenger seat capacity may be increased to eight or nine provided the applicant shows compliance with all the airworthiness requirements of this part in effect on October 18, 1999.

(2) The maximum weight may be increased to greater than 6,000 pounds provided—

(i) The number of passenger seats is not increased above the maximum number certificated on October 18, 1999, or

(ii) The applicant shows compliance with all of the airworthiness requirements of this part in effect on October 18, 1999.

[Doc. No. 26078, 56 FR 41051, Aug. 16, 1991, as amended by Amdt. 27-37, 64 FR 45094, Aug. 18, 1999]

Subpart B—Flight

GENERAL

§ 27.21 Proof of compliance.

Each requirement of this subpart must be met at each appropriate combination of weight and center of gravity within the range of loading conditions for which certification is requested. This must be shown—

(a) By tests upon a rotorcraft of the type for which certification is requested, or by calculations based on, and equal in accuracy to, the results of testing; and

(b) By systematic investigation of each required combination of weight and center of gravity if compliance cannot be reasonably inferred from combinations investigated.

[Doc. No. 5074, 29 FR 15695, Nov. 24, 1964, as amended by Amdt. 27-21, 49 FR 44432, Nov. 6, 1984]

§ 27.25 Weight limits.

(a) *Maximum weight.* The maximum weight (the highest weight at which compliance with each applicable requirement of this part is shown) must be established so that it is—

(1) Not more than—

(i) The highest weight selected by the applicant;

(ii) The design maximum weight (the highest weight at which compliance with each applicable structural loading condition of this part is shown);

(iii) The highest weight at which compliance with each applicable flight requirement of this part is shown; or

(iv) The highest weight in which the provisions of §§ 27.87 or 27.143(c)(1), or combinations thereof, are demonstrated if the weights and operating conditions (altitude and temperature) prescribed by those requirements cannot be met; and

(2) Not less than the sum of—

(i) The empty weight determined under § 27.29; and

(ii) The weight of usable fuel appropriate to the intended operation with full payload;

(iii) The weight of full oil capacity; and

(iv) For each seat, an occupant weight of 170 pounds or any lower weight for which certification is requested.

(b) *Minimum weight.* The minimum weight (the lowest weight at which compliance with each applicable requirement of this part is shown) must be established so that it is—

(1) Not more than the sum of—

(i) The empty weight determined under § 27.29; and

(ii) The weight of the minimum crew necessary to operate the rotorcraft, assuming for each crewmember a weight no more than 170 pounds, or any lower weight selected by the applicant or included in the loading instructions; and

(2) Not less than—

(i) The lowest weight selected by the applicant;

(ii) The design minimum weight (the lowest weight at which compliance with each applicable structural loading condition of this part is shown); or

(iii) The lowest weight at which compliance with each applicable flight requirement of this part is shown.

(c) *Total weight with jettisonable external load.* A total weight for the rotorcraft with a jettisonable external load attached that is greater than the maximum weight established under paragraph (a) of this section may be established for any rotorcraft-load combination if—

(1) The rotorcraft-load combination does not include human external cargo,

(2) Structural component approval for external load operations under either § 27.865 or under equivalent operational standards is obtained,

(3) The portion of the total weight that is greater than the maximum weight established under paragraph (a) of this section is made up only of the weight of all or part of the jettisonable external load,

(4) Structural components of the rotorcraft are shown to comply with the applicable structural requirements of this part under the increased loads and stresses caused by the weight increase over that established under paragraph (a) of this section, and

(5) Operation of the rotorcraft at a total weight greater than the maximum certificated weight established under paragraph (a) of this section is limited by appropriate operating limitations under § 27.865(a) and (d) of this part.

(Secs. 313(a), 601, 603, 604, and 605 of the Federal Aviation Act of 1958 (49 U.S.C. 1354(a), 1421, 1423, 1424, and 1425); and sec. 6(c) of the Dept. of Transportation Act (49 U.S.C. 1655(c)))

[Doc. No. 5074, 29 FR 15695, Nov. 29, 1964, as amended by Amdt. 27-11, 41 FR 55468, Dec. 20, 1976; Amdt. 25-42, 43 FR 2324, Jan. 16, 1978; Amdt. 27-36, 64 FR 43019, Aug. 6, 1999; Amdt. 27-44, 73 FR 10998, Feb. 29, 2008; 73 FR 33876, June 16, 2008]

§ 27.27 Center of gravity limits.

The extreme forward and aft centers of gravity and, where critical, the extreme lateral centers of gravity must be established for each weight established under § 27.25. Such an extreme may not lie beyond—

(a) The extremes selected by the applicant;

(b) The extremes within which the structure is proven; or

(c) The extremes within which compliance with the applicable flight requirements is shown.

[Amdt. 27-2, 33 FR 962, Jan. 26, 1968]

§ 27.29 Empty weight and corresponding center of gravity.

(a) The empty weight and corresponding center of gravity must be determined by weighing the rotorcraft without the crew and payload, but with—

(1) Fixed ballast;

(2) Unusable fuel; and

(3) Full operating fluids, including—

(i) Oil;

(ii) Hydraulic fluid; and

(iii) Other fluids required for normal operation of rotocraft systems, except water intended for injection in the engines.

(b) The condition of the rotorcraft at the time of determining empty weight must be one that is well defined and can be easily repeated, particularly with respect to the weights of fuel, oil, coolant, and installed equipment.

(Secs. 313(a), 601, 603, 604, and 605 of the Federal Aviation Act of 1958 (49 U.S.C. 1354(a), 1421, 1423, 1424, and 1425); and sec. 6(c) of the Dept. of Transportation Act (49 U.S.C. 1655(c)))

[Doc. No. 5074, 29 FR 15695, Nov. 24, 1964, as amended by Amdt. 27-14, 43 FR 2324, Jan. 16, 1978]

§ 27.31 Removable ballast.

Removable ballast may be used in showing compliance with the flight requirements of this subpart.

§ 27.33 Main rotor speed and pitch limits.

(a) *Main rotor speed limits.* A range of main rotor speeds must be established that—

(1) With power on, provides adequate margin to accommodate the variations in rotor speed occurring in any appropriate maneuver, and is consistent with the kind of governor or synchronizer used; and

(2) With power off, allows each appropriate autorotative maneuver to be performed throughout the ranges of airspeed and weight for which certification is requested.

(b) *Normal main rotor high pitch limits (power on).* For rotocraft, except helicopters required to have a main rotor low speed warning under paragraph (e) of this section. It must be shown, with power on and without exceeding approved engine maximum limitations, that main rotor speeds substantially less than the minimum approved main rotor speed will not occur under any sustained flight condition. This must be met by—

(1) Appropriate setting of the main rotor high pitch stop;

(2) Inherent rotorcraft characteristics that make unsafe low main rotor speeds unlikely; or

(3) Adequate means to warn the pilot of unsafe main rotor speeds.

(c) *Normal main rotor low pitch limits (power off).* It must be shown, with power off, that—

(1) The normal main rotor low pitch limit provides sufficient rotor speed, in any autorotative condition, under the most critical combinations of weight and airspeed; and

(2) It is possible to prevent overspeeding of the rotor without exceptional piloting skill.

(d) *Emergency high pitch.* If the main rotor high pitch stop is set to meet paragraph (b)(1) of this section, and if that stop cannot be exceeded inadvertently, additional pitch may be made available for emergency use.

(e) *Main rotor low speed warning for helicopters.* For each single engine helicopter, and each multiengine helicopter that does not have an approved device that automatically increases power on the operating engines when one engine fails, there must be a main rotor low speed warning which meets the following requirements:

(1) The warning must be furnished to the pilot in all flight conditions, including power-on and power-off flight, when the speed of a main rotor approaches a value that can jeopardize safe flight.

(2) The warning may be furnished either through the inherent aerodynamic qualities of the helicopter or by a device.

(3) The warning must be clear and distinct under all conditons, and must be clearly distinguishable from all other warnings. A visual device that requires the attention of the crew within the cockpit is not acceptable by itself.

(4) If a warning device is used, the device must automatically deactivate and reset when the low-speed condition is corrected. If the device has an audible warning, it must also be equipped with a means for the pilot to manually silence the audible warning before the low-speed condition is corrected.

(Secs. 313(a), 601, 603, 604, and 605 of the Federal Aviation Act of 1958 (49 U.S.C. 1354(a), 1421, 1423, 1424, and 1425); and sec. 6(c) of the Dept. of Transportation Act (49 U.S.C. 1655(c)))

[Doc. No. 5074, 29 FR 15695, Nov. 24, 1964, as amended by Amdt. 27-2, 33 FR 962, Jan. 26, 1968; Amdt. 27-14, 43 FR 2324, Jan. 16, 1978]

Performance

§ 27.45 General.

(a) Unless otherwise prescribed, the performance requirements of this subpart must be met for still air and a standard atmosphere.

(b) The performance must correspond to the engine power available under the particular ambient atmospheric conditions, the particular flight condition, and the relative humidity specified in paragraphs (d) or (e) of this section, as appropriate.

(c) The available power must correspond to engine power, not exceeding the approved power, less—

(1) Installation losses; and

(2) The power absorbed by the accessories and services appropriate to the particular ambient atmopheric conditions and the particular flight condition.

(d) For reciprocating engine-powered rotorcraft, the performance, as affected by engine power, must be based on a relative humidity of 80 percent in a standard atmosphere.

(e) For turbine engine-powered rotorcraft, the performance, as affected by engine power, must be based on a relative humidity of—

(1) 80 percent, at and below standard temperature; and

(2) 34 percent, at and above standard temperature plus 50 degrees F. Between these two temperatures, the relative humidity must vary linearly.

(f) For turbine-engine-powered rotorcraft, a means must be provided to permit the pilot to determine prior to takeoff that each engine is capable of developing the power necessary to achieve the applicable rotorcraft performance prescribed in this subpart.

(Secs. 313(a), 601, 603, 604, and 605 of the Federal Aviation Act of 1958 (49 U.S.C. 1354(a), 1421, 1423, 1424, and 1425); and sec. 6(c) of the Dept. of Transportation Act (49 U.S.C. 1655(c)))

[Amdt. 27-14, 43 FR 2324, Jan. 16, 1978, as amended by Amdt. 27-21, 49 FR 44432, Nov. 6, 1984]

§ 27.49 Performance at minimum operating speed.

(a) For helicopters—

(1) The hovering ceiling must be determined over the ranges of weight, altitude, and temperature for which certification is requested, with—

(i) Takeoff power;

(ii) The landing gear extended; and

(iii) The helicopter in-ground effect at a height consistent with normal takeoff procedures; and

(2) The hovering ceiling determined under paragraph (a)(1) of this section must be at least—

(i) For reciprocating engine powered helicopters, 4,000 feet at maximum weight with a standard atmosphere;

(ii) For turbine engine powered helicopters, 2,500 feet pressure altitude at maximum weight at a temperature of standard plus 22°C (standard plus 40°F).

(3) The out-of-ground effect hovering performance must be determined over the ranges of weight, altitude, and temperature for which certification is requested, using takeoff power.

(b) For rotorcraft other than helicopters, the steady rate of climb at the minimum operating speed must be determined over the ranges of weight, altitude, and temperature for which certification is requested, with—

(1) Takeoff power; and

(2) The landing gear extended.

[Amdt. 27-44, 73 FR 10998, Feb. 29, 2008]

§ 27.51 Takeoff.

The takeoff, with takeoff power and r.p.m. at the most critical center of gravity, and with weight from the maximum weight at sea level to the weight for which takeoff certification is requested for each altitude covered by this section—

(a) May not require exceptional piloting skill or exceptionally favorable conditions throughout the ranges of altitude from standard sea level conditions to the maximum altitude for which takeoff and landing certification is requested, and

(b) Must be made in such a manner that a landing can be made safely at any point along the flight path if an engine fails. This must be demonstrated up to the maximum altitude for which takeoff and landing certification is requested or 7,000 feet density altitude, whichever is less.

[Amdt. 27-44, 73 FR 10999, Feb. 29, 2008]

§ 27.65 Climb: all engines operating.

(a) For rotorcraft other than helicopters—

(1) The steady rate of climb, at V_Y, must be determined—

(i) With maximum continuous power on each engine;

(ii) With the landing gear retracted; and

(iii) For the weights, altitudes, and temperatures for which certification is requested; and

(2) The climb gradient, at the rate of climb determined in accordance with paragraph (a)(1) of this section, must be either—

(i) At least 1:10 if the horizontal distance required to take off and climb over a 50-foot obstacle is determined for each weight, altitude, and temperature within the range for which certification is requested; or

(ii) At least 1:6 under standard sea level conditions.

(b) Each helicopter must meet the following requirements:

(1) V_Y must be determined—

(i) For standard sea level conditions;

(ii) At maximum weight; and

(iii) With maximum continuous power on each engine.

(2) The steady rate of climb must be determined—

(i) At the climb speed selected by the applicant at or below V_{NE};

(ii) Within the range from sea level up to the maximum altitude for which certification is requested;

(iii) For the weights and temperatures that correspond to the altitude range set forth in paragraph (b)(2)(ii) of this section and for which certification is requested; and

(iv) With maximum continuous power on each engine.

(Secs. 313(a), 601, 603, 604, and 605 of the Federal Aviation Act of 1958 (49 U.S.C. 1354(a), 1421, 1423, 1424, and 1425); and sec. 6(c) of the Dept. of Transportation Act (49 U.S.C. 1655(c)))

[Doc. No. 5074, 29 FR 15695, Nov. 24, 1964, as amended by Amdt. 27-14, 43 FR 2324, Jan. 16, 1978; Amdt. 27-33, 61 FR 21907, May 10, 1996]

§ 27.67 Climb: one engine inoperative.

For multiengine helicopters, the steady rate of climb (or descent), at V_y (or at the speed for minimum rate of descent), must be determined with—

(a) Maximum weight;

(b) The critical engine inoperative and the remaining engines at either—

(1) Maximum continuous power and, for helicopters for which certification for the use of 30-minute OEI power is requested, at 30-minute OEI power; or

(2) Continuous OEI power for helicopters for which certification for the use of continuous OEI power is requested.

(Secs. 313(a), 601, 603, 604, and 605 of the Federal Aviation Act of 1958 (49 U.S.C. 1354(a), 1421, 1423, 1424, and 1425); and sec. 6(c) of the Dept. of Transportation Act (49 U.S.C. 1655(c)))

[Doc. No 5074, 29 FR 15695, Nov. 24, 1964, as amended by Amdt. 27-23, 53 FR 34210, Sept. 2, 1988]

§ 27.71 Autorotation performance.

For single-engine helicopters and multiengine helicopters that do not meet the Category A engine isolation requirements of Part 29 of this chapter, the minimum rate of descent airspeed and the best angle-of-glide airspeed must be determined in autorotation at—

(a) Maximum weight; and

(b) Rotor speed(s) selected by the applicant.

[Amdt. 27-21, 49 FR 44433, Nov. 6, 1984]

§ 27.75 Landing.

(a) The rotorcraft must be able to be landed with no excessive vertical acceleration, no tendency to bounce, nose over, ground loop, porpoise, or water loop, and without exceptional piloting skill or exceptionally favorable conditions, with—

(1) Approach or autorotation speeds appropriate to the type of rotorcraft and selected by the applicant;

(2) The approach and landing made with—

(i) Power off, for single engine rotorcraft and entered from steady state autorotation; or

(ii) One-engine inoperative (OEI) for multiengine rotorcraft, with each operating engine within approved operating limitations, and entered from an established OEI approach.

(b) Multiengine rotorcraft must be able to be landed safely after complete power failure under normal operating conditions.

[Doc. No. 5074, 29 FR 15695, Nov. 24, 1964, as amended by Amdt. 27-14, 43 FR 2324, Jan. 16, 1978; Amdt. 27-44, 73 FR 10999, Feb. 29, 2008]

§ 27.87 Height-speed envelope.

(a) If there is any combination of height and forward speed (including hover) under which a safe landing cannot be made under the applicable power failure condition in paragraph (b) of this section, a limiting height-speed envelope must be established (including all pertinent information) for that condition, throughout the ranges of—

(1) Altitude, from standard sea level conditions to the maximum altitude capability of the rotorcraft, or 7000 feet density altitude, whichever is less; and

(2) Weight, from the maximum weight at sea level to the weight selected by the applicant for each altitude covered by paragraph (a)(1) of this section. For helicopters, the weight at altitudes above sea level may not be less than the maximum weight or the highest weight allowing hovering out-of-ground effect, whichever is lower.

(b) The applicable power failure conditions are—

(1) For single-engine helicopters, full autorotation;

(2) For multiengine helicopters, OEI (where engine isolation features ensure continued operation of the remaining engines), and the remaining engine(s) within approved limits and at the minimum installed specification power available for the most critical combination of approved ambient temperature and pressure altitude resulting in 7000 feet density altitude or the maximum altitude capability of the helicopter, whichever is less, and

(3) For other rotorcraft, conditions appropriate to the type.

(Secs. 313(a), 601, 603, 604, Federal Aviation Act of 1958 (49 U.S.C. 1354(a), 1421, 1423, 1424), sec. 6(c), Dept. of Transportation Act (49 U.S.C. 1655(c)))

[Doc. No. 5074, 29 FR 15695, Nov. 24, 1964, as amended by Amdt. 27-14, 43 FR 2324, Jan. 16, 1978; Amdt. 27-21, 49 FR 44433, Nov. 6, 1984; Amdt. 27-44, 73 FR 10999, Feb. 29, 2008]

Flight Characteristics

§ 27.141 General.

The rotorcraft must—

(a) Except as specifically required in the applicable section, meet the flight characteristics requirements of this subpart—

(1) At the altitudes and temperatures expected in operation;

(2) Under any critical loading condition within the range of weights and centers of gravity for which certification is requested;

(3) For power-on operations, under any condition of speed, power, and rotor r.p.m. for which certification is requested; and

(4) For power-off operations, under any condition of speed and rotor r.p.m. for which certification is requested that is attainable with the controls rigged in accordance with the approved rigging instructions and tolerances;

(b) Be able to maintain any required flight condition and make a smooth transition from any flight condition to any other flight condition without exceptional piloting skill, alertness, or strength, and without danger of exceeding the limit load factor under any operating condition probable for the type, including—

(1) Sudden failure of one engine, for multiengine rotorcraft meeting Transport Category A engine isolation requirements of Part 29 of this chapter;

(2) Sudden, complete power failure for other rotorcraft; and

(3) Sudden, complete control system failures specified in § 27.695 of this part; and

(c) Have any additional characteristic required for night or instrument operation, if certification for those kinds of operation is requested. Requirements for helicopter instrument flight are contained in appendix B of this part.

[Doc. No. 5074, 29 FR 15695, Nov. 24, 1964, as amended by Amdt. 27-2, 33 FR 962, Jan. 26, 1968; Amdt. 27-11, 41 FR 55468, Dec. 20, 1976; Amdt. 27-19, 48 FR 4389, Jan. 31, 1983; Amdt. 27-21, 49 FR 44433, Nov. 6, 1984]

§ 27.143 Controllability and maneuverability.

(a) The rotorcraft must be safely controllable and maneuverable—

(1) During steady flight; and

(2) During any maneuver appropriate to the type, including—

(i) Takeoff;

(ii) Climb;

(iii) Level flight;

(iv) Turning flight;

(v) Autorotation;

(vi) Landing (power on and power off); and

(vii) Recovery to power-on flight from a balked autorotative approach.

(b) The margin of cyclic control must allow satisfactory roll and pitch control at V_{NE} with—

(1) Critical weight;

(2) Critical center of gravity;

(3) Critical rotor r.p.m.; and

(4) Power off (except for helicopters demonstrating compliance with paragraph (f) of this section) and power on.

(c) Wind velocities from zero to at least 17 knots, from all azimuths, must be established in which the rotorcraft can be operated without loss of control on or near the ground in any maneuver appropriate to the type (such as crosswind takeoffs, sideward flight, and rearward flight)—

(1) With altitude, from standard sea level conditions to the maximum takeoff and landing altitude capability of the rotorcraft or 7000 feet density altitude, whichever is less; with—

(i) Critical Weight;

(ii) Critical center of gravity;

(iii) Critical rotor r.p.m.;

(2) For takeoff and landing altitudes above 7000 feet density altitude with—

(i) Weight selected by the applicant;

(ii) Critical center of gravity; and

(iii) Critical rotor r.p.m.

(d) Wind velocities from zero to at least 17 knots, from all azimuths, must be established in which the rotorcraft can be operated without loss of control out-of-ground-effect, with—

(1) Weight selected by the applicant;

(2) Critical center of gravity;

(3) Rotor r.p.m. selected by the applicant; and

(4) Altitude, from standard sea level conditions to the maximum takeoff and landing altitude capability of the rotorcraft.

(e) The rotorcraft, after (1) failure of one engine in the case of multiengine rotorcraft that meet Transport Category A engine isolation requirements, or (2) complete engine failure in the case of other rotorcraft, must be controllable over the range of speeds and altitudes for which certification is requested when such power failure occurs with maximum continuous power and critical weight. No corrective action time delay for any condition following power failure may be less than—

(i) For the cruise condition, one second, or normal pilot reaction time (whichever is greater); and

(ii) For any other condition, normal pilot reaction time.

(f) For helicopters for which a V_{NE} (power-off) is established under § 27.1505(c), compliance must be demonstrated with the following requirements with critical weight, critical center of gravity, and critical rotor r.p.m.:

(1) The helicopter must be safely slowed to V_{NE} (power-off), without exceptional pilot skill, after the last operating engine is made inoperative at power-on V_{NE}.

(2) At a speed of 1.1 V_{NE} (power-off), the margin of cyclic control must allow satisfactory roll and pitch control with power off.

(Secs. 313(a), 601, 603, 604, and 605 of the Federal Aviation Act of 1958 (49 U.S.C. 1354(a), 1421, 1423, 1424, and 1425); and sec. 6(c) of the Dept. of Transportation Act (49 U.S.C. 1655(c)))

[Doc. No. 5074, 29 FR 15695, Nov. 24, 1964, as amended by Amdt. 27-2, 33 FR 963, Jan. 26, 1968; Amdt. 27-14, 43 FR 2325, Jan. 16, 1978; Amdt. 27-21, 49 FR 44433, Nov. 6, 1984; Amdt. 27-44, 73 FR 10999, Feb. 29, 2008]

§ 27.151 Flight controls.

(a) Longitudinal, lateral, directional, and collective controls may not exhibit excessive breakout force, friction, or preload.

(b) Control system forces and free play may not inhibit a smooth, direct rotorcraft response to control system input.

[Amdt. 27-21, 49 FR 44433, Nov. 6, 1984]

§ 27.161 Trim control.

The trim control—

(a) Must trim any steady longitudinal, lateral, and collective control forces to zero in level flight at any appropriate speed; and

(b) May not introduce any undesirable discontinuities in control force gradients.

[Doc. No. 5074, 29 FR 15695, Nov. 24, 1964, as amended by Amdt. 27-21, 49 FR 44433, Nov. 6, 1984]

§ 27.171 Stability: general.

The rotorcraft must be able to be flown, without undue pilot fatigue or strain, in any normal maneuver for a period of time as long as that expected in normal operation. At least three landings and takeoffs must be made during this demonstration.

§ 27.173 Static longitudinal stability.

(a) The longitudinal control must be designed so that a rearward movement of the control is necessary to obtain an airspeed less than the trim speed, and a forward movement of the control is necessary to obtain an airspeed more than the trim speed.

(b) Throughout the full range of altitude for which certification is requested, with the throttle and collective pitch held constant during the maneuvers specified in § 27.175(a) through (d), the slope of the control position versus airspeed curve must be positive. However, in limited flight conditions or modes of operation determined by the Administrator to be acceptable, the slope of the control position versus airspeed curve may be neutral or negative if the rotorcraft possesses flight characteristics that allow the pilot to maintain airspeed within±5 knots of the desired trim airspeed without exceptional piloting skill or alertness.

[Amdt. 27-21, 49 FR 44433, Nov. 6, 1984, as amended by Amdt. 27-44, 73 FR 10999, Feb. 29, 2008]

§ 27.175 Demonstration of static longitudinal stability.

(a) *Climb.* Static longitudinal stability must be shown in the climb condition at speeds from V_y - 10 kt to V_y+10 kt with—

(1) Critical weight;

(2) Critical center of gravity;

(3) Maximum continuous power;

(4) The landing gear retracted; and

(5) The rotorcraft trimmed at V_Y.

(b) *Cruise.* Static longitudinal stability must be shown in the cruise condition at speeds from 0.8 V_{NE} – 10 kt to 0.8 V_{NE}+10 kt or, if V_H is less than 0.8 V_{NE}, from V_H -10 kt to V_H+10 kt, with—

(1) Critical weight;

(2) Critical center of gravity;

(3) Power for level flight at 0.8 V_{NE} or V_H, whichever is less;

(4) The landing gear retracted; and

(5) The rotorcraft trimmed at 0.8 V_{NE} or V_H, whichever is less.

(c) V_{NE}. Static longitudinal stability must be shown at speeds from V_{NE} – 20 kt to V_{NE} with—

(1) Critical weight;

(2) Critical center of gravity;

(3) Power required for level flight at V_{NE} -10 kt or maximum continuous power, whichever is less;

(4) The landing gear retracted; and

(5) The rotorcraft trimmed at V_{NE} – 10 kt.

(d) *Autorotation.* Static longitudinal stability must be shown in autorotation at—

(1) Airspeeds from the minimum rate of descent airspeed-10 kt to the minimum rate of descent airspeed+10 kt, with—

(i) Critical weight;

(ii) Critical center of gravity;

(iii) The landing gear extended; and

PART 27

FAR

(iv) The rotorcraft trimmed at the minimum rate of descent airspeed.

(2) Airspeeds from best angle-of-glide airspeed−10 kt to the best angle-of-glide airspeed+10 kt, with—

(i) Critical weight;

(ii) Critical center of gravity;

(iii) The landing gear retracted; and

(iv) The rotorcraft trimmed at the best angle-of-glide airspeed.

(Secs. 313(a), 601, 603, 604, and 605 of the Federal Aviation Act of 1958 (49 U.S.C. 1354(a), 1421, 1423, 1424, and 1425); and sec. 6(c) of the Dept. of Transportation Act (49 U.S.C. 1655(c)))

[Doc. No. 5074, 29 FR 15695, Nov. 24, 1964, as amended by Amdt. 27-2, 33 FR 963, Jan. 26, 1968; Amdt. 27-11, 41 FR 55468, Dec. 20, 1976; Amdt. 27-14, 43 FR 2325, Jan. 16, 1978; Amdt. 27-21, 49 FR 44433, Nov. 6, 1984; Amdt. 27-34, 62 FR 46173, Aug. 29, 1997; Amdt. 27-44, 73 FR 10999, Feb. 29, 2008]

§ 27.177 Static directional stability.

(a) The directional controls must operate in such a manner that the sense and direction of motion of the rotorcraft following control displacement are in the direction of the pedal motion with the throttle and collective controls held constant at the trim conditions specified in § 27.175(a), (b), and (c). Sideslip angles must increase with steadily increasing directional control deflection for sideslip angles up to the lesser of—

(1)±25 degrees from trim at a speed of 15 knots less than the speed for minimum rate of descent varying linearly to±10 degrees from trim at V_{NE};

(2) The steady state sideslip angles established by § 27.351;

(3) A sideslip angle selected by the applicant, which corresponds to a sideforce of at least 0.1g; or

(4) The sideslip angle attained by maximum directional control input.

(b) Sufficient cues must accompany the sideslip to alert the pilot when the aircraft is approaching the sideslip limits.

(c) During the maneuver specified in paragraph (a) of this section, the sideslip angle versus directional control position curve may have a negative slope within a small range of angles around trim, provided the desired heading can be maintained without exceptional piloting skill or alertness.

[Amdt. 27-44, 73 FR 11000, Feb. 29, 2008]

Ground and Water Handling Characteristics

§ 27.231 General.

The rotorcraft must have satisfactory ground and water handling characteristics, including freedom from uncontrollable tendencies in any condition expected in operation.

§ 27.235 Taxiing condition.

The rotorcraft must be designed to withstand the loads that would occur when the rotorcraft is taxied over the roughest ground that may reasonably be expected in normal operation.

§ 27.239 Spray characteristics.

If certification for water operation is requested, no spray characteristics during taxiing, takeoff, or landing may obscure the vision of the pilot or damage the rotors, propellers, or other parts of the rotorcraft.

§ 27.241 Ground resonance.

The rotorcraft may have no dangerous tendency to oscillate on the ground with the rotor turning.

Miscellaneous Flight Requirements

§ 27.251 Vibration.

Each part of the rotorcraft must be free from excessive vibration under each appropriate speed and power condition.

Subpart C—Strength Requirements

General

§ 27.301 Loads.

(a) Strength requirements are specified in terms of limit loads (the maximum loads to be expected in service) and ultimate loads (limit loads multiplied by prescribed factors of safety). Unless otherwise provided, prescribed loads are limit loads.

(b) Unless otherwise provided, the specified air, ground, and water loads must be placed in equilibrium with inertia forces, considering each item of mass in the rotorcraft. These loads must be distributed to closely approximate or conservatively represent actual conditions.

(c) If deflections under load would significantly change the distribution of external or internal loads, this redistribution must be taken into account.

§ 27.303 Factor of safety.

Unless otherwise provided, a factor of safety of 1.5 must be used. This factor applies to external and inertia loads unless its application to the resulting internal stresses is more conservative.

§ 27.305 Strength and deformation.

(a) The structure must be able to support limit loads without detrimental or permanent deformation. At any load up to limit loads, the deformation may not interfere with safe operation.

(b) The structure must be able to support ultimate loads without failure. This must be shown by—

(1) Applying ultimate loads to the structure in a static test for at least three seconds; or

(2) Dynamic tests simulating actual load application.

§ 27.307 Proof of structure.

(a) Compliance with the strength and deformation requirements of this subpart must be shown for each critical loading condition accounting for the environment to which the structure will be exposed in operation. Structural analysis (static or fatigue) may be used only if the structure conforms to those structures for which experience has shown this method to be reliable. In other cases, substantiating load tests must be made.

(b) Proof of compliance with the strength requirements of this subpart must include—

(1) Dynamic and endurance tests of rotors, rotor drives, and rotor controls;

(2) Limit load tests of the control system, including control surfaces;

(3) Operation tests of the control system;

(4) Flight stress measurement tests;

(5) Landing gear drop tests; and

(6) Any additional test required for new or unusual design features.

(Secs. 604, 605, 72 Stat. 778, 49 U.S.C. 1424, 1425)

[Doc. No. 5074, 29 FR 15695, Nov. 24, 1964, as amended by Amdt. 27-3, 33 FR 14105, Sept. 18, 1968; Amdt. 27-26, 55 FR 7999, Mar. 6, 1990]

§ 27.309 Design limitations.

The following values and limitations must be established to show compliance with the structural requirements of this subpart:

(a) The design maximum weight.

(b) The main rotor r.p.m. ranges power on and power off.

(c) The maximum forward speeds for each main rotor r.p.m. within the ranges determined under paragraph (b) of this section.

(d) The maximum rearward and sideward flight speeds.

(e) The center of gravity limits corresponding to the limitations determined under paragraphs (b), (c), and (d) of this section.

(f) The rotational speed ratios between each powerplant and each connected rotating component.

(g) The positive and negative limit maneuvering load factors.

Flight Loads

§ 27.321 General.

(a) The flight load factor must be assumed to act normal to the longitudinal axis of the rotorcraft, and to be equal in magnitude and opposite in direction to the rotorcraft inertia load factor at the center of gravity.

(b) Compliance with the flight load requirements of this subpart must be shown—

(1) At each weight from the design minimum weight to the design maximum weight; and

(2) With any practical distribution of disposable load within the operating limitations in the Rotorcraft Flight Manual.

[Doc. No. 5074, 29 FR 15695, Nov. 24, 1964, as amended by Amdt. 27-11, 41 FR 55468, Dec. 20, 1976]

§ 27.337 Limit maneuvering load factor.

The rotorcraft must be designed for—

(a) A limit maneuvering load factor ranging from a positive limit of 3.5 to a negative limit of −1.0; or

(b) Any positive limit maneuvering load factor not less than 2.0 and any negative limit maneuvering load factor of not less than −0.5 for which—

(1) The probability of being exceeded is shown by analysis and flight tests to be extremely remote; and

(2) The selected values are appropriate to each weight condition between the design maximum and design minimum weights.

[Amdt. 27-26, 55 FR 7999, Mar. 6, 1990]

§ 27.339 Resultant limit maneuvering loads.

The loads resulting from the application of limit maneuvering load factors are assumed to act at the center of each rotor hub and at each auxiliary lifting surface, and to act in directions, and with distributions of load among the rotors and auxiliary lifting surfaces, so as to represent each critical maneuvering condition, including power-on and power-off flight with the maximum design rotor tip speed ratio. The rotor tip speed ratio is the ratio of the rotorcraft flight velocity component in the plane of the rotor disc to the rotational tip speed of the rotor blades, and is expressed as follows:

$$\mu = \frac{V \cos \alpha}{\Omega R}$$

where—

V= The airspeed along flight path (f.p.s.);

a= The angle between the projection, in the plane of symmetry, of the axis of no feathering and a line perpendicular to the flight path (radians, positive when axis is pointing aft);

omega= The angular velocity of rotor (radians per second); and

R= The rotor radius (ft).

[Doc. No. 5074, 29 FR 15695, Nov. 24, 1964, as amended by Amdt. 27-11, 41 FR 55469, Dec. 20, 1976]

§ 27.341 Gust loads.

The rotorcraft must be designed to withstand, at each critical airspeed including hovering, the loads resulting from a vertical gust of 30 feet per second.

§ 27.351 Yawing conditions.

(a) Each rotorcraft must be designed for the loads resulting from the maneuvers specified in paragraphs (b) and (c) of this section with—

(1) Unbalanced aerodynamic moments about the center of gravity which the aircraft reacts to in a rational or conservative manner considering the principal masses furnishing the reacting inertia forces; and

(2) Maximum main rotor speed.

(b) To produce the load required in paragraph (a) of this section, in unaccelerated flight with zero yaw, at forward speeds from zero up to 0.6 V_{NE}—

(1) Displace the cockpit directional control suddenly to the maximum deflection limited by the control stops or by the maximum pilot force specified in § 27.397(a);

(2) Attain a resulting sideslip angle or 90°, whichever is less; and

(3) Return the directional control suddenly to neutral.

(c) To produce the load required in paragraph (a) of this section, in unaccelerated flight with zero yaw, at forward speeds from 0.6 V_{NE} up to V_{NE} or V_H, whichever is less—

(1) Displace the cockpit directional control suddenly to the maximum deflection limited by the control stops or by the maximum pilot force specified in § 27.397(a);

(2) Attain a resulting sideslip angle or 15°, whichever is less, at the lesser speed of V_{NE} or V_H;

(3) Vary the sideslip angles of paragraphs (b)(2) and (c)(2) of this section directly with speed; and

(4) Return the directional control suddenly to neutral.

[Amdt. 27-26, 55 FR 7999, Mar. 6, 1990, as amended by Amdt. 27-34, 62 FR 46173, Aug. 29, 1997]

§ 27.361 Engine torque.

(a) For turbine engines, the limit torque may not be less than the highest of—

(1) The mean torque for maximum continuous power multiplied by 1.25;

(2) The torque required by § 27.923;

(3) The torque required by § 27.927; or

(4) The torque imposed by sudden engine stoppage due to malfunction or structural failure (such as compressor jamming).

(b) For reciprocating engines, the limit torque may not be less than the mean torque for maximum continuous power multiplied by—

(1) 1.33, for engines with five or more cylinders; and

(2) Two, three, and four, for engines with four, three, and two cylinders, respectively.

[Amdt. 27-23, 53 FR 34210, Sept. 2, 1988]

Control Surface and System Loads

§ 27.391 General.

Each auxiliary rotor, each fixed or movable stabilizing or control surface, and each system operating any flight control must meet the requirements of §§ 27.395, 27.397, 27.399, 27.411, and 27.427.

[Amdt. 27-26, 55 FR 7999, Mar. 6, 1990, as amended by Amdt. 27-34, 62 FR 46173, Aug. 29, 1997]

§ 27.395 Control system.

(a) The part of each control system from the pilot's controls to the control stops must be designed to withstand pilot forces of not less than—

(1) The forces specified in § 27.397; or

(2) If the system prevents the pilot from applying the limit pilot forces to the system, the maximum forces that the system allows the pilot to apply, but not less than 0.60 times the forces specified in § 27.397.

(b) Each primary control system, including its supporting structure, must be designed as follows:

(1) The system must withstand loads resulting from the limit pilot forces prescribed in § 27.397.

(2) Notwithstanding paragraph (b)(3) of this section, when power-operated actuator controls or power boost controls are used, the system must also withstand the loads resulting from the force output of each normally energized power device, including any single power boost or actuator system failure.

(3) If the system design or the normal operating loads are such that a part of the system cannot react to the limit pilot forces prescribed in § 27.397, that part of the system must be designed to withstand the maximum loads that can be obtained in normal operation. The minimum design loads must, in any case, provide a rugged system for service use, including consideration of fatigue, jamming, ground gusts, control inertia, and friction loads. In the absence of rational analysis, the design loads resulting from 0.60 of the specified limit pilot forces are acceptable minimum design loads.

(4) If operational loads may be exceeded through jamming, ground gusts, control inertia, or friction, the system must withstand the limit pilot forces specified in § 27.397, without yielding.

[Doc. No. 5074, 29 FR 15695, Nov. 24, 1964, as amended by Amdt. 27-26, 55 FR 7999, Mar. 6, 1990]

§ 27.397 Limit pilot forces and torques.

(a) Except as provided in paragraph (b) of this section, the limit pilot forces are as follows:

(1) For foot controls, 130 pounds.

(2) For stick controls, 100 pounds fore and aft, and 67 pounds laterally.

(b) For flap, tab, stabilizer, rotor brake, and landing gear operating controls, the follows apply (R=radius in inches):

(1) Crank, wheel, and lever controls, $[1+R]/3 \times 50$ pounds, but not less than 50 pounds nor more than 100 pounds for hand operated controls or 130 pounds for foot operated controls, applied at any angle within 20 degrees of the plane of motion of the control.

(2) Twist controls, 80R inch-pounds.

[Amdt. 27-11, 41 FR 55469, Dec. 20, 1976, as amended by Amdt. 27-40, 66 FR 23538, May 9, 2001]

§ 27.399 Dual control system.

Each dual primary flight control system must be designed to withstand the loads that result when pilot forces of 0.75 times those obtained under § 27.395 are applied—

(a) In opposition; and

(b) In the same direction.

§ 27.411 Ground clearance: tail rotor guard.

(a) It must be impossible for the tail rotor to contact the landing surface during a normal landing.

(b) If a tail rotor guard is required to show compliance with paragraph (a) of this section—

(1) Suitable design loads must be established for the guard; and

(2) The guard and its supporting structure must be designed to withstand those loads.

§ 27.427 Unsymmetrical loads.

(a) Horizontal tail surfaces and their supporting structure must be designed for unsymmetrical loads arising from yawing and rotor wake effects in combination with the prescribed flight conditions.

(b) To meet the design criteria of paragraph (a) of this section, in the absence of more rational data, both of the following must be met:

(1) One hundred percent of the maximum loading from the symmetrical flight conditions acts on the surface on one side of the plane of symmetry, and no loading acts on the other side.

(2) Fifty percent of the maximum loading from the symmetrical flight conditions acts on the surface on each side of the plane of symmetry but in opposite directions.

(c) For empennage arrangements where the horizontal tail surfaces are supported by the vertical tail surfaces, the vertical tail surfaces and supporting structure must be designed for the combined vertical and horizontal surface loads resulting from each prescribed flight condition, considered separately. The flight conditions must be selected so the maximum design loads are obtained on each surface. In the absence of more rational data, the unsymmetrical horizontal tail surface loading distributions described in this section must be assumed.

[Admt. 27-26, 55 FR 7999, Mar. 6, 1990, as amended by Amdt. 27-27, 55 FR 38966, Sept. 21, 1990]

Ground Loads

§ 27.471 General.

(a) *Loads and equilibrium.* For limit ground loads—

(1) The limit ground loads obtained in the landing conditions in this part must be considered to be external loads that would occur in the rotorcraft structure if it were acting as a rigid body; and

(2) In each specified landing condition, the external loads must be placed in equilibrium with linear and angular inertia loads in a rational or conservative manner.

(b) *Critical centers of gravity.* The critical centers of gravity within the range for which certification is requested must be selected so that the maximum design loads are obtained in each landing gear element.

§ 27.473 Ground loading conditions and assumptions.

(a) For specified landing conditions, a design maximum weight must be used that is not less than the maximum weight. A rotor lift may be assumed to act through the center of gravity throughout the landing impact. This lift may not exceed two-thirds of the design maximum weight.

(b) Unless otherwise prescribed, for each specified landing condition, the rotorcraft must be designed for a limit load factor of not less than the limit inertia load factor substantiated under § 27.725.

[Amdt. 27-2, 33 FR 963, Jan. 26, 1968]

§ 27.475 Tires and shock absorbers.

Unless otherwise prescribed, for each specified landing condition, the tires must be assumed to be in their static position and the shock absorbers to be in their most critical position.

§ 27.477 Landing gear arrangement.

Sections 27.235, 27.479 through 27.485, and 27.493 apply to landing gear with two wheels aft, and one or more wheels forward, of the center of gravity.

§ 27.479 Level landing conditions.

(a) *Attitudes.* Under each of the loading conditions prescribed in paragraph (b) of this section, the rotorcraft is assumed to be in each of the following level landing attitudes:

(1) An attitude in which all wheels contact the ground simultaneously.

(2) An attitude in which the aft wheels contact the ground with the forward wheels just clear of the ground.

(b) *Loading conditions.* The rotorcraft must be designed for the following landing loading conditions:

(1) Vertical loads applied under § 27.471.

(2) The loads resulting from a combination of the loads applied under paragraph (b)(1) of this section with drag loads at each wheel of not less than 25 percent of the vertical load at that wheel.

(3) If there are two wheels forward, a distribution of the loads applied to those wheels under paragraphs (b)(1) and (2) of this section in a ratio of 40:60.

(c) *Pitching moments.* Pitching moments are assumed to be resisted by—

(1) In the case of the attitude in paragraph (a)(1) of this section, the forward landing gear; and

(2) In the case of the attitude in paragraph (a)(2) of this section, the angular inertia forces.

[Doc. No. 5074, 29 FR 15695, Nov. 24, 1964; 29 FR 17885, Dec. 17, 1964]

§ 27.481 Tail-down landing conditions.

(a) The rotorcraft is assumed to be in the maximum nose-up attitude allowing ground clearance by each part of the rotorcraft.

(b) In this attitude, ground loads are assumed to act perpendicular to the ground.

§ 27.483 One-wheel landing conditions.

For the one-wheel landing condition, the rotorcraft is assumed to be in the level attitude and to contact the ground on one aft wheel. In this attitude—

(a) The vertical load must be the same as that obtained on that side under § 27.479(b)(1); and

(b) The unbalanced external loads must be reacted by rotorcraft inertia.

§ 27.485 Lateral drift landing conditions.

(a) The rotorcraft is assumed to be in the level landing attitude, with—

(1) Side loads combined with one-half of the maximum ground reactions obtained in the level landing conditions of § 27.479 (b)(1); and

(2) The loads obtained under paragraph (a)(1) of this section applied—

(i) At the ground contact point; or

(ii) For full-swiveling gear, at the center of the axle.

(b) The rotorcraft must be designed to withstand, at ground contact—

(1) When only the aft wheels contact the ground, side loads of 0.8 times the vertical reaction acting inward on one side, and 0.6 times the vertical reaction acting outward on the other side, all combined with the vertical loads specified in paragraph (a) of this section; and

(2) When all wheels contact the ground simultaneously—

(i) For the aft wheels, the side loads specified in paragraph (b)(1) of this section; and

(ii) For the forward wheels, a side load of 0.8 times the vertical reaction combined with the vertical load specified in paragraph (a) of this section.

§ 27.493 Braked roll conditions.

Under braked roll conditions with the shock absorbers in their static positions—

(a) The limit vertical load must be based on a load factor of at least—

(1) 1.33, for the attitude specified in § 27.479(a)(1); and

(2) 1.0 for the attitude specified in § 27.479(a)(2); and

(b) The structure must be designed to withstand at the ground contact point of each wheel with brakes, a drag load at least the lesser of—

(1) The vertical load multiplied by a coefficient of friction of 0.8; and

(2) The maximum value based on limiting brake torque.

§ 27.497 Ground loading conditions: landing gear with tail wheels.

(a) *General.* Rotorcraft with landing gear with two wheels forward, and one wheel aft, of the center of gravity must be designed for loading conditions as prescribed in this section.

(b) *Level landing attitude with only the forward wheels contacting the ground.* In this attitude—

(1) The vertical loads must be applied under §§ 27.471 through 27.475;

(2) The vertical load at each axle must be combined with a drag load at that axle of not less than 25 percent of that vertical load; and

(3) Unbalanced pitching moments are assumed to be resisted by angular inertia forces.

(c) *Level landing attitude with all wheels contacting the ground simultaneously.* In this attitude, the rotorcraft must be designed for landing loading conditions as prescribed in paragraph (b) of this section.

(d) *Maximum nose-up attitude with only the rear wheel contacting the ground.* The attitude for this condition must be the maximum nose-up attitude expected in normal operation, including autorotative landings. In this attitude—

(1) The appropriate ground loads specified in paragraphs (b)(1) and (2) of this section must be determined and applied, using a rational method to account for the moment arm between the rear wheel ground reaction and the rotorcraft center of gravity; or

(2) The probability of landing with initial contact on the rear wheel must be shown to be extremely remote.

(e) *Level landing attitude with only one forward wheel contacting the ground.* In this attitude, the rotorcraft must be designed for ground loads as specified in paragraphs (b)(1) and (3) of this section.

(f) *Side loads in the level landing attitude.* In the attitudes specified in paragraphs (b) and (c) of this section, the following apply:

(1) The side loads must be combined at each wheel with one-half of the maximum vertical ground reactions obtained for that wheel under paragraphs (b) and (c) of this section. In this condition, the side loads must be—

(i) For the forward wheels, 0.8 times the vertical reaction (on one side) acting inward, and 0.6 times the vertical reaction (on the other side) acting outward; and

(ii) For the rear wheel, 0.8 times the vertical reaction.

(2) The loads specified in paragraph (f)(1) of this section must be applied—

(i) At the ground contact point with the wheel in the trailing position (for non-full swiveling landing gear or for full swiveling landing gear with a lock, steering device, or shimmy damper to keep the wheel in the trailing position); or

(ii) At the center of the axle (for full swiveling landing gear without a lock, steering device, or shimmy damper).

(g) *Braked roll conditions in the level landing attitude.* In the attitudes specified in paragraphs (b) and (c) of this section, and with the shock absorbers in their static positions, the rotorcraft must be designed for braked roll loads as follows:

(1) The limit vertical load must be based on a limit vertical load factor of not less than—

(i) 1.0, for the attitude specified in paragraph (b) of this section; and

(ii) 1.33, for the attitude specified in paragraph (c) of this section.

(2) For each wheel with brakes, a drag load must be applied, at the ground contact point, of not less than the lesser of—

(i) 0.8 times the vertical load; and

(ii) The maximum based on limiting brake torque.

(h) *Rear wheel turning loads in the static ground attitude.* In the static ground attitude, and with the shock absorbers and tires in their static positions, the rotorcraft must be designed for rear wheel turning loads as follows:

(1) A vertical ground reaction equal to the static load on the rear wheel must be combined with an equal sideload.

(2) The load specified in paragraph (h)(1) of this section must be applied to the rear landing gear—

(i) Through the axle, if there is a swivel (the rear wheel being assumed to be swiveled 90 degrees to the longitudinal axis of the rotorcraft); or

(ii) At the ground contact point, if there is a lock, steering device or shimmy damper (the rear wheel being assumed to be in the trailing position).

(i) *Taxiing condition.* The rotorcraft and its landing gear must be designed for loads that would occur when the rotorcraft is taxied over the roughest ground that may reasonably be expected in normal operation.

§ 27.501 Ground loading conditions: landing gear with skids.

(a) *General.* Rotorcraft with landing gear with skids must be designed for the loading conditions specified in this section. In showing compliance with this section, the following apply:

(1) The design maximum weight, center of gravity, and load factor must be determined under §§ 27.471 through 27.475.

(2) Structural yielding of elastic spring members under limit loads is acceptable.

(3) Design ultimate loads for elastic spring members need not exceed those obtained in a drop test of the gear with—

(i) A drop height of 1.5 times that specified in § 27.725; and

(ii) An assumed rotor lift of not more than 1.5 times that used in the limit drop tests prescribed in § 27.725.

(4) Compliance with paragraphs (b) through (e) of this section must be shown with—

(i) The gear in its most critically deflected position for the landing condition being considered; and

(ii) The ground reactions rationally distributed along the bottom of the skid tube.

(b) *Vertical reactions in the level landing attitude.* In the level attitude, and with the rotorcraft contacting the ground along the bottom of both skids, the vertical reactions must be applied as prescribed in paragraph (a) of this section.

(c) *Drag reactions in the level landing attitude.* In the level attitude, and with the rotorcraft contacting the ground along the bottom of both skids, the following apply:

(1) The vertical reactions must be combined with horizontal drag reactions of 50 percent of the vertical reaction applied at the ground.

(2) The resultant ground loads must equal the vertical load specified in paragraph (b) of this section.

(d) *Sideloads in the level landing attitude.* In the level attitude,and with the rotorcraft contacting the ground along the bottom of both skids, the following apply:

(1) The vertical ground reaction must be—

(i) Equal to the vertical loads obtained in the condition specified in paragraph (b) of this section; and

(ii) Divided equally among the skids.

(2) The vertical ground reactions must be combined with a horizontal sideload of 25 percent of their value.

(3) The total sideload must be applied equally between the skids and along the length of the skids.

(4) The unbalanced moments are assumed to be resisted by angular inertia.

(5) The skid gear must be investigated for—

(i) Inward acting sideloads; and

(ii) Outward acting sideloads.

(e) *One-skid landing loads in the level attitude.* In the level attitude, and with the rotorcraft contacting the ground along the bottom of one skid only, the following apply:

(1) The vertical load on the ground contact side must be the same as that obtained on that side in the condition specified in paragraph (b) of this section.

(2) The unbalanced moments are assumed to be resisted by angular inertia.

(f) *Special conditions.* In addition to the conditions specified in paragraphs (b) and (c) of this section, the rotorcraft must be designed for the following ground reactions:

(1) A ground reaction load acting up and aft at an angle of 45 degrees to the longitudinal axis of the rotorcraft. This load must be—

(i) Equal to 1.33 times the maximum weight;

(ii) Distributed symmetrically among the skids;

(iii) Concentrated at the forward end of the straight part of the skid tube; and

(iv) Applied only to the forward end of the skid tube and its attachment to the rotorcraft.

(2) With the rotorcraft in the level landing attitude, a vertical ground reaction load equal to one-half of the vertical load determined under paragraph (b) of this section. This load must be—

(i) Applied only to the skid tube and its attachment to the rotorcraft; and

(ii) Distributed equally over 33.3 percent of the length between the skid tube attachments and centrally located midway between the skid tube attachments.

[Doc. No. 5074, 29 FR 15695, Nov. 24, 1964, as amended by Amdt. 27-2, 33 FR 963, Jan. 26, 1968; Amdt. 27-26, 55 FR 8000, Mar. 6, 1990]

§ 27.505 Ski landing conditions.

If certification for ski operation is requested, the rotorcraft, with skis, must be designed to withstand the following loading conditions (where *P* is the maximum static weight on each ski with the rotorcraft at design maximum weight, and *n* is the limit load factor determined under § 27.473(b).

(a) Up-load conditions in which—

(1) A vertical load of *Pn* and a horizontal load of *Pn*/4 are simultaneously applied at the pedestal bearings; and

(2) A vertical load of 1.33 *P* is applied at the pedestal bearings.

(b) A side-load condition in which a side load of 0.35 *Pn* is applied at the pedestal bearings in a horizontal plane perpendicular to the centerline of the rotorcraft.

(c) A torque-load condition in which a torque load of 1.33 *P* (in foot pounds) is applied to the ski about the vertical axis through the centerline of the pedestal bearings.

Water Loads

§ 27.521 Float landing conditions.

If certification for float operation is requested, the rotorcraft, with floats, must be designed to withstand the following loading conditions (where the limit load factor is determined under § 27.473(b) or assumed to be equal to that determined for wheel landing gear):

(a) Up-load conditions in which—

(1) A load is applied so that, with the rotorcraft in the static level attitude, the resultant water reaction passes vertically through the center of gravity; and

(2) The vertical load prescribed in paragraph (a)(1) of this section is applied simultaneously with an aft component of 0.25 times the vertical component.

(b) A side-load condition in which—

(1) A vertical load of 0.75 times the total vertical load specified in paragraph (a)(1) of this section is divided equally among the floats; and

(2) For each float, the load share determined under paragraph (b)(1) of this section, combined with a total side load of 0.25 times the total vertical load specified in paragraph (b)(1) of this section, is applied to that float only.

Main Component Requirements

§ 27.547 Main rotor structure.

(a) Each main rotor assembly (including rotor hubs and blades) must be designed as prescribed in this section.

(b) [Reserved]

(c) The main rotor structure must be designed to withstand the following loads prescribed in §§ 27.337 through 27.341:

(1) Critical flight loads.

(2) Limit loads occurring under normal conditions of autorotation. For this condition, the rotor r.p.m. must be selected to include the effects of altitude.

(d) The main rotor structure must be designed to withstand loads simulating—

(1) For the rotor blades, hubs, and flapping hinges, the impact force of each blade against its stop during ground operation; and

(2) Any other critical condition expected in normal operation.

(e) The main rotor structure must be designed to withstand the limit torque at any rotational speed, including zero. In addition:

(1) The limit torque need not be greater than the torque defined by a torque limiting device (where provided), and may not be less than the greater of—

(i) The maximum torque likely to be transmitted to the rotor structure in either direction; and

(ii) The limit engine torque specified in § 27.361.

(2) The limit torque must be distributed to the rotor blades in a rational manner.

(Secs. 604, 605, 72 Stat. 778, 49 U.S.C. 1424, 1425)

[Doc. No. 5074, 29 FR 15695, Nov. 24, 1964, as amended by Amdt. 27-3, 33 FR 14105, Sept. 18, 1968]

§ 27.549 Fuselage, landing gear, and rotor pylon structures.

(a) Each fuselage, landing gear, and rotor pylon structure must be designed as prescribed in this section. Resultant rotor forces may be represented as a single force applied at the rotor hub attachment point.

(b) Each structure must be designed to withstand—

(1) The critical loads prescribed in §§ 27.337 through 27.341;

(2) The applicable ground loads prescribed in §§ 27.235, 27.471 through 27.485, 27.493, 27.497, 27.501, 27.505, and 27.521; and

(3) The loads prescribed in § 27.547 (d)(2) and (e).

(c) Auxiliary rotor thrust, and the balancing air and inertia loads occurring under accelerated flight conditions, must be considered.

(d) Each engine mount and adjacent fuselage structure must be designed to withstand the loads occurring under accelerated flight and landing conditions, including engine torque.

(Secs. 604, 605, 72 Stat. 778, 49 U.S.C. 1424, 1425)

[Doc. No. 5074, 29 FR 15695, Nov. 24, 1964, as amended by Amdt. 27-3, 33 FR 14105, Sept. 18, 1968]

Emergency Landing Conditions

§ 27.561 General.

(a) The rotorcraft, although it may be damaged in emergency landing conditions on land or water, must be designed as prescribed in this section to protect the occupants under those conditions.

(b) The structure must be designed to give each occupant every reasonable chance of escaping serious injury in a crash landing when—

(1) Proper use is made of seats, belts, and other safety design provisions;

(2) The wheels are retracted (where applicable); and

(3) Each occupant and each item of mass inside the cabin that could injure an occupant is restrained when subjected to the following ultimate inertial load factors relative to the surrounding structure:

(i) Upward—4g.

(ii) Forward—16g.

(iii) Sideward—8g.

(iv) Downward—20g, after intended displacement of the seat device.

(v) Rearward—1.5g.

(c) The supporting structure must be designed to restrain, under any ultimate inertial load up to those specified in this paragraph, any item of mass above and/or behind the crew and passenger compartment that could injure an occupant if it came loose in an emergency landing. Items of mass to be considered include, but are not limited to, rotors, transmissions, and engines. The items of mass must be restrained for the following ultimate inertial load factors:

(1) Upward—1.5g.

(2) Forward—12g.

(3) Sideward—6g.

(4) Downward—12g.

(5) Rearward—1.5g

(d) Any fuselage structure in the area of internal fuel tanks below the passenger floor level must be designed to resist the following ultimate inertial factors and loads and to protect the fuel tanks from rupture when those loads are applied to that area:

(i) Upward—1.5g.

(ii) Forward—4.0g.

(iii) Sideward—2.0g.

(iv) Downward—4.0g.

[Doc. No. 5074, 29 FR 15695, Nov. 24, 1964, as amended by Amdt. 27-25, 54 FR 47318, Nov. 13, 1989; Amdt. 27-30, 59 FR 50386, Oct. 3, 1994; Amdt. 27-32, 61 FR 10438, Mar. 13, 1996]

§ 27.562 Emergency landing dynamic conditions.

(a) The rotorcraft, although it may be damaged in an emergency crash landing, must be designed to reasonably protect each occupant when—

(1) The occupant properly uses the seats, safety belts, and shoulder harnesses provided in the design; and

(2) The occupant is exposed to the loads resulting from the conditions prescribed in this section.

(b) Each seat type design or other seating device approved for crew or passenger occupancy during takeoff and landing must successfully complete dynamic tests or be demonstrated by rational analysis based on dynamic tests of a similar type seat in accordance with the following criteria. The tests must be conducted with an occupant, simulated by a 170-pound anthropomorphic test dummy (ATD), as defined by 49 CFR 572, subpart B, or its equivalent, sitting in the normal upright position.

(1) A change in downward velocity of not less than 30 feet per second when the seat or other seating device is oriented in its nominal position with respect to the rotorcraft's reference system, the rotorcraft's longitudinal axis is canted upward 60° with respect to the impact velocity vector, and the rotorcraft's lateral axis is perpendicular to a vertical plane containing the impact velocity vector and the rotorcraft's longitudinal axis. Peak floor deceleration must occur in not more than 0.031 seconds after impact and must reach a minimum of 30g's.

(2) A change in forward velocity of not less than 42 feet per second when the seat or other seating device is oriented in its nominal position with respect to the rotorcraft's reference system, the rotorcraft's longitudinal axis is yawed 10° either right or left of the impact velocity vector (whichever would cause the greatest load on the shoulder harness), the rotorcraft's lateral axis is contained in a horizontal plane containing the impact velocity vector, and the rotorcraft's vertical axis is perpendicular to a horizontal plane containing the impact velocity vector. Peak floor deceleration must occur in not more than 0.071 seconds after impact and must reach a minimum of 18.4g's.

(3) Where floor rails or floor or sidewall attachment devices are used to attach the seating devices to the airframe structure for the conditions of this section, the rails or devices must be misaligned with respect to each other by at least 10° vertically (i.e., pitch out of parallel) and by at least a 10° lateral roll, with the directions optional, to account for possible floor warp.

(c) Compliance with the following must be shown:

(1) The seating device system must remain intact although it may experience separation intended as part of its design.

(2) The attachment between the seating device and the airframe structure must remain intact, although the structure may have exceeded its limit load.

(3) The ATD's shoulder harness strap or straps must remain on or in the immediate vicinity of the ATD's shoulder during the impact.

(4) The safety belt must remain on the ATD's pelvis during the impact.

(5) The ATD's head either does not contact any portion of the crew or passenger compartment, or if contact is made, the head impact does not exceed a head injury criteria (HIC) of 1,000 as determined by this equation.

$$HIC = (t_2 - t_1)\left[\frac{1}{(t_2 - t_1)}\int_{t_1}^{t_2} a(t)\,dt\right]^{2.5}$$

Where: a(t) is the resultant acceleration at the center of gravity of the head form expressed as a multiple of g (the acceleration of gravity) and $t_2 - t_1$ is the time duration, in seconds, of major head impact, not to exceed 0.05 seconds.

(6) Loads in individual upper torso harness straps must not exceed 1,750 pounds. If dual straps are used for retaining the upper torso, the total harness strap loads must not exceed 2,000 pounds.

(7) The maximum compressive load measured between the pelvis and the lumbar column of the ATD must not exceed 1,500 pounds.

(d) An alternate approach that achieves an equivalent or greater level of occupant protection, as required by this section, must be substantiated on a rational basis.

[Amdt. 27-25, 54 FR 47318, Nov. 13, 1989]

§ 27.563 Structural ditching provisions.

If certification with ditching provisions is requested, structural strength for ditching must meet the requirements of this section and § 27.801(e).

(a) *Forward speed landing conditions.* The rotorcraft must initially contact the most critical wave for reasonably probable water conditions at forward velocities from zero up to 30 knots in likely pitch, roll, and yaw attitudes. The rotorcraft limit vertical descent velocity may not be less than 5 feet per second relative to the mean water surface. Rotor lift may be used to act through the center of gravity throughout the landing impact. This lift may not exceed two-thirds of the design maximum weight. A maximum forward velocity of less than 30 knots may be used in design if it can be demonstrated that the forward velocity selected would not be exceeded in a normal one-engine-out touchdown.

(b) *Auxiliary or emergency float conditions*—(1) *Floats fixed or deployed before initial water contact.* In addition to the landing loads in paragraph (a) of this section, each auxiliary or emergency float, of its support and attaching structure in the airframe or fuselage, must be designed for the load developed by a fully immersed float unless it can be shown that full immersion is unlikely. If full immersion is unlikely, the highest likely float buoyancy load must be applied. The highest likely buoyancy load must include consideration of a partially immersed float creating restoring moments to compensate the upsetting moments caused by side wind, unsymmetrical rotorcraft loading, water wave action, rotorcraft inertia, and probable structural damage and leakage considered under § 27.801(d). Maximum roll and pitch angles determined from compliance with § 27.801(d) may be used, if significant, to determine the extent of immersion of each float. If the floats are deployed in flight, appropriate air loads derived from the flight limitations with the floats deployed shall be used in substantiation of the floats and their attachment to the rotorcraft. For this purpose, the design airspeed for limit load is the float deployed airspeed operating limit multiplied by 1.11.

(2) *Floats deployed after initial water contact.* Each float must be designed for full or partial immersion perscribed in paragraph (b)(1) of this section. In addition, each float must be designed for combined vertical and drag loads using a relative limit speed of 20 knots between the rotorcraft and the water. The vertical load may not be less than the highest likely buoyancy load determined under paragraph (b)(1) of this section.

[Amdt. 27-26, 55 FR 8000, Mar. 6, 1990]

Fatigue Evaluation

§ 27.571 Fatigue evaluation of flight structure.

(a) *General.* Each portion of the flight structure (the flight structure includes rotors, rotor drive systems between the engines and the rotor hubs, controls, fuselage, landing gear, and their related primary attachments), the failure of which could be catastrophic, must be identified and must be evaluated under paragraph (b), (c), (d), or (e) of this section. The following apply to each fatigue evaluation:

(1) The procedure for the evaluation must be approved.

(2) The locations of probable failure must be determined.

(3) Inflight measurement must be included in determining the following:

(i) Loads or stresses in all critical conditions throughout the range of limitations in § 27.309, except that maneuvering load factors need not exceed the maximum values expected in operation.

(ii) The effect of altitude upon these loads or stresses.

(4) The loading spectra must be as severe as those expected in operation including, but not limited to, external cargo operations, if applicable, and ground-air-ground cycles. The loading spectra must be based on loads or stresses determined under paragraph (a)(3) of this section.

(b) *Fatigue tolerance evaluation.* It must be shown that the fatigue tolerance of the structure ensures that the probability of catastrophic fatigue failure is extremely remote without establishing replacement times, inspection intervals or other procedures under section A27.4 of appendix A.

(c) *Replacement time evaluation.* it must be shown that the probability of catastrophic fatigue failure is extremely remote within a replacement time furnished under section A27.4 of appendix A.

(d) *Fail-safe evaluation.* The following apply to fail-safe evaluation:

(1) It must be shown that all partial failures will become readily detectable under inspection procedures furnished under section A27.4 of appendix A.

(2) The interval between the time when any partial failure becomes readily detectable under paragraph (d)(1) of this section, and the time when any such failure is expected to reduce the remaining strength of the structure to limit or maximum attainable loads (whichever is less), must be determined.

(3) It must be shown that the interval determined under paragraph (d)(2) of this section is long enough, in relation to the inspection intervals and related procedures furnished under section A27.4 of appendix A, to provide a probability of detection great enough to ensure that the probability of catastrophic failure is extremely remote.

(e) *Combination of replacement time and failsafe evaluations.* A component may be evaluated under a combination of paragraphs (c) and (d) of this section. For such component it must be shown that the probability of catastrophic failure is extremely remote with an approved combination of replacement time, inspection intervals, and related procedures furnished under section A27.4 of appendix A.

(Secs. 313(a), 601, 603, 604, and 605, 72 Stat. 752, 775, and 778, (49 U.S.C. 1354(a), 1421, 1423, 1424, and 1425; sec. 6(c), 49 U.S.C. 1655(c)))

[Amdt. 27-3, 33 FR 14106, Sept. 18, 1968, as amended by Amdt. 27-12, 42 FR 15044, Mar. 17, 1977; Amdt. 27-18, 45 FR 60177, Sept. 11 1980; Amdt. 27-26, 55 FR 8000, Mar. 6, 1990]

§ 27.573 Damage Tolerance and Fatigue Evaluation of Composite Rotorcraft Structures.

(a) Each applicant must evaluate the composite rotorcraft structure under the damage tolerance standards of paragraph (d) of this section unless the applicant establishes that a damage tolerance evaluation is impractical within the limits of geometry, inspectability, and good design practice. If an applicant establishes that it is impractical within the limits of geometry, inspectability, and good design practice, the applicant must do a fatigue evaluation in accordance with paragraph (e) of this section.

(b) The methodology used to establish compliance with this section must be submitted to and approved by the Administrator.

(c) Definitions:

(1) *Catastrophic failure* is an event that could prevent continued safe flight and landing.

(2) *Principal Structural Elements (PSEs)* are structural elements that contribute significantly to the carrying of flight or ground loads, the failure of which could result in catastrophic failure of the rotorcraft.

(3) *Threat Assessment* is an assessment that specifies the locations, types, and sizes of damage, considering fatigue, environmental effects, intrinsic and discrete flaws, and impact or other accidental damage (including the discrete source of the accidental damage) that may occur during manufacture or operation.

(d) Damage Tolerance Evaluation:

(1) Each applicant must show that catastrophic failure due to static and fatigue loads, considering the intrinsic or discrete manufacturing defects or accidental damage, is avoided throughout the operational life or prescribed inspection intervals of the rotorcraft by performing damage tolerance evaluations of the strength of composite PSEs and other parts, detail design points, and fabrication techniques. Each applicant must account for the effects of material and process variability along with environmental conditions in the strength and fatigue evaluations. Each applicant must evaluate parts that include PSEs of the airframe, main and tail rotor drive systems, main and tail rotor blades and hubs, rotor controls, fixed and movable control surfaces, engine and transmission mountings, landing gear, other parts, detail design points, and fabrication techniques deemed critical by the FAA. Each damage tolerance evaluation must include:

(i) The identification of all PSEs;

(ii) In-flight and ground measurements for determining the loads or stresses for all PSEs for all critical conditions throughout the range of limits in § 27.309 (including altitude effects), except that maneuvering load factors need not exceed the maximum values expected in service;

(iii) The loading spectra as severe as those expected in service based on loads or stresses determined under paragraph (d)(1)(ii) of this section, including external load operations, if applicable, and other operations including high-torque events;

(iv) A threat assessment for all PSEs that specifies the locations, types, and sizes of damage, considering fatigue, environmental effects, intrinsic and discrete flaws, and impact or other accidental damage (including the discrete source of the accidental damage) that may occur during manufacture or operation; and

(v) An assessment of the residual strength and fatigue characteristics of all PSEs that supports the replacement times and inspection intervals established under paragraph (d)(2) of this section.

(2) Each applicant must establish replacement times, inspections, or other procedures for all PSEs to require the repair or replacement of damaged parts before a catastrophic failure. These replacement times, inspections, or other procedures must be included in the Airworthiness Limitations Section of the Instructions for Continued Airworthiness required by § 27.1529.

(i) Replacement times for PSEs must be determined by tests, or by analysis supported by tests, and must show that the structure is able to withstand the repeated loads of variable magnitude expected in-service. In establishing these replacement times, the following items must be considered:

(A) Damage identified in the threat assessment required by paragraph (d)(1)(iv) of this section;

(B) Maximum acceptable manufacturing defects and in-service damage (*i.e.*, those that do not lower the residual strength below ultimate design loads and those that can be repaired to restore ultimate strength); and

(C) Ultimate load strength capability after applying repeated loads.

(ii) Inspection intervals for PSEs must be established to reveal any damage identified in the threat assessment required by paragraph (d)(1)(iv) of this section that may occur from fatigue or other in-service causes before such damage has grown to the extent that the component cannot sustain the required residual strength capability. In establishing these inspection intervals, the following items must be considered:

(A) The growth rate, including no-growth, of the damage under the repeated loads expected in-service determined by tests or analysis supported by tests;

(B) The required residual strength for the assumed damage established after considering the damage type, inspection interval, detectability of damage, and the techniques adopted for damage detection. The minimum required residual strength is limit load; and

(C) Whether the inspection will detect the damage growth before the minimum residual strength is reached and restored to ultimate load capability, or whether the component will require replacement.

(3) Each applicant must consider the effects of damage on stiffness, dynamic behavior, loads, and functional performance on all PSEs when substantiating the maximum assumed damage size and inspection interval.

(e) Fatigue Evaluation: If an applicant establishes that the damage tolerance evaluation described in paragraph (d) of this section is impractical within the limits of geometry, inspectability, or good design practice, the applicant must do a fatigue evaluation of the particular composite rotorcraft structure and:

(1) Identify all PSEs considered in the fatigue evaluation;

(2) Identify the types of damage for all PSEs considered in the fatigue evaluation;

(3) Establish supplemental procedures to minimize the risk of catastrophic failure associated with the damages identified in paragraph (d) of this section; and

(4) Include these supplemental procedures in the Airworthiness Limitations section of the Instructions for Continued Airworthiness required by § 27.1529.

[Doc. No. FAA-2009-0660, Amdt. 27-47, 76 FR 74663, Dec. 1, 2011]

Subpart D—Design and Construction

General

§ 27.601 Design.

(a) The rotorcraft may have no design features or details that experience has shown to be hazardous or unreliable.

(b) The suitability of each questionable design detail and part must be established by tests.

§ 27.602 Critical parts.

(a) *Critical part.* A critical part is a part, the failure of which could have a catastrophic effect upon the rotocraft, and for which critical characteristics have been identified which must be controlled to ensure the required level of integrity.

(b) If the type design includes crtical parts, a critical parts list shall be established. Procedures shall be established to define the critical design characteristics, identify processes that affect those characteristics, and identify the design change and process change controls necessary for showing compliance with the quality assurance requirements of part 21 of this chapter.

[Doc. No. 29311, 64 FR 46232, Aug. 24, 1999]

§ 27.603 Materials.

The suitability and durability of materials used for parts, the failure of which could adversely affect safety, must—

(a) Be established on the basis of experience or tests;

(b) Meet approved specifications that ensure their having the strength and other properties assumed in the design data; and

(c) Take into account the effects of environmental conditions, such as temperature and humidity, expected in service.

(Secs. 313(a), 601, 603, 604, Federal Aviation Act of 1958 (49 U.S.C. 1354(a), 1421, 1423, 1424); and sec. 6(c) of the Dept. of Transportation Act (49 U.S.C. 1655(c)))

[Doc. No. 5074, 29 FR 15695, Nov. 24, 1964, as amended by Amdt. 27-11, 41 FR 55469, Dec. 20, 1976; Amdt. 27-16, 43 FR 50599, Oct. 30, 1978]

§ 27.605 Fabrication methods.

(a) The methods of fabrication used must produce consistently sound structures. If a fabrication process (such as gluing, spot welding, or heat-treating) requires close control to reach this objective, the process must be performed according to an approved process specification.

(b) Each new aircraft fabrication method must be substantiated by a test program.

(Secs. 313(a), 601, 603, 604, and 605 of the Federal Aviation Act of 1958 (49 U.S.C. 1354(a), 1421, 1423, 1424 and 1425); sec. 6(c) of the Dept. of Transportation Act (49 U.S.C. 1655(c)))

[Doc. No. 5074, 29 FR 15695, Nov. 24, 1964, as amended by Amdt. 27-16, 43 FR 50599, Oct. 30, 1978]

§ 27.607 Fasteners.

(a) Each removable bolt, screw, nut, pin, or other fastener whose loss could jeopardize the safe operation of the rotorcraft must incorporate two separate locking devices. The fastener and its locking devices may not be adversely affected by the environmental conditions associated with the particular installation.

(b) No self-locking nut may be used on any bolt subject to rotation in operation unless a nonfriction locking device is used in addition to the self-locking device.

[Amdt. 27-4, 33 FR 14533, Sept. 27, 1968]

§ 27.609 Protection of structure.

Each part of the structure must—

(a) Be suitably protected against deterioration or loss of strength in service due to any cause, including—

(1) Weathering;

(2) Corrosion; and

(3) Abrasion; and

(b) Have provisions for ventilation and drainage where necessary to prevent the accumulation of corrosive, flammable, or noxious fluids.

§ 27.610 Lightning and static electricity protection.

(a) The rotorcraft must be protected against catastrophic effects from lightning.

(b) For metallic components, compliance with paragraph (a) of this section may be shown by—

(1) Electrically bonding the components properly to the airframe; or

(2) Designing the components so that a strike will not endanger the rotorcraft.

(c) For nonmetallic components, compliance with paragraph (a) of this section may be shown by—

(1) Designing the components to minimize the effect of a strike; or

(2) Incorporating acceptable means of diverting the resulting electrical current so as not to endanger the rotorcraft.

(d) The electrical bonding and protection against lightning and static electricity must—

(1) Minimize the accumulation of electrostatic charge;

(2) Minimize the risk of electric shock to crew, passengers, and service and maintenance personnel using normal precautions;

(3) Provide an electrical return path, under both normal and fault conditions, on rotorcraft having grounded electrical systems; and

(4) Reduce to an acceptable level the effects of static electricity on the functioning of essential electrical and electronic equipment.

[Amdt. 27-21, 49 FR 44433, Nov. 6, 1984, as amended by Amdt. 27-37, 64 FR 45094, Aug. 18, 1999; Amdt. 27-46, 76 FR 33135, June 8, 2011]

§ 27.611 Inspection provisions.

There must be means to allow the close examination of each part that requires—

(a) Recurring inspection;

(b) Adjustment for proper alignment and functioning; or

(c) Lubrication.

§ 27.613 Material strength properties and design values.

(a) Material strength properties must be based on enough tests of material meeting specifications to establish design values on a statistical basis.

(b) Design values must be chosen to minimize the probability of structural failure due to material variability. Except as provided in paragraphs (d) and (e) of this section, compliance with this paragraph must be shown by selecting design values that assure material strength with the following probability—

(1) Where applied loads are eventually distributed through a single member within an assembly, the failure of which would result in loss of structural integrity of the component, 99 percent probability with 95 percent confidence; and

(2) For redundant structure, those in which the failure of individual elements would result in applied loads being safely distributed to other load-carrying members, 90 percent probability with 95 percent confidence.

(c) The strength, detail design, and fabrication of the structure must minimize the probability of disastrous fatigue failure, particularly at points of stress concentration.

(d) Design values may be those contained in the following publications (available from the Naval Publications and Forms Center, 5801 Tabor Avenue, Philadelphia, Pennsylvania 19120) or other values approved by the Administrator:

(1) MIL-HDBK-5, "Metallic Materials and Elements for Flight Vehicle Structure".

(2) MIL-HDBK-17, "Plastics for Flight Vehicles".

(3) ANC-18, "Design of Wood Aircraft Structures".

(4) MIL-HDBK-23, "Composite Construction for Flight Vehicles".

(e) Other design values may be used if a selection of the material is made in which a specimen of each individual item is tested before use and it is determined that the actual strength properties of that particular item will equal or exceed those used in design.

(Secs. 313(a), 601, 603, 604, Federal Aviation Act of 1958 (49 U.S.C. 1354(a), 1421, 1423, 1424), sec. 6(c), Dept. of Transportation Act (49 U.S.C. 1655(c)))

[Doc. No. 5074, 29 FR 15695, Nov. 24, 1964, as amended by Amdt. 27-16, 43 FR 50599, Oct. 30, 1978; Amdt. 27-26, 55 FR 8000, Mar. 6, 1990]

§ 27.619 Special factors.

(a) The special factors prescribed in §§ 27.621 through 27.625 apply to each part of the structure whose strength is—

(1) Uncertain;

(2) Likely to deteriorate in service before normal replacement; or

(3) Subject to appreciable variability due to—

(i) Uncertainties in manufacturing processes; or

(ii) Uncertainties in inspection methods.

(b) For each part to which §§ 27.621 through 27.625 apply, the factor of safety prescribed in § 27.303 must be multiplied by a special factor equal to—

(1) The applicable special factors prescribed in §§ 27.621 through 27.625; or

(2) Any other factor great enough to ensure that the probability of the part being understrength because of the uncertainties specified in paragraph (a) of this section is extremely remote.

§ 27.621 Casting factors.

(a) *General.* The factors, tests, and inspections specified in paragraphs (b) and (c) of this section must be applied in addition to those necessary to establish foundry quality control. The inspections must meet approved specifications. Paragraphs (c) and (d) of this section apply to structural castings except castings that are pressure tested as parts of hydraulic or other fluid systems and do not support structural loads.

(b) *Bearing stresses and surfaces.* The casting factors specified in paragraphs (c) and (d) of this section—

(1) Need not exceed 1.25 with respect to bearing stresses regardless of the method of inspection used; and

(2) Need not be used with respect to the bearing surfaces of a part whose bearing factor is larger than the applicable casting factor.

(c) *Critical castings.* For each casting whose failure would preclude continued safe flight and landing of the rotorcraft or result in serious injury to any occupant, the following apply:

(1) Each critical casting must—

(i) Have a casting factor of not less than 1.25; and

(ii) Receive 100 percent inspection by visual, radiographic, and magnetic particle (for ferromagnetic materials) or penetrant (for nonferromagnetic materials) inspection methods or approved equivalent inspection methods.

(2) For each critical casting with a casting factor less than 1.50, three sample castings must be static tested and shown to meet—

(i) The strength requirements of § 27.305 at an ultimate load corresponding to a casting factor of 1.25; and

(ii) The deformation requirements of § 27.305 at a load of 1.15 times the limit load.

(d) *Noncritical castings.* For each casting other than those specified in paragraph (c) of this section, the following apply:

(1) Except as provided in paragraphs (d)(2) and (3) of this section, the casting factors and corresponding inspections must meet the following table:

Castingfactor	Inspection
2.0 or greater	100percent visual
Less than 2.0,greater than 1.5	100percent visual, and magnetic particle (ferromagnetic materials), penetrant (nonferromagnetic materials), or approved equivalent inspection methods.
1.25 through 1.50	100 percent visual, and magnetic particle (ferromagnetic materials). penetrant (nonferromagnetic materials), and radiographic or approved equivalent inspection methods.

(2) The percentage of castings inspected by nonvisual methods may be reduced below that specified in paragraph (d)(1) of this section when an approved quality control procedure is established.

(3) For castings procured to a specification that guarantees the mechanical properties of the material in the casting and provides for demonstration of these properties by test of coupons cut from the castings on a sampling basis—

(i) A casting factor of 1.0 may be used; and

(ii) The castings must be inspected as provided in paragraph (d)(1) of this section for casting factors of "1.25 through 1.50" and tested under paragraph (c)(2) of this section.

[Doc. No. 5074, 29 FR 15695, Nov. 24, 1964, as amended by Amdt. 27-34, 62 FR 46173, Aug. 29, 1997]

§ 27.623 Bearing factors.

(a) Except as provided in paragraph (b) of this section, each part that has clearance (free fit), and that is subject to pounding or vibration, must have a bearing factor large enough to provide for the effects of normal relative motion.

(b) No bearing factor need be used on a part for which any larger special factor is prescribed.

§ 27.625 Fitting factors.

For each fitting (part or terminal used to join one structural member to another) the following apply:

(a) For each fitting whose strength is not proven by limit and ultimate load tests in which actual stress conditions are simulated in the fitting and surrounding structures, a fitting factor of at least 1.15 must be applied to each part of—

(1) The fitting;

(2) The means of attachment; and

(3) The bearing on the joined members.

(b) No fitting factor need be used—

(1) For joints made under approved practices and based on comprehensive test data (such as continuous joints in metal plating, welded joints, and scarf joints in wood); and

(2) With respect to any bearing surface for which a larger special factor is used.

(c) For each integral fitting, the part must be treated as a fitting up to the point at which the section properties become typical of the member.

(d) Each seat, berth, litter, safety belt, and harness attachment to the structure must be shown by analysis, tests, or both, to be able to withstand the inertia forces prescribed in § 27.561(b)(3) multiplied by a fitting factor of 1.33.

[Doc. No. 5074, 29 FR 15695, Nov. 24, 1964, as amended by Amdt. 27-35, 63 FR 43285, Aug. 12, 1998]

§ 27.629 Flutter.

Each aerodynamic surface of the rotorcraft must be free from flutter under each appropriate speed and power condition.

[Doc. No. 5074, 29 FR 15695, Nov. 24, 1964, as amended by Amdt. 27-26, 55 FR 8000, Mar. 6, 1990]

Rotors

§ 27.653 Pressure venting and drainage of rotor blades.

(a) For each rotor blade—

(1) There must be means for venting the internal pressure of the blade;

(2) Drainage holes must be provided for the blade; and

(3) The blade must be designed to prevent water from becoming trapped in it.

(b) Paragraphs (a)(1) and (2) of this section does not apply to sealed rotor blades capable of withstanding the maximum pressure differentials expected in service.

[Amdt. 27-2, 33 FR 963, Jan. 26, 1968]

§ 27.659 Mass balance.

(a) The rotors and blades must be mass balanced as necessary to—

(1) Prevent excessive vibration; and

(2) Prevent flutter at any speed up to the maximum forward speed.

(b) The structural integrity of the mass balance installation must be substantiated.

[Amdt. 27-2, 33 FR 963, Jan. 26, 1968]

§ 27.661 Rotor blade clearance.

There must be enough clearance between the rotor blades and other parts of the structure to prevent the blades from striking any part of the structure during any operating condition.

[Amdt. 27-2, 33 FR 963, Jan. 26, 1968]

§ 27.663 Ground resonance prevention means.

(a) The reliability of the means for preventing ground resonance must be shown either by analysis and tests, or reliable service experience, or by showing through analysis or tests that malfunction or failure of a single means will not cause ground resonance.

(b) The probable range of variations, during service, of the damping action of the ground resonance prevention means must be established and must be investigated during the test required by § 27.241.

[Amdt. 27-2, 33 FR 963, Jan. 26, 1968, as amended by Amdt. 27-26, 55 FR 8000, Mar. 6, 1990]

Control Systems

§ 27.671 General.

(a) Each control and control system must operate with the ease, smoothness, and positiveness appropriate to its function.

(b) Each element of each flight control system must be designed, or distinctively and permanently marked, to minimize the probability of any incorrect assembly that could result in the malfunction of the system.

§ 27.672 Stability augmentation, automatic, and power-operated systems.

If the functioning of stability augmentation or other automatic or power-operated systems is necessary to show compliance with the flight characteristics requirements of this part, such systems must comply with § 27.671 of this part and the following:

(a) A warning which is clearly distinguishable to the pilot under expected flight conditions without requiring the pilot's attention must be provided for any failure in the stability augmentation system or in any other automatic or power-operated system which could result in an unsafe condition if the pilot is unaware of the failure. Warning systems must not activate the control systems.

(b) The design of the stability augmentation system or of any other automatic or power-operated system must allow initial counteraction of failures without requiring exceptional pilot skill or strength by overriding the failure by movement of the flight controls in the normal sense and deactivating the failed system.

(c) It must be shown that after any single failure of the stability augmentation system or any other automatic or power-operated system—

(1) The rotorcraft is safely controllable when the failure or malfunction occurs at any speed or altitude within the approved operating limitations;

(2) The controllability and maneuverability requirements of this part are met within a practical operational flight envelope (for example, speed, altitude, normal acceleration, and rotorcraft configurations) which is described in the Rotorcraft Flight Manual; and

(3) The trim and stability characteristics are not impaired below a level needed to permit continued safe flight and landing.

[Amdt. 27-21, 49 FR 44433, Nov. 6, 1984; 49 FR 47594, Dec. 6, 1984]

§ 27.673 Primary flight control.

Primary flight controls are those used by the pilot for immediate control of pitch, roll, yaw, and vertical motion of the rotorcraft.

[Amdt. 27-21, 49 FR 44434, Nov. 6, 1984]

§ 27.674 Interconnected controls.

Each primary flight control system must provide for safe flight and landing and operate independently after a malfunction, failure, or jam of any auxiliary interconnected control.

[Amdt. 27-26, 55 FR 8001, Mar. 6, 1990]

§ 27.675 Stops.

(a) Each control system must have stops that positively limit the range of motion of the pilot's controls.

(b) Each stop must be located in the system so that the range of travel of its control is not appreciably affected by—

(1) Wear;

(2) Slackness; or

(3) Takeup adjustments.

(c) Each stop must be able to withstand the loads corresponding to the design conditions for the system.

(d) For each main rotor blade—

(1) Stops that are appropriate to the blade design must be provided to limit travel of the blade about its hinge points; and

(2) There must be means to keep the blade from hitting the droop stops during any operation other than starting and stopping the rotor.

(Secs. 313(a), 601, 603, 604, Federal Aviation Act of 1958 (49 U.S.C. 1354(a), 1421, 1423, 1424), sec. 6(c), Dept. of Transportation Act (49 U.S.C. 1655(c)))

[Doc. No. 5074, 29 FR 15695, Nov. 24, 1964, as amended by Amdt. 27-16, 43 FR 50599, Oct. 30, 1978]

§ 27.679 Control system locks.

If there is a device to lock the control system with the rotorcraft on the ground or water, there must be means to—

(a) Give unmistakable warning to the pilot when the lock is engaged; and

(b) Prevent the lock from engaging in flight.

§ 27.681 Limit load static tests.

(a) Compliance with the limit load requirements of this part must be shown by tests in which—

(1) The direction of the test loads produces the most severe loading in the control system; and

(2) Each fitting, pulley, and bracket used in attaching the system to the main structure is included.

(b) Compliance must be shown (by analyses or individual load tests) with the special factor requirements for control system joints subject to angular motion.

§ 27.683 Operation tests.

It must be shown by operation tests that, when the controls are operated from the pilot compartment with the control system loaded to correspond with loads specified for the system, the system is free from—

(a) Jamming;

(b) Excessive friction; and

(c) Excessive deflection.

§ 27.685 Control system details.

(a) Each detail of each control system must be designed to prevent jamming, chafing, and interference from cargo, passengers, loose objects or the freezing of moisture.

(b) There must be means in the cockpit to prevent the entry of foreign objects into places where they would jam the system.

(c) There must be means to prevent the slapping of cables or tubes against other parts.

(d) Cable systems must be designed as follows:

(1) Cables, cable fittings, turnbuckles, splices, and pulleys must be of an acceptable kind.

(2) The design of the cable systems must prevent any hazardous change in cable tension throughout the range of travel under any operating conditions and temperature variations.

(3) No cable smaller than three thirty-seconds of an inch diameter may be used in any primary control system.

(4) Pulley kinds and sizes must correspond to the cables with which they are used. The pulley cable combinations and strength values which must be used are specified in Military Handbook MIL-HDBK-5C, Vol. 1 & Vol. 2, Metallic Materials and Elements for Flight Vehicle Structures, (Sept. 15, 1976, as amended through December 15, 1978). This incorporation by reference was approved by the Director of the Federal Register in accordance with 5 U.S.C. section 552(a) and 1 CFR part 51. Copies may be obtained from the Naval Publications and Forms Center, 5801 Tabor Avenue, Philadelphia, Pennsylvania, 19120. Copies may be inspected at the FAA, Rotorcraft Standards Staff, 4400 Blue Mount Road, Fort Worth, Texas, or at the National Archives and Records Administration (NARA). For information on the availability of this material at NARA, call 202-741-6030, or go to: *http://www.archives.gov/federal_register/code_of_federal_regulations/ibr_locations.html.*

(5) Pulleys must have close fitting guards to prevent the cables from being displaced or fouled.

(6) Pulleys must lie close enough to the plane passing through the cable to prevent the cable from rubbing against the pulley flange.

(7) No fairlead may cause a change in cable direction of more than 3°.

(8) No clevis pin subject to load or motion and retained only by cotter pins may be used in the control system.

(9) Turnbuckles attached to parts having angular motion must be installed to prevent binding throughout the range of travel.

(10) There must be means for visual inspection at each fairlead, pulley, terminal, and turnbuckle.

(e) Control system joints subject to angular motion must incorporate the following special factors with respect to the ultimate bearing strength of the softest material used as a bearing:

(1) 3.33 for push-pull systems other than ball and roller bearing systems.

(2) 2.0 for cable systems.

(f) For control system joints, the manufacturer's static, non-Brinell rating of ball and roller bearings must not be exceeded.

[Doc. No. 5074, 29 FR 15695, Nov. 24, 1964, as amended by Amdt. 27-11, 41 FR 55469, Dec. 20, 1976; Amdt. 27-26, 55 FR 8001, Mar. 6, 1990; 69 FR 18803, Apr. 9, 2004]

§ 27.687 Spring devices.

(a) Each control system spring device whose failure could cause flutter or other unsafe characteristics must be reliable.

(b) Compliance with paragraph (a) of this section must be shown by tests simulating service conditions.

§ 27.691 Autorotation control mechanism.

Each main rotor blade pitch control mechanism must allow rapid entry into autorotation after power failure.

§ 27.695 Power boost and power-operated control system.

(a) If a power boost or power-operated control system is used, an alternate system must be immediately available that allows continued safe flight and landing in the event of—

(1) Any single failure in the power portion of the system; or

(2) The failure of all engines.

(b) Each alternate system may be a duplicate power portion or a manually operated mechanical system. The power portion includes the power source (such as hydraulic pumps), and such items as valves, lines, and actuators.

(c) The failure of mechanical parts (such as piston rods and links), and the jamming of power cylinders, must be considered unless they are extremely improbable.

LANDING GEAR

§ 27.723 Shock absorption tests.

The landing inertia load factor and the reserve energy absorption capacity of the landing gear must be substantiated by the tests prescribed in §§ 27.725 and 27.727, respectively. These tests must be conducted on the complete rotorcraft or on units consisting of wheel, tire, and shock absorber in their proper relation.

§ 27.725 Limit drop test.

The limit drop test must be conducted as follows:

(a) The drop height must be—

(1) 13 inches from the lowest point of the landing gear to the ground; or

(2) Any lesser height, not less than eight inches, resulting in a drop contact velocity equal to the greatest probable sinking speed likely to occur at ground contact in normal power-off landings.

(b) If considered, the rotor lift specified in § 27.473(a) must be introduced into the drop test by appropriate energy absorbing devices or by the use of an effective mass.

(c) Each landing gear unit must be tested in the attitude simulating the landing condition that is most critical from the standpoint of the energy to be absorbed by it.

(d) When an effective mass is used in showing compliance with paragraph (b) of this section, the following formula may be used instead of more rational computations:

$$W_e = W \times \frac{h+(1-L)d}{h+d}; \text{ and}$$

$$n = n_j \frac{W_e}{W} + L$$

where:

W_e =the effective weight to be used in the drop test (lbs.);

$W = W_M$ for main gear units (lbs.), equal to the static reaction on the particular unit with the rotorcraft in the most critical attitude. A rational method may be used in computing a main gear static reaction, taking into consideration the moment arm between the main wheel reaction and the rotorcraft center of gravity.

$W = W_N$ for nose gear units (lbs.), equal to the vertical component of the static reaction that would exist at the nose wheel, assuming that the mass of the rotorcraft acts at the center of gravity and exerts a force of 1.0g downward and 0.25 *g* forward.

$W = W_T$ for tailwheel units (lbs.), equal to whichever of the following is critical:

(1) The static weight on the tailwheel with the rotorcraft resting on all wheels; or

(2) The vertical component of the ground reaction that would occur at the tailwheel, assuming that the mass of the rotorcraft acts at the center of gravity and exerts a force of 1 *g* downward with the rotorcraft in the maximum nose-up attitude considered in the nose-up landing conditions.

h =specified free drop height (inches).

L =ration of assumed rotor lift to the rotorcraft weight.

d =deflection under impact of the tire (at the proper inflation pressure) plus the vertical component of the axle travels (inches) relative to the drop mass.

n =limit inertia load factor.

n_j =the load factor developed, during impact, on the mass used in the drop test (i.e., the acceleration *dv/dt* in *g*'s recorded in the drop test plus 1.0).

§ 27.727 Reserve energy absorption drop test.

The reserve energy absorption drop test must be conducted as follows:

(a) The drop height must be 1.5 times that specified in § 27.725(a).

(b) Rotor lift, where considered in a manner similar to that prescribed in § 27.725(b), may not exceed 1.5 times the lift allowed under that paragraph.

(c) The landing gear must withstand this test without collapsing. Collapse of the landing gear occurs when a member of the nose, tail, or main gear will not support the rotorcraft in the proper attitude or allows the rotorcraft structure, other than the landing gear and external accessories, to impact the landing surface.

[Doc. No. 5074, 29 FR 15695, Nov. 24, 1964, as amended by Amdt. 27-26, 55 FR 8001, Mar. 6, 1990]

§ 27.729 Retracting mechanism.

For rotorcraft with retractable landing gear, the following apply:

(a) *Loads.* The landing gear, retracting mechansim, wheel-well doors, and supporting structure must be designed for—

(1) The loads occurring in any maneuvering condition with the gear retracted;

(2) The combined friction, inertia, and air loads occurring during retraction and extension at any airspeed up to the design maximum landing gear operating speed; and

(3) The flight loads, including those in yawed flight, occurring with the gear extended at any airspeed up to the design maximum landing gear extended speed.

(b) *Landing gear lock.* A positive means must be provided to keep the gear extended.

(c) *Emergency operation.* When other than manual power is used to operate the gear, emergency means must be provided for extending the gear in the event of—

(1) Any reasonably probable failure in the normal retraction system; or

(2) The failure of any single source of hydraulic, electric, or equivalent energy.

(d) *Operation tests.* The proper functioning of the retracting mechanism must be shown by operation tests.

(e) *Position indicator.* There must be a means to indicate to the pilot when the gear is secured in the extreme positions.

(f) *Control.* The location and operation of the retraction control must meet the requirements of §§ 27.777 and 27.779.

(g) *Landing gear warning.* An aural or equally effective landing gear warning device must be provided that functions continuously when the rotorcraft is in a normal landing mode and the landing gear is not fully extended and locked. A manual shutoff capability must be provided for the warning device and the warning system must automatically reset when the rotorcraft is no longer in the landing mode.

[Amdt. 27-21, 49 FR 44434, Nov. 6, 1984]

§ 27.731 Wheels.

(a) Each landing gear wheel must be approved.

(b) The maximum static load rating of each wheel may not be less than the corresponding static ground reaction with—

(1) Maximum weight; and
(2) Critical center of gravity.
(c) The maximum limit load rating of each wheel must equal or exceed the maximum radial limit load determined under the applicable ground load requirements of this part.

§ 27.733 Tires.
(a) Each landing gear wheel must have a tire—
(1) That is a proper fit on the rim of the wheel; and
(2) Of the proper rating.
(b) The maximum static load rating of each tire must equal or exceed the static ground reaction obtained at its wheel, assuming—
(1) The design maximum weight; and
(2) The most unfavorable center of gravity.
(c) Each tire installed on a retractable landing gear system must, at the maximum size of the tire type expected in service, have a clearance to surrounding structure and systems that is adequate to prevent contact between the tire and any part of the structure or systems.
[Doc. No. 5074, 29 FR 15695, Nov. 24, 1964, as amended by Amdt. 27-11, 41 FR 55469, Dec. 20, 1976]

§ 27.735 Brakes.
For rotorcraft with wheel-type landing gear, a braking device must be installed that is—
(a) Controllable by the pilot;
(b) Usable during power-off landings; and
(c) Adequate to—
(1) Counteract any normal unbalanced torque when starting or stopping the rotor; and
(2) Hold the rotorcraft parked on a 10-degree slope on a dry, smooth pavement.
[Doc. No. 5074, 29 FR 15695, Nov. 24, 1964, as amended by Amdt. 27-21, 49 FR 44434, Nov. 6, 1984]

§ 27.737 Skis.
The maximum limit load rating of each ski must equal or exceed the maximum limit load determined under the applicable ground load requirements of this part.

FLOATS AND HULLS

§ 27.751 Main float buoyancy.
(a) For main floats, the buoyancy necessary to support the maximum weight of the rotorcraft in fresh water must be exceeded by—
(1) 50 percent, for single floats; and
(2) 60 percent, for multiple floats.
(b) Each main float must have enough water-tight compartments so that, with any single main float compartment flooded, the main floats will provide a margin of positive stability great enough to minimize the probability of capsizing.
[Doc. No. 5074, 29 FR 15695, Nov. 24, 1964, as amended by Amdt. 27-2, 33 FR 963, Jan. 26, 1968]

§ 27.753 Main float design.
(a) *Bag floats.* Each bag float must be designed to withstand—
(1) The maximum pressure differential that might be developed at the maximum altitude for which certification with that float is requested; and
(2) The vertical loads prescribed in § 27.521(a), distributed along the length of the bag over three-quarters of its projected area.
(b) *Rigid floats.* Each rigid float must be able to withstand the vertical, horizontal, and side loads prescribed in § 27.521. These loads may be distributed along the length of the float.

§ 27.755 Hulls.
For each rotorcraft, with a hull and auxiliary floats, that is to be approved for both taking off from and landing on water, the hull and auxiliary floats must have enough watertight compartments so that, with any single compartment flooded, the buoyancy of the hull and auxiliary floats (and wheel tires if used) provides a margin of positive stability great enough to minimize the probability of capsizing.

PERSONNEL AND CARGO ACCOMMODATIONS

§ 27.771 Pilot compartment.
For each pilot compartment—
(a) The compartment and its equipment must allow each pilot to perform his duties without unreasonable concentration or fatigue;
(b) If there is provision for a second pilot, the rotorcraft must be controllable with equal safety from either pilot seat; and
(c) The vibration and noise characteristics of cockpit appurtenances may not interfere with safe operation.

§ 27.773 Pilot compartment view.
(a) Each pilot compartment must be free from glare and reflections that could interfere with the pilot's view, and designed so that—
(1) Each pilot's view is sufficiently extensive, clear, and undistorted for safe operation; and
(2) Each pilot is protected from the elements so that moderate rain conditions do not unduly impair his view of the flight path in normal flight and while landing.
(b) If certification for night operation is requested, compliance with paragraph (a) of this section must be shown in night flight tests.

§ 27.775 Windshields and windows.
Windshields and windows must be made of material that will not break into dangerous fragments.
[Amdt. 27-27, 55 FR 38966, Sept. 21, 1990]

§ 27.777 Cockpit controls.
Cockpit controls must be—
(a) Located to provide convenient operation and to prevent confusion and inadvertent operation; and
(b) Located and arranged with respect to the pilots' seats so that there is full and unrestricted movement of each control without interference from the cockpit structure or the pilot's clothing when pilots from 5′2″ to 6′0″ in height are seated.

§ 27.779 Motion and effect of cockpit controls.
Cockpit controls must be designed so that they operate in accordance with the following movements and actuation:
(a) Flight controls, including the collective pitch control, must operate with a sense of motion which corresponds to the effect on the rotorcraft.
(b) Twist-grip engine power controls must be designed so that, for lefthand operation, the motion of the pilot's hand is clockwise to increase power when the hand is viewed from the edge containing the index finger. Other engine power controls, excluding the collective control, must operate with a forward motion to increase power.
(c) Normal landing gear controls must operate downward to extend the landing gear.
[Amdt. 27-21, 49 FR 44434, Nov. 6, 1984]

§ 27.783 Doors.
(a) Each closed cabin must have at least one adequate and easily accessible external door.
(b) Each external door must be located where persons using it will not be endangered by the rotors, propellers, engine intakes, and exhausts when appropriate operating procedures are used. If opening procedures are required, they must be marked inside, on or adjacent to the door opening device.
[Doc. No. 5074, 29 FR 15695, Nov. 24, 1964, as amended by Amdt. 27-26, 55 FR 8001, Mar. 6, 1990]

§ 27.785 Seats, berths, litters, safety belts, and harnesses.
(a) Each seat, safety belt, harness, and adjacent part of the rotorcraft at each station designated for occupancy during takeoff and landing must be free of potentially injurious objects, sharp edges, protuberances, and hard surfaces and must be designed so that a person making proper use of these facilities will not suffer serious injury in an emergency landing as a result of the static inertial load factors specified in § 27.561(b) and dynamic conditions specified in § 27.562.
(b) Each occupant must be protected from serious head injury by a safety belt plus a shoulder harness that will prevent the head from contacting any injurious object except as provided for in § 27.562(c)(5). A shoulder harness (upper torso restraint), in combination with the safety belt, constitutes a torso restraint system as described in TSO-C114.
(c) Each occupant's seat must have a combined safety belt and shoulder harness with a single-point release. Each pilot's combined safety belt and shoulder harness must allow each pilot when seated with safety belt and shoulder harness fastened to perform all functions necessary for flight operations. There must be a means to secure belts and harnesses, when not in use, to prevent interference with the operation of the rotorcraft and with rapid egress in an emergency.
(d) If seat backs do not have a firm handhold, there must be hand grips or rails along each aisle to enable the occupants to steady themselves while using the aisle in moderately rough air.
(e) Each projecting object that could injure persons seated or moving about in the rotorcraft in normal flight must be padded.
(f) Each seat and its supporting structure must be designed for an occupant weight of at least 170 pounds considering the maximum load factors, inertial forces, and reactions between occupant, seat, and safety belt or harness corresponding with the applicable flight and ground load conditions, including the emergency landing conditions of § 27.561(b). In addition—
(1) Each pilot seat must be designed for the reactions resulting from the application of the pilot forces prescribed in § 27.397; and
(2) The inertial forces prescribed in § 27.561(b) must be multiplied by a factor of 1.33 in determining the strength of the attachment of—
(i) Each seat to the structure; and
(ii) Each safety belt or harness to the seat or structure.
(g) When the safety belt and shoulder harness are combined, the rated strength of the safety belt and shoulder harness may not be less than that corresponding to the inertial forces specified in § 27.561(b), considering the occupant weight of at least 170 pounds, considering the dimensional characteristics of the restraint system installation, and using a distribution of at least a 60-percent load to the safety belt and at least a 40-percent load to the shoulder harness. If the safety belt is capable of being used without the shoulder harness, the inertial forces specified must be met by the safety belt alone.
(h) When a headrest is used, the headrest and its supporting structure must be designed to resist the inertia forces specified in § 27.561, with a 1.33 fitting factor and a head weight of at least 13 pounds.
(i) Each seating device system includes the device such as the seat, the cushions, the occupant restraint system, and attachment devices.
(j) Each seating device system may use design features such as crushing or separation of certain parts of the seats to reduce occupant loads for the emergency landing dynamic conditions of § 27.562; otherwise, the system must remain intact and must not interfere with rapid evacuation of the rotorcraft.
(k) For the purposes of this section, a litter is defined as a device designed to carry a nonambulatory person, primarily in a recumbent position, into and on the rotorcraft. Each berth or litter must be designed to withstand the load reaction of an occupant weight of at least 170 pounds when the occupant is subjected to the forward inertial factors specified in § 27.561(b). A berth or litter installed within 15° or less of the longitudinal axis of the rotorcraft must be provided with a padded end-board, cloth diaphram, or equivalent means that can withstand the forward load reaction. A berth or litter oriented greater than 15° with the longitudinal axis of the rotorcraft must be equipped with appropriate restraints, such as straps or safety belts, to withstand the forward load reaction. In addition—
(1) The berth or litter must have a restraint system and must not have corners or other protuberances likely to cause serious injury to a person occupying it during emergency landing conditions; and
(2) The berth or litter attachment and the occupant restraint system attachments to the structure must be designed to withstand the critical loads resulting from flight and ground load conditions and from the conditions prescribed in § 27.561(b). The fitting factor required by § 27.625(d) shall be applied.
[Amdt. 27-21, 49 FR 44434, Nov. 6, 1984, as amended by Amdt. 27-25, 54 FR 47319, Nov. 13, 1989; Amdt. 27-35, 63 FR 43285, Aug. 12, 1998]

§ 27.787 Cargo and baggage compartments.
(a) Each cargo and baggage compartment must be designed for its placarded maximum weight of contents and for the critical load distributions at the appropriate maximum load factors corresponding to the specified flight and ground load conditions, except the emergency landing conditions of § 27.561.
(b) There must be means to prevent the contents of any compartment from becoming a hazard by shifting under the loads specified in paragraph (a) of this section.
(c) Under the emergency landing conditions of § 27.561, cargo and baggage compartments must—

(1) Be positioned so that if the contents break loose they are unlikely to cause injury to the occupants or restrict any of the escape facilities provided for use after an emergency landing; or

(2) Have sufficient strength to withstand the conditions specified in § 27.561 including the means of restraint, and their attachments, required by paragraph (b) of this section. Sufficient strength must be provided for the maximum authorized weight of cargo and baggage at the critical loading distribution.

(d) If cargo compartment lamps are installed, each lamp must be installed so as to prevent contact between lamp bulb and cargo.

[Doc. No. 5074, 29 FR 15695, Nov. 24, 1964, as amended by Amdt. 27-11, 41 FR 55469, Dec. 20, 1976; Amdt. 27-27, 55 FR 38966, Sept. 21, 1990]

§ 27.801 Ditching.

(a) If certification with ditching provisions is requested, the rotorcraft must meet the requirements of this section and §§ 27.807(d), 27.1411 and 27.1415.

(b) Each practicable design measure, compatible with the general characteristics of the rotorcraft, must be taken to minimize the probability that in an emergency landing on water, the behavior of the rotorcraft would cause immediate injury to the occupants or would make it impossible for them to escape.

(c) The probable behavior of the rotorcraft in a water landing must be investigated by model tests or by comparison with rotorcraft of similar configuration for which the ditching characteristics are known. Scoops, flaps, projections, and any other factor likely to affect the hydrodynamic characteristics of the rotorcraft must be considered.

(d) It must be shown that, under reasonably probable water conditions, the flotation time and trim of the rotorcraft will allow the occupants to leave the rotorcraft and enter the life rafts required by § 27.1415. If compliance with this provision is shown by buoyancy and trim computations, appropriate allowances must be made for probable structural damage and leakage. If the rotorcraft has fuel tanks (with fuel jettisoning provisions) that can reasonably be expected to withstand a ditching without leakage, the jettisonable volume of fuel may be considered as buoyancy volume.

(e) Unless the effects of the collapse of external doors and windows are accounted for in the investigation of the probable behavior of the rotorcraft in a water landing (as prescribed in paragraphs (c) and (d) of this section), the external doors and windows must be designed to withstand the probable maximum local pressures.

[Amdt. 27-11, 41 FR 55469, Dec. 20, 1976]

§ 27.805 Flight crew emergency exits.

(a) For rotorcraft with passenger emergency exits that are not convenient to the flight crew, there must be flight crew emergency exits, on both sides of the rotorcraft or as a top hatch in the flight crew area.

(b) Each flight crew emergency exit must be of sufficient size and must be located so as to allow rapid evacuation of the flight crew. This must be shown by test.

(c) Each flight crew emergency exit must not be obstructed by water or flotation devices after an emergency landing on water. This must be shown by test, demonstration, or analysis.

[Doc. No. 29247, 64 FR 45094, Aug. 18, 1999]

§ 27.807 Emergency exits.

(a) *Number and location.* (1) There must be at least one emergency exit on each side of the cabin readily accessible to each passenger. One of these exits must be usable in any probable attitude that may result from a crash;

(2) Doors intended for normal use may also serve as emergency exits, provided that they meet the requirements of this section; and

(3) If emergency flotation devices are installed, there must be an emergency exit accessible to each passenger on each side of the cabin that is shown by test, demonstration, or analysis to;

(i) Be above the waterline; and

(ii) Open without interference from flotation devices, whether stowed or deployed.

(b) *Type and operation.* Each emergency exit prescribed by paragraph (a) of this section must—

(1) Consist of a movable window or panel, or additional external door, providing an unobstructed opening that will admit a 19-by 26-inch ellipse;

(2) Have simple and obvious methods of opening, from the inside and from the outside, which do not require exceptional effort;

(3) Be arranged and marked so as to be readily located and opened even in darkness; and

(4) Be reasonably protected from jamming by fuselage deformation.

(c) *Tests.* The proper functioning of each emergency exit must be shown by test.

(d) *Ditching emergency exits for passengers.* If certification with ditching provisions is requested, the markings required by paragraph (b)(3) of this section must be designed to remain visible if the rotorcraft is capsized and the cabin is submerged.

[Doc. No. 29247, 64 FR 45094, Aug. 18, 1999]

§ 27.831 Ventilation.

(a) The ventilating system for the pilot and passenger compartments must be designed to prevent the presence of excessive quantities of fuel fumes and carbon monoxide.

(b) The concentration of carbon monoxide may not exceed one part in 20,000 parts of air during forward flight or hovering in still air. If the concentration exceeds this value under other conditions, there must be suitable operating restrictions.

§ 27.833 Heaters.

Each combustion heater must be approved.

[Amdt. 27-23, 53 FR 34210, Sept. 2, 1988]

Fire Protection

§ 27.853 Compartment interiors.

For each compartment to be used by the crew or passengers—

(a) The materials must be at least flame-resistant;

(b) [Reserved]

(c) If smoking is to be prohibited, there must be a placard so stating, and if smoking is to be allowed—

(1) There must be an adequate number of self-contained, removable ashtrays; and

(2) Where the crew compartment is separated from the passenger compartment, there must be at least one illuminated sign (using either letters or symbols) notifying all passengers when smoking is prohibited. Signs which notify when smoking is prohibited must—

(i) When illuminated, be legible to each passenger seated in the passenger cabin under all probable lighting conditions; and

(ii) Be so constructed that the crew can turn the illumination on and off.

[Amdt. 27-17, 45 FR 7755, Feb. 4, 1980, as amended by Amdt. 27-37, 64 FR 45095, Aug. 18, 1999]

§ 27.855 Cargo and baggage compartments.

(a) Each cargo and baggage compartment must be constructed of, or lined with, materials that are at least—

(1) Flame resistant, in the case of compartments that are readily accessible to a crewmember in flight; and

(2) Fire resistant, in the case of other compartments.

(b) No compartment may contain any controls, wiring, lines, equipment, or accessories whose damage or failure would affect safe operation, unless those items are protected so that—

(1) They cannot be damaged by the movement of cargo in the compartment; and

(2) Their breakage or failure will not create a fire hazard.

§ 27.859 Heating systems.

(a) *General.* For each heating system that involves the passage of cabin air over, or close to, the exhaust manifold, there must be means to prevent carbon monoxide from entering any cabin or pilot compartment.

(b) *Heat exchangers.* Each heat exchanger must be—

(1) Of suitable materials;

(2) Adequately cooled under all conditions; and

(3) Easily disassembled for inspection.

(c) *Combustion heater fire protection.* Except for heaters which incorporate designs to prevent hazards in the event of fuel leakage in the heater fuel system, fire within the ventilating air passage, or any other heater malfunction, each heater zone must incorporate the fire protection features of the applicable requirements of §§ 27.1183, 27.1185, 27.1189, 27.1191, and be provided with—

(1) Approved, quick-acting fire detectors in numbers and locations ensuring prompt detection of fire in the heater region.

(2) Fire extinguisher systems that provide at least one adequate discharge to all areas of the heater region.

(3) Complete drainage of each part of each zone to minimize the hazards resulting from failure or malfunction of any component containing flammable fluids. The drainage means must be—

(i) Effective under conditions expected to prevail when drainage is needed; and

(ii) Arranged so that no discharged fluid will cause an additional fire hazard.

(4) Ventilation, arranged so that no discharged vapors will cause an additional fire hazard.

(d) *Ventilating air ducts.* Each ventilating air duct passing through any heater region must be fireproof.

(1) Unless isolation is provided by fireproof valves or by equally effective means, the ventilating air duct downstream of each heater must be fireproof for a distance great enough to ensure that any fire originating in the heater can be contained in the duct.

(2) Each part of any ventilating duct passing through any region having a flammable fluid system must be so constructed or isolated from that system that the malfunctioning of any component of that system cannot introduce flammable fluids or vapors into the ventilating airstream.

(e) *Combustion air ducts.* Each combustion air duct must be fireproof for a distance great enough to prevent damage from backfiring or reverse flame propagation.

(1) No combustion air duct may connect with the ventilating airstream unless flames from backfires or reverse burning cannot enter the ventilating airstream under any operating condition, including reverse flow or malfunction of the heater or its associated components.

(2) No combustion air duct may restrict the prompt relief of any backfire that, if so restricted, could cause heater failure.

(f) *Heater control: General.* There must be means to prevent the hazardous accumulation of water or ice on or in any heater control component, control system tubing, or safety control.

(g) *Heater safety controls.* For each combustion heater, safety control means must be provided as follows:

(1) Means independent of the components provided for the normal continuous control of air temperature, airflow, and fuel flow must be provided for each heater to automatically shut off the ignition and fuel supply of that heater at a point remote from that heater when any of the following occurs:

(i) The heat exchanger temperature exceeds safe limits.

(ii) The ventilating air temperature exceeds safe limits.

(iii) The combustion airflow becomes inadequate for safe operation.

(iv) The ventilating airflow becomes inadequate for safe operation.

(2) The means of complying with paragraph (g)(1) of this section for any individual heater must—

(i) Be independent of components serving any other heater, the heat output of which is essential for safe operation; and

(ii) Keep the heater off until restarted by the crew.

(3) There must be means to warn the crew when any heater, the heat output of which is essential for safe operation, has been shut off by the automatic means prescribed in paragraph (g)(1) of this section.

(h) *Air intakes.* Each combustion and ventilating air intake must be located so that no flammable fluids or vapors can enter the heater system—

(1) During normal operation; or

(2) As a result of the malfunction of any other component.

(i) *Heater exhaust.* Each heater exhaust system must meet the requirements of §§ 27.1121 and 27.1123.

(1) Each exhaust shroud must be sealed so that no flammable fluids or hazardous quantities of vapors can reach the exhaust system through joints.

(2) No exhaust system may restrict the prompt relief of any backfire that, if so restricted, could cause heater failure.

(j) *Heater fuel systems.* Each heater fuel system must meet the powerplant fuel system requirements affecting safe heater operation. Each heater fuel system component in the ventilating airstream must be protected by shrouds so that no leakage from those components can enter the ventilating airstream.

(k) *Drains.* There must be means for safe drainage of any fuel that might accumulate in the combustion chamber or the heat exchanger.

(1) Each part of any drain that operates at high temperatures must be protected in the same manner as heater exhausts.

(2) Each drain must be protected against hazardous ice accumulation under any operating condition.

[Doc. No. 5074, 29 FR 15695, Nov. 24, 1964, as amended by Amdt. 27-23, 53 FR 34211, Sept. 2, 1988]

§ 27.861 Fire protection of structure, controls, and other parts.

Each part of the structure, controls, rotor mechanism, and other parts essential to a controlled landing that would be affected by powerplant fires must be fireproof or protected so they can perform their essential functions for at least 5 minutes under any foreseeable powerplant fire conditions.

[Amdt. 27-26, 55 FR 8001, Mar. 6, 1990]

§ 27.863 Flammable fluid fire protection.

(a) In each area where flammable fluids or vapors might escape by leakage of a fluid system, there must be means to minimize the probability of ignition of the fluids and vapors, and the resultant hazards if ignition does occur.

(b) Compliance with paragraph (a) of this section must be shown by analysis or tests, and the following factors must be considered:

(1) Possible sources and paths of fluid leakage, and means of detecting leakage.

(2) Flammability characteristics of fluids, including effects of any combustible or absorbing materials.

(3) Possible ignition sources, including electrical faults, overheating of equipment, and malfunctioning of protective devices.

(4) Means available for controlling or extinguishing a fire, such as stopping flow of fluids, shutting down equipment, fireproof containment, or use of extinguishing agents.

(5) Ability of rotorcraft components that are critical to safety of flight to withstand fire and heat.

(c) If action by the flight crew is required to prevent or counteract a fluid fire (e.g. equipment shutdown or actuation of a fire extinguisher) quick acting means must be provided to alert the crew.

(d) Each area where flammable fluids or vapors might escape by leakage of a fluid system must be identified and defined.

(Secs. 313(a), 601, 603, 604, Federal Aviation Act of 1958 (49 U.S.C. 1354(a), 1421, 1423, 1424), sec. 6(c), Dept. of Transportation Act (49 U.S.C. 1655(c)))

[Amdt. 27-16, 43 FR 50599, Oct. 30, 1978]

External Loads

§ 27.865 External loads.

(a) It must be shown by analysis, test, or both, that the rotorcraft external load attaching means for rotorcraft-load combinations to be used for nonhuman external cargo applications can withstand a limit static load equal to 2.5, or some lower load factor approved under §§ 27.337 through 27.341, multiplied by the maximum external load for which authorization is requested. It must be shown by analysis, test, or both that the rotorcraft external load attaching means and corresponding personnel carrying device system for rotorcraft-load combinations to be used for human external cargo applications can withstand a limit static load equal to 3.5 or some lower load factor, not less than 2.5, approved under §§ 27.337 through 27.341, multiplied by the maximum external load for which authorization is requested. The load for any rotorcraft-load combination class, for any external cargo type, must be applied in the vertical direction. For jettisonable external loads of any applicable external cargo type, the load must also be applied in any direction making the maximum angle with the vertical that can be achieved in service but not less than 30°. However, the 30° angle may be reduced to a lesser angle if—

(1) An operating limitation is established limiting external load operations to such angles for which compliance with this paragraph has been shown; or

(2) It is shown that the lesser angle can not be exceeded in service.

(b) The external load attaching means, for jettisonable rotorcraft-load combinations, must include a quick-release system to enable the pilot to release the external load quickly during flight. The quick-release system must consist of a primary quick release subsystem and a backup quick release subsystem that are isolated from one another. The quick-release system, and the means by which it is controlled, must comply with the following:

(1) A control for the primary quick release subsystem must be installed either on one of the pilot's primary controls or in an equivalently accessible location and must be designed and located so that it may be operated by either the pilot or a crewmember without hazardously limiting the ability to control the rotorcraft during an emergency situation.

(2) A control for the backup quick release subsystem, readily accessible to either the pilot or another crewmember, must be provided.

(3) Both the primary and backup quick release subsystems must—

(i) Be reliable, durable, and function properly with all external loads up to and including the maximum external limit load for which authorization is requested.

(ii) Be protected against electromagnetic interference (EMI) from external and internal sources and against lightning to prevent inadvertent load release.

(A) The minimum level of protection required for jettisonable rotorcraft-load combinations used for nonhuman external cargo is a radio frequency field strength of 20 volts per meter.

(B) The minimum level of protection required for jettisonable rotorcraft-load combinations used for human external cargo is a radio frequency field strength of 200 volts per meter.

(iii) Be protected against any failure that could be induced by a failure mode of any other electrical or mechanical rotorcraft system.

(c) For rotorcraft-load combinations to be used for human external cargo applications, the rotorcraft must—

(1) For jettisonable external loads, have a quick-release system that meets the requirements of paragraph (b) of this section and that—

(i) Provides a dual actuation device for the primary quick release subsystem, and

(ii) Provides a separate dual actuation device for the backup quick release subsystem;

(2) Have a reliable, approved personnel carrying device system that has the structural capability and personnel safety features essential for external occupant safety;

(3) Have placards and markings at all appropriate locations that clearly state the essential system operating instructions and, for the personnel carrying device system, the ingress and egress instructions;

(4) Have equipment to allow direct intercommunication among required crewmembers and external occupants; and

(5) Have the appropriate limitations and procedures incorporated in the flight manual for conducting human external cargo operations.

(d) The critically configured jettisonable external loads must be shown by a combination of analysis, ground tests, and flight tests to be both transportable and releasable throughout the approved operational envelope without hazard to the rotorcraft during normal flight conditions. In addition, these external loads must be shown to be releasable without hazard to the rotorcraft during emergency flight conditions.

(e) A placard or marking must be installed next to the external-load attaching means clearly stating any operational limitations and the maximum authorized external load as demonstrated under § 27.25 and this section.

(f) The fatigue evaluation of § 27.571 of this part does not apply to rotorcraft-load combinations to be used for nonhuman external cargo except for the failure of critical structural elements that would result in a hazard to the rotorcraft. For rotorcraft-load combinations to be used for human external cargo, the fatigue evaluation of § 27.571 of this part applies to the entire quick release and personnel carrying device structural systems and their attachments.

[Amdt. 27-11, 41 FR 55469, Dec. 20, 1976; as amended by Amdt. 27-26, 55 FR 8001, Mar. 6, 1990; Amdt. 27-36, 64 FR 43019, Aug. 6, 1999]

Miscellaneous

§ 27.871 Leveling marks.

There must be reference marks for leveling the rotorcraft on the ground.

§ 27.873 Ballast provisions.

Ballast provisions must be designed and constructed to prevent inadvertent shifting of ballast in flight.

Subpart E—Powerplant

General

§ 27.901 Installation.

(a) For the purpose of this part, the powerplant installation includes each part of the rotorcraft (other than the main and auxiliary rotor structures) that—

(1) Is necessary for propulsion;

(2) Affects the control of the major propulsive units; or

(3) Affects the safety of the major propulsive units between normal inspections or overhauls.

(b) For each powerplant installation—

(1) Each component of the installation must be constructed, arranged, and installed to ensure its continued safe operation between normal inspections or overhauls for the range of temperature and altitude for which approval is requested;

(2) Accessibility must be provided to allow any inspection and maintenance necessary for continued airworthiness;

(3) Electrical interconnections must be provided to prevent differences of potential between major components of the installation and the rest of the rotorcraft;

(4) Axial and radial expansion of turbine engines may not affect the safety of the installation; and

(5) Design precautions must be taken to minimize the possibility of incorrect assembly of components and equipment essential to safe operation of the rotorcraft, except where operation with the incorrect assembly can be shown to be extremely improbable.

(c) The installation must comply with—

(1) The installation instructions provided under § 33.5 of this chapter; and

(2) The applicable provisions of this subpart.

(Secs. 313(a), 601, and 603, 72 Stat. 752, 775, 49 U.S.C. 1354(a), 1421, and 1423; sec. 6(c), 49 U.S.C. 1655(c))

[Doc. No. 5074, 29 FR 15695, Nov. 24, 1964, as amended by Amdt. 27-2, 33 FR 963, Jan. 26, 1968; Amdt. 27-12, 42 FR 15044, Mar. 17, 1977; Amdt. 27-23, 53 FR 34211, Sept. 2, 1988]

§ 27.903 Engines.

(a) *Engine type certification.* Each engine must have an approved type certificate. Reciprocating engines for use in helicopters must be qualified in accordance with § 33.49(d) of this chapter or be otherwise approved for the intended usage.

(b) *Engine or drive system cooling fan blade protection.* (1) If an engine or rotor drive system cooling fan is installed, there must be a means to protect the rotorcraft and allow a safe landing if a fan blade fails. This must be shown by showing that—

(i) The fan blades are contained in case of failure;

(ii) Each fan is located so that a failure will not jeopardize safety; or

(iii) Each fan blade can withstand an ultimate load of 1.5 times the centrifugal force resulting from operation limited by the following:

(A) For fans driven directly by the engine—

(*1*) The terminal engine r.p.m. under uncontrolled conditions; or

(*2*) An overspeed limiting device.

(B) For fans driven by the rotor drive system, the maximum rotor drive system rotational speed to be expected in service, including transients.

(2) Unless a fatigue evaluation under § 27.571 is conducted, it must be shown that cooling fan blades are not operating at resonant conditions within the operating limits of the rotorcraft.

(c) *Turbine engine installation.* For turbine engine installations, the powerplant systems associated with engine control devices, systems, and instrumentation must be designed to give reasonable assurance that those engine operating limitations that adversely affect turbine rotor structural integrity will not be exceeded in service.

(d) *Restart capability:* A means to restart any engine in flight must be provided.

(1) Except for the in-flight shutdown of all engines, engine restart capability must be demonstrated throughout a flight envelope for the rotorcraft.

(2) Following the in-flight shutdown of all engines, in-flight engine restart capability must be provided.

[Doc. No. 5074, 29 FR 15695, Nov. 24, 1964, as amended by Amdt. 27-11, 41 FR 55469, Dec. 20, 1976; Amdt. 27-23, 53 FR 34211, Sept. 2, 1988; Amdt. 27-44, 73 FR 11000, Feb. 29, 2008]

§ 27.907 Engine vibration.

(a) Each engine must be installed to prevent the harmful vibration of any part of the engine or rotorcraft.

(b) The addition of the rotor and the rotor drive system to the engine may not subject the principal rotating parts of the engine to excessive vibration stresses. This must be shown by a vibration investigation.

(c) No part of the rotor drive system may be subjected to excessive vibration stresses.

Rotor Drive System

§ 27.917 Design.

(a) Each rotor drive system must incorporate a unit for each engine to automatically disengage that engine from the main and auxiliary rotors if that engine fails.

(b) Each rotor drive system must be arranged so that each rotor necessary for control in autorotation will continue to be driven by the main rotors after disengagement of the engine from the main and auxiliary rotors.

(c) If a torque limiting device is used in the rotor drive system, it must be located so as to allow continued control of the rotorcraft when the device is operating.

(d) The rotor drive system includes any part necessary to transmit power from the engines to the rotor hubs. This includes gear boxes, shafting, universal joints, couplings, rotor brake assemblies, clutches, supporting bearings for shafting, any attendant accessory pads or drives, and any cooling fans that are a part of, attached to, or mounted on the rotor drive system.

[Doc. No. 5074, 29 FR 15695, Nov. 24, 1964, as amended by Amdt. 27-11, 41 FR 55469, Dec. 20, 1976]

§ 27.921 Rotor brake.

If there is a means to control the rotation of the rotor drive system independently of the engine, any limitations on the use of that means must be specified, and the control for that means must be guarded to prevent inadvertent operation.

§ 27.923 Rotor drive system and control mechanism tests.

(a) Each part tested as prescribed in this section must be in a serviceable condition at the end of the tests. No intervening disassembly which might affect test results may be conducted.

(b) Each rotor drive system and control mechanism must be tested for not less than 100 hours. The test must be conducted on the rotorcraft, and the torque must be absorbed by the rotors to be installed, except that other ground or flight test facilities with other appropriate methods of torque absorption may be used if the conditions of support and vibration closely simulate the conditions that would exist during a test on the rotorcraft.

(c) A 60-hour part of the test prescribed in paragraph (b) of this section must be run at not less than maximum continuous torque and the maximum speed for use with maximum continuous torque. In this test, the main rotor controls must be set in the position that will give maximum longitudinal cyclic pitch change to simulate forward flight. The auxiliary rotor controls must be in the position for normal operation under the conditions of the test.

(d) A 30-hour or, for rotorcraft for which the use of either 30-minute OEI power or continuous OEI power is requested, a 25-hour part of the test prescribed in paragraph (b) of this section must be run at not less than 75 percent of maximum continuous torque and the minimum speed for use with 75 percent of maximum continuous torque. The main and auxiliary rotor controls must be in the position for normal operation under the conditions of the test.

(e) A 10-hour part of the test prescribed in paragraph (b) of this section must be run at not less than takeoff torque and the maximum speed for use with takeoff torque. The main and auxiliary rotor controls must be in the normal position for vertical ascent.

(1) For multiengine rotorcraft for which the use of $2\frac{1}{2}$ minute OEI power is requested, 12 runs during the 10-hour test must be conducted as follows:

(i) Each run must consist of at least one period of $2\frac{1}{2}$ minutes with takeoff torque and the maximum speed for use with takeoff torque on all engines.

(ii) Each run must consist of at least one period for each engine in sequence, during which that engine simulates a power failure and the remaining engines are run at $2\frac{1}{2}$ minute OEI torque and the maximum speed for use with $2\frac{1}{2}$ minute OEI torque for $2\frac{1}{2}$ minutes.

(2) For multiengine turbine-powered rotorcraft for which the use of 30-second and 2-minute OEI power is requested, 10 runs must be conducted as follows:

(i) Immediately following a takeoff run of at least 5 minutes, each power source must simulate a failure, in turn, and apply the maximum torque and the maximum speed for use with 30-second OEI power to the remaining affected drive system power inputs for not less than 30 seconds, followed by application of the maximum torque and the maximum speed for use with 2-minute OEI power for not less than 2 minutes. At least one run sequence must be conducted from a simulated "flight idle" condition. When conducted on a bench test, the test sequence must be conducted following stabilization at takeoff power.

(ii) For the purpose of this paragraph, an affected power input includes all parts of the rotor drive system which can be adversely affected by the application of higher or asymmetric torque and speed prescribed by the test.

(iii) This test may be conducted on a representative bench test facility when engine limitations either preclude repeated use of this power or would result in premature engine removal during the test. The loads, the vibration frequency, and the methods of application to the affected rotor drive system components must be representative of rotorcraft conditions. Test components must be those used to show compliance with the remainder of this section.

(f) The parts of the test prescribed in paragraphs (c) and (d) of this section must be conducted in intervals of not less than 30 minutes and may be accomplished either on the ground or in flight. The part of the test prescribed in paragraph (e) of this section must be conducted in intervals of not less than five minutes.

(g) At intervals of not more than five hours during the tests prescribed in paragraphs (c), (d), and (e) of this section, the engine must be stopped rapidly enough to allow the engine and rotor drive to be automatically disengaged from the rotors.

(h) Under the operating conditions specified in paragraph (c) of this section, 500 complete cycles of lateral control, 500 complete cycles of longitudinal control of the main rotors, and 500 complete cycles of control of each auxiliary rotor must be accomplished. A "complete cycle" involves movement of the controls from the neutral position, through both extreme positions, and back to the neutral position, except that control movements need not produce loads or flapping motions exceeding the maximum loads or motions encountered in flight. The cycling may be accomplished during the testing prescribed in paragraph (c) of this section.

(i) At least 200 start-up clutch engagements must be accomplished—

(1) So that the shaft on the driven side of the clutch is accelerated; and

(2) Using a speed and method selected by the applicant.

(j) For multiengine rotorcraft for which the use of 30-minute OEI power is requested, five runs must be made at 30-minute OEI torque and the maximum speed for use with 30-minute OEI torque, in which each engine, in sequence, is made inoperative and the remaining engine(s) is run for a 30-minute period.

(k) For multiengine rotorcraft for which the use of continuous OEI power is requested, five runs must be made at continuous OEI torque and the maximum speed for use with continuous OEI torque, in which each engine, in sequence, is made inoperative and the remaining engine(s) is run for a 1-hour period.

(Secs. 313(a), 601, and 603, 72 Stat. 752, 775, 49 U.S.C. 1354(a), 1421, and 1423; sec. 6(c), 49 U.S.C. 1655(c))

[Doc. No. 5074, 29 FR 15695, Nov. 24, 1964, as amended by Amdt. 27-2, 33 FR 963, Jan. 26, 1968; Amdt. 27-12, 42 FR 15044, Mar. 17, 1977; Amdt. 27-23, 53 FR 34212, Sept. 2, 1988; Amdt. 27-29, 59 FR 47767, Sept. 16, 1994]

§ 27.927 Additional tests.

(a) Any additional dynamic, endurance, and operational tests, and vibratory investigations necessary to determine that the rotor drive mechanism is safe, must be performed.

(b) If turbine engine torque output to the transmission can exceed the highest engine or transmission torque rating limit, and that output is not directly controlled by the pilot under normal operating conditions (such as where the primary engine power control is accomplished through the flight control), the following test must be made:

(1) Under conditions associated with all engines operating, make 200 applications, for 10 seconds each, or torque that is at least equal to the lesser of—

(i) The maximum torque used in meeting § 27.923 plus 10 percent; or

(ii) The maximum attainable torque output of the engines, assuming that torque limiting devices, if any, function properly.

(2) For multiengine rotorcraft under conditions associated with each engine, in turn, becoming inoperative, apply to the remaining transmission torque inputs the maximum torque attainable under probable operating conditions, assuming that torque limiting devices, if any, function properly. Each transmission input must be tested at this maximum torque for at least 15 minutes.

(3) The tests prescribed in this paragraph must be conducted on the rotorcraft at the maximum rotational speed intended for the power condition of the test and the torque must be absorbed by the rotors to be installed, except that other ground or flight test facilities with other appropriate methods of torque absorption may be used if the conditions of support and vibration closely simulate the conditions that would exist during a test on the rotorcraft.

(c) It must be shown by tests that the rotor drive system is capable of operating under autorotative conditions for 15 minutes after the loss of pressure in the rotor drive primary oil system.

(Secs. 313(a), 601, and 603, 72 Stat. 752, 775, 49 U.S.C. 1354(a), 1421, and 1423; sec. 6(c), 49 U.S.C. 1655(c))

[Amdt. 27-2, 33 FR 963, Jan. 26, 1968, as amended by Amdt. 27-12, 42 FR 15045, Mar. 17, 1977; Amdt. 27-23, 53 FR 34212, Sept. 2, 1988]

§ 27.931 Shafting critical speed.

(a) The critical speeds of any shafting must be determined by demonstration except that analytical methods may be used if reliable methods of analysis are available for the particular design.

(b) If any critical speed lies within, or close to, the operating ranges for idling, power on, and autorotative conditions, the stresses occurring at that speed must be within safe limits. This must be shown by tests.

(c) If analytical methods are used and show that no critical speed lies within the permissible operating ranges, the margins between the calculated critical speeds and the limits of the allowable operating ranges must be adequate to allow for possible variations between the computed and actual values.

§ 27.935 Shafting joints.

Each universal joint, slip joint, and other shafting joints whose lubrication is necessary for operation must have provision for lubrication.

§ 27.939 Turbine engine operating characteristics.

(a) Turbine engine operating characteristics must be investigated in flight to determine that no adverse characteristics (such as stall, surge, or flameout) are present, to a hazardous degree, during normal and emergency operation within the range of operating limitations of the rotorcraft and of the engine.

(b) The turbine engine air inlet system may not, as a result of airflow distortion during normal operation, cause vibration harmful to the engine.

(c) For governor-controlled engines, it must be shown that there exists no hazardous torsional instability of the drive system associated with critical combinations of power, rotational speed, and control displacement.

[Amdt. 27-1, 32 FR 6914, May 5, 1967, as amended by Amdt. 27-11, 41 FR 55469, Dec. 20, 1976]

Fuel System

§ 27.951 General.

(a) Each fuel system must be constructed and arranged to ensure a flow of fuel at a rate and pressure established for proper engine functioning under any likely operating condition, including the maneuvers for which certification is requested.

(b) Each fuel system must be arranged so that—

(1) No fuel pump can draw fuel from more than one tank at a time; or

(2) There are means to prevent introducing air into the system.

(c) Each fuel system for a turbine engine must be capable of sustained operation throughout its flow and pressure range with fuel initially saturated with water at 80°F. and having 0.75cc of free water per gallon added and cooled to the most critical condition for icing likely to be encountered in operation.

[Doc. No. 5074, 29 FR 15695, Nov. 24, 1964, as amended by Amdt. 27-9, 39 FR 35461, Oct. 1, 1974]

§ 27.952 Fuel system crash resistance.

Unless other means acceptable to the Administrator are employed to minimize the hazard of fuel fires to occupants following an otherwise survivable impact (crash landing), the fuel systems must incorporate the design features of this section. These systems must be shown to be capable of sustaining the static and dynamic deceleration loads of this section, considered as ultimate loads acting alone, measured at the system component's center of gravity, without structural damage to system components, fuel tanks, or their attachments that would leak fuel to an ignition source.

(a) *Drop test requirements.* Each tank, or the most critical tank, must be drop-tested as follows:

(1) The drop height must be at least 50 feet.

(2) The drop impact surface must be nondeforming.

(3) The tank must be filled with water to 80 percent of the normal, full capacity.

(4) The tank must be enclosed in a surrounding structure representative of the installation unless it can be established that the surrounding structure is free of projections or other design features likely to contribute to rupture of the tank.

(5) The tank must drop freely and impact in a horizontal position±10°.

(6) After the drop test, there must be no leakage.

(b) *Fuel tank load factors.* Except for fuel tanks located so that tank rupture with fuel release to either significant ignition sources, such as engines, heaters, and auxiliary power units, or occupants is extremely remote, each fuel tank must be designed and installed to retain its contents under the following ultimate inertial load factors, acting alone.

(1) For fuel tanks in the cabin:

(i) Upward—4g.

(ii) Forward—16g.

(iii) Sideward—8g.

(iv) Downward—20g.

(2) For fuel tanks located above or behind the crew or passenger compartment that, if loosened, could injure an occupant in an emergency landing:

(i) Upward—1.5g.

(ii) Forward—8g.

(iii) Sideward—2g.

(iv) Downward—4g.

(3) For fuel tanks in other areas:

(i) Upward—1.5g.

(ii) Forward—4g.

(iii) Sideward—2g.

(iv) Downward—4g.

(c) *Fuel line self-sealing breakaway couplings.* Self-sealing breakaway couplings must be installed unless hazardous relative motion of fuel system components to each other or to local rotorcraft structure is demonstrated to be extremely improbable or unless other means are provided. The couplings or equivalent devices must be installed at all fuel tank-to-fuel line connections, tank-to-tank interconnects, and at other points in the fuel system where local structural deformation could lead to the release of fuel.

(1) The design and construction of self-sealing breakaway couplings must incorporate the following design features:

(i) The load necessary to separate a breakaway coupling must be between 25 to 50 percent of the minimum ultimate failure load (ultimate strength) of the weakest component in the fluid-carrying line. The separation load must in no case be less than 300 pounds, regardless of the size of the fluid line.

(ii) A breakaway coupling must separate whenever its ultimate load (as defined in paragraph (c)(1)(i) of this section) is applied in the failure modes most likely to occur.

(iii) All breakaway couplings must incorporate design provisions to visually ascertain that the coupling is locked together (leak-free) and is open during normal installation and service.

(iv) All breakaway couplings must incorporate design provisions to prevent uncoupling or unintended closing due to operational shocks, vibrations, or accelerations.

(v) No breakaway coupling design may allow the release of fuel once the coupling has performed its intended function.

(2) All individual breakaway couplings, coupling fuel feed systems, or equivalent means must be designed, tested, installed, and maintained so that inadvertent fuel shutoff in flight is improbable in accordance with § 27.955(a) and must comply with the fatigue evaluation requirements of § 27.571 without leaking.

(3) Alternate, equivalent means to the use of breakaway couplings must not create a survivable impact-induced load on the fuel line to which it is installed greater than 25 to 50 percent of the ultimate load (strength) of the weakest component in the line and must comply with the fatigue requirements of § 27.571 without leaking.

(d) *Frangible or deformable structural attachments.* Unless hazardous relative motion of fuel tanks and fuel system components to local rotorcraft structure is demonstrated to be extremely improbable in an otherwise survivable impact, frangible or locally deformable attachments of fuel tanks and fuel system components to local rotorcraft structure must be used. The attachment of fuel tanks and fuel system components to local rotorcraft structure, whether frangible or locally deformable, must be designed such that its separation or relative local deformation will occur without rupture or local tear-out of the fuel tank or fuel system components that will cause fuel leakage. The ultimate strength of frangible or deformable attachments must be as follows:

(1) The load required to separate a frangible attachment from its support structure, or deform a locally deformable attachment relative to its support structure, must be between 25 and 50 percent of the minimum ultimate load (ultimate strength) of the weakest component in the attached system. In no case may the load be less than 300 pounds.

(2) A frangible or locally deformable attachment must separate or locally deform as intended whenever its ultimate load (as defined in paragraph (d)(1) of this section) is applied in the modes most likely to occur.

(3) All frangible or locally deformable attachments must comply with the fatigue requirements of § 27.571.

(e) *Separation of fuel and ignition sources.* To provide maximum crash resistance, fuel must be located as far as practicable from all occupiable areas and from all potential ignition sources.

(f) *Other basic mechanical design criteria.* Fuel tanks, fuel lines, electrical wires, and electrical devices must be designed, constructed, and installed, as far as practicable, to be crash resistant.

(g) *Rigid or semirigid fuel tanks.* Rigid or semirigid fuel tank or bladder walls must be impact and tear resistant.

[Doc. No. 26352, 59 FR 50386, Oct. 3, 1994]

§ 27.953 Fuel system independence.

(a) Each fuel system for multiengine rotorcraft must allow fuel to be supplied to each engine through a system independent of those parts of each system supplying fuel to other engines. However, separate fuel tanks need not be provided for each engine.

(b) If a single fuel tank is used on a multiengine rotorcraft, the following must be provided:

(1) Independent tank outlets for each engine, each incorporating a shutoff valve at the tank. This shutoff valve may also serve as the firewall shutoff valve required by § 27.995 if the line between the valve and the engine compartment does not contain a hazardous amount of fuel that can drain into the engine compartment.

(2) At least two vents arranged to minimize the probability of both vents becoming obstructed simultaneously.

(3) Filler caps designed to minimize the probability of incorrect installation or inflight loss.

(4) A fuel system in which those parts of the system from each tank outlet to any engine are independent of each part of each system supplying fuel to other engines.

§ 27.954 Fuel system lightning protection.

The fuel system must be designed and arranged to prevent the ignition of fuel vapor within the system by—

(a) Direct lightning strikes to areas having a high probability of stroke attachment;

(b) Swept lightning strokes to areas where swept strokes are highly probable; or

(c) Corona and streamering at fuel vent outlets.

[Amdt. 27-23, 53 FR 34212, Sept. 2, 1988]

§ 27.955 Fuel flow.

(a) *General.* The fuel system for each engine must be shown to provide the engine with at least 100 percent of the fuel required under each operating and maneuvering condition to be approved for the rotorcraft including, as applicable, the fuel required to operate the engine(s) under the test conditions required by § 27.927. Unless equivalent methods are used, compliance must be shown by test during which the following provisions are met except that combinations of conditions which are shown to be improbable need not be considered.

(1) The fuel pressure, corrected for critical accelerations, must be within the limits specified by the engine type certificate data sheet.

(2) The fuel level in the tank may not exceed that established as the unusable fuel supply for that tank under § 27.959, plus the minimum additional fuel necessary to conduct the test.

(3) The fuel head between the tank outlet and the engine inlet must be critical with respect to rotorcraft flight attitudes.

(4) The critical fuel pump (for pump-fed systems) is installed to produce (by actual or simulated failure) the critical restriction to fuel flow to be expected from pump failure.

(5) Critical values of engine rotation speed, electrical power, or other sources of fuel pump motive power must be applied.

(6) Critical values of fuel properties which adversely affect fuel flow must be applied.

(7) The fuel filter required by § 27.997 must be blocked to the degree necessary to simulate the accumulation of fuel contamination required to activate the indicator required by § 27.1305(q).

(b) *Fuel transfer systems.* If normal operation of the fuel system requires fuel to be transferred to an engine feed tank, the transfer must occur automatically via a system which has been shown to maintain the fuel level in the engine feed tank within acceptable limits during flight or surface operation of the rotorcraft.

(c) *Multiple fuel tanks.* If an engine can be supplied with fuel from more than one tank, the fuel systems must, in addition to having appropriate manual switching capability, be designed to prevent interruption of fuel flow to that engine, without attention by the flightcrew, when any tank supplying fuel to that engine is depleted of usable fuel during normal operation, and any other tank that normally supplies fuel to the engine alone contains usable fuel.

[Amdt. 27-23, 53 FR 34212, Sept. 2, 1988]

§ 27.959 Unusable fuel supply.

The unusable fuel supply for each tank must be established as not less than the quantity at which the first evidence of malfunction occurs under the most adverse fuel feed condition occurring under any intended operations and flight maneuvers involving that tank.

§ 27.961 Fuel system hot weather operation.

Each suction lift fuel system and other fuel systems with features conducive to vapor formation must be shown by test to operate satisfactorily (within certification limits) when using fuel at a temperature of 110°F under critical operating conditions including, if applicable, the engine operating conditions defined by § 27.927 (b)(1) and (b)(2).

[Amdt. 27-23, 53 FR 34212, Sept. 2, 1988]

§ 27.963 Fuel tanks: general.

(a) Each fuel tank must be able to withstand, without failure, the vibration, inertia, fluid, and structural loads to which it may be subjected in operation.

(b) Each fuel tank of 10 gallons or greater capacity must have internal baffles, or must have external support to resist surging.

(c) Each fuel tank must be separated from the engine compartment by a firewall. At least one-half inch of clear airspace must be provided between the tank and the firewall.

(d) Spaces adjacent to the surfaces of fuel tanks must be ventilated so that fumes cannot accumulate in the tank compartment in case of leakage. If two or more tanks have interconnected outlets, they must be considered as one tank, and the airspaces in those tanks must be interconnected to prevent the flow of fuel from one tank to another as a result of a difference in pressure between those airspaces.

(e) The maximum exposed surface temperature of any component in the fuel tank must be less, by a safe margin as determined by the Administrator, than the lowest expected autoignition temperature of the fuel or fuel vapor in the tank. Compliance with this requirement must be shown under all operating conditions and under all failure or malfunction conditions of all components inside the tank.

(f) Each fuel tank installed in personnel compartments must be isolated by fume-proof and fuel-proof enclosures that are drained and vented to the exterior of the rotorcraft. The design and construction of the enclosures must provide necessary protection for the tank, must be crash resistant during a survivable impact in accordance with § 27.952, and must be adequate to withstand loads and abrasions to be expected in personnel compartments.

(g) Each flexible fuel tank bladder or liner must be approved or shown to be suitable for the particular application and must be puncture resistant. Puncture resistance must be shown by meeting the TSO-C80, paragraph 16.0, requirements using a minimum puncture force of 370 pounds.

(h) Each integral fuel tank must have provisions for inspection and repair of its interior.

[Doc. No. 5074, 29 FR 15695, Nov. 24, 1964, as amended by Amdt. 27-23, 53 FR 34213, Sept. 2, 1988; Amdt. 27-30, 59 FR 50387, Oct. 3, 1994]

§ 27.965 Fuel tank tests.

(a) Each fuel tank must be able to withstand the applicable pressure tests in this section without failure or leakage. If practicable, test pressures may be applied in a manner simulating the pressure distribution in service.

(b) Each conventional metal tank, nonmetallic tank with walls that are not supported by the rotorcraft structure, and integral tank must be subjected to a pressure of 3.5 p.s.i. unless the pressure developed during maximum limit acceleration or emergency deceleration with a full tank exceeds this value, in which case a hydrostatic head, or equivalent test, must be applied to duplicate the acceleration loads as far as possible. However, the pressure need not exceed 3.5 p.s.i. on surfaces not exposed to the acceleration loading.

(c) Each nonmetallic tank with walls supported by the rotorcraft structure must be subjected to the following tests:

(1) A pressure test of at least 2.0 p.s.i. This test may be conducted on the tank alone in conjunction with the test specified in paragraph (c)(2) of this section.

(2) A pressure test, with the tank mounted in the rotorcraft structure, equal to the load developed by the reaction of the contents, with the tank full, during maximum limit acceleration or emergency deceleration. However, the pressure need not exceed 2.0 p.s.i. on surfaces not exposed to the acceleration loading.

(d) Each tank with large unsupported or unstiffened flat areas, or with other features whose failure or deformation could cause leakage, must be subjected to the following test or its equivalent:

(1) Each complete tank assembly and its support must be vibration tested while mounted to simulate the actual installation.

(2) The tank assembly must be vibrated for 25 hours while two-thirds full of any suitable fluid. The amplitude of vibration may not be less than one thirty-second of an inch, unless otherwise substantiated.

(3) The test frequency of vibration must be as follows:

(i) If no frequency of vibration resulting from any r.p.m. within the normal operating range of engine or rotor system speeds is critical, the test frequency of vibration, in number of cycles per minute must, unless a frequency based on a more rational calculation is used, be the number obtained by averaging the maximum and minimum power-on engine speeds (r.p.m.) for reciprocating engine powered rotorcraft or 2,000 c.p.m. for turbine engine powered rotorcraft.

(ii) If only one frequency of vibration resulting from any r.p.m. within the normal operating range of engine or rotor system speeds is critical, that frequency of vibration must be the test frequency.

(iii) If more than one frequency of vibration resulting from any r.p.m. within the normal operating range of engine or rotor system speeds is critical, the most critical of these frequencies must be the test frequency.

(4) Under paragraphs (d)(3)(ii) and (iii) of this section, the time of test must be adjusted to accomplish the same number of vibration cycles as would be accomplished in 25 hours at the frequency specified in paragraph (d)(3)(i) of this section.

(5) During the test, the tank assembly must be rocked at the rate of 16 to 20 complete cycles per minute through an angle of 15 degrees on both sides of the horizontal (30 degrees total), about the most critical axis, for 25 hours. If motion about more than one axis is likely to be critical, the tank must be rocked about each critical axis for 12½ hours.

(Secs. 313(a), 601, and 603, 72 Stat. 752, 775, 49 U.S.C. 1354(a), 1421, and 1423; sec. 6(c), 49 U.S.C. 1655(c))

[Amdt. 27-12, 42 FR 15045, Mar. 17, 1977]

§ 27.967 Fuel tank installation.

(a) Each fuel tank must be supported so that tank loads are not concentrated on unsupported tank surfaces. In addition—

(1) There must be pads, if necessary, to prevent chafing between each tank and its supports;

(2) The padding must be nonabsorbent or treated to prevent the absorption of fuel;

(3) If flexible tank liners are used, they must be supported so that it is not necessary for them to withstand fluid loads; and

(4) Each interior surface of tank compartments must be smooth and free of projections that could cause wear of the liner unless—

(i) There are means for protection of the liner at those points; or

(ii) The construction of the liner itself provides such protection.

(b) Any spaces adjacent to tank surfaces must be adequately ventilated to avoid accumulation of fuel or fumes in those spaces due to minor leakage. If the tank is in a sealed compartment, ventilation may be limited to drain holes that prevent clogging and excessive pressure resulting from altitude changes. If flexible tank liners are installed, the venting arrangement for the spaces between the liner and its container must maintain the proper relationship to tank vent pressures for any expected flight condition.

(c) The location of each tank must meet the requirements of § 27.1185 (a) and (c).

(d) No rotorcraft skin immediately adjacent to a major air outlet from the engine compartment may act as the wall of the integral tank.

[Doc. No. 26352, 59 FR 50387, Oct. 3, 1994]

§ 27.969 Fuel tank expansion space.

Each fuel tank or each group of fuel tanks with interconnected vent systems must have an expansion space of not less than 2 percent of the tank capacity. It must be impossible to fill the fuel tank expansion space inadvertently with the rotorcraft in the normal ground attitude.

[Amdt. 27-23, 53 FR 34213, Sept. 2, 1988]

§ 27.971 Fuel tank sump.

(a) Each fuel tank must have a drainable sump with an effective capacity in any ground attitude to be expected in service of 0.25 percent of the tank capacity or $\frac{1}{16}$ gallon, whichever is greater, unless—

(1) The fuel system has a sediment bowl or chamber that is accessible for preflight drainage and has a minimum capacity of 1 ounce for every 20 gallons of fuel tank capacity; and

(2) Each fuel tank drain is located so that in any ground attitude to be expected in service, water will drain from all parts of the tank to the sediment bowl or chamber.

(b) Each sump, sediment bowl, and sediment chamber drain required by this section must comply with the drain provisions of § 27.999(b).

[Amdt. 27-23, 53 FR 34213, Sept. 2, 1988]

§ 27.973 Fuel tank filler connection.

(a) Each fuel tank filler connection must prevent the entrance of fuel into any part of the rotorcraft other than the tank itself during normal operations and must be crash resistant during a survivable impact in accordance with § 27.952(c). In addition—

(1) Each filler must be marked as prescribed in § 27.1557(c)(1);

(2) Each recessed filler connection that can retain any appreciable quantity of fuel must have a drain that discharges clear of the entire rotorcraft; and

(3) Each filler cap must provide a fuel-tight seal under the fluid pressure expected in normal operation and in a survivable impact.

(b) Each filler cap or filler cap cover must warn when the cap is not fully locked or seated on the filler connection.

[Doc. No. 26352, 59 FR 50387, Oct. 3, 1994]

§ 27.975 Fuel tank vents.

(a) Each fuel tank must be vented from the top part of the expansion space so that venting is effective under all normal flight conditions. Each vent must minimize the probability of stoppage by dirt or ice.

(b) The venting system must be designed to minimize spillage of fuel through the vents to an ignition source in the event of a rollover during landing, ground operation, or a survivable impact.

[Doc. No. 5074, 29 FR 15695, Nov. 24, 1964, as amended by Amdt. 27-23, 53 FR 34213, Sept. 2, 1988; Amdt. 27-30, 59 FR 50387, Oct. 3, 1994; Amdt. 27-35, 63 FR 43285, Aug. 12, 1998]

§ 27.977 Fuel tank outlet.

(a) There must be a fuel stainer for the fuel tank outlet or for the booster pump. This strainer must—

(1) For reciprocating engine powered rotorcraft, have 8 to 16 meshes per inch; and

(2) For turbine engine powered rotorcraft, prevent the passage of any object that could restrict fuel flow or damage any fuel system component.

(b) The clear area of each fuel tank outlet strainer must be at least five times the area of the outlet line.

(c) The diameter of each strainer must be at least that of the fuel tank outlet.

(d) Each finger strainer must be accessible for inspection and cleaning.

[Amdt. 27-11, 41 FR 55470, Dec. 20, 1976]

Fuel System Components

§ 27.991 Fuel pumps.

Compliance with § 27.955 may not be jeopardized by failure of—

(a) Any one pump except pumps that are approved and installed as parts of a type certificated engine; or

(b) Any component required for pump operation except, for engine driven pumps, the engine served by that pump.

[Amdt. 27-23, 53 FR 34213, Sept. 2, 1988]

§ 27.993 Fuel system lines and fittings.

(a) Each fuel line must be installed and supported to prevent excessive vibration and to withstand loads due to fuel pressure and accelerated flight conditions.

(b) Each fuel line connected to components of the rotorcraft between which relative motion could exist must have provisions for flexibility.

(c) Flexible hose must be approved.

(d) Each flexible connection in fuel lines that may be under pressure or subjected to axial loading must use flexible hose assemblies.

(e) No flexible hose that might be adversely affected by high temperatures may be used where excessive temperatures will exist during operation or after engine shutdown.

[Doc. No. 5074, 29 FR 15695, Nov. 24, 1964, as amended by Amdt. 27-2, 33 FR 964, Jan. 26, 1968]

§ 27.995 Fuel valves.

(a) There must be a positive, quick-acting valve to shut off fuel to each engine individually.

(b) The control for this valve must be within easy reach of appropriate crewmembers.

(c) Where there is more than one source of fuel supply there must be means for independent feeding from each source.

(d) No shutoff valve may be on the engine side of any firewall.

§ 27.997 Fuel strainer or filter.

There must be a fuel strainer or filter between the fuel tank outlet and the inlet of the first fuel system component which is susceptible to fuel contamination, including but not limited to the fuel metering device or an engine positive displacement pump, whichever is nearer the fuel tank outlet. This fuel strainer or filter must—

(a) Be accessible for draining and cleaning and must incorporate a screen or element which is easily removable;

(b) Have a sediment trap and drain except that it need not have a drain if the strainer or filter is easily removable for drain purposes;

(c) Be mounted so that its weight is not supported by the connecting lines or by the inlet or outlet connections of the strainer or filter itself, unless adequate strength margins under all loading conditions are provided in the lines and connections; and

(d) Provide a means to remove from the fuel any contaminant which would jeopardize the flow of fuel through rotorcraft or engine fuel system components required for proper rotorcraft fuel system or engine fuel system operation.

[Amdt. 27-9, 39 FR 35461, Oct. 1, 1974, as amended by Amdt. 27-20, 49 FR 6849, Feb. 23, 1984; Amdt. 27-23, 53 FR 34213, Sept. 2, 1988]

§ 27.999 Fuel system drains.

(a) There must be at least one accessible drain at the lowest point in each fuel system to completely drain the system with the rotorcraft in any ground attitude to be expected in service.

(b) Each drain required by paragraph (a) of this section must—

(1) Discharge clear of all parts of the rotorcraft;

(2) Have manual or automatic means to assure positive closure in the off position; and

(3) Have a drain valve—

(i) That is readily accessible and which can be easily opened and closed; and

(ii) That is either located or protected to prevent fuel spillage in the event of a landing with landing gear retracted.

[Doc. No. 574, 29 FR 15695, Nov. 24, 1964, as amended by Amdt. 27-11, 41 FR 55470, Dec. 20, 1976; Amdt. 27-23, 53 FR 34213, Sept. 2, 1988]

Oil System

§ 27.1011 Engines: General.

(a) Each engine must have an independent oil system that can supply it with an appropriate quantity of oil at a temperature not above that safe for continuous operation.

(b) The usable oil capacity of each system may not be less than the product of the endurance of the rotorcraft under critical operating conditions and the maximum oil consumption of the engine under the same conditions, plus a suitable margin to ensure adequate circulation and cooling. Instead of a rational analysis of endurance and consumption, a usable oil capacity of one gallon for each 40 gallons of usable fuel may be used.

(c) The oil cooling provisions for each engine must be able to maintain the oil inlet temperature to that engine at or below the maximum established value. This must be shown by flight tests.

[Doc. No. 5074, 29 FR 15695, Nov. 24, 1964, as amended by Amdt. 27-23, 53 FR 34213, Sept. 2, 1988]

§ 27.1013 Oil tanks.

Each oil tank must be designed and installed so that—

(a) It can withstand, without failure, each vibration, inertia, fluid, and structural load expected in operation;

(b) [Reserved]

(c) Where used with a reciprocating engine, it has an expansion space of not less than the greater of 10 percent of the tank capacity or 0.5 gallon, and where used with a turbine engine, it has an expansion space of not less than 10 percent of the tank capacity.

(d) It is impossible to fill the tank expansion space inadvertently with the rotorcraft in the normal ground attitude;

(e) Adequate venting is provided; and

(f) There are means in the filler opening to prevent oil overflow from entering the oil tank compartment.

[Doc. No. 5074, 29 FR 15695, Nov. 24, 1964, as amended by Amdt. 27-9, 39 FR 35461, Oct. 1, 1974]

§ 27.1015 Oil tank tests.

Each oil tank must be designed and installed so that it can withstand, without leakage, an internal pressure of 5 p.s.i., except that each pressurized oil tank used with a turbine engine must be designed and installed so that it can withstand, without leakage, an internal pressure of 5 p.s.i., plus the maximum operating pressure of the tank.

[Amdt. 27-9, 39 FR 35462, Oct. 1, 1974]

§ 27.1017 Oil lines and fittings.

(a) Each oil line must be supported to prevent excessive vibration.

(b) Each oil line connected to components of the rotorcraft between which relative motion could exist must have provisions for flexibility.

(c) Flexible hose must be approved.

(d) Each oil line must have an inside diameter of not less than the inside diameter of the engine inlet or outlet. No line may have splices between connections.

§ 27.1019 Oil strainer or filter.

(a) Each turbine engine installation must incorporate an oil strainer or filter through which all of the engine oil flows and which meets the following requirements:

(1) Each oil strainer or filter that has a bypass must be constructed and installed so that oil will flow at the normal rate through the rest of the system with the strainer or filter completely blocked.

(2) The oil strainer or filter must have the capacity (with respect to operating limitations established for the engine) to ensure that engine oil system functioning is not impaired when the oil is contaminated to a degree (with respect to particle size and density) that is greater than that established for the engine under Part 33 of this chapter.

(3) The oil strainer or filter, unless it is installed at an oil tank outlet, must incorporate a means to indicate contamination before it reaches the capacity established in accordance with paragraph (a)(2) of this section.

(4) The bypass of a strainer or filter must be constructed and installed so that the release of collected contaminants is minimized by appropriate location of the bypass to ensure that collected contaminants are not in the bypass flow path.

(5) An oil strainer or filter that has no bypass, except one that is installed at an oil tank outlet, must have a means to connect it to the warning system required in § 27.1305(r).

(b) Each oil strainer or filter in a powerplant installation using reciprocating engines must be constructed and installed so that oil will flow at the normal rate through the rest of the system with the strainer or filter element completely blocked.

[Amdt. 27-9, 39 FR 35462, Oct. 1, 1974, as amended by Amdt. 27-20, 49 FR 6849, Feb. 23, 1984; Amdt. 27-23, 53 FR 34213, Sept. 2, 1988]

§ 27.1021 Oil system drains.

A drain (or drains) must be provided to allow safe drainage of the oil system. Each drain must—

(a) Be accessible; and

(b) Have manual or automatic means for positive locking in the closed position.

[Amdt. 27-20, 49 FR 6849, Feb. 23, 1984]

§ 27.1027 Transmissions and gearboxes: General.

(a) The lubrication system for components of the rotor drive system that require continuous lubrication must be sufficiently independent of the lubrication systems of the engine(s) to ensure lubrication during autorotation.

(b) Pressure lubrication systems for transmissions and gearboxes must comply with the engine oil system requirements of §§ 27.1013 (except paragraph (c)), 27.1015, 27.1017, 27.1021, and 27.1337(d).

(c) Each pressure lubrication system must have an oil strainer or filter through which all of the lubricant flows and must—

(1) Be designed to remove from the lubricant any contaminant which may damage transmission and drive system components or impede the flow of lubricant to a hazardous degree;

(2) Be equipped with a means to indicate collection of contaminants on the filter or strainer at or before opening of the bypass required by paragraph (c)(3) of this section; and

(3) Be equipped with a bypass constructed and installed so that—

(i) The lubricant will flow at the normal rate through the rest of the system with the strainer or filter completely blocked; and

(ii) The release of collected contaminants is minimized by appropriate location of the bypass to ensure that collected contaminants are not in the bypass flowpath.

(d) For each lubricant tank or sump outlet supplying lubrication to rotor drive systems and rotor drive system components, a screen must be provided to prevent entrance into the lubrication system of any object that might obstruct the flow of lubricant from the outlet to the filter required by paragraph (c) of this section. The requirements of paragraph (c) do not apply to screens installed at lubricant tank or sump outlets.

(e) Splash-type lubrication systems for rotor drive system gearboxes must comply with §§ 27.1021 and 27.1337(d).

[Amdt. 27-23, 53 FR 34213, Sept. 2, 1988, as amended by Amdt. 27-37, 64 FR 45095, Aug. 18, 1999]

Cooling

§ 27.1041 General.

(a) Each powerplant cooling system must be able to maintain the temperatures of powerplant components within the limits established for these components under critical surface (ground or water) and flight operating conditions for which certification is required and after normal shutdown. Powerplant components to be considered include but may not be limited to engines, rotor drive system components, auxiliary power units, and the cooling or lubricating fluids used with these components.

(b) Compliance with paragraph (a) of this section must be shown in tests conducted under the conditions prescribed in that paragraph.

[Doc. No. 5074, 29 FR 15695, Nov. 24, 1964, as amended by Amdt. 27-23, 53 FR 34213, Sept. 2, 1988]

§ 27.1043 Cooling tests.

(a) *General.* For the tests prescribed in § 27.1041(b), the following apply:

(1) If the tests are conducted under conditions deviating from the maximum ambient atmospheric temperature specified in paragraph (b) of this section, the recorded powerplant temperatures must be corrected under paragraphs (c) and (d) of this section unless a more rational correction method is applicable.

(2) No corrected temperature determined under paragraph (a)(1) of this section may exceed established limits.

(3) For reciprocating engines, the fuel used during the cooling tests must be of the minimum grade approved for the engines, and the mixture settings must be those normally used in the flight stages for which the cooling tests are conducted.

(4) The test procedures must be as prescribed in § 27.1045.

(b) *Maximum ambient atmospheric temperature.* A maximum ambient atmospheric temperature corresponding to sea level conditions of at least 100 degrees F. must be established. The assumed temperature lapse rate is 3.6 degrees F. per thousand feet of altitude above sea level until a temperature of −69.7 degrees F. is reached, above which altitude the temperature is considered constant at −69.7 degrees F. However, for winterization installations, the applicant may select a maximum ambient atmospheric temperature corresponding to sea level conditions of less than 100 degrees F.

(c) *Correction factor (except cylinder barrels).* Unless a more rational correction applies, temperatures of engine fluids and power-plant components (except cylinder barrels) for which temperature limits are established, must be corrected by adding to them the difference between the maximum ambient atmospheric temperature and the temperature of the ambient air at the time of the first occurrence of the maximum component or fluid temperature recorded during the cooling test.

(d) *Correction factor for cylinder barrel temperatures.* Cylinder barrel temperatures must be corrected by adding to them 0.7 times the difference between the maximum ambient atmospheric temperature and the temperature of the ambient air at the time of the first occurrence of the maximum cylinder barrel temperature recorded during the cooling test.

(Secs. 313(a), 601, 603, 604, and 605 of the Federal Aviation Act of 1958 (49 U.S.C. 1354(a), 1421, 1423, 1424, and 1425); and sec. 6(c) of the Dept. of Transportation Act (49 U.S.C. 1655(c)))

[Doc. No. 5074, 29 FR 15695, Nov. 24, 1964, as amended by Amdt. 27-11, 41 FR 55470, Dec. 20, 1976; Amdt. 27-14, 43 FR 2325, Jan. 16, 1978]

§ 27.1045 Cooling test procedures.

(a) *General.* For each stage of flight, the cooling tests must be conducted with the rotorcraft—

(1) In the configuration most critical for cooling; and

(2) Under the conditions most critical for cooling.

(b) *Temperature stabilization.* For the purpose of the cooling tests, a temperature is "stabilized" when its rate of change is less than two degrees F. per minute. The following component and engine fluid temperature stabilization rules apply:

(1) For each rotorcraft, and for each stage of flight—

(i) The temperatures must be stabilized under the conditions from which entry is made into the stage of flight being investigated; or

(ii) If the entry condition normally does not allow temperatures to stabilize, operation through the full entry condition must be conducted before entry into the stage of flight being investigated in order to allow the temperatures to attain their natural levels at the time of entry.

(2) For each helicopter during the takeoff stage of flight, the climb at takeoff power must be preceded by a period of hover during which the temperatures are stabilized.

(c) *Duration of test.* For each stage of flight the tests must be continued until—

(1) The temperatures stabilize or 5 minutes after the occurrence of the highest temperature recorded, as appropriate to the test condition;

(2) That stage of flight is completed; or

(3) An operating limitation is reached.

[Doc. No. 5074, 29 FR 15695, Nov. 24, 1964, as amended by Amdt. 27-23, 53 FR 34214, Sept. 2, 1988]

Induction System

§ 27.1091 Air induction.

(a) The air induction system for each engine must supply the air required by that engine under the operating conditions and maneuvers for which certification is requested.

(b) Each cold air induction system opening must be outside the cowling if backfire flames can emerge.

(c) If fuel can accumulate in any air induction system, that system must have drains that discharge fuel—

(1) Clear of the rotorcraft; and

(2) Out of the path of exhaust flames.

PART 27

FAR

(d) For turbine engine powered rotorcraft—

(1) There must be means to prevent hazardous quantities of fuel leakage or overflow from drains, vents, or other components of flammable fluid systems from entering the engine intake system; and

(2) The air inlet ducts must be located or protected so as to minimize the ingestion of foreign matter during takeoff, landing, and taxiing.

[Doc. No. 5074, 29 FR 15695, Nov. 24, 1964, as amended by Amdt. 27-2, 33 FR 964, Jan. 26, 1968; Amdt. 27-23, 53 FR 34214, Sept. 2, 1988]

§ 27.1093 Induction system icing protection.

(a) *Reciprocating engines.* Each reciprocating engine air induction system must have means to prevent and eliminate icing. Unless this is done by other means, it must be shown that, in air free of visible moisture at a temperature of 30 degrees F., and with the engines at 75 percent of maximum continuous power—

(1) Each rotorcraft with sea level engines using conventional venturi carburetors has a preheater that can provide a heat rise of 90 degrees F.;

(2) Each rotorcraft with sea level engines using carburetors tending to prevent icing has a sheltered alternate source of air, and that the preheat supplied to the alternate air intake is not less than that provided by the engine cooling air downstream of the cylinders;

(3) Each rotorcraft with altitude engines using conventional venturi carburetors has a preheater capable of providing a heat rise of 120 degrees F.; and

(4) Each rotorcraft with altitude engines using carburetors tending to prevent icing has a preheater that can provide a heat rise of—

(i) 100 degrees F.; or

(ii) If a fluid deicing system is used, at least 40 degrees F.

(b) *Turbine engine.* (1) It must be shown that each turbine engine and its air inlet system can operate throughout the flight power range of the engine (including idling)—

(i) Without accumulating ice on engine or inlet system components that would adversely affect engine operation or cause a serious loss of power under the icing conditions specified in appendix C of Part 29 of this chapter; and

(ii) In snow, both falling and blowing, without adverse effect on engine operation, within the limitations established for the rotorcraft.

(2) Each turbine engine must idle for 30 minutes on the ground, with the air bleed available for engine icing protection at its critical condition, without adverse effect, in an atmosphere that is at a temperature between 15° and 30°F (between -9° and -1°C) and has a liquid water content not less than 0.3 gram per cubic meter in the form of drops having a mean effective diameter not less than 20 microns, followed by momentary operation at takeoff power or thrust. During the 30 minutes of idle operation, the engine may be run up periodically to a moderate power or thrust setting in a manner acceptable to the Administrator.

(c) *Supercharged reciprocating engines.* For each engine having superchargers to pressurize the air before it enters the carburetor, the heat rise in the air caused by that supercharging at any altitude may be utilized in determining compliance with paragraph (a) of this section if the heat rise utilized is that which will be available, automatically, for the applicable altitude and operating condition because of supercharging.

(Secs. 313(a), 601, and 603, 72 Stat. 752, 775, 49 U.S.C. 1354(a), 1421, and 1423; sec. 6(c), 49 U.S.C. 1655(c))

[Doc. No. 5074, 29 FR 15695, Nov. 24, 1964, as amended by Amdt. 27-11, 41 FR 55470, Dec. 20, 1976; Amdt. 27-12, 42 FR 15045, Mar. 17, 1977; Amdt. 27-20, 49 FR 6849, Feb. 23, 1984; Amdt. 27-23, 53 FR 34214, Sept. 2, 1988]

Exhaust System

§ 27.1121 General.

For each exhaust system—

(a) There must be means for thermal expansion of manifolds and pipes;

(b) There must be means to prevent local hot spots;

(c) Exhaust gases must discharge clear of the engine air intake, fuel system components, and drains;

(d) Each exhaust system part with a surface hot enough to ignite flammable fluids or vapors must be located or shielded so that leakage from any system carrying flammable fluids or vapors will not result in a fire caused by impingement of the fluids or vapors on any part of the exhaust system including shields for the exhaust system;

(e) Exhaust gases may not impair pilot vision at night due to glare;

(f) If significant traps exist, each turbine engine exhaust system must have drains discharging clear of the rotorcraft, in any normal ground and flight attitudes, to prevent fuel accumulation after the failure of an attempted engine start;

(g) Each exhaust heat exchanger must incorporate means to prevent blockage of the exhaust port after any internal heat exchanger failure.

(Secs. 313(a), 601, and 603, 72 Stat. 752, 775, 49 U.S.C. 1354(a), 1421, and 1423; sec. 6(c), 49 U.S.C. 1655(c))

[Doc. No. 5074, 29 FR 15695, Nov. 24, 1964 as amended by Amdt. 27-12, 42 FR 15045, Mar. 17, 1977]

§ 27.1123 Exhaust piping.

(a) Exhaust piping must be heat and corrosion resistant, and must have provisions to prevent failure due to expansion by operating temperatures.

(b) Exhaust piping must be supported to withstand any vibration and inertia loads to which it would be subjected in operations.

(c) Exhaust piping connected to components between which relative motion could exist must have provisions for flexibility.

[Amdt. 27-11, 41 FR 55470, Dec. 20, 1976]

Powerplant Controls and Accessories

§ 27.1141 Powerplant controls: general.

(a) Powerplant controls must be located and arranged under § 27.777 and marked under § 27.1555.

(b) Each flexible powerplant control must be approved.

(c) Each control must be able to maintain any set position without—

(1) Constant attention; or

(2) Tendency to creep due to control loads or vibration.

(d) Controls of powerplant valves required for safety must have—

(1) For manual valves, positive stops or in the case of fuel valves suitable index provisions, in the open and closed position; and

(2) For power-assisted valves, a means to indicate to the flight crew when the valve—

(i) Is in the fully open or fully closed position; or

(ii) Is moving between the fully open and fully closed position.

(e) For turbine engine powered rotorcraft, no single failure or malfunction, or probable combination thereof, in any powerplant control system may cause the failure of any powerplant function necessary for safety.

(Secs. 313(a), 601, and 603, 72 Stat. 752, 775, 49 U.S.C. 1354(a), 1421, and 1423; sec. 6(c), 49 U.S.C. 1655(c))

[Doc. No. 5074, 29 FR 15695, Nov. 24, 1964, as amended by Amdt. 27-12, 42 FR 15045, Mar. 17, 1977; Amdt. 27-23, 53 FR 34214, Sept. 2, 1988; Amdt. 27-33, 61 FR 21907, May 10, 1996]

§ 27.1143 Engine controls.

(a) There must be a separate power control for each engine.

(b) Power controls must be grouped and arranged to allow—

(1) Separate control of each engine; and

(2) Simultaneous control of all engines.

(c) Each power control must provide a positive and immediately responsive means of controlling its engine.

(d) If a power control incorporates a fuel shutoff feature, the control must have a means to prevent the inadvertent movement of the control into the shutoff position. The means must—

(1) Have a positive lock or stop at the idle position; and

(2) Require a separate and distinct operation to place the control in the shutoff position.

(e) For rotorcraft to be certificated for a 30-second OEI power rating, a means must be provided to automatically activate and control the 30-second OEI power and prevent any engine from exceeding the installed engine limits associated with the 30-second OEI power rating approved for the rotorcraft.

[Doc. No. 5074, 29 FR 15695, Nov. 24, 1964, as amended by Amdt. 27-11, 41 FR 55470, Dec. 20, 1976; Amdt. 27-23, 53 FR 34214, Sept. 2, 1988; Amdt. 27-29, 59 FR 47767, Sept. 16, 1994]

§ 27.1145 Ignition switches.

(a) There must be means to quickly shut off all ignition by the grouping of switches or by a master ignition control.

(b) Each group of ignition switches, except ignition switches for turbine engines for which continuous ignition is not required, and each master ignition control must have a means to prevent its inadvertent operation.

(Secs. 313(a), 601, and 603, 72 Stat. 752, 775, 49 U.S.C. 1354(a), 1421, and 1423; sec. 6(c), 49 U.S.C. 1655(c))

[Doc. No. 5074, 29 FR 15695, Nov. 24, 1964, as amended by Amdt. 27-12, 42 FR 15045, Mar. 17, 1977]

§ 27.1147 Mixture controls.

If there are mixture controls, each engine must have a separate control and the controls must be arranged to allow—

(a) Separate control of each engine; and

(b) Simultaneous control of all engines.

§ 27.1151 Rotor brake controls.

(a) It must be impossible to apply the rotor brake inadvertently in flight.

(b) There must be means to warn the crew if the rotor brake has not been completely released before takeoff.

[Doc. No. 28008, 61 FR 21907, May 10, 1996]

§ 27.1163 Powerplant accessories.

(a) Each engine-mounted accessory must—

(1) Be approved for mounting on the engine involved;

(2) Use the provisions on the engine for mounting; and

(3) Be sealed in such a way as to prevent contamination of the engine oil system and the accessory system.

(b) Unless other means are provided, torque limiting means must be provided for accessory drives located on any component of the transmission and rotor drive system to prevent damage to these components from excessive accessory load.

[Amdt. 27-2, 33 FR 964, Jan. 26, 1968, as amended by Amdt. 27-20, 49 FR 6849, Feb. 23, 1984; Amdt. 27-23, 53 FR 34214, Sept. 2, 1988]

Powerplant Fire Protection

§ 27.1183 Lines, fittings, and components.

(a) Except as provided in paragraph (b) of this section, each line, fitting, and other component carrying flammable fluid in any area subject to engine fire conditions must be fire resistant, except that flammable fluid tanks and supports which are part of and attached to the engine must be fireproof or be enclosed by a fireproof shield unless damage by fire to any non-fireproof part will not cause leakage or spillage of flammable fluid. Components must be shielded or located so as to safeguard against the ignition of leaking flammable fluid. An integral oil sump of less than 25-quart capacity on a reciprocating engine need not be fireproof nor be enclosed by a fireproof shield.

(b) Paragraph (a) does not apply to—

(1) Lines, fittings, and components which are already approved as part of a type certificated engine; and

(2) Vent and drain lines, and their fittings, whose failure will not result in, or add to, a fire hazard.

(c) Each flammable fluid drain and vent must discharge clear of the induction system air inlet.

[Doc. No. 5074, 29 FR 15695, Nov. 24, 1964, as amended by Amdt. 27-1, 32 FR 6914, May 5, 1967; Amdt. 27-9, 39 FR 35462, Oct. 1, 1974; Amdt. 27-20, 49 FR 6849, Feb. 23, 1984]

§ 27.1185 Flammable fluids.

(a) Each fuel tank must be isolated from the engines by a firewall or shroud.

(b) Each tank or reservoir, other than a fuel tank, that is part of a system containing flammable fluids or gases must be isolated from the engine by a firewall or shroud, unless the design of the system, the materials used in the tank and its supports, the shutoff means, and the connections, lines and controls provide a degree of safety equal to that which would exist if the tank or reservoir were isolated from the engines.

(c) There must be at least one-half inch of clear airspace between each tank and each firewall or shroud isolating that tank, unless equivalent means are used to prevent heat transfer from each engine compartment to the flammable fluid.

(d) Absorbent materials close to flammable fluid system components that might leak must be covered or treated to prevent the absorption of hazardous quantities of fluids.

[Doc. No. 5074, 29 FR 15695, Nov. 24, 1964, as amended by Amdt. 27-2, 33 FR 964, Jan. 26, 1968; Amdt. 27-11, 41 FR 55470, Dec. 20, 1976; Amdt. 27-37, 64 FR 45095, Aug. 18, 1999]

§ 27.1187 Ventilation and drainage.

Each compartment containing any part of the powerplant installation must have provision for ventilation and drainage of flammable fluids. The drainage means must be—

(a) Effective under conditions expected to prevail when drainage is needed, and

(b) Arranged so that no discharged fluid will cause an additional fire hazard.

[Doc. No. 29247, 64 FR 45095, Aug. 18, 1999]

§ 27.1189 Shutoff means.

(a) There must be means to shut off each line carrying flammable fluids into the engine compartment, except—

(1) Lines, fittings, and components forming an intergral part of an engine;

(2) For oil systems for which all components of the system, including oil tanks, are fireproof or located in areas not subject to engine fire conditions; and

(3) For reciprocating engine installations only, engine oil system lines in installation using engines of less than 500 cu. in. displacement.

(b) There must be means to guard against inadvertent operation of each shutoff, and to make it possible for the crew to reopen it in flight after it has been closed.

(c) Each shutoff valve and its control must be designed, located, and protected to function properly under any condition likely to result from an engine fire.

[Doc. No. 5074, 29 FR 15695, Nov. 24, 1964, as amended by Amdt. 27-2, 33 FR 964, Jan. 26, 1968; Amdt. 27-20, 49 FR 6850, Feb. 23, 1984; Amdt. 27-23, 53 FR 34214, Sept. 2, 1988]

§ 27.1191 Firewalls.

(a) Each engine, including the combustor, turbine, and tailpipe sections of turbine engines must be isolated by a firewall, shroud, or equivalent means, from personnel compartments, structures, controls, rotor mechanisms, and other parts that are—

(1) Essential to a controlled landing: and

(2) Not protected under § 27.861.

(b) Each auxiliary power unit and combustion heater, and any other combustion equipment to be used in flight, must be isolated from the rest of the rotorcraft by firewalls, shrouds, or equivalent means.

(c) In meeting paragraphs (a) and (b) of this section, account must be taken of the probable path of a fire as affected by the airflow in normal flight and in autorotation.

(d) Each firewall and shroud must be constructed so that no hazardous quantity of air, fluids, or flame can pass from any engine compartment to other parts of the rotorcraft.

(e) Each opening in the firewall or shroud must be sealed with close-fitting, fireproof grommets, bushings, or firewall fittings.

(f) Each firewall and shroud must be fireproof and protected against corrosion.

[Doc. No. 5074, 29 FR 15695, Nov. 24, 1964, as amended by Amdt. 27-2, 22 FR 964, Jan. 26, 1968]

§ 27.1193 Cowling and engine compartment covering.

(a) Each cowling and engine compartment covering must be constructed and supported so that it can resist the vibration, inertia, and air loads to which it may be subjected in operation.

(b) There must be means for rapid and complete drainage of each part of the cowling or engine compartment in the normal ground and flight attitudes.

(c) No drain may discharge where it might cause a fire hazard.

(d) Each cowling and engine compartment covering must be at least fire resistant.

(e) Each part of the cowling or engine compartment covering subject to high temperatures due to its nearness to exhaust system parts or exhaust gas impingement must be fireproof.

(f) A means of retaining each openable or readily removable panel, cowling, or engine or rotor drive system covering must be provided to preclude hazardous damage to rotors or critical control components in the event of structural or mechanical failure of the normal retention means, unless such failure is extremely improbable.

[Doc. No. 5074, 29 FR 15695, Nov. 24, 1964, as amended by Amdt. 27-23, 53 FR 34214, Sept. 2, 1988]

§ 27.1194 Other surfaces.

All surfaces aft of, and near, powerplant compartments, other than tail surfaces not subject to heat, flames, or sparks emanating from a powerplant compartment, must be at least fire resistant.

[Amdt. 27-2, 33 FR 964, Jan. 26, 1968]

§ 27.1195 Fire detector systems.

Each turbine engine powered rotorcraft must have approved quick-acting fire detectors in numbers and locations insuring prompt detection of fire in the engine compartment which cannot be readily observed in flight by the pilot in the cockpit.

[Amdt. 27-5, 36 FR 5493, Mar. 24, 1971]

Subpart F—Equipment

General

§ 27.1301 Function and installation.

Each item of installed equipment must—

(a) Be of a kind and design appropriate to its intended function;

(b) Be labeled as to its identification, function, or operating limitations, or any applicable combination of these factors;

(c) Be installed according to limitations specified for that equipment; and

(d) Function properly when installed.

§ 27.1303 Flight and navigation instruments.

The following are the required flight and navigation instruments:

(a) An airspeed indicator.

(b) An altimeter.

(c) A magnetic direction indicator.

§ 27.1305 Powerplant instruments.

The following are the required powerplant instruments:

(a) A carburetor air temperature indicator, for each engine having a preheater that can provide a heat rise in excess of 60°F.

(b) A cylinder head temperature indicator, for each—

(1) Air cooled engine;

(2) Rotorcraft with cooling shutters; and

(3) Rotorcraft for which compliance with § 27.1043 is shown in any condition other than the most critical flight condition with respect to cooling.

(c) A fuel pressure indicator, for each pump-fed engine.

(d) A fuel quantity indicator, for each fuel tank.

(e) A manifold pressure indicator, for each altitude engine.

(f) An oil temperature warning device to indicate when the temperature exceeds a safe value in each main rotor drive gearbox (including any gearboxes essential to rotor phasing) having an oil system independent of the engine oil system.

(g) An oil pressure warning device to indicate when the pressure falls below a safe value in each pressure-lubricated main rotor drive gearbox (including any gearboxes essential to rotor phasing) having an oil system independent of the engine oil system.

(h) An oil pressure indicator for each engine.

(i) An oil quantity indicator for each oil tank.

(j) An oil temperature indicator for each engine.

(k) At least one tachometer to indicate the r.p.m. of each engine and, as applicable—

(1) The r.p.m. of the single main rotor;

(2) The common r.p.m. of any main rotors whose speeds cannot vary appreciably with respect to each other; or

(3) The r.p.m. of each main rotor whose speed can vary appreciably with respect to that of another main rotor.

(l) A low fuel warning device for each fuel tank which feeds an engine. This device must—

(1) Provide a warning to the flightcrew when approximately 10 minutes of usable fuel remains in the tank; and

(2) Be independent of the normal fuel quantity indicating system.

(m) Means to indicate to the flightcrew the failure of any fuel pump installed to show compliance with § 27.955.

(n) A gas temperature indicator for each turbine engine.

(o) Means to enable the pilot to determine the torque of each turboshaft engine, if a torque limitation is established for that engine under § 27.1521(e).

(p) For each turbine engine, an indicator to indicate the functioning of the powerplant ice protection system.

(q) An indicator for the fuel filter required by § 27.997 to indicate the occurrence of contamination of the filter at the degree established by the applicant in compliance with § 27.955.

(r) For each turbine engine, a warning means for the oil strainer or filter required by § 27.1019, if it has no bypass, to warn the pilot of the occurrence of contamination of the strainer or filter before it reaches the capacity established in accordance with § 27.1019(a)(2).

(s) An indicator to indicate the functioning of any selectable or controllable heater used to prevent ice clogging of fuel system components.

(t) For rotorcraft for which a 30-second/2-minute OEI power rating is requested, a means must be provided to alert the pilot when the engine is at the 30-second and the 2-minute OEI power levels, when the event begins, and when the time interval expires.

(u) For each turbine engine utilizing 30-second/2-minute OEI power, a device or system must be provided for use by ground personnel which—

(1) Automatically records each usage and duration of power at the 30-second and 2-minute OEI levels;

(2) Permits retrieval of the recorded data;

(3) Can be reset only by ground maintenance personnel; and

(4) Has a means to verify proper operation of the system or device.

(v) Warning or caution devices to signal to the flight crew when ferromagnetic particles are detected by the chip detector required by § 27.1337(e).

[Doc. No. 5074, 29 FR 15695, Nov. 24, 1964, as amended by Amdt. 27-9, 39 FR 35462, Oct. 1, 1974; Amdt. 27-23, 53 FR 34214, Sept. 2, 1988; Amdt. 27-29, 59 FR 47767, Sept. 16, 1994; Amdt. 27-37, 64 FR 45095, Aug. 18, 1999; 64 FR 47563, Aug. 31, 1999]

§ 27.1307 Miscellaneous equipment.

The following is the required miscellaneous equipment:

(a) An approved seat for each occupant.

(b) An approved safety belt for each occupant.

(c) A master switch arrangement.

(d) An adequate source of electrical energy, where electrical energy is necessary for operation of the rotorcraft.

(e) Electrical protective devices.

§ 27.1309 Equipment, systems, and installations.

(a) The equipment, systems, and installations whose functioning is required by this subchapter must be designed and installed to ensure that they perform their intended functions under any foreseeable operating condition.

(b) The equipment, systems, and installations of a multiengine rotorcraft must be designed to prevent hazards to the rotorcraft in the event of a probable malfunction or failure.

(c) The equipment, systems, and installations of single-engine rotorcraft must be designed to minimize hazards to the rotorcraft in the event of a probable malfunction or failure.

[Doc. No. 5074, 29 FR 15695, Nov. 24, 1964, as amended by Amdt. 27-21, 49 FR 44435, Nov. 6, 1984; Amdt. 27-46, 76 FR 33135, June 8, 2011]

§ 27.1316 Electrical and electronic system lightning protection.

(a) Each electrical and electronic system that performs a function, for which failure would prevent the continued safe flight and landing of the rotorcraft, must be designed and installed so that—

(1) The function is not adversely affected during and after the time the rotorcraft is exposed to lightning; and

(2) The system automatically recovers normal operation of that function in a timely manner after the rotorcraft is exposed to lightning.

(b) For rotorcraft approved for instrument flight rules operation, each electrical and electronic system that performs a function, for which failure would reduce the capability of the rotorcraft or the ability of the flightcrew to respond to an adverse operating condition, must be designed and installed so that the function recovers normal operation in a timely manner after the rotorcraft is exposed to lightning.

[Doc. No. FAA-2010-0224, Amdt. 27-46, 76 FR 33135, June 8, 2011]

§ 27.1317 High-intensity Radiated Fields (HIRF) Protection.

(a) Except as provided in paragraph (d) of this section, each electrical and electronic system that performs a function whose failure would prevent the continued safe flight and landing of the rotorcraft must be designed and installed so that—

(1) The function is not adversely affected during and after the time the rotorcraft is exposed to HIRF environment I, as described in appendix D to this part;

(2) The system automatically recovers normal operation of that function, in a timely manner, after the rotorcraft is exposed to HIRF environment I, as described in appendix D to this part, unless this conflicts with other operational or functional requirements of that system;

(3) The system is not adversely affected during and after the time the rotorcraft is exposed to HIRF environment II, as described in appendix D to this part; and

(4) Each function required during operation under visual flight rules is not adversely affected during and after the time the rotorcraft is exposed to HIRF environment III, as described in appendix D to this part.

(b) Each electrical and electronic system that performs a function whose failure would significantly reduce the capability of the rotorcraft or the ability of the flightcrew to respond to an adverse operating condition must be designed and installed so the system is not adversely affected when the equipment providing these functions is exposed to equipment HIRF test level 1 or 2, as described in appendix D to this part.

(c) Each electrical and electronic system that performs a function whose failure would reduce the capability of the rotorcraft or the ability of the flightcrew to respond to an adverse operating condition, must be designed and installed so the system is not adversely affected when the equipment providing these functions is exposed to equipment HIRF test level 3, as described in appendix D to this part.

(d) Before December 1, 2012, an electrical or electronic system that performs a function whose failure would prevent the continued safe flight and landing of a rotorcraft may be designed and installed without meeting the provisions of paragraph (a) provided—

(1) The system has previously been shown to comply with special conditions for HIRF, prescribed under § 21.16, issued before December 1, 2007;

(2) The HIRF immunity characteristics of the system have not changed since compliance with the special conditions was demonstrated; and

(3) The data used to demonstrate compliance with the special conditions is provided.

[Doc. No. FAA-2006-23657, 72 FR 44026, Aug. 6, 2007]

Instruments: Installation

§ 27.1321 Arrangement and visibility.

(a) Each flight, navigation, and powerplant instrument for use by any pilot must be easily visible to him.

(b) For each multiengine rotorcraft, identical powerplant instruments must be located so as to prevent confusion as to which engine each instrument relates.

(c) Instrument panel vibration may not damage, or impair the readability or accuracy of, any instrument.

(d) If a visual indicator is provided to indicate malfunction of an instrument, it must be effective under all probable cockpit lighting conditions.

(Secs. 313(a), 601, 603, 604, and 605 of the Federal Aviation Act of 1958 (49 U.S.C. 1354(a), 1421, 1423, 1424, and 1425); and sec. 6(c) of the Dept. of Transportation Act (49 U.S.C. 1655(c)))

[Doc. No. 5074, 29 FR 15695, Nov. 24, 1964; 29 FR 17885, Dec. 17, 1964, as amended by Amdt. 27-13, 42 FR 36971, July 18, 1977]

§ 27.1322 Warning, caution, and advisory lights.

If warning, caution or advisory lights are installed in the cockpit, they must, unless otherwise approved by the Administrator, be—

(a) Red, for warning lights (lights indicating a hazard which may require immediate corrective action):

(b) Amber, for caution lights (lights indicating the possible need for future corrective action);

(c) Green, for safe operation lights; and

(d) Any other color, including white, for lights not described in paragraphs (a) through (c) of this section, provided the color differs sufficiently from the colors prescribed in paragraphs (a) through (c) of this section to avoid possible confusion.

[Amdt. 27-11, 41 FR 55470, Dec. 20, 1976]

§ 27.1323 Airspeed indicating system.

(a) Each airspeed indicating instrument must be calibrated to indicate true airspeed (at sea level with a standard atmosphere) with a minimum practicable instrument calibration error when the corresponding pitot and static pressures are applied.

(b) The airspeed indicating system must be calibrated in flight at forward speeds of 20 knots and over.

(c) At each forward speed above 80 percent of the climbout speed, the airspeed indicator must indicate true airspeed, at sea level with a standard atmosphere, to within an allowable installation error of not more than the greater of—

(1)±3 percent of the calibrated airspeed; or

(2) Five knots.

(Secs. 313(a), 601, 603, 604, and 605 of the Federal Aviation Act of 1958 (49 U.S.C. 1354(a), 1421, 1423, 1424, and 1425); and sec. 6(c) of the Dept. of Transportation Act (49 U.S.C. 1655(c)))

[Doc. No. 5074, 29 FR 15695, Nov. 24, 1964, as amended by Amdt. 27-13, 42 FR 36972, July 18, 1977]

§ 27.1325 Static pressure systems.

(a) Each instrument with static air case connections must be vented so that the influence of rotorcraft speed, the opening and closing of windows, airflow variation, and moisture or other foreign matter does not seriously affect its accuracy.

(b) Each static pressure port must be designed and located in such manner that the correlation between air pressure in the static pressure system and true ambient atmospheric static pressure is not altered when the rotorcraft encounters icing conditions. An anti-icing means or an alternate source of static pressure may be used in showing compliance with this requirement. If the reading of the altimeter, when on the alternate static pressure system, differs from the reading of the altimeter when on the primary static system by more than 50 feet, a correction card must be provided for the alternate static system.

(c) Except as provided in paragraph (d) of this section, if the static pressure system incorporates both a primary and an alternate static pressure source, the means for selecting one or the other source must be designed so that—

(1) When either source is selected, the other is blocked off; and

(2) Both sources cannot be blocked off simultaneously.

(d) For unpressurized rotorcraft, paragraph (c)(1) of this section does not apply if it can be demonstrated that the static pressure system calibration, when either static pressure source is selected is not changed by the other static pressure source being open or blocked.

(Secs. 313(a), 601, 603, 604, and 605 of the Federal Aviation Act of 1958 (49 U.S.C. 1354(a), 1421, 1423, 1424, and 1425); and sec. 6(c) of the Dept. of Transportation Act (49 U.S.C. 1655(c)))

[Doc. No. 5074, 29 FR 15695, Nov. 24, 1964, as amended by Amdt. 27-13, 42 FR 36972, July 18, 1977]

§ 27.1327 Magnetic direction indicator.

(a) Except as provided in paragraph (b) of this section—

(1) Each magnetic direction indicator must be installed so that its accuracy is not excessively affected by the rotorcraft's vibration or magnetic fields; and

(2) The compensated installation may not have a deviation, in level flight, greater than 10 degrees on any heading.

(b) A magnetic nonstabilized direction indicator may deviate more than 10 degrees due to the operation of electrically powered systems such as electrically heated windshields if either a magnetic stabilized direction indicator, which does not have a deviation in level flight greater than 10 degrees on any heading, or a gyroscopic direction indicator, is installed. Deviations of a magnetic nonstabilized direction indicator of more than 10 degrees must be placarded in accordance with § 27.1547(e).

(Secs. 313(a), 601, 603, 604, and 605 of the Federal Aviation Act of 1958 (49 U.S.C. 1354(a), 1421, 1423, 1424, and 1425); and sec. 6(c) of the Dept. of Transportation Act (49 U.S.C. 1655(c)))

[Amdt. 27-13, 42 FR 36972, July 18, 1977]

§ 27.1329 Automatic pilot system.

(a) Each automatic pilot system must be designed so that the automatic pilot can—

(1) Be sufficiently overpowered by one pilot to allow control of the rotorcraft; and

(2) Be readily and positively disengaged by each pilot to prevent it from interfering with control of the rotorcraft.

(b) Unless there is automatic synchronization, each system must have a means to readily indicate to the pilot the alignment of the actuating device in relation to the control system it operates.

(c) Each manually operated control for the system's operation must be readily accessible to the pilots.

(d) The system must be designed and adjusted so that, within the range of adjustment available to the pilot, it cannot produce hazardous loads on the rotorcraft or create hazardous deviations in the flight path under any flight condition appropriate to its use, either during normal operation or in the event of a malfunction, assuming that corrective action begins within a reasonable period of time.

(e) If the automatic pilot integrates signals from auxiliary controls or furnishes signals for operation of other equipment, there must be positive interlocks and sequencing of engagement to prevent improper operation.

(f) If the automatic pilot system can be coupled to airborne navigation equipment, means must be provided to indicate to the pilots the current mode of operation. Selector switch position is not acceptable as a means of indication.

[Amdt. 27-21, 49 FR 44435, Nov. 6, 1984, as amended by Amdt. 27-35, 63 FR 43285, Aug. 12, 1998]

§ 27.1335 Flight director systems.

If a flight director system is installed, means must be provided to indicate to the flight crew its current mode of operation. Selector switch position is not acceptable as a means of indication.

(Secs. 313(a), 601, 603, 604, and 605 of the Federal Aviation Act of 1958 (49 U.S.C. 1354(a), 1421, 1423, 1424, and 1425); and sec. 6(c) of the Dept. of Transportation Act (49 U.S.C. 1655(c)))

[Amdt. 27-13, 42 FR 36972, July 18, 1977]

§ 27.1337 Powerplant instruments.

(a) *Instruments and instrument lines.* (1) Each powerplant instrument line must meet the requirements of §§ 27.- 961 and 27.993.

(2) Each line carrying flammable fluids under pressure must—

(i) Have restricting orifices or other safety devices at the source of pressure to prevent the escape of excessive fluid if the line fails; and

(ii) Be installed and located so that the escape of fluids would not create a hazard.

(3) Each powerplant instrument that utilizes flammable fluids must be installed and located so that the escape of fluid would not create a hazard.

(b) *Fuel quantity indicator.* Each fuel quantity indicator must be installed to clearly indicate to the flight crew the quantity of fuel in each tank in flight. In addition—

(1) Each fuel quantity indicator must be calibrated to read "zero" during level flight when the quantity of fuel remaining in the tank is equal to the unusable fuel supply determined under § 27.959;

(2) When two or more tanks are closely interconnected by a gravity feed system and vented, and when it is impossible to feed from each tank separately, at least one fuel quantity indicator must be installed; and

(3) Each exposed sight gauge used as a fuel quantity indicator must be protected against damage.

(c) *Fuel flowmeter system.* If a fuel flowmeter system is installed, each metering component must have a means for bypassing the fuel supply if malfunction of that component severely restricts fuel flow.

(d) *Oil quantity indicator.* There must be means to indicate the quantity of oil in each tank—

(1) On the ground (including during the filling of each tank); and

(2) In flight, if there is an oil transfer system or reserve oil supply system.

(e) Rotor drive system transmissions and gearboxes utilizing ferromagnetic materials must be equipped with chip detectors designed to indicate the presence of ferromagnetic particles resulting from damage or excessive wear. Chip detectors must—

(1) Be designed to provide a signal to the device required by § 27.1305(v) and be provided with a means to allow crewmembers to check, in flight, the function of each detector electrical circuit and signal.

(2) [Reserved]

(Secs. 313(a), 601, and 603, 72 Stat. 752, 775, 49 U.S.C. 1354(a), 1421, and 1423; sec. 6(c) 49 U.S.C. 1655(c))

[Doc. No. 5074, 29 FR 15695, Nov. 24, 1964, as amended by Amdt. 27-12, 42 FR 15046, Mar. 17, 1977; Amdt. 27-23, 53 FR 34214, Sept. 2, 1988; Amdt. 27-37, 64 FR 45095, Aug. 18, 1999]

Electrical Systems and Equipment

§ 27.1351 General.

(a) *Electrical system capacity.* Electrical equipment must be adequate for its intended use. In addition—

(1) Electric power sources, their transmission cables, and their associated control and protective devices must be

able to furnish the required power at the proper voltage to each load circuit essential for safe operation; and

(2) Compliance with paragraph (a)(1) of this section must be shown by an electrical load analysis, or by electrical measurements that take into account the electrical loads applied to the electrical system, in probable combinations and for probable durations.

(b) *Function.* For each electrical system, the following apply:

(1) Each system, when installed, must be—

(i) Free from hazards in itself, in its method of operation, and in its effects on other parts of the rotorcraft; and

(ii) Protected from fuel, oil, water, other detrimental substances, and mechanical damage.

(2) Electric power sources must function properly when connected in combination or independently.

(3) No failure or malfunction of any source may impair the ability of any remaining source to supply load circuits essential for safe operation.

(4) Each electric power source control must allow the independent operation of each source.

(c) *Generating system.* There must be at least one generator if the system supplies power to load circuits essential for safe operation. In addition—

(1) Each generator must be able to deliver its continuous rated power;

(2) Generator voltage control equipment must be able to dependably regulate each generator output within rated limits;

(3) Each generator must have a reverse current cutout designed to disconnect the generator from the battery and from the other generators when enough reverse current exists to damage that generator; and

(4) Each generator must have an overvoltage control designed and installed to prevent damage to the electrical system, or to equipment supplied by the electrical system, that could result if that generator were to develop an overvoltage condition.

(d) *Instruments.* There must be means to indicate to appropriate crewmembers the electric power system quantities essential for safe operation of the system. In addition—

(1) For direct current systems, an ammeter that can be switched into each generator feeder may be used; and

(2) If there is only one generator, the ammeter may be in the battery feeder.

(e) *External power.* If provisions are made for connecting external power to the rotorcraft, and that external power can be electrically connected to equipment other than that used for engine starting, means must be provided to ensure that no external power supply having a reverse polarity, or a reverse phase sequence, can supply power to the rotorcraft's electrical system.

(Secs. 313(a), 601, 603, 604, and 605 of the Federal Aviation Act of 1958 (49 U.S.C. 1354(a), 1421, 1423, 1424, and 1425); and sec. 6(c) of the Dept. of Transportation Act (49 U.S.C. 1655(c)))

[Doc. No. 5074, 29 FR 15695, Nov. 24, 1964, as amended by Amdt. 27-11, 41 FR 55470, Dec. 20, 1976; Amdt. 27-13, 42 FR 36972, July 18, 1977]

§ 27.1353 Storage battery design and installation.

(a) Each storage battery must be designed and installed as prescribed in this section.

(b) Safe cell temperatures and pressures must be maintained during any probable charging and discharging condition. No uncontrolled increase in cell temperature may result when the battery is recharged (after previous complete discharge)—

(1) At maximum regulated voltage or power;

(2) During a flight of maximum duration; and

(3) Under the most adverse cooling condition likely to occur in service.

(c) Compliance with paragraph (b) of this section must be shown by test unless experience with similar batteries and installations has shown that maintaining safe cell temperatures and pressures presents no problem.

(d) No explosive or toxic gases emitted by any battery in normal operation, or as the result of any probable malfunction in the charging system or battery installation, may accumulate in hazardous quantities within the rotorcraft.

(e) No corrosive fluids or gases that may escape from the battery may damage surrounding structures or adjacent essential equipment.

(f) Each nickel cadmium battery installation capable of being used to start an engine or auxiliary power unit must have provisions to prevent any hazardous effect on structure or essential systems that may be caused by the maximum amount of heat the battery can generate during a short circuit of the battery or of its individual cells.

(g) Nickel cadmium battery installations capable of being used to start an engine or auxiliary power unit must have—

(1) A system to control the charging rate of the battery automatically so as to prevent battery overheating;

(2) A battery temperature sensing and over-temperature warning system with a means for disconnecting the battery from its charging source in the event of an over-temperature condition; or

(3) A battery failure sensing and warning system with a means for disconnecting the battery from its charging source in the event of battery failure.

(Secs. 313(a), 601, 603, 604, and 605 of the Federal Aviation Act of 1958 (49 U.S.C. 1354(a), 1421, 1423, 1424, and 1425); and sec. 6(c) of the Dept. of Transportation Act (49 U.S.C. 1655(c)))

[Doc. No. 5074, 29 FR 15695, Nov. 24, 1964, as amended by Amdt. 27-13, 42 FR 36972, July 18, 1977; Amdt. 27-14, 43 FR 2325, Jan. 16, 1978]

§ 27.1357 Circuit protective devices.

(a) Protective devices, such as fuses or circuit breakers, must be installed in each electrical circuit other than—

(1) The main circuits of starter motors; and

(2) Circuits in which no hazard is presented by their omission.

(b) A protective device for a circuit essential to flight safety may not be used to protect any other circuit.

(c) Each resettable circuit protective device ("trip free" device in which the tripping mechanism cannot be overridden by the operating control) must be designed so that—

(1) A manual operation is required to restore service after trippling; and

(2) If an overload or circuit fault exists, the device will open the circuit regardless of the position of the operating control.

(d) If the ability to reset a circuit breaker or replace a fuse is essential to safety in flight, that circuit breaker or fuse must be located and identified so that it can be readily reset or replaced in flight.

(e) If fuses are used, there must be one spare of each rating, or 50 percent spare fuses of each rating, whichever is greater.

(Secs. 313(a), 601, 603, 604, and 605 of the Federal Aviation Act of 1958 (49 U.S.C. 1354(a), 1421, 1423, 1424, and 1425); and sec. 6(c) of the Dept. of Transportation Act (49 U.S.C. 1655(c)))

[Doc. No. 5074, 29 FR 15695, Nov. 24, 1964; 29 FR 17885, Dec. 17, 1964, as amended by Amdt. 27-13, 42 FR 36972, July 18, 1977]

§ 27.1361 Master switch.

(a) There must be a master switch arrangement to allow ready disconnection of each electric power source from the main bus. The point of disconnection must be adjacent to the sources controlled by the switch.

(b) Load circuits may be connected so that they remain energized after the switch is opened, if they are protected by circuit protective devices, rated at five amperes or less, adjacent to the electric power source.

(c) The master switch or its controls must be installed so that the switch is easily discernible and accessible to a crewmember in flight.

§ 27.1365 Electric cables.

(a) Each electric connecting cable must be of adequate capacity.

(b) Each cable that would overheat in the event of circuit overload or fault must be at least flame resistant and may not emit dangerous quantities of toxic fumes.

(c) Insulation on electrical wire and cable installed in the rotorcraft must be self-extinguishing when tested in accordance with appendix F, part I(a)(3), of part 25 of this chapter.

[Doc. No. 5074, 29 FR 15695, Nov. 24, 1964, as amended by Amdt. 27-35, 63 FR 43285, Aug. 12, 1998]

§ 27.1367 Switches.

Each switch must be—

(a) Able to carry its rated current;

(b) Accessible to the crew; and

(c) Labeled as to operation and the circuit controlled.

Lights

§ 27.1381 Instrument lights.

The instrument lights must—

(a) Make each instrument, switch, and other devices for which they are provided easily readable; and

(b) Be installed so that—

(1) Their direct rays are shielded from the pilot's eyes; and

(2) No objectionable reflections are visible to the pilot.

§ 27.1383 Landing lights.

(a) Each required landing or hovering light must be approved.

(b) Each landing light must be installed so that—

(1) No objectionable glare is visible to the pilot;

(2) The pilot is not adversely affected by halation; and

(3) It provides enough light for night operation, including hovering and landing.

(c) At least one separate switch must be provided, as applicable—

(1) For each separately installed landing light; and

(2) For each group of landing lights installed at a common location.

§ 27.1385 Position light system installation.

(a) *General.* Each part of each position light system must meet the applicable requirements of this section, and each system as a whole must meet the requirements of §§ 27.1387 through 27.1397.

(b) *Forward position lights.* Forward position lights must consist of a red and a green light spaced laterally as far apart as practicable and installed forward on the rotorcraft so that, with the rotorcraft in the normal flying position, the red light is on the left side and the green light is on the right side. Each light must be approved.

(c) *Rear position light.* The rear position light must be a white light mounted as far aft as practicable, and must be approved.

(d) *Circuit.* The two forward position lights and the rear position light must make a single circuit.

(e) *Light covers and color filters.* Each light cover or color filter must be at least flame resistant and may not change color or shape or lose any appreciable light transmission during normal use.

§ 27.1387 Position light system dihedral angles.

(a) Except as provided in paragraph (e) of this section, each forward and rear position light must, as installed, show unbroken light within the dihedral angles described in this section.

(b) Dihedral angle *L* (left) is formed by two intersecting vertical planes, the first parallel to the longitudinal axis of the rotorcraft, and the other at 110 degrees to the left of the first, as viewed when looking forward along the longitudinal axis.

(c) Dihedral angle *R* (right) is formed by two intersecting vertical planes, the first parallel to the longitudinal axis of the rotorcraft, and the other at 110 degrees to the right of the first, as viewed when looking forward along the longitudinal axis.

(d) Dihedral angle *A* (aft) is formed by two intersecting vertical planes making angles of 70 degrees to the right and to the left, respectively, to a vertical plane passing through the longitudinal axis, as viewed when looking aft along the longitudinal axis.

(e) If the rear position light, when mounted as far aft as practicable in accordance with § 25.1385(c), cannot show unbroken light within dihedral angle A (as defined in paragraph (d) of this section), a solid angle or angles of obstructed visibility totaling not more than 0.04 steradians is allowable within that dihedral angle, if such solid angle is within a cone whose apex is at the rear position light and whose elements make an angle of 30° with a vertical line passing through the rear position light.

(49 U.S.C. 1655(c))

[Doc. No. 5074, 29 FR 15695, Nov. 24, 1964, as amended by Amdt. 27-7, 36 FR 21278, Nov. 5, 1971]

§ 27.1389 Position light distribution and intensities.

(a) *General.* the intensities prescribed in this section must be provided by new equipment with light covers and color filters in place. Intensities must be determined with the light source operating at a steady value equal to the average luminous output of the source at the normal operating voltage of the rotorcraft. The light distribution and intensity of each position light must meet the requirements of paragraph (b) of this section.

(b) *Forward and rear position lights.* The light distribution and intensities of forward and rear position lights must be expressed in terms of minimum intensities in the horizontal plane, minimum intensities in any vertical plane, and maximum intensities in overlapping beams, within dihedral angles *L*, *R*, and *A*, and must meet the following requirements:

(1) *Intensities in the horizontal plane.* Each intensity in the horizontal plane (the plane containing the longitudinal axis of the rotorcraft and perpendicular to the plane of symmetry of the rotorcraft) must equal or exceed the values in § 27.1391.

(2) *Intensities in any vertical plane.* Each intensity in any vertical plane (the plane perpendicular to the horizontal plane) must equal or exceed the appropriate value in § 27.1393, where *I* is the minimum intensity prescribed in § 27.1391 for the corresponding angles in the horizontal plane.

(3) *Intensities in overlaps between adjacent signals.* No intensity in any overlap between adjacent signals may exceed the values in § 27.1395, except that higher intensities in overlaps may be used with main beam intensities substantially greater than the minima specified in §§ 27.1391 and 27.1393, if the overlap intensities in relation to the main beam intensities do not adversely affect signal clarity. When the peak intensity of the forward position lights is greater than 100 candles, the maximum overlap intensities between them may exceed the values in § 27.1395 if the overlap intensity in Area A is not more than 10 percent of peak position light intensity and the overlap intensity in Area B is not more than 2.5 percent of peak position light intensity.

§ 27.1391 Minimum intensities in the horizontal plane of forward and rear position lights.

Each position light intensity must equal or exceed the applicable values in the following table:

Dihedral angle (light included)	Angle from right or left of longitudinal axis, measured from dead ahead	Intensity (candles)
L and *R* (forward red and green)	10° to 10° 10° to 20° 20° to 110°	40 30 5
A (rear white)	110° to 180°	20

§ 27.1393 Minimum intensities in any vertical plane of forward and rear position lights.

Each position light intensity must equal or exceed the applicable values in the following table:

Angle above or below the horizontal plane	Intensity, I
0°	1.00
0° to 5°	0.90
5° to 10°	0.80
10° to 15°	0.70
15° to 20°	0.50
20° to 30°	0.30
30° to 40°	0.10
40° to 90°	0.05

§ 27.1395 Maximum intensities in overlapping beams of forward and rear position lights.

No position light intensity may exceed the applicable values in the following table, except as provided in § 27.1389(b)(3).

Overlaps	Maximum Intensity	
	Area A (candles)	Area B (candles)
Green in dihedral angle *L*	10	1
Red in dihedral angle *R*	10	1
Green in dihedral angle *A*	5	1
Red in dihedral angle *A*	5	1
Rear white in dihedral angle *L*	5	1
Rear white in dihedral angle *R*	5	1

Where—

(a) Area A includes all directions in the adjacent dihedral angle that pass through the light source and intersect the common boundary plane at more than 10 degrees but less than 20 degrees, and

(b) Area B includes all directions in the adjacent dihedral angle that pass through the light source and intersect the common boundary plane at more than 20 degrees.

§ 27.1397 Color specifications.

Each position light color must have the applicable International Commission on Illumination chromaticity coordinates as follows:

(a) *Aviation red*—

y is not greater than 0.335; and

z is not greater than 0.002.

(b) *Aviation green*—

x is not greater than 0.440–0.320 *y*;

x is not greater than *y* –0.170; and

y is not less than 0.390–0.170 *x*.

(c) *Aviation white*—

x is not less than 0.300 and not greater than 0.540;

y is not less than *x* –0.040" or y_c –0.010, whichever is the smaller; and

y is not greater than *x*+0.020 nor 0.636–0.400 *x*;

Where y_c is the *y* coordinate of the Planckian radiator for the value of *x* considered.

[Doc. No. 5074, 29 FR 15695, Nov. 24, 1964, as amended by Amdt. 27-6, 36 FR 12972, July 10, 1971]

§ 27.1399 Riding light.

(a) Each riding light required for water operation must be installed so that it can—

(1) Show a white light for at least two nautical miles at night under clear atmospheric conditions; and

(2) Show a maximum practicable unbroken light with the rotorcraft on the water.

(b) Externally hung lights may be used.

[Doc. No. 5074, 29 FR 15695, Nov. 24, 1964, as amended by Amdt. 27-2, 33 FR 964, Jan. 26, 1968]

§ 27.1401 Anticollision light system.

(a) *General.* If certification for night operation is requested, the rotorcraft must have an anticollision light system that—

(1) Consists of one or more approved anticollision lights located so that their emitted light will not impair the crew's vision or detract from the conspicuity of the position lights; and

(2) Meets the requirements of paragraphs (b) through (f) of this section.

(b) *Field of coverage.* The system must consist of enough lights to illuminate the vital areas around the rotorcraft, considering the physical configuration and flight characteristics of the rotorcraft. The field of coverage must extend in each direction within at least 30 degrees below the horizontal plane of the rotorcraft, except that there may be solid angles of obstructed visibility totaling not more than 0.5 steradians.

(c) *Flashing characteristics.* The arrangement of the system, that is, the number of light sources, beam width, speed of rotation, and other characteristics, must give an effective flash frequency of not less than 40, nor more than 100, cycles per minute. The effective flash frequency is the frequency at which the rotorcraft's complete anticollision light system is observed from a distance, and applies to each sector of light including any overlaps that exist when the system consists of more than one light source. In overlaps, flash frequencies may exceed 100, but not 180, cycles per minute.

(d) *Color.* Each anticollision light must be aviation red and must meet the applicable requirements of § 27.1397.

(e) *Light intensity.* The minimum light intensities in any vertical plane, measured with the red filter (if used) and expressed in terms of "effective" intensities, must meet the requirements of paragraph (f) of this section. The following relation must be assumed:

$$I_e = \frac{\int_{t_1}^{t_2} I(t)\,dt}{0.2 + (t_2 - t_1)}$$

where:

I_e =effective intensity (candles).

I(t) =instantaneous intensity as a function of time.

$t_2 - t_1$ =flash time interval (seconds).

Normally, the maximum value of effective intensity is obtained when t_2 and t_1 are chosen so that the effective intensity is equal to the instantaneous intensity at t_2 and t_1.

(f) *Minimum effective intensities for anticollision light.* Each anticollision light effective intensity must equal or exceed the applicable values in the following table:

Angle above or below the horizontal plane	Effective intensity (candles)
0° to 5°	150
5° to 10°	90
10° to 20°	30
20° to 30°	15

[Doc. No. 5074, 29 FR 15695, Nov. 24, 1964, as amended by Amdt. 27-6, 36 FR 12972, July 10, 1971; Amdt. 27-10, 41 FR 5290, Feb. 5, 1976]

Safety Equipment

§ 27.1411 General.

(a) Required safety equipment to be used by the crew in an emergency, such as flares and automatic liferaft releases, must be readily accessible.

(b) Stowage provisions for required safety equipment must be furnished and must—

(1) Be arranged so that the equipment is directly accessible and its location is obvious; and

(2) Protect the safety equipment from damage caused by being subjected to the inertia loads specified in § 27.561.

[Doc. No. 5074, 29 FR 15695, Nov. 24, 1964, as amended by Amdt. 27-11, 41 FR 55470, Dec. 20, 1976]

§ 27.1413 Safety belts.

Each safety belt must be equipped with a metal to metal latching device.

(Secs. 313, 314, and 601 through 610 of the Federal Aviation Act of 1958 (49 U.S.C. 1354, 1355, and 1421 through 1430) and sec. 6(c), Dept. of Transportation Act (49 U.S.C. 1655(c)))

[Doc. No. 5074, 29 FR 15695, Nov. 24, 1964, as amended by Amdt. 27-15, 43 FR 46233, Oct. 5, 1978; Amdt. 27-21, 49 FR 44435, Nov. 6, 1984]

§ 27.1415 Ditching equipment.

(a) Emergency flotation and signaling equipment required by any operating rule in this chapter must meet the requirements of this section.

(b) Each raft and each life preserver must be approved and must be installed so that it is readily available to the crew and passengers. The storage provisions for life preservers must accommodate one life preserver for each occupant for which certification for ditching is requested.

(c) Each raft released automatically or by the pilot must be attached to the rotorcraft by a line to keep it alongside the rotorcraft. This line must be weak enough to break before submerging the empty raft to which it is attached.

(d) Each signaling device must be free from hazard in its operation and must be installed in an accessible location.

[Doc. No. 5074, 29 FR 15695, Nov. 24, 1964, as amended by Amdt. 27-11, 41 FR 55470, Dec. 20, 1976]

§ 27.1419 Ice protection.

(a) To obtain certification for flight into icing conditions, compliance with this section must be shown.

(b) It must be demonstrated that the rotorcraft can be safely operated in the continuous maximum and intermittent maximum icing conditions determined under appendix C of Part 29 of this chapter within the rotorcraft altitude envelope. An analysis must be performed to establish, on the basis of the rotorcraft's operational needs, the adequacy of the ice protection system for the various components of the rotorcraft.

(c) In addition to the analysis and physical evaluation prescribed in paragraph (b) of this section, the effectiveness of the ice protection system and its components must be shown by flight tests of the rotorcraft or its components in measured natural atmospheric icing conditions and by one or more of the following tests as found necessary to determine the adequacy of the ice protection system:

(1) Laboratory dry air or simulated icing tests, or a combination of both, of the components or models of the components.

(2) Flight dry air tests of the ice protection system as a whole, or its individual components.

(3) Flight tests of the rotorcraft or its components in measured simulated icing conditions.

(d) The ice protection provisions of this section are considered to be applicable primarily to the airframe. Powerplant installation requirements are contained in Subpart E of this part.

(e) A means must be indentified or provided for determining the formation of ice on critical parts of the rotorcraft. Unless otherwise restricted, the means must be available for nighttime as well as daytime operation. The rotorcraft flight manual must describe the means of determining ice formation and must contain information necessary for safe operation of the rotorcraft in icing conditions.

[Amdt. 27-19, 48 FR 4389, Jan. 31, 1983]

§ 27.1435 Hydraulic systems.

(a) *Design.* Each hydraulic system and its elements must withstand, without yielding, any structural loads expected in addition to hydraulic loads.

(b) *Tests.* Each system must be substantiated by proof pressure tests. When proof tested, no part of any system may fail, malfunction, or experience a permanent set. The proof load of each system must be at least 1.5 times the maximum operating pressure of that system.

(c) *Accumulators.* No hydraulic accumulator or pressurized reservoir may be installed on the engine side of any firewall unless it is an integral part of an engine.

§ 27.1457 Cockpit voice recorders.

(a) Each cockpit voice recorder required by the operating rules of this chapter must be approved, and must be installed so that it will record the following:

(1) Voice communications transmitted from or received in the rotorcraft by radio.

(2) Voice communications of flight crewmembers on the flight deck.

(3) Voice communications of flight crewmembers on the flight deck, using the rotorcraft's interphone system.

(4) Voice or audio signals identifying navigation or approach aids introduced into a headset or speaker.

(5) Voice communications of flight crewmembers using the passenger loudspeaker system, if there is such a system, and if the fourth channel is available in accordance with the requirements of paragraph (c)(4)(ii) of this section.

(6) If datalink communication equipment is installed, all datalink communications, using an approved data message set. Datalink messages must be recorded as the output signal from the communications unit that translates the signal into usable data.

(b) The recording requirements of paragraph (a)(2) of this section may be met:

(1) By installing a cockpit-mounted area microphone located in the best position for recording voice communications originating at the first and second pilot stations and voice communications of other crewmembers on the flight deck when directed to those stations; or

(2) By installing a continually energized or voice-actuated lip microphone at the first and second pilot stations.

The microphone specified in this paragraph must be so located and, if necessary, the preamplifiers and filters of the recorder must be adjusted or supplemented so that the recorded communications are intelligible when recorded under flight cockpit noise conditions and played back. The level of intelligibility must be approved by the Administrator. Repeated aural or visual playback of the record may be used in evaluating intelligibility.

(c) Each cockpit voice recorder must be installed so that the part of the communication or audio signals specified in paragraph (a) of this section obtained from each of the following sources is recorded on a separate channel:

(1) For the first channel, from each microphone, headset, or speaker used at the first pilot station.

(2) For the second channel, from each microphone, headset, or speaker used at the second pilot station.

(3) For the third channel, from the cockpit-mounted area microphone, or the continually energized or voice-actuated lip microphone at the first and second pilot stations.

(4) For the fourth channel, from:

(i) Each microphone, headset, or speaker used at the stations for the third and fourth crewmembers; or

(ii) If the stations specified in paragraph (c)(4)(i) of this section are not required or if the signal at such a station is picked up by another channel, each microphone on the flight deck that is used with the passenger loudspeaker system if its signals are not picked up by another channel.

(iii) Each microphone on the flight deck that is used with the rotorcraft's loudspeaker system if its signals are not picked up by another channel.

(d) Each cockpit voice recorder must be installed so that:

(1)(i) It receives its electrical power from the bus that provides the maximum reliability for operation of the cockpit voice recorder without jeopardizing service to essential or emergency loads.

(ii) It remains powered for as long as possible without jeopardizing emergency operation of the rotorcraft.

(2) There is an automatic means to simultaneously stop the recorder and prevent each erasure feature from functioning, within 10 minutes after crash impact;

(3) There is an aural or visual means for preflight checking of the recorder for proper operation;

(4) Whether the cockpit voice recorder and digital flight data recorder are installed in separate boxes or in a combination unit, no single electrical failure external to the recorder may disable both the cockpit voice recorder and the digital flight data recorder; and

(5) It has an independent power source—

(i) That provides 10±1 minutes of electrical power to operate both the cockpit voice recorder and cockpit-mounted area microphone;

(ii) That is located as close as practicable to the cockpit voice recorder; and

(iii) To which the cockpit voice recorder and cockpit-mounted area microphone are switched automatically in the event that all other power to the cockpit voice recorder is interrupted either by normal shutdown or by any other loss of power to the electrical power bus.

(e) The record container must be located and mounted to minimize the probability of rupture of the container as a result of crash impact and consequent heat damage to the record from fire.

(f) If the cockpit voice recorder has a bulk erasure device, the installation must be designed to minimize the probability of inadvertent operation and actuation of the device during crash impact.

(g) Each recorder container must be either bright orange or bright yellow.

(h) When both a cockpit voice recorder and a flight data recorder are required by the operating rules, one combination unit may be installed, provided that all other requirements of this section and the requirements for flight data recorders under this part are met.

[Amdt. 27-22, 53 FR 26144, July 11, 1988, as amended by Amdt. 27-43, 73 FR 12563, Mar. 7, 2008; 74 FR 32800, July 9, 2009; Amdt. 27-45, 75 FR 17045, Apr. 5, 2010]

§ 27.1459 Flight data recorders.

(a) Each flight recorder required by the operating rules of Subchapter G of this chapter must be installed so that:

(1) It is supplied with airspeed, altitude, and directional data obtained from sources that meet the accuracy requirements of §§ 27.1323, 27.1325, and 27.1327 of this part, as applicable;

(2) The vertical acceleration sensor is rigidly attached, and located longitudinally within the approved center of gravity limits of the rotorcraft;

(3)(i) It receives its electrical power from the bus that provides the maximum reliability for operation of the flight data recorder without jeopardizing service to essential or emergency loads.

(ii) It remains powered for as long as possible without jeopardizing emergency operation of the rotorcraft.

(4) There is an aural or visual means for preflight checking of the recorder for proper recording of data in the storage medium;

(5) Except for recorders powered solely by the engine-driven electrical generator system, there is an automatic means to simultaneously stop a recorder that has a data erasure feature and prevent each erasure feature from functioning, within 10 minutes after any crash impact; and

(6) Whether the cockpit voice recorder and digital flight data recorder are installed in separate boxes or in a combination unit, no single electrical failure external to the recorder may disable both the cockpit voice recorder and the digital flight data recorder.

(b) Each nonejectable recorder container must be located and mounted so as to minimize the probability of container rupture resulting from crash impact and subsequent damage to the record from fire.

(c) A correlation must be established between the flight recorder readings of airspeed, altitude, and heading and the corresponding readings (taking into account correction factors) of the first pilot's instruments. This correlation must cover the airspeed range over which the aircraft is to be operated, the range of altitude to which the aircraft is limited, and 360 degrees of heading. Correlation may be established on the ground as appropriate.

(d) Each recorder container must:

(1) Be either bright orange or bright yellow;

(2) Have a reflective tape affixed to its external surface to facilitate its location under water; and

(3) Have an underwater locating device, when required by the operating rules of this chapter, on or adjacent to the container which is secured in such a manner that they are not likely to be separated during crash impact.

(e) When both a cockpit voice recorder and a flight data recorder are required by the operating rules, one combination unit may be installed, provided that all other requirements of this section and the requirements for cockpit voice recorders under this part are met.

[Amdt. 27-22, 53 FR 26144, July 11, 1988, as amended by Amdt. 27-43, 73 FR 12564, Mar. 7, 2008; 74 FR 32800, July 9, 2009; Amdt. 27-45, 75 FR 17045, Apr. 5, 2010]

§ 27.1461 Equipment containing high energy rotors.

(a) Equipment containing high energy rotors must meet paragraph (b), (c), or (d) of this section.

(b) High energy rotors contained in equipment must be able to withstand damage caused by malfunctions, vibration, abnormal speeds, and abnormal temperatures. In addition—

(1) Auxiliary rotor cases must be able to contain damage caused by the failure of high energy rotor blades; and

(2) Equipment control devices, systems, and instrumentation must reasonably ensure that no operating limitations affecting the integrity of high energy rotors will be exceeded in service.

(c) It must be shown by test that equipment containing high energy rotors can contain any failure of a high energy rotor that occurs at the highest speed obtainable with the normal speed control devices inoperative.

(d) Equipment containing high energy rotors must be located where rotor failure will neither endanger the occupants nor adversely affect continued safe flight.

[Amdt. 27-2, 33 FR 964, Jan. 26, 1968]

Subpart G—Operating Limitations and Information

§ 27.1501 General.

(a) Each operating limitation specified in §§ 27.1503 through 27.1525 and other limitations and information necessary for safe operation must be established.

(b) The operating limitations and other information necessary for safe operation must be made available to the crewmembers as prescribed in §§ 27.1541 through 27.1589.

(Secs. 313(a), 601, 603, 604, and 605 of the Federal Aviation Act of 1958 (49 U.S.C. 1354(a), 1421, 1423, 1424, and 1425); and sec. 6(c) of the Dept. of Transportation Act (49 U.S.C. 1655(c)))

[Amdt. 27-14, 43 FR 2325, Jan. 16, 1978]

Operating Limitations

§ 27.1503 Airspeed limitations: general.

(a) An operating speed range must be established.

(b) When airspeed limitations are a function of weight, weight distribution, altitude, rotor speed, power, or other factors, airspeed limitations corresponding with the critical combinations of these factors must be established.

§ 27.1505 Never-exceed speed.

(a) The never-exceed speed, V_{NE}, must be established so that it is—

(1) Not less than 40 knots (CAS); and

(2) Not more than the lesser of—

(i) 0.9 times the maximum forward speeds established under § 27.309;

(ii) 0.9 times the maximum speed shown under §§ 27.251 and 27.629; or

(iii) 0.9 times the maximum speed substantiated for advancing blade tip mach number effects.

(b) V_{NE} may vary with altitude, r.p.m., temperature, and weight, if—

(1) No more than two of these variables (or no more than two instruments integrating more than one of these variables) are used at one time; and

(2) The ranges of these variables (or of the indications on instruments integrating more than one of these variables) are large enough to allow an operationally practical and safe variation of V_{NE}.

(c) For helicopters, a stabilized power-off V_{NE} denoted as V_{NE} (power-off) may be established at a speed less than V_{NE} established pursuant to paragraph (a) of this section, if the following conditions are met:

(1) V_{NE} (power-off) is not less than a speed midway between the power-on V_{NE} and the speed used in meeting the requirements of—

(i) § 27.65(b) for single engine helicopters; and

(ii) § 27.67 for multiengine helicopters.

(2) V_{NE} (power-off) is—

(i) A constant airspeed;

(ii) A constant amount less than power-on V_{NE}; or

(iii) A constant airspeed for a portion of the altitude range for which certification is requested, and a constant amount less than power-on V_{NE} for the remainder of the altitude range.

(Secs. 313(a), 601, 603, 604, and 605 of the Federal Aviation Act of 1958 (49 U.S.C. 1354(a), 1421, 1423, 1424, and 1425); and sec. 6(c) of the Dept. of Transportation Act (49 U.S.C. 1655(c)))

[Amdt. 27-2, 33 FR 964, Jan. 26, 1968, and Amdt. 27-14, 43 FR 2325, Jan. 16, 1978; Amdt. 27-21, 49 FR 44435, Nov. 6, 1984]

§ 27.1509 Rotor speed.

(a) *Maximum power-off (autorotation).* The maximum power-off rotor speed must be established so that it does not exceed 95 percent of the lesser of—

(1) The maximum design r.p.m. determined under § 27.309(b); and

(2) The maximum r.p.m. shown during the type tests.

(b) *Minimum power off.* The minimum power-off rotor speed must be established so that it is not less than 105 percent of the greater of—

(1) The minimum shown during the type tests; and

(2) The minimum determined by design substantiation.

(c) *Minimum power on.* The minimum power-on rotor speed must be established so that it is—

(1) Not less than the greater of—

(i) The minimum shown during the type tests; and

(ii) The minimum determined by design substantiation; and

(2) Not more than a value determined under § 27.33(a)(1) and (b)(1).

§ 27.1519 Weight and center of gravity.

The weight and center of gravity limitations determined under §§ 27.25 and 27.27, respectively, must be established as operating limitations.

[Amdt. 27-2, 33 FR 965, Jan. 26, 1968, as amended by Amdt. 27-21, 49 FR 44435, Nov. 6, 1984]

§ 27.1521 Powerplant limitations.

(a) *General.* The powerplant limitations prescribed in this section must be established so that they do not exceed the corresponding limits for which the engines are type certificated.

(b) *Takeoff operation.* The powerplant takeoff operation must be limited by—

(1) The maximum rotational speed, which may not be greater than—

(i) The maximum value determined by the rotor design; or

(ii) The maximum value shown during the type tests;

(2) The maximum allowable manifold pressure (for reciprocating engines);

(3) The time limit for the use of the power corresponding to the limitations established in paragraphs (b)(1) and (2) of this section;

(4) If the time limit in paragraph (b)(3) of this section exceeds two minutes, the maximum allowable cylinder head, coolant outlet, or oil temperatures;

(5) The gas temperature limits for turbine engines over the range of operating and atmospheric conditions for which certification is requested.

(c) *Continuous operation.* The continuous operation must be limited by—

(1) The maximum rotational speed which may not be greater than—

(i) The maximum value determined by the rotor design; or

(ii) The maximum value shown during the type tests;

(2) The minimum rotational speed shown under the rotor speed requirements in § 27.1509(c); and

(3) The gas temperature limits for turbine engines over the range of operating and atmospheric conditions for which certification is requested.

(d) *Fuel grade or designation.* The minimum fuel grade (for reciprocating engines), or fuel designation (for turbine engines), must be established so that it is not less than that required for the operation of the engines within the limitations in paragraphs (b) and (c) of this section.

(e) *Turboshaft engine torque.* For rotorcraft with main rotors driven by turboshaft engines, and that do not have a torque limiting device in the transmission system, the following apply:

(1) A limit engine torque must be established if the maximum torque that the engine can exert is greater than—

(i) The torque that the rotor drive system is designed to transmit; or

(ii) The torque that the main rotor assembly is designed to withstand in showing compliance with § 27.547(e).

(2) The limit engine torque established under paragraph (e)(1) of this section may not exceed either torque specified in paragraph (e)(1)(i) or (ii) of this section.

(f) *Ambient temperature.* For turbine engines, ambient temperature limitations (including limitations for winterization installations, if applicable) must be established as the maximum ambient atmospheric temperature at which compliance with the cooling provisions of §§ 27.1041 through 27.1045 is shown.

(g) *Two and one-half-minute OEI power operation.* Unless otherwise authorized, the use of 2½ -minute OEI power must be limited to engine failure operation of multiengine, turbine-powered rotorcraft for not longer than 2½ minutes after failure of an engine. The use of 2½ -minute OEI power must also be limited by—

(1) The maximum rotational speed, which may not be greater than—

(i) The maximum value determined by the rotor design; or

(ii) The maximum demonstrated during the type tests;

(2) The maximum allowable gas temperature; and

(3) The maximum allowable torque.

(h) *Thirty-minute OEI power operation.* Unless otherwise authorized, the use of 30-minute OEI power must be limited to multiengine, turbine-powered rotorcraft for not longer than 30 minutes after failure of an engine. The use of 30-minute OEI power must also be limited by—

(1) The maximum rotational speed, which may not be greater than—

(i) The maximum value determined by the rotor design; or

(ii) The maximum value demonstrated during the type tests;

(2) The maximum allowable gas temperature; and

(3) The maximum allowable torque.

(i) *Continuous OEI power operation.* Unless otherwise authorized, the use of continuous OEI power must be limited to multiengine, turbine-powered rotorcraft for continued flight after failure of an engine. The use of continuous OEI power must also be limited by—

(1) The maximum rotational speed, which may not be greater than—

(i) The maximum value determined by the rotor design; or

(ii) The maximum value demonstrated during the type tests;

(2) The maximum allowable gas temperature; and

(3) The maximum allowable torque.

(j) *Rated 30-second OEI power operation.* Rated 30-second OEI power is permitted only on multiengine, turbine-powered rotorcraft, also certificated for the use of rated 2-minute OEI power, and can only be used for continued operation of the remaining engine(s) after a failure or precautionary shutdown of an engine. It must be shown that following application of 30-second OEI power, any damage will be readily detectable by the applicable inspections and other related procedures furnished in accordance with Section A27.4 of appendix A of this part and Section A33.4 of appendix A of part 33. The use of 30-second OEI power must be limited to not more than 30 seconds for any period in which that power is used, and by—

(1) The maximum rotational speed, which may not be greater than—

(i) The maximum value determined by the rotor design; or

(ii) The maximum value demonstrated during the type tests;

(2) The maximum allowable gas temperature; and

(3) The maximum allowable torque.

(k) *Rated 2-minute OEI power operation.* Rated 2-minute OEI power is permitted only on multiengine, turbine-powered rotorcraft, also certificated for the use of rated 30-second OEI power, and can only be used for continued operation of the remaining engine(s) after a failure or precautionary shutdown of an engine. It must be shown that following application of 2-minute OEI power, any damage will be readily detectable by the applicable inspections and other related procedures furnished in accordance with Section A27.4 of appendix A of this part and Section A33.4 of appendix A of part 33. The use of 2-minute OEI power must be limited to not more than 2 minutes for any period in which that power is used, and by—

(1) The maximum rotational speed, which may not be greater than—

(i) The maximum value determined by the rotor design; or

(ii) The maximum value demonstrated during the type tests;

(2) The maximum allowable gas temperature; and

(3) The maximum allowable torque.

(Secs. 313(a), 601, 603, 604, and 605 of the Federal Aviation Act of 1958 (49 U.S.C. 1354(a), 1421, 1423, 1424, and 1425); and sec. 6(c) of the Dept. of Transportation Act (49 U.S.C. 1655(c)))

[Doc. No. 5074, 29 FR 15695, Nov. 24, 1964, as amended by Amdt. 27-14, 43 FR 2325, Jan. 16, 1978; Amdt. 27-23, 53 FR 34214, Sept. 2, 1988; Amdt. 27-29, 59 FR 47767, Sept. 16, 1994]

§ 27.1523 Minimum flight crew.

The minimum flight crew must be established so that it is sufficient for safe operation, considering—

(a) The workload on individual crewmembers;

(b) The accessibility and ease of operation of necessary controls by the appropriate crewmember; and

(c) The kinds of operation authorized under § 27.1525.

§ 27.1525 Kinds of operations.

The kinds of operations (such as VFR, IFR, day, night, or icing) for which the rotorcraft is approved are established by demonstrated compliance with the applicable certification requirements and by the installed equipment.

[Amdt. 27-21, 49 FR 44435, Nov. 6, 1984]

§ 27.1527 Maximum operating altitude.

The maximum altitude up to which operation is allowed, as limited by flight, structural, powerplant, functional, or equipment characteristics, must be established.

(Secs. 313(a), 601, 603, 604, and 605 of the Federal Aviation Act of 1958 (49 U.S.C. 1354(a), 1421, 1423, 1424, and 1425); and sec. 6(c) of the Dept. of Transportation Act (49 U.S.C. 1655(c)))

[Amdt. 27-14, 43 FR 2325, Jan. 16, 1978]

§ 27.1529 Instructions for Continued Airworthiness.

The applicant must prepare Instructions for Continued Airworthiness in accordance with appendix A to this part that are acceptable to the Administrator. The instructions may be incomplete at type certification if a program exists to ensure their completion prior to delivery of the first rotorcraft or issuance of a standard certificate of airworthiness, whichever occurs later.

[Amdt. 27-18, 45 FR 60177, Sept. 11, 1980]

Markings and Placards

§ 27.1541 General.

(a) The rotorcraft must contain—

(1) The markings and placards specified in §§ 27.1545 through 27.1565, and

(2) Any additional information, instrument markings, and placards required for the safe operation of rotorcraft with unusual design, operating or handling characteristics.

(b) Each marking and placard prescribed in paragraph (a) of this section—

(1) Must be displayed in a conspicuous place; and

(2) May not be easily erased, disfigured, or obscured.

§ 27.1543 Instrument markings: general.

For each instrument—

(a) When markings are on the cover glass of the instrument, there must be means to maintain the correct alignment of the glass cover with the face of the dial; and

(b) Each arc and line must be wide enough, and located, to be clearly visible to the pilot.

§ 27.1545 Airspeed indicator.

(a) Each airspeed indicator must be marked as specified in paragraph (b) of this section, with the marks located at the corresponding indicated airspeeds.

(b) The following markings must be made:

(1) A red radial line—

(i) For rotocraft other than helicopters, at V_{NE}; and

(ii) For helicopters at V_{NE} (power-on).

(2) A red cross-hatched radial line at V_{NE} (power-off) for helicopters, if V_{NE} (power-off) is less than V_{NE} (power-on).

(3) For the caution range, a yellow arc.

(4) For the safe operating range, a green arc.

(Secs. 313(a), 601, 603, 604, and 605 of the Federal Aviation Act of 1958 (49 U.S.C. 1354(a), 1421, 1423, 1424, and 1425); and sec. 6(c) of the Dept. of Transportation Act (49 U.S.C. 1655(c)))

[Doc. No. 5074, 29 FR 15695, Nov. 24, 1964, as amended by Amdt. 27-14, 43 FR 2325, Jan. 16, 1978; 43 FR 3900, Jan. 30, 1978; Amdt. 27-16, 43 FR 50599, Oct. 30, 1978]

§ 27.1547 Magnetic direction indicator.

(a) A placard meeting the requirements of this section must be installed on or near the magnetic direction indicator.

(b) The placard must show the calibration of the instrument in level flight with the engines operating.

(c) The placard must state whether the calibration was made with radio receivers on or off.

(d) Each calibration reading must be in terms of magnetic heading in not more than 45 degree increments.

(e) If a magnetic nonstabilized direction indicator can have a deviation of more than 10 degrees caused by the operation of electrical equipment, the placard must state which electrical loads, or combination of loads, would cause a deviation of more than 10 degrees when turned on.

(Secs. 313(a), 601, 603, 604, and 605 of the Federal Aviation Act of 1958 (49 U.S.C. 1354(a), 1421, 1423, 1424, and 1425); and sec. 6(c) of the Dept. of Transportation Act (49 U.S.C. 1655(c)))

[Doc. No. 5074, 29 FR 15695, Nov. 24, 1964, as amended by Amdt. 27-13, 42 FR 36972, July 18, 1977]

§ 27.1549 Powerplant instruments.

For each required powerplant instrument, as appropriate to the type of instrument—

(a) Each maximum and, if applicable, minimum safe operating limit must be marked with a red radial or a red line;

(b) Each normal operating range must be marked with a green arc or green line, not extending beyond the maximum and minimum safe limits;

(c) Each takeoff and precautionary range must be marked with a yellow arc or yellow line;

(d) Each engine or propeller range that is restricted because of excessive vibration stresses must be marked with red arcs or red lines; and

(e) Each OEI limit or approved operating range must be marked to be clearly differentiated from the markings of paragraphs (a) through (d) of this section except that no marking is normally required for the 30-second OEI limit.

[Amdt. 27-11, 41 FR 55470, Dec. 20, 1976, as amended by Amdt. 27-23, 53 FR 34215, Sept. 2, 1988; Amdt. 27-29, 59 FR 47768, Sept. 16, 1994]

§ 27.1551 Oil quantity indicator.

Each oil quantity indicator must be marked with enough increments to indicate readily and accurately the quantity of oil.

§ 27.1553 Fuel quantity indicator.

If the unusable fuel supply for any tank exceeds one gallon, or five percent of the tank capacity, whichever is greater, a red arc must be marked on its indicator extending from the calibrated zero reading to the lowest reading obtainable in level flight.

§ 27.1555 Control markings.

(a) Each cockpit control, other than primary flight controls or control whose function is obvious, must be plainly marked as to its function and method of operation.

(b) For powerplant fuel controls—

(1) Each fuel tank selector control must be marked to indicate the position corresponding to each tank and to each existing cross feed position;

(2) If safe operation requires the use of any tanks in a specific sequence, that sequence must be marked on, or adjacent to, the selector for those tanks; and

(3) Each valve control for any engine of a multiengine rotorcraft must be marked to indicate the position corresponding to each engine controlled.

(c) Usable fuel capacity must be marked as follows:

(1) For fuel systems having no selector controls, the usable fuel capacity of the system must be indicated at the fuel quantity indicator.

(2) For fuel systems having selector controls, the usable fuel capacity available at each selector control position must be indicated near the selector control.

(d) For accessory, auxiliary, and emergency controls—

(1) Each essential visual position indicator, such as those showing rotor pitch or landing gear position, must be marked so that each crewmember can determine at any time the position of the unit to which it relates; and

(2) Each emergency control must be red and must be marked as to method of operation.

(e) For rotorcraft incorporating retractable landing gear, the maximum landing gear operating speed must be displayed in clear view of the pilot.

[Doc. No. 5074, 29 FR 15695, Nov. 24, 1964, as amended by Amdt. 27-11, 41 FR 55470, Dec. 20, 1976; Amdt. 27-21, 49 FR 44435, Nov. 6, 1984]

§ 27.1557 Miscellaneous markings and placards.

(a) *Baggage and cargo compartments, and ballast location.* Each baggage and cargo compartment, and each ballast location must have a placard stating any limitations on contents, including weight, that are necessary under the loading requirements.

(b) *Seats.* If the maximum allowable weight to be carried in a seat is less than 170 pounds, a placard stating the lesser weight must be permanently attached to the seat structure.

(c) *Fuel and oil filler openings.* The following apply:

(1) Fuel filler openings must be marked at or near the filler cover with—

(i) The word "fuel";

(ii) For reciprocating engine powered rotorcraft, the minimum fuel grade;

(iii) For turbine engine powered rotorcraft, the permissible fuel designations; and

(iv) For pressure fueling systems, the maximum permissible fueling supply pressure and the maximum permissible defueling pressure.

(2) Oil filler openings must be marked at or near the filler cover with the word "oil".

(d) *Emergency exit placards.* Each placard and operating control for each emergency exit must be red. A placard must be near each emergency exit control and must clearly indicate the location of that exit and its method of operation.

[Doc. No. 5074, 29 FR 15695, Nov. 24, 1964, as amended by Amdt. 27-11, 41 FR 55471, Dec. 20, 1976]

§ 27.1559 Limitations placard.

There must be a placard in clear view of the pilot that specifies the kinds of operations (such as VFR, IFR, day, night, or icing) for which the rotorcraft is approved.

[Amdt. 27-21, 49 FR 44435, Nov. 6, 1984]

§ 27.1561 Safety equipment.

(a) Each safety equipment control to be operated by the crew in emergency, such as controls for automatic liferaft releases, must be plainly marked as to its method of operation.

(b) Each location, such as a locker or compartment, that carries any fire extinguishing, signaling, or other life saving equipment, must be so marked.

§ 27.1565 Tail rotor.

Each tail rotor must be marked so that its disc is conspicuous under normal daylight ground conditions.

[Amdt. 27-2, 33 FR 965, Jan. 26, 1968]

ROTORCRAFT FLIGHT MANUAL AND APPROVED MANUAL MATERIAL

§ 27.1581 General.

(a) *Furnishing information.* A Rotorcraft Flight Manual must be furnished with each rotorcraft, and it must contain the following:

(1) Information required by §§ 27.1583 through 27.1589.

(2) Other information that is necessary for safe operation because of design, operating, or handling characteristics.

(b) *Approved information.* Each part of the manual listed in §§ 27.1583 through 27.1589, that is appropriate to the rotorcraft, must be furnished, verified, and approved, and must be segregated, identified, and clearly distinguished from each unapproved part of that manual.

(c) [Reserved]

(d) *Table of contents.* Each Rotorcraft Flight Manual must include a table of contents if the complexity of the manual indicates a need for it.

(Secs. 313(a), 601, 603, 604, and 605 of the Federal Aviation Act of 1958 (49 U.S.C. 1354(a), 1421, 1423, 1424, and 1425); and sec. 6(c) of the Dept. of Transportation Act (49 U.S.C. 1655(c)))

[Amdt. 27-14, 43 FR 2325, Jan. 16, 1978]

§ 27.1583 Operating limitations.

(a) *Airspeed and rotor limitations.* Information necessary for the marking of airspeed and rotor limitations on, or near, their respective indicators must be furnished. The significance of each limitation and of the color coding must be explained.

(b) *Powerplant limitations.* The following information must be furnished:

(1) Limitations required by § 27.1521.

(2) Explanation of the limitations, when appropriate.

(3) Information necessary for marking the instruments required by §§ 27.1549 through 27.1553.

(c) *Weight and loading distribution.* The weight and center of gravity limits required by §§ 27.25 and 27.27, respectively, must be furnished. If the variety of possible loading conditions warrants, instructions must be included to allow ready observance of the limitations.

(d) *Flight crew.* When a flight crew of more than one is required, the number and functions of the minimum flight crew determined under § 27.1523 must be furnished.

(e) *Kinds of operation.* Each kind of operation for which the rotorcraft and its equipment installations are approved must be listed.

(f) [Reserved]

(g) *Altitude.* The altitude established under § 27.1527 and an explanation of the limiting factors must be furnished.

(Secs. 313(a), 601, 603, 604, and 605 of the Federal Aviation Act of 1958 (49 U.S.C. 1354(a), 1421, 1423, 1424, and 1425); and sec. 6(c) of the Dept. of Transportation Act (49 U.S.C. 1655(c)))

[Doc. No. 5074, 29 FR 15695, Nov. 24, 1964, as amended by Amdt. 27-2, 33 FR 965, Jan. 26, 1968; Amdt. 27-14, 43 FR 2325, Jan. 16, 1978; Amdt. 27-16, 43 FR 50599, Oct. 30, 1978]

§ 27.1585 Operating procedures.

(a) Parts of the manual containing operating procedures must have information concerning any normal and emergency procedures and other information necessary for safe operation, including takeoff and landing procedures and associated airspeeds. The manual must contain any pertinent information including—

(1) The kind of takeoff surface used in the tests and each appropriate climbout speed; and

(2) The kind of landing surface used in the tests and appropriate approach and glide airspeeds.

(b) For multiengine rotorcraft, information identifying each operating condition in which the fuel system independence prescribed in § 27.953 is necessary for safety must be furnished, together with instructions for placing the fuel system in a configuration used to show compliance with that section.

(c) For helicopters for which a V_{NE} (power-off) is established under § 27.1505(c), information must be furnished to explain the V_{NE} (power-off) and the procedures for reducing airspeed to not more than the V_{NE} (power-off) following failure of all engines.

(d) For each rotorcraft showing compliance with § 27.1353 (g)(2) or (g)(3), the operating procedures for disconnecting the battery from its charging source must be furnished.

(e) If the unusable fuel supply in any tank exceeds five percent of the tank capacity, or one gallon, whichever is greater, information must be furnished which indicates that when the fuel quantity indicator reads "zero" in level flight, any fuel remaining in the fuel tank cannot be used safely in flight.

(f) Information on the total quantity of usable fuel for each fuel tank must be furnished.

(g) The airspeeds and rotor speeds for minimum rate of descent and best glide angle as prescribed in § 27.71 must be provided.

(Secs. 313(a), 601, 603, 604, and 605 of the Federal Aviation Act of 1958 (49 U.S.C. 1354(a), 1421, 1423, 1424, and 1425); and sec. 6(c) of the Dept. of Transportation Act (49 U.S.C. 1655(c)))

[Amdt. 27-1, 32 FR 6914, May 5, 1967, as amended by Amdt. 27-14, 43 FR 2326, Jan. 16, 1978; Amdt. 27-16, 43 FR 50599, Oct. 30, 1978; Amdt. 27-21, 49 FR 44435, Nov. 6, 1984]

§ 27.1587 Performance information.

(a) The Rotorcraft Flight Manual must contain the following information, determined in accordance with §§ 27.49 through 27.87 and 27.143(c) and (d):

(1) Enough information to determine the limiting height-speed envelope.

(2) Information relative to—

(i) The steady rates of climb and descent, in-ground effect and out-of-ground effect hovering ceilings, together with the corresponding airspeeds and other pertinent information including the calculated effects of altitude and temperatures;

(ii) The maximum weight for each altitude and temperature condition at which the rotorcraft can safely hover in-ground effect and out-of-ground effect in winds of not less than 17 knots from all azimuths. These data must be clearly referenced to the appropriate hover charts. In addition, if there are other combinations of weight, altitude and temperature for which performance information is provided and at which the rotorcraft cannot land and take off safely with the maximum wind value, those portions of the operating envelope and the appropriate safe wind conditions must be stated in the Rotorcraft Flight Manual;

(iii) For reciprocating engine-powered rotorcraft, the maximum atmospheric temperature at which compliance with the cooling provisions of §§ 27.1041 through 27.1045 is shown; and

(iv) Glide distance as a function of altitude when autorotating at the speeds and conditions for minimum rate of descent and best glide as determined in § 27.71.

(b) The Rotorcraft Flight Manual must contain—

(1) In its performance information section any pertinent information concerning the takeoff weights and altitudes used in compliance with § 27.51; and

(2) The horizontal takeoff distance determined in accordance with § 27.65(a)(2)(i).

(Secs. 313(a), 601, 603, 604, and 605 of the Federal Aviation Act of 1958 (49 U.S.C. 1354(a), 1421, 1423, 1424, and 1425); and sec. 6(c) of the Dept. of Transportation Act (49 U.S.C. 1655(c)))

[Doc. No. 5074, 29 FR 15695, Nov. 24, 1964, as amended by Amdt. 27-14, 43 FR 2326, Jan. 16, 1978; Amdt. 27-21, 49 FR 44435, Nov. 6, 1984; Amdt. 27-44, 73 FR 11000, Feb. 29, 2008; 73 FR 33876, June 16, 2008]

§ 27.1589 Loading information.

There must be loading instructions for each possible loading condition between the maximum and minimum weights determined under § 27.25 that can result in a center of gravity beyond any extreme prescribed in § 27.27, assuming any probable occupant weights.

Appendix A to Part 27—Instructions for Continued Airworthiness

A27.1 *General.*

(a) This appendix specifies requirements for the preparation of Instructions for Continued Airworthiness as required by § 27.1529.

(b) The Instructions for Continued Airworthiness for each rotorcraft must include the Instructions for Continued Airworthiness for each engine and rotor (hereinafter designated'products'), for each appliance required by this chapter, and any required information relating to the interface of those appliances and products with the rotorcraft. If Instructions for Continued Airworthiness are not supplied by the manufacturer of an appliance or product installed in the rotorcraft, the Instructions for Continued Airworthiness for the rotorcraft must include the information essential to the continued airworthiness of the rotorcraft.

(c) The applicant must submit to the FAA a program to show how changes to the Instructions for Continued Airworthiness made by the applicant or by the manufacturers of products and appliances installed in the rotorcraft will be distributed.

A27.2 *Format.*

(a) The Instructions for Continued Airworthiness must be in the form of a manual or manuals as appropriate for the quantity of data to be provided.

(b) The format of the manual or manuals must provide for a practical arrangement.

A27.3 *Content.*

The contents of the manual or manuals must be prepared in the English language. The Instructions for Continued Airworthiness must contain the following manuals or sections, as appropriate, and information:

(a) *Rotorcraft maintenance manual or section.* (1) Introduction information that includes an explanation of the rotorcraft's features and data to the extent necessary for maintenance or preventive maintenance.

(2) A description of the rotorcraft and its systems and installations including its engines, rotors, and appliances.

(3) Basic control and operation information describing how the rotorcraft components and systems are controlled and how they operate, including any special procedures and limitations that apply.

(4) Servicing information that covers details regarding servicing points, capacities of tanks, reservoirs, types of fluids to be used, pressures applicable to the various systems, location of access panels for inspection and servicing, locations of lubrication points, the lubricants to be used, equipment required for servicing, tow instructions and limitations, mooring, jacking, and leveling information.

(b) *Maintenance instructions.* (1) Scheduling information for each part of the rotorcraft and its engines, auxiliary power units, rotors, accessories, instruments and equipment that provides the recommended periods at which they should be cleaned, inspected, adjusted, tested, and lubricated, and the degree of inspection, the applicable wear tolerances, and work recommended at these periods. However, the applicant may refer to an accessory, instrument, or equipment manufacturer as the source of this information if the applicant shows the item has an exceptionally high degree of complexity requiring specialized maintenance techniques, test equipment, or expertise. The recommended overhaul periods and necessary cross references to the Airworthiness Limitations section of the manual must also be included. In addition, the applicant must include an inspection program that includes the frequency and extent of the inspections necessary to provide for the continued airworthiness of the rotorcraft.

(2) Troubleshooting information describing problem malfunctions, how to recognize those malfunctions, and the remedial action for those malfunctions.

(3) Information describing the order and method of removing and replacing products and parts with any necessary precautions to be taken.

(4) Other general procedural instructions including procedures for system testing during ground running, symmetry checks, weighing and determining the center of gravity, lifting and shoring, and storage limitations.

(c) Diagrams of structural access plates and information needed to gain access for inspections when access plates are not provided.

(d) Details for the application of special inspection techniques including radiographic and ultrasonic testing where such processes are specified.

(e) Information needed to apply protective treatments to the structure after inspection.

(f) All data relative to structural fasteners such as identification, discarded recommendations, and torque values.

(g) A list of special tools needed.

A27.4 *Airworthiness Limitations section.*

The Instructions for Continued Airworthiness must contain a section, titled Airworthiness Limitations that is segregated and clearly distinguishable from the rest of the document. This section must set forth each mandatory replacement time, structural inspection interval, and related structural inspection procedure required for type certification. If the Instructions for Continued Airworthiness consist of multiple documents, the section required by this paragraph must be included in the principal manual. This section must contain a legible statement in a prominent location that reads: "The Airworthiness Limitations section is FAA approved and specifies inspections and other maintenance required under §§ 43.16 and 91.403 of the Federal Aviation Regulations unless an alternative program has been FAA approved."

[Amdt. 27-18, 45 FR 60177, Sept. 11, 1980, as amended by Amdt. 27-24, 54 FR 34329, Aug. 18, 1989; Amdt. 27-47, 76 FR 74663, Dec. 1, 2011]

Appendix B to Part 27—Airworthiness Criteria for Helicopter Instrument Flight

I. *General.* A normal category helicopter may not be type certificated for operation under the instrument flight rules (IFR) of this chapter unless it meets the design and installation requirements contained in this appendix.

II. *Definitions.* (a) V_{YI} means instrument climb speed, utilized instead of V_Y for compliance with the climb requirements for instrument flight.

(b) V_{NEI} means instrument flight never exceed speed, utilized instead of V_{NE} for compliance with maximum limit speed requirements for instrument flight.

(c) V_{MINI} means instrument flight minimum speed, utilized in complying with minimum limit speed requirements for instrument flight.

III. *Trim.* It must be possible to trim the cyclic, collective, and directional control forces to zero at all approved IFR airspeeds, power settings, and configurations appropriate to the type.

IV. *Static longitudinal stability.* (a) *General.* The helicopter must possess positive static longitudinal control force stability at critical combinations of weight and center of gravity at the conditions specified in paragraph IV (b) or (c) of this appendix, as appropriate. The stick force must vary with speed so that any substantial speed change results in a stick force clearly perceptible to the pilot. For single-pilot approval, the airspeed must return to within 10 percent of the trim speed when the control force is slowly released for each trim condition specified in paragraph IV(b) of the this appendix.

(b) *For single-pilot approval:*

(1) *Climb.* Stability must be shown in climb throughout the speed range 20 knots either side of trim with—

(i) The helicopter trimmed at V_{YI};

(ii) Landing gear retracted (if retractable); and

(iii) Power required for limit climb rate (at least 1,000 fpm) at V_{YI} or maximum continuous power, whichever is less.

(2) *Cruise.* Stability must be shown throughout the speed range from 0.7 to 1.1 V_H or V_{NEI}, whichever is lower, not to exceed±20 knots from trim with—

(i) The helicopter trimmed and power adjusted for level flight at 0.9 V_H or 0.9 V_{NEI}, whichever is lower; and

(ii) Landing gear retracted (if retractable).

(3) *Slow cruise.* Stability must be shown throughout the speed range from 0.9 V_{MINI} to 1.3 V_{MINI} or 20 knots above trim speed, whichever is greater, with—

(i) the helicopter trimmed and power adjusted for level flight at 1.1 V_{MINI}; and

(ii) Landing gear retracted (if retractable).

(4) *Descent.* Stability must be shown throughout the speed range 20 knots either side of trim with—

(i) The helicopter trimmed at 0.8 V_H or 0.8 V_{NEI} (or 0.8 V_{LE} for the landing gear extended case), whichever is lower;

(ii) Power required for 1,000 fpm descent at trim speed; and

(iii) Landing gear extended and retracted, if applicable.

(5) *Approach.* Stability must be shown throughout the speed range from 0.7 times the minimum recommended approach speed to 20 knots above the maximum recommended approach speed with—

(i) The helicopter trimmed at the recommended approach speed or speeds;

(ii) Landing gear extended and retracted, if applicable; and

(iii) Power required to maintain a 3° glide path and power required to maintain the steepest approach gradient for which approval is requested.

(c) Helicopters approved for a minimum crew of two pilots must comply with the provisions of paragraphs IV(b) (2) and IV(b)(5) of this appendix.

V. *Static Lateral Directional Stability.* (a) Static directional stability must be positive throughout the approved ranges of airspeed, power, and vertical speed. In straight and steady sideslips up to±10° from trim, directional control position must increase without discontinuity with the angle of sideslip, except for a small range of sideslip angles around trim. At greater angles up to the maximum sideslip angle appropriate to the type, increased directional control position must produce an increased angle of sideslip. It must be possible to maintain balanced flight without exceptional pilot skill or alertness.

(b) During sideslips up to±10° from trim throughout the approved ranges of airspeed, power, and vertical speed, there must be no negative dihedral stability perceptible to the pilot through lateral control motion or force. Longitudinal cyclic movement with sideslip must not be excessive.

VI. *Dynamic stability.* (a) For single-pilot approval—

(1) Any oscillation having a period of less than 5 seconds must damp to ½ amplitude in not more than one cycle.

(2) Any oscillation having a period of 5 seconds or more but less than 10 seconds must damp to ½ amplitude in not more than two cycles.

(3) Any oscillation having a period of 10 seconds or more but less than 20 seconds must be damped.

(4) Any oscillation having a period of 20 seconds or more may not achieve double amplitude in less than 20 seconds.

(5) Any aperiodic response may not achieve double amplitude in less than 6 seconds.

(b) For helicopters approved with a minimum crew of two pilots—

(1) Any oscillation having a period of less than 5 seconds must damp to ½ amplitude in not more than two cycles.

(2) Any oscillation having a period of 5 seconds or more but less than 10 seconds must be damped.

(3) Any oscillation having a period of 10 seconds or more may not achieve double amplitude in less than 10 seconds.

VII. *Stability Augmentation System (SAS).*

(a) If a SAS is used, the reliability of the SAS must be related to the effects of its failure. Any SAS failure condition that would prevent continued safe flight and landing must be extremely improbable. It must be shown that, for any failure condition of the SAS that is not shown to be extremely improbable—

(1) The helicopter is safely controllable when the failure or malfunction occurs at any speed or altitude within the approved IFR operating limitations; and

(2) The overall flight characteristics of the helicopter allow for prolonged instrument flight without undue pilot effort. Additional unrelated probable failures affecting the control system must be considered. In addition—

(i) The controllability and maneuverability requirements in Subpart B of this part must be met throughout a practical flight envelope;

(ii) The flight control, trim, and dynamic stability characteristics must not be impaired below a level needed to allow continued safe flight and landing; and

(iii) The static longitudinal and static directional stability requirements of Subpart B must be met throughout a practical flight envelope.

(b) The SAS must be designed so that it cannot create a hazardous deviation in flight path or produce hazardous loads on the helicopter during normal operation or in the event of malfunction or failure, assuming corrective action begins within an appropriate period of time. Where multiple systems are installed, subsequent malfunction conditions must be considered in sequence unless their occurrence is shown to be improbable.

VIII. *Equipment, systems, and installation.* The basic equipment and installation must comply with §§ 29.1303, 29.1431, and 29.1433 through Amendment 29-14, with the following exceptions and additions:

(a) *Flight and Navigation Instruments.* (1) A magnetic gyro-stablized direction indicator instead of a gyroscopic direction indicator required by § 29.1303(h); and

(2) A standby attitude indicator which meets the requirements of §§ 29.1303(g)(1) through (7) instead of a rate-of-turn indicator required by § 29.1303(g). For two-pilot configurations, one pilot's primary indicator may be designated for this purpose. If standby batteries are provided, they may be charged from the aircraft electrical system if adequate isolation is incorporated.

(b) *Miscellaneous requirements.* (1) Instrument systems and other systems essential for IFR flight that could be adversely affected by icing must be adequately protected when exposed to the continuous and intermittent maximum icing conditions defined in appendix C of Part 29 of this chapter, whether or not the rotorcraft is certificated for operation in icing conditions.

(2) There must be means in the generating system to automatically de-energize and disconnect from the main bus any power source developing hazardous overvoltage.

(3) Each required flight instrument using a power supply (electric, vacuum, etc.) must have a visual means integral with the instrument to indicate the adequacy of the power being supplied.

(4) When multiple systems performing like functions are required, each system must be grouped, routed, and spaced so that physical separation between systems is provided to ensure that a single malfunction will not adversely affect more than one system.

(5) For systems that operate the required flight instruments at each pilot's station—

(i) Only the required flight instruments for the first pilot may be connected to that operating system;

(ii) Additional instruments, systems, or equipment may not be connected to an operating system for a second pilot unless provisions are made to ensure the continued normal functioning of the required instruments in the event of any malfunction of the additional instruments, systems, or equipment which is not shown to be extremely improbable;

(iii) The equipment, systems, and installations must be designed so that one display of the information essential to the safety of flight which is provided by the instruments will remain available to a pilot, without additional crewmember action, after any single failure or combination of failures that is not shown to be extremely improbable; and

(iv) For single-pilot configurations, instruments which require a static source must be provided with a means of selecting an alternate source and that source must be calibrated.

IX. *Rotorcraft Flight Manual.* A Rotorcraft Flight Manual or Rotorcraft Flight Manual IFR Supplement must be provided and must contain—

(a) *Limitations.* The approved IFR flight envelope, the IFR flightcrew composition, the revised kinds of operation, and the steepest IFR precision approach gradient for which the helicopter is approved;

(b) *Procedures.* Required information for proper operation of IFR systems and the recommended procedures in the event of stability augmentation or electrical system failures; and

(c) *Performance.* If V_{YI} differs from V_Y, climb performance at V_{YI} and with maximum continuous power throughout the ranges of weight, altitude, and temperature for which approval is requested.

X. Electrical and electronic system lightning protection. For regulations concerning lightning protection for electrical and electronic systems, see § 27.1316.

[Amdt. 27-19, 48 FR 4389, Jan. 31, 1983, as amended by Amdt. 27-44, 73 FR 11000, Feb. 29, 2008; Amdt. 27-46, 76 FR 33135, June 8, 2011]

Appendix C to Part 27—Criteria for Category A

C27.1 General.

A small multiengine rotorcraft may not be type certificated for Category A operation unless it meets the design installation and performance requirements contained in this appendix in addition to the requirements of this part.

C27.2 Applicable part 29 sections.

The following sections of part 29 of this chapter must be met in addition to the requirements of this part:

29.45(a) and (b)(2)—General.
29.49(a)—Performance at minimum operating speed.
29.51—Takeoff data: General.
29.53—Takeoff: Category A.
29.55—Takeoff decision point: Category A.
29.59—Takeoff Path: Category A.
29.60—Elevated heliport takeoff path: Category A.
29.61—Takeoff distance: Category A.
29.62—Rejected takeoff: Category A.
29.64—Climb: General.
29.65(a)—Climb: AEO.
29.67(a)—Climb: OEI.
29.75—Landing: General.
29.77—Landing decision point: Category A.
29.79—Landing: Category A.
29.81—Landing distance (Ground level sites): Category A.
29.85—Balked landing: Category A.
29.87(a)—Height-velocity envelope.
29.547(a) and (b)—Main and tail rotor structure.
29.861(a)—Fire protection of structure, controls, and other parts.
29.901(c)—Powerplant: Installation.
29.903(b) (c) and (e)—Engines.
29.908(a)—Cooling fans.
29.917(b) and (c)(1)—Rotor drive system: Design.
29.927(c)(1)—Additional tests.
29.953(a)—Fuel system independence.
29.1027(a)—Transmission and gearboxes: General.
29.1045(a)(1), (b), (c), (d), and (f)—Climb cooling test procedures.
29.1047(a)—Takeoff cooling test procedures.
29.1181(a)—Designated fire zones: Regions included.
29.1187(e)—Drainage and ventilation of fire zones.
29.1189(c)—Shutoff means.
29.1191(a)(1)—Firewalls.
29.1193(e)—Cowling and engine compartment covering.
29.1195(a) and (d)—Fire extinguishing systems (one shot).
29.1197—Fire extinguishing agents.
29.1199—Extinguishing agent containers.
29.1201—Fire extinguishing system materials.
29.1305(a) (6) and (b)—Powerplant instruments.
29.1309(b)(2) (i) and (d)—Equipment, systems, and installations.
29.1323(c)(1)—Airspeed indicating system.
29.1331(b)—Instruments using a power supply.
29.1351(d)(2)—Electrical systems and equipment: General (operation without normal electrical power).
29.1587(a)—Performance information.

Note: In complying with the paragraphs listed in paragraph C27.2 above, relevant material in the AC "Certification of Transport Category Rotorcraft" should be used.

[Doc. No. 28008, 61 FR 21907, May 10, 1996]

Appendix D to Part 27—HIRF Environments and Equipment HIRF Test Levels

This appendix specifies the HIRF environments and equipment HIRF test levels for electrical and electronic systems under § 27.1317. The field strength values for the HIRF environments and laboratory equipment HIRF test levels are expressed in root-mean-square units measured during the peak of the modulation cycle.

(a) HIRF environment I is specified in the following table:

Table I.—HIRF Environment I

Frequency	Field strength (volts/meter)	
	Peak	Average
10 kHz–2 MHz	50	50
2 MHz–30 MHz	100	100
30 MHz–100 MHz	50	50
100 MHz–400 MHz	100	100
400 MHz–700 MHz	700	50
700 MHz–1 GHz	700	100
1 GHz–2 GHz	2,000	200
2 GHz–6 GHz	3,000	200
6 GHz–8 GHz	1,000	200
8 GHz–12 GHz	3,000	300
12 GHz–18 GHz	2,000	200
18 GHz–40 GHz	600	200

In this table, the higher field strength applies at the frequency band edges.

(b) HIRF environment II is specified in the following table:

Table II.—HIRF Environment II

Frequency	Field strength (volts/meter)	
	Peak	Average
10 kHz–500 kHz	20	20
500 kHz–2 MHz	30	30
2 MHz–30 MHz	100	100
30 MHz–100 MHz	10	10
100 MHz–200 MHz	30	10
200 MHz–400 MHz	10	10
400 MHz–1 GHz	700	40
1 GHz–2 GHz	1,300	160
2 GHz–4 GHz	3,000	120
4 GHz–6 GHz	3,000	160
6 GHz–8 GHz	400	170
8 GHz–12 GHz	1,230	230
12 GHz–18 GHz	730	190
18 GHz–40 GHz	600	150

In this table, the higher field strength applies at the frequency band edges.

(b) HIRF environment II is specified in the following table:

Table III.—HIRF Environment III

Frequency	Field strength (volts/meter)	
	Peak	Average
10 kHz–100 kHz	150	150
100 kHz–400 MHz	200	200
400 MHz–700 MHz	730	200
700 MHz–1 GHz	1,400	240
1 GHz–2 GHz	5,000	250
2 GHz–4 GHz	6,000	490
4 GHz–6 GHz	7,200	400
6 GHz–8 GHz	1,100	170
8 GHz–12 GHz	5,000	330
12 GHz–18 GHz	2,000	330
18 GHz–40 GHz	1,000	420

In this table, the higher field strength applies at the frequency band edges.

(c) HIRF environment III is specified in the following table:

In this table, the higher field strength applies at the frequency band edges.

(d) *Equipment HIRF Test Level 1.*

(1) From 10 kilohertz (kHz) to 400 megahertz (MHz), use conducted susceptibility tests with continuous wave (CW) and 1 kHz square wave modulation with 90 percent depth or greater. The conducted susceptibility current must start at a minimum of 0.6 milliamperes (mA) at 10 kHz, increasing 20 decibels (dB) per frequency decade to a minimum of 30 mA at 500 kHz.

(2) From 500 kHz to 40 MHz, the conducted susceptibility current must be at least 30 mA.

(3) From 40 MHz to 400 MHz, use conducted susceptibility tests, starting at a minimum of 30 mA at 40 MHz, decreasing 20 dB per frequency decade to a minimum of 3 mA at 400 MHz.

(4) From 100 MHz to 400 MHz, use radiated susceptibility tests at a minimum of 20 volts per meter (V/m) peak with CW and 1 kHz square wave modulation with 90 percent depth or greater.

(5) From 400 MHz to 8 gigahertz (GHz), use radiated susceptibility tests at a minimum of 150 V/m peak with pulse modulation of 4 percent duty cycle with a 1 kHz pulse repetition frequency. This signal must be switched on and off at a rate of 1 Hz with a duty cycle of 50 percent.

(e) *Equipment HIRF Test Level 2.* Equipment HIRF test level 2 is HIRF environment II in table II of this appendix reduced by acceptable aircraft transfer function and attenuation curves. Testing must cover the frequency band of 10 kHz to 8 GHz.

(f) *Equipment HIRF Test Level 3.* (1) From 10 kHz to 400 MHz, use conducted susceptibility tests, starting at a minimum of 0.15 mA at 10 kHz, increasing 20 dB per frequency decade to a minimum of 7.5 mA at 500 kHz.

(2) From 500 kHz to 40 MHz, use conducted susceptibility tests at a minimum of 7.5 mA.

(3) From 40 MHz to 400 MHz, use conducted susceptibility tests, starting at a minimum of 7.5 mA at 40 MHz, decreasing 20 dB per frequency decade to a minimum of 0.75 mA at 400 MHz.

(4) From 100 MHz to 8 GHz, use radiated susceptibility tests at a minimum of 5 V/m.

[Doc. No. FAA-2006-23657, 72 FR 44027, Aug. 6, 2007]

PART 29

AIRWORTHINESS STANDARDS: TRANSPORT CATEGORY ROTORCRAFT

Contents

AUTHORITY: 49 U.S.C. 106(g), 40113, 44701-44702, 44704.

SOURCE: Docket No. 5084, 29 FR 16150, Dec. 3, 1964, unless otherwise noted.

Subpart A—General

§ 29.1 Applicability.

(a) This part prescribes airworthiness standards for the issue of type certificates, and changes to those certificates, for transport category rotorcraft.

(b) Transport category rotorcraft must be certificated in accordance with either the Category A or Category B requirements of this part. A multiengine rotorcraft may be type certificated as both Category A and Category B with appropriate and different operating limitations for each category.

(c) Rotorcraft with a maximum weight greater than 20,000 pounds and 10 or more passenger seats must be type certificated as Category A rotorcraft.

(d) Rotorcraft with a maximum weight greater than 20,000 pounds and nine or less passenger seats may be type certificated as Category B rotorcraft provided the Category A requirements of Subparts C, D, E, and F of this part are met.

(e) Rotorcraft with a maximum weight of 20,000 pounds or less but with 10 or more passenger seats may be type certificated as Category B rotorcraft provided the Category A requirements of §§ 29.67(a)(2), 29.87, 29.1517, and subparts C, D, E, and F of this part are met.

(f) Rotorcraft with a maximum weight of 20,000 pounds or less and nine or less passenger seats may be type certificated as Category B rotorcraft.

(g) Each person who applies under Part 21 for a certificate or change described in paragraphs (a) through (f) of this section must show compliance with the applicable requirements of this part.

[Amdt. 29-21, 48 FR 4391, Jan. 31, 1983, as amended by Amdt. 29-39, 61 FR 21898, May 10, 1996; 61 FR 33963, July 1, 1996]

§ 29.2 Special retroactive requirements.

For each rotorcraft manufactured after September 16, 1992, each applicant must show that each occupant's seat is equipped with a safety belt and shoulder harness that meets the requirements of paragraphs (a), (b), and (c) of this section.

(a) Each occupant's seat must have a combined safety belt and shoulder harness with a single-point release. Each pilot's combined safety belt and shoulder harness must allow each pilot, when seated with safety belt and shoulder harness fastened, to perform all functions necessary for flight operations. There must be a means to secure belts and harnesses, when not in use, to prevent interference with the operation of the rotorcraft and with rapid egress in an emergency.

(b) Each occupant must be protected from serious head injury by a safety belt plus a shoulder harness that will prevent the head from contacting any injurious object.

(c) The safety belt and shoulder harness must meet the static and dynamic strength requirements, if applicable, specified by the rotorcraft type certification basis.

(d) For purposes of this section, the date of manufacture is either—

(1) The date the inspection acceptance records, or equivalent, reflect that the rotorcraft is complete and meets the FAA-Approved Type Design Data; or

(2) The date that the foreign civil airworthiness authority certifies the rotorcraft is complete and issues an original standard airworthiness certificate, or equivalent, in that country.

[Doc. No. 26078, 56 FR 41052, Aug. 16, 1991]

Subpart B—Flight

GENERAL

§ 29.21 Proof of compliance.

Each requirement of this subpart must be met at each appropriate combination of weight and center of gravity within the range of loading conditions for which certification is requested. This must be shown—

(a) By tests upon a rotorcraft of the type for which certification is requested, or by calculations based on, and equal in accuracy to, the results of testing; and

(b) By systematic investigation of each required combination of weight and center of gravity, if compliance cannot be reasonably inferred from combinations investigated.

[Doc. No. 5084, 29 FR 16150, Dec. 3, 1964, as amended by Amdt. 29-24, 49 FR 44435, Nov. 6, 1984]

§ 29.25 Weight limits.

(a) *Maximum weight.* The maximum weight (the highest weight at which compliance with each applicable requirement of this part is shown) or, at the option of the applicant, the highest weight for each altitude and for each practicably separable operating condition, such as takeoff, enroute operation, and landing, must be established so that it is not more than—

(1) The highest weight selected by the applicant;

(2) The design maximum weight (the highest weight at which compliance with each applicable structural loading condition of this part is shown); or

(3) The highest weight at which compliance with each applicable flight requirement of this part is shown.

(4) For Category B rotorcraft with 9 or less passenger seats, the maximum weight, altitude, and temperature at which the rotorcraft can safely operate near the ground with the maximum wind velocity determined under § 29.143(c) and may include other demonstrated wind velocities and azimuths. The operating envelopes must be stated in the Limitations section of the Rotorcraft Flight Manual.

(b) *Minimum weight.* The minimum weight (the lowest weight at which compliance with each applicable requirement of this part is shown) must be established so that it is not less than—

(1) The lowest weight selected by the applicant;

(2) The design minimum weight (the lowest weight at which compliance with each structural loading condition of this part is shown); or

(3) The lowest weight at which compliance with each applicable flight requirement of this part is shown.

(c) *Total weight with jettisonable external load.* A total weight for the rotorcraft with a jettisonable external load attached that is greater than the maximum weight established under paragraph (a) of this section may be established for any rotorcraft-load combination if—

(1) The rotorcraft-load combination does not include human external cargo,

(2) Structural component approval for external load operations under either § 29.865 or under equivalent operational standards is obtained,

(3) The portion of the total weight that is greater than the maximum weight established under paragraph (a) of this section is made up only of the weight of all or part of the jettisonable external load,

(4) Structural components of the rotorcraft are shown to comply with the applicable structural requirements of this part under the increased loads and stresses caused by the weight increase over that established under paragraph (a) of this section, and

(5) Operation of the rotorcraft at a total weight greater than the maximum certificated weight established under paragraph (a) of this section is limited by appropriate operating limitations under § 29.865 (a) and (d) of this part.

[Doc. No. 5084, 29 FR 16150, Dec. 3, 1964, as amended by Amdt. 29-12, 41 FR 55471, Dec. 20, 1976; Amdt. 29-43, 64 FR 43020, Aug. 6, 1999; Amdt. 29-51, 73 FR 11001, Feb. 29, 2008]

§ 29.27 Center of gravity limits.

The extreme forward and aft centers of gravity and, where critical, the extreme lateral centers of gravity must be established for each weight established under § 29.25. Such an extreme may not lie beyond—

(a) The extremes selected by the applicant;

(b) The extremes within which the structure is proven; or

(c) The extremes within which compliance with the applicable flight requirements is shown.

[Amdt. 29-3, 33 FR 965, Jan. 26, 1968]

§ 29.29 Empty weight and corresponding center of gravity.

(a) The empty weight and corresponding center of gravity must be determined by weighing the rotorcraft without the crew and payload, but with—

(1) Fixed ballast;

(2) Unusable fuel; and

(3) Full operating fluids, including—

(i) Oil;

(ii) Hydraulic fluid; and

(iii) Other fluids required for normal operation of rotorcraft systems, except water intended for injection in the engines.

(b) The condition of the rotorcraft at the time of determining empty weight must be one that is well defined and can be easily repeated, particularly with respect to the weights of fuel, oil, coolant, and installed equipment.

(Secs. 313(a), 601, 603, 604, and 605 of the Federal Aviation Act of 1958 (49 U.S.C. 1354(a), 1421, 1423, 1424, and 1425); and sec. 6(c) of the Dept. of Transportation Act (49 U.S.C. 1655(c)))

[Doc. No. 5084, 29 FR 16150. Dec. 3, 1964, as amended by Amdt. 29-15, 43 FR 2326, Jan. 16, 1978]

§ 29.31 Removable ballast.

Removable ballast may be used in showing compliance with the flight requirements of this subpart.

§ 29.33 Main rotor speed and pitch limits.

(a) *Main rotor speed limits.* A range of main rotor speeds must be established that—

(1) With power on, provides adequate margin to accommodate the variations in rotor speed occurring in any appropriate maneuver, and is consistent with the kind of governor or synchronizer used; and

(2) With power off, allows each appropriate autorotative maneuver to be performed throughout the ranges of airspeed and weight for which certification is requested.

(b) *Normal main rotor high pitch limit (power on).* For rotorcraft, except helicopters required to have a main rotor low speed warning under paragraph (e) of this section, it must be shown, with power on and without exceeding approved engine maximum limitations, that main rotor speeds substantially less than the minimum approved main rotor speed will not occur under any sustained flight condition. This must be met by—

(1) Appropriate setting of the main rotor high pitch stop;

(2) Inherent rotorcraft characteristics that make unsafe low main rotor speeds unlikely; or

(3) Adequate means to warn the pilot of unsafe main rotor speeds.

(c) *Normal main rotor low pitch limit (power off).* It must be shown, with power off, that—

(1) The normal main rotor low pitch limit provides sufficient rotor speed, in any autorotative condition, under the most critical combinations of weight and airspeed; and

(2) It is possible to prevent overspeeding of the rotor without exceptional piloting skill.

(d) *Emergency high pitch.* If the main rotor high pitch stop is set to meet paragraph (b)(1) of this section, and if that stop cannot be exceeded inadvertently, additional pitch may be made available for emergency use.

(e) *Main rotor low speed warning for helicopters.* For each single engine helicopter, and each multiengine helicopter that does not have an approved device that automatically increases power on the operating engines when one engine fails, there must be a main rotor low speed warning which meets the following requirements:

(1) The warning must be furnished to the pilot in all flight conditions, including power-on and power-off flight, when the speed of a main rotor approaches a value that can jeopardize safe flight.

(2) The warning may be furnished either through the inherent aerodynamic qualities of the helicopter or by a device.

(3) The warning must be clear and distinct under all conditions, and must be clearly distinguishable from all other warnings. A visual device that requires the attention of the crew within the cockpit is not acceptable by itself.

(4) If a warning device is used, the device must automatically deactivate and reset when the low-speed condition is corrected. If the device has an audible warning, it must also be equipped with a means for the pilot to manually silence the audible warning before the low-speed condition is corrected.

(Secs. 313(a), 601, 603, 604, and 605 of the Federal Aviation Act of 1958 (49 U.S.C. 1354(a), 1421, 1423, 1424, and 1425); and sec. 6(c) of the Dept. of Transportation Act (49 U.S.C. 1655(c)))

[Doc. No. 5084, 29 FR 16150, Dec. 3, 1964, as amended by Amdt. 29-3, 33 FR 965, Jan. 26, 1968; Amdt. 29-15, 43 FR 2326, Jan. 16, 1978]

Performance

§ 29.45 General.

(a) The performance prescribed in this subpart must be determined—

(1) With normal piloting skill and;

(2) Without exceptionally favorable conditions.

(b) Compliance with the performance requirements of this subpart must be shown—

(1) For still air at sea level with a standard atmosphere and;

(2) For the approved range of atmospheric variables.

(c) The available power must correspond to engine power, not exceeding the approved power, less—

(1) Installation losses; and

(2) The power absorbed by the accessories and services at the values for which certification is requested and approved.

(d) For reciprocating engine-powered rotorcraft, the performance, as affected by engine power, must be based on a relative humidity of 80 percent in a standard atmosphere.

(e) For turbine engine-powered rotorcraft, the performance, as affected by engine power, must be based on a relative humidity of—

(1) 80 percent, at and below standard temperature; and

(2) 34 percent, at and above standard temperature plus 50°F.

Between these two temperatures, the relative humidity must vary linearly.

(f) For turbine-engine-power rotorcraft, a means must be provided to permit the pilot to detemine prior to takeoff that each engine is capable of developing the power necessary to achieve the applicable rotorcraft performance prescribed in this subpart.

(Secs. 313(a), 601, 603, 604, and 605 of the Federal Aviation Act of 1958 (49 U.S.C. 1354(a), 1421, 1423, 1424, and 1425); and sec. 6(c), Dept. of Transportation Act (49 U.S.C. 1655(c)))

[Doc. No. 5084, 29 FR 16150, Dec. 3, 1964, as amended by Amdt. 29-15, 43 FR 2326, Jan. 16, 1978; Amdt. 29-24, 49 FR 44436, Nov. 6, 1984]

§ 29.49 Performance at minimum operating speed.

(a) For each Category A helicopter, the hovering performance must be determined over the ranges of weight, altitude, and temperature for which takeoff data are scheduled—

(1) With not more than takeoff power;

(2) With the landing gear extended; and

(3) At a height consistent with the procedure used in establishing the takeoff, climbout, and rejected takeoff paths.

(b) For each Category B helicopter, the hovering performance must be determined over the ranges of weight, altitude, and temperature for which certification is requested, with—

(1) Takeoff power;

(2) The landing gear extended; and

(3) The helicopter in ground effect at a height consistent with normal takeoff procedures.

(c) For each helicopter, the out-of-ground effect hovering performance must be determined over the ranges of weight, altitude, and temperature for which certification is requested with takeoff power.

(d) For rotorcraft other than helicopters, the steady rate of climb at the minimum operating speed must be determined over the ranges of weight, altitude, and temperature for which certification is requested with—

(1) Takeoff power; and

(2) The landing gear extended.

[Doc. No. 24802, 61 FR 21898, May 10, 1996; 61 FR 33963, July 1, 1996]

§ 29.51 Takeoff data: general.

(a) The takeoff data required by §§ 29.53, 29.55, 29.59, 29.60, 29.61, 29.62, 29.63, and 29.67 must be determined—

(1) At each weight, altitude, and temperature selected by the applicant; and

(2) With the operating engines within approved operating limitations.

(b) Takeoff data must—

(1) Be determined on a smooth, dry, hard surface; and

(2) Be corrected to assume a level takeoff surface.

(c) No takeoff made to determine the data required by this section may require exceptional piloting skill or alertness, or exceptionally favorable conditions.

[Doc. No. 5084, 29 FR 16150, Dec. 3, 1964, as amended by Amdt. 29-39, 61 FR 21899, May 10, 1996]

§ 29.53 Takeoff: Category A.

The takeoff performance must be determined and scheduled so that, if one engine fails at any time after the start of takeoff, the rotorcraft can—

(a) Return to, and stop safely on, the takeoff area; or

(b) Continue the takeoff and climbout, and attain a configuration and airspeed allowing compliance with § 29.67(a)(2).

[Doc. No. 24802, 61 FR 21899, May 10, 1996; 61 FR 33963, July 1, 1996]

§ 29.55 Takeoff decision point (TDP): Category A.

(a) The TDP is the first point from which a continued takeoff capability is assured under § 29.59 and is the last point in the takeoff path from which a rejected takeoff is assured within the distance determined under § 29.62.

(b) The TDP must be established in relation to the takeoff path using no more than two parameters; e.g., airspeed and height, to designate the TDP.

(c) Determination of the TDP must include the pilot recognition time interval following failure of the critical engine.

[Doc. No. 24802, 61 FR 21899, May 10, 1996]

§ 29.59 Takeoff path: Category A.

(a) The takeoff path extends from the point of commencement of the takeoff procedure to a point at which the rotorcraft is 1,000 feet above the takeoff surface and compliance with § 29.67(a)(2) is shown. In addition—

(1) The takeoff path must remain clear of the height-velocity envelope established in accordance with § 29.87;

(2) The rotorcraft must be flown to the engine failure point; at which point, the critical engine must be made inoperative and remain inoperative for the rest of the takeoff;

(3) After the critical engine is made inoperative, the rotorcraft must continue to the takeoff decision point, and then attain V_{TOSS};

(4) Only primary controls may be used while attaining V_{TOSS} and while establishing a positive rate of climb. Secondary controls that are located on the primary controls may be used after a positive rate of climb and V_{TOSS} are established but in no case less than 3 seconds after the critical engine is made inoperative; and

(5) After attaining V_{TOSS} and a positive rate of a climb, the landing gear may be retracted.

(b) During the takeoff path determination made in accordance with paragraph (a) of this section and after attaining V_{TOSS} and a positive rate of climb, the climb must be continued at a speed as close as practicable to, but not less than, V_{TOSS} until the rotorcraft is 200 feet above the takeoff surface. During this interval, the climb performance must meet or exceed that required by § 29.67(a)(1).

(c) During the continued takeoff, the rotorcraft shall not descend below 15 feet above the takeoff surface when the takeoff decision point is above 15 feet.

(d) From 200 feet above the takeoff surface, the rotorcraft takeoff path must be level or positive until a height 1,000 feet above the takeoff surface is attained with not less than the rate of climb required by § 29.67(a)(2). Any secondary or auxiliary control may be used after attaining 200 feet above the takeoff surface.

(e) Takeoff distance will be determined in accordance with § 29.61.

[Doc. No. 24802, 61 FR 21899, May 10, 1996; 61 FR 33963, July 1, 1996, as amended by Amdt. 29-44, 64 FR 45337, Aug. 19, 1999]

§ 29.60 Elevated heliport takeoff path: Category A.

(a) The elevated heliport takeoff path extends from the point of commencement of the takeoff procedure to a point in the takeoff path at which the rotorcraft is 1,000 feet above the takeoff surface and compliance with § 29.67(a)(2) is shown. In addition—

(1) The requirements of § 29.59(a) must be met;

(2) While attaining V_{TOSS} and a positive rate of climb, the rotorcraft may descend below the level of the takeoff surface if, in so doing and when clearing the elevated heliport edge, every part of the rotorcraft clears all obstacles by at least 15 feet;

(3) The vertical magnitude of any descent below the takeoff surface must be determined; and.

(4) After attaining V_{TOSS} and a positive rate of climb, the landing gear may be retracted.

(b) The scheduled takeoff weight must be such that the climb requirements of § 29.67 (a)(1) and (a)(2) will be met.

(c) Takeoff distance will be determined in accordance with § 29.61.

[Doc. No. 24802, 61 FR 21899, May 10, 1996; 61 FR 33963, July 1, 1996]

§ 29.61 Takeoff distance: Category A.

(a) The normal takeoff distance is the horizontal distance along the takeoff path from the start of the takeoff to the point at which the rotorcraft attains and remains at least 35 feet above the takeoff surface, attains and maintains a speed of at least V_{TOSS}, and establishes a positive rate of climb, assuming the critical engine failure occurs at the engine failure point prior to the takeoff decision point.

(b) For elevated heliports, the takeoff distance is the horizontal distance along the takeoff path from the start of the takeoff to the point at which the rotorcraft attains and maintains a speed of at least V_{TOSS} and establishes a positive rate of climb, assuming the critical engine failure occurs at the engine failure point prior to the takeoff decision point.

[Doc. No. 24802, 61 FR 21899, May 10, 1996]

§ 29.62 Rejected takeoff: Category A.

The rejected takeoff distance and procedures for each condition where takeoff is approved will be established with—

(a) The takeoff path requirements of §§ 29.59 and 29.60 being used up to the TDP where the critical engine failure is recognized and the rotorcraft is landed and brought to a complete stop on the takeoff surface;

(b) The remaining engines operating within approved limits;

(c) The landing gear remaining extended throughout the entire rejected takeoff; and

(d) The use of only the primary controls until the rotorcraft is on the ground. Secondary controls located on the primary control may not be used until the rotorcraft is on the ground. Means other than wheel brakes may be used to stop the rotorcraft if the means are safe and reliable and consistent results can be expected under normal operating conditions.

[Doc. No. 24802, 61 FR 21899, May 10, 1996, as amended by Amdt. 29-44, 64 FR 45337, Aug. 19, 1999]

§ 29.63 Takeoff: Category B.

The horizontal distance required to take off and climb over a 50-foot obstacle must be established with the most unfavorable center of gravity. The takeoff may be begun in any manner if—

(a) The takeoff surface is defined;

(b) Adequate safeguards are maintained to ensure proper center of gravity and control positions; and

(c) A landing can be made safely at any point along the flight path if an engine fails.

[Doc. No. 5084, 29 FR 16150, Dec. 3, 1964, as amended by Amdt. 29-12, 41 FR 55471, Dec. 20, 1976]

§ 29.64 Climb: General.

Compliance with the requirements of §§ 29.65 and 29.67 must be shown at each weight, altitude, and temperature within the operational limits established for the rotorcraft and with the most unfavorable center of gravity for each configuration. Cowl flaps, or other means of controlling the engine-cooling air supply, will be in the position that provides adequate cooling at the temperatures and altitudes for which certification is requested.

[Doc. No. 24802, 61 FR 21900, May 10, 1996]

§ 29.65 Climb: All engines operating.

(a) The steady rate of climb must be determined—

(1) With maximum continuous power;

(2) With the landing gear retracted; and

(3) At V_y for standard sea level conditions and at speeds selected by the applicant for other conditions.

(b) For each Category B rotorcraft except helicopters, the rate of climb determined under paragraph (a) of this section must provide a steady climb gradient of at least 1:6 under standard sea level conditions.

(Secs. 313(a), 601, 603, 604, and 605 of the Federal Aviation Act of 1958 (49 U.S.C. 1354(a), 1421, 1423, 1424, and 1425); and sec. 6(c), Dept. of Transportation Act (49 U.S.C. 1655(c)))

[Doc. No. 5084, 29 FR 16150. Dec. 3, 1964, as amended by Amdt. 29-15, 43 FR 2326, Jan. 16, 1978; Amdt. 29-39, 61 FR 21900, May 10, 1996; 61 FR 33963, July 1, 1996]

§ 29.67 Climb: One engine inoperative (OEI).

(a) For Category A rotorcraft, in the critical takeoff configuration existing along the takeoff path, the following apply:

(1) The steady rate of climb without ground effect, 200 feet above the takeoff surface, must be at least 100 feet per minute for each weight, altitude, and temperature for which takeoff data are to be scheduled with—

(i) The critical engine inoperative and the remaining engines within approved operating limitations, except that for rotorcraft for which the use of 30-second/2-minute OEI power is requested, only the 2-minute OEI power may be used in showing compliance with this paragraph;

(ii) The landing gear extended; and

(iii) The takeoff safety speed selected by the applicant.

(2) The steady rate of climb without ground effect, 1000 feet above the takeoff surface, must be at least 150 feet per minute, for each weight, altitude, and temperature for which takeoff data are to be scheduled with—

(i) The critical engine inoperative and the remaining engines at maximum continuous power including continuous OEI power, if approved, or at 30-minute OEI power for rotorcraft for which certification for use of 30-minute OEI power is requested;

(ii) The landing gear retracted; and

(iii) The speed selected by the applicant.

(3) The steady rate of climb (or descent) in feet per minute, at each altitude and temperature at which the rotorcraft is expected to operate and at any weight within the range of weights for which certification is requested, must be determined with—

(i) The critical engine inoperative and the remaining engines at maximum continuous power including continuous OEI power, if approved, and at 30-minute OEI power for rotorcraft for which certification for the use of 30-minute OEI power is requested;

(ii) The landing gear retracted; and

(iii) The speed selected by the applicant.

(b) For multiengine Category B rotorcraft meeting the Category A engine isolation requirements, the steady rate of climb (or descent) must be determined at the speed for best rate of climb (or minimum rate of descent) at each altitude, temperature, and weight at which the rotorcraft is expected to operate, with the critical engine inoperative and the remaining engines at maximum continuous power including continuous OEI power, if approved, and at 30-minute OEI power for rotorcraft for which certification for the use of 30-minute OEI power is requested.

[Doc. No. 24802, 61 FR 21900, May 10, 1996; 61 FR 33963, July 1, 1996, as amended by Amdt. 29-44, 64 FR 45337, Aug. 19, 1999; 64 FR 47563, Aug. 31, 1999]

§ 29.71 Helicopter angle of glide: Category B.

For each category B helicopter, except multiengine helicopters meeting the requirements of § 29.67(b) and the powerplant installation requirements of category A, the steady angle of glide must be determined in autorotation—

(a) At the forward speed for minimum rate of descent as selected by the applicant;

(b) At the forward speed for best glide angle;

(c) At maximum weight; and

(d) At the rotor speed or speeds selected by the applicant.

[Amdt. 29-12, 41 FR 55471, Dec. 20, 1976]

§ 29.75 Landing: General.

(a) For each rotorcraft—

(1) The corrected landing data must be determined for a smooth, dry, hard, and level surface;

(2) The approach and landing must not require exceptional piloting skill or exceptionally favorable conditions; and

(3) The landing must be made without excessive vertical acceleration or tendency to bounce, nose over, ground loop, porpoise, or water loop.

(b) The landing data required by §§ 29.77, 29.79, 29.81, 29.83, and 29.85 must be determined—

(1) At each weight, altitude, and temperature for which landing data are approved;

(2) With each operating engine within approved operating limitations; and

(3) With the most unfavorable center of gravity.

[Doc. No. 24802, 61 FR 21900, May 10, 1996]

§ 29.77 Landing Decision Point (LDP): Category A.

(a) The LDP is the last point in the approach and landing path from which a balked landing can be accomplished in accordance with § 29.85.

(b) Determination of the LDP must include the pilot recognition time interval following failure of the critical engine.

[Doc. No. 24802, 64 FR 45338, Aug. 19, 1999]

§ 29.79 Landing: Category A.

(a) For Category A rotorcraft—

(1) The landing performance must be determined and scheduled so that if the critical engine fails at any point in the approach path, the rotorcraft can either land and stop safely or climb out and attain a rotorcraft configuration and speed allowing compliance with the climb requirement of § 29.67(a)(2);

(2) The approach and landing paths must be established with the critical engine inoperative so that the transition between each stage can be made smoothly and safely;

(3) The approach and landing speeds must be selected by the applicant and must be appropriate to the type of rotorcraft; and

(4) The approach and landing path must be established to avoid the critical areas of the height-velocity envelope determined in accordance with § 29.87.

(b) It must be possible to make a safe landing on a prepared landing surface after complete power failure occurring during normal cruise.

[Doc. No. 24802, 61 FR 21900, May 10, 1996]

§ 29.81 Landing distance: Category A.

The horizontal distance required to land and come to a complete stop (or to a speed of approximately 3 knots for water landings) from a point 50 ft above the landing surface must be determined from the approach and landing paths established in accordance with § 29.79.

[Doc. No. 24802, 64 FR 45338, Aug. 19, 1999]

§ 29.83 Landing: Category B.

(a) For each Category B rotorcraft, the horizontal distance required to land and come to a complete stop (or to a speed of approximately 3 knots for water landings) from a point 50 feet above the landing surface must be determined with—

(1) Speeds appropriate to the type of rotorcraft and chosen by the applicant to avoid the critical areas of the height-velocity envelope established under § 29.87; and

(2) The approach and landing made with power on and within approved limits.

(b) Each multiengined Category B rotorcraft that meets the powerplant installation requirements for Category A must meet the requirements of—

(1) Sections 29.79 and 29.81; or

(2) Paragraph (a) of this section.

(c) It must be possible to make a safe landing on a prepared landing surface if complete power failure occurs during normal cruise.

[Doc. No. 24802, 61 FR 21900, May 10, 1996; 61 FR 33963, July 1, 1996]

§ 29.85 Balked landing: Category A.

For Category A rotorcraft, the balked landing path with the critical engine inoperative must be established so that—

(a) The transition from each stage of the maneuver to the next stage can be made smoothly and safely;

(b) From the LDP on the approach path selected by the applicant, a safe climbout can be made at speeds allowing compliance with the climb requirements of § 29.67(a)(1) and (2); and

(c) The rotorcraft does not descend below 15 feet above the landing surface. For elevated heliport operations, descent may be below the level of the landing surface provided the deck edge clearance of § 29.60 is maintained and the descent (loss of height) below the landing surface is determined.

[Doc. No. 24802, 64 FR 45338, Aug. 19, 1999]

§ 29.87 Height-velocity envelope.

(a) If there is any combination of height and forward velocity (including hover) under which a safe landing cannot be made after failure of the critical engine and with the remaining engines (where applicable) operating within approved limits, a height-velocity envelope must be established for—

(1) All combinations of pressure altitude and ambient temperature for which takeoff and landing are approved; and

(2) Weight from the maximum weight (at sea level) to the highest weight approved for takeoff and landing at each altitude. For helicopters, this weight need not exceed the highest weight allowing hovering out-of-ground effect at each altitude.

(b) For single-engine or multiengine rotorcraft that do not meet the Category A engine isolation requirements, the height-velocity envelope for complete power failure must be established.

[Doc. No. 24802, 61 FR 21901, May 10, 1996; 61 FR 33963, July 1, 1996]

Flight Characteristics

§ 29.141 General.

The rotorcraft must—

(a) Except as specifically required in the applicable section, meet the flight characteristics requirements of this subpart—

(1) At the approved operating altitudes and temperatures;

(2) Under any critical loading condition within the range of weights and centers of gravity for which certification is requested; and

(3) For power-on operations, under any condition of speed, power, and rotor r.p.m. for which certification is requested; and

(4) For power-off operations, under any condition of speed, and rotor r.p.m. for which certification is requested that is attainable with the controls rigged in accordance with the approved rigging instructions and tolerances;

(b) Be able to maintain any required flight condition and make a smooth transition from any flight condition to any other flight condition without exceptional piloting skill, alertness, or strength, and without danger of exceeding the limit load factor under any operating condition probable for the type, including—

(1) Sudden failure of one engine, for multiengine rotorcraft meeting Transport Category A engine isolation requirements;

(2) Sudden, complete power failure, for other rotorcraft; and

(3) Sudden, complete control system failures specified in § 29.695 of this part; and

(c) Have any additional characteristics required for night or instrument operation, if certification for those kinds of operation is requested. Requirements for helicopter instrument flight are contained in appendix B of this part.

[Doc. No. 5084, 29 FR 16150, Dec. 8, 1964, as amended by Amdt. 29-3, 33 FR 905, Jan. 26, 1968; Amdt. 29-12, 41 FR 55471, Dec. 20, 1976; Amdt. 29-21, 48 FR 4391, Jan. 31, 1983; Amdt. 29-24, 49 FR 44436, Nov. 6, 1984]

§ 29.143 Controllability and maneuverability.

(a) The rotorcraft must be safely controllable and maneuverable—

(1) During steady flight; and

(2) During any maneuver appropriate to the type, including—

(i) Takeoff;

(ii) Climb;

(iii) Level flight;

(iv) Turning flight;

(v) Autorotation; and

(vi) Landing (power on and power off).

(b) The margin of cyclic control must allow satisfactory roll and pitch control at V_{NE} with—

(1) Critical weight;

(2) Critical center of gravity;

(3) Critical rotor r.p.m.; and

(4) Power off (except for helicopters demonstrating compliance with paragraph (f) of this section) and power on.

(c) Wind velocities from zero to at least 17 knots, from all azimuths, must be established in which the rotorcraft can be operated without loss of control on or near the ground in any maneuver appropriate to the type (such as crosswind takeoffs, sideward flight, and rearward flight), with—

(1) Critical weight;

(2) Critical center of gravity;

(3) Critical rotor r.p.m.; and

(4) Altitude, from standard sea level conditions to the maximum takeoff and landing altitude capability of the rotorcraft.

(d) Wind velocities from zero to at least 17 knots, from all azimuths, must be established in which the rotorcraft can be operated without loss of control out-of-ground effect, with—

(1) Weight selected by the applicant;

(2) Critical center of gravity;

(3) Rotor r.p.m. selected by the applicant; and

(4) Altitude, from standard sea level conditions to the maximum takeoff and landing altitude capability of the rotorcraft.

(e) The rotorcraft, after (1) failure of one engine, in the case of multiengine rotorcraft that meet Transport Category A engine isolation requirements, or (2) complete power failure in the case of other rotorcraft, must be controllable over the range of speeds and altitudes for which certification is requested when such power failure occurs with maximum continuous power and critical weight. No corrective action time delay for any condition following power failure may be less than—

(i) For the cruise condition, one second, or normal pilot reaction time (whichever is greater); and

(ii) For any other condition, normal pilot reaction time.

(f) For helicopters for which a V_{NE} (power-off) is established under § 29.1505(c), compliance must be demonstrated with the following requirements with critical weight, critical center of gravity, and critical rotor r.p.m.:

(1) The helicopter must be safely slowed to V_{NE} (power-off), without exceptional pilot skill after the last operating engine is made inoperative at power-on V_{NE}.

(2) At a speed of 1.1 V_{NE} (power-off), the margin of cyclic control must allow satisfactory roll and pitch control with power off.

(Secs. 313(a), 601, 603, 604, and 605 of the Federal Aviation Act of 1958 (49 U.S.C. 1354(a), 1421, 1423, 1424, and 1425); and sec. 6(c) of the Dept. of Transportation Act (49 U.S.C. 1655(c)))

[Doc. No. 5084, 29 FR 16150, Dec. 3, 1964, as amended by Amdt. 29-3, 33 FR 965, Jan. 26, 1968; Amdt. 29-15, 43 FR 2326, Jan. 16, 1978; Amdt. 29-24, 49 FR 44436, Nov. 6, 1984; Amdt. 29-51, 73 FR 11001, Feb. 29, 2008]

§ 29.151 Flight controls.

(a) Longitudinal, lateral, directional, and collective controls may not exhibit excessive breakout force, friction, or preload.

(b) Control system forces and free play may not inhibit a smooth, direct rotorcraft response to control system input.

[Amdt. 29-24, 49 FR 44436, Nov. 6, 1984]

§ 29.161 Trim control.

The trim control—

(a) Must trim any steady longitudinal, lateral, and collective control forces to zero in level flight at any appropriate speed; and

(b) May not introduce any undesirable discontinuities in control force gradients.

[Doc. No. 5084, 29 FR 16150, Dec. 3, 1964, as amended by Amdt. 29-24, 49 FR 44436, Nov. 6, 1984]

§ 29.171 Stability: general.

The rotorcraft must be able to be flown, without undue pilot fatigue or strain, in any normal maneuver for a period of time as long as that expected in normal operation. At least three landings and takeoffs must be made during this demonstration.

§ 29.173 Static longitudinal stability.

(a) The longitudinal control must be designed so that a rearward movement of the control is necessary to obtain an airspeed less than the trim speed, and a forward movement of the control is necessary to obtain an airspeed more than the trim speed.

(b) Throughout the full range of altitude for which certification is requested, with the throttle and collective pitch held constant during the maneuvers specified in § 29.175(a) through (d), the slope of the control position versus airspeed curve must be positive. However, in limited flight conditions or modes of operation determined by the Administrator to be acceptable, the slope of the control position versus airspeed curve may be neutral or negative if the rotorcraft possesses flight characteristics that allow the pilot to maintain airspeed within±5 knots of the desired trim airspeed without exceptional piloting skill or alertness.

[Amdt. 29-24, 49 FR 44436, Nov. 6, 1984, as amended by Amdt. 29-51, 73 FR 11001, Feb. 29, 2008]

§ 29.175 Demonstration of static longitudinal stability.

(a) *Climb.* Static longitudinal stability must be shown in the climb condition at speeds from *Vy* – 10 kt to *Vy*+10 kt with—

(1) Critical weight;

(2) Critical center of gravity;

(3) Maximum continuous power;

(4) The landing gear retracted; and

(5) The rotorcraft trimmed at Vy.

(b) *Cruise.* Static longitudinal stability must be shown in the cruise condition at speeds from 0.8 V_{NE} –10 kt to 0.8 V_{NE}+10 kt or, if V_H is less than 0.8 V_{NE}, from VH – 10 kt to V_H+10 kt, with—

(1) Critical weight;

(2) Critical center of gravity;

(3) Power for level flight at 0.8 V_{NE} or V_H, whichever is less;

(4) The landing gear retracted; and

(5) The rotorcraft trimmed at 0.8 V_{NE} or V_H, whichever is less.

(c) V_{NE}. Static longitudinal stability must be shown at speeds from V_{NE} – 20 kt to V_{NE} with—

(1) Critical weight;

(2) Critical center of gravity;

(3) Power required for level flight at V_{NE} – 10 kt or maximum continuous power, whichever is less;

(4) The landing gear retracted; and

(5) The rotorcraft trimmed at V_{NE} – 10 kt.

(d) *Autorotation.* Static longitudinal stability must be shown in autorotation at—

(1) Airspeeds from the minimum rate of descent airspeed – 10 kt to the minimum rate of descent airspeed+10 kt, with—

(i) Critical weight;

(ii) Critical center of gravity;

(iii) The landing gear extended; and

(iv) The rotorcraft trimmed at the minimum rate of descent airspeed.

(2) Airspeeds from the best angle-of-glide airspeed – 10kt to the best angle-of-glide airspeed+10kt, with—

(i) Critical weight;

(ii) Critical center of gravity;

(iii) The landing gear retracted; and

(iv) The rotorcraft trimmed at the best angle-of-glide airspeed.

[Amdt. 29-51, 73 FR 11001, Feb. 29, 2008]

§ 29.177 Static directional stability.

(a) The directional controls must operate in such a manner that the sense and direction of motion of the rotorcraft following control displacement are in the direction of the pedal motion with throttle and collective controls held constant at the trim conditions specified in § 29.175(a), (b), (c), and (d). Sideslip angles must increase with steadily increasing directional control deflection for sideslip angles up to the lesser of—

(1)±25 degrees from trim at a speed of 15 knots less than the speed for minimum rate of descent varying linearly to±10 degrees from trim at V_{NE};

(2) The steady-state sideslip angles established by § 29.351;

(3) A sideslip angle selected by the applicant, which corresponds to a sideforce of at least 0.1g; or

(4) The sideslip angle attained by maximum directional control input.

(b) Sufficient cues must accompany the sideslip to alert the pilot when approaching sideslip limits.

(c) During the maneuver specified in paragraph (a) of this section, the sideslip angle versus directional control position curve may have a negative slope within a small range of angles around trim, provided the desired heading can be maintained without exceptional piloting skill or alertness.

[Amdt. 29-51, 73 FR 11001, Feb. 29, 2008]

§ 29.181 Dynamic stability: Category A rotorcraft.

Any short-period oscillation occurring at any speed from V_Y to V_{NE} must be positively damped with the primary flight controls free and in a fixed position.

[Amdt. 29-24, 49 FR 44437, Nov. 6, 1984]

Ground and Water Handling Characteristics

§ 29.231 General.

The rotorcraft must have satisfactory ground and water handling characteristics, including freedom from uncontrollable tendencies in any condition expected in operation.

§ 29.235 Taxiing condition.

The rotorcraft must be designed to withstand the loads that would occur when the rotorcraft is taxied over the roughest ground that may reasonably be expected in normal operation.

§ 29.239 Spray characteristics.

If certification for water operation is requested, no spray characteristics during taxiing, takeoff, or landing may obscure the vision of the pilot or damage the rotors, propellers, or other parts of the rotorcraft.

§ 29.241 Ground resonance.

The rotorcraft may have no dangerous tendency to oscillate on the ground with the rotor turning.

Miscellaneous Flight Requirements

§ 29.251 Vibration.

Each part of the rotorcraft must be free from excessive vibration under each appropriate speed and power condition.

Subpart C—Strength Requirements

General

§ 29.301 Loads.

(a) Strength requirements are specified in terms of limit loads (the maximum loads to be expected in service) and ultimate loads (limit loads multiplied by prescribed factors of safety). Unless otherwise provided, prescribed loads are limit loads.

(b) Unless otherwise provided, the specified air, ground, and water loads must be placed in equilibrium with inertia forces, considering each item of mass in the rotorcraft. These loads must be distributed to closely approximate or conservatively represent actual conditions.

(c) If deflections under load would significantly change the distribution of external or internal loads, this redistribution must be taken into account.

§ 29.303 Factor of safety.

Unless otherwise provided, a factor of safety of 1.5 must be used. This factor applies to external and inertia loads unless its application to the resulting internal stresses is more conservative.

§ 29.305 Strength and deformation.

(a) The structure must be able to support limit loads without detrimental or permanent deformation. At any load up to limit loads, the deformation may not interfere with safe operation.

(b) The structure must be able to support ultimate loads without failure. This must be shown by—

(1) Applying ultimate loads to the structure in a static test for at least three seconds; or

(2) Dynamic tests simulating actual load application.

§ 29.307 Proof of structure.

(a) Compliance with the strength and deformation requirements of this subpart must be shown for each critical loading condition accounting for the environment to which the structure will be exposed in operation. Structural analysis (static or fatigue) may be used only if the structure conforms to those structures for which experience has shown this method to be reliable. In other cases, substantiating load tests must be made.

(b) Proof of compliance with the strength requirements of this subpart must include—

(1) Dynamic and endurance tests of rotors, rotor drives, and rotor controls;

(2) Limit load tests of the control system, including control surfaces;

(3) Operation tests of the control system;

(4) Flight stress measurement tests;

(5) Landing gear drop tests; and

(6) Any additional tests required for new or unusual design features.

(Secs. 604, 605, 72 Stat. 778, 49 U.S.C. 1424, 1425)

[Doc. No. 5084, 29 FR 16150, Dec. 3, 1964, as amended by Amdt. 29-4, 33 FR 14106, Sept. 18, 1968; Amdt. 27-26, 55 FR 8001, Mar. 6, 1990]

§ 29.309 Design limitations.

The following values and limitations must be established to show compliance with the structural requirements of this subpart:

(a) The design maximum and design minimum weights.

(b) The main rotor r.p.m. ranges, power on and power off.

(c) The maximum forward speeds for each main rotor r.p.m. within the ranges determined under paragraph (b) of this section.

(d) The maximum rearward and sideward flight speeds.

(e) The center of gravity limits corresponding to the limitations determined under paragraphs (b), (c), and (d) of this section.

(f) The rotational speed ratios between each powerplant and each connected rotating component.

(g) The positive and negative limit maneuvering load factors.

Flight Loads

§ 29.321 General.

(a) The flight load factor must be assumed to act normal to the longitudinal axis of the rotorcraft, and to be equal in magnitude and opposite in direction to the rotorcraft inertia load factor at the center of gravity.

(b) Compliance with the flight load requirements of this subpart must be shown—

(1) At each weight from the design minimum weight to the design maximum weight; and

(2) With any practical distribution of disposable load within the operating limitations in the Rotorcraft Flight Manual.

§ 29.337 Limit maneuvering load factor.

The rotorcraft must be designed for—

(a) A limit maneuvering load factor ranging from a positive limit of 3.5 to a negative limit of –1.0; or

(b) Any positive limit maneuvering load factor not less than 2.0 and any negative limit maneuvering load factor of not less than –0.5 for which—

(1) The probability of being exceeded is shown by analysis and flight tests to be extremely remote; and

(2) The selected values are appropriate to each weight condition between the design maximum and design minimum weights.

[Doc. No. 5084, 29 FR 16150, Dec. 3, 1964, as amended by Amdt. 27-26, 55 FR 8002, Mar. 6, 1990]

§ 29.339 Resultant limit maneuvering loads.

The loads resulting from the application of limit maneuvering load factors are assumed to act at the center of each rotor hub and at each auxiliary lifting surface, and to act in directions and with distributions of load among the rotors and auxiliary lifting surfaces, so as to represent each critical maneuvering condition, including power-on and power-off flight with the maximum design rotor tip speed ratio. The rotor tip speed ratio is the ratio of the rotorcraft flight velocity component in the plane of the rotor disc to the rotational tip speed of the rotor blades, and is expressed as follows:

$$\mu=\frac{V\cos\alpha}{\Omega R}$$

where—

V=The airspeed along the flight path (f.p.s.);

a=The angle between the projection, in the plane of symmetry, of the axis of no feathering and a line perpendicular to the flight path (radians, positive when axis is pointing aft);

Ω=The angular velocity of rotor (radians per second); and

R=The rotor radius (ft.).

§ 29.341 Gust loads.

Each rotorcraft must be designed to withstand, at each critical airspeed including hovering, the loads resulting from vertical and horizontal gusts of 30 feet per second.

§ 29.351 Yawing conditions.

(a) Each rotorcraft must be designed for the loads resulting from the maneuvers specified in paragraphs (b) and (c) of this section, with—

(1) Unbalanced aerodynamic moments about the center of gravity which the aircraft reacts to in a rational or conservative manner considering the principal masses furnishing the reacting inertia forces; and

(2) Maximum main rotor speed.

(b) To produce the load required in paragraph (a) of this section, in unaccelerated flight with zero yaw, at forward speeds from zero up to 0.6 V_{NE}—

(1) Displace the cockpit directional control suddenly to the maximum deflection limited by the control stops or by the maximum pilot force specified in § 29.397(a);

(2) Attain a resulting sideslip angle or 90°, whichever is less; and

(3) Return the directional control suddenly to neutral.

(c) To produce the load required in paragraph (a) of the section, in unaccelerated flight with zero yaw, at forward speeds from 0.6 V_{NE} up to V_{NE} or V_H, whichever is less—

(1) Displace the cockpit directional control suddenly to the maximum deflection limited by the control stops or by the maximum pilot force specified in § 29.397(a);

(2) Attain a resulting sideslip angle or 15°, whichever is less, at the lesser speed of V_{NE} or V_H;

(3) Vary the sideslip angles of paragraphs (b)(2) and (c)(2) of this section directly with speed; and

(4) Return the directional control suddenly to neutral.

[Amdt. 29-26, 55 FR 8002, Mar. 6, 1990, as amended by Amdt. 29-41, 62 FR 46173, Aug. 29, 1997]

§ 29.361 Engine torque.

The limit engine torque may not be less than the following:

(a) For turbine engines, the highest of—

(1) The mean torque for maximum continuous power multiplied by 1.25;

(2) The torque required by § 29.923;

(3) The torque required by § 29.927; or

(4) The torque imposed by sudden engine stoppage due to malfunction or structural failure (such as compressor jamming).

(b) For reciprocating engines, the mean torque for maximum continuous power multiplied by—

(1) 1.33, for engines with five or more cylinders; and

(2) Two, three, and four, for engines with four, three, and two cylinders, respectively.

[Amdt. 29-26, 53 FR 34215, Sept. 2, 1988]

Control Surface and System Loads

§ 29.391 General.

Each auxiliary rotor, each fixed or movable stabilizing or control surface, and each system operating any flight control must meet the requirements of §§ 29.395 through 29.399, 29.411, and 29.427.

[Amdt. 29-26, 55 FR 8002, Mar. 6, 1990, as amended by Amdt. 29-41, 62 FR 46173, Aug. 29, 1997]

§ 29.395 Control system.

(a) The reaction to the loads prescribed in § 29.397 must be provided by—

(1) The control stops only;

(2) The control locks only;

(3) The irreversible mechanism only (with the mechanism locked and with the control surface in the critical positions for the effective parts of the system within its limit of motion);

(4) The attachment of the control system to the rotor blade pitch control horn only (with the control in the critical positions for the affected parts of the system within the limits of its motion); and

(5) The attachment of the control system to the control surface horn (with the control in the critical positions for the affected parts of the system within the limits of its motion).

(b) Each primary control system, including its supporting structure, must be designed as follows:

(1) The system must withstand loads resulting from the limit pilot forces prescribed in § 29.397;

(2) Notwithstanding paragraph (b)(3) of this section, when power-operated actuator controls or power boost controls are used, the system must also withstand the loads resulting from the limit pilot forces prescribed in § 29.397 in conjunction with the forces output of each normally energized power device, including any single power boost or actuator system failure;

(3) If the system design or the normal operating loads are such that a part of the system cannot react to the limit pilot forces prescribed in § 29.397, that part of the system must be designed to withstand the maximum loads that can be obtained in normal operation. The minimum design loads must, in any case, provide a rugged system for service use, including consideration of fatigue, jamming, ground gusts, control inertia, and friction loads. In the absence of a rational analysis, the design loads resulting from 0.60 of the specified limit pilot forces are acceptable minimum design loads; and

(4) If operational loads may be exceeded through jamming, ground gusts, control inertia, or friction, the system must withstand the limit pilot forces specified in § 29.397, without yielding.

[Doc. No. 5084, 29 FR 16150, Dec. 3, 1964, as amended by Amdt. 29-26, 55 FR 8002, Mar. 6, 1990]

§ 29.397 Limit pilot forces and torques.

(a) Except as provided in paragraph (b) of this section, the limit pilot forces are as follows:

(1) For foot controls, 130 pounds.

(2) For stick controls, 100 pounds fore and aft, and 67 pounds laterally.

(b) For flap, tab, stabilizer, rotor brake, and landing gear operating controls, the following apply (R=radius in inches):

(1) Crank wheel, and lever controls, [1+R]/3×50 pounds, but not less than 50 pounds nor more than 100 pounds for hand operated controls or 130 pounds for foot operated controls, applied at any angle within 20 degrees of the plane of motion of the control.

(2) Twist controls, 80R inch-pounds.

[Amdt. 29-12, 41 FR 55471, Dec. 20, 1976, as amended by Amdt. 29-47, 66 FR 23538, May 9, 2001]

§ 29.399 Dual control system.

Each dual primary flight control system must be able to withstand the loads that result when pilot forces not less than 0.75 times those obtained under § 29.395 are applied—

(a) In opposition; and

(b) In the same direction.

§ 29.411 Ground clearance: tail rotor guard.

(a) It must be impossible for the tail rotor to contact the landing surface during a normal landing.

(b) If a tail rotor guard is required to show compliance with paragraph (a) of this section—

(1) Suitable design loads must be established for the guard: and

(2) The guard and its supporting structure must be designed to withstand those loads.

§ 29.427 Unsymmetrical loads.

(a) Horizontal tail surfaces and their supporting structure must be designed for unsymmetrical loads arising from yawing and rotor wake effects in combination with the prescribed flight conditions.

(b) To meet the design criteria of paragraph (a) of this section, in the absence of more rational data, both of the following must be met:

(1) One hundred percent of the maximum loading from the symmetrical flight conditions acts on the surface on one side of the plane of symmetry, and no loading acts on the other side.

(2) Fifty percent of the maximum loading from the symmetrical flight conditions acts on the surface on each side of the plane of symmetry, in opposite directions.

(c) For empennage arrangements where the horizontal tail surfaces are supported by the vertical tail surfaces, the vertical tail surfaces and supporting structure must be designed for the combined vertical and horizontal surface loads resulting from each prescribed flight condition, considered separately. The flight conditions must be selected so that the maximum design loads are obtained on each surface. In the absence of more rational data, the unsymmetrical horizontal tail surface loading distributions described in this section must be assumed.

[Amdt. 27-26, 55 FR 8002, Mar. 6, 1990, as amended by Amdt. 29-31, 55 FR 38966, Sept. 21, 1990]

Ground Loads

§ 29.471 General.

(a) *Loads and equilibrium.* For limit ground loads—

(1) The limit ground loads obtained in the landing conditions in this part must be considered to be external loads that would occur in the rotorcraft structure if it were acting as a rigid body; and

(2) In each specified landing condition, the external loads must be placed in equilibrium with linear and angular inertia loads in a rational or conservative manner.

(b) *Critical centers of gravity.* The critical centers of gravity within the range for which certification is requested must be selected so that the maximum design loads are obtained in each landing gear element.

§ 29.473 Ground loading conditions and assumptions.

(a) For specified landing conditions, a design maximum weight must be used that is not less than the maximum weight. A rotor lift may be assumed to act through the center of gravity throughout the landing impact. This lift may not exceed two-thirds of the design maximum weight.

(b) Unless otherwise prescribed, for each specified landing condition, the rotorcraft must be designed for a limit load factor of not less than the limit inertia load factor substantiated under § 29.725.

(c) Triggering or actuating devices for additional or supplementary energy absorption may not fail under loads established in the tests prescribed in §§ 29.725 and 29.727, but the factor of safety prescribed in § 29.303 need not be used.

[Amdt. 29-3, 33 FR 966, Jan. 26, 1968]

§ 29.475 Tires and shock absorbers.

Unless otherwise prescribed, for each specified landing condition, the tires must be assumed to be in their static position and the shock absorbers to be in their most critical position.

§ 29.477 Landing gear arrangement.

Sections 29.235, 29.479 through 29.485, and 29.493 apply to landing gear with two wheels aft, and one or more wheels forward, of the center of gravity.

§ 29.479 Level landing conditions.

(a) *Attitudes.* Under each of the loading conditions prescribed in paragraph (b) of this section, the rotorcraft is assumed to be in each of the following level landing attitudes:

(1) An attitude in which each wheel contacts the ground simultaneously.

(2) An attitude in which the aft wheels contact the ground with the forward wheels just clear of the ground.

(b) *Loading conditions.* The rotorcraft must be designed for the following landing loading conditions:

(1) Vertical loads applied under § 29.471.

(2) The loads resulting from a combination of the loads applied under paragraph (b)(1) of this section with drag loads at each wheel of not less than 25 percent of the vertical load at that wheel.

(3) The vertical load at the instant of peak drag load combined with a drag component simulating the forces required to accelerate the wheel rolling assembly up to the specified ground speed, with—

(i) The ground speed for determination of the spin-up loads being at least 75 percent of the optimum forward flight speed for minimum rate of descent in autorotation; and

(ii) The loading conditions of paragraph (b) applied to the landing gear and its attaching structure only.

(4) If there are two wheels forward, a distribution of the loads applied to those wheels under paragraphs (b)(1) and (2) of this section in a ratio of 40:60.

(c) *Pitching moments.* Pitching moments are assumed to be resisted by—

(1) In the case of the attitude in paragraph (a)(1) of this section, the forward landing gear; and

(2) In the case of the attitude in paragraph (a)(2) of this section, the angular inertia forces.

§ 29.481 Tail-down landing conditions.

(a) The rotorcraft is assumed to be in the maximum nose-up attitude allowing ground clearance by each part of the rotorcraft.

(b) In this attitude, ground loads are assumed to act perpendicular to the ground.

§ 29.483 One-wheel landing conditions.

For the one-wheel landing condition, the rotorcraft is assumed to be in the level attitude and to contact the ground on one aft wheel. In this attitude—

(a) The vertical load must be the same as that obtained on that side under § 29.479(b)(1); and

(b) The unbalanced external loads must be reacted by rotorcraft inertia.

§ 29.485 Lateral drift landing conditions.

(a) The rotorcraft is assumed to be in the level landing attitude, with—

(1) Side loads combined with one-half of the maximum ground reactions obtained in the level landing conditions of § 29.479(b)(1); and

(2) The loads obtained under paragraph (a)(1) of this section applied—

(i) At the ground contact point; or

(ii) For full-swiveling gear, at the center of the axle.

(b) The rotorcraft must be designed to withstand, at ground contact—

(1) When only the aft wheels contact the ground, side loads of 0.8 times the vertical reaction acting inward on one side and 0.6 times the vertical reaction acting outward on the other side, all combined with the vertical loads specified in paragraph (a) of this section; and

(2) When the wheels contact the ground simultaneously—

(i) For the aft wheels, the side loads specified in paragraph (b)(1) of this section; and

(ii) For the forward wheels, a side load of 0.8 times the vertical reaction combined with the vertical load specified in paragraph (a) of this section.

§ 29.493 Braked roll conditions.

Under braked roll conditions with the shock absorbers in their static positions—

(a) The limit vertical load must be based on a load factor of at least—

(1) 1.33, for the attitude specified in § 29.479(a)(1); and

(2) 1.0, for the attitude specified in § 29.479(a)(2); and

(b) The structure must be designed to withstand, at the ground contact point of each wheel with brakes, a drag load of at least the lesser of—

(1) The vertical load multiplied by a coefficient of friction of 0.8; and

(2) The maximum value based on limiting brake torque.

§ 29.497 Ground loading conditions: landing gear with tail wheels.

(a) *General.* Rotorcraft with landing gear with two wheels forward and one wheel aft of the center of gravity must be designed for loading conditions as prescribed in this section.

(b) *Level landing attitude with only the forward wheels contacting the ground.* In this attitude—

(1) The vertical loads must be applied under §§ 29.471 through 29.475;

(2) The vertical load at each axle must be combined with a drag load at that axle of not less than 25 percent of that vertical load; and

(3) Unbalanced pitching moments are assumed to be resisted by angular inertia forces.

(c) *Level landing attitude with all wheels contacting the ground simultaneously.* In this attitude, the rotorcraft must be designed for landing loading conditions as prescribed in paragraph (b) of this section.

(d) *Maximum nose-up attitude with only the rear wheel contacting the ground.* The attitude for this condition must be the maximum nose-up attitude expected in normal operation, including autorotative landings. In this attitude—

(1) The appropriate ground loads specified in paragraph (b)(1) and (2) of this section must be determined and applied, using a rational method to account for the moment arm between the rear wheel ground reaction and the rotorcraft center of gravity; or

(2) The probability of landing with initial contact on the rear wheel must be shown to be extremely remote.

(e) *Level landing attitude with only one forward wheel contacting the ground.* In this attitude, the rotorcraft must be designed for ground loads as specified in paragraph (b) (1) and (3) of this section.

(f) *Side loads in the level landing attitude.* In the attitudes specified in paragraphs (b) and (c) of this section, the following apply:

(1) The side loads must be combined at each wheel with one-half of the maximum vertical ground reactions obtained for that wheel under paragraphs (b) and (c) of this section. In this condition, the side loads must be—

(i) For the forward wheels, 0.8 times the vertical reaction (on one side) acting inward, and 0.6 times the vertical reaction (on the other side) acting outward; and

(ii) For the rear wheel, 0.8 times the vertical reaction.

(2) The loads specified in paragraph (f)(1) of this section must be applied—

(i) At the ground contact point with the wheel in the trailing position (for non-full swiveling landing gear or for full swiveling landing gear with a lock, steering device, or shimmy damper to keep the wheel in the trailing position); or

(ii) At the center of the axle (for full swiveling landing gear without a lock, steering device, or shimmy damper).

(g) *Braked roll conditions in the level landing attitude.* In the attitudes specified in paragraphs (b) and (c) of this section, and with the shock absorbers in their static positions, the rotorcraft must be designed for braked roll loads as follows:

(1) The limit vertical load must be based on a limit vertical load factor of not less than—

(i) 1.0, for the attitude specified in paragraph (b) of this section; and

(ii) 1.33, for the attitude specified in paragraph (c) of this section.

(2) For each wheel with brakes, a drag load must be applied, at the ground contact point, of not less than the lesser of—

(i) 0.8 times the vertical load; and

(ii) The maximum based on limiting brake torque.

(h) *Rear wheel turning loads in the static ground attitude.* In the static ground attitude, and with the shock absorbers and tires in their static positions, the rotorcraft must be designed for rear wheel turning loads as follows:

(1) A vertical ground reaction equal to the static load on the rear wheel must be combined with an equal side load.

(2) The load specified in paragraph (h)(1) of this section must be applied to the rear landing gear—

(i) Through the axle, if there is a swivel (the rear wheel being assumed to be swiveled 90 degrees to the longitudinal axis of the rotorcraft); or

(ii) At the ground contact point if there is a lock, steering device or shimmy damper (the rear wheel being assumed to be in the trailing position).

(i) *Taxiing condition.* The rotorcraft and its landing gear must be designed for the loads that would occur when the rotorcraft is taxied over the roughest ground that may reasonably be expected in normal operation.

§ 29.501 Ground loading conditions: landing gear with skids.

(a) *General.* Rotorcraft with landing gear with skids must be designed for the loading conditions specified in this section. In showing compliance with this section, the following apply:

(1) The design maximum weight, center of gravity, and load factor must be determined under §§ 29.471 through 29.475.

(2) Structural yielding of elastic spring members under limit loads is acceptable.

(3) Design ultimate loads for elastic spring members need not exceed those obtained in a drop test of the gear with—

(i) A drop height of 1.5 times that specified in § 29.725; and

(ii) An assumed rotor lift of not more than 1.5 times that used in the limit drop tests prescribed in § 29.725.

(4) Compliance with paragraph (b) through (e) of this section must be shown with—

(i) The gear in its most critically deflected position for the landing condition being considered; and

(ii) The ground reactions rationally distributed along the bottom of the skid tube.

(b) *Vertical reactions in the level landing attitude.* In the level attitude, and with the rotorcraft contacting the ground along the bottom of both skids, the vertical reactions must be applied as prescribed in paragraph (a) of this section.

(c) *Drag reactions in the level landing attitude.* In the level attitude, and with the rotorcraft contacting the ground along the bottom of both skids, the following apply:

(1) The vertical reactions must be combined with horizontal drag reactions of 50 percent of the vertical reaction applied at the ground.

(2) The resultant ground loads must equal the vertical load specified in paragraph (b) of this section.

(d) *Sideloads in the level landing attitude.* In the level attitude, and with the rotorcraft contacting the ground along the bottom of both skids, the following apply:

(1) The vertical ground reaction must be—

(i) Equal to the vertical loads obtained in the condition specified in paragraph (b) of this section; and

(ii) Divided equally among the skids.

(2) The vertical ground reactions must be combined with a horizontal sideload of 25 percent of their value.

(3) The total sideload must be applied equally between skids and along the length of the skids.

(4) The unbalanced moments are assumed to be resisted by angular inertia.

(5) The skid gear must be investigated for—

(i) Inward acting sideloads; and

(ii) Outward acting sideloads.

(e) *One-skid landing loads in the level attitude.* In the level attitude, and with the rotorcraft contacting the ground along the bottom of one skid only, the following apply:

(1) The vertical load on the ground contact side must be the same as that obtained on that side in the condition specified in paragraph (b) of this section.

(2) The unbalanced moments are assumed to be resisted by angular inertia.

(f) *Special conditions.* In addition to the conditions specified in paragraphs (b) and (c) of this section, the rotorcraft must be designed for the following ground reactions:

(1) A ground reaction load acting up and aft at an angle of 45 degrees to the longitudinal axis of the rotorcraft. This load must be—

(i) Equal to 1.33 times the maximum weight;

(ii) Distributed symmetrically among the skids;

(iii) Concentrated at the forward end of the straight part of the skid tube; and

(iv) Applied only to the forward end of the skid tube and its attachment to the rotorcraft.

(2) With the rotorcraft in the level landing attitude, a vertical ground reaction load equal to one-half of the vertical load determined under paragraph (b) of this section. This load must be—

(i) Applied only to the skid tube and its attachment to the rotorcraft; and

(ii) Distributed equally over 33.3 percent of the length between the skid tube attachments and centrally located midway between the skid tube attachments.

[Amdt. 29-3, 33 FR 966, Jan. 26, 1968; as amended by Amdt. 27-26, 55 FR 8002, Mar. 6, 1990]

§ 29.505 Ski landing conditions.

If certification for ski operation is requested, the rotorcraft, with skis, must be designed to withstand the following loading conditions (where P is the maximum static weight on each ski with the rotorcraft at design maximum weight, and n is the limit load factor determined under § 29.473(b)):

(a) Up-load conditions in which—

(1) A vertical load of Pn and a horizontal load of $Pn/4$ are simultaneously applied at the pedestal bearings; and

(2) A vertical load of 1.33 P is applied at the pedestal bearings.

(b) A side load condition in which a side load of 0.35 Pn is applied at the pedestal bearings in a horizontal plane perpendicular to the centerline of the rotorcraft.

(c) A torque-load condition in which a torque load of 1.33 P (in foot-pounds) is applied to the ski about the vertical axis through the centerline of the pedestal bearings.

§ 29.511 Ground load: unsymmetrical loads on multiple-wheel units.

(a) In dual-wheel gear units, 60 percent of the total ground reaction for the gear unit must be applied to one wheel and 40 percent to the other.

(b) To provide for the case of one deflated tire, 60 percent of the specified load for the gear unit must be applied to either wheel except that the vertical ground reaction may not be less than the full static value.

(c) In determining the total load on a gear unit, the transverse shift in the load centroid, due to unsymmetrical load distribution on the wheels, may be neglected.

[Amdt. 29-3, 33 FR 966, Jan. 26, 1968]

Water Loads

§ 29.519 Hull type rotorcraft: Water-based and amphibian.

(a) *General.* For hull type rotorcraft, the structure must be designed to withstand the water loading set forth in paragraphs (b), (c), and (d) of this section considering the most severe wave heights and profiles for which approval is desired. The loads for the landing conditions of paragraphs (b) and (c) of this section must be developed and distributed along and among the hull and auxiliary floats, if used, in a rational and conservative manner, assuming a rotor lift not exceeding two-thirds of the rotorcraft weight to act throughout the landing impact.

(b) *Vertical landing conditions.* The rotorcraft must initially contact the most critical wave surface at zero forward speed in likely pitch and roll attitudes which result in critical design loadings. The vertical descent velocity may not be less than 6.5 feet per second relative to the mean water surface.

(c) *Forward speed landing conditions.* The rotorcraft must contact the most critical wave at forward velocities from zero up to 30 knots in likely pitch, roll, and yaw attitudes and with a vertical descent velocity of not less than 6.5 feet per second relative to the mean water surface. A maximum forward velocity of less than 30 knots may be used in design if it can be demonstrated that the forward velocity selected would not be exceeded in a normal one-engine-out landing.

(d) *Auxiliary float immersion condition.* In addition to the loads from the landing conditions, the auxiliary float, and its support and attaching structure in the hull, must be designed for the load developed by a fully immersed float unless it can be shown that full immersion of the float is unlikely, in which case the highest likely float buoyancy load must be applied that considers loading of the float immersed to create restoring moments compensating for upsetting moments caused by side wind, asymmetrical rotorcraft loading, water wave action, and rotorcraft inertia.

[Amdt. 29-3, 33 FR 966, Jan. 26, 196; as amended by Amdt. 27-26, 55 FR 8002, Mar. 6, 1990]

§ 29.521 Float landing conditions.

If certification for float operation (including float amphibian operation) is requested, the rotorcraft, with floats, must be designed to withstand the following loading conditions (where the limit load factor is determined under § 29.473(b) or assumed to be equal to that determined for wheel landing gear):

(a) Up-load conditions in which—

(1) A load is applied so that, with the rotorcraft in the static level attitude, the resultant water reaction passes vertically through the center of gravity; and

(2) The vertical load prescribed in paragraph (a)(1) of this section is applied simultaneously with an aft component of 0.25 times the vertical component

(b) A side load condition in which—

(1) A vertical load of 0.75 times the total vertical load specified in paragraph (a)(1) of this section is divided equally among the floats; and

(2) For each float, the load share determined under paragraph (b)(1) of this section, combined with a total side load of 0.25 times the total vertical load specified in paragraph (b)(1) of this section, is applied to that float only.

[Amdt. 29-3, 33 FR 967, Jan. 26, 1968]

Main Component Requirements

§ 29.547 Main and tail rotor structure.

(a) A rotor is an assembly of rotating components, which includes the rotor hub, blades, blade dampers, the pitch control mechanisms, and all other parts that rotate with the assembly.

(b) Each rotor assembly must be designed as prescribed in this section and must function safely for the critical flight load and operating conditions. A design assessment must be performed, including a detailed failure analysis to identify all failures that will prevent continued safe flight or safe landing, and must identify the means to minimize the likelihood of their occurrence.

(c) The rotor structure must be designed to withstand the following loads prescribed in §§ 29.337 through 29.341 and 29.351:

(1) Critical flight loads.

(2) Limit loads occurring under normal conditions of autorotation.

(d) The rotor structure must be designed to withstand loads simulating—

(1) For the rotor blades, hubs, and flapping hinges, the impact force of each blade against its stop during ground operation; and

(2) Any other critical condition expected in normal operation.

(e) The rotor structure must be designed to withstand the limit torque at any rotational speed, including zero.

In addition:

(1) The limit torque need not be greater than the torque defined by a torque limiting device (where provided), and may not be less than the greater of—

(i) The maximum torque likely to be transmitted to the rotor structure, in either direction, by the rotor drive or by sudden application of the rotor brake; and

(ii) For the main rotor, the limit engine torque specified in § 29.361.

(2) The limit torque must be equally and rationally distributed to the rotor blades.

(Secs. 604, 605, 72 Stat. 778, 49 U.S.C. 1424, 1425)

[Doc. No. 5084, 29 FR 16150, Dec. 3, 1964, as amended by Amdt. 29-4, 33 FR 14106, Sept. 18, 1968; Amdt. 29-40, 61 FR 21907, May 10, 1996]

§ 29.549 Fuselage and rotor pylon structures.

(a) Each fuselage and rotor pylon structure must be designed to withstand—

(1) The critical loads prescribed in §§ 29.337 through 29.341, and 29.351;

(2) The applicable ground loads prescribed in §§ 29.235, 29.471 through 29.485, 29.493, 29.497, 29.505, and 29.521; and

(3) The loads prescribed in § 29.547 (d)(1) and (e)(1)(i).

(b) Auxiliary rotor thrust, the torque reaction of each rotor drive system, and the balancing air and inertia loads occurring under accelerated flight conditions, must be considered.

(c) Each engine mount and adjacent fuselage structure must be designed to withstand the loads occurring under accelerated flight and landing conditions, including engine torque.

(d) [Reserved]

(e) If approval for the use of 2½-minute OEI power is requested, each engine mount and adjacent structure must be designed to withstand the loads resulting from a limit torque equal to 1.25 times the mean torque for 2½-minute OEI power combined with 1g flight loads.

(Secs. 604, 605, 72 Stat. 778, 49 U.S.C. 1424, 1425)

[Doc. No. 5084, 29 FR 16150, Dec. 3, 1964, as amended by Amdt. 29-4, 33 FR 14106, Sept. 18, 1968; Amdt. 29-26, 53 FR 34215, Sept. 2, 1988]

§ 29.551 Auxiliary lifting surfaces.

Each auxiliary lifting surface must be designed to withstand—

(a) The critical flight loads in §§ 29.337 through 29.341, and 29.351;

(b) the applicable ground loads in §§ 29.235, 29.471 through 29.485, 29.493, 29.505, and 29.521; and

(c) Any other critical condition expected in normal operation.

Emergency Landing Conditions

§ 29.561 General.

(a) The rotorcraft, although it may be damaged in emergency landing conditions on land or water, must be designed as prescribed in this section to protect the occupants under those conditions.

(b) The structure must be designed to give each occupant every reasonable chance of escaping serious injury in a crash landing when—

(1) Proper use is made of seats, belts, and other safety design provisions;

(2) The wheels are retracted (where applicable); and

(3) Each occupant and each item of mass inside the cabin that could injure an occupant is restrained when subjected to the following ultimate inertial load factors relative to the surrounding structure:

(i) Upward—4g.

(ii) Forward—16g.

(iii) Sideward—8g.

(iv) Downward—20g, after the intended displacement of the seat device.

(v) Rearward—1.5g.

(c) The supporting structure must be designed to restrain under any ultimate inertial load factor up to those specified in this paragraph, any item of mass above and/or behind the crew and passenger compartment that could injure an occupant if it came loose in an emergency landing. Items of mass to be considered include, but are not limited to, rotors, transmission, and engines. The items of mass must be restrained for the following ultimate inertial load factors:

(1) Upward—1.5g.

(2) Forward—12g.

(3) Sideward—6g.

(4) Downward—12g.

(5) Rearward—1.5g.

(d) Any fuselage structure in the area of internal fuel tanks below the passenger floor level must be designed to resist the following ultimate inertial factors and loads, and to protect the fuel tanks from rupture, if rupture is likely when those loads are applied to that area:

(1) Upward—1.5g.

(2) Forward—4.0g.

(3) Sideward—2.0g.

(4) Downward—4.0g.

[Doc. No. 5084, 29 FR 16150, Dec. 3, 1964, as amended by Amdt. 29-29, 54 FR 47319, Nov. 13, 1989; Amdt. 29-38, 61 FR 10438, Mar. 13, 1996]

§ 29.562 Emergency landing dynamic conditions.

(a) The rotorcraft, although it may be damaged in a crash landing, must be designed to reasonably protect each occupant when—

(1) The occupant properly uses the seats, safety belts, and shoulder harnesses provided in the design; and

(2) The occupant is exposed to loads equivalent to those resulting from the conditions prescribed in this section.

(b) Each seat type design or other seating device approved for crew or passenger occupancy during takeoff and landing must successfully complete dynamic tests or be demonstrated by rational analysis based on dynamic tests of a similar type seat in accordance with the following criteria. The tests must be conducted with an occupant simulated by a 170-pound anthropomorphic test dummy (ATD), as defined by 49 CFR 572, Subpart B, or its equivalent, sitting in the normal upright position.

(1) A change in downward velocity of not less than 30 feet per second when the seat or other seating device is oriented in its nominal position with respect to the rotorcraft's reference system, the rotorcraft's longitudinal axis is canted upward 60° with respect to the impact velocity vector, and the rotorcraft's lateral axis is perpendicular to a vertical plane containing the impact velocity vector and the rotorcraft's longitudinal axis. Peak floor deceleration must occur in not more than 0.031 seconds after impact and must reach a minimum of 30g's.

(2) A change in forward velocity of not less than 42 feet per second when the seat or other seating device is oriented in its nominal position with respect to the rotorcraft's reference system, the rotorcraft's longitudinal axis is yawed 10° either right or left of the impact velocity vector (whichever would cause the greatest load on the shoulder harness), the rotorcraft's lateral axis is contained in a horizontal plane containing the impact velocity vector, and the rotorcraft's vertical axis is perpendicular to a horizontal plane containing the impact velocity vector. Peak floor deceleration must occur in not more than 0.071 seconds after impact and must reach a minimum of 18.4g's.

(3) Where floor rails or floor or sidewall attachment devices are used to attach the seating devices to the airframe structure for the conditions of this section, the rails or devices must be misaligned with respect to each other by at least 10° vertically (i.e., pitch out of parallel) and by at least a 10° lateral roll, with the directions optional, to account for possible floor warp.

(c) Compliance with the following must be shown:

(1) The seating device system must remain intact although it may experience separation intended as part of its design.

(2) The attachment between the seating device and the airframe structure must remain intact although the structure may have exceeded its limit load.

(3) The ATD's shoulder harness strap or straps must remain on or in the immediate vicinity of the ATD's shoulder during the impact.

(4) The safety belt must remain on the ATD's pelvis during the impact.

(5) The ATD's head either does not contact any portion of the crew or passenger compartment or, if contact is made, the head impact does not exceed a head injury criteria (HIC) of 1,000 as determined by this equation.

$$HIC = (t_2 - t_1)\left[\frac{1}{(t_2 - t_1)}\int_{t_1}^{t_2} a(t)\,dt\right]^{2.5}$$

Where: a(t) is the resultant acceleration at the center of gravity of the head form expressed as a multiple of g (the acceleration of gravity) and t_2-t_1 is the time duration, in seconds, of major head impact, not to exceed 0.05 seconds.

(6) Loads in individual shoulder harness straps must not exceed 1,750 pounds. If dual straps are used for retaining the upper torso, the total harness strap loads must not exceed 2,000 pounds.

(7) The maximum compressive load measured between the pelvis and the lumbar column of the ATD must not exceed 1,500 pounds.

(d) An alternate approach that achieves an equivalent or greater level of occupant protection, as required by this section, must be substantiated on a rational basis.

[Amdt. 29-29, 54 FR 47320, Nov. 13, 1989, as amended by Amdt. 29-41, 62 FR 46173, Aug. 29, 1997]

§ 29.563 Structural ditching provisions.

If certification with ditching provisions is requested, structural strength for ditching must meet the requirements of this section and § 29.801(e).

(a) *Forward speed landing conditions.* The rotorcraft must initially contact the most critical wave for reasonably probable water conditions at forward velocities from zero up to 30 knots in likely pitch, roll, and yaw attitudes. The rotorcraft limit vertical descent velocity may not be less than 5 feet per second relative to the mean water surface. Rotor lift may be used to act through the center of gravity throughout the landing impact. This lift may not exceed two-thirds of the design maximum weight. A maximum forward velocity of less than 30 knots may be used in design if it can be demonstrated that the forward velocity selected would not be exceeded in a normal one-engine-out touchdown.

(b) *Auxiliary or emergency float conditions* —(1) *Floats fixed or deployed before initial water contact.* In addition to the landing loads in paragraph (a) of this section, each auxiliary or emergency float, or its support and attaching structure in the airframe or fuselage, must be designed for the load developed by a fully immersed float unless it can be shown that full immersion is unlikely. If full immersion is unlikely, the highest likely float buoyancy load must be applied. The highest likely buoyancy load must include consideration of a partially immersed float creating restoring moments to compensate the upsetting moments caused by side wind, unsymmetrical rotorcraft loading, water wave action, rotorcraft inertia, and probable structural damage and leakage considered under § 29.801(d). Maximum roll and pitch angles determined from compliance with § 29.801(d) may be used, if significant, to determine the extent of immersion of each float. If the floats are deployed in flight, appropriate air loads derived from the flight limitations with the floats deployed shall be used in substantiation of the floats and their attachment to the rotorcraft. For this purpose, the design airspeed for limit load is the float deployed airspeed operating limit multiplied by 1.11.

(2) *Floats deployed after initial water contact.* Each float must be designed for full or partial immersion prescribed in paragraph (b)(1) of this section. In addition, each float must be designed for combined vertical and drag loads using a relative limit speed of 20 knots between the rotorcraft and the water. The vertical load may not be less than the highest likely buoyancy load determined under paragraph (b)(1) of this section.

[Amdt. 27-26, 55 FR 8003, Mar. 6, 1990]

Fatigue Evaluation

§ 29.571 Fatigue Tolerance Evaluation of Metallic Structure.

(a) A fatigue tolerance evaluation of each principal structural element (PSE) must be performed, and appropriate inspections and retirement time or approved equivalent means must be established to avoid catastrophic failure during the operational life of the rotorcraft. The fatigue tolerance evaluation must consider the effects of both fatigue and the damage determined under paragraph (e)(4) of this section. Parts to be evaluated include PSEs of the rotors, rotor drive systems between the engines and rotor hubs, controls, fuselage, fixed and movable control surfaces, engine and transmission mountings, landing gear, and their related primary attachments.

(b) For the purposes of this section, the term—

(1) *Catastrophic failure* means an event that could prevent continued safe flight and landing.

(2) *Principal structural element (PSE)* means a structural element that contributes significantly to the carriage of flight or ground loads, and the fatigue failure of that structural element could result in catastrophic failure of the aircraft.

(c) The methodology used to establish compliance with this section must be submitted to and approved by the Administrator.

(d) Considering all rotorcraft structure, structural elements, and assemblies, each PSE must be identified.

(e) Each fatigue tolerance evaluation required by this section must include:

(1) In-flight measurements to determine the fatigue loads or stresses for the PSEs identified in paragraph (d) of this section in all critical conditions throughout the range of design limitations required by § 29.309 (including altitude effects), except that maneuvering load factors need not exceed the maximum values expected in operations.

(2) The loading spectra as severe as those expected in operations based on loads or stresses determined under paragraph (e)(1) of this section, including external load operations, if applicable, and other high frequency power-cycle operations.

(3) Takeoff, landing, and taxi loads when evaluating the landing gear and other affected PSEs.

(4) For each PSE identified in paragraph (d) of this section, a threat assessment which includes a determination of the probable locations, types, and sizes of damage, taking into account fatigue, environmental effects, intrinsic and discrete flaws, or accidental damage that may occur during manufacture or operation.

(5) A determination of the fatigue tolerance characteristics for the PSE with the damage identified in paragraph (e)(4) of this section that supports the inspection and retirement times, or other approved equivalent means.

(6) Analyses supported by test evidence and, if available, service experience.

(f) A residual strength determination is required that substantiates the maximum damage size assumed in the fatigue tolerance evaluation. In determining inspection intervals based on damage growth, the residual strength evaluation must show that the remaining structure, after damage growth, is able to withstand design limit loads without failure.

(g) The effect of damage on stiffness, dynamic behavior, loads, and functional performance must be considered.

(h) Based on the requirements of this section, inspections and retirement times or approved equivalent means must be established to avoid catastrophic failure. The inspections and retirement times or approved equivalent means must be included in the Airworthiness Limitations Section of the Instructions for Continued Airworthiness required by Section 29.1529 and Section A29.4 of Appendix A of this part.

(i) If inspections for any of the damage types identified in paragraph (e)(4) of this section cannot be established within the limitations of geometry, inspectability, or good design practice, then supplemental procedures, in conjunction with the PSE retirement time, must be established to minimize the risk of occurrence of these types of damage that could result in a catastrophic failure during the operational life of the rotorcraft.

[Doc. No. FAA-2009-0413, Amdt. 29-55, 76 FR 75442, Dec. 2, 2011]

§ 29.573 Damage Tolerance and Fatigue Evaluation of Composite Rotorcraft Structures.

(a) Each applicant must evaluate the composite rotorcraft structure under the damage tolerance standards of paragraph (d) of this section unless the applicant establishes that a damage tolerance evaluation is impractical within the limits of geometry, inspectability, and good design practice. If an applicant establishes that it is impractical within the limits of geometry, inspectability, and good design practice, the applicant must do a fatigue evaluation in accordance with paragraph (e) of this section.

(b) The methodology used to establish compliance with this section must be submitted to and approved by the Administrator.

(c) Definitions:

(1) *Catastrophic failure* is an event that could prevent continued safe flight and landing.

(2) *Principal Structural Elements (PSEs)* are structural elements that contribute significantly to the carrying of flight or ground loads, the failure of which could result in catastrophic failure of the rotorcraft.

(3) *Threat Assessment* is an assessment that specifies the locations, types, and sizes of damage, considering fatigue, environmental effects, intrinsic and discrete flaws, and impact or other accidental damage (including the discrete source of the accidental damage) that may occur during manufacture or operation.

(d) Damage Tolerance Evaluation:

(1) Each applicant must show that catastrophic failure due to static and fatigue loads, considering the intrinsic or discrete manufacturing defects or accidental damage, is avoided throughout the operational life or prescribed inspection intervals of the rotorcraft by performing damage tolerance evaluations of the strength of composite PSEs and other parts, detail design points, and fabrication techniques. Each applicant must account for the effects of material and process variability along with environmental conditions in the strength and fatigue evaluations. Each applicant must evaluate parts that include PSEs of the airframe, main and tail rotor drive systems, main and tail rotor blades and hubs, rotor controls, fixed and movable control surfaces, engine and transmission mountings, landing gear, other parts, detail design points, and fabrication techniques deemed critical by the FAA. Each damage tolerance evaluation must include:

(i) The identification of all PSEs;

(ii) In-flight and ground measurements for determining the loads or stresses for all PSEs for all critical conditions throughout the range of limits in § 29.309 (including altitude effects), except that maneuvering load factors need not exceed the maximum values expected in service;

(iii) The loading spectra as severe as those expected in service based on loads or stresses determined under paragraph (d)(1)(ii) of this section, including external load operations, if applicable, and other operations including high-torque events;

(iv) A threat assessment for all PSEs that specifies the locations, types, and sizes of damage, considering fatigue, environmental effects, intrinsic and discrete flaws, and impact or other accidental damage (including the discrete source of the accidental damage) that may occur during manufacture or operation; and

(v) An assessment of the residual strength and fatigue characteristics of all PSEs that supports the replacement times and inspection intervals established under paragraph (d)(2) of this section.

(2) Each applicant must establish replacement times, inspections, or other procedures for all PSEs to require the repair or replacement of damaged parts before a catastrophic failure. These replacement times, inspections, or other procedures must be included in the Airworthiness Limitations Section of the Instructions for Continued Airworthiness required by § 29.1529.

(i) Replacement times for PSEs must be determined by tests, or by analysis supported by tests, and must show that the structure is able to withstand the repeated loads of variable magnitude expected in-service. In establishing these replacement times, the following items must be considered:

(A) Damage identified in the threat assessment required by paragraph (d)(1)(iv) of this section;

(B) Maximum acceptable manufacturing defects and in-service damage (*i.e.*, those that do not lower the residual strength below ultimate design loads and those that can be repaired to restore ultimate strength); and

(C) Ultimate load strength capability after applying repeated loads.

(ii) Inspection intervals for PSEs must be established to reveal any damage identified in the threat assessment required by paragraph (d)(1)(iv) of this section that may occur from fatigue or other in-service causes before such damage has grown to the extent that the component cannot sustain the required residual strength capability. In establishing these inspection intervals, the following items must be considered:

(A) The growth rate, including no-growth, of the damage under the repeated loads expected in-service determined by tests or analysis supported by tests;

(B) The required residual strength for the assumed damage established after considering the damage type, inspection interval, detectability of damage, and the techniques adopted for damage detection. The minimum required residual strength is limit load; and

(C) Whether the inspection will detect the damage growth before the minimum residual strength is reached and restored to ultimate load capability, or whether the component will require replacement.

(3) Each applicant must consider the effects of damage on stiffness, dynamic behavior, loads, and functional performance on all PSEs when substantiating the maximum assumed damage size and inspection interval.

(e) Fatigue Evaluation: If an applicant establishes that the damage tolerance evaluation described in paragraph (d) of this section is impractical within the limits of geometry, inspectability, or good design practice, the applicant must do a fatigue evaluation of the particular composite rotorcraft structure and:

(1) Identify all PSEs considered in the fatigue evaluation;

(2) Identify the types of damage for all PSEs considered in the fatigue evaluation;

(3) Establish supplemental procedures to minimize the risk of catastrophic failure associated with the damages identified in paragraph (d) of this section; and

(4) Include these supplemental procedures in the Airworthiness Limitations section of the Instructions for Continued Airworthiness required by § 29.1529.

[Doc. No. FAA-2009-0660, Amdt. 29-59, 76 FR 74664, Dec. 1, 2011]

Subpart D—Design and Construction

General

§ 29.601 Design.

(a) The rotorcraft may have no design features or details that experience has shown to be hazardous or unreliable.

(b) The suitability of each questionable design detail and part must be established by tests.

§ 29.602 Critical parts.

(a) *Critical part.* A critical part is a part, the failure of which could have a catastrophic effect upon the rotorcraft, and for which critical characterists have been identified which must be controlled to ensure the required level of integrity.

(b) If the type design includes critical parts, a critical parts list shall be established. Procedures shall be established to define the critical design characteristics, identify processes that affect those characteristics, and identify the design change and process change controls necessary for showing compliance with the quality assurance requirements of part 21 of this chapter.

[Doc. No. 29311, 64 FR 46232, Aug. 24, 1999]

§ 29.603 Materials.

The suitability and durability of materials used for parts, the failure of which could adversely affect safety, must—

(a) Be established on the basis of experience or tests;

(b) Meet approved specifications that ensure their having the strength and other properties assumed in the design data; and

(c) Take into account the effects of environmental conditions, such as temperature and humidity, expected in service.

(Secs. 313(a), 601, 603, 604, and 605 of the Federal Aviation Act of 1958 (49 U.S.C. 1354(a), 1421, 1423, 1424), and sec. 6(c), Dept. of Transportation Act (49 U.S.C. 1655(c)))

[Doc. No. 5084, 29 FR 16150, Dec. 3, 1964, as amended by Amdt. 29-12, 41 FR 55471, Dec. 20, 1976; Amdt. 29-17, 43 FR 50599, Oct. 30, 1978]

§ 29.605 Fabrication methods.

(a) The methods of fabrication used must produce consistently sound structures. If a fabrication process (such as gluing, spot welding, or heat-treating) requires close control to reach this objective, the process must be performed according to an approved process specification.

(b) Each new aircraft fabrication method must be substantiated by a test program.

(Secs. 313(a), 601, 603, 604, Federal Aviation Act of 1958 (49 U.S.C. 1354(a), 1421, 1423, 1424), sec. 6(c), Dept. of Transportation Act (49 U.S.C. 1655(c)))

[Doc. No. 5084, 29 FR 16150. Dec. 3, 1964, as amended by Amdt. 29-17, 43 FR 50599, Oct. 30, 1978]

§ 29.607 Fasteners.

(a) Each removable bolt, screw, nut, pin, or other fastener whose loss could jeopardize the safe operation of the rotorcraft must incorporate two separate locking devices. The fastener and its locking devices may not be adversely affected by the environmental conditions associated with the particular installation.

(b) No self-locking nut may be used on any bolt subject to rotation in operation unless a nonfriction locking device is used in addition to the self-locking device.

[Amdt. 29-5, 33 FR 14533, Sept. 27, 1968]

§ 29.609 Protection of structure.

Each part of the structure must—

(a) Be suitably protected against deterioration or loss of strength in service due to any cause, including—

(1) Weathering;

(2) Corrosion; and

(3) Abrasion; and

(b) Have provisions for ventilation and drainage where necessary to prevent the accumulation of corrosive, flammable, or noxious fluids.

§ 29.610 Lightning and static electricity protection.

(a) The rotorcraft structure must be protected against catastrophic effects from lightning.

(b) For metallic components, compliance with paragraph (a) of this section may be shown by—

(1) Electrically bonding the components properly to the airframe; or

(2) Designing the components so that a strike will not endanger the rotorcraft.

(c) For nonmetallic components, compliance with paragraph (a) of this section may be shown by—

(1) Designing the components to minimize the effect of a strike; or

(2) Incorporating acceptable means of diverting the resulting electrical current to not endanger the rotorcraft.

(d) The electric bonding and protection against lightning and static electricity must—

(1) Minimize the accumulation of electrostatic charge;

(2) Minimize the risk of electric shock to crew, passengers, and service and maintenance personnel using normal precautions;

(3) Provide and electrical return path, under both normal and fault conditions, on rotorcraft having grounded electrical systems; and

(4) Reduce to an acceptable level the effects of static electricity on the functioning of essential electrical and electronic equipment.

[Amdt. 29-24, 49 FR 44437, Nov. 6, 1984; Amdt. 29-40, 61 FR 21907, May 10, 1996; 61 FR 33963, July 1, 1996; Amdt. 29-53, 76 FR 33135, June 8, 2011]

§ 29.611 Inspection provisions.

There must be means to allow close examination of each part that requires—

(a) Recurring inspection;

(b) Adjustment for proper alignment and functioning; or

(c) Lubrication.

§ 29.613 Material strength properties and design values.

(a) Material strength properties must be based on enough tests of material meeting specifications to establish design values on a statistical basis.

(b) Design values must be chosen to minimize the probability of structural failure due to material variability. Except as provided in paragraphs (d) and (e) of this section, compliance with this paragraph must be shown by selecting design values that assure material strength with the following probability—

(1) Where applied loads are eventually distributed through a single member within an assembly, the failure of which would result in loss of structural integrity of the component, 99 percent probability with 95 percent confidence; and

(2) For redundant structures, those in which the failure of individual elements would result in applied loads being safely distributed to other load-carrying members, 90 percent probability with 95 percent confidence.

(c) The strength, detail design, and fabrication of the structure must minimize the probability of disastrous fatigue failure, particularly at points of stress concentration.

(d) Design values may be those contained in the following publications (available from the Naval Publications and Forms Center, 5801 Tabor Avenue, Philadelphia, PA 19120) or other values approved by the Administrator:

(1) MIL—HDBK-5, "Metallic Materials and Elements for Flight Vehicle Structure".

(2) MIL—HDBK-17, "Plastics for Flight Vehicles".

(3) ANC-18, "Design of Wood Aircraft Structures".

(4) MIL—HDBK-23, "Composite Construction for Flight Vehicles".

(e) Other design values may be used if a selection of the material is made in which a specimen of each individual item is tested before use and it is determined that the actual strength properties of that particular item will equal or exceed those used in design.

(Secs. 313(a), 601, 603, 604, Federal Aviation Act of 1958 (49 U.S.C. 1354(a), 1421, 1423, 1424), sec. 6(c), Dept. of Transportation Act (49 U.S.C. 1655(c)))

[Doc. No. 5084, 29 FR 16150, Dec. 3, 1964, as amended by Amdt. 29-17, 43 FR 50599, Oct. 30, 1978; Amdt. 27-26, 55 FR 8003, Mar. 6, 1990]

§ 29.619 Special factors.

(a) The special factors prescribed in §§ 29.621 through 29.625 apply to each part of the structure whose strength is—

(1) Uncertain;

(2) Likely to deteriorate in service before normal replacement; or

(3) Subject to appreciable variability due to—

(i) Uncertainties in manufacturing processes; or

(ii) Uncertainties in inspection methods.

(b) For each part of the rotorcraft to which §§ 29.621 through 29.625 apply, the factor of safety prescribed in § 29.303 must be multiplied by a special factor equal to—

(1) The applicable special factors prescribed in §§ 29.621 through 29.625; or

(2) Any other factor great enough to ensure that the probability of the part being understrength because of the uncertainties specified in paragraph (a) of this section is extremely remote.

§ 29.621 Casting factors.

(a) *General.* The factors, tests, and inspections specified in paragraphs (b) and (c) of this section must be applied in addition to those necessary to establish foundry quality control. The inspections must meet approved specifications. Paragraphs (c) and (d) of this section apply to structural castings except castings that are pressure tested as parts of hydraulic or other fluid systems and do not support structural loads.

(b) *Bearing stresses and surfaces.* The casting factors specified in paragraphs (c) and (d) of this section—

(1) Need not exceed 1.25 with respect to bearing stresses regardless of the method of inspection used; and

(2) Need not be used with respect to the bearing surfaces of a part whose bearing factor is larger than the applicable casting factor.

(c) *Critical castings.* For each casting whose failure would preclude continued safe flight and landing of the rotorcraft or result in serious injury to any occupant, the following apply:

(1) Each critical casting must—

(i) Have a casting factor of not less than 1.25; and

(ii) Receive 100 percent inspection by visual, radiographic, and magnetic particle (for ferromagnetic materials) or penetrant (for nonferromagnetic materials) inspection methods or approved equivalent inspection methods.

(2) For each critical casting with a casting factor less than 1.50, three sample castings must be static tested and shown to meet—

(i) The strength requirements of § 29.305 at an ultimate load corresponding to a casting factor of 1.25; and

(ii) The deformation requirements of § 29.305 at a load of 1.15 times the limit load.

(d) *Noncritical castings.* For each casting other than those specified in paragraph (c) of this section, the following apply:

(1) Except as provided in paragraphs (d)(2) and (3) of this section, the casting factors and corresponding inspections must meet the following table:

Casting factor	Inspection
2.0 or greater	100 percent visual.
Less than 2.0, greater than 1.5	100 percent visual, and magnetic particle (ferromagnetic materials), penetrant (nonferromagnetic materials), or approved equivalent inspection methods.
1.25 through 1.50	100 percent visual, and magnetic particle (ferromagnetic materials), penetrant (nonferromagnetic materials), and radiographic or approved equivalent inspection methods.

(2) The percentage of castings inspected by nonvisual methods may be reduced below that specified in paragraph (d)(1) of this section when an approved quality control procedure is established.

(3) For castings procured to a specification that guarantees the mechanical properties of the material in the casting and provides for demonstration of these properties by test of coupons cut from the castings on a sampling basis—

(i) A casting factor of 1.0 may be used; and

(ii) The castings must be inspected as provided in paragraph (d)(1) of this section for casting factors of "1.25 through 1.50" and tested under paragraph (c)(2) of this section.

[Doc. No. 5084, 29 FR 16150, Dec. 3, 1964, as amended by Amdt. 29-41, 62 FR 46173, Aug. 29, 1997]

§ 29.623 Bearing factors.

(a) Except as provided in paragraph (b) of this section, each part that has clearance (free fit), and that is subject to pounding or vibration, must have a bearing factor large enough to provide for the effects of normal relative motion.

(b) No bearing factor need be used on a part for which any larger special factor is prescribed.

§ 29.625 Fitting factors.

For each fitting (part or terminal used to join one structural member to another) the following apply:

(a) For each fitting whose strength is not proven by limit and ultimate load tests in which actual stress conditions are simulated in the fitting and surrounding structures, a fitting factor of at least 1.15 must be applied to each part of—

(1) The fitting;

(2) The means of attachment; and

(3) The bearing on the joined members.

(b) No fitting factor need be used—

(1) For joints made under approved practices and based on comprehensive test data (such as continuous joints in metal plating, welded joints, and scarf joints in wood); and

(2) With respect to any bearing surface for which a larger special factor is used.

(c) For each integral fitting, the part must be treated as a fitting up to the point at which the section properties become typical of the member.

(d) Each seat, berth, litter, safety belt, and harness attachment to the structure must be shown by analysis, tests, or both, to be able to withstand the inertia forces prescribed in § 29.561(b)(3) multiplied by a fitting factor of 1.33.

[Doc. No. 5084, 29 FR 16150, Dec. 3, 1964, as amended by Amdt. 29-42, 63 FR 43285, Aug. 12, 1998]

§ 29.629 Flutter and divergence.

Each aerodynamic surface of the rotorcraft must be free from flutter and divergence under each appropriate speed and power condition.

[Doc. No. 28008, 61 FR 21907, May 10, 1996]

§ 29.631 Bird strike.

The rotorcraft must be designed to ensure capability of continued safe flight and landing (for Category A) or safe landing (for Category B) after impact with a 2.2-lb (1.0 kg) bird when the velocity of the rotorcraft (relative to the bird along the flight path of the rotorcraft) is equal to V_{NE} or V_H (whichever is the lesser) at altitudes up to 8,000 feet. Compliance must be shown by tests or by analysis based on tests carried out on sufficiently representative structures of similar design.

[Doc. No. 28008, 61 FR 21907, May 10, 1996; 61 FR 33963, July 1, 1996]

Rotors

§ 29.653 Pressure venting and drainage of rotor blades.

(a) For each rotor blade—

(1) There must be means for venting the internal pressure of the blade;

(2) Drainage holes must be provided for the blade; and

(3) The blade must be designed to prevent water from becoming trapped in it.

(b) Paragraphs (a)(1) and (2) of this section does not apply to sealed rotor blades capable of withstanding the maximum pressure differentials expected in service.

[Amdt. 29-3, 33 FR 967, Jan. 26, 1968]

§ 29.659 Mass balance.

(a) The rotor and blades must be mass balanced as necessary to—

(1) Prevent excessive vibration; and

(2) Prevent flutter at any speed up to the maximum forward speed.

(b) The structural integrity of the mass balance installation must be substantiated.

[Amdt. 29-3, 33 FR 967, Jan. 26, 1968]

§ 29.661 Rotor blade clearance.

There must be enough clearance between the rotor blades and other parts of the structure to prevent the blades from striking any part of the structure during any operating condition.

[Amdt. 29-3, 33 FR 967, Jan. 26, 1968]

§ 29.663 Ground resonance prevention means.

(a) The reliability of the means for preventing ground resonance must be shown either by analysis and tests, or reliable service experience, or by showing through analysis or tests that malfunction or failure of a single means will not cause ground resonance.

(b) The probable range of variations, during service, of the damping action of the ground resonance prevention means must be established and must be investigated during the test required by § 29.241.

[Amdt. 27-26, 55 FR 8003, Mar. 6, 1990]

Control Systems

§ 29.671 General.

(a) Each control and control system must operate with the ease, smoothness, and positiveness appropriate to its function.

(b) Each element of each flight control system must be designed, or distinctively and permanently marked, to minimize the probability of any incorrect assembly that could result in the malfunction of the system.

(c) A means must be provided to allow full control movement of all primary flight controls prior to flight, or a means must be provided that will allow the pilot to determine that full control authority is available prior to flight.

[Doc. No. 5084, 29 FR 16150, Dec. 3, 1964, as amended by Amdt. 29-24, 49 FR 44437, Nov. 6, 1984]

§ 29.672 Stability augmentation, automatic, and power-operated systems.

If the functioning of stability augmentation or other automatic or power-operated system is necessary to show compliance with the flight characteristics requirements of this part, the system must comply with § 29.671 of this part and the following:

(a) A warning which is clearly distinguishable to the pilot under expected flight conditions without requiring the pilot's attention must be provided for any failure in the stability augmentation system or in any other automatic or power-operated system which could result in an unsafe condition if the pilot is unaware of the failure. Warning systems must not activate the control systems.

(b) The design of the stability augmentation system or of any other automatic or power-operated system must allow initial counteraction of failures without requiring exceptional pilot skill or strength, by overriding the failure by moving the flight controls in the normal sense, and by deactivating the failed system.

(c) It must be show that after any single failure of the stability augmentation system or any other automatic or power-operated system—

(1) The rotorcraft is safely controllable when the failure or malfunction occurs at any speed or altitude within the approved operating limitations;

(2) The controllability and maneuverability requirements of this part are met within a practical operational flight envelope (for example, speed, altitude, normal acceleration, and rotorcraft configurations) which is described in the Rotorcraft Flight Manual; and

(3) The trim and stability characteristics are not impaired below a level needed to allow continued safe flight and landing.

[Amdt. 29-24, 49 FR 44437, Nov. 6, 1984]

§ 29.673 Primary flight controls.

Primary flight controls are those used by the pilot for immediate control of pitch, roll, yaw, and vertical motion of the rotorcraft.

[Amdt. 29-24, 49 FR 44437, Nov. 6, 1984]

§ 29.674 Interconnected controls.

Each primary flight control system must provide for safe flight and landing and operate independently after a malfunction, failure, or jam of any auxiliary interconnected control.

[Amdt. 27-26, 55 FR 8003, Mar. 6, 1990]

§ 29.675 Stops.

(a) Each control system must have stops that positively limit the range of motionof the pilot's controls.

(b) Each stop must be located in the system so that the range of travel of its control is not appreciably affected by—

(1) Wear;

(2) Slackness; or

(3) Takeup adjustments.

(c) Each stop must be able to withstand the loads corresponding to the design conditions for the system.

(d) For each main rotor blade—

(1) Stops that are appropriate to the blade design must be provided to limit travel of the blade about its hinge points; and

(2) There must be means to keep the blade from hitting the droop stops during any operation other than starting and stopping the rotor.

(Secs. 313(a), 601, 603, 604, Federal Aviation Act of 1958 (49 U.S.C. 1354(a), 1421, 1423, 1424), sec. 6(c), Dept. of Transportation Act (49 U.S.C. 1655(c)))

[Doc. No. 5084, 29 FR 16150. Dec. 3, 1964, as amended by Amdt. 29-17, 43 FR 50599, Oct. 30, 1978]

§ 29.679 Control system locks.

If there is a device to lock the control system with the rotorcraft on the ground or water, there must be means to—

(a) Automatically disengage the lock when the pilot operates the controls in a normal manner, or limit the operation of the rotorcraft so as to give unmistakable warning to the pilot before takeoff; and

(b) Prevent the lock from engaging in flight.

§ 29.681 Limit load static tests.

(a) Compliance with the limit load requirements of this part must be shown by tests in which—

(1) The direction of the test loads produces the most severe loading in the control system; and

(2) Each fitting, pulley, and bracket used in attaching the system to the main structure is included;

(b) Compliance must be shown (by analyses or individual load tests) with the special factor requirements for control system joints subject to angular motion.

§ 29.683 Operation tests.

It must be shown by operation tests that, when the controls are operated from the pilot compartment with the control system loaded to correspond with loads specified for the system, the system is free from—

(a) Jamming;

(b) Excessive friction; and

(c) Excessive deflection.

§ 29.685 Control system details.

(a) Each detail of each control system must be designed to prevent jamming, chafing, and interference from cargo, passengers, loose objects, or the freezing of moisture.

(b) There must be means in the cockpit to prevent the entry of foreign objects into places where they would jam the system.

(c) There must be means to prevent the slapping of cables or tubes against other parts.

(d) Cable systems must be designed as follows:

(1) Cables, cable fittings, turnbuckles, splices, and pulleys must be of an acceptable kind.

(2) The design of cable systems must prevent any hazardous change in cable tension throughout the range of travel under any operating conditions and temperature variations.

(3) No cable smaller than $\frac{1}{8}$ inch diameter may be used in any primary control system.

(4) Pulley kinds and sizes must correspond to the cables with which they are used. The pulley-cable combinations and strength values specified in MIL-HDBK-5 must be used unless they are inapplicable.

(5) Pulleys must have close fitting guards to prevent the cables from being displaced or fouled.

(6) Pulleys must lie close enough to the plane passing through the cable to prevent the cable from rubbing against the pulley flange.

(7) No fairlead may cause a change in cable direction of more than three degrees.

(8) No clevis pin subject to load or motion and retained only by cotter pins may be used in the control system.

(9) Turnbuckles attached to parts having angular motion must be installed to prevent binding throughout the range of travel.

(10) There must be means for visual inspection at each fairlead, pulley, terminal, and turnbuckle.

(e) Control system joints subject to angular motion must incorporate the following special factors with respect to the ultimate bearing strength of the softest material used as a bearing:

(1) 3.33 for push-pull systems other than ball and roller bearing systems.

(2) 2.0 for cable systems.

(f) For control system joints, the manufacturer's static, non-Brinell rating of ball and roller bearings may not be exceeded.

[Doc. No. 5084, 29 FR 16150, Dec. 3, 1964, as amended by Amdt. 29-12, 41 FR 55471, Dec. 20, 1976]

§ 29.687 Spring devices.

(a) Each control system spring device whose failure could cause flutter or other unsafe characteristics must be reliable.

(b) Compliance with paragraph (a) of this section must be shown by tests simulating service conditions.

§ 29.691 Autorotation control mechanism.

Each main rotor blade pitch control mechanism must allow rapid entry into autorotation after power failure.

§ 29.695 Power boost and power-operated control system.

(a) If a power boost or power-operated control system is used, an alternate system must be immediately available that allows continued safe flight and landing in the event of—

(1) Any single failure in the power portion of the system; or

(2) The failure of all engines.

(b) Each alternate system may be a duplicate power portion or a manually operated mechanical system. The power portion includes the power source (such as hydraulic pumps), and such items as valves, lines, and actuators.

(c) The failure of mechanical parts (such as piston rods and links), and the jamming of power cylinders, must be considered unless they are extremely improbable.

Landing Gear

§ 29.723 Shock absorption tests.

The landing inertia load factor and the reserve energy absorption capacity of the landing gear must be substantiated by the tests prescribed in §§ 29.725 and 29.727, respectively. These tests must be conducted on the complete rotorcraft or on units consisting of wheel, tire, and shock absorber in their proper relation.

§ 29.725 Limit drop test.

The limit drop test must be conducted as follows:

(a) The drop height must be at least 8 inches.

(b) If considered, the rotor lift specified in § 29.473(a) must be introduced into the drop test by appropriate energy absorbing devices or by the use of an effective mass.

(c) Each landing gear unit must be tested in the attitude simulating the landing condition that is most critical from the standpoint of the energy to be absorbed by it.

(d) When an effective mass is used in showing compliance with paragraph (b) of this section, the following formulae may be used instead of more rational computations.

$$W_e = W \times \frac{h + (1 - L)d}{h + d}; \text{ and}$$

$$n = n_j \frac{W_e}{W} + L$$

where:

W_e =the effective weight to be used in the drop test (lbs.).

$W=W_M$ for main gear units (lbs.), equal to the static reaction on the particular unit with the rotorcraft in the most critical attitude. A rational method may be used in computing a main gear static reaction, taking into consideration the moment arm between the main wheel reaction and the rotorcraft center of gravity.

$W=W_N$ for nose gear units (lbs.), equal to the vertical component of the static reaction that would exist at the nose wheel, assuming that the mass of the rotorcraft acts at the center of gravity and exerts a force of 1.0 *g* downward and 0.25 *g* forward.

$W=W_t$ for tailwheel units (lbs.) equal to whichever of the following is critical—

(1) The static weight on the tailwheel with the rotorcraft resting on all wheels; or

(2) The vertical component of the ground reaction that would occur at the tailwheel assuming that the mass of the rotorcraft acts at the center of gravity and exerts a force of 1 *g* downward with the rotorcraft in the maximum nose-up attitude considered in the nose-up landing conditions.

h =specified free drop height (inches).

L =ratio of assumed rotor lift to the rotorcraft weight.

d =deflection under impact of the tire (at the proper inflation pressure) plus the vertical component of the axle travel (inches) relative to the drop mass.

n =limit inertia load factor.

n_j =the load factor developed, during impact, on the mass used in the drop test (i.e., the acceleration *dv/dt* in *g*'s recorded in the drop test plus 1.0).

[Doc. No. 5084, 29 FR 16150, Dec. 3, 1964, as amended by Amdt. 29-3, 33 FR 967, Jan. 26. 1968]

§ 29.727 Reserve energy absorption drop test.

The reserve energy absorption drop test must be conducted as follows:

(a) The drop height must be 1.5 times that specified in § 29.725(a).

(b) Rotor lift, where considered in a manner similar to that prescribed in § 29.725(b), may not exceed 1.5 times the lift allowed under that paragraph.

(c) The landing gear must withstand this test without collapsing. Collapse of the landing gear occurs when a member of the nose, tail, or main gear will not support the rotorcraft in the proper attitude or allows the rotorcraft structure, other than landing gear and external accessories, to impact the landing surface.

[Doc. No. 5084, 29 FR 16150, Dec. 3, 1964, as amended by Amdt. 27-26, 55 FR 8003, Mar. 6, 1990]

§ 29.729 Retracting mechanism.

For rotorcraft with retractable landing gear, the following apply:

(a) *Loads.* The landing gear, retracting mechanism, wheel well doors, and supporting structure must be designed for—

(1) The loads occurring in any maneuvering condition with the gear retracted;

(2) The combined friction, inertia, and air loads occurring during retraction and extension at any airspeed up to the design maximum landing gear operating speed; and

(3) The flight loads, including those in yawed flight, occurring with the gear extended at any airspeed up to the design maximum landing gear extended speed.

(b) *Landing gear lock.* A positive means must be provided to keep the gear extended.

(c) *Emergency operation.* When other than manual power is used to operate the gear, emergency means must be provided for extending the gear in the event of—

(1) Any reasonably probable failure in the normal retraction system; or

(2) The failure of any single source of hydraulic, electric, or equivalent energy.

(d) *Operation tests.* The proper functioning of the retracting mechanism must be shown by operation tests.

(e) *Position indicator.* There must be means to indicate to the pilot when the gear is secured in the extreme positions.

(f) *Control.* The location and operation of the retraction control must meet the requirements of §§ 29.777 and 29.779.

(g) *Landing gear warning.* An aural or equally effective landing gear warning device must be provided that functions continuously when the rotorcraft is in a normal landing mode and the landing gear is not fully extended and locked. A manual shutoff capability must be provided for the warning device and the warning system must automatically reset when the rotorcraft is no longer in the landing mode.

[Doc. No. 5084, 29 FR 16150, Dec. 3, 1964, as amended by Amdt. 29-24, 49 FR 44437, Nov. 6, 1984]

§ 29.731 Wheels.

(a) Each landing gear wheel must be approved.

(b) The maximum static load rating of each wheel may not be less than the corresponding static ground reaction with—

(1) Maximum weight; and

(2) Critical center of gravity.

(c) The maximum limit load rating of each wheel must equal or exceed the maximum radial limit load determined under the applicable ground load requirements of this part.

§ 29.733 Tires.

Each landing gear wheel must have a tire—

(a) That is a proper fit on the rim of the wheel; and

(b) Of a rating that is not exceeded under—

(1) The design maximum weight;

(2) A load on each main wheel tire equal to the static ground reaction corresponding to the critical center of gravity; and

(3) A load on nose wheel tires (to be compared with the dynamic rating established for those tires) equal to the reaction obtained at the nose wheel, assuming that the mass of the rotorcraft acts as the most critical center of gravity and exerts a force of 1.0g downward and 0.25g forward, the reactions being distributed to the nose and main wheels according to the principles of statics with the drag reaction at the ground applied only at wheels with brakes.

(c) Each tire installed on a retractable landing gear system must, at the maximum size of the tire type expected in service, have a clearance to surrounding structure and systems that is adequate to prevent contact between the tire and any part of the structure or systems.

[Doc. No. 5084, 29 FR 16150, Dec. 3, 1964, as amended by Amdt. 29-12, 41 FR 55471, Dec. 20, 1976]

§ 29.735 Brakes.

For rotorcraft with wheel-type landing gear, a braking device must be installed that is—

(a) Controllable by the pilot;

(b) Usable during power-off landings; and

(c) Adequate to—

(1) Counteract any normal unbalanced torque when starting or stopping the rotor; and

(2) Hold the rotorcraft parked on a 10-degree slope on a dry, smooth pavement.

[Doc. No. 5084, 29 FR 16150, Dec. 3, 1964, as amended by Amdt. 29-24, 49 FR 44437, Nov. 6, 1984]

§ 29.737 Skis.

(a) The maximum limit load rating of each ski must equal or exceed the maximum limit load determined under the applicable ground load requirements of this part.

(b) There must be a stabilizing means to maintain the ski in an appropriate position during flight. This means must have enough strength to withstand the maximum aerodynamic and inertia loads on the ski.

Floats and Hulls

§ 29.751 Main float buoyancy.

(a) For main floats, the buoyancy necessary to support the maximum weight of the rotorcraft in fresh water must be exceeded by—

(1) 50 percent, for single floats; and

(2) 60 percent, for multiple floats.

(b) Each main float must have enough water-tight compartments so that, with any single main float compartment flooded, the mainfloats will provide a margin of positive stability great enough to minimize the probability of capsizing.

[Doc. No. 5084, 29 FR 16150, Dec. 3, 1964, as amended by Amdt. 29-3, 33 FR 967, Jan. 26, 1968]

§ 29.753 Main float design.

(a) *Bag floats.* Each bag float must be designed to withstand—

(1) The maximum pressure differential that might be developed at the maximum altitude for which certification with that float is requested; and

(2) The vertical loads prescribed in § 29.521(a), distributed along the length of the bag over three-quarters of its projected area.

(b) *Rigid floats.* Each rigid float must be able to withstand the vertical, horizontal, and side loads prescribed in § 29.521. An appropriate load distribution under critical conditions must be used.

§ 29.755 Hull buoyancy.

Water-based and amphibian rotorcraft. The hull and auxiliary floats, if used, must have enough watertight compartments so that, with any single compartment of the hull or auxiliary floats flooded, the buoyancy of the hull and auxiliary floats, and wheel tires if used, provides a margin of positive water stability great enough to minimize the probability of capsizing the rotorcraft for the worst combination of wave heights and surface winds for which approval is desired.

[Amdt. 29-3, 33 FR 967, Jan. 26, 1968; as amended by Amdt. 27-26, 55 FR 8003, Mar. 6, 1990]

§ 29.757 Hull and auxiliary float strength.

The hull, and auxiliary floats if used, must withstand the water loads prescribed by § 29.519 with a rational and conservative distribution of local and distributed water pressures over the hull and float bottom.

[Amdt. 29-3, 33 FR 967, Jan. 26, 1968]

Personnel and Cargo Accommodations

§ 29.771 Pilot compartment.

For each pilot compartment—

(a) The compartment and its equipment must allow each pilot to perform his duties without unreasonable concentration or fatigue;

(b) If there is provision for a second pilot, the rotorcraft must be controllable with equal safety from either pilot position. Flight and powerplant controls must be designed to prevent confusion or inadvertent operation when the rotorcraft is piloted from either position;

(c) The vibration and noise characteristics of cockpit appurtenances may not interfere with safe operation;

(d) Inflight leakage of rain or snow that could distract the crew or harm the structure must be prevented.

[Doc. No. 5084, 29 FR 16150, Dec. 3, 1964, as amended by Amdt. 29-3, 33 FR 967, Jan. 26, 1968; Amdt. 29-24, 49 FR 44437, Nov. 6, 1984]

§ 29.773 Pilot compartment view.

(a) *Nonprecipitation conditions.* For nonprecipitation conditions, the following apply:

(1) Each pilot compartment must be arranged to give the pilots a sufficiently extensive, clear, and undistorted view for safe operation.

(2) Each pilot compartment must be free of glare and reflection that could interfere with the pilot's view. If certification for night operation is requested, this must be shown by night flight tests.

(b) *Precipitation conditions.* For precipitation conditions, the following apply:

(1) Each pilot must have a sufficiently extensive view for safe operation—

(i) In heavy rain at forward speeds up to V_H; and

(ii) In the most severe icing condition for which certification is requested.

(2) The first pilot must have a window that—

(i) Is openable under the conditions prescribed in paragraph (b)(1) of this section; and

(ii) Provides the view prescribed in that paragraph.

[Doc. No. 5084, 29 FR 16150, Dec. 3, 1964, as amended by Amdt. 29-3, 33 FR 967, Jan. 26, 1968]

§ 29.775 Windshields and windows.

Windshields and windows must be made of material that will not break into dangerous fragments.

[Amdt. 29-31, 55 FR 38966, Sept. 21, 1990]

§ 29.777 Cockpit controls.

Cockpit controls must be—

(a) Located to provide convenient operation and to prevent confusion and inadvertent operation; and

(b) Located and arranged with respect to the pilot's seats so that there is full and unrestricted movement of each control without interference from the cockpit structure or the pilot's clothing when pilots from 5′2″ to 6′0″ in height are seated.

§ 29.779 Motion and effect of cockpit controls.

Cockpit controls must be designed so that they operate in accordance with the following movements and actuation:

(a) Flight controls, including the collective pitch control, must operate with a sense of motion which corresponds to the effect on the rotorcraft.

(b) Twist-grip engine power controls must be designed so that, for lefthand operation, the motion of the pilot's hand is clockwise to increase power when the hand is viewed from the edge containing the index finger. Other engine power controls, excluding the collective control, must operate with a forward motion to increase power.

(c) Normal landing gear controls must operate downward to extend the landing gear.

[Amdt. 29-24, 49 FR 44437, Nov. 6, 1984]

§ 29.783 Doors.

(a) Each closed cabin must have at least one adequate and easily accessible external door.

(b) Each external door must be located, and appropriate operating procedures must be established, to ensure that persons using the door will not be endangered by the rotors, propellers, engine intakes, and exhausts when the operating procedures are used.

(c) There must be means for locking crew and external passenger doors and for preventing their opening in flight inadvertently or as a result of mechanical failure. It must be possible to open external doors from inside and outside the cabin with the rotorcraft on the ground even though persons may be crowded against the door on the inside of the rotorcraft. The means of opening must be simple and obvious and so arranged and marked that it can be readily located and operated.

(d) There must be reasonable provisions to prevent the jamming of any external doors in a minor crash as a result of fuselage deformation under the following ultimate inertial forces except for cargo or service doors not suitable for use as an exit in an emergency:

(1) Upward—1.5g.

(2) Forward—4.0g.

(3) Sideward—2.0g.

(4) Downward—4.0g.

(e) There must be means for direct visual inspection of the locking mechanism by crewmembers to determine whether the external doors (including passenger, crew, service, and cargo doors) are fully locked. There must be visual means to signal to appropriate crewmembers when normally used external doors are closed and fully locked.

(f) For outward opening external doors usable for entrance or egress, there must be an auxiliary safety latching device to prevent the door from opening when the primary latching mechanism fails. If the door does not meet the requirements of paragraph (c) of this section with this device in place, suitable operating procedures must be established to prevent the use of the device during takeoff and landing.

(g) If an integral stair is installed in a passenger entry door that is qualified as a passenger emergency exit, the stair must be designed so that under the following conditions the effectiveness of passenger emergency egress will not be impaired:

(1) The door, integral stair, and operating mechanism have been subjected to the inertial forces specified in paragraph (d) of this section, acting separately relative to the surrounding structure.

(2) The rotorcraft is in the normal ground attitude and in each of the attitudes corresponding to collapse of one or more legs, or primary members, as applicable, of the landing gear.

(h) Nonjettisonable doors used as ditching emergency exits must have means to enable them to be secured in the open position and remain secure for emergency egress in sea state conditions prescribed for ditching.

[Doc. No. 5084, 29 FR 16150, Dec. 3, 1964, as amended by Amdt. 29-20, 45 FR 60178, Sept. 11, 1980; Amdt. 29-29, 54 FR 47320, Nov. 13, 1989; Amdt. 27-26, 55 FR 8003, Mar. 6, 1990; Amdt. 29-31, 55 FR 38966, Sept. 21, 1990]

§ 29.785 Seats, berths, litters, safety belts, and harnesses.

(a) Each seat, safety belt, harness, and adjacent part of the rotorcraft at each station designated for occupancy during takeoff and landing must be free of potentially injurious objects, sharp edges, protuberances, and hard surfaces and must be designed so that a person making proper use of these facilities will not suffer serious injury in an emergency landing as a result of the inertial factors specified in § 29.561(b) and dynamic conditions specified in § 29.562.

(b) Each occupant must be protected from serious head injury by a safety belt plus a shoulder harness that will prevent the head from contacting any injurious object, except as provided for in § 29.562(c)(5). A shoulder harness (upper torso restraint), in combination with the safety belt, constitutes a torso restraint system as described in TSO-C114.

(c) Each occupant's seat must have a combined safety belt and shoulder harness with a single-point release. Each pilot's combined safety belt and shoulder harness must allow each pilot when seated with safety belt and shoulder harness fastened to perform all functions necessary for flight operations. There must be a means to secure belt and harness when not in use to prevent interference with the operation of the rotorcraft and with rapid egress in an emergency.

(d) If seat backs do not have a firm handhold, there must be hand grips or rails along each aisle to let the occupants steady themselves while using the aisle in moderately rough air.

(e) Each projecting object that would injure persons seated or moving about in the rotorcraft in normal flight must be padded.

(f) Each seat and its supporting structure must be designed for an occupant weight of at least 170 pounds, considering the maximum load factors, inertial forces, and reactions between the occupant, seat, and safety belt or harness corresponding with the applicable flight and ground-load conditions, including the emergency landing conditions of § 29.561(b). In addition—

(1) Each pilot seat must be designed for the reactions resulting from the application of the pilot forces prescribed in § 29.397; and

(2) The inertial forces prescribed in § 29.561(b) must be multiplied by a factor of 1.33 in determining the strength of the attachment of—

(i) Each seat to the structure; and

(ii) Each safety belt or harness to the seat or structure.

(g) When the safety belt and shoulder harness are combined, the rated strength of the safety belt and shoulder harness may not be less than that corresponding to the inertial forces specified in § 29.561(b), considering the occupant weight of at least 170 pounds, considering the dimensional characteristics of the restraint system installation, and using a distribution of at least a 60-percent load to the safety belt and at least a 40-percent load to the shoulder harness. If the safety belt is capable of being used without the shoulder harness, the inertial forces specified must be met by the safety belt alone.

(h) When a headrest is used, the headrest and its supporting structure must be designed to resist the inertia forces specified in § 29.561, with a 1.33 fitting factor and a head weight of at least 13 pounds.

(i) Each seating device system includes the device such as the seat, the cushions, the occupant restraint system and attachment devices.

(j) Each seating device system may use design features such as crushing or separation of certain parts of the seat in the design to reduce occupant loads for the emergency landing dynamic conditions of § 29.562; otherwise, the system must remain intact and must not interfere with rapid evacuation of the rotorcraft.

(k) For purposes of this section, a litter is defined as a device designed to carry a nonambulatory person, primarily in a recumbent position, into and on the rotorcraft. Each berth or litter must be designed to withstand the load reaction of an occupant weight of at least 170 pounds when the occupant is subjected to the forward inertial factors specified in § 29.561(b). A berth or litter installed within 15° or less of the longitudinal axis of the rotorcraft must be provided with a padded end-board, cloth diaphragm, or equivalent means that can withstand the forward load reaction. A berth or litter oriented greater than 15° with the longitudinal axis of the rotorcraft must be equipped with appropriate restraints, such as straps or safety belts, to withstand the forward reaction. In addition—

(1) The berth or litter must have a restraint system and must not have corners or other protuberances likely to cause serious injury to a person occupying it during emergency landing conditions; and

(2) The berth or litter attachment and the occupant restraint system attachments to the structure must be designed to withstand the critical loads resulting from flight and ground load conditions and from the conditions prescribed in § 29.561(b). The fitting factor required by § 29.625(d) shall be applied.

[Doc. No. 5084, 29 FR 16150, Dec. 3, 1964, as amended by Amdt. 29-24, 49 FR 44437, Nov. 6, 1984; Amdt. 29-29, 54 FR 47320, Nov. 13, 1989; Amdt. 29-42, 63 FR 43285, Aug. 12, 1998]

§ 29.787 Cargo and baggage compartments.

(a) Each cargo and baggage compartment must be designed for its placarded maximum weight of contents and for the critical load distributions at the appropriate maximum load factors corresponding to the specified flight and ground load conditions, except the emergency landing conditions of § 29.561.

(b) There must be means to prevent the contents of any compartment from becoming a hazard by shifting under the loads specified in paragraph (a) of this section.

(c) Under the emergency landing conditions of § 29.561, cargo and baggage compartments must—

(1) Be positioned so that if the contents break loose they are unlikely to cause injury to the occupants or restrict any of the escape facilities provided for use after an emergency landing; or

(2) Have sufficient strength to withstand the conditions specified in § 29.561, including the means of restraint and their attachments required by paragraph (b) of this section. Sufficient strength must be provided for the maximum authorized weight of cargo and baggage at the critical loading distribution.

(d) If cargo compartment lamps are installed, each lamp must be installed so as to prevent contact between lamp bulb and cargo.

[Doc. No. 5084, 29 FR 16150, Dec. 3, 1964, as amended by Amdt. 29-12, 41 FR 55472, Dec. 20, 1976; Amdt. 29-31, 55 FR 38966, Sept. 21, 1990]

§ 29.801 Ditching.

(a) If certification with ditching provisions is requested, the rotorcraft must meet the requirements of this section and §§ 29.807(d), 29.1411 and 29.1415.

(b) Each practicable design measure, compatible with the general characteristics of the rotorcraft, must be taken to minimize the probability that in an emergency landing on water, the behavior of the rotorcraft would cause immediate injury to the occupants or would make it impossible for them to escape.

(c) The probable behavior of the rotorcraft in a water landing must be investigated by model tests or by comparison with rotorcraft of similar configuration for which the ditching characteristics are known. Scoops, flaps, projections, and any other factors likely to affect the hydrodynamic characteristics of the rotorcraft must be considered.

(d) It must be shown that, under reasonably probable water conditions, the flotation time and trim of the rotorcraft will allow the occupants to leave the rotorcraft and enter the liferafts required by § 29.1415. If compliance with this provision is shown by bouyancy and trim computations, appropriate allowances must be made for probable structural damage and leakage. If the rotorcraft has fuel tanks (with fuel jettisoning provisions) that can reasonably be expected to withstand a ditching without leakage, the jettisonable volume of fuel may be considered as bouyancy volume.

(e) Unless the effects of the collapse of external doors and windows are accounted for in the investigation of the probable behavior of the rotorcraft in a water landing (as prescribed in paragraphs (c) and (d) of this section), the external doors and windows must be designed to withstand the probable maximum local pressures.

[Amdt. 29-12, 41 FR 55472, Dec. 20, 1976]

§ 29.803 Emergency evacuation.

(a) Each crew and passenger area must have means for rapid evacuation in a crash landing, with the landing gear (1) extended and (2) retracted, considering the possibility of fire.

(b) Passenger entrance, crew, and service doors may be considered as emergency exits if they meet the requirements of this section and of §§ 29.805 through 29.815.

(c) [Reserved]

(d) Except as provided in paragraph (e) of this section, the following categories of rotorcraft must be tested in accordance with the requirements of appendix D of this part to demonstrate that the maximum seating capacity, including the crewmembers required by the operating rules, can be evacuated from the rotorcraft to the ground within 90 seconds:

(1) Rotorcraft with a seating capacity of more than 44 passengers.

(2) Rotorcraft with all of the following:

(i) Ten or more passengers per passenger exit as determined under § 29.807(b).

(ii) No main aisle, as described in § 29.815, for each row of passenger seats.

(iii) Access to each passenger exit for each passenger by virtue of design features of seats, such as folding or break-over seat backs or folding seats.

(e) A combination of analysis and tests may be used to show that the rotorcraft is capable of being evacuated within 90 seconds under the conditions specified in § 29.803(d) if the Administrator finds that the combination of analysis and tests will provide data, with respect to the emergency evacuation capability of the rotorcraft, equivalent to that which would be obtained by actual demonstration.

[Doc. No. 5084, 29 FR 16150, Dec. 3, 1964, as amended by Amdt. 29-3, 33 FR 967, Jan. 26, 1968; Amdt. 27-26, 55 FR 8004, Mar. 6, 1990]

§ 29.805 Flight crew emergency exits.

(a) For rotorcraft with passenger emergency exits that are not convenient to the flight crew, there must be flight crew emergency exits, on both sides of the rotorcraft or as a top hatch, in the flight crew area.

(b) Each flight crew emergency exit must be of sufficient size and must be located so as to allow rapid evacuation of the flight crew. This must be shown by test.

(c) Each exit must not be obstructed by water or flotation devices after a ditching. This must be shown by test, demonstration, or analysis.

[Amdt. 29-3, 33 FR 968, Jan. 26, 1968; as amended by Amdt. 27-26, 55 FR 8004, Mar. 6, 1990]

§ 29.807 Passenger emergency exits.

(a) *Type.* For the purpose of this part, the types of passenger emergency exit are as follows:

(1) *Type I.* This type must have a rectangular opening of not less than 24 inches wide by 48 inches high, with corner radii not greater than one-third the width of the exit, in the passenger area in the side of the fuselage at floor level and as far away as practicable from areas that might become potential fire hazards in a crash.

(2) *Type II.* This type is the same as Type I, except that the opening must be at least 20 inches wide by 44 inches high.

(3) *Type III.* This type is the same as Type I, except that—

(i) The opening must be at least 20 inches wide by 36 inches high; and

(ii) The exits need not be at floor level.

(4) *Type IV.* This type must have a rectangular opening of not less than 19 inches wide by 26 inches high, with corner radii not greater than one-third the width of the exit, in the side of the fuselage with a step-up inside the rotorcraft of not more than 29 inches.

Openings with dimensions larger than those specified in this section may be used, regardless of shape, if the base of the opening has a flat surface of not less than the specified width.

(b) *Passenger emergency exits; side-of-fuselage.* Emergency exits must be accessible to the passengers and, except as provided in paragraph (d) of this section, must be provided in accordance with the following table:

Passenger seating capacity	Emergency exits for each side of the fuselage			
	Type I	Type II	Type III	Type IV
1 through 10				1
11 through 19			1 or	2
20 through 39		1		1
40 through 59	1			1
60 through 79	1		1 or	2

(c) *Passenger emergency exits; other than side-of-fuselage.* In addition to the requirements of paragraph (b) of this section—

(1) There must be enough openings in the top, bottom, or ends of the fuselage to allow evacuation with the rotorcraft on its side; or

(2) The probability of the rotorcraft coming to rest on its side in a crash landing must be extremely remote.

(d) *Ditching emergency exits for passengers.* If certification with ditching provisions is requested, ditching emergency exits must be provided in accordance with the following requirements and must be proven by test, demonstration, or analysis unless the emergency exits required by paragraph (b) of this section already meet these requirements.

(1) For rotorcraft that have a passenger seating configuration, excluding pilots seats, of nine seats or less, one exit above the waterline in each side of the rotorcraft, meeting at least the dimensions of a Type IV exit.

(2) For rotorcraft that have a passenger seating configuration, excluding pilots seats, of 10 seats or more, one exit above the waterline in a side of the rotorcraft meeting at least the dimensions of a Type III exit, for each unit (or part of a unit) of 35 passenger seats, but no less than two such exits in the passenger cabin, with one on each side of the rotorcraft. However, where it has been shown through analysis, ditching demonstrations, or any other tests found necessary by the Administrator, that the evacuation capability of the rotorcraft during ditching is improved by the use of larger exits, or by other means, the passenger seat to exit ratio may be increased.

(3) Flotation devices, whether stowed or deployed, may not interfere with or obstruct the exits.

(e) *Ramp exits.* One Type I exit only, or one Type II exit only, that is required in the side of the fuselage under paragraph (b) of this section, may be installed instead in the ramp of floor ramp rotorcraft if—

(1) Its installation in the side of the fuselage is impractical; and

(2) Its installation in the ramp meets § 29.813.

(f) *Tests.* The proper functioning of each emergency exit must be shown by test.

[Amdt. 29-3, 33 FR 968, Jan. 26, 1968, as amended by Amdt. 29-12, 41 FR 55472, Dec. 20, 1976; Amdt. 27-26, 55 FR 8004, Mar. 6, 1990]

§ 29.809 Emergency exit arrangement.

(a) Each emergency exit must consist of a movable door or hatch in the external walls of the fuselage and must provide an unobstructed opening to the outside.

(b) Each emergency exit must be openable from the inside and from the outside.

(c) The means of opening each emergency exit must be simple and obvious and may not require exceptional effort.

(d) There must be means for locking each emergency exit and for preventing opening in flight inadvertently or as a result of mechanical failure.

(e) There must be means to minimize the probability of the jamming of any emergency exit in a minor crash landing as a result of fuselage deformation under the ultimate inertial forces in § 29.783(d).

(f) Except as provided in paragraph (h) of this section, each land-based rotorcraft emergency exit must have an approved slide as stated in paragraph (g) of this section, or its equivalent, to assist occupants in descending to the ground from each floor level exit and an approved rope, or its equivalent, for all other exits, if the exit threshold is more that 6 feet above the ground—

(1) With the rotorcraft on the ground and with the landing gear extended;

(2) With one or more legs or part of the landing gear collapsed, broken, or not extended; and

(3) With the rotorcraft resting on its side, if required by § 29.803(d).

(g) The slide for each passenger emergency exit must be a self-supporting slide or equivalent, and must be designed to meet the following requirements:

(1) It must be automatically deployed, and deployment must begin during the interval between the time the exit opening means is actuated from inside the rotorcraft and the time the exit is fully opened. However, each passenger emergency exit which is also a passenger entrance door or a service door must be provided with means to prevent deployment of the slide when the exit is opened from either the inside or the outside under nonemergency conditions for normal use.

(2) It must be automatically erected within 10 seconds after deployment is begun.

(3) It must be of such length after full deployment that the lower end is self-supporting on the ground and provides safe evacuation of occupants to the ground after collapse of one or more legs or part of the landing gear.

(4) It must have the capability, in 25-knot winds directed from the most critical angle, to deploy and, with the assistance of only one person, to remain usable after full deployment to evacuate occupants safely to the ground.

(5) Each slide installation must be qualified by five consecutive deployment and inflation tests conducted (per exit) without failure, and at least three tests of each such five-test series must be conducted using a single representative sample of the device. The sample devices must be deployed and inflated by the system's primary means after being subjected to the inertia forces specified in § 29.561(b). If any part of the system fails or does not function properly during the required tests, the cause of the failure or malfunction must be corrected by positive means and after that, the full series of five consecutive deployment and inflation tests must be conducted without failure.

(h) For rotorcraft having 30 or fewer passenger seats and having an exit threshold more than 6 feet above the ground, a rope or other assist means may be used in place of the slide specified in paragraph (f) of this section, provided an evacuation demonstration is accomplished as prescribed in § 29.803(d) or (e).

(i) If a rope, with its attachment, is used for compliance with paragraph (f), (g), or (h) of this section, it must—

(1) Withstand a 400-pound static load; and

(2) Attach to the fuselage structure at or above the top of the emergency exit opening, or at another approved location if the stowed rope would reduce the pilot's view in flight.

[Amdt. 29-3, 33 FR 968, Jan. 26, 1968, as amended by Amdt. 29-29, 54 FR 47321, Nov. 13, 1989; Amdt. 27-26, 55 FR 8004, Mar. 6, 1990]

§ 29.811 Emergency exit marking.

(a) Each passenger emergency exit, its means of access, and its means of opening must be conspicuously marked for the guidance of occupants using the exits in daylight or in the dark. Such markings must be designed to remain visible for rotorcraft equipped for overwater flights if the rotorcraft is capsized and the cabin is submerged.

(b) The identity and location of each passenger emergency exit must be recognizable from a distance equal to the width of the cabin.

(c) The location of each passenger emergency exit must be indicated by a sign visible to occupants approaching along the main passenger aisle. There must be a locating sign—

(1) Next to or above the aisle near each floor emergency exit, except that one sign may serve two exits if both exists can be seen readily from that sign; and

(2) On each bulkhead or divider that prevents fore and aft vision along the passenger cabin, to indicate emergency exits beyond and obscured by it, except that if this is not possible the sign may be placed at another appropriate location.

(d) Each passenger emergency exit marking and each locating sign must have white letters 1 inch high on a red background 2 inches high, be self or electrically illuminated, and have a minimum luminescence (brightness) of at least 160 microlamberts. The colors may be reversed if this will increase the emergency illumination of the passenger compartment.

(e) The location of each passenger emergency exit operating handle and instructions for opening must be shown—

(1) For each emergency exit, by a marking on or near the exit that is readable from a distance of 30 inches; and

(2) For each Type I or Type II emergency exit with a locking mechanism released by rotary motion of the handle, by—

(i) A red arrow, with a shaft at least three-fourths inch wide and a head twice the width of the shaft, extending along at least 70 degrees of arc at a radius approximately equal to three-fourths of the handle length; and

(ii) The word "open" in red letters 1 inch high, placed horizontally near the head of the arrow.

(f) Each emergency exit, and its means of opening, must be marked on the outside of the rotorcraft. In addition, the following apply:

(1) There must be a 2-inch colored band outlining each passenger emergency exit, except small rotorcraft with a maximum weight of 12,500 pounds or less may have a 2-inch colored band outlining each exit release lever or device of passenger emergency exits which are normally used doors.

(2) Each outside marking, including the band, must have color contrast to be readily distinguishable from the surrounding fuselage surface. The contrast must be such that, if the reflectance of the darker color is 15 percent or less, the reflectance of the lighter color must be at least 45 percent. "Reflectance" is the ratio of the luminous

flux reflected by a body to the luminous flux it receives. When the reflectance of the darker color is greater than 15 percent, at least a 30 percent difference between its reflectance and the reflectance of the lighter color must be provided.

(g) Exits marked as such, though in excess of the required number of exits, must meet the requirements for emergency exits of the particular type. Emergency exits need only be marked with the word "Exit."

[Amdt. 29-3, 33 FR 968, Jan. 26, 1968, as amended by Amdt. 29-24, 49 FR 44438, Nov. 6, 1984; Amdt. 27-26, 55 FR 8004, Mar. 6, 1990; Amdt. 29-31, 55 FR 38967, Sept. 21, 1990]

§ 29.812 Emergency lighting.

For transport Category A rotorcraft, the following apply:

(a) A source of light with its power supply independent of the main lighting system must be installed to—

(1) Illuminate each passenger emergency exit marking and locating sign; and

(2) Provide enough general lighting in the passenger cabin so that the average illumination, when measured at 40-inch intervals at seat armrest height on the center line of the main passenger aisle, is at least 0.05 foot-candle.

(b) Exterior emergency lighting must be provided at each emergency exit. The illumination may not be less than 0.05 foot-candle (measured normal to the direction of incident light) for minimum width on the ground surface, with landing gear extended, equal to the width of the emergency exit where an evacuee is likely to make first contact with the ground outside the cabin. The exterior emergency lighting may be provided by either interior or exterior sources with light intensity measurements made with the emergency exits open.

(c) Each light required by paragraph (a) or (b) of this section must be operable manually from the cockpit station and from a point in the passenger compartment that is readily accessible. The cockpit control device must have an "on," "off," and "armed" position so that when turned on at the cockpit or passenger compartment station or when armed at the cockpit station, the emergency lights will either illuminate or remain illuminated upon interruption of the rotorcraft's normal electric power.

(d) Any means required to assist the occupants in descending to the ground must be illuminated so that the erected assist means is visible from the rotorcraft.

(1) The assist means must be provided with an illumination of not less than 0.03 foot-candle (measured normal to the direction of the incident light) at the ground end of the erected assist means where an evacuee using the established escape route would normally make first contact with the ground, with the rotorcraft in each of the attitudes corresponding to the collapse of one or more legs of the landing gear.

(2) If the emergency lighting subsystem illuminating the assist means is independent of the rotorcraft's main emergency lighting system, it—

(i) Must automatically be activated when the assist means is erected;

(ii) Must provide the illumination required by paragraph (d)(1); and

(iii) May not be adversely affected by stowage.

(e) The energy supply to each emergency lighting unit must provide the required level of illumination for at least 10 minutes at the critical ambient conditions after an emergency landing.

(f) If storage batteries are used as the energy supply for the emergency lighting system, they may be recharged from the rotorcraft's main electrical power system provided the charging circuit is designed to preclude inadvertent battery discharge into charging circuit faults.

[Amdt. 29-24, 49 FR 44438, Nov. 6, 1984]

§ 29.813 Emergency exit access.

(a) Each passageway between passenger compartments, and each passageway leading to Type I and Type II emergency exits, must be—

(1) Unobstructed; and

(2) At least 20 inches wide.

(b) For each emergency exit covered by § 29.809(f), there must be enough space adjacent to that exit to allow a crewmember to assist in the evacuation of passengers without reducing the unobstructed width of the passageway below that required for that exit.

(c) There must be access from each aisle to each Type III and Type IV exit, and

(1) For rotorcraft that have a passenger seating configuration, excluding pilot seats, of 20 or more, the projected opening of the exit provided must not be obstructed by seats, berths, or other protrusions (including seatbacks in any position) for a distance from that exit of not less than the width of the narrowest passenger seat installed on the rotorcraft;

(2) For rotorcraft that have a passenger seating configuration, excluding pilot seats, of 19 or less, there may be minor obstructions in the region described in paragraph (c)(1) of this section, if there are compensating factors to maintain the effectiveness of the exit.

[Doc. No. 5084, 29 FR 16150, Dec. 3, 1964, as amended by Amdt. 29-12, 41 FR 55472, Dec. 20, 1976]

§ 29.815 Main aisle width.

The main passenger aisle width between seats must equal or exceed the values in the following table:

Passenger seating capacity	Minimum main passenger aisle width	
	Less than 25 inches from floor (inches)	25 Inches and more from floor (inches)
10 or less	12	15
11 through 19	12	20
20 or more	15	20

[1] A narrower width not less than 9 inches may be approved when substantiated by tests found necessary by the Administrator.

[Doc. No. 5084, 29 FR 16150, Dec. 3, 1964, as amended by Amdt. 29-12, 41 FR 55472, Dec. 20, 1976]

§ 29.831 Ventilation.

(a) Each passenger and crew compartment must be ventilated, and each crew compartment must have enough fresh air (but not less than 10 cu. ft. per minute per crewmember) to let crewmembers perform their duties without undue discomfort or fatigue.

(b) Crew and passenger compartment air must be free from harmful or hazardous concentrations of gases or vapors.

(c) The concentration of carbon monoxide may not exceed one part in 20,000 parts of air during forward flight. If the concentration exceeds this value under other conditions, there must be suitable operating restrictions.

(d) There must be means to ensure compliance with paragraphs (b) and (c) of this section under any reasonably probable failure of any ventilating, heating, or other system or equipment.

§ 29.833 Heaters.

Each combustion heater must be approved.

Fire Protection

§ 29.851 Fire extinguishers.

(a) *Hand fire extinguishers.* For hand fire extinguishers the following apply:

(1) Each hand fire extinguisher must be approved.

(2) The kinds and quantities of each extinguishing agent used must be appropriate to the kinds of fires likely to occur where that agent is used.

(3) Each extinguisher for use in a personnel compartment must be designed to minimize the hazard of toxic gas concentrations.

(b) *Built-in fire extinguishers.* If a built-in fire extinguishing system is required—

(1) The capacity of each system, in relation to the volume of the compartment where used and the ventilation rate, must be adequate for any fire likely to occur in that compartment.

(2) Each system must be installed so that—

(i) No extinguishing agent likely to enter personnel compartments will be present in a quantity that is hazardous to the occupants; and

(ii) No discharge of the extinguisher can cause structural damage.

§ 29.853 Compartment interiors.

For each compartment to be used by the crew or passengers—

(a) The materials (including finishes or decorative surfaces applied to the materials) must meet the following test criteria as applicable:

(1) Interior ceiling panels, interior wall panels, partitions, galley structure, large cabinet walls, structural flooring, and materials used in the construction of stowage compartments (other than underseat stowage compartments and compartments for stowing small items such as magazines and maps) must be self-extinguishing when tested vertically in accordance with the applicable portions of appendix F of Part 25 of this chapter, or other approved equivalent methods. The average burn length may not exceed 6 inches and the average flame time after removal of the flame source may not exceed 15 seconds. Drippings from the test specimen may not continue to flame for more than an average of 3 seconds after falling.

(2) Floor covering, textiles (including draperies and upholstery), seat cushions, padding, decorative and nondecorative coated fabrics, leather, trays and galley furnishings, electrical conduit, thermal and acoustical insulation and insulation covering, air ducting, joint and edge covering, cargo compartment liners, insulation blankets, cargo covers, and transparencies, molded and thermoformed parts, air ducting joints, and trim strips (decorative and chafing) that are constructed of materials not covered in paragraph (a)(3) of this section, must be self extinguishing when tested vertically in accordance with the applicable portion of appendix F of Part 25 of this chapter, or other approved equivalent methods. The average burn length may not exceed 8 inches and the average flame time after removal of the flame source may not exceed 15 seconds. Drippings from the test specimen may not continue to flame for more than an average of 5 seconds after falling.

(3) Acrylic windows and signs, parts constructed in whole or in part of elastomeric materials, edge lighted instrument assemblies consisting of two or more instruments in a common housing, seat belts, shoulder harnesses, and cargo and baggage tiedown equipment, including containers, bins, pallets, etc., used in passenger or crew compartments, may not have an average burn rate greater than 2.5 inches per minute when tested horizontally in accordance with the applicable portions of appendix F of Part 25 of this chapter, or other approved equivalent methods.

(4) Except for electrical wire and cable insulation, and for small parts (such as knobs, handles, rollers, fasteners, clips, grommets, rub strips, pulleys, and small electrical parts) that the Administrator finds would not contribute significantly to the propagation of a fire, materials in items not specified in paragraphs (a)(1), (a)(2), or (a)(3) of this section may not have a burn rate greater than 4 inches per minute when tested horizontally in accordance with the applicable portions of appendix F of Part 25 of this chapter, or other approved equivalent methods.

(b) In addition to meeting the requirements of paragraph (a)(2), seat cushions, except those on flight crewmember seats, must meet the test requirements of Part II of appendix F of Part 25 of this chapter, or equivalent.

(c) If smoking is to be prohibited, there must be a placard so stating, and if smoking is to be allowed—

(1) There must be an adequate number of self-contained, removable ashtrays; and

(2) Where the crew compartment is separated from the passenger compartment, there must be at least one illuminated sign (using either letters or symbols) notifying all passengers when smoking is prohibited. Signs which notify when smoking is prohibited must—

(i) When illuminated, be legible to each passenger seated in the passenger cabin under all probable lighting conditions; and

(ii) Be so constructed that the crew can turn the illumination on and off.

(d) Each receptacle for towels, paper, or waste must be at least fire-resistant and must have means for containing possible fires;

(e) There must be a hand fire extinguisher for the flight crewmembers; and

(f) At least the following number of hand fire extinguishers must be conveniently located in passenger compartments:

Passenger capacity	Fire extinguishers
7 through 30	1
31 through 60	2
61 or more	3

(Secs. 313(a), 601, 603, 604, Federal Aviation Act of 1958 (49 U.S.C. 1354(a), 1421, 1423, 1424), sec. 6(c), Dept. of Transportation Act (49 U.S.C. 1655(c)))

[Doc. No. 5084, 29 FR 16150, Dec. 3, 1964, as amended by Amdt. 29-3, 33 FR 969, Jan. 26, 1968; Amdt. 29-17, 43

FR 50600, Oct. 30, 1978; Amdt. 29-18, 45 FR 7756, Feb. 4, 1980; Amdt. 29-23, 49 FR 43200, Oct. 26, 1984]

§ 29.855 Cargo and baggage compartments.

(a) Each cargo and baggage compartment must be construced of or lined with materials in accordance with the following:

(1) For accessible and inaccessible compartments not occupied by passengers or crew, the material must be at least fire resistant.

(2) Materials must meet the requirements in § 29.853(a) (1), (a)(2), and (a)(3) for cargo or baggage compartments in which—

(i) The presence of a compartment fire would be easily discovered by a crewmember while at the crewmember's station;

(ii) Each part of the compartment is easily accessible in flight;

(iii) The compartment has a volume of 200 cubic feet or less; and

(iv) Notwithstanding § 29.1439(a), protective breathing equipment is not required.

(b) No compartment may contain any controls, wiring, lines, equipment, or accessories whose damage or failure would affect safe operation, unless those items are protected so that—

(1) They cannot be damaged by the movement of cargo in the compartment; and

(2) Their breakage or failure will not create a fire hazard.

(c) The design and sealing of inaccessible compartments must be adequate to contain compartment fires until a landing and safe evacuation can be made.

(d) Each cargo and baggage compartment that is not sealed so as to contain cargo compartment fires completely without endangering the safety of a rotorcraft or its occupants must be designed, or must have a device, to ensure detection of fires or smoke by a crewmember while at his station and to prevent the accumulation of harmful quantities of smoke, flame, extinguishing agents, and other noxious gases in any crew or passenger compartment. This must be shown in flight.

(e) For rotorcraft used for the carriage of cargo only, the cabin area may be considered a cargo compartment and, in addition to paragraphs (a) through (d) of this section, the following apply:

(1) There must be means to shut off the ventilating airflow to or within the compartment. Controls for this purpose must be accessible to the flight crew in the crew compartment.

(2) Required crew emergency exits must be accessible under all cargo loading conditions.

(3) Sources of heat within each compartment must be shielded and insulated to prevent igniting the cargo.

[Doc. No. 5084, 29 FR 16150, Dec. 3, 1964, as amended by Amdt. 29-3, 33 FR 969, Jan. 26, 1968; Amdt. 29-24, 49 FR 44438, Nov. 6, 1984; Amdt. 27-26, 55 FR 8004, Mar. 6, 1990]

§ 29.859 Combustion heater fire protection.

(a) *Combustion heater fire zones.* The following combustion heater fire zones must be protected against fire under the applicable provisions of §§ 29.1181 through 29.1191, and 29.1195 through 29.1203:

(1) The region surrounding any heater, if that region contains any flammable fluid system components (including the heater fuel system), that could—

(i) Be damaged by heater malfunctioning; or

(ii) Allow flammable fluids or vapors to reach the heater in case of leakage.

(2) Each part of any ventilating air passage that—

(i) Surrounds the combustion chamber; and

(ii) Would not contain (without damage to other rotorcraft components) any fire that may occur within the passage.

(b) *Ventilating air ducts.* Each ventilating air duct passing through any fire zone must be fireproof. In addition—

(1) Unless isolation is provided by fireproof valves or by equally effective means, the ventilating air duct downstream of each heater must be fireproof for a distance great enough to ensure that any fire originating in the heater can be contained in the duct; and

(2) Each part of any ventilating duct passing through any region having a flammable fluid system must be so constructed or isolated from that system that the malfunctioning of any component of that system cannot introduce flammable fluids or vapors into the ventilating airstream.

(c) *Combustion air ducts.* Each combustion air duct must be fireproof for a distance great enough to prevent damage from backfiring or reverse flame propagation. In addition—

(1) No combustion air duct may communicate with the ventilating airstream unless flames from backfires or reverse burning cannot enter the ventilating airstream under any operating condition, including reverse flow or malfunction of the heater or its associated components; and

(2) No combustion air duct may restrict the prompt relief of any backfire that, if so restricted, could cause heater failure.

(d) *Heater controls; general.* There must be means to prevent the hazardous accumulation of water or ice on or in any heater control component, control system tubing, or safety control.

(e) *Heater safety controls.* For each combustion heater, safety control means must be provided as follows:

(1) Means independent of the components provided for the normal continuous control of air temperature, airflow, and fuel flow must be provided, for each heater, to automatically shut off the ignition and fuel supply of that heater at a point remote from that heater when any of the following occurs:

(i) The heat exchanger temperature exceeds safe limits.

(ii) The ventilating air temperature exceeds safe limits.

(iii) The combustion airflow becomes inadequate for safe operation.

(iv) The ventilating airflow becomes inadequate for safe operation.

(2) The means of complying with paragraph (e)(1) of this section for any individual heater must—

(i) Be independent of components serving any other heater whose heat output is essential for safe operation; and

(ii) Keep the heater off until restarted by the crew.

(3) There must be means to warn the crew when any heater whose heat output is essential for safe operation has been shut off by the automatic means prescribed in paragraph (e)(1) of this section.

(f) *Air intakes.* Each combustion and ventilating air intake must be where no flammable fluids or vapors can enter the heater system under any operating condition—

(1) During normal operation; or

(2) As a result of the malfunction of any other component.

(g) *Heater exhaust.* Each heater exhaust system must meet the requirements of §§ 29.1121 and 29.1123. In addition—

(1) Each exhaust shroud must be sealed so that no flammable fluids or hazardous quantities of vapors can reach the exhaust systems through joints; and

(2) No exhaust system may restrict the prompt relief of any backfire that, if so restricted, could cause heater failure.

(h) *Heater fuel systems.* Each heater fuel system must meet the powerplant fuel system requirements affecting safe heater operation. Each heater fuel system component in the ventilating airstream must be protected by shrouds so that no leakage from those components can enter the ventilating airstream.

(i) *Drains.* There must be means for safe drainage of any fuel that might accumulate in the combustion chamber or the heat exchanger. In addition—

(1) Each part of any drain that operates at high temperatures must be protected in the same manner as heater exhausts; and

(2) Each drain must be protected against hazardous ice accumulation under any operating condition.

[Doc. No. 5084, 29 FR 16150, Dec. 3, 1964, as amended by Amdt. 29-2, 32 FR 6914, May 5, 1967]

§ 29.861 Fire protection of structure, controls, and other parts.

Each part of the structure, controls, and the rotor mechanism, and other parts essential to controlled landing and (for category A) flight that would be affected by powerplant fires must be isolated under § 29.1191, or must be—

(a) For category A rotorcraft, fireproof; and

(b) For Category B rotorcraft, fireproof or protected so that they can perform their essential functions for at least 5 minutes under any foreseeable powerplant fire conditions.

[Doc. No. 5084, 29 FR 16150, Dec. 3, 1964, as amended by Amdt. 27-26, 55 FR 8005, Mar. 6, 1990]

§ 29.863 Flammable fluid fire protection.

(a) In each area where flammable fluids or vapors might escape by leakage of a fluid system, there must be means to minimize the probability of ignition of the fluids and vapors, and the resultant hazards if ignition does occur.

(b) Compliance with paragraph (a) of this section must be shown by analysis or tests, and the following factors must be considered:

(1) Possible sources and paths of fluid leakage, and means of detecting leakage.

(2) Flammability characteristics of fluids, including effects of any combustible or absorbing materials.

(3) Possible ignition sources, including electrical faults, overheating of equipment, and malfunctioning of protective devices.

(4) Means available for controlling or extinguishing a fire, such as stopping flow of fluids, shutting down equipment, fireproof containment, or use of extinguishing agents.

(5) Ability of rotorcraft components that are critical to safety of flight to withstand fire and heat.

(c) If action by the flight crew is required to prevent or counteract a fluid fire (e.g. equipment shutdown or actuation of a fire extinguisher), quick acting means must be provided to alert the crew.

(d) Each area where flammable fluids or vapors might escape by leakage of a fluid system must be identified and defined.

(Secs. 313(a), 601, 603, 604, Federal Aviation Act of 1958 (49 U.S.C. 1354(a), 1421, 1423, 1424), sec. 6(c), Dept. of Transportation Act (49 U.S.C. 1655(c)))

[Amdt. 29-17, 43 FR 50600, Oct. 30, 1978]

External Loads

§ 29.865 External loads.

(a) It must be shown by analysis, test, or both, that the rotorcraft external load attaching means for rotorcraft-load combinations to be used for nonhuman external cargo applications can withstand a limit static load equal to 2.5, or some lower load factor approved under §§ 29.337 through 29.341, multiplied by the maximum external load for which authorization is requested. It must be shown by analysis, test, or both that the rotorcraft external load attaching means and corresponding personnel carrying device system for rotorcraft-load combinations to be used for human external cargo applications can withstand a limit static load equal to 3.5 or some lower load factor, not less than 2.5, approved under §§ 29.337 through 29.341, multiplied by the maximum external load for which authorization is requested. The load for any rotorcraft-load combination class, for any external cargo type, must be applied in the vertical direction. For jettisonable external loads of any applicable external cargo type, the load must also be applied in any direction making the maximum angle with the vertical that can be achieved in service but not less than 30°. However, the 30° angle may be reduced to a lesser angle if—

(1) An operating limitation is established limiting external load operations to such angles for which compliance with this paragraph has been shown; or

(2) It is shown that the lesser angle can not be exceeded in service.

(b) The external load attaching means, for jettisonable rotorcraft-load combinations, must include a quick-release system to enable the pilot to release the external load quickly during flight. The quick-release system must consist of a primary quick release subsystem and a backup quick release subsystem that are isolated from one another. The quick release system, and the means by which it is controlled, must comply with the following:

(1) A control for the primary quick release subsystem must be installed either on one of the pilot's primary controls or in an equivalently accessible location and must be designed and located so that it may be operated by either the pilot or a crewmember without hazardously limiting the ability to control the rotorcraft during an emergency situation.

(2) A control for the backup quick release subsystem, readily accessible to either the pilot or another crewmember, must be provided.

(3) Both the primary and backup quick release subsystems must—

(i) Be reliable, durable, and function properly with all external loads up to and including the maximum external limit load for which authorization is requested.

(ii) Be protected against electromagnetic interference (EMI) from external and internal sources and against lightning to prevent inadvertent load release.

(A) The minimum level of protection required for jettisonable rotorcraft-load combinations used for nonhuman external cargo is a radio frequency field strength of 20 volts per meter.

(B) The minimum level of protection required for jettisonable rotorcraft-load combinations used for human external cargo is a radio frequency field strength of 200 volts per meter.

(iii) Be protected against any failure that could be induced by a failure mode of any other electrical or mechanical rotorcraft system.

(c) For rotorcraft-load combinations to be used for human external cargo applications, the rotorcraft must—

(1) For jettisonable external loads, have a quick-release system that meets the requirements of paragraph (b) of this section and that—

(i) Provides a dual actuation device for the primary quick release subsystem, and

(ii) Provides a separate dual actuation device for the backup quick release subsystem;

(2) Have a reliable, approved personnel carrying device system that has the structural capability and personnel safety features essential for external occupant safety;

(3) Have placards and markings at all appropriate locations that clearly state the essential system operating instructions and, for the personnel carrying device system, ingress and egress instructions;

(4) Have equipment to allow direct intercommunication among required crewmembers and external occupants;

(5) Have the appropriate limitations and procedures incorporated in the flight manual for conducting human external cargo operations; and

(6) For human external cargo applications requiring use of Category A rotorcraft, have one-engine-inoperative hover performance data and procedures in the flight manual for the weights, altitudes, and temperatures for which external load approval is requested.

(d) The critically configured jettisonable external loads must be shown by a combination of analysis, ground tests, and flight tests to be both transportable and releasable throughout the approved operational envelope without hazard to the rotorcraft during normal flight conditions. In addition, these external loads—must be shown to be releasable without hazard to the rotorcraft during emergency flight conditions.

(e) A placard or marking must be installed next to the external-load attaching means clearly stating any operational limitations and the maximum authorized external load as demonstrated under § 29.25 and this section.

(f) The fatigue evaluation of § 29.571 of this part does not apply to rotorcraft-load combinations to be used for nonhuman external cargo except for the failure of critical structural elements that would result in a hazard to the rotorcraft. For rotorcraft-load combinations to be used for human external cargo, the fatigue evaluation of § 29.571 of this part applies to the entire quick release and personnel carrying device structural systems and their attachments.

[Amdt. 29-12, 41 FR 55472, Dec. 20, 1976, as amended by Amdt. 27-26, 55 FR 8005, Mar. 6, 1990; Amdt. 29-43, 64 FR 43020, Aug. 6, 1999]

Miscellaneous

§ 29.871 Leveling marks.

There must be reference marks for leveling the rotorcraft on the ground.

§ 29.873 Ballast provisions.

Ballast provisions must be designed and constructed to prevent inadvertent shifting of ballast in flight.

Subpart E—Powerplant

General

§ 29.901 Installation.

(a) For the purpose of this part, the powerplant installation includes each part of the rotorcraft (other than the main and auxiliary rotor structures) that—

(1) Is necessary for propulsion;

(2) Affects the control of the major propulsive units; or

(3) Affects the safety of the major propulsive units between normal inspections or overhauls.

(b) For each powerplant installation—

(1) The installation must comply with—

(i) The installation instructions provided under § 33.5 of this chapter; and

(ii) The applicable provisions of this subpart.

(2) Each component of the installation must be constructed, arranged, and installed to ensure its continued safe operation between normal inspections or overhauls for the range of temperature and altitude for which approval is requested.

(3) Accessibility must be provided to allow any inspection and maintenance necessary for continued airworthiness; and

(4) Electrical interconnections must be provided to prevent differences of potential between major components of the installation and the rest of the rotorcraft.

(5) Axial and radial expansion of turbine engines may not affect the safety of the installation.

(6) Design precautions must be taken to minimize the possibility of incorrect assembly of components and equipment essential to safe operation of the rotorcraft, except where operation with the incorrect assembly can be shown to be extremely improbable.

(c) For each powerplant and auxiliary power unit installation, it must be established that no single failure or malfunction or probable combination of failures will jeopardize the safe operation of the rotorcraft except that the failure of structural elements need not be considered if the probability of any such failure is extremely remote.

(d) Each auxiliary power unit installation must meet the applicable provisions of this subpart.

(Secs. 313(a), 601, 603, 604, Federal Aviation Act of 1958 (49 U.S.C. 1354(a), 1421, 1423, 1424), sec. 6(c), Dept. of Transportation Act (49 U.S.C. 1655(c)))

[Doc. No. 5084, 29 FR 16150, Dec. 3, 1964, as amended by Amdt. 29-3, 33 FR 969, Jan. 26, 1968; Amdt. 29-13, 42 FR 15046, Mar. 17, 1977; Amdt. 29-17, 43 FR 50600, Oct. 30, 1978; Amdt. 29-26, 53 FR 34215, Sept. 2, 1988; Amdt. 29-36, 60 FR 55776, Nov. 2, 1995]

§ 29.903 Engines.

(a) *Engine type certification.* Each engine must have an approved type certificate. Reciprocating engines for use in helicopters must be qualified in accordance with § 33.49(d) of this chapter or be otherwise approved for the intended usage.

(b) *Category A; engine isolation.* For each category A rotorcraft, the powerplants must be arranged and isolated from each other to allow operation, in at least one configuration, so that the failure or malfunction of any engine, or the failure of any system that can affect any engine, will not—

(1) Prevent the continued safe operation of the remaining engines; or

(2) Require immediate action, other than normal pilot action with primary flight controls, by any crewmember to maintain safe operation.

(c) *Category A; control of engine rotation.* For each Category A rotorcraft, there must be a means for stopping the rotation of any engine individually in flight, except that, for turbine engine installations, the means for stopping the engine need be provided only where necessary for safety. In addition—

(1) Each component of the engine stopping system that is located on the engine side of the firewall, and that might be exposed to fire, must be at least fire resistant; or

(2) Duplicate means must be available for stopping the engine and the controls must be where all are not likely to be damaged at the same time in case of fire.

(d) *Turbine engine installation.* For turbine engine installations—

(1) Design precautions must be taken to minimize the hazards to the rotorcraft in the event of an engine rotor failure; and

(2) The powerplant systems associated with engine control devices, systems, and instrumentation must be designed to give reasonable assurance that those engine operating limitations that adversely affect engine rotor structural integrity will not be exceeded in service.

(e) *Restart capability.* (1) A means to restart any engine in flight must be provided.

(2) Except for the in-flight shutdown of all engines, engine restart capability must be demonstrated throughout a flight envelope for the rotorcraft.

(3) Following the in-flight shutdown of all engines, in-flight engine restart capability must be provided.

(Secs. 313(a), 601, and 603, 72 Stat. 752, 775, 49 U.S.C. 1354(a), 1421, and 1423; sec. 6(c), 49 U.S.C. 1655(c))

[Doc. No. 5084, 29 FR 16150, Dec. 3, 1964, as amended by Amdt. 29-12, 41 FR 55472, Dec. 20, 1976; Amdt. 29-26, 53 FR 34215, Sept. 2, 1988; Amdt. 29-31, 55 FR 38967, Sept. 21, 1990; 55 FR 41309, Oct. 10, 1990; Amdt. 29-36, 60 FR 55776, Nov. 2, 1995]

§ 29.907 Engine vibration.

(a) Each engine must be installed to prevent the harmful vibration of any part of the engine or rotorcraft.

(b) The addition of the rotor and the rotor drive system to the engine may not subject the principal rotating parts of the engine to excessive vibration stresses. This must be shown by a vibration investigation.

§ 29.908 Cooling fans.

For cooling fans that are a part of a powerplant installation the following apply:

(a) *Category A.* For cooling fans installed in Category A rotorcraft, it must be shown that a fan blade failure will not prevent continued safe flight either because of damage caused by the failed blade or loss of cooling air.

(b) *Category B.* For cooling fans installed in category B rotorcraft, there must be means to protect the rotorcraft and allow a safe landing if a fan blade fails. It must be shown that—

(1) The fan blade would be contained in the case of a failure;

(2) Each fan is located so that a fan blade failure will not jeopardize safety; or

(3) Each fan blade can withstand an ultimate load of 1.5 times the centrifugal force expected in service, limited by either—

(i) The highest rotational speeds achievable under uncontrolled conditions; or

(ii) An overspeed limiting device.

(c) *Fatigue evaluation.* Unless a fatigue evaluation under § 29.571 is conducted, it must be shown that cooling fan blades are not operating at resonant conditions within the operating limits of the rotorcraft.

(Secs. 313(a), 601, and 603, 72 Stat. 752, 775, 49 U.S.C. 1354(a), 1421, and 1423; sec. 6(c), 49 U.S.C. 1655 (c))

[Amdt. 29-13, 42 FR 15046, Mar. 17, 1977, as amended by Amdt. 29-26, 53 FR 34215, Sept. 2, 1988]

Rotor Drive System

§ 29.917 Design.

(a) *General.* The rotor drive system includes any part necessary to transmit power from the engines to the rotor hubs. This includes gear boxes, shafting, universal joints, couplings, rotor brake assemblies, clutches, supporting bearings for shafting, any attendant accessory pads or drives, and any cooling fans that are a part of, attached to, or mounted on the rotor drive system.

(b) *Design assessment.* A design assessment must be performed to ensure that the rotor drive system functions safely over the full range of conditions for which certification is sought. The design assessment must include a detailed failure analysis to identify all failures that will prevent continued safe flight or safe landing and must identify the means to minimize the likelihood of their occurrence.

(c) *Arrangement.* Rotor drive systems must be arranged as follows:

(1) Each rotor drive system of multiengine rotorcraft must be arranged so that each rotor necessary for operation and control will continue to be driven by the remaining engines if any engine fails.

(2) For single-engine rotorcraft, each rotor drive system must be so arranged that each rotor necessary for control in autorotation will continue to be driven by the main rotors after disengagement of the engine from the main and auxiliary rotors.

(3) Each rotor drive system must incorporate a unit for each engine to automatically disengage that engine from the main and auxiliary rotors if that engine fails.

(4) If a torque limiting device is used in the rotor drive system, it must be located so as to allow continued control of the rotorcraft when the device is operating.

(5) If the rotors must be phased for intermeshing, each system must provide constant and positive phase relationship under any operating condition.

(6) If a rotor dephasing device is incorporated, there must be means to keep the rotors locked in proper phase before operation.

[Doc. No. 5084, 29 FR 16150, Dec. 3, 1964, as amended by Amdt. 29-12, 41 FR 55472, Dec. 20, 1976; Amdt. 29-40, 61 FR 21908, May 10, 1996]

§ 29.921 Rotor brake.

If there is a means to control the rotation of the rotor drive system independently of the engine, any limitations on the use of that means must be specified, and the control for that means must be guarded to prevent inadvertent operation.

§ 29.923 Rotor drive system and control mechanism tests.

(a) *Endurance tests, general.* Each rotor drive system and rotor control mechanism must be tested, as prescribed in paragraphs (b) through (n) and (p) of this section, for at least 200 hours plus the time required to meet the requirements of paragraphs (b)(2), (b)(3), and (k) of this section. These tests must be conducted as follows:

(1) Ten-hour test cycles must be used, except that the test cycle must be extended to include the OEI test of paragraphs (b)(2) and (k), of this section if OEI ratings are requested.

(2) The tests must be conducted on the rotorcraft.

(3) The test torque and rotational speed must be—

(i) Determined by the powerplant limitations; and

(ii) Absorbed by the rotors to be approved for the rotorcraft.

(b) *Endurance tests; takeoff run.* The takeoff run must be conducted as follows:

(1) Except as prescribed in paragraphs (b)(2) and (b)(3) of this section, the takeoff torque run must consist of 1 hour of alternate runs of 5 minutes at takeoff torque and the maximum speed for use with takeoff torque, and 5 minutes at as low an engine idle speed as practicable. The engine must be declutched from the rotor drive system, and the rotor brake, if furnished and so intended, must be applied during the first minute of the idle run. During the remaining 4 minutes of the idle run, the clutch must be engaged so that the engine drives the rotors at the minimum practical r.p.m. The engine and the rotor drive system must be accelerated at the maximum rate. When declutching the engine, it must be decelerated rapidly enough to allow the operation of the overrunning clutch.

(2) For helicopters for which the use of a $2\frac{1}{2}$ -minute OEI rating is requested, the takeoff run must be conducted as prescribed in paragraph (b)(1) of this section, except for the third and sixth runs for which the takeoff torque and the maximum speed for use with takeoff torque are prescribed in that paragraph. For these runs, the following apply:

(i) Each run must consist of at least one period of $2\frac{1}{2}$ minutes with takeoff torque and the maximum speed for use with takeoff torque on all engines.

(ii) Each run must consist of at least one period, for each engine in sequence, during which that engine simulates a power failure and the remaining engines are run at the $2\frac{1}{2}$ -minute OEI torque and the maximum speed for use with $2\frac{1}{2}$ -minute OEI torque for $2\frac{1}{2}$ minutes.

(3) For multiengine, turbine-powered rotorcraft for which the use of 30-second/2-minute OEI power is requested, the takeoff run must be conducted as prescribed in paragraph (b)(1) of this section except for the following:

(i) Immediately following any one 5-minute power-on run required by paragraph (b)(1) of this section, simulate a failure for each power source in turn, and apply the maximum torque and the maximum speed for use with 30-second OEI power to the remaining affected drive system power inputs for not less than 30 seconds. Each application of 30-second OEI power must be followed by two applications of the maximum torque and the maximum speed for use with the 2 minute OEI power for not less than 2 minutes each; the second application must follow a period at stabilized continuous or 30 minute OEI power (whichever is requested by the applicant). At least one run sequence must be conducted from a simulated "flight idle" condition. When conducted on a bench test, the test sequence must be conducted following stabilization at take-off power.

(ii) For the purpose of this paragraph, an affected power input includes all parts of the rotor drive system which can be adversely affected by the application of higher or asymmetric torque and speed prescribed by the test.

(iii) This test may be conducted on a representative bench test facility when engine limitations either preclude repeated use of this power or would result in premature engine removals during the test. The loads, the vibration frequency, and the methods of application to the affected rotor drive system components must be representative of rotorcraft conditions. Test components must be those used to show compliance with the remainder of this section.

(c) *Endurance tests; maximum continuous run.* Three hours of continuous operation at maximum continuous torque and the maximum speed for use with maximum continuous torque must be conducted as follows:

(1) The main rotor controls must be operated at a minimum of 15 times each hour through the main rotor pitch positions of maximum vertical thrust, maximum forward thrust component, maximum aft thrust component, maximum left thrust component, and maximum right thrust component, except that the control movements need not produce loads or blade flapping motion exceeding the maximum loads of motions encountered in flight.

(2) The directional controls must be operated at a minimum of 15 times each hour through the control extremes of maximum right turning torque, neutral torque as required by the power applied to the main rotor, and maximum left turning torque.

(3) Each maximum control position must be held for at least 10 seconds, and the rate of change of control position must be at least as rapid as that for normal operation.

(d) *Endurance tests; 90 percent of maximum continuous run.* One hour of continuous operation at 90 percent of maximum continuous torque and the maximum speed for use with 90 percent of maximum continuous torque must be conducted.

(e) *Endurance tests; 80 percent of maximum continuous run.* One hour of continuous operation at 80 percent of maximum continuous torque and the minimum speed for use with 80 percent of maximum continuous torque must be conducted.

(f) *Endurance tests; 60 percent of maximum continuous run.* Two hours or, for helicopters for which the use of either 30-minute OEI power or continuous OEI power is requested, 1 hour of continuous operation at 60 percent of maximum continuous torque and the minimum speed for use with 60 percent of maximum continuous torque must be conducted.

(g) *Endurance tests; engine malfunctioning run.* It must be determined whether malfunctioning of components, such as the engine fuel or ignition systems, or whether unequal engine power can cause dynamic conditions detrimental to the drive system. If so, a suitable number of hours of operation must be accomplished under those conditions, 1 hour of which must be included in each cycle, and the remaining hours of which must be accomplished at the end of the 20 cycles. If no detrimental condition results, an additional hour of operation in compliance with paragraph (b) of this section must be conducted in accordance with the run schedule of paragraph (b)(1) of this section without consideration of paragraph (b)(2) of this section.

(h) *Endurance tests; overspeed run.* One hour of continuous operation must be conducted at maximum continuous torque and the maximum power-on overspeed expected in service, assuming that speed and torque limiting devices, if any, function properly.

(i) *Endurance tests; rotor control positions.* When the rotor controls are not being cycled during the tie-down tests, the rotor must be operated, using the procedures prescribed in paragraph (c) of this section, to produce each of the maximum thrust positions for the following percentages of test time (except that the control positions need not produce loads or blade flapping motion exceeding the maximum loads or motions encountered in flight):

(1) For full vertical thrust, 20 percent.

(2) For the forward thrust component, 50 percent.

(3) For the right thrust component, 10 percent.

(4) For the left thrust component, 10 percent.

(5) For the aft thrust component, 10 percent.

(j) *Endurance tests, clutch and brake engagements.* A total of at least 400 clutch and brake engagements, including the engagements of paragraph (b) of this section, must be made during the takeoff torque runs and, if necessary, at each change of torque and speed throughout the test. In each clutch engagement, the shaft on the driven side of the clutch must be accelerated from rest. The clutch engagements must be accomplished at the speed and by the method prescribed by the applicant. During deceleration after each clutch engagement, the engines must be stopped rapidly enough to allow the engines to be automatically disengaged from the rotors and rotor drives. If a rotor brake is installed for stopping the rotor, the clutch, during brake engagements, must be disengaged above 40 percent of maximum continuous rotor speed and the rotors allowed to decelerate to 40 percent of maximum continuous rotor speed, at which time the rotor brake must be applied. If the clutch design does not allow stopping the rotors with the engine running, or if no clutch is provided, the engine must be stopped before each application of the rotor brake, and then immediately be started after the rotors stop.

(k) *Endurance tests; OEI power run*—(1) *30-minute OEI power run.* For rotorcraft for which the use of 30-minute OEI power is requested, a run at 30-minute OEI torque and the maximum speed for use with 30-minute OEI torque must be conducted as follows: For each engine, in sequence, that engine must be inoperative and the remaining engines must be run for a 30-minute period.

(2) *Continuous OEI power run.* For rotorcraft for which the use of continuous OEI power is requested, a run at continuous OEI torque and the maximum speed for use with continuous OEI torque must be conducted as follows: For each engine, in sequence, that engine must be inoperative and the remaining engines must be run for 1 hour.

(3) The number of periods prescribed in paragraph (k)(1) or (k)(2) of this section may not be less than the number of engines, nor may it be less than two.

(l) [Reserved]

(m) Any components that are affected by maneuvering and gust loads must be investigated for the same flight conditions as are the main rotors, and their service lives must be determined by fatigue tests or by other acceptable methods. In addition, a level of safety equal to that of the main rotors must be provided for—

(1) Each component in the rotor drive system whose failure would cause an uncontrolled landing;

(2) Each component essential to the phasing of rotors on multirotor rotorcraft, or that furnishes a driving link for the essential control of rotors in autorotation; and

(3) Each component common to two or more engines on multiengine rotorcraft.

(n) *Special tests.* Each rotor drive system designed to operate at two or more gear ratios must be subjected to special testing for durations necessary to substantiate the safety of the rotor drive system.

(o) Each part tested as prescribed in this section must be in a serviceable condition at the end of the tests. No intervening disassembly which might affect test results may be conducted.

(p) *Endurance tests; operating lubricants.* To be approved for use in rotor drive and control systems, lubricants must meet the specifications of lubricants used during the tests prescribed by this section. Additional or alternate lubricants may be qualified by equivalent testing or by comparative analysis of lubricant specifications and rotor drive and control system characteristics. In addition—

(1) At least three 10-hour cycles required by this section must be conducted with transmission and gearbox lubricant temperatures, at the location prescribed for measurement, not lower than the maximum operating temperature for which approval is requested;

(2) For pressure lubricated systems, at least three 10-hour cycles required by this section must be conducted with the lubricant pressure, at the location prescribed for measurement, not higher than the minimum operating pressure for which approval is requested; and

(3) The test conditions of paragraphs (p)(1) and (p)(2) of this section must be applied simultaneously and must be extended to include operation at any one-engine-inoperative rating for which approval is requested.

(Secs. 313(a), 601, 603, 604, Federal Aviation Act of 1958 (49 U.S.C. 1354(a), 1421, 1423, 1424), sec. 6(c), Dept. of Transportation Act (49 U.S.C. 1655(c)))

[Doc. No. 5084, 29 FR 16150, Dec. 3, 1964, as amended by Amdt. 29-1, 30 FR 8778, July 13, 1965; Amdt. 29-17, 43 FR 50600, Oct. 30, 1978; Amdt. 29-26, 53 FR 34215, Sept. 2, 1988; Amdt. 29-31, 55 FR 38967, Sept. 21, 1990; Amdt. 29-34, 59 FR 47768, Sept. 16, 1994; Amdt. 29-40, 61 FR 21908, May 10, 1996; Amdt. 29-42, 63 FR 43285, Aug. 12, 1998]

§ 29.927 Additional tests.

(a) Any additional dynamic, endurance, and operational tests, and vibratory investigations necessary to determine that the rotor drive mechanism is safe, must be performed.

(b) If turbine engine torque output to the transmission can exceed the highest engine or transmission torque limit, and that output is not directly controlled by the pilot under normal operating conditions (such as where the primary engine power control is accomplished through the flight control), the following test must be made:

(1) Under conditions associated with all engines operating, make 200 applications, for 10 seconds each, of torque that is at least equal to the lesser of—

(i) The maximum torque used in meeting § 29.923 plus 10 percent; or

(ii) The maximum torque attainable under probable operating conditions, assuming that torque limiting devices, if any, function properly.

(2) For multiengine rotorcraft under conditions associated with each engine, in turn, becoming inoperative, apply to the remaining transmission torque inputs the maximum torque attainable under probable operating conditions, assuming that torque limiting devices, if any, function properly. Each transmission input must be tested at this maximum torque for at least fifteen minutes.

(c) *Lubrication system failure.* For lubrication systems required for proper operation of rotor drive systems, the following apply:

(1) *Category A.* Unless such failures are extremely remote, it must be shown by test that any failure which results in loss of lubricant in any normal use lubrication

system will not prevent continued safe operation, although not necessarily without damage, at a torque and rotational speed prescribed by the applicant for continued flight, for at least 30 minutes after perception by the flightcrew of the lubrication system failure or loss of lubricant.

(2) *Category B.* The requirements of Category A apply except that the rotor drive system need only be capable of operating under autorotative conditions for at least 15 minutes.

(d) *Overspeed test.* The rotor drive system must be subjected to 50 overspeed runs, each 30±3 seconds in duration, at not less than either the higher of the rotational speed to be expected from an engine control device failure or 105 percent of the maximum rotational speed, including transients, to be expected in service. If speed and torque limiting devices are installed, are independent of the normal engine control, and are shown to be reliable, their rotational speed limits need not be exceeded. These runs must be conducted as follows:

(1) Overspeed runs must be alternated with stabilizing runs of from 1 to 5 minutes duration each at 60 to 80 percent of maximum continuous speed.

(2) Acceleration and deceleration must be accomplished in a period not longer than 10 seconds (except where maximum engine acceleration rate will require more than 10 seconds), and the time for changing speeds may not be deducted from the specified time for the overspeed runs.

(3) Overspeed runs must be made with the rotors in the flattest pitch for smooth operation.

(e) The tests prescribed in paragraphs (b) and (d) of this section must be conducted on the rotorcraft and the torque must be absorbed by the rotors to be installed, except that other ground or flight test facilities with other appropriate methods of torque absorption may be used if the conditions of support and vibration closely simulate the conditions that would exist during a test on the rotorcraft.

(f) Each test prescribed by this section must be conducted without intervening disassembly and, except for the lubrication system failure test required by paragraph (c) of this section, each part tested must be in a serviceable condition at the conclusion of the test.

(Secs. 313(a), 601, 603, 604, Federal Aviation Act of 1958 (49 U.S.C. 1354(a), 1421, 1423 1424), sec. 6(c), Dept. of Transportation Act (49 U.S.C. 1655(c)))

[Amdt. 29-3, 33 FR 969, Jan. 26, 1968, as amended by Amdt. 29-17, 43 FR 50601, Oct. 30, 1978; Amdt. 29-26, 53 FR 34216, Sept. 2, 1988]

§ 29.931 Shafting critical speed.

(a) The critical speeds of any shafting must be determined by demonstration except that analytical methods may be used if reliable methods of analysis are available for the particular design.

(b) If any critical speed lies within, or close to, the operating ranges for idling, power-on, and autorotative conditions, the stresses occurring at that speed must be within safe limits. This must be shown by tests.

(c) If analytical methods are used and show that no critical speed lies within the permissible operating ranges, the margins between the calculated critical speeds and the limits of the allowable operating ranges must be adequate to allow for possible variations between the computed and actual values.

[Amdt. 29-12, 41 FR 55472, Dec. 20, 1976]

§ 29.935 Shafting joints.

Each universal joint, slip joint, and other shafting joints whose lubrication is necessary for operation must have provision for lubrication.

§ 29.939 Turbine engine operating characteristics.

(a) Turbine engine operating characteristics must be investigated in flight to determine that no adverse characteristics (such as stall, surge, of flameout) are present, to a hazardous degree, during normal and emergency operation within the range of operating limitations of the rotorcraft and of the engine.

(b) The turbine engine air inlet system may not, as a result of airflow distortion during normal operation, cause vibration harmful to the engine.

(c) For governor-controlled engines, it must be shown that there exists no hazardous torsional instability of the drive system associated with critical combinations of power, rotational speed, and control displacement.

[Amdt. 29-2, 32 FR 6914, May 5, 1967, as amended by Amdt. 29-12, 41 FR 55473, Dec. 20, 1976]

Fuel System

§ 29.951 General.

(a) Each fuel system must be constructed and arranged to ensure a flow of fuel at a rate and pressure established for proper engine and auxiliary power unit functioning under any likely operating conditions, including the maneuvers for which certification is requested and during which the engine or auxiliary power unit is permitted to be in operation.

(b) Each fuel system must be arranged so that—

(1) No engine or fuel pump can draw fuel from more than one tank at a time; or

(2) There are means to prevent introducing air into the system.

(c) Each fuel system for a turbine engine must be capable of sustained operation throughout its flow and pressure range with fuel initially saturated with water at 80 degrees F. and having 0.75cc of free water per gallon added and cooled to the most critical condition for icing likely to be encountered in operation.

[Doc. No. 5084, 29 FR 16150, Dec. 3, 1964, as amended by Amdt. 29-10, 39 FR 35462, Oct. 1, 1974; Amdt. 29-12, 41 FR 55473, Dec. 20, 1976]

§ 29.952 Fuel system crash resistance.

Unless other means acceptable to the Administrator are employed to minimize the hazard of fuel fires to occupants following an otherwise survivable impact (crash landing), the fuel systems must incorporate the design features of this section. These systems must be shown to be capable of sustaining the static and dynamic deceleration loads of this section, considered as ultimate loads acting alone, measured at the system component's center of gravity without structural damage to the system components, fuel tanks, or their attachments that would leak fuel to an ignition source.

(a) *Drop test requirements.* Each tank, or the most critical tank, must be drop-tested as follows:

(1) The drop height must be at least 50 feet.

(2) The drop impact surface must be nondeforming.

(3) The tanks must be filled with water to 80 percent of the normal, full capacity.

(4) The tank must be enclosed in a surrounding structure representative of the installation unless it can be established that the surrounding structure is free of projections or other design features likely to contribute to upture of the tank.

(5) The tank must drop freely and impact in a horizontal position±10°.

(6) After the drop test, there must be no leakage.

(b) *Fuel tank load factors.* Except for fuel tanks located so that tank rupture with fuel release to either significant ignition sources, such as engines, heaters, and auxiliary power units, or occupants is extremely remote, each fuel tank must be designed and installed to retain its contents under the following ultimate inertial load factors, acting alone.

(1) For fuel tanks in the cabin:

(i) Upward—4g.

(ii) Forward—16g.

(iii) Sideward—8g.

(iv) Downward—20g.

(2) For fuel tanks located above or behind the crew or passenger compartment that, if loosened, could injure an occupant in an emergency landing:

(i) Upward—1.5g.

(ii) Forward—8g.

(iii) Sideward—2g.

(iv) Downward—4g.

(3) For fuel tanks in other areas:

(i) Upward—1.5g.

(ii) Forward—4g.

(iii) Sideward—2g.

(iv) Downward—4g.

(c) *Fuel line self-sealing breakaway couplings.* Self-sealing breakaway couplings must be installed unless hazardous relative motion of fuel system components to each other or to local rotorcraft structure is demonstrated to be extremely improbable or unless other means are provided. The couplings or equivalent devices must be installed at all fuel tank-to-fuel line connections, tank-to-tank interconnects, and at other points in the fuel system where local structural deformation could lead to the release of fuel.

(1) The design and construction of self-sealing breakaway couplings must incorporate the following design features:

(i) The load necessary to separate a breakaway coupling must be between 25 to 50 percent of the minimum ultimate failure load (ultimate strength) of the weakest component in the fluid-carrying line. The separation load must in no case be less than 300 pounds, regardless of the size of the fluid line.

(ii) A breakaway coupling must separate whenever its ultimate load (as defined in paragraph (c)(1)(i) of this section) is applied in the failure modes most likely to occur.

(iii) All breakaway couplings must incorporate design provisions to visually ascertain that the coupling is locked together (leak-free) and is open during normal installation and service.

(iv) All breakaway couplings must incorporate design provisions to prevent uncoupling or unintended closing due to operational shocks, vibrations, or accelerations.

(v) No breakaway coupling design may allow the release of fuel once the coupling has performed its intended function.

(2) All individual breakaway couplings, coupling fuel feed systems, or equivalent means must be designed, tested, installed, and maintained so inadvertent fuel shutoff in flight is improbable in accordance with § 29.955(a) and must comply with the fatigue evaluation requirements of § 29.571 without leaking.

(3) Alternate, equivalent means to the use of breakaway couplings must not create a survivable impact-induced load on the fuel line to which it is installed greater than 25 to 50 percent of the ultimate load (strength) of the weakest component in the line and must comply with the fatigue requirements of § 29.571 without leaking.

(d) *Frangible or deformable structural attachments.* Unless hazardous relative motion of fuel tanks and fuel system components to local rotorcraft structure is demonstrated to be extremely improbable in an otherwise survivable impact, frangible or locally deformable attachments of fuel tanks and fuel system components to local rotorcraft structure must be used. The attachment of fuel tanks and fuel system components to local rotorcraft structure, whether frangible or locally deformable, must be designed such that its separation or relative local deformation will occur without rupture or local tear-out of the fuel tank or fuel system component that will cause fuel leakage. The ultimate strength of frangible or deformable attachments must be as follows:

(1) The load required to separate a frangible attachment from its support structure, or deform a locally deformable attachment relative to its support structure, must be between 25 and 50 percent of the minimum ultimate load (ultimate strength) of the weakest component in the attached system. In no case may the load be less than 300 pounds.

(2) A frangible or locally deformable attachment must separate or locally deform as intended whenever its ultimate load (as defined in paragraph (d)(1) of this section) is applied in the modes most likely to occur.

(3) All frangible or locally deformable attachments must comply with the fatigue requirements of § 29.571.

(e) *Separation of fuel and ignition sources.* To provide maximum crash resistance, fuel must be located as far as practicable from all occupiable areas and from all potential ignition sources.

(f) *Other basic mechanical design criteria.* Fuel tanks, fuel lines, electrical wires, and electrical devices must be designed, constructed, and installed, as far as practicable, to be crash resistant.

(g) *Rigid or semirigid fuel tanks.* Rigid or semirigid fuel tank or bladder walls must be impact and tear resistant.

[Doc. No. 26352, 59 FR 50387, Oct. 3, 1994]

§ 29.953 Fuel system independence.

(a) For category A rotorcraft—

(1) The fuel system must meet the requirements of § 29.903(b); and

(2) Unless other provisions are made to meet paragraph (a)(1) of this section, the fuel system must allow fuel to be supplied to each engine through a system independent of those parts of each system supplying fuel to other engines.

(b) Each fuel system for a multiengine category B rotorcraft must meet the requirements of paragraph (a)(2) of this section. However, separate fuel tanks need not be provided for each engine.

§ 29.954 Fuel system lightning protection.

The fuel system must be designed and arranged to prevent the ignition of fuel vapor within the system by—

(a) Direct lightning strikes to areas having a high probability of stroke attachment;

(b) Swept lightning strokes to areas where swept strokes are highly probable; and

(c) Corona and streamering at fuel vent outlets.

[Amdt. 29-26, 53 FR 34217, Sept. 2, 1988]

§ 29.955 Fuel flow.

(a) *General.* The fuel system for each engine must provide the engine with at least 100 percent of the fuel required under all operating and maneuvering conditions to be approved for the rotorcraft, including, as applicable, the fuel required to operate the engines under the test conditions required by § 29.927. Unless equivalent methods are used, compliance must be shown by test during which the following provisions are met, except that combinations of conditions which are shown to be improbable need not be considered.

(1) The fuel pressure, corrected for accelerations (load factors), must be within the limits specified by the engine type certificate data sheet.

(2) The fuel level in the tank may not exceed that established as the unusable fuel supply for that tank under § 29.959, plus that necessary to conduct the test.

(3) The fuel head between the tank and the engine must be critical with respect to rotorcraft flight attitudes.

(4) The fuel flow transmitter, if installed, and the critical fuel pump (for pump-fed systems) must be installed to produce (by actual or simulated failure) the critical restriction to fuel flow to be expected from component failure.

(5) Critical values of engine rotational speed, electrical power, or other sources of fuel pump motive power must be applied.

(6) Critical values of fuel properties which adversely affect fuel flow are applied during demonstrations of fuel flow capability.

(7) The fuel filter required by § 29.997 is blocked to the degree necessary to simulate the accumulation of fuel contamination required to activate the indicator required by § 29.1305(a)(17).

(b) *Fuel transfer system.* If normal operation of the fuel system requires fuel to be transferred to another tank, the transfer must occur automatically via a system which has been shown to maintain the fuel level in the receiving tank within acceptable limits during flight or surface operation of the rotorcraft.

(c) *Multiple fuel tanks.* If an engine can be supplied with fuel from more than one tank, the fuel system, in addition to having appropriate manual switching capability, must be designed to prevent interruption of fuel flow to that engine, without attention by the flightcrew, when any tank supplying fuel to that engine is depleted of usable fuel during normal operation and any other tank that normally supplies fuel to that engine alone contains usable fuel.

[Amdt. 29-26, 53 FR 34217, Sept. 2, 1988]

§ 29.957 Flow between interconnected tanks.

(a) Where tank outlets are interconnected and allow fuel to flow between them due to gravity or flight accelerations, it must be impossible for fuel to flow between tanks in quantities great enough to cause overflow from the tank vent in any sustained flight condition.

(b) If fuel can be pumped from one tank to another in flight—

(1) The design of the vents and the fuel transfer system must prevent structural damage to tanks from overfilling; and

(2) There must be means to warn the crew before overflow through the vents occurs.

§ 29.959 Unusable fuel supply.

The unusable fuel supply for each tank must be established as not less than the quantity at which the first evidence of malfunction occurs under the most adverse fuel feed condition occurring under any intended operations and flight maneuvers involving that tank.

§ 29.961 Fuel system hot weather operation.

Each suction lift fuel system and other fuel systems conducive to vapor formation must be shown to operate satisfactorily (within certification limits) when using fuel at the most critical temperature for vapor formation under critical operating conditions including, if applicable, the engine operating conditions defined by § 29.927(b)(1) and (b)(2).

[Amdt. 29-26, 53 FR 34217, Sept. 2, 1988]

§ 29.963 Fuel tanks: general.

(a) Each fuel tank must be able to withstand, without failure, the vibration, inertia, fluid, and structural loads to which it may be subjected in operation.

(b) Each flexible fuel tank bladder or liner must be approved or shown to be suitable for the particular application and must be puncture resistant. Puncture resistance must be shown by meeting the TSO-C80, paragraph 16.0, requirements using a minimum puncture force of 370 pounds.

(c) Each integral fuel tank must have facilities for inspection and repair of its interior.

(d) The maximum exposed surface temperature of all components in the fuel tank must be less by a safe margin than the lowest expected autoignition temperature of the fuel or fuel vapor in the tank. Compliance with this requirement must be shown under all operating conditions and under all normal or malfunction conditions of all components inside the tank.

(e) Each fuel tank installed in personnel compartments must be isolated by fume-proof and fuel-proof enclosures that are drained and vented to the exterior of the rotorcraft. The design and construction of the enclosures must provide necessary protection for the tank, must be crash resistant during a survivable impact in accordance with § 29.952, and must be adequate to withstand loads and abrasions to be expected in personnel compartments.

[Doc. No. 5084, 29 FR 16150, Dec. 3, 1964, as amended by Amdt. 29-26, 53 FR 34217, Sept. 2, 1988; Amdt. 29-35, 59 FR 50388, Oct. 3, 1994]

§ 29.965 Fuel tank tests.

(a) Each fuel tank must be able to withstand the applicable pressure tests in this section without failure or leakage. If practicable, test pressures may be applied in a manner simulating the pressure distribution in service.

(b) Each conventional metal tank, each nonmetallic tank with walls that are not supported by the rotorcraft structure, and each integral tank must be subjected to a pressure of 3.5 p.s.i. unless the pressure developed during maximum limit acceleration or emergency deceleration with a full tank exceeds this value, in which case a hydrostatic head, or equivalent test, must be applied to duplicate the acceleration loads as far as possible. However, the pressure need not exceed 3.5 p.s.i. on surfaces not exposed to the acceleration loading.

(c) Each nonmetallic tank with walls supported by the rotorcraft structure must be subjected to the following tests:

(1) A pressure test of at least 2.0 p.s.i. This test may be conducted on the tank alone in conjunction with the test specified in paragraph (c)(2) of this section.

(2) A pressure test, with the tank mounted in the rotorcraft structure, equal to the load developed by the reaction of the contents, with the tank full, during maximum limit acceleration or emergency deceleration. However, the pressure need not exceed 2.0 p.s.i. on surfaces faces not exposed to the acceleration loading.

(d) Each tank with large unsupported or unstiffened flat areas, or with other features whose failure or deformation could cause leakage, must be subjected to the following test or its equivalent:

(1) Each complete tank assembly and its supports must be vibration tested while mounted to simulate the actual installation.

(2) The tank assembly must be vibrated for 25 hours while two-thirds full of any suitable fluid. The amplitude of vibration may not be less than one thirty-second of an inch, unless otherwise substantiated.

(3) The test frequency of vibration must be as follows:

(i) If no frequency of vibration resulting from any r.p.m. within the normal operating range of engine or rotor system speeds is critical, the test frequency of vibration, in number of cycles per minute, must, unless a frequency based on a more rational analysis is used, be the number obtained by averaging the maximum and minimum power-on engine speeds (r.p.m.) for reciprocating engine powered rotorcraft or 2,000 c.p.m. for turbine engine powered rotorcraft.

(ii) If only one frequency of vibration resulting from any r.p.m. within the normal operating range of engine or rotor system speeds is critical, that frequency of vibration must be the test frequency.

(iii) If more than one frequency of vibration resulting from any r.p.m. within the normal operating range of engine or rotor system speeds is critical, the most critical of these frequencies must be the test frequency.

(4) Under paragraph (d)(3)(ii) and (iii), the time of test must be adjusted to accomplish the same number of vibration cycles as would be accomplished in 25 hours at the frequency specified in paragraph (d)(3)(i) of this section.

(5) During the test, the tank assembly must be rocked at the rate of 16 to 20 complete cycles per minute through an angle of 15 degrees on both sides of the horizontal (30 degrees total), about the most critical axis, for 25 hours. If motion about more than one axis is likely to be critical, the tank must be rocked about each critical axis for 12½ hours.

(Secs. 313(a), 601, and 603, 72 Stat. 752, 775, 49 U.S.C. 1354(a), 1421, and 1423; sec. 6(c), 49 U.S.C. 1655 (c))

[Doc. No. 5084, 29 FR 16150, Dec. 3, 1964, as amended by Amdt. 29-13, 42 FR 15046, Mar. 17, 1977]

§ 29.967 Fuel tank installation.

(a) Each fuel tank must be supported so that tank loads are not concentrated on unsupported tank surfaces. In addition—

(1) There must be pads, if necessary, to prevent chafing between each tank and its supports;

(2) The padding must be nonabsorbent or treated to prevent the absorption of fuel;

(3) If flexible tank liners are used, they must be supported so that they are not required to withstand fluid loads; and

(4) Each interior surface of tank compartments must be smooth and free of projections that could cause wear of the liner, unless—

(i) There are means for protection of the liner at those points; or

(ii) The construction of the liner itself provides such protection.

(b) Any spaces adjacent to tank surfaces must be adequately ventilated to avoid accumulation of fuel or fumes in those spaces due to minor leakage. If the tank is in a sealed compartment, ventilation may be limited to drain holes that prevent clogging and that prevent excessive pressure resulting from altitude changes. If flexible tank liners are installed, the venting arrangement for the spaces between the liner and its container must maintain the proper relationship to tank vent pressures for any expected flight condition.

(c) The location of each tank must meet the requirements of § 29.1185(b) and (c).

(d) No rotorcraft skin immediately adjacent to a major air outlet from the engine compartment may act as the wall of an integral tank.

[Doc. No. 5084, 29 FR 16150, Dec. 3, 1964, as amended by Amdt. 29-26, 53 FR 34217, Sept. 2, 1988; Amdt. 29-35, 59 FR 50388, Oct. 3, 1994]

§ 29.969 Fuel tank expansion space.

Each fuel tank or each group of fuel tanks with interconnected vent systems must have an expansion space of not less than 2 percent of the combined tank capacity. It must be impossible to fill the fuel tank expansion space inadvertently with the rotorcraft in the normal ground attitude.

[Amdt. 29-26, 53 FR 34217, Sept. 2, 1988]

§ 29.971 Fuel tank sump.

(a) Each fuel tank must have a sump with a capacity of not less than the greater of—

(1) 0.10 per cent of the tank capacity; or

(2) $\frac{1}{16}$ gallon.

(b) The capacity prescribed in paragraph (a) of this section must be effective with the rotorcraft in any normal attitude, and must be located so that the sump contents cannot escape through the tank outlet opening.

(c) Each fuel tank must allow drainage of hazardous quantities of water from each part of the tank to the sump with the rotorcraft in any ground attitude to be expected in service.

(d) Each fuel tank sump must have a drain that allows complete drainage of the sump on the ground.

[Doc. No. 5084, 29 FR 16150, Dec. 3, 1964, as amended by Amdt. 29-12, 41 FR 55473, Dec. 20, 1976; Amdt. 29-26, 53 FR 34217, Sept. 2, 1988]

§ 29.973 Fuel tank filler connection.

(a) Each fuel tank filler connection must prevent the entrance of fuel into any part of the rotorcraft other than the tank itself during normal operations and must be crash resistant during a survivable impact in accordance with § 29.952(c). In addition—

(1) Each filler must be marked as prescribed in § 29.1557(c)(1);

(2) Each recessed filler connection that can retain any appreciable quantity of fuel must have a drain that discharges clear of the entire rotorcraft; and

(3) Each filler cap must provide a fuel-tight seal under the fluid pressure expected in normal operation and in a survivable impact.

(b) Each filler cap or filler cap cover must warn when the cap is not fully locked or seated on the filler connection.

[Doc. No. 26352, 59 FR 50388, Oct. 3, 1994]

§ 29.975 Fuel tank vents and carburetor vapor vents.

(a) *Fuel tank vents.* Each fuel tank must be vented from the top part of the expansion space so that venting is effective under normal flight conditions. In addition—

(1) The vents must be arranged to avoid stoppage by dirt or ice formation;

(2) The vent arrangement must prevent siphoning of fuel during normal operation;

(3) The venting capacity and vent pressure levels must maintain acceptable differences of pressure between the interior and exterior of the tank, during—

(i) Normal flight operation;

(ii) Maximum rate of ascent and descent; and

(iii) Refueling and defueling (where applicable);

(4) Airspaces of tanks with interconnected outlets must be interconnected;

(5) There may be no point in any vent line where moisture can accumulate with the rotorcraft in the ground attitude or the level flight attitude, unless drainage is provided;

(6) No vent or drainage provision may end at any point—

(i) Where the discharge of fuel from the vent outlet would constitute a fire hazard; or

(ii) From which fumes could enter personnel compartments; and

(7) The venting system must be designed to minimize spillage of fuel through the vents to an ignition source in the event of a rollover during landing, ground operations, or a survivable impact.

(b) *Carburetor vapor vents.* Each carburetor with vapor elimination connections must have a vent line to lead vapors back to one of the fuel tanks. In addition—

(1) Each vent system must have means to avoid stoppage by ice; and

(2) If there is more than one fuel tank, and it is necessary to use the tanks in a definite sequence, each vapor vent return line must lead back to the fuel tank used for takeoff and landing.

[Doc. No. 5084, 29 FR 16150, Dec. 3, 1964, as amended by Amdt. 29-26, 53 FR 34217, Sept. 2, 1988; Amdt. 29-35, 59 FR 50388, Oct. 3, 1994; Amdt. 29-42, 63 FR 43285, Aug. 12, 1998]

§ 29.977 Fuel tank outlet.

(a) There must be a fuel strainer for the fuel tank outlet or for the booster pump. This strainer must—

(1) For reciprocating engine powered airplanes, have 8 to 16 meshes per inch; and

(2) For turbine engine powered airplanes, prevent the passage of any object that could restrict fuel flow or damage any fuel system component.

(b) The clear area of each fuel tank outlet strainer must be at least five times the area of the outlet line.

(c) The diameter of each strainer must be at least that of the fuel tank outlet.

(d) Each finger strainer must be accessible for inspection and cleaning.

[Amdt. 29-12, 41 FR 55473, Dec. 20, 1976]

§ 29.979 Pressure refueling and fueling provisions below fuel level.

(a) Each fueling connection below the fuel level in each tank must have means to prevent the escape of hazardous quantities of fuel from that tank in case of malfunction of the fuel entry valve.

(b) For systems intended for pressure refueling, a means in addition to the normal means for limiting the tank content must be installed to prevent damage to the tank in case of failure of the normal means.

(c) The rotorcraft pressure fueling system (not fuel tanks and fuel tank vents) must withstand an ultimate load that is 2.0 times the load arising from the maximum pressure, including surge, that is likely to occur during fueling. The maximum surge pressure must be established with any combination of tank valves being either intentionally or inadvertently closed.

(d) The rotorcraft defueling system (not including fuel tanks and fuel tank vents) must withstand an ultimate load that is 2.0 times the load arising from the maximum permissible defueling pressure (positive or negative) at the rotorcraft fueling connection.

[Doc. No. 5084, 29 FR 16150, Dec. 3, 1964, as amended by Amdt. 29-12, 41 FR 55473, Dec. 20, 1976]

Fuel System Components

§ 29.991 Fuel pumps.

(a) Compliance with § 29.955 must not be jeopardized by failure of—

(1) Any one pump except pumps that are approved and installed as parts of a type certificated engine; or

(2) Any component required for pump operation except the engine served by that pump.

(b) The following fuel pump installation requirements apply:

(1) When necessary to maintain the proper fuel pressure—

(i) A connection must be provided to transmit the carburetor air intake static pressure to the proper fuel pump relief valve connection; and

(ii) The gauge balance lines must be independently connected to the carburetor inlet pressure to avoid incorrect fuel pressure readings.

(2) The installation of fuel pumps having seals or diaphragms that may leak must have means for draining leaking fuel.

(3) Each drain line must discharge where it will not create a fire hazard.

[Amdt. 29-26, 53 FR 34217, Sept. 2, 1988]

§ 29.993 Fuel system lines and fittings.

(a) Each fuel line must be installed and supported to prevent excessive vibration and to withstand loads due to fuel pressure, valve actuation, and accelerated flight conditions.

(b) Each fuel line connected to components of the rotorcraft between which relative motion could exist must have provisions for flexibility.

(c) Each flexible connection in fuel lines that may be under pressure or subjected to axial loading must use flexible hose assemblies.

(d) Flexible hose must be approved.

(e) No flexible hose that might be adversely affected by high temperatures may be used where excessive temperatures will exist during operation or after engine shutdown.

§ 29.995 Fuel valves.

In addition to meeting the requirements of § 29.1189, each fuel valve must—

(a) [Reserved]

(b) Be supported so that no loads resulting from their operation or from accelerated flight conditions are transmitted to the lines attached to the valve.

(Secs. 313(a), 601, and 603, 72 Stat. 759, 775, 49 U.S.C. 1354(a), 1421, and 1423; sec. 6(c), 49 U.S.C. 1655 (c))

[Doc. No. 5084, 29 FR 16150, Dec. 3, 1964, as amended by Amdt. 29-13, 42 FR 15046, Mar. 17, 1977]

§ 29.997 Fuel strainer or filter.

There must be a fuel strainer or filter between the fuel tank outlet and the inlet of the first fuel system component which is susceptible to fuel contamination, including but not limited to the fuel metering device or an engine positive displacement pump, whichever is nearer the fuel tank outlet. This fuel strainer or filter must—

(a) Be accessible for draining and cleaning and must incorporate a screen or element which is easily removable;

(b) Have a sediment trap and drain, except that it need not have a drain if the strainer or filter is easily removable for drain purposes;

(c) Be mounted so that its weight is not supported by the connecting lines or by the inlet or outlet connections of the strainer or filter inself, unless adequate strengh margins under all loading conditions are provided in the lines and connections; and

(d) Provide a means to remove from the fuel any contaminant which would jeopardize the flow of fuel through rotorcraft or engine fuel system components required for proper rotorcraft or engine fuel system operation.

[Amdt. 29-10, 39 FR 35462, Oct. 1, 1974, as amended by Amdt. 29-22, 49 FR 6850, Feb. 23, 1984; Amdt. 29-26, 53 FR 34217, Sept. 2, 1988]

§ 29.999 Fuel system drains.

(a) There must be at least one accessible drain at the lowest point in each fuel system to completely drain the system with the rotorcraft in any ground attitude to be expected in service.

(b) Each drain required by paragraph (a) of this section including the drains prescribed in § 29.971 must—

(1) Discharge clear of all parts of the rotorcraft;

(2) Have manual or automatic means to ensure positive closure in the off position; and

(3) Have a drain valve—

(i) That is readily accessible and which can be easily opened and closed; and

(ii) That is either located or protected to prevent fuel spillage in the event of a landing with landing gear retracted.

[Doc. No. 5084, 29 FR 16150, Dec. 3, 1964, as amended by Amdt. 29-12, 41 FR 55473, Dec. 20, 1976; Amdt. 29-26, 53 FR 34218, Sept. 2, 1988]

§ 29.1001 Fuel jettisoning.

If a fuel jettisoning system is installed, the following apply:

(a) Fuel jettisoning must be safe during all flight regimes for which jettisoning is to be authorized.

(b) In showing compliance with paragraph (a) of this section, it must be shown that—

(1) The fuel jettisoning system and its operation are free from fire hazard;

(2) No hazard results from fuel or fuel vapors which impinge on any part of the rotorcraft during fuel jettisoning; and

(3) Controllability of the rotorcraft remains satisfactory throughout the fuel jettisoning operation.

(c) Means must be provided to automatically prevent jettisoning fuel below the level required for an all-engine climb at maximum continuous power from sea level to 5,000 feet altitude and cruise thereafter for 30 minutes at maximum range engine power.

(d) The controls for any fuel jettisoning system must be designed to allow flight personnel (minimum crew) to safely interrupt fuel jettisoning during any part of the jettisoning operation.

(e) The fuel jettisoning system must be designed to comply with the powerplant installation requirements of § 29.901(c).

(f) An auxiliary fuel jettisoning system which meets the requirements of paragraphs (a), (b), (d), and (e) of this section may be installed to jettison additional fuel provided it has separate and independent controls.

[Amdt. 29-26, 53 FR 34218, Sept. 2, 1988]

Oil System

§ 29.1011 Engines: general.

(a) Each engine must have an independent oil system that can supply it with an appropriate quantity of oil at a temperature not above that safe for continuous operation.

(b) The usable oil capacity of each system may not be less than the product of the endurance of the rotorcraft under critical operating conditions and the maximum allowable oil consumption of the engine under the same conditions, plus a suitable margin to ensure adequate circulation and cooling. Instead of a rational analysis of endurance and consumption, a usable oil capacity of one gallon for each 40 gallons of usable fuel may be used for reciprocating engine installations.

(c) Oil-fuel ratios lower than those prescribed in paragraph (c) of this section may be used if they are substantiated by data on the oil consumption of the engine.

(d) The ability of the engine and oil cooling provisions to maintain the oil temperature at or below the maximum established value must be shown under the applicable requirements of §§ 29.1041 through 29.1049.

[Doc. No. 5084, 29 FR 16150, Dec. 3, 1964, as amended by Amdt. 29-26, 53 FR 34218, Sept. 2, 1988]

§ 29.1013 Oil tanks.

(a) *Installation.* Each oil tank installation must meet the requirements of § 29.967.

(b) *Expansion space.* Oil tank expansion space must be provided so that—

(1) Each oil tank used with a reciprocating engine has an expansion space of not less than the greater of 10 percent of the tank capacity or 0.5 gallon, and each oil tank used with a turbine engine has an expansion space of not less than 10 percent of the tank capacity;

(2) Each reserve oil tank not directly connected to any engine has an expansion space of not less than two percent of the tank capacity; and

(3) It is impossible to fill the expansion space inadvertently with the rotorcraft in the normal ground attitude.

(c) *Filler connections.* Each recessed oil tank filler connection that can retain any appreciable quantity of oil must have a drain that discharges clear of the entire rotorcraft. In addition—

(1) Each oil tank filler cap must provide an oil-tight seal under the pressure expected in operation;

(2) For category A rotorcraft, each oil tank filler cap or filler cap cover must incorporate features that provide a warning when caps are not fully locked or seated on the filler connection; and

(3) Each oil filler must be marked under § 29.1557(c)(2).

(d) *Vent.* Oil tanks must be vented as follows:

(1) Each oil tank must be vented from the top part of the expansion space to that venting is effective under all normal flight conditions.

(2) Oil tank vents must be arranged so that condensed water vapor that might freeze and obstruct the line cannot accumulate at any point;

(e) *Outlet.* There must be means to prevent entrance into the tank itself, or into the tank outlet, of any object that might obstruct the flow of oil through the system. No oil tank outlet may be enclosed by a screen or guard that would reduce the flow of oil below a safe value at any operating temperature. There must be a shutoff valve at the outlet of each oil tank used with a turbine engine unless the external portion of the oil system (including oil tank supports) is fireproof.

(f) *Flexible liners.* Each flexible oil tank liner must be approved or shown to be suitable for the particular installation.

[Doc. No. 5084, 29 FR 16150, Dec. 3, 1964, as amended by Amdt. 29-10, 39 FR 35462, Oct. 1, 1974]

§ 29.1015 Oil tank tests.

Each oil tank must be designed and installed so that—

(a) It can withstand, without failure, any vibration, inertia, and fluid loads to which it may be subjected in operation; and

(b) It meets the requirements of § 29.965, except that instead of the pressure specified in § 29.965(b)—

(1) For pressurized tanks used with a turbine engine, the test pressure may not be less than 5 p.s.i. plus the maximum operating pressure of the tank; and

(2) For all other tanks, the test pressure may not be less than 5 p.s.i.

[Doc. No. 5084, 29 FR 16150, Dec. 3, 1964, as amended by Amdt. 29-10, 39 FR 35462, Oct. 1, 1974]

§ 29.1017 Oil lines and fittings.

(a) Each oil line must meet the requirements of § 29.993.

(b) Breather lines must be arranged so that—

(1) Condensed water vapor that might freeze and obstruct the line cannot accumulate at any point;

(2) The breather discharge will not constitute a fire hazard if foaming occurs, or cause emitted oil to strike the pilot's windshield; and

(3) The breather does not discharge into the engine air induction system.

§ 29.1019 Oil strainer or filter.

(a) Each turbine engine installation must incorporate an oil strainer or filter through which all of the engine oil flows and which meets the following requirements:

(1) Each oil strainer or filter that has a bypass must be constructed and installed so that oil will flow at the normal rate through the rest of the system with the strainer or filter completely blocked.

(2) The oil strainer or filter must have the capacity (with respect to operating limitations established for the engine) to ensure that engine oil system functioning is not impaired when the oil is contaminated to a degree (with respect to particle size and density) that is greater than that established for the engine under Part 33 of this chapter.

(3) The oil strainer or filter, unless it is installed at an oil tank outlet, must incorporate a means to indicate contamination before it reaches the capacity established in accordance with paragraph (a)(2) of this section.

(4) The bypass of a strainer or filter must be constructed and installed so that the release of collected contaminants is minimized by appropriate location of the bypass to ensure that collected contaminants are not in the bypass flow path.

(5) An oil strainer or filter that has no bypass, except one that is installed at an oil tank outlet, must have a means to connect it to the warning system required in § 29.1305(a)(18).

(b) Each oil strainer or filter in a powerplant installation using reciprocating engines must be constructed and installed so that oil will flow at the normal rate through the rest of the system with the strainer or filter element completely blocked.

[Amdt. 29-10, 39 FR 35463, Oct. 1, 1974, as amended by Amdt. 29-22, 49 FR 6850, Feb. 23, 1984; Amdt. 29-26, 53 FR 34218, Sept. 2, 1988]

§ 29.1021 Oil system drains.

A drain (or drains) must be provided to allow safe drainage of the oil system. Each drain must—

(a) Be accessible; and

(b) Have manual or automatic means for positive locking in the closed position.

[Amdt. 29-22, 49 FR 6850, Feb. 23, 1984]

§ 29.1023 Oil radiators.

(a) Each oil radiator must be able to withstand any vibration, inertia, and oil pressure loads to which it would be subjected in operation.

(b) Each oil radiator air duct must be located, or equipped, so that, in case of fire, and with the airflow as it would be with and without the engine operating, flames cannot directly strike the radiator.

§ 29.1025 Oil valves.

(a) Each oil shutoff must meet the requirements of § 29.1189.

(b) The closing of oil shutoffs may not prevent autorotation.

(c) Each oil valve must have positive stops or suitable index provisions in the "on" and "off" positions and must be supported so that no loads resulting from its operation or from accelerated flight conditions are transmitted to the lines attached to the valve.

§ 29.1027 Transmission and gearboxes: general.

(a) The oil system for components of the rotor drive system that require continuous lubrication must be sufficiently independent of the lubrication systems of the engine(s) to ensure—

(1) Operation with any engine inoperative; and

(2) Safe autorotation.

(b) Pressure lubrication systems for transmissions and gearboxes must comply with the requirements of §§ 29.1013, paragraphs (c), (d), and (f) only, 29.1015, 29.1017, 29.1021, 29.1023, and 29.1337(d). In addition, the system must have—

(1) An oil strainer or filter through which all the lubricant flows, and must—

(i) Be designed to remove from the lubricant any contaminant which may damage transmission and drive system components or impede the flow of lubricant to a hazardous degree; and

(ii) Be equipped with a bypass constructed and installed so that—

(A) The lubricant will flow at the normal rate through the rest of the system with the strainer or filter completely blocked; and

(B) The release of collected contaminants is minimized by appropriate location of the bypass to ensure that collected contaminants are not in the bypass flowpath;

(iii) Be equipped with a means to indicate collection of contaminants on the filter or strainer at or before opening of the bypass;

(2) For each lubricant tank or sump outlet supplying lubrication to rotor drive systems and rotor drive system components, a screen to prevent entrance into the lubrication system of any object that might obstruct the flow of lubricant from the outlet to the filter required by paragraph (b)(1) of this section. The requirements of paragraph (b)(1) of this section do not apply to screens installed at lubricant tank or sump outlets.

(c) Splash type lubrication systems for rotor drive system gearboxes must comply with §§ 29.1021 and 29.1337(d).

[Amdt. 29-26, 53 FR 34218, Sept. 2, 1988]

Cooling

§ 29.1041 General.

(a) The powerplant and auxiliary power unit cooling provisions must be able to maintain the temperatures of powerplant components, engine fluids, and auxiliary power unit components and fluids within the temperature limits established for these components and fluids, under ground, water, and flight operating conditions for which certification is requested, and after normal engine or auxiliary power unit shutdown, or both.

(b) There must be cooling provisions to maintain the fluid temperatures in any power transmission within safe values under any critical surface (ground or water) and flight operating conditions.

(c) Except for ground-use-only auxiliary power units, compliance with paragraphs (a) and (b) of this section must be shown by flight tests in which the temperatures of selected powerplant component and auxiliary power unit component, engine, and transmission fluids are obtained under the conditions prescribed in those paragraphs.

[Doc. No. 5084, 29 FR 16150, Dec. 3, 1964, as amended by Amdt. 29-26, 53 FR 34218, Sept. 2, 1988]

§ 29.1043 Cooling tests.

(a) *General.* For the tests prescribed in § 29.1041(c), the following apply:

(1) If the tests are conducted under conditions deviating from the maximum ambient atmospheric temperature specified in paragraph (b) of this section, the recorded powerplant temperatures must be corrected under paragraphs (c) and (d) of this section, unless a more rational correction method is applicable.

(2) No corrected temperature determined under paragraph (a)(1) of this section may exceed established limits.

(3) The fuel used during the cooling tests must be of the minimum grade approved for the engines, and the mixture settings must be those used in normal operation.

(4) The test procedures must be as prescribed in §§ 29.1045 through 29.1049.

(5) For the purposes of the cooling tests, a temperature is "stabilized" when its rate of change is less than 2°F per minute.

(b) *Maximum ambient atmospheric temperature.* A maximum ambient atmospheric temperature corresponding to sea level conditions of at least 100 degrees F. must be established. The assumed temperature lapse rate is 3.6 degrees F. per thousand feet of altitude above sea level until a temperature of −69.7 degrees F. is reached, above which altitude the temperature is considered constant at −69.7 degrees F. However, for winterization installations, the applicant may select a maximum ambient atmospheric temperature corresponding to sea level conditions of less than 100 degrees F.

(c) *Correction factor (except cylinder barrels).* Unless a more rational correction applies, temperatures of engine fluids and powerplant components (except cylinder barrels) for which temperature limits are established, must be corrected by adding to them the difference between the maximum ambient atmospheric temperature and the temperature of the ambient air at the time of the first occurrence of the maximum component or fluid temperature recorded during the cooling test.

(d) *Correction factor for cylinder barrel temperatures.* Cylinder barrel temperatures must be corrected by adding to them 0.7 times the difference between the maximum ambient atmospheric temperature and the temperature of the ambient air at the time of the first occurrence of the maximum cylinder barrel temperature recorded during the cooling test.

(Secs. 313(a), 601, 603, 604, and 605 of the Federal Aviation Act of 1958 (49 U.S.C. 1354(a), 1421, 1423, 1424, and 1425); and sec. 6(c) of the Dept. of Transportation Act (49 U.S.C. 1655(c)))

[Doc. No. 5084, 29 FR 16150, Dec. 3, 1964, as amended by Amdt. 29-12, 41 FR 55473, Dec. 20, 1976; Amdt. 29-15, 43 FR 2327, Jan. 16, 1978; Amdt. 29-26, 53 FR 34218, Sept. 2, 1988]

§ 29.1045 Climb cooling test procedures.

(a) Climb cooling tests must be conducted under this section for—

(1) Category A rotorcraft; and

(2) Multiengine category B rotorcraft for which certification is requested under the category A powerplant installation requirements, and under the requirements of § 29.861(a) at the steady rate of climb or descent established under § 29.67(b).

(b) The climb or descent cooling tests must be conducted with the engine inoperative that produces the most adverse cooling conditions for the remaining engines and powerplant components.

(c) Each operating engine must—

(1) For helicopters for which the use of 30-minute OEI power is requested, be at 30-minute OEI power for 30 minutes, and then at maximum continuous power (or at full throttle when above the critical altitude);

(2) For helicopters for which the use of continuous OEI power is requested, be at continuous OEI power (or at full throttle when above the critical altitude); and

(3) For other rotorcraft, be at maximum continuous power (or at full throttle when above the critical altitude).

(d) After temperatures have stabilized in flight, the climb must be—

(1) Begun from an altitude not greater than the lower of—

(i) 1,000 feet below the engine critcal altitude; and

PART 29

FAR

(ii) 1,000 feet below the maximum altitude at which the rate of climb is 150 f.p.m; and

(2) Continued for at least five minutes after the occurrence of the highest temperature recorded, or until the rotorcraft reaches the maximum altitude for which certification is requested.

(e) For category B rotorcraft without a positive rate of climb, the descent must begin at the all-engine-critical altitude and end at the higher of—

(1) The maximum altitude at which level flight can be maintained with one engine operative; and

(2) Sea level.

(f) The climb or descent must be conducted at an airspeed representing a normal operational practice for the configuration being tested. However, if the cooling provisions are sensitive to rotorcraft speed, the most critical airspeed must be used, but need not exceed the speeds established under § 29.67(a)(2) or § 29.67(b). The climb cooling test may be conducted in conjunction with the takeoff cooling test of § 29.1047.

[Doc. No. 5084, 29 FR 16150, Dec. 3, 1964, as amended by Amdt. 29-26, 53 FR 34218, Sept. 2, 1988]

§ 29.1047 Takeoff cooling test procedures.

(a) *Category A.* For each category A rotorcraft, cooling must be shown during takeoff and subsequent climb as follows:

(1) Each temperature must be stabilized while hovering in ground effect with—

(i) The power necessary for hovering;

(ii) The appropriate cowl flap and shutter settings; and

(iii) The maximum weight.

(2) After the temperatures have stabilized, a climb must be started at the lowest practicable altitude and must be conducted with one engine inoperative.

(3) The operating engines must be at the greatest power for which approval is sought (or at full throttle when above the critical altitude) for the same period as this power is used in determining the takeoff climbout path under § 29.59.

(4) At the end of the time interval prescribed in paragraph (b)(3) of this section, the power must be changed to that used in meeting § 29.67(a)(2) and the climb must be continued for—

(i) Thirty minutes, if 30-minute OEI power is used; or

(ii) At least 5 minutes after the occurrence of the highest temperature recorded, if continuous OEI power or maximum continuous power is used.

(5) The speeds must be those used in determining the takeoff flight path under § 29.59.

(b) *Category B.* For each category B rotorcraft, cooling must be shown during takeoff and subsequent climb as follows:

(1) Each temperature must be stabilized while hovering in ground effect with—

(i) The power necessary for hovering;

(ii) The appropriate cowl flap and shutter settings; and

(iii) The maximum weight.

(2) After the temperatures have stabilized, a climb must be started at the lowest practicable altitude with takeoff power.

(3) Takeoff power must be used for the same time interval as takeoff power is used in determining the takeoff flight path under § 29.63.

(4) At the end of the time interval prescribed in paragraph (a)(3) of this section, the power must be reduced to maximum continuous power and the climb must be continued for at least five minutes after the occurance of the highest temperature recorded.

(5) The cooling test must be conducted at an airspeed corresponding to normal operating practice for the configuration being tested. However, if the cooling provisions are sensitive to rotorcraft speed, the most critical airspeed must be used, but need not exceed the speed for best rate of climb with maximum continuous power.

[Doc. No. 5084, 29 FR 16150, Dec. 3, 1964, as amended by Amdt. 29-1, 30 FR 8778, July 13, 1965; Amdt. 29-26, 53 FR 34219, Sept. 2, 1988]

§ 29.1049 Hovering cooling test procedures.

The hovering cooling provisions must be shown—

(a) At maximum weight or at the greatest weight at which the rotorcraft can hover (if less), at sea level, with the power required to hover but not more than maximum continuous power, in the ground effect in still air, until at least five minutes after the occurrence of the highest temperature recorded; and

(b) With maximum continuous power, maximum weight, and at the altitude resulting in zero rate of climb for this configuration, until at least five minutes after the occurrence of the highest temperature recorded.

Induction System

§ 29.1091 Air induction.

(a) The air induction system for each engine and auxiliary power unit must supply the air required by that engine and auxiliary power unit under the operating conditions for which certification is requested.

(b) Each engine and auxiliary power unit air induction system must provide air for proper fuel metering and mixture distribution with the induction system valves in any position.

(c) No air intake may open within the engine accessory section or within other areas of any powerplant compartment where emergence of backfire flame would constitute a fire hazard.

(d) Each reciprocating engine must have an alternate air source.

(e) Each alternate air intake must be located to prevent the entrance of rain, ice, or other foreign matter.

(f) For turbine engine powered rotorcraft and rotorcraft incorporating auxiliary power units—

(1) There must be means to prevent hazardous quantities of fuel leakage or overflow from drains, vents, or other components of flammable fluid systems from entering the engine or auxiliary power unit intake system; and

(2) The air inlet ducts must be located or protected so as to minimize the ingestion of foreign matter during takeoff, landing, and taxiing.

(Secs. 313(a), 601, 603, 604, Federal Aviation Act of 1958 (49 U.S.C. 1354(a), 1421, 1423, 1424), sec. 6(c), Dept. of Transportation Act (49 U.S.C. 1655(c)))

[Doc. No. 5084, 29 FR 16150, Dec. 3, 1964, as amended by Amdt. 29-3, 33 FR 969, Jan. 26, 1968; Amdt. 29-17, 43 FR 50601, Oct. 30, 1978]

§ 29.1093 Induction system icing protection.

(a) *Reciprocating engines.* Each reciprocating engine air induction system must have means to prevent and eliminate icing. Unless this is done by other means, it must be shown that, in air free of visible moisture at a temperature of 30°F., and with the engines at 60 percent of maximum continuous power—

(1) Each rotorcraft with sea level engines using conventional venturi carburetors has a preheater that can provide a heat rise of 90°F.;

(2) Each rotorcraft with sea level engines using carburetors tending to prevent icing has a preheater that can provide a heat rise of 70°F.;

(3) Each rotorcraft with altitude engines using conventional venturi carburetors has a preheater that can provide a heat rise of 120°F.; and

(4) Each rotorcraft with altitude engines using carburetors tending to prevent icing has a preheater that can provide a heat rise of 100°F.

(b) *Turbine engines.* (1) It must be shown that each turbine engine and its air inlet system can operate throughout the flight power range of the engine (including idling)—

(i) Without accumulating ice on engine or inlet system components that would adversely affect engine operation or cause a serious loss of power under the icing conditions specified in appendix C of this Part; and

(ii) In snow, both falling and blowing, without adverse effect on engine operation, within the limitations established for the rotorcraft.

(2) Each turbine engine must idle for 30 minutes on the ground, with the air bleed available for engine icing protection at its critical condition, without adverse effect, in an atmosphere that is at a temperature between 15° and 30°F (between -9° and -1°C) and has a liquid water content not less than 0.3 grams per cubic meter in the form of drops having a mean effective diameter not less than 20 microns, followed by momentary operation at takeoff power or thrust. During the 30 minutes of idle operation, the engine may be run up periodically to a moderate power or thrust setting in a manner acceptable to the Administrator.

(c) *Supercharged reciprocating engines.* For each engine having a supercharger to pressurize the air before it enters the carburetor, the heat rise in the air caused by that supercharging at any altitude may be utilized in determining compliance with paragraph (a) of this section if the heat rise utilized is that which will be available, automatically, for the applicable altitude and operation condition because of supercharging.

(Secs. 313(a), 601, and 603, 72 Stat. 752, 775, 49 U.S.C. 1354(a), 1421, and 1423; sec. 6(c), 49 U.S.C. 1655 (c))

[Amdt. 29-3, 33 FR 969, Jan. 26, 1968, as amended by Amdt. 29-12, 41 FR 55473, Dec. 20, 1976; Amdt. 29-13, 42 FR 15046, Mar. 17, 1977; Amdt. 29-22, 49 FR 6850, Feb. 23, 1984; Amdt. 29-26, 53 FR 34219, Sept. 2, 1988]

§ 29.1101 Carburetor air preheater design.

Each carburetor air preheater must be designed and constructed to—

(a) Ensure ventilation of the preheater when the engine is operated in cold air;

(b) Allow inspection of the exhaust manifold parts that it surrounds; and

(c) Allow inspection of critical parts of the preheater itself.

§ 29.1103 Induction systems ducts and air duct systems.

(a) Each induction system duct upstream of the first stage of the engine supercharger and of the auxiliary power unit compressor must have a drain to prevent the hazardous accumulation of fuel and moisture in the ground attitude. No drain may discharge where it might cause a fire hazard.

(b) Each duct must be strong enough to prevent induction system failure from normal backfire conditions.

(c) Each duct connected to components between which relative motion could exist must have means for flexibility.

(d) Each duct within any fire zone for which a fire-extinguishing system is required must be at least—

(1) Fireproof, if it passes through any firewall; or

(2) Fire resistant, for other ducts, except that ducts for auxiliary power units must be fireproof within the auxiliary power unit fire zone.

(e) Each auxiliary power unit induction system duct must be fireproof for a sufficient distance upstream of the auxiliary power unit compartment to prevent hot gas reverse flow from burning through auxiliary power unit ducts and entering any other compartment or area of the rotorcraft in which a hazard would be created resulting from the entry of hot gases. The materials used to form the remainder of the induction system duct and plenum chamber of the auxiliary power unit must be capable of resisting the maximum heat conditions likely to occur.

(f) Each auxiliary power unit induction system duct must be constructed of materials that will not absorb or trap hazardous quantities of flammable fluids that could be ignited in the event of a surge or reverse flow condition.

(Secs. 313(a), 601, 603, 604, Federal Aviation Act of 1958 (49 U.S.C. 1354(a), 1421, 1423, 1424), sec. 6(c), Dept. of Transportation Act (49 U.S.C. 1655(c)))

[Doc. No. 5084, 29 FR 16150, Dec. 3, 1964, as amended by Amdt. 29-17, 43 FR 50602, Oct. 30, 1978]

§ 29.1105 Induction system screens.

If induction system screens are used—

(a) Each screen must be upstream of the carburetor;

(b) No screen may be in any part of the induction system that is the only passage through which air can reach the engine, unless it can be deiced by heated air;

(c) No screen may be deiced by alcohol alone; and

(d) It must be impossible for fuel to strike any screen.

§ 29.1107 Inter-coolers and after-coolers.

Each inter-cooler and after-cooler must be able to withstand the vibration, inertia, and air pressure loads to which it would be subjected in operation.

§ 29.1109 Carburetor air cooling.

It must be shown under § 29.1043 that each installation using two-stage superchargers has means to maintain the air temperature, at the carburetor inlet, at or below the maximum established value.

Exhaust System

§ 29.1121 General.

For powerplant and auxiliary power unit installations the following apply:

(a) Each exhaust system must ensure safe disposal of exhaust gases without fire hazard or carbon monoxide contamination in any personnel compartment.

(b) Each exhaust system part with a surface hot enough to ignite flammable fluids or vapors must be located or shielded so that leakage from any system carrying flammable fluids or vapors will not result in a fire caused by impingement of the fluids or vapors on any part of the exhaust system including shields for the exhaust system.

(c) Each component upon which hot exhaust gases could impinge, or that could be subjected to high temperatures from exhaust system parts, must be fireproof. Each exhaust system component must be separated by a fireproof shield from adjacent parts of the rotorcraft that are outside the engine and auxiliary power unit compartments.

(d) No exhaust gases may discharge so as to cause a fire hazard with respect to any flammable fluid vent or drain.

(e) No exhaust gases may discharge where they will cause a glare seriously affecting pilot vision at night.

(f) Each exhaust system component must be ventilated to prevent points of excessively high temperature.

(g) Each exhaust shroud must be ventilated or insulated to avoid, during normal operation, a temperature high enough to ignite any flammable fluids or vapors outside the shroud.

(h) If significant traps exist, each turbine engine exhaust system must have drains discharging clear of the rotorcraft, in any normal ground and flight attitudes, to prevent fuel accumulation after the failure of an attempted engine start.

(Secs. 313(a), 601, and 603, 72 Stat. 752, 755, 49 U.S.C. 1354(a), 1421, and 1423; sec. 6(c), 49 U.S.C. 1655 (c))

[Doc. No. 5084, 29 FR 16150, Dec. 3, 1964, as amended by Amdt. 29-3, 33 FR 970, Jan. 26, 1968; Amdt. 29-13, 42 FR 15046, Mar. 17, 1977]

§ 29.1123 Exhaust piping.

(a) Exhaust piping must be heat and corrosion resistant, and must have provisions to prevent failure due to expansion by operating temperatures.

(b) Exhaust piping must be supported to withstand any vibration and inertia loads to which it would be subjected in operation.

(c) Exhaust piping connected to components between which relative motion could exist must have provisions for flexibility.

§ 29.1125 Exhaust heat exchangers.

For reciprocating engine powered rotorcraft the following apply:

(a) Each exhaust heat exchanger must be constructed and installed to withstand the vibration, inertia, and other loads to which it would be subjected in operation. In addition—

(1) Each exchanger must be suitable for continued operation at high temperatures and resistant to corrosion from exhaust gases;

(2) There must be means for inspecting the critical parts of each exchanger;

(3) Each exchanger must have cooling provisions wherever it is subject to contact with exhaust gases; and

(4) No exhaust heat exchanger or muff may have stagnant areas or liquid traps that would increase the probability of ignition of flammable fluids or vapors that might be present in case of the failure or malfunction of components carrying flammable fluids.

(b) If an exhaust heat exchanger is used for heating ventilating air used by personnel—

(1) There must be a secondary heat exchanger between the primary exhaust gas heat exchanger and the ventilating air system; or

(2) Other means must be used to prevent harmful contamination of the ventilating air.

[Doc. No. 5084, 29 FR 16150, Dec. 3, 1964, as amended by Amdt. 29-12, 41 FR 55473, Dec. 20, 1976; Amdt. 29-41, 62 FR 46173, Aug. 29, 1997]

Powerplant Controls and Accessories

§ 29.1141 Powerplant controls: general.

(a) Powerplant controls must be located and arranged under § 29.777 and marked under § 29.1555.

(b) Each control must be located so that it cannot be inadvertently operated by persons entering, leaving, or moving normally in the cockpit.

(c) Each flexible powerplant control must be approved.

(d) Each control must be able to maintain any set position without—

(1) Constant attention; or

(2) Tendency to creep due to control loads or vibration.

(e) Each control must be able to withstand operating loads without excessive deflection.

(f) Controls of powerplant valves required for safety must have—

(1) For manual valves, positive stops or in the case of fuel valves suitable index provisions, in the open and closed position; and

(2) For power-assisted valves, a means to indicate to the flight crew when the valve—

(i) Is in the fully open or fully closed position; or

(ii) Is moving between the fully open and fully closed position.

(Secs. 313(a), 601, and 603, 72 Stat. 752, 775, 49 U.S.C. 1354(a), 1421, and 1423; sec. 6(c), 49 U.S.C. 1655(c))

[Doc. No. 5084, 29 FR 16150, Dec. 3, 1964, as amended by Amdt. 29-13, 42 FR 15046, Mar. 17, 1977; Amdt. 29-26, 53 FR 34219, Sept. 2, 1988]

§ 29.1142 Auxiliary power unit controls.

Means must be provided on the flight deck for starting, stopping, and emergency shutdown of each installed auxiliary power unit.

(Secs. 313(a), 601, 603, 604, Federal Aviation Act of 1958 (49 U.S.C. 1354(a), 1421, 1423, 1424), sec. 6(c), Dept. of Transportation Act (49 U.S.C. 1655(c)))

[Amdt. 29-17, 43 FR 50602, Oct. 30, 1978]

§ 29.1143 Engine controls.

(a) There must be a separate power control for each engine.

(b) Power controls must be arranged to allow ready synchronization of all engines by—

(1) Separate control of each engine; and

(2) Simultaneous control of all engines.

(c) Each power control must provide a positive and immediately responsive means of controlling its engine.

(d) Each fluid injection control other than fuel system control must be in the corresponding power control. However, the injection system pump may have a separate control.

(e) If a power control incorporates a fuel shutoff feature, the control must have a means to prevent the inadvertent movement of the control into the shutoff position. The means must—

(1) Have a positive lock or stop at the idle position; and

(2) Require a separate and distinct operation to place the control in the shutoff position.

(f) For rotorcraft to be certificated for a 30-second OEI power rating, a means must be provided to automatically activate and control the 30-second OEI power and prevent any engine from exceeding the installed engine limits associated with the 30-second OEI power rating approved for the rotorcraft.

[Amdt. 29-26, 53 FR 34219, Sept. 2, 1988, as amended by Amdt. 29-34, 59 FR 47768, Sept. 16, 1994]

§ 29.1145 Ignition switches.

(a) Ignition switches must control each ignition circuit on each engine.

(b) There must be means to quickly shut off all ignition by the grouping of switches or by a master ignition control.

(c) Each group of ignition switches, except ignition switches for turbine engines for which continuous ignition is not required, and each master ignition control must have a means to prevent its inadvertent operation.

(Secs. 313(a), 601, and 603, 72 Stat. 759, 775, 49 U.S.C. 1354(a), 1421, and 1423; sec. 6(c), 49 U.S.C. 1655 (c))

[Doc. No. 5084, 29 FR 16150, Dec. 3, 1964, as amended by Amdt. 29-13, 42 FR 15046, Mar. 17, 1977]

§ 29.1147 Mixture controls.

(a) If there are mixture controls, each engine must have a separate control, and the controls must be arranged to allow—

(1) Separate control of each engine; and

(2) Simultaneous control of all engines.

(b) Each intermediate position of the mixture controls that corresponds to a normal operating setting must be identifiable by feel and sight.

§ 29.1151 Rotor brake controls.

(a) It must be impossible to apply the rotor brake inadvertently in flight.

(b) There must be means to warn the crew if the rotor brake has not been completely released before takeoff.

§ 29.1157 Carburetor air temperature controls.

There must be a separate carburetor air temperature control for each engine.

§ 29.1159 Supercharger controls.

Each supercharger control must be accessible to—

(a) The pilots; or

(b) (If there is a separate flight engineer station with a control panel) the flight engineer.

§ 29.1163 Powerplant accessories.

(a) Each engine mounted accessory must—

(1) Be approved for mounting on the engine involved;

(2) Use the provisions on the engine for mounting; and

(3) Be sealed in such a way as to prevent contamination of the engine oil system and the accessory system.

(b) Electrical equipment subject to arcing or sparking must be installed, to minimize the probability of igniting flammable fluids or vapors.

(c) If continued rotation of an engine-driven cabin supercharger or any remote accessory driven by the engine will be a hazard if they malfunction, there must be means to prevent their hazardous rotation without interfering with the continued operation of the engine.

(d) Unless other means are provided, torque limiting means must be provided for accessory drives located on any component of the transmission and rotor drive system to prevent damage to these components from excessive accessory load.

[Doc. No. 5084, 29 FR 16150, Dec. 3, 1964, as amended by Amdt. 29-22, 49 FR 6850, Feb. 23, 1984; Amdt. 29-26, 53 FR 34219, Sept. 2, 1988]

§ 29.1165 Engine ignition systems.

(a) Each battery ignition system must be supplemented with a generator that is automatically available as an alternate source of electrical energy to allow continued engine operation if any battery becomes depleted.

(b) The capacity of batteries and generators must be large enough to meet the simultaneous demands of the engine ignition system and the greatest demands of any electrical system components that draw from the same source.

(c) The design of the engine ignition system must account for—

(1) The condition of an inoperative generator;

(2) The condition of a completely depleted battery with the generator running at its normal operating speed; and

(3) The condition of a completely depleted battery with the generator operating at idling speed, if there is only one battery.

(d) Magneto ground wiring (for separate ignition circuits) that lies on the engine side of any firewall must be installed, located, or protected, to minimize the probability of the simultaneous failure of two or more wires as a result of mechanical damage, electrical fault, or other cause.

(e) No ground wire for any engine may be routed through a fire zone of another engine unless each part of that wire within that zone is fireproof.

(f) Each ignition system must be independent of any electrical circuit that is not used for assisting, controlling, or analyzing the operation of that system.

(g) There must be means to warn appropriate crewmembers if the malfunctioning of any part of the electrical system is causing the continuous discharge of any battery necessary for engine ignition.

[Doc. No. 5084, 29 FR 16150, Dec. 3, 1964, as amended by Amdt. 29-12, 41 FR 55473, Dec. 20, 1976]

Powerplant Fire Protection

§ 29.1181 Designated fire zones: regions included.

(a) Designated fire zones are—

(1) The engine power section of reciprocating engines;

(2) The engine accessory section of reciprocating engines;

(3) Any complete powerplant compartment in which there is no isolation between the engine power section and the engine accessory section, for reciprocating engines;

(4) Any auxiliary power unit compartment;

(5) Any fuel-burning heater and other combustion equipment installation described in § 29.859;

(6) The compressor and accessory sections of turbine engines; and

(7) The combustor, turbine, and tailpipe sections of turbine engine installations except sections that do not contain lines and components carrying flammable fluids or gases and are isolated from the designated fire zone prescribed in paragraph (a)(6) of this section by a firewall that meets § 29.1191.

(b) Each designated fire zone must meet the requirements of §§ 29.1183 through 29.1203.

[Amdt. 29-3, 33 FR 970, Jan. 26, 1968, as amended by Amdt. 29-26, 53 FR 34219, Sept. 2, 1988]

§ 29.1183 Lines, fittings, and components.

(a) Except as provided in paragraph (b) of this section, each line, fitting, and other component carrying flammable fluid in any area subject to engine fire conditions and each component which conveys or contains flammable fluid in a designated fire zone must be fire resistant, except that flammable fluid tanks and supports in a designated fire zone must be fireproof or be enclosed by a fireproof shield unless damage by fire to any non-fireproof part will not cause leakage or spillage of flammable fluid. Components must be shielded or located so as to safeguard against the ignition of leaking flammable fluid. An integral oil sump of less than 25-quart capacity on a reciprocating engine need not be fireproof nor be enclosed by a fireproof shield.

(b) Paragraph (a) of this section does not apply to—

(1) Lines, fittings, and components which are already approved as part of a type certificated engine; and

(2) Vent and drain lines, and their fittings, whose failure will not result in or add to, a fire hazard.

[Doc. No. 5084, 29 FR 16150, Dec. 3, 1964, as amended by Amdt. 29-2, 32 FR 6914, May 5, 1967; Amdt. 29-10, 39 FR 35463, Oct. 1, 1974; Amdt. 29-22, 49 FR 6850, Feb. 23, 1984]

§ 29.1185 Flammable fluids.

(a) No tank or reservoir that is part of a system containing flammable fluids or gases may be in a designated fire zone unless the fluid contained, the design of the system, the materials used in the tank and its supports, the shutoff means, and the connections, lines, and controls provide a degree of safety equal to that which would exist if the tank or reservoir were outside such a zone.

(b) Each fuel tank must be isolated from the engines by a firewall or shroud.

(c) There must be at least one-half inch of clear airspace between each tank or reservoir and each firewall or shroud isolating a designated fire zone, unless equivalent means are used to prevent heat transfer from the fire zone to the flammable fluid.

(d) Absorbent material close to flammable fluid system components that might leak must be covered or treated to prevent the absorption of hazardous quantities of fluids.

§ 29.1187 Drainage and ventilation of fire zones.

(a) There must be complete drainage of each part of each designated fire zone to minimize the hazards resulting from failure or malfunction of any component containing flammable fluids. The drainage means must be—

(1) Effective under conditions expected to prevail when drainage is needed; and

(2) Arranged so that no discharged fluid will cause an additional fire hazard.

(b) Each designated fire zone must be ventilated to prevent the accumulation of flammable vapors.

(c) No ventilation opening may be where it would allow the entry of flammable fluids, vapors, or flame from other zones.

(d) Ventilation means must be arranged so that no discharged vapors will cause an additional fire hazard.

(e) For category A rotorcraft, there must be means to allow the crew to shut off the sources of forced ventilation in any fire zone (other than the engine power section of the powerplant compartment) unless the amount of extinguishing agent and the rate of discharge are based on the maximum airflow through that zone.

§ 29.1189 Shutoff means.

(a) There must be means to shut off or otherwise prevent hazardous quantities of fuel, oil, de-icing fluid, and other flammable fluids from flowing into, within, or through any designated fire zone, except that this means need not be provided—

(1) For lines, fittings, and components forming an integral part of an engine;

(2) For oil systems for turbine engine installations in which all components of the system, including oil tanks, are fireproof or located in areas not subject to engine fire conditions; or

(3) For engine oil systems in category B rotorcraft using reciprocating engines of less than 500 cubic inches displacement.

(b) The closing of any fuel shutoff valve for any engine may not make fuel unavailable to the remaining engines.

(c) For category A rotorcraft, no hazardous quantity of flammable fluid may drain into any designated fire zone after shutoff has been accomplished, nor may the closing of any fuel shutoff valve for an engine make fuel unavailable to the remaining engines.

(d) The operation of any shutoff may not interfere with the later emergency operation of any other equipment, such as the means for declutching the engine from the rotor drive.

(e) Each shutoff valve and its control must be designed, located, and protected to function properly under any condition likely to result from fire in a designated fire zone.

(f) Except for ground-use-only auxiliary power unit installations, there must be means to prevent inadvertent operation of each shutoff and to make it possible to reopen it in flight after it has been closed.

[Doc. No. 5084, 29 FR 16150, Dec. 3, 1964, as amended by Amdt. 29-12, 41 FR 55473, Dec. 20, 1976; Amdt. 29-22, 49 FR 6850, Feb. 23, 1984; Amdt. 29-26, 53 FR 34219, Sept. 2, 1988]

§ 29.1191 Firewalls.

(a) Each engine, including the combustor, turbine, and tailpipe sections of turbine engine installations, must be isolated by a firewall, shroud, or equivalent means, from personnel compartments, structures, controls, rotor mechanisms, and other parts that are—

(1) Essential to controlled flight and landing; and

(2) Not protected under § 29.861.

(b) Each auxiliary power unit, combustion heater, and other combustion equipment to be used in flight, must be isolated from the rest of the rotorcraft by firewalls, shrouds, or equivalent means.

(c) Each firewall or shroud must be constructed so that no hazardous quantity of air, fluid, or flame can pass from any engine compartment to other parts of the rotorcraft.

(d) Each opening in the firewall or shroud must be sealed with close-fitting fireproof grommets, bushings, or firewall fittings.

(e) Each firewall and shroud must be fireproof and protected against corrosion.

(f) In meeting this section, account must be taken of the probable path of a fire as affected by the airflow in normal flight and in autorotation.

[Doc. No. 5084, 29 FR 16150, Dec. 3, 1964, as amended by Amdt. 29-3, 33 FR 970, Jan. 26, 1968]

§ 29.1193 Cowling and engine compartment covering.

(a) Each cowling and engine compartment covering must be constructed and supported so that it can resist the vibration, inertia, and air loads to which it may be subjected in operation.

(b) Cowling must meet the drainage and ventilation requirements of § 29.1187.

(c) On rotorcraft with a diaphragm isolating the engine power section from the engine accessory section, each part of the accessory section cowling subject to flame in case of fire in the engine power section of the powerplant must—

(1) Be fireproof; and

(2) Meet the requirements of § 29.1191.

(d) Each part of the cowling or engine compartment covering subject to high temperatures due to its nearness to exhaust system parts or exhaust gas impingement must be fireproof.

(e) Each rotorcraft must—

(1) Be designated and constructed so that no fire originating in any fire zone can enter, either through openings or by burning through external skin, any other zone or region where it would create additional hazards;

(2) Meet the requirements of paragraph (e)(1) of this section with the landing gear retracted (if applicable); and

(3) Have fireproof skin in areas subject to flame if a fire starts in or burns out of any designated fire zone.

(f) A means of retention for each openable or readily removable panel, cowling, or engine or rotor drive system covering must be provided to preclude hazardous damage to rotors or critical control components in the event of—

(1) Structural or mechanical failure of the normal retention means, unless such failure is extremely improbable; or

(2) Fire in a fire zone, if such fire could adversely affect the normal means of retention.

(Secs. 313(a), 601, and 603, 72 Stat. 759, 775, 49 U.S.C. 1354(a), 1421, and 1423; sec. 6(c), 49 U.S.C. 1655(c))

[Doc. No. 5084, 29 FR 16150, Dec. 3, 1964, as amended by Amdt. 29-3, 33 FR 970, Jan. 26, 1968; Amdt. 29-13, 42 FR 15046, Mar. 17, 1977; Amdt. 29-26, 53 FR 34219, Sept. 2, 1988]

§ 29.1194 Other surfaces.

All surfaces aft of, and near, engine compartments and designated fire zones, other than tail surfaces not subject to heat, flames, or sparks emanating from a designated fire zone or engine compartment, must be at least fire resistant.

[Amdt. 29-3, 33 FR 970, Jan. 26, 1968]

§ 29.1195 Fire extinguishing systems.

(a) Each turbine engine powered rotorcraft and Category A reciprocating engine powered rotorcraft, and each Category B reciprocating engine powered rotorcraft with engines of more than 1,500 cubic inches must have a fire extinguishing system for the designated fire zones. The fire extinguishing system for a powerplant must be able to simultaneously protect all zones of the powerplant compartment for which protection is provided.

(b) For multiengine powered rotorcraft, the fire extinguishing system, the quantity of extinguishing agent, and the rate of discharge must—

(1) For each auxiliary power unit and combustion equipment, provide at least one adequate discharge; and

(2) For each other designated fire zone, provide two adequate discharges.

(c) For single engine rotorcraft, the quantity of extinguishing agent and the rate of discharge must provide at least one adequate discharge for the engine compartment.

(d) It must be shown by either actual or simulated flight tests that under critical airflow conditions in flight the discharge of the extinguishing agent in each designated fire zone will provide an agent concentration capable of extinguishing fires in that zone and of minimizing the probability of reignition.

(Secs. 313(a), 601, 603, 604, Federal Aviation Act of 1958 (49 U.S.C. 1354(a), 1421, 1423, 1424), sec. 6(c), Dept. of Transportation Act (49 U.S.C. 1655(c)))

[Doc. No. 5084, 29 FR 16150, Dec. 3, 1964, as amended by Amdt. 29-3, 33 FR 970, Jan. 26, 1968; Amdt. 29-13, 42 FR 15047, Mar. 17, 1977; Amdt. 29-17, 43 FR 50602, Oct. 30, 1978]

§ 29.1197 Fire extinguishing agents.

(a) Fire extinguishing agents must—

(1) Be capable of extinguishing flames emanating from any burning of fluids or other combustible materials in the area protected by the fire extinguishing system; and

(2) Have thermal stability over the temperature range likely to be experienced in the compartment in which they are stored.

(b) If any toxic extinguishing agent is used, it must be shown by test that entry of harmful concentrations of fluid or fluid vapors into any personnel compartment (due to leakage during normal operation of the rotorcraft, or discharge on the ground or in flight) is prevented, even though a defect may exist in the extinguishing system.

(Secs. 313(a), 601, and 603, 72 Stat. 759, 775, 49 U.S.C. 1354(a), 1421, and 1423; sec. 6(c), 49 U.S.C. 1655(c))

[Doc. No. 5084, 29 FR 16150, Dec. 3, 1964, as amended by Amdt. 29-12, 41 FR 55473, Dec. 20, 1976; Amdt. 29-13, 42 FR 15047, Mar. 17, 1977]

§ 29.1199 Extinguishing agent containers.

(a) Each extinguishing agent container must have a pressure relief to prevent bursting of the container by excessive internal pressures.

(b) The discharge end of each discharge line from a pressure relief connection must be located so that discharge of the fire extinguishing agent would not damage the rotorcraft. The line must also be located or protected to prevent clogging caused by ice or other foreign matter.

(c) There must be a means for each fire extinguishing agent container to indicate that the container has discharged or that the charging pressure is below the established minimum necessary for proper functioning.

(d) The temperature of each container must be maintained, under intended operating conditions, to prevent the pressure in the container from—

(1) Falling below that necessary to provide an adequate rate of discharge; or

(2) Rising high enough to cause premature discharge.

(Secs. 313(a), 601, and 603, 72 Stat. 759, 775, 49 U.S.C. 1354(a), 1421, and 1423; sec. 6(c), 49 U.S.C. 1655 (c))

[Doc. No. 5084, 29 FR 16150, Dec. 3, 1964, as amended by Amdt. 29-13, 42 FR 15047, Mar. 17, 1977]

§ 29.1201 Fire extinguishing system materials.

(a) No materials in any fire extinguishing system may react chemically with any extinguishing agent so as to create a hazard.

(b) Each system component in an engine compartment must be fireproof.

§ 29.1203 Fire detector systems.

(a) For each turbine engine powered rotorcraft and Category A reciprocating engine powered rotorcraft, and for each Category B reciprocating engine powered rotorcraft with engines of more than 900 cubic inches displacement, there must be approved, quick-acting fire detectors in designated fire zones and in the combustor, turbine, and tailpipe sections of turbine installations (whether or not such sections are designated fire zones) in numbers and locations ensuring prompt detection of fire in those zones.

(b) Each fire detector must be constructed and installed to withstand any vibration, inertia, and other loads to which it would be subjected in operation.

(c) No fire detector may be affected by any oil, water, other fluids, or fumes that might be present.

(d) There must be means to allow crewmembers to check, in flight, the functioning of each fire detector system electrical circuit.

(e) The writing and other components of each fire detector system in an engine compartment must be at least fire resistant.

(f) No fire detector system component for any fire zone may pass through another fire zone, unless—

(1) It is protected against the possibility of false warnings resulting from fires in zones through which it passes; or

(2) The zones involved are simultaneously protected by the same detector and extinguishing systems.

[Doc. No. 5084, 29 FR 16150, Dec. 3, 1964, as amended by Amdt. 29-3, 33 FR 970, Jan. 26, 1968]

Subpart F—Equipment

General

§ 29.1301 Function and installation.

Each item of installed equipment must—

(a) Be of a kind and design appropriate to its intended function;

(b) Be labeled as to its identification, function, or operating limitations, or any applicable combination of these factors;

(c) Be installed according to limitations specified for that equipment; and

(d) Function properly when installed.

§ 29.1303 Flight and navigation instruments.

The following are required flight and navigational instruments:

(a) An airspeed indicator. For Category A rotorcraft with V_{NE} less than a speed at which unmistakable pilot cues provide overspeed warning, a maximum allowable airspeed indicator must be provided. If maximum allowable airspeed varies with weight, altitude, temperature, or r.p.m., the indicator must show that variation.

(b) A sensitive altimeter.

(c) A magnetic direction indicator.

(d) A clock displaying hours, minutes, and seconds with a sweep-second pointer or digital presentation.

(e) A free-air temperature indicator.

(f) A non-tumbling gyroscopic bank and pitch indicator.

(g) A gyroscopic rate-of-turn indicator combined with an integral slip-skid indicator (turn-and-bank indicator) except that only a slip-skid indicator is required on rotorcraft with a third attitude instrument system that—

(1) Is usable through flight attitudes of±80 degrees of pitch and±120 degrees of roll;

(2) Is powered from a source independent of the electrical generating system;

(3) Continues reliable operation for a minimum of 30 minutes after total failure of the electrical generating system;

(4) Operates independently of any other attitude indicating system;

(5) Is operative without selection after total failure of the electrical generating system;

(6) Is located on the instrument panel in a position acceptable to the Administrator that will make it plainly visible to and useable by any pilot at his station; and

(7) Is appropriately lighted during all phases of operation.

(h) A gyroscopic direction indicator.

(i) A rate-of-climb (vertical speed) indicator.

(j) For Category A rotorcraft, a speed warning device when V_{NE} is less than the speed at which unmistakable overspeed warning is provided by other pilot cues. The speed warning device must give effective aural warning (differing distinctively from aural warnings used for other purposes) to the pilots whenever the indicated speed exceeds V_{NE} plus 3 knots and must operate satisfactorily throughout the approved range of altitudes and temperatures.

(Secs. 313(a), 601, 603, 604, and 605 of the Federal Aviation Act of 1958 (49 U.S.C. 1354(a), 1421, 1423, 1424, and 1425); and sec. 6(c), Dept. of Transportation Act (49 U.S.C. 1655(c)))

[Doc. No. 5084, 29 FR 16150, Dec. 3, 1964, as amended by Amdt. 29-12, 41 FR 55474, Dec. 20, 1976; Amdt. 29-14, 42 FR 36972, July 18, 1977; Amdt. 29-24, 49 FR 44438, Nov. 6, 1984; 70 FR 2012, Jan. 12, 2005]

§ 29.1305 Powerplant instruments.

The following are required powerplant instruments:

(a) For each rotorcraft—

(1) A carburetor air temperature indicator for each reciprocating engine;

(2) A cylinder head temperature indicator for each air-cooled reciprocating engine, and a coolant temperature indicator for each liquid-cooled reciprocating engine;

(3) A fuel quantity indicator for each fuel tank;

(4) A low fuel warning device for each fuel tank which feeds an engine. This device must—

(i) Provide a warning to the crew when approximately 10 minutes of usable fuel remains in the tank; and

(ii) Be independent of the normal fuel quantity indicating system.

(5) A manifold pressure indicator, for each reciprocating engine of the altitude type;

(6) An oil pressure indicator for each pressure-lubricated gearbox.

(7) An oil pressure warning device for each pressure-lubricated gearbox to indicate when the oil pressure falls below a safe value;

(8) An oil quantity indicator for each oil tank and each rotor drive gearbox, if lubricant is self-contained;

(9) An oil temperature indicator for each engine;

(10) An oil temperature warning device to indicate unsafe oil temperatures in each main rotor drive gearbox, including gearboxes necessary for rotor phasing;

(11) A gas temperature indicator for each turbine engine;

(12) A gas producer rotor tachometer for each turbine engine;

(13) A tachometer for each engine that, if combined with the applicable instrument required by paragraph (a)(14) of this section, indicates rotor r.p.m. during autorotation.

(14) At least one tachometer to indicate, as applicable—

(i) The r.p.m. of the single main rotor;

(ii) The common r.p.m. of any main rotors whose speeds cannot vary appreciably with respect to each other; and

(iii) The r.p.m. of each main rotor whose speed can vary appreciably with respect to that of another main rotor;

(15) A free power turbine tachometer for each turbine engine;

(16) A means, for each turbine engine, to indicate power for that engine;

(17) For each turbine engine, an indicator to indicate the functioning of the powerplant ice protection system;

(18) An indicator for the filter required by § 29.997 to indicate the occurrence of contamination of the filter to the degree established in compliance with § 29.955;

(19) For each turbine engine, a warning means for the oil strainer or filter required by § 29.1019, if it has no bypass, to warn the pilot of the occurrence of contamination of the strainer or filter before it reaches the capacity established in accordance with § 29.1019(a)(2);

(20) An indicator to indicate the functioning of any selectable or controllable heater used to prevent ice clogging of fuel system components;

(21) An individual fuel pressure indicator for each engine, unless the fuel system which supplies that engine does not employ any pumps, filters, or other components subject to degradation or failure which may adversely affect fuel pressure at the engine;

(22) A means to indicate to the flightcrew the failure of any fuel pump installed to show compliance with § 29.955;

(23) Warning or caution devices to signal to the flightcrew when ferromagnetic particles are detected by the chip detector required by § 29.1337(e); and

(24) For auxiliary power units, an individual indicator, warning or caution device, or other means to advise the flightcrew that limits are being exceeded, if exceeding these limits can be hazardous, for—

(i) Gas temperature;

(ii) Oil pressure; and

(iii) Rotor speed.

(25) For rotorcraft for which a 30-second/2-minute OEI power rating is requested, a means must be provided to alert the pilot when the engine is at the 30-second and 2-minute OEI power levels, when the event begins, and when the time interval expires.

(26) For each turbine engine utilizing 30-second/2-minute OEI power, a device or system must be provided for use by ground personnel which—

(i) Automatically records each usage and duration of power at the 30-second and 2-minute OEI levels;

(ii) Permits retrieval of the recorded data;

(iii) Can be reset only by ground maintenance personnel; and

(iv) Has a means to verify proper operation of the system or device.

(b) For category A rotorcraft—

(1) An individual oil pressure indicator for each engine, and either an independent warning device for each engine or a master warning device for the engines with means for isolating the individual warning circuit from the master warning device;

(2) An independent fuel pressure warning device for each engine or a master warning device for all engines with provision for isolating the individual warning device from the master warning device; and

(3) Fire warning indicators.

(c) For category B rotorcraft—

(1) An individual oil pressure indicator for each engine; and

(2) Fire warning indicators, when fire detection is required.

[Doc. No. 5084, 29 FR 16150, Dec. 3, 1964, as amended by Amdt. 29-3, 33 FR 970, Jan. 26, 1968; Amdt. 29-10, 39 FR 35463, Oct. 1, 1974; Amdt. 29-26, 53 FR 34219, Sept. 2, 1988; Amdt. 29-34, 59 FR 47768, Sept. 16, 1994; Amdt. 29-40, 61 FR 21908, May 10, 1996; 61 FR 43952, Aug. 27, 1996]

§ 29.1307 Miscellaneous equipment.

The following is required miscellaneous equipment:

(a) An approved seat for each occupant.

(b) A master switch arrangement for electrical circuits other than ignition.

(c) Hand fire extinguishers.

(d) A windshield wiper or equivalent device for each pilot station.

(e) A two-way radio communication system.

[Amdt. 29-12, 41 FR 55473, Dec. 20, 1976]

§ 29.1309 Equipment, systems, and installations.

(a) The equipment, systems, and installations whose functioning is required by this subchapter must be designed and installed to ensure that they perform their intended functions under any foreseeable operating condition.

(b) The rotorcraft systems and associated components, considered separately and in relation to other systems, must be designed so that—

(1) For Category B rotorcraft, the equipment, systems, and installations must be designed to prevent hazards to the rotorcraft if they malfunction or fail; or

(2) For Category A rotorcraft—

(i) The occurrence of any failure condition which would prevent the continued safe flight and landing of the rotorcraft is extremely improbable; and

(ii) The occurrence of any other failure conditions which would reduce the capability of the rotorcraft or the ability of the crew to cope with adverse operating conditions is improbable.

(c) Warning information must be provided to alert the crew to unsafe system operating conditions and to enable them to take appropriate corrective action. Systems, controls, and associated monitoring and warning means must be designed to minimize crew errors which could create additional hazards.

(d) Compliance with the requirements of paragraph (b)(2) of this section must be shown by analysis and, where necessary, by appropriate ground, flight, or simulator tests. The analysis must consider—

(1) Possible modes of failure, including malfunctions and damage from external sources;

(2) The probability of multiple failures and undetected failures;

(3) The resulting effects on the rotorcraft and occupants, considering the stage of flight and operating conditions; and

(4) The crew warning cues, corrective action required, and the capability of detecting faults.

(e) For Category A rotorcraft, each installation whose functioning is required by this subchapter and which requires a power supply is an "essential load" on the power supply. The power sources and the system must be able to supply the following power loads in probable operating combinations and for probable durations:

(1) Loads connected to the system with the system functioning normally.

(2) Essential loads, after failure of any one prime mover, power converter, or energy storage device.

(3) Essential loads, after failure of—

(i) Any one engine, on rotorcraft with two engines; and

(ii) Any two engines, on rotorcraft with three or more engines.

(f) In determining compliance with paragraphs (e)(2) and (3) of this section, the power loads may be assumed to be reduced under a monitoring procedure consistent with safety in the kinds of operations authorized. Loads not required for controlled flight need not be considered for the two-engine-inoperative condition on rotorcraft with three or more engines.

(g) In showing compliance with paragraphs (a) and (b) of this section with regard to the electrical system and to equipment design and installation, critical environmental conditions must be considered. For electrical generation, distribution, and utilization equipment required by or used in complying with this subchapter, except equipment covered by Technical Standard Orders containing environmental test procedures, the ability to provide continuous, safe service under foreseeable environmental conditions may be shown by environmental tests, design analysis, or reference to previous comparable service experience on other aircraft.

(Secs. 313(a), 601, 603, 604, and 605 of the Federal Aviation Act of 1958 (49 U.S.C. 1354(a), 1421, 1423, 1424, and 1425); and sec. 6(c), Dept. of Transportation Act (49 U.S.C. 1655(c)))

[Doc. No. 5084, 29 FR 16150, Dec. 3, 1964, as amended by Amdt. 29-14, 42 FR 36972, July 18, 1977; Amdt. 29-24, 49 FR 44438, Nov. 6, 1984; Amdt. 29-40, 61 FR 21908, May 10, 1996; Amdt. 29-53, 76 FR 33136, June 8, 2011]

§ 29.1316 Electrical and electronic system lightning protection.

(a) Each electrical and electronic system that performs a function, for which failure would prevent the continued safe flight and landing of the rotorcraft, must be designed and installed so that—

(1) The function is not adversely affected during and after the time the rotorcraft is exposed to lightning; and

(2) The system automatically recovers normal operation of that function in a timely manner after the rotorcraft is exposed to lightning.

(b) Each electrical and electronic system that performs a function, for which failure would reduce the capability of the rotorcraft or the ability of the flightcrew to respond to an adverse operating condition, must be designed and installed so that the function recovers normal operation in a timely manner after the rotorcraft is exposed to lightning.

[Doc. No. FAA-2010-0224, Amdt. 29-53, 76 FR 33136, June 8, 2011]

§ 29.1317 High-intensity Radiated Fields (HIRF) Protection.

(a) Except as provided in paragraph (d) of this section, each electrical and electronic system that performs a function whose failure would prevent the continued safe flight and landing of the rotorcraft must be designed and installed so that—

(1) The function is not adversely affected during and after the time the rotorcraft is exposed to HIRF environment I, as described in appendix E to this part;

(2) The system automatically recovers normal operation of that function, in a timely manner, after the rotorcraft is exposed to HIRF environment I, as described in appendix E to this part, unless this conflicts with other operational or functional requirements of that system;

(3) The system is not adversely affected during and after the time the rotorcraft is exposed to HIRF environment II, as described in appendix E to this part; and

(4) Each function required during operation under visual flight rules is not adversely affected during and after the time the rotorcraft is exposed to HIRF environment III, as described in appendix E to this part.

(b) Each electrical and electronic system that performs a function whose failure would significantly reduce the capability of the rotorcraft or the ability of the flightcrew to respond to an adverse operating condition must be designed and installed so the system is not adversely affected when the equipment providing these functions is exposed to equipment HIRF test level 1 or 2, as described in appendix E to this part.

(c) Each electrical and electronic system that performs such a function whose failure would reduce the capability of the rotorcraft or the ability of the flightcrew to respond to an adverse operating condition must be designed and installed so the system is not adversely affected when the equipment providing these functions is exposed to equipment HIRF test level 3, as described in appendix E to this part.

(d) Before December 1, 2012, an electrical or electronic system that performs a function whose failure would prevent the continued safe flight and landing of a rotorcraft may be designed and installed without meeting the provisions of paragraph (a) provided—

(1) The system has previously been shown to comply with special conditions for HIRF, prescribed under § 21.16, issued before December 1, 2007;

(2) The HIRF immunity characteristics of the system have not changed since compliance with the special conditions was demonstrated; and

(3) The data used to demonstrate compliance with the special conditions is provided.

[Doc. No. FAA-2006-23657, 72 FR 44027, Aug. 6, 2007]

Instruments: Installation

§ 29.1321 Arrangement and visibility.

(a) Each flight, navigation, and powerplant instrument for use by any pilot must be easily visible to him from his station with the minimum practicable deviation from his normal position and line of vision when he is looking forward along the flight path.

(b) Each instrument necessary for safe operation, including the airspeed indicator, gyroscopic direction indicator, gyroscopic bank-and-pitch indicator, slip-skid indicator, altimeter, rate-of-climb indicator, rotor tachometers, and the indicator most representative of engine power, must be grouped and centered as nearly as practicable about the vertical plane of the pilot's forward vision. In addition, for rotorcraft approved for IFR flight—

(1) The instrument that most effectively indicates attitude must be on the panel in the top center position;

(2) The instrument that most effectively indicates direction of flight must be adjacent to and directly below the attitude instrument;

(3) The instrument that most effectively indicates airspeed must be adjacent to and to the left of the attitude instrument; and

(4) The instrument that most effectively indicates altitude or is most frequently utilized in control of altitude must be adjacent to and to the right of the attitude instrument.

(c) Other required powerplant instruments must be closely grouped on the instrument panel.

(d) Identical powerplant instruments for the engines must be located so as to prevent any confusion as to which engine each instrument relates.

(e) Each powerplant instrument vital to safe operation must be plainly visible to appropriate crewmembers.

(f) Instrument panel vibration may not damage, or impair the readability or accuracy of, any instrument.

(g) If a visual indicator is provided to indicate malfunction of an instrument, it must be effective under all probable cockpit lighting conditions.

(Secs. 313(a), 601, 603, 604, and 605 of the Federal Aviation Act of 1958 (49 U.S.C. 1354(a), 1421, 1423, 1424, and 1425); and sec. 6(c), Dept. of Transportation Act (49 U.S.C. 1655(c)))

[Doc. No. 5084, 29 FR 16150, Dec. 3, 1964, as amended by Amdt. 29-14, 42 FR 36972, July 18, 1977; Amdt. 29-21, 48 FR 4391, Jan. 31, 1983]

§ 29.1322 Warning, caution, and advisory lights.

If warning, caution or advisory lights are installed in the cockpit they must, unless otherwise approved by the Administrator, be—

(a) Red, for warning lights (lights indicating a hazard which may require immediate corrective action);

(b) Amber, for caution lights (lights indicating the possible need for future corrective action);

(c) Green, for safe operation lights; and

(d) Any other color, including white, for lights not described in paragraphs (a) through (c) of this section, provided the color differs sufficiently from the colors prescribed in paragraphs (a) through (c) of this section to avoid possible confusion.

[Amdt. 29-12, 41 FR 55474, Dec. 20, 1976]

§ 29.1323 Airspeed indicating system.

For each airspeed indicating system, the following apply:

(a) Each airspeed indicating instrument must be calibrated to indicate true airspeed (at sea level with a standard atmosphere) with a minimum practicable instrument calibration error when the corresponding pitot and static pressures are applied.

(b) Each system must be calibrated to determine system error excluding airspeed instrument error. This calibration must be determined—

(1) In level flight at speeds of 20 knots and greater, and over an appropriate range of speeds for flight conditions of climb and autorotation; and

(2) During takeoff, with repeatable and readable indications that ensure—

(i) Consistent realization of the field lengths specified in the Rotorcraft Flight Manual; and

(ii) Avoidance of the critical areas of the height-velocity envelope as established under § 29.87.

(c) For Category A rotorcraft—

(1) The indication must allow consistent definition of the takeoff decision point; and

(2) The system error, excluding the airspeed instrument calibration error, may not exceed—

(i) Three percent or 5 knots, whichever is greater, in level flight at speeds above 80 percent of takeoff safety speed; and

(ii) Ten knots in climb at speeds from 10 knots below takeoff safety speed to 10 knots above V_Y.

(d) For Category B rotorcraft, the system error, excluding the airspeed instrument calibration error, may not exceed 3 percent or 5 knots, whichever is greater, in level flight at speeds above 80 percent of the climbout speed attained at 50 feet when complying with § 29.63.

(e) Each system must be arranged, so far as practicable, to prevent malfunction or serious error due to the entry of moisture, dirt, or other substances.

(f) Each system must have a heated pitot tube or an equivalent means of preventing malfunction due to icing.

[Doc. No. 5084, 29 FR 16150, Dec. 3, 1964 as amended by Amdt. 29-3, 33 FR 970, Jan. 26, 1968; Amdt. 29-24, 49 FR 44439, Nov. 6, 1984; Amdt. 29-39, 61 FR 21901, May 10, 1996; Amdt. 29-44, 64 FR 45338, Aug. 19, 1999]

§ 29.1325 Static pressure and pressure altimeter systems.

(a) Each instrument with static air case connections must be vented to the outside atmosphere through an appropriate piping system.

(b) Each vent must be located where its orifices are least affected by airflow variation, moisture, or foreign matter.

(c) Each static pressure port must be designed and located in such manner that the correlation between air pressure in the static pressure system and true ambient atmospheric static pressure is not altered when the rotorcraft encounters icing conditions. An anti-icing means or an alternate source of static pressure may be used in showing compliance with this requirement. If the reading of the altimeter, when on the alternate static pressure system, differs from the reading of altimeter when on the primary static system by more than 50 feet, a correction card must be provided for the alternate static system.

(d) Except for the vent into the atmosphere, each system must be airtight.

(e) Each pressure altimeter must be approved and calibrated to indicate pressure altitude in a standard atmosphere with a minimum practicable calibration error when the corresponding static pressures are applied.

(f) Each system must be designed and installed so that an error in indicated pressure altitude, at sea level, with a standard atmosphere, excluding instrument calibration error, does not result in an error of more than±30 feet per 100 knots speed. However, the error need not be less than±30 feet.

(g) Except as provided in paragraph (h) of this section, if the static pressure system incorporates both a primary and an alternate static pressure source, the means for selecting one or the other source must be designed so that—

(1) When either source is selected, the other is blocked off; and

(2) Both sources cannot be blocked off simultaneously.

(h) For unpressurized rotorcraft, paragraph (g)(1) of this section does not apply if it can be demonstrated that the static pressure system calibration, when either static pressure source is selected, is not changed by the other static pressure source being open or blocked.

(Secs. 313(a), 601, 603, 604, and 605 of the Federal Aviation Act of 1958 (49 U.S.C. 1354(a), 1421, 1423,

1424, and 1425); and sec. 6(c), Dept. of Transportation Act (49 U.S.C. 1655(c)))

[Doc. No. 5084, 29 FR 16150, Dec. 3, 1964, as amended by Amdt. 29-14, 42 FR 36972, July 18, 1977; Amdt. 29-24, 49 FR 44439, Nov. 6, 1984]

§ 29.1327 Magnetic direction indicator.

(a) Each magnetic direction indicator must be installed so that its accuracy is not excessively affected by the rotorcraft's vibration or magnetic fields.

(b) The compensated installation may not have a deviation, in level flight, greater than 10 degrees on any heading.

§ 29.1329 Automatic pilot system.

(a) Each automatic pilot system must be designed so that the automatic pilot can—

(1) Be sufficiently overpowered by one pilot to allow control of the rotorcraft; and

(2) Be readily and positively disengaged by each pilot to prevent it from interfering with the control of the rotorcraft.

(b) Unless there is automatic synchronization, each system must have a means to readily indicate to the pilot the alignment of the actuating device in relation to the control system it operates.

(c) Each manually operated control for the system's operation must be readily accessible to the pilots.

(d) The system must be designed and adjusted so that, within the range of adjustment available to the pilot, it cannot produce hazardous loads on the rotorcraft, or create hazardous deviations in the flight path, under any flight condition appropriate to its use, either during normal operation or in the event of a malfunction, assuming that corrective action begins within a reasonable period of time.

(e) If the automatic pilot integrates signals from auxiliary controls or furnishes signals for operation of other equipment, there must be positive interlocks and sequencing of engagement to prevent improper operation.

(f) If the automatic pilot system can be coupled to airborne navigation equipment, means must be provided to indicate to the pilots the current mode of operation. Selector switch position is not acceptable as a means of indication.

[Doc. No. 5084, 29 FR 16150, Dec. 3, 1964, as amended by Amdt. 29-24, 49 FR 44439, Nov. 6, 1984; Amdt. 29-24, 49 FR 47594, Dec. 6, 1984; Amdt. 29-42, 63 FR 43285, Aug. 12, 1998]

§ 29.1331 Instruments using a power supply.

For category A rotorcraft—

(a) Each required flight instrument using a power supply must have—

(1) Two independent sources of power;

(2) A means of selecting either power source; and

(3) A visual means integral with each instrument to indicate when the power adequate to sustain proper instrument performance is not being supplied. The power must be measured at or near the point where it enters the instrument. For electrical instruments, the power is considered to be adequate when the voltage is within the approved limits; and

(b) The installation and power supply system must be such that failure of any flight instrument connected to one source, or of the energy supply from one source, or a fault in any part of the power distribution system does not interfere with the proper supply of energy from any other source.

[Doc. No. 5084, 29 FR 16150, Dec. 3, 1964, as amended by Amdt. 29-24, 49 FR 44439, Nov. 6, 1984]

§ 29.1333 Instrument systems.

For systems that operate the required flight instruments which are located at each pilot's station, the following apply:

(a) Only the required flight instruments for the first pilot may be connected to that operating system.

(b) The equipment, systems, and installations must be designed so that one display of the information essential to the safety of flight which is provided by the flight instruments remains available to a pilot, without additional crewmember action, after any single failure or combination of failures that are not shown to be extremely improbable.

(c) Additional instruments, systems, or equipment may not be connected to the operating system for a second pilot unless provisions are made to ensure the continued normal functioning of the required flight instruments in the event of any malfunction of the additional instruments, systems, or equipment which is not shown to be extremely improbable.

[Amdt. 29-24, 49 FR 44439, Nov. 6, 1984]

§ 29.1335 Flight director systems.

If a flight director system is installed, means must be provided to indicate to the flight crew its current mode of operation. Selector switch position is not acceptable as a means of indication.

(Secs. 313(a), 601, 603, 604, and 605 of the Federal Aviation Act of 1958 (49 U.S.C. 1354(a), 1421, 1423, 1424, and 1425); and sec. 6(c), Dept. of Transportation Act (49 U.S.C. 1655(c)))

[Amdt. 29-14, 42 FR 36973, July 18, 1977]

§ 29.1337 Powerplant instruments.

(a) *Instruments and instrument lines.* (1) Each powerplant and auxiliary power unit instrument line must meet the requirements of §§ 29.993 and 29.1183.

(2) Each line carrying flammable fluids under pressure must—

(i) Have restricting orifices or other safety devices at the source of pressure to prevent the escape of excessive fluid if the line fails; and

(ii) Be installed and located so that the escape of fluids would not create a hazard.

(3) Each powerplant and auxiliary power unit instrument that utilizes flammable fluids must be installed and located so that the escape of fluid would not create a hazard.

(b) *Fuel quantity indicator.* There must be means to indicate to the flight crew members the quantity, in gallons or equivalent units, of usable fuel in each tank during flight. In addition—

(1) Each fuel quantity indicator must be calibrated to read "zero" during level flight when the quantity of fuel remaining in the tank is equal to the unusable fuel supply determined under § 29.959;

(2) When two or more tanks are closely interconnected by a gravity feed system and vented, and when it is impossible to feed from each tank separately, at least one fuel quantity indicator must be installed;

(3) Tanks with interconnected outlets and airspaces may be treated as one tank and need not have separate indicators; and

(4) Each exposed sight gauge used as a fuel quantity indicator must be protected against damage.

(c) *Fuel flowmeter system.* If a fuel flowmeter system is installed, each metering component must have a means for bypassing the fuel supply if malfunction of that component severely restricts fuel flow.

(d) *Oil quantity indicator.* There must be a stick gauge or equivalent means to indicate the quantity of oil—

(1) In each tank; and

(2) In each transmission gearbox.

(e) Rotor drive system transmissions and gearboxes utilizing ferromagnetic materials must be equipped with chip detectors designed to indicate the presence of ferromagnetic particles resulting from damage or excessive wear within the transmission or gearbox. Each chip detector must—

(1) Be designed to provide a signal to the indicator required by § 29.1305(a)(22); and

(2) Be provided with a means to allow crewmembers to check, in flight, the function of each detector electrical circuit and signal.

(Secs. 313(a), 601, and 603, 72 Stat. 759, 775, 49 U.S.C. 1354(a), 1421, and 1423; sec. 6(c), 49 U.S.C. 1655(c))

[Doc. No. 5084, 29 FR 16150, Dec. 3, 1964, as amended by Amdt. 29-13, 42 FR 15047, Mar. 17, 1977; Amdt. 29-26, 53 FR 34219, Sept. 2, 1988]

Electrical Systems and Equipment

§ 29.1351 General.

(a) *Electrical system capacity.* The required generating capacity and the number and kind of power sources must—

(1) Be determined by an electrical load analysis; and

(2) Meet the requirements of § 29.1309.

(b) *Generating system.* The generating system includes electrical power sources, main power busses, transmission cables, and associated control, regulation, and protective devices. It must be designed so that—

(1) Power sources function properly when independent and when connected in combination;

(2) No failure or malfunction of any power source can create a hazard or impair the ability of remaining sources to supply essential loads;

(3) The system voltage and frequency (as applicable) at the terminals of essential load equipment can be maintained within the limits for which the equipment is designed, during any probable operating condition;

(4) System transients due to switching, fault clearing, or other causes do not make essential loads inoperative, and do not cause a smoke or fire hazard;

(5) There are means accessible in flight to appropriate crewmembers for the individual and collective disconnection of the electrical power sources from the main bus; and

(6) There are means to indicate to appropriate crewmembers the generating system quantities essential for the safe operation of the system, such as the voltage and current supplied by each generator.

(c) *External power.* If provisions are made for connecting external power to the rotorcraft, and that external power can be electrically connected to equipment other than that used for engine starting, means must be provided to ensure that no external power supply having a reverse polarity, or a reverse phase sequence, can supply power to the rotorcraft's electrical system.

(d) Operation with the normal electrical power generating system inoperative.

(1) It must be shown by analysis, tests, or both, that the rotorcraft can be operated safely in VFR conditions for a period of not less than 5 minutes, with the normal electrical power generating system (electrical power sources excluding the battery) inoperative, with critical type fuel (from the standpoint of flameout and restart capability), and with the rotorcraft initially at the maximum certificated altitude. Parts of the electrical system may remain on if—

(i) A single malfunction, including a wire bundle or junction box fire, cannot result in loss of the part turned off and the part turned on;

(ii) The parts turned on are electrically and mechanically isolated from the parts turned off; and

(2) Additional requirements for Category A Rotorcraft.

(i) Unless it can be shown that the loss of the normal electrical power generating system is extremely improbable, an emergency electrical power system, independent of the normal electrical power generating system, must be provided, with sufficient capacity to power all systems necessary for continued safe flight and landing.

(ii) Failures, including junction box, control panel, or wire bundle fires, which would result in the loss of the normal and emergency systems, must be shown to be extremely improbable.

(iii) Systems necessary for immediate safety must continue to operate following the loss of the normal electrical power generating system, without the need for flight crew action.

(Secs. 313(a), 601, 603, 604, and 605 of the Federal Aviation Act of 1958 (49 U.S.C. 1354(a), 1421, 1423, 1424, and 1425); and sec. 6(c), Dept. of Transportation Act (49 U.S.C. 1655(c)))

[Doc. No. 5084, 29 FR 16150, Dec. 3, 1964, as amended by Amdt. 29-14, 42 FR 36973, July 18, 1977; Amdt. 29-40, 61 FR 21908, May 10, 1996; Amdt. 29-42, 63 FR 43285, Aug. 12, 1998]

§ 29.1353 Electrical equipment and installations.

(a) Electrical equipment, controls, and wiring must be installed so that operation of any one unit or system of units will not adversely affect the simultaneous operation of any other electrical unit or system essential to safe operation.

(b) Cables must be grouped, routed, and spaced so that damage to essential circuits will be minimized if there are faults in heavy current-carrying cables.

(c) Storage batteries must be designed and installed as follows:

(1) Safe cell temperatures and pressures must be maintained during any probable charging and discharging condition. No uncontrolled increase in cell temperature may result when the battery is recharged (after previous complete discharge)—

(i) At maximum regulated voltage or power;

(ii) During a flight of maximum duration; and

(iii) Under the most adverse cooling condition likely in service.

(2) Compliance with paragraph (a)(1) of this section must be shown by test unless experience with similar batteries and installations has shown that maintaining safe cell temperatures and pressures presents no problem.

(3) No explosive or toxic gases emitted by any battery in normal operation, or as the result of any probable malfunction in the charging system or battery installation, may accumulate in hazardous quantities within the rotorcraft.

(4) No corrosive fluids or gases that may escape from the battery may damage surrounding structures or adjacent essential equipment.

(5) Each nickel cadmium battery installation capable of being used to start an engine or auxiliary power unit must have provisions to prevent any hazardous effect on structure or essential systems that may be caused by the maximum amount of heat the battery can generate during a short circuit of the battery or of its individual cells.

(6) Nickel cadmium battery installations capable of being used to start an engine or auxiliary power unit must have—

(i) A system to control the charging rate of the battery automatically so as to prevent battery overheating;

(ii) A battery temperature sensing and over-temperature warning system with a means for disconnecting the battery from its charging source in the event of an over-temperature condition; or

(iii) A battery failure sensing and warning system with a means for disconnecting the battery from its charging source in the event of battery failure.

(Secs. 313(a), 601, 603, 604, and 605 of the Federal Aviation Act of 1958 (49 U.S.C. 1354(a), 1421, 1423, 1424, and 1425); and sec. 6(c), Dept. of Transportation Act (49 U.S.C. 1655(c)))

[Doc. No. 5084, 29 FR 16150, Dec. 3, 1964, as amended by Amdt. 29-14, 42 FR 36973, July 18, 1977; Amdt. 29-15, 43 FR 2327, Jan. 16, 1978]

§ 29.1355 Distribution system.

(a) The distribution system includes the distribution busses, their associated feeders, and each control and protective device.

(b) If two independent sources of electrical power for particular equipment or systems are required by this chapter, in the event of the failure of one power source for such equipment or system, another power source (including its separate feeder) must be provided automatically or be manually selectable to maintain equipment or system operation.

(Secs. 313(a), 601, 603, 604, and 605 of the Federal Aviation Act of 1958 (49 U.S.C. 1354(a), 1421, 1423, 1424, and 1425); and sec. 6(c), Dept. of Transportation Act (49 U.S.C. 1655(c)))

[Doc. No. 5084, 29 FR 16150, Dec. 3, 1964, as amended by Amdt. 29-14, 42 FR 36973, July 18, 1977; Amdt. 29-24, 49 FR 44439, Nov. 6, 1984]

§ 29.1357 Circuit protective devices.

(a) Automatic protective devices must be used to minimize distress to the electrical system and hazard to the rotorcraft system and hazard to the rotorcraft in the event of wiring faults or serious malfunction of the system or connected equipment.

(b) The protective and control devices in the generating system must be designed to de-energize and disconnect faulty power sources and power transmission equipment from their associated buses with sufficient rapidity to provide protection from hazardous overvoltage and other malfunctioning.

(c) Each resettable circuit protective device must be designed so that, when an overload or circuit fault exists, it will open the circuit regardless of the position of the operating control.

(d) If the ability to reset a circuit breaker or replace a fuse is essential to safety in flight, that circuit breaker or fuse must be located and identified so that it can be readily reset or replaced in flight.

(e) Each essential load must have individual circuit protection. However, individual protection for each circuit in an essential load system (such as each position light circuit in a system) is not required.

(f) If fuses are used, there must be spare fuses for use in flight equal to at least 50 percent of the number of fuses of each rating required for complete circuit protection.

(g) Automatic reset circuit breakers may be used as integral protectors for electrical equipment provided there is circuit protection for the cable supplying power to the equipment.

[Doc. No. 5084, 29 FR 16150, Dec. 3, 1964, as amended by Amdt. 29-24, 49 FR 44440, Nov. 6, 1984]

§ 29.1359 Electrical system fire and smoke protection.

(a) Components of the electrical system must meet the applicable fire and smoke protection provisions of §§ 29.831 and 29.863.

(b) Electrical cables, terminals, and equipment, in designated fire zones, and that are used in emergency procedures, must be at least fire resistant.

(c) Insulation on electrical wire and cable installed in the rotorcraft must be self-extinguishing when tested in accordance with Appendix F, Part I(a)(3), of part 25 of this chapter.

[Doc. No. 5084, 29 FR 16150, Dec. 3, 1964, as amended by Amdt. 29-42, 63 FR 43285, Aug. 12, 1998]

§ 29.1363 Electrical system tests.

(a) When laboratory tests of the electrical system are conducted—

(1) The tests must be performed on a mock-up using the same generating equipment used in the rotorcraft;

(2) The equipment must simulate the electrical characteristics of the distribution wiring and connected loads to the extent necessary for valid test results; and

(3) Laboratory generator drives must simulate the prime movers on the rotorcraft with respect to their reaction to generator loading, including loading due to faults.

(b) For each flight condition that cannot be simulated adequately in the laboratory or by ground tests on the rotorcraft, flight tests must be made.

Lights

§ 29.1381 Instrument lights.

The instrument lights must—

(a) Make each instrument, switch, and other device for which they are provided easily readable; and

(b) Be installed so that—

(1) Their direct rays are shielded from the pilot's eyes; and

(2) No objectionable reflections are visible to the pilot.

§ 29.1383 Landing lights.

(a) Each required landing or hovering light must be approved.

(b) Each landing light must be installed so that—

(1) No objectionable glare is visible to the pilot;

(2) The pilot is not adversely affected by halation; and

(3) It provides enough light for night operation, including hovering and landing.

(c) At least one separate switch must be provided, as applicable—

(1) For each separately installed landing light; and

(2) For each group of landing lights installed at a common location.

§ 29.1385 Position light system installation.

(a) *General.* Each part of each position light system must meet the applicable requirements of this section and each system as a whole must meet the requirements of §§ 29.1387 through 29.1397.

(b) *Forward position lights.* Forward position lights must consist of a red and a green light spaced laterally as far apart as practicable and installed forward on the rotorcraft so that, with the rotorcraft in the normal flying position, the red light is on the left side, and the green light is on the right side. Each light must be approved.

(c) *Rear position light.* The rear position light must be a white light mounted as far aft as practicable, and must be approved.

(d) *Circuit.* The two forward position lights and the rear position light must make a single circuit.

(e) *Light covers and color filters.* Each light cover or color filter must be at least flame resistant and may not change color or shape or lose any appreciable light transmission during normal use.

§ 29.1387 Position light system dihedral angles.

(a) Except as provided in paragraph (e) of this section, each forward and rear position light must, as installed, show unbroken light within the dihedral angles described in this section.

(b) Dihedral angle *L* (left) is formed by two intersecting vertical planes, the first parallel to the longitudinal axis of the rotorcraft, and the other at 110 degrees to the left of the first, as viewed when looking forward along the longitudinal axis.

(c) Dihedral angle *R* (right) is formed by two intersecting vertical planes, the first parallel to the longitudinal axis of the rotorcraft, and the other at 110 degrees to the right of the first, as viewed when looking forward along the longitudinal axis.

(d) Dihedral angle *A* (aft) is formed by two intersecting vertical planes making angles of 70 degrees to the right and to the left, respectively, to a vertical plane passing through the longitudinal axis, as viewed when looking aft along the longitudinal axis.

(e) If the rear position light, when mounted as far aft as practicable in accordance with § 29.1385(c), cannot show unbroken light within dihedral angle A (as defined in paragraph (d) of this section), a solid angle or angles of obstructed visibility totaling not more than 0.04 steradians is allowable within that dihedral angle, if such solid angle is within a cone whose apex is at the rear position light and whose elements make an angle of 30° with a vertical line passing through the rear position light.

(49 U.S.C. 1655(c))

[Doc. No. 5084, 29 FR 16150, Dec. 3, 1964, as amended by Amdt. 29-9, 36 FR 21279, Nov. 5, 1971]

§ 29.1389 Position light distribution and intensities.

(a) *General.* The intensities prescribed in this section must be provided by new equipment with light covers and color filters in place. Intensities must be determined with the light source operating at a steady value equal to the average luminous output of the source at the normal operating voltage of the rotorcraft. The light distribution and intensity of each position light must meet the requirements of paragraph (b) of this section.

(b) *Forward and rear position lights.* The light distribution and intensities of forward and rear position lights must be expressed in terms of minimum intensities in the horizontal plane, minimum intensities in any vertical plane, and maximum intensities in overlapping beams, within dihedral angles, *L*, *R*, and *A*, and must meet the following requirements:

(1) *Intensities in the horizontal plane.* Each intensity in the horizontal plane (the plane containing the longitudinal axis of the rotorcraft and perpendicular to the plane of symmetry of the rotorcraft), must equal or exceed the values in § 29.1391.

(2) *Intensities in any vertical plane.* Each intensity in any vertical plane (the plane perpendicular to the horizontal plane) must equal or exceed the appropriate value in § 29.1393 where *I* is the minimum intensity prescribed in § 29.1391 for the corresponding angles in the horizontal plane.

(3) *Intensities in overlaps between adjacent signals.* No intensity in any overlap between adjacent signals may exceed the values in § 29.1395, except that higher intensities in overlaps may be used with the use of main beam intensities substantially greater than the minima specified in §§ 29.1391 and 29.1393 if the overlap intensities in relation to the main beam intensities do not adversely affect signal clarity.

§ 29.1391 Minimum intensities in the horizontal plane of forward and rear position lights.

Each position light intensity must equal or exceed the applicable values in the following table:

Dihedral angle (light included)	Angle from right or left of longitudinal axis, measured from dead ahead	Intensity (candles)
L and *R* (forward red and green)	0° to 10° 10° to 20° 20° to 110°	40 30 5
A (rear white)	110° to 180°	20

§ 29.1393 Minimum intensities in any vertical plane of forward and rear position lights.

Each position light intensity must equal or exceed the applicable values in the following table:

Angle above or below the horizontal plane	Intensity, I
0°	1.00
0° to 5°	.90
5° to 10°	.80
10° to 15°	.70
15° to 20°	.50
20° to 30°	.30
30° to 40°	.10
40° to 90°	.05

§ 29.1395 Maximum intensities in overlapping beams of forward and rear position lights.

No position light intensity may exceed the applicable values in the following table, except as provided in § 29.1389(b)(3).

Overlaps	Maximum intensity	
	Area A (candles)	Area B (candles)
Green in dihedral angle *L*	10	1
Red in dihedral angle *R*	10	1
Green in dihedral angle *A*	5	1
Red in dihedral angle *A*	5	1
Rear white in dihedral angle *L*	5	1
Rear white in dihedral angle *R*	5	1

Where—

(a) Area A includes all directions in the adjacent dihedral angle that pass through the light source and intersect the common boundary plane at more than 10 degrees but less than 20 degrees; and

(b) Area B includes all directions in the adjacent dihedral angle that pass through the light source and intersect the common boundary plane at more than 20 degrees.

§ 29.1397 Color specifications.

Each position light color must have the applicable International Commission on Illumination chromaticity coordinates as follows:

(a) *Aviation red*—

y is not greater than 0.335; and

z is not greater than 0.002.

(b) *Aviation green*—

x is not greater than 0.440–0.320 *y*;

x is not greater than *y*–0.170; and

y is not less than 0.390–0.170 *x*.

(c) *Aviation white*—

x is not less than 0.300 and not greater than 0.540;

y is not less than *x*–0.040 or y_c–0.010, whichever is the smaller; and

y is not greater than *x*+0.020 nor 0.636–0.400*x*;

Where Y_e is the *y* coordinate of the Planckian radiator for the value of *x* considered.

[Doc. No. 5084, 29 FR 16150, Dec. 3, 1964, as amended by Amdt. 29-7, 36 FR 12972, July 10, 1971]

§ 29.1399 Riding light.

(a) Each riding light required for water operation must be installed so that it can—

(1) Show a white light for at least two miles at night under clear atmospheric conditions; and

(2) Show a maximum practicable unbroken light with the rotorcraft on the water.

(b) Externally hung lights may be used.

§ 29.1401 Anticollision light system.

(a) *General.* If certification for night operation is requested, the rotorcraft must have an anticollision light system that—

(1) Consists of one or more approved anticollision lights located so that their emitted light will not impair the crew's vision or detract from the conspicuity of the position lights; and

(2) Meets the requirements of paragraphs (b) through (f) of this section.

(b) *Field of coverage.* The system must consist of enough lights to illuminate the vital areas around the rotorcraft, considering the physical configuration and flight characteristics of the rotorcraft. The field of coverage must extend in each direction within at least 30 degrees above and 30 degrees below the horizontal plane of the rotorcraft, except that there may be solid angles of obstructed visibility totaling not more than 0.5 steradians.

(c) *Flashing characteristics.* The arrangement of the system, that is, the number of light sources, beam width, speed of rotation, and other characteristics, must give an effective flash frequency of not less than 40, nor more than 100, cycles per minute. The effective flash frequency is the frequency at which the rotorcraft's complete anticollision light system is observed from a distance, and applies to each sector of light including any overlaps that exist when the system consists of more than one light source. In overlaps, flash frequencies may exceed 100, but not 180, cycles per minute.

(d) *Color.* Each anticollision light must be aviation red and must meet the applicable requirements of § 29.1397.

(e) *Light intensity.* The minimum light intensities in any vertical plane, measured with the red filter (if used) and expressed in terms of "effective" intensities must meet the requirements of paragraph (f) of this section. The following relation must be assumed:

$$I_e = \frac{\int_{t_1}^{t_2} I(t)\,dt}{0.2 + (t_2 - t_1)}$$

where:

I_e =effective intensity (candles).

$I(t)$ =instantaneous intensity as a function of time.

t_2–t_1 =flash time interval (seconds).

Normally, the maximum value of effective intensity is obtained when t_2 and t_1 are chosen so that the effective intensity is equal to the instantaneous intensity at t_2 and t_1.

(f) *Minimum effective intensities for anticollision light.* Each anticollision light effective intensity must equal or exceed the applicable values in the following table:

Angle above or below the horizontal plane	Effective intensity (candles)
0° to 5°	150
5° to 10°	90
10° to 20°	30
20° to 30°	15

[Doc. No. 5084, 29 FR 16150, Dec. 3, 1964, as amended by Amdt. 29-7, 36 FR 12972, July 10, 1971; Amdt. 29-11, 41 FR 5290, Feb. 5, 1976]

Safety Equipment

§ 29.1411 General.

(a) *Accessibility.* Required safety equipment to be used by the crew in an emergency, such as automatic liferaft releases, must be readily accessible.

(b) *Stowage provisions.* Stowage provisions for required emergency equipment must be furnished and must—

(1) Be arranged so that the equipment is directly accessible and its location is obvious; and

(2) Protect the safety equipment from inadvertent damage.

(c) *Emergency exit descent device.* The stowage provisions for the emergency exit descent device required by § 29.809(f) must be at the exits for which they are intended.

(d) *Liferafts.* Liferafts must be stowed near exits through which the rafts can be launched during an unplanned ditching. Rafts automatically or remotely released outside the rotorcraft must be attached to the rotorcraft by the static line prescribed in § 29.1415.

(e) *Long-range signaling device.* The stowage provisions for the long-range signaling device required by § 29.1415 must be near an exit available during an unplanned ditching.

(f) *Life preservers.* Each life preserver must be within easy reach of each occupant while seated.

§ 29.1413 Safety belts: passenger warning device.

(a) If there are means to indicate to the passengers when safety belts should be fastened, they must be installed to be operated from either pilot seat.

(b) Each safety belt must be equipped with a metal to metal latching device.

(Secs. 313, 314, and 601 through 610 of the Federal Aviation Act of 1958 (49 U.S.C. 1354, 1355, and 1421 through 1430) and sec. 6(c), Dept. of Transportation Act (49 U.S.C. 1655(c)))

[Doc. No. 5084, 29 FR 16150, Dec. 3, 1964, as amended by Amdt. 29-16 43 FR 46233, Oct. 5, 1978]

§ 29.1415 Ditching equipment.

(a) Emergency flotation and signaling equipment required by any operating rule of this chapter must meet the requirements of this section.

(b) Each liferaft and each life preserver must be approved. In addition—

(1) Provide not less than two rafts, of an approximately equal rated capacity and buoyancy to accommodate the occupants of the rotorcraft; and

(2) Each raft must have a trailing line, and must have a static line designed to hold the raft near the rotorcraft but to release it if the rotorcraft becomes totally submerged.

(c) Approved survival equipment must be attached to each liferaft.

(d) There must be an approved survival type emergency locator transmitter for use in one life raft.

[Doc. No. 5084, 29 FR 16150, Dec. 3, 1964, as amended by Amdt. 29-8, 36 FR 18722, Sept. 21, 1971; Amdt. 29-19, 45 FR 38348, June 9, 1980; Amdt. 27-26, 55 FR 8005, Mar. 6, 1990; Amdt. 29-33, 59 FR 32057, June 21, 1994]

§ 29.1419 Ice protection.

(a) To obtain certification for flight into icing conditions, compliance with this section must be shown.

(b) It must be demonstrated that the rotorcraft can be safely operated in the continuous maximum and intermittent maximum icing conditions determined under appendix C of this part within the rotorcraft altitude envelope. An analysis must be performed to establish, on the basis of the rotorcraft's operational needs, the adequacy of the ice protection system for the various components of the rotorcraft.

(c) In addition to the analysis and physical evaluation prescribed in paragraph (b) of this section, the effectiveness of the ice protection system and its components must be shown by flight tests of the rotorcraft or its components in measured natural atmospheric icing conditions and by one or more of the following tests as found necessary to determine the adequacy of the ice protection system:

(1) Laboratory dry air or simulated icing tests, or a combination of both, of the components or models of the components.

(2) Flight dry air tests of the ice protection system as a whole, or its individual components.

(3) Flight tests of the rotorcraft or its components in measured simulated icing conditions.

(d) The ice protection provisions of this section are considered to be applicable primarily to the airframe. Powerplant installation requirements are contained in Subpart E of this part.

(e) A means must be identified or provided for determining the formation of ice on critical parts of the rotorcraft. Unless otherwise restricted, the means must be available for nighttime as well as daytime operation. The rotorcraft flight manual must describe the means of determining ice formation and must contain information necessary for safe operation of the rotorcraft in icing conditions.

[Amdt. 29-21, 48 FR 4391, Jan. 31, 1983]

Miscellaneous Equipment

§ 29.1431 Electronic equipment.

(a) Radio communication and navigation equipment installations must be free from hazards in themselves, in their method of operation, and in their effects on other components, under any critical environmental conditions.

(b) Radio communication and navigation equipment, controls, and wiring must be installed so that operation of any one unit or system of units will not adversely affect the simultaneous operation of any other radio or electronic unit, or system of units, required by this chapter.

§ 29.1433 Vacuum systems.

(a) There must be means, in addition to the normal pressure relief, to automatically relieve the pressure in the discharge lines from the vacuum air pump when the delivery temperature of the air becomes unsafe.

(b) Each vacuum air system line and fitting on the discharge side of the pump that might contain flammable vapors or fluids must meet the requirements of § 29.1183 if they are in a designated fire zone.

(c) Other vacuum air system components in designated fire zones must be at least fire resistant.

§ 29.1435 Hydraulic systems.

(a) *Design.* Each hydraulic system must be designed as follows:

(1) Each element of the hydraulic system must be designed to withstand, without detrimental, permanent deformation, any structural loads that may be imposed simultaneously with the maximum operating hydraulic loads.

(2) Each element of the hydraulic system must be designed to withstand pressures sufficiently greater than those prescribed in paragraph (b) of this section to show that the system will not rupture under service conditions.

(3) There must be means to indicate the pressure in each main hydraulic power system.

(4) There must be means to ensure that no pressure in any part of the system will exceed a safe limit above the maximum operating pressure of the system, and to prevent excessive pressures resulting from any fluid volumetric change in lines likely to remain closed long enough for such a change to take place. The possibility of detrimental transient (surge) pressures during operation must be considered.

(5) Each hydraulic line, fitting, and component must be installed and supported to prevent excessive vibration and to withstand inertia loads. Each element of the installation must be protected from abrasion, corrosion, and mechanical damage.

(6) Means for providing flexibility must be used to connect points, in a hydraulic fluid line, between which relative motion or differential vibration exists.

(b) *Tests.* Each element of the system must be tested to a proof pressure of 1.5 times the maximum pressure to which that element will be subjected in normal operation, without failure, malfunction, or detrimental deformation of any part of the system.

(c) *Fire protection.* Each hydraulic system using flammable hydraulic fluid must meet the applicable requirements of §§ 29.861, 29.1183, 29.1185, and 29.1189.

§ 29.1439 Protective breathing equipment.

(a) If one or more cargo or baggage compartments are to be accessible in flight, protective breathing equipment must be available for an appropriate crewmember.

(b) For protective breathing equipment required by paragraph (a) of this section or by any operating rule of this chapter—

(1) That equipment must be designed to protect the crew from smoke, carbon dioxide, and other harmful gases while on flight deck duty;

(2) That equipment must include—

(i) Masks covering the eyes, nose, and mouth; or

(ii) Masks covering the nose and mouth, plus accessory equipment to protect the eyes; and

(3) That equipment must supply protective oxygen of 10 minutes duration per crewmember at a pressure altitude of 8,000 feet with a respiratory minute volume of 30 liters per minute BTPD.

§ 29.1457 Cockpit voice recorders.

(a) Each cockpit voice recorder required by the operating rules of this chapter must be approved, and must be installed so that it will record the following:

(1) Voice communications transmitted from or received in the rotorcraft by radio.

(2) Voice communications of flight crewmembers on the flight deck.

(3) Voice communications of flight crewmembers on the flight deck, using the rotorcraft's interphone system.

(4) Voice or audio signals identifying navigation or approach aids introduced into a headset or speaker.

(5) Voice communications of flight crewmembers using the passenger loudspeaker system, if there is such a system, and if the fourth channel is available in accordance with the requirements of paragraph (c)(4)(ii) of this section.

(6) If datalink communication equipment is installed, all datalink communications, using an approved data message set. Datalink messages must be recorded as the output signal from the communications unit that translates the signal into usable data.

(b) The recording requirements of paragraph (a)(2) of this section may be met—

(1) By installing a cockpit-mounted area microphone, located in the best position for recording voice communications originating at the first and second pilot stations and voice communications of other crewmembers on the flight deck when directed to those stations; or

(2) By installing a continually energized or voice-actuated lip microphone at the first and second pilot stations.

The microphone specified in this paragraph must be so located and, if necessary, the preamplifiers and filters of the recorder must be so adjusted or supplemented, that the recorded communications are intelligible when recorded under flight cockpit noise conditions and played back. The level of intelligibility must be approved by the Administrator. Repeated aural or visual playback of the record may be used in evaluating intelligibility.

(c) Each cockpit voice recorder must be installed so that the part of the communication or audio signals specified in paragraph (a) of this section obtained from each of the following sources is recorded on a separate channel:

(1) For the first channel, from each microphone, headset, or speaker used at the first pilot station.

(2) For the second channel, from each microphone, headset, or speaker used at the second pilot station.

(3) For the third channel, from the cockpit-mounted area microphone, or the continually energized or voice-actuated lip microphones at the first and second pilot stations.

(4) For the fourth channel, from—

(i) Each microphone, headset, or speaker used at the stations for the third and fourth crewmembers; or

(ii) If the stations specified in paragraph (c)(4)(i) of this section are not required or if the signal at such a station is picked up by another channel, each microphone on the flight deck that is used with the passenger loudspeaker system if its signals are not picked up by another channel.

(iii) Each microphone on the flight deck that is used with the rotorcraft's loudspeaker system if its signals are not picked up by another channel.

(d) Each cockpit voice recorder must be installed so that—

(1)(i) It receives its electrical power from the bus that provides the maximum reliability for operation of the cockpit voice recorder without jeopardizing service to essential or emergency loads.

(ii) It remains powered for as long as possible without jeopardizing emergency operation of the rotorcraft.

(2) There is an automatic means to simultaneously stop the recorder and prevent each erasure feature from functioning, within 10 minutes after crash impact;

(3) There is an aural or visual means for preflight checking of the recorder for proper operation;

(4) Whether the cockpit voice recorder and digital flight data recorder are installed in separate boxes or in a combination unit, no single electrical failure external to the recorder may disable both the cockpit voice recorder and the digital flight data recorder; and

(5) It has an independent power source—

(i) That provides 10±1 minutes of electrical power to operate both the cockpit voice recorder and cockpit-mounted area microphone;

(ii) That is located as close as practicable to the cockpit voice recorder; and

(iii) To which the cockpit voice recorder and cockpit-mounted area microphone are switched automatically in the event that all other power to the cockpit voice recorder is interrupted either by normal shutdown or by any other loss of power to the electrical power bus.

(e) The record container must be located and mounted to minimize the probability of rupture of the container as a result of crash impact and consequent heat damage to the record from fire.

(f) If the cockpit voice recorder has a bulk erasure device, the installation must be designed to minimize the probability of inadvertent operation and actuation of the device during crash impact.

(g) Each recorder container must be either bright orange or bright yellow.

(h) When both a cockpit voice recorder and a flight data recorder are required by the operating rules, one combination unit may be installed, provided that all other requirements of this section and the requirements for flight data recorders under this part are met.

[Amdt. 29-6, 35 FR 7293, May 9, 1970, as amended by Amdt. 29-50, 73 FR 12564, Mar. 7, 2008; 74 FR 32800, July 9, 2009; Amdt. 29-52, 75 FR 17045, Apr. 5, 2010]

§ 29.1459 Flight data recorders.

(a) Each flight recorder required by the operating rules of Subchapter G of this chapter must be installed so that:

(1) It is supplied with airspeed, altitude, and directional data obtained from sources that meet the accuracy requirements of §§ 29.1323, 29.1325, and 29.1327 of this part, as applicable;

(2) The vertical acceleration sensor is rigidly attached, and located longitudinally within the approved center of gravity limits of the rotorcraft;

(3)(i) It receives its electrical power from the bus that provides the maximum reliability for operation of the flight data recorder without jeopardizing service to essential or emergency loads.

(ii) It remains powered for as long as possible without jeopardizing emergency operation of the rotorcraft.

(4) There is an aural or visual means for perflight checking of the recorder for proper recording of data in the storage medium;

(5) Except for recorders powered solely by the engine-drive electrical generator system, there is an automatic means to simultaneously stop a recorder that has a data erasure feature and prevent each erasure feature from functioning, within 10 minutes after any crash impact; and

(6) Whether the cockpit voice recorder and digital flight data recorder are installed in separate boxes or in a combination unit, no single electrical failure external to the recorder may disable both the cockpit voice recorder and the digital flight data recorder.

(b) Each nonejectable recorder container must be located and mounted so as to minimize the probability of container rupture resulting from crash impact and subsequent damage to the record from fire.

(c) A correlation must be established between the flight recorder readings of airspeed, altitude, and heading and the corresponding readings (taking into account correction factors) of the first pilot's instruments. This correlation must cover the airspeed range over which the aircraft is to be operated, the range of altitude to which the aircraft is limited, and 360 degrees of heading. Correlation may be established on the ground as appropriate.

(d) Each recorder container must:

(1) Be either bright orange or bright yellow;

(2) Have a reflective tape affixed to its external surface to facilitate its location under water; and

(3) Have an underwater locating device, when required by the operating rules of this chapter, on or adjacent to the container which is secured in such a manner that it is not likely to be separated during crash impact.

(e) When both a cockpit voice recorder and a flight data recorder are required by the operating rules, one combination unit may be installed, provided that all other requirements of this section and the requirements for cockpit voice recorders under this part are met.

[Amdt. 29-25, 53 FR 26145, July 11, 1988; 53 FR 26144, July 11, 1988, as amended by Amdt. 29-50, 73 FR 12564, Mar. 7, 2008; 74 FR 32800, July 9, 2009; Amdt. 29-52, 75 FR 17045, Apr. 5, 2010]

§ 29.1461 Equipment containing high energy rotors.

(a) Equipment containing high energy rotors must meet paragraph (b), (c), or (d) of this section.

(b) High energy rotors contained in equipment must be able to withstand damage caused by malfunctions, vibration, abnormal speeds, and abnormal temperatures. In addition—

(1) Auxiliary rotor cases must be able to contain damage caused by the failure of high energy rotor blades; and

(2) Equipment control devices, systems, and instrumentation must reasonably ensure that no operating limitations affecting the integrity of high energy rotors will be exceeded in service.

(c) It must be shown by test that equipment containing high energy rotors can contain any failure of a high energy rotor that occurs at the highest speed obtainable with the normal speed control devices inoperative.

(d) Equipment containing high energy rotors must be located where rotor failure will neither endanger the occupants nor adversely affect continued safe flight.

[Amdt. 29-3, 33 FR 971, Jan. 26, 1968]

Subpart G—Operating Limitations and Information

§ 29.1501 General.

(a) Each operating limitation specified in §§ 29.1503 through 29.1525 and other limitations and information necessary for safe operation must be established.

(b) The operating limitations and other information necessary for safe operation must be made available to the crewmembers as prescribed in §§ 29.1541 through 29.1589.

(Secs. 313(a), 601, 603, 604, and 605 of the Federal Aviation Act of 1958 (49 U.S.C. 1354(a), 1421, 1423, 1424, and 1425); and sec. 6(c), Dept. of Transportation Act (49 U.S.C. 1655(c)))

[Amdt. 29-15, 43 FR 2327, Jan. 16, 1978]

OPERATING LIMITATIONS

§ 29.1503 Airspeed limitations: general.

(a) An operating speed range must be established.

(b) When airspeed limitations are a function of weight, weight distribution, altitude, rotor speed, power, or other factors, airspeed limitations corresponding with the critical combinations of these factors must be established.

§ 29.1505 Never-exceed speed.

(a) The never-exceed speed, V_{NE}, must be established so that it is—

(1) Not less than 40 knots (CAS); and

(2) Not more than the lesser of—

(i) 0.9 times the maximum forward speeds established under § 29.309;

(ii) 0.9 times the maximum speed shown under §§ 29.251 and 29.629; or

(iii) 0.9 times the maximum speed substantiated for advancing blade tip mach number effects under critical altitude conditions.

(b) V_{NE} may vary with altitude, r.p.m., temperature, and weight, if—

(1) No more than two of these variables (or no more than two instruments integrating more than one of these variables) are used at one time; and

(2) The ranges of these variables (or of the indications on instruments integrating more than one of these variables) are large enough to allow an operationally practical and safe variation of V_{NE}.

(c) For helicopters, a stabilized power-off V_{NE} denoted as V_{NE} (power-off) may be established at a speed less than V_{NE} established pursuant to paragraph (a) of this section, if the following conditions are met:

(1) V_{NE} (power-off) is not less than a speed midway between the power-on V_{NE} and the speed used in meeting the requirements of—

(i) § 29.67(a)(3) for Category A helicopters;

(ii) § 29.65(a) for Category B helicopters, except multi-engine helicopters meeting the requirements of § 29.67(b); and

(iii) § 29.67(b) for multi-engine Category B helicopters meeting the requirements of § 29.67(b).

(2) V_{NE} (power-off) is—

(i) A constant airspeed;

(ii) A constant amount less than power-on V_{NE}; or

(iii) A constant airspeed for a portion of the altitude range for which certification is requested, and a constant amount less than power-on V_{NE} for the remainder of the altitude range.

(Secs. 313(a), 601, 603, 604, and 605 of the Federal Aviation Act of 1958 (49 U.S.C. 1354(a), 1421, 1423, 1424, and 1425); and sec. 6(c), Dept. of Transportation Act (49 U.S.C. 1655(c)))

[Amdt. 29-3, 33 FR 971, Jan. 26, 1968, as amended by Amdt. 29-15, 43 FR 2327, Jan. 16, 1978; Amdt. 29-24, 49 FR 44440, Nov. 6, 1984]

§ 29.1509 Rotor speed.

(a) *Maximum power-off (autorotation).* The maximum power-off rotor speed must be established so that it does not exceed 95 percent of the lesser of—

(1) The maximum design r.p.m. determined under § 29.309(b); and

(2) The maximum r.p.m. shown during the type tests.

(b) *Minimum power-off.* The minimum power-off rotor speed must be established so that it is not less than 105 percent of the greater of—

(1) The minimum shown during the type tests; and

(2) The minimum determined by design substantiation.

(c) *Minimum power-on.* The minimum power-on rotor speed must be established so that it is—

(1) Not less than the greater of—

(i) The minimum shown during the type tests; and

(ii) The minimum determined by design substantiation; and

(2) Not more than a value determined under § 29.33 (a)(1) and (c)(1).

§ 29.1517 Limiting height-speed envelope.

For Category A rotorcraft, if a range of heights exists at any speed, including zero, within which it is not possible to make a safe landing following power failure, the range of heights and its variation with forward speed must be established, together with any other pertinent information, such as the kind of landing surface.

[Amdt. 29-21, 48 FR 4391, Jan. 31, 1983]

§ 29.1519 Weight and center of gravity.

The weight and center of gravity limitations determined under §§ 29.25 and 29.27, respectively, must be established as operating limitations.

§ 29.1521 Powerplant limitations.

(a) *General.* The powerplant limitations prescribed in this section must be established so that they do not exceed the corresponding limits for which the engines are type certificated.

(b) *Takeoff operation.* The powerplant takeoff operation must be limited by—

(1) The maximum rotational speed, which may not be greater than—

(i) The maximum value determined by the rotor design; or

(ii) The maximum value shown during the type tests;

(2) The maximum allowable manifold pressure (for reciprocating engines);

(3) The maximum allowable turbine inlet or turbine outlet gas temperature (for turbine engines);

(4) The maximum allowable power or torque for each engine, considering the power input limitations of the transmission with all engines operating;

(5) The maximum allowable power or torque for each engine considering the power input limitations of the transmission with one engine inoperative;

(6) The time limit for the use of the power corresponding to the limitations established in paragraphs (b)(1) through (5) of this section; and

(7) If the time limit established in paragraph (b)(6) of this section exceeds 2 minutes—

(i) The maximum allowable cylinder head or coolant outlet temperature (for reciprocating engines); and

(ii) The maximum allowable engine and transmission oil temperatures.

(c) *Continuous operation.* The continuous operation must be limited by—

(1) The maximum rotational speed, which may not be greater than—

(i) The maximum value determined by the rotor design; or

(ii) The maximum value shown during the type tests;

(2) The minimum rotational speed shown under the rotor speed requirements in § 29.1509(c).

(3) The maximum allowable manifold pressure (for reciprocating engines);

(4) The maximum allowable turbine inlet or turbine outlet gas temperature (for turbine engines);

(5) The maximum allowable power or torque for each engine, considering the power input limitations of the transmission with all engines operating;

(6) The maximum allowable power or torque for each engine, considering the power input limitations of the transmission with one engine inoperative; and

(7) The maximum allowable temperatures for—

(i) The cylinder head or coolant outlet (for reciprocating engines);

(ii) The engine oil; and

(iii) The transmission oil.

(d) *Fuel grade or designation.* The minimum fuel grade (for reciprocating engines) or fuel designation (for turbine engines) must be established so that it is not less than that required for the operation of the engines within the limitations in paragraphs (b) and (c) of this section.

(e) *Ambient temperature.* Ambient temperature limitations (including limitations for winterization installations if applicable) must be established as the maximum ambient atmospheric temperature at which compliance with the cooling provisions of §§ 29.1041 through 29.1049 is shown.

(f) *Two and one-half minute OEI power operation.* Unless otherwise authorized, the use of $2\frac{1}{2}$ -minute OEI power must be limited to engine failure operation of multi-engine, turbine-powered rotorcraft for not longer than $2\frac{1}{2}$ minutes for any period in which that power is used. The use of $2\frac{1}{2}$ -minute OEI power must also be limited by—

(1) The maximum rotational speed, which may not be greater than—

(i) The maximum value determined by the rotor design; or

(ii) The maximum value shown during the type tests;

(2) The maximum allowable gas temperature;

(3) The maximum allowable torque; and

(4) The maximum allowable oil temperature.

(g) *Thirty-minute OEI power operation.* Unless otherwise authorized, the use of 30-minute OEI power must be limited to multiengine, turbine-powered rotorcraft for not longer than 30 minutes after failure of an engine. The use of 30-minute OEI power must also be limited by—

(1) The maximum rotational speed, which may not be greater than—

(i) The maximum value determined by the rotor design; or

(ii) The maximum value shown during the type tests;

(2) The maximum allowable gas temperature;

(3) The maximum allowable torque; and

(4) The maximum allowable oil temperature.

(h) *Continuous OEI power operation.* Unless otherwise authorized, the use of continuous OEI power must be limited to multiengine, turbine-powered rotorcraft for continued flight after failure of an engine. The use of continuous OEI power must also be limited by—

(1) The maximum rotational speed, which may not be greater than—

(i) The maximum value determined by the rotor design; or

(ii) The maximum value shown during the type tests.

(2) The maximum allowable gas temperature;

(3) The maximum allowable torque; and

(4) The maximum allowable oil temperature.

(i) *Rated 30-second OEI power operation.* Rated 30-second OEI power is permitted only on multiengine, turbine-powered rotorcraft, also certificated for the use of rated 2-minute OEI power, and can only be used for continued operation of the remaining engine(s) after a failure or precautionary shutdown of an engine. It must be shown that following application of 30-second OEI power, any damage will be readily detectable by the applicable inspections and other related procedures furnished in accordance with Section A29.4 of appendix A of this part and Section A33.4 of appendix A of part 33. The use of 30-second OEI power must be limited to not more than 30 seconds for any period in which that power is used, and by—

(1) The maximum rotational speed which may not be greater than—

(i) The maximum value determined by the rotor design; or

(ii) The maximum value demonstrated during the type tests;

(2) The maximum allowable gas temperature; and

(3) The maximum allowable torque.

(j) *Rated 2-minute OEI power operation.* Rated 2-minute OEI power is permitted only on multiengine, turbine-powered rotorcraft, also certificated for the use of rated 30-second OEI power, and can only be used for continued operation of the remaining engine(s) after a failure or precautionary shutdown of an engine. It must be shown that following application of 2-minute OEI power, any damage will be readily detectable by the applicable inspections and other related procedures furnished in accordance with Section A29.4 of appendix a of this part and Section A33.4 of appendix A of part 33. The use of 2-minute OEI power must be limited to not more than 2 minutes for any period in which that power is used, and by—

(1) The maximum rotational speed, which may not be greater than—

(i) The maximum value determined by the rotor design; or

(ii) The maximum value demonstrated during the type tests;

(2) The maximum allowable gas temperature; and

(3) The maximum allowable torque.

(Secs. 313(a), 601, 603, 604, and 605 of the Federal Aviation Act of 1958 (49 U.S.C. 1354(a), 1421, 1423, 1424, and 1425); and sec. 6(c), Dept. of Transportation Act (49 U.S.C. 1655(c)))

[Doc. No. 5084, 29 FR 16150, Dec. 3, 1964, as amended by Amdt. 29-1, 30 FR 8778, July 13, 1965; Amdt. 29-3, 33 FR 971, Jan. 26, 1968; Amdt. 29-15, 43 FR 2327, Jan. 16, 1978; Amdt. 29-26, 53 FR 34220, Sept. 2, 1988; Amdt. 29-34, 59 FR 47768, Sept. 16, 1994; Amdt. 29-41, 62 FR 46173, Aug. 29, 1997]

§ 29.1522 Auxiliary power unit limitations.

If an auxiliary power unit that meets the requirements of TSO-C77 is installed in the rotorcraft, the limitations established for that auxiliary power unit under the TSO including the categories of operation must be specified as operating limitations for the rotorcraft.

(Secs. 313(a), 601, 603, 604, Federal Aviation Act of 1958 (49 U.S.C. 1354(a), 1421, 1423), sec. 6(c), Dept. of Transportation Act (49 U.S.C. 1655(c)))

[Amdt. 29-17, 43 FR 50602, Oct. 30, 1978]

§ 29.1523 Minimum flight crew.

The minimum flight crew must be established so that it is sufficient for safe operation, considering—

(a) The workload on individual crewmembers;

(b) The accessibility and ease of operation of necessary controls by the appropriate crewmember; and

(c) The kinds of operation authorized under § 29.1525.

§ 29.1525 Kinds of operations.

The kinds of operations (such as VFR, IFR, day, night, or icing) for which the rotorcraft is approved are established by demonstrated compliance with the applicable certification requirements and by the installed equipment.

[Amdt. 29-24, 49 FR 44440, Nov. 6, 1984]

§ 29.1527 Maximum operating altitude.

The maximum altitude up to which operation is allowed, as limited by flight, structural, powerplant, functional, or equipment characteristics, must be established.

(Secs. 313(a), 601, 603, 604, and 605 of the Federal Aviation Act of 1958 (49 U.S.C. 1354(a), 1421, 1423, 1424, and 1425); and sec. 6(c), Dept. of Transportation Act (49 U.S.C. 1655(c)))

[Amdt. 29-15, 43 FR 2327, Jan. 16, 1978]

§ 29.1529 Instructions for Continued Airworthiness.

The applicant must prepare Instructions for Continued Airworthiness in accordance with appendix A to this part that are acceptable to the Administrator. The instructions may be incomplete at type certification if a program exists to ensure their completion prior to delivery of the first rotorcraft or issuance of a standard certificate of airworthiness, whichever occurs later.

[Amdt. 29-20, 45 FR 60178, Sept. 11, 1980]

Markings and Placards

§ 29.1541 General.

(a) The rotorcraft must contain—

(1) The markings and placards specified in §§ 29.1545 through 29.1565; and

(2) Any additional information, instrument markings, and placards required for the safe operation of the rotorcraft if it has unusual design, operating or handling characteristics.

(b) Each marking and placard prescribed in paragraph (a) of this section—

(1) Must be displayed in a conspicuous place; and

(2) May not be easily erased, disfigured, or obscured.

§ 29.1543 Instrument markings: general.

For each instrument—

(a) When markings are on the cover glass of the instrument there must be means to maintain the correct alignment of the glass cover with the face of the dial; and

(b) Each arc and line must be wide enough, and located to be clearly visible to the pilot.

§ 29.1545 Airspeed indicator.

(a) Each airspeed indicator must be marked as specified in paragraph (b) of this section, with the marks located at the corresponding indicated airspeeds.

(b) The following markings must be made:

(1) A red radial line—

(i) For rotorcraft other than helicopters, at V_{NE}; and

(ii) For helicopters, at a V_{NE} (power-on).

(2) A red, cross-hatched radial line at V_{NE} (power-off) for helicopters, if V_{NE} (power-off) is less than V_{NE} (power-on).

(3) For the caution range, a yellow arc.

(4) For the safe operating range, a green arc.

(Secs. 313(a), 601, 603, 604, and 605 of the Federal Aviation Act of 1958 (49 U.S.C. 1354(a), 1421, 1423, 1424, and 1425); and sec. 6(c), Dept. of Transportation Act (49 U.S.C. 1655(c)))

[Doc. No. 5084, 29 FR 16150, Dec. 3, 1964, as amended by Amdt. 29-15, 43 FR 2327, Jan. 16, 1978; 43 FR 3900, Jan. 30, 1978; Amdt. 29-17, 43 FR 50602, Oct. 30, 1978]

§ 29.1547 Magnetic direction indicator.

(a) A placard meeting the requirements of this section must be installed on or near the magnetic direction indicator.

(b) The placard must show the calibration of the instrument in level flight with the engines operating.

(c) The placard must state whether the calibration was made with radio receivers on or off.

(d) Each calibration reading must be in terms of magnetic heading in not more than 45 degree increments.

§ 29.1549 Powerplant instruments.

For each required powerplant instrument, as appropriate to the type of instruments—

(a) Each maximum and, if applicable, minimum safe operating limit must be marked with a red radial or a red line;

(b) Each normal operating range must be marked with a green arc or green line, not extending beyond the maximum and minimum safe limits;

(c) Each takeoff and precautionary range must be marked with a yellow arc or yellow line;

(d) Each engine or propeller range that is restricted because of excessive vibration stresses must be marked with red arcs or red lines; and

(e) Each OEI limit or approved operating range must be marked to be clearly differentiated from the markings of paragraphs (a) through (d) of this section except that no marking is normally required for the 30-second OEI limit.

[Amdt. 29-12, 41 FR 55474, Dec. 20, 1976, as amended by Amdt. 29-26, 53 FR 34220, Sept. 2, 1988; Amdt. 29-34, 59 FR 47769, Sept. 16, 1994]

§ 29.1551 Oil quantity indicator.

Each oil quantity indicator must be marked with enough increments to indicate readily and accurately the quantity of oil.

§ 29.1553 Fuel quantity indicator.

If the unusable fuel supply for any tank exceeds one gallon, or five percent of the tank capacity, whichever is greater, a red arc must be marked on its indicator extending from the calibrated zero reading to the lowest reading obtainable in level flight.

§ 29.1555 Control markings.

(a) Each cockpit control, other than primary flight controls or control whose function is obvious, must be plainly marked as to its function and method of operation.

(b) For powerplant fuel controls—

(1) Each fuel tank selector valve control must be marked to indicate the position corresponding to each tank and to each existing cross feed position;

(2) If safe operation requires the use of any tanks in a specific sequence, that sequence must be marked on, or adjacent to, the selector for those tanks; and

(3) Each valve control for any engine of a multiengine rotorcraft must be marked to indicate the position corresponding to each engine controlled.

(c) Usable fuel capacity must be marked as follows:

(1) For fuel systems having no selector controls, the usable fuel capacity of the system must be indicated at the fuel quantity indicator.

(2) For fuel systems having selector controls, the usable fuel capacity available at each selector control position must be indicated near the selector control.

(d) For accessory, auxiliary, and emergency controls—

(1) Each essential visual position indicator, such as those showing rotor pitch or landing gear position, must be marked so that each crewmember can determine at any time the position of the unit to which it relates; and

(2) Each emergency control must be red and must be marked as to method of operation.

(e) For rotorcraft incorporating retractable landing gear, the maximum landing gear operating speed must be displayed in clear view of the pilot.

[Doc. No. 5084, 29 FR 16150, Dec. 3, 1964, as amended by Amdt. 29-12, 41 FR 55474, Dec. 20, 1976; Amdt. 29-24, 49 FR 44440, Nov. 6, 1984]

§ 29.1557 Miscellaneous markings and placards.

(a) *Baggage and cargo compartments, and ballast location.* Each baggage and cargo compartment, and each ballast location must have a placard stating any limitations on contents, including weight, that are necessary under the loading requirements.

(b) *Seats.* If the maximum allowable weight to be carried in a seat is less than 170 pounds, a placard stating the lesser weight must be permanently attached to the seat structure.

(c) *Fuel and oil filler openings.* The following apply:

(1) Fuel filler openings must be marked at or near the filler cover with—

(i) The word "fuel";

(ii) For reciprocating engine powered rotorcraft, the minimum fuel grade;

(iii) For turbine-engine-powered rotorcraft, the permissible fuel designations, except that if impractical, this information may be included in the rotorcraft flight manual, and the fuel filler may be marked with an appropriate reference to the flight manual; and

(iv) For pressure fueling systems, the maximum permissible fueling supply pressure and the maximum permissible defueling pressure.

(2) Oil filler openings must be marked at or near the filler cover with the word "oil".

(d) *Emergency exit placards.* Each placard and operating control for each emergency exit must differ in color from the surrounding fuselage surface as prescribed in § 29.811(h)(2). A placard must be near each emergency exit control and must clearly indicate the location of that exit and its method of operation.

[Doc. No. 5084, 29 FR 16150, Dec. 3, 1964, as amended by Amdt. 29-3, 33 FR 971, Jan. 26, 1968; Amdt. 29-12, 41 FR 55474, Dec. 20, 1976; Amdt. 29-26, 53 FR 34220, Sept. 2, 1988]

§ 29.1559 Limitations placard.

There must be a placard in clear view of the pilot that specifies the kinds of operations (VFR, IFR, day, night, or icing) for which the rotorcraft is approved.

[Amdt. 29-24, 49 FR 44440, Nov. 6, 1984]

§ 29.1561 Safety equipment.

(a) Each safety equipment control to be operated by the crew in emergency, such as controls for automatic liferaft releases, must be plainly marked as to its method of operation.

(b) Each location, such as a locker or compartment, that carries any fire extinguishing, signaling, or other life saving equipment, must be so marked.

(c) Stowage provisions for required emergency equipment must be conspicuously marked to identify the contents and facilitate removal of the equipment.

(d) Each liferaft must have obviously marked operating instructions.

(e) Approved survival equipment must be marked for identification and method of operation.

§ 29.1565 Tail rotor.

Each tail rotor must be marked so that its disc is conspicuous under normal daylight ground conditions.

[Amdt. 29-3, 33 FR 971, Jan. 26, 1968]

Rotorcraft Flight Manual

§ 29.1581 General.

(a) *Furnishing information.* A Rotorcraft Flight Manual must be furnished with each rotorcraft, and it must contain the following:

(1) Information required by §§ 29.1583 through 29.1589.

(2) Other information that is necessary for safe operation because of design, operating, or handling characteristics.

(b) *Approved information.* Each part of the manual listed in §§ 29.1583 through 29.1589 that is appropriate to the rotorcraft, must be furnished, verified, and approved, and must be segregated, indentified, and clearly distinguished from each unapproved part of that manual.

(c) [Reserved]

(d) *Table of contents.* Each Rotorcraft Flight Manual must include a table of contents if the complexity of the manual indicates a need for it.

(Secs. 313(a), 601, 603, 604, and 605 of the Federal Aviation Act of 1958 (49 U.S.C. 1354(a), 1421, 1423, 1424, and 1425); and sec. 6(c), Dept. of Transportation Act (49 U.S.C. 1655(c)))

[Amdt. 29-15, 43 FR 2327, Jan. 16, 1978]

§ 29.1583 Operating limitations.

(a) *Airspeed and rotor limitations.* Information necessary for the marking of airspeed and rotor limitations on or near their respective indicators must be furnished. The significance of each limitation and of the color coding must be explained.

(b) *Powerplant limitations.* The following information must be furnished:

(1) Limitations required by § 29.1521.

(2) Explanation of the limitations, when appropriate.

(3) Information necessary for marking the instruments required by §§ 29.1549 through 29.1553.

(c) *Weight and loading distribution.* The weight and center of gravity limits required by §§ 29.25 and 29.27, respectively, must be furnished. If the variety of possible loading conditions warrants, instructions must be included to allow ready observance of the limitations.

(d) *Flight crew.* When a flight crew of more than one is required, the number and functions of the minimum flight crew determined under § 29.1523 must be furnished.

(e) *Kinds of operation.* Each kind of operation for which the rotorcraft and its equipment installations are approved must be listed.

(f) *Limiting heights.* Enough information must be furnished to allow compliance with § 29.1517.

(g) *Maximum allowable wind.* For Category A rotorcraft, the maximum allowable wind for safe operation near the ground must be furnished.

(h) *Altitude.* The altitude established under § 29.1527 and an explanation of the limiting factors must be furnished.

(i) *Ambient temperature.* Maximum and minimum ambient temperature limitations must be furnished.

(Secs. 313(a), 601, 603, 604, and 605 of the Federal Aviation Act of 1958 (49 U.S.C. 1354(a), 1421, 1423, 1424, and 1425); and sec. 6(c), Dept. of Transportation Act (49 U.S.C. 1655(c)))

[Doc. No. 5084, 29 FR 16150, Dec. 3, 1964, as amended by Amdt. 29-3, 33 FR 971, Jan. 26, 1968; Amdt. 29-15, 43 FR 2327, Jan. 16, 1978; Amdt. 29-17, 43 FR 50602, Oct. 30, 1978; Amdt. 29-24, 49 FR 44440, Nov. 6, 1984]

§ 29.1585 Operating procedures.

(a) The parts of the manual containing operating procedures must have information concerning any normal and emergency procedures, and other information necessary for safe operation, including the applicable procedures, such as those involving minimum speeds, to be followed if an engine fails.

(b) For multiengine rotorcraft, information identifying each operating condition in which the fuel system independence prescribed in § 29.953 is necessary for safety must be furnished, together with instructions for placing the fuel system in a configuration used to show compliance with that section.

(c) For helicopters for which a V_{NE} (power-off) is established under § 29.1505(c), information must be furnished to explain the V_{NE} (power-off) and the procedures for reducing airspeed to not more than the V_{NE} (power-off) following failure of all engines.

(d) For each rotorcraft showing compliance with § 29.1353 (c)(6)(ii) or (c)(6)(iii), the operating procedures for disconnecting the battery from its charging source must be furnished.

(e) If the unusable fuel supply in any tank exceeds 5 percent of the tank capacity, or 1 gallon, whichever is greater, information must be furnished which indicates that when the fuel quantity indicator reads "zero" in level flight, any fuel remaining in the fuel tank cannot be used safely in flight.

(f) Information on the total quantity of usable fuel for each fuel tank must be furnished.

(g) For Category B rotorcraft, the airspeeds and corresponding rotor speeds for minimum rate of descent and best glide angle as prescribed in § 29.71 must be provided.

(Secs. 313(a), 601, 603, 604, and 605 of the Federal Aviation Act of 1958 (49 U.S.C. 1354(a), 1421, 1423, 1424, and 1425); and sec. 6(c), Dept. of Transportation Act (49 U.S.C. 1655(c)))

[Amdt. 29-2, 32 FR 6914, May 5, 1967, as amended by Amdt. 29-15, 43 FR 2328, Jan. 16, 1978; Amdt. 29-17, 43 FR 50602, Oct. 30, 1978; Amdt. 29-24, 49 FR 44440, Nov. 6, 1984]

§ 29.1587 Performance information.

Flight manual performance information which exceeds any operating limitation may be shown only to the extent necessary for presentation clarity or to determine the effects of approved optional equipment or procedures. When data beyond operating limits are shown, the limits must be clearly indicated. The following must be provided:

(a) *Category A.* For each category A rotorcraft, the Rotorcraft Flight Manual must contain a summary of the performance data, including data necessary for the application of any operating rule of this chapter, together with descriptions of the conditions, such as airspeeds, under which this data was determined, and must contain—

(1) The indicated airspeeds corresponding with those determined for takeoff, and the procedures to be followed if the critical engine fails during takeoff;

(2) The airspeed calibrations;

(3) The techniques, associated airspeeds, and rates of descent for autorotative landings;

(4) The rejected takeoff distance determined under § 29.62 and the takeoff distance determined under § 29.61;

(5) The landing data determined under § 29.81 and § 29.85;

(6) The steady gradient of climb for each weight, altitude, and temperature for which takeoff data are to be scheduled, along the takeoff path determined in the flight conditions required in § 29.67(a)(1) and (a)(2):

(i) In the flight conditions required in § 29.67(a)(1) between the end of the takeoff distance and the point at which the rotorcraft is 200 feet above the takeoff surface (or 200 feet above the lowest point of the takeoff profile for elevated heliports);

(ii) In the flight conditions required in § 29.67(a)(2) between the points at which the rotorcraft is 200 and 1000 feet above the takeoff surface (or 200 and 1000 feet above the lowest point of the takeoff profile for elevated heliports); and

(7) Out-of-ground effect hover performance determined under § 29.49 and the maximum weight for each altitude and temperature condition at which the rotorcraft can safely hover out-of-ground effect in winds of not less than 17 knots from all azimuths. These data must be clearly referenced to the appropriate hover charts.

(b) *Category B.* For each category B rotorcraft, the Rotorcraft Flight Manual must contain—

(1) The takeoff distance and the climbout speed together with the pertinent information defining the flight path with respect to autorotative landing if an engine fails, including the calculated effects of altitude and temperature;

(2) The steady rates of climb and in-ground-effect hovering ceiling, together with the corresponding airspeeds and other pertinent information, including the calculated effects of altitude and temperature;

(3) The landing distance, appropriate airspeed, and type of landing surface, together with all pertinent information that might affect this distance, including the effects of weight, altitude, and temperature;

(4) The maximum safe wind for operation near the ground;

(5) The airspeed calibrations;

(6) The height-speed envelope except for rotorcraft incorporating this as an operating limitation;

(7) Glide distance as a function of altitude when autorotating at the speeds and conditions for minimum rate of descent and best glide angle, as determined in § 29.71;

(8) Out-of-ground effect hover performance determined under § 29.49 and the maximum safe wind demonstrated under the ambient conditions for data presented. In addition, the maximum weight for each altitude and temperature condition at which the rotorcraft can safely hover out-of-ground-effect in winds of not less than 17 knots from all azimuths. These data must be clearly referenced to the appropriate hover charts; and

(9) Any additional performance data necessary for the application of any operating rule in this chapter.

[Doc. No. 5084, 29 FR 16150, Dec. 3, 1964, as amended by Amdt. 29-21, 48 FR 4392, Jan. 31, 1983; Amdt. 29-24, 49 FR 44440, Nov. 6, 1984; Amdt. 29-39, 61 FR 21901, May 10, 1996; Amdt. 29-40, 61 FR 21908, May 10, 1996; Amdt. 29-44, 64 FR 45338, Aug. 19, 1999; Amdt. 29-51, 73 FR 11001, Feb. 29, 2008]

§ 29.1589 Loading information.

There must be loading instructions for each possible loading condition between the maximum and minimum weights determined under § 29.25 that can result in a center of gravity beyond any extreme prescribed in § 29.27, assuming any probable occupant weights.

Appendix A to Part 29—Instructions for Continued Airworthiness

a29.1 *General*

(a) This appendix specifies requirements for the preparation of Instructions for Continued Airworthiness as required by § 29.1529.

(b) The Instructions for Continued Airworthiness for each rotorcraft must include the Instructions for Continued Airworthiness for each engine and rotor (hereinafter designated "products"), for each applicance required by this chapter, and any required information relating to the interface of those appliances and products with the rotorcraft. If Instructions for Continued Airworthiness are not supplied by the manufacturer of an appliance or product installed in the rotorcraft, the Instructions for Continued Airworthiness for the rotorcraft must include the information essential to the continued airworthiness of the rotorcraft.

(c) The applicant must submit to the FAA a program to show how changes to the Instructions for Continued Airworthiness made by the applicant or by the manufacturers of products and appliances installed in the rotorcraft will be distributed.

a29.2 *Format*

(a) The Instructions for Continued Airworthiness must be in the form of a manual or manuals as appropriate for the quantity of data to be provided.

(b) The format of the manual or manuals must provide for a practical arrangement.

a29.3 *Content*

The contents of the manual or manuals must be prepared in the English language. The Instructions for Continued Airworthiness must contain the following manuals or sections, as appropriate, and information:

(a) *Rotorcraft maintenance manual or section.* (1) Introduction information that includes an explanation of the rotorcraft's features and data to the extent necessary for maintenance or preventive maintenance.

(2) A description of the rotorcraft and its systems and installations including its engines, rotors, and appliances.

(3) Basic control and operation information describing how the rotorcraft components and systems are controlled and how they operate, including any special procedures and limitations that apply.

(4) Servicing information that covers details regarding servicing points, capacities of tanks, reservoirs, types of fluids to be used, pressures applicable to the various systems, location of access panels for inspection and servicing, locations of lubrication points, the lubricants to be used, equipment required for servicing, tow instructions and limitations, mooring, jacking, and leveling information.

(b) *Maintenance Instructions.* (1) Scheduling information for each part of the rotorcraft and its engines, auxiliary power units, rotors, accessories, instruments, and equipment that provides the recommended periods at which they should be cleaned, inspected, adjusted, tested, and lubricated, and the degree of inspection, the applicable wear tolerances, and work recommended at these periods. However, the applicant may refer to an accessory, instrument, or equipment manufacturer as the source of this information if the applicant shows that the item has an exceptionally high degree of complexity requiring specialized maintenance techniques, test equipment, or expertise. The recommended overhaul periods and necessary cross references to the Airworthiness Limitations section of the manual must also be included. In addition, the applicant must include an inspection program that includes the frequency and extent of the inspections necessary to provide for the continued airworthiness of the rotorcraft.

(2) Troubleshooting information describing probable malfunctions, how to recognize those malfunctions, and the remedial action for those malfunctions.

(3) Information describing the order and method of removing and replacing products and parts with any necessary precautions to be taken.

(4) Other general procedural instructions including procedures for system testing during ground running, symmetry checks, weighing and determining the center of gravity, lifting and shoring, and storage limitations.

(c) Diagrams of structural access plates and information needed to gain access for inspections when access plates are not provided.

(d) Details for the application of special inspection techniques including radiographic and ultrasonic testing where such processes are specified.

(e) Information needed to apply protective treatments to the structure after inspection.

(f) All data relative to structural fasteners such as identification, discard recommendations, and torque values.

(g) A list of special tools needed.

a29.4 *Airworthiness Limitations Section*

The Instructions for Continued Airworthiness must contain a section titled Airworthiness Limitations that is segregated and clearly distinguishable from the rest of the document. This section must set forth each mandatory replacement time, structural inspection interval, and related structural inspection procedure required for type certification. If the Instructions for Continued Airworthiness consist of multiple documents, the section required by this paragraph must be included in the principal manual. This section must contain a legible statement in a prominent location that reads: "The Airworthiness Limitations section is FAA approved and specifies maintenance required under §§ 43.16 and 91.403 of the Federal Aviation Regulations unless an alternative program has been FAA approved."

[Amdt. 29-20, 45 FR 60178, Sept. 11, 1980, as amended by Amdt. 29-27, 54 FR 34330, Aug. 18, 1989; Amdt. 29-54, 76 FR 74664, Dec. 1, 2011]

Appendix B to Part 29—Airworthiness Criteria for Helicopter Instrument Flight

I. *General.* A transport category helicopter may not be type certificated for operation under the instrument flight rules (IFR) of this chapter unless it meets the design and installation requirements contained in this appendix.

II. *Definitions.* (a) V_{YI} means instrument climb speed, utilized instead of V_Y for compliance with the climb requirements for instrument flight.

(b) V_{NEI} means instrument flight never exceed speed, utilized instead of V_{NE} for compliance with maximum limit speed requirements for instrument flight.

(c) V_{MINI} means instrument flight minimum speed, utilized in complying with minimum limit speed requirements for instrument flight.

III. *Trim.* It must be possible to trim the cyclic, collective, and directional control forces to zero at all approved IFR airspeeds, power settings, and configurations appropriate to the type.

IV. *Static longitudinal stability.* (a) *General.* The helicopter must possess positive static longitudinal control force stability at critical combinations of weight and center of gravity at the conditions specified in paragraphs IV (b) through (f) of this appendix. The stick force must vary with speed so that any substantial speed change results in a stick force clearly perceptible to the pilot. The airspeed must return to within 10 percent of the trim speed when the control force is slowly released for each trim condition specified in paragraphs IV (b) through (f) of this appendix.

(b) *Climb.* Stability must be shown in climb thoughout the speed range 20 knots either side of trim with—

(1) The helicopter trimmed at V_{YI};

(2) Landing gear retracted (if retractable); and

(3) Power required for limit climb rate (at least 1,000 fpm) at V_{YI} or maximum continuous power, whichever is less.

(c) *Cruise.* Stability must be shown throughout the speed range from 0.7 to 1.1 V_H or V_{NEI}, whichever is lower, not to exceed±20 knots from trim with—

(1) The helicopter trimmed and power adjusted for level flight at 0.9 V_H or 0.9 V_{NEI}, whichever is lower; and

(2) Landing gear retracted (if retractable).

(d) *Slow cruise.* Stability must be shown throughout the speed range from 0.9 V_{MINI} to 1.3 V_{MINI} or 20 knots above trim speed, whichever is greater, with—

(1) The helicopter trimmed and power adjusted for level flight at 1.1 V_{MINI}; and

(2) Landing gear retracted (if retractable).

(e) *Descent.* Stability must be shown throughout the speed range 20 knots either side of trim with—

(1) The helicopter trimmed at 0.8 V_H or 0.8 V_{NEI} (or 0.8 V_{LE} for the landing gear extended case), whichever is lower;

(2) Power required for 1,000 fpm descent at trim speed; and

(3) Landing gear extended and retracted, if applicable.

(f) *Approach.* Stability must be shown throughout the speed range from 0.7 times the minimum recommended approach speed to 20 knots above the maximum recommended approach speed with—

(1) The helicopter trimmed at the recommended approach speed or speeds;

(2) Landing gear extended and retracted, if applicable; and

(3) Power required to maintain a 3° glide path and power required to maintain the steepest approach gradient for which approval is requested.

V. *Static Lateral Directional Stability*

(a) Static directional stability must be positive throughout the approved ranges of airspeed, power, and vertical speed. In straight and steady sideslips up to±10° from trim, directional control position must increase without discontinuity with the angle of sideslip, except for a small range of sideslip angles around trim. At greater angles up to the maximum sideslip angle appropriate to the type, increased directional control position must produce an increased angle of sideslip. It must be possible to maintain balanced flight without exceptional pilot skill or alertness.

(b) During sideslips up to±10° from trim throughout the approved ranges of airspeed, power, and vertical speed there must be no negative dihedral stability perceptible to the pilot through lateral control motion or force. Longitudinal cyclic movement with sideslip must not be excessive.

VI. *Dynamic stability.* (a) Any oscillation having a period of less than 5 seconds must damp to ½ amplitude in not more than one cycle.

(b) Any oscillation having a period of 5 seconds or more but less than 10 seconds must damp to ½ amplitude in not more than two cycles.

(c) Any oscillation having a period of 10 seconds or more but less than 20 seconds must be damped.

(d) Any oscillation having a period of 20 seconds or more may not achieve double amplitude in less than 20 seconds.

(e) Any aperiodic response may not achieve double amplitude in less than 9 seconds.

VII. *Stability Augmentation System (SAS)*

(a) If a SAS is used, the reliability of the SAS must be related to the effects of its failure. Any SAS failure condition that would prevent continued safe flight and landing must be extremely improbable. It must be shown that, for any failure condition of the SAS that is not shown to be extremely improbable—

(1) The helicopter is safely controllable when the failure or malfunction occurs at any speed or altitude within the approved IFR operating limitations; and

(2) The overall flight characteristics of the helicopter allow for prolonged instrument flight without undue pilot effort. Additional unrelated probable failures affecting the control system must be considered. In addition—

(i) The controllability and maneuverability requirements in Subpart B must be met throughout a practical flight envelope;

(ii) The flight control, trim, and dynamic stability characteristics must not be impaired below a level needed to allow continued safe flight and landing;

(iii) For Category A helicopters, the dynamic stability requirements of Subpart B must also be met throughout a practical flight envelope; and

(iv) The static longitudinal and static directional stability requirements of Subpart B must be met throughout a practical flight envelope.

(b) The SAS must be designed so that it cannot create a hazardous deviation in flight path or produce hazardous loads on the helicopter during normal operation or in the event of malfunction or failure, assuming corrective action begins within an appropriate period of time. Where multiple systems are installed, subsequent malfunction conditions must be considered in sequence unless their occurrence is shown to be improbable.

VIII. *Equipment, systems, and installation.* The basic equipment and installation must comply with Subpart F of Part 29 through Amendment 29-14, with the following exceptions and additions:

(a) *Flight and navigation instruments.* (1) A magnetic gyro-stabilized direction indicator instead of the gyroscopic direction indicator required by § 29.1303(h); and

(2) A standby attitude indicator which meets the requirements of §§ 29.1303(g)(1) through (7), instead of a rate-of-turn indicator required by § 29.1303(g). If standby batteries are provided, they may be charged from the aircraft electrical system if adequate isolation is incorporated. The system must be designed so that the standby batteries may not be used for engine starting.

(b) *Miscellaneous requirements.* (1) Instrument systems and other systems essential for IFR flight that could be adversely affected by icing must be provided with adequate ice protection whether or not the rotorcraft is certificated for operation in icing conditions.

(2) There must be means in the generating system to automatically de-energize and disconnect from the main bus any power source developing hazardous overvoltage.

(3) Each required flight instrument using a power supply (electric, vacuum, etc.) must have a visual means integral with the instrument to indicate the adequacy of the power being supplied.

(4) When multiple systems performing like functions are required, each system must be grouped, routed, and spaced so that physical separation between systems is provided to ensure that a single malfunction will not adversely affect more than one system.

(5) For systems that operate the required flight instruments at each pilot's station—

(i) Only the required flight instruments for the first pilot may be connected to that operating system;

(ii) Additional instruments, systems, or equipment may not be connected to an operating system for a second pilot unless provisions are made to ensure the continued normal functioning of the required instruments in the event of any malfunction of the additional instruments, systems, or equipment which is not shown to be extremely improbable;

(iii) The equipment, systems, and installations must be designed so that one display of the information essential to the safety of flight which is provided by the instruments will remain available to a pilot, without additional crew-member action, after any single failure or combination of failures that is not shown to be extremely improbable; and

(iv) For single-pilot configurations, instruments which require a static source must be provided with a means of selecting an alternate source and that source must be calibrated.

(6) In determining compliance with the requirements of § 29.1351(d)(2), the supply of electrical power to all systems necessary for flight under IFR must be included in the evaluation.

(c) *Thunderstorm lights.* In addition to the instrument lights required by § 29.1381(a), thunderstorm lights which provide high intensity white flood lighting to the basic flight instruments must be provided. The thunderstorm lights must be installed to meet the requirements of § 29.1381(b).

IX. *Rotorcraft Flight Manual.* A Rotorcraft Flight Manual or Rotorcraft Flight Manual IFR Supplement must be provided and must contain—

(a) *Limitations.* The approved IFR flight envelope, the IFR flightcrew composition, the revised kinds of operation, and the steepest IFR precision approach gradient for which the helicopter is approved;

(b) *Procedures.* Required information for proper operation of IFR systems and the recommended procedures in the event of stability augmentation or electrical system failures; and

(c) *Performance.* If V_{YI} differs from V_Y, climb performance at V_{YI} and with maximum continuous power throughout the ranges of weight, altitude, and temperature for which approval is requested.

[Amdt. 29-21, 48 FR 4392, Jan. 31, 1983, as amended by Amdt. 29-31, 55 FR 38967, Sept. 21, 1990; 55 FR 41309, Oct. 10, 1990; Amdt. 29-40, 61 FR 21908, May 10, 1996; Amdt. 29-51, 73 FR 11002, Feb. 29, 2008]

Appendix C to Part 29—Icing Certification

(a) *Continuous maximum icing.* The maximum continuous intensity of atmospheric icing conditions (continuous maximum icing) is defined by the variables of the cloud liquid water content, the mean effective diameter of the cloud droplets, the ambient air temperature, and the interrelationship of these three variables as shown in Figure 1 of this appendix. The limiting icing envelope in terms of altitude and temperature is given in Figure 2 of this appendix. The interrelationship of cloud liquid water content with drop diameter and altitude is determined from Figures 1 and 2. The cloud liquid water content for continuous maximum icing conditions of a horizontal extent, other than 17.4 nautical miles, is determined by the value of liquid water content of Figure 1, multiplied by the appropriate factor from Figure 3 of this appendix.

(b) *Intermittent maximum icing.* The intermittent maximum intensity of atmospheric icing conditions (intermittent maximum icing) is defined by the variables of the cloud liquid water content, the mean effective diameter of the cloud droplets, the ambient air temperature, and the interrelationship of these three variables as shown in Figure 4 of this appendix. The limiting icing envelope in terms of altitude and temperature is given in Figure 5 of this appendix. The interrelationship of cloud liquid water content with drop diameter and altitude is determined from Figures 4 and 5. The cloud liquid water content for intermittent maximum icing conditions of a horizontal extent, other than 2.6 nautical miles, is determined by the value of cloud liquid water content of Figure 4 multiplied by the appropriate factor in Figure 6 of this appendix.

[Amdt. 29-21, 48 FR 4393, Jan. 31, 1983]

Figure 1

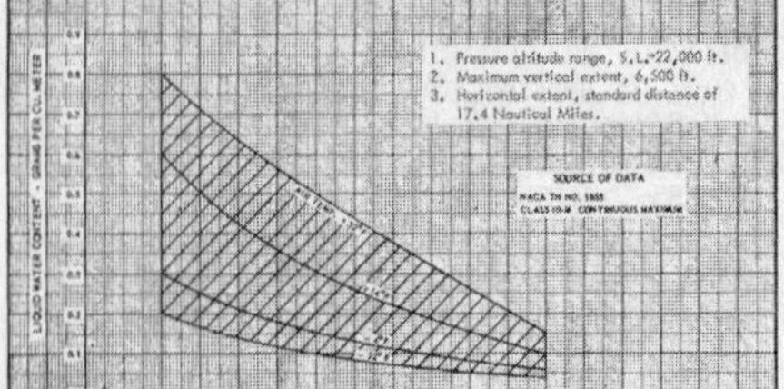

Figure 2

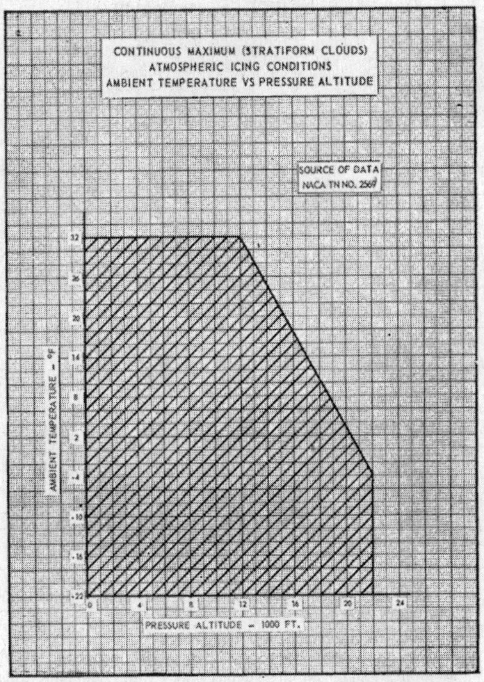

Figure 3

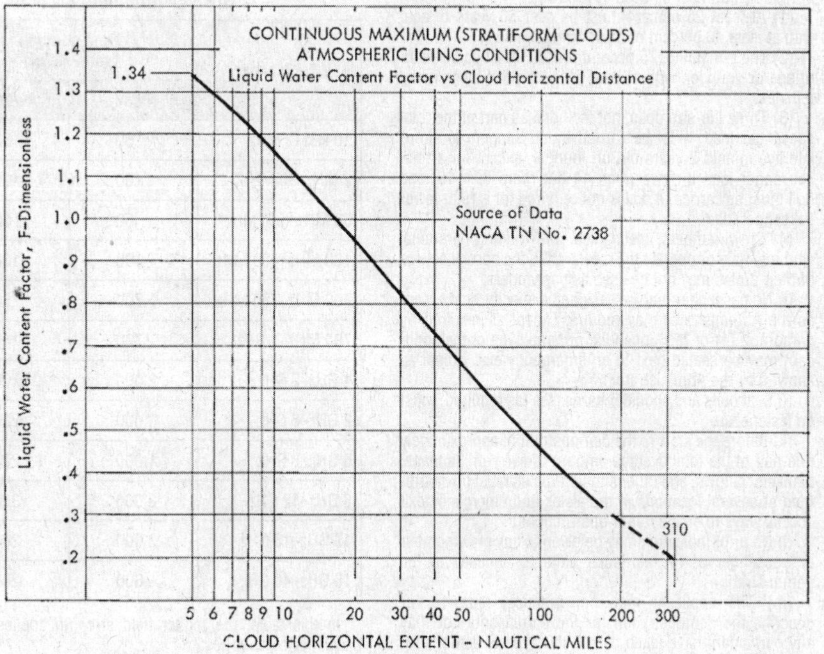

Figure 4

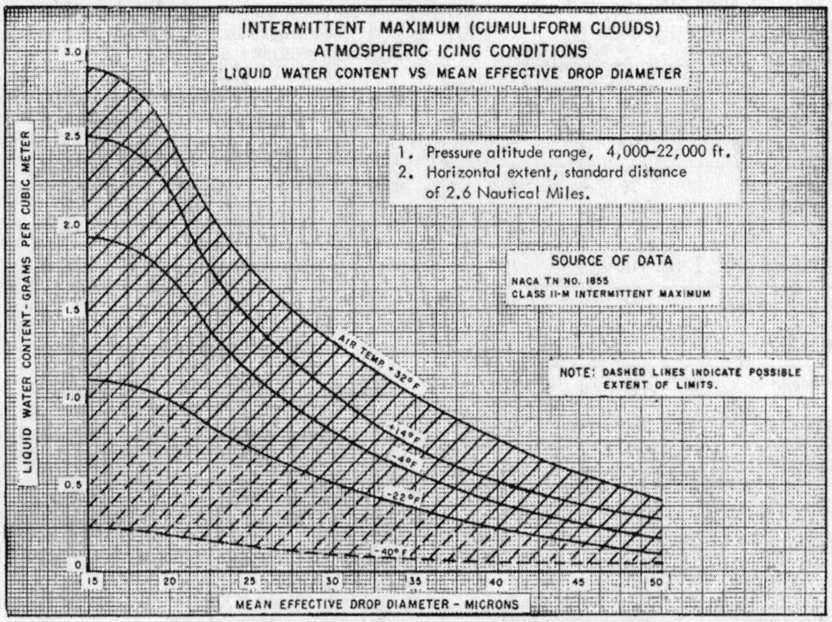

Figure 5

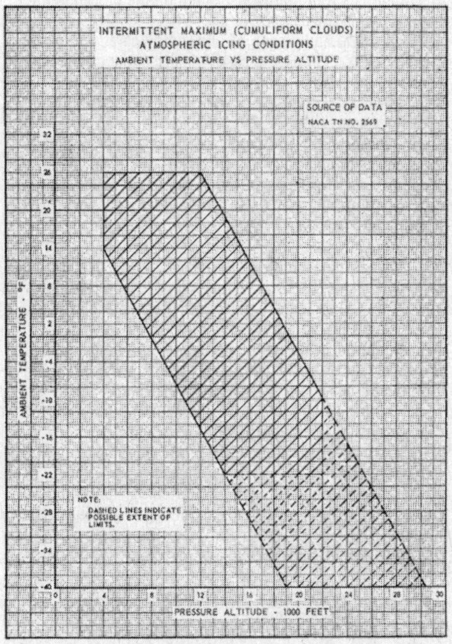

Figure 6

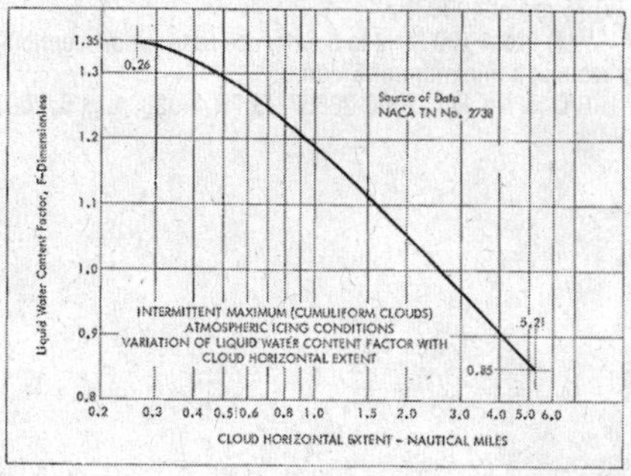

Appendix D to Part 29—Criteria for Demonstration of Emergency Evacuation Procedures Under § 29.803

(a) The demonstration must be conducted either during the dark of the night or during daylight with the dark of night simulated. If the demonstration is conducted indoors during daylight hours, it must be conducted inside a darkened hangar having doors and windows covered. In addition, the doors and windows of the rotorcraft must be covered if the hangar illumination exceeds that of a moonless night. Illumination on the floor or ground may be used, but it must be kept low and shielded against shining into the rotorcraft's windows or doors.

(b) The rotorcraft must be in a normal attitude with landing gear extended.

(c) Safety equipment such as mats or inverted liferafts may be placed on the floor or ground to protect participants. No other equipment that is not part of the rotorcraft's emergency evacuation equipment may be used to aid the participants in reaching the ground.

(d) Except as provided in paragraph (a) of this appendix, only the rotorcraft's emergency lighting system may provide illumination.

(e) All emergency equipment required for the planned operation of the rotorcraft must be installed.

(f) Each external door and exit and each internal door or curtain must be in the takeoff configuration.

(g) Each crewmember must be seated in the normally assigned seat for takeoff and must remain in that seat until receiving the signal for commencement of the demonstration. For compliance with this section, each crewmember must be—

(1) A member of a regularly scheduled line crew; or

(2) A person having knowledge of the operation of exits and emergency equipment.

(h) A representative passenger load of persons in normal health must be used as follows:

(1) At least 25 percent must be over 50 years of age, with at least 40 percent of these being females.

(2) The remaining, 75 percent or less, must be 50 years of age or younger, with at least 30 percent of these being females.

(3) Three life-size dolls, not included as part of the total passenger load, must be carried by passengers to simulate live infants 2 years old or younger, except for a total passenger load of fewer than 44 but more than 19, one doll must be carried. A doll is not required for a 19 or fewer passenger load.

(4) Crewmembers, mechanics, and training personnel who maintain or operate the rotorcraft in the normal course of their duties may not be used as passengers.

(i) No passenger may be assigned a specific seat except as the Administrator may require. Except as required by paragraph (1) of this appendix, no employee of the applicant may be seated next to an emergency exit, except as allowed by the Administrator.

(j) Seat belts and shoulder harnesses (as required) must be fastened.

(k) Before the start of the demonstration, approximately one-half of the total average amount of carry-on baggage, blankets, pillows, and other similar articles must be distributed at several locations in the aisles and emergency exit access ways to create minor obstructions.

(l) No prior indication may be given to any crewmember or passenger of the particular exits to be used in the demonstration.

(m) The applicant may not practice, rehearse, or describe the demonstration for the participants nor may any participant have taken part in this type of demonstration within the preceding 6 months.

(n) A pretakeoff passenger briefing may be given. The passengers may also be advised to follow directions of crewmembers, but not be instructed on the procedures to be followed in the demonstration.

(o) If safety equipment, as allowed by paragraph (c) of this appendix, is provided, either all passenger and cockpit windows must be blacked out or all emergency exits must have safety equipment to prevent disclosure of the available emergency exits.

(p) Not more than 50 percent of the emergency exits in the sides of the fuselage of a rotorcraft that meet all of the requirements applicable to the required emergency exits for that rotorcraft may be used for demonstration. Exits that are not to be used for the demonstration must have the exit handle deactivated or must be indicated by red lights, red tape, or other acceptable means placed outside the exits to indicate fire or other reasons why they are unusable. The exits to be used must be representative of all the emergency exits on the rotorcraft and must be designated by the applicant, subject to approval by the Administrator. If installed, at least one floor level exit (Type I; § 29.807(a)(1)) must be used as required by § 29.807(c).

(q) All evacuees must leave the rotorcraft by a means provided as part of the rotorcraft's equipment.

(r) Approved procedures must be fully utilized during the demonstration.

(s) The evacuation time period is completed when the last occupant has evacuated the rotorcraft and is on the ground.

[Amdt. 27-26, 55 FR 8005, Mar. 6, 1990]

Appendix E to Part 29—HIRF Environments and Equipment HIRF Test Levels

This appendix specifies the HIRF environments and equipment HIRF test levels for electrical and electronic systems under § 29.1317. The field strength values for the HIRF environments and laboratory equipment HIRF test levels are expressed in root-mean-square units measured during the peak of the modulation cycle.

(a) HIRF environment I is specified in the following table:

Table I.
HIRF Environment I

Frequency	Field strength (volts/meter)	
	Peak	Average
10 kHz–2 MHz	50	50
2 MHz–30 MHz	100	100
30 MHz–100 MHz	50	50
100 MHz–400 MHz	100	100
400 MHz–700 MHz	700	50
700 MHz–1 GHz	700	100
1 GHz–2 GHz	2,000	200
2 GHz–6 GHz	3,000	200
6 GHz–8 GHz	1,000	200
8 GHz–12 GHz	3,000	300
12 GHz–18 GHz	2,000	200
18 GHz–40 GHz	600	200

In this table, the higher field strength applies at the frequency band edges.

(b) HIRF environment II is specified in the following table:

Table II.
HIRF Environment II

Frequency	Field strength (volts/meter)	
	Peak	Average
10 kHz–500 kHz	20	20
500 kHz–2 MHz	30	30
2 MHz–30 MHz	100	100
30 MHz–100 MHz	10	10
100 MHz–200 MHz	30	10
200 MHz–400 MHz	10	10
400 MHz–1 GHz	700	40
1 GHz–2 GHz	1,300	160
2 GHz–4 GHz	3,000	120
4 GHz–6 GHz	3,000	160
6 GHz–8 GHz	400	170
8 GHz–12 GHz	1,230	230
12 GHz–18 GHz	730	190
18 GHz–40 GHz	600	150

In this table, the higher field strength applies at the frequency band edges.

(c) HIRF environment III is specified in the following table:

Table III.
HIRF Environment III

Frequency	Field strength (volts/meter)	
	Peak	Average
10 kHz–100 kHz	150	150
100 kHz–400 MHz	200	200
400 MHz–700 MHz	730	200
700 MHz–1 GHz	1,400	240
1 GHz–2 GHz	5,000	250
2 GHz–4 GHz	6,000	490
4 GHz–6 GHz	7,200	400
6 GHz–8 GHz	1,100	170
8 GHz–12 GHz	5,000	330
12 GHz–18 GHz	2,000	330
18 GHz–40 GHz	1,000	420

In this table, the higher field strength applies at the frequency band edges.

(d) *Equipment HIRF Test Level 1.*

(1) From 10 kilohertz (kHz) to 400 megahertz (MHz), use conducted susceptibility tests with continuous wave (CW) and 1 kHz square wave modulation with 90 percent depth or greater. The conducted susceptibility current must start at a minimum of 0.6 milliamperes (mA) at 10 kHz, increasing 20 decibel (dB) per frequency decade to a minimum of 30 mA at 500 kHz.

(2) From 500 kHz to 40 MHz, the conducted susceptibility current must be at least 30 mA.

(3) From 40 MHz to 400 MHz, use conducted susceptibility tests, starting at a minimum of 30 mA at 40 MHz, decreasing 20 dB per frequency decade to a minimum of 3 mA at 400 MHz.

(4) From 100 MHz to 400 MHz, use radiated susceptibility tests at a minimum of 20 volts per meter (V/m) peak with CW and 1 kHz square wave modulation with 90 percent depth or greater.

(5) From 400 MHz to 8 gigahertz (GHz), use radiated susceptibility tests at a minimum of 150 V/m peak with pulse modulation of 4 percent duty cycle with a 1 kHz pulse repetition frequency. This signal must be switched on and off at a rate of 1 Hz with a duty cycle of 50 percent.

(e) *Equipment HIRF Test Level 2.* Equipment HIRF test level 2 is HIRF environment II in table II of this appendix reduced by acceptable aircraft transfer function and attenuation curves. Testing must cover the frequency band of 10 kHz to 8 GHz.

(f) *Equipment HIRF Test Level 3.*

(1) From 10 kHz to 400 MHz, use conducted susceptibility tests, starting at a minimum of 0.15 mA at 10 kHz, increasing 20 dB per frequency decade to a minimum of 7.5 mA at 500 kHz.

(2) From 500 kHz to 40 MHz, use conducted susceptibility tests at a minimum of 7.5 mA.

(3) From 40 MHz to 400 MHz, use conducted susceptibility tests, starting at a minimum of 7.5 mA at 40 MHz, decreasing 20 dB per frequency decade to a minimum of 0.75 mA at 400 MHz.

(4) From 100 MHz to 8 GHz, use radiated susceptibility tests at a minimum of 5 V/m.

[Doc. No. FAA-2006-23657, 72 FR 44028, Aug. 6, 2007]

PART 31

AIRWORTHINESS STANDARDS: MANNED FREE BALLOONS

Contents

Authority: 49 U.S.C. 106(g), 40113, 44701-44702, 44704.

Source: Docket No. 1437, 29 FR 8258, July 1, 1964, as amended by Amdt. 31-1, 29 FR 14563, Oct. 24, 1964, unless otherwise noted.

Subpart A—General

§ 31.1 Applicability.

(a) This part prescribes airworthiness standards for the issue of type certificates and changes to those certificates, for manned free balloons.

(b) Each person who applies under Part 21 for such a certificate or change must show compliance with the applicable requirements of this part.

(c) For purposes of this part—

(1) A captive gas balloon is a balloon that derives its lift from a captive lighter-than-air gas;

(2) A hot air balloon is a balloon that derives its lift from heated air;

(3) The envelope is the enclosure in which the lifting means is contained;

(4) The basket is the container, suspended beneath the envelope, for the balloon occupants;

(5) The trapeze is a harness or is a seat consisting of a horizontal bar or platform suspended beneath the envelope for the balloon occupants; and

(6) The design maximum weight is the maximum total weight of the balloon, less the lifting gas or air.

[Doc. No. 1437, 29 FR 8258, July 1, 1964, as amended by Amdt. 31-3, 41 FR 55474, Dec. 20, 1976]

Subpart B—Flight Requirements

§ 31.12 Proof of compliance.

(a) Each requirement of this subpart must be met at each weight within the range of loading conditions for which certification is requested. This must be shown by—

(1) Tests upon a balloon of the type for which certification is requested or by calculations based on, and equal in accuracy to, the results of testing; and

(2) Systematic investigation of each weight if compliance cannot be reasonably inferred from the weights investigated.

(b) Except as provided in § 31.17(b), allowable weight tolerances during flight testing are +5 percent and −10 percent.

[Amdt. 31-4, 45 FR 60179, Sept. 11, 1980]

§ 31.14 Weight limits.

(a) The range of weights over which the balloon may be safely operated must be established.

(b) *Maximum weight.* The maximum weight is the highest weight at which compliance with each applicable requirement of this part is shown. The maximum weight must be established so that it is not more than—

(1) The highest weight selected by the applicant;

(2) The design maximum weight which is the highest weight at which compliance with each applicable structural loading condition of this part is shown; or

(3) The highest weight at which compliance with each applicable flight requirement of this part is shown.

(c) The information established under paragraphs (a) and (b) of this section must be made available to the pilot in accordance with § 31.81.

[Amdt. 31-3, 41 FR 55474, Dec. 20, 1976]

§ 31.16 Empty weight.

The empty weight must be determined by weighing the balloon with installed equipment but without lifting gas or heater fuel.

[Amdt. 31-4, 45 FR 60179, Sept. 11, 1980]

§ 31.17 Performance: Climb.

(a) Each balloon must be capable of climbing at least 300 feet in the first minute after takeoff with a steady rate of climb. Compliance with the requirements of this section must be shown at each altitude and ambient temperature for which approval is sought.

(b) Compliance with the requirements of paragraph (a) of this section must be shown at the maximum weight with a weight tolerance of +5 percent.

[Amdt. 31-4, 45 FR 60179, Sept. 11, 1980]

§ 31.19 Performance: Uncontrolled descent.

(a) The following must be determined for the most critical uncontrolled descent that can result from any single failure of the heater assembly, fuel cell system, gas value system, or maneuvering vent system, or from any single tear in the balloon envelope between tear stoppers:

(1) The maximum vertical velocity attained.

(2) The altitude loss from the point of failure to the point at which maximum vertical velocity is attained.

(3) The altitude required to achieve level flight after corrective action is inititated, with the balloon descending at the maximum vertical velocity determined in paragraph (a)(1) of this section.

(b) Procedures must be established for landing at the maximum vertical velocity determined in paragraph (a)(1) of this section and for arresting that descent rate in accordance with paragraph (a)(3) of this section.

[Amdt. 31-4, 45 FR 60179, Sept. 11, 1980]

§ 31.20 Controllability.

The applicant must show that the balloon is safely controllable and maneuverable during takeoff, ascent, descent, and landing without requiring exceptional piloting skill.

[Amdt. 31-3, 41 FR 55474, Dec. 20, 1976]

Subpart C—Strength Requirements

§ 31.21 Loads.

Strength requirements are specified in terms of limit loads, that are the maximum load to be expected in service, and ultimate loads, that are limit loads multiplied by prescribed factors of safety. Unless otherwise specified, all prescribed loads are limit loads.

§ 31.23 Flight load factor.

In determining limit load, the limit flight load factor must be at least 1.4.

§ 31.25 Factor of safety.

(a) Except as specified in paragraphs (b) and (c) of this section, the factor of safety is 1.5.

(b) A factor of safety of at least five must be used in envelope design. A reduced factor of safety of at least two may be used if it is shown that the selected factor will preclude failure due to creep or instantaneous rupture from lack of rip stoppers. The selected factor must be applied to the more critical of the maximum operating pressure or envelope stress.

(c) A factor of safety of at least five must be used in the design of all fibrous or non-metallic parts of the rigging and related attachments of the envelope to basket, trapeze, or other means provided for carrying occupants. The primary attachments of the envelope to the basket, trapeze, or other means provided for carrying occupants must be designed so that failure is extremely remote or so that any single failure will not jeopardize safety of flight.

(d) In applying factors of safety, the effect of temperature, and other operating characteristics, or both, that may affect strength of the balloon must be accounted for.

(e) For design purposes, an occupant weight of at least 170 pounds must be assumed.

[Doc. No. 1437, 29 FR 8258, July 1, 1964, as amended by Amdt. 31-2, 30 FR 3377, Mar. 13, 1965]

§ 31.27 Strength.

(a) The structure must be able to support limit loads without detrimental effect.

(b) The structure must be substantiated by test to be able to withstand the ultimate loads for at least three seconds without failure. For the envelope, a test of a representative part is acceptable, if the part tested is large enough to include critical seams, joints, and load attachment points and members.

(c) An ultimate free-fall drop test must be made of the basket, trapeze, or other place provided for occupants. The test must be made at design maximum weight on a horizontal surface, with the basket, trapeze, or other means provided for carrying occupants, striking the surface at angles of 0, 15, and 30 degrees. The weight may be distributed to simulate actual conditions. There must be no distortion or failure that is likely to cause serious injury to the occupants. A drop test height of 36 inches, or a drop test height that produces, upon impact, a velocity equal to the maximum vertical velocity determined in accordance with § 31.19, whichever is higher, must be used.

[Doc. No. 1437, 29 FR 8258, July 1, 1964, as amended by Amdt. 31-4, 45 FR 60179, Sept. 11, 1980]

Subpart D—Design Construction

§ 31.31 General.

The suitability of each design detail or part that bears on safety must be established by tests or analysis.

§ 31.33 Materials.

(a) The suitability and durability of all materials must be established on the basis of experience or tests. Materials must conform to approved specifications that will ensure that they have the strength and other properties assumed in the design data.

(b) Material strength properties must be based on enough tests of material conforming to specifications so as to establish design values on a statistical basis.

§ 31.35 Fabrication methods.

The methods of fabrication used must produce a consistently sound structure. If a fabrication process requires close control to reach this objective, the process must be performed in accordance with an approved process specification.

§ 31.37 Fastenings.

Only approved bolts, pins, screws, and rivets may be used in the structure. Approved locking devices or methods must be used for all these bolts, pins, and screws, unless the installation is shown to be free from vibration. Self-locking nuts may not be used on bolts that are subject to rotation in service.

§ 31.39 Protection.

Each part of the balloon must be suitably protected against deterioration or loss of strength in service due to weathering, corrosion, or other causes.

§ 31.41 Inspection provisions.

There must be a means to allow close examination of each part that require repeated inspection and adjustment.

§ 31.43 Fitting factor.

(a) A fitting factor of at least 1.15 must be used in the analysis of each fitting the strength of which is not proven by limit and ultimate load tests in which the actual stress conditions are simulated in the fitting and surrounding structure. This factor applies to all parts of the fitting, the means of attachment, and the bearing on the members joined.

(b) Each part with an integral fitting must be treated as a fitting up to the point where the section properties become typical of the member.

(c) The fitting factor need not be used if the joint design is made in accordance with approved practices and is based on comprehensive test data.

§ 31.45 Fuel cells.

If fuel cells are used, the fuel cells, their attachments, and related supporting structure must be shown by tests to be capable of withstanding, without detrimental distortion or failure, any inertia loads to which the installation may be subjected, including the drop tests prescribed in § 31.27(c). In the tests, the fuel cells must be loaded to the weight and pressure equivalent to the full fuel quantity condition.

[Amdt. 31-3, 41 FR 55474, Dec. 20, 1976]

§ 31.46 Pressurized fuel systems.

For pressurized fuel systems, each element and its connecting fittings and lines must be tested to an ultimate pressure of at least twice the maximum pressure to which the system will be subjected in normal operation. No part of the system may fail or malfunction during the test. The test configuration must be representative of the normal fuel system installation and balloon configuration.

[Amdt. 31-3, 41 FR 55474, Dec. 20, 1976]

§ 31.47 Burners.

(a) If a burner is used to provide the lifting means, the system must be designed and installed so as not to create a fire hazard.

(b) There must be shielding to protect parts adjacent to the burner flame, and the occupants, from heat effects.

(c) There must be controls, instruments, or other equipment essential to the safe control and operation of the heater. They must be shown to be able to perform their intended functions during normal and emergency operation.

(d) The burner system (including the burner unit, controls, fuel lines, fuel cells, regulators, control valves, and other related elements) must be substantiated by an endurance test of at least 40 hours. Each element of the system must be installed and tested to simulate actual balloon installation and use.

(1) The test program for the main blast valve operation of the burner must include:

(i) Five hours at the maximum fuel pressure for which approval is sought, with a burn time for each one minute cycle of three to ten seconds. The burn time must be established so that each burner is subjected to the maximum thermal shock for temperature affected elements;

(ii) Seven and one-half hours at an intermediate fuel pressure, with a burn time for each one minute cycle of three to ten seconds. An intermediate fuel pressure is 40 to 60 percent of the range between the maximum fuel pressure referenced in paragraph (d)(1)(i) of this section and minimum fuel pressure referenced in paragraph (d)(1)(iii);

(iii) Six hours and fifteen minutes at the minimum fuel pressure for which approval is sought, with a burn time for each one minute cycle of three to ten seconds;

(iv) Fifteen minutes of operation on vapor, with a burn time for each one minute cycle of at least 30 seconds; and

(v) Fifteen hours of normal flight operation.

(2) The test program for the secondary or backup operation of the burner must include six hours of operation with a burn time for each five minute cycle of one minute at an intermediate fuel pressure.

(e) The test must also include at least three flameouts and restarts.

(f) Each element of the system must be serviceable at the end of the test.

[Doc. No. 1437, 29 FR 8258, July 1, 1964, as amended by Amdt. 31-2, 30 FR 3377, Mar. 13, 1965; Amdt. 31-7, 61 FR 18223, Apr. 24, 1996; 61 FR 20877, May 8, 1996]

§ 31.49 Control systems.

(a) Each control must operate easily, smoothly, and positively enough to allow proper performance of its functions. Controls must be arranged and identified to provide for convenience of operation and to prevent the possibility of confusion and subsequent inadvertent operation.

(b) Each control system and operating device must be designed and installed in a manner that will prevent jamming, chafing, or interference from passengers, cargo, or loose objects. Precaution must be taken to prevent foreign objects from jamming the controls. The elements of the control system must have design features or must be distinctly and permanently marked to minimize the possibility of incorrect assembly that could result in malfunctioning of the control system.

(c) Each balloon using a captive gas as the lifting means must have an automatic valve or appendix that is able to release gas automatically at the rate of at least three percent of the total volume per minute when the balloon is at its maximum operating pressure.

(d) Each hot air balloon must have a means to allow the controlled release of hot air during flight.

(e) Each hot air balloon must have a means to indicate the maximum envelope skin temperatures occurring during operation. The indicator must be readily visible to the pilot and marked to indicate the limiting safe temperature of the envelope material. If the markings are on the cover glass of the instrument, there must be provisions to maintain the correct alignment of the glass cover with the face of the dial.

[Doc. No. 1437, 29 FR 8258, July 1, 1964, as amended by Amdt. 31-2, 30 FR 3377, Mar. 13, 1965]

§ 31.51 Ballast.

Each captive gas balloon must have a means for the safe storage and controlled release of ballast. The ballast must consist of material that, if released during flight, is not hazardous to persons on the ground.

§ 31.53 Drag rope.

If a drag rope is used, the end that is released overboard must be stiffened to preclude the probability of the rope becoming entangled with trees, wires, or other objects on the ground.

§ 31.55 Deflation means.

There must be a means to allow emergency deflation of the envelope so as to allow a safe emergency landing. If a system other than a manual system is used, the reliability of the system used must be substantiated.

[Amdt. 31-2, 30 FR 3377, Mar. 13, 1965]

§ 31.57 Rip cords.

(a) If a rip cord is used for emergency deflation, it must be designed and installed to preclude entanglement.

(b) The force required to operate the rip cord may not be less than 25, or more than 75, pounds.

(c) The end of the rip cord to be operated by the pilot must be colored red.

(d) The rip cord must be long enough to allow an increase of at least 10 percent in the vertical dimension of the envelope.

§ 31.59 Trapeze, basket, or other means provided for occupants.

(a) The trapeze, basket, or other means provided for carrying occupants may not rotate independently of the envelope.

(b) Each projecting object on the trapeze, basket, or other means provided for carrying occupants, that could cause injury to the occupants, must be padded.

§ 31.61 Static discharge.

Unless shown not to be necessary for safety, there must be appropriate bonding means in the design of each balloon using flammable gas as a lifting means to ensure that the effects of static discharges will not create a hazard.

[Amdt. 31-2, 30 FR 3377, Mar. 13, 1965]

§ 31.63 Safety belts.

(a) There must be a safety belt, harness, or other restraining means for each occupant, unless the Administrator finds it unnecessary. If installed, the belt, harness, or other restraining means and its supporting structure must meet the strength requirements of subpart C of this part.

(b) This section does not apply to balloons that incorporate a basket or gondola.

[Amdt. 31-2, 30 FR 3377, Mar. 13, 1965, as amended by Amdt. 31-3, 41 FR 55474, Dec. 20, 1976]

§ 31.65 Position lights.

(a) If position lights are installed, there must be one steady aviation white position light and one flashing aviation red (or flashing aviation white) position light with an effective flash frequency of at least 40, but not more than 100, cycles per minute.

(b) Each light must provide 360° horizontal coverage at the intensities prescribed in this paragraph. The following light intensities must be determined with the light source operating at a steady state and with all light covers and color filters in place and at the manufacturer's rated mimimum voltage. For the flashing aviation red light, the measured values must be adjusted to correspond to a red filter temperature of at least 130 °F:

(1) The intensities in the horizontal plane passing through the light unit must equal or exceed the following values:

Position light	Minimum intensity (candles)
Steady white	20
Flashing red or white	40

(2) The intensities in vertical planes must equal or exceed the following values. An intensity of one unit corresponds to the applicable horizontal plane intensity specified in paragraph (b)(1) of this section.

Angles above and below the horizontal in any vertical plane (degrees)	Minimum intensity (units)
0	1.00
0 to 5	0.90
5 to 10	0.80
10 to 15	0.70
15 to 20	0.50
20 to 30	0.30
30 to 40	0.10
40 to 60	0.05

(c) The steady white light must be located not more than 20 feet below the basket, trapeze, or other means for carrying occupants. The flashing red or white light must be located not less than 7, nor more than 10, feet below the steady white light.

(d) There must be a means to retract and store the lights.

(e) Each position light color must have the applicable International Commission on Illumination chromaticity coordinates as follows:

(1) *Aviation red*—

y is not greater than 0.335; and z is not greater than 0.002.

(2) *Aviation white*—

x is not less than 0.300 and not greater than 0.540;

y is not less than x −0.040 or y_o −0.010, whichever is the smaller; and

y is not greater than x +0.020 nor 0.636−0.0400 x;

Where y_o is the y coordinate of the Planckian radiator for the value of x considered.

[Doc. No. 1437, 29 FR 8258, July 1, 1964, as amended by Amdt. 31-1, 29 FR 14563, Oct. 24, 1964; Amdt. 31-4, 45 FR 60179, Sept. 11, 1980]

Subpart E—Equipment

§ 31.71 Function and installation.

(a) Each item of installed equipment must—

(1) Be of a kind and design appropriate to its intended function;

(2) Be permanently and legibly marked or, if the item is too small to mark, tagged as to its identification, function, or operating limitations, or any applicable combination of those factors;

(3) Be installed according to limitations specified for that equipment; and

(4) Function properly when installed.

(b) No item of installed equipment, when performing its function, may affect the function of any other equipment so as to create an unsafe condition.

(c) The equipment, systems, and installations must be designed to prevent hazards to the balloon in the event of a probable malfunction or failure.

[Amdt. 31-4, 45 FR 60180, Sept. 11, 1980]

Subpart F—Operating Limitations and Information

§ 31.81 General.

(a) The following information must be established:

(1) Each operating limitation, including the maximum weight determined under § 31.14.

(2) The normal and emergency procedures.

(3) Other information necessary for safe operation, including—

(i) The empty weight determined under § 31.16;

(ii) The rate of climb determined under § 31.17, and the procedures and conditions used to determine performance;

(iii) The maximum vertical velocity, the altitude drop required to attain that velocity, and altitude drop required to recover from a descent at that velocity, determined under § 31.19, and the procedures and conditions used to determine performance; and

(iv) Pertinent information peculiar to the balloon's operating characteristics.

(b) The information established in compliance with paragraph (a) of this section must be furnished by means of—

(1) A Balloon Flight Manual; or

(2) A placard on the balloon that is clearly visible to the pilot.

[Amdt. 31-4, 45 FR 60180, Sept. 11, 1980]

§ 31.82 Instructions for Continued Airworthiness.

The applicant must prepare Instructions for Continued Airworthiness in accordance with appendix A to this part that are acceptable to the Administrator. The instructions may be incomplete at type certification if a program exists to ensure their completion prior to delivery of the first balloon or issuance of a standard certificate of airworthiness, whichever occurs later.

[Amdt. 31-4, 45 FR 60180, Sept. 11, 1980]

§ 31.83 Conspicuity.

The exterior surface of the envelope must be of a contrasting color or colors so that it will be conspicuous during operation. However, multicolored banners or streamers are acceptable if it can be shown that they are large enough, and there are enough of them of contrasting color, to make the balloon conspicuous during flight.

§ 31.85 Required basic equipment.

In addition to any equipment required by this subchapter for a specific kind of operation, the following equipment is required:

(a) For all balloons:

(1) [Reserved]

(2) An altimeter.

(3) A rate of climb indicator.

(b) For hot air balloons:

(1) A fuel quantity gauge. If fuel cells are used, means must be incorporated to indicate to the crew the quantity of fuel in each cell during flight. The means must be calibrated in appropriate units or in percent of fuel cell capacity.

(2) An envelope temperature indicator.

(c) For captive gas balloons, a compass.

[Amdt. 31-2, 30 FR 3377, Mar. 13, 1965, as amended by Amdt. 31-3, 41 FR 55474, Dec. 20, 1976; Amdt. 31-4, 45 FR 60180, Sept. 11, 1980]

Appendix A to Part 31—Instructions for Continued Airworthiness

a31.1 general

(a) This appendix specifies requirements for the preparation of Instructions for Continued Airworthiness as required by § 31.82.

(b) The Instructions for Continued Airworthiness for each balloon must include the Instructions for Continued Airworthiness for all balloon parts required by this chapter and any required information relating to the interface of those parts with the balloon. If Instructions for Continued Airworthiness are not supplied by the part manufacturer for a balloon part, the Instructions for Continued Airworthiness for the balloon must include the information essential to the continued airworthiness of the balloon.

(c) The applicant must submit to the FAA a program to show how changes to the Instructions for Continued Airworthiness made by the applicant or by the manufacturers of balloon parts will be distributed.

a31.2 format

(a) The Instructions for Continued Airworthiness must be in the form of a manual or manuals as appropriate for the quantity of data to be provided.

(b) The format of the manual or manuals must provide for a practical arrangement.

a31.3 content

The contents of the manual or manuals must be prepared in the English language. The Instructions for Continued Airworthiness must contain the following information:

(a) Introduction information that includes an explanation of the balloon's features and data to the extent necessary for maintenance or preventive maintenance.

(b) A description of the balloon and its systems and installations.

(c) Basic control and operation information for the balloon and its components and systems.

(d) Servicing information that covers details regarding servicing of balloon components, including burner nozzles, fuel tanks, and valves during operations.

(e) Maintenance information for each part of the balloon and its envelope, controls, rigging, basket structure, fuel systems, instruments, and heater assembly that provides the recommended periods at which they should be cleaned, adjusted, tested, and lubricated, the applicable wear tolerances, and the degree of work recommended at these periods. However, the applicant may refer to an accessory, instrument, or equipment manufacturer as the source of this information if the applicant shows that the item has an exceptionally high degree of complexity requiring specialized maintenance techniques, test equipment, or expertise. The recommended overhaul periods and necessary cross references to the Airworthiness Limitations section of the manual must also be included. In addition, the applicant must include an inspection program that includes the frequency and extent of the inspections necessary to provide for the continued airworthiness of the balloon.

(f) Troubleshooting information describing probable malfunctions, how to recognize those malfunctions, and the remedial action for those malfunctions.

(g) Details of what, and how, to inspect after a hard landing.

(h) Instructions for storage preparation including any storage limits.

(i) Instructions for repair on the balloon envelope and its basket or trapeze.

a31.4 airworthiness limitations section

The Instructions for Continued Airworthiness must contain a section titled Airworthiness Limitations that is segregated and clearly distinguishable from the rest of the document. This section must set forth each mandatory replacement time, structural inspection interval, and related structural inspection procedure, including envelope structural integrity, required for type certification. If the Instructions for Continued Airworthiness consist of multiple documents, the section required by this paragraph must be included in the principal manual. This section must contain a legible statement in a prominent location that reads: "The Airworthiness Limitations section is FAA approved and specifies maintenance required under §§ 43.16 and 91.403 of the Federal Aviation Regulations."

[Amdt. 31-4, 45 FR 60180, Sept. 11, 1980, as amended by Amdt. 31-5, 54 FR 34330, Aug. 18, 1989]

PART 33

AIRWORTHINESS STANDARDS: AIRCRAFT ENGINES

Contents

Subpart A—General

Subpart B—Design and Construction; General

Subpart C—Design and Construction; Reciprocating Aircraft Engines

Subpart D—Block Tests; Reciprocating Aircraft Engines

Subpart E—Design and Construction; Turbine Aircraft Engines

Subpart F—Block Tests; Turbine Aircraft Engines

Subpart G—Special Requirements: Turbine Aircraft Engines

Authority: 49 U.S.C. 106(g), 40113, 44701-44702, 44704.

Source: Docket No. 3025, 29 FR 7453, June 10, 1964, unless otherwise noted.

Note: For miscellaneous amendments to cross references in this Part 33, see Amdt. 33-2, 31 FR 9211, July 6, 1966.

Subpart A—General

§ 33.1 Applicability.

(a) This part prescribes airworthiness standards for the issue of type certificates and changes to those certificates, for aircraft engines.

(b) Each person who applies under part 21 for such a certificate or change must show compliance with the applicable requirements of this part and the applicable requirements of part 34 of this chapter.

[Amdt. 33-7, 41 FR 55474, Dec. 20, 1976, as amended by Amdt. 33-14, 55 FR 32861, Aug. 10, 1990]

§ 33.3 General.

Each applicant must show that the aircraft engine concerned meets the applicable requirements of this part.

§ 33.4 Instructions for Continued Airworthiness.

The applicant must prepare Instructions for Continued Airworthiness in accordance with appendix A to this part that are acceptable to the Administrator. The instructions may be incomplete at type certification if a program exists to ensure their completion prior to delivery of the first aircraft with the engine installed, or upon issuance of a standard certificate of airworthiness for the aircraft with the engine installed, whichever occurs later.

[Amdt. 33-9, 45 FR 60181, Sept. 11, 1980]

§ 33.5 Instruction manual for installing and operating the engine.

Each applicant must prepare and make available to the Administrator prior to the issuance of the type certificate, and to the owner at the time of delivery of the engine, approved instructions for installing and operating the engine. The instructions must include at least the following:

(a) *Installation instructions.* (1) The location of engine mounting attachments, the method of attaching the engine to the aircraft, and the maximum allowable load for the mounting attachments and related structure.

(2) The location and description of engine connections to be attached to accessories, pipes, wires, cables, ducts, and cowling.

(3) An outline drawing of the engine including overall dimensions.

(4) A definition of the physical and functional interfaces with the aircraft and aircraft equipment, including the propeller when applicable.

(5) Where an engine system relies on components that are not part of the engine type design, the interface conditions and reliability requirements for those components upon which engine type certification is based must be specified in the engine installation instructions directly or by reference to appropriate documentation.

(6) A list of the instruments necessary for control of the engine, including the overall limits of accuracy and transient response required of such instruments for control of the operation of the engine, must also be stated so that the suitability of the instruments as installed may be assessed.

(b) *Operation instructions.* (1) The operating limitations established by the Administrator.

(2) The power or thrust ratings and procedures for correcting for nonstandard atmosphere.

(3) The recommended procedures, under normal and extreme ambient conditions for—

(i) Starting;

(ii) Operating on the ground; and

(iii) Operating during flight.

(4) For rotorcraft engines having one or more OEI ratings, applicants must provide data on engine performance characteristics and variability to enable the aircraft manufacturer to establish aircraft power assurance procedures.

(5) A description of the primary and all alternate modes, and any back-up system, together with any associated limitations, of the engine control system and its interface with the aircraft systems, including the propeller when applicable.

(c) *Safety analysis assumptions.* The assumptions of the safety analysis as described in § 33.75(d) with respect to the reliability of safety devices, instrumentation, early warning devices, maintenance checks, and similar equipment or procedures that are outside the control of the engine manufacturer.

[Amdt. 33-6, 39 FR 35463, Oct. 1, 1974, as amended by Amdt. 33-9, 45 FR 60181, Sept. 11, 1980; Amdt. 33-24, 47 FR 50867, Sept. 4, 2007; Amdt. 33-25, 73 FR 48123, Aug. 18, 2008; Amdt. 33-26, 73 FR 48284, Aug. 19, 2008]

§ 33.7 Engine ratings and operating limitations.

(a) Engine ratings and operating limitations are established by the Administrator and included in the engine certificate data sheet specified in § 21.41 of this chapter, including ratings and limitations based on the operating conditions and information specified in this section, as applicable, and any other information found necessary for safe operation of the engine.

(b) For reciprocating engines, ratings and operating limitations are established relating to the following:

(1) Horsepower or torque, r.p.m., manifold pressure, and time at critical pressure altitude and sea level pressure altitude for—

(i) Rated maximum continuous power (relating to unsupercharged operation or to operation in each supercharger mode as applicable); and

(ii) Rated takeoff power (relating to unsupercharged operation or to operation in each supercharger mode as applicable).

(2) Fuel grade or specification.

(3) Oil grade or specification.

(4) Temperature of the—

(i) Cylinder;

(ii) Oil at the oil inlet; and

(iii) Turbosupercharger turbine wheel inlet gas.

(5) Pressure of—

(i) Fuel at the fuel inlet; and

(ii) Oil at the main oil gallery.

(6) Accessory drive torque and overhang moment.

(7) Component life.

(8) Turbosupercharger turbine wheel r.p.m.

(c) For turbine engines, ratings and operating limitations are established relating to the following:

(1) Horsepower, torque, or thrust, r.p.m., gas temperature, and time for—

(i) Rated maximum continuous power or thrust (augmented);

(ii) Rated maximum continuous power or thrust (unaugmented);

(iii) Rated takeoff power or thrust (augmented);

(iv) Rated takeoff power or thrust (unaugmented);

(v) Rated 30-minute OEI power;

(vi) Rated $2\frac{1}{2}$-minute OEI power;

(vii) Rated continuous OEI power; and

(viii) Rated 2-minute OEI Power;

(ix) Rated 30-second OEI power; and

(x) Auxiliary power unit (APU) mode of operation.

(2) Fuel designation or specification.

(3) Oil grade or specification.

(4) Hydraulic fluid specification.

(5) Temperature of—

(i) Oil at a location specified by the applicant;

(ii) Induction air at the inlet face of a supersonic engine, including steady state operation and transient over-temperature and time allowed;

(iii) Hydraulic fluid of a supersonic engine;

(iv) Fuel at a location specified by the applicant; and

(v) External surfaces of the engine, if specified by the applicant.

(6) Pressure of—

(i) Fuel at the fuel inlet;

(ii) Oil at a location specified by the applicant;

(iii) Induction air at the inlet face of a supersonic engine, including steady state operation and transient overpressure and time allowed; and

(iv) Hydraulic fluid.

(7) Accessory drive torque and overhang moment.

(8) Component life.

(9) Fuel filtration.

(10) Oil filtration.

(11) Bleed air.

(12) The number of start-stop stress cycles approved for each rotor disc and spacer.

(13) Inlet air distortion at the engine inlet.

(14) Transient rotor shaft overspeed r.p.m., and number of overspeed occurrences.

(15) Transient gas overtemperature, and number of overtemperature occurrences.

(16) Transient engine overtorque, and number of overtorque occurrences.

(17) Maximum engine overtorque for turbopropeller and turboshaft engines incorporating free power turbines.

(18) For engines to be used in supersonic aircraft, engine rotor windmilling rotational r.p.m.

(d) In determining the engine performance and operating limitations, the overall limits of accuracy of the engine control system and of the necessary instrumentation as defined in § 33.5(a)(6) must be taken into account.

[Amdt. 33-6, 39 FR 35463, Oct. 1, 1974, as amended by Amdt. 33-10, 49 FR 6850, Feb. 23, 1984; Amdt. 33-11, 51 FR 10346, Mar. 25, 1986; Amdt. 33-12, 53 FR 34220, Sept. 2, 1988; Amdt. 33-18, 61 FR 31328, June 19, 1996; Amdt. 33-26, 73 FR 48284, Aug. 19, 2008; Amdt. 33-30, 74 FR 45310, Sept. 2, 2009]

§ 33.8 Selection of engine power and thrust ratings.

(a) Requested engine power and thrust ratings must be selected by the applicant.

(b) Each selected rating must be for the lowest power or thrust that all engines of the same type may be expected to produce under the conditions used to determine that rating.

[Amdt. 33-3, 32 FR 3736, Mar. 4, 1967]

Subpart B—Design and Construction; General

§ 33.11 Applicability.

This subpart prescribes the general design and construction requirements for reciprocating and turbine aircraft engines.

§ 33.13 [Reserved]

§ 33.15 Materials.

The suitability and durability of materials used in the engine must—

(a) Be established on the basis of experience or tests; and

(b) Conform to approved specifications (such as industry or military specifications) that ensure their having the strength and other properties assumed in the design data.

(Secs. 313(a), 601, and 603, 72 Stat. 759, 775, 49 U.S.C. 1354(a), 1421, and 1423; sec. 6(c), 49 U.S.C. 1655(c))

[Amdt. 33-8, 42 FR 15047, Mar. 17, 1977, as amended by Amdt. 33-10, 49 FR 6850, Feb. 23, 1984]

§ 33.17 Fire protection.

(a) The design and construction of the engine and the materials used must minimize the probability of the occurrence and spread of fire during normal operation and failure conditions, and must minimize the effect of such a fire. In addition, the design and construction of turbine engines must minimize the probability of the occurrence of an internal fire that could result in structural failure or other hazardous effects.

(b) Except as provided in paragraph (c) of this section, each external line, fitting, and other component, which contains or conveys flammable fluid during normal engine operation, must be fire resistant or fireproof, as determined by the Administrator. Components must be shielded or located to safeguard against the ignition of leaking flammable fluid.

(c) A tank, which contains flammable fluids and any associated shut-off means and supports, which are part of and attached to the engine, must be fireproof either by construction or by protection unless damage by fire will not cause leakage or spillage of a hazardous quantity of flammable fluid. For a reciprocating engine having an integral oil sump of less than 23.7 liters capacity, the oil sump need not be fireproof or enclosed by a fireproof shield.

(d) An engine component designed, constructed, and installed to act as a firewall must be:

(1) Fireproof;

(2) Constructed so that no hazardous quantity of air, fluid or flame can pass around or through the firewall; and,

(3) Protected against corrosion;

(e) In addition to the requirements of paragraphs (a) and (b) of this section, engine control system components that are located in a designated fire zone must be fire resistant or fireproof, as determined by the Administrator.

(f) Unintentional accumulation of hazardous quantities of flammable fluid within the engine must be prevented by draining and venting.

(g) Any components, modules, or equipment, which are susceptible to or are potential sources of static discharges or electrical fault currents must be designed and constructed to be properly grounded to the engine reference, to minimize the risk of ignition in external areas where flammable fluids or vapors could be present.

[Doc. No. FAA-2007-28503, 74 FR 37930, July 30, 2009]

§ 33.19 Durability.

(a) Engine design and construction must minimize the development of an unsafe condition of the engine between overhaul periods. The design of the compressor and turbine rotor cases must provide for the containment of damage from rotor blade failure. Energy levels and trajectories of fragments resulting from rotor blade failure that lie outside the compressor and turbine rotor cases must be defined.

(b) Each component of the propeller blade pitch control system which is a part of the engine type design must meet the requirements of §§ 35.21, 35.23, 35.42 and 35.43 of this chapter.

[Doc. No. 3025, 29 FR 7453, June 10, 1964, as amended by Amdt. 33-9, 45 FR 60181, Sept. 11, 1980; Amdt. 33-10, 49 FR 6851, Feb. 23, 1984; Amdt. 33-28, 73 FR 63346, Oct. 24, 2008]

§ 33.21 Engine cooling.

Engine design and construction must provide the necessary cooling under conditions in which the airplane is expected to operate.

§ 33.23 Engine mounting attachments and structure.

(a) The maximum allowable limit and ultimate loads for engine mounting attachments and related engine structure must be specified.

(b) The engine mounting attachments and related engine structure must be able to withstand—

(1) The specified limit loads without permanent deformation; and

(2) The specified ultimate loads without failure, but may exhibit permanent deformation.

[Amdt. 33-10, 49 FR 6851, Feb. 23, 1984]

§ 33.25 Accessory attachments.

The engine must operate properly with the accessory drive and mounting attachments loaded. Each engine accessory drive and mounting attachment must include provisions for sealing to prevent contamination of, or unacceptable leakage from, the engine interior. A drive and mounting attachment requiring lubrication for external drive splines, or coupling by engine oil, must include provisions for sealing to prevent unacceptable loss of oil and to prevent contamination from sources outside the chamber enclosing the drive connection. The design of the engine must allow for the examination, adjustment, or removal of each accessory required for engine operation.

[Amdt. 33-10, 49 FR 6851, Feb. 23, 1984]

§ 33.27 Turbine, compressor, fan, and turbosupercharger rotor overspeed.

(a) For each fan, compressor, turbine, and turbosupercharger rotor, the applicant must establish by test, analysis, or a combination of both, that each rotor will not burst when operated in the engine for 5 minutes at whichever of the conditions defined in paragraph (b) of this section is the most critical with respect to the integrity of such a rotor.

(1) Test rotors used to demonstrate compliance with this section that do not have the most adverse combination of material properties and dimensional tolerances must be tested at conditions which have been adjusted to ensure the minimum specification rotor possesses the required overspeed capability. This can be accomplished by increasing test speed, temperature, and/or loads.

(2) When an engine test is being used to demonstrate compliance with the overspeed conditions listed in paragraph (b)(3) or (b)(4) of this section and the failure of a component or system is sudden and transient, it may not be possible to operate the engine for 5 minutes after the failure. Under these circumstances, the actual overspeed duration is acceptable if the required maximum overspeed is achieved.

(b) When determining the maximum overspeed condition applicable to each rotor in order to comply with paragraphs (a) and (c) of this section, the applicant must evaluate the following rotor speeds taking into consideration the part's operating temperatures and temperature gradients throughout the engine's operating envelope:

(1) 120 percent of the maximum permissible rotor speed associated with any of the engine ratings except one-engine-inoperative (OEI) ratings of less than 2½ minutes.

(2) 115 percent of the maximum permissible rotor speed associated with any OEI ratings of less than 2½ minutes.

(3) 105 percent of the highest rotor speed that would result from either:

(i) The failure of the component or system which, in a representative installation of the engine, is the most critical with respect to overspeed when operating at any rating condition except OEI ratings of less than 2½ minutes, or

(ii) The failure of any component or system in a representative installation of the engine, in combination with any other failure of a component or system that would not normally be detected during a routine pre-flight check or during normal flight operation, that is the most critical with respect to overspeed, except as provided by paragraph (c) of this section, when operating at any rating condition except OEI ratings of less than 2½ minutes.

(4) 100 percent of the highest rotor speed that would result from the failure of the component or system which, in a representative installation of the engine, is the most critical with respect to overspeed when operating at any OEI rating of less than 2½ minutes.

(c) The highest overspeed that results from a complete loss of load on a turbine rotor, except as provided by paragraph (f) of this section, must be included in the overspeed conditions considered by paragraphs (b)(3)(i), (b)(3)(ii), and (b)(4) of this section, regardless of whether that overspeed results from a failure within the engine or external to the engine. The overspeed resulting from any other single failure must be considered when selecting the most limiting overspeed conditions applicable to each rotor. Overspeeds resulting from combinations of failures must also be considered unless the applicant can show that the probability of occurrence is not greater than extremely remote (probability range of 10^{-7} to 10^{-9} per engine flight hour).

(d) In addition, the applicant must demonstrate that each fan, compressor, turbine, and turbosupercharger rotor complies with paragraphs (d)(1) and (d)(2) of this section for the maximum overspeed achieved when subjected to the conditions specified in paragraphs (b)(3) and (b)(4) of this section. The applicant must use the approach in paragraph (a) of this section which specifies the required test conditions.

(1) Rotor Growth must not cause the engine to:

(i) Catch fire,

(ii) Release high-energy debris through the engine casing or result in a hazardous failure of the engine casing,

(iii) Generate loads greater than those ultimate loads specified in § 33.23(a), or

(iv) Lose the capability of being shut down.

(2) Following an overspeed event and after continued operation, the rotor may not exhibit conditions such as cracking or distortion which preclude continued safe operation.

(e) The design and functioning of engine control systems, instruments, and other methods not covered under § 33.28 must ensure that the engine operating limitations that affect turbine, compressor, fan, and turbosupercharger rotor structural integrity will not be exceeded in service.

(f) Failure of a shaft section may be excluded from consideration in determining the highest overspeed that would result from a complete loss of load on a turbine rotor if the applicant:

(1) Identifies the shaft as an engine life-limited-part and complies with § 33.70.

(2) Uses material and design features that are well understood and that can be analyzed by well-established and validated stress analysis techniques.

(3) Determines, based on an assessment of the environment surrounding the shaft section, that environmental influences are unlikely to cause a shaft failure. This assessment must include complexity of design, corrosion, wear, vibration, fire, contact with adjacent components or structure, overheating, and secondary effects from other failures or combination of failures.

(4) Identifies and declares, in accordance with § 33.5, any assumptions regarding the engine installation in making the assessment described above in paragraph (f)(3) of this section.

(5) Assesses, and considers as appropriate, experience with shaft sections of similar design.

(6) Does not exclude the entire shaft.

(g) If analysis is used to meet the overspeed requirements, then the analytical tool must be validated to prior overspeed test results of a similar rotor. The tool must be validated for each material. The rotor being certified must not exceed the boundaries of the rotors being used to validate the analytical tool in terms of geometric shape, operating stress, and temperature. Validation includes the ability to accurately predict rotor dimensional growth and the burst speed. The predictions must also show that the rotor being certified does not have lower burst and growth margins than rotors used to validate the tool.

[Doc. No. FAA-2010-0398, Amdt. 33-31, 76 FR 42023, July 18, 2011]

§ 33.28 Engine control systems.

(a) *Applicability.* These requirements are applicable to any system or device that is part of engine type design, that controls, limits, or monitors engine operation, and is necessary for the continued airworthiness of the engine.

(b) *Validation* —(1) *Functional aspects.* The applicant must substantiate by tests, analysis, or a combination thereof, that the engine control system performs the intended functions in a manner which:

(i) Enables selected values of relevant control parameters to be maintained and the engine kept within the approved operating limits over changing atmospheric conditions in the declared flight envelope;

(ii) Complies with the operability requirements of §§ 33.51, 33.65 and 33.73, as appropriate, under all likely system inputs and allowable engine power or thrust demands, unless it can be demonstrated that failure of the control function results in a non-dispatchable condition in the intended application;

(iii) Allows modulation of engine power or thrust with adequate sensitivity over the declared range of engine operating conditions; and

(iv) Does not create unacceptable power or thrust oscillations.

(2) *Environmental limits.* The applicant must demonstrate, when complying with §§ 33.53 or 33.91, that the engine control system functionality will not be adversely affected by declared environmental conditions, including electromagnetic interference (EMI), High Intensity Radiated Fields (HIRF), and lightning. The limits to which the system has been qualified must be documented in the engine installation instructions.

(c) *Control transitions.* (1) The applicant must demonstrate that, when fault or failure results in a change from one control mode to another, from one channel to another, or from the primary system to the back-up system, the change occurs so that:

(i) The engine does not exceed any of its operating limitations;

(ii) The engine does not surge, stall, or experience unacceptable thrust or power changes or oscillations or other unacceptable characteristics; and

(iii) There is a means to alert the flight crew if the crew is required to initiate, respond to, or be aware of the control mode change. The means to alert the crew must be described in the engine installation instructions, and the crew action must be described in the engine operating instructions;

(2) The magnitude of any change in thrust or power and the associated transition time must be identified and described in the engine installation instructions and the engine operating instructions.

(d) *Engine control system failures.* The applicant must design and construct the engine control system so that:

(1) The rate for Loss of Thrust (or Power) Control (LOTC/LOPC) events, consistent with the safety objective associated with the intended application can be achieved;

(2) In the full-up configuration, the system is single fault tolerant, as determined by the Administrator, for electrical or electronic failures with respect to LOTC/LOPC events;

(3) Single failures of engine control system components do not result in a hazardous engine effect; and

(4) Foreseeable failures or malfunctions leading to local events in the intended aircraft installation, such as fire, overheat, or failures leading to damage to engine control system components, do not result in a hazardous engine effect due to engine control system failures or malfunctions.

(e) S *ystem safety assessment.* When complying with this section and § 33.75, the applicant must complete a System Safety Assessment for the engine control system. This assessment must identify faults or failures that result in a change in thrust or power, transmission of erroneous data, or an effect on engine operability producing a surge or stall together with the predicted frequency of occurrence of these faults or failures.

(f) *Protection systems.* (1) The design and functioning of engine control devices and systems, together with engine instruments and operating and maintenance instructions, must provide reasonable assurance that those engine operating limitations that affect turbine, compressor, fan, and turbosupercharger rotor structural integrity will not be exceeded in service.

(2) When electronic overspeed protection systems are provided, the design must include a means for testing, at least once per engine start/stop cycle, to establish the availability of the protection function. The means must be such that a complete test of the system can be achieved in the minimum number of cycles. If the test is not fully automatic, the requirement for a manual test must be contained in the engine instructions for operation.

(3) When overspeed protection is provided through hydromechanical or mechanical means, the applicant must demonstrate by test or other acceptable means that the overspeed function remains available between inspection and maintenance periods.

(g) *Software.* The applicant must design, implement, and verify all associated software to minimize the existence of errors by using a method, approved by the FAA, consistent with the criticality of the performed functions.

(h) *Aircraft-supplied data.* Single failures leading to loss, interruption or corruption of aircraft-supplied data (other than thrust or power command signals from the aircraft), or data shared between engines must:

(1) Not result in a hazardous engine effect for any engine; and

(2) Be detected and accommodated. The accommodation strategy must not result in an unacceptable change in thrust or power or an unacceptable change in engine operating and starting characteristics. The applicant must evaluate and document in the engine installation instructions the effects of these failures on engine power or thrust, engine operability, and starting characteristics throughout the flight envelope.

(i) *Aircraft-supplied electrical power.* (1) The applicant must design the engine control system so that the loss, malfunction, or interruption of electrical power supplied from the aircraft to the engine control system will not result in any of the following:

(i) A hazardous engine effect, or

(ii) The unacceptable transmission of erroneous data.

(2) When an engine dedicated power source is required for compliance with paragraph (i)(1) of this section, its capacity should provide sufficient margin to account for engine operation below idle where the engine control system is designed and expected to recover engine operation automatically.

(3) The applicant must identify and declare the need for, and the characteristics of, any electrical power supplied from the aircraft to the engine control system for starting and operating the engine, including transient and steady state voltage limits, in the engine instructions for installation.

(4) Low voltage transients outside the power supply voltage limitations declared in paragraph (i)(3) of this section must meet the requirements of paragraph (i)(1) of this section. The engine control system must be capable of resuming normal operation when aircraft-supplied power returns to within the declared limits.

(j) *Air pressure signal.* The applicant must consider the effects of blockage or leakage of the signal lines on the engine control system as part of the System Safety Assessment of paragraph (e) of this section and must adopt the appropriate design precautions.

(k) *Automatic availability and control of engine power for 30-second OEI rating.* Rotorcraft engines having a 30-second OEI rating must incorporate a means, or a provision for a means, for automatic availability and automatic control of the 30-second OEI power within its operating limitations.

(l) *Engine shut down means.* Means must be provided for shutting down the engine rapidly.

(m) *Programmable logic devices.* The development of programmable logic devices using digital logic or other complex design technologies must provide a level of assurance for the encoded logic commensurate with the hazard associated with the failure or malfunction of the systems in which the devices are located. The applicant must provide evidence that the development of these devices has been done by using a method, approved by the FAA, that is consistent with the criticality of the performed function.

[Amdt. 33-26, 73 FR 48284, Aug. 19, 2008]

§ 33.29 Instrument connection.

(a) Unless it is constructed to prevent its connection to an incorrect instrument, each connection provided for powerplant instruments required by aircraft airworthiness regulations or necessary to insure operation of the engine in compliance with any engine limitation must be marked to identify it with its corresponding instrument.

(b) A connection must be provided on each turbojet engine for an indicator system to indicate rotor system unbalance.

(c) Each rotorcraft turbine engine having a 30-second OEI rating and a 2-minute OEI rating must have a means or a provision for a means to:

(1) Alert the pilot when the engine is at the 30-second OEI and the 2-minute OEI power levels, when the event begins, and when the time interval expires;

(2) Automatically record each usage and duration of power at the 30-second OEI and 2-minute OEI levels;

(3) Alert maintenance personnel in a positive manner that the engine has been operated at either or both of the 30-second and 2-minute OEI power levels, and permit retrieval of the recorded data; and

(4) Enable routine verification of the proper operation of the above means.

(d) The means, or the provision for a means, of paragraphs (c)(2) and (c)(3) of this section must not be capable of being reset in flight.

(e) The applicant must make provision for the installation of instrumentation necessary to ensure operation in compliance with engine operating limitations. Where, in presenting the safety analysis, or complying with any other requirement, dependence is placed on instrumentation that is not otherwise mandatory in the assumed aircraft installation, then the applicant must specify this instrumentation in the engine installation instructions and declare it mandatory in the engine approval documentation.

(f) As part of the System Safety Assessment of § 33.28(e), the applicant must assess the possibility and subsequent effect of incorrect fit of instruments, sensors, or connectors. Where necessary, the applicant must take design precautions to prevent incorrect configuration of the system.

(g) The sensors, together with associated wiring and signal conditioning, must be segregated, electrically and physically, to the extent necessary to ensure that the probability of a fault propagating from instrumentation and monitoring functions to control functions, or vice versa, is consistent with the failure effect of the fault.

(h) The applicant must provide instrumentation enabling the flight crew to monitor the functioning of the turbine cooling system unless appropriate inspections are published in the relevant manuals and evidence shows that:

(1) Other existing instrumentation provides adequate warning of failure or impending failure;

(2) Failure of the cooling system would not lead to hazardous engine effects before detection; or

(3) The probability of failure of the cooling system is extremely remote.

[Amdt. 33-5, 39 FR 1831, Jan. 15, 1974, as amended by Amdt. 33-6, 39 FR 35465, Oct. 1, 1974; Amdt. 33-18, 61 FR 31328, June 19, 1996; Amdt. 33-25, 73 FR 48123, Aug. 18, 2008; Amdt. 33-26, 73 FR 48285, Aug. 19, 2008]

Subpart C—Design and Construction; Reciprocating Aircraft Engines

§ 33.31 Applicability.

This subpart prescribes additional design and construction requirements for reciprocating aircraft engines.

§ 33.33 Vibration.

The engine must be designed and constructed to function throughout its normal operating range of crankshaft rotational speeds and engine powers without inducing excessive stress in any of the engine parts because of vibration and without imparting excessive vibration forces to the aircraft structure.

§ 33.34 Turbocharger rotors.

Each turbocharger case must be designed and constructed to be able to contain fragments of a compressor or turbine that fails at the highest speed that is obtainable with normal speed control devices inoperative.

[Amdt. 33-22, 72 FR 50860, Sept. 4, 2007]

§ 33.35 Fuel and induction system.

(a) The fuel system of the engine must be designed and constructed to supply an appropriate mixture of fuel to the cylinders throughout the complete operating range of the engine under all flight and atmospheric conditions.

(b) The intake passages of the engine through which air or fuel in combination with air passes for combustion purposes must be designed and constructed to minimize the danger of ice accretion in those passages. The engine must be designed and constructed to permit the use of a means for ice prevention.

(c) The type and degree of fuel filtering necessary for protection of the engine fuel system against foreign particles in the fuel must be specified. The applicant must show that foreign particles passing through the prescribed filtering means will not critically impair engine fuel system functioning.

(d) Each passage in the induction system that conducts a mixture of fuel and air must be self-draining, to prevent a liquid lock in the cylinders, in all attitudes that the applicant establishes as those the engine can have when the aircraft in which it is installed is in the static ground attitude.

(e) If provided as part of the engine, the applicant must show for each fluid injection (other than fuel) system and its controls that the flow of the injected fluid is adequately controlled.

[Doc. No. 3025, 29 FR 7453, June 10, 1964, as amended by Amdt. 33-10, 49 FR 6851, Feb. 23, 1984]

§ 33.37 Ignition system.

Each spark ignition engine must have a dual ignition system with at least two spark plugs for each cylinder and two separate electric circuits with separate sources of electrical energy, or have an ignition system of equivalent in-flight reliability.

§ 33.39 Lubrication system.

(a) The lubrication system of the engine must be designed and constructed so that it will function properly in all flight attitudes and atmospheric conditions in which the airplane is expected to operate. In wet sump engines, this requirement must be met when only one-half of the maximum lubricant supply is in the engine.

(b) The lubrication system of the engine must be designed and constructed to allow installing a means of cooling the lubricant.

(c) The crankcase must be vented to the atmosphere to preclude leakage of oil from excessive pressure in the crankcase.

Subpart D—Block Tests; Reciprocating Aircraft Engines

§ 33.41 Applicability.

This subpart prescribes the block tests and inspections for reciprocating aircraft engines.

§ 33.42 General.

Before each endurance test required by this subpart, the adjustment setting and functioning characteristic of each component having an adjustment setting and a functioning characteristic that can be established independent of installation on the engine must be established and recorded.

[Amdt. 33-6, 39 FR 35465, Oct. 1, 1974]

§ 33.43 Vibration test.

(a) Each engine must undergo a vibration survey to establish the torsional and bending vibration characteristics of the crankshaft and the propeller shaft or other output shaft, over the range of crankshaft speed and engine power, under steady state and transient conditions, from idling speed to either 110 percent of the desired maximum continuous speed rating or 103 percent of the maximum desired takeoff speed rating, whichever is higher. The survey must be conducted using, for airplane engines, the same configuration of the propeller type which is used for the endurance test, and using, for other engines, the same configuration of the loading device type which is used for the endurance test.

(b) The torsional and bending vibration stresses of the crankshaft and the propeller shaft or other output shaft may not exceed the endurance limit stress of the material from which the shaft is made. If the maximum stress in the shaft cannot be shown to be below the endurance limit by measurement, the vibration frequency and amplitude must be measured. The peak amplitude must be shown to produce a stress below the endurance limit; if not, the engine must be run at the condition producing the peak amplitude until, for steel shafts, 10 million stress reversals have been sustained without fatigue failure and, for other shafts, until it is shown that fatigue will not occur within the endurance limit stress of the material.

(c) Each accessory drive and mounting attachment must be loaded, with the loads imposed by each accessory used only for an aircraft service being the limit load specified by the applicant for the drive or attachment point.

(d) The vibration survey described in paragraph (a) of this section must be repeated with that cylinder not firing which has the most adverse vibration effect, in order to establish the conditions under which the engine can be operated safely in that abnormal state. However, for this vibration survey, the engine speed range need only extend from idle to the maximum desired takeoff speed, and compliance with paragraph (b) of this section need not be shown.

[Amdt. 33-6, 39 FR 35465, Oct. 1, 1974, as amended by Amdt. 33-10, 49 FR 6851, Feb. 23, 1984]

§ 33.45 Calibration tests.

(a) Each engine must be subjected to the calibration tests necessary to establish its power characteristics and the conditions for the endurance test specified in § 33.49. The results of the power characteristics calibration tests form the basis for establishing the characteristics of the engine over its entire operating range of crankshaft rotational speeds, manifold pressures, fuel/air mixture settings, and altitudes. Power ratings are based upon standard atmospheric conditions with only those accessories installed which are essential for engine functioning.

(b) A power check at sea level conditions must be accomplished on the endurance test engine after the endurance test. Any change in power characteristics which occurs during the endurance test must be determined. Measurements taken during the final portion of the endurance test may be used in showing compliance with the requirements of this paragraph.

[Doc. No. 3025, 29 FR 7453, June 10, 1964, as amended by Amdt. 33-6, 39 FR 35465, Oct. 1, 1974]

§ 33.47 Detonation test.

Each engine must be tested to establish that the engine can function without detonation throughout its range of intended conditions of operation.

§ 33.49 Endurance test.

(a) *General.* Each engine must be subjected to an endurance test that includes a total of 150 hours of operation (except as provided in paragraph (e)(1)(iii) of this section) and, depending upon the type and contemplated use of the engine, consists of one of the series of runs specified in paragraphs (b) through (e) of this section, as applicable. The runs must be made in the order found appropriate by the Administrator for the particular engine being tested. During the endurance test the engine power and the crankshaft rotational speed must be kept within ±3 percent of the rated values. During the runs at rated takeoff power and for at least 35 hours at rated maximum continuous power, one cylinder must be operated at not less than the limiting temperature, the other cylinders must be operated at a temperature not lower than 50 degrees F. below the limiting temperature, and the oil inlet temperature must be maintained within ±10 degrees F. of the limiting temperature. An engine that is equipped with a propeller shaft must be fitted for the endurance test with a propeller that thrust-loads the engine to the maximum thrust which the engine is designed to resist at each applicable operating condition specified in this section. Each accessory drive and mounting attachment must be loaded. During operation at rated takeoff power and rated maximum continuous power, the load imposed by each accessory used only for an aircraft service must be the limit load specified by the applicant for the engine drive or attachment point.

(b) *Unsupercharged engines and engines incorporating a gear-driven single-speed supercharger.* For engines not incorporating a supercharger and for engines incorporating a gear-driven single-speed supercharger the applicant must conduct the following runs:

(1) A 30-hour run consisting of alternate periods of 5 minutes at rated takeoff power with takeoff speed, and 5 minutes at maximum best economy cruising power or maximum recommended cruising power.

(2) A 20-hour run consisting of alternate periods of 1½ hours at rated maximum continuous power with maximum continuous speed, and ½ hour at 75 percent rated maximum continuous power and 91 percent maximum continuous speed.

(3) A 20-hour run consisting of alternate periods of 1½ hours at rated maximum continuous power with maximum continuous speed, and ½ hour at 70 percent rated maximum continuous power and 89 percent maximum continuous speed.

(4) A 20-hour run consisting of alternate periods of 1½ hours at rated maximum continuous power with maximum continuous speed, and ½ hour at 65 percent rated maximum continuous power and 87 percent maximum continuous speed.

(5) A 20-hour run consisting of alternate periods of 1½ hours at rated maximum continuous power with maximum continuous speed, and ½ hour at 60 percent rated maximum continuous power and 84.5 percent maximum continuous speed.

(6) A 20-hour run consisting of alternate periods of 1½ hours at rated maximum continuous power with maximum continuous speed, and ½ hour at 50 percent rated maximum continuous power and 79.5 percent maximum continuous speed.

(7) A 20-hour run consisting of alternate periods of 2½ hours at rated maximum continuous power with maximum continuous speed, and 2½ hours at maximum best economy cruising power or at maximum recommended cruising power.

(c) *Engines incorporating a gear-driven two-speed supercharger.* For engines incorporating a gear-driven two-speed supercharger the applicant must conduct the following runs:

(1) A 30-hour run consisting of alternate periods in the lower gear ratio of 5 minutes at rated takeoff power with takeoff speed, and 5 minutes at maximum best economy cruising power or at maximum recommended cruising power. If a takeoff power rating is desired in the higher gear ratio, 15 hours of the 30-hour run must be made in the higher gear ratio in alternate periods of 5 minutes at the observed horsepower obtainable with the takeoff critical altitude manifold pressure and takeoff speed, and 5 minutes at 70 percent high ratio rated maximum continuous power and 89 percent high ratio maximum continuous speed.

(2) A 15-hour run consisting of alternate periods in the lower gear ratio of 1 hour at rated maximum continuous power with maximum continuous speed, and ½ hour at 75 percent rated maximum continuous power and 91 percent maximum continuous speed.

(3) A 15-hour run consisting of alternate periods in the lower gear ratio of 1 hour at rated maximum continuous power with maximum continuous speed, and ½ hour at 70 percent rated maximum continuous power and 89 percent maximum continuous speed.

(4) A 30-hour run in the higher gear ratio at rated maximum continuous power with maximum continuous speed.

(5) A 5-hour run consisting of alternate periods of 5 minutes in each of the supercharger gear ratios. The first 5 minutes of the test must be made at maximum continuous speed in the higher gear ratio and the observed horsepower obtainable with 90 percent of maximum continuous manifold pressure in the higher gear ratio under sea level conditions. The condition for operation for the alternate 5 minutes in the lower gear ratio must be that obtained by shifting to the lower gear ratio at constant speed.

(6) A 10-hour run consisting of alternate periods in the lower gear ratio of 1 hour at rated maximum continuous power with maximum continuous speed, and 1 hour at 65 percent rated maximum continuous power and 87 percent maximum continuous speed.

(7) A 10-hour run consisting of alternate periods in the lower gear ratio of 1 hour at rated maximum continuous power with maximum continuous speed, and 1 hour at 60 percent rated maximum continuous power and 84.5 percent maximum continuous speed.

(8) A 10-hour run consisting of alternate periods in the lower gear ratio of 1 hour at rated maximum continuous

power with maximum continuous speed, and 1 hour at 50 percent rated maximum continuous power and 79.5 percent maximum continuous speed.

(9) A 20-hour run consisting of alternate periods in the lower gear ratio of 2 hours at rated maximum continuous power with maximum continuous speed, and 2 hours at maximum best economy cruising power and speed or at maximum recommended cruising power.

(10) A 5-hour run in the lower gear ratio at maximum best economy cruising power and speed or at maximum recommended cruising power and speed.

Where simulated altitude test equipment is not available when operating in the higher gear ratio, the runs may be made at the observed horsepower obtained with the critical altitude manifold pressure or specified percentages thereof, and the fuel-air mixtures may be adjusted to be rich enough to suppress detonation.

(d) *Helicopter engines.* To be eligible for use on a helicopter each engine must either comply with paragraphs (a) through (j) of § 29.923 of this chapter, or must undergo the following series of runs:

(1) A 35-hour run consisting of alternate periods of 30 minutes each at rated takeoff power with takeoff speed, and at rated maximum continuous power with maximum continuous speed.

(2) A 25-hour run consisting of alternate periods of 2½ hours each at rated maximum continuous power with maximum continuous speed, and at 70 percent rated maximum continuous power with maximum continuous speed.

(3) A 25-hour run consisting of alternate periods of 2½ hours each at rated maximum continuous power with maximum continuous speed, and at 70 percent rated maximum continuous power with 80 to 90 percent maximum continuous speed.

(4) A 25-hour run consisting of alternate periods of 2½ hours each at 30 percent rated maximum continuous power with takeoff speed, and at 30 percent rated maximum continuous power with 80 to 90 percent maximum continuous speed.

(5) A 25-hour run consisting of alternate periods of 2½ hours each at 80 percent rated maximum continuous power with takeoff speed, and at either rated maximum continuous power with 110 percent maximum continuous speed or at rated takeoff power with 103 percent takeoff speed, whichever results in the greater speed.

(6) A 15-hour run at 105 percent rated maximum continuous power with 105 percent maximum continuous speed or at full throttle and corresponding speed at standard sea level carburetor entrance pressure, if 105 percent of the rated maximum continuous power is not exceeded.

(e) *Turbosupercharged engines.* For engines incorporating a turbosupercharger the following apply except that altitude testing may be simulated provided the applicant shows that the engine and supercharger are being subjected to mechanical loads and operating temperatures no less severe than if run at actual altitude conditions:

(1) For engines used in airplanes the applicant must conduct the runs specified in paragraph (b) of this section, except—

(i) The entire run specified in paragraph (b)(1) of this section must be made at sea level altitude pressure;

(ii) The portions of the runs specified in paragraphs (b)(2) through (7) of this section at rated maximum continuous power must be made at critical altitude pressure, and the portions of the runs at other power must be made at 8,000 feet altitude pressure; and

(iii) The turbosupercharger used during the 150-hour endurance test must be run on the bench for an additional 50 hours at the limiting turbine wheel inlet gas temperature and rotational speed for rated maximum continuous power operation unless the limiting temperature and speed are maintained during 50 hours of the rated maximum continuous power operation.

(2) For engines used in helicopters the applicant must conduct the runs specified in paragraph (d) of this section, except—

(i) The entire run specified in paragraph (d)(1) of this section must be made at critical altitude pressure;

(ii) The portions of the runs specified in paragraphs (d)(2) and (3) of this section at rated maximum continuous power must be made at critical altitude pressure and the portions of the runs at other power must be made at 8,000 feet altitude pressure;

(iii) The entire run specified in paragraph (d)(4) of this section must be made at 8,000 feet altitude pressure;

(iv) The portion of the runs specified in paragraph (d)(5) of this section at 80 percent of rated maximum continuous power must be made at 8,000 feet altitude pressure and the portions of the runs at other power must be made at critical altitude pressure;

(v) The entire run specified in paragraph (d)(6) of this section must be made at critical altitude pressure; and

(vi) The turbosupercharger used during the endurance test must be run on the bench for 50 hours at the limiting turbine wheel inlet gas temperature and rotational speed for rated maximum continuous power operation unless the limiting temperature and speed are maintained during 50 hours of the rated maximum continuous power operation.

[Amdt. 33-3, 32 FR 3736, Mar. 4, 1967, as amended by Amdt. 33-6, 39 FR 35465, Oct. 1, 1974; Amdt. 33-10, 49 FR 6851, Feb. 23, 1984]

§ 33.51 Operation test.

The operation test must include the testing found necessary by the Administrator to demonstrate backfire characteristics, starting, idling, acceleration, overspeeding, functioning of propeller and ignition, and any other operational characteristic of the engine. If the engine incorporates a multispeed supercharger drive, the design and construction must allow the supercharger to be shifted from operation at the lower speed ratio to the higher and the power appropriate to the manifold pressure and speed settings for rated maximum continuous power at the higher supercharger speed ratio must be obtainable within five seconds.

[Doc. No. 3025, 29 FR 7453, June 10, 1964, as amended by Amdt. 33-3, 32 FR 3737, Mar. 4, 1967]

§ 33.53 Engine system and component tests.

(a) For those systems and components that cannot be adequately substantiated in accordance with endurance testing of § 33.49, the applicant must conduct additional tests to demonstrate that systems or components are able to perform the intended functions in all declared environmental and operating conditions.

(b) Temperature limits must be established for each component that requires temperature controlling provisions in the aircraft installation to assure satisfactory functioning, reliability, and durability.

[Docket No. 3025, 29 FR 7453, June 10, 1964, as amended by Amdt. 33-26, 73 FR 48285, Aug. 19, 2008]

§ 33.55 Teardown inspection.

After completing the endurance test—

(a) Each engine must be completely disassembled;

(b) Each component having an adjustment setting and a functioning characteristic that can be established independent of installation on the engine must retain each setting and functioning characteristic within the limits that were established and recorded at the beginning of the test; and

(c) Each engine component must conform to the type design and be eligible for incorporation into an engine for continued operation, in accordance with information submitted in compliance with § 33.4.

[Amdt. 33-6, 39 FR 35466, Oct. 1, 1974, as amended by Amdt. 33-9, 45 FR 60181, Sept. 11, 1980]

§ 33.57 General conduct of block tests.

(a) The applicant may, in conducting the block tests, use separate engines of identical design and construction in the vibration, calibration, detonation, endurance, and operation tests, except that, if a separate engine is used for the endurance test it must be subjected to a calibration check before starting the endurance test.

(b) The applicant may service and make minor repairs to the engine during the block tests in accordance with the service and maintenance instructions submitted in compliance with § 33.4. If the frequency of the service is excessive, or the number of stops due to engine malfunction is excessive, or a major repair, or replacement of a part is found necessary during the block tests or as the result of findings from the teardown inspection, the engine or its parts may be subjected to any additional test the Administrator finds necessary.

(c) Each applicant must furnish all testing facilities, including equipment and competent personnel, to conduct the block tests.

[Doc. No. 3025, 29 FR 7453, June 10, 1964, as amended by Amdt. 33-6, 39 FR 35466, Oct. 1, 1974; Amdt. 33-9, 45 FR 60181, Sept. 11, 1980]

Subpart E—Design and Construction; Turbine Aircraft Engines

§ 33.61 Applicability.

This subpart prescribes additional design and construction requirements for turbine aircraft engines.

§ 33.62 Stress analysis.

A stress analysis must be performed on each turbine engine showing the design safety margin of each turbine engine rotor, spacer, and rotor shaft.

[Amdt. 33-6, 39 FR 35466, Oct. 1, 1974]

§ 33.63 Vibration.

Each engine must be designed and constructed to function throughout its declared flight envelope and operating range of rotational speeds and power/thrust, without inducing excessive stress in any engine part because of vibration and without imparting excessive vibration forces to the aircraft structure.

[Doc. No. 28107, 61 FR 28433, June 4, 1996]

§ 33.64 Pressurized engine static parts.

(a) Strength. The applicant must establish by test, validated analysis, or a combination of both, that all static parts subject to significant gas or liquid pressure loads for a stabilized period of one minute will not:

(1) Exhibit permanent distortion beyond serviceable limits or exhibit leakage that could create a hazardous condition when subjected to the greater of the following pressures:

(i) 1.1 times the maximum working pressure;

(ii) 1.33 times the normal working pressure; or

(iii) 35 kPa (5 p.s.i.) above the normal working pressure.

(2) Exhibit fracture or burst when subjected to the greater of the following pressures:

(i) 1.15 times the maximum possible pressure;

(ii) 1.5 times the maximum working pressure; or

(iii) 35 kPa (5 p.s.i.) above the maximum possible pressure.

(b) Compliance with this section must take into account:

(1) The operating temperature of the part;

(2) Any other significant static loads in addition to pressure loads;

(3) Minimum properties representative of both the material and the processes used in the construction of the part; and

(4) Any adverse geometry conditions allowed by the type design.

[Amdt. 33-27; 73 FR 55437, Sept. 25, 2008; Amdt. 33-27, 73 FR 57235, Oct. 2, 2008]

§ 33.65 Surge and stall characteristics.

When the engine is operated in accordance with operating instructions required by § 33.5(b), starting, a change of power or thrust, power or thrust augmentation, limiting inlet air distortion, or inlet air temperature may not cause surge or stall to the extent that flameout, structural failure, overtemperature, or failure of the engine to recover power or thrust will occur at any point in the operating envelope.

[Amdt. 33-6, 39 FR 35466, Oct. 1, 1974]

§ 33.66 Bleed air system.

The engine must supply bleed air without adverse effect on the engine, excluding reduced thrust or power output, at all conditions up to the discharge flow conditions established as a limitation under § 33.7(c)(11). If bleed air used for engine anti-icing can be controlled, provision must be made for a means to indicate the functioning of the engine ice protection system.

[Amdt. 33-10, 49 FR 6851, Feb. 23, 1984]

§ 33.67 Fuel system.

(a) With fuel supplied to the engine at the flow and pressure specified by the applicant, the engine must function properly under each operating condition required by this part. Each fuel control adjusting means that may not be manipulated while the fuel control device is mounted on the engine must be secured by a locking device and sealed, or otherwise be inaccessible. All other fuel control adjusting means must be accessible and marked to indicate the function of the adjustment unless the function is obvious.

(b) There must be a fuel strainer or filter between the engine fuel inlet opening and the inlet of either the fuel metering device or the engine-driven positive displacement pump whichever is nearer the engine fuel inlet. In addition, the following provisions apply to each strainer or filter required by this paragraph (b):

(1) It must be accessible for draining and cleaning and must incorporate a screen or element that is easily removable.

(2) It must have a sediment trap and drain except that it need not have a drain if the strainer or filter is easily removable for drain purposes.

(3) It must be mounted so that its weight is not supported by the connecting lines or by the inlet or outlet connections of the strainer or filter, unless adequate strength margins under all loading conditions are provided in the lines and connections.

(4) It must have the type and degree of fuel filtering specified as necessary for protection of the engine fuel system against foreign particles in the fuel. The applicant must show:

(i) That foreign particles passing through the specified filtering means do not impair the engine fuel system functioning; and

(ii) That the fuel system is capable of sustained operation throughout its flow and pressure range with the fuel initially saturated with water at 80 °F (27 °C) and having 0.025 fluid ounces per gallon (0.20 milliliters per liter) of free water added and cooled to the most critical condition for icing likely to be encountered in operation. However, this requirement may be met by demonstrating the effectiveness of specified approved fuel anti-icing additives, or that the fuel system incorporates a fuel heater which maintains the fuel temperature at the fuel strainer or fuel inlet above 32 °F (0 °C) under the most critical conditions.

(5) The applicant must demonstrate that the filtering means has the capacity (with respect to engine operating limitations) to ensure that the engine will continue to operate within approved limits, with fuel contaminated to the maximum degree of particle size and density likely to be encountered in service. Operation under these conditions must be demonstrated for a period acceptable to the Administrator, beginning when indication of impending filter blockage is first given by either:

(i) Existing engine instrumentation; or

(ii) Additional means incorporated into the engine fuel system.

(6) Any strainer or filter bypass must be designed and constructed so that the release of collected contaminants is minimized by appropriate location of the bypass to ensure that collected contaminants are not in the bypass flow path.

(c) If provided as part of the engine, the applicant must show for each fluid injection (other than fuel) system and its controls that the flow of the injected fluid is adequately controlled.

[Amdt. 33-6, 39 FR 35466, Oct. 1, 1974, as amended by Amdt. 33-10, 49 FR 6851, Feb. 23, 1984; Amdt. 33-18, 61 FR 31328, June 19, 1996; Amdt. 33-25, 73 FR 48123, Aug. 18, 2008; Amdt. 33-26, 73 FR 48285, Aug. 19, 2008]

§ 33.68 Induction system icing.

Each engine, with all icing protection systems operating, must—

(a) Operate throughout its flight power range (including idling) without the accumulation of ice on the engine components that adversely affects engine operation or that causes a serious loss of power or thrust in continuous maximum and intermittent maximum icing conditions as defined in appendix C of Part 25 of this chapter; and

(b) Idle for 30 minutes on the ground, with the available air bleed for icing protection at its critical condition, without adverse effect, in an atmosphere that is at a temperature between 15° and 30 °F (between -9° and -1 °C) and has a liquid water content not less than 0.3 grams per cubic meter in the form of drops having a mean effective diameter not less than 20 microns, followed by a momentary operation at takeoff power or thrust. During the 30 minutes of idle operation the engine may be run up periodically to a moderate power or thrust setting in a manner acceptable to the Administrator.

[Amdt. 33-6, 39 FR 35466, Oct. 1, 1974, as amended by Amdt. 33-10, 49 FR 6852, Feb. 23, 1984]

§ 33.69 Ignitions system.

Each engine must be equipped with an ignition system for starting the engine on the ground and in flight. An electric ignition system must have at least two igniters and two separate secondary electric circuits, except that only one igniter is required for fuel burning augmentation systems.

[Amdt. 33-6, 39 FR 35466, Oct. 1, 1974]

§ 33.70 Engine life-limited parts.

By a procedure approved by the FAA, operating limitations must be established which specify the maximum allowable number of flight cycles for each engine life-limited part. Engine life-limited parts are rotor and major static structural parts whose primary failure is likely to result in a hazardous engine effect. Typically, engine life-limited parts include, but are not limited to disks, spacers, hubs, shafts, high-pressure casings, and non-redundant mount components. For the purposes of this section, a hazardous engine effect is any of the conditions listed in § 33.75 of this part. The applicant will establish the integrity of each engine life-limited part by:

(a) An engineering plan that contains the steps required to ensure each engine life-limited part is withdrawn from service at an approved life before hazardous engine effects can occur. These steps include validated analysis, test, or service experience which ensures that the combination of loads, material properties, environmental influences and operating conditions, including the effects of other engine parts influencing these parameters, are sufficiently well known and predictable so that the operating limitations can be established and maintained for each engine life-limited part. Applicants must perform appropriate damage tolerance assessments to address the potential for failure from material, manufacturing, and service induced anomalies within the approved life of the part. Applicants must publish a list of the life-limited engine parts and the approved life for each part in the Airworthiness Limitations Section of the Instructions for Continued Airworthiness as required by § 33.4 of this part.

(b) A manufacturing plan that identifies the specific manufacturing constraints necessary to consistently produce each engine life-limited part with the attributes required by the engineering plan.

(c) A service management plan that defines in-service processes for maintenance and the limitations to repair for each engine life-limited part that will maintain attributes consistent with those required by the engineering plan. These processes and limitations will become part of the Instructions for Continued Airworthiness.

[Amdt. 33-22, 72 FR 50860, Sept. 4, 2007]

§ 33.71 Lubrication system.

(a) *General.* Each lubrication system must function properly in the flight attitudes and atmospheric conditions in which an aircraft is expected to operate.

(b) *Oil strainer or filter.* There must be an oil strainer or filter through which all of the engine oil flows. In addition:

(1) Each strainer or filter required by this paragraph that has a bypass must be constructed and installed so that oil will flow at the normal rate through the rest of the system with the strainer or filter element completely blocked.

(2) The type and degree of filtering necessary for protection of the engine oil system against foreign particles in the oil must be specified. The applicant must demonstrate that foreign particles passing through the specified filtering means do not impair engine oil system functioning.

(3) Each strainer or filter required by this paragraph must have the capacity (with respect to operating limitations established for the engine) to ensure that engine oil system functioning is not impaired with the oil contaminated to a degree (with respect to particle size and density) that is greater than that established for the engine in paragraph (b)(2) of this section.

(4) For each strainer or filter required by this paragraph, except the strainer or filter at the oil tank outlet, there must be means to indicate contamination before it reaches the capacity established in accordance with paragraph (b)(3) of this section.

(5) Any filter bypass must be designed and constructed so that the release of collected contaminants is minimized by appropriate location of the bypass to ensure that the collected contaminants are not in the bypass flow path.

(6) Each strainer or filter required by this paragraph that has no bypass, except the strainer or filter at an oil tank outlet or for a scavenge pump, must have provisions for connection with a warning means to warn the pilot of the occurance of contamination of the screen before it reaches the capacity established in accordance with paragraph (b)(3) of this section.

(7) Each strainer or filter required by this paragraph must be accessible for draining and cleaning.

(c) *Oil tanks.* (1) Each oil tank must have an expansion space of not less than 10 percent of the tank capacity.

(2) It must be impossible to inadvertently fill the oil tank expansion space.

(3) Each recessed oil tank filler connection that can retain any appreciable quantity of oil must have provision for fitting a drain.

(4) Each oil tank cap must provide an oil-tight seal. For an applicant seeking eligibility for an engine to be installed on an airplane approved for ETOPS, the oil tank must be designed to prevent a hazardous loss of oil due to an incorrectly installed oil tank cap.

(5) Each oil tank filler must be marked with the word "oil."

(6) Each oil tank must be vented from the top part of the expansion space, with the vent so arranged that condensed water vapor that might freeze and obstruct the line cannot accumulate at any point.

(7) There must be means to prevent entrance into the oil tank or into any oil tank outlet, of any object that might obstruct the flow of oil through the system.

(8) There must be a shutoff valve at the outlet of each oil tank, unless the external portion of the oil system (including oil tank supports) is fireproof.

(9) Each unpressurized oil tank may not leak when subjected to a maximum operating temperature and an internal pressure of 5 p.s.i., and each pressurized oil tank must meet the requirements of § 33.64.

(10) Leaked or spilled oil may not accumulate between the tank and the remainder of the engine.

(11) Each oil tank must have an oil quantity indicator or provisions for one.

(12) If the propeller feathering system depends on engine oil—

(i) There must be means to trap an amount of oil in the tank if the supply becomes depleted due to failure of any part of the lubricating system other than the tank itself;

(ii) The amount of trapped oil must be enough to accomplish the feathering opeation and must be available only to the feathering pump; and

(iii) Provision must be made to prevent sludge or other foreign matter from affecting the safe operation of the propeller feathering system.

(d) *Oil drains.* A drain (or drains) must be provided to allow safe drainage of the oil system. Each drain must—

(1) Be accessible; and

(2) Have manual or automatic means for positive locking in the closed position.

(e) *Oil radiators.* Each oil radiator must withstand, without failure, any vibration, inertia, and oil pressure load to which it is subjected during the block tests.

[Amdt. 33-6, 39 FR 35466, Oct. 1, 1974, as amended by Amdt. 33-10, 49 FR 6852, Feb. 23, 1984; Amdt. 33-21, 72 FR 1877, Jan. 16, 2007; Amdt. 33-27, 73 FR 55437, Sept. 25, 2008; Amdt. 33-27, 73 FR 57235, Oct. 2, 2008]

§ 33.72 Hydraulic actuating systems.

Each hydraulic actuating system must function properly under all conditions in which the engine is expected to operate. Each filter or screen must be accessible for servicing and each tank must meet the design criteria of § 33.71.

[Amdt. 33-6, 39 FR 35467, Oct. 1, 1974]

§ 33.73 Power or thrust response.

The design and construction of the engine must enable an increase—

(a) From minimum to rated takeoff power or thrust with the maximum bleed air and power extraction to be permitted in an aircraft, without overtemperature, surge, stall, or other detrimental factors occurring to the engine whenever the power control lever is moved from the minimum to the maximum position in not more than 1 second, except that the Administrator may allow additional time increments for different regimes of control operation requiring control scheduling; and

(b) From the fixed minimum flight idle power lever position when provided, or if not provided, from not more than 15 percent of the rated takeoff power or thrust available to 95 percent rated takeoff power or thrust in not over 5 seconds. The 5-second power or thrust response must occur from a stabilized static condition using only the bleed air and accessories loads necessary to run the engine. This takeoff rating is specified by the applicant and need not include thrust augmentation.

[Amdt. 33-1, 36 FR 5493, Mar. 24, 1971]

§ 33.74 Continued rotation.

If any of the engine main rotating systems continue to rotate after the engine is shutdown for any reason while in flight, and if means to prevent that continued rotation are not provided, then any continued rotation during the maximum period of flight, and in the flight conditions expected to occur with that engine inoperative, may not result in any condition described in § 33.75(g)(2)(i) through (vi) of this part.

[Amdt. 33-24, 72 FR 50867, Sept. 4, 2007]

§ 33.75 Safety analysis.

(a) (1) The applicant must analyze the engine, including the control system, to assess the likely consequences of all failures that can reasonably be expected to occur. This analysis will take into account, if applicable:

(i) Aircraft-level devices and procedures assumed to be associated with a typical installation. Such assumptions must be stated in the analysis.

(ii) Consequential secondary failures and latent failures.

(iii) Multiple failures referred to in paragraph (d) of this section or that result in the hazardous engine effects defined in paragraph (g)(2) of this section.

(2) The applicant must summarize those failures that could result in major engine effects or hazardous engine effects, as defined in paragraph (g) of this section, and estimate the probability of occurrence of those effects. Any engine part the failure of which could reasonably result in a hazardous engine effect must be clearly identified in this summary.

(3) The applicant must show that hazardous engine effects are predicted to occur at a rate not in excess of that defined as extremely remote (probability range of 10^{-7} to 10^{-9} per engine flight hour). Since the estimated probability for individual failures may be insufficiently precise to enable the applicant to assess the total rate for hazardous engine effects, compliance may be shown by demonstrating that the probability of a hazardous engine effect arising from an individual failure can be predicted to be not greater than 10^{-8} per engine flight hour. In dealing with probabilities of this low order of magnitude, absolute proof is not possible, and compliance may be shown by reliance on engineering judgment and previous experience combined with sound design and test philosophies.

(4) The applicant must show that major engine effects are predicted to occur at a rate not in excess of that defined as remote (probability range of 10^{-5} to 10^{-7} per engine flight hour).

(b) The FAA may require that any assumption as to the effects of failures and likely combination of failures be verified by test.

(c) The primary failure of certain single elements cannot be sensibly estimated in numerical terms. If the failure of such elements is likely to result in hazardous engine effects, then compliance may be shown by reliance on the prescribed integrity requirements of §§ 33.15, 33.27, and 33.70 as applicable. These instances must be stated in the safety analysis.

(d) If reliance is placed on a safety system to prevent a failure from progressing to hazardous engine effects, the possibility of a safety system failure in combination with a basic engine failure must be included in the analysis. Such a safety system may include safety devices, instrumentation, early warning devices, maintenance checks, and other similar equipment or procedures. If items of a safety system are outside the control of the engine manufacturer, the assumptions of the safety analysis with respect to the reliability of these parts must be clearly stated in the analysis and identified in the installation instructions under § 33.5 of this part.

(e) If the safety analysis depends on one or more of the following items, those items must be identified in the analysis and appropriately substantiated.

(1) Maintenance actions being carried out at stated intervals. This includes the verification of the serviceability of items that could fail in a latent manner. When necessary to prevent hazardous engine effects, these maintenance actions and intervals must be published in the instructions for continued airworthiness required under § 33.4 of this part. Additionally, if errors in maintenance of the engine, including the control system, could lead to hazardous engine effects, the appropriate procedures must be included in the relevant engine manuals.

(2) Verification of the satisfactory functioning of safety or other devices at pre-flight or other stated periods. The details of this satisfactory functioning must be published in the appropriate manual.

(3) The provisions of specific instrumentation not otherwise required.

(4) Flight crew actions to be specified in the operating instructions established under § 33.5.

(f) If applicable, the safety analysis must also include, but not be limited to, investigation of the following:

(1) Indicating equipment;

(2) Manual and automatic controls;

(3) Compressor bleed systems;

(4) Refrigerant injection systems;

(5) Gas temperature control systems;

(6) Engine speed, power, or thrust governors and fuel control systems;

(7) Engine overspeed, overtemperature, or topping limiters;

(8) Propeller control systems; and

(9) Engine or propeller thrust reversal systems.

(g) Unless otherwise approved by the FAA and stated in the safety analysis, for compliance with part 33, the following failure definitions apply to the engine:

(1) An engine failure in which the only consequence is partial or complete loss of thrust or power (and associated engine services) from the engine will be regarded as a minor engine effect.

(2) The following effects will be regarded as hazardous engine effects:

(i) Non-containment of high-energy debris;

(ii) Concentration of toxic products in the engine bleed air intended for the cabin sufficient to incapacitate crew or passengers;

(iii) Significant thrust in the opposite direction to that commanded by the pilot;

(iv) Uncontrolled fire;

(v) Failure of the engine mount system leading to inadvertent engine separation;

(vi) Release of the propeller by the engine, if applicable; and

(vii) Complete inability to shut the engine down.

(3) An effect whose severity falls between those effects covered in paragraphs (g)(1) and (g)(2) of this section will be regarded as a major engine effect.

[Amdt. 33-24, 72 FR 50867, Sept. 4, 2007]

§ 33.76 Bird ingestion.

(a) *General.* Compliance with paragraphs (b), (c), and (d) of this section shall be in accordance with the following:

(1) Except as specified in paragraph (d) of this section, all ingestion tests must be conducted with the engine stabilized at no less than 100-percent takeoff power or thrust, for test day ambient conditions prior to the ingestion. In addition, the demonstration of compliance must account for engine operation at sea level takeoff conditions on the hottest day that a minimum engine can achieve maximum rated takeoff thrust or power.

(2) The engine inlet throat area as used in this section to determine the bird quantity and weights will be established by the applicant and identified as a limitation in the installation instructions required under § 33.5.

(3) The impact to the front of the engine from the large single bird, the single largest medium bird which can enter the inlet, and the large flocking bird must be evaluated. Applicants must show that the associated components when struck under the conditions prescribed in paragraphs (b), (c) or (d) of this section, as applicable, will not affect the engine to the extent that the engine cannot comply with the requirements of paragraphs (b)(3), (c)(6) and (d)(4) of this section.

(4) For an engine that incorporates an inlet protection device, compliance with this section shall be established with the device functioning. The engine approval will be endorsed to show that compliance with the requirements has been established with the device functioning.

(5) Objects that are accepted by the Administrator may be substituted for birds when conducting the bird ingestion tests required by paragraphs (b), (c) and (d) of this section.

(6) If compliance with the requirements of this section is not established, the engine type certification documentation will show that the engine shall be limited to aircraft installations in which it is shown that a bird cannot strike the engine, or be ingested into the engine, or adversely restrict airflow into the engine.

(b) *Large single bird.* Compliance with the large bird ingestion requirements shall be in accordance with the following:

(1) The large bird ingestion test shall be conducted using one bird of a weight determined from Table 1 aimed at the most critical exposed location on the first stage rotor blades and ingested at a bird speed of 200-knots for engines to be installed on airplanes, or the maximum airspeed for normal rotorcraft flight operations for engines to be installed on rotorcraft.

(2) Power lever movement is not permitted within 15 seconds following ingestion of the large bird.

(3) Ingestion of a single large bird tested under the conditions prescribed in this section may not result in any condition described in § 33.75(g)(2) of this part.

(4) Compliance with the large bird ingestion requirements of this paragraph may be shown by demonstrating that the requirements of § 33.94(a) constitute a more severe demonstration of blade containment and rotor unbalance than the requirements of this paragraph.

Table 1 to §33.76
Large Bird Weight Requirements

Engine Inlet Throat Area (A)—Square-meters (square-inches)	Bird weight kg. (lb.)
1.35 (2,092)> A	1.85 (4.07) minimum, unless a smaller bird is determined to be a more severe demonstration.
1.35 (2,092)≤ A< 3.90 (6,045)	2.75 (6.05)
3.90 (6,045)≤ A	3.65 (8.03)

(c) *Small and medium flocking bird.* Compliance with the small and medium bird ingestion requirements shall be in accordance with the following:

(1) Analysis or component test, or both, acceptable to the Administrator, shall be conducted to determine the critical ingestion parameters affecting power loss and damage. Critical ingestion parameters shall include, but are not limited to, the effects of bird speed, critical target location, and first stage rotor speed. The critical bird ingestion speed should reflect the most critical condition within the range of airspeeds used for normal flight operations up to 1,500 feet above ground level, but not less than V_1 minimum for airplanes.

(2) Medium bird engine tests shall be conducted so as to simulate a flock encounter, and will use the bird weights and quantities specified in Table 2. When only one bird is specified, that bird will be aimed at the engine core primary flow path; the other critical locations on the engine face area must be addressed, as necessary, by appropriate tests or analysis, or both. When two or more birds are specified in Table 2, the largest of those birds must be aimed at the engine core primary flow path, and a second bird must be aimed at the most critical exposed location on the first stage rotor blades. Any remaining birds must be evenly distributed over the engine face area.

(3) In addition, except for rotorcraft engines, it must also be substantiated by appropriate tests or analysis or both, that when the full fan assembly is subjected to the ingestion of the quantity and weights of bird from Table 3, aimed at the fan assembly's most critical location outboard of the primary core flowpath, and in accordance with the applicable test conditions of this paragraph, that the engine can comply with the acceptance criteria of this paragraph.

(4) A small bird ingestion test is not required if the prescribed number of medium birds pass into the engine rotor blades during the medium bird test.

(5) Small bird ingestion tests shall be conducted so as to simulate a flock encounter using one 85 gram (0.187 lb.) bird for each 0.032 square-meter (49.6 square-inches) of inlet area, or fraction thereof, up to a maximum of 16 birds. The birds will be aimed so as to account for any critical exposed locations on the first stage rotor blades, with any remaining birds evenly distributed over the engine face area.

(6) Ingestion of small and medium birds tested under the conditions prescribed in this paragraph may not cause any of the following:

(i) More than a sustained 25-percent power or thrust loss;

(ii) The engine to be shut down during the required run-on demonstration prescribed in paragraphs (c)(7) or (c)(8) of this section;

(iii) The conditions defined in paragraph (b)(3) of this section.

(iv) Unacceptable deterioration of engine handling characteristics.

(7) Except for rotorcraft engines, the following test schedule shall be used:

(i) Ingestion so as to simulate a flock encounter, with approximately 1 second elapsed time from the moment of the first bird ingestion to the last.

(ii) Followed by 2 minutes without power lever movement after the ingestion.

(iii) Followed by 3 minutes at 75-percent of the test condition.

(iv) Followed by 6 minutes at 60-percent of the test condition.

(v) Followed by 6 minutes at 40-percent of the test condition.

(vi) Followed by 1 minute at approach idle.

(vii) Followed by 2 minutes at 75-percent of the test condition.

(viii) Followed by stabilizing at idle and engine shut down.

(ix) The durations specified are times at the defined conditions with the power being changed between each condition in less than 10 seconds.

(8) For rotorcraft engines, the following test schedule shall be used:

(i) Ingestion so as to simulate a flock encounter within approximately 1 second elapsed time between the first ingestion and the last.

(ii) Followed by 3 minutes at 75-percent of the test condition.

(iii) Followed by 90 seconds at descent flight idle.

(iv) Followed by 30 seconds at 75-percent of the test condition.

(v) Followed by stabilizing at idle and engine shut down.

(vi) The durations specified are times at the defined conditions with the power being changed between each condition in less than 10 seconds.

(9) Engines intended for use in multi-engine rotorcraft are not required to comply with the medium bird ingestion portion of this section, providing that the appropriate type certificate documentation is so endorsed.

(10) If any engine operating limit(s) is exceeded during the initial 2 minutes without power lever movement, as provided by paragraph (c)(7)(ii) of this section, then it shall be established that the limit exceedence will not result in an unsafe condition.

Table 2 to §33.76
Medium Flocking Bird Weight and Quantity Requirements

Engine Inlet Throat Area (A)—Square-meters (square-inches)	Bird quantity	Bird weight kg. (lb.)
0.05 (77.5)> A	none	
0.05 (77.5)≤ A <0.10 (155)	1	0.35 (0.77)
0.10 (155)≤ A <0.20 (310)	1	0.45 (0.99)
0.20 (310)≤ A <0.40 (620)	2	0.45 (0.99)
0.40 (620)≤ A <0.60 (930)	2	0.70 (1.54)
0.60 (930)≤ A <1.00 (1,550)	3	0.70 (1.54)
1.00 (1,550)≤ A <1.35 (2,092)	4	0.70 (1.54)
1.35 (2,092)≤ A <1.70 (2,635)	1	1.15 (2.53)
	plus 3	0.70 (1.54)
1.70 (2,635)≤ A <2.10 (3,255)	1	1.15 (2.53)
	plus 4	0.70 (1.54)
2.10 (3,255)≤ A <2.50 (3,875)	1	1.15 (2.53)
	plus 5	0.70 (1.54)
2.50 (3,875)≤ A <3.90 (6045)	1	1.15 (2.53)
	plus 6	0.70 (1.54)
3.90 (6045)≤ A <4.50 (6975)	3	1.15 (2.53)
4.50 (6975)≤ A	4	1.15 (2.53)

Table 3 to §33.76
Additional Integrity Assessment

Engine Inlet Throat Area (A)—square-meters (square-inches)	Bird quantity	Bird weight kg. (lb.)
1.35 (2,092)> A	none	
1.35 (2,092)≤ A <2.90 (4,495)	1	1.15 (2.53)
2.90 (4,495)≤ A <3.90 (6,045)	2	1.15 (2.53)
3.90 (6,045)≤ A	1	1.15 (2.53)
	plus 6	0.70 (1.54)

(d) *Large flocking bird.* An engine test will be performed as follows:

(1) Large flocking bird engine tests will be performed using the bird mass and weights in Table 4, and ingested at a bird speed of 200 knots.

(2) Prior to the ingestion, the engine must be stabilized at no less than the mechanical rotor speed of the first exposed stage or stages that, on a standard day, would produce 90 percent of the sea level static maximum rated takeoff power or thrust.

(3) The bird must be targeted on the first exposed rotating stage or stages at a blade airfoil height of not less than 50 percent measured at the leading edge.

(4) Ingestion of a large flocking bird under the conditions prescribed in this paragraph must not cause any of the following:

(i) A sustained reduction of power or thrust to less than 50 percent of maximum rated takeoff power or thrust during the run-on segment specified under paragraph (d)(5)(i) of this section.

(ii) Engine shutdown during the required run-on demonstration specified in paragraph (d)(5) of this section.

(iii) The conditions specified in paragraph (b)(3) of this section.

(5) The following test schedule must be used:

(i) Ingestion followed by 1 minute without power lever movement.

(ii) Followed by 13 minutes at not less than 50 percent of maximum rated takeoff power or thrust.

(iii) Followed by 2 minutes between 30 and 35 percent of maximum rated takeoff power or thrust.

(iv) Followed by 1 minute with power or thrust increased from that set in paragraph (d)(5)(iii) of this section, by between 5 and 10 percent of maximum rated takeoff power or thrust.

(v) Followed by 2 minutes with power or thrust reduced from that set in paragraph (d)(5)(iv) of this section, by between 5 and 10 percent of maximum rated takeoff power or thrust.

(vi) Followed by a minimum of 1 minute at ground idle then engine shutdown. The durations specified are times at the defined conditions. Power lever movement between each condition will be 10 seconds or less, except that power lever movements allowed within paragraph (d)(5)(ii) of this section are not limited, and for setting power under paragraph (d)(5)(iii) of this section will be 30 seconds or less.

(6) Compliance with the large flocking bird ingestion requirements of this paragraph (d) may also be demonstrated by:

(i) Incorporating the requirements of paragraph (d)(4) and (d)(5) of this section, into the large single bird test demonstration specified in paragraph (b)(1) of this section; or

(ii) Use of an engine subassembly test at the ingestion conditions specified in paragraph (b)(1) of this section if:

(A) All components critical to complying with the requirements of paragraph (d) of this section are included in the subassembly test;

(B) The components of paragraph (d)(6)(ii)(A) of this section are installed in a representative engine for a run-on demonstration in accordance with paragraphs (d)(4) and (d)(5) of this section; except that section (d)(5)(i) is deleted and section (d)(5)(ii) must be 14 minutes in duration after the engine is started and stabilized; and

(C) The dynamic effects that would have been experienced during a full engine ingestion test can be shown to be negligible with respect to meeting the requirements of paragraphs (d)(4) and (d)(5) of this section.

(7) Applicants must show that an unsafe condition will not result if any engine operating limit is exceeded during the run-on period.

Table 4 to §33.76
Large Flocking Bird Mass and Weight

Engine inlet throat area (square meters/ square inches)	Bird quantity	Bird mass and weight (kg (lbs))
A < 2.50 (3875)	none	
2.50 (3875) ≤ A < 3.50 (5425)	1	1.85 (4.08)
3.50 (5425) ≤ A < 3.90 (6045)	1	2.10 (4.63)
3.90 (6045) ≤ A	1	2.50 (5.51)

[Doc. No. FAA-1998-4815, 65 FR 55854, Sept. 14, 2000, as amended by Amdt. 33-20, 68 FR 75391, Dec. 31, 2003; Amdt. 33-24, 72 FR 50868, Sept. 4, 2007; Amdt. 33-23, 72 FR 58974, Oct. 17, 2007]

§ 33.77 Foreign object ingestion—ice.

(a)-(b) [Reserved]

(c) Ingestion of ice under the conditions of paragraph (e) of this section may not—

(1) Cause a sustained power or thrust loss; or

(2) Require the engine to be shutdown.

(d) For an engine that incorporates a protection device, compliance with this section need not be demonstrated with respect to foreign objects to be ingested under the conditions prescribed in paragraph (e) of this section if it is shown that—

(1) Such foreign objects are of a size that will not pass through the protective device;

(2) The protective device will withstand the impact of the foreign objects; and

(3) The foreign object, or objects, stopped by the protective device will not obstruct the flow of induction air into the engine with a resultant sustained reduction in power or thrust greater than those values required by paragraph (c) of this section.

(e) Compliance with paragraph (c) of this section must be shown by engine test under the following ingestion conditions:

(1) Ice quantity will be the maximum accumulation on a typical inlet cowl and engine face resulting from a 2-minute delay in actuating the anti-icing system; or a slab of ice which is comparable in weight or thickness for that size engine.

(2) The ingestion velocity will simulate ice being sucked into the engine inlet.

(3) Engine operation will be maximum cruise power or thrust.

(4) The ingestion will simulate a continuous maximum icing encounter at 25 degrees Fahrenheit.

[Doc. No. 16919, 49 FR 6852, Feb. 23, 1984, as amended by Amdt. 33-19, 63 FR 14798, Mar. 26, 1998; 63 FR 53278, Oct. 5, 1998; Amdt. 33-20, 65 FR 55856, Sept. 14, 2000]

§ 33.78 Rain and hail ingestion.

(a) *All engines.* (1) The ingestion of large hailstones (0.8 to 0.9 specific gravity) at the maximum true air speed, up to 15,000 feet (4,500 meters), associated with a representative aircraft operating in rough air, with the engine at maximum continuous power, may not cause unacceptable mechanical damage or unacceptable power or thrust loss after the ingestion, or require the engine to be shut down. One-half the number of hailstones shall be aimed randomly over the inlet face area and the other half aimed at the critical inlet face area. The hailstones shall be ingested in a rapid sequence to simulate a hailstone encounter and the number and size of the hailstones shall be determined as follows:

(i) One 1-inch (25 millimeters) diameter hailstone for engines with inlet areas of not more than 100 square inches (0.0645 square meters).

(ii) One 1-inch (25 millimeters) diameter and one 2-inch (50 millimeters) diameter hailstone for each 150 square inches (0.0968 square meters) of inlet area, or fraction thereof, for engines with inlet areas of more than 100 square inches (0.0645 square meters).

(2) In addition to complying with paragraph (a)(1) of this section and except as provided in paragraph (b) of this section, it must be shown that each engine is capable of acceptable operation throughout its specified operating envelope when subjected to sudden encounters with the certification standard concentrations of rain and hail, as defined in appendix B to this part. Acceptable engine operation precludes flameout, run down, continued or non-recoverable surge or stall, or loss of acceleration and deceleration capability, during any three minute continuous period in rain and during any 30 second continuous period in hail. It must also be shown after the ingestion that there is no unacceptable mechanical damage, unacceptable power or thrust loss, or other adverse engine anomalies.

(b) *Engines for rotorcraft.* As an alternative to the requirements specified in paragraph (a)(2) of this section, for rotorcraft turbine engines only, it must be shown that each engine is capable of acceptable operation during and after the ingestion of rain with an overall ratio of water droplet flow to airflow, by weight, with a uniform distribution at the inlet plane, of at least four percent. Acceptable engine operation precludes flameout, run down, continued or non-recoverable surge or stall, or loss of acceleration and deceleration capability. It must also be shown after the ingestion that there is no unacceptable mechanical damage, unacceptable power loss, or other adverse engine anomalies. The rain ingestion must occur under the following static ground level conditions:

(1) A normal stabilization period at take-off power without rain ingestion, followed immediately by the suddenly commencing ingestion of rain for three minutes at takeoff power, then

(2) Continuation of the rain ingestion during subsequent rapid deceleration to minimum idle, then

(3) Continuation of the rain ingestion during three minutes at minimum idle power to be certified for flight operation, then

(4) Continuation of the rain ingestion during subsequent rapid acceleration to takeoff power.

(c) *Engines for supersonic airplanes.* In addition to complying with paragraphs (a)(1) and (a)(2) of this section, a separate test for supersonic airplane engines only, shall be conducted with three hailstones ingested at supersonic cruise velocity. These hailstones shall be aimed at the engine's critical face area, and their ingestion must not cause unacceptable mechanical damage or unacceptable power or thrust loss after the ingestion or require the engine to be shut down. The size of these hailstones shall be determined from the linear variation in diameter from 1-inch (25 millimeters) at 35,000 feet (10,500 meters) to $\frac{1}{4}$ -inch (6 millimeters) at 60,000 feet (18,000 meters) using the diameter corresponding to the lowest expected supersonic cruise altitude. Alternatively, three larger hailstones may be ingested at subsonic velocities such that the kinetic energy of these larger hailstones is equivalent to the applicable supersonic ingestion conditions.

(d) For an engine that incorporates or requires the use of a protection device, demonstration of the rain and hail ingestion capabilities of the engine, as required in paragraphs (a), (b), and (c) of this section, may be waived wholly or in part by the Administrator if the applicant shows that:

(1) The subject rain and hail constituents are of a size that will not pass through the protection device;

(2) The protection device will withstand the impact of the subject rain and hail constituents; and

(3) The subject of rain and hail constituents, stopped by the protection device, will not obstruct the flow of induction air into the engine, resulting in damage, power or thrust loss, or other adverse engine anomalies in excess of what would be accepted in paragraphs (a), (b), and (c) of this section.

[Doc. No. 28652, 63 FR 14799, Mar. 26, 1998]

§ 33.79 Fuel burning thrust augmentor.

Each fuel burning thrust augmentor, including the nozzle, must—

(a) Provide cutoff of the fuel burning thrust augmentor;

(b) Permit on-off cycling;

(c) Be controllable within the intended range of operation;

(d) Upon a failure or malfunction of augmentor combustion, not cause the engine to lose thrust other than that provided by the augmentor; and

(e) Have controls that function compatibly with the other engine controls and automatically shut off augmentor fuel flow if the engine rotor speed drops below the minimum rotational speed at which the augmentor is intended to function.

[Amdt. 33-6, 39 FR 35468, Oct. 1, 1974]

Subpart F—Block Tests; Turbine Aircraft Engines

§ 33.81 Applicability.

This subpart prescribes the block tests and inspections for turbine engines.

[Doc. No. 3025, 29 FR 7453, June 10, 1964, as amended by Amdt. 33-6, 39 FR 35468, Oct. 1, 1974]

§ 33.82 General.

Before each endurance test required by this subpart, the adjustment setting and functioning characteristic of each component having an adjustment setting and a functioning characteristic that can be established independent of installation on the engine must be established and recorded.

[Amdt. 36-6, 39 FR 35468, Oct. 1, 1974]

§ 33.83 Vibration test.

(a) Each engine must undergo vibration surveys to establish that the vibration characteristics of those components that may be subject to mechanically or aerodynamically induced vibratory excitations are acceptable throughout the declared flight envelope. The engine surveys shall be based upon an appropriate combination of experience, analysis, and component test and shall address, as a minimum, blades, vanes, rotor discs, spacers, and rotor shafts.

(b) The surveys shall cover the ranges of power or thrust, and both the physical and corrected rotational speeds for each rotor system, corresponding to operations throughout the range of ambient conditions in the declared flight envelope, from the minimum rotational speed up to 103 percent of the maximum physical and corrected rotational speed permitted for rating periods of two minutes or longer, and up to 100 percent of all other permitted physical and corrected rotational speeds, including those that are overspeeds. If there is any indication of a stress peak arising at the highest of those required physical or corrected rotational speeds, the surveys shall be extended sufficiently to reveal the maximum stress values present, except that the extension need not cover more than a further 2 percentage points increase beyond those speeds.

(c) Evaluations shall be made of the following:

(1) The effects on vibration characteristics of operating with scheduled changes (including tolerances) to variable vane angles, compressor bleeds, accessory loading, the most adverse inlet air flow distortion pattern declared by the manufacturer, and the most adverse conditions in the exhaust duct(s); and

(2) The aerodynamic and aeromechanical factors which might induce or influence flutter in those systems susceptible to that form of vibration.

(d) Except as provided by paragraph (e) of this section, the vibration stresses associated with the vibration characteristics determined under this section, when combined with the appropriate steady stresses, must be less than the endurance limits of the materials concerned, after making due allowances for operating conditions for the permitted variations in properties of the materials. The suitability of these stress margins must be justified for each part evaluated. If it is determined that certain operating conditions, or ranges, need to be limited, operating and installation limitations shall be established.

(e) The effects on vibration characteristics of excitation forces caused by fault conditions (such as, but not limited to, out-of balance, local blockage or enlargement of stator vane passages, fuel nozzle blockage, incorrectly schedule compressor variables, etc.) shall be evaluated by test or analysis, or by reference to previous experience and shall be shown not to create a hazardous condition.

(f) Compliance with this section shall be substantiated for each specific installation configuration that can affect the vibration characteristics of the engine. If these vibration effects cannot be fully investigated during engine certification, the methods by which they can be evaluated and methods by which compliance can be shown shall be substantiated and defined in the installation instructions required by § 33.5.

[Doc. No. 28107, 61 FR 28433, June 4, 1996, as amended by Amdt. 33-33, 77 FR 39624, July 5, 2012; 77 FR 58301, Sept. 20, 2012]

§ 33.84 Engine overtorque test.

(a) If approval of a maximum engine overtorque is sought for an engine incorporating a free power turbine, compliance with this section must be demonstrated by testing.

(1) The test may be run as part of the endurance test requirement of § 33.87. Alternatively, tests may be performed on a complete engine or equivalent testing on individual groups of components.

(2) Upon conclusion of tests conducted to show compliance with this section, each engine part or individual groups of components must meet the requirements of § 33.93(a)(1) and (a)(2).

(b) The test conditions must be as follows:

(1) A total of 15 minutes run at the maximum engine overtorque to be approved. This may be done in separate runs, each being of at least $2\frac{1}{2}$ minutes duration.

(2) A power turbine rotational speed equal to the highest speed at which the maximum overtorque can occur in service. The test speed may not be more than the limit speed of take-off or OEI ratings longer than 2 minutes.

(3) For engines incorporating a reduction gearbox, a gearbox oil temperature equal to the maximum temperature when the maximum engine overtorque could occur in service; and for all other engines, an oil temperature within the normal operating range.

(4) A turbine entry gas temperature equal to the maximum steady state temperature approved for use during periods longer than 20 seconds when operating at conditions not associated with 30-second or 2 minutes OEI ratings. The requirement to run the test at the maximum approved steady state temperature may be waived by the FAA if the applicant can demonstrate that other testing provides substantiation of the temperature effects when considered in combination with the other parameters identified in paragraphs (b)(1), (b)(2) and (b)(3) of this section.

[Doc. No. 2007-28502, 74 FR 45310, Sept. 2, 2009]

§ 33.85 Calibration tests.

(a) Each engine must be subjected to those calibration tests necessary to establish its power characteristics and the conditions for the endurance test specified § 33.87. The results of the power characteristics calibration tests form the basis for establishing the characteristics of the engine over its entire operating range of speeds, pressures, temperatures, and altitudes. Power ratings are based upon standard atmospheric conditions with no airbleed for aircraft services and with only those accessories installed which are essential for engine functioning.

(b) A power check at sea level conditions must be accomplished on the endurance test engine after the endurance test and any change in power characteristics which occurs during the endurance test must be determined. Measurements taken during the final portion of the endurance test may be used in showing compliance with the requirements of this paragraph.

(c) In showing compliance with this section, each condition must stabilize before measurements are taken, except as permitted by paragraph (d) of this section.

(d) In the case of engines having 30-second OEI, and 2-minute OEI ratings, measurements taken during the applicable endurance test prescribed in § 33.87(f) (1) through (8) may be used in showing compliance with the requirements of this section for these OEI ratings.

[Doc. No. 3025, 29 FR 7453, June 10, 1964, as amended by Amdt. 33-6, 39 FR 35468, Oct. 1, 1974; Amdt. 33-18, 61 FR 31328, June 19, 1996]

§ 33.87 Endurance test.

(a) *General.* Each engine must be subjected to an endurance test that includes a total of at least 150 hours of operation and, depending upon the type and contemplated use of the engine, consists of one of the series of runs specified in paragraphs (b) through (g) of this section, as applicable. For engines tested under paragraphs (b), (c), (d), (e) or (g) of this section, the prescribed 6-hour test sequence must be conducted 25 times to complete the required 150 hours of operation. Engines for which the 30-second OEI and 2-minute OEI ratings are desired must be further tested under paragraph (f) of this section. The following test requirements apply:

(1) The runs must be made in the order found appropriate by the FAA for the particular engine being tested.

(2) Any automatic engine control that is part of the engine must control the engine during the endurance test except for operations where automatic control is normally overridden by manual control or where manual control is otherwise specified for a particular test run.

(3) Except as provided in paragraph (a)(5) of this section, power or thrust, gas temperature, rotor shaft rotational speed, and, if limited, temperature of external surfaces of the engine must be at least 100 percent of the value associated with the particular engine operation being tested. More than one test may be run if all parameters cannot be held at the 100 percent level simultaneously.

(4) The runs must be made using fuel, lubricants and hydraulic fluid which conform to the specifications specified in complying with § 33.7(c).

(5) Maximum air bleed for engine and aircraft services must be used during at least one-fifth of the runs, except for the test required under paragraph (f) of this section, provided the validity of the test is not compromised. However, for these runs, the power or thrust or the rotor shaft rotational speed may be less than 100 percent of the value associated with the particular operation being tested if the FAA finds that the validity of the endurance test is not compromised.

(6) Each accessory drive and mounting attachment must be loaded in accordance with paragraphs (a)(6)(i) and (ii) of this section, except as permitted by paragraph (a)(6)(iii) of this section for the test required under paragraph (f) of this section.

(i) The load imposed by each accessory used only for aircraft service must be the limit load specified by the applicant for the engine drive and attachment point during rated maximum continuous power or thrust and higher output.

(ii) The endurance test of any accessory drive and mounting attachment under load may be accomplished on a separate rig if the validity of the test is confirmed by an approved analysis.

(iii) The applicant is not required to load the accessory drives and mounting attachments when running the tests under paragraphs (f)(1) through (f)(8) of this section if the applicant can substantiate that there is no significant effect on the durability of any accessory drive or engine component. However, the applicant must add the equivalent engine output power extraction from the power turbine rotor assembly to the engine shaft output.

(7) During the runs at any rated power or thrust the gas temperature and the oil inlet temperature must be maintained at the limiting temperature except where the test periods are not longer than 5 minutes and do not allow stabilization. At least one run must be made with fuel, oil, and hydraulic fluid at the minimum pressure limit and at least one run must be made with fuel, oil, and hydraulic fluid at the maximum pressure limit with fluid temperature reduced as necessary to allow maximum pressure to be attained.

(8) If the number of occurrences of either transient rotor shaft overspeed, transient gas overtemperature or transient engine overtorque is limited, that number of the accelerations required by paragraphs (b) through (g) of this section must be made at the limiting overspeed, overtemperature or overtorque. If the number of occurrences is not limited, half the required accelerations must be made at the limiting overspeed, overtemperature or overtorque.

(9) For each engine type certificated for use on supersonic aircraft the following additional test requirements apply:

(i) To change the thrust setting, the power control lever must be moved from the initial position to the final position in not more than one second except for movements into the fuel burning thrust augmentor augmentation position if additional time to confirm ignition is necessary.

(ii) During the runs at any rated augmented thrust the hydraulic fluid temperature must be maintained at the limiting temperature except where the test periods are not long enough to allow stabilization.

(iii) During the simulated supersonic runs the fuel temperature and induction air temperature may not be less than the limiting temperature.

(iv) The endurance test must be conducted with the fuel burning thrust augmentor installed, with the primary and secondary exhaust nozzles installed, and with the variable area exhaust nozzles operated during each run according to the methods specified in complying with § 33.5(b).

(v) During the runs at thrust settings for maximum continuous thrust and percentages thereof, the engine must be operated with the inlet air distortion at the limit for those thrust settings.

(b) *Engines other than certain rotorcraft engines.* For each engine except a rotorcraft engine for which a rating is desired under paragraph (c), (d), or (e) of this section, the applicant must conduct the following runs:

(1) *Takeoff and idling.* One hour of alternate five-minute periods at rated takeoff power or thrust and at idling power or thrust. The developed powers or thrusts at takeoff and idling conditions and their corresponding rotor speed and gas temperature conditions must be as established by the power control in accordance with the schedule established by the applicant. The applicant may, during any one period, manually control the rotor speed, power, or thrust while taking data to check performance. For engines with augmented takeoff power ratings that involve increases in turbine inlet temperature, rotor speed, or shaft power, this period of running at takeoff must be at the augmented rating. For engines with augmented takeoff power ratings that do not materially increase operating severity, the amount of running conducted at the augmented rating is determined by the FAA. In changing the power setting after each period, the power-control lever must be moved in the manner prescribed in paragraph (b)(5) of this section.

(2) *Rated maximum continuous and takeoff power or thrust.* Thirty minutes at—

(i) Rated maximum continuous power or thrust during fifteen of the twenty-five 6-hour endurance test cycles; and

(ii) Rated takeoff power or thrust during ten of the twenty-five 6-hour endurance test cycles.

(3) *Rated maximum continuous power or thrust.* One hour and 30 minutes at rated maximum continuous power or thrust.

(4) *Incremental cruise power or thrust.* Two hours and 30 minutes at the successive power lever positions corresponding to at least 15 approximately equal speed and time increments between maximum continuous engine rotational speed and ground or minimum idle rotational speed. For engines operating at constant speed, the thrust and power may be varied in place of speed. If there is significant peak vibration anywhere between ground idle and maximum continuous conditions, the number of increments chosen may be changed to increase the amount of running made while subject to the peak vibrations up to not more than 50 percent of the total time spent in incremental running.

(5) *Acceleration and deceleration runs.* 30 minutes of accelerations and decelerations, consisting of six cycles from idling power or thrust to rated takeoff power or thrust and maintained at the takeoff power lever position for 30 seconds and at the idling power lever position for approximately four and one-half minutes. In complying with this paragraph, the power-control lever must be moved from one extreme position to the other in not more than one second, except that, if different regimes of control operations are incorporated necessitating scheduling of the power-control lever motion in going from one extreme position to the other, a longer period of time is acceptable, but not more than two seconds.

(6) *Starts.* One hundred starts must be made, of which 25 starts must be preceded by at least a two-hour engine shutdown. There must be at least 10 false engine starts, pausing for the applicant's specified minimum fuel drainage time, before attempting a normal start. There must be at least 10 normal restarts with not longer than 15 minutes since engine shutdown. The remaining starts may be made after completing the 150 hours of endurance testing.

(c) *Rotorcraft engines for which a 30-minute OEI power rating is desired.* For each rotorcraft engine for which a 30-minute OEI power rating is desired, the applicant must conduct the following series of tests:

(1) *Takeoff and idling.* One hour of alternate 5-minute periods at rated takeoff power and at idling power. The developed powers at takeoff and idling conditions and their corresponding rotor speed and gas temperature conditions must be as established by the power control in accordance with the schedule established by the applicant. During any one period, the rotor speed and power may be controlled manually while taking data to check performance. For engines with augmented takeoff power ratings that involve increases in turbine inlet temperature, rotor speed, or shaft power, this period of running at rated takeoff power must be at the augmented power rating. In changing the power setting after each period, the power control lever must be moved in the manner prescribed in paragraph (c)(6) of this section.

(2) *Rated maximum continuous and takeoff power.* Thirty minutes at—

(i) Rated maximum continuous power during fifteen of the twenty-five 6-hour endurance test cycles; and

(ii) Rated takeoff power during ten of the twenty-five 6-hour endurance test cycles.

(3) *Rated maximum continuous power.* One hour at rated maximum continuous power.

(4) *Rated 30-minute OEI power.* Thirty minutes at rated 30-minute OEI power.

(5) *Incremental cruise power.* Two hours and 30 minutes at the successive power lever positions corresponding with not less than 15 approximately equal speed and time increments between maximum continuous engine rotational speed and ground or minimum idle rotational speed. For engines operating at constant speed, power may be varied in place of speed. If there are significant peak vibrations anywhere between ground idle and maximum continuous conditions, the number of increments chosen must be changed to increase the amount of running conducted while subject to peak vibrations up to not more than 50 percent of the total time spent in incremental running.

(6) *Acceleration and deceleration runs.* Thirty minutes of accelerations and decelerations, consisting of six cycles from idling power to rated takeoff power and maintained at the takeoff power lever position for 30 seconds and at the idling power lever position for approximately 4½ minutes. In complying with this paragraph, the power control lever must be moved from one extreme position to the other in not more than one second. If, however, different regimes of control operations are incorporated that necessitate scheduling of the power control lever motion from one extreme position to the other, then a longer period of time is acceptable, but not more than two seconds.

(7) *Starts.* One hundred starts, of which 25 starts must be preceded by at least a two-hour engine shutdown. There must be at least 10 false engine starts, pausing for the applicant's specified minimum fuel drainage time, before attempting a normal start. There must be at least 10 normal restarts not more than 15 minutes after engine shutdown. The remaining starts may be made after completing the 150 hours of endurance testing.

(d) *Rotorcraft engines for which a continuous OEI rating is desired.* For each rotorcraft engine for which a continuous OEI power rating is desired, the applicant must conduct the following series of tests:

(1) *Takeoff and idling.* One hour of alternate 5-minute periods at rated takeoff power and at idling power. The developed powers at takeoff and idling conditions and their corresponding rotor speed and gas temperature conditions must be as established by the power control in accordance with the schedule established by the applicant. During any one period the rotor speed and power may be controlled manually while taking data to check performance. For engines with augmented takeoff power ratings that involve increases in turbine inlet temperature, rotor speed, or shaft power, this period of running at rated takeoff power must be at the augmented power rating. In changing the power setting after each period, the power control lever must be moved in the manner prescribed in paragraph (d)(6) of this section.

(2) *Rated maximum continuous and takeoff power.* Thirty minutes at—

(i) Rated maximum continuous power during fifteen of the twenty-five 6-hour endurance test cycles; and

(ii) Rated takeoff power during ten of the twenty-five 6-hour endurance test cycles.

(3) *Rated continuous OEI power.* One hour at rated continuous OEI power.

(4) *Rated maximum continuous power.* One hour at rated maximum continuous power.

(5) *Incremental cruise power.* Two hours at the successive power lever positions corresponding with not less than 12 approximately equal speed and time increments between maximum continuous engine rotational speed and ground or minimum idle rotational speed. For engines operating at constant speed, power may be varied in place of speed. If there are significant peak vibrations anywhere between ground idle and maximum continuous conditions, the number of increments chosen must be changed to increase the amount of running conducted while being subjected to the peak vibrations up to not more than 50 percent of the total time spent in incremental running.

(6) *Acceleration and deceleration runs.* Thirty minutes of accelerations and decelerations, consisting of six cycles from idling power to rated takeoff power and maintained at the takeoff power lever position for 30 seconds and at the idling power lever position for approximately 4½ minutes. In complying with this paragraph, the power control lever must be moved from one extreme position to the other in not more than 1 second, except that if different regimes of control operations are incorporated necessitating scheduling of the power control lever motion in going from one extreme position to the other, a longer period of time is acceptable, but not more than 2 seconds.

(7) *Starts.* One hundred starts, of which 25 starts must be preceded by at least a 2-hour engine shutdown. There must be at least 10 false engine starts, pausing for the applicant's specified minimum fuel drainage time, before attempting a normal start. There must be at least 10 normal restarts with not longer than 15 minutes since engine shutdown. The remaining starts may be made after completing the 150 hours of endurance testing.

(e) *Rotorcraft engines for which a 2½-minute OEI power rating is desired.* For each rotorcraft engine for which a 2½-minute OEI power rating is desired, the applicant must conduct the following series of tests:

(1) *Takeoff, 2½-minute OEI, and idling.* One hour of alternate 5-minute periods at rated takeoff power and at idling power except that, during the third and sixth takeoff power periods, only 2½ minutes need be conducted at rated takeoff power, and the remaining 2½ minutes must be conducted at rated 2½-minute OEI power. The developed powers at takeoff, 2½-minute OEI, and idling conditions and their corresponding rotor speed and gas temperature conditions must be as established by the power control in accordance with the schedule established by the applicant. The applicant may, during any one period, control manually the rotor speed and power while taking data to check performance. For engines with augmented takeoff power ratings that involve increases in turbine inlet temperature, rotor speed, or shaft power, this period of running at rated takeoff power must be at the augmented rating. In changing the power setting after or during each period, the power control lever must be moved in the manner prescribed in paragraph (b)(5), (c)(6), or (d)(6) of this section, as applicable.

(2) The tests required in paragraphs (b)(2) through (b)(6), or (c)(2) through (c)(7), or (d)(2) through (d)(7) of this section, as applicable, except that in one of the 6-hour test sequences, the last 5 minutes of the 30 minutes at takeoff power test period of paragraph (b)(2) of this section, or of the 30 minutes at 30-minute OEI power test period of paragraph (c)(4) of this section, or of the l hour at continuous OEI power test period of paragraph (d)(3) of this section, must be run at 2½-minute OEI power.

(f) *Rotorcraft Engines for which 30-second OEI and 2-minute OEI ratings are desired.* For each rotorcraft engine for which 30-second OEI and 2-minute OEI power ratings are desired, and following completion of the tests under paragraphs (b), (c), (d), or (e) of this section, the applicant may disassemble the tested engine to the extent necessary to show compliance with the requirements of § 33.93(a). The tested engine must then be reassembled using the same parts used during the test runs of paragraphs (b), (c), (d), or (e) of this section, except those parts described as consumables in the Instructions for Continued Airworthiness. Additionally, the tests required in paragraphs (f)(1) through (f)(8) of this section must be run continuously. If a stop occurs during these tests, the interrupted sequence must be repeated unless the applicant shows that the severity of the test would not be reduced if it were continued. The applicant must conduct the following test sequence four times, for a total time of not less than 120 minutes:

(1) *Takeoff power.* Three minutes at rated takeoff power.

(2) *30-second OEI power.* Thirty seconds at rated 30-second OEI power.

(3) *2-minute OEI power.* Two minutes at rated 2-minute OEI power.

(4) *30-minute OEI power, continuous OEI power, or maximum continuous power.* Five minutes at whichever is the greatest of rated 30-minute OEI power, rated continuous OEI power, or rated maximum continuous power, except that, during the first test sequence, this period shall be 65 minutes. However, where the greatest rated power is 30-minute OEI power, that sixty-five minute period shall consist of 30 minutes at 30-minute OEI power followed by 35 minutes at whichever is the greater of continuous OEI power or maximum continuous power.

(5) *50 percent takeoff power.* One minute at 50 percent takeoff power.

(6) *30-second OEI power.* Thirty seconds at rated 30-second OEI power.

(7) *2-minute OEI power.* Two minutes at rated 2-minute OEI power.

(8) *Idle.* One minute at flight idle.

(g) *Supersonic aircraft engines.* For each engine type certificated for use on supersonic aircraft the applicant must conduct the following:

(1) *Subsonic test under sea level ambient atmospheric conditions.* Thirty runs of one hour each must be made, consisting of—

(i) Two periods of 5 minutes at rated takeoff augmented thrust each followed by 5 minutes at idle thrust;

(ii) One period of 5 minutes at rated takeoff thrust followed by 5 minutes at not more than 15 percent of rated takeoff thrust;

(iii) One period of 10 minutes at rated takeoff augmented thrust followed by 2 minutes at idle thrust, except that if rated maximum continuous augmented thrust is lower than rated takeoff augmented thrust, 5 of the 10-minute periods must be at rated maximum continuous augmented thrust; and

(iv) Six periods of 1 minute at rated takeoff augmented thrust each followed by 2 minutes, including acceleration and deceleration time, at idle thrust.

(2) *Simulated supersonic test.* Each run of the simulated supersonic test must be preceded by changing the inlet air temperature and pressure from that attained at subsonic condition to the temperature and pressure attained at supersonic velocity, and must be followed by a return to the temperature attained at subsonic condition. Thirty runs of 4 hours each must be made, consisting of—

(i) One period of 30 minutes at the thrust obtained with the power control lever set at the position for rated maximum continuous augmented thrust followed by 10 minutes at the thrust obtained with the power control lever set at the position for 90 percent of rated maximum continuous augmented thrust. The end of this period in the first five runs must be made with the induction air temperature at the limiting condition of transient overtemperature, but need not be repeated during the periods specified in paragraphs (g)(2)(ii) through (iv) of this section;

(ii) One period repeating the run specified in paragraph (g)(2)(i) of this section, except that it must be followed by 10 minutes at the thrust obtained with the power control lever set at the position for 80 percent of rated maximum continuous augmented thrust;

(iii) One period repeating the run specified in paragraph (g)(2)(i) of this section, except that it must be followed by 10 minutes at the thrust obtained with the power control lever set at the position for 60 percent of rated maximum continuous augmented thrust and then 10 minutes at not more than 15 percent of rated takeoff thrust;

(iv) One period repeating the runs specified in paragraphs (g)(2)(i) and (ii) of this section; and

(v) One period of 30 minutes with 25 of the runs made at the thrust obtained with the power control lever set at the position for rated maximum continuous augmented thrust, each followed by idle thrust and with the remaining 5 runs at the thrust obtained with the power control lever set at the position for rated maximum continuous augmented thrust for 25 minutes each, followed by subsonic operation at not more than 15 percent or rated takeoff thrust and accelerated to rated takeoff thrust for 5 minutes using hot fuel.

(3) *Starts.* One hundred starts must be made, of which 25 starts must be preceded by an engine shutdown of at least 2 hours. There must be at least 10 false engine starts, pausing for the applicant's specified minimum fuel drainage time before attempting a normal start. At least 10 starts must be normal restarts, each made no later than 15 minutes after engine shutdown. The starts may be made at any time, including the period of endurance testing.

[Doc. No. 3025, 29 FR 7453, June 10, 1964, as amended by Amdt. 33-3, 32 FR 3737, Mar. 4, 1967; Amdt. 33-6, 39 FR 35468, Oct. 1, 1974; Amdt. 33-10, 49 FR 6853, Feb. 23, 1984; Amdt. 33-12, 53 FR 34220, Sept. 2, 1988; Amdt. 33-18, 61 FR 31328, June 19, 1996; Amdt. 33-25, 73 FR 48123, Aug. 18, 2008; Amdt. 33-30, 74 FR 45311, Sept. 2, 2009; Amdt. 33-32, 77 FR 22187, Apr. 13, 2012]

§ 33.88 Engine overtemperature test.

(a) Each engine must run for 5 minutes at maximum permissible rpm with the gas temperature at least 75 °F (42 °C) higher than the maximum rating's steady-state operating limit, excluding maximum values of rpm and gas temperature associated with the 30-second OEI and 2-minute OEI ratings. Following this run, the turbine assembly must be within serviceable limits.

(b) In addition to the test requirements in paragraph (a) of this section, each engine for which 30-second OEI and 2-minute OEI ratings are desired, that incorporates a means for automatic temperature control within its operating limitations in accordance with § 33.28(k), must run for a period of 4 minutes at the maximum power-on rpm with the gas temperature at least 35 °F (19 °C) higher than the maximum operating limit at 30-second OEI rating. Following this run, the turbine assembly may exhibit distress beyond the limits for an overtemperature condition provided the engine is shown by analysis or test, as found necessary by the FAA, to maintain the integrity of the turbine assembly.

(c) A separate test vehicle may be used for each test condition.

[Doc. No. 26019, 61 FR 31329, June 19, 1996, as amended by Amdt. 33-25, 73 FR 48124, Aug. 18, 2008; Amdt. 33-26, 73 FR 48285, Aug. 19, 2008]

§ 33.89 Operation test.

(a) The operation test must include testing found necessary by the Administrator to demonstrate—

(1) Starting, idling, acceleration, overspeeding, ignition, functioning of the propeller (if the engine is designated to operate with a propeller);

(2) Compliance with the engine response requirements of § 33.73; and

(3) The minimum power or thrust response time to 95 percent rated takeoff power or thrust, from power lever positions representative of minimum idle and of minimum flight idle, starting from stabilized idle operation, under the following engine load conditions:

(i) No bleed air and power extraction for aircraft use.

(ii) Maximum allowable bleed air and power extraction for aircraft use.

(iii) An intermediate value for bleed air and power extraction representative of that which might be used as a maximum for aircraft during approach to a landing.

(4) If testing facilities are not available, the determination of power extraction required in paragraph (a)(3)(ii) and (iii) of this section may be accomplished through appropriate analytical means.

(b) The operation test must include all testing found necessary by the Administrator to demonstrate that the engine has safe operating characteristics throughout its specified operating envelope.

[Amdt. 33-4, 36 FR 5493, Mar. 24, 1971, as amended by Amdt. 33-6, 39 FR 35469, Oct. 1, 1974; Amdt. 33-10, 49 FR 6853, Feb. 23, 1984]

§ 33.90 Initial maintenance inspection test.

Each applicant, except an applicant for an engine being type certificated through amendment of an existing type certificate or through supplemental type certification procedures, must complete one of the following tests on an engine that substantially conforms to the type design to establish when the initial maintenance inspection is required:

(a) An approved engine test that simulates the conditions in which the engine is expected to operate in service, including typical start-stop cycles.

(b) An approved engine test conducted in accordance with § 33.201 (c) through (f).

[Doc. No. FAA-2002-6717, 72 FR 1877, Jan. 16, 2007]

§ 33.91 Engine system and component tests.

(a) For those systems or components that cannot be adequately substantiated in accordance with endurance testing of § 33.87, the applicant must conduct additional tests to demonstrate that the systems or components are able to perform the intended functions in all declared environmental and operating conditions.

(b) Temperature limits must be established for those components that require temperature controlling provisions in the aircraft installation to assure satisfactory functioning, reliability, and durability.

(c) Each unpressurized hydraulic fluid tank may not fail or leak when subjected to a maximum operating temperature and an internal pressure of 5 p.s.i., and each pressurized hydraulic fluid tank must meet the requirements of § 33.64.

(d) For an engine type certificated for use in supersonic aircraft, the systems, safety devices, and external components that may fail because of operation at maximum and minimum operating temperatures must be identified and tested at maximum and minimum operating temperatures and while temperature and other operating conditions are cycled between maximum and minimum operating values.

[Doc. No. 3025, 29 FR 7453, June 10, 1964, as amended by Amdt. 33-6, 39 FR 35469, Oct. 1, 1974; Amdt. 33-26, 73 FR 48285, Aug. 19, 2008; Amdt. 33-27, 73 FR 55437, Sept. 25, 2008; Amdt. 33-27, 73 FR 57235, Oct. 2, 2008]

§ 33.92 Rotor locking tests.

If continued rotation is prevented by a means to lock the rotor(s), the engine must be subjected to a test that includes 25 operations of this means under the following conditions:

(a) The engine must be shut down from rated maximum continuous thrust or power; and

(b) The means for stopping and locking the rotor(s) must be operated as specified in the engine operating instructions while being subjected to the maximum torque that could result from continued flight in this condition; and

(c) Following rotor locking, the rotor(s) must be held stationary under these conditions for five minutes for each of the 25 operations.

[Doc. No. 28107, 61 FR 28433, June 4, 1996]

§ 33.93 Teardown inspection.

(a) After completing the endurance testing of § 33.87 (b), (c), (d), (e), or (g) of this part, each engine must be completely disassembled, and

(1) Each component having an adjustment setting and a functioning characteristic that can be established independent of installation on the engine must retain each setting and functioning characteristic within the limits that were established and recorded at the beginning of the test; and

(2) Each engine part must conform to the type design and be eligible for incorporation into an engine for continued operation, in accordance with information submitted in compliance with § 33.4.

(b) After completing the endurance testing of § 33.87(f), each engine must be completely disassembled, and

(1) Each component having an adjustment setting and a functioning characteristic that can be established independent of installation on the engine must retain each setting and functioning characteristic within the limits that were established and recorded at the beginning of the test; and

(2) Each engine may exhibit deterioration in excess of that permitted in paragraph (a)(2) of this section, including some engine parts or components that may be unsuitable for further use. The applicant must show by inspection, analysis, test, or by any combination thereof as found necessary by the FAA, that structural integrity of the engine is maintained; or

(c) In lieu of compliance with paragraph (b) of this section, each engine for which the 30-second OEI and 2-minute OEI ratings are desired, may be subjected to the endurance testing of §§ 33.87 (b), (c), (d), or (e) of this part, and followed by the testing of § 33.87(f) without intervening disassembly and inspection. However, the engine must comply with paragraph (a) of this section after completing the endurance testing of § 33.87(f).

[Doc. No. 26019, 61 FR 31329, June 19, 1996, as amended by Amdt. 33-25, 73 FR 48124, Aug. 18, 2008]

§ 33.94 Blade containment and rotor unbalance tests.

(a) Except as provided in paragraph (b) of this section, it must be demonstrated by engine tests that the engine is capable of containing damage without catching fire and without failure of its mounting attachments when operated for at least 15 seconds, unless the resulting engine damage induces a self shutdown, after each of the following events:

(1) Failure of the most critical compressor or fan blade while operating at maximum permissible r.p.m. The blade failure must occur at the outermost retention groove or, for integrally-bladed rotor discs, at least 80 percent of the blade must fail.

(2) Failure of the most critical turbine blade while operating at maximum permissible r.p.m. The blade failure must occur at the outermost retention groove or, for integrally-bladed rotor discs, at least 80 percent of the blade must fail. The most critical turbine blade must be determined by considering turbine blade weight and the strength of the adjacent turbine case at case temperatures and pressures associated with operation at maximum permissible r.p.m.

(b) Analysis based on rig testing, component testing, or service experience may be substitute for one of the engine tests prescribed in paragraphs (a)(1) and (a)(2) of this section if—

(1) That test, of the two prescribed, produces the least rotor unbalance; and

(2) The analysis is shown to be equivalent to the test.

(Secs. 313(a), 601, and 603, Federal Aviation Act of 1958 (49 U.S.C. 1354(a), 1421, and 1423); and 49 U.S.C. 106(g) Revised, Pub. L. 97-449, Jan. 12, 1983)

[Amdt. 33-10, 49 FR 6854, Feb. 23, 1984]

§ 33.95 Engine-propeller systems tests.

If the engine is designed to operate with a propeller, the following tests must be made with a representative propeller installed by either including the tests in the endurance run or otherwise performing them in a manner acceptable to the Administrator:

(a) Feathering operation: 25 cycles.

(b) Negative torque and thrust system operation: 25 cycles from rated maximum continuous power.

(c) Automatic decoupler operation: 25 cycles from rated maximum continuous power (if repeated decoupling and recoupling in service is the intended function of the device).

(d) Reverse thrust operation: 175 cycles from the flight-idle position to full reverse and 25 cycles at rated maximum continuous power from full forward to full reverse thrust. At the end of each cycle the propeller must be operated in reverse pitch for a period of 30 seconds at the maximum rotational speed and power specified by the applicant for reverse pitch operation.

[Doc. No. 3025, 29 FR 7453, June 10, 1964, as amended by Amdt. 33-3, 32 FR 3737, Mar. 4, 1967]

§ 33.96 Engine tests in auxiliary power unit (APU) mode.

If the engine is designed with a propeller brake which will allow the propeller to be brought to a stop while the gas generator portion of the engine remains in operation, and remain stopped during operation of the engine as an auxiliary power unit ("APU mode"), in addition to the requirements of § 33.87, the applicant must conduct the following tests:

(a) Ground locking: A total of 45 hours with the propeller brake engaged in a manner which clearly demonstrates its ability to function without adverse effects on the complete engine while the engine is operating in the APU mode under the maximum conditions of engine speed, torque, temperature, air bleed, and power extraction as specified by the applicant.

(b) Dynamic braking: A total of 400 application-release cycles of brake engagements must be made in a manner which clearly demonstrates its ability to function without adverse effects on the complete engine under the maximum conditions of engine acceleration/deceleration rate, speed, torque, and temperature as specified by the applicant. The propeller must be stopped prior to brake release.

(c) One hundred engine starts and stops with the propeller brake engaged.

(d) The tests required by paragraphs (a), (b), and (c) of this section must be performed on the same engine, but this engine need not be the same engine used for the tests required by § 33.87.

(e) The tests required by paragraphs (a), (b), and (c) of this section must be followed by engine disassembly to the extent necessary to show compliance with the requirements of § 33.93(a) and § 33.93(b).

[Amdt. 33-11, 51 FR 10346, Mar. 25, 1986]

§ 33.97 Thrust reversers.

(a) If the engine incorporates a reverser, the endurance calibration, operation, and vibration tests prescribed in this subpart must be run with the reverser installed. In complying with this section, the power control lever must be moved from one extreme position to the other in not more than one second except, if regimes of control operations are incorporated necessitating scheduling of the power-control lever motion in going from one extreme position to the other, a longer period of time is acceptable but not more than three seconds. In addition, the test prescribed in paragraph (b) of this section must be made. This test may be scheduled as part of the endurance run.

(b) 175 reversals must be made from flight-idle forward thrust to maximum reverse thrust and 25 reversals must be made from rated takeoff thrust to maximum reverse thrust. After each reversal the reverser must be operated at full reverse thrust for a period of one minute, except that, in the case of a reverser intended for use only as a braking means on the ground, the reverser need only be operated at full reverse thrust for 30 seconds.

[Doc. No. 3025, 29 FR 7453, June 10, 1964, as amended by Amdt. 33-3, 32 FR 3737, Mar. 4, 1967]

§ 33.99 General conduct of block tests.

(a) Each applicant may, in making a block test, use separate engines of identical design and construction in the vibration, calibration, endurance, and operation tests, except that, if a separate engine is used for the endurance test it must be subjected to a calibration check before starting the endurance test.

(b) Each applicant may service and make minor repairs to the engine during the block tests in accordance with the service and maintenance instructions submitted in compliance with § 33.4. If the frequency of the service is excessive, or the number of stops due to engine malfunction is excessive, or a major repair, or replacement of a part is found necessary during the block tests or as the result of findings from the teardown inspection, the engine or its parts must be subjected to any additional tests the Administrator finds necessary.

(c) Each applicant must furnish all testing facilities, including equipment and competent personnel, to conduct the block tests.

[Doc. No. 3025, 29 FR 7453, June 10, 1964, as amended by Amdt. 33-6, 39 FR 35470, Oct. 1, 1974; Amdt. 33-9, 45 FR 60181, Sept. 11, 1980]

Subpart G—Special Requirements: Turbine Aircraft Engines

Source: Docket No. FAA-2002-6717, 72 FR 1877, Jan. 16, 2007, unless otherwise noted.

§ 33.201 Design and test requirements for Early ETOPS eligibility.

An applicant seeking type design approval for an engine to be installed on a two-engine airplane approved for ETOPS without the service experience specified in part 25, appendix K, K25.2.1 of this chapter, must comply with the following:

(a) The engine must be designed using a design quality process acceptable to the FAA, that ensures the design features of the engine minimize the occurrence of failures, malfunctions, defects, and maintenance errors that could result in an IFSD, loss of thrust control, or other power loss.

(b) The design features of the engine must address problems shown to result in an IFSD, loss of thrust control, or other power loss in the applicant's other relevant type designs approved within the past 10 years, to the extent that adequate service data is available within that 10-year period. An applicant without adequate service data must show experience with and knowledge of problem mitigating design practices equivalent to that gained from actual service experience in a manner acceptable to the FAA.

(c) Except as specified in paragraph (f) of this section, the applicant must conduct a simulated ETOPS mission cyclic endurance test in accordance with an approved test plan on an engine that substantially conforms to the type design. The test must:

(1) Include a minimum of 3,000 representative service start-stop mission cycles and three simulated diversion cycles at maximum continuous thrust or power for the maximum diversion time for which ETOPS eligibility is sought. Each start-stop mission cycle must include the use of take-off, climb, cruise, descent, approach, and landing thrust or power and the use of thrust reverse (if applicable). The diversions must be evenly distributed over the duration of the test. The last diversion must be conducted within 100 cycles of the completion of the test.

(2) Be performed with the high speed and low speed main engine rotors independently unbalanced to obtain a minimum of 90 percent of the recommended field service maintenance vibration levels. For engines with three main engine rotors, the intermediate speed rotor must be independently unbalanced to obtain a minimum of 90 percent of the recommended production acceptance vibration level. The required peak vibration levels must be verified during a slow acceleration and deceleration run of the test engine covering the main engine rotor operating speed ranges.

(3) Include a minimum of three million vibration cycles for each 60 rpm incremental step of the typical high-speed rotor start-stop mission cycle. The test may be conducted using any rotor speed step increment from 60 to 200 rpm provided the test encompasses the typical service start-stop cycle speed range. For incremental steps greater than 60 rpm, the minimum number of vibration cycles must be linearly increased up to ten million cycles for a 200 rpm incremental step.

(4) Include a minimum of 300,000 vibration cycles for each 60 rpm incremental step of the high-speed rotor approved operational speed range between minimum flight idle and cruise power not covered by paragraph (c)(3) of this section. The test may be conducted using any rotor speed step increment from 60 to 200 rpm provided the test encompasses the applicable speed range. For incremental steps greater than 60 rpm the minimum number of vibration cycles must be linearly increased up to 1 million for a 200 rpm incremental step.

(5) Include vibration surveys at periodic intervals throughout the test. The equivalent value of the peak vibration level observed during the surveys must meet the minimum vibration requirement of § 33.201(c)(2).

(d) Prior to the test required by paragraph (c) of this section, the engine must be subjected to a calibration test to document power and thrust characteristics.

(e) At the conclusion of the testing required by paragraph (c) of this section, the engine must:

(1) Be subjected to a calibration test at sea-level conditions. Any change in power or thrust characteristics must be within approved limits.

(2) Be visually inspected in accordance with the on-wing inspection recommendations and limits contained in the Instructions for Continued Airworthiness submitted in compliance with § 33.4.

(3) Be completely disassembled and inspected—

(i) In accordance with the applicable inspection recommendations and limits contained in the Instructions for Continued Airworthiness submitted in compliance with § 33.4;

(ii) With consideration of the causes of IFSD, loss of thrust control, or other power loss identified by paragraph (b) of this section; and

(iii) In a manner to identify wear or distress conditions that could result in an IFSD, loss of thrust control, or other power loss not specifically identified by paragraph (b) of this section or addressed within the Instructions for Continued Airworthiness.

(4) Not show wear or distress to the extent that could result in an IFSD, loss of thrust control, or other power loss within a period of operation before the component, assembly, or system would likely have been inspected or functionally tested for integrity while in service. Such wear or distress must have corrective action implemented through a design change, a change to maintenance instructions, or operational procedures before ETOPS eligibility is granted. The type and frequency of wear and distress that occurs during the engine test must be consistent with the type and frequency of wear and distress that would be expected to occur on ETOPS eligible engines.

(f) An alternative mission cycle endurance test that provides an equivalent demonstration of the unbalance and vibration specified in paragraph (c) of this section may be used when approved by the FAA.

(g) For an applicant using the simulated ETOPS mission cyclic endurance test to comply with § 33.90, the test may be interrupted so that the engine may be inspected by an on-wing or other method, using criteria acceptable to the FAA, after completion of the test cycles required to comply with § 33.90(a). Following the inspection, the ETOPS test must be resumed to complete the requirements of this section.

Appendix A to Part 33—Instructions for Continued Airworthiness

a33.1 general

(a) This appendix specifies requirements for the preparation of Instructions for Continued Airworthiness as required by § 33.4.

(b) The Instructions for Continued Airworthiness for each engine must include the Instructions for Continued Airworthiness for all engine parts. If Instructions for Continued Airworthiness are not supplied by the engine part manufacturer for an engine part, the Instructions for Continued Airworthiness for the engine must include the information essential to the continued airworthiness of the engine.

(c) The applicant must submit to the FAA a program to show how changes to the Instructions for Continued Airworthiness made by the applicant or by the manufacturers of engine parts will be distributed.

a33.2 format

(a) The Instructions for Continued Airworthiness must be in the form of a manual or manuals as appropriate for the quantity of data to be provided.

(b) The format of the manual or manuals must provide for a practical arrangement.

a33.3 content

The contents of the manual or manuals must be prepared in the English language. The Instructions for Continued Airworthiness must contain the following manuals or sections, as appropriate, and information:

(a) *Engine Maintenance Manual or Section.* (1) Introduction information that includes an explanation of the engine's features and data to the extent necessary for maintenance or preventive maintenance.

(2) A detailed description of the engine and its components, systems, and installations.

(3) Installation instructions, including proper procedures for uncrating, deinhibiting, acceptance checking, lifting, and attaching accessories, with any necessary checks.

(4) Basic control and operating information describing how the engine components, systems, and installations operate, and information describing the methods of starting, running, testing, and stopping the engine and its parts including any special procedures and limitations that apply.

(5) Servicing information that covers details regarding servicing points, capacities of tanks, reservoirs, types of fluids to be used, pressures applicable to the various systems, locations of lubrication points, lubricants to be used, and equipment required for servicing.

(6) Scheduling information for each part of the engine that provides the recommended periods at which it should be cleaned, inspected, adjusted, tested, and lubricated, and the degree of inspection the applicable wear tolerances, and work recommended at these periods. However, the applicant may refer to an accessory, instrument, or equipment manufacturer as the source of this information if the applicant shows that the item has an exceptionally high degree of complexity requiring specialized maintenance techniques, test equipment, or expertise. The recommended overhaul periods and necessary cross references to the Airworthiness Limitations section of the manual must also be included. In addition, the applicant must include an inspection program that includes the frequency and extent of the inspections necessary to provide for the continued airworthiness of the engine.

(7) Troubleshooting information describing probable malfunctions, how to recognize those malfunctions, and the remedial action for those malfunctions.

(8) Information describing the order and method of removing the engine and its parts and replacing parts, with any necessary precautions to be taken. Instructions for proper ground handling, crating, and shipping must also be included.

(9) A list of the tools and equipment necessary for maintenance and directions as to their method of use.

(b) *Engine Overhaul Manual or Section.* (1) Disassembly information including the order and method of disassembly for overhaul.

(2) Cleaning and inspection instructions that cover the materials and apparatus to be used and methods and precautions to be taken during overhaul. Methods of overhaul inspection must also be included.

(3) Details of all fits and clearances relevant to overhaul.

(4) Details of repair methods for worn or otherwise substandard parts and components along with the information necessary to determine when replacement is necessary.

(5) The order and method of assembly at overhaul.

(6) Instructions for testing after overhaul.

(7) Instructions for storage preparation, including any storage limits.

(8) A list of tools needed for overhaul.

(c) *ETOPS Requirements.* For an applicant seeking eligibility for an engine to be installed on an airplane approved for ETOPS, the Instructions for Continued Airworthiness must include procedures for engine condition monitoring. The engine condition monitoring procedures must be able to determine prior to flight, whether an engine is capable of providing, within approved engine operating limits, maximum continuous power or thrust, bleed air, and power extraction required for a relevant engine inoperative diversion. For an engine to be installed on a two-engine airplane approved for ETOPS, the engine condition monitoring procedures must be validated before ETOPS eligibility is granted.

A33.4 airworthiness limitations section

The Instructions for Continued Airworthiness must contain a section titled Airworthiness Limitations that is segregated and clearly distinguishable from the rest of the manual.

(a) For all engines:

(1) The Airworthiness Limitations section must set forth each mandatory replacement time, inspection interval, and related procedure required for type certification. If the Instructions for Continued Airworthiness consist of multiple documents, the section required under this paragraph must be included in the principal manual.

(2) This section must contain a legible statement in a prominent location that reads: "The Airworthiness Limitations section is FAA approved and specifies maintenance required under §§ 43.16 and 91.403 of Title 14 of the Code of Federal Regulations unless an alternative program has been FAA approved."

(b) For rotorcraft engines having 30-second OEI and 2-minute OEI ratings:

(1) The Airworthiness Limitations section must also prescribe the mandatory post-flight inspections and maintenance actions associated with any use of either 30-second OEI or 2-minute OEI ratings.

(2) The applicant must validate the adequacy of the inspections and maintenance actions required under paragraph (b)(1) of this section A33.4.

(3) The applicant must establish an in-service engine evaluation program to ensure the continued adequacy of the instructions for mandatory post-flight inspections and maintenance actions prescribed under paragraph (b)(1) of this section A33.4 and of the data for § 33.5(b)(4) pertaining to power availability. The program must include service engine tests or equivalent service engine test experience on engines of similar design and evaluations of service usage of the 30-second OEI or 2-minute OEI ratings.

[Amdt. 33-9, 45 FR 60181, Sept. 11, 1980, as amended by Amdt. 33-13, 54 FR 34330, Aug. 18, 1989; Amdt. 33-21, 72 FR 1878, Jan. 16, 2007; Amdt. 33-25, 73 FR 48124, Aug. 18, 2008]

Appendix B to Part 33—Certification Standard Atmospheric Concentrations of Rain and Hail

Figure B1, Table B1, Table B2, Table B3, and Table B4 specify the atmospheric concentrations and size distributions of rain and hail for establishing certification, in accordance with the requirements of § 33.78(a)(2). In conducting tests, normally by spraying liquid water to simulate rain conditions and by delivering hail fabricated from ice to simulate hail conditions, the use of water droplets and hail having shapes, sizes and distributions of sizes other than those defined in this appendix B, or the use of a single size or shape for each water droplet or hail, can be accepted, provided that applicant shows that the substitution does not reduce the severity of the test.

FIGURE B1 - Illustration of Rain and Hail Threats. Certification concentrations are obtained using Tables B1 and B2.

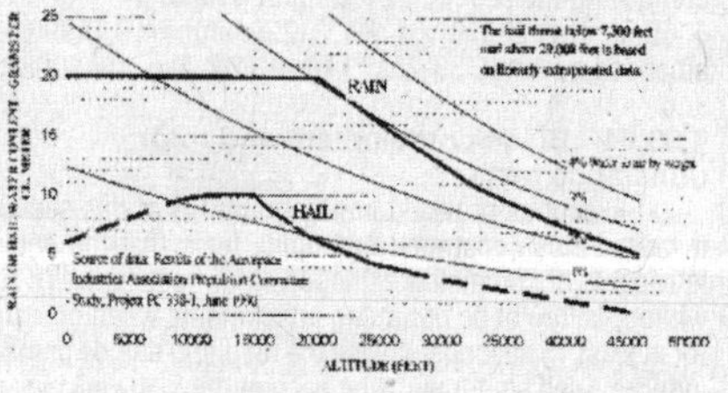

Table B1
Certification Standard Atmospheric Rain Concentrations

Altitude (feet)	Rain water content (RWC) (grams water/meter3air)
0	20.0
20,000	20.0
26,300	15.2
32,700	10.8
39,300	7.7
46,000	5.2

RWC values at other altitudes may be determined by linear interpolation.

Note: Source of data—Results of the Aerospace Industries Association (AIA) Propulsion Committee Study, Project PC 338–1, June 1990.

Table B2
Certification Standard Atmospheric Hail Concentrations

Altitude (feet)	Hail water content (HWC) (grams water/meter3air)
0	6.0
7,300	8.9
8,500	9.4
10,000	9.9
12,000	10.0
15,000	10.0
16,000	8.9
17,700	7.8
19,300	6.6
21,500	5.6
24,300	4.4
29,000	3.3
46,000	0.2

HWC values at other altitudes may be determined by linear interpolation. The hail threat below 7,300 feet and above 29,000 feet is based on linearly extrapolated data.

Note: Source of data—Results of the Aerospace Industries Association (AIA Propulsion Committee (PC) Study, Project PC 338-1, June 1990.

TABLE B3
Certification Standard Atmospheric Rain Droplet Size Distribution

Rain droplet diameter (mm)	Contribution total RWC (%)
0–0.49	0
0.50–0.99	2.25
1.00–1.49	8.75
1.50–1.99	16.25
2.00–2.49	19.00
2.50–2.99	17.75
3.00–3.49	13.50
3.50–3.99	9.50
4.00–4.49	6.00
4.50–4.99	3.00
5.00–5.49	2.00
5.50–5.99	1.25
6.00–6.49	0.50
6.50–7.00	0.25
Total	100.00

Median diameter of rain droplets in 2.66 mm

Note: Source of data—Results of the Aerospace Industries Association (AIA Propulsion Committee (PC) Study, Project PC 338-1, June 1990.

TABLE B4
Certification Standard Atmospheric Hail Size Distribution

Hail diameter (mm)	Contribution total HWC (%)
0–4.9	0
5.0–9.9	17.00
10.0–14.9	25.00
15.0–19.9	22.50
20.0–24.9	16.00
25.0–29.9	9.75
30.0–34.9	4.75
35.0–39.9	2.50
40.0–44.9	1.50
45.0–49.9	0.75
50.0–55.0	0.25
Total	100.00

Median diameter of hail is 16 mm

Note: Source of data—Results of the Aerospace Industries Association (AIA Propulsion Committee (PC) Study, Project PC 338-1, June 1990.

[Doc. No. 28652, 63 FR 14799, Mar. 26, 1998]

PART 33

FAR

PART 34

FUEL VENTING AND EXHAUST EMISSION REQUIREMENTS FOR TURBINE ENGINE POWERED AIRPLANES

Contents

Authority: 42 U.S.C 4321 *et seq.*, 7572l 49 U.S.C. 106(g), 40113, 44701-44702, 44704, 44714

Source: Docket No. 25613, 55 FR 32861, Aug. 10, 1990, unless otherwise noted.

Subpart A—General Provisions

§ 34.1 Definitions.

As used in this part, all terms not defined herein shall have the meaning given them in the Clean Air Act, as amended (42 U.S.C. 7401 et. seq.):

Act means the Clean Air Act, as amended (42 U.S.C. 7401 et. seq.).

Administrator means the Administrator of the Federal Aviation Administration or any person to whom he has delegated his authority in the matter concerned.

Administrator of the EPA means the Administrator of the Environmental Protection Agency and any other officer or employee of the Environmental Protection Agency to whom the authority involved may be delegated.

Aircraft as used in this part means any airplane as defined in 14 CFR part 1 for which a U.S. standard airworthiness certificate or equivalent foreign airworthiness certificate is issued.

Aircraft engine means a propulsion engine which is installed in, or which is manufactured for installation in, an aircraft.

Aircraft gas turbine engine means a turboprop, turbofan, or turbojet aircraft engine.

Characteristic level has the meaning given in Appendix 6 of ICAO Annex 16 as of July 2008. The characteristic level is a calculated emission level for each pollutant based on a statistical assessment of measured emissions from multiple tests.[1]

[1] This incorporation by reference was approved by the Director of the Federal Register in accordance with 5 U.S.C. 552(a) and 1 CFR part 51. This document can be obtained from the ICAO, Document Sales Unit, 999 University Street, Montreal, Quebec H3C 5H7, Canada, phone +1 514-954-8022, or *www.icao.int* or *sales14icao.int.* Copies can be reviewed at the FAA New England Regional Office, 12 New England Executive Park, Burlington, Massachusetts, 781-238-7101, or at the National Archives and Records Administration (NARA). For information on the availability of this material at NARA, call 202-741-6030, or go to: *http://www.archives.gov/federal_register/code_of_federal_regulations/ibr_locations.html.*

Class TP means all aircraft turboprop engines.

Class TF means all turbofan or turbojet aircraft engines or aircraft engines designed for applications that otherwise would have been fulfilled by turbojet and turbofan engines except engines of class T3, T8, and TSS.

Class T3 means all aircraft gas turbine engines of the JT3D model family.

Class T8 means all aircraft gas turbine engines of the JT8D model family.

Class TSS means all aircraft gas turbine engines employed for propulsion of aircraft designed to operate at supersonic flight speeds.

Commercial aircraft engine means any aircraft engine used or intended for use by an "air carrier" (including those engaged in "intrastate air transportation") or a "commercial operator" (including those engaged in "intrastate air transportation") as these terms are defined in Title 49 of the United States Code and Title 14 of the Code of Federal Regulations.

Commercial aircraft gas turbine engine means a turboprop, turbofan, or turbojet commercial aircraft engine.

Date of manufacture of an engine is the date the inspection acceptance records reflect that the engine is complete and meets the FAA approved type design.

Derivative engine for emissions certification purposes means an engine that has the same or similar emissions characteristics as an engine covered by a U.S. type certificate issued under 14 CFR part 33. These characteristics are specified in § 34.48.

Emission measurement system means all of the equipment necessary to transport the emission sample and measure the level of emissions. This includes the sample system and the instrumentation system.

Engine model means all commercial aircraft turbine engines which are of the same general series, displacement, and design characteristics and are approved under the same type certificate.

Excepted, as used in § 34.9, means an engine that may be produced and sold that does not meet otherwise applicable standards. Excepted engines must conform to regulatory conditions specified for an exception in § 34.9. Excepted engines are subject to the standards of this part even though they are not required to comply with the otherwise applicable requirements. Engines excepted with respect to certain standards must comply with other standards from which they are not specifically excepted.

Exempt means an engine that does not meet certain applicable standards but may be produced and sold under the terms allowed by a grant of exemption issued pursuant to § 34.7 of this part and part 11 of this chapter. Exempted engines must conform to regulatory conditions specified in the exemption as well as other applicable regulations. Exempted engines are subject to the standards of this part even though they are not required to comply with the otherwise applicable requirements. Engines exempted with respect to certain standards must comply with other standards as a condition of the exemption.

Exhaust emissions means substances emitted into the atmosphere from the exhaust discharge nozzle of an aircraft or aircraft engine.

Fuel venting emissions means raw fuel, exclusive of hydrocarbons in the exhaust emissions, discharged from aircraft gas turbine engines during all normal ground and flight operations.

In-use aircraft gas turbine engine means an aircraft gas turbine engine which is in service.

Introduction date means the date of manufacture of the first individual production engine of a given engine model or engine type certificate family to be certificated. Neither test engines nor engines not placed into service affect this date.

New aircraft turbine engine means an aircraft gas turbine engine which has never been in service.

Power setting means the power or thrust output of an engine in terms of kilonewtons thrust for turbojet and turbofan engines or shaft power in terms of kilowatts for turboprop engines.

Rated output (rO) means the maximum power/thrust available for takeoff at standard day conditions as approved for the engine by the Federal Aviation Administration, including reheat contribution where applicable, but excluding any contribution due to water injection, expressed in kilowatts or kilonewtons (as applicable), rounded to at least three significant figures.

Rated pressure ratio (rPR) means the ratio between the combustor inlet pressure and the engine inlet pressure achieved by an engine operation at rated output, rounded to at least three significant figures.

Reference day conditions means the reference ambient conditions to which the gaseous emissions (HC and smoke) are to be corrected. The reference day conditions are as follows: Temperature=15 °C, specific humidity=0.00629 kg H_2O/kg of dry air, and pressure=101325 Pa.

Sample system means the system which provides for the transportation of the gaseous emission sample from the sample probe to the inlet of the instrumentation system.

Shaft power means only the measured shaft power output of a turboprop engine.

Smoke means the matter in exhaust emissions which obscures the transmission of light.

Smoke number (SN) means the dimensionless term quantifying smoke emissions.

Standard day conditions means standard ambient conditions as described in the United States Standard Atmosphere 1976, (i.e., temperature=15 °C, specific humidity=0.00 kg H_2O/kg dry air, and pressure= 101325 Pa.)

Taxi/idle (in) means those aircraft operations involving taxi and idle between the time of landing roll-out and final shutdown of all propulsion engines.

Taxi/idle (out) means those aircraft operations involving taxi and idle between the time of initial starting of the propulsion engine(s) used for the taxi and the turn onto the duty runway.

Tier, as used in this part, is a designation related to the NO_x emission standard for the engine as specified in § 34.21 or § 34.23 of this part (e.g., Tier 0).

[Doc. No. 25613, 55 FR 32861, Aug. 10, 1990; 55 FR 37287, Sept. 10, 1990, as amended by Amdt. 34-3, 64 FR 5558, Feb. 3, 1999; Amdt. 34-5, 77 FR 76849, Dec. 31, 2012]

§ 34.2 Abbreviations.

The abbreviations used in this part have the following meanings in both upper and lower case:

CO_2 Carbon dioxide
CO Carbon monoxide
EPA United States Environmental Protection Agency
FAA Federal Aviation Administration, United States Department of Transportation
g Gram(s)
HC Hydrocarbon(s)
HP Horsepower
hr Hour(s)
H_20 water
kg Kilogram(s)
kJ Kilojoule(s)
kN Kilonewton(s)
kW Kilowatt(s)
lb Pound(s)
LTO Landing and takeoff
min Minute(s)
NO_x Oxides of nitrogen
Pa Pascal(s)
rO Rated output
rPR Rated pressure ratio
sec Second(s)
SP Shaft power
SN Smoke number
T Temperature, degrees Kelvin
TIM Time in mode
°C Degrees Celsius
% Percent

[Doc. No. 25613, 55 FR 32861, Aug. 10, 1990, as amended by Amdt. 34-3, 64 FR 5559, Feb. 3, 1999; Amdt. 34-5, 77 FR 76850, Dec. 31, 2012]

§ 34.3 General requirements.

(a) This part provides for the approval or acceptance by the Administrator or the Administrator of the EPA of testing and sampling methods, analytical techniques, and related equipment not identical to those specified in this part. Before either approves or accepts any such alternate, equivalent, or otherwise nonidentical procedures or equipment, the Administrator or the Administrator of the EPA shall consult with the other in determining whether or not the action requires rulemaking under sections 231 and 232 of the Clean Air Act, as amended, consistent with the responsibilities of the Administrator of the EPA and the Secretary of Transportation under sections 231 and 232 of the Clean Air Act.

(b) Under section 232 of the Act, the Secretary of Transportation issues regulations to ensure compliance with 40 CFR part 87. This authority has been delegated to the Administrator of the FAA (49 CFR 1.47).

(c) *U.S. airplanes.* This part applies to civil airplanes that are powered by aircraft gas turbine engines of the classes specified herein and that have U.S. standard airworthiness certificates.

(d) *Foreign airplanes.* Pursuant to the definition of "aircraft" in 40 CFR 87.1, this regulation applies to civil airplanes that are powered by aircraft gas turbine engines of the classes specified herein and that have foreign airworthiness certificates that are equivalent to U.S. standard airworthiness certificates. This regulation applies only to those foreign civil airplanes that, if registered in the United States, would be required by applicable regulations to have a U.S. standard airworthiness certificate in order to conduct the operations intended for the airplane. Pursuant to 40 CFR 87.3(c), this regulation does not apply where it would be inconsistent with an obligation assumed by the United States to a foreign country in a treaty, convention, or agreement.

(e) Reference in this regulation to 40 CFR part 87 refers to title 40 of the Code of Federal Regulations, chapter I—Environmental Protection Agency, part 87, Control of Air Pollution from Aircraft and Aircraft Engines (40 CFR part 87).

(f) This part contains regulations to ensure compliance with certain standards contained in 40 CFR part 87. If EPA takes any action, including the issuance of an exemption or issuance of a revised or alternate procedure, test method, or other regulation, the effect of which is to relax or delay the effective date of any provision of 40 CFR part 87 that is made applicable to an aircraft under this FAR, the Administrator of FAA will grant a general administrative waiver of its more stringent requirements until this FAR is amended to reflect the more relaxed requirements prescribed by EPA.

(g) Unless otherwise stated, all terminology and abbreviations in this FAR that are defined in 40 CFR part 87 have the meaning specified in that part, and all terms in 40 CFR part 87 that are not defined in that part but that are used in this FAR have the meaning given them in the Clean Air Act, as amended by Public Law 91-604.

(h) All interpretations of 40 CFR part 87 that are rendered by the EPA also apply to this FAR.

(i) If the EPA, under 40 CFR 87.3(a), approves or accepts any testing and sampling procedures or methods, analytical techniques, or related equipment not identical to those specified in that part, this FAR requires an applicant to show that such alternate, equivalent, or otherwise nonidentical procedures have been complied with, and that such alternate equipment was used to show compliance, unless the applicant elects to comply with those procedures, methods, techniques, and equipment specified in 40 CFR part 87.

(j) If the EPA, under 40 CFR 87.5, prescribes special test procedures for any aircraft or aircraft engine that is not susceptible to satisfactory testing by the procedures in 40 CFR part 87, the applicant must show the Administrator that those special test procedures have been complied with.

(k) Wherever 40 CFR part 87 requires agreement, acceptance, or approval by the Administrator of the EPA, this FAR requires a showing that such agreement or approval has been obtained.

(l) Pursuant to 42 U.S.C. 7573, no state or political subdivision thereof may adopt or attempt to enforce any standard respecting emissions of any air pollutant from any aircraft or engine thereof unless that standard is identical to a standard made applicable to the aircraft by the terms of this FAR.

(m) If EPA, by regulation or exemption, relaxes a provision of 40 CFR part 87 that is implemented in this FAR, no state or political subdivision thereof may adopt or attempt to enforce the terms of this FAR that are superseded by the relaxed requirement.

(n) If any provision of this FAR is rendered inapplicable to a foreign aircraft as provided in 40 CFR 87.3(c) (international agreements), and § 34.3(d) of this FAR, that provision may not be adopted or enforced against that foreign aircraft by a state or political subdivision thereof.

(o) For exhaust emissions requirements of this FAR that apply beginning February 1, 1974, January 1, 1976, January 1, 1978, January 1, 1984, and August 9, 1985, continued compliance with those requirements is shown for engines for which the type design has been shown to meet those requirements, if the engine is maintained in accordance with applicable maintenance requirements for 14 CFR chapter I. All methods of demonstrating compliance and all model designations previously found acceptable to the Administrator shall be deemed to continue to be an acceptable demonstration of compliance with the specific standards for which they were approved.

(p) Each applicant must allow the Administrator to make, or witness, any test necessary to determine compliance with the applicable provisions of this FAR.

[Doc. No. 25613, 55 FR 32861, Aug. 10, 1990; 55 FR 37287, Sept. 10, 1990; Amdt. 34-5, 77 FR 76850, Dec. 31, 2012]

§ 34.4 [Reserved]

§ 34.5 Special test procedures.

The Administrator or the Administrator of the EPA may, upon written application by a manufacturer or operator of aircraft or aircraft engines, approve test procedures for any aircraft or aircraft engine that is not susceptible to satisfactory testing by the procedures set forth herein. Prior to taking action on any such application, the Administrator or the Administrator of the EPA shall consult with the other.

§ 34.6 Aircraft safety.

(a) The provisions of this part will be revised if at any time the Administrator determines that an emission standard cannot be met within the specified time without creating a safety hazard.

(b) Consistent with 40 CFR 87.6, if the FAA Administrator determines that any emission control regulation in this part cannot be safely applied to an aircraft, that provision may not be adopted or enforced against that aircraft by any state or political subdivision thereof.

§ 34.7 Exemptions.

Notwithstanding part 11 of the Federal Aviation Regulations (14 CFR part 11), all petitions for rulemaking involving either the substance of an emission standard or test procedure prescribed by the EPA that is incorporated in this FAR, or the compliance date for such standard or procedure, must be submitted to the EPA. Information copies of such petitions are invited by the FAA. Petitions for rulemaking or exemption involving provisions of this FAR that do not affect the substance or the compliance date of an emission standard or test procedure that is prescribed by the EPA, and petitions for exemptions under the provisions for which the EPA has specifically granted exemption authority to the Secretary of Transportation are subject to part 11 of the Federal Aviation Regulations (14 CFR part 11). Petitions for rulemaking or exemptions involving these FARs must be submitted to the FAA.

(a) *Exemptions based on flights for short durations at infrequent intervals.* The emission standards of this part do not apply to engines which power aircraft operated in the United States for short durations at infrequent intervals. Such operations are limited to:

(1) Flights of an aircraft for the purpose of export to a foreign country, including any flights essential to demonstrate the integrity of an aircraft prior to a flight to a point outside the United States.

(2) Flights to a base where repairs, alterations or maintenance are to be performed, or to a point of storage, or for the purpose of returning an aircraft to service.

(3) Official visits by representatives of foreign governments.

(4) Other flights the Administrator determines, after consultation with the Administrator of the EPA, to be for short durations at infrequent intervals. A request for such a determination shall be made before the flight takes place.

(b) *Exemptions for very low production engine models.* The emissions standards of this part do not apply to engines of very low production after the date of applicability. For the purpose of this part, "very low production" is limited to a maximum total production for United States civil aviation applications of no more than 200 units covered by the same type certificate after January 1, 1984. Engines manufactured under this provision must be reported to the FAA by serial number on or before the date of manufacture and exemptions granted under this provision are not transferable to any other engine. This exemption is limited to the requirements of § 34.21 only.

(c) *Exemptions for new engines in other categories.* The emissions standards of this part do not apply to engines for which the Administrator determines, with the concurrence of the Administrator of the EPA, that application of any standard under § 34.21 is not justified, based upon consideration of—

(1) Adverse economic impact on the manufacturer;

(2) Adverse economic impact on the aircraft and airline industries at large;

(3) Equity in administering the standards among all economically competing parties;

(4) Public health and welfare effects; and

(5) Other factors which the Administrator, after consultation with the Administrator of the EPA, may deem relevant to the case in question.

(d) *Applicants seeking exemption from other emissions standards of this part and 40 CFR part 87.* Applicants must request exemption from both the FAA and the EPA, even where the underlying regulatory requirements are the same. The FAA and EPA will jointly consider such exemption requests, and will assure consistency in the respective agency determinations.

(e) Applications for exemption from this part shall be submitted in duplicate to the Administrator in accordance with the procedures established by the Administrator in part 11.

(f) The Administrator shall publish in the Federal Register the name of the organization to whom exemptions are granted and the period of such exemptions.

(g) No state or political subdivision thereof may attempt to enforce a standard respecting emissions from an aircraft or engine if such aircraft or engine has been exempted from such standard under this part.

[Doc. No. 25613, 55 FR 32861, Aug. 10, 1990, as amended by Amdt. 34-5, 77 FR 76850, Dec. 31, 2012]

§ 34.9 Exceptions.

(a) *Spare engines.* Certain engines that meet the following description are excepted:

(1) This exception allows production of an engine for installation on an in-service aircraft. A spare engine may not be installed on a new aircraft.

(2) Each spare engine must be identical to a sub-model previously certificated to meet all applicable requirements.

(3) A spare engine may be used only when the emissions of the spare do not exceed the certification requirements of the original engine, for all regulated pollutants.

(4) No separate approval is required to produce spare engines.

(5) The record for each engine excepted under this paragraph (c) must indicate that the engine was produced as an excepted spare engine.

(6) Engines produced under this exception must be labeled "EXCEPTED SPARE" in accordance with § 45.13 of this chapter.

(b) On and after July 18, 2012, and before August 31, 2013, a manufacturer may produce up to six Tier 4 compliant engines that meet the NO_x standards of paragraph (d)(1)(vi) of this section rather than § 34.23(a)(2). No separate approval is required to produce these engines. Engines produced under this exception are to be labeled "COMPLY" in accordance with § 45.13 of this chapter.

[Doc. No. FAA-2012-1333, 77 FR 76850, Dec. 31, 2012]

Subpart B—Engine Fuel Venting Emissions (New and In-Use Aircraft Gas Turbine Engines)

§ 34.10 Applicability.

(a) The provisions of this subpart are applicable to all new aircraft gas turbine engines of classes T3, T8, TSS, and TF equal to or greater than 36 kN (8,090 lb) rated output, manufactured on or after January 1, 1974, and to all in-use aircraft gas turbine engines of classes T3, T8, TSS, and TF equal to or greater than 36 kN (8,090 lb) rated output manufactured after February 1, 1974.

(b) The provisions of this subpart are also applicable to all new aircraft gas turbine engines of class TF less than 36 kN (8,090 lb) rated output and class TP manufactured on or after January 1, 1975, and to all in-use aircraft gas turbine engines of class TF less than 36 kN (8,090 lb) rated output and class TP manufactured after January 1, 1975.

[Doc. No. FAA-2012-1333, 77 FR 76850, Dec. 31, 2012]

§ 34.11 Standard for fuel venting emissions.

(a) No fuel venting emissions shall be discharged into the atmosphere from any new or in-use aircraft gas turbine engine subject to the subpart. This paragraph is directed at the elimination of intentional discharge to the atmosphere of fuel drained from fuel nozzle manifolds after engines are shut down and does not apply to normal fuel seepage from shaft seals, joints, and fittings.

(b) Conformity with the standard set forth in paragraph (a) of this section shall be determined by inspection of the method designed to eliminate these emissions.

(c) As applied to an airframe or an engine, any manufacturer or operator may show compliance with the fuel venting and emissions requirements of this section that were effective beginning February 1, 1974 or January 1, 1975, by any means that prevents the intentional discharge of fuel from fuel nozzle manifolds after the engines are shut down. Acceptable means of compliance include one of the following:

(1) Incorporation of an FAA-approved system that recirculates the fuel back into the fuel system.

(2) Capping or securing the pressurization and drain valve.

(3) Manually draining the fuel from a holding tank into a container.

Subpart C—Exhaust Emissions (New Aircraft Gas Turbine Engines)

§ 34.20 Applicability.

The provisions of this subpart are applicable to all aircraft gas turbine engines of the classes specified beginning on the dates specified in § 34.21.

§ 34.21 Standards for exhaust emissions.

(a) Exhaust emissions of smoke from each new aircraft gas turbine engine of class T8 manufactured on or after February 1, 1974, shall not exceed a smoke number (SN) of 30.

(b) Exhaust emissions of smoke from each new aircraft gas turbine engine of class TF and of rated output of 129 kN (29,000 lb) thrust or greater, manufactured on or after January 1, 1976, shall not exceed

$SN=83.6\ (rO)^{-0.274}$ (rO is in kN).

(c) Exhaust emission of smoke from each new aircraft gas turbine engine of class T3 manufactured on or after January 1, 1978, shall not exceed a smoke number (SN) of 25.

(d) Gaseous exhaust emissions from each new aircraft gas turbine engine shall not exceed:

(1) For Classes TF, T3, T8 engines greater than 26.7 kN (6,000 lb) rated output:

(i) Engines manufactured on or after January 1, 1984:

Hydrocarbons: 19.6 g/kN rO.

(ii) Engines manufactured on or after July 7, 1997:

Carbon Monoxide: 118 g/kN rO.

(iii) Engines of a type or model of which the date of manufacture of the first individual production model was on or before December 31, 1995, and for which the date of manufacture of the individual engine was on or before December 31, 1999 (Tier 2):

Oxides of Nitrogen: (40+2(rPR)) g/kN rO.

(iv) Engines of a type or model of which the date of manufacture of the first individual production model was after December 31, 1995, or for which the date of manufacture of the individual engine was after December 31, 1999 (Tier 2):

Oxides of Nitrogen: (32+1.6(rPR)) g/kN rO.

(v) The emission standards prescribed in paragraphs (d)(1)(iii) and (iv) of this section apply as prescribed beginning July 7, 1997.

(vi) The emission standards of this paragraph apply as prescribed after December 18, 2005. For engines of a type or model of which the first individual production model was manufactured after December 31, 2003 (Tier 4):

(A) That have a rated pressure ratio of 30 or less and a maximum rated output greater than 89 kN:

Oxides of Nitrogen: (19+1.6(rPR)) g/kN rO.

(B) That have a rated pressure ratio of 30 or less and a maximum rated output greater than 26.7 kN but not greater than 89 kN:

Oxides of Nitrogen: (37.572+1.6(rPR) - 0.2087(rO)) g/kN rO.

(C) That have a rated pressure ratio greater than 30 but less than 62.5, and a maximum rated output greater than 89 kN:

Oxides of Nitrogen: (7+2(rPR)) g/kN rO.

(D) That have a rated pressure ratio greater than 30 but less than 62.5, and a maximum rated output greater than 26.7 kN but not greater than 89 kN:

Oxides of Nitrogen: (42.71+1.4286(rPR) - 0.4013(rO) +0.00642(rPR × rO)) g/kN rO.

(E) That have a rated pressure ratio of 62.5 or more:

Oxides of Nitrogen: (32+1.6(rPR)) g/kN rO.

(2) For Class TSS Engines manufactured on or after January 1, 1984:

Hydrocarbons: $140\ (0.92)^{rPR}$ g/kN rO.

(e) Smoke exhaust emissions from each gas turbine engine of the classes specified below shall not exceed:

(1) For Class TF of rated output less than 26.7 kN (6,000 lb) manufactured on or after August 9, 1985:

$SN=83.6(rO)^{-0.274}$ (rO is in kN) not to exceed a maximum of SN=50.

(2) For Classes T3, T8, TSS, and TF of rated output equal to or greater than 26.7 kN (6,000 lb) manufactured on or after January 1, 1984:

$SN=83.6(rO)^{-0.274}$ (rO is in kN) not to exceed a maximum of SN=50.

(3) For Class TP of rated output equal to or greater than 1,000 kW manufactured on or after January 1, 1984:

$SN=187(rO)^{-0.168}$ (rO is in kW).

(f) The standards set forth in paragraphs (a), (b), (c), (d), and (e) of this section refer to a composite gaseous emission sample representing the operation cycles and exhaust smoke emission emitted during operation of the engine as specified in the applicable sections of subpart G of this part, and measured and calculated in accordance with the procedures set forth in subpart G.

(g) Where a gaseous emission standard is specified by a formula, calculate and round the standard to three significant figures or to the nearest 0.1 g/kN (for standards at or above 100 g/kN). Where a smoke standard is specified by a formula, calculate and round the standard to the nearest 0.1 SN. Engines comply with an applicable standard if the testing results show that the engine type certificate family's characteristic level does not exceed the numerical level of that standard, as described in § 34.60.

[Doc. No. 25613, 55 FR 32861, Aug. 10, 1990; 55 FR 37287, Sept. 10, 1990, as amended by Amdt. 34-3, 64 FR 5559, Feb. 3, 1999; Amdt. 34-4, 74 FR 19127, Apr. 28, 2009; Amdt. 34-5, 77 FR 76851, Dec. 31, 2012]

§ 34.23 Exhaust Emission Standards for Engines Manufactured on and after July 18, 2012.

The standards of this section apply to aircraft engines manufactured on and after July 18, 2012, unless otherwise exempted or excepted. Where a gaseous emission standard is specified by a formula, calculate and round the standard to three significant figures or to the nearest 0.1 g/kN (for standards at or above 100 g/kN). Where a smoke standard is specified by a formula, calculate and round the standard to the nearest 0.1 SN. Engines comply with an applicable standard if the testing results show that the engine type certificate family's characteristic level does not exceed the numerical level of that standard, as described in § 34.60.

(a) Gaseous exhaust emissions from each new aircraft gas turbine engine shall not exceed:

(1) For Classes TF, T3 and T8 of rated output less than 26.7 kN (6,000 lb) manufactured on and after July 18, 2012:

$SN=83.6(rO)^{-0.274}$ or 50.0, whichever is smaller

(2) Except as provided in §§ 34.9(b) and 34.21(c), for Classes TF, T3 and T8 engines manufactured on and after July 18, 2012, and for which the first individual production model was manufactured on or before December 31, 2013 (Tier 6):

Tier 6 Oxides of Nitrogen Emission Standards for Subsonic Engines

Class	Rated pressure ratio—rPR	Rated output rO (kN)	NOX(g/kN)
TF, T3, T8	rPR ≤ 30	26.7 < rO < 89.0	38.5486+1.6823 (rPR) - 0.2453 (rO) - (0.00308 (rPR) (rO))
		rO > 89.0	16.72+1.4080 (rPR)
	30 < rPR < 82.6	26.7 < rO ≤ 89.0	46.1600+1.4286 (rPR) - 0.5303 (rO)+(0.00642 (rPR) (rO))
		rO > 89.0	-1.04+2.0 (rPR)
	rPR ≥ 82.6	All	32+1.6 (rPR)

(3) Engines exempted from paragraph (a)(2) of this section produced on or before December 31, 2016 must be labeled "EXEMPT NEW" in accordance with § 45.13 of this chapter. No exemptions to the requirements of paragraph (a)(2) of this section will be granted after December 31, 2016.

(4) For Class TSS Engines manufactured on and after July 18, 2012:

Gaseous Emission Standards for Supersonic Engines

Class	Rated output rO[1] (kN)	NOX (g/kN)	CO (g/kN)
TSS	All	36+2.42 (rPR)	4,550 $(rPR)^{-1.03}$

[1] rO is the rated output with afterburning applied.

(b) Gaseous exhaust emissions from each new aircraft gas turbine engine shall not exceed:

(1) For Classes TF, T3 and T8 engines of a type or model of which the first individual production model was manufactured after December 31, 2013 (Tier 8):

Tier 8 Oxides of Nitrogen Emission Standards for Subsonic Engines

Class	Rated pressure ratio—rPR	Rated output rO (kN)	NOx (g/kN)
TF, T3, T8	rPR ≤ 30	26.7 < rO < 89.0	40.052+1.5681 (rPR) - 0.3615 (rO) - (0.0018 (rPR) (rO))
		rO > 89.0	7.88+1.4080 (rPR)
	30 < rPR < 104.7	26.7 < rO < 89.0	41.9435+1.505 (rPR) - 0.5823 (rO)+(0.005562 (rPR) (rO))
		rO > 89.0	-9.88+2.0 (rPR)
	rPR ≥ 104.7	All	32+1.6 (rPR)

(c) Engines (including engines that are determined to be derivative engines for the purposes of emission certification) type certificated with characteristic levels at or below the NO_x standards of § 34.21(d)(1)(vi) of this part (as applicable based on rated output and rated pressure ratio) and introduced before July 18, 2012, may be produced through December 31, 2012, without meeting the NO_x standard of paragraph (a)(2) of this section.

[Doc. No. 34-5, 77 FR 76851, Dec. 31, 2012]

Subpart D—Exhaust Emissions (In-use Aircraft Gas Turbine Engines)

§ 34.30 Applicability.

The provisions of this subpart are applicable to all in-use aircraft gas turbine engines certificated for operation within the United States of the classes specified, beginning on the dates specified in § 34.31.

§ 34.31 Standards for exhaust emissions.

(a) Exhaust emissions of smoke from each in-use aircraft gas turbine engine of Class T8, beginning February 1, 1974, shall not exceed a smoke number (SN) of 30.

(b) Exhaust emissions of smoke from each in-use aircraft gas turbine engine of Class TF and of rated output of 129 kN (29,000 lb) thrust or greater, beginning January l, 1976, shall not exceed

$SN=83.6(rO)^{-0.274}$ (rO is in kN).

(c) The standards set forth in paragraphs (a) and (b) of this section refer to exhaust smoke emission emitted during operation of the engine as specified in the applicable sections of subpart G of this part, and measured and calculated in accordance with the procedures set forth in subpart G.

[Doc. No. FAA-2012-1333, 77 FR 76852, Dec. 31, 2012]

Subpart E—Certification Provisions

§ 34.48 Derivative engines for emissions certification purposes.

(a) *General.* A derivative engine for emissions certification purposes is an engine configuration that is determined to be similar in design to a previously certificated (original) engine for purposes of compliance with exhaust emissions standards (gaseous and smoke). A type certificate holder may request from the FAA a determination that an engine configuration is considered a derivative engine for emissions certification purposes. To be considered a derivative engine for emission purposes under this part, the configuration must have been derived from the original engine that was certificated to the requirements of part 33 of this chapter and one of the following:

(1) The FAA has determined that a safety issue exists that requires an engine modification.

(2) Emissions from the derivative engines are determined to be similar. In general, this means the emissions must meet the criteria specified in paragraph (b) of this section. The FAA may amend the criteria of paragraph (b) in unusual circumstances, for individual cases, consistent with good engineering judgment.

(3) All of the regulated emissions from the derivative engine are lower than the original engine.

(b) *Emissions similarity.* (1) The type certificate holder must demonstrate that the proposed derivative engine model's emissions meet the applicable standards and differ from the original model's emission rates only within the following ranges:

(i) ±3.0 g/kN for NO_x.

(ii) ±1.0 g/kN for HC.

(iii) ±5.0 g/kN for CO.

(iv) ±2.0 SN for smoke.

(2) If the characteristic level of the original certificated engine model (or any other sub-models within the emission type certificate family tested for certification) before modification is at or above 95% of the applicable standard for any pollutant, an applicant must measure the proposed derivative engine model's emissions for all pollutants to demonstrate that the derivative engine's resulting characteristic levels will not exceed the applicable emission standards. If the characteristic levels of the originally certificated engine model (and all other sub-models within the emission type certificate family tested for certification) are below 95% of the applicable standard for each pollutant, the applicant may use engineering analysis consistent with good engineering judgment to demonstrate that the derivative engine will not exceed the applicable emission standards. The engineering analysis must address all modifications from the original engine, including those approved for previous derivative engines.

(c) *Continued production allowance.* Derivative engines for emissions certification purposes may continue to be produced after the applicability date for new emissions standards when the engines conform to the specifications of this section.

(d) *Non-derivative engines.* If the FAA determines that an engine model does not meet the requirements for a derivative engine for emissions certification purposes, the type certificate holder is required to demonstrate that the engine complies with the emissions standards applicable to a new engine type.

[Doc. No. 34-5, 77 FR 76852, Dec. 31, 2012]

Subpart F [Reserved]

Subpart G—Test Procedures for Engine Exhaust Gaseous Emissions (Aircraft and Aircraft Gas Turbine Engines)

§ 34.60 Introduction.

(a) Use the equipment and procedures specified in Appendix 3, Appendix 5, and Appendix 6 of ICAO Annex 16, as applicable, to demonstrate whether engines meet the applicable gaseous emission standards specified in subpart C of this part. Measure the emissions of all regulated gaseous pollutants. Use the equipment and procedures specified in Appendix 2 and Appendix 6 of ICAO Annex 16 to determine whether engines meet the applicable smoke standard specified in subpart C of this part. The compliance demonstration consists of establishing a mean value from testing the specified number of engines, then calculating a "characteristic level" by applying a set of statistical factors that take into account the number of engines tested. Round each characteristic level to the same number of decimal places as the corresponding emission standard. For turboprop engines, use the procedures specified for turbofan engines, consistent with good engineering judgment.

(b) Use a test fuel that meets the specifications described in Appendix 4 of ICAO Annex 16. The test fuel must not have additives whose purpose is to suppress smoke, such as organometallic compounds.

(c) Prepare test engines by including accessories that are available with production engines if they can reasonably be expected to influence emissions. The test engine may not extract shaft power or bleed service air to provide power to auxiliary gearbox-mounted components required to drive aircraft systems.

(d) Test engines must reach a steady operating temperature before the start of emission measurements.

(e) In consultation with the EPA, the FAA may approve alternative procedures for measuring emissions, including testing and sampling methods, analytical techniques, and equipment specifications that differ from those specified in this part. Manufacturers and operators may request approval of alternative procedures by written request with supporting justification to the FAA Aircraft Certification Office and to the Designated EPA Program Officer. To be approved, one of the following conditions must be met:

(1) The engine cannot be tested using the specified procedures; or

(2) The alternative procedure is shown to be equivalent to, or more accurate or precise than, the specified procedure.

(f) The following landing and takeoff (LTO) cycles apply for emissions testing and for calculating weighted LTO values:

LTO Test Cycles and Time in Mode

Mode	Class					
	TP		TF, T3, T8		TSS	
	TIM (min)	% of rO	TIM (min)	% of rO	TIM (min)	% of rO
Taxi/idle	26.0	7	26.0	7	26.0	5.8
Takeoff	0.5	100	0.7	100	1.2	100
Climbout	2.5	90	2.2	85	2.0	65
Descent	NA	NA	NA	NA	1.2	15
Approach	4.5	30	4.0	30	2.3	34

(g) Engines comply with an applicable standard if the testing results show that the engine type certificate family's characteristic level does not exceed the numerical level of that standard, as described in the applicable appendix of Annex 16.

(h) The system and procedure for sampling and measurement of gaseous emissions shall be as specified by in Appendices 2, 3, 4, 5 and 6 to the International Civil Aviation Organization (ICAO) Annex 16, Environmental Protection, Volume II, Aircraft Engine Emissions, Third Edition, July 2008. This incorporation by reference was approved by the Director of the Federal Register in accordance with 5 U.S.C. 552(a) and 1 CFR part 51. This document can be obtained from the ICAO, Document Sales Unit, 999 University Street, Montreal, Quebec H3C 5H7, Canada, phone +1 514-954-8022, or *www.icao.int* or *sales25icao.int.* Copies can be reviewed at the FAA New England Regional Office, 12 New England Executive Park, Burlington, Massachusetts, 781-238-7101, or at the National Archives and Records Administration (NARA). For information on the availability of this material at NARA, call 202-741-6030, or go to: *http://www.archives.gov/federal_register/code_of_federal_regulations/ibr_locations.html.*

[Doc. No. FAA-2012-1333, 77 FR 76853, Dec. 31, 2012]

PART 35

AIRWORTHINESS STANDARDS: PROPELLERS

Contents

AUTHORITY: 49 U.S.C. 106(g), 40113, 44701-44702, 44704.
SOURCE: Docket No. 2095, 29 FR 7458, June 10, 1964, unless otherwise noted.

Subpart A—General

§ 35.1 Applicability.

(a) This part prescribes airworthiness standards for the issue of type certificates and changes to those certificates, for propellers.

(b) Each person who applies under part 21 for such a certificate or change must show compliance with the applicable requirements of this part.

(c) An applicant is eligible for a propeller type certificate and changes to those certificates after demonstrating compliance with subparts A, B and C of this part. However, the propeller may not be installed on an airplane unless the applicant has shown compliance with either § 23.907 or § 25.907 of this chapter, as applicable, or compliance is not required for installation on that airplane.

(d) For the purposes of this part, the propeller consists of those components listed in the propeller type design, and the propeller system consists of the propeller and all the components necessary for its functioning, but not necessarily included in the propeller type design.

[Amdt. 35-3, 41 FR 55475, Dec. 20, 1976, as amended by Amdt. 35-8, 73 FR 63346, Oct. 24, 2008]

§ 35.2 Propeller configuration.

The applicant must provide a list of all the components, including references to the relevant drawings and software design data, that define the type design of the propeller to be approved under § 21.31 of this chapter.

[Amdt. 35-8, 73 FR 63346, Oct. 24, 2008]

§ 35.3 Instructions for propeller installation and operation.

The applicant must provide instructions that are approved by the Administrator. Those approved instructions must contain:

(a) Instructions for installing the propeller, which:

(1) Include a description of the operational modes of the propeller control system and functional interface of the control system with the airplane and engine systems;

(2) Specify the physical and functional interfaces with the airplane, airplane equipment and engine;

(3) Define the limiting conditions on the interfaces from paragraph (a)(2) of this section;

(4) List the limitations established under § 35.5;

(5) Define the hydraulic fluids approved for use with the propeller, including grade and specification, related operating pressure, and filtration levels; and

(6) State the assumptions made to comply with the requirements of this part.

(b) Instructions for operating the propeller which must specify all procedures necessary for operating the propeller within the limitations of the propeller type design.

[Amdt. 35-8, 73 FR 63346, Oct. 24, 2008]

§ 35.4 Instructions for Continued Airworthiness.

The applicant must prepare Instructions for Continued Airworthiness in accordance with appendix A to this part that are acceptable to the Administrator. The instructions may be incomplete at type certification if a program exists to ensure their completion prior to delivery of the first aircraft with the propeller installed, or upon issuance of a standard certificate of airworthiness for an aircraft with the propeller installed, whichever occurs later.

[Amdt. 35-5, 45 FR 60181, Sept. 11, 1980]

§ 35.5 Propeller ratings and operating limitations.

(a) Propeller ratings and operating limitations must:

(1) Be established by the applicant and approved by the Administrator.

(2) Be included directly or by reference in the propeller type certificate data sheet, as specified in § 21.41 of this chapter.

(3) Be based on the operating conditions demonstrated during the tests required by this part as well as any other information the Administrator requires as necessary for the safe operation of the propeller.

(b) Propeller ratings and operating limitations must be established for the following, as applicable:

(1) Power and rotational speed:

(i) For takeoff.

(ii) For maximum continuous.

(iii) If requested by the applicant, other ratings may also be established.

(2) Overspeed and overtorque limits.

[Amdt. 35-8, 73 FR 63346, Oct. 24, 2008]

§ 35.7 Features and characteristics.

(a) The propeller may not have features or characteristics, revealed by any test or analysis or known to the applicant, that make it unsafe for the uses for which certification is requested.

(b) If a failure occurs during a certification test, the applicant must determine the cause and assess the effect on the airworthiness of the propeller. The applicant must make changes to the design and conduct additional tests that the Administrator finds necessary to establish the airworthiness of the propeller.

[Amdt. 35-8, 73 FR 63346, Oct. 24, 2008]

Subpart B—Design and Construction

§ 35.11 [Reserved]

§ 35.13 [Reserved]

§ 35.15 Safety analysis.

(a)(1) The applicant must analyze the propeller system to assess the likely consequences of all failures that can reasonably be expected to occur. This analysis will take into account, if applicable:

(i) The propeller system in a typical installation. When the analysis depends on representative components, assumed interfaces, or assumed installed conditions, the assumptions must be stated in the analysis.

(ii) Consequential secondary failures and dormant failures.

(iii) Multiple failures referred to in paragraph (d) of this section, or that result in the hazardous propeller effects defined in paragraph (g)(1) of this section.

(2) The applicant must summarize those failures that could result in major propeller effects or hazardous propeller effects defined in paragraph (g) of this section, and estimate the probability of occurrence of those effects.

(3) The applicant must show that hazardous propeller effects are not predicted to occur at a rate in excess of that defined as extremely remote (probability of 10^{-7} or less per propeller flight hour). Since the estimated probability for individual failures may be insufficiently precise to enable the applicant to assess the total rate for hazardous propeller effects, compliance may be shown by demonstrating that the probability of a hazardous propeller effect arising from an individual failure can be predicted to be not greater than 10^{-8} per propeller flight hour. In dealing with probabilities of this low order of magnitude, absolute proof is not possible and reliance must be placed on engineering judgment and previous experience combined with sound design and test philosophies.

(b) If significant doubt exists as to the effects of failures or likely combination of failures, the Administrator may require assumptions used in the analysis to be verified by test.

(c) The primary failures of certain single propeller elements (for example, blades) cannot be sensibly estimated in numerical terms. If the failure of such elements is likely to result in hazardous propeller effects, those elements must be identified as propeller critical parts.

(d) For propeller critical parts, applicants must meet the prescribed integrity specifications of § 35.16. These instances must be stated in the safety analysis.

(e) If the safety analysis depends on one or more of the following items, those items must be identified in the analysis and appropriately substantiated.

(1) Maintenance actions being carried out at stated intervals. This includes verifying that items that could fail in a latent manner are functioning properly. When necessary to prevent hazardous propeller effects, these maintenance actions and intervals must be published in the instructions for continued airworthiness required under § 35.4. Additionally, if errors in maintenance of the propeller system could lead to hazardous propeller effects, the appropriate maintenance procedures must be included in the relevant propeller manuals.

(2) Verification of the satisfactory functioning of safety or other devices at pre-flight or other stated periods. The details of this satisfactory functioning must be published in the appropriate manual.

(3) The provision of specific instrumentation not otherwise required. Such instrumentation must be published in the appropriate documentation.

(4) A fatigue assessment.

(f) If applicable, the safety analysis must include, but not be limited to, assessment of indicating equipment, manual and automatic controls, governors and propeller control systems, synchrophasers, synchronizers, and propeller thrust reversal systems.

(g) Unless otherwise approved by the Administrator and stated in the safety analysis, the following failure definitions apply to compliance with this part.

(1) The following are regarded as hazardous propeller effects:

(i) The development of excessive drag.

(ii) A significant thrust in the opposite direction to that commanded by the pilot.

(iii) The release of the propeller or any major portion of the propeller.

(iv) A failure that results in excessive unbalance.

(2) The following are regarded as major propeller effects for variable pitch propellers:

(i) An inability to feather the propeller for feathering propellers.

(ii) An inability to change propeller pitch when commanded.

(iii) A significant uncommanded change in pitch.

(iv) A significant uncontrollable torque or speed fluctuation.

[Amdt. 35-8, 73 FR 63346, Oct. 24, 2008, as amended by Amdt. 35-9, 78 FR 4041, Jan. 18, 2013]

§ 35.16 Propeller critical parts.

The integrity of each propeller critical part identified by the safety analysis required by § 35.15 must be established by:

(a) A defined engineering process for ensuring the integrity of the propeller critical part throughout its service life,

(b) A defined manufacturing process that identifies the requirements to consistently produce the propeller critical part as required by the engineering process, and

(c) A defined service management process that identifies the continued airworthiness requirements of the propeller critical part as required by the engineering process.

[Amdt. 35-9, 78 FR 4042, Jan. 18, 2013]

§ 35.17 Materials and manufacturing methods.

(a) The suitability and durability of materials used in the propeller must:

(1) Be established on the basis of experience, tests, or both.

(2) Account for environmental conditions expected in service.

(b) All materials and manufacturing methods must conform to specifications acceptable to the Administrator.

(c) The design values of properties of materials must be suitably related to the most adverse properties stated in the material specification for applicable conditions expected in service.

[Amdt. 35-8, 73 FR 63347, Oct. 24, 2008]

§ 35.19 Durability.

Each part of the propeller must be designed and constructed to minimize the development of any unsafe condition of the propeller between overhaul periods.

§ 35.21 Variable and reversible pitch propellers.

(a) No single failure or malfunction in the propeller system will result in unintended travel of the propeller blades to a position below the in-flight low-pitch position. The extent of any intended travel below the in-flight low-pitch position must be documented by the applicant in the appropriate manuals. Failure of structural elements need not be considered if the occurrence of such a failure is shown to be extremely remote under § 35.15.

(b) For propellers incorporating a method to select blade pitch below the in-flight low pitch position, provisions must be made to sense and indicate to the flight crew that the propeller blades are below that position by an amount defined in the installation manual. The method for sensing and indicating the propeller blade pitch position must be such that its failure does not affect the control of the propeller.

[Amdt. 35-8, 73 FR 63347, Oct. 24, 2008]

§ 35.22 Feathering propellers.

(a) Feathering propellers are intended to feather from all flight conditions, taking into account expected wear and leakage. Any feathering and unfeathering limitations must be documented in the appropriate manuals.

(b) Propeller pitch control systems that use engine oil to feather must incorporate a method to allow the propeller to feather if the engine oil system fails.

(c) Feathering propellers must be designed to be capable of unfeathering after the propeller system has stabilized to the minimum declared outside air temperature.

[Amdt. 35-8, 73 FR 63347, Oct. 24, 2008]

§ 35.23 Propeller control system.

The requirements of this section apply to any system or component that controls, limits or monitors propeller functions.

(a) The propeller control system must be designed, constructed and validated to show that:

(1) The propeller control system, operating in normal and alternative operating modes and in transition between operating modes, performs the functions defined by the applicant throughout the declared operating conditions and flight envelope.

(2) The propeller control system functionality is not adversely affected by the declared environmental conditions, including temperature, electromagnetic interference (EMI), high intensity radiated fields (HIRF) and lightning. The environmental limits to which the system has been satisfactorily validated must be documented in the appropriate propeller manuals.

(3) A method is provided to indicate that an operating mode change has occurred if flight crew action is required. In such an event, operating instructions must be provided in the appropriate manuals.

(b) The propeller control system must be designed and constructed so that, in addition to compliance with § 35.15:

(1) No single failure or malfunction of electrical or electronic components in the control system results in a hazardous propeller effect.

(2) Failures or malfunctions directly affecting the propeller control system in a typical airplane, such as structural failures of attachments to the control, fire, or overheat, do not lead to a hazardous propeller effect.

(3) The loss of normal propeller pitch control does not cause a hazardous propeller effect under the intended operating conditions.

(4) The failure or corruption of data or signals shared across propellers does not cause a hazardous propeller effect.

(c) Electronic propeller control system imbedded software must be designed and implemented by a method approved by the Administrator that is consistent with the criticality of the performed functions and that minimizes the existence of software errors.

(d) The propeller control system must be designed and constructed so that the failure or corruption of airplane-supplied data does not result in hazardous propeller effects.

(e) The propeller control system must be designed and constructed so that the loss, interruption or abnormal characteristic of airplane-supplied electrical power does not result in hazardous propeller effects. The power quality requirements must be described in the appropriate manuals.

[Amdt. 35-8, 73 FR 63347, Oct. 24, 2008]

§ 35.24 Strength.

The maximum stresses developed in the propeller may not exceed values acceptable to the Administrator considering the particular form of construction and the most severe operating conditions.

[Amdt. 35-8, 73 FR 63348, Oct. 24, 2008]

Subpart C—Tests and Inspections

§ 35.31 [Reserved]

§ 35.33 General.

(a) Each applicant must furnish test article(s) and suitable testing facilities, including equipment and competent personnel, and conduct the required tests in accordance with part 21 of this chapter.

(b) All automatic controls and safety systems must be in operation unless it is accepted by the Administrator as impossible or not required because of the nature of the test. If needed for substantiation, the applicant may test a different propeller configuration if this does not constitute a less severe test.

(c) Any systems or components that cannot be adequately substantiated by the applicant to the requirements of this part are required to undergo additional tests or analysis to demonstrate that the systems or components are able to perform their intended functions in all declared environmental and operating conditions.

[Amdt. 35-8, 73 FR 63348, Oct. 24, 2008]

§ 35.34 Inspections, adjustments and repairs.

(a) Before and after conducting the tests prescribed in this part, the test article must be subjected to an inspection, and a record must be made of all the relevant parameters, calibrations and settings.

(b) During all tests, only servicing and minor repairs are permitted. If major repairs or part replacement is required, the Administrator must approve the repair or part replacement prior to implementation and may require additional testing. Any unscheduled repair or action on the test article must be recorded and reported.

[Amdt. 35-8, 73 FR 63348, Oct. 24, 2008]

§ 35.35 Centrifugal load tests.

The applicant must demonstrate that a propeller complies with paragraphs (a), (b) and (c) of this section without evidence of failure, malfunction, or permanent deformation that would result in a major or hazardous propeller effect. When the propeller could be sensitive to environmental degradation in service, this must be considered. This section does not apply to fixed-pitch wood or fixed-pitch metal propellers of conventional design.

(a) The hub, blade retention system, and counterweights must be tested for a period of one hour to a load equivalent to twice the maximum centrifugal load to which the propeller would be subjected during operation at the maximum rated rotational speed.

(b) Blade features associated with transitions to the retention system (for example, a composite blade bonded to a metallic retention) must be tested either during the test of paragraph (a) of this section or in a separate component test for a period of one hour to a load equivalent to twice the maximum centrifugal load to which the propeller would be subjected during operation at the maximum rated rotational speed.

(c) Components used with or attached to the propeller (for example, spinners, de-icing equipment, and blade erosion shields) must be subjected to a load equivalent to 159 percent of the maximum centrifugal load to which the component would be subjected during operation at the maximum rated rotational speed. This must be performed by either:

(1) Testing at the required load for a period of 30 minutes; or

(2) Analysis based on test.

[Amdt. 35-8, 73 FR 63348, Oct. 24, 2008]

§ 35.36 Bird impact.

The applicant must demonstrate, by tests or analysis based on tests or experience on similar designs, that the propeller can withstand the impact of a 4-pound bird at the critical location(s) and critical flight condition(s) of a typical installation without causing a major or hazardous propeller effect. This section does not apply to fixed-pitch wood propellers of conventional design.

[Amdt. 35-8, 73 FR 63348, Oct. 24, 2008]

§ 35.37 Fatigue limits and evaluation.

This section does not apply to fixed-pitch wood propellers of conventional design.

(a) Fatigue limits must be established by tests, or analysis based on tests, for propeller:

(1) Hubs.

(2) Blades.

(3) Blade retention components.

(4) Components which are affected by fatigue loads and which are shown under § 35.15 to have a fatigue failure mode leading to hazardous propeller effects.

(b) The fatigue limits must take into account:

(1) All known and reasonably foreseeable vibration and cyclic load patterns that are expected in service; and

(2) Expected service deterioration, variations in material properties, manufacturing variations, and environmental effects.

(c) A fatigue evaluation of the propeller must be conducted to show that hazardous propeller effects due to fatigue will be avoided throughout the intended operational life of the propeller on either:

(1) The intended airplane by complying with § 23.907 or § 25.907 of this chapter, as applicable; or

(2) A typical airplane.

[Amdt. 35-8, 73 FR 63348, Oct. 24, 2008]

§ 35.38 Lightning strike.

The applicant must demonstrate, by tests, analysis based on tests, or experience on similar designs, that the propeller can withstand a lightning strike without causing a major or hazardous propeller effect. The limit to which the propeller has been qualified must be documented in the appropriate manuals. This section does not apply to fixed-pitch wood propellers of conventional design.

[Amdt. 35-8, 73 FR 63348, Oct. 24, 2008]

§ 35.39 Endurance test.

Endurance tests on the propeller system must be made on a representative engine in accordance with paragraph (a) or (b) of this section, as applicable, without evidence of failure or malfunction.

(a) Fixed-pitch and ground adjustable-pitch propellers must be subjected to one of the following tests:

(1) A 50-hour flight test in level flight or in climb. The propeller must be operated at takeoff power and rated rotational speed during at least five hours of this flight test, and at not less than 90 percent of the rated rotational speed for the remainder of the 50 hours.

(2) A 50-hour ground test at takeoff power and rated rotational speed.

(b) Variable-pitch propellers must be subjected to one of the following tests:

(1) A 110-hour endurance test that must include the following conditions:

(i) Five hours at takeoff power and rotational speed and thirty 10-minute cycles composed of:

(A) Acceleration from idle,

(B) Five minutes at takeoff power and rotational speed,

(C) Deceleration, and

(D) Five minutes at idle.

(ii) Fifty hours at maximum continuous power and rotational speed,

(iii) Fifty hours, consisting of ten 5-hour cycles composed of:

(A) Five accelerations and decelerations between idle and takeoff power and rotational speed,

(B) Four and one half hours at approximately even incremental conditions from idle up to, but not including, maximum continuous power and rotational speed, and

(C) Thirty minutes at idle.

(2) The operation of the propeller throughout the engine endurance tests prescribed in part 33 of this chapter.

(c) An analysis based on tests of propellers of similar design may be used in place of the tests of paragraphs (a) and (b) of this section.

[Amdt. 35-8, 73 FR 63348, Oct. 24, 2008]

§ 35.40 Functional test.

The variable-pitch propeller system must be subjected to the applicable functional tests of this section. The same propeller system used in the endurance test (§ 35.39) must be used in the functional tests and must be driven by a representative engine on a test stand or on an airplane. The propeller, must complete these tests without evidence of failure or malfunction. This test may be combined with the endurance test for accumulation of cycles.

(a) Manually-controllable propellers. Five hundred representative flight cycles must be made across the range of pitch and rotational speed.

(b) Governing propellers. Fifteen hundred complete cycles must be made across the range of pitch and rotational speed.

(c) Feathering propellers. Fifty cycles of feather and unfeather operation must be made.

(d) Reversible-pitch propellers. Two hundred complete cycles of control must be made from lowest normal pitch to maximum reverse pitch. During each cycle, the propeller must run for 30 seconds at the maximum power and rotational speed selected by the applicant for maximum reverse pitch.

(e) An analysis based on tests of propellers of similar design may be used in place of the tests of this section.

[Amdt. 35-8, 73 FR 63349, Oct. 24, 2008]

§ 35.41 Overspeed and overtorque.

(a) When the applicant seeks approval of a transient maximum propeller overspeed, the applicant must demonstrate that the propeller is capable of further operation without maintenance action at the maximum propeller overspeed condition. This may be accomplished by:

(1) Performance of 20 runs, each of 30 seconds duration, at the maximum propeller overspeed condition; or

(2) Analysis based on test or service experience.

(b) When the applicant seeks approval of a transient maximum propeller overtorque, the applicant must demonstrate that the propeller is capable of further operation without maintenance action at the maximum propeller overtorque condition. This may be accomplished by:

(1) Performance of 20 runs, each of 30 seconds duration, at the maximum propeller overtorque condition; or

(2) Analysis based on test or service experience.

[Amdt. 35-8, 73 FR 63349, Oct. 24, 2008]

§ 35.42 Components of the propeller control system.

The applicant must demonstrate by tests, analysis based on tests, or service experience on similar components, that each propeller blade pitch control system component, including governors, pitch change assemblies, pitch locks, mechanical stops, and feathering system components, can withstand cyclic operation that simulates the normal load and pitch change travel to which the component would be subjected during the initially declared overhaul period or during a minimum of 1,000 hours of typical operation in service.

[Amdt. 35-8, 73 FR 63349, Oct. 24, 2008]

§ 35.43 Propeller hydraulic components.

Applicants must show by test, validated analysis, or both, that propeller components that contain hydraulic pressure and whose structural failure or leakage from a structural failure could cause a hazardous propeller effect demonstrate structural integrity by:

(a) A proof pressure test to 1.5 times the maximum operating pressure for one minute without permanent deformation or leakage that would prevent performance of the intended function.

(b) A burst pressure test to 2.0 times the maximum operating pressure for one minute without failure. Leakage is permitted and seals may be excluded from the test.

[Amdt. 35-8, 73 FR 63349, Oct. 24, 2008]

§ 35.45 [Reserved]

§ 35.47 [Reserved]

Appendix A to Part 35—Instructions for Continued Airworthiness

a35.1 general

(a) This appendix specifies requirements for the preparation of Instructions for Continued Airworthiness as required by § 35.4.

(b) The Instructions for Continued Airworthiness for each propeller must include the Instructions for Continued Airworthiness for all propeller parts. If Instructions for Continued Airworthiness are not supplied by the propeller part manufacturer for a propeller part, the Instructions for Continued Airworthiness for the propeller must include the information essential to the continued airworthiness of the propeller.

(c) The applicant must submit to the FAA a program to show how changes to the Instructions for Continued Airworthiness made by the applicant or by the manufacturers of propeller parts will be distributed.

a35.2 format

(a) The Instructions for Continued Airworthiness must be in the form of a manual or manuals as appropriate for the quantity of data to be provided.

(b) The format of the manual or manuals must provide for a practical arrangement.

a35.3 content

The contents of the manual must be prepared in the English language. The Instructions for Continued Airworthiness must contain the following sections and information:

(a) *Propeller Maintenance Section.* (1) Introduction information that includes an explanation of the propeller's features and data to the extent necessary for maintenance or preventive maintenance.

(2) A detailed description of the propeller and its systems and installations.

(3) Basic control and operation information describing how the propeller components and systems are controlled and how they operate, including any special procedures that apply.

(4) Instructions for uncrating, acceptance checking, lifting, and installing the propeller.

(5) Instructions for propeller operational checks.

(6) Scheduling information for each part of the propeller that provides the recommended periods at which it should be cleaned, adjusted, and tested, the applicable wear tolerances, and the degree of work recommended at these periods. However, the applicant may refer to an accessory, instrument, or equipment manufacturer as the source of this information if it shows that the item has an exceptionally high degree of complexity requiring specialized maintenance techniques, test equipment, or expertise. The recommended overhaul periods and necessary cross-references to the Airworthiness Limitations section of the manual must also be included. In addition, the applicant must include an inspection program that includes the frequency and extent of the inspections necessary to provide for the continued airworthiness of the propeller.

(7) Troubleshooting information describing probable malfunctions, how to recognize those malfunctions, and the remedial action for those malfunctions.

(8) Information describing the order and method of removing and replacing propeller parts with any necessary precautions to be taken.

(9) A list of the special tools needed for maintenance other than for overhauls.

(b) *Propeller Overhaul Section.* (1) Disassembly information including the order and method of disassembly for overhaul.

(2) Cleaning and inspection instructions that cover the materials and apparatus to be used and methods and precautions to be taken during overhaul. Methods of overhaul inspection must also be included.

(3) Details of all fits and clearances relevant to overhaul.

(4) Details of repair methods for worn or otherwise substandard parts and components along with information necessary to determine when replacement is necessary.

(5) The order and method of assembly at overhaul.

(6) Instructions for testing after overhaul.

(7) Instructions for storage preparation including any storage limits.

(8) A list of tools needed for overhaul.

a35.4 airworthiness limitations section

The Instructions for Continued Airworthiness must contain a section titled Airworthiness Limitations that is segregated and clearly distinguishable from the rest of the document. This section must set forth each mandatory replacement time, inspection interval, and related procedure required for type certification. This section must contain a legible statement in a prominent location that reads: "The Airworthiness Limitations section is FAA approved and specifies maintenance required under §§ 43.16 and 91.403 of the Federal Aviation Regulations unless an alternative program has been FAA approved."

[Amdt. 35-5, 45 FR 60182, Sept. 11, 1980, as amended by Amdt. 35-6, 54 FR 34330, Aug. 18, 1989]

PART 36

NOISE STANDARDS: AIRCRAFT TYPE AND AIRWORTHINESS CERTIFICATION

Contents

Authority: 42 U.S.C. 4321 *et seq.;* 49 U.S.C. 106(g), 40113, 44701-44702, 44704, 44715; sec. 305, Pub. L. 96-193, 94 Stat. 50, 57; E.O. 11514, 35 FR 4247, 3 CFR, 1966-1970 Comp., p. 902.

Source: Docket No. 9337, 34 FR 18364, Nov. 18, 1969, unless otherwise noted.

Subpart A—General

§ 36.1 Applicability and definitions.

(a) This part prescribes noise standards for the issue of the following certificates:

(1) Type certificates, and changes to those certificates, and standard airworthiness certificates, for subsonic transport category large airplanes, and for subsonic jet airplanes regardless of category.

(2) Type certificates and changes to those certificates, standard airworthiness certificates, and restricted category airworthiness certificates, for propeller-driven, small airplanes, and for propeller-driven, commuter category airplanes except those airplanes that are designed for "agricultural aircraft operations" (as defined in § 137.3 of this chapter, as effective on January 1, 1966) or for dispersing fire fighting materials to which § 36.1583 of this part does not apply.

(3) A type certificate and changes to that certificate, and standard airworthiness certificates, for Concorde airplanes.

(4) Type certificates, and changes to those certificates, for helicopters except those helicopters that are designated exclusively for "argicultural aircraft operations" (as defined in § 137.3 of this chapter, as effective on January 1, 1966), for dispensing fire fighting materials, or for carrying external loads (as defined in § 133.1(b) of this chapter, as effective on December 20, 1976).

(5) Type certificates, changes to those certificates, and standard airworthiness certificates, for tiltrotors.

(b) Each person who applies under Part 21 of this chapter for a type of airworthiness certificate specified in this part must show compliance with the applicable requirements of this part, in addition to the applicable airworthiness requirements of this chapter.

(c) Each person who applies under Part 21 of this chapter for approval of an acoustical change described in § 21.93(b) of this chapter must show that the aircraft complies with the applicable provisions of §§ 36.7, 36.9, 36.11 or 36.13 of this part in addition to the applicable airworthiness requirements of this chapter.

(d) Each person who applies for the original issue of a standard airworthiness certificate for a transport category large airplane or for a jet airplane under § 21.183 must, regardless of date of application, show compliance with the following provisions of this part (including appendix B):

(1) The provisions of this part in effect on December 1, 1969, for subsonic airplanes that have not had any flight time before—

(i) December 1, 1973, for airplanes with maximum weights greater than 75,000 pounds, except for airplanes that are powered by Pratt & Whitney Turbo Wasp JT3D series engines;

(ii) December 31, 1974, for airplanes with maximum weights greater than 75,000 pounds and that are powered by Pratt & Whitney Turbo Wasp JT3D series engines; and

(iii) December 31, 1974, for airplanes with maximum weights of 75,000 pounds and less.

(2) The provisions of this part in effect on October 13, 1977, including the stage 2 noise limits, for Concorde airplanes that have not had flight time before January 1, 1980.

(e) Each person who applies for the original issue of a standard airworthiness certificate under § 21.183, or for the original issue of a restricted category airworthiness certificate under § 21.185, for propeller-driven, commuter category airplanes for a propeller driven small airplane that has not had any flight time before January 1, 1980, must show compliance with the applicable provisions of this part.

(f) For the purpose of showing compliance with this part for transport category large airplanes and jet airplanes regardless of category, the following terms have the following meanings:

(1) A "Stage 1 noise level" means a flyover, lateral or approach noise level greater than the Stage 2 noise limits prescribed in section B36.5(b) of appendix B of this part.

(2) A "Stage 1 airplane" means an airplane that has not been shown under this part to comply with the flyover, lateral, and approach noise levels required for Stage 2 or Stage 3 airplanes.

(3) A "Stage 2 noise level" means a noise level at or below the Stage 2 noise limits prescribed in section B36.5(b) of appendix B of this part but higher than the Stage 3 noise limits prescribed in section B36.5(c) of appendix B of this part.

(4) A "Stage 2 airplane" means an airplane that has been shown under this part to comply with Stage 2 noise levels prescribed in section B36.5(b) of appendix B of this part (including use of the applicable tradeoff provisions specified in section B36.6) and that does not comply with the requirements for a Stage 3 airplane.

(5) A "Stage 3 noise level" means a noise level at or below the Stage 3 noise limits prescribed in section B36.5(c) of appendix B of this part.

(6) A "Stage 3 airplane" means an airplane that has been shown under this part to comply with Stage 3 noise levels prescribed in section B36.5(c) of appendix B of this part (including use of the applicable tradeoff provisions specified in section B36.6).

(7) A "subsonic airplane" means an airplane for which the maximum operating limit speed, Mmo , does not exceed a Mach number of 1.

(8) A "supersonic airplane" means an airplane for which the maximum operating limit speed, Mmo , exceeds a Mach number of 1.

(9) A "Stage 4 noise level" means a noise level at or below the Stage 4 noise limit prescribed in section B36.5(d) of appendix B of this part.

(10) A "Stage 4 airplane" means an airplane that has been shown under this part not to exceed the Stage 4 noise limit prescribed in section B36.5(d) of appendix B of this part.

(11) A "Chapter 4 noise level" means a noise level at or below the maximum noise level prescribed in Chapter 4, Paragraph 4.4, Maximum Noise Levels, of the International Civil Aviation Organization (ICAO) Annex 16, Volume I, Amendment 7, effective March 21, 2002. [Incorporated by reference, see § 36.6].

(g) For the purpose of showing compliance with this part for transport category large airplanes and jet airplanes regardless of category, each airplane may not be identified as complying with more than one stage or configuration simultaneously.

(h) For the purpose of showing compliance with this part, for helicopters in the primary, normal, transport, and restricted categories, the following terms have the specified meanings:

(1) *Stage 1 noise level* means a takeoff, flyover, or approach noise level greater than the Stage 2 noise limits prescribed in section H36.305 of appendix H of this part, or a flyover noise level greater than the Stage 2 noise limits prescribed in section J36.305 of appendix J of this part.

(2) *Stage 1 helicopter* means a helicopter that has not been shown under this part to comply with the takeoff, flyover, and approach noise levels required for Stage 2 helicopters as prescribed in section H36.305 of appendix H of this part, or a helicopter that has not been shown under this part to comply with the flyover noise level required for Stage 2 helicopters as prescribed in section J36.305 of appendix J of this part.

(3) *Stage 2 noise level* means a takeoff, flyover, or approach noise level at or below the Stage 2 noise limits prescribed in section H36.305 of appendix H of this part, or a flyover noise level at or below the Stage 2 limit prescribed in section J36.305 of appendix J of this part.

(4) *Stage 2 helicopter* means a helicopter that has been shown under this part to comply with Stage 2 noise limits (including applicable tradeoffs) prescribed in section H36.305 of appendix H of this part, or a helicopter that has been shown under this part to comply with the Stage 2 noise limit prescribed in section J36.305 of appendix J of this part.

(5) *Maximum normal operating RPM* means the highest rotor speed corresponding to the airworthiness limit imposed by the manufacturer and approved by the FAA. Where a tolerance on the highest rotor speed is specified, the maximum normal operating rotor speed is the highest rotor speed for which that tolerance is given. If the rotor speed is automatically linked with flight condition, the maximum normal operating rotor speed corresponding

with that flight condition must be used during the noise certification procedure. If rotor speed can be changed by pilot action, the highest normal operating rotor speed specified in the flight manual limitation section for power-on conditions must be used during the noise certification procedure.

(i) For the purpose of showing compliance with this part for tiltrotors, the following terms have the specified meanings:

Airplane mode means a configuration with nacelles on the down stops (axis aligned horizontally) and rotor speed set to cruise revolutions per minute (RPM).

Airplane mode RPM means the lower range of rotor rotational speed in RPM defined for the airplane mode cruise flight condition.

Fixed operation points mean designated nacelle angle positions selected for airworthiness reference. These are default positions used to refer to normal nacelle positioning operation of the aircraft. The nacelle angle is controlled by a self-centering switch. When the nacelle angle is 0 degrees (airplane mode) and the pilot moves the nacelle switch upwards, the nacelles are programmed to automatically turn to the first default position (for example, 60 degrees) where they will stop. A second upward move of the switch will tilt the nacelle to the second default position (for example, 75 degrees). Above the last default position, the nacelle angle can be set to any angle up to approximately 95 degrees by moving the switch in the up or down direction. The number and position of the fixed operation points may vary on different tiltrotor configurations.

Nacelle angle is defined as the angle between the rotor shaft centerline and the longitudinal axis of the aircraft fuselage.

Tiltrotor means a class of aircraft capable of vertical take-off and landing, within the powered-lift category, with rotors mounted at or near the wing tips that vary in pitch from near vertical to near horizontal configuration relative to the wing and fuselage.

Vertical takeoff and landing (VTOL) mode means the aircraft state or configuration having the rotors orientated with the axis of rotation in a vertical manner (*i.e.*, nacelle angle of approximately 90 degrees) for vertical takeoff and landing operations.

VCON is defined as the maximum authorized speed for any nacelle angle in VTOL/Conversion mode.

VTOL/Conversion mode is all approved nacelle positions where the design operating rotor speed is used for hover operations.

VTOL mode RPM means highest range of RPM that occur for takeoff, approach, hover, and conversion conditions.

[Doc. No. 13243, Amdt. 36-4, 40 FR 1034, Jan. 6, 1975]

Editorial **Note:** For Federal Register citations affecting § 36.1, see the List of CFR Sections Affected, which appears in the Finding Aids section of the printed volume and at *www.fdsys.gov*.

§ 36.2 Requirements as of date of application.

(a) Section 21.17 of this chapter notwithstanding, each person who applies for a type certificate for an aircraft covered by this part, must show that the aircraft meets the applicable requirements of this part that are effective on the date of application for that type certificate. When the time interval between the date of application for the type certificate and the issuance of the type certificate exceeds 5 years, the applicant must show that the aircraft meets the applicable requirements of this part that were effective on a date, to be selected by the applicant, not earlier than 5 years before the issue of the type certificate.

(b) Section 21.101(a) of this chapter notwithstanding, each person who applies for an acoustical change to a type design specified in § 21.93(b) of this chapter must show compliance with the applicable requirements of this part that are effective on the date of application for the change in type design. When the time interval between the date of application for the change in type design and the issuance of the amended or supplemental type certificate exceeds 5 years, the applicant must show that the aircraft meets the applicable requirements of this part that were effective on a date, to be selected by the applicant, not earlier than 5 years before the issue of the amended or supplemental type certificate.

(c) If an applicant elects to comply with a standard in this part that was effective after the filing of the application for a type certificate or change to a type design, the election:

(1) Must be approved by the FAA;

(2) Must include standards adopted between the date of application and the date of the election;

(3) May include other standards adopted after the standard elected by the applicant as determined by the FAA.

[Amdt. 36-54, 67 FR 45211, July 8, 2002; Amdt. 36-24, 67 FR 63195, Oct. 10, 2002]

§ 36.3 Compatibility with airworthiness requirements.

It must be shown that the aircraft meets the airworthiness regulations constituting the type certification basis of the aircraft under all conditions in which compliance with this part is shown, and that all procedures used in complying with this part, and all procedures and information for the flight crew developed under this part, are consistent with the airworthiness regulations constituting the type certification basis of the aircraft.

[Doc. No. 9337, 34 FR 18364, Nov. 18, 1969, as amended by Amdt. 36-14, 53 FR 3540, Feb. 5, 1988]

§ 36.5 Limitation of part.

Pursuant to 49 U.S.C. 1431(b)(4), the noise levels in this part have been determined to be as low as is economically reasonable, technologically practicable, and appropriate to the type of aircraft to which they apply. No determination is made, under this part, that these noise levels are or should be acceptable or unacceptable for operation at, into, or out of, any airport.

§ 36.6 Incorporation by reference.

(a) *General.* This part prescribes certain standards and procedures which are not set forth in full text in the rule. Those standards and procedures are contained in published material which is reasonably available to the class of persons affected and has been approved for incorporation by reference by the Director of the Federal Register under 5 U.S.C. 552 (a) and 1 CFR Part 51.

(b) *Incorporated matter.* (1) Each publication, or part of a publication, which is referenced but not set forth in full-text in this part and which is identified in paragraph (c) of this section is hereby incorporated by reference and made a part of part 36 of this chapter with the approval of the Director of the Federal Register.

(2) Incorporated matter which is subject to subsequent change is incorporated by reference according to the specific reference and to the identification statement. Adoption of any subsequent change in incorporated matter is made under Part 11 of this chapter and 1 CFR Part 51.

(c) *Identification statement.* The complete title or description which identifies each published matter incorporated by reference in this part is as follows:

(1) *International Electrotechnical Commission (IEC) Publications.* (i) IEC Publication No. 179, entitled "Precision Sound Level Meters," dated 1973.

(ii) IEC Publication No. 225, entitled "Octave, Half-Octave, Third Octave Band Filters Intended for the Analysis of Sounds and Vibrations," dated 1966.

(iii) IEC Publication No. 651, entitled "Sound Level Meters," first edition, dated 1979.

(iv) IEC Publication No. 561, entitled "Electro-acoustical Measuring Equipment for Aircraft Noise Certification," first edition, dated 1976.

(v) IEC Publication No. 804, entitled "Integrating-averaging Sound Level Meters," first edition, dated 1985.

(vi) IEC Publication 61094-3, entitled "Measurement Microphones—Part 3: Primary Method for Free-Field Calibration of Laboratory Standard Microphones by the Reciprocity Technique", edition 1.0, dated 1995.

(vii) IEC Publication 61094-4, entitled "Measurement Microphones—Part 4: Specifications for Working Standard Microphones", edition 1.0, dated 1995.

(viii) IEC Publication 61260, entitled "Electroacoustics-Octave-Band and Fractional-Octave-Band filters", edition 1.0, dated 1995.

(ix) IEC Publication 61265, entitled "Instruments for Measurement of Aircraft Noise-Performance Requirements for Systems to Measure One-Third-Octave-Band Sound pressure Levels in Noise Certification of Transport-Category Aeroplanes," edition 1.0, dated 1995.

(x) IEC Publication 60942, entitled "Electroacoustics—Sound Calibrators," edition 2.0, dated 1997.

(2) *Society of Automotive Engineers (SAE) Publications.* (i) SAE ARP 866A, entitled "Standard Values at Atmospheric Absorption as a Function of Temperature and Humidity for Use in Evaluating Aircraft Flyover Noise," dated March 15, 1975.

(3) International Standards and Recommended Practices entitled "Environmental Protection, Annex 16 to the Convention on International Civil Aviation, Volume I, Aircraft Noise", Third Edition, July 1993, Amendment 7, effective March 21, 2002.

(d) *Availability for purchase.* Published material incorporated by reference in this part may be purchased at the price established by the publisher or distributor at the following mailing addresses:

(1) *IEC publications.*

(i) International Electrotechnical Commission, 3, rue de Varembe, Case postale 131, 1211 Geneva 20, Switzerland.

(ii) American National Standard Institute, 11 West 42nd Street, New York City, New York 10036.

(2) *SAE publications.* Society of Automotive Engineers, Inc., 400 Commonwealth Drive, Warrentown, Pennsylvania 15096.

(3) *ICAO publications.* International Civil Aviation Organization (ICAO), Document Sales Unit, 999 University Street, Montreal, Quebec H3C 5H7, Canada.

(e) *Availability for inspection.* A copy of each publication incorporated by reference in this part is available for public inspection at the following locations:

(1) U.S. Department of Transportation, Docket Operations, West Building Ground Floor, Room W12-140, 1200 New Jersey Avenue, SE., Washington, DC 20590.

(2) The National Archives and Records Administration (NARA). For information on the availability of this material at NARA, call 202-741-6030, or go to: *http://www.archives.gov/federal_register/code_of_federal_regulations/ibr_locations.html*.

(3) The respective Region Headquarters of the Federal Aviation Administration as follows:

(i) New England Region Headquarters, 12 New England Executive Park, Burlington, Massachusetts 01803.

(ii) Eastern Region Headquarters, Federal Building, John F. Kennedy (JFK) International Airport, Jamaica, New York 11430.

(iii) Southern Region Headquarters, 1701 Columbia Avenue, College Park, Georgia, 30337.

(iv) Great Lakes Region Headquarters, O'Hare Lake Office Center, 2300 East Devon Avenue, Des Plaines, Illinois 60018.

(v) Central Region Headquarters, Federal Building, 601 East 12th Street, Kanasa City Missouri 64106.

(vi) Southwest Region Headquarters, 2601 Meacham Boulevard, Fort Worth, Texas, 76137-4298.

(vii) Northwest Mountain Region Headquarters, 1601 Lind Avenue, Southwest, Renton, Washington 98055.

(viii) Western-Pacific Region Headquarters, 15000 Aviation Boulevard, Hawthorne, California 92007.

(ix) Alaskan Region Headquarters, 222 West 7th Avenue, #14, Anchorage, Alaska, 99513.

(x) European Office Headquarters, 15, Rue de la Loi (3rd Floor), B-1040 Brussels, Belgium.

[Amdt. 36-9, 43 FR 8739, Mar. 3, 1978, as amended by Amdt. 36-16, 53 FR 47400, Nov. 22, 1988; Amdt. 36-20, 57 FR 42854, Sept. 16, 1992; Amdt. 36-54, 67 FR 45212, July 8, 2002; Amdt. 36-24, 28 FR 1512, Jan. 10, 2003; 68 FR 2402, Jan. 16, 2003; Amdt. 36-26, 70 FR 38748, July 5, 2005; 72 FR 68475, Dec. 5, 2007]

§ 36.7 Acoustical change: Transport category large airplanes and jet airplanes.

(a) *Applicability.* This section applies to all transport category large airplanes and jet airplanes for which an acoustical change approval is applied for under § 21.93(b) of this chapter.

(b) *General requirements.* Except as otherwise specifically provided, for each airplane covered by this section, the acoustical change approval requirements are as follows:

(1) In showing compliance, noise levels must be measured and evaluated in accordance with the applicable procedures and conditions prescribed in Appendix A of this part.

(2) Compliance with the noise limits prescribed in section B36.5 of appendix B must be shown in accordance with the applicable provisions of sections B36.7 and B36.8 of appendix B of this part.

(c) *Stage 1 airplanes.* For each Stage 1 airplane prior to the change in type design, in addition to the provisions of paragraph (b) of this section, the following apply:

(1) If an airplane is a Stage 1 airplane prior to the change in type design, it may not, after the change in type design, exceed the noise levels created prior to the change in type design. The tradeoff provisions of section B36.6 of appendix B of this part may not be used to increase the Stage 1 noise levels, unless the aircraft qualifies as a Stage 2 airplane.

(2) In addition, for an airplane for which application is made after September 17, 1971—

(i) There may be no reduction in power or thrust below the highest airworthiness approved power or thrust, during the tests conducted before and after the change in type design; and

(ii) During the flyover and lateral noise tests conducted before the change in type design, the quietest airworthiness approved configuration available for the highest approved takeoff weight must be used.

(d) *Stage 2 airplanes.* If an airplane is a Stage 2 airplane prior to the change in type design, the following apply, in addition to the provisions of paragraph (b) of this section:

(1) *Airplanes with high bypass ratio jet engines.* For an airplane that has jet engines with a bypass ratio of 2 or more before a change in type design—

(i) The airplane, after the change in type design, may not exceed either (A) each Stage 3 noise limit by more than 3 EPNdB, or (B) each Stage 2 noise limit, whichever is lower:

(ii) The tradeoff provisions of section B36.6 of appendix B of this part may be used in determining compliance under this paragraph with respect to the Stage 2 noise limit or to the Stage 3 plus 3 EPNdB noise limits, as applicable; and

(iii) During the flyover and lateral noise test conducted before the change in type design, the quietest airworthiness approved configuration available for the highest approved takeoff weight must be used.

(2) *Airplanes that do not have high bypass ratio jet engines.* For an airplane that does not have jet engines with a bypass ratio of 2 or more before a change in type design—

(i) The airplane may not be a Stage 1 airplane after the change in type design; and

(ii) During the flyover and lateral noise tests conducted before the change in type design, the quietest airworthiness approved configuration available for the highest approved takeoff weight must be used.

(e) *Stage 3 airplanes.* If an airplane is a Stage 3 airplane prior to the change in type design, the following apply, in addition to the provisions of paragraph (b) of this section:

(1) If compliance with Stage 3 noise levels is not required before the change in type design, the airplane must—

(i) Be a Stage 2 airplane after the change in type design and compliance must be shown under the provisions of paragraph (d)(1) or (d)(2) of this section, as appropriate; or

(ii) Remain a Stage 3 airplane after the change in type design. Compliance must be shown under the provisions of paragraph (e)(2) of this section.

(2) If compliance with Stage 3 noise levels is required before the change in type design, the airplane must be a Stage 3 airplane after the change in type design.

(3) Applications on or after [August 14, 1989.] The airplane must remain a Stage 3 airplane after the change in type design.

(4) If an airplane is a Stage 3 airplane prior to a change in type design, and becomes a Stage 4 after the change in type design, the airplane must remain a Stage 4 airplane.

(f) *Stage 4 airplanes.* If an airplane is a Stage 4 airplane prior to a change in type design, the airplane must remain a Stage 4 airplane after the change in type design.

[Amdt. 36-7, 42 FR 12371, Mar. 3, 1977; Amdt. 36-8, 43 FR 8730, Mar. 2, 1978; Amdt. 36-10, 43 FR 28420, June 29, 1978; Amdt. 36-12, 46 FR 33464, June 29, 1981; Amdt. 36-15, 53 FR 16366, May 6, 1988; 53 FR 18950, May 25, 1988; Amdt. 36-17, 54 FR 21042, May 15, 1989; Amdt. 36-54, 67 FR 45212, July 8, 2002; Amdt. 36-26, 70 FR 38749, July 5, 2005]

§ 36.9 Acoustical change: Propeller-driven small airplanes and propeller-driven commuter category airplanes.

For propeller-driven small airplanes in the primary, normal, utility, acrobatic, transport, and restricted categories and for propeller-driven, commuter category airplanes for which an acoustical change approval is applied for under § 21.93(b) of this chapter after January 1, 1975, the following apply:

(a) If the airplane was type certificated under this part prior to a change in type design, it may not subsequently exceed the noise limits specified in § 36.501 of this part.

(b) If the airplane was not type certificated under this part prior to a change in type design, it may not exceed the higher of the two following values:

(1) The noise limit specified in § 36.501 of this part, or

(2) The noise level created prior to the change in type design, measured and corrected as prescribed in § 36.501 of this part.

[Amdt. 36-16, 53 FR 47400, Nov. 22, 1988; 53 FR 50157, Dec. 13, 1988; Amdt. 36-19, 57 FR 41369, Sept. 9, 1992]

§ 36.11 Acoustical change: Helicopters.

This section applies to all helicopters in the primary, normal, transport, and restricted categories for which an acoustical change approval is applied for under § 21.93(b) of this chapter on or after March 6, 1986. Compliance with the requirements of this section must be demonstrated under appendix H of this part, or, for helicopters having a maximum certificated takeoff weight of not more than 7,000 pounds, compliance with this section may be demonstrated under appendix J of this part.

(a) *General requirements.* Except as otherwise provided, for helicopters covered by this section, the acoustical change approval requirements are as follows:

(1) In showing compliance with the requirements of appendix H of this part, noise levels must be measured, evaluated, and calculated in accordance with the applicable procedures and conditions prescribed in parts B and C of appendix H of this part. For helicopters having a maximum certificated takeoff weight of not more than 7,000 pounds that alternatively demonstrate compliance under appendix J of this part, the flyover noise level prescribed in appendix J of this part must be measured, evaluated, and calculated in accordance with the applicable procedures and conditions prescribed in parts B and C of appendix J of this part.

(2) Compliance with the noise limits prescribed in section H36.305 of appendix H of this part must be shown in accordance with the applicable provisions of part D of appendix H of this part. For those helicopters that demonstrate compliance with the requirements of appendix J of this part, compliance with the noise levels prescribed in section J36.305 of appendix J of this part must be shown in accordance with the applicable provisions of part D of appendix J of this part.

(b) *Stage 1 helicopters.* Except as provided in § 36.805(c), for each Stage 1 helicopter prior to a change in type design, the helicopter noise levels may not, after a change in type design, exceed the noise levels specified in section H36.305(a)(1) of appendix H of this part where the demonstration of compliance is under appendix H of this part. The tradeoff provisions under section H36.305(b) of appendix H of this part may not be used to increase any Stage 1 noise level beyond these limits. If an applicant chooses to demonstrate compliance under appendix J of this part, for each Stage 1 helicopter prior to a change in type design, the helicopter noise levels may not, after a change in type design, exceed the Stage 2 noise levels specified in section J36.305(a) of appendix J of this part.

(c) *Stage 2 helicopters.* For each helicopter that is Stage 2 prior to a change in type design, the helicopter must be a Stage 2 helicopter after a change in type design.

[Doc. No. 26910, 57 FR 42854, Sept. 16, 1992, as amended by Amdt. 36-25, 69 FR 31234, June 2, 2004]

§ 36.13 Acoustical change: Tiltrotor aircraft.

The following requirements apply to tiltrotors in any category for which an acoustical change approval is applied for under § 21.93(b) of this chapter on or after March 11, 2013:

(a) In showing compliance with Appendix K of this part, noise levels must be measured, evaluated, and calculated in accordance with the applicable procedures and conditions prescribed in Appendix K of this part.

(b) Compliance with the noise limits prescribed in section K4 (Noise Limits) of Appendix K of this part must be shown in accordance with the applicable provisions of sections K2 (Noise Evaluation Measure), K3 (Noise Measurement Reference Points), K6 (Noise Certification Reference Procedures), and K7 (Test Procedures) of Appendix K of this part.

(c) After a change in type design, tiltrotor noise levels may not exceed the limits specified in § 36.1103.

[Amdt. 36-29, 78 FR 1139, Jan. 8, 2013]

Subpart B—Transport Category Large Airplanes and Jet Airplanes

§ 36.101 Noise measurement and evaluation.

For transport category large airplanes and jet airplanes, the noise generated by the airplane must be measured and evaluated under appendix A of this part or under an approved equivalent procedure.

[Amdt. 36-54, 67 FR 45212, July 8, 2002]

§ 36.103 Noise limits.

(a) For subsonic transport category large airplanes and subsonic jet airplanes compliance with this section must be shown with noise levels measured and evaluated as prescribed in appendix A of this part, and demonstrated at the measuring points, and in accordance with the test procedures under section B36.8 (or an approved equivalent procedure), stated under appendix B of this part.

(b) Type certification applications between November 5, 1975 and December 31, 2005. If application is made on or after November 5, 1975, and before January 1, 2006, it must be shown that the noise levels of the airplane are no greater than the Stage 3 noise limit prescribed in section B36.5(c) of appendix B of this part.

(c) Type certification applications on or after January 1, 2006. If application is made on or after January 1, 2006, it must be shown that the noise levels of the airplane are no greater than the Stage 4 noise limit prescribed in section B36.5(d) of appendix B of this part. Prior to January 1, 2006, an applicant may seek voluntary certification to Stage 4. If Stage 4 certification is chosen, the requirements of § 36.7(f) of this part will apply.

[Amdt. 36-54, 67 FR 45212, July 8, 2002, as amended by Amdt. 36-26, 70 FR 38749, July 5, 2005]

§ 36.105 Flight Manual Statement of Chapter 4 equivalency.

For each airplane that meets the requirements for Stage 4 certification, the Airplane Flight Manual or operations manual must include the following statement: "The following noise levels comply with part 36, Appendix B, Stage 4 maximum noise level requirements and were obtained by analysis of approved data from noise tests conducted under the provisions of part 36, Amendment 36 (insert part 36 amendment to which the airplane was certificated). The noise measurement and evaluation procedures used to obtain these noise levels are considered by the FAA to be equivalent to the Chapter 4 noise level required by the International Civil Aviation Organization (ICAO) in Annex 16, Volume I, Appendix 2, Amendment 7, effective March 21, 2002." [Incorporated by reference, see § 36.6].

[Amdt. 36-26, 70 FR 38749, July 5, 2005; 70 FR 41610, July 20, 2005]

Subpart C [Reserved]

Subpart D—Noise Limits for Supersonic Transport Category Airplanes

§ 36.301 Noise limits: Concorde.

(a) *General.* For the Concorde airplane, compliance with this subpart must be shown with noise levels measured and evaluated as prescribed in Subpart B of this part, and demonstrated at the measuring points prescribed in appendix B of this part.

(b) *Noise limits.* It must be shown, in accordance with the provisions of this part in effect on October 13, 1977, that the noise levels of the airplane are reduced to the lowest levels that are economically reasonable, technologically practicable, and appropriate for the Concorde type design.

[Amdt. 36-10, 43 FR 28420, June 29, 1978, as amended by Amdt. 36-54, 67 FR 45212, July 8, 2002]

Subpart E [Reserved]

Subpart F—Propeller Driven Small Airplanes and Propeller-Driven, Commuter Category Airplanes

§ 36.501 Noise limits.

(a) Compliance with this subpart must be shown for—

(1) Propeller driven small airplanes for which application for the issuance of a new, amended, or supplemental type certificate in the normal, utility, acrobatic, transport, or restricted category is made on or after October 10, 1973; and propeller-driven, commuter category airplanes for which application for the issuance of a type certificate in the commuter category is made on or after January 15, 1987.

(2) Propeller driven small airplanes and propeller-driven, commuter category airplanes for which application is made for the original issuance of a standard airworthiness certificate or restricted category airworthiness certificate, and that have not had any flight time before January 1, 1980 (regardless of date of application).

(3) Airplanes in the primary category:

(i) Except as provided in paragraph (a)(3)(ii) of this section, for an airplane for which application for a type certificate in the primary category is made, and that was not previously certificated under appendix F of this part, compliance with appendix G of this part must be shown.

(ii) For an airplane in the normal, utility or acrobatic category that (A) has a type certificate issued under this chapter, (B) has a standard airworthiness certificate issued under this chapter, (C) has not undergone an acoustical change from its type design, (D) has not previously been certificated under appendix F or G of this part, and (E) for which application for conversion to the primary category is made, no further showing of compliance with this part is required.

(b) For aircraft covered by this subpart for which certification tests are completed before December 22, 1988, compliance must be shown with noise levels as measured and prescribed in Parts B and C of appendix F, or under approved equivalent procedures. It must be shown that the noise level of the airplane is no greater than the applicable limit set in Part D of appendix F.

(c) For aircraft covered by this subpart for which certification tests are not completed before December 22, 1988, compliance must be shown with noise levels as measured and prescribed in Parts B and C of appendix G, or under approved equivalent procedures. It must be shown that the noise level of the airplane is no greater than the applicable limits set in Part D of appendix G.

[Doc. No. 13243, 40 FR 1034, Jan. 6, 1975, as amended by Amdt. 36-13, 52 FR 1836, Jan. 15, 1987; Amdt. 36-16, 53 FR 47400, Nov. 22, 1988; Amdt. 36-19, 57 FR 41369, Sept. 9, 1992]

Subpart G [Reserved]

Subpart H—Helicopters

Source: Amdt. 36-14, 53 FR 3540, Feb. 5, 1988; 53 FR 7728, Mar. 10, 1988, unless otherwise noted.

§ 36.801 Noise measurement.

For primary, normal, transport, or restricted category helicopters for which certification is sought under appendix H of this part, the noise generated by the helicopter must be measured at the noise measuring points and under the test conditions prescribed in part B of appendix H of this part, or under an FAA-approved equivalent procedure. For those primary, normal, transport, and restricted category helicopters having a maximum certificated takeoff weight of not more than 7,000 pounds for which compliance with appendix J of this part is demonstrated, the noise generated by the helicopter must be measured at the noise measuring point and under the test conditions prescribed in part B of appendix J of this part, or an FAA-approved equivalent procedure.

[Doc. No. 26910, 57 FR 42854, Sept. 16, 1992, as amended by Amdt. 36-25, 69 FR 31234, June 2, 2004]

§ 36.803 Noise evaluation and calculation.

The noise measurement data required under § 36.801 and obtained under appendix H of this part must be corrected to the reference conditions contained in part A of appendix H of this part, and evaluated under the procedures of part C of appendix H of this part, or an FAA-approved equivalent procedure. The noise measurement data required under § 36.801 and obtained under appendix J of this part must be corrected to the reference conditions contained in part A of appendix J of this part, and evaluated under the procedures of part C of appendix J of this part, or an FAA-approved equivalent procedure.

[Doc. No. 26910, 57 FR 42854, Sept. 16, 1992]

§ 36.805 Noise limits.

(a) Compliance with the noise levels prescribed under part D of appendix H of this part, or under part D of appendix J of this part, must be shown for helicopters for which application for issuance of a type certificate in the primary, normal, transport, or restricted category is made on or after March 6, 1986.

(b) For helicopters covered by this section, except as provided in paragraph (c) or (d)(2) of this section, it must be shown either:

(1) For those helicopters demonstrating compliance under appendix H of this part, the noise levels of the helicopter are no greater than the applicable limits prescribed under section H36.305 of appendix H of this part, or

(2) For helicopters demonstrating compliance under appendix J of this part, the noise level of the helicopter is no greater than the limit prescribed under section J36.305 of appendix J of this part.

(c) For helicopters for which application for issuance of an original type certificate in the primary, normal, transport, or restricted category is made on or after March 6, 1986, and which the FAA finds to be the first civil version of a helicopter that was designed and constructed for, and accepted for operational use by, an Armed Force of the United States or the U.S. Coast Guard on or before March 6, 1986, it must be shown that the noise levels of the helicopter are no greater than the noise limits for a change in type design as specified in section H36.305(a)(1)(ii) of appendix H of this part for compliance demonstrated under appendix H of this part, or as specified in section J36.305 of appendix J of this part for compliance demonstrated under appendix J of this part. Subsequent civil versions of any such helicopter must meet the Stage 2 requirements.

(d) Helicopters in the primary category:

(1) Except as provided in paragraph (d)(2) of this section, for a helicopter for which application for a type certificate in the primary category is made, and that was not previously certificated under appendix H of this part, compliance with appendix H of this part must be shown.

(2) For a helicopter that:

(i) Has a normal or transport type certificate issued under this chapter,

(ii) Has a standard airworthiness certificate issued under this chapter,

(iii) Has not undergone an acoustical change from its type design,

(iv) Has not previously been certificated under appendix H of this part, and

(v) For which application for conversion to the primary category is made, no further showing of compliance with this part is required.

[Doc. No. 26910, 57 FR 42855, Sept. 16, 1992]

Subparts I-J [Reserved]

Subpart K—Tiltrotors

Source: 78 FR 1139, Jan. 8, 2013, unless otherwise noted.

§ 36.1101 Noise measurement and evaluation.

For tiltrotors, the noise generated must be measured and evaluated under Appendix K of this part, or under an approved equivalent procedure.

§ 36.1103 Noise limits.

(a) Compliance with the maximum noise levels prescribed in Appendix K of this part must be shown for a tiltrotor for which the application for the issuance of a type certificate is made on or after March 11, 2013.

(b) To demonstrate compliance with this part, noise levels may not exceed the noise limits listed in Appendix K, Section K4, Noise Limits of this part. Appendix K of this part (or an approved equivalent procedure) must also be used to evaluate and demonstrate compliance with the approved test procedures, and at the applicable noise measurement points.

Subparts L-N [Reserved]

Subpart O—Documentation, Operating Limitations and Information

§ 36.1501 Procedures, noise levels and other information.

(a) All procedures, weights, configurations, and other information or data employed for obtaining the certified noise levels prescribed by this part, including equivalent procedures used for flight, testing, and analysis, must be developed and approved. Noise levels achieved during type certification must be included in the approved airplane (rotorcraft) flight manual.

(b) Where supplemental test data are approved for modification or extension of an existing flight data base, such as acoustic data from engine static tests used in the certification of acoustical changes, the test procedures, physical configuration, and other information and procedures that are employed for obtaining the supplemental data must be developed and approved.

[Amdt. 36-15, 53 FR 16366, May 6, 1988]

§ 36.1581 Manuals, markings, and placards.

(a) If an Airplane Flight Manual or Rotorcraft Flight Manual is approved, the approved portion of the Airplane Flight Manual or Rotorcraft Flight Manual must contain the following information, in addition to that specified under § 36.1583 of this part. If an Airplane Flight Manual or Rotorcraft Flight Manual is not approved, the procedures and information must be furnished in any combination of approved manual material, markings, and placards.

(1) For transport category large airplanes and jet airplanes, the noise level information must be one value for each flyover, lateral, and approach as defined and required by appendix B of this part, along with the maximum takeoff weight, maximum landing weight, and configuration.

(2) For propeller driven small airplanes, the noise level information must be one value for takeoff as defined and required by appendix G of this part, along with the maximum takeoff weight and configuration.

(3) For rotorcraft, the noise level information must be one value for each takeoff, flyover, and approach as defined and required by appendix H of this part, or one value for flyover as defined and required by appendix J of this part, at the maximum takeoff weight and configuration.

(b) If supplemental operational noise level information is included in the approved portion of the Airplane Flight Manual, it must be segregated, identified as information in addition to the certificated noise levels, and clearly distinguished from the information required under § 36.1581(a).

(c) The following statement must be furnished near the listed noise levels:

No determination has been made by the Federal Aviation Administration that the noise levels of this aircraft are or should be acceptable or unacceptable for operation at, into, or out of, any airport.

(d) For transport category large airplanes and jet airplanes, for which the weight used in meeting the takeoff or landing noise requirements of this part is less than the maximum weight established under the applicable airworthiness requirements, those lesser weights must be furnished, as operating limitations in the operating limitations section of the Airplane Flight Manual. Further, the maximum takeoff weight must not exceed the takeoff weight that is most critical from a takeoff noise standpoint.

(e) For propeller driven small airplanes and for propeller-driven, commuter category airplanes for which the weight used in meeting the flyover noise requirements of this part is less than the maximum weight by an amount exceeding the amount of fuel needed to conduct the test, that lesser weight must be furnished, as an operating limitation, in the operating limitations section of an approved Airplane Flight Manual, in approved manual material, or on an approved placard.

(f) For primary, normal, transport, and restricted category helicopters, if the weight used in meeting the takeoff, flyover, or approach noise requirements of appendix H of this part, or the weight used in meeting the flyover noise requirement of appendix J of this part, is less than the certificated maximum takeoff weight established under either § 27.25(a) or § 29.25(a) of this chapter, that lesser weight must be furnished as an operating limitation in the operating limitations section of the Rotorcraft Flight Manual, in FAA-approved manual material, or on an FAA-approved placard.

(g) Except as provided in paragraphs (d), (e), and (f) of this section, no operating limitations are furnished under this part.

[Doc. No. 13243, 40 FR 1035, Jan. 6, 1975]

Editorial Note: For Federal Register citations affecting § 36.1581, see the List of CFR Sections Affected, which appears in the Finding Aids section of the printed volume and at *www.fdsys.gov*.

§ 36.1583 Noncomplying agricultural and fire fighting airplanes.

(a) This section applies to propeller-driven, small airplanes that—

(1) Are designed for "agricultural aircraft operations" (as defined in § 137.3 of this chapter, effective on January 1, 1966) or for dispensing fire fighting materials; and

(2) Have not been shown to comply with the noise levels prescribed under appendix F of this part—

(i) For which application is made for the original issue of a standard airworthiness certificate and that do not have any flight time before January 1, 1980; or

(ii) For which application is made for an acoustical change approval, for airplanes which have a standard airworthiness certificate after the change in the type design, and that do not have any flight time in the changed configuration before January 1, 1980.

(b) For airplanes covered by this section an operating limitation reading as follows must be furnished in the manner prescribed in § 36.1581:

Noise abatement: This airplane has not been shown to comply with the noise limits in FAR Part 36 and must be operated in accordance with the noise operating limitation prescribed under FAR § 91.815.

[Amdt. 36-11, 45 FR 67066, Oct. 9, 1980. Redesignated by Amdt. 36-14, 53 FR 3540, Feb. 5, 1988; Amdt. 36-18, 54 FR 34330, Aug. 18, 1989]

Appendix A to Part 36—Aircraft Noise Measurement and Evaluation Under § 36.101

Section A36.1 *Introduction*

A36.1.1 This appendix prescribes the conditions under which airplane noise certification tests must be conducted and states the measurement procedures that must be used to measure airplane noise. The procedures that must be used to determine the noise evaluation quantity designated as effective perceived noise level, EPNL, under §§ 36.101 and 36.803 are also stated.

A36.1.2 The instructions and procedures given are intended to ensure uniformity during compliance tests and to permit comparison between tests of various types of airplanes conducted in various geographical locations.

A36.1.3 A complete list of symbols and units, the mathematical formulation of perceived noisiness, a procedure for determining atmospheric attenuation of sound, and detailed procedures for correcting noise levels from non-reference to reference conditions are included in this appendix.

A36.1.4 For Stage 4 airplanes, an acceptable alternate for noise measurement and evaluation is Appendix 2 to the International Civil Aviation Organization (ICAO) Annex 16, Environmental Protection, Volume I, Aircraft Noise, Third Edition, July 1993, Amendment 7, effective March 21, 2002. [Incorporated by reference, see § 36.6].

Section A36.2 *Noise Certification Test and Measurement Conditions*

A36.2.1 *General.*

A36.2.1.1 This section prescribes the conditions under which noise certification must be conducted and the measurement procedures that must be used.

Note: Many noise certifications involve only minor changes to the airplane type design. The resulting changes in noise can often be established reliably without resorting to a complete test as outlined in this appendix. For this reason, the FAA permits the use of approved equivalent procedures. There are also equivalent procedures that may be used in full certification tests, in the interest of reducing costs and providing reliable results. Guidance material on the use of equivalent procedures in the noise certification of subsonic jet and propeller-driven large airplanes is provided in the current advisory circular for this part.

A36.2.2 *Test environment.*

A36.2.2.1 Locations for measuring noise from an airplane in flight must be surrounded by relatively flat terrain having no excessive sound absorption characteristics such as might be caused by thick, matted, or tall grass, shrubs, or wooded areas. No obstructions that significantly influence the sound field from the airplane must exist within a conical space above the point on the ground vertically below the microphone, the cone being defined by an axis normal to the ground and by a half-angle 80° from this axis.

Note: Those people carrying out the measurements could themselves constitute such obstruction.

A36.2.2.2 The tests must be carried out under the following atmospheric conditions.

(a) No precipitation;

(b) Ambient air temperature not above 95°F (35°C) and not below 14°F (–10°C), and relative humidity not above 95% and not below 20% over the whole noise path between a point 33 ft (10 m) above the ground and the airplane;

Note: Care should be taken to ensure that the noise measuring, airplane flight path tracking, and meteorological instrumentation are also operated within their specific environmental limitations.

(c) Relative humidity and ambient temperature over the whole noise path between a point 33 ft (10 m) above the ground and the airplane such that the sound attenuation in the one-third octave band centered on 8 kHz will not be more than 12 dB/100 m unless:

(1) The dew point and dry bulb temperatures are measured with a device which is accurate to±0.9°F (±0.5°C) and used to obtain relative humidity; in addition layered sections of the atmosphere are used as described in section A36.2.2.3 to compute equivalent weighted sound attenuations in each one-third octave band; or

(2) The peak noy values at the time of PNLT, after adjustment to reference conditions, occur at frequencies less than or equal to 400 Hz.;

(d) If the atmospheric absorption coefficients vary over the PNLTM sound propagation path by more than±1.6 dB/1000 ft (±0.5 dB/100m) in the 3150Hz one-third octave band from the value of the absorption coefficient derived from the meteorological measurement obtained at 33 ft (10 m) above the surface, "layered" sections of the atmosphere must be used as described in section A36.2.2.3 to compute equivalent weighted sound attenuations in each one-third octave band; the FAA will determine whether a sufficient number of layered sections have been used. For each measurement, where multiple layering is not required, equivalent sound attenuations in each one-third octave band must be determined by averaging the atmospheric absorption coefficients for each such band at 33 ft (10 m) above ground level, and at the flight level of the airplane at the time of PNLTM, for each measurement;

(e) Average wind velocity 33 ft (10 m) above ground may not exceed 12 knots and the crosswind velocity for the airplane may not exceed 7 knots. The average wind velocity must be determined using a 30-second averaging period spanning the 10 dB-down time interval. Maximum wind velocity 33 ft (10 m) above ground is not to exceed 15 knots and the crosswind velocity is not to exceed 10 knots during the 10 dB-down time interval;

(f) No anomalous meteorological or wind conditions that would significantly affect the measured noise levels when the noise is recorded at the measuring points specified by the FAA; and

(g) Meteorological measurements must be obtained within 30 minutes of each noise test measurement; meteorological data must be interpolated to actual times of each noise measurement.

A36.2.2.3 When a multiple layering calculation is required by section A36.2.2.2(c) or A36.2.2.2(d) the atmosphere between the airplane and 33 ft (10 m) above the ground must be divided into layers of equal depth. The depth of the layers must be set to not more than the depth of the narrowest layer across which the variation in the atmospheric absorption coefficient of the 3150 Hz one-third octave band is not greater than±1.6 dB/1000 ft (±0.5 dB/100m), with a minimum layer depth of 100 ft (30 m). This requirement must be met for the propagation path at PNLTM. The mean of the values of the atmospheric absorption coefficients at the top and bottom of each layer may be used to characterize the absorption properties of each layer.

A36.2.2.4 The airport control tower or another facility must be aproved by the FAA for use as the central location at which measurements of atmospheric parameters are representative of those conditions existing over the geographical area in which noise measurements are made.

A36.2.3 *Flight path measurement.*

A36.2.3.1 The airplane height and lateral position relative to the flight track must be determined by a method independent of normal flight instrumentation such as radar tracking, theodolite triangulation, or photographic scaling techniques, to be approved by the FAA.

A36.2.3.2 The airplane position along the flight path must be related to the noise recorded at the noise measurement locations by means of synchronizing signals over a distance sufficient to assure adequate data during the period that the noise is within 10 dB of the maximum value of PNLT.

A36.2.3.3 Position and performance data required to make the adjustments referred to in section A36.9 of this appendix must be automatically recorded at an approved sampling rate. Measuring equipment must be approved by the FAA.

Section A36.3 *Measurement of Airplane Noise Received on the Ground*

A36.3.1 *Definitions.*

For the purposes of section A36.3 the following definitions apply:

A36.3.1.1 *Measurement system* means the combination of instruments used for the measurement of sound pressure levels, including a sound calibrator, windscreen, microphone system, signal recording and conditioning devices, and one-third octave band analysis system.

Note: Practical installations may include a number of microphone systems, the outputs from which are recorded simultaneously by a multi-channel recording/analysis device via signal conditioners, as appropriate. For the purpose of this section, each complete measurement channel is considered to be a measurement system to which the requirements apply accordingly.

A36.3.1.2 *Microphone system* means the components of the measurement system which produce an electrical output signal in response to a sound pressure input signal, and which generally include a microphone, a preamplifier, extension cables, and other devices as necessary.

A36.3.1.3 *Sound incidence angle* means in degrees, an angle between the principal axis of the microphone, as defined in IEC 61094-3 and IEC 61094-4, as amended and a line from the sound source to the center of the diaphragm of the microphone.

Note: When the sound incidence angle is 0°, the sound is said to be received at the microphone at "normal (perpendicular) incidence;" when the sound incidence angle is 90°, the sound is said to be received at "grazing incidence."

A36.3.1.4 *Reference direction* means, in degrees, the direction of sound incidence specified by the manufacturer of the microphone, relative to a sound incidence angle of 0°, for which the free-field sensitivity level of the microphone system is within specified tolerance limits.

A36.3.1.5 *Free-field sensitivity of a microphone system* means, in volts per Pascal, for a sinusoidal plane progressive sound wave of specified frequency, at a specified sound incidence angle, the quotient of the root mean square voltage at the output of a microphone system and the root mean square sound pressure that would exist at the position of the microphone in its absence.

A36.3.1.6 *Free-field sensitivity level of a microphone system* means, in decibels, twenty times the logarithm to the base ten of the ratio of the free-field sensitivity of a microphone system and the reference sensitivity of one volt per Pascal.

Note: The free-field sensitivity level of a microphone system may be determined by subtracting the sound pressure level (in decibels re 20 μPa) of the sound incident on the microphone from the voltage level (in decibels re 1 V) at the output of the microphone system, and adding 93.98 dB to the result.

A36.3.1.7 *Time-average band sound pressure level* means in decibels, ten times the logarithm to the base ten, of the ratio of the time mean square of the instantaneous sound pressure during a stated time interval and in a specified one-third octave band, to the square of the reference sound pressure of 20 μPa.

A36.3.1.8 *Level range* means, in decibels, an operating range determined by the setting of the controls that are provided in a measurement system for the recording and one-third octave band analysis of a sound pressure signal. The upper boundary associated with any particular level range must be rounded to the nearest decibel.

A36.3.1.9 *Calibration sound pressure level* means, in decibels, the sound pressure level produced, under reference environmental conditions, in the cavity of the coupler of the sound calibrator that is used to determine the overall acoustical sensitivity of a measurement system.

A36.3.1.10 *Reference level range* means, in decibels, the level range for determining the acoustical sensitivity of the measurement system and containing the calibration sound pressure level.

A36.3.1.11 *Calibration check frequency* means, in hertz, the nominal frequency of the sinusoidal sound pressure signal produced by the sound calibrator.

A36.3.1.12 *Level difference* means, in decibels, for any nominal one-third octave midband frequency, the output signal level measured on any level range minus the level of the corresponding electrical input signal.

A36.3.1.13 *Reference level difference* means, in decibels, for a stated frequency, the level difference measured on a level range for an electrical input signal corresponding to the calibration sound pressure level, adjusted as appropriate, for the level range.

A36.3.1.14 *Level non-linearity* means, in decibels, the level difference measured on any level range, at a stated one-third octave nominal midband frequency, minus the corresponding reference level difference, all input and output signals being relative to the same reference quantity.

A36.3.1.15 *Linear operating range* means, in decibels, for a stated level range and frequency, the range of levels of steady sinusoidal electrical signals applied to the input of the entire measurement system, exclusive of the microphone but including the microphone preamplifier and any other signal-conditioning elements that are considered to be part of the microphone system, extending from a lower to an upper boundary, over which the level non-linearity is within specified tolerance limits.

Note: Microphone extension cables as configured in the field need not be included for the linear operating range determination.

A36.3.1.16 *Windscreen insertion loss* means, in decibels, at a stated nominal one-third octave midband frequency, and for a stated sound incidence angle on the inserted microphone, the indicated sound pressure level without the windscreen installed around the microphone minus the sound pressure level with the windscreen installed.

A36.3.2 *Reference environmental conditions.*

A36.3.2.1 The reference environmental conditions for specifying the performance of a measurement system are:

(a) Air temperature 73.4°F (23°C);

(b) Static air pressure 101.325 kPa; and

(c) Relative humidity 50%.

A36.3.3. *General.*

Note: Measurements of aircraft noise that are made using instruments that conform to the specifications of this section will yield one-third octave band sound pressure levels as a function of time. These one-third octave band levels are to be used for the calculation of effective perceived noise level as described in section A36.4.

A36.3.3.1 The measurement system must consist of equipment approved by the FAA and equivalent to the following:

(a) A windscreen (See A36.3.4.);

(b) A microphone system (See A36.3.5):

(c) A recording and reproducing system to store the measured aircraft noise signals for subsequent analysis (see A36.3.6);

(d) A one-third octave band analysis system (see A36.3.7); and

(e) Calibration systems to maintain the acoustical sensitivity of the above systems within specified tolerance limits (see A36.3.8).

A36.3.3.2. For any component of the measurement system that converts an analog signal to digital form, such conversion must be performed so that the levels of any possible aliases or artifacts of the digitization process will be less than the upper boundary of the linear operating range by at least 50 dB at any frequency less than 12.5 kHz. The sampling rate must be at least 28 kHz. An anti-aliasing filter must be included before the digitization process.

A36.3.4 *Windscreen.*

A36.3.4.1 In the absence of wind and for sinusoidal sounds at grazing incidence, the insertion loss caused by the windscreen of a stated type installed around the microphone must not exceed±1.5 dB at nominal one-third octave midband frequencies from 50 Hz to 10 kHz inclusive.

A36.3.5 *Microphone system.*

A36.3.5.1 The microphone system must meet the specifications in sections A36.3.5.2 to A36.3.5.4. Various microphone systems may be approved by the FAA on the basis of demonstrated equivalent overall electroacoustical performance. Where two or more microphone systems of the same type are used, demonstration that at least one system conforms to the specifications in full is sufficient to demonstrate conformance.

Note: An applicant must still calibrate and check each system as required in section A36.3.9.

A36.3.5.2 The microphone must be mounted with the sensing element 4 ft (1.2 m) above the local ground surface and must be oriented for grazing incidence, *i.e.*, with the sensing element substantially in the plane defined by the predicted reference flight path of the aircraft and the measuring station. The microphone mounting arrangement must minimize the interference of the supports with the sound to be measured. Figure A36-1 illustrates sound incidence angles on a microphone.

A36.3.5.3 The free-field sensitivity level of the microphone and preamplifier in the reference direction, at frequencies over at least the range of one-third-octave nominal midband frequencies from 50 Hz to 5 kHz inclusive, must be within±1.0 dB of that at the calibration check frequency, and within±2.0 dB for nominal midband frequencies of 6.3 kHz, 8 kHz and 10 kHz.

A36.3.5.4 For sinusoidal sound waves at each one-third octave nominal midband frequency over the range from 50 Hz to 10 kHz inclusive, the free-field sensitivity levels of the microphone system at sound incidence angles of 30°, 60°, 90°, 120° and 150°, must not differ from the free-field sensitivity level at a sound incidence angle of 0° ("normal incidence") by more than the values shown in Table A36-1.

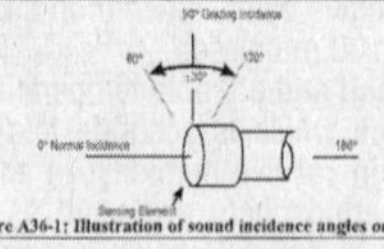

Figure A36-1: Illustration of sound incidence angles on a microphone

Nominal midband frequency kHz	Maximum difference between the free-field sensitivity level of a microphone system at normal incidence and the free-field sensitivity level at specified sound incidence angles dB Sound Incidence angle degrees				
	30	60	90	120	150
0.05 to 1.6	0.5	0.5	1.0	1.0	1.0
2.0	0.5	0.5	1.0	1.0	1.0
2.5	0.5	0.5	1.0	1.5	1.5
3.15	0.5	1.0	1.5	2.0	2.0
4.0	0.5	1.0	2.0	2.5	2.5
5.0	0.5	1.5	2.5	3.0	3.0
6.3	1.0	2.0	3.0	4.0	4.0
8.0	1.5	2.5	4.0	5.5	5.5
10.0	2.0	3.5	5.5	6.5	7.5

Table A36-1 Microphone Directional Response Requirements

The free-field sensitivity level differences at sound incidence angles between any two adjacent sound incidence angles in Table A36-1 must not exceed the tolerance limit for the greater angle.

A36.3.6 *Recording and reproducing systems.*

A36.3.6.1 A recording and reproducing system, such as a digital or analog magnetic tape recorder, a computer-based system or other permanent data storage device, must be used to store sound pressure signals for subsequent analysis. The sound produced by the aircraft must be recorded in such a way that a record of the complete acoustical signal is retained. The recording and reproducing systems must meet the specifications in sections A36.3.6.2 to A36.3.6.9 at the recording speeds and/or data sampling rates used for the noise certification tests. Conformance must be demonstrated for the frequency bandwidths and recording channels selected for the tests.

A36.3.6.2 The recording and reproducing systems must be calibrated as described in section A36.3.9.

(a) For aircraft noise signals for which the high frequency spectral levels decrease rapidly with increasing frequency, appropriate pre-emphasis and complementary de-emphasis networks may be included in the measurement system. If pre-emphasis is included, over the range of nominal one-third octave midband frequencies from 800 Hz to 10 kHz inclusive, the electrical gain provided by the pre-emphasis network must not exceed 20 dB relative to the gain at 800 Hz.

A36.3.6.3 For steady sinusoidal electrical signals applied to the input of the entire measurement system including all parts of the microphone system except the microphone at a selected signal level within 5 dB of that corresponding to the calibration sound pressure level on the reference level range, the time-average signal level indicated by the readout device at any one-third octave nominal midband frequency from 50 Hz to 10 kHz inclusive must be within±1.5 dB of that at the calibration check frequency. The frequency response of a measurement system, which includes components that convert analog signals to digital form, must be within±0.3 dB of the response at 10 kHz over the frequency range from 10 kHz to 11.2 kHz.

Note: Microphone extension cables as configured in the field need not be included for the frequency response determination. This allowance does not eliminate the requirement of including microphone extension cables when performing the pink noise recording in section A36.3.9.5.

A36.3.6.4 For analog tape recordings, the amplitude fluctuations of a 1 kHz sinusoidal signal recorded within 5 dB of the level corresponding to the calibration sound pressure level must not vary by more than±0.5 dB throughout any reel of the type of magnetic tape used. Conformance to this requirement must be demonstrated using a device that has time-averaging properties equivalent to those of the spectrum analyzer.

A36.3.6.5 For all appropriate level ranges and for steady sinusoidal electrical signals applied to the input of the measurement system, including all parts of the microphone system except the microphone, at one-third-octave nominal midband frequencies of 50 Hz, 1 kHz and 10 kHz, and the calibration check frequency, if it is not one of these frequencies, the level non-linearity must not exceed±0.5 dB for a linear operating range of at least 50 dB below the upper boundary of the level range.

Note 1: Level linearity of measurement system components may be tested according to the methods described in IEC 61265 as amended.

Note 2: Microphone extension cables configured in the field need not be included for the level linearity determination.

A36.3.6.6 On the reference level range, the level corresonding to the calibration sound pressure level must be at least 5 dB, but no more than 30 dB less than the upper boundary of the level range.

A36.3.6.7 The linear operating ranges on adjacent level ranges must overlap by at least 50 dB minus the change in attenuation introduced by a change in the level range controls.

Note: It is possible for a measurement system to have level range controls that permit attenuation changes of either 10 dB or 1 dB, for example. With 10 dB steps, the minimum overlap required would be 40 dB, and with 1 dB steps the minimum overlap would be 49 dB.

A36.3.6.8 An overload indicator must be included in the recording and reproducing systems so that an overload indication will occur during an overload condition on any relevant level range.

A36.3.6.9 Attenuators included in the measurement system to permit range changes must operate in known intervals of decibel steps.

A36.3.7 *Analysis systems.*

A36.3.7.1 The analysis system must conform to the specifications in sections A36.3.7.2 to A36.3.7.7 for the frequency bandwidths, channel configurations and gain settings used for analysis.

A36.3.7.2 The output of the analysis system must consist of one-third octave band sound pressure levels as a function of time, obtained by processing the noise signals (preferably recorded) through an analysis system with the following characteristics:

(a) A set of 24 one-third octave band filters, or their equivalent, having nominal midband frequencies from 50 Hz to 10 kHz inclusive;

(b) Response and averaging properties in which, in principle, the output from any one-third octave filter band is squared, averaged and displayed or stored as time-averaged sound pressure levels;

(c) The interval between successive sound pressure level samples must be 500 ms±5 milliseconds(ms) for spectral analysis with or without slow time-weighting, as defined in section A36.3.7.4;

(d) For those analysis systems that do not process the sound pressure signals during the period of time required for readout and/or resetting of the analyzer, the loss of data must not exceed a duration of 5 ms; and

(e) The analysis system must operate in real time from 50 Hz through at least 12 kHz inclusive. This requirement applies to all operating channels of a multi-channel spectral analysis system.

A36.3.7.3 The minimum standard for the one-third octave band analysis system is the class 2 electrical performance requirements of IEC 61260 as amended, over the range of one-third octave nominal midband frequencies from 50 Hz through 10 kHz inclusive.

Note: IEC 61260 specifies procedures for testing of one-third octave band analysis systems for relative attenuation, anti-aliasing filters, real time operation, level linearity, and filter integrated response (effective bandwidth).

A36.3.7.4 When slow time averaging is performed in the analyzer, the response of the one-third octave band analysis system to a sudden onset or interruption of a constant sinusoidal signal at the respective one-third octave nominal midband frequency, must be measured at sampling instants 0.5, 1, 1.5 and 2 seconds(s) after the onset and 0.5 and 1s after interruption. The rising response must be −4±1 dB at 0.5s, −1.75±0.75 dB at 1s, −1±0.5 dB at 1.5s and −0.5±0.5 dB at 2s relative to the steady-state level. The falling response must be such that the sum of the output signal levels, relative to the initial steady-state level, and the corresponding rising response reading is −6.5±1 dB, at both 0.5 and 1s. At subsequent times the sum of the rising and falling responses must be −7.5 dB or less. This equates to an exponential averaging process (slow time-weighting) with a nominal 1s time constant (*i.e.*, 2s averaging time).

A36.3.7.5 When the one-third octave band sound pressure levels are determined from the output of the analyzer without slow time-weighting, slow time-weighting must be simulated in the subsequent processing. Simulated slow time-weighted sound pressure levels can be obtained using a continuous exponential averaging process by the following equation:

$$Ls\,(i,k)=10\log\left[(0.60653)\,10^{0.1\,Ls[i,(k-1)]}+(0.39347)\,10^{0.1\,L(i,k)}\right]$$

where Ls (i,k) is the simulated slow time-weighted sound pressure level and L(i,k) is the as-measured 0.5s time average sound pressure level determined from the output of the analyzer for the k-th instant of time and i-th one-third octave band. For k=1, the slow time-weighted sound pressure Ls [i, (k–1=0)] on the right hand side should be set to 0 dB. An approximation of the continuous exponential averaging is represented by the following equation for a four sample averaging process for k ≥ 4:

$$Ls\ (i,k)=10 \log \left[(0.13)\ 10^{0.1\ L[i,(k-3)]} + (0.21)\ 10^{0.1\ L[i,\ (k-2)]} + (0.27)\ 10^{0.1\ L[i,\ (k-1)]} + (0.39)\ 10^{0.1\ L[i,\ k]}\right]$$

where Ls (i, k) is the simulated slow time-weighted sound pressure level and L (i, k) is the as measured 0.5s time average sound pressure level determined from the output of the analyzer for the k-th instant of time and the i-th one-third octave band.

The sum of the weighting factors is 1.0 in the two equations. Sound pressure levels calculated by means of either equation are valid for the sixth and subsequent 0.5s data samples, or for times greater than 2.5s after initiation of data analysis.

Note: The coefficients in the two equations were calculated for use in determining equivalent slow time-weighted sound pressure levels from samples of 0.5s time average sound pressure levels. The equations do not work with data samples where the averaging time differs from 0.5s.

A36.3.7.6 The instant in time by which a slow time-weighted sound pressure level is characterized must be 0.75s earlier than the actual readout time.

Note: The definition of this instant in time is needed to correlate the recorded noise with the aircraft position when the noise was emitted and takes into account the averaging period of the slow time-weighting. For each 0.5 second data record this instant in time may also be identified as 1.25 seconds after the start of the associated 2 second averaging period.

A36.3.7.7 The resolution of the sound pressure levels, both displayed and stored, must be 0.1 dB or finer.

A36.3.8 *Calibration systems.*

A36.3.8.1 The acoustical sensitivity of the measurement system must be determined using a sound calibrator generating a known sound pressure level at a known frequency. The minimum standard for the sound calibrator is the class 1L requirements of IEC 60942 as amended.

A36.3.9 *Calibration and checking of system.*

A36.3.9.1 Calibration and checking of the measurement system and its constituent components must be carried out to the satisfaction of the FAA by the methods specified in sections A36.3.9.2 through A36.3.9.10. The calibration adjustments, including those for environmental effects on sound calibrator output level, must be reported to the FAA and applied to the measured one-third-octave sound pressure levels determined from the output of the analyzer. Data collected during an overload indication are invalid and may not be used. If the overload condition occurred during recording, the associated test data are invalid, whereas if the overload occurred during analysis, the analysis must be repeated with reduced sensitivity to eliminate the overload.

A36.3.9.2 The free-field frequency response of the microphone system may be determined by use of an electrostatic actuator in combination with manufacturer's data or by tests in an anechoic free-field facility. The correction for frequency response must be determined within 90 days of each test series. The correction for non-uniform frequency response of the microphone system must be reported to the FAA and applied to the measured one-third octave band sound pressure levels determined from the output of the analyzer.

A36.3.9.3 When the angles of incidence of sound emitted from the aircraft are within±30° of grazing incidence at the microphone (see Figure A36-1), a single set of free-field corrections based on grazing incidence is considered sufficient for correction of directional response effects. For other cases, the angle of incidence for each 0.5 second sample must be determined and applied for the correction of incidence effects.

A36.3.9.4 For analog magnetic tape recorders, each reel of magnetic tape must carry at least 30 seconds of pink random or pseudo-random noise at its beginning and end. Data obtained from analog tape-recorded signals will be accepted as reliable only if level differences in the 10 kHz one-third-octave-band are not more than 0.75 dB for the signals recorded at the beginning and end.

A36.3.9.5 The frequency response of the entire measurement system while deployed in the field during the test series, exclusive of the microphone, must be determined at a level within 5 dB of the level corresponding to the calibration sound pressure level on the level range used during the tests for each one-third octave nominal midband frequency from 50 Hz to 10 kHz inclusive, utilizing pink random or pseudo-random noise. Within six months of each test series the output of the noise generator must be determined by a method traceable to the U.S. National Institute of Standards and Technology or to an equivalent national standards laboratory as determined by the FAA. Changes in the relative output from the previous calibration at each one-third octave band may not exceed 0.2 dB. The correction for frequency response must be reported to the FAA and applied to the measured one-third octave sound pressure levels determined from the output of the analyzer.

A36.3.9.6 The performance of switched attenuators in the equipment used during noise certification measurements and calibration must be checked within six months of each test series to ensure that the maximum error does not exceed 0.1 dB.

A36.3.9.7 The sound pressure level produced in the cavity of the coupler of the sound calibrator must be calculated for the test environmental conditions using the manufacturer's supplied information on the influence of atmospheric air pressure and temperature. This sound pressure level is used to establish the acoustical sensitivity of the measurement system. Within six months of each test series the output of the sound calibrator must be determined by a method traceable to the U.S. National Institute of Standards and Technology or to an equivalent national standards laboratory as determined by the FAA. Changes in output from the previous calibration must not exceed 0.2 dB.

A36.3.9.8 Sufficient sound pressure level calibrations must be made during each test day to ensure that the acoustical sensitivity of the measurement system is known at the prevailing environmental conditions corresponding with each test series. The difference between the acoustical sensitivity levels recorded immediately before and immediately after each test series on each day may not exceed 0.5 dB. The 0.5 dB limit applies after any atmospheric pressure corrections have been determined for the calibrator output level. The arithmetic mean of the before and after measurements must be used to represent the acoustical sensitivity level of the measurement system for that test series. The calibration corrections must be reported to the FAA and applied to the measured one-third octave band sound pressure levels determined from the output of the analyzer.

A36.3.9.9 Each recording medium, such as a reel, cartridge, cassette, or diskette, must carry a sound pressure level calibration of at least 10 seconds duration at its beginning and end.

A36.3.9.10 The free-field insertion loss of the windscreen for each one-third octave nominal midband frequency from 50 Hz to 10 kHz inclusive must be determined with sinusoidal sound signals at the incidence angles determined to be applicable for correction of directional response effects per section A36.3.9.3. The interval between angles tested must not exceed 30 degrees. For a windscreen that is undamaged and uncontaminated, the insertion loss may be taken from manufacturer's data. Alternatively, within six months of each test series the insertion loss of the windscreen may be determined by a method traceable to the U.S. National Institute of Standards and Technology or an equivalent national standards laboratory as determined by the FAA. Changes in the insertion loss from the previous calibration at each one-third-octave frequency band must not exceed 0.4 dB. The correction for the free-field insertion loss of the windscreen must be reported to the FAA and applied to the measured one-third octave sound pressure levels determined from the output of the analyzer.

A36.3.10 Adjustments for ambient noise.

A36.3.10.1 Ambient noise, including both an acoustical background and electrical noise of the measurement system, must be recorded for at least 10 seconds at the measurement points with the system gain set at the levels used for the aircraft noise measurements. Ambient noise must be representative of the acoustical background that exists during the flyover test run. The recorded aircraft noise data is acceptable only if the ambient noise levels, when analyzed in the same way, and quoted in PNL (see A36.4.1.3 (a)), are at least 20 dB below the maximum PNL of the aircraft.

A36.3.10.2 Aircraft sound pressure levels within the 10 dB-down points (see A36.4.5.1) must exceed the mean ambient noise levels determined in section A36.3.10.1 by at least 3 dB in each one-third octave band, or must be adjusted using a method approved by the FAA; one method is described in the current advisory circular for this part.

Section A36.4 ***Calculation of Effective Perceived Noise Level From Measured Data***

A36.4.1 *General.*

A36.4.1.1 The basic element for noise certification criteria is the noise evaluation measure known as effective perceived noise level, EPNL, in units of EPNdB, which is a single number evaluator of the subjective effects of airplane noise on human beings. EPNL consists of instantaneous perceived noise level, PNL, corrected for spectral irregularities, and for duration. The spectral irregularity correction, called "tone correction factor", is made at each time increment for only the maximum tone.

A36.4.1.2 Three basic physical properties of sound pressure must be measured: level, frequency distribution, and time variation. To determine EPNL, the instantaneous sound pressure level in each of the 24 one-third octave bands is required for each 0.5 second increment of time during the airplane noise measurement.

A36.4.1.3 The calculation procedure that uses physical measurements of noise to derive the EPNL evaluation measure of subjective response consists of the following five steps:

(a) The 24 one-third octave bands of sound pressure level are converted to perceived noisiness (noy) using the method described in section A36.4.2.1 (a). The noy values are combined and then converted to instantaneous perceived noise levels, PNL(k).

(b) A tone correction factor C(k) is calculated for each spectrum to account for the subjective response to the presence of spectral irregularities.

(c) The tone correction factor is added to the perceived noise level to obtain tone-corrected perceived noise levels PNLT(k), at each one-half second increment:

PNLT(k)=PNL(k) + C(k)

The instantaneous values of tone-corrected perceived noise level are derived and the maximum value, PNLTM, is determined.

(d) A duration correction factor, D, is computed by integration under the curve of tone-corrected perceived noise level versus time.

(e) Effective perceived noise level, EPNL, is determined by the algebraic sum of the maximum tone-corrected perceived noise level and the duration correction factor:

EPNL=PNLTM+D

A36.4.2 *Perceived noise level.*

A36.4.2.1 Instantaneous perceived noise levels, PNL(k), must be calculated from instantaneous one-third octave band sound pressure levels, SPL(i, k) as follows:

(a) Step 1: For each one-third octave band from 50 through 10,000 Hz, convert SPL(i, k) to perceived noisiness n(i, k), by using the mathematical formulation of the noy table given in section A36.4.7.

(b) Step 2: Combine the perceived noisiness values, n(i, k), determined in step 1 by using the following formula:

$$N(k) = n(k) + 0.15\left\{\left[\sum_{I=1}^{24} n(i,k)\right] - n(k)\right\}$$

$$= 0.85n(k) + 0.15\sum_{I=1}^{24} n(i,k)$$

where n(k) is the largest of the 24 values of n(i, k) and N(k) is the total perceived noisiness.

(c) Step 3: Convert the total perceived noisiness, N(k), determined in Step 2 into perceived noise level, PNL(k), using the following formula:

$$PNL(k) = 40.0 + \frac{10}{\log 2}\log N(k)$$

Note: PNL(k) is plotted in the current advisory circular for this part.

A36.4.3 *Correction for spectral irregularities.*

A36.4.3.1 Noise having pronounced spectral irregularities (for example, the maximum discrete frequency components or tones) must be adjusted by the correction factor C(k) calculated as follows:

(a) Step 1: After applying the corrections specified under section A36.3.9, start with the sound pressure level in the 80 Hz one-third octave band (band number 3), calculate the changes in sound pressure level (or "slopes") in the remainder of the one-third octave bands as follows:

s(3, k)=no value
s(4, k)=SPL(4, k)–SPL(3, k)
•
•
s(i, k)=SPL(i, k)–SPL(i–1, k)
•
•
s(24, k)=SPL(24, k)–SPL(23, k)

(b) Step 2: Encircle the value of the slope, s(i, k), where the absolute value of the change in slope is greater than five; that is where:

|Δs(i, k)|=|s(i, k)–s(i–1, k)|>5

(c) Step 3:

(1) If the encircled value of the slope s(i, k) is positive and algebraically greater than the slope s(i–1, k) encircle SPL(i, k).

(2) If the encircled value of the slope s(i, k) is zero or negative and the slope s(i–1, k) is positive, encircle SPL(i–1, k).

(3) For all other cases, no sound pressure level value is to be encircled.

(d) Step 4: Compute new adjusted sound pressure levels SPL′(i, k) as follows:

(1) For non-encircled sound pressure levels, set the new sound pressure levels equal to the original sound pressure levels, SPL′(i, k)=SPL(i, k).

(2) For encircled sound pressure levels in bands 1 through 23 inclusive, set the new sound pressure level equal to the arithmetic average of the preceding and following sound pressure levels as shown below:

SPL′(i, k)= ½ [SPL(i–1, k)+SPL(i+1, k)]

(3) If the sound pressure level in the highest frequency band (i=24) is encircled, set the new sound pressure level in that band equal to:

SPL′(24, k)=SPL(23, k)+ *s* (23, k)

(e) Step 5: Recompute new slope s′(i, k), including one for an imaginary 25th band, as follows:

s′(3, k)= s′(4, k)
s′(4, k)=SPL′(4, k)–SPL′(3, k)
•
•
s′(i, k)=SPL′(i, k)–SPL′(i–1, k)
•
•
s′(24, k)=SPL′(24, k)–SPL′(23, k)
s′(25, k)= s′(24, k)

(f) Step 6: For i, from 3 through 23, compute the arithmetic average of the three adjacent slopes as follows:

s̄(i, k)= ⅓ [s′(i, k)+ s′(i+1, k)+ s′(i+2, k)]

(g) Step 7: Compute final one-third octave-band sound pressure levels, SPL′ (i,k), by beginning with band number 3 and proceeding to band number 24 as follows:

SPL′(3, k)=SPL(3,k)
SPL′(4, k)=SPL′(3,k)+ s̄(3, k)
•
•
SPL′(i, k)=SPL′(i–1, k)+s̄(i–1, k)
•
•
SPL′(24, k)=SPL′(23, k)+s̄(23, k)

(h) Setp 8: Calculate the differences, F(i, k), between the original sound pressure level and the final background sound pressure level as follows:

F(i, k)=SPL(i, k)-SPL′(i, k)

and note only values equal to or greater than 1.5.

(i) Step 9: For each of the relevant one-third octave bands (3 through 24), determine tone correction factors from the sound pressure level differences F(i, k) and Table A36-2.

(j) Step 10: Designate the largest of the tone correction factors, determined in Step 9, as C(k). (An example of the tone correction procedure is given in the current advisory circular for this part). Tone-corrected perceived noise levels PNLT(k) must be determined by adding the C(k) values to corresponding PNL(k) values, that is:

PNLT(k)=PNL(k)+ C(k)

For any i-th one-third octave band, at any k-th increment of time, for which the tone correction factor is suspected to result from something other than (or in addition to) an actual tone (or any spectral irregularity other than airplane noise), an additional analysis may be made using a filter with a bandwidth narrower than one-third of an octave. If the narrow band analysis corroborates these suspicions, then a revised value for the background sound pressure level SPL′(i,k), may be determined from the narrow band analysis and used to compute a revised tone correction factor for that particular one-third octave band. Other methods of rejecting spurious tone corrections may be approved.

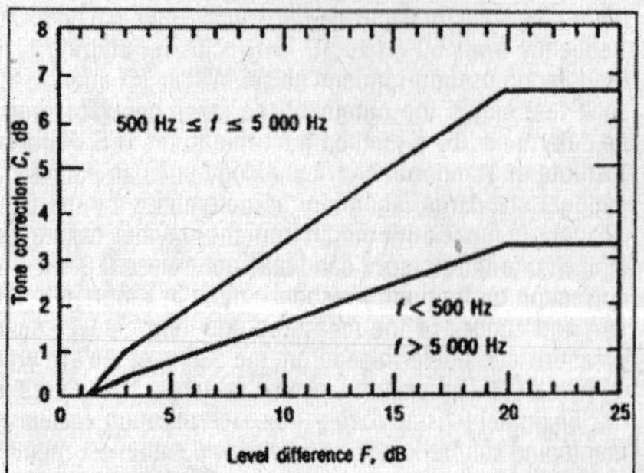

Frequency f, Hz	Level difference F, dB	Tone correction C, dB
50 ≤ f < 500	1½* ≤ F < 3	F/3 — ½
	3 ≤ F < 20	F/6
	20 ≤ F	3⅓
500 ≤ f ≤ 5 000	1½* ≤ F < 3	2 F/3 — 1
	3 ≤ F < 20	F/3
	20 ≤ F	6⅔
5 000 < f ≤ 10 000	1½* ≤ F < 3	F/3 — ½
	3 ≤ F < 20	F/6
	20 ≤ F	3⅓

* See Step 8, 4.3.1.

Table A36-2. Tone correction factor

A36.4.3.2 The tone correction procedure will underestimate EPNL if an important tone is of a frequency such that it is recorded in two adjacent one-third octave bands. An applicant must demonstrate that either:

(a) No important tones are recorded in two adjacent one-third octave bands; or

(b) That if an important tone has occurred, the tone correction has been adjusted to the value it would have had if the tone had been recorded fully in a single one-third octave band.

A36.4.4 Maximum tone-corrected perceived noise level

A36.4.4.1 The maximum tone-corrected perceived noise level, PNLTM, must be the maximum calculated value of the tone-corrected perceived noise level PNLT(k). It must be calculated using the procedure of section A36.4.3. To obtain a satisfactory noise time history, measurements must be made at 0.5 second time intervals.

Note 1: Figure A36-2 is an example of a flyover noise time history where the maximum value is clearly indicated.

Note 2: In the absence of a tone correction factor, PNLTM would equal PNLM.

A36.4.4.2 After the value of PNLTM is obtained, the frequency band for the largest tone correction factor is identified for the two preceding and two succeeding 500 ms data samples. This is performed in order to identity

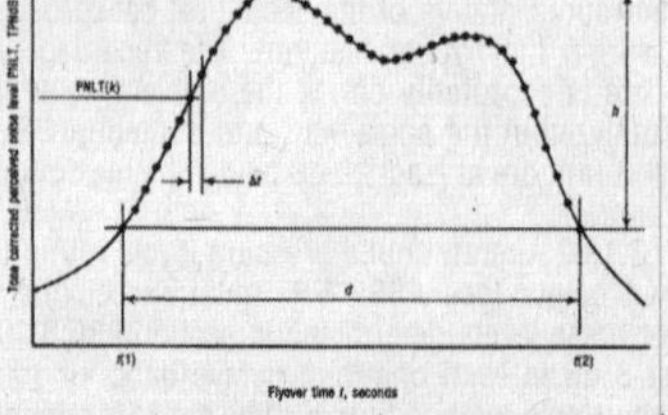

Figure A36-2. Example of perceived noise level corrected for tones as a function of aircraft flyover time

the possibility of tone suppression at PNLTM by one-third octave band sharing of that tone. If the value of the tone correction factor C(k) for PNLTM is less than the average value of C(k) for the five consecutive time intervals, the average value of C(k) must be used to compute a new value for PNLTM.

A36.4.5 *Duration correction.*

A36.4.5.1 The duration correction factor D determined by the integration technique is defined by the expression:

$$D = 10\log\left[\left(\frac{1}{T}\right)\int_{t(1)}^{t(2)} \text{antilog}\frac{PNLT}{10}dt\right] - PNLTM$$

where T is a normalizing time constant, PNLTM is the maximum value of PNLT, t(1) is the first point of time after which PNLT becomes greater than PNLTM-10, and t(2) is the point of time after which PNLT remains constantly less than PNLTM-10.

A36.4.5.2 Since PNLT is calculated from measured values of sound pressure level (SPL), there is no obvious equation for PNLT as a function of time. Consequently, the equation is to be rewritten with a summation sign instead of an integral sign as follows:

$$D = 10\log\left[\left(\frac{1}{T}\right)\int_{k=0}^{d/\Delta t} \Delta t.\text{antilog}\frac{PNLT}{10}dt\right] - PNLTM$$

where Δt is the length of the equal increments of time for which PNLT(k) is calculated and d is the time interval to the nearest 0.5s during which PNLT(k) remains greater or equal to PNLTM-10.

A36.4.5.3 To obtain a satisfactory history of the perceived noise level use one of the following:

(a) Half-Second time intervals for Δt; or

(b) A shorter time interval with approved limits and constants.

A36.4.5.4 The following values for T and Δt must be used in calculating D in the equation given in section A36.4.5.2:

T=10 s, and

Δt=0.5s (or the approved sampling time interval).

Using these values, the equation for D becomes:

$$D = 10\log\left[\sum_{k=0}^{2d} \text{antilog}\frac{PNLT(k)}{10}\right] - PNLTM - 13$$

where d is the duration time defined by the points corresponding to the values PNLTM-10.

A36.4.5.5 If in using the procedures given in section A36.4.5.2, the limits of PNLTM-10 fall between the calculated PNLT(k) values (the usual case), the PNLT(k) values defining the limits of the duration interval must be chosen from the PNLT(k) values closest to PNLTM-10. For those cases with more than one peak value of PNLT(k), the applicable limits must be chosen to yield the largest possible value for the duration time.

A36.4.6 *Effective perceived noise level.*

The total subjective effect of an airplane noise event, designated effective perceived noise level, EPNL, is equal to the algebraic sum of the maximum value of the tone-corrected perceived noise level, PNLTM, and the duration correction D. That is:

EPNL=PNLTM+D

where PNLTM and D are calculated using the procedures given in sections A36.4.2, A36.4.3, A36.4.4. and A36.4.5.

A36.4.7 *Mathematical formulation of noy tables.*

A36.4.7.1 The relationship between sound pressure level (SPL) and the logarithm of perceived noisiness is illustrated in Figure A36-3 and Table A36-3.

A36.4.7.2 The bases of the mathematical formulation are:

(a) The slopes (M(b), M(c), M(d) and M(e)) of the straight lines;

(b) The intercepts (SPL(b) and SPL(c)) of the lines on the SPL axis; and

(c) The coordinates of the discontinuities, SPL(a) and log n(a); SPL(d) and log n=–1.0; and SPL(e) and log n=log (0.3).

A36.4.7.3 Calculate noy values using the following equations:

(a)
SPL ≥ SPL (a)
n=antilog {(c)[SPL–SPL(c)]}
(b)

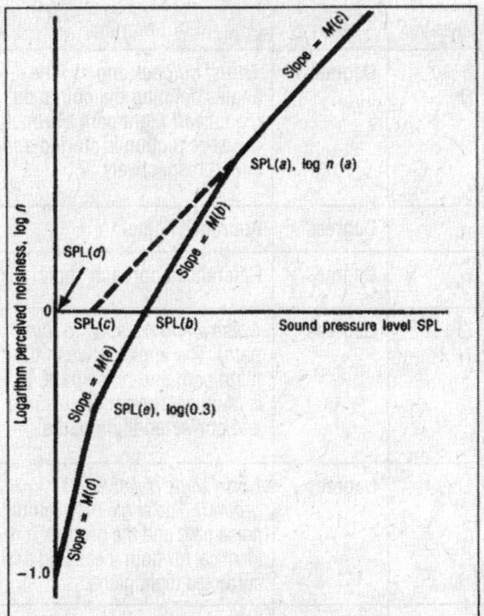

Figure A36-3. Perceived noisiness as a function of sound pressure level

BAND (i)	f HZ	SPL (a)	SPL (b)	SPL (c)	SPL (d)	SPL (e)	M(b)	M(c)	M(d)	M(e)
1	50	91.0	64	52	49	55	0.043478	0.030103	0.079520	0.058098
2	63	85.9	60	51	44	51	0.040570		0.068160	"
3	80	87.3	56	49	39	46	0.036831		"	0.052288
4	100	79.9	53	47	34	42	"		0.059640	0.047534
5	125	79.8	51	46	30	39	0.035336		0.053013	0.043573
6	160	76.0	48	45	27	36	0.033333			"
7	200	74.0	46	43	24	33	"			0.040221
8	250	74.9	44	42	21	30	0.032051			0.037349
9	315	94.6	42	41	18	27	0.030675	0.030103		0.034859
10	400	∞	40	40	16	25	0.030103			
11	500		40	40	16	25				
12	630		40	40	16	25				
13	800		40	40	16	25				
14	1 000		40	40	16	25			0.053013	
15	1 250		38	38	15	23	0.030103		0.059640	0.034859
16	1 600		34	34	12	21	0.029960	NOT APPLICABLE	0.053013	0.040221
17	2 000		32	32	9	18			"	0.037349
18	2 500		30	30	5	15			0.047712	0.034859
19	3 150		29	29	4	14			"	
20	4 000		29	29	5	14			0.053013	
21	5 000		30	30	6	15			"	0.034859
22	6 300	∞	31	31	10	17	0.029960		0.068160	0.037349
23	8 000	44.3	37	34	17	23	0.042285	0.029960	0.079520	"
24	10 000	50.7	41	37	21	29	"	"	0.059640	0.043573

Table A36-3. Constants for mathematically formulated noy values

SPL(b) ≤ SPL < SPL(a)
n=antilog {M(b)[SPL–SPL(b)]}
(c)
SPL(e) ≤ SPL < SPL(b)
n=0.3 antilog {M(e)[SPL–SPL(e)]}
(d)
SPL(d) ≤ SPL < SPL(e)
n=0.1 antilog {M(d)[SPL–SPL(d)]}

A36.4.7.4 Table A36-3 lists the values of the constants necessary to calculate perceived noisiness as a function of sound pressure level.

Section A36.5 Reporting of Data to the FAA

A36.5.1 *General.*

A36.5.1.1 Data representing physical measurements and data used to make corrections to physical measurements must be recorded in an approved permanent form and appended to the record.

A36.5.1.2 All corrections must be reported to and approved by the FAA, including corrections to measurements for equipment response deviations.

A36.5.1.3 Applicants may be required to submit estimates of the individual errors inherent in each of the operations employed in obtaining the final data.

A36.5.2 *Data reporting.*

An applicant is required to submit a noise certification compliance report that includes the following.

A36.5.2.1 The applicant must present measured and corrected sound pressure levels in one-third octave band levels that are obtained with equipment conforming to the standards described in section A36.3 of this appendix.

A36.5.2.2 The applicant must report the make and model of equipment used for measurement and analysis of all acoustic performance and meteorological data.

A36.5.2.3 The applicant must report the following atmospheric environmental data, as measured immediately before, after, or during each test at the observation points prescribed in section A36.2 of this appendix.

(a) Air temperature and relative humidity;

(b) Maximum, minimum and average wind velocities; and

(c) Atmospheric pressure.

A36.5.2.4 The applicant must report conditions of local topography, ground cover, and events that might interfere with sound recordings.

A36.5.2.5 The applicant must report the following:

(a) Type, model and serial numbers (if any) of airplane, engine(s), or propeller(s) (as applicable);

(b) Gross dimensions of airplane and location of engines;

(c) Airplane gross weight for each test run and center of gravity range for each series of test runs;

(d) Airplane configuration such as flap, airbrakes and landing gear positions for each test run;

(e) Whether auxiliary power units (APU), when fitted, are operating for each test run;

(f) Status of pneumatic engine bleeds and engine power take-offs for each test run;

(g) Indicated airspeed in knots or kilometers per hour for each test run;

(h) Engine performance data:

(1) For jet airplanes: engine performance in terms of net thrust, engine pressure ratios, jet exhaust temperatures and fan or compressor shaft rotational speeds as determined from airplane instruments and manufacturer's data for each test run;

(2) For propeller-driven airplanes: engine performance in terms of brake horsepower and residual thrust; or equivalent shaft horsepower; or engine torque and propeller rotational speed; as determined from airplane instruments and manufacturer's data for each test run;

(i) Airplane flight path and ground speed during each test run; and

(j) The applicant must report whether the airplane has any modifications or non-standard equipment likely to affect the noise characteristics of the airplane. The FAA must approve any such modifications or non-standard equipment.

A36.5.3 *Reporting of noise certification reference conditions.*

A36.5.3.1 Airplane position and performance data and the noise measurements must be corrected to the noise certification reference conditions specified in the relevant sections of appendix B of this part. The applicant must report these conditions, including reference parameters, procedures and configurations.

A36.5.4 *Validity of results.*

A36.5.4.1 Three average reference EPNL values and their 90 percent confidence limits must be produced from the test results and reported, each such value being the arithmetical average of the adjusted acoustical measurements for all valid test runs at each measurement point (flyover, lateral, or approach). If more than one acoustic measurement system is used at any single measurement location, the resulting data for each test run must be averaged as a single measurement. The calculation must be performed by:

(a) Computing the arithmetic average for each flight phase using the values from each microphone point; and

(b) Computing the overall arithmetic average for each reference condition (flyover, lateral or approach) using the values in paragraph (a) of this section and the related 90 percent confidence limits.

A36.5.4.2 For each of the three certification measuring points, the minimum sample size is six. The sample size must be large enough to establish statistically for each of the three average noise certification levels a 90 percent confidence limit not exceeding±1.5 EPNdB. No test result may be omitted from the averaging process unless approved by the FAA.

Note: Permitted methods for calculating the 90 percent confidence interval are shown in the current advisory circular for this part.

A36.5.4.3 The average EPNL figures obtained by the process described in section A36.5.4.1 must be those by which the noise performance of the airplane is assessed against the noise certification criteria.

Section A36.6 *Nomenclature: Symbols and Units*

Symbol	Unit	Meaning
antilog		Antilogarithm to the base 10.
C(k)	dB	*Tone correction factor.* The factor to be added to PNL(k) to account for the presence of spectral irregularities such as tones at the k-th increment of time.
d	s	*Duration time.* The time interval between the limits of t(1) and t(2) to the nearest 0.5 second.
D	dB	*Duration correction.* The factor to be added to PNLTM to account for the duration of the noise.
EPNL	EPNdB	*Effective perceived noise level.* The value of PNL adjusted for both spectral irregularities and duration of the noise. (The unit EPNdB is used instead of the unit dB).
$EPNL_r$	EPNdB	Effective perceived noise level adjusted for reference conditions.
f(i)	Hz	*Frequency.* The geometrical mean frequency for the i-th one-third octave band.
F (i, k)	dB	*Delta-dB.* The difference between the original sound pressure level and the final background sound pressure level in the i-th one-third octave band at the k-th interval of time. In this case, background sound pressure level means the broadband noise level that would be present in the one-third octave band in the absence of the tone.
h	dB	*dB-down.* The value to be subtracted from PNLTM that defines the duration of the noise.
H	Percent	*Relative humidity.* The ambient atmospheric relative humidity.
i		*Frequency band index.* The numerical indicator that denotes any one of the 24 one-third octave bands with geometrical mean frequencies from 50 to 10,000 Hz.
k		*Time increment index.* The numerical indicator that denotes the number of equal time increments that have elapsed from a reference zero.
Log		Logarithm to the base 10.
log n(a)		*Noy discontinuity coordinate.* The log n value of the intersection point of the straight lines representing the variation of SPL with log n.
M(b), M(c), etc		*Noy inverse slope.* The reciprocals of the slopes of straight lines representing the variation of SPL with log n.
n	noy	The perceived noisiness at any instant of time that occurs in a specified frequency range.
n(i,k)	noy	The perceived noisiness at the k-th instant of time that occurs in the i-th one-third octave band.

Section A36.7 *Sound Attenuation in Air*

A36.7.1 The atmospheric attenuation of sound must be determined in accordance with the procedure presented in section A36.7.2.

A36.7.2 The relationship between sound attenuation, frequency, temperature, and humidity is expressed by the following equations.

PART 36

FAR

Symbol	Unit	Meaning
n(k)	noy	*Maximum perceived noisiness.* The maximum value of all of the 24 values of n(i) that occurs at the k-th instant of time.
N(k)	noy	*Total perceived noisiness.* The total perceived noisiness at the k-th instant of time calculated from the 24-instantaneous values of n (i, k).
p(b), p(c), etc		*Noy slope.* The slopes of straight lines representing the variation of SPL with log n.
PNL	PNdB	The perceived noise level at any instant of time. (The unit PNdB is used instead of the unit dB).
PNL(k)	PNdB	The perceived noise level calculated from the 24 values of SPL (i, k), at the k-th increment of time. (The unit PNdB is used instead of the unit dB).
PNLM	PNdB	*Maximum perceived noise level.* The maximum value of PNL(k). (The unit PNdB is used instead of the unit dB).
PNLT	TPNdB	*Tone-corrected perceived noise level.* The value of PNL adjusted for the spectral irregularities that occur at any instant of time. (The unit TPNdB is used instead of the unit dB).
PNLT(k)	TPNdB	The tone-corrected perceived noise level that occurs at the k-th increment of time. PNLT(k) is obtained by adjusting the value of PNL(k) for the spectral irregularities that occur at the k-th increment of time. (The unit TPNdB is used instead of the unit dB).
PNLTM	TPNdB	*Maximum tone-corrected perceived noise level.* The maximum value of PNLT(k). (The unit TPNdB is used instead of the unit dB).
$PNLT_r$	TPNdB	Tone-corrected perceived noise level adjusted for reference conditions.
s (i, k)	dB	*Slope of sound pressure level.* The change in level between adjacent one-third octave band sound pressure levels at the i-th band for the k-th instant of time.
Δs (i, k)	dB	Change in slope of sound pressure level.
s' (i, k)	dB	Adjusted slope of sound pressure level. The change in level between adjacent adjusted one-third octave band sound pressure levels at the i-th band for the k-th instant of time.
$\bar{s}$ (i, k)	dB	Average slope of sound pressure level.
SPL	dB re 20 μPa	*Sound pressure level.* The sound pressure level that occurs in a specified frequency range at any instant of time.
SPL(a)	dB re 20 μPa	*Noy discontinuity coordinate.* The SPL value of the intersection point of the straight lines representing the variation of SPL with log n.

Symbol	Unit	Meaning
SPL(b) SPL (c)	dB re 20 μPa	*Noy intercept.* The intercepts on the SPL-axis of the straight lines representing the variation of SPL with log n.
SPL (i, k)	dB re 20 μPa	The sound pressure level at the k-th instant of time that occurs in the i-th one-third octave band.
SPL' (i, k)	dB re 20 μPa	*Adjusted sound pressure level.* The first approximation to background sound pressure level in the i-th one-third octave band for the k-th instant of time.
SPL(i)	dB re 20 μPa	*Maximum sound pressure level.* The sound pressure level that occurs in the i-th one-third octave band of the spectrum for PNLTM.
$SPL(i)_r$	dB re 20 μPa	*Corrected maximum sound pressure level.* The sound pressure level that occurs in the i-th one-third octave band of the spectrum for PNLTM corrected for atmospheric sound absorption.
SPL' (i, k)	dB re 20 μPa	*Final background sound pressure level.* The second and final approximation to background sound pressure level in the i-th one-third octave band for the k-th instant of time.
t	s	*Elapsed time.* The length of time measured from a reference zero.
t(1), t(2)	s	*Time limit.* The beginning and end, respectively, of the noise time history defined by h.
Δt	s	*Time increment.* The equal increments of time for which PNL(k) and PNLT(k) are calculated.
T	s	*Normalizing time constant.* The length of time used as a reference in the integration method for computing duration corrections, where T=10s.
t(°F) (°C)	°F,°C	*Temperature.* The ambient air temperature.
α(i)	dB/1000ft db/100m	*Test atmospheric absorption.* The atmospheric attenuation of sound that occurs in the i-th one-third octave band at the measured air temperature and relative humidity.
$\alpha(i)_0$	dB/1000ft db/100m	*Reference atmospheric absorption.* The atmospheric attenuation of sound that occurs in the i-th one-third octave band at a reference air temperature and relative humidity.
A_1	Degrees	First constant climb angle (Gear up, speed of at least V_2+10 kt (V_2+19 km/h), takeoff thrust).
A_2	Degrees	Second constant climb angle (Gear up, speed of at least V_2+10 kt (V_2+19 km/h), after cut-back).

Symbol	Unit	Meaning
δ ⊠	Degrees	*Thrust cutback angles.* The angles defining the points on the takeoff flight path at which thrust reduction is started and ended respectively.
η	Degrees	Approach angle.
η_r	Degrees	Reference approach angle.
Θ	Degrees	*Noise angle (relative to flight path).* The angle between the flight path and noise path. It is identical for both measured and corrected flight paths.
ψ	Degrees	*Noise angle (relative to ground).* The angle between the noise path and the ground. It is identical for both measured and corrected flight paths.
μ		Engine noise emission parameter.
μ_r		Reference engine noise emission parameter.
Δ_1	EPNdB	*PNLT correction.* The correction to be added to the EPNL calculated from measured data to account for noise level changes due to differences in atmospheric absorption and noise path length between reference and test conditions.
Δ_2	EPNdB	*Adjustment to duration correction.* The adjustment to be made to the EPNL calculated from measured data to account for noise level changes due to the noise duration between reference and test conditions.
Δ_3	EPNdB	*Source noise adjustment.* The adjustment to be made to the EPNL calculated from measured data to account for noise level changes due to differences between reference and test engine operating conditions.

A36.7.2(a) For calculations using the English System of Units:

$$\alpha(i)=10^{\left[2.05\log(f_o/1000)+633\times10^{-4}-1.45325\right]}$$
$$+\eta(\delta)\times10^{[\log(f_o)+8.42994\times10^{-3}-2.4215}$$

and

$$\delta=\sqrt{\frac{1010}{f(0)}}10^{\left(\log H-197274664+2.288074\times10^{-2}g\right)}$$
$$\times10^{\left(-9.589\times10^{-5}g^2+3.0\times10^{-8}g^3\right)}$$

where
η(δ) is listed in Table A36-4 and f0 in Table A36-5;
α(i) is the attenuation coefficient in dB/1000 ft;
θ is the temperature in°F; and
H is the relative humidity, expressed as a percentage.

A36.7.2(b) For calculations using the International System of Units (SI):

$$\alpha(i)=10^{\left[2.05\log(f_o/1000)+1.1394\times10^{-3}g-1.916984\right]}$$
$$+\eta(\delta)\times10^{\left[\log(f_0)+8.42994\times10^{-3}g-2.755624\right]}$$

and

$$\delta=\sqrt{\frac{1010}{f_0}10^{\left(\log H-1328934+3.179768\times10^{-2}g\right)}\times10^{\left(-2.173716\times10^{-4}g^2+1.7496\times10^{-6}g^3\right)}}$$

where

η(δ) is listed in Table A36-4 and f0 in Table A36-5;

α(i) is the attenuation coefficient in dB/100 m;

θ is the temperature in°C; and

H is the relative humidity, expressed as a percentage.

A36.7.3 The values listed in table A36-4 are to be used when calculating the equations listed in section A36.7.2. A term of quadratic interpolation is to be used where necessary.

Section A36.8 *[Reserved]*

Table A36-4. Values of η(δ)

δ	η(δ)	δ	η(δ)
0.00	0.000	2.50	0.450
0.25	0.315	2.80	0.400
0.50	0.700	3.00	0.370
0.60	0.840	3.30	0.330
0.70	0.930	3.60	0.300
0.80	0.975	4.15	0.260
0.90	0.996	4.15	0.245
1.00	1.000	4.80	0.230
1.10	0.970	5.25	0.220
1.20	0.900	5.70	0.210
1.30	0.840	6.05	0.205
1.50	0.750	6.50	0.200
1.70	0.670	7.00	0.200
2.00	0.570	10.00	0.200
2.30	0.495		

Table A36-5. Values of f_0

one-third octave center frequency	f_0 (Hz)	one third octave center frequency	f_0 (Hz)
50	50	800	800
63	63	1000	1000
80	80	1250	1250
100	100	1600	1600
125	125	2000	2000
160	160	2500	2500
200	200	3150	3150
250	250	4000	4000
315	315	5000	4500
400	400	6300	5600
500	500	8000	7100
630	630	10000	9000

Section A36.9 *Adjustment of Airplane Flight Test Results.*

A36.9.1 When certification test conditions are not identical to reference conditions, appropriate adjustments must be made to the measured noise data using the methods described in this section.

A36.9.1.1 Adjustments to the measured noise values must be made using one of the methods described in sections A36.9.3 and A36.9.4 for differences in the following:

(a) Attenuation of the noise along its path as affected by "inverse square" and atmospheric attenuation

(b) Duration of the noise as affected by the distance and the speed of the airplane relative to the measuring point

(c) Source noise emitted by the engine as affected by the differences between test and reference engine operating conditions

(d) Airplane/engine source noise as affected by differences between test and reference airspeeds. In addition to the effect on duration, the effects of airspeed on component noise sources must be accounted for as follows: for conventional airplane configurations, when differences between test and reference airspeeds exceed 15 knots (28 km/h) true airspeed, test data and/or analysis approved by the FAA must be used to quantify the effects of the airspeed adjustment on resulting certification noise levels.

A36.9.1.2 The "integrated" method of adjustment, described in section A36.9.4, must be used on takeoff or approach under the following conditions:

(a) When the amount of the adjustment (using the "simplified" method) is greater than 8 dB on flyover, or 4 dB on approach; or

(b) When the resulting final EPNL value on flyover or approach (using the simplified method) is within 1 dB of the limiting noise levels as prescribed in section B36.5 of this part.

A36.9.2 *Flight profiles.*

As described below, flight profiles for both test and reference conditions are defined by their geometry relative to the ground, together with the associated airplane speed relative to the ground, and the associated engine control parameter(s) used for determining the noise emission of the airplane.

A36.9.2.1 *Takeoff Profile.*

Note: Figure A36-4 illustrates a typical takeoff profile.

(a) The airplane begins the takeoff roll at point A, lifts off at point B and begins its first climb at a constant angle at point C. Where thrust or power (as appropriate) cut-back is used, it is started at point D and completed at point E. From here, the airplane begins a second climb at a constant angle up to point F, the end of the noise certification takeoff flight path.

(b) Position K1 is the takeoff noise measuring station and AK1 is the distance from start of roll to the flyover measuring point. Position K2 is the lateral noise measuring station, which is located on a line parallel to, and the specified distance from, the runway center line where the noise level during takeoff is greatest.

(c) The distance AF is the distance over which the airplane position is measured and synchronized with the noise measurements, as required by section A36.2.3.2 of this part.

A36.9.2.2 *Approach Profile.*

Note: Figure A36-5 illustrates a typical approach profile.

(a) The airplane begins its noise certification approach flight path at point G and touches down on the runway at point J, at a distance OJ from the runway threshold.

(b) Position K3 is the approach noise measuring station and K3 O is the distance from the approach noise measurement point to the runway threshold.

(c) The distance GI is the distance over which the airplane position is measured and synchronized with the noise measurements, as required by section A36.2.3.2 of this part.

The airplane reference point for approach measurements is the instrument landing system (ILS) antenna. If no ILS antenna is installed an alternative reference point must be approved by the FAA.

A36.9.3 *Simplified method of adjustment.*

A36.9.3.1 *General.* As described below, the simplified adjustment method consists of applying adjustments (to the EPNL, which is calculated from the measured data) for the differences between measured and reference conditions at the moment of PNLTM.

A36.9.3.2 *Adjustments to PNL and PNLT.*

(a) The portions of the test flight path and the reference flight path described below, and illustrated in Figure A36-6, include the noise time history that is relevant to the calculation of flyover and approach EPNL. In figure A36-6:

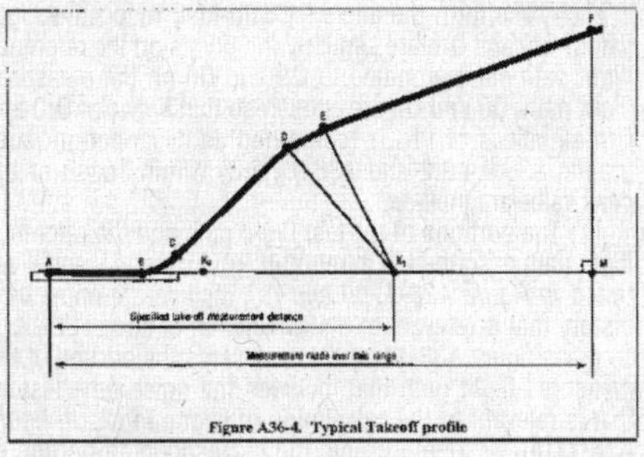

Figure A36-4. Typical Takeoff profile

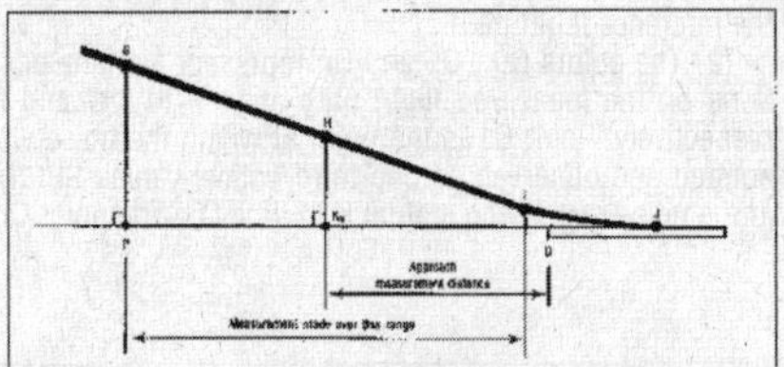

Figure A36-5. Typical Approach profile

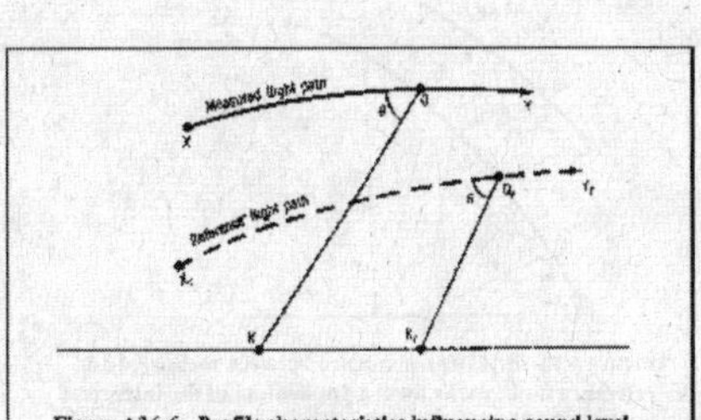

Figure A36-6. Profile characteristics influencing sound level

(1) XY represents the portion of the measured flight path that includes the noise time history relevant to the calculation of flyover and approach EPNL; Xr Yr represents the corresponding portion of the reference flight path.

(2) Q represents the airplane's position on the measured flight path at which the noise was emitted and observed as PNLTM at the noise measuring station K. Qr is the corresponding position on the reference flight path, and Kr the reference measuring station. QK and Qr Kr are, respectively, the measured

and reference noise propagation paths, Qr being determined from the assumption that QK and Qr Kr form the same angle θ with their respective flight paths.

(b) The portions of the test flight path and the reference flight path described in paragraph (b)(1) and (2), and illustrated in Figure A36-7(a) and (b), include the noise time history that is relevant to the calculation of lateral EPNL.

(1) In figure A36-7(a), XY represents the portion of the measured flight path that includes the noise time history that is relevant to the calculation of lateral EPNL; in figure A36-7(b), Xr Yr represents the corresponding portion of the reference flight path.

(2) Q represents the airplane position on the measured flight path at which the noise was emitted and observed as PNLTM at the noise measuring station K. Qr is the corresponding position on the reference flight path, and Kr the reference measuring station. QK and Qr Kr are, respectively, the measured and reference noise propagation paths. In this case Kr is only specified as being on a particular Lateral line; Kr and Qr are therefore determined from the assumptions that QK and Qr Kr :

(i) Form the same angle θ with their respective flight paths; and

(ii) Form the same angle ψ with the ground.

Note: For the lateral noise measurement, sound propagation is affected not only by inverse square and atmospheric attenuation, but also by ground absorption and reflection effects which depend mainly on the angle ψ.

A36.9.3.2.1 The one-third octave band levels SPL(i) comprising PNL (the PNL at the moment of PNLTM observed at K) must be adjusted to reference levels SPL(i) r as follows:

A36.9.3.2.1(a) For calculations using the English System of Units:

$SPL(i)_r = SPL(i) + 0.001[\alpha(i) - \alpha(i)0]QK$

$+0.001\alpha(i)0\,(QK - Q_rK_r)$

$+20\log(QK/Q_rK_r)$

In this expression,

(1) The term $0.001[\alpha(i)-\alpha(i)0]QK$ is the adjustment for the effect of the change in sound attenuation coefficient, and α(i) and α(i)0 are the coefficients for the test and reference atmospheric conditions respectively, determined under section A36.7 of this appendix;

(2) The term $0.001\alpha(i)0\ (QK - Q_rK_r)$ is the adjustment for the effect of the change in the noise path length on the sound attenuation;

(3) The term $20\log(QK/Q_rK_r)$ is the adjustment for the effect of the change in the noise path length due to the "inverse square" law;

(4) QK and Qr Kr are measured in feet and α(i) and α(i)0 are expressed in dB/1000 ft.

A36.9.3.2.1(b) For calculations using the International System of Units:

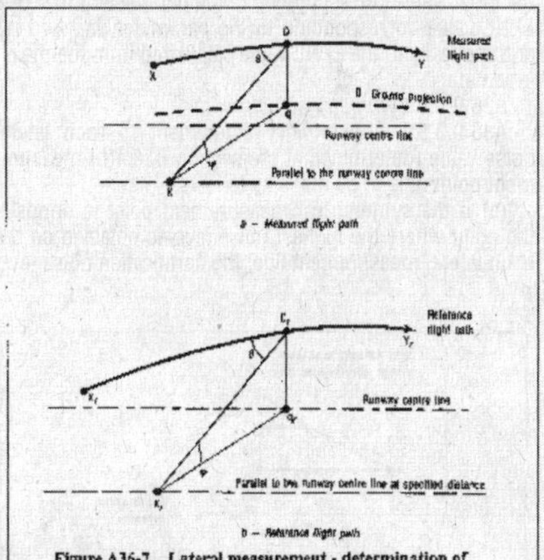

Figure A36-7. Lateral measurement - determination of reference station

$SPL(i)_r = SPL(i) + 0.01[\alpha(i) - \alpha(i)0]QK$
$+0.01\alpha(i)0\ (QK - Q_rK_r)$
$+20 \log(QK/Q_rK_r)$

In this expression,

(1) The term $0.01[\alpha(i)-\alpha(i)0]QK$ is the adjustment for the effect of the change in sound attenuation coefficient, and $\alpha(i)$ and $\alpha(i)0$ are the coefficients for the test and reference atmospheric conditions respectively, determined under section A36.7 of this appendix;

(2) The term $0.01\alpha(i)0\ (QK-Q_rK_r)$ is the adjustment for the effect of the change in the noise path length on the sound attenuation;

(3) The term $20 \log(QK/Q_rK_r)$ is the adjustment for the effect of the change in the noise path length due to the inverse square law;

(4) QK and Qr Kr are measured in meters and $\alpha(i)$ and $\alpha(i)0$ are expressed in dB/100 m.

A36.9.3.2.1.1 *PNLT Correction.*

(a) Convert the corrected values, SPL(i)r , to PNLTr ;

(b) Calculate the correction term Δ1 using the following equation:

Δ1=PNLTr - PNLTM

A36.9.3.2.1.2 Add Δ1 arithmetically to the EPNL calculated from the measured data.

A36.9.3.2.2 If, during a test flight, several peak values of PNLT that are within 2 dB of PNLTM are observed, the procedure defined in section A36.9.3.2.1 must be applied at each peak, and the adjustment term, calculated according to section A36.9.3.2.1, must be added to each peak to give corresponding adjusted peak values of PNLT. If these peak values exceed the value at the moment of PNLTM, the maximum value of such exceedance must be added as a further adjustment to the EPNL calculated from the measured data.

A36.9.3.3 *Adjustments to duration correction.*

A36.9.3.3.1 Whenever the measured flight paths and/or the ground velocities of the test conditions differ from the reference flight paths and/or the ground velocities of the reference conditions, duration adjustments must be applied to the EPNL values calculated from the measured data. The adjustments must be calculated as described below.

A36.9.3.3.2 For the flight path shown in Figure A36-6, the adjustment term is calculated as follows:

Δ2 =-7.5 log(QK/Qr Kr)+10 log(V/Vr)

(a) Add Δ2 arithmetically to the EPNL calculated from the measured data.

A36.9.3.4 *Source noise adjustments.*

A36.9.3.4.1 To account for differences between the parameters affecting engine noise as measured in the certification flight tests, and those calculated or specified in the reference conditions, the source noise adjustment must be calculated and applied. The adjustment is determined from the manufacturer's data approved by the FAA. Typical data used for this adjustment are illustrated in Figure A36-8 that shows a curve of EPNL versus the engine control parameter μ, with the EPNL data being corrected to all the other relevant reference conditions (airplane mass, speed and altitude, air temperature) and for the difference in noise between the test engine and the average engine (as defined in section B36.7(b)(7)). A sufficient number of data points over a range of values of μr is required to calculate the source noise adjustments for lateral, flyover and approach noise measurements.

A36.9.3.4.2 Calculate adjustment term Δ3 by subtracting the EPNL value corresponding to the parameter μ from the EPNL value corresponding to the parameter μr . Add Δ3 arithmetically to the EPNL value calculated from the measured data.

A36.9.3.5 *Symmetry adjustments.*

A36.9.3.5.1 A symmetry adjustment to each lateral noise value (determined at the section B36.4(b) measurement points), is to be made as follows:

(a) If the symmetrical measurement point is opposite the point where the highest noise level is obtained on the main lateral measurement line, the certification noise level

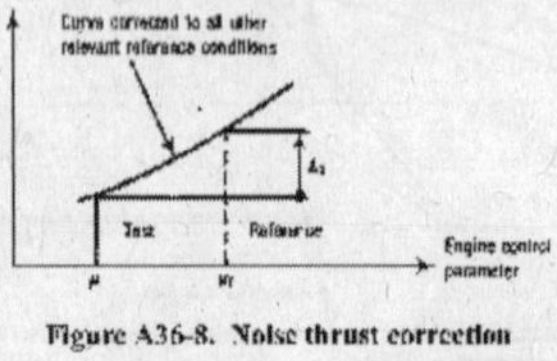

Figure A36-8. Noise thrust correction

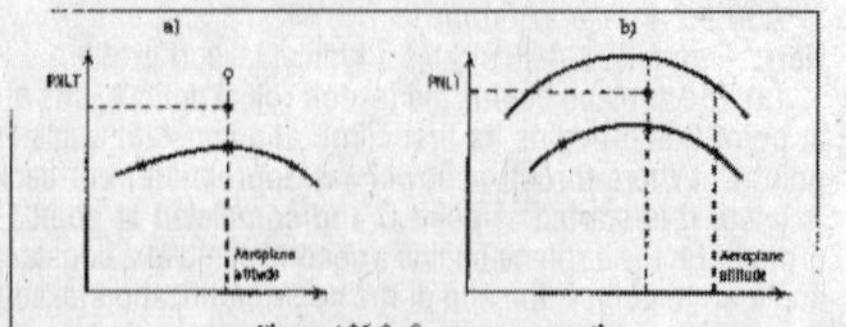

Figure A36-9. Symmetry correction

is the arithmetic mean of the noise levels measured at these two points (see Figure A36-9(a));

(b) If the condition described in paragraph (a) of this section is not met, then it is assumed that the variation of noise with the altitude of the airplane is the same on both sides; there is a constant difference between the lines of noise versus altitude on both sides (see figure A36-9(b)). The certification noise level is the maximum value of the mean between these lines.

A36.9.4 *Integrated method of adjustment*

A36.9.4.1 *General.* As described in this section, the integrated adjustment method consists of recomputing under reference conditions points on the PNLT time history corresponding to measured points obtained during the tests, and computing EPNL directly for the new time history obtained in this way. The main principles are described in sections A36.9.4.2 through A36.9.4.4.1.

A36.9.4.2 *PNLT computations.*

(a) The portions of the test flight path and the reference flight path described in paragraph (a)(1) and (2), and illustrated in Figure A36-10, include the noise time history that is relevant to the calculation of flyover and approach EPNL. In figure A36-10:

(1) XY represents the portion of the measured flight path that includes the noise time history relevant to the calculation of flyover and approach EPNL; Xr Yr represents the corresponding reference flight path.

(2) The points Q0 , Q1 , Qn represent airplane positions on the measured flight path at time t0 , t1 and tn respectively. Point Q1 is the point at which the noise was emitted and observed as one-third octave values SPL(i)1 at the noise measuring station K at time t1 . Point Qr1 represents the corresponding position on the reference flight path for noise observed as SPL(i)r1 at the reference measuring station Kr at time tr1 . Q1 K and Qr1 Kr are respectively the measured and reference noise propagation paths, which in each case form the angle θ1 with their respective flight paths. Qr0 and Qrn are similarly the points on the reference flight path corresponding to Q0 and Qn on the measured flight path. Q0 and Qn are chosen so that between Qr0 and Qrn all values of PNLTr (computed as described in paragraphs A36.9.4.2.2 and A36.9.4.2.3) within 10 dB of the peak value are included.

(b) The portions of the test flight path and the reference flight path described in paragraph (b)(1) and (2), and illustrated *in Figure A36-11(a) and (b),* include the noise time history that is relevant to the calculation of lateral EPNL.

(1) In figure A36-11(a) XY represents the portion of the measured flight path that includes the noise time history that is relevant to the calculation of lateral EPNL; in figure A36-11(b), Xr Yr represents the corresponding portion of the reference flight path.

(2) The points Q0 , Q1 and Qn represent airplane positions on the measured flight path at time t0 , t1 and tn respectively. Point Q1 is the point at which the noise was emitted and observed as one-third octave values SPL(i)1 at the noise measuring station K at time t1 . The point Qr1

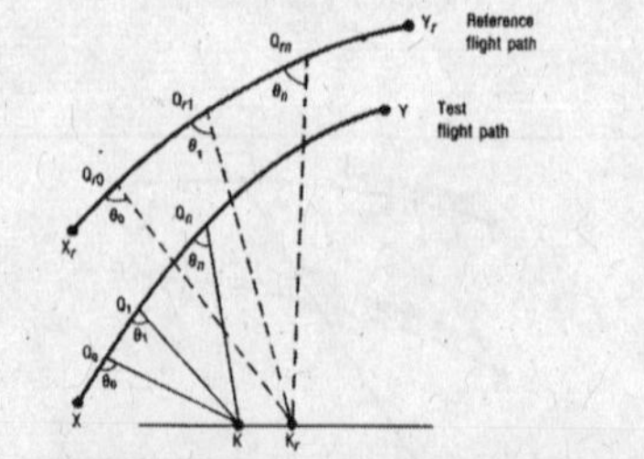

Figure A36-10. Correspondence between measured and reference flight paths for the application of the integrated method of adjustment

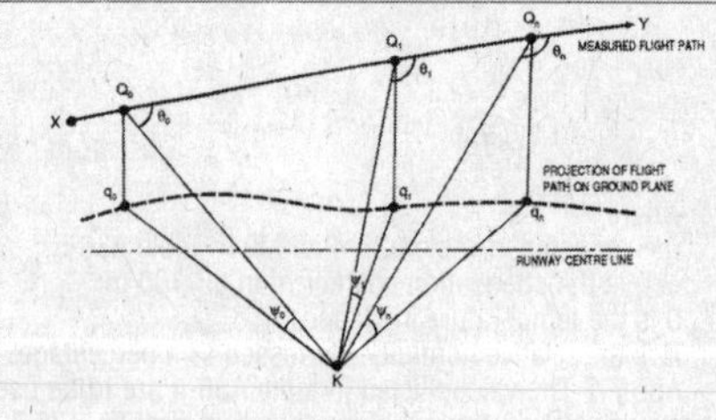

Figure A36-11(a). Measured flight path

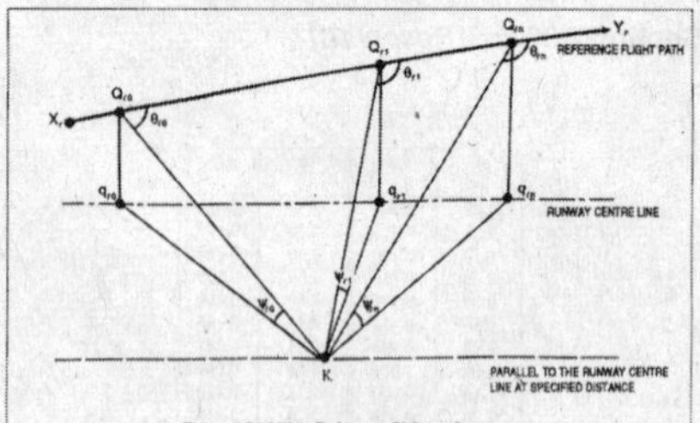

Figure A36-11(b). Reference flight path

represents the corresponding position on the reference flight path for noise observed as SPL(i)r1 at the measuring station Kr at time tr1 . Q1 K and Qr1 Kr are respectively the measured and reference noise propagation paths. Qr0 and Qrn are similarly the points on the reference flight path corresponding to Q0 and Qn on the measured flight path.

Q0 and Qn are chosen to that between Qro and Qrn all values of PNLTr (computed as described in paragraphs A36.9.4.2.2 and A36.9.4.2.3) within 10 dB of the peak value are included. In this case Kr is only specified as being on a particular lateral line. The position of Kr and Qr1 are determined from the following requirements.

(i) Q1 K and Qr1 Kr form the same angle θ1 with their respective flight paths; and

(ii) The differences between the angles1 andr1 must be minimized using a method, approved by the FAA. The differences between the angles are minimized since, for geometrical reasons, it is generally not possible to choose Kr so that the condition described in paragraph A36.9.4.2(b)(2)(i) is met while at the same time keeping1 andr1 equal.

Note: For the lateral noise measurement, sound propagation is affected not only by "inverse square" and atmospheric attenuation, but also by ground absorption and reflection effects which depend mainly on the angle.

A36.9.4.2.1 In paragraphs A36.9.4.2(a)(2) and (b)(2) the time tr1 is later (for Qr1 Kr > Q1 K) than t1 by two separate amounts:

(1) The time taken for the airplane to travel the distance Qr1 Qr0 at a speed Vr less the time taken for it to travel Q1 Q0 at V;

(2) The time taken for sound to travel the distance Qr1 Kr -Q1 K.

Note: For the flight paths described in paragraphs A36.9.4.2(a) and (b), the use of thrust or power cut-back will result in test and reference flight paths at full thrust or power and at cut-back thrust or power. Where the transient region between these thrust or power levels affects the final result, an interpolation must be made between them by an approved method such as that given in the current advisory circular for this part.

A36.9.4.2.2 The measured values of SPL(i)1 must be adjusted to the reference values SPL(i)r1 to account for the differences between measured and reference noise path lengths and between measured and reference atmospheric conditions, using the methods of section A36.9.3.2.1 of this appendix. A corresponding value of PNLr1 must be computed according to the method in section A36.4.2. Values of PNLr must be computed for times t0 through tn .

A36.9.4.2.3 For each value of PNLr1 , a tone correction factor C1 must be determined by analyzing the reference values SPL(i)r using the methods of section A36.4.3 of this appendix, and added to PNLr1 to yield PNLTr1 . Using the process described in this paragraph, values of PNLTr must be computed for times t0 through tn .

A36.9.4.3 *Duration correction.*

A36.9.4.3.1 The values of PNLTr corresponding to those of PNLT at each one-half second interval must be plotted against time (PNLTr1 at time tr1). The duration correction must then be determined using the method of section A36.4.5.1 of this appendix, to yield EPNLr .

A36.9.4.4 *Source Noise Adjustment.*

A36.9.4.4.1 A source noise adjustment, Δ3 , must be determined using the methods of section A36.9.3.4 of this appendix.

A36.9.5 Flight Path Identification Positions

Position	Description
A	Start of Takeoff roll.
B	Lift-off.
C	Start of first constant climb.
D	Start of thrust reduction.
E	Start of second constant climb.
F	End of noise certification Takeoff flight path.
G	Start of noise certification Approach flight path.
H	Position on Approach path directly above noise measuring station.
I	Start of level-off.
J	Touchdown.
K	Noise measurement point.
K_r	Reference measurement point.
K_1	Flyover noise measurement point.
K_2	Lateral noise measurement point.
K_3	Approach noise measurement point.
M	End of noise certification Takeoff flight track.
O	Threshold of Approach end of runway.
P	Start of noise certification Approach flight track.
Q	Position on measured Takeoff flight path corresponding to apparent PNLTM at station K See section A36.9.3.2.
Q_r	Position on corrected Takeoff flight path corresponding to PNLTM at station K. See section A36.9.3.2.
V	Airplane test speed.
V_r	Airplane reference speed.

A36.9.6 Flight Path Distances

Distance	Unit	Meaning
AB	Feet (meters)	Length of takeoff roll. The distance along the runway between the start of takeoff roll and lift off.
AK	Feet (meters)	Takeoff measurement distance. The distance from the start of roll to the takeoff noise measurement station along the extended center line of the runway.
AM	Feet (meters)	Takeoff flight track distance. The distance from the start of roll to the takeoff flight track position along the extended center line of the runway after which the position of the airplane need no longer be recorded.
QK	Feet (meters)	Measured noise path. The distance from the measured airplane position Q to station K.
Q_rK_r	Feet (meters)	Reference noise path. The distance from the reference airplane position Q_r to station K_r.
K_3H	Feet (meters)	Airplane approach height. The height of the airplane above the approach measuring station.
OK_3	Feet (meters)	Approach measurement distance. The distance from the runway threshold to the approach measurement station along the extended center line of the runway.
OP	Feet (meters)	Approach flight track distance. The distance from the runway threshold to the approach flight track position along the extended center line of the runway after which the position of the airplane need no longer be recorded.

[Amdt. 36-54, 67 FR 45212, July 8, 2002; Amdt. 36-24, 67 FR 63195, 63196, Oct. 10, 2002; 68 FR 1512, Jan. 10, 2003; Amdt. 36-26, 70 FR 38749, July 5, 2005]

Appendix B to Part 36—Noise Levels for Transport Category and Jet Airplanes Under § 36.103

Sec.

Section B36.1 *Noise measurement and evaluation*

(a) The procedures of Appendix A of this part, or approved equivalent procedures, must be used to determine noise levels of an airplane. These noise levels must be used to show compliance with the requirements of this appendix.

(b) For Stage 4 airplanes, an acceptable alternative for noise measurement and evaluation is Appendix 2 to the International Civil Aviation Organization (ICAO) Annex 16, Environmental Protection, Volume I, Aircraft Noise, Third Edition, July 1993, Amendment 7, effective March 21, 2002. [Incorporated by reference, see § 36.6].

Section B36.2 *Noise Evaluation Metric*

The noise evaluation metric is the effective perceived noise level expressed in EPNdB, as calculated using the procedures of appendix A of this part.

Section B36.3 *Reference Noise Measurement Points*

When tested using the procedures of this part, except as provided in section B36.6, an airplane may not exceed the noise levels specified in section B36.5 at the following points on level terrain:

(a) Lateral full-power reference noise measurement point:

(1) For jet airplanes: The point on a line parallel to and 1,476 feet (450 m) from the runway centerline, or extended centerline, where the noise level after lift-off is at a maximum during takeoff. For the purpose of showing compliance with Stage 1 or Stage 2 noise limits for an airplane powered by more than three jet engines, the distance from the runway centerline must be 0.35 nautical miles (648 m). For jet airplanes, when approved by the FAA, the maximum lateral noise at takeoff thrust may be assumed to occur at the point (or its approved equivalent) along the extended centerline of the runway where the airplane reaches 985 feet (300 meters) altitude above ground level. A height of 1427 feet (435 meters) may be assumed for Stage 1 or Stage 2 four engine airplanes. The altitude of the airplane as it passes the noise measurement points must be within +328 to −164 feet (+100 to −50 meters) of the target altitude. For airplanes powered by other than jet engines, the altitude for maximum lateral noise must be determined experimentally.

(2) For propeller-driven airplanes: The point on the extended centerline of the runway above which the airplane, at full takeoff power, reaches a height of 2,133 feet (650 meters). For tests conducted before August 7, 2002, an applicant may use the measurement point specified in section B36.3(a)(1) as an alternative.

(b) Flyover reference noise measurement point: The point on the extended centerline of the runway that is 21,325 feet (6,500 m) from the start of the takeoff roll;

(c) Approach reference noise measurement point: The point on the extended centerline of the runway that is 6,562 feet (2,000 m) from the runway threshold. On level ground, this corresponds to a position that is 394 feet (120 m) vertically below the 3° descent path, which originates at a point on the runway 984 feet (300 m) beyond the threshold.

Section B36.4 *Test noise measurement points.*

(a) If the test noise measurement points are not located at the reference noise measurement points, any corrections for the difference in position are to be made using the same adjustment procedures as for the differences between test and reference flight paths.

(b) The applicant must use a sufficient number of lateral test noise measurement points to demonstrate to the FAA that the maximum noise level on the appropriate lateral line has been determined. For jet airplanes, simultaneous measurements must be made at one test noise measurement point at its symmetrical point on the other side of the runway. Propeller-driven airplanes have an inherent asymmetry in lateral noise. Therefore, simultaneous measurements must be made at each and every test noise measurement point at its symmetrical position on the opposite side of the runway. The measurement points are considered to be symmetrical if they are longitudinally within 33 feet (±10 meters) of each other.

Section B36.5 *Maximum Noise Levels*

Except as provided in section B36.6 of this appendix, maximum noise levels, when determined in accordance with the noise evaluation methods of appendix A of this part, may not exceed the following:

(a) For acoustical changes to Stage 1 airplanes, regardless of the number of engines, the noise levels prescribed under § 36.7(c) of this part.

(b) For any Stage 2 airplane regardless of the number of engines:

(1) Flyover: 108 EPNdB for maximum weight of 600,000 pounds or more; for each halving of maximum weight (from 600,000 pounds), reduce the limit by 5 EPNdB; the limit is 93 EPNdB for a maximum weight of 75,000 pounds or less.

(2) Lateral and approach: 108 EPNdB for maximum weight of 600,000 pounds or more; for each halving of maximum weight (from 600,000 pounds), reduce the limit by 2 EPNdB; the limit is 102 EPNdB for a maximum weight of 75,000 pounds or less.

(c) For any Stage 3 airplane:

(1) Flyover.

(i) For airplanes with more than 3 engines: 106 EPNdB for maximum weight of 850,000 pounds or more; for each halving of maximum weight (from 850,000 pounds), reduce the limit by 4 EPNdB; the limit is 89 EPNdB for a maximum weight of 44,673 pounds or less;

(ii) For airplanes with 3 engines: 104 EPNdB for maximum weight of 850,000 pounds or more; for each halving of maximum weight (from 850,000 pounds), reduce the limit by 4 EPNdB; the limit is 89 EPNdB for a maximum weight of 63,177 pounds or less; and

(iii) For airplanes with fewer than 3 engines: 101 EPNdB for maximum weight of 850,000 pounds or more; for each halving of maximum weight (from 850,000 pounds), reduce the limit by 4 EPNdB; the limit is 89 EPNdB for a maximum weight of 106,250 pounds or less.

(2) Lateral, regardless of the number of engines: 103 EPNdB for maximum weight of 882,000 pounds or more; for each halving of maximum weight (from 882,000 pounds), reduce the limit by 2.56 EPNdB; the limit is 94 EPNdB for a maximum weight of 77,200 pounds or less.

(3) Approach, regardless of the number of engines: 105 EPNdB for maximum weight of 617,300 pounds or more; for each halving of maximum weight (from 617,300 pounds), reduce the limit by 2.33 EPNdB; the limit is 98 EPNdB for a maximum weight of 77,200 pounds or less.

(d) For any Stage 4 airplane, the flyover, lateral, and approach maximum noise levels are prescribed in Chapter 4, Paragraph 4.4, Maximum Noise Levels, and Chapter 3, Paragraph 3.4, Maximum Noise Levels, of the International Civil Aviation Organization (ICAO) Annex 16, Environmental Protection, Volume I, Aircraft Noise, Third Edition, July 1993, Amendment 7, effective March 21, 2002. [Incorporated by reference, see § 36.6].

Section B36.6 *Trade-Offs*

Except when prohibited by sections 36.7(c)(1) and 36.7(d)(1)(ii), if the maximum noise levels are exceeded at any one or two measurement points, the following conditions must be met:

(a) The sum of the exceedance(s) may not be greater than 3 EPNdB;

(b) Any exceedance at any single point may not be greater than 2 EPNdB, and

(c) Any exceedance(s) must be offset by a corresponding amount at another point or points.

Section B36.7 *Noise Certification Reference Procedures and Conditions*

(a) General conditions:

(1) All reference procedures must meet the requirements of section 36.3 of this part.

(2) Calculations of airplane performance and flight path must be made using the reference procedures and must be approved by the FAA.

(3) Applicants must use the takeoff and approach reference procedures prescribed in paragraphs (b) and (c) of this section.

(4) [Reserved]

(5) The reference procedures must be determined for the following reference conditions. The reference atmosphere is homogeneous in terms of temperature and relative humidity when used for the calculation of atmospheric absorption coefficients.

(i) Sea level atmospheric pressure of 2116 pounds per square foot (psf) (1013.25 hPa);

(ii) Ambient sea-level air temperature of 77°F (25°C, i.e., ISA+10°C);

(iii) Relative humidity of 70 per cent;

(iv) Zero wind.

(v) In defining the reference takeoff flight path(s) for the takeoff and lateral noise measurements, the runway gradient is zero.

(b) Takeoff reference procedure:

The takeoff reference flight path is to be calculated using the following:

(1) Average engine takeoff thrust or power must be used from the start of takeoff to the point where at least the following height above runway level is reached. The takeoff thrust/power used must be the maximum available for normal operations given in the performance section of the airplane flight manual under the reference atmospheric conditions given in section B36.7(a)(5).

(i) For Stage 1 airplanes and for Stage 2 airplanes that do not have jet engines with a bypass ratio of 2 or more, the following apply:

(A): For airplanes with more than three jet engines—700 feet (214 meters).

(B): For all other airplanes—1,000 feet (305 meters).

(ii) For Stage 2 airplanes that have jet engines with a bypass ratio of 2 or more and for Stage 3 airplanes, the following apply:

(A): For airplanes with more than three engines—689 feet (210 meters).

(B): For airplanes with three engines—853 feet (260 meters).

(C): For airplanes with fewer than three engines—984 feet (300 meters).

(2) Upon reaching the height specified in paragraph (b)(1) of this section, airplane thrust or power must not be reduced below that required to maintain either of the following, whichever is greater:

(i) A climb gradient of 4 per cent; or

(ii) In the case of multi-engine airplanes, level flight with one engine inoperative.

(3) For the purpose of determining the lateral noise level, the reference flight path must be calculated using full takeoff power throughout the test run without a reduction in thrust or power. For tests conducted before August 7, 2002, a single reference flight path that includes thrust cutback in accordance with paragraph (b)(2) of this section, is an acceptable alternative in determining the lateral noise level.

(4) The takeoff reference speed is the all-engine operating takeoff climb speed selected by the applicant for use in normal operation; this speed must be at least V2+10kt (V2+19km/h) but may not be greater than V2+20kt (V2+37km/h). This speed must be attained as soon as practicable after lift-off and be maintained throughout the takeoff noise certification test. For Concorde airplanes, the test day speeds and the acoustic day reference speed are the minimum approved value of V2+35 knots, or the all-engines-operating speed at 35 feet, whichever speed is greater as determined under the regulations constituting the type certification basis of the airplane; this reference speed may not exceed 250 knots. For all airplanes, noise values measured at the test day speeds must be corrected to the acoustic day reference speed.

(5) The takeoff configuration selected by the applicant must be maintained constantly throughout the takeoff reference procedure, except that the landing gear may be retracted. Configuration means the center of gravity position, and the status of the airplane systems that can affect airplane performance or noise. Examples include, the position of lift augmentation devices, whether the APU is operating, and whether air bleeds and engine power take-offs are operating;

(6) The weight of the airplane at the brake release must be the maximum takeoff weight at which the noise certification is requested, which may result in an operating limitation as specified in § 36.1581(d); and

(7) The average engine is defined as the average of all the certification compliant engines used during the airplane flight tests, up to and during certification, when operating within the limitations and according to the procedures given in the Flight Manual. This will determine the relationship of thrust/power to control parameters (e.g., N1 or EPR). Noise measurements made during certification tests must be corrected using this relationship.

(c) Approach reference procedure:

The approach reference flight path must be calculated using the following:

(1) The airplane is stabilized and following a 3° glide path;

(2) For subsonic airplanes, a steady approach speed of Vref+10 kts (Vref+19 km/h) with thrust and power stabilized must be established and maintained over the approach measuring point. Vref is the reference landing speed, which is defined as the speed of the airplane, in a specified landing configuration, at the point where it descends through the landing screen height in the determination of the landing distance for manual landings. For Concorde airplanes, a steady approach speed that is either the landing reference speed + 10 knots or the speed used in establishing the approved landing distance under the airworthiness regulations constituting the type certification basis of the airplane, whichever speed is greater. This speed must be established and maintained over the approach measuring point.

(3) The constant approach configuration used in the airworthiness certification tests, but with the landing gear down, must be maintained throughout the approach reference procedure;

(4) The weight of the airplane at touchdown must be the maximum landing weight permitted in the approach configuration defined in paragraph (c)(3) of this section at which noise certification is requested, except as provided in § 36.1581(d) of this part; and

(5) The most critical configuration must be used; this configuration is defined as that which produces the highest noise level with normal deployment of aerodynamic control surfaces including lift and drag producing devices, at the weight at which certification is requested. This configuration includes all those items listed in section A36.5.2.5 of appendix A of this part that contribute to the noisiest continuous state at the maximum landing weight in normal operation.

Section B36.8 *Noise Certification Test Procedures*

(a) All test procedures must be approved by the FAA.

(b) The test procedures and noise measurements must be conducted and processed in an approved manner to yield the noise evaluation metric EPNL, in units of EPNdB, as described in appendix A of this part.

(c) Acoustic data must be adjusted to the reference conditions specified in this appendix using the methods described in appendix A of this part. Adjustments for speed and thrust must be made as described in section A36.9 of this part.

(d) If the airplane's weight during the test is different from the weight at which noise certification is requested, the required EPNL adjustment may not exceed 2 EPNdB for each takeoff and 1 EPNdB for each approach. Data approved by the FAA must be used to determine the variation of EPNL with weight for both takeoff and approach test conditions. The necessary EPNL adjustment for variations in approach flight path from the reference flight path must not exceed 2 EPNdB.

(e) For approach, a steady glide path angle of 3°±0.5° is acceptable.

(f) If equivalent test procedures different from the reference procedures are used, the test procedures and all methods for adjusting the results to the reference procedures must be approved by the FAA. The adjustments may not exceed 16 EPNdB on takeoff and 8 EPNdB on approach. If the adjustment is more than 8 EPNdB on takeoff, or more than 4 EPNdB on approach, the resulting numbers must be more than 2 EPNdB below the limit noise levels specified in section B36.5.

(g) During takeoff, lateral, and approach tests, the airplane variation in instantaneous indicated airspeed must be maintained within±3% of the average airspeed between the 10 dB-down points. This airspeed is determined by the pilot's airspeed indicator. However, if the instantaneous indicated airspeed exceeds±3 kt (±5.5 km/h) of the average airspeed over the 10 dB-down points, and is determined by the FAA representative on the flight deck to be due to atmospheric turbulence, then the flight so affected must be rejected for noise certification purposes.

Note: Guidance material on the use of equivalent procedures is provided in the current advisory circular for this part.

[Amdt. 36-54, 67 FR 45235, July 8, 2002; Amdt. 36-24, 67 FR 63196, Oct. 10, 2002; 68 FR 1512, Jan. 10, 2003; Amdt. 36-26, 70 FR 38749, July 5, 2005]

Appendixes C-E to Part 36 [Reserved]

Appendix F to Part 36—Flyover Noise Requirements for Propeller-Driven Small Airplane and Propeller-Driven, Commuter Category Airplane Certification Tests Prior to December 22, 1988

part a—general

part a—general

Section F36.1 Scope. This appendix prescribes noise level limits and procedures for measuring and correcting noise data for the propeller driven small airplanes specified in §§ 36.1 and 36.501(b).

part b—noise measurement

Sec. F36.101 *General test conditions.*

(a) The test area must be relatively flat terrain having no excessive sound absorption characteristics such as those caused by thick, matted, or tall grass, by shrubs, or by wooded areas. No obstructions which significantly influence the sound field from the airplane may exist within a conical space above the measurement position, the cone being defined by an axis normal to the ground and by a half-angle 75 degrees from this axis.

(b) The tests must be carried out under the following conditions:

(1) There may be no precipitation.

(2) Relative humidity may not be higher than 90 percent or lower than 30 percent.

(3) Ambient temperature may not be above 86 degrees F. or below 41 degrees F. at 33′ above ground. If the measurement site is within 1 n.m. of an airport thermometer the airport reported temperature may be used.

(4) Reported wind may not be above 10 knots at 33′ above ground. If wind velocities of more than 4 knots are reported, the flight direction must be aligned to within±15 degrees of wind direction and flights with tail wind and head wind must be made in equal numbers. If the measurement site is within 1 n.m. of an airport anemometer, the airport reported wind may be used.

(5) There may be no temperature inversion or anomalous wind conditions that would significantly alter the

noise level of the airplane when the noise is recorded at the required measuring point.

(6) The flight test procedures, measuring equipment, and noise measurement procedures must be approved by the FAA.

(7) Sound pressure level data for noise evaluation purposes must be obtained with acoustical equipment that complies with section F36.103 of this appendix.

Sec. F36.103 Acoustical measurement system. The acoustical measurement system must consist of approved equipment equivalent to the following:

(a) A microphone system with frequency response compatible with measurement and analysis system accuracy as prescribed in section F36.105 of this appendix.

(b) Tripods or similar microphone mountings that minimize interference with the sound being measured.

(c) Recording and reproducing equipment characteristics, frequency response, and dynamic range compatible with the response and accuracy requirements of section F36.105 of this appendix.

(d) Acoustic calibrators using sine wave or broadband noise of known sound pressure level. If broadband noise is used, the signal must be described in terms of its average and maximum root-mean-square (rms) value for nonoverload signal level.

Sec. F36.105 ***Sensing, recording, and reproducing equipment.***

(a) The noise produced by the airplane must be recorded. A magnetic tape recorder is acceptable.

(b) The characteristics of the system must comply with the recommendations in International Electrotechnical Commission (IEC) Publication No. 179, entitled "Precision Sound Level Meters" as incorporated by reference in Part 36 under § 36.6 of this part.

(c) The response of the complete system to a sensibly plane progressive sinusoidal wave of constant amplitude must lie within the tolerance limits specified in IEC Publication No. 179, dated 1973, over the frequency range 45 to 11,200 Hz.

(d) If limitations of the dynamic range of the equipment make it necessary, high frequency pre-emphasis must be added to the recording channel with the converse de-emphasis on playback. The pre-emphasis must be applied such that the instantaneous recorded sound pressure level of the noise signal between 800 and 11,200 Hz does not vary more than 20 dB between the maximum and minimum one-third octave bands.

(e) If requested by the Administrator, the recorded noise signal must be read through an "A" filter with dynamic characteristics designated "slow," as defined in IEC Publication No. 179, dated 1973. The output signal from the filter must be fed to a rectifying circuit with square law rectification, integrated with time constants for charge and discharge of about 1 second or 800 milliseconds.

(f) The equipment must be acoustically calibrated using facilities for acoustic freefield calibration and if analysis of the tape recording is requested by the Administrator, the analysis equipment shall be electronically calibrated by a method approved by the FAA.

(g) A windscreen must be employed with microphone during all measurements of aircraft noise when the wind speed is in excess of 6 knots.

Sec. F36.107 ***Noise measurement procedures.***

(a) The microphones must be oriented in a known direction so that the maximum sound received arrives as nearly as possible in the direction for which the microphones are calibrated. The microphone sensing elements must be approximately 4′ above ground.

(b) Immediately prior to and after each test; a recorded acoustic calibration of the system must be made in the field with an acoustic calibrator for the two purposes of checking system sensitivity and providing an acoustic reference level for the analysis of the sound level data.

(c) The ambient noise, including both acoustical background and electrical noise of the measurement systems, must be recorded and determined in the test area with the system gain set at levels that will be used for aircraft noise measurements. If aircraft sound pressure levels do not exceed the background sound pressure levels by at least 10 dB(A), approved corrections for the contribution of background sound pressure level to the observed sound pressure level must be applied.

Sec. F36.109 ***Data recording, reporting, and approval.***

(a) Data representing physical measurements or corrections to measured data must be recorded in permanent form and appended to the record except that corrections to measurements for normal equipment response deviations need not be reported. All other corrections must be approved. Estimates must be made of the individual errors inherent in each of the operations employed in obtaining the final data.

(b) Measured and corrected sound pressure levels obtained with equipment conforming to the specifications described in section F36.105 of this appendix must be reported.

(c) The type of equipment used for measurement and analysis of all acoustic, airplane performance, and meteorological data must be reported.

(d) The following atmospheric data, measured immediately before, after, or during each test at the observation points prescribed in section F36.101 of this appendix must be reported:

(1) Air temperature and relative humidity.

(2) Maximum, minimum, and average wind velocities.

(e) Comments on local topography, ground cover, and events that might interfere with sound recordings must be reported.

(f) The following airplane information must be reported:

(1) Type, model and serial numbers (if any) of airplanes, engines, and propellers.

(2) Any modifications or nonstandard equipment likely to affect the noise characteristics of the airplane.

(3) Maximum certificated takeoff weights.

(4) Airspeed in knots for each overflight of the measuring point.

(5) Engine performance in terms of revolutions per minute and other relevant parameters for each overflight.

(6) Aircraft height in feet determined by a calibrated altimeter in the aircraft, approved photographic techniques, or approved tracking facilities.

(g) Aircraft speed and position and engine performance parameters must be recorded at an approved sampling rate sufficient to ensure compliance with the test procedures and conditions of this appendix.

Sec. F36.111 ***Flight procedures.***

(a) Tests to demonstrate compliance with the noise level requirements of this appendix must include at least six level flights over the measuring station at a height of 1,000′±30′ and±10 degrees from the zenith when passing overhead.

(b) Each test over flight must be conducted:

(1) At not less than the highest power in the normal operating range provided in an Airplane Flight Manual, or in any combination of approved manual material, approved placard, or approved instrument markings; and

(2) At stabilized speed with propellers synchronized and with the airplane in cruise configuration, except that if the speed at the power setting prescribed in this paragraph would exceed the maximum speed authorized in level flight, accelerated flight is acceptable.

part c—data correction

Sec. F36.201 ***Correction of data.***

(a) Noise data obtained when the temperature is outside the range of 68 degrees F.±9 degrees F., or the relative humidity is below 40 percent, must be corrected to 77 degrees F. and 70 percent relative humidity by a method approved by the FAA.

(b) The performance correction prescribed in paragraph (c) of this section must be used. It must be determined by the method described in this appendix, and must be added algebraically to the measured value. It is limited to 5dB(A).

(c) The performance correction must be computed by using the following formula:

$$\Delta dB = 60 - 20\log_{10}\left\{\left(11{,}430 - D_{50}\frac{R/C}{V_y} + 50\right)\right\}$$

Where:

*D*50 =Takeoff distance to 50 feet at maximum certificated takeoff weight.

R/C=Certificated best rate of climb (fpm).

V_y =Speed for best rate of climb in the same units as rate of climb.

(d) When takeoff distance to 50′ is not listed as approved performance information, the figures of 2000 for single-engine airplanes and 1600′ for multi-engine airplanes must be used.

Sec. F36.203 *Validity of results.*

(a) The test results must produce an average dB(A) and its 90 percent confidence limits, the noise level being the arithmetic average of the corrected acoustical measurements for all valid test runs over the measuring point.

(b) The samples must be large enough to establish statistically a 90 pecent confidence limit not to exceed±1.5 dB(A). No test result may be omitted from the averaging process, unless omission is approved by the FAA.

part d—noise limits

Sec. F36.301 ***Aircraft noise limits.***

(a) Compliance with this section must be shown with noise data measured and corrected as prescribed in Parts B and C of this appendix.

(b) For airplanes for which application for a type certificate is made on or after October 10, 1973, the noise level must not exceed 68 dB(A) up to and including aircraft weights of 1,320 pounds (600 kg.). For weights greater than 1,320 pounds up to and including 3,630 pounds (1.650 kg.) the limit increases at the rate of 1 dB/165 pounds (1 dB/75 kg.) to 82 dB(A) at 3,630 pounds, after which it is constant at 82 dB(A). However, airplanes produced under type certificates covered by this paragraph must also meet paragraph (d) of this section for the original issuance of standard airworthiness certificates or restricted category airworthiness certificates if those airplanes have not had flight time before the date specified in that paragraph.

(c) For airplanes for which application for a type certificate is made on or after January 1, 1975, the noise levels may not exceed the noise limit curve prescribed in paragraph (b) of this section, except that 80 dB(A) may not be exceeded.

(d) For airplanes for which application is made for a standard airworthiness certificate or for a restricted category airworthiness certificate, and that have not had any flight time before January 1, 1980, the requirements of paragraph (c) of this section apply, regardless of date of application, to the original issuance of the certificate for that airplane.

[Doc. No. 13243, 40 FR 1035, Jan. 6, 1975; 40 FR 6347, Feb. 11, 1975, as amended by Amdt. 36-6, 41 FR 56064, Dec. 23, 1976; Amdt. 36-6, 42 FR 4113, Jan. 24, 1977; Amdt. 36-9, 43 FR 8754, Mar. 2, 1978; Amdt. 36-13, 52 FR 1836, Jan. 15, 1987; Amdt. 36-16, 53 FR 47400, Nov. 22, 1988]

Appendix G to Part 36—Takeoff Noise Requirements for Propeller-Driven Small Airplane and Propeller-Driven, Commuter Category Airplane Certification Tests on or After December 22, 1988

part a—general

part a—general

Section G36.1 Scope. This appendix prescribes limiting noise levels and procedures for measuring noise and adjusting these data to standard conditions, for propeller driven small airplanes and propeller-driven, commuter category airplanes specified in §§ 36.1 and 36.501(c).

part b—noise measurement

Sec. G36.101 ***General Test Conditions.***

(a) The test area must be relatively flat terrain having no excessive sound absorption characteristics such as those caused by thick, matted, or tall grass, by shrubs, or by wooded areas. No obstructions which significantly influence the sound field from the airplane may exist within a conical space above the measurement position, the cone being defined by an axis normal to the ground and by a half-angle 75 degrees from the normal ground axis.

(b) The tests must be carried out under the following conditions:

(1) No precipitation;

(2) Ambient air temperature between 36 and 95 degrees F (2.2 and 35 degrees C);

(3) Relative humidity between 20 percent and 95 percent, inclusively;

(4) Wind speed may not exceed 10 knots (19 km/h) and cross wind may not exceed 5 knots (9 km/h), using a 30-second average;

(5) No temperature inversion or anomalous wind condition that would significantly alter the noise level of the airplane when the nose is recorded at the required measuring point, and

(6) The meteorological measurements must be made between 4 ft. (1.2 m) and 33 ft. (10 m) above ground level. If the measurement site is within 1 n.m. of an airport meteorological station, measurements from that station may be used.

(c) The flight test procedures, measuring equipment, and noise measurement procedures must be approved by the FAA.

(d) Sound pressure level data for noise evaluation purposes must be obtained with acoustical equipment that complies with section G36.103 of this appendix.

Sec. G36.103 *Acoustical Measurement System.*

The acoustical measurement system must consist of approved equipment with the following characteristics: (a) A microphone system with frequency response compatible with measurement and analysis system accuracy as prescribed in section G36.105 of this appendix.

(b) Tripods or similar microphone mountings that minimize interference with the sound being measured.

(c) Recording and reproducing equipment characteristics, frequency response, and dynamic range compatible with the response and accuracy requirements of section G36.105 of this appendix.

(d) Acoustic calibrators using sine wave or broadband noise of known sound pressure level. If broadband noise is used, the signal must be described in terms of its average and maximum root-mean-square (rms) value for non-overload signal level.

Sec. G36.105 *Sensing, Recording, and Reproducing Equipment.*

(a) The noise produced by the airplane must be recorded. A magnetic tape recorder, graphic level recorder, or sound level meter is acceptable when approved by the regional certificating authority.

(b) The characteristics of the complete system must comply with the requirements in International Electrotechnical Commission (IEC) Publications No. 651, entitled "Sound Level Meters" and No. 561, entitled "Electro-acoustical Measuring Equipment for Aircraft Noise Certification" as incorporated by reference under § 36.6 of this part. Sound level meters must comply with the requirements for Type 1 sound level meters as specified in IEC Publication No. 651.

(c) The response of the complete system to a sensibly plane progressive sinusoidal wave of constant amplitude must be within the tolerance limits specified in IEC Publication No. 651, over the frequency range 45 to 11,200 Hz.

(d) If equipment dynamic range limitations make it necessary, high frequency pre-emphasis must be added to the recording channel with the converse de-emphasis on playback. The pre-emphasis must be applied such that the instantaneous recorded sound pressure level of the noise signal between 800 and 11,200 Hz does not vary more than 20 dB between the maximum and minimum one-third octave bands.

(e) The output noise signal must be read through an "A" filter with dynamic characteristics designated "slow" as defined in IEC Publication No. 651. A graphic level recorder, sound level meter, or digital equivalent may be used.

(f) The equipment must be acoustically calibrated using facilities for acoustic free-field calibration and if analysis of the tape recording is requested by the Administrator, the analysis equipment shall be electronically calibrated by a method approved by the FAA. Calibrations shall be performed, as appropriate, in accordance with paragraphs A36.3.8 and A36.3.9 of appendix A of this part.

(g) A windscreen must be employed with the microphone during all measurements of aircraft noise when the wind speed is in excess of 5 knots (9 km/hr).

Sec. G36.107 *Noise Measurement Procedures.*

(a) The microphone must be a pressure type, 12.7 mm in diameter, with a protective grid, mounted in an inverted position such that the microphone diaphragm is 7 mm above and parallel to a white-painted metal circular plate. This white-painted metal plate shall be 40 cm in diameter and at least 2.5 mm thick. The plate shall be placed horizontally and flush with the surrounding ground surface with no cavities below the plate. The microphone must be located three-quarters of the distance from the center to the back edge of the plate along a radius normal to the line of flight of the test airplane.

(b) Immediately prior to and after each test, a recorded acoustic calibration of the system must be made in the field with an acoustic calibrator for the purposes of checking system sensitivity and providing an acoustic reference level for the analysis of the sound level data. If a tape recorder or graphic level recorder is used, the frequency response of the electrical system must be determined at a level within 10 dB of the full-scale reading used during the test, utilizing pink or pseudorandom noise.

(c) The ambient noise, including both acoustic background and electrical systems noise, must be recorded and determined in the test area with the system gain set at levels which will be used for aircraft noise measurements. If aircraft sound pressure levels do not exceed the background sound pressure levels by at least 10 dB(A), a takeoff measurement point nearer to the start of the takeoff roll must be used and the results must be adjusted to the reference measurement point by an approved method.

Sec. G36.109 *Data Recording, Reporting, and Approval.*

(a) Data representing physical measurements and adjustments to measured data must be recorded in permanent form and appended to the record, except that corrections to measurements for normal equipment response deviations need not be reported. All other adjustments must be approved. Estimates must be made of the individual errors inherent in each of the operations employed in obtaining the final data.

(b) Measured and corrected sound pressure levels obtained with equipment conforming to the specifications in section G36.105 of this appendix must be reported.

(c) The type of equipment used for measurement and analysis of all acoustical, airplane performance, and meteorological data must be reported.

(d) The following atmospheric data, measured immediately before, after, or during each test at the observation points prescribed in section G36.101 of this appendix must be reported:

(1) Ambient temperature and relative humidity.

(2) Maximum and average wind speeds and directions for each run.

(e) Comments on local topography, ground cover, and events that might interfere with sound recordings must be reported.

(f) The aircraft position relative to the takeoff reference flight path must be determined by an approved method independent of normal flight instrumentation, such as radar tracking, theodolite triangulation, or photographic scaling techniques.

(g) The following airplane information must be reported:

(1) Type, model, and serial numbers (if any) of airplanes, engines, and propellers;

(2) Any modifications or nonstandard equipment likely to affect the noise characteristics of the airplane;

(3) Maximum certificated takeoff weight;

(4) For each test flight, airspeed and ambient temperature at the flyover altitude over the measuring site determined by properly calibrated instruments;

(5) For each test flight, engine performance parameters, such as manifold pressure or power, propeller speed (rpm) and other relevant parameters. Each parameter must be determined by properly calibrated instruments. For instance, propeller RPM must be validated by an independent device accurate to within±1 percent, when the airplane is equipped with a mechanical tachometer.

(6) Airspeed, position, and performance data necessary to make the corrections required in section G36.201 of this appendix must be recorded by an approved method when the airplane is directly over the measuring site.

Sec. G36.111 *Flight Procedures.*

(a) The noise measurement point is on the extended centerline of the runway at a distance of 8200 ft (2500 m) from the start of takeoff roll. The aircraft must pass over the measurement point within±10 degrees from the vertical and within 20% of the reference altitude. The flight test program shall be initiated at the maximum approved takeoff weight and the weight shall be adjusted back to this maximum weight after each hour of flight time. Each flight test must be conducted at the speed for the best rate of climb (V_y)±5 knots (±9 km/hour) indicated airspeed. All test, measurement, and data correction procedures must be approved by the FAA.

(b) The takeoff reference flight path must be calculated for the following atmospheric conditions:

(1) Sea level atmospheric pressure of 1013.25 mb (013.25 hPa);

(2) Ambient air temperature of 59°F (15°C);

(3) Relative humidity of 70 percent; and

(4) Zero wind.

(c) The takeoff reference flight path must be calculated assuming the following two segments:

(1) First segment.

(i) Takeoff power must be used from the brake release point to the point at which the height of 50 ft (15m) above the runway is reached.

(ii) A constant takeoff configuration selected by the applicant must be maintained through this segment.

(iii) The maximum weight of the airplane at brake-release must be the maximum for which noise certification is requested.

(iv) The length of this first segment must correspond to the airworthiness approved value for a takeoff on a level paved runway (or the corresponding value for seaplanes).

(2) Second segment.

(i) The beginning of the second segment corresponds to the end of the first segment.

(ii) The airplane must be in the climb configuration with landing gear up, if retractable, and flap setting corresponding to normal climb position throughout this second segment.

(iii) The airplane speed must be the speed for the best rate of climb (Vy).

(iv) For airplanes equipped with fixed pitch propellers, takeoff power must be maintained throughout the second segment. For airplanes equipped with variable pitch or constant speed propellers, takeoff power and rpm must be maintained throughout the second segment. If airworthiness limitations do not allow the application of takeoff power and rpm up to the reference point, then takeoff power and rpm must be maintained for as long as is permitted by such limitations; thereafter, maximum continuous power and rpm must be maintained. Maximum time allowed at takeoff power under the airworthiness standards must be used in the second segment. The reference height must be calculated assuming climb gradients appropriate to each power setting used.

part c—data corrections

Sec. G36.201 *Corrections to Test Results.*

(a) These corrections account for the effects of:

(1) Differences in atmospheric absorption of sound between meteorological test conditions and reference conditions.

(2) Differences in the noise path length between the actual airplane flight path and the reference flight path.

(3) The change in the helical tip Mach number between test and reference conditions.

(4) The change in the engine power between test and reference conditions.

MEASUREMENT WINDOW FOR NO ABSORPTION CORRECTION Figure G1

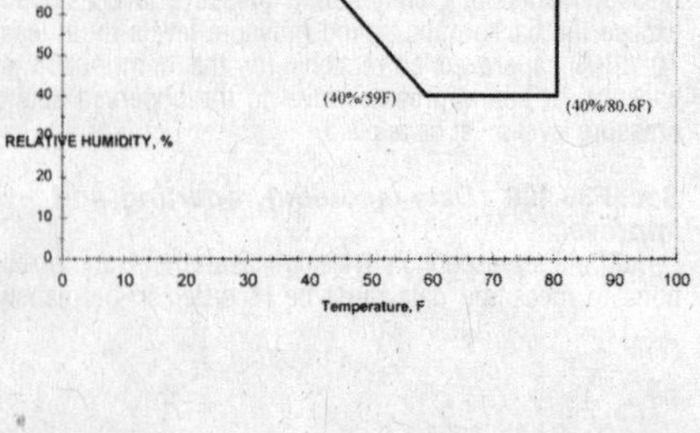

(b) Atmospheric absorption correction is required for noise data obtained when the test conditions are outside those specified in Figure G1. Noise data outside the applicable range must be corrected to 59 F and 70 percent relative humidity by an FAA approved method.

(c) No corrections for helical tip Mach number variation need to be made if the propeller helical tip Mach number is:

(1) At or below 0.70 and the test helical tip Mach number is within 0.014 of the reference helical tip Mach number.

(2) Above 0.70 and at or below 0.80 and the test helical tip Mach number is within 0.007 of the reference helical tip Mach number.

(3) Above 0.80 and the test helical tip Mach number is within 0.005 of the reference helical tip Mach number. For mechanical tachometers, if the helical tip Mach number is above 0.8 and the test helical tip Mach number is within 0.008 of the reference helical tip Mach number.

(d) When the test conditions are outside those specified, corrections must be applied by an approved procedure or by the following simplified procedure:

(1) Measured sound levels must be corrected from test day meteorological conditions to reference conditions by adding an increment equal to

Delta (M)=(HT α—0.7 HR)/1000

where HT is the height in feet under test conditions, HR is the height in feet under reference conditions when the aircraft is directly over the noise measurement point and α is the rate of absorption for the test day conditions at 500 Hz as specified in SAE ARP 866A, entitled "Standard Values of Atmospheric Absorption as a function of Temperature and Humidity for use in Evaluating Aircraft Flyover Noise" as incorporated by reference under § 36.6.

(2) Measured sound levels in decibels must be corrected for height by algebraically adding an increment equal to Delta (1). When test day conditions are within those specified in figure G1:

Delta (1)=22 log (HT/HR)

where HT is the height of the test aircraft when directly over the noise measurement point and HR is the reference height.

When test day conditions are outside those specified in figure G1:

Delta (1)=20 log (HT/HR)

(3) Measured sound levels in decibels must be corrected for helical tip Mach number by algebraically adding an increment equal to:

Delta (2)=k log (MR /MT)

where MT and MR are the test and reference helical tip Mach numbers, respectively. The constant "k" is equal to the slope of the line obtained for measured values of the sound level in dB(A) versus helical tip Mach number. The value of k may be determined from approved data. A nominal value of k=150 may be used when MT is smaller than MR . No correction may be made using the nominal value of k when MT is larger than MR . The reference helical tip Mach number MR is the Mach number corresponding to the reference conditions (RPM, airspeed, temperature) above the measurement point.

(4) Measured sound levels in decibels must be corrected for engine power by algebraically adding an increment equal to

Delta (3)=K3 log (PR /PT)

where PR and PT are the test and reference engine powers respectively obtained from the manifold pressure/torque gauges and engine rpm. The value of K3 shall be determined from approved data from the test airplane. In the absence of flight test data and at the discretion of the Administrator, a value of K3 =17 may be used.

Sec. G36.203 ***Validity of Results.***

(a) The measuring point must be overflown at least six times. The test results must produce an average noise level (LAmax) value within a 90 percent confidence limit. The average noise level is the arithmetic average of the corrected acoustical measurements for all valid test runs over the measuring point.

(b) The samples must be large enough to establish statistically a 90 percent confidence limit not exceeding±1.5 dB(A). No test results may be omitted from the averaging process unless omission is approved by the FAA.

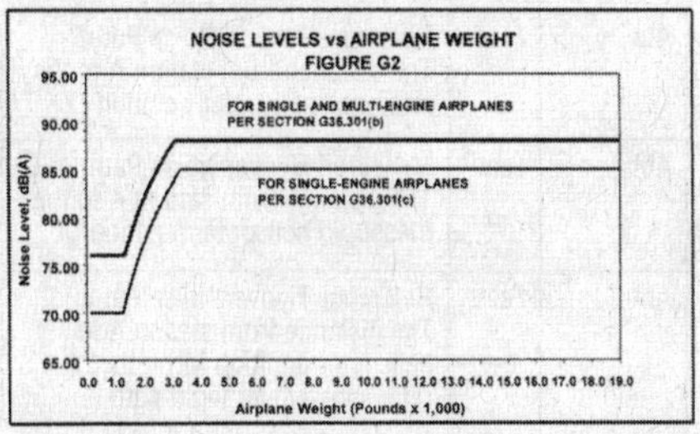

part d—noise limits

Sec. G36.301 ***Aircraft noise limits.***

(a) Compliance with this section must be shown with noise data measured and corrected as prescribed in Parts B and C of this appendix.

(b) For single-engine airplanes for which the original type certification application is received before February 3, 2006 and multi-engine airplanes, the noise level must not exceed 76 dB(A) up to and including aircraft weights of 1,320 pounds (600 kg). For aircraft weights greater than 1,320 pounds, the limit increases from that point with the logarithm of airplane weight at the rate of 9.83 dB (A) per doubling of weight, until the limit of 88 dB (A) is reached, after which the limit is constant up to and including 19,000 pounds (8,618 kg). Figure G2 shows noise level limits vs airplane weight.

(c) For single-engine airplanes for which the original type certification application is received on or after February 3, 2006, the noise level must not exceed 70dB(A) for aircraft having a maximum certificated takeoff weight of 1,257 pounds (570 kg) or less. For aircraft weights greater than 1,257 pounds, the noise limit increases from that point with the logarithm of airplane weight at the rate of 10.75dB(A) per doubling of weight, until the limit of 85dB(A) is reached, after which the limit is constant up to and including 19,000 pounds (8,618 kg). Figure G2 depicts noise level limits for airplane weights for single-engine airplanes.

(Secs. 313(a), 603, and 611(b), Federal Aviation Act of 1958 as amended (49 U.S.C. 1354(a), 1423, and 1431(b)); sec. 6(c), Department of Transportation Act (49 U.S.C. 1655 (c)); Title I, National Environmental Policy Act of 1969 (42 U.S.C. 4321 *et seq*.); E. O. 11514, March 5, 1970 and 14 CFR 11.45).

[Amdt. 36-16, 53 FR 47400, Nov. 22, 1988; 53 FR 50157, Dec. 13, 1988, as amended by Amdt. 36-22, 64 FR 55602, Oct. 13, 1999; Amdt. 36-54, 67 FR 45236, July 8, 2002; Amdt. 36-27, 70 FR 45504, Aug. 5, 2005; Amdt. 36-28, 71 FR 532, Jan. 4, 2006]

Appendix H to Part 36—Noise Requirements For Helicopters Under Subpart H

part a—reference conditions

part b—noise measurement under § 36.801

part c—noise evaluation and calculation under § 36.803

part d—noise limits under § 36.805

part a—reference conditions

Section H36.1 General. This appendix prescribes noise requirements for helicopters specified under § 36.1, including:

(a) The conditions under which helicopter noise certification tests under Part H must be conducted and the measurement procedures that must be used under § 36.801 to measure helicopter noise during each test;

(b) The procedures which must be used under § 36.803 to correct the measured data to the reference conditions and to calculate the noise evaluation quantity designated as Effective Perceived Noise Level (EPNL); and

(c) The noise limits for which compliance must be shown under § 36.805.

Section H36.3 ***Reference Test Conditions.***

(a) *Meteorological conditions.* Aircraft position, performance data and noise measurements must be corrected to the following noise certification reference atmospheric conditions which shall be assumed to exist from the surface to the aircraft altitude:

(1) Sea level pressure of 2,116 psf (1,013.25 hPa).

(2) Ambient temperature of 77 degrees F (25 degrees C).

(3) Relative humidity of 70 percent.

(4) Zero wind.

(b) *Reference test site.* The reference test site is flat and without line-of-sight obstructions across the flight path that encompasses the 10 dB down points.

(c) *Takeoff reference profile.* (1) Figure H1 illustrates a typical takeoff profile, including reference conditions.

(2) The reference flight path is defined as a straight line segment inclined from the starting point (1,640 feet (500 meters) from the center microphone location and 65 feet (20 meters) above ground level) at a constant climb angle β defined by the certificated best rate of climb and Vy for minimum engine performance. The constant climb angle β is derived from the manufacturer's data (approved by the FAA) to define the flight profile for the reference conditions. The constant climb angle β is drawn through Cr and continues, crossing over station A, to the position corresponding to the end of the type certification takeoff path represented by position Ir .

(d) *Level flyover reference profile.* The beginning of the level flyover reference profile is represented by helicopter position Dr (Figure H2). The helicopter approaches position Dr in level flight 492 feet above ground level as measured at Station A. Reference airspeed must be either 0.9VH ; 0.9VNE ; 0.45VH + 65 kts (0.45VH + 120km/h); or 0.45VNE + 65kts (0.45VNE + 120 km/h), whichever of the four speeds is least. The helicopter crosses directly overhead station A in level flight and proceeds to position Jr .

(e) For noise certification purposes, VH is defined as the airspeed in level flight obtained using the minimum specified engine torque corresponding to maximum continuous power available for sea level pressure of 2,116 psf (1,013.25 hPa) at 77°F (25°C) ambient conditions at the relevant maximum certificated weight. The value of VNE is the never-exceed airspeed. The values of VH and VNE that are used for noise certification must be listed in the approved Rotorcraft Flight Manual.

(f) *Approach reference profile.* (1) Figure H3 illustrates approach profile, including reference conditions.

(i) The beginning of the approach profile is represented by helicopter position E. The position of the helicopter is recorded for a sufficient distance (EK) to ensure recording of the entire interval during which the measured helicopter noise level is within 10 dB of Maximum Tone Corrected Perceived Noise Level (PNLTM). The reference flight path, Er Kr represents a stable flight condition in terms of torque, rpm, indicated airspeed, and rate of descent resulting in a 6° approach angle.

(ii) The test approach profile is defined by the approach angle η passing directly over the station A at a height of AH, to position K, which terminates the approach noise certification profile. The test approach angle η must be between 5.5° and 6.5°.

(2) The helicopter approaches position H along a constant 6° approach slope throughout the 10 dB down time period. The helicopter crosses position E and proceeds along the approach slope crossing over station A until it reaches position K.

Section H36.5 Symbols and units. The following symbols and units as used in this appendix for helicopter noise certification have the following meanings.

Flight Profile Identification—Positions

Position	Description
A	Location of the noise measuring point at the flight-track noise measuring station vertically below the reference (takeoff, flyover, or approach) flight path.
C	Start of noise certification takeoff flight path.
C_r	Start of noise certification reference takeoff flight path.
D	Start of noise certification flyover flight path.
D_r	Start of noise certification reference flyover path.
E	Start of noise certification approach flight path.
E_r	Start of noise certification reference approach flight path.
F	Position on takeoff flight path directly above noise measuring station A.
F_r	Position on reference takeoff path directly above noise measuring Station A.
G	Position on flyover flight path directly above noise measuring station A.
G_r	Position on reference flyover path directly above noise measuring Station A.
H	Position on approach flight path directly above noise measuring station A.
H_r	Position on reference path directly above noise measuring Station A.
I	End of noise type certification takeoff flight path.
I_r	End of noise type certification reference takeoff flight path.
J	End of noise type certification flyover flight path.
J_r	End of noise type certification reference flyover flight path.
K	End of noise certification approach type flight path.
K_r	End of noise type certification reference approach flight path.
L	Position on measured takeoff flight path corresponding to PNLTM at station A.
L_r	Position on reference takeoff flight path corresponding to PNLTM of station A.
M	Position on measured flyover flight path corresponding to PNLTM of station A.
M_r	Position on reference flyover flight path corresponding to PNLTM of station A.
N	Position on measured approach flight path corresponding to PNLTM at station A.
N_r	Position on reference approach flight path corresponding to PNLTM at station A.
S	Sideline noise measuring station (note: a subscript denotes the aircraft orientation relative to the direction of flight).

Flight Profile Distances

Distance	Unit	Meaning
AF	Feet	Takeoff Height. The vertical distance between helicopter and station A.
AG	Feet	Flyover Height. The vertical distance between the helicopter and station A.
AH	Feet	Approach Height. The vertical distance between the helicopter and station A.
AL	Feet	Measured Takeoff Noise Path. The distance from station A to the measured helicopter position L.
ALr	Feet	Reference Takeoff Noise Path. The distance from station A to the reference helicopter position Lr.
AM	Feet	Measured Flyover Noise Path. The distance from station A to the measured helicopter position M.
AMr	Feet	Reference Flyover Noise Path. The distance from station A to helicopter position Mron the reference flyover flight path.
AN	Feet	Measured Approach Noise Path. The distance from station A to the measured helicopter noise position N.
ANr	Feet	Reference Approach Noise Path. The distance from station A to the reference helicopter position Nr.
CI	Feet	Takeoff Flight Path Distance. The distance from position C at which the helicopter establishes a constant climb angle on the takeoff flight path passing over station A and continuing to position I at which the position of the helicopter need no longer be recorded.
DJ	Feet	Flyover Flight Path Distance. The distance from position D at which the helicopter is established on the flyover flight path passing over station A and continuing to position J at which the position of the helicopter need no longer be recorded.
EK	Feet	Approach Flight Path Distance. The distance from position E at which the helicopter establishes a constant angle on the approach flight path passing over station A and continuing to position K at which the position of the helicopter need no longer be recorded.

part b—noise measurement under § 36.801

Section H36.101 *Noise certification test and measurement conditions.*

(a) *General.* This section prescribes the conditions under which aircraft noise certification tests must be conducted and the measurement procedures that must be used to measure helicopter noise during each test.

(b) *Test site requirements.* (1) Tests to show compliance with established helicopter noise certification levels must consist of a series of takeoffs, level flyovers, and approaches during which measurement must be taken at noise measuring stations located at the measuring points prescribed in this section.

(2) Each takeoff test, flyover test, and approach test includes simultaneous measurements at the flight-track noise measuring station vertically below the reference flight path and at two sideline noise measuring stations, one on each side of the reference flight track 492 feet (150m) from, and on a line perpendicular to, the flight track of the noise measuring station.

(3) The difference between the elevation of either sideline noise measuring station may not differ from the flight-track noise measuring station by more than 20 feet.

(4) Each noise measuring station must be surrounded by terrain having no excessive sound absorption characteristics, such as might be caused by thick, matted, or tall grass, shrubs, or wooded areas.

(5) During the period when the takeoff, flyover, or approach noise/time record indicates the noise measurement is within 10 dB of PNLTM, no obstruction that significantly influences the sound field from the aircraft may exist—

(i) For any flight-track or sideline noise measuring station, within a conical space above the measuring position (the point on the ground vertically below the microphone), the cone being defined by an axis normal to the ground and by half-angle 80° from this axis; and

(ii) For any sideline noise measuring station, above the line of sight between the microphone and the helicopter.

(6) If a takeoff or flyover test series is conducted at weights other than the maximum takeoff weight for which noise certification is requested, the following additional requirements apply:

(i) At least one takeoff test and one flyover test must be conducted at, or above, the maximum certification weight.

(ii) Each test weight must be within +5 percent or −10 percent of the maximum certification weight.

(7) Each approach test must be conducted with the aircraft stabilized and following a 6.0 degree±0.5 degree approach angle and must meet the requirements of section H36.107 of this part.

(8) If an approach test series is conducted at weights other than the maximum landing weight for which certification is requested, the following additional requirements apply:

(i) At least one approach test must be conducted at a weight at, or above, the maximum landing weight.

(ii) Each test weight must be between +5 percent and −10 percent of the maximum certification weight.

(c) *Weather restrictions.* The tests must be conducted under the following atmospheric conditions:

(1) No rain or other precipitation.

(2) Ambient air temperature between 14°F and 95°F (−10°C and 35°C), inclusively, at a point 33 feet (10 meters) above the ground at the noise measuring station and at the aircraft. The temperature and relative humidity measured at a point 33 feet (10 meters) above the ground at the noise measuring station must be used to adjust for propagation path absorption.

(3) Relative humidity and ambient temperature at a point 33 feet (10 meters) above the ground at the noise measuring station and at the aircraft, is such that the sound attenuation in the one-third octave band centered at 8 kHz is not greater than 12 dB/100 meters and the relative humidity is between 20 percent and 95 percent, inclusively.

(4) Wind velocity as measured at 10 meters above ground does not exceed 10 knots (19 km/h) and the cross-wind component does not exceed 5 knots (9 km/h). The wind shall be determined using a continuous thirty-second averaging period spanning the 10dB down time interval.

(5) No anomalous meteorological conditions (including turbulence) that will significantly affect the noise level of the aircraft when the noise is recorded at each noise measuring station.

(6) The wind velocity, temperature, and relative humidity measurements required under the appendix must be measured in the vicinity of noise measuring stations 10 meters above the ground. The location of the meteorological measurements must be approved by the FAA as representative of those atmospheric conditions existing near the surface over the geographical area which aircraft noise measurements are made. In some cases, a fixed meteorological station (such as those found at airports or other facilities) may meet this requirement.

(7) Temperature and relative humidity measurements must be obtained within 30 minutes of each noise test.

(d) *Aircraft testing procedures.* (1) The aircraft testing procedures and noise measurements must be conducted and processed in a manner that yields the noise evaluation measure designated as Effective Perceived Noise Level (EPNL) in units of EPNdB, as prescribed in Appendix A of this part.

(2) The helicopter height and lateral position relative to the reference flight track (which passes through the flight track noise measuring station) must be determined using an FAA-approved method. The equipment used to make the determination must be independent of normal flight instrumentation. Applicable independent systems are radar tracking, theodolite triangulation, laser trajectography, photo scaling, or differential global positioning system.

(3) The helicopter position along the flight path must be related to the noise recorded at the noise measuring stations by means of synchronized signals recorded at an approved sampling rate. The helicopter position must be recorded relative to the reference flight track during the entire time interval in which the recorded signal is within 10 dB of PNLTM. Measuring and sampling equipment must be approved by the FAA before testing.

(4) Aircraft performance data sufficient to make the corrections required under section H36.205 of this appendix must be recorded at an FAA-approved sampling rate using FAA-approved equipment.

Section H36.103 *Takeoff test conditions.*

(a) This section, in *addition* to the applicable requirements of sections H36.101 and H36.205(b) of this appendix, applies to all takeoff noise tests conducted under this appendix to show compliance with Part 36.

(b) A test series must consist of at least six flights over the flight-track noise measuring station (with simultaneous measurements at all three noise measuring stations) as follows:

(1) An airspeed of either V_y±5 knots or the lowest approved speed±5 knots for the climb after takeoff, whichever speed is greater, must be established and maintained throughout the 10 dB-down time interval.

(2) The horizontal portion of each test flight must be conducted at an altitude of 65 feet (20 meters) above the ground level at the flight-track noise measuring station.

(3) Upon reaching a point 1,640 feet (500 meters) from the noise measuring station, the helicopter must be stabilized at the maximum takeoff power that corresponds to minimum installed engine(s) specification power available for the reference ambient conditions or gearbox torque limit, whichever is lower.

(4) The helicopter must be maintained throughout the 10 dB-down time interval at the best rate of climb speed V_y±5 knots, or the lowest approved speed for climb after takeoff, whichever is greater, for an ambient temperature of 25°C at sea level.

(5) The average rotor speed must not vary from the maximum normal operating rotor RPM by more than±1.0 percent during the 10 dB-down time interval.

(6) The helicopter must stay within±10° or±65 feet (±20 meters), whichever is greater, from the vertical above the reference track throughout the 10dB-down time interval.

(7) A constant takeoff configuration selected by the applicant must be maintained throughout the takeoff reference procedure with the landing gear position consistent with the airworthiness certification tests for establishing best rate-of-climb speed, V_y.

Section H36.105 ***Flyover test conditions.***

(a) This section, in addition to the applicable requirements of sections H36.101 and H36.205(c) of this appendix, applies to all flyover noise tests conducted under this appendix to show compliance with Part 36.

(b) A test series consists of at least six flights. The number of level flights made with a headwind component must be equal to the number of level flights made with a tailwind component with simultaneous measurements at all three noise measuring stations—

(1) In level flight cruise configuration;

(2) At a height of 492 feet±30 feet (150±9 meters) above the ground level at the flight-track noise measuring station; and

(3) The helicopter must fly within±10° or±65 feet (±20 meters), whichever is greater, from the vertical above the reference track throughout the 10 dB-down time interval.

(c) Each flyover noise test must be conducted—

(1) At a speed of 0.9VH ; 0.9VNE ; 0.45VH + 65 kts (0.45VH + 120 km/h); or 0.45VNE + 65 kts (0.45VNE + 120 km/h), whichever speed is least, to be maintained throughout the measured portion of the flyover;

(2) At average rotor speed, which must not vary from the maximum normal operating rotor RPM by more than±1.0 percent during the 10 dB-down time interval.

(3) With the power stabilized during the period when the measured helicopter noise level is within 10 dB of PNLTM.

(d) The airspeed shall not vary from the reference airspeed by more than±5 knots (9 km/hr).

Section H36.107 ***Approach test conditions.***

(a) This section, in addition to the requirements of sections H36.101 and H36.205(d) of this appendix, applies to all approach tests conducted under this appendix to show compliance with Part 36.

(b) A test series must consist of at least six flights over the flight-track noise measuring station (with simultaneous measurements at the three noise measuring stations)—

(1) On an approach slope of 6°±0.5°;

(2) At a height of 394±33 feet (120±10 meters)

(3) The helicopter must fly within±10° or±65 feet (±20 meters) lateral deviation tolerance, whichever is greater, from the vertical above the reference track throughout the 10 dB-down time interval;

(4) At stabilized airspeed equal to the certificated best rate of climb V_y, or the lowest approved speed for approach, whichever is greater, with power stabilized during the approach and over the flight path reference point, and continued to a normal touchdown; and

(5) At average rotor speed, which may not vary from the maximum normal operating rotor RPM by more than±1.0 percent during the 10 dB-down time interval; and

(6) The constant approach configuration used in airworthiness certification tests, with the landing gear extended, must be maintained throughout the approach reference procedure.

(c) The airspeed shall not vary from the reference airspeed by more than±5 knots (±9 km/hr).

Section H36.109 ***Measurement of Helicopter Noise Received on the Ground.***

The measurement system and the measurement, calibration and general analysis procedures to be used are provided in Appendix A, section A36.3 of this part.

Section H36.111 ***Reporting and correcting measured*** *data.*

(a) *General.* Data representing physical measurements, and corrections to measured data, including corrections to measurements for equipment response deviations, must be recorded in permanent form and appended to the record. Each correction must be reported and is subject to FAA approval. An estimate must be made of each individual error inherent in each of the operations employed in obtaining the final data.

(b) *Data reporting.* (1) Measured and corrected sound pressure levels must be presented in one-third octave band levels obtained with equipment conforming to the standards prescribed in section H36.109 of this appendix.

(2) The type of equipment used for measurement and analysis of all acoustic, aircraft performance, and meteorological data must be reported.

(3) The atmospheric environmental data required to demonstrate compliance with this appendix, measured throughout the test period, must be reported.

(4) Conditions of local topography, ground cover, or events which may interfere with sound recording must be reported.

(5) The following aircraft information must be reported:

(i) Type, model, and serial numbers, if any, of aircraft engines and rotors.

(ii) Gross dimensions of aircraft and location of engines.

(iii) Aircraft gross weight for each test run.

(iv) Aircraft configuration, including landing gear positions.

(v) Airspeed in knots.

(vi) Helicopter engine performance as determined from aircraft instruments and manufacturer's data.

(vii) Aircraft flight path, above ground level in feet, determined by an FAA approved method which is independent of normal flight instrumentation, such as radar tracking, theodolite triangulation, laser trajectography, or photographic scaling techniques.

(6) Aircraft speed, and position, and engine performance parameters must be recorded at an approved sampling rate sufficient to correct to the noise certification reference test conditions prescribed in section H36.3 of this appendix. Lateral position relative to the reference flight-track must be reported.

(c) *Data corrections.* (1) Aircraft position, performance data and noise measurement must be corrected to the noise certification reference conditions as prescribed in sections H36.3 and H36.205 of this appendix.

(2) The measured flight path must be corrected by an amount equal to the difference between the applicant's predicted flight path for the certification reference conditions and the measured flight path at the test conditions. Necessary corrections relating to helicopter flight path or performance may be derived from FAA-approved data for the difference between measured and reference conditions, together with appropriate allowances for sound attenuation with distance. The Effective Perceived Noise Level (EPNL) correction may not exceed 2.0 EPNdB except for takeoff flight condition, where the correction may not exceed 4.0 EPNdB, of which the arithmetic sum of Δ1 (described in section H36.205(f)(1)) and the term -7.5 log (AL/ALr) from Δ2 term (described in section H36.205(g)(1)(i)) may not exceed 2.0 EPNdB, for any combination of the following:

(i) The helicopter not passing vertically above the measuring station.

(ii) Any difference between the reference flight track and the actual test flight track; and

(iii) Detailed correction requirements prescribed in section H36.205 of this appendix.

(3) Helicopter sound pressure levels within the 10 dB-down time interval must exceed the mean background sound pressure levels determined under section B36.3.9.11 by at least 3 dB in each one-third octave band, or must be corrected under an FAA-approved method.

(d) *Validity of results.* (1) The test results must produce three average EPNL values within the 90 percent confidence limits, each value consisting of the arithmetic average of the corrected noise measurements for all valid test runs at the takeoff, level flyovers, and approach conditions. The 90 percent confidence limit applies separately to takeoff, flyover, and approach.

(2) The minimum sample size acceptable for each takeoff, approach, and flyover certification measurements is six. The number of samples must be large enough to establish statistically for each of the three average noise certification levels a 90 percent confidence limit which does not exceed±1.5 EPNdB. No test result may be omitted from the averaging process, unless otherwise specified by the FAA.

(3) To comply with this appendix, a minimum of six takeoffs, six approaches, and six level flyovers is required. To be counted toward this requirement, each flight event must be validly recorded at all three noise measuring stations.

(4) The approved values of VH and Vy used in calculating test and reference conditions and flight profiles must be reported along with measured and corrected sound pressure levels.

Section H36.113 ***Atmospheric attenuation of sound.***

(a) The values of the one-third octave band spectra measured during helicopter noise certification tests under this appendix must conform, or be corrected, to the reference conditions prescribed in section H36.3(a). Each correction must account for any differences in the atmospheric attenuation of sound between the test-day conditions and the reference-day conditions along the sound propagation path between the aircraft and the microphone. Unless the meteorological conditions are within the test window prescribed in this appendix, the test data are not acceptable.

(b) *Attenuation rates.* The procedure for determining the atmospheric attenuation rates of sound with distance for each one-third octave bands must be determined in accordance with Society of Automotive Engineering (SAE) ARP 866A. The atmospheric attenuation equations are provided in both the International and English system of units in section A36.7 of this part.

(c) *Correction for atmospheric attenuation.* (1) EPNL values calculated for measured data must be corrected whenever—

(i) The ambient atmospheric conditions of temperature and relative humidity do not conform to the reference conditions, 77°F and 70%, respectively, or

(ii) The measured flight paths do not conform to the reference flight paths.

(iii) The temperature and relative humidity measured at 33 feet (10 meters) above the ground must be used to adjust for propagation path absorption.

(2) The mean attenuation rate over the complete sound propagation path from the aircraft to the microphone must be computed for each one-third octave band from 50 Hz to 10,000 Hz. These rates must be used in computing the corrections required in section H36.111(d) of this appendix.

part c—noise evaluation and calculation under § 36.803

Section H36.201 ***Noise Evaluation in EPNdB.***

(a) Effective Perceived Noise Level (EPNL), in units of effective perceived noise decibels (EPNdB), shall be used for evaluating noise level values under § 36.803 of this part. Except as provided in paragraph (b) of this section, the procedures in appendix A of Part 36 must be used for computing EPNL. appendix A includes requirements governing determination of noise values, including calculations of:

(1) Perceived noise levels;

(2) Corrections for spectral irregularities;

(3) Tone corrections;

(4) Duration corrections;

(5) Effective perceived noise levels; and

(6) Mathematical formulation of noy tables.

(b) Notwithstanding the provisions of section A36.4.3.1(a), for helicopter noise certification, corrections for spectral irregularities shall start with the corrected sound pressure level in the 50 Hz one-third octave band.

Section H36.203 ***Calculation of noise levels.***

(a) To demonstrate compliance with the noise level limits of section H36.305, the noise values measured simultaneously at the three noise measuring points must be arithmetically averaged to obtain a single EPNdB value for each flight.

(b) The calculated noise level for each noise test series, *i.e.* , takeoff, flyover, or approach must be the numerical average of at least six separate flight EPNdB values. The 90 percent confidence limit for all valid test runs under section H36.111(d) of this appendix applies separately to the EPNdB values for each noise test series.

Section H36.205 ***Detailed data correction procedures.***

(a) *General.* If the test conditions do not conform to those prescribed as noise certification reference conditions under section H36.305 of this appendix, the following correction procedure shall apply:

(1) If there is any difference between measured test and reference conditions, an appropriate correction must be made to the EPNL calculated from the measured noise data. Conditions that can result in a different value include:

(i) Atmospheric absorption of sound under measured test conditions that are different from the reference test conditions; or

(ii) Measured flight path that is different from the reference flight path.

(2) The following correction procedures may produce one or more possible correction values which must be added algebraically to the calculated EPNL to bring it to reference conditions:

(i) The flight profiles must be determined for both reference and test conditions. The procedures require noise and flight path recording with a synchronized time signal from which the test profile can be delineated, including the aircraft position for which PNLTM is observed at the noise measuring station. For takeoff, the flight profile corrected to reference conditions may be derived from FAA approved manufacturer's data.

(ii) The sound propagation paths to the microphone from the aircraft position corresponding to PNLTM must be determined for both the test and reference profiles. The SPL values in the spectrum of PNLTM must then be corrected for the effects of—

(A) Change in atmospheric sound absorption;

(B) Atmospheric sound absorption on the linear difference between the two sound path lengths; and

(C) Inverse square law on the difference in sound propagation path length. The corrected values of SPL must then be converted to a reference condition PNLTM value from which PNLTM must be subtracted. The resulting difference represents the correction which must be added algebraically to the EPNL calculated from the measured data.

(iii) As observed at the noise measuring station, the measured PNLTM distance is different from the reference PNLTM distance and therefore the ratio must be calculated and used to determine a noise duration correction factor. Effective perceived noise level, EPNL, is determined by the algebraic sum of the maximum tone corrected perceived noise level (PNLTM) and the duration correction factor.

(iv) For aircraft flyover, alternative source noise corrections require FAA approval and must be determined and adjusted to account for noise level changes caused by the differences between measured test conditions and reference conditions.

(b) *Takeoff profiles.* (1) Figure H1 illustrates a typical takeoff profile, including reference conditions.

(i) The reference takeoff flight path is described in section H36.3(c).

(ii) The test parameters are functions of the helicopter's performance and weight and the atmospheric conditions of temperature, pressure, wind velocity and direction.

(2) For the actual takeoff, the helicopter approaches position C in level flight at 65 feet (20 meters) above ground level at the flight track noise measuring station and at either Vy±5 knots or the lowest approved speed for the climb after takeoff, whichever speed is greater.

(3) Figure H1 illustrates the significant geometrical relationships influencing sound propagation. Position L represents the helicopter location on the measured takeoff path from which PNLTM is observed at station A, and Lr is the corresponding position on the reference sound propagation path. Propagation paths AL and ALr both form the same angle θ (theta) relative to their respective flight paths.

(c) *Level flyover profiles.* (1) The noise type certification level flyover profile is shown in Figure H2. Airspeed must be stabilized within±5 knots of the reference airspeed determined using the procedures in section H36.3(d). The number of level flights made with a headwind component must be equal to the number of level flights made with a tailwind component.

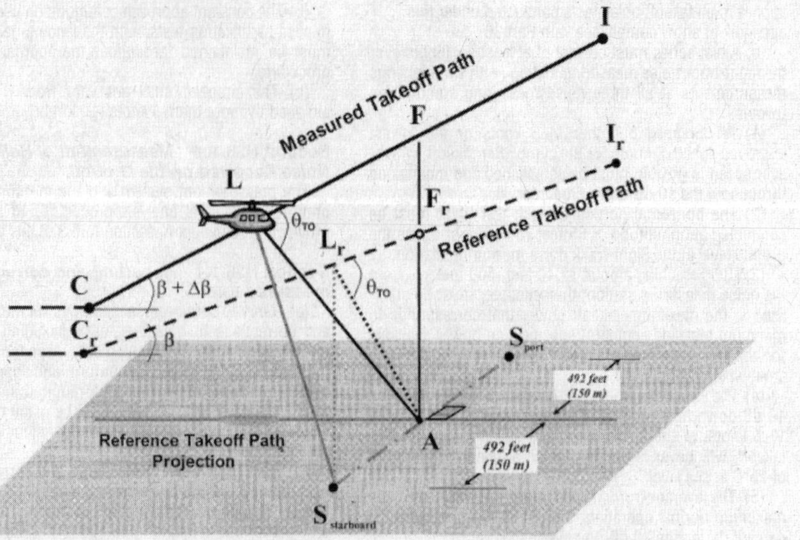

Figure H1.
Comparison of Measured and Reference Takeoff Profiles

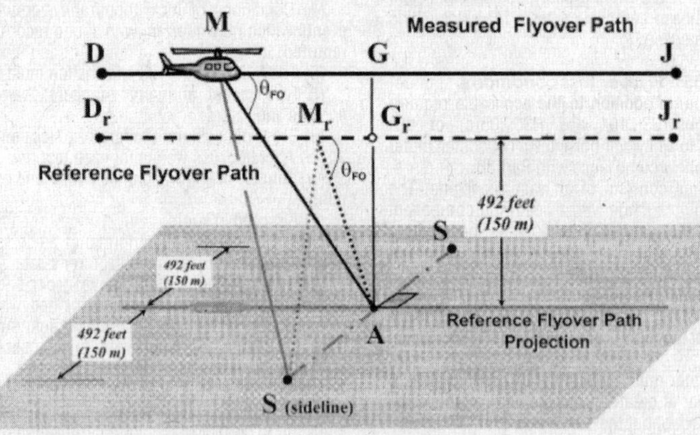

Figure H2.
Comparison of Measured and Reference Flyover Profiles

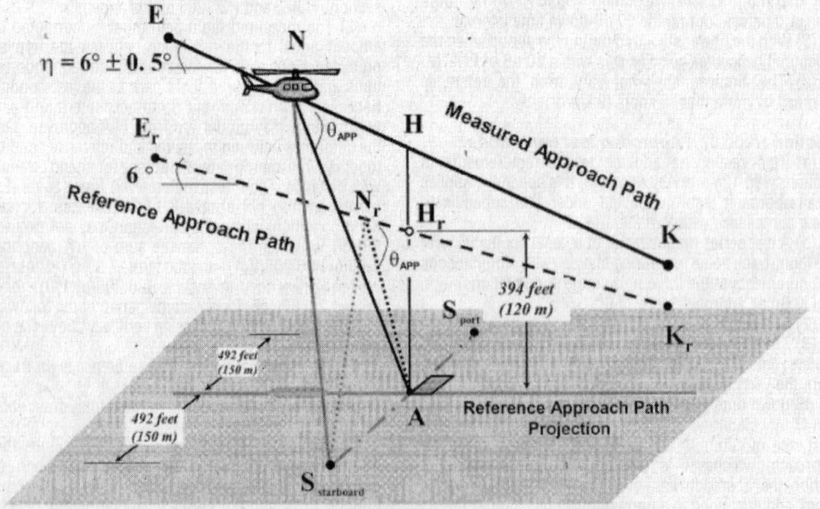

Figure H3.
Comparison of Measured and Reference Approach Profiles

(2) Figure H2 illustrates comparative flyover profiles when test conditions do not conform to prescribed reference conditions. The position of the helicopter shall be recorded for a distance (DJ) sufficient to ensure recording of the entire interval during which the measured helicopter noise level is within 10 dB of PNLTM, as required. The flyover profile is defined by the height AG which is a function of the operating conditions controlled by the pilot. Position M represents the helicopter location on the measured flyover flight path for which PNLTM is observed at station A, and Mr is the corresponding position on the reference flight path.

(d) *Approach profiles.* (1) Figure H3 illustrates a typical approach profile, including reference conditions.

(2) The helicopter approaches position H along a 6° (±0.5°) average approach slope throughout the 10dB-down time interval. Deviation from the 6° average approach slope must be approved by the FAA before testing.

(3) Figure H3 illustrates portions of the measured and reference approach flight paths including the significant geometrical relationships influencing sound propagation. The measured approach path is represented by segment EK with an approach allowable angle θ. Reference positions, Er and Kr , define an idealized reference approach angle of 6°. Position N represents the helicopter location on the measured approach flight path for which PNLTM is observed at measuring station A, and Nr is the corresponding position on the reference approach flight path. The measured and reference noise propagation paths are AN and ANr , respectively, both of which form the same angle, θAPP , corresponding to PNLTM relative to their approach flight paths.

(e) *Correction of noise at source during level flyover.* (1) For level overflight, if any combination of the following three factors, airspeed deviations from reference, rotor speed deviations from reference, and temperature deviations from reference, results in a noise correlating parameter whose value deviates from the reference value of this parameter, then source noise adjustments must be determined from the manufacturer's data that is approved by the FAA.

(2) Off-reference tip Mach number adjustments must be based upon a sensitivity curve of PNLTM versus advancing blade tip Mach number, deduced from overflights performed at different airspeeds surrounding the reference airspeed. If the test aircraft is unable to attain the reference value, then an extrapolation of the sensitivity curve is permitted if data cover at least a range of 0.03 Mach units. The advancing blade tip Mach number must be computed using true airspeed, onboard outside air temperature, and rotor speed. A separate PNLTM versus advancing blade tip Mach number function must be derived for each of the three certification microphone locations, *i.e.*, centerline, sideline left, and sideline right. Sideline left and right are defined relative to the direction of flight for each run. PNLTM adjustments are to be applied to each microphone datum using the appropriate PNLTM function.

(f) *PNLT corrections.* If the measured ambient atmospheric conditions of temperature and relative humidity differ from those prescribed as reference conditions under this appendix (77 degrees F and 70 percent, respectively), corrections to the EPNL values must be calculated from the measured data under paragraph (a) of this section as follows:

(1) *Takeoff flight path.* For the takeoff flight path shown in Figure H1, the spectrum of PNLTM observed at station A for the aircraft at position L is decomposed into its individual SPL(i) values.

(i) Step 1. A set of corrected values are then computed as follows:

$SPL(i)_r = SPL(i) + C [\alpha (i) - \alpha (i)o]AL + C \alpha (i)o (AL - AL_r) + 20 \log (AL/AL_r)$

where SPL(i) and SPL(i)r are the measured and corrected sound pressure levels, respectively, in the i -th one-third octave band. The first correction term adjusts for the effect of change in atmospheric sound absorption where α(i) and α(i)o are the sound attenuation coefficients for the test and reference atmospheric conditions, respectively, for the i -th one-third octave band, and AL is the measured takeoff sound propagation path. The conversion factor constant, *C*, is 0.001 for English System of Units and is 0.01 for International System of Units. The second correction term adjusts for the effects of atmospheric attenuation due to the difference in the sound propagation path length where AL_r is the Reference takeoff sound propagation path. The third correction term, known as the "inverse square" law, adjusts for the effect of the difference in the sound propagation path lengths.

(ii) Step 2. The corrected values of the SPL(i)r are then converted to reference condition PNLT and a correction term calculated as follows:

Δ1= PNLT − PNLTM

which represents the correction to be added algebraically to the EPNL calculated from the measured data.

(2) *Level flyover flight path.* (i) The procedure described in paragraph (f)(1) of this section for takeoff paths is also used for the level flyover paths, with the values of SPL(i) r relating to the flyover sound propagation paths shown in Figure H2 as follows:

$SPL(i)_r = SPL(i) + C [\alpha (i) - \alpha (i)o]AM + C \alpha (i)o (AM - AM_r) + 20 \log (AM/AM_r)$

where the lines AM and AM_r are the measured and reference level flyover sound propagation paths, respectively.

(ii) The remainder of the procedure is the same for the flyover condition as that prescribed in the paragraph (f)(1)(ii) of this section regarding takeoff flight path.

(3) *Approach flight path.* (i) The procedure described in paragraph (f)(1) of this section for takeoff paths is also used for the approach paths, with the values of SPL(i)r relating to the approach sound propagation paths shown in Figure H3 as follows:

$SPL(i)_r = SPL(i) + C [\alpha (i) - \alpha (i)o]AN + C \alpha (i)o (AN - AN_r) + 20 \log (AN/AN_r)$

where the lines AN and ANr are the measured and reference approach sound propagation paths, respectively.

(ii) The remainder of the procedure is the same for the approach condition as that prescribed in the paragraph (f)(1)(ii) of this section regarding takeoff flight path.

(4) *Sideline microphones.* (i) The procedure prescribed in paragraph (f)(1) of this section for takeoff paths is also used for the propagation to the sideline locations, with the values of SPL(i)r relating as follows to the measured sideline sound propagation path shown in Figure H3 as follows:

$SPL(i)r = SPL(i) + C [\alpha (i) - \alpha (i)o]SX + C \alpha (i)o (SX - SX_r) + 20 \log (SX/SX_r)$

where S is the sideline measuring station and, based upon the flight condition, the helicopter positions, X and Xr , correspond to:

$X = L$, and $X_r = Lr$ for takeoff

$X = M$, and $X_r = Mr$ for flyover

$X = N$, and $X_r = Nr$ for approach

(ii) The remainder of the procedure is the same for the sideline paths as that prescribed in the paragraph (f)(1)(ii) of this section regarding takeoff flight paths.

(g) *Duration corrections.* (1) If the measured takeoff and approach flight paths do not conform to those prescribed as the corrected and reference flight paths, respectively, under section A36.5(d)(2) it will be necessary to apply duration corrections to the EPNL values calculated from the measured data. Such corrections must be calculated as follows:

(i) *Takeoff flight path.* For the takeoff path shown in Figure H1, the correction term is calculated using the formula—

$\Delta 2 = -7.5 \log (AL/AL_r) + 10 \log (V/V_r)$

which represents the correction that must be added algebraically to the EPNL calculated from the measured data. The lengths AL and ALr are the measured and reference takeoff distances from the noise measuring station A to the measured and the reference takeoff paths, respectively. A negative sign indicates that, for the particular case of a duration correction, the EPNL calculated from the measured data must be reduced if the measured takeoff path is at greater altitude than the reference takeoff path.

(ii) *Level flyover flight paths.* For the level flyover flight path, the correction term is calculated using the formula—

$\Delta 2 = -7.5 \log (AM/AM_r) + 10 \log (V/V_r)$

where AM is the measured flyover distance from the noise measuring station A to the measured flyover path, and AMr is the reference distance from station A to the reference flyover path.

(iii) *Approach flight path.* For the approach path shown in Figure H3, the correction term is calculated using the formula—

$\Delta 2 = -7.5 \log (AN/AN_r) + 10 \log (V/V_r)$

where AN is the measured approach distance from the noise measuring station A to the measured approach path, and ANr is the reference distance from station A to the reference approach path.

(iv) *Sideline microphones.* For the sideline flight path, the correction term is calculated using the formula—

$\Delta 2 = -7.5 \log (SX/SX_r) + 10 \log (V/V_r)$

where S is the sideline measuring station and based upon the flight condition, the helicopter positions, X and Xr , correspond to:

X = L, and Xr = Lr for takeoff

X = M, and Xr = Mr for flyover

$X = N$, and $X_r = N_r$ for approach

(2) The adjustment procedure described in this section shall apply to the sideline microphones in the take-off, overflight, and approach cases. Although the noise emission is strongly dependent on the directivity pattern, variable from one helicopter type to another, the propagation angle θ shall be the same for test and reference flight paths. The elevation angle ψ shall not be constrained but must be determined and reported. The certification authority shall specify the acceptable limitations on ψ. Corrections to data obtained when these limits are exceeded shall be applied using FAA approved procedures.

part d—noise limits under § 36.805

Section H36.301 ***Noise measurement, evaluation, and calculation.***

Compliance with this part of this appendix must be shown with noise levels measured, evaluated, and calculated as prescribed under Parts B and C of this appendix.

Section H36.303 ***[Reserved]***

Section H36.305 ***Noise levels.***

(a) *Limits.* For compliance with this appendix, it must be shown by flight test that the calculated noise levels of the helicopter, at the measuring points described in Section H36.305(a) of this appendix, do not exceed the following, with appropriate interpolation between weights:

(1) *Stage 1* noise limits for acoustical changes for helicopters are as follows:

(i) For takeoff, flyover, and approach calculated noise levels, the noise levels of each Stage 1 helicopter that exceed the Stage 2 noise limits plus 2 EPNdB may not, after a change in type design, exceed the noise levels created prior to the change in type design.

(ii) For takeoff, flyover, and approach calculated noise levels, the noise levels of each Stage 1 helicopter that do not exceed the Stage 2 noise limits plus 2 EPNdB may not, after the change in type design, exceed the Stage 2 noise limits plus 2 EPNdB.

(2) *Stage 2* noise limits are as follows:

(i) *For takeoff calculated noise levels* —109 EPNdB for maximum takeoff weights of 176,370 pounds (80,000 kg) or more, reduced by 3.01 EPNdB per halving of the weight down to 89 EPNdB, after which the limit is constant.

(ii) *For flyover calculated noise levels* —108 EPNdB for maximum weights of 176,370 pounds (80,000 kg) or more, reduced by 3.01 EPNdB per halving of the weight down to 88 EPNdB, after which the limit is constant.

(iii) *For approach calculated noise levels* —110 EPNdB for maximum weights of 176,370 pounds (80,000 kg) or more, reduced by 3.01 EPNdB per halving of the weight down to 90 EPNdB, after which the limit is constant.

(b) *Tradeoffs.* Except to the extent limited under § 36.11(b) of this part, the noise limits prescribed in paragraph (a) of this section may be exceeded by one or two of the takeoff, flyover, or approach calculated noise levels determined under section H36.203 of this appendix if

(1) The sum of the exceedances is not greater than 4 EPNdB;

(2) No exceedance is greater than 3 EPNdB; and

(3) The exceedances are completely offset by reduction in the other required calculated noise levels.

[Amdt. 36-14, 53 FR 3541, Feb. 5, 1988; 53 FR 4099, Feb. 11, 1988; 53 FR 7728, Mar. 10, 1988, as amended by Amdt. 36-54, 67 FR 45237, July 8, 2002; Amdt. 36-25, 69 FR 31234, June 2, 2004; Amdt. 36-25, 69 FR 41573, July 9, 2004]

Appendix I to Part 36 [Reserved]

Appendix J to Part 36—Alternative Noise Certification Procedure for Helicopters Under Subpart H Having a Maximum Certificated Takeoff Weight of Not More Than 7,000 Pounds

part a—reference conditions

part b—noise measurement procedure under § 36.801

part c—noise evaluation and calculation under § 36.803

part d—noise limits procedure under § 36.805

part a—reference conditions

Section J36.1 *General.*

This appendix prescribes the alternative noise certification requirements identified under § 36.1 of this part and subpart H of this part for helicopters in the primary, normal, transport, and restricted categories having maximum certificated takeoff weight of not more than 7,000 pounds including:

(a) The conditions under which an alternative noise certification test under subpart H of this part must be conducted and the alternative measurement procedure that must be used under § 36.801 of this part to measure the helicopter noise during the test;

(b) The alternative procedures which must be used under § 36.803 of this part to correct the measured data to the reference conditions and to calculate the noise evaluation quantity designated as Sound Exposure Level (SEL); and

(c) The noise limits for which compliance must be shown under § 36.805 of this part.

Section J36.3 *Reference Test Conditions.*

(a) *Meteorological conditions.* The following are the noise certification reference atmospheric conditions which shall be assumed to exist from the surface to the helicopter altitude:

(1) Sea level pressure of 2116 pounds per square foot (76 centimeters mercury);

(2) Ambient temperature of 77 degrees Fahrenheit (25 degrees Celsius);

(3) Relative humidity of 70 percent; and

(4) Zero wind.

(b) *Reference test site.* The reference test site is flat and without line-of-sight obstructions across the flight path that encompasses the 10 dB down points of the A-weighted time history.

(c) *Level flyover reference profile.* The reference flyover profile is a level flight, 492 feet (150 meters) above ground level as measured at the noise measuring station. The reference flyover profile has a linear flight track and passes directly over the noise monitoring station. Airspeed is stabilized at 0.9VH ; 0.9VNE ; 0.45VH + 65 kts (120 km/h); or 0.45VNE + 65 kts (120 km/h), whichever of the four airspeeds is least, and maintained throughout the measured portion of the flyover. Rotor speed is stabilized at the maximum normal operating RPM throughout the 10 dB-down time interval.

(1) For noise certification purposes, VH is defined as the airspeed in level flight obtained using the minimum specification engine power corresponding to maximum continuous power available for sea level pressure of 2,116 psf (1,013.25 hPa) at 77°F (25°C) ambient conditions at the relevant maximum certificated weight. The value of VH and VNE used for noise certification must be included in the Flight Manual.

(2) VNE is the never-exceed airspeed.

(d) The weight of the helicopter shall be the maximum takeoff weight at which noise certification is requested.

Section J36.5 *[Reserved]*

Part B—Noise Measurement Procedure Under § 36.801

Section J36.101 *Noise certification test and measurement conditions.*

(a) *General.* This section prescribes the conditions under which helicopter noise certification tests must be conducted and the measurement procedures that must be used to measure helicopter noise during each test.

(b) *Test site requirements.* (1) The noise measuring station must be surrounded by terrain having no excessive sound absorption characteristics, such as might be caused by thick, matted, or tall grass, shrubs, or wooded areas.

(2) During the period when the flyover noise measurement is within 10 dB of the maximum A-weighted sound level, no obstruction that significantly influences the sound field from the helicopter may exist within a conical space above the noise measuring position (the point on the ground vertically below the microphone), the cone is defined by an axis normal to the ground and by half-angle 80 degrees from this axis.

(c) *Weather restrictions.* The test must be conducted under the following atmospheric conditions:

(1) No rain or other precipitation;

(2) Ambient air temperature between 36 degrees and 95 degrees Fahrenheit (2 degrees and 35 degrees Celsius), inclusively, and relative humidity between 20 percent and 95 percent inclusively, except that testing may not take place where combinations of temperature and relative humidity result in a rate of atmospheric attenuation greater than 10 dB per 100 meters (30.5 dB per 1000 ft) in the one-third octave band centered at 8 kiloHertz.

(3) Wind velocity that does not exceed 10 knots (19 km/h) and a crosswind component that does not exceed 5 knots (9 km/h). The wind shall be determined using a continuous averaging process of no greater than 30 seconds;

(4) Measurements of ambient temperature, relative humidity, wind speed, and wind direction must be made between 4 feet (1.2 meters) and 33 feet (10 meters) above the ground. Unless otherwise approved by the FAA, ambient temperature and relative humidity must be measured at the same height above the ground.

(5) No anomalous wind conditions (including turbulence) or other anomalous meteorological conditions that will significantly affect the noise level of the helicopter when the noise is recorded at the noise measuring station; and

(6) If the measurement site is within 6560 feet (2,000 meters) of a fixed meteorological station (such as those found at airports or other facilities) the weather measurements reported for temperature, relative humidity and wind velocity may be used, if approved by the FAA.

(d) *Helicopter testing procedures.* (1) The helicopter testing procedures and noise measurements must be conducted and processed in a manner which yields the noise evaluation measure designated Sound Exposure Level (SEL) as defined in section J36.109(b) of this appendix.

(2) The helicopter height relative to the noise measurement point sufficient to make corrections required under section J36.205 of this appendix must be determined by an FAA-approved method that is independent of normal flight instrumentation, such as radar tracking, theodolite triangulation, laser trajectography, or photographic scaling techniques.

(3) If an applicant demonstrates that the design characteristics of the helicopter would prevent flight from being conducted in accordance with the reference test conditions prescribed under section J36.3 of this appendix, then with FAA approval, the reference test conditions used under this appendix may vary from the standard reference test conditions, but only to the extent demanded by those design characteristics which make compliance with the reference test conditions impossible.

Section J36.103 *[Reserved]*

Section J36.105 *Flyover test conditions.*

(a) This section prescribes the flight test conditions and allowable random deviations for flyover noise tests conducted under this appendix.

(b) A test series must consist of at least six flights. The number of level flights made with a headwind component must be equal to the number of level flights made with a tailwind component over the noise measurement station:

(1) In level flight and in cruise configuration;

(2) At a height of 492 feet±50 feet (150±15 meters) above the ground level at the noise measuring station; and

(3) Within±10 degrees from the zenith.

(c) Each flyover noise test must be conducted:

(1) At the reference airspeed specified in section J36.3(c) of this appendix, with such airspeed adjusted as necessary to produce the same advancing blade tip Mach number as associated with the reference conditions;

(i) Advancing blade tip Mach number (MAT) is defined as the ratio of the arithmetic sum of blade tip rotational speed (VR) and the helicopter true air speed (VT) over the speed of sound (c) at 77 degrees Fahrenheit (1135.6 ft/sec or 346.13 m/sec) such that MAT =(VR +VT)/c; and

(ii) The airspeed shall not vary from the adjusted reference airspeed by more than±3 knots (±5 km/hr) or an equivalent FAA-approved variation from the reference advancing blade tip Mach number. The adjusted reference airspeed shall be maintained throughout the measured portion of the flyover.

(2) At rotor speed stabilized at the power on maximum normal operating rotor RPM (±1 percent); and

(3) With the power stabilized during the period when the measured helicopter noise level is within 10 dB of the maximum A-weighted sound level (LAMAX).

(d) The helicopter test weight for each flyover test must be within plus 5 percent or minus 10 percent of the maximum takeoff weight for which certification under this part is requested.

(e) The requirements of paragraph (b)(2) of this section notwithstanding, flyovers at an FAA-approved lower height may be used and the results adjusted to the reference measurement point by an FAA-approved method if the ambient noise in the test area, measured in accordance with the requirements prescribed in section J36.109 of this appendix, is found to be within 15 dB(A) of the maximum A-weighted helicopter noise level (LAMAX) measured at the noise measurement station in accordance with section J36.109 of this appendix.

Section J36.107 *[Reserved]*

Section J36.109 *Measurement of helicopter noise received on the ground.*

(a) *General.* (1) The helicopter noise measured under this appendix for noise certification purposes must be obtained with FAA-approved acoustical equipment and measurement practices.

(2) Paragraph (b) of this section identifies and prescribes the specifications for the noise evaluation measurements required under this appendix. Paragraphs (c) and (d) of this section prescribe the required acoustical equipment specifications. Paragraphs (e) and (f) of this section prescribe the calibration and measurement procedures required under this appendix.

(b) *Noise unit definition.* (1) The value of sound exposure level (SEL, or as denoted by symbol, LAE), is defined as the level, in decibels, of the time integral of squared 'A'-weighted sound pressure (PA) over a given time period or event, with reference to the square of the standard reference sound pressure (PO) of 20 micropascals and a reference duration of one second.

(2) This unit is defined by the expression:

$$L_{AE} = 10Log_{10} \frac{1}{T_0} \int_{t_1}^{t_2} \left(\frac{P_A(t)}{P_0} \right)^2 dt \ dB$$

Where TO is the reference integration time of one second and (t2 -t1) is the integration time interval.

(3) The integral equation of paragraph (b)(2) of this section can also be expressed as:

$$L_{AE} = 10Log_{10} \frac{1}{T_0} \int_{t_1}^{t_2} 10^{0.1L_A(t)} dt \ dB$$

Where LA (t) is the time varying A-weighted sound level.

(4) The integration time (t2 -t1) in practice shall not be less than the time interval during which LA (t) first rises to within 10 dB(A) of its maximum value (LAMAX) and last falls below 10 dB(A) of its maximum value.

(5) The SEL may be approximated by the following expression:

LAE =LAMAX + <delta> A

where <delta> A is the duration allowance given by:

<delta> A=10 log10 (T)

where T=(t2 -t1)/2 and LAMAX is defined as the maximum level, in decibels, of the A-weighted sound pressure (slow response) with reference to the square of the standard reference sound pressure (P0).

(c) *Measurement system.* The acoustical measurement system must consist of FAA-approved equipment equivalent to the following:

(1) A microphone system with frequency response that is compatible with the measurement and analysis system accuracy prescribed in paragraph (d) of this section;

(2) Tripods or similar microphone mountings that minimize interference with the sound energy being measured;

(3) Recording and reproducing equipment with characteristics, frequency response, and dynamic range that are compatible with the response and accuracy requirements of paragraph (d) of this section; and

(4) The calibration and checking of measurement systems must use the procedures described in Section A36.3.9.

(d) *Sensing, recording, and reproducing equipment.* (1) The noise levels measured from helicopter flyovers under this appendix may be determined directly by an integrating sound level meter, or the A-weighted sound level time history may be written onto a graphic level recorder set at "slow" response from which the SEL value may be determined. With the approval of the FAA, the noise signal may be tape recorded for subsequent analysis.

(i) The SEL values from each flyover test may be directly determined from an integrating sound level meter complying with the Standards of the International Electrotechnical Commission (IEC) Publication No. 804, "Integrating-averaging Sound Level Meters," as incorporated by reference under § 36.6 of this part, for a Type 1 instrument set at "slow" response.

(ii) The acoustic signal from the helicopter, along with the calibration signals specified under paragraph (e) of this section and the background noise signal required under paragraph (f) of this section may be recorded on a magnetic tape recorder for subsequent analysis by an integrating sound level meter identified in paragraph (d)(1)(i) of this section. The record/playback system (including the audio tape) of the tape recorder must conform to the requirements prescribed in section A36.3.6 of appendix A of this part. The tape recorder shall comply with specifications of IEC Publication No. 561, "Electro-acoustical Measuring Equipment for Aircraft Noise Certification," as incorporated by reference under § 36.6 of this part.

(iii) The characteristics of the complete system shall comply with the recommendations given in IEC Publication No. 651, "Sound Level Meters," as incorporated by reference under § 36.6 of this part, with regard to the specifications concerning microphone, amplifier, and indicating instrument characteristics.

(iv) The response of the complete system to a sensibly plane progressive wave of constant amplitude shall lie within the tolerance limits specified in Table IV and Table V for Type 1 instruments in IEC Publication No. 651, "Sound Level Meters," as incorporated by reference under § 36.6 of this part, for weighting curve "A" over the frequency range of 45 Hz to 11500 Hz.

(v) A windscreen must be used with the microphone during each measurement of the helicopter flyover noise. Correction for any insertion loss produced by the windscreen, as a function of the frequency of the acoustic calibration required under paragraph (e) of this section, must be applied to the measured data and any correction applied must be reported.

(e) *Calibrations.* (1) If the helicopter acoustic signal is tape recorded for subsequent analysis, the measuring system and components of the recording system must be calibrated as prescribed under section A36.3.6 of appendix A of this part.

(2) If the helicopter acoustic signal is directly measured by an integrating sound level meter:

(i) The overall sensitivity of the measuring system shall be checked before and after the series of flyover tests and at intervals (not exceeding one-hour duration) during the flyover tests using an acoustic calibrator using sine wave noise generating a known sound pressure level at a known frequency.

(ii) The performance of equipment in the system will be considered satisfactory if, during each day's testing, the variation in the calibration value does not exceed 0.5 dB. The SEL data collected during the flyover tests shall be adjusted to account for any variation in the calibration value.

(iii) A performance calibration analysis of each piece of calibration equipment, including acoustic calibrators, reference microphones, and voltage insertion devices, must have been made during the six calendar months proceeding the beginning of the helicopter flyover series. Each calibration shall be traceable to the National Institute of Standards and Technology.

(f) *Noise measurement procedures.* (1) The microphone shall be of the pressure-sensitive capacitive type designed for nearly uniform grazing incidence response. The microphone shall be mounted with the center of the sensing element 4 feet (1.2 meters) above the local ground surface and shall be oriented for grazing incidence such that the sensing element, the diaphragm, is substantially in the plane defined by the nominal flight path of the helicopter and the noise measurement station.

(2) If a tape recorder is used, the frequency response of the electrical system must be determined at a level within 10 dB of the full-scale reading used during the test, utilizing pink or pseudorandom noise.

(3) The ambient noise, including both acoustical background and electrical noise of the measurement systems shall be determined in the test area and the system gain set at levels which will be used for helicopter noise measurements. If helicopter sound levels do not exceed the background sound levels by at least 15 dB(A), flyovers at an FAA-approved lower height may be used and the results adjusted to the reference measurement point by an FAA-approved method.

(4) If an integrating sound level meter is used to measure the helicopter noise, the instrument operator shall monitor the continuous A-weighted (slow response) noise levels throughout each flyover to ensure that the SEL integration process includes, at minimum, all of the noise signal between the maximum A-weighted sound level (LAMAX) and the 10 dB down points in the flyover time history. The instrument operator shall note the actual db(A) levels at the start and stop of the SEL integration interval and document these levels along with the value of LAMAX and the integration interval (in seconds) for inclusion in the noise data submitted as part of the reporting requirements under section J36.111(b) of this appendix.

Section J36.111 *Reporting Requirements.*

(a) *General.* Data representing physical measurements, and corrections to measured data, including corrections to measurements for equipment response deviations, must be recorded in permanent form and appended to the record. Each correction is subject to FAA approval.

(b) *Data reporting.* After the completion of the test the following data must be included in the test report furnished to the FAA:

(1) Measured and corrected sound levels obtained with equipment conforming to the standards prescribed in section J36.109 of this appendix;

(2) The type of equipment used for measurement and analysis of all acoustic, aircraft performance and flight path, and meteorological data;

(3) The atmospheric environmental data required to demonstrate compliance with this appendix, measured throughout the test period;

(4) Conditions of local topography, ground cover, or events which may interfere with the sound recording;

(5) The following helicopter information:

(i) Type, model, and serial numbers, if any, of helicopter, engine(s) and rotor(s);

(ii) Gross dimensions of helicopter, location of engines, rotors, type of antitorque system, number of blades for each rotor, and reference operating conditions for each engine and rotor;

(iii) Any modifications of non-standard equipment likely to affect the noise characteristics of the helicopter;

(iv) Maximum takeoff weight for which certification under this appendix is requested;

(v) Aircraft configuration, including landing gear positions;

(vi) VH or VNE (whichever is less) and the adjusted reference airspeed;

(vii) Aircraft gross weight for each test run;

(viii) Indicated and true airspeed for each test run;

(ix) Ground speed, if measured, for each run;

(x) Helicopter engine performance as determined from aircraft instruments and manufacturer's data; and

(xi) Aircraft flight path above ground level, referenced to the elevation of the noise measurement station, in feet, determined by an FAA-approved method which is independent of normal flight instrumentation, such as radar tracking, theodolite triangulation, laser trajectography, or photoscaling techniques; and

(6) Helicopter position and performance data required to make the adjustments prescribed under section J36.205 of this appendix and to demonstrate compliance with the performance and position restrictions prescribed under section J36.105 of this appendix must be recorded at an FAA-approved sampling rate.

Section J36.113 *[Reserved]*

Part C—Noise Evaluation and Calculations Under § 36.803

Section J36.201 *Noise Evaluation in SEL.*

The noise evaluation measure shall be the sound exposure level (SEL) in units of dB(A) as prescribed under section J36.109(b) of this appendix. The SEL value for each flyover may be directly determined by use of an integrating sound level meter. Specifications for the integrating sound level meter and requirements governing the use of such instrumentation are prescribed under section J36.109 of this appendix.

Section J36.203 *Calculation of Noise Levels.*

(a) To demonstrate compliance with the noise level limits specified under section J36.305 of this appendix, the SEL noise levels from each valid flyover, corrected as necessary to reference conditions under section J36.205 of this appendix, must be arithmetically averaged to obtain a single SEL dB(A) mean value for the flyover series. No individual flyover run may be omitted from the averaging process, unless otherwise specified or approved by the FAA.

(b) The minimum sample size acceptable for the helicopter flyover certification measurements is six. The number of samples must be large enough to establish statistically a 90 percent confidence limit that does not exceed±1.5 dB(A).

(c) All data used and calculations performed under this section, including the calculated 90 percent confidence limits, must be documented and provided under the reporting requirements of section J36.111 of this appendix.

Section J36.205 *Detailed Data Correction Procedures.*

(a) When certification test conditions measured under part B of this appendix differ from the reference test conditions prescribed under section J36.3 of this appendix, appropriate adjustments shall be made to the measured noise data in accordance with the methods set out in paragraphs (b) and (c) of this section. At minimum, appropriate adjustments shall be made for off-reference altitude and for the difference between reference airspeed and adjusted reference airspeed.

(b) The adjustment for off-reference altitude may be approximated from:

<delta>J1 =12.5 log10 (HT /492) dB;

where <delta>J1 is the quantity in decibels that must be algebraically added to the measured SEL noise level to correct for an off-reference flight path, HT is the height, in feet, of the test helicopter when directly over the noise measurement point, and the constant (12.5) accounts for the effects on spherical spreading and duration from the off-reference altitude.

(c) The adjustment for the difference between reference airspeed and adjusted reference airspeed is calculated from:

<delta>J3 =10 log10 (VRA /VR) dB;

Where <delta>J3 is the quantity in decibels that must be algebraically added to the measured SEL noise level to correct for the influence of the adjustment of the reference airspeed on the duration of the measured flyover event as perceived at the noise measurement station, VR is the reference airspeed as prescribed under section J36.3.(c) of this appendix, and VRA is the adjusted reference airspeed as prescribed under section J36.105(c) of this appendix.

(d) No correction for source noise during the flyover other than the variation of source noise accounted for by the adjustment of the reference airspeed prescribed for under section J36.105(c) of this appendix need be applied.

(e) No correction for the difference between the reference ground speed and the actual ground speed need be applied.

(f) No correction for off-reference atmospheric attenuation need be applied.

(g) The SEL adjustments must be less than 2.0 dB(A) for differences between test and reference flight procedures prescribed under section J36.105 of this appendix unless a larger adjustment value is approved by the FAA.

(h) All data used and calculations performed under this section must be documented and provided under the reporting requirements specified under section J36.111 of this appendix.

Part D—Noise Limits Procedure Under § 36.805

Section J36.301 *Noise Measurement, Evaluation, and Calculation.*

Compliance with this part of this appendix must be shown with noise levels measured, evaluated, and calculated as prescribed under parts B and C of this appendix.

Section J36.303 *[Reserved]*

Section J36.305 *Noise Limits.*

For compliance with this appendix, the calculated noise levels of the helicopter, at the measuring point described in section J36.101 of this appendix, must be shown to not exceed the following (with appropriate interpolation between weights):

(a) For primary, normal, transport, and restricted category helicopters having a maximum certificated takeoff weight of not more than 7,000 pounds that are noise tested under this appendix, the Stage 2 noise limit is 82 decibels SEL for helicopters up to 1,737 pounds maximum certificated takeoff weight at which the noise certification is requested, and increasing at a rate of 3.0 decibels per doubling of weight thereafter. The limit may be calculated by the equation: LAE (limit) = 82 + 3.0 [log10 (MTOW/1737)/log10 (2)] dB, where MTOW is the maximum takeoff weight, in pounds, for which certification under this appendix is requested.

(b) The procedures required in this amendment shall be done in accordance with the International Electrotechnical Commission IEC Publication No. 804, entitled "Integrating-averaging Sound Level Meters," First Edition, dated 1985. This incorporation by reference was approved by the Director of the Federal Register in accordance with 5 U.S.C. 552(a) and 1 CFR part 51. Copies may be obtained from the Bureau Central de la Commission Electrotechnique Internationale, 1, rue de Varembe, Geneva, Switzerland or the American National Standard Institute, 1430 Broadway, New York City, New York 10018, or at the National Archives and Records Administration (NARA). For information on the availability of this material at NARA, call 202-741-6030, or go to: *http://www.archives.gov/federal_register/code_of_federal_regulations/ibr_locations.html.*

[Doc. No. 26910, 57 FR 42855, Sept. 16, 1992, as amended by Amdt. 36-20, 57 FR 46243, Oct. 7, 1992; 69 FR 18803, Apr. 9, 2004; Amdt. 36-25, 69 FR 31234, June 2, 2004]

Appendix K to Part 36—Noise Requirements for Tiltrotors Under Subpart K

Section K1 *General*

This appendix prescribes noise limits and procedures for measuring noise and adjusting the data to standard conditions for tiltrotors as specified in § 36.1 of this part.

Section K2 *Noise Evaluation Measure*

The noise evaluation measure is the effective perceived noise level in EPNdB, to be calculated in accordance with section A36.4 of Appendix A to this part, except corrections for spectral irregularities must be determined using the 50 Hz sound pressure level found in section H36.201 of Appendix H to this part.

Section K3 *Noise Measurement Reference Points*

The following noise reference points must be used when demonstrating tiltrotor compliance with section K6 (Noise Certification Reference Procedures) and section K7 (Test Procedures) of this appendix:

(a) *Takeoff reference noise measurement points*—

As shown in Figure K1 below:

(1) The centerline noise measurement flight path reference point, designated A, is located on the ground vertically below the reference takeoff flight path. The measurement point is located 1,640 feet (500 m) in the horizontal direction of flight from the point Cr where transition to climbing flight is initiated, as described in section K6.2 of this appendix;

(2) Two sideline noise measurement points, designated as S(starboard) and S(port), are located on the ground perpendicular to and symmetrically stationed at 492 feet (150 m) on each side of the takeoff reference flight path. The measurement points bisect the centerline flight path reference point A.

(b) *Flyover reference noise measurement points*—

As shown in Figure K2 below:

(1) The centerline noise measurement flight path reference point, designated A, is located on the ground 492 feet (150 m) vertically below the reference flyover flight path. The measurement point is defined by the flyover reference procedure in section K6.3 of this appendix;

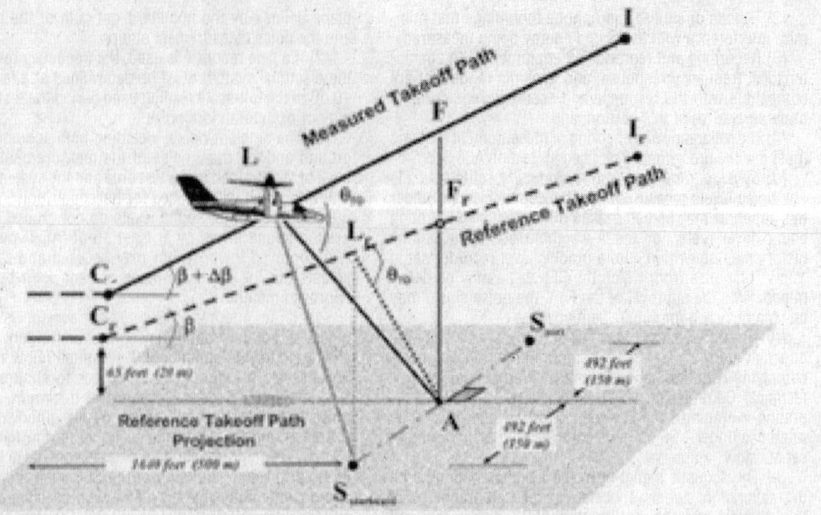

Figure K1.
Comparison of Measured and Reference Takeoff Profiles

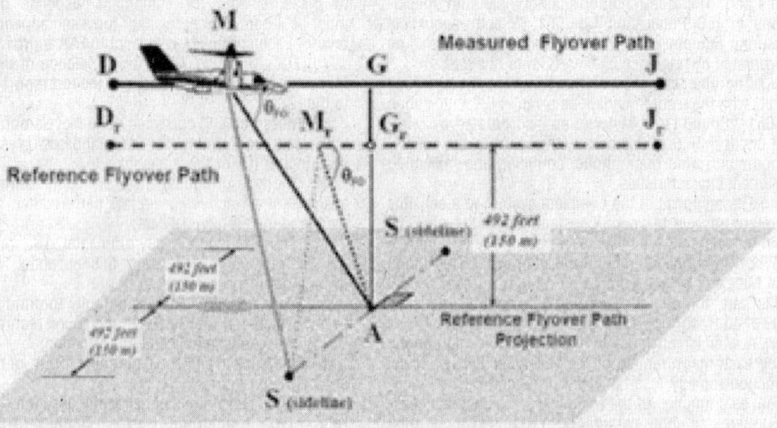

Figure K2.
Comparison of Measured and Reference Flyover Profiles

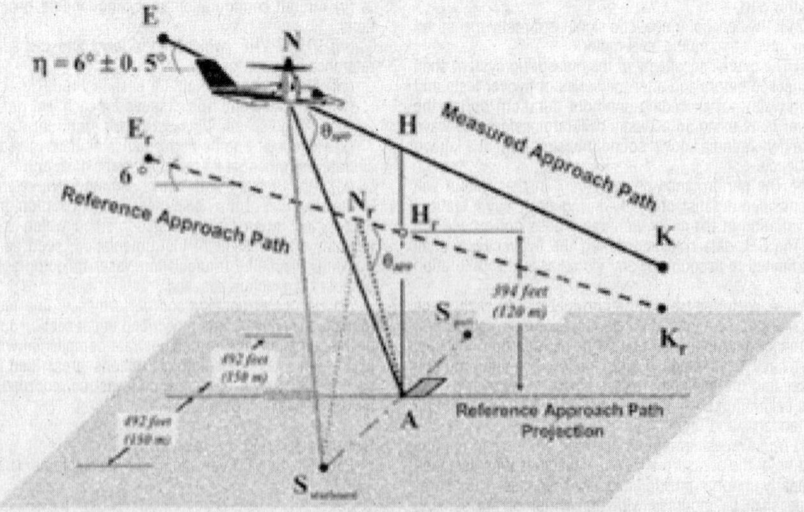

Figure K3.
Comparison of Measured and Reference Approach Profiles

(2) Two sideline noise measurement points, designated as S(sideline), are located on the ground perpendicular to and symmetrically stationed at 492 feet (150 m) on each side of the flyover reference flight path. The measurement points bisect the centerline flight path reference point A.

(c) *Approach reference noise measurement points*—As shown in Figure K3 below:

(1) The centerline noise measurement flight path reference point, designated A, is located on the ground 394 feet (120 m) vertically below the reference approach flight path. The measurement point is defined by the approach reference procedure in section K6.4 of this appendix. On level ground, the measurement point corresponds to a position 3,740 feet (1,140 m) from the intersection of the 6.0 degree approach path with the ground plane;

(2) Two sideline noise measurement points, designated as S(starboard) and S(port), are located on the ground perpendicular to and symmetrically stationed at 492 feet (150 m) on each side of the approach reference flight path. The measurement points bisect the centerline flight path reference point A.

Section K4 *Noise Limits*

For a tiltrotor, the maximum noise levels, as determined in accordance with the noise evaluation in EPNdB and calculation method described in section H36.201 of Appendix H of this part, must not exceed the noise limits as follows:

(a) *At the takeoff flight path reference point:* For a tiltrotor having a maximum certificated takeoff weight (mass) of 176,370 pounds (80,000 kg) or more, in VTOL/Conversion mode, 109 EPNdB, decreasing linearly with the logarithm of the tiltrotor weight (mass) at a rate of 3.0 EPNdB per halving of weight (mass) down to 89 EPNdB, after which the limit is constant. Figure K4 illustrates the takeoff noise limit as a solid line.

(b) *At the Flyover path reference point:* For a tiltrotor having a maximum certificated takeoff weight (mass) of 176,370 pounds (80,000 kg) or more, in VTOL/Conversion mode, 108 EPNdB, decreasing linearly with the logarithm of the tiltrotor weight (mass) at a rate of 3.0 EPNdB per halving of weight (mass) down to 88 EPNdB, after which the limit is constant. Figure K4 illustrates the flyover noise limit as a dashed line.

(c) *At the approach flight path reference point:* For a tiltrotor having a maximum certificated takeoff weight (mass) of 176,370 pounds (80,000 kg) or more, in VTOL/Conversion mode, 110 EPNdB, decreasing linearly with the logarithm of the tiltrotors weight (mass) at a rate of 3.0 EPNdB per halving of weight (mass) down to 90 EPNdB, after which the limit is constant. Figure K4 illustrates the approach noise limit as a dash-dot line.

Section K5 *Trade-Offs*

If the noise evaluation measurement exceeds the noise limits described in K4 of this appendix at one or two measurement points:

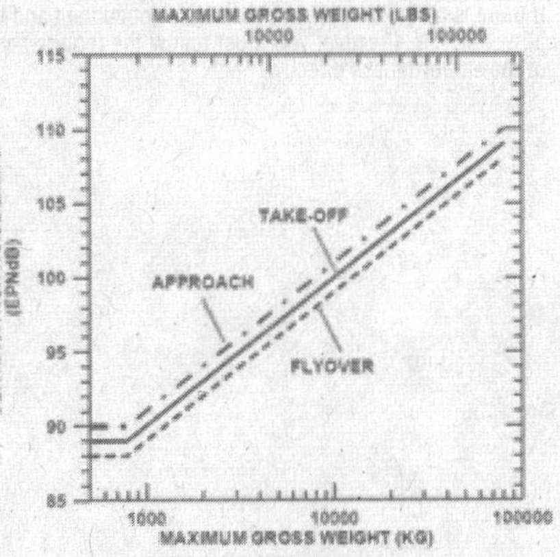

FIGURE K4.
TILTROTOR NOISE LIMITS

(a) The sum of excesses must not be greater than 4 EPNdB;

(b) The excess at any single point must not be greater than 3 EPNdB; and

(c) Any excess must be offset by the remaining noise margin at the other point or points.

Section K6 *Noise Certification Reference Procedures*

K6.1 General Conditions

(a)-(b) [Reserved]

(c) The takeoff, flyover and approach reference procedures must be established in accordance with sections K6.2, K6.3 and K6.4 of this appendix, except as specified in section K6.1(d) of this appendix.

(d) If the design characteristics of the tiltrotor prevent test flights from being conducted in accordance with section K6.2, K6.3 or K6.4 of this appendix, the applicant must revise the test procedures and resubmit the procedures for approval.

(e) The following reference atmospheric conditions must be used to establish the reference procedures:

(1) Sea level atmospheric pressure of 2,116 pounds per square foot (1,013.25 hPa);

(2) Ambient air temperature of 77°Fahrenheit (25°Celsius, i.e. ISA + 10°C);

(3) Relative humidity of 70 percent; and

(4) Zero wind.

(f) For tests conducted in accordance with sections K6.2, K6.3, and K6.4 of this appendix, use the maximum normal operating RPM corresponding to the airworthiness limit imposed by the manufacturer. For configurations for which the rotor speed automatically links with the flight condition, use the maximum normal operating rotor speed corresponding with the reference flight condition. For configurations for which the rotor speed can change by pilot action, use the highest normal rotor speed specified in the flight manual limitation section for the reference conditions.

K6.2 Takeoff Reference Procedure. The takeoff reference flight procedure is as follows:

(a) A constant takeoff configuration must be maintained, including the nacelle angle selected by the applicant;

(b) The tiltrotor power must be stabilized at the maximum takeoff power corresponding to the minimum installed engine(s) specification power available for the reference ambient conditions or gearbox torque limit, whichever is lower. The tiltrotor power must also be stabilized along a path starting from a point located 1,640 feet (500 m) before the flight path reference point, at 65 ft (20 m) above ground level;

(c) The nacelle angle and the corresponding best rate of climb speed, or the lowest approved speed for the climb after takeoff, whichever is the greater, must be maintained throughout the takeoff reference procedure;

(d) The rotor speed must be stabilized at the maximum normal operating RPM certificated for takeoff;

(e) The weight (mass) of the tiltrotors must be the maximum takeoff weight (mass) as requested for noise certification; and

(f) The reference takeoff flight profile is a straight line segment inclined from the starting point 1,640 feet (500 m) before to the center noise measurement point and 65 ft (20 m) above ground level at an angle defined by best rate of climb and the speed corresponding to the selected nacelle angle and for minimum specification engine performance.

K6.3 Flyover Reference Procedure. The flyover reference flight procedure is as follows:

(a) The tiltrotor must be stabilized for level flight along the centerline flyover flight path and over the noise measurement reference point at an altitude of 492 ft (150 m) above ground level;

(b) A constant flyover configuration selected by the applicant must be maintained;

(c) The weight (mass) of the tiltrotor must be the maximum takeoff weight (mass) as requested for noise certification;

(d) In the VTOL/Conversion mode:

(1) The nacelle angle must be at the authorized fixed operation point that is closest to the shallow nacelle angle certificated for zero airspeed;

(2) The airspeed must be 0.9VCON and

(3) The rotor speed must be stabilized at the maximum normal operating RPM certificated for level flight.

K6.4 Approach Reference Procedure. The approach reference procedure is as follows:

(a) The tiltrotor must be stabilized to follow a 6.0 degree approach path;

(b) An approved airworthiness configuration in which maximum noise occurs must be maintained;

(1) An airspeed equal to the best rate of climb speed corresponding to the nacelle angle, or the lowest approved airspeed for the approach, whichever is greater, must be stabilized and maintained; and

(2) The tiltrotor power during the approach must be stabilized over the flight path reference point, and continue as if landing;

(c) The rotor speed must be stabilized at the maximum normal operating RPM certificated for approach;

(d) The constant approach configuration used in airworthiness certification tests, with the landing gear extended, must be maintained; and

(e) The weight (mass) of the tiltrotor at landing must be the maximum landing weight (mass) as requested for noise certification.

Section K7 *Test Procedures*

K7.1 [Reserved]

K7.2 The test procedures and noise measurements must be conducted and processed to yield the noise evaluation measure designated in section K2 of this appendix.

K7.3 If either the test conditions or test procedures do not comply to the applicable noise certification reference conditions or procedures prescribed by this part, the applicant must apply the correction methods described in section H36.205 of Appendix H of this part to the acoustic test data measured.

K7.4 Adjustments for differences between test and reference flight procedures must not exceed:

(a) For takeoff: 4.0 EPNdB, of which the arithmetic sum of delta 1 and the term −7.5 log (QK/QrKr) from delta 2 must not in total exceed 2.0 EPNdB;

(b) For flyover or approach: 2.0 EPNdB.

K7.5 The average rotor RPM must not vary from the normal maximum operating RPM by more than±1.0 percent throughout the 10 dB-down time interval.

K7.6 The tiltrotor airspeed must not vary from the reference airspeed appropriate to the flight demonstration by more than±5 kts (±9 km/h) throughout the 10 dB-down time interval.

K7.7 The number of level flyovers made with a head wind component must be equal to the number of level flyovers made with a tail wind component.

K7.8 The tiltrotor must operate between±10 degrees from the vertical or between±65 feet (±20 m) lateral deviation tolerance, whichever is greater, above the reference track and throughout the 10 dB-down time interval.

K7.9 The tiltrotor altitude must not vary during each flyover by more than±30 ft (±9 m) from the reference altitude throughout the 10 dB-down time interval.

K7.10 During the approach procedure, the tiltrotor must establish a stabilized constant speed approach and fly between approach angles of 5.5 degrees and 6.5 degrees throughout the 10 dB-down time interval.

K7.11 During all test procedures, the tiltrotor weight (mass) must not be less than 90 percent and not more than 105 percent of the maximum certificated weight (mass). For each of the test procedures, complete at least one test at or above this maximum certificated weight (mass).

K7.12 A tiltrotor capable of carrying external loads or external equipment must be noise certificated without such loads or equipment fitted

K7.13 The value of VCON used for noise certification must be included in the approved Flight Manual.

[78 FR 1139, Jan. 8, 2013]

PART 39

AIRWORTHINESS DIRECTIVES

Contents

Authority: 49 U.S.C. 106(g), 40113, 44701.

Source: Doc. No. FAA-2000-8460, 67 FR 48003, July 22, 2002, unless otherwise noted.

§ 39.1 Purpose of this regulation.

The regulations in this part provide a legal framework for FAA's system of Airworthiness Directives.

§ 39.3 Definition of airworthiness directives.

FAA's airworthiness directives are legally enforceable rules that apply to the following products: aircraft, aircraft engines, propellers, and appliances.

§ 39.5 When does FAA issue airworthiness directives?

FAA issues an airworthiness directive addressing a product when we find that:

(a) An unsafe condition exists in the product; and

(b) The condition is likely to exist or develop in other products of the same type design.

§ 39.7 What is the legal effect of failing to comply with an airworthiness directive?

Anyone who operates a product that does not meet the requirements of an applicable airworthiness directive is in violation of this section.

§ 39.9 What if I operate an aircraft or use a product that does not meet the requirements of an airworthiness directive?

If the requirements of an airworthiness directive have not been met, you violate § 39.7 each time you operate the aircraft or use the product.

§ 39.11 What actions do airworthiness directives require?

Airworthiness directives specify inspections you must carry out, conditions and limitations you must comply with, and any actions you must take to resolve an unsafe condition.

§ 39.13 Are airworthiness directives part of the Code of Federal Regulations?

Yes, airworthiness directives are part of the Code of Federal Regulations, but they are not codified in the annual edition. FAA publishes airworthiness directives in full in the Federal Register as amendments to § 39.13.

Editorial Note: For a complete list of citations to airworthiness directives published in the Federal Register, consult the following publications: For airworthiness directives published in the Federal Register since 2001, see the entries for 14 CFR 39.13 in the List of CFR Sections Affected, which appears in the "Finding Aids" section of the printed volume and at *www.fdsys.gov* . For citations to prior amendments, see the entries for 14 CFR 39.13 in the separate publications List of CFR Sections Affected, 1973-1985, List of CFR Sections Affected, 1964-1972, and List of CFR Sections Affected, 1986-2000, and the entries for 14 CFR 507.10 in the List of Sections Affected, 1949-1963. See also the annual editions of the Federal Register Index for subject matter references and citations to FAA airworthiness directives.

§ 39.15 Does an airworthiness directive apply if the product has been changed?

Yes, an airworthiness directive applies to each product identified in the airworthiness directive, even if an individual product has been changed by modifying, altering, or repairing it in the area addressed by the airworthiness directive.

§ 39.17 What must I do if a change in a product affects my ability to accomplish the actions required in an airworthiness directive?

If a change in a product affects your ability to accomplish the actions required by the airworthiness directive in any way, you must request FAA approval of an alternative method of compliance. Unless you can show the change eliminated the unsafe condition, your request should include the specific actions that you propose to address the unsafe condition. Submit your request in the manner described in § 39.19.

§ 39.19 May I address the unsafe condition in a way other than that set out in the airworthiness directive?

Yes, anyone may propose to FAA an alternative method of compliance or a change in the compliance time, if the proposal provides an acceptable level of safety. Unless FAA authorizes otherwise, send your proposal to your principal inspector. Include the specific actions you are proposing to address the unsafe condition. The principal inspector may add comments and will send your request to the manager of the office identified in the airworthiness directive (manager). You may send a copy to the manager at the same time you send it to the principal inspector. If you do not have a principal inspector send your proposal directly to the manager. You may use the alternative you propose only if the manager approves it.

§ 39.21 Where can I get information about FAA-approved alternative methods of compliance?

Each airworthiness directive identifies the office responsible for approving alternative methods of compliance. That office can provide information about alternatives it has already approved.

§ 39.23 May I fly my aircraft to a repair facility to do the work required by an airworthiness directive?

Yes, the operations specifications giving some operators authority to operate include a provision that allow them to fly their aircraft to a repair facility to do the work required by an airworthiness directive. If you do not have this authority, the local Flight Standards District Office of FAA may issue you a special flight permit unless the airworthiness directive states otherwise. To ensure aviation safety, FAA may add special requirements for operating your aircraft to a place where the repairs or modifications can be accomplished. FAA may also decline to issue a special flight permit in particular cases if we determine you cannot move the aircraft safely.

§ 39.25 How do I get a special flight permit?

Apply to FAA for a special flight permit following the procedures in 14 CFR 21.199.

§ 39.27 What do I do if the airworthiness directive conflicts with the service document on which it is based?

In some cases an airworthiness directive incorporates by reference a manufacturer's service document. In these cases, the service document becomes part of the airworthiness directive. In some cases the directions in the service document may be modified by the airworthiness directive. If there is a conflict between the service document and the airworthiness directive, you must follow the requirements of the airworthiness directive.

PART 43

MAINTENANCE, PREVENTIVE MAINTENANCE, REBUILDING, AND ALTERATION

Contents

Authority: 49 U.S.C. 106(g), 40113, 44701, 44703, 44705, 44707, 44711, 44713, 44717, 44725.

Source: Docket No. 1993, 29 FR 5451, Apr. 23, 1964, unless otherwise noted.

Editorial Note: For miscellaneous technical amendments to this part 43, see Amdt. 43-3, 31 FR 3336, Mar. 3, 1966, and Amdt. 43-6, 31 FR 9211, July 6, 1966.

§ 43.1 Applicability.

(a) Except as provided in paragraphs (b) and (d) of this section, this part prescribes rules governing the maintenance, preventive maintenance, rebuilding, and alteration of any—

(1) Aircraft having a U.S. airworthiness certificate;

(2) Foreign-registered civil aircraft used in common carriage or carriage of mail under the provisions of Part 121 or 135 of this chapter; and

(3) Airframe, aircraft engines, propellers, appliances, and component parts of such aircraft.

(b) This part does not apply to—

(1) Any aircraft for which the FAA has issued an experimental certificate, unless the FAA has previously issued a different kind of airworthiness certificate for that aircraft; or

(2) Any aircraft for which the FAA has issued an experimental certificate under the provisions of § 21.191 (i)(3) of this chapter, and the aircraft was previously issued a special airworthiness certificate in the light-sport category under the provisions of § 21.190 of this chapter.

(c) This part applies to all life-limited parts that are removed from a type certificated product, segregated, or controlled as provided in § 43.10.

(d) This part applies to any aircraft issued a special airworthiness certificate in the light-sport category except:

(1) The repair or alteration form specified in §§ 43.5(b) and 43.9(d) is not required to be completed for products not produced under an FAA approval;

(2) Major repairs and major alterations for products not produced under an FAA approval are not required to be recorded in accordance with appendix B of this part; and

(3) The listing of major alterations and major repairs specified in paragraphs (a) and (b) of appendix A of this part is not applicable to products not produced under an FAA approval.

[Doc. No. 1993, 29 FR 5451, Apr. 23, 1964, as amended by Amdt. 43-23, 47 FR 41084, Sept. 16, 1982; Amdt. 43-37, 66 FR 21066, Apr. 27, 2001; Amdt. 43-38, 67 FR 2109, Jan. 15, 2002; Amdt. 43-39, 69 FR 44863, July 27, 2004; Amdt. 43-44, 75 FR 5219, Feb. 1, 2010]

§ 43.2 Records of overhaul and rebuilding.

(a) No person may describe in any required maintenance entry or form an aircraft, airframe, aircraft engine, propeller, appliance, or component part as being overhauled unless—

(1) Using methods, techniques, and practices acceptable to the Administrator, it has been disassembled, cleaned, inspected, repaired as necessary, and reassembled; and

(2) It has been tested in accordance with approved standards and technical data, or in accordance with current standards and technical data acceptable to the Administrator, which have been developed and documented by the holder of the type certificate, supplemental type certificate, or a material, part, process, or appliance approval under part 21 of this chapter.

(b) No person may describe in any required maintenance entry or form an aircraft, airframe, aircraft engine, propeller, appliance, or component part as being rebuilt unless it has been disassembled, cleaned, inspected, repaired as necessary, reassembled, and tested to the same tolerances and limits as a new item, using either new parts or used parts that either conform to new part tolerances and limits or to approved oversized or undersized dimensions.

[Amdt. 43-23, 47 FR 41084, Sept. 16, 1982, as amended by Amdt. 43-43, 74 FR 53394, Oct. 16, 2009]

§ 43.3 Persons authorized to perform maintenance, preventive maintenance, rebuilding, and alterations.

(a) Except as provided in this section and § 43.17, no person may maintain, rebuild, alter, or perform preventive maintenance on an aircraft, airframe, aircraft engine, propeller, appliance, or component part to which this part applies. Those items, the performance of which is a major alteration, a major repair, or preventive maintenance, are listed in appendix A.

(b) The holder of a mechanic certificate may perform maintenance, preventive maintenance, and alterations as provided in Part 65 of this chapter.

(c) The holder of a repairman certificate may perform maintenance, preventive maintenance, and alterations as provided in part 65 of this chapter.

(d) A person working under the supervision of a holder of a mechanic or repairman certificate may perform the maintenance, preventive maintenance, and alterations that his supervisor is authorized to perform, if the supervisor personally observes the work being done to the extent necessary to ensure that it is being done properly and if the supervisor is readily available, in person, for consultation. However, this paragraph does not authorize the performance of any inspection required by Part 91 or Part 125 of this chapter or any inspection performed after a major repair or alteration.

(e) The holder of a repair station certificate may perform maintenance, preventive maintenance, and alterations as provided in Part 145 of this chapter.

(f) The holder of an air carrier operating certificate or an operating certificate issued under Part 121 or 135, may perform maintenance, preventive maintenance, and alterations as provided in Part 121 or 135.

(g) Except for holders of a sport pilot certificate, the holder of a pilot certificate issued under part 61 may perform preventive maintenance on any aircraft owned or operated by that pilot which is not used under part 121, 129, or 135 of this chapter. The holder of a sport pilot certificate may perform preventive maintenance on an aircraft owned or operated by that pilot and issued a special airworthiness certificate in the light-sport category.

(h) Notwithstanding the provisions of paragraph (g) of this section, the Administrator may approve a certificate holder under Part 135 of this chapter, operating rotorcraft in a remote area, to allow a pilot to perform specific preventive maintenance items provided—

(1) The items of preventive maintenance are a result of a known or suspected mechanical difficulty or malfunction that occurred en route to or in a remote area;

(2) The pilot has satisfactorily completed an approved training program and is authorized in writing by the certificate holder for each item of preventive maintenance that the pilot is authorized to perform;

(3) There is no certificated mechanic available to perform preventive maintenance;

(4) The certificate holder has procedures to evaluate the accomplishment of a preventive maintenance item that requires a decision concerning the airworthiness of the rotorcraft; and

(5) The items of preventive maintenance authorized by this section are those listed in paragraph (c) of appendix A of this part.

(i) Notwithstanding the provisions of paragraph (g) of this section, in accordance with an approval issued to the holder of a certificate issued under part 135 of this chapter, a pilot of an aircraft type-certificated for 9 or fewer passenger seats, excluding any pilot seat, may perform the removal and reinstallation of approved aircraft cabin seats, approved cabin-mounted stretchers, and when no tools are required, approved cabin-mounted medical oxygen bottles, provided—

(1) The pilot has satisfactorily completed an approved training program and is authorized in writing by the certificate holder to perform each task; and

(2) The certificate holder has written procedures available to the pilot to evaluate the accomplishment of the task.

(j) A manufacturer may—

(1) Rebuild or alter any aircraft, aircraft engine, propeller, or appliance manufactured by him under a type or production certificate;

(2) Rebuild or alter any appliance or part of aircraft, aircraft engines, propellers, or appliances manufactured by him under a Technical Standard Order Authorization, an FAA-Parts Manufacturer Approval, or Product and Process Specification issued by the Administrator; and

(3) Perform any inspection required by part 91 or part 125 of this chapter on aircraft it manufactured under a type certificate, or currently manufactures under a production certificate.

(k) Updates of databases in installed avionics meeting the conditions of this paragraph are not considered maintenance and may be performed by pilots provided:

(1) The database upload is:

(i) Initiated from the flight deck;

(ii) Performed without disassembling the avionics unit; and

(iii) Performed without the use of tools and/or special equipment.

(2) The pilot must comply with the certificate holder's procedures or the manufacturer's instructions.

(3) The holder of operating certificates must make available written procedures consistent with manufacturer's instructions to the pilot that describe how to:

(i) Perform the database update; and

(ii) Determine the status of the data upload.

[Doc. No. 1993, 29 FR 5451, Apr. 23, 1964, as amended by Amdt. 43-4, 31 FR 5249, Apr. 1, 1966; Amdt. 43-23, 47 FR 41084, Sept. 16, 1982; Amdt. 43-25, 51 FR 40702, Nov. 7, 1986; Amdt. 43-36, 61 FR 19501, May 1, 1996; Amdt. 43-37, 66 FR 21066, Apr. 27, 2001; Amdt. 43-39, 69 FR 44863, July 27, 2004; Amdt. 43-43, 74 FR 53394, Oct. 16, 2009; Amdt. 43-45, 77 FR 71096, Nov. 29, 2012]

§ 43.5 Approval for return to service after maintenance, preventive maintenance, rebuilding, or alteration.

No person may approve for return to service any aircraft, airframe, aircraft engine, propeller, or appliance, that has undergone maintenance, preventive maintenance, rebuilding, or alteration unless—

(a) The maintenance record entry required by § 43.9 or § 43.11, as appropriate, has been made;

(b) The repair or alteration form authorized by or furnished by the Administrator has been executed in a manner prescribed by the Administrator; and

(c) If a repair or an alteration results in any change in the aircraft operating limitations or flight data contained in the approved aircraft flight manual, those operating limitations or flight data are appropriately revised and set forth as prescribed in § 91.9 of this chapter.

[Doc. No. 1993, 29 FR 5451, Apr. 23, 1964, as amended by Amdt. 43-23, 47 FR 41084, Sept. 16, 1982; Amdt. 43-31, 54 FR 34330, Aug. 18, 1989]

§ 43.7 Persons authorized to approve aircraft, airframes, aircraft engines, propellers, appliances, or component parts for return to service after maintenance, preventive maintenance, rebuilding, or alteration.

(a) Except as provided in this section and § 43.17, no person, other than the Administrator, may approve an aircraft, airframe, aircraft engine, propeller, appliance, or component part for return to service after it has undergone maintenance, preventive maintenance, rebuilding, or alteration.

(b) The holder of a mechanic certificate or an inspection authorization may approve an aircraft, airframe, aircraft engine, propeller, appliance, or component part for return to service as provided in Part 65 of this chapter.

(c) The holder of a repair station certificate may approve an aircraft, airframe, aircraft engine, propeller, appliance, or component part for return to service as provided in Part 145 of this chapter.

(d) A manufacturer may approve for return to service any aircraft, airframe, aircraft engine, propeller, appliance, or component part which that manufacturer has worked on under § 43.3(j). However, except for minor alterations, the work must have been done in accordance with technical data approved by the Administrator.

(e) The holder of an air carrier operating certificate or an operating certificate issued under Part 121 or 135, may approve an aircraft, airframe, aircraft engine, propeller, appliance, or component part for return to service as provided in Part 121 or 135 of this chapter, as applicable.

(f) A person holding at least a private pilot certificate may approve an aircraft for return to service after performing preventive maintenance under the provisions of § 43.3(g).

(g) The holder of a repairman certificate (light-sport aircraft) with a maintenance rating may approve an aircraft issued a special airworthiness certificate in light-sport category for return to service, as provided in part 65 of this chapter.

(h) The holder of at least a sport pilot certificate may approve an aircraft owned or operated by that pilot and issued a special airworthiness certificate in the light-sport category for return to service after performing preventive maintenance under the provisions of § 43.3(g).

[Amdt. 43-23, 47 FR 41084, Sept. 16, 1982, as amended by Amdt. 43-36, 61 FR 19501, May 1, 1996; Amdt. 43-37, 66 FR 21066, Apr. 27, 2001; Amdt. 43-39, 69 FR 44863, July 27, 2004]

§ 43.9 Content, form, and disposition of maintenance, preventive maintenance, rebuilding, and alteration records (except inspections performed in accordance with part 91, part 125, § 135.411(a)(1), and § 135.419 of this chapter).

(a) *Maintenance record entries.* Except as provided in paragraphs (b) and (c) of this section, each person who maintains, performs preventive maintenance, rebuilds, or alters an aircraft, airframe, aircraft engine, propeller, appliance, or component part shall make an entry in the maintenance record of that equipment containing the following information:

(1) A description (or reference to data acceptable to the Administrator) of work performed.

(2) The date of completion of the work performed.

(3) The name of the person performing the work if other than the person specified in paragraph (a)(4) of this section.

(4) If the work performed on the aircraft, airframe, aircraft engine, propeller, appliance, or component part has been performed satisfactorily, the signature, certificate number, and kind of certificate held by the person approving the work. The signature constitutes the approval for return to service only for the work performed.

(b) Each holder of an air carrier operating certificate or an operating certificate issued under Part 121 or 135, that is required by its approved operations specifications to provide for a continuous airworthiness maintenance program, shall make a record of the maintenance, preventive maintenance, rebuilding, and alteration, on aircraft, airframes, aircraft engines, propellers, appliances, or component parts which it operates in accordance with the applicable provisions of Part 121 or 135 of this chapter, as appropriate.

(c) This section does not apply to persons performing inspections in accordance with Part 91, 125, § 135.411(a)(1), or § 135.419 of this chapter.

(d) In addition to the entry required by paragraph (a) of this section, major repairs and major alterations shall be entered on a form, and the form disposed of, in the manner prescribed in appendix B, by the person performing the work.

[Amdt. 43-23, 47 FR 41085, Sept. 16, 1982, as amended by Amdt. 43-37, 66 FR 21066, Apr. 27, 2001; Amdt. 43-39, 69 FR 44863, July 27, 2004]

§ 43.10 Disposition of life-limited aircraft parts.

(a) *Definitions used in this section.* For the purposes of this section the following definitions apply.

Life-limited part means any part for which a mandatory replacement limit is specified in the type design, the Instructions for Continued Airworthiness, or the maintenance manual.

Life status means the accumulated cycles, hours, or any other mandatory replacement limit of a life-limited part.

(b) *Temporary removal of parts from type-certificated products.* When a life-limited part is temporarily removed and reinstalled for the purpose of performing maintenance, no disposition under paragraph (c) of this section is required if—

(1) The life status of the part has not changed;

(2) The removal and reinstallation is performed on the same serial numbered product; and

(3) That product does not accumulate time in service while the part is removed.

(c) *Disposition of parts removed from type-certificated products.* Except as provided in paragraph (b) of this section, after April 15, 2002 each person who removes a life-limited part from a type-certificated product must ensure that the part is controlled using one of the methods in this paragraph. The method must deter the installation of the part after it has reached its life limit. Acceptable methods include:

(1) *Record keeping system.* The part may be controlled using a record keeping system that substantiates the part number, serial number, and current life status of the part. Each time the part is removed from a type certificated product, the record must be updated with the current life status. This system may include electronic, paper, or other means of record keeping.

(2) *Tag or record attached to part.* A tag or other record may be attached to the part. The tag or record must include the part number, serial number, and current life status of the part. Each time the part is removed from a type certificated product, either a new tag or record must be created, or the existing tag or record must be updated with the current life status.

(3) *Non-permanent marking.* The part may be legibly marked using a non-permanent method showing its current life status. The life status must be updated each time the part is removed from a type certificated product, or if the mark is removed, another method in this section may be used. The mark must be accomplished in accordance with the instructions under § 45.16 of this chapter in order to maintain the integrity of the part.

(4) *Permanent marking.* The part may be legibly marked using a permanent method showing its current life status. The life status must be updated each time the part is removed from a type certificated product. Unless the part is permanently removed from use on type certificated products, this permanent mark must be accomplished in accordance with the instructions under § 45.16 of this chapter in order to maintain the integrity of the part.

(5) *Segregation.* The part may be segregated using methods that deter its installation on a type-certificated product. These methods must include, at least—

(i) Maintaining a record of the part number, serial number, and current life status, and

(ii) Ensuring the part is physically stored separately from parts that are currently eligible for installation.

(6) *Mutilation.* The part may be mutilated to deter its installation in a type certificated produce. The mutilation must render the part beyond repair and incapable of being reworked to appear to be airworthy.

(7) *Other methods.* Any other method approved or accepted by the FAA.

(d) *Transfer of life-limited parts.* Each person who removes a life-limited part from a type certificated product and later sells or otherwise transfers that part must transfer with the part the mark, tag, or other record used to comply with this section, unless the part is mutilated before it is sold or transferred.

[Doc. No. FAA-2000-8017, 67 FR 2110, Jan. 15, 2002]

§ 43.11 Content, form, and disposition of records for inspections conducted under parts 91 and 125 and §§ 135.411(a)(1) and 135.419 of this chapter.

(a) *Maintenance record entries.* The person approving or disapproving for return to service an aircraft, airframe, aircraft engine, propeller, appliance, or component part after any inspection performed in accordance with part 91, 125, § 135.411(a)(1), or § 135.419 shall make an entry in the maintenance record of that equipment containing the following information:

(1) The type of inspection and a brief description of the extent of the inspection.

(2) The date of the inspection and aircraft total time in service.

(3) The signature, the certificate number, and kind of certificate held by the person approving or disapproving for return to service the aircraft, airframe, aircraft engine, propeller, appliance, component part, or portions thereof.

(4) Except for progressive inspections, if the aircraft is found to be airworthy and approved for return to service, the following or a similarly worded statement—"I certify that this aircraft has been inspected in accordance with (insert type) inspection and was determined to be in airworthy condition."

(5) Except for progressive inspections, if the aircraft is not approved for return to service because of needed maintenance, noncompliance with applicable specifications, airworthiness directives, or other approved data, the following or a similarly worded statement—"I certify that this aircraft has been inspected in accordance with (insert type) inspection and a list of discrepancies and unairworthy items dated (date) has been provided for the aircraft owner or operator."

(6) For progressive inspections, the following or a similarly worded statement—"I certify that in accordance with a progressive inspection program, a routine inspection of (identify whether aircraft or components) and a detailed inspection of (identify components) were performed and the (aircraft or components) are (approved or disapproved) for return to service." If disapproved, the entry will further state "and a list of discrepancies and unairworthy items dated (date) has been provided to the aircraft owner or operator."

(7) If an inspection is conducted under an inspection program provided for in part 91, 125, or § 135.411(a)(1), the entry must identify the inspection program, that part of the inspection program accomplished, and contain a statement that the inspection was performed in accordance with the inspections and procedures for that particular program.

(b) *Listing of discrepancies and placards.* If the person performing any inspection required by part 91 or 125 or § 135.411(a)(1) of this chapter finds that the aircraft is unairworthy or does not meet the applicable type certificate data, airworthiness directives, or other approved data upon which its airworthiness depends, that persons must give the owner or lessee a signed and dated list of those discrepancies. For those items permitted to be inoperative under § 91.213(d)(2) of this chapter, that person shall place a placard, that meets the aircraft's airworthiness certification regulations, on each inoperative instrument and the cockpit control of each item of inoperative equipment, marking it "Inoperative," and shall add the items to the signed and dated list of discrepancies given to the owner or lessee.

[Amdt. 43-23, 47 FR 41085, Sept. 16, 1982, as amended by Amdt. 43-30, 53 FR 50195, Dec. 13, 1988; Amdt. 43-36, 61 FR 19501, May 1, 1996; 71 FR 44188, Aug. 4, 2006]

§ 43.12 Maintenance records: Falsification, reproduction, or alteration.

(a) No person may make or cause to be made:

(1) Any fraudulent or intentionally false entry in any record or report that is required to be made, kept, or used to show compliance with any requirement under this part;

(2) Any reproduction, for fraudulent purpose, of any record or report under this part; or

(3) Any alteration, for fraudulent purpose, of any record or report under this part.

(b) The commission by any person of an act prohibited under paragraph (a) of this section is a basis for suspending or revoking the applicable airman, operator, or production certificate, Technical Standard Order Authorization, FAA-Parts Manufacturer Approval, or Product and Process Specification issued by the Administrator and held by that person.

[Amdt. 43-19, 43 FR 22639, May 25, 1978, as amended by Amdt. 43-23, 47 FR 41085, Sept. 16, 1982]

§ 43.13 Performance rules (general).

(a) Each person performing maintenance, alteration, or preventive maintenance on an aircraft, engine, propeller, or appliance shall use the methods, techniques, and practices prescribed in the current manufacturer's maintenance manual or Instructions for Continued Airworthiness prepared by its manufacturer, or other methods, techniques, and practices acceptable to the Administrator, except as noted in § 43.16. He shall use the tools, equipment, and test apparatus necessary to assure completion of the work in accordance with accepted industry practices. If special equipment or test apparatus is recommended by the manufacturer involved, he must use that equipment or apparatus or its equivalent acceptable to the Administrator.

(b) Each person maintaining or altering, or performing preventive maintenance, shall do that work in such a manner and use materials of such a quality, that the condition of the aircraft, airframe, aircraft engine, propeller, or appliance worked on will be at least equal to its original or properly altered condition (with regard to aerodynamic function, structural strength, resistance to vibration and deterioration, and other qualities affecting airworthiness).

(c) *Special provisions for holders of air carrier operating certificates and operating certificates issued under the provisions of Part 121 or 135 and Part 129 operators holding operations specifications.* Unless otherwise notified by the administrator, the methods, techniques, and practices contained in the maintenance manual or the maintenance part of the manual of the holder of an air carrier operating certificate or an operating certificate under Part 121 or 135 and Part 129 operators holding operations specifications (that is required by its operating specifications to provide a continuous airworthiness maintenance and inspection program) constitute acceptable means of compliance with this section.

[Doc. No. 1993, 29 FR 5451, Apr. 23, 1964, as amended by Amdt. 43-20, 45 FR 60182, Sept. 11, 1980; Amdt. 43-23, 47 FR 41085, Sept. 16, 1982; Amdt. 43-28, 52 FR 20028, June 16, 1987; Amdt. 43-37, 66 FR 21066, Apr. 27, 2001]

§ 43.15 Additional performance rules for inspections.

(a) *General.* Each person performing an inspection required by part 91, 125, or 135 of this chapter, shall—

(1) Perform the inspection so as to determine whether the aircraft, or portion(s) thereof under inspection, meets all applicable airworthiness requirements; and

(2) If the inspection is one provided for in part 125, 135, or § 91.409(e) of this chapter, perform the inspection in accordance with the instructions and procedures set forth in the inspection program for the aircraft being inspected.

(b) *Rotorcraft.* Each person performing an inspection required by Part 91 on a rotorcraft shall inspect the following systems in accordance with the maintenance manual or Instructions for Continued Airworthiness of the manufacturer concerned:

(1) The drive shafts or similar systems.

(2) The main rotor transmission gear box for obvious defects.

(3) The main rotor and center section (or the equivalent area).

(4) The auxiliary rotor on helicopters.

(c) *Annual and 100-hour inspections.* (1) Each person performing an annual or 100-hour inspection shall use a checklist while performing the inspection. The checklist may be of the person's own design, one provided by the manufacturer of the equipment being inspected or one obtained from another source. This checklist must include the scope and detail of the items contained in appendix D to this part and paragraph (b) of this section.

(2) Each person approving a reciprocating-engine-powered aircraft for return to service after an annual or 100-hour inspection shall, before that approval, run the aircraft engine or engines to determine satisfactory performance in accordance with the manufacturer's recommendations of—

(i) Power output (static and idle r.p.m.);

(ii) Magnetos;

(iii) Fuel and oil pressure; and

(iv) Cylinder and oil temperature.

(3) Each person approving a turbine-engine-powered aircraft for return to service after an annual, 100-hour, or progressive inspection shall, before that approval, run the aircraft engine or engines to determine satisfactory performance in accordance with the manufacturer's recommendations.

(d) *Progressive inspection.* (1) Each person performing a progressive inspection shall, at the start of a progressive inspection system, inspect the aircraft completely. After this initial inspection, routine and detailed inspections must be conducted as prescribed in the progressive inspection schedule. Routine inspections consist of visual examination or check of the appliances, the aircraft, and its components and systems, insofar as practicable without disassembly. Detailed inspections consist of a thorough examination of the appliances, the aircraft, and its components and systems, with such disassembly as is necessary. For the purposes of this subparagraph, the overhaul of a component or system is considered to be a detailed inspection.

(2) If the aircraft is away from the station where inspections are normally conducted, an appropriately rated mechanic, a certificated repair station, or the manufacturer of the aircraft may perform inspections in accordance with the procedures and using the forms of the person who would otherwise perform the inspection.

[Doc. No. 1993, 29 FR 5451, Apr. 23, 1964, as amended by Amdt. 43-23, 47 FR 41086, Sept. 16, 1982; Amdt. 43-25, 51 FR 40702, Nov. 7, 1986; Amdt. 43-31, 54 FR 34330, Aug. 18, 1989; 71 FR 44188, Aug. 4, 2006]

§ 43.16 Airworthiness limitations.

Each person performing an inspection or other maintenance specified in an Airworthiness Limitations section of a manufacturer's maintenance manual or Instructions for Continued Airworthiness shall perform the inspection or other maintenance in accordance with that section, or in accordance with operations specifications approved by the Administrator under part 121 or 135, or an inspection program approved under § 91.409(e).

[71 FR 44188, Aug. 4, 2006]

§ 43.17 Maintenance, preventive maintenance, and alterations performed on U.S. aeronautical products by certain Canadian persons.

(a) *Definitions.* For purposes of this section:

Aeronautical product means any civil aircraft or airframe, aircraft engine, propeller, appliance, component, or part to be installed thereon.

Canadian aeronautical product means any aeronautical product under airworthiness regulation by Transport Canada Civil Aviation.

U.S. aeronautical product means any aeronautical product under airworthiness regulation by the FAA.

(b) *Applicability.* This section does not apply to any U.S. aeronautical products maintained or altered under any bilateral agreement made between Canada and any country other than the United States.

(c) *Authorized persons.* (1) A person holding a valid Transport Canada Civil Aviation Maintenance Engineer license and appropriate ratings may, with respect to a U.S.-registered aircraft located in Canada, perform maintenance, preventive maintenance, and alterations in accordance with the requirements of paragraph (d) of this section and approve the affected aircraft for return to service in accordance with the requirements of paragraph (e) of this section.

(2) A Transport Canada Civil Aviation Approved Maintenance Organization (AMO) holding appropriate ratings may, with respect to a U.S.-registered aircraft or other U.S. aeronautical products located in Canada, perform maintenance, preventive maintenance, and alterations in accordance with the requirements of paragraph (d) of this section and approve the affected products for return to service in accordance with the requirements of paragraph (e) of this section.

(d) *Performance requirements.* A person authorized in paragraph (c) of this section may perform maintenance (including any inspection required by Sec. 91.409 of this chapter, except an annual inspection), preventive maintenance, and alterations, provided—

(1) The person performing the work is authorized by Transport Canada Civil Aviation to perform the same type of work with respect to Canadian aeronautical products;

(2) The maintenance, preventive maintenance, or alteration is performed in accordance with a Bilateral Aviation Safety Agreement between the United States and Canada and associated Maintenance Implementation Procedures that provide a level of safety equivalent to that provided by the provisions of this chapter;

(3) The maintenance, preventive maintenance, or alteration is performed such that the affected product complies with the applicable requirements of part 36 of this chapter; and

(4) The maintenance, preventive maintenance, or alteration is recorded in accordance with a Bilateral Aviation Safety Agreement between the United States and Canada and associated Maintenance Implementation Procedures that provide a level of safety equivalent to that provided by the provisions of this chapter.

(e) *Approval requirements.* (1) To return an affected product to service, a person authorized in paragraph (c) of this section must approve (certify) maintenance, preventive maintenance, and alterations performed under this section, except that an Aircraft Maintenance Engineer may not approve a major repair or major alteration.

(2) An AMO whose system of quality control for the maintenance, preventive maintenance, alteration, and inspection of aeronautical products has been approved by Transport Canada Civil Aviation, or an authorized employee performing work for such an AMO, may approve (certify) a major repair or major alteration performed under this section if the work was performed in accordance with technical data approved by the FAA.

(f) No person may operate in air commerce an aircraft, airframe, aircraft engine, propeller, or appliance on which maintenance, preventive maintenance, or alteration has been performed under this section unless it has been approved for return to service by a person authorized in this section.

[Amdt. 43-33, 56 FR 57571, Nov. 12, 1991, as amended by Amdt. 43-40, 71 FR 40877, July 14, 2005]

Appendix A to Part 43—Major Alterations, Major Repairs, and Preventive Maintenance

(a) *Major alterations* —(1) *Airframe major alterations.* Alterations of the following parts and alterations of the following types, when not listed in the aircraft specifications issued by the FAA, are airframe major alterations:

(i) Wings.

(ii) Tail surfaces.

(iii) Fuselage.

(iv) Engine mounts.

(v) Control system.

(vi) Landing gear.

(vii) Hull or floats.

(viii) Elements of an airframe including spars, ribs, fittings, shock absorbers, bracing, cowling, fairings, and balance weights.

(ix) Hydraulic and electrical actuating system of components.

(x) Rotor blades.

(xi) Changes to the empty weight or empty balance which result in an increase in the maximum certificated weight or center of gravity limits of the aircraft.

(xii) Changes to the basic design of the fuel, oil, cooling, heating, cabin pressurization, electrical, hydraulic, de-icing, or exhaust systems.

(xiii) Changes to the wing or to fixed or movable control surfaces which affect flutter and vibration characteristics.

(2) *Powerplant major alterations.* The following alterations of a powerplant when not listed in the engine specifications issued by the FAA, are powerplant major alterations.

(i) Conversion of an aircraft engine from one approved model to another, involving any changes in compression ratio, propeller reduction gear, impeller gear ratios or the substitution of major engine parts which requires extensive rework and testing of the engine.

(ii) Changes to the engine by replacing aircraft engine structural parts with parts not supplied by the original manufacturer or parts not specifically approved by the Administrator.

(iii) Installation of an accessory which is not approved for the engine.

(iv) Removal of accessories that are listed as required equipment on the aircraft or engine specification.

(v) Installation of structural parts other than the type of parts approved for the installation.

(vi) Conversions of any sort for the purpose of using fuel of a rating or grade other than that listed in the engine specifications.

(3) *Propeller major alterations.* The following alterations of a propeller when not authorized in the propeller specifications issued by the FAA are propeller major alterations:

(i) Changes in blade design.

(ii) Changes in hub design.

(iii) Changes in the governor or control design.

(iv) Installation of a propeller governor or feathering system.

(v) Installation of propeller de-icing system.

(vi) Installation of parts not approved for the propeller.

(4) *Appliance major alterations.* Alterations of the basic design not made in accordance with recommendations of the appliance manufacturer or in accordance with an FAA Airworthiness Directive are appliance major alterations. In addition, changes in the basic design of radio communication and navigation equipment approved under type certification or a Technical Standard Order that have an effect on frequency stability, noise level, sensitivity, selectivity, distortion, spurious radiation, AVC characteristics,

or ability to meet environmental test conditions and other changes that have an effect on the performance of the equipment are also major alterations.

(b) *Major repairs* —(1) *Airframe major repairs.* Repairs to the following parts of an airframe and repairs of the following types, involving the strengthening, reinforcing, splicing, and manufacturing of primary structural members or their replacement, when replacement is by fabrication such as riveting or welding, are airframe major repairs.

(i) Box beams.

(ii) Monocoque or semimonocoque wings or control surfaces.

(iii) Wing stringers or chord members.

(iv) Spars.

(v) Spar flanges.

(vi) Members of truss-type beams.

(vii) Thin sheet webs of beams.

(viii) Keel and chine members of boat hulls or floats.

(ix) Corrugated sheet compression members which act as flange material of wings or tail surfaces.

(x) Wing main ribs and compression members.

(xi) Wing or tail surface brace struts.

(xii) Engine mounts.

(xiii) Fuselage longerons.

(xiv) Members of the side truss, horizontal truss, or bulkheads.

(xv) Main seat support braces and brackets.

(xvi) Landing gear brace struts.

(xvii) Axles.

(xviii) Wheels.

(xix) Skis, and ski pedestals.

(xx) Parts of the control system such as control columns, pedals, shafts, brackets, or horns.

(xxi) Repairs involving the substitution of material.

(xxii) The repair of damaged areas in metal or plywood stressed covering exceeding six inches in any direction.

(xxiii) The repair of portions of skin sheets by making additional seams.

(xxiv) The splicing of skin sheets.

(xxv) The repair of three or more adjacent wing or control surface ribs or the leading edge of wings and control surfaces, between such adjacent ribs.

(xxvi) Repair of fabric covering involving an area greater than that required to repair two adjacent ribs.

(xxvii) Replacement of fabric on fabric covered parts such as wings, fuselages, stabilizers, and control surfaces.

(xxviii) Repairing, including rebottoming, of removable or integral fuel tanks and oil tanks.

(2) *Powerplant major repairs.* Repairs of the following parts of an engine and repairs of the following types, are powerplant major repairs:

(i) Separation or disassembly of a crankcase or crankshaft of a reciprocating engine equipped with an integral supercharger.

(ii) Separation or disassembly of a crankcase or crankshaft of a reciprocating engine equipped with other than spur-type propeller reduction gearing.

(iii) Special repairs to structural engine parts by welding, plating, metalizing, or other methods.

(3) *Propeller major repairs.* Repairs of the following types to a propeller are propeller major repairs:

(i) Any repairs to, or straightening of steel blades.

(ii) Repairing or machining of steel hubs.

(iii) Shortening of blades.

(iv) Retipping of wood propellers.

(v) Replacement of outer laminations on fixed pitch wood propellers.

(vi) Repairing elongated bolt holes in the hub of fixed pitch wood propellers.

(vii) Inlay work on wood blades.

(viii) Repairs to composition blades.

(ix) Replacement of tip fabric.

(x) Replacement of plastic covering.

(xi) Repair of propeller governors.

(xii) Overhaul of controllable pitch propellers.

(xiii) Repairs to deep dents, cuts, scars, nicks, etc., and straightening of aluminum blades.

(xiv) The repair or replacement of internal elements of blades.

(4) *Appliance major repairs.* Repairs of the following types to appliances are appliance major repairs:

(i) Calibration and repair of instruments.

(ii) Calibration of radio equipment.

(iii) Rewinding the field coil of an electrical accessory.

(iv) Complete disassembly of complex hydraulic power valves.

(v) Overhaul of pressure type carburetors, and pressure type fuel, oil and hydraulic pumps.

(c) *Preventive maintenance.* Preventive maintenance is limited to the following work, provided it does not involve complex assembly operations:

(1) Removal, installation, and repair of landing gear tires.

(2) Replacing elastic shock absorber cords on landing gear.

(3) Servicing landing gear shock struts by adding oil, air, or both.

(4) Servicing landing gear wheel bearings, such as cleaning and greasing.

(5) Replacing defective safety wiring or cotter keys.

(6) Lubrication not requiring disassembly other than removal of nonstructural items such as cover plates, cowlings, and fairings.

(7) Making simple fabric patches not requiring rib stitching or the removal of structural parts or control surfaces. In the case of balloons, the making of small fabric repairs to envelopes (as defined in, and in accordance with, the balloon manufacturers' instructions) not requiring load tape repair or replacement.

(8) Replenishing hydraulic fluid in the hydraulic reservoir.

(9) Refinishing decorative coating of fuselage, balloon baskets, wings tail group surfaces (excluding balanced control surfaces), fairings, cowlings, landing gear, cabin, or cockpit interior when removal or disassembly of any primary structure or operating system is not required.

(10) Applying preservative or protective material to components where no disassembly of any primary structure or operating system is involved and where such coating is not prohibited or is not contrary to good practices.

(11) Repairing upholstery and decorative furnishings of the cabin, cockpit, or balloon basket interior when the repairing does not require disassembly of any primary structure or operating system or interfere with an operating system or affect the primary structure of the aircraft.

(12) Making small simple repairs to fairings, nonstructural cover plates, cowlings, and small patches and reinforcements not changing the contour so as to interfere with proper air flow.

(13) Replacing side windows where that work does not interfere with the structure or any operating system such as controls, electrical equipment, etc.

(14) Replacing safety belts.

(15) Replacing seats or seat parts with replacement parts approved for the aircraft, not involving disassembly of any primary structure or operating system.

(16) Trouble shooting and repairing broken circuits in landing light wiring circuits.

(17) Replacing bulbs, reflectors, and lenses of position and landing lights.

(18) Replacing wheels and skis where no weight and balance computation is involved.

(19) Replacing any cowling not requiring removal of the propeller or disconnection of flight controls.

(20) Replacing or cleaning spark plugs and setting of spark plug gap clearance.

(21) Replacing any hose connection except hydraulic connections.

(22) Replacing prefabricated fuel lines.

(23) Cleaning or replacing fuel and oil strainers or filter elements.

(24) Replacing and servicing batteries.

(25) Cleaning of balloon burner pilot and main nozzles in accordance with the balloon manufacturer's instructions.

(26) Replacement or adjustment of nonstructural standard fasteners incidental to operations.

(27) The interchange of balloon baskets and burners on envelopes when the basket or burner is designated as interchangeable in the balloon type certificate data and the baskets and burners are specifically designed for quick removal and installation.

(28) The installations of anti-misfueling devices to reduce the diameter of fuel tank filler openings provided the specific device has been made a part of the aircraft type certificiate data by the aircraft manufacturer, the aircraft manufacturer has provided FAA-approved instructions for installation of the specific device, and installation does not involve the disassembly of the existing tank filler opening.

(29) Removing, checking, and replacing magnetic chip detectors.

(30) The inspection and maintenance tasks prescribed and specifically identified as preventive maintenance in a primary category aircraft type certificate or supplemental type certificate holder's approved special inspection and preventive maintenance program when accomplished on a primary category aircraft provided:

(i) They are performed by the holder of at least a private pilot certificate issued under part 61 who is the registered owner (including co-owners) of the affected aircraft and who holds a certificate of competency for the affected aircraft (1) issued by a school approved under § 147.21(e) of this chapter; (2) issued by the holder of the production certificate for that primary category aircraft that has a special training program approved under § 21.24 of this subchapter; or (3) issued by another entity that has a course approved by the Administrator; and

(ii) The inspections and maintenance tasks are performed in accordance with instructions contained by the special inspection and preventive maintenance program approved as part of the aircraft's type design or supplemental type design.

(31) Removing and replacing self-contained, front instrument panel-mounted navigation and communication devices that employ tray-mounted connectors that connect the unit when the unit is installed into the instrument panel, (excluding automatic flight control systems, transponders, and microwave frequency distance measuring equipment (DME)). The approved unit must be designed to be readily and repeatedly removed and replaced, and pertinent instructions must be provided. Prior to the unit's intended use, and operational check must be performed in accordance with the applicable sections of part 91 of this chapter.

(Secs. 313, 601 through 610, and 1102, Federal Aviation Act of 1958 as amended (49 U.S.C. 1354, 1421 through 1430 and 1502); (49 U.S.C. 106(g) (Revised Pub. L. 97-449, Jan. 21, 1983); and 14 CFR 11.45)

[Doc. No. 1993, 29 FR 5451, Apr. 23, 1964, as amended by Amdt. 43-14, 37 FR 14291, June 19, 1972; Amdt. 43-23, 47 FR 41086, Sept. 16, 1982; Amdt. 43-24, 49 FR 44602, Nov. 7, 1984; Amdt. 43-25, 51 FR 40703, Nov. 7, 1986; Amdt. 43-27, 52 FR 17277, May 6, 1987; Amdt. 43-34, 57 FR 41369, Sept. 9, 1992; Amdt. 43-36, 61 FR 19501, May 1, 1996; Amdt. 43-45, 77 FR 71096, Nov. 29, 2012]

Appendix B to Part 43—Recording of Major Repairs and Major Alterations

(a) Except as provided in paragraphs (b), (c), and (d) of this appendix, each person performing a major repair or major alteration shall—

(1) Execute FAA Form 337 at least in duplicate;

(2) Give a signed copy of that form to the aircraft owner; and

(3) Forward a copy of that form to the FAA Aircraft Registration Branch in Oklahoma City, Oklahoma, within 48 hours after the aircraft, airframe, aircraft engine, propeller, or appliance is approved for return to service.

(b) For major repairs made in accordance with a manual or specifications acceptable to the Administrator, a certificated repair station may, in place of the requirements of paragraph (a)—

(1) Use the customer's work order upon which the repair is recorded;

(2) Give the aircraft owner a signed copy of the work order and retain a duplicate copy for at least two years from the date of approval for return to service of the aircraft, airframe, aircraft engine, propeller, or appliance;

(3) Give the aircraft owner a maintenance release signed by an authorized representative of the repair station and incorporating the following information:

(i) Identity of the aircraft, airframe, aircraft engine, propeller or appliance.

(ii) If an aircraft, the make, model, serial number, nationality and registration marks, and location of the repaired area.

(iii) If an airframe, aircraft engine, propeller, or appliance, give the manufacturer's name, name of the part, model, and serial numbers (if any); and

(4) Include the following or a similarly worded statement—

"The aircraft, airframe, aircraft engine, propeller, or appliance identified above was repaired and inspected in accordance with current Regulations of the Federal Aviation Agency and is approved for return to service.

Pertinent details of the repair are on file at this repair station under Order No. ___,

Date

Signed

For signature of authorized representative)

Repair station name) (Certificate No.)

___________."

(Address)

(c) Except as provided in paragraph (d) of this appendix, for a major repair or major alteration made by a person authorized in § 43.17, the person who performs the major repair or major alteration and the person authorized by § 43.17 to approve that work shall execute an FAA Form 337 at least in duplicate. A completed copy of that form shall be—

(1) Given to the aircraft owner; and

(2) Forwarded to the Federal Aviation Administration, Aircraft Registration Branch, AFS-750, Post Office Box 25504, Oklahoma City, OK 73125, within 48 hours after the work is inspected.

(d) For extended-range fuel tanks installed within the passenger compartment or a baggage compartment, the person who performs the work and the person authorized to approve the work by § 43.7 shall execute an FAA Form 337 in at least triplicate. A completed copy of that form shall be—

(1) Placed on board the aircraft as specified in § 91.417 of this chapter;

(2) Given to the aircraft owner; and

(3) Forwarded to the Federal Aviation Administration, Aircraft Registration Branch, AFS-751, Post Office Box 25724, Oklahoma City, OK 73125, within 48 hours after the work is inspected.

(Secs. 101, 610, 72 Stat. 737, 780, 49 U.S.C. 1301, 1430)

[Doc. No. 1993, 29 FR 5451, Apr. 23, 1964, as amended by Amdt. 43-10, 33 FR 15989, Oct. 31, 1968; Amdt. 43-29, 52 FR 34101, Sept. 9, 1987; Amdt. 43-31, 54 FR 34330, Aug. 18, 1989; 71 FR 58495, Oct. 4, 2006; Amdt. 43-41, 72 FR 53680, Sept. 20, 2007]

Appendix C to Part 43 [Reserved]

Appendix D to Part 43—Scope and Detail of Items (as Applicable to the Particular Aircraft) To Be Included in Annual and 100-Hour Inspections

(a) Each person performing an annual or 100-hour inspection shall, before that inspection, remove or open all necessary inspection plates, access doors, fairing, and cowling. He shall thoroughly clean the aircraft and aircraft engine.

(b) Each person performing an annual or 100-hour inspection shall inspect (where applicable) the following components of the fuselage and hull group:

(1) Fabric and skin—for deterioration, distortion, other evidence of failure, and defective or insecure attachment of fittings.

(2) Systems and components—for improper installation, apparent defects, and unsatisfactory operation.

(3) Envelope, gas bags, ballast tanks, and related parts—for poor condition.

(c) Each person performing an annual or 100-hour inspection shall inspect (where applicable) the following components of the cabin and cockpit group:

(1) Generally—for uncleanliness and loose equipment that might foul the controls.

(2) Seats and safety belts—for poor condition and apparent defects.

(3) Windows and windshields—for deterioration and breakage.

(4) Instruments—for poor condition, mounting, marking, and (where practicable) improper operation.

(5) Flight and engine controls—for improper installation and improper operation.

(6) Batteries—for improper installation and improper charge.

(7) All systems—for improper installation, poor general condition, apparent and obvious defects, and insecurity of attachment.

(d) Each person performing an annual or 100-hour inspection shall inspect (where applicable) components of the engine and nacelle group as follows:

(1) Engine section—for visual evidence of excessive oil, fuel, or hydraulic leaks, and sources of such leaks.

(2) Studs and nuts—for improper torquing and obvious defects.

(3) Internal engine—for cylinder compression and for metal particles or foreign matter on screens and sump drain plugs. If there is weak cylinder compression, for improper internal condition and improper internal tolerances.

(4) Engine mount—for cracks, looseness of mounting, and looseness of engine to mount.

(5) Flexible vibration dampeners—for poor condition and deterioration.

(6) Engine controls—for defects, improper travel, and improper safetying.

(7) Lines, hoses, and clamps—for leaks, improper condition and looseness.

(8) Exhaust stacks—for cracks, defects, and improper attachment.

(9) Accessories—for apparent defects in security of mounting.

(10) All systems—for improper installation, poor general condition, defects, and insecure attachment.

(11) Cowling—for cracks, and defects.

(e) Each person performing an annual or 100-hour inspection shall inspect (where applicable) the following components of the landing gear group:

(1) All units—for poor condition and insecurity of attachment.

(2) Shock absorbing devices—for improper oleo fluid level.

(3) Linkages, trusses, and members—for undue or excessive wear fatigue, and distortion.

(4) Retracting and locking mechanism—for improper operation.

(5) Hydraulic lines—for leakage.

(6) Electrical system—for chafing and improper operation of switches.

(7) Wheels—for cracks, defects, and condition of bearings.

(8) Tires—for wear and cuts.

(9) Brakes—for improper adjustment.

(10) Floats and skis—for insecure attachment and obvious or apparent defects.

(f) Each person performing an annual or 100-hour inspection shall inspect (where applicable) all components of the wing and center section assembly for poor general condition, fabric or skin deterioration, distortion, evidence of failure, and insecurity of attachment.

(g) Each person performing an annual or 100-hour inspection shall inspect (where applicable) all components and systems that make up the complete empennage assembly for poor general condition, fabric or skin deterioration, distortion, evidence of failure, insecure attachment, improper component installation, and improper component operation.

(h) Each person performing an annual or 100-hour inspection shall inspect (where applicable) the following components of the propeller group:

(1) Propeller assembly—for cracks, nicks, binds, and oil leakage.

(2) Bolts—for improper torquing and lack of safetying.

(3) Anti-icing devices—for improper operations and obvious defects.

(4) Control mechanisms—for improper operation, insecure mounting, and restricted travel.

(i) Each person performing an annual or 100-hour inspection shall inspect (where applicable) the following components of the radio group:

(1) Radio and electronic equipment—for improper installation and insecure mounting.

(2) Wiring and conduits—for improper routing, insecure mounting, and obvious defects.

(3) Bonding and shielding—for improper installation and poor condition.

(4) Antenna including trailing antenna—for poor condition, insecure mounting, and improper operation.

(j) Each person performing an annual or 100-hour inspection shall inspect (where applicable) each installed miscellaneous item that is not otherwise covered by this listing for improper installation and improper operation.

Appendix E to Part 43—Altimeter System Test and Inspection

Each person performing the altimeter system tests and inspections required by § 91.411 shall comply with the following:

(a) Static pressure system:

(1) Ensure freedom from entrapped moisture and restrictions.

(2) Determine that leakage is within the tolerances established in § 23.1325 or § 25.1325, whichever is applicable.

(3) Determine that the static port heater, if installed, is operative.

(4) Ensure that no alterations or deformations of the airframe surface have been made that would affect the relationship between air pressure in the static pressure system and true ambient static air pressure for any flight condition.

(b) Altimeter:

(1) Test by an appropriately rated repair facility in accordance with the following subparagraphs. Unless otherwise specified, each test for performance may be conducted with the instrument subjected to vibration. When tests are conducted with the temperature substantially different from ambient temperature of approximately 25 degrees C., allowance shall be made for the variation from the specified condition.

(i) *Scale error.* With the barometric pressure scale at 29.92 inches of mercury, the altimeter shall be subjected successively to pressures corresponding to the altitude specified in Table I up to the maximum normally expected operating altitude of the airplane in which the altimeter is to be installed. The reduction in pressure shall be made at a rate not in excess of 20,000 feet per minute to within approximately 2,000 feet of the test point. The test point shall be approached at a rate compatible with the test equipment. The altimeter shall be kept at the pressure corresponding to each test point for at least 1 minute, but not more than 10 minutes, before a reading is taken. The error at all test points must not exceed the tolerances specified in Table I.

(ii) *Hysteresis.* The hysteresis test shall begin not more than 15 minutes after the altimeter's initial exposure to the pressure corresponding to the upper limit of the scale error test prescribed in subparagraph (i); and while the altimeter is at this pressure, the hysteresis test shall commence. Pressure shall be increased at a rate simulating a descent in altitude at the rate of 5,000 to 20,000 feet per minute until within 3,000 feet of the first test point (50 percent of maximum altitude). The test point shall then be approached at a rate of approximately 3,000 feet per minute. The altimeter shall be kept at this pressure for at least 5 minutes, but not more than 15 minutes, before the test reading is taken. After the reading has been taken, the pressure shall be increased further, in the same manner as before, until the pressure corresponding to the second test point (40 percent of maximum altitude) is reached. The altimeter shall be kept at this pressure for at least 1 minute, but not more than 10 minutes, before the test reading is taken. After the reading has been taken, the pressure shall be increased further, in the same manner as before, until atmospheric pressure is reached. The reading of the altimeter at either of the two test points shall not differ by more than the tolerance specified in Table II from the reading of the altimeter for the corresponding altitude recorded during the scale error test prescribed in paragraph (b)(i).

(iii) *After effect.* Not more than 5 minutes after the completion of the hysteresis test prescribed in paragraph (b)(ii), the reading of the altimeter (corrected for any change in atmospheric pressure) shall not differ from the original atmospheric pressure reading by more than the tolerance specified in Table II.

(iv) *Friction.* The altimeter shall be subjected to a steady rate of decrease of pressure approximating 750 feet per minute. At each altitude listed in Table III, the change in reading of the pointers after vibration shall not exceed the corresponding tolerance listed in Table III.

(v) *Case leak.* The leakage of the altimeter case, when the pressure within it corresponds to an altitude of 18,000 feet, shall not change the altimeter reading by more than the tolerance shown in Table II during an interval of 1 minute.

(vi) *Barometric scale error.* At constant atmospheric pressure, the barometric pressure scale shall be set at each of the pressures (falling within its range of adjustment) that are listed in Table IV, and shall cause the pointer to indicate the equivalent altitude difference shown in Table IV with a tolerance of 25 feet.

(2) Altimeters which are the air data computer type with associated computing systems, or which incorporate air data correction internally, may be tested in a manner and to specifications developed by the manufacturer which are acceptable to the Administrator.

(c) Automatic Pressure Altitude Reporting Equipment and ATC Transponder System Integration Test. The test must be conducted by an appropriately rated person under the conditions specified in paragraph (a). Measure the automatic pressure altitude at the output of the installed ATC transponder when interrogated on Mode C at a sufficient number of test points to ensure that the altitude reporting equipment, altimeters, and ATC transponders perform their intended functions as installed in the aircraft. The difference between the automatic reporting output and the altitude displayed at the altimeter shall not exceed 125 feet.

(d) Records: Comply with the provisions of § 43.9 of this chapter as to content, form, and disposition of the records. The person performing the altimeter tests shall record on the altimeter the date and maximum altitude to which the altimeter has been tested and the persons approving the airplane for return to service shall enter that data in the airplane log or other permanent record.

Table I

Altitude	Equivalent pressure (inches of mercury)	Tolerance ±(feet)
-1,000	31.018	20
0	29.921	20
500	29.385	20
1,000	28.856	20
1,500	28.335	25
2,000	27.821	30
3,000	26.817	30
4,000	25.842	35
6,000	23.978	40
8,000	22.225	60
10,000	20.577	80
12,000	19.029	90
14,000	17.577	100
16,000	16.216	110
18,000	14.942	120
20,000	13.750	130
22,000	12.636	140
25,000	11.104	155
30,000	8.885	180
35,000	7.041	205
40,000	5.538	230
45,000	4.355	255
50,000	3.425	280

Table II—Test Tolerances

Test	Tolerance (feet)
Case Leak Test	±100
Hysteresis Test:	
First Test Point (50 percent of maximum altitude)	75
Second Test Point (40 percent of maximum altitude)	75
After Effect Test	30

Table III—Friction

Altitude (feet)	Tolerance (feet)
1,000	±70
2,000	70
3,000	70
5,000	70
10,000	80
15,000	90
20,000	100
25,000	120
30,000	140
35,000	160
40,000	180
50,000	250

Table IV—Pressure-Altitude Difference

Pressure (inches of Hg)	Altitude difference (feet)
28.10	-1,727
28.50	-1,340
29.00	-863
29.50	-392
29.92	0
30.50	+531
30.90	+893
30.99	+974

(Secs. 313, 314, and 601 through 610 of the Federal Aviation Act of 1958 (49 U.S.C. 1354, 1355, and 1421 through 1430) and sec. 6(c), Dept. of Transportation Act (49 U.S.C. 1655(c)))

[Amdt. 43-2, 30 FR 8262, June 29, 1965, as amended by Amdt. 43-7, 32 FR 7587, May 24, 1967; Amdt. 43-19, 43 FR 22639, May 25, 1978; Amdt. 43-23, 47 FR 41086, Sept. 16, 1982; Amdt. 43-31, 54 FR 34330, Aug. 18, 1989]

Appendix F to Part 43—ATC Transponder Tests and Inspections

The ATC transponder tests required by § 91.413 of this chapter may be conducted using a bench check or portable test equipment and must meet the requirements prescribed in paragraphs (a) through (j) of this appendix. If portable test equipment with appropriate coupling to the aircraft antenna system is used, operate the test equipment for ATCRBS transponders at a nominal rate of 235 interrogations per second to avoid possible ATCRBS interference. Operate the test equipment at a nominal rate of 50 Mode S interrogations per second for Mode S. An additional 3 dB loss is allowed to compensate for antenna coupling errors during receiver sensitivity measurements conducted in accordance with paragraph (c)(1) when using portable test equipment.

(a) Radio Reply Frequency:

(1) For all classes of ATCRBS transponders, interrogate the transponder and verify that the reply frequency is 1090 ±3 Megahertz (MHz).

(2) For classes 1B, 2B, and 3B Mode S transponders, interrogate the transponder and verify that the reply frequency is 1090 ±3 MHz.

(3) For classes 1B, 2B, and 3B Mode S transponders that incorporate the optional 1090 ±1 MHz reply frequency, interrogate the transponder and verify that the reply frequency is correct.

(4) For classes 1A, 2A, 3A, and 4 Mode S transponders, interrogate the transponder and verify that the reply frequency is 1090 ±1 MHz.

(b) Suppression: When Classes 1B and 2B ATCRBS Transponders, or Classes 1B, 2B, and 3B Mode S transponders are interrogated Mode 3/A at an interrogation rate between 230 and 1,000 interrogations per second; or when Classes 1A and 2A ATCRBS Transponders, or Classes 1B, 2A, 3A, and 4 Mode S transponders are interrogated at a rate between 230 and 1,200 Mode 3/A interrogations per second:

(1) Verify that the transponder does not respond to more than 1 percent of ATCRBS interrogations when the amplitude of P2 pulse is equal to the P1 pulse.

(2) Verify that the transponder replies to at least 90 percent of ATCRBS interrogations when the amplitude of the P2 pulse is 9 dB less than the P1 pulse. If the test is conducted with a radiated test signal, the interrogation rate shall be 235 ±5 interrogations per second unless a higher rate has been approved for the test equipment used at that location.

(c) Receiver Sensitivity:

(1) Verify that for any class of ATCRBS Transponder, the receiver minimum triggering level (MTL) of the system is -73 ±4 dbm, or that for any class of Mode S transponder the receiver MTL for Mode S format (P6 type) interrogations is -74 ±3 dbm by use of a test set either:

(i) Connected to the antenna end of the transmission line;

(ii) Connected to the antenna terminal of the transponder with a correction for transmission line loss; or

(iii) Utilized radiated signal.

(2) Verify that the difference in Mode 3/A and Mode C receiver sensitivity does not exceed 1 db for either any class of ATCRBS transponder or any class of Mode S transponder.

(d) Radio Frequency (RF) Peak Output Power:

(1) Verify that the transponder RF output power is within specifications for the class of transponder. Use the same conditions as described in (c)(1)(i), (ii), and (iii) above.

(i) For Class 1A and 2A ATCRBS transponders, verify that the minimum RF peak output power is at least 21.0 dbw (125 watts).

(ii) For Class 1B and 2B ATCRBS Transponders, verify that the minimum RF peak output power is at least 18.5 dbw (70 watts).

(iii) For Class 1A, 2A, 3A, and 4 and those Class 1B, 2B, and 3B Mode S transponders that include the optional high RF peak output power, verify that the minimum RF peak output power is at least 21.0 dbw (125 watts).

(iv) For Classes 1B, 2B, and 3B Mode S transponders, verify that the minimum RF peak output power is at least 18.5 dbw (70 watts).

(v) For any class of ATCRBS or any class of Mode S transponders, verify that the maximum RF peak output power does not exceed 27.0 dbw (500 watts).

Note: The tests in (e) through (j) apply only to Mode S transponders.

(e) Mode S Diversity Transmission Channel Isolation: For any class of Mode S transponder that incorporates diversity operation, verify that the RF peak output power transmitted from the selected antenna exceeds the power transmitted from the nonselected antenna by at least 20 db.

(f) Mode S Address: Interrogate the Mode S transponder and verify that it replies only to its assigned address. Use the correct address and at least two incorrect addresses. The interrogations should be made at a nominal rate of 50 interrogations per second.

(g) Mode S Formats: Interrogate the Mode S transponder with uplink formats (UF) for which it is equipped and verify that the replies are made in the correct format. Use the surveillance formats UF=4 and 5. Verify that the altitude reported in the replies to UF=4 are the same as that reported in a valid ATCRBS Mode C reply. Verify that the identity reported in the replies to UF=5 are the same as that reported in a valid ATCRBS Mode 3/A reply. If the transponder is so equipped, use the communication formats UF=20, 21, and 24.

(h) Mode S All-Call Interrogations: Interrogate the Mode S transponder with the Mode S-only all-call format UF=11, and the ATCRBS/Mode S all-call formats (1.6 microsecond P4 pulse) and verify that the correct address and capability are reported in the replies (downlink format DF=11).

(i) ATCRBS-Only All-Call Interrogation: Interrogate the Mode S transponder with the ATCRBS-only all-call interrogation (0.8 microsecond P4 pulse) and verify that no reply is generated.

(j) Squitter: Verify that the Mode S transponder generates a correct squitter approximately once per second.

(k) Records: Comply with the provisions of § 43.9 of this chapter as to content, form, and disposition of the records.

[Amdt. 43-26, 52 FR 3390, Feb. 3, 1987; 52 FR 6651, Mar. 4, 1987, as amended by Amdt. 43-31, 54 FR 34330, Aug. 18, 1989]

PART 61

CERTIFICATION: PILOTS, FLIGHT INSTRUCTORS, AND GROUND INSTRUCTORS

Contents

Authority: 49 U.S.C. 106(f), 106(g), 40113, 44701-44703, 44707, 44709-44711, 45102-45103, 45301-45302.

Source: Docket No. 25910, 62 FR 16298, Apr. 4, 1997, unless otherwise noted.

Special Federal Aviation Regulation No. 73—Robinson R-22/R-44 Special Training and Experience Requirements

Sections

1. Applicability.
2. Required training, aeronautical experience, endorsements, and flight review.
3. Expiration date.

1. Applicability. Under the procedures prescribed herein, this SFAR applies to all persons who seek to manipulate the controls or act as pilot in command of a Robinson model R-22 or R-44 helicopter. The requirements stated in this SFAR are in addition to the current requirements of part 61.

2. Required training, aeronautical experience, endorsements, and flight review.

(a) Awareness Training:

(1) Except as provided in paragraph (a)(2) of this section, no person may manipulate the controls of a Robinson model R-22 or R-44 helicopter after March 27, 1995, for the purpose of flight unless the awareness training specified in paragraph (a)(3) of this section is completed and the person's logbook has been endorsed by a certified flight instructor authorized under paragraph (b)(5) of this section.

(2) A person who holds a rotorcraft category and helicopter class rating on that person's pilot certificate and meets the experience requirements of paragraph (b)(1) or paragraph (b)(2) of this section may not manipulate the controls of a Robinson model R-22 or R-44 helicopter for the purpose of flight after April 26, 1995, unless the awareness training specified in paragraph (a)(3) of this section is completed and the person's logbook has been endorsed by a certified flight instructor authorized under paragraph (b)(5) of this section.

(3) Awareness training must be conducted by a certified flight instructor who has been endorsed under paragraph (b)(5) of this section and consists of instruction in the following general subject areas:

(i) Energy management;

(ii) Mast bumping;

(iii) Low rotor RPM (blade stall);

(iv) Low G hazards; and

(v) Rotor RPM decay.

(4) A person who can show satisfactory completion of the manufacturer's safety course after January 1, 1994, may obtain an endorsement from an FAA aviation safety inspector in lieu of completing the awareness training required in paragraphs (a)(1) and (a)(2) of this section.

(b) Aeronautical Experience:

(1) No person may act as pilot in command of a Robinson model R-22 unless that person:

(i) Has had at least 200 flight hours in helicopters, at least 50 flight hours of which were in the Robinson R-22; or

(ii) Has had at least 10 hours dual instruction in the Robinson R-22 and has received an endorsement from a certified flight instructor authorized under paragraph (b)(5) of this section that the individual has been given the training required by this paragraph and is proficient to act as pilot in command of an R-22. Beginning 12 calendar months after the date of the endorsement, the individual may not act as pilot in command unless the individual has completed a flight review in an R-22 within the preceding 12 calendar months and obtained an endorsement for that flight review. The dual instruction must include at least the following abnormal and emergency procedures flight training:

(A) Enhanced training in autorotation procedures,

(B) Engine rotor RPM control without the use of the governor,

(C) Low rotor RPM recognition and recovery, and

(D) Effects of low G maneuvers and proper recovery procedures.

(2) No person may act as pilot in command of a Robinson R-44 unless that person—

(i) Has had at least 200 flight hours in helicopters, at least 50 flight hours of which were in the Robinson R-44. The pilot in command may credit up to 25 flight hours in the Robinson R-22 toward the 50 hour requirement in the Robinson R-44; or

(ii) Has had at least 10 hours dual instruction in a Robinson helicopter, at least 5 hours of which must have been accomplished in the Robinson R-44 helicopter and has received an endorsement from a certified flight instructor authorized under paragraph (b)(5) of this section that the individual has been given the training required by this paragraph and is proficient to act as pilot in command of an R-44. Beginning 12 calendar months after the date of the endorsement, the individual may not act as pilot in command unless the individual has completed a flight review in a Robinson R-44 within the preceding 12 calendar months and obtained an endorsement for that flight review. The dual instruction must include at least the following abnormal and emergency procedures flight training—

(A) Enhanced training in autorotation procedures;

(B) Engine rotor RPM control without the use of the governor;

(C) Low rotor RPM recognition and recovery; and

(D) Effects of low G maneuvers and proper recovery procedures.

(3) A person who does not hold a rotorcraft category and helicopter class rating must have had at least 20 hours of dual instruction in a Robinson R-22 helicopter prior to operating it in solo flight. In addition, the person must obtain an endorsement from a certified flight instructor authorized under paragraph (b)(5) of this section that instruction has been given in those maneuvers and procedures, and the instructor has found the applicant proficient to solo a Robinson R-22. This endorsement is valid for a period of 90 days. The dual instruction must include at least the following abnormal and emergency procedures flight training:

(i) Enhanced training in autorotation procedures,

(ii) Engine rotor RPM control without the use of the governor,

(iii) Low rotor RPM recognition and recovery, and

(iv) Effects of low G maneuvers and proper recovery procedures.

(4) A person who does not hold a rotorcraft category and helicopter class rating must have had at least 20 hours of dual instruction in a Robinson R-44 helicopter prior to operating it in solo flight. In addition, the person must obtain an endorsement from a certified flight instructor authorized under paragraph (b)(5) of this section that instruction has been given in those maneuvers and procedures, and the instructor has found the applicant proficient to solo a Robinson R-44. This endorsement is valid for a period of 90 days. The dual instruction must include at least the following abnormal and emergency procedures flight training:

(i) Enhanced training in autorotation procedures,

(ii) Engine rotor RPM control without the use of the governor,

(iii) Low rotor RPM recognition and recovery, and

(iv) Effects of low G maneuvers and proper recovery procedures.

(5) No certificated flight instructor may provide instruction or conduct a flight review in a Robinson R-22 or R-44 unless that instructor—

(i) Completes the awareness training in paragraph 2(a) of this SFAR.

(ii) For the Robinson R-22, has had at least 200 flight hours in helicopters, at least 50 flight hours of which were in the Robinson R-22, or for the Robinson R-44, has had at least 200 flight hours in helicopters, 50 flight hours of which were in Robinson helicopters. Up to 25 flight hours of Robinson R-22 flight time may be credited toward the 50 hour requirement.

(iii) Has completed flight training in a Robinson R-22, R-44, or both, on the following abnormal and emergency procedures—

(A) Enhanced training in autorotation procedures;

(B) Engine rotor RPM control without the use of the governor;

(C) Low rotor RPM recognition and recovery; and

(D) Effects of low G maneuvers and proper recovery procedures.

(iv) Has been authorized by endorsement from an FAA aviation safety inspector or authorized designated examiner that the instructor has completed the appropriate training, meets the experience requirements and has satisfactorily demonstrated an ability to provide instruction on the general subject areas of paragraph 2(a)(3) of this SFAR, and the flight training identified in paragraph 2(b)(5)(iii) of this SFAR.

(c) Flight Review:

(1) No flight review completed to satisfy § 61.56 by an individual after becoming eligible to function as pilot in command in a Robinson R-22 helicopter shall be valid for the operation of R-22 helicopter unless that flight review was taken in an R-22.

(2) No flight review completed to satisfy § 61.56 by individual after becoming eligible to function as pilot in command in a Robinson R-44 helicopter shall be valid for the operation of R-44 helicopter unless that flight review was taken in the R-44.

(3) The flight review will include a review of the awareness training subject areas of paragraph 2(a)(3) of this SFAR and the flight training identified in paragraph 2(b) of this SFAR.

(d) Currency Requirements: No person may act as pilot in command of a Robinson model R-22 or R-44 helicopter carrying passengers unless the pilot in command has met the recency of flight experience requirements of § 61.57 in an R-22 or R-44, as appropriate.

3. *Expiration date.* This SFAR No. 73 shall remain in effect until it is revised or rescinded.

[Doc. No. 25910, 62 FR 16298, Apr. 4, 1997, as amended by SFAR 73-1, 63 FR 666, Jan. 7, 1998; 68 FR 43, Jan. 2, 2003; Amdt. 61-120, 73 FR 17246, Apr. 1, 2008; Amdt. SFAR 73-2, 74 FR 25650, May 29, 2009]

Special Federal Aviation Regulation No. 100-2—Relief for U.S. Military and Civilian Personnel Who are Assigned Outside the United States in Support of U.S. Armed Forces Operations

1. *Applicability.* Flight Standards District Offices are authorized to accept from an eligible person, as described in paragraph 2 of this SFAR, the following:

(a) An expired flight instructor certificate to show eligibility for renewal of a flight instructor certificate under § 61.197, or an expired written test report to show eligibility under part 61 to take a practical test;

(b) An expired written test report to show eligibility under §§ 63.33 and 63.57 to take a practical test; and

(c) An expired written test report to show eligibility to take a practical test required under part 65 or an expired inspection authorization to show eligibility for renewal under § 65.93.

2. *Eligibility.* A person is eligible for the relief described in paragraph 1 of this SFAR if:

(a) The person served in a U.S. military or civilian capacity outside the United States in support of the U.S. Armed Forces' operation during some period of time from September 11, 2001, to termination of SFAR 100-2;

(b) The person's flight instructor certificate, airman written test report, or inspection authorization expired some time between September 11, 2001, and 6 calendar months after returning to the United States or termination of SFAR 100-2, whichever is earlier; and

(c) The person complies with § 61.197 or § 65.93 of this chapter, as appropriate, or completes the appropriate practical test within 6 calendar months after returning to the United States, or upon termination of SFAR 100-2, whichever is earlier.

3. *Required documents.* The person must send the Airman Certificate and/or Rating Application (FAA Form 8710-1) to the appropriate Flight Standards District Office. The person must include with the application one of the following documents, which must show the date of assignment outside the United States and the date of return to the United States:

(a) An official U.S. Government notification of personnel action, or equivalent document, showing the person was a civilian on official duty for the U.S. Government outside the United States and was assigned to a U.S. Armed Forces' operation some time between September 11, 2001, to termination of SFAR 100-2;

(b) Military orders showing the person was assigned to duty outside the United States and was assigned to a U.S. Armed Forces' operation some time between September 11, 2001, to termination of SFAR 100-2; or

(c) A letter from the person's military commander or civilian supervisor providing the dates during which the person served outside the United States and was assigned to a U.S. Armed Forces' operation some time between September 11, 2001, to termination of SFAR 100-2.

4. *Expiration date.* This Special Federal Aviation Regulation No. 100-2 is effective until further notice.

[Doc. No. FAA-2009-0923, 75 FR 9766, Mar. 4, 2010]

Special Federal Aviation Regulation No. 108

Note: For the text of SFAR No. 108, see part 91 of this chapter.

Subpart A—General

§ 61.1 Applicability and definitions.

(a) This part prescribes:

(1) The requirements for issuing pilot, flight instructor, and ground instructor certificates and ratings; the conditions under which those certificates and ratings are necessary; and the privileges and limitations of those certificates and ratings.

(2) The requirements for issuing pilot, flight instructor, and ground instructor authorizations; the conditions under which those authorizations are necessary; and the privileges and limitations of those authorizations.

(3) The requirements for issuing pilot, flight instructor, and ground instructor certificates and ratings for persons who have taken courses approved by the Administrator under other parts of this chapter.

(b) For the purpose of this part:

Accredited has the same meaning as defined by the Department of Education in 34 CFR 600.2.

Aeronautical experience means pilot time obtained in an aircraft, flight simulator, or flight training device for meeting the appropriate training and flight time requirements for an airman certificate, rating, flight review, or recency of flight experience requirements of this part.

Authorized instructor means—

(i) A person who holds a ground instructor certificate issued under part 61 of this chapter and is in compliance with § 61.217, when conducting ground training in accordance with the privileges and limitations of his or her ground instructor certificate;

(ii) A person who holds a flight instructor certificate issued under part 61 of this chapter and is in compliance with § 61.197, when conducting ground training or flight training in accordance with the privileges and limitations of his or her flight instructor certificate; or

(iii) A person authorized by the Administrator to provide ground training or flight training under part 61, 121, 135, or 142 of this chapter when conducting ground training or flight training in accordance with that authority.

Complex airplane means an airplane that has a retractable landing gear, flaps, and a controllable pitch propeller, including airplanes equipped with an engine control system consisting of a digital computer and associated accessories for controlling the engine and propeller, such as a full authority digital engine control; or, in the case of a seaplane, flaps and a controllable pitch propeller, including seaplanes equipped with an engine control system consisting of a digital computer and associated accessories for controlling the engine and propeller, such as a full authority digital engine control.

Cross-country time means—

(i) Except as provided in paragraphs (ii) through (vi) of this definition, time acquired during flight—

(A) Conducted by a person who holds a pilot certificate;

(B) Conducted in an aircraft;

(C) That includes a landing at a point other than the point of departure; and

(D) That involves the use of dead reckoning, pilotage, electronic navigation aids, radio aids, or other navigation systems to navigate to the landing point.

(ii) For the purpose of meeting the aeronautical experience requirements (except for a rotorcraft category rating), for a private pilot certificate (except for a powered parachute category rating), a commercial pilot certificate, or an instrument rating, or for the purpose of exercising recreational pilot privileges (except in a rotorcraft) under § 61.101 (c), time acquired during a flight—

(A) Conducted in an appropriate aircraft;

(B) That includes a point of landing that was at least a straight-line distance of more than 50 nautical miles from the original point of departure; and

(C) That involves the use of dead reckoning, pilotage, electronic navigation aids, radio aids, or other navigation systems to navigate to the landing point.

(iii) For the purpose of meeting the aeronautical experience requirements for a sport pilot certificate (except for powered parachute privileges), time acquired during a flight conducted in an appropriate aircraft that—

(A) Includes a point of landing at least a straight line distance of more than 25 nautical miles from the original point of departure; and

(B) Involves, as applicable, the use of dead reckoning; pilotage; electronic navigation aids; radio aids; or other navigation systems to navigate to the landing point.

(iv) For the purpose of meeting the aeronautical experience requirements for a sport pilot certificate with powered parachute privileges or a private pilot certificate with a powered parachute category rating, time acquired during a flight conducted in an appropriate aircraft that—

(A) Includes a point of landing at least a straight line distance of more than 15 nautical miles from the original point of departure; and

(B) Involves, as applicable, the use of dead reckoning; pilotage; electronic navigation aids; radio aids; or other navigation systems to navigate to the landing point.

(v) For the purpose of meeting the aeronautical experience requirements for any pilot certificate with a rotorcraft category rating or an instrument-helicopter rating, or for the purpose of exercising recreational pilot privileges, in a rotorcraft, under § 61.101(c), time acquired during a flight—

(A) Conducted in an appropriate aircraft;

(B) That includes a point of landing that was at least a straight-line distance of more than 25 nautical miles from the original point of departure; and

(C) That involves the use of dead reckoning, pilotage, electronic navigation aids, radio aids, or other navigation systems to navigate to the landing point.

(vi) For the purpose of meeting the aeronautical experience requirements for an airline transport pilot certificate (except with a rotorcraft category rating), time acquired during a flight—

(A) Conducted in an appropriate aircraft;

(B) That is at least a straight-line distance of more than 50 nautical miles from the original point of departure; and

(C) That involves the use of dead reckoning, pilotage, electronic navigation aids, radio aids, or other navigation systems.

(vii) For a military pilot who qualifies for a commercial pilot certificate (except with a rotorcraft category rating) under § 61.73 of this part, time acquired during a flight—

(A) Conducted in an appropriate aircraft;

(B) That is at least a straight-line distance of more than 50 nautical miles from the original point of departure; and

(C) That involves the use of dead reckoning, pilotage, electronic navigation aids, radio aids, or other navigation systems.

Examiner means any person who is authorized by the Administrator to conduct a pilot proficiency test or a practical test for an airman certificate or rating issued under this part, or a person who is authorized to conduct a knowledge test under this part.

Flight training means that training, other than ground training, received from an authorized instructor in flight in an aircraft.

Ground training means that training, other than flight training, received from an authorized instructor.

Institution of higher education has the same meaning as defined by the Department of Education in 34 CFR 600.4.

Instrument approach means an approach procedure defined in part 97 of this chapter.

Instrument training means that time in which instrument training is received from an authorized instructor under actual or simulated instrument conditions.

Knowledge test means a test on the aeronautical knowledge areas required for an airman certificate or rating that can be administered in written form or by a computer.

Nationally recognized accrediting agency has the same meaning as defined by the Department of Education in 34 CFR 600.2.

Night vision goggles means an appliance worn by a pilot that enhances the pilot's ability to maintain visual surface reference at night.

Night vision goggle operation means the portion of a flight that occurs during the time period from 1 hour after sunset to 1 hour before sunrise where the pilot maintains visual surface reference using night vision goggles in an aircraft that is approved for such an operation.

Pilot time means that time in which a person—

(i) Serves as a required pilot flight crewmember;

(ii) Receives training from an authorized instructor in an aircraft, flight simulator, or flight training device; or

(iii) Gives training as an authorized instructor in an aircraft, flight simulator, or flight training device.

Practical test means a test on the areas of operations for an airman certificate, rating, or authorization that is conducted by having the applicant respond to questions and demonstrate maneuvers in flight, in a flight simulator, or in a flight training device.

Set of aircraft means aircraft that share similar performance characteristics, such as similar airspeed and altitude operating envelopes, similar handling characteristics, and the same number and type of propulsion systems.

Student pilot seeking a sport pilot certificate means a person who has received an endorsement—

(i) To exercise student pilot privileges from a certificated flight instructor with a sport pilot rating; or

(ii) That includes a limitation for the operation of a light-sport aircraft specified in § 61.89(c) issued by a certificated flight instructor with other than a sport pilot rating.

Training time means training received—

(i) In flight from an authorized instructor;

(ii) On the ground from an authorized instructor; or

(iii) In a flight simulator or flight training device from an authorized instructor.

[Doc. No. 25910, 62 FR 16298, Apr. 4, 1997; Amdt. 61-103, 62 FR 40893, July 30, 1997; Amdt. 61-110, 69 FR 44864, July 27, 2004; Amdt. 61-124, 74 FR 42546, Aug. 21, 2009; Amdt. 61-128, 76 FR 54105, Aug. 31, 2011; Amdt. 61-130, 78 FR 42372, July 15, 2013]

§ 61.2 Exercise of Privilege.

(a) *Validity.* No person may:

(1) Exercise privileges of a certificate, rating, endorsement, or authorization issued under this part if the certificate, rating or authorization is surrendered, suspended, revoked or expired.

(2) Exercise privileges of a flight instructor certificate if that flight instructor certificate is surrendered, suspended, revoked or expired.

(3) Exercise privileges of a foreign pilot certificate to operate an aircraft of foreign registry under § 61.3(b) if the certificate is surrendered, suspended, revoked or expired.

(4) Exercise privileges of a pilot certificate issued under § 61.75, or an authorization issued under § 61.77, if the foreign pilot certificate relied upon for the issuance of the U.S. pilot certificate or authorization is surrendered, suspended, revoked or expired.

(5) Exercise privileges of a medical certificate issued under part 67 to meet any requirements of part 61 if the medical certificate is surrendered, suspended, revoked or expired according to the duration standards set forth in § 61.23(d).

(6) Use an official government issued driver's license to meet any requirements of part 61 related to holding that driver's license, if the driver's license is surrendered, suspended, revoked or expired.

(b) *Currency.* No person may:

(1) Exercise privileges of an airman certificate, rating, endorsement, or authorization issued under this part unless that person meets the appropriate airman and medical recency requirements of this part, specific to the operation or activity.

(2) Exercise privileges of a foreign pilot license within the United States to conduct an operation described in § 61.3(b), unless that person meets the appropriate airman and medical recency requirements of the country that issued the license, specific to the operation.

[Doc. No. FAA-2006-26661, 74 FR 42546, Aug. 21, 2009]

§ 61.3 Requirement for certificates, ratings, and authorizations.

(a) *Pilot certificate.* No person may serve as a required pilot flight crewmember of a civil aircraft of the United States, unless that person—

(1) Has a pilot certificate or special purpose pilot authorization issued under this part in that person's physical possession or readily accessible in the aircraft when exercising the privileges of that pilot certificate or authorization. However, when the aircraft is operated within a foreign country, a pilot license issued by that country may be used; and

(2) Has a photo identification that is in that person's physical possession or readily accessible in the aircraft when exercising the privileges of that pilot certificate or authorization. The photo identification must be a:

(i) Driver's license issued by a State, the District of Columbia, or territory or possession of the United States;

(ii) Government identification card issued by the Federal government, a State, the District of Columbia, or a territory or possession of the United States;

(iii) U.S. Armed Forces' identification card;

(iv) Official passport;

(v) Credential that authorizes unescorted access to a security identification display area at an airport regulated under 49 CFR part 1542; or

(vi) Other form of identification that the Administrator finds acceptable.

(b) *Required pilot certificate for operating a foreign-registered aircraft.* No person may serve as a required pilot flight crewmember of a civil aircraft of foreign registry within the United States, unless that person's pilot certificate—

(1) Is in that person's physical possession, or readily accessible in the aircraft when exercising the privileges of that pilot certificate; and

(2) Has been issued under this part, or has been issued or validated by the country in which the aircraft is registered.

(c) *Medical certificate.* (1) A person may serve as a required pilot flight crewmember of an aircraft only if that person holds the appropriate medical certificate issued under part 67 of this chapter, or other documentation acceptable to the FAA, that is in that person's physical possession or readily accessible in the aircraft. Paragraph (c)(2) of this section provides certain exceptions to the requirement to hold a medical certificate.

(2) A person is not required to meet the requirements of paragraph (c)(1) of this section if that person—

(i) Is exercising the privileges of a student pilot certificate while seeking a pilot certificate with a glider category rating, a balloon class rating, or glider or balloon privileges;

(ii) Is exercising the privileges of a student pilot certificate while seeking a sport pilot certificate with other than glider or balloon privileges and holds a U.S. driver's license;

(iii) Is exercising the privileges of a student pilot certificate while seeking a pilot certificate with a weight-shift-control aircraft category rating or a powered parachute category rating and holds a U.S. driver's license;

(iv) Is exercising the privileges of a sport pilot certificate with glider or balloon privileges;

(v) Is exercising the privileges of a sport pilot certificate with other than glider or balloon privileges and holds a U.S. driver's license. A person who has applied for or held a medical certificate may exercise the privileges of a sport pilot certificate using a U.S. driver's license only if that person—

(A) Has been found eligible for the issuance of at least a third-class airman medical certificate at the time of his or her most recent application; and

(B) Has not had his or her most recently issued medical certificate suspended or revoked or most recent Authorization for a Special Issuance of a Medical Certificate withdrawn.

(vi) Is holding a pilot certificate with a balloon class rating and is piloting or providing training in a balloon as appropriate;

(vii) Is holding a pilot certificate or a flight instructor certificate with a glider category rating, and is piloting or providing training in a glider, as appropriate;

(viii) Except as provided in paragraph (c)(2)(vii) of this section, is exercising the privileges of a flight instructor certificate, provided the person is not acting as pilot in command or as a required pilot flight crewmember;

(ix) Is exercising the privileges of a ground instructor certificate;

(x) Is operating an aircraft within a foreign country using a pilot license issued by that country and possesses evidence of current medical qualification for that license; or

(xi) Is operating an aircraft with a U.S. pilot certificate, issued on the basis of a foreign pilot license, issued under § 61.75, and holds a medical certificate issued by the foreign country that issued the foreign pilot license, which is in that person's physical possession or readily accessible in the aircraft when exercising the privileges of that airman certificate.

(xii) Is a pilot of the U.S. Armed Forces, has an up-to-date U.S. military medical examination, and holds military pilot flight status.

(d) *Flight instructor certificate.* (1) A person who holds a flight instructor certificate issued under this part must have that certificate, or other documentation acceptable to the Administrator, in that person's physical possession or readily accessible in the aircraft when exercising the privileges of that flight instructor certificate.

(2) Except as provided in paragraph (d)(3) of this section, no person other than the holder of a flight instructor certificate issued under this part with the appropriate rating on that certificate may—

(i) Give training required to qualify a person for solo flight and solo cross-country flight;

(ii) Endorse an applicant for a—

(A) Pilot certificate or rating issued under this part;

(B) Flight instructor certificate or rating issued under this part; or

(C) Ground instructor certificate or rating issued under this part;

(iii) Endorse a pilot logbook to show training given; or

(iv) Endorse a student pilot certificate and logbook for solo operating privileges.

(3) A flight instructor certificate issued under this part is not necessary—

(i) Under paragraph (d)(2) of this section, if the training is given by the holder of a commercial pilot certificate with a lighter-than-air rating, provided the training is given in accordance with the privileges of the certificate in a lighter-than-air aircraft;

(ii) Under paragraph (d)(2) of this section, if the training is given by the holder of an airline transport pilot certificate with a rating appropriate to the aircraft in which the training is given, provided the training is given in accordance with the privileges of the certificate and conducted in accordance with an approved air carrier training program approved under part 121 or part 135 of this chapter;

(iii) Under paragraph (d)(2) of this section, if the training is given by a person who is qualified in accordance with subpart C of part 142 of this chapter, provided the training is conducted in accordance with an approved part 142 training program;

(iv) Under paragraphs (d)(2)(i), (d)(2)(ii)(C), and (d)(2)(iii) of this section, if the training is given by the holder of a ground instructor certificate in accordance with the privileges of the certificate; or

(v) Under paragraph (d)(2)(iii) of this section, if the training is given by an authorized flight instructor under § 61.41 of this part.

(e) *Instrument rating.* No person may act as pilot in command of a civil aircraft under IFR or in weather conditions less than the minimums prescribed for VFR flight unless that person holds:

(1) The appropriate aircraft category, class, type (if required), and instrument rating on that person's pilot certificate for any airplane, helicopter, or powered-lift being flown;

(2) An airline transport pilot certificate with the appropriate aircraft category, class, and type rating (if required) for the aircraft being flown;

(3) For a glider, a pilot certificate with a glider category rating and an airplane instrument rating; or

(4) For an airship, a commercial pilot certificate with a lighter-than-air category rating and airship class rating.

(f) *Category II pilot authorization.* Except for a pilot conducting Category II operations under part 121 or part 135, a person may not:

(1) Act as pilot in command of a civil aircraft during Category II operations unless that person—

(i) Holds a Category II pilot authorization for that category or class of aircraft, and the type of aircraft, if applicable; or

(ii) In the case of a civil aircraft of foreign registry, is authorized by the country of registry to act as pilot in command of that aircraft in Category II operations.

(2) Act as second in command of a civil aircraft during Category II operations unless that person—

(i) Holds a pilot certificate with category and class ratings for that aircraft and an instrument rating for that category aircraft;

(ii) Holds an airline transport pilot certificate with category and class ratings for that aircraft; or

(iii) In the case of a civil aircraft of foreign registry, is authorized by the country of registry to act as second in command of that aircraft during Category II operations.

(g) *Category III pilot authorization.* Except for a pilot conducting Category III operations under part 121 or part 135, a person may not:

(1) Act as pilot in command of a civil aircraft during Category III operations unless that person—

(i) Holds a Category III pilot authorization for that category or class of aircraft, and the type of aircraft, if applicable; or

(ii) In the case of a civil aircraft of foreign registry, is authorized by the country of registry to act as pilot in command of that aircraft in Category III operations.

(2) Act as second in command of a civil aircraft during Category III operations unless that person—

(i) Holds a pilot certificate with category and class ratings for that aircraft and an instrument rating for that category aircraft;

(ii) Holds an airline transport pilot certificate with category and class ratings for that aircraft; or

(iii) In the case of a civil aircraft of foreign registry, is authorized by the country of registry to act as second in command of that aircraft during Category III operations.

(h) *Category A aircraft pilot authorization.* The Administrator may issue a certificate of authorization for a Category II or Category III operation to the pilot of a small aircraft that is a Category A aircraft, as identified in § 97.3(b)(1) of this chapter if:

(1) The Administrator determines that the Category II or Category III operation can be performed safely by that pilot under the terms of the certificate of authorization; and

(2) The Category II or Category III operation does not involve the carriage of persons or property for compensation or hire.

(i) *Ground instructor certificate.* (1) Each person who holds a ground instructor certificate issued under this part or part 143 must have that certificate in that person's physical possession or immediately accessible when exercising the privileges of that certificate.

(2) Except as provided in paragraph (i)(3) of this section, no person other than the holder of a ground instructor certificate, issued under this part or part 143, with the appropriate rating on that certificate may—

(i) Give ground training required to qualify a person for solo flight and solo cross-country flight;

(ii) Endorse an applicant for a knowledge test required for a pilot, flight instructor, or ground instructor certificate or rating issued under this part; or

(iii) Endorse a pilot logbook to show ground training given.

(3) A ground instructor certificate issued under this part is not necessary—

(i) Under paragraph (i)(2) of this section, if the training is given by the holder of a flight instructor certificate issued under this part in accordance with the privileges of that certificate;

(ii) Under paragraph (i)(2) of this section, if the training is given by the holder of a commercial pilot certificate with a lighter-than-air rating, provided the training is given in accordance with the privileges of the certificate in a lighter-than-air aircraft;

(iii) Under paragraph (i)(2) of this section, if the training is given by the holder of an airline transport pilot certificate with a rating appropriate to the aircraft in which the training is given, provided the training is given in accordance with the privileges of the certificate and conducted in accordance with an approved air carrier training program approved under part 121 or part 135 of this chapter;

(iv) Under paragraph (i)(2) of this section, if the training is given by a person who is qualified in accordance with subpart C of part 142 of this chapter, provided the training is conducted in accordance with an approved part 142 training program; or

(v) Under paragraph (i)(2)(iii) of this section, if the training is given by an authorized flight instructor under § 61.41 of this part.

(j) *Age limitation for certain operations* (1) *Age limitation.* No person who holds a pilot certificate issued under this part may serve as a pilot on a civil airplane of U.S. registry in the following operations if the person has reached his or her 65th birthday:

(i) Scheduled international air services carrying passengers in turbojet-powered airplanes;

(ii) Scheduled international air services carrying passengers in airplanes having a passenger-seat configuration of more than nine passenger seats, excluding each crewmember seat;

(iii) Nonscheduled international air transportation for compensation or hire in airplanes having a passenger-seat configuration of more than 30 passenger seats, excluding each crewmember seat; or

(iv) Scheduled international air services, or nonscheduled international air transportation for compensation or hire, in airplanes having a payload capacity of more than 7,500 pounds.

(2) *Age Pairing Requirement.* No person who has attained the age of 60 but who has not attained the age of 65 may serve as a pilot in command in any of the operations described in paragraphs (j)(1)(i) through (iv) of this section unless there is another pilot in the flight deck crew who has not yet attained 60 years of age.

(3) *Definitions.* (i) "International air service," as used in this paragraph (j), means scheduled air service performed in airplanes for the public transport of passengers, mail, or cargo, in which the service passes through the airspace over the territory of more than one country.

(ii) "International air transportation," as used in this paragraph (j), means air transportation performed in airplanes for the public transport of passengers, mail, or cargo, in which the service passes through the airspace over the territory of more than one country.

(k) *Special purpose pilot authorization.* Any person that is required to hold a special purpose pilot authorization, issued in accordance with § 61.77 of this part, must have that authorization and the person's foreign pilot license in that person's physical possession or have it readily accessible in the aircraft when exercising the privileges of that authorization.

(l) *Inspection of certificate.* Each person who holds an airman certificate, medical certificate, authorization, or license required by this part must present it and their photo identification as described in paragraph (a)(2) of this section for inspection upon a request from:

(1) The Administrator;

(2) An authorized representative of the National Transportation Safety Board;

(3) Any Federal, State, or local law enforcement officer; or

(4) An authorized representative of the Transportation Security Administration.

[Doc. No. 25910, 62 FR 16298, Apr. 4, 1997; Amdt. 61-103, 62 FR 40894, July 30, 1997; Amdt. 61-111, 67 FR 65861, Oct. 28, 2002; Amdt. 61-110, 69 FR 44864, July 27, 2004; Amdt. 61-123, 74 FR 34234, July 15, 2009; Amdt. 61-124, 74 FR 42546, Aug. 21, 2009; Amdt. 61-124A, 74 FR 53644, Oct. 20, 2009]

§ 61.4 Qualification and approval of flight simulators and flight training devices.

(a) Except as specified in paragraph (b) or (c) of this section, each flight simulator and flight training device used for training, and for which an airman is to receive credit to satisfy any training, testing, or checking requirement under this chapter, must be qualified and approved by the Administrator for—

(1) The training, testing, and checking for which it is used;

(2) Each particular maneuver, procedure, or crewmember function performed; and

(3) The representation of the specific category and class of aircraft, type of aircraft, particular variation within the type of aircraft, or set of aircraft for certain flight training devices.

(b) Any device used for flight training, testing, or checking that has been determined to be acceptable to or approved by the Administrator prior to August 1, 1996, which can be shown to function as originally designed, is considered to be a flight training device, provided it is used for the same purposes for which it was originally accepted or approved and only to the extent of such acceptance or approval.

(c) The Administrator may approve a device other than a flight simulator or flight training device for specific purposes.

[Doc. No. 25910, 62 FR 16298, Apr. 4, 1997; Amdt. 61-103, 62 FR 40895, July 30, 1997]

§ 61.5 Certificates and ratings issued under this part.

(a) The following certificates are issued under this part to an applicant who satisfactorily accomplishes the training and certification requirements for the certificate sought:

(1) Pilot certificates—

(i) Student pilot.

(ii) Sport pilot.

(iii) Recreational pilot.

(iv) Private pilot.

(v) Commercial pilot.

(vi) Airline transport pilot.

(2) Flight instructor certificates.

(3) Ground instructor certificates.

(b) The following ratings are placed on a pilot certificate (other than student pilot) when an applicant satisfactorily accomplishes the training and certification requirements for the rating sought:

(1) Aircraft category ratings—

(i) Airplane.

(ii) Rotorcraft.

(iii) Glider.

(iv) Lighter-than-air.

(v) Powered-lift.

(vi) Powered parachute.

(vii) Weight-shift-control aircraft.

(2) Airplane class ratings—

(i) Single-engine land.

(ii) Multiengine land.

(iii) Single-engine sea.

(iv) Multiengine sea.

(3) Rotorcraft class ratings—

(i) Helicopter.

(ii) Gyroplane.

(4) Lighter-than-air class ratings—

(i) Airship.

(ii) Balloon.

(5) Weight-shift-control aircraft class ratings—

(i) Weight-shift-control aircraft land.

(ii) Weight-shift-control aircraft sea.

(6) Powered parachute class ratings—

(i) Powered parachute land.

(ii) Powered parachute sea.

(7) Aircraft type ratings—

(i) Large aircraft other than lighter-than-air.

(ii) Turbojet-powered airplanes.

(iii) Other aircraft type ratings specified by the Administrator through the aircraft type certification procedures.

(iv) Second-in-command pilot type rating for aircraft that is certificated for operations with a minimum crew of at least two pilots.

(8) Instrument ratings (on private and commercial pilot certificates only)—

(i) Instrument—Airplane.

(ii) Instrument—Helicopter.

(iii) Instrument—Powered-lift.

(c) The following ratings are placed on a flight instructor certificate when an applicant satisfactorily accomplishes the training and certification requirements for the rating sought:

(1) Aircraft category ratings—

(i) Airplane.

(ii) Rotorcraft.

(iii) Glider.

(iv) Powered-lift.

(2) Airplane class ratings—

(i) Single-engine.

(ii) Multiengine.

(3) Rotorcraft class ratings—

(i) Helicopter.

(ii) Gyroplane.

(4) Instrument ratings—

(i) Instrument—Airplane.

(ii) Instrument—Helicopter.

(iii) Instrument—Powered-lift.

(5) Sport pilot rating.

(d) The following ratings are placed on a ground instructor certificate when an applicant satisfactorily accomplishes the training and certification requirements for the rating sought:

(1) Basic.

(2) Advanced.

(3) Instrument.

[Doc. No. 25910, 62 FR 16298, Apr. 4, 1997, as amended by Amdt. 61-110, 69 FR 44864, July 27, 2004; Amdt. 61-113, 70 FR 45271, Aug. 4, 2005]

§ 61.7 Obsolete certificates and ratings.

(a) The holder of a free-balloon pilot certificate issued before November 1, 1973, may not exercise the privileges of that certificate.

(b) The holder of a pilot certificate that bears any of the following category ratings without an associated class rating may not exercise the privileges of that category rating:

(1) Rotorcraft.

(2) Lighter-than-air.

(3) Helicopter.

(4) Autogyro.

§ 61.9 [Reserved]

§ 61.11 Expired pilot certificates and re-issuance.

(a) No person who holds an expired pilot certificate or rating may act as pilot in command or as a required pilot flight crewmember of an aircraft of the same category or class that is listed on that expired pilot certificate or rating.

(b) The following pilot certificates and ratings have expired and will not be reissued:

(1) An airline transport pilot certificate issued before May 1, 1949, or an airline transport pilot certificate that contains a horsepower limitation.

(2) A private or commercial pilot certificate issued before July 1, 1945.

(3) A pilot certificate with a lighter-than-air or free-balloon rating issued before July 1, 1945.

(c) An airline transport pilot certificate that was issued after April 30, 1949, and that bears an expiration date but does not contain a horsepower limitation, may have that airline transport pilot certificate re-issued without an expiration date.

(d) A private or commercial pilot certificate that was issued after June 30, 1945, and that bears an expiration date, may have that pilot certificate reissued without an expiration date.

(e) A pilot certificate with a lighter-than-air or free-balloon rating that was issued after June 30, 1945, and that bears an expiration date, may have that pilot certificate reissued without an expiration date.

[Doc. No. FAA-2006-26661, 74 FR 42547, Aug. 21, 2009]

§ 61.13 Issuance of airman certificates, ratings, and authorizations.

(a) *Application.* (1) An applicant for an airman certificate, rating, or authorization under this part must make that application on a form and in a manner acceptable to the Administrator.

(2) An applicant—

(i) Must show evidence that the appropriate fee prescribed in appendix A to part 187 of this chapter has been paid when that person applies for airmen certification services administered outside the United States.

(A) Student pilot certificate that is issued outside the United States; or

(B) Knowledge test or practical test for an airman certificate or rating issued under this part, if the test is administered outside the United States.

(ii) May be refused issuance of any U.S. airman certificate, rating, or authorization by the Administrator.

(3) Except as provided in paragraph (a)(2)(ii) of this section, an applicant who satisfactorily accomplishes the training and certification requirements for the certificate, rating, or authorization sought is entitled to receive that airman certificate, rating, or authorization.

(b) *Limitations.* (1) An applicant who cannot comply with certain areas of operation required on the practical test because of physical limitations may be issued an airman certificate, rating, or authorization with the appropriate limitation placed on the applicant's airman certificate provided the—

(i) Applicant is able to meet all other certification requirements for the airman certificate, rating, or authorization sought;

(ii) Physical limitation has been recorded with the FAA on the applicant's medical records; and

(iii) Administrator determines that the applicant's inability to perform the particular area of operation will not adversely affect safety.

(2) A limitation placed on a person's airman certificate may be removed, provided that person demonstrates for an examiner satisfactory proficiency in the area of operation appropriate to the airman certificate, rating, or authorization sought.

(c) *Additional requirements for Category II and Category III pilot authorizations.* (1) A Category II or Category III pilot authorization is issued by a letter of authorization as part of an applicant's instrument rating or airline transport pilot certificate.

(2) Upon original issue, the authorization contains the following limitations:

(i) For Category II operations, the limitation is 1,600 feet RVR and a 150-foot decision height; and

(ii) For Category III operations, each initial limitation is specified in the authorization document.

(3) The limitations on a Category II or Category III pilot authorization may be removed as follows:

(i) In the case of Category II limitations, a limitation is removed when the holder shows that, since the beginning of the sixth preceding month, the holder has made three Category II ILS approaches with a 150-foot decision height to a landing under actual or simulated instrument conditions.

(ii) In the case of Category III limitations, a limitation is removed as specified in the authorization.

(4) To meet the experience requirements of paragraph (c)(3) of this section, and for the practical test required by this part for a Category II or a Category III pilot authorization, a flight simulator or flight training device may be used if it is approved by the Administrator for such use.

(d) *Application during suspension or revocation.* (1) Unless otherwise authorized by the Administrator, a person whose pilot, flight instructor, or ground instructor certificate has been suspended may not apply for any certificate, rating, or authorization during the period of suspension.

(2) Unless otherwise authorized by the Administrator, a person whose pilot, flight instructor, or ground instructor certificate has been revoked may not apply for any certificate, rating, or authorization for 1 year after the date of revocation.

[Doc. No. 25910, 62 FR 40895, July 30, 1997, as amended by Amdt. 61-116, 72 FR 18558, Apr. 12, 2007]

§ 61.14 [Reserved]

§ 61.15 Offenses involving alcohol or drugs.

(a) A conviction for the violation of any Federal or State statute relating to the growing, processing, manufacture, sale, disposition, possession, transportation, or importation of narcotic drugs, marijuana, or depressant or stimulant drugs or substances is grounds for:

(1) Denial of an application for any certificate, rating, or authorization issued under this part for a period of up to 1 year after the date of final conviction; or

(2) Suspension or revocation of any certificate, rating, or authorization issued under this part.

(b) Committing an act prohibited by § 91.17(a) or § 91.19(a) of this chapter is grounds for:

(1) Denial of an application for a certificate, rating, or authorization issued under this part for a period of up to 1 year after the date of that act; or

(2) Suspension or revocation of any certificate, rating, or authorization issued under this part.

(c) For the purposes of paragraphs (d), (e), and (f) of this section, a motor vehicle action means:

(1) A conviction after November 29, 1990, for the violation of any Federal or State statute relating to the operation of a motor vehicle while intoxicated by alcohol or a drug, while impaired by alcohol or a drug, or while under the influence of alcohol or a drug;

(2) The cancellation, suspension, or revocation of a license to operate a motor vehicle after November 29, 1990, for a cause related to the operation of a motor vehicle while intoxicated by alcohol or a drug, while impaired by alcohol or a drug, or while under the influence of alcohol or a drug; or

(3) The denial after November 29, 1990, of an application for a license to operate a motor vehicle for a cause related to the operation of a motor vehicle while intoxicated by alcohol or a drug, while impaired by alcohol or a drug, or while under the influence of alcohol or a drug.

(d) Except for a motor vehicle action that results from the same incident or arises out of the same factual circumstances, a motor vehicle action occurring within 3 years of a previous motor vehicle action is grounds for:

(1) Denial of an application for any certificate, rating, or authorization issued under this part for a period of up to 1 year after the date of the last motor vehicle action; or

(2) Suspension or revocation of any certificate, rating, or authorization issued under this part.

(e) Each person holding a certificate issued under this part shall provide a written report of each motor vehicle action to the FAA, Civil Aviation Security Division (AMC-700), P.O. Box 25810, Oklahoma City, OK 73125, not later than 60 days after the motor vehicle action. The report must include:

(1) The person's name, address, date of birth, and airman certificate number;

(2) The type of violation that resulted in the conviction or the administrative action;

(3) The date of the conviction or administrative action;

(4) The State that holds the record of conviction or administrative action; and

(5) A statement of whether the motor vehicle action resulted from the same incident or arose out of the same factual circumstances related to a previously reported motor vehicle action.

(f) Failure to comply with paragraph (e) of this section is grounds for:

(1) Denial of an application for any certificate, rating, or authorization issued under this part for a period of up to 1 year after the date of the motor vehicle action; or

(2) Suspension or revocation of any certificate, rating, or authorization issued under this part.

§ 61.16 Refusal to submit to an alcohol test or to furnish test results.

A refusal to submit to a test to indicate the percentage by weight of alcohol in the blood, when requested by a law enforcement officer in accordance with § 91.17(c) of this chapter, or a refusal to furnish or authorize the release of the test results requested by the Administrator in accordance with § 91.17(c) or (d) of this chapter, is grounds for:

(a) Denial of an application for any certificate, rating, or authorization issued under this part for a period of up to 1 year after the date of that refusal; or

(b) Suspension or revocation of any certificate, rating, or authorization issued under this part.

§ 61.17 Temporary certificate.

(a) A temporary pilot, flight instructor, or ground instructor certificate or rating is issued for up to 120 days, at which time a permanent certificate will be issued to a person whom the Administrator finds qualified under this part.

(b) A temporary pilot, flight instructor, or ground instructor certificate or rating expires:

(1) On the expiration date shown on the certificate;

(2) Upon receipt of the permanent certificate; or

(3) Upon receipt of a notice that the certificate or rating sought is denied or revoked.

§ 61.18 Security disqualification.

(a) *Eligibility standard.* No person is eligible to hold a certificate, rating, or authorization issued under this part when the Transportation Security Administration (TSA) has notified the FAA in writing that the person poses a security threat.

(b) *Effect of the issuance by the TSA of an Initial Notification of Threat Assessment.* (1) The FAA will hold in abeyance pending the outcome of the TSA's final threat assessment review an application for any certificate, rating, or authorization under this part by any person who has been issued an Initial Notification of Threat Assessment by the TSA.

(2) The FAA will suspend any certificate, rating, or authorization issued under this part after the TSA issues to the holder an Initial Notification of Threat Assessment.

(c) *Effect of the issuance by the TSA of a Final Notification of Threat Assessment.* (1) The FAA will deny an application for any certificate, rating, or authorization under this part to any person who has been issued a Final Notification of Threat Assessment.

(2) The FAA will revoke any certificate, rating, or authorization issued under this part after the TSA has issued to the holder a Final Notification of Threat Assessment.

[Doc. FAA-2003-14293, 68 FR 3774, Jan. 24, 2003]

§ 61.19 Duration of pilot and instructor certificates.

(a) *General.* The holder of a certificate with an expiration date may not, after that date, exercise the privileges of that certificate.

(b) *Student pilot certificate.*

(1) For student pilots who have not reached their 40th birthday, the student pilot certificate does not expire until 60 calendar months after the month of the date of examination shown on the medical certificate.

(2) For student pilots who have reached their 40th birthday, the student pilot certificate does not expire until 24 calendar months after the month of the date of examination shown on the medical certificate.

(3) For student pilots seeking a glider rating, balloon rating, or a sport pilot certificate, the student pilot certificate does not expire until 60 calendar months after the month of the date issued, regardless of the person's age.

(c) *Other pilot certificates.* A pilot certificate (other than a student pilot certificate) issued under this part is issued without a specific expiration date. The holder of a pilot certificate issued on the basis of a foreign pilot license may exercise the privileges of that certificate only while that person's foreign pilot license is effective.

(d) *Flight instructor certificate.* Except as specified in § 61.197(b), a flight instructor certificate expires 24 calendar months from the month in which it was issued, renewed, or reinstated, as appropriate.

(e) *Ground instructor certificate.* A ground instructor certificate is issued without a specific expiration date.

(f) *Return of certificates.* The holder of any airman certificate that is issued under this part, and that has been suspended or revoked, must return that certificate to the FAA when requested to do so by the Administrator.

(g) *Duration of pilot certificates.* Except for a temporary certificate issued under § 61.17 or a student pilot certificate issued under paragraph (b) of this section, the holder of a paper pilot certificate issued under this part may not exercise the privileges of that certificate after March 31, 2010.

[Doc. No. 25910, 62 FR 16298, Apr. 4, 1997, as amended by Amdt. 61-118, 73 FR 10668, Feb. 28, 2008; Amdt. 61-124, 74 FR 42547, Aug. 21, 2009; Amdt. 61-124A, 74 FR 53644, Oct. 20, 2009]

§ 61.21 Duration of a Category II and a Category III pilot authorization (for other than part 121 and part 135 use).

(a) A Category II pilot authorization or a Category III pilot authorization expires at the end of the sixth calendar month after the month in which it was issued or renewed.

(b) Upon passing a practical test for a Category II or Category III pilot authorization, the authorization may be renewed for each type of aircraft for which the authorization is held.

(c) A Category II or Category III pilot authorization for a specific type aircraft for which an authorization is held will not be renewed beyond 12 calendar months from the month the practical test was accomplished in that type aircraft.

(d) If the holder of a Category II or Category III pilot authorization passes the practical test for a renewal in the month before the authorization expires, the holder is considered to have passed it during the month the authorization expired.

§ 61.23 Medical certificates: Requirement and duration.

(a) *Operations requiring a medical certificate.* Except as provided in paragraphs (b) and (c) of this section, a person—

(1) Must hold a first-class medical certificate:

(i) When exercising the pilot-in-command privileges of an airline transport pilot certificate;

(ii) When exercising the second-in-command privileges of an airline transport pilot certificate in a flag or supplemental operation in part 121 of this chapter that requires three or more pilots; or

(iii) When serving as a required pilot flightcrew member in an operation conducted under part 121 of this chapter if the pilot has reached his or her 60th birthday.

(2) Must hold at least a second class medical certificate when exercising:

(i) Second-in-command privileges of an airline transport pilot certificate in part 121 of this chapter (other than operations specified in paragraph (a)(1)(ii) of this section); or

(ii) Privileges of a commercial pilot certificate; or

(3) Must hold at least a third-class medical certificate—

(i) When exercising the privileges of a private pilot certificate;

(ii) When exercising the privileges of a recreational pilot certificate;

(iii) When exercising the privileges of a student pilot certificate;

(iv) When exercising the privileges of a flight instructor certificate and acting as the pilot in command;

(v) When exercising the privileges of a flight instructor certificate and serving as a required pilot flight crewmember;

(vi) When taking a practical test in an aircraft for a recreational pilot, private pilot, commercial pilot, or airline transport pilot certificate, or for a flight instructor certificate; or

(vii) When performing the duties as an Examiner in an aircraft when administering a practical test or proficiency check for an airman certificate, rating, or authorization.

(b) *Operations not requiring a medical certificate.* A person is not required to hold a medical certificate—

(1) When exercising the privileges of a student pilot certificate while seeking—

(i) A sport pilot certificate with glider or balloon privileges; or

(ii) A pilot certificate with a glider category rating or balloon class rating;

(2) When exercising the privileges of a sport pilot certificate with privileges in a glider or balloon;

(3) When exercising the privileges of a pilot certificate with a glider category rating or balloon class rating in a glider or a balloon, as appropriate;

(4) When exercising the privileges of a flight instructor certificate with—

(i) A sport pilot rating in a glider or balloon; or

(ii) A glider category rating;

(5) When exercising the privileges of a flight instructor certificate if the person is not acting as pilot in command or serving as a required pilot flight crewmember;

(6) When exercising the privileges of a ground instructor certificate;

(7) When serving as an Examiner or check airman and administering a practical test or proficiency check for an airman certificate, rating, or authorization conducted in a glider, balloon, flight simulator, or flight training device;

(8) When taking a practical test or a proficiency check for a certificate, rating, authorization or operating privilege conducted in a glider, balloon, flight simulator, or flight training device; or

(9) When a military pilot of the U.S. Armed Forces can show evidence of an up-to-date medical examination authorizing pilot flight status issued by the U.S. Armed Forces and—

(i) The flight does not require higher than a third-class medical certificate; and

(ii) The flight conducted is a domestic flight operation within U.S. airspace.

(c) *Operations requiring either a medical certificate or U.S. driver's license.* (1) A person must hold and possess either a medical certificate issued under part 67 of this chapter or a U.S. driver's license when—

(i) Exercising the privileges of a student pilot certificate while seeking sport pilot privileges in a light-sport aircraft other than a glider or balloon;

(ii) Exercising the privileges of a sport pilot certificate in a light-sport aircraft other than a glider or balloon;

(iii) Exercising the privileges of a flight instructor certificate with a sport pilot rating while acting as pilot in command or serving as a required flight crewmember of a light-sport aircraft other than a glider or balloon; or

(iv) Serving as an Examiner and administering a practical test for the issuance of a sport pilot certificate in a light-sport aircraft other than a glider or balloon.

(2) A person using a U.S. driver's license to meet the requirements of this paragraph must—

(i) Comply with each restriction and limitation imposed by that person's U.S. driver's license and any judicial or administrative order applying to the operation of a motor vehicle;

(ii) Have been found eligible for the issuance of at least a third-class airman medical certificate at the time of his or her most recent application (if the person has applied for a medical certificate);

(iii) Not have had his or her most recently issued medical certificate (if the person has held a medical certificate) suspended or revoked or most recent Authorization for a Special Issuance of a Medical Certificate withdrawn; and

(iv) Not know or have reason to know of any medical condition that would make that person unable to operate a light-sport aircraft in a safe manner.

(d) *Duration of a medical certificate.* Use the following table to determine duration for each class of medical certificate:

[Doc. No. 25910, 62 FR 16298, Apr. 4, 1997; Amdt. 61-103, 62 FR 40895, July 30, 1997; Amdt. 61-110, 69 FR 44864, July 27, 2004, as amended by Amdt. 61-121, 73 FR 43064, July 24, 2008; Amdt. 61-121, 73 FR 48125, Aug. 18, 2008; Amdt. 61-123, 74 FR 34234, July 15, 2009; Amdt. 61-124, 74 FR 42547, Aug. 21, 2009; Amdt. 61-129, 76 FR 78143, Dec. 16, 2011; Amdt. 61-129A, 77 FR 61721, Oct. 11, 2012; Amdt. 61-130, 78 FR 42372, July 15, 2013]

If you hold	And on the date of examination for your most recent medical certificate you were	And you are conducting an operation requiring	Then your medical certificate expires, for that operation, at the end of the last day of the
(1) A first-class medical certificate	(i) Under age 40	an airline transport pilot certificate	12th month after the month of the date of examination shown on the medical certificate.
	(ii) Age 40 or older	an airline transport pilot certificate	6th month after the month of the date of examination shown on the medical certificate.
	(iii) Any age	a commercial pilot certificate or an air traffic control tower operator certificate	12th month after the month of the date of examination shown on the medical certificate.
	(iv) Under age 40	a recreational pilot certificate, a private pilot certificate, a flight instructor certificate (when acting as pilot in command or a required pilot flight crewmember in operations other than glider or balloon), a student pilot certificate, or a sport pilot certificate (when not using a U.S. driver's license as medical qualification)	60th month after the month of the date of examination shown on the medical certificate.
	(v) Age 40 or older	a recreational pilot certificate, a private pilot certificate, a flight instructor certificate (when acting as pilot in command or a required pilot flight crewmember in operations other than glider or balloon), a student pilot certificate, or a sport pilot certificate (when not using a U.S. driver's license as medical qualification)	24th month after the month of the date of examination shown on the medical certificate.
(2) A second-class medical certificate	(i) Any age	a commercial pilot certificate or an air traffic control tower operator certificate	12th month after the month of the date of examination shown on the medical certificate.
	(ii) Under age 40	a recreational pilot certificate, a private pilot certificate, a flight instructor certificate (when acting as pilot in command or a required pilot flight crewmember in operations other than glider or balloon), a student pilot certificate, or a sport pilot certificate (when not using a U.S. driver's license as medical qualification)	60th month after the month of the date of examination shown on the medical certificate.
	(iii) Age 40 or older	a recreational pilot certificate, a private pilot certificate, a flight instructor certificate (when acting as pilot in command or a required pilot flight crewmember in operations other than glider or balloon), a student pilot certificate, or a sport pilot certificate (when not using a U.S. driver's license as medical qualification)	24th month after the month of the date of examination shown on the medical certificate.
(3) A third-class medical certificate	(i) Under age 40	a recreational pilot certificate, a private pilot certificate, a flight instructor certificate (when acting as pilot in command or a required pilot flight crewmember in operations other than glider or balloon), a student pilot certificate, or a sport pilot certificate (when not using a U.S. driver's license as medical qualification)	60th month after the month of the date of examination shown on the medical certificate.
	(ii) Age 40 or older	a recreational pilot certificate, a private pilot certificate, a flight instructor certificate (when acting as pilot in command or a required pilot flight crewmember in operations other than glider or balloon), a student pilot certificate, or a sport pilot certificate (when not using a U.S. driver's license as medical qualification)	24th month after the month of the date of examination shown on the medical certificate.

§ 61.25 Change of name.

(a) An application to change the name on a certificate issued under this part must be accompanied by the applicant's:

(1) Airman certificate; and

(2) A copy of the marriage license, court order, or other document verifying the name change.

(b) The documents in paragraph (a) of this section will be returned to the applicant after inspection.

[Doc. No. 25910, 62 FR 16298, Apr. 4, 1997, as amended by Amdt. 61-124, 74 FR 42548, Aug. 21, 2009]

§ 61.27 Voluntary surrender or exchange of certificate.

(a) The holder of a certificate issued under this part may voluntarily surrender it for:

(1) Cancellation;

(2) Issuance of a lower grade certificate; or

(3) Another certificate with specific ratings deleted.

(b) Any request made under paragraph (a) of this section must include the following signed statement or its equivalent: "This request is made for my own reasons, with full knowledge that my (insert name of certificate or rating, as appropriate) may not be reissued to me unless I again pass the tests prescribed for its issuance."

§ 61.29 Replacement of a lost or destroyed airman or medical certificate or knowledge test report.

(a) A request for the replacement of a lost or destroyed airman certificate issued under this part must be made by letter to the Department of Transportation, FAA, Airman Certification Branch, P.O. Box 25082, Oklahoma City, OK 73125, and must be accompanied by a check or money order for the appropriate fee payable to the FAA.

(b) A request for the replacement of a lost or destroyed medical certificate must be made by letter to the Department of Transportation, FAA, Aerospace Medical Certification Division, P.O. Box 26200, Oklahoma City, OK 73125, and must be accompanied by a check or money order for the appropriate fee payable to the FAA.

(c) A request for the replacement of a lost or destroyed knowledge test report must be made by letter to the Department of Transportation, FAA, Airman Certification Branch, P.O. Box 25082, Oklahoma City, OK 73125, and must be accompanied by a check or money order for the appropriate fee payable to the FAA.

(d) The letter requesting replacement of a lost or destroyed airman certificate, medical certificate, or knowledge test report must state:

(1) The name of the person;

(2) The permanent mailing address (including ZIP code), or if the permanent mailing address includes a post office box number, then the person's current residential address;

(3) The certificate holder's date and place of birth; and

(4) Any information regarding the—

(i) Grade, number, and date of issuance of the airman certificate and ratings, if appropriate;

(ii) Class of medical certificate, the place and date of the medical exam, name of the Airman Medical Examiner (AME), and the circumstances concerning the loss of the original medical certificate, as appropriate; and

(iii) Date the knowledge test was taken, if appropriate.

(e) A person who has lost an airman certificate, medical certificate, or knowledge test report may obtain a facsimile from the FAA Aeromedical Certification Branch or the Airman Certification Branch, as appropriate, confirming that it was issued and the:

(1) Facsimile may be carried as an airman certificate, medical certificate, or knowledge test report, as appropriate, for up to 60 days pending the person's receipt of a duplicate under paragraph (a), (b), or (c) of this section, unless the person has been notified that the certificate has been suspended or revoked.

(2) Request for such a facsimile must include the date on which a duplicate certificate or knowledge test report was previously requested.

[Doc. No. 25910, 62 FR 16298, Apr. 4, 1997; Amdt. 61-103, 62 FR 40896, July 30, 1997; Amdt. 61-121, 73 FR 43065, July 24, 2008; Amdt. 61-124, 74 FR 42548, Aug. 21, 2009]

§ 61.31 Type rating requirements, additional training, and authorization requirements.

(a) *Type ratings required.* A person who acts as a pilot in command of any of the following aircraft must hold a type rating for that aircraft:

(1) Large aircraft (except lighter-than-air).

(2) Turbojet-powered airplanes.

(3) Other aircraft specified by the Administrator through aircraft type certificate procedures.

(b) *Authorization in lieu of a type rating.* A person may be authorized to operate without a type rating for up to 60 days an aircraft requiring a type rating, provided—

(1) The Administrator has authorized the flight or series of flights;

(2) The Administrator has determined that an equivalent level of safety can be achieved through the operating limitations on the authorization;

(3) The person shows that compliance with paragraph (a) of this section is impracticable for the flight or series of flights; and

(4) The flight—

(i) Involves only a ferry flight, training flight, test flight, or practical test for a pilot certificate or rating;

(ii) Is within the United States;

(iii) Does not involve operations for compensation or hire unless the compensation or hire involves payment for the use of the aircraft for training or taking a practical test; and

(iv) Involves only the carriage of flight crewmembers considered essential for the flight.

(5) If the flight or series of flights cannot be accomplished within the time limit of the authorization, the Administrator may authorize an additional period of up to 60 days to accomplish the flight or series of flights.

(c) *Aircraft category, class, and type ratings: Limitations on the carriage of persons, or operating for compensation or hire.* Unless a person holds a category, class, and type rating (if a class and type rating is required) that applies to the aircraft, that person may not act as pilot in command of an aircraft that is carrying another person, or is operated for compensation or hire. That person also may not act as pilot in command of that aircraft for compensation or hire.

(d) *Aircraft category, class, and type ratings: Limitations on operating an aircraft as the pilot in command.* To serve as the pilot in command of an aircraft, a person must—

(1) Hold the appropriate category, class, and type rating (if a class or type rating is required) for the aircraft to be flown; or

(2) Have received training required by this part that is appropriate to the pilot certification level, aircraft category, class, and type rating (if a class or type rating is required) for the aircraft to be flown, and have received an endorsement for solo flight in that aircraft from an authorized instructor.

(e) *Additional training required for operating complex airplanes.* (1) Except as provided in paragraph (e)(2) of this section, no person may act as pilot in command of a complex airplane, unless the person has—

(i) Received and logged ground and flight training from an authorized instructor in a complex airplane, or in a flight simulator or flight training device that is representative of a complex airplane, and has been found proficient in the operation and systems of the airplane; and

(ii) Received a one-time endorsement in the pilot's logbook from an authorized instructor who certifies the person is proficient to operate a complex airplane.

(2) The training and endorsement required by paragraph (e)(1) of this section is not required if the person has logged flight time as pilot in command of a complex airplane, or in a flight simulator or flight training device that is representative of a complex airplane prior to August 4, 1997.

(f) *Additional training required for operating high-performance airplanes.* (1) Except as provided in paragraph (f)(2) of this section, no person may act as pilot in command of a high-performance airplane (an airplane with an engine of more than 200 horsepower), unless the person has—

(i) Received and logged ground and flight training from an authorized instructor in a high-performance airplane, or in a flight simulator or flight training device that is representative of a high-performance airplane, and has been found proficient in the operation and systems of the airplane; and

(ii) Received a one-time endorsement in the pilot's logbook from an authorized instructor who certifies the person is proficient to operate a high-performance airplane.

(2) The training and endorsement required by paragraph (f)(1) of this section is not required if the person has logged flight time as pilot in command of a high-performance airplane, or in a flight simulator or flight training device that is representative of a high-performance airplane prior to August 4, 1997.

(g) *Additional training required for operating pressurized aircraft capable of operating at high altitudes.* (1) Except as provided in paragraph (g)(3) of this section, no person may act as pilot in command of a pressurized aircraft (an aircraft that has a service ceiling or maximum operating altitude, whichever is lower, above 25,000 feet MSL), unless that person has received and logged ground training from an authorized instructor and obtained an endorsement in the person's logbook or training record from an authorized instructor who certifies the person has satisfactorily accomplished the ground training. The ground training must include at least the following subjects:

(i) High-altitude aerodynamics and meteorology;

(ii) Respiration;

(iii) Effects, symptoms, and causes of hypoxia and any other high-altitude sickness;

(iv) Duration of consciousness without supplemental oxygen;

(v) Effects of prolonged usage of supplemental oxygen;

(vi) Causes and effects of gas expansion and gas bubble formation;

(vii) Preventive measures for eliminating gas expansion, gas bubble formation, and high-altitude sickness;

(viii) Physical phenomena and incidents of decompression; and

(ix) Any other physiological aspects of high-altitude flight.

(2) Except as provided in paragraph (g)(3) of this section, no person may act as pilot in command of a pressurized aircraft unless that person has received and logged training from an authorized instructor in a pressurized aircraft, or in a flight simulator or flight training device that is representative of a pressurized aircraft, and obtained an endorsement in the person's logbook or training record from an authorized instructor who found the person proficient in the operation of a pressurized aircraft. The flight training must include at least the following subjects:

(i) Normal cruise flight operations while operating above 25,000 feet MSL;

(ii) Proper emergency procedures for simulated rapid decompression without actually depressurizing the aircraft; and

(iii) Emergency descent procedures.

(3) The training and endorsement required by paragraphs (g)(1) and (g)(2) of this section are not required if that person can document satisfactory accomplishment of any of the following in a pressurized aircraft, or in a flight simulator or flight training device that is representative of a pressurized aircraft:

(i) Serving as pilot in command before April 15, 1991;

(ii) Completing a pilot proficiency check for a pilot certificate or rating before April 15, 1991;

(iii) Completing an official pilot-in-command check conducted by the military services of the United States; or

(iv) Completing a pilot-in-command proficiency check under part 121, 125, or 135 of this chapter conducted by the Administrator or by an approved pilot check airman.

(h) *Additional aircraft type-specific training.* No person may serve as pilot in command of an aircraft that the Administrator has determined requires aircraft type-specific training unless that person has—

(1) Received and logged type-specific training in the aircraft, or in a flight simulator or flight training device that is representative of that type of aircraft; and

(2) Received a logbook endorsement from an authorized instructor who has found the person proficient in the operation of the aircraft and its systems.

(i) *Additional training required for operating tailwheel airplanes.* (1) Except as provided in paragraph (i)(2) of this section, no person may act as pilot in command of a tailwheel airplane unless that person has received and logged flight training from an authorized instructor in a tailwheel airplane and received an endorsement in the person's logbook from an authorized instructor who found the person proficient in the operation of a tailwheel airplane. The flight training must include at least the following maneuvers and procedures:

(i) Normal and crosswind takeoffs and landings;

(ii) Wheel landings (unless the manufacturer has recommended against such landings); and

(iii) Go-around procedures.

(2) The training and endorsement required by paragraph (i)(1) of this section is not required if the person logged pilot-in-command time in a tailwheel airplane before April 15, 1991.

(j) *Additional training required for operating a glider.* (1) No person may act as pilot in command of a glider—

(i) Using ground-tow procedures, unless that person has satisfactorily accomplished ground and flight training on ground-tow procedures and operations, and has received an endorsement from an authorized instructor who certifies in that pilot's logbook that the pilot has been found proficient in ground-tow procedures and operations;

(ii) Using aerotow procedures, unless that person has satisfactorily accomplished ground and flight training on aerotow procedures and operations, and has received an endorsement from an authorized instructor who certifies in that pilot's logbook that the pilot has been found proficient in aerotow procedures and operations; or

(iii) Using self-launch procedures, unless that person has satisfactorily accomplished ground and flight training on self-launch procedures and operations, and has received an endorsement from an authorized instructor who certifies in that pilot's logbook that the pilot has been found proficient in self-launch procedures and operations.

(2) The holder of a glider rating issued prior to August 4, 1997, is considered to be in compliance with the training and logbook endorsement requirements of this paragraph for the specific operating privilege for which the holder is already qualified.

(k) *Additional training required for night vision goggle operations.* (1) Except as provided under paragraph (k)(3) of this section, a person may act as pilot in command of an aircraft using night vision goggles only if that person receives and logs ground training from an authorized instructor and obtains a logbook or training record endorsement from an authorized instructor who certifies the person completed the ground training. The ground training must include the following subjects:

(i) Applicable portions of this chapter that relate to night vision goggle limitations and flight operations;

(ii) Aeromedical factors related to the use of night vision goggles, including how to protect night vision, how the eyes adapt to night, self-imposed stresses that affect night vision, effects of lighting on night vision, cues used to estimate distance and depth perception at night, and visual illusions;

(iii) Normal, abnormal, and emergency operations of night vision goggle equipment;

(iv) Night vision goggle performance and scene interpretation; and

(v) Night vision goggle operation flight planning, including night terrain interpretation and factors affecting terrain interpretation.

(2) Except as provided under paragraph (k)(3) of this section, a person may act as pilot in command of an aircraft using night vision goggles only if that person receives and logs flight training from an authorized instructor and obtains a logbook or training record endorsement from an authorized instructor who found the person proficient in the use of night vision goggles. The flight training must include the following tasks:

(i) Preflight and use of internal and external aircraft lighting systems for night vision goggle operations;

(ii) Preflight preparation of night vision goggles for night vision goggle operations;

(iii) Proper piloting techniques when using night vision goggles during the takeoff, climb, enroute, descent, and landing phases of flight; and

(iv) Normal, abnormal, and emergency flight operations using night vision goggles.

(3) The requirements under paragraphs (k)(1) and (2) of this section do not apply if a person can document satisfactory completion of any of the following pilot proficiency checks using night vision goggles in an aircraft:

(i) A pilot proficiency check on night vision goggle operations conducted by the U.S. Armed Forces.

(ii) A pilot proficiency check on night vision goggle operations under part 135 of this chapter conducted by an Examiner or Check Airman.

(iii) A pilot proficiency check on night vision goggle operations conducted by a night vision goggle manufacturer or authorized instructor, when the pilot—

(A) Is employed by a Federal, State, county, or municipal law enforcement agency; and

(B) Has logged at least 20 hours as pilot in command in night vision goggle operations.

(l) *Exceptions.* (1) This section does not require a category and class rating for aircraft not type-certificated as airplanes, rotorcraft, gliders, lighter-than-air aircraft, powered-lifts, powered parachutes, or weight-shift-control aircraft.

(2) The rating limitations of this section do not apply to—

(i) An applicant when taking a practical test given by an examiner;

(ii) The holder of a student pilot certificate;

(iii) The holder of a pilot certificate when operating an aircraft under the authority of—

(A) A provisional type certificate; or

(B) An experimental certificate, unless the operation involves carrying a passenger;

(iv) The holder of a pilot certificate with a lighter-than-air category rating when operating a balloon;

(v) The holder of a recreational pilot certificate operating under the provisions of § 61.101(h); or

(vi) The holder of a sport pilot certificate when operating a light-sport aircraft.

[Doc. No. 25910, 62 FR 40896, July 30, 1997, as amended by Amdt. 61-104, 63 FR 20286, Apr. 23, 1998; Amdt. 61-110, 69 FR 44865, July 27, 2004; Amdt. 61-124, 74 FR 42548, Aug. 21, 2009; Amdt. 61-128, 76 FR 54105, Aug. 31, 2011]

§ 61.33 Tests: General procedure.

Tests prescribed by or under this part are given at times and places, and by persons designated by the Administrator.

§ 61.35 Knowledge test: Prerequisites and passing grades.

(a) An applicant for a knowledge test must have:

(1) Received an endorsement, if required by this part, from an authorized instructor certifying that the applicant accomplished the appropriate ground-training or a home-study course required by this part for the certificate or rating sought and is prepared for the knowledge test;

(2) After July 31, 2014, for the knowledge test for an airline transport pilot certificate with an airplane category multiengine class rating, a graduation certificate for the airline transport pilot certification training program specified in § 61.156; and

(3) Proper identification at the time of application that contains the applicant's—

(i) Photograph;

(ii) Signature;

(iii) Date of birth, which shows:

(A) For issuance of certificates other than the ATP certificate with an airplane category multiengine class rating, the applicant meets or will meet the age requirements of this part for the certificate sought before the expiration date of the airman knowledge test report; and

(B) For issuance of an ATP certificate with an airplane category multiengine class rating obtained under the aeronautical experience requirements of § 61.159 or § 61.160, the applicant is at least 18 years of age at the time of the knowledge test;

(iv) If the permanent mailing is a post office box number, then the applicant must provide a current residential address.

(b) The Administrator shall specify the minimum passing grade for the knowledge test.

[Doc. No. 25910, 62 FR 16298, Apr. 4, 1997, as amended by Amdt. 61-104, 63 FR 20286, Apr. 23, 1998; Amdt. 61-124, 74 FR 42548, Aug. 21, 2009; Amdt. 61-130, 78 FR 42373, July 15, 2013]

§ 61.37 Knowledge tests: Cheating or other unauthorized conduct.

(a) An applicant for a knowledge test may not:

(1) Copy or intentionally remove any knowledge test;

(2) Give to another applicant or receive from another applicant any part or copy of a knowledge test;

(3) Give assistance on, or receive assistance on, a knowledge test during the period that test is being given;

(4) Take any part of a knowledge test on behalf of another person;

(5) Be represented by, or represent, another person for a knowledge test;

(6) Use any material or aid during the period that the test is being given, unless specifically authorized to do so by the Administrator; and

(7) Intentionally cause, assist, or participate in any act prohibited by this paragraph.

(b) An applicant who the Administrator finds has committed an act prohibited by paragraph (a) of this section is prohibited, for 1 year after the date of committing that act, from:

(1) Applying for any certificate, rating, or authorization issued under this chapter; and

(2) Applying for and taking any test under this chapter.

(c) Any certificate or rating held by an applicant may be suspended or revoked if the Administrator finds that person has committed an act prohibited by paragraph (a) of this section.

§ 61.39 Prerequisites for practical tests.

(a) Except as provided in paragraphs (b), (c), and (e) of this section, to be eligible for a practical test for a certificate or rating issued under this part, an applicant must:

(1) Pass the required knowledge test:

(i) Within the 24-calendar-month period preceding the month the applicant completes the practical test, if a knowledge test is required; or

(ii) Within the 60-calendar month period preceding the month the applicant completes the practical for those applicants who pass the knowledge test after completing the airline transport pilot certification training program in § 61.156;

(2) Present the knowledge test report at the time of application for the practical test, if a knowledge test is required;

(3) Have satisfactorily accomplished the required training and obtained the aeronautical experience prescribed by this part for the certificate or rating sought;

(4) Hold at least a third-class medical certificate, if a medical certificate is required;

(5) Meet the prescribed age requirement of this part for the issuance of the certificate or rating sought;

(6) Have an endorsement, if required by this part, in the applicant's logbook or training record that has been signed by an authorized instructor who certifies that the applicant—

(i) Has received and logged training time within 2 calendar months preceding the month of application in preparation for the practical test;

(ii) Is prepared for the required practical test; and

(iii) Has demonstrated satisfactory knowledge of the subject areas in which the applicant was deficient on the airman knowledge test; and

(7) Have a completed and signed application form.

(b) An applicant for an airline transport pilot certificate with an airplane category multiengine class rating or an airline transport pilot certificate with an airplane type rating may take the practical test with an expired knowledge test only if the applicant passed the knowledge test after July 31, 2014, and is employed:

(1) As a flightcrew member by a part 119 certificate holder conducting operations under parts 125 or 135 of this chapter at the time of the practical test and has satisfactorily accomplished that operator's approved pilot-in-command training or checking program; or

(2) As a flightcrew member by a part 119 certificate holder conducting operations under part 121 of this chapter at the time of the practical test and has satisfactorily accomplished that operator's approved initial training program; or

(3) By the U.S. Armed Forces as a flight crewmember in U.S. military air transport operations at the time of the practical test and has completed the pilot in command aircraft qualification training program that is appropriate to the pilot certificate and rating sought.

(c) An applicant for an airline transport pilot certificate with a rating other than those ratings set forth in paragraph (b) of this section may take the practical test for that certificate or rating with an expired knowledge test report, provided that the applicant is employed:

(1) As a flightcrew member by a part 119 certificate holder conducting operations under parts 125 or 135 of this chapter at the time of the practical test and has satisfactorily accomplished that operator's approved pilot-in-command training or checking program; or

(2) By the U.S. Armed Forces as a flight crewmember in U.S. military air transport operations at the time of the practical test and has completed the pilot in command aircraft qualification training program that is appropriate to the pilot certificate and rating sought.

(d) In addition to the requirements in paragraph (a) of this section, to be eligible for a practical test for an airline transport pilot certificate with an airplane category multiengine class rating or airline transport pilot certificate obtained concurrently with an airplane type rating, an applicant must:

(1) If the applicant passed the knowledge test after July 31, 2014, present the graduation certificate for the airline transport pilot certification training program in § 61.156, at the time of application for the practical test;

(2) If applying for the practical test under the aeronautical experience requirements of § 61.160(a), the applicant must present the documents required by that section to substantiate eligibility; and

(3) If applying for the practical test under the aeronautical experience requirements of § 61.160(b), (c), or (d), the applicant must present an official transcript and certifying document from an institution of higher education that holds a letter of authorization from the Administrator under § 61.169.

(e) A person is not required to comply with the provisions of paragraph (a)(6) of this section if that person:

(1) Holds a foreign pilot license issued by a contracting State to the Convention on International Civil Aviation that authorizes at least the privileges of the pilot certificate sought;

(2) Is only applying for a type rating; or

(3) Is applying for an airline transport pilot certificate or an additional rating to an airline transport pilot certificate in an aircraft that does not require an aircraft type rating practical test.

(f) If all increments of the practical test for a certificate or rating are not completed on the same date, then all the remaining increments of the test must be completed within 2 calendar months after the month the applicant began the test.

(g) If all increments of the practical test for a certificate or rating are not completed within 2 calendar months after the month the applicant began the test, the applicant must retake the entire practical test.

[Doc. No. 25910, 62 FR 16298, Apr. 4, 1997; Amdt. 61-103, 62 FR 40897, July 30, 1997, as amended by Amdt. 61-104, 63 FR 20286, Apr. 23, 1998; Amdt. 61-124, 74 FR 42548, Aug. 21, 2009; Amdt. 61-130, 78 FR 42373, July 15, 2013]

§ 61.41 Flight training received from flight instructors not certificated by the FAA.

(a) A person may credit flight training toward the requirements of a pilot certificate or rating issued under this part, if that person received the training from:

(1) A flight instructor of an Armed Force in a program for training military pilots of either—

(i) The United States; or

(ii) A foreign contracting State to the Convention on International Civil Aviation.

(2) A flight instructor who is authorized to give such training by the licensing authority of a foreign contracting State to the Convention on International Civil Aviation, and the flight training is given outside the United States.

(b) A flight instructor described in paragraph (a) of this section is only authorized to give endorsements to show training given.

§ 61.43 Practical tests: General procedures.

(a) Completion of the practical test for a certificate or rating consists of—

(1) Performing the tasks specified in the areas of operation for the airman certificate or rating sought within the approved practical test standards;

(2) Demonstrating mastery of the aircraft by performing each task successfully;

(3) Demonstrating proficiency and competency within the approved standards; and

(4) Demonstrating sound judgment.

(b) The pilot flight crew complement required during the practical test is based on one of the following requirements that applies to the aircraft being used on the practical test:

(1) If the aircraft's FAA-approved flight manual requires the pilot flight crew complement be a single pilot, then the applicant must demonstrate single pilot proficiency on the practical test.

(2) If the aircraft's type certification data sheet requires the pilot flight crew complement be a single pilot, then the applicant must demonstrate single pilot proficiency on the practical test.

(3) If the FAA Flight Standardization Board report, FAA-approved aircraft flight manual, or aircraft type certification data sheet allows the pilot flight crew complement to be either a single pilot, or a pilot and a copilot, then the applicant may demonstrate single pilot proficiency or have a copilot on the practical test. If the applicant performs the practical test with a copilot, the limitation of "Second in Command Required" will be placed on the applicant's pilot certificate. The limitation may be removed if the applicant passes the practical test by demonstrating single-pilot proficiency in the aircraft in which single-pilot privileges are sought.

(c) If an applicant fails any area of operation, that applicant fails the practical test.

(d) An applicant is not eligible for a certificate or rating sought until all the areas of operation are passed.

(e) The examiner or the applicant may discontinue a practical test at any time:

(1) When the applicant fails one or more of the areas of operation; or

(2) Due to inclement weather conditions, aircraft airworthiness, or any other safety-of-flight concern.

(f) If a practical test is discontinued, the applicant is entitled credit for those areas of operation that were passed, but only if the applicant:

(1) Passes the remainder of the practical test within the 60-day period after the date the practical test was discontinued;

(2) Presents to the examiner for the retest the original notice of disapproval form or the letter of discontinuance form, as appropriate;

(3) Satisfactorily accomplishes any additional training needed and obtains the appropriate instructor endorsements, if additional training is required; and

(4) Presents to the examiner for the retest a properly completed and signed application.

[Doc. No. 25910, 62 FR 16298, Apr. 4, 1997, as amended by Amdt. 61-124, 74 FR 42549, Aug. 21, 2009]

§ 61.45 Practical tests: Required aircraft and equipment.

(a) *General.* Except as provided in paragraph (a)(2) of this section or when permitted to accomplish the entire flight increment of the practical test in a flight simulator or a flight training device, an applicant for a certificate or rating issued under this part must furnish:

(1) An aircraft of U.S. registry for each required test that—

(i) Is of the category, class, and type, if applicable, for which the applicant is applying for a certificate or rating; and

(ii) Has a standard airworthiness certificate or special airworthiness certificate in the limited, primary, or light-sport category.

(2) At the discretion of the examiner who administers the practical test, the applicant may furnish—

(i) An aircraft that has an airworthiness certificate other than a standard airworthiness certificate or special airworthiness certificate in the limited, primary, or light-sport category, but that otherwise meets the requirements of paragraph (a)(1) of this section;

(ii) An aircraft of the same category, class, and type, if applicable, of foreign registry that is properly certificated by the country of registry; or

(iii) A military aircraft of the same category, class, and type, if aircraft class and type are appropriate, for which the applicant is applying for a certificate or rating, and provided—

(A) The aircraft is under the direct operational control of the U.S. Armed Forces;

(B) The aircraft is airworthy under the maintenance standards of the U.S. Armed Forces; and

(C) The applicant has a letter from his or her commanding officer authorizing the use of the aircraft for the practical test.

(b) *Required equipment (other than controls).* (1) Except as provided in paragraph (b)(2) of this section, an aircraft used for a practical test must have—

(i) The equipment for each area of operation required for the practical test;

(ii) No prescribed operating limitations that prohibit its use in any of the areas of operation required for the practical test;

(iii) Except as provided in paragraphs (e) and (f) of this section, at least two pilot stations with adequate visibility for each person to operate the aircraft safely; and

(iv) Cockpit and outside visibility adequate to evaluate the performance of the applicant when an additional jump seat is provided for the examiner.

(2) An applicant for a certificate or rating may use an aircraft with operating characteristics that preclude the applicant from performing all of the tasks required for the practical test. However, the applicant's certificate or rating, as appropriate, will be issued with an appropriate limitation.

(c) *Required controls.* Except for lighter-than-air aircraft, and a glider without an engine, an aircraft used for a practical test must have engine power controls and flight controls that are easily reached and operable in a conventional manner by both pilots, unless the Examiner determines that the practical test can be conducted safely in the aircraft without the controls easily reached by the Examiner.

(d) *Simulated instrument flight equipment.* An applicant for a practical test that involves maneuvering an aircraft solely by reference to instruments must furnish:

(1) Equipment on board the aircraft that permits the applicant to pass the areas of operation that apply to the rating sought; and

(2) A device that prevents the applicant from having visual reference outside the aircraft, but does not prevent the examiner from having visual reference outside the aircraft, and is otherwise acceptable to the Administrator.

(e) *Aircraft with single controls.* A practical test may be conducted in an aircraft having a single set of controls, provided the:

(1) Examiner agrees to conduct the test;

(2) Test does not involve a demonstration of instrument skills; and

(3) Proficiency of the applicant can be observed by an examiner who is in a position to observe the applicant.

(f) *Light-sport aircraft with a single seat.* A practical test for a sport pilot certificate may be conducted in a light-sport aircraft having a single seat provided that the—

(1) Examiner agrees to conduct the test;

(2) Examiner is in a position to observe the operation of the aircraft and evaluate the proficiency of the applicant; and

(3) Pilot certificate of an applicant successfully passing the test is issued a pilot certificate with a limitation "No passenger carriage and flight in a single-seat light-sport aircraft only."

[Doc. No. 25910, 62 FR 16298, Apr. 4, 1997; Amdt. 61-103, 62 FR 40897, July 30, 1997; Amdt. 61-104, 63 FR 20286, Apr. 23, 1998; Amdt. 61-110, 69 FR 44865, July 27, 2004; Amdt. 61-124, 74 FR 42549, Aug. 21, 2009]

§ 61.47 Status of an examiner who is authorized by the Administrator to conduct practical tests.

(a) An examiner represents the Administrator for the purpose of conducting practical tests for certificates and ratings issued under this part and to observe an applicant's ability to perform the areas of operation on the practical test.

(b) The examiner is not the pilot in command of the aircraft during the practical test unless the examiner agrees to act in that capacity for the flight or for a portion of the flight by prior arrangement with:

(1) The applicant; or

(2) A person who would otherwise act as pilot in command of the flight or for a portion of the flight.

(c) Notwithstanding the type of aircraft used during the practical test, the applicant and the examiner (and any other occupants authorized to be on board by the examiner) are not subject to the requirements or limitations for the carriage of passengers that are specified in this chapter.

[Doc. No. 25910, 62 FR 16298, Apr. 4, 1997; Amdt. 61-103, 62 FR 40897, July 30, 1997]

§ 61.49 Retesting after failure.

(a) An applicant for a knowledge or practical test who fails that test may reapply for the test only after the applicant has received:

(1) The necessary training from an authorized instructor who has determined that the applicant is proficient to pass the test; and

(2) An endorsement from an authorized instructor who gave the applicant the additional training.

(b) An applicant for a flight instructor certificate with an airplane category rating or, for a flight instructor certificate with a glider category rating, who has failed the practical test due to deficiencies in instructional proficiency on stall awareness, spin entry, spins, or spin recovery must:

(1) Comply with the requirements of paragraph (a) of this section before being retested;

(2) Bring an aircraft to the retest that is of the appropriate aircraft category for the rating sought and is certificated for spins; and

(3) Demonstrate satisfactory instructional proficiency on stall awareness, spin entry, spins, and spin recovery to an examiner during the retest.

§ 61.51 Pilot logbooks.

(a) *Training time and aeronautical experience.* Each person must document and record the following time in a manner acceptable to the Administrator:

(1) Training and aeronautical experience used to meet the requirements for a certificate, rating, or flight review of this part.

(2) The aeronautical experience required for meeting the recent flight experience requirements of this part.

(b) *Logbook entries.* For the purposes of meeting the requirements of paragraph (a) of this section, each person must enter the following information for each flight or lesson logged:

(1) General—

(i) Date.

(ii) Total flight time or lesson time.

(iii) Location where the aircraft departed and arrived, or for lessons in a flight simulator or flight training device, the location where the lesson occurred.

(iv) Type and identification of aircraft, flight simulator, flight training device, or aviation training device, as appropriate.

(v) The name of a safety pilot, if required by § 91.109 of this chapter.

(2) Type of pilot experience or training—

(i) Solo.

(ii) Pilot in command.

(iii) Second in command.

(iv) Flight and ground training received from an authorized instructor.

(v) Training received in a flight simulator, flight training device, or aviation training device from an authorized instructor.

(3) Conditions of flight—

(i) Day or night.

(ii) Actual instrument.

(iii) Simulated instrument conditions in flight, a flight simulator, flight training device, or aviation training device.

(iv) Use of night vision goggles in an aircraft in flight, in a flight simulator, or in a flight training device.

(c) *Logging of pilot time.* The pilot time described in this section may be used to:

(1) Apply for a certificate or rating issued under this part or a privilege authorized under this part; or

(2) Satisfy the recent flight experience requirements of this part.

(d) *Logging of solo flight time.* Except for a student pilot performing the duties of pilot in command of an airship requiring more than one pilot flight crewmember, a pilot may log as solo flight time only that flight time when the pilot is the sole occupant of the aircraft.

(e) *Logging pilot-in-command flight time.* (1) A sport, recreational, private, commercial, or airline transport pilot may log pilot in command flight time for flights-

(i) When the pilot is the sole manipulator of the controls of an aircraft for which the pilot is rated, or has sport pilot privileges for that category and class of aircraft, if the aircraft class rating is appropriate;

(ii) When the pilot is the sole occupant in the aircraft;

(iii) When the pilot, except for a holder of a sport or recreational pilot certificate, acts as pilot in command of an aircraft for which more than one pilot is required under the type certification of the aircraft or the regulations under which the flight is conducted; or

(iv) When the pilot performs the duties of pilot in command while under the supervision of a qualified pilot in command provided—

(A) The pilot performing the duties of pilot in command holds a commercial or airline transport pilot certificate and aircraft rating that is appropriate to the category and class of aircraft being flown, if a class rating is appropriate;

(B) The pilot performing the duties of pilot in command is undergoing an approved pilot in command training program that includes ground and flight training on the following areas of operation—

(*1*) Preflight preparation;

(*2*) Preflight procedures;

(*3*) Takeoff and departure;

(*4*) In-flight maneuvers;

(*5*) Instrument procedures;

(*6*) Landings and approaches to landings;

(*7*) Normal and abnormal procedures;

(*8*) Emergency procedures; and

(*9*) Postflight procedures;

(C) The supervising pilot in command holds—

(*1*) A commercial pilot certificate and flight instructor certificate, and aircraft rating that is appropriate to the category, class, and type of aircraft being flown, if a class or type rating is required; or

(*2*) An airline transport pilot certificate and aircraft rating that is appropriate to the category, class, and type of aircraft being flown, if a class or type rating is required; and

(D) The supervising pilot in command logs the pilot in command training in the pilot's logbook, certifies the pilot in command training in the pilot's logbook and attests to that certification with his or her signature, and flight instructor certificate number.

(2) If rated to act as pilot in command of the aircraft, an airline transport pilot may log all flight time while acting as pilot in command of an operation requiring an airline transport pilot certificate.

(3) A certificated flight instructor may log pilot in command flight time for all flight time while serving as

the authorized instructor in an operation if the instructor is rated to act as pilot in command of that aircraft.

(4) A student pilot may log pilot-in-command time only when the student pilot—

(i) Is the sole occupant of the aircraft or is performing the duties of pilot of command of an airship requiring more than one pilot flight crewmember;

(ii) Has a solo flight endorsement as required under § 61.87 of this part; and

(iii) Is undergoing training for a pilot certificate or rating.

(f) *Logging second-in-command flight time.* A person may log second-in-command time only for that flight time during which that person:

(1) Is qualified in accordance with the second-in-command requirements of § 61.55 of this part, and occupies a crewmember station in an aircraft that requires more than one pilot by the aircraft's type certificate; or

(2) Holds the appropriate category, class, and instrument rating (if an instrument rating is required for the flight) for the aircraft being flown, and more than one pilot is required under the type certification of the aircraft or the regulations under which the flight is being conducted.

(g) *Logging instrument time.* (1) A person may log instrument time only for that flight time when the person operates the aircraft solely by reference to instruments under actual or simulated instrument flight conditions.

(2) An authorized instructor may log instrument time when conducting instrument flight instruction in actual instrument flight conditions.

(3) For the purposes of logging instrument time to meet the recent instrument experience requirements of § 61.57(c) of this part, the following information must be recorded in the person's logbook—

(i) The location and type of each instrument approach accomplished; and

(ii) The name of the safety pilot, if required.

(4) A person can use time in a flight simulator, flight training device, or aviation training device for acquiring instrument aeronautical experience for a pilot certificate, rating, or instrument recency experience, provided an authorized instructor is present to observe that time and signs the person's logbook or training record to verify the time and the content of the training session.

(h) *Logging training time.* (1) A person may log training time when that person receives training from an authorized instructor in an aircraft, flight simulator, or flight training device.

(2) The training time must be logged in a logbook and must:

(i) Be endorsed in a legible manner by the authorized instructor; and

(ii) Include a description of the training given, the length of the training lesson, and the authorized instructor's signature, certificate number, and certificate expiration date.

(i) *Presentation of required documents.* (1) Persons must present their pilot certificate, medical certificate, logbook, or any other record required by this part for inspection upon a reasonable request by—

(i) The Administrator;

(ii) An authorized representative from the National Transportation Safety Board; or

(iii) Any Federal, State, or local law enforcement officer.

(2) A student pilot must carry the following items in the aircraft on all solo cross-country flights as evidence of the required authorized instructor clearances and endorsements—

(i) Pilot logbook;

(ii) Student pilot certificate; and

(iii) Any other record required by this section.

(3) A sport pilot must carry his or her logbook or other evidence of required authorized instructor endorsements on all flights.

(4) A recreational pilot must carry his or her logbook with the required authorized instructor endorsements on all solo flights—

(i) That exceed 50 nautical miles from the airport at which training was received;

(ii) Within airspace that requires communication with air traffic control;

(iii) Conducted between sunset and sunrise; or

(iv) In an aircraft for which the pilot does not hold an appropriate category or class rating.

(5) A flight instructor with a sport pilot rating must carry his or her logbook or other evidence of required authorized instructor endorsements on all flights when providing flight training.

(j) *Aircraft requirements for logging flight time.* For a person to log flight time, the time must be acquired in an aircraft that is identified as an aircraft under § 61.5(b), and is—

(1) An aircraft of U.S. registry with either a standard or special airworthiness certificate;

(2) An aircraft of foreign registry with an airworthiness certificate that is approved by the aviation authority of a foreign country that is a Member State to the Convention on International Civil Aviation Organization;

(3) A military aircraft under the direct operational control of the U.S. Armed Forces; or

(4) A public aircraft under the direct operational control of a Federal, State, county, or municipal law enforcement agency, if the flight time was acquired by the pilot while engaged on an official law enforcement flight for a Federal, State, County, or Municipal law enforcement agency.

(k) *Logging night vision goggle time.* (1) A person may log night vision goggle time only for the time the person uses night vision goggles as the primary visual reference of the surface and operates:

(i) An aircraft during a night vision goggle operation; or

(ii) A flight simulator or flight training device with the lighting system adjusted to represent the period beginning 1 hour after sunset and ending 1 hour before sunrise.

(2) An authorized instructor may log night vision goggle time when that person conducts training using night vision goggles as the primary visual reference of the surface and operates:

(i) An aircraft during a night goggle operation; or

(ii) A flight simulator or flight training device with the lighting system adjusted to represent the period beginning 1 hour after sunset and ending 1 hour before sunrise.

(3) To log night vision goggle time to meet the recent night vision goggle experience requirements under § 61.57(f), a person must log the information required under § 61.51(b).

[Doc. No. 25910, 62 FR 16298, Apr. 4, 1997; Amdt. 61-103, 62 FR 40897, July 30, 1997; Amdt. 61-104, 63 FR 20286, Apr. 23, 1998; Amdt. 61-110, 69 FR 44865, July 27, 2004; Amdt. 61-124, 74 FR 42549, Aug. 21, 2009; Amdt. 61-128, 76 FR 54105, Aug. 31, 2011]

§ 61.52 Use of aeronautical experience obtained in ultralight vehicles.

(a) Before January 31, 2012, a person may use aeronautical experience obtained in an ultralight vehicle to meet the requirements for the following certificates and ratings issued under this part:

(1) A sport pilot certificate.

(2) A flight instructor certificate with a sport pilot rating;

(3) A private pilot certificate with a weight-shift-control or powered parachute category rating.

(b) Before January 31, 2012, a person may use aeronautical experience obtained in an ultralight vehicle to meet the provisions of § 61.69.

(c) A person using aeronautical experience obtained in an ultralight vehicle to meet the requirements for a certificate or rating specified in paragraph (a) of this section or the requirements of paragraph (b) of this section must—

(1) Have been a registered ultralight pilot with an FAA-recognized ultralight organization when that aeronautical experience was obtained;

(2) Document and log that aeronautical experience in accordance with the provisions for logging aeronautical experience specified by an FAA-recognized ultralight organization and in accordance with the provisions for logging pilot time in aircraft as specified in § 61.51;

(3) Obtain the aeronautical experience in a category and class of vehicle corresponding to the rating or privilege sought; and

(4) Provide the FAA with a certified copy of his or her ultralight pilot records from an FAA-recognized ultralight organization, that —

(i) Document that he or she is a registered ultralight pilot with that FAA-recognized ultralight organization; and

(ii) Indicate that he or she is recognized to operate the category and class of aircraft for which sport pilot privileges are sought.

[Doc. No. FAA-2001-11133, 69 FR 44865, July 27, 2004, as amended by Amdt. 61-125, 75 FR 5220, Feb. 1, 2010]

§ 61.53 Prohibition on operations during medical deficiency.

(a) *Operations that require a medical certificate.* Except as provided for in paragraph (b) of this section, no person who holds a medical certificate issued under part 67 of this chapter may act as pilot in command, or in any other capacity as a required pilot flight crewmember, while that person:

(1) Knows or has reason to know of any medical condition that would make the person unable to meet the requirements for the medical certificate necessary for the pilot operation; or

(2) Is taking medication or receiving other treatment for a medical condition that results in the person being unable to meet the requirements for the medical certificate necessary for the pilot operation.

(b) *Operations that do not require a medical certificate.* For operations provided for in § 61.23(b) of this part, a person shall not act as pilot in command, or in any other capacity as a required pilot flight crewmember, while that person knows or has reason to know of any medical condition that would make the person unable to operate the aircraft in a safe manner.

(c) *Operations requiring a medical certificate or a U.S. driver's license.* For operations provided for in § 61.23(c), a person must meet the provisions of—

(1) Paragraph (a) of this section if that person holds a medical certificate issued under part 67 of this chapter and does not hold a U.S. driver's license.

(2) Paragraph (b) of this section if that person holds a U.S. driver's license.

[Doc. No. 25910, 62 FR 16298, Apr. 4, 1997, as amended by Amdt. 61-110, 69 FR 44866, July 27, 2004; Amdt. 61-124, 74 FR 42550, Aug. 21, 2009]

§ 61.55 Second-in-command qualifications.

(a) A person may serve as a second-in-command of an aircraft type certificated for more than one required pilot flight crewmember or in operations requiring a second-in-command pilot flight crewmember only if that person holds:

(1) At least a private pilot certificate with the appropriate category and class rating; and

(2) An instrument rating or privilege that applies to the aircraft being flown if the flight is under IFR; and

(3) At least a pilot type rating for the aircraft being flown unless the flight will be conducted as domestic flight operations within the United States airspace.

(b) Except as provided in paragraph (e) of this section, no person may serve as a second-in-command of an aircraft type certificated for more than one required pilot flight crewmember or in operations requiring a second-in-command unless that person has within the previous 12 calendar months:

(1) Become familiar with the following information for the specific type aircraft for which second-in-command privileges are requested—

(i) Operational procedures applicable to the powerplant, equipment, and systems.

(ii) Performance specifications and limitations.

(iii) Normal, abnormal, and emergency operating procedures.

(iv) Flight manual.

(v) Placards and markings.

(2) Except as provided in paragraph (g) of this section, performed and logged pilot time in the type of aircraft or in a flight simulator that represents the type of aircraft for which second-in-command privileges are requested, which includes—

(i) Three takeoffs and three landings to a full stop as the sole manipulator of the flight controls;

(ii) Engine-out procedures and maneuvering with an engine out while executing the duties of pilot in command; and

(iii) Crew resource management training.

(c) If a person complies with the requirements in paragraph (b) of this section in the calendar month before or the calendar month after the month in which compliance with this section is required, then that person is considered to have accomplished the training and practice in the month it is due.

(d) A person may receive a second-in-command pilot type rating for an aircraft after satisfactorily completing the second-in-command familiarization training requirements under paragraph (b) of this section in that type of aircraft provided the training was completed within the 12 calendar months before the month of application for the SIC pilot type rating. The person must comply with the following application and pilot certification procedures:

(1) The person who provided the training must sign the applicant's logbook or training record after each lesson in accordance with § 61.51(h)(2) of this part. In lieu of the trainer, it is permissible for a qualified management official within the organization to sign the applicant's training records or logbook and make the required endorsement. The qualified management official must hold the position of Chief Pilot, Director of Training, Director of Operations,

or another comparable management position within the organization that provided the training and must be in a position to verify the applicant's training records and that the training was given.

(2) The trainer or qualified management official must make an endorsement in the applicant's logbook that states "[Applicant's Name and Pilot Certificate Number] has demonstrated the skill and knowledge required for the safe operation of the [Type of Aircraft], relevant to the duties and responsibilities of a second in command."

(3) If the applicant's flight experience and/or training records are in an electronic form, the applicant must present a paper copy of those records containing the signature of the trainer or qualified management official to an FAA Flight Standards District Office or Examiner.

(4) The applicant must complete and sign an Airman Certificate and/or Rating Application, FAA Form 8710-1, and present the application to an FAA Flight Standards District Office or to an Examiner.

(5) The person who provided the ground and flight training to the applicant must sign the "Instructor's Recommendation" section of the Airman Certificate and/or Rating Application, FAA Form 8710-1. In lieu of the trainer, it is permissible for a qualified management official within the organization to sign the applicant's FAA Form 8710-1.

(6) The applicant must appear in person at a FAA Flight Standards District Office or to an Examiner with his or her logbook/training records and with the completed and signed FAA Form 8710-1.

(7) There is no practical test required for the issuance of the "SIC Privileges Only" pilot type rating.

(e) A person may receive a second-in-command pilot type rating for the type of aircraft after satisfactorily completing an approved second-in-command training program, proficiency check, or competency check under subpart K of part 91, part 125, or part 135, as appropriate, in that type of aircraft provided the training was completed within the 12 calendar months before the month of application for the SIC pilot type rating. The person must comply with the following application and pilot certification procedures:

(1) The person who provided the training must sign the applicant's logbook or training record after each lesson in accordance with § 61.51(h)(2) of this part. In lieu of the trainer, it is permissible for a qualified management official within the organization to sign the applicant's training records or logbook and make the required endorsement. The qualified management official must hold the position of Chief Pilot, Director of Training, Director of Operations, or another comparable management position within the organization that provided the training and must be in a position to verify the applicant's training records and that the training was given.

(2) The trainer or qualified management official must make an endorsement in the applicant's logbook that states "[Applicant's Name and Pilot Certificate Number] has demonstrated the skill and knowledge required for the safe operation of the [Type of Aircraft], relevant to the duties and responsibilities of a second in command."

(3) If the applicant's flight experience and/or training records are in an electronic form, the applicant must provide a paper copy of those records containing the signature of the trainer or qualified management official to an FAA Flight Standards District Office, an Examiner, or an Aircrew Program Designee.

(4) The applicant must complete and sign an Airman Certificate and/or Rating Application, FAA Form 8710-1, and present the application to an FAA Flight Standards District Office or to an Examiner or to an authorized Aircrew Program Designee.

(5) The person who provided the ground and flight training to the applicant must sign the "Instructor's Recommendation" section of the Airman Certificate and/or Rating Application, FAA Form 8710-1. In lieu of the trainer, it is permissible for a qualified management official within the organization to sign the applicant's FAA Form 8710-1.

(6) The applicant must appear in person at an FAA Flight Standards District Office or to an Examiner or to an authorized Aircrew Program Designee with his or her logbook/training records and with the completed and signed FAA Form 8710-1.

(7) There is no practical test required for the issuance of the "SIC Privileges Only" pilot type rating.

(f) The familiarization training requirements of paragraph (b) of this section do not apply to a person who is:

(1) Designated and qualified as pilot in command under subpart K of part 91, part 121, 125, or 135 of this chapter in that specific type of aircraft;

(2) Designated as the second in command under subpart K of part 91, part 121, 125, or 135 of this chapter in that specific type of aircraft;

(3) Designated as the second in command in that specific type of aircraft for the purpose of receiving flight training required by this section, and no passengers or cargo are carried on the aircraft; or

(4) Designated as a safety pilot for purposes required by § 91.109 of this chapter.

(g) The holder of a commercial or airline transport pilot certificate with the appropriate category and class rating is not required to meet the requirements of paragraph (b)(2) of this section, provided the pilot:

(1) Is conducting a ferry flight, aircraft flight test, or evaluation flight of an aircraft's equipment; and

(2) Is not carrying any person or property on board the aircraft, other than necessary for conduct of the flight.

(h) For the purpose of meeting the requirements of paragraph (b) of this section, a person may serve as second in command in that specific type aircraft, provided:

(1) The flight is conducted under day VFR or day IFR; and

(2) No person or property is carried on board the aircraft, other than necessary for conduct of the flight.

(i) The training under paragraphs (b) and (d) of this section and the training, proficiency check, and competency check under paragraph (e) of this section may be accomplished in a flight simulator that is used in accordance with an approved training course conducted by a training center certificated under part 142 of this chapter or under subpart K of part 91, part 121 or part 135 of this chapter.

(j) When an applicant for an initial second-in-command qualification for a particular type of aircraft receives all the training in a flight simulator, that applicant must satisfactorily complete one takeoff and one landing in an aircraft of the same type for which the qualification is sought. This requirement does not apply to an applicant who completes a proficiency check under part 121 or competency check under subpart K, part 91, part 125, or part 135 for the particular type of aircraft.

[Doc. No. 25910, 62 FR 16298, Apr. 4, 1997; Amdt. 61-103, 62 FR 40898, July 30, 1997; Amdt. 61-109, 68 FR 54559, Sept. 17, 2003; Amdt. 61-113, 70 FR 45271, Aug. 4, 2005; Amdt. 61-109, 70 FR 61890, Oct. 27, 2005; Amdt. 61-124, 74 FR 42550, Aug. 21, 2009; Amdt. 61-128, 76 FR 54105, Aug. 31, 2011; Amdt. 61-130, 78 FR 42374, July 15, 2013]

§ 61.56 Flight review.

(a) Except as provided in paragraphs (b) and (f) of this section, a flight review consists of a minimum of 1 hour of flight training and 1 hour of ground training. The review must include:

(1) A review of the current general operating and flight rules of part 91 of this chapter; and

(2) A review of those maneuvers and procedures that, at the discretion of the person giving the review, are necessary for the pilot to demonstrate the safe exercise of the privileges of the pilot certificate.

(b) Glider pilots may substitute a minimum of three instructional flights in a glider, each of which includes a flight to traffic pattern altitude, in lieu of the 1 hour of flight training required in paragraph (a) of this section.

(c) Except as provided in paragraphs (d), (e), and (g) of this section, no person may act as pilot in command of an aircraft unless, since the beginning of the 24th calendar month before the month in which that pilot acts as pilot in command, that person has—

(1) Accomplished a flight review given in an aircraft for which that pilot is rated by an authorized instructor and

(2) A logbook endorsed from an authorized instructor who gave the review certifying that the person has satisfactorily completed the review.

(d) A person who has, within the period specified in paragraph (c) of this section, passed a pilot proficiency check conducted by an examiner, an approved pilot check airman, or a U.S. Armed Force, for a pilot certificate, rating, or operating privilege need not accomplish the flight review required by this section.

(e) A person who has, within the period specified in paragraph (c) of this section, satisfactorily accomplished one or more phases of an FAA-sponsored pilot proficiency award program need not accomplish the flight review required by this section.

(f) A person who holds a flight instructor certificate and who has, within the period specified in paragraph (c) of this section, satisfactorily completed a renewal of a flight instructor certificate under the provisions in § 61.197 need not accomplish the one hour of ground training specified in paragraph (a) of this section.

(g) A student pilot need not accomplish the flight review required by this section provided the student pilot is undergoing training for a certificate and has a current solo flight endorsement as required under § 61.87 of this part.

(h) The requirements of this section may be accomplished in combination with the requirements of § 61.57 and other applicable recent experience requirements at the discretion of the authorized instructor conducting the flight review.

(i) A flight simulator or flight training device may be used to meet the flight review requirements of this section subject to the following conditions:

(1) The flight simulator or flight training device must be used in accordance with an approved course conducted by a training center certificated under part 142 of this chapter.

(2) Unless the flight review is undertaken in a flight simulator that is approved for landings, the applicant must meet the takeoff and landing requirements of § 61.57(a) or § 61.57(b) of this part.

(3) The flight simulator or flight training device used must represent an aircraft or set of aircraft for which the pilot is rated.

[Doc. No. 25910, 62 FR 16298, Apr. 4, 1997; Amdt. 61-103, 62 FR 40898, July 30, 1997; Amdt. 61-104, 63 FR 20287, Apr. 23, 1998; Amdt. 61-124, 74 FR 42550, Aug. 21, 2009]

§ 61.57 Recent flight experience: Pilot in command.

(a) *General experience.* (1) Except as provided in paragraph (e) of this section, no person may act as a pilot in command of an aircraft carrying passengers or of an aircraft certificated for more than one pilot flight crewmember unless that person has made at least three takeoffs and three landings within the preceding 90 days, and—

(i) The person acted as the sole manipulator of the flight controls; and

(ii) The required takeoffs and landings were performed in an aircraft of the same category, class, and type (if a type rating is required), and, if the aircraft to be flown is an airplane with a tailwheel, the takeoffs and landings must have been made to a full stop in an airplane with a tailwheel.

(2) For the purpose of meeting the requirements of paragraph (a)(1) of this section, a person may act as a pilot in command of an aircraft under day VFR or day IFR, provided no persons or property are carried on board the aircraft, other than those necessary for the conduct of the flight.

(3) The takeoffs and landings required by paragraph (a)(1) of this section may be accomplished in a flight simulator or flight training device that is—

(i) Approved by the Administrator for landings; and

(ii) Used in accordance with an approved course conducted by a training center certificated under part 142 of this chapter.

(b) *Night takeoff and landing experience.* (1) Except as provided in paragraph (e) of this section, no person may act as pilot in command of an aircraft carrying passengers during the period beginning 1 hour after sunset and ending 1 hour before sunrise, unless within the preceding 90 days that person has made at least three takeoffs and three landings to a full stop during the period beginning 1 hour after sunset and ending 1 hour before sunrise, and—

(i) That person acted as sole manipulator of the flight controls; and

(ii) The required takeoffs and landings were performed in an aircraft of the same category, class, and type (if a type rating is required).

(2) The takeoffs and landings required by paragraph (b)(1) of this section may be accomplished in a flight simulator that is—

(i) Approved by the Administrator for takeoffs and landings, if the visual system is adjusted to represent the period described in paragraph (b)(1) of this section; and

(ii) Used in accordance with an approved course conducted by a training center certificated under part 142 of this chapter.

(c) *Instrument experience.* Except as provided in paragraph (e) of this section, a person may act as pilot in command under IFR or weather conditions less than the minimums prescribed for VFR only if:

(1) *Use of an airplane, powered-lift, helicopter, or airship for maintaining instrument experience.* Within the 6 calendar months preceding the month of the flight, that person performed and logged at least the following tasks

and iterations in an airplane, powered-lift, helicopter, or airship, as appropriate, for the instrument rating privileges to be maintained in actual weather conditions, or under simulated conditions using a view-limiting device that involves having performed the following—

(i) Six instrument approaches.

(ii) Holding procedures and tasks.

(iii) Intercepting and tracking courses through the use of navigational electronic systems.

(2) *Use of a flight simulator or flight training device for maintaining instrument experience.* Within the 6 calendar months preceding the month of the flight, that person performed and logged at least the following tasks and iterations in a flight simulator or flight training device, provided the flight simulator or flight training device represents the category of aircraft for the instrument rating privileges to be maintained and involves having performed the following—

(i) Six instrument approaches.

(ii) Holding procedures and tasks.

(iii) Intercepting and tracking courses through the use of navigational electronic systems.

(3) *Use of an aviation training device for maintaining instrument experience.* Within the 2 calendar months preceding the month of the flight, that person performed and logged at least the following tasks, iterations, and time in an aviation training device and has performed the following—

(i) Three hours of instrument experience.

(ii) Holding procedures and tasks.

(iii) Six instrument approaches.

(iv) Two unusual attitude recoveries while in a descending, Vne airspeed condition and two unusual attitude recoveries while in an ascending, stall speed condition.

(v) Interception and tracking courses through the use of navigational electronic systems.

(4) *Combination of completing instrument experience in an aircraft and a flight simulator, flight training device, and aviation training device.* A person who elects to complete the instrument experience with a combination of an aircraft, flight simulator or flight training device, and aviation training device must have performed and logged the following within the 6 calendar months preceding the month of the flight—

(i) Instrument experience in an airplane, powered-lift, helicopter, or airship, as appropriate, for the instrument rating privileges to be maintained, performed in actual weather conditions, or under simulated weather conditions while using a view-limiting device, on the following instrument currency tasks:

(A) Instrument approaches.

(B) Holding procedures and tasks.

(C) Interception and tracking courses through the use of navigational electronic systems.

(ii) Instrument experience in a flight simulator or flight training device that represents the category of aircraft for the instrument rating privileges to be maintained and involves performing at least the following tasks—

(A) Instrument approaches.

(B) Holding procedures and tasks.

(C) Interception and tracking courses through the use of navigational electronic systems.

(iii) Instrument experience in an aviation training device that represents the category of aircraft for the instrument rating privileges to be maintained and involves performing at least the following tasks—

(A) Six instrument approaches.

(B) Holding procedures and tasks.

(C) Interception and tracking courses through the use of navigational electronic systems.

(5) *Combination of completing instrument experience in a flight simulator or flight training device, and an aviation training device.* A person who elects to complete the instrument experience with a combination of a flight simulator, flight training device, and aviation training device must have performed the following within the 6 calendar months preceding the month of the flight—

(i) Instrument recency experience in a flight simulator or flight training device that represents the category of aircraft for the instrument rating privileges to be maintained and involves having performed the following tasks:

(A) Six instrument approaches.

(B) Holding procedures and tasks.

(C) Interception and tracking courses through the use of navigational electronic systems.

(ii) Three hours of instrument experience in an aviation training device that represents the category of aircraft for the instrument rating privileges to be maintained and involves performing at least the following tasks—

(A) Six instrument approaches.

(B) Holding procedures and tasks.

(C) Interception and tracking courses through the use of navigational electronic systems.

(D) Two unusual attitude recoveries while in a descending, Vne airspeed condition and two unusual attitude recoveries while in an ascending, stall speed condition.

(6) Maintaining instrument recent experience in a glider.

(i) Within the 6 calendar months preceding the month of the flight, that person must have performed and logged at least the following instrument currency tasks, iterations, and flight time, and the instrument currency must have been performed in actual weather conditions or under simulated weather conditions—

(A) One hour of instrument flight time in a glider or in a single engine airplane using a view-limiting device while performing interception and tracking courses through the use of navigation electronic systems.

(B) Two hours of instrument flight time in a glider or a single engine airplane with the use of a view-limiting device while performing straight glides, turns to specific headings, steep turns, flight at various airspeeds, navigation, and slow flight and stalls.

(ii) Before a pilot is allowed to carry a passenger in a glider under IFR or in weather conditions less than the minimums prescribed for VFR, that pilot must—

(A) Have logged and performed 2 hours of instrument flight time in a glider within the 6 calendar months preceding the month of the flight.

(B) Use a view-limiting-device while practicing performance maneuvers, performance airspeeds, navigation, slow flight, and stalls.

(d) *Instrument proficiency check.* Except as provided in paragraph (e) of this section, a person who has failed to meet the instrument experience requirements of paragraph (c) for more than six calendar months may reestablish instrument currency only by completing an instrument proficiency check. The instrument proficiency check must consist of the areas of operation and instrument tasks required in the instrument rating practical test standards.

(1) The instrument proficiency check must be—

(i) In an aircraft that is appropriate to the aircraft category;

(ii) For other than a glider, in a flight simulator or flight training device that is representative of the aircraft category; or

(iii) For a glider, in a single-engine airplane or a glider.

(2) The instrument proficiency check must be given by—

(i) An examiner;

(ii) A person authorized by the U.S. Armed Forces to conduct instrument flight tests, provided the person being tested is a member of the U.S. Armed Forces;

(iii) A company check pilot who is authorized to conduct instrument flight tests under part 121, 125, or 135 of this chapter or subpart K of part 91 of this chapter, and provided that both the check pilot and the pilot being tested are employees of that operator or fractional ownership program manager, as applicable;

(iv) An authorized instructor; or

(v) A person approved by the Administrator to conduct instrument practical tests.

(e) *Exceptions.* (1) Paragraphs (a) and (b) of this section do not apply to a pilot in command who is employed by a certificate holder under part 125 and engaged in a flight operation for that certificate holder if the pilot is in compliance with §§ 125.281 and 125.285 of this chapter.

(2) This section does not apply to a pilot in command who is employed by an air carrier certificated under part 121 or 135 and is engaged in a flight operation under part 91, 121, or 135 for that air carrier if the pilot is in compliance with §§ 121.435 or 121.436, as applicable, and § 121.439, or §§ 135.243 and 135.247 of this chapter, as appropriate.

(3) Paragraph (b) of this section does not apply to a pilot in command of a turbine-powered airplane that is type certificated for more than one pilot crewmember, provided that pilot has complied with the requirements of paragraph (e)(3)(i) or (ii) of this section:

(i) The pilot in command must hold at least a commercial pilot certificate with the appropriate category, class, and type rating for each airplane that is type certificated for more than one pilot crewmember that the pilot seeks to operate under this alternative, and:

(A) That pilot must have logged at least 1,500 hours of aeronautical experience as a pilot;

(B) In each airplane that is type certificated for more than one pilot crewmember that the pilot seeks to operate under this alternative, that pilot must have accomplished and logged the daytime takeoff and landing recent flight experience of paragraph (a) of this section, as the sole manipulator of the flight controls;

(C) Within the preceding 90 days prior to the operation of that airplane that is type certificated for more than one pilot crewmember, the pilot must have accomplished and logged at least 15 hours of flight time in the type of airplane that the pilot seeks to operate under this alternative; and

(D) That pilot has accomplished and logged at least 3 takeoffs and 3 landings to a full stop, as the sole manipulator of the flight controls, in a turbine-powered airplane that requires more than one pilot crewmember. The pilot must have performed the takeoffs and landings during the period beginning 1 hour after sunset and ending 1 hour before sunrise within the preceding 6 months prior to the month of the flight.

(ii) The pilot in command must hold at least a commercial pilot certificate with the appropriate category, class, and type rating for each airplane that is type certificated for more than one pilot crewmember that the pilot seeks to operate under this alternative, and:

(A) That pilot must have logged at least 1,500 hours of aeronautical experience as a pilot;

(B) In each airplane that is type certificated for more than one pilot crewmember that the pilot seeks to operate under this alternative, that pilot must have accomplished and logged the daytime takeoff and landing recent flight experience of paragraph (a) of this section, as the sole manipulator of the flight controls;

(C) Within the preceding 90 days prior to the operation of that airplane that is type certificated for more than one pilot crewmember, the pilot must have accomplished and logged at least 15 hours of flight time in the type of airplane that the pilot seeks to operate under this alternative; and

(D) Within the preceding 12 months prior to the month of the flight, the pilot must have completed a training program that is approved under part 142 of this chapter. The approved training program must have required and the pilot must have performed, at least 6 takeoffs and 6 landings to a full stop as the sole manipulator of the controls in a flight simulator that is representative of a turbine-powered airplane that requires more than one pilot crewmember. The flight simulator's visual system must have been adjusted to represent the period beginning 1 hour after sunset and ending 1 hour before sunrise.

(f) *Night vision goggle operating experience.* (1) A person may act as pilot in command in a night vision goggle operation with passengers on board only if, within 2 calendar months preceding the month of the flight, that person performs and logs the following tasks as the sole manipulator of the controls on a flight during a night vision goggle operation—

(i) Three takeoffs and three landings, with each takeoff and landing including a climbout, cruise, descent, and approach phase of flight (only required if the pilot wants to use night vision goggles during the takeoff and landing phases of the flight).

(ii) Three hovering tasks (only required if the pilot wants to use night vision goggles when operating helicopters or powered-lifts during the hovering phase of flight).

(iii) Three area departure and area arrival tasks.

(iv) Three tasks of transitioning from aided night flight (*aided night flight* means that the pilot uses night vision goggles to maintain visual surface reference) to unaided night flight (*unaided night flight* means that the pilot does not use night vision goggles) and back to aided night flight.

(v) Three night vision goggle operations, or when operating helicopters or powered-lifts, six night vision goggle operations.

(2) A person may act as pilot in command using night vision goggles only if, within the 4 calendar months preceding the month of the flight, that person performs and logs the tasks listed in paragraph (f)(1)(i) through (v) of this section as the sole manipulator of the controls during a night vision goggle operation.

(g) *Night vision goggle proficiency check.* A person must either meet the night vision goggle experience requirements of paragraphs (f)(1) or (f)(2) of this section or pass a night vision goggle proficiency check to act as pilot in command using night vision goggles. The proficiency check must be performed in the category of aircraft that is appropriate to the night vision goggle operation for which the person is seeking the night vision goggle privilege or in a flight simulator or flight training device that is representative of that category of aircraft. The check must consist of the tasks listed in § 61.31(k), and the check must be performed by:

(1) An Examiner who is qualified to perform night vision goggle operations in that same aircraft category and class;

(2) A person who is authorized by the U.S. Armed Forces to perform night vision goggle proficiency checks, provided the person being administered the check is also a member of the U.S. Armed Forces;

(3) A company check pilot who is authorized to perform night vision goggle proficiency checks under parts 121, 125, or 135 of this chapter, provided that both the check pilot and the pilot being tested are employees of that operator;

(4) An authorized flight instructor who is qualified to perform night vision goggle operations in that same aircraft category and class;

(5) A person who is qualified as pilot in command for night vision goggle operations in accordance with paragraph (f) of this section; or

(6) A person approved by the FAA to perform night vision goggle proficiency checks.

[Doc. No. 25910, 62 FR 16298, Apr. 4, 1997; Amdt. 61-103, 62 FR 40898, July 30, 1997; Amdt. 61-106, 64 FR 23529, Apr. 30, 1999; Amdt. 61-109, 68 FR 54559, Sept. 17, 2003; Amdt. 61-124, 74 FR 42550, Aug. 21, 2009; Amdt. 61-127, 76 FR 19267, Apr. 7, 2011; Amdt. 61-129, 76 FR 78143, Dec. 16, 2011; Amdt. 61-130, 78 FR 42374, July 15, 2013]

§ 61.58 Pilot-in-command proficiency check: Operation of an aircraft that requires more than one pilot flight crewmember or is turbojet-powered.

(a) Except as otherwise provided in this section, to serve as pilot in command of an aircraft that is type certificated for more than one required pilot flight crewmember or is turbojet-powered, a person must—

(1) Within the preceding 12 calendar months, complete a pilot-in-command proficiency check in an aircraft that is type certificated for more than one required pilot flight crewmember or is turbojet-powered; and

(2) Within the preceding 24 calendar months, complete a pilot-in-command proficiency check in the particular type of aircraft in which that person will serve as pilot in command, that is type certificated for more than one required pilot flight crewmember or is turbojet-powered.

(b) This section does not apply to persons conducting operations under subpart K of part 91, part 121, 125, 133, 135, or 137 of this chapter, or persons maintaining continuing qualification under an Advanced Qualification program approved under subpart Y of part 121 of this chapter.

(c) The pilot-in-command proficiency check given in accordance with the provisions of subpart K of part 91, part 121, 125, or 135 of this chapter may be used to satisfy the requirements of this section.

(d) The pilot-in-command proficiency check required by paragraph (a) of this section may be accomplished by satisfactory completion of one of the following:

(1) A pilot-in-command proficiency check conducted by a person authorized by the Administrator, consisting of the aeronautical knowledge areas, areas of operations, and tasks required for a type rating, in an aircraft that is type certificated for more than one pilot flight crewmember or is turbojet-powered;

(2) The practical test required for a type rating, in an aircraft that is type certificated for more than one required pilot flight crewmember or is turbojet-powered;

(3) The initial or periodic practical test required for the issuance of a pilot examiner or check airman designation, in an aircraft that is type certificated for more than one required pilot flight crewmember or is turbojet-powered;

(4) A pilot proficiency check administered by a U.S. Armed Force that qualifies the military pilot for pilot-in-command designation with instrument privileges, and was performed in a military aircraft that the military requires to be operated by more than one pilot flight crewmember or is turbojet-powered;

(5) For a pilot authorized by the Administrator to operate an experimental turbojet-powered aircraft that possesses, by original design or through modification, more than a single seat, the required proficiency check for all of the experimental turbojet-powered aircraft for which the pilot holds an authorization may be accomplished by completing any one of the following:

(i) A single proficiency check, conducted by an examiner authorized by the Administrator, in any one of the experimental turbojet-powered aircraft for which the airman holds an authorization to operate if conducted within the prior 12 months;

(ii) A single proficiency check, conducted by an examiner authorized by the Administrator, in any experimental turbojet-powered aircraft (*e.g.*, if a pilot acquires a new authorization to operate an additional experimental turbojet-powered aircraft, the check for that new authorization will meet the intent), if conducted within the prior 12 months;

(iii) Current qualification under an Advanced Qualification Program (AQP) under subpart Y of part 121 of this chapter;

(iv) Any proficiency check conducted under subpart K of part 91, part 121, or part 135 of this chapter within the prior 12 months if conducted in a turbojet-powered aircraft; or

(v) Any other § 61.58 proficiency check conducted within the prior 12 months if conducted in a turbojet-powered aircraft.

(e) The pilot of a multi-seat experimental turbojet-powered aircraft who has not received a proficiency check within the prior 12 months in accordance with this section may continue to operate such aircraft in accordance with the pilot's authorizations. However, the pilot is prohibited from carriage of any persons in any experimental turbojet-powered aircraft with the exception of those individuals authorized by the Administrator to conduct training, conduct flight checks, or perform pilot certification functions in such aircraft, and only during flights specifically related to training, flight checks, or certification in such aircraft.

(f) This section will not apply to a pilot authorized by the Administrator to serve as pilot in command in experimental turbojet-powered aircraft that possesses, by original design, a single seat, when operating such single-seat aircraft.

(g) A check or test described in paragraphs (d)(1) through (5) of this section may be accomplished in a flight simulator under part 142 of this chapter, subject to the following:

(1) Except as provided for in paragraphs (g)(2) and (3) of this section, if an otherwise qualified and approved flight simulator used for a pilot-in-command proficiency check is not qualified and approved for a specific required maneuver—

(i) The training center must annotate, in the applicant's training record, the maneuver or maneuvers omitted; and

(ii) Prior to acting as pilot in command, the pilot must demonstrate proficiency in each omitted maneuver in an aircraft or flight simulator qualified and approved for each omitted maneuver.

(2) If the flight simulator used pursuant to paragraph (g) of this section is not qualified and approved for circling approaches—

(i) The applicant's record must include the statement, "Proficiency in circling approaches not demonstrated"; and

(ii) The applicant may not perform circling approaches as pilot in command when weather conditions are less than the basic VFR conditions described in § 91.155 of this chapter, until proficiency in circling approaches has been successfully demonstrated in a flight simulator qualified and approved for circling approaches or in an aircraft to a person authorized by the Administrator to conduct the check required by this section.

(3) If the flight simulator used pursuant to paragraph (g) of this section is not qualified and approved for landings, the applicant must—

(i) Hold a type rating in the airplane represented by the simulator; and

(ii) Have completed within the preceding 90 days at least three takeoffs and three landings (one to a full stop) as the sole manipulator of the flight controls in the type airplane for which the pilot-in-command proficiency check is sought.

(h) For the purpose of meeting the pilot-in-command proficiency check requirements of paragraph (a) of this section, a person may act as pilot in command of a flight under day VFR conditions or day IFR conditions if no person or property is carried, other than as necessary to demonstrate compliance with this part.

(i) If a pilot takes the pilot-in-command proficiency check required by this section in the calendar month before or the calendar month after the month in which it is due, the pilot is considered to have taken it in the month in which it was due for the purpose of computing when the next pilot-in-command proficiency check is due.

(j) A pilot-in-command of a turbojet powered aircraft that is type certificated for one pilot does not have to comply with the pilot-in-command proficiency check requirements in paragraphs (a)(1) and (a)(2) of this section until October 31, 2012.

(k) Unless required by the aircraft's operating limitations, a pilot-in-command of an experimental turbojet-powered aircraft does not have to comply with the pilot-in-command proficiency check requirements in paragraphs (a)(1) and (a)(2) of this section until October 31, 2012.

[Doc. No. 25910, 62 FR 40899, July 30, 1997, as amended by Amdt. 61-109, 68 FR 54559, Sept. 17, 2003; Amdt. 61-112, 70 FR 54814, Sept. 16, 2005; Amdt. 61-128, 76 FR 54106, Aug. 31, 2011; 76 FR 63184, Oct. 12, 2011]

§ 61.59 Falsification, reproduction, or alteration of applications, certificates, logbooks, reports, or records.

(a) No person may make or cause to be made:

(1) Any fraudulent or intentionally false statement on any application for a certificate, rating, authorization, or duplicate thereof, issued under this part;

(2) Any fraudulent or intentionally false entry in any logbook, record, or report that is required to be kept, made, or used to show compliance with any requirement for the issuance or exercise of the privileges of any certificate, rating, or authorization under this part;

(3) Any reproduction for fraudulent purpose of any certificate, rating, or authorization, under this part; or

(4) Any alteration of any certificate, rating, or authorization under this part.

(b) The commission of an act prohibited under paragraph (a) of this section is a basis for suspending or revoking any airman certificate, rating, or authorization held by that person.

§ 61.60 Change of address.

The holder of a pilot, flight instructor, or ground instructor certificate who has made a change in permanent mailing address may not, after 30 days from that date, exercise the privileges of the certificate unless the holder has notified in writing the FAA, Airman Certification Branch, P.O. Box 25082, Oklahoma City, OK 73125, of the new permanent mailing address, or if the permanent mailing address includes a post office box number, then the holder's current residential address.

Subpart B—Aircraft Ratings and Pilot Authorizations

§ 61.61 Applicability.

This subpart prescribes the requirements for the issuance of additional aircraft ratings after a pilot certificate is issued, issuance of a type rating concurrently with a pilot certificate, and the requirements for and limitations of pilot authorizations issued by the Administrator.

[Doc. No. FAA-2006-26661, 76 FR 78143, Dec. 16, 2011]

§ 61.63 Additional aircraft ratings (other than for ratings at the airline transport pilot certification level).

(a) *General.* For an additional aircraft rating on a pilot certificate, other than for an airline transport pilot certificate, a person must meet the requirements of this section appropriate to the additional aircraft rating sought.

(b) *Additional aircraft category rating.* A person who applies to add a category rating to a pilot certificate:

(1) Must complete the training and have the applicable aeronautical experience.

(2) Must have a logbook or training record endorsement from an authorized instructor attesting that the person was found competent in the appropriate aeronautical knowledge areas and proficient in the appropriate areas of operation.

(3) Must pass the practical test.

(4) Need not take an additional knowledge test, provided the applicant holds an airplane, rotorcraft, powered-lift, weight-shift-control aircraft, powered parachute, or airship rating at that pilot certificate level.

(c) *Additional aircraft class rating.* A person who applies for an additional class rating on a pilot certificate:

(1) Must have a logbook or training record endorsement from an authorized instructor attesting that the person was found competent in the appropriate aeronautical knowledge areas and proficient in the appropriate areas of operation.

(2) Must pass the practical test.

(3) Need not meet the specified training time requirements prescribed by this part that apply to the pilot certificate for the aircraft class rating sought; unless, the person only holds a lighter-than-air category rating with a balloon class rating and is seeking an airship class rating, then that person must receive the specified training time

requirements and possess the appropriate aeronautical experience.

(4) Need not take an additional knowledge test, provided the applicant holds an airplane, rotorcraft, powered-lift, weight-shift-control aircraft, powered parachute, or airship rating at that pilot certificate level.

(d) *Additional aircraft type rating.* Except as provided under paragraph (d)(6) of this section, a person who applies for an aircraft type rating or an aircraft type rating to be completed concurrently with an aircraft category or class rating—

(1) Must hold or concurrently obtain an appropriate instrument rating, except as provided in paragraph (e) of this section.

(2) Must have a logbook or training record endorsement from an authorized instructor attesting that the person is competent in the appropriate aeronautical knowledge areas and proficient in the appropriate areas of operation at the airline transport pilot certification level.

(3) Must pass the practical test at the airline transport pilot certification level.

(4) Must perform the practical test in actual or simulated instrument conditions, except as provided in paragraph (e) of this section.

(5) Need not take an additional knowledge test if the applicant holds an airplane, rotorcraft, powered-lift, or airship rating on the pilot certificate.

(6) In the case of a pilot employee of a part 121 or part 135 certificate holder or of a fractional ownership program manager under subpart K of part 91 of this chapter, the pilot must—

(i) Meet the appropriate requirements under paragraphs (d)(1), (d)(3), and (d)(4) of this section; and

(ii) Receive a flight training record endorsement from the certificate holder attesting that the person completed the certificate holder's approved ground and flight training program.

(e) *Aircraft not capable of instrument maneuvers and procedures.* (1) An applicant for a type rating or a type rating in addition to an aircraft category and/or class rating who provides an aircraft that is not capable of the instrument maneuvers and procedures required on the practical test:

(i) May apply for the type rating, but the rating will be limited to "VFR only."

(ii) May have the "VFR only" limitation removed for that aircraft type after the applicant:

(A) Passes a practical test in that type of aircraft in actual or simulated instrument conditions;

(B) Passes a practical test in that type of aircraft on the appropriate instrument maneuvers and procedures in § 61.157; or

(C) Becomes qualified under § 61.73(d) for that type of aircraft.

(2) When an instrument rating is issued to a person who holds one or more type ratings, the amended pilot certificate must bear the "VFR only" limitation for each aircraft type rating that the person did not demonstrate instrument competency.

(f) *Multiengine airplane with a single-pilot station.* An applicant for a type rating, at other than the ATP certification level, in a multiengine airplane with a single-pilot station must perform the practical test in the multi-seat version of that airplane, or the practical test may be performed in the single-seat version of that airplane if the Examiner is in a position to observe the applicant during the practical test and there is no multi-seat version of that multiengine airplane.

(g) *Single engine airplane with a single-pilot station.* An applicant for a type rating, at other than the ATP certification level, in a single engine airplane with a single-pilot station must perform the practical test in the multi-seat version of that single engine airplane, or the practical test may be performed in the single-seat version of that airplane if the Examiner is in a position to observe the applicant during the practical test and there is no multi-seat version of that single engine airplane.

(h) *Aircraft category and class rating for the operation of aircraft with an experimental certificate.* A person holding a recreational, private, or commercial pilot certificate may apply for a category and class rating limited to a specific make and model of experimental aircraft, provided—

(1) The person logged 5 hours flight time while acting as pilot in command in the same category, class, make, and model of aircraft.

(2) The person received a logbook endorsement from an authorized instructor who determined the pilot's proficiency to act as pilot in command of the same category, class, make, and model of aircraft.

(3) The flight time specified under paragraph (h)(1) of this section was logged between September 1, 2004 and August 31, 2005.

(i) *Waiver authority.* An Examiner who conducts a practical test may waive any task for which the FAA has provided waiver authority.

[Doc. No. FAA-2006-26661, 74 FR 42552, Aug. 21, 2009, as amended by Amdt. 61-125, 75 FR 5220, Feb. 1, 2010]

§ 61.64 Use of a flight simulator and flight training device.

(a) *Use of a flight simulator or flight training device.* If an applicant for a certificate or rating uses a flight simulator or flight training device for training or any portion of the practical test, the flight simulator and flight training device—

(1) Must represent the category, class, and type (if a type rating is applicable) for the rating sought; and

(2) Must be qualified and approved by the Administrator and used in accordance with an approved course of training under part 141 or part 142 of this chapter; or under part 121 or part 135 of this chapter, provided the applicant is a pilot employee of that air carrier operator.

(b) Except as provided in paragraph (f) of this section, if an airplane is not used during the practical test for a type rating for a turbojet airplane (except for preflight inspection), an applicant must accomplish the entire practical test in a Level C or higher flight simulator and the applicant must—

(1) Hold a type rating in a turbojet airplane of the same class of airplane for which the type rating is sought, and that type rating may not contain a supervised operating experience limitation;

(2) Have 1,000 hours of flight time in two different turbojet airplanes of the same class of airplane for which the type rating is sought;

(3) Have been appointed by the U.S. Armed Forces as pilot in command in a turbojet airplane of the same class of airplane for which the type rating is sought;

(4) Have 500 hours of flight time in the same type of airplane for which the type rating is sought; or

(5) Have logged at least 2,000 hours of flight time, of which 500 hours were in turbine-powered airplanes of the same class of airplane for which the type rating is sought.

(c) Except as provided in paragraph (f) of this section, if an airplane is not used during the practical test for a type rating for a turbo-propeller airplane (except for preflight inspection), an applicant must accomplish the entire practical test in a Level C or higher flight simulator and the applicant must—

(1) Hold a type rating in a turbo-propeller airplane of the same class of airplane for which the type rating is sought, and that type rating may not contain a supervised operating experience limitation;

(2) Have 1,000 hours of flight time in two different turbo-propeller airplanes of the same class of airplane for which the type rating is sought;

(3) Have been appointed by the U.S. Armed Forces as pilot in command in a turbo-propeller airplane of the same class of airplane for which the type rating is sought;

(4) Have 500 hours of flight time in the same type of airplane for which the type rating is sought; or

(5) Have logged at least 2,000 hours of flight time, of which 500 hours were in turbine-powered airplanes of the same class of airplane for which the type rating is sought.

(d) Except as provided in paragraph (f) of this section, if a helicopter is not used during the practical test for a type rating in a helicopter (except for preflight inspection), an applicant must accomplish the entire practical test in a Level C or higher flight simulator and the applicant must meet one of the following requirements—

(1) Hold a type rating in a helicopter and that type rating may not contain the supervised operating experience limitation;

(2) Have been appointed by the U.S. Armed Forces as pilot in command of a helicopter;

(3) Have 500 hours of flight time in the type of helicopter; or

(4) Have 1,000 hours of flight time in two different types of helicopters.

(e) Except as provided in paragraph (f) of this section, if a powered-lift is not used during the practical test for a type rating in a powered-lift (except for preflight inspection), an applicant must accomplish the entire practical test in a Level C or higher flight simulator and the applicant must meet one of the following requirements—

(1) Hold a type rating in a powered-lift without a supervised operating experience limitation;

(2) Have been appointed by the U.S. Armed Forces as pilot in command of a powered-lift;

(3) Have 500 hours of flight time in the type of powered-lift for which the rating is sought; or

(4) Have 1,000 hours of flight time in two different types of powered-lifts.

(f) If the applicant does not meet one of the experience requirements of paragraphs (b)(1) through (5), (c)(1) through (5), (d)(1) through (4) or (e)(1) through (4) of this section, as appropriate to the type rating sought, then—

(1) The applicant must complete the following tasks on the practical test in an aircraft appropriate to category, class, and type for the rating sought: Preflight inspection, normal takeoff, normal instrument landing system approach, missed approach, and normal landing; or

(2) The applicant's pilot certificate will be issued with a limitation that states: "The [name of the additional type rating] is subject to pilot in command limitations," and the applicant is restricted from serving as pilot in command in an aircraft of that type.

(g) The limitation described under paragraph (f)(2) of this section may be removed from the pilot certificate if the applicant complies with the following—

(1) Performs 25 hours of flight time in an aircraft of the category, class, and type for which the limitation applies under the direct observation of the pilot in command who holds a category, class, and type rating, without limitations, for the aircraft;

(2) Logs each flight and the pilot in command who observed the flight attests in writing to each flight;

(3) Obtains the flight time while performing the duties of pilot in command; and

(4) Presents evidence of the supervised operating experience to any Examiner or FAA Flight Standards District Office to have the limitation removed.

[Doc. No. FAA-2006-26661, 76 FR 78143, Dec. 16, 2011]

§ 61.65 Instrument rating requirements.

(a) *General.* A person who applies for an instrument rating must:

(1) Hold at least a current private pilot certificate, or be concurrently applying for a private pilot certificate, with an airplane, helicopter, or powered-lift rating appropriate to the instrument rating sought;

(2) Be able to read, speak, write, and understand the English language. If the applicant is unable to meet any of these requirements due to a medical condition, the Administrator may place such operating limitations on the applicant's pilot certificate as are necessary for the safe operation of the aircraft;

(3) Receive and log ground training from an authorized instructor or accomplish a home-study course of training on the aeronautical knowledge areas of paragraph (b) of this section that apply to the instrument rating sought;

(4) Receive a logbook or training record endorsement from an authorized instructor certifying that the person is prepared to take the required knowledge test;

(5) Receive and log training on the areas of operation of paragraph (c) of this section from an authorized instructor in an aircraft, flight simulator, or flight training device that represents an airplane, helicopter, or powered-lift appropriate to the instrument rating sought;

(6) Receive a logbook or training record endorsement from an authorized instructor certifying that the person is prepared to take the required practical test;

(7) Pass the required knowledge test on the aeronautical knowledge areas of paragraph (b) of this section; however, an applicant is not required to take another knowledge test when that person already holds an instrument rating; and

(8) Pass the required practical test on the areas of operation in paragraph (c) of this section in—

(i) An airplane, helicopter, or powered-lift appropriate to the rating sought; or

(ii) A flight simulator or a flight training device appropriate to the rating sought and for the specific maneuver or instrument approach procedure performed. If an approved flight training device is used for the practical test, the instrument approach procedures conducted in that flight training device are limited to one precision and one nonprecision approach, provided the flight training device is approved for the procedure performed.

(b) *Aeronautical knowledge.* A person who applies for an instrument rating must have received and logged ground training from an authorized instructor or accomplished a home-study course on the following aeronautical knowledge areas that apply to the instrument rating sought:

(1) Federal Aviation Regulations of this chapter that apply to flight operations under IFR;

(2) Appropriate information that applies to flight operations under IFR in the "Aeronautical Information Manual;"

(3) Air traffic control system and procedures for instrument flight operations;

(4) IFR navigation and approaches by use of navigation systems;

(5) Use of IFR en route and instrument approach procedure charts;

(6) Procurement and use of aviation weather reports and forecasts and the elements of forecasting weather trends based on that information and personal observation of weather conditions;

(7) Safe and efficient operation of aircraft under instrument flight rules and conditions;

(8) Recognition of critical weather situations and windshear avoidance;

(9) Aeronautical decision making and judgment; and

(10) Crew resource management, including crew communication and coordination.

(c) *Flight proficiency.* A person who applies for an instrument rating must receive and log training from an authorized instructor in an aircraft, or in a flight simulator or flight training device, in accordance with paragraph (g) of this section, that includes the following areas of operation:

(1) Preflight preparation;

(2) Preflight procedures;

(3) Air traffic control clearances and procedures;

(4) Flight by reference to instruments;

(5) Navigation systems;

(6) Instrument approach procedures;

(7) Emergency operations; and

(8) Postflight procedures.

(d) *Aeronautical experience for the instrument-airplane rating.* A person who applies for an instrument-airplane rating must have logged:

(1) Except as provided in paragraph (g) of this section, 50 hours of cross-country flight time as pilot in command, of which 10 hours must have been in an airplane; and

(2) Forty hours of actual or simulated instrument time in the areas of operation listed in paragraph (c) of this section, of which 15 hours must have been received from an authorized instructor who holds an instrument-airplane rating, and the instrument time includes:

(i) Three hours of instrument flight training from an authorized instructor in an airplane that is appropriate to the instrument-airplane rating within 2 calendar months before the date of the practical test; and

(ii) Instrument flight training on cross country flight procedures, including one cross country flight in an airplane with an authorized instructor, that is performed under instrument flight rules, when a flight plan has been filed with an air traffic control facility, and that involves—

(A) A flight of 250 nautical miles along airways or by directed routing from an air traffic control facility;

(B) An instrument approach at each airport; and

(C) Three different kinds of approaches with the use of navigation systems.

(e) *Aeronautical experience for the instrument-helicopter rating.* A person who applies for an instrument-helicopter rating must have logged:

(1) Except as provided in paragraph (g) of this section, 50 hours of cross-country flight time as pilot in command, of which 10 hours must have been in a helicopter; and

(2) Forty hours of actual or simulated instrument time in the areas of operation listed under paragraph (c) of this section, of which 15 hours must have been with an authorized instructor who holds an instrument-helicopter rating, and the instrument time includes:

(i) Three hours of instrument flight training from an authorized instructor in a helicopter that is appropriate to the instrument-helicopter rating within 2 calendar months before the date of the practical test; and

(ii) Instrument flight training on cross country flight procedures, including one cross country flight in a helicopter with an authorized instructor that is performed under instrument flight rules and a flight plan has been filed with an air traffic control facility, and involves—

(A) A flight of 100 nautical miles along airways or by directed routing from an air traffic control facility;

(B) An instrument approach at each airport; and

(C) Three different kinds of approaches with the use of navigation systems.

(f) *Aeronautical experience for the instrument-powered-lift rating.* A person who applies for an instrument-powered-lift rating must have logged:

(1) Except as provided in paragraph (g) of this section, 50 hours of cross-country flight time as pilot in command, of which 10 hours must have been in a powered-lift; and

(2) Forty hours of actual or simulated instrument time in the areas of operation listed under paragraph (c) of this section, of which 15 hours must have been received from an authorized instructor who holds an instrument-powered-lift rating, and the instrument time includes:

(i) Three hours of instrument flight training from an authorized instructor in a powered-lift that is appropriate to the instrument-powered-lift rating within 2 calendar months before the date of the practical test; and

(ii) Instrument flight training on cross country flight procedures, including one cross country flight in a powered-lift with an authorized instructor that is performed under instrument flight rules, when a flight plan has been filed with an air traffic control facility, that involves—

(A) A flight of 250 nautical miles along airways or by directed routing from an air traffic control facility;

(B) An instrument approach at each airport; and

(C) Three different kinds of approaches with the use of navigation systems.

(g) An applicant for a combined private pilot certificate with an instrument rating may satisfy the cross-country flight time requirements of this section by crediting:

(1) For an instrument-airplane rating or an instrument-powered-lift rating, up to 45 hours of cross-country flight time performing the duties of pilot in command with an authorized instructor; or

(2) For an instrument-helicopter rating, up to 47 hours of cross-country flight time performing the duties of pilot in command with an authorized instructor.

(h) *Use of flight simulators or flight training devices.* If the instrument time was provided by an authorized instructor in a flight simulator or flight training device—

(1) A maximum of 30 hours may be performed in that flight simulator or flight training device if the instrument time was completed in accordance with part 142 of this chapter; or

(2) A maximum of 20 hours may be performed in that flight simulator or flight training device if the instrument time was not completed in accordance with part 142 of this chapter.

(i) *Use of an aviation training device.* A maximum of 10 hours of instrument time received in an aviation training device may be credited for the instrument time requirements of this section if—

(1) The device is approved and authorized by the FAA;

(2) An authorized instructor provides the instrument time in the device;

(3) No more than 10 hours of instrument time in a flight simulator or flight training device was credited for the instrument time requirements of this section;

(4) A view-limiting device was worn by the applicant when logging instrument time in the device; and

(5) The FAA approved the instrument training and instrument tasks performed in the device.

[Doc. No. 25910, 62 FR 16298, Apr. 4, 1997; Amdt. 61-103, 62 FR 40900, July 30, 1997; Amdt. 61-124, 74 FR 42554, Aug. 21, 2009; Amdt. 61-127, 76 FR 19267, Apr. 7, 2011; Amdt. 61-128, 76 FR 54106, Aug. 31, 2011]

§ 61.67 Category II pilot authorization requirements.

(a) *General.* A person who applies for a Category II pilot authorization must hold:

(1) At least a private or commercial pilot certificate with an instrument rating or an airline transport pilot certificate;

(2) A type rating for the aircraft for which the authorization is sought if that aircraft requires a type rating; and

(3) A category and class rating for the aircraft for which the authorization is sought.

(b) *Experience requirements.* An applicant for a Category II pilot authorization must have at least—

(1) 50 hours of night flight time as pilot in command.

(2) 75 hours of instrument time under actual or simulated instrument conditions that may include not more than—

(i) A combination of 25 hours of simulated instrument flight time in a flight simulator or flight training device; or

(ii) 40 hours of simulated instrument flight time if accomplished in an approved course conducted by an appropriately rated training center certificated under part 142 of this chapter.

(3) 250 hours of cross-country flight time as pilot in command.

(c) *Practical test requirements.* (1) A practical test must be passed by a person who applies for—

(i) Issuance or renewal of a Category II pilot authorization; and

(ii) The addition of another type aircraft to the applicant's Category II pilot authorization.

(2) To be eligible for the practical test for an authorization under this section, an applicant must—

(i) Meet the requirements of paragraphs (a) and (b) of this section; and

(ii) If the applicant has not passed a practical test for this authorization during the 12 calendar months preceding the month of the test, then that person must—

(A) Meet the requirements of § 61.57(c); and

(B) Have performed at least six ILS approaches during the 6 calendar months preceding the month of the test, of which at least three of the approaches must have been conducted without the use of an approach coupler.

(3) The approaches specified in paragraph (c)(2)(ii)(B) of this section—

(i) Must be conducted under actual or simulated instrument flight conditions;

(ii) Must be conducted to the decision height for the ILS approach in the type aircraft in which the practical test is to be conducted;

(iii) Need not be conducted to the decision height authorized for Category II operations;

(iv) Must be conducted to the decision height authorized for Category II operations only if conducted in a flight simulator or flight training device; and

(v) Must be accomplished in an aircraft of the same category and class, and type, as applicable, as the aircraft in which the practical test is to be conducted or in a flight simulator that—

(A) Represents an aircraft of the same category and class, and type, as applicable, as the aircraft in which the authorization is sought; and

(B) Is used in accordance with an approved course conducted by a training center certificated under part 142 of this chapter.

(4) The flight time acquired in meeting the requirements of paragraph (c)(2)(ii)(B) of this section may be used to meet the requirements of paragraph (c)(2)(ii)(A) of this section.

(d) *Practical test procedures.* The practical test consists of an oral increment and a flight increment.

(1) *Oral increment.* In the oral increment of the practical test an applicant must demonstrate knowledge of the following:

(i) Required landing distance;

(ii) Recognition of the decision height;

(iii) Missed approach procedures and techniques using computed or fixed attitude guidance displays;

(iv) Use and limitations of RVR;

(v) Use of visual clues, their availability or limitations, and altitude at which they are normally discernible at reduced RVR readings;

(vi) Procedures and techniques related to transition from nonvisual to visual flight during a final approach under reduced RVR;

(vii) Effects of vertical and horizontal windshear;

(viii) Characteristics and limitations of the ILS and runway lighting system;

(ix) Characteristics and limitations of the flight director system, auto approach coupler (including split axis type if equipped), auto throttle system (if equipped), and other required Category II equipment;

(x) Assigned duties of the second in command during Category II approaches, unless the aircraft for which authorization is sought does not require a second in command; and

(xi) Instrument and equipment failure warning systems.

(2) *Flight increment.* The following requirements apply to the flight increment of the practical test:

(i) The flight increment must be conducted in an aircraft of the same category, class, and type, as applicable, as the aircraft in which the authorization is sought or in a flight simulator that—

(A) Represents an aircraft of the same category and class, and type, as applicable, as the aircraft in which the authorization is sought; and

(B) Is used in accordance with an approved course conducted by a training center certificated under part 142 of this chapter.

(ii) The flight increment must consist of at least two ILS approaches to 100 feet AGL including at least one landing and one missed approach.

(iii) All approaches performed during the flight increment must be made with the use of an approved flight control guidance system, except if an approved auto approach coupler is installed, at least one approach must be hand flown using flight director commands.

(iv) If a multiengine airplane with the performance capability to execute a missed approach with one engine inoperative is used for the practical test, the flight increment must include the performance of one missed approach with an engine, which shall be the most critical engine, if applicable, set at idle or zero thrust before reaching the middle marker.

(v) If a multiengine flight simulator or multiengine flight training device is used for the practical test, the applicant must execute a missed approach with the most critical engine, if applicable, failed.

(vi) For an authorization for an aircraft that requires a type rating, the practical test must be performed in coordination with a second in command who holds a type rating in the aircraft in which the authorization is sought.

(vii) Oral questioning may be conducted at any time during a practical test.

[Doc. No. 25910, 62 FR 16298, Apr. 4, 1997; Amdt. 61-103, 62 FR 40900, July 30, 1997]

§ 61.68 Category III pilot authorization requirements.

(a) *General.* A person who applies for a Category III pilot authorization must hold:

(1) At least a private pilot certificate or commercial pilot certificate with an instrument rating or an airline transport pilot certificate;

(2) A type rating for the aircraft for which the authorization is sought if that aircraft requires a type rating; and

(3) A category and class rating for the aircraft for which the authorization is sought.

(b) *Experience requirements.* An applicant for a Category III pilot authorization must have at least—

(1) 50 hours of night flight time as pilot in command.

(2) 75 hours of instrument flight time during actual or simulated instrument conditions that may include not more than—

(i) A combination of 25 hours of simulated instrument flight time in a flight simulator or flight training device; or

(ii) 40 hours of simulated instrument flight time if accomplished in an approved course conducted by an appropriately rated training center certificated under part 142 of this chapter.

(3) 250 hours of cross-country flight time as pilot in command.

(c) *Practical test requirements.* (1) A practical test must be passed by a person who applies for—

(i) Issuance or renewal of a Category III pilot authorization; and

(ii) The addition of another type of aircraft to the applicant's Category III pilot authorization.

(2) To be eligible for the practical test for an authorization under this section, an applicant must—

(i) Meet the requirements of paragraphs (a) and (b) of this section; and

(ii) If the applicant has not passed a practical test for this authorization during the 12 calendar months preceding the month of the test, then that person must—

(A) Meet the requirements of § 61.57(c); and

(B) Have performed at least six ILS approaches during the 6 calendar months preceding the month of the test, of which at least three of the approaches must have been conducted without the use of an approach coupler.

(3) The approaches specified in paragraph (c)(2)(ii)(B) of this section—

(i) Must be conducted under actual or simulated instrument flight conditions;

(ii) Must be conducted to the alert height or decision height for the ILS approach in the type aircraft in which the practical test is to be conducted;

(iii) Need not be conducted to the decision height authorized for Category III operations;

(iv) Must be conducted to the alert height or decision height, as applicable, authorized for Category III operations only if conducted in a flight simulator or flight training device; and

(v) Must be accomplished in an aircraft of the same category and class, and type, as applicable, as the aircraft in which the practical test is to be conducted or in a flight simulator that—

(A) Represents an aircraft of the same category and class, and type, as applicable, as the aircraft for which the authorization is sought; and

(B) Is used in accordance with an approved course conducted by a training center certificated under part 142 of this chapter.

(4) The flight time acquired in meeting the requirements of paragraph (c)(2)(ii)(B) of this section may be used to meet the requirements of paragraph (c)(2)(ii)(A) of this section.

(d) *Practical test procedures.* The practical test consists of an oral increment and a flight increment.

(1) *Oral increment.* In the oral increment of the practical test an applicant must demonstrate knowledge of the following:

(i) Required landing distance;

(ii) Determination and recognition of the alert height or decision height, as applicable, including use of a radar altimeter;

(iii) Recognition of and proper reaction to significant failures encountered prior to and after reaching the alert height or decision height, as applicable;

(iv) Missed approach procedures and techniques using computed or fixed attitude guidance displays and expected height loss as they relate to manual go-around or automatic go-around, and initiation altitude, as applicable;

(v) Use and limitations of RVR, including determination of controlling RVR and required transmissometers;

(vi) Use, availability, or limitations of visual cues and the altitude at which they are normally discernible at reduced RVR readings including—

(A) Unexpected deterioration of conditions to less than minimum RVR during approach, flare, and rollout;

(B) Demonstration of expected visual references with weather at minimum conditions;

(C) The expected sequence of visual cues during an approach in which visibility is at or above landing minima; and

(D) Procedures and techniques for making a transition from instrument reference flight to visual flight during a final approach under reduced RVR.

(vii) Effects of vertical and horizontal windshear;

(viii) Characteristics and limitations of the ILS and runway lighting system;

(ix) Characteristics and limitations of the flight director system auto approach coupler (including split axis type if equipped), auto throttle system (if equipped), and other Category III equipment;

(x) Assigned duties of the second in command during Category III operations, unless the aircraft for which authorization is sought does not require a second in command;

(xi) Recognition of the limits of acceptable aircraft position and flight path tracking during approach, flare, and, if applicable, rollout; and

(xii) Recognition of, and reaction to, airborne or ground system faults or abnormalities, particularly after passing alert height or decision height, as applicable.

(2) *Flight increment.* The following requirements apply to the flight increment of the practical test—

(i) The flight increment may be conducted in an aircraft of the same category and class, and type, as applicable, as the aircraft for which the authorization is sought, or in a flight simulator that—

(A) Represents an aircraft of the same category and class, and type, as applicable, as the aircraft in which the authorization is sought; and

(B) Is used in accordance with an approved course conducted by a training center certificated under part 142 of this chapter.

(ii) The flight increment must consist of at least two ILS approaches to 100 feet AGL, including one landing and one missed approach initiated from a very low altitude that may result in a touchdown during the go-around maneuver;

(iii) All approaches performed during the flight increment must be made with the approved automatic landing system or an equivalent landing system approved by the Administrator;

(iv) If a multiengine aircraft with the performance capability to execute a missed approach with one engine inoperative is used for the practical test, the flight increment must include the performance of one missed approach with the most critical engine, if applicable, set at idle or zero thrust before reaching the middle or outer marker;

(v) If a multiengine flight simulator or multiengine flight training device is used, a missed approach must be executed with an engine, which shall be the most critical engine, if applicable, failed;

(vi) For an authorization for an aircraft that requires a type rating, the practical test must be performed in coordination with a second in command who holds a type rating in the aircraft in which the authorization is sought;

(vii) Oral questioning may be conducted at any time during the practical test;

(viii) Subject to the limitations of this paragraph, for Category IIIb operations predicated on the use of a fail-passive rollout control system, at least one manual rollout using visual reference or a combination of visual and instrument references must be executed. The maneuver required by this paragraph shall be initiated by a fail-passive disconnect of the rollout control system—

(A) After main gear touchdown;

(B) Prior to nose gear touchdown;

(C) In conditions representative of the most adverse lateral touchdown displacement allowing a safe landing on the runway; and

(D) In weather conditions anticipated in Category IIIb operations.

[Doc. No. 25910, 62 FR 16298, Apr. 4, 1997; Amdt. 61-103, 62 FR 40900, July 30, 1997]

§ 61.69 Glider and unpowered ultralight vehicle towing: Experience and training requirements.

(a) No person may act as pilot in command for towing a glider or unpowered ultralight vehicle unless that person—

(1) Holds a private, commercial or airline transport pilot certificate with a category rating for powered aircraft;

(2) Has logged at least 100 hours of pilot-in-command time in the aircraft category, class and type, if required, that the pilot is using to tow a glider or unpowered ultralight vehicle;

(3) Has a logbook endorsement from an authorized instructor who certifies that the person has received ground and flight training in gliders or unpowered ultralight vehicles and is proficient in—

(i) The techniques and procedures essential to the safe towing of gliders or unpowered ultralight vehicles, including airspeed limitations;

(ii) Emergency procedures;

(iii) Signals used; and

(iv) Maximum angles of bank.

(4) Except as provided in paragraph (b) of this section, has logged at least three flights as the sole manipulator of the controls of an aircraft while towing a glider or unpowered ultralight vehicle, or has simulated towing flight procedures in an aircraft while accompanied by a pilot who meets the requirements of paragraphs (c) and (d) of this section.

(5) Except as provided in paragraph (b) of this section, has received a logbook endorsement from the pilot, described in paragraph (a)(4) of this section, certifying that the person has accomplished at least 3 flights in an aircraft while towing a glider or unpowered ultralight vehicle, or while simulating towing flight procedures; and

(6) Within 24 calendar months before the flight has—

(i) Made at least three actual or simulated tows of a glider or unpowered ultralight vehicle while accompanied by a qualified pilot who meets the requirements of this section; or

(ii) Made at least three flights as pilot in command of a glider or unpowered ultralight vehicle towed by an aircraft.

(b) Any person who, before May 17, 1967, has made and logged 10 or more flights as pilot in command of an aircraft towing a glider or unpowered ultralight vehicle in accordance with a certificate of waiver need not comply with paragraphs (a)(4) and (a)(5) of this section.

(c) The pilot, described in paragraph (a)(4) of this section, who endorses the logbook of a person seeking towing privileges must have—

(1) Met the requirements of this section prior to endorsing the logbook of the person seeking towing privileges; and

(2) Logged at least 10 flights as pilot in command of an aircraft while towing a glider or unpowered ultralight vehicle.

(d) If the pilot described in paragraph (a)(4) of this section holds only a private pilot certificate, then that pilot must have—

(1) Logged at least 100 hours of pilot-in-command time in airplanes, or 200 hours of pilot-in-command time in a combination of powered and other-than-powered aircraft; and

(2) Performed and logged at least three flights within the 12 calendar months preceding the month that pilot accompanies or endorses the logbook of a person seeking towing privileges—

(i) In an aircraft while towing a glider or unpowered ultralight vehicle accompanied by another pilot who meets the requirements of this section; or

(ii) As pilot in command of a glider or unpowered ultralight vehicle being towed by another aircraft.

[Doc. No. FAA-2001-11133, 69 FR 44866, July 27, 2004, as amended by Amdt. 61-124, 74 FR 42555, Aug. 21, 2009]

§ 61.71 Graduates of an approved training program other than under this part: Special rules.

(a) A person who graduates from an approved training program under part 141 or part 142 of this chapter is considered to have met the applicable aeronautical experience, aeronautical knowledge, and areas of operation requirements of this part if that person presents the graduation certificate and passes the required practical test within the 60-day period after the date of graduation.

(b) A person may apply for an airline transport pilot certificate, type rating, or both under this part, and will be considered to have met the applicable requirements under § 61.157, except for the airline transport pilot certification training program required by § 61.156, for that certificate and rating, if that person has:

(1) Satisfactorily accomplished an approved training program and a proficiency check for that airplane type that includes all the tasks and maneuvers required to serve as pilot in command in accordance with the requirements of subparts N and O of part 121 of this chapter; and

(2) Applied for an airline transport pilot certificate, type rating, or both within the 60-day period from the date the person satisfactorily accomplished the requirements of paragraph (b)(1) for that airplane type.

(c) A person who holds a foreign pilot license and is applying for an equivalent U.S. pilot certificate on the basis of a Bilateral Aviation Safety Agreement and associated Implementation Procedures for Licensing may be considered to have met the applicable aeronautical experience, aeronautical knowledge, and areas of operation requirements of this part.

[Doc. No. 25910, 62 FR 16298, Apr. 4, 1997; Amdt. 61-103, 62 FR 40901, July 30, 1997; Amdt. 61-128, 76 FR 54107, Aug. 31, 2011; Amdt. 61-130, 78 FR 42374, July 15, 2013]

§ 61.73 Military pilots or former military pilots: Special rules.

(a) *General.* Except for a person who has been removed from flying status for lack of proficiency or because of a disciplinary action involving aircraft operations, a U.S. military pilot or former military pilot who meets the requirements of this section may apply, on the basis of his or her military pilot qualifications, for:

(1) A commercial pilot certificate with the appropriate aircraft category and class rating.

(2) An instrument rating with the appropriate aircraft rating.

(3) A type rating.

(b) *Military pilots and former military pilots in the U.S. Armed Forces.* A person who qualifies as a military pilot or former military pilot in the U.S. Armed Forces may apply for a pilot certificate and ratings under paragraph (a) of this section if that person—

(1) Presents evidentiary documents described under paragraphs (h)(1), (2), and (3) of this section that show the person's status in the U.S. Armed Forces.

(2) Has passed the military competency aeronautical knowledge test on the appropriate parts of this chapter for commercial pilot privileges and limitations, air traffic and general operating rules, and accident reporting rules.

(3) Presents official U.S. military records that show compliance with one of the following requirements—

(i) Before the date of the application, passing an official U.S. military pilot and instrument proficiency check in a military aircraft of the kind of aircraft category, class, and type, if class or type of aircraft is applicable, for the ratings sought; or

(ii) Before the date of application, logging 10 hours of pilot time as a military pilot in a U.S. military aircraft in the kind of aircraft category, class, and type, if a class rating or type rating is applicable, for the aircraft rating sought.

(c) *A military pilot in the Armed Forces of a foreign contracting State to the Convention on International Civil Aviation.* A person who is a military pilot in the Armed Forces of a foreign contracting State to the Convention on International Civil Aviation and is assigned to pilot duties in the U.S. Armed Forces, for purposes other than receiving flight training, may apply for a commercial pilot certificate and ratings under paragraph (a) of this section, provided that person—

(1) Presents evidentiary documents described under paragraph (h)(4) of this section that show the person is a military pilot in the Armed Forces of a foreign contracting State to the Convention on International Civil Aviation, and is assigned to pilot duties in the U.S. Armed Forces, for purposes other than receiving flight training.

(2) Has passed the military competency aeronautical knowledge test on the appropriate parts of this chapter for commercial pilot privileges and limitations, air traffic and general operating rules, and accident reporting rules.

(3) Presents official U.S. military records that show compliance with one of the following requirements:

(i) Before the date of the application, passed an official U.S. military pilot and instrument proficiency check in a military aircraft of the kind of aircraft category, class, or type, if class or type of aircraft is applicable, for the ratings; or

(ii) Before the date of the application, logged 10 hours of pilot time as a military pilot in a U.S. military aircraft of the kind of category, class, and type of aircraft, if a class rating or type rating is applicable, for the aircraft rating.

(d) *Instrument rating.* A person who is qualified as a U.S. military pilot or former military pilot may apply for an instrument rating to be added to a pilot certificate if that person—

(1) Has passed an instrument proficiency check in the U.S. Armed Forces in the aircraft category for the instrument rating sought; and

(2) Has an official U.S. Armed Forces record that shows the person is instrument pilot qualified by the U.S. Armed Forces to conduct instrument flying on Federal airways in that aircraft category and class for the instrument rating sought.

(e) *Aircraft type rating.* An aircraft type rating may only be issued for a type of aircraft that has a comparable civilian type designation by the Administrator.

(f) *Aircraft type rating placed on an airline transport pilot certificate.* A person who is a military pilot or former military pilot of the U.S. Armed Forces and requests an aircraft type rating to be placed on an existing U.S. airline transport pilot certificate may be issued the rating at the airline transport pilot certification level, provided that person:

(1) Holds a category and class rating for that type of aircraft at the airline transport pilot certification level; and

(2) Has passed an official U.S. military pilot check and instrument proficiency check in that type of aircraft.

(g) *Flight instructor certificate and ratings.* A person who can show official U.S. military documentation of being a U.S. military instructor pilot or U.S. military pilot examiner, or a former instructor pilot or pilot examiner may apply for and be issued a flight instructor certificate with the appropriate ratings if that person:

(1) Holds a commercial or airline transport pilot certificate with the appropriate aircraft category and class rating, if a class rating is appropriate, for the flight instructor rating sought;

(2) Holds an instrument rating, or has instrument privileges, on the pilot certificate that is appropriate to the flight instructor rating sought; and

(3) Presents the following documents:

(i) A knowledge test report that shows the person passed a knowledge test on the aeronautical knowledge areas listed under § 61.185(a) appropriate to the flight instructor rating sought and the knowledge test was passed within the preceding 24 calendar months prior to the month of application. If the U.S. military instructor pilot or pilot examiner already holds a flight instructor certificate, holding of a flight instructor certificate suffices for the knowledge test report.

(ii) An official U.S. Armed Forces record or order that shows the person is or was qualified as a U.S. Armed Forces military instructor pilot or pilot examiner for the flight instructor rating sought.

(iii) An official U.S. Armed Forces record or order that shows the person completed a U.S. Armed Forces' instructor pilot or pilot examiner training course and received an aircraft rating qualification as a military instructor pilot or pilot examiner that is appropriate to the flight instructor rating sought.

(iv) An official U.S. Armed Forces record or order that shows the person passed a U.S. Armed Forces instructor pilot or pilot examiner proficiency check in an aircraft as a military instructor pilot or pilot examiner that is appropriate to the flight instructor rating sought.

(h) *Documents for qualifying for a pilot certificate and rating.* The following documents are required for a person to apply for a pilot certificate and rating:

(1) An official U.S. Armed Forces record that shows the person is or was a military pilot.

(2) An official U.S. Armed Forces record that shows the person graduated from a U.S. Armed Forces undergraduate pilot training school and received a rating qualification as a military pilot.

(3) An official U.S. Armed Forces record that shows the pilot passed a pilot proficiency check and instrument proficiency check in an aircraft as a military pilot.

(4) If a person is a military pilot in the Armed Forces from a foreign contracting State to the Convention on International Civil Aviation and is applying for a pilot certificate and rating, that person must present the following:

(i) An official U.S. Armed Forces record that shows the person is a military pilot in the U.S. Armed Forces;

(ii) An official U.S. Armed Forces record that shows the person is assigned as a military pilot in the U.S. Armed Forces for purposes other than receiving flight training;

(iii) An official record that shows the person graduated from a military undergraduate pilot training school from the Armed Forces from a foreign contracting State to the Convention on International Civil Aviation or from the U.S. Armed Forces, and received a qualification as a military pilot; and

(iv) An official U.S. Armed Forces record that shows that the person passed a pilot proficiency check and instrument proficiency check in an aircraft as a military pilot in the U.S. Armed Forces.

[Doc. No. FAA-2006-26661, 74 FR 42555, Aug. 21, 2009]

§ 61.75 Private pilot certificate issued on the basis of a foreign pilot license.

(a) *General.* A person who holds a foreign pilot license at the private pilot level or higher that was issued by a contracting State to the Convention on International Civil Aviation may apply for and be issued a U.S. private pilot certificate with the appropriate ratings if the foreign pilot license meets the requirements of this section.

(b) *Certificate issued.* A U.S. private pilot certificate issued under this section must specify the person's foreign license number and country of issuance. A person who holds a foreign pilot license issued by a contracting State to the Convention on International Civil Aviation may be issued a U.S. private pilot certificate based on the foreign pilot license without any further showing of proficiency, provided the applicant:

(1) Meets the requirements of this section;

(2) Holds a foreign pilot license, at the private pilot license level or higher, that does not contain a limitation stating that the applicant has not met all of the standards of ICAO for that license;

(3) Does not hold a U.S. pilot certificate other than a U.S. student pilot certificate;

(4) Holds a medical certificate issued under part 67 of this chapter or a medical license issued by the country that issued the person's foreign pilot license; and

(5) Is able to read, speak, write, and understand the English language. If the applicant is unable to meet one of these requirements due to medical reasons, then the Administrator may place such operating limitations on that applicant's pilot certificate as are necessary for the safe operation of the aircraft.

(c) *Aircraft ratings issued.* Aircraft ratings listed on a person's foreign pilot license, in addition to any issued after testing under the provisions of this part, may be placed on that person's U.S. pilot certificate for private pilot privileges only.

(d) *Instrument ratings issued.* A person who holds an instrument rating on the foreign pilot license issued by a contracting State to the Convention on International Civil Aviation may be issued an instrument rating on a U.S. pilot certificate provided:

(1) The person's foreign pilot license authorizes instrument privileges;

(2) Within 24 months preceding the month in which the person applies for the instrument rating, the person passes the appropriate knowledge test; and

(3) The person is able to read, speak, write, and understand the English language. If the applicant is unable to meet one of these requirements due to medical reasons, then the Administrator may place such operating limitations on that applicant's pilot certificate as are necessary for the safe operation of the aircraft.

(e) *Operating privileges and limitations.* A person who receives a U.S. private pilot certificate that has been issued under the provisions of this section:

(1) May act as pilot in command of a civil aircraft of the United States in accordance with the pilot privileges authorized by this part and the limitations placed on that U.S. pilot certificate;

(2) Is limited to the privileges placed on the certificate by the Administrator;

(3) Is subject to the limitations and restrictions on the person's U.S. certificate and foreign pilot license when exercising the privileges of that U.S. pilot certificate in an aircraft of U.S. registry operating within or outside the United States; and

(f) *Limitation on licenses used as the basis for a U.S. certificate.* A person may use only one foreign pilot license as a basis for the issuance of a U.S. pilot certificate. The foreign pilot license and medical certification used as a basis for issuing a U.S. pilot certificate under this section must be written in English or accompanied by an English transcription that has been signed by an official or representative of the foreign aviation authority that issued the foreign pilot license.

(g) *Limitation placed on a U.S. pilot certificate.* A U.S. pilot certificate issued under this section can only be exercised when the pilot has the foreign pilot license, upon which the issuance of the U.S. pilot certificate was based, in the holder's possession or readily accessible in the aircraft.

[Doc. No. 25910, 62 FR 16298, Apr. 4, 1997, as amended by Amdt. 61-124, 74 FR 42556, Aug. 21, 2009]

§ 61.77 Special purpose pilot authorization: Operation of a civil aircraft of the United States and leased by a non-U.S. citizen.

(a) *General.* The holder of a foreign pilot license issued by a contracting State to the Convention on International Civil Aviation who meets the requirements of this section may be issued a special purpose pilot authorization by the Administrator for the purpose of performing pilot duties—

(1) On a civil aircraft of U.S. registry that is leased to a person who is not a citizen of the United States, and

(2) For carrying persons or property for compensation or hire for operations in—

(i) Scheduled international air services in turbojet-powered airplanes of U.S. registry;

(ii) Scheduled international air services in airplanes of U.S. registry having a configuration of more than nine passenger seats, excluding crewmember seats;

(iii) Nonscheduled international air transportation in airplanes of U.S. registry having a configuration of more than 30 passenger seats, excluding crewmember seats; or

(iv) Scheduled international air services, or nonscheduled international air transportation, in airplanes of U.S. registry having a payload capacity of more than 7,500 pounds.

(b) *Eligibility.* To be eligible for the issuance or renewal of a special purpose pilot authorization, an applicant must present the following to an FAA Flight Standards District Office:

(1) A foreign pilot license issued by the aeronautical authority of a contracting State to the Convention on International Civil Aviation that contains the appropriate aircraft category, class, type rating, if appropriate, and instrument rating for the aircraft to be flown;

(2) A certification by the lessee of the aircraft—

(i) Stating that the applicant is employed by the lessee;

(ii) Specifying the aircraft type on which the applicant will perform pilot duties; and

(iii) Stating that the applicant has received ground and flight instruction that qualifies the applicant to perform the duties to be assigned on the aircraft.

(3) Documentation showing when the applicant will reach the age of 65 years (an official copy of the applicant's birth certificate or other official documentation);

(4) Documentation the applicant meets the medical standards for the issuance of the foreign pilot license from the aeronautical authority of that contracting State to the Convention on International Civil Aviation; and

(5) A statement that the applicant does not already hold a special purpose pilot authorization; however, if the applicant already holds a special purpose pilot authorization, then that special purpose pilot authorization must be surrendered to either the FAA Flight Standards District Office that issued it, or the FAA Flight Standards District Office processing the application for the authorization, prior to being issued another special purpose pilot authorization.

(c) *Privileges.* A person issued a special purpose pilot authorization under this section—

(1) May exercise the privileges prescribed on the special purpose pilot authorization; and

(2) Must comply with the limitations specified in this section and any additional limitations specified on the special purpose pilot authorization.

(d) *General limitations.* A special purpose pilot authorization may be used only—

(1) For flights between foreign countries or for flights in foreign air commerce within the time period allotted on the authorization.

(2) If the foreign pilot license required by paragraph (b)(1) of this section, the medical documentation required by paragraph (b)(4) of this section, and the special purpose pilot authorization issued under this section are in the holder's physical possession or immediately accessible in the aircraft.

(3) While the holder is employed by the person to whom the aircraft described in the certification required by paragraph (b)(2) of this section is leased.

(4) While the holder is performing pilot duties on the U.S.-registered aircraft described in the certification required by paragraph (b)(2) of this section.

(5) If the holder has only one special purpose pilot authorization as provided in paragraph (b)(5) of this section.

(e) *Age limitation.* No person who holds a special purpose pilot authorization issued under this part, may serve as a pilot on a civil airplane of U.S. registry if the person has reached his or her 65th birthday, in the following operations:

(1) Scheduled international air services carrying passengers in turbojet-powered airplanes;

(2) Scheduled international air services carrying passengers in airplanes having a passenger-seat configuration of more than nine passenger seats, excluding each crewmember seat;

(3) Nonscheduled international air transportation for compensation or hire in airplanes having a passenger-seat configuration of more than 30 passenger seats, excluding each crewmember seat; or

(4) Scheduled international air services, or nonscheduled international air transportation for compensation or hire, in airplanes having a payload capacity of more than 7,500 pounds.

(f) *Definitions.* (1) *International air service,* as used in paragraph (e) of this section, means scheduled air service performed in airplanes for the public transport of passengers, mail, or cargo, in which the service passes through the air space over the territory of more than one country.

(2) *International air transportation,* as used in paragraph (e) of this section, means air transportation performed in airplanes for the public transport of passengers, mail, or cargo, in which service passes through the air space over the territory of more than one country.

(g) *Age Pairing Requirement.* No person who has attained the age of 60 but who has not attained the age of 65 may serve as a pilot in command in any of the operations described in § 61.3(j)(1)(i) through (iv) unless there is another pilot in the flight deck crew who has not yet attained 60 years of age.

(h) *Expiration date.* Each special purpose pilot authorization issued under this section expires—

(1) 60 calendar months from the month it was issued, unless sooner suspended or revoked;

(2) When the lease agreement for the aircraft expires or the lessee terminates the employment of the person who holds the special purpose pilot authorization;

(3) Whenever the person's foreign pilot license has been suspended, revoked, or is no longer valid; or

(4) When the person no longer meets the medical standards for the issuance of the foreign pilot license.

(i) *Renewal.* A person exercising the privileges of a special purpose pilot authorization may apply for a 60-calendar-month extension of that authorization, provided the person—

(1) Continues to meet the requirements of this section; and

(2) Surrenders the expired special purpose pilot authorization upon receipt of the new authorization.

(j) *Surrender.* The holder of a special purpose pilot authorization must surrender the authorization to the Administrator within 7 days after the date the authorization terminates.

[Doc. No. 25910, 62 FR 40901, July 30, 1997, as amended by Amdt. 61-123, 74 FR 34234, July 15, 2009; Amdt. 61-124, 74 FR 42557, Aug. 21, 2009]

Subpart C—Student Pilots

§ 61.81 Applicability.

This subpart prescribes the requirements for the issuance of student pilot certificates, the conditions under which those certificates are necessary, and the general operating rules and limitations for the holders of those certificates.

§ 61.83 Eligibility requirements for student pilots.

To be eligible for a student pilot certificate, an applicant must:

(a) Be at least 16 years of age for other than the operation of a glider or balloon.

(b) Be at least 14 years of age for the operation of a glider or balloon.

(c) Be able to read, speak, write, and understand the English language. If the applicant is unable to meet one of these requirements due to medical reasons, then the Administrator may place such operating limitations on that applicant's pilot certificate as are necessary for the safe operation of the aircraft.

§ 61.85 Application.

An application for a student pilot certificate is made on a form and in a manner provided by the Administrator and is submitted to:

(a) A designated aviation medical examiner if applying for an FAA medical certificate under part 67 of this chapter;

(b) An examiner; or

(c) A Flight Standards District Office.

§ 61.87 Solo requirements for student pilots.

(a) *General.* A student pilot may not operate an aircraft in solo flight unless that student has met the requirements of this section. The term "solo flight" as used in this subpart means that flight time during which a student pilot is the sole occupant of the aircraft or that flight time during which the student performs the duties of a pilot in command of a gas balloon or an airship requiring more than one pilot flight crewmember.

(b) *Aeronautical knowledge.* A student pilot must demonstrate satisfactory aeronautical knowledge on a knowledge test that meets the requirements of this paragraph:

(1) The test must address the student pilot's knowledge of—

(i) Applicable sections of parts 61 and 91 of this chapter;

(ii) Airspace rules and procedures for the airport where the solo flight will be performed; and

(iii) Flight characteristics and operational limitations for the make and model of aircraft to be flown.

(2) The student's authorized instructor must—

(i) Administer the test; and

(ii) At the conclusion of the test, review all incorrect answers with the student before authorizing that student to conduct a solo flight.

(c) *Pre-solo flight training.* Prior to conducting a solo flight, a student pilot must have:

(1) Received and logged flight training for the maneuvers and procedures of this section that are appropriate to the make and model of aircraft to be flown; and

(2) Demonstrated satisfactory proficiency and safety, as judged by an authorized instructor, on the maneuvers and procedures required by this section in the make and model of aircraft or similar make and model of aircraft to be flown.

(d) *Maneuvers and procedures for pre-solo flight training in a single-engine airplane.* A student pilot who is receiving training for a single-engine airplane rating or privileges must receive and log flight training for the following maneuvers and procedures:

(1) Proper flight preparation procedures, including preflight planning and preparation, powerplant operation, and aircraft systems;

(2) Taxiing or surface operations, including runups;

(3) Takeoffs and landings, including normal and crosswind;

(4) Straight and level flight, and turns in both directions;

(5) Climbs and climbing turns;

(6) Airport traffic patterns, including entry and departure procedures;

(7) Collision avoidance, windshear avoidance, and wake turbulence avoidance;

(8) Descents, with and without turns, using high and low drag configurations;

(9) Flight at various airspeeds from cruise to slow flight;

(10) Stall entries from various flight attitudes and power combinations with recovery initiated at the first indication of a stall, and recovery from a full stall;

(11) Emergency procedures and equipment malfunctions;

(12) Ground reference maneuvers;

(13) Approaches to a landing area with simulated engine malfunctions;

(14) Slips to a landing; and

(15) Go-arounds.

(e) *Maneuvers and procedures for pre-solo flight training in a multiengine airplane.* A student pilot who is receiving training for a multiengine airplane rating must receive and log flight training for the following maneuvers and procedures:

(1) Proper flight preparation procedures, including preflight planning and preparation, powerplant operation, and aircraft systems;

(2) Taxiing or surface operations, including runups;

(3) Takeoffs and landings, including normal and crosswind;
(4) Straight and level flight, and turns in both directions;
(5) Climbs and climbing turns;
(6) Airport traffic patterns, including entry and departure procedures;
(7) Collision avoidance, windshear avoidance, and wake turbulence avoidance;
(8) Descents, with and without turns, using high and low drag configurations;
(9) Flight at various airspeeds from cruise to slow flight;
(10) Stall entries from various flight attitudes and power combinations with recovery initiated at the first indication of a stall, and recovery from a full stall;
(11) Emergency procedures and equipment malfunctions;
(12) Ground reference maneuvers;
(13) Approaches to a landing area with simulated engine malfunctions; and
(14) Go-arounds.

(f) *Maneuvers and procedures for pre-solo flight training in a helicopter.* A student pilot who is receiving training for a helicopter rating must receive and log flight training for the following maneuvers and procedures:
(1) Proper flight preparation procedures, including preflight planning and preparation, powerplant operation, and aircraft systems;
(2) Taxiing or surface operations, including runups;
(3) Takeoffs and landings, including normal and crosswind;
(4) Straight and level flight, and turns in both directions;
(5) Climbs and climbing turns;
(6) Airport traffic patterns, including entry and departure procedures;
(7) Collision avoidance, windshear avoidance, and wake turbulence avoidance;
(8) Descents with and without turns;
(9) Flight at various airspeeds;
(10) Emergency procedures and equipment malfunctions;
(11) Ground reference maneuvers;
(12) Approaches to the landing area;
(13) Hovering and hovering turns;
(14) Go-arounds;
(15) Simulated emergency procedures, including autorotational descents with a power recovery and power recovery to a hover;
(16) Rapid decelerations; and
(17) Simulated one-engine-inoperative approaches and landings for multiengine helicopters.

(g) *Maneuvers and procedures for pre-solo flight training in a gyroplane.* A student pilot who is receiving training for a gyroplane rating or privileges must receive and log flight training for the following maneuvers and procedures:
(1) Proper flight preparation procedures, including preflight planning and preparation, powerplant operation, and aircraft systems;
(2) Taxiing or surface operations, including runups;
(3) Takeoffs and landings, including normal and crosswind;
(4) Straight and level flight, and turns in both directions;
(5) Climbs and climbing turns;
(6) Airport traffic patterns, including entry and departure procedures;
(7) Collision avoidance, windshear avoidance, and wake turbulence avoidance;
(8) Descents with and without turns;
(9) Flight at various airspeeds;
(10) Emergency procedures and equipment malfunctions;
(11) Ground reference maneuvers;
(12) Approaches to the landing area;
(13) High rates of descent with power on and with simulated power off, and recovery from those flight configurations;
(14) Go-arounds; and
(15) Simulated emergency procedures, including simulated power-off landings and simulated power failure during departures.

(h) *Maneuvers and procedures for pre-solo flight training in a powered-lift.* A student pilot who is receiving training for a powered-lift rating must receive and log flight training in the following maneuvers and procedures:
(1) Proper flight preparation procedures, including preflight planning and preparation, powerplant operation, and aircraft systems;
(2) Taxiing or surface operations, including runups;
(3) Takeoffs and landings, including normal and crosswind;
(4) Straight and level flight, and turns in both directions;
(5) Climbs and climbing turns;
(6) Airport traffic patterns, including entry and departure procedures;
(7) Collision avoidance, windshear avoidance, and wake turbulence avoidance;
(8) Descents with and without turns;
(9) Flight at various airspeeds from cruise to slow flight;
(10) Stall entries from various flight attitudes and power combinations with recovery initiated at the first indication of a stall, and recovery from a full stall;
(11) Emergency procedures and equipment malfunctions;
(12) Ground reference maneuvers;
(13) Approaches to a landing with simulated engine malfunctions;
(14) Go-arounds;
(15) Approaches to the landing area;
(16) Hovering and hovering turns; and
(17) For multiengine powered-lifts, simulated one-engine-inoperative approaches and landings.

(i) *Maneuvers and procedures for pre-solo flight training in a glider.* A student pilot who is receiving training for a glider rating or privileges must receive and log flight training for the following maneuvers and procedures:
(1) Proper flight preparation procedures, including preflight planning, preparation, aircraft systems, and, if appropriate, powerplant operations;
(2) Taxiing or surface operations, including runups, if applicable;
(3) Launches, including normal and crosswind;
(4) Straight and level flight, and turns in both directions, if applicable;
(5) Airport traffic patterns, including entry procedures;
(6) Collision avoidance, windshear avoidance, and wake turbulence avoidance;
(7) Descents with and without turns using high and low drag configurations;
(8) Flight at various airspeeds;
(9) Emergency procedures and equipment malfunctions;
(10) Ground reference maneuvers, if applicable;
(11) Inspection of towline rigging and review of signals and release procedures, if applicable;
(12) Aerotow, ground tow, or self-launch procedures;
(13) Procedures for disassembly and assembly of the glider;
(14) Stall entry, stall, and stall recovery;
(15) Straight glides, turns, and spirals;
(16) Landings, including normal and crosswind;
(17) Slips to a landing;
(18) Procedures and techniques for thermalling; and
(19) Emergency operations, including towline break procedures.

(j) *Maneuvers and procedures for pre-solo flight training in an airship.* A student pilot who is receiving training for an airship rating or privileges must receive and log flight training for the following maneuvers and procedures:
(1) Proper flight preparation procedures, including preflight planning and preparation, powerplant operation, and aircraft systems;
(2) Taxiing or surface operations, including runups;
(3) Takeoffs and landings, including normal and crosswind;
(4) Straight and level flight, and turns in both directions;
(5) Climbs and climbing turns;
(6) Airport traffic patterns, including entry and departure procedures;
(7) Collision avoidance, windshear avoidance, and wake turbulence avoidance;
(8) Descents with and without turns;
(9) Flight at various airspeeds from cruise to slow flight;
(10) Emergency procedures and equipment malfunctions;
(11) Ground reference maneuvers;
(12) Rigging, ballasting, and controlling pressure in the ballonets, and superheating; and
(13) Landings with positive and with negative static trim.

(k) *Maneuvers and procedures for pre-solo flight training in a balloon.* A student pilot who is receiving training in a balloon must receive and log flight training for the following maneuvers and procedures:
(1) Layout and assembly procedures;
(2) Proper flight preparation procedures, including preflight planning and preparation, and aircraft systems;
(3) Ascents and descents;
(4) Landing and recovery procedures;
(5) Emergency procedures and equipment malfunctions;
(6) Operation of hot air or gas source, ballast, valves, vents, and rip panels, as appropriate;
(7) Use of deflation valves or rip panels for simulating an emergency;
(8) The effects of wind on climb and approach angles; and
(9) Obstruction detection and avoidance techniques.

(l) *Maneuvers and procedures for pre-solo flight training in a powered parachute.* A student pilot who is receiving training for a powered parachute rating or privileges must receive and log flight training for the following maneuvers and procedures:
(1) Proper flight preparation procedures, including preflight planning and preparation, preflight assembly and rigging, aircraft systems, and powerplant operations.
(2) Taxiing or surface operations, including run-ups.
(3) Takeoffs and landings, including normal and crosswind.
(4) Straight and level flight, and turns in both directions.
(5) Climbs, and climbing turns in both directions.
(6) Airport traffic patterns, including entry and departure procedures.
(7) Collision avoidance, windshear avoidance, and wake turbulence avoidance.
(8) Descents, and descending turns in both directions.
(9) Emergency procedures and equipment malfunctions.
(10) Ground reference maneuvers.
(11) Straight glides, and gliding turns in both directions.
(12) Go-arounds.
(13) Approaches to landing areas with a simulated engine malfunction.
(14) Procedures for canopy packing and aircraft disassembly.

(m) *Maneuvers and procedures for pre-solo flight training in a weight-shift-control aircraft.* A student pilot who is receiving training for a weight-shift-control aircraft rating or privileges must receive and log flight training for the following maneuvers and procedures:
(1) Proper flight preparation procedures, including preflight planning and preparation, preflight assembly and rigging, aircraft systems, and powerplant operations.
(2) Taxiing or surface operations, including run-ups.
(3) Takeoffs and landings, including normal and crosswind.
(4) Straight and level flight, and turns in both directions.
(5) Climbs, and climbing turns in both directions.
(6) Airport traffic patterns, including entry and departure procedures.
(7) Collision avoidance, windshear avoidance, and wake turbulence avoidance.
(8) Descents, and descending turns in both directions.
(9) Flight at various airspeeds from maximum cruise to slow flight.
(10) Emergency procedures and equipment malfunctions.
(11) Ground reference maneuvers.
(12) Stall entry, stall, and stall recovery.
(13) Straight glides, and gliding turns in both directions.
(14) Go-arounds.
(15) Approaches to landing areas with a simulated engine malfunction.
(16) Procedures for disassembly.

(n) *Limitations on student pilots operating an aircraft in solo flight.* A student pilot may not operate an aircraft in solo flight unless that student pilot has received:
(1) An endorsement from an authorized instructor on his or her student pilot certificate for the specific make and model aircraft to be flown; and
(2) An endorsement in the student's logbook for the specific make and model aircraft to be flown by an authorized instructor, who gave the training within the 90 days preceding the date of the flight.

(o) *Limitations on student pilots operating an aircraft in solo flight at night.* A student pilot may not operate an aircraft in solo flight at night unless that student pilot has received:
(1) Flight training at night on night flying procedures that includes takeoffs, approaches, landings, and go-arounds at night at the airport where the solo flight will be conducted;
(2) Navigation training at night in the vicinity of the airport where the solo flight will be conducted; and
(3) An endorsement in the student's logbook for the specific make and model aircraft to be flown for night solo flight by an authorized instructor who gave the training within the 90-day period preceding the date of the flight.

(p) *Limitations on flight instructors authorizing solo flight.* No instructor may authorize a student pilot to perform a solo flight unless that instructor has—

(1) Given that student pilot training in the make and model of aircraft or a similar make and model of aircraft in which the solo flight is to be flown;

(2) Determined the student pilot is proficient in the maneuvers and procedures prescribed in this section;

(3) Determined the student pilot is proficient in the make and model of aircraft to be flown;

(4) Ensured that the student pilot's certificate has been endorsed by an instructor authorized to provide flight training for the specific make and model aircraft to be flown; and

(5) Endorsed the student pilot's logbook for the specific make and model aircraft to be flown, and that endorsement remains current for solo flight privileges, provided an authorized instructor updates the student's logbook every 90 days thereafter.

[Doc. No. 25910, 62 FR 16298, Apr. 4, 1997; Amdt. 61-103, 62 FR 40902, July 30, 1997; Amdt. 61-104, 63 FR 20287, Apr. 23, 1998; Amdt. 61-110, 69 FR 44866, July 27, 2004; Amdt. 61-124, 74 FR 42557, Aug. 21, 2009]

§ 61.89 General limitations.

(a) A student pilot may not act as pilot in command of an aircraft:

(1) That is carrying a passenger;

(2) That is carrying property for compensation or hire;

(3) For compensation or hire;

(4) In furtherance of a business;

(5) On an international flight, except that a student pilot may make solo training flights from Haines, Gustavus, or Juneau, Alaska, to White Horse, Yukon, Canada, and return over the province of British Columbia;

(6) With a flight or surface visibility of less than 3 statute miles during daylight hours or 5 statute miles at night;

(7) When the flight cannot be made with visual reference to the surface; or

(8) In a manner contrary to any limitations placed in the pilot's logbook by an authorized instructor.

(b) A student pilot may not act as a required pilot flight crewmember on any aircraft for which more than one pilot is required by the type certificate of the aircraft or regulations under which the flight is conducted, except when receiving flight training from an authorized instructor on board an airship, and no person other than a required flight crewmember is carried on the aircraft.

(c) A student pilot seeking a sport pilot certificate must comply with the provisions of paragraphs (a) and (b) of this section and may not act as pilot in command—

(1) Of an aircraft other than a light-sport aircraft;

(2) At night;

(3) At an altitude of more than 10,000 feet MSL or 2,000 feet AGL, whichever is higher;

(4) In Class B, C, and D airspace, at an airport located in Class B, C, or D airspace, and to, from, through, or on an airport having an operational control tower without having received the ground and flight training specified in § 61.94 and an endorsement from an authorized instructor;

(5) Of a light-sport aircraft without having received the applicable ground training, flight training, and instructor endorsements specified in § 61.327 (a) and (b).

[Doc. No. 25910, 62 FR 16298, Apr. 4, 1997, as amended by Amdt. 61-110, 69 FR 44867, July 27, 2004; Amdt. 61-125, 75 FR 5220, Feb. 1, 2010]

§ 61.91 [Reserved]

§ 61.93 Solo cross-country flight requirements.

(a) *General.* (1) Except as provided in paragraph (b) of this section, a student pilot must meet the requirements of this section before—

(i) Conducting a solo cross-country flight, or any flight greater than 25 nautical miles from the airport from where the flight originated.

(ii) Making a solo flight and landing at any location other than the airport of origination.

(2) Except as provided in paragraph (b) of this section, a student pilot who seeks solo cross-country flight privileges must:

(i) Have received flight training from an instructor authorized to provide flight training on the maneuvers and procedures of this section that are appropriate to the make and model of aircraft for which solo cross-country privileges are sought;

(ii) Have demonstrated cross-country proficiency on the appropriate maneuvers and procedures of this section to an authorized instructor;

(iii) Have satisfactorily accomplished the pre-solo flight maneuvers and procedures required by § 61.87 of this part in the make and model of aircraft or similar make and model of aircraft for which solo cross-country privileges are sought; and

(iv) Comply with any limitations included in the authorized instructor's endorsement that are required by paragraph (c) of this section.

(3) A student pilot who seeks solo cross-country flight privileges must have received ground and flight training from an authorized instructor on the cross-country maneuvers and procedures listed in this section that are appropriate to the aircraft to be flown.

(b) *Authorization to perform certain solo flights and cross-country flights.* A student pilot must obtain an endorsement from an authorized instructor to make solo flights from the airport where the student pilot normally receives training to another location. A student pilot who receives this endorsement must comply with the requirements of this paragraph.

(1) Solo flights may be made to another airport that is within 25 nautical miles from the airport where the student pilot normally receives training, provided—

(i) An authorized instructor has given the student pilot flight training at the other airport, and that training includes flight in both directions over the route, entering and exiting the traffic pattern, and takeoffs and landings at the other airport;

(ii) The authorized instructor who gave the training endorses the student pilot's logbook authorizing the flight;

(iii) The student pilot has a solo flight endorsement in accordance with § 61.87 of this part;

(iv) The authorized instructor has determined that the student pilot is proficient to make the flight; and

(v) The purpose of the flight is to practice takeoffs and landings at that other airport.

(2) Repeated specific solo cross-country flights may be made to another airport that is within 50 nautical miles of the airport from which the flight originated, provided—

(i) The authorized instructor has given the student flight training in both directions over the route, including entering and exiting the traffic patterns, takeoffs, and landings at the airports to be used;

(ii) The authorized instructor who gave the training has endorsed the student's logbook certifying that the student is proficient to make such flights;

(iii) The student has a solo flight endorsement in accordance with § 61.87 of this part; and

(iv) The student has a solo cross country flight endorsement in accordance with paragraph (c) of this section; however, for repeated solo cross country flights to another airport within 50 nautical miles from which the flight originated, separate endorsements are not required to be made for each flight.

(c) *Endorsements for solo cross-country flights.* Except as specified in paragraph (b)(2) of this section, a student pilot must have the endorsements prescribed in this paragraph for each cross-country flight:

(1) *Student pilot certificate endorsement.* A student pilot must have a solo cross-country endorsement from the authorized instructor who conducted the training, and that endorsement must be placed on that person's student pilot certificate for the specific category of aircraft to be flown.

(2) *Logbook endorsement.* (i) A student pilot must have a solo cross-country endorsement from an authorized instructor that is placed in the student pilot's logbook for the specific make and model of aircraft to be flown.

(ii) For each cross-country flight, the authorized instructor who reviews the cross-country planning must make an endorsement in the person's logbook after reviewing that person's cross-country planning, as specified in paragraph (d) of this section. The endorsement must—

(A) Specify the make and model of aircraft to be flown;

(B) State that the student's preflight planning and preparation is correct and that the student is prepared to make the flight safely under the known conditions; and

(C) State that any limitations required by the student's authorized instructor are met.

(d) *Limitations on authorized instructors to permit solo cross-country flights.* An authorized instructor may not permit a student pilot to conduct a solo cross-country flight unless that instructor has:

(1) Determined that the student's cross-country planning is correct for the flight;

(2) Reviewed the current and forecast weather conditions and has determined that the flight can be completed under VFR;

(3) Determined that the student is proficient to conduct the flight safely;

(4) Determined that the student has the appropriate solo cross-country endorsement for the make and model of aircraft to be flown; and

(5) Determined that the student's solo flight endorsement is current for the make and model aircraft to be flown.

(e) *Maneuvers and procedures for cross-country flight training in a single-engine airplane.* A student pilot who is receiving training for cross-country flight in a single-engine airplane must receive and log flight training in the following maneuvers and procedures:

(1) Use of aeronautical charts for VFR navigation using pilotage and dead reckoning with the aid of a magnetic compass;

(2) Use of aircraft performance charts pertaining to cross-country flight;

(3) Procurement and analysis of aeronautical weather reports and forecasts, including recognition of critical weather situations and estimating visibility while in flight;

(4) Emergency procedures;

(5) Traffic pattern procedures that include area departure, area arrival, entry into the traffic pattern, and approach;

(6) Procedures and operating practices for collision avoidance, wake turbulence precautions, and windshear avoidance;

(7) Recognition, avoidance, and operational restrictions of hazardous terrain features in the geographical area where the cross-country flight will be flown;

(8) Procedures for operating the instruments and equipment installed in the aircraft to be flown, including recognition and use of the proper operational procedures and indications;

(9) Use of radios for VFR navigation and two-way communication, except that a student pilot seeking a sport pilot certificate must only receive and log flight training on the use of radios installed in the aircraft to be flown;

(10) Takeoff, approach, and landing procedures, including short-field, soft-field, and crosswind takeoffs, approaches, and landings;

(11) Climbs at best angle and best rate; and

(12) Control and maneuvering solely by reference to flight instruments, including straight and level flight, turns, descents, climbs, use of radio aids, and ATC directives. For student pilots seeking a sport pilot certificate, the provisions of this paragraph only apply when receiving training for cross-country flight in an airplane that has a VH greater than 87 knots CAS.

(f) *Maneuvers and procedures for cross-country flight training in a multiengine airplane.* A student pilot who is receiving training for cross-country flight in a multiengine airplane must receive and log flight training in the following maneuvers and procedures:

(1) Use of aeronautical charts for VFR navigation using pilotage and dead reckoning with the aid of a magnetic compass;

(2) Use of aircraft performance charts pertaining to cross-country flight;

(3) Procurement and analysis of aeronautical weather reports and forecasts, including recognition of critical weather situations and estimating visibility while in flight;

(4) Emergency procedures;

(5) Traffic pattern procedures that include area departure, area arrival, entry into the traffic pattern, and approach;

(6) Procedures and operating practices for collision avoidance, wake turbulence precautions, and windshear avoidance;

(7) Recognition, avoidance, and operational restrictions of hazardous terrain features in the geographical area where the cross-country flight will be flown;

(8) Procedures for operating the instruments and equipment installed in the aircraft to be flown, including recognition and use of the proper operational procedures and indications;

(9) Use of radios for VFR navigation and two-way communications;

(10) Takeoff, approach, and landing procedures, including short-field, soft-field, and crosswind takeoffs, approaches, and landings;

(11) Climbs at best angle and best rate; and

(12) Control and maneuvering solely by reference to flight instruments, including straight and level flight, turns, descents, climbs, use of radio aids, and ATC directives.

(g) *Maneuvers and procedures for cross-country flight training in a helicopter.* A student pilot who is receiving training for cross-country flight in a helicopter must receive and log flight training for the following maneuvers and procedures:

(1) Use of aeronautical charts for VFR navigation using pilotage and dead reckoning with the aid of a magnetic compass;

(2) Use of aircraft performance charts pertaining to cross-country flight;

(3) Procurement and analysis of aeronautical weather reports and forecasts, including recognition of critical weather situations and estimating visibility while in flight;

(4) Emergency procedures;

(5) Traffic pattern procedures that include area departure, area arrival, entry into the traffic pattern, and approach;

(6) Procedures and operating practices for collision avoidance, wake turbulence precautions, and windshear avoidance;

(7) Recognition, avoidance, and operational restrictions of hazardous terrain features in the geographical area where the cross-country flight will be flown;

(8) Procedures for operating the instruments and equipment installed in the aircraft to be flown, including recognition and use of the proper operational procedures and indications;

(9) Use of radios for VFR navigation and two-way communications; and

(10) Takeoff, approach, and landing procedures.

(h) *Maneuvers and procedures for cross-country flight training in a gyroplane.* A student pilot who is receiving training for cross-country flight in a gyroplane must receive and log flight training in the following maneuvers and procedures:

(1) Use of aeronautical charts for VFR navigation using pilotage and dead reckoning with the aid of a magnetic compass;

(2) Use of aircraft performance charts pertaining to cross-country flight;

(3) Procurement and analysis of aeronautical weather reports and forecasts, including recognition of critical weather situations and estimating visibility while in flight;

(4) Emergency procedures;

(5) Traffic pattern procedures that include area departure, area arrival, entry into the traffic pattern, and approach;

(6) Procedures and operating practices for collision avoidance, wake turbulence precautions, and windshear avoidance;

(7) Recognition, avoidance, and operational restrictions of hazardous terrain features in the geographical area where the cross-country flight will be flown;

(8) Procedures for operating the instruments and equipment installed in the aircraft to be flown, including recognition and use of the proper operational procedures and indications;

(9) Use of radios for VFR navigation and two-way communication, except that a student pilot seeking a sport pilot certificate must only receive and log flight training on the use of radios installed in the aircraft to be flown; and

(10) Takeoff, approach, and landing procedures, including short-field and soft-field takeoffs, approaches, and landings.

(i) *Maneuvers and procedures for cross-country flight training in a powered-lift.* A student pilot who is receiving training for cross-country flight training in a powered-lift must receive and log flight training in the following maneuvers and procedures:

(1) Use of aeronautical charts for VFR navigation using pilotage and dead reckoning with the aid of a magnetic compass;

(2) Use of aircraft performance charts pertaining to cross-country flight;

(3) Procurement and analysis of aeronautical weather reports and forecasts, including recognition of critical weather situations and estimating visibility while in flight;

(4) Emergency procedures;

(5) Traffic pattern procedures that include area departure, area arrival, entry into the traffic pattern, and approach;

(6) Procedures and operating practices for collision avoidance, wake turbulence precautions, and windshear avoidance;

(7) Recognition, avoidance, and operational restrictions of hazardous terrain features in the geographical area where the cross-country flight will be flown;

(8) Procedures for operating the instruments and equipment installed in the aircraft to be flown, including recognition and use of the proper operational procedures and indications;

(9) Use of radios for VFR navigation and two-way communications;

(10) Takeoff, approach, and landing procedures that include high-altitude, steep, and shallow takeoffs, approaches, and landings; and

(11) Control and maneuvering solely by reference to flight instruments, including straight and level flight, turns, descents, climbs, use of radio aids, and ATC directives.

(j) *Maneuvers and procedures for cross-country flight training in a glider.* A student pilot who is receiving training for cross-country flight in a glider must receive and log flight training in the following maneuvers and procedures:

(1) Use of aeronautical charts for VFR navigation using pilotage and dead reckoning with the aid of a magnetic compass;

(2) Use of aircraft performance charts pertaining to cross-country flight;

(3) Procurement and analysis of aeronautical weather reports and forecasts, including recognition of critical weather situations and estimating visibility while in flight;

(4) Emergency procedures;

(5) Traffic pattern procedures that include area departure, area arrival, entry into the traffic pattern, and approach;

(6) Procedures and operating practices for collision avoidance, wake turbulence precautions, and windshear avoidance;

(7) Recognition, avoidance, and operational restrictions of hazardous terrain features in the geographical area where the cross-country flight will be flown;

(8) Procedures for operating the instruments and equipment installed in the aircraft to be flown, including recognition and use of the proper operational procedures and indications;

(9) Landings accomplished without the use of the altimeter from at least 2,000 feet above the surface; and

(10) Recognition of weather and upper air conditions favorable for cross-country soaring, ascending and descending flight, and altitude control.

(k) *Maneuvers and procedures for cross-country flight training in an airship.* A student pilot who is receiving training for cross-country flight in an airship must receive and log flight training for the following maneuvers and procedures:

(1) Use of aeronautical charts for VFR navigation using pilotage and dead reckoning with the aid of a magnetic compass;

(2) Use of aircraft performance charts pertaining to cross-country flight;

(3) Procurement and analysis of aeronautical weather reports and forecasts, including recognition of critical weather situations and estimating visibility while in flight;

(4) Emergency procedures;

(5) Traffic pattern procedures that include area departure, area arrival, entry into the traffic pattern, and approach;

(6) Procedures and operating practices for collision avoidance, wake turbulence precautions, and windshear avoidance;

(7) Recognition, avoidance, and operational restrictions of hazardous terrain features in the geographical area where the cross-country flight will be flown;

(8) Procedures for operating the instruments and equipment installed in the aircraft to be flown, including recognition and use of the proper operational procedures and indications;

(9) Use of radios for VFR navigation and two-way communication, except that a student pilot seeking a sport pilot certificate must only receive and log flight training on the use of radios installed in the aircraft to be flown;

(10) Control of air pressure with regard to ascending and descending flight and altitude control;

(11) Control of the airship solely by reference to flight instruments, except for a student pilot seeking a sport pilot certificate; and

(12) Recognition of weather and upper air conditions conducive for the direction of cross-country flight.

(l) *Maneuvers and procedures for cross-country flight training in a powered parachute.* A student pilot who is receiving training for cross-country flight in a powered parachute must receive and log flight training in the following maneuvers and procedures:

(1) Use of aeronautical charts for VFR navigation using pilotage and dead reckoning with the aid of a magnetic compass, as appropriate.

(2) Use of aircraft performance charts pertaining to cross-country flight.

(3) Procurement and analysis of aeronautical weather reports and forecasts, including recognizing critical weather situations and estimating visibility while in flight.

(4) Emergency procedures.

(5) Traffic pattern procedures that include area departure, area arrival, entry into the traffic pattern, and approach.

(6) Procedures and operating practices for collision avoidance, wake turbulence precautions, and windshear avoidance.

(7) Recognition, avoidance, and operational restrictions of hazardous terrain features in the geographical area where the cross-country flight will be flown.

(8) Procedures for operating the instruments and equipment installed in the aircraft to be flown, including recognition and use of the proper operational procedures and indications.

(9) If equipped for flight with navigation radios, the use of radios for VFR navigation.

(10) Recognition of weather and upper air conditions favorable for the cross-country flight.

(11) Takeoff, approach and landing procedures.

(m) *Maneuvers and procedures for cross-country flight training in a weight-shift-control aircraft.* A student pilot who is receiving training for cross-country flight in a weight-shift-control aircraft must receive and log flight training for the following maneuvers and procedures:

(1) Use of aeronautical charts for VFR navigation using pilotage and dead reckoning with the aid of a magnetic compass, as appropriate.

(2) Use of aircraft performance charts pertaining to cross-country flight.

(3) Procurement and analysis of aeronautical weather reports and forecasts, including recognizing critical weather situations and estimating visibility while in flight.

(4) Emergency procedures.

(5) Traffic pattern procedures that include area departure, area arrival, entry into the traffic pattern, and approach.

(6) Procedures and operating practices for collision avoidance, wake turbulence precautions, and windshear avoidance.

(7) Recognition, avoidance, and operational restrictions of hazardous terrain features in the geographical area where the cross-country flight will be flown.

(8) Procedures for operating the instruments and equipment installed in the aircraft to be flown, including recognition and use of the proper operational procedures and indications.

(9) If equipped for flight using navigation radios, the use of radios for VFR navigation.

(10) Recognition of weather and upper air conditions favorable for the cross-country flight.

(11) Takeoff, approach and landing procedures, including crosswind approaches and landings.

[Doc. No. 25910, 62 FR 16298, Apr. 4, 1997; Amdt. 61-103, 62 FR 40902, July 30, 1997; Amdt. 61-110, 69 FR 44867, July 27, 2004; Amdt. 61-124, 74 FR 42557, Aug. 21, 2009; Amdt. 61-125, 75 FR 5220, Feb. 1, 2010]

§ 61.94 Student pilot seeking a sport pilot certificate or a recreational pilot certificate: Operations at airports within, and in airspace located within, Class B, C, and D airspace, or at airports with an operational control tower in other airspace.

(a) A student pilot seeking a sport pilot certificate or a recreational pilot certificate who wants to obtain privileges to operate in Class B, C, and D airspace, at an airport located in Class B, C, or D airspace, and to, from, through, or at an airport having an operational control tower, must receive and log ground and flight training from an authorized instructor in the following aeronautical knowledge areas and areas of operation:

(1) The use of radios, communications, navigation systems and facilities, and radar services.

(2) Operations at airports with an operating control tower, to include three takeoffs and landings to a full stop, with each landing involving a flight in the traffic pattern, at an airport with an operating control tower.

(3) Applicable flight rules of part 91 of this chapter for operations in Class B, C, and D airspace and air traffic control clearances.

(4) Ground and flight training for the specific Class B, C, or D airspace for which the solo flight is authorized, if applicable, within the 90-day period preceding the date of the flight in that airspace. The flight training must be received in the specific airspace area for which solo flight is authorized.

(5) Ground and flight training for the specific airport located in Class B, C, or D airspace for which the solo flight is authorized, if applicable, within the 90-day period preceding the date of the flight at that airport. The flight and ground training must be received at the specific airport for which solo flight is authorized.

(b) The authorized instructor who provides the training specified in paragraph (a) of this section must provide a logbook endorsement that certifies the student has received that training and is proficient to conduct solo flight in that specific airspace or at that specific airport and in those aeronautical knowledge areas and areas of operation specified in this section.

[Doc. No. FAA-2001-11133, 69 FR 44867, July 27, 2004]

§ 61.95 Operations in Class B airspace and at airports located within Class B airspace.

(a) A student pilot may not operate an aircraft on a solo flight in Class B airspace unless:

(1) The student pilot has received both ground and flight training from an authorized instructor on that Class B airspace area, and the flight training was received in the specific Class B airspace area for which solo flight is authorized;

(2) The logbook of that student pilot has been endorsed by the authorized instructor who gave the student pilot flight training, and the endorsement is dated within the 90-day period preceding the date of the flight in that Class B airspace area; and

(3) The logbook endorsement specifies that the student pilot has received the required ground and flight training, and has been found proficient to conduct solo flight in that specific Class B airspace area.

(b) A student pilot may not operate an aircraft on a solo flight to, from, or at an airport located within Class B airspace pursuant to § 91.131(b) of this chapter unless:

(1) The student pilot has received both ground and flight training from an instructor authorized to provide training to operate at that airport, and the flight and ground training has been received at the specific airport for which the solo flight is authorized;

(2) The logbook of that student pilot has been endorsed by an authorized instructor who gave the student pilot flight training, and the endorsement is dated within the 90-day period preceding the date of the flight at that airport; and

(3) The logbook endorsement specifies that the student pilot has received the required ground and flight training, and has been found proficient to conduct solo flight operations at that specific airport.

(c) This section does not apply to a student pilot seeking a sport pilot certificate or a recreational pilot certificate.

[Doc. No. 25910, 62 FR 16298, Apr. 4, 1997; Amdt. 61-103, 62 FR 40902, July 30, 1997; Amdt. 61-110, 69 FR 44868, July 27, 2004]

Subpart D—Recreational Pilots

§ 61.96 Applicability and eligibility requirements: General.

(a) This subpart prescribes the requirement for the issuance of recreational pilot certificates and ratings, the conditions under which those certificates and ratings are necessary, and the general operating rules for persons who hold those certificates and ratings.

(b) To be eligible for a recreational pilot certificate, a person who applies for that certificate must:

(1) Be at least 17 years of age;

(2) Be able to read, speak, write, and understand the English language. If the applicant is unable to meet one of these requirements due to medical reasons, then the Administrator may place such operating limitations on that applicant's pilot certificate as are necessary for the safe operation of the aircraft;

(3) Receive a logbook endorsement from an authorized instructor who—

(i) Conducted the training or reviewed the applicant's home study on the aeronautical knowledge areas listed in § 61.97(b) of this part that apply to the aircraft category and class rating sought; and

(ii) Certified that the applicant is prepared for the required knowledge test.

(4) Pass the required knowledge test on the aeronautical knowledge areas listed in § 61.97(b) of this part;

(5) Receive flight training and a logbook endorsement from an authorized instructor who—

(i) Conducted the training on the areas of operation listed in § 61.98(b) of this part that apply to the aircraft category and class rating sought; and

(ii) Certified that the applicant is prepared for the required practical test.

(6) Meet the aeronautical experience requirements of § 61.99 of this part that apply to the aircraft category and class rating sought before applying for the practical test;

(7) Pass the practical test on the areas of operation listed in § 61.98(b) that apply to the aircraft category and class rating;

(8) Comply with the sections of this part that apply to the aircraft category and class rating; and

(9) Hold either a student pilot certificate or sport pilot certificate.

[Doc. No. 25910, 62 FR 16298, Apr. 4, 1997; Amdt. 61-103, 62 FR 40902, July 30, 1997; Amdt. 61-124, 74 FR 42558, Aug. 21, 2009]

§ 61.97 Aeronautical knowledge.

(a) *General.* A person who applies for a recreational pilot certificate must receive and log ground training from an authorized instructor or complete a home-study course on the aeronautical knowledge areas of paragraph (b) of this section that apply to the aircraft category and class rating sought.

(b) *Aeronautical knowledge areas.* (1) Applicable Federal Aviation Regulations of this chapter that relate to recreational pilot privileges, limitations, and flight operations;

(2) Accident reporting requirements of the National Transportation Safety Board;

(3) Use of the applicable portions of the "Aeronautical Information Manual" and FAA advisory circulars;

(4) Use of aeronautical charts for VFR navigation using pilotage with the aid of a magnetic compass;

(5) Recognition of critical weather situations from the ground and in flight, windshear avoidance, and the procurement and use of aeronautical weather reports and forecasts;

(6) Safe and efficient operation of aircraft, including collision avoidance, and recognition and avoidance of wake turbulence;

(7) Effects of density altitude on takeoff and climb performance;

(8) Weight and balance computations;

(9) Principles of aerodynamics, powerplants, and aircraft systems;

(10) Stall awareness, spin entry, spins, and spin recovery techniques, if applying for an airplane single-engine rating;

(11) Aeronautical decision making and judgment; and

(12) Preflight action that includes—

(i) How to obtain information on runway lengths at airports of intended use, data on takeoff and landing distances, weather reports and forecasts, and fuel requirements; and

(ii) How to plan for alternatives if the planned flight cannot be completed or delays are encountered.

[Doc. No. 25910, 62 FR 16298, Apr. 4, 1997; Amdt. 61-103, 62 FR 40902, July 30, 1997]

§ 61.98 Flight proficiency.

(a) *General.* A person who applies for a recreational pilot certificate must receive and log ground and flight training from an authorized instructor on the areas of operation of this section that apply to the aircraft category and class rating sought.

(b) *Areas of operation.* (1) *For a single-engine airplane rating:* (i) Preflight preparation;

(ii) Preflight procedures;

(iii) Airport operations;

(iv) Takeoffs, landings, and go-arounds;

(v) Performance maneuvers;

(vi) Ground reference maneuvers;

(vii) Navigation;

(viii) Slow flight and stalls;

(ix) Emergency operations; and

(x) Postflight procedures.

(2) *For a helicopter rating:* (i) Preflight preparation;

(ii) Preflight procedures;

(iii) Airport and heliport operations;

(iv) Hovering maneuvers;

(v) Takeoffs, landings, and go-arounds;

(vi) Performance maneuvers;

(vii) Ground reference maneuvers;

(viii) Navigation;

(ix) Emergency operations; and

(x) Postflight procedures.

(3) *For a gyroplane rating:* (i) Preflight preparation;

(ii) Preflight procedures;

(iii) Airport operations;

(iv) Takeoffs, landings, and go-arounds;

(v) Performance maneuvers;

(vi) Ground reference maneuvers;

(vii) Navigation;

(viii) Flight at slow airspeeds;

(ix) Emergency operations; and

(x) Postflight procedures.

[Doc. No. 25910, 62 FR 16298, Apr. 4, 1997; Amdt. 61-103, 62 FR 40902, July 30, 1997]

§ 61.99 Aeronautical experience.

A person who applies for a recreational pilot certificate must receive and log at least 30 hours of flight time that includes at least—

(a) 15 hours of flight training from an authorized instructor on the areas of operation listed in § 61.98 of this part that consists of at least:

(1) Except as provided in § 61.100 of this part, 2 hours of flight training en route to an airport that is located more than 25 nautical miles from the airport where the applicant normally trains, which includes at least three takeoffs and three landings at the airport located more than 25 nautical miles from the airport where the applicant normally trains; and

(2) Three hours of flight training with an authorized instructor in the aircraft for the rating sought in preparation for the practical test within the preceding 2 calendar months from the month of the test.

(b) 3 hours of solo flying in the aircraft for the rating sought, on the areas of operation listed in § 61.98 of this part that apply to the aircraft category and class rating sought.

[Doc. No. 25910, 62 FR 16298, Apr. 4, 1997, as amended by Amdt. 61-110, 69 FR 44868, July 27, 2004; Amdt. 61-124A, 74 FR 53645, Oct. 20, 2009]

§ 61.100 Pilots based on small islands.

(a) An applicant located on an island from which the flight training required in § 61.99(a)(1) of this part cannot be accomplished without flying over water for more than 10 nautical miles from the nearest shoreline need not comply with the requirements of that section. However, if other airports that permit civil operations are available to which a flight may be made without flying over water for more than 10 nautical miles from the nearest shoreline, the applicant must show completion of a dual flight between two airports, which must include three landings at the other airport.

(b) An applicant who complies with paragraph (a) of this section and meets all requirements for the issuance of a recreational pilot certificate, except the requirements of § 61.99(a)(1) of this part, will be issued a pilot certificate with an endorsement containing the following limitation, "Passenger carrying prohibited on flights more than 10 nautical miles from (the appropriate island)." The limitation may be subsequently amended to include another island if the applicant complies with the requirements of paragraph (a) of this section for another island.

(c) Upon meeting the requirements of § 61.99(a)(1) of this part, the applicant may have the limitation(s) in paragraph (b) of this section removed.

§ 61.101 Recreational pilot privileges and limitations.

(a) A person who holds a recreational pilot certificate may:

(1) Carry no more than one passenger; and

(2) Not pay less than the pro rata share of the operating expenses of a flight with a passenger, provided the expenses involve only fuel, oil, airport expenses, or aircraft rental fees.

(b) A person who holds a recreational pilot certificate may act as pilot in command of an aircraft on a flight within 50 nautical miles from the departure airport, provided that person has—

(1) Received ground and flight training for takeoff, departure, arrival, and landing procedures at the departure airport;

(2) Received ground and flight training for the area, terrain, and aids to navigation that are in the vicinity of the departure airport;

(3) Been found proficient to operate the aircraft at the departure airport and the area within 50 nautical miles from that airport; and

(4) Received from an authorized instructor a logbook endorsement, which is carried in the person's possession in the aircraft, that permits flight within 50 nautical miles from the departure airport.

(c) A person who holds a recreational pilot certificate may act as pilot in command of an aircraft on a flight that exceeds 50 nautical miles from the departure airport, provided that person has—

(1) Received ground and flight training from an authorized instructor on the cross-country training requirements of subpart E of this part that apply to the aircraft rating held;

(2) Been found proficient in cross-country flying; and

(3) Received from an authorized instructor a logbook endorsement, which is carried on the person's possession in the aircraft, that certifies the person has received and been found proficient in the cross-country training requirements of subpart E of this part that apply to the aircraft rating held.

(d) A person who holds a recreational pilot certificate may act as pilot in command of an aircraft in Class B, C, and D airspace, at an airport located in Class B, C, or D airspace, and to, from, through, or at an airport having an operational control tower, provided that person has—

(1) Received and logged ground and flight training from an authorized instructor on the following aeronautical knowledge areas and areas of operation, as appropriate to the aircraft rating held:

(i) The use of radios, communications, navigation system and facilities, and radar services.

(ii) Operations at airports with an operating control tower to include three takeoffs and landings to a full stop, with each landing involving a flight in the traffic pattern at an airport with an operating control tower.

(iii) Applicable flight rules of part 91 of this chapter for operations in Class B, C, and D airspace and air traffic control clearances;

(2) Been found proficient in those aeronautical knowledge areas and areas of operation specified in paragraph (d)(1) of this section; and

(3) Received from an authorized instructor a logbook endorsement, which is carried on the person's possession or readily accessible in the aircraft, that certifies the person has received and been found proficient in those aeronautical knowledge areas and areas of operation specified in paragraph (d)(1) of this section.

(e) Except as provided in paragraphs (d) and (i) of this section, a recreational pilot may not act as pilot in command of an aircraft—

(1) That is certificated—

(i) For more than four occupants;

(ii) With more than one powerplant;

(iii) With a powerplant of more than 180 horsepower, except aircraft certificated in the rotorcraft category; or

(iv) With retractable landing gear;

(2) That is classified as a multiengine airplane, powered-lift, glider, airship, balloon, powered parachute, or weight-shift-control aircraft;

(3) That is carrying a passenger or property for compensation or hire;

(4) For compensation or hire;

(5) In furtherance of a business;

(6) Between sunset and sunrise;

(7) In Class A, B, C, and D airspace, at an airport located in Class B, C, or D airspace, or to, from, through, or at an airport having an operational control tower;

(8) At an altitude of more than 10,000 feet MSL or 2,000 feet AGL, whichever is higher;

(9) When the flight or surface visibility is less than 3 statute miles;

(10) Without visual reference to the surface;

(11) On a flight outside the United States, unless authorized by the country in which the flight is conducted;

(12) To demonstrate that aircraft in flight as an aircraft salesperson to a prospective buyer;

(13) That is used in a passenger-carrying airlift and sponsored by a charitable organization; and

(14) That is towing any object.

(f) A recreational pilot may not act as a pilot flight crewmember on any aircraft for which more than one pilot is required by the type certificate of the aircraft or the regulations under which the flight is conducted, except when:

(1) Receiving flight training from a person authorized to provide flight training on board an airship; and

(2) No person other than a required flight crewmember is carried on the aircraft.

(g) A person who holds a recreational pilot certificate, has logged fewer than 400 flight hours, and has not logged pilot-in-command time in an aircraft within the 180 days preceding the flight shall not act as pilot in command of an aircraft until the pilot receives flight training and a logbook endorsement from an authorized instructor, and the instructor certifies that the person is proficient to act as pilot in command of the aircraft. This requirement can be met in combination with the requirements of §§ 61.56 and 61.57 of this part, at the discretion of the authorized instructor.

(h) A recreational pilot certificate issued under this subpart carries the notation, "Holder does not meet ICAO requirements."

(i) For the purpose of obtaining additional certificates or ratings while under the supervision of an authorized instructor, a recreational pilot may fly as the sole occupant of an aircraft:

(1) For which the pilot does not hold an appropriate category or class rating;

(2) Within airspace that requires communication with air traffic control; or

(3) Between sunset and sunrise, provided the flight or surface visibility is at least 5 statute miles.

(j) In order to fly solo as provided in paragraph (i) of this section, the recreational pilot must meet the appropriate aeronautical knowledge and flight training requirements of § 61.87 for that aircraft. When operating an aircraft under the conditions specified in paragraph (i) of this section, the recreational pilot shall carry the logbook that has been endorsed for each flight by an authorized instructor who:

(1) Has given the recreational pilot training in the make and model of aircraft in which the solo flight is to be made;

(2) Has found that the recreational pilot has met the applicable requirements of § 61.87; and

(3) Has found that the recreational pilot is competent to make solo flights in accordance with the logbook endorsement.

[Doc. No. 25910, 62 FR 16298, Apr. 4, 1997, as amended by Amdt. 61-110, 69 FR 44868, July 27, 2004; Amdt. 61-124, 74 FR 42558, Aug. 21, 2009]

Subpart E—Private Pilots

§ 61.102 Applicability.

This subpart prescribes the requirements for the issuance of private pilot certificates and ratings, the conditions under which those certificates and ratings are necessary, and the general operating rules for persons who hold those certificates and ratings.

§ 61.103 Eligibility requirements: General.

To be eligible for a private pilot certificate, a person must:

(a) Be at least 17 years of age for a rating in other than a glider or balloon.

(b) Be at least 16 years of age for a rating in a glider or balloon.

(c) Be able to read, speak, write, and understand the English language. If the applicant is unable to meet one of these requirements due to medical reasons, then the Administrator may place such operating limitations on that applicant's pilot certificate as are necessary for the safe operation of the aircraft.

(d) Receive a logbook endorsement from an authorized instructor who:

(1) Conducted the training or reviewed the person's home study on the aeronautical knowledge areas listed in § 61.105(b) of this part that apply to the aircraft rating sought; and

(2) Certified that the person is prepared for the required knowledge test.

(e) Pass the required knowledge test on the aeronautical knowledge areas listed in § 61.105(b) of this part.

(f) Receive flight training and a logbook endorsement from an authorized instructor who:

(1) Conducted the training in the areas of operation listed in § 61.107(b) of this part that apply to the aircraft rating sought; and

(2) Certified that the person is prepared for the required practical test.

(g) Meet the aeronautical experience requirements of this part that apply to the aircraft rating sought before applying for the practical test.

(h) Pass a practical test on the areas of operation listed in § 61.107(b) of this part that apply to the aircraft rating sought.

(i) Comply with the appropriate sections of this part that apply to the aircraft category and class rating sought.

(j) Hold a U.S. student pilot certificate, sport pilot certificate, or recreational pilot certificate.

[Doc. No. 25910, 62 FR 16298, Apr. 4, 1997, as amended by Amdt. 61-124, 74 FR 42558, Aug. 21, 2009]

§ 61.105 Aeronautical knowledge.

(a) *General.* A person who is applying for a private pilot certificate must receive and log ground training from an authorized instructor or complete a home-study course on the aeronautical knowledge areas of paragraph (b) of this section that apply to the aircraft category and class rating sought.

(b) *Aeronautical knowledge areas.* (1) Applicable Federal Aviation Regulations of this chapter that relate to private pilot privileges, limitations, and flight operations;

(2) Accident reporting requirements of the National Transportation Safety Board;

(3) Use of the applicable portions of the "Aeronautical Information Manual" and FAA advisory circulars;

(4) Use of aeronautical charts for VFR navigation using pilotage, dead reckoning, and navigation systems;

(5) Radio communication procedures;

(6) Recognition of critical weather situations from the ground and in flight, windshear avoidance, and the procurement and use of aeronautical weather reports and forecasts;

(7) Safe and efficient operation of aircraft, including collision avoidance, and recognition and avoidance of wake turbulence;

(8) Effects of density altitude on takeoff and climb performance;

(9) Weight and balance computations;

(10) Principles of aerodynamics, powerplants, and aircraft systems;

(11) Stall awareness, spin entry, spins, and spin recovery techniques for the airplane and glider category ratings;

(12) Aeronautical decision making and judgment; and

(13) Preflight action that includes—

(i) How to obtain information on runway lengths at airports of intended use, data on takeoff and landing distances, weather reports and forecasts, and fuel requirements; and

(ii) How to plan for alternatives if the planned flight cannot be completed or delays are encountered.

[Doc. No. 25910, 62 FR 16298, Apr. 4, 1997; Amdt. 61-103, 62 FR 40902, July 30, 1997]

§ 61.107 Flight proficiency.

(a) *General.* A person who applies for a private pilot certificate must receive and log ground and flight training from an authorized instructor on the areas of operation of this section that apply to the aircraft category and class rating sought.

(b) *Areas of operation.* (1) For an airplane category rating with a single-engine class rating:

(i) Preflight preparation;

(ii) Preflight procedures;

(iii) Airport and seaplane base operations;

(iv) Takeoffs, landings, and go-arounds;

(v) Performance maneuvers;

(vi) Ground reference maneuvers;

(vii) Navigation;

(viii) Slow flight and stalls;

(ix) Basic instrument maneuvers;

(x) Emergency operations;

(xi) Night operations, except as provided in § 61.110 of this part; and

(xii) Postflight procedures.

(2) For an airplane category rating with a multiengine class rating:

(i) Preflight preparation;

(ii) Preflight procedures;

(iii) Airport and seaplane base operations;

(iv) Takeoffs, landings, and go-arounds;

(v) Performance maneuvers;

(vi) Ground reference maneuvers;

(vii) Navigation;

(viii) Slow flight and stalls;

(ix) Basic instrument maneuvers;

(x) Emergency operations;

(xi) Multiengine operations;

(xii) Night operations, except as provided in § 61.110 of this part; and

(xiii) Postflight procedures.

(3) For a rotorcraft category rating with a helicopter class rating:

(i) Preflight preparation;

(ii) Preflight procedures;

(iii) Airport and heliport operations;

(iv) Hovering maneuvers;

(v) Takeoffs, landings, and go-arounds;

(vi) Performance maneuvers;

(vii) Navigation;

(viii) Emergency operations;

(ix) Night operations, except as provided in § 61.110 of this part; and

(x) Postflight procedures.

(4) For a rotorcraft category rating with a gyroplane class rating:

(i) Preflight preparation;

(ii) Preflight procedures;

(iii) Airport operations;

(iv) Takeoffs, landings, and go-arounds;

(v) Performance maneuvers;

(vi) Ground reference maneuvers;

(vii) Navigation;

(viii) Flight at slow airspeeds;

(ix) Emergency operations;

(x) Night operations, except as provided in § 61.110 of this part; and

(xi) Postflight procedures.

(5) For a powered-lift category rating:

(i) Preflight preparation;

(ii) Preflight procedures;

(iii) Airport and heliport operations;

(iv) Hovering maneuvers;

(v) Takeoffs, landings, and go-arounds;

(vi) Performance maneuvers;

(vii) Ground reference maneuvers;

(viii) Navigation;

(ix) Slow flight and stalls;

(x) Basic instrument maneuvers;

(xi) Emergency operations;

(xii) Night operations, except as provided in § 61.110 of this part; and

(xiii) Postflight procedures.

(6) For a glider category rating:

(i) Preflight preparation;

(ii) Preflight procedures;

(iii) Airport and gliderport operations;

(iv) Launches and landings;

(v) Performance speeds;

(vi) Soaring techniques;

(vii) Performance maneuvers;

(viii) Navigation;

(ix) Slow flight and stalls;

(x) Emergency operations; and

(xi) Postflight procedures.

(7) For a lighter-than-air category rating with an airship class rating:

(i) Preflight preparation;

(ii) Preflight procedures;

(iii) Airport operations;

(iv) Takeoffs, landings, and go-arounds;

(v) Performance maneuvers;

(vi) Ground reference maneuvers;

(vii) Navigation;

(viii) Emergency operations; and

(ix) Postflight procedures.

(8) For a lighter-than-air category rating with a balloon class rating:

(i) Preflight preparation;

(ii) Preflight procedures;

(iii) Airport operations;

(iv) Launches and landings;

(v) Performance maneuvers;

(vi) Navigation;

(vii) Emergency operations; and

(viii) Postflight procedures.

(9) For a powered parachute category rating—

(i) Preflight preparation;

(ii) Preflight procedures;

(iii) Airport and seaplane base operations, as applicable;

(iv) Takeoffs, landings, and go-arounds;

(v) Performance maneuvers;

(vi) Ground reference maneuvers;

(vii) Navigation;

(viii) Night operations, except as provided in § 61.110;

(ix) Emergency operations; and

(x) Post-flight procedures.

(10) For a weight-shift-control aircraft category rating—

(i) Preflight preparation;

(ii) Preflight procedures;

(iii) Airport and seaplane base operations, as applicable;

(iv) Takeoffs, landings, and go-arounds;

(v) Performance maneuvers;

(vi) Ground reference maneuvers;

(vii) Navigation;

(viii) Slow flight and stalls;

(ix) Night operations, except as provided in § 61.110;

(x) Emergency operations; and

(xi) Post-flight procedures.

[Doc. No. 25910, 62 FR 16298, Apr. 4, 1997, as amended by Amdt. 61-110, 69 FR 44868, July 27, 2004]

§ 61.109 Aeronautical experience.

(a) *For an airplane single-engine rating.* Except as provided in paragraph (k) of this section, a person who applies for a private pilot certificate with an airplane category and single-engine class rating must log at least 40 hours of flight time that includes at least 20 hours of flight training from an authorized instructor and 10 hours of solo flight training in the areas of operation listed in § 61.107(b)(1) of this part, and the training must include at least—

(1) 3 hours of cross-country flight training in a single-engine airplane;

(2) Except as provided in § 61.110 of this part, 3 hours of night flight training in a single-engine airplane that includes—

(i) One cross-country flight of over 100 nautical miles total distance; and

(ii) 10 takeoffs and 10 landings to a full stop (with each landing involving a flight in the traffic pattern) at an airport.

(3) 3 hours of flight training in a single-engine airplane on the control and maneuvering of an airplane solely by reference to instruments, including straight and level flight, constant airspeed climbs and descents, turns to a heading, recovery from unusual flight attitudes, radio communications, and the use of navigation systems/facilities and radar services appropriate to instrument flight;

(4) 3 hours of flight training with an authorized instructor in a single-engine airplane in preparation for the practical test, which must have been performed within the preceding 2 calendar months from the month of the test; and

(5) 10 hours of solo flight time in a single-engine airplane, consisting of at least—

(i) 5 hours of solo cross-country time;

(ii) One solo cross country flight of 150 nautical miles total distance, with full-stop landings at three points, and one segment of the flight consisting of a straight-line distance of more than 50 nautical miles between the takeoff and landing locations; and

(iii) Three takeoffs and three landings to a full stop (with each landing involving a flight in the traffic pattern) at an airport with an operating control tower.

(b) *For an airplane multiengine rating.* Except as provided in paragraph (k) of this section, a person who applies for a private pilot certificate with an airplane category and multiengine class rating must log at least 40 hours of flight time that includes at least 20 hours of flight training from an authorized instructor and 10 hours of solo flight training in the areas of operation listed in § 61.107(b)(2) of this part, and the training must include at least—

(1) 3 hours of cross-country flight training in a multiengine airplane;

(2) Except as provided in § 61.110 of this part, 3 hours of night flight training in a multiengine airplane that includes—

(i) One cross-country flight of over 100 nautical miles total distance; and

(ii) 10 takeoffs and 10 landings to a full stop (with each landing involving a flight in the traffic pattern) at an airport.

(3) 3 hours of flight training in a multiengine airplane on the control and maneuvering of an airplane solely by reference to instruments, including straight and level flight, constant airspeed climbs and descents, turns to a heading, recovery from unusual flight attitudes, radio communications, and the use of navigation systems/facilities and radar services appropriate to instrument flight;

(4) 3 hours of flight training with an authorized instructor in a multiengine airplane in preparation for the practical test, which must have been performed within the preceding 2 calendar months from the month of the test; and

(5) 10 hours of solo flight time in an airplane consisting of at least—

(i) 5 hours of solo cross-country time;

(ii) One solo cross country flight of 150 nautical miles total distance, with full-stop landings at three points, and one segment of the flight consisting of a straight-line distance of more than 50 nautical miles between the takeoff and landing locations; and

(iii) Three takeoffs and three landings to a full stop (with each landing involving a flight in the traffic pattern) at an airport with an operating control tower.

(c) *For a helicopter rating.* Except as provided in paragraph (k) of this section, a person who applies for a private pilot certificate with rotorcraft category and helicopter class rating must log at least 40 hours of flight time that includes at least 20 hours of flight training from an authorized instructor and 10 hours of solo flight training in the areas of operation listed in § 61.107(b)(3) of this part, and the training must include at least—

(1) 3 hours of cross-country flight training in a helicopter;

(2) Except as provided in § 61.110 of this part, 3 hours of night flight training in a helicopter that includes—

(i) One cross-country flight of over 50 nautical miles total distance; and

(ii) 10 takeoffs and 10 landings to a full stop (with each landing involving a flight in the traffic pattern) at an airport.

(3) 3 hours of flight training with an authorized instructor in a helicopter in preparation for the practical test, which must have been performed within the preceding 2 calendar months from the month of the test; and

(4) 10 hours of solo flight time in a helicopter, consisting of at least—

(i) 3 hours cross-country time;

(ii) One solo cross country flight of 100 nautical miles total distance, with landings at three points, and one segment of the flight being a straight-line distance of more than 25 nautical miles between the takeoff and landing locations; and

(iii) Three takeoffs and three landings to a full stop (with each landing involving a flight in the traffic pattern) at an airport with an operating control tower.

(d) *For a gyroplane rating.* Except as provided in paragraph (k) of this section, a person who applies for a private pilot certificate with rotorcraft category and gyroplane class rating must log at least 40 hours of flight time that includes at least 20 hours of flight training from an authorized instructor and 10 hours of solo flight training in the areas of operation listed in § 61.107(b)(4) of this part, and the training must include at least—

(1) 3 hours of cross-country flight training in a gyroplane;

(2) Except as provided in § 61.110 of this part, 3 hours of night flight training in a gyroplane that includes—

(i) One cross-country flight of over 50 nautical miles total distance; and

(ii) 10 takeoffs and 10 landings to a full stop (with each landing involving a flight in the traffic pattern) at an airport.

(3) 3 hours of flight training with an authorized instructor in a gyroplane in preparation for the practical test, which must have been performed within the preceding 2 calendar months from the month of the test; and

(4) 10 hours of solo flight time in a gyroplane, consisting of at least—

(i) 3 hours of cross-country time;

(ii) One solo cross country flight of 100 nautical miles total distance, with landings at three points, and one segment of the flight being a straight-line distance of more than 25 nautical miles between the takeoff and landing locations; and

(iii) Three takeoffs and three landings to a full stop (with each landing involving a flight in the traffic pattern) at an airport with an operating control tower.

(e) *For a powered-lift rating.* Except as provided in paragraph (k) of this section, a person who applies for a private pilot certificate with a powered-lift category rating must log at least 40 hours of flight time that includes at least 20 hours of flight training from an authorized instructor and 10 hours of solo flight training in the areas of operation listed in § 61.107(b)(5) of this part, and the training must include at least—

(1) 3 hours of cross-country flight training in a powered-lift;

(2) Except as provided in § 61.110 of this part, 3 hours of night flight training in a powered-lift that includes—

(i) One cross-country flight of over 100 nautical miles total distance; and

(ii) 10 takeoffs and 10 landings to a full stop (with each landing involving a flight in the traffic pattern) at an airport.

(3) 3 hours of flight training in a powered-lift on the control and maneuvering of a powered-lift solely by reference to instruments, including straight and level flight, constant airspeed climbs and descents, turns to a heading, recovery from unusual flight attitudes, radio communications, and the use of navigation systems/facilities and radar services appropriate to instrument flight;

(4) 3 hours of flight training with an authorized instructor in a powered-lift in preparation for the practical test, which must have been performed within the preceding 2 calendar months from the month of the test; and

(5) 10 hours of solo flight time in an airplane or powered-lift consisting of at least—

(i) 5 hours cross-country time;

(ii) One solo cross country flight of 150 nautical miles total distance, with full-stop landings at three points, and one segment of the flight consisting of a straight-line distance of more than 50 nautical miles between the takeoff and landing locations; and

(iii) Three takeoffs and three landings to a full stop (with each landing involving a flight in the traffic pattern) at an airport with an operating control tower.

(f) *For a glider category rating.* (1) If the applicant for a private pilot certificate with a glider category rating has not logged at least 40 hours of flight time as a pilot in a heavier-than-air aircraft, the applicant must log at least 10 hours of flight time in a glider in the areas of operation listed in § 61.107(b)(6) of this part, and that flight time must include at least—

(i) 20 flights in a glider in the areas of operations listed in § 61.107(b)(6) of this part, including at least 3 training flights with an authorized instructor in a glider in preparation for the practical test that must have been performed within the preceding 2 calendar months from the month of the test; and

(ii) 2 hours of solo flight time in a glider in the areas of operation listed in § 61.107(b)(6) of this part, with not less than 10 launches and landings being performed.

(2) If the applicant has logged at least 40 hours of flight time in a heavier-than-air aircraft, the applicant must log at least 3 hours of flight time in a glider in the areas of operation listed in § 61.107(b)(6) of this part, and that flight time must include at least—

(i) 10 solo flights in a glider in the areas of operation listed in § 61.107(b)(6) of this part; and

(ii) 3 training flights with an authorized instructor in a glider in preparation for the practical test that must have been performed within the preceding 2 calendar months from the month of the test.

(g) *For an airship rating.* A person who applies for a private pilot certificate with a lighter-than-air category and airship class rating must log at least:

(1) 25 hours of flight training in airships on the areas of operation listed in § 61.107(b)(7) of this part, which consists of at least:

(i) 3 hours of cross-country flight training in an airship;

(ii) Except as provided in § 61.110 of this part, 3 hours of night flight training in an airship that includes:

(A) A cross-country flight of over 25 nautical miles total distance; and

(B) Five takeoffs and five landings to a full stop (with each landing involving a flight in the traffic pattern) at an airport.

(2) 3 hours of flight training in an airship on the control and maneuvering of an airship solely by reference to instruments, including straight and level flight, constant airspeed climbs and descents, turns to a heading, recovery from unusual flight attitudes, radio communications, and the use of navigation systems/facilities and radar services appropriate to instrument flight;

(3) Three hours of flight training with an authorized instructor in an airship in preparation for the practical test within the preceding 2 calendar months from the month of the test; and

(4) 5 hours performing the duties of pilot in command in an airship with an authorized instructor.

(h) *For a balloon rating.* A person who applies for a private pilot certificate with a lighter-than-air category and balloon class rating must log at least 10 hours of flight training that includes at least six training flights with an authorized instructor in the areas of operation listed in § 61.107(b)(8) of this part, that includes—

(1) *Gas balloon.* If the training is being performed in a gas balloon, at least two flights of 2 hours each that consists of—

(i) At least one training flight with an authorized instructor in a gas balloon in preparation for the practical test within the preceding 2 calendar months from the month of the test;

(ii) At least one flight performing the duties of pilot in command in a gas balloon with an authorized instructor; and

(iii) At least one flight involving a controlled ascent to 3,000 feet above the launch site.

(2) *Balloon with an airborne heater.* If the training is being performed in a balloon with an airborne heater, at least—

(i) At least two training flights of 1 hour each with an authorized instructor in a balloon with an airborne heater in preparation for the practical test within the preceding 2 calendar months from the month of the test;

(ii) One solo flight in a balloon with an airborne heater; and

(iii) At least one flight involving a controlled ascent to 2,000 feet above the launch site.

(i) *For a powered parachute rating.* A person who applies for a private pilot certificate with a powered parachute category rating must log at least 25 hours of flight time in a powered parachute that includes at least 10 hours of flight training with an authorized instructor, including 30 takeoffs and landings, and 10 hours of solo flight training in the areas of operation listed in § 61.107 (b)(9) and the training must include at least—

(1) One hour of cross-country flight training in a powered parachute that includes a 1-hour cross-country flight with a landing at an airport at least 25 nautical miles from the airport of departure;

(2) Except as provided in § 61.110, 3 hours of night flight training in a powered parachute that includes 10 takeoffs and landings (with each landing involving a flight in the traffic pattern) at an airport;

(3) Three hours of flight training with an authorized instructor in a powered parachute in preparation for the practical test, which must have been performed within the preceding 2 calendar months from the month of the test;

(4) Three hours of solo flight time in a powered parachute, consisting of at least—

(i) One solo cross-country flight with a landing at an airport at least 25 nautical miles from the departure airport; and

(ii) Twenty solo takeoffs and landings to a full stop (with each landing involving a flight in a traffic pattern) at an airport; and

(5) Three takeoffs and landings (with each landing involving a flight in the traffic pattern) in an aircraft at an airport with an operating control tower.

(j) *For a weight-shift-control aircraft rating.* A person who applies for a private pilot certificate with a weight-shift-control rating must log at least 40 hours of flight time that includes at least 20 hours of flight training with an authorized instructor and 10 hours of solo flight training in the areas of operation listed in § 61.107(b)(10) and the training must include at least—

(1) Three hours of cross-country flight training in a weight-shift-control aircraft;

(2) Except as provided in § 61.110, 3 hours of night flight training in a weight-shift-control aircraft that includes—

(i) One cross-country flight of over 75 nautical miles total distance that includes a point of landing that is a straight-line distance of more than 50 nautical miles from the original point of departure; and

(ii) Ten takeoffs and landings (with each landing involving a flight in the traffic pattern) at an airport;

(3) Three hours of flight training with an authorized instructor in a weight-shift-control aircraft in preparation for the practical test, which must have been performed within the preceding 2 calendar months from the month of the test;

(4) Ten hours of solo flight time in a weight-shift-control aircraft, consisting of at least—

(i) Five hours of solo cross-country time; and

(ii) One solo cross-country flight over 100 nautical miles total distance, with landings at a minimum of three points, and one segment of the flight being a straight line distance of at least 50 nautical miles between takeoff and landing locations; and

(5) Three takeoffs and landings (with each landing involving a flight in the traffic pattern) in an aircraft at an airport with an operating control tower.

(k) *Permitted credit for use of a flight simulator or flight training device.* (1) Except as provided in paragraphs (k)(2) of this section, a maximum of 2.5 hours of training in a flight simulator or flight training device representing the category, class, and type, if applicable, of aircraft appropriate to the rating sought, may be credited toward the flight training time required by this section, if received from an authorized instructor.

(2) A maximum of 5 hours of training in a flight simulator or flight training device representing the category, class, and type, if applicable, of aircraft appropriate to the rating sought, may be credited toward the flight training time required by this section if the training is accomplished in a course conducted by a training center certificated under part 142 of this chapter.

(3) Except when fewer hours are approved by the Administrator, an applicant for a private pilot certificate with an airplane, rotorcraft, or powered-lift rating, who has satisfactorily completed an approved private pilot course conducted by a training center certificated under part 142 of this chapter, need only have a total of 35 hours of aeronautical experience to meet the requirements of this section.

[Doc. No. 25910, 62 FR 40902, July 30, 1997, as amended by Amdt. 61-104, 63 FR 20287, Apr. 23, 1998; Amdt. 61-110, 69 FR 44868, July 27, 2004; Amdt. 61-124, 74 FR 42558, Aug. 21, 2009; Amdt. 61-124A, 74 FR 53645, Oct. 20, 2009; Amdt. 61-125, 75 FR 5220, Feb. 1, 2010]

§ 61.110 Night flying exceptions.

(a) Subject to the limitations of paragraph (b) of this section, a person is not required to comply with the night flight training requirements of this subpart if the person receives flight training in and resides in the State of Alaska.

(b) A person who receives flight training in and resides in the State of Alaska but does not meet the night flight training requirements of this section:

(1) May be issued a pilot certificate with a limitation "Night flying prohibited"; and

(2) Must comply with the appropriate night flight training requirements of this subpart within the 12-calendar-month period after the issuance of the pilot certificate. At the end of that period, the certificate will become invalid for use until the person complies with the appropriate night training requirements of this subpart. The person may have the "Night flying prohibited" limitation removed if the person—

(i) Accomplishes the appropriate night flight training requirements of this subpart; and

(ii) Presents to an examiner a logbook or training record endorsement from an authorized instructor that verifies accomplishment of the appropriate night flight training requirements of this subpart.

(c) A person who does not meet the night flying requirements in § 61.109(d)(2), (i)(2), or (j)(2) may be issued a private pilot certificate with the limitation "Night flying prohibited." This limitation may be removed by an examiner if the holder complies with the requirements of § 61.109(d)(2), (i)(2), or (j)(2), as appropriate.

[Doc. No. 25910, 62 FR 16298, Apr. 4, 1997; Amdt. 61-103, 62 FR 40904, July 30, 1997; Amdt. 61-110, 69 FR 44869, July 27, 2004]

§ 61.111 Cross-country flights: Pilots based on small islands.

(a) Except as provided in paragraph (b) of this section, an applicant located on an island from which the cross-country flight training required in § 61.109 of this part cannot be accomplished without flying over water for more than 10 nautical miles from the nearest shoreline need not comply with the requirements of that section.

(b) If other airports that permit civil operations are available to which a flight may be made without flying over water for more than 10 nautical miles from the nearest shoreline, the applicant must show completion of two round-trip solo flights between those two airports that are farthest apart, including a landing at each airport on both flights.

(c) An applicant who complies with paragraph (a) or paragraph (b) of this section, and meets all requirements for the issuance of a private pilot certificate, except the cross-country training requirements of § 61.109 of this part, will be issued a pilot certificate with an endorsement containing the following limitation, "Passenger carrying prohibited on flights more than 10 nautical miles from (the appropriate island)." The limitation may be subsequently amended to include another island if the applicant complies with the requirements of paragraph (b) of this section for another island.

(d) Upon meeting the cross-country training requirements of § 61.109 of this part, the applicant may have the limitation in paragraph (c) of this section removed.

[Doc. No. 25910, 62 FR 16298, Apr. 4, 1997; Amdt. 61-103, 62 FR 40904, July 30, 1997]

§ 61.113 Private pilot privileges and limitations: Pilot in command.

(a) Except as provided in paragraphs (b) through (h) of this section, no person who holds a private pilot certificate may act as pilot in command of an aircraft that is carrying passengers or property for compensation or hire; nor may that person, for compensation or hire, act as pilot in command of an aircraft.

(b) A private pilot may, for compensation or hire, act as pilot in command of an aircraft in connection with any business or employment if:

(1) The flight is only incidental to that business or employment; and

(2) The aircraft does not carry passengers or property for compensation or hire.

(c) A private pilot may not pay less than the pro rata share of the operating expenses of a flight with passengers, provided the expenses involve only fuel, oil, airport expenditures, or rental fees.

(d) A private pilot may act as pilot in command of a charitable, nonprofit, or community event flight described in § 91.146, if the sponsor and pilot comply with the requirements of § 91.146.

(e) A private pilot may be reimbursed for aircraft operating expenses that are directly related to search and location operations, provided the expenses involve only fuel, oil, airport expenditures, or rental fees, and the operation is sanctioned and under the direction and control of:

(1) A local, State, or Federal agency; or

(2) An organization that conducts search and location operations.

(f) A private pilot who is an aircraft salesman and who has at least 200 hours of logged flight time may demonstrate an aircraft in flight to a prospective buyer.

(g) A private pilot who meets the requirements of § 61.69 may act as a pilot in command of an aircraft towing a glider or unpowered ultralight vehicle.

(h) A private pilot may act as pilot in command for the purpose of conducting a production flight test in a light-sport aircraft intended for certification in the light-sport category under § 21.190 of this chapter, provided that—

(1) The aircraft is a powered parachute or a weight-shift-control aircraft;

(2) The person has at least 100 hours of pilot-in-command time in the category and class of aircraft flown; and

(3) The person is familiar with the processes and procedures applicable to the conduct of production flight testing, to include operations conducted under a special flight permit and any associated operating limitations.

[Doc. No. 25910, 62 FR 16298, Apr. 4, 1997, as amended by Amdt. 61-110, 69 FR 44869, July 27, 2004; Amdt. 61-115, 72 FR 6910, Feb. 13, 2007; Amdt. 61-125, 75 FR 5220, Feb. 1, 2010]

§ 61.115 Balloon rating: Limitations.

(a) If a person who applies for a private pilot certificate with a balloon rating takes a practical test in a balloon with an airborne heater:

(1) The pilot certificate will contain a limitation restricting the exercise of the privileges of that certificate to a balloon with an airborne heater; and

(2) The limitation may be removed when the person obtains the required aeronautical experience in a gas balloon and receives a logbook endorsement from an authorized instructor who attests to the person's accomplishment of the required aeronautical experience and ability to satisfactorily operate a gas balloon.

(b) If a person who applies for a private pilot certificate with a balloon rating takes a practical test in a gas balloon:

(1) The pilot certificate will contain a limitation restricting the exercise of the privilege of that certificate to a gas balloon; and

(2) The limitation may be removed when the person obtains the required aeronautical experience in a balloon with an airborne heater and receives a logbook endorsement from an authorized instructor who attests to the person's accomplishment of the required aeronautical experience and ability to satisfactorily operate a balloon with an airborne heater.

§ 61.117 Private pilot privileges and limitations: Second in command of aircraft requiring more than one pilot.

Except as provided in § 61.113 of this part, no private pilot may, for compensation or hire, act as second in command of an aircraft that is type certificated for more than one pilot, nor may that pilot act as second in command of such an aircraft that is carrying passengers or property for compensation or hire.

[Doc. No. 25910, 62 FR 16298, Apr. 4, 1997; Amdt. 61-103, 62 FR 40904, July 30, 1997]

§ 61.118-61.120 [Reserved]

Subpart F—Commercial Pilots

§ 61.121 Applicability.

This subpart prescribes the requirements for the issuance of commercial pilot certificates and ratings, the conditions under which those certificates and ratings are necessary, and the general operating rules for persons who hold those certificates and ratings.

§ 61.123 Eligibility requirements: General.

To be eligible for a commercial pilot certificate, a person must:

(a) Be at least 18 years of age;

(b) Be able to read, speak, write, and understand the English language. If the applicant is unable to meet one of these requirements due to medical reasons, then the Administrator may place such operating limitations on that applicant's pilot certificate as are necessary for the safe operation of the aircraft.

(c) Receive a logbook endorsement from an authorized instructor who:

(1) Conducted the required ground training or reviewed the person's home study on the aeronautical knowledge areas listed in § 61.125 of this part that apply to the aircraft category and class rating sought; and

(2) Certified that the person is prepared for the required knowledge test that applies to the aircraft category and class rating sought.

(d) Pass the required knowledge test on the aeronautical knowledge areas listed in § 61.125 of this part;

(e) Receive the required training and a logbook endorsement from an authorized instructor who:

(1) Conducted the training on the areas of operation listed in § 61.127(b) of this part that apply to the aircraft category and class rating sought; and

(2) Certified that the person is prepared for the required practical test.

(f) Meet the aeronautical experience requirements of this subpart that apply to the aircraft category and class rating sought before applying for the practical test;

(g) Pass the required practical test on the areas of operation listed in § 61.127(b) of this part that apply to the aircraft category and class rating sought;

(h) Hold at least a private pilot certificate issued under this part or meet the requirements of § 61.73; and

(i) Comply with the sections of this part that apply to the aircraft category and class rating sought.

§ 61.125 Aeronautical knowledge.

(a) *General.* A person who applies for a commercial pilot certificate must receive and log ground training from an authorized instructor, or complete a home-study course, on the aeronautical knowledge areas of paragraph (b) of this section that apply to the aircraft category and class rating sought.

(b) *Aeronautical knowledge areas.* (1) Applicable Federal Aviation Regulations of this chapter that relate to commercial pilot privileges, limitations, and flight operations;

(2) Accident reporting requirements of the National Transportation Safety Board;

(3) Basic aerodynamics and the principles of flight;

(4) Meteorology to include recognition of critical weather situations, windshear recognition and avoidance, and the use of aeronautical weather reports and forecasts;

(5) Safe and efficient operation of aircraft;

(6) Weight and balance computations;

(7) Use of performance charts;

(8) Significance and effects of exceeding aircraft performance limitations;

(9) Use of aeronautical charts and a magnetic compass for pilotage and dead reckoning;

(10) Use of air navigation facilities;

(11) Aeronautical decision making and judgment;

(12) Principles and functions of aircraft systems;

(13) Maneuvers, procedures, and emergency operations appropriate to the aircraft;

(14) Night and high-altitude operations;

(15) Procedures for operating within the National Airspace System; and

(16) Procedures for flight and ground training for lighter-than-air ratings.

§ 61.127 Flight proficiency.

(a) *General.* A person who applies for a commercial pilot certificate must receive and log ground and flight training from an authorized instructor on the areas of operation of this section that apply to the aircraft category and class rating sought.

(b) *Areas of operation.* (1) For an airplane category rating with a single-engine class rating:

(i) Preflight preparation;
(ii) Preflight procedures;
(iii) Airport and seaplane base operations;
(iv) Takeoffs, landings, and go-arounds;
(v) Performance maneuvers;
(vi) Ground reference maneuvers;
(vii) Navigation;
(viii) Slow flight and stalls;
(ix) Emergency operations;
(x) High-altitude operations; and
(xi) Postflight procedures.

(2) For an airplane category rating with a multiengine class rating:

(i) Preflight preparation;
(ii) Preflight procedures;
(iii) Airport and seaplane base operations;
(iv) Takeoffs, landings, and go-arounds;
(v) Performance maneuvers;
(vi) Navigation;
(vii) Slow flight and stalls;
(viii) Emergency operations;
(ix) Multiengine operations;
(x) High-altitude operations; and
(xi) Postflight procedures.

(3) For a rotorcraft category rating with a helicopter class rating:

(i) Preflight preparation;
(ii) Preflight procedures;
(iii) Airport and heliport operations;
(iv) Hovering maneuvers;
(v) Takeoffs, landings, and go-arounds;
(vi) Performance maneuvers;
(vii) Navigation;
(viii) Emergency operations;
(ix) Special operations; and
(x) Postflight procedures.

(4) For a rotorcraft category rating with a gyroplane class rating:

(i) Preflight preparation;
(ii) Preflight procedures;
(iii) Airport operations;
(iv) Takeoffs, landings, and go-arounds;
(v) Performance maneuvers;
(vi) Ground reference maneuvers;
(vii) Navigation;
(viii) Flight at slow airspeeds;
(ix) Emergency operations; and
(x) Postflight procedures.

(5) For a powered-lift category rating:

(i) Preflight preparation;
(ii) Preflight procedures;
(iii) Airport and heliport operations;
(iv) Hovering maneuvers;
(v) Takeoffs, landings, and go-arounds;
(vi) Performance maneuvers;
(vii) Navigation;
(viii) Slow flight and stalls;
(ix) Emergency operations;
(x) High-altitude operations;
(xi) Special operations; and
(xii) Postflight procedures.

(6) For a glider category rating:

(i) Preflight preparation;
(ii) Preflight procedures;
(iii) Airport and gliderport operations;
(iv) Launches and landings;
(v) Performance speeds;
(vi) Soaring techniques;
(vii) Performance maneuvers;
(viii) Navigation;
(ix) Slow flight and stalls;
(x) Emergency operations; and
(xi) Postflight procedures.

(7) For a lighter-than-air category rating with an airship class rating:

(i) Fundamentals of instructing;
(ii) Technical subjects;
(iii) Preflight preparation;
(iv) Preflight lesson on a maneuver to be performed in flight;
(v) Preflight procedures;
(vi) Airport operations;
(vii) Takeoffs, landings, and go-arounds;
(viii) Performance maneuvers;
(ix) Navigation;
(x) Emergency operations; and
(xi) Postflight procedures.

(8) For a lighter-than-air category rating with a balloon class rating:

(i) Fundamentals of instructing;
(ii) Technical subjects;
(iii) Preflight preparation;
(iv) Preflight lesson on a maneuver to be performed in flight;
(v) Preflight procedures;
(vi) Airport operations;
(vii) Launches and landings;
(viii) Performance maneuvers;
(ix) Navigation;
(x) Emergency operations; and
(xi) Postflight procedures.

[Doc. No. 25910, 62 FR 16298, Apr. 4, 1997, as amended by Amdt. 61-124, 74 FR 42558, Aug. 21, 2009]

§ 61.129 Aeronautical experience.

(a) *For an airplane single-engine rating.* Except as provided in paragraph (i) of this section, a person who applies for a commercial pilot certificate with an airplane category and single-engine class rating must log at least 250 hours of flight time as a pilot that consists of at least:

(1) 100 hours in powered aircraft, of which 50 hours must be in airplanes.

(2) 100 hours of pilot-in-command flight time, which includes at least—

(i) 50 hours in airplanes; and

(ii) 50 hours in cross-country flight of which at least 10 hours must be in airplanes.

(3) 20 hours of training on the areas of operation listed in § 61.127(b)(1) of this part that includes at least—

(i) Ten hours of instrument training using a view-limiting device including attitude instrument flying, partial panel skills, recovery from unusual flight attitudes, and intercepting and tracking navigational systems. Five hours of the 10 hours required on instrument training must be in a single engine airplane;

(ii) 10 hours of training in an airplane that has a retractable landing gear, flaps, and a controllable pitch propeller, or is turbine-powered, or for an applicant seeking a single-engine seaplane rating, 10 hours of training in a seaplane that has flaps and a controllable pitch propeller;

(iii) One 2-hour cross country flight in a single engine airplane in daytime conditions that consists of a total straight-line distance of more than 100 nautical miles from the original point of departure;

(iv) One 2-hour cross country flight in a single engine airplane in nighttime conditions that consists of a total straight-line distance of more than 100 nautical miles from the original point of departure; and

(v) Three hours in a single-engine airplane with an authorized instructor in preparation for the practical test within the preceding 2 calendar months from the month of the test.

(4) Ten hours of solo flight time in a single engine airplane or 10 hours of flight time performing the duties of pilot in command in a single engine airplane with an authorized instructor on board (either of which may be credited towards the flight time requirement under paragraph (a)(2) of this section), on the areas of operation listed under § 61.127(b)(1) that include—

(i) One cross-country flight of not less than 300 nautical miles total distance, with landings at a minimum of three points, one of which is a straight-line distance of at least 250 nautical miles from the original departure point. However, if this requirement is being met in Hawaii, the longest segment need only have a straight-line distance of at least 150 nautical miles; and

(ii) 5 hours in night VFR conditions with 10 takeoffs and 10 landings (with each landing involving a flight in the traffic pattern) at an airport with an operating control tower.

(b) *For an airplane multiengine rating.* Except as provided in paragraph (i) of this section, a person who applies for a commercial pilot certificate with an airplane category and multiengine class rating must log at least 250 hours of flight time as a pilot that consists of at least:

(1) 100 hours in powered aircraft, of which 50 hours must be in airplanes.

(2) 100 hours of pilot-in-command flight time, which includes at least—

(i) 50 hours in airplanes; and

(ii) 50 hours in cross-country flight of which at least 10 hours must be in airplanes.

(3) 20 hours of training on the areas of operation listed in § 61.127(b)(2) of this part that includes at least—

(i) Ten hours of instrument training using a view-limiting device including attitude instrument flying, partial panel skills, recovery from unusual flight attitudes, and intercepting and tracking navigational systems. Five hours of the 10 hours required on instrument training must be in a multiengine airplane;

(ii) 10 hours of training in a multiengine airplane that has a retractable landing gear, flaps, and controllable pitch propellers, or is turbine-powered, or for an applicant seeking a multiengine seaplane rating, 10 hours of training in a multiengine seaplane that has flaps and a controllable pitch propeller;

(iii) One 2-hour cross country flight in a multiengine airplane in daytime conditions that consists of a total straight-line distance of more than 100 nautical miles from the original point of departure;

(iv) One 2-hour cross country flight in a multiengine airplane in nighttime conditions that consists of a total straight-line distance of more than 100 nautical miles from the original point of departure; and

(v) Three hours in a multiengine airplane with an authorized instructor in preparation for the practical test within the preceding 2 calendar months from the month of the test.

(4) 10 hours of solo flight time in a multiengine airplane or 10 hours of flight time performing the duties of pilot in command in a multiengine airplane with an authorized instructor (either of which may be credited towards the flight time requirement in paragraph (b)(2) of this section), on the areas of operation listed in § 61.127(b)(2) of this part that includes at least—

(i) One cross-country flight of not less than 300 nautical miles total distance with landings at a minimum of three points, one of which is a straight-line distance of at least 250 nautical miles from the original departure point. However, if this requirement is being met in Hawaii, the longest segment need only have a straight-line distance of at least 150 nautical miles; and

(ii) 5 hours in night VFR conditions with 10 takeoffs and 10 landings (with each landing involving a flight with a traffic pattern) at an airport with an operating control tower.

(c) *For a helicopter rating.* Except as provided in paragraph (i) of this section, a person who applies for a commercial pilot certificate with a rotorcraft category and helicopter class rating must log at least 150 hours of flight time as a pilot that consists of at least:

(1) 100 hours in powered aircraft, of which 50 hours must be in helicopters.

(2) 100 hours of pilot-in-command flight time, which includes at least—

(i) 35 hours in helicopters; and

(ii) 10 hours in cross-country flight in helicopters.

(3) 20 hours of training on the areas of operation listed in § 61.127(b)(3) of this part that includes at least—

(i) Five hours on the control and maneuvering of a helicopter solely by reference to instruments using a view-limiting device including attitude instrument flying, partial panel skills, recovery from unusual flight attitudes, and intercepting and tracking navigational systems. This aeronautical experience may be performed in an aircraft, flight simulator, flight training device, or an aviation training device;

(ii) One 2-hour cross country flight in a helicopter in daytime conditions that consists of a total straight-line distance of more than 50 nautical miles from the original point of departure;

(iii) One 2-hour cross country flight in a helicopter in nighttime conditions that consists of a total straight-line distance of more than 50 nautical miles from the original point of departure; and

(iv) Three hours in a helicopter with an authorized instructor in preparation for the practical test within the preceding 2 calendar months from the month of the test.

(4) Ten hours of solo flight time in a helicopter or 10 hours of flight time performing the duties of pilot in command in a helicopter with an authorized instructor on board (either of which may be credited towards the flight time requirement under paragraph (c)(2) of this section), on the areas of operation listed under § 61.127(b)(3) that includes—

(i) One cross-country flight with landings at a minimum of three points, with one segment consisting of a straight-line distance of at least 50 nautical miles from the original point of departure; and

(ii) 5 hours in night VFR conditions with 10 takeoffs and 10 landings (with each landing involving a flight in the traffic pattern).

(d) *For a gyroplane rating.* A person who applies for a commercial pilot certificate with a rotorcraft category and gyroplane class rating must log at least 150 hours of flight time as a pilot (of which 5 hours may have been accomplished in a flight simulator or flight training device that is representative of a gyroplane) that consists of at least:

(1) 100 hours in powered aircraft, of which 25 hours must be in gyroplanes.

(2) 100 hours of pilot-in-command flight time, which includes at least—

(i) 10 hours in gyroplanes; and

(ii) 3 hours in cross-country flight in gyroplanes.

(3) 20 hours of training on the areas of operation listed in § 61.127(b)(4) of this part that includes at least—

(i) 2.5 hours on the control and maneuvering of a gyroplane solely by reference to instruments using a view-limiting device including attitude instrument flying, partial panel skills, recovery from unusual flight attitudes, and intercepting and tracking navigational systems. This aeronautical experience may be performed in an aircraft, flight simulator, flight training device, or an aviation training device;

(ii) One 2-hour cross country flight in a gyroplane in daytime conditions that consists of a total straight-line distance of more than 50 nautical miles from the original point of departure;

(iii) Two hours of flight training during nighttime conditions in a gyroplane at an airport, that includes 10 takeoffs and 10 landings to a full stop (with each landing involving a flight in the traffic pattern); and

(iv) Three hours in a gyroplane with an authorized instructor in preparation for the practical test within the preceding 2 calendar months from the month of the test.

(4) Ten hours of solo flight time in a gyroplane or 10 hours of flight time performing the duties of pilot in command in a gyroplane with an authorized instructor on board (either of which may be credited towards the flight time requirement under paragraph (d)(2) of this section), on the areas of operation listed in § 61.127(b)(4) that includes—

(i) One cross-country flight with landings at a minimum of three points, with one segment consisting of a straight-line distance of at least 50 nautical miles from the original point of departure; and

(ii) 5 hours in night VFR conditions with 10 takeoffs and 10 landings (with each landing involving a flight in the traffic pattern).

(e) *For a powered-lift rating.* Except as provided in paragraph (i) of this section, a person who applies for a commercial pilot certificate with a powered-lift category rating must log at least 250 hours of flight time as a pilot that consists of at least:

(1) 100 hours in powered aircraft, of which 50 hours must be in a powered-lift.

(2) 100 hours of pilot-in-command flight time, which includes at least—

(i) 50 hours in a powered-lift; and

(ii) 50 hours in cross-country flight of which 10 hours must be in a powered-lift.

(3) 20 hours of training on the areas of operation listed in § 61.127(b)(5) of this part that includes at least—

(i) Ten hours of instrument training using a view-limiting device including attitude instrument flying, partial panel skills, recovery from unusual flight attitudes, and intercepting and tracking navigational systems. Five hours of the 10 hours required on instrument training must be in a powered-lift;

(ii) One 2-hour cross country flight in a powered-lift in daytime conditions that consists of a total straight-line distance of more than 100 nautical miles from the original point of departure;

(iii) One 2-hour cross country flight in a powered-lift in nighttime conditions that consists of a total straight-line distance of more than 100 nautical miles from the original point of departure; and

(iv) 3 hours in a powered-lift with an authorized instructor in preparation for the practical test within the preceding 2 calendar months from the month of the test.

(4) Ten hours of solo flight time in a powered-lift or 10 hours of flight time performing the duties of pilot in command in a powered-lift with an authorized instructor on board (either of which may be credited towards the flight time requirement under paragraph (e)(2) of this section, on the areas of operation listed in § 61.127(b)(5) that includes—

(i) One cross-country flight of not less than 300 nautical miles total distance with landings at a minimum of three points, one of which is a straight-line distance of at least 250 nautical miles from the original departure point. However, if this requirement is being met in Hawaii the longest segment need only have a straight-line distance of at least 150 nautical miles; and

(ii) 5 hours in night VFR conditions with 10 takeoffs and 10 landings (with each landing involving a flight in the traffic pattern) at an airport with an operating control tower.

(f) *For a glider rating.* A person who applies for a commercial pilot certificate with a glider category rating must log at least—

(1) 25 hours of flight time as a pilot in a glider and that flight time must include at least 100 flights in a glider as pilot in command, including at least—

(i) Three hours of flight training in a glider with an authorized instructor or 10 training flights in a glider with an authorized instructor on the areas of operation listed in § 61.127(b)(6) of this part, including at least 3 training flights in a glider with an authorized instructor in preparation for the practical test within the preceding 2 calendar months from the month of the test; and

(ii) 2 hours of solo flight that include not less than 10 solo flights in a glider on the areas of operation listed in § 61.127(b)(6) of this part; or

(2) 200 hours of flight time as a pilot in heavier-than-air aircraft and at least 20 flights in a glider as pilot in command, including at least—

(i) Three hours of flight training in a glider or 10 training flights in a glider with an authorized instructor on the areas of operation listed in § 61.127(b)(6) of this part including at least 3 training flights in a glider with an authorized instructor in preparation for the practical test within the preceding 2 calendar months from the month of the test; and

(ii) 5 solo flights in a glider on the areas of operation listed in § 61.127(b)(6) of this part.

(g) *For an airship rating.* A person who applies for a commercial pilot certificate with a lighter-than-air category and airship class rating must log at least 200 hours of flight time as a pilot, which includes at least the following hours:

(1) 50 hours in airships.

(2) Thirty hours of pilot in command flight time in airships or performing the duties of pilot in command in an airship with an authorized instructor aboard, which consists of—

(i) 10 hours of cross-country flight time in airships; and

(ii) 10 hours of night flight time in airships.

(3) Forty hours of instrument time to include—

(i) Instrument training using a view-limiting device for attitude instrument flying, partial panel skills, recovery from unusual flight attitudes, and intercepting and tracking navigational systems; and

(ii) Twenty hours of instrument flight time, of which 10 hours must be in flight in airships.

(4) 20 hours of flight training in airships on the areas of operation listed in § 61.127(b)(7) of this part, which includes at least—

(i) Three hours in an airship with an authorized instructor in preparation for the practical test within the preceding 2 calendar months from the month of the test;

(ii) One hour cross country flight in an airship in daytime conditions that consists of a total straight-line distance of more than 25 nautical miles from the point of departure; and

(iii) One hour cross country flight in an airship in nighttime conditions that consists of a total straight-line distance of more than 25 nautical miles from the point of departure.

(5) 10 hours of flight training performing the duties of pilot in command with an authorized instructor on the areas of operation listed in § 61.127(b)(7) of this part, which includes at least—

(i) One cross-country flight with landings at a minimum of three points, with one segment consisting of a straight-line distance of at least 25 nautical miles from the original point of departure; and

(ii) 5 hours in night VFR conditions with 10 takeoffs and 10 landings (with each landing involving a flight in the traffic pattern).

(h) *For a balloon rating.* A person who applies for a commercial pilot certificate with a lighter-than-air category and a balloon class rating must log at least 35 hours of flight time as a pilot, which includes at least the following requirements:

(1) 20 hours in balloons;

(2) 10 flights in balloons;

(3) Two flights in balloons as the pilot in command; and

(4) 10 hours of flight training that includes at least 10 training flights with an authorized instructor in balloons on the areas of operation listed in § 61.127(b)(8) of this part, which consists of at least—

(i) For a gas balloon—

(A) Two training flights of 2 hours each in a gas balloon with an authorized instructor in preparation for the practical test within the preceding 2 calendar months from the month of the test;

(B) 2 flights performing the duties of pilot in command in a gas balloon with an authorized instructor on the appropriate areas of operation; and

(C) One flight involving a controlled ascent to 5,000 feet above the launch site.

(ii) For a balloon with an airborne heater—

(A) Two training flights of 1 hour each in a balloon with an airborne heater with an authorized instructor in preparation for the practical test within the preceding 2 calendar months from the month of the test;

(B) Two solo flights in a balloon with an airborne heater on the appropriate areas of operation; and

(C) One flight involving a controlled ascent to 3,000 feet above the launch site.

(i) *Permitted credit for use of a flight simulator or flight training device.* (1) Except as provided in paragraph (i)(2) of this section, an applicant who has not accomplished the training required by this section in a course conducted by a training center certificated under part 142 of this chapter may:

(i) Credit a maximum of 50 hours toward the total aeronautical experience requirements for an airplane or powered-lift rating, provided the aeronautical experience was obtained from an authorized instructor in a flight simulator or flight training device that represents that class of airplane or powered-lift category and type, if applicable, appropriate to the rating sought; and

(ii) Credit a maximum of 25 hours toward the total aeronautical experience requirements of this section for a helicopter rating, provided the aeronautical experience was obtained from an authorized instructor in a flight simulator or flight training device that represents a helicopter and type, if applicable, appropriate to the rating sought.

(2) An applicant who has accomplished the training required by this section in a course conducted by a training center certificated under part 142 of this chapter may:

(i) Credit a maximum of 100 hours toward the total aeronautical experience requirements of this section for an airplane and powered-lift rating, provided the aeronautical experience was obtained from an authorized instructor in a flight simulator or flight training device that represents that class of airplane or powered-lift category and type, if applicable, appropriate to the rating sought; and

(ii) Credit a maximum of 50 hours toward the total aeronautical experience requirements of this section for a helicopter rating, provided the aeronautical experience was obtained from an authorized instructor in a flight simulator or flight training device that represents a helicopter and type, if applicable, appropriate to the rating sought.

(3) Except when fewer hours are approved by the FAA, an applicant for the commercial pilot certificate with the airplane or powered-lift rating who has completed 190 hours of aeronautical experience is considered to have met the total aeronautical experience requirements of this section, provided the applicant satisfactorily completed an approved commercial pilot course under part 142 of this chapter and the approved course was appropriate to the commercial pilot certificate and aircraft rating sought.

[Doc. No. 25910, 62 FR 16298, Apr. 4, 1997; Amdt. 61-101, 62 FR 16892, Apr. 8, 1997; Amdt. 61-103, 62 FR 40904, July 30, 1997; Amdt. 61-104, 63 FR 20288, Apr. 23, 1998; Amdt. 61-124, 74 FR 42558, Aug. 21, 2009; Amdt. 61-124A, 74 FR 53645, Oct. 20, 2009]

§ 61.131 Exceptions to the night flying requirements.

(a) Subject to the limitations of paragraph (b) of this section, a person is not required to comply with the night flight training requirements of this subpart if the person receives flight training in and resides in the State of Alaska.

(b) A person who receives flight training in and resides in the State of Alaska but does not meet the night flight training requirements of this section:

(1) May be issued a pilot certificate with the limitation "night flying prohibited."

(2) Must comply with the appropriate night flight training requirements of this subpart within the 12-calendar-month period after the issuance of the pilot certificate. At the end of that period, the certificate will become invalid for use until the person complies with the appropriate night flight training requirements of this subpart. The person may have the "night flying prohibited" limitation removed if the person—

(i) Accomplishes the appropriate night flight training requirements of this subpart; and

(ii) Presents to an examiner a logbook or training record endorsement from an authorized instructor that verifies accomplishment of the appropriate night flight training requirements of this subpart.

[Doc. No. 25910, 62 FR 16298, Apr. 4, 1997; Amdt. 61-103, 62 FR 40905, July 30, 1997]

§ 61.133 Commercial pilot privileges and limitations.

(a) *Privileges* —(1) *General.* A person who holds a commercial pilot certificate may act as pilot in command of an aircraft—

(i) Carrying persons or property for compensation or hire, provided the person is qualified in accordance with this part and with the applicable parts of this chapter that apply to the operation; and

(ii) For compensation or hire, provided the person is qualified in accordance with this part and with the applicable parts of this chapter that apply to the operation.

(2) *Commercial pilots with lighter-than-air category ratings.* A person with a commercial pilot certificate with a lighter-than-air category rating may—

(i) *For an airship* —(A) Give flight and ground training in an airship for the issuance of a certificate or rating;

(B) Give an endorsement for a pilot certificate with an airship rating;

(C) Endorse a student pilot certificate or logbook for solo operating privileges in an airship;

(D) Act as pilot in command of an airship under IFR or in weather conditions less than the minimum prescribed for VFR flight; and

(E) Give flight and ground training and endorsements that are required for a flight review, an operating privilege or recency-of-experience requirements of this part.

(ii) *For a balloon* —(A) Give flight and ground training in a balloon for the issuance of a certificate or rating;

(B) Give an endorsement for a pilot certificate with a balloon rating;

(C) Endorse a student pilot certificate or logbook for solo operating privileges in a balloon; and

(D) Give ground and flight training and endorsements that are required for a flight review, an operating privilege, or recency-of-experience requirements of this part.

(b) *Limitations.* (1) A person who applies for a commercial pilot certificate with an airplane category or powered-lift category rating and does not hold an instrument rating in the same category and class will be issued a commercial pilot certificate that contains the limitation, "The carriage of passengers for hire in (airplanes) (powered-lifts) on cross-country flights in excess of 50 nautical miles or at night is prohibited." The limitation may be removed when the person satisfactorily accomplishes the requirements listed in § 61.65 of this part for an instrument rating in the same category and class of aircraft listed on the person's commercial pilot certificate.

(2) If a person who applies for a commercial pilot certificate with a balloon rating takes a practical test in a balloon with an airborne heater—

(i) The pilot certificate will contain a limitation restricting the exercise of the privileges of that certificate to a balloon with an airborne heater.

(ii) The limitation specified in paragraph (b)(2)(i) of this section may be removed when the person obtains the required aeronautical experience in a gas balloon and receives a logbook endorsement from an authorized instructor who attests to the person's accomplishment of the required aeronautical experience and ability to satisfactorily operate a gas balloon.

(3) If a person who applies for a commercial pilot certificate with a balloon rating takes a practical test in a gas balloon—

(i) The pilot certificate will contain a limitation restricting the exercise of the privileges of that certificate to a gas balloon.

(ii) The limitation specified in paragraph (b)(3)(i) of this section may be removed when the person obtains the required aeronautical experience in a balloon with an airborne heater and receives a logbook endorsement from an authorized instructor who attests to the person's accomplishment of the required aeronautical experience and ability to satisfactorily operate a balloon with an airborne heater.

[Doc. No. 25910, 62 FR 16298, Apr. 4, 1997; Amdt. 61-103, 62 FR 40905, July 30, 1997]

§§ 61.135-61.141 [Reserved]

Subpart G—Airline Transport Pilots

§ 61.151 Applicability.

This subpart prescribes the requirements for the issuance of airline transport pilot certificates and ratings, the conditions under which those certificates and ratings are necessary, and the general operating rules for persons who hold those certificates and ratings.

§ 61.153 Eligibility requirements: General.

To be eligible for an airline transport pilot certificate, a person must:

(a) Meet the following age requirements:

(1) For an airline transport pilot certificate obtained under the aeronautical experience requirements of §§ 61.159, 61.161, or 61.163, be at least 23 years of age; or

(2) For an airline transport pilot certificate obtained under the aeronautical experience requirements of § 61.160, be at least 21 years of age.

(b) Be able to read, speak, write, and understand the English language. If the applicant is unable to meet one of these requirements due to medical reasons, then the Administrator may place such operating limitations on that applicant's pilot certificate as are necessary for the safe operation of the aircraft;

(c) Be of good moral character;

(d) Meet at least one of the following requirements:

(1) Holds a commercial pilot certificate with an instrument rating issued under this part;

(2) Meet the military experience requirements under § 61.73 of this part to qualify for a commercial pilot certificate, and an instrument rating if the person is a rated military pilot or former rated military pilot of an Armed Force of the United States; or

(3) Holds either a foreign airline transport pilot license with instrument privileges, or a foreign commercial pilot license with an instrument rating, that—

(i) Was issued by a contracting State to the Convention on International Civil Aviation; and

(ii) Contains no geographical limitations.

(e) After July 31, 2014, for an airline transport pilot certificate with an airplane category multiengine class rating or an airline transport pilot certificate obtained concurrently with an airplane type rating, receive a graduation certificate from an authorized training provider certifying completion of the airline transport pilot certification training program specified in § 61.156 before applying for the knowledge test required by paragraph (g) of this section;

(f) Meet the aeronautical experience requirements of this subpart that apply to the aircraft category and class rating sought before applying for the practical test;

(g) Pass a knowledge test on the aeronautical knowledge areas of § 61.155(c) of this part that apply to the aircraft category and class rating sought;

(h) Pass the practical test on the areas of operation listed in § 61.157(e) of this part that apply to the aircraft category and class rating sought; and

(i) Comply with the sections of this subpart that apply to the aircraft category and class rating sought.

[Doc. No. 25910, 62 FR 16298, Apr. 4, 1997; Amdt. 61-103, 62 FR 40905, July 30, 1997; Amdt. 61-124, 74 FR 42559, Aug. 21, 2009; Amdt. 61-130, 78 FR 42374, July 15, 2013]

§ 61.155 Aeronautical knowledge.

(a) *General.* The knowledge test for an airline transport pilot certificate is based on the aeronautical knowledge areas listed in paragraph (c) of this section that are appropriate to the aircraft category and class rating sought.

(b) *Aircraft type rating.* A person who is applying for an additional aircraft type rating to be added to an airline transport pilot certificate is not required to pass a knowledge test if that person's airline transport pilot certificate lists the aircraft category and class rating that is appropriate to the type rating sought.

(c) *Aeronautical knowledge areas.* (1) Applicable Federal Aviation Regulations of this chapter that relate to airline transport pilot privileges, limitations, and flight operations;

(2) Meteorology, including knowledge of and effects of fronts, frontal characteristics, cloud formations, icing, and upper-air data;

(3) General system of weather and NOTAM collection, dissemination, interpretation, and use;

(4) Interpretation and use of weather charts, maps, forecasts, sequence reports, abbreviations, and symbols;

(5) National Weather Service functions as they pertain to operations in the National Airspace System;

(6) Windshear and microburst awareness, identification, and avoidance;

(7) Principles of air navigation under instrument meteorological conditions in the National Airspace System;

(8) Air traffic control procedures and pilot responsibilities as they relate to en route operations, terminal area and radar operations, and instrument departure and approach procedures;

(9) Aircraft loading, weight and balance, use of charts, graphs, tables, formulas, and computations, and their effect on aircraft performance;

(10) Aerodynamics relating to an aircraft's flight characteristics and performance in normal and abnormal flight regimes;

(11) Human factors;

(12) Aeronautical decision making and judgment;

(13) Crew resource management to include crew communication and coordination; and

(14) After July 31, 2014, for airplane category multiengine class rating or airplane type rating, the content of the airline transport pilot certification training program in § 61.156.

(d) An applicant who successfully completes the knowledge test for an airline transport pilot certificate prior to August 1, 2014, must successfully complete the practical test within 24 months from the month in which the knowledge test was successfully completed. An applicant who passes the knowledge test prior to August 1, 2014, but fails to successfully complete the practical test within 24 months must complete the airline transport pilot certification training program specified in § 61.156 and retake the knowledge test prior to applying for the practical test.

[Docket No. 25910, 62 FR 16298, Apr. 4, 1997, as amended by Amdt. 61-130, 78 FR 42374, July 15, 2013]

§ 61.156 Training requirements: Airplane category—multiengine class rating or airplane type rating concurrently with airline transport pilot certificate.

After July 31, 2014, a person who applies for the knowledge test for an airline transport pilot certificate with an airplane category multiengine class rating must present a graduation certificate from an authorized training provider under part 121, 135, 141, or 142 of this chapter certifying the applicant has completed the following training in a course approved by the Administrator.

(a) *Academic training.* The applicant for the knowledge test must receive at least 30 hours of classroom instruction that includes the following:

(1) At least 8 hours of instruction on aerodynamics including high altitude operations;

(2) At least 2 hours of instruction on meteorology, including adverse weather phenomena and weather detection systems; and

(3) At least 14 hours of instruction on air carrier operations, including the following areas:

(i) Physiology;

(ii) Communications;

(iii) Checklist philosophy;

(iv) Operational control;

(v) Minimum equipment list/configuration deviation list;

(vi) Ground operations;

(vii) Turbine engines;

(viii) Transport category aircraft performance;

(ix) Automation, navigation, and flight path warning systems.

(4) At least 6 hours of instruction on leadership, professional development, crew resource management, and safety culture.

(b) *FSTD training.* The applicant for the knowledge test must receive at least 10 hours of training in a flight simulation training device qualified under part 60 of this chapter that represents a multiengine turbine airplane. The training must include the following:

(1) At least 6 hours of training in a Level C or higher full flight simulator qualified under part 60 of this chapter that represents a multiengine turbine airplane with a maximum takeoff weight of 40,000 pounds or greater. The training must include the following areas:

(i) Low energy states/stalls;

(ii) Upset recovery techniques; and

(iii) Adverse weather conditions, including icing, thunderstorms, and crosswinds with gusts.

(2) The remaining FSTD training may be completed in a Level 4 or higher flight simulation training device. The training must include the following areas:

(i) Navigation including flight management systems; and

(ii) Automation including autoflight.

(c) *Deviation authority.* The Administrator may issue deviation authority from the weight requirement in paragraph (b)(1) of this section upon a determination that the objectives of the training can be met in an alternative device.

[Doc. No. FAA-2010-0100, 78 FR 42375, July 15, 2013]

§ 61.157 Flight proficiency.

(a) *General.* (1) The practical test for an airline transport pilot certificate is given for—

(i) An airplane category and single engine class rating.

(ii) An airplane category and multiengine class rating.

(iii) A rotorcraft category and helicopter class rating.

(iv) A powered-lift category rating.

(v) An aircraft type rating.

(2) A person who is applying for an airline transport pilot practical test must meet—

(i) The eligibility requirements of § 61.153; and

(ii) The aeronautical knowledge and aeronautical experience requirements of this subpart that apply to the aircraft category and class rating sought.

(b) *Aircraft type rating.* Except as provided in paragraph (c) of this section, a person who applies for an aircraft type rating to be added to an airline transport pilot certificate or applies for a type rating to be concurrently completed with an airline transport pilot certificate:

(1) Must receive and log ground and flight training from an authorized instructor on the areas of operation under this section that apply to the aircraft type rating;

(2) Must receive a logbook endorsement from an authorized instructor that certifies the applicant completed the training on the areas of operation listed under paragraph (e) of this section that apply to the aircraft type rating; and

(3) Must perform the practical test in actual or simulated instrument conditions, except as provided under paragraph (g) of this section.

(c) *Exceptions.* A person who applies for an aircraft type rating to be added to an airline transport pilot certificate or an aircraft type rating concurrently with an airline transport pilot certificate, and who is an employee of a certificate holder operating under part 121 or part 135 of this chapter, does not need to comply with the requirements of paragraph (b) of this section if the applicant presents a training record that shows completion of that certificate holder's approved training program for the aircraft type rating.

(d) *Upgrading type ratings.* Any type rating(s) and limitations on a pilot certificate of an applicant who completes an airline transport pilot practical test will be included at the airline transport pilot certification level, provided the applicant passes the practical test in the same category and class of aircraft for which the applicant holds the type rating(s).

(e) *Areas of operation.* (1) For an airplane category—single engine class rating:

(i) Preflight preparation;

(ii) Preflight procedures;

(iii) Takeoff and departure phase;

(iv) In-flight maneuvers;

(v) Instrument procedures;

(vi) Landings and approaches to landings;

(vii) Normal and abnormal procedures;

(viii) Emergency procedures; and

(ix) Postflight procedures.

(2) For an airplane category—multiengine class rating:

(i) Preflight preparation;

(ii) Preflight procedures;

(iii) Takeoff and departure phase;

(iv) In-flight maneuvers;

(v) Instrument procedures;

(vi) Landings and approaches to landings;

(vii) Normal and abnormal procedures;

(viii) Emergency procedures; and

(ix) Postflight procedures.

(3) For a powered-lift category rating:

(i) Preflight preparation;

(ii) Preflight procedures;

(iii) Takeoff and departure phase;

(iv) In-flight maneuvers;

(v) Instrument procedures;

(vi) Landings and approaches to landings;

(vii) Normal and abnormal procedures;

(viii) Emergency procedures; and

(ix) Postflight procedures.

(4) For a rotorcraft category—helicopter class rating:

(i) Preflight preparation;

(ii) Preflight procedures;

(iii) Takeoff and departure phase;

(iv) In-flight maneuvers;

(v) Instrument procedures;

(vi) Landings and approaches to landings;

(vii) Normal and abnormal procedures;

(viii) Emergency procedures; and

(ix) Postflight procedures.

(f) *Proficiency and competency checks conducted under part 121, part 135, or subpart K of part 91.* (1) Successful completion of any of the following checks satisfies the flight proficiency requirements of this section for the issuance of an airline transport pilot certificate and/or the appropriate aircraft rating:

(i) A proficiency check under § 121.441 of this chapter.

(ii) Both a competency check under § 135.293(a)(2) and § 135.293(b) of this chapter and pilot-in-command instrument proficiency check under § 135.297 of this chapter.

(iii) Both a competency check under § 91.1065 of this chapter and a pilot-in-command instrument proficiency check under § 91.1069 of this chapter.

(2) The checks specified in paragraph (f)(1) of this section must be conducted by one of the following:

(i) An FAA Aviation Safety Inspector.

(ii) An Aircrew Program Designee who is authorized to perform proficiency and/or competency checks for the air carrier whose approved training program has been satisfactorily completed by the pilot applicant.

(iii) A Training Center Evaluator with appropriate certification authority who is also authorized to perform the portions of the competency and/or proficiency checks required by paragraph (f)(1) of this section for the air carrier whose approved training program has been satisfactorily completed by the pilot applicant.

(g) *Aircraft not capable of instrument maneuvers and procedures.* An applicant may add a type rating to an airline transport pilot certificate with an aircraft that is not capable of the instrument maneuvers and procedures required on the practical test under the following circumstances—

(1) The rating is limited to "VFR only."

(2) The type rating is added to an airline transport pilot certificate that has instrument privileges in that category and class of aircraft.

(3) The "VFR only" limitation may be removed for that aircraft type after the applicant:

(i) Passes a practical test in that type of aircraft on the appropriate instrument maneuvers and procedures in § 61.157; or

(ii) Becomes qualified in § 61.73(d) for that type of aircraft.

(h) *Multiengine airplane with a single-pilot station.* An applicant for a type rating, at the ATP certification level, in a multiengine airplane with a single-pilot station must perform the practical test in the multi-seat version of that airplane. The practical test may be performed in the single-seat version of that airplane if the Examiner is in a position to observe the applicant during the practical test in the case where there is no multi-seat version of that multiengine airplane.

(i) *Single engine airplane with a single-pilot station.* An applicant for a type rating, at the ATP certification level, in a single engine airplane with a single-pilot station must perform the practical test in the multi-seat version of that single engine airplane. The practical test may be performed in the single-seat version of that airplane if the Examiner is in a position to observe the applicant during the practical test in the case where there is no multi-seat version of that single engine airplane.

(j) *Waiver authority.* An Examiner who conducts a practical test may waive any task for which the FAA has provided waiver authority.

[Doc. No. FAA-2006-26661, 74 FR 42560, Aug. 21, 2009; Amdt. 61-124A, 74 FR 53647, Oct. 20, 2009; Amdt. 61-130, 78 FR 42375, July 15, 2013]

§ 61.158 [Reserved]

§ 61.159 Aeronautical experience: Airplane category rating.

(a) Except as provided in paragraphs (b), (c), and (d) of this section, a person who is applying for an airline transport pilot certificate with an airplane category and class rating must have at least 1,500 hours of total time as a pilot that includes at least:

(1) 500 hours of cross-country flight time.

(2) 100 hours of night flight time.

(3) 50 hours of flight time in the class of aircraft for which the rating is sought. A maximum of 25 hours of training in a full flight simulator representing a multiengine airplane may be credited toward the flight time requirement of this paragraph if the training was accomplished as part of an approved training course in parts 121, 135, 141, or 142 of this chapter. A flight training device or aviation training device may not be used to satisfy this requirement.

(4) 75 hours of instrument flight time, in actual or simulated instrument conditions, subject to the following:

(i) Except as provided in paragraph (a)(4)(ii) of this section, an applicant may not receive credit for more than a total of 25 hours of simulated instrument time in a flight simulator or flight training device.

(ii) A maximum of 50 hours of training in a flight simulator or flight training device may be credited toward the instrument flight time requirements of paragraph (a)(4) of this section if the training was accomplished in a course conducted by a training center certificated under part 142 of this chapter.

(iii) Training in a flight simulator or flight training device must be accomplished in a flight simulator or flight training device, representing an airplane.

(5) Not more than 100 hours of the total aeronautical experience requirements of paragraph (a) of this section may be obtained in a full flight simulator or flight training device that represents an airplane, provided the aeronautical experience was accomplished as part of an approved training course in parts 121, 135, 141, or 142 of this chapter.

(6) Not more than 100 hours of the total aeronautical experience requirements of paragraph (a) of this section may be obtained in a flight simulator or flight training device that represents an airplane, provided the aeronautical experience was obtained in an approved course conducted by a training center certificated under part 142 of this chapter.

(b) A person who has performed at least 20 night takeoffs and landings to a full stop may substitute each additional night takeoff and landing to a full stop for 1 hour of night flight time to satisfy the requirements of paragraph (a)(2) of this section; however, not more than 25 hours of night flight time may be credited in this manner.

(c) A commercial pilot may credit the following second-in-command flight time or flight-engineer flight time toward the 1,500 hours of total time as a pilot required by paragraph (a) of this section:

(1) Second-in-command time, provided the time is acquired in an airplane—

(i) Required to have more than one pilot flight crewmember by the airplane's flight manual, type certificate, or the regulations under which the flight is being conducted;

(ii) Engaged in operations under subpart K of part 91, part 121, or part 135 of this chapter for which a second in command is required; or

(iii) That is required by the operating rules of this chapter to have more than one pilot flight crewmember.

(2) Flight-engineer time, provided the time—

(i) Is acquired in an airplane required to have a flight engineer by the airplane's flight manual or type certificate;

(ii) Is acquired while engaged in operations under part 121 of this chapter for which a flight engineer is required;

(iii) Is acquired while the person is participating in a pilot training program approved under part 121 of this chapter; and

(iv) Does not exceed more than 1 hour for each 3 hours of flight engineer flight time for a total credited time of no more than 500 hours.

(3) Flight-engineer time, provided the flight time—

(i) Is acquired as a U.S. Armed Forces' flight engineer crewmember in an airplane that requires a flight engineer crewmember by the flight manual;

(ii) Is acquired while the person is participating in a flight engineer crewmember training program for the U.S. Armed Forces; and

(iii) Does not exceed 1 hour for each 3 hours of flight engineer flight time for a total credited time of no more than 500 hours.

(d) An applicant is issued an airline transport pilot certificate with the limitation, "Holder does not meet the pilot in command aeronautical experience requirements of ICAO," as prescribed under Article 39 of the Convention on International Civil Aviation, if the applicant does not meet the ICAO requirements contained in Annex 1 "Personnel Licensing" to the Convention on International Civil Aviation, but otherwise meets the aeronautical experience requirements of this section.

(e) An applicant is entitled to an airline transport pilot certificate without the ICAO limitation specified under paragraph (d) of this section when the applicant presents satisfactory evidence of having met the ICAO requirements under paragraph (d) of this section and otherwise meets the aeronautical experience requirements of this section.

[Doc. No. 25910, 62 FR 16298, Apr. 4, 1997; Amdt. 61-103, 62 FR 40906, July 30, 1997; Amdt. 61-104, 63 FR 20288, Apr. 23, 1998; Amdt. 61-109, 68 FR 54560, Sept. 17, 2003; Amdt. 61-124, 74 FR 42561, Aug. 21, 2009; Amdt. 61-130, 78 FR 42375, July 15, 2013]

§ 61.160 Aeronautical experience—airplane category restricted privileges.

(a) Except for a person who has been removed from flying status for lack of proficiency or because of a disciplinary action involving aircraft operations, a U.S. military pilot or former U.S. military pilot may apply for an airline transport pilot certificate with an airplane category multiengine class rating or an airline transport pilot certificate concurrently with an airplane type rating with a minimum of 750 hours of total time as a pilot if the pilot presents:

(1) An official Form DD-214 (Certificate of Release or Discharge from Active Duty) indicating that the person was honorably discharged from the U.S. Armed Forces or an official U.S. Armed Forces record that shows the pilot is currently serving in the U.S. Armed Forces; and

(2) An official U.S. Armed Forces record that shows the person graduated from a U.S. Armed Forces undergraduate pilot training school and received a rating qualification as a military pilot.

(b) A person may apply for an airline transport pilot certificate with an airplane category multiengine class rating or an airline transport pilot certificate concurrently with an airplane type rating with a minimum of 1,000 hours of total time as a pilot if the person:

(1) Holds a Bachelor's degree with an aviation major from an institution of higher education, as defined in § 61.1, that has been issued a letter of authorization by the Administrator under § 61.169;

(2) Completes 60 semester credit hours of aviation and aviation-related coursework that has been recognized by the Administrator as coursework designed to improve and enhance the knowledge and skills of a person seeking a career as a professional pilot;

(3) Holds a commercial pilot certificate with an airplane category and instrument rating if:

(i) The required ground training was completed as part of an approved part 141 curriculum at the institution of higher education; and

(ii) The required flight training was completed as part of an approved part 141 curriculum at the institution of higher education or at a part 141 pilot school that has a training agreement under § 141.26 of this chapter with the institution of higher education; and

(4) Presents official transcripts or other documentation acceptable to the Administrator from the institution of higher education certifying that the graduate has satisfied the requirements in paragraphs (b)(1) through (3) of this section.

(c) A person may apply for an airline transport pilot certificate with an airplane category multiengine class rating or an airline transport pilot certificate concurrently with an airplane type rating with a minimum of 1,250 hours of total time as a pilot if the person:

(1) Holds an Associate's degree with an aviation major from an institution of higher education, as defined in § 61.1, that has been issued a letter of authorization by the Administrator under § 61.169;

(2) Completes at least 30 semester credit hours of aviation and aviation-related coursework that has been recognized by the Administrator as coursework designed to improve and enhance the knowledge and skills of a person seeking a career as a professional pilot;

(3) Holds a commercial pilot certificate with an airplane category and instrument rating if:

(i) The required ground training was completed as part of an approved part 141 curriculum at the institution of higher education; and

(ii) The required flight training was completed as part of an approved part 141 curriculum at the institution of higher education or at a part 141 pilot school that has a written training agreement under § 141.26 of this chapter with the institution of higher education; and

(4) Presents official transcripts or other documentation acceptable to the Administrator from the institution of higher education certifying that the graduate has satisfied the requirements in paragraphs (c)(1) through (3) of this section.

(d) A graduate of an institution of higher education who completes fewer than 60 semester credit hours but at least 30 credit hours and otherwise satisfies the requirements of paragraph (b) may apply for airline transport pilot certificate with an airplane category multiengine class rating or an airline transport pilot certificate concurrently with an airplane type rating with a minimum of 1,250 hours of total time as a pilot.

(e) A person who applies for an airline transport pilot certificate under the total flight times listed in paragraphs (a), (b), and (c) of this section must otherwise meet the aeronautical experience requirements of § 61.159, except that the person may apply for an airline transport pilot certificate with 200 hours of cross-country flight time.

(f) A person who has 1,500 hours total time as a pilot, 200 hours of cross-country flight time, and otherwise meets the aeronautical experience requirements of § 61.159 may apply for an airline transport pilot certificate under this section.

(g) An airline transport pilot certificate obtained under this section is subject to the pilot in command limitations set forth in § 61.167(b) and must contain the following limitation, "Restricted in accordance with 14 CFR 61.167." The pilot is entitled to an airline transport pilot certificate without the limitation specified in this paragraph when the applicant presents satisfactory evidence of having met the aeronautical experience requirements of § 61.159 and the age requirement of § 61.153(a)(1).

(h) An applicant who meets the aeronautical experience requirements of paragraphs (a), (b), (c), and (d) of this section is issued an airline transport pilot certificate with the limitation, "Holder does not meet the pilot in command aeronautical experience requirements of ICAO," as prescribed under Article 39 of the Convention on International Civil Aviation if the applicant does not meet the ICAO requirements contained in Annex 1 "Personnel Licensing" to the Convention on International Civil Aviation. An applicant is entitled to an airline transport pilot certificate without the ICAO limitation specified under this paragraph when the applicant presents satisfactory evidence of having met the ICAO requirements and otherwise meets the aeronautical experience requirements of § 61.159.

[Doc. No. FAA-2010-0100, 78 FR 42375, July 15, 2013]

§ 61.161 Aeronautical experience: Rotorcraft category and helicopter class rating.

(a) A person who is applying for an airline transport pilot certificate with a rotorcraft category and helicopter class rating, must have at least 1,200 hours of total time as a pilot that includes at least:

(1) 500 hours of cross-country flight time;

(2) 100 hours of night flight time, of which 15 hours are in helicopters;

(3) 200 hours of flight time in helicopters, which includes at least 75 hours as a pilot in command, or as second in command performing the duties of a pilot in command under the supervision of a pilot in command, or any combination thereof; and

(4) 75 hours of instrument flight time in actual or simulated instrument meteorological conditions, of which at least 50 hours are obtained in flight with at least 25 hours in helicopters as a pilot in command, or as second in command performing the duties of a pilot in command under the supervision of a pilot in command, or any combination thereof.

(b) Training in a flight simulator or flight training device may be credited toward the instrument flight time requirements of paragraph (a)(4) of this section, subject to the following:

(1) Training in a flight simulator or a flight training device must be accomplished in a flight simulator or flight training device that represents a rotorcraft.

(2) Except as provided in paragraph (b)(3) of this section, an applicant may receive credit for not more than a total of 25 hours of simulated instrument time in a flight simulator and flight training device.

(3) A maximum of 50 hours of training in a flight simulator or flight training device may be credited toward the instrument flight time requirements of paragraph (a)(4) of this section if the aeronautical experience is accomplished in an approved course conducted by a training center certificated under part 142 of this chapter.

[Doc. No. 25910, 62 FR 16298, Apr. 4, 1997; Amdt. 61-103, 62 FR 40906, July 30, 1997; Amdt. 61-104, 63 FR 20289, Apr. 23, 1998]

§ 61.163 Aeronautical experience: Powered-lift category rating.

(a) A person who is applying for an airline transport pilot certificate with a powered-lift category rating must have at least 1,500 hours of total time as a pilot that includes at least:

(1) 500 hours of cross-country flight time;

(2) 100 hours of night flight time;

(3) 250 hours in a powered-lift as a pilot in command, or as a second in command performing the duties of a pilot in command under the supervision of a pilot in command, or any combination thereof, which includes at least—

(i) 100 hours of cross-country flight time; and

(ii) 25 hours of night flight time.

(4) 75 hours of instrument flight time in actual or simulated instrument conditions, subject to the following:

(i) Except as provided in paragraph (a)(4)(ii) of this section, an applicant may not receive credit for more than a total of 25 hours of simulated instrument time in a flight simulator or flight training device.

(ii) A maximum of 50 hours of training in a flight simulator or flight training device may be credited toward the instrument flight time requirements of paragraph (a)(4) of this section if the training was accomplished in a course conducted by a training center certificated under part 142 of this chapter.

(iii) Training in a flight simulator or flight training device must be accomplished in a flight simulator or flight training device that represents a powered-lift.

(b) Not more than 100 hours of the total aeronautical experience requirements of paragraph (a) of this section may be obtained in a flight simulator or flight training device that represents a powered-lift, provided the aeronautical experience was obtained in an approved course conducted by a training center certificated under part 142 of this chapter.

[Doc. No. 25910, 62 FR 16298, Apr. 4, 1997; Amdt. 61-103, 62 FR 40906, July 30, 1997; Amdt. 61-104, 63 FR 20289, Apr. 23, 1998]

§ 61.165 Additional aircraft category and class ratings.

(a) *Rotorcraft category and helicopter class rating.* A person applying for an airline transport certificate with a rotorcraft category and helicopter class rating who holds an airline transport pilot certificate with another aircraft category rating must:

(1) Meet the eligibility requirements of § 61.153 of this part;

(2) Pass a knowledge test on the aeronautical knowledge areas of § 61.155(c) of this part;

(3) Comply with the requirements in § 61.157(b) of this part, if appropriate;

(4) Meet the applicable aeronautical experience requirements of § 61.161 of this part; and

(5) Pass the practical test on the areas of operation of § 61.157(e)(4) of this part.

(b) *Airplane category rating with a single-engine class rating.* A person applying for an airline transport certificate with an airplane category and single-engine class rating who holds an airline transport pilot certificate with another aircraft category rating must:

(1) Meet the eligibility requirements of § 61.153 of this part;

(2) Pass a knowledge test on the aeronautical knowledge areas of § 61.155(c) of this part;

(3) Comply with the requirements in § 61.157(b) of this part, if appropriate;

(4) Meet the applicable aeronautical experience requirements of § 61.159 of this part; and

(5) Pass the practical test on the areas of operation of § 61.157(e)(1) of this part.

(c) *Airplane category rating with a multiengine class rating.* A person applying for an airline transport certificate with an airplane category and multiengine class rating who holds an airline transport certificate with another aircraft category rating must:

(1) Meet the eligibility requirements of § 61.153 of this part;

(2) After July 31, 2014, successfully complete the airline transport pilot certification training program specified in § 61.156;

(3) Pass a knowledge test for an airplane category multiengine class rating or type rating on the aeronautical knowledge areas of § 61.155(c);

(4) Comply with the requirements in § 61.157(b) of this part, if appropriate;

(5) Meet the aeronautical experience requirements of § 61.159 or § 61.160; and

(6) Pass the practical test on the areas of operation of § 61.157(e)(2) of this part.

(d) *Powered-lift category.* A person applying for an airline transport pilot certificate with a powered-lift category rating who holds an airline transport certificate with another aircraft category rating must:

(1) Meet the eligibility requirements of § 61.153 of this part;

(2) Pass a required knowledge test on the aeronautical knowledge areas of § 61.155(c) of this part;

(3) Comply with the requirements in § 61.157(b) of this part, if appropriate;

(4) Meet the applicable aeronautical experience requirements of § 61.163 of this part; and

(5) Pass the required practical test on the areas of operation of § 61.157(e)(3) of this part.

(e) *Additional class rating within the same aircraft category.* Except as provided in paragraph (f) of this section, a person applying for an airline transport pilot certificate with an additional class rating who holds an airline transport certificate in the same aircraft category must—

(1) Meet the eligibility requirements of § 61.153, except paragraph (g) of that section;

(2) Comply with the requirements in § 61.157(b) of this part, if applicable;

(3) Meet the applicable aeronautical experience requirements of subpart G of this part; and

(4) Pass a practical test on the areas of operation of § 61.157(e) appropriate to the aircraft rating sought.

(f) *Adding a multiengine class rating or airplane type rating to an airline transport pilot certificate with a single engine class rating.* A person applying to add a multiengine class rating or airplane type rating to an airline transport pilot certificate with an airplane category single engine class rating must—

(1) Meet the eligibility requirements of § 61.153;

(2) Pass a required knowledge test on the aeronautical knowledge areas of § 61.155(c), as applicable to multiengine airplanes;

(3) Comply with the requirements in § 61.157(b), if applicable;

(4) Meet the applicable aeronautical experience requirements of § 61.159; and

(5) Pass a practical test on the areas of operation of § 61.157(e)(2).

(g) *Category class ratings for the operation of aircraft with experimental certificates.* Notwithstanding the provisions of paragraphs (a) through (f) of this section, a person holding an airline transport certificate may apply for a category and class rating limited to a specific make and model of experimental aircraft, provided—

(1) The person has logged at least 5 hours flight time while acting as pilot in command in the same category, class, make, and model of aircraft that has been issued an experimental certificate;

(2) The person has received a logbook endorsement from an authorized instructor who has determined that he or she is proficient to act as pilot in command of the same category, class, make, and model of aircraft for which application is made; and

(3) The flight time specified in paragraph (g)(1) of this section must be logged between September 1, 2004 and August 31, 2005.

[Doc. No. 25910, 62 FR 16298, Apr. 4, 1997; Amdt. 61-103, 62 FR 40906, July 30, 1997; Amdt. 61-110, 69 FR 44869, July 27, 2004; Amdt. 61-130, 78 FR 42376, July 15, 2013]

§ 61.167 Airline transport pilot privileges and limitations.

(a) *Privileges.* (1) A person who holds an airline transport pilot certificate is entitled to the same privileges as a person who holds a commercial pilot certificate with an instrument rating.

(2) A person who holds an airline transport pilot certificate and has met the aeronautical experience requirements of § 61.159 and the age requirements of § 61.153(a)(1) of this part may instruct—

(i) Other pilots in air transportation service in aircraft of the category, class, and type, as applicable, for which the airline transport pilot is rated and endorse the logbook or other training record of the person to whom training has been given;

(ii) In flight simulators, and flight training devices representing the aircraft referenced in paragraph (b)(1) of this section, when instructing under the provisions of this section and endorse the logbook or other training record of the person to whom training has been given;

(iii) Only as provided in this section, except that an airline transport pilot who also holds a flight instructor certificate can exercise the instructor privileges under subpart H of this part for which he or she is rated; and

(iv) In an aircraft, only if the aircraft has functioning dual controls, when instructing under the provisions of this section.

(3) Excluding briefings and debriefings, an airline transport pilot may not instruct in aircraft, flight simulators, and flight training devices under this section—

(i) For more than 8 hours in any 24-consecutive-hour period; or

(ii) For more than 36 hours in any 7-consecutive-day period.

(4) An airline transport pilot may not instruct in Category II or Category III operations unless he or she has been trained and successfully tested under Category II or Category III operations, as applicable.

(b) *Limitations.* A person who holds an airline transport pilot certificate and has not satisfied the age requirement of § 61.153(a)(1) and the aeronautical experience requirements of § 61.159 may not:

(1) Act as pilot in command in operations conducted under part 121, § 91.1053(a)(2)(i), or § 135.243(a)(1) of this chapter, or

(2) Serve as second in command in flag or supplemental operations in part 121 of this chapter requiring three or more pilots.

[Doc. No. FAA-2010-0100, 78 FR 42376, July 15, 2013]

§ 61.169 Letters of authorization for institutions of higher education.

(a) An institution of higher education that is accredited, as defined in § 61.1, may apply for a letter of authorization for the purpose of certifying its graduates for an airline transport pilot certificate under the academic and aeronautical experience requirements in § 61.160. The application must be in a form and manner acceptable to the Administrator.

(b) An institution of higher education must comply with the provisions of the letter of authorization and may not certify a graduate unless it determines that the graduate has satisfied the requirements of § 61.160, as appropriate.

(c) The Administrator may rescind or amend a letter of authorization if the Administrator determines that the institution of higher education is not complying or is unable to comply with the provisions of the letter of authorization.

[Doc. No. FAA-2010-0100, 78 FR 42377, July 15, 2013]

§§ 61.170-69.171 [Reserved]

Subpart H—Flight Instructors Other than Flight Instructors With a Sport Pilot Rating

§ 61.181 Applicability.

This subpart prescribes the requirements for the issuance of flight instructor certificates and ratings (except for flight instructor certificates with a sport pilot rating), the conditions under which those certificates and ratings are necessary, and the limitations on those certificates and ratings.

[Doc. No. FAA-2001-11133, 69 FR 44869, July 27, 2004]

§ 61.183 Eligibility requirements.

To be eligible for a flight instructor certificate or rating a person must:

(a) Be at least 18 years of age;

(b) Be able to read, speak, write, and understand the English language. If the applicant is unable to meet one of these requirements due to medical reasons, then the Administrator may place such operating limitations on that applicant's flight instructor certificate as are necessary;

(c) Hold either a commercial pilot certificate or airline transport pilot certificate with:

(1) An aircraft category and class rating that is appropriate to the flight instructor rating sought; and

(2) An instrument rating, or privileges on that person's pilot certificate that are appropriate to the flight instructor rating sought, if applying for—

(i) A flight instructor certificate with an airplane category and single-engine class rating;

(ii) A flight instructor certificate with an airplane category and multiengine class rating;

(iii) A flight instructor certificate with a powered-lift rating; or

(iv) A flight instructor certificate with an instrument rating.

(d) Receive a logbook endorsement from an authorized instructor on the fundamentals of instructing listed in § 61.185 of this part appropriate to the required knowledge test;

(e) Pass a knowledge test on the areas listed in § 61.185(a)(1) of this part, unless the applicant:

(1) Holds a flight instructor certificate or ground instructor certificate issued under this part;

(2) Holds a teacher's certificate issued by a State, county, city, or municipality that authorizes the person to teach at an educational level of the 7th grade or higher; or

(3) Is employed as a teacher at an accredited college or university.

(f) Pass a knowledge test on the aeronautical knowledge areas listed in § 61.185(a)(2) and (a)(3) of this part that are appropriate to the flight instructor rating sought;

(g) Receive a logbook endorsement from an authorized instructor on the areas of operation listed in § 61.187(b) of this part, appropriate to the flight instructor rating sought;

(h) Pass the required practical test that is appropriate to the flight instructor rating sought in an:

(1) Aircraft that is representative of the category and class of aircraft for the aircraft rating sought; or

(2) Flight simulator or approved flight training device that is representative of the category and class of aircraft for the rating sought, and used in accordance with a course at a training center certificated under part 142 of this chapter.

(i) Accomplish the following for a flight instructor certificate with an airplane or a glider rating:

(1) Receive a logbook endorsement from an authorized instructor indicating that the applicant is competent and possesses instructional proficiency in stall awareness, spin entry, spins, and spin recovery procedures after providing the applicant with flight training in those training areas in an airplane or glider, as appropriate, that is certificated for spins; and

(2) Demonstrate instructional proficiency in stall awareness, spin entry, spins, and spin recovery procedures. However, upon presentation of the endorsement specified in paragraph (i)(1) of this section an examiner may accept that endorsement as satisfactory evidence of instructional proficiency in stall awareness, spin entry, spins, and spin recovery procedures for the practical test, provided that the practical test is not a retest as a result of the applicant failing the previous test for deficiencies in the knowledge or skill of stall awareness, spin entry, spins, or spin recovery instructional procedures. If the retest is a result of deficiencies in the ability of an applicant to demonstrate knowledge or skill of stall awareness, spin entry, spins, or spin recovery instructional procedures, the examiner must test the person on stall awareness, spin entry, spins, and spin recovery instructional procedures in an airplane or glider, as appropriate, that is certificated for spins;

(j) Log at least 15 hours as pilot in command in the category and class of aircraft that is appropriate to the flight instructor rating sought; and

(k) Comply with the appropriate sections of this part that apply to the flight instructor rating sought.

[Doc. No. 25910, 62 FR 16298, Apr. 4, 1997; Amdt. 61-103, 62 FR 40907, July 30, 1997; Amdt. 61-124, 74 FR 42561, Aug. 21, 2009]

§ 61.185 Aeronautical knowledge.

(a) A person who is applying for a flight instructor certificate must receive and log ground training from an authorized instructor on:

(1) Except as provided in paragraph (b) of this section, the fundamentals of instructing, including:

(i) The learning process;

(ii) Elements of effective teaching;

(iii) Student evaluation and testing;

(iv) Course development;

(v) Lesson planning; and

(vi) Classroom training techniques.

(2) The aeronautical knowledge areas for a recreational, private, and commercial pilot certificate applicable to the aircraft category for which flight instructor privileges are sought; and

(3) The aeronautical knowledge areas for the instrument rating applicable to the category for which instrument flight instructor privileges are sought.

(b) The following applicants do not need to comply with paragraph (a)(1) of this section:

(1) The holder of a flight instructor certificate or ground instructor certificate issued under this part;

(2) The holder of a current teacher's certificate issued by a State, county, city, or municipality that authorizes the person to teach at an educational level of the 7th grade or higher; or

(3) A person employed as a teacher at an accredited college or university.

[Doc. No. 25910, 62 FR 16298, Apr. 4, 1997; Amdt. 61-103, 62 FR 40907, July 30, 1997]

§ 61.187 Flight proficiency.

(a) *General.* A person who is applying for a flight instructor certificate must receive and log flight and ground training from an authorized instructor on the areas of operation listed in this section that apply to the flight instructor rating sought. The applicant's logbook must contain an endorsement from an authorized instructor certifying that the person is proficient to pass a practical test on those areas of operation.

(b) *Areas of operation.* (1) For an airplane category rating with a single-engine class rating:

(i) Fundamentals of instructing;

(ii) Technical subject areas;

(iii) Preflight preparation;

(iv) Preflight lesson on a maneuver to be performed in flight;

(v) Preflight procedures;

(vi) Airport and seaplane base operations;

(vii) Takeoffs, landings, and go-arounds;

(viii) Fundamentals of flight;

(ix) Performance maneuvers;

(x) Ground reference maneuvers;

(xi) Slow flight, stalls, and spins;

(xii) Basic instrument maneuvers;

(xiii) Emergency operations; and

(xiv) Postflight procedures.

(2) For an airplane category rating with a multiengine class rating:

(i) Fundamentals of instructing;

(ii) Technical subject areas;

(iii) Preflight preparation;

(iv) Preflight lesson on a maneuver to be performed in flight;

(v) Preflight procedures;

(vi) Airport and seaplane base operations;

(vii) Takeoffs, landings, and go-arounds;

(viii) Fundamentals of flight;

(ix) Performance maneuvers;

(x) Ground reference maneuvers;

(xi) Slow flight and stalls;

(xii) Basic instrument maneuvers;

(xiii) Emergency operations;

(xiv) Multiengine operations; and

(xv) Postflight procedures.

(3) For a rotorcraft category rating with a helicopter class rating:

(i) Fundamentals of instructing;

(ii) Technical subject areas;

(iii) Preflight preparation;

(iv) Preflight lesson on a maneuver to be performed in flight;

(v) Preflight procedures;

(vi) Airport and heliport operations;

(vii) Hovering maneuvers;

(viii) Takeoffs, landings, and go-arounds;

(ix) Fundamentals of flight;

(x) Performance maneuvers;

(xi) Emergency operations;

(xii) Special operations; and

(xiii) Postflight procedures.

(4) For a rotorcraft category rating with a gyroplane class rating:

(i) Fundamentals of instructing;

(ii) Technical subject areas;

(iii) Preflight preparation;

(iv) Preflight lesson on a maneuver to be performed in flight;

(v) Preflight procedures;

(vi) Airport operations;

(vii) Takeoffs, landings, and go-arounds;

(viii) Fundamentals of flight;

(ix) Performance maneuvers;

(x) Flight at slow airspeeds;

(xi) Ground reference maneuvers;

(xii) Emergency operations; and

(xiii) Postflight procedures.

(5) For a powered-lift category rating:

(i) Fundamentals of instructing;

(ii) Technical subject areas;

(iii) Preflight preparation;

(iv) Preflight lesson on a maneuver to be performed in flight;

(v) Preflight procedures;

(vi) Airport and heliport operations;

(vii) Hovering maneuvers;

(viii) Takeoffs, landings, and go-arounds;

(ix) Fundamentals of flight;

(x) Performance maneuvers;

(xi) Ground reference maneuvers;

(xii) Slow flight and stalls;

(xiii) Basic instrument maneuvers;

(xiv) Emergency operations;

(xv) Special operations; and

(xvi) Postflight procedures.

(6) For a glider category rating:

(i) Fundamentals of instructing;

(ii) Technical subject areas;

(iii) Preflight preparation;

(iv) Preflight lesson on a maneuver to be performed in flight;

(v) Preflight procedures;

(vi) Airport and gliderport operations;

(vii) Launches and landings;

(viii) Fundamentals of flight;

(ix) Performance speeds;

(x) Soaring techniques;

(xi) Performance maneuvers;

(xii) Slow flight, stalls, and spins;

(xiii) Emergency operations; and

(xiv) Postflight procedures.

(7) For an instrument rating with the appropriate aircraft category and class rating:

(i) Fundamentals of instructing;

(ii) Technical subject areas;

(iii) Preflight preparation;

(iv) Preflight lesson on a maneuver to be performed in flight;

(v) Air traffic control clearances and procedures;

(vi) Flight by reference to instruments;
(vii) Navigation aids;
(viii) Instrument approach procedures;
(ix) Emergency operations; and
(x) Postflight procedures.

(c) The flight training required by this section may be accomplished:

(1) In an aircraft that is representative of the category and class of aircraft for the rating sought; or

(2) In a flight simulator or flight training device representative of the category and class of aircraft for the rating sought, and used in accordance with an approved course at a training center certificated under part 142 of this chapter.

[Doc. No. 25910, 62 FR 16298, Apr. 4, 1997; Amdt. 61-103, 62 FR 40907, July 30, 1997; Amdt. 61-124, 74 FR 42561, Aug. 21, 2009]

§ 61.189 Flight instructor records.

(a) A flight instructor must sign the logbook of each person to whom that instructor has given flight training or ground training.

(b) A flight instructor must maintain a record in a logbook or a separate document that contains the following:

(1) The name of each person whose logbook or student pilot certificate that instructor has endorsed for solo flight privileges, and the date of the endorsement; and

(2) The name of each person that instructor has endorsed for a knowledge test or practical test, and the record shall also indicate the kind of test, the date, and the results.

(c) Each flight instructor must retain the records required by this section for at least 3 years.

§ 61.191 Additional flight instructor ratings.

(a) A person who applies for an additional flight instructor rating on a flight instructor certificate must meet the eligibility requirements listed in § 61.183 of this part that apply to the flight instructor rating sought.

(b) A person who applies for an additional rating on a flight instructor certificate is not required to pass the knowledge test on the areas listed in § 61.185(a)(1) of this part.

[Doc. No. 25910, 62 FR 16298, Apr. 4, 1997; Amdt. 61-103, 62 FR 40907, July 30, 1997]

§ 61.193 Flight instructor privileges.

A person who holds a flight instructor certificate is authorized within the limitations of that person's flight instructor certificate and ratings to train and issue endorsements that are required for:

(a) A student pilot certificate;
(b) A pilot certificate;
(c) A flight instructor certificate;
(d) A ground instructor certificate;
(e) An aircraft rating;
(f) An instrument rating;
(g) A flight review, operating privilege, or recency of experience requirement of this part;
(h) A practical test; and
(i) A knowledge test.

[Doc. No. 25910, 62 FR 16298, Apr. 4, 1997; Amdt. 61-103, 62 FR 40907, July 30, 1997, as amended by Amdt. 61-124, 74 FR 42561, Aug. 21, 2009]

§ 61.195 Flight instructor limitations and qualifications.

A person who holds a flight instructor certificate is subject to the following limitations:

(a) *Hours of training.* In any 24-consecutive-hour period, a flight instructor may not conduct more than 8 hours of flight training.

(b) *Aircraft Ratings.* A flight instructor may not conduct flight training in any aircraft for which the flight instructor does not hold:

(1) A pilot certificate and flight instructor certificate with the applicable category and class rating; and

(2) If appropriate, a type rating.

(c) *Instrument Rating.* A flight instructor who provides instrument training for the issuance of an instrument rating, a type rating not limited to VFR, or the instrument training required for commercial pilot and airline transport pilot certificates must hold an instrument rating on his or her pilot certificate and flight instructor certificate that is appropriate to the category and class of aircraft used for the training provided.

(d) *Limitations on endorsements.* A flight instructor may not endorse a:

(1) Student pilot's certificate or logbook for solo flight privileges, unless that flight instructor has—

(i) Given that student the flight training required for solo flight privileges required by this part; and

(ii) Determined that the student is prepared to conduct the flight safely under known circumstances, subject to any limitations listed in the student's logbook that the instructor considers necessary for the safety of the flight.

(2) Student pilot's certificate and logbook for a solo cross-country flight, unless that flight instructor has determined the student's flight preparation, planning, equipment, and proposed procedures are adequate for the proposed flight under the existing conditions and within any limitations listed in the logbook that the instructor considers necessary for the safety of the flight;

(3) Student pilot's logbook for solo flight in a Class B airspace area or at an airport within Class B airspace unless that flight instructor has—

(i) Given that student ground and flight training in that Class B airspace or at that airport; and

(ii) Determined that the student is proficient to operate the aircraft safely.

(4) Logbook of a recreational pilot, unless that flight instructor has—

(i) Given that pilot the ground and flight training required by this part; and

(ii) Determined that the recreational pilot is proficient to operate the aircraft safely.

(5) Logbook of a pilot for a flight review, unless that instructor has conducted a review of that pilot in accordance with the requirements of § 61.56(a) of this part; or

(6) Logbook of a pilot for an instrument proficiency check, unless that instructor has tested that pilot in accordance with the requirements of § 61.57(d) of this part.

(e) *Training in an aircraft that requires a type rating.* A flight instructor may not give flight training in an aircraft that requires the pilot in command to hold a type rating unless the flight instructor holds a type rating for that aircraft on his or her pilot certificate.

(f) *Training received in a multiengine airplane, a helicopter, or a powered-lift.* A flight instructor may not give training required for the issuance of a certificate or rating in a multiengine airplane, a helicopter, or a powered-lift unless that flight instructor has at least 5 flight hours of pilot-in-command time in the specific make and model of multiengine airplane, helicopter, or powered-lift, as appropriate.

(g) *Position in aircraft and required pilot stations for providing flight training.* (1) A flight instructor must perform all training from in an aircraft that complies with the requirements of § 91.109 of this chapter.

(2) A flight instructor who provides flight training for a pilot certificate or rating issued under this part must provide that flight training in an aircraft that meets the following requirements—

(i) The aircraft must have at least two pilot stations and be of the same category, class, and type, if appropriate, that applies to the pilot certificate or rating sought.

(ii) For single-place aircraft, the pre-solo flight training must have been provided in an aircraft that has two pilot stations and is of the same category, class, and type, if appropriate.

(h) *Qualifications of the flight instructor for training first-time flight instructor applicants.* (1) The ground training provided to an initial applicant for a flight instructor certificate must be given by an authorized instructor who—

(i) Holds a ground or flight instructor certificate with the appropriate rating, has held that certificate for at least 24 calendar months, and has given at least 40 hours of ground training; or

(ii) Holds a ground or flight instructor certificate with the appropriate rating, and has given at least 100 hours of ground training in an FAA-approved course.

(2) Except for an instructor who meets the requirements of paragraph (h)(3)(ii) of this section, a flight instructor who provides training to an initial applicant for a flight instructor certificate must—

(i) Meet the eligibility requirements prescribed in § 61.183 of this part;

(ii) Hold the appropriate flight instructor certificate and rating;

(iii) Have held a flight instructor certificate for at least 24 months;

(iv) For training in preparation for an airplane, rotorcraft, or powered-lift rating, have given at least 200 hours of flight training as a flight instructor; and

(v) For training in preparation for a glider rating, have given at least 80 hours of flight training as a flight instructor.

(3) A flight instructor who serves as a flight instructor in an FAA-approved course for the issuance of a flight instructor rating must hold a flight instructor certificate with the appropriate rating and pass the required initial and recurrent flight instructor proficiency tests, in accordance with the requirements of the part under which the FAA-approved course is conducted, and must—

(i) Meet the requirements of paragraph (h)(2) of this section; or

(ii) Have trained and endorsed at least five applicants for a practical test for a pilot certificate, flight instructor certificate, ground instructor certificate, or an additional rating, and at least 80 percent of those applicants passed that test on their first attempt; and

(A) Given at least 400 hours of flight training as a flight instructor for training in an airplane, a rotorcraft, or for a powered-lift rating; or

(B) Given at least 100 hours of flight training as a flight instructor, for training in a glider rating.

(i) *Prohibition against self-endorsements.* A flight instructor shall not make any self-endorsement for a certificate, rating, flight review, authorization, operating privilege, practical test, or knowledge test that is required by this part.

(j) *Additional qualifications required to give training in Category II or Category III operations.* A flight instructor may not give training in Category II or Category III operations unless the flight instructor has been trained and tested in Category II or Category III operations, pursuant to § 61.67 or § 61.68 of this part, as applicable.

(k) *Training for night vision goggle operations.* A flight instructor may not conduct training for night vision goggle operations unless the flight instructor:

(1) Has a pilot and flight instructor certificate with the applicable category and class rating for the training;

(2) If appropriate, has a type rating on his or her pilot certificate for the aircraft;

(3) Is pilot in command qualified for night vision goggle operations, in accordance with § 61.31(k);

(4) Has logged 100 night vision goggle operations as the sole manipulator of the controls;

(5) Has logged 20 night vision goggle operations as the sole manipulator of the controls in the category and class, and type of aircraft, if aircraft class and type is appropriate, that the training will be given in;

(6) Is qualified to act as pilot in command in night vision goggle operations under § 61.57(f) or (g); and

(7) Has a logbook endorsement from an FAA Aviation Safety Inspector or a person who is authorized by the FAA to provide that logbook endorsement that states the flight instructor is authorized to perform the night vision goggle pilot in command qualification and recent flight experience requirements under § 61.31(k) and § 61.57(f) and (g).

[Doc. No. 25910, 62 FR 16298, Apr. 4, 1997; Amdt. 61-103, 62 FR 40907, July 30, 1997; Amdt. 61-124, 74 FR 42561, Aug. 21, 2009]

§ 61.197 Renewal requirements for flight instructor certification.

(a) A person who holds a flight instructor certificate that has not expired may renew that flight instructor certificate by—

(1) Passing a practical test for—

(i) One of the ratings listed on the current flight instructor certificate; or

(ii) An additional flight instructor rating; or

(2) Submitting a completed and signed application with the FAA and satisfactorily completing one of the following renewal requirements—

(i) A record of training students showing that, during the preceding 24 calendar months, the flight instructor has endorsed at least 5 students for a practical test for a certificate or rating and at least 80 percent of those students passed that test on the first attempt.

(ii) A record showing that, within the preceding 24 calendar months, the flight instructor has served as a company check pilot, chief flight instructor, company check airman, or flight instructor in a part 121 or part 135 operation, or in a position involving the regular evaluation of pilots.

(iii) A graduation certificate showing that, within the preceding 3 calendar months, the person has successfully completed an approved flight instructor refresher course consisting of ground training or flight training, or a combination of both.

(iv) A record showing that, within the preceding 12 months from the month of application, the flight instructor passed an official U.S. Armed Forces military instructor pilot proficiency check.

(b) The expiration month of a renewed flight instructor certificate shall be 24 calendar months from—

(1) The month the renewal requirements of paragraph (a) of this section are accomplished; or

(2) The month of expiration of the current flight instructor certificate provided—

(i) The renewal requirements of paragraph (a) of this section are accomplished within the 3 calendar months preceding the expiration month of the current flight instructor certificate, and

(ii) If the renewal is accomplished under paragraph (a)(2)(iii) of this section, the approved flight instructor refresher course must be completed within the 3 calendar months preceding the expiration month of the current flight instructor certificate.

(c) The practical test required by paragraph (a)(1) of this section may be accomplished in a flight simulator or flight training device if the test is accomplished pursuant to an approved course conducted by a training center certificated under part 142 of this chapter.

[Doc. No. 25910, 63 FR 20289, Apr. 23, 1998, as amended by Amdt. 61-124, 74 FR 42562, Aug. 21, 2009]

§ 61.199 Reinstatement requirements of an expired flight instructor certificate.

(a) *Flight instructor certificates.* The holder of an expired flight instructor certificate who has not complied with the flight instructor renewal requirements of § 61.197 may reinstate that flight instructor certificate and ratings by filing a completed and signed application with the FAA and satisfactorily completing one of the following reinstatement requirements:

(1) A flight instructor certification practical test, as prescribed by § 61.183(h), for one of the ratings held on the expired flight instructor certificate.

(2) A flight instructor certification practical test for an additional rating.

(b) *Flight instructor ratings.* (1) A flight instructor rating or a limited flight instructor rating on a pilot certificate is no longer valid and may not be exchanged for a similar rating or a flight instructor certificate.

(2) The holder of a flight instructor rating or a limited flight instructor rating on a pilot certificate may be issued a flight instructor certificate with the current ratings, but only if the person passes the required knowledge and practical test prescribed in this subpart for the issuance of the current flight instructor certificate and rating.

[Doc. No. 25910, 62 FR 16298, Apr. 4, 1997, as amended by Amdt. 61-104, 63 FR 20289, Apr. 23, 1998; Amdt. 61-124, 74 FR 42562, Aug. 21, 2009]

§ 61.201 [Reserved]

Subpart I—Ground Instructors

§ 61.211 Applicability.

This subpart prescribes the requirements for the issuance of ground instructor certificates and ratings, the conditions under which those certificates and ratings are necessary, and the limitations upon those certificates and ratings.

§ 61.213 Eligibility requirements.

(a) To be eligible for a ground instructor certificate or rating a person must:

(1) Be at least 18 years of age;

(2) Be able to read, write, speak, and understand the English language. If the applicant is unable to meet one of these requirements due to medical reasons, then the Administrator may place such operating limitations on that applicant's ground instructor certificate as are necessary;

(3) Except as provided in paragraph (b) of this section, pass a knowledge test on the fundamentals of instructing to include—

(i) The learning process;

(ii) Elements of effective teaching;

(iii) Student evaluation and testing;

(iv) Course development;

(v) Lesson planning; and

(vi) Classroom training techniques.

(4) Pass a knowledge test on the aeronautical knowledge areas in—

(i) For a basic ground instructor rating §§ 61.97, 61.105, and 61.309;

(ii) For an advanced ground instructor rating §§ 61.97, 61.105, 61.125, 61.155, and 61.309; and

(iii) For an instrument ground instructor rating, § 61.65.

(b) The knowledge test specified in paragraph (a)(3) of this section is not required if the applicant:

(1) Holds a ground instructor certificate or flight instructor certificate issued under this part;

(2) Holds a teacher's certificate issued by a State, county, city, or municipality that authorizes the person to teach at an educational level of the 7th grade or higher; or

(3) Is employed as a teacher at an accredited college or university.

[Doc. No. 25910, 62 FR 16298, Apr. 4, 1997, as amended by Amdt. 61-110, 69 FR 44869, July 27, 2004; Amdt. 61-124, 74 FR 42562, Aug. 21, 2009]

§ 61.215 Ground instructor privileges.

(a) A person who holds a basic ground instructor rating is authorized to provide—

(1) Ground training in the aeronautical knowledge areas required for the issuance of a sport pilot certificate, recreational pilot certificate, private pilot certificate, or associated ratings under this part;

(2) Ground training required for a sport pilot, recreational pilot, and private pilot flight review; and

(3) A recommendation for a knowledge test required for the issuance of a sport pilot certificate, recreational pilot certificate, or private pilot certificate under this part.

(b) A person who holds an advanced ground instructor rating is authorized to provide:

(1) Ground training on the aeronautical knowledge areas required for the issuance of any certificate or rating under this part except for the aeronautical knowledge areas required for an instrument rating.

(2) The ground training required for any flight review except for the training required for an instrument rating.

(3) A recommendation for a knowledge test required for the issuance of any certificate or rating under this part except for an instrument rating.

(c) A person who holds an instrument ground instructor rating is authorized to provide:

(1) Ground training in the aeronautical knowledge areas required for the issuance of an instrument rating under this part;

(2) Ground training required for an instrument proficiency check; and

(3) A recommendation for a knowledge test required for the issuance of an instrument rating under this part.

(d) A person who holds a ground instructor certificate is authorized, within the limitations of the ratings on the ground instructor certificate, to endorse the logbook or other training record of a person to whom the holder has provided the training or recommendation specified in paragraphs (a) through (c) of this section.

[Doc. No. 25910, 62 FR 16298, Apr. 4, 1997, as amended by Amdt. 61-110, 69 FR 44869, July 27, 2004; Amdt. 61-124, 74 FR 42562, Aug. 21, 2009]

§ 61.217 Recent experience requirements.

The holder of a ground instructor certificate may not perform the duties of a ground instructor unless the person can show that one of the following occurred during the preceding 12 calendar months:

(a) Employment or activity as a ground instructor giving pilot, flight instructor, or ground instructor training;

(b) Employment or activity as a flight instructor giving pilot, flight instructor, or ground instructor ground or flight training;

(c) Completion of an approved flight instructor refresher course and receipt of a graduation certificate for that course; or

(d) An endorsement from an authorized instructor certifying that the person has demonstrated knowledge in the subject areas prescribed under § 61.213(a)(3) and (a)(4), as appropriate.

[Doc. No. FAA-2006-26661, 74 FR 42562, Aug. 21, 2009]

Subpart J—Sport Pilots

Source: Docket No. FAA-2001-11133, 69 FR 44869, July 27, 2004, unless otherwise noted.

If you hold	And you hold	Then you may operate	And
(1) A medical certificate	(i) A sport pilot certificate,	(A) Any light-sport aircraft for which you hold the endorsements required for its category and class	(*1*) You must hold any other endorsements required by this subpart, and comply with the limitations in §61.315.
	(ii) At least a recreational pilot certificate with a category and class rating,	(A) Any light-sport aircraft in that category and class,	(*1*) You do not have to hold any of the endorsements required by this subpart, nor do you have to comply with the limitations in §61.315.
	(iii) At least a recreational pilot certificate but not a rating for the category and class of light sport aircraft you operate,	(A) That light-sport aircraft, only if you hold the endorsements required in §61.321 for its category and class,	(*1*) You must comply with the limitations in §61.315, except §61.315(c)(14) and, if a private pilot or higher, §61.315(c)(7).
(2) Only a U.S. driver's license	(i) A sport pilot certificate,	(A) Any light-sport aircraft for which you hold the endorsements required for its category and class.	(*1*) You must hold any other endorsements required by this subpart, and comply with the limitations in §61.315.
	(ii) At least a recreational pilot certificate with a category and class rating,	(A) Any light-sport aircraft in that category and class,	(*1*) You do not have to hold any of the endorsements required by this subpart, but you must comply with the limitations in §61.315.
	(iii) At least a recreational pilot certificate but not a rating for the category and class of light-sport aircraft you operate,	(A) That light-sport aircraft, only if you hold the endorsements required in §61.321 for its category and class,	(*1*) You must comply with the limitations in §61.315, except §61.315(c)(14) and, if a private pilot or higher, §61.315(c)(7).
(3) Neither a medical certificate nor a U.S. driver's license	(i) A sport pilot certificate,	(A) Any light-sport glider or balloon for which you hold the endorsements required for its category and class	(*1*) You must hold any other endorsements required by this subpart, and comply with the limitations in §61.315.
	(ii) At least a private pilot certificate with a category and class rating for glider or balloon,	(A) Any light-sport glider or balloon in that category and class	(*1*) You do not have to hold any of the endorsements required by this subpart, nor do you have to comply with the limitations in §61.315.
	(iii) At least a private pilot certificate but not a rating for glider or balloon,	(A) Any light-sport glider or balloon, only if you hold the endorsements required in §61.321 for its category and class	(*1*) You must comply with the limitations in §61.315, except §61.315(c)(14) and, if a private pilot or higher, §61.315(c)(7).

PART 61

FAR

§ 61.301 What is the purpose of this subpart and to whom does it apply?

(a) This subpart prescribes the following requirements that apply to a sport pilot certificate:

(1) Eligibility.
(2) Aeronautical knowledge.
(3) Flight proficiency.
(4) Aeronautical experience.
(5) Endorsements.
(6) Privileges and limits.

(b) Other provisions of this part apply to the logging of flight time and testing.

(c) This subpart applies to applicants for, and holders of, sport pilot certificates. It also applies to holders of recreational pilot certificates and higher, as provided in § 61.303.

[Docket No. FAA-2001-11133, 69 FR 44869, July 27, 2004, as amended by Amdt. 61-125, 75 FR 5221, Feb. 1, 2010]

§ 61.303 If I want to operate a light-sport aircraft, what operating limits and endorsement requirements in this subpart must I comply with?

(a) Use the following table to determine what operating limits and endorsement requirements in this subpart, if any, apply to you when you operate a light-sport aircraft. The medical certificate specified in this table must be in compliance with § 61.2 in regards to currency and validity. If you hold a recreational pilot certificate, but not a medical certificate, you must comply with cross country requirements in § 61.101 (c), even if your flight does not exceed 50 nautical miles from your departure airport. You must also comply with requirements in other subparts of this part that apply to your certificate and the operation you conduct.

(b) A person using a U.S. driver's license to meet the requirements of this paragraph must—

(1) Comply with each restriction and limitation imposed by that person's U.S. driver's license and any judicial or administrative order applying to the operation of a motor vehicle;

(2) Have been found eligible for the issuance of at least a third-class airman medical certificate at the time of his or her most recent application (if the person has applied for a medical certificate);

(3) Not have had his or her most recently issued medical certificate (if the person has held a medical certificate) suspended or revoked or most recent Authorization for a Special Issuance of a Medical Certificate withdrawn; and

(4) Not know or have reason to know of any medical condition that would make that person unable to operate a light-sport aircraft in a safe manner.

[Doc. No. FAA-2001-11133, 69 FR 44869, July 27, 2004, as amended by Amdt. 61-124, 74 FR 42562, Aug. 21, 2009; Amdt. 61-125, 75 FR 5221, Feb. 1, 2010]

§ 61.305 What are the age and language requirements for a sport pilot certificate?

(a) To be eligible for a sport pilot certificate you must:

(1) Be at least 17 years old (or 16 years old if you are applying to operate a glider or balloon).

(2) Be able to read, speak, write, and understand English. If you cannot read, speak, write, and understand English because of medical reasons, the FAA may place limits on your certificate as are necessary for the safe operation of light-sport aircraft.

§ 61.307 What tests do I have to take to obtain a sport pilot certificate?

To obtain a sport pilot certificate, you must pass the following tests:

(a) *Knowledge test.* You must pass a knowledge test on the applicable aeronautical knowledge areas listed in § 61.309. Before you may take the knowledge test for a sport pilot certificate, you must receive a logbook endorsement from the authorized instructor who trained you or reviewed and evaluated your home-study course on the aeronautical knowledge areas listed in § 61.309 certifying you are prepared for the test.

(b) *Practical test.* You must pass a practical test on the applicable areas of operation listed in §§ 61.309 and 61.311. Before you may take the practical test for a sport pilot certificate, you must receive a logbook endorsement from the authorized instructor who provided you with flight training on the areas of operation specified in §§ 61.309 and 61.311 in preparation for the practical test. This endorsement certifies that you meet the applicable aeronautical knowledge and experience requirements and are prepared for the practical test.

If you are applying for a sport pilot certificate with . . .	Then you must log at least . . .	Which must include at least . . .
(a) Airplane category and single-engine land or sea class privileges,	(1) 20 hours of flight time, including at least 15 hours of flight training from an authorized instructor in a single-engine airplane and at least 5 hours of solo flight training in the areas of operation listed in §61.311,	(i) 2 hours of cross-country flight training, (ii) 10 takeoffs and landings to a full stop (with each landing involving a flight in the traffic pattern) at an airport, (iii) One solo cross-country flight of at least 75 nautical miles total distance, with a full-stop landing at a minimum of two points and one segment of the flight consisting of a straight-line distance of at least 25 nautical miles between the takeoff and landing locations, and (iv) 2 hours of flight training with an authorized instructor on those areas of operation specified in §61.311 in preparation for the practical test within the preceding 2 calendar months from the month of the test.
(b) Glider category privileges, and you have not logged at least 20 hours of flight time in a heavier-than-air aircraft,	(1) 10 hours of flight time in a glider, including 10 flights in a glider receiving flight training from an authorized instructor and at least 2 hours of solo flight training in the areas of operation listed in §61.311,	(i) Five solo launches and landings, and (ii) at least 3 training flights with an authorized instructor on those areas of operation specified in §61.311 in preparation for the practical test within the preceding 2 calendar months from the month of the test.
(c) Glider category privileges, and you have logged 20 hours flight time in a heavier-than-air aircraft,	(1) 3 hours of flight time in a glider, including five flights in a glider while receiving flight training from an authorized instructor and at least 1 hour of solo flight training in the areas of operation listed in §61.311,	(i) Three solo launches and landings, and (ii) at least 3 training flights with an authorized instructor on those areas of operation specified in §61.311 in preparation for the practical test within the preceding 2 calendar months from the month of the test.
(d) Rotorcraft category and gyroplane class privileges,	(1) 20 hours of flight time, including 15 hours of flight training from an authorized instructor in a gyroplane and at least 5 hours of solo flight training in the areas of operation listed in §61.311,	(i) 2 hours of cross-country flight training, (ii) 10 takeoffs and landings to a full stop (with each landing involving a flight in the traffic pattern) at an airport, (iii) One solo cross-country flight of at least 50 nautical miles total distance, with a full-stop landing at a minimum of two points, and one segment of the flight consisting of a straight-line distance of at least 25 nautical miles between the takeoff and landing locations, and (iv) 2 hours of flight training with an authorized instructor on those areas of operation specified in §61.311 in preparation for the practical test within the preceding 2 calendar months from the month of the test.
(e) Lighter-than-air category and airship class privileges,	(1) 20 hours of flight time, including 15 hours of flight training from an authorized instructor in an airship and at least 3 hours performing the duties of pilot in command in an airship with an authorized instructor in the areas of operation listed in §61.311,	(i) 2 hours of cross-country flight training, (ii) Three takeoffs and landings to a full stop (with each landing involving a flight in the traffic pattern) at an airport, (iii) One cross-country flight of at least 25 nautical miles between the takeoff and landing locations, and (iv) 2 hours of flight training with an authorized instructor on those areas of operation specified in §61.311 in preparation for the practical test within the preceding 2 calendar months from the month of the test.
(f) Lighter-than-air category and balloon class privileges,	(1) 7 hours of flight time in a balloon, including three flights with an authorized instructor and one flight performing the duties of pilot in command in a balloon with an authorized instructor in the areas of operation listed in §61.311,	(i) 2 hours of cross-country flight training, and (ii) 1 hours of flight training with an authorized instructor on those areas of operation specified in §61.311 in preparation for the practical test within the preceding 2 calendar months from the month of the test.
(g) Powered parachute category land or sea class privileges,	(1) 12 hours of flight time in a powered parachute, including 10 hours of flight training from an authorized instructor in a powered parachute, and at least 2 hours of solo flight training in the areas of operation listed in §61.311	(i) 1 hour of cross-country flight training, (ii) 20 takeoffs and landings to a full stop in a powered parachute with each landing involving flight in the traffic pattern at an airport; (iii) 10 solo takeoffs and landings to a full stop (with each landing involving a flight in the traffic pattern) at an airport, (iv) One solo flight with a landing at a different airport and one segment of the flight consisting of a straight-line distance of at least 10 nautical miles between takeoff and landing locations, and (v) 1 hours of flight training with an authorized instructor on those areas of operation specified in §61.311 in preparation for the practical test within the preceding 2 calendar months from the month of the test.
(h) Weight-shift-control aircraft category land or sea class privileges,	(1) 20 hours of light time, including 15 hours of flight training from an authorized instructor in a weight-shift-control aircraft and at least 5 hours of solo flight training in the areas of operation listed in §61.311,	(i) 2 hours of cross-country flight training, (ii) 10 takeoffs and landings to a full stop (with each landing involving a flight in the traffic pattern) at an airport, (iii) One solo cross-country flight of at least 50 nautical miles total distance, with a full-stop landing at a minimum of two points, and one segment of the flight consisting of a straight-line distance of at least 25 nautical miles between takeoff and landing locations, and (iv) 2 hours of flight training with an authorized instructor on those areas of operation specified in §61.311 in preparation for the practical test within the preceding 2 calendar months from the month of the test.

§ 61.309 What aeronautical knowledge must I have to apply for a sport pilot certificate?

To apply for a sport pilot certificate you must receive and log ground training from an authorized instructor or complete a home-study course on the following aeronautical knowledge areas:

(a) Applicable regulations of this chapter that relate to sport pilot privileges, limits, and flight operations.

(b) Accident reporting requirements of the National Transportation Safety Board.

(c) Use of the applicable portions of the aeronautical information manual and FAA advisory circulars.

(d) Use of aeronautical charts for VFR navigation using pilotage, dead reckoning, and navigation systems, as appropriate.

(e) Recognition of critical weather situations from the ground and in flight, windshear avoidance, and the procurement and use of aeronautical weather reports and forecasts.

(f) Safe and efficient operation of aircraft, including collision avoidance, and recognition and avoidance of wake turbulence.

(g) Effects of density altitude on takeoff and climb performance.

(h) Weight and balance computations.

(i) Principles of aerodynamics, powerplants, and aircraft systems.

(j) Stall awareness, spin entry, spins, and spin recovery techniques, as applicable.

(k) Aeronautical decision making and risk management.

(l) Preflight actions that include—

(1) How to get information on runway lengths at airports of intended use, data on takeoff and landing distances, weather reports and forecasts, and fuel requirements; and

(2) How to plan for alternatives if the planned flight cannot be completed or if you encounter delays.

[Docket No. FAA-2001-11133, 69 FR 44869, July 27, 2004, as amended by Amdt. 61-125, 75 FR 5221, Feb. 1, 2010]

§ 61.311 What flight proficiency requirements must I meet to apply for a sport pilot certificate?

To apply for a sport pilot certificate you must receive and log ground and flight training from an authorized instructor on the following areas of operation, as appropriate, for airplane single-engine land or sea, glider, gyroplane, airship, balloon, powered parachute land or sea, and weight-shift-control aircraft land or sea privileges:

(a) Preflight preparation.

(b) Preflight procedures.

(c) Airport, seaplane base, and gliderport operations, as applicable.

(d) Takeoffs (or launches), landings, and go-arounds.

(e) Performance maneuvers, and for gliders, performance speeds.

(f) Ground reference maneuvers (not applicable to gliders and balloons).

(g) Soaring techniques (applicable only to gliders).

(h) Navigation.

(i) Slow flight (not applicable to lighter-than-air aircraft and powered parachutes).

(j) Stalls (not applicable to lighter-than-air aircraft, gyroplanes, and powered parachutes).

(k) Emergency operations.

(l) Post-flight procedures.

[Docket No. FAA-2001-11133, 69 FR 44869, July 27, 2004, as amended by Amdt. 61-125, 75 FR 5221, Feb. 1, 2010]

§ 61.313 What aeronautical experience must I have to apply for a sport pilot certificate?

Use the following table to determine the aeronautical experience you must have to apply for a sport pilot certificate:

[Doc. No. FAA-2001-11133, 69 FR 44869, July 27, 2004; Amdt. 61-124A, 74 FR 53647, Oct. 20, 2009; Amdt. 61-125, 75 FR 5221, Feb. 1, 2010]

§ 61.315 What are the privileges and limits of my sport pilot certificate?

(a) If you hold a sport pilot certificate you may act as pilot in command of a light-sport aircraft, except as specified in paragraph (c) of this section.

(b) You may share the operating expenses of a flight with a passenger, provided the expenses involve only fuel, oil, airport expenses, or aircraft rental fees. You must pay at least half the operating expenses of the flight.

(c) You may not act as pilot in command of a light-sport aircraft:

(1) That is carrying a passenger or property for compensation or hire.

(2) For compensation or hire.

(3) In furtherance of a business.

(4) While carrying more than one passenger.

(5) At night.

(6) In Class A airspace.

(7) In Class B, C, and D airspace, at an airport located in Class B, C, or D airspace, and to, from, through, or at an airport having an operational control tower unless you have met the requirements specified in § 61.325.

(8) Outside the United States, unless you have prior authorization from the country in which you seek to operate. Your sport pilot certificate carries the limit "Holder does not meet ICAO requirements."

(9) To demonstrate the aircraft in flight to a prospective buyer if you are an aircraft salesperson.

(10) In a passenger-carrying airlift sponsored by a charitable organization.

(11) At an altitude of more than 10,000 feet MSL or 2,000 feet AGL, whichever is higher.

(12) When the flight or surface visibility is less than 3 statute miles.

(13) Without visual reference to the surface.

(14) If the aircraft:

(i) Has a VH greater than 87 knots CAS, unless you have met the requirements of § 61.327(b).

(ii) Has a VH less than or equal to 87 knots CAS, unless you have met the requirements of § 61.327(a) or have logged flight time as pilot in command of an airplane with a VH less than or equal to 87 knots CAS before April 2, 2010.

(15) Contrary to any operating limitation placed on the airworthiness certificate of the aircraft being flown.

(16) Contrary to any limit on your pilot certificate or airman medical certificate, or any other limit or endorsement from an authorized instructor.

(17) Contrary to any restriction or limitation on your U.S. driver's license or any restriction or limitation imposed by judicial or administrative order when using your driver's license to satisfy a requirement of this part.

(18) While towing any object.

(19) As a pilot flight crewmember on any aircraft for which more than one pilot is required by the type certificate of the aircraft or the regulations under which the flight is conducted.

[Docket No. FAA-2001-11133, 69 FR 44869, July 27, 2004, as amended by Amdt. 61-125, 75 FR 5221, Feb. 1, 2010; Amdt. 61-125A, 75 FR 15610, Mar. 30, 2010]

§ 61.317 Is my sport pilot certificate issued with aircraft category and class ratings?

Your sport pilot certificate does not list aircraft category and class ratings. When you successfully pass the practical test for a sport pilot certificate, regardless of the light-sport aircraft privileges you seek, the FAA will issue you a sport pilot certificate without any category and class ratings. The FAA will provide you with a logbook endorsement for the category and class of aircraft in which you are authorized to act as pilot in command.

[Docket No. FAA-2001-11133, 69 FR 44869, July 27, 2004, as amended by Amdt. 61-125, 75 FR 5222, Feb. 1, 2010; Amdt. 61-125A, 75 FR 15610, Mar. 30, 2010]

§ 61.319 [Reserved]

§ 61.321 How do I obtain privileges to operate an additional category or class of light-sport aircraft?

If you hold a sport pilot certificate and seek to operate an additional category or class of light-sport aircraft, you must—

(a) Receive a logbook endorsement from the authorized instructor who trained you on the applicable aeronautical knowledge areas specified in § 61.309 and areas of operation specified in § 61.311. The endorsement certifies you have met the aeronautical knowledge and flight proficiency requirements for the additional light-sport aircraft privilege you seek;

(b) Successfully complete a proficiency check from an authorized instructor other than the instructor who trained you on the aeronautical knowledge areas and areas of operation specified in §§ 61.309 and 61.311 for the additional light-sport aircraft privilege you seek;

(c) Complete an application for those privileges on a form and in a manner acceptable to the FAA and present this application to the authorized instructor who conducted the proficiency check specified in paragraph (b) of this section; and

(d) Receive a logbook endorsement from the instructor who conducted the proficiency check specified in paragraph (b) of this section certifying you are proficient in the applicable areas of operation and aeronautical knowledge areas, and that you are authorized for the additional category and class light-sport aircraft privilege.

§ 61.323 [Reserved]

§ 61.325 How do I obtain privileges to operate a light-sport aircraft at an airport within, or in airspace within, Class B, C, and D airspace, or in other airspace with an airport having an operational control tower?

If you hold a sport pilot certificate and seek privileges to operate a light-sport aircraft in Class B, C, or D airspace, at an airport located in Class B, C, or D airspace, or to, from, through, or at an airport having an operational control tower, you must receive and log ground and flight training. The authorized instructor who provides this training must provide a logbook endorsement that certifies you are proficient in the following aeronautical knowledge areas and areas of operation:

(a) The use of radios, communications, navigation system/facilities, and radar services.

(b) Operations at airports with an operating control tower to include three takeoffs and landings to a full stop, with each landing involving a flight in the traffic pattern, at an airport with an operating control tower.

(c) Applicable flight rules of part 91 of this chapter for operations in Class B, C, and D airspace and air traffic control clearances.

§ 61.327 Are there specific endorsement requirements to operate a light-sport aircraft based on VH ?

(a) Except as specified in paragraph (c) of this section, if you hold a sport pilot certificate and you seek to operate a light-sport aircraft that is an airplane with a VH less than or equal to 87 knots CAS you must—

(1) Receive and log ground and flight training from an authorized instructor in an airplane that has a VH less than or equal to 87 knots CAS; and

(2) Receive a logbook endorsement from the authorized instructor who provided the training specified in paragraph (a)(1) of this section certifying that you are proficient in the operation of light-sport aircraft that is an airplane with a VH less than or equal to 87 knots CAS.

(b) If you hold a sport pilot certificate and you seek to operate a light-sport aircraft that has a VH greater than 87 knots CAS you must—

(1) Receive and log ground and flight training from an authorized instructor in an aircraft that has a VH greater than 87 knots CAS; and

(2) Receive a logbook endorsement from the authorized instructor who provided the training specified in paragraph (b)(1) of this section certifying that you are proficient in the operation of light-sport aircraft with a VH greater than 87 knots CAS.

(c) The training and endorsements required by paragraph (a) of this section are not required if you have logged flight time as pilot in command of an airplane with a VH less than or equal to 87 knots CAS prior to April 2, 2010.

[Docket No. FAA-2007-29015, 75 FR 5222, Feb. 1, 2010; Amdt. 61-125A, 75 FR 15610, Mar. 30, 2010]

Subpart K—Flight Instructors With a Sport Pilot Rating

Source: Doc. No. FAA-2001-11133, 69 FR 44875, July 27, 2004, unless otherwise noted.

§ 61.401 What is the purpose of this subpart?

(a) This part prescribes the following requirements that apply to a flight instructor certificate with a sport pilot rating:

(1) Eligibility.

(2) Aeronautical knowledge.

(3) Flight proficiency.

(4) Endorsements.

(5) Privileges and limits.

(b) Other provisions of this part apply to the logging of flight time and testing.

[Doc. No. FAA-2001-11133, 69 FR 44875, July 27, 2004, as amended by Amdt. 61-125, 75 FR 5222, Feb. 1, 2010]

§ 61.403 What are the age, language, and pilot certificate requirements for a flight instructor certificate with a sport pilot rating?

To be eligible for a flight instructor certificate with a sport pilot rating you must:

(a) Be at least 18 years old.

(b) Be able to read, speak, write, and understand English. If you cannot read, speak, write, and understand English because of medical reasons, the FAA may place limits on your certificate as are necessary for the safe operation of light-sport aircraft.

(c) Hold at least a sport pilot certificate with category and class ratings or privileges, as applicable, that are appropriate to the flight instructor privileges sought.

[Doc. No. FAA-2001-11133, 69 FR 44875, July 27, 2004, as amended by Amdt. 61-124, 74 FR 42562, Aug. 21, 2009]

§ 61.405 What tests do I have to take to obtain a flight instructor certificate with a sport pilot rating?

To obtain a flight instructor certificate with a sport pilot rating you must pass the following tests:

(a) *Knowledge test.* Before you take a knowledge test, you must receive a logbook endorsement certifying you are prepared for the test from an authorized instructor who trained you or evaluated your home-study course on the aeronautical knowledge areas listed in § 61.407. You must pass knowledge tests on—

(1) The fundamentals of instructing listed in § 61.407(a), unless you meet the requirements of § 61.407(c); and

(2) The aeronautical knowledge areas for a sport pilot certificate applicable to the aircraft category and class for which flight instructor privileges are sought.

(b) *Practical test.* (1) Before you take the practical test, you must—

(i) Receive a logbook endorsement from the authorized instructor who provided you with flight training on the areas of operation specified in § 61.409 that apply to the category and class of aircraft privileges you seek. This endorsement certifies you meet the applicable aeronautical knowledge and experience requirements and are prepared for the practical test;

(ii) If you are seeking privileges to provide instruction in an airplane or glider, receive a logbook endorsement from an authorized instructor indicating that you are competent and possess instructional proficiency in stall awareness, spin entry, spins, and spin recovery procedures after you have received flight training in those training areas in an airplane or glider, as appropriate, that is certificated for spins;

(2) You must pass a practical test—

(i) On the areas of operation listed in § 61.409 that are appropriate to the category and class of aircraft privileges you seek;

(ii) In an aircraft representative of the category and class of aircraft for the privileges you seek;

(iii) In which you demonstrate that you are able to teach stall awareness, spin entry, spins, and spin recovery procedures if you are seeking privileges to provide instruction in an airplane or glider. If you have not failed a practical test based on deficiencies in your ability to demonstrate knowledge or skill in these areas and you provide the endorsement required by paragraph (b)(1)(ii) of this section, an examiner may accept the endorsement instead of the demonstration required by this paragraph. If you are taking a test because you previously failed a test based on not meeting the requirements of this paragraph, you must pass a practical test on stall awareness, spin entry, spins, and spin recovery instructional competency and proficiency in the applicable category and class of aircraft that is certificated for spins.

§ 61.407 What aeronautical knowledge must I have to apply for a flight instructor certificate with a sport pilot rating?

(a) Except as specified in paragraph (c) of this section you must receive and log ground training from an authorized instructor on the fundamentals of instruction that includes:

(1) The learning process.
(2) Elements of effective teaching.
(3) Student evaluation and testing.
(4) Course development.
(5) Lesson planning.
(6) Classroom training techniques.

(b) You must receive and log ground training from an authorized instructor on the aeronautical knowledge areas applicable to a sport pilot certificate for the aircraft category and class in which you seek flight instructor privileges.

If you are applying for a flight instructor certificate with a sport pilot rating for . . .	Then you must log at least . . .	Which must include at least . . .
(a) Airplane category and single-engine class privileges,	(1) 150 hours of flight time as a pilot,	(i) 100 hours of flight time as pilot in command in powered aircraft, (ii) 50 hours of flight time in a single-engine airplane, (iii) 25 hours of cross-country flight time, (iv) 10 hours of cross-country flight time in a single-engine airplane, and
		(v) 15 hours of flight time as pilot in command in a single-engine airplane that is a light-sport aircraft.
(b) Glider category privileges,	(1) 25 hours of flight time as pilot in command in a glider, 100 flights in a glider, and 15 flights as pilot in command in a glider that is a light-sport aircraft, or (2) 100 hours in heavier-than-air aircraft, 20 flights in a glider, and 15 flights as pilot in command in a glider that is a light-sport aircraft	
(c) Rotorcraft category and gyroplane class privileges,	(1) 125 hours of flight time as a pilot,	(i) 100 hours of flight time as pilot in command in powered aircraft, (ii) 50 hours of flight time in a gyroplane,
		(iii) 10 hours of cross-country flight time,
		(iv) 3 hours of cross-country flight time in a gyroplane, and
		(v) 15 hours of flight time as pilot in command in a gyroplane that is a light-sport aircraft.
(d) Lighter-than-air category and airship class privileges,	(1) 100 hours of flight time as a pilot,	(i) 40 hours of flight time in an airship, (ii) 20 hours of pilot in command time in an airship,
		(iii) 10 hours of cross-country flight time,
		(iv) 5 hours of cross-country flight time in an airship, and
		(v) 15 hours of flight time as pilot in command in an airship that is a light-sport aircraft.
(e) Lighter-than-air category and balloon class privileges,	(1) 35 hours of flight time as pilot-in-command,	(i) 20 hours of flight time in a balloon, (ii) 10 flights in a balloon, and
		(iii) 5 flights as pilot in command in a balloon that is a light-sport aircraft.
(f) Weight-shift-control aircraft category privileges,	(1) 150 hours of flight time as a pilot,	(i) 100 hours of flight time as pilot in command in powered aircraft, (ii) 50 hours of flight time in a weight-shift-control aircraft,
		(iii) 25 hours of cross-country flight time,
		(iv) 10 hours of cross-country flight time in a weight-shift-control aircraft, and
		(v) 15 hours of flight time as pilot in command in a weight-shift-control aircraft that is a light-sport aircraft.
(g) Powered-parachute category privileges,	(1) 100 hours of flight time as a pilot,	(i) 75 hours of flight time as pilot in command in powered aircraft, (ii) 50 hours of flight time in a powered parachute,
		(iii) 15 hours of cross-country flight time,
		(iv) 5 hours of cross-country flight time in a powered parachute, and
		(v) 15 hours of flight time as pilot in command in a powered parachute that is a light-sport aircraft.

(c) You do not have to meet the requirements of paragraph (a) of this section if you—

(1) Hold a flight instructor certificate or ground instructor certificate issued under this part;

(2) Hold a teacher's certificate issued by a State, county, city, or municipality; or

(3) Are employed as a teacher at an accredited college or university.

[Doc. No. FAA-2001-11133, 69 FR 44875, July 27, 2004, as amended by Amdt. 61-124, 74 FR 42562, Aug. 21, 2009]

§ 61.409 What flight proficiency requirements must I meet to apply for a flight instructor certificate with a sport pilot rating?

You must receive and log ground and flight training from an authorized instructor on the following areas of operation for the aircraft category and class in which you seek flight instructor privileges:

(a) Technical subject areas.

(b) Preflight preparation.

(c) Preflight lesson on a maneuver to be performed in flight.

(d) Preflight procedures.

(e) Airport, seaplane base, and gliderport operations, as applicable.

(f) Takeoffs (or launches), landings, and go-arounds.

(g) Fundamentals of flight.

(h) Performance maneuvers and for gliders, performance speeds.

(i) Ground reference maneuvers (except for gliders and lighter-than-air).

(j) Soaring techniques.

(k) Slow flight (not applicable to lighter-than-air and powered parachutes).

(l) Stalls (not applicable to lighter-than-air, powered parachutes, and gyroplanes).

(m) Spins (applicable to airplanes and gliders).

(n) Emergency operations.

(o) Tumble entry and avoidance techniques (applicable to weight-shift-control aircraft).

(p) Post-flight procedures.

§ 61.411 What aeronautical experience must I have to apply for a flight instructor certificate with a sport pilot rating?

Use the following table to determine the experience you must have for each aircraft category and class:

§ 61.413 What are the privileges of my flight instructor certificate with a sport pilot rating?

If you hold a flight instructor certificate with a sport pilot rating, you are authorized, within the limits of your certificate and rating, to provide training and endorsements that are required for, and relate to—

(a) A student pilot seeking a sport pilot certificate;

(b) A sport pilot certificate;

(c) A flight instructor certificate with a sport pilot rating;

(d) A powered parachute or weight-shift-control aircraft rating;

(e) Sport pilot privileges;

(f) A flight review or operating privilege for a sport pilot;

(g) A practical test for a sport pilot certificate, a private pilot certificate with a powered parachute or weight-shift-control aircraft rating or a flight instructor certificate with a sport pilot rating;

(h) A knowledge test for a sport pilot certificate, a private pilot certificate with a powered parachute or weight-shift-control aircraft rating or a flight instructor certificate with a sport pilot rating; and

(i) A proficiency check for an additional category or class privilege for a sport pilot certificate or a flight instructor certificate with a sport pilot rating.

[Doc. No. FAA-2001-11133, 69 FR 44875, July 27, 2004, as amended by Amdt. 61-125, 75 FR 5222, Feb. 1, 2010]

§ 61.415 What are the limits of a flight instructor certificate with a sport pilot rating?

If you hold a flight instructor certificate with a sport pilot rating, you may only provide flight training in a light-sport aircraft and are subject to the following limits:

(a) You may not provide ground or flight training in any aircraft for which you do not hold:

(1) A sport pilot certificate with applicable category and class privileges or a pilot certificate with the applicable category and class rating; and

(2) Applicable category and class privileges for your flight instructor certificate with a sport pilot rating.

(b) You may not provide ground or flight training for a private pilot certificate with a powered parachute or weight-shift-control aircraft rating unless you hold:

(1) At least a private pilot certificate with the applicable category and class rating; and

(2) Applicable category and class privileges for your flight instructor certificate with a sport pilot rating.

(c) You may not conduct more than 8 hours of flight training in any 24-consecutive-hour period.

(d) You may not endorse a:

(1) Student pilot's certificate or logbook for solo flight privileges, unless you have—

(i) Given that student the flight training required for solo flight privileges required by this part; and

(ii) Determined that the student is prepared to conduct the flight safely under known circumstances, subject to any limitations listed in the student's logbook that you consider necessary for the safety of the flight.

(2) Student pilot's certificate and logbook for a solo cross-country flight, unless you have determined the student's flight preparation, planning, equipment, and proposed procedures are adequate for the proposed flight under the existing conditions and within any limitations listed in the logbook that you consider necessary for the safety of the flight.

(3) Student pilot's certificate and logbook for solo flight in Class B, C, and D airspace areas, at an airport within Class B, C, or D airspace and to from, through or on an airport having an operational control tower, unless that you have—

(i) Given that student ground and flight training in that airspace or at that airport; and

(ii) Determined that the student is proficient to operate the aircraft safely.

(4) Logbook of a pilot for a flight review, unless you have conducted a review of that pilot in accordance with the requirements of § 61.56.

(e) You may not provide training to operate a light-sport aircraft in Class B, C, and D airspace, at an airport located in Class B, C, or D airspace, and to, from, through, or at an airport having an operational control tower, unless you have the endorsement specified in § 61.325, or are otherwise authorized to conduct operations in this airspace and at these airports.

(f) You may not provide training in a light-sport aircraft that is an airplane with a VH less than or equal to 87 knots CAS unless you have the endorsement specified in § 61.327 (a), or are otherwise authorized to operate that light-sport aircraft.

(g) You may not provide training in a light-sport aircraft with a VH greater than 87 knots CAS unless you have the endorsement specified in § 61.327 (b), or are otherwise authorized to operate that light-sport aircraft.

(h) You must perform all training in an aircraft that complies with the requirements of § 91.109 of this chapter.

(i) If you provide flight training for a certificate, rating or privilege, you must provide that flight training in an aircraft that meets the following:

(1) The aircraft must have at least two pilot stations and be of the same category and class appropriate to the certificate, rating or privilege sought.

(2) For single place aircraft, pre-solo flight training must be provided in an aircraft that has two pilot stations and is of the same category and class appropriate to the certificate, rating, or privilege sought.

[Doc. No. FAA-2001-11133, 69 FR 44875, July 27, 2004, as amended by Amdt. 61-125, 75 FR 5222, Feb. 1, 2010; Amdt. 61-125A, 75 FR 15610, Mar. 30, 2010]

§ 61.417 Will my flight instructor certificate with a sport pilot rating list aircraft category and class ratings?

Your flight instructor certificate does not list aircraft category and class ratings. When you successfully pass the practical test for a flight instructor certificate with a sport pilot rating, regardless of the light-sport aircraft privileges you seek, the FAA will issue you a flight instructor certificate with a sport pilot rating without any category and class ratings. The FAA will provide you with a logbook endorsement for the category and class of light-sport aircraft you are authorized to provide training in.

§ 61.419 How do I obtain privileges to provide training in an additional category or class of light-sport aircraft?

If you hold a flight instructor certificate with a sport pilot rating and seek to provide training in an additional category or class of light-sport aircraft you must—

(a) Receive a logbook endorsement from the authorized instructor who trained you on the applicable areas of operation specified in § 61.409 certifying you have met the aeronautical knowledge and flight proficiency requirements for the additional category and class flight instructor privilege you seek;

(b) Successfully complete a proficiency check from an authorized instructor other than the instructor who trained you on the areas specified in § 61.409 for the additional category and class flight instructor privilege you seek;

(c) Complete an application for those privileges on a form and in a manner acceptable to the FAA and present this application to the authorized instructor who conducted the proficiency check specified in paragraph (b) of this section; and

(d) Receive a logbook endorsement from the instructor who conducted the proficiency check specified in paragraph (b) of this section certifying you are proficient in the areas of operation and authorized for the additional category and class flight instructor privilege.

§ 61.421 May I give myself an endorsement?

No. If you hold a flight instructor certificate with a sport pilot rating, you may not give yourself an endorsement for any certificate, privilege, rating, flight review, authorization, practical test, knowledge test, or proficiency check required by this part.

§ 61.423 What are the recordkeeping requirements for a flight instructor with a sport pilot rating?

(a) As a flight instructor with a sport pilot rating you must:

(1) Sign the logbook of each person to whom you have given flight training or ground training.

(2) Keep a record of the name, date, and type of endorsement for:

(i) Each person whose logbook or student pilot certificate you have endorsed for solo flight privileges.

(ii) Each person for whom you have provided an endorsement for a knowledge test, practical test, or proficiency check, and the record must indicate the kind of test or check, and the results.

(iii) Each person whose logbook you have endorsed as proficient to operate—

(A) An additional category or class of light-sport aircraft;

(B) In Class B, C, and D airspace; at an airport located in Class B, C, or D airspace; and to, from, through, or at an airport having an operational control tower;

(C) A light-sport aircraft that is an airplane with a VH less than or equal to 87 knots CAS; and

(D) A light-sport aircraft with a VH greater than 87 knots CAS.

(iv) Each person whose logbook you have endorsed as proficient to provide flight training in an additional category or class of light-sport aircraft.

(b) Within 10 days after providing an endorsement for a person to operate or provide training in an additional category and class of light-sport aircraft you must—

(1) Complete, sign, and submit to the FAA the application presented to you to obtain those privileges; and

(2) Retain a copy of the form.

(c) You must keep the records listed in this section for 3 years. You may keep these records in a logbook or a separate document.

[Doc. No. FAA-2001-11133, 69 FR 44875, July 27, 2004, as amended by Amdt. 61-125, 75 FR 5222, Feb. 1, 2010; Amdt. 61-125A, 75 FR 15610, Mar. 30, 2010]

§ 61.425 How do I renew my flight instructor certificate?

If you hold a flight instructor certificate with a sport pilot rating you may renew your certificate in accordance with the provisions of § 61.197.

§ 61.427 What must I do if my flight instructor certificate with a sport pilot rating expires?

You may exchange your expired flight instructor certificate with a sport pilot rating for a new certificate with a sport pilot rating and any other rating on that certificate by passing a practical test as prescribed in § 61.405(b) or § 61.183(h) for one of the ratings listed on the expired flight instructor certificate. The FAA will reinstate any privilege authorized by the expired certificate.

§ 61.429 May I exercise the privileges of a flight instructor certificate with a sport pilot rating if I hold a flight instructor certificate with another rating?

If you hold a flight instructor certificate, a commercial pilot certificate with an airship rating, or a commercial pilot certificate with a balloon rating issued under this part, and you seek to exercise the privileges of a flight instructor certificate with a sport pilot rating, you may do so without any further showing of proficiency, subject to the following limits:

(a) You are limited to the aircraft category and class ratings listed on your flight instructor certificate, commercial pilot certificate with an airship rating, or commercial pilot certificate with a balloon rating, as appropriate, when exercising your flight instructor privileges and the privileges specified in § 61.413.

(b) You must comply with the limits specified in § 61.415 and the recordkeeping requirements of § 61.423.

(c) If you want to exercise the privileges of your flight instructor certificate in a category or class of light-sport aircraft for which you are not currently rated, you must meet all applicable requirements to provide training in an additional category or class of light-sport aircraft specified in § 61.419.

[Doc. No. FAA-2001-11133, 69 FR 44875, July 27, 2004, as amended by Amdt. 61-124, 74 FR 42562, Aug. 21, 2009; Amdt. 61-125, 75 FR 5222, Feb. 1, 2010]

PART 63

CERTIFICATION: FLIGHT CREWMEMBERS OTHER THAN PILOTS

Contents

Authority: 49 U.S.C. 106(g), 40113, 44701-44703, 44707, 44709-44711, 45102-45103, 45301-45302.

Special Federal Aviation Regulation No. 100-2

Editorial Note: For the text of SFAR No. 100-2, see part 61 of this chapter.

Subpart A—General

Source: Docket No. 1179, 27 FR 7969, Aug. 10, 1962, unless otherwise noted.

§ 63.1 Applicability.

This part prescribes the requirements for issuing flight engineer and flight navigator certificates and the general operating rules for holders of those certificates.

§ 63.2 Certification of foreign flight crewmembers other than pilots.

A person who is neither a United States citizen nor a resident alien is issued a certificate under this part (other than under § 63.23 or § 63.42) outside the United States only when the Administrator finds that the certificate is needed for the operation of a U.S.-registered civil aircraft.

(Secs. 313, 601, 602, Federal Aviation Act of 1958, as amended (49 U.S.C. 1354, 1421, and 1422); sec. 6(c), Department of Transportation Act (49 U.S.C. 1655(c)); Title V, Independent Offices Appropriations Act of 1952 (31 U.S.C. 483(a)); sec. 28, International Air Transportation Competition Act of 1979 (49 U.S.C. 1159(b)))

[Doc. No. 22052, 47 FR 35693, Aug. 18, 1982]

§ 63.3 Certificates and ratings required.

(a) No person may act as a flight engineer of a civil aircraft of U.S. registry unless he has in his personal possession a current flight engineer certificate with appropriate ratings issued to him under this part and a second-class (or higher) medical certificate issued to him under part 67 of this chapter within the preceding 12 months. However, when the aircraft is operated within a foreign country, a current flight engineer certificate issued by the country in which the aircraft is operated, with evidence of current medical qualification for that certificate, may be used. Also, in the case of a flight engineer certificate issued under § 63.42, evidence of current medical qualification accepted for the issue of that certificate is used in place of a medical certificate.

(b) No person may act as a flight navigator of a civil aircraft of U.S. registry unless he has in his personal possession a current flight navigator certificate issued to him under this part and a second-class (or higher) medical certificate issued to him under part 67 of this chapter within the preceding 12 months. However, when the aircraft is operated within a foreign country, a current flight navigator certificate issued by the country in which the aircraft is operated, with evidence of current medical qualification for that certificate, may be used.

(c) Each person who holds a flight engineer or flight navigator certificate, or medical certificate, shall present either or both for inspection upon the request of the Administrator or an authorized representative of the National Transportation Safety Board, or of any Federal, State, or local law enforcement officer.

(Secs. 3, 6, 9, 80 Stat. 931, 49 U.S.C. 1652, 1655, 1657)

[Doc. No. 1179, 27 FR 7969, Aug. 10, 1962, as amended by Amdt. 63-1, 27 FR 10410, Oct. 25, 1962; Amdt. 63-3, 30 FR 14559, Nov. 23, 1965; Amdt. 63-7, 31 FR 13523, Oct. 20, 1966; Doc. No. 8084, 32 FR 5769, Apr. 11, 1967; Amdt. 63-9, 33 FR 18613, Dec. 17, 1968; Amdt. 63-11, 35 FR 5320, Mar. 31, 1970]

§ 63.11 Application and issue.

(a) An application for a certificate and appropriate class rating, or for an additional rating, under this part must be made on a form and in a manner prescribed by the Administrator. Each person who applies for airmen certification services to be administered outside the United States for any certificate or rating issued under this part must show evidence that the fee prescribed in appendix A of part 187 of this chapter has been paid.

(b) An applicant who meets the requirements of this part is entitled to an appropriate certificate and appropriate class ratings.

(c) Unless authorized by the Administrator, a person whose flight engineer certificate is suspended may not apply for any rating to be added to that certificate during the period of suspension.

(d) Unless the order of revocation provides otherwise, a person whose flight engineer or flight navigator certificate is revoked may not apply for the same kind of certificate for 1 year after the date of revocation.

(Secs. 313, 601, 602, Federal Aviation Act of 1958, as amended (49 U.S.C. 1354, 1421, and 1422); sec. 6(c), Department of Transportation Act (49 U.S.C. 1655(c)); Title V, Independent Offices Appropriations Act of 1952 (31 U.S.C. 483(a)); sec. 28, International Air Transportation Competition Act of 1979 (49 U.S.C. 1159(b)))

[Doc. No. 1179, 27 FR 7969, Aug. 10, 1962, as amended by Amdt. 63-3, 30 FR 14559, Nov. 23, 1965; Amdt. 63-7, 31 FR 13523, Oct. 20, 1966; Amdt. 63-22, 47 FR 35693, Aug. 16, 1982; Amdt. 63-35, 72 FR 18558, Apr. 12, 2007]

§ 63.12 Offenses involving alcohol or drugs.

(a) A conviction for the violation of any Federal or state statute relating to the growing, processing, manufacture, sale, disposition, possession, transportation, or importation of narcotic drugs, marihuana, or depressant or stimulant drugs or substances is grounds for—

(1) Denial of an application for any certificate or rating issued under this part for a period of up to 1 year after the date of final conviction; or

(2) Suspension or revocation of any certificate or rating issued under this part.

(b) The commission of an act prohibited by § 91.17(a) or § 91.19(a) of this chapter is grounds for—

(1) Denial of an application for a certificate or rating issued under this part for a period of up to 1 year after the date of that act; or

(2) Suspension or revocation of any certificate or rating issued under this part.

[Doc. No. 21956, 50 FR 15379, Apr. 17, 1985, as amended by Amdt. 63-27, 54 FR 34330, Aug. 18, 1989]

§ 63.12a Refusal to submit to an alcohol test or to furnish test results.

A refusal to submit to a test to indicate the percentage by weight of alcohol in the blood, when requested by a law enforcement officer in accordance with § 91.11(c) of this chapter, or a refusal to furnish or authorize the release of the test results when requested by the Administrator in accordance with § 91.17 (c) or (d) of this chapter, is grounds for—

(a) Denial of an application for any certificate or rating issued under this part for a period of up to 1 year after the date of that refusal; or

(b) Suspension or revocation of any certificate or rating issued under this part.

[Docket No. 21956, 51 FR 1229, Jan. 9, 1986, as amended by Amdt. 63-27, 54 FR 34330, Aug. 18, 1989]

§ 63.12b [Reserved]

§ 63.13 Temporary certificate.

A certificate effective for a period of not more than 120 days may be issued to a qualified applicant, pending review of his application and supplementary documents and the issue of the certificate for which he applied.

[Doc. No. 1179, 27 FR 7969, Aug. 10, 1962, as amended by Amdt. 63-19, 43 FR 22639, May 25, 1978]

§ 63.14 Security disqualification.

(a) *Eligibility standard.* No person is eligible to hold a certificate, rating, or authorization issued under this part when the Transportation Security Administration (TSA) has notified the FAA in writing that the person poses a security threat.

(b) *Effect of the issuance by the TSA of an Initial Notification of Threat Assessment.* (1) The FAA will hold in abeyance pending the outcome of the TSA's final threat assessment review an application for any certificate, rating, or authorization under this part by any person who has been issued an Initial Notification of Threat Assessment by the TSA.

(2) The FAA will suspend any certificate, rating, or authorization issued under this part after the TSA issues to the holder an Initial Notification of Threat Assessment.

(c) *Effect of the issuance by the TSA of a Final Notification of Threat Assessment.* (1) The FAA will deny an application for any certificate, rating, or authorization under this part to any person who has been issued a Final Notification of Threat Assessment.

(2) The FAA will revoke any certificate, rating, or authorization issued under this part after the TSA has issued to the holder a Final Notification of Threat Assessment.

[Doc. No. FAA-2003-14293, 68 FR 3774, Jan. 24, 2003]

§ 63.15 Duration of certificates.

(a) Except as provided in § 63.23 and paragraph (b) of this section, a certificate or rating issued under this part is effective until it is surrendered, suspended, or revoked.

(b) A flight engineer certificate (with any amendment thereto) issued under § 63.42 expires at the end of the 24th month after the month in which the certificate was issued or renewed. However, the holder may exercise the privileges of that certificate only while the foreign flight engineer license on which that certificate is based is effective.

(c) Any certificate issued under this part ceases to be effective if it is surrendered, suspended, or revoked. The holder of any certificate issued under this part that is suspended or revoked shall, upon the Administrator's request, return it to the Administrator.

(d) Except for temporary certificate issued under § 63.13, the holder of a paper certificate issued under this part may not exercise the privileges of that certificate after March 31, 2013.

(Sec. 6, 80 Stat. 937, 49 U.S.C. 1655; secs. 313, 601, 602, Federal Aviation Act of 1958, as amended (49 U.S.C. 1354, 1421, and 1422); sec. 6(c), Department of Transportation Act (49 U.S.C. 1655(c)); Title V, Independent Offices Appropriations Act of 1952 (31 U.S.C. 483(a)); sec. 28,

International Air Transportation Competition Act of 1979 (49 U.S.C. 1159(b)))

[Doc. No. 8846, 33 FR 18613, Dec. 17, 1968, as amended by Amdt. 63-22, 47 FR 35693, Aug. 16, 1982; Amdt. 63-36, 73 FR 10668, Feb. 28, 2008]

§ 63.15a [Reserved]

§ 63.16 Change of name; replacement of lost or destroyed certificate.

(a) An application for a change of name on a certificate issued under this part must be accompanied by the applicant's current certificate and the marriage license, court order, or other document verifying the change. The documents are returned to the applicant after inspection.

(b) An application for a replacement of a lost or destroyed certificate is made by letter to the Department of Transportation, Federal Aviation Administration, Airman Certification Branch, Post Office Box 25082, Oklahoma City, OK 73125. The letter must—

(1) Contain the name in which the certificate was issued, the permanent mailing address (including zip code), social security number (if any), and date and place of birth of the certificate holder, and any available information regarding the grade, number, and date of issue of the certificate, and the ratings on it; and

(2) Be accompanied by a check or money order for $2, payable to the Federal Aviation Administration.

(c) An application for a replacement of a lost or destroyed medical certificate is made by letter to the Department of Transportation, Federal Aviation Administration, Civil Aeromedical Institute, Aeromedical Certification Branch, Post Office Box 25082, Oklahoma City, OK 73125, accompanied by a check or money order for $2.00.

(d) A person whose certificate issued under this part or medical certificate, or both, has been lost may obtain a telegram from the Federal Aviation Administration confirming that it was issued. The telegram may be carried as a certificate for a period not to exceed 60 days pending his receiving a duplicate under paragraph (b) or (c) of this section, unless he has been notified that the certificate has been suspended or revoked. The request for such a telegram may be made by prepaid telegram, stating the date upon which a duplicate certificate was requested, or including the request for a duplicate and a money order for the necessary amount. The request for a telegraphic certificate should be sent to the office prescribed in paragraph (b) or (c) of this section, as appropriate. However, a request for both at the same time should be sent to the office prescribed in paragraph (b) of this section.

[Doc. No. 7258, 31 FR 13523, Oct. 20, 1966, as amended by Doc. No. 8084, 32 FR 5769, Apr. 11, 1967; Amdt. 63-12, 35 FR 14075, Sept. 4, 1970; Amdt. 63-13, 36 FR 28654, Feb. 11, 1971]

§ 63.17 Tests: General procedure.

(a) Tests prescribed by or under this part are given at times and places, and by persons, designated by the Administrator.

(b) The minimum passing grade for each test is 70 percent.

§ 63.18 Written tests: Cheating or other unauthorized conduct.

(a) Except as authorized by the Administrator, no person may—

(1) Copy, or intentionally remove, a written test under this part;

(2) Give to another, or receive from another, any part or copy of that test;

(3) Give help on that test to, or receive help on that test from, any person during the period that test is being given.

(4) Take any part of that test in behalf of another person;

(5) Use any material or aid during the period that test is being given; or

(6) Intentionally cause, assist, or participate in any act prohibited by this paragraph.

(b) No person who commits an act prohibited by paragraph (a) of this section is eligible for any airman or ground instructor certificate or rating under this chapter for a period of 1 year after the date of that act. In addition, the commission of that act is a basis for suspending or revoking any airman or ground instructor certificate or rating held by that person.

[Doc. No. 4086, 30 FR 2196, Feb. 18, 1965]

§ 63.19 Operations during physical deficiency.

No person may serve as a flight engineer or flight navigator during a period of known physical deficiency, or increase in physical deficiency, that would make him unable to meet the physical requirements for his current medical certificate.

§ 63.20 Applications, certificates, logbooks, reports, and records; falsification, reproduction, or alteration.

(a) No person may make or cause to be made—

(1) Any fraudulent or intentionally false statement on any application for a certificate or rating under this part;

(2) Any fraudulent or intentionally false entry in any logbook, record, or report that is required to be kept, made, or used, to show compliance with any requirement for any certificate or rating under this part;

(3) Any reproduction, for fraudulent purpose, of any certificate or rating under this part; or

(4) Any alteration of any certificate or rating under this part.

(b) The commission by any person of an act prohibited under paragraph (a) of this section is a basis for suspending or revoking any airman or ground instructor certificate or rating held by that person.

[Doc. No. 4086, 30 FR 2196, Feb. 18, 1965]

§ 63.21 Change of address.

Within 30 days after any change in his permanent mailing address, the holder of a certificate issued under this part shall notify the Department of Transportation, Federal Aviation Administration, Airman Certification Branch, Post Office Box 25082, Oklahoma City, OK 73125, in writing, of his new address.

[Doc. No. 10536, 35 FR 14075, Sept. 4, 1970]

§ 63.23 Special purpose flight engineer and flight navigator certificates: Operation of U.S.-registered civil airplanes leased by a person not a U.S. citizen.

(a) *General.* The holder of a current foreign flight engineer or flight navigator certificate, license, or authorization issued by a foreign contracting State to the Convention on International Civil Aviation, who meets the requirements of this section, may hold a special purpose flight engineer or flight navigator certificate, as appropriate, authorizing the holder to perform flight engineer or flight navigator duties on a civil airplane of U.S. registry, leased to a person not a citizen of the United States, carrying persons or property for compensation or hire. Special purpose flight engineer and flight navigator certificates are issued under this section only for airplane types that can have a maximum passenger seating configuration, excluding any flight crewmember seat, of more than 30 seats or a maximum payload capacity (as defined in § 135.2(e) of this chapter) of more than 7,500 pounds.

(b) *Eligibility.* To be eligible for the issuance, or renewal, of a certificate under this section, an applicant must present the following to the Administrator:

(1) A current foreign flight engineer or flight navigator certificate, license, or authorization issued by the aeronautical authority of a foreign contracting State to the Convention on International Civil Aviation or a facsimile acceptable to the Administrator. The certificate or license must authorize the applicant to perform the flight engineer or flight navigator duties to be authorized by a certificate issued under this section on the same airplane type as the leased airplane.

(2) A current certification by the lessee of the airplane—

(i) Stating that the applicant is employed by the lessee;

(ii) Specifying the airplane type on which the applicant will perform flight engineer or flight navigator duties; and

(iii) Stating that the applicant has received ground and flight instruction which qualifies the applicant to perform the duties to be assigned on the airplane.

(3) Documentation showing that the applicant currently meets the medical standards for the foreign flight engineer or flight navigator certificate, license, or authorization required by paragraph (b)(1) of this section, except that a U.S. medical certificate issued under part 67 of this chapter is not evidence that the applicant meets those standards unless the State which issued the applicant's foreign flight engineer or flight navigator certificate, license, or authorization accepts a U.S. medical certificate as evidence of medical fitness for a flight engineer or flight navigator certificate, license, or authorization.

(c) *Privileges.* The holder of a special purpose flight engineer or flight navigator certificate issued under this section may exercise the same privileges as those shown on the certificate, license, or authorization specified in paragraph (b)(1) of this section, subject to the limitations specified in this section.

(d) *Limitations.* Each certificate issued under this section is subject to the following limitations:

(1) It is valid only—

(i) For flights between foreign countries and for flights in foreign air commerce;

(ii) While it and the certificate, license, or authorization required by paragraph (b)(1) of this section are in the certificate holder's personal possession and are current;

(iii) While the certificate holder is employed by the person to whom the airplane described in the certification required by paragraph (b)(2) of this section is leased;

(iv) While the certificate holder is performing flight engineer or flight navigator duties on the U.S.-registered civil airplane described in the certification required by paragraph (b)(2) of this section; and

(v) While the medical documentation required by paragraph (b)(3) of this section is in the certificate holder's personal possession and is currently valid.

(2) Each certificate issued under this section contains the following:

(i) The name of the person to whom the U.S.-registered civil airplane is leased.

(ii) The type of airplane.

(iii) The limitation: "Issued under, and subject to, § 63.23 of the Federal Aviation Regulations."

(iv) The limitation: "Subject to the privileges and limitations shown on the holder's foreign flight (engineer or navigator) certificate, license, or authorization."

(3) Any additional limitations placed on the certificate which the Administrator considers necessary.

(e) *Termination.* Each special purpose flight engineer or flight navigator certificate issued under this section terminates—

(1) When the lease agreement for the airplane described in the certification required by paragraph (b)(2) of this section terminates;

(2) When the foreign flight engineer or flight navigator certificate, license, or authorization, or the medical documentation required by paragraph (b) of this section is suspended, revoked, or no longer valid; or

(3) After 24 months after the month in which the special purpose flight engineer or flight navigator certificate was issued.

(f) *Surrender of certificate.* The certificate holder shall surrender the special purpose flight engineer or flight navigator certificate to the Administrator within 7 days after the date it terminates.

(g) *Renewal.* The certificate holder may have the certificate renewed by complying with the requirements of paragraph (b) of this section at the time of application for renewal.

(Secs. 313(a), 601, and 602, Federal Aviation Act of 1958; as amended (49 U.S.C. 1354(a), 1421, and 1422); sec. 6(c), Department of Transportation Act (49 U.S.C. 1655(c)))

[Doc. No. 19300, 45 FR 5672, Jan. 24, 1980]

Subpart B—Flight Engineers

Authority: Secs. 313(a), 601, and 602, Federal Aviation Act of 1958; 49 U.S.C. 1354, 1421, 1422.

Source: Docket No. 6458, 30 FR 14559, Nov. 23, 1965, unless otherwise noted.

§ 63.31 Eligibility requirements; general.

To be eligible for a flight engineer certificate, a person must—

(a) Be at least 21 years of age;

(b) Be able to read, speak, and understand the English language, or have an appropriate limitation placed on his flight engineer certificate;

(c) Hold at least a second-class medical certificate issued under part 67 of this chapter within the 12 months before the date he applies, or other evidence of medical qualification accepted for the issue of a flight engineer certificate under § 63.42; and

(d) Comply with the requirements of this subpart that apply to the rating he seeks.

(Sec. 6, 80 Stat. 937, 49 U.S.C. 1655)

[Doc. No. 6458, 30 FR 14559, Nov. 23, 1965, as amended by Amdt. 63-9, 33 FR 18614, Dec. 17, 1968]

§ 63.33 Aircraft ratings.

(a) The aircraft class ratings to be placed on flight engineer certificates are—

(1) Reciprocating engine powered;

(2) Turbopropeller powered; and

(3) Turbojet powered.

(b) To be eligible for an additional aircraft class rating after his flight engineer certificate with a class rating is issued to him, an applicant must pass the written test that is appropriate to the class of airplane for which an additional rating is sought, and—

(1) Pass the flight test for that class of aircraft; or

(2) Satisfactorily complete an approved flight engineer training program that is appropriate to the additional class rating sought.

§ 63.35 Knowledge requirements.

(a) An applicant for a flight engineer certificate must pass a written test on the following:

(1) The regulations of this chapter that apply to the duties of a flight engineer.

(2) The theory of flight and aerodynamics.

(3) Basic meteorology with respect to engine operations.

(4) Center of gravity computations.

(b) An applicant for the original or additional issue of a flight engineer class rating must pass a written test for that airplane class on the following:

(1) Preflight.

(2) Airplane equipment.

(3) Airplane systems.

(4) Airplane loading.

(5) Airplane procedures and engine operations with respect to limitations.

(6) Normal operating procedures.

(7) Emergency procedures.

(8) Mathematical computation of engine operations and fuel consumption.

(c) Before taking the written tests prescribed in paragraphs (a) and (b) of this section, an applicant for a flight engineer certificate must present satisfactory evidence of having completed one of the experience requirements of § 63.37. However, he may take the written tests before acquiring the flight training required by § 63.37.

(d) An applicant for a flight engineer certificate or rating must have passed the written tests required by paragraphs (a) and (b) of this section since the beginning of the 24th calendar month before the month in which the flight is taken. However, this limitation does not apply to an applicant for a flight engineer certificate or rating if—

(1) The applicant—

(i) Within the period ending 24 calendar months after the month in which the applicant passed the written test, is employed as a flight crewmember or mechanic by a U.S. air carrier or commercial operator operating either under part 121 or as a commuter air carrier under part 135 (as defined in part 298 of this title) and is employed by such a certificate holder at the time of the flight test;

(ii) If employed as a flight crewmember, has completed initial training, and, if appropriate, transition or upgrade training; and

(iii) Meets the recurrent training requirements of the applicable part or, for mechanics, meets the recency of experience requirements of part 65; or

(2) Within the period ending 24 calendar months after the month in which the applicant passed the written test, the applicant participated in a flight engineer or maintenance training program of a U.S. scheduled military air transportation service and is currently participating in that program.

(e) An air carrier or commercial operator with an approved training program under part 121 of this chapter may, when authorized by the Administrator, provide as part of that program a written test that it may administer to satisfy the test required for an additional rating under paragraph (b) of this section.

(Sec. 6, 80 Stat. 937, 49 U.S.C. 1655; secs. 313(a), 601 through 605 of the Federal Aviation Act of 1958 (49 U.S.C. 1354(a), 1421 through 1425); sec. 6(c), Department of Transportation Act (49 U.S.C. 1655(c)); and 14 CFR 11.49)

[Doc. No. 1179, 27 FR 7969, Aug. 10, 1962, as amended by Amdt. 63-17, 40 FR 32830, Aug. 5, 1975; Doc. 63-21, 47 FR 13316, Mar. 29, 1982]

§ 63.37 Aeronautical experience requirements.

(a) Except as otherwise specified therein, the flight time used to satisfy the aeronautical experience requirements of paragraph (b) of this section must have been obtained on an airplane—

(1) On which a flight engineer is required by this chapter; or

(2) That has at least three engines that are rated at least 800 horsepower each or the equivalent in turbine-powered engines.

(b) An applicant for a flight engineer certificate with a class rating must present, for the class rating sought, satisfactory evidence of one of the following:

(1) At least 3 years of diversified practical experience in aircraft and aircraft engine maintenance (of which at least 1 year was in maintaining multiengine aircraft with engines rated at least 800 horsepower each, or the equivalent in turbine engine powered aircraft), and at least 5 hours of flight training in the duties of a flight engineer.

(2) Graduation from at least a 2-year specialized aeronautical training course in maintaining aircraft and aircraft engines (of which at least 6 calendar months were in maintaining multiengine aircraft with engines rated at least 800 horsepower each or the equivalent in turbine engine powered aircraft), and at least 5 hours of flight training in the duties of a flight engineer.

(3) A degree in aeronautical, electrical, or mechanical engineering from a recognized college, university, or engineering school; at least 6 calendar months of practical experience in maintaining multiengine aircraft with engines rated at least 800 horsepower each, or the equivalent in turbine engine powered aircraft; and at least 5 hours of flight training in the duties of a flight engineer.

(4) At least a commercial pilot certificate with an instrument rating and at least 5 hours of flight training in the duties of a flight engineer.

(5) At least 200 hours of flight time in a transport category airplane (or in a military airplane with at least two engines and at least equivalent weight and horsepower) as pilot in command or second in command performing the functions of a pilot in command under the supervision of a pilot in command.

(6) At least 100 hours of flight time as a flight engineer.

(7) Within the 90-day period before he applies, successful completion of an approved flight engineer ground and flight course of instruction as provided in appendix C of this part.

(Sec. 6, 80 Stat. 937, 49 U.S.C. 1655)

[Doc. No. 6458, 30 FR 14559, Nov. 23, 1965, as amended by Amdt. 63-5, 31 FR 9047, July 1, 1966; Amdt. 63-17, 40 FR 32830, Aug. 5, 1975]

§ 63.39 Skill requirements.

(a) An applicant for a flight engineer certificate with a class rating must pass a practical test on the duties of a flight engineer in the class of airplane for which a rating is sought. The test may only be given on an airplane specified in § 63.37(a).

(b) The applicant must—

(1) Show that he can satisfactorily perform preflight inspection, servicing, starting, pretakeoff, and postlanding procedures;

(2) In flight, show that he can satisfactorily perform the normal duties and procedures relating to the airplane, airplane engines, propellers (if appropriate), systems, and appliances; and

(3) In flight, in an airplane simulator, or in an approved flight engineer training device, show that he can satisfactorily perform emergency duties and procedures and recognize and take appropriate action for malfunctions of the airplane, engines, propellers (if appropriate), systems and appliances.

§ 63.41 Retesting after failure.

An applicant for a flight engineer certificate who fails a written test or practical test for that certificate may apply for retesting—

(a) After 30 days after the date he failed that test; or

(b) After he has received additional practice or instruction (flight, synthetic trainer, or ground training, or any combination thereof) that is necessary, in the opinion of the Administrator or the applicant's instructor (if the Administrator has authorized him to determine the additional instruction necessary) to prepare the applicant for retesting.

§ 63.42 Flight engineer certificate issued on basis of a foreign flight engineer license.

(a) *Certificates issued.* The holder of a current foreign flight engineer license issued by a contracting State to the Convention on International Civil Aviation, who meets the requirements of this section, may have a flight engineer certificate issued to him for the operation of civil aircraft of U.S. registry. Each flight engineer certificate issued under this section specifies the number and State of issuance of the foreign flight engineer license on which it is based. If the holder of the certificate cannot read, speak, or understand the English language, the Administrator may place any limitation on the certificate that he considers necessary for safety.

(b) *Medical standards and certification.* An applicant must submit evidence that he currently meets the medical standards for the foreign flight engineer license on which the application for a certificate under this section is based. A current medical certificate issued under part 67 of this chapter will be excepted as evidence that the applicant meets those standards. However, a medical certificate issued under part 67 of this chapter is not evidence that the applicant meets those standards outside the United States unless the State that issued the applicant's foreign flight engineer license also accepts that medical certificate as evidence of the applicant's physical fitness for his foreign flight engineer license.

(c) *Ratings issued.* Aircraft class ratings listed on the applicant's foreign flight engineer license, in addition to any issued to him after testing under the provisions of this part, are placed on the applicant's flight engineer certificate. An applicant without an aircraft class rating on his foreign flight engineer license may be issued a class rating if he shows that he currently meets the requirements for exercising the privileges of his foreign flight engineer license on that class of aircraft.

(d) *Privileges and limitations.* The holder of a flight engineer certificate issued under this section may act as a flight engineer of a civil aircraft of U.S. registry subject to the limitations of this part and any additional limitations placed on his certificate by the Administrator. He is subject to these limitations while he is acting as a flight engineer of the aircraft within or outside the United States. However, he may not act as flight engineer or in any other capacity as a required flight crewmember, of a civil aircraft of U.S. registry that is carrying persons or property for compensation or hire.

(e) *Renewal of certificate and ratings.* The holder of a certificate issued under this section may have that certificate and the ratings placed thereon renewed if, at the time of application for renewal, the foreign flight engineer license on which that certificate is based is in effect. Application for the renewal of the certificate and ratings thereon must be made before the expiration of the certificate.

(Sec. 6, 80 Stat. 937, 49 U.S.C. 1655)

[Doc. No. 8846, 33 FR 18614, Dec. 17, 1968, as amended by Amdt. 63-20, 45 FR 5673, Jan. 24, 1980]

§ 63.43 Flight engineer courses.

An applicant for approval of a flight engineer course must submit a letter to the Administrator requesting approval, and must also submit three copies of each course outline, a description of the facilities and equipment, and a list of the instructors and their qualifications. An air carrier or commercial operator with an approved flight engineer training course under part 121 of this chapter may apply for approval of a training course under this part by letter without submitting the additional information required by this paragraph. Minimum requirements for obtaining approval of a flight engineer course are set forth in appendix C of this part.

Subpart C—Flight Navigators

Authority: Secs. 313(a), 314, 601, and 607; 49 U.S.C. 1354(a), 1355, 1421, and 1427.

Source: Docket No. 1179, 27 FR 7970, Aug. 10, 1962, unless otherwise noted.

§ 63.51 Eligibility requirements; general.

To be eligible for a flight navigator certificate, a person must—

(a) Be at least 21 years of age;

(b) Be able to read, write, speak, and understand the English language;

(c) Hold at least a second-class medical certificate issued under part 67 of this chapter within the 12 months before the date he applies; and

(d) Comply with §§ 63.53, 63.55, and 63.57.

§ 63.53 Knowledge requirements.

(a) An applicant for a flight navigator certificate must pass a written test on—

(1) The regulations of this chapter that apply to the duties of a flight navigator;

(2) The fundamentals of flight navigation, including flight planning and cruise control;

(3) Practical meteorology, including analysis of weather maps, weather reports, and weather forecasts; and weather sequence abbreviations, symbols, and nomenclature;

(4) The types of air navigation facilities and procedures in general use;

(5) Calibrating and using air navigation instruments;

(6) Navigation by dead reckoning;

(7) Navigation by celestial means;

(8) Navigation by radio aids;

(9) Pilotage and map reading; and

(10) Interpretation of navigation aid identification signals.

(b) A report of the test is mailed to the applicant. A passing grade is evidence, for a period of 24 months after the test, that the applicant has complied with this section.

[Doc. No. 1179, 27 FR 7970, Aug. 10 1962, as amended by Amdt. 63-19, 43 FR 22639, May 25, 1978]

§ 63.55 Experience requirements.

(a) An applicant for a flight navigator certificate must be a graduate of a flight navigator course approved by the Administrator or present satisfactory documentary evidence of—

(1) Satisfactory determination of his position in flight at least 25 times by night by celestial observations and at least 25 times by day by celestial observations in conjunction with other aids; and

(2) At least 200 hours of satisfactory flight navigation including celestial and radio navigation and dead reckoning.

A pilot who has logged 500 hours of cross-country flight time, of which at least 100 hours were at night, may be credited with not more than 100 hours for the purposes of paragraph (a)(2) of this section.

(b) Flight time used exclusively for practicing long-range navigation methods, with emphasis on celestial navigation and dead reckoning, is considered to be satisfactory navigation experience for the purposes of paragraph (a) of this section. It must be substantiated by a logbook, by records of an armed force or a certificated air carrier, or by a letter signed by a certificated flight navigator and attached to the application.

§ 63.57 Skill requirements.

(a) An applicant for a flight navigator certificate must pass a practical test in navigating aircraft by—

(1) Dead reckoning;

(2) Celestial means; and

(3) Radio aids to navigation.

(b) An applicant must pass the written test prescribed by § 63.53 before taking the test under this section. However, if a delay in taking the test under this section would inconvenience the applicant or an air carrier, he may take it before he receives the result of the written test, or after he has failed the written test.

(c) The test requirements for this section are set forth in appendix A of this part.

[Doc. No. 1179, 27 FR 7970, Aug. 10, 1962, as amended by Amdt. 63-19, 43 FR 22639, May 25, 1978]

§ 63.59 Retesting after failure.

(a) An applicant for a flight navigator certificate who fails a written or practical test for that certificate may apply for retesting—

(1) After 30 days after the date he failed that test; or

(2) Before the 30 days have expired if the applicant presents a signed statement from a certificated flight navigator, certificated ground instructor, or any other qualified person approved by the Administrator, certifying that that person has given the applicant additional instruction in each of the subjects failed and that person considers the applicant ready for retesting.

(b) A statement from a certificated flight navigator, or from an operations official of an approved navigator course, is acceptable, for the purposes of paragraph (a)(2) of this section, for the written test and for the flight test. A statement from a person approved by the Administrator is acceptable for the written tests. A statement from a supervising or check navigator with the United States Armed Forces is acceptable for the written test and for the practical test.

(c) If the applicant failed the flight test, the additional instruction must have been administered in flight.

[Doc. No. 1179, 27 FR 7970, Aug. 10, 1962, as amended by Amdt. 63-19, 43 FR 22640, May 25, 1978]

§ 63.61 Flight navigator courses.

An applicant for approval of a flight navigator course must submit a letter to the Administrator requesting approval, and must also submit three copies of the course outline, a description of his facilities and equipment, and a list of the instructors and their qualifications. Requirements for the course are set forth in appendix B to this part.

Appendix A to Part 63—Test Requirements for Flight Navigator Certificate

(a) *Demonstration of skill.* An applicant will be required to pass practical tests on the prescribed subjects. These tests may be given by FAA inspectors and designated flight navigator examiners.

(b) *The examination.* The practical examination consists of a ground test and a flight test as itemized on the examination check sheet. Each item must be completed satisfactorily in order for the applicant to obtain a passing grade. Items 5, 6, 7 of the ground test may be completed orally, and items 17, 22, 23, 34, 36, 37, 38, and 39 of the flight test may be completed by an oral examination when a lack of ground facilities or navigation equipment makes such procedure necessary. In these cases a notation to that effect shall be made in the "Remarks" space on the check sheet.

(c) *Examination procedure.* (1) An applicant will provide an aircraft in which celestial observations can be taken in all directions. Minimum equipment shall include a table for plotting, a drift meter or absolute altimeter, an instrument for taking visual bearings, and a radio direction finder.

(2) More than one flight may be used to complete the flight test and any type of flight pattern may be used. The test will be conducted chiefly over water whenever practicable, and without regard to radio range legs or radials. If the test is conducted chiefly over land, a chart should be used which shows very little or no topographical and aeronautical data. The total flight time will cover a period of at least four hours. Only one applicant may be examined at one time, and no applicant may perform other than navigator duties during the examination.

(3) When the test is conducted with an aircraft belonging to an air carrier, the navigation procedures should conform with those set forth in the carrier's operations manual. Items of the flight test which are not performed during the routine navigation of the flight will be completed by oral examination after the flight or at times during flight which the applicant indicates may be used for tests on those items. Since in-flight weather conditions, the reliability of the weather forecast, and the stability of the aircraft will have considerable effect on an applicant's performance, good judgment must be used by the agent or examiner in evaluating the tests.

(d) *Ground test.* For the ground test, in the order of the numbered items on the examination check sheet, an applicant will be required to:

(1) Identify without a star identifier, at least six navigational stars and all planets available for navigation at the time of the examination and explain the method of identification.

(2) Identify two additional stars with a star identifier or sky diagrams and explain identification procedure.

(3) Precompute a time-altitude curve for a period of about 20 minutes and take 10 single observations of a celestial body which is rising or setting rapidly. The intervals between observations should be at least one minute. Mark each observation on the graph to show accuracy. All observations, after corrections, shall plot within 8 minutes of arc from the time-altitude curve, and the average error shall not exceed 5 minutes of arc.

(4) Take and plot one 3-star fix and 3 LOP's of the sun. Plotted fix or an average of LOP's must fall within 5 miles of the actual position of the observer.

(5) Demonstrate or explain the compensation and swinging of a liquid-type magnetic compass.

(6) Demonstrate or explain a method of aligning one type of drift meter.

(7) Demonstrate or explain a method of aligning an astro-compass or periscopic sextant.

(e) *Flight test.* For the flight test, in the order of the numbered items on the examination check sheet, an applicant will be required to:

(1) Demonstrate his ability to read weather symbols and interpret synoptic surface and upper air weather maps with particular emphasis being placed on winds.

(2) Prepare a flight plan by zones from the forecast winds or pressure data of an upper air chart and the operator's data.

(3) Compute from the operator's data the predicted fuel consumption for each zone of the flight, including the alternate.

(4) Determine the point-of-no-return for the flight with all engines running and the equitime point with one engine inoperative. Graphical methods which are part of the company's operations manual may be used for these computations.

(5) Prepare a cruise control (howgozit) chart from the operator's data.

(6) Enter actual fuel consumed on the cruise control chart and interpret the variations of the actual curve from the predicted curve.

(7) Check the presence on board and operating condition of all navigation equipment. Normally a check list will be used. This check will include a time tick or chronometer comparison. Any lack of thoroughness during this check will justify this item being graded unsatisfactory.

(8) Locate emergency equipment, such as, the nearest fire extinguisher, life preserver, life rafts, exits, axe, first aid kits, etc.

(9) Recite the navigator's duties and stations during emergencies for the type of aircraft used for the test.

(10) Demonstrate the proper use of a flux gate compass or gyrosyn compass (when available), with special emphasis on the caging methods and the location of switches, circuit breakers, and fuses. If these compasses are not part of the aircraft's equipment, an oral examination will be given.

(11) Be accurate and use good judgment when setting and altering headings. Erroneous application of variation, deviation, or drift correction, or incorrect measurement of course on the chart will be graded as unsatisfactory.

(12) Demonstrate or explain the use of characteristics of various chart projections used in long-range air navigation, including the plotting of courses and bearings, and the measuring of distances.

(13) Demonstrate ability to identify designated landmarks by the use of a sectional or WAC chart.

(14) Use a computer with facility and accuracy for the computation of winds, drift correction and drift angles, ground speeds, ETA's, fuel loads, etc.

(15) Determine track, ground speed, and wind by the double drift method. When a drift meter is not part of the aircraft's equipment, an oral examination on the use of the drift meter and a double drift problem shall be completed.

(16) Determine ground speed and wind by the timing method with a drift meter. When a drift meter is not part of the aircraft's equipment, an oral examination on the procedure and a problem shall be completed.

(17) Demonstrate the use of air plot for determining wind between fixes and for plotting pressure lines of position when using pressure and absolute altimeter comparisons.

(18) Give ETA's to well defined check points at least once each hour after the second hour of flight. The average error shall not be more than 5 percent of the intervening time intervals, and the maximum error of any one ETA shall not be more than 10 percent.

(19) Demonstrate knowledge and use of D/F equipment and radio facility information. Grading on this item will be based largely on the applicant's selection of those radio aids which will be of most value to his navigation, the manner with which he uses equipment, including filter box controls, and the precision with which he reads bearings. The aircraft's compass heading and all compass corrections must be considered for each bearing.

(20) Use care in tuning to radio stations to insure maximum reception of signal and check for interference signals. Receiver will be checked to ascertain that antenna and BFO (Voice-CW) switches are in correct positions.

(21) Identify at least three radio stations using International Morse code only for identification. The agent or examiner will tune in these stations so that the applicant will have no knowledge of the direction, distance, or frequency of the stations.

(22) Take at least one radio bearing by manual use of the loop. The agent or examiner will check the applicant's bearing by taking a manual bearing on the same station immediately after the applicant.

(23) Show the use of good judgment in evaluating radio bearings, and explain why certain bearings may be of doubtful value.

(24) Determine and apply correctly the correction required to be made to radio bearings before plotting them on a Mercator chart, and demonstrate the ability to plot bearings accurately on charts of the Mercator and Lambert conformal projections.

(25) Compute the compass heading, ETA, and fuel remaining if it is assumed that the flight would be diverted to an alternate airport at a time specified by the agent or examiner.

(26) Check the counter scales of a Loran receiver for accuracy, and explain the basic (face) adjustments which

affect tuning and counter alignment. A guide sheet may be used for this test.

(27) Demonstrate a knowledge of the basic principle of Loran and the ability to tune a Loran receiver, to match signals, to read time differences, to plot Loran LOP's, and to identify and use sky waves.

(28) Take and plot bearings from a consol station and explain the precautions which must be taken when tuning a radio receiver for consol signals. Also, discuss those conditions which affect the reliability of consol bearings.

(29) Demonstrate the ability to properly operate and read an absolute altimeter.

(30) Determine the "D" factors for a series of compared readings of an absolute altimeter and a pressure altimeter.

(31) Determine drift angle or lateral displacement from the true headingline by application of Bellamy's formula or a variation thereof.

(32) Interpret the altimeter comparison data with respect to the pressure system found at flight level. From this data evaluate the accuracy of the prognostic weather map used for flight planning and apply this analysis to the navigation of the flight.

(33) Interpret single LOP's for most probable position, and show how a series of single LOP's of the same body may be used to indicate the probable track and ground speed. Also, show how a series of single LOP's (celestial or radio) from the same celestial body or radio station may be used to determine position when the change of azimuth or bearing is 30° or more between observations.

(34) Select one of the celestial LOP's used during the flight and explain how to make a single line of position approach to a point selected by the agent or examiner, giving headings, times, and ETA's.

(35) Demonstrate the proper use of an astro-compass or periscopic sextant for taking bearings.

(36) Determine compass deviation as soon as possible after reaching cruising altitude and whenever there is a change of compass heading of 15° or more.

(37) Take celestial fixes at hourly intervals when conditions permit. The accuracy of these fixes shall be checked by means of a Loran, radio, or visual fix whenever practicable. After allowing for the probable error of a Loran, radio, or visual fix, a celestial fix under favorable conditions should plot within 10 miles of the actual position.

(38) Select celestial bodies for observation, when possible, whose azimuths will differ by approximately 120° for a 3-body fix and will differ by approximately 90° for a 2-body fix. The altitudes of the selected bodies should be between 25° and 75° whenever practicable.

(39) Have POMAR and any other required reports ready for transmission at time of schedule, and be able to inform the pilot in command promptly with regard to the aircraft's position and progress in comparison with the flight plan.

(40) Keep a log with sufficient legible entries to provide a record from which the flight could be retraced.

(41) Note significant weather changes which might influence the drift or ground speed of the aircraft, such as, temperature, "D" factors, frontal conditions, turbulence, etc.

(42) Determine the wind between fixes as a regular practice.

(43) Estimate the time required and average ground speed during a letdown, under conditions specified by the pilot in command.

(44) Work with sufficient speed to determine the aircraft's position hourly by celestial means and also make all other observations and records pertinent to the navigation. The applicant should be able to take the observation, compute, and plot a celestial LOP within a time limit of 8 minutes; take and plot a Loran LOP within a time limit of 3 minutes for ground waves and 4 minutes for sky waves; observe the absolute and pressure altimeters and compute the drift or lateral displacement within a time limit of 3 minutes.

(45) Be accurate in reading instruments and making computations. Errors which are made and corrected without affecting the navigation will be disregarded unless they cause considerable loss of time.

An uncorrected error in computation (including reading instruments and books) which will affect the reported position more than 25 miles, the heading more than 3°, or any ETA more than 15 minutes will cause this item to be graded unsatisfactory.

(46) Be alert to changing weather or other conditions during flight which might affect the navigation. An applicant should not fail to take celestial observations just prior to encountering a broken or overcast sky condition; and he should not fail to take a bearing on a radio station, which operates at scheduled intervals and which would be a valuable aid to the navigation.

(47) Show a logical choice and sequence in using the various navigation methods according to time and accuracy, and check the positions determined by one method against positions determined by other methods.

(48) Use a logical sequence in performing the various duties of a navigator and plan work according to a schedule. The more important duties should not be neglected for others of less importance.

Appendix B to Part 63—Flight Navigator Training Course Requirements

(a) *Training course outline* —(1) *Format.* The ground course outline and the flight course outline shall be combined in one looseleaf binder and shall include a table of contents, divided into two parts—ground course and flight course. Each part of the table of contents must contain a list of the major subjects, together with hours allotted to each subject and the total classroom and flight hours.

(2) *Ground course outline.* (i) It is not mandatory that a course outline have the subject headings arranged exactly as listed in this paragraph. Any arrangement of general headings and subheadings will be satisfactory provided all the subject material listed here is included and the acceptable minimum number of hours is assigned to each subject. Each general subject shall be broken down into detail showing items to be covered.

(ii) If any agency desires to include additional subjects in the ground training curriculum, such as international law, flight hygiene, or others which are not required, the hours allotted these additional subjects may not be included in the minimum classroom hours.

(iii) The following subjects with classroom hours are considered the minimum coverage for a ground training course for flight navigators:

Subject	Classroom hours
Federal Aviation Administration	5
To include Parts 63, 91, and 121 of this chapter.	
Meteorology	40
To include:	
Basic weather principles.	
Temperature.	
Pressure.	
Winds.	
Moisture in the atmosphere.	
Stability.	
Clouds.	
Hazards.	
Air masses.	
Front weather.	
Fog.	
Thunderstorms.	
Icing.	
World weather and climate.	
Weather maps and weather reports.	
Forecasting.	
International Morse code:	
Ability to receive code groups of letters and numerals at a speed of eight words per minute	
Navigation instruments (exclusive of radio and radar)	20
To include:	
Compasses.	
Pressure altimeters.	
Airspeed indicators.	
Driftmeters.	
Bearing indicators.	
Aircraft octants.	
Instrument calibration and alignment.	
Charts and pilotage	15
To include:	
Chart projections.	
Chart symbols.	
Principles of pilotage.	
Dead reckoning	30
To include:	
Air plot.	
Ground plot.	
Calculation of ETA.	
Vector analysis.	
Use of computer.	
Search.	
Absolute altimeter with:	
Applications	15
To include:	
Principles of construction.	
Operating instructions.	
Use of Bellamy's formula.	
Flight planning with single drift correction.	
Radio and long-range navigational aids	35
To include:	
Principles of radio transmission and reception.	
Radio aids to navigation.	
Government publications.	
Airborne D/F equipment.	
Errors of radio bearings.	
Quadrantal correction.	
Plotting radio bearings.	
ICAO Q code for direction finding.	
Loran.	
Consol.	
Celestial navigation	150
To include:	
The solar system.	
The celestial sphere.	
The astronomical triangle.	
Theory of lines of position.	
Use of the Air Almanac.	
Time and its applications.	
Navigation tables.	
Precomputation.	
Celestial line of position approach.	
Star identification.	
Corrections to celestial observations.	
Flight planning and cruise control	25
To include:	
The flight plan.	
Fuel consumption charts.	

Subject	Classroom hours
Methods of cruise control.	
Flight progress chart.	
Point-of-no-return.	
Equitime point.	
Long-range flight problems	15
Total (exclusive of final examinations)	350

(3) *Flight course outline.* (i) A minimum of 150 hours of supervised flight training shall be given, of which at least 50 hours of flight training must be given at night, and celestial navigation must be used during flights which total at least 125 hours.

(ii) A maximum of 50 hours of the required flight training may be obtained in acceptable types of synthetic flight navigator training devices.

(iii) Flights should be at least four hours in length and should be conducted off civil airways. Some training on long-range flights is desirable, but is not required. There is no limit to the number of students that may be trained on one flight, but at least one astrodrome or one periscopic sextant mounting must be provided for each group of four students.

(iv) Training must be given in dead reckoning, pilotage, radio navigation, celestial navigation, and the use of the absolute altimeter.

(b) *Equipment.* (1) Classroom equipment shall include one table at least 24″ × 32″ in dimensions for each student.

(2) Aircraft suitable for the flight training must be available to the approved course operator to insure that the flight training may be completed without undue delay.

The approved course operator may contract or obtain written agreements with aircraft operators for the use of suitable aircraft. A copy of the contract or written agreement with an aircraft operator shall be attached to each of the three copies of the course outline submitted for approval. In all cases, the approved course operator is responsible for the nature and quality of instruction given during flight.

(c) *Instructors.* (1) Sufficient classroom instructors must be available to prevent an excessive ratio of students to instructors. Any ratio in excess of 20 to 1 will be considered unsatisfactory.

(2) At least one ground instructor must hold a valid flight navigator certificate, and be utilized to coordinate instruction of ground school subjects.

(3) Each instructor who conducts flight training must hold a valid flight navigator certificate.

(d) *Revision of training course.* (1) Requests for revisions to course outlines, facilities, and equipment shall follow procedures for original approval of the course. Revisions should be submitted in such form that an entire page or pages of the approved outline can be removed and replaced by the revisions.

(2) The list of instructors may be revised at any time without request for approval, provided the minimum requirement of paragraph (e) of this section is maintained.

(e) *Credit for previous training and experience.* (1) Credit may be granted by an operator to students for previous training and experience which is provable and comparable to portions of the approved curriculum. When granting such credit, the approved course operator should be fully cognizant of the fact that he is responsible for the proficiency of his graduates in accordance with subdivision (i) of paragraph (3) of this section.

(2) Where advanced credit is allowed, the operator shall evaluate the student's previous training and experience in accordance with the normal practices of accredited technical schools. Before credit is given for any ground school subject or portion thereof, the student must pass an appropriate examination given by the operator. The results of the examination, the basis for credit allowance, and the hours credited shall be incorporated as a part of the student's records.

(3) Credit up to a maximum of 50 hours toward the flight training requirement may be given to pilots who have logged at least 500 hours while a member of a flight crew which required a certificated flight navigator or the Armed Forces equivalent. A similar credit may also be given to a licensed deck officer of the Maritime Service who has served as such for at least one year on ocean-going vessels. One-half of the flight time credited under the terms of this paragraph may be applied toward the 50 hours of flight training required at night.

(f) *Students records and reports.* Approval of a course shall not be continued in effect unless the course operator keeps an accurate record of each student, including a chronological log of all instruction, subjects covered and course examinations and grades, and unless he prepares and transmits to the local Flight Standards District Office not later than January 31 of each year, a report containing the following information for the previous calendar year:

(1) The names of all students graduated, together with their school grades for ground and flight subjects.

(2) The names of all students failed or dropped, together with their school grades and reasons for dropping.

(g) *Quality of instruction.* Approval of a course shall not be continued in effect unless at least 80 percent of the students who apply within 90 days after graduation are able to qualify on the first attempt for certification as flight navigators.

(h) *Statement of graduation.* Each student who successfully completes an approved flight navigator course shall be given a statement of graduation.

(i) *Inspections.* Approved course operations will be inspected by authorized representatives of the Administrator as often as deemed necessary to insure that instruction is maintained at the required standards, but the period between inspections shall not exceed 12 months.

(j) *Change of ownership, name, or location*—(1) *Change of ownership.* Approval of a flight navigator course shall not be continued in effect after the course has changed ownership. The new owner must obtain a new approval by following the procedure prescribed for original approval.

(2) *Change in name.* An approved course changed in name but not changed in ownership shall remain valid if the change is reported by the approved course operator to the local Flight Standards District Office. A letter of approval under the new name will be issued by the regional office.

(3) *Change in location.* An approved course shall remain in effect even though the approved course operator changes location if the change is reported without delay by the operator to the local Flight Standards District Office, which will inspect the facilities to be used. If they are found to be adequate, a letter of approval showing the new location will be issued by the regional office.

(k) *Cancellation of approval.* (1) Failure to meet or maintain any of the requirements set forth in this section for the approval or operation of an approved flight navigator course shall be considered sufficient reason for cancellation of the approval.

(2) If an operator should desire voluntary cancellation of his approved course, he should submit the effective letter of approval and a written request for cancellation to the Administrator through the local Flight Standards District Office.

(l) *Duration.* The authority to operate an approved flight navigator course shall expire 24 months after the last day of the month of issuance.

(m) *Renewal.* Application for renewal of authority to operate an approved flight navigator course may be made by letter to the local Flight Standards District Office at any time within 60 days before to the expiration date. Renewal of approval will depend upon the course operator meeting the current conditions for approval and having a satisfactory record as an operator.

[Doc. No. 1179, 27 FR 7970, Aug. 10, 1962, as amended by Amdt. 63-6, 31 FR 9211, July 6, 1966; Amdt. 63-28, 54 FR 39291, Sept. 25, 1989]

Appendix C to Part 63—Flight Engineer Training Course Requirements

(a) *Training course outline* —(1) *Format.* The ground course outline and the flight course outline are independent. Each must be contained in a looseleaf binder to include a table of contents. If an applicant desires approval of both a ground school course and a flight school course, they must be combined in one looseleaf binder that includes a separate table of contents for each course. Separate course outlines are required for each type of airplane.

(2) *Ground course outline.* (i) It is not mandatory that the subject headings be arranged exactly as listed in this paragraph. Any arrangement of subjects is satisfactory if all the subject material listed here is included and at least the minimum programmed hours are assigned to each subject. Each general subject must be broken down into detail showing the items to be covered.

(ii) If any course operator desires to include additional subjects in the ground course curriculum, such as international law, flight hygiene, or others that are not required, the hours allotted these additional subjects may not be included in the minimum programmed classroom hours.

(iii) The following subjects and classroom hours are the minimum programmed coverage for the initial approval of a ground training course for flight engineers. Subsequent to initial approval of a ground training course an applicant may apply to the Administrator for a reduction in the programmed hours. Approval of a reduction in the approved programmed hours is based on improved training effectiveness due to improvements in methods, training aids, quality of instruction, or any combination thereof.

Subject	Classroom hours
Federal Aviation Regulations	10
To include the regulations of this chapter that apply to flight engineers	
Theory of Flight and Aerodynamics	10
Airplane Familiarization	90
To include as appropriate:	
Specifications.	
Construction features.	
Flight controls.	
Hydraulic systems.	
Pneumatic systems.	
Electrical systems.	
Anti-icing and de-icing systems.	
Pressurization and air-conditioning systems.	
Vacuum systems.	
Pilot static systems.	
Instrument systems.	
Fuel and oil systems.	
Emergency equipment.	
Engine Familiarization	45
To include as appropriate:	
Specifications.	
Construction features.	
Lubrication.	
Ignition.	
Carburetor and induction, supercharging and fuel control systems	
Accessories.	
Propellers.	
Instrumentation.	
Emergency equipment.	
Normal Operations (Ground and Flight)	50
To include as appropriate:	
Servicing methods and procedures.	
Operation of all the airplane systems.	
Operation of all the engine systems.	
Loading and center of gravity computations.	
Cruise control (normal, long range, maximum endurance)	
Power and fuel computation.	
Meteorology as applicable to engine operation	

Subject	Classroom hours
Emergency Operations	80
To include as appropriate:	
Landing gear, brakes, flaps, speed brakes, and leading edge devices	
Pressurization and air-conditioning.	
Portable fire extinguishers.	
Fuselage fire and smoke control.	
Loss of electrical power.	
Engine fire control.	
Engine shut-down and restart.	
Oxygen.	
Total (exclusive of final tests)	235

The above subjects, except Theory of Flight and Aerodynamics, and Regulations must apply to the same type of airplane in which the student flight engineer is to receive flight training.

(3) Flight Course Outline. (i) The flight training curriculum must include at least 10 hours of flight instruction in an airplane specified in § 63.37(a). The flight time required for the practical test may not be credited as part of the required flight instruction.

(ii) All of the flight training must be given in the same type airplane.

(iii) As appropriate to the airplane type, the following subjects must be taught in the flight training course:

Subject
normal duties, procedures and operations
To include as appropriate:
Airplane preflight.
Engine starting, power checks, pretakeoff, postlanding and shut-down procedures.
Power control.
Temperature control.
Engine operation analysis.
Operation of all systems.
Fuel management.
Logbook entries.
Pressurization and air conditioning.
recognition and correction of in-flight malfunctions
To include:
Analysis of abnormal engine operation.
Analysis of abnormal operation of all systems.
Corrective action.
emergency operations in flight
To include as appropriate:
Engine fire control.
Fuselage fire control.
Smoke control.
Loss of power or pressure in each system.
Engine overspeed.
Fuel dumping.
Landing gear, spoilers, speed brakes, and flap extension and retraction.
Engine shut-down and restart.
Use of oxygen.

(iv) If the Administrator finds a simulator or flight engineer training device to accurately reproduce the design, function, and control characteristics, as pertaining to the duties and responsibilities of a flight engineer on the type of airplane to be flown, the flight training time may be reduced by a ratio of 1 hour of flight time to 2 hours of airplane simulator time, or 3 hours of flight engineer training device time, as the case may be, subject to the following limitations:

(*a*) Except as provided in subdivision (*b*) of this paragraph, the required flight instruction time in an airplane may not be less than 5 hours.

(*b*) As to a flight engineer student holding at least a commercial pilot certificate with an instrument rating, airplane simulator or a combination of airplane simulator and flight engineer training device time may be submitted for up to all 10 hours of the required flight instruction time in an airplane. However, not more than 15 hours of flight engineer training device time may be substituted for flight instruction time.

(v) To obtain credit for flight training time, airplane simulator time, or flight engineer training device time, the student must occupy the flight engineer station and operate the controls.

(b) *Classroom equipment.* Classroom equipment should consist of systems and procedural training devices, satisfactory to the Administrator, that duplicate the operation of the systems of the airplane in which the student is to receive his flight training.

(c) *Contracts or agreements.* (1) An approved flight engineer course operator may contract with other persons to obtain suitable airplanes, airplane simulators, or other training devices or equipment.

(2) An operator who is approved to conduct both the flight engineer ground course and the flight engineer flight course may contract with others to conduct one course or the other in its entirety but may not contract with others to conduct both courses for the same airplane type.

(3) An operator who has approval to conduct a flight engineer ground course or flight course for a type of airplane, but not both courses, may not contract with another person to conduct that course in whole or in part.

(4) An operator who contracts with another to conduct a flight engineer course may not authorize or permit the course to be conducted in whole or in part by a third person.

(5) In all cases, the course operator who is approved to operate the course is responsible for the nature and quality of the instruction given.

(6) A copy of each contract authorized under this paragraph must be attached to each of the 3 copies of the course outline submitted for approval.

(d) *Instructors.* (1) Only certificated flight engineers may give the flight instruction required by this appendix in an airplane, simulator, or flight engineer training device.

(2) There must be a sufficient number of qualified instructors available to prevent an excess ratio of students to instructors.

(e) *Revisions.* (1) Requests for revisions of the course outlines, facilities or equipment must follow the procedures for original approval of the course. Revisions must be submitted in such form that an entire page or pages of the approved outline can be removed and replaced by the revisions.

(2) The list of instructors may be revised at any time without request for approval, if the requirements of paragraph (d) of this appendix are maintained.

(f) *Ground school credits.* (1) Credit may be granted a student in the ground school course by the course operator for comparable previous training or experience that the student can show by written evidence: however, the course operator must still meet the quality of instruction as described in paragraph (h) of this appendix.

(2) Before credit for previous training or experience may be given, the student must pass a test given by the course operator on the subject for which the credit is to be given. The course operator shall incorporate results of the test, the basis for credit allowance, and the hours credited as part of the student's records.

(g) *Records and reports.* (1) The course operator must maintain, for at least two years after a student graduates, fails, or drops from a course, a record of the student's training, including a chronological log of the subject course, attendance examinations, and grades.

(2) Except as provided in paragraph (3) of this section, the course operator must submit to the Administrator, not later than January 31 of each year, a report for the previous calendar year's training, to include:

(i) Name, enrollment and graduation date of each student;

(ii) Ground school hours and grades of each student;

(iii) Flight, airplane simulator, flight engineer training device hours, and grades of each student; and

(iv) Names of students failed or dropped, together with their school grades and reasons for dropping.

(3) Upon request, the Administrator may waive the reporting requirements of paragraph (2) of this section for an approved flight engineer course that is part of an approved training course under subpart N of part 121 of this chapter.

(h) *Quality of instruction.* (1) Approval of a ground course is discontinued whenever less than 80 percent of the students pass the FAA written test on the first attempt.

(2) Approval of a flight course is discontinued whenever less than 80 percent of the students pass the FAA practical test on the first attempt.

(3) Notwithstanding paragraphs (1) and (2) of this section, approval of a ground or flight course may be continued when the Administrator finds—

(i) That the failure rate was based on less than a representative number of students; or

(ii) That the course operator has taken satisfactory means to improve the effectiveness of the training.

(i) *Time limitation.* Each student must apply for the written test and the flight test within 90 days after completing the ground school course.

(j) *Statement of course completion.* (1) The course operator shall give to each student who successfully completes an approved flight engineer ground school training course, and passes the FAA written test, a statement of successful completion of the course that indicates the date of training, the type of airplane on which the ground course training was based, and the number of hours received in the ground school course.

(2) The course operator shall give each student who successfully completes an approved flight engineer flight course, and passed the FAA practical test, a statement of successful completion of the flight course that indicates the dates of the training, the type of airplane used in the flight course, and the number of hours received in the flight course.

(3) A course operator who is approved to conduct both the ground course and the flight course may include both courses in a single statement of course completion if the provisions of paragraphs (1) and (2) of this section are included.

(4) The requirements of this paragraph do not apply to an air carrier or commercial operator with an approved training course under part 121 of this chapter providing the student receives a flight engineer certificate upon completion of that course.

(k) *Inspections.* Each course operator shall allow the Administrator at any time or place, to make any inspection necessary to ensure that the quality and effectiveness of the instruction are maintained at the required standards.

(l) *Change of ownership, name, or location.* (1) Approval of a flight engineer ground course or flight course is discontinued if the ownership of the course changes. The new owner must obtain a new approval by following the procedure prescribed for original approval.

(2) Approval of a flight engineer ground course or flight course does not terminate upon a change in the name of the course that is reported to the Administrator within 30 days. The Administrator issues a new letter of approval, using the new name, upon receipt of notice within that time.

(3) Approval of a flight engineer ground course or flight course does not terminate upon a change in location of the course that is reported to the Administrator within 30 days. The Administrator issues a new letter of approval, showing the new location, upon receipt of notice within that time, if he finds the new facilities to be adequate.

(m) *Cancellation of approval.* (1) Failure to meet or maintain any of the requirements of this appendix for the approval of a flight engineer ground course or flight course is reason for cancellation of the approval.

(2) If a course operator desires to voluntarily terminate the course, he should notify the Administrator in writing and return the last letter of approval.

(n) *Duration.* Except for a course operated as part of an approved training course under subpart N of part 121 of this chapter, the approval to operate a flight engineer ground course or flight course terminates 24 months after the last day of the month of issue.

(o) *Renewal.* (1) Renewal of approval to operate a flight engineer ground course or flight course is conditioned upon the course operator's meeting the requirements of this appendix.

(2) Application for renewal may be made to the Administrator at any time after 60 days before the termination date.

(p) *Course operator approvals.* An applicant for approval of a flight engineer ground course, or flight course, or both, must meet all of the requirements of this appendix concerning application, approval, and continuing approval of that course or courses.

(q) *Practical test eligibility.* An applicant for a flight engineer certificate and class rating under the provisions of § 63.37(b)(6) is not eligible to take the practical test unless he has successfully completed an approved flight engineer ground school course in the same type of airplane for which he has completed an approved flight engineer flight course.

[Doc. No. 6458, 30 FR 14560, Nov. 23, 1965, as amended by Amdt. 63-15, 37 FR 9758, May 17, 1972]

PART 63

FAR

PART 65

CERTIFICATION: AIRMEN OTHER THAN FLIGHT CREWMEMBERS

Contents

Authority: 49 U.S.C. 106(g). 40113, 44701-44703, 44707, 44709-44711, 45102-45103, 45301-45302.

Source: Docket No. 1179, 27 FR 7973, Aug. 10, 1962, unless otherwise noted.

Special Federal Aviation Regulation No. 100-2

Editorial Note: For the text of SFAR No. 100-2, see part 61 of this chapter.

Special Federal Aviation Regulation No. 103—Process for Requesting Waiver of Mandatory Separation Age for a Federal Aviation Administration Air Traffic Control Specialist In Flight Service Stations, Enroute or Terminal Facilities, and the David J. Hurley Air Traffic Control System Command Center

1. *To whom does this SFAR apply?* This Special Federal Aviation Regulation (SFAR) applies to you if you are an air traffic control specialist (ATCS) employed by the FAA in flight service stations, enroute facilities, terminal facilities, or at the David J. Hurley Air Traffic Control System Command Center who wishes to obtain a waiver of the mandatory separation age as provided by 5 U.S.C. section 8335(a).

2. *When must I file for a waiver?* No earlier than the beginning of the twelfth month before, but no later than the beginning of the sixth month before, the month in which you turn 56, your official chain-of-command must receive your written request asking for a waiver of mandatory separation.

3. *What if I do not file a request before six months before the month in which I turn 56?* If your official chain-of-command does not receive your written request for a waiver of mandatory separation before the beginning of the sixth month before the month in which you turn 56, your request will be denied.

4. *How will the FAA determine if my request meets the filing time requirements of this SFAR?*

a. We consider your request to be filed in a timely manner under this SFAR if your official chain-of-command receives it or it is postmarked:

i. After 12 a.m. on the first day of the twelfth month before the month in which you turn 56; and

ii. Before 12 a.m. of the first day of the sixth month before the month in which you turn 56.

b. If you file your request by mail and the postmark is not legible, we will consider it to comply with paragraph a.2 of this section if we receive it by 12 p.m. of the fifth day of the sixth month before the month in which you turn 56.

c. If the last day of the time period specified in paragraph a.2 or paragraph b falls on a Saturday, Sunday, or Federal holiday, we will consider the time period to end at 12 p.m. of the next business day.

5. *Where must I file my request for waiver and what must it include?*

a. You must file your request for waiver of mandatory separation in writing with the Air Traffic Manager in flight service stations, enroute facilities, terminal facilities, or the David J. Hurley Air Traffic Control System Command Center in which you are employed.

b. Your request for waiver must include all of the following:

i. Your name.
ii. Your current facility.
iii. Your starting date at the facility.
iv. A list of positions at the facility that you are certified in and how many hours it took to achieve certification at the facility.
v. Your area of specialty at the facility.
vi. Your shift schedule.
vii. [Reserved]
viii. A list of all facilities where you have worked as a certified professional controller (CPC) including facility level and dates at each facility;
ix. Evidence of your exceptional skills and experience as a controller; and
x. Your signature.

6. *How will my waiver request be reviewed?*

a. Upon receipt of your request for waiver, the Air Traffic Manager of your facility will make a written recommendation that the Administrator either approve or deny your request. If the manager recommends approval of your request, he or she will certify in writing the accuracy of the information you provided as evidence of your exceptional skills and experience as a controller.

b. The Air Traffic Manager will then forward the written recommendation with a copy of your request to the senior executive manager in the Air Traffic Manager's regional chain-of-command.

c. The senior executive manager in the regional chain-of-command will make a written recommendation that the Administrator either approve or deny your request. If the senior executive manager recommends approval of your request, he or she will certify in writing the accuracy of the information you have provided as evidence of exceptional skills and experience.

d. The senior executive manager in the regional chain-of-command will then forward his or her recommendation with a copy of your request to the appropriate Vice President at FAA Headquarters. Depending on the facility in which you are employed, the request will be forwarded to either the Vice President for Flight Services, the Vice President for Enroute and Oceanic Services, the Vice President for Terminal Services or the Vice President for Systems Operations. For example, if you work at a flight service station at the time that you request a waiver, the request will be forwarded to the Vice President for Flight Services.

e. The appropriate Vice President will review your request and make a written recommendation that the Administrator either approve or deny your request, which will be forwarded to the Administrator.

f. The Administrator will issue the final decision on your request.

7. *If I am granted a waiver, when will it expire?*

a. Waivers will be granted for a period of one year.

b. No later than 90-days prior to expiration of a waiver, you may request that the waiver be extended using the same process identified in section 6.

c. If you timely request an extension of the waiver and it is denied, you will receive a 60-day advance notice of your separation date simultaneously with notification of the denial.

d. If you do not request an extension of the waiver granted, you will receive a 60-day advance notice of your separation date.

e. Action to separate you from your covered position becomes effective on the last day of the month in which the 60-day notice expires.

8. *Under what circumstances may my waiver be terminated?*

a. The FAA/DOT may terminate your waiver under the following circumstances:

i. The needs of the FAA; or

ii. If you are identified as a primary contributor to an operational error/deviation or runway incursion.

b. If the waiver is terminated for either of the reasons identified in paragraph 1 of this section, the air traffic control specialist will receive a 60-day advance notice.

c. Action to separate you from your covered position becomes effective on the last day of the month in which the 60-day notice expires.

9. *Appeal of denial or termination of waiver request:* The denial or termination of a waiver of mandatory separation request is neither appealable nor grievable.

[Doc. No. FAA-2004-17334, 70 FR 1636, Jan. 7, 2005, as amended by Amdt. 65-55, 76 FR 12, Jan. 3, 2011]

Subpart A—General

§ 65.1 Applicability.

This part prescribes the requirements for issuing the following certificates and associated ratings and the general operating rules for the holders of those certificates and ratings:

(a) Air-traffic control-tower operators.

(b) Aircraft dispatchers.

(c) Mechanics.

(d) Repairmen.

(e) Parachute riggers.

§ 65.3 Certification of foreign airmen other than flight crewmembers.

A person who is neither a U.S. citizen nor a resident alien is issued a certificate under subpart D of this part, outside the United States, only when the Administrator finds that the certificate is needed for the operation or continued airworthiness of a U.S.-registered civil aircraft.

[Doc. 65-28, 47 FR 35693, Aug. 16, 1982]

§ 65.11 Application and issue.

(a) Application for a certificate and appropriate class rating, or for an additional rating, under this part must be made on a form and in a manner prescribed by the Administrator. Each person who applies for airmen certification services to be administered outside the United States or for any certificate or rating issued under this part must show evidence that the fee prescribed in appendix A of part 187 of this chapter has been paid.

(b) An applicant who meets the requirements of this part is entitled to an appropriate certificate and rating.

(c) Unless authorized by the Administrator, a person whose air traffic control tower operator, mechanic, or parachute rigger certificate is suspended may not apply for any rating to be added to that certificate during the period of suspension.

(d) Unless the order of revocation provides otherwise—

(1) A person whose air traffic control tower operator, aircraft dispatcher, or parachute rigger certificate is revoked may not apply for the same kind of certificate for 1 year after the date of revocation; and

(2) A person whose mechanic or repairman certificate is revoked may not apply for either of those kinds of certificates for 1 year after the date of revocation.

[Doc. No. 1179, 27 FR 7973, Aug. 10, 1962, as amended by Amdt. 65-9, 31 FR 13524, Oct. 20, 1966; Amdt. 65-28, 47 FR 35693, Aug. 16, 1982; Amdt. 65-49, 72 FR 18559, Apr. 12, 2007]

§ 65.12 Offenses involving alcohol or drugs.

(a) A conviction for the violation of any Federal or state statute relating to the growing, processing, manufacture, sale, disposition, possession, transportation, or importation of narcotic drugs, marihuana, or depressant or stimulant drugs or substances is grounds for—

(1) Denial of an application for any certificate or rating issued under this part for a period of up to 1 year after the date of final conviction; or

(2) Suspension or revocation of any certificate or rating issued under this part.

(b) The commission of an act prohibited by § 91.19(a) of this chapter is grounds for—

(1) Denial of an application for a certificate or rating issued under this part for a period of up to 1 year after the date of that act; or

(2) Suspension or revocation of any certificate or rating issued under this part.

[Doc. No. 21956, 50 FR 15379, Apr. 17, 1985, as amended by Amdt. 65-34, 54 FR 34330, Aug. 18, 1989]

§ 65.13 Temporary certificate.

A certificate and ratings effective for a period of not more than 120 days may be issued to a qualified applicant, pending review of his application and supplementary documents and the issue of the certificate and ratings for which he applied.

[Doc. No. 1179, 27 FR 7973, Aug. 10, 1962, as amended by Amdt. 65-23, 43 FR 22640, May 25, 1978]

§ 65.14 Security disqualification.

(a) *Eligibility standard.* No person is eligible to hold a certificate, rating, or authorization issued under this part when the Transportation Security Administration (TSA) has notified the FAA in writing that the person poses a security threat.

(b) *Effect of the issuance by the TSA of an Initial Notification of Threat Assessment.* (1) The FAA will hold in abeyance pending the outcome of the TSA's final threat assessment review an application for any certificate, rating, or authorization under this part by any person who has been issued an Initial Notification of Threat Assessment by the TSA.

(2) The FAA will suspend any certificate, rating, or authorization issued under this part after the TSA issues to the holder an Initial Notification of Threat Assessment.

(c) *Effect of the issuance by the TSA of a Final Notification of Threat Assessment.* (1) The FAA will deny an application for any certificate, rating, or authorization under this part to any person who has been issued a Final Notification of Threat Assessment.

(2) The FAA will revoke any certificate, rating, or authorization issued under this part after the TSA has issued to the holder a Final Notification of Threat Assessment.

[Doc. No. FAA-2003-14293, 68 FR 3775, Jan. 24, 2003]

§ 65.15 Duration of certificates.

(a) Except for repairman certificates, a certificate or rating issued under this part is effective until it is surrendered, suspended, or revoked.

(b) Unless it is sooner surrendered, suspended, or revoked, a repairman certificate is effective until the holder is relieved from the duties for which the holder was employed and certificated.

(c) The holder of a certificate issued under this part that is suspended, revoked, or no longer effective shall return it to the Administrator.

(d) Except for temporary certificates issued under § 65.13, the holder of a paper certificate issued under this part may not exercise the privileges of that certificate after March 31, 2013.

[Doc. No. 22052, 47 FR 35693, Aug. 16, 1982, as amended by Amdt. 65-51, 73 FR 10668, Feb. 28, 2008]

§ 65.16 Change of name: Replacement of lost or destroyed certificate.

(a) An application for a change of name on a certificate issued under this part must be accompanied by the applicant's current certificate and the marriage license, court order, or other document verifying the change. The documents are returned to the applicant after inspection.

(b) An application for a replacement of a lost or destroyed certificate is made by letter to the Department of Transportation, Federal Aviation Administration, Airman Certification Branch, Post Office Box 25082, Oklahoma City, OK 73125. The letter must—

(1) Contain the name in which the certificate was issued, the permanent mailing address (including zip code), social security number (if any), and date and place of birth of the certificate holder, and any available information regarding the grade, number, and date of issue of the certificate, and the ratings on it; and

(2) Be accompanied by a check or money order for $2, payable to the Federal Aviation Administration.

(c) An application for a replacement of a lost or destroyed medical certificate is made by letter to the Department of Transportation, Federal Aviation Administration, Aerospace Medical Certification Division, Post Office Box 26200, Oklahoma City, OK 73125, accompanied by a check or money order for $2.00.

(d) A person whose certificate issued under this part or medical certificate, or both, has been lost may obtain a telegram from the FAA confirming that it was issued. The telegram may be carried as a certificate for a period not to exceed 60 days pending his receiving a duplicate certificate under paragraph (b) or (c) of this section, unless he has been notified that the certificate has been suspended or revoked. The request for such a telegram may be made by prepaid telegram, stating the date upon which a duplicate certificate was requested, or including the request for a duplicate and a money order for the necessary amount. The request for a telegraphic certificate should be sent to the office prescribed in paragraph (b) or (c) of this section, as appropriate. However, a request for both at the same time should be sent to the office prescribed in paragraph (b) of this section.

[Doc. No. 7258, 31 FR 13524, Oct. 20, 1966, as amended by Doc. No. 8084, 32 FR 5769, Apr. 11, 1967; Amdt. 65-16, 35 FR 14075, Sept. 4, 1970; Amdt. 65-17, 36 FR 2865, Feb. 11, 1971; Amdt. 65-52, 73 FR 43065, July 24, 2008]

§ 65.17 Tests: General procedure.

(a) Tests prescribed by or under this part are given at times and places, and by persons, designated by the Administrator.

(b) The minimum passing grade for each test is 70 percent.

§ 65.18 Written tests: Cheating or other unauthorized conduct.

(a) Except as authorized by the Administrator, no person may—

(1) Copy, or intentionally remove, a written test under this part;

(2) Give to another, or receive from another, any part or copy of that test;

(3) Give help on that test to, or receive help on that test from, any person during the period that test is being given;

(4) Take any part of that test in behalf of another person;

(5) Use any material or aid during the period that test is being given; or

(6) Intentionally cause, assist, or participate in any act prohibited by this paragraph.

(b) No person who commits an act prohibited by paragraph (a) of this section is eligible for any airman or ground instructor certificate or rating under this chapter for a period of 1 year after the date of that act. In addition, the commission of that act is a basis for suspending or revoking any airman or ground instructor certificate or rating held by that person.

[Doc. No. 4086, 30 FR 2196, Feb. 18, 1965]

§ 65.19 Retesting after failure.

An applicant for a written, oral, or practical test for a certificate and rating, or for an additional rating under this part, may apply for retesting—

(a) After 30 days after the date the applicant failed the test; or

(b) Before the 30 days have expired if the applicant presents a signed statement from an airman holding the certificate and rating sought by the applicant, certifying that the airman has given the applicant additional instruction in each of the subjects failed and that the airman considers the applicant ready for retesting.

[Doc. No. 16383, 43 FR 22640, May 25, 1978]

§ 65.20 Applications, certificates, logbooks, reports, and records: Falsification, reproduction, or alteration.

(a) No person may make or cause to be made—

(1) Any fraudulent or intentionally false statement on any application for a certificate or rating under this part;

(2) Any fraudulent or intentionally false entry in any logbook, record, or report that is required to be kept, made, or used, to show compliance with any requirement for any certificate or rating under this part;

(3) Any reproduction, for fraudulent purpose, of any certificate or rating under this part; or

(4) Any alteration of any certificate or rating under this part.

(b) The commission by any person of an act prohibited under paragraph (a) of this section is a basis for suspending or revoking any airman or ground instructor certificate or rating held by that person.

[Doc. No. 4086, 30 FR 2196, Feb. 18, 1965]

§ 65.21 Change of address.

Within 30 days after any change in his permanent mailing address, the holder of a certificate issued under this part shall notify the Department of Transportation, Federal Aviation Administration, Airman Certification Branch, Post Office Box 25082, Oklahoma City, OK 73125, in writing, of his new address.

[Doc. No. 10536, 35 FR 14075, Sept. 4, 1970]

§ 65.23 [Reserved]

Subpart B—Air Traffic Control Tower Operators

Source: Docket No. 10193, 35 FR 12326, Aug. 1, 1970, unless otherwise noted.

§ 65.31 Required certificates, and rating or qualification.

No person may act as an air traffic control tower operator at an air traffic control tower in connection with civil aircraft unless he—

(a) Holds an air traffic control tower operator certificate issued to him under this subpart;

(b) Holds a facility rating for that control tower issued to him under this subpart, or has qualified for the operating position at which he acts and is under the supervision of the holder of a facility rating for that control tower; and

For the purpose of this subpart, *operating position* means an air traffic control function performed within or directly associated with the control tower;

(c) Except for a person employed by the FAA or employed by, or on active duty with, the Department of the Air Force, Army, or Navy or the Coast Guard, holds at least a second-class medical certificate issued under part 67 of this chapter.

[Doc. No. 10193, 35 FR 12326, Aug. 1, 1970, as amended by Amdt. 65-25, 45 FR 18911, Mar. 24, 1980; Amdt. 65-31, 52 FR 17518, May 8, 1987]

§ 65.33 Eligibility requirements: General.

To be eligible for an air traffic control tower operator certificate a person must—

(a) Be at least 18 years of age;

(b) Be of good moral character;

(c) Be able to read, write, and understand the English language and speak it without accent or impediment of speech that would interfere with two-way radio conversation;

(d) Except for a person employed by the FAA or employed by, or on active duty with, the Department of the Air Force, Army, or Navy or the Coast Guard, hold at least a second-class medical certificate issued under part 67 of this chapter within the 12 months before the date application is made; and

(e) Comply with § 65.35.

[Doc. No. 10193, 35 FR 12326, Aug. 1, 1970, as amended by Amdt. 65-25, 45 FR 18911, Mar. 24, 1980; Amdt. 65-31, 52 FR 17518, May 8, 1987]

§ 65.35 Knowledge requirements.

Each applicant for an air traffic control tower operator certificate must pass a written test on—

(a) The flight rules in part 91 of this chapter:

(b) Airport traffic control procedures, and this subpart:

(c) En route traffic control procedures;

(d) Communications operating procedures;

(e) Flight assistance service;

(f) Air navigation, and aids to air navigation; and

(g) Aviation weather.

§ 65.37 Skill requirements: Operating positions.

No person may act as an air traffic control tower operator at any operating position unless he has passed a practical test on—

(a) Control tower equipment and its use;

(b) Weather reporting procedures and use of reports;

(c) Notices to Airmen, and use of the Airman's Information Manual;

(d) Use of operational forms;

(e) Performance of noncontrol operational duties; and

(f) Each of the following procedures that is applicable to that operating position and is required by the person performing the examination:

(1) The airport, including rules, equipment, runways, taxiways, and obstructions.

(2) The terrain features, visual checkpoints, and obstructions within the lateral boundaries of the surface areas of Class B, Class C, Class D, or Class E airspace designated for the airport.

(3) Traffic patterns and associated procedures for use of preferential runways and noise abatement.

(4) Operational agreements.

(5) The center, alternate airports, and those airways, routes, reporting points, and air navigation aids used for terminal air traffic control.

(6) Search and rescue procedures.

(7) Terminal air traffic control procedures and phraseology.

(8) Holding procedures, prescribed instrument approach, and departure procedures.

(9) Radar alignment and technical operation.

(10) The application of the prescribed radar and nonradar separation standard, as appropriate.

[Doc. No. 10193, 35 FR 12326, Aug. 1, 1991, as amended by Amdt. 65-36, 56 FR 65653, Dec. 17, 1991]

§ 65.39 Practical experience requirements: Facility rating.

Each applicant for a facility rating at any air traffic control tower must have satisfactorily served—

(a) As an air traffic control tower operator at that control tower without a facility rating for at least 6 months; or

(b) As an air traffic control tower operator with a facility rating at a different control tower for at least 6 months before the date he applies for the rating.

However, an applicant who is a member of an Armed Force of the United States meets the requirements of this section if he has satisfactorily served as an air traffic control tower operator for at least 6 months.

[Doc. No. 1179, 27 FR 7973, Aug. 10, 1962, as amended by Amdt. 65-19, 36 FR 21280, Nov. 5, 1971]

§ 65.41 Skill requirements: Facility ratings.

Each applicant for a facility rating at an air traffic control tower must have passed a practical test on each item listed in § 65.37 of this part that is applicable to each operating position at the control tower at which the rating is sought.

§ 65.43 Rating privileges and exchange.

(a) The holder of a senior rating on August 31, 1970, may at any time after that date exchange his rating for a facility rating at the same air traffic control tower. However, if he does not do so before August 31, 1971, he may not thereafter exercise the privileges of his senior rating at the control tower concerned until he makes the exchange.

(b) The holder of a junior rating on August 31, 1970, may not control air traffic, at any operating position at the control tower concerned, until he has met the applicable requirements of § 65.37 of this part. However, before meeting those requirements he may control air traffic under the supervision, where required, of an operator with a senior rating (or facility rating) in accordance with § 65.41 of this part in effect before August 31, 1970.

§ 65.45 Performance of duties.

(a) An air traffic control tower operator shall perform his duties in accordance with the limitations on his certificate and the procedures and practices prescribed in air traffic control manuals of the FAA, to provide for the safe, orderly, and expeditious flow of air traffic.

(b) An operator with a facility rating may control traffic at any operating position at the control tower at which he holds a facility rating. However, he may not issue an air traffic clearance for IFR flight without authorization from the appropriate facility exercising IFR control at that location.

(c) An operator who does not hold a facility rating for a particular control tower may act at each operating position for which he has qualified, under the supervision of an operator holding a facility rating for that control tower.

[Doc. No. 10193, 35 FR 12326, Aug. 1, 1970, as amended by Amdt. 65-16, 35 FR 14075, Sept. 4, 1970]

§§ 65.46-65.46b [Reserved]

§ 65.47 Maximum hours.

Except in an emergency, a certificated air traffic control tower operator must be relieved of all duties for at least 24 consecutive hours at least once during each 7 consecutive days. Such an operator may not serve or be required to serve—

(a) For more than 10 consecutive hours; or

(b) For more than 10 hours during a period of 24 consecutive hours, unless he has had a rest period of at least 8 hours at or before the end of the 10 hours of duty.

§ 65.49 General operating rules.

(a) Except for a person employed by the FAA or employed by, or on active duty with, the Department of the Air Force, Army, or Navy, or the Coast Guard, no person may act as an air traffic control tower operator under a certificate issued to him or her under this part unless he or she has in his or her personal possession an appropriate current medical certificate issued under part 67 of this chapter.

(b) Each person holding an air traffic control tower operator certificate shall keep it readily available when performing duties in an air traffic control tower, and shall present that certificate or his medical certificate or both for inspection upon the request of the Administrator or an authorized representative of the National Transportation Safety Board, or of any Federal, State, or local law enforcement officer.

(c) A certificated air traffic control tower operator who does not hold a facility rating for a particular control tower may not act at any operating position at the control tower concerned unless there is maintained at that control tower, readily available to persons named in paragraph (b) of this section, a current record of the operating positions at which he has qualified.

(d) An air traffic control tower operator may not perform duties under his certificate during any period of known physical deficiency that would make him unable to meet the physical requirements for his current medical certificate. However, if the deficiency is temporary, he may perform duties that are not affected by it whenever another certificated and qualified operator is present and on duty.

(e) A certificated air traffic control tower operator may not control air traffic with equipment that the Administrator has found to be inadequate.

(f) The holder of an air traffic control tower operator certificate, or an applicant for one, shall, upon the reasonable request of the Administrator, cooperate fully in any test that is made of him.

[Doc. No. 1179, 27 FR 7973, Aug. 10, 1962, as amended by Amdt. 65-31, 52 FR 17519, May 8, 1987]

§ 65.50 Currency requirements.

The holder of an air traffic control tower operator certificate may not perform any duties under that certificate unless—

(a) He has served for at least three of the preceding 6 months as an air traffic control tower operator at the control tower to which his facility rating applies, or at the operating positions for which he has qualified; or

(b) He has shown that he meets the requirements for his certificate and facility rating at the control tower concerned, or for operating at positions for which he has previously qualified.

Subpart C—Aircraft Dispatchers

SOURCE: Docket No. FAA-1998-4553, 64 FR 68923, Dec. 8, 1999, unless otherwise noted.

§ 65.51 Certificate required.

(a) No person may act as an aircraft dispatcher (exercising responsibility with the pilot in command in the operational control of a flight) in connection with any civil aircraft in air commerce unless that person has in his or her personal possession an aircraft dispatcher certificate issued under this subpart.

(b) Each person who holds an aircraft dispatcher certificate must present it for inspection upon the request of the Administrator or an authorized representative of the National Transportation Safety Board, or of any Federal, State, or local law enforcement officer.

§ 65.53 Eligibility requirements: General.

(a) To be eligible to take the aircraft dispatcher knowledge test, a person must be at least 21 years of age.

(b) To be eligible for an aircraft dispatcher certificate, a person must—

(1) Be at least 23 years of age;

(2) Be able to read, speak, write, and understand the English language;

(3) Pass the required knowledge test prescribed by § 65.55 of this part;

(4) Pass the required practical test prescribed by § 65.59 of this part; and

(5) Comply with the requirements of § 65.57 of this part.

§ 65.55 Knowledge requirements.

(a) A person who applies for an aircraft dispatcher certificate must pass a knowledge test on the following aeronautical knowledge areas:

(1) Applicable Federal Aviation Regulations of this chapter that relate to airline transport pilot privileges, limitations, and flight operations;

(2) Meteorology, including knowledge of and effects of fronts, frontal characteristics, cloud formations, icing, and upper-air data;

(3) General system of weather and NOTAM collection, dissemination, interpretation, and use;

(4) Interpretation and use of weather charts, maps, forecasts, sequence reports, abbreviations, and symbols;

(5) National Weather Service functions as they pertain to operations in the National Airspace System;

(6) Windshear and microburst awareness, identification, and avoidance;

(7) Principles of air navigation under instrument meteorological conditions in the National Airspace System;

(8) Air traffic control procedures and pilot responsibilities as they relate to enroute operations, terminal area and radar operations, and instrument departure and approach procedures;

(9) Aircraft loading, weight and balance, use of charts, graphs, tables, formulas, and computations, and their effect on aircraft performance;

(10) Aerodynamics relating to an aircraft's flight characteristics and performance in normal and abnormal flight regimes;

(11) Human factors;

(12) Aeronautical decision making and judgment; and

(13) Crew resource management, including crew communication and coordination.

(b) The applicant must present documentary evidence satisfactory to the administrator of having passed an aircraft dispatcher knowledge test within the preceding 24 calendar months.

§ 65.57 Experience or training requirements.

An applicant for an aircraft dispatcher certificate must present documentary evidence satisfactory to the Administrator that he or she has the experience prescribed in paragraph (a) of this section or has accomplished the training described in paragraph (b) of this section as follows:

(a) A total of at least 2 years experience in the 3 years before the date of application, in any one or in any combination of the following areas:

(1) In military aircraft operations as a—

(i) Pilot;

(ii) Flight navigator; or

(iii) Meteorologist.

(2) In aircraft operations conducted under part 121 of this chapter as—

(i) An assistant in dispatching air carrier aircraft, under the direct supervision of a dispatcher certificated under this subpart;

(ii) A pilot;

(iii) A flight engineer; or

(iv) A meteorologist.

(3) In aircraft operations as—

(i) An Air Traffic Controller; or

(ii) A Flight Service Specialist.

(4) In aircraft operations, performing other duties that the Administrator finds provide equivalent experience.

(b) A statement of graduation issued or revalidated in accordance with § 65.70(b) of this part, showing that the person has successfully completed an approved aircraft dispatcher course.

§ 65.59 Skill requirements.

An applicant for an aircraft dispatcher certificate must pass a practical test given by the Administrator, with respect to any one type of large aircraft used in air carrier operations. The practical test must be based on the aircraft dispatcher practical test standards, as published by the FAA, on the items outlined in appendix A of this part.

§ 65.61 Aircraft dispatcher certification courses: Content and minimum hours.

(a) An approved aircraft dispatcher certification course must:

(1) Provide instruction in the areas of knowledge and topics listed in appendix A of this part;

(2) Include a minimum of 200 hours of instruction.

(b) An applicant for approval of an aircraft dispatcher course must submit an outline that describes the major topics and subtopics to be covered and the number of hours proposed for each.

(c) Additional subject headings for an aircraft dispatcher certification course may also be included, however the hours proposed for any subjects not listed in appendix A of this part must be in addition to the minimum 200 course hours required in paragraph (a) of this section.

(d) For the purpose of completing an approved course, a student may substitute previous experience or training for a portion of the minimum 200 hours of training. The course operator determines the number of hours of credit based on an evaluation of the experience or training to determine if it is comparable to portions of the approved course curriculum. The credit allowed, including the total hours and the basis for it, must be placed in the student's record required by § 65.70(a) of this part.

§ 65.63 Aircraft dispatcher certification courses: Application, duration, and other general requirements.

(a) *Application.* Application for original approval of an aircraft dispatcher certification course or the renewal of approval of an aircraft dispatcher certification course under this part must be:

(1) Made in writing to the Administrator;

(2) Accompanied by two copies of the course outline required under § 65.61(b) of this part, for which approval is sought;

(3) Accompanied by a description of the equipment and facilities to be used; and

(4) Accompanied by a list of the instructors and their qualifications.

(b) *Duration.* Unless withdrawn or canceled, an approval of an aircraft dispatcher certification course of study expires:

(1) On the last day of the 24th month from the month the approval was issued; or

(2) Except as provided in paragraph (f) of this section, on the date that any change in ownership of the school occurs.

(c) *Renewal.* Application for renewal of an approved aircraft dispatcher certification course must be made within 30 days preceding the month the approval expires, provided the course operator meets the following requirements:

(1) At least 80 percent of the graduates from that aircraft dispatcher certification course, who applied for the practical test required by § 65.59 of this part, passed the practical test on their first attempt; and

(2) The aircraft dispatcher certification course continues to meet the requirements of this subpart for course approval.

(d) *Course revisions.* Requests for approval of a revision of the course outline, facilities, or equipment must be in accordance with paragraph (a) of this section. Proposed revisions of the course outline or the description of facilities and equipment must be submitted in a format that will allow an entire page or pages of the approved outline or description to be removed and replaced by any approved revision. The list of instructors may be revised at any time without request for approval, provided the minimum requirements of § 65.67 of this part are maintained and the Administrator is notified in writing.

(e) *Withdrawal or cancellation of approval.* Failure to continue to meet the requirements of this subpart for the approval or operation of an approved aircraft dispatcher certification course is grounds for withdrawal of approval of the course. A course operator may request cancellation of course approval by a letter to the Administrator. The operator must forward any records to the FAA as requested by the Administrator.

(f) *Change in ownership.* A change in ownership of a part 65, appendix A-approved course does not terminate that aircraft dispatcher certification course approval if, within 10 days after the date that any change in ownership of the school occurs:

(1) Application is made for an appropriate amendment to the approval; and

(2) No change in the facilities, personnel, or approved aircraft dispatcher certification course is involved.

(g) *Change in name or location.* A change in name or location of an approved aircraft dispatcher certification course does not invalidate the approval if, within 10 days after the date that any change in name or location occurs, the course operator of the part 65, appendix A-approved course notifies the Administrator, in writing, of the change.

§ 65.65 Aircraft dispatcher certification courses: Training facilities.

An applicant for approval of authority to operate an aircraft dispatcher course of study must have facilities, equipment, and materials adequate to provide each student the theoretical and practical aspects of aircraft dispatching. Each room, training booth, or other space used for instructional purposes must be temperature controlled, lighted, and ventilated to conform to local building, sanitation, and health codes. In addition, the training facility must be so located that the students in that facility are not distracted by the instruction conducted in other rooms.

§ 65.67 Aircraft dispatcher certification courses: Personnel.

(a) Each applicant for an aircraft dispatcher certification course must meet the following personnel requirements:

(1) Each applicant must have adequate personnel, including one instructor who holds an aircraft dispatcher certificate and is available to coordinate all training course instruction.

(2) Each applicant must not exceed a ratio of 25 students for one instructor.

(b) The instructor who teaches the practical dispatch applications area of the appendix A course must hold an aircraft dispatchers certificate

§ 65.70 Aircraft dispatcher certification courses: Records.

(a) The operator of an aircraft dispatcher course must maintain a record for each student, including a chronological log of all instructors, subjects covered, and course examinations and results. The record must be retained for at least 3 years after graduation. The course operator also must prepare, for its records, and transmit to the Administrator not later than January 31 of each year, a report containing the following information for the previous year:

(1) The names of all students who graduated, together with the results of their aircraft dispatcher certification courses.

(2) The names of all the students who failed or withdrew, together with the results of their aircraft dispatcher certification courses or the reasons for their withdrawal.

(b) Each student who successfully completes the approved aircraft dispatcher certification course must be given a written statement of graduation, which is valid for 90 days. After 90 days, the course operator may revalidate the graduation certificate for an additional 90 days if the course operator determines that the student remains proficient in the subject areas listed in appendix A of this part.

Subpart D—Mechanics

§ 65.71 Eligibility requirements: General.

(a) To be eligible for a mechanic certificate and associated ratings, a person must—

(1) Be at least 18 years of age;

(2) Be able to read, write, speak, and understand the English language, or in the case of an applicant who does not meet this requirement and who is employed outside of the United States by a U.S. air carrier, have his certificate endorsed "Valid only outside the United States";

(3) Have passed all of the prescribed tests within a period of 24 months; and

(4) Comply with the sections of this subpart that apply to the rating he seeks.

(b) A certificated mechanic who applies for an additional rating must meet the requirements of § 65.77 and, within a period of 24 months, pass the tests prescribed by §§ 65.75 and 65.79 for the additional rating sought.

[Doc. No. 1179, 27 FR 7973, Aug. 10, 1962, as amended by Amdt. 65-6, 31 FR 5950, Apr. 19, 1966]

§ 65.73 Ratings.

(a) The following ratings are issued under this subpart:

(1) Airframe.

(2) Powerplant.

(b) A mechanic certificate with an aircraft or aircraft engine rating, or both, that was issued before, and was valid on, June 15, 1952, is equal to a mechanic certificate with an airframe or powerplant rating, or both, as the case may be, and may be exchanged for such a corresponding certificate and rating or ratings.

§ 65.75 Knowledge requirements.

(a) Each applicant for a mechanic certificate or rating must, after meeting the applicable experience requirements of § 65.77, pass a written test covering the construction and maintenance of aircraft appropriate to the rating he seeks, the regulations in this subpart, and the applicable provisions of parts 43 and 91 of this chapter. The basic principles covering the installation and maintenance of propellers are included in the powerplant test.

(b) The applicant must pass each section of the test before applying for the oral and practical tests prescribed by § 65.79. A report of the written test is sent to the applicant.

[Doc. No. 1179, 27 FR 7973, Aug. 10, 1962, as amended by Amdt. 65-1, 27 FR 10410, Oct. 25, 1962; Amdt. 65-6, 31 FR 5950, Apr. 19, 1966]

§ 65.77 Experience requirements.

Each applicant for a mechanic certificate or rating must present either an appropriate graduation certificate or certificate of completion from a certificated aviation maintenance technician school or documentary evidence, satisfactory to the Administrator, of—

(a) At least 18 months of practical experience with the procedures, practices, materials, tools, machine tools, and equipment generally used in constructing, maintaining, or altering airframes, or powerplants appropriate to the rating sought; or

(b) At least 30 months of practical experience concurrently performing the duties appropriate to both the airframe and powerplant ratings.

[Doc. No. 1179, 27 FR, 7973, Aug. 10, 1962, as amended by Amdt. 65-14, 35 FR, 5533, Apr. 3, 1970]

§ 65.79 Skill requirements.

Each applicant for a mechanic certificate or rating must pass an oral and a practical test on the rating he seeks. The tests cover the applicant's basic skill in performing practical projects on the subjects covered by the written test for that rating. An applicant for a powerplant rating must show his ability to make satisfactory minor repairs to, and minor alterations of, propellers.

§ 65.80 Certificated aviation maintenance technician school students.

Whenever an aviation maintenance technician school certificated under part 147 of this chapter shows to an FAA inspector that any of its students has made satisfactory progress at the school and is prepared to take the oral and practical tests prescribed by § 65.79, that student may take those tests during the final subjects of his training in the approved curriculum, before he meets the applicable experience requirements of § 65.77 and before he passes each section of the written test prescribed by § 65.75.

[Doc. No. 9444, 35 FR 5533, Apr. 3, 1970]

§ 65.81 General privileges and limitations.

(a) A certificated mechanic may perform or supervise the maintenance, preventive maintenance or alteration of an aircraft or appliance, or a part thereof, for which he is rated (but excluding major repairs to, and major alterations of, propellers, and any repair to, or alteration of, instruments), and may perform additional duties in accordance with §§ 65.85, 65.87, and 65.95. However, he may not supervise the maintenance, preventive maintenance, or alteration of, or approve and return to service, any aircraft or appliance, or part thereof, for which he is rated unless he has satisfactorily performed the work concerned at an earlier date. If he has not so performed that work at an earlier date, he may show his ability to do it by performing it to the satisfaction of the Administrator or under the direct supervision of a certificated and appropriately rated mechanic, or a certificated repairman, who has had previous experience in the specific operation concerned.

(b) A certificated mechanic may not exercise the privileges of his certificate and rating unless he understands the current instructions of the manufacturer, and the maintenance manuals, for the specific operation concerned.

[Doc. No. 1179, 27 FR 7973, Aug. 10, 1962, as amended by Amdt. 65-2, 29 FR 5451, Apr. 23, 1964; Amdt. 65-26, 45 FR 46737, July 10, 1980]

§ 65.83 Recent experience requirements.

A certificated mechanic may not exercise the privileges of his certificate and rating unless, within the preceding 24 months—

(a) The Administrator has found that he is able to do that work; or

(b) He has, for at least 6 months—

(1) Served as a mechanic under his certificate and rating;

(2) Technically supervised other mechanics;

(3) Supervised, in an executive capacity, the maintenance or alteration of aircraft; or

(4) Been engaged in any combination of paragraph (b) (1), (2), or (3) of this section.

§ 65.85 Airframe rating; additional privileges.

(a) Except as provided in paragraph (b) of this section, a certificated mechanic with an airframe rating may approve and return to service an airframe, or any related part or appliance, after he has performed, supervised, or inspected its maintenance or alteration (excluding major repairs and major alterations). In addition, he may perform the 100-hour inspection required by part 91 of this chapter on an airframe, or any related part or appliance, and approve and return it to service.

(b) A certificated mechanic with an airframe rating can approve and return to service an airframe, or any related part or appliance, of an aircraft with a special airworthiness certificate in the light-sport category after performing and inspecting a major repair or major alteration for products that are not produced under an FAA approval provided the work was performed in accordance with instructions developed by the manufacturer or a person acceptable to the FAA.

[Doc. No. 1179, 27 FR 7973, Aug. 10, 1962, as amended by Amdt. 65-10, 32 FR 5770, Apr. 11, 1967; Amdt. 65-45, 69 FR 44879, July 27, 2004]

§ 65.87 Powerplant rating; additional privileges.

(a) Except as provided in paragraph (b) of this section, a certificated mechanic with a powerplant rating may approve and return to service a powerplant or propeller or any related part or appliance, after he has performed, supervised, or inspected its maintenance or alteration (excluding major repairs and major alterations). In addition, he may perform the 100-hour inspection required by part 91 of this chapter on a powerplant or propeller, or any part thereof, and approve and return it to service.

(b) A certificated mechanic with a powerplant rating can approve and return to service a powerplant or propeller, or any related part or appliance, of an aircraft with a special airworthiness certificate in the light-sport category after performing and inspecting a major repair or major alteration for products that are not produced under an FAA approval, provided the work was performed in accordance with instructions developed by the manufacturer or a person acceptable to the FAA.

[Doc. No. 1179, 27 FR 7973, Aug. 10, 1962, as amended by Amdt. 65-10, 32 FR 5770, Apr. 11, 1967; Amdt. 65-45, 69 FR 44879, July 27, 2004]

§ 65.89 Display of certificate.

Each person who holds a mechanic certificate shall keep it within the immediate area where he normally exercises the privileges of the certificate and shall present it for inspection upon the request of the Administrator or an authorized representative of the National Transportation Safety Board, or of any Federal, State, or local law enforcement officer.

[Doc. No. 7258, 31 FR 13524, Oct. 20, 1966, as amended by Doc. No. 8084, 32 FR 5769, Apr. 11, 1967]

§ 65.91 Inspection authorization.

(a) An application for an inspection authorization is made on a form and in a manner prescribed by the Administrator.

(b) An applicant who meets the requirements of this section is entitled to an inspection authorization.

(c) To be eligible for an inspection authorization, an applicant must—

(1) Hold a currently effective mechanic certificate with both an airframe rating and a powerplant rating, each of which is currently effective and has been in effect for a total of at least 3 years;

(2) Have been actively engaged, for at least the 2-year period before the date he applies, in maintaining aircraft certificated and maintained in accordance with this chapter;

(3) Have a fixed base of operations at which he may be located in person or by telephone during a normal working week but it need not be the place where he will exercise his inspection authority;

(4) Have available to him the equipment, facilities, and inspection data necessary to properly inspect airframes, powerplants, propellers, or any related part or appliance; and

(5) Pass a written test on his ability to inspect according to safety standards for returning aircraft to service after major repairs and major alterations and annual and progressive inspections performed under part 43 of this chapter.

An applicant who fails the test prescribed in paragraph (c)(5) of this section may not apply for retesting until at least 90 days after the date he failed the test.

[Doc. No. 1179, 27 FR 7973, Aug. 10, 1962, as amended by Amdt. 65-5, 31 FR 3337, Mar. 3, 1966; Amdt. 65-22, 42 FR 46279, Sept. 15, 1977; Amdt. 65-30, 50 FR 15700, Apr. 19, 1985]

§ 65.92 Inspection authorization: Duration.

(a) Each inspection authorization expires on March 31 of each odd-numbered year. However, the holder may exercise the privileges of that authorization only while he holds a currently effective mechanic certificate with both a currently effective airframe rating and a currently effective powerplant rating.

(b) An inspection authorization ceases to be effective whenever any of the following occurs:

(1) The authorization is surrendered, suspended, or revoked.

(2) The holder no longer has a fixed base of operation.

(3) The holder no longer has the equipment, facilities, and inspection data required by § 65.91(c) (3) and (4) for issuance of his authorization.

(c) The holder of an inspection authorization that is suspended or revoked shall, upon the Administrator's request, return it to the Administrator.

[Doc. No. 12537, 42 FR 46279, Sept. 15, 1977, as amended by Amdt. 65-50, 72 FR 4404, Jan. 30, 2007]

§ 65.93 Inspection authorization: Renewal.

(a) To be eligible for renewal of an inspection authorization for a 2-year period an applicant must present evidence during the month of March of each odd-numbered year, at an FAA Flight Standards District Office or an International Field Office, that the applicant still meets the requirements of § 65.91(c) (1) through (4). In addition, during the time the applicant held the inspection authorization, the applicant must show completion of one of the activities in § 65.93(a) (1) through (5) below by March 31 of the first year of the 2-year inspection authorization period, and completion of one of the five activities during the second year of the 2-year period:

(1) Performed at least one annual inspection for each 90 days that the applicant held the current authority; or

(2) Performed at least two major repairs or major alterations for each 90 days that the applicant held the current authority; or

(3) Performed or supervised and approved at least one progressive inspection in accordance with standards prescribed by the Administrator; or

(4) Attended and successfully completed a refresher course, acceptable to the Administrator, of not less than 8 hours of instruction; or

(5) Passed an oral test by an FAA inspector to determine that the applicant's knowledge of applicable regulations and standards is current.

(b) The holder of an inspection authorization that has been in effect:

(1) for less than 90 days before the expiration date need not comply with paragraphs (a)(1) through (5) of this section.

(2) for less than 90 days before March 31 of an even-numbered year need not comply with paragraphs (a)(1) through (5) of this section for the first year of the 2-year inspection authorization period.

(c) An inspection authorization holder who does not complete one of the activities set forth in § 65.93(a) (1) through (5) of this section by March 31 of the first year of the 2-year inspection authorization period may not exercise inspection authorization privileges after March 31 of the first year. The inspection authorization holder may resume exercising inspection authorization privileges after passing an oral test from an FAA inspector to determine that the applicant's knowledge of the applicable regulations and standards is current. An inspection authorization holder who passes this oral test is deemed to have completed the requirements of § 65.93(a) (1) through (5) by March 31 of the first year.

[Docket No. FAA-2007-27108, 72 FR 4404, Jan. 30, 2007]

§ 65.95 Inspection authorization: Privileges and limitations.

(a) The holder of an inspection authorization may—

(1) Inspect and approve for return to service any aircraft or related part or appliance (except any aircraft maintained in accordance with a continuous airworthiness program under part 121 of this chapter) after a major repair or major alteration to it in accordance with part 43 [New] of this chapter, if the work was done in accordance with technical data approved by the Administrator; and

(2) Perform an annual, or perform or supervise a progressive inspection according to §§ 43.13 and 43.15 of this chapter.

(b) When he exercises the privileges of an inspection authorization the holder shall keep it available for inspection by the aircraft owner, the mechanic submitting the aircraft, repair, or alteration for approval (if any), and shall present it upon the request of the Administrator or an authorized representative of the National Transportation Safety Board, or of any Federal, State, or local law enforcement officer.

(c) If the holder of an inspection authorization changes his fixed base of operation, he may not exercise the privileges of the authorization until he has notified the FAA Flight Standards District Office or International Field Office for the area in which the new base is located, in writing, of the change.

[Doc. No. 1179, 27 FR 7973, Aug. 10, 1962, as amended by Amdt. 65-2, 29 FR 5451, Apr. 23, 1964; Amdt. 65-4, 30 FR 3638, Mar. 14, 1965; Amdt. 65-5, 31 FR 3337, Mar. 3, 1966; Amdt. 65-9, 31 FR 13524, Oct. 20, 1966; 32 FR 5769, Apr. 11, 1967; Amdt. 65-35, 54 FR 39292, Sept. 25, 1989; Amdt. 65-41, 66 FR 21066, Apr. 27, 2001]

Subpart E—Repairmen

§ 65.101 Eligibility requirements: General.

(a) To be eligible for a repairman certificate a person must—

(1) Be at least 18 years of age;

(2) Be specially qualified to perform maintenance on aircraft or components thereof, appropriate to the job for which he is employed;

(3) Be employed for a specific job requiring those special qualifications by a certificated repair station, or by a certificated commercial operator or certificated air carrier, that is required by its operating certificate or approved operations specifications to provide a continuous airworthiness maintenance program according to its maintenance manuals;

(4) Be recommended for certification by his employer, to the satisfaction of the Administrator, as able to satisfactorily maintain aircraft or components, appropriate to the job for which he is employed;

(5) Have either—

(i) At least 18 months of practical experience in the procedures, practices, inspection methods, materials, tools, machine tools, and equipment generally used in the maintenance duties of the specific job for which the person is to be employed and certificated; or

(ii) Completed formal training that is acceptable to the Administrator and is specifically designed to qualify the applicant for the job on which the applicant is to be employed; and

(6) Be able to read, write, speak, and understand the English language, or, in the case of an applicant who does not meet this requirement and who is employed outside the United States by a certificated repair station, a certificated U.S. commercial operator, or a certificated U.S. air carrier, described in paragraph (a)(3) of this section, have this certificate endorsed "Valid only outside the United States."

(b) This section does not apply to the issuance of a repairman certificate (experimental aircraft builder) under § 65.104 or to a repairman certificate (light-sport aircraft) under § 65.107.

[Doc. No. 1179, 27 FR 7973, Aug. 10, 1962, as amended by Amdt. 65-11, 32 FR 13506, Sept. 27, 1967; Amdt. 65-24, 44 FR 46781, Aug. 9, 1979; Amdt. 65-27, 47 FR 13316, Mar. 29, 1982; Amdt. 65-45, 69 FR 44879, July 27, 2004; 72 FR 7739, Feb. 20, 2007]

§ 65.103 Repairman certificate: Privileges and limitations.

(a) A certificated repairman may perform or supervise the maintenance, preventive maintenance, or alteration of aircraft or aircraft components appropriate to the job for which the repairman was employed and certificated, but only in connection with duties for the certificate holder by whom the repairman was employed and recommended.

(b) A certificated repairman may not perform or supervise duties under the repairman certificate unless the repairman understands the current instructions of the certificate holder by whom the repairman is employed and the manufacturer's instructions for continued airworthiness relating to the specific operations concerned.

(c) This section does not apply to the holder of a repairman certificate (light-sport aircraft) while that repairman is performing work under that certificate.

[Doc. No. 18241, 45 FR 46738, July 10, 1980, as amended by Amdt. 65-45, 69 FR 44879, July 27, 2004]

§ 65.104 Repairman certificate—experimental aircraft builder—Eligibility, privileges and limitations.

(a) To be eligible for a repairman certificate (experimental aircraft builder), an individual must—

(1) Be at least 18 years of age;

(2) Be the primary builder of the aircraft to which the privileges of the certificate are applicable;

(3) Show to the satisfaction of the Administrator that the individual has the requisite skill to determine whether the aircraft is in a condition for safe operations; and

(4) Be a citizen of the United States or an individual citizen of a foreign country who has lawfully been admitted for permanent residence in the United States.

(b) The holder of a repairman certificate (experimental aircraft builder) may perform condition inspections on the aircraft constructed by the holder in accordance with the operating limitations of that aircraft.

(c) Section 65.103 does not apply to the holder of a repairman certificate (experimental aircraft builder) while performing under that certificate.

[Doc. No. 18739, 44 FR 46781, Aug. 9, 1979]

§ 65.105 Display of certificate.

Each person who holds a repairman certificate shall keep it within the immediate area where he normally exercises the privileges of the certificate and shall present it for inspection upon the request of the Administrator or an authorized representative of the National Transportation Safety Board, or of any Federal, State, or local law enforcement officer.

[Doc. No. 7258, 31 FR 13524, Oct. 20, 1966, as amended by Doc. No. 8084, 32 FR 5769, Apr. 11, 1967]

§ 65.107 Repairman certificate (light-sport aircraft): Eligibility, privileges, and limits.

(a) Use the following table to determine your eligibility for a repairman certificate (light-sport aircraft) and appropriate rating:

To be eligible for	You must
(1) A repairman certificate (light-sport aircraft)	(i) Be at least 18 years old, (ii) Be able to read, speak, write, and understand English. If for medical reasons you cannot meet one of these requirements, the FAA may place limits on your repairman certificate necessary to safely perform the actions authorized by the certificate and rating,
	(iii) Demonstrate the requisite skill to determine whether a light-sport aircraft is in a condition for safe operation, and
	(iv) Be a citizen of the United States, or a citizen of a foreign country who has been lawfully admitted for permanent residence in the United States.
(2) A repairman certificate (light-sport aircraft) with an inspection rating	(i) Meet the requirements of paragraph (a)(1) of this section, and (ii) Complete a 16-hour training course acceptable to the FAA on inspecting the particular class of experimental light-sport aircraft for which you intend to exercise the privileges of this rating.
(3) A repairman certificate (light-sport aircraft) with a maintenance rating	(i) Meet the requirements of paragraph (a)(1) of this section, and
	(ii) Complete a training course acceptable to the FAA on maintaining the particular class of light-sport aircraft for which you intend to exercise the privileges of this rating. The training course must, at a minimum, provide the following number of hours of instruction:
	(A) For airplane class privileges—120-hours,
	(B) For weight-shift control aircraft class privileges—104 hours,
	(C) For powered parachute class privileges—104 hours,
	(D) For lighter than air class privileges—80 hours,
	(E) For glider class privileges—80 hours.

(b) The holder of a repairman certificate (light-sport aircraft) with an inspection rating may perform the annual condition inspection on a light-sport aircraft:

(1) That is owned by the holder;

(2) That has been issued an experimental certificate for operating a light-sport aircraft under § 21.191(i) of this chapter; and

(3) That is in the same class of light-sport-aircraft for which the holder has completed the training specified in paragraph (a)(2)(ii) of this section.

(c) The holder of a repairman certificate (light-sport aircraft) with a maintenance rating may—

(1) Approve and return to service an aircraft that has been issued a special airworthiness certificate in the light-sport category under § 21.190 of this chapter, or any part thereof, after performing or inspecting maintenance (to include the annual condition inspection and the 100-hour inspection required by § 91.327 of this chapter), preventive maintenance, or an alteration (excluding a major repair or a major alteration on a product produced under an FAA approval);

(2) Perform the annual condition inspection on a light-sport aircraft that has been issued an experimental certificate for operating a light-sport aircraft under § 21.191(i) of this chapter; and

(3) Only perform maintenance, preventive maintenance, and an alteration on a light-sport aircraft that is in the same class of light-sport aircraft for which the holder has completed the training specified in paragraph (a)(3)(ii) of this section. Before performing a major repair, the holder must complete additional training acceptable to the FAA and appropriate to the repair performed.

(d) The holder of a repairman certificate (light-sport aircraft) with a maintenance rating may not approve for return to service any aircraft or part thereof unless that person has previously performed the work concerned satisfactorily. If that person has not previously performed that work, the person may show the ability to do the work by performing it to the satisfaction of the FAA, or by performing it under the direct supervision of a certificated and appropriately rated mechanic, or a certificated repairman, who has had previous experience in the specific operation concerned. The repairman may not exercise the privileges of the certificate unless the repairman understands the current instructions of the manufacturer and the maintenance manuals for the specific operation concerned.

[Doc. No. FAA-2001-11133, 69 FR 44879, July 27, 2004]

Subpart F—Parachute Riggers

§ 65.111 Certificate required.

(a) No person may pack, maintain, or alter any personnel-carrying parachute intended for emergency use in connection with civil aircraft of the United States (including the reserve parachute of a dual parachute system to be used for intentional parachute jumping) unless that person holds an appropriate current certificate and type rating issued under this subpart and complies with §§ 65.127 through 65.133.

(b) No person may pack any main parachute of a dual-parachute system to be used for intentional parachute jumping in connection with civil aircraft of the United States unless that person—

(1) Has an appropriate current certificate issued under this subpart;

(2) Is under the supervision of a current certificated parachute rigger;

(3) Is the person making the next parachute jump with that parachute in accordance with § 105.43(a) of this chapter; or

(4) Is the parachutist in command making the next parachute jump with that parachute in a tandem parachute operation conducted under § 105.45(b)(1) of this chapter.

(c) No person may maintain or alter any main parachute of a dual-parachute system to be used for intentional parachute jumping in connection with civil aircraft of the United States unless that person—

(1) Has an appropriate current certificate issued under this subpart; or

(2) Is under the supervision of a current certificated parachute rigger;

(d) Each person who holds a parachute rigger certificate shall present it for inspection upon the request of the Administrator or an authorized representative of the National Transportation Safety Board, or of any Federal, State, or local law enforcement officer.

(e) The following parachute rigger certificates are issued under this part:

(1) Senior parachute rigger.

(2) Master parachute rigger.

(f) Sections 65.127 through 65.133 do not apply to parachutes packed, maintained, or altered for the use of the armed forces.

[Doc. No. 1179, 27 FR 7973, Aug. 10, 1962, as amended by Amdt. 65-9, 31 FR 13524, Oct. 20, 1966; 32 FR 5769, Apr. 11, 1967; Amdt. 65-42, 66 FR 23553, May 9, 2001; Amdt. 65-54, 75 FR 31285, June 3, 2010]

§ 65.113 Eligibility requirements: General.

(a) To be eligible for a parachute rigger certificate, a person must—

(1) Be at least 18 years of age;

(2) Be able to read, write, speak, and understand the English language, or, in the case of a citizen of Puerto Rico, or a person who is employed outside of the United States

by a U.S. air carrier, and who does not meet this requirement, be issued a certificate that is valid only in Puerto Rico or while he is employed outside of the United States by that air carrier, as the case may be; and

(3) Comply with the sections of this subpart that apply to the certificate and type rating he seeks.

(b) Except for a master parachute rigger certificate, a parachute rigger certificate that was issued before, and was valid on, October 31, 1962, is equal to a senior parachute rigger certificate, and may be exchanged for such a corresponding certificate.

§ 65.115 Senior parachute rigger certificate: Experience, knowledge, and skill requirements.

Except as provided in § 65.117, an applicant for a senior parachute rigger certificate must—

(a) Present evidence satisfactory to the Administrator that he has packed at least 20 parachutes of each type for which he seeks a rating, in accordance with the manufacturer's instructions and under the supervision of a certificated parachute rigger holding a rating for that type or a person holding an appropriate military rating;

(b) Pass a written test, with respect to parachutes in common use, on—

(1) Their construction, packing, and maintenance;

(2) The manufacturer's instructions;

(3) The regulations of this subpart; and

(c) Pass an oral and practical test showing his ability to pack and maintain at least one type of parachute in common use, appropriate to the type rating he seeks.

[Doc. No. 10468, 37 FR 13251, July 6, 1972]

§ 65.117 Military riggers or former military riggers: Special certification rule.

In place of the procedure in § 65.115, an applicant for a senior parachute rigger certificate is entitled to it if he passes a written test on the regulations of this subpart and presents satisfactory documentary evidence that he—

(a) Is a member or civilian employee of an Armed Force of the United States, is a civilian employee of a regular armed force of a foreign country, or has, within the 12 months before he applies, been honorably discharged or released from any status covered by this paragraph;

(b) Is serving, or has served within the 12 months before he applies, as a parachute rigger for such an Armed Force; and

(c) Has the experience required by § 65.115(a).

§ 65.119 Master parachute rigger certificate: Experience, knowledge, and skill requirements.

An applicant for a master parachute rigger certificate must meet the following requirements:

(a) Present evidence satisfactory to the Administrator that he has had at least 3 years of experience as a parachute rigger and has satisfactorily packed at least 100 parachutes of each of two types in common use, in accordance with the manufacturer's instructions—

(1) While a certificated and appropriately rated senior parachute rigger; or

(2) While under the supervision of a certificated and appropriately rated parachute rigger or a person holding appropriate military ratings.

An applicant may combine experience specified in paragraphs (a) (1) and (2) of this section to meet the requirements of this paragraph.

(b) If the applicant is not the holder of a senior parachute rigger certificate, pass a written test, with respect to parachutes in common use, on—

(1) Their construction, packing, and maintenance;

(2) The manufacturer's instructions; and

(3) The regulations of this subpart.

(c) Pass an oral and practical test showing his ability to pack and maintain two types of parachutes in common use, appropriate to the type ratings he seeks.

[Doc. No. 10468, 37 FR 13252, July 6, 1972]

§ 65.121 Type ratings.

(a) The following type ratings are issued under this subpart:

(1) Seat.

(2) Back.

(3) Chest.

(4) Lap.

(b) The holder of a senior parachute rigger certificate who qualifies for a master parachute rigger certificate is entitled to have placed on his master parachute rigger certificate the ratings that were on his senior parachute rigger certificate.

§ 65.123 Additional type ratings: Requirements.

A certificated parachute rigger who applies for an additional type rating must—

(a) Present evidence satisfactory to the Administrator that he has packed at least 20 parachutes of the type for which he seeks a rating, in accordance with the manufacturer's instructions and under the supervision of a certificated parachute rigger holding a rating for that type or a person holding an appropriate military rating; and

(b) Pass a practical test, to the satisfaction of the Administrator, showing his ability to pack and maintain the type of parachute for which he seeks a rating.

[Doc. No. 1179, 27 FR 7973, Aug. 10, 1962, as amended by Amdt. 65-20, 37 FR 13251, July 6, 1972]

§ 65.125 Certificates: Privileges.

(a) A certificated senior parachute rigger may—

(1) Pack or maintain (except for major repair) any type of parachute for which he is rated; and

(2) Supervise other persons in packing any type of parachute for which that person is rated in accordance with § 105.43(a) or § 105.45(b)(1) of this chapter.

(b) A certificated master parachute rigger may—

(1) Pack, maintain, or alter any type of parachute for which he is rated; and

(2) Supervise other persons in packing, maintaining, or altering any type of parachute for which the certificated parachute rigger is rated in accordance with § 105.43(a) or § 105.45(b)(1) of this chapter.

(c) A certificated parachute rigger need not comply with §§ 65.127 through 65.133 (relating to facilities, equipment, performance standards, records, recent experience, and seal) in packing, maintaining, or altering (if authorized) the main parachute of a dual parachute pack to be used for intentional jumping.

[Doc. No. 1179, 27 FR 7973, Aug. 10, 1962, as amended by Amdt. 65-20, 37 FR 13252, July 6, 1972; Amdt. 65-42, 66 FR 23553, May 9, 2001]

§ 65.127 Facilities and equipment.

No certificated parachute rigger may exercise the privileges of his certificate unless he has at least the following facilities and equipment available to him:

(a) A smooth top table at least three feet wide by 40 feet long.

(b) Suitable housing that is adequately heated, lighted, and ventilated for drying and airing parachutes.

(c) Enough packing tools and other equipment to pack and maintain the types of parachutes that he services.

(d) Adequate housing facilities to perform his duties and to protect his tools and equipment.

[Doc. No. 1179, 27 FR 7973, Aug. 10, 1962, as amended by Amdt. 65-27, 47 FR 13316, Mar. 29, 1982]

§ 65.129 Performance standards.

No certificated parachute rigger may—

(a) Pack, maintain, or alter any parachute unless he is rated for that type;

(b) Pack a parachute that is not safe for emergency use;

(c) Pack a parachute that has not been thoroughly dried and aired;

(d) Alter a parachute in a manner that is not specifically authorized by the Administrator or the manufacturer;

(e) Pack, maintain, or alter a parachute in any manner that deviates from procedures approved by the Administrator or the manufacturer of the parachute; or

(f) Exercise the privileges of his certificate and type rating unless he understands the current manufacturer's instructions for the operation involved and has—

(1) Performed duties under his certificate for at least 90 days within the preceding 12 months; or

(2) Shown the Administrator that he is able to perform those duties.

§ 65.131 Records.

(a) Each certificated parachute rigger shall keep a record of the packing, maintenance, and alteration of parachutes performed or supervised by him. He shall keep in that record, with respect to each parachute worked on, a statement of—

(1) Its type and make;

(2) Its serial number;

(3) The name and address of its owner;

(4) The kind and extent of the work performed;

(5) The date when and place where the work was performed; and

(6) The results of any drop tests made with it.

(b) Each person who makes a record under paragraph (a) of this section shall keep it for at least 2 years after the date it is made.

(c) Each certificated parachute rigger who packs a parachute shall write, on the parachute packing record attached to the parachute, the date and place of the packing and a notation of any defects he finds on inspection. He shall sign that record with his name and the number of his certificate.

§ 65.133 Seal.

Each certificated parachute rigger must have a seal with an identifying mark prescribed by the Administrator, and a seal press. After packing a parachute he shall seal the pack with his seal in accordance with the manufacturer's recommendation for that type of parachute.

Appendix A to Part 65—Aircraft Dispatcher Courses

Overview

This appendix sets forth the areas of knowledge necessary to perform dispatcher functions. The items listed below indicate the minimum set of topics that must be covered in a training course for aircraft dispatcher certification. The order of coverage is at the discretion of the approved school. For the latest technological advancements refer to the Practical Test Standards as published by the FAA.

I. Regulations

A. Subpart C of this part;

B. Parts 1, 25, 61, 71, 91, 121, 139, and 175, of this chapter;

C. 49 CFR part 830;

D. General Operating Manual.

II. Meteorology

A. Basic Weather Studies

(1) The earth's motion and its effects on weather.

(2) Analysis of the following regional weather types, characteristics, and structures, or combinations thereof:

(a) Maritime.

(b) Continental.

(c) Polar.

(d) Tropical.

(3) Analysis of the following local weather types, characteristics, and structures or combinations thereof:

(a) Coastal.

(b) Mountainous.

(c) Island.

(d) Plains.

(4) The following characteristics of the atmosphere:

(a) Layers.

(b) Composition.

(c) Global Wind Patterns.

(d) Ozone.

(5) Pressure:

(a) Units of Measure.

(b) Weather Systems Characteristics.

(c) Temperature Effects on Pressure.

(d) Altimeters.

(e) Pressure Gradient Force.

(f) Pressure Pattern Flying Weather.

(6) Wind:

(a) Major Wind Systems and Coriolis Force.

(b) Jetstreams and their Characteristics.

(c) Local Wind and Related Terms.

(7) States of Matter:

(a) Solids, Liquid, and Gases.

(b) Causes of change of state.

(8) Clouds:

(a) Composition, Formation, and Dissipation.

(b) Types and Associated Precipitation.

(c) Use of Cloud Knowledge in Forecasting.

(9) Fog:

(a) Causes, Formation, and Dissipation.

(b) Types.

(10) Ice:

(a) Causes, Formation, and Dissipation.

(b) Types.

(11) Stability/Instability:

(a) Temperature Lapse Rate, Convection.

(b) Adiabatic Processes.

(c) Lifting Processes.

(d) Divergence.

(e) Convergence.

(12) Turbulence:

(a) Jetstream Associated.

(b) Pressure Pattern Recognition.

(c) Low Level Windshear.

(d) Mountain Waves.
(e) Thunderstorms.
(f) Clear Air Turbulence.
(13) Airmasses:
(a) Classification and Characteristics.
(b) Source Regions.
(c) Use of Airmass Knowledge in Forecasting.
(14) Fronts:
(a) Structure and Characteristics, Both Vertical and Horizontal.
(b) Frontal Types.
(c) Frontal Weather Flying.
(15) Theory of Storm Systems:
(a) Thunderstorms.
(b) Tornadoes.
(c) Hurricanes and Typhoons.
(d) Microbursts.
(e) Causes, Formation, and Dissipation.
B. Weather, Analysis, and Forecasts
(1) Observations:
(a) Surface Observations.
(i) Observations made by certified weather observer.
(ii) Automated Weather Observations.
(b) Terminal Forecasts.
(c) Significant En route Reports and Forecasts.
(i) Pilot Reports.
(ii) Area Forecasts.
(iii) Sigmets, Airmets.
(iv) Center Weather Advisories.
(d) Weather Imagery.
(i) Surface Analysis.
(ii) Weather Depiction.
(iii) Significant Weather Prognosis.
(iv) Winds and Temperature Aloft.
(v) Tropopause Chart.
(vi) Composite Moisture Stability Chart.
(vii) Surface Weather Prognostic Chart.
(viii) Radar Meteorology.
(ix) Satellite Meteorology.
(x) Other charts as applicable.
(e) Meteorological Information Data Collection Systems.
(2) Data Collection, Analysis, and Forecast Facilities.
(3) Service Outlets Providing Aviation Weather Products.
C. Weather Related Aircraft Hazards
(1) Crosswinds and Gusts.
(2) Contaminated Runways.
(3) Restrictions to Surface Visibility.
(4) Turbulence and Windshear.
(5) Icing.
(6) Thunderstorms and Microburst.
(7) Volcanic Ash.
III. Navigation
A. Study of the Earth
(1) Time reference and location (0 Longitude, UTC).
(2) Definitions.
(3) Projections.
(4) Charts.
B. Chart Reading, Application, and Use.
C. National Airspace Plan.
D. Navigation Systems.
E. Airborne Navigation Instruments.
F. Instrument Approach Procedures.
(1) Transition Procedures.
(2) Precision Approach Procedures.
(3) Non-precision Approach Procedures.
(4) Minimums and the relationship to weather.
G. Special Navigation and Operations.
(1) North Atlantic.
(2) Pacific.
(3) Global Differences.
IV. AIRCRAFT
A. Aircraft Flight Manual.
B. Systems Overview.
(1) Flight controls.
(2) Hydraulics.
(3) Electrical.
(4) Air Conditioning and Pressurization.
(5) Ice and Rain protection.
(6) Avionics, Communication, and Navigation.
(7) Powerplants and Auxiliary Power Units.
(8) Emergency and Abnormal Procedures.
(9) Fuel Systems and Sources.
C. Minimum Equipment List/Configuration Deviation List (MEL/CDL) and Applications.
D. Performance.
(1) Aircraft in general.
(2) Principles of flight:
(a) Group one aircraft.
(b) Group two aircraft.
(3) Aircraft Limitations.
(4) Weight and Balance.
(5) Flight instrument errors.
(6) Aircraft performance:
(a) Take-off performance.
(b) En route performance.
(c) Landing performance.
V. Communications
A. Regulatory requirements.
B. Communication Protocol.
C. Voice and Data Communications.
D. Notice to Airmen (NOTAMS).
E. Aeronautical Publications.
F. Abnormal Procedures.
VI. Air Traffic Control
A. Responsibilities.
B. Facilities and Equipment.
C. Airspace classification and route structure.
D. Flight Plans.
(1) Domestic.
(2) International.
E. Separation Minimums.
F. Priority Handling.
G. Holding Procedures.
H. Traffic Management.
VII. Emergency and Abnormal Procedures
A. Security measures on the ground.
B. Security measures in the air.
C. FAA responsibility and services.
D. Collection and dissemination of information on overdue or missing aircraft.
E. Means of declaring an emergency.
F. Responsibility for declaring an emergency.
G. Required reporting of an emergency.
H. NTSB reporting requirements.
VIII. Practical Dispatch Applications
A. Human Factors.
(1) Decisionmaking:
(a) Situation Assessment.
(b) Generation and Evaluation of Alternatives.
(i) Tradeoffs and Prioritization.
(ii) Contingency Planning.
(c) Support Tools and Technologies.
(2) Human Error:
(a) Causes.
(i) Individual and Organizational Factors.
(ii) Technology-Induced Error.
(b) Prevention.
(c) Detection and Recovery.
(3) Teamwork:
(a) Communication and Information Exchange.
(b) Cooperative and Distributed Problem-Solving.
(c) Resource Management.
(i) Air Traffic Control (ATC) activities and workload.
(ii) Flightcrew activities and workload.
(iii) Maintenance activities and workload.
(iv) Operations Control Staff activities and workload.
B. Applied Dispatching.
(1) Briefing techniques, Dispatcher, Pilot.
(2) Preflight:
(a) Safety.
(b) Weather Analysis.
(i) Satellite imagery.
(ii) Upper and lower altitude charts.
(iii) Significant en route reports and forecasts.
(iv) Surface charts.
(v) Surface observations.
(vi) Terminal forecasts and orientation to Enhanced Weather Information System (EWINS).
(c) NOTAMS and airport conditions.
(d) Crew.
(i) Qualifications.
(ii) Limitations.
(e) Aircraft.
(i) Systems.
(ii) Navigation instruments and avionics systems.
(iii) Flight instruments.
(iv) Operations manuals and MEL/CDL.
(v) Performance and limitations.
(f) Flight Planning.
(i) Route of flight.
1. Standard Instrument Departures and Standard Terminal Arrival Routes.
2. En route charts.
3. Operational altitude.
4. Departure and arrival charts.
(ii) Minimum departure fuel.
1. Climb.
2. Cruise.
3. Descent.
(g) Weight and balance.
(h) Economics of flight overview (Performance, Fuel Tankering).
(i) Decision to operate the flight.
(j) ATC flight plan filing.
(k) Flight documentation.
(i) Flight plan.
(ii) Dispatch release.
(3) Authorize flight departure with concurrence of pilot in command.
(4) In-flight operational control:
(a) Current situational awareness.
(b) Information exchange.
(c) Amend original flight release as required.
(5) Post-Flight:
(a) Arrival verification.
(b) Weather debrief.
(c) Flight irregularity reports as required.

[Doc. No. FAA-1998-4553, 64 FR 68925, Dec. 8, 1999]

PART 65

FAR

PART 67

MEDICAL STANDARDS AND CERTIFICATION

Contents

Authority: 49 U.S.C. 106(g), 40113, 44701-44703, 44707, 44709-44711, 45102-45103, 45301-45303.

Source: Docket No. 27940, 61 FR 11256, Mar. 19, 1996, unless otherwise noted.

Subpart A—General

§ 67.1 Applicability.

This part prescribes the medical standards and certification procedures for issuing medical certificates for airmen and for remaining eligible for a medical certificate.

§ 67.3 Issue.

A person who meets the medical standards prescribed in this part, based on medical examination and evaluation of the person's history and condition, is entitled to an appropriate medical certificate.

[Doc. No. FAA-2007-27812, 73 FR 43065, July 24, 2008]

§ 67.4 Application.

An applicant for first-, second- and third-class medical certification must:

(a) Apply on a form and in a manner prescribed by the Administrator;

(b) Be examined by an aviation medical examiner designated in accordance with part 183 of this chapter. An applicant may obtain a list of aviation medical examiners from the FAA Office of Aerospace Medicine homepage on the FAA Web site, from any FAA Regional Flight Surgeon, or by contacting the Manager of the Aerospace Medical Education Division, P.O. Box 26200, Oklahoma City, Oklahoma 73125.

(c) Show proof of age and identity by presenting a government-issued photo identification (such as a valid U.S. driver's license, identification card issued by a driver's license authority, military identification, or passport). If an applicant does not have government-issued identification, he or she may use non-photo, government-issued identification (such as a birth certificate or voter registration card) in conjunction with photo identification (such as a work identification card or a student identification card).

[Doc. No. FAA-2007-27812, 73 FR 43065, July 24, 2008]

§ 67.7 Access to the National Driver Register.

At the time of application for a certificate issued under this part, each person who applies for a medical certificate shall execute an express consent form authorizing the Administrator to request the chief driver licensing official of any state designated by the Administrator to transmit information contained in the National Driver Register about the person to the Administrator. The Administrator shall make information received from the National Driver Register, if any, available on request to the person for review and written comment.

Subpart B—First-Class Airman Medical Certificate

§ 67.101 Eligibility.

To be eligible for a first-class airman medical certificate, and to remain eligible for a first-class airman medical certificate, a person must meet the requirements of this subpart.

§ 67.103 Eye.

Eye standards for a first-class airman medical certificate are:

(a) Distant visual acuity of 20/20 or better in each eye separately, with or without corrective lenses. If corrective lenses (spectacles or contact lenses) are necessary for 20/20 vision, the person may be eligible only on the condition that corrective lenses are worn while exercising the privileges of an airman certificate.

(b) Near vision of 20/40 or better, Snellen equivalent, at 16 inches in each eye separately, with or without corrective lenses. If age 50 or older, near vision of 20/40 or better, Snellen equivalent, at both 16 inches and 32 inches in each eye separately, with or without corrective lenses.

(c) Ability to perceive those colors necessary for the safe performance of airman duties.

(d) Normal fields of vision.

(e) No acute or chronic pathological condition of either eye or adnexa that interferes with the proper function of an eye, that may reasonably be expected to progress to that degree, or that may reasonably be expected to be aggravated by flying.

(f) Bifoveal fixation and vergence-phoria relationship sufficient to prevent a break in fusion under conditions that may reasonably be expected to occur in performing airman duties. Tests for the factors named in this paragraph are not required except for persons found to have more than 1 prism diopter of hyperphoria, 6 prism diopters of esophoria, or 6 prism diopters of exophoria. If any of these values are exceeded, the Federal Air Surgeon may require the person to be examined by a qualified eye specialist to determine if there is bifoveal fixation and an adequate vergence-phoria relationship. However, if otherwise eligible, the person is issued a medical certificate pending the results of the examination.

§ 67.105 Ear, nose, throat, and equilibrium.

Ear, nose, throat, and equilibrium standards for a first-class airman medical certificate are:

(a) The person shall demonstrate acceptable hearing by at least one of the following tests:

(1) Demonstrate an ability to hear an average conversational voice in a quiet room, using both ears, at a distance of 6 feet from the examiner, with the back turned to the examiner.

(2) Demonstrate an acceptable understanding of speech as determined by audiometric speech discrimination testing to a score of at least 70 percent obtained in one ear or in a sound field environment.

(3) Provide acceptable results of pure tone audiometric testing of unaided hearing acuity according to the following table of worst acceptable thresholds, using the calibration standards of the American National Standards Institute, 1969 (11 West 42d Street, New York, NY 10036):

Frequency (Hz)	500 Hz	1000 Hz	2000 Hz	3000 Hz
Better ear (Db)	35	30	30	40
Poorer ear (Db)	35	50	50	60

(b) No disease or condition of the middle or internal ear, nose, oral cavity, pharynx, or larynx that—

(1) Interferes with, or is aggravated by, flying or may reasonably be expected to do so; or

(2) Interferes with, or may reasonably be expected to interfere with, clear and effective speech communication.

(c) No disease or condition manifested by, or that may reasonably be expected to be manifested by, vertigo or a disturbance of equilibrium.

§ 67.107 Mental.

Mental standards for a first-class airman medical certificate are:

(a) No established medical history or clinical diagnosis of any of the following:

(1) A personality disorder that is severe enough to have repeatedly manifested itself by overt acts.

(2) A psychosis. As used in this section, "psychosis" refers to a mental disorder in which:

(i) The individual has manifested delusions, hallucinations, grossly bizarre or disorganized behavior, or other commonly accepted symptoms of this condition; or

(ii) The individual may reasonably be expected to manifest delusions, hallucinations, grossly bizarre or disorganized behavior, or other commonly accepted symptoms of this condition.

(3) A bipolar disorder.

(4) Substance dependence, except where there is established clinical evidence, satisfactory to the Federal Air Surgeon, of recovery, including sustained total abstinence from the substance(s) for not less than the preceding 2 years. As used in this section—

(i) "Substance" includes: Alcohol; other sedatives and hypnotics; anxiolytics; opioids; central nervous system stimulants such as cocaine, amphetamines, and similarly acting sympathomimetics; hallucinogens; phencyclidine or similarly acting arylcyclohexylamines; cannabis; inhalants; and other psychoactive drugs and chemicals; and

(ii) "Substance dependence" means a condition in which a person is dependent on a substance, other than tobacco or ordinary xanthine-containing (e.g., caffeine) beverages, as evidenced by—

(A) Increased tolerance;

(B) Manifestation of withdrawal symptoms;

(C) Impaired control of use; or

(D) Continued use despite damage to physical health or impairment of social, personal, or occupational functioning.

(b) No substance abuse within the preceding 2 years defined as:

(1) Use of a substance in a situation in which that use was physically hazardous, if there has been at any other time an instance of the use of a substance also in a situation in which that use was physically hazardous;

(2) A verified positive drug test result, an alcohol test result of 0.04 or greater alcohol concentration, or a refusal to submit to a drug or alcohol test required by the U.S. Department of Transportation or an agency of the U.S. Department of Transportation; or

(3) Misuse of a substance that the Federal Air Surgeon, based on case history and appropriate, qualified medical judgment relating to the substance involved, finds—

(i) Makes the person unable to safely perform the duties or exercise the privileges of the airman certificate applied for or held; or

(ii) May reasonably be expected, for the maximum duration of the airman medical certificate applied for or held, to make the person unable to perform those duties or exercise those privileges.

(c) No other personality disorder, neurosis, or other mental condition that the Federal Air Surgeon, based on the case history and appropriate, qualified medical judgment relating to the condition involved, finds—

(1) Makes the person unable to safely perform the duties or exercise the privileges of the airman certificate applied for or held; or

(2) May reasonably be expected, for the maximum duration of the airman medical certificate applied for or held, to make the person unable to perform those duties or exercise those privileges.

[Doc. No. 27940, 61 FR 11256, Mar. 19, 1996, as amended by Amdt. 67-19, 71 FR 35764, June 21, 2006]

§ 67.109 Neurologic.

Neurologic standards for a first-class airman medical certificate are:

(a) No established medical history or clinical diagnosis of any of the following:

(1) Epilepsy;

(2) A disturbance of consciousness without satisfactory medical explanation of the cause; or

(3) A transient loss of control of nervous system function(s) without satisfactory medical explanation of the cause.

(b) No other seizure disorder, disturbance of consciousness, or neurologic condition that the Federal Air Surgeon, based on the case history and appropriate, qualified medical judgment relating to the condition involved, finds—

(1) Makes the person unable to safely perform the duties or exercise the privileges of the airman certificate applied for or held; or

(2) May reasonably be expected, for the maximum duration of the airman medical certificate applied for or held, to make the person unable to perform those duties or exercise those privileges.

§ 67.111 Cardiovascular.

Cardiovascular standards for a first-class airman medical certificate are:

(a) No established medical history or clinical diagnosis of any of the following:

(1) Myocardial infarction;

(2) Angina pectoris;

(3) Coronary heart disease that has required treatment or, if untreated, that has been symptomatic or clinically significant;

(4) Cardiac valve replacement;

(5) Permanent cardiac pacemaker implantation; or

(6) Heart replacement;

(b) A person applying for first-class medical certification must demonstrate an absence of myocardial infarction and other clinically significant abnormality on electrocardiographic examination:

(1) At the first application after reaching the 35th birthday; and

(2) On an annual basis after reaching the 40th birthday.

(c) An electrocardiogram will satisfy a requirement of paragraph (b) of this section if it is dated no earlier than 60 days before the date of the application it is to accompany and was performed and transmitted according to acceptable standards and techniques.

§ 67.113 General medical condition.

The general medical standards for a first-class airman medical certificate are:

(a) No established medical history or clinical diagnosis of diabetes mellitus that requires insulin or any other hypoglycemic drug for control.

(b) No other organic, functional, or structural disease, defect, or limitation that the Federal Air Surgeon, based on the case history and appropriate, qualified medical judgment relating to the condition involved, finds—

(1) Makes the person unable to safely perform the duties or exercise the privileges of the airman certificate applied for or held; or

(2) May reasonably be expected, for the maximum duration of the airman medical certificate applied for or held, to make the person unable to perform those duties or exercise those privileges.

(c) No medication or other treatment that the Federal Air Surgeon, based on the case history and appropriate, qualified medical judgment relating to the medication or other treatment involved, finds—

(1) Makes the person unable to safely perform the duties or exercise the privileges of the airman certificate applied for or held; or

(2) May reasonably be expected, for the maximum duration of the airman medical certificate applied for or held, to make the person unable to perform those duties or exercise those privileges.

§ 67.115 Discretionary issuance.

A person who does not meet the provisions of §§ 67.103 through 67.113 may apply for the discretionary issuance of a certificate under § 67.401.

Subpart C—Second-Class Airman Medical Certificate

§ 67.201 Eligibility.

To be eligible for a second-class airman medical certificate, and to remain eligible for a second-class airman medical certificate, a person must meet the requirements of this subpart.

§ 67.203 Eye.

Eye standards for a second-class airman medical certificate are:

(a) Distant visual acuity of 20/20 or better in each eye separately, with or without corrective lenses. If corrective lenses (spectacles or contact lenses) are necessary for 20/20 vision, the person may be eligible only on the condition that corrective lenses are worn while exercising the privileges of an airman certificate.

(b) Near vision of 20/40 or better, Snellen equivalent, at 16 inches in each eye separately, with or without corrective lenses. If age 50 or older, near vision of 20/40 or better, Snellen equivalent, at both 16 inches and 32 inches in each eye separately, with or without corrective lenses.

(c) Ability to perceive those colors necessary for the safe performance of airman duties.

(d) Normal fields of vision.

(e) No acute or chronic pathological condition of either eye or adnexa that interferes with the proper function of an eye, that may reasonably be expected to progress to that degree, or that may reasonably be expected to be aggravated by flying.

(f) Bifoveal fixation and vergence-phoria relationship sufficient to prevent a break in fusion under conditions that may reasonably be expected to occur in performing airman duties. Tests for the factors named in this paragraph are not required except for persons found to have more than 1 prism diopter of hyperphoria, 6 prism diopters of esophoria, or 6 prism diopters of exophoria. If any of these values are exceeded, the Federal Air Surgeon may require the person to be examined by a qualified eye specialist to determine if there is bifoveal fixation and an adequate vergence-phoria relationship. However, if otherwise eligible, the person is issued a medical certificate pending the results of the examination.

§ 67.205 Ear, nose, throat, and equilibrium.

Ear, nose, throat, and equilibrium standards for a second-class airman medical certificate are:

(a) The person shall demonstrate acceptable hearing by at least one of the following tests:

(1) Demonstrate an ability to hear an average conversational voice in a quiet room, using both ears, at a distance of 6 feet from the examiner, with the back turned to the examiner.

(2) Demonstrate an acceptable understanding of speech as determined by audiometric speech discrimination testing to a score of at least 70 percent obtained in one ear or in a sound field environment.

(3) Provide acceptable results of pure tone audiometric testing of unaided hearing acuity according to the following table of worst acceptable thresholds, using the calibration standards of the American National Standards Institute, 1969:

Frequency (Hz)	500 Hz	1000 Hz	2000 Hz	3000 Hz
Better ear (Db)	35	30	30	40
Poorer ear (Db)	35	50	50	60

(b) No disease or condition of the middle or internal ear, nose, oral cavity, pharynx, or larynx that—

(1) Interferes with, or is aggravated by, flying or may reasonably be expected to do so; or

(2) Interferes with, or may reasonably be expected to interfere with, clear and effective speech communication.

(c) No disease or condition manifested by, or that may reasonably be expected to be manifested by, vertigo or a disturbance of equilibrium.

§ 67.207 Mental.

Mental standards for a second-class airman medical certificate are:

(a) No established medical history or clinical diagnosis of any of the following:

(1) A personality disorder that is severe enough to have repeatedly manifested itself by overt acts.

(2) A psychosis. As used in this section, "psychosis" refers to a mental disorder in which:

(i) The individual has manifested delusions, hallucinations, grossly bizarre or disorganized behavior, or other commonly accepted symptoms of this condition; or

(ii) The individual may reasonably be expected to manifest delusions, hallucinations, grossly bizarre or disorganized behavior, or other commonly accepted symptoms of this condition.

(3) A bipolar disorder.

(4) Substance dependence, except where there is established clinical evidence, satisfactory to the Federal Air Surgeon, of recovery, including sustained total abstinence from the substance(s) for not less than the preceding 2 years. As used in this section—

(i) "Substance" includes: Alcohol; other sedatives and hypnotics; anxiolytics; opioids; central nervous system stimulants such as cocaine, amphetamines, and similarly acting sympathomimetics; hallucinogens; phencyclidine or similarly acting arylcyclohexylamines; cannabis; inhalants; and other psychoactive drugs and chemicals; and

(ii) "Substance dependence" means a condition in which a person is dependent on a substance, other than tobacco or ordinary xanthine-containing (e.g., caffeine) beverages, as evidenced by—

(A) Increased tolerance;

(B) Manifestation of withdrawal symptoms;

(C) Impaired control of use; or

(D) Continued use despite damage to physical health or impairment of social, personal, or occupational functioning.

(b) No substance abuse within the preceding 2 years defined as:

(1) Use of a substance in a situation in which that use was physically hazardous, if there has been at any other time an instance of the use of a substance also in a situation in which that use was physically hazardous;

(2) A verified positive drug test result, an alcohol test result of 0.04 or greater alcohol concentration, or a refusal to submit to a drug or alcohol test required by the U.S. Department of Transportation or an agency of the U.S. Department of Transportation; or

(3) Misuse of a substance that the Federal Air Surgeon, based on case history and appropriate, qualified medical judgment relating to the substance involved, finds—

(i) Makes the person unable to safely perform the duties or exercise the privileges of the airman certificate applied for or held; or

(ii) May reasonably be expected, for the maximum duration of the airman medical certificate applied for or held, to make the person unable to perform those duties or exercise those privileges.

(c) No other personality disorder, neurosis, or other mental condition that the Federal Air Surgeon, based on the case history and appropriate, qualified medical judgment relating to the condition involved, finds—

(1) Makes the person unable to safely perform the duties or exercise the privileges of the airman certificate applied for or held; or

(2) May reasonably be expected, for the maximum duration of the airman medical certificate applied for or held, to make the person unable to perform those duties or exercise those privileges.

[Doc. No. 27940, 61 FR 11256, Mar. 19, 1996, as amended by Amdt. 67-19, 71 FR 35764, June 21, 2006]

§ 67.209 Neurologic.

Neurologic standards for a second-class airman medical certificate are:

(a) No established medical history or clinical diagnosis of any of the following:

(1) Epilepsy;

(2) A disturbance of consciousness without satisfactory medical explanation of the cause; or

(3) A transient loss of control of nervous system function(s) without satisfactory medical explanation of the cause;

(b) No other seizure disorder, disturbance of consciousness, or neurologic condition that the Federal Air Surgeon, based on the case history and appropriate, qualified medical judgment relating to the condition involved, finds—

(1) Makes the person unable to safely perform the duties or exercise the privileges of the airman certificate applied for or held; or

(2) May reasonably be expected, for the maximum duration of the airman medical certificate applied for or held, to make the person unable to perform those duties or exercise those privileges.

§ 67.211 Cardiovascular.

Cardiovascular standards for a second-class medical certificate are no established medical history or clinical diagnosis of any of the following:

(a) Myocardial infarction;

(b) Angina pectoris;

(c) Coronary heart disease that has required treatment or, if untreated, that has been symptomatic or clinically significant;

(d) Cardiac valve replacement;

(e) Permanent cardiac pacemaker implantation; or

(f) Heart replacement.

§ 67.213 General medical condition.

The general medical standards for a second-class airman medical certificate are:

(a) No established medical history or clinical diagnosis of diabetes mellitus that requires insulin or any other hypoglycemic drug for control.

(b) No other organic, functional, or structural disease, defect, or limitation that the Federal Air Surgeon, based on the case history and appropriate, qualified medical judgment relating to the condition involved, finds—

(1) Makes the person unable to safely perform the duties or exercise the privileges of the airman certificate applied for or held; or

(2) May reasonably be expected, for the maximum duration of the airman medical certificate applied for or held, to make the person unable to perform those duties or exercise those privileges.

(c) No medication or other treatment that the Federal Air Surgeon, based on the case history and appropriate, qualified medical judgment relating to the medication or other treatment involved, finds—

(1) Makes the person unable to safely perform the duties or exercise the privileges of the airman certificate applied for or held; or

(2) May reasonably be expected, for the maximum duration of the airman medical certificate applied for or held, to make the person unable to perform those duties or exercise those privileges.

§ 67.215 Discretionary issuance.

A person who does not meet the provisions of §§ 67.203 through 67.213 may apply for the discretionary issuance of a certificate under § 67.401.

Subpart D—Third-Class Airman Medical Certificate

§ 67.301 Eligibility.

To be eligible for a third-class airman medical certificate, or to remain eligible for a third-class airman medical certificate, a person must meet the requirements of this subpart.

§ 67.303 Eye.

Eye standards for a third-class airman medical certificate are:

(a) Distant visual acuity of 20/40 or better in each eye separately, with or without corrective lenses. If corrective lenses (spectacles or contact lenses) are necessary for 20/40 vision, the person may be eligible only on the condition that corrective lenses are worn while exercising the privileges of an airman certificate.

(b) Near vision of 20/40 or better, Snellen equivalent, at 16 inches in each eye separately, with or without corrective lenses.

(c) Ability to perceive those colors necessary for the safe performance of airman duties.

(d) No acute or chronic pathological condition of either eye or adnexa that interferes with the proper function of an eye, that may reasonably be expected to progress to that degree, or that may reasonably be expected to be aggravated by flying.

§ 67.305 Ear, nose, throat, and equilibrium.

Ear, nose, throat, and equilibrium standards for a third-class airman medical certificate are:

(a) The person shall demonstrate acceptable hearing by at least one of the following tests:

(1) Demonstrate an ability to hear an average conversational voice in a quiet room, using both ears, at a distance of 6 feet from the examiner, with the back turned to the examiner.

(2) Demonstrate an acceptable understanding of speech as determined by audiometric speech discrimination testing to a score of at least 70 percent obtained in one ear or in a sound field environment.

(3) Provide acceptable results of pure tone audiometric testing of unaided hearing acuity according to the following table of worst acceptable thresholds, using the calibration standards of the American National Standards Institute, 1969:

Frequency (Hz)	500 Hz	1000 Hz	2000 Hz	3000 Hz
Better ear (Db)	35	30	30	40
Poorer ear (Db)	35	50	50	60

(b) No disease or condition of the middle or internal ear, nose, oral cavity, pharynx, or larynx that—

(1) Interferes with, or is aggravated by, flying or may reasonably be expected to do so; or

(2) Interferes with clear and effective speech communication.

(c) No disease or condition manifested by, or that may reasonably be expected to be manifested by, vertigo or a disturbance of equilibrium.

§ 67.307 Mental.

Mental standards for a third-class airman medical certificate are:

(a) No established medical history or clinical diagnosis of any of the following:

(1) A personality disorder that is severe enough to have repeatedly manifested itself by overt acts.

(2) A psychosis. As used in this section, "psychosis" refers to a mental disorder in which—

(i) The individual has manifested delusions, hallucinations, grossly bizarre or disorganized behavior, or other commonly accepted symptoms of this condition; or

(ii) The individual may reasonably be expected to manifest delusions, hallucinations, grossly bizarre or disorganized behavior, or other commonly accepted symptoms of this condition.

(3) A bipolar disorder.

(4) Substance dependence, except where there is established clinical evidence, satisfactory to the Federal Air Surgeon, of recovery, including sustained total abstinence from the substance(s) for not less than the preceding 2 years. As used in this section—

(i) "Substance" includes: alcohol; other sedatives and hypnotics; anxiolytics; opioids; central nervous system stimulants such as cocaine, amphetamines, and similarly acting sympathomimetics; hallucinogens; phencyclidine or similarly acting arylcyclohexylamines; cannabis; inhalants; and other psychoactive drugs and chemicals; and

(ii) "Substance dependence" means a condition in which a person is dependent on a substance, other than tobacco or ordinary xanthine-containing (e.g., caffeine) beverages, as evidenced by—

(A) Increased tolerance;

(B) Manifestation of withdrawal symptoms;

(C) Impaired control of use; or

(D) Continued use despite damage to physical health or impairment of social, personal, or occupational functioning.

(b) No substance abuse within the preceding 2 years defined as:

(1) Use of a substance in a situation in which that use was physically hazardous, if there has been at any other time an instance of the use of a substance also in a situation in which that use was physically hazardous;

(2) A verified positive drug test result, an alcohol test result of 0.04 or greater alcohol concentration, or a refusal to submit to a drug or alcohol test required by the U.S. Department of Transportation or an agency of the U.S. Department of Transportation; or

(3) Misuse of a substance that the Federal Air Surgeon, based on case history and appropriate, qualified medical judgment relating to the substance involved, finds—

(i) Makes the person unable to safely perform the duties or exercise the privileges of the airman certificate applied for or held; or

(ii) May reasonably be expected, for the maximum duration of the airman medical certificate applied for or held, to make the person unable to perform those duties or exercise those privileges.

(c) No other personality disorder, neurosis, or other mental condition that the Federal Air Surgeon, based on the case history and appropriate, qualified medical judgment relating to the condition involved, finds—

(1) Makes the person unable to safely perform the duties or exercise the privileges of the airman certificate applied for or held; or

(2) May reasonably be expected, for the maximum duration of the airman medical certificate applied for or held, to make the person unable to perform those duties or exercise those privileges.

[Doc. No. 27940, 61 FR 11256, Mar. 19, 1996, as amended by Amdt. 67-19, 71 FR 35764, June 21, 2006]

§ 67.309 Neurologic.

Neurologic standards for a third-class airman medical certificate are:

(a) No established medical history or clinical diagnosis of any of the following:

(1) Epilepsy;

(2) A disturbance of consciousness without satisfactory medical explanation of the cause; or

(3) A transient loss of control of nervous system function(s) without satisfactory medical explanation of the cause.

(b) No other seizure disorder, disturbance of consciousness, or neurologic condition that the Federal Air Surgeon, based on the case history and appropriate, qualified medical judgment relating to the condition involved, finds—

(1) Makes the person unable to safely perform the duties or exercise the privileges of the airman certificate applied for or held; or

(2) May reasonably be expected, for the maximum duration of the airman medical certificate applied for or held, to make the person unable to perform those duties or exercise those privileges.

§ 67.311 Cardiovascular.

Cardiovascular standards for a third-class airman medical certificate are no established medical history or clinical diagnosis of any of the following:

(a) Myocardial infarction;

(b) Angina pectoris;

(c) Coronary heart disease that has required treatment or, if untreated, that has been symptomatic or clinically significant;

(d) Cardiac valve replacement;

(e) Permanent cardiac pacemaker implantation; or

(f) Heart replacement.

§ 67.313 General medical condition.

The general medical standards for a third-class airman medical certificate are:

(a) No established medical history or clinical diagnosis of diabetes mellitus that requires insulin or any other hypoglycemic drug for control.

(b) No other organic, functional, or structural disease, defect, or limitation that the Federal Air Surgeon, based on the case history and appropriate, qualified medical judgment relating to the condition involved, finds—

(1) Makes the person unable to safely perform the duties or exercise the privileges of the airman certificate applied for or held; or

(2) May reasonably be expected, for the maximum duration of the airman medical certificate applied for or held, to make the person unable to perform those duties or exercise those privileges.

(c) No medication or other treatment that the Federal Air Surgeon, based on the case history and appropriate, qualified medical judgment relating to the medication or other treatment involved, finds—

(1) Makes the person unable to safely perform the duties or exercise the privileges of the airman certificate applied for or held; or

(2) May reasonably be expected, for the maximum duration of the airman medical certificate applied for or held, to make the person unable to perform those duties or exercise those privileges.

§ 67.315 Discretionary issuance.

A person who does not meet the provisions of §§ 67.303 through 67.313 may apply for the discretionary issuance of a certificate under § 67.401.

Subpart E—Certification Procedures

§ 67.401 Special issuance of medical certificates.

(a) At the discretion of the Federal Air Surgeon, an Authorization for Special Issuance of a Medical Certificate (Authorization), valid for a specified period, may be granted to a person who does not meet the provisions of subparts B, C, or D of this part if the person shows to the satisfaction of the Federal Air Surgeon that the duties authorized by the class of medical certificate applied for can be performed without endangering public safety during the period in which the Authorization would be in force. The Federal Air Surgeon may authorize a special medical flight test, practical test, or medical evaluation for this purpose. A medical certificate of the appropriate class may be issued to a person who does not meet the provisions of subparts B, C, or D of this part if that person possesses a valid Authorization and is otherwise eligible. An airman medical certificate issued in accordance with this section shall expire no later than the end of the validity period or upon the withdrawal of the Authorization upon which it is based. At the end of its specified validity period, for grant of a new Authorization, the person must again show to the satisfaction of the Federal Air Surgeon that the duties authorized by the class of medical certificate applied for can be performed without endangering public safety during the period in which the Authorization would be in force.

(b) At the discretion of the Federal Air Surgeon, a Statement of Demonstrated Ability (SODA) may be granted, instead of an Authorization, to a person whose disqualifying condition is static or nonprogressive and who has been found capable of performing airman duties without endangering public safety. A SODA does not expire and authorizes a designated aviation medical examiner to issue a medical certificate of a specified class if the examiner finds that the condition described on its face has not adversely changed.

(c) In granting an Authorization or SODA, the Federal Air Surgeon may consider the person's operational experience and any medical facts that may affect the ability of the person to perform airman duties including—

(1) The combined effect on the person of failure to meet more than one requirement of this part; and

(2) The prognosis derived from professional consideration of all available information regarding the person.

(d) In granting an Authorization or SODA under this section, the Federal Air Surgeon specifies the class of medical certificate authorized to be issued and may do any or all of the following:

(1) Limit the duration of an Authorization;

(2) Condition the granting of a new Authorization on the results of subsequent medical tests, examinations, or evaluations;

(3) State on the Authorization or SODA, and any medical certificate based upon it, any operational limitation needed for safety; or

(4) Condition the continued effect of an Authorization or SODA, and any second- or third-class medical certificate based upon it, on compliance with a statement of functional limitations issued to the person in coordination with the Director of Flight Standards or the Director's designee.

(e) In determining whether an Authorization or SODA should be granted to an applicant for a third-class medical certificate, the Federal Air Surgeon considers the freedom of an airman, exercising the privileges of a private pilot certificate, to accept reasonable risks to his or her person and property that are not acceptable in the exercise of commercial or airline transport pilot privileges, and, at the same time, considers the need to protect the safety of persons and property in other aircraft and on the ground.

(f) An Authorization or SODA granted under the provisions of this section to a person who does not meet the applicable provisions of subparts B, C, or D of this part may be withdrawn, at the discretion of the Federal Air Surgeon, at any time if—

(1) There is adverse change in the holder's medical condition;

(2) The holder fails to comply with a statement of functional limitations or operational limitations issued as a condition of certification under this section;

(3) Public safety would be endangered by the holder's exercise of airman privileges;

(4) The holder fails to provide medical information reasonably needed by the Federal Air Surgeon for certification under this section; or

(5) The holder makes or causes to be made a statement or entry that is the basis for withdrawal of an Authorization or SODA under § 67.403.

(g) A person who has been granted an Authorization or SODA under this section based on a special medical flight or practical test need not take the test again during later physical examinations unless the Federal Air Surgeon determines or has reason to believe that the physical deficiency has or may have degraded to a degree to require another special medical flight test or practical test.

(h) The authority of the Federal Air Surgeon under this section is also exercised by the Manager, Aeromedical Certification Division, and each Regional Flight Surgeon.

(i) If an Authorization or SODA is withdrawn under paragraph (f) of this section the following procedures apply:

(1) The holder of the Authorization or SODA will be served a letter of withdrawal, stating the reason for the action;

(2) By not later than 60 days after the service of the letter of withdrawal, the holder of the Authorization or SODA may request, in writing, that the Federal Air Surgeon provide for review of the decision to withdraw. The request for review may be accompanied by supporting medical evidence;

(3) Within 60 days of receipt of a request for review, a written final decision either affirming or reversing the decision to withdraw will be issued; and

(4) A medical certificate rendered invalid pursuant to a withdrawal, in accordance with paragraph (a) of this section, shall be surrendered to the Administrator upon request.

[Docket No. 27940, 61 FR 11256, Mar. 19, 1996, as amended by Amdt. 67-20, 73 FR 43066, July 24, 2008; Amdt. 67-21, 77 FR 16668, Mar. 22, 2012]

§ 67.403 Applications, certificates, logbooks, reports, and records: Falsification, reproduction, or alteration; incorrect statements.

(a) No person may make or cause to be made—

(1) A fraudulent or intentionally false statement on any application for a medical certificate or on a request for any Authorization for Special Issuance of a Medical Certificate (Authorization) or Statement of Demonstrated Ability (SODA) under this part;

(2) A fraudulent or intentionally false entry in any logbook, record, or report that is kept, made, or used, to show compliance with any requirement for any medical certificate or for any Authorization or SODA under this part;

(3) A reproduction, for fraudulent purposes, of any medical certificate under this part; or

(4) An alteration of any medical certificate under this part.

(b) The commission by any person of an act prohibited under paragraph (a) of this section is a basis for—

(1) Suspending or revoking all airman, ground instructor, and medical certificates and ratings held by that person;

(2) Withdrawing all Authorizations or SODA's held by that person; and

(3) Denying all applications for medical certification and requests for Authorizations or SODA's.

(c) The following may serve as a basis for suspending or revoking a medical certificate; withdrawing an Authorization or SODA; or denying an application for a medical certificate or request for an authorization or SODA:

(1) An incorrect statement, upon which the FAA relied, made in support of an application for a medical certificate or request for an Authorization or SODA.

(2) An incorrect entry, upon which the FAA relied, made in any logbook, record, or report that is kept, made, or used to show compliance with any requirement for a medical certificate or an Authorization or SODA.

§ 67.405 Medical examinations: Who may perform?

(a) *First-class.* Any aviation medical examiner who is specifically designated for the purpose may perform examinations for the first-class medical certificate.

(b) *Second- and third-class.* Any aviation medical examiner may perform examinations for the second-or third-class medical certificate.

[Doc. No. FAA-2007-27812, 73 FR 43066, July 24, 2008]

§ 67.407 Delegation of authority.

(a) The authority of the Administrator under 49 U.S.C. 44703 to issue or deny medical certificates is delegated to the Federal Air Surgeon to the extent necessary to—

(1) Examine applicants for and holders of medical certificates to determine whether they meet applicable medical standards; and

(2) Issue, renew, and deny medical certificates, and issue, renew, deny, and withdraw Authorizations for Special Issuance of a Medical Certificate and Statements of Demonstrated Ability to a person based upon meeting or failing to meet applicable medical standards.

(b) Subject to limitations in this chapter, the delegated functions of the Federal Air Surgeon to examine applicants for and holders of medical certificates for compliance with applicable medical standards and to issue, renew, and deny medical certificates are also delegated to aviation medical examiners and to authorized representatives of the Federal Air Surgeon within the FAA.

(c) The authority of the Administrator under 49 U.S.C. 44702, to reconsider the action of an aviation medical examiner is delegated to the Federal Air Surgeon; the Manager, Aeromedical Certification Division; and each Regional Flight Surgeon. Where the person does not meet the standards of §§ 67.107(b)(3) and (c), 67.109(b), 67.113(b) and (c), 67.207(b)(3) and (c), 67.209(b), 67.213(b) and (c), 67.307(b)(3) and (c), 67.309(b), or 67.313(b) and (c), any action taken under this paragraph other than by the Federal Air Surgeon is subject to reconsideration by the Federal Air Surgeon. A certificate issued by an aviation medical examiner is considered to be affirmed as issued unless an FAA official named in this paragraph (authorized official) reverses that issuance within 60 days after the date of issuance. However, if within 60 days after the date of issuance an authorized official requests the certificate holder to submit additional medical information, an authorized official may reverse the issuance within 60 days after receipt of the requested information.

(d) The authority of the Administrator under 49 U.S.C. 44709 to re-examine any civil airman to the extent necessary to determine an airman's qualification to continue to hold an airman medical certificate, is delegated to the Federal Air Surgeon and his or her authorized representatives within the FAA.

§ 67.409 Denial of medical certificate.

(a) Any person who is denied a medical certificate by an aviation medical examiner may, within 30 days after the date of the denial, apply in writing and in duplicate to the Federal Air Surgeon, Attention: Manager, Aeromedical Certification Division, AAM-300, Federal Aviation Administration, P.O. Box 26080, Oklahoma City, Oklahoma 73126, for reconsideration of that denial. If the person does not ask for reconsideration during the 30-day period after the date of the denial, he or she is considered to have withdrawn the application for a medical certificate.

(b) The denial of a medical certificate—

(1) By an aviation medical examiner is not a denial by the Administrator under 49 U.S.C. 44703.

(2) By the Federal Air Surgeon is considered to be a denial by the Administrator under 49 U.S.C. 44703.

(3) By the Manager, Aeromedical Certification Division, or a Regional Flight Surgeon is considered to be a denial by the Administrator under 49 U.S.C. 44703 except where the person does not meet the standards of §§ 67.107(b)(3) and (c), 67.109(b), or 67.113(b) and (c); 67.207(b)(3) and (c), 67.209(b), or 67.213(b) and (c); or 67.307(b)(3) and (c), 67.309(b), or 67.313(b) and (c).

(c) Any action taken under § 67.407(c) that wholly or partly reverses the issue of a medical certificate by an aviation medical examiner is the denial of a medical certificate under paragraph (b) of this section.

(d) If the issue of a medical certificate is wholly or partly reversed by the Federal Air Surgeon; the Manager, Aeromedical Certification Division; or a Regional Flight Surgeon, the person holding that certificate shall surrender it, upon request of the FAA.

§ 67.411 [Reserved]

§ 67.413 Medical records.

(a) Whenever the Administrator finds that additional medical information or history is necessary to determine whether you meet the medical standards required to hold a medical certificate, you must:

(1) Furnish that information to the FAA; or

(2) Authorize any clinic, hospital, physician, or other person to release to the FAA all available information or records concerning that history.

(b) If you fail to provide the requested medical information or history or to authorize its release, the FAA may suspend, modify, or revoke your medical certificate or, in the case of an applicant, deny the application for a medical certificate.

(c) If your medical certificate is suspended, modified, or revoked under paragraph (b) of this section, that suspension or modification remains in effect until you provide the requested information, history, or authorization to the FAA and until the FAA determines that you meet the medical standards set forth in this part.

[Doc. No. FAA-2007-27812, 73 FR 43066, July 24, 2008]

§ 67.415 Return of medical certificate after suspension or revocation.

The holder of any medical certificate issued under this part that is suspended or revoked shall, upon the Administrator's request, return it to the Administrator.

PART 71

DESIGNATION OF CLASS A, B, C, D, AND E AIRSPACE AREAS; AIR TRAFFIC SERVICE ROUTES; AND REPORTING POINTS

Contents

Authority: 49 U.S.C. 106(g), 40103, 40113, 40120; E.O. 10854, 24 FR 9565, 3 CFR, 1959-1963 Comp., p. 389.

Source: Amdt. 71-14, 56 FR 65654, Dec. 17, 1991, unless otherwise noted.

Special Federal Aviation Regulation No. 97

Editorial Note: For the text of SFAR No. 97, see part 91 of this chapter.

§ 71.1 Applicability.

A listing for Class A, B, C, D, and E airspace areas; air traffic service routes; and reporting points can be found in FAA Order 7400.9W, Airspace Designations and Reporting Points, dated August 8, 2012. This incorporation by reference was approved by the Director of the Federal Register in accordance with 5 U.S.C. 552 (a) and 1 CFR part 51. The approval to incorporate by reference FAA Order 7400.9W is effective September 15, 2012, through September 15, 2013. During the incorporation by reference period, proposed changes to the listings of Class A, B, C, D, and E airspace areas; air traffic service routes; and reporting points will be published in full text as proposed rule documents in the Federal Register. Amendments to the listings of Class A, B, C, D, and E airspace areas; air traffic service routes; and reporting points will be published in full text as final rules in the Federal Register. Periodically, the final rule amendments will be integrated into a revised edition of the Order and submitted to the Director of the Federal Register for approval for incorporation by reference in this section. Copies of FAA Order 7400.9W may be obtained from Airspace, Regulations and ATC Procedures Group, Federal Aviation Administration, 800 Independence Avenue SW., Washington, DC 20591, (202) 267-8783. An electronic version of the Order is available on the FAA Web site at *http://www.faa.gov/air_traffic/publications.* Copies of FAA Order 7400.9W may be inspected in Docket No. FAA-2012-0842; Amendment No. 71-44 on *http://www.regulations.gov.* A copy of AFF Order 7400.9W may be inspected at the National Archives and Records Administration (NARA). For information on the availability of this material at NARA, call 202-741-6030, or go to: *http://www.archives.gov/federal-register/cfr/ibr-locations.html.*

[Doc. No. FAA-2012-0842, 77 FR 50908, Aug. 23, 2012]

Effective Date Note: By Doc. No. FAA-2012-0842, 77 FR 50908, Aug. 23, 2012, § 71.1 was revised, effective Sept. 15, 2012 through Sept. 15, 2013.

§ 71.3 [Reserved]

§ 71.5 Reporting points.

The reporting points listed in subpart H of FAA Order 7400.9W (incorporated by reference, see § 71.1) consist of geographic locations at which the position of an aircraft must be reported in accordance with part 91 of this chapter.

[Doc. No. 29334, 73 FR 54495, Sept. 22, 2008, as amended by Amdt. 71-40, 73 FR 60940, Oct. 15, 2008; Amdt. 71-41, 74 FR 46490, Sept. 10, 2009; Amdt. 71-42, 75 FR 55268, Sept. 10, 2010; Amdt. 71-43, 76 FR 53329, Aug. 26, 2011; Amdt. 71-44, 77 FR 50908, Aug. 23, 2012]

Effective Date Note: By Amdt. 71-44, 77 FR 50908, Aug. 23, 2012, § 71.5 was amended by removing the words "FAA Order 7400.9V" and adding, in their place, the words "FAA Order 7400.9W", effective Sept. 15, 2012 through Sept. 15, 2013.

§ 71.7 Bearings, radials, and mileages.

All bearings and radials in this part are true and are applied from point of origin and all mileages in this part are stated as nautical miles.

§ 71.9 Overlapping airspace designations.

(a) When overlapping airspace designations apply to the same airspace, the operating rules associated with the more restrictive airspace designation apply.

(b) For the purpose of this section—

(1) Class A airspace is more restrictive than Class B, Class C, Class D, Class E, or Class G airspace;

(2) Class B airspace is more restrictive than Class C, Class D, Class E, or Class G airspace;

(3) Class C airspace is more restrictive than Class D, Class E, or Class G airspace;

(4) Class D airspace is more restrictive than Class E or Class G airspace; and

(5) Class E is more restrictive than Class G airspace.

§ 71.11 Air Traffic Service (ATS) routes.

Unless otherwise specified, the following apply:

(a) An Air Traffic Service (ATS) route is based on a centerline that extends from one navigation aid, fix, or intersection, to another navigation aid, fix, or intersection (or through several navigation aids, fixes, or intersections) specified for that route.

(b) An ATS route does not include the airspace of a prohibited area.

[Doc. No. FAA-2003-14698, 68 FR 16947, Apr. 8, 2003, as amended by Amdt. 71-33, 70 FR 23004, May 3, 2005]

§ 71.13 Classification of Air Traffic Service (ATS) routes.

Unless otherwise specified, ATS routes are classified as follows:

(a) In subpart A of this part:

(1) Jet routes.

(2) Area navigation (RNAV) routes.

(b) In subpart E of this part:

(1) VOR Federal airways.

(2) Colored Federal airways.

(i) Green Federal airways.

(ii) Amber Federal airways.

(iii) Red Federal airways.

(iv) Blue Federal airways.

(3) Area navigation (RNAV) routes.

[Doc. No. FAA-2003-14698, 68 FR 16947, Apr. 8, 2003]

§ 71.15 Designation of jet routes and VOR Federal airways.

Unless otherwise specified, the place names appearing in the descriptions of airspace areas designated as jet routes in subpart A of FAA Order 7400.9W, and as VOR Federal airways in subpart E of FAA Order 7400.9W, are the names of VOR or VORTAC navigation aids. FAA Order 7400.9W is incorporated by reference in § 71.1.

[Doc. No. 29334, 73 FR 54495, Sept. 22, 2008, as amended by Amdt. 71-40, 73 FR 60940, Oct. 15, 2008; Amdt. 71-41, 74 FR 46490, Sept. 10, 2009; Amdt. 71-42, 75 FR 55268, Sept. 10, 2010; Amdt. 71-43, 76 FR 53329, Aug. 26, 2011; Amdt. 71-44, 77 FR 50908, Aug. 23, 2012]

Effective Date Note: By Amdt. 71-44, 77 FR 50908, Aug. 23, 2012, § 71.15 was amended by removing the words "FAA Order 7400.9V" and adding, in their place, the words "FAA Order 7400.9W", effective Sept. 15, 2012 through Sept. 15, 2013.

Subpart A—Class A Airspace

§ 71.31 Class A airspace.

The airspace descriptions contained in § 71.33 and the routes contained in subpart A of FAA Order 7400.9W (incorporated by reference, see § 71.1) are designated as Class A airspace within which all pilots and aircraft are subject to the rating requirements, operating rules, and equipment requirements of part 91 of this chapter.

[Doc. No. 29334, 73 FR 54495, Sept. 22, 2008, as amended by Amdt. 71-40, 73 FR 60940, Oct. 15, 2008; Amdt. 71-41, 74 FR 46490, Sept. 10, 2009; Amdt. 71-42, 75 FR 55268, Sept. 10, 2010; Amdt. 71-43, 76 FR 53329, Aug. 26, 2011; Amdt. 71-44, 77 FR 50908, Aug. 23, 2012]

Effective Date Note: By Amdt. 71-44, 77 FR 50908, Aug. 23, 2012, § 71.31 was amended by removing the words "FAA Order 7400.9V" and adding, in their place, the words "FAA Order 7400.9W", effective Sept. 15, 2012 through Sept. 15, 2013.

§ 71.33 Class A airspace areas.

(a) That airspace of the United States, including that airspace overlying the waters within 12 nautical miles of the coast of the 48 contiguous States, from 18,000 feet MSL to and including FL600 excluding the states of Alaska and Hawaii, Santa Barbara Island, Farallon Island, and the airspace south of latitude 25°04′00″ North.

(b) That airspace of the State of Alaska, including that airspace overlying the waters within 12 nautical miles of the coast, from 18,000 feet MSL to and including FL600 but not including the airspace less than 1,500 feet above the surface of the earth and the Alaska Peninsula west of longitude 160°00′00″ West.

(c) The airspace areas listed as offshore airspace areas in subpart A of FAA Order 7400.9W (incorporated by reference, see § 71.1) that are designated in international airspace within areas of domestic radio navigational signal or ATC radar coverage, and within which domestic ATC procedures are applied.

[Amdt. 71-14, 56 FR 65654, Dec. 17, 1991]

Editorial Note: For Federal Register citations affecting § 71.33, see the List of CFR Sections Affected, which appears in the Finding Aids section of the printed volume and at *www.fdsys.gov.*

Effective Date Note: By Amdt. 71-44, 77 FR 50908, Aug. 23, 2012, § 71.33(c) was amended by removing the words "FAA Order 7400.9V" and adding, in their place, the words "FAA Order 7400.9W", effective Sept. 15, 2012 through Sept. 15, 2013.

Subpart B—Class B Airspace

§ 71.41 Class B airspace.

The Class B airspace areas listed in subpart B of FAA Order 7400.9W (incorporated by reference, see § 71.1) consist of specified airspace within which all aircraft operators are subject to the minimum pilot qualification requirements, operating rules, and aircraft equipment requirements of part 91 of this chapter. Each Class B airspace area designated for an airport in subpart B of FAA Order 7400.9W (incorporated by reference, see § 71.1) contains at least one primary airport around which the airspace is designated.

[Doc. No. 29334, 73 FR 54495, Sept. 22, 2008, as amended by Amdt. 71-40, 73 FR 60940, Oct. 15, 2008; Amdt. 71-41, 74 FR 46490, Sept. 10, 2009; Amdt. 71-42, 75 FR 55268, Sept. 10, 2010; Amdt. 71-43, 76 FR 53329, Aug. 26, 2011; Amdt. 71-44, 77 FR 50908, Aug. 23, 2012]

Effective Date Note: By Amdt. 71-44, 77 FR 50908, Aug. 23, 2012, § 71.41 was amended by removing the words "FAA Order 7400.9V" and adding, in their place, the words "FAA Order 7400.9W", effective Sept. 15, 2012, through Sept. 15, 2013.

Subpart C—Class C Airspace

§ 71.51 Class C airspace.

The Class C airspace areas listed in subpart C of FAA Order 7400.9W (incorporated by reference, see § 71.1) consist of specified airspace within which all aircraft operators are subject to operating rules and equipment requirements specified in part 91 of this chapter. Each Class C airspace area designated for an airport in subpart C of FAA Order 7400.9W (incorporated by reference, see § 71.1) contains at least one primary airport around which the airspace is designated.

[Doc. No. 29334, 73 FR 54495, Sept. 22, 2008, as amended by Amdt. 71-40, 73 FR 60940, Oct. 15, 2008; Amdt. 71-41, 74 FR 46490, Sept. 10, 2009; Amdt. 71-42, 75 FR 55269, Sept. 10, 2010; Amdt. 71-43, 76 FR 53329, Aug. 26, 2011; Amdt. 71-44, 77 FR 50908, Aug. 23, 2012]

Effective Date Note: By Amdt. 71-44, 77 FR 50908, Aug. 23, 2012, § 71.51 was amended by removing the words "FAA Order 7400.9V" and adding, in their place, the words "FAA Order 7400.9W", effective Sept. 15, 2012, through Sept. 15, 2013.

Subpart D—Class D Airspace

§ 71.61 Class D airspace.

The Class D airspace areas listed in subpart D of FAA Order 7400.9W (incorporated by reference, see § 71.1) consist of specified airspace within which all aircraft operators are subject to operating rules and equipment requirements specified in part 91 of this chapter. Each Class D airspace area designated for an airport in subpart D of FAA Order 7400.9W (incorporated by reference, see § 71.1) contains at least one primary airport around which the airspace is designated.

[Doc. No. 29334, 73 FR 54495, Sept. 22, 2008, as amended by Amdt. 71-40, 73 FR 60940, Oct. 15, 2008; Amdt. 71-41, 74 FR 46490, Sept. 10, 2009; Amdt. 71-42, 75 FR 55269, Sept. 10, 2010; Amdt. 71-43, 76 FR 53329, Aug. 26, 2011; Amdt. 71-44, 77 FR 50908, Aug. 23, 2012]

Effective Date Note: By Amdt. 71-44, 77 FR 50908, Aug. 23, 2012, § 71.61 was amended by removing the words "FAA Order 7400.9V" and adding, in their place, the words "FAA Order 7400.9W", effective Sept. 15, 2012, through Sept. 15, 2013.

Subpart E—Class E Airspace

§ 71.71 Class E airspace.

Class E Airspace consists of:

(a) The airspace of the United States, including that airspace overlying the waters within 12 nautical miles of the coast of the 48 contiguous states and Alaska, extending upward from 14,500 feet MSL up to, but not including 18,000 feet MSL, and the airspace above FL600, excluding—

(1) The Alaska peninsula west of longitude 160°00′00″ W.; and

(2) The airspace below 1,500 feet above the surface of the earth.

(b) The airspace areas designated for an airport in subpart E of FAA Order 7400.9W (incorporated by reference, see § 71.1) within which all aircraft operators are subject to the operating rules specified in part 91 of this chapter.

(c) The airspace areas listed as domestic airspace areas in subpart E of FAA Order 7400.9W (incorporated by reference, see § 71.1) which extend upward from 700 feet or more above the surface of the earth when designated in conjunction with an airport for which an approved instrument approach procedure has been prescribed, or from 1,200 feet or more above the surface of the earth for the purpose of transitioning to or from the terminal or en route environment. When such areas are designated in conjunction with airways or routes, the extent of such designation has the lateral extent identical to that of a Federal airway and extends upward from 1,200 feet or higher. Unless otherwise specified, the airspace areas in the paragraph extend upward from 1,200 feet or higher above the surface to, but not including, 14,500 feet MSL.

(d) The Federal airways described in subpart E of FAA Order 7400.9W (incorporated by reference, see § 71.1).

(e) The airspace areas listed as en route domestic airspace areas in subpart E of FAA Order 7400.9W (incorporated by reference, see § 71.1). Unless otherwise specified, each airspace area has a lateral extent identical to that of a Federal airway and extends upward from 1,200 feet above the surface of the earth to the overlying or adjacent controlled airspace.

(f) The airspace areas listed as offshore airspace areas in subpart E of FAA Order 7400.9W (incorporated by reference, see § 71.1) that are designated in international airspace within areas of domestic radio navigational signal or ATC radar coverage, and within which domestic ATC procedures are applied. Unless otherwise specified, each airspace area extends upward from a specified altitude up to, but not including, 18,000 feet MSL.

[Doc. No. 29334, 73 FR 54495, Sept. 22, 2008, as amended by Amdt. 71-40, 73 FR 60940, Oct. 15, 2008; Amdt. 71-41, 74 FR 46490, Sept. 10, 2009; Amdt. 71-42, 75 FR 55269, Sept. 10, 2010; Amdt. 71-43, 76 FR 53329, Aug. 26, 2011; Amdt. 71-44, 77 FR 50908, Aug. 23, 2012]

Effective Date Note: By Amdt. 71-44, 77 FR 50908, Aug. 23, 2012, § 71.71(b),(c),(d), and (f) was amended by removing the words "FAA Order 7400.9V" and adding, in their place, the words "FAA Order 7400.9W", effective Sept. 15, 2012, through Sept. 15, 2013.

Subparts F-G [Reserved]

Subpart H—Reporting Points

§ 71.901 Applicability.

Unless otherwise designated:

(a) Each reporting point listed in subpart H of FAA Order 7400.9W (incorporated by reference, see § 71.1) applies to all directions of flight. In any case where a geographic location is designated as a reporting point for less than all airways passing through that point, or for a particular direction of flight along an airway only, it is so indicated by including the airways or direction of flight in the designation of geographical location.

(b) Place names appearing in the reporting point descriptions indicate VOR or VORTAC facilities identified by those names.

[Doc. No. 29334, 73 FR 54495, Sept. 22, 2008, as amended by Amdt. 71-40, 73 FR 60940, Oct. 15, 2008; Amdt. 71-41, 74 FR 46490, Sept. 10, 2009; Amdt. 71-42, 75 FR 55269, Sept. 10, 2010; Amdt. 71-43, 76 FR 53329, Aug. 26, 2011; Amdt. 71-44, 77 FR 50908, Aug. 23, 2012]

Effective Date Note: By Amdt. 71-44, 77 FR 50908, Aug. 23, 2012, § 71.901 was amended by removing the words "FAA Order 7400.9V" and adding, in their place, the words "FAA Order 7400.9W", effective Sept. 15, 2012, through Sept. 15, 2013.

PART 73

SPECIAL USE AIRSPACE

Contents

AUTHORITY: 49 U.S.C. 106(g), 40103, 40113, 40120; E.O. 10854, 24 FR 9565, 3 CFR, 1959-1963 Comp., p. 389.

SOURCE: 46 FR 779, Jan. 2, 1981, unless otherwise noted.

Subpart A—General

§ 73.1 Applicability.

The airspace that is described in subpart B and subpart C of this part is designated as special use airspace. These parts prescribe the requirements for the use of that airspace.

§ 73.3 Special use airspace.

(a) Special use airspace consists of airspace of defined dimensions identified by an area on the surface of the earth wherein activities must be confined because of their nature, or wherein limitations are imposed upon aircraft operations that are not a part of those activities, or both.

(b) The vertical limits of special use airspace are measured by designated altitude floors and ceilings expressed as flight levels or as feet above mean sea level. Unless otherwise specified, the word "to" (an altitude or flight level) means "to and including" (that altitude or flight level).

(c) The horizontal limits of special use airspace are measured by boundaries described by geographic coordinates or other appropriate references that clearly define their perimeter.

(d) The period of time during which a designation of special use airspace is in effect is stated in the designation.

§ 73.5 Bearings; radials; miles.

(a) All bearings and radials in this part are true from point of origin.

(b) Unless otherwise specified, all mileages in this part are stated as statute miles.

Subpart B—Restricted Areas

§ 73.11 Applicability.

This subpart designates restricted areas and prescribes limitations on the operation of aircraft within them.

§ 73.13 Restrictions.

No person may operate an aircraft within a restricted area between the designated altitudes and during the time of designation, unless he has the advance permission of

(a) The using agency described in § 73.15; or

(b) The controlling agency described in § 73.17.

§ 73.15 Using agency.

(a) For the purposes of this subpart, the following are using agencies;

(1) The agency, organization, or military command whose activity within a restricted area necessitated the area being so designated.

(b) Upon the request of the FAA, the using agency shall execute a letter establishing procedures for joint use of a restricted area by the using agency and the controlling agency, under which the using agency would notify the controlling agency whenever the controlling agency may grant permission for transit through the restricted area in accordance with the terms of the letter.

(c) The using agency shall—

(1) Schedule activities within the restricted area;

(2) Authorize transit through, or flight within, the restricted area as feasible; and

(3) Contain within the restricted area all activities conducted therein in accordance with the purpose for which it was designated.

§ 73.17 Controlling agency.

For the purposes of this part, the controlling agency is the FAA facility that may authorize transit through or flight within a restricted area in accordance with a joint-use letter issued under § 73.15.

§ 73.19 Reports by using agency.

(a) Each using agency shall prepare a report on the use of each restricted area assigned thereto during any part of the preceding 12-month period ended September 30, and transmit it by the following January 31 of each year to the Manager, Air Traffic Division in the regional office of the Federal Aviation Administration having jurisdiction over the area in which the restricted area is located, with a copy to the Program Director for Air Traffic Airspace Management, Federal Aviation Administration, Washington, DC 20591.

(b) In the report under this section the using agency shall:

(1) State the name and number of the restricted area as published in this part, and the period covered by the report.

(2) State the activities (including average daily number of operations if appropriate) conducted in the area, and any other pertinent information concerning current and future electronic monitoring devices.

(3) State the number of hours daily, the days of the week, and the number of weeks during the year that the area was used.

(4) For restricted areas having a joint-use designation, also state the number of hours daily, the days of the week, and the number of weeks during the year that the restricted area was released to the controlling agency for public use.

(5) State the mean sea level altitudes or flight levels (whichever is appropriate) used in aircraft operations and the maximum and average ordinate of surface firing (expressed in feet, mean sea level altitude) used on a daily, weekly, and yearly basis.

(6) Include a chart of the area (of optional scale and design) depicting, if used, aircraft operating areas, flight patterns, ordnance delivery areas, surface firing points, and target, fan, and impact areas. After once submitting an appropriate chart, subsequent annual charts are not required unless there is a change in the area, activity or altitude (or flight levels) used, which might alter the depiction of the activities originally reported. If no change is to be submitted, a statement indicating "no change" shall be included in the report.

(7) Include any other information not otherwise required under this part which is considered pertinent to activities carried on in the restricted area.

(c) If it is determined that the information submitted under paragraph (b) of this section is not sufficient to evaluate the nature and extent of the use of a restricted area, the FAA may request the using agency to submit supplementary reports. Within 60 days after receiving a request for additional information, the using agency shall submit such information as the Program Director for Air Traffic Airspace Management considers appropriate. Supplementary reports must be sent to the FAA officials designated in paragraph (a) of this section.

(Secs. 307 and 313(a), Federal Aviation Act of 1958 (49 U.S.C. 1348 and 1354(a)))

[Doc. No. 15379, 42 FR 54798, Oct. 11, 1977, as amended by Amdt. 73-5, 54 FR 39292, Sept. 25, 1989; Amdt. 73-6, 58 FR 42001, Aug. 6, 1993; Amdt. 73-8, 61 FR 26435, May 28, 1996; Amdt. 73-8, 63 FR 16890, Apr. 7, 1998]

EDITORIAL NOTE: The restricted areas formerly carried as §§ 608.21 to 608.72 of this title were transferred to part 73 as §§ 73.21 to 73.72 under subpart B but are not carried in the Code of Federal Regulations. For FEDERAL REGISTER citations affecting these restricted areas, see the List of CFR Sections Affected, which appears in the Finding Aids section of the printed volume and at *www.fdsys.gov.*

Subpart C—Prohibited Areas

§ 73.81 Applicability.

This subpart designates prohibited areas and prescribes limitations on the operation of aircraft therein.

§ 73.83 Restrictions.

No person may operate an aircraft within a prohibited area unless authorization has been granted by the using agency.

§ 73.85 Using agency.

For the purpose of this subpart, the using agency is the agency, organization or military command that established the requirements for the prohibited area.

EDITORIAL NOTE: Sections 73.87 through 73.99 are reserved for descriptions of designated prohibited areas. For FEDERAL REGISTER citations affecting these prohibited areas, see the List of CFR Sections Affected, which appears in the Finding Aids section of the printed volume and at *www.fdsys.gov.*

FEDERAL AVIATION REGULATIONS (FAR)

PART 91

GENERAL OPERATING AND FLIGHT RULES

Contents

Subpart I—Operating Noise Limits

Subpart J—Waivers

Subpart K—Fractional Ownership Operations

Subpart L—Continued Airworthiness and Safety Improvements

Subpart M—Special Federal Aviation Regulations

Authority: 49 U.S.C. 106(g), 1155, 40103, 40113, 40120, 44101, 44111, 44701, 44704, 44709, 44711, 44712, 44715, 44716, 44717, 44722, 46306, 46315, 46316, 46504, 46506-46507, 47122, 47508, 47528-47531, articles 12 and 29 of the Convention on International Civil Aviation (61 Stat. 1180).

Link to an amendment published at 78 FR 39582, July 2, 2013.

Special Federal Aviation Regulation No. 50-2—Special Flight Rules in the Vicinity of the Grand Canyon National Park, AZ

Section 1. Applicability. This rule prescribes special operating rules for all persons operating aircraft in the following airspace, designated as the Grand Canyon National Park Special Flight Rules Area:

That airspace extending upward from the surface up to but not including 14,500 feet MSL within an area bounded by a line beginning at lat. 36°09′30″ N., long. 114°03′00″ W.; northeast to lat. 36°14′00″ N., long. 113°09′50″ W.; thence northeast along the boundary of the Grand Canyon National Park to lat. 36°24′47″ N., long. 112°52′00″ W.; to lat. 36°30′30″ N., long. 112°36′15″ W. to lat. 36°21′30″ N., long. 112°00′00″ W. to lat. 36°35′30″ N., long. 111°53′10″ W., to lat. 36°53′00″ N., long. 111°36′45″ W. to lat. 36°53′00″ N., long. 111°33′00″ W.; to lat. 36°19′00″ N., long. 111°50′50″ W.; to lat. 36°17′00″ N., long. 111°42′00″ W.; to lat. 35°59′30″ N., long. 111°42′00″ W.; to lat. 35°57′30″ N., long. 112°03′55″ W.; thence counterclockwise via the 5 statute mile radius of the Grand Canyon Airport airport reference point (lat. 35°57′09″ N., long. 112°08′47″ W.) to lat. 35°57′30″ N., long. 112°14′00″ W.; to lat. 35°57′30″ N., long. 113°11′00″ W.; to lat. 35°42′30″ N., long. 113°11′00″ W.; to 35°38′30″ N.; long. 113°27′30″ W.; thence counterclockwise via the 5 statute mile radius of the Peach Springs VORTAC to lat. 35°41′20″ N., long. 113°36′00″ W.; to lat. 35°55′25″ N., long. 113°49′10″ W.; to lat. 35°57′45″ N., 113°45′20″ W.; thence northwest along the park boundary to lat. 36°02′20″ N., long. 113°50′15″ W.; to 36°00′10″ N., long. 113°53′45″ W.; thence to the point of beginning.

Section 3. Aircraft operations: general. Except in an emergency, no person may operate an aircraft in the Special Flight Rules, Area under VFR on or after September 22, 1988, or under IFR on or after April 6, 1989, unless the operation—(a) Is conducted in accordance with the following procedures:

Note: The following procedures do not relieve the pilot from see-and-avoid responsibility or compliance with FAR 91.119.

(1) Unless necessary to maintain a safe distance from other aircraft or terrain—

(i) Remain clear of the areas described in Section 4; and

(ii) Remain at or above the following altitudes in each sector of the canyon:

Eastern section from Lees Ferry to North Canyon and North Canyon to Boundary Ridge: as prescribed in Section 5.

Boundary Ridge to Supai Point (Yumtheska Point): 10,000 feet MSL.

Western section from Diamond Creek to the Grant Wash Cliffs: 8,000 feet MSL.

(2) Proceed through the four flight corridors describe in Section 4 at the following altitudes unless otherwise authorized in writing by the Flight Standards District Office:

Northbound

11,500 or
13,500 feet MSL

Southbound

>10,500 or
>12,500 feet MSL

(b) Is authorized in writing by the Flight Standards District Office and is conducted in compliance with the conditions contained in that authorization. Normally authorization will be granted for operation in the areas described in Section 4 or below the altitudes listed in Section 5 only for operations of aircraft necessary for law enforcement, firefighting, emergency medical treatment/evacuation of persons in the vicinity of the Park; for support of Park maintenance or activities; or for aerial access to and maintenance of other property located within the Special Flight Rules Area. Authorization may be issued on a continuing basis.

(c)(1) Prior to November 1, 1988, is conducted in accordance with a specific authorization to operate in that airspace incorporated in the operator's part 135 operations specifications in accordance with the provisions of SFAR 50-1, notwithstanding the provisions of Sections 4 and 5; and

(2) On or after November 1, 1988, is conducted in accordance with a specific authorization to operate in that airspace incorporated in the operated in the operator's operations specifications and approved by the Flight Standards District Office in accordance with the provisions of SFAR 50-2.

(d) Is a search and rescue mission directed by the U.S. Air Force Rescue Coordination Center.

(e) Is conducted within 3 nautical miles of Whitmore Airstrip, Pearce Ferry Airstrip, North Rim Airstrip, Cliff Dwellers Airstrip, or Marble Canyon Airstrip at an altitudes less than 3,000 feet above airport elevation, for the purpose of landing at or taking off from that facility. Or

(f) Is conducted under an IFR clearance and the pilot is acting in accordance with ATC instructions. An IFR flight plan may not be filed on a route or at an altitude that would require operation in an area described in Section 4.

Section 4. Flight-free zones. Except in an emergency or if otherwise necessary for safety of flight, or unless otherwise authorized by the Flight Standards District Office for a purpose listed in Section 3(b), no person may operate an aircraft in the Special Flight Rules Area within the following areas:

(a) Desert View Flight-Free Zone. Within an area bounded by a line beginning at Lat. 35°59′30″ N., Long. 111°46′20″ W. to 35°59′30″ N., Long. 111°52′45″ W.; to Lat. 36°04′50″ N., Long. 111°52′00″ W.; to Lat. 36°06′00″ N., Long. 111°46′20″ W.; to the point of origin; but not including the airspace at and above 10,500 feet MSL within 1 mile of the western boundary of the zone. The area between the Desert View and Bright Angel Flight-Free Zones is designated the "Zuni Point Corridor."

(b) Bright Angel Flight-Free Zone. Within an area bounded by a line beginning at Lat. 35°59′30″ N., Long. 111°55′30″ W.; to Lat. 35°59′30″ N., Long. 112°04′00″ W.; thence counterclockwise via the 5 statute mile radius of the Grand Canyon Airport point (Lat. 35°57′09″ N., Long. 112°08′47″ W.) to Lat. 36°01′30″ N., Long. 112°11′00″ W.; to Lat. 36°06′15″ N., Long. 112°12′50″ W.; to Lat. 36°14′40″ N., Long. 112°08′50″ W.; to Lat. 36°14′40″ N., Long. 111°57′30″ W.; to Lat. 36°12′30″ N., Long. 111°53′50″ W.; to the point of origin; but not including the airspace at and above 10,500 feet MSL within 1 mile of the eastern boundary between the southern boundary and Lat. 36°04′50″ N. or the airspace at and above 10,500 feet MSL within 2 miles of the northwest boundary. The area bounded by the Bright Angel and Shinumo Flight-Free Zones is designated the "Dragon Corridor."

(c) Shinumo Flight-Free Zone. Within an area bounded by a line beginning at Lat. 36°04′00″ N., Long. 112°16′40″ W.; northwest along the park boundary to a point at Lat. 36°12′47″ N., Long. 112°30′53″ W.; to Lat. 36°21′15″ N., Long. 112°20′20″ W.; east along the park boundary to Lat. 36°21′15″ N., Long. 112°13′55″ W.; to Lat. 36°14′40″ N., Long. 112°11′25″ W.; to the point of origin. The area between the Thunder River/Toroweap and Shinumo Flight Free Zones is designated the "Fossil Canyon Corridor."

(d) Toroweap/Thunder River Flight-Free Zone. Within an area bounded by a line beginning at Lat. 36°22′45″ N., Long. 112°20′35″ W.; thence northwest along the boundary of the Grand Canyon National Park to Lat. 36°17′48″ N., Long. 113°03′15″ W.; to Lat. 36°15′00″ N., Long. 113°07′10″ W.; to Lat. 36°10′30″ N., Long. 113°07′10″ W.; thence east along the Colorado River to the confluence of Havasu Canyon (Lat. 36°18′40″ N., Long. 112°45′45″ W.;) including that area within a 1.5 nautical mile radius of Toroweap Overlook (Lat. 36°12′45″ N., Long. 113°03′30″ W.); to the point of origin; but not including the following airspace designated as the "Tuckup Corridor": at or above 10,500 feet MSL within 2 nautical miles either side of a line extending between Lat. 36°24′47″ N., Long. 112°48′50″ W. and Lat. 36°17′10″ N., Long. 112°48′50″ W.; to the point of origin.

Section 5. Minimum flight altitudes. Except in an emergency or if otherwise necessary for safety of flight, or unless otherwise authorized by the Flight Standards District Office for a purpose listed in Section 3(b), no person may operate an aircraft in the Special Flight Rules Area at an altitude lower than the following:

(a) Eastern section from Lees Ferry to North Canyon: 5,000 feet MSL.

(b) Eastern section from North Canyon to Boundary Ridge: 6,000 feet MSL.

(c) Boundary Ridge to Supai (Yumtheska) Point: 7,500 feet MSL.

(d) Supai Point to Diamond Creek: 6,500 feet MSL.

(e) Western section from Diamond Creek to the Grand Wash Cliffs: 5,000 feet MSL.

Section 9. Termination date. Section 1. Applicability, Section 4, Flight-free zones, and Section 5. Minimum flight altitudes, expire on April 19, 2001.

Note: [Removed]

[66 FR 1003, Jan. 4, 2001, as amended at 66 FR 16584, Mar. 26, 2001; 72 FR 9846, Mar. 6, 2007]

Special Federal Aviation Regulation No. 60—Air Traffic Control System Emergency Operation

1. Each person shall, before conducting any operation under the Federal Aviation Regulations (14 CFR chapter I), be familiar with all available information concerning that operation, including Notices to Airmen issued under § 91.139 and, when activated, the provisions of the National Air Traffic Reduced Complement Operations Plan available for inspection at operating air traffic facilities and Regional air traffic division offices, and the General Aviation Reservation Program. No operator may change the designated airport of intended operation for any flight contained in the October 1, 1990, OAG.

2. Notwithstanding any provision of the Federal Aviation Regulations to the contrary, no person may operate an aircraft in the Air Traffic Control System:

a. Contrary to any restriction, prohibition, procedure or other action taken by the Director of the Office of Air Traffic Systems Management (Director) pursuant to paragraph 3 of this regulation and announced in a Notice to Airmen pursuant to § 91.139 of the Federal Aviation Regulations.

b. When the National Air Traffic Reduced Complement Operations Plan is activated pursuant to paragraph 4 of this regulation, except in accordance with the pertinent provisions of the National Air Traffic Reduced Complement Operations Plan.

3. Prior to or in connection with the implementation of the RCOP, and as conditions warrant, the Director is authorized to:

a. Restrict, prohibit, or permit VFR and/or IFR operations at any airport, Class B airspace area, Class C airspace area, or other class of controlled airspace.

b. Give priority at any airport to flights that are of military necessity, or are medical emergency flights, Presidential flights, and flights transporting critical Government employees.

c. Implement, at any airport, traffic management procedures, that may include reduction of flight operations. Reduction of flight operations will be accomplished, to the extent practical, on a pro rata basis among and between air carrier, commercial operator, and general aviation operations. Flights cancelled under this SFAR at a high density traffic airport will be considered to have been operated for purposes of part 93 of the Federal Aviation Regulations.

4. The Director may activate the National Air Traffic Reduced Complement Operations Plan at any time he finds that it is necessary for the safety and efficiency of the National Airspace System. Upon activation of the RCOP and notwithstanding any provision of the FAR to the contrary, the Director is authorized to suspend or modify any airspace designation.

5. Notice of restrictions, prohibitions, procedures and other actions taken by the Director under this regulation with respect to the operation of the Air Traffic Control system will be announced in Notices to Airmen issued pursuant to § 91.139 of the Federal Aviation Regulations.

6. The Director may delegate his authority under this regulation to the extent he considers necessary for the safe and efficient operation of the National Air Traffic Control System.

(**Authority:** 49 U.S.C. app. 1301(7), 1303, 1344, 1348, 1352 through 1355, 1401, 1421 through 1431, 1471, 1472, 1502, 1510, 1522, and 2121 through 2125; articles 12, 29, 31, and 32(a) of the Convention on International Civil Aviation (61 stat. 1180); 42 U.S.C. 4321 *et seq.;* E.O. 11514, 35 FR 4247, 3 CFR, 1966-1970 Comp., p. 902; 49 U.S.C. 106(g))

[Doc. No. 26351, 55 FR 40760, Oct. 4, 1990, as amended by Amdt. 91-227, 56 FR 65652, Dec. 17, 1991]

Special Federal Aviation Regulation No. 79—Prohibition Against Certain Flights Within the Flight Information Region (FIR) of the Democratic People's Republic of Korea (DPRK)

1. *Applicability.* This rule applies to the following persons:

(a) All U.S. air carriers or commercial operators.

(b) All persons exercising the privileges of an airman certificate issued by the FAA, except such persons operating U.S.-registered aircraft for a foreign air carrier.

(c) All operators of aircraft registered in the United States except where the operator of such aircraft is a foreign air carrier.

2. *Flight Prohibition.* Except as provided in paragraphs 3 and 4 of this SFAR, no person described in paragraph 1 may conduct flight operations through the Pyongyang FIR west of 132 degrees east longitude.

3. *Permitted Operations.* This SFAR does not prohibit persons described in paragraph 1 from conducting flight operations within the Pyongyang FIR west of 132 degrees east longitude where such operations are authorized either by exemption issued by the Administrator or by another agency of the United States Government with FAA approval.

4. *Emergency situations.* In an emergency that requires immediate decision and action for the safety of the flight, the pilot in command on an aircraft may deviate from this SFAR to the extent required by that emergency. Except for U.S. air carriers and commercial operators that are subject to the requirements of 14 CFR parts 121, 125, or 135, each person who deviates from this rule shall, within ten (10) days of the deviation, excluding Saturdays, Sundays, and Federal holidays, submit to the nearest FAA Flight Standards District Office a complete report of the operations of the aircraft involved in the deviation, including a description of the deviation and the reasons therefore.

5. *Expiration.* This Special Federal Aviation Regulation No. 79 will remain in effect until further notice.

[Doc. No. 28831, 62 FR 20078, Apr. 24, 1997, as amended at 63 FR 8017, Feb. 17, 1998; 63 FR 19286, Apr. 17, 1998]

Special Federal Aviation Regulation No. 87—Prohibition Against Certain Flights Within the Territory and Airspace of Ethiopia

1. *Applicability.* This Special Federal Aviation Regulation (SFAR) No. 87 applies to all U.S. air carriers or commercial operators, all persons exercising the privileges of an airman certificate issued by the FAA unless that person is engaged in the operation of a U.S.-registered aircraft for a foreign air carrier, and all operators using aircraft registered in the United States except where the operator of such aircraft is a foreign air carrier.

2. *Flight prohibition.* Except as provided in paragraphs 3 and 4 of this SFAR, no person described in paragraph 1 may conduct flight operations within the territory and airspace of Ethiopia north of 12 degrees north latitude.

3. *Permitted operations.* This SFAR does not prohibit persons described in paragraph 1 from conducting flight operations within the territory and airspace of Ethiopia where such operations are authorized either by exemption

issued by the Administrator or by an authorization issued by another agency of the United States Government with the approval of the FAA.

4. *Emergency situations.* In an emergency that requires immediate decision and action for the safety of the flight, the pilot in command of an aircraft may deviate from this SFAR to the extent required by that emergency. Except for U.S. air carriers and commercial operators that are subject to the requirements of 14 CFR 121.557, 121.559, or 135.19, each person who deviates from this rule shall, within ten (10) days of the deviation, excluding Saturdays, Sundays, and Federal holidays, submit to the nearest FAA Flight Standards District Office a complete report of the operations of the aircraft involved in the deviation, including a description of the deviation and the reasons therefor.

5. *Expiration.* This Special Federal Aviation Regulation shall remain in effect until further notice.

[Doc. No. FAA-2000-7360; 65 FR 31215, May 16, 2000]

Special Federal Aviation Regulation No. 97—Special Operating Rules for the Conduct of Instrument Flight Rules (IFR) Area Navigation (RNAV) Operations using Global Positioning Systems (GPS) in Alaska

Those persons identified in Section 1 may conduct IFR en route RNAV operations in the State of Alaska and its airspace on published air traffic routes using TSO C145a/C146a navigation systems as the only means of IFR navigation. Despite contrary provisions of parts 71, 91, 95, 121, 125, and 135 of this chapter, a person may operate aircraft in accordance with this SFAR if the following requirements are met.

Section 1. *Purpose, use, and limitations*

a. This SFAR permits TSO C145a/C146a GPS (RNAV) systems to be used for IFR en route operations in the United States airspace over and near Alaska (as set forth in paragraph c of this section) at Special Minimum En Route Altitudes (MEA) that are outside the operational service volume of ground-based navigation aids, if the aircraft operation also meets the requirements of sections 3 and 4 of this SFAR.

b. Certificate holders and part 91 operators may operate aircraft under this SFAR provided that they comply with the requirements of this SFAR.

c. Operations conducted under this SFAR are limited to United States Airspace within and near the State of Alaska as defined in the following area description:

From 62°00′00.000″N, Long. 141°00′00.00″W.; to
Lat. 59°47′54.11″N., Long. 135°28′38.34″W.; to
Lat. 56°00′04.11″N., Long. 130°00′07.80″W.; to
Lat. 54°43′00.00″N., Long. 130°37′00.00″W.; to
Lat. 51°24′00.00″N., Long. 167°49′00.00″W.; to
Lat. 50°08′00.00″N., Long. 176°34′00.00″W.; to
Lat. 45°42′00.00″N., Long. -162°55′00.00″E.; to
Lat. 50°05′00.00″N., Long. -159°00′00.00″E.; to
Lat. 54°00′00.00″N., Long. -169°00′00.00″E.; to
Lat. 60°00 00.00″N., Long. -180°00′ 00.00″E; to
Lat. 65°00′00.00″N., Long. 168°58′23.00″W.; to
Lat. 90°00′00.00″N., Long. 00°00′0.00″W.; to
Lat. 62°00′00.000″N, Long. 141°00′00.00″W.

(d) No person may operate an aircraft under IFR during the en route portion of flight below the standard MEA or at the special MEA unless the operation is conducted in accordance with sections 3 and 4 of this SFAR.

Section 2. *Definitions and abbreviations*

For the purposes of this SFAR, the following definitions and abbreviations apply.

Area navigation (RNAV). RNAV is a method of navigation that permits aircraft operations on any desired flight path.

Area navigation (RNAV) route. RNAV route is a published route based on RNAV that can be used by suitably equipped aircraft.

Certificate holder. A certificate holder means a person holding a certificate issued under part 119 or part 125 of this chapter or holding operations specifications issued under part 129 of this chapter.

Global Navigation Satellite System (GNSS). GNSS is a world-wide position and time determination system that uses satellite ranging signals to determine user location. It encompasses all satellite ranging technologies, including GPS and additional satellites. Components of the GNSS include GPS, the Global Orbiting Navigation Satellite System, and WAAS satellites.

Global Positioning System (GPS). GPS is a satellite-based radio navigational, positioning, and time transfer system. The system provides highly accurate position and velocity information and precise time on a continuous global basis to properly equipped users.

Minimum crossing altitude (MCA). The minimum crossing altitude (MCA) applies to the operation of an aircraft proceeding to a higher minimum en route altitude when crossing specified fixes.

Required navigation system. Required navigation system means navigation equipment that meets the performance requirements of TSO C145a/C146a navigation systems certified for IFR en route operations.

Route segment. Route segment is a portion of a route bounded on each end by a fix or NAVAID.

Special MEA. Special MEA refers to the minimum en route altitudes, using required navigation systems, on published routes outside the operational service volume of ground-based navigation aids and are depicted on the published Low Altitude and High Altitude En Route Charts using the color blue and with the suffix "G." For example, a GPS MEA of 4000 feet MSL would be depicted using the color blue, as 4000G.

Standard MEA. Standard MEA refers to the minimum en route IFR altitude on published routes that uses ground-based navigation aids and are depicted on the published Low Altitude and High Altitude En Route Charts using the color black.

Station referenced. Station referenced refers to radio navigational aids or fixes that are referenced by ground based navigation facilities such as VOR facilities.

Wide Area Augmentation System (WAAS). WAAS is an augmentation to GPS that calculates GPS integrity and correction data on the ground and uses geo-stationary satellites to broadcast GPS integrity and correction data to GPS/WAAS users and to provide ranging signals. It is a safety critical system consisting of a ground network of reference and integrity monitor data processing sites to assess current GPS performance, as well as a space segment that broadcasts that assessment to GNSS users to support en route through precision approach navigation. Users of the system include all aircraft applying the WAAS data and ranging signal.

Section 3. *Operational Requirements*

To operate an aircraft under this SFAR, the following requirements must be met:

a. Training and qualification for operations and maintenance personnel on required navigation equipment used under this SFAR.

b. Use authorized procedures for normal, abnormal, and emergency situations unique to these operations, including degraded navigation capabilities, and satellite system outages.

c. For certificate holders, training of flight crewmembers and other personnel authorized to exercise operational control on the use of those procedures specified in paragraph b of this section.

d. Part 129 operators must have approval from the State of the operator to conduct operations in accordance with this SFAR.

e. In order to operate under this SFAR, a certificate holder must be authorized in operations specifications.

Section 4. *Equipment Requirements*

a. The certificate holder must have properly installed, certificated, and functional dual required navigation systems as defined in section 2 of this SFAR for the en route operations covered under this SFAR.

b. When the aircraft is being operated under part 91, the aircraft must be equipped with at least one properly installed, certificated, and functional required navigation system as defined in section 2 of this SFAR for the en route operations covered under this SFAR.

Section 5. *Expiration date*

This Special Federal Aviation Regulation will remain in effect until rescinded.

[Doc. No. FAA-2003-14305, 68 FR 14077, Mar. 21, 2003]

Special Federal Aviation Regulation No. 104—Prohibition Against Certain Flights by Syrian Air Carriers to the United States

1. *Applicability.* This Special Federal Aviation Regulation (SFAR) No. 104 applies to any air carrier owned or controlled by Syria that is engaged in scheduled international air services.

2. *Special flight restrictions.* Except as provided in paragraphs 3 and 4 of this SFAR No. 104, no air carrier described in paragraph 1 may take off from or land in the territory of the United States.

3. *Permitted operations.* This SFAR does not prohibit overflights of the territory of the United States by any air carrier described in paragraph 1.

4. *Emergency situations.* In an emergency that requires immediate decision and action for the safety of the flight, the pilot in command of an aircraft of any air carrier described in paragraph 1 may deviate from this SFAR to the extent required by that emergency. Each person who deviates from this rule must, within 10 days of the deviation, excluding Saturdays, Sundays, and Federal holidays, submit to the nearest FAA Flight Standards District Office a complete report of the operations or the aircraft involved in the deviation, including a description of the deviation and the reasons therefor.

5. *Duration.* This SFAR No. 104 will remain in effect until further notice.

[Doc. No. FAA-2004-17763, 69 FR 31719, June 4, 2004]

Special Federal Aviation Regulation No. 107—Prohibition Against Certain Flights Within the Territory and Airspace of Somalia

1. *Applicability.* This rule applies to the following persons:

(a) All U.S. air carriers or commercial operators;

(b) All persons exercising the privileges of an airman certificate issued by the FAA except such persons operating U.S.-registered aircraft for a foreign air carrier; and

(c) All operators of aircraft registered in the United States except where the operator of such aircraft is a foreign air carrier.

2. *Flight prohibition.* Except as provided below, or in paragraphs 3 and 4 of this SFAR, no person described in paragraph 1 may conduct flight operations within the territory and airspace of Somalia below flight level (FL) 200.

(a) Overflights of Somalia may be conducted above FL 200 subject to the approval of, and in accordance with the conditions established by, the appropriate authorities of Somalia.

(b) Flights departing from countries adjacent to Somalia whose climb performance will not permit operation above FL 200 prior to entering Somali airspace may operate at altitudes below FL 200 within Somalia to the extent necessary to permit a climb above FL 200, subject to the approval of, and in accordance with the conditions established by, the appropriate authorities of Somalia.

3. *Permitted operations.* This SFAR does not prohibit persons described in section 1 from conducting flight operations within the territory and airspace below FL 200 of Somalia when such operations are authorized either by another agency of the United States Government with the approval of the FAA or by an exemption issued by the Administrator.

4. *Emergency situations.* In an emergency that requires immediate decision and action for the safety of the flight, the pilot in command of an aircraft may deviate from this SFAR to the extent required by that emergency. Except for U.S. air carriers and commercial operators that are subject to the requirements of Title 14 CFR parts 119, 121, or 135, each person who deviates from this rule must, within 10 days of the deviation, excluding Saturdays, Sundays, and Federal holidays, submit to the nearest FAA Flight Standards District Office a complete report of the operations of the aircraft involved in the deviation, including a description of the deviation and the reasons for it.

5. *Expiration.* This Special Federal Aviation Regulation will remain in effect until further notice.

[Doc. No. FAA-2007-27602, 72 FR 16712, Apr. 5, 2007]

Special Federal Aviation Regulation No. 108—Mitsubishi MU-2B Series Special Training, Experience, and Operating Requirements

1. *Applicability.* After February 5, 2009, this Special Federal Aviation Regulation (SFAR) applies to all persons who operate the Mitsubishi MU-2B series airplane including those who act as pilot-in-command, act as second-in-command, or other persons who manipulate the controls while under the supervision of a pilot-in-command. This SFAR also applies to those persons who provide pilot training for the Mitsubishi MU-2B series airplane. The requirements in this SFAR are in addition to the requirements of 14 CFR parts 61, 91, and 135 of this chapter.

2. *Compliance and Eligibility.* (a) Except as provided in paragraph (b) of this section, no person may manipulate the controls, act as pilot-in-command, act as second-in-command, or provide pilot training for the Mitsubishi MU-2B series airplane unless that person meets the applicable requirements of this SFAR.

(b) A person, who does not meet the requirements of this SFAR, may manipulate the controls of the Mitsubishi MU-2B series airplane if a pilot-in-command meeting the applicable requirements of this SFAR is occupying a pilot station, and the flight is being conducted for one of the following reasons—

(1) The pilot-in-command is providing pilot training to the manipulator of the controls, and no passengers or cargo are carried on board the airplane;

(2) The pilot-in-command is conducting a maintenance test flight with a second pilot or certificated mechanic, and no passengers or cargo are carried on board the airplane; or

(3) The pilot-in-command is conducting a simulated instrument flight and is using a safety pilot other than the pilot-in-command who manipulates the controls for the purposes of 14 CFR 91.109, and no passengers or cargo are carried on board the airplane.

(c) A person is required to complete *Initial/transition training* if that person has fewer than—

(1) 50 hours of documented flight time manipulating the controls while serving as pilot-in-command of a Mitsubishi MU-2B series airplane in the preceding 24 months; or

(2) 500 hours of documented flight time manipulating the controls while serving as pilot-in-command of a Mitsubishi MU-2B series airplane.

(d) A person is eligible to receive *Requalification training* in lieu of Initial/transition training if that person has at least—

(1) 50 hours of documented flight time manipulating the controls while serving as pilot-in-command of a Mitsubishi MU-2B series airplane in the preceding 24 months; or

(2) 500 hours of documented flight time manipulating the controls while serving as pilot-in-command of a Mitsubishi MU-2B series airplane.

(e) A person is required to complete *Recurrent training* within the preceding 12 months. Successful completion of Initial/transition or Requalification training within the preceding 12 months satisfies the requirement of Recurrent training. A person must successfully complete Initial/transition training or Requalification training before being eligible to receive Recurrent training.

(f) Successful completion of Initial/transition training or Requalification training is a one-time requirement. A person may elect to retake Initial/transition training or Requalification training in lieu of Recurrent training.

(g) A person is required to complete Differences training if that person operates more than one MU-2B model. Differences training between the K and M models of the MU-2B airplane, and the J and L models of the MU-2B airplane, may be accomplished with Level A training. All other Differences training must be accomplished with Level B training. Persons that are operating two models of the MU-2B airplane are required to receive 1.5 hours of Differences training. Persons that are operating three or more models of the MU-2B airplane are required to receive 3.0 hours of Differences training. An additional 1.5 hours of Differences training is required for each model added at a later date. Differences Training is not a recurring annual requirement. Once a person has received Differences training between the applicable different models, no additional Differences training between those models is required.

3. *Required Pilot Training.* (a) Except as provided in section 2 paragraph (b) of this SFAR, no person may manipulate the controls, act as pilot-in-command, or act as second-in-command of a Mitsubishi MU-2B series airplane for the purpose of flight unless—

(1) The applicable requirements for ground and flight training on Initial/transition, Requalification, Recurrent, and Differences training have been completed, as specified in this SFAR, including Appendices A through D of this SFAR; and

(2) That person's logbook has been endorsed in accordance with paragraph (f) of this section.

(b) No person may manipulate the controls, act as pilot-in-command, or act as second-in-command, of a Mitsubishi MU-2B series airplane for the purpose of flight unless—

(1) That person satisfactorily completes, if applicable, annual Recurrent pilot training on the *Special Emphasis Items,* and all items listed in the *Training Course Final Phase Check* as specified in Appendix C of this SFAR; and

(2) That person's logbook has been endorsed in accordance with paragraph (f) of this section.

(c) Satisfactory completion of the competency check required by 14 CFR 135.293 within the preceding 12 calendar months may not be substituted for the Mitsubishi MU-2B series airplane annual recurrent flight training of this section.

(d) Satisfactory completion of a Federal Aviation Administration sponsored pilot proficiency award program, as described in 14 CFR 61.56(e) may not be substituted for the Mitsubishi MU-2B series airplane annual recurrent flight training of this section.

(e) If a person complies with the requirements of paragraph (a) or (b) of this section in the calendar month before or the calendar month after the month in which compliance with these paragraphs are required, that person is considered to have accomplished the training requirement in the month the training is due.

(f) The endorsement required under paragraph (a) and (b) of this section must be made by—

(1) A certificated flight instructor meeting the qualifications of section 5 of this SFAR; or

(2) For persons operating the Mitsubishi MU-2B series airplane for a part 119 certificate holder within the last 12 calendar months, the 14 CFR part 119 certificate holder's flight instructor if authorized by the FAA and if that flight instructor meets the requirements of section 5 of this SFAR.

(g) All training conducted for the Mitsubishi MU-2B series airplane must be completed in accordance with the applicable MU-2B series checklist listed in table 1 of this SFAR or an MU-2B series airplane checklist that has been accepted by the Federal Aviation Administration's MU-2B Flight Standardization Board.

Table 1 to SFAR 108—MU-2B Series Airplane Manufacturer's Checklists

Model	Type certificate	Cockpit checklist MHI document No.	Date the checklist was accepted by the FSB
MU-2B-60	A10SW	YET06220C	2/12/2007
MU-2B-40	A10SW	YET06256A	2/12/2007
MU-2B-36A	A10SW	YET06257B	2/12/2007
MU-2B-36	A2PC	YET06252B	2/12/2007
MU-2B-35	A2PC	YET06251B	2/12/2007
MU-2B-30	A2PC	YET06250A	3/2/2007
MU-2B-26A	A10SW	YET06255A	2/12/2007
MU-2B-26	A2PC	YET06249A	3/2/2007
MU-2B-26	A10SW	YET06254A	3/2/2007
MU-2B-25	A10SW	YET06253A	3/2/2007
MU-2B-25	A2PC	YET06248A	3/2/2007
MU-2B-20	A2PC	YET06247A	2/12/2007
MU-2B-15	A2PC	YET06246A	3/2/2007
MU-2B-10	A2PC	YET06245A	3/2/2007
MU-2B	A2PC	YET06244A	3/2/2007

4. *Aeronautical Experience.* No person may act as pilot-in-command of a Mitsubishi MU-2B series airplane for the purpose of flight unless that person holds an airplane category and multi-engine land class rating, and has logged a minimum of 100 flight hours of pilot-in-command time in multi-engine airplanes.

5. *Instruction, Checking and Evaluation.* (a) *Flight Instructor (Airplane).* No flight instructor may provide instruction or conduct a flight review in a Mitsubishi MU-2B series airplane unless that flight instructor meets the requirements of this paragraph.

(1) Each flight instructor who provides flight training in the Mitsubishi MU-2B series airplane must meet the pilot training and documentation requirements of section 3 of this SFAR before giving flight instruction in the Mitsubishi MU-2B series airplane.

(2) Each flight instructor who provides flight training in the Mitsubishi MU-2B series airplane must meet the currency requirements of paragraphs (a) and (c) of section 6 of this SFAR before giving flight instruction in the Mitsubishi MU-2B series airplane.

(3) Each flight instructor who provides flight training in the Mitsubishi MU-2B series airplane must have a minimum total pilot time of 2,000 pilot-in-command hours, 800 pilot-in-command hours in multiengine airplanes.

(4) Each flight instructor who provides flight training in the Mitsubishi MU-2B series airplane must have—

(i) 300 pilot-in-command hours in the Mitsubishi MU-2B series airplane, 50 hours of which must have been within the preceding 12 months; or

(ii) 100 pilot-in-command hours in the Mitsubishi MU-2B series airplane, 25 hours of which must have been within the preceding 12 months, and 300 hours providing instruction in a FAA-approved Mitsubishi MU-2B simulator or FAA-approved Mitsubishi MU-2B flight training device, 25 hours of which must have been within the preceding 12 months.

(b) *Flight Instructor (Simulator/ Flight Training Device).* No flight instructor may provide instruction for the Mitsubishi MU-2B series airplane unless that instructor meets the requirements of this paragraph.

(1) Each flight instructor who provides flight training for the Mitsubishi MU-2B series airplane must meet the pilot training and documentation requirements of section 3 of this SFAR before giving flight instruction for the Mitsubishi MU-2B series airplane.

(2) Each flight instructor who provides flight training for the Mitsubishi MU-2B series airplane must meet the currency requirements of paragraph (c) of section 6 of this SFAR before giving flight instruction for the Mitsubishi MU-2B series airplane.

(3) Each flight instructor who provides flight training for the Mitsubishi MU-2B series airplane must have—

(i) A minimum total pilot time of 2000 pilot-in-command hours and 800 pilot-in-command hours in multiengine airplanes; and

(ii) Within the preceding 12 months, either 50 hours of Mitsubishi MU-2B series airplane pilot-in-command experience or 50 hours providing simulator or flight training device instruction for the Mitsubishi MU-2B.

(c) *Checking and Evaluation.* No person may provide checking or evaluation for the Mitsubishi MU-2B series airplane unless that person meets the requirements of this paragraph.

(1) For the purpose of checking, designated pilot examiners, training center evaluators, and check airmen must have completed the appropriate training in the Mitsubishi MU-2B series airplane in accordance with section 3 of this SFAR.

(2) For checking conducted in the Mitsubishi MU-2B series airplane, each designated pilot examiner and check airman must have 100 hours pilot-in-command flight time in the Mitsubishi MU-2B series airplane and maintain currency in accordance with section 6 of this SFAR.

6. *Currency Requirements and Flight Review.* (a) The takeoff and landing currency requirements of 14 CFR 61.57 must be maintained in the Mitsubishi MU-2B series airplane. Takeoff and landings in other multiengine airplanes do not meet the takeoff landing currency requirements for the Mitsubishi MU-2B series airplane. Takeoff and landings in either the short-body or long-body Mitsubishi MU-2B model airplane may be credited toward takeoff and landing currency for both Mitsubishi MU-2B model groups.

(b) Instrument experience obtained in other category and class of aircraft may be used to satisfy the instrument currency requirements of 14 CFR 61.57 for the Mitsubishi MU-2B series airplane.

(c) Satisfactory completion of a flight review to satisfy the requirements of 14 CFR 61.56 is valid for operation of a Mitsubishi MU-2B series airplane only if that flight review is conducted in a Mitsubishi MU-2B series airplane. The flight review for Mitsubishi MU-2B series airplanes must include the *Special Emphasis Items,* and all items listed in the *Training Course Final Phase Check* of Appendix C of this SFAR.

(d) A person who successfully completes the Initial/transition, Requalification, or Recurrent training requirements, as described in section 3 of this SFAR, also meets the requirements of 14 CFR 61.56 and need not accomplish a separate flight review provided that at least 1 hour of the flight training was conducted in the Mitsubishi MU-2B series airplane.

7. *Operating Requirements.* (a) Except as provided in paragraph (b) of this section, no person may operate a Mitsubishi MU-2B airplane in single pilot operations unless that airplane has a functional autopilot.

(b) A person may operate a Mitsubishi MU-2B airplane in single pilot operations without a functional autopilot when—

(1) Operating under day visual flight rule requirements; or

(2) Authorized under a FAA approved minimum equipment list for that airplane, operating under instrument flight rule requirements in daytime visual meteorological conditions.

(c) No person may operate a Mitsubishi MU-2B series airplane unless a copy of the appropriate Mitsubishi Heavy Industries MU-2B Airplane Flight Manual is carried on board the airplane and is accessible during each flight at the pilot station.

(d) No person may operate a Mitsubishi MU-2B series airplane unless an MU-2B series airplane checklist, appropriate for the model being operated and accepted by the Federal Aviation Administration MU-2B Flight Standardization Board, is accessible for each flight at the pilot station and is used by the flight crewmembers when operating the airplane.

(e) No person may operate a Mitsubishi MU-2B series airplane contrary to the MU-2B training program in the Appendices of this SFAR.

(f) If there are any differences between the training and operating requirements of this SFAR and the MU-2B Airplane Flight Manual's procedures sections (Normal, Abnormal, and Emergency) and the MU-2B airplane series checklist specified in section 3(g), table 1, the person operating the airplane must operate the airplane in accordance with the training specified in section 3(g), table 1.

8. *Credit for Prior Training.* Initial/transition or requalification training conducted between July 27, 2006, and April 7, 2008, using Mitsubishi Heavy Industries MU-2B Training Program, Part number YET 05301, Revision Original, dated July 27, 2006, or Revision 1, dated September 19, 2006, is considered to be compliant with this SFAR, if the student met the eligibility requirements for the applicable category of training and the student's instructor met the experience requirements of this SFAR.

9. *Incorporation by Reference.* You must proceed in accordance with the Mitsubishi Heavy Industries MU-2B Checklists as listed in Table 1 of this SFAR which are incorporated by reference. The Director of the Federal Register approved this incorporation by reference in accordance with 5 U.S.C. section 552(a) and 1 CFR part 51. The Mitsubishi Heavy Industries MU-2B Checklists are distributed by Turbine Aircraft Services, Inc. You may obtain a copy from Turbine Aircraft Services Inc., 4550 Jimmy Doolittle Drive, Addison, Texas 75001, USA. You may inspect a copy at U.S. Department of Transportation, Docket Management Facility, Room W 12-140, West Building Ground Floor, 1200 New Jersey Ave., SE., Washington, DC 20590-0001, or at the National Archives and Records Administration at NARA, call 202-741-6030, or go to: *http://www.archives.gov/federal_register/code_of_federal_regulations/ibr_locations.html.*

10. *Expiration.* This SFAR will remain in effect until further notice.

Appendix A to SFAR 108—MU-2B General Training Requirements

(a) The Mitsubishi MU-2B Training Program consists of both ground and flight training. The minimum pilot training requirement hours are shown in Table 1 of this appendix for ground instruction and Table 2 of this appendix for flight instruction. An additional ground training requirement for Differences Training is shown in Table 3.

(b) The MU-2B is certificated by the Federal Aviation Administration (FAA) as a single pilot airplane. No training credit is given for second in command (SIC) training and no credit is given for right seat time under this program. Only the sole manipulator of the controls of the MU-2B airplane, Flight Training Device (FTD), or Level C or D simulator can receive training credit under this program.

(c) The training program references the applicable MU-2B airplane flight manual (AFM) in several sections. There may be differences between sequencing of procedures found in the AFM's procedures sections and the checklists, procedures, and techniques found within this training program. The FAA's Mitsubishi MU-2B SFAR requires that if there are any differences between the AFM's procedures sections (Normal, Abnormal, and Emergency) and the training and operating requirements of the Mitsubishi MU-2B SFAR, the person operating the airplane must operate the airplane in accordance with the training specified in the SFAR and this MU-2B training program.

(d) Minimum Programmed Training Hours

Table 1 to Appendix A of SFAR 108

Ground instruction		
Initial/transition	Requalificaton	Recurrent
20 hours	12 hours	8 hours.

Table 2 to Appendix A of SFAR 108

Flight instruction		
Initial/transition	Requalification	Recurrent
12 hours with a minimum of 6 hours at Level E	8 hours Level C or Level E	4 hours at Level E, or 6 hours at Level C.

Table 3 to Appendix A of SFAR 108

Differences training	
2 models currently	1.5 hours at Level A or B.
More than 2 models currently	3 hours at Level A or B.
Each additional model added	1.5 hours at Level A or B.

(e) Definitions of Levels of Training as Used in This Appendix

(1) LEVEL A Training—Training that is conducted through self instruction by the pilot.

(2) LEVEL B Training—Training that is conducted in the classroom environment with the aid of a qualified instructor who meets the requirements of this SFAR.

(3) LEVEL C Training—Training that is accomplished in an FAA-approved Level 5, 6, or 7 Flight Training Device (FTD). In addition to the basic FTD requirements, the FTD must be representative of the MU-2B cockpit controls and be specifically approved by the FAA for the MU-2B airplane.

(4) LEVEL E Training—Training that must be accomplished in the MU-2B airplane, Level C simulator, or Level D simulator.

Appendix B to SFAR 108—MU-2B Ground Training Curriculum Contents

All items in the ground training curriculum must be covered. The order of presentation is at the discretion of the instructor. The student must satisfactorily complete a written or oral exam given by the training provider based on this MU-2B Training Program.

I. Aircraft General
A. Introduction
B. Airplane (Structures/Aerodynamics/Engines) Overview
1. Fuselage
2. Wing
3. Empennage
4. Doors
5. Windshield and Windows
C. Airplane Systems
1. Electrical Power
2. Lighting
3. Fuel System
4. Powerplant
5. Environmental
6. Fire Protection
7. Ice and Rain Protection
8. Landing Gear and Brakes
9. Flight Controls and Trim
10. Pilot Static System/Flight Instruments
11. Oxygen System
D. Operating Limitations
1. Weights
2. Center of Gravity and Loading
3. Airspeeds
4. Maneuvering Load Factors
5. Takeoff And Landing Operations
6. Enroute Operations
E. Required Placards
F. Instrument Markings
G. Flight Characteristics
1. Control System
2. Stability and Stall Characteristics
3. Single Engine Operation
4. Maneuvering and Trim
5. Takeoff and Landing
II. Electrical Power
A. General Description
B. DC Electrical System
1. DC Power Generation
2. DC Power Distribution
3. Battery System
4. External Power System
C. AC Electrical System
1. AC Power Generation
2. Controls and Indicators
3. AC Power Distribution
D. Limitations
1. General Limitations
2. Instrument Markings
III. Lighting
A. Exterior Lighting System
1. Navigation Lights
2. Anti-Collision Lights
3. Wing Inspection Lights
4. Taxi Lights
5. Landing Lights
6. Rotating Beacon
7. Operation
B. Interior Lighting System
1. Flight Compartment Lights
2. Passenger Compartment Lights
C. Emergency Lighting System
1. Cockpit Emergency Lighting
2. Aircraft Emergency Lighting
D. Procedures
1. Normal
2. Abnormal
3. Emergency
IV. Master Caution System
A. System Description and Operation
1. Master Caution Light and Reset Switch
2. Annunciator and Indicator Panels
3. Operation Lights
4. System Tests
B. Procedures
V. Fuel System
A. Fuel Storage
1. Refueling/Balancing
2. De-Fueling and Draining
3. Tank Vent System
B. Fuel Distribution
1. Fuel Transfer
2. Fuel Balancing
3. Boost Pump Operation
C. Fuel Indicating
1. Fuel Quantity
2. Low Fuel Warning
D. Fuel System Limitations
1. Approved Fuels
2. Fuel Anti-Icing Additives
3. Fuel Temperature Limitations
4. Fuel Transfer and Fuel Imbalance
5. Fuel Pumps
6. Refueling
7. Capacity
8. Unusable Fuel
VI. Powerplant
A. Engine Description
1. Major Sections
2. Cockpit Controls
3. Instrumentation
4. Operation
B. Engine Systems
1. Lubrication
2. Fuel
3. Ignition
4. Engine Starting
5. Anti-Ice
C. Propeller System
1. Ground Operations
2. In-Flight Operations
3. Synchronization
4. De-Ice
D. Ground Checks
1. Overspeed Governor
2. SRL and Delta P/P
3. NTS and Feather Valve
4. Supplementary NTS
E. In Flight Post Maintenance Checks
1. NTS In-Flight
2. Flight Idle Fuel Flow
F. Limitations
1. Powerplant
2. Engine Starting Conditions
3. Airstart Envelope
4. Engine Starting
5. Oil
6. Fuel
7. Starter/Generator
8. External Power
9. Instrument Markings (as applicable)
a. TPE331-10-511M
b. TPE331-5/6-252/251M
c. TPE331-1-151M
G. Engine Malfunctions and Failures
1. Propeller Coupling
2. Torque Sensor
3. Engine Overspeed
4. Fuel Control Spline
VII. Fire Protection
A. Introduction
B. Engine Fire Detection
1. System Description
2. Annunciator
C. Portable Fire Extinguishers
VIII. Pneumatics
A. System Description
B. System Operation
1. Air Sources
2. Limitations
C. Wing and Tail De-Ice
1. System Description
2. Controls
D. Entrance and Baggage Door Seal

1. Air Source
2. Operation
IX. Ice and Rain Protection
A. General Description
B. Wing De-Ice
1. System Description
2. Operation
3. Controls and Indications
C. Engine Anti-Ice
1. System Description
2. Operation
3. Controls and Indications
D. Window Defog
1. Controls
2. Operation
E. Tail De-Ice
1. Horizontal Stabilizer De-Ice
2. Vertical Stabilizer De-Ice
F. Pitot Static System Anti-Icing
1. Pitot Tube Heating
2. Static Port Heating
3. AOA Transmitter Heating
G. Windshield De-Ice/Anti-Ice
1. System Description
2. Controls and Indications
H. Windshield Wiper
1. System Description
2. Control and Operation
I. Propeller De-Ice
1. System Description
2. Controls and Indications
J. Ice Detector
1. System Description
2. Controls and Indications
3. Operation
K. Limitations
1. Temperatures
2. Cycling
X. Air Conditioning
A. System Description and Operation
1. Refrigeration Unit (ACM)
2. Air Distribution
3. Ventilation
4. Temperature Control
5. Water Separator
B. Limitations
XI. Pressurization
A. General
B. Component Description
1. Cabin Pressure Controller
2. Altitude Pressure Regulator
3. Ram Air
4. Outflow Safety Valves
5. Air Filters
6. Manual Control Valve
7. Pneumatic Relays
8. Venturi
C. System Operation
1. Ground Operation
2. Takeoff Mode
3. In-Flight Operation
4. Landing Operation
D. Emergency Operation
1. High Altitude
2. Low Altitude
E. Limitations
1. Maximum Differential
2. Landing Limitations
XII. Landing Gear and Brakes
A. General Description
1. Landing Gear Doors
2. Controls and Indicators
3. Warning Systems
4. Emergency Extension
B. Nosewheel Steering
C. Landing Gear/Brakes/Tires
D. Limitations
1. Airspeed (with flaps)
2. Emergency Extension
3. Tire Speed
4. Brake Energy
XIII. Flight Controls
A. Primary Flight Controls (Elevator/Rudder/Spoilers)
1. Description
2. Operations
B. Trim Systems
1. System Description
2. Roll Trim
a. Normal Operation
b. Emergency Operation
3. Rudder Trim
4. Pitch Trim
a. General
b. Operations
c. Trim-in-Motion Alert System
C. Secondary Flight Controls
1. System Description
2. Flaps
D. Limitations
1. Instrument Markings
2. Placards
E. Flight Characteristics
1. Control Systems
2. Stability and Stall Characteristics
3. Single Engine Operation
5. Maneuvering and Trim
6. Takeoff and Landing
XIV. Avionics
A. Pitot-Static System
1. System Description
2. Pilot's System
3. Co-Pilot's System
4. Alternate Static
B. Air Data Computer
C. Attitude Instrument Displays (EFIS and Standard)
1. EADI
2. Standard Attitude Gyro
D. AHRS
1. System Description
2. Controls and Indications
E. Navigation
1. Nav Systems Descriptions
2. Compass System Descriptions
3. Display Systems
4. Terrain Awareness System
5. Traffic Avoidance System
F. Communications
1. VHF Communications Systems
2. Audio Control
G. Standby Flight Instruments
1. System Description
2. Controls and Indications
H. Automatic Flight Control System
1. Controls and Indications
2. Yaw Damper
3. Trim-in-Motion Alert System
4. Autopilot Automatic Disconnect
5. Aural Alert System
I. Angle of Attack (AOA) System
1. System Description
2. Controls and Indications
J. Limitations
XV. Oxygen System
A. System Description
B. Crew Oxygen
1. Oxygen Cylinder Assembly
2. Pressure Gauge
3. Outlet Valves
4. Duration
C. Passenger Oxygen
1. System Description
2. Duration
D. Limitations
XVI. Performance and Planning
A. Takeoff Performance Charts
1. Runway Requirements
2. Normal and with One Engine Inoperative
B. Climb Performance
1. Normal and with One Engine Inoperative
2. Obstacle Clearance
3. Power Assurance Charts
C. Cruise Performance
1. Power Charts
2. Maximum Practical Altitude
3. Cruise Speeds/Engine Health
4. Buffet Boundary
D. Landing Performance
1. Runway Requirements
a. Dry Runway
b. Wet Runway
2. Go-Around
a. One Engine Inoperative
b. All Engines
XVII. Weight and Balance
A. Aircraft Loading Procedures
B. Limitations
1. Weight Limits
2. C.G. Limits
C. Plotter
1. Description
2. Use
D. Calculations
1. AFM Procedures
2. Examples
XVIII. General Subjects
A. Controlled Flight into Terrain Awareness
B. CRM/SPRM
1. Crew Resource Management
2. Single Pilot Resource Management
C. MU-2B Flight Standardization Board Report

Appendix C to SFAR 108—MU-2B Final Phase Check and Flight Training Requirements

(I) *MU-2B Final Phase Check Requirements*

(A) Completion of the MU-2B Training Program in this appendix requires successful completion of a final phase check taken in the MU-2B airplane or a Level C or D simulator for Initial/Transition training. The final phase check for Requalification or Recurrent Training may be taken in the MU-2B airplane, a Level C or D simulator, or in a Level 5, 6, or 7 FAA-approved MU-2B Flight Training Device (FTD). The final phase check must be conducted by a qualified flight instructor who meets the requirements of the MU-2B SFAR. Simultaneous training and checking is not allowed for Initial/Transition training.

(B) For pilots operating under 14 CFR part 135, checking must be done in accordance with applicable regulations. For the purpose of recurrent testing in 14 CFR 135.293(b), the MU-2B is considered a separate type of aircraft.

(C) The final phase check must be conducted using the standards contained in the FAA Commercial Pilot—Airplane Multi-Engine Land, and Instrument Rating—Airplane Practical Test Standards (PTS).

(D) The final phase check portion of the training is comprised of the following tasks for *all* airmen (instrument rated and non instrument rated). An (*) indicates those maneuvers for Initial/Transition training which must be completed in the MU-2B airplane, or a Level C or D simulator.

(1) Preflight Check.
(2) Start and Taxi Procedures.
(3) *Normal Takeoff (X-Wind) (Two Engine).
(4) *Takeoff Engine Failure.
(5) Rejected Takeoff.
(6) *Steep Turns.
(7) *Approach to Stalls (3) (must include Accelerated Stalls).
(8) *Maneuvering with One Engine Inoperative—Loss of Directional Control (Vmc).
(9) Abnormal and Emergency Procedures—To include MU-2B operation in icing conditions without the autopilot or without trim-in-motion or automatic autopilot disconnect.
(10) *Precision Approach (One Engine Inoperative).
(11) Go Around/Rejected Landing.
(12) Normal Landing (X-Wind).
(13) *Landing with One Engine Inoperative.
(14) *Landing with Non-Standard Flap Configuration (0 or 5 degrees).
(15) Postflight Procedures.

(E) The following additional tasks are required for those airmen who possess an instrument rating. An (*) indicates those maneuvers for Initial/Transition training which must be completed in the MU-2B airplane, or a Level C or D simulator.

(1) Preflight Check.
(2) Unusual Attitudes.
(3) Abnormal and Emergency Procedures.
(4) Basic Instrument Flight Maneuvers.
(5) Area Arrival and Departure.
(6) Holding.
(7) Precision Approach (Two Engine).
(8) *Non-Precision Approaches (2)—Must include a Non-Precision Approach with One Engine Inoperative.
(9) Missed Approach from either Precision or Non Precision Instrument Approach (Two Engine).
(10) Landing from a Straight-In or Circling Approach.
(11) Circling Approach.
(12) Postflight Procedures.

(F) A form titled "Training Course Final Phase Check" has been included in this appendix for use in creating a training and final check record for the student and the training provider.

(II) *MU-2B Required Flight Training Tasks*

(A) General Flight Training Requirements: All flight training maneuvers must be consistent with this training program and the applicable MU-2B checklist accepted by the FAA. The maneuver profiles shown in Appendix D to this SFAR No. 108 are presented to show the required training scenarios. Profiles conducted in flight require planning and care on the part of both the instructor and student in order to provide the highest level of safety possible. The maneuver profiles shown in Appendix D to this SFAR No. 108 do not account for local geographic and flight conditions. The instructor and student must consider local conditions when performing these maneuvers in flight.

(B) Special Emphasis Items: Certain aspects of pilot knowledge, skills and abilities must be emphasized and evaluated during the training and checking process of the MU-2B Training Program.

(1) Accelerated stall awareness and recovery procedures with emphasis on configuration management. Awareness of the margin to stall in all flight operations and configurations must be emphasized throughout training.

(2) Vmc awareness and early recognition must be trained and checked. Minimum airspeeds for one engine inoperative must be emphasized in all configurations.

(3) Airspeed management and recognition of airspeed deterioration below recommended speeds and recovery methods in this training program must be emphasized throughout training and checking.

(4) Knowledge of icing conditions and encounters must be emphasized throughout training and checking including: Equipment requirements, certification standards, minimum airspeeds, and the use of the autopilot and other applicable AFM procedures.

(5) Airplane performance characteristics with all engines operating and with one engine inoperative must be emphasized.

(C) MU-2B Flight Training Program Proficiency Standards.

(1) Each pilot, regardless of the level of pilot certificate held, must be trained to and maintain the proficiency standards described below.

(a) General VFR/IFR.

(i) Bank Angle—±5 degrees of prescribed bank angle

(ii) Heading—±10 degrees

(iii) Altitude—±100 feet

(iv) Airspeed—±10 knots

(b) Instrument Approach—Final Approach Segment.

Precision Approach

(i) Heading—±10 degrees

(ii) Altitude—±100 feet

(iii) Airspeed—±10 knots prior to final

(iv) Airspeed—±10 knots after established on final

(v) Glide Slope (GS)/Localizer Deviation—Within 3/4 scale—not below GS

Non-Precision Approach

Straight In

(vi) Initial Approach Altitude—±100 feet

(vii) Heading—±10 degrees

(viii) Altitude (MDA)—+ 100, -0 feet

(ix) Airspeed—+ 10 knots

(x) Course Deviation Indicator—Within 3/4 scale or ±10 degrees on RMI

Circling Approach

(xi) Maximum Bank—30 degrees

(xii) Heading—Within 10 degrees

(xiii) Altitude—+100, -0 feet

(xiv) Airspeed—Within 10 knots but not less than Vref

(c) In all cases, a pilot must show complete mastery of the aircraft with the outcome of each maneuver or procedure never seriously in doubt.

(D) Maneuvers and Procedures. All flight training maneuvers and procedures must be conducted as they are applicable to the MU-2B and each type of operations involved.

Preflight

(1) Preflight Inspection—The pilot must—

(a) Conduct an actual visual inspection of the exterior and interior of the airplane, locating each item and explaining briefly the purpose of inspecting it; and

(b) Demonstrate the use of the appropriate checklist, appropriate control system checks, starting procedures, radio and electronic equipment checks, and the selection of proper navigation and communications radio facilities and frequencies prior to flight.

(2) Taxiing—this maneuver includes taxiing in compliance with instructions issued by the appropriate ATC facility or by the person conducting the check.

(3) Pre-Takeoff Checks—The pilot must satisfactorily complete all pre-takeoff aircraft systems and powerplant checks before takeoff.

Takeoff and Departure

(1) Normal—One normal takeoff, which for the purpose of this maneuver, begins when the airplane is taxied into position on the runway to be used.

(2) Instrument Takeoff—Takeoff with simulated instrument conditions at or before reaching an altitude of 200 feet above the airport elevation and visibility of 1800 RVR.

(3) Crosswind—One crosswind takeoff, if practical, under the existing meteorological, airport and traffic conditions.

(4) Powerplant Failure—One takeoff with a simulated failure of the most critical powerplant at a point after Vlof. In the MU-2B airplane, all simulated powerplant failures must only be initiated when the person conducting the training or checking determines that it is safe under the prevailing conditions. The instructor must assure that the power lever does not move beyond the flight idle gate.

(5) Rejected Takeoff—A rejected takeoff performed in an airplane during a normal takeoff run after reaching a reasonable speed determined by giving due consideration to aircraft characteristics, runway length, surface conditions, wind direction and velocity, brake heat energy, and any other pertinent factors that may adversely affect safety or the airplane.

(6) Area departure—Demonstrate adequate knowledge of departure procedures, establishing appropriate ATC communications and following clearances.

Flight Maneuvers and Procedures

(1) Steep bank turns—Each steep turn must involve a bank angle of 50 degrees with a heading change of at least 180 degrees but no more than 360 degrees.

(2) Approaches to stalls—Must be performed in each of the following configurations; takeoff, clean, and landing. One approach to a stall must be performed in either the takeoff, clean, or landing configuration while in a turn with a bank angle between 15 degrees and 30 degrees.

(3) Accelerated stalls—must be done in the flaps 20 and flaps 0 configurations.

(4) Recovery procedures must be initiated at the first indication of a stall.

Normal and Abnormal Procedures and Operations

(1) Runway trim.

(2) Normal and abnormal operations of the following systems:

(a) Pressurization.

(b) Pneumatic.

(c) Air conditioning.

(d) Fuel.

(e) Electrical.

(f) Flight control.

(g) Anti-icing and de-icing.

(h) Autopilot.

(i) Stall warning devices, as applicable.

(j) Airborne radar and weather detection devices.

(k) Other systems, devices or aids available.

(l) Electrical, flight control and flight instrument system malfunction or failure.

(m) Landing gear and flap system malfunction or failure.

(n) Failure of navigation or communications equipment.

Flight Emergency Procedures

(1) Powerplant failure.

(2) Powerplant, cabin, flight deck, wing and electrical fires.

(3) Smoke control.

(4) Fuel jettisoning, as applicable.

(5) Any other emergency procedures outlined in the appropriate AFM or FAA-accepted checklist.

Instrument Procedures

(1) Area departure.

(2) Use of navigation systems including adherence to assigned course and/or radial.

(3) Holding procedures.

(4) Aircraft approach category airspeeds.

(5) Approach procedures: Each instrument approach must be performed according to all procedures and limitations approved for that facility. An instrument approach procedure begins when the airplane is over the initial approach fix for the approach procedure being used and ends when the airplane touches down on the runway or when transition to missed approach configuration is completed.

(a) ILS, ILS/DME, approach.

(i) A manually controlled ILS with a powerplant inoperative; occurring before initiating the final approach course and continuing to full stop or through the missed approach procedure.

(ii) A manually controlled ILS utilizing raw data to 200 feet or decision height (DH).

(iii) An ILS with the autopilot coupled.

(b) Non-precision approaches.

(i) NDB, NDB/DME approach, straight in or circle.

(ii) VOR, VOR/DME, straight in or circle.

(iii) LOC, LOC/DME, LOC backcourse.

(iv) GPS approach (If the aircraft/FTD/flight simulator has a GPS installed, the applicant must demonstrate GPS approach proficiency.)

(v) ASR approach.

(c) Missed approach procedure: One missed approach procedure must be a complete approved missed approach procedure as published or as assigned by ATC.

(i) From a precision approach.

(ii) From a non-precision approach.

(iii) With a simulated powerplant failure.

(d) Circling approach.

(i) The circling approach must be made to the authorized MDA and followed by a change in heading and the necessary maneuvering (by visual reference) to maintain a flight path that permits a normal landing on the runway.

(ii) The circling approach must be performed without excessive maneuvering and without exceeding the normal operating limits of the airplane and the angle of bank must not exceed 30°.

Landings and Approaches to Landings

(1) Airport orientation.

(2) Normal landings with stabilized approach.

(3) Crosswind landings.

(4) From a precision instrument approach.

(5) From a precision instrument approach with a powerplant inoperative.

(6) From a non-precision instrument approach.

(7) From a non-precision instrument approach with a powerplant inoperative.

(8) From a circling approach or VFR traffic pattern.

(9) Go Around/Rejected landings—a normal missed approach procedure or a visual go-around after the landing is rejected. The landing should be rejected at approximately 50 feet and approximately over the runway threshold.

(10) Zero flap landing.

(a) Runway requirements.

(b) Airspeeds.

Appendix D to SFAR 108—MU-2B
Maneuver Profiles

(A) The Maneuver Profiles are provided to develop pilot proficiency with the procedures and techniques contained within this MU-2B Flight Training Program.

(B) Though constructed for use in the airplane they may also be used in the Flight Training Device (FTD). When an FTD is used, a maneuver may be performed at lower altitudes or carried to its completion. When training is conducted in the MU-2B airplane, all maneuvers must be performed in a manner sufficient to evaluate the performance of the student while never jeopardizing the safety of the flight.

(C) The maneuvers profiles are broken down into three sections by similar aircraft model groups. The three sections of this program are:

(1) Marquise (-60), Solitaire (-40), N (-36A), P (-26A)—Figures A-1 through A-28

(2) J (-35), K (-25), L (-;36), M (-26)—Figures B-1 through B-28

(3) B, D (-10), F (-20), G (-30)—Figures C-1 through C-28

(D) Each MU-2B profile in its respective section follows the outline below.

(1) Normal Takeoff (5- and 20-degrees flaps).

(2) Takeoff Engine Failure (5- and 20-degrees flaps).

(3) Takeoff Engine Failure on Runway or Rejected Takeoff.

(4) Takeoff Engine Failure after Liftoff—Unable to Climb (Classroom or FTD only).

(5) Steep Turns.

(6) Slow Flight Maneuvers.

(7) One Engine Inoperative Maneuvering/Loss of Directional Control.

TRAINING COURSE FINAL PHASE CHECK		
NAME OF AIRMAN *(last, first, middle initial)*	GRADE OF CERTIFICATE	CERTIFICATE NUMBER

DATE OF CHECK	LOCATION OF CHECK	TYPE OF CHECK	MU-2B MODEL	FTD MODEL

SCHOOL NAME	INSTRUCTOR NAME	CFI NUMBER	EXPIRES

FLIGHT MANEUVERS GRADE *(S-Satisfactory U-Unsatisfactory)*

MANEUVERS REQUIRED FOR ALL AIRMEN	A/C	FTD
PREFLIGHT CHECK		
START AND TAXI PROCEDURES		
*NORMAL TAKEOFF (X WIND) (TWO ENGINE)		
*TAKEOFF ENGINE FAILURE		
REJECTED TAKEOFF		
*STEEP TURNS		
*APPROACH TO STALL (3)		
*MANEUVERING WITH ONE ENGINE INOP (VMC)		
ABNORMAL AND EMERGENCY PROCEDURES - TO INCLUDE THE MU-2 OPERATIOIN IN ICING CONDITIONS WITHOUT THE AUTOPILOT OR WITHOUT TRIM-IN-MOTION/AUTOMATIC AUTOPILOT DISCONNECT.		
*PRECISION APROACH (ONE ENGINE INOPERATIVE)		
GO AROUND / REJECTED LANDING		
NORMAL LANDING (X WIND)		
*LANDING WITH ONE ENGINE INOPERATIVE		
*LANDING WITH NON-STANDARD FLAP CONFIG		
POST FLIGHT PROCEDURES		
ADDITIONAL MANEUVERS REQUIRED FOR INSTRUMENT RATED AIRMEN	**A/C**	**FTD**
PREFLIGHT CHECK		
UNUSUAL ATTITUDES		
ABNORMAL AND EMERGENCY PROCEDURES		
BASIC INSTRUMENT FLIGHT MANEUVERS		
AREA ARRIVAL AND DEPARTURE		
HOLDING		
PRECISION APPROACH (TWO ENGINE)		
*NON-PRECISION APPROACHES (2)		
MISSED APPROACH FROM EITHER PRECISION OR NON-PRECISION APPROACH (TWO ENGINE) MUST INCLUDE AN APPROACH WITH ONE ENGINE INOP		
LANDING FROM A STRAIGHT-IN/CIRCLING APPROACH		
CIRCLING APPROACH		
POST FLIGHT PROCEDURES		

RESULTS OF CHECK	SATISFACTORY / UNSATISFACTORY	FLIGHT TIMES	AIRCRAFT	FTD
INSTRUCTOR SIGNATURE		AIRMAN SIGNATURE		

(8) Approach to Stall (clean configuration/wings level).
(9) Approach to Stall (takeoff configuration/15- to 30-degrees bank).
(10) Approach to Stall (landing configuration/gear down/40-degrees flaps).
(11) Accelerated Stall (no flaps).
(12) Emergency Descent (low speed).
(13) Emergency Descent (high speed).
(14) Unusual Altitude Recovery (nose high).
(15) Unusual Altitude Recovery (nose low).
(16) Normal Landing (20- and 40-degrees flaps).
(17) Go Around/Rejected Landing.
(18) No Flap or 5-degrees flaps Landing.
(19) One Engine Inoperative Landing (5- and 20-degrees flaps).
(20) Crosswind Landing.
(21) ILS and Missed Approach.
(22) Two Engine Missed Approach.
(23) One Engine Inoperative ILS and Missed Approach.
(24) One Engine Inoperative Missed Approach.
(25) Non-Precision and Missed Approach.
(26) One Engine Inoperative Non-Precision and Missed Approach.
(27) Circling Approach at Weather Minimums.
(28) One Engine Inoperative Circling Approach at Weather Minimums.

Engine Performance

(A) The following should be considered in reference to power settings and airspeeds:

(1) Power settings shown in *italics* are provided as guidance only during training and are not referenced in the AFM. Power setting guidance is provided to show the approximate power setting that will produce the desired airspeed or flight condition. Actual power settings may be different from those stated and should be noted by the instructor and student for reference during other maneuvers. Power settings in the profiles are stated in torque or PSI and will vary with aircraft model, engine model, weight, and density altitude. Power settings are based on standard atmospheric conditions.

(2) Some pilots prefer to set power initially using fuel flow, because the fuel flow system is not field adjustable. Fuel flow settings refer to engine operations only. If fuel flow is used to set power for takeoff, check torque and temperature after setting fuel flow and adjust torque or temperature, whichever is limiting, for maximum takeoff power prior to liftoff.

(3) Improperly adjusted torque or improperly calibrated temperatures are a safety of flight issue and must be checked and corrected prior to conducting flight training.

(4) The pilot should refer to the performance section of the airplane flight manual to determine actual speeds required for his/her particular model and specific weight for any given operation.

MU-2B MARQUISE (-60), SOLITAIRE (-40), N (-36A), P (-26A)
NORMAL TAKE-OFF, 5° OR 20° FLAPS

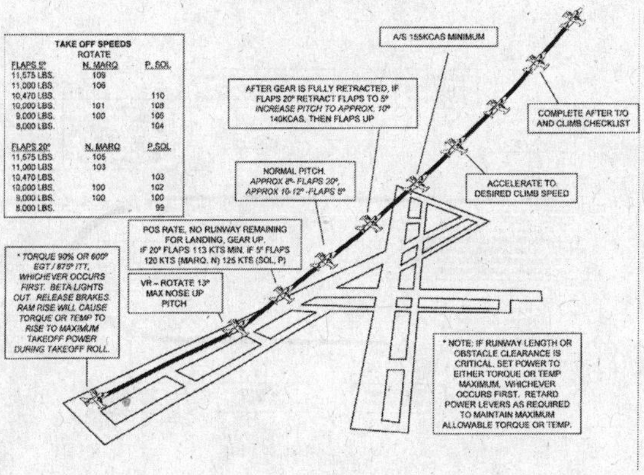

MU-2B MARQUISE (-60), SOLITAIRE (-40), N (-36A), P (-26A)
TAKE-OFF ENGINE FAILURE ON RUNWAY

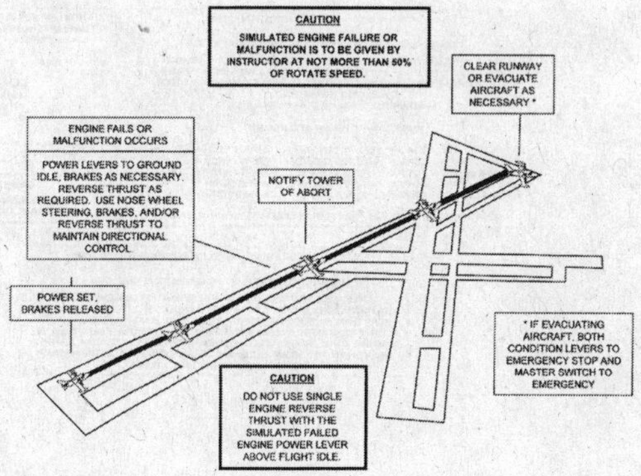

A-3

MU-2B MARQUISE (-60), SOLITAIRE (-40), N (-36A), P (-26A)
TAKE-OFF ENGINE FAILURE – FLAPS 5° OR 20°

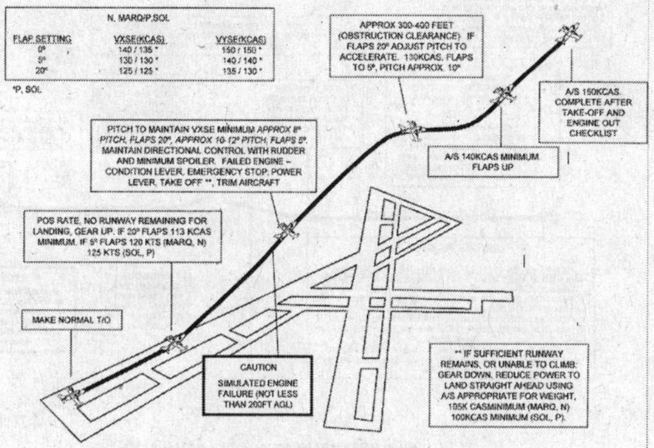

MU-2B MARQUISE (-60), SOLITAIRE (-40), N (-36A), P (-26A)
TAKE-OFF ENGINE FAILURE - UNABLE TO CLIMB
CLASSROOM DISCUSSION OR FTD USE ONLY

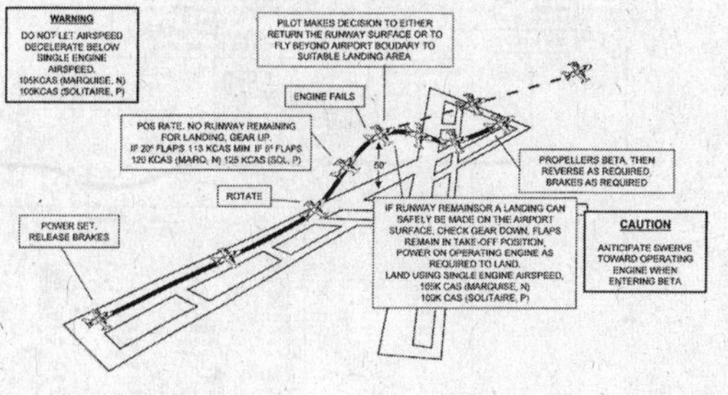

A-4

MU-2B MARQUISE (-60), SOLITAIRE (-40), N (-36A), P (-26A)
STEEP TURNS

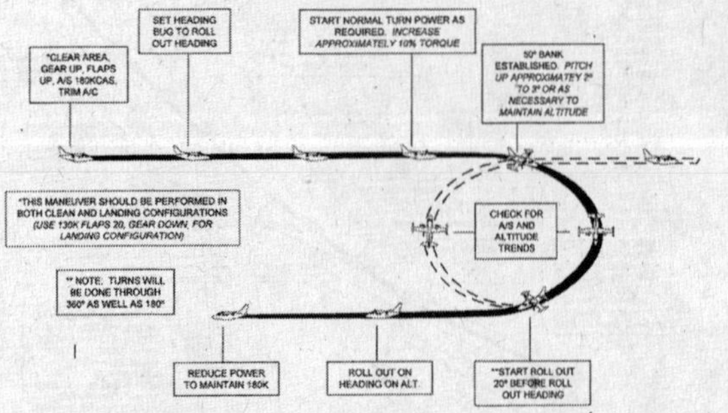

MU-2B MARQUISE (-60), SOLITAIRE (-40), N (-36A), P (-26A)
SLOW FLIGHT MANEUVERING
MINIMUM CONTROLLABLE AIRSPEED

SLOW FLIGHT MANEUVERING IS CONDUCTED AS FOLLOWS.

CLEAR THE AREA PRIOR TO BEGINNING THE MANEUVER.

START WITH CLEAN CONFIGURATION AND CHANGE AIRCRAFT CONFIGURATION FROM CLEAN TO FULL FLAP AND GEAR IN STAGES. USE A MAXIMUM OF 15° BANK AND PERFORM HEADING CHANGES OF 90° LEFT AND RIGHT. CONSTANT ALTITUDE IS REQUIRED THROUGHOUT.
MAINTAIN 115KCAS IN ALL CONFIGURATIONS.

**APPROXIMATE POWER SETTINGS ARE:

CLEAN	*TORQUE (35%) PER ENGINE*	*APPROX PITCH +12*
5° FLAP	*TORQUE (32%) PER ENGINE*	*APPROX PITCH +8*
5° FLAP & GEAR	*TORQUE (44%) PER ENGINE*	*APPROX PITCH +9*
20° FLAP & GEAR	*TORQUE (42%) PER ENGINE*	*APPROX PITCH +4*
40° FLAP & GEAR	*TORQUE (54%) PER ENGINE*	*APPROX PITCH 0*

** NOTE: POWER SETTINGS WILL VARY WITH AIRCRAFT WEIGHT AND ALTITUDE.

STALL SPEEDS (APPROXIMATE)
AT MAXIMUM GROSS TAKEOFF WEIGHT
N, MARQUISE / P, SOLITIARE

ANGLE OF BANK / FLAPS	0°	15°
UP	106/104*	108/106*
5°	99/ 98*	100/ 99*
20°	87/ 85*	88/ 88*
40°	81/ 78*	83/ 79*

*P, SOL

Vmc FLAPS 5° 99K (MARQ. N), 100KCAS (SOL. P)
FLAPS 20° 99K (MARQ. N), 93KCAS (SOL. P)

CAUTION
STALL WARNING MAY ACTIVATE 4 TO 9 KCAS ABOVE STALL

MINIMUM CONTROLLABLE AIRSPEED IS CONDUCTED AS FOLLOWS:

CLEAR THE AREA PRIOR TO BEGINNING THE MANEUVER.

THE MANEUVER MAY BE DONE IN ANY COMBINATION OF GEAR OR FLAP CONFIGURATIONS. IF BANK IS TO BE USED, IT SHOULD BE DONE AT BANK OF NOT MORE THAN 15°. BEGIN THE MANEUVER BY CONFIGURING THE AIRCRAFT IN THE DESIRED GEAR AND FLAP CONFIGURATION. SLOW THE AIRCRAFT UNTIL THE STALL WARNING (STICK SHAKER) IS ACTIVATED AND ADD POWER TO MAINTAIN ALTITUDE AND A SPEED JUST ABOVE AERODYNAMIC STALL. DO NOT ALLOW THE AIRCRAFT TO REACH AERODYNAMIC STALL BUFFET.

A-6

MU-2B MARQUISE (-60), SOLITAIRE (-40), N (-36A), P (-26A)
ONE ENGINE INOPERATIVE MANEUVERING LOSS OF DIRECTIONAL CONTROL

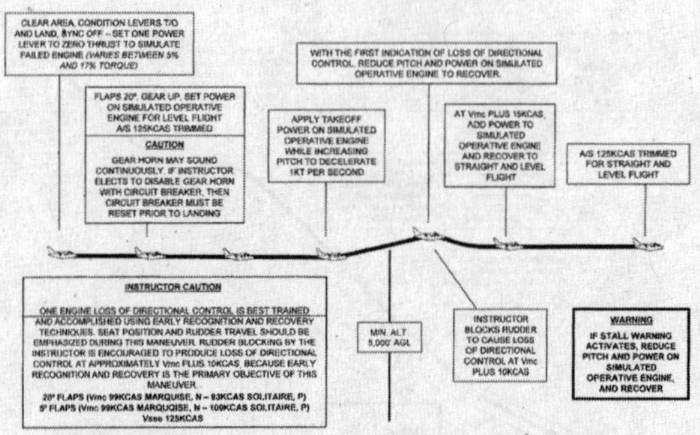

A-7

MU-2B MARQUISE (-60), SOLITAIRE (-40), N (-36A), P (-26A)
APPROACH TO STALL CLEAN CONFIGURATION / WINGS LEVEL

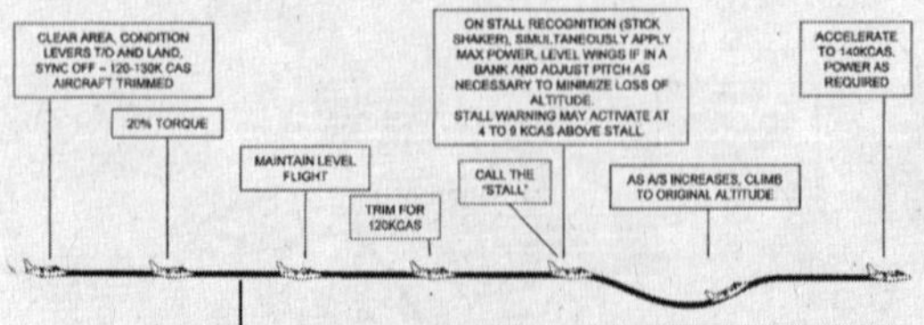

STALL SPEEDS

FLAPS SET / GR.WT.	UP	5	20	40
7,000	/85 *	/80 *	/72 *	/64 *
7,500	/88 *	/82 *	/74 *	/66 *
8,000	/91 *	/85 *	/76 *	/68 *
8,500	/93 *	/88 *	/79 *	/70 *
9,000	93/96 *	87/90 *	76/81 *	72/72 *
9,500	95/99 *	90/93 *	79/83 *	74/74 *
10,000	98/101*	92/95 *	81/85 *	75/76 *
10,470	/104 *	/98 *	/88 *	/78 *
10,500	101/	94/	83/	77/
11,000	103/	96/	85/	79/
11,575	106/	99/	87/	81/

*P, SOL

A-8

MU-2B MARQUISE (-60), SOLITAIRE (-40), N (-36A), P (-26A)
APPROACH TO STALL
TAKEOFF CONFIGURATION 15-30° BANK

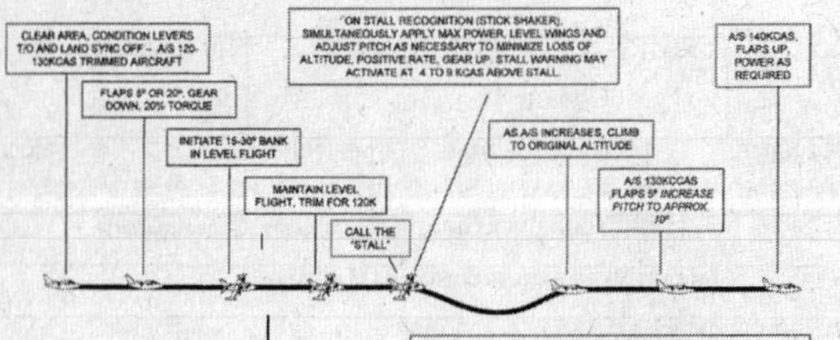

STALL SPEEDS (APPROXIMATE)
AT MAXIMUM GROSS TAKEOFF WEIGHT
N, MARQUISE / P, SOLITIARE

ANGLE OF BANK / FLAPS	10	20	30	40	50	60
UP	107/104*	109/108*	113/112*	120/119*	131/130*	148/146*
5°	99/ 98*	102/101*	106/105*	113/112*	123/122*	138/138*
20°	87/ 88*	89/ 90*	93/ 94*	98/100*	108/109*	122/123*
40°	82/ 79*	84/ 80*	87/ 84*	92/ 90*	101/ 98*	113/110*

*P, SOL

MU-2B MARQUISE (-60), SOLITAIRE (-40), N (-36A), P (-26A)
APPROACH TO STALL
GEAR DOWN – FULL FLAPS

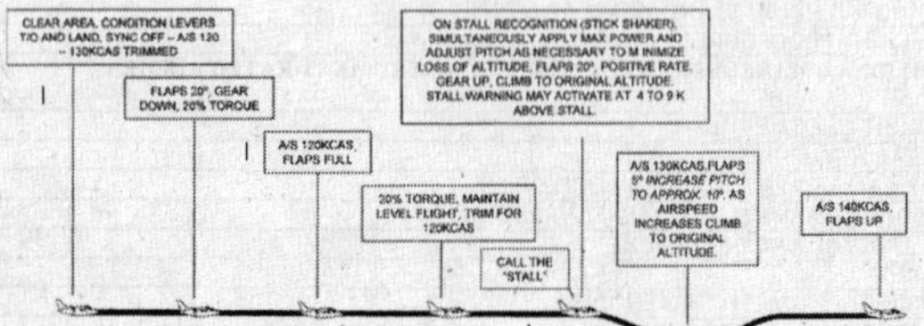

STALL SPEEDS

FLAPS SET / GR.WT.	UP	5	20	40
7,000	/85 *	/80 *	/72 *	/64 *
7,500	/88 *	/82 *	/74 *	/66 *
8,000	/91 *	/85 *	/76 *	/68 *
8,500	/93 *	/88 *	/79 *	/70 *
9,000	93/96 *	87/90 *	76/81 *	72/72 *
9,500	95/99 *	90/93 *	79/83 *	74/74 *
10,000	98/101*	92/95 *	81/85 *	75/76 *
10,470	/104 *	/98 *	/88 *	/78 *
10,500	101/	94/	83/	77/
11,000	103/	96/	85/	79/
11,575	106/	99/	87/	81/

*P, SOL

A-10

MU-2B MARQUISE (-60), SOLITAIRE (-40), N (-36A), P (-26A)
ACCELERATED STALLS

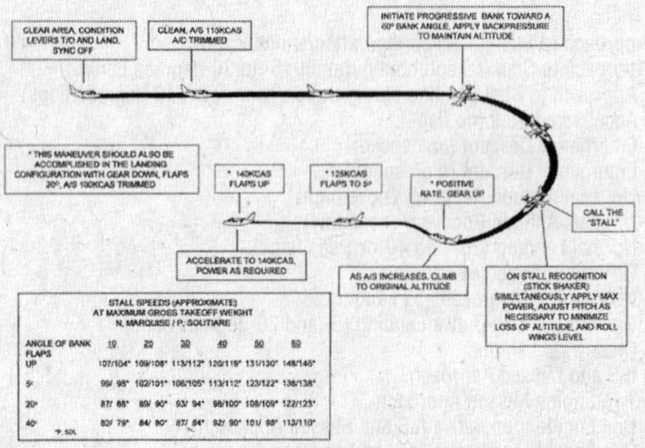

STALL SPEEDS (APPROXIMATE)
AT MAXIMUM GROSS TAKEOFF WEIGHT
N, MARQUISE / P, SOLITIARE

ANGLE OF BANK / FLAPS	10	20	30	40	50	60
UP	107/104*	109/108*	113/112*	120/119*	131/130*	148/146*
5°	99/ 98*	102/101*	106/105*	113/112*	123/122*	138/138*
20°	87/ 88*	89/ 90*	93/ 94*	98/100*	108/109*	122/123*
40°	82/ 79*	84/ 80*	87/ 84*	92/ 90*	101/ 98*	113/110*

*P, SOL

A-11

MU-2B MARQUISE (-60), SOLITAIRE (-40), N (-36A), P (-26A)
UNUSUAL ATTITUDE RECOVERY (NOSE HIGH)

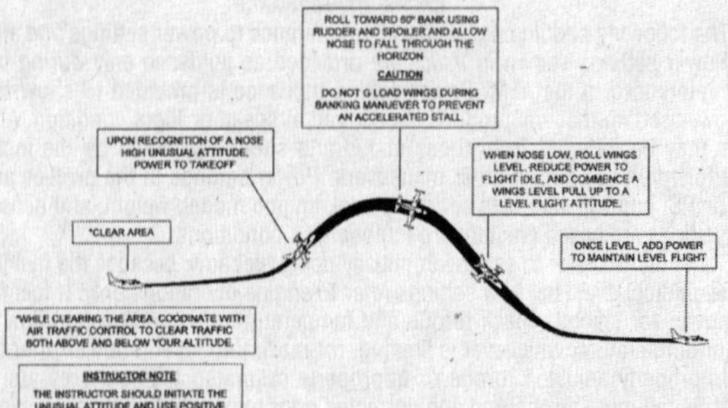

A-14

MU-2B MARQUISE (-60), SOLITAIRE (-40), N (-36A), P (-26A)

EMERGENCY DESCENT (HIGH SPEED)

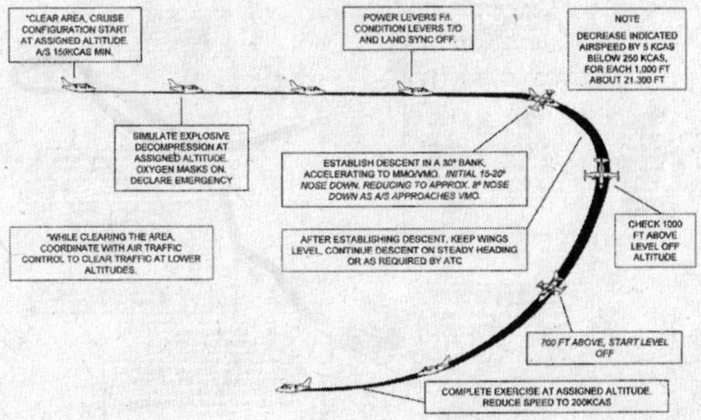

A-13

MU-2B MARQUISE (-60), SOLITAIRE (-40), N (-36A), P (-26A)

EMERGENCY DESCENT (LOW SPEED)

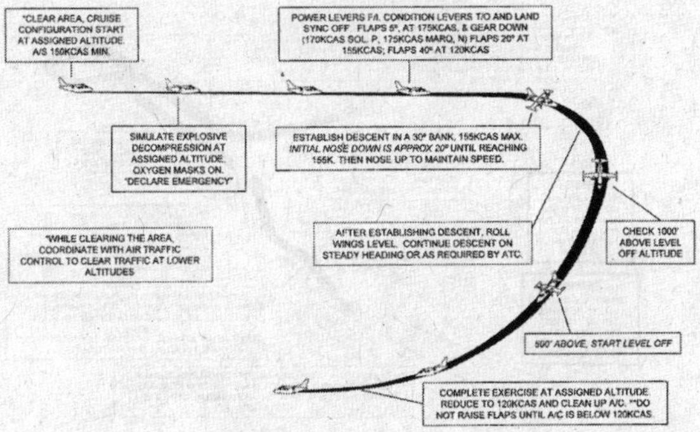

MU-2B MARQUISE (-60), SOLITAIRE (-40), N (-36A), P (-26A)

UNUSUAL ATTITUDE RECOVERY (NOSE LOW)

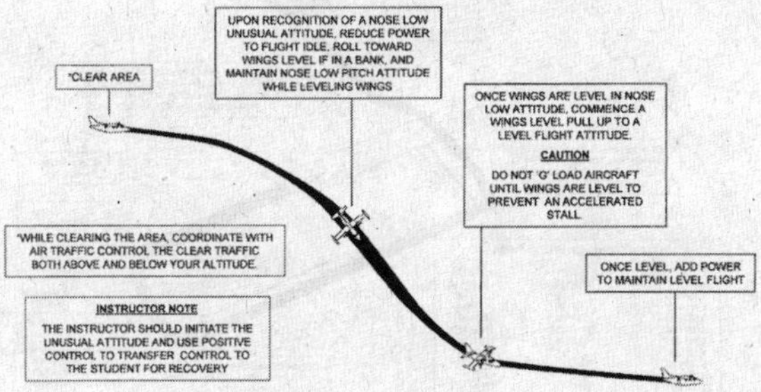

MU-2B MARQUISE (-60), SOLITAIRE (-40), N (-36A), P (-26A)

NORMAL LANDING (20°or 40° FLAPS)

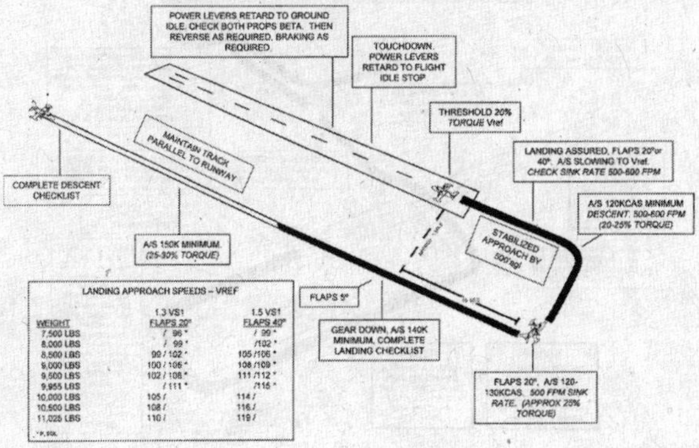

LANDING APPROACH SPEEDS – VREF

WEIGHT	1.3 VS1 FLAPS 20°	1.5 VS1 FLAPS 40°
7,500 LBS	/ 96 *	/ 99 *
8,000 LBS	/ 99 *	/102 *
8,500 LBS	99 / 102 *	105 /106 *
9,000 LBS	100 / 106 *	108 /109 *
9,500 LBS	102 / 108 *	111 /112 *
9,955 LBS	/ 111 *	/115 *
10,000 LBS	105 /	114 /
10,500 LBS	108 /	116 /
11,025 LBS	110 /	119 /

* P, SOL

A-16

MU-2B MARQUISE (-60), SOLITAIRE (-40), N (-36A), P (-26A)

GO AROUND - REJECTED LANDING

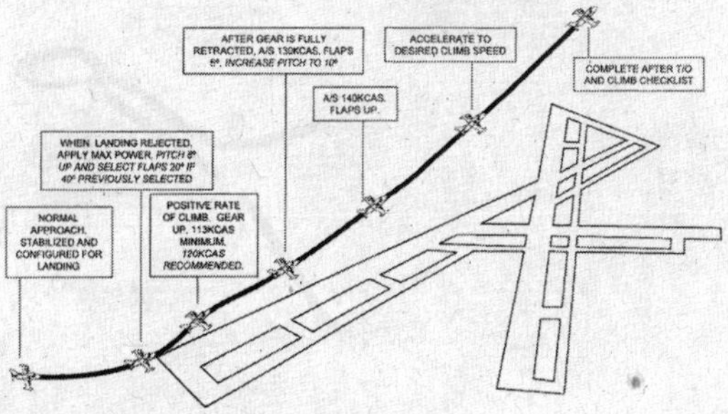

A-17

MU-2B MARQUISE (-60), SOLITAIRE (-40), N (-36A), P (-26A)

NO FLAP OR 5° FLAP LANDING

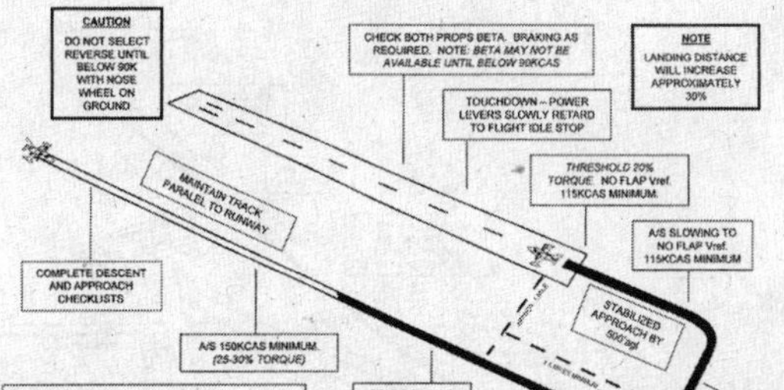

NO FLAP VREF 1.25 VS1 (USE FOR FLAPS UP OR 5°)

	UP	5°
7,500	/ 115*	/ 115*
8,000	/ 115*	/ 115*
8,500	/ 116*	/ 115*
9,000	116 / 120*	115 / 115*
9,500	119 / 123*	115 / 117*
9,995	/ 126*	/ 120*
10,000	123 /	115 /
10,500	127 /	116 /
11,025	129 /	121 /

*SOL, P

A-18

MU-2B MARQUISE (-60), SOLITAIRE (-40), N (-36A), P (-26A)

ONE ENGINE INOPERATIVE LANDING

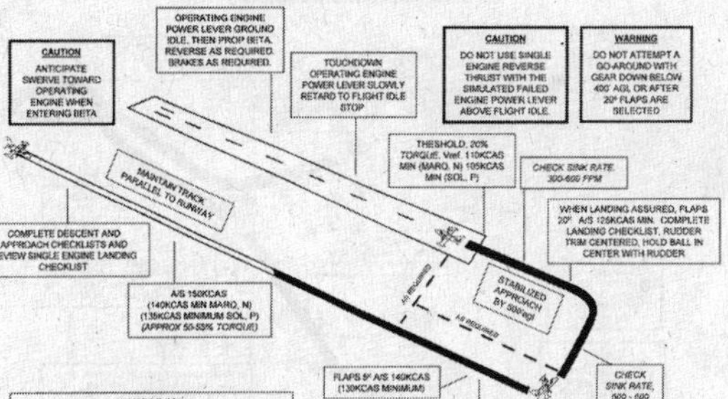

N, MARQ/P, SOL

FLAP SETTING	VXSE(KCAS)	VYSE(KCAS)
UP	140 / 135 *	150 / 150 *
5°	130 / 130 *	140 / 140 *
20°	125 / 125 *	135 / 130 *

*P, SOL

A-19

MU-2B MARQUISE (-60), SOLITAIRE (-40), N (-36A), P (-26A)

CROSSWIND LANDING

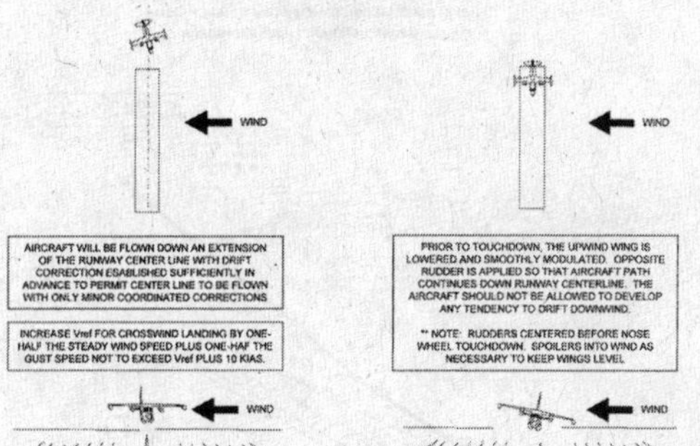

A-20

MU-2B MARQUISE (-60), SOLITAIRE (-40), N (-36A), P (-26A)

ILS AND MISSED APPROACH

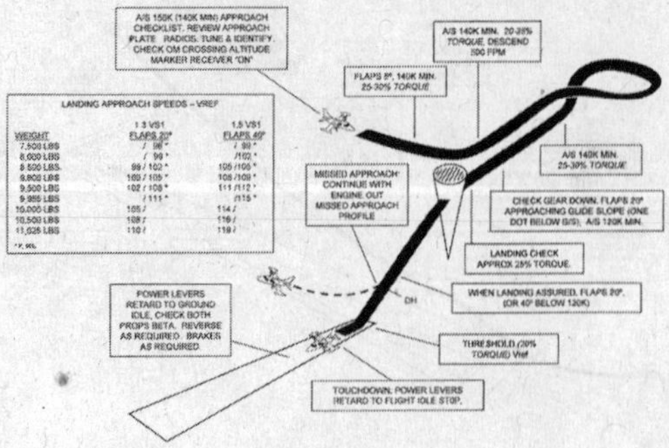

A-21

MU-2B MARQUISE (-60), SOLITAIRE (-40), N (-36A), P (-26A)

TWO ENGINE MISSED APPROACH

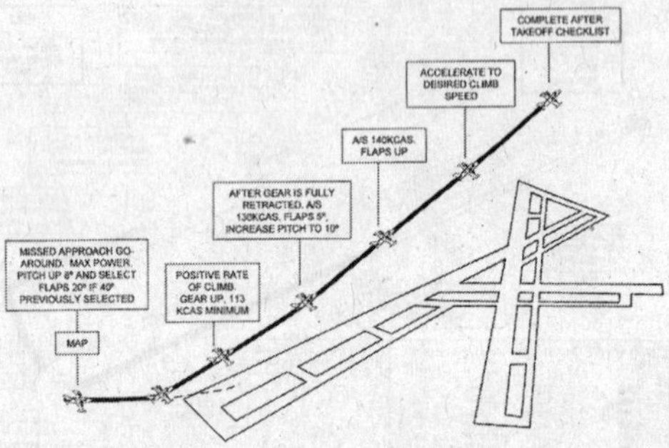

A-22

MU-2B MARQUISE (-60), SOLITAIRE (-40), N (-36A), P (-26A)

ONE ENGINE INOPERATIVE ILS AND MISSED APPROACH

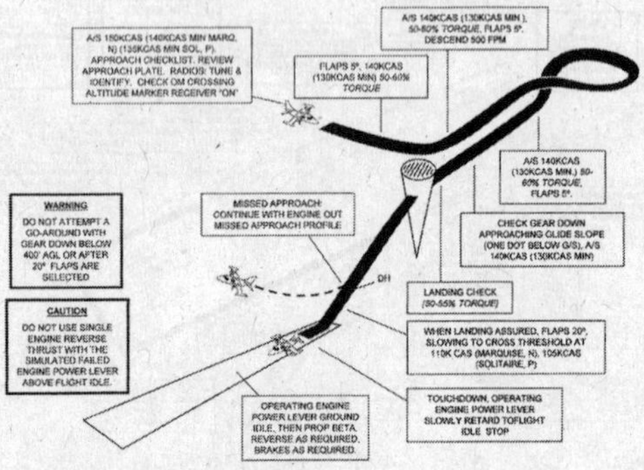

A-23

MU-2B MARQUISE (-60), SOLITAIRE (-40), N (-36A), P (-26A)

ONE ENGINE INOPERATIVE MISSED APPROACH

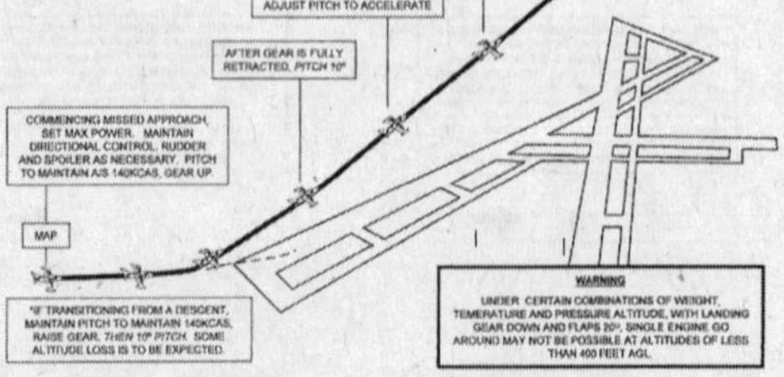

A-24

MU-2B MARQUISE (-60), SOLITAIRE (-40), N (-36A), P (-26A)

NON-PRECISION AND MISSED APPROACH

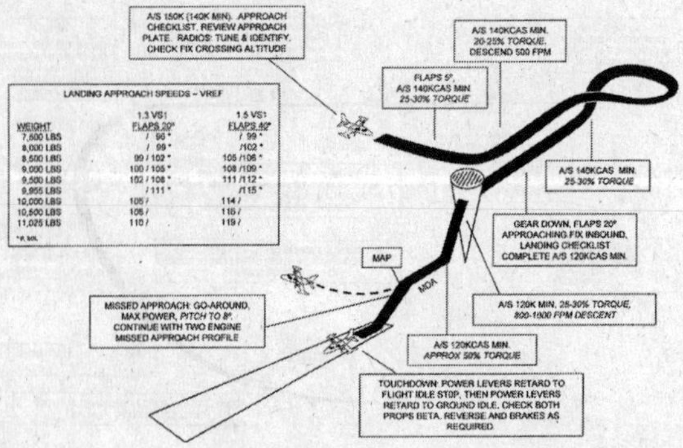

A-25

MU-2B MARQUISE (-60), SOLITAIRE (-40), N (-36A), P (-26A)

ONE ENGINE INOPERATIVE NON-PRECISION AND MISSED APPROACH

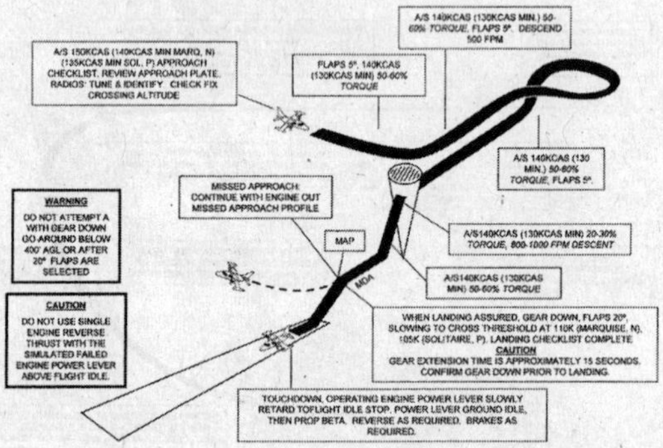

A-26

MU-2B MARQUISE (-60), SOLITAIRE (-40), N (-36A), P (-26A)

CIRCLING APPROACH AT WEATHER MINIMUMS

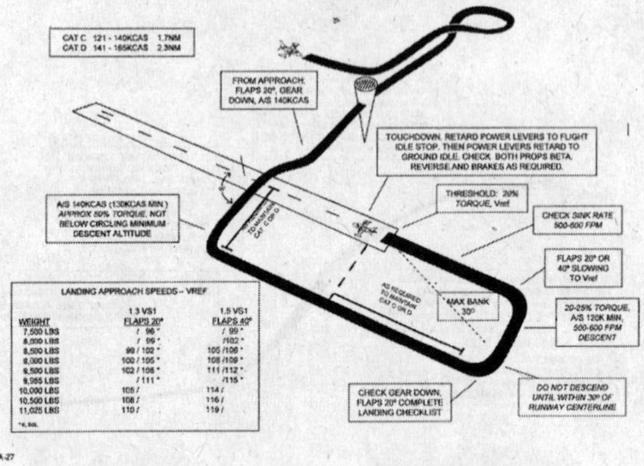

A-27

MU-2B MARQUISE (-60), SOLITAIRE (-40), N (-36A), P (-26A)

ONE ENGINE INOPERATIVE CIRCLING APPROACH AT WEATHER MINIMUMS

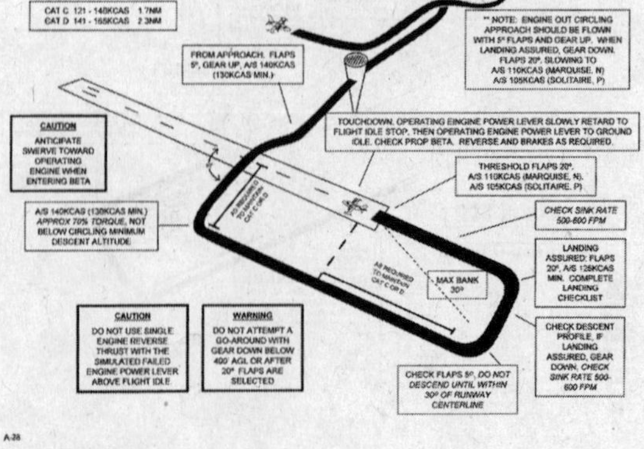

A-28

MU-2B J (-35), K (-25), L (-36), M (-26)
NORMAL TAKE-OFF, 5° OR 20° FLAPS

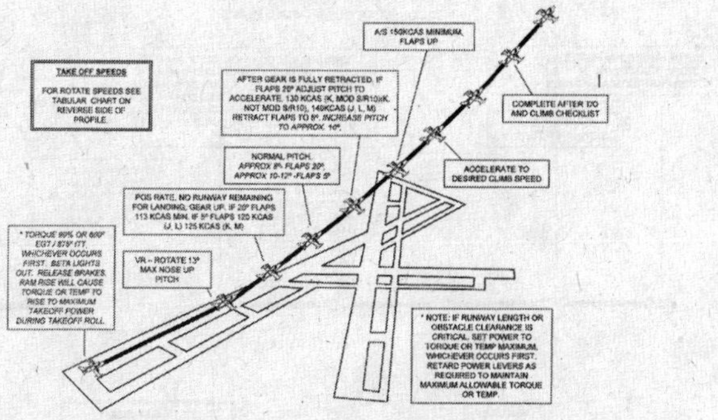

B-1

MU-2B J (-35), K (-25), L (-36), M (-26)
TAKE-OFF ENGINE FAILURE - UNABLE TO CLIMB
CLASSROOM DISCUSSION OR FTD USE ONLY

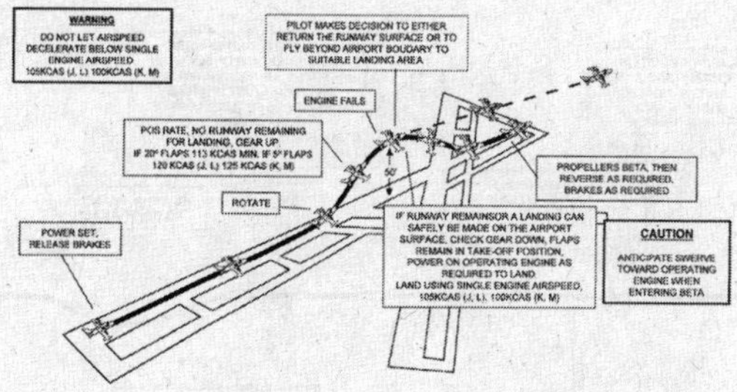

B-4

TAKE OFF SPEEDS
ROTATE

FLAPS 5°	K	M	J	L
11,575 LBS				109
11,000 LBS				106
10,800 LBS			109	105
10,470 LBS		110		104
10,000 LBS		108	105	101
9,920 LBS	108			101
9,500 LBS	107	107	103	101
9,000 LBS	106	106	101	100
8,000 LBS	104	104	100	
7,500 LBS	102			

FLAPS 20°	K	M	J	L
11,575 LBS				105
11,000 LBS				103
10,800 LBS			105	
10,470 LBS		103		
10,000 LBS		102	102	100
9,920 LBS	102			
9,500 LBS	101	101	101	100
9,000 LBS	100	100	100	100
8,000 LBS	99	99	100	
7,500 LBS	98			

MU-2B J (-35), K (-25), L (-36), M (-26)
STEEP TURNS

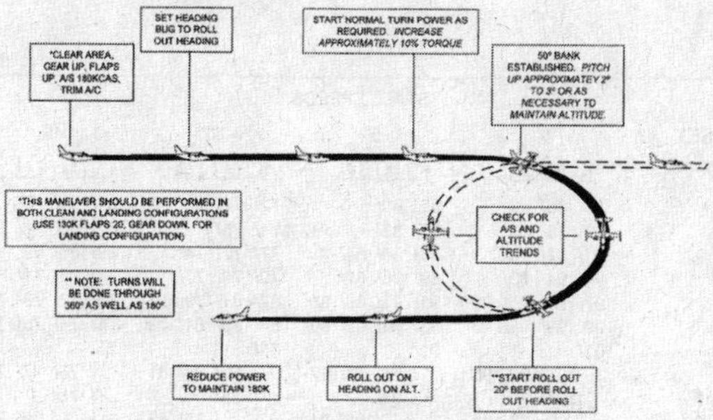

B-5

MU-2B J (-35), K (-25), L (-36), M (-26)
TAKE-OFF ENGINE FAILURE – FLAPS 5° OR 20°

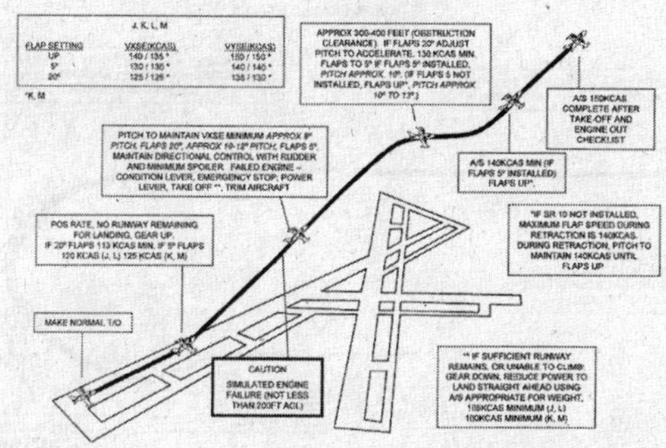

B-2

MU-2B J (-35), K (-25), L (-36), M (-26)
SLOW FLIGHT MANEUVERING
MINIMUM CONTROLLABLE AIRSPEED

SLOW FLIGHT MANEUVERING IS CONDUCTED AS FOLLOWS:

CLEAR THE AREA PRIOR TO BEGINNING THE MANEUVER.

START WITH CLEAN CONFIGURATION AND CHANGE AIRCRAFT CONFIGURATION FROM CLEAN TO FULL FLAP AND GEAR IN STAGES. USE A MAXIMUM OF 15° BANK AND PERFORM HEADING CHANGES OF 90° LEFT AND RIGHT. CONSTANT ALTITUDE IS REQUIRED THROUGHOUT.
MAINTAIN 115KCAS IN ALL CONFIGURATIONS.

**APPROXIMATE POWER SETTINGS ARE:

CLEAN	*TORQUE (35%) PER ENGINE*	*APPROX PITCH +12*
5° FLAP	*TORQUE (32%) PER ENGINE*	*APPROX PITCH +8*
5° FLAP & GEAR	*TORQUE (44%) PER ENGINE*	*APPROX PITCH +9*
20° FLAP & GEAR	*TORQUE (42%) PER ENGINE*	*APPROX PITCH +4*
40° FLAP & GEAR	*TORQUE (54%) PER ENGINE*	*APPROX PITCH 0*

** NOTE: POWER SETTINGS WILL VARY WITH AIRCRAFT WEIGHT AND ALTITUDE.

STALL SPEEDS (APPROXIMATE)
AT MAXIMUM GROSS TAKEOFF WEIGHT
J, K, L, M

ANGLE OF BANK	J / L / K / M 0°	J / L / K / M 15°
FLAPS		
UP	104/106/101/104	107/108/103/106
5°	98/ 99/ 95/ 98	100/101/ 97/100
20°	86/ 87/ 85/ 87	88/ 89/ 87/ 89
40°	79/ 81/ 76/ 78	82/ 83/ 78/ 80

CAUTION
STALL WARNING MAY ACTIVATE 4 TO 9 KTS ABOVE STALL

MINIMUM CONTROLLABLE AIRSPEED IS CONDUCTED AS FOLLOWS:

CLEAR THE AREA PRIOR TO BEGINNING THE MANEUVER.

THE MANEUVER MAY BE DONE IN ANY COMBINATION OF GEAR OR FLAP CONFIGURATIONS. IF BANK IS TO BE USED, IT SHOULD BE DONE AT BANK OF NOT MORE THAN 10°. BEGIN THE MANEUVER BY CONFIGURING THE AIRCRAFT IN THE DESIRED GEAR AND FLAP CONFIGURATION. SLOW THE AIRCRAFT UNTIL THE STALL WARNING (STICK SHAKER) IS ACTIVATED AND ADD POWER TO MAINTAIN ALTITUDE AND A SPEED JUST ABOVE AERODYNAMIC STALL. DO NOT ALLOW THE AIRCRAFT TO REACH AERODYNAMIC STALL BUFFET.

B-6

MU-2B J (-35), K (-25), L (-36), M (-26)
TAKE-OFF ENGINE FAILURE ON RUNWAY

CAUTION
SIMULATED ENGINE FAILURE OR MALFUNCTION IS TO BE GIVEN BY INSTRUCTOR AT NOT MORE THAN 80% OF ROTATE SPEEDS.

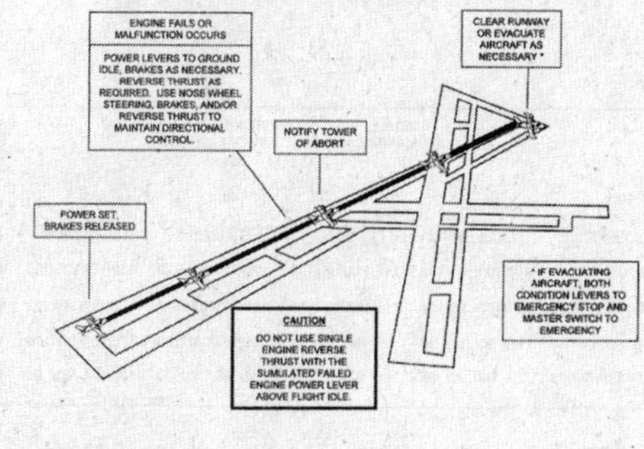

B-3

MU-2B J (-35), K (-25), L (-36), M (-26)
ONE ENGINE INOPERATIVE MANEUVERING
LOSS OF DIRECTIONAL CONTROL

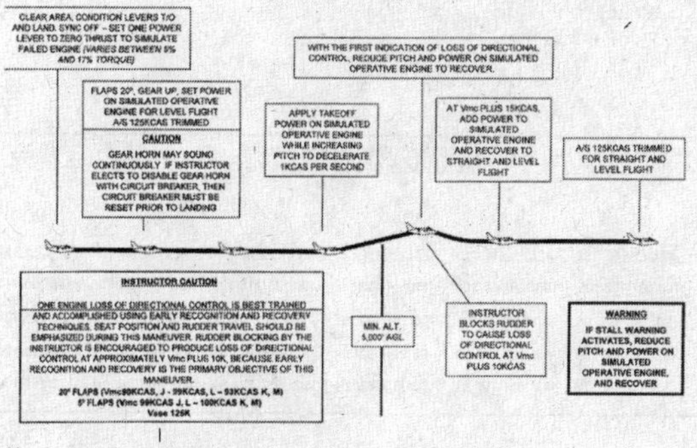

B-7

MU-2B J (-35), K (-25), L (-36), M (-26)

APPROACH TO STALL CLEAN CONFIGURATION / WINGS LEVEL

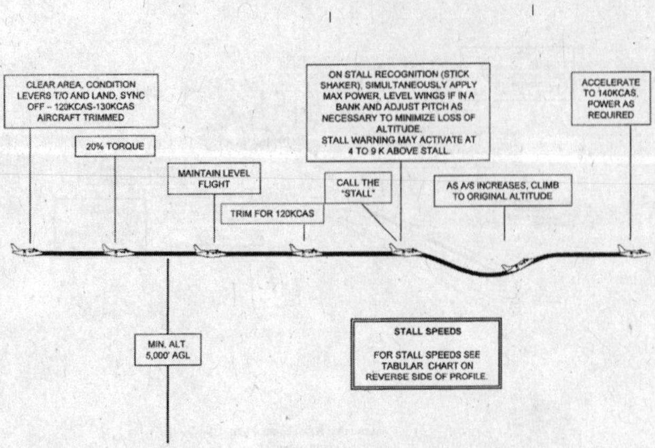

B-8

STALL SPEEDS

FLAPS SET	0	5	20	40
GR.WT.	K / M / J / L	K / M / J / L	K / M / J / L	K / M / J / L
7,000	85/ 85/	80/ 80/	72/ 72/	64/ 64/
7,500	88 / 88/	83/ 83/	74/ 75/	67/ 66/
8,000	91/ 91/ 90/	86/ 85/ 84/	77/ 77/ 74/	69/ 68/ 69
8,500	94/ 94/ 93/	89/ 88/ 87/	79/ 79/ 77/	71/ 70/ 71/
9,000	97/ 96/ 95/ 93	91/ 91/ 89/ 88	82/ 81/ 79/ 77	73/ 72/ 73/ 72
9,500	99 /99/ 98/ 96	93/ 93/ 92/ 90	84/ 83/ 81/ 79	75/ 74/ 75/ 74
9,920	101/	95/	85/	76/
10,000	/102/100/ 98	/ 96/ 94/ 92	/ 86/ 84/ 81	/ 76/ 77/ 76
10,470	/104/	/ 98/	/ 88/	/ 78/
10,500	/103/101	/ 96/ 94	/ 85/ 83	/ 79/ 77
10,800	/104/	/ 98/	/ 86/	/ 80/ 78
11,000	/103	/ 97	/ 85	/ 79
11,500	/106	/ 99	/ 87	/ 81

MU-2B J (-35), K (-25), L (-36), M (-26)

APPROACH TO STALL

TAKEOFF CONFIGURATION 15-30° BANK

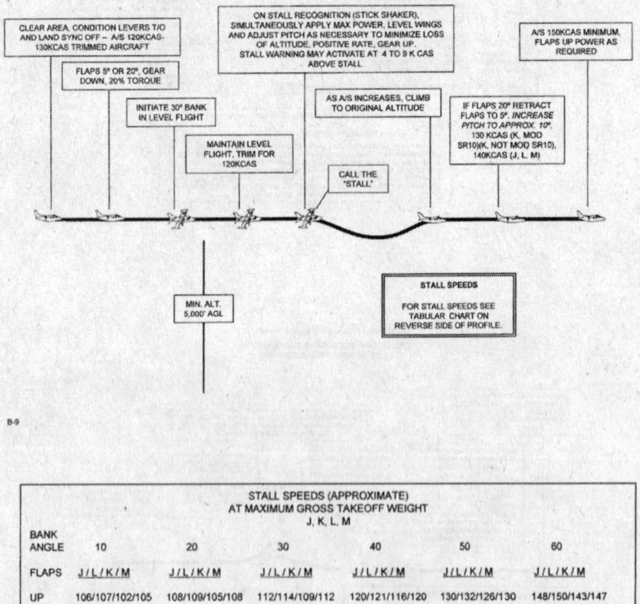

B-9

STALL SPEEDS (APPROXIMATE)
AT MAXIMUM GROSS TAKEOFF WEIGHT
J, K, L, M

BANK ANGLE	10	20	30	40	50	60
FLAPS	J / L / K / M	J / L / K / M	J / L / K / M	J / L / K / M	J / L / K / M	J / L / K / M
UP	106/107/102/105	108/109/105/108	112/114/109/112	120/121/116/120	130/132/126/130	148/150/143/147
5°	99/100/ 96/ 98	101/102/ 98/101	105/107/102/105	112/113/109/112	122/123/119/122	138/140/134/138
20°	87/ 88/ 86/ 88	89/ 90/ 88/ 90	92 /94/ 92/ 94	98/100/ 97/100	108/109/107/109	122/123/120/123
40°	81/ 82/ 77/ 79	83/ 84/ 79/ 81	86/ 87/ 82/ 84	92/ 93/ 87/ 90	100/102/ 96/ 98	112/115/108/110

MU-2B J (-35), K (-25), L (-36), M (-26)

APPROACH TO STALL

GEAR DOWN – FULL FLAPS

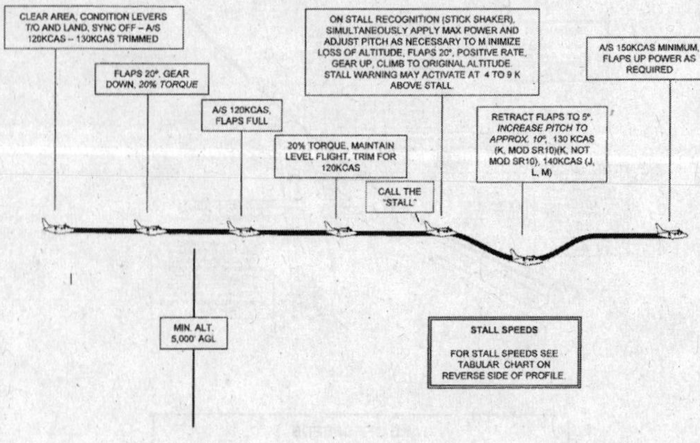

B-10

STALL SPEEDS

FLAPS SET	0	5	20	40
GR.WT.	K / M / J / L	K / M / J / L	K / M / J / L	K / M / J / L
7,000	85/ 85/	80/ 80/	72/ 72/	64/ 64/
7,500	88 / 88/	83/ 83/	74/ 75/	67/ 66/
8,000	91/ 91/ 90/	86/ 85/ 84/	77/ 77/ 74/	69/ 68/ 69
8,500	94/ 94/ 93/	89/ 88/ 87/	79/ 79/ 77/	71/ 70/ 71/
9,000	97/ 96/ 95/ 93	91/ 91/ 89/ 88	82/ 81/ 79/ 77	73/ 72/ 73/ 72
9,500	99 /99/ 98/ 96	93/ 93/ 92/ 90	84/ 83/ 81/ 79	75/ 74/ 75/ 74
9,920	101/	95/	85/	76/
10,000	/102/100/ 98	/ 96/ 94/ 92	/ 86/ 84/ 81	/ 76/ 77/ 76
10,470	/104/	/ 98/	/ 88/	/ 78/
10,500	/103/101	/ 96/ 94	/ 85/ 83	/ 79/ 77
10,800	/104/	/ 97/	/ 86/	/ 80/ 78
11,000	/103	/ 97	/ 85	/ 79
11,500	/106	/ 99	/ 87	/ 81

MU-2B J (-35), K (-25), L (-36), M (-26)

ACCELERATED STALLS

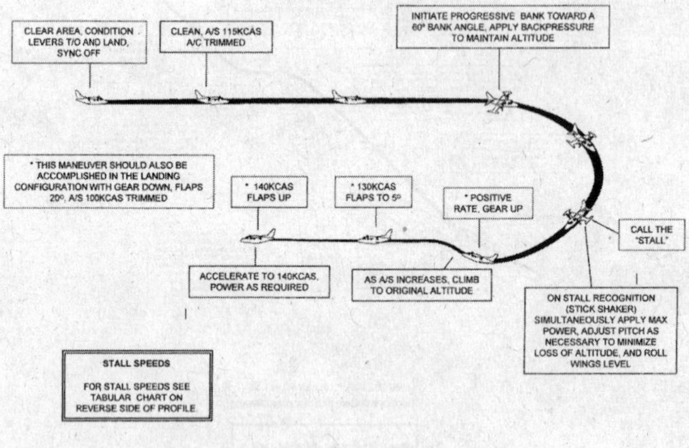

B-11

STALL SPEEDS (APPROXIMATE)
AT MAXIMUM GROSS TAKEOFF WEIGHT
J, K, L, M

BANK ANGLE	10	20	30	40	50	60
FLAPS	J / L / K / M	J / L / K / M	J / L / K / M	J / L / K / M	J / L / K / M	J / L / K / M
UP	106/107/102/105	108/109/105/108	112/114/109/112	120/121/116/120	130/132/126/130	148/150/143/147
5°	99/100/ 96/ 98	101/102/ 98/101	105/107/102/105	112/113/109/112	122/123/119/122	138/140/134/138
20°	87/ 88/ 86/ 88	89/ 90/ 88/ 90	92 /94/ 92/ 94	98/100/ 97/100	108/109/107/109	122/123/120/123
40°	81/ 82/ 77/ 79	83/ 84/ 79/ 81	86/ 87/ 82/ 84	92/ 93/ 87/ 90	100/102/ 96/ 98	112/115/108/110

MU-2B J (-35), K (-25), L (-36), M (-26)
EMERGENCY DESCENT (LOW SPEED)

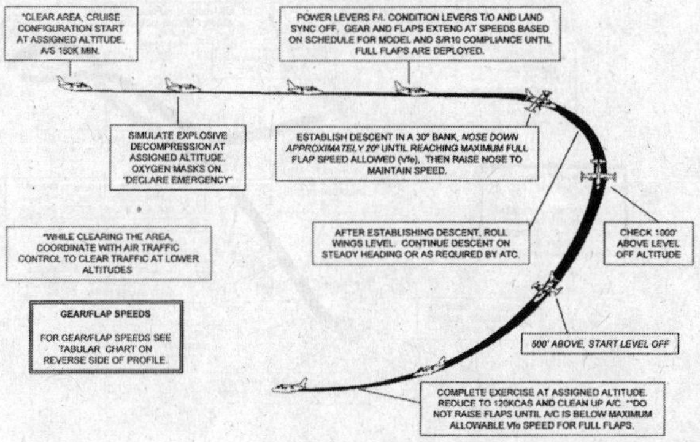

B-12

GEAR AND FLAP EXTEND SCHEDULE
(K+ AND J+ ARE MODIFIED BY S/R10)

GEAR			
K, K+:	**160KCAS**		
M, J, J+:	**170KCAS**		
L:	**175KCAS**		
FLAPS	**5º**	**20º**	**40º**
J: S/N 548 – 609 NOT MODIFIED BY S/R10	**146KCAS**	**146KCAS**	**120KCAS**
J+: S/N 548 – 609 MODIFIED BY S/R10 AND S/N 610 - 654	**175KCAS**	**146KCAS**	**120KCAS**
K: S/N 239 – 279 NOT MODIFIED BY S/R10	**140KCAS**	**140KCAS**	**120KCAS**
K+: S/N 239 – 279 MODIFIED BY S/R10 AND S/N 280 - 318	**175KCAS**	**140KCAS**	**120KCAS**
L / M	**175KCAS**	**155KCAS**	**120KCAS**

MU-2B J (-35), K (-25), L (-36), M (-26)
EMERGENCY DESCENT (HIGH SPEED)

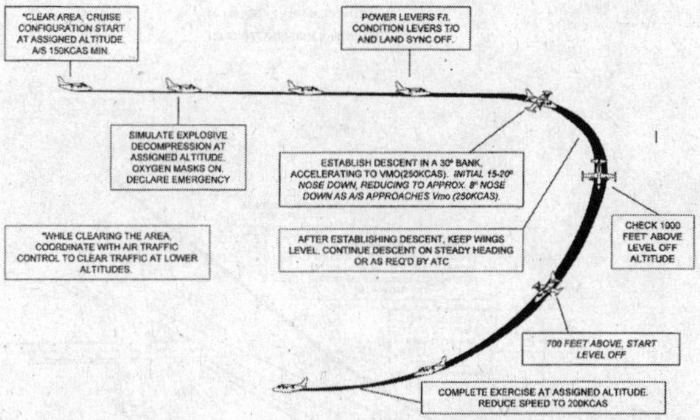

B-13

MU-2B J (-35), K (-25), L (-36), M (-26)
UNUSUAL ATTITUDE RECOVERY (NOSE HIGH)

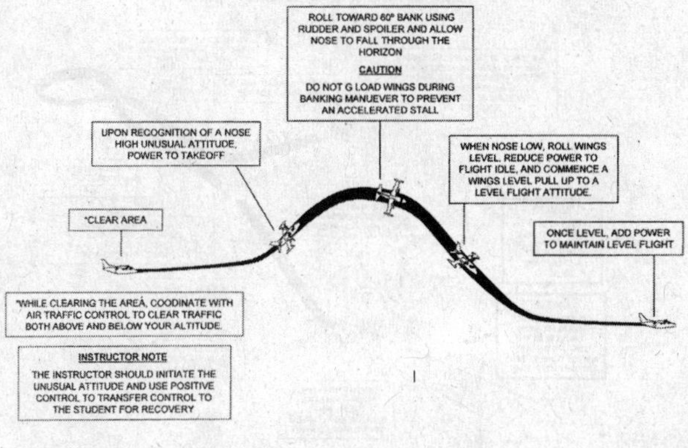

B-14

MU-2B J (-35), K (-25), L (-36), M (-26)
UNUSUAL ATTITUDE RECOVERY (NOSE LOW)

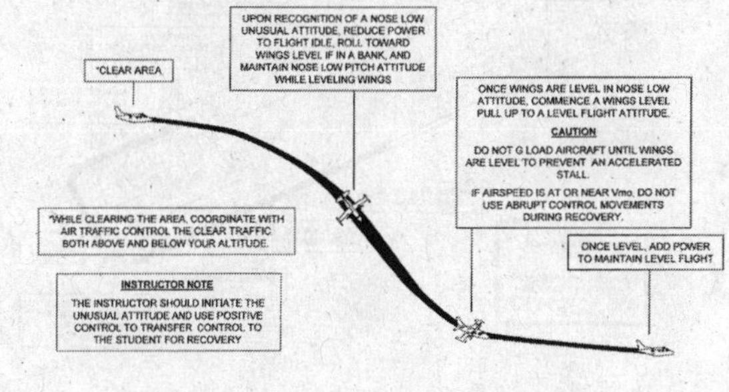

B-15

MU-2B J (-35), K (-25), L (-36), M (-26)
NORMAL LANDING (20°or 40° FLAPS)

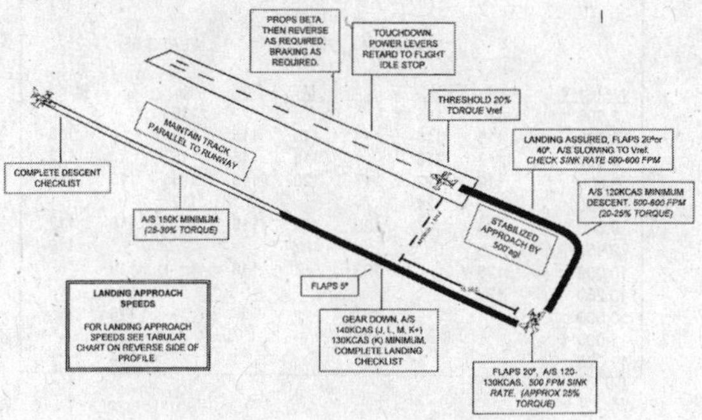

B-16

LANDING APPROACH SPEEDS Vref
J, K, L, M

	FLAPS 20º (1.3 VS1)				FLAPS 40º (1.5 VS1)			
WEIGHT	K	M	J	L	K	M	J	L
7,000	93	96				96		
7,500	96	100	93		99	100	100	
8,000	100	103	96		103	103	103	
8,500	103	106	100	99	106	106	106	105
9,000	106	109	103	99	109	109	109	108
9,435	108				112			
9,500		112	106	103		112	112	111
9,955		115				115		
10,000			109	105			115	114
10,260			110				117	
10,500				108				117
11,000				110				119
11,025				110				119

MU-2B J (-35), K (-25), L (-36), M (-26)
GO AROUND - REJECTED LANDING

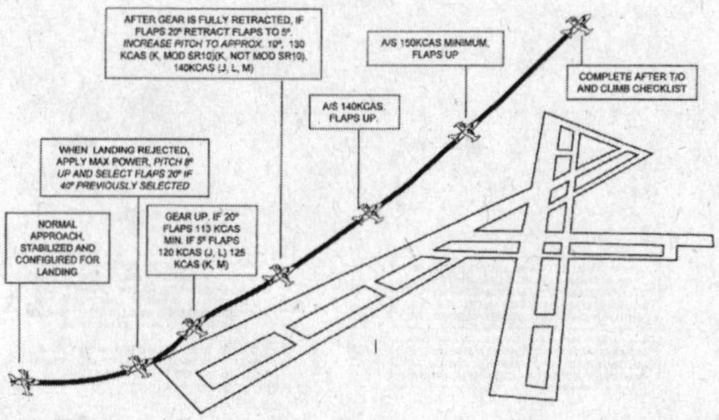

B-17

MU-2B J (-35), K (-25), L (-36), M (-26)

NO FLAP OR 5° FLAP LANDING

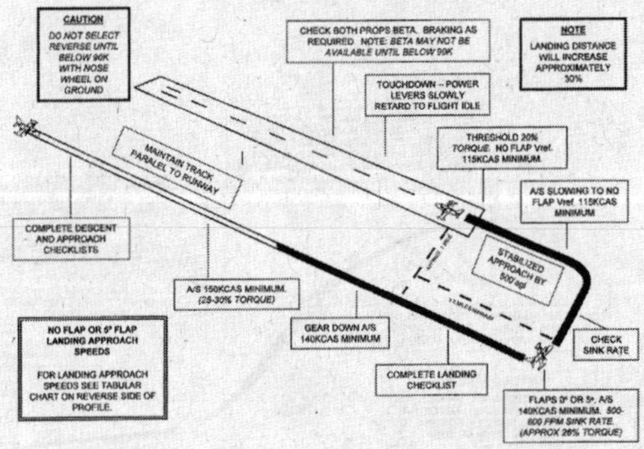

B-18

NO FLAP Vref 1.25 VS1
(BUT NOT BELOW 115KCAS)
USE FOR FLAP UP OR 5°

J, K, L, M

	FLAPS UP				FLAPS 5°			
WEIGHT	J	K	L	M	J	K	L	M
7,500		115				115		
8.000	115	115		115	115	115		115
8.500	117	118		118	115	115		115
9.000	119	122	117	120	115	115	115	115
9,435		124				117		
9.500	123		120	124	115		115	117
9.955				127				119
10.000	125		123		118		115	
10.260	128							
10.500			127				118	
11,000			129				124	
11.025			129				124	

MU-2B J (-35), K (-25), L (-36), M (-26)

ONE ENGINE INOPERATIVE LANDING

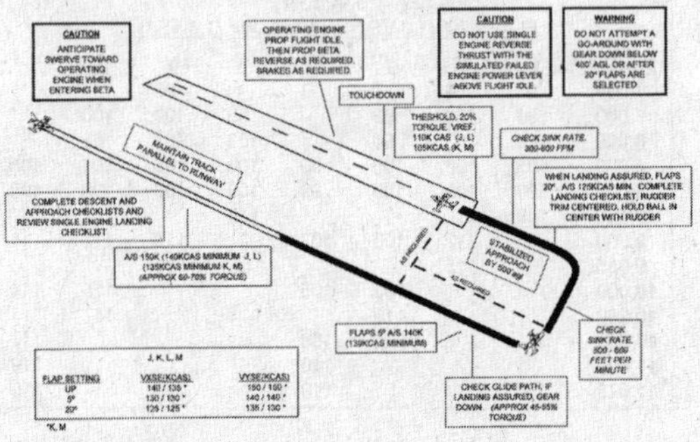

B-19

MU-2B J (-35), K (-25), L (-36), M (-26)

CROSSWIND LANDING

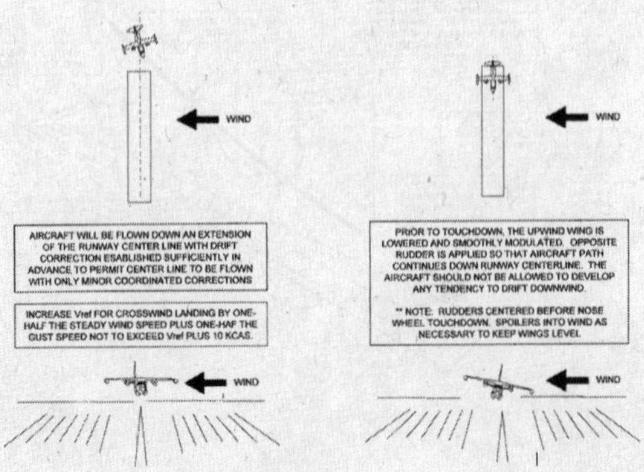

MU-2B J (-35), K (-25), L (-36), M (-26)

ILS AND MISSED APPROACH

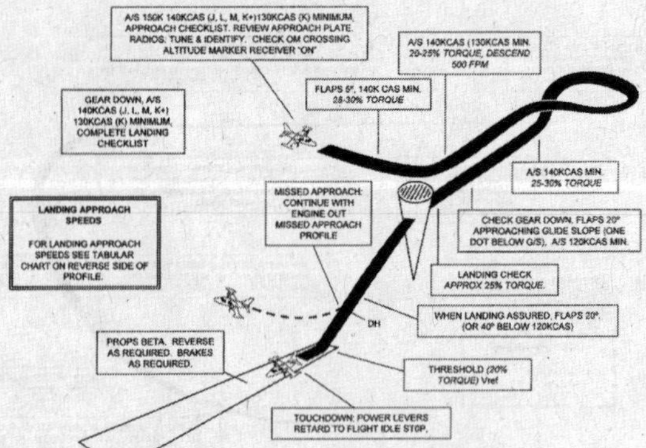

B-21

LANDING APPROACH SPEEDS Vref
J, K, L, M

	FLAPS 20° (1.3 VS1)				FLAPS 40° (1.5 VS1)			
WEIGHT	K	M	J	L	K	M	J	L
7,000	93	96				96		
7,500	96	100	93		99	100	100	
8,000	100	103	96		103	103	103	
8,500	103	106	100	99	106	106	106	105
9,000	106	109	103	99	109	109	109	108
9,435	108				112			
9,500		112	106	103		112	112	111
9,955		115				115		
10,000			109	105			115	114
10,260			110				117	
10,500				108				117
11,000				110				119
11,025				110				119

MU-2B J (-35), K (-25), L (-36), M (-26)

TWO ENGINE MISSED APPROACH

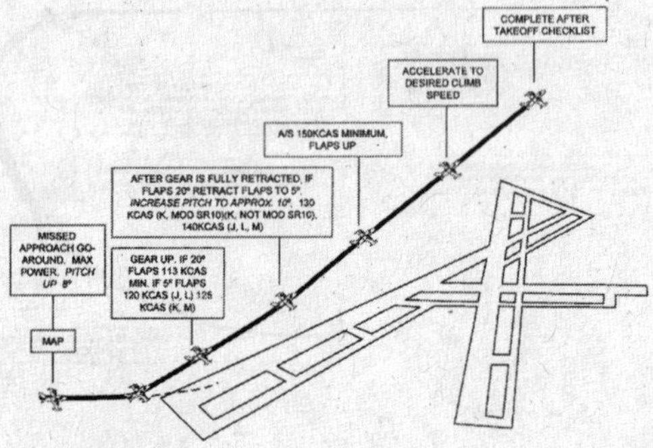

B-22

MU-2B J (-35), K (-25), L (-36), M (-26)

ONE ENGINE INOPERATIVE ILS AND MISSED APPROACH

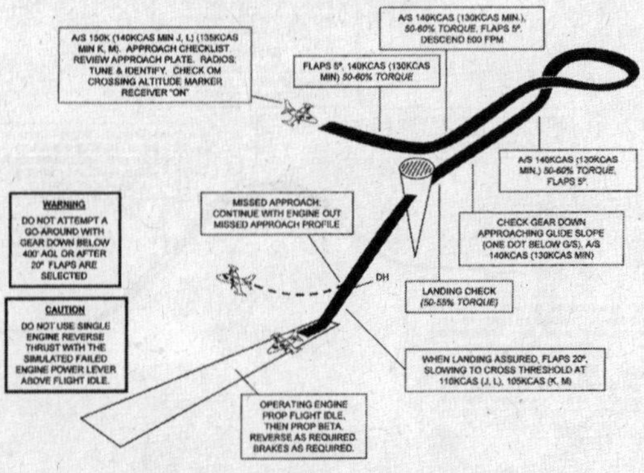

B-23

MU-2B J (-35), K (-25), L (-36), M (-26)
ONE ENGINE INOPERATIVE MISSED APPROACH

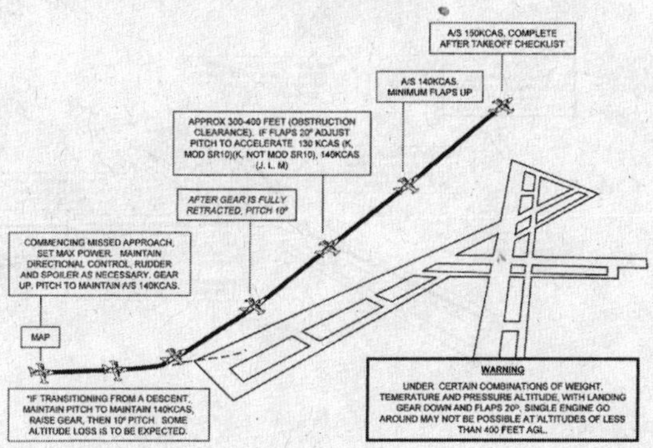

B-24

MU-2B J (-35), K (-25), L (-36), M (-26)
NON-PRECISION AND MISSED APPROACH

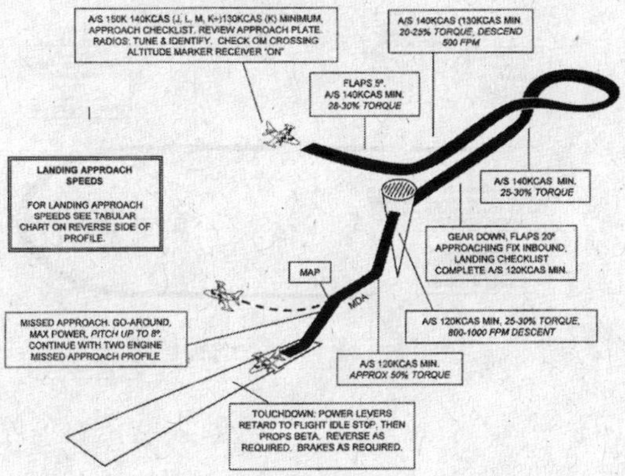

B-25

LANDING APPROACH SPEEDS Vref J, K, L, M	FLAPS 20° (1.3 VS1)				FLAPS 40° (1.5 VS1)			
WEIGHT	K	M	J	L	K	M	J	L
7,000	93	96				96		
7,500	96	100	93		99	100	100	
8,000	100	103	96		103	103	103	
8,500	103	106	100	99	106	106	106	105
9,000	106	109	103	99	109	109	109	108
9,435	108				112			
9,500		112	106	103		112	112	111
9,955		115				115		
10,000			109	105			115	114
10,260			110				117	
10,500				108				117
11,000				110				119
11,025				110				119

MU-2B J (-35), K (-25), L (-36), M (-26)
ONE ENGINE INOPERATIVE NON-PRECISION AND MISSED APPROACH

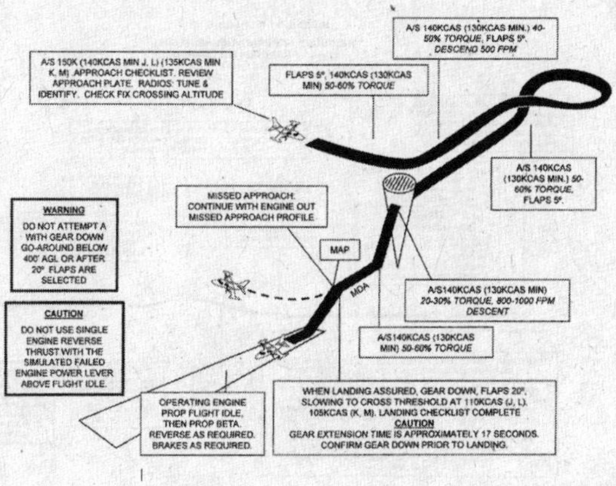

B-26

MU-2B J (-35), K (-25), L (-36), M (-26)
CIRCLING APPROACH AT WEATHER MINIMUMS

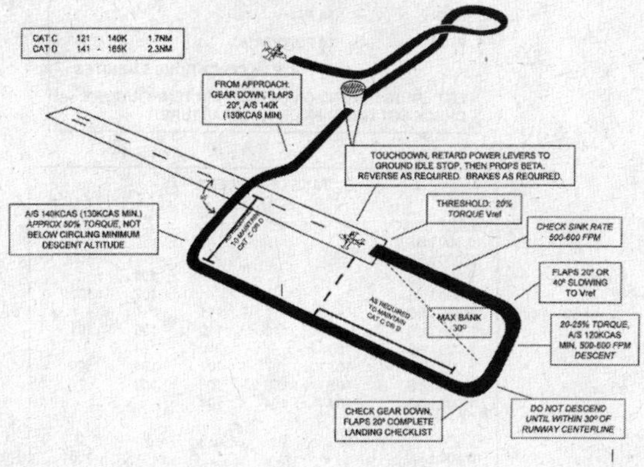

B-27

LANDING APPROACH SPEEDS Vref J, K, L, M	FLAPS 20° (1.3 VS1)				FLAPS 40° (1.5 VS1)			
WEIGHT	K	M	J	L	K	M	J	L
7,000	93	96				96		
7,500	96	100	93		99	100	100	
8,000	100	103	96		103	103	103	
8,500	103	106	100	99	106	106	106	105
9,000	106	109	103	99	109	109	109	108
9,435	108				112			
9,500		112	106	103		112	112	111
9,955		115				115		
10,000			109	105			115	114
10,260			110				117	
10,500				108				117
11,000				110				119
11,025				110				119

MU-2B J (-35), K (-25), L (-36), M (-26)
ONE ENGINE INOPERATIVE CIRCLING APPROACH AT WEATHER MINIMUMS

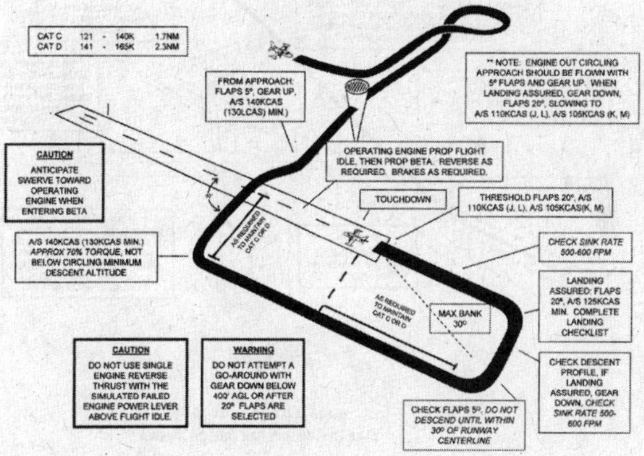

B-28

MU-2B B, D (-10), F (-20), G (-30)
NORMAL TAKE-OFF, 5° OR 20° FLAPS

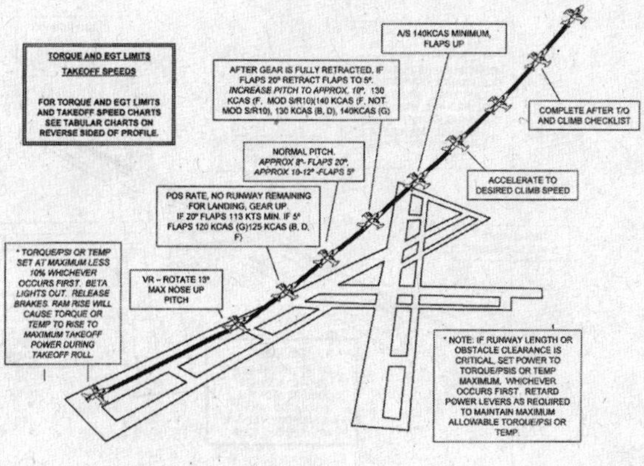

C-1

TORQUE LIMITS	
B, D	64 PSI
F, G	60 PSI (STATIC)
	64 PSI (RAM CONDITIONS 5 MINUTES

EGT LIMITS DEPEND ON OUTSIDE AIR TEMPERATURE, CHECK EGT LIMITS PRIOR TO DEPARTURE.

TAKE OFF SPEEDS
ROTATE

FLAPS 5°	B	B+	D	F	G
10,800 LBS					109
10,000 LBS					105
9,920 LBS				108	
9,500 LBS				107	103
9,350 LBS		111	111		
9,000 LBS		110	110	106	101
8,930 LBS	109				
8,000 LBS	107	107	107	104	100
7,500 LBS	106	106	106	102	
7,000 LBS	104	104	104		

FLAPS 20°	B	B+	D	F	G
10,800 LBS					105
10,000 LBS					102
9,920 LBS				102	
9,500 LBS				101	101
9,350 LBS		104	104		
9,000 LBS		103	103	100	100
8,930 LBS	103				
8,000 LBS	101	101	101	99	100
7,500 LBS	100	100	100	98	
7,000 LBS	99	99	99		

B: NOT MODIFIED BYH S/B 036 AND S/B 092
B+: MODIFIED BY S/B 036 AND S/B 092

MU-2B B, D (-10), F (-20), G (-30)
TAKE-OFF ENGINE FAILURE – FLAPS 5° OR 20°

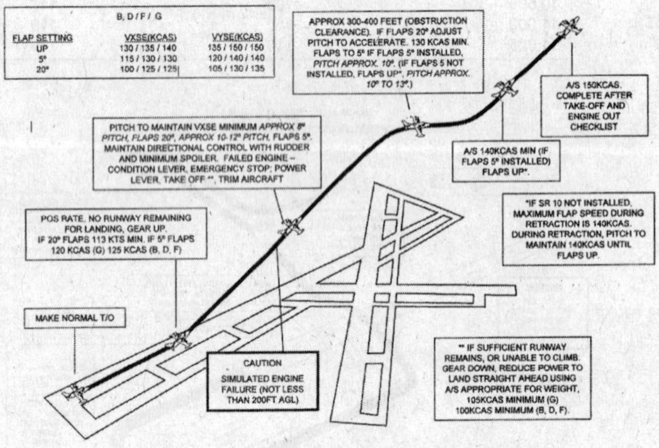

C-2

MU-2B B, D (-10), F (-20), G (-30)
TAKE-OFF ENGINE FAILURE ON RUNWAY

CAUTION
SIMULATED ENGINE FAILURE OR MALFUNCTION IS TO BE GIVEN BY INSTRUCTOR AT NOT MORE THAN 50% OF ROTATE SPEEDS.

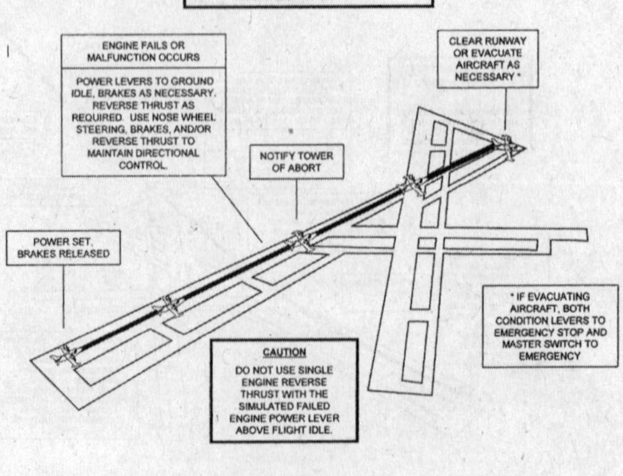

C-3

MU-2B B, D (-10), F (-20), G (-30)
TAKE-OFF ENGINE FAILURE - UNABLE TO CLIMB
CLASSROOM DISCUSSION OR FTD USE ONLY

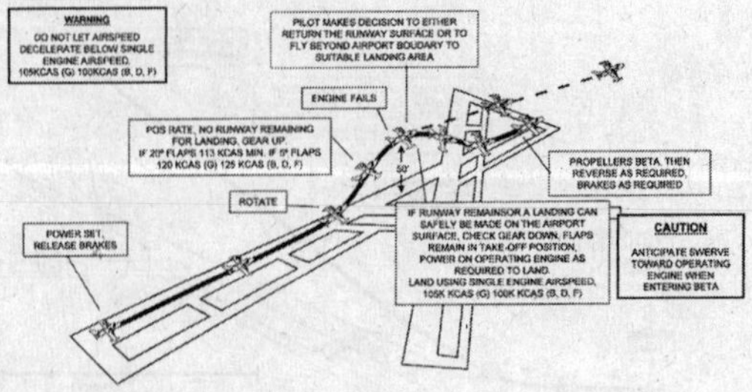

C-4

MU-2B B, D (-10), F (-20), G (-30)
STEEP TURNS

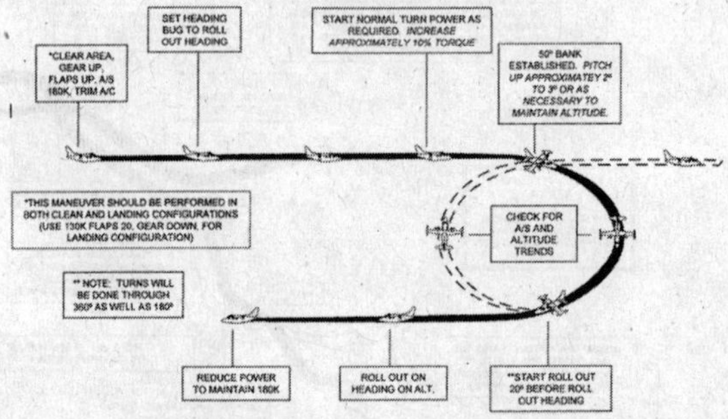

C-5

MU-2B B, D (-10), F (-20), G (-30)
SLOW FLIGHT MANEUVERING
MINIMUM CONTROLLABLE AIRSPEED

SLOW FLIGHT MANEUVERING IS CONDUCTED AS FOLLOWS:

CLEAR THE AREA PRIOR TO BEGINNING THE MANEUVER.

START WITH CLEAN CONFIGURATION AND CHANGE AIRCRAFT CONFIGURATION FROM CLEAN TO FULL FLAP AND GEAR IN STAGES. USE A MAXIMUM OF 15° BANK AND PERFORM HEADING CHANGES OF 90° LEFT AND RIGHT. CONSTANT ALTITUDE IS REQUIRED THROUGHOUT.
MAINTAIN 115K IN ALL CONFIGURATIONS.

**APPROXIMATE POWER SETTINGS ARE:

CLEAN	*TORQUE (35%) OR PSI (23) PER ENGINE*	*APPROX PITCH +12*
5° FLAP	*TORQUE (32%) OR PSI (21) PER ENGINE*	*APPROX PITCH +8*
5° FLAP & GEAR	*TORQUE (44%) OR PSI (29) PER ENGINE*	*APPROX PITCH +9*
20° FLAP & GEAR	*TORQUE (42%) OR PSI (27) PER ENGINE*	*APPROX PITCH +4*
40° FLAP & GEAR	*TORQUE (54%) OR PSI (35) PER ENGINE*	*APPROX PITCH 0*

** NOTE: POWER SETTINGS WILL VARY WITH AIRCRAFT WEIGHT AND ALTITUDE.

STALL SPEEDS (APPROXIMATE)
AT MAXIMUM GROSS TAKEOFF WEIGHT
B, B+, D, F, G

	B/ B+ / D / F / G	B / B+ / D / F / G
ANGLE OF BANK	0°	15°
FLAPS		
UP	95/ 98 / 98/102/104	98/ 99/ 99/104/106
5°	85/ 88/ 88/ 95/ 98	88/ 89/ 89/ 97/100
20°	80/ 81/ 81/ 85/ 88	81/ 83/ 83/ 87/ 88
40°	72/ 73/ 73/ 77/ 80	73/ 74/ 74/ 78/ 81

Vmc: 20° FLAPS (90KCAS G, 93KCAS F, 89KCAS D, 89/91KCAS B)
5° FLAPS (99KCAS G, 100KCAS F, 97KCAS D, 97/99KCAS B)
(FOR B MODEL Vmc SPEED CONSULT SERIAL NUMBER APPLICABILITY IN AFM)

CAUTION
STALL WARNING MAY ACTIVATE 4 TO 9 KTS ABOVE STALL

MINIMUM CONTROLLABLE AIRSPEED IS CONDUCTED AS FOLLOWS:

CLEAR THE AREA PRIOR TO BEGINNING THE MANEUVER.

THE MANEUVER MAY BE DONE IN ANY COMBINATION OF GEAR OR FLAP CONFIGURATIONS. IF BANK IS TO BE USED, IT SHOULD BE DONE AT BANK OF NOT MORE THAN 10°. BEGIN THE MANEUVER BY CONFIGURING THE AIRCRAFT IN THE DESIRED GEAR AND FLAP CONFIGURATION. SLOW THE AIRCRAFT UNTIL THE STALL WARNING (STICK SHAKER) IS ACTIVATED AND ADD POWER TO MAINTAIN ALTITUDE AND A SPEED JUST ABOVE AERODYNAMIC STALL. DO NOT ALLOW THE AIRCRAFT TO REACH AERODYNAMIC STALL BUFFET.

C-6

MU-2B B, D (-10), F (-20), G (-30)
ONE ENGINE INOPERATIVE MANEUVERING
LOSS OF DIRECTIONAL CONTROL

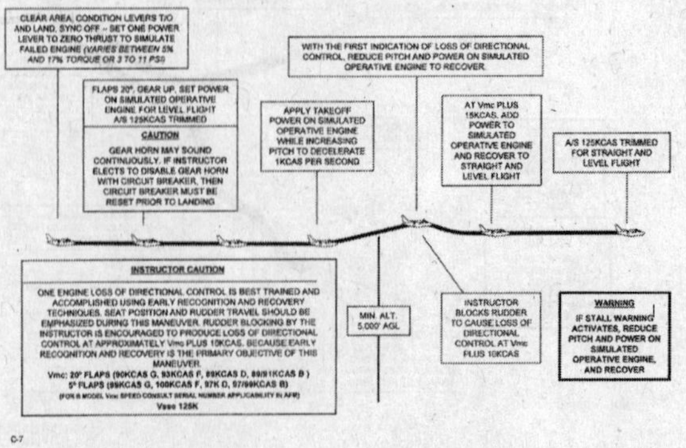

C-7

MU-2B B, D (-10), F (-20), G (-30)

APPROACH TO STALL CLEAN CONFIGURATION / WINGS LEVEL

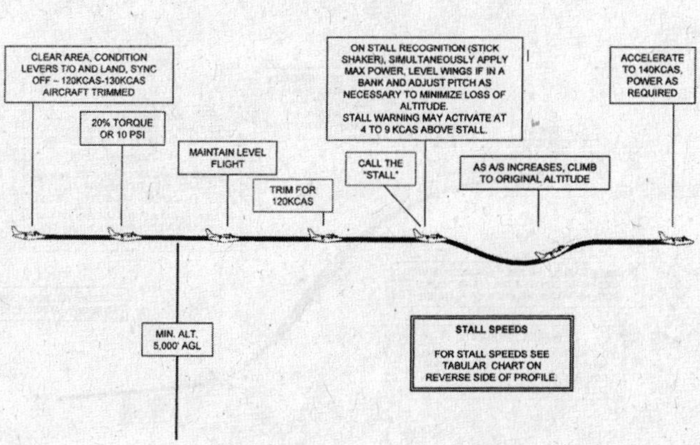

C-8

STALL SPEEDS

FLAPS SET	0	5	20	40
GR.WT.	B/ B+, D/ F/ G	B/ B+, D/ F/ G	B/ B+, D/ F/ G	B/ B+, D/ F/ G
7,000	85/ 85/ 85	76/ 76/ 80	70/ 70/ 72	63/ 63/ 64
7,500	88/ 88/ 85/	78/ 78/ 83/	73/ 73/ 74/	66/ 63/ 67/
8,000	90/ 90/ 91/ 90	81/ 81/ 86/ 84	75/ 75/ 77/ 74	68/ 68/ 69/ 69
8,500	93/ 93/ 94/ 93	83/ 83/ 88/ 87	78/ 78/ 79/ 77	70/ 70/ 71/ 71
8,930	95/	85/	79/	72/
9,000	/ 95/ 97/ 95	/ 86/ 91/ 90	/ 80/ 81/ 79	/ 72/ 73/ 73
9,350	/ 97/	/ 87/	/ 81/	/ 73/
9,500	/ 99/ 98	93/ 92	/ 83/ 81	/ 75/ 75
9,920	/101/	95/	/ 85/	/ 76/
10,000	/101	/ 94	/ 83	/ 77
10,500	/103	/ 97	/ 85	/ 79
10,800	/105	/ 98	/. 87	/ 81

MU-2B B, D (-10), F (-20), G (-30)

APPROACH TO STALL

TAKEOFF CONFIGURATION 15-30° BANK

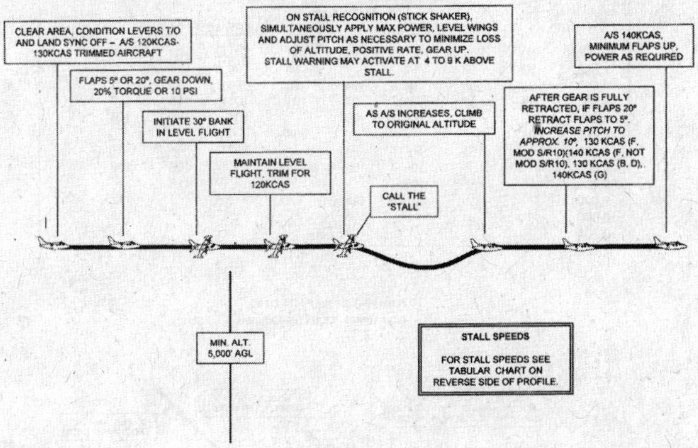

C-9

STALL SPEEDS (APPROXIMATE)
AT MAXIMUM GROSS TAKEOFF WEIGHT
B, B+, D, F, G

BANK ANGLE	10	20	30	40	50	60
FLAPS	B/ B+, D/ F/ G	B/ B+, D/ F/ G	B/ B+, D/ F/ G	B/ B+, D/ F/ G	B/ B+, D/ F/ G	B/ B+, D/ F/ G
UP	96/ 99/102/106	99/ 101/105/108	103/105/109/112	109/111/116/120	120/122/126/130	136/138/143/148
5°	87/ 88/ 96/ 99	89/ 90/ 98/101	92/ 94/102/105	98/100/109/112	107/109/119/122	120/124/135/138
20°	80/ 82/ 86/ 87	82/ 84/ 88/ 89	86/ 87/ 92/ 93	91/ 93/ 97/ 98	99/101/107/108	113/114/120/122
40°	72/ 74/ 77/ 81	74/ 75/ 79/ 82	77/ 79/ 82/ 86	82/ 83/ 87/ 91	90/ 91/ 95/100	102/103/108/113

MU-2B B, D (-10), F (-20), G (-30)

APPROACH TO STALL

GEAR DOWN – FULL FLAPS

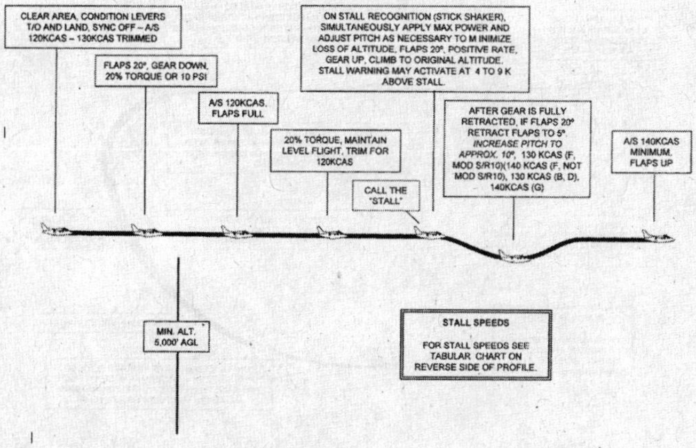

C-10

STALL SPEEDS

FLAPS SET	0	5	20	40
GR.WT.	B/ B+, D/ F/ G	B/ B+, D/ F/ G	B/ B+, D/ F/ G	B/ B+, D/ F/ G
7,000	85/ 85/ 85	76/ 76/ 80	70/ 70/ 72	63/ 63/ 64
7,500	88/ 88/ 85/	78/ 78/ 83/	73/ 73/ 74/	66/ 63/ 67/
8,000	90/ 90/ 91/ 90	81/ 81/ 86/ 84	75/ 75/ 77/ 74	68/ 68/ 69/ 69
8,500	93/ 93/ 94/ 93	83/ 83/ 88/ 87	78/ 78/ 79/ 77	70/ 70/ 71/ 71
8,930	95/	85/	79/	72/
9,000	/ 95/ 97/ 95	/ 86/ 91/ 90	/ 80/ 81/ 79	/ 72/ 73/ 73
9,350	/ 97/	/ 87/	/ 81/	/ 73/
9,500	/ 99/ 98	93/ 92	/ 83/ 81	/ 75/ 75
9,920	/101/	95/	/ 85/	/ 76/
10,000	/101	/ 94	/ 83	/ 77
10,500	/103	/ 97	/ 85	/ 79
10,800	/105	/ 98	/ 87	/ 81

MU-2B B, D (-10), F (-20), G (-30)

ACCELERATED STALLS

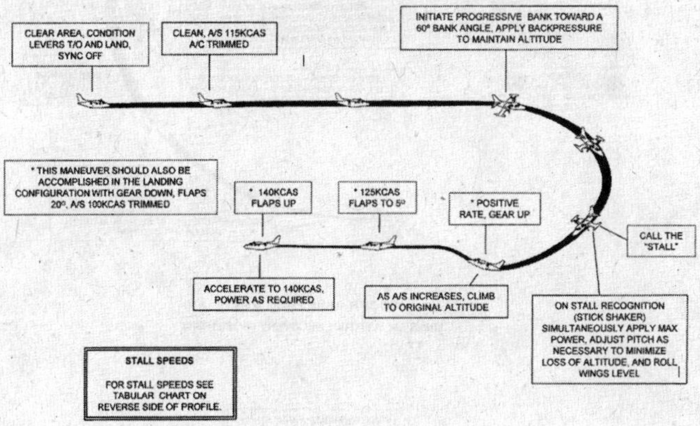

C-11

STALL SPEEDS (APPROXIMATE)
AT MAXIMUM GROSS TAKEOFF WEIGHT
B, B+, D, F, G

BANK ANGLE	10	20	30	40	50	60
FLAPS	B/ B+, D/ F/ G	B/ B+, D/ F/ G	B/ B+, D/ F/ G	B/ B+, D/ F/ G	B/ B+, D/ F/ G	B/ B+, D/ F/ G
UP	96/ 99/102/106	99/ 101/105/108	103/105/109/112	109/111/116/120	120/122/126/130	136/138/143/148
5°	87/ 88/ 96/ 99	89/ 90/ 98/101	92/ 94/102/105	98/100/109/112	107/109/119/122	120/124/135/138
20°	80/ 82/ 86/ 87	82/ 84/ 88/ 89	86/ 87/ 92/ 93	91/ 93/ 97/ 98	99/101/107/108	113/114/120/122
40°	72/ 74/ 77/ 81	74/ 75/ 79/ 82	77/ 79/ 82/ 86	82/ 83/ 87/ 91	90/ 91/ 95/100	102/103/108/113

MU-2B B, D (-10), F (-20), G (-30)

EMERGENCY DESCENT (LOW SPEED)

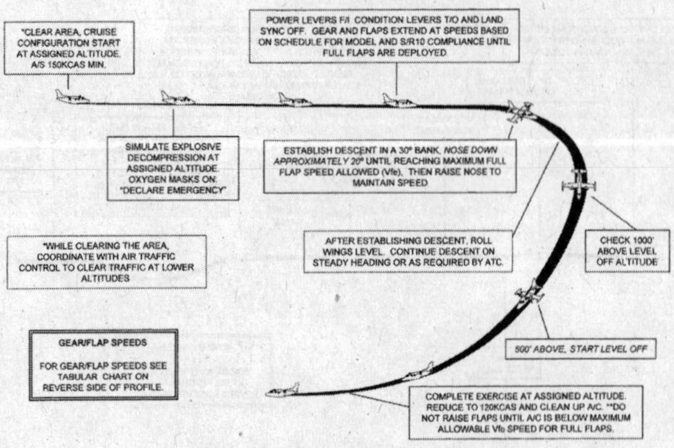

MU-2B B, D (-10), F (-20), G (-30)

UNUSUAL ATTITUDE RECOVERY (NOSE LOW)

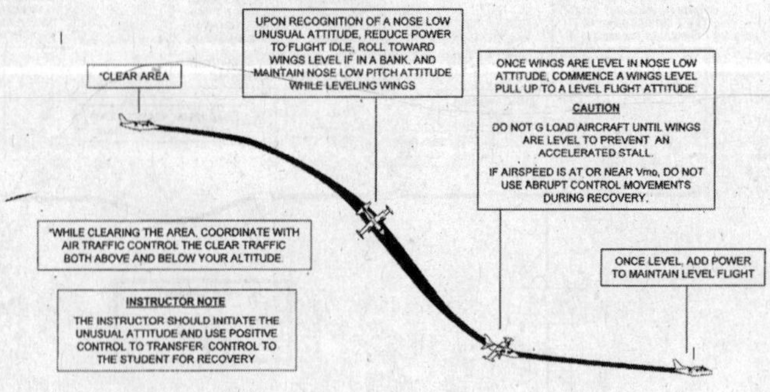

GEAR AND FLAP EXTEND SCHEDULE

(F+ AND G+ ARE MODIFIED BY S/R10)

GEAR			
B, D, F, F+:	**160KCAS**		
G, G+:	**170KCAS**		
FLAPS	**5°**	**20°**	**40°**
G: NOT MODIFIED BY S/R10	**146KCAS**	**146KCAS**	**120KCAS**
G+: MODIFIED BY S/R10 AND	**175KCAS**	**146KCAS**	**120KCAS**
F: NOT MODIFIED BY S/R10	**140KCAS**	**140KCAS**	**120KCAS**
F+: MODIFIED BY S/R10 AND	**175KCAS**	**140KCAS**	**120KCAS**
B, D, F	**140KCAS**	**140KCAS**	**120KCAS**

C-15

MU-2B B, D (-10), F (-20), G (-30)

NORMAL LANDING (20° or 40° FLAPS)

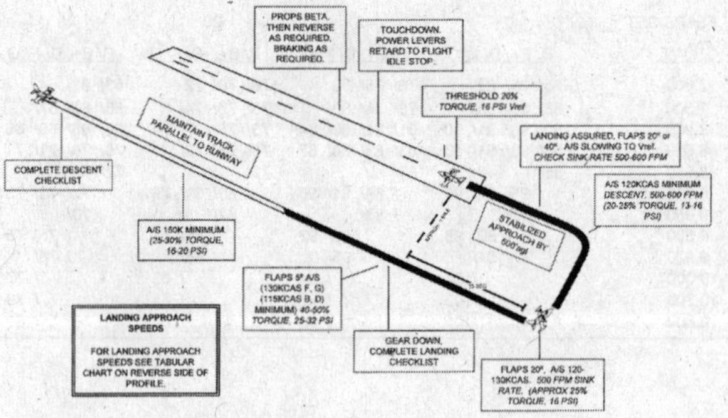

C-16

MU-2B B, D (-10), F (-20), G (-30)

EMERGENCY DESCENT (HIGH SPEED)

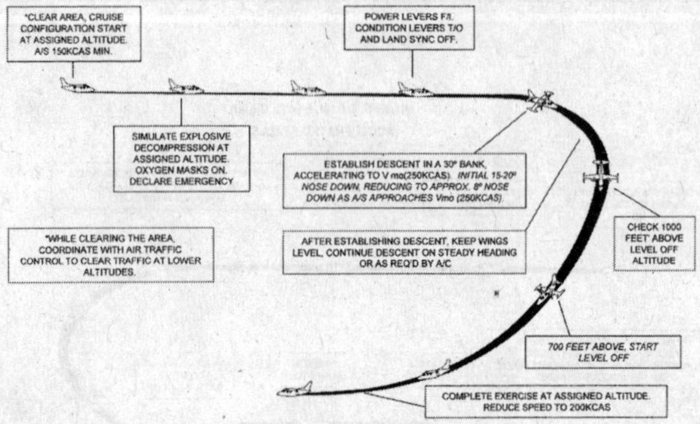

LANDING APPROACH SPEEDS Vref B, B+, D, F, G								
	FLAPS 20° (1.3 VSI)				FLAPS 40° (1.5 VSI)			
WEIGHT	B	B+, D	F	G	B	B+, D	F	G
7,000	92	92	93		94	94		
7,500	95	95	96	94	98	98	99	100
8,000	98	98	100	97	101	101	103	103
8,500		101	103	100		104	106	106
8,490	101				104			
8,930		103				107		
9,000			106	103			109	109
9,435			108				112	
9,500				105				112
10,000				108				115
10,260				109				117

MU-2B B, D (-10), F (-20), G (-30)

UNUSUAL ATTITUDE RECOVERY (NOSE HIGH)

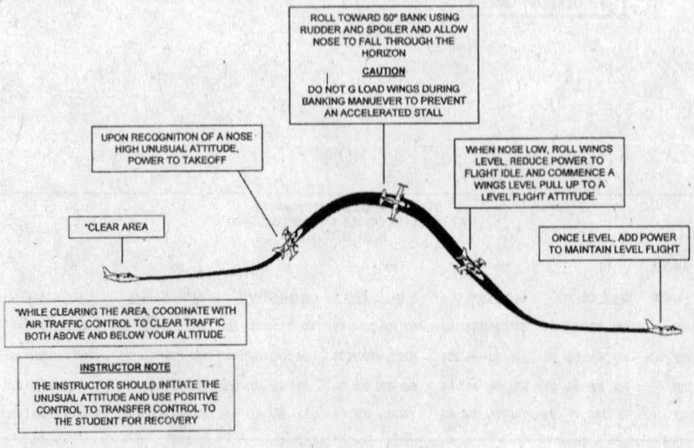

MU-2B B, D (-10), F (-20), G (-30)

GO AROUND - REJECTED LANDING

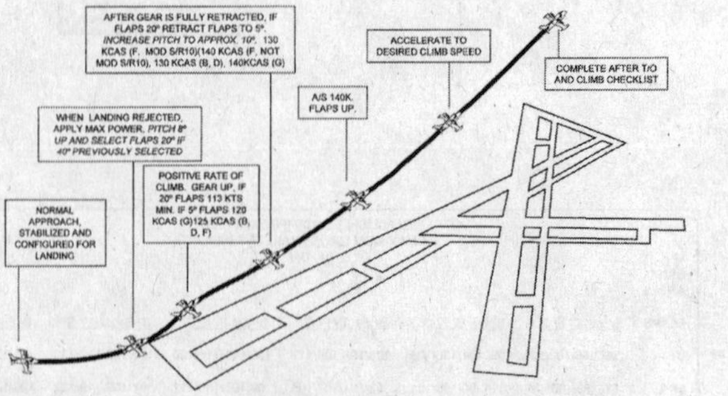

C-17

MU-2B B, D (-10), F (-20), G (-30)

NO FLAP OR 5° FLAP LANDING

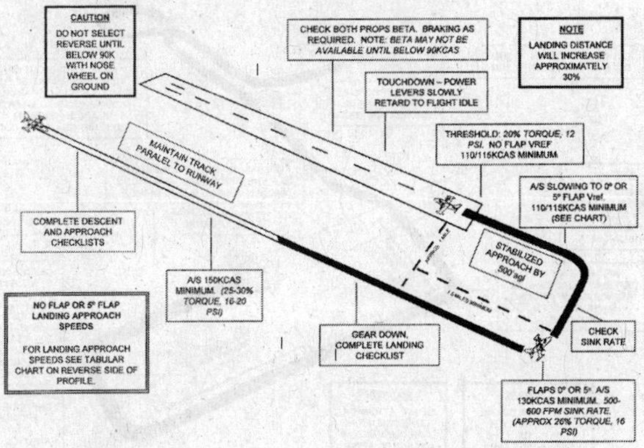

A-18

NO FLAP Vref 1.25 VS1 (BUT NOT BELOW 110KCAS (B, B+, D, F) 115KCAS (G)) USE FOR FLAP UP OR 5° B, B+, D, F, G										
	FLAPS UP					FLAPS 5°				
WEIGHT	B	B+	D	F	G	B	B+	D	F	G
7,500	110	110	110	110		110	110	110	110	
8.000	113	113	113	114	115	110	110	110	110	115
8,490	117					110				
8.500		117	117	118	117		110	110	110	115
8,930		119	119				110	110		
9.000				122	119				114	115
9,435				124					117	
9.500					123					115
10.000					127					118
10.260					128					120

MU-2B B, D (-10), F (-20), G (-30)

ONE ENGINE INOPERATIVE LANDING

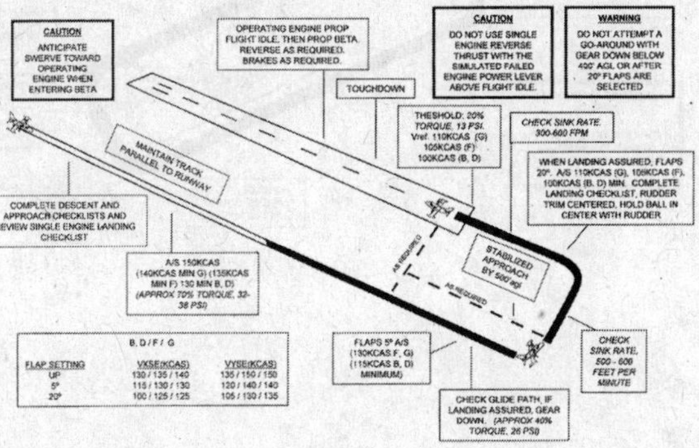

C-19

MU-2B B, D (-10), F (-20), G (-30)

CROSSWIND LANDING

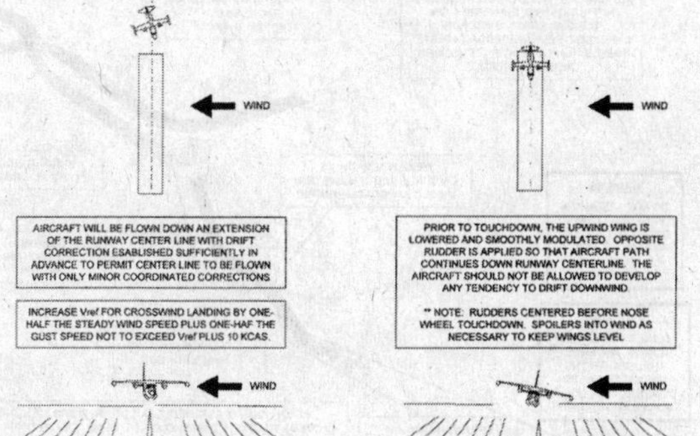

C-20

MU-2B B, D (-10), F (-20), G (-30)

ILS AND MISSED APPROACH

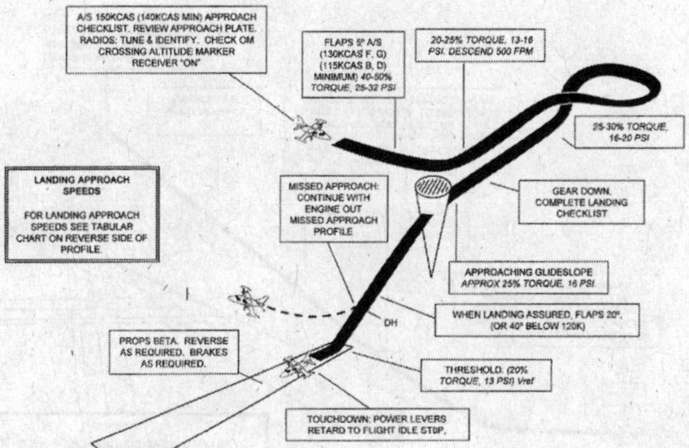

C-21

LANDING APPROACH SPEEDS Vref B, B+, D, F, G								
	FLAPS 20° (1.3 VSI)				FLAPS 40° (1.5 VSI)			
WEIGHT	B	B+, D	F	G	B	B+, D	F	G
7,000	92	92	93		94	94		
7,500	95	95	96	94	98	98	99	100
8,000	98	98	100	97	101	101	103	103
8,500		101	103	100		104	106	106
8,490	101				104			
8,930		103				107		
9,000			106	103			109	109
9,435			108				112	
9,500				105				112
10,000				108				115
10,260				109				117

MU-2B B, D (-10), F (-20), G (-30)

TWO ENGINE MISSED APPROACH

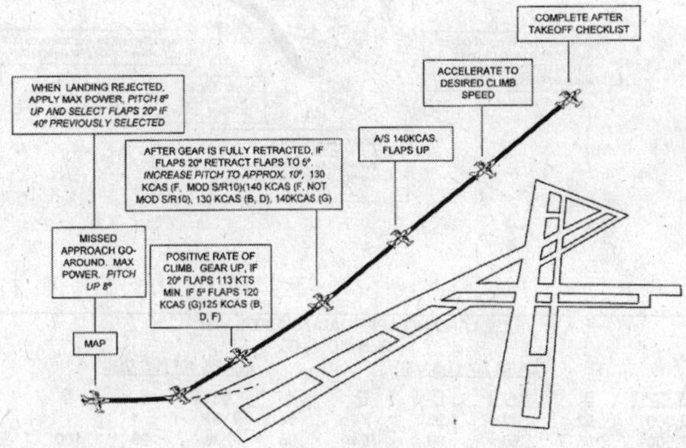

C-22

MU-2B B, D (-10), F (-20), G (-30)

ONE ENGINE INOPERATIVE ILS AND MISSED APPROACH

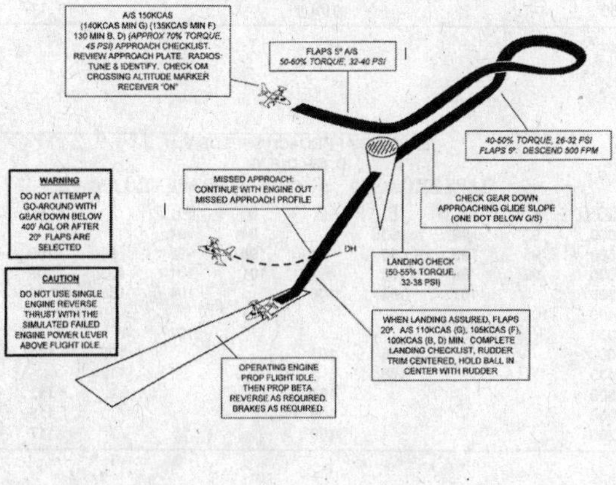

C-23

MU-2B B, D (-10), F (-20), G (-30)

ONE ENGINE INOPERATIVE MISSED APPROACH

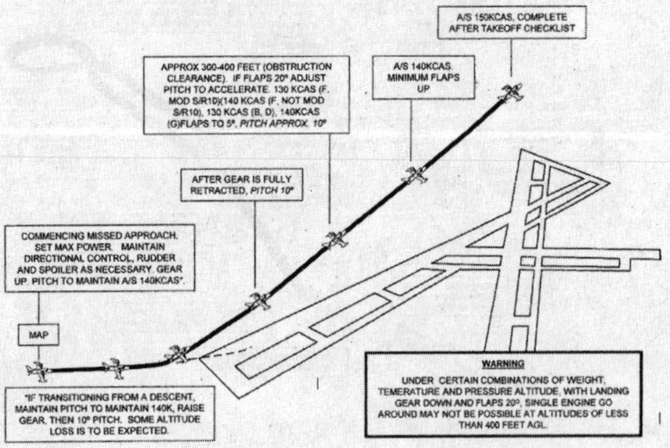

MU-2B B, D (-10), F (-20), G (-30)

NON-PRECISION AND MISSED APPROACH

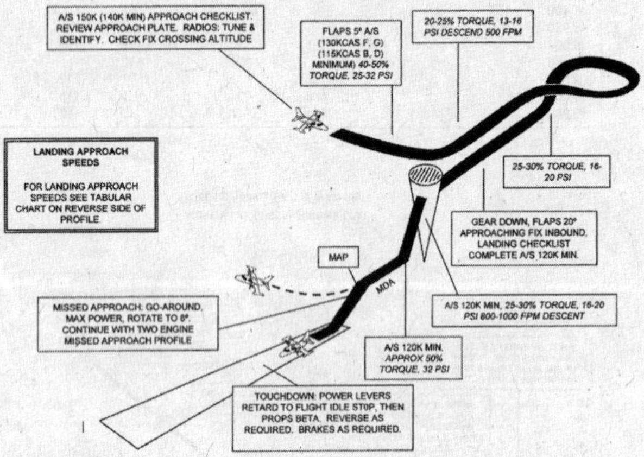

LANDING APPROACH SPEEDS Vref
B, B+, D, F, G

	FLAPS 20° (1.3 VSI)				FLAPS 40° (1.5 VSI)			
WEIGHT	B	B+, D	F	G	B	B+, D	F	G
7,000	92	92	93		94	94		
7,500	95	95	96	94	98	98	99	100
8,000	98	98	100	97	101	101	103	103
8,500		101	103	100		104	106	106
8,490	101				104			
8,930		103				107		
9,000			106	103			109	109
9,435			108				112	
9,500				105				112
10,000				108				115
10,260				109				117

LANDING APPROACH SPEEDS Vref
B, B+, D, F, G

	FLAPS 20° (1.3 VSI)				FLAPS 40° (1.5 VSI)			
WEIGHT	B	B+, D	F	G	B	B+, D	F	G
7,000	92	92	93		94	94		
7,500	95	95	96	94	98	98	99	100
8,000	98	98	100	97	101	101	103	103
8,500		101	103	100		104	106	106
8,490	101				104			
8,930		103				107		
9,000			106	103			109	109
9,435			108				112	
9,500				105				112
10,000				108				115
10,260				109				117

MU-2B B, D (-10), F (-20), G (-30)

ONE ENGINE INOPERATIVE CIRCLING APPROACH AT WEATHER MINIMUMS

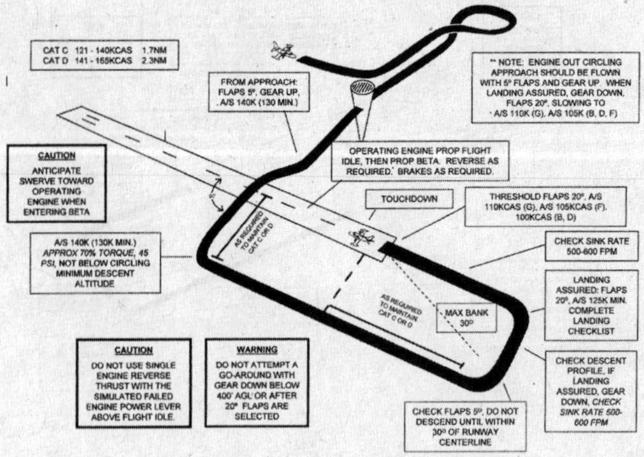

MU-2B B, D (-10), F (-20), G (-30)

CIRCLING APPROACH AT WEATHER MINIMUMS

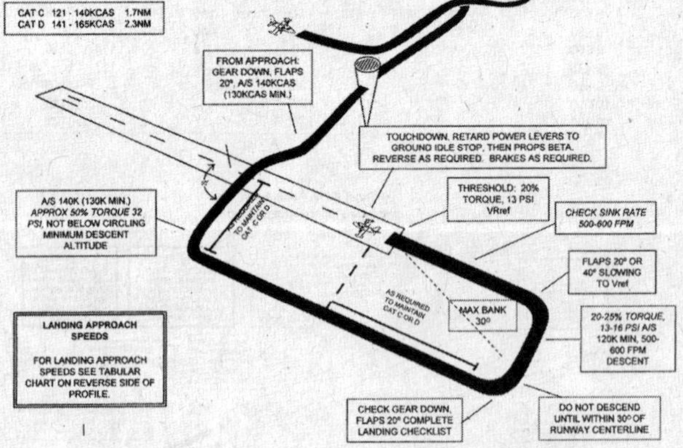

MU-2B B, D (-10), F (-20), G (-30)

ONE ENGINE INOPERATIVE NON-PRECISION AND MISSED APPROACH

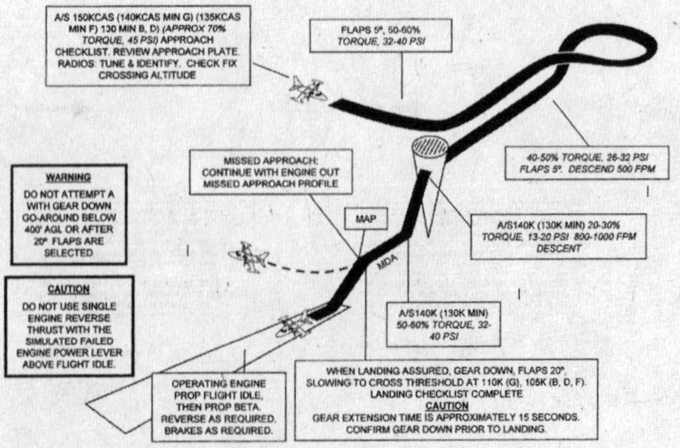

In Flight Maneuvering

(A) Maneuvers conducted at altitude such as stalls and steep turns must always be preceded by clearing turns and at least one crew member must continually clear the flying area during the maneuver. The instructor must emphasize the importance of clearing the area, even if the maneuvers are being done in an FTD or simulator. This will create the habit pattern in the pilot to clear the area before practicing maneuvers.

(B) During stalling maneuvers and upon recognition of the indication of a stall, the pilot must call the "stall" to the instructor and then proceed with the recovery. In addition, during training, the pilot must announce the completion of the stall recovery maneuver. Instructors must exercise caution when conducting stall maneuvers and be prepared to take the controls if the safe outcome of the maneuver is in doubt.

(C) During accelerated stall maneuvers, it is important that the instructor pay close attention to the position of the ball throughout the maneuver and recovery so as to maintain coordinated flight. Stall recognition and recovery is the completion criteria, and it is not necessary to continue the stall beyond the stick shaker to aerodynamic buffet.

(D) When demonstrating a loss of directional control with one engine inoperative, the engine failure must only be simulated. During the slowing of the aircraft to demonstrate loss of directional control, the instructor should use the rudder block method to allow the student to experience the loss of directional control associated with VMC, at a speed of approximately 10 knots above actual VMC.

Note: To accurately simulate single engine operations, zero thrust must be established. The zero thrust torque setting will vary greatly from model to model. It is important to establish to zero thrust torque setting for your aircraft. This requires that the aircraft be flown on one engine to establish the zero thrust setting. This is accomplished by establishing single engine flight with one propeller feathered and noting the performance with the operating engine at maximum torque or temperature. It is suggested that two airspeeds be established for zero thrust power settings. They are 120 kts, flaps 20, gear up for takeoff and 140 knots, flaps 5, gear up for in-flight and approach maneuvering. Once performance has been established and recorded for each airspeed, restart the other engine and find the torque setting that duplicates the performance (climb or descent rate, airspeed) as was recorded with that propeller feathered. This torque setting will be zero thrust for the simulated inoperative engine. The student/pilot should note that the performance experienced with one engine operating at flight idle, may produce greater performance than if the engine were stopped and the propeller feathered.

Pre-maneuver briefings for any maneuver that requires either an actual engine shutdown or a simulated engine failure must be undertaken when using an aircraft. In the case of an actual engine shutdown, a minimum altitude of 3,000 ft above ground level (agl) must be used and done in a position where a safe landing can be made at an airport in the event of difficulty.

Takeoff and Landing

(A) When using the profiles to establish the procedure for configuring the aircraft for takeoff or landing, it is important to understand that each task for the procedure, as noted on the procedure diagram, establishes the point at which each task should have been completed and not the exact point at which the task should be accomplished unless otherwise stated in the task box. Numbers which represent performance such as descent rates or other maneuvering information that is not contained in the aircraft flight manual are shown in *italics.*

(B) In all takeoff profiles the prompt for the gear to be retracted is "No Runway Remaining, Gear Up". This should set the decision point for making a landback after an engine failure and should normally be reached at altitudes of less than 100 ft AGL. It is impractical to attempt a landback from above 100 ft AGL, because it can require distances up to 10,000 ft from the beginning of the takeoff run to bring the aircraft to a stop. But, even on very long runways, landback will not be necessary above 100 ft AGL and above Vyse for the flap configurations, if the single engine climb capability found in the POM charts, with the gear up, is positive (250 fpm or better) and obstacles clearance is not an issue.

(C) The manufacturers FAA-accepted checklists and checklist in Appendix C to this SFAR No. 108 describe a procedure for the discontinuance of flight following an engine failure after takeoff and the realization that the aircraft cannot climb. The corresponding flight profile in this training program is "Takeoff Engine Failure, Unable to Climb". This maneuver must not be attempted in the aircraft, but must be the subject of a classroom discussion or be demonstrated in the FTD.

(D) The focus of all landing procedures, whether two engine or engine out, is on a stabilized approach from an altitude of 500 feet. This will not be possible for all approach procedure maneuvering, especially during non-precision or circle to land approaches. Approach procedures for these two approaches should be stabilized from the point at which the pilot leaves the Minimum Descent Altitude for the landing.

(E) When performing one engine inoperative approaches, landings or missed approaches, the instructor must be prepared to add power to the simulated failed engine at the first sign of deteriorating airspeed or other situation that indicates the student's inability to correctly perform the maneuver.

(F) While maneuvering in the pattern or during instrument approach procedures with one engine inoperative, a 30° bank angle must not be exceeded. This will become especially important when executing non-precision and circle to land approaches.

Emergency and Abnormal Procedures

(A) During training, either in the FTD or in the aircraft, the performance of emergency and abnormal procedures is critical to the completion of the training program. All emergency and abnormal procedures should be simulated when training in the MU-2B airplane.

(B) When presenting emergency scenarios to the student, the instructor must not introduce multiple emergencies concurrently.

Scenario Based Training (SBT)

SBT flight training creates an environment of realism. The SBT programs utilize a highly structured flight operation scenario to simulate the overall flight environment. The pilot is required to plan a routine, point-to-point flight and initiate the flight. During the conduct of the flight, "reality-based" abnormal or emergency events are introduced without warning. Because the pilot is constantly operating in the world of unknowns, this type of training also builds in the "startle factor", and just as in the real-world, the consequences of the pilot's actions (decisions, judgment, airmanship, tactile skills, etc.) will continue to escalate and affect the outcome of the planned flight. Although flying skills are an integral part of this type of training, SBT enables the pilot to gain experience in dealing with unexpected events and more importantly further enhances the development of good judgment and decisionmaking.

[Doc. No. FAA-2006-24981, 73 FR 7051, Feb. 6, 2008, as amended by Amdt. 91-324, 76 FR 54107, Aug. 31, 2011]

Subpart A—General

Source: Docket No. 18334, 54 FR 34292, Aug. 18, 1989, unless otherwise noted.

§ 91.1 Applicability.

(a) Except as provided in paragraphs (b) and (c) of this section and §§ 91.701 and 91.703, this part prescribes rules governing the operation of aircraft (other than moored balloons, kites, unmanned rockets, and unmanned free balloons, which are governed by part 101 of this chapter, and ultralight vehicles operated in accordance with part 103 of this chapter) within the United States, including the waters within 3 nautical miles of the U.S. coast.

(b) Each person operating an aircraft in the airspace overlying the waters between 3 and 12 nautical miles from the coast of the United States must comply with §§ 91.1 through 91.21; §§ 91.101 through 91.143; §§ 91.151 through 91.159; §§ 91.167 through 91.193; § 91.203; § 91.205; §§ 91.209 through 91.217; § 91.221, § 91.225; §§ 91.303 through 91.319; §§ 91.323 through 91.327; § 91.605; § 91.609; §§ 91.703 through 91.715; and § 91.903.

(c) This part applies to each person on board an aircraft being operated under this part, unless otherwise specified.

(d) This part also establishes requirements for operators to take actions to support the continued airworthiness of each airplane.

[Doc. No. 18334, 54 FR 34292, Aug. 18, 1989, as amended by Amdt. 91-257, 64 FR 1079, Jan. 7, 1999; Amdt. 91-282, 69 FR 44880, July 27, 2004; Amdt. 91-297, 72 FR 63410, Nov. 8, 2007; Amdt. 91-314, 75 FR 30193, May 28, 2010]

§ 91.3 Responsibility and authority of the pilot in command.

(a) The pilot in command of an aircraft is directly responsible for, and is the final authority as to, the operation of that aircraft.

(b) In an in-flight emergency requiring immediate action, the pilot in command may deviate from any rule of this part to the extent required to meet that emergency.

(c) Each pilot in command who deviates from a rule under paragraph (b) of this section shall, upon the request of the Administrator, send a written report of that deviation to the Administrator.

(Approved by the Office of Management and Budget under control number 2120-0005)

§ 91.5 Pilot in command of aircraft requiring more than one required pilot.

No person may operate an aircraft that is type certificated for more than one required pilot flight crewmember unless the pilot in command meets the requirements of § 61.58 of this chapter.

§ 91.7 Civil aircraft airworthiness.

(a) No person may operate a civil aircraft unless it is in an airworthy condition.

(b) The pilot in command of a civil aircraft is responsible for determining whether that aircraft is in condition for safe flight. The pilot in command shall discontinue the flight when unairworthy mechanical, electrical, or structural conditions occur.

§ 91.9 Civil aircraft flight manual, marking, and placard requirements.

(a) Except as provided in paragraph (d) of this section, no person may operate a civil aircraft without complying with the operating limitations specified in the approved Airplane or Rotorcraft Flight Manual, markings, and placards, or as otherwise prescribed by the certificating authority of the country of registry.

(b) No person may operate a U.S.-registered civil aircraft—

(1) For which an Airplane or Rotorcraft Flight Manual is required by § 21.5 of this chapter unless there is available in the aircraft a current, approved Airplane or Rotorcraft Flight Manual or the manual provided for in § 121.141(b); and

(2) For which an Airplane or Rotorcraft Flight Manual is not required by § 21.5 of this chapter, unless there is available in the aircraft a current approved Airplane or Rotorcraft Flight Manual, approved manual material, markings, and placards, or any combination thereof.

(c) No person may operate a U.S.-registered civil aircraft unless that aircraft is identified in accordance with part 45 of this chapter.

(d) Any person taking off or landing a helicopter certificated under part 29 of this chapter at a heliport constructed over water may make such momentary flight as is necessary for takeoff or landing through the prohibited range of the limiting height-speed envelope established for the helicopter if that flight through the prohibited range takes place over water on which a safe ditching can be accomplished and if the helicopter is amphibious or is equipped with floats or other emergency flotation gear adequate to accomplish a safe emergency ditching on open water.

§ 91.11 Prohibition on interference with crewmembers.

No person may assault, threaten, intimidate, or interfere with a crewmember in the performance of the crewmember's duties aboard an aircraft being operated.

§ 91.13 Careless or reckless operation.

(a) *Aircraft operations for the purpose of air navigation.* No person may operate an aircraft in a careless or reckless manner so as to endanger the life or property of another.

(b) *Aircraft operations other than for the purpose of air navigation.* No person may operate an aircraft, other than for the purpose of air navigation, on any part of the surface of an airport used by aircraft for air commerce (including areas used by those aircraft for receiving or discharging persons or cargo), in a careless or reckless manner so as to endanger the life or property of another.

§ 91.15 Dropping objects.

No pilot in command of a civil aircraft may allow any object to be dropped from that aircraft in flight that creates a hazard to persons or property. However, this section does not prohibit the dropping of any object if reasonable precautions are taken to avoid injury or damage to persons or property.

§ 91.17 Alcohol or drugs.

(a) No person may act or attempt to act as a crewmember of a civil aircraft—

(1) Within 8 hours after the consumption of any alcoholic beverage;

(2) While under the influence of alcohol;

(3) While using any drug that affects the person's faculties in any way contrary to safety; or

(4) While having an alcohol concentration of 0.04 or greater in a blood or breath specimen. Alcohol concentration means grams of alcohol per deciliter of blood or grams of alcohol per 210 liters of breath.

(b) Except in an emergency, no pilot of a civil aircraft may allow a person who appears to be intoxicated or who demonstrates by manner or physical indications that the individual is under the influence of drugs (except a medical patient under proper care) to be carried in that aircraft.

(c) A crewmember shall do the following:

(1) On request of a law enforcement officer, submit to a test to indicate the alcohol concentration in the blood or breath, when—

(i) The law enforcement officer is authorized under State or local law to conduct the test or to have the test conducted; and

(ii) The law enforcement officer is requesting submission to the test to investigate a suspected violation of State or local law governing the same or substantially similar conduct prohibited by paragraph (a)(1), (a)(2), or (a)(4) of this section.

(2) Whenever the FAA has a reasonable basis to believe that a person may have violated paragraph (a)(1), (a)(2), or (a)(4) of this section, on request of the FAA, that person must furnish to the FAA the results, or authorize any clinic, hospital, or doctor, or other person to release to the FAA, the results of each test taken within 4 hours after acting or attempting to act as a crewmember that indicates an alcohol concentration in the blood or breath specimen.

(d) Whenever the Administrator has a reasonable basis to believe that a person may have violated paragraph (a)(3) of this section, that person shall, upon request by the Administrator, furnish the Administrator, or authorize any clinic, hospital, doctor, or other person to release to the Administrator, the results of each test taken within 4 hours after acting or attempting to act as a crewmember that indicates the presence of any drugs in the body.

(e) Any test information obtained by the Administrator under paragraph (c) or (d) of this section may be evaluated in determining a person's qualifications for any airman certificate or possible violations of this chapter and may be used as evidence in any legal proceeding under section 602, 609, or 901 of the Federal Aviation Act of 1958.

[Doc. No. 18334, 54 FR 34292, Aug. 18, 1989, as amended by Amdt. 91-291, June 21, 2006]

§ 91.19 Carriage of narcotic drugs, marihuana, and depressant or stimulant drugs or substances.

(a) Except as provided in paragraph (b) of this section, no person may operate a civil aircraft within the United States with knowledge that narcotic drugs, marihuana, and depressant or stimulant drugs or substances as defined in Federal or State statutes are carried in the aircraft.

(b) Paragraph (a) of this section does not apply to any carriage of narcotic drugs, marihuana, and depressant or stimulant drugs or substances authorized by or under any Federal or State statute or by any Federal or State agency.

§ 91.21 Portable electronic devices.

(a) Except as provided in paragraph (b) of this section, no person may operate, nor may any operator or pilot in command of an aircraft allow the operation of, any portable electronic device on any of the following U.S.-registered civil aircraft:

(1) Aircraft operated by a holder of an air carrier operating certificate or an operating certificate; or

(2) Any other aircraft while it is operated under IFR.

(b) Paragraph (a) of this section does not apply to—

(1) Portable voice recorders;

(2) Hearing aids;

(3) Heart pacemakers;

(4) Electric shavers; or

(5) Any other portable electronic device that the operator of the aircraft has determined will not cause interference with the navigation or communication system of the aircraft on which it is to be used.

(c) In the case of an aircraft operated by a holder of an air carrier operating certificate or an operating certificate, the determination required by paragraph (b)(5) of this section shall be made by that operator of the aircraft on which the particular device is to be used. In the case of other aircraft, the determination may be made by the pilot in command or other operator of the aircraft.

§ 91.23 Truth-in-leasing clause requirement in leases and conditional sales contracts.

(a) Except as provided in paragraph (b) of this section, the parties to a lease or contract of conditional sale involving a U.S.-registered large civil aircraft and entered into after January 2, 1973, shall execute a written lease or contract and include therein a written truth-in-leasing clause as a concluding paragraph in large print, immediately preceding the space for the signature of the parties, which contains the following with respect to each such aircraft:

(1) Identification of the Federal Aviation Regulations under which the aircraft has been maintained and inspected during the 12 months preceding the execution of the lease or contract of conditional sale, and certification by the parties thereto regarding the aircraft's status of compliance with applicable maintenance and inspection requirements in this part for the operation to be conducted under the lease or contract of conditional sale.

(2) The name and address (printed or typed) and the signature of the person responsible for operational control of the aircraft under the lease or contract of conditional sale, and certification that each person understands that person's responsibilities for compliance with applicable Federal Aviation Regulations.

(3) A statement that an explanation of factors bearing on operational control and pertinent Federal Aviation Regulations can be obtained from the nearest FAA Flight Standards district office.

(b) The requirements of paragraph (a) of this section do not apply—

(1) To a lease or contract of conditional sale when—

(i) The party to whom the aircraft is furnished is a foreign air carrier or certificate holder under part 121, 125, 135, or 141 of this chapter, or

(ii) The party furnishing the aircraft is a foreign air carrier or a person operating under part 121, 125, and 141 of this chapter, or a person operating under part 135 of this chapter having authority to engage in on-demand operations with large aircraft.

(2) To a contract of conditional sale, when the aircraft involved has not been registered anywhere prior to the execution of the contract, except as a new aircraft under a dealer's aircraft registration certificate issued in accordance with § 47.61 of this chapter.

(c) No person may operate a large civil aircraft of U.S. registry that is subject to a lease or contract of conditional sale to which paragraph (a) of this section applies, unless—

(1) The lessee or conditional buyer, or the registered owner if the lessee is not a citizen of the United States, has mailed a copy of the lease or contract that complies with the requirements of paragraph (a) of this section, within 24 hours of its execution, to the Aircraft Registration Branch, Attn: Technical Section, P.O. Box 25724, Oklahoma City, OK 73125;

(2) A copy of the lease or contract that complies with the requirements of paragraph (a) of this section is carried in the aircraft. The copy of the lease or contract shall be made available for review upon request by the Administrator, and

(3) The lessee or conditional buyer, or the registered owner if the lessee is not a citizen of the United States, has notified by telephone or in person the FAA Flight Standards district office nearest the airport where the flight will originate. Unless otherwise authorized by that office, the notification shall be given at least 48 hours before takeoff in the case of the first flight of that aircraft under that lease or contract and inform the FAA of—

(i) The location of the airport of departure;

(ii) The departure time; and

(iii) The registration number of the aircraft involved.

(d) The copy of the lease or contract furnished to the FAA under paragraph (c) of this section is commercial or financial information obtained from a person. It is, therefore, privileged and confidential and will not be made available by the FAA for public inspection or copying under 5 U.S.C. 552(b)(4) unless recorded with the FAA under part 49 of this chapter.

(e) For the purpose of this section, a lease means any agreement by a person to furnish an aircraft to another person for compensation or hire, whether with or without flight crewmembers, other than an agreement for the sale of an aircraft and a contract of conditional sale under section 101 of the Federal Aviation Act of 1958. The person furnishing the aircraft is referred to as the lessor, and the person to whom it is furnished the lessee.

(Approved by the Office of Management and Budget under control number 2120-0005)

[Doc. No. 18334, 54 FR 34292, Aug. 18, 1989, as amended by Amdt. 91-212, 54 FR 39293, Sept. 25, 1989; Amdt. 91-253, 62 FR 13253, Mar. 19, 1997; Amdt. 91-267, 66 FR 21066, Apr. 27, 2001]

§ 91.25 Aviation Safety Reporting Program: Prohibition against use of reports for enforcement purposes.

The Administrator of the FAA will not use reports submitted to the National Aeronautics and Space Administration under the Aviation Safety Reporting Program (or information derived therefrom) in any enforcement action except information concerning accidents or criminal offenses which are wholly excluded from the Program.

§§ 91.27-91.99 [Reserved]

Subpart B—Flight Rules

Source: Docket No. 18334, 54 FR 34294, Aug. 18, 1989, unless otherwise noted.

General

§ 91.101 Applicability.

This subpart prescribes flight rules governing the operation of aircraft within the United States and within 12 nautical miles from the coast of the United States.

§ 91.103 Preflight action.

Each pilot in command shall, before beginning a flight, become familiar with all available information concerning that flight. This information must include—

(a) For a flight under IFR or a flight not in the vicinity of an airport, weather reports and forecasts, fuel requirements, alternatives available if the planned flight cannot be completed, and any known traffic delays of which the pilot in command has been advised by ATC;

(b) For any flight, runway lengths at airports of intended use, and the following takeoff and landing distance information:

(1) For civil aircraft for which an approved Airplane or Rotorcraft Flight Manual containing takeoff and landing distance data is required, the takeoff and landing distance data contained therein; and

(2) For civil aircraft other than those specified in paragraph (b)(1) of this section, other reliable information appropriate to the aircraft, relating to aircraft performance under expected values of airport elevation and runway slope, aircraft gross weight, and wind and temperature.

§ 91.105 Flight crewmembers at stations.

(a) During takeoff and landing, and while en route, each required flight crewmember shall—

(1) Be at the crewmember station unless the absence is necessary to perform duties in connection with the operation of the aircraft or in connection with physiological needs; and

(2) Keep the safety belt fastened while at the crewmember station.

(b) Each required flight crewmember of a U.S.-registered civil aircraft shall, during takeoff and landing, keep his or her shoulder harness fastened while at his or her assigned duty station. This paragraph does not apply if—

(1) The seat at the crewmember's station is not equipped with a shoulder harness; or

(2) The crewmember would be unable to perform required duties with the shoulder harness fastened.

[Doc. No. 18334, 54 FR 34294, Aug. 18, 1989, as amended by Amdt. 91-231, 57 FR 42671, Sept. 15, 1992]

§ 91.107 Use of safety belts, shoulder harnesses, and child restraint systems.

(a) Unless otherwise authorized by the Administrator—

(1) No pilot may take off a U.S.-registered civil aircraft (except a free balloon that incorporates a basket or gondola,

or an airship type certificated before November 2, 1987) unless the pilot in command of that aircraft ensures that each person on board is briefed on how to fasten and unfasten that person's safety belt and, if installed, shoulder harness.

(2) No pilot may cause to be moved on the surface, take off, or land a U.S.-registered civil aircraft (except a free balloon that incorporates a basket or gondola, or an airship type certificated before November 2, 1987) unless the pilot in command of that aircraft ensures that each person on board has been notified to fasten his or her safety belt and, if installed, his or her shoulder harness.

(3) Except as provided in this paragraph, each person on board a U.S.-registered civil aircraft (except a free balloon that incorporates a basket or gondola or an airship type certificated before November 2, 1987) must occupy an approved seat or berth with a safety belt and, if installed, shoulder harness, properly secured about him or her during movement on the surface, takeoff, and landing. For seaplane and float equipped rotorcraft operations during movement on the surface, the person pushing off the seaplane or rotorcraft from the dock and the person mooring the seaplane or rotorcraft at the dock are excepted from the preceding seating and safety belt requirements. Notwithstanding the preceding requirements of this paragraph, a person may:

(i) Be held by an adult who is occupying an approved seat or berth, provided that the person being held has not reached his or her second birthday and does not occupy or use any restraining device;

(ii) Use the floor of the aircraft as a seat, provided that the person is on board for the purpose of engaging in sport parachuting; or

(iii) Notwithstanding any other requirement of this chapter, occupy an approved child restraint system furnished by the operator or one of the persons described in paragraph (a)(3)(iii)(A) of this section provided that:

(A) The child is accompanied by a parent, guardian, or attendant designated by the child's parent or guardian to attend to the safety of the child during the flight;

(B) Except as provided in paragraph (a)(3)(iii)(B)(*4*) of this action, the approved child restraint system bears one or more labels as follows:

(*1*) Seats manufactured to U.S. standards between January 1, 1981, and February 25, 1985, must bear the label: "This child restraint system conforms to all applicable Federal motor vehicle safety standards";

(*2*) Seats manufactured to U.S. standards on or after February 26, 1985, must bear two labels:

(*i*) "This child restraint system conforms to all applicable Federal motor vehicle safety standards"; and

(*ii*) "THIS RESTRAINT IS CERTIFIED FOR USE IN MOTOR VEHICLES AND AIRCRAFT" in red lettering;

(*3*) Seats that do not qualify under paragraphs (a)(3)(iii)(B)(*1*) and (a)(3)(iii)(B)(*2*) of this section must bear a label or markings showing:

(*i*) That the seat was approved by a foreign government;

(*ii*) That the seat was manufactured under the standards of the United Nations; or

(*iii*) That the seat or child restraint device furnished by the operator was approved by the FAA through Type Certificate or Supplemental Type Certificate.

(*iv*) That the seat or child restraint device furnished by the operator, or one of the persons described in paragraph (a)(3)(iii)(A) of this section, was approved by the FAA in accordance with § 21.8 or Technical Standard Order C-100b, or a later version.

(*4*) Except as provided in § 91.107(a)(3)(iii)(B)(*3*)(*iii*) and § 91.107(a)(3)(iii)(B)(*3*)(*iv*), booster-type child restraint systems (as defined in Federal Motor Vehicle Safety Standard No. 213 (49 CFR 571.213)), vest- and harness-type child restraint systems, and lap held child restraints are not approved for use in aircraft; and

(C) The operator complies with the following requirements:

(*1*) The restraint system must be properly secured to an approved forward-facing seat or berth;

(*2*) The child must be properly secured in the restraint system and must not exceed the specified weight limit for the restraint system; and

(*3*) The restraint system must bear the appropriate label(s).

(b) Unless otherwise stated, this section does not apply to operations conducted under part 121, 125, or 135 of this chapter. Paragraph (a)(3) of this section does not apply to persons subject to § 91.105.

[Doc. No. 26142, 57 FR 42671, Sept. 15, 1992, as amended by Amdt. 91-250, 61 FR 28421, June 4, 1996; Amdt. 91-289, 70 FR 50906, Aug. 26, 2005; Amdt. 91-292, 71 FR 40009, July 14, 2006; Amdt. 91-317, 75 FR 48857, Aug. 12, 2010]

§ 91.109 Flight instruction; Simulated instrument flight and certain flight tests.

(a) No person may operate a civil aircraft (except a manned free balloon) that is being used for flight instruction unless that aircraft has fully functioning dual controls. However, instrument flight instruction may be given in an airplane that is equipped with a single, functioning throwover control wheel that controls the elevator and ailerons, in place of fixed, dual controls, when—

(1) The instructor has determined that the flight can be conducted safely; and

(2) The person manipulating the controls has at least a private pilot certificate with appropriate category and class ratings.

(b) An airplane equipped with a single, functioning throwover control wheel that controls the elevator and ailerons, in place of fixed, dual controls may be used for flight instruction to conduct a flight review required by § 61.56 of this chapter, or to obtain recent flight experience or an instrument proficiency check required by § 61.57 when—

(1) The airplane is equipped with operable rudder pedals at both pilot stations;

(2) The pilot manipulating the controls is qualified to serve and serves as pilot in command during the entire flight;

(3) The instructor is current and qualified to serve as pilot in command of the airplane, meets the requirements of § 61.195(b), and has logged at least 25 hours of pilot-in-command flight time in the make and model of airplane; and

(4) The pilot in command and the instructor have determined the flight can be conducted safely.

(c) No person may operate a civil aircraft in simulated instrument flight unless—

(1) The other control seat is occupied by a safety pilot who possesses at least a private pilot certificate with category and class ratings appropriate to the aircraft being flown.

(2) The safety pilot has adequate vision forward and to each side of the aircraft, or a competent observer in the aircraft adequately supplements the vision of the safety pilot; and

(3) Except in the case of lighter-than-air aircraft, that aircraft is equipped with fully functioning dual controls. However, simulated instrument flight may be conducted in a single-engine airplane, equipped with a single, functioning, throwover control wheel, in place of fixed, dual controls of the elevator and ailerons, when—

(i) The safety pilot has determined that the flight can be conducted safely; and

(ii) The person manipulating the controls has at least a private pilot certificate with appropriate category and class ratings.

(d) No person may operate a civil aircraft that is being used for a flight test for an airline transport pilot certificate or a class or type rating on that certificate, or for a part 121 proficiency flight test, unless the pilot seated at the controls, other than the pilot being checked, is fully qualified to act as pilot in command of the aircraft.

[Doc. No. 18334, 54 FR 34294, Aug. 18, 1989, as amended by Amdt. 91-324, 76 FR 54107, Aug. 31, 2011]

§ 91.111 Operating near other aircraft.

(a) No person may operate an aircraft so close to another aircraft as to create a collision hazard.

(b) No person may operate an aircraft in formation flight except by arrangement with the pilot in command of each aircraft in the formation.

(c) No person may operate an aircraft, carrying passengers for hire, in formation flight.

§ 91.113 Right-of-way rules: Except water operations.

(a) *Inapplicability.* This section does not apply to the operation of an aircraft on water.

(b) *General.* When weather conditions permit, regardless of whether an operation is conducted under instrument flight rules or visual flight rules, vigilance shall be maintained by each person operating an aircraft so as to see and avoid other aircraft. When a rule of this section gives another aircraft the right-of-way, the pilot shall give way to that aircraft and may not pass over, under, or ahead of it unless well clear.

(c) *In distress.* An aircraft in distress has the right-of-way over all other air traffic.

(d) *Converging.* When aircraft of the same category are converging at approximately the same altitude (except head-on, or nearly so), the aircraft to the other's right has the right-of-way. If the aircraft are of different categories—

(1) A balloon has the right-of-way over any other category of aircraft;

(2) A glider has the right-of-way over an airship, powered parachute, weight-shift-control aircraft, airplane, or rotorcraft.

(3) An airship has the right-of-way over a powered parachute, weight-shift-control aircraft, airplane, or rotorcraft.

However, an aircraft towing or refueling other aircraft has the right-of-way over all other engine-driven aircraft.

(e) *Approaching head-on.* When aircraft are approaching each other head-on, or nearly so, each pilot of each aircraft shall alter course to the right.

(f) *Overtaking.* Each aircraft that is being overtaken has the right-of-way and each pilot of an overtaking aircraft shall alter course to the right to pass well clear.

(g) *Landing.* Aircraft, while on final approach to land or while landing, have the right-of-way over other aircraft in flight or operating on the surface, except that they shall not take advantage of this rule to force an aircraft off the runway surface which has already landed and is attempting to make way for an aircraft on final approach. When two or more aircraft are approaching an airport for the purpose of landing, the aircraft at the lower altitude has the right-of-way, but it shall not take advantage of this rule to cut in front of another which is on final approach to land or to overtake that aircraft.

[Doc. No. 18334, 54 FR 34294, Aug. 18, 1989, as amended by Amdt. 91-282, 69 FR 44880, July 27, 2004]

§ 91.115 Right-of-way rules: Water operations.

(a) *General.* Each person operating an aircraft on the water shall, insofar as possible, keep clear of all vessels and avoid impeding their navigation, and shall give way to any vessel or other aircraft that is given the right-of-way by any rule of this section.

(b) *Crossing.* When aircraft, or an aircraft and a vessel, are on crossing courses, the aircraft or vessel to the other's right has the right-of-way.

(c) *Approaching head-on.* When aircraft, or an aircraft and a vessel, are approaching head-on, or nearly so, each shall alter its course to the right to keep well clear.

(d) *Overtaking.* Each aircraft or vessel that is being overtaken has the right-of-way, and the one overtaking shall alter course to keep well clear.

(e) *Special circumstances.* When aircraft, or an aircraft and a vessel, approach so as to involve risk of collision, each aircraft or vessel shall proceed with careful regard to existing circumstances, including the limitations of the respective craft.

§ 91.117 Aircraft speed.

(a) Unless otherwise authorized by the Administrator, no person may operate an aircraft below 10,000 feet MSL at an indicated airspeed of more than 250 knots (288 m.p.h.).

(b) Unless otherwise authorized or required by ATC, no person may operate an aircraft at or below 2,500 feet above the surface within 4 nautical miles of the primary airport of a Class C or Class D airspace area at an indicated airspeed of more than 200 knots (230 mph.). This paragraph (b) does not apply to any operations within a Class B airspace area. Such operations shall comply with paragraph (a) of this section.

(c) No person may operate an aircraft in the airspace underlying a Class B airspace area designated for an airport or in a VFR corridor designated through such a Class B airspace area, at an indicated airspeed of more than 200 knots (230 mph).

(d) If the minimum safe airspeed for any particular operation is greater than the maximum speed prescribed in this section, the aircraft may be operated at that minimum speed.

[Doc. No. 18334, 54 FR 34292, Aug. 18, 1989, as amended by Amdt. 91-219, 55 FR 34708, Aug. 24, 1990; Amdt. 91-227, 56 FR 65657, Dec. 17, 1991; Amdt. 91-233, 58 FR 43554, Aug. 17, 1993]

§ 91.119 Minimum safe altitudes: General.

Except when necessary for takeoff or landing, no person may operate an aircraft below the following altitudes:

(a) *Anywhere.* An altitude allowing, if a power unit fails, an emergency landing without undue hazard to persons or property on the surface.

(b) *Over congested areas.* Over any congested area of a city, town, or settlement, or over any open air assembly of persons, an altitude of 1,000 feet above the highest obstacle within a horizontal radius of 2,000 feet of the aircraft.

(c) *Over other than congested areas.* An altitude of 500 feet above the surface, except over open water or sparsely populated areas. In those cases, the aircraft may not be operated closer than 500 feet to any person, vessel, vehicle, or structure.

(d) *Helicopters, powered parachutes, and weight-shift-control aircraft.* If the operation is conducted without hazard to persons or property on the surface—

(1) A helicopter may be operated at less than the minimums prescribed in paragraph (b) or (c) of this section, provided each person operating the helicopter complies with any routes or altitudes specifically prescribed for helicopters by the FAA; and

(2) A powered parachute or weight-shift-control aircraft may be operated at less than the minimums prescribed in paragraph (c) of this section.

[Docket No. 18334, 54 FR 34294, Aug. 18, 1989, as amended by Amdt. 91-311, 75 FR 5223, Feb. 1, 2010]

§ 91.121 Altimeter settings.

(a) Each person operating an aircraft shall maintain the cruising altitude or flight level of that aircraft, as the case may be, by reference to an altimeter that is set, when operating—

(1) Below 18,000 feet MSL, to—

(i) The current reported altimeter setting of a station along the route and within 100 nautical miles of the aircraft;

(ii) If there is no station within the area prescribed in paragraph (a)(1)(i) of this section, the current reported altimeter setting of an appropriate available station; or

(iii) In the case of an aircraft not equipped with a radio, the elevation of the departure airport or an appropriate altimeter setting available before departure; or

(2) At or above 18,000 feet MSL, to 29.92″ Hg.

(b) The lowest usable flight level is determined by the atmospheric pressure in the area of operation as shown in the following table:

Current altimeter setting	Lowest usable flight level
29.92 (or higher)	180
29.91 through 29.42	185
29.41 through 28.92	190
28.91 through 28.42	195
28.41 through 27.92	200
27.91 through 27.42	205
27.41 through 26.92	210

(c) To convert minimum altitude prescribed under §§ 91.119 and 91.177 to the minimum flight level, the pilot shall take the flight level equivalent of the minimum altitude in feet and add the appropriate number of feet specified below, according to the current reported altimeter setting:

Current altimeter setting	Adjustment factor
29.92 (or higher)	None
29.91 through 29.42	500
29.41 through 28.92	1,000
28.91 through 28.42	1,500
28.41 through 27.92	2,000
27.91 through 27.42	2,500
27.41 through 26.92	3,000

§ 91.123 Compliance with ATC clearances and instructions.

(a) When an ATC clearance has been obtained, no pilot in command may deviate from that clearance unless an amended clearance is obtained, an emergency exists, or the deviation is in response to a traffic alert and collision avoidance system resolution advisory. However, except in Class A airspace, a pilot may cancel an IFR flight plan if the operation is being conducted in VFR weather conditions. When a pilot is uncertain of an ATC clearance, that pilot shall immediately request clarification from ATC.

(b) Except in an emergency, no person may operate an aircraft contrary to an ATC instruction in an area in which air traffic control is exercised.

(c) Each pilot in command who, in an emergency, or in response to a traffic alert and collision avoidance system resolution advisory, deviates from an ATC clearance or instruction shall notify ATC of that deviation as soon as possible.

(d) Each pilot in command who (though not deviating from a rule of this subpart) is given priority by ATC in an emergency, shall submit a detailed report of that emergency within 48 hours to the manager of that ATC facility, if requested by ATC.

(e) Unless otherwise authorized by ATC, no person operating an aircraft may operate that aircraft according to any clearance or instruction that has been issued to the pilot of another aircraft for radar air traffic control purposes.

(Approved by the Office of Management and Budget under control number 2120-0005)

[Doc. No. 18834, 54 FR 34294, Aug. 18, 1989, as amended by Amdt. 91-227, 56 FR 65658, Dec. 17, 1991; Amdt. 91-244, 60 FR 50679, Sept. 29, 1995]

§ 91.125 ATC light signals.

ATC light signals have the meaning shown in the following table:

Color and type of signal	Meaning with respect to aircraft on the surface	Meaning with respect to aircraft in flight
Steady green	Cleared for takeoff	Cleared to land.
Flashing green	Cleared to taxi	Return for landing (to be followed by steady green at proper time).
Steady red	Stop	Give way to other aircraft and continue circling.
Flashing red	Taxi clear of runway in use	Airport unsafe—do not land.
Flashing white	Return to starting point on airport	Not applicable.
Alternating red and green	Exercise extreme caution	Exercise extreme caution.

§ 91.126 Operating on or in the vicinity of an airport in Class G airspace.

(a) *General.* Unless otherwise authorized or required, each person operating an aircraft on or in the vicinity of an airport in a Class G airspace area must comply with the requirements of this section.

(b) *Direction of turns.* When approaching to land at an airport without an operating control tower in Class G airspace—

(1) Each pilot of an airplane must make all turns of that airplane to the left unless the airport displays approved light signals or visual markings indicating that turns should be made to the right, in which case the pilot must make all turns to the right; and

(2) Each pilot of a helicopter or a powered parachute must avoid the flow of fixed-wing aircraft.

(c) *Flap settings.* Except when necessary for training or certification, the pilot in command of a civil turbojet-powered aircraft must use, as a final flap setting, the minimum certificated landing flap setting set forth in the approved performance information in the Airplane Flight Manual for the applicable conditions. However, each pilot in command has the final authority and responsibility for the safe operation of the pilot's airplane, and may use a different flap setting for that airplane if the pilot determines that it is necessary in the interest of safety.

(d) *Communications with control towers.* Unless otherwise authorized or required by ATC, no person may operate an aircraft to, from, through, or on an airport having an operational control tower unless two-way radio communications are maintained between that aircraft and the control tower. Communications must be established prior to 4 nautical miles from the airport, up to and including 2,500 feet AGL. However, if the aircraft radio fails in flight, the pilot in command may operate that aircraft and land if weather conditions are at or above basic VFR weather minimums, visual contact with the tower is maintained, and a clearance to land is received. If the aircraft radio fails while in flight under IFR, the pilot must comply with § 91.185.

[Doc. No. 24458, 56 FR 65658, Dec. 17, 1991, as amended by Amdt. 91-239, 59 FR 11693, Mar. 11, 1994; Amdt. 91-282, 69 FR 44880, July 27, 2004]

§ 91.127 Operating on or in the vicinity of an airport in Class E airspace.

(a) Unless otherwise required by part 93 of this chapter or unless otherwise authorized or required by the ATC facility having jurisdiction over the Class E airspace area, each person operating an aircraft on or in the vicinity of an airport in a Class E airspace area must comply with the requirements of § 91.126.

(b) *Departures.* Each pilot of an aircraft must comply with any traffic patterns established for that airport in part 93 of this chapter.

(c) *Communications with control towers.* Unless otherwise authorized or required by ATC, no person may operate an aircraft to, from, through, or on an airport having an operational control tower unless two-way radio communications are maintained between that aircraft and the control tower. Communications must be established prior to 4 nautical miles from the airport, up to and including 2,500 feet AGL. However, if the aircraft radio fails in flight, the pilot in command may operate that aircraft and land if weather conditions are at or above basic VFR weather minimums, visual contact with the tower is maintained, and a clearance to land is received. If the aircraft radio fails while in flight under IFR, the pilot must comply with § 91.185.

[Doc. No. 24458, 56 FR 65658, Dec. 17, 1991, as amended by Amdt. 91-239, 59 FR 11693, Mar. 11, 1994]

§ 91.129 Operations in Class D airspace.

(a) *General.* Unless otherwise authorized or required by the ATC facility having jurisdiction over the Class D airspace area, each person operating an aircraft in Class D airspace must comply with the applicable provisions of this section. In addition, each person must comply with §§ 91.126 and 91.127. For the purpose of this section, the primary airport is the airport for which the Class D airspace area is designated. A satellite airport is any other airport within the Class D airspace area.

(b) *Deviations.* An operator may deviate from any provision of this section under the provisions of an ATC authorization issued by the ATC facility having jurisdiction over the airspace concerned. ATC may authorize a deviation on a continuing basis or for an individual flight, as appropriate.

(c) *Communications.* Each person operating an aircraft in Class D airspace must meet the following two-way radio communications requirements:

(1) *Arrival or through flight.* Each person must establish two-way radio communications with the ATC facility (including foreign ATC in the case of foreign airspace designated in the United States) providing air traffic services prior to entering that airspace and thereafter maintain those communications while within that airspace.

(2) *Departing flight.* Each person—

(i) From the primary airport or satellite airport with an operating control tower must establish and maintain two-way radio communications with the control tower, and thereafter as instructed by ATC while operating in the Class D airspace area; or

(ii) From a satellite airport without an operating control tower, must establish and maintain two-way radio communications with the ATC facility having jurisdiction over the Class D airspace area as soon as practicable after departing.

(d) *Communications failure.* Each person who operates an aircraft in a Class D airspace area must maintain two-way radio communications with the ATC facility having jurisdiction over that area.

(1) If the aircraft radio fails in flight under IFR, the pilot must comply with § 91.185 of the part.

(2) If the aircraft radio fails in flight under VFR, the pilot in command may operate that aircraft and land if—

(i) Weather conditions are at or above basic VFR weather minimums;

(ii) Visual contact with the tower is maintained; and

(iii) A clearance to land is received.

(e) *Minimum altitudes when operating to an airport in Class D airspace.* (1) Unless required by the applicable distance-from-cloud criteria, each pilot operating a large or turbine-powered airplane must enter the traffic pattern at an altitude of at least 1,500 feet above the elevation of the airport and maintain at least 1,500 feet until further descent is required for a safe landing.

(2) Each pilot operating a large or turbine-powered airplane approaching to land on a runway served by an instrument approach procedure with vertical guidance, if the airplane is so equipped, must:

(i) Operate that airplane at an altitude at or above the glide path between the published final approach fix and the decision altitude (DA), or decision height (DH), as applicable; or

(ii) If compliance with the applicable distance-from-cloud criteria requires glide path interception closer in, operate that airplane at or above the glide path, between the point of interception of glide path and the DA or the DH.

(3) Each pilot operating an airplane approaching to land on a runway served by a visual approach slope indicator must maintain an altitude at or above the glide path until a lower altitude is necessary for a safe landing.

(4) Paragraphs (e)(2) and (e)(3) of this section do not prohibit normal bracketing maneuvers above or below the glide path that are conducted for the purpose of remaining on the glide path.

(f) *Approaches.* Except when conducting a circling approach under part 97 of this chapter or unless otherwise required by ATC, each pilot must—

(1) Circle the airport to the left, if operating an airplane; or

(2) Avoid the flow of fixed-wing aircraft, if operating a helicopter.

(g) *Departures.* No person may operate an aircraft departing from an airport except in compliance with the following:

(1) Each pilot must comply with any departure procedures established for that airport by the FAA.

(2) Unless otherwise required by the prescribed departure procedure for that airport or the applicable distance from clouds criteria, each pilot of a turbine-powered airplane and each pilot of a large airplane must climb to an altitude of 1,500 feet above the surface as rapidly as practicable.

(h) *Noise abatement.* Where a formal runway use program has been established by the FAA, each pilot of a large or turbine-powered airplane assigned a noise abatement runway by ATC must use that runway. However, consistent with the final authority of the pilot in command concerning the safe operation of the aircraft as prescribed in § 91.3(a), ATC may assign a different runway if requested by the pilot in the interest of safety.

(i) *Takeoff, landing, taxi clearance.* No person may, at any airport with an operating control tower, operate an aircraft on a runway or taxiway, or take off or land an aircraft, unless an appropriate clearance is received from ATC.

[Doc. No. 24458, 56 FR 65658, Dec. 17, 1991, as amended by Amdt. 91-234, 58 FR 48793, Sept. 20, 1993; Amdt. 91-296, 72 FR 31678, June 7, 2007; 77 FR 28250, May 14, 2012]

§ 91.130 Operations in Class C airspace.

(a) *General.* Unless otherwise authorized by ATC, each aircraft operation in Class C airspace must be conducted in compliance with this section and § 91.129. For the purpose of this section, the primary airport is the airport for which the Class C airspace area is designated. A satellite airport is any other airport within the Class C airspace area.

(b) *Traffic patterns.* No person may take off or land an aircraft at a satellite airport within a Class C airspace area except in compliance with FAA arrival and departure traffic patterns.

(c) *Communications.* Each person operating an aircraft in Class C airspace must meet the following two-way radio communications requirements:

(1) *Arrival or through flight.* Each person must establish two-way radio communications with the ATC facility (including foreign ATC in the case of foreign airspace designated in the United States) providing air traffic services prior to entering that airspace and thereafter maintain those communications while within that airspace.

(2) *Departing flight.* Each person—

(i) From the primary airport or satellite airport with an operating control tower must establish and maintain two-way radio communications with the control tower, and thereafter as instructed by ATC while operating in the Class C airspace area; or

(ii) From a satellite airport without an operating control tower, must establish and maintain two-way radio communications with the ATC facility having jurisdiction over the Class C airspace area as soon as practicable after departing.

(d) *Equipment requirements.* Unless otherwise authorized by the ATC having jurisdiction over the Class C airspace area, no person may operate an aircraft within a Class C airspace area designated for an airport unless that aircraft is equipped with the applicable equipment specified in § 91.215, and after January 1, 2020, § 91.225.

(e) *Deviations.* An operator may deviate from any provision of this section under the provisions of an ATC authorization issued by the ATC facility having jurisdiction over the airspace concerned. ATC may authorize a deviation on a continuing basis or for an individual flight, as appropriate.

[Doc. No. 24458, 56 FR 65659, Dec. 17, 1991, as amended by Amdt. 91-232, 58 FR 40736, July 30, 1993; Amdt. 91-239, 59 FR 11693, Mar. 11, 1994; Amdt. 91-314, 75 FR 30193, May 28, 2010]

§ 91.131 Operations in Class B airspace.

(a) *Operating rules.* No person may operate an aircraft within a Class B airspace area except in compliance with § 91.129 and the following rules:

(1) The operator must receive an ATC clearance from the ATC facility having jurisdiction for that area before operating an aircraft in that area.

(2) Unless otherwise authorized by ATC, each person operating a large turbine engine-powered airplane to or from a primary airport for which a Class B airspace area is designated must operate at or above the designated floors of the Class B airspace area while within the lateral limits of that area.

(3) Any person conducting pilot training operations at an airport within a Class B airspace area must comply with any procedures established by ATC for such operations in that area.

(b) *Pilot requirements.* (1) No person may take off or land a civil aircraft at an airport within a Class B airspace area or operate a civil aircraft within a Class B airspace area unless—

(i) The pilot in command holds at least a private pilot certificate;

(ii) The pilot in command holds a recreational pilot certificate and has met—

(A) The requirements of § 61.101(d) of this chapter; or

(B) The requirements for a student pilot seeking a recreational pilot certificate in § 61.94 of this chapter;

(iii) The pilot in command holds a sport pilot certificate and has met—

(A) The requirements of § 61.325 of this chapter; or

(B) The requirements for a student pilot seeking a recreational pilot certificate in § 61.94 of this chapter; or

(iv) The aircraft is operated by a student pilot who has met the requirements of § 61.94 or § 61.95 of this chapter, as applicable.

(2) Notwithstanding the provisions of paragraphs (b)(1)(ii), (b)(1)(iii) and (b)(1)(iv) of this section, no person may take off or land a civil aircraft at those airports listed in section 4 of appendix D to this part unless the pilot in command holds at least a private pilot certificate.

(c) *Communications and navigation equipment requirements.* Unless otherwise authorized by ATC, no person may operate an aircraft within a Class B airspace area unless that aircraft is equipped with—

(1) *For IFR operation.* An operable VOR or TACAN receiver or an operable and suitable RNAV system; and

(2) *For all operations.* An operable two-way radio capable of communications with ATC on appropriate frequencies for that Class B airspace area.

(d) *Other equipment requirements.* No person may operate an aircraft in a Class B airspace area unless the aircraft is equipped with—

(1) The applicable operating transponder and automatic altitude reporting equipment specified in § 91.215 (a), except as provided in § 91.215 (e), and

(2) After January 1, 2020, the applicable Automatic Dependent Surveillance-Broadcast Out equipment specified in § 91.225.

[Doc. No. 24458, 56 FR 65658, Dec. 17, 1991, as amended by Amdt. 91-282, 69 FR 44880, July 27, 2004; Amdt. 91-296, 72 FR 31678, June 7, 2007; Amdt. 91-314, 75 FR 30193, May 28, 2010]

§ 91.133 Restricted and prohibited areas.

(a) No person may operate an aircraft within a restricted area (designated in part 73) contrary to the restrictions imposed, or within a prohibited area, unless that person has the permission of the using or controlling agency, as appropriate.

(b) Each person conducting, within a restricted area, an aircraft operation (approved by the using agency) that creates the same hazards as the operations for which the restricted area was designated may deviate from the rules of this subpart that are not compatible with the operation of the aircraft.

§ 91.135 Operations in Class A airspace.

Except as provided in paragraph (d) of this section, each person operating an aircraft in Class A airspace must conduct that operation under instrument flight rules (IFR) and in compliance with the following:

(a) *Clearance.* Operations may be conducted only under an ATC clearance received prior to entering the airspace.

(b) *Communications.* Unless otherwise authorized by ATC, each aircraft operating in Class A airspace must be equipped with a two-way radio capable of communicating with ATC on a frequency assigned by ATC. Each pilot must maintain two-way radio communications with ATC while operating in Class A airspace.

(c) *Equipment requirements.* Unless otherwise authorized by ATC, no person may operate an aircraft within Class A airspace unless that aircraft is equipped with the applicable equipment specified in § 91.215, and after January 1, 2020, § 91.225.

(d) *ATC authorizations.* An operator may deviate from any provision of this section under the provisions of an ATC authorization issued by the ATC facility having jurisdiction of the airspace concerned. In the case of an inoperative transponder, ATC may immediately approve an operation within a Class A airspace area allowing flight to continue, if desired, to the airport of ultimate destination, including any intermediate stops, or to proceed to a place where suitable repairs can be made, or both. Requests for deviation from any provision of this section must be submitted in writing, at least 4 days before the proposed operation. ATC may authorize a deviation on a continuing basis or for an individual flight.

[Doc. No. 24458, 56 FR 65659, Dec. 17, 1991, as amended by Amdt. 91-314, 75 FR 30193, May 28, 2010]

§ 91.137 Temporary flight restrictions in the vicinity of disaster/hazard areas.

(a) The Administrator will issue a Notice to Airmen (NOTAM) designating an area within which temporary flight restrictions apply and specifying the hazard or condition requiring their imposition, whenever he determines it is necessary in order to—

(1) Protect persons and property on the surface or in the air from a hazard associated with an incident on the surface;

(2) Provide a safe environment for the operation of disaster relief aircraft; or

(3) Prevent an unsafe congestion of sightseeing and other aircraft above an incident or event which may generate a high degree of public interest.

The Notice to Airmen will specify the hazard or condition that requires the imposition of temporary flight restrictions.

(b) When a NOTAM has been issued under paragraph (a)(1) of this section, no person may operate an aircraft within the designated area unless that aircraft is participating in the hazard relief activities and is being operated under the direction of the official in charge of on scene emergency response activities.

(c) When a NOTAM has been issued under paragraph (a)(2) of this section, no person may operate an aircraft within the designated area unless at least one of the following conditions are met:

(1) The aircraft is participating in hazard relief activities and is being operated under the direction of the official in charge of on scene emergency response activities.

(2) The aircraft is carrying law enforcement officials.

(3) The aircraft is operating under the ATC approved IFR flight plan.

(4) The operation is conducted directly to or from an airport within the area, or is necessitated by the impracticability of VFR flight above or around the area due to weather, or terrain; notification is given to the Flight Service Station (FSS) or ATC facility specified in the NOTAM to receive advisories concerning disaster relief aircraft operations; and the operation does not hamper or endanger relief activities and is not conducted for the purpose of observing the disaster.

(5) The aircraft is carrying properly accredited news representatives, and, prior to entering the area, a flight plan is filed with the appropriate FAA or ATC facility specified in the Notice to Airmen and the operation is conducted above the altitude used by the disaster relief aircraft, unless otherwise authorized by the official in charge of on scene emergency response activities.

(d) When a NOTAM has been issued under paragraph (a)(3) of this section, no person may operate an aircraft within the designated area unless at least one of the following conditions is met:

(1) The operation is conducted directly to or from an airport within the area, or is necessitated by the impracticability of VFR flight above or around the area due to weather or terrain, and the operation is not conducted for the purpose of observing the incident or event.

(2) The aircraft is operating under an ATC approved IFR flight plan.

(3) The aircraft is carrying incident or event personnel, or law enforcement officials.

(4) The aircraft is carrying properly accredited news representatives and, prior to entering that area, a flight plan is filed with the appropriate FSS or ATC facility specified in the NOTAM.

(e) Flight plans filed and notifications made with an FSS or ATC facility under this section shall include the following information:

(1) Aircraft identification, type and color.

(2) Radio communications frequencies to be used.

(3) Proposed times of entry of, and exit from, the designated area.

(4) Name of news media or organization and purpose of flight.

(5) Any other information requested by ATC.

§ 91.138 Temporary flight restrictions in national disaster areas in the State of Hawaii.

(a) When the Administrator has determined, pursuant to a request and justification provided by the Governor of the State of Hawaii, or the Governor's designee, that an inhabited area within a declared national disaster area in the State of Hawaii is in need of protection for humanitarian reasons, the Administrator will issue a Notice to Airmen (NOTAM) designating an area within which temporary flight restrictions apply. The Administrator will designate the extent and duration of the temporary flight restrictions necessary to provide for the protection of persons and property on the surface.

(b) When a NOTAM has been issued in accordance with this section, no person may operate an aircraft within the designated area unless at least one of the following conditions is met:

(1) That person has obtained authorization from the official in charge of associated emergency or disaster relief response activities, and is operating the aircraft under the conditions of that authorization.

(2) The aircraft is carrying law enforcement officials.

(3) The aircraft is carrying persons involved in an emergency or a legitimate scientific purpose.

(4) The aircraft is carrying properly accredited newspersons, and that prior to entering the area, a flight plan is filed with the appropriate FAA or ATC facility specified in the NOTAM and the operation is conducted in compliance with the conditions and restrictions established by the official in charge of on-scene emergency response activities.

(5) The aircraft is operating in accordance with an ATC clearance or instruction.

(c) A NOTAM issued under this section is effective for 90 days or until the national disaster area designation is terminated, whichever comes first, unless terminated by notice or extended by the Administrator at the request of the Governor of the State of Hawaii or the Governor's designee.

[Doc. No. 26476, 56 FR 23178, May 20, 1991, as amended by Amdt. 91-270, 66 FR 47377, Sept. 11, 2001]

§ 91.139 Emergency air traffic rules.

(a) This section prescribes a process for utilizing Notices to Airmen (NOTAMs) to advise of the issuance and operations under emergency air traffic rules and regulations and designates the official who is authorized to issue NOTAMs on behalf of the Administrator in certain matters under this section.

(b) Whenever the Administrator determines that an emergency condition exists, or will exist, relating to the FAA's ability to operate the air traffic control system and during which normal flight operations under this chapter cannot be conducted consistent with the required levels of safety and efficiency—

(1) The Administrator issues an immediately effective air traffic rule or regulation in response to that emergency condition; and

(2) The Administrator or the Associate Administrator for Air Traffic may utilize the NOTAM system to provide notification of the issuance of the rule or regulation.

Those NOTAMs communicate information concerning the rules and regulations that govern flight operations, the use of navigation facilities, and designation of that airspace in which the rules and regulations apply.

(c) When a NOTAM has been issued under this section, no person may operate an aircraft, or other device governed by the regulation concerned, within the designated airspace except in accordance with the authorizations, terms, and conditions prescribed in the regulation covered by the NOTAM.

§ 91.141 Flight restrictions in the proximity of the Presidential and other parties.

No person may operate an aircraft over or in the vicinity of any area to be visited or traveled by the President, the Vice President, or other public figures contrary to the restrictions established by the Administrator and published in a Notice to Airmen (NOTAM).

§ 91.143 Flight limitation in the proximity of space flight operations.

When a Notice to Airmen (NOTAM) is issued in accordance with this section, no person may operate any aircraft of U.S. registry, or pilot any aircraft under the authority of an airman certificate issued by the Federal Aviation Administration, within areas designated in a NOTAM for space flight operation except when authorized by ATC.

[Doc. No. FAA-2004-19246, 69 FR 59753, Oct. 5, 2004]

§ 91.144 Temporary restriction on flight operations during abnormally high barometric pressure conditions.

(a) *Special flight restrictions.* When any information indicates that barometric pressure on the route of flight currently exceeds or will exceed 31 inches of mercury, no person may operate an aircraft or initiate a flight contrary to the requirements established by the Administrator and published in a Notice to Airmen issued under this section.

(b) *Waivers.* The Administrator is authorized to waive any restriction issued under paragraph (a) of this section to permit emergency supply, transport, or medical services to be delivered to isolated communities, where the operation can be conducted with an acceptable level of safety.

[Amdt. 91-240, 59 FR 17452, Apr. 12, 1994; 59 FR 37669, July 25, 1994]

§ 91.145 Management of aircraft operations in the vicinity of aerial demonstrations and major sporting events.

(a) The FAA will issue a Notice to Airmen (NOTAM) designating an area of airspace in which a temporary flight restriction applies when it determines that a temporary flight restriction is necessary to protect persons or property on the surface or in the air, to maintain air safety and efficiency, or to prevent the unsafe congestion of aircraft in the vicinity of an aerial demonstration or major sporting event. These demonstrations and events may include:

(1) United States Naval Flight Demonstration Team (Blue Angels);

(2) United States Air Force Air Demonstration Squadron (Thunderbirds);

(3) United States Army Parachute Team (Golden Knights);

(4) Summer/Winter Olympic Games;

(5) Annual Tournament of Roses Football Game;

(6) World Cup Soccer;

(7) Major League Baseball All-Star Game;

(8) World Series;

(9) Kodak Albuquerque International Balloon Fiesta;

(10) Sandia Classic Hang Gliding Competition;

(11) Indianapolis 500 Mile Race;

(12) Any other aerial demonstration or sporting event the FAA determines to need a temporary flight restriction in accordance with paragraph (b) of this section.

(b) In deciding whether a temporary flight restriction is necessary for an aerial demonstration or major sporting event not listed in paragraph (a) of this section, the FAA considers the following factors:

(1) Area where the event will be held.

(2) Effect flight restrictions will have on known aircraft operations.

(3) Any existing ATC airspace traffic management restrictions.

(4) Estimated duration of the event.

(5) Degree of public interest.

(6) Number of spectators.

(7) Provisions for spectator safety.

(8) Number and types of participating aircraft.

(9) Use of mixed high and low performance aircraft.

(10) Impact on non-participating aircraft.

(11) Weather minimums.

(12) Emergency procedures that will be in effect.

(c) A NOTAM issued under this section will state the name of the aerial demonstration or sporting event and specify the effective dates and times, the geographic features or coordinates, and any other restrictions or procedures governing flight operations in the designated airspace.

(d) When a NOTAM has been issued in accordance with this section, no person may operate an aircraft or device, or engage in any activity within the designated airspace area, except in accordance with the authorizations, terms, and conditions of the temporary flight restriction published in the NOTAM, unless otherwise authorized by:

(1) Air traffic control; or

(2) A Flight Standards Certificate of Waiver or Authorization issued for the demonstration or event.

(e) For the purpose of this section:

(1) *Flight restricted airspace area for an aerial demonstration*— The amount of airspace needed to protect persons and property on the surface or in the air, to maintain air safety and efficiency, or to prevent the unsafe congestion of aircraft will vary depending on the aerial demonstration and the factors listed in paragraph (b) of this section. The restricted airspace area will normally be limited to a 5 nautical mile radius from the center of the demonstration and an altitude 17000 mean sea level (for high performance aircraft) or 13000 feet above the surface (for certain parachute operations), but will be no greater than the minimum airspace necessary for the management of aircraft operations in the vicinity of the specified area.

(2) *Flight restricted area for a major sporting event*— The amount of airspace needed to protect persons and property on the surface or in the air, to maintain air safety and efficiency, or to prevent the unsafe congestion of aircraft will vary depending on the size of the event and the factors listed in paragraph (b) of this section. The restricted airspace will normally be limited to a 3 nautical mile radius from the center of the event and 2500 feet above the surface but will not be greater than the minimum airspace necessary for the management of aircraft operations in the vicinity of the specified area.

(f) A NOTAM issued under this section will be issued at least 30 days in advance of an aerial demonstration or a major sporting event, unless the FAA finds good cause for a shorter period and explains this in the NOTAM.

(g) When warranted, the FAA Administrator may exclude the following flights from the provisions of this section:

(1) Essential military.

(2) Medical and rescue.

(3) Presidential and Vice Presidential.

(4) Visiting heads of state.

(5) Law enforcement and security.

(6) Public health and welfare.

[Doc. No. FAA-2000-8274, 66 FR 47378, Sept. 11, 2001]

§ 91.146 Passenger-carrying flights for the benefit of a charitable, nonprofit, or community event.

(a) *Definitions.* For purposes of this section, the following definitions apply:

Charitable event means an event that raises funds for the benefit of a charitable organization recognized by the Department of the Treasury whose donors may deduct contributions under section 170 of the Internal Revenue Code (26 U.S.C. Section 170).

Community event means an event that raises funds for the benefit of any local or community cause that is not a charitable event or non-profit event.

Non-profit event means an event that raises funds for the benefit of a non-profit organization recognized under State or Federal law, as long as one of the organization's purposes is the promotion of aviation safety.

(b) Passenger carrying flights for the benefit of a charitable, nonprofit, or community event identified in paragraph (c) of this section are not subject to the certification requirements of part 119 or the drug and alcohol testing requirements in part 120 of this chapter, provided the following conditions are satisfied and the limitations in paragraphs (c) and (d) are not exceeded:

(1) The flight is nonstop and begins and ends at the same airport and is conducted within a 25-statute mile radius of that airport;

(2) The flight is conducted from a public airport that is adequate for the airplane or helicopter used, or from another location the FAA approves for the operation;

(3) The airplane or helicopter has a maximum of 30 seats, excluding each crewmember seat, and a maximum payload capacity of 7,500 pounds;

(4) The flight is not an aerobatic or a formation flight;

(5) Each airplane or helicopter holds a standard airworthiness certificate, is airworthy, and is operated in compli-

ance with the applicable requirements of subpart E of this part;

(6) Each flight is made during day VFR conditions;

(7) Reimbursement of the operator of the airplane or helicopter is limited to that portion of the passenger payment for the flight that does not exceed the pro rata cost of owning, operating, and maintaining the aircraft for that flight, which may include fuel, oil, airport expenditures, and rental fees;

(8) The beneficiary of the funds raised is not in the business of transportation by air;

(9) A private pilot acting as pilot in command has at least 500 hours of flight time;

(10) Each flight is conducted in accordance with the safety provisions of part 136, subpart A of this chapter; and

(11) Flights are not conducted over a national park, unit of a national park, or abutting tribal lands, unless the operator has secured a letter of agreement from the FAA, as specified under subpart B of part 136 of this chapter, and is operating in accordance with that agreement during the flights.

(c) (1) Passenger-carrying flights or series of flights are limited to a total of four charitable events or non-profit events per year, with no event lasting more than three consecutive days.

(2) Passenger-carrying flights or series of flights are limited to one community event per year, with no event lasting more than three consecutive days.

(d) Pilots and sponsors of events described in this section are limited to no more than 4 events per calendar year.

(e) At least seven days before the event, each sponsor of an event described in this section must furnish to the FAA Flight Standards District Office with jurisdiction over the geographical area where the event is scheduled:

(1) A signed letter detailing the name of the sponsor, the purpose of the event, the date and time of the event, the location of the event, all prior events under this section participated in by the sponsor in the current calendar year;

(2) A photocopy of each pilot in command's pilot certificate, medical certificate, and logbook entries that show the pilot is current in accordance with §§ 61.56 and 61.57 of this chapter and that any private pilot has at least 500 hours of flight time; and

(3) A signed statement from each pilot that lists all prior events under this section in which the pilot has participated during the current calendar year.

[Doc. No. FAA-1998-4521, 72 FR 6910, Feb. 13, 2007, as amended by Amdt. 91-308, 74 FR 32804, July 9, 2009]

§ 91.147 Passenger carrying flights for compensation or hire.

Each Operator conducting passenger-carrying flights for compensation or hire must meet the following requirements unless all flights are conducted under § 91.146.

(a) For the purposes of this section and for drug and alcohol testing, *Operator* means any person conducting nonstop passenger-carrying flights in an airplane or helicopter for compensation or hire in accordance with §§ 119.1(e)(2), 135.1(a)(5), or 121.1(d), of this chapter that begin and end at the same airport and are conducted within a 25-statute mile radius of that airport.

(b) An Operator must comply with the safety provisions of part 136, subpart A of this chapter, and apply for and receive a Letter of Authorization from the Flight Standards District Office nearest to its principal place of business.

(c) Each application for a Letter of Authorization must include the following information:

(1) Name of Operator, agent, and any d/b/a (doing-business-as) under which that Operator does business;

(2) Principal business address and mailing address;

(3) Principal place of business (if different from business address);

(4) Name of person responsible for management of the business;

(5) Name of person responsible for aircraft maintenance;

(6) Type of aircraft, registration number(s), and make/model/series; and

(7) An Antidrug and Alcohol Misuse Prevention Program registration.

(d) The Operator must register and implement its drug and alcohol testing programs in accordance with part 120 of this chapter.

(e) The Operator must comply with the provisions of the Letter of Authorization received.

[Doc. No. FAA-1998-4521, 72 FR 6911, Feb. 13, 2007, as amended by Amdt. 91-307, 74 FR 22652, May 14, 2009; Amdt. 91-320, 76 FR 8893, Feb. 16, 2011]

§§ 91.148-91.149 [Reserved]

Visual Flight Rules

§ 91.151 Fuel requirements for flight in VFR conditions.

(a) No person may begin a flight in an airplane under VFR conditions unless (considering wind and forecast weather conditions) there is enough fuel to fly to the first point of intended landing and, assuming normal cruising speed—

(1) During the day, to fly after that for at least 30 minutes; or

(2) At night, to fly after that for at least 45 minutes.

(b) No person may begin a flight in a rotorcraft under VFR conditions unless (considering wind and forecast weather conditions) there is enough fuel to fly to the first point of intended landing and, assuming normal cruising speed, to fly after that for at least 20 minutes.

§ 91.153 VFR flight plan: Information required.

(a) *Information required.* Unless otherwise authorized by ATC, each person filing a VFR flight plan shall include in it the following information:

(1) The aircraft identification number and, if necessary, its radio call sign.

(2) The type of the aircraft or, in the case of a formation flight, the type of each aircraft and the number of aircraft in the formation.

(3) The full name and address of the pilot in command or, in the case of a formation flight, the formation commander.

(4) The point and proposed time of departure.

(5) The proposed route, cruising altitude (or flight level), and true airspeed at that altitude.

(6) The point of first intended landing and the estimated elapsed time until over that point.

(7) The amount of fuel on board (in hours).

(8) The number of persons in the aircraft, except where that information is otherwise readily available to the FAA.

(9) Any other information the pilot in command or ATC believes is necessary for ATC purposes.

(b) *Cancellation.* When a flight plan has been activated, the pilot in command, upon canceling or completing the flight under the flight plan, shall notify an FAA Flight Service Station or ATC facility.

§ 91.155 Basic VFR weather minimums.

(a) Except as provided in paragraph (b) of this section and § 91.157, no person may operate an aircraft under VFR when the flight visibility is less, or at a distance from clouds that is less, than that prescribed for the corresponding altitude and class of airspace in the following table:

Airspace	Flight visibility	Distance from clouds
Class A	Not Applicable	Not Applicable.
Class B	3 statute miles	Clear of Clouds.
Class C	3 statute miles	500 feet below. 1,000 feet above. 2,000 feet horizontal.
Class D	3 statute miles	500 feet below. 1,000 feet above. 2,000 feet horizontal.
Class E:		
Less than 10,000 feet MSL	3 statute miles	500 feet below. 1,000 feet above. 2,000 feet horizontal
At or above 10,000 feet MSL	5 statute miles	1,000 feet below. 1,000 feet above. 1 statute mile horizontal.
Class G:		
1,200 feet or less above the surface (regardless of MSL altitude)		
Day, except as provided in §91.155(b)	1 statute mile	Clear of clouds.
Night, except as provided in §91.155(b)	3 statute miles	500 feet below. 1,000 feet above. 2,000 feet horizontal.
More than 1,200 feet above the surface but less than 10,000 feet MSL		
Day	1 statute mile	500 feet below. 1,000 feet above. 2,000 feet horizontal.
Night	3 statute miles	500 feet below. 1,000 feet above. 2,000 feet horizontal.
More than 1,200 feet above the surface and at or above 10,000 feet MSL	5 statute miles	1,000 feet below. 1,000 feet above. 1 statute mile horizontal.

(b) *Class G Airspace.* Notwithstanding the provisions of paragraph (a) of this section, the following operations may be conducted in Class G airspace below 1,200 feet above the surface:

(1) *Helicopter.* A helicopter may be operated clear of clouds if operated at a speed that allows the pilot adequate opportunity to see any air traffic or obstruction in time to avoid a collision.

(2) *Airplane, powered parachute, or weight-shift-control aircraft.* If the visibility is less than 3 statute miles but not less than 1 statute mile during night hours and you are operating in an airport traffic pattern within 1/2 mile of the runway, you may operate an airplane, powered parachute, or weight-shift-control aircraft clear of clouds.

(c) Except as provided in § 91.157, no person may operate an aircraft beneath the ceiling under VFR within the lateral boundaries of controlled airspace designated to the surface for an airport when the ceiling is less than 1,000 feet.

(d) Except as provided in § 91.157 of this part, no person may take off or land an aircraft, or enter the traffic pattern of an airport, under VFR, within the lateral boundaries of the surface areas of Class B, Class C, Class D, or Class E airspace designated for an airport—

(1) Unless ground visibility at that airport is at least 3 statute miles; or

(2) If ground visibility is not reported at that airport, unless flight visibility during landing or takeoff, or while operating in the traffic pattern is at least 3 statute miles.

(e) For the purpose of this section, an aircraft operating at the base altitude of a Class E airspace area is considered to be within the airspace directly below that area.

[Doc. No. 24458, 56 FR 65660, Dec. 17, 1991, as amended by Amdt. 91-235, 58 FR 51968, Oct. 5, 1993; Amdt. 91-282, 69 FR 44880, July 27, 2004]

§ 91.157 Special VFR weather minimums.

(a) Except as provided in appendix D, section 3, of this part, special VFR operations may be conducted under the weather minimums and requirements of this section, instead of those contained in § 91.155, below 10,000 feet MSL within the airspace contained by the upward extension of the lateral boundaries of the controlled airspace designated to the surface for an airport.

(b) Special VFR operations may only be conducted—

(1) With an ATC clearance;

(2) Clear of clouds;

(3) Except for helicopters, when flight visibility is at least 1 statute mile; and

(4) Except for helicopters, between sunrise and sunset (or in Alaska, when the sun is 6 degrees or more below the horizon) unless—

(i) The person being granted the ATC clearance meets the applicable requirements for instrument flight under part 61 of this chapter; and

(ii) The aircraft is equipped as required in § 91.205(d).

(c) No person may take off or land an aircraft (other than a helicopter) under special VFR—

(1) Unless ground visibility is at least 1 statute mile; or

(2) If ground visibility is not reported, unless flight visibility is at least 1 statute mile. For the purposes of this paragraph, the term flight visibility includes the visibility from the cockpit of an aircraft in takeoff position if:

(i) The flight is conducted under this part 91; and

(ii) The airport at which the aircraft is located is a satellite airport that does not have weather reporting capabilities.

(d) The determination of visibility by a pilot in accordance with paragraph (c)(2) of this section is not an official weather report or an official ground visibility report.

[Amdt. 91-235, 58 FR 51968, Oct. 5, 1993, as amended by Amdt. 91-247, 60 FR 66874, Dec. 27, 1995; Amdt. 91-262, 65 FR 16116, Mar. 24, 2000]

§ 91.159 VFR cruising altitude or flight level.

Except while holding in a holding pattern of 2 minutes or less, or while turning, each person operating an aircraft under VFR in level cruising flight more than 3,000 feet above the surface shall maintain the appropriate altitude or flight level prescribed below, unless otherwise authorized by ATC:

(a) When operating below 18,000 feet MSL and—

(1) On a magnetic course of zero degrees through 179 degrees, any odd thousand foot MSL altitude +500 feet (such as 3,500, 5,500, or 7,500); or

(2) On a magnetic course of 180 degrees through 359 degrees, any even thousand foot MSL altitude +500 feet (such as 4,500, 6,500, or 8,500).

(b) When operating above 18,000 feet MSL, maintain the altitude or flight level assigned by ATC.

[Doc. No. 18334, 54 FR 34294, Aug. 18, 1989, as amended by Amdt. 91-276, 68 FR 61321, Oct. 27, 2003; 68 FR 70133, Dec. 17, 2003]

§ 91.161 Special awareness training required for pilots flying under visual flight rules within a 60-nautical mile radius of the Washington, DC VOR/DME.

(a) *Operations within a 60-nautical mile radius of the Washington, DC VOR/DME under visual flight rules (VFR).* Except as provided under paragraph (e) of this section, no person may serve as a pilot in command or as second in command of an aircraft while flying within a 60-nautical mile radius of the DCA VOR/DME, under VFR, unless that pilot has completed Special Awareness Training and holds a certificate of training completion.

(b) *Special Awareness Training.* The Special Awareness Training consists of information to educate pilots about the procedures for flying in the Washington, DC area and, more generally, in other types of special use airspace. This free training is available on the FAA's Web site. Upon completion of the training, each person will need to print out a copy of the certificate of training completion.

(c) *Inspection of certificate of training completion.* Each person who holds a certificate for completing the Special Awareness Training must present it for inspection upon request from:

(1) An authorized representative of the FAA;

(2) An authorized representative of the National Transportation Safety Board;

(3) Any Federal, State, or local law enforcement officer; or

(4) An authorized representative of the Transportation Security Administration.

(d) *Emergency declared.* The failure to complete the Special Awareness Training course on flying in and around the Washington, DC Metropolitan Area is not a violation of this section if an emergency is declared by the pilot, as described under § 91.3(b), or there was a failure of two-way radio communications when operating under IFR as described under § 91.185.

(e) *Exceptions.* The requirements of this section do not apply if the flight is being performed in an aircraft of an air ambulance operator certificated to conduct part 135 operations under this chapter, the U.S. Armed Forces, or a law enforcement agency.

[Doc. No. FAA-2006-25250, 73 FR 46803, Aug. 12, 2008]

§§ 91.162-91.165 [Reserved]

Instrument Flight Rules

§ 91.167 Fuel requirements for flight in IFR conditions.

(a) No person may operate a civil aircraft in IFR conditions unless it carries enough fuel (considering weather reports and forecasts and weather conditions) to—

(1) Complete the flight to the first airport of intended landing;

(2) Except as provided in paragraph (b) of this section, fly from that airport to the alternate airport; and

(3) Fly after that for 45 minutes at normal cruising speed or, for helicopters, fly after that for 30 minutes at normal cruising speed.

(b) Paragraph (a)(2) of this section does not apply if:

(1) Part 97 of this chapter prescribes a standard instrument approach procedure to, or a special instrument approach procedure has been issued by the Administrator to the operator for, the first airport of intended landing; and

(2) Appropriate weather reports or weather forecasts, or a combination of them, indicate the following:

(i) *For aircraft other than helicopters.* For at least 1 hour before and for 1 hour after the estimated time of arrival, the ceiling will be at least 2,000 feet above the airport elevation and the visibility will be at least 3 statute miles.

(ii) *For helicopters.* At the estimated time of arrival and for 1 hour after the estimated time of arrival, the ceiling will be at least 1,000 feet above the airport elevation, or at least 400 feet above the lowest applicable approach minima, whichever is higher, and the visibility will be at least 2 statute miles.

[Doc. No. 98-4390, 65 FR 3546, Jan. 21, 2000]

§ 91.169 IFR flight plan: Information required.

(a) *Information required.* Unless otherwise authorized by ATC, each person filing an IFR flight plan must include in it the following information:

(1) Information required under § 91.153 (a) of this part;

(2) Except as provided in paragraph (b) of this section, an alternate airport.

(b) Paragraph (a)(2) of this section does not apply if:

(1) Part 97 of this chapter prescribes a standard instrument approach procedure to, or a special instrument approach procedure has been issued by the Administrator to the operator for, the first airport of intended landing; and

(2) Appropriate weather reports or weather forecasts, or a combination of them, indicate the following:

(i) *For aircraft other than helicopters.* For at least 1 hour before and for 1 hour after the estimated time of arrival, the ceiling will be at least 2,000 feet above the airport elevation and the visibility will be at least 3 statute miles.

(ii) *For helicopters.* At the estimated time of arrival and for 1 hour after the estimated time of arrival, the ceiling will be at least 1,000 feet above the airport elevation, or at least 400 feet above the lowest applicable approach minima, whichever is higher, and the visibility will be at least 2 statute miles.

(c) *IFR alternate airport weather minima.* Unless otherwise authorized by the Administrator, no person may include an alternate airport in an IFR flight plan unless appropriate weather reports or weather forecasts, or a combination of them, indicate that, at the estimated time of arrival at the alternate airport, the ceiling and visibility at that airport will be at or above the following weather minima:

(1) If an instrument approach procedure has been published in part 97 of this chapter, or a special instrument approach procedure has been issued by the Administrator to the operator, for that airport, the following minima:

(i) *For aircraft other than helicopters:* The alternate airport minima specified in that procedure, or if none are specified the following standard approach minima:

(A) *For a precision approach procedure.* Ceiling 600 feet and visibility 2 statute miles.

(B) *For a nonprecision approach procedure.* Ceiling 800 feet and visibility 2 statute miles.

(ii) *For helicopters:* Ceiling 200 feet above the minimum for the approach to be flown, and visibility at least 1 statute mile but never less than the minimum visibility for the approach to be flown, and

(2) If no instrument approach procedure has been published in part 97 of this chapter and no special instrument approach procedure has been issued by the Administrator to the operator, for the alternate airport, the ceiling and visibility minima are those allowing descent from the MEA, approach, and landing under basic VFR.

(d) *Cancellation.* When a flight plan has been activated, the pilot in command, upon canceling or completing the flight under the flight plan, shall notify an FAA Flight Service Station or ATC facility.

[Doc. No. 18334, 54 FR 34294, Aug. 18, 1989, as amended by Amdt. 91-259, 65 FR 3546, Jan. 21, 2000]

§ 91.171 VOR equipment check for IFR operations.

(a) No person may operate a civil aircraft under IFR using the VOR system of radio navigation unless the VOR equipment of that aircraft—

(1) Is maintained, checked, and inspected under an approved procedure; or

(2) Has been operationally checked within the preceding 30 days, and was found to be within the limits of the permissible indicated bearing error set forth in paragraph (b) or (c) of this section.

(b) Except as provided in paragraph (c) of this section, each person conducting a VOR check under paragraph (a)(2) of this section shall—

(1) Use, at the airport of intended departure, an FAA-operated or approved test signal or a test signal radiated by a certificated and appropriately rated radio repair station or, outside the United States, a test signal operated or approved by an appropriate authority to check the VOR equipment (the maximum permissible indicated bearing error is plus or minus 4 degrees); or

(2) Use, at the airport of intended departure, a point on the airport surface designated as a VOR system checkpoint by the Administrator, or, outside the United States, by an appropriate authority (the maximum permissible bearing error is plus or minus 4 degrees);

(3) If neither a test signal nor a designated checkpoint on the surface is available, use an airborne checkpoint designated by the Administrator or, outside the United States, by an appropriate authority (the maximum permissible bearing error is plus or minus 6 degrees); or

(4) If no check signal or point is available, while in flight—

(i) Select a VOR radial that lies along the centerline of an established VOR airway;

(ii) Select a prominent ground point along the selected radial preferably more than 20 nautical miles from the VOR ground facility and maneuver the aircraft directly over the point at a reasonably low altitude; and

(iii) Note the VOR bearing indicated by the receiver when over the ground point (the maximum permissible variation between the published radial and the indicated bearing is 6 degrees).

(c) If dual system VOR (units independent of each other except for the antenna) is installed in the aircraft, the person checking the equipment may check one system against the other in place of the check procedures specified in paragraph (b) of this section. Both systems shall be tuned to the same VOR ground facility and note the indicated bearings to that station. The maximum permissible variation between the two indicated bearings is 4 degrees.

(d) Each person making the VOR operational check, as specified in paragraph (b) or (c) of this section, shall enter the date, place, bearing error, and sign the aircraft log or other record. In addition, if a test signal radiated by a repair station, as specified in paragraph (b)(1) of this section, is used, an entry must be made in the aircraft log or other record by the repair station certificate holder or the certificate holder's representative certifying to the bearing transmitted by the repair station for the check and the date of transmission.

(Approved by the Office of Management and Budget under control number 2120-0005)

§ 91.173 ATC clearance and flight plan required.

No person may operate an aircraft in controlled airspace under IFR unless that person has—

(a) Filed an IFR flight plan; and

(b) Received an appropriate ATC clearance.

§ 91.175 Takeoff and landing under IFR.

(a) *Instrument approaches to civil airports.* Unless otherwise authorized by the FAA, when it is necessary to use an instrument approach to a civil airport, each person operating an aircraft must use a standard instrument approach procedure prescribed in part 97 of this chapter for that airport. This paragraph does not apply to United States military aircraft.

(b) *Authorized DA/DH or MDA.* For the purpose of this section, when the approach procedure being used provides for and requires the use of a DA/DH or MDA, the authorized DA/DH or MDA is the highest of the following:

(1) The DA/DH or MDA prescribed by the approach procedure.

(2) The DA/DH or MDA prescribed for the pilot in command.

(3) The DA/DH or MDA appropriate for the aircraft equipment available and used during the approach.

(c) *Operation below DA/ DH or MDA.* Except as provided in paragraph (l) of this section, where a DA/DH or MDA is applicable, no pilot may operate an aircraft, except a military aircraft of the United States, below the authorized

MDA or continue an approach below the authorized DA/DH unless—

(1) The aircraft is continuously in a position from which a descent to a landing on the intended runway can be made at a normal rate of descent using normal maneuvers, and for operations conducted under part 121 or part 135 unless that descent rate will allow touchdown to occur within the touchdown zone of the runway of intended landing;

(2) The flight visibility is not less than the visibility prescribed in the standard instrument approach being used; and

(3) Except for a Category II or Category III approach where any necessary visual reference requirements are specified by the Administrator, at least one of the following visual references for the intended runway is distinctly visible and identifiable to the pilot:

(i) The approach light system, except that the pilot may not descend below 100 feet above the touchdown zone elevation using the approach lights as a reference unless the red terminating bars or the red side row bars are also distinctly visible and identifiable.

(ii) The threshold.

(iii) The threshold markings.

(iv) The threshold lights.

(v) The runway end identifier lights.

(vi) The visual approach slope indicator.

(vii) The touchdown zone or touchdown zone markings.

(viii) The touchdown zone lights.

(ix) The runway or runway markings.

(x) The runway lights.

(d) *Landing.* No pilot operating an aircraft, except a military aircraft of the United States, may land that aircraft when—

(1) For operations conducted under paragraph (l) of this section, the requirements of (l)(4) of this section are not met; or

(2) For all other part 91 operations and parts 121, 125, 129, and 135 operations, the flight visibility is less than the visibility prescribed in the standard instrument approach procedure being used.

(e) *Missed approach procedures.* Each pilot operating an aircraft, except a military aircraft of the United States, shall immediately execute an appropriate missed approach procedure when either of the following conditions exist:

(1) Whenever operating an aircraft pursuant to paragraph (c) or (l) of this section and the requirements of that paragraph are not met at either of the following times:

(i) When the aircraft is being operated below MDA; or

(ii) Upon arrival at the missed approach point, including a DA/DH where a DA/DH is specified and its use is required, and at any time after that until touchdown.

(2) Whenever an identifiable part of the airport is not distinctly visible to the pilot during a circling maneuver at or above MDA, unless the inability to see an identifiable part of the airport results only from a normal bank of the aircraft during the circling approach.

(f) *Civil airport takeoff minimums.* This paragraph applies to persons operating an aircraft under part 121, 125, 129, or 135 of this chapter.

(1) Unless otherwise authorized by the FAA, no pilot may takeoff from a civil airport under IFR unless the weather conditions at time of takeoff are at or above the weather minimums for IFR takeoff prescribed for that airport under part 97 of this chapter.

(2) If takeoff weather minimums are not prescribed under part 97 of this chapter for a particular airport, the following weather minimums apply to takeoffs under IFR:

(i) For aircraft, other than helicopters, having two engines or less—1 statute mile visibility.

(ii) For aircraft having more than two engines— 1/2 statute mile visibility.

(iii) For helicopters— 1/2 statute mile visibility.

(3) Except as provided in paragraph (f)(4) of this section, no pilot may takeoff under IFR from a civil airport having published obstacle departure procedures (ODPs) under part 97 of this chapter for the takeoff runway to be used, unless the pilot uses such ODPs or an alternative procedure or route assigned by air traffic control.

(4) Notwithstanding the requirements of paragraph (f) (3) of this section, no pilot may takeoff from an airport under IFR unless:

(i) For part 121 and part 135 operators, the pilot uses a takeoff obstacle clearance or avoidance procedure that ensures compliance with the applicable airplane performance operating limitations requirements under part 121, subpart I or part 135, subpart I for takeoff at that airport; or

(ii) For part 129 operators, the pilot uses a takeoff obstacle clearance or avoidance procedure that ensures compliance with the airplane performance operating limitations prescribed by the State of the operator for takeoff at that airport.

(g) *Military airports.* Unless otherwise prescribed by the Administrator, each person operating a civil aircraft under IFR into or out of a military airport shall comply with the instrument approach procedures and the takeoff and landing minimum prescribed by the military authority having jurisdiction of that airport.

(h) *Comparable values of RVR and ground visibility.* (1) Except for Category II or Category III minimums, if RVR minimums for takeoff or landing are prescribed in an instrument approach procedure, but RVR is not reported for the runway of intended operation, the RVR minimum shall be converted to ground visibility in accordance with the table in paragraph (h)(2) of this section and shall be the visibility minimum for takeoff or landing on that runway.

(2)

RVR (feet)	Visibility (statute miles)
1,600	1/4
2,400	1/2
3,200	5/8
4,000	3/4
4,500	7/8
5,000	1
6,000	1 1/4

(i) *Operations on unpublished routes and use of radar in instrument approach procedures.* When radar is approved at certain locations for ATC purposes, it may be used not only for surveillance and precision radar approaches, as applicable, but also may be used in conjunction with instrument approach procedures predicated on other types of radio navigational aids. Radar vectors may be authorized to provide course guidance through the segments of an approach to the final course or fix. When operating on an unpublished route or while being radar vectored, the pilot, when an approach clearance is received, shall, in addition to complying with § 91.177, maintain the last altitude assigned to that pilot until the aircraft is established on a segment of a published route or instrument approach procedure unless a different altitude is assigned by ATC. After the aircraft is so established, published altitudes apply to descent within each succeeding route or approach segment unless a different altitude is assigned by ATC. Upon reaching the final approach course or fix, the pilot may either complete the instrument approach in accordance with a procedure approved for the facility or continue a surveillance or precision radar approach to a landing.

(j) *Limitation on procedure turns.* In the case of a radar vector to a final approach course or fix, a timed approach from a holding fix, or an approach for which the procedure specifies "No PT," no pilot may make a procedure turn unless cleared to do so by ATC.

(k) *ILS components.* The basic components of an ILS are the localizer, glide slope, and outer marker, and, when installed for use with Category II or Category III instrument approach procedures, an inner marker. The following means may be used to substitute for the outer marker: Compass locator; precision approach radar (PAR) or airport surveillance radar (ASR); DME, VOR, or nondirectional beacon fixes authorized in the standard instrument approach procedure; or a suitable RNAV system in conjunction with a fix identified in the standard instrument approach procedure. Applicability of, and substitution for, the inner marker for a Category II or III approach is determined by the appropriate 14 CFR part 97 approach procedure, letter of authorization, or operations specifications issued to an operator.

(l) *Approach to straight-in landing operations below DH, or MDA using an enhanced flight vision system (EFVS).* For straight-in instrument approach procedures other than Category II or Category III, no pilot operating under this section or §§ 121.651, 125.381, and 135.225 of this chapter may operate an aircraft at any airport below the authorized MDA or continue an approach below the authorized DH and land unless—

(1) The aircraft is continuously in a position from which a descent to a landing on the intended runway can be made at a normal rate of descent using normal maneuvers, and, for operations conducted under part 121 or part 135 of this chapter, the descent rate will allow touchdown to occur within the touchdown zone of the runway of intended landing;

(2) The pilot determines that the enhanced flight visibility observed by use of a certified enhanced flight vision system is not less than the visibility prescribed in the standard instrument approach procedure being used;

(3) The following visual references for the intended runway are distinctly visible and identifiable to the pilot using the enhanced flight vision system:

(i) The approach light system (if installed); or

(ii) The following visual references in both paragraphs (l)(3)(ii)(A) and (B) of this section:

(A) The runway threshold, identified by at least one of the following:

(*1*) The beginning of the runway landing surface;

(*2*) The threshold lights; or

(*3*) The runway end identifier lights.

(B) The touchdown zone, identified by at least one of the following:

(*1*) The runway touchdown zone landing surface;

(*2*) The touchdown zone lights;

(*3*) The touchdown zone markings; or

(*4*) The runway lights.

(4) At 100 feet above the touchdown zone elevation of the runway of intended landing and below that altitude, the flight visibility must be sufficient for the following to be distinctly visible and identifiable to the pilot without reliance on the enhanced flight vision system to continue to a landing:

(i) The lights or markings of the threshold; or

(ii) The lights or markings of the touchdown zone;

(5) The pilot(s) is qualified to use an EFVS as follows—

(i) For parts 119 and 125 certificate holders, the applicable training, testing and qualification provisions of parts 121, 125, and 135 of this chapter;

(ii) For foreign persons, in accordance with the requirements of the civil aviation authority of the State of the operator; or

(iii) For persons conducting any other operation, in accordance with the applicable currency and proficiency requirements of part 61 of this chapter;

(6) For parts 119 and 125 certificate holders, and part 129 operations specifications holders, their operations specifications authorize use of EFVS; and

(7) The aircraft is equipped with, and the pilot uses, an enhanced flight vision system, the display of which is suitable for maneuvering the aircraft and has either an FAA type design approval or, for a foreign-registered aircraft, the EFVS complies with all of the EFVS requirements of this chapter.

(m) For purposes of this section, "enhanced flight vision system" (EFVS) is an installed airborne system comprised of the following features and characteristics:

(1) An electronic means to provide a display of the forward external scene topography (the natural or manmade features of a place or region especially in a way to show their relative positions and elevation) through the use of imaging sensors, such as a forward-looking infrared, millimeter wave radiometry, millimeter wave radar, and low-light level image intensifying;

(2) The EFVS sensor imagery and aircraft flight symbology (*i.e.,* at least airspeed, vertical speed, aircraft attitude, heading, altitude, command guidance as appropriate for the approach to be flown, path deviation indications, and flight path vector, and flight path angle reference cue) are presented on a head-up display, or an equivalent display, so that they are clearly visible to the pilot flying in his or her normal position and line of vision and looking forward along the flight path, to include:

(i) The displayed EFVS imagery, attitude symbology, flight path vector, and flight path angle reference cue, and other cues, which are referenced to this imagery and external scene topography, must be presented so that they are aligned with and scaled to the external view; and

(ii) The flight path angle reference cue must be displayed with the pitch scale, selectable by the pilot to the desired descent angle for the approach, and suitable for monitoring the vertical flight path of the aircraft on approaches without vertical guidance; and

(iii) The displayed imagery and aircraft flight symbology do not adversely obscure the pilot's outside view or field of view through the cockpit window;

(3) The EFVS includes the display element, sensors, computers and power supplies, indications, and controls. It may receive inputs from an airborne navigation system or flight guidance system; and

(4) The display characteristics and dynamics are suitable for manual control of the aircraft.

[Doc. No. 18334, 54 FR 34294, Aug. 18, 1989, as amended by Amdt. 91-267, 66 FR 21066, Apr. 27, 2001; Amdt. 91-281, 69 FR 1640, Jan. 9, 2004; Amdt. 91-296, 72 FR 31678, June 7, 2007; Amdt. 91-306, 74 FR 20205, May 1, 2009]

§ 91.177 Minimum altitudes for IFR operations.

(a) *Operation of aircraft at minimum altitudes.* Except when necessary for takeoff or landing, or unless otherwise authorized by the FAA, no person may operate an aircraft under IFR below—

(1) The applicable minimum altitudes prescribed in parts 95 and 97 of this chapter. However, if both a MEA and a MOCA are prescribed for a particular route or route segment, a person may operate an aircraft below the MEA down to, but not below, the MOCA, provided the applicable navigation signals are available. For aircraft using VOR for navigation, this applies only when the aircraft is within 22 nautical miles of that VOR (based on the reasonable estimate by the pilot operating the aircraft of that distance); or

(2) If no applicable minimum altitude is prescribed in parts 95 and 97 of this chapter, then—

(i) In the case of operations over an area designated as a mountainous area in part 95 of this chapter, an altitude of 2,000 feet above the highest obstacle within a horizontal distance of 4 nautical miles from the course to be flown; or

(ii) In any other case, an altitude of 1,000 feet above the highest obstacle within a horizontal distance of 4 nautical miles from the course to be flown.

(b) *Climb.* Climb to a higher minimum IFR altitude shall begin immediately after passing the point beyond which that minimum altitude applies, except that when ground obstructions intervene, the point beyond which that higher minimum altitude applies shall be crossed at or above the applicable MCA.

[Doc. No. 18334, 54 FR 34294, Aug. 18, 1989, as amended by Amdt. 91-296, 72 FR 31678, June 7, 2007; Amdt. 91-315, 75 FR 30690, June 2, 2010]

§ 91.179 IFR cruising altitude or flight level.

Unless otherwise authorized by ATC, the following rules apply—

(a) *In controlled airspace.* Each person operating an aircraft under IFR in level cruising flight in controlled airspace shall maintain the altitude or flight level assigned that aircraft by ATC. However, if the ATC clearance assigns "VFR conditions on-top," that person shall maintain an altitude or flight level as prescribed by § 91.159.

(b) *In uncontrolled airspace.* Except while in a holding pattern of 2 minutes or less or while turning, each person operating an aircraft under IFR in level cruising flight in uncontrolled airspace shall maintain an appropriate altitude as follows:

(1) When operating below 18,000 feet MSL and—

(i) On a magnetic course of zero degrees through 179 degrees, any odd thousand foot MSL altitude (such as 3,000, 5,000, or 7,000); or

(ii) On a magnetic course of 180 degrees through 359 degrees, any even thousand foot MSL altitude (such as 2,000, 4,000, or 6,000).

(2) When operating at or above 18,000 feet MSL but below flight level 290, and—

(i) On a magnetic course of zero degrees through 179 degrees, any odd flight level (such as 190, 210, or 230); or

(ii) On a magnetic course of 180 degrees through 359 degrees, any even flight level (such as 180, 200, or 220).

(3) When operating at flight level 290 and above in non-RVSM airspace, and—

(i) On a magnetic course of zero degrees through 179 degrees, any flight level, at 4,000-foot intervals, beginning at and including flight level 290 (such as flight level 290, 330, or 370); or

(ii) On a magnetic course of 180 degrees through 359 degrees, any flight level, at 4,000-foot intervals, beginning at and including flight level 310 (such as flight level 310, 350, or 390).

(4) When operating at flight level 290 and above in airspace designated as Reduced Vertical Separation Minimum (RVSM) airspace and—

(i) On a magnetic course of zero degrees through 179 degrees, any odd flight level, at 2,000-foot intervals beginning at and including flight level 290 (such as flight level 290, 310, 330, 350, 370, 390, 410); or

(ii) On a magnetic course of 180 degrees through 359 degrees, any even flight level, at 2000-foot intervals beginning at and including flight level 300 (such as 300, 320, 340, 360, 380, 400).

[Doc. No. 18334, 54 FR 34294, Aug. 18, 1989, as amended by Amdt. 91-276, 68 FR 61321, Oct. 27, 2003; 68 FR 70133, Dec. 17, 2003; Amdt. 91-296, 72 FR 31679, June 7, 2007]

§ 91.180 Operations within airspace designated as Reduced Vertical Separation Minimum airspace.

(a) Except as provided in paragraph (b) of this section, no person may operate a civil aircraft in airspace designated as Reduced Vertical Separation Minimum (RVSM) airspace unless:

(1) The operator and the operator's aircraft comply with the minimum standards of appendix G of this part; and

(2) The operator is authorized by the Administrator or the country of registry to conduct such operations.

(b) The Administrator may authorize a deviation from the requirements of this section.

[Amdt. 91-276, 68 FR 70133, Dec. 17, 2003]

§ 91.181 Course to be flown.

Unless otherwise authorized by ATC, no person may operate an aircraft within controlled airspace under IFR except as follows:

(a) On an ATS route, along the centerline of that airway.

(b) On any other route, along the direct course between the navigational aids or fixes defining that route. However, this section does not prohibit maneuvering the aircraft to pass well clear of other air traffic or the maneuvering of the aircraft in VFR conditions to clear the intended flight path both before and during climb or descent.

[Doc. No. 18334, 54 FR 34294, Aug. 18, 1989, as amended by Amdt. 91-296, 72 FR 31679, June 7, 2007]

§ 91.183 IFR communications.

Unless otherwise authorized by ATC, the pilot in command of each aircraft operated under IFR in controlled airspace must ensure that a continuous watch is maintained on the appropriate frequency and must report the following as soon as possible—

(a) The time and altitude of passing each designated reporting point, or the reporting points specified by ATC, except that while the aircraft is under radar control, only the passing of those reporting points specifically requested by ATC need be reported;

(b) Any unforecast weather conditions encountered; and

(c) Any other information relating to the safety of flight.

[Doc. No. 18334, 54 FR 34294, Aug. 18, 1989, as amended by Amdt. 91-296, 72 FR 31679, June 7, 2007]

§ 91.185 IFR operations: Two-way radio communications failure.

(a) *General.* Unless otherwise authorized by ATC, each pilot who has two-way radio communications failure when operating under IFR shall comply with the rules of this section.

(b) *VFR conditions.* If the failure occurs in VFR conditions, or if VFR conditions are encountered after the failure, each pilot shall continue the flight under VFR and land as soon as practicable.

(c) *IFR conditions.* If the failure occurs in IFR conditions, or if paragraph (b) of this section cannot be complied with, each pilot shall continue the flight according to the following:

(1) *Route.* (i) By the route assigned in the last ATC clearance received;

(ii) If being radar vectored, by the direct route from the point of radio failure to the fix, route, or airway specified in the vector clearance;

(iii) In the absence of an assigned route, by the route that ATC has advised may be expected in a further clearance; or

(iv) In the absence of an assigned route or a route that ATC has advised may be expected in a further clearance, by the route filed in the flight plan.

(2) *Altitude.* At the highest of the following altitudes or flight levels for the route segment being flown:

(i) The altitude or flight level assigned in the last ATC clearance received;

(ii) The minimum altitude (converted, if appropriate, to minimum flight level as prescribed in § 91.121(c)) for IFR operations; or

(iii) The altitude or flight level ATC has advised may be expected in a further clearance.

(3) *Leave clearance limit.* (i) When the clearance limit is a fix from which an approach begins, commence descent or descent and approach as close as possible to the expect-further-clearance time if one has been received, or if one has not been received, as close as possible to the estimated time of arrival as calculated from the filed or amended (with ATC) estimated time en route.

(ii) If the clearance limit is not a fix from which an approach begins, leave the clearance limit at the expect-further-clearance time if one has been received, or if none has been received, upon arrival over the clearance limit, and proceed to a fix from which an approach begins and commence descent or descent and approach as close as possible to the estimated time of arrival as calculated from the filed or amended (with ATC) estimated time en route.

[Doc. No. 18334, 54 FR 34294, Aug. 18, 1989; Amdt. 91-211, 54 FR 41211, Oct. 5, 1989]

§ 91.187 Operation under IFR in controlled airspace: Malfunction reports.

(a) The pilot in command of each aircraft operated in controlled airspace under IFR shall report as soon as practical to ATC any malfunctions of navigational, approach, or communication equipment occurring in flight.

(b) In each report required by paragraph (a) of this section, the pilot in command shall include the—

(1) Aircraft identification;

(2) Equipment affected;

(3) Degree to which the capability of the pilot to operate under IFR in the ATC system is impaired; and

(4) Nature and extent of assistance desired from ATC.

§ 91.189 Category II and III operations: General operating rules.

(a) No person may operate a civil aircraft in a Category II or III operation unless—

(1) The flight crew of the aircraft consists of a pilot in command and a second in command who hold the appropriate authorizations and ratings prescribed in § 61.3 of this chapter;

(2) Each flight crewmember has adequate knowledge of, and familiarity with, the aircraft and the procedures to be used; and

(3) The instrument panel in front of the pilot who is controlling the aircraft has appropriate instrumentation for the type of flight control guidance system that is being used.

(b) Unless otherwise authorized by the Administrator, no person may operate a civil aircraft in a Category II or Category III operation unless each ground component required for that operation and the related airborne equipment is installed and operating.

(c) *Authorized DA/DH.* For the purpose of this section, when the approach procedure being used provides for and requires the use of a DA/DH, the authorized DA/DH is the highest of the following:

(1) The DA/DH prescribed by the approach procedure.

(2) The DA/DH prescribed for the pilot in command.

(3) The DA/DH for which the aircraft is equipped.

(d) Unless otherwise authorized by the Administrator, no pilot operating an aircraft in a Category II or Category III approach that provides and requires use of a DA/DH may continue the approach below the authorized decision height unless the following conditions are met:

(1) The aircraft is in a position from which a descent to a landing on the intended runway can be made at a normal rate of descent using normal maneuvers, and where that descent rate will allow touchdown to occur within the touchdown zone of the runway of intended landing.

(2) At least one of the following visual references for the intended runway is distinctly visible and identifiable to the pilot:

(i) The approach light system, except that the pilot may not descend below 100 feet above the touchdown zone elevation using the approach lights as a reference unless the red terminating bars or the red side row bars are also distinctly visible and identifiable.

(ii) The threshold.

(iii) The threshold markings.

(iv) The threshold lights.

(v) The touchdown zone or touchdown zone markings.

(vi) The touchdown zone lights.

(e) Unless otherwise authorized by the Administrator, each pilot operating an aircraft shall immediately execute an appropriate missed approach whenever, prior to touchdown, the requirements of paragraph (d) of this section are not met.

(f) No person operating an aircraft using a Category III approach without decision height may land that aircraft except in accordance with the provisions of the letter of authorization issued by the Administrator.

(g) Paragraphs (a) through (f) of this section do not apply to operations conducted by certificate holders operating under part 121, 125, 129, or 135 of this chapter, or holders of management specifications issued in accordance with subpart K of this part. Holders of operations specifications or management specifications may operate a civil aircraft in a Category II or Category III operation only in accordance with their operations specifications or management specifications, as applicable.

[Doc. No. 18334, 54 FR 34294, Aug. 18, 1989, as amended by Amdt. 91-280, 68 FR 54560, Sept. 17, 2003; Amdt. 91-296, 72 FR 31679, June 7, 2007]

§ 91.191 Category II and Category III manual.

(a) Except as provided in paragraph (c) of this section, after August 4, 1997, no person may operate a U.S.-registered civil aircraft in a Category II or a Category III operation unless—

(1) There is available in the aircraft a current and approved Category II or Category III manual, as appropriate, for that aircraft;

(2) The operation is conducted in accordance with the procedures, instructions, and limitations in the appropriate manual; and

(3) The instruments and equipment listed in the manual that are required for a particular Category II or Category III operation have been inspected and maintained in accordance with the maintenance program contained in the manual.

(b) Each operator must keep a current copy of each approved manual at its principal base of operations and must make each manual available for inspection upon request by the Administrator.

(c) This section does not apply to operations conducted by a certificate holder operating under part 121 or part 135 of this chapter or a holder of management specifications issued in accordance with subpart K of this part.

[Doc. No. 26933, 61 FR 34560, July 2, 1996, as amended by Amdt. 91-280, 68 FR 54560, Sept. 17, 2003]

§ 91.193 Certificate of authorization for certain Category II operations.

The Administrator may issue a certificate of authorization authorizing deviations from the requirements of §§ 91.189, 91.191, and 91.205(f) for the operation of small aircraft identified as Category A aircraft in § 97.3 of this chapter in Category II operations if the Administrator finds that the proposed operation can be safely conducted under the terms of the certificate. Such authorization does not permit operation of the aircraft carrying persons or property for compensation or hire.

§§ 91.195-91.199 [Reserved]

Subpart C—Equipment, Instrument, and Certificate Requirements

Source: Docket No. 18334, 54 FR 34304, Aug. 18, 1989, unless otherwise noted.

§ 91.201 [Reserved]

§ 91.203 Civil aircraft: Certifications required.

(a) Except as provided in § 91.715, no person may operate a civil aircraft unless it has within it the following:

(1) An appropriate and current airworthiness certificate. Each U.S. airworthiness certificate used to comply with this subparagraph (except a special flight permit, a copy of the applicable operations specifications issued under § 21.197(c) of this chapter, appropriate sections of the air carrier manual required by parts 121 and 135 of this chapter containing that portion of the operations specifications issued under § 21.197(c), or an authorization under § 91.611) must have on it the registration number assigned to the aircraft under part 47 of this chapter. However, the airworthiness certificate need not have on it an assigned special identification number before 10 days after that number is first affixed to the aircraft. A revised airworthiness certificate having on it an assigned special identification number, that has been affixed to an aircraft, may only be obtained upon application to an FAA Flight Standards district office.

(2) An effective U.S. registration certificate issued to its owner or, for operation within the United States, the second copy of the Aircraft registration Application as provided for in § 47.31(c), or a registration certification issued under the laws of a foreign country.

(b) No person may operate a civil aircraft unless the airworthiness certificate required by paragraph (a) of this section or a special flight authorization issued under § 91.715 is displayed at the cabin or cockpit entrance so that it is legible to passengers or crew.

(c) No person may operate an aircraft with a fuel tank installed within the passenger compartment or a baggage compartment unless the installation was accomplished pursuant to part 43 of this chapter, and a copy of FAA Form 337 authorizing that installation is on board the aircraft.

(d) No person may operate a civil airplane (domestic or foreign) into or out of an airport in the United States unless it complies with the fuel venting and exhaust emissions requirements of part 34 of this chapter.

[Doc. No. 18334, 54 FR 34292, Aug. 18, 1989, as amended by Amdt. 91-218, 55 FR 32861, Aug. 10, 1990; Amdt. 91-318, 75 FR 41983, July 20, 2010]

§ 91.205 Powered civil aircraft with standard category U.S. airworthiness certificates: Instrument and equipment requirements.

(a) *General.* Except as provided in paragraphs (c)(3) and (e) of this section, no person may operate a powered civil aircraft with a standard category U.S. airworthiness certificate in any operation described in paragraphs (b) through (f) of this section unless that aircraft contains the instruments and equipment specified in those paragraphs (or FAA-approved equivalents) for that type of operation, and those instruments and items of equipment are in operable condition.

(b) *Visual-flight rules (day).* For VFR flight during the day, the following instruments and equipment are required:

(1) Airspeed indicator.

(2) Altimeter.

(3) Magnetic direction indicator.

(4) Tachometer for each engine.

(5) Oil pressure gauge for each engine using pressure system.

(6) Temperature gauge for each liquid-cooled engine.

(7) Oil temperature gauge for each air-cooled engine.

(8) Manifold pressure gauge for each altitude engine.

(9) Fuel gauge indicating the quantity of fuel in each tank.

(10) Landing gear position indicator, if the aircraft has a retractable landing gear.

(11) For small civil airplanes certificated after March 11, 1996, in accordance with part 23 of this chapter, an approved aviation red or aviation white anticollision light system. In the event of failure of any light of the anticollision light system, operation of the aircraft may continue to a location where repairs or replacement can be made.

(12) If the aircraft is operated for hire over water and beyond power-off gliding distance from shore, approved flotation gear readily available to each occupant and, unless the aircraft is operating under part 121 of this subchapter, at least one pyrotechnic signaling device. As used in this section, "shore" means that area of the land adjacent to the water which is above the high water mark and excludes land areas which are intermittently under water.

(13) An approved safety belt with an approved metal-to-metal latching device for each occupant 2 years of age or older.

(14) For small civil airplanes manufactured after July 18, 1978, an approved shoulder harness for each front seat. The shoulder harness must be designed to protect the occupant from serious head injury when the occupant experiences the ultimate inertia forces specified in § 23.561(b)(2) of this chapter. Each shoulder harness installed at a flight crewmember station must permit the crewmember, when seated and with the safety belt and shoulder harness fastened, to perform all functions necessary for flight operations. For purposes of this paragraph—

(i) The date of manufacture of an airplane is the date the inspection acceptance records reflect that the airplane is complete and meets the FAA-approved type design data; and

(ii) A front seat is a seat located at a flight crewmember station or any seat located alongside such a seat.

(15) An emergency locator transmitter, if required by § 91.207.

(16) For normal, utility, and acrobatic category airplanes with a seating configuration, excluding pilot seats, of 9 or less, manufactured after December 12, 1986, a shoulder harness for—

(i) Each front seat that meets the requirements of § 23.785 (g) and (h) of this chapter in effect on December 12, 1985;

(ii) Each additional seat that meets the requirements of § 23.785(g) of this chapter in effect on December 12, 1985.

(17) For rotorcraft manufactured after September 16, 1992, a shoulder harness for each seat that meets the requirements of § 27.2 or § 29.2 of this chapter in effect on September 16, 1991.

(c) *Visual flight rules (night).* For VFR flight at night, the following instruments and equipment are required:

(1) Instruments and equipment specified in paragraph (b) of this section.

(2) Approved position lights.

(3) An approved aviation red or aviation white anticollision light system on all U.S.-registered civil aircraft. Anticollision light systems initially installed after August 11, 1971, on aircraft for which a type certificate was issued or applied for before August 11, 1971, must at least meet the anticollision light standards of part 23, 25, 27, or 29 of this chapter, as applicable, that were in effect on August 10, 1971, except that the color may be either aviation red or aviation white. In the event of failure of any light of the anticollision light system, operations with the aircraft may be continued to a stop where repairs or replacement can be made.

(4) If the aircraft is operated for hire, one electric landing light.

(5) An adequate source of electrical energy for all installed electrical and radio equipment.

(6) One spare set of fuses, or three spare fuses of each kind required, that are accessible to the pilot in flight.

(d) *Instrument flight rules.* For IFR flight, the following instruments and equipment are required:

(1) Instruments and equipment specified in paragraph (b) of this section, and, for night flight, instruments and equipment specified in paragraph (c) of this section.

(2) Two-way radio communication and navigation equipment suitable for the route to be flown.

(3) Gyroscopic rate-of-turn indicator, except on the following aircraft:

(i) Airplanes with a third attitude instrument system usable through flight attitudes of 360 degrees of pitch and roll and installed in accordance with the instrument requirements prescribed in § 121.305(j) of this chapter; and

(ii) Rotorcraft with a third attitude instrument system usable through flight attitudes of ±80 degrees of pitch and ±120 degrees of roll and installed in accordance with § 29.1303(g) of this chapter.

(4) Slip-skid indicator.

(5) Sensitive altimeter adjustable for barometric pressure.

(6) A clock displaying hours, minutes, and seconds with a sweep-second pointer or digital presentation.

(7) Generator or alternator of adequate capacity.

(8) Gyroscopic pitch and bank indicator (artificial horizon).

(9) Gyroscopic direction indicator (directional gyro or equivalent).

(e) *Flight at and above 24,000 feet MSL (FL 240).* If VOR navigation equipment is required under paragraph (d)(2) of this section, no person may operate a U.S.-registered civil aircraft within the 50 states and the District of Columbia at or above FL 240 unless that aircraft is equipped with approved DME or a suitable RNAV system. When the DME or RNAV system required by this paragraph fails at and above FL 240, the pilot in command of the aircraft must notify ATC immediately, and then may continue operations at and above FL 240 to the next airport of intended landing where repairs or replacement of the equipment can be made.

(f) *Category II operations.* The requirements for Category II operations are the instruments and equipment specified in—

(1) Paragraph (d) of this section; and

(2) Appendix A to this part.

(g) *Category III operations.* The instruments and equipment required for Category III operations are specified in paragraph (d) of this section.

(h) *Night vision goggle operations.* For night vision goggle operations, the following instruments and equipment must be installed in the aircraft, functioning in a normal manner, and approved for use by the FAA:

(1) Instruments and equipment specified in paragraph (b) of this section, instruments and equipment specified in paragraph (c) of this section;

(2) Night vision goggles;

(3) Interior and exterior aircraft lighting system required for night vision goggle operations;

(4) Two-way radio communications system;

(5) Gyroscopic pitch and bank indicator (artificial horizon);

(6) Generator or alternator of adequate capacity for the required instruments and equipment; and

(7) Radar altimeter.

(i) *Exclusions.* Paragraphs (f) and (g) of this section do not apply to operations conducted by a holder of a certificate issued under part 121 or part 135 of this chapter.

[Doc. No. 18334, 54 FR 34292, Aug. 18, 1989, as amended by Amdt. 91-220, 55 FR 43310, Oct. 26, 1990; Amdt. 91-223, 56 FR 41052, Aug. 16, 1991; Amdt. 91-231,

57 FR 42672, Sept. 15, 1992; Amdt. 91-248, 61 FR 5171, Feb. 9, 1996; Amdt. 91-251, 61 FR 34560, July 2, 1996; Amdt. 91-285, 69 FR 77599, Dec. 27, 2004; Amdt. 91-296, 72 FR 31679, June 7, 2007; Amdt. 91-309, 74 FR 42563, Aug. 21, 2009]

§ 91.207 Emergency locator transmitters.

(a) Except as provided in paragraphs (e) and (f) of this section, no person may operate a U.S.-registered civil airplane unless—

(1) There is attached to the airplane an approved automatic type emergency locator transmitter that is in operable condition for the following operations, except that after June 21, 1995, an emergency locator transmitter that meets the requirements of TSO-C91 may not be used for new installations:

(i) Those operations governed by the supplemental air carrier and commercial operator rules of parts 121 and 125;

(ii) Charter flights governed by the domestic and flag air carrier rules of part 121 of this chapter; and

(iii) Operations governed by part 135 of this chapter; or

(2) For operations other than those specified in paragraph (a)(1) of this section, there must be attached to the airplane an approved personal type or an approved automatic type emergency locator transmitter that is in operable condition, except that after June 21, 1995, an emergency locator transmitter that meets the requirements of TSO-C91 may not be used for new installations.

(b) Each emergency locator transmitter required by paragraph (a) of this section must be attached to the airplane in such a manner that the probability of damage to the transmitter in the event of crash impact is minimized. Fixed and deployable automatic type transmitters must be attached to the airplane as far aft as practicable.

(c) Batteries used in the emergency locator transmitters required by paragraphs (a) and (b) of this section must be replaced (or recharged, if the batteries are rechargeable)—

(1) When the transmitter has been in use for more than 1 cumulative hour; or

(2) When 50 percent of their useful life (or, for rechargeable batteries, 50 percent of their useful life of charge) has expired, as established by the transmitter manufacturer under its approval.

The new expiration date for replacing (or recharging) the battery must be legibly marked on the outside of the transmitter and entered in the aircraft maintenance record. Paragraph (c)(2) of this section does not apply to batteries (such as water-activated batteries) that are essentially unaffected during probable storage intervals.

(d) Each emergency locator transmitter required by paragraph (a) of this section must be inspected within 12 calendar months after the last inspection for—

(1) Proper installation;

(2) Battery corrosion;

(3) Operation of the controls and crash sensor; and

(4) The presence of a sufficient signal radiated from its antenna.

(e) Notwithstanding paragraph (a) of this section, a person may—

(1) Ferry a newly acquired airplane from the place where possession of it was taken to a place where the emergency locator transmitter is to be installed; and

(2) Ferry an airplane with an inoperative emergency locator transmitter from a place where repairs or replacements cannot be made to a place where they can be made.

No person other than required crewmembers may be carried aboard an airplane being ferried under paragraph (e) of this section.

(f) Paragraph (a) of this section does not apply to—

(1) Before January 1, 2004, turbojet-powered aircraft;

(2) Aircraft while engaged in scheduled flights by scheduled air carriers;

(3) Aircraft while engaged in training operations conducted entirely within a 50-nautical mile radius of the airport from which such local flight operations began;

(4) Aircraft while engaged in flight operations incident to design and testing;

(5) New aircraft while engaged in flight operations incident to their manufacture, preparation, and delivery;

(6) Aircraft while engaged in flight operations incident to the aerial application of chemicals and other substances for agricultural purposes;

(7) Aircraft certificated by the Administrator for research and development purposes;

(8) Aircraft while used for showing compliance with regulations, crew training, exhibition, air racing, or market surveys;

(9) Aircraft equipped to carry not more than one person.

(10) An aircraft during any period for which the transmitter has been temporarily removed for inspection, repair, modification, or replacement, subject to the following:

(i) No person may operate the aircraft unless the aircraft records contain an entry which includes the date of initial removal, the make, model, serial number, and reason for removing the transmitter, and a placard located in view of the pilot to show "ELT not installed."

(ii) No person may operate the aircraft more than 90 days after the ELT is initially removed from the aircraft; and

(11) On and after January 1, 2004, aircraft with a maximum payload capacity of more than 18,000 pounds when used in air transportation.

[Doc. No. 18334, 54 FR 34304, Aug. 18, 1989, as amended by Amdt. 91-242, 59 FR 32057, June 21, 1994; 59 FR 34578, July 6, 1994; Amdt. 91-265, 65 FR 81319, Dec. 22, 2000; 66 FR 16316, Mar. 23, 2001]

§ 91.209 Aircraft lights.

No person may:

(a) During the period from sunset to sunrise (or, in Alaska, during the period a prominent unlighted object cannot be seen from a distance of 3 statute miles or the sun is more than 6 degrees below the horizon)—

(1) Operate an aircraft unless it has lighted position lights;

(2) Park or move an aircraft in, or in dangerous proximity to, a night flight operations area of an airport unless the aircraft—

(i) Is clearly illuminated;

(ii) Has lighted position lights; or

(iii) is in an area that is marked by obstruction lights;

(3) Anchor an aircraft unless the aircraft—

(i) Has lighted anchor lights; or

(ii) Is in an area where anchor lights are not required on vessels; or

(b) Operate an aircraft that is equipped with an anticollision light system, unless it has lighted anticollision lights. However, the anticollision lights need not be lighted when the pilot-in-command determines that, because of operating conditions, it would be in the interest of safety to turn the lights off.

[Doc. No. 27806, 61 FR 5171, Feb. 9, 1996]

§ 91.211 Supplemental oxygen.

(a) *General.* No person may operate a civil aircraft of U.S. registry—

(1) At cabin pressure altitudes above 12,500 feet (MSL) up to and including 14,000 feet (MSL) unless the required minimum flight crew is provided with and uses supplemental oxygen for that part of the flight at those altitudes that is of more than 30 minutes duration;

(2) At cabin pressure altitudes above 14,000 feet (MSL) unless the required minimum flight crew is provided with and uses supplemental oxygen during the entire flight time at those altitudes; and

(3) At cabin pressure altitudes above 15,000 feet (MSL) unless each occupant of the aircraft is provided with supplemental oxygen.

(b) *Pressurized cabin aircraft.* (1) No person may operate a civil aircraft of U.S. registry with a pressurized cabin—

(i) At flight altitudes above flight level 250 unless at least a 10-minute supply of supplemental oxygen, in addition to any oxygen required to satisfy paragraph (a) of this section, is available for each occupant of the aircraft for use in the event that a descent is necessitated by loss of cabin pressurization; and

(ii) At flight altitudes above flight level 350 unless one pilot at the controls of the airplane is wearing and using an oxygen mask that is secured and sealed and that either supplies oxygen at all times or automatically supplies oxygen whenever the cabin pressure altitude of the airplane exceeds 14,000 feet (MSL), except that the one pilot need not wear and use an oxygen mask while at or below flight level 410 if there are two pilots at the controls and each pilot has a quick-donning type of oxygen mask that can be placed on the face with one hand from the ready position within 5 seconds, supplying oxygen and properly secured and sealed.

(2) Notwithstanding paragraph (b)(1)(ii) of this section, if for any reason at any time it is necessary for one pilot to leave the controls of the aircraft when operating at flight altitudes above flight level 350, the remaining pilot at the controls shall put on and use an oxygen mask until the other pilot has returned to that crewmember's station.

§ 91.213 Inoperative instruments and equipment.

(a) Except as provided in paragraph (d) of this section, no person may take off an aircraft with inoperative instruments or equipment installed unless the following conditions are met:

(1) An approved Minimum Equipment List exists for that aircraft.

(2) The aircraft has within it a letter of authorization, issued by the FAA Flight Standards district office having jurisdiction over the area in which the operator is located, authorizing operation of the aircraft under the Minimum Equipment List. The letter of authorization may be obtained by written request of the airworthiness certificate holder. The Minimum Equipment List and the letter of authorization constitute a supplemental type certificate for the aircraft.

(3) The approved Minimum Equipment List must—

(i) Be prepared in accordance with the limitations specified in paragraph (b) of this section; and

(ii) Provide for the operation of the aircraft with the instruments and equipment in an inoperable condition.

(4) The aircraft records available to the pilot must include an entry describing the inoperable instruments and equipment.

(5) The aircraft is operated under all applicable conditions and limitations contained in the Minimum Equipment List and the letter authorizing the use of the list.

(b) The following instruments and equipment may not be included in a Minimum Equipment List:

(1) Instruments and equipment that are either specifically or otherwise required by the airworthiness requirements under which the aircraft is type certificated and which are essential for safe operations under all operating conditions.

(2) Instruments and equipment required by an airworthiness directive to be in operable condition unless the airworthiness directive provides otherwise.

(3) Instruments and equipment required for specific operations by this part.

(c) A person authorized to use an approved Minimum Equipment List issued for a specific aircraft under subpart K of this part, part 121, 125, or 135 of this chapter must use that Minimum Equipment List to comply with the requirements in this section.

(d) Except for operations conducted in accordance with paragraph (a) or (c) of this section, a person may takeoff an aircraft in operations conducted under this part with inoperative instruments and equipment without an approved Minimum Equipment List provided—

(1) The flight operation is conducted in a—

(i) Rotorcraft, non-turbine-powered airplane, glider, lighter-than-air aircraft, powered parachute, or weight-shift-control aircraft, for which a master minimum equipment list has not been developed; or

(ii) Small rotorcraft, nonturbine-powered small airplane, glider, or lighter-than-air aircraft for which a Master Minimum Equipment List has been developed; and

(2) The inoperative instruments and equipment are not—

(i) Part of the VFR-day type certification instruments and equipment prescribed in the applicable airworthiness regulations under which the aircraft was type certificated;

(ii) Indicated as required on the aircraft's equipment list, or on the Kinds of Operations Equipment List for the kind of flight operation being conducted;

(iii) Required by § 91.205 or any other rule of this part for the specific kind of flight operation being conducted; or

(iv) Required to be operational by an airworthiness directive; and

(3) The inoperative instruments and equipment are—

(i) Removed from the aircraft, the cockpit control placarded, and the maintenance recorded in accordance with § 43.9 of this chapter; or

(ii) Deactivated and placarded "Inoperative." If deactivation of the inoperative instrument or equipment involves maintenance, it must be accomplished and recorded in accordance with part 43 of this chapter; and

(4) A determination is made by a pilot, who is certificated and appropriately rated under part 61 of this chapter, or by a person, who is certificated and appropriately rated to perform maintenance on the aircraft, that the inoperative instrument or equipment does not constitute a hazard to the aircraft.

An aircraft with inoperative instruments or equipment as provided in paragraph (d) of this section is considered to be in a properly altered condition acceptable to the Administrator.

(e) Notwithstanding any other provision of this section, an aircraft with inoperable instruments or equipment may be operated under a special flight permit issued in accordance with §§ 21.197 and 21.199 of this chapter.

[Doc. No. 18334, 54 FR 34304, Aug. 18, 1989, as amended by Amdt. 91-280, 68 FR 54560, Sept. 17, 2003; Amdt. 91-282, 69 FR 44880, July 27, 2004]

§ 91.215 ATC transponder and altitude reporting equipment and use.

(a) *All airspace: U.S.-registered civil aircraft.* For operations not conducted under part 121 or 135 of this chapter, ATC transponder equipment installed must meet the performance and environmental requirements of any class of TSO-C74b (Mode A) or any class of TSO-C74c (Mode A with altitude reporting capability) as appropriate, or the appropriate class of TSO-C112 (Mode S).

(b) *All airspace.* Unless otherwise authorized or directed by ATC, no person may operate an aircraft in the airspace described in paragraphs (b)(1) through (b)(5) of this section, unless that aircraft is equipped with an operable coded radar beacon transponder having either Mode 3/A 4096 code capability, replying to Mode 3/A interrogations with the code specified by ATC, or a Mode S capability, replying to Mode 3/A interrogations with the code specified by ATC and intermode and Mode S interrogations in accordance with the applicable provisions specified in TSO C-112, and that aircraft is equipped with automatic pressure altitude reporting equipment having a Mode C capability that automatically replies to Mode C interrogations by transmitting pressure altitude information in 100-foot increments. This requirement applies—

(1) *All aircraft.* In Class A, Class B, and Class C airspace areas;

(2) *All aircraft.* In all airspace within 30 nautical miles of an airport listed in appendix D, section 1 of this part from the surface upward to 10,000 feet MSL;

(3) Notwithstanding paragraph (b)(2) of this section, any aircraft which was not originally certificated with an engine-driven electrical system or which has not subsequently been certified with such a system installed, balloon or glider may conduct operations in the airspace within 30 nautical miles of an airport listed in appendix D, section 1 of this part provided such operations are conducted—

(i) Outside any Class A, Class B, or Class C airspace area; and

(ii) Below the altitude of the ceiling of a Class B or Class C airspace area designated for an airport or 10,000 feet MSL, whichever is lower; and

(4) All aircraft in all airspace above the ceiling and within the lateral boundaries of a Class B or Class C airspace area designated for an airport upward to 10,000 feet MSL; and

(5) All aircraft except any aircraft which was not originally certificated with an engine-driven electrical system or which has not subsequently been certified with such a system installed, balloon, or glider—

(i) In all airspace of the 48 contiguous states and the District of Columbia at and above 10,000 feet MSL, excluding the airspace at and below 2,500 feet above the surface; and

(ii) In the airspace from the surface to 10,000 feet MSL within a 10-nautical-mile radius of any airport listed in appendix D, section 2 of this part, excluding the airspace below 1,200 feet outside of the lateral boundaries of the surface area of the airspace designated for that airport.

(c) *Transponder-on operation.* While in the airspace as specified in paragraph (b) of this section or in all controlled airspace, each person operating an aircraft equipped with an operable ATC transponder maintained in accordance with § 91.413 of this part shall operate the transponder, including Mode C equipment if installed, and shall reply on the appropriate code or as assigned by ATC.

(d) *ATC authorized deviations.* Requests for ATC authorized deviations must be made to the ATC facility having jurisdiction over the concerned airspace within the time periods specified as follows:

(1) For operation of an aircraft with an operating transponder but without operating automatic pressure altitude reporting equipment having a Mode C capability, the request may be made at any time.

(2) For operation of an aircraft with an inoperative transponder to the airport of ultimate destination, including any intermediate stops, or to proceed to a place where suitable repairs can be made or both, the request may be made at any time.

(3) For operation of an aircraft that is not equipped with a transponder, the request must be made at least one hour before the proposed operation.

(Approved by the Office of Management and Budget under control number 2120-0005)

[Doc. No. 18334, 54 FR 34304, Aug. 18, 1989, as amended by Amdt. 91-221, 56 FR 469, Jan. 4, 1991; Amdt. 91-227, 56 FR 65660, Dec. 17, 1991; Amdt. 91-227, 7 FR 328, Jan. 3, 1992; Amdt. 91-229, 57 FR 34618, Aug. 5, 1992; Amdt. 91-267, 66 FR 21066, Apr. 27, 2001]

§ 91.217 Data correspondence between automatically reported pressure altitude data and the pilot's altitude reference.

(a) No person may operate any automatic pressure altitude reporting equipment associated with a radar beacon transponder—

(1) When deactivation of that equipment is directed by ATC;

(2) Unless, as installed, that equipment was tested and calibrated to transmit altitude data corresponding within 125 feet (on a 95 percent probability basis) of the indicated or calibrated datum of the altimeter normally used to maintain flight altitude, with that altimeter referenced to 29.92 inches of mercury for altitudes from sea level to the maximum operating altitude of the aircraft; or

(3) Unless the altimeters and digitizers in that equipment meet the standards of TSO-C10b and TSO-C88, respectively.

(b) No person may operate any automatic pressure altitude reporting equipment associated with a radar beacon transponder or with ADS-B Out equipment unless the pressure altitude reported for ADS-B Out and Mode C/S is derived from the same source for aircraft equipped with both a transponder and ADS-B Out.

[Docket No. 18334, 54 FR 34304, Aug. 18, 1989, as amended by Amdt. 91-314, 75 FR 30193, May 28, 2010]

§ 91.219 Altitude alerting system or device: Turbojet-powered civil airplanes.

(a) Except as provided in paragraph (d) of this section, no person may operate a turbojet-powered U.S.-registered civil airplane unless that airplane is equipped with an approved altitude alerting system or device that is in operable condition and meets the requirements of paragraph (b) of this section.

(b) Each altitude alerting system or device required by paragraph (a) of this section must be able to—

(1) Alert the pilot—

(i) Upon approaching a preselected altitude in either ascent or descent, by a sequence of both aural and visual signals in sufficient time to establish level flight at that preselected altitude; or

(ii) Upon approaching a preselected altitude in either ascent or descent, by a sequence of visual signals in sufficient time to establish level flight at that preselected altitude, and when deviating above and below that preselected altitude, by an aural signal;

(2) Provide the required signals from sea level to the highest operating altitude approved for the airplane in which it is installed;

(3) Preselect altitudes in increments that are commensurate with the altitudes at which the aircraft is operated;

(4) Be tested without special equipment to determine proper operation of the alerting signals; and

(5) Accept necessary barometric pressure settings if the system or device operates on barometric pressure. However, for operation below 3,000 feet AGL, the system or device need only provide one signal, either visual or aural, to comply with this paragraph. A radio altimeter may be included to provide the signal if the operator has an approved procedure for its use to determine DA/DH or MDA, as appropriate.

(c) Each operator to which this section applies must establish and assign procedures for the use of the altitude alerting system or device and each flight crewmember must comply with those procedures assigned to him.

(d) Paragraph (a) of this section does not apply to any operation of an airplane that has an experimental certificate or to the operation of any airplane for the following purposes:

(1) Ferrying a newly acquired airplane from the place where possession of it was taken to a place where the altitude alerting system or device is to be installed.

(2) Continuing a flight as originally planned, if the altitude alerting system or device becomes inoperative after the airplane has taken off; however, the flight may not depart from a place where repair or replacement can be made.

(3) Ferrying an airplane with any inoperative altitude alerting system or device from a place where repairs or replacements cannot be made to a place where it can be made.

(4) Conducting an airworthiness flight test of the airplane.

(5) Ferrying an airplane to a place outside the United States for the purpose of registering it in a foreign country.

(6) Conducting a sales demonstration of the operation of the airplane.

(7) Training foreign flight crews in the operation of the airplane before ferrying it to a place outside the United States for the purpose of registering it in a foreign country.

[Doc. No. 18334, 54 FR 34304, Aug. 18, 1989, as amended by Amdt. 91-296, 72 FR 31679, June 7, 2007]

§ 91.221 Traffic alert and collision avoidance system equipment and use.

(a) *All airspace: U.S.-registered civil aircraft.* Any traffic alert and collision avoidance system installed in a U.S.-registered civil aircraft must be approved by the Administrator.

(b) *Traffic alert and collision avoidance system, operation required.* Each person operating an aircraft equipped with an operable traffic alert and collision avoidance system shall have that system on and operating.

§ 91.223 Terrain awareness and warning system.

(a) *Airplanes manufactured after March 29, 2002.* Except as provided in paragraph (d) of this section, no person may operate a turbine-powered U.S.-registered airplane configured with six or more passenger seats, excluding any pilot seat, unless that airplane is equipped with an approved terrain awareness and warning system that as a minimum meets the requirements for Class B equipment in Technical Standard Order (TSO)-C151.

(b) *Airplanes manufactured on or before March 29, 2002.* Except as provided in paragraph (d) of this section, no person may operate a turbine-powered U.S.-registered airplane configured with six or more passenger seats, excluding any pilot seat, after March 29, 2005, unless that airplane is equipped with an approved terrain awareness and warning system that as a minimum meets the requirements for Class B equipment in Technical Standard Order (TSO)-C151.

(Approved by the Office of Management and Budget under control number 2120-0631)

(c) *Airplane Flight Manual.* The Airplane Flight Manual shall contain appropriate procedures for—

(1) The use of the terrain awareness and warning system; and

(2) Proper flight crew reaction in response to the terrain awareness and warning system audio and visual warnings.

(d) *Exceptions.* Paragraphs (a) and (b) of this section do not apply to—

(1) Parachuting operations when conducted entirely within a 50 nautical mile radius of the airport from which such local flight operations began.

(2) Firefighting operations.

(3) Flight operations when incident to the aerial application of chemicals and other substances.

[Doc. No. 29312, 65 FR 16755, Mar. 29, 2000]

§ 91.225 Automatic Dependent Surveillance-Broadcast (ADS-B) Out equipment and use.

(a) After January 1, 2020, and unless otherwise authorized by ATC, no person may operate an aircraft in Class A airspace unless the aircraft has equipment installed that—

(1) Meets the requirements in TSO-C166b, Extended Squitter Automatic Dependent Surveillance-Broadcast (ADS-B) and Traffic Information Service-Broadcast (TIS-B) Equipment Operating on the Radio Frequency of 1090 Megahertz (MHz); and

(2) Meets the requirements of § 91.227.

(b) After January 1, 2020, and unless otherwise authorized by ATC, no person may operate an aircraft below 18,000 feet MSL and in airspace described in paragraph (d) of this section unless the aircraft has equipment installed that—

(1) Meets the requirements in—

(i) TSO-C166b; or

(ii) TSO-C154c, Universal Access Transceiver (UAT) Automatic Dependent Surveillance-Broadcast (ADS-B) Equipment Operating on the Frequency of 978 MHz;

(2) Meets the requirements of § 91.227.

(c) Operators with equipment installed with an approved deviation under § 21.618 of this chapter also are in compliance with this section.

(d) After January 1, 2020, and unless otherwise authorized by ATC, no person may operate an aircraft in the following airspace unless the aircraft has equipment installed that meets the requirements in paragraph (b) of this section:

(1) Class B and Class C airspace areas;

(2) Except as provided for in paragraph (e) of this section, within 30 nautical miles of an airport listed in

appendix D, section 1 to this part from the surface upward to 10,000 feet MSL;

(3) Above the ceiling and within the lateral boundaries of a Class B or Class C airspace area designated for an airport upward to 10,000 feet MSL;

(4) Except as provided in paragraph (e) of this section, Class E airspace within the 48 contiguous states and the District of Columbia at and above 10,000 feet MSL, excluding the airspace at and below 2,500 feet above the surface; and

(5) Class E airspace at and above 3,000 feet MSL over the Gulf of Mexico from the coastline of the United States out to 12 nautical miles.

(e) The requirements of paragraph (b) of this section do not apply to any aircraft that was not originally certificated with an electrical system, or that has not subsequently been certified with such a system installed, including balloons and gliders. These aircraft may conduct operations without ADS-B Out in the airspace specified in paragraphs (d)(2) and (d)(4) of this section. Operations authorized by this section must be conducted—

(1) Outside any Class B or Class C airspace area; and

(2) Below the altitude of the ceiling of a Class B or Class C airspace area designated for an airport, or 10,000 feet MSL, whichever is lower.

(f) Each person operating an aircraft equipped with ADS-B Out must operate this equipment in the transmit mode at all times.

(g) Requests for ATC authorized deviations from the requirements of this section must be made to the ATC facility having jurisdiction over the concerned airspace within the time periods specified as follows:

(1) For operation of an aircraft with an inoperative ADS-B Out, to the airport of ultimate destination, including any intermediate stops, or to proceed to a place where suitable repairs can be made or both, the request may be made at any time.

(2) For operation of an aircraft that is not equipped with ADS-B Out, the request must be made at least 1 hour before the proposed operation.

(h) The standards required in this section are incorporated by reference with the approval of the Director of the Office of the Federal Register under 5 U.S.C. 552(a) and 1 CFR part 51. All approved materials are available for inspection at the FAA's Office of Rulemaking (ARM-1), 800 Independence Avenue, SW., Washington, DC 20590 (telephone 202-267-9677), or at the National Archives and Records Administration (NARA). For information on the availability of this material at NARA, call 202-741-6030, or go to *http://www.archives.gov/federal_register/code_of_federal_regulations/ibr_locations.html.* This material is also available from the sources indicated in paragraphs (h)(1) and (h)(2) of this section.

(1) Copies of Technical Standard Order (TSO)-C166b, Extended Squitter Automatic Dependent Surveillance-Broadcast (ADS-B) and Traffic Information Service-Broadcast (TIS-B) Equipment Operating on the Radio Frequency of 1090 Megahertz (MHz) (December 2, 2009) and TSO-C154c, Universal Access Transceiver (UAT) Automatic Dependent Surveillance-Broadcast (ADS-B) Equipment Operating on the Frequency of 978 MHz (December 2, 2009) may be obtained from the U.S. Department of Transportation, Subsequent Distribution Office, DOT Warehouse M30, Ardmore East Business Center, 3341 Q 75th Avenue, Landover, MD 20785; telephone (301) 322-5377. Copies of TSO -C166B and TSO-C154c are also available on the FAA's Web site, at *http://www.faa.gov/aircraft/air_cert/design_approvals/tso/.* Select the link "Search Technical Standard Orders."

(2) Copies of Section 2, Equipment Performance Requirements and Test Procedures, of RTCA DO-260B, Minimum Operational Performance Standards for 1090 MHz Extended Squitter Automatic Dependent Surveillance-Broadcast (ADS-B) and Traffic Information Services-Broadcast (TIS-B), December 2, 2009 (referenced in TSO-C166b) and Section 2, Equipment Performance Requirements and Test Procedures, of RTCA DO-282B, Minimum Operational Performance Standards for Universal Access Transceiver (UAT) Automatic Dependent Surveillance-Broadcast (ADS-B), December 2, 2009 (referenced in TSO C-154c) may be obtained from RTCA, Inc., 1828 L Street, NW., Suite 805, Washington, DC 20036-5133, telephone 202-833-9339. Copies of RTCA DO-260B and RTCA DO-282B are also available on RTCA Inc.'s Web site, at *http://www.rtca.org/onlinecart/allproducts.cfm.*

[Doc. No. FAA-2007-29305, 75 FR 30193, May 28, 2010; Amdt. 91-314-A, 75 FR 37712, June 30, 2010; Amdt. 91-316, 75 FR 37712, June 30, 2010]

§ 91.227 Automatic Dependent Surveillance-Broadcast (ADS-B) Out equipment performance requirements.

(a) *Definitions.* For the purposes of this section:

ADS-B Out is a function of an aircraft's onboard avionics that periodically broadcasts the aircraft's state vector (3-dimensional position and 3-dimensional velocity) and other required information as described in this section.

Navigation Accuracy Category for Position (NAC_P) specifies the accuracy of a reported aircraft's position, as defined in TSO-C166b and TSO-C154c.

Navigation Accuracy Category for Velocity (NAC_V) specifies the accuracy of a reported aircraft's velocity, as defined in TSO-C166b and TSO-C154c.

Navigation Integrity Category (NIC) specifies an integrity containment radius around an aircraft's reported position, as defined in TSO-C166b and TSO-C154c.

Position Source refers to the equipment installed onboard an aircraft used to process and provide aircraft position (for example, latitude, longitude, and velocity) information.

Source Integrity Level (SIL) indicates the probability of the reported horizontal position exceeding the containment radius defined by the NIC on a per sample or per hour basis, as defined in TSO-C166b and TSO-C154c.

System Design Assurance (SDA) indicates the probability of an aircraft malfunction causing false or misleading information to be transmitted, as defined in TSO-C166b and TSO-C154c.

Total latency is the total time between when the position is measured and when the position is transmitted by the aircraft.

Uncompensated latency is the time for which the aircraft does not compensate for latency.

(b) *1090 MHz ES and UAT Broadcast Links and Power Requirements*—

(1) Aircraft operating in Class A airspace must have equipment installed that meets the antenna and power output requirements of Class A1, A1S, A2, A3, B1S, or B1 equipment as defined in TSO-C166b, Extended Squitter Automatic Dependent Surveillance-Broadcast (ADS-B) and Traffic Information Service-Broadcast (TIS-B) Equipment Operating on the Radio Frequency of 1090 Megahertz (MHz).

(2) Aircraft operating in airspace designated for ADS-B Out, but outside of Class A airspace, must have equipment installed that meets the antenna and output power requirements of either:

(i) Class A1, A1S, A2, A3, B1S, or B1 as defined in TSO-C166b; or

(ii) Class A1H, A1S, A2, A3, B1S, or B1 equipment as defined in TSO-C154c, Universal Access Transceiver (UAT) Automatic Dependent Surveillance-Broadcast (ADS-B) Equipment Operating on the Frequency of 978 MHz.

(c) *ADS-B Out Performance Requirements for NACP, NACV, NIC, SDA, and SIL*—

(1) For aircraft broadcasting ADS-B Out as required under § 91.225 (a) and (b)—

(i) The aircraft's NACP must be less than 0.05 nautical miles;

(ii) The aircraft's NACV must be less than 10 meters per second;

(iii) The aircraft's NIC must be less than 0.2 nautical miles;

(iv) The aircraft's SDA must be 2; and

(v) The aircraft's SIL must be 3.

(2) Changes in NACP, NACV, SDA, and SIL must be broadcast within 10 seconds.

(3) Changes in NIC must be broadcast within 12 seconds.

(d) *Minimum Broadcast Message Element Set for ADS-B Out.* Each aircraft must broadcast the following information, as defined in TSO-C166b or TSO-C154c. The pilot must enter information for message elements listed in paragraphs (d)(7) through (d)(10) of this section during the appropriate phase of flight.

(1) The length and width of the aircraft;

(2) An indication of the aircraft's latitude and longitude;

(3) An indication of the aircraft's barometric pressure altitude;

(4) An indication of the aircraft's velocity;

(5) An indication if TCAS II or ACAS is installed and operating in a mode that can generate resolution advisory alerts;

(6) If an operable TCAS II or ACAS is installed, an indication if a resolution advisory is in effect;

(7) An indication of the Mode 3/A transponder code specified by ATC;

(8) An indication of the aircraft's call sign that is submitted on the flight plan, or the aircraft's registration number, except when the pilot has not filed a flight plan, has not requested ATC services, and is using a TSO-C154c self-assigned temporary 24-bit address;

(9) An indication if the flightcrew has identified an emergency, radio communication failure, or unlawful interference;

(10) An indication of the aircraft's "IDENT" to ATC;

(11) An indication of the aircraft assigned ICAO 24-bit address, except when the pilot has not filed a flight plan, has not requested ATC services, and is using a TSO-C154c self-assigned temporary 24-bit address;

(12) An indication of the aircraft's emitter category;

(13) An indication of whether an ADS-B In capability is installed;

(14) An indication of the aircraft's geometric altitude;

(15) An indication of the Navigation Accuracy Category for Position (NACP);

(16) An indication of the Navigation Accuracy Category for Velocity (NACV);

(17) An indication of the Navigation Integrity Category (NIC);

(18) An indication of the System Design Assurance (SDA); and

(19) An indication of the Source Integrity Level (SIL).

(e) *ADS-B Latency Requirements*—

(1) The aircraft must transmit its geometric position no later than 2.0 seconds from the time of measurement of the position to the time of transmission.

(2) Within the 2.0 total latency allocation, a maximum of 0.6 seconds can be uncompensated latency. The aircraft must compensate for any latency above 0.6 seconds up to the maximum 2.0 seconds total by extrapolating the geometric position to the time of message transmission.

(3) The aircraft must transmit its position and velocity at least once per second while airborne or while moving on the airport surface.

(4) The aircraft must transmit its position at least once every 5 seconds while stationary on the airport surface.

(f) *Equipment with an approved deviation.* Operators with equipment installed with an approved deviation under § 21.618 of this chapter also are in compliance with this section.

(g) *Incorporation by Reference.* The standards required in this section are incorporated by reference with the approval of the Director of the Office of the Federal Register under 5 U.S.C. 552(a) and 1 CFR part 51. All approved materials are available for inspection at the FAA's Office of Rulemaking (ARM-1), 800 Independence Avenue, SW., Washington, DC 20590 (telephone 202-267-9677), or at the National Archives and Records Administration (NARA). For information on the availability of this material at NARA, call 202-741-6030, or go to *http://www.archives.gov/federal_register/code_of_federal_regulations/ibr_locations.html.* This material is also available from the sources indicated in paragraphs (g)(1) and (g)(2) of this section.

(1) Copies of Technical Standard Order (TSO)-C166b, Extended Squitter Automatic Dependent Surveillance-Broadcast (ADS-B) and Traffic Information Service-Broadcast (TIS-B) Equipment Operating on the Radio Frequency of 1090 Megahertz (MHz) (December 2, 2009) and TSO-C154c, Universal Access Transceiver (UAT) Automatic Dependent Surveillance-Broadcast (ADS-B) Equipment Operating on the Frequency of 978 MHz (December 2, 2009) may be obtained from the U.S. Department of Transportation, Subsequent Distribution Office, DOT Warehouse M30, Ardmore East Business Center, 3341 Q 75th Avenue, Landover, MD 20785; telephone (301) 322-5377. Copies of TSO -C166B and TSO-C154c are also available on the FAA's Web site, at *http://www.faa.gov/aircraft/air_cert/design_approvals/tso/.* Select the link "Search Technical Standard Orders."

(2) Copies of Section 2, Equipment Performance Requirements and Test Procedures, of RTCA DO-260B, Minimum Operational Performance Standards for 1090 MHz Extended Squitter Automatic Dependent Surveillance-Broadcast (ADS-B) and Traffic Information Services-Broadcast (TIS-B), December 2, 2009 (referenced in TSO-C166b) and Section 2, Equipment Performance Requirements and Test Procedures, of RTCA DO-282B, Minimum Operational Performance Standards for Universal Access Transceiver (UAT) Automatic Dependent Surveillance-Broadcast (ADS-B), December 2, 2009 (referenced in TSO C-154c) may be obtained from RTCA, Inc., 1828 L Street, NW., Suite 805, Washington, DC 20036-5133, telephone 202-833-9339. Copies of RTCA DO-260B and RTCA DO-282B are also available on RTCA Inc.'s Web site, at *http://www.rtca.org/onlinecart/allproducts.cfm.*

[Doc. No. FAA-2007-29305, 75 FR 30194, May 28, 2010; Amdt. 91-314-A, 75 FR 37712, June 30, 2010; Amdt. 91-316, 75 FR 37712, June 30, 2010]

§§ 91.228-91.299 [Reserved]

Subpart D—Special Flight Operations

SOURCE: Docket No. 18334, 54 FR 34308, Aug. 18, 1989, unless otherwise noted.

§ 91.301 [Reserved]

§ 91.303 Aerobatic flight.

No person may operate an aircraft in aerobatic flight—

(a) Over any congested area of a city, town, or settlement;

(b) Over an open air assembly of persons;

(c) Within the lateral boundaries of the surface areas of Class B, Class C, Class D, or Class E airspace designated for an airport;

(d) Within 4 nautical miles of the center line of any Federal airway;

(e) Below an altitude of 1,500 feet above the surface; or

(f) When flight visibility is less than 3 statute miles.

For the purposes of this section, aerobatic flight means an intentional maneuver involving an abrupt change in an aircraft's attitude, an abnormal attitude, or abnormal acceleration, not necessary for normal flight.

[Doc. No. 18834, 54 FR 34308, Aug. 18, 1989, as amended by Amdt. 91-227, 56 FR 65661, Dec. 17, 1991]

§ 91.305 Flight test areas.

No person may flight test an aircraft except over open water, or sparsely populated areas, having light air traffic.

§ 91.307 Parachutes and parachuting.

(a) No pilot of a civil aircraft may allow a parachute that is available for emergency use to be carried in that aircraft unless it is an approved type and has been packed by a certificated and appropriately rated parachute rigger—

(1) Within the preceding 180 days, if its canopy, shrouds, and harness are composed exclusively of nylon, rayon, or other similar synthetic fiber or materials that are substantially resistant to damage from mold, mildew, or other fungi and other rotting agents propagated in a moist environment; or

(2) Within the preceding 60 days, if any part of the parachute is composed of silk, pongee, or other natural fiber or materials not specified in paragraph (a)(1) of this section.

(b) Except in an emergency, no pilot in command may allow, and no person may conduct, a parachute operation from an aircraft within the United States except in accordance with part 105 of this chapter.

(c) Unless each occupant of the aircraft is wearing an approved parachute, no pilot of a civil aircraft carrying any person (other than a crewmember) may execute any intentional maneuver that exceeds—

(1) A bank of 60 degrees relative to the horizon; or

(2) A nose-up or nose-down attitude of 30 degrees relative to the horizon.

(d) Paragraph (c) of this section does not apply to—

(1) Flight tests for pilot certification or rating; or

(2) Spins and other flight maneuvers required by the regulations for any certificate or rating when given by—

(i) A certificated flight instructor; or

(ii) An airline transport pilot instructing in accordance with § 61.67 of this chapter.

(e) For the purposes of this section, *approved parachute* means—

(1) A parachute manufactured under a type certificate or a technical standard order (C-23 series); or

(2) A personnel-carrying military parachute identified by an NAF, AAF, or AN drawing number, an AAF order number, or any other military designation or specification number.

[Doc. No. 18334, 54 FR 34308, Aug. 18, 1989, as amended by Amdt. 91-255, 62 FR 68137, Dec. 30, 1997; Amdt. 91-268, 66 FR 23553, May 9, 2001; Amdt. 91-305, 73 FR 69530, Nov. 19, 2008]

§ 91.309 Towing: Gliders and unpowered ultralight vehicles.

(a) No person may operate a civil aircraft towing a glider or unpowered ultralight vehicle unless—

(1) The pilot in command of the towing aircraft is qualified under § 61.69 of this chapter;

(2) The towing aircraft is equipped with a tow-hitch of a kind, and installed in a manner, that is approved by the Administrator;

(3) The towline used has breaking strength not less than 80 percent of the maximum certificated operating weight of the glider or unpowered ultralight vehicle and not more than twice this operating weight. However, the towline used may have a breaking strength more than twice the maximum certificated operating weight of the glider or unpowered ultralight vehicle if—

(i) A safety link is installed at the point of attachment of the towline to the glider or unpowered ultralight vehicle with a breaking strength not less than 80 percent of the maximum certificated operating weight of the glider or unpowered ultralight vehicle and not greater than twice this operating weight;

(ii) A safety link is installed at the point of attachment of the towline to the towing aircraft with a breaking strength greater, but not more than 25 percent greater, than that of the safety link at the towed glider or unpowered ultralight vehicle end of the towline and not greater than twice the maximum certificated operating weight of the glider or unpowered ultralight vehicle;

(4) Before conducting any towing operation within the lateral boundaries of the surface areas of Class B, Class C, Class D, or Class E airspace designated for an airport, or before making each towing flight within such controlled airspace if required by ATC, the pilot in command notifies the control tower. If a control tower does not exist or is not in operation, the pilot in command must notify the FAA flight service station serving that controlled airspace before conducting any towing operations in that airspace; and

(5) The pilots of the towing aircraft and the glider or unpowered ultralight vehicle have agreed upon a general course of action, including takeoff and release signals, airspeeds, and emergency procedures for each pilot.

(b) No pilot of a civil aircraft may intentionally release a towline, after release of a glider or unpowered ultralight vehicle, in a manner that endangers the life or property of another.

[Doc. No. 18834, 54 FR 34308, Aug. 18, 1989, as amended by Amdt. 91-227, 56 FR 65661, Dec. 17, 1991; Amdt. 91-282, 69 FR 44880, July 27, 2004]

§ 91.311 Towing: Other than under § 91.309.

No pilot of a civil aircraft may tow anything with that aircraft (other than under § 91.309) except in accordance with the terms of a certificate of waiver issued by the Administrator.

§ 91.313 Restricted category civil aircraft: Operating limitations.

(a) No person may operate a restricted category civil aircraft—

(1) For other than the special purpose for which it is certificated; or

(2) In an operation other than one necessary to accomplish the work activity directly associated with that special purpose.

(b) For the purpose of paragraph (a) of this section, operating a restricted category civil aircraft to provide flight crewmember training in a special purpose operation for which the aircraft is certificated is considered to be an operation for that special purpose.

(c) No person may operate a restricted category civil aircraft carrying persons or property for compensation or hire. For the purposes of this paragraph, a special purpose operation involving the carriage of persons or material necessary to accomplish that operation, such as crop dusting, seeding, spraying, and banner towing (including the carrying of required persons or material to the location of that operation), and operation for the purpose of providing flight crewmember training in a special purpose operation, are not considered to be the carriage of persons or property for compensation or hire.

(d) No person may be carried on a restricted category civil aircraft unless that person—

(1) Is a flight crewmember;

(2) Is a flight crewmember trainee;

(3) Performs an essential function in connection with a special purpose operation for which the aircraft is certificated; or

(4) Is necessary to accomplish the work activity directly associated with that special purpose.

(e) Except when operating in accordance with the terms and conditions of a certificate of waiver or special operating limitations issued by the Administrator, no person may operate a restricted category civil aircraft within the United States—

(1) Over a densely populated area;

(2) In a congested airway; or

(3) Near a busy airport where passenger transport operations are conducted.

(f) This section does not apply to nonpassenger-carrying civil rotorcraft external-load operations conducted under part 133 of this chapter.

(g) No person may operate a small restricted-category civil airplane manufactured after July 18, 1978, unless an approved shoulder harness is installed for each front seat. The shoulder harness must be designed to protect each occupant from serious head injury when the occupant experiences the ultimate inertia forces specified in § 23.561(b)(2) of this chapter. The shoulder harness installation at each flight crewmember station must permit the crewmember, when seated and with the safety belt and shoulder harness fastened, to perform all functions necessary for flight operation. For purposes of this paragraph—

(1) The date of manufacture of an airplane is the date the inspection acceptance records reflect that the airplane is complete and meets the FAA-approved type design data; and

(2) A front seat is a seat located at a flight crewmember station or any seat located alongside such a seat.

§ 91.315 Limited category civil aircraft: Operating limitations.

No person may operate a limited category civil aircraft carrying persons or property for compensation or hire.

§ 91.317 Provisionally certificated civil aircraft: Operating limitations.

(a) No person may operate a provisionally certificated civil aircraft unless that person is eligible for a provisional airworthiness certificate under § 21.213 of this chapter.

(b) No person may operate a provisionally certificated civil aircraft outside the United States unless that person has specific authority to do so from the Administrator and each foreign country involved.

(c) Unless otherwise authorized by the Director, Flight Standards Service, no person may operate a provisionally certificated civil aircraft in air transportation.

(d) Unless otherwise authorized by the Administrator, no person may operate a provisionally certificated civil aircraft except—

(1) In direct conjunction with the type or supplemental type certification of that aircraft;

(2) For training flight crews, including simulated air carrier operations;

(3) Demonstration flight by the manufacturer for prospective purchasers;

(4) Market surveys by the manufacturer;

(5) Flight checking of instruments, accessories, and equipment that do not affect the basic airworthiness of the aircraft; or

(6) Service testing of the aircraft.

(e) Each person operating a provisionally certificated civil aircraft shall operate within the prescribed limitations displayed in the aircraft or set forth in the provisional aircraft flight manual or other appropriate document. However, when operating in direct conjunction with the type or supplemental type certification of the aircraft, that person shall operate under the experimental aircraft limitations of § 21.191 of this chapter and when flight testing, shall operate under the requirements of § 91.305 of this part.

(f) Each person operating a provisionally certificated civil aircraft shall establish approved procedures for—

(1) The use and guidance of flight and ground personnel in operating under this section; and

(2) Operating in and out of airports where takeoffs or approaches over populated areas are necessary. No person may operate that aircraft except in compliance with the approved procedures.

(g) Each person operating a provisionally certificated civil aircraft shall ensure that each flight crewmember is properly certificated and has adequate knowledge of, and familiarity with, the aircraft and procedures to be used by that crewmember.

(h) Each person operating a provisionally certificated civil aircraft shall maintain it as required by applicable regulations and as may be specially prescribed by the Administrator.

(i) Whenever the manufacturer, or the Administrator, determines that a change in design, construction, or operation is necessary to ensure safe operation, no person may operate a provisionally certificated civil aircraft until that change has been made and approved. Section 21.99 of this chapter applies to operations under this section.

(j) Each person operating a provisionally certificated civil aircraft—

(1) May carry in that aircraft only persons who have a proper interest in the operations allowed by this section or who are specifically authorized by both the manufacturer and the Administrator; and

(2) Shall advise each person carried that the aircraft is provisionally certificated.

(k) The Administrator may prescribe additional limitations or procedures that the Administrator considers necessary, including limitations on the number of persons who may be carried in the aircraft.

(Approved by the Office of Management and Budget under control number 2120-0005)

[Doc. No. 18334, 54 FR 34308, Aug. 18, 1989, as amended by Amdt. 91-212, 54 FR 39293, Sept. 25, 1989]

§ 91.319 Aircraft having experimental certificates: Operating limitations.

(a) No person may operate an aircraft that has an experimental certificate—

(1) For other than the purpose for which the certificate was issued; or

(2) Carrying persons or property for compensation or hire.

(b) No person may operate an aircraft that has an experimental certificate outside of an area assigned by the Administrator until it is shown that—

(1) The aircraft is controllable throughout its normal range of speeds and throughout all the maneuvers to be executed; and

(2) The aircraft has no hazardous operating characteristics or design features.

(c) Unless otherwise authorized by the Administrator in special operating limitations, no person may operate an aircraft that has an experimental certificate over a densely populated area or in a congested airway. The Administrator may issue special operating limitations for particular aircraft to permit takeoffs and landings to be conducted over a densely populated area or in a congested airway, in accordance with terms and conditions specified in the authorization in the interest of safety in air commerce.

(d) Each person operating an aircraft that has an experimental certificate shall—

(1) Advise each person carried of the experimental nature of the aircraft;

(2) Operate under VFR, day only, unless otherwise specifically authorized by the Administrator; and

(3) Notify the control tower of the experimental nature of the aircraft when operating the aircraft into or out of airports with operating control towers.

(e) No person may operate an aircraft that is issued an experimental certificate under § 21.191(i) of this chapter for compensation or hire, except a person may operate an aircraft issued an experimental certificate under § 21.191(i)(1) for compensation or hire to—

(1) Tow a glider that is a light-sport aircraft or unpowered ultralight vehicle in accordance with § 91.309; or

(2) Conduct flight training in an aircraft which that person provides prior to January 31, 2010.

(f) No person may lease an aircraft that is issued an experimental certificate under § 21.191(i) of this chapter, except in accordance with paragraph (e)(1) of this section.

(g) No person may operate an aircraft issued an experimental certificate under § 21.191(i)(1) of this chapter to tow a glider that is a light-sport aircraft or unpowered ultralight vehicle for compensation or hire or to conduct flight training for compensation or hire in an aircraft which that persons provides unless within the preceding 100 hours of time in service the aircraft has—

(1) Been inspected by a certificated repairman (light-sport aircraft) with a maintenance rating, an appropriately rated mechanic, or an appropriately rated repair station in accordance with inspection procedures developed by the aircraft manufacturer or a person acceptable to the FAA; or

(2) Received an inspection for the issuance of an airworthiness certificate in accordance with part 21 of this chapter.

(h) The FAA may issue deviation authority providing relief from the provisions of paragraph (a) of this section for the purpose of conducting flight training. The FAA will issue this deviation authority as a letter of deviation authority.

(1) The FAA may cancel or amend a letter of deviation authority at any time.

(2) An applicant must submit a request for deviation authority to the FAA at least 60 days before the date of intended operations. A request for deviation authority must contain a complete description of the proposed operation and justification that establishes a level of safety equivalent to that provided under the regulations for the deviation requested.

(i) The Administrator may prescribe additional limitations that the Administrator considers necessary, including limitations on the persons that may be carried in the aircraft.

(Approved by the Office of Management and Budget under control number 2120-0005)

[Doc. No. 18334, 54 FR 34308, Aug. 18, 1989, as amended by Amdt. 91-282, 69 FR 44881, July 27, 2004]

§ 91.321 Carriage of candidates in elections.

(a) As an aircraft operator, you may receive payment for carrying a candidate, agent of a candidate, or person traveling on behalf of a candidate, running for Federal, State, or local election, without having to comply with the rules in parts 121, 125 or 135 of this chapter, under the following conditions:

(1) Your primary business is not as an air carrier or commercial operator;

(2) You carry the candidate, agent, or person traveling on behalf of a candidate, under the rules of part 91; and

(3) By Federal, state or local law, you are required to receive payment for carrying the candidate, agent, or person traveling on behalf of a candidate. For federal elections, the payment may not exceed the amount required by the Federal Election Commission. For a state or local election, the payment may not exceed the amount required under the applicable state or local law.

(b) For the purposes of this section, for Federal elections, the terms *candidate* and *election* have the same meaning as set forth in the regulations of the Federal Election Commission. For State or local elections, the terms *candidate* and *election* have the same meaning as provided by the applicable State or local law and those terms relate to candidates for election to public office in State and local government elections.

[Doc. No. FAA-2005-20168, 70 FR 4982, Jan. 31, 2005]

§ 91.323 Increased maximum certificated weights for certain airplanes operated in Alaska.

(a) Notwithstanding any other provision of the Federal Aviation Regulations, the Administrator will approve, as provided in this section, an increase in the maximum certificated weight of an airplane type certificated under Aeronautics Bulletin No. 7-A of the U.S. Department of Commerce dated January 1, 1931, as amended, or under the normal category of part 4a of the former Civil Air Regulations (14 CFR part 4a, 1964 ed.) if that airplane is operated in the State of Alaska by—

(1) A certificate holder conducting operations under part 121 or part 135 of this chapter; or

(2) The U.S. Department of Interior in conducting its game and fish law enforcement activities or its management, fire detection, and fire suppression activities concerning public lands.

(b) The maximum certificated weight approved under this section may not exceed—

(1) 12,500 pounds;

(2) 115 percent of the maximum weight listed in the FAA aircraft specifications;

(3) The weight at which the airplane meets the positive maneuvering load factor requirement for the normal category specified in § 23.337 of this chapter; or

(4) The weight at which the airplane meets the climb performance requirements under which it was type certificated.

(c) In determining the maximum certificated weight, the Administrator considers the structural soundness of the airplane and the terrain to be traversed.

(d) The maximum certificated weight determined under this section is added to the airplane's operation limitations and is identified as the maximum weight authorized for operations within the State of Alaska.

[Doc. No. 18334, 54 FR 34308, Aug. 18, 1989; Amdt. 91-211, 54 FR 41211, Oct. 5, 1989, as amended by Amdt. 91-253, 62 FR 13253, Mar. 19, 1997]

§ 91.325 Primary category aircraft: Operating limitations.

(a) No person may operate a primary category aircraft carrying persons or property for compensation or hire.

(b) No person may operate a primary category aircraft that is maintained by the pilot-owner under an approved special inspection and maintenance program except—

(1) The pilot-owner; or

(2) A designee of the pilot-owner, provided that the pilot-owner does not receive compensation for the use of the aircraft.

[Doc. No. 23345, 57 FR 41370, Sept. 9, 1992]

§ 91.327 Aircraft having a special airworthiness certificate in the light-sport category: Operating limitations.

(a) No person may operate an aircraft that has a special airworthiness certificate in the light-sport category for compensation or hire except—

(1) To tow a glider or an unpowered ultralight vehicle in accordance with § 91.309 of this chapter; or

(2) To conduct flight training.

(b) No person may operate an aircraft that has a special airworthiness certificate in the light-sport category unless—

(1) The aircraft is maintained by a certificated repairman with a light-sport aircraft maintenance rating, an appropriately rated mechanic, or an appropriately rated repair station in accordance with the applicable provisions of part 43 of this chapter and maintenance and inspection procedures developed by the aircraft manufacturer or a person acceptable to the FAA;

(2) A condition inspection is performed once every 12 calendar months by a certificated repairman (light-sport aircraft) with a maintenance rating, an appropriately rated mechanic, or an appropriately rated repair station in accordance with inspection procedures developed by the aircraft manufacturer or a person acceptable to the FAA;

(3) The owner or operator complies with all applicable airworthiness directives;

(4) The owner or operator complies with each safety directive applicable to the aircraft that corrects an existing unsafe condition. In lieu of complying with a safety directive an owner or operator may—

(i) Correct the unsafe condition in a manner different from that specified in the safety directive provided the person issuing the directive concurs with the action; or

(ii) Obtain an FAA waiver from the provisions of the safety directive based on a conclusion that the safety directive was issued without adhering to the applicable consensus standard;

(5) Each alteration accomplished after the aircraft's date of manufacture meets the applicable and current consensus standard and has been authorized by either the manufacturer or a person acceptable to the FAA;

(6) Each major alteration to an aircraft product produced under a consensus standard is authorized, performed and inspected in accordance with maintenance and inspection procedures developed by the manufacturer or a person acceptable to the FAA; and

(7) The owner or operator complies with the requirements for the recording of major repairs and major alterations performed on type-certificated products in accordance with § 43.9(d) of this chapter, and with the retention requirements in § 91.417.

(c) No person may operate an aircraft issued a special airworthiness certificate in the light-sport category to tow a glider or unpowered ultralight vehicle for compensation or hire or conduct flight training for compensation or hire in an aircraft which that persons provides unless within the preceding 100 hours of time in service the aircraft has—

(1) Been inspected by a certificated repairman with a light-sport aircraft maintenance rating, an appropriately rated mechanic, or an appropriately rated repair station in accordance with inspection procedures developed by the aircraft manufacturer or a person acceptable to the FAA and been approved for return to service in accordance with part 43 of this chapter; or

(2) Received an inspection for the issuance of an airworthiness certificate in accordance with part 21 of this chapter.

(d) Each person operating an aircraft issued a special airworthiness certificate in the light-sport category must operate the aircraft in accordance with the aircraft's operating instructions, including any provisions for necessary operating equipment specified in the aircraft's equipment list.

(e) Each person operating an aircraft issued a special airworthiness certificate in the light-sport category must advise each person carried of the special nature of the aircraft and that the aircraft does not meet the airworthiness requirements for an aircraft issued a standard airworthiness certificate.

(f) The FAA may prescribe additional limitations that it considers necessary.

[Doc. No. FAA-2001-11133, 69 FR 44881, July 27, 2004]

§§ 91.328-91.399 [Reserved]

Subpart E—Maintenance, Preventive Maintenance, and Alterations

SOURCE: Docket No. 18334, 54 FR 34311, Aug. 18, 1989, unless otherwise noted.

§ 91.401 Applicability.

(a) This subpart prescribes rules governing the maintenance, preventive maintenance, and alterations of U.S.-registered civil aircraft operating within or outside of the United States.

(b) Sections 91.405, 91.409, 91.411, 91.417, and 91.419 of this subpart do not apply to an aircraft maintained in accordance with a continuous airworthiness maintenance program as provided in part 121, 129, or §§ 91.1411 or 135.411(a)(2) of this chapter.

(c) Sections 91.405 and 91.409 of this part do not apply to an airplane inspected in accordance with part 125 of this chapter.

[Doc. No. 18334, 54 FR 34311, Aug. 18, 1989, as amended by Amdt. 91-267, 66 FR 21066, Apr. 27, 2001; Amdt. 91-280, 68 FR 54560, Sept. 17, 2003]

§ 91.403 General.

(a) The owner or operator of an aircraft is primarily responsible for maintaining that aircraft in an airworthy condition, including compliance with part 39 of this chapter.

(b) No person may perform maintenance, preventive maintenance, or alterations on an aircraft other than as prescribed in this subpart and other applicable regulations, including part 43 of this chapter.

(c) No person may operate an aircraft for which a manufacturer's maintenance manual or instructions for continued airworthiness has been issued that contains an airworthiness limitations section unless the mandatory replacement times, inspection intervals, and related procedures specified in that section or alternative inspection intervals and related procedures set forth in an operations specification approved by the Administrator under part 121 or 135 of this chapter or in accordance with an inspection program approved under § 91.409(e) have been complied with.

(d) A person must not alter an aircraft based on a supplemental type certificate unless the owner or operator of the aircraft is the holder of the supplemental type certificate, or has written permission from the holder.

[Doc. No. 18334, 54 FR 34311, Aug. 18, 1989, as amended by Amdt. 91-267, 66 FR 21066, Apr. 27, 2001; Amdt. 91-293, 71 FR 56005, Sept. 26, 2006]

§ 91.405 Maintenance required.

Each owner or operator of an aircraft—

(a) Shall have that aircraft inspected as prescribed in subpart E of this part and shall between required inspections, except as provided in paragraph (c) of this section, have discrepancies repaired as prescribed in part 43 of this chapter;

(b) Shall ensure that maintenance personnel make appropriate entries in the aircraft maintenance records indicating the aircraft has been approved for return to service;

(c) Shall have any inoperative instrument or item of equipment, permitted to be inoperative by § 91.213(d)(2) of this part, repaired, replaced, removed, or inspected at the next required inspection; and

(d) When listed discrepancies include inoperative instruments or equipment, shall ensure that a placard has been installed as required by § 43.11 of this chapter.

§ 91.407 Operation after maintenance, preventive maintenance, rebuilding, or alteration.

(a) No person may operate any aircraft that has undergone maintenance, preventive maintenance, rebuilding, or alteration unless—

(1) It has been approved for return to service by a person authorized under § 43.7 of this chapter; and

(2) The maintenance record entry required by § 43.9 or § 43.11, as applicable, of this chapter has been made.

(b) No person may carry any person (other than crewmembers) in an aircraft that has been maintained, rebuilt, or altered in a manner that may have appreciably changed its flight characteristics or substantially affected its operation in flight until an appropriately rated pilot with at least a private pilot certificate flies the aircraft, makes an operational check of the maintenance performed or alteration made, and logs the flight in the aircraft records.

(c) The aircraft does not have to be flown as required by paragraph (b) of this section if, prior to flight, ground tests, inspection, or both show conclusively that the maintenance, preventive maintenance, rebuilding, or alteration has not appreciably changed the flight characteristics or substantially affected the flight operation of the aircraft.

(Approved by the Office of Management and Budget under control number 2120-0005)

§ 91.409 Inspections.

(a) Except as provided in paragraph (c) of this section, no person may operate an aircraft unless, within the preceding 12 calendar months, it has had—

(1) An annual inspection in accordance with part 43 of this chapter and has been approved for return to service by a person authorized by § 43.7 of this chapter; or

(2) An inspection for the issuance of an airworthiness certificate in accordance with part 21 of this chapter.

No inspection performed under paragraph (b) of this section may be substituted for any inspection required by this paragraph unless it is performed by a person authorized to perform annual inspections and is entered as an "annual" inspection in the required maintenance records.

(b) Except as provided in paragraph (c) of this section, no person may operate an aircraft carrying any person (other than a crewmember) for hire, and no person may give flight instruction for hire in an aircraft which that person provides, unless within the preceding 100 hours of time in service the aircraft has received an annual or 100-hour inspection and been approved for return to service in accordance with part 43 of this chapter or has received an inspection for the issuance of an airworthiness certificate in accordance with part 21 of this chapter. The 100-hour limitation may be exceeded by not more than 10 hours while en route to reach a place where the inspection can be done. The excess time used to reach a place where the inspection can be done must be included in computing the next 100 hours of time in service.

(c) Paragraphs (a) and (b) of this section do not apply to—

(1) An aircraft that carries a special flight permit, a current experimental certificate, or a light-sport or provisional airworthiness certificate;

(2) An aircraft inspected in accordance with an approved aircraft inspection program under part 125 or 135 of this chapter and so identified by the registration number in the operations specifications of the certificate holder having the approved inspection program;

(3) An aircraft subject to the requirements of paragraph (d) or (e) of this section; or

(4) Turbine-powered rotorcraft when the operator elects to inspect that rotorcraft in accordance with paragraph (e) of this section.

(d) *Progressive inspection.* Each registered owner or operator of an aircraft desiring to use a progressive inspection program must submit a written request to the FAA Flight Standards district office having jurisdiction over the area in which the applicant is located, and shall provide—

(1) A certificated mechanic holding an inspection authorization, a certificated airframe repair station, or the manufacturer of the aircraft to supervise or conduct the progressive inspection;

(2) A current inspection procedures manual available and readily understandable to pilot and maintenance personnel containing, in detail—

(i) An explanation of the progressive inspection, including the continuity of inspection responsibility, the making of reports, and the keeping of records and technical reference material;

(ii) An inspection schedule, specifying the intervals in hours or days when routine and detailed inspections will be performed and including instructions for exceeding an inspection interval by not more than 10 hours while en route and for changing an inspection interval because of service experience;

(iii) Sample routine and detailed inspection forms and instructions for their use; and

(iv) Sample reports and records and instructions for their use;

(3) Enough housing and equipment for necessary disassembly and proper inspection of the aircraft; and

(4) Appropriate current technical information for the aircraft.

The frequency and detail of the progressive inspection shall provide for the complete inspection of the aircraft within each 12 calendar months and be consistent with the manufacturer's recommendations, field service experience, and the kind of operation in which the aircraft is engaged. The progressive inspection schedule must ensure that the aircraft, at all times, will be airworthy and will conform to all applicable FAA aircraft specifications, type certificate data sheets, airworthiness directives, and other approved data. If the progressive inspection is discontinued, the owner or operator shall immediately notify the local FAA Flight Standards district office, in writing, of the discontinuance. After the discontinuance, the first annual inspection under § 91.409(a)(1) is due within 12 calendar months after the last complete inspection of the aircraft under the progressive inspection. The 100-hour inspection under § 91.409(b) is due within 100 hours after that complete inspection. A complete inspection of the aircraft, for the purpose of determining when the annual and 100-hour inspections are due, requires a detailed inspection of the aircraft and all its components in accordance with the progressive inspection. A routine inspection of the aircraft and a detailed inspection of several components is not considered to be a complete inspection.

(e) *Large airplanes (to which part 125 is not applicable), turbojet multiengine airplanes, turbopropeller-powered multiengine airplanes, and turbine-powered rotorcraft.* No person may operate a large airplane, turbojet multiengine airplane, turbopropeller-powered multiengine airplane, or turbine-powered rotorcraft unless the replacement times for life-limited parts specified in the aircraft specifications, type data sheets, or other documents approved by the Administrator are complied with and the airplane or turbine-powered rotorcraft, including the airframe, engines, propellers, rotors, appliances, survival equipment, and emergency equipment, is inspected in accordance with an inspection program selected under the provisions of paragraph (f) of this section, except that, the owner or operator of a turbine-powered rotorcraft may elect to use the inspection provisions of § 91.409(a), (b), (c), or (d) in lieu of an inspection option of § 91.409(f).

(f) *Selection of inspection program under paragraph (e) of this section.* The registered owner or operator of each airplane or turbine-powered rotorcraft described in paragraph (e) of this section must select, identify in the aircraft maintenance records, and use one of the following programs for the inspection of the aircraft:

(1) A continuous airworthiness inspection program that is part of a continuous airworthiness maintenance program currently in use by a person holding an air carrier operating certificate or an operating certificate issued under part 121 or 135 of this chapter and operating that make and model aircraft under part 121 of this chapter or operating that make and model under part 135 of this chapter and maintaining it under § 135.411(a)(2) of this chapter.

(2) An approved aircraft inspection program approved under § 135.419 of this chapter and currently in use by a person holding an operating certificate issued under part 135 of this chapter.

(3) A current inspection program recommended by the manufacturer.

(4) Any other inspection program established by the registered owner or operator of that airplane or turbine-powered rotorcraft and approved by the Administrator under paragraph (g) of this section. However, the Administrator may require revision of this inspection program in accordance with the provisions of § 91.415.

Each operator shall include in the selected program the name and address of the person responsible for scheduling the inspections required by the program and make a copy of that program available to the person performing inspections on the aircraft and, upon request, to the Administrator.

(g) *Inspection program approved under paragraph (e) of this section.* Each operator of an airplane or turbine-powered rotorcraft desiring to establish or change an approved inspection program under paragraph (f)(4) of this section must submit the program for approval to the local FAA Flight Standards district office having jurisdiction over the area in which the aircraft is based. The program must be in writing and include at least the following information:

(1) Instructions and procedures for the conduct of inspections for the particular make and model airplane or turbine-powered rotorcraft, including necessary tests and checks. The instructions and procedures must set forth in detail the parts and areas of the airframe, engines, propellers, rotors, and appliances, including survival and emergency equipment required to be inspected.

(2) A schedule for performing the inspections that must be performed under the program expressed in terms of the time in service, calendar time, number of system operations, or any combination of these.

(h) *Changes from one inspection program to another.* When an operator changes from one inspection program under paragraph (f) of this section to another, the time in service, calendar times, or cycles of operation accumulated under the previous program must be applied in determining inspection due times under the new program.

(Approved by the Office of Management and Budget under control number 2120-0005)

[Doc. No. 18334, 54 FR 34311, Aug. 18, 1989; Amdt. 91-211, 54 FR 41211, Oct. 5, 1989; Amdt. 91-267, 66 FR 21066, Apr. 27, 2001; Amdt. 91-282, 69 FR 44882, July 27, 2004]

§ 91.410 [Reserved]

§ 91.411 Altimeter system and altitude reporting equipment tests and inspections.

(a) No person may operate an airplane, or helicopter, in controlled airspace under IFR unless—

(1) Within the preceding 24 calendar months, each static pressure system, each altimeter instrument, and each automatic pressure altitude reporting system has been tested and inspected and found to comply with appendices E and F of part 43 of this chapter;

(2) Except for the use of system drain and alternate static pressure valves, following any opening and closing of the static pressure system, that system has been tested and inspected and found to comply with paragraph (a), appendix E, of part 43 of this chapter; and

(3) Following installation or maintenance on the automatic pressure altitude reporting system of the ATC transponder where data correspondence error could be introduced, the integrated system has been tested, inspected, and found to comply with paragraph (c), appendix E, of part 43 of this chapter.

(b) The tests required by paragraph (a) of this section must be conducted by—

(1) The manufacturer of the airplane, or helicopter, on which the tests and inspections are to be performed;

(2) A certificated repair station properly equipped to perform those functions and holding—

(i) An instrument rating, Class I;

(ii) A limited instrument rating appropriate to the make and model of appliance to be tested;

(iii) A limited rating appropriate to the test to be performed;

(iv) An airframe rating appropriate to the airplane, or helicopter, to be tested; or

(3) A certificated mechanic with an airframe rating (static pressure system tests and inspections only).

(c) Altimeter and altitude reporting equipment approved under Technical Standard Orders are considered to be tested and inspected as of the date of their manufacture.

(d) No person may operate an airplane, or helicopter, in controlled airspace under IFR at an altitude above the maximum altitude at which all altimeters and the automatic altitude reporting system of that airplane, or helicopter, have been tested.

[Docket No. 18334, 54 FR 34308, Aug. 18, 1989, as amended by Amdt. 91-269, 66 FR 41116, Aug. 6, 2001; 72 FR 7739, Feb. 20, 2007]

§ 91.413 ATC transponder tests and inspections.

(a) No persons may use an ATC transponder that is specified in 91.215(a), 121.345(c), or § 135.143(c) of this chapter unless, within the preceding 24 calendar months, the ATC transponder has been tested and inspected and found to comply with appendix F of part 43 of this chapter; and

(b) Following any installation or maintenance on an ATC transponder where data correspondence error could be introduced, the integrated system has been tested, inspected, and found to comply with paragraph (c), appendix E, of part 43 of this chapter.

(c) The tests and inspections specified in this section must be conducted by—

(1) A certificated repair station properly equipped to perform those functions and holding—

(i) A radio rating, Class III;

(ii) A limited radio rating appropriate to the make and model transponder to be tested;

(iii) A limited rating appropriate to the test to be performed;

(2) A holder of a continuous airworthiness maintenance program as provided in part 121 or § 135.411(a)(2) of this chapter; or

(3) The manufacturer of the aircraft on which the transponder to be tested is installed, if the transponder was installed by that manufacturer.

[Doc. No. 18334, 54 FR 34311, Aug. 18, 1989, as amended by Amdt. 91-267, 66 FR 21066, Apr. 27, 2001; Amdt. 91-269, 66 FR 41116, Aug. 6, 2001]

§ 91.415 Changes to aircraft inspection programs.

(a) Whenever the Administrator finds that revisions to an approved aircraft inspection program under § 91.409(f)(4) or § 91.1109 are necessary for the continued adequacy of the program, the owner or operator must, after notification by the Administrator, make any changes in the program found to be necessary by the Administrator.

(b) The owner or operator may petition the Administrator to reconsider the notice to make any changes in a program in accordance with paragraph (a) of this section.

(c) The petition must be filed with the Director, Flight Standards Service within 30 days after the certificate holder or fractional ownership program manager receives the notice.

(d) Except in the case of an emergency requiring immediate action in the interest of safety, the filing of the petition stays the notice pending a decision by the Administrator.

[Doc. No. 18334, 54 FR 34311, Aug. 18, 1989, as amended by Amdt. 91-280, 68 FR 54560, Sept. 17, 2003]

§ 91.417 Maintenance records.

(a) Except for work performed in accordance with §§ 91.411 and 91.413, each registered owner or operator shall keep the following records for the periods specified in paragraph (b) of this section:

(1) Records of the maintenance, preventive maintenance, and alteration and records of the 100-hour, annual, progressive, and other required or approved inspections, as appropriate, for each aircraft (including the airframe) and each engine, propeller, rotor, and appliance of an aircraft. The records must include—

(i) A description (or reference to data acceptable to the Administrator) of the work performed; and

(ii) The date of completion of the work performed; and

(iii) The signature, and certificate number of the person approving the aircraft for return to service.

(2) Records containing the following information:

(i) The total time in service of the airframe, each engine, each propeller, and each rotor.

(ii) The current status of life-limited parts of each airframe, engine, propeller, rotor, and appliance.

(iii) The time since last overhaul of all items installed on the aircraft which are required to be overhauled on a specified time basis.

(iv) The current inspection status of the aircraft, including the time since the last inspection required by the inspection program under which the aircraft and its appliances are maintained.

(v) The current status of applicable airworthiness directives (AD) and safety directives including, for each, the method of compliance, the AD or safety directive number and revision date. If the AD or safety directive involves recurring action, the time and date when the next action is required.

(vi) Copies of the forms prescribed by § 43.9(d) of this chapter for each major alteration to the airframe and currently installed engines, rotors, propellers, and appliances.

(b) The owner or operator shall retain the following records for the periods prescribed:

(1) The records specified in paragraph (a)(1) of this section shall be retained until the work is repeated or superseded by other work or for 1 year after the work is performed.

(2) The records specified in paragraph (a)(2) of this section shall be retained and transferred with the aircraft at the time the aircraft is sold.

(3) A list of defects furnished to a registered owner or operator under § 43.11 of this chapter shall be retained until the defects are repaired and the aircraft is approved for return to service.

(c) The owner or operator shall make all maintenance records required to be kept by this section available for inspection by the Administrator or any authorized representative of the National Transportation Safety Board (NTSB). In addition, the owner or operator shall present Form 337 described in paragraph (d) of this section for inspection upon request of any law enforcement officer.

(d) When a fuel tank is installed within the passenger compartment or a baggage compartment pursuant to part 43 of this chapter, a copy of FAA Form 337 shall be kept on board the modified aircraft by the owner or operator.

(Approved by the Office of Management and Budget under control number 2120-0005)

[Docket No. 18334, 54 FR 34311, Aug. 18, 1989, as amended by Amdt. 91-311, 75 FR 5223, Feb. 1, 2010; Amdt. 91-323, 76 FR 39260, July 6, 2011]

§ 91.419 Transfer of maintenance records.

Any owner or operator who sells a U.S.-registered aircraft shall transfer to the purchaser, at the time of sale, the following records of that aircraft, in plain language form or in coded form at the election of the purchaser, if the coded form provides for the preservation and retrieval of information in a manner acceptable to the Administrator:

(a) The records specified in § 91.417(a)(2).

(b) The records specified in § 91.417(a)(1) which are not included in the records covered by paragraph (a) of this section, except that the purchaser may permit the seller to keep physical custody of such records. However, custody of records by the seller does not relieve the purchaser of the responsibility under § 91.417(c) to make the records available for inspection by the Administrator or any authorized representative of the National Transportation Safety Board (NTSB).

§ 91.421 Rebuilt engine maintenance records.

(a) The owner or operator may use a new maintenance record, without previous operating history, for an aircraft engine rebuilt by the manufacturer or by an agency approved by the manufacturer.

(b) Each manufacturer or agency that grants zero time to an engine rebuilt by it shall enter in the new record—

(1) A signed statement of the date the engine was rebuilt;

(2) Each change made as required by airworthiness directives; and

(3) Each change made in compliance with manufacturer's service bulletins, if the entry is specifically requested in that bulletin.

(c) For the purposes of this section, a rebuilt engine is a used engine that has been completely disassembled, inspected, repaired as necessary, reassembled, tested, and approved in the same manner and to the same tolerances and limits as a new engine with either new or used parts. However, all parts used in it must conform to the production drawing tolerances and limits for new parts or be of approved oversized or undersized dimensions for a new engine.

§§ 91.423-91.499 [Reserved]

Subpart F—Large and Turbine-Powered Multiengine Airplanes and Fractional Ownership Program Aircraft

Source: Doc. No. 18334, 54 FR 34314, Aug. 18, 1989, unless otherwise noted.

§ 91.501 Applicability.

(a) This subpart prescribes operating rules, in addition to those prescribed in other subparts of this part, governing the operation of large airplanes of U.S. registry, turbojet-powered multiengine civil airplanes of U.S. registry, and fractional ownership program aircraft of U.S. registry that are operating under subpart K of this part in operations not involving common carriage. The operating rules in this subpart do not apply to those aircraft when they are required to be operated under parts 121, 125, 129, 135, and 137 of this chapter. (Section 91.409 prescribes an inspection program for large and for turbine-powered (turbojet and turboprop) multiengine airplanes and turbine-powered rotorcraft of U.S. registry when they are operated under this part or part 129 or 137.)

(b) Operations that may be conducted under the rules in this subpart instead of those in parts 121, 129, 135, and 137 of this chapter when common carriage is not involved, include—

(1) Ferry or training flights;

(2) Aerial work operations such as aerial photography or survey, or pipeline patrol, but not including fire fighting operations;

(3) Flights for the demonstration of an airplane to prospective customers when no charge is made except for those specified in paragraph (d) of this section;

(4) Flights conducted by the operator of an airplane for his personal transportation, or the transportation of his guests when no charge, assessment, or fee is made for the transportation;

(5) Carriage of officials, employees, guests, and property of a company on an airplane operated by that company, or the parent or a subsidiary of the company or a subsidiary of the parent, when the carriage is within the scope of, and incidental to, the business of the company (other than transportation by air) and no charge, assessment or fee is made for the carriage in excess of the cost of owning, operating, and maintaining the airplane, except that no charge of any kind may be made for the carriage of a guest of a company, when the carriage is not within the scope of, and incidental to, the business of that company;

(6) The carriage of company officials, employees, and guests of the company on an airplane operated under a time sharing, interchange, or joint ownership agreement as defined in paragraph (c) of this section;

(7) The carriage of property (other than mail) on an airplane operated by a person in the furtherance of a business or employment (other than transportation by air) when the carriage is within the scope of, and incidental to, that business or employment and no charge, assessment, or fee is made for the carriage other than those specified in paragraph (d) of this section;

(8) The carriage on an airplane of an athletic team, sports group, choral group, or similar group having a common purpose or objective when there is no charge, assessment, or fee of any kind made by any person for that carriage; and

(9) The carriage of persons on an airplane operated by a person in the furtherance of a business other than transportation by air for the purpose of selling them land, goods, or property, including franchises or distributorships, when the carriage is within the scope of, and incidental to, that business and no charge, assessment, or fee is made for that carriage.

(10) Any operation identified in paragraphs (b)(1) through (b)(9) of this section when conducted—

(i) By a fractional ownership program manager, or

(ii) By a fractional owner in a fractional ownership program aircraft operated under subpart K of this part, except that a flight under a joint ownership arrangement under paragraph (b)(6) of this section may not be conducted. For a flight under an interchange agreement under paragraph (b)(6) of this section, the exchange of equal time for the operation must be properly accounted for as part of the total hours associated with the fractional owner's share of ownership.

(c) As used in this section—

(1) A *time sharing agreement* means an arrangement whereby a person leases his airplane with flight crew to another person, and no charge is made for the flights conducted under that arrangement other than those specified in paragraph (d) of this section;

(2) An *interchange agreement* means an arrangement whereby a person leases his airplane to another person in exchange for equal time, when needed, on the other person's airplane, and no charge, assessment, or fee is made, except that a charge may be made not to exceed the difference between the cost of owning, operating, and maintaining the two airplanes;

(3) A *joint ownership agreement* means an arrangement whereby one of the registered joint owners of an airplane employs and furnishes the flight crew for that airplane and each of the registered joint owners pays a share of the charge specified in the agreement.

(d) The following may be charged, as expenses of a specific flight, for transportation as authorized by paragraphs (b) (3) and (7) and (c)(1) of this section:

(1) Fuel, oil, lubricants, and other additives.

(2) Travel expenses of the crew, including food, lodging, and ground transportation.

(3) Hangar and tie-down costs away from the aircraft's base of operation.

(4) Insurance obtained for the specific flight.

(5) Landing fees, airport taxes, and similar assessments.

(6) Customs, foreign permit, and similar fees directly related to the flight.

(7) In flight food and beverages.

(8) Passenger ground transportation.

(9) Flight planning and weather contract services.

(10) An additional charge equal to 100 percent of the expenses listed in paragraph (d)(1) of this section.

[Doc. No. 18334, 54 FR 34314, Aug. 18, 1989, as amended by Amdt. 91-280, 68 FR 54560, Sept. 17, 2003]

§ 91.503 Flying equipment and operating information.

(a) The pilot in command of an airplane shall ensure that the following flying equipment and aeronautical charts and data, in current and appropriate form, are accessible for each flight at the pilot station of the airplane:

(1) A flashlight having at least two size "D" cells, or the equivalent, that is in good working order.

(2) A cockpit checklist containing the procedures required by paragraph (b) of this section.

(3) Pertinent aeronautical charts.

(4) For IFR, VFR over-the-top, or night operations, each pertinent navigational en route, terminal area, and approach and letdown chart.

(5) In the case of multiengine airplanes, one-engine inoperative climb performance data.

(b) Each cockpit checklist must contain the following procedures and shall be used by the flight crewmembers when operating the airplane:

(1) Before starting engines.

(2) Before takeoff.

(3) Cruise.

(4) Before landing.

(5) After landing.

(6) Stopping engines.

(7) Emergencies.

(c) Each emergency cockpit checklist procedure required by paragraph (b)(7) of this section must contain the following procedures, as appropriate:

(1) Emergency operation of fuel, hydraulic, electrical, and mechanical systems.

(2) Emergency operation of instruments and controls.

(3) Engine inoperative procedures.

(4) Any other procedures necessary for safety.

(d) The equipment, charts, and data prescribed in this section shall be used by the pilot in command and other members of the flight crew, when pertinent.

§ 91.505 Familiarity with operating limitations and emergency equipment.

(a) Each pilot in command of an airplane shall, before beginning a flight, become familiar with the Airplane Flight Manual for that airplane, if one is required, and with any placards, listings, instrument markings, or any combination thereof, containing each operating limitation prescribed for that airplane by the Administrator, including those specified in § 91.9(b).

(b) Each required member of the crew shall, before beginning a flight, become familiar with the emergency equipment installed on the airplane to which that crewmember is assigned and with the procedures to be followed for the use of that equipment in an emergency situation.

§ 91.507 Equipment requirements: Over-the-top or night VFR operations.

No person may operate an airplane over-the-top or at night under VFR unless that airplane is equipped with the instruments and equipment required for IFR operations under § 91.205(d) and one electric landing light for night operations. Each required instrument and item of equipment must be in operable condition.

§ 91.509 Survival equipment for overwater operations.

(a) No person may take off an airplane for a flight over water more than 50 nautical miles from the nearest shore unless that airplane is equipped with a life preserver or an approved flotation means for each occupant of the airplane.

(b) Except as provided in paragraph (c) of this section, no person may take off an airplane for flight over water more than 30 minutes flying time or 100 nautical miles from the nearest shore, whichever is less, unless it has on board the following survival equipment:

(1) A life preserver, equipped with an approved survivor locator light, for each occupant of the airplane.

(2) Enough liferafts (each equipped with an approved survival locator light) of a rated capacity and buoyancy to accommodate the occupants of the airplane.

(3) At least one pyrotechnic signaling device for each liferaft.

(4) One self-buoyant, water-resistant, portable emergency radio signaling device that is capable of transmission on the appropriate emergency frequency or frequencies and not dependent upon the airplane power supply.

(5) A lifeline stored in accordance with § 25.1411(g) of this chapter.

(c) A fractional ownership program manager under subpart K of this part may apply for a deviation from paragraphs (b)(2) through (5) of this section for a particular over water operation or the Administrator may amend the management specifications to require the carriage of all or any specific items of the equipment listed in paragraphs (b)(2) through (5) of this section.

(d) The required life rafts, life preservers, and signaling devices must be installed in conspicuously marked locations and easily accessible in the event of a ditching without appreciable time for preparatory procedures.

(e) A survival kit, appropriately equipped for the route to be flown, must be attached to each required life raft.

(f) As used in this section, the term shore means that area of the land adjacent to the water that is above the high water mark and excludes land areas that are intermittently under water.

[Doc. No. 18334, 54 FR 34314, Aug. 18, 1989, as amended by Amdt. 91-280, 68 FR 54561, Sept. 17, 2003]

§ 91.511 Communication and navigation equipment for overwater operations.

(a) Except as provided in paragraphs (c), (d), and (f) of this section, no person may take off an airplane for a flight over water more than 30 minutes flying time or 100 nautical miles from the nearest shore unless it has at least the following operable equipment:

(1) Radio communication equipment appropriate to the facilities to be used and able to transmit to, and receive from, at least one communication facility from any place along the route:

(i) Two transmitters.

(ii) Two microphones.

(iii) Two headsets or one headset and one speaker.

(iv) Two independent receivers.

(2) Appropriate electronic navigational equipment consisting of at least two independent electronic navigation units capable of providing the pilot with the information necessary to navigate the airplane within the airspace assigned by air traffic control. However, a receiver that can receive both communications and required navigational signals may be used in place of a separate communications receiver and a separate navigational signal receiver or unit.

(b) For the purposes of paragraphs (a)(1)(iv) and (a)(2) of this section, a receiver or electronic navigation unit is independent if the function of any part of it does not depend on the functioning of any part of another receiver or electronic navigation unit.

(c) Notwithstanding the provisions of paragraph (a) of this section, a person may operate an airplane on which no passengers are carried from a place where repairs or replacement cannot be made to a place where they can be made, if not more than one of each of the dual items of radio communication and navigational equipment specified in paragraphs (a)(1) (i) through (iv) and (a)(2) of this section malfunctions or becomes inoperative.

(d) Notwithstanding the provisions of paragraph (a) of this section, when both VHF and HF communications equipment are required for the route and the airplane has two VHF transmitters and two VHF receivers for communications, only one HF transmitter and one HF receiver is required for communications.

(e) As used in this section, the term *shore* means that area of the land adjacent to the water which is above the high-water mark and excludes land areas which are intermittently under water.

(f) Notwithstanding the requirements in paragraph (a)(2) of this section, a person may operate in the Gulf of Mexico, the Caribbean Sea, and the Atlantic Ocean west of a line which extends from 44°47′00″ N/67°00′00″ W to 39°00′00″ N/67°00′00″ W to 38°30′00″ N/60°00′00″ W south along the 60°00′00″ W longitude line to the point where the line intersects with the northern coast of South America, when:

(1) A single long-range navigation system is installed, operational, and appropriate for the route; and

(2) Flight conditions and the aircraft's capabilities are such that no more than a 30-minute gap in two-way radio very high frequency communications is expected to exist.

[Doc. No. 18334, 54 FR 34314, Aug. 18, 1989, as amended by Amdt. 91-249, 61 FR 7190, Feb. 26, 1996; Amdt. 91-296, 72 FR 31679, June 7, 2007]

§ 91.513 Emergency equipment.

(a) No person may operate an airplane unless it is equipped with the emergency equipment listed in this section.

(b) Each item of equipment—

(1) Must be inspected in accordance with § 91.409 to ensure its continued serviceability and immediate readiness for its intended purposes;

(2) Must be readily accessible to the crew;

(3) Must clearly indicate its method of operation; and

(4) When carried in a compartment or container, must have that compartment or container marked as to contents and date of last inspection.

(c) Hand fire extinguishers must be provided for use in crew, passenger, and cargo compartments in accordance with the following:

(1) The type and quantity of extinguishing agent must be suitable for the kinds of fires likely to occur in the compartment where the extinguisher is intended to be used.

(2) At least one hand fire extinguisher must be provided and located on or near the flight deck in a place that is readily accessible to the flight crew.

(3) At least one hand fire extinguisher must be conveniently located in the passenger compartment of each airplane accommodating more than six but less than 31 passengers, and at least two hand fire extinguishers must be conveniently located in the passenger compartment of each airplane accommodating more than 30 passengers.

(4) Hand fire extinguishers must be installed and secured in such a manner that they will not interfere with the safe operation of the airplane or adversely affect the safety of the crew and passengers. They must be readily accessible and, unless the locations of the fire extinguishers are obvious, their stowage provisions must be properly identified.

(d) First aid kits for treatment of injuries likely to occur in flight or in minor accidents must be provided.

(e) Each airplane accommodating more than 19 passengers must be equipped with a crash axe.

(f) Each passenger-carrying airplane must have a portable battery-powered megaphone or megaphones readily accessible to the crewmembers assigned to direct emergency evacuation, installed as follows:

(1) One megaphone on each airplane with a seating capacity of more than 60 but less than 100 passengers, at the most rearward location in the passenger cabin where it would be readily accessible to a normal flight attendant seat. However, the Administrator may grant a deviation from the requirements of this subparagraph if the Administrator finds that a different location would be more useful for evacuation of persons during an emergency.

(2) On each airplane with a seating capacity of 100 or more passengers, one megaphone installed at the forward end and one installed at the most rearward location where it would be readily accessible to a normal flight attendant seat.

§ 91.515 Flight altitude rules.

(a) Notwithstanding § 91.119, and except as provided in paragraph (b) of this section, no person may operate an airplane under VFR at less than—

(1) One thousand feet above the surface, or 1,000 feet from any mountain, hill, or other obstruction to flight, for day operations; and

(2) The altitudes prescribed in § 91.177, for night operations.

(b) This section does not apply—

(1) During takeoff or landing;

(2) When a different altitude is authorized by a waiver to this section under subpart J of this part; or

(3) When a flight is conducted under the special VFR weather minimums of § 91.157 with an appropriate clearance from ATC.

§ 91.517 Passenger information.

(a) Except as provided in paragraph (b) of this section, no person may operate an airplane carrying passengers unless it is equipped with signs that are visible to passengers and flight attendants to notify them when smoking is prohibited and when safety belts must be fastened. The signs must be so constructed that the crew can turn them on and off. They must be turned on during airplane movement on the surface, for each takeoff, for each landing, and when otherwise considered to be necessary by the pilot in command.

(b) The pilot in command of an airplane that is not required, in accordance with applicable aircraft and equipment requirements of this chapter, to be equipped as provided in paragraph (a) of this section shall ensure that the passengers are notified orally each time that it is necessary to fasten their safety belts and when smoking is prohibited.

(c) If passenger information signs are installed, no passenger or crewmember may smoke while any "no smoking" sign is lighted nor may any passenger or crewmember smoke in any lavatory.

(d) Each passenger required by § 91.107(a)(3) to occupy a seat or berth shall fasten his or her safety belt about him or her and keep it fastened while any "fasten seat belt" sign is lighted.

(e) Each passenger shall comply with instructions given him or her by crewmembers regarding compliance with paragraphs (b), (c), and (d) of this section.

[Doc. No. 26142, 57 FR 42672, Sept. 15, 1992]

§ 91.519 Passenger briefing.

(a) Before each takeoff the pilot in command of an airplane carrying passengers shall ensure that all passengers have been orally briefed on—

(1) *Smoking.* Each passenger shall be briefed on when, where, and under what conditions smoking is prohibited. This briefing shall include a statement, as appropriate, that the Federal Aviation Regulations require passenger compliance with lighted passenger information signs and no smoking placards, prohibit smoking in lavatories, and require compliance with crewmember instructions with regard to these items;

(2) *Use of safety belts and shoulder harnesses.* Each passenger shall be briefed on when, where, and under what conditions it is necessary to have his or her safety belt and, if installed, his or her shoulder harness fastened about him or her. This briefing shall include a statement, as appropriate, that Federal Aviation Regulations require passenger compliance with the lighted passenger sign and/or crewmember instructions with regard to these items;

(3) Location and means for opening the passenger entry door and emergency exits;

(4) Location of survival equipment;

(5) Ditching procedures and the use of flotation equipment required under § 91.509 for a flight over water; and

(6) The normal and emergency use of oxygen equipment installed on the airplane.

(b) The oral briefing required by paragraph (a) of this section shall be given by the pilot in command or a member of the crew, but need not be given when the pilot in command determines that the passengers are familiar with the contents of the briefing. It may be supplemented by printed cards for the use of each passenger containing—

(1) A diagram of, and methods of operating, the emergency exits; and

(2) Other instructions necessary for use of emergency equipment.

(c) Each card used under paragraph (b) must be carried in convenient locations on the airplane for the use of each passenger and must contain information that is pertinent only to the type and model airplane on which it is used.

(d) For operations under subpart K of this part, the passenger briefing requirements of § 91.1035 apply, instead of the requirements of paragraphs (a) through (c) of this section.

[Doc. No. 18334, 54 FR 34314, Aug. 18, 1989, as amended by Amdt. 91-231, 57 FR 42672, Sept. 15, 1992; Amdt. 91-280, 68 FR 54561, Sept. 17, 2003]

§ 91.521 Shoulder harness.

(a) No person may operate a transport category airplane that was type certificated after January 1, 1958, unless it is equipped at each seat at a flight deck station with a combined safety belt and shoulder harness that meets the applicable requirements specified in § 25.785 of this chapter, except that—

(1) Shoulder harnesses and combined safety belt and shoulder harnesses that were approved and installed before March 6, 1980, may continue to be used; and

(2) Safety belt and shoulder harness restraint systems may be designed to the inertia load factors established under the certification basis of the airplane.

(b) No person may operate a transport category airplane unless it is equipped at each required flight attendant seat in the passenger compartment with a combined safety belt and shoulder harness that meets the applicable requirements specified in § 25.785 of this chapter, except that—

(1) Shoulder harnesses and combined safety belt and shoulder harnesses that were approved and installed before March 6, 1980, may continue to be used; and

(2) Safety belt and shoulder harness restraint systems may be designed to the inertia load factors established under the certification basis of the airplane.

§ 91.523 Carry-on baggage.

No pilot in command of an airplane having a seating capacity of more than 19 passengers may permit a passenger to stow baggage aboard that airplane except—

(a) In a suitable baggage or cargo storage compartment, or as provided in § 91.525; or

(b) Under a passenger seat in such a way that it will not slide forward under crash impacts severe enough to induce the ultimate inertia forces specified in § 25.561(b)(3) of this chapter, or the requirements of the regulations under which the airplane was type certificated. Restraining devices must also limit sideward motion of under-seat baggage and be designed to withstand crash impacts severe enough to induce sideward forces specified in § 25.561(b)(3) of this chapter.

§ 91.525 Carriage of cargo.

(a) No pilot in command may permit cargo to be carried in any airplane unless—

(1) It is carried in an approved cargo rack, bin, or compartment installed in the airplane;

(2) It is secured by means approved by the Administrator; or

(3) It is carried in accordance with each of the following:

(i) It is properly secured by a safety belt or other tiedown having enough strength to eliminate the possibility of shifting under all normally anticipated flight and ground conditions.

(ii) It is packaged or covered to avoid possible injury to passengers.

(iii) It does not impose any load on seats or on the floor structure that exceeds the load limitation for those components.

(iv) It is not located in a position that restricts the access to or use of any required emergency or regular exit, or the use of the aisle between the crew and the passenger compartment.

(v) It is not carried directly above seated passengers.

(b) When cargo is carried in cargo compartments that are designed to require the physical entry of a crewmember to extinguish any fire that may occur during flight, the cargo must be loaded so as to allow a crewmember to effectively reach all parts of the compartment with the contents of a hand fire extinguisher.

§ 91.527 Operating in icing conditions.

(a) No pilot may take off an airplane that has frost, ice, or snow adhering to any propeller, windshield, stabilizing or control surface; to a powerplant installation; or to an airspeed, altimeter, rate of climb, or flight attitude instrument system or wing, except that takeoffs may be made with frost under the wing in the area of the fuel tanks if authorized by the FAA.

(b) No pilot may fly under IFR into known or forecast light or moderate icing conditions, or under VFR into known light or moderate icing conditions, unless—

(1) The aircraft has functioning deicing or anti-icing equipment protecting each rotor blade, propeller, windshield, wing, stabilizing or control surface, and each airspeed, altimeter, rate of climb, or flight attitude instrument system;

(2) The airplane has ice protection provisions that meet section 34 of Special Federal Aviation Regulation No. 23; or

(3) The airplane meets transport category airplane type certification provisions, including the requirements for certification for flight in icing conditions.

(c) Except for an airplane that has ice protection provisions that meet the requirements in section 34 of Special Federal Aviation Regulation No. 23, or those for transport category airplane type certification, no pilot may fly an airplane into known or forecast severe icing conditions.

(d) If current weather reports and briefing information relied upon by the pilot in command indicate that the forecast icing conditions that would otherwise prohibit the flight will not be encountered during the flight because of changed weather conditions since the forecast, the restrictions in paragraphs (b) and (c) of this section based on forecast conditions do not apply.

[Doc. No. 18334, 54 FR 34314, Aug. 18, 1989, as amended by Amdt. 91-310, 74 FR 62696, Dec. 1, 2009]

§ 91.529 Flight engineer requirements.

(a) No person may operate the following airplanes without a flight crewmember holding a current flight engineer certificate:

(1) An airplane for which a type certificate was issued before January 2, 1964, having a maximum certificated takeoff weight of more than 80,000 pounds.

(2) An airplane type certificated after January 1, 1964, for which a flight engineer is required by the type certification requirements.

(b) No person may serve as a required flight engineer on an airplane unless, within the preceding 6 calendar months, that person has had at least 50 hours of flight time as a flight engineer on that type airplane or has been checked by the Administrator on that type airplane and is found to be familiar and competent with all essential current information and operating procedures.

§ 91.531 Second in command requirements.

(a) Except as provided in paragraph (b) and (d) of this section, no person may operate the following airplanes without a pilot who is designated as second in command of that airplane:

(1) A large airplane, except that a person may operate an airplane certificated under SFAR 41 without a pilot who is designated as second in command if that airplane is certificated for operation with one pilot.

(2) A turbojet-powered multiengine airplane for which two pilots are required under the type certification requirements for that airplane.

(3) A commuter category airplane, except that a person may operate a commuter category airplane notwithstanding paragraph (a)(1) of this section, that has a passenger seating configuration, excluding pilot seats, of nine or less without a pilot who is designated as second in command if that airplane is type certificated for operations with one pilot.

(b) The Administrator may issue a letter of authorization for the operation of an airplane without compliance with the requirements of paragraph (a) of this section if that airplane is designed for and type certificated with only one pilot station. The authorization contains any conditions that the Administrator finds necessary for safe operation.

(c) No person may designate a pilot to serve as second in command, nor may any pilot serve as second in command, of an airplane required under this section to have two pilots unless that pilot meets the qualifications for second in command prescribed in § 61.55 of this chapter.

(d) No person may operate an aircraft under subpart K of this part without a pilot who is designated as second in command of that aircraft in accordance with § 91.1049(d). The second in command must meet the experience requirements of § 91.1053.

[Doc. No. 18334, 54 FR 34314, Aug. 18, 1989, as amended by Amdt. 91-280, 68 FR 54561, Sept. 17, 2003]

§ 91.533 Flight attendant requirements.

(a) No person may operate an airplane unless at least the following number of flight attendants are on board the airplane:

(1) For airplanes having more than 19 but less than 51 passengers on board, one flight attendant.

(2) For airplanes having more than 50 but less than 101 passengers on board, two flight attendants.

(3) For airplanes having more than 100 passengers on board, two flight attendants plus one additional flight attendant for each unit (or part of a unit) of 50 passengers above 100.

(b) No person may serve as a flight attendant on an airplane when required by paragraph (a) of this section unless that person has demonstrated to the pilot in command familiarity with the necessary functions to be performed in an emergency or a situation requiring emergency evacuation and is capable of using the emergency equipment installed on that airplane.

§ 91.535 Stowage of food, beverage, and passenger service equipment during aircraft movement on the surface, takeoff, and landing.

(a) No operator may move an aircraft on the surface, take off, or land when any food, beverage, or tableware furnished by the operator is located at any passenger seat.

(b) No operator may move an aircraft on the surface, take off, or land unless each food and beverage tray and seat back tray table is secured in its stowed position.

(c) No operator may permit an aircraft to move on the surface, take off, or land unless each passenger serving cart is secured in its stowed position.

(d) No operator may permit an aircraft to move on the surface, take off, or land unless each movie screen that extends into the aisle is stowed.

(e) Each passenger shall comply with instructions given by a crewmember with regard to compliance with this section.

[Doc. No. 26142, 57 FR 42672, Sept. 15, 1992]

§§ 91.536-91.599 [Reserved]

Subpart G—Additional Equipment and Operating Requirements for Large and Transport Category Aircraft

Source: Docket No. 18334, 54 FR 34318, Aug. 18, 1989, unless otherwise noted.

§ 91.601 Applicability.

This subpart applies to operation of large and transport category U.S.-registered civil aircraft.

§ 91.603 Aural speed warning device.

No person may operate a transport category airplane in air commerce unless that airplane is equipped with an aural speed warning device that complies with § 25.1303(c)(1).

§ 91.605 Transport category civil airplane weight limitations.

(a) No person may take off any transport category airplane (other than a turbine-engine-powered airplane certificated after September 30, 1958) unless—

(1) The takeoff weight does not exceed the authorized maximum takeoff weight for the elevation of the airport of takeoff;

(2) The elevation of the airport of takeoff is within the altitude range for which maximum takeoff weights have been determined;

(3) Normal consumption of fuel and oil in flight to the airport of intended landing will leave a weight on arrival not in excess of the authorized maximum landing weight for the elevation of that airport; and

(4) The elevations of the airport of intended landing and of all specified alternate airports are within the altitude range for which the maximum landing weights have been determined.

(b) No person may operate a turbine-engine-powered transport category airplane certificated after September 30, 1958, contrary to the Airplane Flight Manual, or take off that airplane unless—

(1) The takeoff weight does not exceed the takeoff weight specified in the Airplane Flight Manual for the elevation of the airport and for the ambient temperature existing at the time of takeoff;

(2) Normal consumption of fuel and oil in flight to the airport of intended landing and to the alternate airports will leave a weight on arrival not in excess of the landing weight specified in the Airplane Flight Manual for the elevation of each of the airports involved and for the ambient temperatures expected at the time of landing;

(3) The takeoff weight does not exceed the weight shown in the Airplane Flight Manual to correspond with the minimum distances required for takeoff, considering the elevation of the airport, the runway to be used, the effective runway gradient, the ambient temperature and wind component at the time of takeoff, and, if operating limitations exist for the minimum distances required for takeoff from wet runways, the runway surface condition (dry or wet). Wet runway distances associated with grooved or porous friction course runways, if provided in the Airplane Flight Manual, may be used only for runways that are grooved or treated with a porous friction course (PFC) overlay, and that the operator determines are designed, constructed, and maintained in a manner acceptable to the Administrator.

(4) Where the takeoff distance includes a clearway, the clearway distance is not greater than one-half of—

(i) The takeoff run, in the case of airplanes certificated after September 30, 1958, and before August 30, 1959; or

(ii) The runway length, in the case of airplanes certificated after August 29, 1959.

(c) No person may take off a turbine-engine-powered transport category airplane certificated after August 29, 1959, unless, in addition to the requirements of paragraph (b) of this section—

(1) The accelerate-stop distance is no greater than the length of the runway plus the length of the stopway (if present); and

(2) The takeoff distance is no greater than the length of the runway plus the length of the clearway (if present); and

(3) The takeoff run is no greater than the length of the runway.

[Doc. No. 18334, 54 FR 34318, Aug. 18, 1989, as amended by Amdt. 91-256, 63 FR 8321, Feb. 18, 1998]

§ 91.607 Emergency exits for airplanes carrying passengers for hire.

(a) Notwithstanding any other provision of this chapter, no person may operate a large airplane (type certificated under the Civil Air Regulations effective before April 9, 1957) in passenger-carrying operations for hire, with more than the number of occupants—

(1) Allowed under Civil Air Regulations § 4b.362 (a), (b), and (c) as in effect on December 20, 1951; or

(2) Approved under Special Civil Air Regulations SR-387, SR-389, SR-389A, or SR-389B, or under this section as in effect.

However, an airplane type listed in the following table may be operated with up to the listed number of occupants (including crewmembers) and the corresponding number of exits (including emergency exits and doors) approved for the emergency exit of passengers or with an occupant-exit configuration approved under paragraph (b) or (c) of this section.

Airplane type	Maximum number of occupants including all crewmembers	Corresponding number of exits authorized for passenger use
B–307	61	4
B–377	96	9
C–46	67	4
CV–240	53	6
CV–340 and CV–440	53	6
DC–3	35	4
DC–3 (Super)	39	5
DC–4	86	5
DC–6	87	7
DC–6B	112	11
L–18	17	3
L–049, L–649, L–749	87	7
L–1049 series	96	9
M–202	53	6
M–404	53	7
Viscount 700 series	53	7

(b) Occupants in addition to those authorized under paragraph (a) of this section may be carried as follows:

(1) For each additional floor-level exit at least 24 inches wide by 48 inches high, with an unobstructed 20-inch-wide access aisleway between the exit and the main passenger aisle, 12 additional occupants.

(2) For each additional window exit located over a wing that meets the requirements of the airworthiness standards under which the airplane was type certificated or that is large enough to inscribe an ellipse 19×26 inches, eight additional occupants.

(3) For each additional window exit that is not located over a wing but that otherwise complies with paragraph (b)(2) of this section, five additional occupants.

(4) For each airplane having a ratio (as computed from the table in paragraph (a) of this section) of maximum number of occupants to number of exits greater than 14:1, and for each airplane that does not have at least one full-size, door-type exit in the side of the fuselage in the rear part of the cabin, the first additional exit must be a floor-level exit that complies with paragraph (b)(1) of this section and must be located in the rear part of the cabin on the opposite side of the fuselage from the main entrance door. However, no person may operate an airplane under this section carrying more than 115 occupants unless there is such an exit on each side of the fuselage in the rear part of the cabin.

(c) No person may eliminate any approved exit except in accordance with the following:

(1) The previously authorized maximum number of occupants must be reduced by the same number of additional occupants authorized for that exit under this section.

(2) Exits must be eliminated in accordance with the following priority schedule: First, non-over-wing window exits; second, over-wing window exits; third, floor-level exits located in the forward part of the cabin; and fourth, floor-level exits located in the rear of the cabin.

(3) At least one exit must be retained on each side of the fuselage regardless of the number of occupants.

(4) No person may remove any exit that would result in a ratio of maximum number of occupants to approved exits greater than 14:1.

(d) This section does not relieve any person operating under part 121 of this chapter from complying with § 121.291.

§ 91.609 Flight data recorders and cockpit voice recorders.

(a) No holder of an air carrier operating certificate or an operating certificate may conduct any operation under this part with an aircraft listed in the holder's operations specifications or current list of aircraft used in air transportation unless that aircraft complies with any applicable flight recorder and cockpit voice recorder requirements of the part under which its certificate is issued except that the operator may—

(1) Ferry an aircraft with an inoperative flight recorder or cockpit voice recorder from a place where repair or replacement cannot be made to a place where they can be made;

(2) Continue a flight as originally planned, if the flight recorder or cockpit voice recorder becomes inoperative after the aircraft has taken off;

(3) Conduct an airworthiness flight test during which the flight recorder or cockpit voice recorder is turned off to test it or to test any communications or electrical equipment installed in the aircraft; or

(4) Ferry a newly acquired aircraft from the place where possession of it is taken to a place where the flight recorder or cockpit voice recorder is to be installed.

(b) Notwithstanding paragraphs (c) and (e) of this section, an operator other than the holder of an air carrier or a commercial operator certificate may—

(1) Ferry an aircraft with an inoperative flight recorder or cockpit voice recorder from a place where repair or replacement cannot be made to a place where they can be made;

(2) Continue a flight as originally planned if the flight recorder or cockpit voice recorder becomes inoperative after the aircraft has taken off;

(3) Conduct an airworthiness flight test during which the flight recorder or cockpit voice recorder is turned off to test it or to test any communications or electrical equipment installed in the aircraft;

(4) Ferry a newly acquired aircraft from a place where possession of it was taken to a place where the flight recorder or cockpit voice recorder is to be installed; or

(5) Operate an aircraft:

(i) For not more than 15 days while the flight recorder and/or cockpit voice recorder is inoperative and/or removed for repair provided that the aircraft maintenance records contain an entry that indicates the date of failure, and a placard is located in view of the pilot to show that the flight recorder or cockpit voice recorder is inoperative.

(ii) For not more than an additional 15 days, provided that the requirements in paragraph (b)(5)(i) are met and that a certificated pilot, or a certificated person authorized to return an aircraft to service under § 43.7 of this chapter, certifies in the aircraft maintenance records that additional time is required to complete repairs or obtain a replacement unit.

(c)(1) No person may operate a U.S. civil registered, multiengine, turbine-powered airplane or rotorcraft having a passenger seating configuration, excluding any pilot seats of 10 or more that has been manufactured after October 11, 1991, unless it is equipped with one or more approved flight recorders that utilize a digital method of recording and storing data and a method of readily retrieving that data from the storage medium, that are capable of recording the data specified in appendix E to this part, for an airplane, or appendix F to this part, for a rotorcraft, of this part within the range, accuracy, and recording interval specified, and that are capable of retaining no less than 8 hours of aircraft operation.

(2) All airplanes subject to paragraph (c)(1) of this section that are manufactured before April 7, 2010, by April 7, 2012, must meet the requirements of § 23.1459(a)(7) or § 25.1459(a)(8) of this chapter, as applicable.

(3) All airplanes and rotorcraft subject to paragraph (c)(1) of this section that are manufactured on or after April 7, 2010, must meet the flight data recorder requirements of § 23.1459, § 25.1459, § 27.1459, or § 29.1459 of this chapter, as applicable, and retain at least the last 25 hours of recorded information using a recorder that meets the standards of TSO-C124a, or later revision.

(d) Whenever a flight recorder, required by this section, is installed, it must be operated continuously from the instant the airplane begins the takeoff roll or the rotorcraft begins lift-off until the airplane has completed the landing roll or the rotorcraft has landed at its destination.

(e) Unless otherwise authorized by the Administrator, after October 11, 1991, no person may operate a U.S. civil registered multiengine, turbine-powered airplane or rotorcraft having a passenger seating configuration of six passengers or more and for which two pilots are required by type certification or operating rule unless it is equipped with an approved cockpit voice recorder that:

(1) Is installed in compliance with § 23.1457(a)(1) and (2), (b), (c), (d)(1)(i), (2) and (3), (e), (f), and (g); § 25.1457(a)(1) and (2), (b), (c), (d)(1)(i), (2) and (3), (e), (f), and (g); § 27.1457(a)(1) and (2), (b), (c), (d)(1)(i), (2) and (3), (e), (f), and (g); or § 29.1457(a)(1) and (2), (b), (c), (d)(1)(i), (2) and (3), (e), (f), and (g) of this chapter, as applicable; and

(2) Is operated continuously from the use of the checklist before the flight to completion of the final checklist at the end of the flight.

(f) In complying with this section, an approved cockpit voice recorder having an erasure feature may be used, so that at any time during the operation of the recorder, information recorded more than 15 minutes earlier may be erased or otherwise obliterated.

(g) In the event of an accident or occurrence requiring immediate notification to the National Transportation Safety Board under part 830 of its regulations that results in the termination of the flight, any operator who has installed approved flight recorders and approved cockpit voice recorders shall keep the recorded information for at least 60 days or, if requested by the Administrator or the Board, for a longer period. Information obtained from the record is used to assist in determining the cause of accidents or occurrences in connection with the investigation under part 830. The Administrator does not use the cockpit voice recorder record in any civil penalty or certificate action.

(h) All airplanes required by this section to have a cockpit voice recorder and a flight data recorder, that are manufactured before April 7, 2010, must by April 7, 2012, have a cockpit voice recorder that also—

(1) Meets the requirements of § 23.1457(d)(6) or § 25.1457(d)(6) of this chapter, as applicable; and

(2) If transport category, meets the requirements of § 25.1457(a)(3), (a)(4), and (a)(5) of this chapter.

(i) All airplanes or rotorcraft required by this section to have a cockpit voice recorder and flight data recorder, that are manufactured on or after April 7, 2010, must have a cockpit voice recorder installed that also—

(1) Is installed in accordance with the requirements of § 23.1457 (except for paragraphs (a)(6) and (d)(5)); § 25.1457 (except for paragraphs (a)(6) and (d)(5)); § 27.1457 (except for paragraphs (a)(6) and (d)(5)); or § 29.1457 (except for paragraphs (a)(6) and (d)(5)) of this chapter, as applicable; and

(2) Retains at least the last 2 hours of recorded information using a recorder that meets the standards of TSO-C123a, or later revision.

(3) For all airplanes or rotorcraft manufactured on or after April 6, 2012, also meets the requirements of § 23.1457(a)(6) and (d)(5); § 25.1457(a)(6) and (d)(5); § 27.1457(a)(6) and (d)(5); or § 29.1457(a)(6) and (d)(5) of this chapter, as applicable.

(j) All airplanes or rotorcraft required by this section to have a cockpit voice recorder and a flight data recorder, that install datalink communication equipment on or after April 6, 2012, must record all datalink messages as required by the certification rule applicable to the aircraft.

(k) An aircraft operated under this part under deviation authority from part 125 of this chapter must comply with all of the applicable flight data recorder requirements of part 125 applicable to the aircraft, notwithstanding such deviation authority.

[Doc. No. 18334, 54 FR 34318, Aug. 18, 1989, as amended by Amdt. 91-226, 56 FR 51621, Oct. 11, 1991; Amdt. 91-228, 57 FR 19353, May 5, 1992; Amdt. 91-300, 73 FR 12564, Mar. 7, 2008; Amdt. 91-304, 73 FR 73178, Dec. 2, 2008; Amdt. 91-300, 74 FR 32800, July 9, 2009; Amdt. 91-313, 75 FR 17045, Apr. 5, 2010]

§ 91.611 Authorization for ferry flight with one engine inoperative.

(a) *General.* The holder of an air carrier operating certificate or an operating certificate issued under part 125 may conduct a ferry flight of a four-engine airplane or a turbine-engine-powered airplane equipped with three engines, with one engine inoperative, to a base for the purpose of repairing that engine subject to the following:

(1) The airplane model has been test flown and found satisfactory for safe flight in accordance with paragraph (b) or (c) of this section, as appropriate. However, each operator who before November 19, 1966, has shown that a model of airplane with an engine inoperative is satisfactory for safe flight by a test flight conducted in accordance with performance data contained in the applicable Airplane Flight Manual under paragraph (a)(2) of this section need not repeat the test flight for that model.

(2) The approved Airplane Flight Manual contains the following performance data and the flight is conducted in accordance with that data:

(i) Maximum weight.

(ii) Center of gravity limits.

(iii) Configuration of the inoperative propeller (if applicable).

(iv) Runway length for takeoff (including temperature accountability).

(v) Altitude range.

(vi) Certificate limitations.

(vii) Ranges of operational limits.

(viii) Performance information.

(ix) Operating procedures.

(3) The operator has FAA approved procedures for the safe operation of the airplane, including specific requirements for—

(i) Limiting the operating weight on any ferry flight to the minimum necessary for the flight plus the necessary reserve fuel load;

(ii) A limitation that takeoffs must be made from dry runways unless, based on a showing of actual operating takeoff techniques on wet runways with one engine inoperative, takeoffs with full controllability from wet runways have been approved for the specific model aircraft and included in the Airplane Flight Manual:

(iii) Operations from airports where the runways may require a takeoff or approach over populated areas; and

(iv) Inspection procedures for determining the operating condition of the operative engines.

(4) No person may take off an airplane under this section if—

(i) The initial climb is over thickly populated areas; or

(ii) Weather conditions at the takeoff or destination airport are less than those required for VFR flight.

(5) Persons other than required flight crewmembers shall not be carried during the flight.

(6) No person may use a flight crewmember for flight under this section unless that crewmember is thoroughly familiar with the operating procedures for one-engine inoperative ferry flight contained in the certificate holder's manual and the limitations and performance information in the Airplane Flight Manual.

(b) *Flight tests: reciprocating-engine-powered airplanes.* The airplane performance of a reciprocating-engine-powered airplane with one engine inoperative must be determined by flight test as follows:

(1) A speed not less than 1.3 VS1 must be chosen at which the airplane may be controlled satisfactorily in a climb with the critical engine inoperative (with its propeller removed or in a configuration desired by the operator and with all other engines operating at the maximum power determined in paragraph (b)(3) of this section.

(2) The distance required to accelerate to the speed listed in paragraph (b)(1) of this section and to climb to 50 feet must be determined with—

(i) The landing gear extended;

(ii) The critical engine inoperative and its propeller removed or in a configuration desired by the operator; and

(iii) The other engines operating at not more than maximum power established under paragraph (b)(3) of this section.

(3) The takeoff, flight and landing procedures, such as the approximate trim settings, method of power application, maximum power, and speed must be established.

(4) The performance must be determined at a maximum weight not greater than the weight that allows a rate of climb of at least 400 feet per minute in the en route configuration set forth in § 25.67(d) of this chapter in effect on January 31, 1977, at an altitude of 5,000 feet.

(5) The performance must be determined using temperature accountability for the takeoff field length, computed in accordance with § 25.61 of this chapter in effect on January 31, 1977.

(c) *Flight tests: Turbine-engine-powered airplanes.* The airplane performance of a turbine-engine-powered airplane with one engine inoperative must be determined by flight tests, including at least three takeoff tests, in accordance with the following:

(1) Takeoff speeds VR and V2, not less than the corresponding speeds under which the airplane was type certificated under § 25.107 of this chapter, must be chosen at which the airplane may be controlled satisfactorily with the critical engine inoperative (with its propeller removed or in a configuration desired by the operator, if applicable) and with all other engines operating at not more than the power selected for type certification as set forth in § 25.101 of this chapter.

(2) The minimum takeoff field length must be the horizontal distance required to accelerate and climb to the 35-foot height at V2 speed (including any additional speed increment obtained in the tests) multiplied by 115 percent and determined with—

(i) The landing gear extended;

(ii) The critical engine inoperative and its propeller removed or in a configuration desired by the operator (if applicable); and

(iii) The other engine operating at not more than the power selected for type certification as set forth in § 25.101 of this chapter.

(3) The takeoff, flight, and landing procedures such as the approximate trim setting, method of power application, maximum power, and speed must be established. The airplane must be satisfactorily controllable during the entire takeoff run when operated according to these procedures.

(4) The performance must be determined at a maximum weight not greater than the weight determined under § 25.121(c) of this chapter but with—

(i) The actual steady gradient of the final takeoff climb requirement not less than 1.2 percent at the end of the takeoff path with two critical engines inoperative; and

(ii) The climb speed not less than the two-engine inoperative trim speed for the actual steady gradient of the final takeoff climb prescribed by paragraph (c)(4)(i) of this section.

(5) The airplane must be satisfactorily controllable in a climb with two critical engines inoperative. Climb performance may be shown by calculations based on, and equal in accuracy to, the results of testing.

(6) The performance must be determined using temperature accountability for takeoff distance and final takeoff climb computed in accordance with § 25.101 of this chapter.

For the purpose of paragraphs (c)(4) and (5) of this section, *two critical engines* means two adjacent engines on one side of an airplane with four engines, and the center engine and one outboard engine on an airplane with three engines.

§ 91.613 Materials for compartment interiors.

(a) No person may operate an airplane that conforms to an amended or supplemental type certificate issued in accordance with SFAR No. 41 for a maximum certificated takeoff weight in excess of 12,500 pounds unless within 1 year after issuance of the initial airworthiness certificate under that SFAR the airplane meets the compartment interior requirements set forth in § 25.853 (a), (b), (b-1), (b-2), and (b-3) of this chapter in effect on September 26, 1978.

(b) Thermal/acoustic insulation materials. For transport category airplanes type certificated after January 1, 1958:

(1) For airplanes manufactured before September 2, 2005, when thermal/acoustic insulation is installed in the fuselage as replacements after September 2, 2005, the insulation must meet the flame propagation requirements of § 25.856 of this chapter, effective September 2, 2003, if it is:

(i) Of a blanket construction or

(ii) Installed around air ducting.

(2) For airplanes manufactured after September 2, 2005, thermal/acoustic insulation materials installed in the fuselage must meet the flame propagation requirements of § 25.856 of this chapter, effective September 2, 2003.

[Doc. No. 18334, 54 FR 34318, Aug. 18, 1989, as amended by Amdt. 91-279, 68 FR 45083, July 31, 2003; Amdt. 91-290, 70 FR 77752, Dec. 30, 2005]

§§ 91.615-91.699 [Reserved]

Subpart H—Foreign Aircraft Operations and Operations of U.S.-Registered Civil Aircraft Outside of the United States; and Rules Governing Persons on Board Such Aircraft

Source: Docket No. 18334, 54 FR 34320, Aug. 18, 1989, unless otherwise noted.

§ 91.701 Applicability.

(a) This subpart applies to the operations of civil aircraft of U.S. registry outside of the United States and the operations of foreign civil aircraft within the United States.

(b) Section 91.702 of this subpart also applies to each person on board an aircraft operated as follows:

(1) A U.S. registered civil aircraft operated outside the United States;

(2) Any aircraft operated outside the United States—

(i) That has its next scheduled destination or last place of departure in the United States if the aircraft next lands in the United States; or

(ii) If the aircraft lands in the United States with the individual still on the aircraft regardless of whether it was a scheduled or otherwise planned landing site.

[Doc. No. FAA-1998-4954, 64 FR 1079, Jan. 7, 1999]

§ 91.702 Persons on board.

Section 91.11 of this part (Prohibitions on interference with crewmembers) applies to each person on board an aircraft.

[Doc. No. FAA-1998-4954, 64 FR 1079, Jan. 7, 1999]

§ 91.703 Operations of civil aircraft of U.S. registry outside of the United States.

(a) Each person operating a civil aircraft of U.S. registry outside of the United States shall—

(1) When over the high seas, comply with annex 2 (Rules of the Air) to the Convention on International Civil Aviation and with §§ 91.117(c), 91.127, 91.129, and 91.131;

(2) When within a foreign country, comply with the regulations relating to the flight and maneuver of aircraft there in force;

(3) Except for §§ 91.117(a), 91.307(b), 91.309, 91.323, and 91.711, comply with this part so far as it is not inconsistent with applicable regulations of the foreign country where the aircraft is operated or annex 2 of the Convention on International Civil Aviation; and

(4) When operating within airspace designated as Minimum Navigation Performance Specifications (MNPS) airspace, comply with § 91.705. When operating within airspace designated as Reduced Vertical Separation Minimum (RVSM) airspace, comply with § 91.706.

(5) For aircraft subject to ICAO Annex 16, carry on board the aircraft documents that summarize the noise operating characteristics and certifications of the aircraft that demonstrate compliance with this part and part 36 of this chapter.

(b) Annex 2 to the Convention on International Civil Aviation, Ninth Edition—July 1990, with Amendments through Amendment 32 effective February 19, 1996, to which reference is made in this part, is incorporated into this part and made a part hereof as provided in 5 U.S.C. § 552 and pursuant to 1 CFR part 51. Annex 2 (including a complete historic file of changes thereto) is available for public inspection at the Rules Docket, AGC-200, Federal Aviation Administration, 800 Independence Avenue SW., Washington, DC 20591; or at the National Archives and Records Administration (NARA). For information on the availability of this material at NARA, call 202-741-6030, or go to: *http://www.archives.gov/federal_register/code_of_federal_regulations/ibr_locations.html*. In addition, Annex 2 may be purchased from the International Civil Aviation Organization (Attention: Distribution Officer), P.O. Box 400, Succursale, Place de L'Aviation Internationale, 1000 Sherbrooke Street West, Montreal, Quebec, Canada H3A 2R2.

[Doc. No. 18834, 54 FR 34320, Aug. 18, 1989, as amended by Amdt. 91-227, 56 FR 65661, Dec. 17, 1991; Amdt. 91-254, 62 FR 17487, Apr. 9, 1997; 69 FR 18803, Apr. 9, 2004; Amdt. 91-299, 73 FR 10143, Feb. 26, 2008; Amdt. 91-312, 75 FR 9333, Mar. 2, 2010]

§ 91.705 Operations within airspace designated as Minimum Navigation Performance Specification Airspace.

(a) Except as provided in paragraph (b) of this section, no person may operate a civil aircraft of U.S. registry in airspace designated as Minimum Navigation Performance Specifications airspace unless—

(1) The aircraft has approved navigation performance capability that complies with the requirements of appendix C of this part; and

(2) The operator is authorized by the Administrator to perform such operations.

(b) The Administrator may authorize a deviation from the requirements of this section in accordance with Section 3 of appendix C to this part.

[Doc. No. 28870, 62 FR 17487, Apr. 9, 1997]

§ 91.706 Operations within airspace designed as Reduced Vertical Separation Minimum Airspace.

(a) Except as provided in paragraph (b) of this section, no person may operate a civil aircraft of U.S. registry in airspace designated as Reduced Vertical Separation Minimum (RVSM) airspace unless:

(1) The operator and the operator's aircraft comply with the requirements of appendix G of this part; and

(2) The operator is authorized by the Administrator to conduct such operations.

(b) The Administrator may authorize a deviation from the requirements of this section in accordance with Section 5 of appendix G to this part.

[Doc. No. 28870, 62 FR 17487, Apr. 9, 1997]

§ 91.707 Flights between Mexico or Canada and the United States.

Unless otherwise authorized by ATC, no person may operate a civil aircraft between Mexico or Canada and the United States without filing an IFR or VFR flight plan, as appropriate.

§ 91.709 Operations to Cuba.

No person may operate a civil aircraft from the United States to Cuba unless—

(a) Departure is from an international airport of entry designated in § 6.13 of the Air Commerce Regulations of the Bureau of Customs (19 CFR 6.13); and

(b) In the case of departure from any of the 48 contiguous States or the District of Columbia, the pilot in command of the aircraft has filed—

(1) A DVFR or IFR flight plan as prescribed in § 99.11 or § 99.13 of this chapter; and

(2) A written statement, within 1 hour before departure, with the Office of Immigration and Naturalization Service at the airport of departure, containing—

(i) All information in the flight plan;

(ii) The name of each occupant of the aircraft;

(iii) The number of occupants of the aircraft; and

(iv) A description of the cargo, if any.

This section does not apply to the operation of aircraft by a scheduled air carrier over routes authorized in operations specifications issued by the Administrator.

(Approved by the Office of Management and Budget under control number 2120-0005)

§ 91.711 Special rules for foreign civil aircraft.

(a) *General.* In addition to the other applicable regulations of this part, each person operating a foreign civil aircraft within the United States shall comply with this section.

(b) *VFR.* No person may conduct VFR operations which require two-way radio communications under this part unless at least one crewmember of that aircraft is able to conduct two-way radio communications in the English language and is on duty during that operation.

(c) *IFR.* No person may operate a foreign civil aircraft under IFR unless—

(1) That aircraft is equipped with—

(i) Radio equipment allowing two-way radio communication with ATC when it is operated in controlled airspace; and

(ii) Navigation equipment suitable for the route to be flown.

(2) Each person piloting the aircraft—

(i) Holds a current United States instrument rating or is authorized by his foreign airman certificate to pilot under IFR; and

(ii) Is thoroughly familiar with the United States en route, holding, and letdown procedures; and

(3) At least one crewmember of that aircraft is able to conduct two-way radiotelephone communications in the English language and that crewmember is on duty while the aircraft is approaching, operating within, or leaving the United States.

(d) *Over water.* Each person operating a foreign civil aircraft over water off the shores of the United States shall give flight notification or file a flight plan in accordance with the Supplementary Procedures for the ICAO region concerned.

(e) *Flight at and above FL 240.* If VOR navigation equipment is required under paragraph (c)(1)(ii) of this section, no person may operate a foreign civil aircraft within the 50 States and the District of Columbia at or above FL 240, unless the aircraft is equipped with approved DME or a suitable RNAV system. When the DME or RNAV system required by this paragraph fails at and above FL 240, the pilot in command of the aircraft must notify ATC immediately and may then continue operations at and above FL 240 to the next airport of intended landing where repairs or replacement of the equipment can be made. A foreign civil aircraft may be operated within the 50 States and the District of Columbia at or above FL 240 without DME or an

RNAV system when operated for the following purposes, and ATC is notified before each takeoff:

(1) Ferry flights to and from a place in the United States where repairs or alterations are to be made.

(2) Ferry flights to a new country of registry.

(3) Flight of a new aircraft of U.S. manufacture for the purpose of—

(i) Flight testing the aircraft;

(ii) Training foreign flight crews in the operation of the aircraft; or

(iii) Ferrying the aircraft for export delivery outside the United States.

(4) Ferry, demonstration, and test flight of an aircraft brought to the United States for the purpose of demonstration or testing the whole or any part thereof.

[Doc. No. 18834, 54 FR 34320, Aug. 18, 1989, as amended by Amdt. 91-227, 56 FR 65661, Dec. 17, 1991; Amdt. 91-296, 72 FR 31679, June 7, 2007]

§ 91.713 Operation of civil aircraft of Cuban registry.

No person may operate a civil aircraft of Cuban registry except in controlled airspace and in accordance with air traffic clearance or air traffic control instructions that may require use of specific airways or routes and landings at specific airports.

§ 91.715 Special flight authorizations for foreign civil aircraft.

(a) Foreign civil aircraft may be operated without airworthiness certificates required under § 91.203 if a special flight authorization for that operation is issued under this section. Application for a special flight authorization must be made to the Flight Standards Division Manager or Aircraft Certification Directorate Manager of the FAA region in which the applicant is located or to the region within which the U.S. point of entry is located. However, in the case of an aircraft to be operated in the U.S. for the purpose of demonstration at an airshow, the application may be made to the Flight Standards Division Manager or Aircraft Certification Directorate Manager of the FAA region in which the airshow is located.

(b) The Administrator may issue a special flight authorization for a foreign civil aircraft subject to any conditions and limitations that the Administrator considers necessary for safe operation in the U.S. airspace.

(c) No person may operate a foreign civil aircraft under a special flight authorization unless that operation also complies with part 375 of the Special Regulations of the Department of Transportation (14 CFR part 375).

(Approved by the Office of Management and Budget under control number 2120-0005)

[Doc. No. 18334, 54 FR 34320, Aug. 18, 1989, as amended by Amdt. 91-212, 54 FR 39293, Sept. 25, 1989]

§§ 91.717-91.799 [Reserved]

Subpart I—Operating Noise Limits

SOURCE: Docket No. 18334, 54 FR 34321, Aug. 18, 1989, unless otherwise noted.

§ 91.801 Applicability: Relation to part 36.

Link to an amendment published at 78 FR 39583, July 2, 2013.

(a) This subpart prescribes operating noise limits and related requirements that apply, as follows, to the operation of civil aircraft in the United States.

(1) Sections 91.803, 91.805, 91.807, 91.809, and 91.811 apply to civil subsonic jet (turbojet) airplanes with maximum weights of more than 75,000 pounds and—

(i) If U.S. registered, that have standard airworthiness certificates; or

(ii) If foreign registered, that would be required by this chapter to have a U.S. standard airworthiness certificate in order to conduct the operations intended for the airplane were it registered in the United States. Those sections apply to operations to or from airports in the United States under this part and parts 121, 125, 129, and 135 of this chapter.

(2) Section 91.813 applies to U.S. operators of civil subsonic jet (turbojet) airplanes covered by this subpart. This section applies to operators operating to or from airports in the United States under this part and parts 121, 125, and 135, but not to those operating under part 129 of this chapter.

(3) Sections 91.803, 91.819, and 91.821 apply to U.S.-registered civil supersonic airplanes having standard airworthiness certificates and to foreign-registered civil supersonic airplanes that, if registered in the United States, would be required by this chapter to have U.S. standard airworthiness certificates in order to conduct the operations intended for the airplane. Those sections apply to operations under this part and under parts 121, 125, 129, and 135 of this chapter.

(b) Unless otherwise specified, as used in this subpart "part 36" refers to 14 CFR part 36, including the noise levels under appendix C of that part, notwithstanding the provisions of that part excepting certain airplanes from the specified noise requirements. For purposes of this subpart, the various stages of noise levels, the terms used to describe airplanes with respect to those levels, and the terms "subsonic airplane" and "supersonic airplane" have the meanings specified under part 36 of this chapter. For purposes of this subpart, for subsonic airplanes operated in foreign air commerce in the United States, the Administrator may accept compliance with the noise requirements under annex 16 of the International Civil Aviation Organization when those requirements have been shown to be substantially compatible with, and achieve results equivalent to those achievable under, part 36 for that airplane. Determinations made under these provisions are subject to the limitations of § 36.5 of this chapter as if those noise levels were part 36 noise levels.

(c) Sections 91.851 through 91.877 of this subpart prescribe operating noise limits and related requirements that apply to any civil subsonic jet (turbojet) airplane (for which an airworthiness certificate other than an experimental certificate has been issued by the Administrator) with a maximum certificated takeoff weight of more than 75,000 pounds operating to or from an airport in the 48 contiguous United States and the District of Columbia under this part, parts 121, 125, 129, or 135 of this chapter on and after September 25, 1991.

(d) Section 91.877 prescribes reporting requirements that apply to any civil subsonic jet (turbojet) airplane with a maximum weight of more than 75,000 pounds operated by an air carrier or foreign air carrier between the contiguous United States and the State of Hawaii, between the State of Hawaii and any point outside of the 48 contiguous United States, or between the islands of Hawaii in turnaround service, under part 121 or 129 of this chapter on or after November 5, 1990.

[Doc. No. 18334, 54 FR 34321, Aug. 18, 1989; Amdt. 91-211, 54 FR 41211, Oct. 5, 1989, as amended by Amdt. 91-225, 56 FR 48658, Sept. 25, 1991; Amdt. 91-252, 61 FR 66185, Dec. 16, 1996; Amdt. 91-275, 67 FR 45237, July 8, 2002; Amdt. 91-276, 67 FR 46571, July 15, 2002]

§ 91.803 Part 125 operators: Designation of applicable regulations.

For airplanes covered by this subpart and operated under part 125 of this chapter, the following regulations apply as specified:

(a) For each airplane operation to which requirements prescribed under this subpart applied before November 29, 1980, those requirements of this subpart continue to apply.

(b) For each subsonic airplane operation to which requirements prescribed under this subpart did not apply before November 29, 1980, because the airplane was not operated in the United States under this part or part 121, 129, or 135 of this chapter, the requirements prescribed under § 91.805 of this subpart apply.

(c) For each supersonic airplane operation to which requirements prescribed under this subpart did not apply before November 29, 1980, because the airplane was not operated in the United States under this part or part 121, 129, or 135 of this chapter, the requirements of §§ 91.819 and 91.821 of this subpart apply.

(d) For each airplane required to operate under part 125 for which a deviation under that part is approved to operate, in whole or in part, under this part or part 121, 129, or 135 of this chapter, notwithstanding the approval, the requirements prescribed under paragraphs (a), (b), and (c) of this section continue to apply.

[Docket No. 18334, 54 FR 34321, Aug. 18, 1989, as amended by Amdt. 91-276, 67 FR 46571, July 15, 2002]

§ 91.805 Final compliance: Subsonic airplanes.

Except as provided in §§ 91.809 and 91.811, on and after January 1, 1985, no person may operate to or from an airport in the United States any subsonic airplane covered by this subpart unless that airplane has been shown to comply with Stage 2 or Stage 3 noise levels under part 36 of this chapter.

§§ 91.807-91.813 [Reserved]

§ 91.815 Agricultural and fire fighting airplanes: Noise operating limitations.

(a) This section applies to propeller-driven, small airplanes having standard airworthiness certificates that are designed for "agricultural aircraft operations" (as defined in § 137.3 of this chapter, as effective on January 1, 1966) or for dispensing fire fighting materials.

(b) If the Airplane Flight Manual, or other approved manual material information, markings, or placards for the airplane indicate that the airplane has not been shown to comply with the noise limits under part 36 of this chapter, no person may operate that airplane, except—

(1) To the extent necessary to accomplish the work activity directly associated with the purpose for which it is designed;

(2) To provide flight crewmember training in the special purpose operation for which the airplane is designed; and

(3) To conduct "nondispensing aerial work operations" in accordance with the requirements under § 137.29(c) of this chapter.

§ 91.817 Civil aircraft sonic boom.

(a) No person may operate a civil aircraft in the United States at a true flight Mach number greater than 1 except in compliance with conditions and limitations in an authorization to exceed Mach 1 issued to the operator under appendix B of this part.

(b) In addition, no person may operate a civil aircraft for which the maximum operating limit speed MM0 exceeds a Mach number of 1, to or from an airport in the United States, unless—

(1) Information available to the flight crew includes flight limitations that ensure that flights entering or leaving the United States will not cause a sonic boom to reach the surface within the United States; and

(2) The operator complies with the flight limitations prescribed in paragraph (b)(1) of this section or complies with conditions and limitations in an authorization to exceed Mach 1 issued under appendix B of this part.

(Approved by the Office of Management and Budget under control number 2120-0005)

§ 91.819 Civil supersonic airplanes that do not comply with part 36.

(a) *Applicability.* This section applies to civil supersonic airplanes that have not been shown to comply with the Stage 2 noise limits of part 36 in effect on October 13, 1977, using applicable trade-off provisions, and that are operated in the United States, after July 31, 1978.

(b) *Airport use.* Except in an emergency, the following apply to each person who operates a civil supersonic airplane to or from an airport in the United States:

(1) Regardless of whether a type design change approval is applied for under part 21 of this chapter, no person may land or take off an airplane covered by this section for which the type design is changed, after July 31, 1978, in a manner constituting an "acoustical change" under § 21.93 unless the acoustical change requirements of part 36 are complied with.

(2) No flight may be scheduled, or otherwise planned, for takeoff or landing after 10 p.m. and before 7 a.m. local time.

§ 91.821 Civil supersonic airplanes: Noise limits.

Except for Concorde airplanes having flight time before January 1, 1980, no person may operate in the United States, a civil supersonic airplane that does not comply with Stage 2 noise limits of part 36 in effect on October 13, 1977, using applicable trade-off provisions.

§§ 91.823-91.849 [Reserved]

§ 91.851 Definitions.

For the purposes of §§ 91.851 through 91.877 of this subpart:

Chapter 4 noise level means a noise level at or below the maximum noise level prescribed in Chapter 4, Paragraph 4.4, Maximum Noise Levels, of the International Civil Aviation Organization (ICAO) Annex 16, Volume I, Amendment 7, effective March 21, 2002. The Director of the Federal Register in accordance with 5 U.S.C. 552(a) and 1 CFR part 51 approved the incorporation by reference of this document, which can be obtained from the International Civil Aviation Organization (ICAO), Document Sales Unit, 999 University Street, Montreal, Quebec H3C 5H7, Canada. Also, you may obtain documents on the Internet at *http://*

www.ICAO.int/eshop/index.cfm. Copies may be reviewed at the U.S. Department of Transportation, Docket Operations, West Building Ground Floor, Room W12-140, 1200 New Jersey Avenue, SE., Washington, DC 20590 or at the National Archives and Records Administration (NARA). For information on the availability of this material at NARA, call 202-741-6030, or go to: *http://www.archives.gov/federal_register/code_of_federal_regulations/ibr_locations.html.*

Contiguous United States means the area encompassed by the 48 contiguous United States and the District of Columbia.

Fleet means those civil subsonic jet (turbojet) airplanes with a maximum certificated weight of more than 75,000 pounds that are listed on an operator's operations specifications as eligible for operation in the contiguous United States.

Import means a change in ownership of an airplane from a non-U.S. person to a U.S. person when the airplane is brought into the United States for operation.

Operations specifications means an enumeration of airplanes by type, model, series, and serial number operated by the operator or foreign air carrier on a given day, regardless of how or whether such airplanes are formally listed or designated by the operator.

Owner means any person that has indicia of ownership sufficient to register the airplane in the United States pursuant to part 47 of this chapter.

New entrant means an air carrier or foreign air carrier that, on or before November 5, 1990, did not conduct operations under part 121 or 129 of this chapter using an airplane covered by this subpart to or from any airport in the contiguous United States, but that initiates such operation after that date.

Stage 2 noise levels mean the requirements for Stage 2 noise levels as defined in part 36 of this chapter in effect on November 5, 1990.

Stage 3 noise levels mean the requirements for Stage 3 noise levels as defined in part 36 of this chapter in effect on November 5, 1990.

Stage 4 noise level means a noise level at or below the Stage 4 noise limit prescribed in part 36 of this chapter.

Stage 2 airplane means a civil subsonic jet (turbojet) airplane with a maximum certificated weight of 75,000 pounds or more that complies with Stage 2 noise levels as defined in part 36 of this chapter.

Stage 3 airplane means a civil subsonic jet (turbojet) airplane with a maximum certificated weight of 75,000 pounds or more that complies with Stage 3 noise levels as defined in part 36 of this chapter.

Stage 4 airplane means an airplane that has been shown not to exceed the Stage 4 noise limit prescribed in part 36 of this chapter. A Stage 4 airplane complies with all of the noise operating rules of this part.

[Doc. No. 26433, 56 FR 48658, Sept. 25, 1991, as amended by Amdt. 91-252, 61 FR 66185, Dec. 16, 1996; Amdt. 91-275, 67 FR 45237, July 8, 2002; Amdt. 91-288, 70 FR 38749, July 5, 2005; 72 FR 68475, Dec. 5, 2007]

§ 91.853 Final compliance: Civil subsonic airplanes.

Except as provided in § 91.873, after December 31, 1999, no person shall operate to or from any airport in the contiguous United States any airplane subject to § 91.801(c) of this subpart, unless that airplane has been shown to comply with Stage 3 or Stage 4 noise levels.

[Doc. No. FAA-2003-16526, 70 FR 38749, July 5, 2005]

§ 91.855 Entry and nonaddition rule.

No person may operate any airplane subject to § 91.801(c) of this subpart to or from an airport in the contiguous United States unless one or more of the following apply:

(a) The airplane complies with Stage 3 or Stage 4 noise levels.

(b) The airplane complies with Stage 2 noise levels and was owned by a U.S. person on and since November 5, 1990. Stage 2 airplanes that meet these criteria and are leased to foreign airlines are also subject to the return provisions of paragraph (e) of this section.

(c) The airplane complies with Stage 2 noise levels, is owned by a non-U.S. person, and is the subject of a binding lease to a U.S. person effective before and on September 25, 1991. Any such airplane may be operated for the term of the lease in effect on that date, and any extensions thereof provided for in that lease.

(d) The airplane complies with Stage 2 noise levels and is operated by a foreign air carrier.

(e) The airplane complies with Stage 2 noise levels and is operated by a foreign operator other than for the purpose of foreign air commerce.

(f) The airplane complies with Stage 2 noise levels and—

(1) On November 5, 1990, was owned by:

(i) A corporation, trust, or partnership organized under the laws of the United States or any State (including individual States, territories, possessions, and the District of Columbia);

(ii) An individual who is a citizen of the United States; or

(iii) An entity owned or controlled by a corporation, trust, partnership, or individual described in paragraph (f)(1) (i) or (ii) of this section; and

(2) Enters into the United States not later than 6 months after the expiration of a lease agreement (including any extensions thereof) between an owner described in paragraph (f)(1) of this section and a foreign airline.

(g) The airplane complies with Stage 2 noise levels and was purchased by the importer under a written contract executed before November 5, 1990.

(h) Any Stage 2 airplane described in this section is eligible for operation in the contiguous United States only as provided under § 91.865 or 91.867.

[Doc. No. 26433, 56 FR 48658, Sept. 25, 1991; 56 FR 51167, Oct. 10, 1991, as amended by Amdt. 91-288, 70 FR 38750, July 5, 2005]

§ 91.857 Stage 2 operations outside of the 48 contiguous United States.

An operator of a Stage 2 airplane that is operating only between points outside the contiguous United States on or after November 5, 1990, must include in its operations specifications a statement that such airplane may not be used to provide air transportation to or from any airport in the contiguous United States.

[Doc. No. FAA-2002-12771, 67 FR 46571, July 15, 2002]

§ 91.858 Special flight authorizations for non-revenue Stage 2 operations.

(a) After December 31, 1999, any operator of a Stage 2 airplane over 75,000 pounds may operate that airplane in nonrevenue service in the contiguous United States only for the following purposes:

(1) Sell, lease, or scrap the airplane;

(2) Obtain modifications to meet Stage 3 noise levels;

(3) Obtain scheduled heavy maintenance or significant modifications;

(4) Deliver the airplane to a lessee or return it to a lessor;

(5) Park or store the airplane; and

(6) Prepare the airplane for any of the purposes listed in paragraph (a)(1) thru (a)(5) of this section.

(b) An operator of a Stage 2 airplane that needs to operate in the contiguous United States for any of the purposes listed above may apply to FAA's Office of Environment and Energy for a special flight authorization. The applicant must file in advance. Applications are due 30 days in advance of the planned flight and must provide the information necessary for the FAA to determine that the planned flight is within the limits prescribed in the law.

[Doc. No. FAA-2002-12771, 67 FR 46571, July 15, 2002]

§ 91.859 Modification to meet Stage 3 or Stage 4 noise levels.

For an airplane subject to § 91.801(c) of this subpart and otherwise prohibited from operation to or from an airport in the contiguous United States by § 91.855, any person may apply for a special flight authorization for that airplane to operate in the contiguous United States for the purpose of obtaining modifications to meet Stage 3 or Stage 4 noise levels.

[Doc. No. FAA-2003-16526, 70 FR 38750, July 5, 2005]

§ 91.861 Base level.

(a) *U.S. Operators.* The base level of a U.S. operator is equal to the number of owned or leased Stage 2 airplanes subject to § 91.801(c) of this subpart that were listed on that operator's operations specifications for operations to or from airports in the contiguous United States on any one day selected by the operator during the period January 1, 1990, through July 1, 1991, plus or minus adjustments made pursuant to paragraphs (a) (1) and (2).

(1) The base level of a U.S. operator shall be increased by a number equal to the total of the following—

(i) The number of Stage 2 airplanes returned to service in the United States pursuant to § 91.855(f);

(ii) The number of Stage 2 airplanes purchased pursuant to § 91.855(g); and

(iii) Any U.S. operator base level acquired with a Stage 2 airplane transferred from another person under § 91.863.

(2) The base level of a U.S. operator shall be decreased by the amount of U.S. operator base level transferred with the corresponding number of Stage 2 airplanes to another person under § 91.863.

(b) Foreign air carriers. The base level of a foreign air carrier is equal to the number of owned or leased Stage 2 airplanes that were listed on that carrier's U.S. operations specifications on any one day during the period January 1, 1990, through July 1, 1991, plus or minus any adjustments to the base levels made pursuant to paragraphs (b)(1) and (2).

(1) The base level of a foreign air carrier shall be increased by the amount of foreign air carrier base level acquired with a Stage 2 airplane from another person under § 91.863.

(2) The base level of a foreign air carrier shall be decreased by the amount of foreign air carrier base level transferred with a Stage 2 airplane to another person under § 91.863.

(c) New entrants do not have a base level.

[Doc. No. 26433, 56 FR 48659, Sept. 25, 1991; 56 FR 51167, Oct. 10, 1991]

§ 91.863 Transfers of Stage 2 airplanes with base level.

(a) Stage 2 airplanes may be transferred with or without the corresponding amount of base level. Base level may not be transferred without the corresponding number of Stage 2 airplanes.

(b) No portion of a U.S. operator's base level established under § 91.861(a) may be used for operations by a foreign air carrier. No portion of a foreign air carrier's base level established under § 91.861(b) may be used for operations by a U.S. operator.

(c) Whenever a transfer of Stage 2 airplanes with base level occurs, the transferring and acquiring parties shall, within 10 days, jointly submit written notification of the transfer to the FAA, Office of Environment and Energy. Such notification shall state:

(1) The names of the transferring and acquiring parties;

(2) The name, address, and telephone number of the individual responsible for submitting the notification on behalf of the transferring and acquiring parties;

(3) The total number of Stage 2 airplanes transferred, listed by airplane type, model, series, and serial number;

(4) The corresponding amount of base level transferred and whether it is U.S. operator or foreign air carrier base level; and

(5) The effective date of the transaction.

(d) If, taken as a whole, a transaction or series of transactions made pursuant to this section does not produce an increase or decrease in the number of Stage 2 airplanes for either the acquiring or transferring operator, such transaction or series of transactions may not be used to establish compliance with the requirements of § 91.865.

[Doc. No. 26433, 56 FR 48659, Sept. 25, 1991]

§ 91.865 Phased compliance for operators with base level.

Except as provided in paragraph (a) of this section, each operator that operates an airplane under part 91, 121, 125, 129, or 135 of this chapter, regardless of the national registry of the airplane, shall comply with paragraph (b) or (d) of this section at each interim compliance date with regard to its subsonic airplane fleet covered by § 91.801(c) of this subpart.

(a) This section does not apply to new entrants covered by § 91.867 or to foreign operators not engaged in foreign air commerce.

(b) Each operator that chooses to comply with this paragraph pursuant to any interim compliance requirement shall reduce the number of Stage 2 airplanes it operates that are eligible for operation in the contiguous United States to a maximum of:

(1) After December 31, 1994, 75 percent of the base level held by the operator;

(2) After December 31, 1996, 50 percent of the base level held by the operator;

(3) After December 31, 1998, 25 percent of the base level held by the operator.

(c) Except as provided under § 91.871, the number of Stage 2 airplanes that must be reduced at each compliance date contained in paragraph (b) of this section shall be determined by reference to the amount of base level held by the operator on that compliance date, as calculated under § 91.861.

(d) Each operator that chooses to comply with this paragraph pursuant to any interim compliance requirement shall operate a fleet that consists of:

(1) After December 31, 1994, not less than 55 percent Stage 3 airplanes;

(2) After December 31, 1996, not less than 65 percent Stage 3 airplanes;

(3) After December 31, 1998, not less than 75 percent Stage 3 airplanes.

(e) Calculations resulting in fractions may be rounded to permit the continued operation of the next whole number of Stage 2 airplanes.

[Doc. No. 26433, 56 FR 48659, Sept. 25, 1991]

§ 91.867 Phased compliance for new entrants.

(a) New entrant U.S. air carriers.

(1) A new entrant initiating operations under part 121 of this chapter on or before December 31, 1994, may initiate service without regard to the percentage of its fleet composed of Stage 3 airplanes.

(2) After December 31, 1994, at least 25 percent of the fleet of a new entrant must comply with Stage 3 noise levels.

(3) After December 31, 1996, at least 50 percent of the fleet of a new entrant must comply with Stage 3 noise levels.

(4) After December 31, 1998, at least 75 percent of the fleet of a new entrant must comply with Stage 3 noise levels.

(b) New entrant foreign air carriers.

(1) A new entrant foreign air carrier initiating part 129 operations on or before December 31, 1994, may initiate service without regard to the percentage of its fleet composed of Stage 3 airplanes.

(2) After December 31, 1994, at least 25 percent of the fleet on U.S. operations specifications of a new entrant foreign air carrier must comply with Stage 3 noise levels.

(3) After December 31, 1996, at least 50 percent of the fleet on U.S. operations specifications of a new entrant foreign air carrier must comply with Stage 3 noise levels.

(4) After December 31, 1998, at least 75 percent of the fleet on U.S. operations specifications of a new entrant foreign air carrier must comply with Stage 3 noise levels.

(c) Calculations resulting in fractions may be rounded to permit the continued operation of the next whole number of Stage 2 airplanes.

[Doc. No. 26433, 56 FR 48659, Sept. 25, 1991, as amended by Amdt. 91-252, 61 FR 66185, Dec. 16, 1996]

§ 91.869 Carry-forward compliance.

(a) Any operator that exceeds the requirements of paragraph (b) of § 91.865 of this part on or before December 31, 1994, or on or before December 31, 1996, may claim a credit that may be applied at a subsequent interim compliance date.

(b) Any operator that eliminates or modifies more Stage 2 airplanes pursuant to § 91.865(b) than required as of December 31, 1994, or December 31, 1996, may count the number of additional Stage 2 airplanes reduced as a credit toward—

(1) The number of Stage 2 airplanes it would otherwise be required to reduce following a subsequent interim compliance date specified in § 91.865(b); or

(2) The number of Stage 3 airplanes it would otherwise be required to operate in its fleet following a subsequent interim compliance date to meet the percentage requirements specified in § 91.865(d).

[Doc. No. 26433, 56 FR 48659, Sept. 25, 1991; 56 FR 65783, Dec. 18, 1991]

§ 91.871 Waivers from interim compliance requirements.

(a) Any U.S. operator or foreign air carrier subject to the requirements of § 91.865 or 91.867 of this subpart may request a waiver from any individual compliance requirement.

(b) Applications must be filed with the Secretary of Transportation at least 120 days prior to the compliance date from which the waiver is requested.

(c) Applicants must show that a grant of waiver would be in the public interest, and must include in its application its plans and activities for modifying its fleet, including evidence of good faith efforts to comply with the requirements of § 91.865 or § 91.867. The application should contain all information the applicant considers relevant, including, as appropriate, the following:

(1) The applicant's balance sheet and cash flow positions;

(2) The composition of the applicant's current fleet; and

(3) The applicant's delivery position with respect to new airplanes or noise-abatement equipment.

(d) Waivers will be granted only upon a showing by the applicant that compliance with the requirements of § 91.865 or 91.867 at a particular interim compliance date is financially onerous, physically impossible, or technologically infeasible, or that it would have an adverse effect on competition or on service to small communities.

(e) The conditions of any waiver granted under this section shall be determined by the circumstances presented in the application, but in no case may the term extend beyond the next interim compliance date.

(f) A summary of any request for a waiver under this section will be published in the Federal Register, and public comment will be invited. Unless the Secretary finds that circumstances require otherwise, the public comment period will be at least 14 days.

[Doc. No. 26433, 56 FR 48660, Sept. 25, 1991]

§ 91.873 Waivers from final compliance.

(a) A U.S. air carrier or a foreign air carrier may apply for a waiver from the prohibition contained in § 91.853 of this part for its remaining Stage 2 airplanes, provided that, by July 1, 1999, at least 85 percent of the airplanes used by the carrier to provide service to or from an airport in the contiguous United States will comply with the Stage 3 noise levels.

(b) An application for the waiver described in paragraph (a) of this section must be filed with the Secretary of Transportation no later than January 1, 1999, or, in the case of a foreign air carrier, no later than April 20, 2000. Such application must include a plan with firm orders for replacing or modifying all airplanes to comply with Stage 3 noise levels at the earliest practicable time.

(c) To be eligible to apply for the waiver under this section, a new entrant U.S. air carrier must initiate service no later than January 1, 1999, and must comply fully with all provisions of this section.

(d) The Secretary may grant a waiver under this section if the Secretary finds that granting such waiver is in the public interest. In making such a finding, the Secretary shall include consideration of the effect of granting such waiver on competition in the air carrier industry and the effect on small community air service, and any other information submitted by the applicant that the Secretary considers relevant.

(e) The term of any waiver granted under this section shall be determined by the circumstances presented in the application, but in no case will the waiver permit the operation of any Stage 2 airplane covered by this subchapter in the contiguous United States after December 31, 2003.

(f) A summary of any request for a waiver under this section will be published in the Federal Register, and public comment will be invited. Unless the secretary finds that circumstances require otherwise, the public comment period will be at least 14 days.

[Doc. No. 26433, 56 FR 48660, Sept. 25, 1991; 56 FR 51167 Oct. 10, 1991; Amdt. 91-276, 67 FR 46571, July 15, 2002]

§ 91.875 Annual progress reports.

(a) Each operator subject to § 91.865 or § 91.867 of this chapter shall submit an annual report to the FAA, Office of Environment and Energy, on the progress it has made toward complying with the requirements of that section. Such reports shall be submitted no later than 45 days after the end of a calendar year. All progress reports must provide the information through the end of the calendar year, be certified by the operator as true and complete (under penalty of 18 U.S.C. 1001), and include the following information:

(1) The name and address of the operator;

(2) The name, title, and telephone number of the person designated by the operator to be responsible for ensuring the accuracy of the information in the report;

(3) The operator's progress during the reporting period toward compliance with the requirements of § 91.853, § 91.865 or § 91.867. For airplanes on U.S. operations specifications, each operator shall identify the airplanes by type, model, series, and serial number.

(i) Each Stage 2 airplane added or removed from operation or U.S. operations specifications (grouped separately by those airplanes acquired with and without base level);

(ii) Each Stage 2 airplane modified to Stage 3 noise levels (identifying the manufacturer and model of noise abatement retrofit equipment;

(iii) Each Stage 3 airplane on U.S. operations specifications as of the last day of the reporting period; and

(iv) For each Stage 2 airplane transferred or acquired, the name and address of the recipient or transferor; and, if base level was transferred, the person to or from whom base level was transferred or acquired pursuant to Section 91.863 along with the effective date of each base level transaction, and the type of base level transferred or acquired.

(b) Each operator subject to § 91.865 or § 91.867 of this chapter shall submit an initial progress report covering the period from January 1, 1990, through December 31, 1991, and provide:

(1) For each operator subject to § 91.865:

(i) The date used to establish its base level pursuant to § 91.861(a); and

(ii) A list of those Stage 2 airplanes (by type, model, series and serial number) in its base level, including adjustments made pursuant to § 91.861 after the date its base level was established.

(2) For each U.S. operator:

(i) A plan to meet the compliance schedules in § 91.865 or § 91.867 and the final compliance date of § 91.853, including the schedule for delivery of replacement Stage 3 airplanes or the installation of noise abatement retrofit equipment; and

(ii) A separate list (by type, model, series, and serial number) of those airplanes included in the operator's base level, pursuant to § 91.861(a)(1) (i) and (ii), under the categories "returned" or "purchased," along with the date each was added to its operations specifications.

(c) Each operator subject to § 91.865 or § 91.867 of this chapter shall submit subsequent annual progress reports covering the calendar year preceding the report and including any changes in the information provided in paragraphs (a) and (b) of this section; including the use of any carry-forward credits pursuant to § 91.869.

(d) An operator may request, in any report, that specific planning data be considered proprietary.

(e) If an operator's actions during any reporting period cause it to achieve compliance with § 91.853, the report should include a statement to that effect. Further progress reports are not required unless there is any change in the information reported pursuant to paragraph (a) of this section.

(f) For each U.S. operator subject to § 91.865, progress reports submitted for calendar years 1994, 1996, and 1998, shall also state how the operator achieved compliance with the requirements of that section, i.e.—

(1) By reducing the number of Stage 2 airplanes in its fleet to no more than the maximum permitted percentage of its base level under § 91.865(b), or

(2) By operating a fleet that consists of at least the minimum required percentage of Stage 3 airplanes under § 91.865(d).

(Approved by the Office of Management and Budget under control number 2120-0553)

[Doc. No. 26433, 56 FR 48660, Sept. 25, 1991; 56 FR 51168, Oct. 10, 1991, as amended by 57 FR 5977, Feb. 19, 1992]

§ 91.877 Annual reporting of Hawaiian operations.

(a) Each air carrier or foreign air carrier subject to § 91.865 or § 91.867 of this part that conducts operations between the contiguous United States and the State of Hawaii, between the State of Hawaii and any point outside of the contiguous United States, or between the islands of Hawaii in turnaround service, on or since November 5, 1990, shall include in its annual report the information described in paragraph (c) of this section.

(b) Each air carrier or foreign air carrier not subject to § 91.865 or § 91.867 of this part that conducts operations between the contiguous U.S. and the State of Hawaii, between the State of Hawaii and any point outside of the contiguous United States, or between the islands of Hawaii in turnaround service, on or since November 5, 1990, shall submit an annual report to the FAA, Office of Environment and Energy, on its compliance with the Hawaiian operations provisions of 49 U.S.C. 47528. Such reports shall be submitted no later than 45 days after the end of a calendar year. All progress reports must provide the information through the end of the calendar year, be certified by the operator as true and complete (under penalty of 18 U.S.C. 1001), and include the following information—

(1) The name and address of the air carrier or foreign air carrier;

(2) The name, title, and telephone number of the person designated by the air carrier or foreign air carrier to be responsible for ensuring the accuracy of the information in the report; and

(3) The information specified in paragraph (c) of this section.

(c) The following information must be included in reports filed pursuant to this section—

(1) For operations conducted between the contiguous United States and the State of Hawaii—

(i) The number of Stage 2 airplanes used to conduct such operations as of November 5, 1990;

(ii) Any change to that number during the calendar year being reported, including the date of such change;

(2) For air carriers that conduct inter-island turnaround service in the State of Hawaii—

(i) The number of Stage 2 airplanes used to conduct such operations as of November 5, 1990;

(ii) Any change to that number during the calendar year being reported, including the date of such change;

(iii) For an air carrier that provided inter-island trunaround service within the state of Hawaii on November 5, 1990, the number reported under paragraph (c)(2)(i) of this section may include all Stage 2 airplanes with a maximum certificated takeoff weight of more than 75,000 pounds that were owned or leased by the air carrier on November 5, 1990, regardless of whether such airplanes were operated by that air carrier or foreign air carrier on that date.

(3) For operations conducted between the State of Hawaii and a point outside the contiguous United States—

(i) The number of Stage 2 airplanes used to conduct such operations as of November 5, 1990; and

(ii) Any change to that number during the calendar year being reported, including the date of such change.

(d) Reports or amended reports for years predating this regulation are required to be filed concurrently with the next annual report.

[Doc. No. 28213, 61 FR 66185, Dec. 16, 1996]

§§ 91.879-91.880 [Reserved]

§ 91.881 xxx

Link to an amendment published at 78 FR 39583, July 2, 2013.

§ 91.883 xxx

Link to an amendment published at 78 FR 39583, July 2, 2013.

§§ 91.884-91.899 [Reserved]

Subpart J—Waivers

§ 91.901 [Reserved]

§ 91.903 Policy and procedures.

(a) The Administrator may issue a certificate of waiver authorizing the operation of aircraft in deviation from any rule listed in this subpart if the Administrator finds that the proposed operation can be safely conducted under the terms of that certificate of waiver.

(b) An application for a certificate of waiver under this part is made on a form and in a manner prescribed by the Administrator and may be submitted to any FAA office.

(c) A certificate of waiver is effective as specified in that certificate of waiver.

[Doc. No. 18334, 54 FR 34325, Aug. 18, 1989]

§ 91.905 List of rules subject to waivers.

Sec.

91.107 Use of safety belts.

91.111 Operating near other aircraft.

91.113 Right-of-way rules: Except water operations.

91.115 Right-of-way rules: Water operations.

91.117 Aircraft speed.

91.119 Minimum safe altitudes: General.

91.121 Altimeter settings.

91.123 Compliance with ATC clearances and instructions.

91.125 ATC light signals.

91.126 Operating on or in the vicinity of an airport in Class G airspace.

91.127 Operating on or in the vicinity of an airport in Class E airspace.

91.129 Operations in Class D airspace.

91.130 Operations in Class C airspace.

91.131 Operations in Class B airspace.

91.133 Restricted and prohibited areas.

91.135 Operations in Class A airspace.

91.137 Temporary flight restrictions.

91.141 Flight restrictions in the proximity of the Presidential and other parties.

91.143 Flight limitation in the proximity of space flight operations.

91.153 VFR flight plan: Information required.

91.155 Basic VFR weather minimums

91.157 Special VFR weather minimums.

91.159 VFR cruising altitude or flight level.

91.169 IFR flight plan: Information required.

91.173 ATC clearance and flight plan required.

91.175 Takeoff and landing under IFR.

91.177 Minimum altitudes for IFR operations.

91.179 IFR cruising altitude or flight level.

91.181 Course to be flown.

91.183 IFR radio communications.

91.185 IFR operations: Two-way radio communications failure.

91.187 Operation under IFR in controlled airspace: Malfunction reports.

91.209 Aircraft lights.

91.303 Aerobatic flights.

91.305 Flight test areas.

91.311 Towing: Other than under § 91.309.

91.313(e) Restricted category civil aircraft: Operating limitations.

91.515 Flight altitude rules.

91.705 Operations within the North Atlantic Minimum Navigation Performance Specifications Airspace.

91.707 Flights between Mexico or Canada and the United States.

91.713 Operation of civil aircraft of Cuban registry.

[Doc. No. 18334, 54 FR 34325, Aug. 18, 1989, as amended by Amdt. 91-227, 56 FR 65661, Dec. 17, 1991]

§§ 91.907-91.999 [Reserved]

Subpart K—Fractional Ownership Operations

SOURCE: Docket No. FAA-2001-10047, 68 FR 54561, Sept. 17, 2003, unless otherwise noted.

§ 91.1001 Applicability.

(a) This subpart prescribes rules, in addition to those prescribed in other subparts of this part, that apply to fractional owners and fractional ownership program managers governing—

(1) The provision of program management services in a fractional ownership program;

(2) The operation of a fractional ownership program aircraft in a fractional ownership program; and

(3) The operation of a program aircraft included in a fractional ownership program managed by an affiliate of the manager of the program to which the owner belongs.

(b) As used in this part—

(1) *Affiliate of a program manager* means a manager that, directly, or indirectly, through one or more intermediaries, controls, is controlled by, or is under common control with, another program manager. The holding of at least forty percent (40 percent) of the equity and forty percent (40 percent) of the voting power of an entity will be presumed to constitute control for purposes of determining an affiliation under this subpart.

(2) A *dry-lease aircraft exchange* means an arrangement, documented by the written program agreements, under which the program aircraft are available, on an as needed basis without crew, to each fractional owner.

(3) A *fractional owner or owner* means an individual or entity that possesses a minimum fractional ownership interest in a program aircraft and that has entered into the applicable program agreements; provided, however, that in the case of the flight operations described in paragraph (b)(6)(ii) of this section, and solely for purposes of requirements pertaining to those flight operations, the fractional owner operating the aircraft will be deemed to be a fractional owner in the program managed by the affiliate.

(4) A *fractional ownership interest* means the ownership of an interest or holding of a multi-year leasehold interest and/or a multi-year leasehold interest that is convertible into an ownership interest in a program aircraft.

(5) A *fractional ownership program or program* means any system of aircraft ownership and exchange that consists of all of the following elements:

(i) The provision for fractional ownership program management services by a single fractional ownership program manager on behalf of the fractional owners.

(ii) Two or more airworthy aircraft.

(iii) One or more fractional owners per program aircraft, with at least one program aircraft having more than one owner.

(iv) Possession of at least a minimum fractional ownership interest in one or more program aircraft by each fractional owner.

(v) A dry-lease aircraft exchange arrangement among all of the fractional owners.

(vi) Multi-year program agreements covering the fractional ownership, fractional ownership program management services, and dry-lease aircraft exchange aspects of the program.

(6) A *fractional ownership program aircraft or program aircraft* means:

(i) An aircraft in which a fractional owner has a minimal fractional ownership interest and that has been included in the dry-lease aircraft exchange pursuant to the program agreements, or

(ii) In the case of a fractional owner from one program operating an aircraft in a different fractional ownership program managed by an affiliate of the operating owner's program manager, the aircraft being operated by the fractional owner, so long as the aircraft is:

(A) Included in the fractional ownership program managed by the affiliate of the operating owner's program manager, and

(B) Included in the operating owner's program's dry-lease aircraft exchange pursuant to the program agreements of the operating owner's program.

(iii) An aircraft owned in whole or in part by the program manager that has been included in the dry-lease aircraft exchange and is used to supplement program operations.

(7) A *Fractional Ownership Program Flight or Program Flight* means a flight under this subpart when one or more passengers or property designated by a fractional owner are on board the aircraft.

(8) *Fractional ownership program management services or program management services* mean administrative and aviation support services furnished in accordance with the applicable requirements of this subpart or provided by the program manager on behalf of the fractional owners, including, but not limited to, the—

(i) Establishment and implementation of program safety guidelines;

(ii) Employment, furnishing, or contracting of pilots and other crewmembers;

(iii) Training and qualification of pilots and other crewmembers and personnel;

(iv) Scheduling and coordination of the program aircraft and crews;

(v) Maintenance of program aircraft;

(vi) Satisfaction of recordkeeping requirements;

(vii) Development and use of a program operations manual and procedures; and

(viii) Application for and maintenance of management specifications and other authorizations and approvals.

(9) A *fractional ownership program manager or program manager* means the entity that offers fractional ownership program management services to fractional owners, and is designated in the multi-year program agreements referenced in paragraph (b)(1)(v) of this section to fulfill the requirements of this chapter applicable to the manager of the program containing the aircraft being flown. When a fractional owner is operating an aircraft in a fractional ownership program managed by an affiliate of the owner's program manager, the references in this subpart to the flight-related responsibilities of the program manager apply, with respect to that particular flight, to the affiliate of the owner's program manager rather than to the owner's program manager.

(10) A *minimum fractional ownership interest* means—

(i) A fractional ownership interest equal to, or greater than, one-sixteenth (1/16) of at least one subsonic, fixed-wing or powered-lift program aircraft; or

(ii) A fractional ownership interest equal to, or greater than, one-thirty-second (1/32) of at least one rotorcraft program aircraft.

(c) The rules in this subpart that refer to a fractional owner or a fractional ownership program manager also apply to any person who engages in an operation governed by this subpart without the management specifications required by this subpart.

§ 91.1002 Compliance date.

No person that conducted flights before November 17, 2003 under a program that meets the definition of fractional ownership program in § 91.1001 may conduct such flights after February 17, 2005 unless it has obtained management specifications under this subpart.

[Doc. No. FAA-2001-10047, 68 FR 54561, Sept. 17, 2003; 69 FR 74413, Dec. 14, 2004]

§ 91.1003 Management contract between owner and program manager.

Each owner must have a contract with the program manager that—

(a) Requires the program manager to ensure that the program conforms to all applicable requirements of this chapter.

(b) Provides the owner the right to inspect and to audit, or have a designee of the owner inspect and audit, the records of the program manager pertaining to the operational safety of the program and those records required to show compliance with the management specifications and all applicable regulations. These records include, but are not limited to, the management specifications, authorizations, approvals, manuals, log books, and maintenance records maintained by the program manager.

(c) Designates the program manager as the owner's agent to receive service of notices pertaining to the program that the FAA seeks to provide to owners and authorizes the FAA to send such notices to the program manager in its capacity as the agent of the owner for such service.

(d) Acknowledges the FAA's right to contact the owner directly if the Administrator determines that direct contact is necessary.

§ 91.1005 Prohibitions and limitations.

(a) Except as provided in § 91.321 or § 91.501, no owner may carry persons or property for compensation or hire on a program flight.

(b) During the term of the multi-year program agreements under which a fractional owner has obtained a minimum fractional ownership interest in a program aircraft, the flight hours used during that term by the owner on program aircraft must not exceed the total hours associated with the fractional owner's share of ownership.

(c) No person may sell or lease an aircraft interest in a fractional ownership program that is smaller than that prescribed in the definition of "minimum fractional ownership interest" in § 91.1001(b)(10) unless flights associated with that interest are operated under part 121 or 135 of this chapter and are conducted by an air carrier or commercial operator certificated under part 119 of this chapter.

§ 91.1007 Flights conducted under part 121 or part 135 of this chapter.

(a) Except as provided in § 91.501(b), when a nonprogram aircraft is used to substitute for a program flight, the flight must be operated in compliance with part 121 or part 135 of this chapter, as applicable.

(b) A program manager who holds a certificate under part 119 of this chapter may conduct a flight for the use of a fractional owner under part 121 or part 135 of this chapter if the aircraft is listed on that certificate holder's operations specifications for part 121 or part 135, as applicable.

(c) The fractional owner must be informed when a flight is being conducted as a program flight or is being conducted under part 121 or part 135 of this chapter.

Operational Control

§ 91.1009 Clarification of operational control.

(a) An owner is in operational control of a program flight when the owner—

(1) Has the rights and is subject to the limitations set forth in §§ 91.1003 through 91.1013;

(2) Has directed that a program aircraft carry passengers or property designated by that owner; and

(3) The aircraft is carrying those passengers or property.

(b) An owner is not in operational control of a flight in the following circumstances:

(1) A program aircraft is used for a flight for administrative purposes such as demonstration, positioning, ferrying, maintenance, or crew training, and no passengers or property designated by such owner are being carried; or

(2) The aircraft being used for the flight is being operated under part 121 or 135 of this chapter.

§ 91.1011 Operational control responsibilities and delegation.

(a) Each owner in operational control of a program flight is ultimately responsible for safe operations and for complying with all applicable requirements of this chapter, including those related to airworthiness and operations in connection with the flight. Each owner may delegate some or all of the performance of the tasks associated with carrying out this responsibility to the program manager, and may rely on the program manager for aviation expertise and program management services. When the owner delegates performance of tasks to the program manager or relies on the program manager's expertise, the owner and the program manager are jointly and individually responsible for compliance.

(b) The management specifications, authorizations, and approvals required by this subpart are issued to, and in the sole name of, the program manager on behalf of the fractional owners collectively. The management specifications, authorizations, and approvals will not be affected by any change in ownership of a program aircraft, as long as the aircraft remains a program aircraft in the identified program.

§ 91.1013 Operational control briefing and acknowledgment.

(a) Upon the signing of an initial program management services contract, or a renewal or extension of a program management services contract, the program manager must brief the fractional owner on the owner's operational control responsibilities, and the owner must review and sign an acknowledgment of these operational control responsibilities. The acknowledgment must be included with the program management services contract. The acknowledgment must define when a fractional owner is in operational control and the owner's responsibilities and liabilities under the program. These include:

(1) Responsibility for compliance with the management specifications and all applicable regulations.

(2) Enforcement actions for any noncompliance.

(3) Liability risk in the event of a flight-related occurrence that causes personal injury or property damage.

(b) The fractional owner's signature on the acknowledgment will serve as the owner's affirmation that the owner has read, understands, and accepts the operational control responsibilities described in the acknowledgment.

(c) Each program manager must ensure that the fractional owner or owner's representatives have access to the acknowledgments for such owner's program aircraft. Each program manager must ensure that the FAA has access to the acknowledgments for all program aircraft.

Program Management

§ 91.1014 Issuing or denying management specifications.

(a) A person applying to the Administrator for management specifications under this subpart must submit an application—

(1) In a form and manner prescribed by the Administrator; and

(2) Containing any information the Administrator requires the applicant to submit.

(b) Management specifications will be issued to the program manager on behalf of the fractional owners if, after investigation, the Administrator finds that the applicant:

(1) Meets the applicable requirements of this subpart; and

(2) Is properly and adequately equipped in accordance with the requirements of this chapter and is able to conduct safe operations under appropriate provisions of part 91 of this chapter and management specifications issued under this subpart.

(c) An application for management specifications will be denied if the Administrator finds that the applicant is not properly or adequately equipped or is not able to conduct safe operations under this part.

§ 91.1015 Management specifications.

(a) Each person conducting operations under this subpart or furnishing fractional ownership program management services to fractional owners must do so in accordance with management specifications issued by the Administrator to the fractional ownership program manager under this subpart. Management specifications must include:

(1) The current list of all fractional owners and types of aircraft, registration markings and serial numbers;

(2) The authorizations, limitations, and certain procedures under which these operations are to be conducted,

(3) Certain other procedures under which each class and size of aircraft is to be operated;

(4) Authorization for an inspection program approved under § 91.1109, including the type of aircraft, the registration markings and serial numbers of each aircraft to be operated under the program. No person may conduct any program flight using any aircraft not listed.

(5) Time limitations, or standards for determining time limitations, for overhauls, inspections, and checks for airframes, engines, propellers, rotors, appliances, and emergency equipment of aircraft.

(6) The specific location of the program manager's principal base of operations and, if different, the address that will serve as the primary point of contact for correspondence between the FAA and the program manager and the name and mailing address of the program manager's agent for service;

(7) Other business names the program manager may use;

(8) Authorization for the method of controlling weight and balance of aircraft;

(9) Any authorized deviation and exemption granted from any requirement of this chapter; and

(10) Any other information the Administrator determines is necessary.

(b) The program manager may keep the current list of all fractional owners required by paragraph (a)(1) of this section at its principal base of operation or other location approved by the Administrator and referenced in its management specifications. Each program manager shall make this list of owners available for inspection by the Administrator.

(c) Management specifications issued under this subpart are effective unless—

(1) The management specifications are amended as provided in § 91.1017; or

(2) The Administrator suspends or revokes the management specifications.

(d) At least 30 days before it proposes to establish or change the location of its principal base of operations, its main operations base, or its main maintenance base, a program manager must provide written notification to the Flight Standards District Office that issued the program manager's management specifications.

(e) Each program manager must maintain a complete and separate set of its management specifications at its principal base of operations, or at a place approved by the Administrator, and must make its management specifications available for inspection by the Administrator and the fractional owner(s) to whom the program manager furnishes its services for review and audit.

(f) Each program manager must insert pertinent excerpts of its management specifications, or references thereto, in its program manual and must—

(1) Clearly identify each such excerpt as a part of its management specifications; and

(2) State that compliance with each management specifications requirement is mandatory.

(g) Each program manager must keep each of its employees and other persons who perform duties material to its operations informed of the provisions of its management specifications that apply to that employee's or person's duties and responsibilities.

§ 91.1017 Amending program manager's management specifications.

(a) The Administrator may amend any management specifications issued under this subpart if—

(1) The Administrator determines that safety and the public interest require the amendment of any management specifications; or

(2) The program manager applies for the amendment of any management specifications, and the Administrator determines that safety and the public interest allows the amendment.

(b) Except as provided in paragraph (e) of this section, when the Administrator initiates an amendment of a program manager's management specifications, the following procedure applies:

(1) The Flight Standards District Office that issued the program manager's management specifications will notify the program manager in writing of the proposed amendment.

(2) The Flight Standards District Office that issued the program manager's management specifications will set a reasonable period (but not less than 7 days) within which the program manager may submit written information, views, and arguments on the amendment.

(3) After considering all material presented, the Flight Standards District Office that issued the program manager's

management specifications will notify the program manager of—

(i) The adoption of the proposed amendment,

(ii) The partial adoption of the proposed amendment, or

(iii) The withdrawal of the proposed amendment.

(4) If the Flight Standards District Office that issued the program manager's management specifications issues an amendment of the management specifications, it becomes effective not less than 30 days after the program manager receives notice of it unless—

(i) The Flight Standards District Office that issued the program manager's management specifications finds under paragraph (e) of this section that there is an emergency requiring immediate action with respect to safety; or

(ii) The program manager petitions for reconsideration of the amendment under paragraph (d) of this section.

(c) When the program manager applies for an amendment to its management specifications, the following procedure applies:

(1) The program manager must file an application to amend its management specifications—

(i) At least 90 days before the date proposed by the applicant for the amendment to become effective, unless a shorter time is approved, in cases such as mergers, acquisitions of operational assets that require an additional showing of safety (for example, proving tests or validation tests), and resumption of operations following a suspension of operations as a result of bankruptcy actions.

(ii) At least 15 days before the date proposed by the applicant for the amendment to become effective in all other cases.

(2) The application must be submitted to the Flight Standards District Office that issued the program manager's management specifications in a form and manner prescribed by the Administrator.

(3) After considering all material presented, the Flight Standards District Office that issued the program manager's management specifications will notify the program manager of—

(i) The adoption of the applied for amendment;

(ii) The partial adoption of the applied for amendment; or

(iii) The denial of the applied for amendment. The program manager may petition for reconsideration of a denial under paragraph (d) of this section.

(4) If the Flight Standards District Office that issued the program manager's management specifications approves the amendment, following coordination with the program manager regarding its implementation, the amendment is effective on the date the Administrator approves it.

(d) When a program manager seeks reconsideration of a decision of the Flight Standards District Office that issued the program manager's management specifications concerning the amendment of management specifications, the following procedure applies:

(1) The program manager must petition for reconsideration of that decision within 30 days of the date that the program manager receives a notice of denial of the amendment of its management specifications, or of the date it receives notice of an FAA-initiated amendment of its management specifications, whichever circumstance applies.

(2) The program manager must address its petition to the Director, Flight Standards Service.

(3) A petition for reconsideration, if filed within the 30-day period, suspends the effectiveness of any amendment issued by the Flight Standards District Office that issued the program manager's management specifications unless that District Office has found, under paragraph (e) of this section, that an emergency exists requiring immediate action with respect to safety.

(4) If a petition for reconsideration is not filed within 30 days, the procedures of paragraph (c) of this section apply.

(e) If the Flight Standards District Office that issued the program manager's management specifications finds that an emergency exists requiring immediate action with respect to safety that makes the procedures set out in this section impracticable or contrary to the public interest—

(1) The Flight Standards District Office amends the management specifications and makes the amendment effective on the day the program manager receives notice of it; and

(2) In the notice to the program manager, the Flight Standards District Office will articulate the reasons for its finding that an emergency exists requiring immediate action with respect to safety or that makes it impracticable or contrary to the public interest to stay the effectiveness of the amendment.

§ 91.1019 Conducting tests and inspections.

(a) At any time or place, the Administrator may conduct an inspection or test, other than an en route inspection, to determine whether a program manager under this subpart is complying with title 49 of the United States Code, applicable regulations, and the program manager's management specifications.

(b) The program manager must—

(1) Make available to the Administrator at the program manager's principal base of operations, or at a place approved by the Administrator, the program manager's management specifications; and

(2) Allow the Administrator to make any test or inspection, other than an en route inspection, to determine compliance respecting any matter stated in paragraph (a) of this section.

(c) Each employee of, or person used by, the program manager who is responsible for maintaining the program manager's records required by or necessary to demonstrate compliance with this subpart must make those records available to the Administrator.

(d) The Administrator may determine a program manager's continued eligibility to hold its management specifications on any grounds listed in paragraph (a) of this section, or any other appropriate grounds.

(e) Failure by any program manager to make available to the Administrator upon request, the management specifications, or any required record, document, or report is grounds for suspension of all or any part of the program manager's management specifications.

§ 91.1021 Internal safety reporting and incident/accident response.

(a) Each program manager must establish an internal anonymous safety reporting procedure that fosters an environment of safety without any potential for retribution for filing the report.

(b) Each program manager must establish procedures to respond to an aviation incident/accident.

§ 91.1023 Program operating manual requirements.

(a) Each program manager must prepare and keep current a program operating manual setting forth procedures and policies acceptable to the Administrator. The program manager's management, flight, ground, and maintenance personnel must use this manual to conduct operations under this subpart. However, the Administrator may authorize a deviation from this paragraph if the Administrator finds that, because of the limited size of the operation, part of the manual is not necessary for guidance of management, flight, ground, or maintenance personnel.

(b) Each program manager must maintain at least one copy of the manual at its principal base of operations.

(c) No manual may be contrary to any applicable U.S. regulations, foreign regulations applicable to the program flights in foreign countries, or the program manager's management specifications.

(d) The program manager must make a copy of the manual, or appropriate portions of the manual (and changes and additions), available to its maintenance and ground operations personnel and must furnish the manual to—

(1) Its crewmembers; and

(2) Representatives of the Administrator assigned to the program manager.

(e) Each employee of the program manager to whom a manual or appropriate portions of it are furnished under paragraph (d)(1) of this section must keep it up-to-date with the changes and additions furnished to them.

(f) Except as provided in paragraph (h) of this section, the appropriate parts of the manual must be carried on each aircraft when away from the principal operations base. The appropriate parts must be available for use by ground or flight personnel.

(g) For the purpose of complying with paragraph (d) of this section, a program manager may furnish the persons listed therein with all or part of its manual in printed form or other form, acceptable to the Administrator, that is retrievable in the English language. If the program manager furnishes all or part of the manual in other than printed form, it must ensure there is a compatible reading device available to those persons that provides a legible image of the maintenance information and instructions, or a system that is able to retrieve the maintenance information and instructions in the English language.

(h) If a program manager conducts aircraft inspections or maintenance at specified facilities where the approved aircraft inspection program is available, the program manager is not required to ensure that the approved aircraft inspection program is carried aboard the aircraft en route to those facilities.

(i) Program managers that are also certificated to operate under part 121 or 135 of this chapter may be authorized to use the operating manual required by those parts to meet the manual requirements of subpart K, provided:

(1) The policies and procedures are consistent for both operations, or

(2) When policies and procedures are different, the applicable policies and procedures are identified and used.

§ 91.1025 Program operating manual contents.

Each program operating manual must have the date of the last revision on each revised page. Unless otherwise authorized by the Administrator, the manual must include the following:

(a) Procedures for ensuring compliance with aircraft weight and balance limitations;

(b) Copies of the program manager's management specifications or appropriate extracted information, including area of operations authorized, category and class of aircraft authorized, crew complements, and types of operations authorized;

(c) Procedures for complying with accident notification requirements;

(d) Procedures for ensuring that the pilot in command knows that required airworthiness inspections have been made and that the aircraft has been approved for return to service in compliance with applicable maintenance requirements;

(e) Procedures for reporting and recording mechanical irregularities that come to the attention of the pilot in command before, during, and after completion of a flight;

(f) Procedures to be followed by the pilot in command for determining that mechanical irregularities or defects reported for previous flights have been corrected or that correction of certain mechanical irregularities or defects have been deferred;

(g) Procedures to be followed by the pilot in command to obtain maintenance, preventive maintenance, and servicing of the aircraft at a place where previous arrangements have not been made by the program manager or owner, when the pilot is authorized to so act for the operator;

(h) Procedures under § 91.213 for the release of, and continuation of flight if any item of equipment required for the particular type of operation becomes inoperative or unserviceable en route;

(i) Procedures for refueling aircraft, eliminating fuel contamination, protecting from fire (including electrostatic protection), and supervising and protecting passengers during refueling;

(j) Procedures to be followed by the pilot in command in the briefing under § 91.1035.

(k) Procedures for ensuring compliance with emergency procedures, including a list of the functions assigned each category of required crewmembers in connection with an emergency and emergency evacuation duties;

(l) The approved aircraft inspection program, when applicable;

(m) Procedures for the evacuation of persons who may need the assistance of another person to move expeditiously to an exit if an emergency occurs;

(n) Procedures for performance planning that take into account take off, landing and en route conditions;

(o) An approved Destination Airport Analysis, when required by § 91.1037(c), that includes the following elements, supported by aircraft performance data supplied by the aircraft manufacturer for the appropriate runway conditions—

(1) Pilot qualifications and experience;

(2) Aircraft performance data to include normal, abnormal and emergency procedures as supplied by the aircraft manufacturer;

(3) Airport facilities and topography;

(4) Runway conditions (including contamination);

(5) Airport or area weather reporting;

(6) Appropriate additional runway safety margins, if required;

(7) Airplane inoperative equipment;

(8) Environmental conditions; and

(9) Other criteria that affect aircraft performance.

(p) A suitable system (which may include a coded or electronic system) that provides for preservation and retrieval of maintenance recordkeeping information required by § 91.1113 in a manner acceptable to the Administrator that provides—

(1) A description (or reference to date acceptable to the Administrator) of the work performed:

(2) The name of the person performing the work if the work is performed by a person outside the organization of the program manager; and

(3) The name or other positive identification of the individual approving the work.

(q) Flight locating and scheduling procedures; and

(r) Other procedures and policy instructions regarding program operations that are issued by the program manager or required by the Administrator.

§ 91.1027 Recordkeeping.

(a) Each program manager must keep at its principal base of operations or at other places approved by the Administrator, and must make available for inspection by the Administrator all of the following:

(1) The program manager's management specifications.

(2) A current list of the aircraft used or available for use in operations under this subpart, the operations for which each is equipped (for example, MNPS, RNP5/10, RVSM.).

(3) An individual record of each pilot used in operations under this subpart, including the following information:

(i) The full name of the pilot.

(ii) The pilot certificate (by type and number) and ratings that the pilot holds.

(iii) The pilot's aeronautical experience in sufficient detail to determine the pilot's qualifications to pilot aircraft in operations under this subpart.

(iv) The pilot's current duties and the date of the pilot's assignment to those duties.

(v) The effective date and class of the medical certificate that the pilot holds.

(vi) The date and result of each of the initial and recurrent competency tests and proficiency checks required by this subpart and the type of aircraft flown during that test or check.

(vii) The pilot's flight time in sufficient detail to determine compliance with the flight time limitations of this subpart.

(viii) The pilot's check pilot authorization, if any.

(ix) Any action taken concerning the pilot's release from employment for physical or professional disqualification; and

(x) The date of the satisfactory completion of initial, transition, upgrade, and differences training and each recurrent training phase required by this subpart.

(4) An individual record for each flight attendant used in operations under this subpart, including the following information:

(i) The full name of the flight attendant, and

(ii) The date and result of training required by § 91.1063, as applicable.

(5) A current list of all fractional owners and associated aircraft. This list or a reference to its location must be included in the management specifications and should be of sufficient detail to determine the minimum fractional ownership interest of each aircraft.

(b) Each program manager must keep each record required by paragraph (a)(2) of this section for at least 6 months, and must keep each record required by paragraphs (a)(3) and (a)(4) of this section for at least 12 months. When an employee is no longer employed or affiliated with the program manager or fractional owner, each record required by paragraphs (a)(3) and (a)(4) of this section must be retained for at least 12 months.

(c) Each program manager is responsible for the preparation and accuracy of a load manifest in duplicate containing information concerning the loading of the aircraft. The manifest must be prepared before each takeoff and must include—

(1) The number of passengers;

(2) The total weight of the loaded aircraft;

(3) The maximum allowable takeoff weight for that flight;

(4) The center of gravity limits;

(5) The center of gravity of the loaded aircraft, except that the actual center of gravity need not be computed if the aircraft is loaded according to a loading schedule or other approved method that ensures that the center of gravity of the loaded aircraft is within approved limits. In those cases, an entry must be made on the manifest indicating that the center of gravity is within limits according to a loading schedule or other approved method;

(6) The registration number of the aircraft or flight number;

(7) The origin and destination; and

(8) Identification of crewmembers and their crew position assignments.

(d) The pilot in command of the aircraft for which a load manifest must be prepared must carry a copy of the completed load manifest in the aircraft to its destination. The program manager must keep copies of completed load manifest for at least 30 days at its principal operations base, or at another location used by it and approved by the Administrator.

(e) Each program manager is responsible for providing a written document that states the name of the entity having operational control on that flight and the part of this chapter under which the flight is operated. The pilot in command of the aircraft must carry a copy of the document in the aircraft to its destination. The program manager must keep a copy of the document for at least 30 days at its principal operations base, or at another location used by it and approved by the Administrator.

(f) Records may be kept either in paper or other form acceptable to the Administrator.

(g) Program managers that are also certificated to operate under part 121 or 135 of this chapter may satisfy the recordkeeping requirements of this section and of § 91.1113 with records maintained to fulfill equivalent obligations under part 121 or 135 of this chapter.

§ 91.1029 Flight scheduling and locating requirements.

(a) Each program manager must establish and use an adequate system to schedule and release program aircraft.

(b) Except as provided in paragraph (d) of this section, each program manager must have adequate procedures established for locating each flight, for which a flight plan is not filed, that—

(1) Provide the program manager with at least the information required to be included in a VFR flight plan;

(2) Provide for timely notification of an FAA facility or search and rescue facility, if an aircraft is overdue or missing; and

(3) Provide the program manager with the location, date, and estimated time for reestablishing radio or telephone communications, if the flight will operate in an area where communications cannot be maintained.

(c) Flight locating information must be retained at the program manager's principal base of operations, or at other places designated by the program manager in the flight locating procedures, until the completion of the flight.

(d) The flight locating requirements of paragraph (b) of this section do not apply to a flight for which an FAA flight plan has been filed and the flight plan is canceled within 25 nautical miles of the destination airport.

§ 91.1031 Pilot in command or second in command: Designation required.

(a) Each program manager must designate a—

(1) Pilot in command for each program flight; and

(2) Second in command for each program flight requiring two pilots.

(b) The pilot in command, as designated by the program manager, must remain the pilot in command at all times during that flight.

§ 91.1033 Operating information required.

(a) Each program manager must, for all program operations, provide the following materials, in current and appropriate form, accessible to the pilot at the pilot station, and the pilot must use them—

(1) A cockpit checklist;

(2) For multiengine aircraft or for aircraft with retractable landing gear, an emergency cockpit checklist containing the procedures required by paragraph (c) of this section, as appropriate;

(3) At least one set of pertinent aeronautical charts; and

(4) For IFR operations, at least one set of pertinent navigational en route, terminal area, and instrument approach procedure charts.

(b) Each cockpit checklist required by paragraph (a)(1) of this section must contain the following procedures:

(1) Before starting engines;

(2) Before takeoff;

(3) Cruise;

(4) Before landing;

(5) After landing; and

(6) Stopping engines.

(c) Each emergency cockpit checklist required by paragraph (a)(2) of this section must contain the following procedures, as appropriate:

(1) Emergency operation of fuel, hydraulic, electrical, and mechanical systems.

(2) Emergency operation of instruments and controls.

(3) Engine inoperative procedures.

(4) Any other emergency procedures necessary for safety.

§ 91.1035 Passenger awareness.

(a) Prior to each takeoff, the pilot in command of an aircraft carrying passengers on a program flight must ensure that all passengers have been orally briefed on—

(1) *Smoking:* Each passenger must be briefed on when, where, and under what conditions smoking is prohibited. This briefing must include a statement, as appropriate, that the regulations require passenger compliance with lighted passenger information signs and no smoking placards, prohibit smoking in lavatories, and require compliance with crewmember instructions with regard to these items;

(2) *Use of safety belts, shoulder harnesses, and child restraint systems:* Each passenger must be briefed on when, where and under what conditions it is necessary to have his or her safety belt and, if installed, his or her shoulder harness fastened about him or her, and if a child is being transported, the appropriate use of child restraint systems, if available. This briefing must include a statement, as appropriate, that the regulations require passenger compliance with the lighted passenger information sign and/or crewmember instructions with regard to these items;

(3) The placement of seat backs in an upright position before takeoff and landing;

(4) Location and means for opening the passenger entry door and emergency exits;

(5) Location of survival equipment;

(6) Ditching procedures and the use of flotation equipment required under § 91.509 for a flight over water;

(7) The normal and emergency use of oxygen installed in the aircraft; and

(8) Location and operation of fire extinguishers.

(b) Prior to each takeoff, the pilot in command of an aircraft carrying passengers on a program flight must ensure that each person who may need the assistance of another person to move expeditiously to an exit if an emergency occurs and that person's attendant, if any, has received a briefing as to the procedures to be followed if an evacuation occurs. This paragraph does not apply to a person who has been given a briefing before a previous leg of that flight in the same aircraft.

(c) Prior to each takeoff, the pilot in command must advise the passengers of the name of the entity in operational control of the flight.

(d) The oral briefings required by paragraphs (a), (b), and (c) of this section must be given by the pilot in command or another crewmember.

(e) The oral briefing required by paragraph (a) of this section may be delivered by means of an approved recording playback device that is audible to each passenger under normal noise levels.

(f) The oral briefing required by paragraph (a) of this section must be supplemented by printed cards that must be carried in the aircraft in locations convenient for the use of each passenger. The cards must—

(1) Be appropriate for the aircraft on which they are to be used;

(2) Contain a diagram of, and method of operating, the emergency exits; and

(3) Contain other instructions necessary for the use of emergency equipment on board the aircraft.

§ 91.1037 Large transport category airplanes: Turbine engine powered; Limitations; Destination and alternate airports.

(a) No program manager or any other person may permit a turbine engine powered large transport category airplane on a program flight to take off that airplane at a weight that (allowing for normal consumption of fuel and oil in flight to the destination or alternate airport) the weight of the airplane on arrival would exceed the landing weight in the Airplane Flight Manual for the elevation of the destination or alternate airport and the ambient temperature expected at the time of landing.

(b) Except as provided in paragraph (c) of this section, no program manager or any other person may permit a turbine engine powered large transport category airplane on a program flight to take off that airplane unless its weight on arrival, allowing for normal consumption of fuel and oil in flight (in accordance with the landing distance in the Airplane Flight Manual for the elevation of the destination airport and the wind conditions expected there at the time of landing), would allow a full stop landing at the intended destination airport within 60 percent of the effective length of each runway described below from a point 50 feet above the intersection of the obstruction clearance plane and the runway. For the purpose of determining the allowable landing weight at the destination airport, the following is assumed:

(1) The airplane is landed on the most favorable runway and in the most favorable direction, in still air.

(2) The airplane is landed on the most suitable runway considering the probable wind velocity and direction and the ground handling characteristics of that airplane, and considering other conditions such as landing aids and terrain.

(c) A program manager or other person flying a turbine engine powered large transport category airplane on a program flight may permit that airplane to take off at a weight in excess of that allowed by paragraph (b) of this section if all of the following conditions exist:

(1) The operation is conducted in accordance with an approved Destination Airport Analysis in that person's program operating manual that contains the elements listed in § 91.1025(o).

(2) The airplane's weight on arrival, allowing for normal consumption of fuel and oil in flight (in accordance with the landing distance in the Airplane Flight Manual for the elevation of the destination airport and the wind conditions expected there at the time of landing), would allow a full stop landing at the intended destination airport within 80 percent of the effective length of each runway described below from a point 50 feet above the intersection of the obstruction clearance plane and the runway. For the purpose of determining the allowable landing weight at the destination airport, the following is assumed:

(i) The airplane is landed on the most favorable runway and in the most favorable direction, in still air.

(ii) The airplane is landed on the most suitable runway considering the probable wind velocity and direction and the ground handling characteristics of that airplane, and considering other conditions such as landing aids and terrain.

(3) The operation is authorized by management specifications.

(d) No program manager or other person may select an airport as an alternate airport for a turbine engine powered large transport category airplane unless (based on the assumptions in paragraph (b) of this section) that airplane, at the weight expected at the time of arrival, can be brought to a full stop landing within 80 percent of the effective length of the runway from a point 50 feet above the intersection of the obstruction clearance plane and the runway.

(e) Unless, based on a showing of actual operating landing techniques on wet runways, a shorter landing distance (but never less than that required by paragraph (b) or (c) of this section) has been approved for a specific type and model airplane and included in the Airplane Flight Manual, no person may take off a turbojet airplane when the appropriate weather reports or forecasts, or any combination of them, indicate that the runways at the destination or alternate airport may be wet or slippery at the estimated time of arrival unless the effective runway length at the destination airport is at least 115 percent of the runway length required under paragraph (b) or (c) of this section.

§ 91.1039 IFR takeoff, approach and landing minimums.

(a) No pilot on a program aircraft operating a program flight may begin an instrument approach procedure to an airport unless—

(1) Either that airport or the alternate airport has a weather reporting facility operated by the U.S. National Weather Service, a source approved by the U.S. National Weather Service, or a source approved by the Administrator; and

(2) The latest weather report issued by the weather reporting facility includes a current local altimeter setting for the destination airport. If no local altimeter setting is available at the destination airport, the pilot must obtain the current local altimeter setting from a source provided by the facility designated on the approach chart for the destination airport.

(b) For flight planning purposes, if the destination airport does not have a weather reporting facility described in paragraph (a)(1) of this section, the pilot must designate as an alternate an airport that has a weather reporting facility meeting that criteria.

(c) The MDA or Decision Altitude and visibility landing minimums prescribed in part 97 of this chapter or in the program manager's management specifications are increased by 100 feet and 1/2 mile respectively, but not to exceed the ceiling and visibility minimums for that airport when used as an alternate airport, for each pilot in command of a turbine-powered aircraft who has not served at least 100 hours as pilot in command in that type of aircraft.

(d) No person may take off an aircraft under IFR from an airport where weather conditions are at or above takeoff minimums but are below authorized IFR landing minimums unless there is an alternate airport within one hour's flying time (at normal cruising speed, in still air) of the airport of departure.

(e) Each pilot making an IFR takeoff or approach and landing at an airport must comply with applicable instrument approach procedures and take off and landing weather minimums prescribed by the authority having jurisdiction over the airport. In addition, no pilot may, at that airport take off when the visibility is less than 600 feet.

§ 91.1041 Aircraft proving and validation tests.

(a) No program manager may permit the operation of an aircraft, other than a turbojet aircraft, for which two pilots are required by the type certification requirements of this chapter for operations under VFR, if it has not previously proved such an aircraft in operations under this part in at least 25 hours of proving tests acceptable to the Administrator including—

(1) Five hours of night time, if night flights are to be authorized;

(2) Five instrument approach procedures under simulated or actual conditions, if IFR flights are to be authorized; and

(3) Entry into a representative number of en route airports as determined by the Administrator.

(b) No program manager may permit the operation of a turbojet airplane if it has not previously proved a turbojet airplane in operations under this part in at least 25 hours of proving tests acceptable to the Administrator including—

(1) Five hours of night time, if night flights are to be authorized;

(2) Five instrument approach procedures under simulated or actual conditions, if IFR flights are to be authorized; and

(3) Entry into a representative number of en route airports as determined by the Administrator.

(c) No program manager may carry passengers in an aircraft during proving tests, except those needed to make the tests and those designated by the Administrator to observe the tests. However, pilot flight training may be conducted during the proving tests.

(d) Validation testing is required to determine that a program manager is capable of conducting operations safely and in compliance with applicable regulatory standards. Validation tests are required for the following authorizations:

(1) The addition of an aircraft for which two pilots are required for operations under VFR or a turbojet airplane, if that aircraft or an aircraft of the same make or similar design has not been previously proved or validated in operations under this part.

(2) Operations outside U.S. airspace.

(3) Class II navigation authorizations.

(4) Special performance or operational authorizations.

(e) Validation tests must be accomplished by test methods acceptable to the Administrator. Actual flights may not be required when an applicant can demonstrate competence and compliance with appropriate regulations without conducting a flight.

(f) Proving tests and validation tests may be conducted simultaneously when appropriate.

(g) The Administrator may authorize deviations from this section if the Administrator finds that special circumstances make full compliance with this section unnecessary.

§ 91.1043 [Reserved]

§ 91.1045 Additional equipment requirements.

No person may operate a program aircraft on a program flight unless the aircraft is equipped with the following—

(a) Airplanes having a passenger-seat configuration of more than 30 seats or a payload capacity of more than 7,500 pounds:

(1) A cockpit voice recorder as required by § 121.359 of this chapter as applicable to the aircraft specified in that section.

(2) A flight recorder as required by § 121.343 or § 121.344 of this chapter as applicable to the aircraft specified in that section.

(3) A terrain awareness and warning system as required by § 121.354 of this chapter as applicable to the aircraft specified in that section.

(4) A traffic alert and collision avoidance system as required by § 121.356 of this chapter as applicable to the aircraft specified in that section.

(5) Airborne weather radar as required by § 121.357 of this chapter, as applicable to the aircraft specified in that section.

(b) Airplanes having a passenger-seat configuration of 30 seats or fewer, excluding each crewmember, and a payload capacity of 7,500 pounds or less, and any rotorcraft (as applicable):

(1) A cockpit voice recorder as required by § 135.151 of this chapter as applicable to the aircraft specified in that section.

(2) A flight recorder as required by § 135.152 of this chapter as applicable to the aircraft specified in that section.

(3) A terrain awareness and warning system as required by § 135.154 of this chapter as applicable to the aircraft specified in that section.

(4) A traffic alert and collision avoidance system as required by § 135.180 of this chapter as applicable to the aircraft specified in that section.

(5) As applicable to the aircraft specified in that section, either:

(i) Airborne thunderstorm detection equipment as required by § 135.173 of this chapter; or

(ii) Airborne weather radar as required by § 135.175 of this chapter.

§ 91.1047 Drug and alcohol misuse education program.

(a) Each program manager must provide each direct employee performing flight crewmember, flight attendant, flight instructor, or aircraft maintenance duties with drug and alcohol misuse education.

(b) No program manager may use any contract employee to perform flight crewmember, flight attendant, flight instructor, or aircraft maintenance duties for the program manager unless that contract employee has been provided with drug and alcohol misuse education.

(c) Program managers must disclose to their owners and prospective owners the existence of a company drug and alcohol misuse testing program. If the program manager has implemented a company testing program, the program manager's disclosure must include the following:

(1) Information on the substances that they test for, for example, alcohol and a list of the drugs;

(2) The categories of employees tested, the types of tests, for example, pre-employment, random, reasonable cause/suspicion, post accident, return to duty and follow-up; and

(3) The degree to which the program manager's company testing program is comparable to the federally mandated drug and alcohol testing program required under part 120 of this chapter regarding the information in paragraphs (c)(1) and (c)(2) of this section.

(d) If a program aircraft is operated on a program flight into an airport at which no maintenance personnel are available that are subject to the requirements of paragraphs (a) or (b) of this section and emergency maintenance is required, the program manager may use persons not meeting the requirements of paragraphs (a) or (b) of this section to provide such emergency maintenance under both of the following conditions:

(1) The program manager must notify the Drug Abatement Program Division, AAM-800, 800 Independence Avenue, SW., Washington, DC 20591 in writing within 10 days after being provided emergency maintenance in accordance with this paragraph. The program manager must retain copies of all such written notifications for two years.

(2) The aircraft must be reinspected by maintenance personnel who meet the requirements of paragraph (a) or (b) of this section when the aircraft is next at an airport where such maintenance personnel are available.

(e) For purposes of this section, emergency maintenance means maintenance that—

(1) Is not scheduled, and

(2) Is made necessary by an aircraft condition not discovered prior to the departure for that location.

(f) Notwithstanding paragraphs (a) and (b) of this section, drug and alcohol misuse education conducted under an FAA-approved drug and alcohol misuse prevention program may be used to satisfy these requirements.

[Docket No. FAA-2001-10047, 68 FR 54561, Sept. 17, 2003, as amended by Amdt. 91-307, 74 FR 22653, May 14, 2009]

§ 91.1049 Personnel.

(a) Each program manager and each fractional owner must use in program operations on program aircraft flight crews meeting § 91.1053 criteria and qualified under the appropriate regulations. The program manager must provide oversight of those crews.

(b) Each program manager must employ (either directly or by contract) an adequate number of pilots per program aircraft. Flight crew staffing must be determined based on the following factors, at a minimum:

(1) Number of program aircraft.

(2) Program manager flight, duty, and rest time considerations, and in all cases within the limits set forth in §§ 91.1057 through 91.1061.

(3) Vacations.

(4) Operational efficiencies.

(5) Training.

(6) Single pilot operations, if authorized by deviation under paragraph (d) of this section.

(c) Each program manager must publish pilot and flight attendant duty schedules sufficiently in advance to follow the flight, duty, and rest time limits in §§ 91.1057 through 91.1061 in program operations.

(d) Unless otherwise authorized by the Administrator, when any program aircraft is flown in program operations with passengers onboard, the crew must consist of at least two qualified pilots employed or contracted by the program manager or the fractional owner.

(e) The program manager must ensure that trained and qualified scheduling or flight release personnel are on duty to schedule and release program aircraft during all hours that such aircraft are available for program operations.

§ 91.1050 Employment of former FAA employees.

(a) Except as specified in paragraph (c) of this section, no fractional owner or fractional ownership program manager may knowingly employ or make a contractual arrangement which permits an individual to act as an agent or representative of the fractional owner or fractional ownership program manager in any matter before the Federal Aviation Administration if the individual, in the preceding 2 years—

(1) Served as, or was directly responsible for the oversight of, a Flight Standards Service aviation safety inspector; and

(2) Had direct responsibility to inspect, or oversee the inspection of, the operations of the fractional owner or fractional ownership program manager.

(b) For the purpose of this section, an individual shall be considered to be acting as an agent or representative of a fractional owner or fractional ownership program manager in a matter before the agency if the individual makes any written or oral communication on behalf of the fractional owner or fractional ownership program manager to the agency (or any of its officers or employees) in connection with a particular matter, whether or not involving a specific party and without regard to whether the individual has participated in, or had responsibility for, the particular matter while serving as a Flight Standards Service aviation safety inspector.

(c) The provisions of this section do not prohibit a fractional owner or fractional ownership program manager from knowingly employing or making a contractual arrangement which permits an individual to act as an agent or representative of the fractional owner or fractional ownership program manager in any matter before the Federal Aviation Administration if the individual was employed by the fractional owner or fractional ownership program manager before October 21, 2011.

[Doc. No. FAA-2008-1154, 76 FR 52235, Aug. 22, 2011]

§ 91.1051 Pilot safety background check.

Within 90 days of an individual beginning service as a pilot, the program manager must request the following information:

(a) FAA records pertaining to—

(1) Current pilot certificates and associated type ratings.

(2) Current medical certificates.

(3) Summaries of legal enforcement actions resulting in a finding by the Administrator of a violation.

(b) Records from all previous employers during the five years preceding the date of the employment application where the applicant worked as a pilot. If any of these firms are in bankruptcy, the records must be requested from the trustees in bankruptcy for those employees. If the previous employer is no longer in business, a documented good faith effort must be made to obtain the records. Records from previous employers must include, as applicable—

(1) Crew member records.

(2) Drug testing—collection, testing, and rehabilitation records pertaining to the individual.

(3) Alcohol misuse prevention program records pertaining to the individual.

(4) The applicant's individual record that includes certifications, ratings, aeronautical experience, effective date and class of the medical certificate.

§ 91.1053 Crewmember experience.

(a) No program manager or owner may use any person, nor may any person serve, as a pilot in command or second in command of a program aircraft, or as a flight attendant on a program aircraft, in program operations under this subpart unless that person has met the applicable requirements of part 61 of this chapter and has the following experience and ratings:

(1) Total flight time for all pilots:

(i) Pilot in command—A minimum of 1,500 hours.

(ii) Second in command—A minimum of 500 hours.

(2) For multi-engine turbine-powered fixed-wing and powered-lift aircraft, the following FAA certification and ratings requirements:

(i) Pilot in command—Airline transport pilot and applicable type ratings.

(ii) Second in command—Commercial pilot and instrument ratings.

(iii) Flight attendant (if required or used)—Appropriately trained personnel.

(3) For all other aircraft, the following FAA certification and rating requirements:

(i) Pilot in command—Commercial pilot and instrument ratings.

(ii) Second in command—Commercial pilot and instrument ratings.

(iii) Flight attendant (if required or used)—Appropriately trained personnel.

(b) The Administrator may authorize deviations from paragraph (a)(1) of this section if the Flight Standards District Office that issued the program manager's management specifications finds that the crewmember has comparable experience, and can effectively perform the functions associated with the position in accordance with the requirements of this chapter. Grants of deviation under this paragraph may be granted after consideration of the size and scope of the operation, the qualifications of the intended personnel and the circumstances set forth in § 91.1055(b)(1) through (3). The Administrator may, at any time, terminate any grant of deviation authority issued under this paragraph.

§ 91.1055 Pilot operating limitations and pairing requirement.

(a) If the second in command of a fixed-wing program aircraft has fewer than 100 hours of flight time as second in command flying in the aircraft make and model and, if a type rating is required, in the type aircraft being flown, and the pilot in command is not an appropriately qualified check pilot, the pilot in command shall make all takeoffs and landings in any of the following situations:

(1) Landings at the destination airport when a Destination Airport Analysis is required by § 91.1037(c); and

(2) In any of the following conditions:

(i) The prevailing visibility for the airport is at or below 3⁄4 mile.

(ii) The runway visual range for the runway to be used is at or below 4,000 feet.

(iii) The runway to be used has water, snow, slush, ice or similar contamination that may adversely affect aircraft performance.

(iv) The braking action on the runway to be used is reported to be less than "good."

(v) The crosswind component for the runway to be used is in excess of 15 knots.

(vi) Windshear is reported in the vicinity of the airport.

(vii) Any other condition in which the pilot in command determines it to be prudent to exercise the pilot in command's authority.

(b) No program manager may release a program flight under this subpart unless, for that aircraft make or model and, if a type rating is required, for that type aircraft, either the pilot in command or the second in command has at least 75 hours of flight time, either as pilot in command or second in command. The Administrator may, upon application by the program manager, authorize deviations from the requirements of this paragraph by an appropriate amendment to the management specifications in any of the following circumstances:

(1) A newly authorized program manager does not employ any pilots who meet the minimum requirements of this paragraph.

(2) An existing program manager adds to its fleet a new category and class aircraft not used before in its operation.

(3) An existing program manager establishes a new base to which it assigns pilots who will be required to become qualified on the aircraft operated from that base.

(c) No person may be assigned in the capacity of pilot in command in a program operation to more than two aircraft types that require a separate type rating.

§ 91.1057 Flight, duty and rest time requirements: All crewmembers.

(a) For purposes of this subpart—

Augmented flight crew means at least three pilots.

Calendar day means the period of elapsed time, using Coordinated Universal Time or local time that begins at midnight and ends 24 hours later at the next midnight.

Duty period means the period of elapsed time between reporting for an assignment involving flight time and release from that assignment by the program manager. All time between these two points is part of the duty period, even if flight time is interrupted by nonflight-related duties. The time is calculated using either Coordinated Universal Time or local time to reflect the total elapsed time.

Extension of flight time means an increase in the flight time because of circumstances beyond the control of the program manager or flight crewmember (such as adverse weather) that are not known at the time of departure and that prevent the flightcrew from reaching the destination within the planned flight time.

Flight attendant means an individual, other than a flight crewmember, who is assigned by the program manager, in accordance with the required minimum crew complement under the program manager's management specifications or in addition to that minimum complement, to duty in an aircraft during flight time and whose duties include but are not necessarily limited to cabin-safety-related responsibilities.

Multi-time zone flight means an easterly or westerly flight or multiple flights in one direction in the same duty period that results in a time zone difference of 5 or more hours and is conducted in a geographic area that is south of 60 degrees north latitude and north of 60 degrees south latitude.

Reserve status means that status in which a flight crewmember, by arrangement with the program manager: Holds himself or herself fit to fly to the extent that this is within the control of the flight crewmember; remains within a reasonable response time of the aircraft as agreed between the flight crewmember and the program manager; and maintains a ready means whereby the flight crewmember may be contacted by the program manager. Reserve status is not part of any duty period or rest period.

Rest period means a period of time required pursuant to this subpart that is free of all responsibility for work or duty prior to the commencement of, or following completion of, a duty period, and during which the flight crewmember or flight attendant cannot be required to receive contact from the program manager. A rest period does not include any time during which the program manager imposes on a flight crewmember or flight attendant any duty or restraint, including any actual work or present responsibility for work should the occasion arise.

Standby means that portion of a duty period during which a flight crewmember is subject to the control of the program manager and holds himself or herself in a condition of readiness to undertake a flight. Standby is not part of any rest period.

(b) A program manager may assign a crewmember and a crewmember may accept an assignment for flight time only when the applicable requirements of this section and §§ 91.1059-91.1062 are met.

(c) No program manager may assign any crewmember to any duty during any required rest period.

(d) Time spent in transportation, not local in character, that a program manager requires of a crewmember and provides to transport the crewmember to an airport at which he or she is to serve on a flight as a crewmember, or from an airport at which he or she was relieved from duty to return to his or her home station, is not considered part of a rest period.

(e) A flight crewmember may continue a flight assignment if the flight to which he or she is assigned would normally terminate within the flight time limitations, but because of circumstances beyond the control of the program manager or flight crewmember (such as adverse weather conditions), is not at the time of departure expected to reach its destination within the planned flight time. The extension of flight time under this paragraph may not exceed the maximum time limits set forth in § 91.1059.

(f) Each flight assignment must provide for at least 10 consecutive hours of rest during the 24-hour period that precedes the completion time of the assignment.

(g) The program manager must provide each crewmember at least 13 rest periods of at least 24 consecutive hours each in each calendar quarter.

(h) A flight crewmember may decline a flight assignment if, in the flight crewmember's determination, to do so would not be consistent with the standard of safe operation required under this subpart, this part, and applicable provisions of this title.

(i) Any rest period required by this subpart may occur concurrently with any other rest period.

(j) If authorized by the Administrator, a program manager may use the applicable unscheduled flight time limitations, duty period limitations, and rest requirements of part 121 or part 135 of this chapter instead of the flight time limitations, duty period limitations, and rest requirements of this subpart.

§ 91.1059 Flight time limitations and rest requirements: One or two pilot crews.

(a) No program manager may assign any flight crewmember, and no flight crewmember may accept an assignment, for flight time as a member of a one- or two-pilot crew if that crewmember's total flight time in all commercial flying will exceed—

(1) 500 hours in any calendar quarter;

(2) 800 hours in any two consecutive calendar quarters;

(3) 1,400 hours in any calendar year.

(b) Except as provided in paragraph (c) of this section, during any 24 consecutive hours the total flight time of the assigned flight, when added to any commercial flying by that flight crewmember, may not exceed—

(1) 8 hours for a flight crew consisting of one pilot; or

(2) 10 hours for a flight crew consisting of two pilots qualified under this subpart for the operation being conducted.

(c) No program manager may assign any flight crewmember, and no flight crewmember may accept an assignment, if that crewmember's flight time or duty period will exceed, or rest time will be less than—

	Normal duty	Extension of flight time
(1) Minimum Rest Immediately Before Duty	10 Hours	10 Hours.
(2) Duty Period	Up to 14 Hours	Up to 14 Hours.
(3) Flight Time For 1 Pilot	Up to 8 Hours	Exceeding 8 Hours up to 9 Hours.
(4) Flight Time For 2 Pilots	Up to 10 Hours	Exceeding 10 Hours up to 12 Hours.
(5) Minimum After Duty Rest	10 Hours	12 Hours.
(6) Minimum After Duty Rest Period for Multi-Time Zone Flights	14 Hours	18 Hours.

§ 91.1061 Augmented flight crews.

(a) No program manager may assign any flight crewmember, and no flight crewmember may accept an assignment, for flight time as a member of an augmented crew if that crewmember's total flight time in all commercial flying will exceed—

(1) 500 hours in any calendar quarter;

(2) 800 hours in any two consecutive calendar quarters;

(3) 1,400 hours in any calendar year.

(b) No program manager may assign any pilot to an augmented crew, unless the program manager ensures:

(1) Adequate sleeping facilities are installed on the aircraft for the pilots.

(2) No more than 8 hours of flight deck duty is accrued in any 24 consecutive hours.

(3) For a three-pilot crew, the crew must consist of at least the following:

(i) A pilot in command (PIC) who meets the applicable flight crewmember requirements of this subpart and § 61.57 of this chapter.

(ii) A PIC qualified pilot who meets the applicable flight crewmember requirements of this subpart and § 61.57(c) and (d) of this chapter.

(iii) A second in command (SIC) who meets the SIC qualifications of this subpart. For flight under IFR, that person must also meet the recent instrument experience requirements of part 61 of this chapter.

(4) For a four-pilot crew, at least three pilots who meet the conditions of paragraph (b)(3) of this section, plus a fourth pilot who meets the SIC qualifications of this subpart. For flight under IFR, that person must also meet the recent instrument experience requirements of part 61 of this chapter.

(c) No program manager may assign any flight crewmember, and no flight crewmember may accept an assignment, if that crewmember's flight time or duty period will exceed, or rest time will be less than—

	3-Pilot crew	4-Pilot crew
(1) Minimum Rest Immediately Before Duty	10 Hours	10 Hours
(2) Duty Period	Up to 16 Hours	Up to 18 Hours
(3) Flight Time	Up to 12 Hours	Up to 16 Hours
(4) Minimum After Duty Rest	12 Hours	18 Hours
(5) Minimum After Duty Rest Period for Multi-Time Zone Flights	18 hours	24 hours

§ 91.1062 Duty periods and rest requirements: Flight attendants.

(a) Except as provided in paragraph (b) of this section, a program manager may assign a duty period to a flight attendant only when the assignment meets the applicable duty period limitations and rest requirements of this paragraph.

(1) Except as provided in paragraphs (a)(4), (a)(5), and (a)(6) of this section, no program manager may assign a flight attendant to a scheduled duty period of more than 14 hours.

(2) Except as provided in paragraph (a)(3) of this section, a flight attendant scheduled to a duty period of 14 hours or less as provided under paragraph (a)(1) of this section must be given a scheduled rest period of at least 9 consecutive hours. This rest period must occur between the completion of the scheduled duty period and the commencement of the subsequent duty period.

(3) The rest period required under paragraph (a)(2) of this section may be scheduled or reduced to 8 consecutive hours if the flight attendant is provided a subsequent rest period of at least 10 consecutive hours; this subsequent rest period must be scheduled to begin no later than 24 hours after the beginning of the reduced rest period and must occur between the completion of the scheduled duty period and the commencement of the subsequent duty period.

(4) A program manager may assign a flight attendant to a scheduled duty period of more than 14 hours, but no more than 16 hours, if the program manager has assigned to the flight or flights in that duty period at least one flight attendant in addition to the minimum flight attendant complement required for the flight or flights in that duty period under the program manager's management specifications.

(5) A program manager may assign a flight attendant to a scheduled duty period of more than 16 hours, but no more than 18 hours, if the program manager has assigned to the flight or flights in that duty period at least two flight attendants in addition to the minimum flight attendant complement required for the flight or flights in that duty period under the program manager's management specifications.

(6) A program manager may assign a flight attendant to a scheduled duty period of more than 18 hours, but no more than 20 hours, if the scheduled duty period includes one or more flights that land or take off outside the 48 contiguous states and the District of Columbia, and if the program manager has assigned to the flight or flights in that duty period at least three flight attendants in addition to the minimum flight attendant complement required for the flight or flights in that duty period under the program manager's management specifications.

(7) Except as provided in paragraph (a)(8) of this section, a flight attendant scheduled to a duty period of more than 14 hours but no more than 20 hours, as provided in paragraphs (a)(4), (a)(5), and (a)(6) of this section, must be given a scheduled rest period of at least 12 consecutive hours. This rest period must occur between the completion of the scheduled duty period and the commencement of the subsequent duty period.

(8) The rest period required under paragraph (a)(7) of this section may be scheduled or reduced to 10 consecutive hours if the flight attendant is provided a subsequent rest period of at least 14 consecutive hours; this subsequent rest period must be scheduled to begin no later than 24 hours after the beginning of the reduced rest period and must occur between the completion of the scheduled duty period and the commencement of the subsequent duty period.

(9) Notwithstanding paragraphs (a)(4), (a)(5), and (a)(6) of this section, if a program manager elects to reduce the rest period to 10 hours as authorized by paragraph (a)(8) of this section, the program manager may not schedule a flight attendant for a duty period of more than 14 hours during the 24-hour period commencing after the beginning of the reduced rest period.

(b) Notwithstanding paragraph (a) of this section, a program manager may apply the flight crewmember flight time and duty limitations and rest requirements of this part to flight attendants for all operations conducted under this part provided that the program manager establishes written procedures that—

(1) Apply to all flight attendants used in the program manager's operation;

(2) Include the flight crewmember rest and duty requirements of §§ 91.1057, 91.1059, and 91.1061, as appropriate to the operation being conducted, except that rest facilities on board the aircraft are not required;

(3) Include provisions to add one flight attendant to the minimum flight attendant complement for each flight crewmember who is in excess of the minimum number required in the aircraft type certificate data sheet and who is assigned to the aircraft under the provisions of § 91.1061; and

(4) Are approved by the Administrator and described or referenced in the program manager's management specifications.

§ 91.1063 Testing and training: Applicability and terms used.

(a) Sections 91.1065 through 91.1107:

(1) Prescribe the tests and checks required for pilots and flight attendant crewmembers and for the approval of check pilots in operations under this subpart;

(2) Prescribe the requirements for establishing and maintaining an approved training program for crewmembers, check pilots and instructors, and other operations personnel employed or used by the program manager in program operations;

(3) Prescribe the requirements for the qualification, approval and use of aircraft simulators and flight training devices in the conduct of an approved training program; and

(4) Permits training center personnel authorized under part 142 of this chapter who meet the requirements of § 91.1075 to conduct training, testing and checking under contract or other arrangements to those persons subject to the requirements of this subpart.

(b) If authorized by the Administrator, a program manager may comply with the applicable training and testing sections of subparts N and O of part 121 of this chapter instead of §§ 91.1065 through 91.1107, except for the operating experience requirements of § 121.434 of this chapter.

(c) If authorized by the Administrator, a program manager may comply with the applicable training and testing sections of subparts G and H of part 135 of this chapter instead of §§ 91.1065 through 91.1107, except for the operating experience requirements of § 135.244 of this chapter.

(d) For the purposes of this subpart, the following terms and definitions apply:

(1) *Initial training.* The training required for crewmembers who have not qualified and served in the same capacity on an aircraft.

(2) *Transition training.* The training required for crewmembers who have qualified and served in the same capacity on another aircraft.

(3) *Upgrade training.* The training required for crewmembers who have qualified and served as second in command on a particular aircraft type, before they serve as pilot in command on that aircraft.

(4) *Differences training.* The training required for crewmembers who have qualified and served on a particular type aircraft, when the Administrator finds differences training is necessary before a crewmember serves in the same capacity on a particular variation of that aircraft.

(5) *Recurrent training.* The training required for crewmembers to remain adequately trained and currently proficient for each aircraft crewmember position, and type of operation in which the crewmember serves.

(6) *In flight.* The maneuvers, procedures, or functions that will be conducted in the aircraft.

(7) *Training center.* An organization governed by the applicable requirements of part 142 of this chapter that conducts training, testing, and checking under contract or other arrangement to program managers subject to the requirements of this subpart.

(8) *Requalification training.* The training required for crewmembers previously trained and qualified, but who have become unqualified because of not having met within the required period any of the following:

(i) Recurrent crewmember training requirements of § 91.1107.

(ii) Instrument proficiency check requirements of § 91.1069.

(iii) Testing requirements of § 91.1065.

(iv) Recurrent flight attendant testing requirements of § 91.1067.

§ 91.1065 Initial and recurrent pilot testing requirements.

(a) No program manager or owner may use a pilot, nor may any person serve as a pilot, unless, since the beginning of the 12th month before that service, that pilot has passed either a written or oral test (or a combination), given by the Administrator or an authorized check pilot, on that pilot's knowledge in the following areas—

(1) The appropriate provisions of parts 61 and 91 of this chapter and the management specifications and the operating manual of the program manager;

(2) For each type of aircraft to be flown by the pilot, the aircraft powerplant, major components and systems, major appliances, performance and operating limitations, standard and emergency operating procedures, and the contents of the accepted operating manual or equivalent, as applicable;

(3) For each type of aircraft to be flown by the pilot, the method of determining compliance with weight and balance limitations for takeoff, landing and en route operations;

(4) Navigation and use of air navigation aids appropriate to the operation or pilot authorization, including, when applicable, instrument approach facilities and procedures;

(5) Air traffic control procedures, including IFR procedures when applicable;

(6) Meteorology in general, including the principles of frontal systems, icing, fog, thunderstorms, and windshear, and, if appropriate for the operation of the program manager, high altitude weather;

(7) Procedures for—

(i) Recognizing and avoiding severe weather situations;

(ii) Escaping from severe weather situations, in case of inadvertent encounters, including low-altitude windshear (except that rotorcraft aircraft pilots are not required to be tested on escaping from low-altitude windshear); and

(iii) Operating in or near thunderstorms (including best penetration altitudes), turbulent air (including clear air turbulence), icing, hail, and other potentially hazardous meteorological conditions; and

(8) New equipment, procedures, or techniques, as appropriate.

(b) No program manager or owner may use a pilot, nor may any person serve as a pilot, in any aircraft unless, since the beginning of the 12th month before that service, that pilot has passed a competency check given by the Administrator or an authorized check pilot in that class of aircraft, if single-engine aircraft other than turbojet, or that type of aircraft, if rotorcraft, multiengine aircraft, or turbojet airplane, to determine the pilot's competence in practical skills and techniques in that aircraft or class of aircraft. The extent of the competency check will be determined by the Administrator or authorized check pilot conducting the competency check. The competency check may include any of the maneuvers and procedures currently required for the original issuance of the particular pilot certificate required for the operations authorized and appropriate to the category, class and type of aircraft involved. For the purposes of this paragraph, type, as to an airplane, means any one of a group of airplanes determined by the Administrator to have a similar means of propulsion, the same manufacturer, and no significantly different handling or flight characteristics. For the purposes of this paragraph, type, as to a rotorcraft, means a basic make and model.

(c) The instrument proficiency check required by § 91.1069 may be substituted for the competency check required by this section for the type of aircraft used in the check.

(d) For the purpose of this subpart, competent performance of a procedure or maneuver by a person to be used as a pilot requires that the pilot be the obvious master of the aircraft, with the successful outcome of the maneuver never in doubt.

(e) The Administrator or authorized check pilot certifies the competency of each pilot who passes the knowledge or flight check in the program manager's pilot records.

(f) All or portions of a required competency check may be given in an aircraft simulator or other appropriate training device, if approved by the Administrator.

§ 91.1067 Initial and recurrent flight attendant crewmember testing requirements.

No program manager or owner may use a flight attendant crewmember, nor may any person serve as a flight attendant crewmember unless, since the beginning of the 12th month before that service, the program manager has determined by appropriate initial and recurrent testing that the person is knowledgeable and competent in the following areas as appropriate to assigned duties and responsibilities—

(a) Authority of the pilot in command;

(b) Passenger handling, including procedures to be followed in handling deranged persons or other persons whose conduct might jeopardize safety;

(c) Crewmember assignments, functions, and responsibilities during ditching and evacuation of persons who may need the assistance of another person to move expeditiously to an exit in an emergency;

(d) Briefing of passengers;

(e) Location and operation of portable fire extinguishers and other items of emergency equipment;

(f) Proper use of cabin equipment and controls;

(g) Location and operation of passenger oxygen equipment;

(h) Location and operation of all normal and emergency exits, including evacuation slides and escape ropes; and

(i) Seating of persons who may need assistance of another person to move rapidly to an exit in an emergency as prescribed by the program manager's operations manual.

§ 91.1069 Flight crew: Instrument proficiency check requirements.

(a) No program manager or owner may use a pilot, nor may any person serve, as a pilot in command of an aircraft under IFR unless, since the beginning of the 6th month before that service, that pilot has passed an instrument proficiency check under this section administered by the Administrator or an authorized check pilot.

(b) No program manager or owner may use a pilot, nor may any person serve, as a second command pilot of an aircraft under IFR unless, since the beginning of the 12th month before that service, that pilot has passed an instrument proficiency check under this section administered by the Administrator or an authorized check pilot.

(c) No pilot may use any type of precision instrument approach procedure under IFR unless, since the beginning of the 6th month before that use, the pilot satisfactorily demonstrated that type of approach procedure. No pilot may use any type of nonprecision approach procedure under IFR unless, since the beginning of the 6th month before that use, the pilot has satisfactorily demonstrated either that type of approach procedure or any other two different types of nonprecision approach procedures. The instrument approach procedure or procedures must include at least one straight-in approach, one circling approach, and one missed approach. Each type of approach procedure demonstrated must be conducted to published minimums for that procedure.

(d) The instrument proficiency checks required by paragraphs (a) and (b) of this section consists of either an oral or written equipment test (or a combination) and a flight check under simulated or actual IFR conditions. The equipment test includes questions on emergency procedures, engine operation, fuel and lubrication systems, power settings, stall speeds, best engine-out speed, propeller and supercharger operations, and hydraulic, mechanical, and electrical systems, as appropriate. The flight check includes navigation by instruments, recovery from simulated emergencies, and standard instrument approaches involving navigational facilities which that pilot is to be authorized to use.

(e) Each pilot taking the instrument proficiency check must show that standard of competence required by § 91.1065(d).

(1) The instrument proficiency check must—

(i) For a pilot in command of an aircraft requiring that the PIC hold an airline transport pilot certificate, include the procedures and maneuvers for an airline transport pilot certificate in the particular type of aircraft, if appropriate; and

(ii) For a pilot in command of a rotorcraft or a second in command of any aircraft requiring that the SIC hold a commercial pilot certificate include the procedures and maneuvers for a commercial pilot certificate with an instrument rating and, if required, for the appropriate type rating.

(2) The instrument proficiency check must be given by an authorized check pilot or by the Administrator.

(f) If the pilot is assigned to pilot only one type of aircraft, that pilot must take the instrument proficiency check required by paragraph (a) of this section in that type of aircraft.

(g) If the pilot in command is assigned to pilot more than one type of aircraft, that pilot must take the instrument proficiency check required by paragraph (a) of this section in each type of aircraft to which that pilot is assigned, in rotation, but not more than one flight check during each period described in paragraph (a) of this section.

(h) If the pilot in command is assigned to pilot both single-engine and multiengine aircraft, that pilot must initially take the instrument proficiency check required by paragraph (a) of this section in a multiengine aircraft, and each succeeding check alternately in single-engine and multiengine aircraft, but not more than one flight check during each period described in paragraph (a) of this section.

(i) All or portions of a required flight check may be given in an aircraft simulator or other appropriate training device, if approved by the Administrator.

§ 91.1071 Crewmember: Tests and checks, grace provisions, training to accepted standards.

(a) If a crewmember who is required to take a test or a flight check under this subpart, completes the test or flight check in the month before or after the month in which it is required, that crewmember is considered to have completed the test or check in the month in which it is required.

(b) If a pilot being checked under this subpart fails any of the required maneuvers, the person giving the check may give additional training to the pilot during the course of the check. In addition to repeating the maneuvers failed, the person giving the check may require the pilot being checked to repeat any other maneuvers that are necessary to determine the pilot's proficiency. If the pilot being checked is unable to demonstrate satisfactory performance to the person conducting the check, the program manager may not use the pilot, nor may the pilot serve, as a flight crewmember in operations under this subpart until the pilot has satisfactorily completed the check. If a pilot who demonstrates unsatisfactory performance is employed as a pilot for a certificate holder operating under part 121, 125, or 135 of this chapter, he or she must notify that certificate holder of the unsatisfactory performance.

§ 91.1073 Training program: General.

(a) Each program manager must have a training program and must:

(1) Establish, obtain the appropriate initial and final approval of, and provide a training program that meets this subpart and that ensures that each crewmember, including each flight attendant if the program manager uses a flight attendant crewmember, flight instructor, check pilot, and each person assigned duties for the carriage and handling of hazardous materials (as defined in 49 CFR 171.8) is adequately trained to perform these assigned duties.

(2) Provide adequate ground and flight training facilities and properly qualified ground instructors for the training required by this subpart.

(3) Provide and keep current for each aircraft type used and, if applicable, the particular variations within the aircraft type, appropriate training material, examinations, forms, instructions, and procedures for use in conducting the training and checks required by this subpart.

(4) Provide enough flight instructors, check pilots, and simulator instructors to conduct required flight training and flight checks, and simulator training courses allowed under this subpart.

(b) Whenever a crewmember who is required to take recurrent training under this subpart completes the training in the month before, or the month after, the month in which that training is required, the crewmember is considered to have completed it in the month in which it was required.

(c) Each instructor, supervisor, or check pilot who is responsible for a particular ground training subject, segment of flight training, course of training, flight check, or competence check under this subpart must certify as to the proficiency and knowledge of the crewmember, flight instructor, or check pilot concerned upon completion of that training or check. That certification must be made a part of the crewmember's record. When the certification required by this paragraph is made by an entry in a computerized recordkeeping system, the certifying instructor, supervisor, or check pilot, must be identified with that entry. However, the signature of the certifying instructor, supervisor, or check pilot is not required for computerized entries.

(d) Training subjects that apply to more than one aircraft or crewmember position and that have been satisfactorily completed during previous training while employed by the program manager for another aircraft or another crewmember position, need not be repeated during subsequent training other than recurrent training.

(e) Aircraft simulators and other training devices may be used in the program manager's training program if approved by the Administrator.

(f) Each program manager is responsible for establishing safe and efficient crew management practices for all phases of flight in program operations including crew resource management training for all crewmembers used in program operations.

(g) If an aircraft simulator has been approved by the Administrator for use in the program manager's training program, the program manager must ensure that each pilot annually completes at least one flight training session in an approved simulator for at least one program aircraft. The training session may be the flight training portion of any of the pilot training or check requirements of this subpart, including the initial, transition, upgrade, requalification, differences, or recurrent training, or the accomplishment of a competency check or instrument proficiency check. If there is no approved simulator for that aircraft type in operation, then all flight training and checking must be accomplished in the aircraft.

§ 91.1075 Training program: Special rules.

Other than the program manager, only the following are eligible under this subpart to conduct training, testing, and checking under contract or other arrangement to those persons subject to the requirements of this subpart.

(a) Another program manager operating under this subpart:

(b) A training center certificated under part 142 of this chapter to conduct training, testing, and checking required by this subpart if the training center—

(1) Holds applicable training specifications issued under part 142 of this chapter;

(2) Has facilities, training equipment, and courseware meeting the applicable requirements of part 142 of this chapter;

(3) Has approved curriculums, curriculum segments, and portions of curriculum segments applicable for use in training courses required by this subpart; and

(4) Has sufficient instructors and check pilots qualified under the applicable requirements of §§ 91.1089 through 91.1095 to conduct training, testing, and checking to persons subject to the requirements of this subpart.

(c) A part 119 certificate holder operating under part 121 or part 135 of this chapter.

(d) As authorized by the Administrator, a training center that is not certificated under part 142 of this chapter.

§ 91.1077 Training program and revision: Initial and final approval.

(a) To obtain initial and final approval of a training program, or a revision to an approved training program, each program manager must submit to the Administrator—

(1) An outline of the proposed or revised curriculum, that provides enough information for a preliminary evaluation of the proposed training program or revision; and

(2) Additional relevant information that may be requested by the Administrator.

(b) If the proposed training program or revision complies with this subpart, the Administrator grants initial approval in writing after which the program manager may conduct the training under that program. The Administrator then evaluates the effectiveness of the training program and advises the program manager of deficiencies, if any, that must be corrected.

(c) The Administrator grants final approval of the proposed training program or revision if the program manager shows that the training conducted under the initial approval in paragraph (b) of this section ensures that each person who successfully completes the training is adequately trained to perform that person's assigned duties.

(d) Whenever the Administrator finds that revisions are necessary for the continued adequacy of a training program that has been granted final approval, the program manager must, after notification by the Administrator, make any changes in the program that are found necessary by the Administrator. Within 30 days after the program manager receives the notice, it may file a petition to reconsider the notice with the Administrator. The filing of a petition to reconsider stays the notice pending a decision by the Administrator. However, if the Administrator finds that there is an emergency that requires immediate action in the interest of safety, the Administrator may, upon a statement of the reasons, require a change effective without stay.

§ 91.1079 Training program: Curriculum.

(a) Each program manager must prepare and keep current a written training program curriculum for each type of aircraft for each crewmember required for that type aircraft. The curriculum must include ground and flight training required by this subpart.

(b) Each training program curriculum must include the following:

(1) A list of principal ground training subjects, including emergency training subjects, that are provided.

(2) A list of all the training devices, mock-ups, systems trainers, procedures trainers, or other training aids that the program manager will use.

(3) Detailed descriptions or pictorial displays of the approved normal, abnormal, and emergency maneuvers, procedures and functions that will be performed during each flight training phase or flight check, indicating those maneuvers, procedures and functions that are to be performed during the inflight portions of flight training and flight checks.

§ 91.1081 Crewmember training requirements.

(a) Each program manager must include in its training program the following initial and transition ground training as appropriate to the particular assignment of the crewmember:

(1) Basic indoctrination ground training for newly hired crewmembers including instruction in at least the—

(i) Duties and responsibilities of crewmembers as applicable;

(ii) Appropriate provisions of this chapter;

(iii) Contents of the program manager's management specifications (not required for flight attendants); and

(iv) Appropriate portions of the program manager's operating manual.

(2) The initial and transition ground training in §§ 91.1101 and 91.1105, as applicable.

(3) Emergency training in § 91.1083.

(b) Each training program must provide the initial and transition flight training in § 91.1103, as applicable.

(c) Each training program must provide recurrent ground and flight training as provided in § 91.1107.

(d) Upgrade training in §§ 91.1101 and 91.1103 for a particular type aircraft may be included in the training program for crewmembers who have qualified and served as second in command on that aircraft.

(e) In addition to initial, transition, upgrade and recurrent training, each training program must provide ground and flight training, instruction, and practice necessary to ensure that each crewmember—

(1) Remains adequately trained and currently proficient for each aircraft, crewmember position, and type of operation in which the crewmember serves; and

(2) Qualifies in new equipment, facilities, procedures, and techniques, including modifications to aircraft.

§ 91.1083 Crewmember emergency training.

(a) Each training program must provide emergency training under this section for each aircraft type, model, and configuration, each crewmember, and each kind of operation conducted, as appropriate for each crewmember and the program manager.

(b) Emergency training must provide the following:

(1) Instruction in emergency assignments and procedures, including coordination among crewmembers.

(2) Individual instruction in the location, function, and operation of emergency equipment including—

(i) Equipment used in ditching and evacuation;

(ii) First aid equipment and its proper use; and

(iii) Portable fire extinguishers, with emphasis on the type of extinguisher to be used on different classes of fires.

(3) Instruction in the handling of emergency situations including—

(i) Rapid decompression;

(ii) Fire in flight or on the surface and smoke control procedures with emphasis on electrical equipment and related circuit breakers found in cabin areas;

(iii) Ditching and evacuation;

(iv) Illness, injury, or other abnormal situations involving passengers or crewmembers; and

(v) Hijacking and other unusual situations.

(4) Review and discussion of previous aircraft accidents and incidents involving actual emergency situations.

(c) Each crewmember must perform at least the following emergency drills, using the proper emergency equipment and procedures, unless the Administrator finds that, for a particular drill, the crewmember can be adequately trained by demonstration:

(1) Ditching, if applicable.

(2) Emergency evacuation.

(3) Fire extinguishing and smoke control.

(4) Operation and use of emergency exits, including deployment and use of evacuation slides, if applicable.

(5) Use of crew and passenger oxygen.

(6) Removal of life rafts from the aircraft, inflation of the life rafts, use of lifelines, and boarding of passengers and crew, if applicable.

(7) Donning and inflation of life vests and the use of other individual flotation devices, if applicable.

(d) Crewmembers who serve in operations above 25,000 feet must receive instruction in the following:

(1) Respiration.

(2) Hypoxia.

(3) Duration of consciousness without supplemental oxygen at altitude.

(4) Gas expansion.

(5) Gas bubble formation.

(6) Physical phenomena and incidents of decompression.

§ 91.1085 Hazardous materials recognition training.

No program manager may use any person to perform, and no person may perform, any assigned duties and responsibilities for the handling or carriage of hazardous materials (as defined in 49 CFR 171.8), unless that person has received training in the recognition of hazardous materials.

§ 91.1087 Approval of aircraft simulators and other training devices.

(a) Training courses using aircraft simulators and other training devices may be included in the program manager's training program if approved by the Administrator.

(b) Each aircraft simulator and other training device that is used in a training course or in checks required under this subpart must meet the following requirements:

(1) It must be specifically approved for—

(i) The program manager; and

(ii) The particular maneuver, procedure, or crewmember function involved.

(2) It must maintain the performance, functional, and other characteristics that are required for approval.

(3) Additionally, for aircraft simulators, it must be—

(i) Approved for the type aircraft and, if applicable, the particular variation within type for which the training or check is being conducted; and

(ii) Modified to conform with any modification to the aircraft being simulated that changes the performance, functional, or other characteristics required for approval.

(c) A particular aircraft simulator or other training device may be used by more than one program manager.

(d) In granting initial and final approval of training programs or revisions to them, the Administrator considers the training devices, methods, and procedures listed in the program manager's curriculum under § 91.1079.

§ 91.1089 Qualifications: Check pilots (aircraft) and check pilots (simulator).

(a) For the purposes of this section and § 91.1093:

(1) A check pilot (aircraft) is a person who is qualified to conduct flight checks in an aircraft, in a flight simulator, or in a flight training device for a particular type aircraft.

(2) A check pilot (simulator) is a person who is qualified to conduct flight checks, but only in a flight simulator, in a flight training device, or both, for a particular type aircraft.

(3) Check pilots (aircraft) and check pilots (simulator) are those check pilots who perform the functions described in § 91.1073(a)(4) and (c).

(b) No program manager may use a person, nor may any person serve as a check pilot (aircraft) in a training program established under this subpart unless, with respect to the aircraft type involved, that person—

(1) Holds the pilot certificates and ratings required to serve as a pilot in command in operations under this subpart;

(2) Has satisfactorily completed the training phases for the aircraft, including recurrent training, that are required to serve as a pilot in command in operations under this subpart;

(3) Has satisfactorily completed the proficiency or competency checks that are required to serve as a pilot in command in operations under this subpart;

(4) Has satisfactorily completed the applicable training requirements of § 91.1093;

(5) Holds at least a Class III medical certificate unless serving as a required crewmember, in which case holds a Class I or Class II medical certificate as appropriate; and

(6) Has been approved by the Administrator for the check pilot duties involved.

(c) No program manager may use a person, nor may any person serve as a check pilot (simulator) in a training program established under this subpart unless, with respect to the aircraft type involved, that person meets the provisions of paragraph (b) of this section, or—

(1) Holds the applicable pilot certificates and ratings, except medical certificate, required to serve as a pilot in command in operations under this subpart;

(2) Has satisfactorily completed the appropriate training phases for the aircraft, including recurrent training, that are required to serve as a pilot in command in operations under this subpart;

(3) Has satisfactorily completed the appropriate proficiency or competency checks that are required to serve as a pilot in command in operations under this subpart;

(4) Has satisfactorily completed the applicable training requirements of § 91.1093; and

(5) Has been approved by the Administrator for the check pilot (simulator) duties involved.

(d) Completion of the requirements in paragraphs (b) (2), (3), and (4) or (c)(2), (3), and (4) of this section, as applicable, must be entered in the individual's training record maintained by the program manager.

(e) A check pilot who does not hold an appropriate medical certificate may function as a check pilot (simulator), but may not serve as a flightcrew member in operations under this subpart.

(f) A check pilot (simulator) must accomplish the following—

(1) Fly at least two flight segments as a required crewmember for the type, class, or category aircraft involved within the 12-month period preceding the performance of any check pilot duty in a flight simulator; or

(2) Before performing any check pilot duty in a flight simulator, satisfactorily complete an approved line-observation program within the period prescribed by that program.

(g) The flight segments or line-observation program required in paragraph (f) of this section are considered to be completed in the month required if completed in the month before or the month after the month in which they are due.

§ 91.1091 Qualifications: Flight instructors (aircraft) and flight instructors (simulator).

(a) For the purposes of this section and § 91.1095:

(1) A flight instructor (aircraft) is a person who is qualified to instruct in an aircraft, in a flight simulator, or in a flight training device for a particular type, class, or category aircraft.

(2) A flight instructor (simulator) is a person who is qualified to instruct in a flight simulator, in a flight training device, or in both, for a particular type, class, or category aircraft.

(3) Flight instructors (aircraft) and flight instructors (simulator) are those instructors who perform the functions described in § 91.1073(a)(4) and (c).

(b) No program manager may use a person, nor may any person serve as a flight instructor (aircraft) in a training program established under this subpart unless, with respect to the type, class, or category aircraft involved, that person—

(1) Holds the pilot certificates and ratings required to serve as a pilot in command in operations under this subpart or part 121 or 135 of this chapter;

(2) Has satisfactorily completed the training phases for the aircraft, including recurrent training, that are required to serve as a pilot in command in operations under this subpart;

(3) Has satisfactorily completed the proficiency or competency checks that are required to serve as a pilot in command in operations under this subpart;

(4) Has satisfactorily completed the applicable training requirements of § 91.1095; and

(5) Holds at least a Class III medical certificate.

(c) No program manager may use a person, nor may any person serve as a flight instructor (simulator) in a training program established under this subpart, unless, with respect to the type, class, or category aircraft involved, that person meets the provisions of paragraph (b) of this section, or—

(1) Holds the pilot certificates and ratings, except medical certificate, required to serve as a pilot in command in operations under this subpart or part 121 or 135 of this chapter;

(2) Has satisfactorily completed the appropriate training phases for the aircraft, including recurrent training, that are required to serve as a pilot in command in operations under this subpart;

(3) Has satisfactorily completed the appropriate proficiency or competency checks that are required to serve as a pilot in command in operations under this subpart; and

(4) Has satisfactorily completed the applicable training requirements of § 91.1095.

(d) Completion of the requirements in paragraphs (b) (2), (3), and (4) or (c)(2), (3), and (4) of this section, as applicable, must be entered in the individual's training record maintained by the program manager.

(e) A pilot who does not hold a medical certificate may function as a flight instructor in an aircraft if functioning as a non-required crewmember, but may not serve as a flightcrew member in operations under this subpart.

(f) A flight instructor (simulator) must accomplish the following—

(1) Fly at least two flight segments as a required crewmember for the type, class, or category aircraft involved within the 12-month period preceding the performance of any flight instructor duty in a flight simulator; or

(2) Satisfactorily complete an approved line-observation program within the period prescribed by that program preceding the performance of any flight instructor duty in a flight simulator.

(g) The flight segments or line-observation program required in paragraph (f) of this section are considered completed in the month required if completed in the month before, or in the month after, the month in which they are due.

[Doc. No. FAA-2001-10047, 68 FR 54561, Sept. 17, 2003, as amended by Amdt. 91-322, 76 FR 31823, June 2, 2011]

§ 91.1093 Initial and transition training and checking: Check pilots (aircraft), check pilots (simulator).

(a) No program manager may use a person nor may any person serve as a check pilot unless—

(1) That person has satisfactorily completed initial or transition check pilot training; and

(2) Within the preceding 24 months, that person satisfactorily conducts a proficiency or competency check under the observation of an FAA inspector or an aircrew designated examiner employed by the program manager. The observation check may be accomplished in part or in full in an aircraft, in a flight simulator, or in a flight training device.

(b) The observation check required by paragraph (a)(2) of this section is considered to have been completed in the month required if completed in the month before or the month after the month in which it is due.

(c) The initial ground training for check pilots must include the following:

(1) Check pilot duties, functions, and responsibilities.

(2) The applicable provisions of the Code of Federal Regulations and the program manager's policies and procedures.

(3) The applicable methods, procedures, and techniques for conducting the required checks.

(4) Proper evaluation of student performance including the detection of—

(i) Improper and insufficient training; and

(ii) Personal characteristics of an applicant that could adversely affect safety.

(5) The corrective action in the case of unsatisfactory checks.

(6) The approved methods, procedures, and limitations for performing the required normal, abnormal, and emergency procedures in the aircraft.

(d) The transition ground training for a check pilot must include the approved methods, procedures, and limitations for performing the required normal, abnormal, and emergency procedures applicable to the aircraft to which the check pilot is in transition.

(e) The initial and transition flight training for a check pilot (aircraft) must include the following—

(1) The safety measures for emergency situations that are likely to develop during a check;

(2) The potential results of improper, untimely, or nonexecution of safety measures during a check;

(3) Training and practice in conducting flight checks from the left and right pilot seats in the required normal, abnormal, and emergency procedures to ensure competence to conduct the pilot flight checks required by this subpart; and

(4) The safety measures to be taken from either pilot seat for emergency situations that are likely to develop during checking.

(f) The requirements of paragraph (e) of this section may be accomplished in full or in part in flight, in a flight simulator, or in a flight training device, as appropriate.

(g) The initial and transition flight training for a check pilot (simulator) must include the following:

(1) Training and practice in conducting flight checks in the required normal, abnormal, and emergency procedures to ensure competence to conduct the flight checks required by this subpart. This training and practice must be accomplished in a flight simulator or in a flight training device.

(2) Training in the operation of flight simulators, flight training devices, or both, to ensure competence to conduct the flight checks required by this subpart.

§ 91.1095 Initial and transition training and checking: Flight instructors (aircraft), flight instructors (simulator).

(a) No program manager may use a person nor may any person serve as a flight instructor unless—

(1) That person has satisfactorily completed initial or transition flight instructor training; and

(2) Within the preceding 24 months, that person satisfactorily conducts instruction under the observation of an FAA inspector, a program manager check pilot, or an aircrew designated examiner employed by the program manager. The observation check may be accomplished in part or in full in an aircraft, in a flight simulator, or in a flight training device.

(b) The observation check required by paragraph (a)(2) of this section is considered to have been completed in the month required if completed in the month before, or the month after, the month in which it is due.

(c) The initial ground training for flight instructors must include the following:

(1) Flight instructor duties, functions, and responsibilities.

(2) The applicable Code of Federal Regulations and the program manager's policies and procedures.

(3) The applicable methods, procedures, and techniques for conducting flight instruction.

(4) Proper evaluation of student performance including the detection of—

(i) Improper and insufficient training; and

(ii) Personal characteristics of an applicant that could adversely affect safety.

(5) The corrective action in the case of unsatisfactory training progress.

(6) The approved methods, procedures, and limitations for performing the required normal, abnormal, and emergency procedures in the aircraft.

(7) Except for holders of a flight instructor certificate—

(i) The fundamental principles of the teaching-learning process;

(ii) Teaching methods and procedures; and

(iii) The instructor-student relationship.

(d) The transition ground training for flight instructors must include the approved methods, procedures, and limitations for performing the required normal, abnormal, and emergency procedures applicable to the type, class, or category aircraft to which the flight instructor is in transition.

(e) The initial and transition flight training for flight instructors (aircraft) must include the following—

(1) The safety measures for emergency situations that are likely to develop during instruction;

(2) The potential results of improper or untimely safety measures during instruction;

(3) Training and practice from the left and right pilot seats in the required normal, abnormal, and emergency maneuvers to ensure competence to conduct the flight instruction required by this subpart; and

(4) The safety measures to be taken from either the left or right pilot seat for emergency situations that are likely to develop during instruction.

(f) The requirements of paragraph (e) of this section may be accomplished in full or in part in flight, in a flight simulator, or in a flight training device, as appropriate.

(g) The initial and transition flight training for a flight instructor (simulator) must include the following:

(1) Training and practice in the required normal, abnormal, and emergency procedures to ensure competence to conduct the flight instruction required by this subpart. These maneuvers and procedures must be accomplished in full or in part in a flight simulator or in a flight training device.

(2) Training in the operation of flight simulators, flight training devices, or both, to ensure competence to conduct the flight instruction required by this subpart.

§ 91.1097 Pilot and flight attendant crewmember training programs.

(a) Each program manager must establish and maintain an approved pilot training program, and each program manager who uses a flight attendant crewmember must establish and maintain an approved flight attendant training program, that is appropriate to the operations to which each pilot and flight attendant is to be assigned, and will ensure that they are adequately trained to meet the applicable knowledge and practical testing requirements of §§ 91.1065 through 91.1071.

(b) Each program manager required to have a training program by paragraph (a) of this section must include in that program ground and flight training curriculums for—

(1) Initial training;

(2) Transition training;

(3) Upgrade training;

(4) Differences training;

(5) Recurrent training; and

(6) Requalification training.

(c) Each program manager must provide current and appropriate study materials for use by each required pilot and flight attendant.

(d) The program manager must furnish copies of the pilot and flight attendant crewmember training program, and all changes and additions, to the assigned representative of the Administrator. If the program manager uses training facilities of other persons, a copy of those training programs or appropriate portions used for those facilities must also be furnished. Curricula that follow FAA published curricula may be cited by reference in the copy of the training program furnished to the representative of the Administrator and need not be furnished with the program.

§ 91.1099 Crewmember initial and recurrent training requirements.

No program manager may use a person, nor may any person serve, as a crewmember in operations under this subpart unless that crewmember has completed the appropriate initial or recurrent training phase of the training program appropriate to the type of operation in which the crewmember is to serve since the beginning of the 12th month before that service.

§ 91.1101 Pilots: Initial, transition, and upgrade ground training.

Initial, transition, and upgrade ground training for pilots must include instruction in at least the following, as applicable to their duties:

(a) General subjects—

(1) The program manager's flight locating procedures;

(2) Principles and methods for determining weight and balance, and runway limitations for takeoff and landing;

(3) Enough meteorology to ensure a practical knowledge of weather phenomena, including the principles of frontal systems, icing, fog, thunderstorms, windshear and, if appropriate, high altitude weather situations;

(4) Air traffic control systems, procedures, and phraseology;

(5) Navigation and the use of navigational aids, including instrument approach procedures;

(6) Normal and emergency communication procedures;

(7) Visual cues before and during descent below Decision Altitude or MDA; and

(8) Other instructions necessary to ensure the pilot's competence.

(b) For each aircraft type—

(1) A general description;

(2) Performance characteristics;

(3) Engines and propellers;

(4) Major components;

(5) Major aircraft systems (that is, flight controls, electrical, and hydraulic), other systems, as appropriate, principles of normal, abnormal, and emergency operations, appropriate procedures and limitations;

(6) Knowledge and procedures for—

(i) Recognizing and avoiding severe weather situations;

(ii) Escaping from severe weather situations, in case of inadvertent encounters, including low-altitude windshear (except that rotorcraft pilots are not required to be trained in escaping from low-altitude windshear);

(iii) Operating in or near thunderstorms (including best penetration altitudes), turbulent air (including clear air turbulence), inflight icing, hail, and other potentially hazardous meteorological conditions; and

(iv) Operating airplanes during ground icing conditions, (that is, any time conditions are such that frost, ice, or snow may reasonably be expected to adhere to the aircraft), if the program manager expects to authorize takeoffs in ground icing conditions, including:

(A) The use of holdover times when using deicing/anti-icing fluids;

(B) Airplane deicing/anti-icing procedures, including inspection and check procedures and responsibilities;

(C) Communications;

(D) Airplane surface contamination (that is, adherence of frost, ice, or snow) and critical area identification, and knowledge of how contamination adversely affects airplane performance and flight characteristics;

(E) Types and characteristics of deicing/anti-icing fluids, if used by the program manager;

(F) Cold weather preflight inspection procedures;

(G) Techniques for recognizing contamination on the airplane;

(7) Operating limitations;

(8) Fuel consumption and cruise control;

(9) Flight planning;

(10) Each normal and emergency procedure; and

(11) The approved Aircraft Flight Manual or equivalent.

§ 91.1103 Pilots: Initial, transition, upgrade, requalification, and differences flight training.

(a) Initial, transition, upgrade, requalification, and differences training for pilots must include flight and practice in each of the maneuvers and procedures contained in each of the curriculums that are a part of the approved training program.

(b) The maneuvers and procedures required by paragraph (a) of this section must be performed in flight, except to the extent that certain maneuvers and procedures may be performed in an aircraft simulator, or an appropriate training device, as allowed by this subpart.

(c) If the program manager's approved training program includes a course of training using an aircraft simulator or other training device, each pilot must successfully complete—

(1) Training and practice in the simulator or training device in at least the maneuvers and procedures in this subpart that are capable of being performed in the aircraft simulator or training device; and

(2) A flight check in the aircraft or a check in the simulator or training device to the level of proficiency of a pilot in command or second in command, as applicable, in at least the maneuvers and procedures that are capable of being performed in an aircraft simulator or training device.

§ 91.1105 Flight attendants: Initial and transition ground training.

Initial and transition ground training for flight attendants must include instruction in at least the following—

(a) General subjects—

(1) The authority of the pilot in command; and

(2) Passenger handling, including procedures to be followed in handling deranged persons or other persons whose conduct might jeopardize safety.

(b) For each aircraft type—

(1) A general description of the aircraft emphasizing physical characteristics that may have a bearing on ditching, evacuation, and inflight emergency procedures and on other related duties;

(2) The use of both the public address system and the means of communicating with other flight crewmembers, including emergency means in the case of attempted hijacking or other unusual situations; and

(3) Proper use of electrical galley equipment and the controls for cabin heat and ventilation.

§ 91.1107 Recurrent training.

(a) Each program manager must ensure that each crewmember receives recurrent training and is adequately trained and currently proficient for the type aircraft and crewmember position involved.

(b) Recurrent ground training for crewmembers must include at least the following:

(1) A quiz or other review to determine the crewmember's knowledge of the aircraft and crewmember position involved.

(2) Instruction as necessary in the subjects required for initial ground training by this subpart, as appropriate, including low-altitude windshear training and training on operating during ground icing conditions, as prescribed in § 91.1097 and described in § 91.1101, and emergency training.

(c) Recurrent flight training for pilots must include, at least, flight training in the maneuvers or procedures in this subpart, except that satisfactory completion of the check required by § 91.1065 within the preceding 12 months may be substituted for recurrent flight training.

§ 91.1109 Aircraft maintenance: Inspection program.

Each program manager must establish an aircraft inspection program for each make and model program aircraft and ensure each aircraft is inspected in accordance with that inspection program.

(a) The inspection program must be in writing and include at least the following information:

(1) Instructions and procedures for the conduct of inspections for the particular make and model aircraft, including necessary tests and checks. The instructions and procedures must set forth in detail the parts and areas of the airframe, engines, propellers, rotors, and appliances, including survival and emergency equipment required to be inspected.

(2) A schedule for performing the inspections that must be accomplished under the inspection program expressed in terms of the time in service, calendar time, number of system operations, or any combination thereof.

(3) The name and address of the person responsible for scheduling the inspections required by the inspection program. A copy of the inspection program must be made available to the person performing inspections on the aircraft and, upon request, to the Administrator.

(b) Each person desiring to establish or change an approved inspection program under this section must submit the inspection program for approval to the Flight Standards District Office that issued the program manager's management specifications. The inspection program must be derived from one of the following programs:

(1) An inspection program currently recommended by the manufacturer of the aircraft, aircraft engines, propellers, appliances, and survival and emergency equipment;

(2) An inspection program that is part of a continuous airworthiness maintenance program currently in use by a person holding an air carrier or operating certificate issued under part 119 of this chapter and operating that make and model aircraft under part 121 or 135 of this chapter;

(3) An aircraft inspection program approved under § 135.419 of this chapter and currently in use under part 135 of this chapter by a person holding a certificate issued under part 119 of this chapter; or

(4) An airplane inspection program approved under § 125.247 of this chapter and currently in use under part 125 of this chapter.

(5) An inspection program that is part of the program manager's continuous airworthiness maintenance program under §§ 91.1411 through 91.1443.

(c) The Administrator may require revision of the inspection program approved under this section in accordance with the provisions of § 91.415.

§ 91.1111 Maintenance training.

The program manager must ensure that all employees who are responsible for maintenance related to program aircraft undergo appropriate initial and annual recurrent training and are competent to perform those duties.

§ 91.1113 Maintenance recordkeeping.

Each fractional ownership program manager must keep (using the system specified in the manual required in § 91.1025) the records specified in § 91.417(a) for the periods specified in § 91.417(b).

§ 91.1115 Inoperable instruments and equipment.

(a) No person may take off an aircraft with inoperable instruments or equipment installed unless the following conditions are met:

(1) An approved Minimum Equipment List exists for that aircraft.

(2) The program manager has been issued management specifications authorizing operations in accordance with an approved Minimum Equipment List. The flight crew must have direct access at all times prior to flight to all of the information contained in the approved Minimum Equipment List through printed or other means approved by the Administrator in the program manager's management specifications. An approved Minimum Equipment List, as authorized by the management specifications, constitutes an approved change to the type design without requiring recertification.

(3) The approved Minimum Equipment List must:

(i) Be prepared in accordance with the limitations specified in paragraph (b) of this section.

(ii) Provide for the operation of the aircraft with certain instruments and equipment in an inoperable condition.

(4) Records identifying the inoperable instruments and equipment and the information required by (a)(3)(ii) of this section must be available to the pilot.

(5) The aircraft is operated under all applicable conditions and limitations contained in the Minimum Equipment List and the management specifications authorizing use of the Minimum Equipment List.

(b) The following instruments and equipment may not be included in the Minimum Equipment List:

(1) Instruments and equipment that are either specifically or otherwise required by the airworthiness requirements under which the airplane is type certificated and that are essential for safe operations under all operating conditions.

(2) Instruments and equipment required by an airworthiness directive to be in operable condition unless the airworthiness directive provides otherwise.

(3) Instruments and equipment required for specific operations by this part.

(c) Notwithstanding paragraphs (b)(1) and (b)(3) of this section, an aircraft with inoperable instruments or equipment may be operated under a special flight permit under §§ 21.197 and 21.199 of this chapter.

(d) A person authorized to use an approved Minimum Equipment List issued for a specific aircraft under part 121, 125, or 135 of this chapter must use that Minimum Equipment List to comply with this section.

§ 91.1411 Continuous airworthiness maintenance program use by fractional ownership program manager.

Fractional ownership program aircraft may be maintained under a continuous airworthiness maintenance program (CAMP) under §§ 91.1413 through 91.1443. Any program manager who elects to maintain the program aircraft using a continuous airworthiness maintenance program must comply with §§ 91.1413 through 91.1443.

§ 91.1413 CAMP: Responsibility for airworthiness.

(a) For aircraft maintained in accordance with a Continuous Airworthiness Maintenance Program, each program manager is primarily responsible for the following:

(1) Maintaining the airworthiness of the program aircraft, including airframes, aircraft engines, propellers, rotors, appliances, and parts.

(2) Maintaining its aircraft in accordance with the requirements of this chapter.

(3) Repairing defects that occur between regularly scheduled maintenance required under part 43 of this chapter.

(b) Each program manager who maintains program aircraft under a CAMP must—

(1) Employ a Director of Maintenance or equivalent position. The Director of Maintenance must be a certificated mechanic with airframe and powerplant ratings who has responsibility for the maintenance program on all program aircraft maintained under a continuous airworthiness maintenance program. This person cannot also act as Chief Inspector.

(2) Employ a Chief Inspector or equivalent position. The Chief Inspector must be a certificated mechanic with airframe and powerplant ratings who has overall responsibility for inspection aspects of the CAMP. This person cannot also act as Director of Maintenance.

(3) Have the personnel to perform the maintenance of program aircraft, including airframes, aircraft engines, propellers, rotors, appliances, emergency equipment and parts, under its manual and this chapter; or make arrangements with another person for the performance of maintenance. However, the program manager must ensure that any maintenance, preventive maintenance, or alteration that is performed by another person is performed under the program manager's operating manual and this chapter.

§ 91.1415 CAMP: Mechanical reliability reports.

(a) Each program manager who maintains program aircraft under a CAMP must report the occurrence or detection of each failure, malfunction, or defect in an aircraft concerning—

(1) Fires during flight and whether the related fire-warning system functioned properly;

(2) Fires during flight not protected by related fire-warning system;

(3) False fire-warning during flight;

(4) An exhaust system that causes damage during flight to the engine, adjacent structure, equipment, or components;

(5) An aircraft component that causes accumulation or circulation of smoke, vapor, or toxic or noxious fumes in the crew compartment or passenger cabin during flight;

(6) Engine shutdown during flight because of flameout;

(7) Engine shutdown during flight when external damage to the engine or aircraft structure occurs;

(8) Engine shutdown during flight because of foreign object ingestion or icing;

(9) Shutdown of more than one engine during flight;

(10) A propeller feathering system or ability of the system to control overspeed during flight;

(11) A fuel or fuel-dumping system that affects fuel flow or causes hazardous leakage during flight;

(12) An unwanted landing gear extension or retraction or opening or closing of landing gear doors during flight;

(13) Brake system components that result in loss of brake actuating force when the aircraft is in motion on the ground;

(14) Aircraft structure that requires major repair;

(15) Cracks, permanent deformation, or corrosion of aircraft structures, if more than the maximum acceptable to the manufacturer or the FAA; and

(16) Aircraft components or systems that result in taking emergency actions during flight (except action to shut down an engine).

(b) For the purpose of this section, *during flight* means the period from the moment the aircraft leaves the surface of the earth on takeoff until it touches down on landing.

(c) In addition to the reports required by paragraph (a) of this section, each program manager must report any other failure, malfunction, or defect in an aircraft that occurs or is detected at any time if, in the manager's opinion, the failure, malfunction, or defect has endangered or may endanger the safe operation of the aircraft.

(d) Each program manager must send each report required by this section, in writing, covering each 24-hour period beginning at 0900 hours local time of each day and ending at 0900 hours local time on the next day to the Flight Standards District Office that issued the program manager's management specifications. Each report of occurrences during a 24-hour period must be mailed or transmitted to that office within the next 72 hours. However, a report that is due on Saturday or Sunday may be mailed or transmitted on the following Monday and one that is due on a holiday may be mailed or transmitted on the next workday. For aircraft operated in areas where mail is not collected, reports may be mailed or transmitted within 72 hours after the aircraft returns to a point where the mail is collected.

(e) The program manager must transmit the reports required by this section on a form and in a manner prescribed by the Administrator, and must include as much of the following as is available:

(1) The type and identification number of the aircraft.

(2) The name of the program manager.

(3) The date.

(4) The nature of the failure, malfunction, or defect.

(5) Identification of the part and system involved, including available information pertaining to type designation of the major component and time since last overhaul, if known.

(6) Apparent cause of the failure, malfunction or defect (for example, wear, crack, design deficiency, or personnel error).

(7) Other pertinent information necessary for more complete identification, determination of seriousness, or corrective action.

(f) A program manager that is also the holder of a type certificate (including a supplemental type certificate), a Parts Manufacturer Approval, or a Technical Standard Order Authorization, or that is the licensee of a type certificate need not report a failure, malfunction, or defect under this section if the failure, malfunction, or defect has been reported by it under § 21.3 of this chapter or under the accident reporting provisions of part 830 of the regulations of the National Transportation Safety Board.

(g) No person may withhold a report required by this section even when not all information required by this section is available.

(h) When the program manager receives additional information, including information from the manufacturer or other agency, concerning a report required by this section, the program manager must expeditiously submit it as a supplement to the first report and reference the date and place of submission of the first report.

§ 91.1417 CAMP: Mechanical interruption summary report.

Each program manager who maintains program aircraft under a CAMP must mail or deliver, before the end of the 10th day of the following month, a summary report of the following occurrences in multiengine aircraft for the preceding month to the Flight Standards District Office that issued the management specifications:

(a) Each interruption to a flight, unscheduled change of aircraft en route, or unscheduled stop or diversion from a route, caused by known or suspected mechanical difficulties or malfunctions that are not required to be reported under § 91.1415.

(b) The number of propeller featherings in flight, listed by type of propeller and engine and aircraft on which it was installed. Propeller featherings for training, demonstration, or flight check purposes need not be reported.

§ 91.1423 CAMP: Maintenance organization.

(a) Each program manager who maintains program aircraft under a CAMP that has its personnel perform any of its maintenance (other than required inspections), preventive maintenance, or alterations, and each person with whom it arranges for the performance of that work, must have an organization adequate to perform the work.

(b) Each program manager who has personnel perform any inspections required by the program manager's manual under § 91.1427(b) (2) or (3), (in this subpart referred to as required inspections), and each person with whom the program manager arranges for the performance of that work, must have an organization adequate to perform that work.

(c) Each person performing required inspections in addition to other maintenance, preventive maintenance, or alterations, must organize the performance of those functions so as to separate the required inspection functions from the other maintenance, preventive maintenance, or alteration functions. The separation must be below the level of administrative control at which overall responsibility for the required inspection functions and other maintenance, preventive maintenance, or alterations is exercised.

§ 91.1425 CAMP: Maintenance, preventive maintenance, and alteration programs.

Each program manager who maintains program aircraft under a CAMP must have an inspection program and a program covering other maintenance, preventive maintenance, or alterations that ensures that—

(a) Maintenance, preventive maintenance, or alterations performed by its personnel, or by other persons, are performed under the program manager's manual;

(b) Competent personnel and adequate facilities and equipment are provided for the proper performance of maintenance, preventive maintenance, or alterations; and

(c) Each aircraft released to service is airworthy and has been properly maintained for operation under this part.

§ 91.1427 CAMP: Manual requirements.

(a) Each program manager who maintains program aircraft under a CAMP must put in the operating manual the chart or description of the program manager's organization required by § 91.1423 and a list of persons with whom it has arranged for the performance of any of its required inspections, and other maintenance, preventive maintenance, or alterations, including a general description of that work.

(b) Each program manager must put in the operating manual the programs required by § 91.1425 that must be followed in performing maintenance, preventive maintenance, or alterations of that program manager's aircraft, including airframes, aircraft engines, propellers, rotors, appliances, emergency equipment, and parts, and must include at least the following:

(1) The method of performing routine and nonroutine maintenance (other than required inspections), preventive maintenance, or alterations.

(2) A designation of the items of maintenance and alteration that must be inspected (required inspections) including at least those that could result in a failure, malfunction, or defect endangering the safe operation of the aircraft, if not performed properly or if improper parts or materials are used.

(3) The method of performing required inspections and a designation by occupational title of personnel authorized to perform each required inspection.

(4) Procedures for the reinspection of work performed under previous required inspection findings (buy-back procedures).

(5) Procedures, standards, and limits necessary for required inspections and acceptance or rejection of the items required to be inspected and for periodic inspection and calibration of precision tools, measuring devices, and test equipment.

(6) Procedures to ensure that all required inspections are performed.

(7) Instructions to prevent any person who performs any item of work from performing any required inspection of that work.

(8) Instructions and procedures to prevent any decision of an inspector regarding any required inspection from being countermanded by persons other than supervisory personnel of the inspection unit, or a person at the level of administrative control that has overall responsibility for the management of both the required inspection functions and the other maintenance, preventive maintenance, or alterations functions.

(9) Procedures to ensure that maintenance (including required inspections), preventive maintenance, or alterations that are not completed because of work interruptions are properly completed before the aircraft is released to service.

(c) Each program manager must put in the manual a suitable system (which may include an electronic or coded system) that provides for the retention of the following information—

(1) A description (or reference to data acceptable to the Administrator) of the work performed;

(2) The name of the person performing the work if the work is performed by a person outside the organization of the program manager; and

(3) The name or other positive identification of the individual approving the work.

(d) For the purposes of this part, the program manager must prepare that part of its manual containing maintenance information and instructions, in whole or in part, in a format acceptable to the Administrator, that is retrievable in the English language.

§ 91.1429 CAMP: Required inspection personnel.

(a) No person who maintains an aircraft under a CAMP may use any person to perform required inspections unless the person performing the inspection is appropriately certificated, properly trained, qualified, and authorized to do so.

(b) No person may allow any person to perform a required inspection unless, at the time the work was performed, the person performing that inspection is under the supervision and control of the chief inspector.

(c) No person may perform a required inspection if that person performed the item of work required to be inspected.

(d) Each program manager must maintain, or must ensure that each person with whom it arranges to perform required inspections maintains, a current listing of persons who have been trained, qualified, and authorized to conduct required inspections. The persons must be identified by name, occupational title, and the inspections that they are authorized to perform. The program manager (or person with whom it arranges to perform its required inspections) must give written information to each person so authorized, describing the extent of that person's responsibilities, authorities, and inspectional limitations. The list must be made available for inspection by the Administrator upon request.

§ 91.1431 CAMP: Continuing analysis and surveillance.

(a) Each program manager who maintains program aircraft under a CAMP must establish and maintain a system for the continuing analysis and surveillance of the performance and effectiveness of its inspection program and the program covering other maintenance, preventive maintenance, and alterations and for the correction of any deficiency in those programs, regardless of whether those programs are carried out by employees of the program manager or by another person.

(b) Whenever the Administrator finds that the programs described in paragraph (a) of this section does not contain adequate procedures and standards to meet this part, the program manager must, after notification by the Administrator, make changes in those programs requested by the Administrator.

(c) A program manager may petition the Administrator to reconsider the notice to make a change in a program. The petition must be filed with the Director, Flight Standards Service, within 30 days after the program manager receives the notice. Except in the case of an emergency requiring immediate action in the interest of safety, the filing of the petition stays the notice pending a decision by the Administrator.

§ 91.1433 CAMP: Maintenance and preventive maintenance training program.

Each program manager who maintains program aircraft under a CAMP or a person performing maintenance or preventive maintenance functions for it must have a training program to ensure that each person (including inspection personnel) who determines the adequacy of work done is fully informed about procedures and techniques and new equipment in use and is competent to perform that person's duties.

§ 91.1435 CAMP: Certificate requirements.

(a) Except for maintenance, preventive maintenance, alterations, and required inspections performed by repair stations located outside the United States certificated under the provisions of part 145 of this chapter, each person who is directly in charge of maintenance, preventive maintenance, or alterations for a CAMP, and each person performing required inspections for a CAMP must hold an appropriate airman certificate.

(b) For the purpose of this section, a person "directly in charge" is each person assigned to a position in which that person is responsible for the work of a shop or station that performs maintenance, preventive maintenance, alterations, or other functions affecting airworthiness. A person who is directly in charge need not physically observe and direct each worker constantly but must be available for consultation and decision on matters requiring instruction or decision from higher authority than that of the person performing the work.

§ 91.1437 CAMP: Authority to perform and approve maintenance.

A program manager who maintains program aircraft under a CAMP may employ maintenance personnel, or make arrangements with other persons to perform maintenance and preventive maintenance as provided in its maintenance manual. Unless properly certificated, the program manager may not perform or approve maintenance for return to service.

§ 91.1439 CAMP: Maintenance recording requirements.

(a) Each program manager who maintains program aircraft under a CAMP must keep (using the system specified in the manual required in § 91.1427) the following records for the periods specified in paragraph (b) of this section:

(1) All the records necessary to show that all requirements for the issuance of an airworthiness release under § 91.1443 have been met.

(2) Records containing the following information:

(i) The total time in service of the airframe, engine, propeller, and rotor.

(ii) The current status of life-limited parts of each airframe, engine, propeller, rotor, and appliance.

(iii) The time since last overhaul of each item installed on the aircraft that are required to be overhauled on a specified time basis.

(iv) The identification of the current inspection status of the aircraft, including the time since the last inspections required by the inspection program under which the aircraft and its appliances are maintained.

(v) The current status of applicable airworthiness directives, including the date and methods of compliance, and, if the airworthiness directive involves recurring action, the time and date when the next action is required.

(vi) A list of current major alterations and repairs to each airframe, engine, propeller, rotor, and appliance.

(b) Each program manager must retain the records required to be kept by this section for the following periods:

(1) Except for the records of the last complete overhaul of each airframe, engine, propeller, rotor, and appliance the records specified in paragraph (a)(1) of this section must be retained until the work is repeated or superseded by other work or for one year after the work is performed.

(2) The records of the last complete overhaul of each airframe, engine, propeller, rotor, and appliance must be retained until the work is superseded by work of equivalent scope and detail.

(3) The records specified in paragraph (a)(2) of this section must be retained as specified unless transferred with the aircraft at the time the aircraft is sold.

(c) The program manager must make all maintenance records required to be kept by this section available for inspection by the Administrator or any representative of the National Transportation Safety Board.

§ 91.1441 CAMP: Transfer of maintenance records.

When a U.S.-registered fractional ownership program aircraft maintained under a CAMP is removed from the list of program aircraft in the management specifications, the program manager must transfer to the purchaser, at the time of the sale, the following records of that aircraft, in plain language form or in coded form that provides for the preservation and retrieval of information in a manner acceptable to the Administrator:

(a) The records specified in § 91.1439(a)(2).

(b) The records specified in § 91.1439(a)(1) that are not included in the records covered by paragraph (a) of this section, except that the purchaser may allow the program manager to keep physical custody of such records. However, custody of records by the program manager does not relieve the purchaser of its responsibility under § 91.1439(c) to make the records available for inspection by the Administrator or any representative of the National Transportation Safety Board.

§ 91.1443 CAMP: Airworthiness release or aircraft maintenance log entry.

(a) No program aircraft maintained under a CAMP may be operated after maintenance, preventive maintenance, or alterations are performed unless qualified, certificated personnel employed by the program manager prepare, or cause the person with whom the program manager arranges for the performance of the maintenance, preventive maintenance, or alterations, to prepare—

(1) An airworthiness release; or

(2) An appropriate entry in the aircraft maintenance log.

(b) The airworthiness release or log entry required by paragraph (a) of this section must—

(1) Be prepared in accordance with the procedure in the program manager's manual;

(2) Include a certification that—

(i) The work was performed in accordance with the requirements of the program manager's manual;

(ii) All items required to be inspected were inspected by an authorized person who determined that the work was satisfactorily completed;

(iii) No known condition exists that would make the aircraft unairworthy;

(iv) So far as the work performed is concerned, the aircraft is in condition for safe operation; and

(3) Be signed by an authorized certificated mechanic.

(c) Notwithstanding paragraph (b)(3) of this section, after maintenance, preventive maintenance, or alterations performed by a repair station certificated under the provisions of part 145 of this chapter, the approval for return to service or log entry required by paragraph (a) of this section may be signed by a person authorized by that repair station.

(d) Instead of restating each of the conditions of the certification required by paragraph (b) of this section, the program manager may state in its manual that the signature of an authorized certificated mechanic or repairman constitutes that certification.

Subpart L—Continued Airworthiness and Safety Improvements

SOURCE: Amdt. 91-297, 72 FR 63410, Nov. 8, 2007, unless otherwise noted.

§ 91.1501 Purpose and definition.

(a) This subpart requires operators to support the continued airworthiness of each airplane. These requirements may include, but are not limited to, revising the inspection program, incorporating design changes, and incorporating revisions to Instructions for Continued Airworthiness.

(b) For purposes of this subpart, the "FAA Oversight Office" is the aircraft certification office or office of the Transport Airplane Directorate with oversight responsibility for the relevant type certificate or supplemental type certificate, as determined by the Administrator.

§ 91.1503 [Reserved]

§ 91.1505 Repairs assessment for pressurized fuselages.

(a) No person may operate an Airbus Model A300 (excluding the -600 series), British Aerospace Model BAC 1-11, Boeing Model, 707, 720, 727, 737 or 747, McDonnell Douglas Model DC-8, DC-9/MD-80 or DC-10, Fokker Model F28, or Lockheed Model L-1011 airplane beyond applicable flight cycle implementation time specified below, or May 25, 2001, whichever occurs later, unless repair assessment guidelines applicable to the fuselage pressure boundary (fuselage skin, door skin, and bulkhead webs) that have been approved by the FAA Aircraft Certification Office (ACO), or office of the Transport Airplane Directorate, having cognizance over the type certificate for the affected airplane are incorporated within its inspection program:

(1) For the Airbus Model A300 (excluding the -600 series), the flight cycle implementation time is:

(i) Model B2: 36,000 flights.

(ii) Model B4-100 (including Model B4-2C): 30,000 flights above the window line, and 36,000 flights below the window line.

(iii) Model B4-200: 25,500 flights above the window line, and 34,000 flights below the window line.

(2) For all models of the British Aerospace BAC 1-11, the flight cycle implementation time is 60,000 flights.

(3) For all models of the Boeing 707, the flight cycle implementation time is 15,000 flights.

(4) For all models of the Boeing 720, the flight cycle implementation time is 23,000 flights.

(5) For all models of the Boeing 727, the flight cycle implementation time is 45,000 flights.

(6) For all models of the Boeing 737, the flight cycle implementation time is 60,000 flights.

(7) For all models of the Boeing 747, the flight cycle implementation time is 15,000 flights.

(8) For all models of the McDonnell Douglas DC-8, the flight cycle implementation time is 30,000 flights.

(9) For all models of the McDonnell Douglas DC-9/MD-80, the flight cycle implementation time is 60,000 flights.

(10) For all models of the McDonnell Douglas DC-10, the flight cycle implementation time is 30,000 flights.

(11) For all models of the Lockheed L-1011, the flight cycle implementation time is 27,000 flights.

(12) For the Fokker F-28 Mark 1000, 2000, 3000, and 4000, the flight cycle implementation time is 60,000 flights.

(b) [Reserved]

[Doc. No. 29104, 65 FR 24125, Apr. 25, 2000; 65 FR 35703, June 5, 2000; 65 FR 50744, Aug. 21, 2000, as amended by Amdt. 91-266, 66 FR 23130, May 7, 2001; Amdt. 91-277, 67 FR 72834, Dec. 9, 2002; Amdt. 91-283, 69 FR 45941, July 30, 2004. Redesignated and amended by Amdt. 91-297, 72 FR 63410, Nov. 8, 2007]

§ 91.1507 Fuel tank system inspection program.

(a) Except as provided in paragraph (g) of this section, this section applies to transport category, turbine-powered airplanes with a type certificate issued after January 1, 1958, that, as a result of original type certification or later increase in capacity, have—

(1) A maximum type-certificated passenger capacity of 30 or more, or

(2) A maximum payload capacity of 7,500 pounds or more.

(b) For each airplane on which an auxiliary fuel tank is installed under a field approval, before June 16, 2008, the operator must submit to the FAA Oversight Office proposed maintenance instructions for the tank that meet the requirements of Special Federal Aviation Regulation No. 88 (SFAR 88) of this chapter.

(c) After December 16, 2008, no operator may operate an airplane identified in paragraph (a) of this section unless the inspection program for that airplane has been revised to include applicable inspections, procedures, and limitations for fuel tank systems.

(d) The proposed fuel tank system inspection program revisions specified in paragraph (c) of this section must be based on fuel tank system Instructions for Continued Airworthiness (ICA) that have been developed in accordance with the applicable provisions of SFAR 88 of this chapter or § 25.1529 and part 25, Appendix H, of this chapter, in effect on June 6, 2001 (including those developed for auxiliary fuel tanks, if any, installed under supplemental type certificates or other design approval) and that have been approved by the FAA Oversight Office.

(e) After December 16, 2008, before returning an airplane to service after any alterations for which fuel tank ICA are developed under SFAR 88, or under § 25.1529 in effect on June 6, 2001, the operator must include in the inspection program for the airplane inspections and procedures for the fuel tank system based on those ICA.

(f) The fuel tank system inspection program changes identified in paragraphs (d) and (e) of this section and any later fuel tank system revisions must be submitted to the Flight Standards District Office (FSDO) responsible for review and approval.

(g) This section does not apply to the following airplane models:

(1) Bombardier CL-44

(2) Concorde

(3) deHavilland D.H. 106 Comet 4C

(4) VFW-Vereinigte Flugtechnische Werk VFW-614

(5) Illyushin Aviation IL 96T

(6) Bristol Aircraft Britannia 305

(7) Handley Page Herald Type 300

(8) Avions Marcel Dassault—Breguet Aviation Mercure 100C

(9) Airbus Caravelle

(10) Lockheed L-300

Subpart M—Special Federal Aviation Regulations

§ 91.1603 Special Federal Aviation Regulation No. 112—Prohibition Against Certain Flights Within the Tripoli (HLLL) Flight Information Region (FIR).

(a) *Applicability.* This section applies to the following persons:

(1) All U.S. air carriers and U.S. commercial operators;

(2) All persons exercising the privileges of an airman certificate issued by the FAA, except when such persons are operating a U.S.-registered aircraft for a foreign air carrier; and

(3) All operators of U.S.-registered civil aircraft, except operators of such aircraft that are foreign air carriers.

(b) *Flight prohibition.* Except as provided in paragraphs (c) and (d) of this section, no person described in paragraph (a) of this section may conduct flight operations within the Tripoli (HLLL) FIR.

(c) *Permitted operations.* This section does not prohibit persons described in paragraph (a) of this section from conducting flight operations within the Tripoli (HLLL) FIR under the following conditions:

(1) Flight operations are conducted under a contract, grant or cooperative agreement with another department, agency, or instrumentality of the United States Government with the approval of the FAA, or by an exemption issued by the Administrator. The FAA will process requests for approval or exemption in a timely manner, with an order of preference first for those operations in support of U.S. government-sponsored activities, second for those operations in support of government-sponsored activities of another State with the support of a U.S. government agency, and third for all other operations.

(2) Flight operations are coordinated with any mechanism established by paragraph 8 of U.N. Security Council Resolution 1973 (2011).

(d) *Emergency situations.* In an emergency that requires immediate decision and action for the safety of the flight, the pilot in command of an aircraft may deviate from this section to the extent required by that emergency. Except for U.S. air carriers and commercial operators that are subject to the requirements of 14 CFR parts 119, 121, 125, or 135, each person who deviates from this section must, within 10 days of the deviation, excluding Saturdays, Sundays, and Federal holidays, submit to the nearest FAA Flight Standards District Office a complete report of the operations of the aircraft involved in the deviation, including a description of the deviation and the reasons for it.

(e) *Expiration.* This Special Federal Aviation Regulation will remain in effect for 3 years from the effective date. The FAA may amend, rescind, or extend this Special Federal Aviation Regulation as necessary.

[Doc. No. FAA-2011-0246, 76 FR 16238, Mar. 23, 2011]

§ 91.1605 Special Federal Aviation Regulation No. 77—Prohibition Against Certain Flights Within the Territory and Airspace of Iraq.

(a) *Applicability.* This rule applies to the following persons:

(1) All U.S. air carriers or commercial operators;

(2) All persons exercising the privileges of an airman certificate issued by the FAA except such persons operating U.S.-registered aircraft for a foreign air carrier; or

(3) All operators of aircraft registered in the United States except where the operator of such aircraft is a foreign air carrier.

(b) *Flight prohibition.* No person may conduct flight operations over or within the territory of Iraq, except as provided in paragraphs (c) and (d) of this section or except as follows:

(1) Overflights of Iraq may be conducted above flight level (FL) 200 subject to the approval of, and in accordance with the conditions established by, the appropriate authorities of Iraq.

(2) Flights departing from the countries adjacent to Iraq whose climb performance will not permit operations above FL200 prior to entering Iraqi airspace may operate at altitudes below FL200 within Iraq to the extent necessary to permit a climb above FL200, subject to the approval of, and in accordance with the conditions established by, the appropriate authorities of Iraq.

(3) Flights originating from or destined to areas outside of Iraq may be operated to or from Erbil International Airport (ORER) or Sulaymaniyah International Airport (ORSU) within the territory of Iraq north of 34°30′ North latitude. Such flights may operate below FL200 only when initiating an arrival to or departure from Erbil International Airport (ORER) or Sulaymaniyah International Airport (ORSU).

(4) Flights departing Erbil and Sulaymaniyah whose climb performance will not permit operation above FL200 prior to entering Iraqi airspace south of the 34°30′ North latitude may operate at altitudes below FL 200 to the extent necessary to permit a climb above FL200.

(5) Prior to conducting the flight operations described in paragraphs (b)(3) and (4) of this section, the operator must obtain a letter of authorization or operations specification, as appropriate, from the Director, Flight Standards Service, AFS-1, which will specify the limitations and conditions under which the operation must be conducted. All flights conducted under paragraphs (b)(3) and (4) of this section are subject to the approval of, and must be conducted in accordance with the conditions established by the appropriate authorities of Iraq.

(c) *Permitted Operations.* This SFAR does not prohibit persons described in paragraph (a) of this section from conducting flight operations within the territory and airspace of Iraq when such operations are authorized either by another agency of the United States Government with the approval of the FAA, or by an exemption granted by the Administrator.

(d) *Emergency situations.* In an emergency that requires immediate decision and action for the safety of the flight, the pilot in command of an aircraft may deviate from this SFAR to the extent required by that emergency. Except for U.S. air carriers or commercial operators that are subject to the requirements of parts 119, 121, or 135, each person who deviates from this rule shall, within ten (10) days of the deviation, excluding Saturdays, Sundays, and Federal holidays, submit to the Flight Standards Service Air Transportation Division (AFS-200) a complete report of the operations of the aircraft involved in the deviation including a description of the deviation and the reasons therefore.

[Doc. No. FAA-2003-14766, 77 FR 72712, Dec. 6, 2012]

Appendix A to Part 91—Category II Operations: Manual, Instruments, Equipment, and Maintenance

1. Category II Manual

(a) *Application for approval.* An applicant for approval of a Category II manual or an amendment to an approved Category II manual must submit the proposed manual or amendment to the Flight Standards District Office having jurisdiction of the area in which the applicant is located. If the application requests an evaluation program, it must include the following:

(1) The location of the aircraft and the place where the demonstrations are to be conducted; and

(2) The date the demonstrations are to commence (at least 10 days after filing the application).

(b) *Contents.* Each Category II manual must contain:

(1) The registration number, make, and model of the aircraft to which it applies;

(2) A maintenance program as specified in section 4 of this appendix; and

(3) The procedures and instructions related to recognition of decision height, use of runway visual range information, approach monitoring, the decision region (the region between the middle marker and the decision height), the maximum permissible deviations of the basic ILS indicator within the decision region, a missed approach, use of airborne low approach equipment, minimum altitude for the use of the autopilot, instrument and equipment failure warning systems, instrument failure, and other procedures, instructions, and limitations that may be found necessary by the Administrator.

2. Required Instruments and Equipment

The instruments and equipment listed in this section must be installed in each aircraft operated in a Category II operation. This section does not require duplication of instruments and equipment required by § 91.205 or any other provisions of this chapter.

(a) *Group I.* (1) Two localizer and glide slope receiving systems. Each system must provide a basic ILS display and each side of the instrument panel must have a basic ILS display. However, a single localizer antenna and a single glide slope antenna may be used.

(2) A communications system that does not affect the operation of at least one of the ILS systems.

(3) A marker beacon receiver that provides distinctive aural and visual indications of the outer and the middle markers.

(4) Two gyroscopic pitch and bank indicating systems.

(5) Two gyroscopic direction indicating systems.

(6) Two airspeed indicators.

(7) Two sensitive altimeters adjustable for barometric pressure, each having a placarded correction for altimeter scale error and for the wheel height of the aircraft. After June 26, 1979, two sensitive altimeters adjustable for barometric pressure, having markings at 20-foot intervals and each having a placarded correction for altimeter scale error and for the wheel height of the aircraft.

(8) Two vertical speed indicators.

(9) A flight control guidance system that consists of either an automatic approach coupler or a flight director system. A flight director system must display computed information as steering command in relation to an ILS localizer and, on the same instrument, either computed information as pitch command in relation to an ILS glide slope or basic ILS glide slope information. An automatic approach coupler must provide at least automatic steering in relation to an ILS localizer. The flight control guidance system may be operated from one of the receiving systems required by subparagraph (1) of this paragraph.

(10) For Category II operations with decision heights below 150 feet either a marker beacon receiver providing aural and visual indications of the inner marker or a radio altimeter.

(b) *Group II.* (1) Warning systems for immediate detection by the pilot of system faults in items (1), (4), (5), and (9) of Group I and, if installed for use in Category III operations, the radio altimeter and autothrottle system.

(2) Dual controls.

(3) An externally vented static pressure system with an alternate static pressure source.

(4) A windshield wiper or equivalent means of providing adequate cockpit visibility for a safe visual transition by either pilot to touchdown and rollout.

(5) A heat source for each airspeed system pitot tube installed or an equivalent means of preventing malfunctioning due to icing of the pitot system.

3. Instruments and Equipment Approval

(a) *General.* The instruments and equipment required by section 2 of this appendix must be approved as provided in this section before being used in Category II operations. Before presenting an aircraft for approval of the instruments and equipment, it must be shown that since the beginning of the 12th calendar month before the date of submission—

(1) The ILS localizer and glide slope equipment were bench checked according to the manufacturer's instructions and found to meet those standards specified in RTCA Paper 23-63/DO-117 dated March 14, 1963, "Standard Adjustment Criteria for Airborne Localizer and Glide Slope Receivers," which may be obtained from the RTCA Secretariat, 1425 K St., NW., Washington, DC 20005.

(2) The altimeters and the static pressure systems were tested and inspected in accordance with appendix E to part 43 of this chapter; and

(3) All other instruments and items of equipment specified in section 2(a) of this appendix that are listed in the proposed maintenance program were bench checked and found to meet the manufacturer's specifications.

(b) *Flight control guidance system.* All components of the flight control guidance system must be approved as installed by the evaluation program specified in paragraph (e) of this section if they have not been approved for Category III operations under applicable type or supplemental type certification procedures. In addition, subsequent changes to make, model, or design of the components must be approved under this paragraph. Related systems or devices, such as the autothrottle and computed missed approach guidance system, must be approved in the same manner if they are to be used for Category II operations.

(c) *Radio altimeter.* A radio altimeter must meet the performance criteria of this paragraph for original approval and after each subsequent alteration.

(1) It must display to the flight crew clearly and positively the wheel height of the main landing gear above the terrain.

(2) It must display wheel height above the terrain to an accuracy of plus or minus 5 feet or 5 percent, whichever is greater, under the following conditions:

(i) Pitch angles of zero to plus or minus 5 degrees about the mean approach attitude.

(ii) Roll angles of zero to 20 degrees in either direction.

(iii) Forward velocities from minimum approach speed up to 200 knots.

(iv) Sink rates from zero to 15 feet per second at altitudes from 100 to 200 feet.

(3) Over level ground, it must track the actual altitude of the aircraft without significant lag or oscillation.

(4) With the aircraft at an altitude of 200 feet or less, any abrupt change in terrain representing no more than 10 percent of the aircraft's altitude must not cause the altimeter to unlock, and indicator response to such changes must not exceed 0.1 seconds and, in addition, if the system unlocks for greater changes, it must reacquire the signal in less than 1 second.

(5) Systems that contain a push-to-test feature must test the entire system (with or without an antenna) at a simulated altitude of less than 500 feet.

(6) The system must provide to the flight crew a positive failure warning display any time there is a loss of power or an absence of ground return signals within the designed range of operating altitudes.

(d) *Other instruments and equipment.* All other instruments and items of equipment required by § 2 of this appendix must be capable of performing as necessary for Category II operations. Approval is also required after each subsequent alteration to these instruments and items of equipment.

(e) *Evaluation program* —(1) *Application.* Approval by evaluation is requested as a part of the application for approval of the Category II manual.

(2) *Demonstrations.* Unless otherwise authorized by the Administrator, the evaluation program for each aircraft requires the demonstrations specified in this paragraph. At least 50 ILS approaches must be flown with at least five approaches on each of three different ILS facilities and no more than one half of the total approaches on any one ILS facility. All approaches shall be flown under simulated instrument conditions to a 100-foot decision height and 90 percent of the total approaches made must be successful. A successful approach is one in which—

(i) At the 100-foot decision height, the indicated airspeed and heading are satisfactory for a normal flare and landing (speed must be plus or minus 5 knots of programmed airspeed, but may not be less than computed threshold speed if autothrottles are used);

(ii) The aircraft at the 100-foot decision height, is positioned so that the cockpit is within, and tracking so as to remain within, the lateral confines of the runway extended;

(iii) Deviation from glide slope after leaving the outer marker does not exceed 50 percent of full-scale deflection as displayed on the ILS indicator;

(iv) No unusual roughness or excessive attitude changes occur after leaving the middle marker; and

(v) In the case of an aircraft equipped with an approach coupler, the aircraft is sufficiently in trim when the approach coupler is disconnected at the decision height to allow for the continuation of a normal approach and landing.

(3) *Records.* During the evaluation program the following information must be maintained by the applicant for the aircraft with respect to each approach and made available to the Administrator upon request:

(i) Each deficiency in airborne instruments and equipment that prevented the initiation of an approach.

(ii) The reasons for discontinuing an approach, including the altitude above the runway at which it was discontinued.

(iii) Speed control at the 100-foot decision height if auto throttles are used.

(iv) Trim condition of the aircraft upon disconnecting the auto coupler with respect to continuation to flare and landing.

(v) Position of the aircraft at the middle marker and at the decision height indicated both on a diagram of the basic ILS display and a diagram of the runway extended to the middle marker. Estimated touchdown point must be indicated on the runway diagram.

(vi) Compatibility of flight director with the auto coupler, if applicable.

(vii) Quality of overall system performance.

(4) *Evaluation.* A final evaluation of the flight control guidance system is made upon successful completion of the demonstrations. If no hazardous tendencies have been displayed or are otherwise known to exist, the system is approved as installed.

4. Maintenance program

(a) Each maintenance program must contain the following:

(1) A list of each instrument and item of equipment specified in § 2 of this appendix that is installed in the aircraft and approved for Category II operations, including the make and model of those specified in § 2(a).

(2) A schedule that provides for the performance of inspections under subparagraph (5) of this paragraph within 3 calendar months after the date of the previous inspection. The inspection must be performed by a person authorized by part 43 of this chapter, except that each alternate inspection may be replaced by a functional flight check. This functional flight check must be performed by a pilot holding a Category II pilot authorization for the type aircraft checked.

(3) A schedule that provides for the performance of bench checks for each listed instrument and item of equipment that is specified in section 2(a) within 12 calendar months after the date of the previous bench check.

(4) A schedule that provides for the performance of a test and inspection of each static pressure system in accordance with appendix E to part 43 of this chapter within 12 calendar months after the date of the previous test and inspection.

(5) The procedures for the performance of the periodic inspections and functional flight checks to determine the ability of each listed instrument and item of equipment specified in section 2(a) of this appendix to perform as approved for Category II operations including a procedure for recording functional flight checks.

(6) A procedure for assuring that the pilot is informed of all defects in listed instruments and items of equipment.

(7) A procedure for assuring that the condition of each listed instrument and item of equipment upon which maintenance is performed is at least equal to its Category II approval condition before it is returned to service for Category II operations.

(8) A procedure for an entry in the maintenance records required by § 43.9 of this chapter that shows the date, airport, and reasons for each discontinued Category II operation because of a malfunction of a listed instrument or item of equipment.

(b) *Bench check.* A bench check required by this section must comply with this paragraph.

(1) It must be performed by a certificated repair station holding one of the following ratings as appropriate to the equipment checked:

(i) An instrument rating.

(ii) A radio rating.

(2) It must consist of removal of an instrument or item of equipment and performance of the following:

(i) A visual inspection for cleanliness, impending failure, and the need for lubrication, repair, or replacement of parts;

(ii) Correction of items found by that visual inspection; and

(iii) Calibration to at least the manufacturer's specifications unless otherwise specified in the approved Category II manual for the aircraft in which the instrument or item of equipment is installed.

(c) *Extensions.* After the completion of one maintenance cycle of 12 calendar months, a request to extend the period for checks, tests, and inspections is approved if it is shown that the performance of particular equipment justifies the requested extension.

[Doc. No. 18334, 54 FR 34325, Aug. 18, 1989, as amended by Amdt. 91-269, 66 FR 41116, Aug. 6, 2001]

Appendix B to Part 91—Authorizations To Exceed Mach 1 (§ 91.817)

Section 1. Application

(a) An applicant for an authorization to exceed Mach 1 must apply in a form and manner prescribed by the Administrator and must comply with this appendix.

(b) In addition, each application for an authorization to exceed Mach 1 covered by section 2(a) of this appendix must contain all information requested by the Administrator necessary to assist him in determining whether the designation of a particular test area or issuance of a particular authorization is a "major Federal action significantly affecting the quality of the human environment" within the meaning of the National Environmental Policy Act of 1969 (42 U.S.C. 4321 *et seq.*), and to assist him in complying with that act and with related Executive Orders, guidelines, and orders prior to such action.

(c) In addition, each application for an authorization to exceed Mach 1 covered by section 2(a) of this appendix must contain—

(1) Information showing that operation at a speed greater than Mach 1 is necessary to accomplish one or more of the purposes specified in section 2(a) of this appendix, including a showing that the purpose of the test cannot be safely or properly accomplished by overocean testing;

(2) A description of the test area proposed by the applicant, including an environmental analysis of that area meeting the requirements of paragraph (b) of this section; and

(3) Conditions and limitations that will ensure that no measurable sonic boom overpressure will reach the surface outside of the designated test area.

(d) An application is denied if the Administrator finds that such action is necessary to protect or enhance the environment.

Section 2. Issuance

(a) For a flight in a designated test area, an authorization to exceed Mach 1 may be issued when the Administrator has taken the environmental protective actions specified in section 1(b) of this appendix and the applicant shows one or more of the following:

(1) The flight is necessary to show compliance with airworthiness requirements.

(2) The flight is necessary to determine the sonic boom characteristics of the airplane or to establish means of reducing or eliminating the effects of sonic boom.

(3) The flight is necessary to demonstrate the conditions and limitations under which speeds greater than a true flight Mach number of 1 will not cause a measurable sonic boom overpressure to reach the surface.

(b) For a flight outside of a designated test area, an authorization to exceed Mach 1 may be issued if the applicant shows conservatively under paragraph (a)(3) of this section that—

(1) The flight will not cause a measurable sonic boom overpressure to reach the surface when the aircraft is operated under conditions and limitations demonstrated under paragraph (a)(3) of this section; and

(2) Those conditions and limitations represent all foreseeable operating conditions.

Section 3. Duration

(a) An authorization to exceed Mach 1 is effective until it expires or is surrendered, or until it is suspended or terminated by the Administrator. Such an authorization may be amended or suspended by the Administrator at any time if the Administrator finds that such action is necessary to protect the environment. Within 30 days of notification of amendnent, the holder of the authorization must request reconsideration or the amendment becomes final. Within 30 days of notification of suspension, the holder of the authorization must request reconsideration or the authorization is automatically terminated. If reconsideration is requested within the 30-day period, the amendment or suspension continues until the holder shows why the authorization should not be amended or terminated. Upon such showing, the Administrator may terminate or amend the authorization if the Administrator finds that such action is necessary to protect the environment, or he may reinstate the authorization without amendment if he finds that termination or amendnent is not necessary to protect the environment.

(b) Findings and actions by the Administrator under this section do not affect any certificate issued under title VI of the Federal Aviation Act of 1958.

[Doc. No. 18334, 54 FR 34327, Aug. 18, 1989]

Appendix C to Part 91—Operations in the North Atlantic (NAT) Minimum Navigation Performance Specifications (MNPS) Airspace

Section 1

NAT MNPS airspace is that volume of airspace between FL 285 and FL 420 extending between latitude 27 degrees north and the North Pole, bounded in the east by the eastern boundaries of control areas Santa Maria Oceanic, Shanwick Oceanic, and Reykjavik Oceanic and in the west by the western boundary of Reykjavik Oceanic Control Area, the western boundary of Gander Oceanic Control Area, and the western boundary of New York Oceanic Control Area, excluding the areas west of 60 degrees west and south of 38 degrees 30 minutes north.

Section 2

The navigation performance capability required for aircraft to be operated in the airspace defined in section 1 of this appendix is as follows:

(a) The standard deviation of lateral track errors shall be less than 6.3 NM (11.7 Km). Standard deviation is a statistical measure of data about a mean value. The mean is zero nautical miles. The overall form of data is such that the plus and minus 1 standard deviation about the mean encompasses approximately 68 percent of the data and plus or minus 2 deviations encompasses approximately 95 percent.

(b) The proportion of the total flight time spent by aircraft 30 NM (55.6 Km) or more off the cleared track shall be less than 5.3×10^{-4} (less than 1 hour in 1,887 flight hours).

(c) The proportion of the total flight time spent by aircraft between 50 NM and 70 NM (92.6 Km and 129.6 Km) off the cleared track shall be less than 13×10^{-5} (less than 1 hour in 7,693 flight hours.)

Section 3

Air traffic control (ATC) may authorize an aircraft operator to deviate from the requirements of § 91.705 for a specific flight if, at the time of flight plan filing for that flight, ATC determines that the aircraft may be provided appropriate separation and that the flight will not interfere with, or impose a burden upon, the operations of other aircraft which meet the requirements of § 91.705.

[Doc. No. 18334, 54 FR 34327, Aug. 18, 1989, as amended by Amdt. 91-254, 62 FR 17487, Apr. 9, 1997]

Appendix D to Part 91—Airports/Locations: Special Operating Restrictions

Section 1. Locations at which the requirements of § 91.215(b)(2) and § 91.225(d)(2) apply. The requirements of §§ 91.215(b)(2) and 91.225(d)(2) apply below 10,000 feet MSL within a 30-nautical-mile radius of each location in the following list.

Atlanta, GA (The William B. Hartsfield Atlanta International Airport)
Baltimore, MD (Baltimore Washington International Airport)
Boston, MA (General Edward Lawrence Logan International Airport)
Chantilly, VA (Washington Dulles International Airport)
Charlotte, NC (Charlotte/Douglas International Airport)
Chicago, IL Chicago-O'Hare International Airport)
Cleveland, OH (Cleveland-Hopkins International Airport)
Covington, KY (Cincinnati Northern Kentucky International Airport)
Dallas, TX (Dallas/Fort Worth Regional Airport)
Denver, CO (Denver International Airport)
Detroit, MI (Metropolitan Wayne County Airport)
Honolulu, HI (Honolulu International Airport)
Houston, TX (George Bush Intercontinental Airport/Houston)
Kansas City, KS (Mid-Continent International Airport)
Las Vegas, NV (McCarran International Airport)
Los Angeles, CA (Los Angeles International Airport)
Memphis, TN (Memphis International Airport)
Miami, FL (Miami International Airport)
Minneapolis, MN (Minneapolis-St. Paul International Airport)
Newark, NJ (Newark International Airport)
New Orleans, LA (New Orleans International Airport-Moisant Field)
New York, NY (John F. Kennedy International Airport)
New York, NY (LaGuardia Airport)
Orlando, FL (Orlando International Airport)
Philadelphia, PA (Philadelphia International Airport)
Phoenix, AZ (Phoenix Sky Harbor International Airport)
Pittsburgh, PA (Greater Pittsburgh International Airport)
St. Louis, MO (Lambert-St. Louis International Airport)
Salt Lake City, UT (Salt Lake City International Airport)
San Diego, CA (San Diego International Airport)
San Francisco, CA (San Francisco International Airport)
Seattle, WA (Seattle-Tacoma International Airport)
Tampa, FL (Tampa International Airport)
Washington, DC (Ronald Reagan Washington National Airport and Andrews Air Force Base, MD)

Section 2. Airports at which the requirements of § 91.215(b)(5)(ii) apply. [Reserved]

Section 3. Locations at which fixed-wing Special VFR operations are prohibited.

The Special VFR weather minimums of § 91.157 do not apply to the following airports:

Atlanta, GA (The William B. Hartsfield Atlanta International Airport)
Baltimore, MD (Baltimore/Washington International Airport)
Boston, MA (General Edward Lawrence Logan International Airport)
Buffalo, NY (Greater Buffalo International Airport)
Chicago, IL (Chicago-O'Hare International Airport)
Cleveland, OH (Cleveland-Hopkins International Airport)
Columbus, OH (Port Columbus International Airport)
Covington, KY (Cincinnati Northern Kentucky International Airport)
Dallas, TX (Dallas/Fort Worth Regional Airport)
Dallas, TX (Love Field)
Denver, CO (Denver International Airport)
Detroit, MI (Metropolitan Wayne County Airport)
Honolulu, HI (Honolulu International Airport)
Houston, TX (George Bush Intercontinental Airport/Houston)
Indianapolis, IN (Indianapolis International Airport)
Los Angeles, CA (Los Angeles International Airport)
Louisville, KY (Standiford Field)
Memphis, TN (Memphis International Airport)
Miami, FL (Miami International Airport)
Minneapolis, MN (Minneapolis-St. Paul International Airport)
Newark, NJ (Newark International Airport)
New York, NY (John F. Kennedy International Airport)
New York, NY (LaGuardia Airport)
New Orleans, LA (New Orleans International Airport-Moisant Field)
Philadelphia, PA (Philadelphia International Airport)
Pittsburgh, PA (Greater Pittsburgh International Airport)
Portland, OR (Portland International Airport)
San Francisco, CA (San Francisco International Airport)
Seattle, WA (Seattle-Tacoma International Airport)
St. Louis, MO (Lambert-St. Louis International Airport)
Tampa, FL (Tampa International Airport)
Washington, DC (Ronald Reagan Washington National Airport and Andrews Air Force Base, MD)

Section 4. Locations at which solo student, sport, and recreational pilot activity is not permitted.

Pursuant to § 91.131(b)(2), solo student, sport, and recreational pilot operations are not permitted at any of the following airports.

Atlanta, GA (The William B. Hartsfield Atlanta International Airport)
Boston, MA (General Edward Lawrence Logan International Airport)
Chicago, IL (Chicago-O'Hare International Airport)
Dallas, TX (Dallas/Fort Worth Regional Airport)
Los Angeles, CA (Los Angeles International Airport)
Miami, FL (Miami International Airport)
Newark, NJ (Newark International Airport)
New York, NY (John F. Kennedy International Airport)
New York, NY (LaGuardia Airport)
San Francisco, CA (San Francisco International Airport)
Washington, DC (Ronald Reagan Washington National Airport)
Andrews Air Force Base, MD

[Amdt. 91-227, 56 FR 65661, Dec. 17, 1991, as amended by Amdt. 91-235, 58 FR 51968, Oct. 5, 1993; Amdt. 91-236, 59 FR 2918, Jan. 19, 1994; Amdt. 91-237, 59 FR 6547, Feb. 11, 1994; 59 FR 37667, July 25, 1994; Amdt. 91-258, 64 FR 66769, Nov. 30, 1999; Amdt. 91-278, 68 FR 9795, Feb. 28, 2003; Amdt. 91-282, 69 FR 44882, July 27, 2004; Amdt. 91-314, 75 FR 30195, May 28, 2010; Amdt. 91-319, 75 FR 61613, Oct. 6, 2010]

Effective Date Note: By Amdt. 91-236, 59 FR 2918, Jan. 19, 1994, as corrected by Amdt. 91-237, 59 FR 6547, Feb. 11, 1994, appendix D to part 91 was amended in sections 1 and 3 in the Denver, CO, entry by revising "Stapleton" to read "Denver" effective Mar. 9, 1994. By Amdt. 91-238, 59 FR 10958, Mar. 9, 1994, the effective date was delayed to May 15, 1994. By Amdt. 91-241, 59 FR 24916, May 13, 1994, the effective date was suspended indefinitely.

Appendix E to Part 91—Airplane Flight Recorder Specifications

Link to an amendment published at 78 FR 39971, July 3, 2013.

Parameters	Range	Installed system1minimum accuracy (to recovered data)	Sampling interval (per second)	Resolution[4] read out
Relative Time (From Recorded on Prior to Takeoff)	8 hr minimum	±0.125% per hour	1	1 sec.
Indicated Airspeed	Vso to VD (KIAS)	±5% or ±10 kts., whichever is greater. Resolution 2 kts. below 175 KIAS	1	1%[3]
Altitude	-1,000 ft. to max cert. alt. of A/C	±100 to ±700 ft. (see Table 1, TSO C51-a)	11	25 to 150 ft.
Magnetic Heading	360°	±5°	1	1°
Vertical Acceleration	-3g to +6g	±0.2g in addition to ±0.3g maximum datum	4 (or 1 per second where peaks, ref. to 1g are recorded)	0.03g.
Longitudinal Acceleration	±1.0g	±1.5% max. range excluding datum error of ±5%	2	0.01g.
Pitch Attitude	100% of usable	±2°	1	0.8°
Roll Attitude	±60° or 100% of usable range, whichever is greater	±2°	1	0.8°
Stabilizer Trim Position, or	Full Range	±3% unless higher uniquely required	1	1%[3]
Pitch Control Position[5]				
Engine Power, Each Engine:	Full Range	±3% unless higher uniquely required	1	1%[3]
Fan or N[1]Speed or EPR or Cockpit indications Used for Aircraft Certification OR	Maximum Range	±5%	1	1%[3]
Prop. speed and Torque (Sample Once/Sec as Close together as Practicable)			1 (prop Speed) 1 (torque)	1%[3] 1%[3]
Altitude Rate[2](need depends on altitude resolution)	±8,000 fpm	±10%. Resolution 250 fpm below 12,000 ft. indicated	1	250 fpm. below 12,000
Angle of Attack[2](need depends on altitude resolution)	-20° to 40° or 100% of usable range	±2°	1	0.8%[3]
Radio Transmitter Keying (Discrete)	On/Off		1	
TE Flaps (Discrete or Analog)	Each discrete position (U, D, T/O, AAP) OR		1	
LE Flaps (Discrete or Analog)	Analog 0–100% range	±3%	1	1%[3]
	Each discrete position (U, D, T/O, AAP) OR		1	
Thrust Reverser, Each Engine (Discrete)	Analog 0–100% range	±3°	1	1%[3]
	Stowed or full reverse			
Spoiler/ Speedbrake (Discrete)	Stowed or out		1	
Autopilot Engaged (Discrete)	Engaged or Disengaged		1	

[1] When data sources are aircraft instruments (except altimeters) of acceptable quality to fly the aircraft the recording system excluding these sensors (but including all other characteristics of the recording system) shall contribute no more than half of the values in this column.

[2] If data from the altitude encoding altimeter (100 ft. resolution) is used, then either one of these parameters should also be recorded. If however, altitude is recorded at a minimum resolution of 25 feet, then these two parameters can be omitted.

[3] Per cent of full range.

[4] This column applies to aircraft manufactured after October 11, 1991.

[5] For Pitch Control Position only, for all aircraft manufactured on or after April 6, 2012, the sampling interval (per second) is 8. Each input must be recorded at this rate. Alternately sampling inputs (interleaving) to meet this sampling interval is prohibited.

[Doc. No. 18334, 54 FR 34327, Aug. 18, 1989, as amended by Amdt. 91-300, 73 FR 12565, Mar. 7, 2008; 73 FR 15280, Mar. 21, 2008; Amdt. 91-313, 75 FR 17046, Apr. 5, 2010]

Appendix F to Part 91—Helicopter Flight Recorder Specifications

Parameters	Range	Installed system1minimum accuracy (to recovered data)	Sampling interval (per second)	Resolution 3 read out
Relative Time (From Recorded on Prior to Takeoff)	4 hr minimum	±0.125% per hour	1	1 sec.
Indicated Airspeed	VM in to VD (KIAS) (minimum airspeed signal attainable with installed pilot-static system)	±5% or ±10 kts., whichever is greater	1	1 kt.
Altitude	-1,000 ft. to 20,000 ft. pressure altitude	±100 to ±700 ft. (see Table 1, TSO C51–a)	1	25 to 150 ft.
Magnetic Heading	360°	±5°	1	1°
Vertical Acceleration	-3g to +6g	±0.2g in addition to ±0.3g maximum datum	4 (or 1 per second where peaks, ref. to 1g are recorded)	0.05g.
Longitudinal Acceleration	±1.0g	±1.5% max. range excluding datum error of ±5%	2	0.03g.
Pitch Attitude	100% of usable range	±2°	1	0.8°
Roll Attitude	±60 or 100% of usable range, whichever is greater	±2°	1	0.8°
Altitude Rate	±8,000 fpm	±10% Resolution 250 fpm below 12,000 ft. indicated	1	250 fpm below 12,000.
Engine Power, Each Engine				
Main Rotor Speed	Maximum Range	±5%	1	1%2.
Free or Power Turbine	Maximum Range	±5%	1	1%2.
Engine Torque	Maximum Range	±5%	1	1%2.

Parameters	Range	Installed system1minimum accuracy (to recovered data)	Sampling interval (per second)	Resolution 3 read out
Flight Control Hydraulic Pressure				
Primary (Discrete)	High/Low		1	
Secondary—if applicable (Discrete)	High/Low		1	
Radio Transmitter Keying (Discrete)	On/Off		1	
Autopilot Engaged (Discrete)	Engaged or Disengaged		1	
SAS Status-Engaged (Discrete)	Engaged or Disengaged		1	
SAS Fault Status (Discrete)	Fault/OK		1	
Flight Controls				
Collective[4]	Full range	±3%	2	1%2.
Pedal Position[4]	Full range	±3%	2	1%2.
Lat. Cyclic[4]	Full range	±3%	2	1%2.
Long. Cyclic[4]	Full range	±3%	2	1%2.
Controllable Stabilator Position[4]	Full range	±3%	2	1%2.

[1] When data sources are aircraft instruments (except altimeters) of acceptable quality to fly the aircraft the recording system excluding these sensors (but including all other characteristics of the recording system) shall contribute no more than half of the values in this column.

[2] Per cent of full range.

[3] This column applies to aircraft manufactured after October 11, 1991.

[4] For all aircraft manufactured on or after April 6, 2012, the sampling interval per second is 4.

[Doc. No. 18334, 54 FR 34328, Aug. 18, 1989; 54 FR 41211, Oct. 5, 1989; 54 FR 53036, Dec. 26, 1989; Amdt. 91-300, 73 FR 12565, Mar. 7, 2008; 73 FR 15280, Mar. 21, 2008; Amdt. 91-313, 75 FR 17046, Apr. 5, 2010]

Appendix G to Part 91—Operations in Reduced Vertical Separation Minimum (RVSM) Airspace

Section 1. Definitions

Reduced Vertical Separation Minimum (RVSM) Airspace. Within RVSM airspace, air traffic control (ATC) separates aircraft by a minimum of 1,000 feet vertically between flight level (FL) 290 and FL 410 inclusive. RVSM airspace is special qualification airspace; the operator and the aircraft used by the operator must be approved by the Administrator. Air-traffic control notifies operators of RVSM by providing route planning information. Section 8 of this appendix identifies airspace where RVSM may be applied.

RVSM Group Aircraft. Aircraft within a group of aircraft, approved as a group by the Administrator, in which each of the aircraft satisfy each of the following:

(a) The aircraft have been manufactured to the same design, and have been approved under the same type certificate, amended type certificate, or supplemental type certificate.

(b) The static system of each aircraft is installed in a manner and position that is the same as those of the other aircraft in the group. The same static source error correction is incorporated in each aircraft of the group.

(c) The avionics units installed in each aircraft to meet the minimum RVSM equipment requirements of this appendix are:

(1) Manufactured to the same manufacturer specification and have the same part number; or

(2) Of a different manufacturer or part number, if the applicant demonstrates that the equipment provides equivalent system performance.

RVSM Nongroup Aircraft. An aircraft that is approved for RVSM operations as an individual aircraft.

RVSM Flight envelope. An RVSM flight envelope includes the range of Mach number, weight divided by atmospheric pressure ratio, and altitudes over which an aircraft is approved to be operated in cruising flight within RVSM airspace. RVSM flight envelopes are defined as follows:

(a) The *full RVSM flight envelope* is bounded as follows:

(1) The altitude flight envelope extends from FL 290 upward to the lowest altitude of the following:

(i) FL 410 (the RVSM altitude limit);

(ii) The maximum certificated altitude for the aircraft; or

(iii) The altitude limited by cruise thrust, buffet, or other flight limitations.

(2) The airspeed flight envelope extends:

(i) From the airspeed of the slats/flaps-up maximum endurance (holding) airspeed, or the maneuvering airspeed, whichever is lower;

(ii) To the maximum operating airspeed (Vmo /Mmo), or airspeed limited by cruise thrust buffet, or other flight limitations, whichever is lower.

(3) All permissible gross weights within the flight envelopes defined in paragraphs (1) and (2) of this definition.

(b) The *basic RVSM flight envelope* is the same as the full RVSM flight envelope except that the airspeed flight envelope extends:

(1) From the airspeed of the slats/flaps-up maximum endurance (holding) airspeed, or the maneuver airspeed, whichever is lower;

(2) To the upper Mach/airspeed boundary defined for the full RVSM flight envelope, or a specified lower value not less than the long-range cruise Mach number plus .04 Mach, unless further limited by available cruise thrust, buffet, or other flight limitations.

Section 2. Aircraft Approval

(a) An operator may be authorized to conduct RVSM operations if the Administrator finds that its aircraft comply with this section.

(b) The applicant for authorization shall submit the appropriate data package for aircraft approval. The package must consist of at least the following:

(1) An identification of the RVSM aircraft group or the nongroup aircraft;

(2) A definition of the RVSM flight envelopes applicable to the subject aircraft;

(3) Documentation that establishes compliance with the applicable RVSM aircraft requirements of this section; and

(4) The conformity tests used to ensure that aircraft approved with the data package meet the RVSM aircraft requirements.

(c) *Altitude-keeping equipment: All aircraft.* To approve an aircraft group or a nongroup aircraft, the Administrator must find that the aircraft meets the following requirements:

(1) The aircraft must be equipped with two operational independent altitude measurement systems.

(2) The aircraft must be equipped with at least one automatic altitude control system that controls the aircraft altitude—

(i) Within a tolerance band of ±65 feet about an acquired altitude when the aircraft is operated in straight and level flight under nonturbulent, nongust conditions; or

(ii) Within a tolerance band of ±130 feet under nonturbulent, nongust conditions for aircraft for which application for type certification occurred on or before April 9, 1997 that are equipped with an automatic altitude control system with flight management/performance system inputs.

(3) The aircraft must be equipped with an altitude alert system that signals an alert when the altitude displayed to the flight crew deviates from the selected altitude by more than:

(i) ±300 feet for aircraft for which application for type certification was made on or before April 9, 1997; or

(ii) ±200 feet for aircraft for which application for type certification is made after April 9, 1997.

(d) *Altimetry system error containment: Group aircraft for which application for type certification was made on or before April 9, 1997.* To approve group aircraft for which application for type certification was made on or before April 9, 1997, the Administrator must find that the altimetry system error (ASE) is contained as follows:

(1) At the point in the basic RVSM flight envelope where mean ASE reaches its largest absolute value, the absolute value may not exceed 80 feet.

(2) At the point in the basic RVSM flight envelope where mean ASE plus three standard deviations reaches its largest absolute value, the absolute value may not exceed 200 feet.

(3) At the point in the full RVSM flight envelope where mean ASE reaches its largest absolute value, the absolute value may not exceed 120 feet.

(4) At the point in the full RVSM flight envelope where mean ASE plus three standard deviations reaches its largest absolute value, the absolute value may not exceed 245 feet.

(5) *Necessary operating restrictions.* If the applicant demonstrates that its aircraft otherwise comply with the ASE containment requirements, the Administrator may establish an operating restriction on that applicant's aircraft to restrict the aircraft from operating in areas of the basic RVSM flight envelope where the absolute value of mean ASE exceeds 80 feet, and/or the absolute value of mean ASE plus three standard deviations exceeds 200 feet; or from operating in areas of the full RVSM flight envelope where the absolute value of the mean ASE exceeds 120 feet and/or the absolute value of the mean ASE plus three standard deviations exceeds 245 feet.

(e) *Altimetry system error containment: Group aircraft for which application for type certification is made after April 9, 1997.* To approve group aircraft for which application for type certification is made after April 9, 1997, the Administrator must find that the altimetry system error (ASE) is contained as follows:

(1) At the point in the full RVSM flight envelope where mean ASE reaches its largest absolute value, the absolute value may not exceed 80 feet.

(2) At the point in the full RVSM flight envelope where mean ASE plus three standard deviations reaches its largest absolute value, the absolute value may not exceed 200 feet.

(f) *Altimetry system error containment: Nongroup aircraft.* To approve a nongroup aircraft, the Administrator must find that the altimetry system error (ASE) is contained as follows:

(1) For each condition in the basic RVSM flight envelope, the largest combined absolute value for residual static source error plus the avionics error may not exceed 160 feet.

(2) For each condition in the full RVSM flight envelope, the largest combined absolute value for residual static source error plus the avionics error may not exceed 200 feet.

(g) Traffic Alert and Collision Avoidance System (TCAS) Compatibility With RVSM Operations: All aircraft. After March 31, 2002, unless otherwise authorized by the Administrator, if you operate an aircraft that is equipped with TCAS II in RVSM airspace, it must be a TCAS II that meets TSO C-119b (Version 7.0), or a later version.

(h) If the Administrator finds that the applicant's aircraft comply with this section, the Administrator notifies the applicant in writing.

Section 3. Operator Authorization

(a) Authority for an operator to conduct flight in airspace where RVSM is applied is issued in operations specifications, a Letter of Authorization, or management specifications issued under subpart K of this part, as appropriate. To issue an RVSM authorization, the Administrator must find that the operator's aircraft have been approved in accordance with

Section 2 of this appendix and the operator complies with this section.

(b) An applicant for authorization to operate within RVSM airspace shall apply in a form and manner prescribed by the Administrator. The application must include the following:

(1) An approved RVSM maintenance program outlining procedures to maintain RVSM aircraft in accordance with the requirements of this appendix. Each program must contain the following:

(i) Periodic inspections, functional flight tests, and maintenance and inspection procedures, with acceptable maintenance practices, for ensuring continued compliance with the RVSM aircraft requirements.

(ii) A quality assurance program for ensuring continuing accuracy and reliability of test equipment used for testing aircraft to determine compliance with the RVSM aircraft requirements.

(iii) Procedures for returning noncompliant aircraft to service.

(2) For an applicant who operates under part 121 or 135 of this chapter or under subpart K of this part, initial and recurring pilot training requirements.

(3) Policies and procedures: An applicant who operates under part 121 or 135 of this chapter or under subpart K of this part must submit RVSM policies and procedures that will enable it to conduct RVSM operations safely.

(c) Validation and Demonstration. In a manner prescribed by the Administrator, the operator must provide evidence that:

(1) It is capable to operate and maintain each aircraft or aircraft group for which it applies for approval to operate in RVSM airspace; and

(2) Each pilot has an adequate knowledge of RVSM requirements, policies, and procedures.

Section 4. RVSM Operations

(a) Each person requesting a clearance to operate within RVSM airspace shall correctly annotate the flight plan filed with air traffic control with the status of the operator and aircraft with regard to RVSM approval. Each operator shall verify RVSM applicability for the flight planned route through the appropriate flight planning information sources.

(b) No person may show, on the flight plan filed with air traffic control, an operator or aircraft as approved for RVSM operations, or operate on a route or in an area where RVSM approval is required, unless:

(1) The operator is authorized by the Administrator to perform such operations; and

(2) The aircraft has been approved and complies with the requirements of Section 2 of this appendix.

Section 5. Deviation Authority Approval

The Administrator may authorize an aircraft operator to deviate from the requirements of § 91.180 or § 91.706 for a specific flight in RVSM airspace if that operator has not been approved in accordance with section 3 of this appendix if:

(a) The operator submits a request in a time and manner acceptable to the Administrator; and

(b) At the time of filing the flight plan for that flight, ATC determines that the aircraft may be provided appropriate separation and that the flight will not interfere with, or impose a burden on, the operations of operators who have been approved for RVSM operations in accordance with Section 3 of this appendix.

Section 6. Reporting Altitude-Keeping Errors

Each operator shall report to the Administrator each event in which the operator's aircraft has exhibited the following altitude-keeping performance:

(a) Total vertical error of 300 feet or more;

(b) Altimetry system error of 245 feet or more; or

(c) Assigned altitude deviation of 300 feet or more.

Section 7. Removal or Amendment of Authority

The Administrator may amend operations specifications or management specifications issued under subpart K of this part to revoke or restrict an RVSM authorization, or may revoke or restrict an RVSM letter of authorization, if the Administrator determines that the operator is not complying, or is unable to comply, with this appendix or subpart H of this part. Examples of reasons for amendment, revocation, ore restriction include, but are not limited to, an operator's:

(a) Committing one or more altitude-keeping errors in RVSM airspace;

(b) Failing to make an effective and timely response to identify and correct an altitude-keeping error; or

(c) Failing to report an altitude-keeping error.

Section 8. Airspace Designation

(a) *RVSM in the North Atlantic.* (1) RVSM may be applied in the NAT in the following ICAO Flight Information Regions (FIRs): New York Oceanic, Gander Oceanic, Sondrestrom FIR, Reykjavik Oceanic, Shanwick Oceanic, and Santa Maria Oceanic.

(2) RVSM may be effective in the Minimum Navigation Performance Specification (MNPS) airspace within the NAT. The MNPS airspace within the NAT is defined by the volume of airspace between FL 285 and FL 420 (inclusive) extending between latitude 27 degrees north and the North Pole, bounded in the east by the eastern boundaries of control areas Santa Maria Oceanic, Shanwick Oceanic, and Reykjavik Oceanic and in the west by the western boundaries of control areas Reykjavik Oceanic, Gander Oceanic, and New York Oceanic, excluding the areas west of 60 degrees west and south of 38 degrees 30 minutes north.

(b) *RVSM in the Pacific.* (1) RVSM may be applied in the Pacific in the following ICAO Flight Information Regions (FIRs): Anchorage Arctic, Anchorage Continental, Anchorage Oceanic, Auckland Oceanic, Brisbane, Edmonton, Honiara, Los Angeles, Melbourne, Nadi, Naha, Nauru, New Zealand, Oakland, Oakland Oceanic, Port Moresby, Seattle, Tahiti, Tokyo, Ujung Pandang and Vancouver.

(c) *RVSM in the West Atlantic Route System (WATRS).* RVSM may be applied in the New York FIR portion of the West Atlantic Route System (WATRS). The area is defined as beginning at a point 38°30′ N/60°00′W direct to 38°30′N/69°15′ W direct to 38°20′ N/69°57′ W direct to 37°31′ N/71°41′ W direct to 37°13′ N/72°40′ W direct to 35°05′ N/72°40′ W direct to 34°54′ N/72°57′ W direct to 34°29′ N/73°34′ W direct to 34°33′ N/73°41′ W direct to 34°19′ N/74°02′ W direct to 34°14′ N/73°57′ W direct to 32°12′ N/76°49′ W direct to 32°20′ N/77°00′ W direct to 28°08′ N/77°00′ W direct to 27°50′ N/76°32′ W direct to 27°50′ N/74°50′ W direct to 25°00′ N/73°21′ W direct to 25°00′05′ N/69°13′06′ W direct to 25°00′ N/69°07′ W direct to 23°30′ N/68°40′ W direct to 23°30′ N/60°00′ W to the point of beginning.

(d) *RVSM in the United States.* RVSM may be applied in the airspace of the 48 contiguous states, District of Columbia, and Alaska, including that airspace overlying the waters within 12 nautical miles of the coast.

(e) *RVSM in the gulf of Mexico.* RVSM may be applied in the Gulf of Mexico in the following areas: Gulf of Mexico High Offshore Airspace, Houston Oceanic ICAO FIR and Miami Oceanic ICAO FIR.

(f) *RVSM in Atlantic High Offshore Airspace and the San Juan FIR.* RVSM may be applied in Atlantic High Offshore Airspace and in the San Juan ICAO FIR.

[Doc. No. 28870, 62 FR 17487, Apr. 9, 1997, as amended by Amdt. 91-261, 65 FR 5942, Feb. 7, 2000; Amdt. 91-271, 66 FR 63895, Dec. 10, 2001; Amdt. 91-274, 68 FR 54584, Sept. 17, 2003; Amdt. 91-276, 68 FR 70133, Dec. 17, 2003]

PART 93

SPECIAL AIR TRAFFIC RULES

Contents

Authority: 49 U.S.C. 106(g), 40103, 40106, 40109, 40113, 44502, 44514, 44701, 44719, 46301.

Special Federal Aviation Regulation No. 60

Editorial Note: For the text of SFAR No. 60, see part 91 of this chapter.

Subpart A—General

§ 93.1 Applicability.

This part prescribes special air traffic rules for operating aircraft in certain areas described in this part, unless otherwise authorized by air traffic control.

[Doc. No. FAA-2002-13235, 68 FR 9795, Feb. 28, 2003]

Subparts B-C [Reserved]

Subpart D—Anchorage, Alaska, Terminal Area

Source: Docket No. 29029, 64 FR 14976, Mar. 29, 1999, unless otherwise noted.

§ 93.51 Applicability.

This subpart prescribes special air traffic rules for aircraft operating in the Anchorage, Alaska, Terminal Area.

[Doc. No. FAA-2002-13235, 68 FR 9795, Feb. 28, 2003]

§ 93.53 Description of area.

The Anchorage, Alaska, Terminal Area is designated as that airspace extending upward from the surface to the upper limit of each of the segments described in § 93.55. It is bounded by a line beginning at Point MacKenzie, extending westerly along the bank of Knik Arm to a point intersecting the 350° bearing from the Anchorage International ATCT; thence north to intercept the 5.2-mile arc centered on the geographical center of Anchorage, Alaska, ATCT; thence counterclockwise along that arc to its intersection with a line bearing 180° from the intersection of the new Seward Highway and International Airport Road; thence due north to O'Malley Road; thence east along O'Malley Road to its intersection with Lake Otis Parkway; thence northerly along Lake Otis Parkway to its intersection with Abbott Road; thence east along Abbott Road to its intersection with Abbott Loop Road; thence north to its intersection with Tudor Road; thence easterly along Tudor Road to its intersection with Muldoon Road; thence northerly along Muldoon Road to the intersection of the Glenn Highway; thence north and east along the Glenn Highway to Ski Bowl Road; thence southeast along the Ski Bowl Road to a point one-half mile south of the Glenn Highway; thence north and east one-half mile south of and parallel to the Glenn Highway to its intersection with a line one-half mile east of and parallel to the Bryant Airport Runway 16/34 extended centerline; thence northeast along a line one-half mile east of and parallel to Bryant Airport Runway 16/34 extended centerline to lat. 61°17′13″N., long. 149°37′35″W.; thence west along lat. 61°17′13″N., to long. 149°43′08″W.; thence north along long. 149°43′08″W., to lat. 61°17′30″N.; thence to lat. 61°17′58″N., long 149°44′08″W.; thence to lat. 61°19′10″N., long. 149°46′44″W.; thence north along long. 149°46′44″W., to intercept the 4.7-mile radius arc centered on Elmendorf Air Force Base (AFB), Alaska; thence counterclockwise along the 4.7-mile radius arc to its intersection with the west bank of Knik Arm; thence southerly along the west bank of Knik Arm to the point of beginning.

[Doc. No. 29029, 64 FR 14976, Mar. 29, 1999; Amdt. 93-77, 64 FR 17439, Apr. 9, 1999]

§ 93.55 Subdivision of Terminal Area.

The Anchorage, Alaska, Terminal Area is subdivided as follows:

(a) *International segment.* That area from the surface to and including 4,100 feet MSL, within a 5.2-mile radius of the Anchorage International ATCT; excluding that airspace east of the 350° bearing from the Anchorage International ATCT and north of the 090° bearing from the Anchorage International ATCT and east of a line bearing 180° and 360° from the intersection of the new Seward Highway and International Airport Road and the airspace extending upward from the surface to but not including 600 feet MSL, south of lat. 61°08′28″N.

(b) *Merrill segment.* That area from the surface to and including 2,500 feet MSL, within a line beginning at Point Noname; thence direct to the mouth of Ship Creek; thence direct to the intersection of the Glenn Highway and Muldoon Road; thence south along Muldoon Road to Tudor Road; thence west along Tudor Road to the new Seward Highway; thence direct to West Anchorage High School;

thence direct to Point MacKenzie; thence via the north bank of Knik Arm to the point of beginning.

(c) *Lake Hood segment.* That area from the surface to and including 2,500 feet MSL, within a line beginning at Point MacKenzie; thence direct to West Anchorage High School; thence direct to the intersection of Tudor Road and the new Seward Highway; thence south along the new Seward Highway to the 090° bearing from the Anchorage International ATCT; thence west direct to the Anchorage International ATCT; thence north along the 350° bearing from the Anchorage International ATCT to the north bank of Knik arm; thence via the north bank of Knik Arm to the point of beginning.

(d) *Elmendorf segment.* That area from the surface to and including 3,000 feet MSL, within a line beginning at Point Noname; thence via the north bank of Knik Arm to the intersection of the 4.7-mile radius of Elmendorf AFB; thence clockwise along the 4.7-mile radius of Elmendorf AFB to long. 149°46′44″W.; thence south along long. 149°46′44″W. to lat. 61°19′10″N.; thence to lat. 61°17′58″N., long. 149°44′08″W.; thence to lat. 61°17′30″N., long. 149°43′08″W.; thence south along long. 149°43′08″W. to the Glenn Highway; thence south and west along the Glenn Highway to Muldoon Road; thence direct to the mouth of Ship Creek; thence direct to the point of beginning.

(e) *Bryant segment.* That area from the surface to and including 2,000 feet MSL, within a line beginning at lat. 61°17′13″N., long. 149°37′35″W.; thence west along lat. 61°17′13″N., to long. 149°43′08″W.; thence south along long. 149°43′08″W., to the Glenn Highway; thence north and east along the Glenn Highway to Ski Bowl Road; thence southeast along the Ski Bowl Road to a point one-half mile south of the Glenn Highway; thence north and east one-half mile south of and parallel to the Glenn Highway to its intersection with a line one-half mile east of and parallel to the Bryant Airport Runway 16/34 extended centerline; thence northeast along a line one-half mile east of and parallel to Bryant Airport runway 16/34 extended centerline to the point of beginning.

(f) *Seward Highway segment.* That area from the surface to and including 4,100 feet MSL, within a line beginning at the intersection of a line bearing 180° from the intersection of the new Seward Highway and International Airport Road, and O'Malley Road; thence east along O'Malley Road to its intersection with Lake Otis Park Way, lat. 61°07′23″N., long 149°50′03″W.; thence northerly along Lake Otis Park Way to its intersection with Abbott Road, lat. 61°08′14″N., long. 149°50′03″W.; thence east along Abbott Road to its intersection with Abbott Loop Road, lat. 61°08′14″N., long. 149°48′16″W.; thence due north to intersect with Tudor Road, lat. 61°10′51″N., long. 149°48′16″W.; thence west along Tudor Road to its intersection with the new Seward Highway, lat. 61°10′51″N., long. 149°51′38″W.; thence south along the new Seward Highway to its intersection with a line bearing 180° and 360° from the intersection of the new Seward Highway and International Airport Road; thence south to the point of beginning.

[Doc. No. 29029, 64 FR 14976, Mar. 29, 1999; Amdt. 93-77, 64 FR 17439, Apr. 9, 1999]

§ 93.57 General rules: All segments.

(a) Each person operating an aircraft to, from, or on an airport within the Anchorage, Alaska, Terminal Area shall operate that aircraft according to the rules set forth in this section and §§ 93.59, 93.61, 93.63, 93.65, 93.67, or 93.68 as applicable, unless otherwise authorized or required by ATC.

(b) Each person operating an airplane within the Anchorage, Alaska Terminal Area shall conform to the flow of traffic depicted on the appropriate aeronautical charts.

(c) Each person operating a helicopter shall operate it in a manner so as to avoid the flow of airplanes.

(d) Except as provided in § 93.65 (d) and (e), and § 93.67(b), each person operating an aircraft in the Anchorage, Alaska, Terminal Area shall operate that aircraft only within the designated segment containing the arrival or departure airport.

(e) Except as provided in §§ 93.63(d) and 93.67(b), each person operating an aircraft in the Anchorage, Alaska, Terminal Area shall maintain two-way radio communications with the ATCT serving the segment containing the arrival or departure airport.

§ 93.59 General rules: International segment.

(a) No person may operate an aircraft at an altitude between 1,200 feet MSL and 2,000 feet MSL in that portion of this segment lying north of the midchannel of Knik Arm.

(b) Each person operating an airplane at a speed of more than 105 knots within this segment (except that part described in paragraph (a) of this section) shall operate that airplane at an altitude of at least 1,600 feet MSL until maneuvering for a safe landing requires further descent.

(c) Each person operating an airplane at a speed of 105 knots or less within this segment (except that part described in paragraph (a) of this section) shall operate that airplane at an altitude of at least 900 feet MSL until maneuvering for a safe landing requires further descent.

§ 93.61 General rules: Lake Hood segment.

(a) No person may operate an aircraft at an altitude between 1,200 feet MSL and 2,000 feet MSL in that portion of this segment lying north of the midchannel of Knik Arm.

(b) Each person operating an airplane within this segment (except that part described in paragraph (a) of this section) shall operate that airplane at an altitude of at least 600 feet MSL until maneuvering for a safe landing requires further descent.

§ 93.63 General rules: Merrill segment.

(a) No person may operate an aircraft at an altitude between 600 feet MSL and 2,000 feet MSL in that portion of this segment lying north of the midchannel of Knik Arm.

(b) Each person operating an airplane at a speed of more than 105 knots within this segment (except for that part described in paragraph (a) of this section) shall operate that airplane at an altitude of at least 1,200 feet MSL until maneuvering for a safe landing requires further descent.

(c) Each person operating an airplane at a speed of 105 knots or less within this segment (except for that part described in paragraph (a) of this section) shall operate that airplane at an altitude of at least 900 feet MSL until maneuvering for a safe landing requires further descent.

(d) Whenever the Merrill ATCT is not operating, each person operating an aircraft either in that portion of the Merrill segment north of midchannel of Knik Arm, or in the Seward Highway segment at or below 1200 feet MSL, shall contact Anchorage Approach Control for wake turbulence and other advisories. Aircraft operating within the remainder of the segment should self-announce intentions on the Merrill Field CTAF.

§ 93.65 General rules: Elmendorf segment.

(a) Each person operating a turbine-powered aircraft within this segment shall operate that aircraft at an altitude of at least 1,700 feet MSL until maneuvering for a safe landing requires further descent.

(b) Each person operating an airplane (other than turbine-powered aircraft) at a speed of more than 105 knots within this segment shall operate that airplane at an altitude of at least 1,200 feet MSL until maneuvering for a safe landing requires further descent.

(c) Each person operating an airplane (other than turbine-powered aircraft) at a speed of 105 knots or less within the segment shall operate that airplane at an altitude of at least 800 feet MSL until maneuvering for a safe landing requires further descent.

(d) A person landing or departing from Elmendorf AFB, may operate that aircraft at an altitude between 1,500 feet MSL and 1,700 feet MSL within that portion of the International and Lake Hood segments lying north of the midchannel of Knik Arm.

(e) A person landing or departing from Elmendorf AFB, may operate that aircraft at an altitude between 900 feet MSL and 1,700 feet MSL within that portion of the Merrill segment lying north of the midchannel of Knik Arm.

(f) A person operating in VFR conditions, at or below 600 feet MSL, north of a line beginning at the intersection of Farrell Road and the long. 149°43′08″W.; thence west along Farrell Road to the east end of Sixmile Lake; thence west along a line bearing on the middle of Lake Lorraine to the northwest bank of Knik Arm; is not required to establish two-way radio communications with ATC.

[Doc. No. 29029, 64 FR 14977, Mar. 29, 1999; Amdt. 93-77, 64 FR 17439, Apr. 9, 1999]

§ 93.67 General rules: Bryant segment.

(a) Each person operating an airplane to or from the Bryant Airport shall conform to the flow of traffic shown on the appropriate aeronautical charts, and while in the traffic pattern, shall operate that airplane at an altitude of at least 1,000 feet MSL until maneuvering for a safe landing requires further descent.

(b) Each person operating an aircraft within the Bryant segment should self-announce intentions on the Bryant Airport CTAF.

§ 93.68 General rules: Seward Highway segment.

(a) Each person operating an airplane in the Seward Highway segment shall operate that airplane at an altitude of at least 1,000 feet MSL unless maneuvering for a safe landing requires further descent.

(b) Each person operating an aircraft at or below 1,200 feet MSL that will transition to or from the Lake Hood or Merrill segment shall contact the appropriate ATCT prior to entering the Seward Highway segment. All other persons operating an airplane at or below 1,200 feet MSL in this segment shall contact Anchorage Approach Control.

(c) At all times, each person operating an aircraft above 1,200 MSL shall contact Anchorage Approach Control prior to entering the Seward Highway segment.

§ 93.69 Special requirements, Lake Campbell and Sixmile Lake Airports.

Each person operating an aircraft to or from Lake Campbell or Sixmile Lake Airport shall conform to the flow of traffic for the Lake operations that are depicted on the appropriate aeronautical charts.

Subpart E—Flight Restrictions in the Vicinity of Niagara Falls, New York

§ 93.71 General operating procedures.

(a) Flight restrictions are in effect below 3,500 feet MSL in the airspace above Niagara Falls, New York, west of a line from latitude 43°06′33″ N., longitude 79°03′30″ W. (the Whirlpool Rapids Bridge) to latitude 43°04′47″ N., longitude 79°02′44″ W. (the Niagara River Inlet) to latitude 43°04′29″ N., longitude 79°03′30″ W. (the International Control Dam) to the United States/Canadian Border and thence along the border to the point of origin.

(b) No flight is authorized below 3,500 feet MSL in the area described in paragraph (a) of this section, except for aircraft operations conducted directly to or from an airport/heliport within the area, aircraft operating on an ATC-approved IFR flight plan, aircraft operating the Scenic Falls Route pursuant to approval of Transport Canada, aircraft carrying law enforcement officials, or aircraft carrying properly accredited news representatives for which a flight plan has been filed with Buffalo NY (BUF) Automated Flight Service Station (AFSS).

(c) Check with Transport Canada for flight restrictions in Canadian airspace. Commercial air tour operations approved by Transport Canada will be conducting a north/south orbit of the Niagara Falls area below 3,500 feet MSL over the Niagara River.

(d) The minimum altitude for VFR flight over the Scenic Falls area is 3,500 feet MSL.

(e) Comply with the following procedures when conducting flight over the area described in paragraph (a) of this section:

(1) Fly a clockwise pattern;

(2) Do not proceed north of the Rainbow Bridge;

(3) Prior to joining the pattern, broadcast flight intentions on frequency 122.05 Mhz, giving altitude and position, and monitor the frequency while in the pattern;

(4) Use the Niagara Falls airport altimeter setting. Contact Niagara Falls Airport Traffic Control Tower to obtain the current altimeter setting, to facilitate the exchange of traffic advisories/restrictions, and to reduce the risk of midair collisions between aircraft operating in the vicinity of the Falls. If the Control Tower is closed, use the appropriate Automatic Terminal Information Service (ATIS) Frequency;

(5) Do not exceed 130 knots;

(6) Anticipate heavy congestion of VFR traffic at or above 3,500 feet MSL; and

(7) Use caution to avoid high-speed civil and military aircraft transiting the area to or from Niagara Falls Airport.

(f) These procedures do not relieve pilots from the requirements of § 91.113 of this chapter to see and avoid other aircraft.

(g) Flight following, to and from the area, is available through Buffalo Approach.

[Doc. No. FAA-2002-13235, 68 FR 9795, Feb. 28, 2003]

Subpart F—Valparaiso, Florida, Terminal Area

§ 93.80 Applicability.

This subpart prescribes special air traffic rules for aircraft operating in the Valparaiso, Florida, Terminal Area.

[Doc. No. FAA-2002-13235, 68 FR 9795, Feb. 28, 2003]

§ 93.81 Applicability and description of area.

The Valparaiso, Florida Terminal Area is designated as follows:

(a) North-South Corridor. The North-South Corridor includes the airspace extending upward from the surface up to, but not including, 18,000 feet MSL, bounded by a line beginning at:

Latitude 30°42′51″ N., Longitude 86°38′02″ W.; to
Latitude 30°43′18″ N., Longitude 86°27′37″ W.; to
Latitude 30°37′01″ N., Longitude 86°27′37″ W.; to
Latitude 30°37′01″ N., Longitude 86°25′30″ W.; to
Latitude 30°33′01″ N., Longitude 86°25′30″ W.; to
Latitude 30°33′01″ N., Longitude 86°25′00″ W.; to
Latitude 30°25′01″ N., Longitude 86°25′00″ W.; to
Latitude 30°25′01″ N., Longitude 86°38′12″ W.; to
Latitude 30°29′02″ N., Longitude 86°38′02″ W.; to point of beginning.

(b) East-West Corridor—The East-West Corridor is divided into three sections to accommodate the different altitudes as portions of the corridor underlie restricted areas R-2915C, R-2919B, and R-2914B.

(1) The west section would include that airspace extending upward from the surface to but not including 8,500 feet MSL, bounded by a line beginning at: Latitude 30°22′47″ N., Longitude 86°51′30″ W.: then along the shoreline to Latitude 30°23′46″ N., Longitude 86°38′15″ W.; to Latitude 30°20′51″ N., Longitude 86°38′50″ W.; then 3 NM from and parallel to the shoreline to Latitude 30°19′31″ N., Longitude 86°51′30″ W.; to the beginning.

(2) The center section would include that airspace extending upward from the surface to but not including 18,000 feet MSL, bounded by a line beginning at:

Latitude 30°25′01″ N., Longitude 86°38′12″ W.; to
Latitude 30°25′01″ N., Longitude 86°25′00″ W.; to
Latitude 30°25′01″ N., Longitude 86°22′26″ W.; to
Latitude 30°19′46″ N., Longitude 86°23′45″ W.; then 3 NM from and parallel to the shoreline to Latitude 30°20′51″ N., Longitude 86°38′50″ W.; to Latitude 30°23′46″ N., Longitude 86°38′15″ W.; to the beginning.

(3) The east section would include that airspace extending upward from the surface to but not including 8,500 feet MSL, bounded by a line beginning at:

Latitude 30°25′01″ N., Longitude 86°22′26″ W.; to
Latitude 30°22′01″ N., Longitude 86°08′00″ W.; to
Latitude 30°19′16″ N., Longitude 85°56′00″ W.; to
Latitude 30°11′01″ N., Longitude 85°56′00″ W.; then 3 NM from and parallel to the shoreline to Latitude 30°19′46″ N., Longitude 86°23′45″ W.; to the beginning.

[Amdt. 93-70, 59 FR 46154, Sept. 6, 1994 as amended by Amdt. 93-82, 68 FR 9795, Feb. 28, 2003]

§ 93.83 Aircraft operations.

(a) *North-South Corridor.* Unless otherwise authorized by ATC (including the Eglin Radar Control Facility), no person may operate an aircraft in flight within the North-South Corridor designated in § 93.81(b)(1) unless—

(1) Before operating within the corridor, that person obtains a clearance from the Eglin Radar Control Facility or an appropriate FAA ATC facility; and

(2) That person maintains two-way radio communication with the Eglin Radar Control Facility or an appropriate FAA ATC facility while within the corridor.

(b) *East-West Corridor.* Unless otherwise authorized by ATC (including the Eglin Radar Control Facility), no person may operate an aircraft in flight within the East-West Corridor designated in § 93.81(b)(2) unless—

(1) Before operating within the corridor, that person establishes two-way radio communications with Eglin Radar Control Facility or an appropriate FAA ATC facility and receives an ATC advisory concerning operations being conducted therein; and

(2) That person maintains two-way radio communications with the Eglin Radar Control Facility or an appropriate FAA ATC facility while within the corridor.

[Amdt. 93-70, 59 FR 46155, Sept. 6, 1994]

Subpart G—Special Flight Rules in the Vicinity of Los Angeles International Airport

Source: Doc. No. FAA-2002-14149, 68 FR 41214, July 10, 2003, unless otherwise noted.

§ 93.91 Applicability.

This subpart prescribes special air traffic rules for aircraft conducting VFR operations in the Los Angeles, California Special Flight Rules Area.

§ 93.93 Description of area.

The Los Angeles Special Flight Rules Area is designated as that part of Area A of the Los Angeles Class B airspace area at 3,500 feet above mean sea level (MSL) and at 4,500 feet MSL, beginning at Ballona Creek/Pacific Ocean (lat. 33°57′42″ N, long. 118°27′23″ W), then eastbound along Manchester Blvd. to the intersection of Manchester/405 Freeway (lat. 33°57′42″ N, long. 118°22′10″ W), then southbound along the 405 Freeway to the intersection of the 405 Freeway/Imperial Highway (lat. 33°55′51″ N, long. 118° 22′06″ W), then westbound along Imperial Highway to the intersection of Imperial Highway/Pacific Ocean (lat. 33°55′51″ N, long. 118°26′05″ W), then northbound along the shoreline to the point of beginning.

§ 93.95 General operating procedures.

Unless otherwise authorized by the Administrator, no person may operate an aircraft in the airspace described in § 93.93 unless the operation is conducted in accordance with the following procedures:

(a) The flight must be conducted under VFR and only when operation may be conducted in compliance with § 91.155(a) of this chapter.

(b) The aircraft must be equipped as specified in § 91.215(b) of this chapter replying on code 1201 prior to entering and while operating in this area.

(c) The pilot shall have a current Los Angeles Terminal Area Chart in the aircraft.

(d) The pilot shall operate on the Santa Monica very high frequency omni-directional radio range (VOR) 132° radial.

(e) Aircraft navigating in a southeasterly direction shall be in level flight at 3,500 feet MSL.

(f) Aircraft navigating in a northwesterly direction shall be in level flight at 4,500 feet MSL.

(g) Indicated airspeed shall not exceed 140 knots.

(h) Anti-collision lights and aircraft position/navigation lights shall be on. Use of landing lights is recommended.

(i) Turbojet aircraft are prohibited from VFR operations in this area.

§ 93.97 Operations in the SFRA.

Notwithstanding the provisions of § 91.131(a) of this chapter, an air traffic control authorization is not required in the Los Angeles Special Flight Rules Area for operations in compliance with § 93.95. All other provisions of § 91.131 of this chapter apply to operations in the Los Angeles Special Flight Rules Area.

Subpart H—Mandatory Use of the New York North Shore Helicopter Route

Source: Doc. No. FAA-2010-0302, 77 FR 39921, July 6, 2012, unless otherwise noted.

Effective Date **Note:** By Doc. No. FAA-2010-0302, 77 FR 39921, July 6, 2012, subpart H was added, effective Aug. 6, 2012 through Aug. 6, 2014.

§ 93.101 Applicability.

This subpart prescribes a special air traffic rule for civil helicopters operating VFR along the North Shore, Long Island, New York, between August 6, 2012 and August 6, 2014.

§ 93.103 Helicopter operations.

(a) Unless otherwise authorized, each person piloting a helicopter along Long Island, New York's northern shoreline between the VPLYD waypoint and Orient Point, shall utilize the North Shore Helicopter route and altitude, as published.

(b) Pilots may deviate from the route and altitude requirements of paragraph (a) of this section when necessary for safety, weather conditions or transitioning to or from a destination or point of landing.

Subpart I [Reserved]

Subpart J—Lorain County Regional Airport Traffic Rule

§ 93.117 Applicability.

This subpart prescribes a special air traffic rule for aircraft operating at the Lorain County Regional Airport, Lorain County, Ohio.

[Doc. No. FAA-2002-13235, 68 FR 9795, Feb. 28, 2003]

§ 93.119 Aircraft operations.

Each person piloting an airplane landing at the Lorain County Regional Airport shall enter the traffic pattern north of the airport and shall execute a right traffic pattern for a landing to the southwest or a left traffic pattern for a landing to the northeast. Each person taking off from the airport shall execute a departure turn to the north as soon as practicable after takeoff.

[Doc. No. 8669, 33 FR 11749, Aug. 20, 1968]

Subpart K—High Density Traffic Airports

§ 93.121 Applicability.

This subpart designates high density traffic airports and prescribes air traffic rules for operating aircraft, other than helicopters, to or from those airports.

[Doc. No. 9974, 35 FR 16592, Oct. 24, 1970, as amended by Amdt. 93-27, 38 FR 29464, Oct. 25, 1973]

§ 93.123 High density traffic airports.

(a) Each of the following airports is designated as a high density traffic airport and, except as provided in § 93.129 and paragraph (b) of this section, or unless otherwise authorized by ATC, is limited to the hourly number of allocated IFR operations (takeoffs and landings) that may be reserved for the specified classes of users for that airport:

IFR Operations per Hour

Airport

Class of user	LaGuardia 4,5	Newark	O'Hare 2,3,5	Ronald Reagan National1
Air carriers	48	40	120	37
Commuters	14	10	25	11
Other	6	10	10	12

John F. Kennedy

	Air carriers	Commuters	Other
1500	69	15	2
1600	74	12	2
1700	80	13	0
1800	75	10	2
1900	63	12	2

[1] Washington National Airport operations are subject to modifications per Section 93.124.

[2] The hour period in effect at O'Hare begins at 6:45 a.m. and continues in 30-minute increments until 9:15 p.m.

[3] Operations at O'Hare International Airport shall not—

(a) Except as provided in paragraph (c) of the note, exceed 62 for air carriers and 13 for commuters and 5 for "other" during any 30-minute period beginning at 6:45 a.m. and continuing every 30 minutes thereafter.

(b) Except as provided in paragraph (c) of the note, exceed more than 120 for air carriers, 25 for commuters, and 10 for "other" in any two consecutive 30-minute periods.

(c) For the hours beginning at 6:45 a.m., 7:45 a.m., 11:45 a.m., 7:45 p.m. and 8:45 p.m., the hourly limitations shall be 105 for air carriers, 40 for commuters and 10 for "other," and the 30-minute limitations shall be 55 for air carriers, 20 for commuters and 5 for "other." For the hour beginning at 3:45 p.m., the hourly limitations shall be 115 for air carriers, 30 for commuters and 10 for "others", and the 30-minute limitations shall be 60 for air carriers, 15 for commuters and 5 for "other."

[4] Operations at LaGuardia Airport shall not—

(a) Exceed 26 for air carriers, 7 for commuters and 3 for "other" during any 30-minute period.

(b) Exceed 48 for air carriers, 14 for commuters, and 6 for "other" in any two consecutive 30-minute periods.

[5] Pursuant to bilateral agreement, 14 slots at LaGuardia and 24 slots at O'Hare are allocated to the Canadian carriers. These slots are excluded from the hourly quotas set forth in § 93.123 above.

(b) The following exceptions apply to the allocations of reservations prescribed in paragraph (a) Of this section.

(1) The allocations of reservations among the several classes of users do not apply from 12 midnight to 6 a.m. local time, but the total hourly limitation remains applicable.

(2) [Reserved]

(3) The allocation of 37 IFR reservations per hour for air carriers except commuters at Washington National Airport does not include charter flights, or other nonscheduled flights of scheduled or supplemental air carriers. These flights may be conducted without regard to the limitation of 37 IFR reservations per hour.

(4) The allocation of IFR reservations for air carriers except commuters at LaGuardia, Newark, O'Hare, and Washington National Airports does not include extra sections of scheduled flights. The allocation of IFR reservations for scheduled commuters at Washington National Airport does not include extra sections of scheduled flights. These flights may be conducted without regard to the limitation upon the hourly IFR reservations at those airports.

(5) Any reservation allocated to, but not taken by, air carrier operations (except commuters) is available for a scheduled commuter operation.

(6) Any reservation allocated to, but not taken by, air carrier operations (except commuters) or scheduled commuter operations is available for other operations.

(c) For purposes of this subpart—

(1) The number of operations allocated to *air carriers except commuters,* as used in paragraph (a) of this section refers to the number of operations conducted by air carriers with turboprop and reciprocating engine aircraft having a certificated maximum passenger seating capacity of 75 or more or with turbojet powered aircraft having a certificated maximum passenger seating capacity of 56 or more, or, if used for cargo service in air transportation, with any aircraft having a maximum payload capacity of 18,000 pounds or more.

(2) The number of operations allocated to scheduled commuters, as used in paragraph (a) of this section, refers to the number of operations conducted by air carriers with turboprop and reciprocating engine aircraft having a certificated maximum passenger seating capacity of less than 75 or by turbojet aircraft having a certificated maximum passenger seating capacity of less than 56, or if used for cargo service in air transportation, with any aircraft having a maximum payload capacity of less than 18,000 pounds. For purposes of aircraft operations at Ronald Reagan Washington National Airport, the term "commuters" means aircraft operations using aircraft having a certificated maximum seating capacity of 76 or less.

(3) Notwithstanding the provisions of paragraph (c)(2) of this section, a limited number of operations allocated for "scheduled commuters" under paragraph (a) of this section may be conducted with aircraft described in § 93.221(e) of this part pursuant to the requirements of § 93.221(e).

[Doc. No. 9113, 34 FR 2603, Feb. 26, 1969, as amended by Amdt. 93-37, 45 FR 62408, Sept. 18, 1980; Amdt. 93-44, 46 FR 58048, Nov. 27, 1981; Amdt. 93-46, 49 FR 8244, Mar. 6, 1984; Amdt. 93-57, 54 FR 34906, Aug. 22, 1989; 54 FR 37303, Sept. 8, 1989; Amdt. 93-59, 54 FR 39843, Sept. 28, 1989; Amdt. 93-62, 56 FR 41207, Aug. 19, 1991; Amdt. 93-78, 64 FR 53564, Oct. 1, 1999; Amdt. 93-84, 70 FR 29063, May 19, 2005]

§ 93.125 Arrival or departure reservation.

Except between 12 Midnight and 6 a.m. local time, no person may operate an aircraft to or from an airport designated as a high density traffic airport unless he has received, for that operation, an arrival or departure reservation from ATC.

[Doc. No. 9974, 37 FR 22794, Oct. 25, 1972]

§ 93.129 Additional operations.

(a) *IFR.* The operator of an aircraft may take off or land the aircraft under IFR at a designated high density traffic airport without regard to the maximum number of operations allocated for that airport if the operation is not a scheduled operation to or from a high density airport and he obtains a departure or arrival reservation, as appropriate, from ATC. The reservation is granted by ATC whenever the aircraft may be accommodated without significant additional delay to the operations allocated for the airport for which the reservations is requested.

(b) *VFR.* The operator of an aircraft may take off and land the aircraft under VFR at a designated high density traffic airport without regard to the maximum number of operations allocated for that airport if the operation is not a scheduled operation to or from a high density airport and he obtains a departure or arrival reservation, as appropriate, from ATC. The reservation is granted by ATC whenever the aircraft may be accommodated without significant additional delay to the operations allocated for the airport for which the reservation is requested and the ceiling reported at the airport is at least 1,000 feet and the ground visibility reported at the airport is at least 3 miles.

(c) For the purpose of this section a *scheduled operation to or from the high* density airport is any operation regularly conducted by an air carrier or commuter between a high density airport and another point regularly served by that operator unless the service is conducted pursuant to irregular charter or hiring of aircraft or is a nonpassenger flight.

(d) An aircraft operator must obtain an IFR reservation in accordance with procedures established by the Administrator. For IFR flights to or from a high density airport, reservations for takeoff and arrival shall be obtained prior to takeoff.

[Doc. No. 9113, 34 FR 2603, Feb. 26, 1969, as amended by Amdt. 93-25, 37 FR 22794, Oct. 25, 1972; Amdt. 93-44, 46 FR 58049, Nov. 27, 1981; Amdt. 93-46, 49 FR 8244, Mar. 6, 1984]

§ 93.130 Suspension of allocations.

The Administrator may suspend the effectiveness of any allocation prescribed in § 93.123 and the reservation requirements prescribed in § 93.125 if he finds such action to be consistent with the efficient use of the airspace. Such suspension may be terminated whenever the Administrator determines that such action is necessary for the efficient use of the airspace.

[Doc. No. 9974, 35 FR 16592, Oct. 24, 1970, as amended by Amdt. 93-21, 35 FR 16636, Oct. 27, 1970; Amdt. 93-27, 38 FR 29464, Oct. 25, 1973]

§ 93.133 Exceptions.

Except as provided in § 93.130, the provisions of §§ 93.123 and 93.125 do not apply to—

(a) The Newark Airport, Newark, NJ;

(b) The Kennedy International Airport, New York, NY, except during the hours from 3 p.m. through 7:59 p.m., local time; and

(c) O'Hare International Airport from 9:15 p.m. to 6:44 a.m., local time.

[Doc. No. 24471, 49 FR 8244, Mar. 6, 1984]

Subpart L [Reserved]

Subpart M—Ketchikan International Airport Traffic Rule

Source: Docket No. 14687, 41 FR 14879, Apr. 8, 1976, unless otherwise noted.

§ 93.151 Applicability.

This subpart prescribes a special air traffic rule for aircraft conducting VFR operations in the vicinity of the Ketchikan International Airport or Ketchikan Harbor, Alaska.

[Doc. No. FAA-2002-13235, 68 FR 9795, Feb. 28, 2003]

§ 93.152 Description of area.

Within that airspace below 3,000 feet MSL within the lateral boundary of the surface area of the Ketchikan Class E airspace regardless of whether that airspace is in effect.

[Doc. No. FAA-2002-13235, 68 FR 9795, Feb. 28, 2003]

§ 93.153 Communications.

(a) When the Ketchikan Flight Service Station is in operation, no person may operate an aircraft within the airspace specified in § 93.151, or taxi onto the runway at Ketchikan International Airport, unless that person has established two-way radio communications with the Ketchikan Flight Service Station for the purpose of receiving traffic advisories and continues to monitor the advisory frequency at all times while operating within the specified airspace.

(b) When the Ketchikan Flight Service Station is not in operation, no person may operate an aircraft within the airspace specified in § 93.151, or taxi onto the runway at Ketchikan International Airport, unless that person continuously monitors and communicates, as appropriate, on the designated common traffic advisory frequency as follows:

(1) *For inbound flights.* Announces position and intentions when no less than 10 miles from Ketchikan International Airport, and monitors the designated frequency until clear of the movement area on the airport or Ketchikan Harbor.

(2) *For departing flights.* Announces position and intentions prior to taxiing onto the active runway on the airport or onto the movement area of Ketchikan Harbor and monitors the designated frequency until outside the airspace described in § 93.151 and announces position and intentions upon departing that airspace.

(c) Notwithstanding the provisions of paragraphs (a) and (b) of this section, if two-way radio communications failure occurs in flight, a person may operate an aircraft within the airspace specified in § 93.151, and land, if weather conditions are at or above basic VFR weather minimums.

[Doc. No. 26653, 56 FR 48094, Sept. 23, 1991]

§ 93.155 Aircraft operations.

(a) When an advisory is received from the Ketchikan Flight Service Station stating that an aircraft is on final approach to the Ketchikan International Airport, no person may taxi onto the runway of that airport until the approaching aircraft has landed and has cleared the runway.

(b) Unless otherwise authorized by ATC, each person operating a large airplane or a turbine engine powered airplane shall—

(1) When approaching to land at the Ketchikan International Airport, maintain an altitude of at least 900 feet MSL until within three miles of the airport; and

(2) After takeoff from the Ketchikan International Airport, maintain runway heading until reaching an altitude of 900 feet MSL.

Subpart N [Reserved]

Subpart O—Special Flight Rules in the Vicinity of Luke AFB, AZ

Source: 74 FR 69278, Dec. 31, 2009, unless otherwise noted.

§ 93.175 Applicability.

This subpart prescribes a Special Air Traffic Rule for aircraft conducting VFR operations in the vicinity of Luke Air Force Base, AZ.

§ 93.176 Description of area.

The Luke Air Force Base, Arizona Terminal Area is designated during official daylight hours Monday through Friday while Luke pilot flight training is underway, as broadcast on the local Automatic Terminal Information Service (ATIS), and other times by Notice to Airmen (NOTAM), as follows:

(a) East Sector:

(1) South section includes airspace extending from 3,000 feet MSL to the base of the overlaying Phoenix Class B airspace bounded by a line beginning at: Lat. 33°23′56″ N; Long. 112°28′37″ W; to Lat. 33°22′32″ N; Long. 112°37′14″ W; to Lat. 33°25′39″ N; Long. 112°37′29″ W; to Lat. 33°31′55″ N; Long. 112°30′32″ W; to Lat. 33°28′00″ N; Long. 112°28′41″ W; to point of beginning.

(2) South section lower includes airspace extending from 2,100 feet MSL to the base of the overlaying Phoenix Class B airspace, excluding the Luke Class D airspace area bounded by a line beginning at: Lat. 33°28′00″ N; Long. 112°28′41″ W; to Lat. 33°23′56″ N; Long. 112°28′37″ W; to Lat. 33°27′53″ N; Long. 112°24′12″ W; to point of beginning.

(3) Center section includes airspace extending from surface to the base of the overlaying Phoenix Class B airspace, excluding the Luke Class D airspace area bounded by a line beginning at: Lat. 33°42′22″ N; Long. 112°19′16″ W; to Lat. 33°38′40″ N; Long. 112°14′03″ W; to Lat. 33°27′53″ N; Long. 112°24′12″ W; to Lat. 33°28′00″ N; Long. 112°28′41″ W; to Lat. 33°31′55″ N; Long. 112°30′32″ W; to point of beginning.

(4) The north section includes that airspace extending upward from 3,000 feet MSL to 4,000 feet MSL, bounded by a line beginning at: Lat. 33°42′22″ N; Long. 112°19′16″ W; to Lat. 33°46′58″ N; Long. 112°16′41″ W; to Lat. 33°44′48″ N; Long. 112°10′59″ W; to Lat. 33°38′40″ N; Long. 112°14′03″ W; to point of beginning.

(b) West Sector:

(1) The north section includes that airspace extending upward from 3,000 feet MSL to 6,000 feet MSL, bounded by a line beginning at: Lat. 33°51′52″ N; Long. 112°37′54″ W; to Lat. 33°49′34″ N; Long. 112°23′34″ W; to Lat. 33°46′58″ N; Long. 112°16′41″ W; to Lat. 33°42′22″ N; Long. 112°19′16″ W; to Lat. 33°39′27″ N; Long. 112°22′27″ W; to point of beginning.

(2) The south section includes that airspace extending upward from the surface to 6,000 feet MSL, bounded by a line beginning at: Lat. 33°39′27″ N; Long. 112°22′27″ W; to Lat. 33°38′06″ N; Long. 112°23′51″ W; to Lat. 33°38′07″ N; Long. 112°28′50″ W; to Lat. 33°39′34″ N; Long. 112°31′39″ W; to Lat. 33°39′32″ N; Long. 112°37′36″ W; to Lat. 33°51′52″ N; Long. 112°37′54″ W; to point of beginning.

§ 93.177 Operations in the Special Air Traffic Rule Area.

(a) Unless otherwise authorized by Air Traffic Control (ATC), no person may operate an aircraft in flight within the Luke Terminal Area designated in § 93.176 unless—

(1) Before operating within the Luke Terminal area, that person establishes radio contact with the Luke RAPCON; and

(2) That person maintains two-way radio communication with the Luke RAPCON or an appropriate ATC facility while within the designated area.

(b) Requests for deviation from the provisions of this section apply only to aircraft not equipped with an operational radio. The request must be submitted at least 24 hours before the proposed operation to Luke RAPCON.

Subparts P-R [Reserved]

Subpart S—Allocation of Commuter and Air Carrier IFR Operations at High Density Traffic Airports

Source: Docket No. 24105, 50 FR 52195, Dec. 20, 1985, unless otherwise noted.

§ 93.211 Applicability.

(a) This subpart prescribes rules applicable to the allocation and withdrawal of IFR operational authority (takeoffs and landings) to individual air carriers and commuter operators at the High Density Traffic Airports identified in subpart K of this part except for Newark Airport.

(b) This subpart also prescribes rules concerning the transfer of allocated IFR operational authority and the use of that authority once allocated.

§ 93.213 Definitions and general provisions.

(a) For purposes of this subpart—

(1) *New entrant carrier* means a commuter operator or air carrier which does not hold a slot at a particular airport and has never sold or given up a slot at that airport after December 16, 1985.

(2) *Slot* means the operational authority to conduct one IFR landing or takeoff operation each day during a specific hour or 30 minute period at one of the High Density Traffic Airports, as specified in subpart K of this part.

(3) *Summer season* means the period of time from the first Sunday in April until the last Sunday in October.

(4) *Winter season* means the period of time from the last Sunday in October until the first Sunday in April.

(5) *Limited incumbent carrier* means an air carrier or commuter operator that holds or operates fewer than 12 air carrier or commuter slots, in any combination, at a particular airport, not including international slots, Essential Air Service Program slots, or slots between the hours of 2200 and 0659 at Washington National Airport or LaGuardia Airport. However, for the purposes of this paragraph (a)(5), the carrier is considered to hold the number of slots at that airport that the carrier has, since December 16, 1985:

(i) Returned to the FAA;

(ii) Had recalled by the FAA under § 93.227(a); or

(iii) Transferred to another party other than by trade for one or more slots at the same airport.

(b) The definitions specified in subpart K of this part also apply to this subpart.

(c) For purposes of this subpart, if an air carrier, commuter operator, or other person has more than a 50-percent ownership or control of one or more other air carriers, commuter operators, or other persons, they shall be considered to be a single air carrier, commuter operator, or person. In addition, if a single company has more than a 50-percent ownership or control of two or more air carriers and/or commuter operators or any combination thereof, those air carriers and/or commuter operators shall be considered to be a single operator. A single operator may be considered to be both an air carrier and commuter operator for purposes of this subpart.

[Doc. No. 24105, 50 FR 52195, Dec. 20, 1985, as amended by Amdt. 93-52, 51 FR 21717, June 13, 1986; Amdt. 93-57, 54 FR 34906, Aug. 22, 1989; 54 FR 37303, Sept. 8, 1989; Amdt. 93-65, 57 FR 37314, Aug. 18, 1992]

§ 93.215 Initial allocation of slots.

(a) Each air carrier and commuter operator holding a permanent slot on December 16, 1985, as evidenced by the records of the air carrier and commuter operator scheduling committees, shall be allocated those slots subject to withdrawal under the provisions of this subpart. The Chief Counsel of the FAA shall be the final decisionmaker for initial allocation determinations.

(b) Any permanent slot whose use on December 16, 1985 is divided among different operators, by day of the week, or otherwise, as evidenced by records of the scheduling committees, shall be allocated in conformity with those records. The Chief Counsel of the FAA shall be the final decisionmaker for these determinations.

(c) A carrier may permanently designate a slot it holds at Kennedy International Airport as a seasonal slot, to be held by the carrier only during the corresponding season in future years, if it notifies the FAA (at the address specified in § 93.225(e)), in writing, the preceding winter seasons or by October 15 of the preceding year for summer seasons.

(d) Within 30 days after December 16, 1985, each U.S. air carrier and commuter operator must notify the office specified in § 93.221(a)(1), in writing, of those slots used for operations described in § 93.217(a)(1) on December 16, 1985.

(e) Any slot not held by an operator on December 16, 1985 shall be allocated in accordance with the provisions of §§ 93.217, 93.219 or 93.225 of this subpart.

[Doc. No. 24105, 50 FR 52195, Dec. 20, 1985, as amended by Amdt. 93-52, 51 FR 21717, June 13, 1986]

§ 93.217 Allocation of slots for international operations and applicable limitations.

(a) Any air carrier of commuter operator having the authority to conduct international operations shall be provided slots for those operations, excluding transborder service solely between HDR airports and Canada, subject to the following conditions and the other provisions of this section:

(1) The slot may be used only for a flight segment in which either the takeoff or landing is at a foreign point or, for foreign operators, the flight segment is a continuation of a flight that begins or ends at a foreign point. Slots may be obtained and used under this section only for operations at Kennedy and O'Hare airports unless otherwise required by bilateral agreement and only for scheduled service unless the requesting carrier qualifies for the slot on the basis of historic seasonal operations, under § 93.217(a)(5).

(2) Slots used for an operation described in paragraph (a)(1) of this section may not be bought, sold, leased, or otherwise transferred, except that such a slot may be traded to another slot-holder on a one-for-one basis for a slot at the same airport in a different hour or half-hour period if the trade is for the purpose of conducting such an operation in a different hour or half-hour period.

(3) Slots used for operations described in paragraph (a)(1) of this section must be returned to the FAA if the slot will not be used for such operations for more than a 2-week period.

(4) Each air carrier or commuter operator having a slot that is used for operations described in paragraph (a)(1) of this section but is not used every day of the week shall notify the office specified in § 93.221(a)(1) in writing of those days on which the slots will not be used.

(5) Except as provided in paragraph (a)(10) of this section, at Kennedy and O'Hare Airports, a slot shall be allocated, upon request, for seasonal international operations, including charter operations, if the Chief Counsel of the FAA determines that the slot had been permanently allocated to and used by the requesting carrier in the same hour and for the same time period during the corresponding season of the preceding year. Requests for such slots must be submitted to the office specified in § 93.221(a)(1), by the deadline published in a Federal Register notice for each season. For operations during the 1986 summer season, requests under this paragraph must have been submitted to the FAA on or before February 1, 1986. Each carrier requesting a slot under this paragraph must submit its entire international schedule at the relevant airport for the particular season, noting which requests are in addition to or changes from the previous year.

(6) Except as provided in paragraph (a)(10) of this section, additional slots shall be allocated at O'Hare Airport for international scheduled air carrier and commuter operations (beyond those slots allocated under §§ 93.215 and 93.217(a)(5) if a request is submitted to the office specified in § 93.221(a)(1) and filed by the deadline published in a Federal Register notice for each season. These slots will be allocated at the time requested unless a slot is available within one hour of the requested time, in which case the unallocated slots will be used to satisfy the request.

(7) If required by bilateral agreement, additional slots shall be allocated at LaGuardia Airport for international scheduled passenger operations within the hour requested.

(8) To the extent vacant slots are available, additional slots during the high density hours shall be allocated at Kennedy Airport for new international scheduled air carrier and commuter operations (beyond those operations for which slots have been allocated under §§ 93.215 and 93.217(a)(5)), if a request is submitted to the office specified in § 93.221(a)(1) by the deadline published in a Federal Register notice for each season. In addition, slots may be withdrawn from domestic operations for operations at Kennedy Airport under this paragraph if required by international obligations.

(9) In determining the hour in which a slot request under §§ 93.217(a)(6) and 93.217(a)(8) will be granted, the following will be taken into consideration, among other things:

(i) The availability of vacant slot times;

(ii) International obligations;

(iii) Airport terminal capacity, including facilities and personnel of the U.S. Customs Service and the U.S. Immigration and Naturalization Service;

(iv) The extent and regularity of intended use of a slot; and

(v) Schedule constraints of carriers requesting slots.

(10) At O'Hare Airport, a slot will not be allocated under this section to a carrier holding or operating 100 or more permanent slots on the previous May 15 for a winter season or October 15 for a summer season unless:

(i) Allocation of the slot does not result in a total allocation to that carrier under this section that exceeds the number of slots allocated to and scheduled by that carrier under this section on February 23, 1990, and as reduced by the number of slots reclassified under § 93.218, and does not exceed by more than 2 the number of slots allocated to and scheduled by that carrier during any half hour of that day, or

(ii) Notwithstanding the number of slots allocated under paragraph (a)(10)(i) of this section, a slot is available for allocation without withdrawal of a permanent slot from any carrier.

(b) If a slot allocated under § 93.215 was scheduled for an operation described in paragraph (a)(1) of this section on December 16, 1985, its use shall be subject to the requirements of paragraphs (a)(1) through (a)(4) of this section. The requirements also apply to slots used for international operations at LaGuardia Airport.

(c) If a slot is offered to a carrier in other than the hour requested, the carrier shall have 14 days after the date of the offer to accept the newly offered slot. Acceptance must be in writing and sent to the office specified in § 93.221(a)(1) and must repeat the certified statements required by paragraph (e) of this section.

(d) The Office of the Secretary of Transportation reserves the right not to apply the provisions of this section, concerning the allocation of slots, to any foreign air carrier or commuter operator of a country that provides slots to U.S. air carriers and commuter operators on a basis more restrictive than provided by this subpart. Decisions not to apply the provisions of this section will be made by the Office of the Secretary of Transportation.

(e) Each request for slots under this section shall state the airport, days of the week and time of the day of the desired slots and the period of time the slots are to be used. Each request shall identify whether the slot is requested under paragraph (a)(5), (6), or (8) and identify any changes from the previous year if requested under both paragraphs. The request must be accompanied by a certified statement signed by an officer of the operator indicating that the operator has or has contracted for aircraft capable of being utilized in using the slots requested and that the operator has bona fide plans to use the requested slots for operations described in paragraph (a).

[Doc. No. 24105, 51 FR 21717, June 13, 1986, as amended by Amdt. 93-61, 55 FR 53243, Dec. 27, 1990; 56 FR 1059, Jan. 10, 1991; Amdt. 93-78, 64 FR 53565, Oct. 1, 1999]

§ 93.218 Slots for transborder service to and from Canada.

(a) Except as otherwise provided in this subpart, international slots identified by U.S. carriers for international operations in December 1985 and the equivalent number of international slots held as of February 24, 1998, will be domestic slots. The Chief Counsel of the FAA shall be the final decisionmaker for these determinations.

(b) Canadian carriers shall have a guaranteed base level of slots of 42 slots at LaGuardia, 36 slots at O'Hare for the Sumner season, and 32 slots at O'Hare in the Winter season.

(c) Any modification to the slot base by the Government of Canada or the Canadian carriers that results in a decrease of the guaranteed base in paragraph (b) of this section shall permanently modify the base number of slots.

[Doc. No. FAA-1999-4971, 64 FR 53565, Oct. 1, 1999]

§ 93.219 Allocation of slots for essential air service operations and applicable limitations.

Whenever the Office of the Secretary of Transportation determines that slots are needed for operations to or from a High Density Traffic Airport under the Department of Transportation's Essential Air Service (EAS) Program, those slots shall be provided to the designated air carrier or commuter operator subject to the following limitations:

(a) Slots obtained under this section may not be bought, sold, leased or otherwise transferred, except that such slots may be traded for other slots on a one-for-one basis at the same airport.

(b) Any slot obtained under this section must be returned to the FAA if it will not be used for EAS purposes for more than a 2-week period. A slot returned under this paragraph may be reallocated to the operator which returned it upon request to the FAA office specified in § 93.221(a)(1) if that slot has not been reallocated to an operator to provide substitute essential air service.

(c) Slots shall be allocated for EAS purposes in a time period within 90 minutes of the time period requested.

(d) The Department will not honor requests for slots for EAS purposes to a point if the requesting carrier has previously traded away or sold slots it had used or obtained for use in providing essential air service to that point.

(e) Slots obtained under Civil Aeronautics Board Order No. 84-11-40 shall be considered to have been obtained under this section.

§ 93.221 Transfer of slots.

(a) Except as otherwise provided in this subpart, effective April 1, 1986, slots may be bought, sold or leased for any consideration and any time period and they may be traded in any combination for slots at the same airport or any other high density traffic airport. Transfers, including leases, shall comply with the following conditions:

(1) Requests for confirmation must be submitted in writing to Slot Administration Office, AGC-230, Office of the Chief Counsel, Federal Aviation Administration, 800 Independence Ave., SW., Washington, DC 20591, in a format to be prescribed by the Administrator. Requests will provide the names of the transferor and recipient; business address and telephone number of the persons representing the transferor and recipient; whether the slot is to be used for an arrival or departure; the date the slot was acquired by the transferor; the section of this subpart under which the slot was allocated to the transferor; whether the slot has been used by the transferor for international or essential air service operations; and whether the slot will be used by the recipient for international or essential air service operations. After withdrawal priorities have been established under § 93.223 of this part, the requests must include the slot designations of the transferred slots as described in § 93.223(b)(5).

(2) The slot transferred must come from the transferor's then-current FAA-approved base.

(3) Written evidence of each transferor's consent to the transfer must be provided to the FAA.

(4) The recipient of a transferred slot may not use the slot until written confirmation has been received from the FAA.

(5)(i) Until a slot obtained by a new entrant or limited incumbent carrier in a lottery held under § 93.225 after June 1, 1991, has been used by the carrier that obtained it for a continuous 24-month period after the lottery in accordance with § 93.227(a), that slot may be transferred only by trade for one or more slots at the same airport or to other new entrant or limited incumbent carriers under § 93.221(a)(5)(iii). This transfer restriction shall apply to the same extent to any slot or slots acquired by trading the slot obtained in a lottery. To remove the transfer restriction, documentation of 24 months' continuous use must be submitted to the FAA Office of the Chief Counsel.

(ii) Failure to use a slot acquired by trading a slot obtained in a lottery for a continuous 24-month period after the lottery, shall void all trades involving the lottery slot, which shall be returned to the FAA. All use of the lottery slot shall be counted toward fulfilling the minimum use requirements under § 93.227(a) applicable to the slot or slots for which the lottery slot was traded, including subsequent trades.

(iii) Slots obtained by new entrant or limited incumbent carriers in a lottery may be sold, leased, or otherwise transferred to another entrant or limited incumbent carrier after a minimum of 60 days of use by the obtaining carrier. The transfer restrictions of § 93.221(a)(5)(i) shall continue to apply to the slot until documentation of 24 months' continuous use has been submitted and the transfer restriction removed.

(6) The Office of the Secretary of Transportation must determine that the transfer will not be injurious to the essential air service program.

(b) A record of each slot transfer shall be kept on file by the office specified in paragraph (a)(1) of this section and will be made available to the public upon request.

(c) Any person may buy or sell slots and any air carrier or commuter may use them. Notwithstanding § 93.123, air carrier slots may be used with aircraft of the kind described in § 93.123 (c)(1) or (c)(2) but commuter slots may only be used with aircraft of the kind described in § 93.0123(c)(2).

(d) Air carriers and commuter operators considered to be a single operator under the provisions of § 93.213(c) of this subpart but operating under separate names shall report transfers of slots between them.

(e) Notwithstanding § 93.123(c)(2) of this part, a commuter slot at O'Hare International Airport may be used with an aircraft described in § 93.123(c)(1) of this part on the following conditions:

(1) Air carrier aircraft that may be operated under this paragraph are limited to aircraft:

(i) Having an actual seating configuration of 110 or fewer passengers; and

(ii) Having a maximum certificated takeoff weight of less than 126,000 pounds.

(2) No more than 50 percent of the total number of commuter slots held by a slot holder at O'Hare International Airport may be used with aircraft described in paragraph (e)(1) of this section.

(3) An air carrier or commuter operator planning to operate an aircraft described in paragraph (e)(1) of this section in a commuter slot shall notify ATC at least 75 days in advance of the planned start date of such operation. The notice shall include the slot number, proposed time of operation, aircraft type, aircraft series, actual aircraft seating configuration, and planned start date. ATC will approve or disapprove the proposed operation no later than 45 days prior to the planned start date. If an operator does not initiate operation of a commuter slot under this section within 30 days of the planned start date first submitted to the FAA, the ATC approval for that operation will expire. That operator may file a new or revised notice for the same half-hour slot time.

(4) An operation may not be conducted under paragraph (e)(1) of this section unless a gate is available for that operation without planned waiting time.

(5) For the purposes of this paragraph (e), notice to ATC shall be submitted in writing to: Director, Air Traffic System Management, ATM-1, Federal Aviation Administration, 800 Independence Avenue SW., Washington, DC 20591.

[Doc. No. 24105, 50 FR 52195, Dec. 20, 1985, as amended by Amdt. 93-52, 51 FR 21717, June 13, 1986; Amdt. 93-58, 54 FR 39293, Sept. 25, 1989; Amdt. 93-62, 56 FR 41208, Aug. 19, 1991; Amdt. 93-65, 57 FR 37314, Aug. 18, 1992; Amdt. 93-68, 58 FR 39616, July 23, 1993]

§ 93.223 Slot withdrawal.

(a) Slots do not represent a property right but represent an operating privilege subject to absolute FAA control. Slots may be withdrawn at any time to fulfill the Department's operational needs, such as providing slots for international or essential air service operations or eliminating slots. Before withdrawing any slots under this section to provide them for international operations, essential air services or other operational needs, those slots returned under § 93.224 of this part and those recalled by the agency under § 93.227 will be allocated.

(b) Separate slot pools shall be established for air carriers and commuter operators at each airport. The FAA shall assign, by random lottery, withdrawal priority numbers for the recall priority of slots at each airport. Each additional permanent slot, if any, will be assigned the next higher number for air carrier or commuter slots, as appropriate, at each airport. Each slot shall be assigned a designation consisting of the applicable withdrawal priority number; the airport code; a code indicating whether the slot is an air carrier or commuter operator slot; and the time period of the slot. The designation shall also indicate, as appropriate, if the slot is daily or for certain days of the week only; is limited to arrivals or departures; is allocated for international operations or for EAS purposes; and, at Kennedy International Airport, is a summer or winter slot.

(c) Whenever slots must be withdrawn, they will be withdrawn in accordance with the priority list established under paragraph (b) of this section, except:

(1) Slots obtained in a lottery held pursuant to § 93.225 of this part shall be subject to withdrawal pursuant to paragraph (i) of that section, and

(2) Slots necessary for international and essential air service operations shall be exempt from withdrawal for use for other international or essential air service operations.

(3) Except as provided in § 93.227(a), the FAA shall not withdraw slots held at an airport by an air carrier or commuter operator holding and operating 12 or fewer slots at that airport (excluding slots used for operations described in § 93.212(a)(1)), if withdrawal would reduce the number of slots held below the number of slots operated.

(4) No slot comprising the guaranteed base of slots, as defined in section 93.318(b), shall be withdrawn for use for international operations or for new entrants.

(d) The following withdrawal priority rule shall be used to permit application of the one-for-one trade provisions for international and essential air service slots and the slot withdrawal provisions where the slots are needed for other than international or essential air service operations. If an operator has more than one slot in a specific time period in which it also has a slot being used for international or essential air service operations, the international and essential air service slots will be considered to be those with the lowest withdrawal priority.

(e) The operator(s) using each slot to be withdrawn shall be notified by the FAA of the withdrawal and shall cease operations using that slot on the date indicated in the notice. Generally, the FAA will provide at least 30 days after notification for the operator to cease operations unless exigencies require a shorter time period.

(f) For 24 months following a lottery held after June 1, 1991, a slot acquired in that lottery shall be withdrawn by the FAA upon the sale, merger, or acquisition of more than 50 percent ownership or control of the carrier using that slot or one acquired by trade of that slot, if the resulting total of slots held or operated at the airport by the surviving entity would exceed 12 slots.

[Doc. No. 24105, 50 FR 52195, Dec. 20, 1985, as amended by Amdt. 93-52, 51 FR 21718, June 13, 1986; Amdt. 93-57, 54 FR 34906, Aug. 22, 1989; Amdt. 93-65, 57 FR 37314, Aug. 18, 1992; Amdt. 93-78, 64 FR 53565, Oct. 1, 1999]

§ 93.224 Return of slots.

(a) Whenever a slot is required to be returned under this subpart, the holder must notify the office specified in § 93.221(a)(1) in writing of the date after which the slot will not be used.

(b) Slots may be voluntarily returned for use by other operators by notifying the office specified in § 93.221(a)(1) in writing.

§ 93.225 Lottery of available slots.

(a) Whenever the FAA determines that sufficient slots have become available for distribution for purposes other than international or essential air service operations, but generally not more than twice a year, they shall be allocated in accordance with the provisions of this section.

(b) A random lottery shall be held to determine the order of slot selection.

(c) Slot allocation lotteries shall be held on an airport-by-airport basis with separate lotteries for air carrier and commuter operator slots. The slots to be allocated in each lottery will be each unallocated slot not necessary for international or Essential Air Service Program operations, including any slot created by an increase in the operating limits set forth in § 93.123(a).

(d) The FAA shall publish a notice in the Federal Register announcing any lottery dates. The notice may include special procedures to be in effect for the lotteries.

(e) Participation in a lottery is open to each U.S. air carrier or commuter operator operating at the airport and providing scheduled passenger service at the airport, as well as where provided for by bilateral agreement. Any U.S. carrier, or foreign air carrier where provided for by bilateral agreement, that is not operating scheduled service at the airport and has not failed to operate slots obtained in the previous lottery, or slots traded for those obtained by lottery, but wishes to initiate scheduled passenger service at the airport, shall be included in the lottery if that operator notifies, in writing, the Slot Administration Office, AGC-230, Office of the Chief Counsel, Federal Aviation Administration, 800 Independence Avenue, SW.,

Washington, DC 20591. The notification must be received 15 days prior to the lottery date and state whether there is any common ownership or control of, by, or with any other air carrier or commuter operator as defined in § 93.213(c). New entrant and limited incumbent carriers will be permitted to complete their selections before participation by other incumbent carriers is initiated.

(f) At the lottery, each operator must make its selection within 5 minutes after being called or it shall lose its turn. If capacity still remains after each operator has had an opportunity to select slots, the allocation sequence will be repeated in the same order. An operator may select any two slots available at the airport during each sequence, except that new entrant carriers may select four slots, if available, in the first sequence.

(g) To select slots during a slot lottery session, a carrier must have appropriate economic authority for scheduled passenger service under Title IV of the Federal Aviation Act of 1958, as amended (49 U.S.C. App. 1371 *et seq.*), and must hold FAA operating authority under part 121 or part 135 of this chapter as appropriate for the slots the operator seeks to select.

(h) During the first selection sequence, 25 percent of the slots available but no less than two slots shall be reserved for selection by new entrant carriers. If new entrant carriers do not select all of the slots set aside for new entrant carriers, limited incumbent carriers may select the remaining slots. If every participating new entrant carrier and limited incumbent carrier has ceased selection of available slots or has obtained 12 slots at that airport, other incumbent carriers may participate in selecting the remaining slots; however, slots selected by non-limited incumbent carriers will be allocated only until the date of the next lottery.

(i) Slots obtained under this section shall retain their withdrawal priority as established under § 93.223. If the slot is newly created, a withdrawal priority shall be assigned. That priority number shall be higher than any other slot assigned a withdrawal number previously.

[Doc. No. 24105, 50 FR 52195, Dec. 20, 1985, as amended by Amdt. 93-52, 51 FR 21718, June 13, 1986; Amdt. 93-58, 54 FR 39293, Sept. 25, 1989; Amdt. 93-65, 57 FR 37314, Aug. 18, 1992; 57 FR 47993, Oct. 21, 1992; Amdt. 93-78, 64 FR 53565, Oct. 1, 1999]

§ 93.226 Allocation of slots in low-demand periods.

(a) If there are available slots in the following time periods and there are no pending requests for international or EAS operations at these times, FAA will allocate slots upon request on a first-come, first-served basis, as set forth in this section:

(1) Any period for which a slot is available less than 5 days per week.

(2) Any time period for which a slot is available for less than a full season.

(3) For LaGuardia and Washington National Airports:

(i) 6:00 a.m.-6:59 a.m.

(ii) 10:00 p.m.-midnight.

(b) Slots will be allocated only to operators with the economic and operating authority and aircraft required to use the slots.

(c) Requests for allocations under this section shall be submitted in writing to the address listed in § 93.221(a)(1) and shall identify the request as made under this section.

(d) The FAA may deny requests made under this section after a determination that all remaining slots in a particular category should be distributed by lottery.

(e) Slots may be allocated on a seasonal or temporary basis under this provision.

[Doc. No. 24105, 51 FR 21718, June 13, 1986]

§ 93.227 Slot use and loss.

(a) Except as provided in paragraphs (b), (c), (d), (g), and (l) of this section, any slot not utilized 80 percent of the time over a 2-month period shall be recalled by the FAA.

(b) Paragraph (a) of this section does not apply to slots obtained under § 93.225 of this part during:

(1) The first 90 days after they are allocated to a new entrant carrier; or

(2) The first 60 days after they are allocated to a limited incumbent or other incumbent carrier.

(c) Paragraph (a) of this section does not apply to slots of an operator forced by a strike to cease operations using those slots.

(d) In the case of a carrier that files for protection under the Federal bankruptcy laws and has not received a Notice of Withdrawal from the FAA for the subject slot or slots, paragraph (a) of this section does not apply:

(1) During a period after the initial petition in bankruptcy, to any slot held or operated by that carrier, for:

(i) 60 days after the carrier files the initial petition in bankruptcy; and

(ii) 30 days after the carrier, in anticipation of transferring slots, submits information to a Federal government agency in connection with a statutory antitrust, economic impact, or similar review of the transfer, provided that the information is submitted more than 30 days after filing the initial petition in bankruptcy, and provided further that any slot to be transferred has not become subject to withdrawal under any other provision of this § 93.227; and

(2) During a period after a carrier ceases operations at an airport, to any slot held or operated by that carrier at that airport, for:

(i) 30 days after the carrier ceases operations at that airport, provided that the slot has not become subject to withdrawal under any other provision of this § 93.227; and

(ii) 30 days after the parties to a proposed transfer of any such slot comply with requests for additional information by a Federal government agency in connection with an antitrust, economic impact, or similar investigation of the transfer, provided that—

(A) The original notice of the transfer is filed with the Federal agency within 30 days after the carrier ceases operation at the airport;

(B) The request for additional information is made within 10 days of the filing of the notice by the carrier;

(C) The carrier submits the additional information to the Federal agency within 15 days of the request by such agency; and

(D) Any slot to be transferred has not become subject to withdrawal under any other provision of this § 93.227.

(e) Persons having slots withdrawn pursuant to paragraph (a) of this section must cease all use of those slots upon receipt of notice from the FAA.

(f) Persons holding slots but not using them pursuant to the provisions of paragraphs (b), (c) and (d) may lease those slots for use by others. A slot obtained in a lottery may not be leased after the expiration of the applicable time period specified in paragraph (b) of this section unless it has been operated for a 2-month period at least 65 percent of the time by the operator which obtained it in the lottery.

(g) This section does not apply to slots used for the operations described in § 93.217(a)(1) except that a U.S. air carrier or commuter operator required to file a report under paragraph (i) of this section shall include all slots operated at the airport, including slots described in § 93.217(a)(1).

(h) Within 30 days after an operator files for protection under the Federal bankruptcy laws, the FAA shall recall any slots of that operator, if—(1) the slots were formerly used for essential air service and (2) the Office of the Secretary of Transportation determines those slots are required to provide substitute essential air service to or from the same points.

(i) Every air carrier and commuter operator or other person holding a slot at a high density airport shall, within 14 days after the last day of the 2-month period beginning January 1, 1986, and every 2 months thereafter, forward, in writing, to the address identified in § 93.221(a)(1), a list of all slots held by the air carrier, commuter operator or other person along with a listing of which air carrier or commuter operator actually operated the slot for each day of the 2-month period. The report shall identify the flight number for which the slot was used and the equipment used, and shall identify the flight as an arrival or departure. The report shall identify any common ownership or control of, by, or with any other carrier as defined in § 93.213(c) of this subpart. The report shall be signed by a senior official of the air carrier or commuter operator. If the slot is held by an "other person," the report must be signed by an official representative.

(j) The Chief Counsel of the FAA may waive the requirements of paragraph (a) of this section in the event of a highly unusual and unpredictable condition which is beyond the control of the slot-holder and which exists for a period of 9 or more days. Examples of conditions which could justify waiver under this paragraph are weather conditions which result in the restricted operation of an airport for an extended period of time or the grounding of an aircraft type.

(k) The Chief Counsel of the FAA may, upon request, grant a waiver from the requirements of paragraph (a) of this section for a slot used for the domestic segment of an intercontinental all-cargo flight. To qualify for a waiver, a carrier must operate the slot a substantial percentage of the time and must return the slot to the FAA in advance for the time periods it will not be used.

(l) The FAA will treat as used any slot held by a carrier at a High Density Traffic Airport on Thanksgiving Day, the Friday following Thanksgiving Day, and the period from December 24 through the first Saturday in January.

[Doc. No. 24105, 50 FR 52195, Dec. 20, 1985, as amended by Amdt. 93-52, 51 FR 21718, June 13, 1986; Amdt. 93-65, 57 FR 37315, Aug. 18, 1992; Amdt. 93-71, 59 FR 58771, Nov. 15, 1994]

Subpart T—Ronald Reagan Washington National Airport Traffic Rules

SOURCE: Docket No. 25143, 51 FR 43587, Dec. 3, 1986; Amdt. 93-82, 68 FR 9795, Feb. 28, 2003, unless otherwise noted.

§ 93.251 Applicability.

This subpart prescribes rules applicable to the operation of aircraft to or from Ronald Reagan Washington National Airport.

§ 93.253 Nonstop operations.

No person may operate an aircraft nonstop in air transportation between Ronald Reagan Washington National Airport and another airport that is more than 1,250 miles away from Ronald Reagan Washington National Airport.

Subpart U—Special Flight Rules in the Vicinity of Grand Canyon National Park, AZ

SOURCE: Doc. No. 28537, 61 FR 69330, Dec. 31, 1996, unless otherwise noted.

§ 93.301 Applicability.

This subpart prescribes special operating rules for all persons operating aircraft in the following airspace, designated as the Grand Canyon National Park Special Flight Rules Area: That airspace extending from the surface up to but not including 18,000 feet MSL within an area bounded by a line beginning at Lat. 35°55′12″ N., Long. 112°04′05″ W.; east to Lat. 35°55′30″ N., Long. 111°45′00″ W.; to Lat. 35°59′02″ N., Long. 111°36′03″ W.; north to Lat. 36°15′30″ N., Long. 111°36′06″ W.; to Lat. 36°24′49″ N., Long. 111°47′45″ W.; to Lat. 36°52′23″ N., Long. 111°33′10″ W.; west-northwest to Lat. 36°53′37″ N., Long. 111°38′29″ W.; southwest to Lat. 36°35′02″ N., Long. 111°53′28″ W.; to Lat. 36°21′30″ N., Long. 112°00′03″ W.; west-northwest to Lat. 36°30′30″ N., Long. 112°35′59″ W.; southwest to Lat. 36°24′46″ N., Long. 112°51′10″ W., thence west along the boundary of Grand Canyon National Park (GCNP) to Lat. 36°14′08″ N., Long. 113°10′07″ W.; west-southwest to Lat. 36°09′30″ N., Long. 114°03′03″ W.; southeast to Lat. 36°05′11″ N., Long. 113°58′46″ W.; thence south along the boundary of GCNP to Lat. 35°58′23″ N., Long. 113°54′14″ W.; north to Lat. 36°00′10″ N., Long. 113°53′48″ W.; northeast to Lat. 36°02′14″ N., Long. 113°50′16″ W.; to Lat. 36°02′17″ N., Long. 113°53′48″ W.; northeast to Lat. 36°02′14″ N., Long. 113°50′16″ W.; to Lat. 36°02′17″ N., Long. 113°49′11″ W.; southeast to Lat. 36°01′22″ N., Long. 113°48′21″ W.; to Lat. 35°59′15″ N., Long. 113°47′13″ W.; to Lat. 35°57′51″ N., Long. 113°46′01″ W.; to Lat. 35°57′45″ N., Long. 113°45′23″ W.; southwest to Lat. 35°54′48″ N., Long. 113°50′24″ W.; southeast to Lat. 35°41′01″ N., Long. 113°35′27″ W.; thence clockwise via the 4.2-nautical mile radius of the Peach Springs VORTAC to Lat. 36°38′53″ N., Long. 113°27′49″ W.; northeast to Lat. 35°42′58″ N., Long. 113°10′57″ W.; north to Lat. 35°57′51″ N., Long. 113°11′06″ W.; east to Lat. 35°57′44″ N., Long. 112°14′04″ W.; thence clockwise via the 4.3-nautical mile radius of the Grand Canyon National Park Airport reference point (Lat. 35°57′08″ N., Long. 112°08′49″ W.) to the point of origin.

[Doc. No. 5926, 65 FR 17742, Apr. 4, 2000]

§ 93.303 Definitions.

For the purposes of this subpart:

Allocation means authorization to conduct a commercial air tour in the Grand Canyon National Park (GCNP) Special Flight Rules Area (SFRA).

Commercial air tour means any flight conducted for compensation or hire in a powered aircraft where a purpose of the flight is sightseeing. If the operator of a flight asserts that the flight is not a commercial air tour, factors that can be considered by the Administrator in making a determination of whether the flight is a commercial air tour include, but are not limited to—

(1) Whether there was a holding out to the public of willingness to conduct a sightseeing flight for compensation or hire;

(2) Whether a narrative was provided that referred to areas or points of interest on the surface;

(3) The area of operation;

(4) The frequency of flights;

(5) The route of flight;

(6) The inclusion of sightseeing flights as part of any travel arrangement package; or

(7) Whether the flight in question would or would not have been canceled based on poor visibility of the surface.

Commercial Special Flight Rules Area Operation means any portion of any flight within the Grand Canyon National Park Special Flight Rules Area that is conducted by a certificate holder that has operations specifications authorizing flights within the Grand Canyon National Park Special Flight Rules Area. This term does not include operations conducted under an FAA Form 7711-1, Certificate of Waiver or Authorization. The types of flights covered by this definition are set forth in the "Las Vegas Flight Standards District Office Grand Canyon National Park Special Flight Rules Area Procedures Manual" which is available from the Las Vegas Flight Standards District Office.

Flight Standards District Office means the FAA Flight Standards District Office with jurisdiction for the geographical area containing the Grand Canyon.

GCNP quiet aircraft technology designation means an aircraft that is subject to § 93.301 and has been shown to comply with the noise limit specified in appendix A of this part.

Number of passenger seats means the number of passenger seats for which an individual aircraft is configured.

Park means Grand Canyon National Park.

Special Flight Rules Area means the Grand Canyon National Park Special Flight Rules Area.

[65 FR 17732, Apr. 4, 2000, as amended at 70 FR 16092, Mar. 29, 2005]

§ 93.305 Flight-free zones and flight corridors.

Except in an emergency or if otherwise necessary for safety of flight, or unless otherwise authorized by the Flight Standards District Office for a purpose listed in 93.309, no person may operate an aircraft in the Special Flight Rules Area within the following flight-free zones:

(a) *Desert View Flight-free Zone.* That airspace extending from the surface up to but not including 14,500 feet MSL within an area bounded by a line beginning at Lat. 35°59′58″ N., Long. 111°52′47″ W.; thence east to Lat. 36°00′00″ N., Long. 111°51′04″ W.; thence north to 36°00′24″ N., Long. 111°51′04″ W.; thence east to 36°00′24″ N., Long. 111°45′44″ W.; thence north along the GCNP boundary to Lat. 36°14′05″ N., Long. 111°48′34″ W.; thence southwest to Lat. 36°12′06″ N., Long. 111°51′14″ W.; to the point of origin; but not including the airspace at and above 10,500 feet MSL within 1 nautical mile of the western boundary of the zone. The corridor to the west between the Desert View and Bright Angel Flight-free Zones, is designated the "Zuni Point Corridor." This corridor is 2 nautical miles wide for commercial air tour flights and 4 nautical miles wide for transient and general aviation operations.

(b) *Bright Angel Flight-free Zone.* That airspace extending from the surface up to but not including 14,500 feet MSL within an area bounded by a line beginning at Lat. 35°58′39″ N., Long. 111°55′43″ W.; north to Lat. 36°12′41″ N., Long. 111°53′54″ W.; northwest to Lat. 36°18′18″ N., Long. 111°58′15″ W.; thence west along the GCNP boundary to Lat. 36°20′11″ N., Long. 112°06′25″ W.; south-southwest to Lat. 36°09′31″ N., Long. 112°11′15″ W.; to Lat. 36°04′16″ N., Long. 112°17′20″ W.; thence southeast along the GCNP boundary to Lat. 36°01′54″ N., Long. 112°11′24″ W.; thence clockwise via the 4.3-nautical mile radius of the Grand Canyon National Park Airport reference point (Lat. 35°57′08″ N., Long. 112°08′49″ W.) to Lat. 35°59′37″ N., Long. 112°04′29″ W.; thence east along the GCNP boundary to the point of origin; but not including the airspace at and above 10,500 feet MSL within 1 nautical mile of the eastern boundary or the airspace at and above 10,500 feet MSL within 2 nautical miles of the northwestern boundary. The corridor to the east, between this flight-free zone and the Desert View Flight-free Zone, is designated the "Zuni Point Corridor." The corridor to the west, between the Bright Angel and Toroweap/Shinumo Flight-free Zones, is designated the "Dragon Corridor." This corridor is 2 nautical miles wide for commercial air tour flights and 4 nautical miles wide for transient and general aviation operations. The Bright Angel Flight-free Zone does not include the following airspace designated as the Bright Angel Corridor: That airspace one-half nautical mile on either side of a line extending from Lat. 36°14′57″ N., Long. 112°08′45″ W. and Lat. 36°15′01″ N., Long. 111°55′39″ W.

(c) *Toroweap/Shinumo Flight-free Zone.* That airspace extending from the surface up to but not including 14,500 feet MSL within an area bounded by a line beginning at Lat. 36°05′44″ N., Long. 112°19′27″ W.; north-northeast to Lat. 36°10′49″ N., Long. 112°13′19″ W.; to Lat. 36°21′02″ N., Long. 112°08′47″ W.; thence west and south along the GCNP boundary to Lat 36°10′58″ N., Long. 113°08′35″ W.; south to Lat. 36°10′12″ N., Long. 113°08′34″ W.; thence in an easterly direction along the park boundary to the point of origin; but not including the following airspace designated as the "Tuckup Corridor": at or above 10,500 feet MSL within 2 nautical miles either side of a line extending between Lat. 36°24′42″ N., Long. 112°48′47″ W. and Lat. 36°14′17″ N., Long. 112°48′31″ W. The airspace designated as the "Fossil Canyon Corridor" is also excluded from the Toroweap/Shinumo Flight-free Zone at or above 10,500 feet MSL within 2 nautical miles either side of a line extending between Lat. 36°16′26″ N., Long. 112°34′35″ W. and Lat. 36°22′51″ N., Long. 112°18′18″ W. The Fossil Canyon Corridor is to be used for transient and general aviation operations only.

(d) *Sanup Flight-free Zone.* That airspace extending from the surface up to but not including 8,000 feet MSL within an area bounded by a line beginning at Lat. 35°59′32″ N., Long. 113°20′28″ W.; west to Lat. 36°00′55″ N., Long. 113°42′09″ W.; southeast to Lat. 35°59′57″ N., Long. 113°41′09″ W.; to Lat. 35°59′09″ N., Long. 113°40′53″ W.; to Lat. 35°58′45″ N., Long. 113°40′15″ W.; to Lat. 35°57′52″ N., Long. 113°39′34″ W.; to Lat. 35°56′44″ N., Long. 113°39′07″ W.; to Lat. 35°56′04″ N., Long. 113°39′20″ W.; to Lat. 35°55′02″ N., Long. 113°40′43″ W.; to Lat. 35°54′47″ N., Long. 113°40′51″ W.; southeast to Lat. 35°50′16″ N., Long. 113°37′13″ W.; thence along the park boundary to the point of origin.

[Doc. No. 28537, 61 FR 69330, Dec. 31, 1996, as amended by Amdt. 93-80, 65 FR 17742, Apr. 4, 2000]

§ 93.307 Minimum flight altitudes.

Except in an emergency, or if otherwise necessary for safety of flight, or unless otherwise authorized by the Flight Standards District Office for a purpose listed in 93.309, no person may operate an aircraft in the Special Flight Rules Area at an altitude lower than the following:

(a) *Minimum sector altitudes* —(1) *Commercial air tours* —(i) *Marble Canyon Sector.* Lees Ferry to Boundary Ridge: 6,000 feet MSL.

(ii) *Supai Sector.* Boundary Ridge to Supai Point: 7,500 feet MSL.

(iii) *Diamond Creek Sector.* Supai Point to Diamond Creek: 6,500 feet MSL.

(iv) *Pearce Ferry Sector.* Diamond Creek to the Grand Wash Cliffs: 5,000 feet MSL.

(2) *Transient and general aviation operations* —(i) *Marble Canyon Sector.* Lees Ferry to Boundary Ridge: 8,000 feet MSL.

(ii) *Supai Sector.* Boundary Ridge to Supai Point: 10,000 feet MSL.

(iii) *Diamond Creek Sector.* Supai Point to Diamond Creek: 9,000 feet MSL.

(iv) *Pearce Ferry Sector.* Diamond Creek to the Grand Wash Cliffs: 8,000 feet MSL.

(b) *Minimum corridor altitudes* —(1) *Commercial air tours* —(i) *Zuni Point Corridors.* 7,500 feet MSL.

(ii) *Dragon Corridor.* 7,500 feet MSL.

(2) *Transient and general aviation operations* —(i) *Zuni Point Corridor.* 10,500 feet MSL.

(ii) *Dragon Corridor.* 10,500 feet MSL.

(iii) *Tuckup Corridor.* 10,500 feet MSL.

(iv) Fossil Canyon Corridor. 10,500 feet

[Doc. No. 28537, 61 FR 69330, Dec. 31, 1996, as amended by Amdt. 93-80, 65 FR 17742, 17743, Apr. 4, 2000]

§ 93.309 General operating procedures.

Except in an emergency, no person may operate an aircraft in the Special Flight Rules Area unless the operation is conducted in accordance with the following procedures.

(NOTE: The following procedures do not relieve the pilot from see-and-avoid responsibility or compliance with the minimum safe altitude requirements specified in § 91.119 of this chapter.):

(a) Unless necessary to maintain a safe distance from other aircraft or terrain remain clear of the flight-free zones described in § 93.305;

(b) Unless necessary to maintain a safe distance from other aircraft or terrain, proceed through the Zuni Point, Dragon, Tuckup, and Fossil Canyon Flight Corridors described in § 93.305 at the following altitudes unless otherwise authorized in writing by the Flight Standards District Office:

(1) *Northbound.* 11,500 or 13,500 feet MSL.

(2) *Southbound.* 10,500 or 12,500 feet MSL.

(c) For operation in the flight-free zones described in § 93.305, or flight below the altitudes listed in § 93.307, is authorized in writing by the Flight Standards District Office and is conducted in compliance with the conditions contained in that authorization. Normally authorization will be granted for operation in the areas described in § 93.305 or below the altitudes listed in § 93.307 only for operations of aircraft necessary for law enforcement, firefighting, emergency medical treatment/evacuation of persons in the vicinity of the Park; for support of Park maintenance or activities; or for aerial access to and maintenance of other property located within the Special Flight Rules Area. Authorization may be issued on a continuing basis;

(d) Is conducted in accordance with a specific authorization to operate in that airspace incorporated in the operator's operations specifications and approved by the Flight Standards District Office in accordance with the provisions of this subpart;

(e) Is a search and rescue mission directed by the U.S. Air Force Rescue Coordination Center;

(f) Is conducted within 3 nautical miles of Grand Canyon Bar Ten Airstrip, Pearce Ferry Airstrip, Cliff Dwellers Airstrip, Marble Canyon Airstrip, or Tuweep Airstrip at an altitude less than 3,000 feet above airport elevation, for the purpose of landing at or taking off from that facility; or

(g) Is conducted under an instrument flight rules (IFR) clearance and the pilot is acting in accordance with ATC instructions. An IFR flight plan may not be filed on a route or at an altitude that would require operation in an area described in § 93.305.

[Doc. No. 28537, 61 FR 69330, Dec. 31, 1996, as amended by Amdt. 93-80, 65 FR 17742, 17743, Apr. 4, 2000]

§ 93.311 Minimum terrain clearance.

Except in an emergency, when necessary for takeoff or landing, or unless otherwise authorized by the Flight Standards District Office for a purpose listed in § 93.309(c), no person may operate an aircraft within 500 feet of any terrain or structure located between the north and south rims of the Grand Canyon.

§ 93.313 Communications.

Except when in contact with the Grand Canyon National Park Airport Traffic Control Tower during arrival or departure or on a search and rescue mission directed by the U.S. Air Force Rescue Coordination Center, no person may operate an aircraft in the Special Flight Rules Area unless he monitors the appropriate frequency continuously while in that airspace.

§ 93.315 Requirements for commercial Special Flight Rules Area operations.

Each person conducting commercial Special Flight Rules Area operations must be certificated in accordance with Part 119 for Part 135 or 121 operations and hold appropriate Grand Canyon National Park Special Flight Rules Area operations specifications.

[65 FR 17732, Apr. 4, 2000]

§ 93.316 [Reserved]

§ 93.317 Commercial Special Flight Rules Area operation curfew.

Unless otherwise authorized by the Flight Standards District Office, no person may conduct a commercial Special Flight Rules Area operation in the Dragon and Zuni Point corridors during the following flight-free periods:

(a) Summer season (May 1-September 30)-6 p.m. to 8 a.m. daily; and

(b) Winter season (October 1-April 30)-5 p.m. to 9 a.m. daily.

[65 FR 17732, Apr. 4, 2000]

§ 93.319 Commercial air tour limitations.

(a) Unless excepted under paragraph (f) or (g) of this section, no certificate holder certificated in accordance with part 119 for part 121 or 135 operations may conduct more commercial air tours in the Grand Canyon National

Park in any calendar year than the number of allocations specified on the certificate holder's operations specifications.

(b) The Administrator determines the number of initial allocations for each certificate holder based on the total number of commercial air tours conducted by the certificate holder and reported to the FAA during the period beginning on May 1, 1997 and ending on April 30, 1998, unless excepted under paragraph (g).

(c) Certificate holders who conducted commercial air tours during the base year and reported them to the FAA receive an initial allocation.

(d) A certificate holder must use one allocation for each flight that is a commercial air tour, unless excepted under paragraph (f) or (g) of this section.

(e) Each certificate holder's operation specifications will identify the following information, as applicable:

(1) Total SFRA allocations; and

(2) Dragon corridor and Zuni Point corridor allocations.

(f) Certificate holders satisfying the requirements of § 93.315 of this subpart are not required to use a commercial air tour allocation for each commercial air tour flight in the GCNP SFRA provided the following conditions are satisfied:

(1) The certificate holder conducts its operations in conformance with the routes and airspace authorizations as specified in its Grand Canyon National Park Special Flight Rules Area operations specifications;

(2) The certificate holder must have executed a written contract with the Hualapai Indian Nation which grants the certificate holder a trespass permit and specifies the maximum number of flights to be permitted to land at Grand Canyon West Airport and at other sites located in the vicinity of that airport and operates in compliance with that contract; and

(3) The certificate holder must have a valid operations specification that authorizes the certificate holder to conduct the operations specified in the contract with the Hualapai Indian Nation and specifically approves the number of operations that may transit the Grand Canyon National Park Special Flight Rules Area under this exception.

(g) Certificate holders conducting commercial air tours at or above 14,500 feet MSL but below 18,000 feet MSL who did not receive initial allocations in 1999 because they were not required to report during the base year may operate without an allocation when conducting air tours at those altitudes. Certificate holders conducting commercial air tours in the area affected by the eastward shift of the SFRA who did not receive initial allocations in 1999 because they were not required to report during the base year may continue to operate on the specified routes without an allocation in the area bounded by longitude line 111 degrees 42 minutes east and longitude line 111 degrees 36 minutes east. This exception does not include operation in the Zuni Point corridor.

[65 FR 17732, Apr. 4, 2000]

§ 93.321 Transfer and termination of allocations.

(a) Allocations are not a property interest; they are an operating privilege subject to absolute FAA control.

(b) Allocations are subject to the following conditions:

(1) The Administrator will re-authorize and re-distribute allocations no earlier than two years from the effective date of this rule.

(2) Allocations that are held by the FAA at the time of reallocation may be distributed among remaining certificate holders, proportionate to the size of each certificate holder's allocation.

(3) The aggregate SFRA allocations will not exceed the number of operations reported to the FAA for the base year beginning on May 1, 1997 and ending on April 30, 1998, except as adjusted to incorporate operations occurring for the base year of April 1, 2000 and ending on March 31, 2001, that operate at or above 14,500 feet MSL and below 18,000 feet MSL and operations in the area affected by the eastward shift of the SFRA bounded by longitude line 111 degrees 42 minutes east to longitude 111 degrees 36 minutes east.

(4) Allocations may be transferred among Part 135 or Part 121 certificate holders, subject to all of the following:

(i) Such transactions are subject to all other applicable requirements of this chapter.

(ii) Allocations authorizing commercial air tours outside the Dragon and Zuni Point corridors may not be transferred into the Dragon and Zuni Point corridors. Allocations authorizing commercial air tours within the Dragon and Zuni Point corridors may be transferred outside of the Dragon and Zuni Point corridors.

(iii) A certificate holder must notify in writing the Las Vegas Flight Standards District Office within 10 calendar days of a transfer of allocations. This notification must identify the parties involved, the type of transfer (permanent or temporary) and the number of allocations transferred. Permanent transfers are not effective until the Flight Standards District Office reissues the operations specifications reflecting the transfer. Temporary transfers are effective upon notification.

(5) An allocation will revert to the FAA upon voluntary cessation of commercial air tours within the SFRA for any consecutive 180-day period unless the certificate holder notifies the FSDO in writing, prior to the expiration of the 180-day time period, of the following: the reason why the certificate holder has not conducted any commercial air tours during the consecutive 180-day period; and the date the certificate holder intends on resuming commercial air tours operations. The FSDO will notify the certificate holder of any extension to the consecutive 180-days. A certificate holder may be granted one extension.

(6) The FAA retains the right to re-distribute, reduce, or revoke allocations based on:

(i) Efficiency of airspace;

(ii) Voluntary surrender of allocations;

(iii) Involuntary cessation of operations; and

(iv) Aviation safety.

[65 FR 17733, Apr. 4, 2000]

§ 93.323 Flight plans.

Each certificate holder conducting a commercial SFRA operation must file a visual flight rules (VFR) flight plan in accordance with § 91.153. This section does not apply to operations conducted in accordance with § 93.309(g). The flight plan must be on file with a FAA Flight Service Station prior to each flight. Each VFR flight plan must identify the purpose of the flight in the "remarks" section according to one of the types set forth in the "Las Vegas Flight Standards District Office Grand Canyon National Park Special Flight Rules Area Procedures Manual" which is available from the Las Vegas Flight Standards District Office.

[65 FR 17733, Apr. 4, 2000]

§ 93.325 Quarterly reporting.

(a) Each certificate holder must submit in writing, within 30 days of the end of each calendar quarter, the total number of commercial SFRA operations conducted for that quarter. Quarterly reports must be filed with the Las Vegas Flight Standards District Office.

(b) Each quarterly report must contain the following information.

(1) Make and model of aircraft;

(2) Identification number (registration number) for each aircraft;

(3) Departure airport for each segment flown;

(4) Departure date and actual Universal Coordinated Time, as applicable for each segment flown;

(5) Type of operation; and

(6) Route(s) flown.

[65 FR 17733, Apr. 4, 2000]

Appendix to Subpart U of Part 93—Special Flight Rules in the Vicinity of the Grand Canyon National Park, AZ

PART 93 - SUBPART U
SPECIAL FLIGHT RULES IN THE VICINITY OF THE GRAND CANYON NATIONAL PARK, AZ.

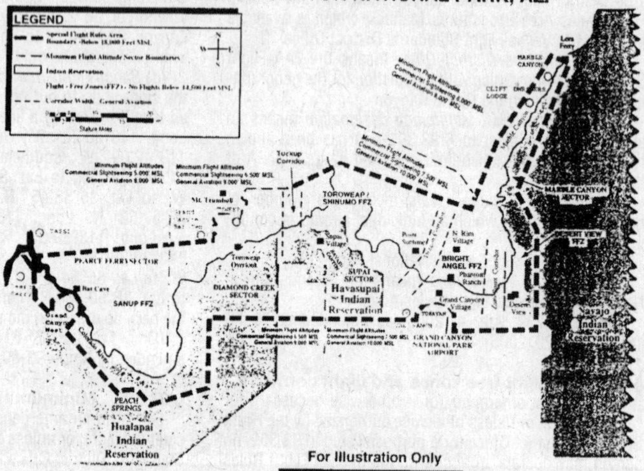

For Illustration Only

NOT FOR USE IN NAVIGATION

Appendix A to Subpart U of Part 93—GCNP Quiet Aircraft Technology Designation

This appendix contains procedures for determining the GCNP quiet aircraft technology designation status for each aircraft subject to § 93.301 determined during the noise certification process as prescribed under part 36 of this chapter. Where no certificated noise level is available, the Administrator may approve an alternative measurement procedure.

Aircraft Noise Limit for GCNP Quiet Aircraft Technology Designation

A. For helicopters with a flyover noise level obtained in accordance with the measurement procedures prescribed in Appendix H of 14 CFR part 36, the limit is 80 dB for helicopters having a seating configuration of two or fewer passenger seats, increasing at 3 dB per doubling of the number of passenger seats for helicopters having a seating configuration of three or more passenger seats. The noise limit for helicopters with three or more passenger seats can be calculated by the formula:

EPNL(H) = 80 +10log(# PAX seats/2) dB

B. For helicopters with a flyover noise level obtained in accordance with the measurement procedures prescribed in Appendix J of 14 CFR part 36, the limit is 77 dB for helicopters having a seating configuration of two or fewer passenger seats, increasing at 3 dB per doubling of the number of passenger seats for helicopters having a seating configuration of three or more passenger seats. The noise limit for helicopters with three or more passenger seats can be calculated by the formula:

SEL(J) = 77 + 10log(# PAX seats/2) dB

C. For propeller-driven airplanes with a measured flyover noise level obtained in accordance with the measurement procedures prescribed in Appendix F of 14 CFR part 36 without the performance correction defined in Sec. F35.201(c), the limit is 69 dB for airplanes having a seating configuration of two or fewer passenger seats, increasing at 3 dB per doubling of the number of passenger seats for airplanes having a seating configuration of three or more passenger seats. The noise limit for propeller-driven airplanes with three or more passenger seats can be calculated by the formula:

LAmax(F) = 69 + 10log(# PAX seats/2) dB

D. In the event that a flyover noise level is not available in accordance with Appendix F of 14 CFR part 36, the noise limit for propeller-driven airplanes with a takeoff noise level obtained in accordance with the measurement procedures prescribed in Appendix G is 74 dB or 77 dB, depending on 14 CFR part 36 amendment level, for airplanes having a seating configuration of two or fewer passenger seats, increasing at 3 dB per doubling of the number of passenger seats for airplanes having a seating configuration of three or more passenger seats. The noise limit for propeller-driven airplanes with three or more passenger seats can be calculated by the formula:

LAmax(G) = 74 + 10log(# PAX seats/2) dB for certifications obtained under 14 CFR part 36, Amendment 21 or earlier;

LAmax(G)=77+10log(# PAX seats/2) dB for certifications obtained under 14 CFR part 36, Amendment 22 or later.

[FAA-2003-14715, 70 FR 16092, Mar. 29, 2005]

Subpart V—Washington, DC Metropolitan Area Special Flight Rules Area

SOURCE: Doc. No. FAA-2004-17005, 73 FR 76213, Dec. 16, 2008, unless otherwise noted.

§ 93.331 Purpose and applicability of this subpart.

This subpart prescribes special air traffic rules for aircraft operating in the Washington, DC Metropolitan Area. Because identification and control of aircraft is required for reasons of national security, the areas described in this subpart constitute national defense airspace. The purpose of establishing this area is to facilitate the tracking of, and communication with, aircraft to deter persons who would use an aircraft as a weapon, or as a means of delivering weapons, to conduct an attack on persons, property, or buildings in the area. This subpart applies to pilots conducting any type of flight operations in the airspace designated as the Washington, DC Metropolitan Area Special Flight Rules Area (DC SFRA) (as defined in § 93.335), which includes the airspace designated as the Washington, DC Metropolitan Area Flight Restricted Zone (DC FRZ) (as defined in § 93.335).

§ 93.333 Failure to comply with this subpart.

(a) *Any violation.* The FAA may take civil enforcement action against a pilot for violations, whether inadvertent or intentional, including imposition of civil penalties and suspension or revocation of airmen's certificates.

(b) *Knowing or willful violations.* The DC FRZ and DC SFRA were established for reasons of national security under the provisions of 49 U.S.C. 40103(b)(3). Areas established by the FAA under that authority constitute "national defense airspace" as that term is used in 49 U.S.C. 46307. In addition to being subject to the provisions of paragraph (a) of this section, persons who knowingly or willfully violate national defense airspace established pursuant to 49 U.S.C. 40103(b)(3) may be subject to criminal prosecution.

§ 93.335 Definitions.

For purposes of this subpart—

DC FRZ flight plan is a flight plan filed for the sole purpose of complying with the requirements for VFR operations into, out of, and through the DC FRZ. This flight plan is separate and distinct from a standard VFR flight plan, and does not include search and rescue services.

DC SFRA flight plan is a flight plan filed for the sole purpose of complying with the requirements for VFR operations into, out of, and through the DC SFRA. This flight plan is separate and distinct from a standard VFR flight plan, and does not include search and rescue services.

Fringe airports are the following airports located near the outer boundary of the Washington, DC Metropolitan Area Special Flight Rules Area: Barnes (MD47), Flying M Farms (MD77), Mountain Road (MD43), Robinson (MD14), and Skyview (51VA).

Washington, DC Metropolitan Area Flight Restricted Zone (DC FRZ) is an area bounded by a line beginning at the Washington VOR/DME (DCA) 311° radial at 15 nautical miles (NM) (Lat. 38°59′31″ N., Long. 077°18′30″ W.); then clockwise along the DCA 15 nautical mile arc to the DCA 002° radial at 15 NM (Lat. 39°06′28″ N., Long 077°04′32″ W.); then southeast via a line drawn to the DCA 049° radial at 14 NM (Lat. 39°02′18″ N., Long. 076°50′38″ W.); thence south via a line drawn to the DCA 064° radial at 13 NM (Lat. 38°59′01″ N., Long. 076°48′32″ W.); thence clockwise along the 13 NM arc to the DCA 276° radial at 13 NM (Lat.38°50′53″ N., Long 077°18′48″ W.); thence north to the point of beginning, excluding the airspace within a one nautical mile radius of the Freeway Airport, W00, Mitchellville, MD from the surface up to but not including flight level (FL) 180. The DC FRZ is within and part of the Washington, DC Metropolitan Area SFRA.

Washington, DC Metropolitan Area Special Flight Rules Area (DC SFRA) is an area of airspace over the surface of the earth where the ready identification, location, and control of aircraft is required in the interests of national security. Specifically, the DC SFRA is that airspace, from the surface to, but not including, FL 180, within a 30-mile radius of Lat. 38°51′34″ N., Long. 077°02′11″ W., or the DCA VOR/DME. The DC SFRA includes the DC FRZ.

[Doc. No. FAA-2004-17005, 73 FR 76213, Dec. 16, 2008; Amdt. 93-91, 73 FR 79314, Dec. 29, 2008]

§ 93.337 Requirements for operating in the DC SFRA.

A pilot conducting any type of flight operation in the DC SFRA must comply with the restrictions listed in this subpart and all special instructions issued by the FAA in the interest of national security. Those special instructions may be issued in any manner the FAA considers appropriate, including a NOTAM. Additionally, a pilot must comply with all of the applicable requirements of this chapter.

§ 93.339 Requirements for operating in the DC SFRA, including the DC FRZ.

(a) Except as provided in paragraphs (b) and (c) of this section and in § 93.345, or unless authorized by Air Traffic Control, no pilot may operate an aircraft, including an ultralight vehicle or any civil aircraft or public aircraft, in the DC SFRA, including the DC FRZ, unless—

(1) The aircraft is equipped with an operable two-way radio capable of communicating with Air Traffic Control on appropriate radio frequencies;

(2) Before operating an aircraft in the DC SFRA, including the DC FRZ, the pilot establishes two-way radio communications with the appropriate Air Traffic Control facility and maintains such communications while operating the aircraft in the DC SFRA, including the DC FRZ;

(3) The aircraft is equipped with an operating automatic altitude reporting transponder;

(4) Before operating an aircraft in the DC SFRA, including the DC FRZ, the pilot obtains and transmits a discrete transponder code from Air Traffic Control, and the aircraft's transponder continues to transmit the assigned code while operating within the DC SFRA;

(5) For VFR operations, the pilot must file and activate a DC FRZ or DC SFRA flight plan by obtaining a discrete transponder code. The flight plan is closed upon landing at an airport within the DC SFRA or when the aircraft exits the DC SFRA;

(6) Before operating the aircraft into, out of, or through the Washington, DC Tri-Area Class B Airspace Area, the pilot receives a specific Air Traffic Control clearance to operate in the Class B airspace area; and

(7) Before operating the aircraft into, out of, or through Class D airspace area that is within the DC SFRA, the pilot complies with § 91.129 of this chapter.

(b) Paragraph (a)(5) of this section does not apply to operators of Department of Defense aircraft, law enforcement operations, or lifeguard or air ambulance operations under an FAA/TSA airspace authorization, if the flight crew is in contact with Air Traffic Control and is transmitting an Air Traffic Control-assigned discrete transponder code.

(c) When operating an aircraft in the VFR traffic pattern at an airport within the DC SFRA (but not within the DC FRZ) that does not have an airport traffic control tower, a pilot must—

(1) File a DC SFRA flight plan for traffic pattern work;

(2) Communicate traffic pattern position via the published Common Traffic Advisory Frequency (CTAF);

(3) Monitor VHF frequency 121.5 or UHF frequency 243.0, if the aircraft is suitably equipped;

(4) Obtain and transmit the Air Traffic Control-assigned discrete transponder code; and

(5) When exiting the VFR traffic pattern, comply with paragraphs (a)(1) through (a)(7) of this section.

(d) When operating an aircraft in the VFR traffic pattern at an airport within the DC SFRA (but not within the DC FRZ) that has an operating airport traffic control tower, a pilot must—

(1) Before departure or before entering the traffic pattern, request to remain in the traffic pattern;

(2) Remain in two-way radio communications with the tower. If the aircraft is suitably equipped, the pilot must also monitor VHF frequency 121.5 or UHF frequency 243.0;

(3) Continuously operate the aircraft transponder on code 1234 unless Air Traffic Control assigns a different code; and

(4) Before exiting the traffic pattern, comply with paragraphs (a)(1) through (a)(7) of this section.

(e) Pilots must transmit the assigned transponder code. No pilot may use transponder code 1200 while in the DC SFRA.

§ 93.341 Aircraft operations in the DC FRZ.

(a) Except as provided in paragraph (b) of this section, no pilot may conduct any flight operation under part 91, 101, 103, 105, 125, 133, 135, or 137 of this chapter in the DC FRZ, unless the specific flight is operating under an FAA/TSA authorization.

(b) Department of Defense (DOD) operations, law enforcement operations, and lifeguard or air ambulance operations under an FAA/TSA airspace authorization are excepted from the prohibition in paragraph (a) of this section if the pilot is in contact with Air Traffic Control and operates the aircraft transponder on an Air Traffic Control-assigned beacon code.

(c) The following aircraft operations are permitted in the DC FRZ:

(1) Aircraft operations under the DCA Access Standard Security Program (DASSP) (49 CFR part 1562) with a Transportation Security Administration (TSA) flight authorization.

(2) Law enforcement and other U.S. Federal aircraft operations with prior FAA approval.

(3) Foreign-operated military and state aircraft operations with a State Department-authorized diplomatic clearance, with State Department notification to the FAA and TSA.

(4) Federal, State, Federal DOD contract, local government agency aircraft operations and part 121, 129 or 135 air carrier flights with TSA-approved full aircraft operator standard security programs/procedures, if operating with DOD permission and notification to the FAA and the National Capital Regional Coordination Center (NCRCC). These flights may land and depart Andrews Air Force Base, MD, with prior permission, if required.

(5) Aircraft operations maintaining radio contact with Air Traffic Control and continuously transmitting an Air Traffic Control-assigned discrete transponder code. The pilot must monitor VHF frequency 121.5 or UHF frequency 243.0.

(d) Before departing from an airport within the DC FRZ, or before entering the DC FRZ, all aircraft, except DOD, law enforcement, and lifeguard or air ambulance aircraft operating under an FAA/TSA airspace authorization must file and activate an IFR or a DC FRZ or a DC SFRA flight plan and transmit a discrete transponder code assigned by an Air Traffic Control facility. Aircraft must transmit the discrete transponder code at all times while in the DC FRZ or DC SFRA.

§ 93.343 Requirements for aircraft operations to or from College Park Airport, Potomac Airfield, or Washington Executive/Hyde Field Airport.

(a) A pilot may not operate an aircraft to or from College Park Airport, MD, Potomac Airfield, MD, or Washington Executive/Hyde Field Airport, MD unless—

(1) The aircraft and its crew and passengers comply with security rules issued by the TSA in 49 CFR part 1562, subpart A;

(2) Before departing, the pilot files an IFR or DC FRZ or DC SFRA flight plan with the Washington Hub Flight Service Station (FSS) for each departure and arrival from/to College Park, Potomac Airfield, and Washington Executive/Hyde Field airports, whether or not the aircraft makes an intermediate stop;

(3) When filing a flight plan with the Washington Hub FSS, the pilot identifies himself or herself by providing the assigned pilot identification code. The Washington Hub FSS will accept the flight plan only after verifying the code; and

(4) The pilot complies with the applicable IFR or VFR egress procedures in paragraph (b), (c) or (d) of this section.

(b) If using IFR procedures, a pilot must—

(1) Obtain an Air Traffic Control clearance from the Potomac TRACON; and

(2) Comply with Air Traffic Control departure instructions from Washington Executive/Hyde Field, Potomac Airport, or College Park Airport. The pilot must then proceed on the Air Traffic Control-assigned course and remain clear of the DC FRZ.

(c) If using VFR egress procedures, a pilot must—

(1) Depart as instructed by Air Traffic Control and expect a heading directly out of the DC FRZ until the pilot establishes two-way radio communication with Potomac Approach; and

(2) Operate as assigned by Air Traffic Control until clear of the DC FRZ, the DC SFRA, and the Class B or Class D airspace area.

(d) If using VFR ingress procedures, the aircraft must remain outside the DC SFRA until the pilot establishes communications with Air Traffic Control and receives authorization for the aircraft to enter the DC SFRA.

(e) VFR arrivals:

(1) If landing at College Park Airport a pilot may receive routing via the vicinity of Freeway Airport; or

(2) If landing at Washington Executive/Hyde Field or Potomac Airport, the pilot may receive routing via the vicinity of Maryland Airport or the Nottingham VORTAC.

§ 93.345 VFR outbound procedures for fringe airports.

(a) A pilot may depart from a fringe airport as defined in § 93.335 without filing a flight plan or communicating with Air Traffic Control, unless requested, provided:

(1) The aircraft's transponder transmits code 1205;

(2) The pilot exits the DC SFRA by the most direct route before proceeding on course; and

(3) The pilot monitors VHF frequency 121.5 or UHF frequency 243.0.

(b) No pilot may operate an aircraft arriving at a fringe airport or transit the DC SFRA unless that pilot complies with the DC SFRA operating procedures in this subpart.

Subpart W—New York Class B Airspace Hudson River and East River Exclusion Special Flight Rules Area

SOURCE: 74 FR 59910, Nov. 19, 2009, unless otherwise noted.

§ 93.350 Definitions.

For the purposes of this subpart only the following definitions apply:

(a) *Local operation.* Any aircraft within the Hudson River Exclusion that is conducting an operation other than as described in paragraph (b) of this section. Local operations include but are not limited to operations for sightseeing, electronic news gathering, and law enforcement.

(b) *Transient operation.* Aircraft transiting the entire length of the Hudson River Class B Exclusion, as defined in paragraph (d) of this section, from one end to the other.

(c) *New York Class B airspace East River Exclusion* is that airspace below 1,500 feet MSL between the east and west banks of, and overlying, the East River beginning at lat. 40°38′39″N., long. 74°02′03″W., thence north along a line drawn direct to the southwestern tip of Governors Island, thence north along a line direct to the southwest tip of Manhattan Island, thence north along the west bank of the East River to the LGA VOR/DME 6-mile arc, thence counterclockwise along the 6-mile arc to the east bank of the East River, thence south along the east bank of the East River to the point of beginning at lat. 40°38′39″N., long 74°02′03″W.; and that airspace 1,100 feet MSL and below between the east and west banks of, and overlying the East River, from the LGA VOR/DME 6-mile arc to the north tip of Roosevelt Island.

(d) *New York Class B airspace Hudson River Exclusion* is that area from the surface up to but not including the overlying floor of the New York Class B airspace area, between the east and west banks of, and overlying, the Hudson River within the area beginning north of LaGuardia Airport on the west bank of the Hudson River at lat. 40°57′45″N., long. 73°54′48″W. (near Alpine Tower), thence south along the west bank of the Hudson River to intersect the Colts Neck VOR/DME 012° radial, thence southwest along the Colts Neck 012° radial to the Hudson River shoreline, thence south along the shoreline to the Verrazano-Narrows Bridge, thence east along the Bridge to the east bank of the Hudson River, thence north along the east bank of the Hudson River to lat. 40°38′39″N., long. 74°02′03″W., thence north along a line drawn direct to the southwesternmost point of Governors Island, thence north along a line drawn direct to the southwest tip of Manhattan Island, thence north along the east bank of the Hudson River to the LGA VOR/DME 11-mile arc, north of LaGuardia Airport, thence counterclockwise along the 11-mile arc to lat. 40°57′54″N., long. 73°54′23″W., thence to the point of beginning.

§ 93.351 General requirements for operating in the East River and/or Hudson River Exclusions.

Pilots must adhere to the following requirements:

(a) Maintain an indicated airspeed not to exceed 140 knots.

(b) Anti-collision lights and aircraft position/navigation lights shall be on, if equipped. Use of landing lights is recommended.

(c) Self announce position on the appropriate radio frequency for the East River or Hudson River as depicted on the New York VFR Terminal Area Chart (TAC) and/or New York Helicopter Route Chart.

(d) Have a current New York TAC chart and/or New York Helicopter Route Chart in the aircraft and be familiar with the information contained therein.

§ 93.352 Hudson River Exclusion specific operating procedures.

In addition to the requirements in § 93.351, the following procedures apply:

(a) Pilots must self announce, at the charted mandatory reporting points, the following information: aircraft type, current position, direction of flight, and altitude.

(b) Pilots must fly along the west shoreline of the Hudson River when southbound, and along the east shoreline of the Hudson River when northbound; while remaining within the boundaries of the Hudson River Exclusion as defined in § 93.350(d).

(c) Aircraft transiting the area within the Hudson River Exclusion in accordance with § 93.350(b) must transit the Hudson River Exclusion at or above an altitude of 1,000 feet MSL up to, but not including, the floor of the overlying Class B airspace.

§ 93.353 East River Exclusion specific operating procedures.

No person may operate an airplane in the East River Exclusion extending from the southwestern tip of Governors Island to the north tip of Roosevelt Island except:

(a) Seaplanes landing on or taking off from the river; or

(b) Airplanes authorized by ATC. Pilots must contact LaGuardia Airport Traffic Control Tower prior to Governors Island for authorization.

PART 95

IFR ALTITUDES

Contents

Authority: 49 U.S.C. 106(g), 40103, 40113, and 14 CFR 11.49(b)(2).

Special Federal Aviation Regulation No. 97

Editorial Note: For the text of SFAR No. 97, see part 91 of this chapter.

Subpart A—General

§ 95.1 Applicability.

(a) This part prescribes altitudes governing the operation of aircraft under IFR on ATS routes, or other direct routes for which an MEA is designated in this part. In addition, it designates mountainous areas and changeover points.

(b) The MAA is the highest altitude on an ATS route, or other direct route for which an MEA is designated, at which adequate reception of VOR signals is assured.

(c) The MCA applies to the operation of an aircraft proceeding to a higher minimum en route altitude when crossing specified fixes.

(d) The MEA is the minimum en route IFR altitude on an ATS route, ATS route segment, or other direct route. The MEA applies to the entire width of the ATS route, ATS route segment, or other direct route between fixes defining that route. Unless otherwise specified, an MEA prescribed for an off airway route or route segment applies to the airspace 4 nautical miles on each side of a direct course between the navigation fixes defining that route or route segment.

(e) The MOCA assures obstruction clearance on an ATS route, ATS route segment, or other direct route, and adequate reception of VOR navigation signals within 22 nautical miles of a VOR station used to define the route.

(f) The MRA applies to the operation of an aircraft over an intersection defined by ground-based navigation aids. The MRA is the lowest altitude at which the intersection can be determined using the ground-based navigation aids.

(g) The changeover point (COP) applies to operation of an aircraft along a Federal airway, jet route, or other direct route; for which an MEA is designated in this part. It is the point for transfer of the airborne navigation reference from the ground-based navigation aid behind the aircraft to the next appropriate ground-based navigation aid to ensure continuous reception of signals.

[Doc. No. FAA-2003-14698, 68 FR 16947, Apr. 8, 2003]

§ 95.3 Symbols.

For the purposes of this part—

(a) *COP* means changeover point.

(b) *L* means compass locator;

(c) *LF/MF* means low frequency, medium frequency;

(d) *LFR* means low frequency radio range;

(e) *VOR-E* means VOR and distance measuring equipment; and

(f) *Z* means a very high frequency location marker.

[Doc. No. 1580, 28 FR 6718, June 29, 1963, as amended by Amdt. 95-118, 29 FR 13166, Sept. 23, 1964]

Subpart B—Designated Mountainous Areas

§ 95.11 General.

The areas described in this subpart are designated mountainous areas.

[Doc. No. 1580, 28 FR 6718, June 29, 1963]

§ 95.13 Eastern United States Mountainous Area.

All of the following area excluding those portions specified in the exceptions.

(a) *Area.*

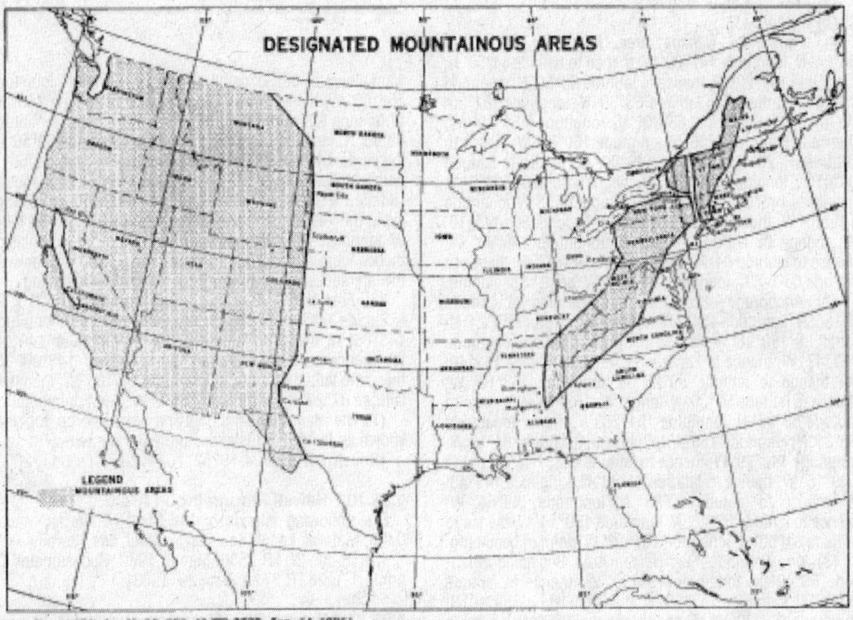

[Doc. No. 13384, Amdt. 95-255, 40 FR 2578, Jan. 14, 1975]

Beginning at latitude 47°10′ N., longitude 67°55′ W.; thence west and south along the Canadian Border to latitude 45°00′ N., longitude 74°15′ W.; thence to latitude 44°20′ N., longitude 75°30′ W.; thence to latitude 43°05′ N., longitude 75°30′ W.; thence to latitude 42°57′ N., longitude 77°30′ W.; thence to latitude 42°52′ N., longitude 78°42′ W.; thence to latitude 42°26′ N., longitude 79°13′ W.; thence to latitude 42°05′ N., longitude 80°00′ W.; thence to latitude 40°50′ N., longitude 80°00′ W.; thence to latitude 40°26′ N., longitude 79°54′ W.; thence to latitude 38°25′ N., longitude 81°46′ W.; thence to latitude 36°00′ N., longitude 86°00′ W.; thence to latitude 33°37′ N., longitude 86°45′ W.; thence to latitude 32°30′ N., longitude 86°25′ W.; thence to latitude 33°22′ N., longitude 85°00′ W.; thence to latitude 36°35′ N., longitude 79°20′ W.; thence to latitude 40°11′ N., longitude 76°24′ W.; thence to latitude 41°24′ N., longitude 74°30′ W.; thence to latitude 41°43′ N., longitude 72°40′ W.; thence to latitude 42°13′ N., longitude 72°44′ W.; thence to latitude 43°12′ N., longitude 71°30′ W.; thence to latitude 43°45′ N., longitude 70°30′ W.; thence to latitude 45°00′ N., longitude 69°30′ W.; thence to latitude 47°10′ N., longitude 67°55′ W., point of beginning.

(b) *Exceptions.* The area bounded by the following coordinates:

Beginning at latitude 45°00′ N., longitude 73°26′ W.; thence to latitude 44°32′ N., longitude 73°04′ W.; thence to latitude 42°51′ N., longitude 73°41′ W.; thence to latitude 41°38′ N., longitude 73°46′ W.; thence to latitude 41°16′ N., longitude 73°50′ W.; thence to latitude 41°17′ N., longitude 74°00′ W.; thence to latitude 41°25′ N., longitude 73°58′ W.; thence to latitude 41°26′ N., longitude 74°01′ W.; thence to latitude 41°37′ N., longitude 73°58′ W.; thence to latitude 42°41′ N., longitude 73°55′ W.; thence to latitude 43°02′ N., longitude 76°15′ W.; thence to latitude 43°17′ N., longitude 75°21′ W.; thence to latitude 42°59′ N., longitude 74°43′ W.; thence to latitude 42°52′ N., longitude 73°53′ W.; thence to latitude 44°30′ N., longitude 73°18′ W.; thence to latitude 45°00′ N., longitude 73°39′ W.; thence to latitude 45°00′ N., longitude 73°26′ W., point of beginning.

[21 FR 2750, Apr. 28, 1956. Redesignated by Amdt. 1-1, 28 FR 6718, June 29, 1963, as amended at 73 FR 63885, Oct. 28, 2008]

§ 95.15 Western United States Mountainous Area.

All of the following area excluding that portion specified in the exceptions:

(a) *Area.* From the Pacific coastline of the United States, eastward along the Canadian and Mexican borders, to the following coordinates:

Beginning at latitude 49°00′ N., longitude 108°00′ W.; thence to latitude 46°45′ N., longitude 104°00′ W.; thence to latitude 44°06′ N., longitude 103°15′ W.; thence to latitude 43°00′ N., longitude 103°15′ W.; thence to latitude 41°52′ N., longitude 103°39′ W.; thence to latitude 35°11′ N., longitude 103°39′ W.; thence to latitude 33°17′ N., longitude 104°27′ W.; thence to latitude 32°17′ N., longitude 104°14′ W.; thence to latitude 29°48′ N., longitude 102°00′ W.

(b) *Exceptions.*

(1) Beginning at latitude 35°25′ N., longitude 119°09′ W.; thence to latitude 35°29′ N., longitude 118°58′ W.; thence to latitude 36°49′ N., longitude 119°37′ W.; thence to latitude 38°30′ N., longitude 121°24′ W.; thence to latitude 39°30′ N., longitude 121°32′ W.; thence to latitude 40°08′ N., longitude 122°08′ W.; thence to latitude 40°06′ N., longitude 122°20′ W.; thence to latitude 39°05′ N., longitude 122°12′ W.; thence to latitude 38°01′ N., longitude 121°51′ W.; thence to latitude 37°37′ N., longitude 121°12′ W.; thence to latitude 37°00′ N., longitude 120°58′ W.; thence to latitude 36°14′ N., longitude 120°11′ W., point of beginning.

(2) Beginning at latitude 49°00′ N., longitude 122°21′ W.; thence to latitude 48°34′ N., longitude 122°21′ W.; thence to latitude 48°08′ N., longitude 122°00′ W.; thence to latitude 47°12′ N., longitude 122°00′ W.; thence to latitude 46°59′ N., longitude 122°13′ W.; thence to latitude 46°52′ N., longitude 122°16′ W.; thence to latitude 46°50′ N., longitude 122°40′ W.; thence to latitude 46°35′ N., longitude 122°48′ W.; thence to latitude 46°35′ N., longitude 123°17′ W.; thence to latitude

47°15′ N., longitude 123°17′ W.; thence to latitude 47°41′ N., longitude 122°54′ W.; thence to latitude 48°03′ N., longitude 122°48′ W.; thence to latitude 48°17′ N., longitude 123°15′ W.; thence North and East along the United States and Canada Boundary to latitude 49°00′ N., longitude 122°21′ W., point of beginning.

[21 FR 2750, Apr. 28, 1956. Redesignated by Amdt. 1-1, 28 FR 6718, June 29, 1963, and amended by Amdt. 95-255, 40 FR 2579, Jan. 14, 1975]

§ 95.17 Alaska Mountainous Area.

All of the following area excluding those portions specified in the exceptions:

(a) *Area.* The State of Alaska.

(b) *Exceptions;*

(1) *Fairbanks—Nenana Area.* Beginning at latitude 64°54′ N, longitude 147°00′ W; thence to latitude 64°50′ N, longitude 151°22′ W, thence to latitude 63°50′ N, longitude 152°50′ W; thence to latitude 63°30′ N, longitude 152°30′ W; thence to latitude 63°30′ N, longitude 151°30′ W; thence to latitude 64°05′ N, longitude 150°30′ W; thence to latitude 64°20′ N, longitude 149°00′ W; thence to latitude 64°07′ N, longitude 146°30′ W; thence to latitude 63°53′ N, longitude 146°00′ W; thence to latitude 63°53′ N, longitude 145°00′ W; thence to latitude 64°09′ N, longitude 145°16′ W; thence to latitude 64°12′ N, longitude 146°00′ W; thence to latitude 64°25′ N, longitude 146°37′ W; thence to latitude 64°54′ N, longitude 147°00′ W, point of beginning.

(2) *Anchorage—Homer Area.* Beginning at latitude 61°50′ N, longitude 151°12′ W; thence to latitude 61°24′ N, longitude 150°28′ W; thence to latitude 61°08′ N, longitude 151°47′ W; thence to latitude 59°49′ N, longitude 152°40′ W; thence to latitude 59°25′ N, longitude 153°10′ W; thence to latitude 59°00′ N, longitude 153°10′ W; thence to latitude 59°33′ N, longitude 151°28′ W; thence to latitude 60°31′ N, longitude 150°43′ W; thence to latitude 61°13′ N, longitude 149°39′ W; thence to latitude 61°37′ N, longitude 149°15′ W; thence to latitude 61°44′ N, longitude 149°48′ W; thence to latitude 62°23′ N, longitude 149°54′ W; thence to latitude 62°23′ N, longitude 150°14′ W; thence to latitude 61°50′ N, longitude 151°12′ W, point of beginning.

(3) *King Salmon—Port Heiden Area.* Beginning at latitude 58°49′ N, longitude 159°30′ W; thence to latitude 59°40′ N, longitude 157°00′ W; thence to latitude 59°40′ N, longitude 155°30′ W; thence to latitude 59°50′ N, longitude 154°50′ W; thence to latitude 59°35′ N, longitude 154°40′ W; thence to latitude 58°57′ N, longitude 156°05′ W; thence to latitude 58°00′ N, longitude 156°20′ W; thence to latitude 57°00′ N, longitude 158°20′ W; thence to latitude 56°43′ N, longitude 158°39′ W; thence to latitude 56°27′ N, longitude 160°00′ W; thence along the shoreline to latitude 58°49′ N, longitude 159°30′ W, point of beginning.

(4) *Bethel—Aniak Area.* Beginning at latitude 63°28′ N, longitude 161°30′ W; thence to latitude 62°40′ N, longitude 163°03′ W; thence to latitude 62°05′ N, longitude 162°38′ W; thence to latitude 61°51′ N, longitude 160°43′ W; thence to latitude 62°55′ N, longitude 160°30′ W; thence to latitude 63°00′ N, longitude 158°00′ W; thence to latitude 61°45′ N, longitude 159°30′ W; thence to latitude 61°34′ N, longitude 159°15′ W; thence to latitude 61°07′ N, longitude 160°20′ W; thence to latitude 60°25′ N, longitude 160°40′ W; thence to latitude 59°36′ N, longitude 161°49′ W; thence along the shoreline to latitude 63°28′ N, longitude 161°30′ W; point of beginning; and Nunivak Island.

(5) *North Slope Area.* Beginning at a point where latitude 69°30′ N intersects the northwest coast of Alaska and east-

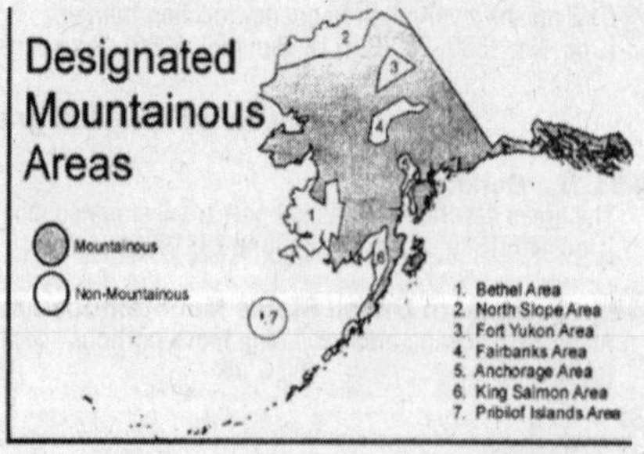

ward along the 69°30′ parallel to latitude 69°30′ N, longitude 156°00′ W; thence to latitude 69°10′ N, longitude 153°00′ W; thence eastward along the 69°10′ N parallel to latitude 69°10′ N, longitude 149°00′ W; thence to latitude 69°50′ N, longitude 146°00′ W; thence eastward along the 69°50′ N parallel to latitude 69°50′ N, longitude 145°00′ W; thence to latitude 69°35′ N, longitude 141°00′ W; thence northward along the 141°00′ W Meridian to a point where the 141°00′ W Meridian intersects the northeast coastline of Alaska; thence westward along the northern coastline of Alaska to the intersection of latitude 69°30′ N; point of beginning .

(6) *Fort Yukon Area.* Beginning at latitude 67°20′ N, longitude 144°00′ W; thence to latitude 66°00′ N, longitude 143°00′ W; thence to latitude 66°05′ N, longitude 149°00′ W; thence to latitude 66°45′ N, longitude 148°00′ W; thence to latitude 67°00′ N, longitude 147°00′ W; thence to latitude 67°20′ N, longitude 144°00′ W; point of beginning.

(7) The islands of Saint Paul and Saint George, together known as the Pribilof Islands, in the Bering Sea.

[Doc. No. FAA-2004-19352, 70 FR 7360, Feb. 11, 2005]

§ 95.19 Hawaii Mountainous Area.

The following islands of the State of Hawaii: Kauai, Oahu, Molokai, Lanai, Kehoolawe, Maui, and Hawaii.

[Amdt. 88, 27 FR 4536, May 8, 1962. Redesignated by Amdt. 1-1, 28 FR 6719, June 29, 1963]

§ 95.21 Puerto Rico Mountainous Area.

The area bounded by the following coordinates:

Beginning at latitude 18°22′ N., longitude 66°58′ W., thence to latitude 18°19′ N., longitude 66°06′ W.; thence

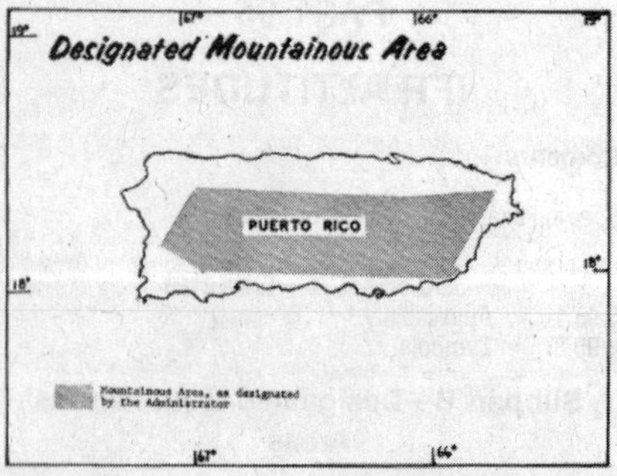

to latitude 18°20′ N., longitude 65°50′ W.; thence to latitude 18°20′ N., longitude 65°42′ W.; thence to latitude 18°03′ N., longitude 65°52′ W.; thence to latitude 18°02′ N., longitude 65°51′ W.; thence to latitude 17°59′ N., longitude 65°55′ W.; thence to latitude 18°05′ N., longitude 66°57′ W.; thence to latitude 18°11′ N., longitude 67°07′ W.; thence to latitude 18°22′ N., longitude 66°58′ W.; the point of beginning.

[Amdt. 88, 27 FR 4536, May 8, 1962; 27 FR 5603, June 13, 1962. Redesignated by Amdt. 1-1, 28 FR 6719, June 29, 1963]

Subpart C—En Route IFR Altitudes Over Particular Routes and Intersections

Editorial Note: The prescribed IFR altitudes for flights over particular routes and intersections in this subpart were formerly carried as §§ 610.11 through 610.6887 of this title and were transferred to part 95 as §§ 95.41 through 95.6887, respectively, but are not carried in the Code of Federal Regulations. For Federal Register citations affecting these routes, see the List of CFR Sections Affected, which appears in the Finding Aids section of the printed volume and at *www.fdsys.gov*.

§ 95.31 General.

This subpart prescribes IFR altitudes for flights along particular routes or route segments and over additional intersections not listed as a part of a route or route segment.

[Doc. No. 1580, 28 FR 6719, June 29, 1963]

Subpart D—Changeover Points

Editorial Note: The prescribed COP's for Federal airways, jet routes, or other direct routes for which an MEA is designated in this part are not carried in the Code of Federal Regulations. For Federal Register citations affecting these routes see the List of CFR Sections Affected, which appears in the Finding Aids section of the printed volume and at *www.fdsys.gov*.

§ 95.8001 General.

This subpart prescribes COP's for Federal airways, jet routes, area navigation routes, or other direct routes for which an MEA is designated in this part. Unless otherwise specified the COP is midway between the navigation facilities or way points for straight route segments, or at the intersection of radials or courses forming a dogleg in the case of dogleg route segments.

[Doc. No. 10580, 35 FR 14610, Sept. 18, 1970]

PART 97

STANDARD INSTRUMENT PROCEDURES

Contents

Subpart A—General

Subpart B—Procedures

Subpart C—TERPS Procedures

Authority: 49 U.S.C. 106(g), 40103, 40106, 40113, 40114, 40120, 44502, 44514, 44701, 44719, and 44721-44722.

Source: Docket No. 1580, 28 FR 6719, June 29, 1963, unless otherwise noted.

Subpart A—General

§ 97.1 Applicability.

(a) This part prescribes standard instrument approach procedures to civil airports in the United States and the weather minimums that apply to landings under IFR at those airports.

(b) This part also prescribes obstacle departure procedures (ODPs) for certain civil airports in the United States and the weather minimums that apply to takeoffs under IFR at civil airports in the United States.

[Doc. No. FAA-2002-14002, 72 FR 31679, June 7, 2007]

§ 97.3 Symbols and terms used in procedures.

As used in the standard instrument procedures prescribed in this part—

Aircraft approach category means a grouping of aircraft based on a speed of VREF, if specified, or if VREF is not specified, 1.3 V_{so} at the maximum certificated landing weight. VREF, V_{so}, and the maximum certificated landing weight are those values as established for the aircraft by the certification authority of the country of registry. The categories are as follows—

(1) Category A: Speed less than 91 knots.

(2) Category B: Speed 91 knots or more but less than 121 knots.

(3) Category C: Speed 121 knots or more but less than 141 knots.

(4) Category D: Speed 141 knots or more but less than 166 knots.

(5) Category E: Speed 166 knots or more.

Approach procedure segments for which altitudes (minimum altitudes, unless otherwise specified) and paths are prescribed in procedures, are as follows—

(1) Initial approach is the segment between the initial approach fix and the intermediate fix or the point where the aircraft is established on the intermediate course or final approach course.

(2) Initial approach altitude is the altitude (or altitudes, in high altitude procedure) prescribed for the initial approach segment of an instrument approach.

(3) Intermediate approach is the segment between the intermediate fix or point and the final approach fix.

(4) Final approach is the segment between the final approach fix or point and the runway, airport, or missed approach point.

(5) Missed approach is the segment between the missed approach point, or point of arrival at decision altitude or decision height (DA/DH), and the missed approach fix at the prescribed altitude.

Ceiling means the minimum ceiling, expressed in feet above the airport elevation, required for takeoff or required for designating an airport as an alternate airport.

Copter procedures means helicopter procedures, with applicable minimums as prescribed in § 97.35. Helicopters may also use other procedures prescribed in subpart C of this part and may use the Category A minimum descent altitude (MDA), or decision altitude or decision height (DA/DH). For other than "copter-only" approaches, the required visibility minimum for Category I approaches may be reduced to one-half the published visibility minimum for Category A aircraft, but in no case may it be reduced to less than one-quarter mile prevailing visibility, or, if reported, 1,200 feet RVR. Reduction of visibility minima on Category II instrument approach procedures is prohibited.

FAF means final approach fix.

HAA means height above airport and is expressed in feet.

HAL means height above landing and is the height of the DA/MDA above a designated helicopter landing area elevation used for helicopter instrument approach procedures and is expressed in feet.

HAS means height above the surface and is the height of the DA/MDA above the highest terrain/surface within a 5,200-foot radius of the missed approach point used in helicopter instrument approach procedures and is expressed in feet above ground level (AGL).

HAT means height above touchdown.

HCH means helipoint crossing height and is the computed height of the vertical guidance path above the helipoint elevation at the helipoint expressed in feet.

Helipoint means the aiming point for the final approach course. It is normally the center point of the touchdown and lift-off area (TLOF).

Hold in lieu of PT means a holding pattern established under applicable FAA criteria, and used in lieu of a procedure turn to execute a course reversal.

MAP means missed approach point.

More than 65 knots means an aircraft that has a stalling speed of more than 65 knots (as established in an approved flight manual) at maximum certificated landing weight with full flaps, landing gear extended, and power off.

MSA means minimum safe altitude, expressed in feet above mean sea level, depicted on an approach chart that provides at least 1,000 feet of obstacle clearance for emergency use within a certain distance from the specified navigation facility or fix.

NA means not authorized.

NOPT means no procedure turn required. Altitude prescribed applies only if procedure turn is not executed.

Procedure turn means the maneuver prescribed when it is necessary to reverse direction to establish the aircraft on an intermediate or final approach course. The outbound course, direction of turn, distance within which the turn must be completed, and minimum altitude are specified in the procedure. However, the point at which the turn may be begun, and the type and rate of turn, is left to the discretion of the pilot.

RA means radio altimeter setting height.

RVV means runway visibility value.

SIAP means standard instrument approach procedure.

65 knots or less means an aircraft that has a stalling speed of 65 knots or less (as established in an approved flight manual) at maximum certificated landing weight with full flaps, landing gear extended, and power off.

T means nonstandard takeoff minimums or specified departure routes/procedures or both.

TDZ means touchdown zone.

Visibility minimum means the minimum visibility specified for approach, landing, or takeoff, expressed in statute miles, or in feet where RVR is reported.

[Doc. No. FAA-2002-14002, 72 FR 31679, June 7, 2007]

§ 97.5 Bearings, courses, tracks, headings, radials, miles.

(a) All bearings, courses, tracks, headings, and radials in this part are magnetic, unless otherwise designated.

(b) RVR values are stated in feet. Other visibility values are stated in statute miles. All other mileages are stated in nautical miles.

[Doc. No. 561, 32 FR 13912, Oct. 6, 1967, as amended by Amdt. 97-1336, 72 FR 31680, June 7, 2007]

Subpart B—Procedures

Editorial Note: The procedures set forth in this subpart were formerly carried as §§ 609.100 through 609.500 of this title and were transferred to part 97 as §§ 97.11 through 97.19, respectively, but are not carried in the Code of Federal Regulations. For Federal Register citations affecting these procedures, see the List of CFR Sections Affected, which appears in the Finding Aids section of the printed volume and at *www.fdsys.gov*.

§ 97.10 [Reserved]

Subpart C—TERPS Procedures

Source: Docket No. 8130, 32 FR 13912, Oct. 6, 1967, unless otherwise noted.

Editorial Note: The procedures for §§ 97.21 through 97.35, respectively, are not carried in the Code of Federal Regulations. For Federal Register citations affecting these procedures, see the List of CFR Sections Affected, which appears in the Finding Aids section of the printed volume and at *www.fdsys.gov*.

§ 97.20 General.

(a) This subpart prescribes standard instrument approach procedures and takeoff minimums and obstacle departure procedures (ODPs) based on the criteria contained in FAA Order 8260.3, U.S. Standard for Terminal Instrument Procedures (TERPs), and other related Orders in the 8260 series that also address instrument procedure design criteria.

(b) Standard instrument approach procedures and associated supporting data adopted by the FAA are documented on FAA Forms 8260-3, 8260-4, 8260-5. Takeoff minimums and obstacle departure procedures (ODPs) are documented on FAA Form 8260-15A. These forms are incorporated by reference. The Director of the Federal Register approved this incorporation by reference pursuant to 5 U.S.C. 552(a) and 1 CFR part 51. The standard instrument approach procedures and takeoff minimums and obstacle departure procedures (ODPs) are available for examination at the FAA's Rules Docket (AGC-200) and at the National Flight Data Center, 800 Independence Avenue, SW., Washington, DC 20590, or at the National Archives and Records Administration (NARA). For information on the availability of this material at NARA, call 202-741-6030, or go to *http://www.archives.gov/federal_register/code_of_federal_regulations/ibr_locations.html*.

(c) Standard instrument approach procedures and takeoff minimums and obstacle departure procedures (ODPs) are depicted on aeronautical charts published by the FAA National Aeronautical Charting Office. These charts are available for purchase from the FAA's National Aeronautical Charting Office, Distribution Division, 6303 Ivy Lane, Suite 400, Greenbelt, MD 20770.

[Doc. No. FAA-2002-14002, 72 FR 31680, June 7, 2007]

PART 99

SECURITY CONTROL OF AIR TRAFFIC

Contents

Subpart A—General

Subpart B—Designated Air Defense Identification Zones

Authority: 49 U.S.C. 106(g), 40101, 40103, 40106, 40113, 40120, 44502, 44721.

Source: Docket No. 25113, 53 FR 18217, May 20, 1988, unless otherwise noted.

Subpart A—General

§ 99.1 Applicability.

(a) This subpart prescribes rules for operating all aircraft (except for Department of Defense and law enforcement aircraft) in a defense area, or into, within, or out of the United States through an Air Defense Identification Zone (ADIZ) designated in subpart B.

(b) Except for §§ 99.7, 99.13, and 99.15 this subpart does not apply to the operation of any aircraft-

(1) Within the 48 contiguous States and the District of Columbia, or within the State of Alaska, on a flight which remains within 10 nautical miles of the point of departure;

(2) Operating at true airspeed of less than 180 knots in the Hawaii ADIZ or over any island, or within 12 nautical miles of the coastline of any island, in the Hawaii ADIZ;

(3) Operating at true airspeed of less than 180 knots in the Alaska ADIZ while the pilot maintains a continuous listening watch on the appropriate frequency; or

(4) Operating at true airspeed of less than 180 knots in the Guam ADIZ.

(c) An FAA ATC center may exempt the following operations from this subpart (except § 99.7) on a local basis only, with the concurrence of the U.S. military commanders concerned, or pursuant to an agreement with a U.S. Federal security or intelligence agency:

(1) Aircraft operations that are conducted wholly within the boundaries of an ADIZ and are not currently significant to the air defense system.

(2) Aircraft operations conducted in accordance with special procedures prescribed by a U.S. military authority, or a U.S. Federal security or intelligence agency concerned.

[Doc. No. 25113, 53 FR 18217, May 20, 1988, as amended by Amdt. 99-14, 53 FR 44182, Nov. 2, 1988; 66 FR 49822, Sept. 28, 2001; 69 FR 16755, Mar. 30, 2004]

§ 99.3 Definitions.

Aeronautical facility means, for the purposes of this subpart, a communications facility where flight plans or position reports are normally filed during flight operations.

Air defense identification zone (ADIZ) means an area of airspace over land or water in which the ready identification, location, and control of all aircraft (except for Department of Defense and law enforcement aircraft) is required in the interest of national security.

Defense area means any airspace of the contiguous United States that is not an ADIZ in which the control of aircraft is required for reasons of national security.

Defense visual flight rules (DVFR) means, for the purposes of this subpart, a flight within an ADIZ conducted by any aircraft (except for Department of Defense and law enforcement aircraft) in accordance with visual flight rules in part 91 of this title.

[Doc. No. FAA-2001-10693, 66 FR 49822, Sept. 28, 2001, as amended at 69 FR 16755, Mar. 30, 2004]

§ 99.5 Emergency situations.

In an emergency that requires immediate decision and action for the safety of the flight, the pilot in command of an aircraft may deviate from the rules in this part to the extent required by that emergency. He shall report the reasons for the deviation to the communications facility where flight plans or position reports are normally filed (referred to in this part as "an appropriate aeronautical facility") as soon as possible.

§ 99.7 Special security instructions.

Each person operating an aircraft in an ADIZ or Defense Area must, in addition to the applicable rules of this part, comply with special security instructions issued by the Administrator in the interest of national security, pursuant to agreement between the FAA and the Department of Defense, or between the FAA and a U.S. Federal security or intelligence agency.

[69 FR 16756, Mar. 30, 2004]

§ 99.9 Radio requirements.

(a) A person who operates a civil aircraft into an ADIZ must have a functioning two-way radio, and the pilot must maintain a continuous listening watch on the appropriate aeronautical facility's frequency.

(b) No person may operate an aircraft into, within, or whose departure point is within an ADIZ unless—

(1) The person files a DVFR flight plan containing the time and point of ADIZ penetration, and

(2) The aircraft departs within five minutes of the estimated departure time contained in the flight plan.

(c) If the pilot operating an aircraft under DVFR in an ADIZ cannot maintain two-way radio communications, the pilot may proceed, in accordance with original DVFR flight plan, or land as soon as practicable. The pilot must report the radio failure to an appropriate aeronautical facility as soon as possible.

(d) If a pilot operating an aircraft under IFR in an ADIZ cannot maintain two-way radio communications, the pilot must proceed in accordance with § 91.185 of this chapter.

[Doc. No. FAA-2001-10693, 66 FR 49822, Sept. 28, 2001, as amended at 69 FR 16756, Mar. 30, 2004]

§ 99.11 ADIZ flight plan requirements.

(a) No person may operate an aircraft into, within, or from a departure point within an ADIZ, unless the person files, activates, and closes a flight plan with the appropriate aeronautical facility, or is otherwise authorized by air traffic control.

(b) Unless ATC authorizes an abbreviated flight plan—

(1) A flight plan for IFR flight must contain the information specified in § 91.169; and

(2) A flight plan for VFR flight must contain the information specified in § 91.153(a) (1) through (6).

(3) If airport of departure is within the Alaskan ADIZ and there is no facility for filing a flight plan then:

(i) Immediately after takeoff or when within range of an appropriate aeronautical facility, comply with provisions of paragraph (b)(1) or (b)(2) as appropriate.

(ii) Proceed according to the instructions issued by the appropriate aeronautical facility.

(c) The pilot shall designate a flight plan for VFR flight as a DVFR flight plan.

(d) The pilot in command of an aircraft for which a flight plan has been filed must file an arrival or completion notice with an appropriate aeronautical facility.

[Doc. No. 25113, 53 FR 18217, May 20, 1988; 53 FR 44182, Nov. 2, 1988, as amended by Amdt. 99-15, 54 FR 34331, Aug. 18, 1989; 66 FR 49822, Sept. 28, 2001; 69 FR 16756, Mar. 30, 2004]

§ 99.12 [Reserved]

§ 99.13 Transponder-on requirements.

(a) *Aircraft transponder-on operation.* Each person operating an aircraft into or out of the United States into, within, or across an ADIZ designated in subpart B of this part, if that aircraft is equipped with an operable radar beacon transponder, shall operate the transponder, including altitude encoding equipment if installed, and shall reply on the appropriate code or as assigned by ATC.

(b) *ATC transponder equipment and use.* Effective September 7, 1990, unless otherwise authorized by ATC, no person may operate a civil aircraft into or out of the United States into, within, or across the contiguous U.S. ADIZ designated in subpart B of this part unless that aircraft is equipped with a coded radar beacon transponder.

(c) *ATC transponder and altitude reporting equipment and use.* Effective December 30, 1990, unless otherwise authorized by ATC, no person may operate a civil aircraft into or out of the United States into, within, or across the contiguous U.S. ADIZ unless that aircraft is equipped with a coded radar beacon transponder and automatic pressure altitude reporting equipment having altitude reporting capability that automatically replies to interrogations by transmitting pressure altitude information in 100-foot increments.

(d) Paragraphs (b) and (c) of this section do not apply to the operation of an aircraft which was not originally certificated with an engine-driven electrical system and which has not subsequently been certified with such a system installed, a balloon, or a glider.

[Doc. No. 24903, 55 FR 8395, Mar. 7, 1990. Redesignated at 69 FR 16756, Mar. 30, 2004]

§ 99.15 Position reports.

(a) The pilot of an aircraft operating in or penetrating an ADIZ under IFR—

(1) In controlled airspace, must make the position reports required in § 91.183; and

(2) In uncontrolled airspace, must make the position reports required in this section.

(b) No pilot may operate an aircraft penetrating an ADIZ under DVFR unless—

(1) The pilot reports to an appropriate aeronautical facility before penetration: the time, position, and altitude at which the aircraft passed the last reporting point before penetration and the estimated time of arrival over the next appropriate reporting point along the flight route;

(2) If there is no appropriate reporting point along the flight route, the pilot reports at least 15 minutes before penetration: The estimated time, position, and altitude at which the pilot will penetrate; or

(3) If the departure airport is within an ADIZ or so close to the ADIZ boundary that it prevents the pilot from complying with paragraphs (b)(1) or (2) of this section, the pilot must report immediately after departure: the time of departure, the altitude, and the estimated time of arrival over the first reporting point along the flight route.

(c) In addition to any other reports as ATC may require, no pilot in command of a foreign civil aircraft may enter the United States through an ADIZ unless that pilot makes the reports required in this section or reports the position of the aircraft when it is not less that one hour and not more that 2 hours average direct cruising distance from the United States.

[69 FR 16756, Mar. 30, 2004]

§ 99.17 Deviation from flight plans and ATC clearances and instructions.

(a) No pilot may deviate from the provisions of an ATC clearance or ATC instruction except in accordance with § 91.123 of this chapter.

(b) No pilot may deviate from the filed IFR flight plan when operating an aircraft in uncontrolled airspace unless that pilot notifies an appropriate aeronautical facility before deviating.

(c) No pilot may deviate from the filed DVFR flight plan unless that pilot notifies an appropriate aeronautical facility before deviating.

[69 FR 16756, Mar. 30, 2004]

§§ 99.19-99.31 [Reserved]

Subpart B—Designated Air Defense Identification Zones

§ 99.41 General.

The airspace above the areas described in this subpart is established as an ADIZ. The lines between points described in this subpart are great circles except that the lines joining adjacent points on the same parallel of latitude are rhumb lines.

[69 FR 16756, Mar. 30, 2004]

§ 99.43 Contiguous U.S. ADIZ.

The area bounded by a line from 43°15′N, 65°55′W; 44°21′N; 67°16′W; 43°10′N; 69°40′W; 41°05′N; 69°40′W; 40°32′N; 72°15′W; 39°55′N; 73°00′W; 39°38′N; 73°00′W; 39°36′N; 73°40′W; 37°00′N; 75°30′W; 36°10′N; 75°10′W; 35°10′N; 75°10′W; 32°00′N; 80°30′W; 30°30′N; 81°00′W; 26°40′N; 79°40′W; 25°00′N; 80°05′W; 24°25′N; 81°15′W;

24°20′N; 81°45′W; 24°30′N; 82°06′W; 24°41′N; 82°06′W; 24°43′N; 82°00′W; 25°00′N; 81°30′W; 25°10′N; 81°23′W; 25°35′N; 81°30′W; 26°15′N 82°20′W; 27°50′N; 83°05′W; 28°55′N; 83°30′W; 29°42′N; 84°00′W; 29°20′N; 85°00′W; 30°00′N; 87°10′W; 30°00′N; 88°30′W; 28°45′N; 88°55′W; 28°45′N; 90°00′W; 29°25′N; 94°00′W; 28°20′N; 96°00′W; 27°30′N; 97°00′W; 26°00′N; 97°00′W; 25°58′N; 97°07′W; westward along the U.S./Mexico border to 32°32′03″N, 117°07′25″W; 32°30′N; 117°25′W; 32°35′N; 118°30′W; 33°05′N; 119°45′W; 33°55′N; 120°40′W; 34°50′N; 121°10′W; 38°50′N; 124°00′W; 40°00′N; 124°35′W; 40°25′N; 124°40′W; 42°50′N; 124°50′W; 46°15′N; 124°30′W; 48°30′N; 125°00′W; 48°20′N; 128°00′W; 48°20′N; 132°00′W; 37°42′N; 130°40′W; 29°00′N; 124°00′W; 30°45′N; 120°50′W; 32°00′N; 118°24′W; 32°30′N; 117°20′W; 32°32′03″N; 117°07′25″W; eastward along the U.S./Mexico border to 25°58′N, 97°07′W; 26°00′N; 97°00′W; 26°00′N; 95°00′W; 26°30′N; 95°00′W; then via 26°30′N; parallel to 26°30′N; 84°00′W; 24°00′N; 83°00′W; then Via 24°00′N; parallel to 24°00′N; 79°25′W; 25°40′N; 79°25′W; 27°30′N; 78°50′W; 30°45′N; 74°00′W; 39°30′N; 63°45′W; 43°00′N; 65°48′W; to point of beginning.

[Doc. No. FAA-2001-10693, 66 FR 49822, Sept. 28, 2001. Redesignated at 69 FR 16756, Mar. 30, 2004]

§ 99.45 Alaska ADIZ.

The area is bounded by a line from 54°00′N; 136°00′W; 56°57′N; 144°00′W; 57°00′N; 145°00′W; 53°00′N; 158°00′W; 50°00′N; 169°00′W; 50°00′N; 180°00′; 50°00′N; 170°00′E; 53°00′N; 170°00′E; 60°00′00″N; 180°00′; 65°00′N; 169°00′W; then along 169°00′W; to 75°00′N; 169°00′W; then along the 75°00′N; parallel to 75°00′N; 141°00′W; 69°50′N; 141°00′W 71°18′N; 156°44′W; 68°40′N; 167°10′W; 67°00′N; 165°00′W; 65°40′N; 168°15′W; 63°45′N; 165°30′W; 61°20′N; 166°40′W; 59°00′N; 163°00′W; then south along 163°00′W to 54°00′N, 163°00′W; 56°30′N; 154°00′W; 59°20′N; 146°00′W; 59°30′N; 140°00′W; 57°00′N; 136°00′W; 54°35′N, 133°00′W; to point of beginning.

[Doc. No. FAA-2001-10693, 66 FR 49822, Sept. 28, 2001. Redesignated at 69 FR 16756, Mar. 30, 2004]

§ 99.47 Guam ADIZ.

(a) *Inner boundary.* From a point 13°52′07″ N, 143°59′16″ E, counterclockwise along the 50-nautical-mile radius arc of the NIMITZ VORTAC (located at 13°27′11″ N, 144°43′51″ E); to a point 13°02′08″ N, 145°28′17″ E; then to a point 14°49′07″ N, 146°13′58″ E; counterclockwise along the 35-nautical-mile radius arc of the SAIPAN NDB (located at 15°06′46″ N, 145°42′42″ E); to a point 15°24′21″ N, 145°11′21″ E; then to the point of origin.

(b) *Outer boundary.* The area bounded by a circle with a radius of 250 NM centered at latitude 13°32′41″ N, longitude 144°50′30″ E.

[Doc. No. 25113, 53 FR 18217, May 20, 1988. Redesignated at 69 FR 16756, Mar. 30, 2004]

§ 99.49 Hawaii ADIZ.

(a) *Outer boundary.* The area included in the irregular octagonal figure formed by a line connecting 26°30′ N, 156°00′ W; 26°30′ N, 161°00′ W; 24°00′ N, 164°00′ W; 20°00′ N, 164°00′ W; 17°00′ N, 160°00′ W; 17°00′ N, 156°00′ W; 20°00′ N, 153°00′ W; 22°00′ N, 153°00′ W; to point of beginning.

(b) *Inner boundary.* The inner boundary to follow a line connecting 22°30′ N, 157°00′ W; 22°30′ N, 160°00′ W; 22°00′ N, 161°00′ W; 21°00′ N, 161°00′ W; 20°00′ N, 160°00′ W; 20°00′ N, 156°30′ W; 21°00′ N, 155°30′ W; to point of beginning.

[Doc. No. 25113, 53 FR 18217, May 20, 1988. Redesignated at 69 FR 16756, Mar. 30, 2004]

PART 99

FAR

PART 103

ULTRALIGHT VEHICLES

Contents

Authority: 49 U.S.C. 106(g), 40103-40104, 40113, 44701.

Source: Docket No. 21631, 47 FR 38776, Sept. 2, 1982, unless otherwise noted.

Subpart A—General

§ 103.1 Applicability.

This part prescribes rules governing the operation of ultralight vehicles in the United States. For the purposes of this part, an ultralight vehicle is a vehicle that:

(a) Is used or intended to be used for manned operation in the air by a single occupant;

(b) Is used or intended to be used for recreation or sport purposes only;

(c) Does not have any U.S. or foreign airworthiness certificate; and

(d) If unpowered, weighs less than 155 pounds; or

(e) If powered:

(1) Weighs less than 254 pounds empty weight, excluding floats and safety devices which are intended for deployment in a potentially catastrophic situation;

(2) Has a fuel capacity not exceeding 5 U.S. gallons;

(3) Is not capable of more than 55 knots calibrated airspeed at full power in level flight; and

(4) Has a power-off stall speed which does not exceed 24 knots calibrated airspeed.

§ 103.3 Inspection requirements.

(a) Any person operating an ultralight vehicle under this part shall, upon request, allow the Administrator, or his designee, to inspect the vehicle to determine the applicability of this part.

(b) The pilot or operator of an ultralight vehicle must, upon request of the Administrator, furnish satisfactory evidence that the vehicle is subject only to the provisions of this part.

§ 103.5 Waivers.

No person may conduct operations that require a deviation from this part except under a written waiver issued by the Administrator.

§ 103.7 Certification and registration.

(a) Notwithstanding any other section pertaining to certification of aircraft or their parts or equipment, ultralight vehicles and their component parts and equipment are not required to meet the airworthiness certification standards specified for aircraft or to have certificates of airworthiness.

(b) Notwithstanding any other section pertaining to airman certification, operators of ultralight vehicles are not required to meet any aeronautical knowledge, age, or experience requirements to operate those vehicles or to have airman or medical certificates.

(c) Notwithstanding any other section pertaining to registration and marking of aircraft, ultralight vehicles are not required to be registered or to bear markings of any type.

Subpart B—Operating Rules

§ 103.9 Hazardous operations.

(a) No person may operate any ultralight vehicle in a manner that creates a hazard to other persons or property.

(b) No person may allow an object to be dropped from an ultralight vehicle if such action creates a hazard to other persons or property.

§ 103.11 Daylight operations.

(a) No person may operate an ultralight vehicle except between the hours of sunrise and sunset.

(b) Notwithstanding paragraph (a) of this section, ultralight vehicles may be operated during the twilight periods 30 minutes before official sunrise and 30 minutes after official sunset or, in Alaska, during the period of civil twilight as defined in the Air Almanac, if:

(1) The vehicle is equipped with an operating anticollision light visible for at least 3 statute miles; and

(2) All operations are conducted in uncontrolled airspace.

§ 103.13 Operation near aircraft; right-of-way rules.

(a) Each person operating an ultralight vehicle shall maintain vigilance so as to see and avoid aircraft and shall yield the right-of-way to all aircraft.

(b) No person may operate an ultralight vehicle in a manner that creates a collision hazard with respect to any aircraft.

(c) Powered ultralights shall yield the right-of-way to unpowered ultralights.

§ 103.15 Operations over congested areas.

No person may operate an ultralight vehicle over any congested area of a city, town, or settlement, or over any open air assembly of persons.

§ 103.17 Operations in certain airspace.

No person may operate an ultralight vehicle within Class A, Class B, Class C, or Class D airspace or within the lateral boundaries of the surface area of Class E airspace designated for an airport unless that person has prior authorization from the ATC facility having jurisdiction over that airspace.

[Amdt. 103-17, 56 FR 65662, Dec. 17, 1991]

§ 103.19 Operations in prohibited or restricted areas.

No person may operate an ultralight vehicle in prohibited or restricted areas unless that person has permission from the using or controlling agency, as appropriate.

§ 103.20 Flight restrictions in the proximity of certain areas designated by notice to airmen.

No person may operate an ultralight vehicle in areas designated in a Notice to Airmen under § 91.137, § 91.138, § 91.141, § 91.143 or § 91.145 of this chapter, unless authorized by:

(a) Air Traffic Control (ATC); or

(b) A Flight Standards Certificate of Waiver or Authorization issued for the demonstration or event.

[Doc. No. FAA-2000-8274, 66 FR 47378, Sept. 11, 2001]

§ 103.21 Visual reference with the surface.

No person may operate an ultralight vehicle except by visual reference with the surface.

§ 103.23 Flight visibility and cloud clearance requirements.

No person may operate an ultralight vehicle when the flight visibility or distance from clouds is less than that in the table found below. All operations in Class A, Class B, Class C, and Class D airspace or Class E airspace designated for an airport must receive prior ATC authorization as required in § 103.17 of this part.

[Amdt. 103-17, 56 FR 65662, Dec. 17, 1991]

Airspace	Flight visibility	Distance from clouds
Class A	Not applicable	Not Applicable.
Class B	3 statute miles	Clear of Clouds.
Class C	3 statute miles	500 feet below. 1,000 feet above. 2,000 feet horizontal.
Class D	3 statute miles	500 feet below. 1,000 feet above. 2,000 feet horizontal.
Class E:		
Less than 10,000 feet MSL	3 statute miles	500 feet below. 1,000 feet above. 2,000 feet horizontal.
At or above 10,000 feet MSL	5 statute miles	1,000 feet below. 1,000 feet above. 1 statute mile horizontal.
Class G:		
1,200 feet or less above the surface (regardless of MSL altitude)	1 statute mile	Clear of clouds.
More than 1,200 feet above the surface but less than 10,000 feet MSL	1 statute mile	500 feet below. 1,000 feet above. 2,000 feet horizontal.
More than 1,200 feet above the surface and at or above 10,000 feet MSL	5 statute miles	1,000 feet below. 1,000 feet above. 1 statute mile horizontal.

[Amdt. 103–17, 56 FR 65662, Dec. 17, 1991]

PART 105

PARACHUTE OPERATIONS

Contents

AUTHORITY: 49 U.S.C. 106(g), 40113-40114, 44701-44702, 44721.

SOURCE: Doc. No. FAA-1999-5483, 66 FR 23553, May 9, 2001, unless otherwise noted.

§ 105.1 Applicability.

(a) Except as provided in paragraphs (b) and (c) of this section, this part prescribes rules governing parachute operations conducted in the United States.

(b) This part does not apply to a parachute operation conducted—

(1) In response to an in-flight emergency, or

(2) To meet an emergency on the surface when it is conducted at the direction or with the approval of an agency of the United States, or of a State, Puerto Rico, the District of Columbia, or a possession of the United States, or an agency or political subdivision thereof.

(c) Sections 105.5, 105.9, 105.13, 105.15, 105.17, 105.19 through 105.23, 105.25(a)(1) and 105.27 of this part do not apply to a parachute operation conducted by a member of an Armed Force—

(1) Over or within a restricted area when that area is under the control of an Armed Force.

(2) During military operations in uncontrolled airspace.

§ 105.3 Definitions.

For the purposes of this part—

Approved parachute means a parachute manufactured under a type certificate or a Technical Standard Order (C-23 series), or a personnel-carrying U.S. military parachute (other than a high altitude, high speed, or ejection type) identified by a Navy Air Facility, an Army Air Field, and Air Force-Navy drawing number, an Army Air Field order number, or any other military designation or specification number.

Automatic Activation Device means a self-contained mechanical or electro-mechanical device that is attached to the interior of the reserve parachute container, which automatically initiates parachute deployment of the reserve parachute at a pre-set altitude, time, percentage of terminal velocity, or combination thereof.

Direct Supervision means that a certificated rigger personally observes a non-certificated person packing a main parachute to the extent necessary to ensure that it is being done properly, and takes responsibility for that packing.

Drop Zone means any pre-determined area upon which parachutists or objects land after making an intentional parachute jump or drop. The center-point target of a drop zone is expressed in nautical miles from the nearest VOR facility when 30 nautical miles or less; or from the nearest airport, town, or city depicted on the appropriate Coast and Geodetic Survey World Aeronautical Chart or Sectional Aeronautical Chart, when the nearest VOR facility is more than 30 nautical miles from the drop zone.

Foreign parachutist means a parachutist who is neither a U.S. citizen or a resident alien and is participating in parachute operations within the United States using parachute equipment not manufctured in the United States.

Freefall means the portion of a parachute jump or drop between aircraft exit and parachute deployment in which the parachute is activated manually by the parachutist at the parachutist's discretion or automatically, or, in the case of an object, is activated automatically.

Main parachute means a parachute worn as the primary parachute used or intended to be used in conjunction with a reserve parachute.

Object means any item other than a person that descends to the surface from an aircraft in flight when a parachute is used or is intended to be used during all or part of the descent.

Parachute drop means the descent of an object to the surface from an aircraft in flight when a parachute is used or intended to be used during all or part of that descent.

Parachute jump means a parachute operation that involves the descent of one or more persons to the surface from an aircraft in flight when an aircraft is used or intended to be used during all or part of that descent.

Parachute operation means the performance of all activity for the purpose of, or in support of, a parachute jump or a parachute drop. This parachute operation can involve, but is not limited to, the following persons: parachutist, parachutist in command and passenger in tandem parachute operations, drop zone or owner or operator, jump master, certificated parachute rigger, or pilot.

Parachutist means a person who intends to exit an aircraft while in flight using a single-harness, dual parachute system to descend to the surface.

Parachutist in command means the person responsible fro the operation and safety of a tandem parachute operation.

Passenger parachutist means a person who boards an aircraft, acting as other than the parachutist in command of a tandem parachute operation, with the intent of exiting the aircraft while in-flight using the forward harness of a dual harness tandem parachute system to descend to the surface.

Pilot chute means a small parachute used to initiate and/or accelerate deployment of a main or reserve parachute.

Ram-air parachute means a parachute with a canopy consisting of an upper and lower surface that is inflated by ram air entering through specially designed openings in the front of the canopy to form a gliding airfoil.

Reserve parachute means an approved parachute worn for emergency use to be activated only upon failure of the main parachute or in any other emergency where use of the main parachute is impractical or use of the main parachute would increase risk.

Single-harness, dual parachute system: means the combination of a main parachute, approved reserve parachute, and approved single person harness and dual-parachute container. This parachute system may have an operational automatic activation device installed.

Tandem parachute operation: means a parachute operation in which more than one person simultaneously uses the same tandem parachute system while descending to the surface from an aircraft in flight.

Tandem parachute system: means the combination of a main parachute, approved reserve parachute, and approved harness and dual parachute container, and a separate approved forward harness for a passenger parachutist. This parachute system must have an operational automatic activation device installed.

§ 105.5 General.

No person may conduct a parachute operation, and no pilot in command of an aircraft may allow a parachute operation to be conducted from an aircraft, if that operation creates a hazard to air traffic or to persons or property on the surface.

§ 105.7 Use of alcohol and drugs.

No person may conduct a parachute operation, and no pilot in command of an aircraft may allow a person to conduct a parachute operation from that aircraft, if that person is or appears to be under the influence of—

(a) Alcohol, or

(b) Any drug that affects that person's faculties in any way contrary to safety.

§ 105.9 Inspections.

The Administrator may inspect any parachute operation to which this part applies (including inspections at the site where the parachute operation is being conducted) to determine compliance with the regulations of this part.

Subpart B—Operating Rules

§ 105.13 Radio equipment and use requirements.

(a) Except when otherwise authorized by air traffic control—

(1) No person may conduct a parachute operation, and no pilot in command of an aircraft may allow a parachute operation to be conducted from that aircraft, in or into controlled airspace unless, during that flight—

(i) The aircraft is equipped with a functioning two-way radio communication system appropriate to the air traffic control facilities being used; and

(ii) Radio communications have been established between the aircraft and the air traffic control facility having jurisdiction over the affected airspace of the first intended exit altitude at least 5 minutes before the parachute operation begins. The pilot in command must establish radio communications to receive information regarding air traffic activity in the vicinity of the parachute operation.

(2) The pilot in command of an aircraft used for any parachute operation in or into controlled airspace must, during each flight—

(i) Continuously monitor the appropriate frequency of the aircraft's radio communications system from the time radio communications are first established between the aircraft and air traffic control, until the pilot advises air traffic control that the parachute operation has ended for that flight.

(ii) Advise air traffic control when the last parachutist or object leaves the aircraft.

(b) Parachute operations must be aborted if, prior to receipt of a required air traffic control authorization, or during any parachute operation in or into controlled airspace, the required radio communications system is or becomes inoperative.

§ 105.15 Information required and notice of cancellation or postponement of a parachute operation.

(a) Each person requesting an authorization under §§ 105.21(b) and 105.25(a)(2) of this part and each person submitting a notification under § 105.25(a)(3) of this part must provide the following information (on an individual or group basis):

(1) The date and time the parachute operation will begin.

(2) The radius of the drop zone around the target expressed in nautical miles.

(3) The location of the center of the drop zone in relation to—

(i) The nearest VOR facility in terms of the VOR radial on which it is located and its distance in nautical miles from the VOR facility when that facility is 30 nautical miles or less from the drop zone target; or

(ii) the nearest airport, town, or city depicted on the appropriate Coast and Geodetic Survey World Aeronautical Chart or Sectional Aeronautical Chart, when the nearest VOR facility is more than 30 nautical miles from the drop zone target.

(4) Each altitude above mean sea level at which the aircraft will be operated when parachutists or objects exist the aircraft.

(5) The duration of the intended parachute operation.

(6) The name, address, and telephone number of the person who requests the authorization or gives notice of the parachute operation.

(7) The registration number of the aircraft to be used.

(8) The name of the air traffic control facility with jurisdiction of the airspace at the first intended exit altitude to be used for the parachute operation.

(b) Each holder of a certificate of authorization issued under §§ 105.21(b) and 105.25(b) of this part must present that certificate for inspection upon the request of the Administrator or any Federal, State, or local official.

(c) Each person requesting an authorization under §§ 105.21(b) and 105.25(a)(2) of this part and each person submitting a notice under § 105.25(a)(3) of this part must promptly notify the air traffic control facility having jurisdiction over the affected airspace if the proposed or scheduled parachute operation is canceled or postponed.

§ 105.17 Flight visibility and clearance from cloud requirements.

No person may conduct a parachute operation, and no pilot in command of an aircraft may allow a parachute operation to be conducted from that aircraft—

(a) Into or through a cloud, or

(b) When the flight visibility or the distance from any cloud is less than that prescribed in the following table:

Altitude	Flight visibility (statute miles)	Distance from clouds
1,200 feet or less above the surface regardless of the MSL altitude	3	500 feet below, 1,000 feet above, 2,000 feet horizontal.
More than 1,200 feet above the surface but less than 10,000 feet MSL	3	500 feet below, 1,000 feet above, 2,000 feet horizontal.
More than 1,200 feet above the surface and at or above 10,000 feet MSL	5	1,000 feet below, 1,000 feet above, 1 mile horizontal.

§ 105.19 Parachute operations between sunset and sunrise.

(a) No person may conduct a parachute operation, and no pilot in command of an aircraft may allow a person to conduct a parachute operation from an aircraft between sunset and sunrise, unless the person or object descending from the aircraft displays a light that is visible for at least 3 statute miles.

(b) The light required by paragraph (a) of this section must be displayed from the time that the person or object is under a properly functioning open parachute until that person or object reaches the surface.

§ 105.21 Parachute operations over or into a congested area or an open-air assembly of persons.

(a) No person may conduct a parachute operation, and no pilot in command of an aircraft may allow a parachute operation to be conducted from that aircraft, over or into a congested area of a city, town, or settlement, or an open-air assembly of persons unless a certificate of authorization for that parachute operation has been issued under this section. However, a parachutist may drift over a congested area or an open-air assembly of persons with a fully deployed and properly functioning parachute if that parachutist is at a sufficient altitude to avoid creating a hazard to persons or property on the surface.

(b) An application for a certificate of authorization issued under this section must—

(1) Be made in the form and manner prescribed by the Administrator, and

(2) Contain the information required in § 105.15(a) of this part.

(c) Each holder of, and each person named as a participant in a certificate of authorization issued under this section must comply with all requirements contained in the certificate of authorization.

(d) Each holder of a certificate of authorization issued under this section must present that certificate for inspection upon the request of the Administrator, or any Federal, State, or local official.

§ 105.23 Parachute operations over or onto airports.

No person may conduct a parachute operation, and no pilot in command of an aircraft may allow a parachute operation to be conducted from that aircraft, over or onto any airport unless—

(a) For airports with an operating control tower:

(1) Prior approval has been obtained from the management of the airport to conduct parachute operations over or on that airport.

(2) Approval has been obtained from the control tower to conduct parachute operations over or onto that airport.

(3) Two-way radio communications are maintained between the pilot of the aircraft involved in the parachute operation and the control tower of the airport over or onto which the parachute operation is being conducted.

(b) For airports without an operating control tower, prior approval has been obtained from the management of the airport to conduct parachute operations over or on that airport.

(c) A parachutist may drift over that airport with a fully deployed and properly functioning parachute if the parachutist is at least 2,000 feet above that airport's traffic pattern, and avoids creating a hazard to air traffic or to persons and property on the ground.

§ 105.25 Parachute operations in designated airspace.

(a) No person may conduct a parachute operation, and no pilot in command of an aircraft may allow a parachute operation to be conducted from that aircraft—

(1) Over or within a restricted area or prohibited area unless the controlling agency of the area concerned has authorized that parachute operation;

(2) Within or into a Class A, B, C, D airspace area without, or in violation of the requirements of, an air traffic control authorization issued under this section;

(3) Except as provided in paragraph (c) and (d) of this section, within or into Class E or G airspace area unless the air traffic control facility having jurisdiction over the airspace at the first intended exit altitude is notified of the parachute operation no earlier than 24 hours before or no later than 1 hour before the parachute operation begins.

(b) Each request for a parachute operation authorization or notification required under this section must be submitted to the air traffic control facility having jurisdiction over the airspace at the first intended exit altitude and must include the information prescribed by § 105.15(a) of this part.

(c) For the purposes of paragraph (a)(3) of this section, air traffic control facilities may accept a written notification from an organization that conducts parachute operations and lists the scheduled series of parachute operations to be conducted over a stated period of time not longer than 12 calendar months. The notification must contain the information prescribed by § 105.15(a) of this part, identify the responsible persons associated with that parachute operation, and be submitted at least 15 days, but not more than 30 days, before the parachute operation begins. The FAA may revoke the acceptance of the notification for any failure of the organization conducting the parachute operations to comply with its requirements.

(d) Paragraph (a)(3) of this section does not apply to a parachute operation conducted by a member of an Armed Force within a restricted area that extends upward from the surface when that area is under the control of an Armed Force.

Subpart C—Parachute Equipment and Packing

§ 105.41 Applicability.

This subpart prescribed rules governing parachute equipment used in civil parachute operations.

§ 105.43 Use of single-harness, dual-parachute systems.

No person may conduct a parachute operation using a single-harness, dual-parachute system, and no pilot in command of an aircraft may allow any person to conduct a parachute operation from that aircraft using a single-harness, dual-parachute system, unless that system has at least one main parachute, one approved reserve parachute, and one approved single person harness and container that are packed as follows:

(a) The main parachute must have been packed within 180 days before the date of its use by a certificated parachute rigger, the person making the next jump with that parachute, or a non-certificated person under the direct supervision of a certificated parachute rigger.

(b) The reserve parachute must have been packed by a certificated parachute rigger—

(1) Within 180 days before the date of its use, if its canopy, shroud, and harness are composed exclusively of nylon, rayon, or similar synthetic fiber or material that is substantially resistant to damage from mold, mildew, and other fungi, and other rotting agents propagated in a moist environment; or

(2) Within 60 days before the date of its use, if it is composed of any amount of silk, pongee, or other natural fiber, or material not specified in paragraph (b)(1) of this section.

(c) If installed, the automatic activation device must be maintained in accordance with manufacturer instructions for that automatic activation device.

[Doc. No. FAA-1999-5483, 66 FR 23553, May 9, 2001, as amended by Amdt. 105-13, 73 FR 69531, Nov. 19, 2008]

§ 105.45 Use of tandem parachute systems.

(a) No person may conduct a parachute operation using a tandem parachute system, and no pilot in command of an aircraft may allow any person to conduct a parachute operation from that aircraft using a tandem parachute system, unless—

(1) One of the parachutists using the tandem parachute system is the parachutist in command, and meets the following requirements:

(i) Has a minimum of 3 years of experience in parachuting, and must provide documentation that the parachutist—

(ii) Has completed a minimum of 500 freefall parachute jumps using a ram-air parachute, and

(iii) Holds a master parachute license issued by an organization recognized by the FAA, and

(iv) Has successfully completed a tandem instructor course given by the manufacturer of the tandem parachute system used in the parachute operation or a course acceptable to the Administrator.

(v) Has been certified by the appropriate parachute manufacturer or tandem course provider as being properly trained on the use of the specific tandem parachute system to be used.

(2) The person acting as parachutist in command:

(i) Has briefed the passenger parachutist before boarding the aircraft. The briefing must include the procedures to be used in case of an emergency with the aircraft or after exiting the aircraft, while preparing to exit and exiting the aircraft, freefall, operating the parachute after freefall, landing approach, and landing.

(ii) Uses the harness position prescribed by the manufacturer of the tandem parachute equipment.

(b) No person may make a parachute jump with a tandem parachute system unless—

(1) The main parachute has been packed by a certificated parachute rigger, the parachutist in command making the next jump with that parachute, or a person under the direct supervision of a certificated parachute rigger.

(2) The reserve parachute has been packed by a certificated parachute rigger in accordance with § 105.43(b) of this part.

(3) The tandem parachute system contains an operational automatic activation device for the reserve parachute, approved by the manufacturer of that tandem parachute system. The device must—

(i) Have been maintained in accordance with manufacturer instructions, and

(ii) Be armed during each tandem parachute operation.

(4) The passenger parachutist is provided with a manual main parachute activation device and instructed on the use of that device, if required by the owner/operator.

(5) The main parachute is equipped with a single-point release system.

(6) The reserve parachute meets Technical Standard Order C23 specifications.

§ 105.47 Use of static lines.

(a) Except as provided in paragraph (c) of this section, no person may conduct a parachute operation using a static line attached to the aircraft and the main parachute unless an assist device, described and attached as follows, is used to aid the pilot chute in performing its function, or, if no pilot chute is used, to aid in the direct deployment of the main parachute canopy. The assist device must—

(1) Be long enough to allow the main parachute container to open before a load is placed on the device.

(2) Have a static load strength of—

(i) At least 28 pounds but not more than 160 pounds if it is used to aid the pilot chute in performing its function; or

(ii) At least 56 pounds but not more than 320 pounds if it is used to aid in the direct deployment of the main parachute canopy; and

(3) Be attached as follows:

(i) At one end, to the static line above the static-line pins or, if static-line pins are not used, above the static-line ties to the parachute cone.

(ii) At the other end, to the pilot chute apex, bridle cord, or bridle loop, or, if no pilot chute is used, to the main parachute canopy.

(b) No person may attach an assist device required by paragraph (a) of this section to any main parachute unless that person is a certificated parachute rigger or that person makes the next parachute jump with that parachute.

(c) An assist device is not required for parachute operations using direct-deployed, ram-air parachutes.

§ 105.49 Foreign parachutists and equipment.

(a) No person may conduct a parachute operation, and no pilot in command of an aircraft may allow a parachute operation to be conducted from that aircraft with an unapproved foreign parachute system unless—

(1) The parachute system is worn by a foreign parachutist who is the owner of that system.

(2) The parachute system is of a single-harness dual parachute type.

(3) The parachute system meets the civil aviation authority requirements of the foreign parachutist's country.

(4) All foreign non-approved parachutes deployed by a foreign parachutist during a parachute operation conducted under this section shall be packed as follows—

(i) The main parachute must be packed by the foreign parachutist making the next parachute jump with that parachute, a certificated parachute rigger, or any other person acceptable to the Administrator.

(ii) The reserve parachute must be packed in accordance with the foreign parachutist's civil aviation authority requirements, by a certificated parachute rigger, or any other person acceptable to the Administrator.

PART 110

GENERAL REQUIREMENTS

Contents

Authority: 49 U.S.C. 106(g), 1153, 40101, 40102, 40103, 40113, 44105, 44106, 44111, 44701-44717, 44722, 44901, 44903, 44904, 44906, 44912, 44914, 44936, 44938, 46103, 46105.

Source: Doc. No. FAA-2009-0140, 76 FR 7486, Feb. 10, 2011, unless otherwise noted.

§ 110.1 Applicability.

This part governs all operations conducted under subchapter G of this chapter.

§ 110.2 Definitions

For the purpose of this subchapter, the term—

All-cargo operation means any operation for compensation or hire that is other than a passenger-carrying operation or, if passengers are carried, they are only those specified in § 121.583(a) or § 135.85 of this chapter.

Certificate-holding district office means the Flight Standards District Office that has responsibility for administering the certificate and is charged with the overall inspection of the certificate holder's operations.

Commercial air tour means a flight conducted for compensation or hire in an airplane or helicopter where a purpose of the flight is sightseeing. The FAA may consider the following factors in determining whether a flight is a commercial air tour:

(1) Whether there was a holding out to the public of willingness to conduct a sightseeing flight for compensation or hire;

(2) Whether the person offering the flight provided a narrative that referred to areas or points of interest on the surface below the route of the flight;

(3) The area of operation;

(4) How often the person offering the flight conducts such flights;

(5) The route of flight;

(6) The inclusion of sightseeing flights as part of any travel arrangement package;

(7) Whether the flight in question would have been canceled based on poor visibility of the surface below the route of the flight; and

(8) Any other factors that the FAA considers appropriate.

Commuter operation means any scheduled operation conducted by any person operating one of the following types of aircraft with a frequency of operations of at least five round trips per week on at least one route between two or more points according to the published flight schedules:

(1) Airplanes, other than turbojet-powered airplanes, having a maximum passenger-seat configuration of 9 seats or less, excluding each crewmember seat, and a maximum payload capacity of 7,500 pounds or less; or

(2) Rotorcraft.

Direct air carrier means a person who provides or offers to provide air transportation and who has control over the operational functions performed in providing that transportation.

DOD commercial air carrier evaluator means a qualified Air Mobility Command, Survey and Analysis Office cockpit evaluator performing the duties specified in Public Law 99-661 when the evaluator is flying on an air carrier that is contracted or pursuing a contract with the U.S. Department of Defense (DOD).

Domestic operation means any scheduled operation conducted by any person operating any airplane described in paragraph (1) of this definition at locations described in paragraph (2) of this definition:

(1) Airplanes:

(i) Turbojet-powered airplanes;

(ii) Airplanes having a passenger-seat configuration of more than 9 passenger seats, excluding each crewmember seat; or

(iii) Airplanes having a payload capacity of more than 7,500 pounds.

(2) Locations:

(i) Between any points within the 48 contiguous States of the United States or the District of Columbia; or

(ii) Operations solely within the 48 contiguous States of the United States or the District of Columbia; or

(iii) Operations entirely within any State, territory, or possession of the United States; or

(iv) When specifically authorized by the Administrator, operations between any point within the 48 contiguous States of the United States or the District of Columbia and any specifically authorized point located outside the 48 contiguous States of the United States or the District of Columbia.

Empty weight means the weight of the airframe, engines, propellers, rotors, and fixed equipment. Empty weight excludes the weight of the crew and payload, but includes the weight of all fixed ballast, unusable fuel supply, undrainable oil, total quantity of engine coolant, and total quantity of hydraulic fluid.

Flag operation means any scheduled operation conducted by any person operating any airplane described in paragraph (1) of this definition at the locations described in paragraph (2) of this definition:

(1) Airplanes:

(i) Turbojet-powered airplanes;

(ii) Airplanes having a passenger-seat configuration of more than 9 passenger seats, excluding each crewmember seat; or

(iii) Airplanes having a payload capacity of more than 7,500 pounds.

(2) Locations:

(i) Between any point within the State of Alaska or the State of Hawaii or any territory or possession of the United States and any point outside the State of Alaska or the State of Hawaii or any territory or possession of the United States, respectively; or

(ii) Between any point within the 48 contiguous States of the United States or the District of Columbia and any point outside the 48 contiguous States of the United States and the District of Columbia.

(iii) Between any point outside the U.S. and another point outside the U.S.

Justifiable aircraft equipment means any equipment necessary for the operation of the aircraft. It does not include equipment or ballast specifically installed, permanently or otherwise, for the purpose of altering the empty weight of an aircraft to meet the maximum payload capacity.

Kind of operation means one of the various operations a certificate holder is authorized to conduct, as specified in its operations specifications, *i.e.,* domestic, flag, supplemental, commuter, or on-demand operations.

Maximum payload capacity means:

(1) For an aircraft for which a maximum zero fuel weight is prescribed in FAA technical specifications, the maximum zero fuel weight, less empty weight, less all justifiable aircraft equipment, and less the operating load (consisting of minimum flightcrew, foods and beverages, and supplies and equipment related to foods and beverages, but not including disposable fuel or oil).

(2) For all other aircraft, the maximum certificated takeoff weight of an aircraft, less the empty weight, less all justifiable aircraft equipment, and less the operating load (consisting of minimum fuel load, oil, and flightcrew). The allowance for the weight of the crew, oil, and fuel is as follows:

(i) Crew—for each crewmember required by the Federal Aviation Regulations—

(A) For male flightcrew members—180 pounds.

(B) For female flightcrew members—140 pounds.

(C) For male flight attendants—180 pounds.

(D) For female flight attendants—130 pounds.

(E) For flight attendants not identified by gender—140 pounds.

(ii) Oil—350 pounds or the oil capacity as specified on the Type Certificate Data Sheet.

(iii) Fuel—the minimum weight of fuel required by the applicable Federal Aviation Regulations for a flight between domestic points 174 nautical miles apart under VFR weather conditions that does not involve extended overwater operations.

Maximum zero fuel weight means the maximum permissible weight of an aircraft with no disposable fuel or oil. The zero fuel weight figure may be found in either the aircraft type certificate data sheet, the approved Aircraft Flight Manual, or both.

Noncommon carriage means an aircraft operation for compensation or hire that does not involve a holding out to others.

On-demand operation means any operation for compensation or hire that is one of the following:

(1) Passenger-carrying operations conducted as a public charter under part 380 of this chapter or any operations in which the departure time, departure location, and arrival location are specifically negotiated with the customer or the customer's representative that are any of the following types of operations:

(i) Common carriage operations conducted with airplanes, including turbojet-powered airplanes, having a passenger-seat configuration of 30 seats or fewer, excluding each crewmember seat, and a payload capacity of 7,500 pounds or less, except that operations using a specific airplane that is also used in domestic or flag operations and that is so listed in the operations specifications as required by § 119.49(a)(4) of this chapter for those operations are considered supplemental operations;

(ii) Noncommon or private carriage operations conducted with airplanes having a passenger-seat configuration of less than 20 seats, excluding each crewmember seat, and a payload capacity of less than 6,000 pounds; or

(iii) Any rotorcraft operation.

(2) Scheduled passenger-carrying operations conducted with one of the following types of aircraft with a frequency of operations of less than five round trips per week on at least one route between two or more points according to the published flight schedules:

(i) Airplanes, other than turbojet powered airplanes, having a maximum passenger-seat configuration of 9 seats or less, excluding each crewmember seat, and a maximum payload capacity of 7,500 pounds or less; or

(ii) Rotorcraft.

(3) All-cargo operations conducted with airplanes having a payload capacity of 7,500 pounds or less, or with rotorcraft.

Passenger-carrying operation means any aircraft operation carrying any person, unless the only persons on the aircraft are those identified in §§ 121.583(a) or 135.85 of this chapter, as applicable. An aircraft used in a passenger-carrying operation may also carry cargo or mail in addition to passengers.

Principal base of operations means the primary operating location of a certificate holder as established by the certificate holder.

Provisional airport means an airport approved by the Administrator for use by a certificate holder for the purpose of providing service to a community when the regular airport used by the certificate holder is not available.

Regular airport means an airport used by a certificate holder in scheduled operations and listed in its operations specifications.

Scheduled operation means any common carriage passenger-carrying operation for compensation or hire conducted by an air carrier or commercial operator for which the certificate holder or its representative offers in advance the departure location, departure time, and arrival location. It does not include any passenger-carrying operation that is conducted as a public charter operation under part 380 of this chapter.

Supplemental operation means any common carriage operation for compensation or hire conducted with any airplane described in paragraph (1) of this definition that is a type of operation described in paragraph (2) of this definition:

(1) Airplanes:

(i) Airplanes having a passenger-seat configuration of more than 30 seats, excluding each crewmember seat;

(ii) Airplanes having a payload capacity of more than 7,500 pounds; or

(iii) Each propeller-powered airplane having a passenger-seat configuration of more than 9 seats and less than 31 seats, excluding each crewmember seat, that is also used in domestic or flag operations and that is so listed in the operations specifications as required by § 119.49(a)(4) of this chapter for those operations; or

(iv) Each turbojet powered airplane having a passenger seat configuration of 1 or more and less than 31 seats, excluding each crewmember seat, that is also used in domestic or flag operations and that is so listed in the operations specifications as required by § 119.49(a)(4) of this chapter for those operations.

(2) Types of operation:

(i) Operations for which the departure time, departure location, and arrival location are specifically negotiated with the customer or the customer's representative;

(ii) All-cargo operations; or

(iii) Passenger-carrying public charter operations conducted under part 380 of this chapter.

Wet lease means any leasing arrangement whereby a person agrees to provide an entire aircraft and at least one crewmember. A wet lease does not include a code-sharing arrangement.

When common carriage is not involved or operations not involving common carriage means any of the following:

(1) Noncommon carriage.

(2) Operations in which persons or cargo are transported without compensation or hire.

(3) Operations not involving the transportation of persons or cargo.

(4) Private carriage.

Years in service means the calendar time elapsed since an aircraft was issued its first U.S. or first foreign airworthiness certificate.

PART 117

FLIGHT AND DUTY LIMITATIONS AND REST REQUIREMENTS: FLIGHTCREW MEMBERS

Sec.

Authority: 49 U.S.C. 106(g), 40113, 40119, 44101, 44701-44702, 44705, 44709-44711, 44713, 44716-44717, 44722, 46901, 44903-44904, 44912, 46105.

§ 117.1 Applicability.

(a) This part prescribes flight and duty limitations and rest requirements for all flightcrew members and certificate holders conducting passenger operations under part 121 of this chapter.

(b) This part applies to all operations directed by part 121 certificate holders under part 91, other than subpart K, of this chapter if any segment is conducted as a domestic passenger, flag passenger, or supplemental passenger operation.

(c) This part applies to all flightcrew members when participating in an operation under part 91, other than subpart K of this chapter, on behalf of the part 121 certificate holder if any flight segment is conducted as a domestic passenger, flag passenger, or supplemental passenger operation

(d) Notwithstanding paragraphs (a), (b) and (c) of this section, a certificate holder may conduct under part 117 its part 121 operations pursuant to 121.470, 121.480, or 121.500.

§ 117.3 Definitions.

In addition to the definitions in §§ 1.1 and 110.2 of this chapter, the following definitions apply to this part. In the event there is a conflict in definitions, the definitions in this part control.

Acclimated means a condition in which a flightcrew member has been in a theater for 72 hours or has been given at least 36 consecutive hours free from duty.

Airport/standby reserve means a defined duty period during which a flightcrew member is required by a certificate holder to be at an airport for a possible assignment.

Augmented flightcrew means a flightcrew that has more than the minimum number of flightcrew members required by the airplane type certificate to operate the aircraft to allow a flightcrew member to be replaced by another qualified flightcrew member for in-flight rest.

Calendar day means a 24-hour period from 0000 through 2359 using Coordinated Universal Time or local time.

Certificate holder means a person who holds or is required to hold an air carrier certificate or operating certificate issued under part 119 of this chapter.

Deadhead transportation means transportation of a flightcrew member as a passenger or non-operating flightcrew member, by any mode of transportation, as required by a certificate holder, excluding transportation to or from a suitable accommodation. All time spent in deadhead transportation is duty and is not rest. For purposes of determining the maximum flight duty period in Table B of this part, deadhead transportation is not considered a flight segment.

Duty means any task that a flightcrew member performs as required by the certificate holder, including but not limited to flight duty period, flight duty, pre- and post-flight duties, administrative work, training, deadhead transportation, aircraft positioning on the ground, aircraft loading, and aircraft servicing.

Fatigue means a physiological state of reduced mental or physical performance capability resulting from lack of sleep or increased physical activity that can reduce a flightcrew member's alertness and ability to safely operate an aircraft or perform safety-related duties.

Fatigue risk management system (FRMS) means a management system for a certificate holder to use to mitigate the effects of fatigue in its particular operations. It is a data-driven process and a systematic method used to continuously monitor and manage safety risks associated with fatigue-related error.

Fit for duty means physiologically and mentally prepared and capable of performing assigned duties at the highest degree of safety.

Flight duty period (FDP) means a period that begins when a flightcrew member is required to report for duty with the intention of conducting a flight, a series of flights, or positioning or ferrying flights, and ends when the aircraft is parked after the last flight and there is no intention for further aircraft movement by the same flightcrew member. A flight duty period includes the duties performed by the flightcrew member on behalf of the certificate holder that occur before a flight segment or between flight segments without a required intervening rest period. Examples of tasks that are part of the flight duty period include deadhead transportation, training conducted in an aircraft or flight simulator, and airport/standby reserve, if the above tasks occur before a flight segment or between flight segments without an intervening required rest period:

Home base means the location designated by a certificate holder where a flightcrew member normally begins and ends his or her duty periods.

Lineholder means a flightcrew member who has an assigned flight duty period and is not acting as a reserve flightcrew member.

Long-call reserve means that, prior to beginning the rest period required by § 117.25, the flightcrew member is notified by the certificate holder to report for a flight duty period following the completion of the rest period.

Physiological night's rest means 10 hours of rest that encompasses the hours of 0100 and 0700 at the flightcrew member's home base, unless the individual has acclimated to a different theater. If the flightcrew member has acclimated to a different theater, the rest must encompass the hours of 0100 and 0700 at the acclimated location.

Report time means the time that the certificate holder requires a flightcrew member to report for an assignment.

Reserve availability period means a duty period during which a certificate holder requires a flightcrew member on short call reserve to be available to receive an assignment for a flight duty period.

Reserve flightcrew member means a flightcrew member who a certificate holder requires to be available to receive an assignment for duty.

Rest facility means a bunk or seat accommodation installed in an aircraft that provides a flightcrew member with a sleep opportunity.

(1) *Class 1 rest facility* means a bunk or other surface that allows for a flat sleeping position and is located separate from both the flight deck and passenger cabin in an area that is temperature-controlled, allows the flightcrew member to control light, and provides isolation from noise and disturbance.

(2) *Class 2 rest facility* means a seat in an aircraft cabin that allows for a flat or near flat sleeping position; is separated from passengers by a minimum of a curtain to provide darkness and some sound mitigation; and is reasonably free from disturbance by passengers or flightcrew members.

(3) *Class 3 rest facility* means a seat in an aircraft cabin or flight deck that reclines at least 40 degrees and provides leg and foot support.

Rest period means a continuous period determined prospectively during which the flightcrew member is free from all restraint by the certificate holder, including freedom from present responsibility for work should the occasion arise.

Scheduled means to appoint, assign, or designate for a fixed time.

Short-call reserve means a period of time in which a flightcrew member is assigned to a reserve availability period.

Split duty means a flight duty period that has a scheduled break in duty that is less than a required rest period.

Suitable accommodation means a temperature-controlled facility with sound mitigation and the ability to control light that provides a flightcrew member with the ability to sleep either in a bed, bunk or in a chair that allows for flat or near flat sleeping position. Suitable accommodation only applies to ground facilities and does not apply to aircraft onboard rest facilities.

Theater means a geographical area where local time at the flightcrew member's flight duty period departure point and arrival point differ by more than 60 degrees longitude.

Unforeseen operational circumstance means an unplanned event of insufficient duration to allow for adjustments to schedules, including unforecast weather, equipment malfunction, or air traffic delay that is not reasonably expected.

Window of circadian low means a period of maximum sleepiness that occurs between 0200 and 0559 during a physiological night.

§ 117.5 Fitness for duty.

(a) Each flightcrew member must report for any flight duty period rested and prepared to perform his or her assigned duties.

(b) No certificate holder may assign and no flightcrew member may accept assignment to a flight duty period if the flightcrew member has reported for a flight duty period too fatigued to safely perform his or her assigned duties.

(c) No certificate holder may permit a flightcrew member to continue a flight duty period if the flightcrew member has reported him or herself too fatigued to continue the assigned flight duty period.

(d) As part of the dispatch or flight release, as applicable, each flightcrew member must affirmatively state he or she is fit for duty prior to commencing flight.

§ 117.7 Fatigue risk management system.

(a) No certificate holder may exceed any provision of this part unless approved by the FAA under a Fatigue Risk Management System that provides at least an equivalent level of safety against fatigue-related accidents or incidents as the other provisions of this part.

(b) The Fatigue Risk Management System must include:

(1) A fatigue risk management policy.

(2) An education and awareness training program.

(3) A fatigue reporting system.

(4) A system for monitoring flightcrew fatigue.

(5) An incident reporting process.

(6) A performance evaluation.

§ 117.9 Fatigue education and awareness training program.

(a) Each certificate holder must develop and implement an education and awareness training program, approved by the Administrator. This program must provide annual education and awareness training to all employees of the certificate holder responsible for administering the provisions of this rule including flightcrew members, dispatchers, individuals directly involved in the scheduling of flightcrew members, individuals directly involved in operational control, and any employee providing direct management oversight of those areas.

(b) The fatigue education and awareness training program must be designed to increase awareness of:

(1) Fatigue;

(2) The effects of fatigue on pilots; and

(3) Fatigue countermeasures

(c) (1) Each certificate holder must update its fatigue education and awareness training program every two years and submit the update to the Administrator for review and acceptance.

(2) Not later than 12 months after the date of submission of the fatigue education and awareness training program required by (c)(1) of this section, the Administrator shall review and accept or reject the update. If the Administrator rejects an update, the Administrator shall provide suggested modifications for resubmission of the update.

§ 117.11 Flight time limitation.

(a) No certificate holder may schedule and no flightcrew member may accept an assignment or continue an assigned flight duty period if the total flight time:

(1) Will exceed the limits specified in Table A of this part if the operation is conducted with the minimum required flightcrew.

(2) Will exceed 13 hours if the operation is conducted with a 3-pilot flightcrew.

(3) Will exceed 17 hours if the operation is conducted with a 4-pilot flightcrew.

(b) If unforeseen operational circumstances arise after takeoff that are beyond the certificate holder's control, a flightcrew member may exceed the maximum flight time specified in paragraph (a) of this section and the cumulative flight time limits in 117.23(b) to the extent necessary to safely land the aircraft at the next destination airport or alternate, as appropriate.

(c) Each certificate holder must report to the Administrator within 10 days any flight time that exceeded the maximum flight time limits permitted by this section. The report must contain the following:

(1) A description of the extended flight time limitation and the circumstances surrounding the need for the extension; and

(2) If the circumstances giving rise to the extension were within the certificate holder's control, the corrective action(s) that the certificate holder intends to take to minimize the need for future extensions.

(d) Each certificate holder must implement the corrective action(s) reported in paragraph (c)(2) of this section within 30 days from the date of the extended flight time limitation.

§ 117.13 Flight duty period: Unaugmented operations.

(a) Except as provided for in § 117.15, no certificate holder may assign and no flightcrew member may accept an assignment for an unaugmented flight operation if the scheduled flight duty period will exceed the limits in Table B of this part.

(b) If the flightcrew member is not acclimated:

(1) The maximum flight duty period in Table B of this part is reduced by 30 minutes.

(2) The applicable flight duty period is based on the local time at the theater in which the flightcrew member was last acclimated.

§ 117.15 Flight duty period: Split duty.

For an unaugmented operation only, if a flightcrew member is provided with a rest opportunity (an opportunity to sleep) in a suitable accommodation during his or her flight duty period, the time that the flightcrew member spends in the suitable accommodation is not part of that flightcrew member's flight duty period if all of the following conditions are met:

(a) The rest opportunity is provided between the hours of 22:00 and 05:00 local time.

(b) The time spent in the suitable accommodation is at least 3 hours, measured from the time that the flightcrew member reaches the suitable accommodation.

(c) The rest opportunity is scheduled before the beginning of the flight duty period in which that rest opportunity is taken.

(d) The rest opportunity that the flightcrew member is actually provided may not be less than the rest opportunity that was scheduled.

(e) The rest opportunity is not provided until the first segment of the flight duty period has been completed.

(f) The combined time of the flight duty period and the rest opportunity provided in this section does not exceed 14 hours.

§ 117.17 Flight duty period: Augmented flightcrew.

(a) For flight operations conducted with an acclimated augmented flightcrew, no certificate holder may assign and no flightcrew member may accept an assignment if the scheduled flight duty period will exceed the limits specified in Table C of this part.

(b) If the flightcrew member is not acclimated:

(1) The maximum flight duty period in Table C of this part is reduced by 30 minutes.

(2) The applicable flight duty period is based on the local time at the theater in which the flightcrew member was last acclimated.

(c) No certificate holder may assign and no flightcrew member may accept an assignment under this section unless during the flight duty period:

(1) Two consecutive hours in the second half of the flight duty period are available for in-flight rest for the pilot flying the aircraft during landing.

(2) Ninety consecutive minutes are available for in-flight rest for the pilot performing monitoring duties during landing.

(d) No certificate holder may assign and no flightcrew member may accept an assignment involving more than three flight segments under this section.

(e) At all times during flight, at least one flightcrew member qualified in accordance with § 121.543(b)(3)(i) of this chapter must be at the flight controls.

§ 117.19 Flight duty period extensions.

(a) For augmented and unaugmented operations, if unforeseen operational circumstances arise prior to takeoff:

(1) The pilot in command and the certificate holder may extend the maximum flight duty period permitted in Tables B or C of this part up to 2 hours.

(2) An extension in the flight duty period under paragraph (a)(1) of this section of more than 30 minutes may occur only once prior to receiving a rest period described in § 117.25(b).

(3) A flight duty period cannot be extended under paragraph (a)(1) of this section if it causes a flightcrew member to exceed the cumulative flight duty period limits specified in 117.23(c).

(4) Each certificate holder must report to the Administrator within 10 days any flight duty period that exceeded the maximum flight duty period permitted in Tables B or C of this part by more than 30 minutes. The report must contain the following:

(i) A description of the extended flight duty period and the circumstances surrounding the need for the extension; and

(ii) If the circumstances giving rise to the extension were within the certificate holder's control, the corrective action(s) that the certificate holder intends to take to minimize the need for future extensions.

(5) Each certificate holder must implement the corrective action(s) reported in paragraph (a)(4) of this section within 30 days from the date of the extended flight duty period.

(b) For augmented and unaugmented operations, if unforeseen operational circumstances arise after takeoff:

(1) The pilot in command and the certificate holder may extend maximum flight duty periods specified in Tables B or C of this part to the extent necessary to safely land the aircraft at the next destination airport or alternate airport, as appropriate.

(2) An extension of the flight duty period under paragraph (b)(1) of this section of more than 30 minutes may occur only once prior to receiving a rest period described in § 117.25(b).

(3) An extension taken under paragraph (b) of this section may exceed the cumulative flight duty period limits specified in 117.23(c).

(4) Each certificate holder must report to the Administrator within 10 days any flight duty period that exceeded the maximum flight duty period limits permitted by Tables B or C of this part. The report must contain a description of the circumstances surrounding the affected flight duty period.

§ 117.21 Reserve status.

(a) Unless specifically designated as airport/standby or short-call reserve by the certificate holder, all reserve is considered long-call reserve.

(b) Any reserve that meets the definition of airport/standby reserve must be designated as airport/standby reserve. For airport/standby reserve, all time spent in a reserve status is part of the flightcrew member's flight duty period.

(c) For short call reserve,

(1) The reserve availability period may not exceed 14 hours.

(2) For a flightcrew member who has completed a reserve availability period, no certificate holder may schedule and no flightcrew member may accept an assignment of a reserve availability period unless the flightcrew member receives the required rest in § 117.25(e).

(3) For an unaugmented operation, the total number of hours a flightcrew member may spend in a flight duty period and a reserve availability period may not exceed the lesser of the maximum applicable flight duty period in Table B of this part plus 4 hours, or 16 hours, as measured from the beginning of the reserve availability period.

(4) For an augmented operation, the total number of hours a flightcrew member may spend in a flight duty period and a reserve availability period may not exceed the flight duty period in Table C of this part plus 4 hours, as measured from the beginning of the reserve availability period.

(d) For long call reserve, if a certificate holder contacts a flightcrew member to assign him or her to a flight duty period that will begin before and operate into the flightcrew member's window of circadian low, the flightcrew member must receive a 12 hour notice of report time from the certificate holder.

(e) A certificate holder may shift a reserve flightcrew member's reserve status from long-call to short-call only if the flightcrew member receives a rest period as provided in § 117.25(e).

§ 117.23 Cumulative limitations.

(a) The limitations of this section include all flying by flightcrew members on behalf of any certificate holder or 91K Program Manager during the applicable periods.

(b) No certificate holder may schedule and no flightcrew member may accept an assignment if the flightcrew member's total flight time will exceed the following:

(1) 100 hours in any 672 consecutive hours and

(2) 1,000 hours in any 365 consecutive calendar day period.

(c) No certificate holder may schedule and no flightcrew member may accept an assignment if the flightcrew member's total Flight Duty Period will exceed:

(1) 60 flight duty period hours in any 168 consecutive hours and

(2) 190 flight duty period hours in any 672 consecutive hours.

§ 117.25 Rest period.

(a) No certificate holder may assign and no flightcrew member may accept assignment to any reserve or duty with the certificate holder during any required rest period.

(b) Before beginning any reserve or flight duty period a flightcrew member must be given at least 30 consecutive hours free from all duty in any 168 consecutive hour period.

(c) If a flightcrew member operating in a new theater has received 36 consecutive hours of rest, that flightcrew member is acclimated and the rest period meets the requirements of paragraph (b) of this section.

(d) If a flightcrew member travels more than 60° longitude during a flight duty period or a series of flight duty periods that require him or her to be away from home base for more than 168 consecutive hours, the flightcrew member must be given a minimum of 56 consecutive hours rest upon return to home base. This rest must encompass three physiological nights' rest based on local time.

(e) No certificate holder may schedule and no flightcrew member may accept an assignment for any reserve or flight duty period unless the flightcrew member is given a rest period of at least 10 consecutive hours immediately before beginning the reserve or flight duty period measured from the time the flightcrew member is released from duty. The 10 hour rest period must provide the flightcrew member with a minimum of 8 uninterrupted hours of sleep opportunity.

(f) If a flightcrew member determines that a rest period under paragraph (e) of this section will not provide eight uninterrupted hours of sleep opportunity, the flightcrew member must notify the certificate holder. The flightcrew member cannot report for the assigned flight duty period until he or she receives a rest period specified in paragraph (e) of this section.

(g) If a flightcrew member engaged in deadhead transportation exceeds the applicable flight duty period in Table B of this part, the flightcrew member must be given a rest period equal to the length of the deadhead transportation but not less than the required rest in paragraph (e) of this section before beginning a flight duty period.

§ 117.27 Consecutive nighttime operations.

A certificate holder may schedule and a flightcrew member may accept up to five consecutive flight duty periods that infringe on the window of circadian low if the certificate holder provides the flightcrew member with an opportunity to rest in a suitable accommodation during each of the consecutive nighttime flight duty periods. The rest opportunity must be at least 2 hours, measured from the time that the flightcrew member reaches the suitable accommodation, and must comply with the conditions specified in § 117.15(a), (c), (d), and (e). Otherwise, no certificate holder may schedule and no flightcrew member may accept more than three consecutive flight duty periods that infringe on the window of circadian low. For purposes of this section, any split duty rest that is provided in accordance with § 117.15 counts as part of a flight duty period.

§ 117.29 Emergency and government sponsored operations.

(a) This section applies to operations conducted pursuant to contracts with the U.S. Government and operations conducted pursuant to a deviation under § 119.57 of this chapter that cannot otherwise be conducted under this part because of circumstances that could prevent flightcrew members from being relieved by another crew or safely provided with the rest required under § 117.25 at the end of the applicable flight duty period.

(b) The pilot-in-command may determine that maximum applicable flight duty periods must be exceeded to the extent necessary to allow the flightcrew to fly to the closest destination where they can safely be relieved from duty by another flightcrew or can receive the requisite amount of rest prior to commencing their next flight duty period.

(c) A flight duty period may not be extended for an operation conducted pursuant to a contract with the U.S. Government if it causes a flightcrew member to exceed the cumulative flight time limits in § 117.23(b) and the cumulative flight duty period limits in § 117.23(c).

(d) The flightcrew shall be given a rest period immediately after reaching the destination described in paragraph (b) of this section equal to the length of the actual flight duty period or 24 hours, whichever is less.

(e) Each certificate holder must report within 10 days:

(1) Any flight duty period that exceeded the maximum flight duty period permitted in Tables B or C of this part, as applicable, by more than 30 minutes; and

(2) Any flight time that exceeded the maximum flight time limits permitted in Table A of this part and § 117.11, as applicable.

(f) The report must contain the following:

(1) A description of the extended flight duty period and flight time limitation, and the circumstances surrounding the need for the extension; and

(2) If the circumstances giving rise to the extension(s) were within the certificate holder's control, the corrective action(s) that the certificate holder intends to take to minimize the need for future extensions.

(g) Each certificate holder must implement the corrective action(s) reported pursuant to paragraph (e)(2) of this section within 30 days from the date of the extended flight duty period.

Table A to Part 117_Maximum Flight Time Limits for Unaugmented

Time of report (acclimated)	Maximum flight time (hours)
0000-0459	8
0500-1959	9
2000-2359	8

Table B to Part 117_Flight Duty Period: Unaugmented Operations

Scheduled time of start (acclimated time)	Maximum flight duty period (hours) for lineholders based on number of flight segments: 1	2	3	4	5	6	7+
0000-0359	9	9	9	9	9	9	9
0400-0459	10	10	10	10	9	9	9
0500-0559	12	12	12	12	11.5	11	10.5
0600-0659	13	13	12	12	11.5	11	10.5
0700-1159	14	14	13	13	12.5	12	11.5
1200-1259	13	13	13	13	12.5	12	11.5
1300-1659	12	12	12	12	11.5	11	10.5
1700-2159	12	12	11	11	10	9	9
2200-2259	11	11	10	10	9	9	9
2300-2359	10	10	10	9	9	9	9

Table C to Part 117_Flight Duty Period: Augmented Operations

Scheduled time of start (acclimated time)	Maximum flight duty period (hours) based on rest facility and number of pilots: Class 1 rest facility		Class 2 rest facility		Class 3 rest facility	
	3 pilots	4 pilots	3 pilots	4 pilots	3 pilots	4 pilots
0000-0559	15	17	14	15.5	13	13.5
0600-0659	16	18.5	15	16.5	14	14.5
0700-1259	17	19	16.5	18	15	15.5
1300-1659	16	18.5	15	16.5	14	14.5
1700-2359	15	17	14	15.5	13	13.5

PART 119

CERTIFICATION: AIR CARRIERS AND COMMERCIAL OPERATORS

Contents

Subpart A—General

Subpart B—Applicability of Operating Requirements to Different Kinds of Operations Under Parts 121, 125, and 135 of This Chapter

Subpart C—Certification, Operations Specifications, and Certain Other Requirements for Operations Conducted Under Part 121 or Part 135 of This Chapter

Authority: 49 U.S.C. 106(g), 1153, 40101, 40102, 40103, 40113, 44105, 44106, 44111, 44701-44717, 44722, 44901, 44903, 44904, 44906, 44912, 44914, 44936, 44938, 46103, 46105.

Source: Doc. No. 28154, 60 FR 65913, Dec. 20, 1995, unless otherwise noted.

Subpart A—General

§ 119.1 Applicability.

(a) This part applies to each person operating or intending to operate civil aircraft—

(1) As an air carrier or commercial operator, or both, in air commerce; or

(2) When common carriage is not involved, in operations of U.S.-registered civil airplanes with a seat configuration of 20 or more passengers, or a maximum payload capacity of 6,000 pounds or more.

(b) This part prescribes—

(1) The types of air operator certificates issued by the Federal Aviation Administration, including air carrier certificates and operating certificates;

(2) The certification requirements an operator must meet in order to obtain and hold a certificate authorizing operations under part 121, 125, or 135 of this chapter and operations specifications for each kind of operation to be conducted and each class and size of aircraft to be operated under part 121 or 135 of this chapter;

(3) The requirements an operator must meet to conduct operations under part 121, 125, or 135 of this chapter and in operating each class and size of aircraft authorized in its operations specifications;

(4) Requirements affecting wet leasing of aircraft and other arrangements for transportation by air;

(5) Requirements for obtaining deviation authority to perform operations under a military contract and obtaining deviation authority to perform an emergency operation; and

(6) Requirements for management personnel for operations conducted under part 121 or part 135 of this chapter.

(c) Persons subject to this part must comply with the other requirements of this chapter, except where those requirements are modified by or where additional requirements are imposed by part 119, 121, 125, or 135 of this chapter.

(d) This part does not govern operations conducted under part 91, subpart K (when common carriage is not involved) nor does it govern operations conducted under part 129, 133, 137, or 139 of this chapter.

(e) Except for operations when common carriage is not involved conducted with airplanes having a passenger-seat configuration of 20 seats or more, excluding any required crewmember seat, or a payload capacity of 6,000 pounds or more, this part does not apply to—

(1) Student instruction;

(2) Nonstop Commercial Air Tours conducted after September 11, 2007, in an airplane or helicopter having a standard airworthiness certificate and passenger-seat configuration of 30 seats or fewer and a maximum payload capacity of 7,500 pounds or less that begin and end at the same airport, and are conducted within a 25-statute mile radius of that airport, in compliance with the Letter of Authorization issued under § 91.147 of this chapter. For nonstop Commercial Air Tours conducted in accordance with part 136, subpart B of this chapter, National Parks Air Tour Management, the requirements of part 119 of this chapter apply unless excepted in § 136.37(g)(2). For Nonstop Commercial Air Tours conducted in the vicinity of the Grand Canyon National Park, Arizona, the requirements of SFAR 50-2, part 93, subpart U, and part 119 of this chapter, as applicable, apply.

(3) Ferry or training flights;

(4) Aerial work operations, including—

(i) Crop dusting, seeding, spraying, and bird chasing;

(ii) Banner towing;

(iii) Aerial photography or survey;

(iv) Fire fighting;

(v) Helicopter operations in construction or repair work (but it does apply to transportation to and from the site of operations); and

(vi) Powerline or pipeline patrol;

(5) Sightseeing flights conducted in hot air balloons;

(6) Nonstop flights conducted within a 25-statute-mile radius of the airport of takeoff carrying persons or objects for the purpose of conducting intentional parachute operations.

(7) Helicopter flights conducted within a 25 statute mile radius of the airport of takeoff if—

(i) Not more than two passengers are carried in the helicopter in addition to the required flightcrew;

(ii) Each flight is made under day VFR conditions;

(iii) The helicopter used is certificated in the standard category and complies with the 100-hour inspection requirements of part 91 of this chapter;

(iv) The operator notifies the FAA Flight Standards District Office responsible for the geographic area concerned at least 72 hours before each flight and furnishes any essential information that the office requests;

(v) The number of flights does not exceed a total of six in any calendar year;

(vi) Each flight has been approved by the Administrator; and

(vii) Cargo is not carried in or on the helicopter;

(8) Operations conducted under part 133 of this chapter or 375 of this title;

(9) Emergency mail service conducted under 49 U.S.C. 41906; or

(10) Operations conducted under the provisions of § 91.321 of this chapter.

[Docket No. 28154, 60 FR 65913, Dec. 20, 1995, as amended by Amdt. 119-4, 66 FR 23557, May 9, 2001; Amdt. 119-5, 67 FR 9554, Mar. 1, 2002; Amdt. 119-7, 68 FR 54584, Sept. 17, 2003; 72 FR 6911, Feb. 13, 2007]

§ 119.3 [Reserved]

§ 119.5 Certifications, authorizations, and prohibitions.

(a) A person authorized by the Administrator to conduct operations as a direct air carrier will be issued an Air Carrier Certificate.

(b) A person who is not authorized to conduct direct air carrier operations, but who is authorized by the Administrator to conduct operations as a U.S. commercial operator, will be issued an Operating Certificate.

(c) A person who is not authorized to conduct direct air carrier operations, but who is authorized by the Administrator to conduct operations when common carriage is not involved as an operator of U.S.-registered civil airplanes with a seat configuration of 20 or more passengers, or a maximum payload capacity of 6,000 pounds or more, will be issued an Operating Certificate.

(d) A person authorized to engage in common carriage under part 121 or part 135 of this chapter, or both, shall be issued only one certificate authorizing such common carriage, regardless of the kind of operation or the class or size of aircraft to be operated.

(e) A person authorized to engage in noncommon or private carriage under part 125 or part 135 of this chapter, or both, shall be issued only one certificate authorizing such carriage, regardless of the kind of operation or the class or size of aircraft to be operated.

(f) A person conducting operations under more than one paragraph of §§ 119.21, 119.23, or 119.25 shall conduct those operations in compliance with—

(1) The requirements specified in each paragraph of those sections for the kind of operation conducted under that paragraph; and

(2) The appropriate authorizations, limitations, and procedures specified in the operations specifications for each kind of operation.

(g) No person may operate as a direct air carrier or as a commercial operator without, or in violation of, an appropriate certificate and appropriate operations specifications. No person may operate as a direct air carrier or as a commercial operator in violation of any deviation or exemption authority, if issued to that person or that person's representative.

(h) A person holding an Operating Certificate authorizing noncommon or private carriage operations shall not conduct any operations in common carriage. A person holding an Air Carrier Certificate or Operating Certificate authorizing common carriage operations shall not conduct any operations in noncommon carriage.

(i) No person may operate as a direct air carrier without holding appropriate economic authority from the Department of Transportation.

(j) A certificate holder under this part may not operate aircraft under part 121 or part 135 of this chapter in a geographical area unless its operations specifications specifically authorize the certificate holder to operate in that area.

(k) No person may advertise or otherwise offer to perform an operation subject to this part unless that person is authorized by the Federal Aviation Administration to conduct that operation.

(l) No person may operate an aircraft under this part, part 121 of this chapter, or part 135 of this chapter in violation of an air carrier operating certificate, operating certificate, or appropriate operations specifications issued under this part.

[Doc. No. 28154, 60 FR 65913, Dec. 20, 1995, as amended by Amdt. 119-3, 62 FR 13253, Mar. 19, 1997; 62 FR 15570, Apr. 1, 1997]

§ 119.7 Operations specifications.

(a) Each certificate holder's operations specifications must contain—

(1) The authorizations, limitations, and certain procedures under which each kind of operation, if applicable, is to be conducted; and

(2) Certain other procedures under which each class and size of aircraft is to be operated.

(b) Except for operations specifications paragraphs identifying authorized kinds of operations, operations specifications are not a part of a certificate.

§ 119.9 Use of business names.

(a) A certificate holder under this part may not operate an aircraft under part 121 or part 135 of this chapter using a business name other than a business name appearing in the certificate holder's operations specifications.

(b) No person may operate an aircraft under part 121 or part 135 of this chapter unless the name of the certificate holder who is operating the aircraft, or the air carrier or operating certificate number of the certificate holder who is operating the aircraft, is legibly displayed on the aircraft and is clearly visible and readable from the outside of the aircraft to a person standing on the ground at any time except during flight time. The means of displaying the name on the aircraft and its readability must be acceptable to the Administrator.

[Doc. No. 28154, 60 FR 65913, Dec. 20, 1995, as amended by Amdt. 119-3, 62 FR 13253, Mar. 19, 1997]

Subpart B—Applicability of Operating Requirements to Different Kinds of Operations Under Parts 121, 125, and 135 of This Chapter

§ 119.21 Commercial operators engaged in intrastate common carriage and direct air carriers.

(a) Each person who conducts airplane operations as a commercial operator engaged in intrastate common carriage of persons or property for compensation or hire in air commerce, or as a direct air carrier, shall comply with the certification and operations specifications requirements in subpart C of this part, and shall conduct its:

(1) Domestic operations in accordance with the applicable requirements of part 121 of this chapter, and shall be issued operations specifications for those operations in accordance with those requirements. However, based on a showing of safety in air commerce, the Administrator may permit persons who conduct domestic operations between any point located within any of the following Alaskan islands and any point in the State of Alaska to comply with the requirements applicable to flag operations contained in subpart U of part 121 of this chapter:

(i) The Aleutian Islands.

(ii) The Pribilof Islands.

(iii) The Shumagin Islands.

(2) Flag operations in accordance with the applicable requirements of part 121 of this chapter, and shall be issued operations specifications for those operations in accordance with those requirements.

(3) Supplemental operations in accordance with the applicable requirements of part 121 of this chapter, and shall be issued operations specifications for those operations in accordance with those requirements. However, based on a determination of safety in air commerce, the Administrator may authorize or require those operations to be conducted under paragraph (a)(1) or (a)(2) of this section.

(4) Commuter operations in accordance with the applicable requirements of part 135 of this chapter, and shall be issued operations specifications for those operations in accordance with those requirements.

(5) On-demand operations in accordance with the applicable requirements of part 135 of this chapter, and shall be issued operations specifications for those operations in accordance with those requirements.

(b) Persons who are subject to the requirements of paragraph (a)(4) of this section may conduct those operations in accordance with the requirements of paragraph (a)(1) or (a)(2) of this section, provided they obtain authorization from the Administrator.

(c) Persons who are subject to the requirements of paragraph (a)(5) of this section may conduct those operations in accordance with the requirements of paragraph (a)(3) of this section, provided they obtain authorization from the Administrator.

[Doc. No. 28154, 60 FR 65913, Dec. 20, 1995, as amended by Amdt. 119-2, 61 FR 30433, June 14, 1996; Amdt. 119-3, 62 FR 13254, Mar. 19, 1997]

§ 119.23 Operators engaged in passenger-carrying operations, cargo operations, or both with airplanes when common carriage is not involved.

(a) Each person who conducts operations when common carriage is not involved with airplanes having a passenger-seat configuration of 20 seats or more, excluding each crewmember seat, or a payload capacity of 6,000 pounds or more, shall, unless deviation authority is issued—

(1) Comply with the certification and operations specifications requirements of part 125 of this chapter;

(2) Conduct its operations with those airplanes in accordance with the requirements of part 125 of this chapter; and

(3) Be issued operations specifications in accordance with those requirements.

(b) Each person who conducts noncommon carriage (except as provided in § 91.501(b) of this chapter) or private carriage operations for compensation or hire with airplanes having a passenger-seat configuration of less than 20 seats, excluding each crewmember seat, and a payload capacity of less than 6,000 pounds shall—

(1) Comply with the certification and operations specifications requirements in subpart C of this part;

(2) Conduct those operations in accordance with the requirements of part 135 of this chapter, except for those requirements applicable only to commuter operations; and

(3) Be issued operations specifications in accordance with those requirements.

[Doc. No. 28154, 60 FR 65913, Dec. 20, 1995, as amended by Amdt. 119-2, 61 FR 30434, June 14, 1996]

§ 119.25 Rotorcraft operations: Direct air carriers and commercial operators.

Each person who conducts rotorcraft operations for compensation or hire must comply with the certification and operations specifications requirements of Subpart C of this part, and shall conduct its:

(a) Commuter operations in accordance with the applicable requirements of part 135 of this chapter, and shall be issued operations specifications for those operations in accordance with those requirements.

(b) On-demand operations in accordance with the applicable requirements of part 135 of this chapter, and shall be issued operations specifications for those operations in accordance with those requirements.

Subpart C—Certification, Operations Specifications, and Certain Other Requirements for Operations Conducted Under Part 121 or Part 135 of This Chapter

§ 119.31 Applicability.

This subpart sets out certification requirements and prescribes the content of operations specifications and certain other requirements for operations conducted under part 121 or part 135 of this chapter.

§ 119.33 General requirements.

(a) A person may not operate as a direct air carrier unless that person—

(1) Is a citizen of the United States;

(2) Obtains an Air Carrier Certificate; and

(3) Obtains operations specifications that prescribe the authorizations, limitations, and procedures under which each kind of operation must be conducted.

(b) A person other than a direct air carrier may not conduct any commercial passenger or cargo aircraft operation for compensation or hire under part 121 or part 135 of this chapter unless that person—

(1) Is a citizen of the United States;

(2) Obtains an Operating Certificate; and

(3) Obtains operations specifications that prescribe the authorizations, limitations, and procedures under which each kind of operation must be conducted.

(c) Each applicant for a certificate under this part and each applicant for operations specifications authorizing a new kind of operation that is subject to § 121.163 or § 135.145 of this chapter shall conduct proving tests as authorized by the Administrator during the application process for authority to conduct operations under part 121 or part 135 of this chapter. All proving tests must be conducted in a manner acceptable to the Administrator. All proving tests must be conducted under the appropriate operating and maintenance requirements of part 121 or 135 of this chapter that would apply if the applicant were fully certificated. The Administrator will issue a letter of authorization to each person stating the various authorities under which the proving tests shall be conducted.

[Doc. No. 28154, 60 FR 65913, Dec. 20, 1995, as amended by Amdt. 119-2, 61 FR 30434, June 14, 1996]

§ 119.35 Certificate application requirements for all operators.

(a) A person applying to the Administrator for an Air Carrier Certificate or Operating Certificate under this part (applicant) must submit an application—

(1) In a form and manner prescribed by the Administrator; and

(2) Containing any information the Administrator requires the applicant to submit.

(b) Each applicant must submit the application to the Administrator at least 90 days before the date of intended operation.

[Doc. No. 28154, 62 FR 13254, Mar. 19, 1997; 62 FR 15570, Apr. 1, 1997]

§ 119.36 Additional certificate application requirements for commercial operators.

(a) Each applicant for the original issue of an operating certificate for the purpose of conducting intrastate common carriage operations under part 121 or part 135 of this chapter must submit an application in a form and manner prescribed by the Administrator to the Flight Standards District Office in whose area the applicant proposes to establish or has established his or her principal base of operations.

(b) Each application submitted under paragraph (a) of this section must contain a signed statement showing the following:

(1) For corporate applicants:

(i) The name and address of each stockholder who owns 5 percent or more of the total voting stock of the corporation, and if that stockholder is not the sole beneficial owner of the stock, the name and address of each beneficial owner. An individual is considered to own the stock owned, directly or indirectly, by or for his or her spouse, children, grandchildren, or parents.

(ii) The name and address of each director and each officer and each person employed or who will be employed in a management position described in §§ 119.65 and 119.69, as applicable.

(iii) The name and address of each person directly or indirectly controlling or controlled by the applicant and each person under direct or indirect control with the applicant.

(2) For non-corporate applicants:

(i) The name and address of each person having a financial interest therein and the nature and extent of that interest.

(ii) The name and address of each person employed or who will be employed in a management position described in §§ 119.65 and 119.69, as applicable.

(c) In addition, each applicant for the original issue of an operating certificate under paragraph (a) of this section must submit with the application a signed statement showing—

(1) The nature and scope of its intended operation, including the name and address of each person, if any, with whom the applicant has a contract to provide services as a commercial operator and the scope, nature, date, and duration of each of those contracts; and

(2) For applicants intending to conduct operations under part 121 of this chapter, the financial information listed in paragraph (e) of this section.

(d) Each applicant for, or holder of, a certificate issued under paragraph (a) of this section, shall notify the Administrator within 10 days after—

(1) A change in any of the persons, or the names and addresses of any of the persons, submitted to the Administrator under paragraph (b)(1) or (b)(2) of this section; or

(2) For applicants intending to conduct operations under part 121 of this chapter, a change in the financial information submitted to the Administrator under paragraph (e) of this section that occurs while the application for the issue is pending before the FAA and that would make the applicant's financial situation substantially less favorable than originally reported.

(e) Each applicant for the original issue of an operating certificate under paragraph (a) of this section who intends to conduct operations under part 121 of this chapter must submit the following financial information:

(1) A balance sheet that shows assets, liabilities, and net worth, as of a date not more than 60 days before the date of application.

(2) An itemization of liabilities more than 60 days past due on the balance sheet date, if any, showing each creditor's name and address, a description of the liability, and the amount and due date of the liability.

(3) An itemization of claims in litigation, if any, against the applicant as of the date of application showing each claimant's name and address and a description and the amount of the claim.

(4) A detailed projection of the proposed operation covering 6 complete months after the month in which the certificate is expected to be issued including—

(i) Estimated amount and source of both operating and nonoperating revenue, including identification of its existing and anticipated income producing contracts and estimated revenue per mile or hour of operation by aircraft type;

(ii) Estimated amount of operating and nonoperating expenses by expense objective classification; and

(iii) Estimated net profit or loss for the period.

(5) An estimate of the cash that will be needed for the proposed operations during the first 6 months after the month in which the certificate is expected to be issued, including—

(i) Acquisition of property and equipment (explain);

(ii) Retirement of debt (explain);

(iii) Additional working capital (explain);

(iv) Operating losses other than depreciation and amortization (explain); and

(v) Other (explain).

(6) An estimate of the cash that will be available during the first 6 months after the month in which the certificate is expected to be issued, from—

(i) Sale of property or flight equipment (explain);

(ii) New debt (explain);

(iii) New equity (explain);

(iv) Working capital reduction (explain);

(v) Operations (profits) (explain);

(vi) Depreciation and amortization (explain); and

(vii) Other (explain).

(7) A schedule of insurance coverage in effect on the balance sheet date showing insurance companies; policy numbers; types, amounts, and period of coverage; and special conditions, exclusions, and limitations.

(8) Any other financial information that the Administrator requires to enable him or her to determine that the applicant has sufficient financial resources to conduct his or her operations with the degree of safety required in the public interest.

(f) Each financial statement containing financial information required by paragraph (e) of this section must be based on accounts prepared and maintained on an accrual basis in accordance with generally accepted accounting principles applied on a consistent basis, and must contain the name and address of the applicant's public accounting firm, if any. Information submitted must be signed by an officer, owner, or partner of the applicant or certificate holder.

[Doc. No. 28154, 62 FR 13254, Mar. 19, 1997; 62 FR 15570, Apr. 1, 1997]

§ 119.37 Contents of an Air Carrier Certificate or Operating Certificate.

The Air Carrier Certificate or Operating Certificate includes—

(a) The certificate holder's name;

(b) The location of the certificate holder's principal base of operations;

(c) The certificate number;

(d) The certificate's effective date; and

(e) The name or the designator of the certificate-holding district office.

§ 119.39 Issuing or denying a certificate.

(a) An applicant may be issued an Air Carrier Certificate or Operating Certificate if, after investigation, the Administrator finds that the applicant—

(1) Meets the applicable requirements of this part;

(2) Holds the economic authority applicable to the kinds of operations to be conducted, issued by the Department of Transportation, if required; and

(3) Is properly and adequately equipped in accordance with the requirements of this chapter and is able to conduct a safe operation under appropriate provisions of part 121 or part 135 of this chapter and operations specifications issued under this part.

(b) An application for a certificate may be denied if the Administrator finds that—

(1) The applicant is not properly or adequately equipped or is not able to conduct safe operations under this subchapter;

(2) The applicant previously held an Air Carrier Certificate or Operating Certificate which was revoked;

(3) The applicant intends to or fills a key management position listed in § 119.65(a) or § 119.69(a), as applicable, with an individual who exercised control over or who held the same or a similar position with a certificate holder whose certificate was revoked, or is in the process of being revoked, and that individual materially contributed to the circumstances causing revocation or causing the revocation process;

(4) An individual who will have control over or have a substantial ownership interest in the applicant had the same or similar control or interest in a certificate holder whose certificate was revoked, or is in the process of being revoked, and that individual materially contributed to the circumstances causing revocation or causing the revocation process; or

(5) In the case of an applicant for an Operating Certificate for intrastate common carriage, that for financial reasons the applicant is not able to conduct a safe operation.

§ 119.41 Amending a certificate.

(a) The Administrator may amend any certificate issued under this part if—

(1) The Administrator determines, under 49 U.S.C. 44709 and part 13 of this chapter, that safety in air commerce and the public interest requires the amendment; or

(2) The certificate holder applies for the amendment and the certificate-holding district office determines that safety in air commerce and the public interest allows the amendment.

(b) When the Administrator proposes to issue an order amending, suspending, or revoking all or part of any certificate, the procedure in § 13.19 of this chapter applies.

(c) When the certificate holder applies for an amendment of its certificate, the following procedure applies:

(1) The certificate holder must file an application to amend its certificate with the certificate-holding district office at least 15 days before the date proposed by the applicant for the amendment to become effective, unless the administrator approves filing within a shorter period; and

(2) The application must be submitted to the certificate-holding district office in the form and manner prescribed by the Administrator.

(d) When a certificate holder seeks reconsideration of a decision from the certificate-holding district office concerning amendments of a certificate, the following procedure applies:

(1) The petition for reconsideration must be made within 30 days after the certificate holder receives the notice of denial; and

(2) The certificate holder must petition for reconsideration to the Director, Flight Standards Service.

§ 119.43 Certificate holder's duty to maintain operations specifications.

(a) Each certificate holder shall maintain a complete and separate set of its operations specifications at its principal base of operations.

(b) Each certificate holder shall insert pertinent excerpts of its operations specifications, or references thereto, in its manual and shall—

(1) Clearly identify each such excerpt as a part of its operations specifications; and

(2) State that compliance with each operations specifications requirement is mandatory.

(c) Each certificate holder shall keep each of its employees and other persons used in its operations informed of the provisions of its operations specifications that apply to that employee's or person's duties and responsibilities.

§ 119.45 [Reserved]

§ 119.47 Maintaining a principal base of operations, main operations base, and main maintenance base; change of address.

(a) Each certificate holder must maintain a principal base of operations. Each certificate holder may also establish a main operations base and a main maintenance base which may be located at either the same location as the principal base of operations or at separate locations.

(b) At least 30 days before it proposes to establish or change the location of its principal base of operations, its main operations base, or its main maintenance base, a certificate holder must provide written notification to its certificate-holding district office.

§ 119.49 Contents of operations specifications.

(a) Each certificate holder conducting domestic, flag, or commuter operations must obtain operations specifications containing all of the following:

(1) The specific location of the certificate holder's principal base of operations and, if different, the address that shall serve as the primary point of contact for correspondence between the FAA and the certificate holder and the name and mailing address of the certificate holder's agent for service.

(2) Other business names under which the certificate holder may operate.

(3) Reference to the economic authority issued by the Department of Transportation, if required.

(4) Type of aircraft, registration markings, and serial numbers of each aircraft authorized for use, each regular and alternate airport to be used in scheduled operations, and, except for commuter operations, each provisional and refueling airport.

(i) Subject to the approval of the Administrator with regard to form and content, the certificate holder may incorporate by reference the items listed in paragraph (a)(4) of this section into the certificate holder's operations specifications by maintaining a current listing of those items and by referring to the specific list in the applicable paragraph of the operations specifications.

(ii) The certificate holder may not conduct any operation using any aircraft or airport not listed.

(5) Kinds of operations authorized.

(6) Authorization and limitations for routes and areas of operations.

(7) Airport limitations.

(8) Time limitations, or standards for determining time limitations, for overhauling, inspecting, and checking airframes, engines, propellers, rotors, appliances, and emergency equipment.

(9) Authorization for the method of controlling weight and balance of aircraft.

(10) Interline equipment interchange requirements, if relevant.

(11) Aircraft wet lease information required by § 119.53(c).

(12) Any authorized deviation and exemption granted from any requirement of this chapter.

(13) An authorization permitting, or a prohibition against, accepting, handling, and transporting materials regulated as hazardous materials in transport under 49 CFR parts 171 through 180.

(14) Any other item the Administrator determines is necessary.

(b) Each certificate holder conducting supplemental operations must obtain operations specifications containing all of the following:

(1) The specific location of the certificate holder's principal base of operations, and, if different, the address that shall serve as the primary point of contact for correspondence between the FAA and the certificate holder and the name and mailing address of the certificate holder's agent for service.

(2) Other business names under which the certificate holder may operate.

(3) Reference to the economic authority issued by the Department of Transportation, if required.

(4) Type of aircraft, registration markings, and serial number of each aircraft authorized for use.

(i) Subject to the approval of the Administrator with regard to form and content, the certificate holder may incorporate by reference the items listed in paragraph (b)(4) of this section into the certificate holder's operations specifications by maintaining a current listing of those items and by referring to the specific list in the applicable paragraph of the operations specifications.

(ii) The certificate holder may not conduct any operation using any aircraft not listed.

(5) Kinds of operations authorized.

(6) Authorization and limitations for routes and areas of operations.

(7) Special airport authorizations and limitations.

(8) Time limitations, or standards for determining time limitations, for overhauling, inspecting, and checking airframes, engines, propellers, appliances, and emergency equipment.

(9) Authorization for the method of controlling weight and balance of aircraft.

(10) Aircraft wet lease information required by § 119.53(c).

(11) Any authorization or requirement to conduct supplemental operations as provided by § 119.21(a)(3).

(12) Any authorized deviation or exemption from any requirement of this chapter.

(13) An authorization permitting, or a prohibition against, accepting, handling, and transporting materials regulated as hazardous materials in transport under 49 CFR parts 171 through 180.

(14) Any other item the Administrator determines is necessary.

(c) Each certificate holder conducting on-demand operations must obtain operations specifications containing all of the following:

(1) The specific location of the certificate holder's principal base of operations, and if different, the address that shall serve as the primary point of contact for correspondence between the FAA and the name and mailing address of the certificate holder's agent for service.

(2) Other business names under which the certificate holder may operate.

(3) Reference to the economic authority issued by the Department of Transportation, if required.

(4) Kind and area of operations authorized.

(5) Category and class of aircraft that may be used in those operations.

(6) Type of aircraft, registration markings, and serial number of each aircraft that is subject to an airworthiness maintenance program required by § 135.411(a)(2) of this chapter.

(i) Subject to the approval of the Administrator with regard to form and content, the certificate holder may incorporate by reference the items listed in paragraph (c)(6) of this section into the certificate holder's operations specifications by maintaining a current listing of those items and by referring to the specific list in the applicable paragraph of the operations specifications.

(ii) The certificate holder may not conduct any operation using any aircraft not listed.

(7) Registration markings of each aircraft that is to be inspected under an approved aircraft inspection program under § 135.419 of this chapter.

(8) Time limitations or standards for determining time limitations, for overhauls, inspections, and checks for airframes, engines, propellers, rotors, appliances, and emergency equipment of aircraft that are subject to an airworthiness maintenance program required by § 135.411(a)(2) of this chapter.

(9) Additional maintenance items required by the Administrator under § 135.421 of this chapter.

(10) Aircraft wet lease information required by § 119.53(c).

(11) Any authorized deviation or exemption from any requirement of this chapter.

(12) An authorization permitting, or a prohibition against, accepting, handling, and transporting materials regulated as hazardous materials in transport under 49 CFR parts 171 through 180.

(13) Any other item the Administrator determines is necessary.

[Docket No. 28154, 60 FR 65913, Dec. 20, 1995, as amended by Amdt. 119-10, 70 FR 58823, Oct. 7, 2005; Amdt. 119-13, 75 FR 26645, May 12, 2010]

§ 119.51 Amending operations specifications.

(a) The Administrator may amend any operations specifications issued under this part if—

(1) The Administrator determines that safety in air commerce and the public interest require the amendment; or

(2) The certificate holder applies for the amendment, and the Administrator determines that safety in air commerce and the public interest allows the amendment.

(b) Except as provided in paragraph (e) of this section, when the Administrator initiates an amendment to a certificate holder's operations specifications, the following procedure applies:

(1) The certificate-holding district office notifies the certificate holder in writing of the proposed amendment.

(2) The certificate-holding district office sets a reasonable period (but not less than 7 days) within which the certificate holder may submit written information, views, and arguments on the amendment.

(3) After considering all material presented, the certificate-holding district office notifies the certificate holder of—

(i) The adoption of the proposed amendment;

(ii) The partial adoption of the proposed amendment; or

(iii) The withdrawal of the proposed amendment.

(4) If the certificate-holding district office issues an amendment to the operations specifications, it becomes effective not less than 30 days after the certificate holder receives notice of it unless—

(i) The certificate-holding district office finds under paragraph (e) of this section that there is an emergency requiring immediate action with respect to safety in air commerce; or

(ii) The certificate holder petitions for reconsideration of the amendment under paragraph (d) of this section.

(c) When the certificate holder applies for an amendment to its operations specifications, the following procedure applies:

(1) The certificate holder must file an application to amend its operations specifications—

(i) At least 90 days before the date proposed by the applicant for the amendment to become effective, unless a shorter time is approved, in cases of mergers; acquisitions of airline operational assets that require an additional showing of safety (e.g., proving tests); changes in the kind of operation as defined in § 110.2; resumption of operations following a suspension of operations as a result of bankruptcy actions; or the initial introduction of aircraft not before proven for use in air carrier or commercial operator operations.

(ii) At least 15 days before the date proposed by the applicant for the amendment to become effective in all other cases.

(2) The application must be submitted to the certificate-holding district office in a form and manner prescribed by the Administrator.

(3) After considering all material presented, the certificate-holding district office notifies the certificate holder of—

(i) The adoption of the applied for amendment;

(ii) The partial adoption of the applied for amendment; or

(iii) The denial of the applied for amendment. The certificate holder may petition for reconsideration of a denial under paragraph (d) of this section.

(4) If the certificate-holding district office approves the amendment, following coordination with the certificate holder regarding its implementation, the amendment is effective on the date the Administrator approves it.

(d) When a certificate holder seeks reconsideration of a decision from the certificate-holding district office concerning the amendment of operations specifications, the following procedure applies:

(1) The certificate holder must petition for reconsideration of that decision within 30 days of the date that the certificate holder receives a notice of denial of the amendment to its operations specifications, or of the date it receives notice of an FAA-initiated amendment to its operations specifications, whichever circumstance applies.

(2) The certificate holder must address its petition to the Director, Flight Standards Service.

(3) A petition for reconsideration, if filed within the 30-day period, suspends the effectiveness of any amendment issued by the certificate-holding district office unless the certificate-holding district office has found, under paragraph (e) of this section, that an emergency exists requiring immediate action with respect to safety in air transportation or air commerce.

(4) If a petition for reconsideration is not filed within 30 days, the procedures of paragraph (c) of this section apply.

(e) If the certificate-holding district office finds that an emergency exists requiring immediate action with respect to safety in air commerce or air transportation that makes the procedures set out in this section impracticable or contrary to the public interest:

(1) The certificate-holding district office amends the operations specifications and makes the amendment effective on the day the certificate holder receives notice of it.

(2) In the notice to the certificate holder, the certificate-holding district office articulates the reasons for its finding that an emergency exists requiring immediate action with respect to safety in air transportation or air commerce or that makes it impracticable or contrary to the public interest to stay the effectiveness of the amendment.

[Doc. No. 28154, 60 FR 65913, Dec. 20, 1995, as amended by Amdt. 119-14, 76 FR 7488, Feb. 10, 2011]

§ 119.53 Wet leasing of aircraft and other arrangements for transportation by air.

(a) Unless otherwise authorized by the Administrator, prior to conducting operations involving a wet lease, each certificate holder under this part authorized to conduct common carriage operations under this subchapter shall provide the Administrator with a copy of the wet lease to be executed which would lease the aircraft to any other person engaged in common carriage operations under this subchapter, including foreign air carriers, or to any other foreign person engaged in common carriage wholly outside the United States.

(b) No certificate holder under this part may wet lease from a foreign air carrier or any other foreign person or any person not authorized to engage in common carriage.

(c) Upon receiving a copy of a wet lease, the Administrator determines which party to the agreement has operational control of the aircraft and issues amendments to the operations specifications of each party to the agreement, as needed. The lessor must provide the following information to be incorporated into the operations specifications of both parties, as needed.

(1) The names of the parties to the agreement and the duration thereof.

(2) The nationality and registration markings of each aircraft involved in the agreement.

(3) The kind of operation (e.g., domestic, flag, supplemental, commuter, or on-demand).

(4) The airports or areas of operation.

(5) A statement specifying the party deemed to have operational control and the times, airports, or areas under which such operational control is exercised.

(d) In making the determination of paragraph (c) of this section, the Administrator will consider the following:

(1) Crewmembers and training.

(2) Airworthiness and performance of maintenance.

(3) Dispatch.

(4) Servicing the aircraft.

(5) Scheduling.

(6) Any other factor the Administrator considers relevant.

(e) Other arrangements for transportation by air: Except as provided in paragraph (f) of this section, a certificate holder under this part operating under part 121 or 135 of this chapter may not conduct any operation for another certificate holder under this part or a foreign air carrier under part 129 of this chapter or a foreign person engaged in common carriage wholly outside the United States unless it holds applicable Department of Transportation economic authority, if required, and is authorized under its operations specifications to conduct the same kinds of operations (as defined in § 110.2). The certificate holder conducting the substitute operation must conduct that operation in accordance with the same operations authority held by the certificate holder arranging for the substitute operation. These substitute operations must be conducted between airports for which the substitute certificate holder holds authority for scheduled operations or within areas of operations for which the substitute certificate holder has authority for supplemental or on-demand operations.

(f) A certificate holder under this part may, if authorized by the Department of Transportation under § 380.3 of this title and the Administrator in the case of interstate commuter, interstate domestic, and flag operations, or the Administrator in the case of scheduled intrastate common carriage operations, conduct one or more flights for passengers who are stranded because of the cancellation of their scheduled flights. These flights must be conducted under the rules of part 121 or part 135 of this chapter applicable to supplemental or on-demand operations.

[Doc. No. 28154, 60 FR 65913, Dec. 20, 1995, as amended by Amdt. 119-14, 76 FR 7488, Feb. 10, 2011]

§ 119.55 Obtaining deviation authority to perform operations under a U.S. military contract.

Link to an amendment published at 77 FR 402, Jan. 4, 2012.

Link to a correction published at 77 FR 28764, May 16, 2012.

(a) The Administrator may authorize a certificate holder that is authorized to conduct supplemental or on-demand operations to deviate from the applicable requirements of this part, part 121, or part 135 of this chapter in order to perform operations under a U.S. military contract.

(b) A certificate holder that has a contract with the U.S. Department of Defense's Air Mobility Command (AMC) must submit a request for deviation authority to AMC. AMC will review the requests, then forward the carriers' consolidated requests, along with AMC's recommendations, to the FAA for review and action.

(c) The Administrator may authorize a deviation to perform operations under a U.S. military contract under the following conditions—

(1) The Department of Defense certifies to the Administrator that the operation is essential to the national defense;

(2) The Department of Defense further certifies that the certificate holder cannot perform the operation without deviation authority;

(3) The certificate holder will perform the operation under a contract or subcontract for the benefit of a U.S. armed service; and

(4) The Administrator finds that the deviation is based on grounds other than economic advantage either to the certificate holder or to the United States.

(d) In the case where the Administrator authorizes a deviation under this section, the Administrator will issue an appropriate amendment to the certificate holder's operations specifications.

(e) The Administrator may, at any time, terminate any grant of deviation authority issued under this section.

§ 119.57 Obtaining deviation authority to perform an emergency operation.

(a) In emergency conditions, the Administrator may authorize deviations if—

(1) Those conditions necessitate the transportation of persons or supplies for the protection of life or property; and

(2) The Administrator finds that a deviation is necessary for the expeditious conduct of the operations.

(b) When the Administrator authorizes deviations for operations under emergency conditions—

(1) The Administrator will issue an appropriate amendment to the certificate holder's operations specifications; or

(2) If the nature of the emergency does not permit timely amendment of the operations specifications—

(i) The Administrator may authorize the deviation orally; and

(ii) The certificate holder shall provide documentation describing the nature of the emergency to the certificate-holding district office within 24 hours after completing the operation.

§ 119.59 Conducting tests and inspections.

(a) At any time or place, the Administrator may conduct an inspection or test to determine whether a certificate holder under this part is complying with title 49 of the United States Code, applicable regulations, the certificate, or the certificate holder's operations specifications.

(b) The certificate holder must—

(1) Make available to the Administrator at the certificate holder's principal base of operations—

(i) The certificate holder's Air Carrier Certificate or the certificate holder's Operating Certificate and the certificate holder's operations specifications; and

(ii) A current listing that will include the location and persons responsible for each record, document, and report required to be kept by the certificate holder under title 49 of the United States Code applicable to the operation of the certificate holder.

(2) Allow the Administrator to make any test or inspection to determine compliance respecting any matter stated in paragraph (a) of this section.

(c) Each employee of, or person used by, the certificate holder who is responsible for maintaining the certificate holder's records must make those records available to the Administrator.

(d) The Administrator may determine a certificate holder's continued eligibility to hold its certificate and/or operations specifications on any grounds listed in paragraph (a) of this section, or any other appropriate grounds.

(e) Failure by any certificate holder to make available to the Administrator upon request, the certificate, operations specifications, or any required record, document, or report is grounds for suspension of all or any part of the certificate holder's certificate and operations specifications.

(f) In the case of operators conducting intrastate common carriage operations, these inspections and tests include inspections and tests of financial books and records.

§ 119.61 Duration and surrender of certificate and operations specifications.

(a) An Air Carrier Certificate or Operating Certificate issued under this part is effective until—

(1) The certificate holder surrenders it to the Administrator; or

(2) The Administrator suspends, revokes, or otherwise terminates the certificate.

(b) Operations specifications issued under this part, part 121, or part 135 of this chapter are effective unless—

(1) The Administrator suspends, revokes, or otherwise terminates the certificate;

(2) The operations specifications are amended as provided in § 119.51;

(3) The certificate holder does not conduct a kind of operation for more than the time specified in § 119.63 and fails to follow the procedures of § 119.63 upon resuming that kind of operation; or

(4) The Administrator suspends or revokes the operations specifications for a kind of operation.

(c) Within 30 days after a certificate holder terminates operations under part 135 of this chapter, the operating certificate and operations specifications must be surrendered by the certificate holder to the certificate-holding district office.

§ 119.63 Recency of operation.

(a) Except as provided in paragraph (b) of this section, no certificate holder may conduct a kind of operation for which it holds authority in its operations specifications unless the certificate holder has conducted that kind of operation within the preceding number of consecutive calendar days specified in this paragraph:

(1) For domestic, flag, or commuter operations—30 days.

(2) For supplemental or on-demand operations—90 days, except that if the certificate holder has authority to conduct domestic, flag, or commuter operations, and has conducted domestic, flag or commuter operations within the previous 30 days, this paragraph does not apply.

(b) If a certificate holder does not conduct a kind of operation for which it is authorized in its operations specifications within the number of calendar days specified in paragraph (a) of this section, it shall not conduct such kind of operation unless—

(1) It advises the Administrator at least 5 consecutive calendar days before resumption of that kind of operation; and

(2) It makes itself available and accessible during the 5 consecutive calendar day period in the event that the FAA decides to conduct a full inspection reexamination to determine whether the certificate holder remains properly and adequately equipped and able to conduct a safe operation.

[Doc. No. 28154, 60 FR 65913, Dec. 20, 1995, as amended by Amdt. 119-2, 61 FR 30434, June 14, 1996]

§ 119.65 Management personnel required for operations conducted under part 121 of this chapter.

(a) Each certificate holder must have sufficient qualified management and technical personnel to ensure the highest degree of safety in its operations. The certificate holder must have qualified personnel serving full-time in the following or equivalent positions:

(1) Director of Safety.

(2) Director of Operations.

(3) Chief Pilot.

(4) Director of Maintenance.

(5) Chief Inspector.

(b) The Administrator may approve positions or numbers of positions other than those listed in paragraph (a) of this section for a particular operation if the certificate holder shows that it can perform the operation with the highest degree of safety under the direction of fewer or different categories of management personnel due to—

(1) The kind of operation involved;

(2) The number and type of airplanes used; and

(3) The area of operations.

(c) The title of the positions required under paragraph (a) of this section or the title and number of equivalent positions approved under paragraph (b) of this section shall be set forth in the certificate holder's operations specifications.

(d) The individuals who serve in the positions required or approved under paragraph (a) or (b) of this section and anyone in a position to exercise control over operations conducted under the operating certificate must—

(1) Be qualified through training, experience, and expertise;

(2) To the extent of their responsibilities, have a full understanding of the following materials with respect to the certificate holder's operation—

(i) Aviation safety standards and safe operating practices;

(ii) 14 CFR Chapter I (Federal Aviation Regulations);

(iii) The certificate holder's operations specifications;

(iv) All appropriate maintenance and airworthiness requirements of this chapter (e.g., parts 1, 21, 23, 25, 43, 45, 47, 65, 91, and 121 of this chapter); and

(v) The manual required by § 121.133 of this chapter; and

(3) Discharge their duties to meet applicable legal requirements and to maintain safe operations.

(e) Each certificate holder must:

(1) State in the general policy provisions of the manual required by § 121.133 of this chapter, the duties, responsibilities, and authority of personnel required under paragraph (a) of this section;

(2) List in the manual the names and business addresses of the individuals assigned to those positions; and

(3) Notify the certificate-holding district office within 10 days of any change in personnel or any vacancy in any position listed.

§ 119.67 Management personnel: Qualifications for operations conducted under part 121 of this chapter.

(a) To serve as Director of Operations under § 119.65(a) a person must—

(1) Hold an airline transport pilot certificate;

(2) Have at least 3 years supervisory or managerial experience within the last 6 years in a position that exercised operational control over any operations conducted with large airplanes under part 121 or part 135 of this chapter, or if the certificate holder uses only small airplanes in its operations, the experience may be obtained in large or small airplanes; and

(3) In the case of a person becoming a Director of Operations—

(i) For the first time ever, have at least 3 years experience, within the past 6 years, as pilot in command of a large airplane operated under part 121 or part 135 of this chapter, if the certificate holder operates large airplanes. If the certificate holder uses only small airplanes in its operation, the experience may be obtained in either large or small airplanes.

(ii) In the case of a person with previous experience as a Director of Operations, have at least 3 years experience as pilot in command of a large airplane operated under part 121 or part 135 of this chapter, if the certificate holder operates large airplanes. If the certificate holder uses only small airplanes in its operation, the experience may be obtained in either large or small airplanes.

(b) To serve as Chief Pilot under § 119.65(a) a person must hold an airline transport pilot certificate with appropriate ratings for at least one of the airplanes used in the certificate holder's operation and:

(1) In the case of a person becoming a Chief Pilot for the first time ever, have at least 3 years experience, within the past 6 years, as pilot in command of a large airplane operated under part 121 or part 135 of this chapter, if the certificate holder operates large airplanes. If the certificate holder uses only small airplanes in its operation, the experience may be obtained in either large or small airplanes.

(2) In the case of a person with previous experience as a Chief Pilot, have at least 3 years experience, as pilot in command of a large airplane operated under part 121 or part 135 of this chapter, if the certificate holder operates large airplanes. If the certificate holder uses only small airplanes in its operation, the experience may be obtained in either large or small airplanes.

(c) To serve as Director of Maintenance under § 119.65(a) a person must—

(1) Hold a mechanic certificate with airframe and powerplant ratings;

(2) Have 1 year of experience in a position responsible for returning airplanes to service;

(3) Have at least 1 year of experience in a supervisory capacity under either paragraph (c)(4)(i) or (c)(4)(ii) of this section maintaining the same category and class of airplane as the certificate holder uses; and

(4) Have 3 years experience within the past 6 years in one or a combination of the following—

(i) Maintaining large airplanes with 10 or more passenger seats, including at the time of appointment as Director of Maintenance, experience in maintaining the same category and class of airplane as the certificate holder uses; or

(ii) Repairing airplanes in a certificated airframe repair station that is rated to maintain airplanes in the same category and class of airplane as the certificate holder uses.

(d) To serve as Chief Inspector under § 119.65(a) a person must—

(1) Hold a mechanic certificate with both airframe and powerplant ratings, and have held these ratings for at least 3 years;

(2) Have at least 3 years of maintenance experience on different types of large airplanes with 10 or more passenger seats with an air carrier or certificated repair station, 1 year of which must have been as maintenance inspector; and

(3) Have at least 1 year of experience in a supervisory capacity maintaining the same category and class of aircraft as the certificate holder uses.

(e) A certificate holder may request a deviation to employ a person who does not meet the appropriate airman experience, managerial experience, or supervisory experience requirements of this section if the Manager of the Air Transportation Division, AFS-200, or the Manager of the Aircraft Maintenance Division, AFS-300, as appropriate, finds that the person has comparable experience, and can effectively perform the functions associated with the position in accordance with the requirements of this chapter and the procedures outlined in the certificate holder's manual. Grants of deviation under this paragraph may be granted after consideration of the size and scope of the operation and the qualifications of the intended personnel. The Administrator may, at any time, terminate any grant of deviation authority issued under this paragraph.

[Doc. No. 28154, 60 FR 65913, Dec. 20, 1995, as amended by Amdt. 119-2, 61 FR 30434, June 14, 1996; Amdt. 119-3, 62 FR 13255, Mar. 19, 1997]

§ 119.69 Management personnel required for operations conducted under part 135 of this chapter.

(a) Each certificate holder must have sufficient qualified management and technical personnel to ensure the safety of its operations. Except for a certificate holder using only one pilot in its operations, the certificate holder must have qualified personnel serving in the following or equivalent positions:

(1) Director of Operations.

(2) Chief Pilot.

(3) Director of Maintenance.

(b) The Administrator may approve positions or numbers of positions other than those listed in paragraph (a) of this section for a particular operation if the certificate holder shows that it can perform the operation with the highest degree of safety under the direction of fewer or different categories of management personnel due to—

(1) The kind of operation involved;

(2) The number and type of aircraft used; and

(3) The area of operations.

(c) The title of the positions required under paragraph (a) of this section or the title and number of equivalent positions approved under paragraph (b) of this section shall be set forth in the certificate holder's operations specifications.

(d) The individuals who serve in the positions required or approved under paragraph (a) or (b) of this section and anyone in a position to exercise control over operations conducted under the operating certificate must—

(1) Be qualified through training, experience, and expertise;

(2) To the extent of their responsibilities, have a full understanding of the following material with respect to the certificate holder's operation—

(i) Aviation safety standards and safe operating practices;

(ii) 14 CFR Chapter I (Federal Aviation Regulations);

(iii) The certificate holder's operations specifications;

(iv) All appropriate maintenance and airworthiness requirements of this chapter (e.g., parts 1, 21, 23, 25, 43, 45, 47, 65, 91, and 135 of this chapter); and

(v) The manual required by § 135.21 of this chapter; and

(3) Discharge their duties to meet applicable legal requirements and to maintain safe operations.

(e) Each certificate holder must—

(1) State in the general policy provisions of the manual required by § 135.21 of this chapter, the duties, responsibilities, and authority of personnel required or approved under paragraph (a) or (b), respectively, of this section;

(2) List in the manual the names and business addresses of the individuals assigned to those positions; and

(3) Notify the certificate-holding district office within 10 days of any change in personnel or any vacancy in any position listed.

§ 119.71 Management personnel: Qualifications for operations conducted under part 135 of this chapter.

(a) To serve as Director of Operations under § 119.69(a) for a certificate holder conducting any operations for which the pilot in command is required to hold an airline transport pilot certificate a person must hold an airline transport pilot certificate and either:

(1) Have at least 3 years supervisory or managerial experience within the last 6 years in a position that exercised operational control over any operations conducted under part 121 or part 135 of this chapter; or

(2) In the case of a person becoming Director of Operations—

(i) For the first time ever, have at least 3 years experience, within the past 6 years, as pilot in command of an aircraft operated under part 121 or part 135 of this chapter.

(ii) In the case of a person with previous experience as a Director of Operations, have at least 3 years experience, as pilot in command of an aircraft operated under part 121 or part 135 of this chapter.

(b) To serve as Director of Operations under § 119.69(a) for a certificate holder that only conducts operations for which the pilot in command is required to hold a commercial pilot certificate, a person must hold at least a commercial pilot certificate. If an instrument rating is required for any pilot in command for that certificate holder, the Director of Operations must also hold an instrument rating. In addition, the Director of Operations must either—

(1) Have at least 3 years supervisory or managerial experience within the last 6 years in a position that exercised operational control over any operations conducted under part 121 or part 135 of this chapter; or

(2) In the case of a person becoming Director of Operations—

(i) For the first time ever, have at least 3 years experience, within the past 6 years, as pilot in command of an aircraft operated under part 121 or part 135 of this chapter.

(ii) In the case of a person with previous experience as a Director of Operations, have at least 3 years experience as pilot in command of an aircraft operated under part 121 or part 135 of this chapter.

(c) To serve as Chief Pilot under § 119.69(a) for a certificate holder conducting any operation for which the pilot in command is required to hold an airline transport pilot certificate a person must hold an airline transport pilot certificate with appropriate ratings and be qualified to serve as pilot in command in at least one aircraft used in the certificate holder's operation and:

(1) In the case of a person becoming a Chief Pilot for the first time ever, have at least 3 years experience, within the past 6 years, as pilot in command of an aircraft operated under part 121 or part 135 of this chapter.

(2) In the case of a person with previous experience as a Chief Pilot, have at least 3 years experience as pilot in command of an aircraft operated under part 121 or part 135 of this chapter.

(d) To serve as Chief Pilot under § 119.69(a) for a certificate holder that only conducts operations for which the pilot in command is required to hold a commercial pilot certificate, a person must hold at least a commercial pilot certificate. If an instrument rating is required for any pilot in command for that certificate holder, the Chief Pilot must also hold an instrument rating. The Chief Pilot must be qualified to serve as pilot in command in at least one aircraft used in the certificate holder's operation. In addition, the Chief Pilot must:

(1) In the case of a person becoming a Chief Pilot for the first time ever, have at least 3 years experience, within the past 6 years, as pilot in command of an aircraft operated under part 121 or part 135 of this chapter.

(2) In the case of a person with previous experience as a Chief Pilot, have at least 3 years experience as pilot in command of an aircraft operated under part 121 or part 135 of this chapter.

(e) To serve as Director of Maintenance under § 119.69(a) a person must hold a mechanic certificate with airframe and powerplant ratings and either:

(1) Have 3 years of experience within the past 6 years maintaining aircraft as a certificated mechanic, including, at the time of appointment as Director of Maintenance, experience in maintaining the same category and class of aircraft as the certificate holder uses; or

(2) Have 3 years of experience within the past 6 years repairing aircraft in a certificated airframe repair station, including 1 year in the capacity of approving aircraft for return to service.

(f) A certificate holder may request a deviation to employ a person who does not meet the appropriate airmen experience requirements, managerial experience requirements, or supervisory experience requirements of this section if the Manager of the Air Transportation Division, AFS-200, or the Manager of the Aircraft Maintenance Division, AFS-300, as appropriate, find that the person has comparable experience, and can effectively perform the functions associated with the position in accordance with the requirements of this chapter and the procedures outlined in the certificate holder's manual. The Administrator may, at any time, terminate any grant of deviation authority issued under this paragraph.

[Doc. No. 28154, 60 FR 65913, Dec. 20, 1995, as amended by Amdt. 119-3, 62 FR 13255, Mar. 19, 1997; Amdt. 119-12, 72 FR 54816, Sept. 27, 2007]

§ 119.73 Employment of former FAA employees.

(a) Except as specified in paragraph (c) of this section, no certificate holder conducting operations under part 121 or 135 of this chapter may knowingly employ or make a contractual arrangement which permits an individual to act as an agent or representative of the certificate holder in any matter before the Federal Aviation Administration if the individual, in the preceding 2 years—

(1) Served as, or was directly responsible for the oversight of, a Flight Standards Service aviation safety inspector; and

(2) Had direct responsibility to inspect, or oversee the inspection of, the operations of the certificate holder.

(b) For the purpose of this section, an individual shall be considered to be acting as an agent or representative of a certificate holder in a matter before the agency if the individual makes any written or oral communication on behalf of the certificate holder to the agency (or any of its officers or employees) in connection with a particular matter, whether or not involving a specific party and without regard to whether the individual has participated in, or had responsibility for, the particular matter while serving as a Flight Standards Service aviation safety inspector.

(c) The provisions of this section do not prohibit a certificate holder from knowingly employing or making a contractual arrangement which permits an individual to act as an agent or representative of the certificate holder in any matter before the Federal Aviation Administration if the individual was employed by the certificate holder before October 21, 2011.

[Doc. No. FAA-2008-1154, 76 FR 52235, Aug. 22, 2011]

PART 120

DRUG AND ALCOHOL TESTING PROGRAM

Contents

Authority: 49 U.S.C. 106(g), 40101-40103, 40113, 40120, 41706, 41721, 44106, 44701, 44702, 44703, 44709, 44710, 44711, 45101-45105, 46105, 46306.

Link to an amendment published at 78 FR 42003, July 15, 2013.

Source: Doc. No. FAA-2008-0937, 74 FR 22653, May 14, 2009, unless otherwise noted.

Subpart A—General

§ 120.1 Applicability.

This part applies to the following persons:

(a) All air carriers and operators certificated under part 119 of this chapter authorized to conduct operations under part 121 or part 135 of this chapter, all air traffic control facilities not operated by the FAA or by or under contract to the U.S. military; and all operators as defined in 14 CFR 91.147.

(b) All individuals who perform, either directly or by contract, a safety-sensitive function listed in subpart E or subpart F of this part.

(c) All part 145 certificate holders who perform safety-sensitive functions and elect to implement a drug and alcohol testing program under this part.

(d) All contractors who elect to implement a drug and alcohol testing program under this part.

§ 120.3 Purpose.

The purpose of this part is to establish a program designed to help prevent accidents and injuries resulting from the use of prohibited drugs or the misuse of alcohol by employees who perform safety-sensitive functions in aviation.

§ 120.5 Procedures.

Each employer having a drug and alcohol testing program under this part must ensure that all drug and alcohol testing conducted pursuant to this part complies with the procedures set forth in 49 CFR part 40.

§ 120.7 Definitions.

For the purposes of this part, the following definitions apply:

(a) *Accident* means an occurrence associated with the operation of an aircraft which takes place between the time any individual boards the aircraft with the intention of flight and all such individuals have disembarked, and in which any individual suffers death or serious injury, or in which the aircraft receives substantial damage.

(b) *Alcohol* means the intoxicating agent in beverage alcohol, ethyl alcohol, or other low molecular weight alcohols, including methyl or isopropyl alcohol.

(c) *Alcohol concentration (or content)* means the alcohol in a volume of breath expressed in terms of grams of alcohol per 210 liters of breath as indicated by an evidential breath test under subpart F of this part.

(d) *Alcohol* use means the consumption of any beverage, mixture, or preparation, including any medication, containing alcohol.

(e) *Contractor* is an individual or company that performs a safety-sensitive function by contract for an employer or another contractor.

(f) *Covered employee* means an individual who performs, either directly or by contract, a safety-sensitive function listed in §§ 120.105 and 120.215 for an employer (as defined in paragraph (i) of this section). For purposes of pre-employment testing only, the term "covered employee" includes an individual applying to perform a safety-sensitive function.

(g) *DOT agency* means an agency (or "operating administration") of the United States Department of Transportation administering regulations requiring drug and alcohol testing (14 CFR parts 61, 65, 121, and 135; 46 CFR part 16; 49 CFR parts 199, 219, and 382) in accordance with 49 CFR part 40.

(h) *Employee* is an individual who is hired, either directly or by contract, to perform a safety-sensitive function for an employer, as defined in paragraph (i) of this section. An employee is also an individual who transfers into a position to perform a safety-sensitive function for an employer.

(i) *Employer* is a part 119 certificate holder with authority to operate under parts 121 and/or 135 of this chapter, an operator as defined in § 91.147 of this chapter, or an air traffic control facility not operated by the FAA or by or under contract to the U.S. Military. An employer may use a contract employee who is not included under that employer's FAA-mandated drug and alcohol testing program to perform a safety-sensitive function only if that contract employee is included under the contractor's FAA-mandated drug and alcohol testing program and is performing a safety-sensitive function on behalf of that contractor (i.e., within the scope of employment with the contractor.)

(j) *Hire* means retaining an individual for a safety-sensitive function as a paid employee, as a volunteer, or through barter or other form of compensation.

(k) *Performing* (a safety-sensitive function): an employee is considered to be performing a safety-sensitive function during any period in which he or she is actually performing, ready to perform, or immediately available to perform such function.

(l) *Positive rate for random drug testing* means the number of verified positive results for random drug tests conducted under subpart E of this part, plus the number of refusals of random drug tests required by subpart E of this part, divided by the total number of random drug test results (*i.e.*, positives, negatives, and refusals) under subpart E of this part.

(m) *Prohibited drug* means marijuana, cocaine, opiates, phencyclidine (PCP), and amphetamines, as specified in 49 CFR 40.85.

(n) *Refusal to submit to alcohol test* means that a covered employee has engaged in conduct including but not limited to that described in 49 CFR 40.261, or has failed to remain readily available for post-accident testing as required by subpart F of this part.

(o) *Refusal to submit to drug test* means that an employee engages in conduct including but not limited to that described in 49 CFR 40.191.

(p) *Safety-sensitive function* means a function listed in §§ 120.105 and 120.215.

(q) *Verified negative drug test result* means a drug test result from an HHS-certified laboratory that has undergone review by an MRO and has been determined by the MRO to be a negative result.

(r) *Verified positive drug test result* means a drug test result from an HHS-certified laboratory that has undergone review by an MRO and has been determined by the MRO to be a positive result.

(s) *Violation rate for random alcohol testing* means the number of 0.04, and above, random alcohol confirmation test results conducted under subpart F of this part, plus the number of refusals of random alcohol tests required by subpart F of this part, divided by the total number of random alcohol screening tests (including refusals) conducted under subpart F of this part.

[Doc. No. FAA-2008-0937, 74 FR 22653, May 14, 2009; Amdt. 120-0A, 75 FR 3153, Jan. 20, 2010]

Subpart B—Individuals Certificated Under Parts 61, 63, and 65

§ 120.11 Refusal to submit to a drug or alcohol test by a Part 61 certificate holder.

(a) This section applies to all individuals who hold a certificate under part 61 of this chapter and who are subject to drug and alcohol testing under this part.

(b) Refusal by the holder of a certificate issued under part 61 of this chapter to take a drug or alcohol test required under the provisions of this part is grounds for:

(1) Denial of an application for any certificate, rating, or authorization issued under part 61 of this chapter for a period of up to 1 year after the date of such refusal; and

(2) Suspension or revocation of any certificate, rating, or authorization issued under part 61 of this chapter.

§ 120.13 Refusal to submit to a drug or alcohol test by a Part 63 certificate holder.

(a) This section applies to all individuals who hold a certificate under part 63 of this chapter and who are subject to drug and alcohol testing under this part.

(b) Refusal by the holder of a certificate issued under part 63 of this chapter to take a drug or alcohol test required under the provisions of this part is grounds for:

(1) Denial of an application for any certificate or rating issued under part 63 of this chapter for a period of up to 1 year after the date of such refusal; and

(2) Suspension or revocation of any certificate or rating issued under part 63 of this chapter.

[Doc. No. FAA-2008-0937, 74 FR 22653, May 14, 2009; Amdt. 120-0A, 75 FR 3153, Jan. 20, 2010]

§ 120.15 Refusal to submit to a drug or alcohol test by a Part 65 certificate holder.

(a) This section applies to all individuals who hold a certificate under part 65 of this chapter and who are subject to drug and alcohol testing under this part.

(b) Refusal by the holder of a certificate issued under part 65 of this chapter to take a drug or alcohol test required under the provisions of this part is grounds for:

(1) Denial of an application for any certificate or rating issued under part 65 of this chapter for a period of up to 1 year after the date of such refusal; and

(2) Suspension or revocation of any certificate or rating issued under part 65 of this chapter.

[Doc. No. FAA-2008-0937, 74 FR 22653, May 14, 2009; Amdt. 120-0A, 75 FR 3153, Jan. 20, 2010]

Subpart C—Air Traffic Controllers

§ 120.17 Use of prohibited drugs.

(a) Each employer shall provide each employee performing a function listed in subpart E of this part, and his or her supervisor, with the training specified in that subpart. No employer may use any contractor to perform an air traffic control function unless that contractor provides each of its employees performing that function for the employer, and his or her supervisor, with the training specified in subpart E of this part.

(b) No employer may knowingly use any individual to perform, nor may any individual perform for an employer, either directly or by contract, any air traffic control function while that individual has a prohibited drug, as defined in this part, in his or her system.

(c) No employer shall knowingly use any individual to perform, nor may any individual perform for an employer, either directly or by contract, any air traffic control function if the individual has a verified positive drug test result on, or has refused to submit to, a drug test required by subpart E of this part and the individual has not met the requirements of subpart E of this part for returning to the performance of safety-sensitive duties.

(d) Each employer shall test each of its employees who perform any air traffic control function in accordance with subpart E of this part. No employer may use any contractor to perform any air traffic control function unless that contractor tests each employee performing such a function for the employer in accordance with subpart E of this part.

[Doc. No. FAA-2008-0937, 74 FR 22653, May 14, 2009; Amdt. 120-0A, 75 FR 3153, Jan. 20, 2010]

§ 120.19 Misuse of alcohol.

(a) This section applies to covered employees who perform air traffic control duties directly or by contract for an employer that is an air traffic control facility not operated by the FAA or the US military.

(b) *Alcohol concentration.* No covered employee shall report for duty or remain on duty requiring the performance of safety-sensitive functions while having an alcohol concentration of 0.04 or greater. No employer having actual knowledge that an employee has an alcohol concentration of 0.04 or greater shall permit the employee to perform or continue to perform safety-sensitive functions.

(c) *On-duty use.* No covered employee shall use alcohol while performing safety-sensitive functions. No employer having actual knowledge that a covered employee is using alcohol while performing safety-sensitive functions shall permit the employee to perform or continue to perform safety-sensitive functions.

(d) *Pre-duty use.* No covered employee shall perform air traffic control duties within 8 hours after using alcohol. No employer having actual knowledge that such an employee has used alcohol within 8 hours shall permit the employee to perform or continue to perform air traffic control duties.

(e) *Use following an accident.* No covered employee who has actual knowledge of an accident involving an aircraft for which he or she performed a safety-sensitive function at or near the time of the accident shall use alcohol for 8 hours following the accident, unless he or she has been given a post-accident test under subpart F of this part or the employer has determined that the employee's performance could not have contributed to the accident.

(f) *Refusal to submit to a required alcohol test.* A covered employee may not refuse to submit to any alcohol test required under subpart F of this part. An employer may not permit an employee who refuses to submit to such a test to perform or continue to perform safety-sensitive functions.

§ 120.21 Testing for alcohol.

(a) Each air traffic control facility not operated by the FAA or the U.S. military must establish an alcohol testing program in accordance with the provisions of subpart F of this part.

(b) No employer shall use any individual who meets the definition of covered employee in subpart A of this part to perform a safety-sensitive function listed in subpart F of this part unless that individual is subject to testing for alcohol misuse in accordance with the provisions of that subpart.

Subpart D—Part 119 Certificate Holders Authorized To Conduct Operations under Part 121 or Part 135 or Operators Under § 91.147 of This Chapter and Safety-Sensitive Employees

§ 120.31 Prohibited drugs.

(a) Each certificate holder or operator shall provide each employee performing a function listed in subpart E of this part, and his or her supervisor, with the training specified in that subpart.

(b) No certificate holder or operator may use any contractor to perform a function listed in subpart E of this part unless that contractor provides each of its employees performing that function for the certificate holder or operator, and his or her supervisor, with the training specified in that subpart.

§ 120.33 Use of prohibited drugs.

(a) This section applies to individuals who perform a function listed in subpart E of this part for a certificate holder or operator. For the purpose of this section, an individual who performs such a function pursuant to a contract with the certificate holder or the operator is considered to be performing that function for the certificate holder or the operator.

(b) No certificate holder or operator may knowingly use any individual to perform, nor may any individual perform for a certificate holder or an operator, either directly or by contract, any function listed in subpart E of this part while that individual has a prohibited drug, as defined in this part, in his or her system.

(c) No certificate holder or operator shall knowingly use any individual to perform, nor shall any individual perform for a certificate holder or operator, either directly or by contract, any safety-sensitive function if that individual has a verified positive drug test result on, or has refused to submit to, a drug test required by subpart E of this part and the individual has not met the requirements of that subpart for returning to the performance of safety-sensitive duties.

[Doc. No. FAA-2008-0937, 74 FR 22653, May 14, 2009; Amdt. 120-0A, 75 FR 3153, Jan. 20, 2010]

§ 120.35 Testing for prohibited drugs.

(a) Each certificate holder or operator shall test each of its employees who perform a function listed in subpart E of this part in accordance with that subpart.

(b) Except as provided in paragraph (c) of this section, no certificate holder or operator may use any contractor to perform a function listed in subpart E of this part unless that contractor tests each employee performing such a function for the certificate holder or operator in accordance with that subpart.

(c) If a certificate holder conducts an on-demand operation into an airport at which no maintenance providers are available that are subject to the requirements of subpart E of this part and emergency maintenance is required, the certificate holder may use individuals not meeting the requirements of paragraph (b) of this section to provide such emergency maintenance under both of the following conditions:

(1) The certificate holder must give written notification of the emergency maintenance to the Drug Abatement Program Division, AAM-800, 800 Independence Avenue, SW., Washington, DC 20591, within 10 days after being provided same in accordance with this paragraph. A certificate holder must retain copies of all such written notifications for two years.

(2) The aircraft must be reinspected by maintenance personnel who meet the requirements of paragraph (b) of this section when the aircraft is next at an airport where such maintenance personnel are available.

(d) For purposes of this section, emergency maintenance means maintenance that—

(1) Is not scheduled and

(2) Is made necessary by an aircraft condition not discovered prior to the departure for that location.

§ 120.37 Misuse of alcohol.

(a) *General.* This section applies to covered employees who perform a function listed in subpart F of this part for a certificate holder. For the purpose of this section, an individual who meets the definition of covered employee in subpart F of this part is considered to be performing the function for the certificate holder.

(b) *Alcohol concentration.* No covered employee shall report for duty or remain on duty requiring the performance of safety-sensitive functions while having an alcohol concentration of 0.04 or greater. No certificate holder having actual knowledge that an employee has an alcohol concentration of 0.04 or greater shall permit the employee to perform or continue to perform safety-sensitive functions.

(c) *On-duty use.* No covered employee shall use alcohol while performing safety-sensitive functions. No certificate holder having actual knowledge that a covered employee is using alcohol while performing safety-sensitive functions shall permit the employee to perform or continue to perform safety-sensitive functions.

(d) *Pre-duty use.* (1) No covered employee shall perform flight crewmember or flight attendant duties within 8 hours after using alcohol. No certificate holder having actual knowledge that such an employee has used alcohol within 8 hours shall permit the employee to perform or continue to perform the specified duties.

(2) No covered employee shall perform safety-sensitive duties other than those specified in paragraph (d)(1) of this section within 4 hours after using alcohol. No certificate holder having actual knowledge that such an employee has used alcohol within 4 hours shall permit the employee to perform or to continue to perform safety-sensitive functions.

(e) *Use following an accident.* No covered employee who has actual knowledge of an accident involving an aircraft for which he or she performed a safety-sensitive function at or near the time of the accident shall use alcohol for 8 hours following the accident, unless he or she has been given a post-accident test under subpart F of this part, or the employer has determined that the employee's performance could not have contributed to the accident.

(f) *Refusal to submit to a required alcohol test.* A covered employee must not refuse to submit to any alcohol test required under subpart F of this part. A certificate holder must not permit an employee who refuses to submit to such a test to perform or continue to perform safety-sensitive functions.

§ 120.39 Testing for alcohol.

(a) Each certificate holder must establish an alcohol testing program in accordance with the provisions of subpart F of this part.

(b) Except as provided in paragraph (c) of this section, no certificate holder or operator may use any individual who meets the definition of covered employee in subpart A of this part to perform a safety-sensitive function listed in that subpart F of this part unless that individual is subject to testing for alcohol misuse in accordance with the provisions of that subpart.

(c) If a certificate holder conducts an on-demand operation into an airport at which no maintenance providers are available that are subject to the requirements of subpart F of this part and emergency maintenance is required, the certificate holder may use individuals not meeting the requirements of paragraph (b) of this section to provide such emergency maintenance under both of the following conditions:

(1) The certificate holder must give written notification of the emergency maintenance to the Drug Abatement Program Division, AAM-800, 800 Independence Avenue, SW., Washington, DC 20591, within 10 days after being provided same in accordance with this paragraph. A certificate holder must retain copies of all such written notifications for two years.

(2) The aircraft must be reinspected by maintenance personnel who meet the requirements of paragraph (b) of this section when the aircraft is next at an airport where such maintenance personnel are available.

(d) For purposes of this section, emergency maintenance means maintenance that—

(1) Is not scheduled and

(2) Is made necessary by an aircraft condition not discovered prior to the departure for that location.

Subpart E—Drug Testing Program Requirements

§ 120.101 Scope.

This subpart contains the standards and components that must be included in a drug testing program required by this part.

§ 120.103 General.

(a) *Purpose.* The purpose of this subpart is to establish a program designed to help prevent accidents and injuries resulting from the use of prohibited drugs by employees who perform safety-sensitive functions.

(b) *DOT procedures.* (1) Each employer shall ensure that drug testing programs conducted pursuant to 14 CFR parts 65, 91, 121, and 135 comply with the requirements of this subpart and the "Procedures for Transportation Workplace Drug Testing Programs" published by the Department of Transportation (DOT) (49 CFR part 40).

(2) An employer may not use or contract with any drug testing laboratory that is not certified by the Department of Health and Human Services (HHS) under the National Laboratory Certification Program.

(c) *Employer responsibility.* As an employer, you are responsible for all actions of your officials, representatives, and service agents in carrying out the requirements of this subpart and 49 CFR part 40.

(d) *Applicable Federal Regulations.* The following applicable regulations appear in 49 CFR or 14 CFR:

(1) 49 CFR Part 40—Procedures for Transportation Workplace Drug Testing Programs

(2) 14 CFR:

(i) § 67.107—First-Class Airman Medical Certificate, Mental.

(ii) § 67.207—Second-Class Airman Medical Certificate, Mental.

(iii) § 67.307—Third-Class Airman Medical Certificate, Mental.

(iv) § 91.147—Passenger carrying flight for compensation or hire.

(v) § 135.1—Applicability

(e) Falsification. No individual may make, or cause to be made, any of the following:

(1) Any fraudulent or intentionally false statement in any application of a drug testing program.

(2) Any fraudulent or intentionally false entry in any record or report that is made, kept, or used to show compliance with this part.

(3) Any reproduction or alteration, for fraudulent purposes, of any report or record required to be kept by this part.

[Doc. No. FAA-2008-0937, 74 FR 22653, May 14, 2009; Amdt. 120-0A, 75 FR 3153, Jan. 20, 2010]

§ 120.105 Employees who must be tested.

Each employee, including any assistant, helper, or individual in a training status, who performs a safety-sensitive function listed in this section directly or by contract (including by subcontract at any tier) for an employer as defined in this subpart must be subject to drug testing under a drug testing program implemented in accordance with this subpart. This includes full-time, part-time, temporary, and intermittent employees regardless of the degree of supervision. The safety-sensitive functions are:

(a) Flight crewmember duties.

(b) Flight attendant duties.

(c) Flight instruction duties.

(d) Aircraft dispatcher duties.

(e) Aircraft maintenance and preventive maintenance duties.

(f) Ground security coordinator duties.

(g) Aviation screening duties.

(h) Air traffic control duties.

§ 120.107 Substances for which testing must be conducted.

Each employer shall test each employee who performs a safety-sensitive function for evidence of marijuana, cocaine, opiates, phencyclidine (PCP), and amphetamines during each test required by § 120.109.

§ 120.109 Types of drug testing required.

Each employer shall conduct the types of testing described in this section in accordance with the procedures set forth in this subpart and the DOT "Procedures for Transportation Workplace Drug Testing Programs" (49 CFR part 40).

(a) *Pre-employment drug testing.* (1) No employer may hire any individual for a safety-sensitive function listed in § 120.105 unless the employer first conducts a pre-employment test and receives a verified negative drug test result for that individual.

(2) No employer may allow an individual to transfer from a nonsafety-sensitive to a safety-sensitive function unless the employer first conducts a pre-employment test and receives a verified negative drug test result for the individual.

(3) Employers must conduct another pre-employment test and receive a verified negative drug test result before hiring or transferring an individual into a safety-sensitive function if more than 180 days elapse between conducting the pre-employment test required by paragraphs (a)(1) or (2) of this section and hiring or transferring the individual into a safety-sensitive function, resulting in that individual being brought under an FAA drug testing program.

(4) If the following criteria are met, an employer is permitted to conduct a pre-employment test, and if such a test is conducted, the employer must receive a negative test result before putting the individual into a safety-sensitive function:

(i) The individual previously performed a safety-sensitive function for the employer and the employer is not required to pre-employment test the individual under paragraphs (a)(1) or (2) of this section before putting the individual to work in a safety-sensitive function;

(ii) The employer removed the individual from the employer's random testing program conducted under this subpart for reasons other than a verified positive test result on an FAA-mandated drug test or a refusal to submit to such testing; and

(iii) The individual will be returning to the performance of a safety-sensitive function.

(5) Before hiring or transferring an individual to a safety-sensitive function, the employer must advise each individual that the individual will be required to undergo pre-employment testing in accordance with this subpart, to determine the presence of marijuana, cocaine, opiates, phencyclidine (PCP), and amphetamines, or a metabolite of those drugs in the individual's system. The employer shall provide this same notification to each individual required by the employer to undergo pre-employment testing under paragraph (a)(4) of this section.

(b) *Random drug testing.* (1) Except as provided in paragraphs (b)(2) through (b)(4) of this section, the minimum annual percentage rate for random drug testing shall be 50 percent of covered employees.

(2) The Administrator's decision to increase or decrease the minimum annual percentage rate for random drug testing is based on the reported positive rate for the entire industry. All information used for this determination is drawn from the statistical reports required by § 120.119. In order to ensure reliability of the data, the Administrator considers the quality and completeness of the reported data, may obtain additional information or reports from employers, and may make appropriate modifications in calculating the industry positive rate. Each year, the Administrator will publish in the Federal Register the minimum annual percentage rate for random drug testing of covered employees. The new minimum annual percentage rate for random drug testing will be applicable starting January 1 of the calendar year following publication.

(3) When the minimum annual percentage rate for random drug testing is 50 percent, the Administrator may lower this rate to 25 percent of all covered employees if the Administrator determines that the data received under the reporting requirements of this subpart for two consecutive calendar years indicate that the reported positive rate is less than 1.0 percent.

(4) When the minimum annual percentage rate for random drug testing is 25 percent, and the data received under the reporting requirements of this subpart for any calendar year indicate that the reported positive rate is equal to or greater than 1.0 percent, the Administrator will increase the minimum annual percentage rate for random drug testing to 50 percent of all covered employees.

(5) The selection of employees for random drug testing shall be made by a scientifically valid method, such as a random-number table or a computer-based random number generator that is matched with employees' Social Security numbers, payroll identification numbers, or other comparable identifying numbers. Under the selection process used, each covered employee shall have an equal chance of being tested each time selections are made.

(6) As an employer, you must select and test a percentage of employees at least equal to the minimum annual percentage rate each year.

(i) As an employer, to determine whether you have met the minimum annual percentage rate, you must divide the number of random testing results for safety-sensitive employees by the average number of safety-sensitive employees eligible for random testing.

(A) To calculate whether you have met the annual minimum percentage rate, count all random positives, random negatives, and random refusals as your "random testing results."

(B) To calculate the average number of safety-sensitive employees eligible for random testing throughout the year, add the total number of safety-sensitive employees eligible for testing during each random testing period for the year and divide that total by the number of random testing periods. Only safety-sensitive employees are to be in an employer's random testing pool, and all safety-sensitive employees must be in the random pool. If you are an employer conducting random testing more often than once per month (*e.g.*, you select daily, weekly, bi-weekly) you do not need to compute this total number of safety-sensitive employees more than on a once per month basis.

(ii) As an employer, you may use a service agent to perform random selections for you, and your safety-sensitive employees may be part of a larger random testing pool of safety-sensitive employees. However, you must ensure that the service agent you use is testing at the appropriate percentage established for your industry and that only safety-sensitive employees are in the random testing pool. For example:

(A) If the service agent has your employees in a random testing pool for your company alone, you must ensure that the testing is conducted at least at the minimum annual percentage rate under this part.

(B) If the service agent has your employees in a random testing pool combined with other FAA-regulated companies, you must ensure that the testing is conducted at least at the minimum annual percentage rate under this part.

(C) If the service agent has your employees in a random testing pool combined with other DOT-regulated companies, you must ensure that the testing is conducted at least at the highest rate required for any DOT-regulated company in the pool.

(7) Each employer shall ensure that random drug tests conducted under this subpart are unannounced and that the dates for administering random tests are spread reasonably throughout the calendar year.

(8) Each employer shall require that each safety-sensitive employee who is notified of selection for random drug testing proceeds to the collection site immediately; provided, however, that if the employee is performing a safety-sensitive function at the time of the notification, the employer shall instead ensure that the employee ceases to perform the safety-sensitive function and proceeds to the collection site as soon as possible.

(9) If a given covered employee is subject to random drug testing under the drug testing rules of more than one DOT agency, the employee shall be subject to random drug testing at the percentage rate established for the calendar year by the DOT agency regulating more than 50 percent of the employee's function.

(10) If an employer is required to conduct random drug testing under the drug testing rules of more than one DOT agency, the employer may—

(i) Establish separate pools for random selection, with each pool containing the covered employees who are subject to testing at the same required rate; or

(ii) Randomly select covered employees for testing at the highest percentage rate established for the calendar year by any DOT agency to which the employer is subject.

(11) An employer required to conduct random drug testing under the anti-drug rules of more than one DOT agency shall provide each such agency access to the employer's records of random drug testing, as determined to be necessary by the agency to ensure the employer's compliance with the rule.

(c) *Post-accident drug testing.* Each employer shall test each employee who performs a safety-sensitive function for the presence of marijuana, cocaine, opiates, phencyclidine (PCP), and amphetamines, or a metabolite of those drugs in the employee's system if that employee's performance either contributed to an accident or can not be completely discounted as a contributing factor to the accident. The employee shall be tested as soon as possible but not later than 32 hours after the accident. The decision not to administer a test under this section must be based on a determination, using the best information available at the time of the determination, that the employee's performance could not have contributed to the accident. The employee shall submit to post-accident testing under this section.

(d) *Drug testing based on reasonable cause.* Each employer must test each employee who performs a safety-sensitive function and who is reasonably suspected of having used a prohibited drug. The decision to test must be based on a reasonable and articulable belief that the employee is using a prohibited drug on the basis of specific contemporaneous physical, behavioral, or performance indicators of probable drug use. At least two of the employee's supervisors, one of whom is trained in detection of the symptoms of possible drug use, must substantiate and concur in the decision to test an employee who is reasonably suspected of drug use; except that in the case of an employer, other than a part 121 certificate holder, who employs 50 or fewer employees who perform safety-sensitive functions, one supervisor who is trained in detection of symptoms of possible drug use must substantiate the decision to test an employee who is reasonably suspected of drug use.

(e) *Return to duty drug testing.* Each employer shall ensure that before an individual is returned to duty to perform a safety-sensitive function after refusing to submit to a drug test required by this subpart or receiving a verified positive drug test result on a test conducted under this subpart the individual shall undergo a return-to-duty drug test. No employer shall allow an individual required to undergo return-to-duty testing to perform a safety-sensitive function unless the employer has received a verified negative drug test result for the individual. The test cannot occur until after the SAP has determined that the employee has successfully complied with the prescribed education and/or treatment.

(f) *Follow-up drug testing.* (1) Each employer shall implement a reasonable program of unannounced testing of each individual who has been hired to perform or who has been returned to the performance of a safety-sensitive function after refusing to submit to a drug test required by this subpart or receiving a verified positive drug test result on a test conducted under this subpart.

(2) The number and frequency of such testing shall be determined by the employer's Substance Abuse Professional conducted in accordance with the provisions of 49 CFR part 40, but shall consist of at least six tests in the first 12 months following the employee's return to duty.

(3) The employer must direct the employee to undergo testing for alcohol in accordance with subpart F of this part, in addition to drugs, if the Substance Abuse Professional determines that alcohol testing is necessary for the particular employee. Any such alcohol testing shall be conducted in accordance with the provisions of 49 CFR part 40.

(4) Follow-up testing shall not exceed 60 months after the date the individual begins to perform or returns to the performance of a safety-sensitive function. The Substance Abuse Professional may terminate the requirement for follow-up testing at any time after the first six tests have been conducted, if the Substance Abuse Professional determines that such testing is no longer necessary.

§ 120.111 Administrative and other matters.

(a) *MRO record retention requirements.* (1) Records concerning drug tests confirmed positive by the laboratory shall be maintained by the MRO for 5 years. Such records include the MRO copies of the custody and control form, medical interviews, documentation of the basis for verifying as negative test results confirmed as positive by the laboratory, any other documentation concerning the MRO's verification process.

(2) Should the employer change MRO's for any reason, the employer shall ensure that the former MRO forwards all records maintained pursuant to this rule to the new MRO within ten working days of receiving notice from the employer of the new MRO's name and address.

(3) Any employer obtaining MRO services by contract, including a contract through a C/TPA, shall ensure that the contract includes a recordkeeping provision that is consistent with this paragraph, including requirements for transferring records to a new MRO.

(b) *Access to records.* The employer and the MRO shall permit the Administrator or the Administrator's representative to examine records required to be kept under this subpart and 49 CFR part 40. The Administrator or the Administrator's representative may require that all records maintained by the service agent for the employer must be produced at the employer's place of business.

(c) *Release of drug testing information.* An employer shall release information regarding an employee's drug testing results, evaluation, or rehabilitation to a third party in accordance with 49 CFR part 40. Except as required by law, this subpart, or 49 CFR part 40, no employer shall release employee information.

(d) *Refusal to submit to testing.* Each employer must notify the FAA within 2 working days of any employee who holds a certificate issued under part 61, part 63, or part 65 of this chapter who has refused to submit to a drug test required under this subpart. Notification must be sent to: Federal Aviation Administration, Office of Aerospace Medicine, Drug Abatement Division (AAM-800), 800 Independence Avenue, SW., Washington, DC 20591, or by fax to (202) 267-5200.

(e) *Permanent disqualification from service.* (1) An employee who has verified positive drug test results on two drug tests required by this subpart of this chapter, and conducted after September 19, 1994, is permanently precluded from performing for an employer the safety-sensitive duties the employee performed prior to the second drug test.

(2) An employee who has engaged in prohibited drug use during the performance of a safety-sensitive function after September 19, 1994 is permanently precluded from performing that safety-sensitive function for an employer.

(f) *DOT management information system annual reports.* Copies of any annual reports submitted to the FAA under this subpart must be maintained by the employer for a minimum of 5 years.

§ 120.113 Medical Review Officer, Substance Abuse Professional, and Employer Responsibilities.

(a) The employer shall designate or appoint a Medical Review Officer (MRO) who shall be qualified in accordance with 49 CFR part 40 and shall perform the functions set forth in 49 CFR part 40 and this subpart. If the employer does not have a qualified individual on staff to serve as MRO, the employer may contract for the provision of MRO services as part of its drug testing program.

(b) *Medical Review Officer (MRO).* The MRO must perform the functions set forth in subpart G of 49 CFR part 40, and subpart E of this part. The MRO shall not delay verification of the primary test result following a request for a split specimen test unless such delay is based on reasons other than the fact that the split specimen test result is pending. If the primary test result is verified as positive, actions required under this rule (*e.g.,* notification to the Federal Air Surgeon, removal from safety-sensitive position) are not stayed during the 72-hour request period or pending receipt of the split specimen test result.

(c) *Substance Abuse Professional (SAP).* The SAP must perform the functions set forth in 49 CFR part 40, subpart O.

(d) *Additional Medical Review Officer, Substance Abuse Professional, and Employer Responsibilities Regarding 14 CFR part 67 Airman Medical Certificate Holders.* (1) As part of verifying a confirmed positive test result or refusal to submit to a test, the MRO must ask and the individual must answer whether he or she holds an airman medical certificate issued under 14 CFR part 67 or would be required to hold an airman medical certificate to perform a safety-sensitive function for the employer. If the individual answers in the affirmative to either question, in addition to notifying the employer in accordance with 49 CFR part 40, the MRO must forward to the Federal Air Surgeon, at the address listed in paragraph (d)(5) of this section, the name of the individual, along with identifying information and supporting documentation, within 2 working days after verifying a positive drug test result or refusal to submit to a test.

(2) During the SAP interview required for a verified positive test result or a refusal to submit to a test, the SAP must ask and the individual must answer whether he or she holds or would be required to hold an airman medical certificate issued under 14 CFR part 67 to perform a safety-sensitive function for the employer. If the individual answers in the affirmative, the individual must obtain an airman medical certificate issued by the Federal Air Surgeon dated after the verified positive drug test result date or refusal to test date. After the individual obtains this airman medical certificate, the SAP may recommend to the employer that the individual may be returned to a safety-sensitive position. The receipt of an airman medical certificate does not alter any obligations otherwise required by 49 CFR part 40 or this subpart.

(3) An employer must forward to the Federal Air Surgeon within 2 working days of receipt, copies of all reports provided to the employer by a SAP regarding the following:

(i) An individual who the MRO has reported to the Federal Air Surgeon under § 120.113 (d)(1); or

(ii) An individual who the employer has reported to the Federal Air Surgeon under § 120.111(d).

(4) The employer must not permit an employee who is required to hold an airman medical certificate under 14 CFR part 67 to perform a safety-sensitive duty to resume that duty until the employee has:

(i) Been issued an airman medical certificate from the Federal Air Surgeon after the date of the verified positive drug test result or refusal to test; and

(ii) Met the return to duty requirements in accordance with 49 CFR part 40.

(5) Reports required under this section shall be forwarded to the Federal Air Surgeon, Federal Aviation Administration, Office of Aerospace Medicine, Attn: Drug Abatement Division (AAM-800), 800 Independence Avenue, SW., Washington, DC 20591.

(6) MROs, SAPs, and employers who send reports to the Federal Air Surgeon must keep a copy of each report for 5 years.

§ 120.115 Employee Assistance Program (EAP).

Link to an amendment published at 78 FR 42003, July 15, 2013.

(a) The employer shall provide an EAP for employees. The employer may establish the EAP as a part of its internal personnel services or the employer may contract with an entity that will provide EAP services to an employee. Each EAP must include education and training on drug use for employees and training for supervisors making determinations for testing of employees based on reasonable cause.

(b) *EAP education program.* (1) Each EAP education program must include at least the following elements:

(i) Display and distribution of informational material;

(ii) Display and distribution of a community service hot-line telephone number for employee assistance; and

(iii) Display and distribution of the employer's policy regarding drug use in the workplace.

(2) The employer's policy shall include information regarding the consequences under the rule of using drugs while performing safety-sensitive functions, receiving a verified positive drug test result, or refusing to submit to a drug test required under the rule.

(c) *EAP training program.* (1) Each employer shall implement a reasonable program of initial training for employees. The employee training program must include at least the following elements:

(i) The effects and consequences of drug use on individual health, safety, and work environment;

(ii) The manifestations and behavioral cues that may indicate drug use and abuse; and

(iii) Documentation of training given to employees and employer's supervisory personnel.

(2) The employer's supervisory personnel who will determine when an employee is subject to testing based on reasonable cause shall receive specific training on specific, contemporaneous physical, behavioral, and performance indicators of probable drug use in addition to the training specified in § 120.115 (c).

(3) The employer shall ensure that supervisors who will make reasonable cause determinations receive at least 60 minutes of initial training.

(4) The employer shall implement a reasonable recurrent training program for supervisory personnel making reasonable cause determinations during subsequent years.

(5) The employer shall identify the employee and supervisor for EAP training in the employer's drug testing plan submitted to the FAA for approval.

§ 120.117 Implementing a drug testing program.

Link to an amendment published at 78 FR 42003, July 15, 2013.

(a) Each company must meet the requirements of this subpart. Use the following chart to determine whether your company must obtain an Antidrug and Alcohol Misuse Prevention Program Operations Specification or whether you must register with the FAA:

If you are . . .	You must . . .
(1) A part 119 certificate holder with authority to operate under parts 121 and/or 135	Obtain an Antidrug and Alcohol Misuse Prevention Program Operations Specification by contacting your FAA Principal Operations Inspector.
(2) An operator as defined in §91.147 of this chapter	Register with the FAA by contacting the Flight Standards District Office nearest to your principal place of business.
(3) An air traffic control facility not operated by the FAA or by or under contract to the U.S. Military	Register with the FAA, Office of Aerospace Medicine, Drug Abatement Division (AAM–800), 800 Independence Avenue, SW., Washington, DC 20591.
(4) A part 145 certificate holder who has your own drug testing program	Obtain an Antidrug and Alcohol Misuse Prevention Program Operations Specification by contacting your Principal Maintenance Inspector or register with the FAA, Office of Aerospace Medicine, Drug Abatement Division (AAM–800), 800 Independence Avenue, SW., Washington, DC 20591, if you opt to conduct your own drug testing program.

If you are . . .	You must . . .
(5) A contractor who has your own drug testing program	Register with the FAA, Office of Aerospace Medicine, Drug Abatement Division (AAM–800), 800 Independence Avenue, SW., Washington, DC 20591, if you opt to conduct your own drug testing program.

(b) Use the following chart for implementing a drug testing program if you are applying for a part 119 certificate with authority to operate under parts 121 or 135 of this chapter, if you intend to begin operations as defined in § 91.147 of this chapter, or if you intend to begin air traffic control operations (not operated by the FAA or by or under contract to the U.S. Military). Use it to determine whether you need to have an Antidrug and Alcohol Misuse Prevention Program Operations Specification, or whether you need to register with the FAA. Your employees who perform safety-sensitive functions must be tested in accordance with this subpart. The chart follows:

If you . . .	You must . . .
(1) Apply for a part 119 certificate with authority to operate under parts 121 or 135	(i) Have an Antidrug and Alcohol Misuse Prevention Program Operations Specification. (ii) Implement an FAA drug testing program no later than the date you start operations, and
	(iii) Meet the requirements of this subpart.
(2) Intend to begin operations as defined in §91.147 of this chapter	(i) Register with the FAA, by contacting the Flight Standards District Office nearest to your principal place of business prior to starting operations.
	(ii) Implement an FAA drug testing program no later than the date you start operations, and
	(iii) Meet the requirements of this subpart.
(3) Intend to begin air traffic control operations (at an air traffic control facility not operated by the FAA or by or under contract to the U.S. military)	(i) Register with the FAA, Office of Aerospace Medicine, Drug Abatement Division (AAM–800), 800 Independence Avenue, SW., Washington, DC 20591 prior to starting operations. (ii) Implement an FAA drug testing program no later than the date you start operations, and
	(iii) Meet the requirements of this subpart.

(c) If you are an individual or company that intends to provide safety-sensitive services by contract to a part 119 certificate holder with authority to operate under parts 121 and/or 135 of this chapter, an operation as defined in § 91.147 of this chapter, or an air traffic control facility not operated by the FAA or by or under contract to the U.S. military, use the following chart to determine what you must do if you opt to have your own drug testing program.

If you . . .	And you opt to conduct your own drug program, you must . . .
(1) Are a part 145 certificate holder	(i) Have an Antidrug and Alcohol Misuse Prevention Program Operations Specification or register with the FAA, Office of Aerospace Medicine, Drug Abatement Division (AAM–800), 800 Independence Avenue, SW., Washington, DC 20591,
	(ii) Implement an FAA drug testing program no later than the date you start performing safety-sensitive functions for a part 119 certificate holder with authority to operate under parts 121 or 135, or operator as defined in §91.147 of this chapter, and
	(iii) Meet the requirements of this subpart as if you were an employer.
(2) Are a contractor	(i) Register with the FAA, Office of Aerospace Medicine, Drug Abatement Division (AAM–800), 800 Independence Avenue, SW., Washington, DC 20591,
	(ii) Implement an FAA drug testing program no later than the date you start performing safety-sensitive functions for a part 119 certificate holder with authority to operate under parts 121 or 135, or operator as defined in §91.147 of this chapter, or an air traffic control facility not operated by the FAA or by or under contract to the U.S. Military, and
	(iii) Meet the requirements of this subpart as if you were an employer.

(d) *Obtaining an Antidrug and Alcohol Misuse Prevention Program Operations Specification.* (1) To obtain an Antidrug and Alcohol Misuse Prevention Program Operations Specification, you must contact your FAA Principal Operations Inspector or Principal Maintenance Inspector. Provide him/her with the following information:

(i) Company name.

(ii) Certificate number.

(iii) Telephone number.

(iv) Address where your drug and alcohol testing program records are kept.

(v) Whether you have 50 or more safety-sensitive employees, or 49 or fewer safety-sensitive employees. (Part 119 certificate holders with authority to operate only under part 121 of this chapter are not required to provide this information.)

(2) You must certify on your Antidrug and Alcohol Misuse Prevention Program Operations Specification issued by your FAA Principal Operations Inspector or Principal Maintenance Inspector that you will comply with this part and 49 CFR part 40.

(3) You are required to obtain only one Antidrug and Alcohol Misuse Prevention Program Operations Specification to satisfy this requirement under this part.

(4) You must update the Antidrug and Alcohol Misuse Prevention Program Operations Specification when any changes to the information contained in the Operation Specification occur.

(e) *Registering a drug and alcohol testing program with the FAA.* (1) To register with the FAA, submit the following information:

(i) Company name.

(ii) Telephone number.

(iii) Address where your drug and alcohol testing program records are kept.

(iv) Type of safety-sensitive functions you perform for an employer (such as flight instruction duties, aircraft dispatcher duties, maintenance or preventive maintenance duties, ground security coordinator duties, aviation screening duties, air traffic control duties).

(v) Whether you have 50 or more safety-sensitive employees, or 49 or fewer covered employees.

(vi) A signed statement indicating that: your company will comply with this part and 49 CFR part 40; and, if you are a contractor, you intend to provide safety-sensitive functions by contract to a part 119 certificate holder with authority to operate under part 121 and/or part 135 of this chapter, an operator as defined in § 91.147 of this chapter, or an air traffic control facility not operated by the FAA or by or under contract to the U.S. military.

(2) Send this information in the form and manner prescribed by the Administrator, in duplicate to the appropriate address below:

(i) For § 91.147 operators: The Flight Standards District Office nearest to your principal place of business.

(ii) For all others: The Federal Aviation Administration, Office of Aerospace Medicine, Drug Abatement Division (AAM-800), 800 Independence Avenue, SW., Washington, DC 20591.

(3) Update the registration information as changes occur. Send the updates in duplicate to the address specified in paragraph (e)(2) of this section.

(4) This registration will satisfy the registration requirements for both your drug testing program under this subpart and your alcohol testing program under subpart F of this part.

[Doc. No. FAA-2008-0937, 74 FR 22653, May 14, 2009; Amdt. 120-0A, 75 FR 3154, Jan. 20, 2010]

§ 120.119 Annual reports.

(a) Annual reports of testing results must be submitted to the FAA by March 15 of the succeeding calendar year for the prior calendar year (January 1 through December 31) in accordance with the following provisions:

(1) Each part 121 certificate holder shall submit an annual report each year.

(2) Each entity conducting a drug testing program under this part, other than a part 121 certificate holder, that has 50 or more employees performing a safety-sensitive function on January 1 of any calendar year shall submit an annual report to the FAA for that calendar year.

(3) The Administrator reserves the right to require that aviation employers not otherwise required to submit annual reports prepare and submit such reports to the FAA. Employers that will be required to submit annual reports under this provision will be notified in writing by the FAA.

(b) As an employer, you must use the Management Information System (MIS) form and instructions as required by 49 CFR part 40 (at 49 CFR 40.26 and appendix H to 49 CFR part 40). You may also use the electronic version of the MIS form provided by DOT. The Administrator may designate means (*e.g.*, electronic program transmitted via the Internet) other than hard-copy, for MIS form submission. For information on where to submit MIS forms and for the electronic version of the form, *see: http://www.faa.gov/about/office_org/headquarters_offices/avs/offices/aam/drug_alcohol*.

(c) A service agent may prepare the MIS report on behalf of an employer. However, a company official (*e.g.*, Designated Employer Representative as defined in 49 CFR part 40) must certify the accuracy and completeness of the MIS report, no matter who prepares it.

[Doc. No. FAA-2008-0937, 74 FR 22653, May 14, 2009; Amdt. 120-0A, 75 FR 3154, Jan. 20, 2010]

§ 120.121 Preemption.

(a) The issuance of 14 CFR parts 65, 91, 121, and 135 by the FAA preempts any State or local law, rule, regulation, order, or standard covering the subject matter of 14 CFR parts 65, 91, 121, and 135, including but not limited to, drug testing of aviation personnel performing safety-sensitive functions.

(b) The issuance of 14 CFR parts 65, 91, 121, and 135 does not preempt provisions of state criminal law that impose sanctions for reckless conduct of an individual that leads to actual loss of life, injury, or damage to property whether such provisions apply specifically to aviation employees or generally to the public.

§ 120.123 Drug testing outside the territory of the United States.

(a) No part of the testing process (including specimen collection, laboratory processing, and MRO actions) shall be conducted outside the territory of the United States.

(1) Each employee who is assigned to perform safety-sensitive functions solely outside the territory of the United States shall be removed from the random testing pool upon the inception of such assignment.

(2) Each covered employee who is removed from the random testing pool under this section shall be returned to the random testing pool when the employee resumes the performance of safety-sensitive functions wholly or partially within the territory of the United States.

(b) The provisions of this subpart shall not apply to any individual who performs a function listed in § 120.105 by contract for an employer outside the territory of the United States.

§ 120.125 Waivers from 49 CFR 40.21.

An employer subject to this part may petition the Drug Abatement Division, Office of Aerospace Medicine, for a waiver allowing the employer to stand down an employee following a report of a laboratory confirmed positive drug test or refusal, pending the outcome of the verification process.

(a) Each petition for a waiver must be in writing and include substantial facts and justification to support the waiver. Each petition must satisfy the substantive requirements for obtaining a waiver, as provided in 49 CFR 40.21.

(b) Each petition for a waiver must be submitted to the Federal Aviation Administration, Office of Aerospace Medicine, Drug Abatement Division (AAM-800), 800 Independence Avenue, SW., Washington, DC 20591.

(c) The Administrator may grant a waiver subject to 49 CFR 40.21(d).

Subpart F—Alcohol Testing Program Requirements

§ 120.201 Scope.

This subpart contains the standards and components that must be included in an alcohol testing program required by this part.

§ 120.203 General.

(a) *Purpose*. The purpose of this subpart is to establish programs designed to help prevent accidents and injuries resulting from the misuse of alcohol by employees who perform safety-sensitive functions in aviation.

(b) *Alcohol testing procedures*. Each employer shall ensure that all alcohol testing conducted pursuant to this subpart complies with the procedures set forth in 49 CFR part 40. The provisions of 49 CFR part 40 that address alcohol testing are made applicable to employers by this subpart.

(c) *Employer responsibility*. As an employer, you are responsible for all actions of your officials, representatives, and service agents in carrying out the requirements of the DOT agency regulations.

§ 120.205 Preemption of State and local laws.

(a) Except as provided in paragraph (a)(2) of this section, these regulations preempt any State or local law, rule, regulation, or order to the extent that:

(1) Compliance with both the State or local requirement and this subpart is not possible; or

(2) Compliance with the State or local requirement is an obstacle to the accomplishment and execution of any requirement in this subpart.

(b) The alcohol testing requirements of this title shall not be construed to preempt provisions of State criminal law that impose sanctions for reckless conduct leading to actual loss of life, injury, or damage to property, whether the provisions apply specifically to transportation employees or employers or to the general public.

§ 120.207 Other requirements imposed by employers.

Except as expressly provided in these alcohol testing requirements, nothing in this subpart shall be construed to affect the authority of employers, or the rights of employees, with respect to the use or possession of alcohol, including any authority and rights with respect to alcohol testing and rehabilitation.

§ 120.209 Requirement for notice.

Before performing an alcohol test under this subpart, each employer shall notify a covered employee that the alcohol test is required by this subpart. No employer shall falsely represent that a test is administered under this subpart.

§ 120.211 Applicable Federal regulations.

The following applicable regulations appear in 49 CFR and 14 CFR:

(a) 49 CFR Part 40—Procedures for Transportation Workplace Drug Testing Programs

(b) 14 CFR:

(1) § 67.107—First-Class Airman Medical Certificate, Mental.

(2) § 67.207—Second-Class Airman Medical Certificate, Mental.

(3) § 67.307—Third-Class Airman Medical Certificate, Mental.

(4) § 91.147—Passenger carrying flights for compensation or hire.

(5) § 135.1—Applicability

[Doc. No. FAA-2008-0937, 74 FR 22653, May 14, 2009; Amdt. 120-0A, 75 FR 3154, Jan. 20, 2010]

§ 120.213 Falsification.

No individual may make, or cause to be made, any of the following:

(a) Any fraudulent or intentionally false statement in any application of an alcohol testing program.

(b) Any fraudulent or intentionally false entry in any record or report that is made, kept, or used to show compliance with this subpart.

(c) Any reproduction or alteration, for fraudulent purposes, of any report or record required to be kept by this subpart.

§ 120.215 Covered employees.

(a) Each employee, including any assistant, helper, or individual in a training status, who performs a safety-sensitive function listed in this section directly or by contract (including by subcontract at any tier) for an employer as defined in this subpart must be subject to alcohol testing under an alcohol testing program implemented in accordance with this subpart. This includes full-time, part-time, temporary, and intermittent employees regardless of the degree of supervision. The safety-sensitive functions are:

(1) Flight crewmember duties.

(2) Flight attendant duties.

(3) Flight instruction duties.

(4) Aircraft dispatcher duties.

(5) Aircraft maintenance or preventive maintenance duties.

(6) Ground security coordinator duties.

(7) Aviation screening duties.

(8) Air traffic control duties.

(b) Each employer must identify any employee who is subject to the alcohol testing regulations of more than one DOT agency. Prior to conducting any alcohol test on a covered employee subject to the alcohol testing regulations of more than one DOT agency, the employer must determine which DOT agency authorizes or requires the test.

§ 120.217 Tests required.

(a) *Pre-employment alcohol testing*. As an employer, you may, but are not required to, conduct pre-employment alcohol testing under this subpart. If you choose to conduct pre-employment alcohol testing, you must comply with the following requirements:

(1) You must conduct a pre-employment alcohol test before the first performance of safety-sensitive functions by every covered employee (whether a new employee or someone who has transferred to a position involving the performance of safety-sensitive functions).

(2) You must treat all safety-sensitive employees performing safety-sensitive functions the same for the purpose of pre-employment alcohol testing (*i.e.*, you must not test some covered employees and not others).

(3) You must conduct the pre-employment tests after making a contingent offer of employment or transfer, subject to the employee passing the pre-employment alcohol test.

(4) You must conduct all pre-employment alcohol tests using the alcohol testing procedures of 49 CFR part 40.

(5) You must not allow a covered employee to begin performing safety-sensitive functions unless the result of the employee's test indicates an alcohol concentration of less than 0.04. If a pre-employment test result under this paragraph indicates an alcohol concentration of 0.02 or greater but less than 0.04, the provisions of § 120.221(f) apply.

(b) *Post-accident alcohol testing*. (1) As soon as practicable following an accident, each employer shall test each surviving covered employee for alcohol if that employee's performance of a safety-sensitive function either contributed to the accident or cannot be completely discounted as a contributing factor to the accident. The decision not to administer a test under this section shall be based on the employer's determination, using the best available information at the time of the determination, that the covered employee's performance could not have contributed to the accident.

(2) If a test required by this section is not administered within 2 hours following the accident, the employer shall prepare and maintain on file a record stating the reasons the test was not promptly administered. If a test required by this section is not administered within 8 hours following the accident, the employer shall cease attempts to administer an alcohol test and shall prepare and maintain the same record. Records shall be submitted to the FAA upon request of the Administrator or his or her designee.

(3) A covered employee who is subject to post-accident testing shall remain readily available for such testing or may be deemed by the employer to have refused to submit to testing. Nothing in this section shall be construed to require the delay of necessary medical attention for injured people following an accident or to prohibit a covered employee from leaving the scene of an accident for the period necessary to obtain assistance in responding to the accident or to obtain necessary emergency medical care.

(c) *Random alcohol testing*. (1) Except as provided in paragraphs (c)(2) through (c)(4) of this section, the minimum annual percentage rate for random alcohol testing will be 25 percent of the covered employees.

(2) The Administrator's decision to increase or decrease the minimum annual percentage rate for random alcohol testing is based on the violation rate for the entire industry. All information used for this determination is drawn from MIS reports required by this subpart. In order to ensure reliability of the data, the Administrator considers the quality and completeness of the reported data, may obtain additional information or reports from employers, and may make appropriate modifications in calculating the industry violation rate. Each year, the Administrator will publish in the FEDERAL REGISTER the minimum annual percentage rate for random alcohol testing of covered employees. The new minimum annual percentage rate for random alcohol testing will be applicable starting January 1 of the calendar year following publication.

(3)(i) When the minimum annual percentage rate for random alcohol testing is 25 percent or more, the Administrator may lower this rate to 10 percent of all covered employees if the Administrator determines that the data received under the reporting requirements of this subpart for two consecutive calendar years indicate that the violation rate is less than 0.5 percent.

(ii) When the minimum annual percentage rate for random alcohol testing is 50 percent, the Administrator may lower this rate to 25 percent of all covered employees if the Administrator determines that the data received under the reporting requirements of this subpart for two consecutive calendar years indicate that the violation rate is less than 1.0 percent but equal to or greater than 0.5 percent.

(4)(i) When the minimum annual percentage rate for random alcohol testing is 10 percent, and the data received under the reporting requirements of this subpart for that calendar year indicate that the violation rate is equal to or greater than 0.5 percent but less than 1.0 percent, the Administrator will increase the minimum annual percentage rate for random alcohol testing to 25 percent of all covered employees.

(ii) When the minimum annual percentage rate for random alcohol testing is 25 percent or less, and the data received under the reporting requirements of this subpart for that calendar year indicate that the violation rate is equal to or greater than 1.0 percent, the Administrator will increase the minimum annual percentage rate for random alcohol testing to 50 percent of all covered employees.

(5) The selection of employees for random alcohol testing shall be made by a scientifically valid method, such as a random-number table or a computer-based random number generator that is matched with employees' Social Security numbers, payroll identification numbers, or other comparable identifying numbers. Under the selection process used, each covered employee shall have an equal chance of being tested each time selections are made.

(6) As an employer, you must select and test a percentage of employees at least equal to the minimum annual percentage rate each year.

(i) As an employer, to determine whether you have met the minimum annual percentage rate, you must divide the number of random alcohol screening test results for safety-sensitive employees by the average number of safety-sensitive employees eligible for random testing.

(A) To calculate whether you have met the annual minimum percentage rate, count all random screening test results below 0.02 breath alcohol concentration, random screening test results of 0.02 or greater breath alcohol concentration, and random refusals as your "random alcohol screening test results."

(B) To calculate the average number of safety-sensitive employees eligible for random testing throughout the year, add the total number of safety-sensitive employees eligible for testing during each random testing period for the year and divide that total by the number of random testing periods. Only safety-sensitive employees are to be in an employer's random testing pool, and all safety-sensitive employees must be in the random pool. If you are an employer conducting random testing more often than once per month (*e.g.*, you select daily, weekly, bi-weekly) you do not need to compute this total number of safety-sensitive employees more than on a once per month basis.

(ii) As an employer, you may use a service agent to perform random selections for you, and your safety-sensitive employees may be part of a larger random testing pool of safety-sensitive employees. However, you must ensure that the service agent you use is testing at the appropriate percentage established for your industry and that only safety-sensitive employees are in the random testing pool. For example:

(A) If the service agent has your employees in a random testing pool for your company alone, you must ensure that the testing is conducted at least at the minimum annual percentage rate under this part.

(B) If the service agent has your employees in a random testing pool combined with other FAA-regulated companies, you must ensure that the testing is conducted at least at the minimum annual percentage rate under this part.

(C) If the service agent has your employees in a random testing pool combined with other DOT-regulated companies, you must ensure that the testing is conducted at least at the highest rate required for any DOT-regulated company in the pool.

(7) Each employer shall ensure that random alcohol tests conducted under this subpart are unannounced and that the dates for administering random tests are spread reasonably throughout the calendar year.

(8) Each employer shall require that each covered employee who is notified of selection for random testing proceeds to the testing site immediately; provided, however, that if the employee is performing a safety-sensitive function at the time of the notification, the employer shall instead ensure that the employee ceases to perform the safety-sensitive function and proceeds to the testing site as soon as possible.

(9) A covered employee shall only be randomly tested while the employee is performing safety-sensitive functions; just before the employee is to perform safety-sensitive functions; or just after the employee has ceased performing such functions.

(10) If a given covered employee is subject to random alcohol testing under the alcohol testing rules of more than one DOT agency, the employee shall be subject to random alcohol testing at the percentage rate established for the calendar year by the DOT agency regulating more than 50 percent of the employee's functions.

(11) If an employer is required to conduct random alcohol testing under the alcohol testing rules of more than one DOT agency, the employer may—

(i) Establish separate pools for random selection, with each pool containing the covered employees who are subject to testing at the same required rate; or

(ii) Randomly select such employees for testing at the highest percentage rate established for the calendar year by any DOT agency to which the employer is subject.

(d) *Reasonable suspicion alcohol testing.* (1) An employer shall require a covered employee to submit to an alcohol test when the employer has reasonable suspicion to believe that the employee has violated the alcohol misuse prohibitions in §§ 120.19 or 120.37.

(2) The employer's determination that reasonable suspicion exists to require the covered employee to undergo an alcohol test shall be based on specific, contemporaneous, articulable observations concerning the appearance, behavior, speech or body odors of the employee. The required observations shall be made by a supervisor who is trained in detecting the symptoms of alcohol misuse. The supervisor who makes the determination that reasonable suspicion exists shall not conduct the breath alcohol test on that employee.

(3) Alcohol testing is authorized by this section only if the observations required by paragraph (d)(2) of this section are made during, just preceding, or just after the period of the work day that the covered employee is required to be in compliance with this rule. An employee may be directed by the employer to undergo reasonable suspicion testing for alcohol only while the employee is performing safety-sensitive functions; just before the employee is to perform safety-sensitive functions; or just after the employee has ceased performing such functions.

(4)(i) If a test required by this section is not administered within 2 hours following the determination made under paragraph (d)(2) of this section, the employer shall prepare and maintain on file a record stating the reasons the test was not promptly administered. If a test required by this section is not administered within 8 hours following the determination made under paragraph (d)(2) of this section, the employer shall cease attempts to administer an alcohol test and shall state in the record the reasons for not administering the test.

(ii) Notwithstanding the absence of a reasonable suspicion alcohol test under this section, no covered employee shall report for duty or remain on duty requiring the performance of safety-sensitive functions while the employee is under the influence of, or impaired by, alcohol, as shown by the behavioral, speech, or performance indicators of alcohol misuse, nor shall an employer permit the covered employee to perform or continue to perform safety-sensitive functions until:

(A) An alcohol test is administered and the employee's alcohol concentration measures less than 0.02; or

(B) The start of the employee's next regularly scheduled duty period, but not less than 8 hours following the determination made under paragraph (d)(2) of this section that there is reasonable suspicion that the employee has violated the alcohol misuse provisions in §§ 120.19 or 120.37.

(iii) No employer shall take any action under this subpart against a covered employee based solely on the employee's behavior and appearance in the absence of an alcohol test. This does not prohibit an employer with authority independent of this subpart from taking any action otherwise consistent with law.

(e) *Return-to-duty alcohol testing.* Each employer shall ensure that before a covered employee returns to duty requiring the performance of a safety-sensitive function after engaging in conduct prohibited in §§ 120.19 or 120.37 the employee shall undergo a return-to-duty alcohol test with a result indicating an alcohol concentration of less than 0.02. The test cannot occur until after the SAP has determined that the employee has successfully complied with the prescribed education and/or treatment.

(f) *Follow-up alcohol testing.* (1) Each employer shall ensure that the employee who engages in conduct prohibited by §§ 120.19 or 120.37, is subject to unannounced follow-up alcohol testing as directed by a SAP.

(2) The number and frequency of such testing shall be determined by the employer's SAP, but must consist of at least six tests in the first 12 months following the employee's return to duty.

(3) The employer must direct the employee to undergo testing for drugs in accordance with subpart E of this part, in addition to alcohol, if the SAP determines that drug testing is necessary for the particular employee. Any such drug testing shall be conducted in accordance with the provisions of 49 CFR part 40.

(4) Follow-up testing shall not exceed 60 months after the date the individual begins to perform, or returns to the performance of, a safety-sensitive function. The SAP may terminate the requirement for follow-up testing at any time after the first six tests have been conducted, if the SAP determines that such testing is no longer necessary.

(5) A covered employee shall be tested for alcohol under this section only while the employee is performing safety-sensitive functions, just before the employee is to perform safety-sensitive functions, or just after the employee has ceased performing such functions.

(g) *Retesting of covered employees with an alcohol concentration of 0.02 or greater but less than 0.04.* Each employer shall retest a covered employee to ensure compliance with the provisions of § 120.221(f) if the employer chooses to permit the employee to perform a safety-sensitive function within 8 hours following the administration of an alcohol test indicating an alcohol concentration of 0.02 or greater but less than 0.04.

§ 120.219 Handling of test results, record retention, and confidentiality.

(a) *Retention of records.* (1) *General requirement.* In addition to the records required to be maintained under 49 CFR part 40, employers must maintain records required by this subpart in a secure location with controlled access.

(2) *Period of retention*.

(i) *Five years*.

(A) Copies of any annual reports submitted to the FAA under this subpart for a minimum of 5 years.

(B) Records of notifications to the Federal Air Surgeon of refusals to submit to testing and violations of the alcohol misuse prohibitions in this chapter by covered employees who hold medical certificates issued under part 67 of this chapter.

(C) Documents presented by a covered employee to dispute the result of an alcohol test administered under this subpart.

(D) Records related to other violations of §§ 120.19 or 120.37.

(ii) *Two years.* Records related to the testing process and training required under this subpart.

(A) Documents related to the random selection process.

(B) Documents generated in connection with decisions to administer reasonable suspicion alcohol tests.

(C) Documents generated in connection with decisions on post-accident tests.

(D) Documents verifying existence of a medical explanation of the inability of a covered employee to provide adequate breath for testing.

(E) Materials on alcohol misuse awareness, including a copy of the employer's policy on alcohol misuse.

(F) Documentation of compliance with the requirements of § 120.223(a).

(G) Documentation of training provided to supervisors for the purpose of qualifying the supervisors to make a determination concerning the need for alcohol testing based on reasonable suspicion.

(H) Certification that any training conducted under this subpart complies with the requirements for such training.

(b) *Annual reports.* (1) Annual reports of alcohol testing program results must be submitted to the FAA by March 15 of the succeeding calendar year for the prior calendar year (January 1 through December 31) in accordance with the provisions of paragraphs (b)(1)(i) through (iii) of this section.

(i) Each part 121 certificate holder shall submit an annual report each year.

(ii) Each entity conducting an alcohol testing program under this part, other than a part 121 certificate holder, that has 50 or more employees performing a safety-sensitive function on January 1 of any calendar year shall submit an annual report to the FAA for that calendar year.

(iii) The Administrator reserves the right to require that aviation employers not otherwise required to submit annual reports prepare and submit such reports to the FAA. Employers that will be required to submit annual reports under this provision will be notified in writing by the FAA.

(2) As an employer, you must use the Management Information System (MIS) form and instructions as required by 49 CFR part 40 (at 49 CFR 40.26 and appendix H to 49 CFR part 40). You may also use the electronic version of the MIS form provided by the DOT. The Administrator may designate means (*e.g.*, electronic program transmitted via the Internet) other than hard-copy, for MIS form submission. For information on where to submit MIS forms and for the electronic version of the form, see: *http://www.faa.gov/about/office_org/headquarters_offices/avs/offices/aam/drug_alcohol/*.

(3) A service agent may prepare the MIS report on behalf of an employer. However, a company official (*e.g.*, Designated Employer Representative as defined in 49 CFR part 40) must certify the accuracy and completeness of the MIS report, no matter who prepares it.

(c) *Access to records and facilities*.

(1) Except as required by law or expressly authorized or required in this subpart, no employer shall release covered employee information that is contained in records required to be maintained under this subpart.

(2) A covered employee is entitled, upon written request, to obtain copies of any records pertaining to the employee's use of alcohol, including any records pertaining to his or her alcohol tests in accordance with 49 CFR part 40. The employer shall promptly provide the records requested by the employee. Access to an employee's records shall not be contingent upon payment for records other than those specifically requested.

(3) Each employer shall permit access to all facilities utilized in complying with the requirements of this subpart to the Secretary of Transportation or any DOT agency with regulatory authority over the employer or any of its covered employees.

§ 120.221 Consequences for employees engaging in alcohol-related conduct.

Link to an amendment published at 78 FR 42004, July 15, 2013.

(a) *Removal from safety-sensitive function.* (1) Except as provided in 49 CFR part 40, no covered employee shall perform safety-sensitive functions if the employee has engaged in conduct prohibited by §§ 120.19 or 120.37, or an alcohol misuse rule of another DOT agency.

(2) No employer shall permit any covered employee to perform safety-sensitive functions if the employer has determined that the employee has violated this section.

(b) *Permanent disqualification from service.* An employee who violates §§ 120.19 or 120.37, or who engages in alcohol use that violates another alcohol misuse provision of §§ 120.19 or 120.37 and who had previously engaged in alcohol use that violated the provisions of §§ 120.19 or 120.37 after becoming subject to such prohibitions is permanently precluded from performing for an employer the safety-sensitive duties the employee performed before such violation.

(c) *Notice to the Federal Air Surgeon.* (1) An employer who determines that a covered employee who holds an airman medical certificate issued under part 67 of this chapter has engaged in alcohol use that violated the alcohol misuse provisions of §§ 120.19 or 120.37 shall notify the Federal Air Surgeon within 2 working days.

(2) Each such employer shall forward to the Federal Air Surgeon a copy of the report of any evaluation performed under the provisions of § 120.223(c) within 2 working days of the employer's receipt of the report.

(3) All documents must be sent to the Federal Air Surgeon, Federal Aviation Administration, Office of Aerospace Medicine, Attn: Drug Abatement Division (AAM-800), 800 Independence Avenue, SW., Washington, DC 20591.

(4) No covered employee who is required to hold an airman medical certificate in order to perform a safety-sensitive duty may perform that duty following a violation of this subpart until the covered employee obtains an airman medical certificate issued by the Federal Air Surgeon dated after the alcohol test result or refusal to test date. After the covered employee obtains this airman medical certificate, the SAP may recommend to the employer that the covered employee may be returned to a safety-sensitive position. The receipt of an airman medical certificate does not alter any obligations otherwise required by 49 CFR part 40 or this subpart.

(5) Once the Federal Air Surgeon has recommended under paragraph (c)(4) of this section that the employee be permitted to perform safety-sensitive duties, the employer cannot permit the employee to perform those safety-sensitive duties until the employer has ensured that the employee meets the return to duty requirements in accordance with 49 CFR part 40.

(d) *Notice of refusals.* Each covered employer must notify the FAA within 2 working days of any employee who holds a certificate issued under part 61, part 63, or part 65 of this chapter who has refused to submit to an alcohol test required under this subpart. Notification must be sent to: Federal Aviation Administration, Office of Aerospace Medicine, Drug Abatement Division (AAM-800), 800 Independence Avenue, SW., Washington, DC 20591, or by fax to (202) 267-5200.

(e) *Required evaluation and alcohol testing.* No covered employee who has engaged in conduct prohibited by §§ 120.19 or 120.37 shall perform safety-sensitive functions unless the employee has met the requirements of 49 CFR part 40. No employer shall permit a covered employee who has engaged in such conduct to perform safety-sensitive functions unless the employee has met the requirements of 49 CFR part 40.

(f) *Other alcohol-related conduct.* (1) No covered employee tested under this subpart who is found to have an alcohol concentration of 0.02 or greater but less than 0.04 shall perform or continue to perform safety-sensitive functions for an employer, nor shall an employer permit the employee to perform or continue to perform safety-sensitive functions, until:

(i) The employee's alcohol concentration measures less than 0.02; or

(ii) The start of the employee's next regularly scheduled duty period, but not less than 8 hours following administration of the test.

(2) Except as provided in paragraph (f)(1) of this section, no employer shall take any action under this rule against an employee based solely on test results showing an alcohol concentration less than 0.04. This does not prohibit an employer with authority independent of this rule from taking any action otherwise consistent with law.

§ 120.223 Alcohol misuse information, training, and substance abuse professionals.

(a) *Employer obligation to promulgate a policy on the misuse of alcohol.* (1) *General requirements.* Each employer shall provide educational materials that explain these alcohol testing requirements and the employer's policies and procedures with respect to meeting those requirements.

(i) The employer shall ensure that a copy of these materials is distributed to each covered employee prior to the start of alcohol testing under the employer's FAA-mandated alcohol testing program and to each individual subsequently hired for or transferred to a covered position.

(ii) Each employer shall provide written notice to representatives of employee organizations of the availability of this information.

(2) *Required content.* The materials to be made available to employees shall include detailed discussion of at least the following:

(i) The identity of the individual designated by the employer to answer employee questions about the materials.

(ii) The categories of employees who are subject to the provisions of these alcohol testing requirements.

(iii) Sufficient information about the safety-sensitive functions performed by those employees to make clear what period of the work day the covered employee is required to be in compliance with these alcohol testing requirements.

(iv) Specific information concerning employee conduct that is prohibited by this chapter.

(v) The circumstances under which a covered employee will be tested for alcohol under this subpart.

(vi) The procedures that will be used to test for the presence of alcohol, protect the employee and the integrity of the breath testing process, safeguard the validity of the test results, and ensure that those results are attributed to the correct employee.

(vii) The requirement that a covered employee submit to alcohol tests administered in accordance with this subpart.

(viii) An explanation of what constitutes a refusal to submit to an alcohol test and the attendant consequences.

(ix) The consequences for covered employees found to have violated the prohibitions in this chapter, including the requirement that the employee be removed immediately from performing safety-sensitive functions, and the process in 49 CFR part 40, subpart O.

(x) The consequences for covered employees found to have an alcohol concentration of 0.02 or greater but less than 0.04.

(xi) Information concerning the effects of alcohol misuse on an individual's health, work, and personal life; signs and symptoms of an alcohol problem; available methods of evaluating and resolving problems associated with the misuse of alcohol; and intervening when an alcohol problem is suspected, including confrontation, referral to any available employee assistance program, and/or referral to management.

(xii) Optional provisions. The materials supplied to covered employees may also include information on additional employer policies with respect to the use or possession of alcohol, including any consequences for an employee found to have a specified alcohol level, that are based on the employer's authority independent of this subpart. Any such additional policies or consequences must be clearly and obviously described as being based on independent authority.

(b) *Training for supervisors.* Each employer shall ensure that persons designated to determine whether reasonable suspicion exists to require a covered employee to undergo alcohol testing under § 120.217(d) of this subpart receive at least 60 minutes of training on the physical, behavioral, speech, and performance indicators of probable alcohol misuse.

(c) *Substance abuse professional (SAP) duties.* The SAP must perform the functions set forth in 49 CFR part 40, subpart O, and this subpart.

§ 120.225 How to implement an alcohol testing program.

Link to an amendment published at 78 FR 42005, July 15, 2013.

(a) Each company must meet the requirements of this subpart. Use the following chart to determine whether your company must obtain an Antidrug and Alcohol Misuse Prevention Program Operations Specification or whether you must register with the FAA:

If you are . . .	You must . . .
(1) A part 119 certificate holder with authority to operate under parts 121 and/or 135	Obtain an Antidrug and Alcohol Misuse Prevention Program Operations Specification by contacting your FAA Principal Operations Inspector.
(2) An operator as defined in §91.147 of this chapter	Register with the FAA, by contacting the Flight Standards District Office nearest to your principal place of business.
(3) An air traffic control facility not operated by the FAA or by or under contract to the U.S. Military	Register with the FAA, Office of Aerospace Medicine, Drug Abatement Division (AAM–800), 800 Independence Avenue, SW., Washington, DC 20591.
(4) A part 145 certificate holder who has your own alcohol testing program	Obtain an Antidrug and Alcohol Misuse Prevention Program Operations Specification by contacting your Principal Maintenance Inspector or register with the FAA Office of Aerospace Medicine, Drug Abatement Division (AAM–800), 800 Independence Avenue, SW., Washington, DC 20591 if you opt to conduct your own alcohol testing program.
(5) A contractor who has your own alcohol testing program	Register with the FAA, Office of Aerospace Medicine, Drug Abatement Division (AAM–800), 800 Independence Avenue, SW., Washington, DC 20591 if you opt to conduct your own alcohol testing program.

(b) Use the following chart for implementing an alcohol testing program if you are applying for a part 119 certificate with authority to operate under parts 121 and/or 135 of this chapter, if you intend to begin operations as defined in § 91.147 of this chapter, or if you intend to begin operations as defined air traffic control operations (not operated by the FAA or by or under contract to the U.S. Military). Use it to determine whether you need to have an Antidrug and Alcohol Misuse Prevention Program Operations Specification, or whether you need to register with the FAA. Your employees who perform safety-sensitive duties must be tested in accordance with this subpart. The chart follows:

If you . . .	You must . . .
(1) Apply for a part 119 certificate with authority to operate under parts 121 and/ or 135	(i) Have an Antidrug and Alcohol Misuse Prevention Program Operations Specification, (ii) Implement an FAA alcohol testing program no later than the date you start operations, and
	(iii) Meet the requirements of this subpart.
(2) Intend to begin operations as defined in §91.147 of this chapter	(i) Register with the FAA by contacting the Flight Standards District Office nearest your principal place of business prior to starting operations,
	(ii) Implement an FAA alcohol testing program no later than the date you start operations, and
	(iii) Meet the requirements of this subpart.
(3) Intend to begin air traffic control operations (at an air traffic control facility not operated by the FAA or by or under contract to the U.S. military)	(i) Register with the FAA, Office of Aerospace Medicine, Drug Abatement Division (AAM–800), 800 Independence Avenue, SW., Washington, DC 20591 prior to starting operations, (ii) Implement an FAA alcohol testing program no later than the date you start operations, and (iii) Meet the requirements of this subpart.

(c) If you are an individual or a company that intends to provide safety-sensitive services by contract to a part 119 certificate holder with authority to operate under parts 121 and/or 135 of this chapter or an operator as defined in § 91.147 of this chapter, use the following chart to determine what you must do if you opt to have your own alcohol testing program.

If you . . .	And you opt to conduct your own Alcohol Testing Program, you must . . .
(1) Are a part 145 certificate holder	(i) Have an Antidrug and Alcohol Misuse Prevention Program Operations Specifications or register with the FAA, Office of Aerospace Medicine, Drug Abatement Division (AAM–800), 800 Independence Avenue, SW., Washington, DC 20591,
	(ii) Implement an FAA alcohol testing program no later than the date you start performing safety-sensitive functions for a part 119 certificate holder with the authority to operate under parts 121 and/ or 135, or operator as defined in §91.147 of this chapter, and

If you . . .	And you opt to conduct your own Alcohol Testing Program, you must . . .
	(iii) Meet the requirements of this subpart as if you were an employer.
(2) Are a contractor	(i) Register with the FAA, Office of Aerospace Medicine, Drug Abatement Division (AAM–800), 800 Independence Avenue, SW., Washington, DC 20591,
	(ii) Implement an FAA alcohol testing program no later than the date you start performing safety-sensitive functions for a part 119 certificate holder with authority to operate under parts 121 and/or 135, or operator as defined in §91.147 of this chapter, and
	(iii) Meet the requirements of this subpart as if you were an employer.

(d)(1) To obtain an Antidrug and Alcohol Misuse Prevention Program Operations Specification, you must contact your FAA Principal Operations Inspector or Principal Maintenance Inspector. Provide him/her with the following information:

(i) Company name.

(ii) Certificate number.

(iii) Telephone number.

(iv) Address where your drug and alcohol testing program records are kept.

(v) Whether you have 50 or more covered employees, or 49 or fewer covered employees. (Part 119 certificate holders with authority to operate only under part 121 of this chapter are not required to provide this information.)

(2) You must certify on your Antidrug and Alcohol Misuse Prevention Program Operations Specification, issued by your FAA Principal Operations Inspector or Principal Maintenance Inspector, that you will comply with this part and 49 CFR part 40.

(3) You are required to obtain only one Antidrug and Alcohol Misuse Prevention Program Operations Specification to satisfy this requirement under this part.

(4) You must update the Antidrug and Alcohol Misuse Prevention Program Operations Specification when any changes to the information contained in the Operation Specification occur.

(e)(1) To register with the FAA, submit the following information:

(i) Company name.

(ii) Telephone number.

(iii) Address where your drug and alcohol testing program records are kept.

(iv) Type of safety-sensitive functions you perform for an employer (such as flight instruction duties, aircraft dispatcher duties, maintenance or preventive maintenance duties, ground security coordinator duties, aviation screening duties, air traffic control duties).

(v) Whether you have 50 or more covered employees, or 49 or fewer covered employees.

(vi) A signed statement indicating that: Your company will comply with this part and 49 CFR part 40; and, if you are a contractor, you intend to provide safety-sensitive functions by contract to a part 119 certificate holder with authority to operate under part 121 and/or 135 of this chapter, an operator as defined by § 91.147 of this chapter, or an air traffic control facility not operated by the FAA or by or under contract to the U.S. Military.

(2) Send this information in the form and manner prescribed by the Administrator, in duplicate to the appropriate address below:

(i) For § 91.147 operators: The Flight Standards District Office nearest to your principal place of business.

(ii) For all others: The Federal Aviation Administration, Office of Aerospace Medicine, Drug Abatement Division (AAM-800), 800 Independence Avenue, SW., Washington, DC 20591.

(3) Update the registration information as changes occur. Send the updates in duplicate to the address specified in paragraph (e)(2) of this section.

(4) This registration will satisfy the registration requirements for both your drug testing program under subpart E of this part and your alcohol testing program under this subpart.

[Doc. No. FAA-2008-0937, 74 FR 22653, May 14, 2009; Amdt. 120-0A, 75 FR 3154, Jan. 20, 2010]

§ 120.227 Employees located outside the U.S.

(a) No covered employee shall be tested for alcohol misuse while located outside the territory of the United States.

(1) Each covered employee who is assigned to perform safety-sensitive functions solely outside the territory of the United States shall be removed from the random testing pool upon the inception of such assignment.

(2) Each covered employee who is removed from the random testing pool under this paragraph shall be returned to the random testing pool when the employee resumes the performance of safety-sensitive functions wholly or partially within the territory of the United States.

(b) The provisions of this subpart shall not apply to any person who performs a safety-sensitive function by contract for an employer outside the territory of the United States.

PART 125

CERTIFICATION AND OPERATIONS: AIRPLANES HAVING A SEATING CAPACITY OF 20 OR MORE PASSENGERS OR A MAXIMUM PAYLOAD CAPACITY OF 6,000 POUNDS OR MORE; AND RULES GOVERNING PERSONS ON BOARD SUCH AIRCRAFT

Contents

Appendix E to Part 125—Airplane Flight Recorder Specifications

Authority: 49 U.S.C. 106(g), 40113, 44701-44702, 44705, 44710-44711, 44713, 44716-44717, 44722.

Source: Docket No. 19779, 45 FR 67235, Oct. 9, 1980, unless otherwise noted.

Special Federal Aviation Regulation No. 89

Editorial Note: For the text of SFAR No. 89, see part 121 of this chapter.

Special Federal Aviation Regulation No. 97

Editorial Note: For the text of SFAR No. 97, see part 91 of this chapter.

Special Federal Aviation Regulation No. 106

Editorial Note: For the text of SFAR No. 106, see part 121 of this chapter.

Subpart A—General

§ 125.1 Applicability.

(a) Except as provided in paragraphs (b), (c) and (d) of this section, this part prescribes rules governing the operations of U.S.-registered civil airplanes which have a seating configuration of 20 or more passengers or a maximum payload capacity of 6,000 pounds or more when common carriage is not involved.

(b) The rules of this part do not apply to the operations of airplanes specified in paragraph (a) of this section, when—

(1) They are required to be operated under part 121, 129, 135, or 137 of this chapter;

(2) They have been issued restricted, limited, or provisional airworthiness certificates, special flight permits, or experimental certificates;

(3) They are being operated by a part 125 certificate holder without carrying passengers or cargo under part 91 for training, ferrying, positioning, or maintenance purposes;

(4) They are being operated under part 91 by an operator certificated to operate those airplanes under the rules of parts 121, 135, or 137 of this chapter, they are being operated under the applicable rules of part 121 or part 135 of this chapter by an applicant for a certificate under part 119 of this chapter or they are being operated by a foreign air carrier or a foreign person engaged in common carriage solely outside the United States under part 91 of this chapter;

(5) They are being operated under a deviation authority issued under § 125.3;

(6) They are being operated under part 91, subpart K by a fractional owner as defined in § 91.1001 of this chapter; or

(7) They are being operated by a fractional ownership program manager as defined in § 91.1001 of this chapter, for training, ferrying, positioning, maintenance, or demonstration purposes under part 91 of this chapter and without carrying passengers or cargo for compensation or hire except as permitted for demonstration flights under § 91.501(b)(3) of this chapter.

(c) The rules of this part, except § 125.247, do not apply to the operation of airplanes specified in paragraph (a) when they are operated outside the United States by a person who is not a citizen of the United States.

(d) The provisions of this part apply to each person on board an aircraft being operated under this part, unless otherwise specified.

(e) This part also establishes requirements for operators to take actions to support the continued airworthiness of each airplane.

[Doc. No. 19779, 45 FR 67235, Oct. 9, 1980, as amended by Amdt. 125-4, 47 FR 44719, Oct. 12, 1982; Amdt. 125-5, 49 FR 34816, Sept. 4, 1984; Amdt. 125-6, 51 FR 873, Jan. 8, 1986; Amdt. 125-9, 52 FR 20028, May 28, 1987; Amdt. 121-251, 60 FR 65937, Dec. 20, 1995; Amdt. 125-31, 64 FR 1080, Jan. 7, 1999; Amdt. 125-44, 68 FR 54585, Sept. 17, 2003; Amdt. 125-53, 72 FR 63412, Nov. 8, 2007]

§ 125.3 Deviation authority.

(a) The Administrator may, upon consideration of the circumstances of a particular operation, issue deviation authority providing relief from specified sections of part 125. This deviation authority will be issued as a Letter of Deviation Authority.

(b) A Letter of Deviation Authority may be terminated or amended at any time by the Administrator.

(c) A request for deviation authority must be submitted to the nearest Flight Standards District Office, not less than 60 days prior to the date of intended operations. A request for deviation authority must contain a complete statement of the circumstances and justification for the deviation requested.

(d) After February 2, 2012, no deviation authority from the flight data recorder requirements of this part will be granted. Any previously issued deviation from the flight data recorder requirements of this part is no longer valid.

[Doc. No. 19779, 45 FR 67235, Oct. 9, 1980, as amended by Amdt. 125-13, 54 FR 39294, Sept. 25, 1989; Amdt. 125-56, 73 FR 73179, Dec. 2, 2008]

§ 125.5 Operating certificate and operations specifications required.

(a) After February 3, 1981, no person may engage in operations governed by this part unless that person holds a certificate and operations specification or appropriate deviation authority.

(b) Applicants who file an application before June 1, 1981 shall continue to operate under the rules applicable to their operations on February 2, 1981 until the application for an operating certificate required by this part has been denied or the operating certificate and operations specifications required by this part have been issued.

(c) The rules of this part which apply to a certificate holder also apply to any person who engages in any operation governed by this part without an appropriate certificate and operations specifications required by this part or a Letter of Deviation Authority issued under § 125.3.

[Doc. No. 19779, 45 FR 67235, Oct. 9, 1980, as amended by Amdt. 125-1A, 46 FR 10903, Feb. 5, 1981]

§ 125.7 Display of certificate.

(a) The certificate holder must display a true copy of the certificate in each of its aircraft.

(b) Each operator holding a Letter of Deviation Authority issued under this part must carry a true copy in each of its airplanes.

§ 125.9 Definitions.

(a) For the purposes of this part, *maximum payload capacity* means:

(1) For an airplane for which a maximum zero fuel weight is prescribed in FAA technical specifications, the maximum zero fuel weight, less empty weight, less all justifiable airplane equipment, and less the operating load (consisting of minimum flightcrew, foods and beverages and supplies and equipment related to foods and beverages, but not including disposable fuel or oil):

(2) For all other airplanes, the maximum certificated takeoff weight of an airplane, less the empty weight, less all justifiable airplane equipment, and less the operating load (consisting of minimum fuel load, oil, and flightcrew). The allowance for the weight of the crew, oil, and fuel is as follows:

(i) Crew—200 pounds for each crewmember required under this chapter

(ii) Oil—350 pounds.

(iii) Fuel—the minimum weight of fuel required under this chapter for a flight between domestic points 174 nautical miles apart under VFR weather conditions that does not involve extended overwater operations.

(b) For the purposes of this part, *empty weight* means the weight of the airframe, engines, propellers, and fixed equipment. Empty weight excludes the weight of the crew and payload, but includes the weight of all fixed ballast, unusable fuel supply, undrainable oil, total quantity of engine coolant, and total quantity of hydraulic fluid.

(c) For the purposes of this part, *maximum zero fuel weight* means the maximum permissible weight of an airplane with no disposable fuel or oil. The zero fuel weight figure may be found in either the airplane type certificate data sheet or the approved Airplane Flight Manual, or both.

(d) For the purposes of this section, *justifiable airplane equipment* means any equipment necessary for the operation of the airplane. It does not include equipment or ballast specifically installed, permanently or otherwise, for the purpose of altering the empty weight of an airplane to meet the maximum payload capacity.

§ 125.11 Certificate eligibility and prohibited operations.

(a) No person is eligible for a certificate or operations specifications under this part if the person holds the appropriate operating certificate and/or operations specifications necessary to conduct operations under part 121, 129 or 135 of this chapter.

(b) No certificate holder may conduct any operation which results directly or indirectly from any person's holding out to the public to furnish transportation.

(c) No person holding operations specifications under this part may operate or list on its operations specifications any aircraft listed on any operations specifications or other required aircraft listing under part 121, 129, or 135 of this chapter.

[Doc. No. 19779, 45 FR 67235, Oct. 9, 1980 as amended by Amdt. 125-9, 52 FR 20028, May 28, 1987]

Subpart B—Certification Rules and Miscellaneous Requirements

§ 125.21 Application for operating certificate.

(a) Each applicant for the issuance of an operating certificate must submit an application in a form and manner prescribed by the Administrator to the FAA Flight Standards district office in whose area the applicant proposes to establish or has established its principal operations base. The application must be submitted at least 60 days before the date of intended operations.

(b) Each application submitted under paragraph (a) of this section must contain a signed statement showing the following:

(1) The name and address of each director and each officer or person employed or who will be employed in a management position described in § 125.25.

(2) A list of flight crewmembers with the type of airman certificate held, including ratings and certificate numbers.

§ 125.23 Rules applicable to operations subject to this part.

Each person operating an airplane in operations under this part shall—

(a) While operating inside the United States, comply with the applicable rules in part 91 of this chapter; and

(b) While operating outside the United States, comply with Annex 2, Rules of the Air, to the Convention on International Civil Aviation or the regulations of any foreign country, whichever applies, and with any rules of parts 61 and 91 of this chapter and this part that are more restrictive than that Annex or those regulations and that can be complied with without violating that Annex or those regulations. Annex 2 is incorporated by reference in § 91.703(b) of this chapter.

[Doc. No. 19779, 45 FR 67235, Oct. 9, 1980, as amended by Amdt. 125-12, 54 FR 34331, Aug. 18, 1989]

§ 125.25 Management personnel required.

(a) Each applicant for a certificate under this part must show that it has enough management personnel, including at least a director of operations, to assure that its operations are conducted in accordance with the requirements of this part.

(b) Each applicant shall—

(1) Set forth the duties, responsibilities, and authority of each of its management personnel in the general policy section of its manual;

(2) List in the manual the names and addresses of each of its management personnel;

(3) Designate a person as responsible for the scheduling of inspections required by the manual and for the updating of the approved weight and balance system on all airplanes.

(c) Each certificate holder shall notify the FAA Flight Standards district office charged with the overall inspection of the certificate holder of any change made in the assignment of persons to the listed positions within 10 days, excluding Saturdays, Sundays, and Federal holidays, of such change.

§ 125.26 Employment of former FAA employees.

(a) Except as specified in paragraph (c) of this section, no certificate holder may knowingly employ or make a contractual arrangement which permits an individual to act as an agent or representative of the certificate holder in any matter before the Federal Aviation Administration if the individual, in the preceding 2 years—

(1) Served as, or was directly responsible for the oversight of, a Flight Standards Service aviation safety inspector; and

(2) Had direct responsibility to inspect, or oversee the inspection of, the operations of the certificate holder.

(b) For the purpose of this section, an individual shall be considered to be acting as an agent or representative of a certificate holder in a matter before the agency if the individual makes any written or oral communication on behalf of the certificate holder to the agency (or any of its officers or employees) in connection with a particular matter, whether or not involving a specific party and without regard to whether the individual has participated in, or had responsibility for, the particular matter while serving as a Flight Standards Service aviation safety inspector.

(c) The provisions of this section do not prohibit a certificate holder from knowingly employing or making a contractual arrangement which permits an individual to act as an agent or representative of the certificate holder in any matter before the Federal Aviation Administration if the individual was employed by the certificate holder before October 21, 2011.

[Doc. No. FAA-2008-1154, 76 FR 52235, Aug. 22, 2011]

§ 125.27 Issue of certificate.

(a) An applicant for a certificate under this subpart is entitled to a certificate if the Administrator finds that the applicant is properly and adequately equipped and able to conduct a safe operation in accordance with the requirements of this part and the operations specifications provided for in this part.

(b) The Administrator may deny an application for a certificate under this subpart if the Administrator finds—

(1) That an operating certificate required under this part or part 121, 123, or 135 of this chapter previously issued to the applicant was revoked; or

(2) That a person who was employed in a management position under § 125.25 of this part with (or has exercised control with respect to) any certificate holder under part 121, 123, 125, or 135 of this chapter whose operating certificate has been revoked, will be employed in any of those positions or a similar position with the applicant and that the person's employment or control contributed materially to the reasons for revoking that certificate.

§ 125.29 Duration of certificate.

(a) A certificate issued under this part is effective until surrendered, suspended, or revoked.

(b) The Administrator may suspend or revoke a certificate under section 609 of the Federal Aviation Act of 1958 and the applicable procedures of part 13 of this chapter for any cause that, at the time of suspension or revocation, would have been grounds for denying an application for a certificate.

(c) If the Administrator suspends or revokes a certificate or it is otherwise terminated, the holder of that certificate shall return it to the Administrator.

§ 125.31 Contents of certificate and operations specifications.

(a) Each certificate issued under this part contains the following:

(1) The holder's name.

(2) A description of the operations authorized.

(3) The date it is issued.

(b) The operations specifications issued under this part contain the following:

(1) The kinds of operations authorized.

(2) The types and registration numbers of airplanes authorized for use.

(3) Approval of the provisions of the operator's manual relating to airplane inspections, together with necessary conditions and limitations.

(4) Registration numbers of airplanes that are to be inspected under an approved airplane inspection program under § 125.247.

(5) Procedures for control of weight and balance of airplanes.

(6) Any other item that the Administrator determines is necessary to cover a particular situation.

§ 125.33 Operations specifications not a part of certificate.

Operations specifications are not a part of an operating certificate.

§ 125.35 Amendment of operations specifications.

(a) The FAA Flight Standards district office charged with the overall inspection of the certificate holder may amend any operations specifications issued under this part if—

(1) It determines that safety in air commerce requires that amendment; or

(2) Upon application by the holder, that district office determines that safety in air commerce allows that amendment.

(b) The certificate holder must file an application to amend operations specifications at least 15 days before the date proposed by the applicant for the amendment to become effective, unless a shorter filing period is approved. The application must be on a form and in a manner prescribed by the Administrator and be submitted to the FAA Flight Standards district office charged with the overall inspection of the certificate holder.

(c) Within 30 days after a notice of refusal to approve a holder's application for amendment is received, the holder may petition the Director, Flight Standards Service, to reconsider the refusal to amend.

(d) When the FAA Flight Standards district office charged with the overall inspection of the certificate holder amends operations specifications, that district office gives notice in writing to the holder of a proposed amendment to the operations specifications, fixing a period of not less than 7 days within which the holder may submit written information, views, and arguments concerning the proposed amendment. After consideration of all relevant matter presented, that district office notifies the holder of any amendment adopted, or a rescission of the notice. That amendment becomes effective not less than 30 days after the holder receives notice of the adoption of the amendment, unless the holder petitions the Director, Flight Standards Service, for reconsideration of the amendment. In that case, the effective date of the amendment is stayed pending a decision by the Director. If the Director finds there is an emergency requiring immediate action as to safety in air commerce that makes the provisions of this paragraph impracticable or contrary to the public interest, the Director notifies the certificate holder that the amendment is effective on the date of receipt, without previous notice.

[Doc. No. 19779, 45 FR 67235, Oct. 9, 1980, as amended by Amdt. 125-13, 54 FR 39294, Sept. 25, 1989]

§ 125.37 Duty period limitations.

(a) Each flight crewmember and flight attendant must be relieved from all duty for at least 8 consecutive hours during any 24-hour period.

(b) The Administrator may specify rest, flight time, and duty time limitations in the operations specifications that are other than those specified in paragraph (a) of this section.

[Doc. No. 19779, 45 FR 67235, Oct. 9, 1980, as amended by Amdt. 125-21, 59 FR 42993, Aug. 19, 1994]

§ 125.39 Carriage of narcotic drugs, marihuana, and depressant or stimulant drugs or substances.

If the holder of a certificate issued under this part permits any airplane owned or leased by that holder to be engaged in any operation that the certificate holder knows to be in violation of § 91.19(a) of this chapter, that operation is a basis for suspending or revoking the certificate.

[Doc. No. 19779, 45 FR 67235, Oct. 9, 1980, as amended by Amdt. 125-12, 54 FR 34331, Aug. 18, 1989]

§ 125.41 Availability of certificate and operations specifications.

Each certificate holder shall make its operating certificate and operations specifications available for inspection by the Administrator at its principal operations base.

§ 125.43 Use of operations specifications.

(a) Each certificate holder shall keep each of its employees informed of the provisions of its operations specifications that apply to the employee's duties and responsibilities.

(b) Each certificate holder shall maintain a complete and separate set of its operations specifications. In addition, each certificate holder shall insert pertinent excerpts of its operations specifications, or reference thereto, in its manual in such a manner that they retain their identity as operations specifications.

§ 125.45 Inspection authority.

Each certificate holder shall allow the Administrator, at any time or place, to make any inspections or tests to determine its compliance with the Federal Aviation Act of 1958, the Federal Aviation Regulations, its operating certificate and operations specifications, its letter of deviation authority, or its eligibililty to continue to hold its certificate or its letter of deviation authority.

§ 125.47 Change of address.

Each certificate holder shall notify the FAA Flight Standards district office charged with the overall inspection of its operations, in writing, at least 30 days in advance, of any change in the address of its principal business office, its principal operations base, or its principal maintenance base.

§ 125.49 Airport requirements.

(a) No certificate holder may use any airport unless it is adequate for the proposed operation, considering such items as size, surface, obstructions, and lighting.

(b) No pilot of an airplane carrying passengers at night may take off from, or land on, an airport unless—

(1) That pilot has determined the wind direction from an illuminated wind direction indicator or local ground communications, or, in the case of takeoff, that pilot's personal observations; and

(2) The limits of the area to be used for landing or takeoff are clearly shown by boundary or runway marker lights.

(c) For the purposes of paragraph (b) of this section, if the area to be used for takeoff or landing is marked by flare pots or lanterns, their use must be approved by the Administrator.

§ 125.51 En route navigation facilities.

(a) Except as provided in paragraph (b) of this section, no certificate holder may conduct any operation over a route (including to any destination, refueling or alternate airports) unless suitable navigation aids are available over the route to navigate the airplane along the route within the degree of accuracy required for ATC. Navigation aids required for routes outside of controlled airspace are listed in the certificate holder's operations specifications except for those aids required for routes to alternate airports.

(b) Navigation aids are not required for any of the following operations—

(1) Day VFR operations that the certificate holder shows can be conducted safely by pilotage because of the characteristics of the terrain;

(2) Night VFR operations on routes that the certificate holder shows have reliably lighted landmarks adequate for safe operations; and

(3) Other operations approved by the certificate holding district office.

[Doc. No. FAA-2002-14002, 72 FR 31682, June 7, 2007]

§ 125.53 Flight locating requirements.

(a) Each certificate holder must have procedures established for locating each flight for which an FAA flight plan is not filed that—

(1) Provide the certificate holder with at least the information required to be included in a VFR flight plan;

(2) Provide for timely notification of an FAA facility or search and rescue facility, if an airplane is overdue or missing; and

(3) Provide the certificate holder with the location, date, and estimated time for reestablishing radio or telephone communications, if the flight will operate in an area where communications cannot be maintained.

(b) Flight locating information shall be retained at the certificate holder's principal operations base, or at other places designated by the certificate holder in the flight locating procedures, until the completion of the flight.

(c) Each certificate holder shall furnish the representative of the Administrator assigned to it with a copy of its flight locating procedures and any changes or additions, unless those procedures are included in a manual required under this part.

Subpart C—Manual Requirements

§ 125.71 Preparation.

(a) Each certificate holder shall prepare and keep current a manual setting forth the certificate holder's procedures and policies acceptable to the Administrator. This manual must be used by the certificate holder's flight, ground, and maintenance personnel in conducting its operations. However, the Administrator may authorize a deviation from this paragraph if the Administrator finds that, because of the limited size of the operation, all or part of the manual is not necessary for guidance of flight, ground, or maintenance personnel.

(b) Each certificate holder shall maintain at least one copy of the manual at its principal operations base.

(c) The manual must not be contrary to any applicable Federal regulations, foreign regulation applicable to the certificate holder's operations in foreign countries, or the certificate holder's operating certificate or operations specifications.

(d) A copy of the manual, or appropriate portions of the manual (and changes and additions) shall be made available to maintenance and ground operations personnel by the certificate holder and furnished to—

(1) Its flight crewmembers; and

(2) The FAA Flight Standards district office charged with the overall inspection of its operations.

(e) Each employee of the certificate holder to whom a manual or appropriate portions of it are furnished under paragraph (d)(1) of this section shall keep it up to date with the changes and additions furnished to them.

(f) For the purpose of complying with paragraph (d) of this section, a certificate holder may furnish the persons listed therein with the maintenance part of its manual in printed form or other form, acceptable to the Administrator, that is retrievable in the English language. If the certificate holder furnishes the maintenance part of the manual in other than printed form, it must ensure there is a compatible reading device available to those persons that provides a legible image of the maintenance information and instructions or a system that is able to retrieve the maintenance information and instructions in the English language.

(g) If a certificate holder conducts airplane inspections or maintenance at specified stations where it keeps the approved inspection program manual, it is not required to carry the manual aboard the airplane en route to those stations.

[Doc. No. 19779, 45 FR 67235, Oct. 9, 1980, as amended by Amdt. 125-28, 62 FR 13257, Mar. 19, 1997]

§ 125.73 Contents.

Each manual shall have the date of the last revision and revision number on each revised page. The manual must include—

(a) The name of each management person who is authorized to act for the certificate holder, the person's assigned area of responsibility, and the person's duties, responsibilities, and authority;

(b) Procedures for ensuring compliance with airplane weight and balance limitations;

(c) Copies of the certificate holder's operations specifications or appropriate extracted information, including area of operations authorized, category and class of airplane authorized, crew complements, and types of operations authorized;

(d) Procedures for complying with accident notification requirements;

(e) Procedures for ensuring that the pilot in command knows that required airworthiness inspections have been made and that the airplane has been approved for return to service in compliance with applicable maintenance requirements;

(f) Procedures for reporting and recording mechanical irregularities that come to the attention of the pilot in command before, during, and after completion of a flight;

(g) Procedures to be followed by the pilot in command for determining that mechanical irregularities or defects reported for previous flights have been corrected or that correction has been deferred;

(h) Procedures to be followed by the pilot in command to obtain maintenance, preventive maintenance, and servicing of the airplane at a place where previous arrangements have not been made by the operator, when the pilot is authorized to so act for the operator;

(i) Procedures for the release for, or continuation of, flight if any item of equipment required for the particular type of operation becomes inoperative or unserviceable en route;

(j) Procedures for refueling airplanes, eliminating fuel contamination, protecting from fire (including electrostatic protection), and supervising and protecting passengers during refueling;

(k) Procedures to be followed by the pilot in command in the briefing under § 125.327;

(l) Flight locating procedures, when applicable;

(m) Procedures for ensuring compliance with emergency procedures, including a list of the functions assigned each category of required crewmembers in connection with an emergency and emergency evacuation;

(n) The approved airplane inspection program;

(o) Procedures and instructions to enable personnel to recognize hazardous materials, as defined in title 49 CFR, and if these materials are to be carried, stored, or handled, procedures and instructions for—

(1) Accepting shipment of hazardous material required by title 49 CFR, to assure proper packaging, marking, labeling, shipping documents, compatibility of articles, and instructions on their loading, storage, and handling;

(2) Notification and reporting hazardous material incidents as required by title 49 CFR; and

(3) Notification of the pilot in command when there are hazardous materials aboard, as required by title 49 CFR;

(p) Procedures for the evacuation of persons who may need the assistance of another person to move expeditiously to an exit if an emergency occurs;

(q) The identity of each person who will administer tests required by this part, including the designation of the tests authorized to be given by the person; and

(r) Other procedures and policy instructions regarding the certificate holder's operations that are issued by the certificate holder.

§ 125.75 Airplane flight manual.

(a) Each certificate holder shall keep a current approved Airplane Flight Manual or approved equivalent for each type airplane that it operates.

(b) Each certificate holder shall carry the approved Airplane Flight Manual or the approved equivalent aboard each airplane it operates. A certificate holder may elect to carry a combination of the manuals required by this section and § 125.71. If it so elects, the certificate holder may revise the operating procedures sections and modify the presentation of performance from the applicable Airplane Flight Manual if the revised operating procedures and modified performance data presentation are approved by the Administrator.

Subpart D—Airplane Requirements

§ 125.91 Airplane requirements: General.

(a) No certificate holder may operate an airplane governed by this part unless it—

(1) Carries an appropriate current airworthiness certificate issued under this chapter; and

(2) Is in an airworthy condition and meets the applicable airworthiness requirements of this chapter, including those relating to identification and equipment.

(b) No person may operate an airplane unless the current empty weight and center of gravity are calculated from the values established by actual weighing of the airplane within the preceding 36 calendar months.

(c) Paragraph (b) of this section does not apply to airplanes issued an original airworthiness certificate within the preceding 36 calendar months.

§ 125.93 Airplane limitations.

No certificate holder may operate a land airplane (other than a DC-3, C-46, CV-240, CV-340, CV-440, CV-580, CV-600, CV-640, or Martin 404) in an extended overwater operation unless it is certificated or approved as adequate for ditching under the ditching provisions of part 25 of this chapter.

Subpart E—Special Airworthiness Requirements

§ 125.111 General.

(a) Except as provided in paragraph (b) of this section, no certificate holder may use an airplane powered by airplane engines rated at more than 600 horsepower each for maximum continuous operation unless that airplane meets the requirements of §§ 125.113 through 125.181.

(b) If the Administrator determines that, for a particular model of airplane used in cargo service, literal compliance with any requirement under paragraph (a) of this section would be extremely difficult and that compliance would not contribute materially to the objective sought, the Administrator may require compliance with only those requirements that are necessary to accomplish the basic objectives of this part.

(c) This section does not apply to any airplane certificated under—

(1) Part 4b of the Civil Air Regulations in effect after October 31, 1946;

(2) Part 25 of this chapter; or

(3) Special Civil Air Regulation 422, 422A, or 422B.

§ 125.113 Cabin interiors.

(a) Upon the first major overhaul of an airplane cabin or refurbishing of the cabin interior, all materials in each compartment used by the crew or passengers that do not meet the following requirements must be replaced with materials that meet these requirements:

(1) For an airplane for which the application for the type certificate was filed prior to May 1, 1972, § 25.853 in effect on April 30, 1972.

(2) For an airplane for which the application for the type certificate was filed on or after May 1, 1972, the materials requirement under which the airplane was type certificated.

(b) Except as provided in paragraph (a) of this section, each compartment used by the crew or passengers must meet the following requirements:

(1) Materials must be at least flash resistant.

(2) The wall and ceiling linings and the covering of upholstering, floors, and furnishings must be flame resistant.

(3) Each compartment where smoking is to be allowed must be equipped with self-contained ash trays that are completely removable and other compartments must be placarded against smoking.

(4) Each receptacle for used towels, papers, and wastes must be of fire-resistant material and must have a cover or other means of containing possible fires started in the receptacles.

(c) Thermal/acoustic insulation materials. For transport category airplanes type certificated after January 1, 1958:

(1) For airplanes manufactured before September 2, 2005, when thermal/acoustic insulation is installed in the fuselage as replacements after September 2, 2005, the insulation must meet the flame propagation requirements of § 25.856 of this chapter, effective September 2, 2003, if it is:

(i) of a blanket construction or

(ii) Installed around air ducting.

(2) For airplanes manufactured after September 2, 2005, thermal/acoustic insulation materials installed in the fuselage must meet the flame propagation requirements of § 25.856 of this chapter, effective September 2, 2003.

[Doc. No. 19799, 45 FR 67235, Oct. 9, 1980, as amended by Amdt. 125-43, 68 FR 45084, July 31, 2003; Amdt. 125-50, 70 FR 77752, Dec. 30, 2005]

§ 125.115 Internal doors.

In any case where internal doors are equipped with louvres or other ventilating means, there must be a means convenient to the crew for closing the flow of air through the door when necessary.

§ 125.117 Ventilation.

Each passenger or crew compartment must be suitably ventilated. Carbon monoxide concentration may not be more than one part in 20,000 parts of air, and fuel fumes may not be present. In any case where partitions between compartments have louvres or other means allowing air to flow between compartments, there must be a means convenient to the crew for closing the flow of air through the partitions when necessary.

§ 125.119 Fire precautions.

(a) Each compartment must be designed so that, when used for storing cargo or baggage, it meets the following requirements:

(1) No compartment may include controls, wiring, lines, equipment, or accessories that would upon damage or failure, affect the safe operation of the airplane unless the item is adequately shielded, isolated, or otherwise protected so that it cannot be damaged by movement of cargo in the compartment and so that damage to or failure of the item would not create a fire hazard in the compartment.

(2) Cargo or baggage may not interfere with the functioning of the fire-protective features of the compartment.

(3) Materials used in the construction of the compartments, including tie-down equipment, must be at least flame resistant.

(4) Each compartment must include provisions for safeguarding against fires according to the classifications set forth in paragraphs (b) through (f) of this section.

(b) *Class A.* Cargo and baggage compartments are classified in the "A" category if a fire therein would be readily discernible to a member of the crew while at that crewmember's station, and all parts of the compartment are easily accessible in flight. There must be a hand fire extinguisher available for each Class A compartment.

(c) *Class B.* Cargo and baggage compartments are classified in the "B" category if enough access is provided while in flight to enable a member of the crew to effectively reach all of the compartment and its contents with a hand fire extinguisher and the compartment is so designed that, when the access provisions are being used, no hazardous amount of smoke, flames, or extinguishing agent enters any compartment occupied by the crew or passengers. Each Class B compartment must comply with the following:

(1) It must have a separate approved smoke or fire detector system to give warning at the pilot or flight engineer station.

(2) There must be a hand-held fire extinguisher available for the compartment.

(3) It must be lined with fire-resistant material, except that additional service lining of flame-resistant material may be used.

(d) *Class C.* Cargo and baggage compartments are classified in the "C" category if they do not conform with the requirements for the "A", "B", "D", or "E" categories. Each Class C compartment must comply with the following:

(1) It must have a separate approved smoke or fire detector system to give warning at the pilot or flight engineer station.

(2) It must have an approved built-in fire-extinguishing system controlled from the pilot or flight engineer station.

(3) It must be designed to exclude hazardous quantities of smoke, flames, or extinguishing agents from entering into any compartment occupied by the crew or passengers.

(4) It must have ventilation and draft control so that the extinguishing agent provided can control any fire that may start in the compartment.

(5) It must be lined with fire-resistant material, except that additional service lining of flame-resistant material may be used.

(e) *Class D.* Cargo and baggage compartments are classified in the "D" category if they are so designed and constructed that a fire occurring therein will be completely confined without endangering the safety of the airplane or the occupants. Each Class D compartment must comply with the following:

(1) It must have a means to exclude hazardous quantities of smoke, flames, or noxious gases from entering any compartment occupied by the crew or passengers.

(2) Ventilation and drafts must be controlled within each compartment so that any fire likely to occur in the compartment will not progress beyond safe limits.

(3) It must be completely lined with fire-resistant material.

(4) Consideration must be given to the effect of heat within the compartment on adjacent critical parts of the airplane.

(f) *Class E.* On airplanes used for the carriage of cargo only, the cabin area may be classified as a Class "E" compartment. Each Class E compartment must comply with the following:

(1) It must be completely lined with fire-resistant material.

(2) It must have a separate system of an approved type smoke or fire detector to give warning at the pilot or flight engineer station.

(3) It must have a means to shut off the ventilating air flow to or within the compartment and the controls for that means must be accessible to the flightcrew in the crew compartment.

(4) It must have a means to exclude hazardous quantities of smoke, flames, or noxious gases from entering the flightcrew compartment.

(5) Required crew emergency exits must be accessible under all cargo loading conditions.

§ 125.121 Proof of compliance with § 125.119.

Compliance with those provisions of § 125.119 that refer to compartment accessibility, the entry of hazardous quantities of smoke or extinguishing agent into compartment occupied by the crew or passengers, and the dissipation of the extinguishing agent in Class "C" compartments must be shown by tests in flight. During these tests it must be shown that no inadvertent operation of smoke or fire detectors in other compartments within the airplane would occur as a result of fire contained in any one compartment, either during the time it is being extinguished, or thereafter, unless the extinguishing system floods those compartments simultaneously.

§ 125.123 Propeller deicing fluid.

If combustible fluid is used for propeller deicing, the certificate holder must comply with § 125.153.

§ 125.125 Pressure cross-feed arrangements.

(a) Pressure cross-feed lines may not pass through parts of the airplane used for carrying persons or cargo unless there is a means to allow crewmembers to shut off the supply of fuel to these lines or the lines are enclosed in a fuel and fume-proof enclosure that is ventilated and drained to the exterior of the airplane. However, such an enclosure need not be used if those lines incorporate no fittings on or within the personnel or cargo areas and are suitably routed or protected to prevent accidental damage.

(b) Lines that can be isolated from the rest of the fuel system by valves at each end must incorporate provisions for relieving excessive pressures that may result from exposure of the isolated line to high temperatures.

§ 125.127 Location of fuel tanks.

(a) Fuel tanks must be located in accordance with § 125.153.

(b) No part of the engine nacelle skin that lies immediately behind a major air outlet from the engine compartment may be used as the wall of an integral tank.

(c) Fuel tanks must be isolated from personnel compartments by means of fume- and fuel-proof enclosures.

§ 125.129 Fuel system lines and fittings.

(a) Fuel lines must be installed and supported so as to prevent excessive vibration and so as to be adequate to withstand loads due to fuel pressure and accelerated flight conditions.

(b) Lines connected to components of the airplane between which there may be relative motion must incorporate provisions for flexibility.

(c) Flexible connections in lines that may be under pressure and subject to axial loading must use flexible hose assemblies rather than hose clamp connections.

(d) Flexible hoses must be of an acceptable type or proven suitable for the particular application.

§ 125.131 Fuel lines and fittings in designated fire zones.

Fuel lines and fittings in each designated fire zone must comply with § 125.157.

§ 125.133 Fuel valves.

Each fuel valve must—

(a) Comply with § 125.155;

(b) Have positive stops or suitable index provisions in the "on" and "off" positions; and

(c) Be supported so that loads resulting from its operation or from accelerated flight conditions are not transmitted to the lines connected to the valve.

§ 125.135 Oil lines and fittings in designated fire zones.

Oil lines and fittings in each designated fire zone must comply with § 125.157.

§ 125.137 Oil valves.

(a) Each oil valve must—

(1) Comply with § 125.155;

(2) Have positive stops or suitable index provisions in the "on" and "off" positions; and

(3) Be supported so that loads resulting from its operation or from accelerated flight conditions are not transmitted to the lines attached to the valve.

(b) The closing of an oil shutoff means must not prevent feathering the propeller, unless equivalent safety provisions are incorporated.

§ 125.139 Oil system drains.

Accessible drains incorporating either a manual or automatic means for positive locking in the closed position must be provided to allow safe drainage of the entire oil system.

§ 125.141 Engine breather lines.

(a) Engine breather lines must be so arranged that condensed water vapor that may freeze and obstruct the line cannot accumulate at any point.

(b) Engine breathers must discharge in a location that does not constitute a fire hazard in case foaming occurs and so that oil emitted from the line does not impinge upon the pilots' windshield.

(c) Engine breathers may not discharge into the engine air induction system.

§ 125.143 Firewalls.

Each engine, auxiliary power unit, fuel-burning heater, or other item of combusting equipment that is intended for operation in flight must be isolated from the rest of the airplane by means of firewalls or shrouds, or by other equivalent means.

§ 125.145 Firewall construction.

Each firewall and shroud must—

(a) Be so made that no hazardous quantity of air, fluids, or flame can pass from the engine compartment to other parts of the airplane;

(b) Have all openings in the firewall or shroud sealed with close-fitting fireproof grommets, bushings, or firewall fittings;

(c) Be made of fireproof material; and

(d) Be protected against corrosion.

§ 125.147 Cowling.

(a) Cowling must be made and supported so as to resist the vibration, inertia, and air loads to which it may be normally subjected.

(b) Provisions must be made to allow rapid and complete drainage of the cowling in normal ground and flight attitudes. Drains must not discharge in locations constituting a fire hazard. Parts of the cowling that are subjected to high temperatures because they are near exhaust system parts or because of exhaust gas impingement must be made of fireproof material. Unless otherwise specified in these regulations, all other parts of the cowling must be made of material that is at least fire resistant.

§ 125.149 Engine accessory section diaphragm.

Unless equivalent protection can be shown by other means, a diaphragm that complies with § 125.145 must be provided on air-cooled engines to isolate the engine power section and all parts of the exhaust system from the engine accessory compartment.

§ 125.151 Powerplant fire protection.

(a) Designated fire zones must be protected from fire by compliance with §§ 125.153 through 125.159.

(b) Designated fire zones are—

(1) Engine accessory sections;

(2) Installations where no isolation is provided between the engine and accessory compartment; and

(3) Areas that contain auxiliary power units, fuel-burning heaters, and other combustion equipment.

§ 125.153 Flammable fluids.

(a) No tanks or reservoirs that are a part of a system containing flammable fluids or gases may be located in designated fire zones, except where the fluid contained, the design of the system, the materials used in the tank, the shutoff means, and the connections, lines, and controls provide equivalent safety.

(b) At least one-half inch of clear airspace must be provided between any tank or reservior and a firewall or shroud isolating a designated fire zone.

§ 125.155 Shutoff means.

(a) Each engine must have a means for shutting off or otherwise preventing hazardous amounts of fuel, oil, deicer, and other flammable fluids from flowing into, within, or through any designated fire zone. However, means need not be provided to shut off flow in lines that are an integral part of an engine.

(b) The shutoff means must allow an emergency operating sequence that is compatible with the emergency operation of other equipment, such as feathering the propeller, to facilitate rapid and effective control of fires.

(c) Shutoff means must be located outside of designated fire zones, unless equivalent safety is provided, and it must be shown that no hazardous amount of flammable fluid will drain into any designated fire zone after a shutoff.

(d) Adequate provisions must be made to guard against inadvertent operation of the shutoff means and to make it possible for the crew to reopen the shutoff means after it has been closed.

§ 125.157 Lines and fittings.

(a) Each line, and its fittings, that is located in a designated fire zone, if it carries flammable fluids or gases under pressure, or is attached directly to the engine, or is subject to relative motion between components (except lines and fittings forming an integral part of the engine), must be flexible and fire-resistant with fire-resistant, factory-fixed, detachable, or other approved fire-resistant ends.

(b) Lines and fittings that are not subject to pressure or to relative motion between components must be of fire-resistant materials.

§ 125.159 Vent and drain lines.

All vent and drain lines, and their fittings, that are located in a designated fire zone must, if they carry flammable fluids or gases, comply with § 125.157, if the Administrator finds that the rupture or breakage of any vent or drain line may result in a fire hazard.

§ 125.161 Fire-extinguishing systems.

(a) Unless the certificate holder shows that equivalent protection against destruction of the airplane in case of fire is provided by the use of fireproof materials in the nacelle and other components that would be subjected to flame, fire-extinguishing systems must be provided to serve all designated fire zones.

(b) Materials in the fire-extinguishing system must not react chemically with the extinguishing agent so as to be a hazard.

§ 125.163 Fire-extinguishing agents.

Only methyl bromide, carbon dioxide, or another agent that has been shown to provide equivalent extinguishing action may be used as a fire-extinguishing agent. If methyl bromide or any other toxic extinguishing agent is used, provisions must be made to prevent harmful concentrations of fluid or fluid vapors from entering any personnel compartment either because of leakage during normal operation of the airplane or because of discharging the fire extinguisher on the ground or in flight when there is a defect in the extinguishing system. If a methyl bromide system is used, the containers must be charged with dry agent and sealed by the fire-extinguisher manufacturer or some other person using satisfactory recharging equipment. If carbon dioxide is used, it must not be possible to discharge enough gas into the personnel compartments to create a danger of suffocating the occupants.

§ 125.165 Extinguishing agent container pressure relief.

Extinguishing agent containers must be provided with a pressure relief to prevent bursting of the container because of excessive internal pressures. The discharge line from the relief connection must terminate outside the airplane in a place convenient for inspection on the ground. An indicator must be provided at the discharge end of the line to provide a visual indication when the container has discharged.

§ 125.167 Extinguishing agent container compartment temperature.

Precautions must be taken to ensure that the extinguishing agent containers are installed in places where reasonable temperatures can be maintained for effective use of the extinguishing system.

§ 125.169 Fire-extinguishing system materials.

(a) Except as provided in paragraph (b) of this section, each component of a fire-extinguishing system that is in a designated fire zone must be made of fireproof materials.

(b) Connections that are subject to relative motion between components of the airplane must be made of flexible materials that are at least fire-resistant and be located so as to minimize the probability of failure.

§ 125.171 Fire-detector systems.

Enough quick-acting fire detectors must be provided in each designated fire zone to assure the detection of any fire that may occur in that zone.

§ 125.173 Fire detectors.

Fire detectors must be made and installed in a manner that assures their ability to resist, without failure, all vibration, inertia, and other loads to which they may be normally subjected. Fire detectors must be unaffected by exposure to fumes, oil, water, or other fluids that may be present.

§ 125.175 Protection of other airplane components against fire.

(a) Except as provided in paragraph (b) of this section, all airplane surfaces aft of the nacelles in the area of one nacelle diameter on both sides of the nacelle centerline must be made of material that is at least fire resistant.

(b) Paragraph (a) of this section does not apply to tail surfaces lying behind nacelles unless the dimensional configuration of the airplane is such that the tail surfaces could be affected readily by heat, flames, or sparks emanating from a designated fire zone or from the engine from a designated fire zone or from the engine compartment of any nacelle.

§ 125.177 Control of engine rotation.

(a) Except as provided in paragraph (b) of this section, each airplane must have a means of individually stopping and restarting the rotation of any engine in flight.

(b) In the case of turbine engine installations, a means of stopping rotation need be provided only if the Administrator finds that rotation could jeopardize the safety of the airplane.

§ 125.179 Fuel system independence.

(a) Each airplane fuel system must be arranged so that the failure of any one component does not result in the irrecoverable loss of power of more than one engine.

(b) A separate fuel tank need not be provided for each engine if the certificate holder shows that the fuel system incorporates features that provide equivalent safety.

§ 125.181 Induction system ice prevention.

A means for preventing the malfunctioning of each engine due to ice accumulation in the engine air induction system must be provided for each airplane.

§ 125.183 Carriage of cargo in passenger compartments.

(a) Except as provided in paragraph (b) or (c) of this section, no certificate holder may carry cargo in the passenger compartment of an airplane.

(b) Cargo may be carried aft of the foremost seated passengers if it is carried in an approved cargo bin that meets the following requirements:

(1) The bin must withstand the load factors and emergency landing conditions applicable to the passenger seats of the airplane in which the bin is installed, multiplied by a factor of 1.15, using the combined weight of the bin and the maximum weight of cargo that may be carried in the bin.

(2) The maximum weight of cargo that the bin is approved to carry and any instructions necessary to ensure proper weight distribution within the bin must be conspicuously marked on the bin.

(3) The bin may not impose any load on the floor or other structure of the airplane that exceeds the load limitations of that structure.

(4) The bin must be attached to the seat tracks or to the floor structure of the airplane, and its attachment must withstand the load factors and emergency landing conditions applicable to the passenger seats of the airplane in which the bin is installed, multiplied by either the factor 1.15 or the seat attachment factor specified for the airplane, whichever is greater, using the combined weight of the bin and the maximum weight of cargo that may be carried in the bin.

(5) The bin may not be installed in a position that restricts access to or use of any required emergency exit, or of the aisle in the passenger compartment.

(6) The bin must be fully enclosed and made of material that is at least flame-resistant.

(7) Suitable safeguards must be provided within the bin to prevent the cargo from shifting under emergency landing conditions.

(8) The bin may not be installed in a position that obscures any passenger's view of the "seat belt" sign, "no smoking" sign, or any required exit sign, unless an auxiliary sign or other approved means for proper notification of the passenger is provided.

(c) All cargo may be carried forward of the foremost seated passengers and carry-on baggage may be carried alongside the foremost seated passengers if the cargo (including carry-on baggage) is carried either in approved bins as specified in paragraph (b) of this section or in accordance with the following:

(1) It is properly secured by a safety belt or other tie down having enough strength to eliminate the possibility of shifting under all normally anticipated flight and ground conditions.

(2) It is packaged or covered in a manner to avoid possible injury to passengers.

(3) It does not impose any load on seats or the floor structure that exceeds the load limitation for those components.

(4) Its location does not restrict access to or use of any required emergency or regular exit, or of the aisle in the passenger compartment.

(5) Its location does not obscure any passenger's view of the "seat belt" sign, "no smoking" sign, or required exit sign, unless an auxiliary sign or other approved means for proper notification of the passenger is provided.

§ 125.185 Carriage of cargo in cargo compartments.

When cargo is carried in cargo compartments that are designed to require the physical entry of a crewmember to extinguish any fire that may occur during flight, the cargo must be loaded so as to allow a crewmember to effectively reach all parts of the compartment with the contents of a hand-held fire extinguisher.

§ 125.187 Landing gear: Aural warning device.

(a) Except for airplanes that comply with the requirements of § 25.729 of this chapter on or after January 6, 1992, each airplane must have a landing gear aural warning device that functions continuously under the following conditions:

(1) For airplanes with an established approach wing-flap position, whenever the wing flaps are extended beyond the maximum certificated approach climb configuration position in the Airplane Flight Manual and the landing gear is not fully extended and locked.

(2) For airplanes without an established approach climb wing-flap position, whenever the wing flaps are extended beyond the position at which landing gear extension is normally performed and the landing gear is not fully extended and locked.

(b) The warning system required by paragraph (a) of this section—

(1) May not have a manual shutoff;

(2) Must be in addition to the throttle-actuated device installed under the type certification airworthiness requirements; and

(3) May utilize any part of the throttle-actuated system including the aural warning device.

(c) The flap position sensing unit may be installed at any suitable place in the airplane.

[Doc. No. 19779, 45 FR 67235, Oct. 9, 1980, as amended by Amdt. 125-16, 56 FR 63762, Dec. 5, 1991]

§ 125.189 Demonstration of emergency evacuation procedures.

(a) Each certificate holder must show, by actual demonstration conducted in accordance with paragraph (a) of appendix B of this part, that the emergency evacuation procedures for each type and model of airplane with a seating of more than 44 passengers, that is used in its passenger-carrying operations, allow the evacuation of the full seating capacity, including crewmembers, in 90 seconds or less, in each of the following circumstances:

(1) A demonstration must be conducted by the certificate holder upon the initial introduction of a type and model of airplane into passenger-carrying operations. However, the demonstration need not be repeated for any airplane type or model that has the same number and type of exits, the same cabin configuration, and the same emergency equipment as any other airplane used by the certificate holder in successfully demonstrating emergency evacuation in compliance with this paragraph.

(2) A demonstration must be conducted—

(i) Upon increasing by more than 5 percent the passenger seating capacity for which successful demonstration has been conducted; or

(ii) Upon a major change in the passenger cabin interior configuration that will affect the emergency evacuation of passengers.

(b) If a certificate holder has conducted a successful demonstration required by § 121.291(a) in the same type airplane as a part 121 or part 123 certificate holder, it need not conduct a demonstration under this paragraph in that type airplane to achieve certification under part 125.

(c) Each certificate holder operating or proposing to operate one or more landplanes in extended overwater operations, or otherwise required to have certain equipment under § 125.209, must show, by a simulated ditching conducted in accordance with paragraph (b) of appendix B of this part, that it has the ability to efficiently carry out its ditching procedures.

(d) If a certificate holder has conducted a successful demonstration required by § 121.291(b) in the same type airplane as a part 121 or part 123 certificate holder, it need not conduct a demonstration under this paragraph in that type airplane to achieve certification under part 125.

Subpart F—Instrument and Equipment Requirements

§ 125.201 Inoperable instruments and equipment.

(a) No person may take off an airplane with inoperable instruments or equipment installed unless the following conditions are met:

(1) An approved Minimum Equipment List exists for that airplane.

(2) The Flight Standards District Office having certification responsibility has issued the certificate holder operations specifications authorizing operations in accordance with an approved Minimum Equipment List. The flight crew shall have direct access at all times prior to flight to

all of the information contained in the approved Minimum Equipment List through printed or other means approved by the Administrator in the certificate holders operations specifications. An approved Minimum Equipment List, as authorized by the operations specifications, constitutes an approved change to the type design without requiring recertification.

(3) The approved Minimum Equipment List must:

(i) Be prepared in accordance with the limitations specified in paragraph (b) of this section.

(ii) Provide for the operation of the airplane with certain instruments and equipment in an inoperable condition.

(4) Records identifying the inoperable instruments and equipment and the information required by paragraph (a)(3)(ii) of this section must be available to the pilot.

(5) The airplane is operated under all applicable conditions and limitations contained in the Minimum Equipment List and the operations specifications authorizing use of the Minimum Equipment List.

(b) The following instruments and equipment may not be included in the Minimum Equipment List:

(1) Instruments and equipment that are either specifically or otherwise required by the airworthiness requirements under which the airplane is type certificated and which are essential for safe operations under all operating conditions.

(2) Instruments and equipment required by an airworthiness directive to be in operable condition unless the airworthiness directive provides otherwise.

(3) Instruments and equipment required for specific operations by this part.

(c) Notwithstanding paragraphs (b)(1) and (b)(3) of this section, an airplane with inoperable instruments or equipment may be operated under a special flight permit under §§ 21.197 and 21.199 of this chapter.

[Doc. No. 25780, 56 FR 12310, Mar. 22, 1991]

§ 125.203 Communication and navigation equipment.

(a) *Communication equipment—general.* No person may operate an airplane unless it has two-way radio communication equipment able, at least in flight, to transmit to, and receive from, appropriate facilities 22 nautical miles away.

(b) *Navigation equipment for operations over the top.* No person may operate an airplane over the top unless it has navigation equipment suitable for the route to be flown.

(c) *Communication and navigation equipment for IFR or extended over-water operations—General.* Except as provided in paragraph (f) of this section, no person may operate an airplane carrying passengers under IFR or in extended over-water operations unless—

(1) The en route navigation aids necessary for navigating the airplane along the route (e.g., ATS routes, arrival and departure routes, and instrument approach procedures, including missed approach procedures if a missed approach routing is specified in the procedure) are available and suitable for use by the aircraft navigation systems required by this section;

(2) The airplane used in those operations is equipped with at least the following equipment—

(i) Except as provided in paragraph (d) of this section, two approved independent navigation systems suitable for navigating the airplane along the route within the degree of accuracy required for ATC;

(ii) One marker beacon receiver providing visual and aural signals;

(iii) One ILS receiver;

(iv) Two transmitters;

(v) Two microphones;

(vi) Two headsets or one headset and one speaker; and

(vii) Two independent communication systems, one of which must have two-way voice communication capability, capable of transmitting to, and receiving from, at least one appropriate facility from any place on the route to be flown; and

(3) Any RNAV system used to meet the navigation equipment requirements of this section is authorized in the certificate holder's operations specifications.

(d) *Use of a single independent navigation system for operations under IFR—not for extended overwater operations.* Notwithstanding the requirements of paragraph (c)(2)(i) of this section, the airplane may be equipped with a single independent navigation system suitable for navigating the airplane along the route to be flown within the degree of accuracy required for ATC if—

(1) It can be shown that the airplane is equipped with at least one other independent navigation system suitable, in the event of loss of the navigation capability of the single independent navigation system permitted by this paragraph at any point along the route, for proceeding safely to a suitable airport and completing an instrument approach; and

(2) The airplane has sufficient fuel so that the flight may proceed safely to a suitable airport by use of the remaining navigation system, and complete an instrument approach and land.

(e) *Use of VOR navigation equipment.* If VOR navigation equipment is required by paragraph (c) or (d) of this section, no person may operate an airplane unless it is equipped with at least one approved DME or a suitable RNAV system.

(f) *Extended over-water operations.* Notwithstanding the requirements of paragraph (c) of this section, installation and use of a single long-range navigation system and a single long-range communication system for extended over-water operations in certain geographic areas may be authorized by the Administrator and approved in the certificate holder's operations specifications. The following are among the operational factors the Administrator may consider in granting an authorization:

(1) The ability of the flight crew to navigate the airplane along the route to be flown within the degree of accuracy required for ATC;

(2) The length of the route being flown; and

(3) The duration of the very high frequency communications gap.

[Doc. No. FAA-2002-14002, 72 FR 31682, June 7, 2007]

§ 125.204 Portable electronic devices.

(a) Except as provided in paragraph (b) of this section, no person may operate, nor may any operator or pilot in command of an aircraft allow the operation of, any portable electronic device on any U.S.-registered civil aircraft operating under this part.

(b) Paragraph (a) of this section does not apply to—

(1) Portable voice recorders;

(2) Hearing aids;

(3) Heart pacemakers;

(4) Electric shavers; or

(5) Any other portable electronic device that the Part 125 certificate holder has determined will not cause interference with the navigation or communication system of the aircraft on which it is to be used.

(c) The determination required by paragraph (b)(5) of this section shall be made by that Part 125 certificate holder operating the particular device to be used.

[Doc. No. FAA-1998-4954, 64 FR 1080, Jan. 7, 1999]

§ 125.205 Equipment requirements: Airplanes under IFR.

No person may operate an airplane under IFR unless it has—

(a) A vertical speed indicator;

(b) A free-air temperature indicator;

(c) A heated pitot tube for each airspeed indicator;

(d) A power failure warning device or vacuum indicator to show the power available for gyroscopic instruments from each power source;

(e) An alternate source of static pressure for the altimeter and the airspeed and vertical speed indicators;

(f) At least two generators each of which is on a separate engine, or which any combination of one-half of the total number are rated sufficiently to supply the electrical loads of all required instruments and equipment necessary for safe emergency operation of the airplane; and

(g) Two independent sources of energy (with means of selecting either), of which at least one is an engine-driven pump or generator, each of which is able to drive all gyroscopic instruments and installed so that failure of one instrument or source does not interfere with the energy supply to the remaining instruments or the other energy source. For the purposes of this paragraph, each engine-driven source of energy must be on a different engine.

(h) For the purposes of paragraph (f) of this section, a continuous inflight electrical load includes one that draws current continuously during flight, such as radio equipment, electrically driven instruments, and lights, but does not include occasional intermittent loads.

(i) An airspeed indicating system with heated pitot tube or equivalent means for preventing malfunctioning due to icing.

(j) A sensitive altimeter.

(k) Instrument lights providing enough light to make each required instrument, switch, or similar instrument easily readable and installed so that the direct rays are shielded from the flight crewmembers' eyes and that no objectionable reflections are visible to them. There must be a means of controlling the intensity of illumination unless it is shown that nondimming instrument lights are satisfactory.

§ 125.206 Pitot heat indication systems.

(a) Except as provided in paragraph (b) of this section, after April 12, 1981, no person may operate a transport category airplane equipped with a flight instrument pitot heating system unless the airplane is equipped with an operable pitot heat indication system that complies with § 25.1326 of this chapter in effect on April 12, 1978.

(b) A certificate holder may obtain an extension of the April 12, 1981, compliance date specified in paragraph (a) of this section, but not beyond April 12, 1983, from the Director, Flight Standards Service if the certificate holder—

(1) Shows that due to circumstances beyond its control it cannot comply by the specified compliance date; and

(2) Submits by the specified compliance date a schedule for compliance acceptable to the Director, indicating that compliance will be achieved at the earliest practicable date.

[Doc. No. 18904, 46 FR 43806, Aug. 31, 1981, as amended by Amdt. 125-13, 54 FR 39294, Sept. 25, 1989]

§ 125.207 Emergency equipment requirements.

(a) No person may operate an airplane having a seating capacity of 20 or more passengers unless it is equipped with the following emergency equipment:

(1) One approved first aid kit for treatment of injuries likely to occur in flight or in a minor accident, which meets the following specifications and requirements:

(i) Each first aid kit must be dust and moisture proof and contain only materials that either meet Federal Specifications GGK-391a, as revised, or as approved by the Administrator.

(ii) Required first aid kits must be readily accessible to the cabin flight attendants.

(iii) Except as provided in paragraph (a)(1)(iv) of this section, at time of takeoff, each first aid kit must contain at least the following or other contents approved by the Administrator:

Contents	Quantity
Adhesive bandage compressors, 1 in	16
Antiseptic swabs	20
Ammonia inhalants	10
Bandage compressors, 4 in	8
Triangular bandage compressors, 40 in	5
Arm splint, noninflatable	1
Leg splint, noninflatable	1
Roller bandage, 4 in	4
Adhesive tape, 1-in standard roll	2
Bandage scissors	1
Protective latex gloves or equivalent nonpermeable gloves	[1]1

[1] Pair.

(iv) Protective latex gloves or equivalent nonpermeable gloves may be placed in the first aid kit or in a location that is readily accessible to crewmembers.

(2) A crash axe carried so as to be accessible to the crew but inaccessible to passengers during normal operations.

(3) Signs that are visible to all occupants to notify them when smoking is prohibited and when safety belts should be fastened. The signs must be so constructed that they can be turned on and off by a crewmember. They must be turned on for each takeoff and each landing and when otherwise considered to be necessary by the pilot in command.

(4) The additional emergency equipment specified in appendix A of this part.

(b) *Megaphones.* Each passenger-carrying airplane must have a portable battery-powered megaphone or megaphones readily accessible to the crewmembers assigned to direct emergency evacuation, installed as follows:

(1) One megaphone on each airplane with a seating capacity of more than 60 and less than 100 passengers, at the most rearward location in the passenger cabin where it would be readily accessible to a normal flight attendant seat. However, the Administrator may grant a deviation from the requirements of this paragraph if the Administrator finds that a different location would be more useful for evacuation of persons during an emergency.

(2) Two megaphones in the passenger cabin on each airplane with a seating capacity of more than 99 and less than 200 passengers, one installed at the forward end and the other at the most rearward location where it would be readily accessible to a normal flight attendant seat.

(3) Three megaphones in the passenger cabin on each airplane with a seating capacity of more than 199 passengers, one installed at the forward end, one installed at the most rearward location where it would be readily accessible to a normal flight attendant seat, and one installed in a readily accessible location in the mid-section of the airplane.

[Doc. No. 19779, 45 FR 67235, Oct. 9, 1980, as amended by Amdt. 125-19, 59 FR 1781, Jan. 12, 1994; Amdt. 125-22, 59 FR 52643, Oct. 18, 1994; 59 FR 55208, Nov. 4, 1994]

§ 125.209 Emergency equipment: Extended overwater operations.

(a) No person may operate an airplane in extended overwater operations unless it carries, installed in conspicuously marked locations easily accessible to the occupants if a ditching occurs, the following equipment:

(1) An approved life preserver equipped with an approved survivor locator light, or an approved flotation means, for each occupant of the aircraft. The life preserver or other flotation means must be easily accessible to each seated occupant. If a flotation means other than a life preserver is used, it must be readily removable from the airplane.

(2) Enough approved life rafts (with proper buoyancy) to carry all occupants of the airplane, and at least the following equipment for each raft clearly marked for easy identification—

(i) One canopy (for sail, sunshade, or rain catcher);

(ii) One radar reflector (or similar device);

(iii) One life raft repair kit;

(iv) One bailing bucket;

(v) One signaling mirror;

(vi) One police whistle;

(vii) One raft knife;

(viii) One CO_2 bottle for emergency inflation;

(ix) One inflation pump;

(x) Two oars;

(xi) One 75-foot retaining line;

(xii) One magnetic compass;

(xiii) One dye marker;

(xiv) One flashlight having at least two size "D" cells or equivalent;

(xv) At least one approved pyrotechnic signaling device;

(xvi) A 2-day supply of emergency food rations supplying at least 1,000 calories a day for each person;

(xvii) One sea water desalting kit for each two persons that raft is rated to carry, or two pints of water for each person the raft is rated to carry;

(xviii) One fishing kit; and

(xix) One book on survival appropriate for the area in which the airplane is operated.

(b) No person may operate an airplane in extended overwater operations unless there is attached to one of the life rafts required by paragraph (a) of this section, an approved survival type emergency locator transmitter. Batteries used in this transmitter must be replaced (or recharged, if the batteries are rechargeable) when the transmitter has been in use for more than one cumulative hour, or, when 50 percent of their useful life (or for rechargeable batteries, 50 percent of their useful life of charge) has expired, as established by the transmitter manufacturer under its approval. The new expiration date for replacing (or recharging) the battery must be legibly marked on the outside of the transmitter. The battery useful life (or useful life of charge) requirements of this paragraph do not apply to batteries (such as water-activated batteries) that are essentially unaffected during probable storage intervals.

[Doc. No. 19779, 45 FR 67235, Oct. 9, 1980, as amended by Amdt. 125-20, 59 FR 32058, June 21, 1994]

§ 125.211 Seat and safety belts.

(a) No person may operate an airplane unless there are available during the takeoff, en route flight, and landing—

(1) An approved seat or berth for each person on board the airplane who is at least 2 years old; and

(2) An approved safety belt for separate use by each person on board the airplane who is at least 2 years old, except that two persons occupying a berth may share one approved safety belt and two persons occupying a multiple lounge or divan seat may share one approved safety belt during en route flight only.

(b) Except as provided in paragraphs (b)(1) and (b)(2) of this section, each person on board an airplane operated under this part shall occupy an approved seat or berth with a separate safety belt properly secured about him or her during movement on the surface, takeoff, and landing. A safety belt provided for the occupant of a seat may not be used for more than one person who has reached his or her second birthday. Notwithstanding the preceding requirements, a child may:

(1) Be held by an adult who is occupying an approved seat or berth, provided the child has not reached his or her second birthday and the child does not occupy or use any restraining device; or

(2) Notwithstanding any other requirement of this chapter, occupy an approved child restraint system furnished by the certificate holder or one of the persons described in paragraph (b)(2)(i) of this section, provided:

(i) The child is accompanied by a parent, guardian, or attendant designated by the child's parent or guardian to attend to the safety of the child during the flight;

(ii) Except as provided in paragraph (b)(2)(ii)(D) of this section, the approved child restraint system bears one or more labels as follows:

(A) Seats manufactured to U.S. standards between January 1, 1981, and February 25, 1985, must bear the label: "This child restraint system conforms to all applicable Federal motor vehicle safety standards";

(B) Seats manufactured to U.S. standards on or after February 26, 1985, must bear two labels:

(*1*) "This child restraint system conforms to all applicable Federal motor vehicle safety standards"; and

(*2*) "THIS RESTRAINT IS CERTIFIED FOR USE IN MOTOR VEHICLES AND AIRCRAFT" in red lettering;

(C) Seats that do not qualify under paragraphs (b)(2)(ii)(A) and (b)(2)(ii)(B) of this section must bear a label or markings showing:

(*1*) That the seat was approved by a foreign government;

(*2*) That the seat was manufactured under the standards of the United Nations; or

(*3*) That the seat or child restraint device furnished by the certificate holder was approved by the FAA through Type Certificate or Supplemental Type Certificate.

(*4*) That the seat or child restraint device furnished by the certificate holder, or one of the persons described in paragraph (b)(2)(i) of this section, was approved by the FAA in accordance with § 21.305(d) or Technical Standard Order C-100b, or a later version.

(D) Except as provided in § 125.211(b)(2)(ii)(C)(*3*) and § 125.211(b)(2)(ii)(C)(*4*), booster-type child restraint systems (as defined in Federal Motor Vehicle Safety Standard No. 213 (49 CFR 571.213)), vest- and harness-type child restraint systems, and lap held child restraints are not approved for use in aircraft; and

(iii) The certificate holder complies with the following requirements:

(A) The restraint system must be properly secured to an approved forward-facing seat or berth;

(B) The child must be properly secured in the restraint system and must not exceed the specified weight limit for the restraint system; and

(C) The restraint system must bear the appropriate label(s).

(c) Except as provided in paragraph (c)(3) of this section, the following prohibitions apply to certificate holders:

(1) Except as provided in § 125.211(b)(2)(ii)(C)(*3*) and § 125.211(b)(2)(ii)(C)(*4*), no certificate holder may permit a child, in an aircraft, to occupy a booster-type child restraint system, a vest-type child restraint system, a harness-type child restraint system, or a lap held child restraint system during take off, landing, and movement on the surface.

(2) Except as required in paragraph (c)(1) of this section, no certificate holder may prohibit a child, if requested by the child's parent, guardian, or designated attendant, from occupying a child restraint system furnished by the child's parent, guardian, or designated attendant provided:

(i) The child holds a ticket for an approved seat or berth or such seat or berth is otherwise made available by the certificate holder for the child's use;

(ii) The requirements of paragraph (b)(2)(i) of this section are met;

(iii) The requirements of paragraph (b)(2)(iii) of this section are met; and

(iv) The child restraint system has one or more of the labels described in paragraphs (b)(2)(ii)(A) through (b)(2)(ii)(C) of this section.

(3) This section does not prohibit the certificate holder from providing child restraint systems authorized by this section or, consistent with safe operating practices, determining the most appropriate passenger seat location for the child restraint system.

(d) Each sideward facing seat must comply with the applicable requirements of § 25.785(c) of this chapter.

(e) No certificate holder may take off or land an airplane unless each passenger seat back is in the upright position. Each passenger shall comply with instructions given by a crewmember in compliance with this paragraph. This paragraph does not apply to seats on which cargo or persons who are unable to sit erect for a medical reason are carried in accordance with procedures in the certificate holder's manual if the seat back does not obstruct any passenger's access to the aisle or to any emergency exit.

(f) Each occupant of a seat equipped with a shoulder harness must fasten the shoulder harness during takeoff and landing, except that, in the case of crewmembers, the shoulder harness need not be fastened if the crewmember cannot perform his required duties with the shoulder harness fastened.

[Doc. No. 19799, 45 FR 67235, Oct. 9, 1980, as amended by Amdt. 125-17, 57 FR 42674, Sept. 15, 1992; Amdt. 125-26, 61 FR 28422, June 4, 1996; Amdt. 125-48, 70 FR 50907, Aug. 26, 2005; Amdt. 125-51, 71 FR 40009, July 14, 2006; 71 FR 59373, Oct. 10, 2006]

§ 125.213 Miscellaneous equipment.

No person may conduct any operation unless the following equipment is installed in the airplane:

(a) If protective fuses are installed on an airplane, the number of spare fuses approved for the airplane and appropriately described in the certificate holder's manual.

(b) A windshield wiper or equivalent for each pilot station.

(c) A power supply and distribution system that meets the requirements of §§ 25.1309, 25.1331, 25.1351 (a) and (b) (1) through (4), 25.1353, 25.1355, and 25.1431(b) or that is able to produce and distribute the load for the required instruments and equipment, with use of an external power supply if any one power source or component of the power distribution system fails. The use of common elements in the system may be approved if the Administrator finds that they are designed to be reasonably protected against malfunctioning. Engine-driven sources of energy, when used, must be on separate engines.

(d) A means for indicating the adequacy of the power being supplied to required flight instruments.

(e) Two independent static pressure systems, vented to the outside atmospheric pressure so that they will be least affected by air flow variation or moisture or other foreign matter, and installed so as to be airtight except for the vent. When a means is provided for transferring an instrument from its primary operating system to an alternative system, the means must include a positive positioning control and must be marked to indicate clearly which system is being used.

(f) A placard on each door that is the means of access to a required passenger emergency exit to indicate that it must be open during takeoff and landing.

(g) A means for the crew, in an emergency, to unlock each door that leads to a compartment that is normally accessible to passengers and that can be locked by passengers.

§ 125.215 Operating information required.

(a) The operator of an airplane must provide the following materials, in current and appropriate form, accessible to the pilot at the pilot station, and the pilot shall use them:

(1) A cockpit checklist.

(2) An emergency cockpit checklist containing the procedures required by paragraph (c) of this section, as appropriate.

(3) Pertinent aeronautical charts.

(4) For IFR operations, each pertinent navigational en route, terminal area, and approach and letdown chart;

(5) One-engine-inoperative climb performance data and, if the airplane is approved for use in IFR or over-the-top operations, that data must be sufficient to enable the pilot to determine that the airplane is capable of carrying passengers over-the-top or in IFR conditions at a weight that will allow it to climb, with the critical engine inoperative, at least 50 feet a minute when operating at the MEA's of the route to be flown or 5,000 feet MSL, whichever is higher.

(b) Each cockpit checklist required by paragraph (a)(1) of this section must contain the following procedures:

(1) Before starting engines;

(2) Before take-off;

(3) Cruise;
(4) Before landing;
(5) After landing;
(6) Stopping engines.

(c) Each emergency cockpit checklist required by paragraph (a)(2) of this section must contain the following procedures, as appropriate:

(1) Emergency operation of fuel, hydraulic, electrical, and mechanical systems.

(2) Emergency operation of instruments and controls.

(3) Engine inoperative procedures.

(4) Any other emergency procedures necessary for safety.

§ 125.217 Passenger information.

(a) Except as provided in paragraph (b) of this section, no person may operate an airplane carrying passengers unless it is equipped with signs that meet the requirements of § 25.791 of this chapter and that are visible to passengers and flight attendants to notify them when smoking is prohibited and when safety belts must be fastened. The signs must be so constructed that the crew can turn them on and off. They must be turned on during airplane movement on the surface, for each takeoff, for each landing, and when otherwise considered to be necessary by the pilot in command.

(b) No passenger or crewmember may smoke while any "No Smoking" sign is lighted nor may any passenger or crewmember smoke in any lavatory.

(c) Each passenger required by § 125.211(b) to occupy a seat or berth shall fasten his or her safety belt about him or her and keep it fastened while any "Fasten Seat Belt" sign is lighted.

(d) Each passenger shall comply with instructions given him or her by crewmembers regarding compliance with paragraphs (b) and (c) of this section.

[Doc. No. 26142, 57 FR 42675, Sept. 15, 1992]

§ 125.219 Oxygen for medical use by passengers.

(a) Except as provided in paragraphs (d) and (e) of this section, no certificate holder may allow the carriage or operation of equipment for the storage, generation or dispensing of medical oxygen unless the unit to be carried is constructed so that all valves, fittings, and gauges are protected from damage during that carriage or operation and unless the following conditions are met:

(1) The equipment must be—

(i) Of an approved type or in conformity with the manufacturing, packaging, marking, labeling, and maintenance requirements of title 49 CFR parts 171, 172, and 173, except § 173.24(a)(1);

(ii) When owned by the certificate holder, maintained under the certificate holder's approved maintenance program;

(iii) Free of flammable contaminants on all exterior surfaces; and

(iv) Appropriately secured.

(2) When the oxygen is stored in the form of a liquid, the equipment must have been under the certificate holder's approved maintenance program since its purchase new or since the storage container was last purged.

(3) When the oxygen is stored in the form of a compressed gas as defined in title 49 CFR 173.300(a)—

(i) When owned by the certificate holder, it must be maintained under its approved maintenance program; and

(ii) The pressure in any oxygen cylinder must not exceed the rated cylinder pressure.

(4) The pilot in command must be advised when the equipment is on board and when it is intended to be used.

(5) The equipment must be stowed, and each person using the equipment must be seated so as not to restrict access to or use of any required emergency or regular exit or of the aisle in the passenger compartment.

(b) When oxygen is being used, no person may smoke and no certificate holder may allow any person to smoke within 10 feet of oxygen storage and dispensing equipment carried under paragraph (a) of this section.

(c) No certificate holder may allow any person other than a person trained in the use of medical oxygen equipment to connect or disconnect oxygen bottles or any other ancillary component while any passenger is aboard the airplane.

(d) Paragraph (a)(1)(i) of this section does not apply when that equipment is furnished by a professional or medical emergency service for use on board an airplane in a medical emergency when no other practical means of transportation (including any other properly equipped certificate holder) is reasonably available and the person carried under the medical emergency is accompanied by a person trained in the use of medical oxygen.

(e) Each certificate holder who, under the authority of paragraph (d) of this section, deviates from paragraph (a)(1)(i) of this section under a medical emergency shall, within 10 days, excluding Saturdays, Sundays, and Federal holidays, after the deviation, send to the FAA Flight Standards district office charged with the overall inspection of the certificate holder a complete report of the operation involved, including a description of the deviation and the reasons for it.

§ 125.221 Icing conditions: Operating limitations.

(a) No pilot may take off an airplane that has frost, ice, or snow adhering to any propeller, windshield, stabilizing or control surface; to a powerplant installation; or to an airspeed, altimeter, rate of climb, flight attitude instrument system, or wing, except that takeoffs may be made with frost under the wing in the area of the fuel tanks if authorized by the FAA.

(b) No certificate holder may authorize an airplane to take off and no pilot may take off an airplane any time conditions are such that frost, ice, or snow may reasonably be expected to adhere to the airplane unless the pilot has completed the testing required under § 125.287(a)(9) and unless one of the following requirements is met:

(1) A pretakeoff contamination check, that has been established by the certificate holder and approved by the Administrator for the specific airplane type, has been completed within 5 minutes prior to beginning takeoff. A pretakeoff contamination check is a check to make sure the wings and control surfaces are free of frost, ice, or snow.

(2) The certificate holder has an approved alternative procedure and under that procedure the airplane is determined to be free of frost, ice, or snow.

(3) The certificate holder has an approved deicing/anti-icing program that complies with § 121.629(c) of this chapter and the takeoff complies with that program.

(c) No pilot may fly under IFR into known or forecast light or moderate icing conditions, or under VFR into known light or moderate icing conditions, unless—

(1) The aircraft has functioning deicing or anti-icing equipment protecting each propeller, windshield, wing, stabilizing or control surface, and each airspeed, altimeter, rate of climb, or flight attitude instrument system;

(2) The airplane has ice protection provisions that meet appendix C of this part; or

(3) The airplane meets transport category airplane type certification provisions, including the requirements for certification for flight in icing conditions.

(d) Except for an airplane that has ice protection provisions that meet appendix C of this part or those for transport category airplane type certification, no pilot may fly an airplane into known or forecast severe icing conditions.

(e) If current weather reports and briefing information relied upon by the pilot in command indicate that the forecast icing condition that would otherwise prohibit the flight will not be encountered during the flight because of changed weather conditions since the forecast, the restrictions in paragraphs (b) and (c) of this section based on forecast conditions do not apply.

[45 FR 67235, Oct. 9, 1980, as amended by Amdt. 125-18, 58 FR 69629, Dec. 30, 1993; Amdt. 125-58, 74 FR 62696, Dec. 1, 2009]

§ 125.223 Airborne weather radar equipment requirements.

(a) No person may operate an airplane governed by this part in passenger-carrying operations unless approved airborne weather radar equipment is installed in the airplane.

(b) No person may begin a flight under IFR or night VFR conditions when current weather reports indicate that thunderstorms, or other potentially hazardous weather conditions that can be detected with airborne weather radar equipment, may reasonably be expected along the route to be flown, unless the airborne weather radar equipment required by paragraph (a) of this section is in satisfactory operating condition.

(c) If the airborne weather radar equipment becomes inoperative en route, the airplane must be operated under the instructions and procedures specified for that event in the manual required by § 125.71.

(d) This section does not apply to airplanes used solely within the State of Hawaii, within the State of Alaska, within that part of Canada west of longitude 130 degrees W, between latitude 70 degrees N, and latitude 53 degrees N, or during any training, test, or ferry flight.

(e) Without regard to any other provision of this part, an alternate electrical power supply is not required for airborne weather radar equipment.

§ 125.224 Collision avoidance system.

Effective January 1, 2005, any airplane you operate under this part 125 must be equipped and operated according to the following table:

Collision Avoidance Systems

If you operate any . . .	Then you must operate that airplane with:
(a) Turbine-powered airplane of more than 33,000 pounds maximum certificated takeoff weight	(1) An appropriate class of Mode S transponder that meets Technical Standard Order (TSO) C–112, or a later version, and one of the following approved units: (i) TCAS II that meets TSO C–119b (version 7.0), or a later version.
	(ii) TCAS II that meets TSO C–119a (version 6.04A Enhanced) that was installed in that airplane before May 1, 2003. If that TCAS II version 6.04A Enhanced no longer can be repaired to TSO C–119a standards, it must be replaced with a TCAS II that meets TSO C–119b (version 7.0), or a later version. (iii) A collision avoidance system equivalent to TSO C–119b (version 7.0), or a later version, capable of coordinating with units that meet TSO C–119a (version 6.04A Enhanced), or a later version.
(b) Piston-powered airplane of more than 33,000 pounds maximum certificated takeoff weight	(1) TCAS I that meets TSO C–118, or a later version, or (2) A collision avoidance system equivalent to TSO C–118, or a later version, or (1)(3) A collision avoidance system and Mode S transponder that meet paragraph (a)(1) of this section.

[Doc. No. FAA-2001-10910, 68 FR 15903, Apr. 1, 2003]

§ 125.225 Flight data recorders.

(a) Except as provided in paragraph (d) of this section, after October 11, 1991, no person may operate a large airplane type certificated before October 1, 1969, for operations above 25,000 feet altitude, nor a multiengine, turbine powered airplane type certificated before October 1, 1969, unless it is equipped with one or more approved flight recorders that utilize a digital method of recording and storing data and a method of readily retrieving that data from the storage medium. The following information must be able to be determined within the ranges, accuracies, resolution, and recording intervals specified in appendix D of this part:

(1) Time;
(2) Altitude;
(3) Airspeed;
(4) Vertical acceleration;
(5) Heading;
(6) Time of each radio transmission to or from air traffic control;
(7) Pitch attitude;
(8) Roll attitude;
(9) Longitudinal acceleration;
(10) Control column or pitch control surface position; and
(11) Thrust of each engine.

(b) Except as provided in paragraph (d) of this section, after October 11, 1991, no person may operate a large airplane type certificated after September 30, 1969, for operations above 25,000 feet altitude, nor a multiengine, turbine powered airplane type certificated after September 30, 1969, unless it is equipped with one or more approved flight recorders that utilize a digital method of recording and storing data and a method of readily retrieving that data from the storage medium. The following information must be able to be determined within the ranges, accuracies, resolutions, and recording intervals specified in appendix D of this part:

(1) Time;

(2) Altitude;

(3) Airspeed;

(4) Vertical acceleration;

(5) Heading;

(6) Time of each radio transmission either to or from air traffic control;

(7) Pitch attitude;

(8) Roll attitude;

(9) Longitudinal acceleration;

(10) Pitch trim position;

(11) Control column or pitch control surface position;

(12) Control wheel or lateral control surface position;

(13) Rudder pedal or yaw control surface position;

(14) Thrust of each engine;

(15) Position of each trust reverser;

(16) Trailing edge flap or cockpit flap control position; and

(17) Leading edge flap or cockpit flap control position.

(c) After October 11, 1991, no person may operate a large airplane equipped with a digital data bus and ARINC 717 digital flight data acquisition unit (DFDAU) or equivalent unless it is equipped with one or more approved flight recorders that utilize a digital method of recording and storing data and a method of readily retrieving that data from the storage medium. Any parameters specified in appendix D of this part that are available on the digital data bus must be recorded within the ranges, accuracies, resolutions, and sampling intervals specified.

(d) No person may operate under this part an airplane that is manufactured after October 11, 1991, unless it is equipped with one or more approved flight recorders that utilize a digital method of recording and storing data and a method of readily retrieving that data from the storage medium. The parameters specified in appendix D of this part must be recorded within the ranges, accuracies, resolutions and sampling intervals specified. For the purpose of this section, "manufactured" means the point in time at which the airplane inspection acceptance records reflect that the airplane is complete and meets the FAA-approved type design data.

(e) Whenever a flight recorder required by this section is installed, it must be operated continuously from the instant the airplane begins the takeoff roll until it has completed the landing roll at an airport.

(f) Except as provided in paragraph (g) of this section, and except for recorded data erased as authorized in this paragraph, each certificate holder shall keep the recorded data prescribed in paragraph (a), (b), (c), or (d) of this section, as applicable, until the airplane has been operated for at least 25 hours of the operating time specified in § 125.227(a) of this chapter. A total of 1 hour of recorded data may be erased for the purpose of testing the flight recorder or the flight recorder system. Any erasure made in accordance with this paragraph must be of the oldest recorded data accumulated at the time of testing. Except as provided in paragraph (g) of this section, no record need be kept more than 60 days.

(g) In the event of an accident or occurrence that requires immediate notification of the National Transportation Safety Board under 49 CFR part 830 and that results in termination of the flight, the certificate holder shall remove the recording media from the airplane and keep the recorded data required by paragraph (a), (b), (c), or (d) of this section, as applicable, for at least 60 days or for a longer period upon the request of the Board or the Administrator.

(h) Each flight recorder required by this section must be installed in accordance with the requirements of § 25.1459 of this chapter in effect on August 31, 1977. The correlation required by § 25.1459(c) of this chapter need be established only on one airplane of any group of airplanes.

(1) That are of the same type;

(2) On which the flight recorder models and their installations are the same; and

(3) On which there are no differences in the type design with respect to the installation of the first pilot's instruments associated with the flight recorder. The most recent instrument calibration, including the recording medium from which this calibration is derived, and the recorder correlation must be retained by the certificate holder.

(i) Each flight recorder required by this section that records the data specified in paragraphs (a), (b), (c), or (d) of this section must have an approved device to assist in locating that recorder under water.

(j) After August 20, 2001, this section applies only to the airplane models listed in § 125.226(l)(2). All other airplanes must comply with the requirements of § 125.226.

[Doc. No. 25530, 53 FR 26148, July 11, 1988; 53 FR 30906, Aug. 16, 1988; Amdt. 125-54, 73 FR 12568, Mar. 7, 2008]

§ 125.226 Digital flight data recorders.

(a) Except as provided in paragraph (l) of this section, no person may operate under this part a turbine-engine-powered transport category airplane unless it is equipped with one or more approved flight recorders that use a digital method of recording and storing data and a method of readily retrieving that data from the storage medium. The operational parameters required to be recorded by digital flight data recorders required by this section are as follows: the phrase "when an information source is installed" following a parameter indicates that recording of that parameter is not intended to require a change in installed equipment:

(1) Time;

(2) Pressure altitude;

(3) Indicated airspeed;

(4) Heading—primary flight crew reference (if selectable, record discrete, true or magnetic);

(5) Normal acceleration (Vertical);

(6) Pitch attitude;

(7) Roll attitude;

(8) Manual radio transmitter keying, or CVR/DFDR synchronization reference;

(9) Thrust/power of each engine—primary flight crew reference;

(10) Autopilot engagement status;

(11) Longitudinal acceleration;

(12) Pitch control input;

(13) Lateral control input;

(14) Rudder pedal input;

(15) Primary pitch control surface position;

(16) Primary lateral control surface position;

(17) Primary yaw control surface position;

(18) Lateral acceleration;

(19) Pitch trim surface position or parameters of paragraph (a)(82) of this section if currently recorded;

(20) Trailing edge flap or cockpit flap control selection (except when parameters of paragraph (a)(85) of this section apply);

(21) Leading edge flap or cockpit flap control selection (except when parameters of paragraph (a)(86) of this section apply);

(22) Each Thrust reverser position (or equivalent for propeller airplane);

(23) Ground spoiler position or speed brake selection (except when parameters of paragraph (a)(87) of this section apply);

(24) Outside or total air temperature;

(25) Automatic Flight Control System (AFCS) modes and engagement status, including autothrottle;

(26) Radio altitude (when an information source is installed);

(27) Localizer deviation, MLS Azimuth;

(28) Glideslope deviation, MLS Elevation;

(29) Marker beacon passage;

(30) Master warning;

(31) Air/ground sensor (primary airplane system reference nose or main gear);

(32) Angle of attack (when information source is installed);

(33) Hydraulic pressure low (each system);

(34) Ground speed (when an information source is installed);

(35) Ground proximity warning system;

(36) Landing gear position or landing gear cockpit control selection;

(37) Drift angle (when an information source is installed);

(38) Wind speed and direction (when an information source is installed);

(39) Latitude and longitude (when an information source is installed);

(40) Stick shaker/pusher (when an information source is installed);

(41) Windshear (when an information source is installed);

(42) Throttle/power lever position;

(43) Additional engine parameters (as designed in appendix E of this part);

(44) Traffic alert and collision avoidance system;

(45) DME 1 and 2 distances;

(46) Nav 1 and 2 selected frequency;

(47) Selected barometric setting (when an information source is installed);

(48) Selected altitude (when an information source is installed);

(49) Selected speed (when an information source is installed);

(50) Selected mach (when an information source is installed);

(51) Selected vertical speed (when an information source is installed);

(52) Selected heading (when an information source is installed);

(53) Selected flight path (when an information source is installed);

(54) Selected decision height (when an information source is installed);

(55) EFIS display format;

(56) Multi-function/engine/alerts display format;

(57) Thrust command (when an information source is installed);

(58) Thrust target (when an information source is installed);

(59) Fuel quantity in CG trim tank (when an information source is installed);

(60) Primary Navigation System Reference;

(61) Icing (when an information source is installed);

(62) Engine warning each engine vibration (when an information source is installed);

(63) Engine warning each engine over temp. (when an information source is installed);

(64) Engine warning each engine oil pressure low (when an information source is installed);

(65) Engine warning each engine over speed (when an information source is installed);

(66) Yaw trim surface position;

(67) Roll trim surface position;

(68) Brake pressure (selected system);

(69) Brake pedal application (left and right);

(70) Yaw of sideslip angle (when an information source is installed);

(71) Engine bleed valve position (when an information source is installed);

(72) De-icing or anti-icing system selection (when an information source is installed);

(73) Computed center of gravity (when an information source is installed);

(74) AC electrical bus status;

(75) DC electrical bus status;

(76) APU bleed valve position (when an information source is installed);

(77) Hydraulic pressure (each system);

(78) Loss of cabin pressure;

(79) Computer failure;

(80) Heads-up display (when an information source is installed);

(81) Para-visual display (when an information source is installed);

(82) Cockpit trim control input position-pitch;

(83) Cockpit trim control input position—roll;

(84) Cockpit trim control input position—yaw;

(85) Trailing edge flap and cockpit flap control position;

(86) Leading edge flap and cockpit flap control position;

(87) Ground spoiler position and speed brake selection;

(88) All cockpit flight control input forces (control wheel, control column, rudder pedal);

(89) Yaw damper status;

(90) Yaw damper command; and

(91) Standby rudder valve status.

(b) For all turbine-engine powered transport category airplanes manufactured on or before October 11, 1991, by August 20, 2001—

(1) For airplanes not equipped as of July 16, 1996, with a flight data acquisition unit (FDAU), the parameters listed in paragraphs (a)(1) through (a)(18) of this section must be recorded within the ranges and accuracies specified in Appendix D of this part, and—

(i) For airplanes with more than two engines, the parameter described in paragraph (a)(18) is not required unless sufficient capacity is available on the existing recorder to record that parameter.

(ii) Parameters listed in paragraphs (a)(12) through (a)(17) each may be recorded from a single source.

(2) For airplanes that were equipped as of July 16, 1996, with a flight data acquisition unit (FDAU), the parameters listed in paragraphs (a)(1) through (a)(22) of this section must be recorded within the ranges, accuracies, and recording intervals specified in Appendix E of this part. Parameters listed in paragraphs (a)(12) through (a)(17) each may be recorded from a single source.

(3) The approved flight recorder required by this section must be installed at the earliest time practicable, but no later than the next heavy maintenance check after August 18, 1999 and no later than August 20, 2001. A heavy maintenance check is considered to be any time an airplane is scheduled to be out of service for 4 or more days and is scheduled to include access to major structural components.

(c) For all turbine-engine-powered transport category airplanes manufactured on or before October 11, 1991—

(1) That were equipped as of July 16, 1996, with one or more digital data bus(es) and an ARINC 717 digital flight data acquisition unit (DFDAU) or equivalent, the parameters specified in paragraphs (a)(1) through (a)(22) of this section must be recorded within the ranges, accuracies, resolutions, and sampling intervals specified in Appendix E of this part by August 20, 2001. Parameters listed in paragraphs (a)(12) through (a)(14) each may be recorded from a single source.

(2) Commensurate with the capacity of the recording system (DFDAU or equivalent and the DFDR), all additional parameters for which information sources are installed and which are connected to the recording system must be recorded within the ranges, accuracies, resolutions, and sampling intervals specified in Appendix E of this part by August 20, 2001.

(3) That were subject to § 125.225(e) of this part, all conditions of § 125.225(c) must continue to be met until compliance with paragraph (c)(1) of this section is accomplished.

(d) For all turbine-engine-powered transport category airplanes that were manufactured after October 11, 1991—

(1) The parameters listed in paragraphs (a)(1) through (a)(34) of this section must be recorded within the ranges, accuracies, resolutions, and recording intervals specified in Appendix E of this part by August 20, 2001. Paramaters listed in paragraphs (a)(12) through (a)(14) each may be recorded from a single source.

(2) Commensurate with the capacity of the recording system, all additional parameters for which information sources are installed and which are connected to the recording system, must be recorded within the ranges, accuracies, resolutions, and sampling intervals specified in Appendix E of this part by August 20, 2001.

(e) For all turbine-engine-powered transport category airplanes that are manufactured after August 18, 2000—

(1) The parameters listed in paragraph (a) (1) through (57) of this section must be recorded within the ranges, accuracies, resolutions, and recording intervals specified in Appendix E of this part.

(2) Commensurate with the capacity of the recording system, all additional parameters for which information sources are installed and which are connected to the recording system, must be recorded within the ranges, accuracies, resolutions, and sampling intervals specified in Appendix E of this part.

(3) In addition to the requirements of paragraphs (e)(1) and (e)(2) of this section, all Boeing 737 model airplanes must also comply with the requirements of paragraph (n) of this section, as applicable.

(f) For all turbine-engine-powered transport category airplanes manufactured after August 19, 2002—

(1) The parameters listed in paragraphs (a)(1) through (a)(88) of this section must be recorded within the ranges, accuracies, resolutions, and recording intervals specified in Appendix E to this part.

(2) In addition to the requirements of paragraphs (f) (1) of this section, all Boeing 737 model airplanes must also comply with the requirements of paragraph (n) of this section.

(g) Whenever a flight data recorder required by this section is installed, it must be operated continuously from the instant the airplane begins its takeoff roll until it has completed its landing roll.

(h) Except as provided in paragraph (i) of this section, and except for recorded data erased as authorized in this paragraph, each certificate holder shall keep the recorded data prescribed by this section, as appropriate, until the airplane has been operated for at least 25 hours of the operating time specified in § 121.359(a) of this part. A total of 1 hour of recorded data may be erased for the purpose of testing the flight recorder or the flight recorder system. Any erasure made in accordance with this paragraph must be of the oldest recorded data accumulated at the time of testing. Except as provided in paragraph (i) of this section, no record need to be kept more than 60 days.

(i) In the event of an accident or occurrence that requires immediate notification of the National Transportation Safety Board under 49 CFR 830 of its regulations and that results in termination of the flight, the certificate holder shall remove the recorder from the airplane and keep the recorder data prescribed by this section, as appropriate, for at least 60 days or for a longer period upon the request of the Board or the Administrator.

(j) Each flight data recorder system required by this section must be installed in accordance with the requirements of § 25.1459(a) (except paragraphs (a)(3)(ii) and (7)), (b), (d) and (e) of this chapter. A correlation must be established between the values recorded by the flight data recorder and the corresponding values being measured. The correlation must contain a sufficient number of correlation points to accurately establish the conversion from the recorded values to engineering units or discrete state over the full operating range of the parameter. Except for airplanes having separate altitude and airspeed sensors that are an integral part of the flight data recorder system, a single correlation may be established for any group of airplanes—

(1) That are of the same type;

(2) On which the flight recorder system and its installation are the same; and

(3) On which there is no difference in the type design with respect to the installation of those sensors associated with the flight data recorder system. Documentation sufficient to convert recorded data into the engineering units and discrete values specified in the applicable appendix must be maintained by the certificate holder.

(k) Each flight data recorder required by this section must have an approved device to assist in locating that recorder under water.

(l) The following airplanes that were manufactured before August 18, 1997 need not comply with this section, but must continue to comply with applicable paragraphs of § 125.225 of this chapter, as appropriate:

(1) Airplanes that meet the Stage 2 noise levels of part 36 of this chapter and are subject to § 91.801(c) of this chapter, until January 1, 2000. On and after January 1, 2000, any Stage 2 airplane otherwise allowed to be operated under Part 91 of this chapter must comply with the applicable flight data recorder requirements of this section for that airplane.

(2) British Aerospace 1-11, General Dynamics Convair 580, General Dynamics Convair 600, General Dynamics Convair 640, deHavilland Aircraft Company Ltd. DHC-7, Fairchild Industries FH 227, Fokker F-27 (except Mark 50), F-28 Mark 1000 and Mark 4000, Gulfstream Aerospace G-159, Jetstream 4100 Series, Lockheed Aircraft Corporation Electra 10-A, Lockheed Aircraft Corporation Electra 10-B, Lockheed Aircraft Corporation Electra 10-E, Lockheed Aircraft Corporation Electra L-188, Lockheed Martin Model 382 (L-100) Hercules, Maryland Air Industries, Inc. F27, Mitsubishi Heavy Industries, Ltd. YS-11, Short Bros. Limited SD3-30, Short Bros. Limited SD3-60.

(m) All aircraft subject to the requirements of this section that are manufactured on or after April 7, 2010, must have a flight data recorder installed that also—

(1) Meets the requirements in § 25.1459(a)(3), (a)(7), and (a)(8) of this chapter; and

(2) Retains the 25 hours of recorded information required in paragraph (f) of this section using a recorder that meets the standards of TSO-C124a, or later revision.

(n) In addition to all other applicable requirements of this section, all Boeing 737 model airplanes manufactured after August 18, 2000 must record the parameters listed in paragraphs (a)(88) through (a)(91) of this section within the ranges, accuracies, resolutions, and recording intervals specified in Appendix E to this part. Compliance with this paragraph is required no later than February 2, 2011.

[Doc. No. 28109, 62 FR 38387, July 17, 1997; 62 FR 48135, Sept. 12, 1997, as amended by Amdt. 125-42, 68 FR 42937, July 18, 2003; 68 FR 50069, Aug. 20, 2003; Amdt. 125-54, 73 FR 12568, Mar. 7, 2008; Amdt. 125-56, 73 FR 73179, Dec. 2, 2008; Amdt. 125-54, 74 FR 32801, 32804, July 9, 2009]

§ 125.227 Cockpit voice recorders.

(a) No certificate holder may operate a large turbine engine powered airplane or a large pressurized airplane with four reciprocating engines unless an approved cockpit voice recorder is installed in that airplane and is operated continuously from the start of the use of the checklist (before starting engines for the purpose of flight) to completion of the final checklist at the termination of the flight.

(b) Each certificate holder shall establish a schedule for completion, before the prescribed dates, of the cockpit voice recorder installations required by paragraph (a) of this section. In addition, the certificate holder shall identify any airplane specified in paragraph (a) of this section he intends to discontinue using before the prescribed dates.

(c) The cockpit voice recorder required by this section must also meet the following standards:

(1) The requirements of part 25 of this chapter in effect after October 11, 1991.

(2) After September 1, 1980, each recorder container must—

(i) Be either bright orange or bright yellow;

(ii) Have reflective tape affixed to the external surface to facilitate its location under water; and

(iii) Have an approved underwater locating device on or adjacent to the container which is secured in such a manner that it is not likely to be separated during crash impact, unless the cockpit voice recorder and the flight recorder, required by § 125.225 of this chapter, are installed adjacent to each other in such a manner that they are not likely to be separated during crash impact.

(d) In complying with this section, an approved cockpit voice recorder having an erasure feature may be used so that, at any time during the operation of the recorder, information recorded more than 30 minutes earlier may be erased or otherwise obliterated.

(e) For those aircraft equipped to record the uninterrupted audio signals received by a boom or a mask microphone the flight crewmembers are required to use the boom microphone below 18,000 feet mean sea level. No person may operate a large turbine engine powered airplane or a large pressurized airplane with four reciprocating engines manufactured after October 11, 1991, or on which a cockpit voice recorder has been installed after October 11, 1991, unless it is equipped to record the uninterrupted audio signal received by a boom or mask microphone in accordance with § 25.1457(c)(5) of this chapter.

(f) In the event of an accident or occurrence requiring immediate notification of the National Transportation Safety Board under 49 CFR part 830 of its regulations, which results in the termination of the flight, the certificate holder shall keep the recorded information for at least 60 days or, if requested by the Administrator or the Board, for a longer period. Information obtained from the record is used to assist in determining the cause of accidents or occurrences in connection with investigations under 49 CFR part 830. The Administrator does not use the record in any civil penalty or certificate action.

(g) By April 7, 2012, all turbine engine-powered airplanes subject to this section that are manufactured before April 7, 2010, must have a cockpit voice recorder installed that also—

(1) Meets the requirements of § 25.1457(a)(3), (a)(4), (a)(5), and (d)(6) of this chapter;

(2) Retains at least the last 2 hours of recorded information using a recorder that meets the standards of TSO-C123a, or later revision; and

(3) Is operated continuously from the start of the use of the checklist (before starting the engines for the purpose of flight), to the completion of the final checklist at the termination of the flight.

(h) All turbine engine-powered airplanes subject to this section that are manufactured on or after April 7, 2010, must have a cockpit voice recorder installed that also—

(1) Is installed in accordance with the requirements of § 25.1457 (except for paragraph (a)(6)) of this chapter;

(2) Retains at least the last 2 hours of recorded information using a recorder that meets the standards of TSO-C123a, or later revision; and

(3) Is operated continuously from the start of the use of the checklist (before starting the engines for the purpose of flight), to the completion of the final checklist at the termination of the flight.

(4) For all airplanes manufactured on or after December 6, 2010, also meets the requirements of § 25.1457(a)(6) of this chapter.

(i) All airplanes required by this part to have a cockpit voice recorder and a flight data recorder, that install datalink communication equipment on or after December 6, 2010, must record all datalink messages as required by the certification rule applicable to the airplane.

[Doc. No. 25530, 53 FR 26149, July 11, 1988, as amended by Amdt. 125-54, 73 FR 12568, Mar. 7, 2008; Amdt. 125-54, 74 FR 32801, July 9, 2009; Amdt. 125-60, 75 FR 17046; Apr. 5, 2010]

§ 125.228 Flight data recorders: filtered data.

(a) A flight data signal is filtered when an original sensor signal has been changed in any way, other than changes necessary to:

(1) Accomplish analog to digital conversion of the signal;

(2) Format a digital signal to be DFDR compatible; or

(3) Eliminate a high frequency component of a signal that is outside the operational bandwidth of the sensor.

(b) An original sensor signal for any flight recorder parameter required to be recorded under § 125.226 may be filtered only if the recorded signal value continues to meet the requirements of Appendix D or E of this part, as applicable.

(c) For a parameter described in § 125.226(a) (12) through (17), (42), or (88), or the corresponding parameter in Appendix D of this part, if the recorded signal value is filtered and does not meet the requirements of Appendix D or E of this part, as applicable, the certificate holder must:

(1) Remove the filtering and ensure that the recorded signal value meets the requirements of Appendix D or E of this part, as applicable; or

(2) Demonstrate by test and analysis that the original sensor signal value can be reconstructed from the recorded data. This demonstration requires that:

(i) The FAA determine that the procedure and the test results submitted by the certificate holder as its compliance with paragraph (c)(2) of this section are repeatable; and

(ii) The certificate holder maintains documentation of the procedure required to reconstruct the original sensor signal value. This documentation is also subject to the requirements of § 125.226(i).

(d) *Compliance.* Compliance is required as follows:

(1) No later than October 20, 2011, each operator must determine, for each airplane it operates, whether the airplane's DFDR system is filtering any of the parameters listed in paragraph (c) of this section. The operator must create a record of this determination for each airplane it operates, and maintain it as part of the correlation documentation required by § 125.226(j)(3) of this part.

(2) For airplanes that are not filtering any listed parameter, no further action is required unless the airplane's DFDR system is modified in a manner that would cause it to meet the definition of filtering on any listed parameter.

(3) For airplanes found to be filtering a parameter listed in paragraph (c) of this section, the operator must either:

(i) No later than April 21, 2014, remove the filtering; or

(ii) No later than April 22, 2013, submit the necessary procedure and test results required by paragraph (c)(2) of this section.

(4) After April 21, 2014, no aircraft flight data recording system may filter any parameter listed in paragraph (c) of this section that does not meet the requirements of Appendix D or E of this part, unless the certificate holder possesses test and analysis procedures and the test results that have been approved by the FAA. All records of tests, analysis and procedures used to comply with this section must be maintained as part of the correlation documentation required by § 125.226(j)(3) of this part.

[Doc. No. FAA-2006-26135, 75 FR 7356, Feb. 19, 2010]

Subpart G—Maintenance

§ 125.241 Applicability.

This subpart prescribes rules, in addition to those prescribed in other parts of this chapter, for the maintenance of airplanes, airframes, aircraft engines, propellers, appliances, each item of survival and emergency equipment, and their component parts operated under this part.

§ 125.243 Certificate holder's responsibilities.

(a) With regard to airplanes, including airframes, aircraft engines, propellers, appliances, and survival and emergency equipment, operated by a certificate holder, that certificate holder is primarily responsible for—

(1) Airworthiness;

(2) The performance of maintenance, preventive maintenance, and alteration in accordance with applicable regulations and the certificate holder's manual;

(3) The scheduling and performance of inspections required by this part; and

(4) Ensuring that maintenance personnel make entries in the airplane maintenance log and maintenance records which meet the requirements of part 43 of this chapter and the certificate holder's manual, and which indicate that the airplane has been approved for return to service after maintenance, preventive maintenance, or alteration has been performed.

§ 125.245 Organization required to perform maintenance, preventive maintenance, and alteration.

The certificate holder must ensure that each person with whom it arranges for the performance of maintenance, preventive maintenance, alteration, or required inspection items identified in the certificate holder's manual in accordance with § 125.249(a)(3)(ii) must have an organization adequate to perform that work.

§ 125.247 Inspection programs and maintenance.

(a) No person may operate an airplane subject to this part unless

(1) The replacement times for life-limited parts specified in the aircraft type certificate data sheets, or other documents approved by the Administrator, are complied with;

(2) Defects disclosed between inspections, or as a result of inspection, have been corrected in accordance with part 43 of this chapter; and

(3) The airplane, including airframe, aircraft engines, propellers, appliances, and survival and emergency equipment, and their component parts, is inspected in accordance with an inspection program approved by the Administrator.

(b) The inspection program specified in paragraph (a)(3) of this section must include at least the following:

(1) Instructions, procedures, and standards for the conduct of inspections for the particular make and model of airplane, including necessary tests and checks. The instructions and procedures must set forth in detail the parts and areas of the airframe, aircraft engines, propellers, appliances, and survival and emergency equipment required to be inspected.

(2) A schedule for the performance of inspections that must be performed under the program, expressed in terms of the time in service, calendar time, number of system operations, or any combination of these.

(c) No person may be used to perform the inspections required by this part unless that person is authorized to perform maintenance under part 43 of this chapter.

(d) No person may operate an airplane subject to this part unless—

(1) The installed engines have been maintained in accordance with the overhaul periods recommended by the manufacturer or a program approved by the Administrator; and

(2) The engine overhaul periods are specified in the inspection programs required by § 125.247(a)(3).

(e) Inspection programs which may be approved for use under this part include, but are not limited to—

(1) A continuous inspection program which is a part of a current continuous airworthiness program approved for use by a certificate holder under part 121 or part 135 of this chapter;

(2) Inspection programs currently recommended by the manufacturer of the airplane, aircraft engines, propellers, appliances, or survival and emergency equipment; or

(3) An inspection program developed by a certificate holder under this part.

[Doc. No. 19779, 45 FR 67235, Oct. 9, 1980, as amended by Amdt. 125-2, 46 FR 24409, Apr. 30, 1981]

§ 125.248 [Reserved]

§ 125.249 Maintenance manual requirements.

(a) Each certificate holder's manual required by § 125.71 of this part shall contain, in addition to the items required by § 125.73 of this part, at least the following:

(1) A description of the certificate holders maintenance organization, when the certificate holder has such an organization.

(2) A list of those persons with whom the certificate holder has arranged for performance of inspections under this part. The list shall include the persons' names and addresses.

(3) The inspection programs required by § 125.247 of this part to be followed in the performance of inspections under this part including—

(i) The method of performing routine and nonroutine inspections (other than required inspections);

(ii) The designation of the items that must be inspected (required inspections), including at least those which if improperly accomplished could result in a failure, malfunction, or defect endangering the safe operation of the airplane;

(iii) The method of performing required inspections;

(iv) Procedures for the inspection of work performed under previously required inspection findings ("buy-back procedures");

(v) Procedures, standards, and limits necessary for required inspections and acceptance or rejection of the items required to be inspected;

(vi) Instructions to prevent any person who performs any item of work from performing any required inspection of that work; and

(vii) Procedures to ensure that work interruptions do not adversely affect required inspections and to ensure required inspections are properly completed before the airplane is released to service.

(b) In addition, each certificate holder's manual shall contain a suitable system which may include a coded system that provides for the retention of the following:

(1) A description (or reference to data acceptable to the Administrator) of the work performed.

(2) The name of the person performing the work and the person's certificate type and number.

(3) The name of the person approving the work and the person's certificate type and number.

§ 125.251 Required inspection personnel.

(a) No person may use any person to perform required inspections unless the person performing the inspection is appropriately certificated, properly trained, qualified, and authorized to do so.

(b) No person may perform a required inspection if that person performed the item of work required to be inspected.

Subpart H—Airman and Crewmember Requirements

§ 125.261 Airman: Limitations on use of services.

(a) No certificate holder may use any person as an airman nor may any person serve as an airman unless that person—

(1) Holds an appropriate current airman certificate issued by the FAA;

(2) Has any required appropriate current airman and medical certificates in that person's possession while engaged in operations under this part; and

(3) Is otherwise qualified for the operation for which that person is to be used.

(b) Each airman covered by paragraph (a) of this section shall present the certificates for inspection upon the request of the Administrator.

§ 125.263 Composition of flightcrew.

(a) No certificate holder may operate an airplane with less than the minimum flightcrew specified in the type certificate and the Airplane Flight Manual approved for that type airplane and required by this part for the kind of operation being conducted.

(b) In any case in which this part requires the performance of two or more functions for which an airman certificate is necessary, that requirement is not satisfied by the performance of multiple functions at the same time by one airman.

(c) On each flight requiring a flight engineer, at least one flight crewmember, other than the flight engineer, must be qualified to provide emergency performance of the flight engineer's functions for the safe completion of the flight if the flight engineer becomes ill or is otherwise incapacitated. A pilot need not hold a flight engineer's certificate to perform the flight engineer's functions in such a situation.

§ 125.265 Flight engineer requirements.

(a) No person may operate an airplane for which a flight engineer is required by the type certification requirements without a flight crewmember holding a current flight engineer certificate.

(b) No person may serve as a required flight engineer on an airplane unless, within the preceding 6 calendar months, that person has had at least 50 hours of flight time as a flight engineer on that type airplane, or the Administrator has checked that person on that type airplane and determined that person is familiar and competent with all essential current information and operating procedures.

§ 125.267 Flight navigator and long-range navigation equipment.

(a) No certificate holder may operate an airplane outside the 48 conterminous States and the District of Columbia when its position cannot be reliably fixed for a period of more than 1 hour, without—

(1) A flight crewmember who holds a current flight navigator certificate; or

(2) Two independent, properly functioning, and approved long-range means of navigation which enable

a reliable determination to be made of the position of the airplane by each pilot seated at that person's duty station.

(b) Operations where a flight navigator or long-range navigation equipment, or both, are required are specified in the operations specifications of the operator.

§ 125.269 Flight attendants.

(a) Each certificate holder shall provide at least the following flight attendants on each passenger-carrying airplane used:

(1) For airplanes having more than 19 but less than 51 passengers—one flight attendant.

(2) For airplanes having more than 50 but less than 101 passengers—two flight attendants.

(3) For airplanes having more than 100 passengers—two flight attendants plus one additional flight attendant for each unit (or part of a unit) of 50 passengers above 100 passengers.

(b) The number of flight attendants approved under paragraphs (a) and (b) of this section are set forth in the certificate holder's operations specifications.

(c) During takeoff and landing, flight attendants required by this section shall be located as near as practicable to required floor level exits and shall be uniformly distributed throughout the airplane to provide the most effective egress of passengers in event of an emergency evacuation.

§ 125.271 Emergency and emergency evacuation duties.

(a) Each certificate holder shall, for each type and model of airplane, assign to each category of required crewmember, as appropriate, the necessary functions to be performed in an emergency or a situation requiring emergency evacuation. The certificate holder shall show those functions are realistic, can be practically accomplished, and will meet any reasonably anticipated emergency, including the possible incapacitation of individual crewmembers or their inability to reach the passenger cabin because of shifting cargo in combination cargo-passenger airplanes.

(b) The certificate holder shall describe in its manual the functions of each category of required crewmembers under paragraph (a) of this section.

Subpart I—Flight Crewmember Requirements

§ 125.281 Pilot-in-command qualifications.

No certificate holder may use any person, nor may any person serve, as pilot in command of an airplane unless that person—

(a) Holds at least a commercial pilot certificate, an appropriate category, class, and type rating, and an instrument rating; and

(b) Has had at least 1,200 hours of flight time as a pilot, including 500 hours of cross-country flight time, 100 hours of night flight time, including at least 10 night takeoffs and landings, and 75 hours of actual or simulated instrument flight time, at least 50 hours of which were actual flight.

§ 125.283 Second-in-command qualifications.

No certificate holder may use any person, nor may any person serve, as second in command of an airplane unless that person—

(a) Holds at least a commercial pilot certificate with appropriate category and class ratings, and an instrument rating; and

(b) For flight under IFR, meets the recent instrument experience requirements prescribed for a pilot in command in part 61 of this chapter.

§ 125.285 Pilot qualifications: Recent experience.

(a) No certificate holder may use any person, nor may any person serve, as a required pilot flight crewmember unless within the preceding 90 calendar days that person has made at least three takeoffs and landings in the type airplane in which that person is to serve. The takeoffs and landings required by this paragraph may be performed in a flight simulator if the flight simulator is qualified and approved by the Administrator for such purpose. However, any person who fails to qualify for a 90-consecutive-day period following the date of that person's last qualification under this paragraph must reestablish recency of experience as provided in paragraph (b) of this section.

(b) A required pilot flight crewmember who has not met the requirements of paragraph (a) of this section may reestablish recency of experience by making at least three takeoffs and landings under the supervision of an authorized check airman, in accordance with the following:

(1) At least one takeoff must be made with a simulated failure of the most critical powerplant.

(2) At least one landing must be made from an ILS approach to the lowest ILS minimums authorized for the certificate holder.

(3) At least one landing must be made to a complete stop.

(c) A required pilot flight crewmember who performs the maneuvers required by paragraph (b) of this section in a qualified and approved flight simulator, as prescribed in paragraph (a) of this section, must—

(1) Have previously logged 100 hours of flight time in the same type airplane in which the pilot is to serve; and

(2) Be observed on the first two landings made in operations under this part by an authorized check airman who acts as pilot in command and occupies a pilot seat. The landings must be made in weather minimums that are not less than those contained in the certificate holder's operations specifications for Category I operations and must be made within 45 days following completion of simulator testing.

(d) An authorized check airman who observes the takeoffs and landings prescribed in paragraphs (b) and (c)(3) of this section shall certify that the person being observed is proficient and qualified to perform flight duty in operations under this part, and may require any additional maneuvers that are determined necessary to make this certifying statement.

[Doc. No. 19779, 45 FR 67235, Oct. 9, 1980, as amended by Amdt. 125-27, 61 FR 34561, July 2, 1996]

§ 125.287 Initial and recurrent pilot testing requirements.

(a) No certificate holder may use any person, nor may any person serve as a pilot, unless, since the beginning of the 12th calendar month before that service, that person has passed a written or oral test, given by the Administrator or an authorized check airman on that person's knowledge in the following areas—

(1) The appropriate provisions of parts 61, 91, and 125 of this chapter and the operations specifications and the manual of the certificate holder;

(2) For each type of airplane to be flown by the pilot, the airplane powerplant, major components and systems, major appliances, performance and operating limitations, standard and emergency operating procedures, and the contents of the approved Airplane Flight Manual or approved equivalent, as applicable;

(3) For each type of airplane to be flown by the pilot, the method of determining compliance with weight and balance limitations for takeoff, landing, and en route operations;

(4) Navigation and use of air navigation aids appropriate to the operation of pilot authorization, including, when applicable, instrument approach facilities and procedures;

(5) Air traffic control procedures, including IFR procedures when applicable;

(6) Meteorology in general, including the principles of frontal systems, icing, fog, thunderstorms, and windshear, and, if appropriate for the operation of the certificate holder, high altitude weather;

(7) Procedures for avoiding operations in thunderstorms and hail, and for operating in turbulent air or in icing conditions;

(8) New equipment, procedures, or techniques, as appropriate;

(9) Knowledge and procedures for operating during ground icing conditions, (i.e., any time conditions are such that frost, ice, or snow may reasonably be expected to adhere to the airplane), if the certificate holder expects to authorize takeoffs in ground icing conditions, including:

(i) The use of holdover times when using deicing/anti-icing fluids.

(ii) Airplane deicing/anti-icing procedures, including inspection and check procedures and responsibilities.

(iii) Communications.

(iv) Airplane surface contamination (i.e., adherence of frost, ice, or snow) and critical area identification, and knowledge of how contamination adversely affects airplane performance and flight characteristics.

(v) Types and characteristics of deicing/anti-icing fluids, if used by the certificate holder.

(vi) Cold weather preflight inspection procedures.

(vii) Techniques for recognizing contamination on the airplane.

(b) No certificate holder may use any person, nor may any person serve, as a pilot in any airplane unless, since the beginning of the 12th calendar month before that service, that person has passed a competency check given by the Administrator or an authorized check airman in that type of airplane to determine that person's competence in practical skills and techniques in that airplane or type of airplane. The extent of the competency check shall be determined by the Administrator or authorized check airman conducting the competency check. The competency check may include any of the maneuvers and procedures currently required for the original issuance of the particular pilot certificate required for the operations authorized and appropriate to the category, class, and type of airplane involved. For the purposes of this paragraph, type, as to an airplane, means any one of a group of airplanes determined by the Administrator to have a similar means of propulsion, the same manufacturer, and no significantly different handling or flight characteristics.

(c) The instrument proficiency check required by § 125.291 may be substituted for the competency check required by this section for the type of airplane used in the check.

(d) For the purposes of this part, competent performance of a procedure or maneuver by a person to be used as a pilot requires that the pilot be the obvious master of the airplane with the successful outcome of the maneuver never in doubt.

(e) The Administrator or authorized check airman certifies the competency of each pilot who passes the knowledge or flight check in the certificate holder's pilot records.

(f) Portions of a required competency check may be given in an airplane simulator or other appropriate training device, if approved by the Administrator.

[45 FR 67235, Oct. 9, 1980, as amended by Amdt. 125-18, 58 FR 69629, Dec. 30, 1993]

§ 125.289 Initial and recurrent flight attendant crewmember testing requirements.

No certificate holder may use any person, nor may any person serve, as a flight attendant crewmember, unless, since the beginning of the 12th calendar month before that service, the certificate holder has determined by appropriate initial and recurrent testing that the person is knowledgeable and competent in the following areas as appropriate to assigned duties and responsibilities:

(a) Authority of the pilot in command;

(b) Passenger handling, including procedures to be followed in handling deranged persons or other persons whose conduct might jeopardize safety;

(c) Crewmember assignments, functions, and responsibilities during ditching and evacuation of persons who may need the assistance of another person to move expeditiously to an exit in an emergency;

(d) Briefing of passengers;

(e) Location and operation of portable fire extinguishers and other items of emergency equipment;

(f) Proper use of cabin equipment and controls;

(g) Location and operation of passenger oxygen equipment;

(h) Location and operation of all normal and emergency exits, including evacuation chutes and escape ropes; and

(i) Seating of persons who may need assistance of another person to move rapidly to an exit in an emergency as prescribed by the certificate holder's operations manual.

§ 125.291 Pilot in command: Instrument proficiency check requirements.

(a) No certificate holder may use any person, nor may any person serve, as a pilot in command of an airplane under IFR unless, since the beginning of the sixth calendar month before that service, that person has passed an instrument proficiency check and the Administrator or an authorized check airman has so certified in a letter of competency.

(b) No pilot may use any type of precision instrument approach procedure under IFR unless, since the beginning of the sixth calendar month before that use, the pilot has satisfactorily demonstrated that type of approach procedure and has been issued a letter of competency under paragraph (g) of this section. No pilot may use any type of nonprecision approach procedure under IFR unless, since the beginning of the sixth calendar month before that use, the pilot has satisfactorily demonstrated either that type of approach procedure or any other two different types of nonprecision approach procedures and has been issued a letter of competency under paragraph (g) of this section. The instrument approach procedure or procedures must include at least one straight-in approach, one circling approach, and one missed approach. Each type of approach procedure demonstrated must be conducted to published minimums for that procedure.

(c) The instrument proficiency check required by paragraph (a) of this section consists of an oral or written equipment test and a flight check under simulated or actual IFR conditions. The equipment test includes questions on emergency procedures, engine operation, fuel and lubrication systems, power settings, stall speeds, best engine-out speed, propeller and supercharge operations, and hydraulic, mechanical, and electrical systems, as appropriate. The flight check includes navigation by instruments, recovery from simulated emergencies, and standard instrument approaches involving navigational facilities which that pilot is to be authorized to use.

(1) For a pilot in command of an airplane, the instrument proficiency check must include the procedures and maneuvers for a commercial pilot certificate with an instrument rating and, if required, for the appropriate type rating.

(2) The instrument proficiency check must be given by an authorized check airman or by the Administrator.

(d) If the pilot in command is assigned to pilot only one type of airplane, that pilot must take the instrument proficiency check required by paragraph (a) of this section in that type of airplane.

(e) If the pilot in command is assigned to pilot more than one type of airplane, that pilot must take the instrument proficiency check required by paragraph (a) of this section in each type of airplane to which that pilot is assigned, in rotation, but not more than one flight check during each period described in paragraph (a) of this section.

(f) Portions of a required flight check may be given in an airplane simulator or other appropriate training device, if approved by the Administrator.

(g) The Administrator or authorized check airman issues a letter of competency to each pilot who passes the instrument proficiency check. The letter of competency contains a list of the types of instrument approach procedures and facilities authorized.

§ 125.293 Crewmember: Tests and checks, grace provisions, accepted standards.

(a) If a crewmember who is required to take a test or a flight check under this part completes the test or flight check in the calendar month before or after the calendar month in which it is required, that crewmember is considered to have completed the test or check in the calendar month in which it is required.

(b) If a pilot being checked under this subpart fails any of the required maneuvers, the person giving the check may give additional training to the pilot during the course of the check. In addition to repeating the maneuvers failed, the person giving the check may require the pilot being checked to repeat any other maneuvers that are necessary to determine the pilot's proficiency. If the pilot being checked is unable to demonstrate satisfactory performance to the person conducting the check, the certificate holder may not use the pilot, nor may the pilot serve, in the capacity for which the pilot is being checked in operations under this part until the pilot has satisfactorily completed the check.

§ 125.295 Check airman authorization: Application and issue.

Each certificate holder desiring FAA approval of a check airman shall submit a request in writing to the FAA Flight Standards district office charged with the overall inspection of the certificate holder. The Administrator may issue a letter of authority to each check airman if that airman passes the appropriate oral and flight test. The letter of authority lists the tests and checks in this part that the check airman is qualified to give, and the category, class and type airplane, where appropriate, for which the check airman is qualified.

§ 125.296 Training, testing, and checking conducted by training centers: Special rules.

A crewmember who has successfully completed training, testing, or checking in accordance with an approved training program that meets the requirements of this part and that is conducted in accordance with an approved course conducted by a training center certificated under part 142 of this chapter, is considered to meet applicable requirements of this part.

[Doc. No. 26933, 61 FR 34561, July 2, 1996]

§ 125.297 Approval of flight simulators and flight training devices.

(a) Flight simulators and flight training devices approved by the Administrator may be used in training, testing, and checking required by this subpart.

(b) Each flight simulator and flight training device that is used in training, testing, and checking required under this subpart must be used in accordance with an approved training course conducted by a training center certificated under part 142 of this chapter, or meet the following requirements:

(1) It must be specifically approved for—

(i) The certificate holder;

(ii) The type airplane and, if applicable, the particular variation within type for which the check is being conducted; and

(iii) The particular maneuver, procedure, or crewmember function involved.

(2) It must maintain the performance, functional, and other characteristics that are required for approval.

(3) It must be modified to conform with any modification to the airplane being simulated that changes the performance, functional, or other characteristics required for approval.

[Doc. No. 19779, 45 FR 67235, Oct. 9, 1980, as amended by Amdt. 125-27, 61 FR 34561, July 2, 1996]

Subpart J—Flight Operations

§ 125.311 Flight crewmembers at controls.

(a) Except as provided in paragraph (b) of this section, each required flight crewmember on flight deck duty must remain at the assigned duty station with seat belt fastened while the airplane is taking off or landing and while it is en route.

(b) A required flight crewmember may leave the assigned duty station—

(1) If the crewmember's absence is necessary for the performance of duties in connection with the operation of the airplane;

(2) If the crewmember's absence is in connection with physiological needs; or

(3) If the crewmember is taking a rest period and relief is provided—

(i) In the case of the assigned pilot in command, by a pilot qualified to act as pilot in command.

(ii) In the case of the assigned second in command, by a pilot qualified to act as second in command of that airplane during en route operations. However, the relief pilot need not meet the recent experience requirements of § 125.285.

§ 125.313 Manipulation of controls when carrying passengers.

No pilot in command may allow any person to manipulate the controls of an airplane while carrying passengers during flight, nor may any person manipulate the controls while carrying passengers during flight, unless that person is a qualified pilot of the certificate holder operating that airplane.

§ 125.315 Admission to flight deck.

(a) No person may admit any person to the flight deck of an airplane unless the person being admitted is—

(1) A crewmember;

(2) An FAA inspector or an authorized representative of the National Transportation Safety Board who is performing official duties; or

(3) Any person who has the permission of the pilot in command.

(b) No person may admit any person to the flight deck unless there is a seat available for the use of that person in the passenger compartment, except—

(1) An FAA inspector or an authorized representative of the Administrator or National Transportation Safety Board who is checking or observing flight operations; or

(2) A certificated airman employed by the certificate holder whose duties require an airman certificate.

§ 125.317 Inspector's credentials: Admission to pilots' compartment: Forward observer's seat.

(a) Whenever, in performing the duties of conducting an inspection, an FAA inspector presents an Aviation Safety Inspector credential, FAA Form 110A, to the pilot in command of an airplane operated by the certificate holder, the inspector must be given free and uninterrupted access to the pilot compartment of that airplane. However, this paragraph does not limit the emergency authority of the pilot in command to exclude any person from the pilot compartment in the interest of safety.

(b) A forward observer's seat on the flight deck, or forward passenger seat with headset or speaker, must be provided for use by the Administrator while conducting en route inspections. The suitability of the location of the seat and the headset or speaker for use in conducting en route inspections is determined by the Administrator.

§ 125.319 Emergencies.

(a) In an emergency situation that requires immediate decision and action, the pilot in command may take any action considered necessary under the circumstances. In such a case, the pilot in command may deviate from prescribed operations, procedures and methods, weather minimums, and this chapter, to the extent required in the interests of safety.

(b) In an emergency situation arising during flight that requires immediate decision and action by appropriate management personnel in the case of operations conducted with a flight following service and which is known to them, those personnel shall advise the pilot in command of the emergency, shall ascertain the decision of the pilot in command, and shall have the decision recorded. If they cannot communicate with the pilot, they shall declare an emergency and take any action that they consider necessary under the circumstances.

(c) Whenever emergency authority is exercised, the pilot in command or the appropriate management personnel shall keep the appropriate ground radio station fully informed of the progress of the flight. The person declaring the emergency shall send a written report of any deviation, through the operator's director of operations, to the Administrator within 10 days, exclusive of Saturdays, Sundays, and Federal holidays, after the flight is completed or, in the case of operations outside the United States, upon return to the home base.

§ 125.321 Reporting potentially hazardous meteorological conditions and irregularities of ground and navigation facilities.

Whenever the pilot in command encounters a meteorological condition or an irregularity in a ground facility or navigation aid in flight, the knowledge of which the pilot in command considers essential to the safety of other flights, the pilot in command shall notify an appropriate ground station as soon as practicable.

[Doc. No. 19779, 45 FR 67235, Oct. 9, 1980, as amended by Amdt. 125-52, 72 FR 31683, June 7, 2007]

§ 125.323 Reporting mechanical irregularities.

The pilot in command shall ensure that all mechanical irregularities occurring during flight are entered in the maintenance log of the airplane at the next place of landing. Before each flight, the pilot in command shall ascertain the status of each irregularity entered in the log at the end of the preceding flight.

§ 125.325 Instrument approach procedures and IFR landing minimums.

No person may make an instrument approach at an airport except in accordance with IFR weather minimums and unless the type of instrument approach procedure to be used is listed in the certificate holder's operations specifications.

§ 125.327 Briefing of passengers before flight.

(a) Before each takeoff, each pilot in command of an airplane carrying passengers shall ensure that all passengers have been orally briefed on—

(1) *Smoking.* Each passenger shall be briefed on when, where, and under what conditions smoking is prohibited. This briefing shall include a statement that the Federal Aviation Regulations require passenger compliance with the lighted passenger information signs, posted placards, areas designated for safety purposes as no smoking areas, and crewmember instructions with regard to these items.

(2) *The use of safety belts, including instructions on how to fasten and unfasten the safety belts.* Each passenger shall be briefed on when, where, and under what conditions the safety belt must be fastened about him or her. This briefing shall include a statement that the Federal Aviation Regulations require passenger compliance with lighted passenger information signs and crewmember instructions concerning the use of safety belts.

(3) The placement of seat backs in an upright position before takeoff and landing;

(4) Location and means for opening the passenger entry door and emergency exits;

(5) Location of survival equipment;

(6) If the flight involves extended overwater operation, ditching procedures and the use of required flotation equipment;

(7) If the flight involves operations above 12,000 feet MSL, the normal and emergency use of oxygen; and

(8) Location and operation of fire extinguishers.

(b) Before each takeoff, the pilot in command shall ensure that each person who may need the assistance of another person to move expeditiously to an exit if an emergency occurs and that person's attendant, if any, has received a briefing as to the procedures to be followed if an evacuation occurs. This paragraph does not apply to a person who has been given a briefing before a previous leg of a flight in the same airplane.

(c) The oral briefing required by paragraph (a) of this section shall be given by the pilot in command or a member of the crew. It shall be supplemented by printed cards for the use of each passenger containing—

(1) A diagram and method of operating the emergency exits; and

(2) Other instructions necessary for the use of emergency equipment on board the airplane.

Each card used under this paragraph must be carried in the airplane in locations convenient for the use of each passenger and must contain information that is appropriate to the airplane on which it is to be used.

(d) The certificate holder shall describe in its manual the procedure to be followed in the briefing required by paragraph (a) of this section.

(e) If the airplane does not proceed directly over water after takeoff, no part of the briefing required by paragraph (a)(6) of this section has to be given before takeoff but the briefing required by paragraph (a)(6) must be given before reaching the overwater part of the flight.

[Doc. No. 19779, 45 FR 67235, Oct. 9, 1980, as amended by Amdt. 125-17, 57 FR 42675, Sept. 15, 1992]

§ 125.328 Prohibition on crew interference.

No person may assault, threaten, intimidate, or interfere with a crewmember in the performance of the crewmember's duties aboard an aircraft being operated under this part.

[Doc. No. FAA-1998-4954, 64 FR 1080, Jan. 7, 1999]

§ 125.329 Minimum altitudes for use of autopilot.

(a) Except as provided in paragraphs (b), (c), (d), and (e) of this section, no person may use an autopilot at an altitude above the terrain which is less than 500 feet or less than twice the maximum altitude loss specified in the approved Airplane Flight Manual or equivalent for a malfunction of the autopilot, whichever is higher.

(b) When using an instrument approach facility other than ILS, no person may use an autopilot at an altitude above the terrain that is less than 50 feet below the approved minimum descent altitude for that procedure, or less than twice the maximum loss specified in the approved Airplane Flight Manual or equivalent for a malfunction of the autopilot under approach conditions, whichever is higher.

(c) For ILS approaches when reported weather conditions are less than the basic weather conditions in § 91.155 of this chapter, no person may use an autopilot with an approach coupler at an altitude above the terrain that is less than 50 feet above the terrain, or the maximum altitude loss specified in the approved Airplane Flight Manual or equivalent for the malfunction of the autopilot with approach coupler, whichever is higher.

(d) Without regard to paragraph (a), (b), or (c) of this section, the Administrator may issue operations specifications to allow the use, to touchdown, of an approved flight control guidance system with automatic capability, if—

(1) The system does not contain any altitude loss (above zero) specified in the approved Airplane Flight Manual or equivalent for malfunction of the autopilot with approach coupler; and

(2) The Administrator finds that the use of the system to touchdown will not otherwise adversely affect the safety standards of this section.

(e) Notwithstanding paragraph (a) of this section, the Administrator issues operations specifications to allow the use of an approved autopilot system with automatic capability during the takeoff and initial climb phase of flight provided:

(1) The Airplane Flight Manual specifies a minimum altitude engagement certification restriction;

(2) The system is not engaged prior to the minimum engagement certification restriction specified in the Airplane Flight Manual or an altitude specified by the Administrator, whichever is higher; and

(3) The Administrator finds that the use of the system will not otherwise affect the safety standards required by this section.

[Doc. No. 19779, 45 FR 67325, Oct. 9, 1980, as amended by Amdt. 125-12, 54 FR 34332, Aug. 18, 1989; Amdt. 125-29, 62 FR 27922, May 21, 1997]

§ 125.331 Carriage of persons without compliance with the passenger-carrying provisions of this part.

The following persons may be carried aboard an airplane without complying with the passenger-carrying requirements of this part:

(a) A crewmember.

(b) A person necessary for the safe handling of animals on the airplane.

(c) A person necessary for the safe handling of hazardous materials (as defined in subchapter C of title 49 CFR).

(d) A person performing duty as a security or honor guard accompanying a shipment made by or under the authority of the U.S. Government.

(e) A military courier or a military route supervisor carried by a military cargo contract operator if that carriage is specifically authorized by the appropriate military service.

(f) An authorized representative of the Administrator conducting an en route inspection.

(g) A person authorized by the Administrator.

§ 125.333 Stowage of food, beverage, and passenger service equipment during airplane movement on the surface, takeoff, and landing.

(a) No certificate holder may move an airplane on the surface, take off, or land when any food, beverage, or tableware furnished by the certificate holder is located at any passenger seat.

(b) No certificate holder may move an airplane on the surface, take off, or land unless each food and beverage tray and seat back tray table is secured in its stowed position.

(c) No certificate holder may permit an airplane to move on the surface, take off, or land unless each passenger serving cart is secured in its stowed position.

(d) Each passenger shall comply with instructions given by a crewmember with regard to compliance with this section.

[Doc. No. 26142, 57 FR 42675, Sept. 15, 1992]

Subpart K—Flight Release Rules

§ 125.351 Flight release authority.

(a) No person may start a flight without authority from the person authorized by the certificate holder to exercise operational control over the flight.

(b) No person may start a flight unless the pilot in command or the person authorized by the cetificate holder to exercise operational control over the flight has executed a flight release setting forth the conditions under which the flight will be conducted. The pilot in command may sign the flight release only when both the pilot in command and the person authorized to exercise operational control believe the flight can be made safely, unless the pilot in command is authorized by the certificate holder to exercise operational control and execute the flight release without the approval of any other person.

(c) No person may continue a flight from an intermediate airport without a new flight release if the airplane has been on the ground more than 6 hours.

§ 125.353 Facilities and services.

During a flight, the pilot in command shall obtain any additional available information of meteorological conditions and irregularities of facilities and services that may affect the safety of the flight.

§ 125.355 Airplane equipment.

No person may release an airplane unless it is airworthy and is equipped as prescribed.

§ 125.357 Communication and navigation facilities.

No person may release an airplane over any route or route segment unless communication and navigation facilities equal to those required by § 125.51 are in satisfactory operating condition.

§ 125.359 Flight release under VFR.

No person may release an airplane for VFR operation unless the ceiling and visibility en route, as indicated by available weather reports or forecasts, or any combination thereof, are and will remain at or above applicable VFR minimums until the airplane arrives at the airport or airports specified in the flight release.

§ 125.361 Flight release under IFR or over-the-top.

Except as provided in § 125.363, no person may release an airplane for operations under IFR or over-the-top unless appropriate weather reports or forecasts, or any combination thereof, indicate that the weather conditions will be at or above the authorized minimums at the estimated time of arrival at the airport or airports to which released.

§ 125.363 Flight release over water.

(a) No person may release an airplane for a flight that involves extended overwater operation unless appropriate weather reports or forecasts, or any combination thereof, indicate that the weather conditions will be at or above the authorized minimums at the estimated time of arrival at any airport to which released or to any required alternate airport.

(b) Each certificate holder shall conduct extended overwater operations under IFR unless it shows that operating under IFR is not necessary for safety.

(c) Each certificate holder shall conduct other overwater operations under IFR if the Administrator determines that operation under IFR is necessary for safety.

(d) Each authorization to conduct extended overwater operations under VFR and each requirement to conduct other overwater operations under IFR will be specified in the operations specifications.

§ 125.365 Alternate airport for departure.

(a) If the weather conditions at the airport of takeoff are below the landing minimums in the certificate holder's operations specifications for that airport, no person may release an airplane from that airport unless the flight release specifies an alternate airport located within the following distances from the airport of takeoff:

(1) *Airplanes having two engines.* Not more than 1 hour from the departure airport at normal cruising speed in still air with one engine inoperative.

(2) *Airplanes having three or more engines.* Not more than 2 hours from the departure airport at normal cruising speed in still air with one engine inoperative.

(b) For the purposes of paragraph (a) of this section, the alternate airport weather conditions must meet the requirements of the certificate holder's operations specifications.

(c) No person may release an airplane from an airport unless that person lists each required alternate airport in the flight release.

§ 125.367 Alternate airport for destination: IFR or over-the-top.

(a) Except as provided in paragraph (b) of this section, each person releasing an airplane for operation under IFR or over-the-top shall list at least one alternate airport for each destination airport in the flight release.

(b) An alternate airport need not be designated for IFR or over-the-top operations where the airplane carries enough fuel to meet the requirements of §§ 125.375 and 125.377 for flights outside the 48 conterminous States and the District of Columbia over routes without an available alternate airport for a particular airport of destination.

(c) For the purposes of paragraph (a) of this section, the weather requirements at the alternate airport must meet the requirements of the operator's operations specifications.

(d) No person may release a flight unless that person lists each required alternate airport in the flight release.

§ 125.369 Alternate airport weather minimums.

No person may list an airport as an alternate airport in the flight release unless the appropriate weather reports or forecasts, or any combination thereof, indicate that the weather conditions will be at or above the alternate weather minimums specified in the certificate holder's operations specifications for that airport when the flight arrives.

§ 125.371 Continuing flight in unsafe conditions.

(a) No pilot in command may allow a flight to continue toward any airport to which it has been released if, in the opinion of the pilot in command, the flight cannot be completed safely, unless, in the opinion of the pilot in command, there is no safer procedure. In that event, continuation toward that airport is an emergency situation.

§ 125.373 Original flight release or amendment of flight release.

(a) A certificate holder may specify any airport authorized for the type of airplane as a destination for the purpose of original release.

(b) No person may allow a flight to continue to an airport to which it has been released unless the weather conditions at an alternate airport that was specified in the flight release are forecast to be at or above the alternate minimums specified in the operations specifications for that airport at the time the airplane would arrive at the alternate airport. However, the flight release may be amended en route to include any alternate airport that is within the fuel range of the airplane as specified in § 125.375 or § 125.377.

(c) No person may change an original destination or alternate airport that is specified in the original flight release to another airport while the airplane is en route unless the other airport is authorized for that type of airplane.

(d) Each person who amends a flight release en route shall record that amendment.

§ 125.375 Fuel supply: Nonturbine and turbopropeller-powered airplanes.

(a) Except as provided in paragraph (b) of this section, no person may release for flight or take off a nonturbine or turbopropeller-powered airplane unless, considering the wind and other weather conditions expected, it has enough fuel—

(1) To fly to and land at the airport to which it is released;

(2) Thereafter, to fly to and land at the most distant alternate airport specified in the flight release; and

(3) Thereafter, to fly for 45 minutes at normal crusing fuel consumption.

(b) If the airplane is released for any flight other than from one point in the conterminous United States to another point in the conterminous United States, it must carry enough fuel to meet the requirements of paragraphs (a) (1) and (2) of this section and thereafter fly for 30 minutes plus 15 percent of the total time required to fly at normal cruising fuel consumption to the airports specified in paragraphs (a) (1) and (2) of this section, or fly for 90 minutes at normal cruising fuel consumption, whichever is less.

(c) No person may release a nonturbine or turbopropeller-powered airplane to an airport for which an alternate is not specified under § 125.367(b) unless it has enough fuel, considering wind and other weather conditions expected, to fly to that airport and thereafter to fly for 3 hours at normal cruising fuel consumption.

§ 125.377 Fuel supply: Turbine-engine-powered airplanes other than turbopropeller.

(a) Except as provided in paragraph (b) of this section, no person may release for flight or takeoff a turbine-powered airplane (other than a turbopropeller-powered airplane) unless, considering the wind and other weather conditions expected, it has enough fuel—

(1) To fly to and land at the airport to which it is released;

(2) Thereafter, to fly to and land at the most distant alternate airport specified in the flight release; and

(3) Thereafter, to fly for 45 minutes at normal cruising fuel consumption.

(b) For any operation outside the 48 conterminous United States and the District of Columbia, unless authorized by the Administrator in the operations specifications, no person may release for flight or take off a turbine-engine powered airplane (other than a turbopropeller-powered airplane) unless, considering wind and other weather conditions expected, it has enough fuel—

(1) To fly and land at the airport to which it is released;

(2) After that, to fly for a period of 10 percent of the total time required to fly from the airport of departure and land at the airport to which it was released;

(3) After that, to fly to and land at the most distant alternate airport specified in the flight release, if an alternate is required; and

(4) After that, to fly for 30 minutes at holding speed at 1,500 feet above the alternate airport (or the destination airport if no alternate is required) under standard temperature conditions.

(c) No person may release a turbine-engine-powered airplane (other than a turbopropeller airplane) to an airport for which an alternate is not specified under § 125.367(b) unless it has enough fuel, considering wind and other weather conditions expected, to fly to that airport and thereafter to fly for at least 2 hours at normal cruising fuel consumption.

(d) The Administrator may amend the operations specifications of a certificate holder to require more fuel than any of the minimums stated in paragraph (a) or (b) of this section if the Administrator finds that additional fuel is necessary on a particular route in the interest of safety.

§ 125.379 Landing weather minimums: IFR.

(a) If the pilot in command of an airplane has not served 100 hours as pilot in command in the type of airplane being operated, the MDA or DA/DH and visibility landing minimums in the certificate holder's operations specification are increased by 100 feet and one-half mile (or the RVR equivalent). The MDA or DA/DH and visibility minimums need not be increased above those applicable to the airport when used as an alternate airport, but in no event may the landing minimums be less than a 300-foot ceiling and 1 mile of visibility.

(b) The 100 hours of pilot-in-command experience required by paragraph (a) may be reduced (not to exceed 50 percent) by substituting one landing in operations under this part in the type of airplane for 1 required hour of pilot-in-command experience if the pilot has at least 100 hours as pilot in command of another type airplane in operations under this part.

(c) Category II minimums, when authorized in the certificate holder's operations specifications, do not apply until the pilot in command subject to paragraph (a) of this section meets the requirements of that paragraph in the type of airplane the pilot is operating.

[Doc. No. 19779, 45 FR 67235, Oct. 9, 1980, as amended by Amdt. 125-52, 72 FR 31683, June 7, 2007]

§ 125.381 Takeoff and landing weather minimums: IFR.

(a) Regardless of any clearance from ATC, if the reported weather conditions are less than that specified in the certificate holder's operations specifications, no pilot may—

(1) Take off an airplane under IFR; or

(2) Except as provided in paragraph (c) of this section, land an airplane under IFR.

(b) Except as provided in paragraph (c) of this section, no pilot may execute an instrument approach procedure if the latest reported visibility is less than the landing minimums specified in the certificate holder's operations specifications.

(c) If a pilot initiates an instrument approach procedure based on a weather report that indicates that the specified visibility minimums exist and subsequently receives another weather report that indicates that conditions are below the minimum requirements, then the pilot may continue with the approach only if, the requirements of § 91.175(l) of this chapter, or both of the following conditions are met—

(1) The later weather report is received when the airplane is in one of the following approach phases:

(i) The airplane is on a ILS approach and has passed the final approach fix;

(ii) The airplane is on an ASR or PAR final approach and has been turned over to the final approach controller; or

(iii) The airplane is on a nonprecision final approach and the airplane—

(A) Has passed the appropriate facility or final approach fix; or

(B) Where a final approach fix is not specified, has completed the procedure turn and is established inbound toward the airport on the final approach course within the distance prescribed in the procedure; and

(2) The pilot in command finds, on reaching the authorized MDA, or DA/DH, that the actual weather conditions are at or above the minimums prescribed for the procedure being used.

[Doc. No. 19779, 45 FR 67235, Oct. 9, 1980, as amended by Amdt. 125-2, 46 FR 24409, Apr. 30, 1981; Amdt. 125-45, 69 FR 1641, Jan. 9, 2004; Amdt. 125-52, 72 FR 31683, June 7, 2007]

§ 125.383 Load manifest.

(a) Each certificate holder is responsible for the preparation and accuracy of a load manifest in duplicate containing information concerning the loading of the airplane. The manifest must be prepared before each takeoff and must include—

(1) The number of passengers;

(2) The total weight of the loaded airplane;

(3) The maximum allowable takeoff and landing weights for that flight;

(4) The center of gravity limits;

(5) The center of gravity of the loaded airplane, except that the actual center of gravity need not be computed if the airplane is loaded according to a loading schedule or other approved method that ensures that the center of gravity of the loaded airplane is within approved limits. In those cases, an entry shall be made on the manifest indicating that the center of gravity is within limits according to a loading schedule or other approved method:

(6) The registration number of the airplane;

(7) The origin and destination; and

(8) Names of passengers.

(b) The pilot in command of an airplane for which a load manifest must be prepared shall carry a copy of the completed load manifest in the airplane to its destination. The certificate holder shall keep copies of completed load manifests for at least 30 days at its principal operations base, or at another location used by it and approved by the Administrator.

Subpart L—Records and Reports

§ 125.401 Crewmember record.

(a) Each certificate holder shall—

(1) Maintain current records of each crewmember that show whether or not that crewmember complies with this chapter (e.g., proficiency checks, airplane qualifications, any required physical examinations, and flight time records); and

(2) Record each action taken concerning the release from employment or physical or professional disqualification of any flight crewmember and keep the record for at least 6 months thereafter.

(b) Each certificate holder shall maintain the records required by paragraph (a) of this section at its principal operations base, or at another location used by it and approved by the Administrator.

(c) Computer record systems approved by the Administrator may be used in complying with the requirements of paragraph (a) of this section.

§ 125.403 Flight release form.

(a) The flight release may be in any form but must contain at least the following information concerning each flight:

(1) Company or organization name.

(2) Make, model, and registration number of the airplane being used.

(3) Date of flight.

(4) Name and duty assignment of each crewmember.

(5) Departure airport, destination airports, alternate airports, and route.

(6) Minimum fuel supply (in gallons or pounds).

(7) A statement of the type of operation (e.g., IFR, VFR).

(b) The airplane flight release must contain, or have attached to it, weather reports, available weather forecasts, or a combination thereof.

§ 125.405 Disposition of load manifest, flight release, and flight plans.

(a) The pilot in command of an airplane shall carry in the airplane to its destination the original or a signed copy of the—

(1) Load manifest required by § 125.383;

(2) Flight release;

(3) Airworthiness release; and

(4) Flight plan, including route.

(b) If a flight originates at the principal operations base of the certificate holder, it shall retain at that base a signed copy of each document listed in paragraph (a) of this section.

(c) Except as provided in paragraph (d) of this section, if a flight originates at a place other than the principal operations base of the certificate holder, the pilot in command (or another person not aboard the airplane who is authorized by the operator) shall, before or immediately after departure of the flight, mail signed copies of the documents listed in paragraph (a) of this section to the principal operations base.

(d) If a flight originates at a place other than the principal operations base of the certificate holder and there is at that place a person to manage the flight departure for the operator who does not depart on the airplane, signed copies of the documents listed in paragraph (a) of this section may be retained at that place for not more than 30 days before being sent to the principal operations base of the certificate holder. However, the documents for a particular flight need not be further retained at that place or be sent to the principal operations base, if the originals or other copies of them have been previously returned to the principal operations base.

(e) The certificate holder shall:

(1) Identify in its operations manual the person having custody of the copies of documents retained in accordance with paragraph (d) of this section; and

(2) Retain at its principal operations base either the original or a copy of the records required by this section for at least 30 days.

§ 125.407 Maintenance log: Airplanes.

(a) Each person who takes corrective action or defers action concerning a reported or observed failure or malfunction of an airframe, aircraft engine, propeller, or appliance shall record the action taken in the airplane maintenance log in accordance with part 43 of this chapter.

(b) Each certificate holder shall establish a procedure for keeping copies of the airplane maintenance log required by this section in the airplane for access by appropriate personnel and shall include that procedure in the manual required by § 125.249.

§ 125.409 Service difficulty reports.

(a) Each certificate holder shall report the occurrence or detection of each failure, malfunction, or defect, in a form and manner prescribed by the Administrator.

(b) Each certificate holder shall submit each report required by this section, covering each 24-hour period beginning at 0900 local time of each day and ending at 0900 local time on the next day, to the FAA office in Oklahoma City, Oklahoma. Each report of occurrences during a 24-hour period shall be submitted to the collection point within the next 96 hours. However, a report due on Saturday or Sunday may be submitted on the following Monday, and a report due on a holiday may be submitted on the next work day.

[Doc. No. 19779, 45 FR 67235, Oct. 9, 1980, as amended by Amdt. 125-49, 70 FR 76979, Dec. 29, 2005]

§ 125.411 Airworthiness release or maintenance record entry.

(a) No certificate holder may operate an airplane after maintenance, preventive maintenance, or alteration is performed on the airplane unless the person performing that maintenance, preventive maintenance, or alteration prepares or causes to be prepared—

(1) An airworthiness release; or

(2) An entry in the aircraft maintenance records in accordance with the certificate holder's manual.

(b) The airworthiness release or maintenance record entry required by paragraph (a) of this section must—

(1) Be prepared in accordance with the procedures set forth in the certificate holder's manual;

(2) Include a certification that—

(i) The work was performed in accordance with the requirements of the certificate holder's manual;

(ii) All items required to be inspected were inspected by an authorized person who determined that the work was satisfactorily completed;

(iii) No known condition exists that would make the airplane unairworthy; and

(iv) So far as the work performed is concerned, the airplane is in condition for safe operation; and

(3) Be signed by a person authorized in part 43 of this chapter to perform maintenance, preventive maintenance, and alteration.

(c) When an airworthiness release form is prepared, the certificate holder must give a copy to the pilot in command and keep a record of it for at least 60 days.

(d) Instead of restating each of the conditions of the certification required by paragraph (b) of this section, the certificate holder may state in its manual that the signature of a person authorized in part 43 of this chapter constitutes that certification.

Subpart M—Continued Airworthiness and Safety Improvements

Source: Amdt. 125-53, 72 FR 63412, Nov. 8, 2007, unless otherwise noted.

§ 125.501 Purpose and definition.

(a) This subpart requires operators to support the continued airworthiness of each airplane. These requirements may include, but are not limited to, revising the inspection program, incorporating design changes, and incorporating revisions to Instructions for Continued Airworthiness.

(b) For purposes of this subpart, the "FAA Oversight Office" is the aircraft certification office or office of the Transport Airplane Directorate with oversight responsibility for the relevant type certificate or supplemental type certificate, as determined by the Administrator.

§ 125.503 [Reserved]

§ 125.505 Repairs assessment for pressurized fuselages.

(a) No person may operate an Airbus Model A300 (excluding the -600 series), British Aerospace Model BAC 1-11, Boeing Model 707, 720, 727, 737 or 747, McDonnell Douglas Model DC-8, DC-9/MD-80 or DC-10, Fokker Model F28, or Lockheed Model L-1011 beyond the applicable flight cycle implementation time specified below, or May 25, 2001, whichever occurs later, unless operations specifications have been issued to reference repair assessment guidelines applicable to the fuselage pressure boundary (fuselage skin, door skin, and bulkhead webs), and those guidelines are incorporated in its maintenance program. The repair assessment guidelines must be approved by the FAA Aircraft Certification Office (ACO), or office of the Transport Airplane Directorate, having cognizance over the type certificate for the affected airplane.

(1) For the Airbus Model A300 (excluding the -600 series), the flight cycle implementation time is:

(i) Model B2: 36,000 flights.

(ii) Model B4-100 (including Model B4-2C): 30,000 flights above the window line, and 36,000 flights below the window line.

(iii) Model B4-200: 25,500 flights above the window line, and 34,000 flights below the window line.

(2) For all models of the British Aerospace BAC 1-11, the flight cycle implementation time is 60,000 flights.

(3) For all models of the Boeing 707, the flight cycle implementation time is 15,000 flights.

(4) For all models of the Boeing 720, the flight cycle implementation time is 23,000 flights.

(5) For all models of the Boeing 727, the flight cycle implementation time is 45,000 flights.

(6) For all models of the Boeing 737, the flight cycle implementation time is 60,000 flights.

(7) For all models of the Boeing 747, the flight cycle implementation time is 15,000 flights.

(8) For all models of the McDonnell Douglas DC-8, the flight cycle implementation time is 30,000 flights.

(9) For all models of the McDonnell Douglas DC-9/MD-80, the flight cycle implementation time is 60,000 flights.

(10) For all models of the McDonnell Douglas DC-10, the flight cycle implementation time is 30,000 flights.

(11) For all models of the Lockheed L-1011, the flight cycle implementation time is 27,000 flights.

(12) For the Fokker F-28 Mark, 1000, 2000, 3000, and 4000, the flight cycle implementation time is 60,000 flights.

(b) [Reserved]

[Doc. No. 29104, 65 FR 24126, Apr. 25, 2000; 65 FR 50744, Aug. 21, 2000, as amended by Amdt. 125-36, 66 FR 23131, May 7, 2001; Amdt. 125-40, 67 FR 72834, Dec. 9, 2002; Amdt. 125-46, 69 FR 45942, July 30, 2004. Redesignated by Amdt. 125-53, 72 FR 63412, Nov. 8, 2007]

§ 125.507 Fuel tank system inspection program.

(a) Except as provided in paragraph (g) of this section, this section applies to transport category, turbine-powered airplanes with a type certificate issued after January 1, 1958, that, as a result of original type certification or later increase in capacity, have—

(1) A maximum type-certificated passenger capacity of 30 or more, or

(2) A maximum payload capacity of 7500 pounds or more.

(b) For each airplane on which an auxiliary fuel tank is installed under a field approval, before June 16, 2008, the certificate holder must submit to the FAA Oversight Office proposed maintenance instructions for the tank that meet the requirements of Special Federal Aviation Regulation No. 88 (SFAR 88) of this chapter.

(c) After December 16, 2008, no certificate holder may operate an airplane identified in paragraph (a) of this section unless the inspection program for that airplane has been revised to include applicable inspections, procedures, and limitations for fuel tank systems.

(d) The proposed fuel tank system inspection program revisions must be based on fuel tank system Instructions for Continued Airworthiness (ICA) that have been developed in accordance with the applicable provisions of SFAR 88 of this chapter or § 25.1529 and part 25, Appendix H, of this chapter, in effect on June 6, 2001 (including those developed for auxiliary fuel tanks, if any, installed under supplemental type certificates or other design approval) and that have been approved by the FAA Oversight Office.

(e) After December 16, 2008, before returning an aircraft to service after any alteration for which fuel tank ICA are developed under SFAR 88, or under § 25.1529 in effect on June 6, 2001, the certificate holder must include in the inspection program for the airplane inspections and procedures for the fuel tank system based on those ICA.

(f) The fuel tank system inspection program changes identified in paragraphs (d) and (e) of this section and any later fuel tank system revisions must be submitted to the Principal Inspector for review and approval.

(g) This section does not apply to the following airplane models:

(1) Bombardier CL-44

(2) Concorde

(3) deHavilland D.H. 106 Comet 4C

(4) VFW-Vereinigte Flugtechnische Werk VFW-614

(5) Illyushin Aviation IL 96T

(6) Bristol Aircraft Britannia 305

(7) Handley Page Herald Type 300

(8) Avions Marcel Dassault—Breguet Aviation Mercure 100C

(9) Airbus Caravelle

(10) Lockheed L-300

§ 125.509 Flammability reduction means.

(a) *Applicability.* Except as provided in paragraph (m) of this section, this section applies to transport category, turbine-powered airplanes with a type certificate issued after January 1, 1958, that, as a result of original type certification or later increase in capacity have:

(1) A maximum type-certificated passenger capacity of 30 or more, or

(2) A maximum payload capacity of 7,500 pounds or more.

(b) *New Production Airplanes.* Except in accordance with § 125.201, no person may operate an airplane identified in Table 1 of this section (including all-cargo airplanes) for which the State of Manufacture issued the original certificate of airworthiness or export airworthiness approval after December 27, 2010 unless an Ignition Mitigation Means (IMM) or Flammability Reduction Means (FRM) meeting the requirements of § 26.33 of this chapter is operational.

Table 1

Model—Boeing	Model—Airbus
747 Series	A318, A319, A320, A321 Series
737 Series	A330, A340 Series
777 Series	
767 Series	

(c) *Auxiliary Fuel Tanks.* After the applicable date stated in paragraph (e) of this section, no person may operate any airplane subject to § 26.33 of this chapter that has an Auxiliary Fuel Tank installed pursuant to a field approval, unless the following requirements are met:

(1) The person complies with 14 CFR 26.35 by the applicable date stated in that section.

(2) The person installs Flammability Impact Mitigation Means (FIMM), if applicable, that is approved by the FAA Oversight Office.

(3) Except in accordance with § 125.201, the FIMM, if applicable, are operational.

(d) *Retrofit.* Except as provided in paragraph (j) of this section, after the dates specified in paragraph (e) of this section, no person may operate an airplane to which this section applies unless the requirements of paragraphs (d)(1) and (d)(2) of this section are met.

(1) Ignition Mitigation Means (IMM), Flammability Reduction Means (FRM), or FIMM, if required by §§ 26.33, 26.35, or 26.37 of this chapter, that are approved by the FAA Oversight Office, are installed within the compliance times specified in paragraph (e) of this section.

(2) Except in accordance with § 125.201 of this part, the IMM, FRM or FIMM, as applicable, are operational.

(e) *Compliance Times.* The installations required by paragraph (d) of this section must be accomplished no later than the applicable dates specified in paragraph (e)(1), (e)(2) or (e)(3) of this section.

(1) Fifty percent of each person's fleet of airplanes subject to paragraph (d)(1) of this section must be modified no later than December 26, 2014.

(2) One hundred percent of each person's fleet of airplanes subject to paragraph (d)(1) of this section must be modified no later than December 26, 2017.

(3) For those persons that have only one airplane of a model identified in Table 1 of this section, the airplane must be modified no later than December 26, 2017.

(f) *Compliance after Installation.* Except in accordance with § 125.201, no person may—

(1) Operate an airplane on which IMM or FRM has been installed before the dates specified in paragraph (e) of this section unless the IMM or FRM is operational, or

(2) Deactivate or remove an IMM or FRM once installed unless it is replaced by a means that complies with paragraph (d) of this section.

(g) *Inspection Program Revisions.* No person may operate an airplane for which airworthiness limitations have been approved by the FAA Oversight Office in accordance with §§ 26.33, 26.35, or 26.37 of this chapter after the airplane is modified in accordance with paragraph (d) of this section unless the inspection program for that airplane is revised to include those applicable airworthiness limitations.

(h) After the inspection program is revised as required by paragraph (g) of this section, before returning an airplane to service after any alteration for which airworthiness limitations are required by §§ 25.981, 26.33, 26.35, or 26.37 of this chapter, the person must revise the inspection program for the airplane to include those airworthiness limitations.

(i) The inspection program changes identified in paragraphs (g) and (h) of this section must be submitted to the operator's assigned Flight Standards Office responsible for review and approval prior to incorporation.

(j) The requirements of paragraph (d) of this section do not apply to airplanes operated in all-cargo service, but those airplanes are subject to paragraph (f) of this section.

(k) After the date by which any person is required by this section to modify 100 percent of the affected fleet, no person may operate in passenger service any airplane model specified in Table 2 of this section unless the airplane has been modified to comply with § 26.33(c) of this chapter.

Table 2

Model—Boeing	Model—Airbus
747 Series	A318, A319, A320, A321 Series.
737 Series	A300, A310 Series.
777 Series	A330, A340 Series.
767 Series	
757 Series	

(l) No person may operate any airplane on which an auxiliary fuel tank is installed after December 26, 2017 unless the FAA has certified the tank as compliant with § 25.981 of this chapter, in effect on December 26, 2008.

(m) *Exclusions.* The requirements of this section do not apply to the following airplane models:

(1) Convair CV-240, 340, 440, including turbine powered conversions.

(2) Lockheed L-188 Electra.

(3) Vickers VC-10.

(4) Douglas DC-3, including turbine powered conversions.

(5) Bombardier CL-44.

(6) Mitsubishi YS-11.

(7) BAC 1-11.

(8) Concorde.

(9) deHavilland D.H. 106 Comet 4C.

(10) VFW—Vereinigte Flugtechnische VFW-614.

(11) Illyushin Aviation IL 96T.

(12) Bristol Aircraft Britannia 305.

(13) Handley Page Herald Type 300.

(14) Avions Marcel Dassault—Breguet Aviation Mercure 100C.

(15) Airbus Caravelle.

(16) Fokker F-27/Fairchild Hiller FH-227.

(17) Lockheed L-300.

[Doc. No. FAA-2005-22997, 73 FR 42502, July 21, 2008, as amended by Amdt. 125-57, 74 FR 31619, July 2, 2009]

Appendix A to Part 125—Additional Emergency Equipment

(a) *Means for emergency evacuation.* Each passenger-carrying landplane emergency exit (other than over-the-wing) that is more that 6 feet from the ground with the airplane on the ground and the landing gear extended must have an approved means to assist the occupants in descending to the ground. The assisting means for a floor level emergency exit must meet the requirements of § 25.809(f)(1) of this chapter in effect on April 30, 1972, except that, for any airplane for which the application for the type certificate was filed after that date, it must meet the requirements under which the airplane was type certificated. An assisting means that deploys automatically must be armed during taxiing, takeoffs, and landings. However, if the Administrator finds that the design of the exit makes compliance impractical, the Administrator may grant a deviation from the requirement of automatic deployment if the assisting means automatically erects upon deployment and, with respect to required emergency exits, if an emergency evacuation demonstration is conducted in accordance with § 125.189. This paragraph does not apply to the rear window emergency exit of DC-3 airplanes operated with less than 36 occupants, including crewmembers, and less than five exits authorized for passenger use.

(b) *Interior emergency exit marking.* The following must be complied with for each passenger-carrying airplane:

(1) Each passenger emergency exit, its means of access, and means of opening must be conspicuously marked. The identity and location of each passenger emergency exit must be recognizable from a distance equal to the width of the cabin. The location of each passenger emergency exit must be indicated by a sign visible to occupants approaching along the main passenger aisle. There must be a locating sign—

(i) Above the aisle near each over-the-wing passenger emergency exit, or at another ceiling location if it is more practical because of low headroom;

(ii) Next to each floor level passenger emergency exit, except that one sign may serve two such exits if they both can be seen readily from that sign; and

(iii) On each bulkhead or divider that prevents fore and aft vision along the passenger cabin, to indicate emergency exits beyond and obscured by it, except that if this is not possible the sign may be placed at another appropriate location.

(2) Each passenger emergency exit marking and each locating sign must meet the following:

(i) For an airplane for which the application for the type certificate was filed prior to May 1, 1972, each passenger emergency exit marking and each locating sign must be manufactured to meet the requirements of § 25.812(b) of this chapter in effect on April 30, 1972. On these airplanes, no sign may continue to be used if its luminescence (brightness) decreases to below 100 microlamberts. The colors may be reversed if it increases the emergency illumination of the passenger compartment. However, the Administrator may authorize deviation from the 2-inch background requirements if the Administrator finds that special circumstances exist that make compliance impractical and that the proposed deviation provides an equivalent level of safety.

(ii) For an airplane for which the application for the type certificate was filed on or after May 1, 1972, each passenger emergency exit marking and each locating sign must be manufactured to meet the interior emergency exit marking requirements under which the airplane was type certificated. On these airplanes, no sign may continue to be used if its luminescence (brightness) decreases to below 250 microlamberts.

(c) *Lighting for interior emergency exit markings.* Each passenger-carrying airplane must have an emergency lighting system, independent of the main lighting system. However, sources of general cabin illumination may be common to both the emergency and the main lighting systems if the power supply to the emergency lighting system is independent of the power supply to the main lighting system. The emergency lighting system must—

(1) Illuminate each passenger exit marking and locating sign; and

(2) Provide enough general lighting in the passenger cabin so that the average illumination, when measured at 40-inch intervals at seat armrest height, on the centerline of the main passenger aisle, is at least 0.05 foot-candles.

(d) *Emergency light operation.* Except for lights forming part of emergency lighting subsystems provided in compliance with § 25.812(g) of this chapter (as prescribed in paragraph (h) of this section) that serve no more than one assist means, are independent of the airplane's main emergency lighting systems, and are automatically activated when the assist means is deployed, each light required by paragraphs (c) and (h) must comply with the following:

(1) Each light must be operable manually and must operate automatically from the independent lighting system—

(i) In a crash landing; or

(ii) Whenever the airplane's normal electric power to the light is interrupted.

(2) Each light must—

(i) Be operable manually from the flightcrew station and from a point in the passenger compartment that is readily accessible to a normal flight attendant seat;

(ii) Have a means to prevent inadvertent operation of the manual controls; and

(iii) When armed or turned on at either station, remain lighted or become lighted upon interruption of the airplane's normal electric power.

Each light must be armed or turned on during taxiing, takeoff, and landing. In showing compliance with this paragraph, a transverse vertical separation of the fuselage need not be considered.

(3) Each light must provide the required level of illumination for at least 10 minutes at the critical ambient conditions after emergency landing.

(e) *Emergency exit operating handles.* (1) For a passenger-carrying airplane for which the application for the type certificate was filed prior to May 1, 1972, the location of each passenger emergency exit operating handle and instructions for opening the exit must be shown by a marking on or near the exit that is readable from a distance of 30 inches. In addition, for each Type I and Type II emergency exit with a locking mechanism released by rotary motion of the handle, the instructions for opening must be shown by—

(i) A red arrow with a shaft at least $^3/_4$ inch wide and a head twice the width of the shaft, extending along at least 70 degrees of arc at a radius approximately equal to $^3/_4$ of the handle length; and

(ii) The word "open" in red letters 1 inch high placed horizontally near the head of the arrow.

(2) For a passenger-carrying airplane for which the application for the type certificate was filed on or after May 1, 1972, the location of each passenger emergency exit operating handle and instructions for opening the exit must be shown in accordance with the requirements under which the airplane was type certificated. On these airplanes, no operating handle or operating handle cover may continue to be used if its luminescence (brightness) decreases to below 100 microlamberts.

(f) *Emergency exit access.* Access to emergency exits must be provided as follows for each passenger-carrying airplane:

(1) Each passageway between individual passenger areas, or leading to a Type I or Type II emergency exit, must be unobstructed and at least 20 inches wide.

(2) There must be enough space next to each Type I or Type II emergency exit to allow a crewmember to assist in the evacuation of passengers without reducing the unobstructed width of the passageway below that required in paragraph (f)(1) of this section. However, the Administrator may authorize deviation from this requirement for an airplane certificated under the provisions of part 4b of the Civil Air Regulations in effect before December 20, 1951, if the Administrator finds that special circumstances exist that provide an equivalent level of safety.

(3) There must be access from the main aisle to each Type III and Type IV exit. The access from the aisle to these exits must not be obstructed by seats, berths, or other protrusions in a manner that would reduce the effectiveness of the exit. In addition—

(i) For an airplane for which the application for the type certificate was filed prior to May 1, 1972, the access must meet the requirements of § 25.813(c) of this chapter in effect on April 30, 1972; and

(ii) For an airplane for which the application for the type certificate was filed on or after May 1, 1972, the access must meet the emergency exit access requirements under which the airplane was certificated.

(4) If it is necessary to pass through a passageway between passenger compartments to reach any required emergency exit from any seat in the passenger cabin, the passageway must not be obstructed. However, curtains may be used if they allow free entry through the passageway.

(5) No door may be installed in any partition between passenger compartments.

(6) If it is necessary to pass through a doorway separating the passenger cabin from other areas to reach any required emergency exit from any passenger seat, the door must have a means to latch it in open position, and the door must be latched open during each takeoff and landing. The latching means must be able to withstand the loads imposed upon it when the door is subjected to the ultimate interia forces, relative to the surrounding structure, listed in § 25.561(b) of this chapter.

(g) *Exterior exit markings.* Each passenger emergency exit and the means of opening that exit from the outside must be marked on the outside of the airplane. There must be a 2-inch colored band outlining each passenger emergency exit on the side of the fuselage. Each outside marking, including the band, must be readily distinguishable from the surrounding fuselage area by contrast in color. The markings must comply with the following:

(1) If the reflectance of the darker color is 15 percent or less, the reflectance of the lighter color must be at least 45 percent. "Reflectance" is the ratio of the luminous flux reflected by a body to the luminous flux it receives.

(2) If the reflectance of the darker color is greater than 15 percent, at least a 30 percent difference between its reflectance and the reflectance of the lighter color must be provided.

(3) Exits that are not in the side of the fuselage must have the external means of opening and applicable instructions marked conspicuously in red or, if red is inconspicuous against the background color, in bright chrome yellow and, when the opening means for such an exit is located on only one side of the fuselage, a conspicuous marking to that effect must be provided on the other side.

(h) *Exterior emergency lighting and escape route.* (1) Each passenger-carrying airplane must be equipped with exterior lighting that meets the following requirements:

(i) For an airplane for which the application for the type certificate was filed prior to May 1, 1972, the requirements of § 25.812(f) and (g) of this chapter in effect on April 30, 1972.

(ii) For an airplane for which the application for the type certificate was filed on or after May 1, 1972, the exterior emergency lighting requirements under which the airplane was type certificated.

(2) Each passenger-carrying airplane must be equipped with a slip-resistant escape route that meets the following requirements:

(i) For an airplane for which the application for the type certificate was filed prior to May 1, 1972, the requirements of § 25.803(e) of this chapter in effect on April 30, 1972.

(ii) For an airplane for which the application for the type certificate was filed on or after May 1, 1972, the slip-resistant escape route requirements under which the airplane was type certificated.

(i) *Floor level exits.* Each floor level door or exit in the side of the fuselage (other than those leading into a cargo or baggage compartment that is not accessible from the passenger cabin) that is 44 or more inches high and 20 or more inches wide, but not wider than 46 inches, each passenger ventral exit (except the ventral exits on M-404 and CV-240 airplanes) and each tail cone exit must meet the requirements of this section for floor level emergency exits. However, the Administrator may grant a deviation from this paragraph if the Administrator finds that circumstances make full compliance impractical and that an acceptable level of safety has been achieved.

(j) *Additional emergency exits.* Approved emergency exits in the passenger compartments that are in excess of the minimum number of required emergency exits must meet all of the applicable provisions of this section except paragraph (f), (1), (2), and (3) and must be readily accessible.

(k) On each large passenger-carrying turbojet-powered airplane, each ventral exit and tailcone exit must be—

(1) Designed and constructed so that it cannot be opened during flight; and

(2) Marked with a placard readable from a distance of 30 inches and installed at a conspicuous location near the means of opening the exit, stating that the exit has been designed and constructed so that it cannot be opened during flight.

Appendix B to Part 125—Criteria for Demonstration of Emergency Evacuation Procedures Under § 125.189

(a) *Aborted takeoff demonstration.* (1) The demonstration must be conducted either during the dark of the night or during daylight with the dark of the night simulated. If the demonstration is conducted indoors during daylight hours, it must be conducted with each window covered and each door closed to minimize the daylight effect. Illumination on the floor or ground may be used, but it must be kept low and shielded against shining into the airplane's windows or doors.

(2) The airplane must be in a normal ground attitude with landing gear extended.

(3) Stands or ramps may be used for descent from the wing to the ground. Safety equipment such as mats or inverted life rafts may be placed on the ground to protect participants. No other equipment that is not part of the airplane's emergency evacuation equipment may be used to aid the participants in reaching the ground.

(4) The airplane's normal electric power sources must be deenergized.

(5) All emergency equipment for the type of passenger-carrying operation involved must be installed in accordance with the certificate holder's manual.

(6) Each external door and exit and each internal door or curtain must be in position to simulate a normal takeoff.

(7) A representative passenger load of persons in normal health must be used. At least 30 percent must be females. At least 5 percent must be over 60 years of age with a proportionate number of females. At least 5 percent, but not more than 10 percent, must be children under 12 years of age, prorated through that age group. Three life-size dolls, not included as part of the total passenger load, must be carried by passengers to simulate live infants 2 years old or younger. Crewmembers, mechanics, and training personnel who maintain or operate the airplane in the normal course of their duties may not be used as passengers.

(8) No passenger may be assigned a specific seat except as the Administrator may require. Except as required by item (12) of this paragraph, no employee of the certificate holder may be seated next to an emergency exit.

(9) Seat belts and shoulder harnesses (as required) must be fastened.

(10) Before the start of the demonstration, approximately one-half of the total average amount of carry-on baggage, blankets, pillows, and other similar articles must be distributed at several locations in the aisles and emergency exit access ways to create minor obstructions.

(11) The seating density and arrangement of the airplane must be representative of the highest capacity passenger version of that airplane the certificate holder operates or proposes to operate.

(12) Each crewmember must be a member of a regularly scheduled line crew, must be seated in that crewmember's normally assigned seat for takeoff, and must remain in that seat until the signal for commencement of the demonstration is received.

(13) No crewmember or passenger may be given prior knowledge of the emergency exits available for the demonstration.

(14) The certificate holder may not practice, rehearse, or describe the demonstration for the participants nor may any participant have taken part in this type of demonstration within the preceding 6 months.

(15) The pretakeoff passenger briefing required by § 125.327 may be given in accordance with the certificate holder's manual. The passengers may also be warned to follow directions of crewmembers, but may not be instructed on the procedures to be followed in the demonstration.

(16) If safety equipment as allowed by item (3) of this section is provided, either all passenger and cockpit windows must be blacked out or all of the emergency exits must have safety equipment to prevent disclosure of the available emergency exits.

(17) Not more than 50 percent of the emergency exits in the sides of the fuselage of an airplane that meet all of the requirements applicable to the required emergency exits for that airplane may be used for the demonstration. Exits that are not to be used in the demonstration must have the exit handle deactivated or must be indicated by red lights, red tape or other acceptable means, placed outside the exits to indicate fire or other reason that they are unusable. The exits to be used must be representative of all of the emergency exits on the airplane and must be designated by the certificate holder, subject to approval by the Administrator. At least one floor level exit must be used.

(18) All evacuees, except those using an over-the-wing exit, must leave the airplane by a means provided as part of the airplane's equipment.

(19) The certificate holder's approved procedures and all of the emergency equipment that is normally available, including slides, ropes, lights, and megaphones, must be fully utilized during the demonstration.

(20) The evacuation time period is completed when the last occupant has evacuated the airplane and is on the ground. Evacuees using stands or ramps allowed by item (3) above are considered to be on the ground when they are on the stand or ramp: *Provided,* That the acceptance rate of the stand or ramp is no greater than the acceptance rate of the means available on the airplane for descent from the wing during an actual crash situation.

(b) *Ditching demonstration.* The demonstration must assume that daylight hours exist outside the airplane and that all required crewmembers are available for the demonstration.

(1) If the certificate holder's manual requires the use of passengers to assist in the launching of liferafts, the needed passengers must be aboard the airplane and participate in the demonstration according to the manual.

(2) A stand must be placed at each emergency exit and wing with the top of the platform at a height simulating the water level of the airplane following a ditching.

(3) After the ditching signal has been received, each evacuee must don a life vest according to the certificate holder's manual.

(4) Each liferaft must be launched and inflated according to the certificate holder's manual and all other required emergency equipment must be placed in rafts.

(5) Each evacuee must enter a liferaft and the crewmembers assigned to each liferaft must indicate the location of emergency equipment aboard the raft and describe its use.

(6) Either the airplane, a mockup of the airplane, or a floating device simulating a passenger compartment must be used.

(i) If a mockup of the airplane is used, it must be a life-size mockup of the interior and representative of the airplane currently used by or proposed to be used by the certificate holder and must contain adequate seats for use of the evacuees. Operation of the emergency exits and the doors must closely simulate that on the airplane. Sufficient wing area must be installed outside the over-the-wing exits to demonstrate the evacuation.

(ii) If a floating device simulating a passenger compartment is used, it must be representative, to the extent possible, of the passenger compartment of the airplane used in operations. Operation of the emergency exits and the doors must closely simulate operation on that airplane. Sufficient wing area must be installed outside the over-the-wing exits to demonstrate the evacuation. The device must be equipped with the same survival equipment as is installed on the airplane, to accommodate all persons participating in the demonstration.

Appendix C to Part 125—Ice Protection

If certification with ice protection provisions is desired, compliance with the following must be shown:

(a) The recommended procedures for the use of the ice protection equipment must be set forth in the Airplane Flight Manual.

(b) An analysis must be performed to establish, on the basis of the airplane's operational needs, the adequacy of the ice protection system for the various components of the airplane. In addition, tests of the ice protection system must be conducted to demonstrate that the airplane is capable of operating safely in continuous maximum and intermittent maximum icing conditions as described in appendix C of part 25 of this chapter.

(c) Compliance with all or portions of this section may be accomplished by reference, where applicable because of similarity of the designs, to analyses and tests performed by the applicant for a type certificated model.

APPENDIX D TO PART 125—AIRPLANE FLIGHT RECORDER SPECIFICATION

Parameters	Range	Accuracy sensor input to DFDR readout	Sampling interval (per second)	Resolution4read out
Time (GMT or Frame Counter) (range 0 to 4095, sampled 1 per frame)	24 Hrs	±0.125% Per Hour	0.25 (1 per 4 seconds)	1 sec.
Altitude	–1,000 ft to max certificated altitude of aircraft	±100 to ±700 ft (See Table 1, TSO-C51a)	1	5' to 35'[1]
Airspeed	50 KIAS to V_{so}, and V_{so} to 1.2 V_D	±5%, ±3%	1	1 kt.
Heading	360°	±2°	1	0.5°
Normal Acceleration (Vertical)	–3g to +6g	±1% of max range excluding datum error of ±5%	8	0.01g.
Pitch Attitude	±75°	±2°	1	0.5°.
Roll Attitude	±180°	±2°	1	0.5°.
Radio Transmitter Keying	On-Off (Discrete)		1	
Thrust/Power on Each Engine	Full range forward	±2%	1	0.2%[2]
Trailing Edge Flap or Cockpit Control Selection	Full range or each discrete position	±3° or as pilot's Indicator	0.5	0.5%[2]
Leading Edge Flap or Cockpit Control Selection	Full range or each discrete position	±3° or as pilot's indicator	0.5	0.5%[2]
Thrust Reverser Position	Stowed, in transit, and reverse (Discrete)		1 (per 4 seconds per engine)	
Ground Spoiler Position/Speed Brake Selection	Full range or each discrete position	±2% unless higher accuracy uniquely required	1	0.2%[2].
Marker Beacon Passage	Discrete		1	
Autopilot Engagement	Discrete		1	
Longitudinal Acceleration	±1g	±1.5% max range excluding datum error of ±5%	4	0.01g
Pilot Input and/or Surface Position-Primary Controls (Pitch, Roll, Yaw)[3]	Full range	±2° unless higher accuracy uniquely required	1	0.2%[2].
Lateral Acceleration	±1g	±1.5% max range excluding datum error of ±5%	4	0.01g.
Pitch Trim Position	Full range	±3% unless higher accuracy uniquely required	1	0.3%[2]
Glideslope Deviation	±400 Microamps	±3%	1	0.3%[2]
Localizer Deviation	±400 Microamps	±3%	1	0.3%[2].
AFCS Mode and Engagement Status	Discrete		1	
Radio Altitude	–20 ft to 2,500 ft	±2 Ft or ±3% Whichever is Greater Below 500 Ft and ±5% Above 500 Ft		1 ft + 5%[2] above 500'.
Master Warning	Discrete		1	
Main Gear Squat Switch Status	Discrete		1	
Angle of Attack (if recorded directly)	As installed	As installed	2	0.3%[2].
Outside Air Temperature or Total Air Temperature	–50 °C to +90 °C	±2 °C	0.5	0.3 °C
Hydraulics, Each System Low Pressure	Discrete		0.5	or 0.5%[2].
Groundspeed	As Installed	Most Accurate Systems Installed (IMS Equipped Aircraft Only)	1	0.2%[2].
If additional recording capacity is available, recording of the following parameters is recommended. The parameters are listed in order of significance:				
Drift Angle	When available. As installed	As installed	4	
Wind Speed and Direction	When available. As installed	As installed	4	
Latitude and Longitude	When available. As installed	As installed	4	
Brake pressure/Brake pedal position	As installed	As installed	1	
Additional engine parameters:				
EPR	As installed	As installed	1 (per engine)	
N^1	As installed	As installed	1 (per engine)	
N^2	As installed	As installed	1 (per engine)	
EGT	As installed	As installed	1 (per engine)	

Parameters	Range	Accuracy sensor input to DFDR readout	Sampling interval (per second)	Resolution4read out
Throttle Lever Position	As installed	As installed	1 (per engine)	
Fuel Flow	As installed	As installed	1 (per engine)	
TCAS:				
TA	As installed	As installed	1	
RA	As installed	As installed	1	
Sensitivity level (as selected by crew)	As installed	As installed	2	
GPWS (ground proximity warning system)	Discrete		1	
Landing gear or gear selector position	Discrete		0.25 (1 per 4 seconds)	
DME 1 and 2 Distance	0–200 NM;	As installed	0.25	1 mi.
Nav 1 and 2 Frequency Selection	Full range	As installed	0.25	

[1] When altitude rate is recorded. Altitude rate must have sufficient resolution and sampling to permit the derivation of altitude to 5 feet.

[2] Percent of full range.

[3] For airplanes that can demonstrate the capability of deriving either the control input on control movement (one from the other) for all modes of operation and flight regimes, the "or" applies. For airplanes with non-mechanical control systems (fly-by-wire) the "and" applies. In airplanes with split surfaces, suitable combination of inputs is acceptable in lieu of recording each surface separately.

[4] This column applies to aircraft manufactured after October 11, 1991.

[Doc. No. 25530, 53 FR 26150, July 11, 1988; 53 FR 30906, Aug. 16, 1988]

Appendix E to Part 125—Airplane Flight Recorder Specifications

Link to an amendment published at 78 FR 39971, July 3, 2013.

The recorded values must meet the designated range, resolution and accuracy requirements during static and dynamic conditions. Dynamic condition means the parameter is experiencing change at the maximum rate attainable, including the maximum rate of reversal. All data recorded must be correlated in time to within one second.

Parameters	Range	Accuracy (sensor input)	Seconds per sampling interval	Resolution	Remarks
1. Time or Relative Times Counts.[1]	24 Hrs, 0 to 4095	±0.125% Per Hour	4	1 sec	UTC time preferred when available. Count increments each 4 seconds of system operation.
2. Pressure Altitude	−1000 ft to max certificated altitude of aircraft. +5000 ft	±100 to ±700 ft (see table, TSO C124a or TSO C51a)	1	5' to 35'	Data should be obtained from the air data computer when practicable.
3. Indicated airspeed or Calibrated airspeed	50 KIAS or minimum value to Max V_{so}, to 1.2 V_D	±5% and ±3%	1	1 kt	Data should be obtained from the air data computer when practicable.
4, Heading (Primary flight crew reference)	0–360° and Discrete "true" or "mag"	±2°	1	0.5°	When true or magnetic heading can be selected as the primary heading reference, a discrete indicating selection must be recorded.
5. Normal Acceleration (Vertical)[9]	−3g to +6g	±1% of max range excluding datum error of ±5%	0.125	0.004g.	
6. Pitch Attitude	±75°	±2°	1 or 0.25 for airplanes operated under §125.226(f)	0.5°	A sampling rate of 0.25 is recommended.
7. Roll Attitude[2]	±180°	±2°	1 or 0.5 for airplanes operated under §121.344(f)	0.5°	A sampling rate of 0.5 is recommended.
8. Manual Radio Transmitter Keying or CVR/DFDR synchronization reference	On-Off (Discrete) None.		1		Preferably each crew member but one discrete acceptable for all transmission provided the CVR/FDR system complies with TSO C124a CVR synchronization requirements (paragraph 4.2.1 ED–55).
9. Thrust/Power on each engine—primary flight crew reference	Full Range Forward	±2%	1 (per engine)	0.3% of full range	Sufficient parameters (e.g., EPR, N1 or Torque, NP) as appropriate to the particular engine being recorded to determine power in forward and reverse thrust, including potential overspeed condition.
10. Autopilot Engagement	Discrete "on" or "off"		1.		
11. Longitudinal Acceleration	±1g	±1.5% max. range excluding datum error of ±5%	0.25	0.004g.	
12a. Pitch control(s) position (nonfly-by-wire systems)[18]	Full range	±2° unless higher accuracy uniquely required	0.5 or 0.25 for airplanes operated under §125.226(f)	0.5% of full range	For airplanes that have a flight control breakaway capability that allows either pilot to operate the controls independently, record both control inputs. The control inputs may be sampled alternately once per second to produce the sampling interval of 0.5 or 0.25, as applicable.

Parameters	Range	Accuracy (sensor input)	Seconds per sampling interval	Resolution	Remarks
12b. Pitch control(s) position (fly-by-wire systems)[3 18]	Full range	±2° unless higher accuracy uniquely required	0.5 or 0.25 for airplanes operated under §125.226(f)	0.2% of full range	
13a. Lateral control position(s) (nonfly-by-wire)[18]	Full range	±2° unless higher accuracy uniquely required	0.5 or 0.25 for airplanes operated under §125.226(f)	0.2% of full range	For airplanes that have a flight control break away capability that allows either pilot to operate the controls independently, record both control inputs. The control inputs may be sampled alternately once per second to produce the sampling interval of 0.5 or 0.25, as applicable.
13b. Lateral control position(s) (fly-by-wire)[4 18]	Full range	±2° unless higher accuracy uniquely required	0.5 or 0.25 for airplanes operated under §125.226(f)	0.2% of full range	
14a.Yaw control position(s) (nonfly-by-wire)[5 18]	Full range	±2° unless higher accuracy uniquely required	0.5	0.3% of full range	For airplanes that have a flight control breakaway capability that allows either pilot to operate the controls independently, record both control inputs. The control inputs may be sampled alternately once per second to produce the sampling interval of 0.5.
14b. Yaw control position(s) (fly-by-wire)[18]	Full range	±2° unless higher accuracy uniquely required	0.5	0.2% of full range	
15. Pitch control surface(s) position[6 18]	Full range	±2° unless higher accuracy uniquely required	0.5 or 0.25 for airplanes operated under §125.226(f)	0.3% of full range	For airplanes fitted with multiple or split surfaces, a suitable combination of inputs is acceptable in lieu of recording each surface separately. The control surfaces may be sampled alternately to produce the sampling interval of 0.5 or 0.25, as applicable.
16. Lateral control surface(s) position[7 18]	Full Range	±2° unless higher accuracy uniquely required	0.5 or 0.25 for airplanes operated under §125.226(f)	0.2% of full range	A suitable combination of surface position sensors is acceptable in lieu of recording each surface separately. The control surfaces may be sampled alternately to produce the sampling interval of 0.5 or 0.25, as applicable.
17. Yaw control surface(s) position[8 18]	Full range	±2° unless higher accuracy uniquely required	0.5	0.2% of full range	For airplanes fitted with multiple or split surfaces, a suitable combination of surface position sensors is acceptable in lieu of recording each surface separately. The control surfaces may be sampled alternately to produce the sampling interval of 0.5.
18. Lateral Acceleration	±1g	±1.5% max. range excluding datum error of ±5%	0.25	0.004g.	
19. Pitch Trim Surface Position	Full Range	±3° Unless Higher Accuracy Uniquely Required	1	0.6% of full range	
20. Trailing Edge Flap or Cockpit Control Selection.[10]	Full Range or Each Position (discrete)	±3° or as Pilot's indicator	2	0.5% of full range	Flap position and cockpit control may each be sampled at 4 second intervals, to give a data point every 2 seconds.
21. Leading Edge Flap or Cockpit Control Selection.[11]	Full Range or Each Discrete Position	±3° or as Pilot's indicator and sufficient to determine each discrete position	2	0.5% of full range	Left and right sides, or flap position and cockpit control may each be sampled at 4 second intervals, so as to give a data point every 2 seconds.
22. Each Thrust Reverser Position (or equivalent for propeller airplane)	Stowed, In Transit, and Reverse (Discrete)		1 (per engine).		Turbo-jet—2 discretes enable the 3 states to be determined. Turbo-prop—1 discrete.
23. Ground Spoiler Position or Speed Brake Selection[12]	Full Range or Each Position (discrete)	±2° Unless higher accuracy uniquely required	1 or 0.5 for airplanes operated under §125.226(f)	0.2% of full range	
24. Outside Air Temperature or Total Air Temperature.[13]	−50 °C to +90 °C	±2 °C	2	0.3 °C.	
25. Autopilot/Autothrottle/ AFCS Mode and Engagement Status	A suitable combination of discretes		1		Discretes should show which systems are engaged and which primary modes are controlling the flight path and speed of the aircraft.
26. Radio Altitude[14]	−20 ft to 2,500 ft	±2 ft or ±3% Whichever is Greater Below 500 ft and ±5% above 500 ft	1	1 ft +5% Above 500 ft	For autoland/category 3 operations. Each radio altimeter should be recorded, but arranged so that at least one is recorded each second.

Parameters	Range	Accuracy (sensor input)	Seconds per sampling interval	Resolution	Remarks
27. Localizer Deviation, MLS Azimuth, or GPS Lateral Deviation	±400 Microamps or available sensor range as installed ±62°	As installed. ±3% recommended	1	0.3% of full range	For autoland/category 3 operations. each system should be recorded but arranged so that at least one is recorded each second. It is not necessary to record ILS and MLS at the same time, only the approach aid in use need be recorded.
28. Glideslope Deviation, MLS Elevation, or GPS Vertical Deviation	±400 Microamps or available sensor range as installed. 0.9 to + 30°	As installed ±3% recommended	1	0.3% of full range	For autoland/category 3 operations. each system should be recorded but arranged so that at least one is recorded each second. It is not necessary to record ILS and MLS at the same time, only the approach aid in use need be recorded.
29. Marker Beacon Passage	Discrete "on" or "off"		1		A single discrete is acceptable for all markers.
30. Master Warning	Discrete		1		Record the master warning and record each 'red' warning that cannot be determined from other parameters or from the cockpit voice recorder.
31. Air/ground sensor (primary airplane system reference nose or main gear)	Discrete "air" or "ground"		1 (0.25 recommended).		
32. Angle of Attack (If measured directly)	As installed	As Installed	2 or 0.5 for airplanes operated under §125.226(f)	0.3% of full range	If left and right sensors are available, each may be recorded at 4 or 1 second intervals, as appropriate, so as to give a data point at 2 seconds or 0.5 second, as required.
33. Hydraulic Pressure Low, Each System	Discrete or available sensor range, "low" or "normal"	±5%	2	0.5% of full range.	
34. Groundspeed	As Installed	Most Accurate Systems Installed	1	0.2% of full range.	
35. GPWS (ground proximity warning system)	Discrete "warning" or "off"		1		A suitable combination of discretes unless recorder capacity is limited in which case a single discrete for all modes is acceptable.
36. Landing Gear Position or Landing gear cockpit control selection	Discrete		4		A suitable combination of discretes should be recorded.
37. Drift Angle.[15]	As installed	As installed	4	0.1%.	
38. Wind Speed and Direction	As installed	As installed	4	1 knot, and 1.0°.	
39. Latitude and Longitude	As installed	As installed	4	0.002°, or as installed	Provided by the Primary Navigation System Reference. Where capacity permits Latitude/ longtitude resolution should be 0.0002°.
40. Stick shaker and pusher activation	Discrete(s) "on" or "off"		1		A suitable combination of discretes to determine activation.
41. WIndshear Detection	Discrete "warning" or "off"		1		
42. Throttle/power lever position.[16]	Full Range	±2%	1 for each lever	2% of full range	For airplanes with non-mechanically linked cockpit engine controls.
43. Additional Engine Parameters	As installed	As installed	Each engine each second	2% of full range	Where capacity permits, the preferred priority is indicated vibration level, N2, EGT, Fuel Flow, Fuel Cut-off lever position and N3, unless engine manufacturer recommends otherwise.
44. Traffic Alert and Collision Avoidance System (TCAS)	Discretes	As installed	1		A suitable combination of discretes should be recorded to determine the status of-Combined Control, Vertical Control, Up Advisory, and Down Advisory. (ref. ARINC Characteristic 735 Attachment 6E, TCAS VERTICAL RA DATA OUTPUT WORD.)
45. DME 1 and 2 Distance	0–200 NM	As installed	4	1 NM	1 mile.
46. Nav 1 and 2 Selected Frequency	Full range	As installed	4		Sufficient to determine selected frequency
47. Selected barometric setting	Full range	±5%	(1 per 64 sec.)	0.2% of full range.	
48. Selected Altitude	Full range	±5%	1	100 ft.	
49. Selected speed	Full range	±5%	1	1 knot.	
50. Selected Mach	Full range	±5%	1	.01.	
51. Selected vertical speed	Full range	±5%	1	100 ft/min.	

Parameters	Range	Accuracy (sensor input)	Seconds per sampling interval	Resolution	Remarks
52. Selected heading	Full range	±5%	1	1°.	
53. Selected flight path	Full range	±5%	1	1°.	
54. Selected decision height	Full range	±5%	64	1 ft.	
55. EFIS display format	Discrete(s)		4		Discretes should show the display system status (e.g., off, normal, fail, composite, sector, plan, nav aids, weather radar, range, copy).
56. Multi-function/Engine Alerts Display format	Discrete(s)		4		Discretes should show the display system status (e.g., off, normal, fail, and the identity of display pages for emergency procedures, need not be recorded).
57. Thrust command.[17]	Full Range	±2%	2	2% of full range	
58. Thrust target	Full range	±2%	4	2% of full range.	
59. Fuel quantity in CG trim tank	Full range	±5%	(1 per 64 sec.)	1% of full range.	
60. Primary Navigation System Reference	Discrete GPS, INS, VOR/DME, MLS, Loran C, Omega, Localizer Glideslope		4		A suitable combination of discrete to determine the Primary Navigation System reference.
61. Ice Detection	Discrete "ice" or "no ice"		4		
62. Engine warning each engine vibration	Discrete		1		
63. Engine warning each engine over temp	Discrete		1		
64. Engine warning each engine oil pressure low	Discrete		1		
65. Engine warning each engine over speed	Discrete		1		
66. Yaw Trim Surface Position	Full Range	±3% Unless Higher Accuracy Uniquely Required	2	0.3% of full range.	
67. Roll Trim Surface Position	Full Range	±3% Unless Higher Accuracy Uniquely Required	2	0.3% of full range.	
68. Brake Pressure (left and right)	As installed	±5%	1		To determine braking effort applied by pilots or by autobrakes.
69. Brake Pedal Application (left and right)	Discrete or Analog "applied" or "off"	±5% (Analog)	1		To determine braking applied by pilots.
70. Yaw or sideslip angle	Full Range	±5%	1	0,5°.	
71. Engine bleed valve position	Decrete "open" or "closed"		4		
72. De-icing or anti-icing system selection	Discrete "on" or "off"		4		
73. Computed center of gravity	Full Range	±5%	(1 per 64 sec.)	1% of full range.	
74. AC electrical bus status	Discrete "power" or "off"		4		Each bus.
75. DC electrical bus status	Discrete "power" or "off"		4		Each bus.
76. APU bleed valve position	Discrete "open" or "closed		4.		
77. Hydraulic Pressure (each system)	Full range	±5%	2	100 psi.	
78. Loss of cabin pressure	Discrete "loss" or "normal"		1.		
79. Computer failure (critical flight and engine control systems)	Discrete "fail" or "normal"		4.		
80. Heads-up display (when an information source is installed)	Discrete(s) "on" or "off"		4.		
81. Para-visual display (when an information source is installed)	Discrete(s) "on" or "off"		1.		

Parameters	Range	Accuracy (sensor input)	Seconds per sampling interval	Resolution	Remarks
82. Cockpit trim control input position—pitch	Full Range	±5%	1	0.2% of full range	Where mechanical means for control inputs are not available, cockpit display trim positions should be recorded.
83. Cockpit trim control input position—roll	Full Range	±5%	1	0.7% of full range	Where mechanical means for control inputs are not available, cockpit display trim position should be recorded.
84. Cockpit trim control input position—yaw	Full Range	±5%	1	0.3% of full range	Where mechanical means for control input are not available, cockpit display trim positions should be recorded.
85. Trailing edge flap and cockpit flap control position	Full Range	±5%	2	0.5% of full range	Trailing edge flaps and cockpit flap control position may each be sampled alternately at 4 second intervals to provide a sample each 0.5 second.
86. Leading edge flap and cockpit flap control position	Full Range or Discrete	±5%	1	0.5% of full range.	
87. Ground spoiler position and speed brake selection	Full Range or Discrete	±5%	0.5	0.3% of full range	
88. All cockpit flight control input forces (control wheel, control column, rudder pedal)[18 19]	Full range Control wheel ±70 lbs Control column ±85 lbs Rudder pedal ±165 lbs	±5%	1	0.3% of full range	For fly-by-wire flight control systems, where flight control surface position is a function of the displacement of the control input device only, it is not necessary to record this parameter. For airplanes that have a flight control break away capability that allows either pilot to operate the control independently, record both control force inputs. The control force inputs may be sampled alternately once per 2 seconds to produce the sampling interval of 1.
89. Yaw damper status	Discrete (on/off)	0.5			
90. Yaw damper command	Full range	As installed	0.5	1% of full range	
91. Standby rudder valve status	Discrete	0.5			

[1] For A300 B2/B4 airplanes, resolution = 6 seconds.
[2] For A330/A340 series airplanes, resolution = 0.703°.
[3] For A318/A319/A320/A321 series airplanes, resolution = 0.275% (0.088°>0.064°)
For A330/A340 series airplanes, resolution = 2.20% (0.703°>0.064°)
[4] For A318/A319/A320/A321 series airplanes, resolution = 0.22% (0.088°>0.080°)
For A330/A340 series airplanes, resolution = 1.76% (0.703°>0.080°)
[5] For A330/A340 series airplanes, resolution = 1.18% (0.703°>0.120°).
[6] For A330/A340 series airplanes, resolution = 0.783% (0.352°>0.090°)
[7] For A330/A340 series airplanes, aileron resolution = 0.704% (0.352°>0.100°). For A330/A340 series airplanes, spoiler resolution = 1.406% (0.703°>0.100°).
[8] For A330/A340 series airplanes, resolution = 0.30% (0.176°>0.12°)
For A330/A340 series airplanes, seconds per sampling interval = 1
[9] For B-717 series airplanes, resolution = .005g. For Dassault F900C/F900EX airplanes, resolution = .007g.
[10] For A330/A340 series airplanes, resolution = 1.05% (0.250°>0.120°)
[11] For A330/A340 series airplanes, resolution = 1.05% (0.250°>0.120°). For A330 B2/B4 series airplanes, resolution = 0.92% (0.230°>0.125°).
[12] For A330/A340 series airplanes, spoiler resolution = 1.406% (0.703°>0.100°).
[13] For A330/A340 series airplanes, resolution = 0.5°C.
[14] For Dassault F900C/F900EX airplanes, Radio Altitude resolution = 1.25 ft.
[15] For A330/A340 series airplanes, resolution = 0.352 degrees.
[16] For A318/A319/A320/A321 series airplanes, resolution = 4.32%. For A330/A340 series airplanes, resolution is 3.27% of full range for throttle lever angle (TLA); for reverse thrust, reverse throttle lever angle (RLA) resolution is nonlinear over the active reverse thrust range, which is 51.54 degrees to 96.14 degrees. The resolved element is 2.8 degrees uniformly over the entire active reverse thrust range, or 2.9% of the full range value of 96.14 degrees.
[17] For A318/A319/A320/A321 series airplanes, with IAE engines, resolution = 2.58%.
[18] For all aircraft manufactured on or after December 6, 2010, the seconds per sampling interval is 0.125. Each input must be recorded at this rate. Alternately sampling inputs (interleaving) to meet this sampling interval is prohibited.
[19] For all 737 model airplanes manufactured between August 19, 2000, and April 6, 2010: The seconds per sampling interval is 0.5 per control input; the remarks regarding the sampling rate do not apply; a single control wheel force transducer installed on the left cable control is acceptable provided the left and right control wheel positions also are recorded.

[Doc. No. 28109, 62 FR 38390, July 17, 1997; 62 FR 48135, Sept. 12, 1997, as amended by Amdt. 125-32, 64 FR 46121, Aug. 24, 1999; 65 FR 2295, Jan. 14, 2000; Amdt. 125-32, 65 FR 2295, Jan. 14, 2000; Amdt. 125-34, 65 FR 51745, Aug. 24, 2000; 65 FR 81735, Dec. 27, 2000; Amdt. 125-39, 67 FR 54323, Aug. 21, 2002; Amdt. 125-42, 68 FR 42937, July 18, 2003; 68 FR 50069, Aug. 20, 2003; 68 FR 53877, Sept. 15, 2003; Amdt. 125-54, 73 FR 12568, Mar. 7, 2008; Amdt. 125-56, 73 FR 73180, Dec. 2, 2008; Amdt. 125-60, 75 FR 17046, Apr. 5, 2010; Amdt. 125-59, 75 FR 7357, Feb. 19, 2010]

PART 129

OPERATIONS: FOREIGN AIR CARRIERS AND FOREIGN OPERATORS OF U.S.-REGISTERED AIRCRAFT ENGAGED IN COMMON CARRIAGE

Contents

Authority: 49 U.S.C. 1372, 40113, 40119, 44101, 44701-44702, 44705, 44709-44711, 44713, 44716-44717, 44722, 44901-44904, 44906, 44912, 46105, Pub. L. 107-71 sec. 104.

Source: Docket No. 1994, 29 FR 1720, Feb. 5, 1964, unless otherwise noted.

Special Federal Aviation Regulation No. 97

Editorial **Note:** For the text of SFAR No. 97, see part 91 of this chapter.

Subpart A—General

§ 129.1 Applicability and definitions.

(a) *Foreign air carrier operations in the United States.* This part prescribes rules governing the operation within the United States of each foreign air carrier holding the following:

(1) A permit issued by the U.S. Department of Transportation under 49 U.S.C. 41301 through 41306, or

(2) Other appropriate economic or exemption authority issued by the U.S. Department of Transportation.

(b) *Operations of U.S.-registered aircraft solely outside the United States.* In addition to the operations specified under paragraph (a) of this section, §§ 129.5, 129.7, 129.9, 129.11, 129.14, 129.20 and 129.24, and subpart B of this part also apply to operations of U.S.-registered aircraft operated solely outside the United States in common carriage by a foreign person or foreign air carrier.

(c) *Definitions.* For the purpose of this part—

(1) *Foreign person* means any person who is not a citizen of the United States and who operates a U.S.-registered aircraft in common carriage solely outside the United States.

(2) *Years in service* means the calendar time elapsed since an aircraft was issued its first U.S. or first foreign airworthiness certificate.

[Doc. No. FAA-1999-5401, 67 FR 72762, Dec. 6, 2002, as amended by Amdt. 129-43, 72 FR 63413, Nov. 8, 2007; Amdt. 129-45, 73 FR 12570, Mar. 7, 2008; Amdt. 129-45, 74 FR 32801, July 9, 2009; Amdt. 129-49, 76 FR 7489, Feb. 10, 2011]

§ 129.5 Operations specifications.

(a) Each foreign air carrier conducting operations within the United States, and each foreign air carrier or foreign person operating U.S.-registered aircraft solely outside the United States in common carriage must conduct its operations in accordance with operations specifications issued by the Administrator under this part.

(b) Each foreign air carrier conducting operations within the United States must conduct its operations in accordance with the Standards contained in Annex 1 (Personnel Licensing), Annex 6 (Operation of Aircraft), Part I (International Commercial Air Transport—Aeroplanes) or Part III (International Operations—Helicopters), as appropriate, and in Annex 8 (Airworthiness of Aircraft) to the Convention on International Civil Aviation.

(c) No foreign air carrier may operate to or from locations within the United States without, or in violation of, appropriate operations specifications.

(d) No foreign air carrier or foreign person shall operate U.S.-registered aircraft solely outside the United States in common carriage without, or in violation of, appropriate operations specifications.

(e) Each foreign air carrier must keep each of its employees and other persons used in its operations informed of the provisions of its operations specifications that apply to that employee's or person's duties and responsibilities.

(f) Operations specifications issued under this part are effective until—

(1) The foreign air carrier or foreign person surrenders them to the FAA;

(2) The Administrator suspends or terminates the operations specifications; or

(3) The operations specifications are amended as provided in § 129.11.

(g) Within 30 days after a foreign air carrier or foreign person terminates operations under part 129 of this subchapter, the operations specifications must be surrendered by the foreign air carrier or foreign person to the responsible Flight Standards District Office.

(h) No person operating under this part may operate or list on its operations specifications any airplane listed on operations specifications issued under part 125 of this chapter.

[Doc. No. FAA-2009-0140; 76 FR 7489, Feb. 10, 2011]

§ 129.7 Application, issuance, or denial of operations specifications.

(a) A foreign air carrier or foreign person applying to the FAA for operations specifications under this part must submit an application—

(1) In a form and manner prescribed by the Administrator; and

(2) At least 90 days before the intended date of operation.

(b) An authorized officer or employee of the applicant, having knowledge of the matters stated in the application, must sign the application and certify in writing that the statements in the application are true. The application must include two copies of the appropriate written authority issued to that officer or employee by the applicant.

(c) A foreign applicant may be issued operations specifications, if after review, the Administrator finds the applicant—

(1) Meets the applicable requirements of this part;

(2) Holds the economic or exemption authority required by the Department of Transportation, applicable to the operations to be conducted;

(3) Complies with the applicable security requirements of 49 CFR chapter XII;

(4) Is properly and adequately equipped to conduct the operations described in the operations specifications; and

(5) Holds a valid air operator certificate issued by the State of the Operator.

(d) An application may be denied if the Administrator finds that the applicant is not properly or adequately equipped to conduct the operations to be described in the operations specifications.

[Doc. No. FAA-2009-0140; 76 FR 7489, Feb. 10, 2011]

§ 129.9 Contents of operations specifications.

(a) The contents of operations specifications issued to a foreign air carrier conducting operations within the United States under § 129.1(a) shall include:

(1) The specific location and mailing address of the applicant's principal place of business in the State of the Operator and, if different, the address that will serve as the primary point of contact for correspondence between the FAA and the foreign air carrier;

(2) Within 1 year after February 10, 2011, the designation of an agent for service within the United States, including the agent's full name and office address or usual place of residence;

(3) The certificate number and validity of the foreign air carrier's Air Operator Certificate issued by the State of the Operator;

(4) Each regular and alternate airport to be used in scheduled operations;

(5) The type of aircraft and registration markings of each aircraft;

(6) The approved maintenance program and minimum equipment list for United States registered aircraft authorized for use; and

(7) Any other item the Administrator determines is necessary.

(b) The contents of operations specifications issued to a foreign air carrier or foreign person operating U.S.-registered aircraft solely outside the United States in common carriage in accordance with § 129.1(b) shall include—

(1) The specific location and mailing address of the principal place of business in the State of the Operator and, if different, the address that will serve as the primary point of contact for correspondence between the FAA and the foreign air carrier or foreign person;

(2) Within 1 year after February 10, 2011, the designation of an agent for service within the United States, including the agent's full name and office address or usual place of residence;

(3) In the case of a foreign air carrier, the certificate number and validity of the foreign air carrier's Air Operator Certificate issued by the State of the Operator;

(4) Any other business names under which the foreign air carrier or foreign person may operate;

(5) The type, registration markings, and serial number of each United States registered aircraft authorized for use;

(6) The approved maintenance program and minimum equipment list for United States registered aircraft authorized for use; and

(7) Any other item the Administrator determines is necessary.

[Doc. No. FAA-2009-0140; 76 FR 7489, Feb. 10, 2011; Amdt. 129-49-A, 76 FR 15212, Mar. 21, 2011]

§ 129.11 Amendment, suspension and termination of operations specifications.

(a) The Administrator may amend any operations specifications issued under this part if—

(1) The Administrator determines that safety in air commerce and the public interest require the amendment; or

(2) The foreign air carrier or foreign person applies for an amendment, and the Administrator determines that safety in air commerce and the public interest allows the amendment.

(b) The Administrator may suspend or terminate any operations specifications issued under this part if the Administrator determines that safety in air commerce and the public interest require the suspension or termination;

(c) Except as provided in paragraphs (f) and (g) of this section, when the Administrator initiates an action to amend, suspend or terminate a foreign air carrier or foreign person's operations specifications, the following procedure applies:

(1) The responsible Flight Standards District Office notifies the foreign air carrier or foreign person in writing of the proposed amendment, suspension or termination.

(2) The responsible Flight Standards District Office sets a reasonable period (but not less than 7 days) within which the foreign air carrier or foreign person may submit written information, views, and arguments on the amendment, suspension or termination.

(3) After considering all material presented, the responsible Flight Standards District Office notifies the foreign air carrier or foreign person of—

(i) The adoption of the proposed amendment, suspension or termination;

(ii) The partial adoption of the proposed amendment, suspension or termination; or

(iii) The withdrawal of the proposed amendment, suspension or termination.

(4) If the responsible Flight Standards District Office issues an action to amend, suspend or terminate the operations specifications, it becomes effective not less than 30 days after the foreign air carrier or foreign person receives notice of it unless—

(i) The responsible Flight Standards District Office finds under paragraph (g) of this section that there is an emergency requiring immediate action with respect to safety in air commerce; or

(ii) The foreign air carrier or foreign person petitions for reconsideration of the amendment, suspension or termination under paragraph (e) of this section.

(d) When the foreign air carrier or foreign person applies for an amendment to its operations specifications, the following procedure applies:

(1) The foreign air carrier or foreign person must file an application to amend its operations specifications—

(i) At least 90 days before the date proposed by the applicant for the amendment to become effective in cases of mergers; acquisitions of airline operational assets that require an additional showing to Department of Transportation for economic authority; major changes in the type of operation; and resumption of operations following a suspension of operations as a result of bankruptcy actions, unless a shorter time is approved by the Administrator.

(ii) At least 30 days before the date proposed by the applicant for the amendment to become effective in all other cases.

(2) The application must be submitted to the responsible Flight Standards District Office in a form and manner prescribed by the Administrator.

(3) After considering all material presented, the responsible Flight Standards District Office notifies the foreign air carrier or foreign person of—

(i) The adoption of the applied for amendment;

(ii) The partial adoption of the applied for amendment; or

(iii) The denial of the applied for amendment.

(4) If the responsible Flight Standards District Office approves the amendment, following coordination with the foreign air carrier or foreign person regarding its implementation, the amendment is effective on the date the responsible Flight Standards District Office approves it.

(e) The foreign air carrier or foreign person may petition for reconsideration of a full or partial adoption of an amendment, a denial of an amendment or a suspension or termination of operations specifications.

(f) When a foreign air carrier or foreign person seeks reconsideration of a decision from the responsible Flight Standards District Office concerning the amendment, suspension or termination of operations specifications, the following procedure applies:

(1) The foreign air carrier or foreign person must petition for reconsideration of that decision within 30 days after the date that the foreign air carrier or foreign person receives a notice of the decision.

(2) The foreign air carrier or foreign person must address its petition to the Director, Flight Standards Service.

(3) A petition for reconsideration, if filed within the 30-day period, suspends the effectiveness of any amendment, suspension or termination issued by the responsible Flight Standards District Office unless the responsible Flight Standards District Office has found, under paragraph (g) of this section, that an emergency exists requiring immediate action with respect to safety in air transportation or air commerce.

(g) If the responsible Flight Standards District Office finds that an emergency exists requiring immediate action with respect to safety in air commerce or air transportation that makes the procedures set out in this section impracticable or contrary to the public interest, that office may make the amendment, suspension or termination effective on the day the foreign air carrier or foreign person receives notice of it. In the notice to the foreign air carrier or foreign person, the responsible Flight Standards District Office will articulate the reasons for its finding that an emergency exists requiring immediate action with respect to safety in air transportation or air commerce or that makes it impracticable or contrary to the public interest to stay the effectiveness of the amendment, suspension or termination.

[Doc. No. FAA-2009-0140, 76 FR 7490, Feb. 10, 2011]

§ 129.13 Airworthiness and registration certificates.

(a) No foreign air carrier may operate any aircraft within the United States unless that aircraft carries a current registration certificate and displays the nationality and registration markings of the State of Registry, and an airworthiness certificate issued or validated by:

(1) The State of Registry; or

(2) The State of the Operator, provided that the State of the Operator and the State of Registry have entered into an agreement under Article 83 *bis* of the Convention on International Civil Aviation that covers the aircraft.

(b) No foreign air carrier may operate a foreign aircraft within the United States except in accordance with the limitations on maximum certificated weights prescribed for that aircraft and that operation by the country of manufacture of the aircraft.

[Docket No. 1994, 29 FR 1720, Feb. 5, 1964, as amended by Amdt. 129-33, 67 FR 42455, June 21, 2002; Amdt. 129-49, 76 FR 7490, Feb. 10, 2011]

§ 129.14 Maintenance program and minimum equipment list requirements for U.S.-registered aircraft.

(a) Each foreign air carrier and each foreign person operating a U.S.-registered aircraft within or outside the United States in common carriage must ensure that each aircraft is maintained in accordance with a program approved by the Administrator in the operations specifications.

(b) No foreign air carrier or foreign person may operate a U.S.-registered aircraft with inoperable instruments or equipment unless the following conditions are met:

(1) A master minimum equipment list exists for the aircraft type.

(2) The foreign operator submits for review and approval its aircraft minimum equipment list based on the master minimum equipment list, to the FAA Flight Standards District Office having geographic responsibility for the operator. The foreign operator must show, before minimum equipment list approval can be obtained, that the maintenance procedures used under its maintenance program are adequate to support the use of its minimum equipment list.

(3) For leased aircraft maintained and operated under a U.S. operator's continuous airworthiness maintenance program and FAA-approved minimum equipment list, the foreign operator submits the U.S. operator's approved continuous airworthiness maintenance program and approved aircraft minimum equipment list to the FAA office prescribed in paragraph (b)(2) of this section for review and evaluation. The foreign operator must show that it is capable of operating under the lessor's approved maintenance program and that it is also capable of meeting the maintenance and operational requirements specified in the lessor's approved minimum equipment list.

(4) The FAA operations specification permitting the operator to use an approved minimum equipment list is carried aboard the aircraft. An approved minimum equipment list, as authorized by the operations specifications, constitutes an approved change to the type design without requiring recertification.

(5) The approved minimum equipment list provides for the operation of the aircraft with certain instruments and equipment in an inoperable condition.

(6) The aircraft records available to the pilot must include an entry describing the inoperable instruments and equipment.

(7) The aircraft is operated under all applicable conditions and limitations contained in the minimum equipment list and the operations specification authorizing the use of the list.

[Doc. No. 24856, 52 FR 20029, May 28, 1987, as amended by Amdt. 129-49, 76 FR 7490, Feb. 10, 2011]

§ 129.15 Flightcrew member certificates.

Each person acting as a flightcrew member must hold a certificate or license that shows the person's ability to perform duties in connection with the operation of the aircraft. The certificate or license must have been issued or rendered valid by:

(a) The State in which the aircraft is registered; or

(b) The State of the Operator, provided that the State of the Operator and the State of Registry have entered into an agreement under Article 83 *bis* of the Convention on International Civil Aviation that covers the aircraft.

[Doc. No. FAA-2009-0140; 76 FR 7491, Feb. 10, 2011]

§ 129.17 Aircraft communication and navigation equipment for operations under IFR or over the top.

(a) *Aircraft navigation equipment requirements—General.* No foreign air carrier may conduct operations under IFR or over the top unless—

(1) The en route navigation aids necessary for navigating the aircraft along the route (e.g., ATS routes, arrival and departure routes, and instrument approach procedures, including missed approach procedures if a missed approach routing is specified in the procedure) are available and suitable for use by the aircraft navigation equipment required by this section;

(2) The aircraft used in those operations is equipped with at least the following—

(i) Except as provided in paragraph (c) of this section, two approved independent navigation systems suitable for navigating the aircraft along the route to be flown within the degree of accuracy required for ATC;

(ii) One marker beacon receiver providing visual and aural signals; and

(iii) One ILS receiver; and

(3) Any RNAV system used to meet the navigation equipment requirements of this section is authorized in the foreign air carrier's operations specifications.

(b) *Aircraft communication equipment requirements.* No foreign air carrier may operate an aircraft under IFR or over the top, unless it is equipped with—

(1) At least two independent communication systems necessary under normal operating conditions to fulfill the functions specified in § 121.347(a) of this chapter; and

(2) At least one of the communication systems required by paragraph (b)(1) of this section must have two-way voice communication capability.

(c) *Use of a single independent navigation system for operations under IFR or over the top.* Notwithstanding the requirements of paragraph (a)(2)(i) of this section, the aircraft may be equipped with a single independent navigation system suitable for navigating the aircraft along the route to be flown within the degree of accuracy required for ATC if:

(1) It can be shown that the aircraft is equipped with at least one other independent navigation system suitable, in the event of loss of the navigation capability of the single independent navigation system permitted by this paragraph at any point along the route, for proceeding safely to a suitable airport and completing an instrument approach; and

(2) The aircraft has sufficient fuel so that the flight may proceed safely to a suitable airport by use of the remaining navigation system, and complete an instrument approach and land.

(d) *VOR navigation equipment.* If VOR navigation equipment is required by paragraph (a) or (c) of this section, no foreign air carrier may operate an aircraft unless it is equipped with at least one approved DME or suitable RNAV system.

[Doc. No. FAA-2002-14002, 72 FR 31683, June 7, 2007]

§ 129.18 Collision avoidance system.

Effective January 1, 2005, any airplane you, as a foreign air carrier, operate under part 129 must be equipped and operated according to the following table:

Collision Avoidance Systems

If you operate in the United States any . . .	Then you must operate that airplane with:
(a) Turbine-powered airplane of more than 33,000 pounds maximum certificated takeoff weight	(1) An appropriate class of Mode S transponder that meets Technical Standard Order (TSO) C–112, or a later version, and one of the followign approved units; (i) TCAS II that meets TSO C–119b (version 7.0), or takeoff weight a later version.

If you operate in the United States any . . .	Then you must operate that airplane with:
	(ii) TCAS II that meets TSO C–119a (version 6.04A Enhanced) that was installed in that airplane before May 1, 2003. If that TCAS II version 6.04A Enhanced no longer can be repaired to TSO C–119a standards, it must be replaced with a TCAS II that meets TSO C–119b (version 7.0), or a later version. (iii) A collision avoidance system equivalent to TSO C–119b (version 7.0), or a later version, capable of coordinating with units that meet TSO C–119a (version 6.04A Enhanced), or a later version.
(b) Turbine-powered airplane with a passenger-seat configuration, excluding any pilot seat, or 10–30 seats	(1) TCAS I that meets TSO C–118, or a later version, or (2) A collision avoidance system equivalent to excluding any TSO C–118, or a later version, or (3) A collision avoidance system and Mode S transponder that meet paragraph (a)(1) of this section.

[Doc. No. FAA-2001-10910, 68 FR 15903, Apr. 1, 2003]

§ 129.19 Air traffic rules and procedures.

(a) Each pilot must be familiar with the applicable rules, the navigational and communications facilities, and the air traffic control and other procedures, of the areas to be traversed by him within the United States.

(b) Each foreign air carrier shall establish procedures to assure that each of its pilots has the knowledge required by paragraph (a) of this section and shall check the ability of each of its pilots to operate safely according to applicable rules and procedures.

(c) Each foreign air carrier shall conform to the practices, procedures, and other requirements prescribed by the Administrator for U.S. air carriers for the areas to be operated in.

§ 129.20 Digital flight data recorders.

No person may operate an aircraft under this part that is registered in the United States unless it is equipped with one or more approved flight recorders that use a digital method of recording and storing data and a method of readily retrieving that data from the storage medium. The flight data recorder must record the parameters that would be required to be recorded if the aircraft were operated under part 121, 125, or 135 of this chapter, and must be installed by the compliance times required by those parts, as applicable to the aircraft.

[Doc. No. 28109, 62 FR 38396, July 17, 1997]

§ 129.21 Control of traffic.

(a) Subject to applicable immigration laws and regulations, each foreign air carrier must furnish sufficient personnel necessary to provide two-way voice communications between its aircraft and stations at places where the FAA finds that communication is necessary but cannot be maintained in a language with which station operators are familiar.

(b) Each person furnished by a foreign air carrier under paragraph (a) of this section must be able to speak English and the language necessary to maintain communications with its aircraft and must assist station operators in directing traffic.

[Doc. No. FAA-2002-14002, 72 FR 31683, June 7, 2007]

§ 129.22 Communication and navigation equipment for rotorcraft operations under VFR over routes navigated by pilotage.

(a) No foreign air carrier may operate a rotorcraft under VFR over routes that can be navigated by pilotage unless the rotorcraft is equipped with the radio communication equipment necessary under normal operating conditions to fulfill the following:

(1) Communicate with at least one appropriate station from any point on the route;

(2) Communicate with appropriate air traffic control facilities from any point within Class B, Class C, or Class D airspace, or within a Class E surface area designated for an airport in which flights are intended; and

(3) Receive meteorological information from any point en route.

(b) No foreign air carrier may operate a rotorcraft at night under VFR over routes that can be navigated by pilotage unless that rotorcraft is equipped with—

(1) Radio communication equipment necessary under normal operating conditions to fulfill the functions specified in paragraph (a) of this section; and

(2) Navigation equipment suitable for the route to be flown.

[Doc. No. FAA-2002-14002, 72 FR 31683, June 7, 2007]

§ 129.23 Transport category cargo service airplanes: Increased zero fuel and landing weights.

(a) Notwithstanding the applicable structural provisions of the transport category airworthiness regulations, but subject to paragraphs (b) through (g) of this section, a foreign air carrier may operate (for cargo service only) any of the following transport category airplanes (certificated under part 4b of the Civil Air Regulations effective before March 13, 1956) at increased zero fuel and landing weights—

(1) DC-6A, DC-6B, DC-7B, and DC-7C; and

(2) L-1049 B, C, D, E, F, G, and H, and the L-1649A when modified in accordance with supplemental type certificate SA 4-1402.

(b) The zero fuel weight (maximum weight of the airplane with no disposable fuel and oil) and the structural landing weight may be increased beyond the maximum approved in full compliance with applicable rules only if the Administrator finds that—

(1) The increase is not likely to reduce seriously the structural strength;

(2) The probability of sudden fatigue failure is not noticeably increased;

(3) The flutter, deformation, and vibration characteristics do not fall below those required by applicable regulations; and

(4) All other applicable weight limitations will be met.

(c) No zero fuel weight may be increased by more than five percent, and the increase in the structural landing weight may not exceed the amount, in pounds, of the increase in zero fuel weight.

(d) Each airplane must be inspected in accordance with the approved special inspection procedures, for operations at increased weights, established and issued by the manufacturer of the type of airplane.

(e) A foreign air carrier may not operate an airplane under this section unless the country of registry requires the airplane to be operated in accordance with the passenger-carrying transport category performance operating limitations in part 121 or the equivalent.

(f) The Airplane Flight Manual for each airplane operated under this section must be appropriately revised to include the operating limitations and information needed for operation at the increased weights.

(g) Each airplane operated at an increased weight under this section must, before it is used in passenger service, be inspected under the special inspection procedures for return to passenger service established and issued by the manufacturer and approved by the Administrator.

[Doc. No. 6403, 29 FR 19098, Dec. 30, 1964]

§ 129.24 Cockpit voice recorders.

No person may operate an aircraft under this part that is registered in the United States unless it is equipped with an approved cockpit voice recorder that meets the standards of TSO-C123a, or later revision. The cockpit voice recorder must record the information that would be required to be recorded if the aircraft were operated under part 121, 125, or 135 of this chapter, and must be installed by the compliance times required by that part, as applicable to the aircraft.

[Doc. No. FAA-2005-20245, 73 FR 12570, Mar. 7, 2008]

§ 129.25 Airplane security.

Foreign air carriers conducting operations under this part must comply with the applicable security requirements in 49 CFR chapter XII.

[67 FR 8350, Feb. 22, 2002]

§ 129.28 Flightdeck security.

(a) After August 20, 2002, except for a newly manufactured airplane on a non-revenue delivery flight, no foreign air carrier covered by § 129.1(a), may operate:

(1) A passenger carrying transport category airplane within the United States, except for overflights, unless the airplane is equipped with a door between the passenger and pilot compartment that incorporates features to restrict the unwanted entry of persons into the flightdeck that are operable from the flightdeck only; or

(2) A transport category all-cargo airplane within the United States, except for overflights, that has a door installed between the pilot compartment and any other occupied compartment on or after June 21, 2002, unless the door incorporates features to restrict the unwanted entry of persons into the flightdeck that are operable from the flightdeck only.

(b) To the extent necessary to meet the requirements of paragraph (a) of this section, the requirements of § 129.13(a) to maintain airworthiness certification are waived until April 9, 2003. After that date, the requirements of § 129.13(a) apply in full.

(c) After April 9, 2003, except for a newly manufactured airplane on a non-revenue delivery flight, no foreign air carrier covered by § 129.1(a) may operate a passenger carrying transport category airplane, or a transport category all-cargo airplane that has a door installed between the pilot compartment and any other occupied compartment on or after June 21, 2002, within the United States, except for overflights, unless the airplane's flightdeck door installation meets the requirements of paragraphs (c)(1) and(2) of this section or an alternative standard found acceptable to the Administrator.

(1) Except for a newly manufactured airplane on a non-revenue delivery flight, no foreign air carrier covered by § 129.1(a) may operate:

(i) After April 9, 2003, a passenger carrying transport category airplane within the United States, except on overflights, unless the airplane's flightdeck door installation meets the requirements of paragraphs (c)(2) and (c)(3) of this section or an alternative standard found acceptable to the Administrator.

(ii) After October 1, 2003, a transport category all-cargo airplane that had a door installed between the pilot compartment and any other occupied compartment on or after June 21, 2002, within the United States, except on overflights, unless the airplane's flightdeck door installation meets the requirements of paragraphs (c)(2) and (c)(3) of this section or an alternative standard found acceptable to the Administrator; or the operator must implement a security program approved by the Transportation Security Administration (TSA) for the operation of all airplanes in that operator's fleet.

(2) The door must resist forcible intrusion by unauthorized persons and be capable of withstanding impacts of 300 joules (221.3 foot-pounds) at the critical locations on the door, as well as a 1,113-newton (250 pounds) constant tensile load on the knob or handle, and

(3) The door must resist penetration by small arms fire and fragmentation devices to a level equivalent to Level IIIa of the National Institute of Justice Standard (NIJ) 0101.04.

(d) After August 20, 2002, no foreign air carrier covered by § 129.1 may operate a passenger carrying transport category airplane, or a transport category all-cargo airplane that has a door installed between the pilot compartment and any other occupied compartment on or after June 21, 2002, within the United States, except for overflights, unless the carrier has procedures in place that are acceptable to the civil aviation authority responsible for oversight of the foreign air carriers operating under this part to prevent access to the flightdeck except as authorized as follows:

(1) No person other than a person who is assigned to perform duty on the flight deck may have a key to the flight deck door that will provide access to the flightdeck.

(2) Except when it is necessary to permit access and egress by persons authorized in accordance with paragraph (d)(3) of this section, a pilot in command of an airplane that has a lockable flight deck door in accordance with § 129.28(a) and that is carrying passengers shall ensure that the door separating the flight crew compartment from the passenger compartment is closed and locked at all times when the airplane is being operated.

(3) No person may admit any person to the flight deck of an airplane unless the person being admitted is—

(i) A crewmember,

(ii) An inspector of the civil aviation authority responsible for oversight of the part 129 operator, or

(iii) Any other person authorized by the civil aviation authority responsible for oversight of the part 129 operator.

(e) The requirements of paragraph (a) through (d) except (d)(3), do not apply to transport category passenger carrying airplanes originally type certificated with a maximum passenger seating configuration of 19 seats or less, or to all-cargo airplanes with a payload capacity of 7,500 pounds or less.

[Doc. No. FAA-2002-12504, 67 FR 79824, Dec. 30, 2002, as amended by Amdt. 129-38, 68 FR 42882, July 18, 2003]

§ 129.29 Smoking prohibitions.

(a) No person may smoke and no operator may permit smoking in any aircraft lavatory.

(b) Unless otherwise authorized by the Secretary of Transportation, no person may smoke and no operator may permit smoking anywhere on the aircraft (including the passenger cabin and the flight deck) during scheduled passenger foreign air transportation or during any scheduled passenger interstate or intrastate air transportation.

[Doc. No. FAA-2000-7467, 65 FR 36780, June 9, 2000]

Subpart B—Continued Airworthiness and Safety Improvements

§ 129.101 Purpose and definition.

(a) This subpart requires a foreign person or foreign air carrier operating a U.S. registered airplane in common carriage to support the continued airworthiness of each airplane. These requirements may include, but are not limited to, revising the maintenance program, incorporating design changes, and incorporating revisions to Instructions for Continued Airworthiness.

(b) For purposes of this subpart, the "FAA Oversight Office" is the aircraft certification office or office of the Transport Airplane Directorate with oversight responsibility for the relevant type certificate or supplemental type certificate, as determined by the Administrator.

[Amdt. 129-43, 72 FR 63413, Nov. 8, 2007]

§ 129.103 [Reserved]

§ 129.105 Aging airplane inspections and records reviews for U.S.-registered multiengine aircraft.

(a) *Operation after inspection and records review.* After the dates specified in this paragraph, a foreign air carrier or foreign person may not operate a U.S.-registered multiengine airplane under this part unless the Administrator has notified the foreign air carrier or foreign person that the Administrator has completed the aging airplane inspection and records review required by this section. During the inspection and records review, the foreign air carrier or foreign person must demonstrate to the Administrator that the maintenance of age sensitive parts and components of the airplane has been adequate and timely enough to ensure the highest degree of safety.

(1) *Airplanes exceeding 24 years in service on* December 8, 2003; *initial and repetitive inspections and records reviews.* For an airplane that has exceeded 24 years in service on December 8, 2003, no later than December 5, 2007, and thereafter at intervals not to exceed 7 years.

(2) *Airplanes exceeding 14 years in service but not 24 years in service on* December 8, 2003; *initial and repetitive inspections and records reviews.* For an airplane that has exceeded 14 years in service, but not 24 years in service, on December 8, 2003, no later than December 4, 2008, and thereafter at intervals not to exceed 7 years.

(3) *Airplanes not exceeding 14 years in service on* December 8, 2003; *initial and repetitive inspections and records reviews.* For an airplane that has not exceeded 14 years in service on December 8, 2003, no later than 5 years after the start of the airplane's 15th year in service and thereafter at intervals not to exceed 7 years.

(b) *Unforeseen schedule conflict.* In the event of an unforeseen scheduling conflict for a specific airplane, the Administrator may approve an extension of up to 90 days beyond an interval specified in paragraph (b) of this section.

(c) *Airplane and records availability.* The foreign air carrier or foreign person must make available to the Administrator each U.S.-registered multiengine airplane for which an inspection and records review is required under this section, in a condition for inspection specified by the Administrator, together with the records containing the following information:

(1) Total years in service of the airplane;

(2) Total time in service of the airframe;

(3) Total flight cycles of the airframe;

(4) Date of the last inspection and records review required by this section;

(5) Current status of life-limited parts of the airframe;

(6) Time since the last overhaul of all structural components required to be overhauled on a specific time basis;

(7) Current inspection status of the airplane, including the time since the last inspection required by the inspection program under which the airplane is maintained;

(8) Current status of applicable airworthiness directives, including the date and methods of compliance, and if the airworthiness directive involves recurring action, the time and date when the next action is required;

(9) A list of major structural alterations; and

(10) A report of major structural repairs and the current inspection status for those repairs.

(d) *Notification to Administrator.* Each foreign air carrier or foreign person must notify the Administrator at least 60 days before the date on which the airplane and airplane records will be made available for the inspection and records review.

[Doc. No. FAA-1999-5401, 67 FR 72763, Dec. 6, 2002, as amended by Amdt. 129-34, 70 FR 5533, Feb. 2, 2005; Amdt. 129-41, 70 FR 23936, May 6, 2005. Redesignated by Amdt. 129-43, 72 FR 63413, Nov. 8, 2007]

§ 129.107 Repairs assessment for pressurized fuselages.

(a) No foreign air carrier or foreign persons operating a U.S. registered airplane may operate an Airbus Model A300 (excluding –600 series), British Aerospace Model BAC 1-11, Boeing Model 707, 720, 727, 737, or 747, McDonnell Douglas Model DC-8, DC-9/MD-80 or DC-10, Fokker Model F28, or Lockheed Model L-1011 beyond the applicable flight cycle implementation time specified below, or May 25, 2001, whichever occurs later, unless operations specifications have been issued to reference repair assessment guidelines applicable to the fuselage pressure boundary (fuselage skin, door skin, and bulkhead webs), and those guidelines are incorporated in its maintenance program. The repair assessment guidelines must be approved by the FAA Aircraft Certification Office (ACO), or office of the Transport Airplane Directorate, having cognizance over the type certificate for the affected airplane.

(1) For the Airbus Model A300 (excluding the -600 series), the flight cycle implementation time is:

(i) Model B2: 36,000 flights.

(ii) Model B4-100 (including Model B4-2C): 30,000 flights above the window line, and 36,000 flights below the window line.

(iii) Model B4-200: 25,500 flights above the window line, and 34,000 flights below the window line.

(2) For all models of the British Aerospace BAC 1-11, the flight cycle implementation time is 60,000 flights.

(3) For all models of the Boeing 707, the flight cycle implementation time is 15,000 flights.

(4) For all models of the Boeing 720, the flight cycle implementation time is 23,000 flights.

(5) For all models of the Boeing 727, the flight cycle implementation time is 45,000 flights.

(6) For all models of the Boeing 737, the flight cycle implementation time is 60,00 flights.

(7) For all models of the Boeing 747, the flight cycle implementation time is 15,000 flights.

(8) For all models of the McDonnell Douglas DC-8, the flight cycle implementation time is 30,000 flights.

(9) For all models of the McDonnell Douglas DC-9/MD-80, the flight cycle implementation time is 60,000 flights.

(10) For all models of the McDonnell Douglas DC-10, the flight cycle implementation time is 30,000 flights.

(11) For all models of the Lockheed L-1011, the flight cycle implementation time is 27,000 flights.

(12) For the Fokker F-28 Mark 1000, 2000, 3000, and 4000, the flight cycle implementation time is 60,000 flights.

(b) [Reserved]

[Doc. No. 29104, 65 FR 24126, Apr. 25, 2000; 65 FR 35703, June 5, 2000, as amended by Amdt. 129-30, 66 FR 23131, May 7, 2001; Amdt. 129-35, 67 FR 72834, Dec. 9, 2002; Amdt. 129-39, 69 FR 45942, July 30, 2004. Redesignated and amended by Amdt. 129-43, 72 FR 63413, Nov. 8, 2007]

§ 129.109 Supplemental inspections for U.S.-registered aircraft.

(a) *Applicability.* This section applies to U.S.-registered, transport category, turbine powered airplanes with a type certificate issued after January 1, 1958 that as a result of original type certification or later increase in capacity have—

(1) A maximum type certificated passenger seating capacity of 30 or more; or

(2) A maximum payload capacity of 7,500 pounds or more.

(b) *General requirements.* After December 20, 2010, a certificate holder may not operate an airplane under this part unless the following requirements have been met:

(1) *Baseline Structure.* The certificate holder's maintenance program for the airplane includes FAA-approved damage-tolerance-based inspections and procedures for airplane structure susceptible to fatigue cracking that could contribute to a catastrophic failure. For the purpose of this section, this structure is termed "fatigue critical structure."

(2) *Adverse effects of repairs, alterations, and modifications.* The maintenance program for the airplane includes a means for addressing the adverse effects repairs, alterations, and modifications may have on fatigue critical structure and on inspections required by paragraph (b)(1) of this section. The means for addressing these adverse effects must be approved by the FAA Oversight Office.

(3) *Changes to maintenance program.* The changes made to the maintenance program required by paragraph (b)(1) and (b)(2) of this section, and any later revisions to these changes, must be submitted to the Principal Maintenance Inspector for review and approval.

[Doc. No. FAA-1999-5401, 70 FR 5532, Feb. 2, 2005. Redesignated by Amdt. 129-43, 72 FR 63413, Nov. 8, 2007; Amdt. 129-44, 72 FR 70508, Dec. 12, 2007]

§ 129.111 Electrical wiring interconnection systems (EWIS) maintenance program.

(a) Except as provided in paragraph (f) of this section, this section applies to transport category, turbine-powered airplanes with a type certificate issued after January 1, 1958, that, as a result of original type certification or later increase in capacity, have—

(1) A maximum type-certificated passenger capacity of 30 or more, or

(2) A maximum payload capacity of 7500 pounds or more.

(b) After March 10, 2011, no foreign person or foreign air carrier may operate a U.S.-registered airplane identified in paragraph (a) of this section unless the maintenance program for that airplane includes inspections and procedures for EWIS.

(c) The proposed EWIS maintenance program changes must be based on EWIS Instructions for Continued Airworthiness (ICA) that have been developed in accordance with the provisions of Appendix H of part 25 of this chapter applicable to each affected airplane (including those ICA developed for supplemental type certificates installed on each airplane) and that have been approved by the FAA Oversight Office.

(1) For airplanes subject to § 26.11 of this chapter, the EWIS ICA must comply with paragraphs H25.5(a)(1) and (b).

(2) For airplanes subject to § 25.1729 of this chapter, the EWIS ICA must comply with paragraph H25.4 and all of paragraph H25.5.

(d) After March 10, 2011, before returning a U.S.-registered airplane to service after any alterations for which EWIS ICA are developed, the foreign person or foreign air carrier must include in the maintenance program for that airplane inspections and procedures for EWIS based on those ICA.

(e) The EWIS maintenance program changes identified in paragraphs (c) and (d) of this section and any later EWIS revisions must be submitted to the Principal Inspector or Flight Standards International Field Office responsible for review and approval.

(f) This section does not apply to the following airplane models:

(1) Lockheed L-188

(2) Bombardier CL-44

(3) Mitsubishi YS-11

(4) British Aerospace BAC 1-11

(5) Concorde

(6) deHavilland D.H. 106 Comet 4C

(7) VFW-Vereinigte Flugtechnische Werk VFW-614

(8) Illyushin Aviation IL 96T

(9) Bristol Aircraft Britannia 305

(10) Handley Page Herald Type 300

(11) Avions Marcel Dassault—Breguet Aviation Mercure 100C

(12) Airbus Caravelle

(13) Lockheed L-300

[Amdt. 129-43, 72 FR 63413, Nov. 8, 2007]

§ 129.113 Fuel tank system maintenance program.

(a) Except as provided in paragraph (g) of this section, this section applies to transport category, turbine-powered airplanes with a type certificate issued after January 1, 1958, that, as a result of original type certification or later increase in capacity, have—

(1) A maximum type-certificated passenger capacity of 30 or more, or

(2) A maximum payload capacity of 7500 pounds or more.

(b) For each U.S.-registered airplane on which an auxiliary fuel tank is installed under a field approval, before June 16, 2008, the foreign person or foreign air carrier operating the airplane must submit to the FAA Oversight Office proposed maintenance instructions for the tank that meet the requirements of Special Federal Aviation Regulation No. 88 (SFAR 88) of this chapter.

(c) After December 16, 2008, no foreign person or foreign air carrier may operate a U.S.-registered airplane identified in paragraph (a) of this section unless the maintenance program for that airplane has been revised to include applicable inspections, procedures, and limitations for fuel tank systems.

(d) The proposed fuel tank system maintenance program revisions must be based on fuel tank system Instructions for Continued Airworthiness (ICA) that have been developed in accordance with the applicable provisions of SFAR 88 of this chapter or § 25.1529 and part 25, Appendix H, of this chapter, in effect on June 6, 2001 (including those developed for auxiliary fuel tanks, if any, installed under supplemental type certificates or other design approval) and that have been approved by the FAA Oversight Office.

(e) After December 16, 2008, before returning a U.S.-registered airplane to service after any alteration for which fuel tank ICA are developed under SFAR 88, or under § 25.1529 in effect on June 6, 2001, the foreign person or foreign air carrier must include in the maintenance program for the airplane inspections and procedures for the fuel tank system based on those ICA.

(f) The fuel tank system maintenance program changes identified in paragraphs (d) and (e) of this section and any later fuel tank system revisions must be submitted to the Principal Inspector or Flight Standards International Field Office responsible for review and approval.

(g) This section does not apply to the following airplane models:

(1) Bombardier CL-44
(2) Concorde
(3) deHavilland D.H. 106 Comet 4C
(4) VFW-Vereinigte Flugtechnische Werk VFW-614
(5) Illyushin Aviation IL 96T
(6) Bristol Aircraft Britannia 305
(7) Handley Page Herald Type 300
(8) Avions Marcel Dassault—Breguet Aviation Mercure 100C
(9) Airbus Caravelle
(10) Lockheed L-300

[Amdt. 129-43, 72 FR 63413, Nov. 8, 2007]

§ 129.115 Limit of validity.

(a) *Applicability.* This section applies to foreign air carriers or foreign persons operating any U.S.-registered transport category, turbine-powered airplane with a maximum takeoff gross weight greater than 75,000 pounds and a type certificate issued after January 1, 1958, regardless of whether the maximum takeoff gross weight is a result of an original type certificate or a later design change. This section also applies to foreign air carriers or foreign persons operating any other U.S.-registered transport category, turbine-powered airplane with a type certificate issued after January 1, 1958, regardless of the maximum takeoff gross weight, for which a limit of validity of the engineering data that supports the structural maintenance program (hereafter referred to as LOV) is required in accordance with § 25.571 or § 26.21 of this chapter after January 14, 2011.

(b) *Limit of validity.* No foreign air carrier or foreign person may operate a U.S.-registered airplane identified in paragraph (a) of this section after the applicable date identified in Table 1 of this section, unless an Airworthiness Limitations section (ALS) approved under Appendix H to part 25 or § 26.21 of this chapter is incorporated into its maintenance program. The ALS must—

(1) Include an LOV approved under § 25.571 or § 26.21 of this chapter, as applicable, except as provided in paragraph (f) of this section; and

(2) Be clearly distinguishable within its maintenance program.

(c) *Operation of airplanes excluded from § 26.21.* No certificate holder may operate an airplane identified in § 26.21(g) of this chapter after July 14, 2013, unless an ALS approved under Appendix H to part 25 or § 26.21 of this chapter is incorporated into its maintenance program. The ALS must—

(1) Include an LOV approved under § 25.571 or § 26.21 of this chapter, as applicable, except as provided in paragraph (f) of this section; and

(2) Be clearly distinguishable within its maintenance program

(d) *Extended limit of validity.* No foreign air carrier or foreign person may operate an airplane beyond the LOV or extended LOV specified in paragraph (b)(1), (c), (d), or (f) of this section, as applicable, unless the following conditions are met:

(1) An ALS must be incorporated into its maintenance program that—

(i) Includes an extended LOV and any widespread fatigue damage airworthiness limitation items (ALIs) approved under § 26.23 of this chapter; and

(ii) Is approved under § 26.23 of this chapter;

(2) The extended LOV and the airworthiness limitation items pertaining to widespread fatigue damage must be clearly distinguishable within its maintenance program.

(e) *Principal Maintenance Inspector approval.* Foreign air carriers or foreign persons must submit the maintenance program revisions required by paragraphs (b), (c), and (d) of this section to the Principal Maintenance Inspector or Flight Standards International Field Office for review and approval.

(f) *Exception.* For any airplane for which an LOV has not been approved as of the applicable compliance date specified in paragraph (c) or Table 1 of this section, instead of including an approved LOV in the ALS, an operator must include the applicable default LOV specified in Table 1 or Table 2 of this section, as applicable, in the ALS.

Table 1—Airplanes Subject to §26.21

Airplane model	Compliance date—months after January 14, 2011	Default LOV [flight cycles (FC) or flight hours (FH)]
Airbus—Existing1Models Only:		
A300 B2–1A, B2–1C, B2K–3C, B2–203	30	48,000 FC
A300 B4–2C, B4–103	30	40,000 FC
A300 B4–203	30	34,000 FC
A300–600 Series	30	30,000 FC/67,500 FH
A310–200 Series	30	40,000 FC/60,000 FH
A310–300 Series	30	35,000 FC/60,000 FH
A318 Series	60	48,000 FC/60,000 FH
A319 Series	60	48,000 FC/60,000 FH
A320–100 Series	60	48,000 FC/48,000 FH
A320–200 Series	60	48,000 FC/60,000 FH
A321 Series	60	48,000 FC/60,000 FH
A330–200, –300 Series (except WV050 family) (non enhanced)	60	40,000 FC/60,000 FH
A330–200, –300 Series WV050 family (enhanced)	60	33,000 FC/100,000 FH
A330–200 Freighter Series	60	See NOTE.
A340–200, –300 Series (except WV 027 and WV050 family) (non enhanced)	60	20,000 FC/80,000 FH
A340–200, –300 Series WV 027 (non enhanced)	60	30,000 FC/60,000 FH
A340–300 Series WV050 family (enhanced)	60	20,000 FC/100,000 FH
A340–500, –600 Series	60	16,600 FC/100,000 FH
A380–800 Series	72	See NOTE.
Boeing—Existing1Models Only:		
717	60	60,000 FC/60,000 FH
727 (all series)	30	60,000 FC
737 (Classics): 737–100, –200, –200C, –300, –400, –500	30	75,000 FC
737 (NG): 737–600, –700, –700C, –800, –900, –900ER	60	75,000 FC
747 (Classics): 747–100, –100B, –100B SUD, –200B, –200C, –200F, –300, 747SP, 747SR	30	20,000 FC
747–400: 747–400, –400D, –400F	60	20,000 FC
757	60	50,000 FC
767	60	50,000 FC
777–200, –300	60	40,000 FC
777–200LR, 777–300ER	72	40,000 FC
777F	72	11,000 FC
Bombardier—Existing1Models Only:		
CL–600: 2D15 (Regional Jet Series 705), 2D24 (Regional Jet Series 900)	72	60,000 FC
Embraer—Existing1Models Only:		
ERJ 170	72	See NOTE.
ERJ 190	72	See NOTE.
Fokker—Existing1 Models Only:		
F.28 Mark 0070, Mark 0100	30	90,000 FC
Lockheed—Existing1Models Only:		

Airplane model	Compliance date—months after January 14, 2011	Default LOV [flight cycles (FC) or flight hours (FH)]
L–1011	30	36,000 FC
188	30	26,600 FC
382 (all series)	30	20,000 FC/50,000 FH
McDonnell Douglas—Existing1Models Only:		
DC–8, –8F	30	50,000 FC/50,000 FH
DC–9 (except for MD–80 series)	30	100,000 FC/100,000 FH
MD–80 (DC–9–81, –82, –83, –87, MD–88)	30	50,000 FC/50,000 FH
MD–90	60	60,000 FC/90,000 FH
DC–10–10, –15	30	42,000 FC/60,000 FH
DC–10–30, –40, –10F, –30F, –40F	30	30,000 FC/60,000 FH
MD–10–10F	60	42,000 FC/60,000 FH
MD–10–30F	60	30,000 FC/60,000 FH
MD–11, MD–11F	60	20,000 FC/60,000 FH
Maximum Takeoff Gross Weight Changes	30, or within 12 months after the LOV is approved, or before operating the airplane, whichever occurs latest	Not applicable.
All airplanes whose maximum takeoff gross weight has been decreased to 75,000 pounds or below after January 14, 2011 or increased to greater than 75,000 pounds at any time by an amended type certificate or supplemental type certificate		
All Other Airplane Models (TCs and amended TCs) not Listed in Table 2	72, or within 12 months after the LOV is approved, or before operating the airplane, whichever occurs latest	Not applicable.

[1] Type certificated as of January 14, 2011.

Note: Airplane operation limitation is stated in the Airworthiness Limitation section.

Note: Airplane operation limitation is stated in the Airworthiness Limitation section.

Table 2—Airplanes Excluded from §26.21

Airplane model	Default LOV [flight cycles (FC) or flight hours (FH)]
Airbus:	
Caravelle	15,000 FC/24,000 FH
Avions Marcel Dassault:	
Breguet Aviation Mercure 100C	20,000 FC/16,000 FH
Boeing:	
Boeing 707 (–100 Series and –200 Series)	20,000 FC
Boeing 707 (–300 Series and –400 Series)	20,000 FC
Boeing 720	30,000 FC
Bombardier:	
CL–44D4 and CL–44J	20,000 FC
BD–700	15,000 FH
Bristol Aeroplane Company:	
Britannia 305	10,000 FC
British Aerospace Airbus, Ltd.:	
BAC 1–11 (all models)	85,000 FC
British Aerospace (Commercial Aircraft) Ltd.:	
Armstrong Whitworth Argosy A.W. 650 Series 101	20,000 FC
BAE Systems (Operations) Ltd.:	
BAe 146–100A (all models)	50,000 FC
BAe 146–200–07	50,000 FC
BAe 146–200–07 Dev	50,000 FC
BAe 146–200–11	50,000 FC
BAe 146–200–07A	47,000 FC
BAe 146–200–11 Dev	43,000 FC
BAe 146–300 (all models)	40,000 FC
Avro 146–RJ70A (all models)	40,000 FC
Avro 146–RJ85A and 146–RJ100A (all models)	50,000 FC
D & R Nevada, LLC:	
Convair Model 22	1,000 FC/1,000 FH
Convair Model 23M	1,000 FC/1,000 FH
deHavilland Aircraft Company, Ltd.:	
D.H. 106 Comet 4C	8,000 FH
Gulfstream:	
GV	40,000 FH
GV–SP	40,000 FH
Ilyushin Aviation Complex:	
IL–96T	10,000 FC/30,000 FH
Lockheed:	
300–50A01 (USAF C 141A)	20,000 FC

[Doc. No. FAA-2006-24281, 75 FR 69787, Nov. 15, 2010, as amended by Amdt. 129-51, 77 FR 30878, May 24, 2012; Amdt. 129-51A, 77 FR 55107, Sept. 7, 2012]

§ 129.117 Flammability reduction means.

(a) *Applicability.* Except as provided in paragraph (o) of this section, this section applies to U.S.-registered transport category, turbine-powered airplanes with a type certificate issued after January 1, 1958, that as a result of original type certification or later increase in capacity have:

(1) A maximum type-certificated passenger capacity of 30 or more, or

(2) A maximum payload capacity of 7,500 pounds or more.

(b) *New Production Airplanes.* Except in accordance with § 129.14, no foreign air carrier or foreign person may operate an airplane identified in Table 1 of this section (including all-cargo airplanes) for which application is made for original certificate of airworthiness or export airworthiness approval after December 27, 2010 unless an Ignition Mitigation Means (IMM) or Flammability Reduction Means (FRM) meeting the requirements of § 26.33 of this chapter is operational.

Table 1

Model—Boeing	Model—Airbus
747 Series	A318, A319, A320, A321 Series
737 Series	A330, A340 Series
777 Series	
767 Series	

(c) *Auxiliary Fuel Tanks.* After the applicable date stated in paragraph (e) of this section, no foreign air carrier or foreign person may operate any airplane subject § 26.33 of this chapter that has an Auxiliary Fuel Tank installed pursuant to a field approval, unless the following requirements are met:

(1) The foreign air carrier or foreign person complies with 14 CFR 26.35 by the applicable date stated in that section.

(2) The foreign air carrier or foreign person installs Flammability Impact Mitigation Means (FIMM), if applicable, that are approved by the FAA Oversight Office.

(3) Except in accordance with § 129.14, the FIMM, if applicable, are operational.

(d) *Retrofit.* After the dates specified in paragraph (e) of this section, no foreign air carrier or foreign person may operate an airplane to which this section applies unless the requirements of paragraphs (d)(1) and (d)(2) of this section are met.

(1) IMM, FRM or FIMM, if required by §§ 26.33, 26.35, or 26.37 of this chapter, that are approved by the FAA Oversight Office, are installed within the compliance times specified in paragraph (e) of this section.

(2) Except in accordance with § 129.14, the IMM, FRM or FIMM, as applicable, are operational.

(e) *Compliance Times.* Except as provided in paragraphs (k) and (l) of this section, the installations required by paragraph (d) of this section must be accomplished no later than the applicable dates specified in paragraph (e)(1) or (e)(2) of this section.

(1) Fifty percent of each foreign air carrier or foreign person's fleet identified in paragraph (d)(1) of this section must be modified no later than December 26, 2014.

(2) One hundred percent of each foreign air carrier or foreign person's fleet of airplanes subject to paragraph (d)(1) or this section must be modified no later than December 26, 2017.

(3) For those foreign air carriers or foreign persons that have only one airplane for a model identified in Table 1, the airplane must be modified no later than December 26, 2017.

(f) *Compliance after Installation.* Except in accordance with § 129.14, no person may—

(1) Operate an airplane on which IMM or FRM has been installed before the dates specified in paragraph (e) of this section unless the IMM or FRM is operational.

(2) Deactivate or remove an IMM or FRM once installed unless it is replaced by a means that complies with paragraph (d) of this section.

(g) *Maintenance Program Revisions.* No foreign air carrier or foreign person may operate an airplane for which airworthiness limitations have been approved by the FAA Oversight Office in accordance with §§ 26.33, 26.35, or 26.37 of this chapter after the airplane is modified in accordance with paragraph (d) of this section unless the maintenance program for that airplane is revised to include those applicable airworthiness limitations.

(h) After the maintenance program is revised as required by paragraph (g) of this section, before returning an airplane to service after any alteration for which airworthiness limitations are required by §§ 25.981, 26.33, 26.35, or 26.37 of this chapter, the foreign person or foreign air carrier must revise the maintenance program for the airplane to include those airworthiness limitations.

(i) The maintenance program changes identified in paragraphs (g) and (h) of this section must be submitted to the operator's assigned Flight Standards Office or Principal Inspector for review and approval prior to incorporation.

(j) The requirements of paragraph (d) of this section do not apply to airplanes operated in all-cargo service, but those airplanes are subject to paragraph (f) of this section.

(k) The compliance dates specified in paragraph (e) of this section may be extended by one year, provided that—

(1) No later than March 26, 2009, the foreign air carrier or foreign person notifies its assigned Flight Standards Office or Principal Inspector that it intends to comply with this paragraph;

(2) No later than June 24, 2009, the foreign air carrier or foreign person applies for an amendment to its operations specifications in accordance with § 129.11 to include a requirement for the airplane models specified in Table 2 of this section to use ground air conditioning systems for actual gate times of more than 30 minutes, when available at the gate and operational, whenever the ambient temperature exceeds 60 degrees Fahrenheit; and

(3) Thereafter, the certificate holder uses ground air conditioning systems as described in paragraph (k)(2) of this section on each airplane subject to the extension.

Table 2

Model—Boeing	Model—Airbus
747 Series	A318, A319, A320, A321 Series
737 Series	A300, A310 Series
777 Series	A330, A340 Series
767 Series	
757 Series	

(l) For any foreign air carrier or foreign person for which the operating certificate is issued after December 26, 2008, the compliance date specified in paragraph (e) of this section may be extended by one year, provided that the foreign air carrier or foreign person meets the requirements of paragraph (k)(2) of this section when its initial operations specifications are issued and, thereafter, uses ground air conditioning systems as described in paragraph (k)(2) of this section on each airplane subject to the extension.

(m) After the date by which any person is required by this section to modify 100 percent of the affected fleet, no person may operate in passenger service any airplane model specified in Table 2 of this section unless the airplane has been modified to comply with § 26.33(c) of this chapter.

Table 3

Model—Boeing	Model—Airbus
747 Series	A318, A319, A320, A321 Series
737 Series	A300, A310 Series
777 Series	A330, A340 Series
767 Series	
757 Series	

(n) No foreign air carrier or foreign person may operate any airplane on which an auxiliary fuel tank is installed after December 26, 2017 unless the FAA has certified the tank as compliant with § 25.981 of this chapter, in effect on December 26, 2008.

(o) *Exclusions.* The requirements of this section do not apply to the following airplane models:

(1) Convair CV-240, 340, 440, including turbine powered conversions.

(2) Lockheed L-188 Electra.

(3) Vickers VC-10.

(4) Douglas DC-3, including turbine powered conversions.

(5) Bombardier CL-44.

(6) Mitsubishi YS-11.

(7) BAC 1-11.

(8) Concorde.

(9) deHavilland D.H. 106 Comet 4C.

(10) VFW—Vereinigte Flugtechnische VFW-614.

(11) Illyushin Aviation IL 96T.

(12) Bristol Aircraft Britannia 305.

(13) Handley Page Herald Type 300.

(14) Avions Marcel Dassault—Breguet Aviation Mercure 100C.

(15) Airbus Caravelle.

(16) Fokker F-27/Fairchild Hiller FH-227.

(17) Lockheed L-300.

[Doc. No. FAA-2005-22997, 73 FR 42503, July 21, 2008, as amended by Amdt. 129-47, 74 FR 31620, July 2, 2009]

Subpart C—Special Federal Aviation Regulations

§ 129.201 SFAR No. 111—Lavatory Oxygen Systems.

The requirements of § 121.1500 of this chapter also apply to this part.

[Doc. No. FAA-2011-0186, 76 FR 12556, Mar. 8, 2011]

PART 133

ROTORCRAFT EXTERNAL-LOAD OPERATIONS

Contents

Authority: 49 U.S.C. 106(g), 40113, 44701-44702.

Source: Docket No. 1529, 29 FR 603, Jan. 24, 1964, unless otherwise noted.

Subpart A—Applicability

§ 133.1 Applicability.

This part prescribes—

(a) Airworthiness certification rules for rotorcraft used in; and

(b) Operating and certification rules governing the conduct of rotorcraft external-load operations in the United States by any person.

(c) The certification rules of this part do not apply to—

(1) Rotorcraft manufacturers when developing external-load attaching means;

(2) Rotorcraft manufacturers demonstrating compliance of equipment utilized under this part or appropriate portions of part 27 or 29 of this chapter;

(3) Operations conducted by a person demonstrating compliance for the issuance of a certificate or authorization under this part;

(4) Training flights conducted in preparation for the demonstration of compliance with this part; or

(5) A Federal, State, or local government conducting operations with public aircraft.

(d) For the purpose of this part, a person other than a crewmember or a person who is essential and directly connected with the external-load operation may be carried only in approved Class D rotorcraft-load combinations.

[Doc. No. 15176, 42 FR 24198, May 12, 1977, as amended by Amdt. 133-9, 51 FR 40707, Nov. 7, 1986]

Subpart B—Certification Rules

§ 133.11 Certificate required.

(a) No person subject to this part may conduct rotorcraft external-load operations within the United States without, or in violation of the terms of, a Rotorcraft External-Load Operator Certificate issued by the Administrator under § 133.17.

(b) No person holding a Rotorcraft External-Load Operator Certificate may conduct rotorcraft external-load operations subject to this part under a business name that is not on that certificate.

[Doc. No. 15176, 42 FR 24198, May 12, 1977, as amended by Amdt. 133-7, 42 FR 32531, June 27, 1977; Amdt. 133-9, 51 FR 40707, Nov. 7, 1986]

§ 133.13 Duration of certificate.

Unless sooner surrendered, suspended, or revoked, a Rotorcraft External-Load Operator Certificate expires at the end of the twenty-fourth month after the month in which it is issued or renewed.

[Doc. No. 15176, 42 FR 24198, May 12, 1977, as amended by Amdt. 133-7, 42 FR 32531, June 27, 1977; Amdt. 133-9, 51 FR 40707, Nov. 7, 1986]

§ 133.14 Carriage of narcotic drugs, marihuana, and depressant or stimulant drugs or substances.

If the holder of a certificate issued under this part permits any aircraft owned or leased by that holder to be engaged in any operation that the certificate holder knows to be in violation of § 91.19(a) of this chapter, that operation is a basis for suspending or revoking the certificate.

[Doc. No. 12035, 38 FR 17493, July 2, 1973, as amended by Amdt. 133-10, 54 FR 34332, Aug. 18, 1989]

§ 133.15 Application for certificate issuance or renewal.

Application for an original certificate or renewal of a certificate issued under this part is made on a form, and in a manner, prescribed by the Administrator. The form may be obtained from an FAA Flight Standards District Office. The completed application is sent to the district office that has jurisdiction over the area in which the applicant's home base of operation is located.

[Doc. No. 15176, 42 FR 24198, May 12, 1977, as amended by Amdt. 133-11, 54 FR 39294, Sept. 25, 1989]

§ 133.17 Requirements for issuance of a rotorcraft external-load operator certificate.

If an applicant shows that he complies with §§ 133.19, 133.21, and 133.23, the Administrator issues a Rotorcraft External-Load Operator Certificate to him with an authorization to operate specified rotorcraft with those classes of rotorcraft-load combinations for which he complies with the applicable provisions of subpart D of this part.

§ 133.19 Rotorcraft.

(a) The applicant must have the exclusive use of at least one rotorcraft that—

(1) Was type certificated under, and meets the requirements of, part 27 or 29 of this chapter (but not necessarily with external-load-carrying attaching means installed) or of § 21.25 of this chapter for the special purpose of rotorcraft external-load operations;

(2) Complies with the certification provisions in subpart D of this part that apply to the rotorcraft-load combinations for which authorization is requested; and

(3) Has a valid standard or restricted category airworthiness certificate.

(b) For the purposes of paragraph (a) of this section, a person has exclusive use of a rotorcraft if he has the sole possession, control, and use of it for flight, as owner, or has a written agreement (including arrangements for the performance of required maintenance) giving him that possession, control, and use for at least six consecutive months.

[Doc. No. 15176, 42 FR 24198, May 12, 1977]

§ 133.21 Personnel.

(a) The applicant must hold, or have available the services of at least one person who holds, a current commercial or airline transport pilot certificate, with a rating appropriate for the rotorcraft prescribed in § 133.19, issued by the Administrator.

(b) The applicant must designate one pilot, who may be the applicant, as chief pilot for rotorcraft external-load operations. The applicant also may designate qualified pilots as assistant chief pilots to perform the functions of the chief pilot when the chief pilot is not readily available. The chief pilot and assistant chief pilots must be acceptable to the Administrator and each must hold a current Commercial or Airline Transport Pilot Certificate, with a rating appropriate for the rotorcraft prescribed in § 133.19.

(c) The holder of a Rotorcraft External-Load Operator Certificate shall report any change in designation of chief pilot or assistant chief pilot immediately to the FAA certificate-holding office. The new chief pilot must be designated and must comply with § 133.23 within 30 days or the operator may not conduct further operations under the Rotorcraft External-Load Operator Certificate unless otherwise authorized by the FAA certificate-holding office.

[Doc. No. 1529, 29 FR 603, Jan. 24, 1964, as amended by Amdt. 133-9, 51 FR 40707, Nov. 7, 1986]

§ 133.22 Employment of former FAA employees.

(a) Except as specified in paragraph (c) of this section, no certificate holder may knowingly employ or make a contractual arrangement which permits an individual to act as an agent or representative of the certificate holder in any matter before the Federal Aviation Administration if the individual, in the preceding 2 years—

(1) Served as, or was directly responsible for the oversight of, a Flight Standards Service aviation safety inspector; and

(2) Had direct responsibility to inspect, or oversee the inspection of, the operations of the certificate holder.

(b) For the purpose of this section, an individual shall be considered to be acting as an agent or representative of a certificate holder in a matter before the agency if the individual makes any written or oral communication on behalf of the certificate holder to the agency (or any of its officers or employees) in connection with a particular matter, whether or not involving a specific party and without regard to whether the individual has participated in, or had responsibility for, the particular matter while serving as a Flight Standards Service aviation safety inspector.

(c) The provisions of this section do not prohibit a certificate holder from knowingly employing or making a contractual arrangement which permits an individual to act as an agent or representative of the certificate holder in any matter before the Federal Aviation Administration if the individual was employed by the certificate holder before October 21, 2011.

[Doc. No. FAA-2008-1154, 76 FR 52236, Aug. 22, 2011]

§ 133.23 Knowledge and skill.

(a) Except as provided in paragraph (d) of this section, the applicant, or the chief pilot designated in accordance with § 133.21(b), must demonstrate to the Administrator satisfactory knowledge and skill regarding rotorcraft external-load operations as set forth in paragraphs (b) and (c) of this section.

(b) The test of knowledge (which may be oral or written, at the option of the applicant) covers the following subjects:

(1) Steps to be taken before starting operations, including a survey of the flight area.

(2) Proper method of loading, rigging, or attaching the external load.

(3) Performance capabilities, under approved operating procedures and limitations, of the rotorcraft to be used.

(4) Proper instructions of flight crew and ground workers.

(5) Appropriate rotorcraft-load combination flight manual.

(c) The test of skill requires appropriate maneuvers for each class requested. The appropriate maneuvers for each load class must be demonstrated in the rotorcraft prescribed in § 133.19.

(1) Takeoffs and landings.

(2) Demonstration of directional control while hovering.

(3) Acceleration from a hover.

(4) Flight at operational airspeeds.

(5) Approaches to landing or working area.

(6) Maneuvering the external load into the release position.

(7) Demonstration of winch operation, if a winch is installed to hoist the external load.

(d) Compliance with paragraphs (b) and (c) of this section need not be shown if the Administrator finds, on the basis of the applicant's (or his designated chief pilot's) previous experience and safety record in rotorcraft external-load operations, that his knowledge and skill are adequate.

[Doc. No. 1529, 29 FR 603, Jan. 24, 1964, as amended by Amdt. 133-9, 51 FR 40707, Nov. 7, 1986]

§ 133.25 Amendment of certificate.

(a) The holder of a Rotorcraft External-Load Certificate may apply to the FAA Flight Standards District Office having jurisdiction over the area in which the applicant's home base of operation is located, or to the Flight Standards District Office nearest the area in which operations are to be conducted, for an amendment of the applicant's certificate, to add or delete a rotorcraft-load combination authorization, by executing the appropriate portion of the form used in applying for a Rotorcraft External-Load Operator Certificate. If the applicant for the amendment shows compliance with §§ 133.19, and 133.49, the Flight Standards District Office issues an amended Rotorcraft External-Load Operator Certificate to the applicant with authorization to operate with those classes of rotorcraft-load combinations for which the applicant complies with the applicable provisions of subpart D of this part.

(b) The holder of a rotorcraft external-load certificate may apply for an amendment to add or delete a rotorcraft authorization by submitting to the certificate-holding FAA Flight Standards District Office a new list of rotorcraft, by registration number, with the classes of rotorcraft-load combinations for which authorization is requested.

[Doc. No. 18434, 43 FR 52206, Nov. 9, 1978, as amended by Amdt. 133-9, 51 FR 40707, Nov. 7, 1986; Amdt. 133-11, 54 FR 39294, Sept. 25, 1989]

§ 133.27 Availability, transfer, and surrender of certificate.

(a) Each holder of a rotorcraft external-load operator certificate shall keep that certificate and a list of authorized rotorcraft at the home base of operations and shall make it available for inspection by the Administrator upon request.

(b) Each person conducting a rotorcraft external-load operation shall carry a facsimile of the Rotorcraft External-Load Operator Certificate in each rotorcraft used in the operation.

(c) If the Administrator suspends or revokes a Rotorcraft External-Load Operator Certificate, the holder of that certificate shall return it to the Administrator. If the certificate holder, for any other reason, discontinues operations under his certificate, and does not resume operations within two years, he shall return the certificate to the FAA Flight Standards District Office having jurisdiction over the area in which his home base of operations is located.

[Doc. No. 1529, 29 FR 603, Jan. 24, 1964, as amended by Amdt. 133-9, 51 FR 40708, Nov. 7, 1986; Amdt. 133-11, 54 FR 39294, Sept. 25, 1989]

Subpart C—Operating Rules and Related Requirements

§ 133.31 Emergency operations.

(a) In an emergency involving the safety of persons or property, the certificate holder may deviate from the rules of this part to the extent required to meet that emergency.

(b) Each person who, under the authority of this section, deviates from a rule of this part shall notify the Administrator within 10 days after the deviation. Upon the request of the Administrator, that person shall provide the certificate-holding FAA Flight Standards District Office a complete report of the aircraft operation involved, including a description of the deviation and reasons for it.

[Doc. No. 24550, 51 FR 40708, Nov. 7, 1986, as amended by Amdt. 133-11, 54 FR 39294, Sept. 25, 1989]

§ 133.33 Operating rules.

(a) No person may conduct a rotorcraft external-load operation without, or contrary to, the Rotorcraft-Load Combination Flight Manual prescribed in § 133.47.

(b) No person may conduct a rotorcraft external-load operation unless—

(1) The rotorcraft complies with § 133.19; and

(2) The rotorcraft and rotorcraft-load combination is authorized under the Rotorcraft External-Load Operator Certificate.

(c) Before a person may operate a rotorcraft with an external-load configuration that differs substantially from any that person has previously carried with that type of rotorcraft (whether or not the rotorcraft-load combination is of the same class), that person must conduct, in a manner that will not endanger persons or property on the surface, such of the following flight-operational checks as the Administrator determines are appropriate to the rotorcraft-load combination:

(1) A determination that the weight of the rotorcraft-load combination and the location of its center of gravity are within approved limits, that the external load is securely fastened, and that the external load does not interfere with devices provided for its emergency release.

(2) Make an initial liftoff and verify that controllability is satisfactory.

(3) While hovering, verify that directional control is adequate.

(4) Accelerate into forward flight to verify that no attitude (whether of the rotorcraft or of the external load) is encountered in which the rotorcraft is uncontrollable or which is otherwise hazardous.

(5) In forward flight, check for hazardous oscillations of the external load, but if the external load is not visible to the pilot, other crewmembers or ground personnel may make this check and signal the pilot.

(6) Increase the forward airspeed and determine an operational airspeed at which no hazardous oscillation or hazardous aerodynamic turbulence is encountered.

(d) Notwithstanding the provisions of part 91 of this chapter, the holder of a Rotorcraft External-Load Operator Certificate may conduct (in rotorcraft type certificated under and meeting the requirements of part 27 or 29 of this chapter, including the external-load attaching means) rotorcraft external-load operations over congested areas if those operations are conducted without hazard to persons or property on the surface and comply with the following:

(1) The operator must develop a plan for each complete operation, coordinate this plan with the FAA Flight Standards District Office having jurisdiction over the area in which the operation will be conducted, and obtain approval for the operation from that district office. The plan must include an agreement with the appropriate political subdivision that local officials will exclude unauthorized persons from the area in which the operation will be conducted, coordination with air traffic control, if necessary, and a detailed chart depicting the flight routes and altitudes.

(2) Each flight must be conducted at an altitude, and on a route, that will allow a jettisonable external load to be released, and the rotorcraft landed, in an emergency without hazard to persons or property on the surface.

(e) Notwithstanding the provisions of part 91 of this chapter, and except as provided in § 133.45(d), the holder of a Rotorcraft External-Load Operator Certificate may conduct external-load operations, including approaches, departures, and load positioning maneuvers necessary for the operation, below 500 feet above the surface and closer than 500 feet to persons, vessels, vehicles, and structures, if the operations are conducted without creating a hazard to persons or property on the surface.

(f) No person may conduct rotorcraft external-load operations under IFR unless specifically approved by the Administrator. However, under no circumstances may a person be carried as part of the external-load under IFR.

[Doc. No. 24550, 51 FR 40708, Nov. 7, 1986, as amended by Amdt. 133-11, 54 FR 39294, Sept. 25, 1989]

§ 133.35 Carriage of persons.

(a) No certificate holder may allow a person to be carried during rotorcraft external-load operations unless that person—

(1) Is a flight crewmember;

(2) Is a flight crewmember trainee;

(3) Performs an essential function in connection with the external-load operation; or

(4) Is necessary to accomplish the work activity directly associated with that operation.

(b) The pilot in command shall ensure that all persons are briefed before takeoff on all pertinent procedures to be followed (including normal, abnormal, and emergency procedures) and equipment to be used during the external-load operation.

[Doc. No. 24550, 51 FR 40708, Nov. 7, 1986]

§ 133.37 Crewmember training, currency, and testing requirements.

(a) No certificate holder may use, nor may any person serve, as a pilot in operations conducted under this part unless that person—

(1) Has successfully demonstrated, to the Administrator knowledge and skill with respect to the rotorcraft-load combination in accordance with § 133.23 (in the case of a pilot other than the chief pilot or an assistant chief pilot who has been designated in accordance with § 133.21(b), this demonstration may be made to the chief pilot or assistant chief pilot); and

(2) Has in his or her personal possession a letter of competency or an appropriate logbook entry indicating compliance with paragraph (a)(1) of this section.

(b) No certificate holder may use, nor may any person serve as, a crewmember or other operations personnel in Class D operations conducted under this part unless, within the preceding 12 calendar months, that person has successfully completed either an approved initial or a recurrent training program.

(c) Notwithstanding the provisions of paragraph (b) of this section, a person who has performed a rotorcraft external-load operation of the same class and in an aircraft of the same type within the past 12 calendar months need not undergo recurrent training.

[Doc. No. 24550, 51 FR 40708, Nov. 7, 1986]

§ 133.39 Inspection authority.

Each person conducting an operation under this part shall allow the Administrator to make any inspections or tests that he considers necessary to determine compliance with the Federal Aviation Regulations and the Rotorcraft External-Load Operator Certificate.

[Doc. No. 1529, 29 FR 603, Jan. 24, 1964. Redesignated by Amdt. 133-9, 51 FR 40708, Nov. 7, 1986]

Subpart D—Airworthiness Requirements

§ 133.41 Flight characteristics requirements.

(a) The applicant must demonstrate to the Administrator, by performing the operational flight checks prescribed in paragraphs (b), (c), and (d) of this section, as applicable, that the rotorcraft-load combination has satisfactory flight characteristics, unless these operational flight checks have been demonstrated previously and the rotorcraft-load combination flight characteristics were satisfactory. For the purposes of this demonstration, the external-load weight (including the external-load attaching means) is the maximum weight for which authorization is requested.

(b) Class A rotorcraft-load combinations: The operational flight check must consist of at least the following maneuvers:

(1) Take off and landing.

(2) Demonstration of adequate directional control while hovering.

(3) Acceleration from a hover.

(4) Horizontal flight at airspeeds up to the maximum airspeed for which authorization is requested.

(c) *Class B and D rotorcraft-load combinations:* The operational flight check must consist of at least the following maneuvers:

(1) Pickup of the external load.

(2) Demonstration of adequate directional control while hovering.

(3) Acceleration from a hover.

(4) Horizontal flight at airspeeds up to the maximum airspeed for which authorization is requested.

(5) Demonstrating appropriate lifting device operation.

(6) Maneuvering of the external load into release position and its release, under probable flight operation conditions, by means of each of the quick-release controls installed on the rotorcraft.

(d) Class C rotorcraft-load combinations: For Class C rotorcraft-load combinations used in wire-stringing, cable-laying, or similar operations, the operational flight check must consist of the maneuvers, as applicable, prescribed in paragraph (c) of this section.

[Doc. No. 1529, 29 FR 603, Jan. 24, 1964, as amended by Amdt. 133-5, 41 FR 55475, Dec. 20, 1976; Amdt. 133-9, 51 FR 40709, Nov. 7, 1986]

§ 133.43 Structures and design.

(a) *External-load attaching means.* Each external-load attaching means must have been approved under—

(1) Part 8 of the Civil Air Regulations on or before January 17, 1964;

(2) Part 133, before February 1, 1977;

(3) Part 27 or 29 of this chapter, as applicable, irrespective of the date of approval; or

(4) Section 21.25 of this chapter.

(b) *Quick release devices.* Each quick release device must have been approved under—

(1) Part 27 or 29 of this chapter, as applicable;

(2) Part 133, before February 1, 1977; or

(3) Section 21.25 of this chapter, except the device must comply with §§ 27.865(b) and 29.865(b), as applicable, of this chapter.

(c) *Weight and center of gravity*—

(1) *Weight.* The total weight of the rotorcraft-load combination must not exceed the total weight approved for the rotorcraft during its type certification.

(2) *Center of gravity.* The location of the center of gravity must, for all loading conditions, be within the range established for the rotorcraft during its type certification. For Class C rotorcraft-load combinations, the magnitude and direction of the loading force must be established at those values for which the effective location of the center of gravity remains within its established range.

[Doc. No. 14324, 41 FR 55475, Dec. 20, 1976, as amended by Amdt. 133-12, 55 FR 8006, Mar. 6, 1990]

§ 133.45 Operating limitations.

In addition to the operating limitations set forth in the approved Rotorcraft Flight Manual, and to any other limitations the Administrator may prescribe, the operator shall establish at least the following limitations and set them forth in the Rotorcraft-Load Combination Flight Manual for rotorcraft-load combination operations:

(a) The rotorcraft-load combination may be operated only within the weight and center of gravity limitations established in accordance with § 133.43(c).

(b) The rotorcraft-load combination may not be operated with an external load weight exceeding that used in showing compliance with §§ 133.41 and 133.43.

(c) The rotorcraft-load combination may not be operated at airspeeds greater than those established in accordance with § 133.41 (b), (c), and (d).

(d) No person may conduct an external-load operation under this part with a rotorcraft type certificated in the restricted category under § 21.25 of this chapter over a densely populated area, in a congested airway, or near a busy airport where passenger transport operations are conducted.

(e) The rotorcraft-load combination of Class D may be conducted only in accordance with the following:

(1) The rotorcraft to be used must have been type certificated under transport Category A for the operating weight and provide hover capability with one engine inoperative at that operating weight and altitude.

(2) The rotorcraft must be equipped to allow direct radio intercommunication among required crewmembers.

(3) The personnel lifting device must be FAA approved.

(4) The lifting device must have an emergency release requiring two distinct actions.

[Doc. No. 1529, 29 FR 603, Jan. 24, 1964, as amended by Amdt. 133-1, 30 FR 883, Jan. 28, 1965; Amdt. 133-5, 41 FR 55476, Dec. 20, 1976; Amdt. 133-6, 42 FR 24198, May 12, 1977; Amdt. 133-9, 51 FR 40709, Nov. 7, 1986]

§ 133.47 Rotorcraft-load combination flight manual.

The applicant must prepare a Rotorcraft-Load Combination Flight Manual and submit it for approval by the Administrator. The manual must be prepared in accordance with the rotorcraft flight manual provisions of subpart G of part 27 or 29 of this chapter, whichever is applicable. The limiting height-speed envelope data need not be listed as operating limitations. The manual must set forth—

(a) Operating limitations, procedures (normal and emergency), performance, and other information established under this subpart;

(b) The class of rotorcraft-load combinations for which the airworthiness of the rotorcraft has been demonstrated in accordance with §§ 133.41 and 133.43; and

(c) In the information section of the Rotorcraft-Load Combination Flight Manual—

(1) Information on any peculiarities discovered when operating particular rotorcraft-load combinations;

(2) Precautionary advice regarding static electricity discharges for Class B, Class C, and Class D rotorcraft-load combinations; and

(3) Any other information essential for safe operation with external loads.

[Doc. No. 1529, 29 FR 603, Jan. 24, 1964, as amended by Amdt. 133-9, 51 FR 40709, Nov. 7, 1986]

§ 133.49 Markings and placards.

The following markings and placards must be displayed conspicuously and must be such that they cannot be easily erased, disfigured, or obscured:

(a) A placard (displayed in the cockpit or cabin) stating the class of rotorcraft-load combination for which the rotorcraft has been approved and the occupancy limitation prescribed in § 133.45(a).

(b) A placard, marking, or instruction (displayed next to the external-load attaching means) stating the maximum external load prescribed as an operating limitation in § 133.45(c).

§ 133.51 Airworthiness certification.

A Rotorcraft External-Load Operator Certificate is a current and valid airworthiness certificate for each rotorcraft type certificated under part 27 or 29 of this chapter (or their predecessor parts) and listed by registration number on a list attached to the certificate, when the rotorcraft is being used in operations conducted under this part.

[Doc. No. 24550, 51 FR 40709, Nov. 7, 1986]

PART 135

OPERATING REQUIREMENTS: COMMUTER AND ON DEMAND OPERATIONS AND RULES GOVERNING PERSONS ON BOARD SUCH AIRCRAFT

Contents

Subpart J—Maintenance, Preventive Maintenance, and Alterations

Subpart K—Hazardous Materials Training Program

Authority: 49 U.S.C. 106(f), 106(g), 41706, 40113, 44701-44702, 44705, 44709, 44711-44713, 44715-44717, 44722, 45101-45105.

Source: Docket No. 16097, 43 FR 46783, Oct. 10, 1978, unless otherwise noted.

Special Federal Aviation Regulation No. 50-2

Editorial Note: For the text of SFAR No. 50-2, see part 91 of this chapter.

Special Federal Aviation Regulation No. 71

Editorial Note: For the text of SFAR No. 71, see part 91 of this chapter.

Special Federal Aviation Regulation No. 89

Editorial Note: For the text of SFAR No. 89, see part 121 of this chapter.

Special Federal Aviation Regulation No. 97

Editorial Note: For the text of SFAR No. 97, see part 91 of this chapter.

Special Federal Aviation Regulation No. 106

Editorial Note: For the text of SFAR No. 106, see part 121 of this chapter.

Special Federal Aviation Regulation No. 108

Editorial Note: For the text of SFAR No. 108, see part 91 of this chapter.

Subpart A—General

§ 135.1 Applicability.

(a) This part prescribes rules governing—

(1) The commuter or on-demand operations of each person who holds or is required to hold an Air Carrier Certificate or Operating Certificate under part 119 of this chapter.

(2) Each person employed or used by a certificate holder conducting operations under this part including the maintenance, preventative maintenance and alteration of an aircraft.

(3) The transportation of mail by aircraft conducted under a postal service contract awarded under 39 U.S.C. 5402c.

(4) Each person who applies for provisional approval of an Advanced Qualification Program curriculum, curriculum segment, or portion of a curriculum segment under subpart Y of part 121 of this chapter of 14 CFR part 121 and each person employed or used by an air carrier or commercial operator under this part to perform training, qualification, or evaluation functions under an Advanced Qualification Program under subpart Y of part 121 of this chapter of 14 CFR part 121.

(5) Nonstop Commercial Air Tour flights conducted for compensation or hire in accordance with § 119.1(e)(2) of this chapter that begin and end at the same airport and are conducted within a 25-statute-mile radius of that airport; provided further that these operations must comply only with the drug and alcohol testing requirements in §§ 120.31, 120.33, 120.35, 120.37, and 120.39 of this chapter; and with the provisions of part 136, subpart A, and § 91.147 of this chapter by September 11, 2007.

(6) Each person who is on board an aircraft being operated under this part.

(7) Each person who is an applicant for an Air Carrier Certificate or an Operating Certificate under 119 of this chapter, when conducting proving tests.

(8) Commercial Air tours conducted by holders of operations specifications issued under this part must comply with the provisions of part 136, Subpart A of this chapter by September 11, 2007.

(b) [Reserved]

(c) An operator who does not hold a part 119 certificate and who operates under the provisions of § 91.147 of this chapter is permitted to use a person who is otherwise authorized to perform aircraft maintenance or preventive maintenance duties and who is not subject to anti-drug and alcohol misuse prevent programs to perform—

(1) Aircraft maintenance or preventive maintenance on the operator's aircraft if the operator would otherwise be required to transport the aircraft more than 50 nautical miles further than the repair point closest to operator's principal place of operation to obtain these services; or

(2) Emergency repairs on the operator's aircraft if the aircraft cannot be safely operated to a location where an employee subject to FAA-approved programs can perform the repairs.

[Doc. No. 16097, 43 FR 46783, Oct. 10, 1978]

Editorial Note: For Federal Register citations affecting § 135.1, see the List of CFR Sections Affected, which appears in the Finding Aids section of the printed volume and at *www.fdsys.gov.*

§ 135.2 Compliance schedule for operators that transition to part 121 of this chapter; certain new entrant operators.

(a) *Applicability.* This section applies to the following:

(1) Each certificate holder that was issued an air carrier or operating certificate and operations specifications under the requirements of part 135 of this chapter or under SFAR No. 38-2 of 14 CFR part 121 before January 19, 1996, and that conducts scheduled passenger-carrying operations with:

(i) Nontransport category turbopropeller powered airplanes type certificated after December 31, 1964, that have a passenger seat configuration of 10-19 seats;

(ii) Transport category turbopropeller powered airplanes that have a passenger seat configuration of 20-30 seats; or

(iii) Turbojet engine powered airplanes having a passenger seat configuration of 1-30 seats.

(2) Each person who, after January 19, 1996, applies for or obtains an initial air carrier or operating certificate and operations specifications to conduct scheduled passenger-carrying operations in the kinds of airplanes described in paragraphs (a)(1)(i), (a)(1)(ii), or paragraph (a)(1)(iii) of this section.

(b) *Obtaining operations specifications.* A certificate holder described in paragraph (a)(1) of this section may not, after March 20, 1997, operate an airplane described in paragraphs (a)(1)(i), (a)(1)(ii), or (a)(1)(iii) of this section in scheduled passenger-carrying operations, unless it obtains operations specifications to conduct its scheduled operations under part 121 of this chapter on or before March 20, 1997.

(c) *Regular or accelerated compliance.* Except as provided in paragraphs (d), and (e) of this section, each certificate holder described in paragraph (a)(1) of this section shall comply with each applicable requirement of part 121 of this chapter on and after March 20, 1997 or on and after the date on which the certificate holder is issued operations specifications under this part, whichever occurs first. Except as provided in paragraphs (d) and (e) of this section, each person described in paragraph (a)(2) of this section shall comply with each applicable requirement of part 121 of this chapter on and after the date on which that person is issued a certificate and operations specifications under part 121 of this chapter.

(d) *Delayed compliance dates.* Unless paragraph (e) of this section specifies an earlier compliance date, no certificate holder that is covered by paragraph (a) of this section may operate an airplane in 14 CFR part 121 operations on or after a date listed in this paragraph unless that airplane meets the applicable requirement of this paragraph:

(1) *Nontransport category turbopropeller powered airplanes type certificated after December 31, 1964, that have a passenger seat configuration of 10-19 seats.* No certificate holder may operate under this part an airplane that is described in paragraph (a)(1)(i) of this section on or after a date listed in paragraph (d)(1) of this section unless that airplane meets the applicable requirement listed in paragraph (d)(1) of this section:

(i) December 20, 1997:

(A) Section 121.289, Landing gear aural warning.

(B) Section 121.308, Lavatory fire protection.

(C) Section 121.310(e), Emergency exit handle illumination.

(D) Section 121.337(b)(8), Protective breathing equipment.

(E) Section 121.340, Emergency flotation means.

(ii) December 20, 1999: Section 121.342, Pitot heat indication system.

(iii) December 20, 2010:

(A) For airplanes described in § 121.157(f), the Airplane Performance Operating Limitations in §§ 121.189 through 121.197.

(B) Section 121.161(b), Ditching approval.

(C) Section 121.305(j), Third attitude indicator.

(D) Section 121.312(c), Passenger seat cushion flammability.

(iv) March 12, 1999: Section 121.310(b)(1), Interior emergency exit locating sign.

(2) *Transport category turbopropeller powered airplanes that have a passenger seat configuration of 20-30 seats.* No certificate holder may operate under this part an airplane that is described in paragraph (a)(1)(ii) of this section on or after a date listed in paragraph (d)(2) of this section unless that airplane meets the applicable requirement listed in paragraph (d)(2) of this section:

(i) December 20, 1997:

(A) Section 121.308, Lavatory fire protection.

(B) Section 121.337(b) (8) and (9), Protective breathing equipment.

(C) Section 121.340, Emergency flotation means.

(ii) December 20, 2010: Section 121.305(j), Third attitude indicator.

(e) *Newly manufactured airplanes.* No certificate holder that is described in paragraph (a) of this section may operate under part 121 of this chapter an airplane manufactured on or after a date listed in this paragraph (e) unless that airplane meets the applicable requirement listed in this paragraph (e).

(1) For nontransport category turbopropeller powered airplanes type certificated after December 31, 1964, that have a passenger seat configuration of 10-19 seats:

(i) Manufactured on or after March 20, 1997:

(A) Section 121.305(j), Third attitude indicator.

(B) Section 121.311(f), Safety belts and shoulder harnesses.

(ii) Manufactured on or after December 20, 1997: Section 121.317(a), Fasten seat belt light.

(iii) Manufactured on or after December 20, 1999: Section 121.293, Takeoff warning system.

(iv) Manufactured on or after March 12, 1999: Section 121.310(b)(1), Interior emergency exit locating sign.

(2) For transport category turbopropeller powered airplanes that have a passenger seat configuration of 20-30 seats manufactured on or after March 20, 1997: Section 121.305(j), Third attitude indicator.

(f) *New type certification requirements.* No person may operate an airplane for which the application for a type certificate was filed after March 29, 1995, in 14 CFR part 121 operations unless that airplane is type certificated under part 25 of this chapter.

(g) *Transition plan.* Before March 19, 1996 each certificate holder described in paragraph (a)(1) of this section must submit to the FAA a transition plan (containing a calendar of events) for moving from conducting its scheduled operations under the commuter requirements of part 135 of this chapter to the requirements for domestic or flag operations under part 121 of this chapter. Each transition plan must contain details on the following:

(1) Plans for obtaining new operations specifications authorizing domestic or flag operations;

(2) Plans for being in compliance with the applicable requirements of part 121 of this chapter on or before March 20, 1997; and

(3) Plans for complying with the compliance date schedules contained in paragraphs (d) and (e) of this section.

[Doc. No. 28154, 60 FR 65938, Dec. 20, 1995, as amended by Amdt. 135-65, 61 FR 30435, June 14, 1996; Amdt. 135-66, 62 FR 13257, Mar. 19, 1997]

§ 135.3 Rules applicable to operations subject to this part.

(a) Each person operating an aircraft in operations under this part shall—

(1) While operating inside the United States, comply with the applicable rules of this chapter; and

(2) While operating outside the United States, comply with Annex 2, Rules of the Air, to the Convention on International Civil Aviation or the regulations of any foreign country, whichever applies, and with any rules of parts 61 and 91 of this chapter and this part that are more restrictive than that Annex or those regulations and that can be complied with without violating that Annex or those regulations. Annex 2 is incorporated by reference in § 91.703(b) of this chapter.

(b) After March 19, 1997, each certificate holder that conducts commuter operations under this part with airplanes in which two pilots are required by the type certification rules of this chapter shall comply with subparts N and O of part 121 of this chapter instead of the requirements of subparts E, G, and H of this part. Each affected certificate holder must submit to the Administrator and obtain approval of a transition plan (containing a calendar of events) for moving from its present part 135 training, checking, testing, and qualification requirements to the requirements of part 121 of this chapter. Each transition plan must be submitted by March 19, 1996, and must contain details on how the certificate holder plans to be in compliance with subparts N and O of part 121 on or before March 19, 1997.

(c) If authorized by the Administrator upon application, each certificate holder that conducts operations under this part to which paragraph (b) of this section does not apply, may comply with the applicable sections of subparts N and O of part 121 instead of the requirements of subparts E, G, and H of this part, except that those authorized certificate holders may choose to comply with the operating experience requirements of § 135.244, instead of the requirements of § 121.434 of this chapter.

[Doc. No. 27993, 60 FR 65949, Dec. 20, 1995, as amended by Amdt. 135-65, 61 FR 30435, June 14, 1996]

§ 135.4 Applicability of rules for eligible on-demand operations.

(a) An "eligible on-demand operation" is an on-demand operation conducted under this part that meets the following requirements:

(1) *Two-pilot crew.* The flightcrew must consist of at least two qualified pilots employed or contracted by the certificate holder.

(2) *Flight crew experience.* The crewmembers must have met the applicable requirements of part 61 of this chapter and have the following experience and ratings:

(i) Total flight time for all pilots:

(A) Pilot in command—A minimum of 1,500 hours.

(B) Second in command—A minimum of 500 hours.

(ii) For multi-engine turbine-powered fixed-wing and powered-lift aircraft, the following FAA certification and ratings requirements:

(A) Pilot in command—Airline transport pilot and applicable type ratings.

(B) Second in command—Commercial pilot and instrument ratings.

(iii) For all other aircraft, the following FAA certification and rating requirements:

(A) Pilot in command—Commercial pilot and instrument ratings.

(B) Second in command—Commercial pilot and instrument ratings.

(3) *Pilot operating limitations.* If the second in command of a fixed-wing aircraft has fewer than 100 hours of flight time as second in command flying in the aircraft make and model and, if a type rating is required, in the type aircraft being flown, and the pilot in command is not an appropriately qualified check pilot, the pilot in command shall make all takeoffs and landings in any of the following situations:

(i) Landings at the destination airport when a Destination Airport Analysis is required by § 135.385(f); and

(ii) In any of the following conditions:

(A) The prevailing visibility for the airport is at or below $^3/_4$ mile.

(B) The runway visual range for the runway to be used is at or below 4,000 feet.

(C) The runway to be used has water, snow, slush, ice, or similar contamination that may adversely affect aircraft performance.

(D) The braking action on the runway to be used is reported to be less than "good."

(E) The crosswind component for the runway to be used is in excess of 15 knots.

(F) Windshear is reported in the vicinity of the airport.

(G) Any other condition in which the pilot in command determines it to be prudent to exercise the pilot in command's authority.

(4) *Crew pairing.* Either the pilot in command or the second in command must have at least 75 hours of flight time in that aircraft make or model and, if a type rating is required, for that type aircraft, either as pilot in command or second in command.

(b) The Administrator may authorize deviations from paragraphs (a)(2)(i) or (a)(4) of this section if the Flight Standards District Office that issued the certificate holder's operations specifications finds that the crewmember has comparable experience, and can effectively perform the functions associated with the position in accordance with the requirements of this chapter. The Administrator may, at any time, terminate any grant of deviation authority issued under this paragraph. Grants of deviation under this paragraph may be granted after consideration of the size and scope of the operation, the qualifications of the intended personnel and the following circumstances:

(1) A newly authorized certificate holder does not employ any pilots who meet the minimum requirements of paragraphs (a)(2)(i) or (a)(4) of this section.

(2) An existing certificate holder adds to its fleet a new category and class aircraft not used before in its operation.

(3) An existing certificate holder establishes a new base to which it assigns pilots who will be required to become qualified on the aircraft operated from that base.

(c) An eligible on-demand operation may comply with alternative requirements specified in §§ 135.225(b), 135.385(f), and 135.387(b) instead of the requirements that apply to other on-demand operations.

[Doc. No. FAA-2001-10047, 68 FR 54585, Sept. 17, 2003]

§ 135.7 Applicability of rules to unauthorized operators.

The rules in this part which apply to a person certificated under part 119 of this chapter also apply to a person who engages in any operation governed by this part without an appropriate certificate and operations specifications required by part 119 of this chapter.

[Doc. No. 16097, 43 FR 46783, Oct. 10, 1978, as amended by Amdt. 135-58, 60 FR 65939, Dec. 20, 1995]

§ 135.12 Previously trained crewmembers.

A certificate holder may use a crewmember who received the certificate holder's training in accordance with subparts E, G, and H of this part before March 19, 1997 without complying with initial training and qualification requirements of subparts N and O of part 121 of this chapter. The crewmember must comply with the applicable recurrent training requirements of part 121 of this chapter.

[Doc. No. 27993, 60 FR 65950, Dec. 20, 1995]

§ 135.19 Emergency operations.

(a) In an emergency involving the safety of persons or property, the certificate holder may deviate from the rules of this part relating to aircraft and equipment and weather minimums to the extent required to meet that emergency.

(b) In an emergency involving the safety of persons or property, the pilot in command may deviate from the rules of this part to the extent required to meet that emergency.

(c) Each person who, under the authority of this section, deviates from a rule of this part shall, within 10 days, excluding Saturdays, Sundays, and Federal holidays, after the deviation, send to the FAA Flight Standards District Office charged with the overall inspection of the certificate holder a complete report of the aircraft operation involved, including a description of the deviation and reasons for it.

§ 135.21 Manual requirements.

(a) Each certificate holder, other than one who uses only one pilot in the certificate holder's operations, shall prepare and keep current a manual setting forth the certificate holder's procedures and policies acceptable to the Administrator. This manual must be used by the certificate holder's flight, ground, and maintenance personnel in conducting its operations. However, the Administrator may authorize a deviation from this paragraph if the Administrator finds that, because of the limited size of the operation, all or part of the manual is not necessary for guidance of flight, ground, or maintenance personnel.

(b) Each certificate holder shall maintain at least one copy of the manual at its principal base of operations.

(c) The manual must not be contrary to any applicable Federal regulations, foreign regulation applicable to the certificate holder's operations in foreign countries, or the certificate holder's operating certificate or operations specifications.

(d) A copy of the manual, or appropriate portions of the manual (and changes and additions) shall be made available to maintenance and ground operations personnel by the certificate holder and furnished to—

(1) Its flight crewmembers; and

(2) Representatives of the Administrator assigned to the certificate holder.

(e) Each employee of the certificate holder to whom a manual or appropriate portions of it are furnished under paragraph (d)(1) of this section shall keep it up to date with the changes and additions furnished to them.

(f) Except as provided in paragraph (h) of this section, each certificate holder must carry appropriate parts of the manual on each aircraft when away from the principal operations base. The appropriate parts must be available for use by ground or flight personnel.

(g) For the purpose of complying with paragraph (d) of this section, a certificate holder may furnish the persons listed therein with all or part of its manual in printed form or other form, acceptable to the Administrator, that is retrievable in the English language. If the certificate holder furnishes all or part of the manual in other than printed form, it must ensure there is a compatible reading device available to those persons that provides a legible image of the information and instructions, or a system that is able to retrieve the information and instructions in the English language.

(h) If a certificate holder conducts aircraft inspections or maintenance at specified stations where it keeps the approved inspection program manual, it is not required to carry the manual aboard the aircraft en route to those stations.

[Doc. No. 16097, 43 FR 46783, Oct. 10, 1978, as amended by Amdt. 135-18, 47 FR 33396, Aug. 2, 1982; Amdt. 135-58, 60 FR 65939, Dec. 20, 1995; Amdt. 135-66, 62 FR 13257, Mar. 19, 1997; Amdt. 135-91, 68 FR 54585, Sept. 17, 2003]

§ 135.23 Manual contents.

Each manual shall have the date of the last revision on each revised page. The manual must include—

(a) The name of each management person required under § 119.69(a) of this chapter who is authorized to act for the certificate holder, the person's assigned area of responsibility, the person's duties, responsibilities, and authority, and the name and title of each person authorized to exercise operational control under § 135.77;

(b) Procedures for ensuring compliance with aircraft weight and balance limitations and, for multiengine aircraft, for determining compliance with § 135.185;

(c) Copies of the certificate holder's operations specifications or appropriate extracted information, including area of operations authorized, category and class of aircraft authorized, crew complements, and types of operations authorized;

(d) Procedures for complying with accident notification requirements;

(e) Procedures for ensuring that the pilot in command knows that required airworthiness inspections have been made and that the aircraft has been approved for return to service in compliance with applicable maintenance requirements;

(f) Procedures for reporting and recording mechanical irregularities that come to the attention of the pilot in command before, during, and after completion of a flight;

(g) Procedures to be followed by the pilot in command for determining that mechanical irregularities or defects reported for previous flights have been corrected or that correction has been deferred;

(h) Procedures to be followed by the pilot in command to obtain maintenance, preventive maintenance, and servicing of the aircraft at a place where previous arrangements have not been made by the operator, when the pilot is authorized to so act for the operator;

(i) Procedures under § 135.179 for the release for, or continuation of, flight if any item of equipment required for the particular type of operation becomes inoperative or unserviceable en route;

(j) Procedures for refueling aircraft, eliminating fuel contamination, protecting from fire (including electrostatic protection), and supervising and protecting passengers during refueling;

(k) Procedures to be followed by the pilot in command in the briefing under § 135.117;

(l) Flight locating procedures, when applicable;

(m) Procedures for ensuring compliance with emergency procedures, including a list of the functions assigned each category of required crewmembers in connection with an emergency and emergency evacuation duties under § 135.123;

(n) En route qualification procedures for pilots, when applicable;

(o) The approved aircraft inspection program, when applicable;

(p)(1) Procedures and information, as described in paragraph (p)(2) of this section, to assist each crewmember and person performing or directly supervising the following job functions involving items for transport on an aircraft:

(i) Acceptance;

(ii) Rejection;

(iii) Handling;

(iv) Storage incidental to transport;

(v) Packaging of company material; or

(vi) Loading.

(2) Ensure that the procedures and information described in this paragraph are sufficient to assist a person in identifying packages that are marked or labeled as containing hazardous materials or that show signs of containing undeclared hazardous materials. The procedures and information must include:

(i) Procedures for rejecting packages that do not conform to the Hazardous Materials Regulations in 49 CFR parts 171 through 180 or that appear to contain undeclared hazardous materials;

(ii) Procedures for complying with the hazardous materials incident reporting requirements of 49 CFR 171.15 and 171.16 and discrepancy reporting requirements of 49 CFR 175.31.

(iii) The certificate holder's hazmat policies and whether the certificate holder is authorized to carry, or is prohibited from carrying, hazardous materials; and

(iv) If the certificate holder's operations specifications permit the transport of hazardous materials, procedures and information to ensure the following:

(A) That packages containing hazardous materials are properly offered and accepted in compliance with 49 CFR parts 171 through 180;

(B) That packages containing hazardous materials are properly handled, stored, packaged, loaded and carried on board an aircraft in compliance with 49 CFR parts 171 through 180;

(C) That the requirements for Notice to the Pilot in Command (49 CFR 175.33) are complied with; and

(D) That aircraft replacement parts, consumable materials or other items regulated by 49 CFR parts 171 through 180 are properly handled, packaged, and transported.

(q) Procedures for the evacuation of persons who may need the assistance of another person to move expeditiously to an exit if an emergency occurs; and

(r) If required by § 135.385, an approved Destination Airport Analysis establishing runway safety margins at destination airports, taking into account the following factors as supported by published aircraft performance data supplied by the aircraft manufacturer for the appropriate runway conditions—

(1) Pilot qualifications and experience;

(2) Aircraft performance data to include normal, abnormal and emergency procedures as supplied by the aircraft manufacturer;

(3) Airport facilities and topography;

(4) Runway conditions (including contamination);

(5) Airport or area weather reporting;

(6) Appropriate additional runway safety margins, if required;

(7) Airplane inoperative equipment;

(8) Environmental conditions; and

(9) Other criteria affecting aircraft performance.

(s) Other procedures and policy instructions regarding the certificate holder's operations issued by the certificate holder.

[Doc. No. 16097, 43 FR 46783, Oct. 10, 1978, as amended by Amdt. 135-20, 51 FR 40709, Nov. 7, 1986; Amdt. 135-58, 60 FR 65939, Dec. 20, 1995; Amdt. 135-91, 68 FR 54586, Sept. 17, 2003; Amdt. 135-101, 70 FR 58829, Oct. 7, 2005]

§ 135.25 Aircraft requirements.

(a) Except as provided in paragraph (d) of this section, no certificate holder may operate an aircraft under this part unless that aircraft—

(1) Is registered as a civil aircraft of the United States and carries an appropriate and current airworthiness certificate issued under this chapter; and

(2) Is in an airworthy condition and meets the applicable airworthiness requirements of this chapter, including those relating to identification and equipment.

(b) Each certificate holder must have the exclusive use of at least one aircraft that meets the requirements for at least one kind of operation authorized in the certificate holder's operations specifications. In addition, for each kind of operation for which the certificate holder does not have the exclusive use of an aircraft, the certificate holder must have available for use under a written agreement (including arrangements for performing required maintenance) at least one aircraft that meets the requirements for that kind of operation. However, this paragraph does not prohibit the operator from using or authorizing the use of the aircraft for other than operations under this part and does not require the certificate holder to have exclusive use of all aircraft that the certificate holder uses.

(c) For the purposes of paragraph (b) of this section, a person has exclusive use of an aircraft if that person has the sole possession, control, and use of it for flight, as owner, or has a written agreement (including arrangements for performing required maintenance), in effect when the aircraft is operated, giving the person that possession, control, and use for at least 6 consecutive months.

(d) A certificate holder may operate in common carriage, and for the carriage of mail, a civil aircraft which is leased or chartered to it without crew and is registered in a country which is a party to the Convention on International Civil Aviation if—

(1) The aircraft carries an appropriate airworthiness certificate issued by the country of registration and meets the registration and identification requirements of that country;

(2) The aircraft is of a type design which is approved under a U.S. type certificate and complies with all of the requirements of this chapter (14 CFR chapter I) that would be applicable to that aircraft were it registered in the United States, including the requirements which must be met for issuance of a U.S. standard airworthiness certificate (including type design conformity, condition for safe operation, and the noise, fuel venting, and engine emission requirements of this chapter), except that a U.S. registration certificate and a U.S. standard airworthiness certificate will not be issued for the aircraft;

(3) The aircraft is operated by U.S.-certificated airmen employed by the certificate holder; and

(4) The certificate holder files a copy of the aircraft lease or charter agreement with the FAA Aircraft Registry, Department of Transportation, 6400 South MacArthur Boulevard, Oklahoma City, OK (Mailing address: P.O. Box 25504, Oklahoma City, OK 73125).

[Doc. No. 16097, 43 FR 46783, Oct. 10, 1978, as amended by Amdt. 135-8, 45 FR 68649, Oct. 16, 1980; Amdt. 135-66, 62 FR 13257, Mar. 19, 1997]

§ 135.41 Carriage of narcotic drugs, marihuana, and depressant or stimulant drugs or substances.

If the holder of a certificate operating under this part allows any aircraft owned or leased by that holder to be engaged in any operation that the certificate holder knows to be in violation of § 91.19(a) of this chapter, that operation is a basis for suspending or revoking the certificate.

[Doc. No. 28154, 60 FR 65939, Dec. 20, 1995]

§ 135.43 Crewmember certificates: International operations.

(a) This section describes the certificates that were issued to United States citizens who were employed by air carriers at the time of issuance as flight crewmembers on United States registered aircraft engaged in international air commerce. The purpose of the certificate is to facilitate the entry and clearance of those crewmembers into ICAO contracting states. They were issued under Annex 9, as amended, to the Convention on International Civil Aviation.

(b) The holder of a certificate issued under this section, or the air carrier by whom the holder is employed, shall surrender the certificate for cancellation at the nearest FAA Flight Standards District Office at the termination of the holder's employment with that air carrier.

[Doc. No. 28154, 61 FR 30435, June 14, 1996]

Subpart B—Flight Operations

§ 135.61 General.

This subpart prescribes rules, in addition to those in part 91 of this chapter, that apply to operations under this part.

§ 135.63 Recordkeeping requirements.

(a) Each certificate holder shall keep at its principal business office or at other places approved by the Administrator, and shall make available for inspection by the Administrator the following—

(1) The certificate holder's operating certificate;

(2) The certificate holder's operations specifications;

(3) A current list of the aircraft used or available for use in operations under this part and the operations for which each is equipped;

(4) An individual record of each pilot used in operations under this part, including the following information:

(i) The full name of the pilot.

(ii) The pilot certificate (by type and number) and ratings that the pilot holds.

(iii) The pilot's aeronautical experience in sufficient detail to determine the pilot's qualifications to pilot aircraft in operations under this part.

(iv) The pilot's current duties and the date of the pilot's assignment to those duties.

(v) The effective date and class of the medical certificate that the pilot holds.

(vi) The date and result of each of the initial and recurrent competency tests and proficiency and route checks required by this part and the type of aircraft flown during that test or check.

(vii) The pilot's flight time in sufficient detail to determine compliance with the flight time limitations of this part.

(viii) The pilot's check pilot authorization, if any.

(ix) Any action taken concerning the pilot's release from employment for physical or professional disqualification.

(x) The date of the completion of the initial phase and each recurrent phase of the training required by this part; and

(5) An individual record for each flight attendant who is required under this part, maintained in sufficient detail to determine compliance with the applicable portions of § 135.273 of this part.

(b) Each certificate holder must keep each record required by paragraph (a)(3) of this section for at least 6 months, and must keep each record required by paragraphs (a)(4) and (a)(5) of this section for at least 12 months.

(c) For multiengine aircraft, each certificate holder is responsible for the preparation and accuracy of a load manifest in duplicate containing information concerning the loading of the aircraft. The manifest must be prepared before each takeoff and must include:

(1) The number of passengers;

(2) The total weight of the loaded aircraft;

(3) The maximum allowable takeoff weight for that flight;

(4) The center of gravity limits;

(5) The center of gravity of the loaded aircraft, except that the actual center of gravity need not be computed if the aircraft is loaded according to a loading schedule or other approved method that ensures that the center of gravity of the loaded aircraft is within approved limits. In those cases, an entry shall be made on the manifest indicating that the center of gravity is within limits according to a loading schedule or other approved method;

(6) The registration number of the aircraft or flight number;

(7) The origin and destination; and

(8) Identification of crew members and their crew position assignments.

(d) The pilot in command of an aircraft for which a load manifest must be prepared shall carry a copy of the completed load manifest in the aircraft to its destination. The certificate holder shall keep copies of completed load manifests for at least 30 days at its principal operations base, or at another location used by it and approved by the Administrator.

[Doc. No. 16097, 43 FR 46783, Oct. 10, 1978, as amended by Amdt. 135-52, 59 FR 42993, Aug. 19, 1994]

§ 135.64 Retention of contracts and amendments: Commercial operators who conduct intrastate operations for compensation or hire.

Each commercial operator who conducts intrastate operations for compensation or hire shall keep a copy of each written contract under which it provides services as a commercial operator for a period of at least one year after the date of execution of the contract. In the case of an oral contract, it shall keep a memorandum stating its elements, and of any amendments to it, for a period of at least one year after the execution of that contract or change.

[Doc. No. 28154, 60 FR 65939, Dec. 20, 1995, as amended by Amdt. 135-65, 61 FR 30435, June 14, 1996; Amdt. 135-66, 62 FR 13257, Mar. 19, 1997]

§ 135.65 Reporting mechanical irregularities.

(a) Each certificate holder shall provide an aircraft maintenance log to be carried on board each aircraft for recording or deferring mechanical irregularities and their correction.

(b) The pilot in command shall enter or have entered in the aircraft maintenance log each mechanical irregularity that comes to the pilot's attention during flight time. Before each flight, the pilot in command shall, if the pilot does not already know, determine the status of each irregularity entered in the maintenance log at the end of the preceding flight.

(c) Each person who takes corrective action or defers action concerning a reported or observed failure or malfunction of an airframe, powerplant, propeller, rotor, or appliance, shall record the action taken in the aircraft maintenance log under the applicable maintenance requirements of this chapter.

(d) Each certificate holder shall establish a procedure for keeping copies of the aircraft maintenance log required by this section in the aircraft for access by appropriate personnel and shall include that procedure in the manual required by § 135.21.

§ 135.67 Reporting potentially hazardous meteorological conditions and irregularities of ground facilities or navigation aids.

Whenever a pilot encounters a potentially hazardous meteorological condition or an irregularity in a ground facility or navigation aid in flight, the knowledge of which the pilot considers essential to the safety of other flights, the pilot shall notify an appropriate ground radio station as soon as practicable.

[Doc. No. 16097, 43 FR 46783, Oct. 1, 1978, as amended at Amdt. 135-1, 44 FR 26737, May 7, 1979; Amdt. 135-110, 72 FR 31684, June 7, 2007]

§ 135.69 Restriction or suspension of operations: Continuation of flight in an emergency.

(a) During operations under this part, if a certificate holder or pilot in command knows of conditions, including airport and runway conditions, that are a hazard to safe operations, the certificate holder or pilot in command, as the case may be, shall restrict or suspend operations as necessary until those conditions are corrected.

(b) No pilot in command may allow a flight to continue toward any airport of intended landing under the conditions set forth in paragraph (a) of this section, unless, in the opinion of the pilot in command, the conditions that are a hazard to safe operations may reasonably be expected to be corrected by the estimated time of arrival or, unless there is no safer procedure. In the latter event, the continuation toward that airport is an emergency situation under § 135.19.

§ 135.71 Airworthiness check.

The pilot in command may not begin a flight unless the pilot determines that the airworthiness inspections required by § 91.409 of this chapter, or § 135.419, whichever is applicable, have been made.

[Doc. No. 16097, 43 FR 46783, Oct. 10, 1978, as amended by Amdt. 135-32, 54 FR 34332, Aug. 18, 1989]

§ 135.73 Inspections and tests.

Each certificate holder and each person employed by the certificate holder shall allow the Administrator, at any time or place, to make inspections or tests (including en route inspections) to determine the holder's compliance with the Federal Aviation Act of 1958, applicable regulations, and the certificate holder's operating certificate, and operations specifications.

§ 135.75 Inspectors credentials: Admission to pilots' compartment: Forward observer's seat.

(a) Whenever, in performing the duties of conducting an inspection, an FAA inspector presents an Aviation Safety Inspector credential, FAA Form 110A, to the pilot in command of an aircraft operated by the certificate holder, the inspector must be given free and uninterrupted access to the pilot compartment of that aircraft. However, this paragraph does not limit the emergency authority of the pilot in command to exclude any person from the pilot compartment in the interest of safety.

(b) A forward observer's seat on the flight deck, or forward passenger seat with headset or speaker must be provided for use by the Administrator while conducting en route inspections. The suitability of the location of the seat and the headset or speaker for use in conducting en route inspections is determined by the Administrator.

§ 135.76 DOD Commercial Air Carrier Evaluator's Credentials: Admission to pilots compartment: Forward observer's seat.

(a) Whenever, in performing the duties of conducting an evaluation, a DOD commercial air carrier evaluator presents S&A Form 110B, "DOD Commercial Air Carrier Evaluator's Credential," to the pilot in command of an aircraft operated by the certificate holder, the evaluator must be given free and uninterrupted access to the pilot's compartment of that aircraft. However, this paragraph does not limit the emergency authority of the pilot in command to exclude any person from the pilot compartment in the interest of safety.

(b) A forward observer's seat on the flight deck or forward passenger seat with headset or speaker must be provided for use by the evaluator while conducting en route evaluations. The suitability of the location of the seat and the headset or speaker for use in conducting en route evaluations is determined by the FAA.

[Doc. No. FAA-2003-15571, 68 FR 41218, July 10, 2003]

§ 135.77 Responsibility for operational control.

Each certificate holder is responsible for operational control and shall list, in the manual required by § 135.21, the name and title of each person authorized by it to exercise operational control.

§ 135.78 Instrument approach procedures and IFR landing minimums.

No person may make an instrument approach at an airport except in accordance with IFR weather minimums and instrument approach procedures set forth in the certificate holder's operations specifications.

[Doc. No. FAA-2002-14002, 72 FR 31684, June 7, 2007]

§ 135.79 Flight locating requirements.

(a) Each certificate holder must have procedures established for locating each flight, for which an FAA flight plan is not filed, that—

(1) Provide the certificate holder with at least the information required to be included in a VFR flight plan;

(2) Provide for timely notification of an FAA facility or search and rescue facility, if an aircraft is overdue or missing; and

(3) Provide the certificate holder with the location, date, and estimated time for reestablishing communications, if the flight will operate in an area where communications cannot be maintained.

(b) Flight locating information shall be retained at the certificate holder's principal place of business, or at other places designated by the certificate holder in the flight locating procedures, until the completion of the flight.

(c) Each certificate holder shall furnish the representative of the Administrator assigned to it with a copy of its flight locating procedures and any changes or additions, unless those procedures are included in a manual required under this part.

[Doc. No. 16097, 43 FR 46783, Oct. 10, 1978, as amended by Amdt. 135-110, 72 FR 31684, June 7, 2007]

§ 135.81 Informing personnel of operational information and appropriate changes.

Each certificate holder shall inform each person in its employment of the operations specifications that apply to that person's duties and responsibilities and shall make available to each pilot in the certificate holder's employ the following materials in current form:

(a) Airman's Information Manual (Alaska Supplement in Alaska and Pacific Chart Supplement in Pacific-Asia Regions) or a commercial publication that contains the same information.

(b) This part and part 91 of this chapter.

(c) Aircraft Equipment Manuals, and Aircraft Flight Manual or equivalent.

(d) For foreign operations, the International Flight Information Manual or a commercial publication that contains the same information concerning the pertinent operational and entry requirements of the foreign country or countries involved.

§ 135.83 Operating information required.

(a) The operator of an aircraft must provide the following materials, in current and appropriate form, accessible to the pilot at the pilot station, and the pilot shall use them:

(1) A cockpit checklist.

(2) For multiengine aircraft or for aircraft with retractable landing gear, an emergency cockpit checklist containing the procedures required by paragraph (c) of this section, as appropriate.

(3) Pertinent aeronautical charts.

(4) For IFR operations, each pertinent navigational en route, terminal area, and approach and letdown chart.

(5) For multiengine aircraft, one-engine-inoperative climb performance data and if the aircraft is approved for use in IFR or over-the-top operations, that data must be sufficient to enable the pilot to determine compliance with § 135.181(a)(2).

(b) Each cockpit checklist required by paragraph (a)(1) of this section must contain the following procedures:

(1) Before starting engines;

(2) Before takeoff;

(3) Cruise;

(4) Before landing;

(5) After landing;

(6) Stopping engines.

(c) Each emergency cockpit checklist required by paragraph (a)(2) of this section must contain the following procedures, as appropriate:

(1) Emergency operation of fuel, hydraulic, electrical, and mechanical systems.

(2) Emergency operation of instruments and controls.

(3) Engine inoperative procedures.

(4) Any other emergency procedures necessary for safety.

§ 135.85 Carriage of persons without compliance with the passenger-carrying provisions of this part.

The following persons may be carried aboard an aircraft without complying with the passenger-carrying requirements of this part:

(a) A crewmember or other employee of the certificate holder.

(b) A person necessary for the safe handling of animals on the aircraft.

(c) A person necessary for the safe handling of hazardous materials (as defined in subchapter C of title 49 CFR).

(d) A person performing duty as a security or honor guard accompanying a shipment made by or under the authority of the U.S. Government.

(e) A military courier or a military route supervisor carried by a military cargo contract air carrier or commercial operator in operations under a military cargo contract, if that carriage is specifically authorized by the appropriate military service.

(f) An authorized representative of the Administrator conducting an en route inspection.

(g) A person, authorized by the Administrator, who is performing a duty connected with a cargo operation of the certificate holder.

(h) A DOD commercial air carrier evaluator conducting an en route evaluation.

[Docket No. 16097, 43 FR 46783, Oct. 10, 1978, as amended by Amdt. 135-88, 68 FR 41218, July 10, 2003]

§ 135.87 Carriage of cargo including carry-on baggage.

No person may carry cargo, including carry-on baggage, in or on any aircraft unless—

(a) It is carried in an approved cargo rack, bin, or compartment installed in or on the aircraft;

(b) It is secured by an approved means; or

(c) It is carried in accordance with each of the following:

(1) For cargo, it is properly secured by a safety belt or other tie-down having enough strength to eliminate the possibility of shifting under all normally anticipated flight and ground conditions, or for carry-on baggage, it is restrained so as to prevent its movement during air turbulence.

(2) It is packaged or covered to avoid possible injury to occupants.

(3) It does not impose any load on seats or on the floor structure that exceeds the load limitation for those components.

(4) It is not located in a position that obstructs the access to, or use of, any required emergency or regular exit, or the use of the aisle between the crew and the passenger compartment, or located in a position that obscures any passenger's view of the "seat belt" sign, "no smoking" sign, or any required exit sign, unless an auxiliary sign or other approved means for proper notification of the passengers is provided.

(5) It is not carried directly above seated occupants.

(6) It is stowed in compliance with this section for takeoff and landing.

(7) For cargo only operations, paragraph (c)(4) of this section does not apply if the cargo is loaded so that at least one emergency or regular exit is available to provide all occupants of the aircraft a means of unobstructed exit from the aircraft if an emergency occurs.

(d) Each passenger seat under which baggage is stowed shall be fitted with a means to prevent articles of baggage stowed under it from sliding under crash impacts severe enough to induce the ultimate inertia forces specified in the emergency landing condition regulations under which the aircraft was type certificated.

(e) When cargo is carried in cargo compartments that are designed to require the physical entry of a crewmember to extinguish any fire that may occur during flight, the cargo must be loaded so as to allow a crewmember to effectively reach all parts of the compartment with the contents of a hand fire extinguisher.

§ 135.89 Pilot requirements: Use of oxygen.

(a) *Unpressurized aircraft.* Each pilot of an unpressurized aircraft shall use oxygen continuously when flying—

(1) At altitudes above 10,000 feet through 12,000 feet MSL for that part of the flight at those altitudes that is of more than 30 minutes duration; and

(2) Above 12,000 feet MSL.

(b) *Pressurized aircraft.* (1) Whenever a pressurized aircraft is operated with the cabin pressure altitude more than 10,000 feet MSL, each pilot shall comply with paragraph (a) of this section.

(2) Whenever a pressurized aircraft is operated at altitudes above 25,000 feet through 35,000 feet MSL, unless each pilot has an approved quick-donning type oxygen mask—

(i) At least one pilot at the controls shall wear, secured and sealed, an oxygen mask that either supplies oxygen at all times or automatically supplies oxygen whenever the cabin pressure altitude exceeds 12,000 feet MSL; and

(ii) During that flight, each other pilot on flight deck duty shall have an oxygen mask, connected to an oxygen supply, located so as to allow immediate placing of the mask on the pilot's face sealed and secured for use.

(3) Whenever a pressurized aircraft is operated at altitudes above 35,000 feet MSL, at least one pilot at the controls shall wear, secured and sealed, an oxygen mask required by paragraph (b)(2)(i) of this section.

(4) If one pilot leaves a pilot duty station of an aircraft when operating at altitudes above 25,000 feet MSL, the remaining pilot at the controls shall put on and use an approved oxygen mask until the other pilot returns to the pilot duty station of the aircraft.

§ 135.91 Oxygen for medical use by passengers.

(a) Except as provided in paragraphs (d) and (e) of this section, no certificate holder may allow the carriage or operation of equipment for the storage, generation or dispensing of medical oxygen unless the unit to be carried is constructed so that all valves, fittings, and gauges are protected from damage during that carriage or operation and unless the following conditions are met—

(1) The equipment must be—

(i) Of an approved type or in conformity with the manufacturing, packaging, marking, labeling, and maintenance requirements of title 49 CFR parts 171, 172, and 173, except § 173.24(a)(1);

(ii) When owned by the certificate holder, maintained under the certificate holder's approved maintenance program;

(iii) Free of flammable contaminants on all exterior surfaces; and

(iv) Appropriately secured.

(2) When the oxygen is stored in the form of a liquid, the equipment must have been under the certificate holder's approved maintenance program since its purchase new or since the storage container was last purged.

(3) When the oxygen is stored in the form of a compressed gas as defined in title 49 CFR 173.300(a)—

(i) When owned by the certificate holder, it must be maintained under its approved maintenance program; and

(ii) The pressure in any oxygen cylinder must not exceed the rated cylinder pressure.

(4) The pilot in command must be advised when the equipment is on board, and when it is intended to be used.

(5) The equipment must be stowed, and each person using the equipment must be seated, so as not to restrict access to or use of any required emergency or regular exit, or of the aisle in the passenger compartment.

(b) No person may smoke and no certificate holder may allow any person to smoke within 10 feet of oxygen storage and dispensing equipment carried under paragraph (a) of this section.

(c) No certificate holder may allow any person other than a person trained in the use of medical oxygen equipment to connect or disconnect oxygen bottles or any other ancillary component while any passenger is aboard the aircraft.

(d) Paragraph (a)(1)(i) of this section does not apply when that equipment is furnished by a professional or medical emergency service for use on board an aircraft in a medical emergency when no other practical means of transportation (including any other properly equipped certificate holder) is reasonably available and the person carried under the medical emergency is accompanied by a person trained in the use of medical oxygen.

(e) Each certificate holder who, under the authority of paragraph (d) of this section, deviates from paragraph (a)(1)(i) of this section under a medical emergency shall, within 10 days, excluding Saturdays, Sundays, and Federal holidays, after the deviation, send to the certificate-holding district office a complete report of the operation involved, including a description of the deviation and the reasons for it.

[Doc. No. 16097, 43 FR 46783, Oct. 10, 1978, as amended by Amdt. 135-60, 61 FR 2616, Jan. 26, 1996]

§ 135.93 Autopilot: Minimum altitudes for use.

(a) Except as provided in paragraphs (b), (c), (d), and (e) of this section, no person may use an autopilot at an altitude above the terrain which is less than 500 feet or less than twice the maximum altitude loss specified in the approved Aircraft Flight Manual or equivalent for a malfunction of the autopilot, whichever is higher.

(b) When using an instrument approach facility other than ILS, no person may use an autopilot at an altitude above the terrain that is less than 50 feet below the approved minimum descent altitude for that procedure, or less than twice the maximum loss specified in the approved Airplane Flight Manual or equivalent for a malfunction of the autopilot under approach conditions, whichever is higher.

(c) For ILS approaches, when reported weather conditions are less than the basic weather conditions in § 91.155 of this chapter, no person may use an autopilot with an approach coupler at an altitude above the terrain that is less than 50 feet above the terrain, or the maximum altitude loss specified in the approved Airplane Flight Manual or equivalent for the malfunction of the autopilot with approach coupler, whichever is higher.

(d) Without regard to paragraph (a), (b), or (c) of this section, the Administrator may issue operations specifications to allow the use, to touchdown, of an approved flight control guidance system with automatic capability, if—

(1) The system does not contain any altitude loss (above zero) specified in the approved Aircraft Flight Manual or equivalent for malfunction of the autopilot with approach coupler; and

(2) The Administrator finds that the use of the system to touchdown will not otherwise adversely affect the safety standards of this section.

(e) Notwithstanding paragraph (a) of this section, the Administrator issues operations specifications to allow the use of an approved autopilot system with automatic capability during the takeoff and initial climb phase of flight provided:

(1) The Airplane Flight Manual specifies a minimum altitude engagement certification restriction;

(2) The system is not engaged prior to the minimum engagement certification restriction specified in the Airplane Flight Manual, or an altitude specified by the Administrator, whichever is higher; and

(3) The Administrator finds that the use of the system will not otherwise affect the safety standards required by this section.

(f) This section does not apply to operations conducted in rotorcraft.

[Doc. No. 16097, 43 FR 46783, Oct. 10, 1978, as amended by Amdt. 135-32, 54 FR 34332, Aug. 18, 1989; Amdt. 135-68, 62 FR 27923, May 21, 1997]

§ 135.95 Airmen: Limitations on use of services.

No certificate holder may use the services of any person as an airman unless the person performing those services—

(a) Holds an appropriate and current airman certificate; and

(b) Is qualified, under this chapter, for the operation for which the person is to be used.

§ 135.97 Aircraft and facilities for recent flight experience.

Each certificate holder shall provide aircraft and facilities to enable each of its pilots to maintain and demonstrate the pilot's ability to conduct all operations for which the pilot is authorized.

§ 135.98 Operations in the North Polar Area.

After August 13, 2008, no certificate holder may operate an aircraft in the region north of 78° N latitude ("North Polar Area"), other than intrastate operations wholly within the state of Alaska, unless authorized by the FAA. The certificate holder's operation specifications must include the following:

(a) The designation of airports that may be used for en-route diversions and the requirements the airports must meet at the time of diversion.

(b) Except for all-cargo operations, a recovery plan for passengers at designated diversion airports.

(c) A fuel-freeze strategy and procedures for monitoring fuel freezing for operations in the North Polar Area.

(d) A plan to ensure communication capability for operations in the North Polar Area.

(e) An MEL for operations in the North Polar Area.

(f) A training plan for operations in the North Polar Area.

(g) A plan for mitigating crew exposure to radiation during solar flare activity.

(h) A plan for providing at least two cold weather anti-exposure suits in the aircraft, to protect crewmembers during outside activity at a diversion airport with extreme climatic conditions. The FAA may relieve the certificate holder from this requirement if the season of the year makes the equipment unnecessary.

[Doc. No. FAA-2002-6717, 72 FR 1885, Jan. 16, 2007, as amended by Amdt. 135-112, 73 FR 8798, Feb. 15, 2008]

§ 135.99 Composition of flight crew.

(a) No certificate holder may operate an aircraft with less than the minimum flight crew specified in the aircraft operating limitations or the Aircraft Flight Manual for that aircraft and required by this part for the kind of operation being conducted.

(b) No certificate holder may operate an aircraft without a second in command if that aircraft has a passenger seating configuration, excluding any pilot seat, of ten seats or more.

§ 135.100 Flight crewmember duties.

(a) No certificate holder shall require, nor may any flight crewmember perform, any duties during a critical phase of flight except those duties required for the safe operation of the aircraft. Duties such as company required calls made for such nonsafety related purposes as ordering galley supplies and confirming passenger connections, announcements made to passengers promoting the air carrier or pointing out sights of interest, and filling out company payroll and related records are not required for the safe operation of the aircraft.

(b) No flight crewmember may engage in, nor may any pilot in command permit, any activity during a critical phase of flight which could distract any flight crewmember from the performance of his or her duties or which could interfere in any way with the proper conduct of those duties. Activities such as eating meals, engaging in nonessential conversations within the cockpit and nonessential communications between the cabin and cockpit crews, and reading publications not related to the proper conduct of the flight are not required for the safe operation of the aircraft.

(c) For the purposes of this section, critical phases of flight includes all ground operations involving taxi, takeoff and landing, and all other flight operations conducted below 10,000 feet, except cruise flight.

NOTE: Taxi is defined as "movement of an airplane under its own power on the surface of an airport."

[Doc. No. 20661, 46 FR 5502, Jan. 19, 1981]

§ 135.101 Second in command required under IFR.

Except as provided in § 135.105, no person may operate an aircraft carrying passengers under IFR unless there is a second in command in the aircraft.

[Doc. No. 28743, 62 FR 42374, Aug. 6, 1997]

§ 135.103 [Reserved]

§ 135.105 Exception to second in command requirement: Approval for use of autopilot system.

(a) Except as provided in §§ 135.99 and 135.111, unless two pilots are required by this chapter for operations under VFR, a person may operate an aircraft without a second in command, if it is equipped with an operative approved autopilot system and the use of that system is authorized by appropriate operations specifications. No certificate holder may use any person, nor may any person serve, as a pilot in command under this section of an aircraft operated in a commuter operation, as defined in part 119 of this chapter unless that person has at least 100 hours pilot in command flight time in the make and model of aircraft to be flown and has met all other applicable requirements of this part.

(b) The certificate holder may apply for an amendment of its operations specifications to authorize the use of an autopilot system in place of a second in command.

(c) The Administrator issues an amendment to the operations specifications authorizing the use of an autopilot system, in place of a second in command, if—

(1) The autopilot is capable of operating the aircraft controls to maintain flight and maneuver it about the three axes; and

(2) The certificate holder shows, to the satisfaction of the Administrator, that operations using the autopilot system can be conducted safely and in compliance with this part.

The amendment contains any conditions or limitations on the use of the autopilot system that the Administrator determines are needed in the interest of safety.

[Doc. No. 16097, 43 FR 46783, Oct. 10, 1978, as amended by Amdt. 135-3, 45 FR 7542, Feb. 4, 1980; Amdt. 135-58, 60 FR 65939, Dec. 20, 1995]

§ 135.107 Flight attendant crewmember requirement.

No certificate holder may operate an aircraft that has a passenger seating configuration, excluding any pilot seat, of more than 19 unless there is a flight attendant crewmember on board the aircraft.

§ 135.109 Pilot in command or second in command: Designation required.

(a) Each certificate holder shall designate a—

(1) Pilot in command for each flight; and

(2) Second in command for each flight requiring two pilots.

(b) The pilot in command, as designated by the certificate holder, shall remain the pilot in command at all times during that flight.

§ 135.111 Second in command required in Category II operations.

No person may operate an aircraft in a Category II operation unless there is a second in command of the aircraft.

§ 135.113 Passenger occupancy of pilot seat.

No certificate holder may operate an aircraft type certificated after October 15, 1971, that has a passenger seating configuration, excluding any pilot seat, of more than eight seats if any person other than the pilot in command, a second in command, a company check airman, or an authorized representative of the Administrator, the National Transportation Safety Board, or the United States Postal Service occupies a pilot seat.

§ 135.115 Manipulation of controls.

No pilot in command may allow any person to manipulate the flight controls of an aircraft during flight conducted under this part, nor may any person manipulate the controls during such flight unless that person is—

(a) A pilot employed by the certificate holder and qualified in the aircraft; or

(b) An authorized safety representative of the Administrator who has the permission of the pilot in command, is qualified in the aircraft, and is checking flight operations.

§ 135.117 Briefing of passengers before flight.

(a) Before each takeoff each pilot in command of an aircraft carrying passengers shall ensure that all passengers have been orally briefed on—

(1) *Smoking.* Each passenger shall be briefed on when, where, and under what conditions smoking is prohibited (including, but not limited to, any applicable requirements of part 252 of this title). This briefing shall include a statement that the Federal Aviation Regulations require passenger compliance with the lighted passenger information signs (if such signs are required), posted placards, areas designated for safety purposes as no smoking areas, and crewmember instructions with regard to these items. The briefing shall also include a statement (if the aircraft is equipped with a lavatory) that Federal law prohibits: tampering with, disabling, or destroying any smoke detector installed in an aircraft lavatory; smoking in lavatories; and, when applicable, smoking in passenger compartments.

(2) The use of safety belts, including instructions on how to fasten and unfasten the safety belts. Each passenger shall be briefed on when, where, and under what conditions the safety belt must be fastened about that passenger. This briefing shall include a statement that the Federal Aviation Regulations require passenger compliance with lighted passenger information signs and crewmember instructions concerning the use of safety belts.

(3) The placement of seat backs in an upright position before takeoff and landing;

(4) Location and means for opening the passenger entry door and emergency exits;

(5) Location of survival equipment;

(6) If the flight involves extended overwater operation, ditching procedures and the use of required flotation equipment;

(7) If the flight involves operations above 12,000 feet MSL, the normal and emergency use of oxygen; and

(8) Location and operation of fire extinguishers.

(b) Before each takeoff the pilot in command shall ensure that each person who may need the assistance of another person to move expeditiously to an exit if an emergency occurs and that person's attendant, if any, has received a briefing as to the procedures to be followed if an evacuation occurs. This paragraph does not apply to a person who has been given a briefing before a previous leg of a flight in the same aircraft.

(c) The oral briefing required by paragraph (a) of this section shall be given by the pilot in command or a crewmember.

(d) Notwithstanding the provisions of paragraph (c) of this section, for aircraft certificated to carry 19 passengers or less, the oral briefing required by paragraph (a) of this section shall be given by the pilot in command, a crewmember, or other qualified person designated by the certificate holder and approved by the Administrator.

(e) The oral briefing required by paragraph (a) of this section must be supplemented by printed cards which must be carried in the aircraft in locations convenient for the use of each passenger. The cards must—

(1) Be appropriate for the aircraft on which they are to be used;

(2) Contain a diagram of, and method of operating, the emergency exits;

(3) Contain other instructions necessary for the use of emergency equipment on board the aircraft; and

(4) No later than June 12, 2005, for scheduled Commuter passenger-carrying flights, include the sentence, "Final assembly of this aircraft was completed in [INSERT NAME OF COUNTRY]."

(f) The briefing required by paragraph (a) may be delivered by means of an approved recording playback device that is audible to each passenger under normal noise levels.

[Doc. No. 16097, 43 FR 46783, Oct. 10, 1978, as amended by Amdt. 135-9, 51 FR 40709, Nov. 7, 1986; Amdt. 135-25, 53 FR 12362, Apr. 13, 1988; Amdt. 135-44, 57 FR 42675, Sept. 15, 1992; 57 FR 43776, Sept. 22, 1992; 69 FR 39294, June 29, 2004]

§ 135.119 Prohibition against carriage of weapons.

No person may, while on board an aircraft being operated by a certificate holder, carry on or about that person a deadly or dangerous weapon, either concealed or unconcealed. This section does not apply to—

(a) Officials or employees of a municipality or a State, or of the United States, who are authorized to carry arms; or

(b) Crewmembers and other persons authorized by the certificate holder to carry arms.

§ 135.120 Prohibition on interference with crewmembers.

No person may assault, threaten, intimidate, or interfere with a crewmember in the performance of the crewmember's duties aboard an aircraft being operated under this part.

[Doc. No. FAA-1998-4954, 64 FR 1080, Jan. 7, 1999]

§ 135.121 Alcoholic beverages.

(a) No person may drink any alcoholic beverage aboard an aircraft unless the certificate holder operating the aircraft has served that beverage.

(b) No certificate holder may serve any alcoholic beverage to any person aboard its aircraft if that person appears to be intoxicated.

(c) No certificate holder may allow any person to board any of its aircraft if that person appears to be intoxicated.

§ 135.122 Stowage of food, beverage, and passenger service equipment during aircraft movement on the surface, takeoff, and landing.

(a) No certificate holder may move an aircraft on the surface, take off, or land when any food, beverage, or tableware furnished by the certificate holder is located at any passenger seat.

(b) No certificate holder may move an aircraft on the surface, take off, or land unless each food and beverage tray and seat back tray table is secured in its stowed position.

(c) No certificate holder may permit an aircraft to move on the surface, take off, or land unless each passenger serving cart is secured in its stowed position.

(d) Each passenger shall comply with instructions given by a crewmember with regard to compliance with this section.

[Doc. No. 26142, 57 FR 42675, Sept. 15, 1992]

§ 135.123 Emergency and emergency evacuation duties.

(a) Each certificate holder shall assign to each required crewmember for each type of aircraft as appropriate, the necessary functions to be performed in an emergency or in a situation requiring emergency evacuation. The certificate holder shall ensure that those functions can be practicably accomplished, and will meet any reasonably anticipated emergency including incapacitation of individual crewmembers or their inability to reach the passenger cabin because of shifting cargo in combination cargo-passenger aircraft.

(b) The certificate holder shall describe in the manual required under § 135.21 the functions of each category of required crewmembers assigned under paragraph (a) of this section.

§ 135.125 Aircraft security.

Certificate holders conducting operators conducting operations under this part must comply with the applicable security requirements in 49 CFR chapter XII.

[67 FR 8350, Feb. 22, 2002]

§ 135.127 Passenger information requirements and smoking prohibitions.

(a) No person may conduct a scheduled flight on which smoking is prohibited by part 252 of this title unless the "No Smoking" passenger information signs are lighted during the entire flight, or one or more "No Smoking" placards meeting the requirements of § 25.1541 of this chapter are posted during the entire flight. If both the lighted signs and the placards are used, the signs must remain lighted during the entire flight segment.

(b) No person may smoke while a "No Smoking" sign is lighted or while "No Smoking" placards are posted, except as follows:

(1) *On-demand operations.* The pilot in command of an aircraft engaged in an on-demand operation may authorize smoking on the flight deck (if it is physically separated from any passenger compartment), except in any of the following situations:

(i) During aircraft movement on the surface or during takeoff or landing;

(ii) During scheduled passenger-carrying public charter operations conducted under part 380 of this title;

(iii) During on-demand operations conducted interstate that meet paragraph (2) of the definition "On-demand operation" in § 110.2 of this chapter, unless permitted under paragraph (b)(2) of this section; or

(iv) During any operation where smoking is prohibited by part 252 of this title or by international agreement.

(2) *Certain intrastate commuter operations and certain intrastate on-demand operations.* Except during aircraft movement on the surface or during takeoff or landing, a pilot in command of an aircraft engaged in a commuter operation or an on-demand operation that meets paragraph (2) of the definition of "On-demand operation" in § 110.2 of this chapter may authorize smoking on the flight deck (if it is physically separated from the passenger compartment, if any) if—

(i) Smoking on the flight deck is not otherwise prohibited by part 252 of this title;

(ii) The flight is conducted entirely within the same State of the United States (a flight from one place in Hawaii to another place in Hawaii through the airspace over a place outside Hawaii is not entirely within the same State); and

(iii) The aircraft is either not turbojet-powered or the aircraft is not capable of carrying at least 30 passengers.

(c) No person may smoke in any aircraft lavatory.

(d) No person may operate an aircraft with a lavatory equipped with a smoke detector unless there is in that lavatory a sign or placard which reads: "Federal law provides for a penalty of up to $2,000 for tampering with the smoke detector installed in this lavatory."

(e) No person may tamper with, disable, or destroy any smoke detector installed in any aircraft lavatory.

(f) On flight segments other than those described in paragraph (a) of this section, the "No Smoking" sign required by § 135.177(a)(3) of this part must be turned on during any movement of the aircraft on the surface, for each takeoff or landing, and at any other time considered necessary by the pilot in command.

(g) The passenger information requirements prescribed in § 91.517 (b) and (d) of this chapter are in addition to the requirements prescribed in this section.

(h) Each passenger shall comply with instructions given him or her by crewmembers regarding compliance with paragraphs (b), (c), and (e) of this section.

[Doc. No. 25590, 55 FR 8367, Mar. 7, 1990, as amended by Amdt. 135-35, 55 FR 20135, May 15, 1990; Amdt. 135-44, 57 FR 42675, Sept. 15, 1992; Amdt. 135-60, 61 FR 2616, Jan. 26, 1996; Amdt. 135-76, 65 FR 36780, June 9, 2000; Amdt. 135-124, 76 FR 7491, Feb. 10, 2011]

§ 135.128 Use of safety belts and child restraint systems.

(a) Except as provided in this paragraph, each person on board an aircraft operated under this part shall occupy an approved seat or berth with a separate safety belt properly secured about him or her during movement on the surface, takeoff, and landing. For seaplane and float equipped rotorcraft operations during movement on the surface, the person pushing off the seaplane or rotorcraft from the dock and the person mooring the seaplane or rotorcraft at the dock are excepted from the preceding seating and safety belt requirements. A safety belt provided for the occupant of a seat may not be used by more than one person who has reached his or her second birthday. Notwithstanding the preceding requirements, a child may:

(1) Be held by an adult who is occupying an approved seat or berth, provided the child has not reached his or her second birthday and the child does not occupy or use any restraining device; or

(2) Notwithstanding any other requirement of this chapter, occupy an approved child restraint system furnished by the certificate holder or one of the persons described in paragraph (a)(2)(i) of this section, provided:

(i) The child is accompanied by a parent, guardian, or attendant designated by the child's parent or guardian to attend to the safety of the child during the flight;

(ii) Except as provided in paragraph (a)(2)(ii)(D) of this section, the approved child restraint system bears one or more labels as follows:

(A) Seats manufactured to U.S. standards between January 1, 1981, and February 25, 1985, must bear the label: "This child restraint system conforms to all applicable Federal motor vehicle safety standards";

(B) Seats manufactured to U.S. standards on or after February 26, 1985, must bear two labels:

(*1*) "This child restraint system conforms to all applicable Federal motor vehicle safety standards"; and

(*2*) "THIS RESTRAINT IS CERTIFIED FOR USE IN MOTOR VEHICLES AND AIRCRAFT" in red lettering;

(C) Seats that do not qualify under paragraphs (a)(2)(ii)(A) and (a)(2)(ii)(B) of this section must bear a label or markings showing:

(*1*) That the seat was approved by a foreign government;

(*2*) That the seat was manufactured under the standards of the United Nations;

(*3*) That the seat or child restraint device furnished by the certificate holder was approved by the FAA through Type Certificate or Supplemental Type Certificate.

(*4*) That the seat or child restraint device furnished by the certificate holder, or one of the persons described in paragraph (b)(2)(i) of this section, was approved by the FAA in accordance with § 21.305(d) or Technical Standard Order C-100b, or a later version.

(D) Except as provided in § 135.128(a)(2)(ii)(C)(*3*) and § 135.128(a)(2)(ii)(C)(*4*), booster-type child restraint systems (as defined in Federal Motor Vehicle Safety Standard No. 213 (49 CFR 571.213)), vest- and harness-type child restraint systems, and lap held child restraints are not approved for use in aircraft; and

(iii) The certificate holder complies with the following requirements:

(A) The restraint system must be properly secured to an approved forward-facing seat or berth;

(B) The child must be properly secured in the restraint system and must not exceed the specified weight limit for the restraint system; and

(C) The restraint system must bear the appropriate label(s).

(b) Except as provided in paragraph (b)(3) of this section, the following prohibitions apply to certificate holders:

(1) Except as provided in § 135.128 (a)(2)(ii)(C)(*3*) and § 135.128 (a)(2)(ii)(C)(*4*), no certificate holder may permit a child, in an aircraft, to occupy a booster-type child restraint system, a vest-type child restraint system, a harness-type child restraint system, or a lap held child restraint system during take off, landing, and movement on the surface.

(2) Except as required in paragraph (b)(1) of this section, no certificate holder may prohibit a child, if requested by the child's parent, guardian, or designated attendant, from occupying a child restraint system furnished by the child's parent, guardian, or designated attendant provided:

(i) The child holds a ticket for an approved seat or berth or such seat or berth is otherwise made available by the certificate holder for the child's use;

(ii) The requirements of paragraph (a)(2)(i) of this section are met;

(iii) The requirements of paragraph (a)(2)(iii) of this section are met; and

(iv) The child restraint system has one or more of the labels described in paragraphs (a)(2)(ii)(A) through (a)(2)(ii)(C) of this section.

(3) This section does not prohibit the certificate holder from providing child restraint systems authorized by this or, consistent with safe operating practices, determining the most appropriate passenger seat location for the child restraint system.

[Doc. No. 26142, 57 FR 42676, Sept. 15, 1992, as amended by Amdt. 135-62, 61 FR 28422, June 4, 1996; Amdt. 135-100, 70 FR 50907, Aug. 26, 2005; Amdt. 135-106, 71 FR 40010, July 14, 2006; 71 FR 59374, Oct. 10, 2006]

§ 135.129 Exit seating.

(a)(1) *Applicability.* This section applies to all certificate holders operating under this part, except for on-demand operations with aircraft having 19 or fewer passenger seats and commuter operations with aircraft having 9 or fewer passenger seats.

(2) *Duty to make determination of suitability.* Each certificate holder shall determine, to the extent necessary to perform the applicable functions of paragraph (d) of this section, the suitability of each person it permits to occupy an exit seat. For the purpose of this section—

(i) *Exit seat* means—

(A) Each seat having direct access to an exit; and

(B) Each seat in a row of seats through which passengers would have to pass to gain access to an exit, from the first seat inboard of the exit to the first aisle inboard of the exit.

(ii) A passenger seat having *direct access* means a seat from which a passenger can proceed directly to the exit without entering an aisle or passing around an obstruction.

(3) *Persons designated to make determination.* Each certificate holder shall make the passenger exit seating determinations required by this paragraph in a non-discriminatory manner consistent with the requirements of this section, by persons designated in the certificate holder's required operations manual.

(4) *Submission of designation for approval.* Each certificate holder shall designate the exit seats for each passenger seating configuration in its fleet in accordance with the definitions in this paragraph and submit those designations for approval as part of the procedures required to be submitted for approval under paragraphs (n) and (p) of this section.

(b) No certificate holder may seat a person in a seat affected by this section if the certificate holder determines that it is likely that the person would be unable to perform one or more of the applicable functions listed in paragraph (d) of this section because—

(1) The person lacks sufficient mobility, strength, or dexterity in both arms and hands, and both legs:

(i) To reach upward, sideways, and downward to the location of emergency exit and exit-slide operating mechanisms;

(ii) To grasp and push, pull, turn, or otherwise manipulate those mechanisms;

(iii) To push, shove, pull, or otherwise open emergency exits;

(iv) To lift out, hold, deposit on nearby seats, or maneuver over the seatbacks to the next row objects the size and weight of over-wing window exit doors;

(v) To remove obstructions of size and weight similar over-wing exit doors;

(vi) To reach the emergency exit expeditiously;

(vii) To maintain balance while removing obstructions;

(viii) To exit expeditiously;

(ix) To stabilize an escape slide after deployment; or

(x) To assist others in getting off an escape slide;

(2) The person is less than 15 years of age or lacks the capacity to perform one or more of the applicable functions listed in paragraph (d) of this section without the assistance of an adult companion, parent, or other relative;

(3) The person lacks the ability to read and understand instructions required by this section and related to emergency evacuation provided by the certificate holder in printed or graphic form or the ability to understand oral crew commands.

(4) The person lacks sufficient visual capacity to perform one or more of the applicable functions in paragraph (d) of this section without the assistance of visual aids beyond contact lenses or eyeglasses;

(5) The person lacks sufficient aural capacity to hear and understand instructions shouted by flight attendants, without assistance beyond a hearing aid;

(6) The person lacks the ability adequately to impart information orally to other passengers; or,

(7) The person has:

(i) A condition or responsibilities, such as caring for small children, that might prevent the person from performing one or more of the applicable functions listed in paragraph (d) of this section; or

(ii) A condition that might cause the person harm if he or she performs one or more of the applicable functions listed in paragraph (d) of this section.

(c) Each passenger shall comply with instructions given by a crewmember or other authorized employee of the certificate holder implementing exit seating restrictions established in accordance with this section.

(d) Each certificate holder shall include on passenger information cards, presented in the language in which briefings and oral commands are given by the crew, at each exit seat affected by this section, information that, in the event of an emergency in which a crewmember is not available to assist, a passenger occupying an exit seat may use if called upon to perform the following functions:

(1) Locate the emergency exit;

(2) Recognize the emergency exit opening mechanism;

(3) Comprehend the instructions for operating the emergency exit;

(4) Operate the emergency exit;

(5) Assess whether opening the emergency exit will increase the hazards to which passengers may be exposed;

(6) Follow oral directions and hand signals given by a crewmember;

(7) Stow or secure the emergency exit door so that it will not impede use of the exit;

(8) Assess the condition of an escape slide, activate the slide, and stabilize the slide after deployment to assist others in getting off the slide;

(9) Pass expeditiously through the emergency exit; and

(10) Assess, select, and follow a safe path away from the emergency exit.

(e) Each certificate holder shall include on passenger information cards, at each exit seat—

(1) In the primary language in which emergency commands are given by the crew, the selection criteria set forth in paragraph (b) of this section, and a request that a passenger identify himself or herself to allow reseating if he or she—

(i) Cannot meet the selection criteria set forth in paragraph (b) of this section;

(ii) Has a nondiscernible condition that will prevent him or her from performing the applicable functions listed in paragraph (d) of this section;

(iii) May suffer bodily harm as the result of performing one or more of those functions; or

(iv) Does not wish to perform those functions; and,

(2) In each language used by the certificate holder for passenger information cards, a request that a passenger identify himself or herself to allow reseating if he or she lacks the ability to read, speak, or understand the language or the graphic form in which instructions required by this section and related to emergency evacuation are provided by the certificate holder, or the ability to understand the specified language in which crew commands will be given in an emergency;

(3) May suffer bodily harm as the result of performing one or more of those functions; or,

(4) Does not wish to perform those functions.

A certificate holder shall not require the passenger to disclose his or her reason for needing reseating.

(f) Each certificate holder shall make available for inspection by the public at all passenger loading gates and ticket counters at each airport where it conducts passenger operations, written procedures established for making determinations in regard to exit row seating.

(g) No certificate holder may allow taxi or pushback unless at least one required crewmember has verified that no exit seat is occupied by a person the crewmember determines is likely to be unable to perform the applicable functions listed in paragraph (d) of this section.

(h) Each certificate holder shall include in its passenger briefings a reference to the passenger information cards, required by paragraphs (d) and (e), the selection criteria set forth in paragraph (b), and the functions to be performed, set forth in paragraph (d) of this section.

(i) Each certificate holder shall include in its passenger briefings a request that a passenger identify himself or herself to allow reseating if he or she—

(1) Cannot meet the selection criteria set forth in paragraph (b) of this section;

(2) Has a nondiscernible condition that will prevent him or her from performing the applicable functions listed in paragraph (d) of this section;

(3) May suffer bodily harm as the result of performing one or more of those functions; or,

(4) Does not wish to perform those functions.

A certificate holder shall not require the passenger to disclose his or her reason for needing reseating.

(j) [Reserved]

(k) In the event a certificate holder determines in accordance with this section that it is likely that a passenger assigned to an exit seat would be unable to perform the functions listed in paragraph (d) of this section or a passenger requests a non-exit seat, the certificate holder shall expeditiously relocate the passenger to a non-exit seat.

(l) In the event of full booking in the non-exit seats and if necessary to accommodate a passenger being relocated from an exit seat, the certificate holder shall move a passenger who is willing and able to assume the evacuation functions that may be required, to an exit seat.

(m) A certificate holder may deny transportation to any passenger under this section only because—

(1) The passenger refuses to comply with instructions given by a crewmember or other authorized employee of the certificate holder implementing exit seating restrictions established in accordance with this section, or

(2) The only seat that will physically accommodate the person's handicap is an exit seat.

(n) In order to comply with this section certificate holders shall—

(1) Establish procedures that address:

(i) The criteria listed in paragraph (b) of this section;

(ii) The functions listed in paragraph (d) of this section;

(iii) The requirements for airport information, passenger information cards, crewmember verification of appropriate seating in exit seats, passenger briefings, seat assignments, and denial of transportation as set forth in this section;

(iv) How to resolve disputes arising from implementation of this section, including identification of the certificate holder employee on the airport to whom complaints should be addressed for resolution; and,

(2) Submit their procedures for preliminary review and approval to the principal operations inspectors assigned to them at the certificate-holding district office.

(o) Certificate holders shall assign seats prior to boarding consistent with the criteria listed in paragraph (b) and the functions listed in paragraph (d) of this section, to the maximum extent feasible.

(p) The procedures required by paragraph (n) of this section will not become effective until final approval is granted by the Director, Flight Standards Service, Washington, DC. Approval will be based solely upon the safety aspects of the certificate holder's procedures.

[Doc. No. 25821, 55 FR 8073, Mar. 6, 1990, as amended by Amdt. 135-45, 57 FR 48664, Oct. 27, 1992; Amdt. 135-50, 59 FR 33603, June 29, 1994; Amdt. 135-60, 61 FR 2616, Jan. 26, 1996]

Subpart C—Aircraft and Equipment

§ 135.141 Applicability.

This subpart prescribes aircraft and equipment requirements for operations under this part. The requirements of this subpart are in addition to the aircraft and equipment requirements of part 91 of this chapter. However, this part does not require the duplication of any equipment required by this chapter.

§ 135.143 General requirements.

(a) No person may operate an aircraft under this part unless that aircraft and its equipment meet the applicable regulations of this chapter.

(b) Except as provided in § 135.179, no person may operate an aircraft under this part unless the required instruments and equipment in it have been approved and are in an operable condition.

(c) ATC transponder equipment installed within the time periods indicated below must meet the performance and environmental requirements of the following TSO's:

(1) *Through January 1, 1992:* (i) Any class of TSO-C74b or any class of TSO-C74c as appropriate, provided that the equipment was manufactured before January 1, 1990; or

(ii) The appropriate class of TSO-C112 (Mode S).

(2) *After January 1, 1992:* The appropriate class of TSO-C112 (Mode S). For purposes of paragraph (c)(2) of this section, "installation" does not include—

(i) Temporary installation of TSO-C74b or TSO-C74c substitute equipment, as appropriate, during maintenance of the permanent equipment;

(ii) Reinstallation of equipment after temporary removal for maintenance; or

(iii) For fleet operations, installation of equipment in a fleet aircraft after removal of the equipment for maintenance from another aircraft in the same operator's fleet.

[Doc. No. 16097, 43 FR 46783, Oct. 10, 1978, as amended by Amdt. 135-22, 52 FR 3392, Feb. 3, 1987]

§ 135.144 Portable electronic devices.

(a) Except as provided in paragraph (b) of this section, no person may operate, nor may any operator or pilot in command of an aircraft allow the operation of, any portable electronic device on any of the following U.S.-registered civil aircraft operating under this part.

(b) Paragraph (a) of this section does not apply to—

(1) Portable voice recorders;

(2) Hearing aids;

(3) Heart pacemakers;

(4) Electric shavers; or

(5) Any other portable electronic device that the part 119 certificate holder has determined will not cause interference with the navigation or communication system of the aircraft on which it is to be used.

(c). The determination required by paragraph (b)(5) of this section shall be made by that part 119 certificate holder operating the aircraft on which the particular device is to be used.

[Doc. No. FAA-1998-4954, 64 FR 1080, Jan. 7, 1999]

§ 135.145 Aircraft proving and validation tests.

(a) No certificate holder may operate an aircraft, other than a turbojet aircraft, for which two pilots are required by this chapter for operations under VFR, if it has not previously proved such an aircraft in operations under this part in at least 25 hours of proving tests acceptable to the Administrator including—

(1) Five hours of night time, if night flights are to be authorized;

(2) Five instrument approach procedures under simulated or actual conditions, if IFR flights are to be authorized; and

(3) Entry into a representative number of en route airports as determined by the Administrator.

(b) No certificate holder may operate a turbojet airplane if it has not previously proved a turbojet airplane in operations under this part in at least 25 hours of proving tests acceptable to the Administrator including—

(1) Five hours of night time, if night flights are to be authorized;

(2) Five instrument approach procedures under simulated or actual conditions, if IFR flights are to be authorized; and

(3) Entry into a representative number of en route airports as determined by the Administrator.

(c) No certificate holder may carry passengers in an aircraft during proving tests, except those needed to make the tests and those designated by the Administrator to observe the tests. However, pilot flight training may be conducted during the proving tests.

(d) Validation testing is required to determine that a certificate holder is capable of conducting operations safely and in compliance with applicable regulatory standards. Validation tests are required for the following authorizations:

(1) The addition of an aircraft for which two pilots are required for operations under VFR or a turbojet airplane, if that aircraft or an aircraft of the same make or similar design has not been previously proved or validated in operations under this part.

(2) Operations outside U.S. airspace.

(3) Class II navigation authorizations.

(4) Special performance or operational authorizations.

(e) Validation tests must be accomplished by test methods acceptable to the Administrator. Actual flights may not be required when an applicant can demonstrate competence and compliance with appropriate regulations without conducting a flight.

(f) Proving tests and validation tests may be conducted simultaneously when appropriate.

(g) The Administrator may authorize deviations from this section if the Administrator finds that special circumstances make full compliance with this section unnecessary.

[Doc. No. FAA-2001-10047, 68 FR 54586, Sept. 17, 2003]

§ 135.147 Dual controls required.

No person may operate an aircraft in operations requiring two pilots unless it is equipped with functioning dual controls. However, if the aircraft type certification operating limitations do not require two pilots, a throwover control wheel may be used in place of two control wheels.

§ 135.149 Equipment requirements: General.

No person may operate an aircraft unless it is equipped with—

(a) A sensitive altimeter that is adjustable for barometric pressure;

(b) Heating or deicing equipment for each carburetor or, for a pressure carburetor, an alternate air source;

(c) For turbojet airplanes, in addition to two gyroscopic bank-and-pitch indicators (artificial horizons) for use at the pilot stations, a third indicator that is installed in accordance with the instrument requirements prescribed in § 121.305(j) of this chapter.

(d) [Reserved]

(e) For turbine powered aircraft, any other equipment as the Administrator may require.

[Doc. No. 16097, 43 FR 46783, Oct. 10, 1978, as amended at Amdt. 135-1, 44 FR 26737, May 7, 1979; Amdt. 135-34, 54 FR 43926, Oct. 27, 1989; Amdt. 135-38, 55 FR 43310, Oct. 26, 1990]

§ 135.150 Public address and crewmember interphone systems.

No person may operate an aircraft having a passenger seating configuration, excluding any pilot seat, of more than 19 unless it is equipped with—

(a) A public address system which—

(1) Is capable of operation independent of the crewmember interphone system required by paragraph (b) of this section, except for handsets, headsets, microphones, selector switches, and signaling devices;

(2) Is approved in accordance with § 21.305 of this chapter;

(3) Is accessible for immediate use from each of two flight crewmember stations in the pilot compartment;

(4) For each required floor-level passenger emergency exit which has an adjacent flight attendant seat, has a microphone which is readily accessible to the seated flight attendant, except that one microphone may serve more than one exit, provided the proximity of the exits allows unassisted verbal communication between seated flight attendants;

(5) Is capable of operation within 10 seconds by a flight attendant at each of those stations in the passenger compartment from which its use is accessible;

(6) Is audible at all passenger seats, lavatories, and flight attendant seats and work stations; and

(7) For transport category airplanes manufactured on or after November 27, 1990, meets the requirements of § 25.1423 of this chapter.

(b) A crewmember interphone system which—

(1) Is capable of operation independent of the public address system required by paragraph (a) of this section, except for handsets, headsets, microphones, selector switches, and signaling devices;

(2) Is approved in accordance with § 21.305 of this chapter;

(3) Provides a means of two-way communication between the pilot compartment and—

(i) Each passenger compartment; and

(ii) Each galley located on other than the main passenger deck level;

(4) Is accessible for immediate use from each of two flight crewmember stations in the pilot compartment;

(5) Is accessible for use from at least one normal flight attendant station in each passenger compartment;

(6) Is capable of operation within 10 seconds by a flight attendant at each of those stations in each passenger compartment from which its use is accessible; and

(7) For large turbojet-powered airplanes—

(i) Is accessible for use at enough flight attendant stations so that all floor-level emergency exits (or entryways to those exits in the case of exits located within galleys) in each passenger compartment are observable from one or more of those stations so equipped;

(ii) Has an alerting system incorporating aural or visual signals for use by flight crewmembers to alert flight attendants and for use by flight attendants to alert flight crewmembers;

(iii) For the alerting system required by paragraph (b)(7)(ii) of this section, has a means for the recipient of a call to determine whether it is a normal call or an emergency call; and

(iv) When the airplane is on the ground, provides a means of two-way communication between ground personnel and either of at least two flight crewmembers in the pilot compartment. The interphone system station for use by ground personnel must be so located that personnel using the system may avoid visible detection from within the airplane.

[Doc. No. 24995, 54 FR 43926, Oct. 27, 1989]

§ 135.151 Cockpit voice recorders.

(a) No person may operate a multiengine, turbine-powered airplane or rotorcraft having a passenger seating configuration of six or more and for which two pilots are required by certification or operating rules unless it is equipped with an approved cockpit voice recorder that:

(1) Is installed in compliance with § 23.1457(a)(1) and (2), (b), (c), (d)(1)(i), (2) and (3), (e), (f), and (g); § 25.1457(a)(1) and (2), (b), (c), (d)(1)(i), (2) and (3), (e), (f), and (g), § 27.1457(a)(1) and (2), (b), (c), (d)(1)(i), (2) and (3), (e), (f), and (g); or § 29.1457(a)(1) and (2), (b), (c), (d)(1)(i), (2) and (3), (e), (f), and (g) of this chapter, as applicable; and

(2) Is operated continuously from the use of the check list before the flight to completion of the final check list at the end of the flight.

(b) No person may operate a multiengine, turbine-powered airplane or rotorcraft having a passenger seating configuration of 20 or more seats unless it is equipped with an approved cockpit voice recorder that—

(1) Is installed in accordance with the requirements of § 23.1457 (except paragraphs (a)(6), (d)(1)(ii), (4), and (5)); § 25.1457 (except paragraphs (a)(6), (d)(1)(ii), (4), and (5)); § 27.1457 (except paragraphs (a)(6), (d)(1)(ii), (4), and (5)); or § 29.1457 (except paragraphs (a)(6), (d)(1)(ii), (4), and (5)) of this chapter, as applicable; and

(2) Is operated continuously from the use of the check list before the flight to completion of the final check list at the end of the flight.

(c) In the event of an accident, or occurrence requiring immediate notification of the National Transportation Safety Board which results in termination of the flight, the certificate holder shall keep the recorded information for at least 60 days or, if requested by the Administrator or the Board, for a longer period. Information obtained from the record may be used to assist in determining the cause of accidents or occurrences in connection with investigations. The Administrator does not use the record in any civil penalty or certificate action.

(d) For those aircraft equipped to record the uninterrupted audio signals received by a boom or a mask microphone the flight crewmembers are required to use the boom microphone below 18,000 feet mean sea level. No person may operate a large turbine engine powered airplane manufactured after October 11, 1991, or on which a cockpit voice recorder has been installed after October 11, 1991, unless it is equipped to record the uninterrupted audio signal received by a boom or mask microphone in accordance with § 25.1457(c)(5) of this chapter.

(e) In complying with this section, an approved cockpit voice recorder having an erasure feature may be used, so that during the operation of the recorder, information:

(1) Recorded in accordance with paragraph (a) of this section and recorded more than 15 minutes earlier; or

(2) Recorded in accordance with paragraph (b) of this section and recorded more than 30 minutes earlier; may be erased or otherwise obliterated.

(f) By April 7, 2012, all airplanes subject to paragraph (a) or paragraph (b) of this section that are manufactured before April 7, 2010, and that are required to have a flight data recorder installed in accordance with § 135.152, must have a cockpit voice recorder that also—

(1) Meets the requirements in § 23.1457(d)(6) or § 25.1457(d)(6) of this chapter, as applicable; and

(2) If transport category, meet the requirements in § 25.1457(a)(3), (a)(4), and (a)(5) of this chapter.

(g)(1) No person may operate a multiengine, turbine-powered airplane or rotorcraft that is manufactured on or after April 7, 2010, that has a passenger seating configuration of six or more seats, for which two pilots are required by certification or operating rules, and that is required to have a flight data recorder under § 135.152, unless it is equipped with an approved cockpit voice recorder that also—

(i) Is installed in accordance with the requirements of § 23.1457 (except for paragraph (a)(6)); § 25.1457 (except for paragraph (a)(6)); § 27.1457 (except for paragraph (a)(6)); or § 29.1457 (except for paragraph (a)(6)) of this chapter, as applicable; and

(ii) Is operated continuously from the use of the check list before the flight, to completion of the final check list at the end of the flight; and

(iii) Retains at least the last 2 hours of recorded information using a recorder that meets the standards of TSO-C123a, or later revision.

(iv) For all airplanes or rotorcraft manufactured on or after December 6, 2010, also meets the requirements of § 23.1457(a)(6); § 25.1457(a)(6); § 27.1457(a)(6); or § 29.457(a)(6) of this chapter, as applicable.

(2) No person may operate a multiengine, turbine-powered airplane or rotorcraft that is manufactured on or after April 7, 2010, has a passenger seating configuration of 20 or more seats, and that is required to have a flight data recorder under § 135.152, unless it is equipped with an approved cockpit voice recorder that also—

(i) Is installed in accordance with the requirements of § 23.1457 (except for paragraph (a)(6)); § 25.1457 (except for paragraph (a)(6)); § 27.1457 (except for paragraph (a)(6)); or § 29.1457 (except for paragraph (a)(6)) of this chapter, as applicable; and

(ii) Is operated continuously from the use of the check list before the flight, to completion of the final check list at the end of the flight; and

(iii) Retains at least the last 2 hours of recorded information using a recorder that meets the standards of TSO-C123a, or later revision.

(iv) For all airplanes or rotorcraft manufactured on or after December 6, 2010, also meets the requirements of § 23.1457(a)(6); § 25.1457(a)(6); § 27.1457(a)(6); or § 29.457(a)(6) of this chapter, as applicable.

(h) All airplanes or rotorcraft required by this part to have a cockpit voice recorder and a flight data recorder, that install datalink communication equipment on or after December 6, 2010, must record all datalink messages as required by the certification rule applicable to the aircraft.

[Doc. No. 16097, 43 FR 46783, Oct. 10, 1978, as amended by Amdt. 135-23, 52 FR 9637, Mar. 25, 1987; Amdt. 135-26, 53 FR 26151, July 11, 1988; Amdt. 135-60, 61 FR 2616, Jan. 26, 1996; Amdt. 135-113, 73 FR 12570, Mar. 7, 2008; Amdt. 135-113, 74 FR 32801, July 9, 2009; Amdt. 135-121, 75 FR 17046, Apr. 5, 2010]

§ 135.152 Flight data recorders.

(a) Except as provided in paragraph (k) of this section, no person may operate under this part a multi-engine, turbine-engine powered airplane or rotorcraft having a passenger seating configuration, excluding any required crewmember seat, of 10 to 19 seats, that was either brought onto the U.S. register after, or was registered outside the United States and added to the operator's U.S. operations specifications after, October 11, 1991, unless it is equipped with one or more approved flight recorders that use a digital method of recording and storing data and a method of readily retrieving that data from the storage medium. The parameters specified in either Appendix B or C of this part, as applicable must be recorded within the range, accuracy, resolution, and recording intervals as specified. The recorder shall retain no less than 25 hours of aircraft operation.

(b) After October 11, 1991, no person may operate a multiengine, turbine-powered airplane having a passenger seating configuration of 20 to 30 seats or a multiengine, turbine-powered rotorcraft having a passenger seating configuration of 20 or more seats unless it is equipped with one or more approved flight recorders that utilize a digital method of recording and storing data, and a method of readily retrieving that data from the storage medium. The parameters in appendix D or E of this part, as applicable, that are set forth below, must be recorded within the ranges, accuracies, resolutions, and sampling intervals as specified.

(1) Except as provided in paragraph (b)(3) of this section for aircraft type certificated before October 1, 1969, the following parameters must be recorded:

(i) Time;
(ii) Altitude;
(iii) Airspeed;
(iv) Vertical acceleration;
(v) Heading;
(vi) Time of each radio transmission to or from air traffic control;
(vii) Pitch attitude;
(viii) Roll attitude;
(ix) Longitudinal acceleration;
(x) Control column or pitch control surface position; and
(xi) Thrust of each engine.

(2) Except as provided in paragraph (b)(3) of this section for aircraft type certificated after September 30, 1969, the following parameters must be recorded:

(i) Time;
(ii) Altitude;
(iii) Airspeed;
(iv) Vertical acceleration;
(v) Heading;
(vi) Time of each radio transmission either to or from air traffic control;
(vii) Pitch attitude;
(viii) Roll attitude;
(ix) Longitudinal acceleration;
(x) Pitch trim position;
(xi) Control column or pitch control surface position;
(xii) Control wheel or lateral control surface position;
(xiii) Rudder pedal or yaw control surface position;
(xiv) Thrust of each engine;
(xv) Position of each thrust reverser;
(xvi) Trailing edge flap or cockpit flap control position; and
(xvii) Leading edge flap or cockpit flap control position.

(3) For aircraft manufactured after October 11, 1991, all of the parameters listed in appendix D or E of this part, as applicable, must be recorded.

(c) Whenever a flight recorder required by this section is installed, it must be operated continuously from the instant the airplane begins the takeoff roll or the rotorcraft begins the lift-off until the airplane has completed the landing roll or the rotorcraft has landed at its destination.

(d) Except as provided in paragraph (c) of this section, and except for recorded data erased as authorized in this paragraph, each certificate holder shall keep the recorded data prescribed in paragraph (a) of this section until the aircraft has been operating for at least 25 hours of the operating time specified in paragraph (c) of this section. In addition, each certificate holder shall keep the recorded data prescribed in paragraph (b) of this section for an airplane until the airplane has been operating for at least 25 hours, and for a rotorcraft until the rotorcraft has been operating for at least 10 hours, of the operating time specified in paragraph (c) of this section. A total of 1 hour of recorded data may be erased for the purpose of testing the flight recorder or the flight recorder system. Any erasure made in accordance with this paragraph must be of the oldest recorded data accumulated at the time of testing. Except as provided in paragraph (c) of this section, no record need be kept more than 60 days.

(e) In the event of an accident or occurrence that requires the immediate notification of the National Transportation Safety Board under 49 CFR part 830 of its regulations and that results in termination of the flight, the certificate holder shall remove the recording media from the aircraft and keep the recorded data required by paragraphs (a) and (b) of this section for at least 60 days or for a longer period upon request of the Board or the Administrator.

(f)(1) For airplanes manufactured on or before August 18, 2000, and all other aircraft, each flight recorder required by this section must be installed in accordance with the requirements of § 23.1459 (except paragraphs (a)(3)(ii) and (6)), § 25.1459 (except paragraphs (a)(3)(ii) and (7)), § 27.1459 (except paragraphs (a)(3)(ii) and (6)), or § 29.1459 (except paragraphs (a)(3)(ii) and (6)), as appropriate, of this chapter. The correlation required by paragraph (c) of §§ 23.1459, 25.1459, 27.1459, or 29.1459 of this chapter, as appropriate, need be established only on one aircraft of a group of aircraft:

(i) That are of the same type;
(ii) On which the flight recorder models and their installations are the same; and
(iii) On which there are no differences in the type designs with respect to the installation of the first pilot's instruments associated with the flight recorder. The most recent instrument calibration, including the recording medium from which this calibration is derived, and the recorder correlation must be retained by the certificate holder.

(2) For airplanes manufactured after August 18, 2000, each flight data recorder system required by this section must be installed in accordance with the requirements of § 23.1459(a) (except paragraphs (a)(3)(ii) and (6)), (b), (d) and (e), or § 25.1459(a) (except paragraphs (a)(3)(ii) and (7)), (b), (d) and (e) of this chapter. A correlation must be established between the values recorded by the flight data recorder and the corresponding values being measured. The correlation must contain a sufficient number of correlation points to accurately establish the conversion from the recorded values to engineering units or discrete state over the full operating range of the parameter. Except for airplanes having separate altitude and airspeed sensors that are an integral part of the flight data recorder system, a single correlation may be established for any group of airplanes—

(i) That are of the same type;
(ii) On which the flight recorder system and its installation are the same; and
(iii) On which there is no difference in the type design with respect to the installation of those sensors associated with the flight data recorder system. Documentation sufficient to convert recorded data into the engineering units and discrete values specified in the applicable appendix must be maintained by the certificate holder.

(g) Each flight recorder required by this section that records the data specified in paragraphs (a) and (b) of this section must have an approved device to assist in locating that recorder under water.

(h) The operational parameters required to be recorded by digital flight data recorders required by paragraphs (i) and (j) of this section are as follows, the phrase "when an information source is installed" following a parameter indicates that recording of that parameter is not intended to require a change in installed equipment.

(1) Time;
(2) Pressure altitude;
(3) Indicated airspeed;
(4) Heading—primary flight crew reference (if selectable, record discrete, true or magnetic);
(5) Normal acceleration (Vertical);
(6) Pitch attitude;
(7) Roll attitude;
(8) Manual radio transmitter keying, or CVR/DFDR synchronization reference;
(9) Thrust/power of each engine—primary flight crew reference;
(10) Autopilot engagement status;
(11) Longitudinal acceleration;
(12) Pitch control input;
(13) Lateral control input;
(14) Rudder pedal input;
(15) Primary pitch control surface position;
(16) Primary lateral control surface position;
(17) Primary yaw control surface position;
(18) Lateral acceleration;
(19) Pitch trim surface position or parameters of paragraph (h)(82) of this section if currently recorded;
(20) Trailing edge flap or cockpit flap control selection (except when parameters of paragraph (h)(85) of this section apply);
(21) Leading edge flap or cockpit flap control selection (except when parameters of paragraph (h)(86) of this section apply);
(22) Each Thrust reverser position (or equivalent for propeller airplane);
(23) Ground spoiler position or speed brake selection (except when parameters of paragraph (h)(87) of this section apply);
(24) Outside or total air temperature;
(25) Automatic Flight Control System (AFCS) modes and engagement status, including autothrottle;
(26) Radio altitude (when an information source is installed);
(27) Localizer deviation, MLS Azimuth;
(28) Glideslope deviation, MLS Elevation;
(29) Marker beacon passage;
(30) Master warning;
(31) Air/ground sensor (primary airplane system reference nose or main gear);
(32) Angle of attack (when information source is installed);
(33) Hydraulic pressure low (each system);
(34) Ground speed (when an information source is installed);
(35) Ground proximity warning system;
(36) Landing gear position or landing gear cockpit control selection;
(37) Drift angle (when an information source is installed);
(38) Wind speed and direction (when an information source is installed);
(39) Latitude and longitude (when an information source is installed);
(40) Stick shaker/pusher (when an information source is installed);
(41) Windshear (when an information source is installed);
(42) Throttle/power lever position;
(43) Additional engine parameters (as designated in appendix F of this part);
(44) Traffic alert and collision avoidance system;
(45) DME 1 and 2 distances;
(46) Nav 1 and 2 selected frequency;
(47) Selected barometric setting (when an information source is installed);
(48) Selected altitude (when an information source is installed);
(49) Selected speed (when an information source is installed);
(50) Selected mach (when an information source is installed);
(51) Selected vertical speed (when an information source is installed);
(52) Selected heading (when an information source is installed);
(53) Selected flight path (when an information source is installed);
(54) Selected decision height (when an information source is installed);
(55) EFIS display format;
(56) Multi-function/engine/alerts display format;
(57) Thrust command (when an information source is installed);
(58) Thrust target (when an information source is installed);
(59) Fuel quantity in CG trim tank (when an information source is installed);
(60) Primary Navigation System Reference;
(61) Icing (when an information source is installed);
(62) Engine warning each engine vibration (when an information source is installed);
(63) Engine warning each engine over temp. (when an information source is installed);
(64) Engine warning each engine oil pressure low (when an information source is installed);
(65) Engine warning each engine over speed (when an information source is installed;
(66) Yaw trim surface position;
(67) Roll trim surface position;
(68) Brake pressure (selected system);
(69) Brake pedal application (left and right);
(70) Yaw or sideslip angle (when an information source is installed);
(71) Engine bleed valve position (when an information source is installed);
(72) De-icing or anti-icing system selection (when an information source is installed);
(73) Computed center of gravity (when an information source is installed);
(74) AC electrical bus status;
(75) DC electrical bus status;
(76) APU bleed valve position (when an information source is installed);
(77) Hydraulic pressure (each system);
(78) Loss of cabin pressure;
(79) Computer failure;
(80) Heads-up display (when an information source is installed);
(81) Para-visual display (when an information source is installed);
(82) Cockpit trim control input position—pitch;
(83) Cockpit trim control input position—roll;

(84) Cockpit trim control input position—yaw;

(85) Trailing edge flap and cockpit flap control position;

(86) Leading edge flap and cockpit flap control position;

(87) Ground spoiler position and speed brake selection; and

(88) All cockpit flight control input forces (control wheel, control column, rudder pedal).

(i) For all turbine-engine powered airplanes with a seating configuration, excluding any required crewmember seat, of 10 to 30 passenger seats, manufactured after August 18, 2000—

(1) The parameters listed in paragraphs (h)(1) through (h)(57) of this section must be recorded within the ranges, accuracies, resolutions, and recording intervals specified in Appendix F of this part.

(2) Commensurate with the capacity of the recording system, all additional parameters for which information sources are installed and which are connected to the recording system must be recorded within the ranges, accuracies, resolutions, and sampling intervals specified in Appendix F of this part.

(j) For all turbine-engine-powered airplanes with a seating configuration, excluding any required crewmember seat, of 10 to 30 passenger seats, that are manufactured after August 19, 2002 the parameters listed in paragraph (a)(1) through (a)(88) of this section must be recorded within the ranges, accuracies, resolutions, and recording intervals specified in Appendix F of this part.

(k) For aircraft manufactured before August 18, 1997, the following aircraft types need not comply with this section: Bell 212, Bell 214ST, Bell 412, Bell 412SP, Boeing Chinook (BV-234), Boeing/Kawasaki Vertol 107 (BV/KV-107-II), deHavilland DHC-6, Eurocopter Puma 330J, Sikorsky 58, Sikorsky 61N, Sikorsky 76A.

(l) By April 7, 2012, all aircraft manufactured before April 7, 2010, must also meet the requirements in § 23.1459(a)(7), § 25.1459(a)(8), § 27.1459(e), or § 29.1459(e) of this chapter, as applicable.

(m) All aircraft manufactured on or after April 7, 2010, must have a flight data recorder installed that also—

(1) Meets the requirements of § 23.1459(a)(3), (a)(6), and (a)(7), § 25.1459(a)(3), (a)(7), and (a)(8), § 27.1459(a)(3), (a)(6), and (e), or § 29.1459(a)(3), (a)(6), and (e) of this chapter, as applicable; and

(2) Retains the 25 hours of recorded information required in paragraph (d) of this section using a recorder that meets the standards of TSO-C124a, or later revision.

[Doc. No. 25530, 53 FR 26151, July 11, 1988, as amended by Amdt. 135-69, 62 FR 38396, July 17, 1997; 62 FR 48135, Sept. 12, 1997; Amdt. 135-89, 68 FR 42939, July 18, 2003; Amdt. 135-113, 73 FR 12570, Mar. 7, 2008; Amdt. 135-113, 74 FR 32801, July 9, 2009]

§ 135.153 [Reserved]

§ 135.154 Terrain awareness and warning system.

(a) *Airplanes manufactured after March 29, 2002:*

(1) No person may operate a turbine-powered airplane configured with 10 or more passenger seats, excluding any pilot seat, unless that airplane is equipped with an approved terrain awareness and warning system that meets the requirements for Class A equipment in Technical Standard Order (TSO)-C151. The airplane must also include an approved terrain situational awareness display.

(2) No person may operate a turbine-powered airplane configured with 6 to 9 passenger seats, excluding any pilot seat, unless that airplane is equipped with an approved terrain awareness and warning system that meets as a minimum the requirements for Class B equipment in Technical Standard Order (TSO)-C151.

(b) *Airplanes manufactured on or before March 29, 2002:*

(1) No person may operate a turbine-powered airplane configured with 10 or more passenger seats, excluding any pilot seat, after March 29, 2005, unless that airplane is equipped with an approved terrain awareness and warning system that meets the requirements for Class A equipment in Technical Standard Order (TSO)-C151. The airplane must also include an approved terrain situational awareness display.

(2) No person may operate a turbine-powered airplane configured with 6 to 9 passenger seats, excluding any pilot seat, after March 29, 2005, unless that airplane is equipped with an approved terrain awareness and warning system that meets as a minimum the requirements for Class B equipment in Technical Standard Order (TSO)-C151.

(Approved by the Office of Management and Budget under control number 2120-0631)

(c) *Airplane Flight Manual.* The Airplane Flight Manual shall contain appropriate procedures for—

(1) The use of the terrain awareness and warning system; and

(2) Proper flight crew reaction in response to the terrain awareness and warning system audio and visual warnings.

[Doc. No. 29312, 65 FR 16755, Mar. 29, 2000]

§ 135.155 Fire extinguishers: Passenger-carrying aircraft.

No person may operate an aircraft carrying passengers unless it is equipped with hand fire extinguishers of an approved type for use in crew and passenger compartments as follows—

(a) The type and quantity of extinguishing agent must be suitable for the kinds of fires likely to occur;

(b) At least one hand fire extinguisher must be provided and conveniently located on the flight deck for use by the flight crew; and

(c) At least one hand fire extinguisher must be conveniently located in the passenger compartment of each aircraft having a passenger seating configuration, excluding any pilot seat, of at least 10 seats but less than 31 seats.

§ 135.156 Flight data recorders: filtered data.

(a) A flight data signal is filtered when an original sensor signal has been changed in any way, other than changes necessary to:

(1) Accomplish analog to digital conversion of the signal;

(2) Format a digital signal to be DFDR compatible; or

(3) Eliminate a high frequency component of a signal that is outside the operational bandwidth of the sensor.

(b) An original sensor signal for any flight recorder parameter required to be recorded under § 135.152 may be filtered only if the recorded signal value continues to meet the requirements of Appendix D or F of this part, as applicable.

(c) For a parameter described in § 135.152(h)(12) through (17), (42), or (88), or the corresponding parameter in Appendix D of this part, if the recorded signal value is filtered and does not meet the requirements of Appendix D or F of this part, as applicable, the certificate holder must:

(1) Remove the filtering and ensure that the recorded signal value meets the requirements of Appendix D or F of this part, as applicable; or

(2) Demonstrate by test and analysis that the original sensor signal value can be reconstructed from the recorded data. This demonstration requires that:

(i) The FAA determine that the procedure and test results submitted by the certificate holder as its compliance with paragraph (c)(2) of this section are repeatable; and

(ii) The certificate holder maintains documentation of the procedure required to reconstruct the original sensor signal value. This documentation is also subject to the requirements of § 135.152(e).

(d) *Compliance.* Compliance is required as follows:

(1) No later than October 20, 2011, each operator must determine, for each aircraft on its operations specifications, whether the aircraft's DFDR system is filtering any of the parameters listed in paragraph (c) of this section. The operator must create a record of this determination for each aircraft it operates, and maintain it as part of the correlation documentation required by § 135.152 (f)(1)(iii) or (f)(2)(iii) of this part as applicable.

(2) For aircraft that are not filtering any listed parameter, no further action is required unless the aircraft's DFDR system is modified in a manner that would cause it to meet the definition of filtering on any listed parameter.

(3) For aircraft found to be filtering a parameter listed in paragraph (c) of this section the operator must either:

(i) No later than April 21, 2014, remove the filtering; or

(ii) No later than April 22, 2013, submit the necessary procedure and test results required by paragraph (c)(2) of this section.

(4) After April 21, 2014, no aircraft flight data recording system may filter any parameter listed in paragraph (c) of this section that does not meet the requirements of Appendix D or F of this part, unless the certificate holder possesses test and analysis procedures and the test results that have been approved by the FAA. All records of tests, analysis and procedures used to comply with this section must be maintained as part of the correlation documentation required by § 135.152 (f)(1)(iii) or (f)(2)(iii) of this part as applicable.

[Doc. No. FAA-2006-26135, 75 FR 7357, Feb. 19, 2010]

§ 135.157 Oxygen equipment requirements.

(a) *Unpressurized aircraft.* No person may operate an unpressurized aircraft at altitudes prescribed in this section unless it is equipped with enough oxygen dispensers and oxygen to supply the pilots under § 135.89(a) and to supply, when flying—

(1) At altitudes above 10,000 feet through 15,000 feet MSL, oxygen to at least 10 percent of the occupants of the aircraft, other than the pilots, for that part of the flight at those altitudes that is of more than 30 minutes duration; and

(2) Above 15,000 feet MSL, oxygen to each occupant of the aircraft other than the pilots.

(b) *Pressurized aircraft.* No person may operate a pressurized aircraft—

(1) At altitudes above 25,000 feet MSL, unless at least a 10-minute supply of supplemental oxygen is available for each occupant of the aircraft, other than the pilots, for use when a descent is necessary due to loss of cabin pressurization; and

(2) Unless it is equipped with enough oxygen dispensers and oxygen to comply with paragraph (a) of this section whenever the cabin pressure altitude exceeds 10,000 feet MSL and, if the cabin pressurization fails, to comply with § 135.89 (a) or to provide a 2-hour supply for each pilot, whichever is greater, and to supply when flying—

(i) At altitudes above 10,000 feet through 15,000 feet MSL, oxygen to at least 10 percent of the occupants of the aircraft, other than the pilots, for that part of the flight at those altitudes that is of more than 30 minutes duration; and

(ii) Above 15,000 feet MSL, oxygen to each occupant of the aircraft, other than the pilots, for one hour unless, at all times during flight above that altitude, the aircraft can safely descend to 15,000 feet MSL within four minutes, in which case only a 30-minute supply is required.

(c) The equipment required by this section must have a means—

(1) To enable the pilots to readily determine, in flight, the amount of oxygen available in each source of supply and whether the oxygen is being delivered to the dispensing units; or

(2) In the case of individual dispensing units, to enable each user to make those determinations with respect to that person's oxygen supply and delivery; and

(3) To allow the pilots to use undiluted oxygen at their discretion at altitudes above 25,000 feet MSL.

§ 135.158 Pitot heat indication systems.

(a) Except as provided in paragraph (b) of this section, after April 12, 1981, no person may operate a transport category airplane equipped with a flight instrument pitot heating system unless the airplane is also equipped with an operable pitot heat indication system that complies with § 25.1326 of this chapter in effect on April 12, 1978.

(b) A certificate holder may obtain an extension of the April 12, 1981, compliance date specified in paragraph (a) of this section, but not beyond April 12, 1983, from the Director, Flight Standards Service if the certificate holder—

(1) Shows that due to circumstances beyond its control it cannot comply by the specified compliance date; and

(2) Submits by the specified compliance date a schedule for compliance, acceptable to the Director, indicating that compliance will be achieved at the earliest practicable date.

[Doc. No. 18094, Amdt. 135-17, 46 FR 48306, Aug. 31, 1981, as amended by Amdt. 135-33, 54 FR 39294, Sept. 25, 1989]

§ 135.159 Equipment requirements: Carrying passengers under VFR at night or under VFR over-the-top conditions.

No person may operate an aircraft carrying passengers under VFR at night or under VFR over-the-top, unless it is equipped with—

(a) A gyroscopic rate-of-turn indicator except on the following aircraft:

(1) Airplanes with a third attitude instrument system usable through flight attitudes of 360 degrees of pitch-and-roll and installed in accordance with the instrument requirements prescribed in § 121.305(j) of this chapter.

(2) Helicopters with a third attitude instrument system usable through flight attitudes of ±80 degrees of pitch and ±120 degrees of roll and installed in accordance with § 29.1303(g) of this chapter.

(3) Helicopters with a maximum certificated takeoff weight of 6,000 pounds or less.

(b) A slip skid indicator.

(c) A gyroscopic bank-and-pitch indicator.

(d) A gyroscopic direction indicator.

(e) A generator or generators able to supply all probable combinations of continuous in-flight electrical loads for required equipment and for recharging the battery.

(f) For night flights—

(1) An anticollision light system;

(2) Instrument lights to make all instruments, switches, and gauges easily readable, the direct rays of which are shielded from the pilots' eyes; and

(3) A flashlight having at least two size "D" cells or equivalent.

(g) For the purpose of paragraph (e) of this section, a continuous in-flight electrical load includes one that draws current continuously during flight, such as radio equipment and electrically driven instruments and lights, but does not include occasional intermittent loads.

(h) Notwithstanding provisions of paragraphs (b), (c), and (d), helicopters having a maximum certificated takeoff weight of 6,000 pounds or less may be operated until January 6, 1988, under visual flight rules at night without a slip skid indicator, a gyroscopic bank-and-pitch indicator, or a gyroscopic direction indicator.

[Doc. No. 24550, 51 FR 40709, Nov. 7, 1986, as amended by Amdt. 135-38, 55 FR 43310, Oct. 26, 1990]

§ 135.161 Communication and navigation equipment for aircraft operations under VFR over routes navigated by pilotage.

(a) No person may operate an aircraft under VFR over routes that can be navigated by pilotage unless the aircraft is equipped with the two-way radio communication equipment necessary under normal operating conditions to fulfill the following:

(1) Communicate with at least one appropriate station from any point on the route, except in remote locations and areas of mountainous terrain where geographical constraints make such communication impossible.

(2) Communicate with appropriate air traffic control facilities from any point within Class B, Class C, or Class D airspace, or within a Class E surface area designated for an airport in which flights are intended; and

(3) Receive meteorological information from any point en route, except in remote locations and areas of mountainous terrain where geographical constraints make such communication impossible.

(b) No person may operate an aircraft at night under VFR over routes that can be navigated by pilotage unless that aircraft is equipped with—

(1) Two-way radio communication equipment necessary under normal operating conditions to fulfill the functions specified in paragraph (a) of this section; and

(2) Navigation equipment suitable for the route to be flown.

[Doc. No. FAA-2002-14002, 72 FR 31684, June 7, 2007, as amended by Amdt. 135-116, 74 FR 20205, May 1, 2009]

§ 135.163 Equipment requirements: Aircraft carrying passengers under IFR.

No person may operate an aircraft under IFR, carrying passengers, unless it has—

(a) A vertical speed indicator;

(b) A free-air temperature indicator;

(c) A heated pitot tube for each airspeed indicator;

(d) A power failure warning device or vacuum indicator to show the power available for gyroscopic instruments from each power source;

(e) An alternate source of static pressure for the altimeter and the airspeed and vertical speed indicators;

(f) For a single-engine aircraft:

(1) Two independent electrical power generating sources each of which is able to supply all probable combinations of continuous inflight electrical loads for required instruments and equipment; or

(2) In addition to the primary electrical power generating source, a standby battery or an alternate source of electric power that is capable of supplying 150% of the electrical loads of all required instruments and equipment necessary for safe emergency operation of the aircraft for at least one hour;

(g) For multi-engine aircraft, at least two generators or alternators each of which is on a separate engine, of which any combination of one-half of the total number are rated sufficiently to supply the electrical loads of all required instruments and equipment necessary for safe emergency operation of the aircraft except that for multi-engine helicopters, the two required generators may be mounted on the main rotor drive train; and

(h) Two independent sources of energy (with means of selecting either) of which at least one is an engine-driven pump or generator, each of which is able to drive all required gyroscopic instruments powered by, or to be powered by, that particular source and installed so that failure of one instrument or source, does not interfere with the energy supply to the remaining instruments or the other energy source unless, for single-engine aircraft in all cargo operations only, the rate of turn indicator has a source of energy separate from the bank and pitch and direction indicators. For the purpose of this paragraph, for multi-engine aircraft, each engine-driven source of energy must be on a different engine.

(i) For the purpose of paragraph (f) of this section, a continuous inflight electrical load includes one that draws current continuously during flight, such as radio equipment, electrically driven instruments, and lights, but does not include occasional intermittent loads.

[Doc. No. 16097, 43 FR 46783, Oct. 10, 1978, as amended by Amdt. 135-70, 62 FR 42374, Aug. 6, 1997; Amdt. 135-72, 63 FR 25573, May 8, 1998]

§ 135.165 Communication and navigation equipment: Extended over-water or IFR operations.

(a) *Aircraft navigation equipment requirements — General.* Except as provided in paragraph (g) of this section, no person may conduct operations under IFR or extended over-water unless—

(1) The en route navigation aids necessary for navigating the aircraft along the route (e.g., ATS routes, arrival and departure routes, and instrument approach procedures, including missed approach procedures if a missed approach routing is specified in the procedure) are available and suitable for use by the navigation systems required by this section:

(2) The aircraft used in extended over-water operations is equipped with at least two-approved independent navigation systems suitable for navigating the aircraft along the route to be flown within the degree of accuracy required for ATC.

(3) The aircraft used for IFR operations is equipped with at least—

(i) One marker beacon receiver providing visual and aural signals; and

(ii) One ILS receiver.

(4) Any RNAV system used to meet the navigation equipment requirements of this section is authorized in the certificate holder's operations specifications.

(b) *Use of a single independent navigation system for IFR operations.* The aircraft may be equipped with a single independent navigation system suitable for navigating the aircraft along the route to be flown within the degree of accuracy required for ATC if:

(1) It can be shown that the aircraft is equipped with at least one other independent navigation system suitable, in the event of loss of the navigation capability of the single independent navigation system permitted by this paragraph at any point along the route, for proceeding safely to a suitable airport and completing an instrument approach; and

(2) The aircraft has sufficient fuel so that the flight may proceed safely to a suitable airport by use of the remaining navigation system, and complete an instrument approach and land.

(c) *VOR navigation equipment.* Whenever VOR navigation equipment is required by paragraph (a) or (b) of this section, no person may operate an aircraft unless it is equipped with at least one approved DME or suitable RNAV system.

(d) *Airplane communication equipment requirements.* Except as permitted in paragraph (e) of this section, no person may operate a turbojet airplane having a passenger seat configuration, excluding any pilot seat, of 10 seats or more, or a multiengine airplane in a commuter operation, as defined in part 119 of this chapter, under IFR or in extended over-water operations unless the airplane is equipped with—

(1) At least two independent communication systems necessary under normal operating conditions to fulfill the functions specified in § 121.347(a) of this chapter; and

(2) At least one of the communication systems required by paragraph (d)(1) of this section must have two-way voice communication capability.

(e) *IFR or extended over-water communications equipment requirements.* A person may operate an aircraft other than that specified in paragraph (d) of this section under IFR or in extended over-water operations if it meets all of the requirements of this section, with the exception that only one communication system transmitter is required for operations other than extended over-water operations.

(f) *Additional aircraft communication equipment requirements.* In addition to the requirements in paragraphs (d) and (e) of this section, no person may operate an aircraft under IFR or in extended over-water operations unless it is equipped with at least:

(1) Two microphones; and

(2) Two headsets or one headset and one speaker.

(g) *Extended over-water exceptions.* Notwithstanding the requirements of paragraphs (a), (d), and (e) of this section, installation and use of a single long-range navigation system and a single long-range communication system for extended over-water operations in certain geographic areas may be authorized by the Administrator and approved in the certificate holder's operations specifications. The following are among the operational factors the Administrator may consider in granting an authorization:

(1) The ability of the flight crew to navigate the airplane along the route within the degree of accuracy required for ATC;

(2) The length of the route being flown; and

(3) The duration of the very high frequency communications gap.

[Doc. No. FAA-2002-14002, 72 FR 31684, June 7, 2007]

§ 135.167 Emergency equipment: Extended overwater operations.

(a) Except where the Administrator, by amending the operations specifications of the certificate holder, requires the carriage of all or any specific items of the equipment listed below for any overwater operation, or, upon application of the certificate holder, the Administrator allows deviation for a particular extended overwater operation, no person may operate an aircraft in extended overwater operations unless it carries, installed in conspicuously marked locations easily accessible to the occupants if a ditching occurs, the following equipment:

(1) An approved life preserver equipped with an approved survivor locator light for each occupant of the aircraft. The life preserver must be easily accessible to each seated occupant.

(2) Enough approved liferafts of a rated capacity and buoyancy to accommodate the occupants of the aircraft.

(b) Each liferaft required by paragraph (a) of this section must be equipped with or contain at least the following:

(1) One approved survivor locator light.

(2) One approved pyrotechnic signaling device.

(3) Either—

(i) One survival kit, appropriately equipped for the route to be flown; or

(ii) One canopy (for sail, sunshade, or rain catcher);

(iii) One radar reflector;

(iv) One liferaft repair kit;

(v) One bailing bucket;

(vi) One signaling mirror;

(vii) One police whistle;

(viii) One raft knife;

(ix) One CO_2 bottle for emergency inflation;

(x) One inflation pump;

(xi) Two oars;

(xii) One 75-foot retaining line;

(xiii) One magnetic compass;

(xiv) One dye marker;

(xv) One flashlight having at least two size "D" cells or equivalent;

(xvi) A 2-day supply of emergency food rations supplying at least 1,000 calories per day for each person;

(xvii) For each two persons the raft is rated to carry, two pints of water or one sea water desalting kit;

(xviii) One fishing kit; and

(xix) One book on survival appropriate for the area in which the aircraft is operated.

(c) No person may operate an airplane in extended overwater operations unless there is attached to one of the life rafts required by paragraph (a) of this section, an approved survival type emergency locator transmitter. Batteries used in this transmitter must be replaced (or recharged, if the batteries are rechargeable) when the transmitter has been in use for more than 1 cumulative hour, or, when 50 percent of their useful life (or for rechargeable batteries, 50 percent of their useful life of charge) has expired, as established by the transmitter manufacturer under its approval. The new expiration date for replacing (or recharging) the battery must be legibly marked on the outside of the transmitter. The battery useful life (or useful life of charge) requirements of this paragraph do not apply to batteries (such as water-activated batteries) that are essentially unaffected during probable storage intervals.

[Doc. No. 16097, 43 FR 46783, Oct. 10, 1978, as amended by Amdt. 135-4, 45 FR 38348, June 30, 1980; Amdt. 135-20, 51 FR 40710, Nov. 7, 1986; Amdt. 135-49, 59 FR 32058, June 21, 1994; Amdt. 135-91, 68 FR 54586, Sept. 17, 2003]

§ 135.168 [Reserved]

§ 135.169 Additional airworthiness requirements.

(a) Except for commuter category airplanes, no person may operate a large airplane unless it meets the additional airworthiness requirements of §§ 121.213 through 121.283 and 121.307 of this chapter.

(b) No person may operate a reciprocating-engine or turbopropeller-powered small airplane that has a passenger seating configuration, excluding pilot seats, of 10 seats or more unless it is type certificated—

(1) In the transport category;

(2) Before July 1, 1970, in the normal category and meets special conditions issued by the Administrator for airplanes intended for use in operations under this part;

(3) Before July 19, 1970, in the normal category and meets the additional airworthiness standards in Special Federal Aviation Regulation No. 23;

(4) In the normal category and meets the additional airworthiness standards in appendix A;

(5) In the normal category and complies with section 1.(a) of Special Federal Aviation Regulation No. 41;

(6) In the normal category and complies with section 1.(b) of Special Federal Aviation Regulation No. 41; or

(7) In the commuter category.

(c) No person may operate a small airplane with a passenger seating configuration, excluding any pilot seat, of 10 seats or more, with a seating configuration greater than the maximum seating configuration used in that type airplane in operations under this part before August 19, 1977. This paragraph does not apply to—

(1) An airplane that is type certificated in the transport category; or

(2) An airplane that complies with—

(i) Appendix A of this part provided that its passenger seating configuration, excluding pilot seats, does not exceed 19 seats; or

(ii) Special Federal Aviation Regulation No. 41.

(d) Cargo or baggage compartments:

(1) After March 20, 1991, each Class C or D compartment, as defined in § 25.857 of part 25 of this chapter, greater than 200 cubic feet in volume in a transport category airplane type certificated after January 1, 1958, must have ceiling and sidewall panels which are constructed of:

(i) Glass fiber reinforced resin;

(ii) Materials which meet the test requirements of part 25, appendix F, part III of this chapter; or

(iii) In the case of liner installations approved prior to March 20, 1989, aluminum.

(2) For compliance with this paragraph, the term "liner" includes any design feature, such as a joint or fastener, which would affect the capability of the liner to safely contain a fire.

[Doc. No. 16097, 43 FR 46783, Oct. 10, 1978, as amended by Amdt. 135-2, 44 FR 53731, Sept. 17, 1979; Amdt. 135-21, 52 FR 1836, Jan. 15, 1987; 52 FR 34745, Sept. 14, 1987; Amdt. 135-31, 54 FR 7389, Feb. 17, 1989; Amdt. 135-55, 60 FR 6628, Feb. 2, 1995]

§ 135.170 Materials for compartment interiors.

(a) No person may operate an airplane that conforms to an amended or supplemental type certificate issued in accordance with SFAR No. 41 for a maximum certificated takeoff weight in excess of 12,500 pounds unless within one year after issuance of the initial airworthiness certificate under that SFAR, the airplane meets the compartment interior requirements set forth in § 25.853(a) in effect March 6, 1995 (formerly § 25.853 (a), (b), (b-1), (b-2), and (b-3) of this chapter in effect on September 26, 1978).

(b) Except for commuter category airplanes and airplanes certificated under Special Federal Aviation Regulation No. 41, no person may operate a large airplane unless it meets the following additional airworthiness requirements:

(1) Except for those materials covered by paragraph (b)(2) of this section, all materials in each compartment used by the crewmembers or passengers must meet the requirements of § 25.853 of this chapter in effect as follows or later amendment thereto:

(i) Except as provided in paragraph (b)(1)(iv) of this section, each airplane with a passenger capacity of 20 or more and manufactured after August 19, 1988, but prior to August 20, 1990, must comply with the heat release rate testing provisions of § 25.853(d) in effect March 6, 1995 (formerly § 25.853(a-1) in effect on August 20, 1986), except that the total heat release over the first 2 minutes of sample exposure rate must not exceed 100 kilowatt minutes per square meter and the peak heat release rate must not exceed 100 kilowatts per square meter.

(ii) Each airplane with a passenger capacity of 20 or more and manufactured after August 19, 1990, must comply with the heat release rate and smoke testing provisions of § 25.853(d) in effect March 6, 1995 (formerly § 25.83(a-1) in effect on September 26, 1988).

(iii) Except as provided in paragraph (b)(1) (v) or (vi) of this section, each airplane for which the application for type certificate was filed prior to May 1, 1972, must comply with the provisions of § 25.853 in effect on April 30, 1972, regardless of the passenger capacity, if there is a substantially complete replacement of the cabin interior after April 30, 1972.

(iv) Except as provided in paragraph (b)(1) (v) or (vi) of this section, each airplane for which the application for type certificate was filed after May 1, 1972, must comply with the material requirements under which the airplane was type certificated regardless of the passenger capacity if there is a substantially complete replacement of the cabin interior after that date.

(v) Except as provided in paragraph (b)(1)(vi) of this section, each airplane that was type certificated after January 1, 1958, must comply with the heat release testing provisions of § 25.853(d) in effect March 6, 1995 (formerly § 25.853(a-1) in effect on August 20, 1986), if there is a substantially complete replacement of the cabin interior components identified in that paragraph on or after that date, except that the total heat release over the first 2 minutes of sample exposure shall not exceed 100 kilowatt-minutes per square meter and the peak heat release rate shall not exceed 100 kilowatts per square meter.

(vi) Each airplane that was type certificated after January 1, 1958, must comply with the heat release rate and smoke testing provisions of § 25.853(d) in effect March 6, 1995 (formerly § 25.853(a-1) in effect on August 20, 1986), if there is a substantially complete replacement of the cabin interior components identified in that paragraph after August 19, 1990.

(vii) Contrary provisions of this section notwithstanding, the Manager of the Transport Airplane Directorate, Aircraft Certification Service, Federal Aviation Administration, may authorize deviation from the requirements of paragraph (b)(1)(i), (b)(1)(ii), (b)(1)(v), or (b)(1)(vi) of this section for specific components of the cabin interior that do not meet applicable flammability and smoke emission requirements, if the determination is made that special circumstances exist that make compliance impractical. Such grants of deviation will be limited to those airplanes manufactured within 1 year after the applicable date specified in this section and those airplanes in which the interior is replaced within 1 year of that date. A request for such grant of deviation must include a thorough and accurate analysis of each component subject to § 25.853(d) in effect March 6, 1995 (formerly § 25.853(a-1) in effect on August 20, 1986), the steps being taken to achieve compliance, and, for the few components for which timely compliance will not be achieved, credible reasons for such noncompliance.

(viii) Contrary provisions of this section notwithstanding, galley carts and standard galley containers that do not meet the flammability and smoke emission requirements of § 25.853(d) in effect March 6, 1995 (formerly § 25.853(a-1) in effect on August 20, 1986), may be used in airplanes that must meet the requirements of paragraph (b)(1)(i), (b)(1)(ii), (b)(1)(iv) or (b)(1)(vi) of this section provided the galley carts or standard containers were manufactured prior to March 6, 1995.

(2) For airplanes type certificated after January 1, 1958, seat cushions, except those on flight crewmember seats, in any compartment occupied by crew or passengers must comply with the requirements pertaining to fire protection of seat cushions in § 25.853(c) effective November 26, 1984.

(c) Thermal/acoustic insulation materials. For transport category airplanes type certificated after January 1, 1958:

(1) For airplanes manufactured before September 2, 2005, when thermal/acoustic insulation is installed in the fuselage as replacements after September 2, 2005, the insulation must meet the flame propagation requirements of § 25.856 of this chapter, effective September 2, 2003, if it is:

(i) Of a blanket construction, or

(ii) Installed around air ducting.

(2) For airplanes manufactured after September 2, 2005, thermal/acoustic insulation materials installed in the fuselage must meet the flame propagation requirements of § 25.856 of this chapter, effective September 2, 2003.

[Doc. No. 26192, 60 FR 6628, Feb. 2, 1995; Amdt. 135-55, 60 FR 11194, Mar. 1, 1995; Amdt. 135-56, 60 FR 13011, Mar. 9, 1995; Amdt. 135-90, 68 FR 45084, July 31, 2003; Amdt. 135-103, 70 FR 77752, Dec. 30, 2005]

§ 135.171 Shoulder harness installation at flight crewmember stations.

(a) No person may operate a turbojet aircraft or an aircraft having a passenger seating configuration, excluding any pilot seat, of 10 seats or more unless it is equipped with an approved shoulder harness installed for each flight crewmember station.

(b) Each flight crewmember occupying a station equipped with a shoulder harness must fasten the shoulder harness during takeoff and landing, except that the shoulder harness may be unfastened if the crewmember cannot perform the required duties with the shoulder harness fastened.

§ 135.173 Airborne thunderstorm detection equipment requirements.

(a) No person may operate an aircraft that has a passenger seating configuration, excluding any pilot seat, of 10 seats or more in passenger-carrying operations, except a helicopter operating under day VFR conditions, unless the aircraft is equipped with either approved thunderstorm detection equipment or approved airborne weather radar equipment.

(b) No person may operate a helicopter that has a passenger seating configuration, excluding any pilot seat, of 10 seats or more in passenger-carrying operations, under night VFR when current weather reports indicate that thunderstorms or other potentially hazardous weather conditions that can be detected with airborne thunderstorm detection equipment may reasonably be expected along the route to be flown, unless the helicopter is equipped with either approved thunderstorm detection equipment or approved airborne weather radar equipment.

(c) No person may begin a flight under IFR or night VFR conditions when current weather reports indicate that thunderstorms or other potentially hazardous weather conditions that can be detected with airborne thunderstorm detection equipment, required by paragraph (a) or (b) of this section, may reasonably be expected along the route to be flown, unless the airborne thunderstorm detection equipment is in satisfactory operating condition.

(d) If the airborne thunderstorm detection equipment becomes inoperative en route, the aircraft must be operated under the instructions and procedures specified for that event in the manual required by § 135.21.

(e) This section does not apply to aircraft used solely within the State of Hawaii, within the State of Alaska, within that part of Canada west of longitude 130 degrees W, between latitude 70 degrees N, and latitude 53 degrees N, or during any training, test, or ferry flight.

(f) Without regard to any other provision of this part, an alternate electrical power supply is not required for airborne thunderstorm detection equipment.

[Doc. No. 16097, 43 FR 46783, Oct. 10, 1978, as amended by Amdt. 135-20, 51 FR 40710, Nov. 7, 1986; Amdt. 135-60, 61 FR 2616, Jan. 26, 1996]

§ 135.175 Airborne weather radar equipment requirements.

(a) No person may operate a large, transport category aircraft in passenger-carrying operations unless approved airborne weather radar equipment is installed in the aircraft.

(b) No person may begin a flight under IFR or night VFR conditions when current weather reports indicate that thunderstorms, or other potentially hazardous weather conditions that can be detected with airborne weather radar equipment, may reasonably be expected along the route to be flown, unless the airborne weather radar equipment required by paragraph (a) of this section is in satisfactory operating condition.

(c) If the airborne weather radar equipment becomes inoperative en route, the aircraft must be operated under the instructions and procedures specified for that event in the manual required by § 135.21.

(d) This section does not apply to aircraft used solely within the State of Hawaii, within the State of Alaska, within that part of Canada west of longitude 130 degrees W, between latitude 70 degrees N, and latitude 53 degrees N, or during any training, test, or ferry flight.

(e) Without regard to any other provision of this part, an alternate electrical power supply is not required for airborne weather radar equipment.

§ 135.177 Emergency equipment requirements for aircraft having a passenger seating configuration of more than 19 passengers.

(a) No person may operate an aircraft having a passenger seating configuration, excluding any pilot seat, of more than 19 seats unless it is equipped with the following emergency equipment:

(1) At least one approved first-aid kit for treatment of injuries likely to occur in flight or in a minor accident that must:

(i) Be readily accessible to crewmembers.

(ii) Be stored securely and kept free from dust, moisture, and damaging temperatures.

(iii) Contain at least the following appropriately maintained contents in the specified quantities:

Contents	Quantity
Adhesive bandage compresses, 1-inch	16
Antiseptic swabs	20
Ammonia inhalants	10
Bandage compresses, 4-inch	8
Triangular bandage compresses, 40-inch	5
Arm splint, noninflatable	1
Leg splint, noninflatable	1
Roller bandage, 4-inch	4
Adhesive tape, 1-inch standard roll	2
Bandage scissors	1
Protective nonpermeable gloves or equivalent	1 pair

(2) A crash axe carried so as to be accessible to the crew but inaccessible to passengers during normal operations.

(3) Signs that are visible to all occupants to notify them when smoking is prohibited and when safety belts must be fastened. The signs must be constructed so that they can be turned on during any movement of the aircraft on the surface, for each takeoff or landing, and at other times considered necessary by the pilot in command. "No smoking" signs shall be turned on when required by § 135.127.

(4) [Reserved]

(b) Each item of equipment must be inspected regularly under inspection periods established in the operations specifications to ensure its condition for continued serviceability and immediate readiness to perform its intended emergency purposes.

[Doc. No. 16097, 43 FR 46783, Oct. 10, 1978, as amended by Amdt. 135-25, 53 FR 12362, Apr. 13, 1988; Amdt. 135-43, 57 FR 19245, May 4, 1992; Amdt. 135-44, 57 FR 42676, Sept. 15, 1992; Amdt. 135-47, 59 FR 1781, Jan. 12, 1994; Amdt. 135-53, 59 FR 52643, Oct. 18, 1994; 59 FR 55208, Nov. 4, 1994; Amdt. 121-281, 66 FR 19045, Apr. 12, 2001]

§ 135.178 Additional emergency equipment.

No person may operate an airplane having a passenger seating configuration of more than 19 seats, unless it has the additional emergency equipment specified in paragraphs (a) through (l) of this section.

(a) *Means for emergency evacuation.* Each passenger-carrying landplane emergency exit (other than over-the-wing) that is more than 6 feet from the ground, with the airplane on the ground and the landing gear extended, must have an approved means to assist the occupants in descending to the ground. The assisting means for a floor-level emergency exit must meet the requirements of § 25.809(f)(1) of this chapter in effect on April 30, 1972, except that, for any airplane for which the application for the type certificate was filed after that date, it must meet the requirements under which the airplane was type certificated. An assisting means that deploys automatically must be armed during taxiing, takeoffs, and landings; however, the Administrator may grant a deviation from the requirement of automatic deployment if he finds that the design of the exit makes compliance impractical, if the assisting means automatically erects upon deployment and, with respect to required emergency exits, if an emergency evacuation demonstration is conducted in accordance with § 121.291(a) of this chapter. This paragraph does not apply to the rear window emergency exit of Douglas DC-3 airplanes operated with fewer than 36 occupants, including crewmembers, and fewer than five exits authorized for passenger use.

(b) *Interior emergency exit marking.* The following must be complied with for each passenger-carrying airplane:

(1) Each passenger emergency exit, its means of access, and its means of opening must be conspicuously marked. The identity and locating of each passenger emergency exit must be recognizable from a distance equal to the width of the cabin. The location of each passenger emergency exit must be indicated by a sign visible to occupants approaching along the main passenger aisle. There must be a locating sign—

(i) Above the aisle near each over-the-wing passenger emergency exit, or at another ceiling location if it is more practical because of low headroom;

(ii) Next to each floor level passenger emergency exit, except that one sign may serve two such exits if they both can be seen readily from that sign; and

(iii) On each bulkhead or divider that prevents fore and aft vision along the passenger cabin, to indicate emergency exits beyond and obscured by it, except that if this is not possible, the sign may be placed at another appropriate location.

(2) Each passenger emergency exit marking and each locating sign must meet the following:

(i) For an airplane for which the application for the type certificate was filed prior to May 1, 1972, each passenger emergency exit marking and each locating sign must be manufactured to meet the requirements of § 25.812(b) of this chapter in effect on April 30, 1972. On these airplanes, no sign may continue to be used if its luminescence (brightness) decreases to below 100 microlamberts. The colors may be reversed if it increases the emergency illumination of the passenger compartment. However, the Administrator may authorize deviation from the 2-inch background requirements if he finds that special circumstances exist that make compliance impractical and that the proposed deviation provides an equivalent level of safety.

(ii) For an airplane for which the application for the type certificate was filed on or after May 1, 1972, each passenger emergency exit marking and each locating sign must be manufactured to meet the interior emergency exit marking requirements under which the airplane was type certificated. On these airplanes, no sign may continue to be used if its luminescence (brightness) decreases to below 250 microlamberts.

(c) *Lighting for interior emergency exit markings.* Each passenger-carrying airplane must have an emergency lighting system, independent of the main lighting system; however, sources of general cabin illumination may be common to both the emergency and the main lighting systems if the power supply to the emergency lighting system is independent of the power supply to the main lighting system. The emergency lighting system must—

(1) Illuminate each passenger exit marking and locating sign;

(2) Provide enough general lighting in the passenger cabin so that the average illumination when measured at 40-inch intervals at seat armrest height, on the centerline of the main passenger aisle, is at least 0.05 foot-candles; and

(3) For airplanes type certificated after January 1, 1958, include floor proximity emergency escape path marking which meets the requirements of § 25.812(e) of this chapter in effect on November 26, 1984.

(d) *Emergency light operation.* Except for lights forming part of emergency lighting subsystems provided in compliance with § 25.812(h) of this chapter (as prescribed in paragraph (h) of this section) that serve no more than one assist means, are independent of the airplane's main emergency lighting systems, and are automatically activated when the assist means is deployed, each light required by paragraphs (c) and (h) of this section must:

(1) Be operable manually both from the flightcrew station and from a point in the passenger compartment that is readily accessible to a normal flight attendant seat;

(2) Have a means to prevent inadvertent operation of the manual controls;

(3) When armed or turned on at either station, remain lighted or become lighted upon interruption of the airplane's normal electric power;

(4) Be armed or turned on during taxiing, takeoff, and landing. In showing compliance with this paragraph, a transverse vertical separation of the fuselage need not be considered;

(5) Provide the required level of illumination for at least 10 minutes at the critical ambient conditions after emergency landing; and

(6) Have a cockpit control device that has an "on," "off," and "armed" position.

(e) *Emergency exit operating handles.* (1) For a passenger-carrying airplane for which the application for the type certificate was filed prior to May 1, 1972, the location of each passenger emergency exit operating handle, and instructions for opening the exit, must be shown by a marking on or near the exit that is readable from a distance of 30 inches. In addition, for each Type I and Type II emergency exit with a locking mechanism released by rotary motion of the handle, the instructions for opening must be shown by—

(i) A red arrow with a shaft at least three-fourths inch wide and a head twice the width of the shaft, extending along at least 70° of arc at a radius approximately equal to three-fourths of the handle length; and

(ii) The word "open" in red letters 1 inch high placed horizontally near the head of the arrow.

(2) For a passenger-carrying airplane for which the application for the type certificate was filed on or after May 1, 1972, the location of each passenger emergency exit operating handle and instructions for opening the exit must be shown in accordance with the requirements under which the airplane was type certificated. On these airplanes, no operating handle or operating handle cover may continue to be used if its luminescence (brightness) decreases to below 100 microlamberts.

(f) *Emergency exit access.* Access to emergency exits must be provided as follows for each passenger-carrying airplane:

(1) Each passageway between individual passenger areas, or leading to a Type I or Type II emergency exit, must be unobstructed and at least 20 inches wide.

(2) There must be enough space next to each Type I or Type II emergency exit to allow a crewmember to assist in the evacuation of passengers without reducing the unobstructed width of the passageway below that required in paragraph (f)(1) of this section; however, the Administrator may authorize deviation from this requirement for an airplane certificated under the provisions of part 4b of the Civil Air Regulations in effect before December 20, 1951, if he finds that special circumstances exist that provide an equivalent level of safety.

(3) There must be access from the main aisle to each Type III and Type IV exit. The access from the aisle to these exits must not be obstructed by seats, berths, or other protrusions in a manner that would reduce the effectiveness of the exit. In addition, for a transport category airplane type certificated after January 1, 1958, there must be placards installed in accordance with § 25.813(c)(3) of this chapter for each Type III exit after December 3, 1992.

(4) If it is necessary to pass through a passageway between passenger compartments to reach any required emergency exit from any seat in the passenger cabin, the passageway must not be obstructed. Curtains may, however, be used if they allow free entry through the passageway.

(5) No door may be installed in any partition between passenger compartments.

(6) If it is necessary to pass through a doorway separating the passenger cabin from other areas to reach a required emergency exit from any passenger seat, the door must have a means to latch it in the open position, and the door must be latched open during each takeoff and landing. The latching means must be able to withstand the loads imposed upon it when the door is subjected to the ultimate inertia forces, relative to the surrounding structure, listed in § 25.561(b) of this chapter.

(g) *Exterior exit markings.* Each passenger emergency exit and the means of opening that exit from the outside must be marked on the outside of the airplane. There must be a 2-inch colored band outlining each passenger emergency exit on the side of the fuselage. Each outside marking, including the band, must be readily distinguishable from the surrounding fuselage area by contrast in color. The markings must comply with the following:

(1) If the reflectance of the darker color is 15 percent or less, the reflectance of the lighter color must be at least 45 percent.

(2) If the reflectance of the darker color is greater than 15 percent, at least a 30 percent difference between its reflectance and the reflectance of the lighter color must be provided.

(3) Exits that are not in the side of the fuselage must have the external means of opening and applicable instructions marked conspicuously in red or, if red is inconspicuous against the background color, in bright chrome yellow and, when the opening means for such an exit is located on only one side of the fuselage, a conspicuous marking to that effect must be provided on the other side. "Reflectance" is the ratio of the luminous flux reflected by a body to the luminous flux it receives.

(h) *Exterior emergency lighting and escape route.* (1) Each passenger-carrying airplane must be equipped with exterior lighting that meets the following requirements:

(i) For an airplane for which the application for the type certificate was filed prior to May 1, 1972, the requirements of § 25.812 (f) and (g) of this chapter in effect on April 30, 1972.

(ii) For an airplane for which the application for the type certificate was filed on or after May 1, 1972, the exterior emergency lighting requirements under which the airplane was type certificated.

(2) Each passenger-carrying airplane must be equipped with a slip-resistant escape route that meets the following requirements:

(i) For an airplane for which the application for the type certificate was filed prior to May 1, 1972, the requirements of § 25.803(e) of this chapter in effect on April 30, 1972.

(ii) For an airplane for which the application for the type certificate was filed on or after May 1, 1972, the slip-resistant escape route requirements under which the airplane was type certificated.

(i) *Floor level exits.* Each floor level door or exit in the side of the fuselage (other than those leading into a cargo or baggage compartment that is not accessible from the passenger cabin) that is 44 or more inches high and 20 or more inches wide, but not wider than 46 inches, each passenger ventral exit (except the ventral exits on Martin 404 and Convair 240 airplanes), and each tail cone exit, must meet the requirements of this section for floor level emergency exits. However, the Administrator may grant a deviation from this paragraph if he finds that circumstances make full compliance impractical and that an acceptable level of safety has been achieved.

(j) *Additional emergency exits.* Approved emergency exits in the passenger compartments that are in excess of the minimum number of required emergency exits must meet all of the applicable provisions of this section, except paragraphs (f) (1), (2), and (3) of this section, and must be readily accessible.

(k) On each large passenger-carrying turbojet-powered airplane, each ventral exit and tailcone exit must be—

(1) Designed and constructed so that it cannot be opened during flight; and

(2) Marked with a placard readable from a distance of 30 inches and installed at a conspicuous location near the means of opening the exit, stating that the exit has been designed and constructed so that it cannot be opened during flight.

(l) *Portable lights.* No person may operate a passenger-carrying airplane unless it is equipped with flashlight stowage provisions accessible from each flight attendant seat.

[Doc. No. 26530, 57 FR 19245, May 4, 1992; 57 FR 29120, June 30, 1992, as amended at 57 FR 34682, Aug. 6, 1992]

§ 135.179 Inoperable instruments and equipment.

(a) No person may take off an aircraft with inoperable instruments or equipment installed unless the following conditions are met:

(1) An approved Minimum Equipment List exists for that aircraft.

(2) The certificate-holding district office has issued the certificate holder operations specifications authorizing operations in accordance with an approved Minimum Equipment List. The flight crew shall have direct access at all times prior to flight to all of the information contained in the approved Minimum Equipment List through printed or other means approved by the Administrator in the certificate holders operations specifications. An approved Minimum Equipment List, as authorized by the operations specifications, constitutes an approved change to the type design without requiring recertification.

(3) The approved Minimum Equipment List must:

(i) Be prepared in accordance with the limitations specified in paragraph (b) of this section.

(ii) Provide for the operation of the aircraft with certain instruments and equipment in an inoperable condition.

(4) Records identifying the inoperable instruments and equipment and the information required by (a)(3)(ii) of this section must be available to the pilot.

(5) The aircraft is operated under all applicable conditions and limitations contained in the Minimum Equipment List and the operations specifications authorizing use of the Minimum Equipment List.

(b) The following instruments and equipment may not be included in the Minimum Equipment List:

(1) Instruments and equipment that are either specifically or otherwise required by the airworthiness requirements under which the airplane is type certificated and which are essential for safe operations under all operating conditions.

(2) Instruments and equipment required by an airworthiness directive to be in operable condition unless the airworthiness directive provides otherwise.

(3) Instruments and equipment required for specific operations by this part.

(c) Notwithstanding paragraphs (b)(1) and (b)(3) of this section, an aircraft with inoperable instruments or equipment may be operated under a special flight permit under §§ 21.197 and 21.199 of this chapter.

[Doc. No. 25780, 56 FR 12311, Mar. 22, 1991; 56 FR 14920, Apr. 8, 1991, as amended by Amdt. 135-60, 61 FR 2616, Jan. 26, 1996; Amdt. 135-91, 68 FR 54586, Sept. 17, 2003]

§ 135.180 Traffic Alert and Collision Avoidance System.

(a) Unless otherwise authorized by the Administrator, after December 31, 1995, no person may operate a turbine powered airplane that has a passenger seat configuration, excluding any pilot seat, of 10 to 30 seats unless it is equipped with an approved traffic alert and collision avoidance system. If a TCAS II system is installed, it must be capable of coordinating with TCAS units that meet TSO C-119.

(b) The airplane flight manual required by § 135.21 of this part shall contain the following information on the TCAS I system required by this section:

(1) Appropriate procedures for—

(i) The use of the equipment; and

(ii) Proper flightcrew action with respect to the equipment operation.

(2) An outline of all input sources that must be operating for the TCAS to function properly.

[Doc. No. 25355, 54 FR 951, Jan. 10, 1989, as amended by Amdt. 135-54, 59 FR 67587, Dec. 29, 1994]

§ 135.181 Performance requirements: Aircraft operated over-the-top or in IFR conditions.

(a) Except as provided in paragraphs (b) and (c) of this section, no person may—

(1) Operate a single-engine aircraft carrying passengers over-the-top; or

(2) Operate a multiengine aircraft carrying passengers over-the-top or in IFR conditions at a weight that will not allow it to climb, with the critical engine inoperative, at least 50 feet a minute when operating at the MEAs of the route to be flown or 5,000 feet MSL, whichever is higher.

(b) Notwithstanding the restrictions in paragraph (a)(2) of this section, multiengine helicopters carrying passengers offshore may conduct such operations in over-the-top or in IFR conditions at a weight that will allow the helicopter to climb at least 50 feet per minute with the critical engine inoperative when operating at the MEA of the route to be flown or 1,500 feet MSL, whichever is higher.

(c) Without regard to paragraph (a) of this section, if the latest weather reports or forecasts, or any combination of them, indicate that the weather along the planned route (including takeoff and landing) allows flight under VFR under the ceiling (if a ceiling exists) and that the weather is forecast to remain so until at least 1 hour after the estimated time of arrival at the destination, a person may operate an aircraft over-the-top.

(d) Without regard to paragraph (a) of this section, a person may operate an aircraft over-the-top under conditions allowing—

(1) For multiengine aircraft, descent or continuance of the flight under VFR if its critical engine fails; or

(2) For single-engine aircraft, descent under VFR if its engine fails.

[Doc. No. 16097, 43 FR 46783, Oct. 10, 1978, as amended by Amdt. 135-20, 51 FR 40710, Nov. 7, 1986; Amdt. 135-70, 62 FR 42374, Aug. 6, 1997]

§ 135.183 Performance requirements: Land aircraft operated over water.

No person may operate a land aircraft carrying passengers over water unless—

(a) It is operated at an altitude that allows it to reach land in the case of engine failure;

(b) It is necessary for takeoff or landing;

(c) It is a multiengine aircraft operated at a weight that will allow it to climb, with the critical engine inoperative, at least 50 feet a minute, at an altitude of 1,000 feet above the surface; or

(d) It is a helicopter equipped with helicopter flotation devices.

§ 135.185 Empty weight and center of gravity: Currency requirement.

(a) No person may operate a multiengine aircraft unless the current empty weight and center of gravity are calculated from values established by actual weighing of the aircraft within the preceding 36 calendar months.

(b) Paragraph (a) of this section does not apply to—

(1) Aircraft issued an original airworthiness certificate within the preceding 36 calendar months; and

(2) Aircraft operated under a weight and balance system approved in the operations specifications of the certificate holder.

Subpart D—VFR/IFR Operating Limitations and Weather Requirements

§ 135.201 Applicability.

This subpart prescribes the operating limitations for VFR/IFR flight operations and associated weather requirements for operations under this part.

§ 135.203 VFR: Minimum altitudes.

Except when necessary for takeoff and landing, no person may operate under VFR—

(a) An airplane—

(1) During the day, below 500 feet above the surface or less than 500 feet horizontally from any obstacle; or

(2) At night, at an altitude less than 1,000 feet above the highest obstacle within a horizontal distance of 5 miles from the course intended to be flown or, in designated mountainous terrain, less than 2,000 feet above the highest obstacle within a horizontal distance of 5 miles from the course intended to be flown; or

(b) A helicopter over a congested area at an altitude less than 300 feet above the surface.

§ 135.205 VFR: Visibility requirements.

(a) No person may operate an airplane under VFR in uncontrolled airspace when the ceiling is less than 1,000 feet unless flight visibility is at least 2 miles.

(b) No person may operate a helicopter under VFR in Class G airspace at an altitude of 1,200 feet or less above the surface or within the lateral boundaries of the surface areas of Class B, Class C, Class D, or Class E airspace designated for an airport unless the visibility is at least—

(1) During the day— ½ mile; or

(2) At night—1 mile.

[Doc. No. 16097, 43 FR 46783, Oct. 10, 1978, as amended by Amdt. 135-41, 56 FR 65663, Dec. 17, 1991]

§ 135.207 VFR: Helicopter surface reference requirements.

No person may operate a helicopter under VFR unless that person has visual surface reference or, at night, visual surface light reference, sufficient to safely control the helicopter.

§ 135.209 VFR: Fuel supply.

(a) No person may begin a flight operation in an airplane under VFR unless, considering wind and forecast weather conditions, it has enough fuel to fly to the first point of intended landing and, assuming normal cruising fuel consumption—

(1) During the day, to fly after that for at least 30 minutes; or

(2) At night, to fly after that for at least 45 minutes.

(b) No person may begin a flight operation in a helicopter under VFR unless, considering wind and forecast weather conditions, it has enough fuel to fly to the first point of intended landing and, assuming normal cruising fuel consumption, to fly after that for at least 20 minutes.

§ 135.211 VFR: Over-the-top carrying passengers: Operating limitations.

Subject to any additional limitations in § 135.181, no person may operate an aircraft under VFR over-the-top carrying passengers, unless—

(a) Weather reports or forecasts, or any combination of them, indicate that the weather at the intended point of termination of over-the-top flight—

(1) Allows descent to beneath the ceiling under VFR and is forecast to remain so until at least 1 hour after the estimated time of arrival at that point; or

(2) Allows an IFR approach and landing with flight clear of the clouds without reaching the prescribed initial approach altitude over the final approach facility, unless the approach is made with the use of radar under § 91.175(i) of this chapter; or

(b) It is operated under conditions allowing—

(1) For multiengine aircraft, descent or continuation of the flight under VFR if its critical engine fails; or

(2) For single-engine aircraft, descent under VFR if its engine fails.

[Doc. No. 16097, 43 FR 46783, Oct. 10, 1978, as amended by Amdt. 135-32, 54 FR 34332, Aug. 18, 1989; 73 FR 20164, Apr. 15, 2008]

§ 135.213 Weather reports and forecasts.

(a) Whenever a person operating an aircraft under this part is required to use a weather report or forecast, that person shall use that of the U.S. National Weather Service, a source approved by the U.S. National Weather Service, or a source approved by the Administrator. However, for operations under VFR, the pilot in command may, if such a report is not available, use weather information based on that pilot's own observations or on those of other persons competent to supply appropriate observations.

(b) For the purposes of paragraph (a) of this section, weather observations made and furnished to pilots to conduct IFR operations at an airport must be taken at the airport where those IFR operations are conducted, unless the Administrator issues operations specifications allowing the use of weather observations taken at a location not at the airport where the IFR operations are conducted. The Administrator issues such operations specifications when, after investigation by the U.S. National Weather Service and the certificate-holding district office, it is found that the standards of safety for that operation would allow the deviation from this paragraph for a particular operation for which an air carrier operating certificate or operating certificate has been issued.

[Doc. No. 16097, 43 FR 46783, Oct. 10, 1978, as amended by Amdt. 135-60, 61 FR 2616, Jan. 26, 1996]

§ 135.215 IFR: Operating limitations.

(a) Except as provided in paragraphs (b), (c) and (d) of this section, no person may operate an aircraft under IFR outside of controlled airspace or at any airport that does not have an approved standard instrument approach procedure.

(b) The Administrator may issue operations specifications to the certificate holder to allow it to operate under IFR over routes outside controlled airspace if—

(1) The certificate holder shows the Administrator that the flight crew is able to navigate, without visual reference to the ground, over an intended track without deviating more than 5 degrees or 5 miles, whichever is less, from that track; and

(2) The Administrator determines that the proposed operations can be conducted safely.

(c) A person may operate an aircraft under IFR outside of controlled airspace if the certificate holder has been approved for the operations and that operation is necessary to—

(1) Conduct an instrument approach to an airport for which there is in use a current approved standard or special instrument approach procedure; or

(2) Climb into controlled airspace during an approved missed approach procedure; or

(3) Make an IFR departure from an airport having an approved instrument approach procedure.

(d) The Administrator may issue operations specifications to the certificate holder to allow it to depart at an airport that does not have an approved standard instrument approach procedure when the Administrator determines that it is necessary to make an IFR departure from that airport and that the proposed operations can be conducted safely. The approval to operate at that airport does not include an approval to make an IFR approach to that airport.

§ 135.217 IFR: Takeoff limitations.

No person may takeoff an aircraft under IFR from an airport where weather conditions are at or above takeoff minimums but are below authorized IFR landing minimums unless there is an alternate airport within 1 hour's flying time (at normal cruising speed, in still air) of the airport of departure.

§ 135.219 IFR: Destination airport weather minimums.

No person may take off an aircraft under IFR or begin an IFR or over-the-top operation unless the latest weather reports or forecasts, or any combination of them, indicate that weather conditions at the estimated time of arrival at the next airport of intended landing will be at or above authorized IFR landing minimums.

§ 135.221 IFR: Alternate airport weather minimums.

No person may designate an alternate airport unless the weather reports or forecasts, or any combination of them, indicate that the weather conditions will be at or above authorized alternate airport landing minimums for that airport at the estimated time of arrival.

§ 135.223 IFR: Alternate airport requirements.

(a) Except as provided in paragraph (b) of this section, no person may operate an aircraft in IFR conditions unless it carries enough fuel (considering weather reports or forecasts or any combination of them) to—

(1) Complete the flight to the first airport of intended landing;

(2) Fly from that airport to the alternate airport; and

(3) Fly after that for 45 minutes at normal cruising speed or, for helicopters, fly after that for 30 minutes at normal cruising speed.

(b) Paragraph (a)(2) of this section does not apply if part 97 of this chapter prescribes a standard instrument approach procedure for the first airport of intended landing and, for at least one hour before and after the estimated time of arrival, the appropriate weather reports or forecasts, or any combination of them, indicate that—

(1) The ceiling will be at least 1,500 feet above the lowest circling approach MDA; or

(2) If a circling instrument approach is not authorized for the airport, the ceiling will be at least 1,500 feet above the lowest published minimum or 2,000 feet above the airport elevation, whichever is higher; and

(3) Visibility for that airport is forecast to be at least three miles, or two miles more than the lowest applicable visibility minimums, whichever is the greater, for the instrument approach procedure to be used at the destination airport.

[Doc. No. 16097, 43 FR 46783, Oct. 10, 1978, as amended by Amdt. 135-20, 51 FR 40710, Nov. 7, 1986]

§ 135.225 IFR: Takeoff, approach and landing minimums.

(a) Except to the extent permitted by paragraph (b) of this section, no pilot may begin an instrument approach procedure to an airport unless—

(1) That airport has a weather reporting facility operated by the U.S. National Weather Service, a source approved by U.S. National Weather Service, or a source approved by the Administrator; and

(2) The latest weather report issued by that weather reporting facility indicates that weather conditions are at or above the authorized IFR landing minimums for that airport.

(b) A pilot conducting an eligible on-demand operation may begin an instrument approach procedure to an airport that does not have a weather reporting facility operated by the U.S. National Weather Service, a source approved by the U.S. National Weather Service, or a source approved by the Administrator if—

(1) The alternate airport has a weather reporting facility operated by the U.S. National Weather Service, a source approved by the U.S. National Weather Service, or a source approved by the Administrator; and

(2) The latest weather report issued by the weather reporting facility includes a current local altimeter setting for the destination airport. If no local altimeter setting for the destination airport is available, the pilot may use the current altimeter setting provided by the facility designated on the approach chart for the destination airport.

(c) If a pilot has begun the final approach segment of an instrument approach to an airport under paragraph (b) of this section, and the pilot receives a later weather report indicating that conditions have worsened to below the minimum requirements, then the pilot may continue the approach only if the requirements of § 91.175(l) of this chapter, or both of the following conditions, are met—

(1) The later weather report is received when the aircraft is in one of the following approach phases:

(i) The aircraft is on an ILS final approach and has passed the final approach fix;

(ii) The aircraft is on an ASR or PAR final approach and has been turned over to the final approach controller; or

(iii) The aircraft is on a nonprecision final approach and the aircraft—

(A) Has passed the appropriate facility or final approach fix; or

(B) Where a final approach fix is not specified, has completed the procedure turn and is established inbound toward the airport on the final approach course within the distance prescribed in the procedure; and

(2) The pilot in command finds, on reaching the authorized MDA or DA/DH, that the actual weather conditions are at or above the minimums prescribed for the procedure being used.

(d) If a pilot has begun the final approach segment of an instrument approach to an airport under paragraph (c) of this section and a later weather report indicating below minimum conditions is received after the aircraft is—

(1) On an ILS final approach and has passed the final approach fix; or

(2) On an ASR or PAR final approach and has been turned over to the final approach controller; or

(3) On a final approach using a VOR, NDB, or comparable approach procedure; and the aircraft—

(i) Has passed the appropriate facility or final approach fix; or

(ii) Where a final approach fix is not specified, has completed the procedure turn and is established inbound toward the airport on the final approach course within the distance prescribed in the procedure; the approach may be continued and a landing made if the pilot finds, upon reaching the authorized MDA or DH, that actual weather conditions are at least equal to the minimums prescribed for the procedure.

(e) The MDA or DA/DH and visibility landing minimums prescribed in part 97 of this chapter or in the operator's operations specifications are increased by 100 feet and ½ mile respectively, but not to exceed the ceiling and visibility minimums for that airport when used as an alternate airport, for each pilot in command of a turbine-powered airplane who has not served at least 100 hours as pilot in command in that type of airplane.

(f) Each pilot making an IFR takeoff or approach and landing at a military or foreign airport shall comply with applicable instrument approach procedures and weather minimums prescribed by the authority having jurisdiction over that airport. In addition, unless authorized by the certificate holder's operations specifications, no pilot may, at that airport—

(1) Take off under IFR when the visibility is less than 1 mile; or

(2) Make an instrument approach when the visibility is less than ½ mile.

(g) If takeoff minimums are specified in part 97 of this chapter for the take- off airport, no pilot may take off an aircraft under IFR when the weather conditions reported by the facility described in paragraph (a)(1) of this section are less than the takeoff minimums specified for the takeoff airport in part 97 or in the certificate holder's operations specifications.

(h) Except as provided in paragraph (i) of this section, if takeoff minimums are not prescribed in part 97 of this chapter for the takeoff airport, no pilot may takeoff an aircraft under IFR when the weather conditions reported by the facility described in paragraph (a)(1) of this section are less than that prescribed in part 91 of this chapter or in the certificate holder's operations specifications.

(i) At airports where straight-in instrument approach procedures are authorized, a pilot may takeoff an aircraft under IFR when the weather conditions reported by the facility described in paragraph (a)(1) of this section are equal to or better than the lowest straight-in landing minimums, unless otherwise restricted, if—

(1) The wind direction and velocity at the time of takeoff are such that a straight-in instrument approach can be made to the runway served by the instrument approach;

(2) The associated ground facilities upon which the landing minimums are predicated and the related airborne equipment are in normal operation; and

(3) The certificate holder has been approved for such operations.

[Doc. No. 16097, 43 FR 46783, Oct. 10, 1978, as amended by Amdt. 135-91, 68 FR 54586, Sept. 17, 2003; Amdt. 135-93, 69 FR 1641, Jan. 9, 2004; Amdt. 135-110, 72 FR 31685, June 7, 2007; Amdt. 135-126, 77 FR 1632, Jan. 11, 2012]

§ 135.227 Icing conditions: Operating limitations.

(a) No pilot may take off an aircraft that has frost, ice, or snow adhering to any rotor blade, propeller, windshield, stabilizing or control surface; to a powerplant installation; or to an airspeed, altimeter, rate of climb, flight attitude instrument system, or wing, except that takeoffs may be made with frost under the wing in the area of the fuel tanks if authorized by the FAA.

(b) No certificate holder may authorize an airplane to take off and no pilot may take off an airplane any time conditions are such that frost, ice, or snow may reasonably be expected to adhere to the airplane unless the pilot has completed all applicable training as required by § 135.341 and unless one of the following requirements is met:

(1) A pretakeoff contamination check, that has been established by the certificate holder and approved by the Administrator for the specific airplane type, has been completed within 5 minutes prior to beginning takeoff. A pretakeoff contamination check is a check to make sure the wings and control surfaces are free of frost, ice, or snow.

(2) The certificate holder has an approved alternative procedure and under that procedure the airplane is determined to be free of frost, ice, or snow.

(3) The certificate holder has an approved deicing/anti-icing program that complies with § 121.629(c) of this chapter and the takeoff complies with that program.

(c) No pilot may fly under IFR into known or forecast light or moderate icing conditions or under VFR into known light or moderate icing conditions, unless—

(1) The aircraft has functioning deicing or anti-icing equipment protecting each rotor blade, propeller, windshield, wing, stabilizing or control surface, and each airspeed, altimeter, rate of climb, or flight attitude instrument system;

(2) The airplane has ice protection provisions that meet section 34 of appendix A of this part; or

(3) The airplane meets transport category airplane type certification provisions, including the requirements for certification for flight in icing conditions.

(d) No pilot may fly a helicopter under IFR into known or forecast icing conditions or under VFR into known icing conditions unless it has been type certificated and appropriately equipped for operations in icing conditions.

(e) Except for an airplane that has ice protection provisions that meet section 34 of appendix A, or those for transport category airplane type certification, no pilot may fly an aircraft into known or forecast severe icing conditions.

(f) If current weather reports and briefing information relied upon by the pilot in command indicate that the forecast icing condition that would otherwise prohibit the flight will not be encountered during the flight because of changed weather conditions since the forecast, the restrictions in paragraphs (c), (d), and (e) of this section based on forecast conditions do not apply.

[Doc. No. 16097, 43 FR 46783, Oct. 10, 1978, as amended by Amdt. 133-20, 51 FR 40710, Nov. 7, 1986; Amdt. 135-46, 58 FR 69629, Dec. 30, 1993; Amdt. 135-60, 61 FR 2616, Jan. 26, 1996; Amdt. 135-119, 74 FR 62696, Dec. 1, 2009]

§ 135.229 Airport requirements.

(a) No certificate holder may use any airport unless it is adequate for the proposed operation, considering such items as size, surface, obstructions, and lighting.

(b) No pilot of an aircraft carrying passengers at night may takeoff from, or land on, an airport unless—

(1) That pilot has determined the wind direction from an illuminated wind direction indicator or local ground communications or, in the case of takeoff, that pilot's personal observations; and

(2) The limits of the area to be used for landing or takeoff are clearly shown—

(i) For airplanes, by boundary or runway marker lights;

(ii) For helicopters, by boundary or runway marker lights or reflective material.

(c) For the purpose of paragraph (b) of this section, if the area to be used for takeoff or landing is marked by flare pots or lanterns, their use must be approved by the Administrator.

Subpart E—Flight Crewmember Requirements

§ 135.241 Applicability.

Except as provided in § 135.3, this subpart prescribes the flight crewmember requirements for operations under this part.

[Doc. No. 16097, 43 FR 46783, Oct. 10, 1978, as amended by Amdt. 121-250, 60 FR 65950, Dec. 20, 1995]

§ 135.243 Pilot in command qualifications.

(a) No certificate holder may use a person, nor may any person serve, as pilot in command in passenger-carrying operations—

(1) Of a turbojet airplane, of an airplane having a passenger-seat configuration, excluding each crewmember seat, of 10 seats or more, or of a multiengine airplane in a commuter operation as defined in part 119 of this chapter, unless that person holds an airline transport pilot certificate with appropriate category and class ratings and, if required, an appropriate type rating for that airplane.

(2) Of a helicopter in a scheduled interstate air transportation operation by an air carrier within the 48 contiguous states unless that person holds an airline transport pilot certificate, appropriate type ratings, and an instrument rating.

(b) Except as provided in paragraph (a) of this section, no certificate holder may use a person, nor may any person serve, as pilot in command of an aircraft under VFR unless that person—

(1) Holds at least a commercial pilot certificate with appropriate category and class ratings and, if required, an appropriate type rating for that aircraft; and

(2) Has had at least 500 hours time as a pilot, including at least 100 hours of cross-country flight time, at least 25 hours of which were at night; and

(3) For an airplane, holds an instrument rating or an airline transport pilot certificate with an airplane category rating; or

(4) For helicopter operations conducted VFR over-the-top, holds a helicopter instrument rating, or an airline transport pilot certificate with a category and class rating for that aircraft, not limited to VFR.

(c) Except as provided in paragraph (a) of this section, no certificate holder may use a person, nor may any person serve, as pilot in command of an aircraft under IFR unless that person—

(1) Holds at least a commercial pilot certificate with appropriate category and class ratings and, if required, an appropriate type rating for that aircraft; and

(2) Has had at least 1,200 hours of flight time as a pilot, including 500 hours of cross country flight time, 100 hours of night flight time, and 75 hours of actual or simulated instrument time at least 50 hours of which were in actual flight; and

(3) For an airplane, holds an instrument rating or an airline transport pilot certificate with an airplane category rating; or

(4) For a helicopter, holds a helicopter instrument rating, or an airline transport pilot certificate with a category and class rating for that aircraft, not limited to VFR.

(d) Paragraph (b)(3) of this section does not apply when—

(1) The aircraft used is a single reciprocating-engine-powered airplane;

(2) The certificate holder does not conduct any operation pursuant to a published flight schedule which specifies five or more round trips a week between two or more points and places between which the round trips are performed, and does not transport mail by air under a contract or contracts with the United States Postal Service having total amount estimated at the beginning of any semiannual reporting period (January 1-June 30; July 1-December 31) to be in excess of $20,000 over the 12 months commencing with the beginning of the reporting period;

(3) The area, as specified in the certificate holder's operations specifications, is an isolated area, as determined by the Flight Standards district office, if it is shown that—

(i) The primary means of navigation in the area is by pilotage, since radio navigational aids are largely ineffective; and

(ii) The primary means of transportation in the area is by air;

(4) Each flight is conducted under day VFR with a ceiling of not less than 1,000 feet and visibility not less than 3 statute miles;

(5) Weather reports or forecasts, or any combination of them, indicate that for the period commencing with the planned departure and ending 30 minutes after the planned arrival at the destination the flight may be conducted under VFR with a ceiling of not less than 1,000 feet and visibility of not less than 3 statute miles, except that if weather reports and forecasts are not available, the pilot in command may use that pilot's observations or those of other persons competent to supply weather observations if those observations indicate the flight may be conducted under VFR with the ceiling and visibility required in this paragraph;

(6) The distance of each flight from the certificate holder's base of operation to destination does not exceed 250 nautical miles for a pilot who holds a commercial pilot certificate with an airplane rating without an instrument rating, provided the pilot's certificate does not contain any limitation to the contrary; and

(7) The areas to be flown are approved by the certificate-holding FAA Flight Standards district office and are listed in the certificate holder's operations specifications.

[Doc. No. 16097, 43 FR 46783, Oct. 10, 1978; Amdt. 135-1, 43 FR 49975, Oct. 26, 1978, as amended by Amdt. 135-15, 46 FR 30971, June 11, 1981; Amdt. 135-58, 60 FR 65939, Dec. 20, 1995]

§ 135.244 Operating experience.

(a) No certificate holder may use any person, nor may any person serve, as a pilot in command of an aircraft operated in a commuter operation, as defined in part 119 of this chapter unless that person has completed, prior to designation as pilot in command, on that make and basic model aircraft and in that crewmember position, the following operating experience in each make and basic model of aircraft to be flown:

(1) Aircraft, single engine—10 hours.

(2) Aircraft multiengine, reciprocating engine-powered—15 hours.

(3) Aircraft multiengine, turbine engine-powered—20 hours.

(4) Airplane, turbojet-powered—25 hours.

(b) In acquiring the operating experience, each person must comply with the following:

(1) The operating experience must be acquired after satisfactory completion of the appropriate ground and flight training for the aircraft and crewmember position. Approved provisions for the operating experience must be included in the certificate holder's training program.

(2) The experience must be acquired in flight during commuter passenger-carrying operations under this part. However, in the case of an aircraft not previously used by the certificate holder in operations under this part, operating experience acquired in the aircraft during proving flights or ferry flights may be used to meet this requirement.

(3) Each person must acquire the operating experience while performing the duties of a pilot in command under the supervision of a qualified check pilot.

(4) The hours of operating experience may be reduced to not less than 50 percent of the hours required by this section by the substitution of one additional takeoff and landing for each hour of flight.

[Doc. No. 20011, 45 FR 7541, Feb. 4, 1980, as amended by Amdt. 135-9, 45 FR 80461, Dec. 14, 1980; Amdt. 135-58, 60 FR 65940, Dec. 20, 1995]

§ 135.245 Second in command qualifications.

(a) Except as provided in paragraph (b), no certificate holder may use any person, nor may any person serve, as second in command of an aircraft unless that person holds at least a commercial pilot certificate with appropriate category and class ratings and an instrument rating. For flight under IFR, that person must meet the recent instrument experience requirements of part 61 of this chapter.

(b) A second in command of a helicopter operated under VFR, other than over-the-top, must have at least a commercial pilot certificate with an appropriate aircraft category and class rating.

[44 FR 26738, May 7, 1979]

§ 135.247 Pilot qualifications: Recent experience.

(a) No certificate holder may use any person, nor may any person serve, as pilot in command of an aircraft carrying passengers unless, within the preceding 90 days, that person has—

(1) Made three takeoffs and three landings as the sole manipulator of the flight controls in an aircraft of the same category and class and, if a type rating is required, of the same type in which that person is to serve; or

(2) For operation during the period beginning 1 hour after sunset and ending 1 hour before sunrise (as published in the Air Almanac), made three takeoffs and three landings during that period as the sole manipulator of the flight controls in an aircraft of the same category and class and, if a type rating is required, of the same type in which that person is to serve.

A person who complies with paragraph (a)(2) of this section need not comply with paragraph (a)(1) of this section.

(3) Paragraph (a)(2) of this section does not apply to a pilot in command of a turbine-powered airplane that is type certificated for more than one pilot crewmember, provided that pilot has complied with the requirements of paragraph (a)(3)(i) or (ii) of this section:

(i) The pilot in command must hold at least a commercial pilot certificate with the appropriate category, class, and type rating for each airplane that is type certificated for more than one pilot crewmember that the pilot seeks to operate under this alternative, and:

(A) That pilot must have logged at least 1,500 hours of aeronautical experience as a pilot;

(B) In each airplane that is type certificated for more than one pilot crewmember that the pilot seeks to operate under this alternative, that pilot must have accomplished and logged the daytime takeoff and landing recent flight experience of paragraph (a) of this section, as the sole manipulator of the flight controls;

(C) Within the preceding 90 days prior to the operation of that airplane that is type certificated for more than one pilot crewmember, the pilot must have accomplished and logged at least 15 hours of flight time in the type of airplane that the pilot seeks to operate under this alternative; and

(D) That pilot has accomplished and logged at least 3 takeoffs and 3 landings to a full stop, as the sole manipulator of the flight controls, in a turbine-powered airplane that requires more than one pilot crewmember. The pilot must have performed the takeoffs and landings during the period beginning 1 hour after sunset and ending 1 hour before sunrise within the preceding 6 months prior to the month of the flight.

(ii) The pilot in command must hold at least a commercial pilot certificate with the appropriate category, class, and type rating for each airplane that is type certificated for more than one pilot crewmember that the pilot seeks to operate under this alternative, and:

(A) That pilot must have logged at least 1,500 hours of aeronautical experience as a pilot;

(B) In each airplane that is type certificated for more than one pilot crewmember that the pilot seeks to operate under this alternative, that pilot must have accomplished and logged the daytime takeoff and landing recent flight experience of paragraph (a) of this section, as the sole manipulator of the flight controls;

(C) Within the preceding 90 days prior to the operation of that airplane that is type certificated for more than one pilot crewmember, the pilot must have accomplished and logged at least 15 hours of flight time in the type of airplane that the pilot seeks to operate under this alternative; and

(D) Within the preceding 12 months prior to the month of the flight, the pilot must have completed a training program that is approved under part 142 of this chapter. The approved training program must have required and the pilot must have performed, at least 6 takeoffs and 6 landings to a full stop as the sole manipulator of the controls in a flight simulator that is representative of a turbine-powered airplane that requires more than one pilot crewmember. The flight simulator's visual system must have been adjusted to represent the period beginning 1 hour after sunset and ending 1 hour before sunrise.

(b) For the purpose of paragraph (a) of this section, if the aircraft is a tailwheel airplane, each takeoff must be made in a tailwheel airplane and each landing must be made to a full stop in a tailwheel airplane.

[Doc. No. 16097, 43 FR 46783, Oct. 10, 1978, as amended by Amdt. 135-91, 68 FR 54587, Sept. 17, 2003]

§§ 135.249-135.255 [Reserved]

Subpart F—Crewmember Flight Time and Duty Period Limitations and Rest Requirements

SOURCE: Docket No. 23634, 50 FR 29320, July 18, 1985, unless otherwise noted.

§ 135.261 Applicability.

Sections 135.263 through 135.273 of this part prescribe flight time limitations, duty period limitations, and rest requirements for operations conducted under this part as follows:

(a) Section 135.263 applies to all operations under this subpart.

(b) Section 135.265 applies to:

(1) Scheduled passenger-carrying operations except those conducted solely within the state of Alaska. "Scheduled passenger-carrying operations" means passenger-carrying operations that are conducted in accordance with a published schedule which covers at least five round trips per week on at least one route between two or more points, includes dates or times (or both), and is openly advertised or otherwise made readily available to the general public, and

(2) Any other operation under this part, if the operator elects to comply with § 135.265 and obtains an appropriate operations specification amendment.

(c) Sections 135.267 and 135.269 apply to any operation that is not a scheduled passenger-carrying operation and to any operation conducted solely within the State of Alaska, unless the operator elects to comply with § 135.265 as authorized under paragraph (b)(2) of this section.

(d) Section 135.271 contains special daily flight time limits for operations conducted under the helicopter emergency medical evacuation service (HEMES).

(e) Section 135.273 prescribes duty period limitations and rest requirements for flight attendants in all operations conducted under this part.

[Doc. No. 23634, 50 FR 29320, July 18, 1985, as amended by Amdt. 135-52, 59 FR 42993, Aug. 19, 1994]

§ 135.263 Flight time limitations and rest requirements: All certificate holders.

(a) A certificate holder may assign a flight crewmember and a flight crewmember may accept an assignment for flight time only when the applicable requirements of §§ 135.263 through 135.271 are met.

(b) No certificate holder may assign any flight crewmember to any duty with the certificate holder during any required rest period.

(c) Time spent in transportation, not local in character, that a certificate holder requires of a flight crewmember and provides to transport the crewmember to an airport at which he is to serve on a flight as a crewmember, or from an airport at which he was relieved from duty to return to his home station, is not considered part of a rest period.

(d) A flight crewmember is not considered to be assigned flight time in excess of flight time limitations if the flights to which he is assigned normally terminate within the limitations, but due to circumstances beyond the control of the certificate holder or flight crewmember (such as adverse weather conditions), are not at the time of departure expected to reach their destination within the planned flight time.

§ 135.265 Flight time limitations and rest requirements: Scheduled operations.

(a) No certificate holder may schedule any flight crewmember, and no flight crewmember may accept an assignment, for flight time in scheduled operations or in other commercial flying if that crewmember's total flight time in all commercial flying will exceed—

(1) 1,200 hours in any calendar year.

(2) 120 hours in any calendar month.

(3) 34 hours in any 7 consecutive days.

(4) 8 hours during any 24 consecutive hours for a flight crew consisting of one pilot.

(5) 8 hours between required rest periods for a flight crew consisting of two pilots qualified under this part for the operation being conducted.

(b) Except as provided in paragraph (c) of this section, no certificate holder may schedule a flight crewmember, and no flight crewmember may accept an assignment, for flight time during the 24 consecutive hours preceding the scheduled completion of any flight segment without a scheduled rest period during that 24 hours of at least the following:

(1) 9 consecutive hours of rest for less than 8 hours of scheduled flight time.

(2) 10 consecutive hours of rest for 8 or more but less than 9 hours of scheduled flight time.

(3) 11 consecutive hours of rest for 9 or more hours of scheduled flight time.

(c) A certificate holder may schedule a flight crewmember for less than the rest required in paragraph (b) of this section or may reduce a scheduled rest under the following conditions:

(1) A rest required under paragraph (b)(1) of this section may be scheduled for or reduced to a minimum of 8 hours if the flight crewmember is given a rest period of at least 10 hours that must begin no later than 24 hours after the commencement of the reduced rest period.

(2) A rest required under paragraph (b)(2) of this section may be scheduled for or reduced to a minimum of 8 hours if the flight crewmember is given a rest period of at least 11 hours that must begin no later than 24 hours after the commencement of the reduced rest period.

(3) A rest required under paragraph (b)(3) of this section may be scheduled for or reduced to a minimum of 9 hours if the flight crewmember is given a rest period of at least 12 hours that must begin no later than 24 hours after the commencement of the reduced rest period.

(d) Each certificate holder shall relieve each flight crewmember engaged in scheduled air transportation from all further duty for at least 24 consecutive hours during any 7 consecutive days.

§ 135.267 Flight time limitations and rest requirements: Unscheduled one- and two-pilot crews.

(a) No certificate holder may assign any flight crewmember, and no flight crewmember may accept an assignment, for flight time as a member of a one- or two-pilot crew if that crewmember's total flight time in all commercial flying will exceed—

(1) 500 hours in any calendar quarter.

(2) 800 hours in any two consecutive calendar quarters.

(3) 1,400 hours in any calendar year.

(b) Except as provided in paragraph (c) of this section, during any 24 consecutive hours the total flight time of the assigned flight when added to any other commercial flying by that flight crewmember may not exceed—

(1) 8 hours for a flight crew consisting of one pilot; or

(2) 10 hours for a flight crew consisting of two pilots qualified under this part for the operation being conducted.

(c) A flight crewmember's flight time may exceed the flight time limits of paragraph (b) of this section if the assigned flight time occurs during a regularly assigned duty period of no more than 14 hours and—

(1) If this duty period is immediately preceded by and followed by a required rest period of at least 10 consecutive hours of rest;

(2) If flight time is assigned during this period, that total flight time when added to any other commercial flying by the flight crewmember may not exceed—

(i) 8 hours for a flight crew consisting of one pilot; or

(ii) 10 hours for a flight crew consisting of two pilots; and

(3) If the combined duty and rest periods equal 24 hours.

(d) Each assignment under paragraph (b) of this section must provide for at least 10 consecutive hours of rest during the 24-hour period that precedes the planned completion time of the assignment.

(e) When a flight crewmember has exceeded the daily flight time limitations in this section, because of circumstances beyond the control of the certificate holder or flight crewmember (such as adverse weather conditions), that flight crewmember must have a rest period before being assigned or accepting an assignment for flight time of at least—

(1) 11 consecutive hours of rest if the flight time limitation is exceeded by not more than 30 minutes;

(2) 12 consecutive hours of rest if the flight time limitation is exceeded by more than 30 minutes, but not more than 60 minutes; and

(3) 16 consecutive hours of rest if the flight time limitation is exceeded by more than 60 minutes.

(f) The certificate holder must provide each flight crewmember at least 13 rest periods of at least 24 consecutive hours each in each calendar quarter.

[Doc. No. 23634, 50 FR 29320, July 18, 1985, as amended by Amdt. 135-33, 54 FR 39294, Sept. 25, 1989; Amdt. 135-60, 61 FR 2616, Jan. 26, 1996]

§ 135.269 Flight time limitations and rest requirements: Unscheduled three- and four-pilot crews.

(a) No certificate holder may assign any flight crewmember, and no flight crewmember may accept an assignment, for flight time as a member of a three- or four-pilot crew if that crewmember's total flight time in all commercial flying will exceed—

(1) 500 hours in any calendar quarter.

(2) 800 hours in any two consecutive calendar quarters.

(3) 1,400 hours in any calendar year.

(b) No certificate holder may assign any pilot to a crew of three or four pilots, unless that assignment provides—

(1) At least 10 consecutive hours of rest immediately preceding the assignment;

(2) No more than 8 hours of flight deck duty in any 24 consecutive hours;

(3) No more than 18 duty hours for a three-pilot crew or 20 duty hours for a four-pilot crew in any 24 consecutive hours;

(4) No more than 12 hours aloft for a three-pilot crew or 16 hours aloft for a four-pilot crew during the maximum duty hours specified in paragraph (b)(3) of this section;

(5) Adequate sleeping facilities on the aircraft for the relief pilot;

(6) Upon completion of the assignment, a rest period of at least 12 hours;

(7) For a three-pilot crew, a crew which consists of at least the following:

(i) A pilot in command (PIC) who meets the applicable flight crewmember requirements of subpart E of part 135;

(ii) A PIC who meets the applicable flight crewmember requirements of subpart E of part 135, except those prescribed in §§ 135.244 and 135.247; and

(iii) A second in command (SIC) who meets the SIC qualifications of § 135.245.

(8) For a four-pilot crew, at least three pilots who meet the conditions of paragraph (b)(7) of this section, plus a fourth pilot who meets the SIC qualifications of § 135.245.

(c) When a flight crewmember has exceeded the daily flight deck duty limitation in this section by more than 60 minutes, because of circumstances beyond the control of the certificate holder or flight crewmember, that flight crewmember must have a rest period before the next duty period of at least 16 consecutive hours.

(d) A certificate holder must provide each flight crewmember at least 13 rest periods of at least 24 consecutive hours each in each calendar quarter.

§ 135.271 Helicopter hospital emergency medical evacuation service (HEMES).

(a) No certificate holder may assign any flight crewmember, and no flight crewmember may accept an assignment for flight time if that crewmember's total flight time in all commercial flight will exceed—

(1) 500 hours in any calendar quarter.

(2) 800 hours in any two consecutive calendar quarters.

(3) 1,400 hours in any calendar year.

(b) No certificate holder may assign a helicopter flight crewmember, and no flight crewmember may accept an assignment, for hospital emergency medical evacuation service helicopter operations unless that assignment provides for at least 10 consecutive hours of rest immediately preceding reporting to the hospital for availability for flight time.

(c) No flight crewmember may accrue more than 8 hours of flight time during any 24-consecutive hour period of a HEMES assignment, unless an emergency medical evacuation operation is prolonged. Each flight crewmember who exceeds the daily 8 hour flight time limitation in this paragraph must be relieved of the HEMES assignment immediately upon the completion of that emergency medical evacuation operation and must be given a rest period in compliance with paragraph (h) of this section.

(d) Each flight crewmember must receive at least 8 consecutive hours of rest during any 24 consecutive hour period of a HEMES assignment. A flight crewmember must be relieved of the HEMES assignment if he or she has not or cannot receive at least 8 consecutive hours of rest during any 24 consecutive hour period of a HEMES assignment.

(e) A HEMES assignment may not exceed 72 consecutive hours at the hospital.

(f) An adequate place of rest must be provided at, or in close proximity to, the hospital at which the HEMES assignment is being performed.

(g) No certificate holder may assign any other duties to a flight crewmember during a HEMES assignment.

(h) Each pilot must be given a rest period upon completion of the HEMES assignment and prior to being assigned any further duty with the certificate holder of—

(1) At least 12 consecutive hours for an assignment of less than 48 hours.

(2) At least 16 consecutive hours for an assignment of more than 48 hours.

(i) The certificate holder must provide each flight crewmember at least 13 rest periods of at least 24 consecutive hours each in each calendar quarter.

§ 135.273 Duty period limitations and rest time requirements.

(a) For purposes of this section—

Calendar day means the period of elapsed time, using Coordinated Universal Time or local time, that begins at midnight and ends 24 hours later at the next midnight.

Duty period means the period of elapsed time between reporting for an assignment involving flight time and release from that assignment by the certificate holder. The time is calculated using either Coordinated Universal Time or local time to reflect the total elapsed time.

Flight attendant means an individual, other than a flight crewmember, who is assigned by the certificate holder, in accordance with the required minimum crew complement under the certificate holder's operations specifications or in addition to that minimum complement, to duty in an aircraft during flight time and whose duties include but are not necessarily limited to cabin-safety-related responsibilities.

Rest period means the period free of all responsibility for work or duty should the occasion arise.

(b) Except as provided in paragraph (c) of this section, a certificate holder may assign a duty period to a flight attendant only when the applicable duty period limitations and rest requirements of this paragraph are met.

(1) Except as provided in paragraphs (b)(4), (b)(5), and (b)(6) of this section, no certificate holder may assign a flight attendant to a scheduled duty period of more than 14 hours.

(2) Except as provided in paragraph (b)(3) of this section, a flight attendant scheduled to a duty period of 14 hours or less as provided under paragraph (b)(1) of this section must be given a scheduled rest period of at least 9 consecutive hours. This rest period must occur between the completion of the scheduled duty period and the commencement of the subsequent duty period.

(3) The rest period required under paragraph (b)(2) of this section may be scheduled or reduced to 8 consecutive hours if the flight attendant is provided a subsequent rest period of at least 10 consecutive hours; this subsequent rest period must be scheduled to begin no later than 24 hours after the beginning of the reduced rest period and must occur between the completion of the scheduled duty period and the commencement of the subsequent duty period.

(4) A certificate holder may assign a flight attendant to a scheduled duty period of more than 14 hours, but no more than 16 hours, if the certificate holder has assigned to the flight or flights in that duty period at least one flight attendant in addition to the minimum flight attendant complement required for the flight or flights in that duty period under the certificate holder's operations specifications.

(5) A certificate holder may assign a flight attendant to a scheduled duty period of more than 16 hours, but no more than 18 hours, if the certificate holder has assigned to the flight or flights in that duty period at least two flight attendants in addition to the minimum flight attendant complement required for the flight or flights in that duty period under the certificate holder's operations specifications.

(6) A certificate holder may assign a flight attendant to a scheduled duty period of more than 18 hours, but no more than 20 hours, if the scheduled duty period includes one or more flights that land or take off outside the 48 contiguous states and the District of Columbia, and if the certificate holder has assigned to the flight or flights in that duty period at least three flight attendants in addition to the minimum flight attendant complement required for the flight or flights in that duty period under the certificate holder's operations specifications.

(7) Except as provided in paragraph (b)(8) of this section, a flight attendant scheduled to a duty period of more than 14 hours but no more than 20 hours, as provided in paragraphs (b)(4), (b)(5), and (b)(6) of this section, must be given a scheduled rest period of at least 12 consecutive hours. This rest period must occur between the completion of the scheduled duty period and the commencement of the subsequent duty period.

(8) The rest period required under paragraph (b)(7) of this section may be scheduled or reduced to 10 consecutive hours if the flight attendant is provided a subsequent rest period of at least 14 consecutive hours; this subsequent rest period must be scheduled to begin no later than 24 hours after the beginning of the reduced rest period and must occur between the completion of the scheduled duty period and the commencement of the subsequent duty period.

(9) Notwithstanding paragraphs (b)(4), (b)(5), and (b)(6) of this section, if a certificate holder elects to reduce the rest period to 10 hours as authorized by paragraph (b)(8) of this section, the certificate holder may not schedule a flight attendant for a duty period of more than 14 hours during the 24-hour period commencing after the beginning of the reduced rest period.

(10) No certificate holder may assign a flight attendant any duty period with the certificate holder unless the flight attendant has had at least the minimum rest required under this section.

(11) No certificate holder may assign a flight attendant to perform any duty with the certificate holder during any required rest period.

(12) Time spent in transportation, not local in character, that a certificate holder requires of a flight attendant and provides to transport the flight attendant to an airport at which that flight attendant is to serve on a flight as a crewmember, or from an airport at which the flight attendant was relieved from duty to return to the flight attendant's home station, is not considered part of a rest period.

(13) Each certificate holder must relieve each flight attendant engaged in air transportation from all further duty for at least 24 consecutive hours during any 7 consecutive calendar days.

(14) A flight attendant is not considered to be scheduled for duty in excess of duty period limitations if the flights to which the flight attendant is assigned are scheduled and normally terminate within the limitations but due to circumstances beyond the control of the certificate holder (such as adverse weather conditions) are not at the time of departure expected to reach their destination within the scheduled time.

(c) Notwithstanding paragraph (b) of this section, a certificate holder may apply the flight crewmember flight time and duty limitations and rest requirements of this part to flight attendants for all operations conducted under this part provided that—

(1) The certificate holder establishes written procedures that—

(i) Apply to all flight attendants used in the certificate holder's operation;

(ii) Include the flight crewmember requirements contained in subpart F of this part, as appropriate to the operation being conducted, except that rest facilities on board the aircraft are not required; and

(iii) Include provisions to add one flight attendant to the minimum flight attendant complement for each flight crewmember who is in excess of the minimum number required in the aircraft type certificate data sheet and who is assigned to the aircraft under the provisions of subpart F of this part, as applicable.

(iv) Are approved by the Administrator and described or referenced in the certificate holder's operations specifications; and

(2) Whenever the Administrator finds that revisions are necessary for the continued adequacy of duty period limitation and rest requirement procedures that are required by paragraph (c)(1) of this section and that had been granted final approval, the certificate holder must, after notification by the Administrator, make any changes in the procedures that are found necessary by the Administrator. Within 30 days after the certificate holder receives such notice, it may file a petition to reconsider the notice with the certificate-holding district office. The filing of a petition to reconsider stays the notice, pending decision by the Administrator. However, if the Administrator finds that there is an emergency that requires immediate action in the interest of safety, the Administrator may, upon a statement of the reasons, require a change effective without stay.

[Amdt. 135-52, 59 FR 42993, Aug. 19, 1994, as amended by Amdt. 135-60, 61 FR 2616, Jan. 26, 1996]

Subpart G—Crewmember Testing Requirements

§ 135.291 Applicability.

Except as provided in § 135.3, this subpart—

(a) Prescribes the tests and checks required for pilot and flight attendant crewmembers and for the approval of check pilots in operations under this part; and

(b) Permits training center personnel authorized under part 142 of this chapter who meet the requirements of §§ 135.337 and 135.339 to conduct training, testing, and checking under contract or other arrangement to those persons subject to the requirements of this subpart.

[Doc. No. 26933, 61 FR 34561, July 2, 1996, as amended by Amdt. 135-91, 68 FR 54587, Sept. 17, 2003]

§ 135.293 Initial and recurrent pilot testing requirements.

(a) No certificate holder may use a pilot, nor may any person serve as a pilot, unless, since the beginning of the 12th calendar month before that service, that pilot has passed a written or oral test, given by the Administrator or an authorized check pilot, on that pilot's knowledge in the following areas—

(1) The appropriate provisions of parts 61, 91, and 135 of this chapter and the operations specifications and the manual of the certificate holder;

(2) For each type of aircraft to be flown by the pilot, the aircraft powerplant, major components and systems, major appliances, performance and operating limitations, standard and emergency operating procedures, and the contents of the approved Aircraft Flight Manual or equivalent, as applicable;

(3) For each type of aircraft to be flown by the pilot, the method of determining compliance with weight and balance limitations for takeoff, landing and en route operations;

(4) Navigation and use of air navigation aids appropriate to the operation or pilot authorization, including, when applicable, instrument approach facilities and procedures;

(5) Air traffic control procedures, including IFR procedures when applicable;

(6) Meteorology in general, including the principles of frontal systems, icing, fog, thunderstorms, and windshear, and, if appropriate for the operation of the certificate holder, high altitude weather;

(7) Procedures for—

(i) Recognizing and avoiding severe weather situations;

(ii) Escaping from severe weather situations, in case of inadvertent encounters, including low-altitude windshear (except that rotorcraft pilots are not required to be tested on escaping from low-altitude windshear); and

(iii) Operating in or near thunderstorms (including best penetrating altitudes), turbulent air (including clear air turbulence), icing, hail, and other potentially hazardous meteorological conditions; and

(8) New equipment, procedures, or techniques, as appropriate.

(b) No certificate holder may use a pilot, nor may any person serve as a pilot, in any aircraft unless, since the beginning of the 12th calendar month before that service, that pilot has passed a competency check given by the Administrator or an authorized check pilot in that class of aircraft, if single-engine airplane other than turbojet, or that type of aircraft, if helicopter, multiengine airplane, or turbojet airplane, to determine the pilot's competence in practical skills and techniques in that aircraft or class of aircraft. The extent of the competency check shall be determined by the Administrator or authorized check pilot conducting the competency check. The competency check may include any of the maneuvers and procedures currently required for the original issuance of the particular pilot certificate required for the operations authorized and appropriate to the category, class and type of aircraft involved. For the purposes of this paragraph, type, as to an airplane, means any one of a group of airplanes determined by the Administrator to have a similar means of propulsion, the same manufacturer, and no significantly different handling or flight characteristics. For the purposes of this paragraph, type, as to a helicopter, means a basic make and model.

(c) The instrument proficiency check required by § 135.297 may be substituted for the competency check required by this section for the type of aircraft used in the check.

(d) For the purpose of this part, competent performance of a procedure or maneuver by a person to be used as a pilot requires that the pilot be the obvious master of the aircraft, with the successful outcome of the maneuver never in doubt.

(e) The Administrator or authorized check pilot certifies the competency of each pilot who passes the knowledge or flight check in the certificate holder's pilot records.

(f) Portions of a required competency check may be given in an aircraft simulator or other appropriate training device, if approved by the Administrator.

[Doc. No. 16097, 43 FR 46783, Oct. 10, 1978, as amended by Amdt. 135-27, 53 FR 37697, Sept. 27, 1988]

§ 135.295 Initial and recurrent flight attendant crewmember testing requirements.

No certificate holder may use a flight attendant crewmember, nor may any person serve as a flight attendant crewmember unless, since the beginning of the 12th calendar month before that service, the certificate holder has determined by appropriate initial and recurrent testing that the person is knowledgeable and competent in the following areas as appropriate to assigned duties and responsibilities—

(a) Authority of the pilot in command;

(b) Passenger handling, including procedures to be followed in handling deranged persons or other persons whose conduct might jeopardize safety;

(c) Crewmember assignments, functions, and responsibilities during ditching and evacuation of persons who may need the assistance of another person to move expeditiously to an exit in an emergency;

(d) Briefing of passengers;

(e) Location and operation of portable fire extinguishers and other items of emergency equipment;

(f) Proper use of cabin equipment and controls;

(g) Location and operation of passenger oxygen equipment;

(h) Location and operation of all normal and emergency exits, including evacuation chutes and escape ropes; and

(i) Seating of persons who may need assistance of another person to move rapidly to an exit in an emergency as prescribed by the certificate holder's operations manual.

§ 135.297 Pilot in command: Instrument proficiency check requirements.

(a) No certificate holder may use a pilot, nor may any person serve, as a pilot in command of an aircraft under IFR unless, since the beginning of the 6th calendar month before that service, that pilot has passed an instrument proficiency check under this section administered by the Administrator or an authorized check pilot.

(b) No pilot may use any type of precision instrument approach procedure under IFR unless, since the beginning of the 6th calendar month before that use, the pilot satisfactorily demonstrated that type of approach procedure. No pilot may use any type of nonprecision approach procedure under IFR unless, since the beginning of the 6th calendar month before that use, the pilot has satisfactorily demonstrated either that type of approach procedure or any other two different types of nonprecision approach procedures. The instrument approach procedure or procedures must include at least one straight-in approach, one circling approach, and one missed approach. Each type of approach procedure demonstrated must be conducted to published minimums for that procedure.

(c) The instrument proficiency check required by paragraph (a) of this section consists of an oral or written equipment test and a flight check under simulated or actual IFR conditions. The equipment test includes questions on emergency procedures, engine operation, fuel and lubrication systems, power settings, stall speeds, best engine-out speed, propeller and supercharger operations, and hydraulic, mechanical, and electrical systems, as appropriate. The flight check includes navigation by instruments, recovery from simulated emergencies, and standard instrument approaches involving navigational facilities which that pilot is to be authorized to use. Each pilot taking the instrument proficiency check must show that standard of competence required by § 135.293(d).

(1) The instrument proficiency check must—

(i) For a pilot in command of an airplane under § 135.243(a), include the procedures and maneuvers for an airline transport pilot certificate in the particular type of airplane, if appropriate; and

(ii) For a pilot in command of an airplane or helicopter under § 135.243(c), include the procedures and maneuvers for a commercial pilot certificate with an instrument rating and, if required, for the appropriate type rating.

(2) The instrument proficiency check must be given by an authorized check airman or by the Administrator.

(d) If the pilot in command is assigned to pilot only one type of aircraft, that pilot must take the instrument proficiency check required by paragraph (a) of this section in that type of aircraft.

(e) If the pilot in command is assigned to pilot more than one type of aircraft, that pilot must take the instrument proficiency check required by paragraph (a) of this section in each type of aircraft to which that pilot is assigned, in rotation, but not more than one flight check during each period described in paragraph (a) of this section.

(f) If the pilot in command is assigned to pilot both single-engine and multiengine aircraft, that pilot must initially take the instrument proficiency check required by paragraph (a) of this section in a multiengine aircraft, and each succeeding check alternately in single-engine and multiengine aircraft, but not more than one flight check during each period described in paragraph (a) of this section. Portions of a required flight check may be given in an aircraft simulator or other appropriate training device, if approved by the Administrator.

(g) If the pilot in command is authorized to use an autopilot system in place of a second in command, that pilot must show, during the required instrument proficiency check, that the pilot is able (without a second in command) both with and without using the autopilot to—

(1) Conduct instrument operations competently; and

(2) Properly conduct air-ground communications and comply with complex air traffic control instructions.

(3) Each pilot taking the autopilot check must show that, while using the autopilot, the airplane can be operated as proficiently as it would be if a second in command were present to handle air-ground communications and air traffic control instructions. The autopilot check need only be demonstrated once every 12 calendar months during the instrument proficiency check required under paragraph (a) of this section.

[Doc. No. 16097, 43 FR 46783, Oct. 10, 1978, as amended by Amdt. 135-15, 46 FR 30971, June 11, 1981]

§ 135.299 Pilot in command: Line checks: Routes and airports.

(a) No certificate holder may use a pilot, nor may any person serve, as a pilot in command of a flight unless, since the beginning of the 12th calendar month before that service, that pilot has passed a flight check in one of the types of aircraft which that pilot is to fly. The flight check shall—

(1) Be given by an approved check pilot or by the Administrator;

(2) Consist of at least one flight over one route segment; and

(3) Include takeoffs and landings at one or more representative airports. In addition to the requirements of this paragraph, for a pilot authorized to conduct IFR operations, at least one flight shall be flown over a civil airway, an approved off-airway route, or a portion of either of them.

(b) The pilot who conducts the check shall determine whether the pilot being checked satisfactorily performs the duties and responsibilities of a pilot in command in operations under this part, and shall so certify in the pilot training record.

(c) Each certificate holder shall establish in the manual required by § 135.21 a procedure which will ensure that each pilot who has not flown over a route and into an airport within the preceding 90 days will, before beginning the flight, become familiar with all available information required for the safe operation of that flight.

§ 135.301 Crewmember: Tests and checks, grace provisions, training to accepted standards.

(a) If a crewmember who is required to take a test or a flight check under this part, completes the test or flight check in the calendar month before or after the calendar month in which it is required, that crewmember is considered to have completed the test or check in the calendar month in which it is required.

(b) If a pilot being checked under this subpart fails any of the required maneuvers, the person giving the check may give additional training to the pilot during the course of the check. In addition to repeating the maneuvers failed, the person giving the check may require the pilot being checked to repeat any other maneuvers that are necessary to determine the pilot's proficiency. If the pilot being checked is unable to demonstrate satisfactory performance to the person conducting the check, the certificate holder may not use the pilot, nor may the pilot serve, as a flight crewmember in operations under this part until the pilot has satisfactorily completed the check.

Subpart H—Training

§ 135.321 Applicability and terms used.

(a) Except as provided in § 135.3, this subpart prescribes the requirements applicable to—

(1) A certificate holder under this part which contracts with, or otherwise arranges to use the services of a training center certificated under part 142 to perform training, testing, and checking functions;

(2) Each certificate holder for establishing and maintaining an approved training program for crewmembers, check airmen and instructors, and other operations personnel employed or used by that certificate holder; and

(3) Each certificate holder for the qualification, approval, and use of aircraft simulators and flight training devices in the conduct of the program.

(b) For the purposes of this subpart, the following terms and definitions apply:

(1) *Initial training.* The training required for crewmembers who have not qualified and served in the same capacity on an aircraft.

(2) *Transition training.* The training required for crewmembers who have qualified and served in the same capacity on another aircraft.

(3) *Upgrade training.* The training required for crewmembers who have qualified and served as second in command on a particular aircraft type, before they serve as pilot in command on that aircraft.

(4) *Differences training.* The training required for crewmembers who have qualified and served on a particular type aircraft, when the Administrator finds differences training is necessary before a crewmember serves in the same capacity on a particular variation of that aircraft.

(5) *Recurrent training.* The training required for crewmembers to remain adequately trained and currently proficient for each aircraft, crewmember position, and type of operation in which the crewmember serves.

(6) *In flight.* The maneuvers, procedures, or functions that must be conducted in the aircraft.

(7) *Training center.* An organization governed by the applicable requirements of part 142 of this chapter that conducts training, testing, and checking under contract or other arrangement to certificate holders subject to the requirements of this part.

(8) *Requalification training.* The training required for crewmembers previously trained and qualified, but who have become unqualified due to not having met within the required period the—

(i) Recurrent pilot testing requirements of § 135.293;

(ii) Instrument proficiency check requirements of § 135.297; or

(iii) Line checks required by § 135.299.

[Doc. No. 16097, 43 FR 46783, Oct. 10, 1978, as amended by Amdt. 121-250, 60 FR 65950, Dec. 20, 1995; Amdt. 135-63, 61 FR 34561, July 2, 1996; Amdt. 135-91, 68 FR 54588, Sept. 17, 2003]

§ 135.323 Training program: General.

(a) Each certificate holder required to have a training program under § 135.341 shall:

(1) Establish and implement a training program that satisfies the requirements of this subpart and that ensures that each crewmember, aircraft dispatcher, flight instructor and check airman is adequately trained to perform his or her assigned duties. Prior to implementation, the certificate holder must obtain initial and final FAA approval of the training program.

(2) Provide adequate ground and flight training facilities and properly qualified ground instructors for the training required by this subpart.

(3) Provide and keep current for each aircraft type used and, if applicable, the particular variations within the aircraft type, appropriate training material, examinations, forms, instructions, and procedures for use in conducting the training and checks required by this subpart.

(4) Provide enough flight instructors, check airmen, and simulator instructors to conduct required flight training and flight checks, and simulator training courses allowed under this subpart.

(b) Whenever a crewmember who is required to take recurrent training under this subpart completes the training in the calendar month before, or the calendar month after, the month in which that training is required, the crewmember is considered to have completed it in the calendar month in which it was required.

(c) Each instructor, supervisor, or check airman who is responsible for a particular ground training subject, segment of flight training, course of training, flight check, or competence check under this part shall certify as to the proficiency and knowledge of the crewmember, flight instructor, or check airman concerned upon completion of that training or check. That certification shall be made a part of the crewmember's record. When the certification required by this paragraph is made by an entry in a computerized recordkeeping system, the certifying instructor, supervisor, or check airman, must be identified with that entry. However, the signature of the certifying instructor, supervisor, or check airman, is not required for computerized entries.

(d) Training subjects that apply to more than one aircraft or crewmember position and that have been satisfactorily completed during previous training while employed by the certificate holder for another aircraft or another crewmember position, need not be repeated during subsequent training other than recurrent training.

(e) Aircraft simulators and other training devices may be used in the certificate holder's training program if approved by the Administrator.

[Doc. No. 16097, 43 FR 46783, Oct. 10, 1978, as amended by Amdt. 135-101, 70 FR 58829, Oct. 7, 2005]

§ 135.324 Training program: Special rules.

(a) Other than the certificate holder, only another certificate holder certificated under this part or a training center certificated under part 142 of this chapter is eligible under this subpart to conduct training, testing, and checking under contract or other arrangement to those persons subject to the requirements of this subpart.

(b) A certificate holder may contract with, or otherwise arrange to use the services of, a training center certificated under part 142 of this chapter to conduct training, testing, and checking required by this part only if the training center—

(1) Holds applicable training specifications issued under part 142 of this chapter;

(2) Has facilities, training equipment, and courseware meeting the applicable requirements of part 142 of this chapter;

(3) Has approved curriculums, curriculum segments, and portions of curriculum segments applicable for use in training courses required by this subpart; and

(4) Has sufficient instructor and check airmen qualified under the applicable requirements of §§ 135.337 through 135.340 to provide training, testing, and checking to persons subject to the requirements of this subpart.

[Doc. No. 26933, 61 FR 34562, July 2, 1996, as amended by Amdt. 135-67, 62 FR 13791, Mar. 21, 1997; Amdt. 135-91, 68 FR 54588, Sept. 17, 2003]

§ 135.325 Training program and revision: Initial and final approval.

(a) To obtain initial and final approval of a training program, or a revision to an approved training program, each certificate holder must submit to the Administrator—

(1) An outline of the proposed or revised curriculum, that provides enough information for a preliminary evaluation of the proposed training program or revision; and

(2) Additional relevant information that may be requested by the Administrator.

(b) If the proposed training program or revision complies with this subpart, the Administrator grants initial approval in writing after which the certificate holder may conduct the training under that program. The Administrator then evaluates the effectiveness of the training program and advises the certificate holder of deficiencies, if any, that must be corrected.

(c) The Administrator grants final approval of the proposed training program or revision if the certificate holder shows that the training conducted under the initial approval in paragraph (b) of this section ensures that each person who successfully completes the training is adequately trained to perform that person's assigned duties.

(d) Whenever the Administrator finds that revisions are necessary for the continued adequacy of a training program that has been granted final approval, the certificate holder shall, after notification by the Administrator, make any changes in the program that are found necessary by the Administrator. Within 30 days after the certificate holder receives the notice, it may file a petition to reconsider the notice with the Administrator. The filing of a petition to reconsider stays the notice pending a decision by the Administrator. However, if the Administrator finds that there is an emergency that requires immediate action in the interest of safety, the Administrator may, upon a statement of the reasons, require a change effective without stay.

§ 135.327 Training program: Curriculum.

(a) Each certificate holder must prepare and keep current a written training program curriculum for each type of aircraft for each crewmember required for that type aircraft. The curriculum must include ground and flight training required by this subpart.

(b) Each training program curriculum must include the following:

(1) A list of principal ground training subjects, including emergency training subjects, that are provided.

(2) A list of all the training devices, mockups, systems trainers, procedures trainers, or other training aids that the certificate holder will use.

(3) Detailed descriptions or pictorial displays of the approved normal, abnormal, and emergency maneuvers, procedures and functions that will be performed during each flight training phase or flight check, indicating those maneuvers, procedures and functions that are to be performed during the inflight portions of flight training and flight checks.

§ 135.329 Crewmember training requirements.

(a) Each certificate holder must include in its training program the following initial and transition ground training as appropriate to the particular assignment of the crewmember:

(1) Basic indoctrination ground training for newly hired crewmembers including instruction in at least the—

(i) Duties and responsibilities of crewmembers as applicable;

(ii) Appropriate provisions of this chapter;

(iii) Contents of the certificate holder's operating certificate and operations specifications (not required for flight attendants); and

(iv) Appropriate portions of the certificate holder's operating manual.

(2) The initial and transition ground training in §§ 135.345 and 135.349, as applicable.

(3) Emergency training in § 135.331.

(4) Crew resource management training in § 135.330.

(b) Each training program must provide the initial and transition flight training in § 135.347, as applicable.

(c) Each training program must provide recurrent ground and flight training in § 135.351.

(d) Upgrade training in §§ 135.345 and 135.347 for a particular type aircraft may be included in the training program for crewmembers who have qualified and served as second in command on that aircraft.

(e) In addition to initial, transition, upgrade and recurrent training, each training program must provide ground and flight training, instruction, and practice necessary to ensure that each crewmember—

(1) Remains adequately trained and currently proficient for each aircraft, crewmember position, and type of operation in which the crewmember serves; and

(2) Qualifies in new equipment, facilities, procedures, and techniques, including modifications to aircraft.

[Docket No. 16097, 43 FR 46783, Oct. 10, 1978, as amended by Amdt. 135-122, 76 FR 3837, Jan. 21, 2011]

§ 135.330 Crew resource management training.

(a) Each certificate holder must have an approved crew resource management training program that includes initial and recurrent training. The training program must include at least the following:

(1) Authority of the pilot in command;

(2) Communication processes, decisions, and coordination, to include communication with Air Traffic Control, personnel performing flight locating and other operational functions, and passengers;

(3) Building and maintenance of a flight team;

(4) Workload and time management;

(5) Situational awareness;

(6) Effects of fatigue on performance, avoidance strategies and countermeasures;

(7) Effects of stress and stress reduction strategies; and

(8) Aeronautical decision-making and judgment training tailored to the operator's flight operations and aviation environment.

(b) After March 22, 2013, no certificate holder may use a person as a flightcrew member or flight attendant unless that person has completed approved crew resource management initial training with that certificate holder.

(c) For flightcrew members and flight attendants, the Administrator, at his or her discretion, may credit crew resource management training completed with that certificate holder before March 22, 2013, toward all or part of the initial CRM training required by this section.

(d) In granting credit for initial CRM training, the Administrator considers training aids, devices, methods and procedures used by the certificate holder in a voluntary CRM program included in a training program required by § 135.341, § 135.345, or § 135.349.

[Docket No. FAA-2009-0023, 76 FR 3837, Jan. 21, 2011]

§ 135.331 Crewmember emergency training.

(a) Each training program must provide emergency training under this section for each aircraft type, model, and configuration, each crewmember, and each kind of operation conducted, as appropriate for each crewmember and the certificate holder.

(b) Emergency training must provide the following:

(1) Instruction in emergency assignments and procedures, including coordination among crewmembers.

(2) Individual instruction in the location, function, and operation of emergency equipment including—

(i) Equipment used in ditching and evacuation;

(ii) First aid equipment and its proper use; and

(iii) Portable fire extinguishers, with emphasis on the type of extinguisher to be used on different classes of fires.

(3) Instruction in the handling of emergency situations including—

(i) Rapid decompression;

(ii) Fire in flight or on the surface and smoke control procedures with emphasis on electrical equipment and related circuit breakers found in cabin areas;

(iii) Ditching and evacuation;

(iv) Illness, injury, or other abnormal situations involving passengers or crewmembers; and

(v) Hijacking and other unusual situations.

(4) Review of the certificate holder's previous aircraft accidents and incidents involving actual emergency situations.

(c) Each crewmember must perform at least the following emergency drills, using the proper emergency equipment and procedures, unless the Administrator finds that, for a particular drill, the crewmember can be adequately trained by demonstration:

(1) Ditching, if applicable.

(2) Emergency evacuation.

(3) Fire extinguishing and smoke control.

(4) Operation and use of emergency exits, including deployment and use of evacuation chutes, if applicable.

(5) Use of crew and passenger oxygen.

(6) Removal of life rafts from the aircraft, inflation of the life rafts, use of life lines, and boarding of passengers and crew, if applicable.

(7) Donning and inflation of life vests and the use of other individual flotation devices, if applicable.

(d) Crewmembers who serve in operations above 25,000 feet must receive instruction in the following:

(1) Respiration.

(2) Hypoxia.

(3) Duration of consciousness without supplemental oxygen at altitude.

(4) Gas expansion.

(5) Gas bubble formation.

(6) Physical phenomena and incidents of decompression.

§ 135.335 Approval of aircraft simulators and other training devices.

(a) Training courses using aircraft simulators and other training devices may be included in the certificate holder's training program if approved by the Administrator.

(b) Each aircraft simulator and other training device that is used in a training course or in checks required under this subpart must meet the following requirements:

(1) It must be specifically approved for—

(i) The certificate holder; and

(ii) The particular maneuver, procedure, or crewmember function involved.

(2) It must maintain the performance, functional, and other character- istics that are required for approval.

(3) Additionally, for aircraft simulators, it must be—

(i) Approved for the type aircraft and, if applicable, the particular variation within type for which the training or check is being conducted; and

(ii) Modified to conform with any modification to the aircraft being simulated that changes the performance, functional, or other characteristics required for approval.

(c) A particular aircraft simulator or other training device may be used by more than one certificate holder.

(d) In granting initial and final approval of training programs or revisions to them, the Administrator considers the training devices, methods and procedures listed in the certificate holder's curriculum under § 135.327.

[Doc. No. 16907, 43 FR 46783, Oct. 10, 1978, as amended by Amdt. 135-1, 44 FR 26738, May 7, 1979]

§ 135.336 Airline transport pilot certification training program.

(a) A certificate holder may obtain approval to establish and implement a training program to satisfy the requirements of § 61.156 of this chapter. The training program must be separate from the air carrier training program required by this part.

(b) No certificate holder may use a person nor may any person serve as an instructor in a training program approved to meet the requirements of § 61.156 of this chapter unless the instructor:

(1) Holds an airline transport pilot certificate with an airplane category multiengine class rating;

(2) Has at least 2 years of experience as a pilot in command in operations conducted under § 91.1053(a)(2)(i) of this chapter, § 135.243(a)(1) of this part, or as a pilot in command or second in command in any operation conducted under part 121 of this chapter;

(3) Except for the holder of a flight instructor certificate, receives initial training on the following topics:

(i) The fundamental principles of the learning process;

(ii) Elements of effective teaching, instruction methods, and techniques;

(iii) Instructor duties, privileges, responsibilities, and limitations;

(iv) Training policies and procedures; and

(v) Evaluation.

(4) If providing training in a flight simulation training device, holds an aircraft type rating for the aircraft represented by the flight simulation training device utilized in the training program and have received training and evaluation within the preceding 12 months from the certificate holder on:

(i) Proper operation of flight simulator and flight training device controls and systems;

(ii) Proper operation of environmental and fault panels;

(iii) Data and motion limitations of simulation;

(iv) Minimum equipment requirements for each curriculum; and

(v) The maneuvers that will be demonstrated in the flight simulation training device.

(c) A certificate holder may not issue a graduation certificate to a student unless that student has completed all the curriculum requirements of the course.

(d) A certificate holder must conduct evaluations to ensure that training techniques, procedures, and standards are acceptable to the Administrator.

[Doc. No. FAA-2010-0100, 78 FR 42379, July 15, 2013]

§ 135.337 Qualifications: Check airmen (aircraft) and check airmen (simulator).

(a) For the purposes of this section and § 135.339:

(1) A check airman (aircraft) is a person who is qualified to conduct flight checks in an aircraft, in a flight simulator, or in a flight training device for a particular type aircraft.

(2) A check airman (simulator) is a person who is qualified to conduct flight checks, but only in a flight simulator, in a flight training device, or both, for a particular type aircraft.

(3) Check airmen (aircraft) and check airmen (simulator) are those check airmen who perform the functions described in §§ 135.321 (a) and 135.323(a)(4) and (c).

(b) No certificate holder may use a person, nor may any person serve as a check airman (aircraft) in a training program established under this subpart unless, with respect to the aircraft type involved, that person—

(1) Holds the airman certificates and ratings required to serve as a pilot in command in operations under this part;

(2) Has satisfactorily completed the training phases for the aircraft, including recurrent training, that are required to serve as a pilot in command in operations under this part;

(3) Has satisfactorily completed the proficiency or competency checks that are required to serve as a pilot in command in operations under this part;

(4) Has satisfactorily completed the applicable training requirements of § 135.339;

(5) Holds at least a Class III medical certificate unless serving as a required crewmember, in which case holds a Class I or Class II medical certificate as appropriate.

(6) Has satisfied the recency of experience requirements of § 135.247; and

(7) Has been approved by the Administrator for the check airman duties involved.

(c) No certificate holder may use a person, nor may any person serve as a check airman (simulator) in a training program established under this subpart unless, with respect to the aircraft type involved, that person meets the provisions of paragraph (b) of this section, or—

(1) Holds the applicable airman certificates and ratings, except medical certificate, required to serve as a pilot in command in operations under this part;

(2) Has satisfactorily completed the appropriate training phases for the aircraft, including recurrent training, that are required to serve as a pilot in command in operations under this part;

(3) Has satisfactorily completed the appropriate proficiency or competency checks that are required to serve as a pilot in command in operations under this part;

(4) Has satisfactorily completed the applicable training requirements of § 135.339; and

(5) Has been approved by the Administrator for the check airman (simulator) duties involved.

(d) Completion of the requirements in paragraphs (b) (2), (3), and (4) or (c) (2), (3), and (4) of this section, as applicable, shall be entered in the individual's training record maintained by the certificate holder.

(e) Check airmen who do not hold an appropriate medical certificate may function as check airmen (simulator), but may not serve as flightcrew members in operations under this part.

(f) A check airman (simulator) must accomplish the following—

(1) Fly at least two flight segments as a required crewmember for the type, class, or category aircraft involved within the 12-month preceding the performance of any check airman duty in a flight simulator; or

(2) Satisfactorily complete an approved line-observation program within the period prescribed by that program and that must precede the performance of any check airman duty in a flight simulator.

(g) The flight segments or line-observation program required in paragraph (f) of this section are considered to be completed in the month required if completed in the calendar month before or the calendar month after the month in which they are due.

[Doc. No. 28471, 61 FR 30744, June 17, 1996]

§ 135.338 Qualifications: Flight instructors (aircraft) and flight instructors (simulator).

(a) For the purposes of this section and § 135.340:

(1) A flight instructor (aircraft) is a person who is qualified to instruct in an aircraft, in a flight simulator, or in a flight training device for a particular type, class, or category aircraft.

(2) A flight instructor (simulator) is a person who is qualified to instruct in a flight simulator, in a flight training device, or in both, for a particular type, class, or category aircraft.

(3) Flight instructors (aircraft) and flight instructors (simulator) are those instructors who perform the functions described in § 135.321(a) and 135.323 (a)(4) and (c).

(b) No certificate holder may use a person, nor may any person serve as a flight instructor (aircraft) in a training program established under this subpart unless, with respect to the type, class, or category aircraft involved, that person—

(1) Holds the airman certificates and ratings required to serve as a pilot in command in operations under this part;

(2) Has satisfactorily completed the training phases for the aircraft, including recurrent training, that are required to serve as a pilot in command in operations under this part;

(3) Has satisfactorily completed the proficiency or competency checks that are required to serve as a pilot in command in operations under this part;

(4) Has satisfactorily completed the applicable training requirements of § 135.340;

(5) Holds at least a Class III medical certificate; and

(6) Has satisfied the recency of experience requirements of § 135.247.

(c) No certificate holder may use a person, nor may any person serve as a flight instructor (simulator) in a training program established under this subpart, unless, with respect to the type, class, or category aircraft involved, that person meets the provisions of paragraph (b) of this section, or—

(1) Holds the airman certificates and ratings, except medical certificate, required to serve as a pilot in command in operations under this part except before March 19, 1997 that person need not hold a type rating for the type, class, or category of aircraft involved.

(2) Has satisfactorily completed the appropriate training phases for the aircraft, including recurrent training, that are required to serve as a pilot in command in operations under this part;

(3) Has satisfactorily completed the appropriate proficiency or competency checks that are required to serve as a pilot in command in operations under this part; and

(4) Has satisfactorily completed the applicable training requirements of § 135.340.

(d) Completion of the requirements in paragraphs (b) (2), (3), and (4) or (c) (2), (3), and (4) of this section, as applicable, shall be entered in the individual's training record maintained by the certificate holder.

(e) An airman who does not hold a medical certificate may function as a flight instructor in an aircraft if functioning as a non-required crewmember, but may not serve as a flightcrew member in operations under this part.

(f) A flight instructor (simulator) must accomplish the following—

(1) Fly at least two flight segments as a required crewmember for the type, class, or category aircraft involved within the 12-month period preceding the performance of any flight instructor duty in a flight simulator; or

(2) Satisfactorily complete an approved line-observation program within the period prescribed by that program preceding the performance of any flight instructor duty in a flight simulator.

(g) The flight segments or line-observation program required in paragraph (f) of this section are considered completed in the month required if completed in the calendar month before, or in the calendar month after, the month in which they are due.

[Doc. No. 28471, 61 FR 30744, June 17, 1996; 62 FR 3739, Jan. 24, 1997, as amended by Amdt. 135-125, 76 FR 35104, June 16, 2011]

§ 135.339 Initial and transition training and checking: Check airmen (aircraft), check airmen (simulator).

(a) No certificate holder may use a person nor may any person serve as a check airman unless—

(1) That person has satisfactorily completed initial or transition check airman training; and

(2) Within the preceding 24 calendar months, that person satisfactorily conducts a proficiency or competency check under the observation of an FAA inspector or an aircrew designated examiner employed by the operator. The observation check may be accomplished in part or in full in an aircraft, in a flight simulator, or in a flight training device. This paragraph applies after March 19, 1997.

(b) The observation check required by paragraph (a)(2) of this section is considered to have been completed in the month required if completed in the calendar month before or the calendar month after the month in which it is due.

(c) The initial ground training for check airmen must include the following:

(1) Check airman duties, functions, and responsibilities.

(2) The applicable Code of Federal Regulations and the certificate holder's policies and procedures.

(3) The applicable methods, procedures, and techniques for conducting the required checks.

(4) Proper evaluation of student performance including the detection of—

(i) Improper and insufficient training; and

(ii) Personal characteristics of an applicant that could adversely affect safety.

(5) The corrective action in the case of unsatisfactory checks.

(6) The approved methods, procedures, and limitations for performing the required normal, abnormal, and emergency procedures in the aircraft.

(d) The transition ground training for check airmen must include the approved methods, procedures, and limitations for performing the required normal, abnormal, and emergency procedures applicable to the aircraft to which the check airman is in transition.

(e) The initial and transition flight training for check airmen (aircraft) must include the following—

(1) The safety measures for emergency situations that are likely to develop during a check;

(2) The potential results of improper, untimely, or nonexecution of safety measures during a check;

(3) Training and practice in conducting flight checks from the left and right pilot seats in the required normal, abnormal, and emergency procedures to ensure competence to conduct the pilot flight checks required by this part; and

(4) The safety measures to be taken from either pilot seat for emergency situations that are likely to develop during checking.

(f) The requirements of paragraph (e) of this section may be accomplished in full or in part in flight, in a flight simulator, or in a flight training device, as appropriate.

(g) The initial and transition flight training for check airmen (simulator) must include the following:

(1) Training and practice in conducting flight checks in the required normal, abnormal, and emergency procedures to ensure competence to conduct the flight checks required by this part. This training and practice must be accomplished in a flight simulator or in a flight training device.

(2) Training in the operation of flight simulators, flight training devices, or both, to ensure competence to conduct the flight checks required by this part.

[Doc. No. 28471, 61 FR 30745, June 17, 1996; 62 FR 3739, Jan. 24, 1997]

§ 135.340 Initial and transition training and checking: Flight instructors (aircraft), flight instructors (simulator).

(a) No certificate holder may use a person nor may any person serve as a flight instructor unless—

(1) That person has satisfactorily completed initial or transition flight instructor training; and

(2) Within the preceding 24 calendar months, that person satisfactorily conducts instruction under the observation of an FAA inspector, an operator check airman, or an aircrew designated examiner employed by the operator. The observation check may be accomplished in part or in full in an aircraft, in a flight simulator, or in a flight training device. This paragraph applies after March 19, 1997.

(b) The observation check required by paragraph (a)(2) of this section is considered to have been completed in the month required if completed in the calendar month before, or the calendar month after, the month in which it is due.

(c) The initial ground training for flight instructors must include the following:

(1) Flight instructor duties, functions, and responsibilities.

(2) The applicable Code of Federal Regulations and the certificate holder's policies and procedures.

(3) The applicable methods, procedures, and techniques for conducting flight instruction.

(4) Proper evaluation of student performance including the detection of—

(i) Improper and insufficient training; and

(ii) Personal characteristics of an applicant that could adversely affect safety.

(5) The corrective action in the case of unsatisfactory training progress.

(6) The approved methods, procedures, and limitations for performing the required normal, abnormal, and emergency procedures in the aircraft.

(7) Except for holders of a flight instructor certificate—

(i) The fundamental principles of the teaching-learning process;

(ii) Teaching methods and procedures; and

(iii) The instructor-student relationship.

(d) The transition ground training for flight instructors must include the approved methods, procedures, and limitations for performing the required normal, abnormal, and emergency procedures applicable to the type, class, or category aircraft to which the flight instructor is in transition.

(e) The initial and transition flight training for flight instructors (aircraft) must include the following—

(1) The safety measures for emergency situations that are likely to develop during instruction;

(2) The potential results of improper or untimely safety measures during instruction;

(3) Training and practice from the left and right pilot seats in the required normal, abnormal, and emergency maneuvers to ensure competence to conduct the flight instruction required by this part; and

(4) The safety measures to be taken from either the left or right pilot seat for emergency situations that are likely to develop during instruction.

(f) The requirements of paragraph (e) of this section may be accomplished in full or in part in flight, in a flight simulator, or in a flight training device, as appropriate.

(g) The initial and transition flight training for a flight instructor (simulator) must include the following:

(1) Training and practice in the required normal, abnormal, and emergency procedures to ensure competence to conduct the flight instruction required by this part. These maneuvers and procedures must be accomplished in full or in part in a flight simulator or in a flight training device.

(2) Training in the operation of flight simulators, flight training devices, or both, to ensure competence to conduct the flight instruction required by this part.

[Doc. No. 28471, 61 FR 30745, June 17, 1996; 61 FR 34927, July 3, 1996; 62 FR 3739, Jan. 24, 1997]

§ 135.341 Pilot and flight attendant crewmember training programs.

(a) Each certificate holder, other than one who uses only one pilot in the certificate holder's operations, shall establish and maintain an approved pilot training program, and each certificate holder who uses a flight attendant crewmember shall establish and maintain an approved flight attendant training program, that is appropriate to the operations to which each pilot and flight attend- ant is to be assigned, and will ensure that they are adequately trained to meet the applicable knowledge and practical testing requirements of §§ 135.293 through 135.301. However, the Administrator may authorize a deviation from this section if the Administrator finds that, because of the limited size and scope of the operation, safety will allow a deviation from these requirements. This deviation authority does not extend to the training provided under paragraph (c) of this section.

(b) Each certificate holder required to have a training program by paragraph (a) of this section shall include in that program ground and flight training curriculums for—

(1) Initial training;

(2) Transition training;

(3) Upgrade training;

(4) Differences training; and

(5) Recurrent training.

(c) Each certificate holder required to have a training program by paragraph (a) of this section shall provide current and appropriate study materials for use by each required pilot and flight attendant.

(d) The certificate holder shall furnish copies of the pilot and flight attendant crewmember training program, and all changes and additions, to the assigned representative of the Administrator. If the certificate holder uses training facilities of other persons, a copy of those training programs or appropriate portions used for those facilities shall also be furnished. Curricula that follow FAA published curricula may be cited by reference in the copy of the training program furnished to the representative of the Administrator and need not be furnished with the program.

[Doc. No. 16097, 43 FR 46783, Oct. 10, 1978, as amended by Amdt. 135-18, 47 FR 33396, Aug. 2, 1982; Amdt. 135-127, 78 FR 42379, July 15, 2013]

§ 135.343 Crewmember initial and recurrent training requirements.

No certificate holder may use a person, nor may any person serve, as a crewmember in operations under this part unless that crewmember has completed the appropriate initial or recurrent training phase of the training program appropriate to the type of operation in which the crewmember is to serve since the beginning of the 12th calendar month before that service. This section does not apply to a certificate holder that uses only one pilot in the certificate holder's operations.

[Doc. No. 16097, 43 FR 46783, Oct. 10, 1978, as amended by Amdt. 135-18, 47 FR 33396, Aug. 2, 1982]

§ 135.345 Pilots: Initial, transition, and upgrade ground training.

Initial, transition, and upgrade ground training for pilots must include instruction in at least the following, as applicable to their duties:

(a) General subjects—

(1) The certificate holder's flight locating procedures;

(2) Principles and methods for determining weight and balance, and runway limitations for takeoff and landing;

(3) Enough meteorology to ensure a practical knowledge of weather phenomena, including the principles of frontal systems, icing, fog, thunderstorms, windshear and, if appropriate, high altitude weather situations;

(4) Air traffic control systems, procedures, and phraseology;

(5) Navigation and the use of navigational aids, including instrument approach procedures;

(6) Normal and emergency communication procedures;

(7) Visual cues before and during descent below DA/DH or MDA;

(8) ETOPS, if applicable;

(9) After August 13, 2008, passenger recovery plan for any passenger-carrying operation (other than intrastate operations wholly within the state of Alaska) in the North Polar area; and

(10) Other instructions necessary to ensure the pilot's competence.

(b) For each aircraft type—

(1) A general description;

(2) Performance characteristics;

(3) Engines and propellers;

(4) Major components;

(5) Major aircraft systems (i.e., flight controls, electrical, and hydraulic), other systems, as appropriate, principles of normal, abnormal, and emergency operations, appropriate procedures and limitations;

(6) Knowledge and procedures for—

(i) Recognizing and avoiding severe weather situations;

(ii) Escaping from severe weather situations, in case of inadvertent encounters, including low-altitude windshear (except that rotorcraft pilots are not required to be trained in escaping from low-altitude windshear);

(iii) Operating in or near thunderstorms (including best penetrating altitudes), turbulent air (including clear air turbulence), icing, hail, and other potentially hazardous meteorological conditions; and

(iv) Operating airplanes during ground icing conditions, (i.e., any time conditions are such that frost, ice, or snow may reasonably be expected to adhere to the airplane), if the certificate holder expects to authorize takeoffs in ground icing conditions, including:

(A) The use of holdover times when using deicing/anti-icing fluids;

(B) Airplane deicing/anti-icing procedures, including inspection and check procedures and responsibilities;

(C) Communications;

(D) Airplane surface contamination (i.e., adherence of frost, ice, or snow) and critical area identification, and knowledge of how contamination adversely affects airplane performance and flight characteristics;

(E) Types and characteristics of deicing/anti-icing fluids, if used by the certificate holder;

(F) Cold weather preflight inspection procedures;

(G) Techniques for recognizing contamination on the airplane;

(7) Operating limitations;

(8) Fuel consumption and cruise control;

(9) Flight planning;

(10) Each normal and emergency procedure; and

(11) The approved Aircraft Flight Manual, or equivalent.

[Doc. No. 16097, 43 FR 46783, Oct. 10, 1978, as amended by Amdt. 135-27, 53 FR 37697, Sept. 27, 1988; Amdt. 135-46, 58 FR 69630, Dec. 30, 1993; Amdt. 135-108, 72 FR 1885, Jan. 16, 2007; Amdt. 135-110, 72 FR 31685, June 7, 2007; Amdt. 135-112, 73 FR 8798, Feb. 15, 2008]

§ 135.347 Pilots: Initial, transition, upgrade, and differences flight training.

(a) Initial, transition, upgrade, and differences training for pilots must include flight and practice in each of the maneuvers and procedures in the approved training program curriculum.

(b) The maneuvers and procedures required by paragraph (a) of this section must be performed in flight, except to the extent that certain maneuvers and procedures may be performed in an aircraft simulator, or an appropriate training device, as allowed by this subpart.

(c) If the certificate holder's approved training program includes a course of training using an aircraft simulator or other training device, each pilot must successfully complete—

(1) Training and practice in the simulator or training device in at least the maneuvers and procedures in this subpart that are capable of being performed in the aircraft simulator or training device; and

(2) A flight check in the aircraft or a check in the simulator or training device to the level of proficiency of a pilot in command or second in command, as applicable, in at least the maneuvers and procedures that are capable of being performed in an aircraft simulator or training device.

§ 135.349 Flight attendants: Initial and transition ground training.

Initial and transition ground training for flight attendants must include instruction in at least the following—

(a) General subjects—

(1) The authority of the pilot in command; and

(2) Passenger handling, including procedures to be followed in handling deranged persons or other persons whose conduct might jeopardize safety.

(b) For each aircraft type—

(1) A general description of the aircraft emphasizing physical characteristics that may have a bearing on ditching, evacuation, and inflight emergency procedures and on other related duties;

(2) The use of both the public address system and the means of communicating with other flight crewmembers, including emergency means in the case of attempted hijacking or other unusual situations; and

(3) Proper use of electrical galley equipment and the controls for cabin heat and ventilation.

§ 135.351 Recurrent training.

(a) Each certificate holder must ensure that each crewmember receives recurrent training and is adequately trained and currently proficient for the type aircraft and crewmember position involved.

(b) Recurrent ground training for crewmembers must include at least the following:

(1) A quiz or other review to determine the crewmember's knowledge of the aircraft and crewmember position involved.

(2) Instruction as necessary in the subjects required for initial ground training by this subpart, as appropriate, including low-altitude windshear training and training on operating during ground icing conditions as prescribed in § 135.341 and described in § 135.345, crew resource management training as prescribed in § 135.330, and emergency training as prescribed in § 135.331.

(c) Recurrent flight training for pilots must include, at least, flight training in the maneuvers or procedures in this subpart, except that satisfactory completion of the check required by § 135.293 within the preceding 12 calendar months may be substituted for recurrent flight training.

[Doc. No. 16097, 43 FR 46783, Oct. 10, 1978, as amended by Amdt. 135-27, 53 FR 37698, Sept. 27, 1988; Amdt. 135-46, 58 FR 69630, Dec. 30, 1993; Amdt. 135-122, 76 FR 3837, Jan. 21, 2011]

§ 135.353 [Reserved]

Subpart I—Airplane Performance Operating Limitations

§ 135.361 Applicability.

(a) This subpart prescribes airplane performance operating limitations applicable to the operation of the categories of airplanes listed in § 135.363 when operated under this part.

(b) For the purpose of this subpart, *effective length of the runway,* for landing means the distance from the point at which the obstruction clearance plane associated with the approach end of the runway intersects the centerline of the runway to the far end of the runway.

(c) For the purpose of this subpart, *obstruction clearance plane* means a plane sloping upward from the runway at a slope of 1:20 to the horizontal, and tangent to or clearing all obstructions within a specified area surrounding the runway as shown in a profile view of that area. In the plan view, the centerline of the specified area coincides with the centerline of the runway, beginning at the point where the obstruction clearance plane intersects the centerline of the runway and proceeding to a point at least 1,500 feet from the beginning point. After that the centerline coincides with the takeoff path over the ground for the runway (in the case of takeoffs) or with the instrument approach counterpart (for landings), or, where the applicable one of these paths has not been established, it proceeds consistent with turns of at least 4,000-foot radius until a point is reached beyond which the obstruction clearance plane clears all obstructions. This area extends laterally 200 feet on each side of the centerline at the point where the obstruction clearance plane intersects the runway and continues at this width to the end of the runway; then it increases uniformly to 500 feet on each side of the centerline at a point 1,500 feet from the intersection of the obstruction clearance plane with the runway; after that it extends laterally 500 feet on each side of the centerline.

§ 135.363 General.

(a) Each certificate holder operating a reciprocating engine powered large transport category airplane shall comply with §§ 135.365 through 135.377.

(b) Each certificate holder operating a turbine engine powered large transport category airplane shall comply with §§ 135.379 through 135.387, except that when it operates a turbopropeller-powered large transport category airplane certificated after August 29, 1959, but previously type certificated with the same number of reciprocating engines, it may comply with §§ 135.365 through 135.377.

(c) Each certificate holder operating a large nontransport category airplane shall comply with §§ 135.389 through 135.395 and any determination of compliance must be based only on approved performance data. For the purpose of this subpart, a large nontrans- port category airplane is an airplane that was type certificated before July 1, 1942.

(d) Each certificate holder operating a small transport category airplane shall comply with § 135.397.

(e) Each certificate holder operating a small nontransport category airplane shall comply with § 135.399.

(f) The performance data in the Airplane Flight Manual applies in determining compliance with §§ 135.365 through 135.387. Where conditions are different from those on which the performance data is based, compliance is determined by interpolation or by computing the effects of change in the specific variables, if the results of the interpolation or computations are substantially as accurate as the results of direct tests.

(g) No person may take off a reciprocating engine powered large transport category airplane at a weight that is more than the allowable weight for the runway being used (determined under the runway takeoff limitations of the transport category operating rules of this subpart) after taking into account the temperature operating correction factors in section 4a.749a-T or section 4b.117 of the Civil Air Regulations in effect on January 31, 1965, and in the applicable Airplane Flight Manual.

(h) The Administrator may author- ize in the operations specifications deviations from this subpart if special circumstances make a literal observ- ance of a requirement unnecessary for safety.

(i) The 10-mile width specified in §§ 135.369 through 135.373 may be reduced to 5 miles, for not more than 20 miles, when operating under VFR or where navigation facilities furnish reliable and accurate identification of high ground and obstructions located outside of 5 miles, but within 10 miles, on each side of the intended track.

(j) Each certificate holder operating a commuter category airplane shall comply with § 135.398.

[Doc. No. 16097, 43 FR 46783, Oct. 10, 1978, as amended by Amdt. 135-21, 52 FR 1836, Jan. 15, 1987]

§ 135.364 Maximum flying time outside the United States.

After August 13, 2008, no certificate holder may operate an airplane, other than an all-cargo airplane with more than two engines, on a planned route that exceeds 180 minutes flying time (at the one-engine-inoperative cruise speed under standard conditions in still air) from an Adequate Airport outside the continental United States unless the operation is approved by the FAA in accordance with Appendix G of this part, Extended Operations (ETOPS).

[Doc. No. FAA-1999-6717, 73 FR 8798, Feb. 15, 2008]

§ 135.365 Large transport category airplanes: Reciprocating engine powered: Weight limitations.

(a) No person may take off a reciprocating engine powered large transport category airplane from an airport located at an elevation outside of the range for which maximum takeoff weights have been determined for that airplane.

(b) No person may take off a reciprocating engine powered large transport category airplane for an airport of intended destination that is located at an elevation outside of the range for which maximum landing weights have been determined for that airplane.

(c) No person may specify, or have specified, an alternate airport that is located at an elevation outside of the range for which maximum landing weights have been determined for the reciprocating engine powered large transport category airplane concerned.

(d) No person may take off a reciprocating engine powered large transport category airplane at a weight more than the maximum authorized takeoff weight for the elevation of the airport.

(e) No person may take off a reciprocating engine powered large transport category airplane if its weight on arrival at the airport of destination will be more than the maximum authorized landing weight for the elevation of that airport, allowing for normal consumption of fuel and oil en route.

§ 135.367 Large transport category airplanes: Reciprocating engine powered: Takeoff limitations.

(a) No person operating a reciprocating engine powered large transport category airplane may take off that airplane unless it is possible—

(1) To stop the airplane safely on the runway, as shown by the accelerate-stop distance data, at any time during takeoff until reaching critical-engine failure speed;

(2) If the critical engine fails at any time after the airplane reaches critical-engine failure speed V_1, to continue the takeoff and reach a height of 50 feet, as indicated by the takeoff path data, before passing over the end of the runway; and

(3) To clear all obstacles either by at least 50 feet vertically (as shown by the takeoff path data) or 200 feet horizontally within the airport boundaries and 300 feet horizontally beyond the boundaries, without banking before reaching a height of 50 feet (as shown by the takeoff path data) and after that without banking more than 15 degrees.

(b) In applying this section, corrections must be made for any runway gradient. To allow for wind effect, takeoff data based on still air may be corrected by taking into account not more than 50 percent of any reported headwind component and not less than 150 percent of any reported tailwind component.

§ 135.369 Large transport category airplanes: Reciprocating engine powered: En route limitations: All engines operating.

(a) No person operating a reciprocating engine powered large transport category airplane may take off that airplane at a weight, allowing for normal consumption of fuel and oil, that does not allow a rate of climb (in feet per minute), with all engines operating, of at least 6.90 Vs_0 (that is, the number of feet per minute obtained by multiplying the number of knots by 6.90) at an altitude of a least 1,000 feet above the highest ground or obstruction within ten miles of each side of the intended track.

(b) This section does not apply to large transport category airplanes certificated under part 4a of the Civil Air Regulations.

§ 135.371 Large transport category airplanes: Reciprocating engine powered: En route limitations: One engine inoperative.

(a) Except as provided in paragraph (b) of this section, no person operating a reciprocating engine powered large transport category airplane may take off that airplane at a weight, allowing for normal consumption of fuel and oil, that does not allow a rate of climb (in feet per minute), with one engine inoperative, of at least (0.079–0.106/N) Vs_o 2 (where N is the number of engines installed and Vs_o is expressed in knots) at an altitude of least 1,000 feet above the highest ground or obstruction within 10 miles of each side of the intended track. However, for the purposes of this paragraph the rate of climb for transport category airplanes certificated under part 4a of the Civil Air Regulations is 0.026 Vs_o 2.

(b) In place of the requirements of paragraph (a) of this section, a person may, under an approved procedure, operate a reciprocating engine powered large transport category airplane at an all-engines-operating altitude that allows the airplane to continue, after an engine failure, to an alternate airport where a landing can be made under § 135.377, allowing for normal consumption of fuel and oil. After the assumed failure, the flight path must clear the ground and any obstruction within five miles on each side of the intended track by at least 2,000 feet.

(c) If an approved procedure under paragraph (b) of this section is used, the certificate holder shall comply with the following:

(1) The rate of climb (as prescribed in the Airplane Flight Manual for the appropriate weight and altitude) used in calculating the airplane's flight path shall be diminished by an amount in feet per minute, equal to (0.079–0.106/N) Vs_o 2 (when N is the number of engines installed and Vs_o is expressed in knots) for airplanes certificated under part 25 of this chapter and by 0.026 Vs_o 2 for airplanes certificated under part 4a of the Civil Air Regulations.

(2) The all-engines-operating altitude shall be sufficient so that in the event the critical engine becomes inoperative at any point along the route, the flight will be able to proceed to a predetermined alternate airport by use of this procedure. In determining the takeoff weight, the airplane is assumed to pass over the critical obstruction following engine failure at a point no closer to the critical obstruction than the nearest approved navigational fix, unless the Administrator approves a procedure established on a different basis upon finding that adequate operational safeguards exist.

(3) The airplane must meet the provisions of paragraph (a) of this section at 1,000 feet above the airport used as an alternate in this procedure.

(4) The procedure must include an approved method of accounting for winds and temperatures that would otherwise adversely affect the flight path.

(5) In complying with this procedure, fuel jettisoning is allowed if the certificate holder shows that it has an adequate training program, that proper instructions are given to the flight crew, and all other precautions are taken to ensure a safe procedure.

(6) The certificate holder and the pilot in command shall jointly elect an alternate airport for which the appropriate weather reports or forecasts, or any combination of them, indicate that weather conditions will be at or above the alternate weather minimum specified in the certificate holder's operations specifications for that airport when the flight arrives.

[Docket No. 16097, 43 FR 46783, Oct. 10, 1978, as amended by Amdt. 135-110, 72 FR 31685, June 7, 2007]

§ 135.373 Part 25 transport category airplanes with four or more engines: Reciprocating engine powered: En route limitations: Two engines inoperative.

(a) No person may operate an airplane certificated under part 25 and having four or more engines unless—

(1) There is no place along the intended track that is more than 90 minutes (with all engines operating at cruising power) from an airport that meets § 135.377; or

(2) It is operated at a weight allowing the airplane, with the two critical engines inoperative, to climb at 0.013 Vs_o 2 feet per minute (that is, the number of feet per minute obtained by multiplying the number of knots squared by 0.013) at an altitude of 1,000 feet above the highest ground or obstruction within 10 miles on each side of the intended track, or at an altitude of 5,000 feet, whichever is higher.

(b) For the purposes of paragraph (a)(2) of this section, it is assumed that—

(1) The two engines fail at the point that is most critical with respect to the takeoff weight;

(2) Consumption of fuel and oil is normal with all engines operating up to the point where the two engines fail with two engines operating beyond that point;

(3) Where the engines are assumed to fail at an altitude above the prescribed minimum altitude, compliance with the prescribed rate of climb at the prescribed minimum altitude need not be shown during the descent from the cruising altitude to the prescribed minimum altitude, if those requirements can be met once the prescribed minimum altitude is reached, and assuming descent to be along a net flight path and the rate of descent to be 0.013 Vs_o 2 greater than the rate in the approved performance data; and

(4) If fuel jettisoning is provided, the airplane's weight at the point where the two engines fail is considered to be not less than that which would include enough fuel to proceed to an airport meeting § 135.377 and to arrive at an altitude of at least 1,000 feet directly over that airport.

§ 135.375 Large transport category airplanes: Reciprocating engine powered: Landing limitations: Destination airports.

(a) Except as provided in paragraph (b) of this section, no person operating a reciprocating engine powered large transport category airplane may take off that airplane, unless its weight on arrival, allowing for normal consumption of fuel and oil in flight, would allow a full stop landing at the intended destination within 60 percent of the effective length of each runway described below from a point 50 feet directly above the intersection of the obstruction clearance plane and the runway. For the purposes of determining the allowable landing weight at the destination airport the following is assumed:

(1) The airplane is landed on the most favorable runway and in the most favorable direction in still air.

(2) The airplane is landed on the most suitable runway considering the probable wind velocity and direction (forecast for the expected time of arrival), the ground handling characteristics of the type of airplane, and other conditions such as landing aids and terrain, and allowing for the effect of the landing path and roll of not more than 50 percent of the headwind component or not less than 150 percent of the tailwind component.

(b) An airplane that would be prohibited from being taken off because it could not meet paragraph (a)(2) of this section may be taken off if an alternate airport is selected that meets all of this section except that the airplane can accomplish a full stop landing within 70 percent of the effective length of the runway.

§ 135.377 Large transport category airplanes: Reciprocating engine powered: Landing limitations: Alternate airports.

No person may list an airport as an alternate airport in a flight plan unless the airplane (at the weight anticipated at the time of arrival at the airport), based on the assumptions in § 135.375(a) (1) and (2), can be brought to a full stop landing within 70 percent of the effective length of the runway.

§ 135.379 Large transport category airplanes: Turbine engine powered: Takeoff limitations.

(a) No person operating a turbine engine powered large transport category airplane may take off that airplane at a weight greater than that listed in the Airplane Flight Manual for the elevation of the airport and for the ambient temperature existing at take- off.

(b) No person operating a turbine engine powered large transport category airplane certificated after August 26, 1957, but before August 30, 1959 (SR422, 422A), may take off that airplane at a weight greater than that listed in the Airplane Flight Manual for the minumum distance required for takeoff. In the case of an airplane certificated after September 30, 1958 (SR422A, 422B), the takeoff distance may include a clearway distance but the clearway distance included may not be greater than one-half of the takeoff run.

(c) No person operating a turbine engine powered large transport category airplane certificated after August 29, 1959 (SR422B), may take off that airplane at a weight greater than that listed in the Airplane Flight Manual at which compliance with the following may be shown:

(1) The accelerate-stop distance, as defined in § 25.109 of this chapter, must not exceed the length of the runway plus the length of any stopway.

(2) The takeoff distance must not exceed the length of the runway plus the length of any clearway except that the length of any clearway included must not be greater than one-half the length of the runway.

(3) The takeoff run must not be greater than the length of the runway.

(d) No person operating a turbine engine powered large transport category airplane may take off that airplane at a weight greater than that listed in the Airplane Flight Manual—

(1) For an airplane certificated after August 26, 1957, but before October 1, 1958 (SR422), that allows a takeoff path that clears all obstacles either by at least (35+0.01 D) feet vertically (D is the distance along the intended flight path from the end of the runway in feet), or by at least 200 feet horizontally within the airport boundaries and by at least 300 feet horizontally after passing the boundaries; or

(2) For an airplane certificated after September 30, 1958 (SR422A, 422B), that allows a net takeoff flight path that clears all obstacles either by a height of at least 35 feet vertically, or by at least 200 feet horizontally within the airport boundaries and by at least 300 feet horizontally after passing the boundaries.

(e) In determining maximum weights, minimum distances, and flight paths under paragraphs (a) through (d) of this section, correction must be made for the runway to be used, the elevation of the airport, the effective runway gradient, the ambient temperature and wind component at the time of takeoff, and, if operating limitations exist for the minimum distances required for takeoff from wet runways, the runway surface condition (dry or wet). Wet runway distances associated with grooved or porous friction course runways, if provided in the Airplane Flight Manual, may be used only for runways that are grooved or treated with a porous friction course (PFC) overlay, and that the operator determines are designed, constructed, and maintained in a manner acceptable to the Administrator.

(f) For the purposes of this section, it is assumed that the airplane is not banked before reaching a height of 50 feet, as shown by the takeoff path or net takeoff flight path data (as appropriate) in the Airplane Flight Manual, and after that the maximum bank is not more than 15 degrees.

(g) For the purposes of this section, the terms, *takeoff distance, takeoff run, net takeoff flight path,* have the same meanings as set forth in the rules under which the airplane was certificated.

[Doc. No. 16097, 43 FR 46783, Oct. 10, 1978, as amended by Amdt. 135-71, 63 FR 8321, Feb. 18, 1998]

§ 135.381 Large transport category airplanes: Turbine engine powered: En route limitations: One engine inoperative.

(a) No person operating a turbine engine powered large transport category airplane may take off that airplane at a weight, allowing for normal consumption of fuel and oil, that is greater than that which (under the approved, one engine inoperative, en route net flight path data in the Airplane Flight Manual for that airplane) will allow compliance with paragraph (a) (1) or (2) of this section, based on the ambient temperatures expected en route.

(1) There is a positive slope at an altitude of at least 1,000 feet above all terrain and obstructions within five statute miles on each side of the intended track, and, in addition, if that airplane was certificated after August 29, 1958 (SR422B), there is a positive slope at 1,500 feet above the airport where the airplane is assumed to land after an engine fails.

(2) The net flight path allows the airplane to continue flight from the cruising altitude to an airport where a landing can be made under § 135.387 clearing all terrain and obstructions within five statute miles of the intended track by at least 2,000 feet vertically and with a positive slope at 1,000 feet above the airport where the airplane lands after an engine fails, or, if that airplane was certificated after September 30, 1958 (SR422A, 422B), with a positive slope at 1,500 feet above the airport where the airplane lands after an engine fails.

(b) For the purpose of paragraph (a)(2) of this section, it is assumed that—

(1) The engine fails at the most critical point en route;

(2) The airplane passes over the critical obstruction, after engine failure at a point that is no closer to the obstruction than the approved navigation fix, unless the Administrator authorizes a different procedure based on adequate operational safeguards;

(3) An approved method is used to allow for adverse winds;

(4) Fuel jettisoning will be allowed if the certificate holder shows that the crew is properly instructed, that the training program is adequate, and that all other precautions are taken to ensure a safe procedure;

(5) The alternate airport is selected and meets the prescribed weather minimums; and

(6) The consumption of fuel and oil after engine failure is the same as the consumption that is allowed for in the approved net flight path data in the Airplane Flight Manual.

[Docket No. 16097, 43 FR 46783, Oct. 10, 1978, as amended by Amdt. 135-110, 72 FR 31685, June 7, 2007]

§ 135.383 Large transport category airplanes: Turbine engine powered: En route limitations: Two engines inoperative.

(a) Airplanes certificated after August 26, 1957, but before October 1, 1958 (SR422). No person may operate a turbine engine powered large transport category airplane along an intended route unless that person complies with either of the following:

(1) There is no place along the intended track that is more than 90 minutes (with all engines operating at cruising power) from an airport that meets § 135.387.

(2) Its weight, according to the two-engine-inoperative, en route, net flight path data in the Airplane Flight Manual, allows the airplane to fly from the point where the two engines are assumed to fail simultaneously to an airport that meets § 135.387, with a net flight path (considering the ambient temperature anticipated along the track) having a positive slope at an altitude of at least 1,000 feet above all terrain and obstructions within five statute miles on each side of the intended track, or at an altitude of 5,000 feet, whichever is higher.

For the purposes of paragraph (a)(2) of this section, it is assumed that the two engines fail at the most critical point en route, that if fuel jettisoning is provided, the airplane's weight at the point where the engines fail includes enough fuel to continue to the airport and to arrive at an altitude of at least 1,000 feet directly over the airport, and that the fuel and oil consumption after engine failure is the same as the consumption allowed for in the net flight path data in the Airplane Flight Manual.

(b) Airplanes certificated after September 30, 1958, but before August 30, 1959 (SR422A). No person may operate a turbine engine powered large transport category airplane along an intended route unless that person complies with either of the following:

(1) There is no place along the intended track that is more than 90 minutes (with all engines operating at cruising power) from an airport that meets § 135.387.

(2) Its weight, according to the two-engine-inoperative, en route, net flight path data in the Airplane Flight Manual allows the airplane to fly from the point where the two engines are assumed to fail simultaneously to an airport that meets § 135.387 with a net flight path (considering the ambient temperatures anticipated along the track) having a positive slope at an altitude of at least 1,000 feet above all terrain and obstructions within five statute miles on each side of the intended track, or at an altitude of 2,000 feet, whichever is higher.

For the purpose of paragraph (b)(2) of this section, it is assumed that the two engines fail at the most critical point en route, that the airplane's weight at the point where the engines fail includes enough fuel to continue to the airport, to arrive at an altitude of at least 1,500 feet directly over the airport, and after that to fly for 15 minutes at cruise power or thrust, or both, and that the consumption of fuel and oil after engine failure is the same as the consumption allowed for in the net flight path data in the Airplane Flight Manual.

(c) Aircraft certificated after August 29, 1959 (SR422B). No person may operate a turbine engine powered large transport category airplane along an intended route unless that person complies with either of the following:

(1) There is no place along the intended track that is more than 90 minutes (with all engines operating at cruising power) from an airport that meets § 135.387.

(2) Its weight, according to the two-engine-inoperative, en route, net flight path data in the Airplane Flight Manual, allows the airplane to fly from the point where the two engines are assumed to fail simultaneously to an airport that meets § 135.387, with the net flight path (considering the ambient temperatures anticipated along the track) clearing vertically by at least 2,000 feet all terrain and obstructions within five statute miles on each side of the intended track. For the purposes of this paragraph, it is assumed that—

(i) The two engines fail at the most critical point en route;

(ii) The net flight path has a positive slope at 1,500 feet above the airport where the landing is assumed to be made after the engines fail;

(iii) Fuel jettisoning will be approved if the certificate holder shows that the crew is properly instructed, that the training program is adequate, and that all other precautions are taken to ensure a safe procedure;

(iv) The airplane's weight at the point where the two engines are assumed to fail provides enough fuel to continue to the airport, to arrive at an altitude of at least 1,500 feet directly over the airport, and after that to fly for 15 minutes at cruise power or thrust, or both; and

(v) The consumption of fuel and oil after the engines fail is the same as the consumption that is allowed for in the net flight path data in the Airplane Flight Manual.

§ 135.385 Large transport category airplanes: Turbine engine powered: Landing limitations: Destination airports.

(a) No person operating a turbine engine powered large transport category airplane may take off that airplane at a weight that (allowing for normal consumption of fuel and oil in flight to the destination or alternate airport) the weight of the airplane on arrival would exceed the landing weight in the Airplane Flight Manual for the elevation of the destination or alternate airport and the ambient temperature anticipated at the time of landing.

(b) Except as provided in paragraph (c), (d), (e), or (f) of this section, no person operating a turbine engine powered large transport category airplane may take off that airplane unless its weight on arrival, allowing for normal consumption of fuel and oil in flight (in accordance with the landing distance in the Airplane Flight Manual for the elevation of the destination airport and the wind conditions expected there at the time of landing), would allow a full stop landing at the intended destination airport within 60 percent of the effective length of each runway described below from a point 50 feet above the intersection of the obstruction clearance plane and the runway. For the purpose of determining the allowable landing weight at the destination airport the following is assumed:

(1) The airplane is landed on the most favorable runway and in the most favorable direction, in still air.

(2) The airplane is landed on the most suitable runway considering the probable wind velocity and direction and the ground handling characteristics of the airplane, and considering other conditions such as landing aids and terrain.

(c) A turbopropeller powered airplane that would be prohibited from being taken off because it could not meet paragraph (b)(2) of this section, may be taken off if an alternate airport is selected that meets all of this section except that the airplane can accomplish a full stop landing within 70 percent of the effective length of the runway.

(d) Unless, based on a showing of actual operating landing techniques on wet runways, a shorter landing distance (but never less than that required by paragraph (b) of this section) has been approved for a specific type and model airplane and included in the Airplane Flight Manual, no person may take off a turbojet airplane when the appropriate weather reports or forecasts, or any combination of them, indicate that the runways at the destination airport may be wet or slippery at the estimated time of arrival unless the effective runway length at the destination airport is at least 115 percent of the runway length required under paragraph (b) of this section.

(e) A turbojet airplane that would be prohibited from being taken off because it could not meet paragraph (b)(2) of this section may be taken off if an alternate airport is selected that meets all of paragraph (b) of this section.

(f) An eligible on-demand operator may take off a turbine engine powered large transport category airplane on an on-demand flight if all of the following conditions exist:

(1) The operation is permitted by an approved Destination Airport Analysis in that person's operations manual.

(2) The airplane's weight on arrival, allowing for normal consumption of fuel and oil in flight (in accordance with the landing distance in the Airplane Flight Manual for the elevation of the destination airport and the wind conditions expected there at the time of landing), would allow a full stop landing at the intended destination airport within 80 percent of the effective length of each runway described below from a point 50 feet above the intersection of the obstruction clearance plane and the runway. For the purpose of determining the allowable landing weight at the destination airport, the following is assumed:

(i) The airplane is landed on the most favorable runway and in the most favorable direction, in still air.

(ii) The airplane is landed on the most suitable runway considering the probable wind velocity and direction and the ground handling characteristics of the airplane, and considering other conditions such as landing aids and terrain.

(3) The operation is authorized by operations specifications.

[Doc. No. 16097, 43 FR 46783, Oct. 10, 1978, as amended by Amdt. 135-91, 68 FR 54588, Sept. 17, 2003]

§ 135.387 Large transport category airplanes: Turbine engine powered: Landing limitations: Alternate airports.

(a) Except as provided in paragraph (b) of this section, no person may select an airport as an alternate airport for a turbine engine powered large transport category airplane unless (based on the assumptions in § 135.385(b)) that airplane, at the weight expected at the time of arrival, can be brought to a full stop landing within 70 percent of the effective length of the runway for turbo-propeller-powered airplanes and 60 percent of the effective length of the runway for turbojet airplanes, from a point 50 feet above the intersection of the obstruction clearance plane and the runway.

(b) Eligible on-demand operators may select an airport as an alternate airport for a turbine engine powered large transport category airplane if (based on the assumptions in § 135.385(f)) that airplane, at the weight expected at the time of arrival, can be brought to a full stop landing within 80 percent of the effective length of the runway from a point 50 feet above the intersection of the obstruction clearance plane and the runway.

[Doc. No. FAA-2001-10047, 68 FR 54588, Sept. 17, 2003]

§ 135.389 Large nontransport category airplanes: Takeoff limitations.

(a) No person operating a large nontransport category airplane may take off that airplane at a weight greater than the weight that would allow the airplane to be brought to a safe stop within the effective length of the runway, from any point during the takeoff before reaching 105 percent of minimum control speed (the minimum speed at which an airplane can be safely controlled in flight after an engine becomes inoperative) or 115 percent of the power off stalling speed in the takeoff configuration, whichever is greater.

(b) For the purposes of this section—

(1) It may be assumed that takeoff power is used on all engines during the acceleration;

(2) Not more than 50 percent of the reported headwind component, or not less than 150 percent of the reported tailwind component, may be taken into account;

(3) The average runway gradient (the difference between the elevations of the endpoints of the runway divided by the total length) must be considered if it is more than one-half of one percent;

(4) It is assumed that the airplane is operating in standard atmosphere; and

(5) For takeoff, *effective length of the runway* means the distance from the end of the runway at which the takeoff is started to a point at which the obstruction clearance plane associated with the other end of the runway intersects the runway centerline.

§ 135.391 Large nontransport category airplanes: En route limitations: One engine inoperative.

(a) Except as provided in paragraph (b) of this section, no person operating a large nontransport category airplane may take off that airplane at a weight that does not allow a rate of climb of at least 50 feet a minute, with the critical engine inoperative, at an altitude of at least 1,000 feet above the highest obstruction within five miles on each side of the intended track, or 5,000 feet, whichever is higher.

(b) Without regard to paragraph (a) of this section, if the Administrator finds that safe operations are not impaired, a person may operate the airplane at an altitude that allows the airplane, in case of engine failure, to clear all obstructions within five miles on each side of the intended track by 1,000 feet. If this procedure is used, the rate of descent for the appropriate weight and altitude is assumed to be 50 feet a minute greater than the rate in the approved performance data. Before approving such a procedure, the Administrator considers the following for the route, route segement, or area concerned:

(1) The reliability of wind and weather forecasting.

(2) The location and kinds of navigation aids.

(3) The prevailing weather conditions, particularly the frequency and amount of turbulence normally encountered.

(4) Terrain features.

(5) Air traffic problems.

(6) Any other operational factors that affect the operations.

(c) For the purposes of this section, it is assumed that—

(1) The critical engine is inoperative;

(2) The propeller of the inoperative engine is in the minimum drag position;

(3) The wing flaps and landing gear are in the most favorable position;

(4) The operating engines are operating at the maximum continuous power available;

(5) The airplane is operating in standard atmosphere; and

(6) The weight of the airplane is progressively reduced by the anticipated consumption of fuel and oil.

§ 135.393 Large nontransport category airplanes: Landing limitations: Destination airports.

(a) No person operating a large nontransport category airplane may take off that airplane at a weight that—

(1) Allowing for anticipated consumption of fuel and oil, is greater than the weight that would allow a full stop landing within 60 percent of the effective length of the most suitable runway at the destination airport; and

(2) Is greater than the weight allowable if the landing is to be made on the runway—

(i) With the greatest effective length in still air; and

(ii) Required by the probable wind, taking into account not more than 50 percent of the headwind component or not less than 150 percent of the tailwind component.

(b) For the purpose of this section, it is assumed that—

(1) The airplane passes directly over the intersection of the obstruction clearance plane and the runway at a height of 50 feet in a steady gliding approach at a true indicated airspeed of at least 1.3 V_{SO};

(2) The landing does not require exceptional pilot skill; and

(3) The airplane is operating in standard atmosphere.

§ 135.395 Large nontransport category airplanes: Landing limitations: Alternate airports.

No person may select an airport as an alternate airport for a large nontransport category airplane unless that airplane (at the weight anticipated at the time of arrival), based on the assumptions in § 135.393(b), can be brought to a full stop landing within 70 percent of the effective length of the runway.

§ 135.397 Small transport category airplane performance operating limitations.

(a) No person may operate a reciprocating engine powered small transport category airplane unless that person complies with the weight limitations in § 135.365, the takeoff limitations in § 135.367 (except paragraph (a)(3)), and the landing limitations in §§ 135.375 and 135.377.

(b) No person may operate a turbine engine powered small transport category airplane unless that person complies with the takeoff limitations in § 135.379 (except paragraphs (d) and (f)) and the landing limitations in §§ 135.385 and 135.387.

§ 135.398 Commuter category airplanes performance operating limitations.

(a) No person may operate a commuter category airplane unless that person complies with the takeoff weight limitations in the approved Airplane Flight Manual.

(b) No person may take off an airplane type certificated in the commuter category at a weight greater than that listed in the Airplane Flight Manual that allows a net takeoff flight path that clears all obstacles either by a height of at least 35 feet vertically, or at least 200 feet horizontally within the airport boundaries and by at least 300 feet horizontally after passing the boundaries.

(c) No person may operate a commuter category airplane unless that person complies with the landing limitations prescribed in §§ 135.385 and 135.387 of this part. For purposes of this paragraph, §§ 135.385 and 135.387 are applicable to all commuter category airplanes notwithstanding their stated applicability to turbine-engine-powered large transport category airplanes.

(d) In determining maximum weights, minimum distances and flight paths under paragraphs (a) through (c) of this section, correction must be made for the runway to be used, the elevation of the airport, the effective runway gradient, and ambient temperature, and wind component at the time of takeoff.

(e) For the purposes of this section, the assumption is that the airplane is not banked before reaching a height of 50 feet as shown by the net takeoff flight path data in the Airplane Flight Manual and thereafter the maximum bank is not more than 15 degrees.

[Doc. No. 23516, 52 FR 1836, Jan. 15, 1987]

§ 135.399 Small nontransport category airplane performance operating limitations.

(a) No person may operate a reciprocating engine or turbopropeller-powered small airplane that is certificated under § 135.169(b) (2), (3), (4), (5), or (6) unless that person complies with the takeoff weight limitations in the approved Airplane Flight Manual or equivalent for operations under this part, and, if the airplane is certificated under § 135.169(b) (4) or (5) with the landing weight limitations in the Approved Airplane Flight Manual or equivalent for operations under this part.

(b) No person may operate an airplane that is certificated under § 135.169(b)(6) unless that person complies with the landing limitations prescribed in §§ 135.385 and 135.387 of this part. For purposes of this paragraph, §§ 135.385 and 135.387 are applicable to reciprocating and turbopropeller-powered small airplanes notwithstanding their stated applicability to turbine engine powered large transport category airplanes.

[44 FR 53731, Sept. 17, 1979]

Subpart J—Maintenance, Preventive Maintenance, and Alterations

§ 135.411 Applicability.

(a) This subpart prescribes rules in addition to those in other parts of this chapter for the maintenance, preventive maintenance, and alterations for each certificate holder as follows:

(1) Aircraft that are type certificated for a passenger seating configuration, excluding any pilot seat, of nine seats or less, shall be maintained under parts 91 and 43 of this chapter and §§ 135.415, 135.417, 135.421 and 135.422. An approved aircraft inspection program may be used under § 135.419.

(2) Aircraft that are type certificated for a passenger seating configuration, excluding any pilot seat, of ten seats or more, shall be maintained under a maintenance program in §§ 135.415, 135.417, 135.423 through 135.443.

(b) A certificate holder who is not otherwise required, may elect to maintain its aircraft under paragraph (a)(2) of this section.

(c) Single engine aircraft used in passenger-carrying IFR operations shall also be maintained in accordance with § 135.421 (c), (d), and (e).

(d) A certificate holder who elects to operate in accordance with § 135.364 must maintain its aircraft under paragraph (a)(2) of this section and the additional requirements of Appendix G of this part.

[Doc. No. 16097, 43 FR 46783, Oct. 10, 1978, as amended by Amdt. 135-70, 62 FR 42374, Aug. 6, 1997; Amdt. 135-78, 65 FR 60556, Oct. 11, 2000; Amdt. 135-92, 68 FR 69308, Dec. 12, 2003; Amdt. 135-81, 70 FR 5533, Feb. 2, 2005; Amdt. 135-108, 72 FR 1885, Jan. 16, 2007; 72 FR 53114, Sept. 18, 2007]

§ 135.413 Responsibility for airworthiness.

(a) Each certificate holder is primarily responsible for the airworthiness of its aircraft, including airframes, aircraft engines, propellers, rotors, appliances, and parts, and shall have its aircraft maintained under this chapter, and shall have defects repaired between required maintenance under part 43 of this chapter.

(b) Each certificate holder who maintains its aircraft under § 135.411(a)(2) shall—

(1) Perform the maintenance, preventive maintenance, and alteration of its aircraft, including airframe, aircraft engines, propellers, rotors, appliances, emergency equipment and parts, under its manual and this chapter; or

(2) Make arrangements with another person for the performance of maintenance, preventive maintenance, or alteration. However, the certificate holder shall ensure that any maintenance, preventive maintenance, or alteration that is performed by another person is performed under the certificate holder's manual and this chapter.

§ 135.415 Service difficulty reports.

(a) Each certificate holder shall report the occurrence or detection of each failure, malfunction, or defect in an aircraft concerning—

(1) Fires during flight and whether the related fire-warning system functioned properly;

(2) Fires during flight not protected by related fire-warning system;

(3) False fire-warning during flight;

(4) An exhaust system that causes damage during flight to the engine, adjacent structure, equipment, or components;

(5) An aircraft component that causes accumulation or circulation of smoke, vapor, or toxic or noxious fumes in the crew compartment or passenger cabin during flight;

(6) Engine shutdown during flight because of flameout;

(7) Engine shutdown during flight when external damage to the engine or aircraft structure occurs;

(8) Engine shutdown during flight due to foreign object ingestion or icing;

(9) Shutdown of more than one engine during flight;

(10) A propeller feathering system or ability of the system to control overspeed during flight;

(11) A fuel or fuel-dumping system that affects fuel flow or causes hazardous leakage during flight;

(12) An unwanted landing gear extension or retraction or opening or closing of landing gear doors during flight;

(13) Brake system components that result in loss of brake actuating force when the aircraft is in motion on the ground;

(14) Aircraft structure that requires major repair;

(15) Cracks, permanent deformation, or corrosion of aircraft structures, if more than the maximum acceptable to the manufacturer or the FAA; and

(16) Aircraft components or systems that result in taking emergency actions during flight (except action to shut-down an engine).

(b) For the purpose of this section, *during flight* means the period from the moment the aircraft leaves the surface of the earth on takeoff until it touches down on landing.

(c) In addition to the reports required by paragraph (a) of this section, each certificate holder shall report any other failure, malfunction, or defect in an aircraft that occurs or is detected at any time if, in its opinion, the failure, malfunction, or defect has endangered or may endanger the safe operation of the aircraft.

(d) Each certificate holder shall submit each report required by this section, covering each 24-hour period beginning at 0900 local time of each day and ending at 0900 local time on the next day, to the FAA offices in Oklahoma City, Oklahoma. Each report of occurrences during a 24-hour period shall be submitted to the collection point within the next 96 hours. However, a report due on Saturday or Sunday may be submitted on the following Monday, and a report due on a holiday may be submitted on the next workday.

(e) The certificate holder shall transmit the reports required by this section on a form and in a manner prescribed by the Administrator, and shall include as much of the following as is available:

(1) The type and identification number of the aircraft.

(2) The name of the operator.

(3) The date.

(4) The nature of the failure, malfunction, or defect.

(5) Identification of the part and system involved, including available information pertaining to type designation of the major component and time since last overhaul, if known.

(6) Apparent cause of the failure, malfunction or defect (e.g., wear, crack, design deficiency, or personnel error).

(7) Other pertinent information necessary for more complete identification, determination of seriousness, or corrective action.

(f) A certificate holder that is also the holder of a type certificate (including a supplemental type certificate), a Parts Manufacturer Approval, or a Technical Standard Order Authorization, or that is the licensee of a type certificate need not report a failure, malfunction, or defect under this section if the failure, malfunction, or defect has been reported by it under § 21.3 or § 37.17 of this chapter or under the accident reporting provisions of part 830 of the regulations of the National Transportation Safety Board.

(g) No person may withhold a report required by this section even though all information required by this section is not available.

(h) When the certificate holder gets additional information, including information from the manufacturer or other agency, concerning a report required by this section, it shall expeditiously submit it as a supplement to the first report and reference the date and place of submission of the first report.

[Doc. No. 16097, 43 FR 46783, Oct. 10, 1978, as amended by Amdt. 135-102, 70 FR 76979, Dec. 29, 2005]

§ 135.417 Mechanical interruption summary report.

Each certificate holder shall mail or deliver, before the end of the 10th day of the following month, a summary report of the following occurrences in multiengine aircraft for the preceding month to the certificate-holding district office:

(a) Each interruption to a flight, unscheduled change of aircraft en route, or unscheduled stop or diversion from a route, caused by known or suspected mechanical difficulties or malfunctions that are not required to be reported under § 135.415.

(b) The number of propeller featherings in flight, listed by type of propeller and engine and aircraft on which it was installed. Propeller featherings for training, demonstration, or flight check purposes need not be reported.

[Doc. No. 16097, 43 FR 46783, Oct. 10, 1978, as amended by Amdt. 135-60, 61 FR 2616, Jan. 26, 1996]

§ 135.419 Approved aircraft inspection program.

(a) Whenever the Administrator finds that the aircraft inspections required or allowed under part 91 of this chapter are not adequate to meet this part, or upon application by a certificate holder, the Administrator may amend the certificate holder's operations specifications under § 119.51, to require or allow an approved aircraft inspection program for any make and model aircraft of which the certificate holder has the exclusive use of at least one aircraft (as defined in § 135.25(b)).

(b) A certificate holder who applies for an amendment of its operations specifications to allow an approved aircraft inspection program must submit that program with its application for approval by the Administrator.

(c) Each certificate holder who is required by its operations specifications to have an approved aircraft inspection program shall submit a program for approval by the Administrator within 30 days of the amendment of its operations specifications or within any other period that the Administrator may prescribe in the operations specifications.

(d) The aircraft inspection program submitted for approval by the Administrator must contain the following:

(1) Instructions and procedures for the conduct of aircraft inspections (which must include necessary tests and checks), setting forth in detail the parts and areas of the airframe, engines, propellers, rotors, and appliances, including emergency equipment, that must be inspected.

(2) A schedule for the performance of the aircraft inspections under paragraph (d)(1) of this section expressed in terms of the time in service, calendar time, number of system operations, or any combination of these.

(3) Instructions and procedures for recording discrepancies found during inspections and correction or deferral of discrepancies including form and disposition of records.

(e) After approval, the certificate holder shall include the approved aircraft inspection program in the manual required by § 135.21.

(f) Whenever the Administrator finds that revisions to an approved aircraft inspection program are necessary for the continued adequacy of the program, the certificate holder shall, after notification by the Administrator, make any changes in the program found by the Administrator to be necessary. The certificate holder may petition the Administrator to reconsider the notice to make any changes in a program. The petition must be filed with the representatives of the Administrator assigned to it within 30 days after the certificate holder receives the notice. Except in the case of an emergency requiring immediate action in the interest of safety, the filing of the petition stays the notice pending a decision by the Administrator.

(g) Each certificate holder who has an approved aircraft inspection program shall have each aircraft that is subject to the program inspected in accordance with the program.

(h) The registration number of each aircraft that is subject to an approved aircraft inspection program must be included in the operations specifications of the certificate holder.

[Doc. No. 16097, 43 FR 46783, Oct. 10, 1978, as amended by Amdt. 135-104, 71 FR 536, Jan. 4, 2006]

§ 135.421 Additional maintenance requirements.

(a) Each certificate holder who operates an aircraft type certificated for a passenger seating configuration, excluding any pilot seat, of nine seats or less, must comply with the manufacturer's recommended maintenance programs, or a program approved by the Administrator, for each aircraft engine, propeller, rotor, and each item of emergency equipment required by this chapter.

(b) For the purpose of this section, a manufacturer's maintenance program is one which is contained in the maintenance manual or maintenance instructions set forth by the manufacturer as required by this chapter for the aircraft, aircraft engine, propeller, rotor or item of emergency equipment.

(c) For each single engine aircraft to be used in passenger-carrying IFR operations, each certificate holder must incorporate into its maintenance program either:

(1) The manufacturer's recommended engine trend monitoring program, which includes an oil analysis, if appropriate, or

(2) An FAA approved engine trend monitoring program that includes an oil analysis at each 100 hour interval or at the manufacturer's suggested interval, whichever is more frequent.

(d) For single engine aircraft to be used in passenger-carrying IFR operations, written maintenance instructions containing the methods, techniques, and practices necessary to maintain the equipment specified in §§ 135.105, and 135.163 (f) and (h) are required.

(e) No certificate holder may operate a single engine aircraft under IFR, carrying passengers, unless the certificate holder records and maintains in the engine maintenance records the results of each test, observation, and inspection required by the applicable engine trend monitoring program specified in (c) (1) and (2) of this section.

[Doc. No. 16097, 43 FR 46783, Oct. 10, 1978, as amended by Amdt. 135-70, 62 FR 42374, Aug. 6, 1997]

§ 135.422 Aging airplane inspections and records reviews for multiengine airplanes certificated with nine or fewer passenger seats.

(a) *Applicability.* This section applies to multiengine airplanes certificated with nine or fewer passenger seats, operated by a certificate holder in a scheduled operation under this part, except for those airplanes operated by a certificate holder in a scheduled operation between any point within the State of Alaska and any other point within the State of Alaska.

(b) *Operation after inspections and records review.* After the dates specified in this paragraph, a certificate holder may not operate a multiengine airplane in a scheduled operation under this part unless the Administrator has notified the certificate holder that the Administrator has completed the aging airplane inspection and records review required by this section. During the inspection and records review, the certificate holder must demonstrate to the Administrator that the maintenance of age-sensitive parts and components of the airplane has been adequate and timely enough to ensure the highest degree of safety.

(1) *Airplanes exceeding 24 years in service on December 8, 2003; initial and repetitive inspections and records reviews.* For an airplane that has exceeded 24 years in service on December 8, 2003, no later than December 5, 2007, and thereafter at intervals not to exceed 7 years.

(2) *Airplanes exceeding 14 years in service but not 24 years in service on December 8, 2003; initial and repetitive inspections and records reviews.* For an airplane that has exceeded 14 years in service, but not 24 years in service, on December 8, 2003, no later than December 4, 2008, and thereafter at intervals not to exceed 7 years.

(3) *Airplanes not exceeding 14 years in service on December 8, 2003; initial and repetitive inspections and records reviews.* For an airplane that has not exceeded 14 years in service on December 8, 2003, no later than 5 years after the start of the airplane's 15th year in service and thereafter at intervals not to exceed 7 years.

(c) *Unforeseen schedule conflict.* In the event of an unforeseen scheduling conflict for a specific airplane, the Administrator may approve an extension of up to 90 days beyond an interval specified in paragraph (b) of this section.

(d) *Airplane and records availability.* The certificate holder must make available to the Administrator each airplane for which an inspection and records review is required under this section, in a condition for inspection specified by the Administrator, together with the records containing the following information:

(1) Total years in service of the airplane;

(2) Total time in service of the airframe;

(3) Date of the last inspection and records review required by this section;

(4) Current status of life-limited parts of the airframe;

(5) Time since the last overhaul of all structural components required to be overhauled on a specific time basis;

(6) Current inspection status of the airplane, including the time since the last inspection required by the inspection program under which the airplane is maintained;

(7) Current status of applicable airworthiness directives, including the date and methods of compliance, and, if the airworthiness directive involves recurring action, the time and date when the next action is required;

(8) A list of major structural alterations; and

(9) A report of major structural repairs and the current inspection status for these repairs.

(e) *Notification to the Administrator.* Each certificate holder must notify the Administrator at least 60 days before the date on which the airplane and airplane records will be made available for the inspection and records review.

[Doc. No. FAA-1999-5401, 70 FR 5533, Feb. 2, 2005]

§ 135.423 Maintenance, preventive maintenance, and alteration organization.

(a) Each certificate holder that performs any of its maintenance (other than required inspections), preventive maintenance, or alterations, and each person with whom it arranges for the performance of that work, must have an organization adequate to perform the work.

(b) Each certificate holder that performs any inspections required by its manual under § 135.427(b) (2) or (3), (in this subpart referred to as *required inspections*), and each person with whom it arranges for the performance of that work, must have an organization adequate to perform that work.

(c) Each person performing required inspections in addition to other maintenance, preventive maintenance, or alterations, shall organize the performance of those functions so as to separate the required inspection functions from the other maintenance, preventive maintenance, and alteration functions. The separation shall be below the level of administrative control at which overall responsibility for the required inspection functions and other maintenance, preventive maintenance, and alteration functions is exercised.

[Doc. No. 16097, 43 FR 46783, Oct. 10, 1978. Redesignated by Amdt. 135-81, 67 FR 72765, Dec. 6, 2002. Redesignated by Amdt. 135-81, 70 FR 5533, Feb. 2, 2005]

§ 135.425 Maintenance, preventive maintenance, and alteration programs.

Each certificate holder shall have an inspection program and a program covering other maintenance, preventive maintenance, and alterations, that ensures that—

(a) Maintenance, preventive maintenance, and alterations performed by it, or by other persons, are performed under the certificate holder's manual;

(b) Competent personnel and adequate facilities and equipment are provided for the proper performance of maintenance, preventive maintenance, and alterations; and

(c) Each aircraft released to service is airworthy and has been properly maintained for operation under this part.

§ 135.427 Manual requirements.

(a) Each certificate holder shall put in its manual the chart or description of the certificate holder's organization required by § 135.423 and a list of persons with whom it has arranged for the performance of any of its required inspections, other maintenance, preventive maintenance, or alterations, including a general description of that work.

(b) Each certificate holder shall put in its manual the programs required by § 135.425 that must be followed in performing maintenance, preventive maintenance, and alterations of that certificate holder's aircraft, including airframes, aircraft engines, propellers, rotors, appliances, emergency equipment, and parts, and must include at least the following:

(1) The method of performing routine and nonroutine maintenance (other than required inspections), preventive maintenance, and alterations.

(2) A designation of the items of maintenance and alteration that must be inspected (required inspections) including at least those that could result in a failure, malfunction, or defect endangering the safe operation of the aircraft, if not performed properly or if improper parts or materials are used.

(3) The method of performing required inspections and a designation by occupational title of personnel authorized to perform each required inspection.

(4) Procedures for the reinspection of work performed under previous required inspection findings (*buy-back procedures*).

(5) Procedures, standards, and limits necessary for required inspections and acceptance or rejection of the items required to be inspected and for periodic inspection and calibration of precision tools, measuring devices, and test equipment.

(6) Procedures to ensure that all required inspections are performed.

(7) Instructions to prevent any person who performs any item of work from performing any required inspection of that work.

(8) Instructions and procedures to prevent any decision of an inspector regarding any required inspection from being countermanded by persons other than supervisory personnel of the inspection unit, or a person at the level of administrative control that has overall responsibility for the management of both the required inspection functions and the other maintenance, preventive maintenance, and alterations functions.

(9) Procedures to ensure that required inspections, other maintenance, preventive maintenance, and alterations that are not completed as a result of work interruptions are properly completed before the aircraft is released to service.

(c) Each certificate holder shall put in its manual a suitable system (which may include a coded system) that provides for the retention of the following information—

(1) A description (or reference to data acceptable to the Administrator) of the work performed;

(2) The name of the person performing the work if the work is performed by a person outside the organization of the certificate holder; and

(3) The name or other positive identification of the individual approving the work.

(d) For the purposes of this part, the certificate holder must prepare that part of its manual containing maintenance information and instructions, in whole or in part, in printed form or other form, acceptable to the Administrator, that is retrievable in the English language.

[Doc. No. 16097, 43 FR 46783, Oct. 10, 1978, as amended by Amdt. 135-66, 62 FR 13257, Mar. 19, 1997; 69 FR 18472, Apr. 8, 2004; Amdt. 135-118, 74 FR 38522, Aug. 4, 2009]

§ 135.429 Required inspection personnel.

(a) No person may use any person to perform required inspections unless the person performing the inspection is appropriately certificated, properly trained, qualified, and authorized to do so.

(b) No person may allow any person to perform a required inspection unless, at the time, the person performing that inspection is under the supervision and control of an inspection unit.

(c) No person may perform a required inspection if that person performed the item of work required to be inspected.

(d) In the case of rotorcraft that operate in remote areas or sites, the Administrator may approve procedures for the performance of required inspection items by a pilot when no other qualified person is available, provided—

(1) The pilot is employed by the certificate holder;

(2) It can be shown to the satisfaction of the Administrator that each pilot authorized to perform required inspections is properly trained and qualified;

(3) The required inspection is a result of a mechanical interruption and is not a part of a certificate holder's continuous airworthiness maintenance program;

(4) Each item is inspected after each flight until the item has been inspected by an appropriately certificated mechanic other than the one who originally performed the item of work; and

(5) Each item of work that is a required inspection item that is part of the flight control system shall be flight tested and reinspected before the aircraft is approved for return to service.

(e) Each certificate holder shall maintain, or shall determine that each person with whom it arranges to perform its required inspections maintains, a current listing of persons who have been trained, qualified, and authorized to conduct required inspections. The persons must be identified by name, occupational title and the inspections that they are authorized to perform. The certificate holder (or person with whom it arranges to perform its required inspections) shall give written information to each person so authorized, describing the extent of that person's responsibilities, authorities, and inspectional limitations. The list shall be made available for inspection by the Administrator upon request.

[Doc. No. 16097, 43 FR 46783, Oct. 10, 1978, as amended by Amdt. 135-20, 51 FR 40710, Nov. 7, 1986]

§ 135.431 Continuing analysis and surveillance.

(a) Each certificate holder shall establish and maintain a system for the continuing analysis and surveillance of the performance and effectiveness of its inspection program and the program covering other maintenance, preventive maintenance, and alterations and for the correction of any deficiency in those programs, regardless of whether those programs are carried out by the certificate holder or by another person.

(b) Whenever the Administrator finds that either or both of the programs described in paragraph (a) of this section does not contain adequate procedures and standards to meet this part, the certificate holder shall, after notification by the Administrator, make changes in those programs requested by the Administrator.

(c) A certificate holder may petition the Administrator to reconsider the notice to make a change in a program. The petition must be filed with the certificate-holding district office within 30 days after the certificate holder receives the notice. Except in the case of an emergency requiring immediate action in the interest of safety, the filing of the petition stays the notice pending a decision by the Administrator.

[Doc. No. 16097, 43 FR 46783, Oct. 10, 1978, as amended by Amdt. 135-60, 61 FR 2617, Jan. 26, 1996]

§ 135.433 Maintenance and preventive maintenance training program.

Each certificate holder or a person performing maintenance or preventive maintenance functions for it shall have a training program to ensure that each person (including inspection personnel) who determines the adequacy of work done is fully informed about procedures and techniques and new equipment in use and is competent to perform that person's duties.

§ 135.435 Certificate requirements.

(a) Except for maintenance, preventive maintenance, alterations, and required inspections performed by a certificated repair station that is located outside the United States, each person who is directly in charge of maintenance, preventive maintenance, or alterations, and each person performing required inspections must hold an appropriate airman certificate.

(b) For the purpose of this section, a person *directly in charge* is each person assigned to a position in which that person is responsible for the work of a shop or station that performs maintenance, preventive maintenance, alterations, or other functions affecting airworthiness. A person who is *directly in charge* need not physically observe and direct each worker constantly but must be available for consultation and decision on matters requiring instruction or decision from higher authority than that of the person performing the work.

[Doc. No. 16097, 43 FR 46783, Oct. 10, 1978, as amended by Amdt. 135-82, 66 FR 41117, Aug. 6, 2001]

§ 135.437 Authority to perform and approve maintenance, preventive maintenance, and alterations.

(a) A certificate holder may perform or make arrangements with other persons to perform maintenance, preventive maintenance, and alterations as provided in its maintenance manual. In addition, a certificate holder may perform these functions for another certificate holder as provided in the maintenance manual of the other certificate holder.

(b) A certificate holder may approve any airframe, aircraft engine, propeller, rotor, or appliance for return to service after maintenance, preventive maintenance, or alterations that are performed under paragraph (a) of this section. However, in the case of a major repair or alteration, the work must have been done in accordance with technical data approved by the Administrator.

§ 135.439 Maintenance recording requirements.

(a) Each certificate holder shall keep (using the system specified in the manual required in § 135.427) the following records for the periods specified in paragraph (b) of this section:

(1) All the records necessary to show that all requirements for the issuance of an airworthiness release under § 135.443 have been met.

(2) Records containing the following information:

(i) The total time in service of the airframe, engine, propeller, and rotor.

(ii) The current status of life-limited parts of each airframe, engine, propeller, rotor, and appliance.

(iii) The time since last overhaul of each item installed on the aircraft which are required to be overhauled on a specified time basis.

(iv) The identification of the current inspection status of the aircraft, including the time since the last inspections required by the inspection program under which the aircraft and its appliances are maintained.

(v) The current status of applicable airworthiness directives, including the date and methods of compliance, and, if the airworthiness directive involves recurring action, the time and date when the next action is required.

(vi) A list of current major alterations and repairs to each airframe, engine, propeller, rotor, and appliance.

(b) Each certificate holder shall retain the records required to be kept by this section for the following periods:

(1) Except for the records of the last complete overhaul of each airframe, engine, propeller, rotor, and appliance the records specified in paragraph (a)(1) of this section shall be retained until the work is repeated or superseded by other work or for one year after the work is performed.

(2) The records of the last complete overhaul of each airframe, engine, propeller, rotor, and appliance shall be retained until the work is superseded by work of equivalent scope and detail.

(3) The records specified in paragraph (a)(2) of this section shall be retained and transferred with the aircraft at the time the aircraft is sold.

(c) The certificate holder shall make all maintenance records required to be kept by this section available for inspection by the Administrator or any representative of the National Transportation Safety Board.

[Doc. No. 16097, 43 FR 46783, Oct. 10, 1978; 43 FR 49975, Oct. 26, 1978]

§ 135.441 Transfer of maintenance records.

Each certificate holder who sells a United States registered aircraft shall transfer to the purchaser, at the time of the sale, the following records of that aircraft, in plain language form or in coded form which provides for the preservation and retrieval of information in a manner acceptable to the Administrator:

(a) The records specified in § 135.439(a)(2).

(b) The records specified in § 135.439(a)(1) which are not included in the records covered by paragraph (a) of this section, except that the purchaser may allow the seller to keep physical custody of such records. However, custody of records by the seller does not relieve the purchaser of its responsibility under § 135.439(c) to make the records available for inspection by the Administrator or any representative of the National Transportation Safety Board.

§ 135.443 Airworthiness release or aircraft maintenance log entry.

(a) No certificate holder may operate an aircraft after maintenance, preventive maintenance, or alterations are performed on the aircraft unless the certificate holder prepares, or causes the person with whom the certificate holder arranges for the performance of the maintenance, preventive maintenance, or alterations, to prepare—

(1) An airworthiness release; or

(2) An appropriate entry in the aircraft maintenance log.

(b) The airworthiness release or log entry required by paragraph (a) of this section must—

(1) Be prepared in accordance with the procedure in the certificate holder's manual;

(2) Include a certification that—

(i) The work was performed in accordance with the requirements of the certificate holder's manual;

(ii) All items required to be inspected were inspected by an authorized person who determined that the work was satisfactorily completed;

(iii) No known condition exists that would make the aircraft unairworthy; and

(iv) So far as the work performed is concerned, the aircraft is in condition for safe operation; and

(3) Be signed by an authorized certificated mechanic or repairman, except that a certificated repairman may sign the release or entry only for the work for which that person is employed and for which that person is certificated.

(c) Notwithstanding paragraph (b)(3) of this section, after maintenance, preventive maintenance, or alterations performed by a repair station located outside the United States, the airworthiness release or log entry required by paragraph (a) of this section may be signed by a person authorized by that repair station.

(d) Instead of restating each of the conditions of the certification required by paragraph (b) of this section, the certificate holder may state in its manual that the signature of an authorized certificated mechanic or repairman constitutes that certification.

[Doc. No. 16097, 43 FR 46783, Oct. 10, 1978, as amended by Amdt. 135-29, 53 FR 47375, Nov. 22, 1988; Amdt. 135-82, 66 FR 41117, Aug. 6, 2001]

Subpart K—Hazardous Materials Training Program

SOURCE: Doc. No. FAA-2003-15085, 70 FR 58829, Oct. 7, 2005, unless otherwise noted.

§ 135.501 Applicability and definitions.

(a) This subpart prescribes the requirements applicable to each certificate holder for training each crewmember and person performing or directly supervising any of the following job functions involving any item for transport on board an aircraft:

(1) Acceptance;

(2) Rejection;

(3) Handling;

(4) Storage incidental to transport;

(5) Packaging of company material; or

(6) Loading.

(b) *Definitions.* For purposes of this subpart, the following definitions apply:

(1) *Company material (COMAT)* —Material owned or used by a certificate holder.

(2) *Initial hazardous materials training* —The basic training required for each newly hired person, or each person changing job functions, who performs or directly supervises any of the job functions specified in paragraph (a) of this section.

(3) *Recurrent hazardous materials training* —The training required every 24 months for each person who has satisfactorily completed the certificate holder's approved initial hazardous materials training program and performs or directly supervises any of the job functions specified in paragraph (a) of this section.

§ 135.503 Hazardous materials training: General.

(a) Each certificate holder must establish and implement a hazardous materials training program that:

(1) Satisfies the requirements of Appendix O of part 121 of this part;

(2) Ensures that each person performing or directly supervising any of the job functions specified in § 135.501(a) is trained to comply with all applicable parts of 49 CFR parts 171 through 180 and the requirements of this subpart; and

(3) Enables the trained person to recognize items that contain, or may contain, hazardous materials regulated by 49 CFR parts 171 through 180.

(b) Each certificate holder must provide initial hazardous materials training and recurrent hazardous materials training to each crewmember and person performing or directly supervising any of the job functions specified in § 135.501(a).

(c) Each certificate holder's hazardous materials training program must be approved by the FAA prior to implementation.

§ 135.505 Hazardous materials training required.

(a) *Training requirement.* Except as provided in paragraphs (b), (c) and (f) of this section, no certificate holder may use any crewmember or person to perform any of the job functions or direct supervisory responsibilities, and no person may perform any of the job functions or direct supervisory responsibilities, specified in § 135.501(a) unless that person has satisfactorily completed the certificate holder's FAA-approved initial or recurrent hazardous materials training program within the past 24 months.

(b) *New hire or new job function.* A person who is a new hire and has not yet satisfactorily completed the required initial hazardous materials training, or a person who is changing job functions and has not received initial or recurrent training for a job function involving storage incidental to transport, or loading of items for transport on an aircraft, may perform those job functions for not more than 30 days from the date of hire or a change in job function, if the person is under the direct visual supervision of a person who is authorized by the certificate holder to supervise that person and who has successfully completed the certificate holder's FAA-approved initial or recurrent training program within the past 24 months.

(c) *Persons who work for more than one certificate holder.* A certificate holder that uses or assigns a person to perform or directly supervise a job function specified in § 135.501(a), when that person also performs or directly supervises the same job function for another certificate holder, need only train that person in its own policies and procedures regarding those job functions, if all of the following are met:

(1) The certificate holder using this exception receives written verification from the person designated to hold the training records representing the other certificate holder that the person has satisfactorily completed hazardous materials training for the specific job function under the other certificate holder's FAA approved hazardous material training program under appendix O of part 121 of this chapter; and

(2) The certificate holder who trained the person has the same operations specifications regarding the acceptance, handling, and transport of hazardous materials as the certificate holder using this exception.

(d) *Recurrent hazardous materials training—Completion date.* A person who satisfactorily completes recurrent hazardous materials training in the calendar month before, or the calendar month after, the month in which the recurrent training is due, is considered to have taken that training during the month in which it is due. If the person completes this training earlier than the month before it is due, the month of the completion date becomes his or her new anniversary month.

(e) *Repair stations.* A certificate holder must ensure that each repair station performing work for, or on the certificate holder's behalf is notified in writing of the certificate holder's policies and operations specification authorization permitting or prohibition against the acceptance, rejection, handling, storage incidental to transport, and transportation of hazardous materials, including company material. This notification requirement applies only to repair stations that are regulated by 49 CFR parts 171 through 180.

(f) *Certificate holders operating at foreign locations.* This exception applies if a certificate holder operating at a foreign location where the country requires the certificate holder to use persons working in that country to load aircraft. In such a case, the certificate holder may use those persons even if they have not been trained in accordance with the certificate holder's FAA approved hazardous materials training program. Those persons, however, must be under the direct visual supervision of someone who has successfully completed the certificate holder's approved initial or recurrent hazardous materials training program in accordance with this part. This exception applies only to those persons who load aircraft.

§ 135.507 Hazardous materials training records.

(a) *General requirement.* Each certificate holder must maintain a record of all training required by this part received within the preceding three years for each person who performs or directly supervises a job function specified in § 135.501(a). The record must be maintained during the time that the person performs or directly supervises any of those job functions, and for 90 days thereafter. These training records must be kept for direct employees of the certificate holder, as well as independent contractors, subcontractors, and any other person who performs or directly supervises these job functions for the certificate holder.

(b) *Location of records.* The certificate holder must retain the training records required by paragraph (a) of this section for all initial and recurrent training received within the preceding 3 years for all persons performing or directly supervising the job functions listed in Appendix O of part 121 of this chapter at a designated location. The records must be available upon request at the location where the trained person performs or directly supervises the job function specified in § 135.501(a). Records may be maintained electronically and provided on location electronically. When the person ceases to perform or directly supervise a hazardous materials job function, the certificate holder must retain the hazardous materials training records for an additional 90 days and make them available upon request at the last location where the person worked.

(c) *Content of records.* Each record must contain the following:

(1) The individual's name;

(2) The most recent training completion date;

(3) A description, copy or reference to training materials used to meet the training requirement;

(4) The name and address of the organization providing the training; and

(5) A copy of the certification issued when the individual was trained, which shows that a test has been completed satisfactorily.

(d) *New hire or new job function.* Each certificate holder using a person under the exception in § 135.505(b) must maintain a record for that person. The records must be available upon request at the location where the trained person performs or directly supervises the job function specified in § 135.501(a). Records may be maintained electronically and provided on location electronically. The record must include the following:

(1) A signed statement from an authorized representative of the certificate holder authorizing the use of the person in accordance with the exception;

(2) The date of hire or change in job function;

(3) The person's name and assigned job function;

(4) The name of the supervisor of the job function; and

(5) The date the person is to complete hazardous materials training in accordance with Appendix O of part 121 of this chapter.

Appendix A to Part 135—Additional Airworthiness Standards for 10 or More Passenger Airplanes

APPLICABILITY

1. *Applicability.* This appendix prescribes the additional airworthiness standards required by § 135.169.

2. *References.* Unless otherwise provided, references in this appendix to specific sections of part 23 of the Federal Aviation Regulations (FAR part 23) are to those sections of part 23 in effect on March 30, 1967.

FLIGHT REQUIREMENTS

3. *General.* Compliance must be shown with the applicable requirements of subpart B of FAR part 23, as supplemented or modified in §§ 4 through 10.

PERFORMANCE

4. *General.* (a) Unless otherwise prescribed in this appendix, compliance with each applicable performance requirement in sections 4 through 7 must be shown for ambient atmospheric conditions and still air.

(b) The performance must correspond to the propulsive thrust available under the particular ambient atmospheric conditions and the particular flight condition. The available propulsive thrust must correspond to engine power or thrust, not exceeding the approved power or thrust less—

(1) Installation losses; and

(2) The power or equivalent thrust absorbed by the accessories and services appropriate to the particular ambient atmospheric conditions and the particular flight condition.

(c) Unless otherwise prescribed in this appendix, the applicant must select the take-off, en route, and landing configurations for the airplane.

(d) The airplane configuration may vary with weight, altitude, and temperature, to the extent they are compatible with the operating procedures required by paragraph (e) of this section.

(e) Unless otherwise prescribed in this appendix, in determining the critical engine inoperative takeoff performance, the accelerate-stop distance, takeoff distance, changes in the airplane's configuration, speed, power, and thrust must be made under procedures established by the applicant for operation in service.

(f) Procedures for the execution of balked landings must be established by the applicant and included in the Airplane Flight Manual.

(g) The procedures established under paragraphs (e) and (f) of this section must—

(1) Be able to be consistently executed in service by a crew of average skill;

(2) Use methods or devices that are safe and reliable; and

(3) Include allowance for any time delays, in the execution of the procedures, that may reasonably be expected in service.

5. *Takeoff.* (a) *General.* Takeoff speeds, the accelerate-stop distance, the takeoff distance, and the one-engine-inoperative takeoff flight path data (described in paragraphs (b), (c), (d), and (f) of this section), must be determined for—

(1) Each weight, altitude, and ambient temperature within the operational limits selected by the applicant;

(2) The selected configuration for takeoff;

(3) The center of gravity in the most unfavorable position;

(4) The operating engine within approved operating limitations; and

(5) Takeoff data based on smooth, dry, hard-surface runway.

PART 135

FAR

(b) *Takeoff speeds.* (1) The decision speed V_1 is the calibrated airspeed on the ground at which, as a result of engine failure or other reasons, the pilot is assumed to have made a decision to continue or discontinue the takeoff. The speed V_1 must be selected by the applicant but may not be less than—

(i) 1.10 V_{S1};

(ii) 1.10 V_{MC};

(iii) A speed that allows acceleration to V_1 and stop under paragraph (c) of this section; or

(iv) A speed at which the airplane can be rotated for takeoff and shown to be adequate to safely continue the takeoff, using normal piloting skill, when the critical engine is suddenly made inoperative.

(2) The initial climb out speed V_2, in terms of calibrated airspeed, must be selected by the applicant so as to allow the gradient of climb required in section 6(b)(2), but it must not be less than V_1 or less than 1.2 V_{S1}.

(3) Other essential take off speeds necessary for safe operation of the airplane.

(c) *Accelerate-stop distance.* (1) The accelerate-stop distance is the sum of the distances necessary to—

(i) Accelerate the airplane from a standing start to V_1; and

(ii) Come to a full stop from the point at which V_1 is reached assuming that in the case of engine failure, failure of the critical engine is recognized by the pilot at the speed V_1.

(2) Means other than wheel brakes may be used to determine the accelerate-stop distance if that means is available with the critical engine inoperative and—

(i) Is safe and reliable;

(ii) Is used so that consistent results can be expected under normal operating conditions; and

(iii) Is such that exceptional skill is not required to control the airplane.

(d) *All engines operating takeoff distance.* The all engine operating takeoff distance is the horizontal distance required to takeoff and climb to a height of 50 feet above the takeoff surface under the procedures in FAR 23.51(a).

(e) *One-engine-inoperative takeoff.* Determine the weight for each altitude and temperature within the operational limits established for the airplane, at which the airplane has the capability, after failure of the critical engine at V_1 determined under paragraph (b) of this section, to take off and climb at not less than V_2, to a height 1,000 feet above the takeoff surface and attain the speed and configuration at which compliance is shown with the en route one-engine-inoperative gradient of climb specified in section 6(c).

(f) *One-engine-inoperative takeoff flight path data.* The one-engine-inoperative takeoff flight path data consist of takeoff flight paths extending from a standing start to a point in the takeoff at which the airplane reaches a height 1,000 feet above the takeoff surface under paragraph (e) of this section.

6. *Climb.* (a) *Landing climb: All-engines-operating.* The maximum weight must be determined with the airplane in the landing configuration, for each altitude, and ambient temperature within the operational limits established for the airplane, with the most unfavorable center of gravity, and out-of-ground effect in free air, at which the steady gradient of climb will not be less than 3.3 percent, with:

(1) The engines at the power that is available 8 seconds after initiation of movement of the power or thrust controls from the minimum flight idle to the takeoff position.

(2) A climb speed not greater than the approach speed established under section 7 and not less than the greater of 1.05 V_{MC} or 1.10 V_{S1}.

(b) *Takeoff climb: one-engine-inoperative.* The maximum weight at which the airplane meets the minimum climb performance specified in paragraphs (1) and (2) of this paragraph must be determined for each altitude and ambient temperature within the operational limits established for the airplane, out of ground effect in free air, with the airplane in the takeoff configuration, with the most unfavorable center of gravity, the critical engine inoperative, the remaining engines at the maximum takeoff power or thrust, and the propeller of the inoperative engine windmilling with the propeller controls in the normal position except that, if an approved automatic feathering system is installed, the propellers may be in the feathered position:

(1) *Takeoff: landing gear extended.* The minimum steady gradient of climb must be measurably positive at the speed V_1.

(2) *Takeoff: landing gear retracted.* The minimum steady gradient of climb may not be less than 2 percent at speed V_2. For airplanes with fixed landing gear this requirement must be met with the landing gear extended.

(c) *En route climb: one-engine-inoperative.* The maximum weight must be determined for each altitude and ambient temperature within the operational limits established for the airplane, at which the steady gradient of climb is not less 1.2 percent at an altitude 1,000 feet above the takeoff surface, with the airplane in the en route configuration, the critical engine inoperative, the remaining engine at the maximum continuous power or thrust, and the most unfavorable center of gravity.

7. *Landing.* (a) The landing field length described in paragraph (b) of this section must be determined for standard atmosphere at each weight and altitude within the operational limits established by the applicant.

(b) The landing field length is equal to the landing distance determined under FAR 23.75(a) divided by a factor of 0.6 for the destination airport and 0.7 for the alternate airport. Instead of the gliding approach specified in FAR 23.75(a)(1), the landing may be preceded by a steady approach down to the 50-foot height at a gradient of descent not greater than 5.2 percent (3°) at a calibrated airspeed not less than 1.3 V_{S1}.

Trim

8. *Trim.* (a) *Lateral and directional trim.* The airplane must maintain lateral and directional trim in level flight at a speed of V_H or V_{MO}/M_{MO}, whichever is lower, with landing gear and wing flaps retracted.

(b) *Longitudinal trim.* The airplane must maintain longitudinal trim during the following conditions, except that it need not maintain trim at a speed greater than V_{MO}/M_{MO}:

(1) In the approach conditions specified in FAR 23.161(c) (3) through (5), except that instead of the speeds specified in those paragraphs, trim must be maintained with a stick force of not more than 10 pounds down to a speed used in showing compliance with section 7 or 1.4 V_{S1} whichever is lower.

(2) In level flight at any speed from V_H or V_{MO}/M_{MO}, whichever is lower, to either V_x or 1.4 V_{S1}, with the landing gear and wing flaps retracted.

Stability

9. *Static longitudinal stability.* (a) In showing compliance with FAR 23.175(b) and with paragraph (b) of this section, the airspeed must return to within ±7½ percent of the trim speed.

(b) *Cruise stability.* The stick force curve must have a stable slope for a speed range of ±50 knots from the trim speed except that the speeds need not exceed V_{FC}/M_{FC} or be less than 1.4 V_{S1}. This speed range will be considered to begin at the outer extremes of the friction band and the stick force may not exceed 50 pounds with—

(1) Landing gear retracted;

(2) Wing flaps retracted;

(3) The maximum cruising power as selected by the applicant as an operating limitation for turbine engines or 75 percent of maximum continuous power for reciprocating engines except that the power need not exceed that required at V_{MO}/M_{MO};

(4) Maximum takeoff weight; and

(5) The airplane trimmed for level flight with the power specified in paragraph (3) of this paragraph.

V_{FC}/M_{FC} may not be less than a speed midway between V_{MO}/M_{MO} and V_{DF}/M_{DF}, except that, for altitudes where Mach number is the limiting factor, M_{FC} need not exceed the Mach number at which effective speed warning occurs.

(c) *Climb stability (turbopropeller powered airplanes only).* In showing compliance with FAR 23.175(a), an applicant must, instead of the power specified in FAR 23.175(a)(4), use the maximum power or thrust selected by the applicant as an operating limitation for use during climb at the best rate of climb speed, except that the speed need not be less than 1.4 V_{S1}.

Stalls

10. *Stall warning.* If artificial stall warning is required to comply with FAR 23.207, the warning device must give clearly distinguishable indications under expected conditions of flight. The use of a visual warning device that requires the attention of the crew within the cockpit is not acceptable by itself.

Control Systems

11. *Electric trim tabs.* The airplane must meet FAR 23.677 and in addition it must be shown that the airplane is safely controllable and that a pilot can perform all the maneuvers and operations necessary to effect a safe landing following any probable electric trim tab runaway which might be reasonably expected in service allowing for appropriate time delay after pilot recognition of the runaway. This demonstration must be conducted at the critical airplane weights and center of gravity positions.

Instruments: Installation

12. *Arrangement and visibility.* Each instrument must meet FAR 23.1321 and in addition:

(a) Each flight, navigation, and powerplant instrument for use by any pilot must be plainly visible to the pilot from the pilot's station with the minimum practicable deviation from the pilot's normal position and line of vision when the pilot is looking forward along the flight path.

(b) The flight instruments required by FAR 23.1303 and by the applicable operating rules must be grouped on the instrument panel and centered as nearly as practicable about the vertical plane of each pilot's forward vision. In addition—

(1) The instrument that most effectively indicates the attitude must be in the panel in the top center position;

(2) The instrument that most effectively indicates the airspeed must be on the panel directly to the left of the instrument in the top center position;

(3) The instrument that most effectively indicates altitude must be adjacent to and directly to the right of the instrument in the top center position; and

(4) The instrument that most effectively indicates direction of flight must be adjacent to and directly below the instrument in the top center position.

13. *Airspeed indicating system.* Each airspeed indicating system must meet FAR 23.1323 and in addition:

(a) Airspeed indicating instruments must be of an approved type and must be calibrated to indicate true airspeed at sea level in the standard atmosphere with a minimum practicable instrument calibration error when the corresponding pitot and static pressures are supplied to the instruments.

(b) The airspeed indicating system must be calibrated to determine the system error, i.e., the relation between IAS and CAS, in flight and during the accelerate-takeoff ground run. The ground run calibration must be obtained between 0.8 of the minimum value of V_1 and 1.2 times the maximum value of V_1, considering the approved ranges of altitude and weight. The ground run calibration is determined assuming an engine failure at the minimum value of V_1.

(c) The airspeed error of the installation excluding the instrument calibration error, must not exceed 3 percent or 5 knots whichever is greater, throughout the speed range from V_{MO} to 1.3 V_{S1} with flaps retracted and from 1.3 V_{SO} to V_{FE} with flaps in the landing position.

(d) Information showing the relationship between IAS and CAS must be shown in the Airplane Flight manual.

14. *Static air vent system.* The static air vent system must meet FAR 23.1325. The altimeter system calibration must be determined and shown in the Airplane Flight Manual.

Operating Limitations and Information

15. *Maximum operating limit speed V_{MO}/M_{MO}.* Instead of establishing operating limitations based on V_{NE} and V_{NO}, the applicant must establish a maximum operating limit speed V_{MO}/M_{MO} as follows:

(a) The maximum operating limit speed must not exceed the design cruising speed V_C and must be sufficiently below V_D/M_D or V_{DF}/M_{DF} to make it highly improbable that the latter speeds will be inadvertently exceeded in flight.

(b) The speed V_{MO} must not exceed 0.8 V_D/M_D or 0.8 V_{DF}/M_{DF} unless flight demonstrations involving upsets as specified by the Administrator indicates a lower speed margin will not result in speeds exceeding V_D/M_D or V_{DF}. Atmospheric variations, horizontal gusts, system and equipment errors, and airframe production variations are taken into account.

16. *Minimum flight crew.* In addition to meeting FAR 23.1523, the applicant must establish the minimum number and type of qualified flight crew personnel sufficient for safe operation of the airplane considering—

(a) Each kind of operation for which the applicant desires approval;

(b) The workload on each crewmember considering the following:

(1) Flight path control.

(2) Collision avoidance.

(3) Navigation.

(4) Communications.

(5) Operation and monitoring of all essential aircraft systems.

(6) Command decisions; and

(c) The accessibility and ease of operation of necessary controls by the appropriate crewmember during all normal and emergency operations when at the crewmember flight station.

17. *Airspeed indicator.* The airspeed indicator must meet FAR 23.1545 except that, the airspeed notations and markings in terms of V_{NO} and V_{NH} must be replaced by the V_{MO}/M_{MO} notations. The airspeed indicator markings must be easily read and understood by the pilot. A placard adjacent to the airspeed indicator is an acceptable means of showing compliance with FAR 23.1545(c).

Airplane Flight Manual

18. *General.* The Airplane Flight Manual must be prepared under FARs 23.1583 and 23.1587, and in addition the operating limitations and performance information in sections 19 and 20 must be included.

19. *Operating limitations.* The Airplane Flight Manual must include the following limitations—

(a) *Airspeed limitations.* (1) The maximum operating limit speed V_{MO}/M_{MO} and a statement that this speed limit may not be deliberately exceeded in any regime of flight (climb, cruise, or descent) unless a higher speed is authorized for flight test or pilot training;

(2) If an airspeed limitation is based upon compressibility effects, a statement to this effect and information as to any symptoms, the probable behavior of the airplane, and the recommended recovery procedures; and

(3) The airspeed limits, shown in terms of V_{MO}/M_{MO} instead of V_{NO} and V_{NE}.

(b) *Takeoff weight limitations.* The maximum takeoff weight for each airport elevation, ambient temperature, and available takeoff runway length within the range selected by the applicant may not exceed the weight at which—

(1) The all-engine-operating takeoff distance determined under section 5(b) or the accelerate-stop distance determined under section 5(c), whichever is greater, is equal to the available runway length;

(2) The airplane complies with the one-engine-inoperative takeoff requirements specified in section 5(e); and

(3) The airplane complies with the one-engine-inoperative takeoff and en route climb requirements specified in sections 6 (b) and (c).

(c) *Landing weight limitations.* The maximum landing weight for each airport elevation (standard temperature) and available landing runway length, within the range selected by the applicant. This weight may not exceed the weight at which the landing field length determined under section 7(b) is equal to the available runway length. In showing compliance with this operating limitation, it is acceptable to assume that the landing weight at the destination will be equal to the takeoff weight reduced by the normal consumption of fuel and oil en route.

20. *Performance information.* The Airplane Flight Manual must contain the performance information determined under the performance requirements of this appendix. The information must include the following:

(a) Sufficient information so that the takeoff weight limits specified in section 19(b) can be determined for all temperatures and altitudes within the operation limitations selected by the applicant.

(b) The conditions under which the performance information was obtained, including the airspeed at the 50-foot height used to determine landing distances.

(c) The performance information (determined by extrapolation and computed for the range of weights between the maximum landing and takeoff weights) for—

(1) Climb in the landing configuration; and

(2) Landing distance.

(d) Procedure established under section 4 related to the limitations and information required by this section in the form of guidance material including any relevant limitations or information.

(e) An explanation of significant or unusual flight or ground handling characteristics of the airplane.

(f) Airspeeds, as indicated airspeeds, corresponding to those determined for takeoff under section 5(b).

21. *Maximum operating altitudes.* The maximum operating altitude to which operation is allowed, as limited by flight, structural, powerplant, functional, or equipment characteristics, must be specified in the Airplane Flight Manual.

22. *Stowage provision for airplane flight manual.* Provision must be made for stowing the Airplane Flight Manual in a suitable fixed container which is readily accessible to the pilot.

23. *Operating procedures.* Procedures for restarting turbine engines in flight (including the effects of altitude) must be set forth in the Airplane Flight Manual.

Airframe Requirements

Flight Loads

24. *Engine torque.* (a) Each turbopropeller engine mount and its supporting structure must be designed for the torque effects of:

(1) The conditions in FAR 23.361(a).

(2) The limit engine torque corresponding to takeoff power and propeller speed multiplied by a factor accounting for propeller control system malfunction, including quick feathering action, simultaneously with 1 g level flight loads. In the absence of a ration- al analysis, a factor of 1.6 must be used.

(b) The limit torque is obtained by multiplying the mean torque by a factor of 1.25.

25. *Turbine engine gyroscopic loads.* Each turbopropeller engine mount and its supporting structure must be designed for the gyroscopic loads that result, with the engines at maximum continuous r.p.m., under either—

(a) The conditions in FARs 23.351 and 23.423; or

(b) All possible combinations of the following:

(1) A yaw velocity of 2.5 radians per second.

(2) A pitch velocity of 1.0 radians per second.

(3) A normal load factor of 2.5.

(4) Maximum continuous thrust.

26. *Unsymmetrical loads due to engine failure.* (a) Turbopropeller powered airplanes must be designed for the unsymmet- rical loads resulting from the failure of the critical engine including the following conditions in combination with a single malfunction of the propeller drag limiting system, considering the probable pilot corrective action on the flight controls:

(1) At speeds between V_{MO} and V_D, the loads resulting from power failure because of fuel flow interruption are considered to be limit loads.

(2) At speeds between V_{MO} and V_C, the loads resulting from the disconnection of the engine compressor from the turbine or from loss of the turbine blades are considered to be ultimate loads.

(3) The time history of the thrust decay and drag buildup occurring as a result of the prescribed engine failures must be substantiated by test or other data applicable to the particular engine-propeller combination.

(4) The timing and magnitude of the probable pilot corrective action must be conserv- atively estimated, considering the characteristics of the particular engine-propeller-airplane combination.

(b) Pilot corrective action may be assumed to be initiated at the time maximum yawing velocity is reached, but not earlier than 2 seconds after the engine failure. The magnitude of the corrective action may be based on the control forces in FAR 23.397 except that lower forces may be assumed where it is shown by analysis or test that these forces can control the yaw and roll resulting from the prescribed engine failure conditions.

Ground Loads

27. *Dual wheel landing gear units.* Each dual wheel landing gear unit and its supporting structure must be shown to comply with the following:

(a) *Pivoting.* The airplane must be assumed to pivot about one side of the main gear with the brakes on that side locked. The limit vertical load factor must be 1.0 and the coefficient of friction 0.8. This condition need apply only to the main gear and its supporting structure.

(b) *Unequal tire inflation.* A 60-40 percent distribution of the loads established under FAR 23.471 through FAR 23.483 must be applied to the dual wheels.

(c) *Flat tire.* (1) Sixty percent of the loads in FAR 23.471 through FAR 23.483 must be applied to either wheel in a unit.

(2) Sixty percent of the limit drag and side loads and 100 percent of the limit vertical load established under FARs 23.493 and 23.485 must be applied to either wheel in a unit except that the vertical load need not exceed the maximum vertical load in paragraph (c)(1) of this section.

Fatigue Evaluation

28. *Fatigue evaluation of wing and associated structure.* Unless it is shown that the structure, operating stress levels, materials and expected use are comparable from a fatigue standpoint to a similar design which has had substantial satisfactory service experience, the strength, detail design, and the fabrication of those parts of the wing, wing carrythrough, and attaching structure whose failure would be catastrophic must be evaluated under either—

(a) A fatigue strength investigation in which the structure is shown by analysis, tests, or both to be able to withstand the repeated loads of variable magnitude expected in service; or

(b) A fail-safe strength investigation in which it is shown by analysis, tests, or both that catastrophic failure of the structure is not probable after fatigue, or obvious partial failure, of a principal structural element, and that the remaining structure is able to withstand a static ultimate load factor of 75 percent of the critical limit load factor at V_C. These loads must be multiplied by a factor of 1.15 unless the dynamic effects of failure under static load are otherwise considered.

Design and Construction

29. *Flutter.* For multiengine turbopropeller powered airplanes, a dynamic evaluation must be made and must include—

(a) The significant elastic, inertia, and aerodynamic forces associated with the rotations and displacements of the plane of the propeller; and

(b) Engine-propeller-nacelle stiffness and damping variations appropriate to the particular configuration.

Landing Gear

30. *Flap operated landing gear warning device.* Airplanes having retractable landing gear and wing flaps must be equipped with a warning device that functions continuously when the wing flaps are extended to a flap position that activates the warning device to give adequate warning before landing, using normal landing procedures, if the landing gear is not fully extended and locked. There may not be a manual shut off for this warning device. The flap position sensing unit may be installed at any suitable location. The system for this device may use any part of the system (including the aural warning device) provided for other landing gear warning devices.

Personnel and Cargo Accommodations

31. *Cargo and baggage compartments.* Cargo and baggage compartments must be designed to meet FAR 23.787 (a) and (b), and in addition means must be provided to protect passengers from injury by the contents of any cargo or baggage compartment when the ultimate forward inertia force is 9 g.

32. *Doors and exits.* The airplane must meet FAR 23.783 and FAR 23.807 (a)(3), (b), and (c), and in addition:

(a) There must be a means to lock and safeguard each external door and exit against opening in flight either inadvertently by persons, or as a result of mechanical failure. Each external door must be operable from both the inside and the outside.

(b) There must be means for direct visual inspection of the locking mechanism by crewmembers to determine whether external doors and exits, for which the initial opening movement is outward, are fully locked. In addition, there must be a visual means to signal to crewmembers when normally used external doors are closed and fully locked.

(c) The passenger entrance door must qualify as a floor level emergency exit. Each additional required emergency exit except floor level exits must be located over the wing or must be provided with acceptable means to assist the occupants in descending to the ground. In addition to the passenger entrance door:

(1) For a total seating capacity of 15 or less, an emergency exit as defined in FAR 23.807(b) is required on each side of the cabin.

(2) For a total seating capacity of 16 through 23, three emergency exits as defined in FAR 23.807(b) are required with one on the same side as the door and two on the side opposite the door.

(d) An evacuation demonstration must be conducted utilizing the maximum number of occupants for which certification is desired. It must be conducted under simulated night conditions utilizing only the emergency exits on the most critical side of the aircraft. The participants must be representative of average airline passengers with no previous practice or rehearsal for the demonstration. Evacuation must be completed within 90 seconds.

(e) Each emergency exit must be marked with the word "Exit" by a sign which has white letters 1 inch high on a red background 2 inches high, be self-illuminated or independently internally electrically illuminated, and have a minimum luminescence (brightness) of at least 160 microlamberts. The colors may be reversed if the passenger compartment illumination is essentially the same.

(f) Access to window type emergency exits must not be obstructed by seats or seat backs.

(g) The width of the main passenger aisle at any point between seats must equal or exceed the values in the following table:

Total seating capacity	Minimum main passenger aisle width	
	Less than 25 inches from floor	25 inches and more from floor
10 through 23	9 inches	15 inches.

Miscellaneous

33. *Lightning strike protection.* Parts that are electrically insulated from the basic airframe must be connected to it through lightning arrestors unless a lightning strike on the insulated part—

(a) Is improbable because of shielding by other parts; or

(b) Is not hazardous.

34. *Ice protection.* If certification with ice protection provisions is desired, compliance with the following must be shown:

(a) The recommended procedures for the use of the ice protection equipment must be set forth in the Airplane Flight Manual.

(b) An analysis must be performed to establish, on the basis of the airplane's operational needs, the adequacy of the ice protection system for the various components of the airplane. In addition, tests of the ice protection system must be conducted to demonstrate that the airplane is capable of operating safely in continuous maximum and intermittent maximum icing conditions as described in appendix C of part 25 of this chapter.

(c) Compliance with all or portions of this section may be accomplished by reference, where applicable because of similarity of the designs, to analysis and tests performed by the applicant for a type certificated model.

35. *Maintenance information.* The applicant must make available to the owner at the time of delivery of the airplane the information the applicant considers essential for the proper maintenance of the airplane. That information must include the following:

(a) Description of systems, including electrical, hydraulic, and fuel controls.

(b) Lubrication instructions setting forth the frequency and the lubricants and fluids which are to be used in the various systems.

(c) Pressures and electrical loads applicable to the various systems.

(d) Tolerances and adjustments necessary for proper functioning.

(e) Methods of leveling, raising, and towing.

(f) Methods of balancing control surfaces.

(g) Identification of primary and secondary structures.

(h) Frequency and extent of inspections necessary to the proper operation of the airplane.

(i) Special repair methods applicable to the airplane.

(j) Special inspection techniques, such as X-ray, ultrasonic, and magnetic particle inspection.

(k) List of special tools.

Propulsion

General

36. *Vibration characteristics.* For turbopropeller powered airplanes, the engine installation must not result in vibration characteristics of the engine exceeding those established during the type certification of the engine.

37. *In flight restarting of engine.* If the engine on turbopropeller powered airplanes cannot be restarted at the maximum cruise altitude, a determination must be made of the altitude below which restarts can be consistently accomplished. Restart information must be provided in the Airplane Flight Manual.

38. *Engines.* (a) *For turbopropeller powered airplanes.* The engine installation must comply with the following:

(1) *Engine isolation.* The powerplants must be arranged and isolated from each other to allow operation, in at least one configuration, so that the failure or malfunction of any engine, or of any system that can affect the engine, will not—

(i) Prevent the continued safe operation of the remaining engines; or

(ii) Require immediate action by any crewmember for continued safe operation.

(2) *Control of engine rotation.* There must be a means to individually stop and restart the rotation of any engine in flight except that engine rotation need not be stopped if continued rotation could not jeopardize the safety of the airplane. Each component of the stopping and restarting system on the engine side of the firewall, and that might be exposed to fire, must be at least fire resistant. If hydraulic propeller feathering systems are used for this purpose, the feathering lines must be at least fire resistant under the operating conditions that may be expected to exist during feathering.

(3) *Engine speed and gas temperature control devices.* The powerplant systems associated with engine control devices, systems, and instrumentation must provide reasonable assurance that those engine operating limitations that adversely affect turbine rotor structural integrity will not be exceeded in service.

(b) *For reciprocating engine powered airplanes.* To provide engine isolation, the powerplants must be arranged and isolated from each other to allow operation, in at least one configuration, so that the failure or malfunction of any engine, or of any system that can affect that engine, will not—

(1) Prevent the continued safe operation of the remaining engines; or

(2) Require immediate action by any crewmember for continued safe operation.

39. *Turbopropeller reversing systems.* (a) Turbopropeller reversing systems intended for ground operation must be designed so that no single failure or malfunction of the system will result in unwanted reverse thrust under any expected operating condition. Failure of structural elements need not be considered if the probability of this kind of failure is extremely remote.

(b) Turbopropeller reversing systems intended for in flight use must be designed so that no unsafe condition will result during normal operation of the system, or from any failure (or reasonably likely combination of failures) of the reversing system, under any anticipated condition of operation of the airplane. Failure of structural elements need not be considered if the probability of this kind of failure is extremely remote.

(c) Compliance with this section may be shown by failure analysis, testing, or both for propeller systems that allow propeller blades to move from the flight low-pitch position to a position that is substantially less than that at the normal flight low-pitch stop position. The analysis may include or be supported by the analysis made to show compliance with the type certification of the propeller and associated installation components. Credit will be given for pertinent analysis and testing completed by the engine and propeller manufacturers.

40. *Turbopropeller drag-limiting systems.* Turbopropeller drag-limiting systems must be designed so that no single failure or malfunction of any of the systems during normal or emergency operation results in propeller drag in excess of that for which the airplane was designed. Failure of structural elements of the drag-limiting systems need not be considered if the probability of this kind of failure is extremely remote.

41. *Turbine engine powerplant operating characteristics.* For turbopropeller powered airplanes, the turbine engine powerplant operating characteristics must be investigated in flight to determine that no adverse characteristics (such as stall, surge, or flameout) are present to a hazardous degree, during normal and emergency operation within the range of operating limitations of the airplane and of the engine.

42. *Fuel flow.* (a) For turbopropeller powered airplanes—

(1) The fuel system must provide for continuous supply of fuel to the engines for normal operation without interruption due to depletion of fuel in any tank other than the main tank; and

(2) The fuel flow rate for turbopropeller engine fuel pump systems must not be less than 125 percent of the fuel flow required to develop the standard sea level atmospheric conditions takeoff power selected and included as an operating limitation in the Airplane Flight Manual.

(b) For reciprocating engine powered airplanes, it is acceptable for the fuel flow rate for each pump system (main and reserve supply) to be 125 percent of the takeoff fuel consumption of the engine.

Fuel System Components

43. *Fuel pumps.* For turbopropeller powered airplanes, a reliable and independent power source must be provided for each pump used with turbine engines which do not have provisions for mechanically driving the main pumps. It must be demonstrated that the pump installations provide a reliability and durability equivalent to that in FAR 23.991(a).

44. *Fuel strainer or filter.* For turbopropeller powered airplanes, the following apply:

(a) There must be a fuel strainer or filter between the tank outlet and the fuel metering device of the engine. In addition, the fuel strainer or filter must be—

(1) Between the tank outlet and the engine-driven positive displacement pump inlet, if there is an engine-driven positive displacement pump;

(2) Accessible for drainage and cleaning and, for the strainer screen, easily removable; and

(3) Mounted so that its weight is not supported by the connecting lines or by the inlet or outlet connections of the strainer or filter itself.

(b) Unless there are means in the fuel system to prevent the accumulation of ice on the filter, there must be means to automatically maintain the fuel-flow if ice-clogging of the filter occurs; and

(c) The fuel strainer or filter must be of adequate capacity (for operating limitations established to ensure proper service) and of appropriate mesh to insure proper engine operation, with the fuel contaminated to a degree (for particle size and density) that can be reasonably expected in service. The degree of fuel filtering may not be less than that established for the engine type certification.

45. *Lightning strike protection.* Protection must be provided against the ignition of flammable vapors in the fuel vent system due to lightning strikes.

Cooling

46. *Cooling test procedures for turbopropeller powered airplanes.* (a) Turbopropeller powered airplanes must be shown to comply with FAR 23.1041 during takeoff, climb, en route, and landing stages of flight that correspond to the applicable performance requirements. The cooling tests must be conducted with the airplane in the configuration, and operating under the conditions that are critical relative to cooling during each stage of flight. For the cooling tests a temperature is "stabilized" when its rate of change is less than 2 °F. per minute.

(b) Temperatures must be stabilized under the conditions from which entry is made into each stage of flight being investigated unless the entry condition is not one during which component and engine fluid temperatures would stabilize, in which case, operation through the full entry condition must be conducted before entry into the stage of flight being investigated to allow temperatures to reach their natural levels at the time of entry. The takeoff cooling test must be preceded by a period during which the powerplant component and engine fluid temperatures are stabilized with the engines at ground idle.

(c) Cooling tests for each stage of flight must be continued until—

(1) The component and engine fluid temperatures stabilize;

(2) The stage of flight is completed; or

(3) An operating limitation is reached.

Induction System

47. *Air induction.* For turbopropeller powered airplanes—

(a) There must be means to prevent hazardous quantities of fuel leakage or overflow from drains, vents, or other components of flammable fluid systems from entering the engine intake systems; and

(b) The air inlet ducts must be located or protected so as to minimize the ingestion of foreign matter during takeoff, landing, and taxiing.

48. *Induction system icing protection.* For turbopropeller powered airplanes, each turbine engine must be able to operate throughout its flight power range without adverse effect on engine operation or serious loss of power or thrust, under the icing conditions specified in appendix C of part 25 of this chapter. In addition, there must be means to indicate to appropriate flight crewmembers the functioning of the powerplant ice protection system.

49. *Turbine engine bleed air systems.* Turbine engine bleed air systems of turbopropeller powered airplanes must be investigated to determine—

(a) That no hazard to the airplane will result if a duct rupture occurs. This condition must consider that a failure of the duct can occur anywhere between the engine port and the airplane bleed service; and

(b) That, if the bleed air system is used for direct cabin pressurization, it is not possible for hazardous contamination of the cabin air system to occur in event of lubrication system failure.

Exhaust System

50. *Exhaust system drains.* Turbopropeller engine exhaust systems having low spots or pockets must incorporate drains at those locations. These drains must discharge clear of the airplane in normal and ground attitudes to prevent the accumulation of fuel after the failure of an attempted engine start.

Powerplant Controls and Accessories

51. *Engine controls.* If throttles or power levers for turbopropeller powered airplanes are such that any position of these controls will reduce the fuel flow to the engine(s) below that necessary for satisfactory and safe idle operation of the engine while the airplane is in flight, a means must be provided to prevent inadvertent movement of the control into this position. The means provided must incorporate a positive lock or stop at this idle position and must require a separate and distinct operation by the crew to displace the control from the normal engine operating range.

52. *Reverse thrust controls.* For turbopropeller powered airplanes, the propeller reverse thrust controls must have a means to prevent their inadvertent operation. The means must have a positive lock or stop at the idle position and must require a separate and distinct operation by the crew to displace the control from the flight regime.

53. *Engine ignition systems.* Each turbopropeller airplane ignition system must be considered an essential electrical load.

54. *Powerplant accessories.* The powerplant accessories must meet FAR 23.1163, and if the continued rotation of any accessory remotely driven by the engine is hazardous when malfunctioning occurs, there must be means to prevent rotation without interfering with the continued operation of the engine.

Powerplant Fire Protection

55. *Fire detector system.* For turbopropeller powered airplanes, the following apply:

(a) There must be a means that ensures prompt detection of fire in the engine compartment. An overtemperature switch in each engine cooling air exit is an acceptable method of meeting this requirement.

(b) Each fire detector must be constructed and installed to withstand the vibration, inertia, and other loads to which it may be subjected in operation.

(c) No fire detector may be affected by any oil, water, other fluids, or fumes that might be present.

(d) There must be means to allow the flight crew to check, in flight, the functioning of each fire detector electric circuit.

(e) Wiring and other components of each fire detector system in a fire zone must be at least fire resistant.

56. *Fire protection, cowling and nacelle skin.* For reciprocating engine powered airplanes, the engine cowling must be designed and constructed so that no fire originating in the engine compartment can enter either through openings or by burn through, any other region where it would create additional hazards.

57. *Flammable fluid fire protection.* If flammable fluids or vapors might be liberated by the leakage of fluid systems in areas other than engine compartments, there must be means to—

(a) Prevent the ignition of those fluids or vapors by any other equipment; or

(b) Control any fire resulting from that ignition.

Equipment

58. *Powerplant instruments.* (a) The following are required for turbopropeller airplanes:

(1) The instruments required by FAR 23.1305 (a) (1) through (4), (b) (2) and (4).

(2) A gas temperature indicator for each engine.

(3) Free air temperature indicator.

(4) A fuel flowmeter indicator for each engine.

(5) Oil pressure warning means for each engine.

(6) A torque indicator or adequate means for indicating power output for each engine.

(7) Fire warning indicator for each engine.

(8) A means to indicate when the propeller blade angle is below the low-pitch position corresponding to idle operation in flight.

(9) A means to indicate the functioning of the ice protection system for each engine.

(b) For turbopropeller powered airplanes, the turbopropeller blade position indicator must begin indicating when the blade has moved below the flight low-pitch position.

(c) The following instruments are required for reciprocating engine powered airplanes:

(1) The instruments required by FAR 23.1305.

(2) A cylinder head temperature indicator for each engine.

(3) A manifold pressure indicator for each engine.

Systems and Equipments

General

59. *Function and installation.* The systems and equipment of the airplane must meet FAR 23.1301, and the following:

(a) Each item of additional installed equipment must—

(1) Be of a kind and design appropriate to its intended function;

(2) Be labeled as to its identification, function, or operating limitations, or any applicable combination of these factors, unless misuse or inadvertent actuation cannot create a hazard;

(3) Be installed according to limitations specified for that equipment; and

(4) Function properly when installed.

(b) Systems and installations must be designed to safeguard against hazards to the aircraft in the event of their malfunction or failure.

(c) Where an installation, the functioning of which is necessary in showing compliance with the applicable requirements, requires a power supply, that installation must be considered an essential load on the power supply, and the power sources and the distribution system must be capable of supplying the following power loads in probable operation combinations and for probable durations:

(1) All essential loads after failure of any prime mover, power converter, or energy storage device.

(2) All essential loads after failure of any one engine on two-engine airplanes.

(3) In determining the probable operating combinations and durations of essential loads for the power failure conditions described in paragraphs (1) and (2) of this paragraph, it is permissible to assume that the power loads are reduced in accordance with a monitoring procedure which is consistent with safety in the types of operations authorized.

60. *Ventilation.* The ventilation system of the airplane must meet FAR 23.831, and in addition, for pressurized aircraft, the ventilating air in flight crew and passenger compartments must be free of harmful or hazardous concentrations of gases and vapors in normal operation and in the event of reasonably probable failures or malfunctioning of the ventilating, heating, pressurization, or other systems, and equipment. If accumulation of hazardous quantities of smoke in the cockpit area is reasonably probable, smoke evacuation must be readily accomplished.

Electrical Systems and Equipment

61. *General.* The electrical systems and equipment of the airplane must meet FAR 23.1351, and the following:

(a) *Electrical system capacity.* The required generating capacity, and number and kinds of power sources must—

(1) Be determined by an electrical load analysis; and

(2) Meet FAR 23.1301.

(b) *Generating system.* The generating system includes electrical power sources, main power busses, transmission cables, and associated control, regulation and protective devices. It must be designed so that—

(1) The system voltage and frequency (as applicable) at the terminals of all essential load equipment can be maintained within the limits for which the equipment is designed, during any probable operating conditions;

(2) System transients due to switching, fault clearing, or other causes do not make essential loads inoperative, and do not cause a smoke or fire hazard;

(3) There are means, accessible in flight to appropriate crewmembers, for the individual and collective disconnection of the electrical power sources from the system; and

(4) There are means to indicate to appropriate crewmembers the generating system quantities essential for the safe operation of the system, including the voltage and current supplied by each generator.

62. *Electrical equipment and installation.* Electrical equipment, controls, and wiring must be installed so that operation of any one unit or system of units will not adversely affect the simultaneous operation of any other electrical unit or system essential to the safe operation.

Appendix B to Part 135—Airplane Flight Recorder Specifications

Parameters	Range	Installed system[1] minimum accuracy (to recovered data)	Sampling interval (per second)	Resolution[4] read out
Relative time (from recorded on prior to takeoff)	25 hr minimum	±0.125% per hour	1	1 sec.
Indicated airspeed	V_{so} to V_D (KIAS)	±5% or ±10 kts., whichever is greater. Resolution 2 kts. below 175 KIAS	1	1%[3].
Altitude	−1,000 ft. to max cert. alt. of A/C	±100 to ±700 ft. (see Table 1, TSO C51–a)	1	25 to 150
Magnetic heading	360°	±5°	1	1°
Vertical acceleration	−3g to +6g	±0.2g in addition to ±0.3g maximum datum	4 (or 1 per second where peaks, ref. to 1g are recorded)	0.03g.
Longitudinal acceleration	±1.0g	±1.5% max. range excluding datum error of ±5%	2	0.01g.
Pitch attitude	100% of usable	±2°	1	0.8°
Roll attitude	±60° or 100% of usable range, whichever is greater	±2°	1	0.8°
Stabilizer trim position	Full range	±3% unless higher uniquely required	1	1%[3].

Or				
Pitch control position	Full range	±3% unless higher uniquely required	1	1%[3].
Engine Power, Each Engine				
Fan or N_1 speed or EPR or cockpit indications used for aircraft certification	Maximum range	±5%	1	1%[3].
Or				
Prop. speed and torque (sample once/sec as close together as practicable)			1 (prop speed), 1 (torque)	
Altitude rate[2] (need depends on altitude resolution)	±8,000 fpm	±10%. Resolution 250 fpm below 12,000 ft. indicated	1	250 fpm Below 12,000
Angle of attack[2] (need depends on altitude resolution)	−20° to 40° or of usable range	±2°	1	0.8%[3]
Radio transmitter keying (discrete)	On/off		1	
TE flaps (discrete or analog)	Each discrete position (U, D, T/O, AAP)		1	
	Or			
	Analog 0–100% range	±3°	1	1%[3]
LE flaps (discrete or analog)	Each discrete position (U, D, T/O, AAP)		1	
	Or			
	Analog 0–100% range	±3°	1	1%[3].
Thrust reverser, each engine (Discrete)	Stowed or full reverse		1	
Spoiler/speedbrake (discrete)	Stowed or out		1	
Autopilot engaged (discrete)	Engaged or disengaged		1	

[1] When data sources are aircraft instruments (except altimeters) of acceptable quality to fly the aircraft the recording system excluding these sensors (but including all other characteristics of the recording system) shall contribute no more than half of the values in this column.

[2] If data from the altitude encoding altimeter (100 ft. resolution) is used, then either one of these parameters should also be recorded. If however, altitude is recorded at a minimum resolution of 25 feet, then these two parameters can be omitted.

[3] Per cent of full range.

[4] This column applies to aircraft manufacturing after October 11, 1991.

[Doc. No. 25530, 53 FR 26152, July 11, 1988; 53 FR 30906, Aug. 16, 1988, as amended by Amdt. 135-69, 62 FR 38397, July 17, 1997]

Appendix C to Part 135—Helicopter Flight Recorder Specifications

Parameters	Range	Installed system[1] minimum accuracy (to recovered data)	Sampling interval (per second)	Resolution[3] read out
Relative time (from recorded on prior to takeoff)	25 hr minimum	±0.125% per hour	1	1 sec.
Indicated airspeed	V_{min} to V_D (KIAS) (minimum airspeed signal attainable with installed pilot-static system)	±5% or ±10 kts., whichever is greater	1	1 kt.
Altitude	−1,000 ft. to 20,000 ft. pressure altitude	±100 to ±700 ft. (see Table 1, TSO C51–a)	1	25 to 150 ft.
Magnetic heading	360°	±5°	1	1°.
Vertical acceleration	−3g to +6g	±0.2g in addition to ±0.3g maximum datum	4 (or 1 per second where peaks, ref. to 1g are recorded)	0.05g.
Longitudinal acceleration	±1.0g	±1.5% max. range excluding datum error of ±5%	2	0.03g.
Pitch attitude	100% of usable range	±2°	1	0.8°.
Roll attitude	±60° or 100% of usable range, whichever is greater	±2°	1	0.8°.
Altitude rate	±8,000 fpm	±10% Resolution 250 fpm below 12,000 ft. indicated	1	250 fpm below 12,000.
Engine Power, Each Engine				
Main rotor speed	Maximum range	±5%	1	1%[2]
Free or power turbine	Maximum range	+5%	1	1%[2]
Engine torque	Maximum range	±5%	1	1%[2]
Flight Control—Hydraulic Pressure				
Primary (discrete)	High/low		1	
Secondary—if applicable (discrete)	High/low		1	

Parameters	Range	Installed system1minimum accuracy (to recovered data)	Sampling interval (per second)	Resolution3read out
Radio transmitter keying (discrete)	On/off		1	
Autopilot engaged (discrete)	Engaged or disengaged		1	
SAS status—engaged (discrete)	Engaged/disengaged		1	
SAS fault status (discrete)	Fault/OK		1	
Flight Controls				
Collective[4]	Full range	±3%	2	1%[2]
Pedal Position[4]	Full range	±3%	2	1%[2]
Lat. Cyclic[4]	Full range	±3%	2	1%[2]
Long. Cyclic[4]	Full range	±3%	2	1%[2]
Controllable Stabilator Position[4]	Full range	±3%	2	1%[2]

[1] When data sources are aircraft instruments (except altimeters) of acceptable quality to fly the aircraft the recording system excluding these sensors (but including all other characteristics of the recording system) shall contribute no more than half of the values in this column.

[2] Per cent of full range.

[3] This column applies to aircraft manufactured after October 11, 1991.

[4] For all aircraft manufactured on or after December 6, 2010, the sampling interval per second is 4.

[Doc. No. 25530, 53 FR 26152, July 11, 1988; 53 FR 30906, Aug. 16, 1988, as amended by Amdt. 135-69, 62 FR 38397, July 17, 1997; Amdt. 135-113, 73 FR 12570, Mar. 7, 2008; 73 FR 15281, Mar. 21, 2008; Amdt. 135-121, 75 FR 17047, Apr. 5, 2010]

Appendix D to Part 135—Airplane Flight Recorder Specification

Parameters	Range	Installed system1minimum accuracy (to recovered data)	Sampling interval (per second)	Resolution3read out
Relative time (from recorded on prior to takeoff)	25 hr minimum	±0.125% per hour	1	1 sec.
Indicated airspeed	$V_m$in to V_D(KIAS) (minimum airspeed signal attainable with installed pilot-static system)	±5% or ±10 kts., whichever is greater	1	1 kt.
Altitude	–1,000 ft. to 20,000 ft. pressure altitude	±100 to ±700 ft. (see Table 1, TSO C51–a)	1	25 to 150 ft.
Magnetic heading	360°	±5°	1	1°.
Vertical acceleration	–3g to +6g	±0.2g in addition to ±0.3g maximum datum	4 (or 1 per second where peaks, ref. to 1g are recorded)	0.05g.
Longitudinal acceleration	±1.0g	±1.5% max. range excluding datum error of ±5%	2	0.03g.
Pitch attitude	100% of usable range	±2°	1	0.8°.
Roll attitude	±60° or 100% of usable range, whichever is greater	±2°	1	0.8°.
Altitude rate	±8,000 fpm	±10% Resolution 250 fpm below 12,000 ft. indicated	1	250 fpm below 12,000.
Engine Power, Each Engine				
Main rotor speed	Maximum range	±5%	1	1%[2]
Free or power turbine	Maximum range	+5%	1	1%[2]
Engine torque	Maximum range	±5%	1	1%[2]
Flight Control—Hydraulic Pressure				
Primary (discrete)	High/low		1	
Secondary—if applicable (discrete)	High/low		1	
Radio transmitter keying (discrete)	On/off		1	
Autopilot engaged (discrete)	Engaged or disengaged		1	
SAS status—engaged (discrete)	Engaged/disengaged		1	
SAS fault status (discrete)	Fault/OK		1	

Parameters	Range	Installed system1minimum accuracy (to recovered data)	Sampling interval (per second)	Resolution3read out
Flight Controls				
Collective[4]	Full range	±3%	2	1%[2]
Pedal Position[4]	Full range	±3%	2	1%[2]
Lat. Cyclic[4]	Full range	±3%	2	1%[2]
Long. Cyclic[4]	Full range	±3%	2	1%[2]
Controllable Stabilator Position[4]	Full range	±3%	2	1%[2]

[1] When altitude rate is recorded. Altitude rate must have sufficient resolution and sampling to permit the derivation of altitude to 5 feet.
[2] Per cent of full range.
[3] For airplanes that can demonstrate the capability of deriving either the control input on control movement (one from the other) for all modes of operation and flight regimes, the "or" applies. For airplanes with non-mechanical control systems (fly-by-wire) the "and" applies. In airplanes with split surfaces, suitable combination of inputs is acceptable in lieu of recording each surface separately.
[4] This column applies to aircraft manufactured after October 11, 1991.
[Doc. No. 25530, 53 FR 26153, July 11, 1988; 53 FR 30906, Aug. 16, 1988]

Appendix E to Part 135—Helicopter Flight Recorder Specifications

Parameters	Range	Accuracy sensor input to DFDR readout	Sampling interval (per second)	Resolution2read out
Time (GMT)	24 Hrs	±0.125% Per Hour	0.25 (1 per 4 seconds)	1 sec
Altitude	–1,000 ft to max certificated altitude of aircraft	±100 to ±700 ft (See Table 1, TSO-C51a)	1	5' to 30'.
Airspeed	As the installed measuring system	±3%	1	1 kt
Heading	360°	±2°	1	0.5°.
Normal Acceleration (Vertical)	–3g to +6g	±1% of max range excluding datum error of ±5%	8	0.01g
Pitch Attitude	±75°	±2°	2	0.5°
Roll Attitude	±180°	±2°	2	0.5°.
Radio Transmitter Keying	On-Off (Discrete)		1	0.25 sec
Power in Each Engine: Free Power Turbine Speed *and*Engine Torque	0–130% (power Turbine Speed) Full range (Torque)	±2%	1 speed 1 torque (per engine)	0.2%[1]to 0.4%[1]
Main Rotor Speed	0–130%	±2%	2	0.3%[1]
Altitude Rate	±6,000 ft/min	As installed	2	0.2%[1]
Pilot Input—Primary Controls (Collective, Longitudinal Cyclic, Lateral Cyclic, Pedal)[3]	Full range	±3%	2	0.5%[1]
Flight Control Hydraulic Pressure Low	Discrete, each circuit		1	
Flight Control Hydraulic Pressure Selector Switch Position, 1st and 2nd stage	Discrete		1	
AFCS Mode and Engagement Status	Discrete (5 bits necessary)		1	
Stability Augmentation System Engage	Discrete		1	
SAS Fault Status	Discrete		0.25	
Main Gearbox Temperature Low	As installed	As installed	0.25	0.5%[1]
Main Gearbox Temperature High	As installed	As installed	0.5	0.5%[1]
Controllable Stabilator Position	Full Range	±3%	2	0.4%[1].
Longitudinal Acceleration	±1g	±1.5% max range excluding datum error of ±5%	4	0.01g.
Lateral Acceleration	±1g	±1.5% max range excluding datum of ±5%	4	0.01g.
Master Warning	Discrete		1	
Nav 1 and 2 Frequency Selection	Full range	As installed	0.25	

Parameters	Range	Accuracy sensor input to DFDR readout	Sampling interval (per second)	Resolution2read out
Outside Air Temperature	-50 °C to +90 °C	±2° c	0.5	0.3° c

[1] Per cent of full range.

[2] This column applies to aircraft manufactured after October 11, 1991.

[3] For all aircraft manufactured on or after December 6, 2010, the sampling interval per second is 4.

[Doc. No. 25530, 53 FR 26154, July 11, 1988; 53 FR 30906, Aug. 16, 1988; Amdt. 135-113, 73 FR 12571, Mar. 7, 2008; 73 FR 15281, Mar. 21, 2008; Amdt. 135-121, 75 FR 17047, Apr. 5, 2010]

Appendix F to Part 135—Airplane Flight Recorder Specification

The recorded values must meet the designated range, resolution and accuracy requirements during static and dynamic conditions. Dynamic condition means the parameter is experiencing change at the maximum rate attainable, including the maximum rate of reversal. All data recorded must be correlated in time to within one second.

Parameters	Range	Accuracy (sensor input)	Seconds per sampling interval	Resolution	Remarks
1. Time or Relative Time Counts[1]	24 Hrs, 0 to 4095	±0.125% Per Hour	4	1 sec	UTC time preferred when available. Counter increments each 4 seconds of system operation.
2. Pressure Altitude	-1000 ft to max certificated altitude of aircraft. +5000 ft	±100 to ±700 ft (see table, TSO C124a or TSO C51a)	1	5' to 35&inch;	Data should be obtained from the air data computer when practicable.
3. Indicated airspeed or Calibrated airspeed	50 KIAS or minimum value to Max V_{so} and V_{so} to 1.2 V_D	±5% and ±3%	1	1 kt	Data should be obtained from the air data computer when practicable.
4. Heading (Primary flight crew reference)	0–360° and Discrete "true" or "mag"	±2°	1	0.5°	When true or magnetic heading can be selected as the primary heading reference, a discrete indicating selection must be recorded.
5. Normal Acceleration (Vertical)[9]	-3g to +6g	±1% of max range excluding datum error of ±5%	0.125	0.004g	
6. Pitch Attitude	±75%	±2°	1 or 0.25 for airplanes operated under §135.152(j)	0.5°	A sampling rate of 0.25 is recommended.
7. Roll Attitude[2]	±180°	±2°	1 or 0.5 0.5 airplanes operated under §135.152(j)	0.5°	A sampling rate of 0.5 is recommended.
8. Manual Radio Transmitter Keying or CVR/DFDR synchronization reference	On-Off (Discrete) None		1		Preferably each crew member but one discrete acceptable for all transmission provided the CVR/FDR system complies with TSO C124a CVR synchronization requirements (paragraph 4.2.1 ED-55).
9. Thrust/Power on each engine—primary flight crew reference	Full Range Forward	±2%	1 (per engine)	0.3% of full range	Sufficient parameters (e.g. EPR, N1 or Torque, NP) as appropriate to the particular engine being recorded to determine power in forward and reverse thrust, including potential overspeed condition.
10. Autopilot Engagement	Discrete "on" or "off"		1		
11. Longitudinal Acceleration	±1g	±1.5% max. range excluding datum error of ±5%	0.25	0.004g.	
12a. Pitch control(s) position (nonfly-by-wire systems)[18]	Full Range	±2° unless higher accuracy uniquely required	0.5 or 0.25 for airplanes operated under §135.152(j)	0.5% of full range	For airplanes that have a flight control breakaway capability that allows either pilot to operate the controls independently, record both control inputs. The control inputs may be sampled alternately once per second to produce the sampling interval of 0.5 or 0.25, as applicable.
12b. Pitch control(s) position (fly-by-wire systems)[3] [18]	Full Range	±2° unless higher accuracy uniquely required	0.5 or 0.25 for airplanes operated under §135.152(j)	0.2% of full range	
13a. Lateral control position(s) (nonfly-by-wire)[18]	Full Range	±2° unless higher accuracy uniquely required	0.5 or 0.25 for airplanes operated under §135.152(j)	0.2% of full range	For airplanes that have a flight control breakaway capability that allows either pilot to operate the controls independently, record both control inputs. The control inputs may be sampled alternately once per second to produce the sampling interval of 0.5 or 0.25, as applicable.
13b. Lateral control position(s) (fly-by-wire)[4] [18]	Full Range	±2° unless higher accuracy uniquely required	0.5 or 0.25 for airplanes operated under §135.152(j)	0.2% of full range	
14a. Yaw control position(s) (nonfly-by-wire)[5] [18]	Full Range	±2° unless higher accuracy uniquely required	0.5	0.3% of full range	For airplanes that have a flight control breakaway capability that allows either pilot to operate the controls independently, record both control inputs. The control inputs may be sampled alternately once per second to produce the sampling of 0.5 or 0.25, as applicable.
14b. Yaw control position(s) (fly-by-wire)[18]	Full Range	±2° unless higher accuracy uniquely required	0.5	0.2% of full range	

Parameters	Range	Accuracy (sensor input)	Seconds per sampling interval	Resolution	Remarks
15. Pitch control surface(s) position[6 18]	Full Range	±2° unless higher accuracy uniquely required	0.5 or 0.25 for airplanes operated under §135.152(j).	0.3% of full range	For airplanes fitted with multiple or split surfaces, a suitable combination of inputs is acceptable in lieu of recording each surface separately. The control surfaces may be sampled alternately to produce the sampling interval of 0.5 or 0.25, as applicable.
16. Lateral control surface(s) position[7 18]	Full Range	±2° unless higher accuracy uniquely required	0.5 or 0.25 for airplanes operated under §135.152(j)	0.2% of full range	A suitable combination of surface position sensors is acceptable in lieu of recording each surface separately. The control surfaces may be sampled alternately to produce the sampling interval of 0.5 or 0.25, as applicable.
17. Yaw control surface(s) position[8 18]	Full Range	±2° unless higher accuracy uniquely required	0.5	0.2% of full range	For airplanes with multiple or split surfaces, a suitable combination of surface position sensors is acceptable in lieu of recording each surface separately. The control surfaces may be sampled alternately to produce the sampling interval of 0.5.
18. Lateral Acceleration	±1g	±1.5% max. range excluding datum error of ±5%	0.25	0.004g	
19. Pitch Trim Surface Position	Full Range	±3° Unless Higher Accuracy Uniquely Required	1	0.6% of full range	
20. Trailing Edge Flap or Cockpit Control Selection[10]	Full Range or Each Position (discrete)	±3° or as Pilot's Indicator	2	0.5% of full range	Flap position and cockpit control may each be sampled alternately at 4 second intervals, to give a data point every 2 seconds.
21. Leading Edge Flap or Cockpit Control Selection[11]	Full Range or Each Discrete Position	±3° or as Pilot's Indicator and sufficient to determine each discrete position	2	0.5% of full range	Left and right sides, of flap position and cockpit control may each be sampled at 4 second intervals, so as to give a data point to every 2 seconds.
22. Each Thrust reverser Position (or equivalent for propeller airplane)	Stowed, In Transit, and reverse (Discrete)		1 (per engine		Turbo-jet—2 discretes enable the 3 states to be determined Turbo-prop—1 discrete
23. Ground Spoiler Position or Speed Brake Selection[12]	Full Range or Each Position (discrete)	±2° Unless Higher Accuracy Uniquely Required	1 or 0.5 for airplanes operated under §135.152(j)	0.5% of full range	
24. Outside Air Temperature or Total Air Temperature[13]	-50 °C to +90 °C	±2 °C	2	0.3 °C	
25. Autopilot/Autothrottle/AFCS Mode and Engagement Status	A suitable combination of discretes		1		Discretes should show which systems are engaged and which primary modes are controlling the flight path and speed of the aircraft.
26. Radio Altitude[14]	-20 ft to 2,500 ft	±2 ft or ±3% Whichever is Greater Below 500 ft and ±5% Above 500 ft	1	1 ft +5% above 500 ft	For autoland/category 3 operations. Each radio altimeter should be recorded, but arranged so that at least one is recorded each second.
27. Localizer Deviation, MLS Azimuth, or GPS Lateral Deviation	±400 Microamps or available sensor range as installed ±62°	As installed ±3% recommended.	1	0.3% of full range	For autoland/category 3 operations. Each system should be recorded but arranged so that at least one is recorded each second. It is not necessary to record ILS and MLS at the same time, only the approach aid in use need be recorded.
28. Glideslope Deviation, MLS Elevation, or GPS Vertical Deviation	±400 Microamps or available sensor range as installed 0.9 to + 30°	As installed ±3% recommended	1	0.3% of full range	For autoland/category 3 operations. Each system should be recorded but arranged so that at least one is recorded each second. It is not necessary to record ILS and MLS at the same time, only the approach aid in use need be recorded.
29. Marker Beacon Passage	Discrete "on" or "off"		1		A single discrete is acceptable for all markers.
30. Master Warning	Discrete		1		Record the master warning and record each "red" warning that cannot be determined from other parameters or from the cockpit voice recorder.
31. Air/ground sensor (primary airplane system reference nose or main gear)	Discrete "air" or "ground"		1 (0.25 recommended.)		
32. Angle of Attack (If measured directly)	As installed	As installed	2 or 0.5 for airplanes operated under §135.152(j)	0.3% of full range	If left and right sensors are available, each may be recorded at 4 or 1 second intervals, as appropriate, so as to give a data point at 2 seconds or 0.5 second, as required.
33. Hydraulic Pressure Low, Each System	Discrete or available sensor range, "low" or "normal"	±5%	2	0.5% of full range.	
34. Groundspeed	As installed	Most Accurate Systems Installed	1	0.2% of full range.	
35. GPWS (ground proximity warning system)	Discrete "warning" or "off"		1		A suitable combination of discretes unless recorder capacity is limited in which case a single discrete for all modes is acceptable.

Parameters	Range	Accuracy (sensor input)	Seconds per sampling interval	Resolution	Remarks
36. Landing Gear Position or Landing gear cockpit control selection	Discrete		4		A suitable combination of discretes should be recorded.
37. Drift Angle[15]	As installed	As installed	4	0.1°	
38. Wind Speed and Direction	As installed	As installed	4	1 knot, and 1.0°.	
39. Latitude and Longitude	As installed	As installed	4	0.002°, or as installed	Provided by the Primary Navigation System Reference. Where capacity permits latitude/longitude resolution should be 0.0002°.
40. Stick shaker and pusher activation	Discrete(s) "on" or "off"		1		A suitable combination of discretes to determine activation.
41. Windshear Detection	Discrete "warning" or "off"		1.		
42. Throttle/power lever position[16]	Full Range	±2%	1 for each lever	2% of full range	For airplanes with non-mechanically linked cockpit engine controls.
43. Additional Engine Parameters	As installed	As installed	Each engine each second	2% of full range	Where capacity permits, the preferred priority is indicated vibration level, N2, EGT, Fuel Flow, Fuel Cut-off lever position and N3, unless engine manufacturer recommends otherwise.
44. Traffic Alert and Collision Avoidance System (TCAS)	Discretes	As installed	1		A suitable combination of discretes should be recorded to determine the status of—Combined Control, Vertical Control, Up Advisory, and down advisory. (ref. ARINC Characteristic 735 Attachment 6E, TCAS VERTICAL RA DATA OUTPUT WORD.)
45. DME 1 and 2 Distance	0–200 NM;	As installed	4	1 NM	1 mile.
46. Nav 1 and 2 Selected Frequency	Full range	As installed	4		Sufficient to determine selected frequency.
47. Selected barometric setting	Full Range	±5%	(1 per 64 sec.)	0.2% of full range.	
48. Selected altitude	Full Range	±5%	1	100 ft.	
49. Selected speed	Full Range	±5%	1	1 knot.	
50. Selected Mach	Full Range	±5%	1	.01.	
51. Selected vertical speed	Full Range	±5%	1	100 ft./min.	
52. Selected heading	Full Range	±5%	1	1°.	
53. Selected flight path	Full Range	±5%	1	1°.	
54. Selected decision height	Full Range	±5%	64	1 ft.	
55. EFIS display format	Discrete(s)		4		Discretes should show the display system status (e.g., off, normal, fail, composite, sector, plan, nav aids, weather radar, range, copy.
56. Multi-function/Engine Alerts Display format	Discrete(s)		4		Discretes should show the display system status (e.g., off, normal, fail, and the identity of display pages for emergency procedures, need not be recorded.
57. Thrust comand[17]	Full Range	±2%	2	2% of full range	
58. Thrust target	Full Range	±2%	4	2% of full range.	
59. Fuel quantity in CG trim tank	Full Range	±5%	(1 per 64 sec.)	1% of full range.	
60. Primary Navigation System Reference	Discrete GPS, INS, VOR/DME, MLS, Loran C, Omega, Localizer Glidescope		4		A suitable combination of discretes to determine the Primary Navigation System reference.
61. Ice Detection	Discrete "ice" or "no ice"		4.		
62. Engine warning each engine vibration	Discrete		1.		
63. Engine warning each engine over temp.	Discrete		1.		
64. Engine warning each engine oil pressure low	Discrete		1.		
65. Engine warning each engine over speed	Discrete		1.		
66. Yaw Trim Surface Position	Full Range	±3% Unless Higher Accuracy Uniquely Required	2	0.3% of full range.	
67. Roll Trim Surface Position	Full Range	±3% Unless Higher Accuracy Uniquely Required	2	0.3% of full range.	

PART 135

FAR

Parameters	Range	Accuracy (sensor input)	Seconds per sampling interval	Resolution	Remarks
68. Brake Pressure (left and right)	As installed	±5%	1		To determine braking effort applied by pilots or by autobrakes.
69. Brake Pedal Application (left and right)	Discrete or Analog "applied" or "off"	±5% (Analog)	1		To determine braking applied by pilots.
70. Yaw or sideslip angle	Full Range	±5%	1	0.5°.	
71. Engine bleed valve position	Discrete "open" or "closed"		4.		
72. De-icing or anti-icing system selection	Discrete "on" or "off"		4.		
73. Computed center of gravity	Full Range	±5%	(1 per 64 sec.)	1% of full range.	
74. AC electrical bus status	Discrete "power" or "off"		4		Each bus.
75. DC electrical bus status	Discrete "power" or "off"		4		Each bus.
76. APU bleed valve position	Discrete "open" or "closed"		4.		
77. Hydraulic Pressure (each system)	Full range	±5%	2	100 psi.	
78. Loss of cabin pressure	Discrete "loss" or "normal"		1.		
79. Computer failure (critical flight and engine control systems)	Discrete "fail" or "normal"		4.		
80. Heads-up display (when an information source is installed)	Discrete(s) "on" or "off"		4.		
81. Para-visual display (when an information source is installed)	Discrete(s) "on" or "off"		1.		
82. Cockpit trim control input position—pitch	Full Range	±5%	1	0.2% of full range	Where mechanical means for control inputs are not available, cockpit display trim positions should be recorded.
83. Cockpit trim control input position—roll	Full Range	±5%	1	0.7% of full range	Where mechanical means for control inputs are not available, cockpit display trim position should be recorded.
84. Cockpit trim control input position—yaw	Full Range	±5%	1	0.3% of full range	Where mechanical means for control input are not available, cockpit display trim positions should be recorded.
85. Trailing edge flap and cockpit flap control position	Full Range	±5%	2	0.5% of full range	Trailing edge flaps and cockpit flap control position may each be sampled alternately at 4 second intervals to provide a sample each 0.5 second.
86. Leading edge flap and cockpit flap control position	Full Range or Discrete	±5%	1	0.5% of full range.	
87. Ground spoiler position and speed brake selection	Full Range or Discrete	±5%	0.5	0.3% of full range	
88. All cockpit flight control input forces (control wheel, control column, rudder pedal)[18]	Full Range Control wheel ±70 lbs. Control column ±85 lbs. Rudder pedal ±165 lbs	±5°	1	0.3% of full range	For fly-by-wire flight control systems, where flight control surface position is a function of the displacement of the control input device only, it is not necessary to record this parameter. For airplanes that have a flight control breakaway capability that allows either pilot to operate the control independently, record both control force inputs. The control force inputs may be sampled alternately once per 2 seconds to produce the sampling interval of 1.

[1] For A300 B2/B4 airplanes, resolution = 6 seconds.
[2] For A330/A340 series airplanes, resolution = 0.703°.
[3] For A318/A319/A320/A321 series airplanes, resolution = 0.275% (0.088°>0.064°). For A330/A340 series airplanes, resolution = 2.20% (0.703°>0.064°).
[4] For A318/A319/A320/A321 series airplanes, resolution = 0.22% (0.088°>0.080°). For A330/A340 series airplanes, resolution = 1.76% (0.703°>0.080°).
[5] For A330/A340 series airplanes, resolution = 1.18% (0.703°>0.120°).
[6] For A330/A340 series airplanes, resolution = 0.783% (0.352°>0.090°).
[7] For A330/A340 series airplanes, aileron resolution = 0.704% (0.352°>0.100°). For A330/A340 series airplanes, spoiler resolution = 1.406% (0.703°>0.100°).
[8] For A330/A340 series airplanes, resolution = 0.30% (0.176°>0.12°). For A330/A340 series airplanes, seconds per sampling interval = 1.
[9] For B-717 series airplanes, resolution =.005g. For Dassault F900C/F900EX airplanes, resolution =.007g.
[10] For A330/A340 series airplanes, resolution = 1.05% (0.250°>0.120°).
[11] For A330/A340 series airplanes, resolution = 1.05% (0.250°>0.120°). For A300 B2/B4 series airplanes, resolution = 0.92% (0.230°>0.125°).
[12] For A330/A340 series airplanes, spoiler resolution = 1.406% (0.703°>0.100°).
[13] For A330/A340 series airplanes, resolution = 0.5 °C.
[14] For Dassault F900C/F900EX airplanes, Radio Altitude resolution = 1.25 ft.
[15] For A330/A340 series airplanes, resolution = 0.352 degrees.
[16] For A318/A319/A320/A321 series airplanes, resolution = 4.32%. For A330/A340 series airplanes, resolution is 3.27% of full range for throttle lever angle (TLA); for reverse thrust, reverse throttle lever angle (RLA) resolution is nonlinear over the active reverse thrust range, which is 51.54 degrees to 96.14 degrees. The resolved element is 2.8 degrees uniformly over the entire active reverse thrust range, or 2.9% of the full range value of 96.14 degrees.
[17] For A318/A319/A320/A321 series airplanes, with IAE engines, resolution = 2.58%.
[18]- For all aircraft manufactured on or after December 6, 2010, the seconds per sampling interval is 0.125. Each input must be recorded at this rate. Alternately sampling inputs (interleaving) to meet this sampling interval is prohibited.

[Doc. No. 28109, 62 FR 38398, July 17, 1997; 62 FR 48135, Sept. 12, 1997; Amdt. 135-85, 67 FR 54323, Aug. 21, 2002; Amdt. 135-89, 68 FR 42939, July 18, 2003; 68 FR 50069, Aug. 20, 2003; Amdt. 135-113, 73 FR 12570, Mar. 7, 2008; Amdt. 135-121, 75 FR 17047, Apr. 5, 2010; Amdt. 135-120, 75 FR 7357, Feb. 19, 2010]

63. *Distribution system.* (a) For the purpose of complying with this section, the distribution system includes the distribution busses, their associated feeders, and each control and protective device.

(b) Each system must be designed so that essential load circuits can be supplied in the event of reasonably probable faults or open circuits, including faults in heavy current carrying cables.

(c) If two independent sources of electrical power for particular equipment or systems are required under this appendix, their electrical energy supply must be ensured by means such as duplicate electrical equipment, throw-over switching, or multichannel or loop circuits separately routed.

64. *Circuit protective devices.* The circuit protective devices for the electrical circuits of the airplane must meet FAR 23.1357, and in addition circuits for loads which are essential to safe operation must have individual and exclusive circuit protection.

Appendix G to Part 135—Extended Operations (ETOPS)

G135.1 Definitions.

G135.1.1 *Adequate Airport* means an airport that an airplane operator may list with approval from the FAA because that airport meets the landing limitations of § 135.385 or is a military airport that is active and operational.

G135.1.2 *ETOPS Alternate Airport* means an adequate airport that is designated in a dispatch or flight release for use in the event of a diversion during ETOPS. This definition applies to flight planning and does not in any way limit the authority of the pilot in command during flight.

G135.1.3 *ETOPS Entry Point* means the first point on the route of an ETOPS flight, determined using a one-engine inoperative cruise speed under standard conditions in still air, that is more than 180 minutes from an adequate airport.

G135.1.4 *ETOPS Qualified Person* means a person, performing maintenance for the certificate holder, who has satisfactorily completed the certificate holder's ETOPS training program.

G135.2 Requirements.

G135.2.1 *General.* After August 13, 2008, no certificate holder may operate an airplane, other than an all-cargo airplane with more than two engines, outside the continental United States more than 180 minutes flying time (at the one-engine-inoperative cruise speed under standard conditions in still air) from an airport described in § 135.364 unless—

(a) The certificate holder receives ETOPS approval from the FAA;

(b) The operation is conducted in a multi-engine transport category turbine-powered airplane;

(c) The operation is planned to be no more than 240 minutes flying time (at the one engine inoperative cruise speed under standard conditions in still air) from an airport described in § 135.364; and

(d) The certificate holder meets the requirements of this appendix.

G135.2.2 *Required certificate holder experience prior to conducting ETOPS.*

Before applying for ETOPS approval, the certificate holder must have at least 12 months experience conducting international operations (excluding Canada and Mexico) with multi-engine transport category turbine-engine powered airplanes. The certificate holder may consider the following experience as international operations:

(a) Operations to or from the State of Hawaii.

(b) For certificate holders granted approval to operate under part 135 or part 121 before February 15, 2007, up to 6 months of domestic operating experience and operations in Canada and Mexico in multi-engine transport category turbojet-powered airplanes may be credited as part of the required 12 months of international experience required by paragraph G135.2.2(a) of this appendix.

(c) ETOPS experience with other aircraft types to the extent authorized by the FAA.

G135.2.3 *Airplane requirements.* No certificate holder may conduct ETOPS in an airplane that was manufactured after February 17, 2015 unless the airplane meets the standards of § 25.1535.

G135.2.4 *Crew information requirements.* The certificate holder must ensure that flight crews have in-flight access to current weather and operational information needed to comply with § 135.83, § 135.225, and § 135.229. This includes information on all ETOPS Alternate Airports, all destination alternates, and the destination airport proposed for each ETOPS flight.

G135.2.5 *Operational Requirements.*

(a) No person may allow a flight to continue beyond its ETOPS Entry Point unless—

(1) The weather conditions at each ETOPS Alternate Airport are forecast to be at or above the operating minima in the certificate holder's operations specifications for that airport when it might be used (from the earliest to the latest possible landing time), and

(3) All ETOPS Alternate Airports within the authorized ETOPS maximum diversion time are reviewed for any changes in conditions that have occurred since dispatch.

(b) In the event that an operator cannot comply with paragraph G135.2.5(a)(1) of this appendix for a specific airport, another ETOPS Alternate Airport must be substituted within the maximum ETOPS diversion time that could be authorized for that flight with weather conditions at or above operating minima.

(c) Pilots must plan and conduct ETOPS under instrument flight rules.

(d) *Time-Limited Systems.*

(1) Except as provided in paragraph G135.2.5(d)(3) of this appendix, the time required to fly the distance to each ETOPS Alternate Airport (at the all-engines-operating cruise speed, corrected for wind and temperature) may not exceed the time specified in the Airplane Flight Manual for the airplane's most limiting fire suppression system time required by regulation for any cargo or baggage compartments (if installed), minus 15 minutes.

(2) Except as provided in G135.2.5(d)(3) of this appendix, the time required to fly the distance to each ETOPS Alternate Airport (at the approved one-engine-inoperative cruise speed, corrected for wind and temperature) may not exceed the time specified in the Airplane Flight Manual for the airplane's most time limited system time (other than the airplane's most limiting fire suppression system time required by regulation for any cargo or baggage compartments), minus 15 minutes.

(3) A certificate holder operating an airplane without the Airplane Flight Manual information needed to comply with paragraphs G135.2.5(d)(1) and (d)(2) of this appendix, may continue ETOPS with that airplane until February 17, 2015.

G135.2.6 *Communications Requirements.*

(a) No person may conduct an ETOPS flight unless the following communications equipment, appropriate to the route to be flown, is installed and operational:

(1) Two independent communication transmitters, at least one of which allows voice communication.

(2) Two independent communication receivers, at least one of which allows voice communication.

(3) Two headsets, or one headset and one speaker.

(b) In areas where voice communication facilities are not available, or are of such poor quality that voice communication is not possible, communication using an alternative system must be substituted.

G135.2.7 *Fuel Requirements.* No person may dispatch or release for flight an ETOPS flight unless, considering wind and other weather conditions expected, it has the fuel otherwise required by this part and enough fuel to satisfy each of the following requirements:

(a) *Fuel to fly to an ETOPS Alternate Airport.* (1) Fuel to account for rapid decompression and engine failure. The airplane must carry the greater of the following amounts of fuel:

(i) Fuel sufficient to fly to an ETOPS Alternate Airport assuming a rapid decompression at the most critical point followed by descent to a safe altitude in compliance with the oxygen supply requirements of § 135.157;

(ii) Fuel sufficient to fly to an ETOPS Alternate Airport (at the one-engine-inoperative cruise speed under standard conditions in still air) assuming a rapid decompression and a simultaneous engine failure at the most critical point followed by descent to a safe altitude in compliance with the oxygen requirements of § 135.157; or

(iii) Fuel sufficient to fly to an ETOPS Alternate Airport (at the one-engine-inoperative cruise speed under standard conditions in still air) assuming an engine failure at the most critical point followed by descent to the one engine inoperative cruise altitude.

(2) Fuel to account for errors in wind forecasting. In calculating the amount of fuel required by paragraph G135.2.7(a)(1) of this appendix, the certificate holder must increase the actual forecast wind speed by 5% (resulting in an increase in headwind or a decrease in tailwind) to account for any potential errors in wind forecasting. If a certificate holder is not using the actual forecast wind based on a wind model accepted by the FAA, the airplane must carry additional fuel equal to 5% of the fuel required by paragraph G135.2.7(a) of this appendix, as reserve fuel to allow for errors in wind data.

(3) Fuel to account for icing. In calculating the amount of fuel required by paragraph G135.2.7(a)(1) of this appendix, (after completing the wind calculation in G135.2.7(a)(2) of this appendix), the certificate holder must ensure that the airplane carries the greater of the following amounts of fuel in anticipation of possible icing during the diversion:

(i) Fuel that would be burned as a result of airframe icing during 10 percent of the time icing is forecast (including the fuel used by engine and wing anti-ice during this period).

(ii) Fuel that would be used for engine anti-ice, and if appropriate wing anti-ice, for the entire time during which icing is forecast.

(4) Fuel to account for engine deterioration. In calculating the amount of fuel required by paragraph G135.2.7(a)(1) of this appendix (after completing the wind calculation in paragraph G135.2.7(a)(2) of this appendix), the certificate holder must ensure the airplane also carries fuel equal to 5% of the fuel specified above, to account for deterioration in cruise fuel burn performance unless the certificate holder has a program to monitor airplane in-service deterioration to cruise fuel burn performance.

(b) *Fuel to account for holding, approach, and landing.* In addition to the fuel required by paragraph G135.2.7 (a) of this appendix, the airplane must carry fuel sufficient to hold at 1500 feet above field elevation for 15 minutes upon reaching the ETOPS Alternate Airport and then conduct an instrument approach and land.

(c) *Fuel to account for APU use.* If an APU is a required power source, the certificate holder must account for its fuel consumption during the appropriate phases of flight.

G135.2.8 *Maintenance Program Requirements.* In order to conduct an ETOPS flight under § 135.364, each certificate holder must develop and comply with the ETOPS maintenance program as authorized in the certificate holder's operations specifications for each two-engine airplane-engine combination used in ETOPS. This provision does not apply to operations using an airplane with more than two engines. The certificate holder must develop this ETOPS maintenance program to supplement the maintenance program currently approved for the operator. This ETOPS maintenance program must include the following elements:

(a) *ETOPS maintenance document.* The certificate holder must have an ETOPS maintenance document for use by each person involved in ETOPS. The document must—

(1) List each ETOPS Significant System,

(2) Refer to or include all of the ETOPS maintenance elements in this section,

(3) Refer to or include all supportive programs and procedures,

(4) Refer to or include all duties and responsibilities, and

(5) Clearly state where referenced material is located in the certificate holder's document system.

(b) *ETOPS pre-departure service check.* The certificate holder must develop a pre-departure check tailored to their specific operation.

(1) The certificate holder must complete a pre-departure service check immediately before each ETOPS flight.

(2) At a minimum, this check must:

(i) Verify the condition of all ETOPS Significant Systems;

(ii) Verify the overall status of the airplane by reviewing applicable maintenance records; and

(iii) Include an interior and exterior inspection to include a determination of engine and APU oil levels and consumption rates.

(3) An appropriately trained maintenance person, who is ETOPS qualified must accomplish and certify by signature ETOPS specific tasks. Before an ETOPS flight may commence, an ETOPS pre-departure service check (PDSC) Signatory Person, who has been authorized by the certificate holder, must certify by signature, that the ETOPS PDSC has been completed.

(4) For the purposes of this paragraph (b) only, the following definitions apply:

(i) ETOPS qualified person: A person is ETOPS qualified when that person satisfactorily completes the operator's ETOPS training program and is authorized by the certificate holder.

(ii) ETOPS PDSC Signatory Person: A person is an ETOPS PDSC Signatory Person when that person is ETOPS Qualified and that person:

(A) When certifying the completion of the ETOPS PDSC in the United States:

(*1*) Works for an operator authorized to engage in part 135 or 121 operation or works for a part 145 repair station; and

(*2*) Holds a U.S. Mechanic's Certificate with airframe and powerplant ratings.

(B) When certifying the completion of the ETOPS PDSC outside of the U.S. holds a certificate in accordance with § 43.17(c)(1) of this chapter; or

(C) When certifying the completion of the ETOPS PDSC outside the U.S. holds the certificates needed or has the requisite experience or training to return aircraft to service on behalf of an ETOPS maintenance entity.

(iii) ETOPS maintenance entity: An entity authorized to perform ETOPS maintenance and complete ETOPS predeparture service checks and that entity is:

(A) Certificated to engage in part 135 or 121 operations;

(B) Repair station certificated under part 145 of this title; or

(C) Entity authorized pursuant to § 43.17(c)(2) of this chapter.

(c) *Limitations on dual maintenance.* (1) Except as specified in paragraph G135.2.8(c)(2) of this appendix, the certificate holder may not perform scheduled or unscheduled dual maintenance during the same maintenance visit on the same or a substantially similar ETOPS Significant System listed in the ETOPS maintenance document, if the improper maintenance could result in the failure of an ETOPS Significant System.

(2) In the event dual maintenance as defined in paragraph G135.2.8(c)(1) of this appendix cannot be avoided, the certificate holder may perform maintenance provided:

(i) The maintenance action on each affected ETOPS Significant System is performed by a different technician, or

(ii) The maintenance action on each affected ETOPS Significant System is performed by the same technician under the direct supervision of a second qualified individual; and

(iii) For either paragraph G135.2.8(c)(2)(i) or (ii) of this appendix, a qualified individual conducts a ground verification test and any in-flight verification test required under the program developed pursuant to paragraph G135.2.8(d) of this appendix.

(d) *Verification program.* The certificate holder must develop a program for the resolution of discrepancies that will ensure the effectiveness of maintenance actions taken on ETOPS Significant Systems. The verification program must identify potential problems and verify satisfactory corrective action. The verification program must include ground verification and in-flight verification policy and procedures. The certificate holder must establish procedures to clearly indicate who is going to initiate the verification action and what action is necessary. The verification action may be performed on an ETOPS revenue flight provided the verification action is documented as satisfactorily completed upon reaching the ETOPS entry point.

(e) *Task identification.* The certificate holder must identify all ETOPS-specific tasks. An ETOPS qualified person must accomplish and certify by signature that the ETOPS-specific task has been completed.

(f) *Centralized maintenance control procedures.* The certificate holder must develop procedures for centralized maintenance control for ETOPS.

(g) *ETOPS parts control program.* The certificate holder must develop an ETOPS parts control program to ensure the proper identification of parts used to maintain the configuration of airplanes used in ETOPS.

(h) *Enhanced Continuing Analysis and Surveillance System (E-CASS) program.* A certificate holder's existing CASS must be enhanced to include all elements of the ETOPS maintenance program. In addition to the reporting requirements of § 135.415 and § 135.417, the program includes reporting procedures, in the form specified in § 135.415(e), for the following significant events detrimental to ETOPS within 96 hours of the occurrence to the certificate holding district office (CHDO):

(1) IFSDs, except planned IFSDs performed for flight training.

(2) Diversions and turnbacks for failures, malfunctions, or defects associated with any airplane or engine system.

(3) Uncommanded power or thrust changes or surges.

(4) Inability to control the engine or obtain desired power or thrust.

(5) Inadvertent fuel loss or unavailability, or uncorrectable fuel imbalance in flight.

(6) Failures, malfunctions or defects associated with ETOPS Significant Systems.

(7) Any event that would jeopardize the safe flight and landing of the airplane on an ETOPS flight.

(i) *Propulsion system monitoring.*

The certificate holder, in coordination with the CHDO, must—

(1) Establish criteria as to what action is to be taken when adverse trends in propulsion system conditions are detected, and

(2) Investigate common cause effects or systemic errors and submit the findings to the CHDO within 30 days.

(j) *Engine condition monitoring.*

(1) The certificate holder must establish an engine-condition monitoring program to detect deterioration at an early stage and to allow for corrective action before safe operation is affected.

(2) This program must describe the parameters to be monitored, the method of data collection, the method of analyzing data, and the process for taking corrective action.

(3) The program must ensure that engine limit margins are maintained so that a prolonged engine-inoperative diversion may be conducted at approved power levels and in all expected environmental conditions without exceeding approved engine limits. This includes approved limits for items such as rotor speeds and exhaust gas temperatures.

(k) *Oil consumption monitoring.* The certificate holder must develop an engine oil consumption monitoring program to ensure that there is enough oil to complete each ETOPS flight. APU oil consumption must be included if an APU is required for ETOPS. The operator's consumption limit may not exceed the manufacturer's recommendation. Monitoring must be continuous and include oil added at each ETOPS departure point. The program must compare the amount of oil added at each ETOPS departure point with the running average consumption to identify sudden increases.

(l) *APU in-flight start program.* If an APU is required for ETOPS, but is not required to run during the ETOPS portion of the flight, the certificate holder must have a program acceptable to the FAA for cold soak in-flight start and run reliability.

(m) *Maintenance training.* For each airplane-engine combination, the certificate holder must develop a maintenance training program to ensure that it provides training adequate to support ETOPS. It must include ETOPS specific training for all persons involved in ETOPS maintenance that focuses on the special nature of ETOPS. This training must be in addition to the operator's maintenance training program used to qualify individuals for specific airplanes and engines.

(n) *Configuration, maintenance, and procedures (CMP) document.* The certificate holder must use a system to ensure compliance with the minimum requirements set forth in the current version of the CMP document for each airplane-engine combination that has a CMP.

(o) *Reporting.* The certificate holder must report quarterly to the CHDO and the airplane and engine manufacturer for each airplane authorized for ETOPS. The report must provide the operating hours and cycles for each airplane.

G135.2.9 *Delayed compliance date for all airplanes.* A certificate holder need not comply with this appendix for any airplane until August 13, 2008.

[Doc. No. FAA-2002-6717, 72 FR 1885, Jan. 16, 2007, as amended by Amdt. 135-108, 72 FR 7348, Feb. 15, 2007; 72 FR 26542, May 10, 2007; Amdt. 135-112, 73 FR 8798, Feb. 15, 2008; Amdt. 135-115, 73 FR 33882, June 16, 2008]

PART 136

COMMERCIAL AIR TOURS AND NATIONAL PARKS AIR TOUR MANAGEMENT

Contents

Subpart A—National Air Tour Safety Standards

Subpart B—National Parks Air Tour Management

Subpart C—Grand Canyon National Park

Authority: 49 U.S.C. 106(g), 40113, 40119, 44101, 44701, 44701-44702, 44705, 44709-44711, 44713, 44716-44717, 44722, 44901, 44903-44904, 44912, 46105.

Source: Docket No. FAA-2001-8690, 67 FR 65667, Oct. 25, 2002, unless otherwise noted.

Subpart A—National Air Tour Safety Standards

Source: Docket No. FAA-1998-4521, 72 FR 6912, Feb. 13, 2007, unless otherwise noted.

§ 136.1 Applicability and definitions.

(a) This subpart applies to each person operating or intending to operate a commercial air tour in an airplane or helicopter and, when applicable, to all occupants of the airplane or helicopter engaged in a commercial air tour. When any requirement of this subpart is more stringent than any other requirement of this chapter, the person operating the commercial air tour must comply with the requirement in this subpart.

(b) As of September 11, 2007, this subpart is applicable to:

(1) Part 121 or 135 operators conducting a commercial air tour and holding a part 119 certificate;

(2) Part 91 operators conducting flights as described in § 119.1(e)(2); and

(3) Part 91 operators conducting flights as described in 14 CFR 91.146

(c) This subpart is not applicable to operations conducted in balloons, gliders (powered or un-powered), parachutes (powered or un-powered), gyroplanes, or airships.

(d) For the purposes of this subpart the following definitions apply:

Commercial Air Tour means a flight conducted for compensation or hire in an airplane or helicopter where a purpose of the flight is sightseeing. The FAA may consider the following factors in determining whether a flight is a commercial air tour for purposes of this subpart:

(1) Whether there was a holding out to the public of willingness to conduct a sightseeing flight for compensation or hire;

(2) Whether the person offering the flight provided a narrative that referred to areas or points of interest on the surface below the route of the flight;

(3) The area of operation;

(4) How often the person offering the flight conducts such flights;

(5) The route of the flight;

(6) The inclusion of sightseeing flights as part of any travel arrangement package;

(7) Whether the flight in question would have been canceled based on poor visibility of the surface below the route of the flight; and

(8) Any other factors that the FAA considers appropriate.

Commercial Air Tour operator means any person who conducts a commercial air tour.

Life preserver means a flotation device used by an aircraft occupant if the aircraft ditches in water. If an inflatable device, it must be un-inflated and ready for its intended use once inflated. In evaluating whether a non-inflatable life preserver is acceptable to the FAA, the operator must demonstrate to the FAA that such a preserver can be used during an evacuation and will allow all passengers to exit the aircraft without blocking the exit. Each occupant must have the physical capacity to wear and inflate the type of device used once briefed by the commercial air tour operator. Seat cushions do not meet this definition.

Raw terrain means any area on the surface, including water, devoid of any person, structure, vehicle, or vessel.

Shoreline means that area of the land adjacent to the water of an ocean, sea, lake, pond, river or tidal basin that is above the high water mark and excludes land areas unsuitable for landing such as vertical cliffs or land intermittently under water during the particular flight.

Suitable landing area for helicopters means an area that provides the operator reasonable capability to land without damage to equipment or injury to persons. Suitable landing areas must be site-specific, designated by the operator, and accepted by the FAA. These site-specific areas would provide an emergency landing area for a single-engine helicopter or a multiengine helicopter that does not have the capability to reach a safe landing area after an engine power loss.

(e) In an in-flight emergency requiring immediate action, the pilot in command may deviate from any rule of this subpart to the extent required to meet that emergency.

§ 136.3 Letters of Authorization.

Operators subject to this subpart who have Letters of Authorization may use the procedures described in 14 CFR 119.51 to amend or have the FAA reconsider those Letters of Authorization.

§ 136.5 Additional requirements for Hawaii.

No person may conduct a commercial air tour in the State of Hawaii unless they comply with the additional requirements and restrictions in appendix A to part 136.

§ 136.7 Passenger briefings.

(a) Before takeoff each pilot in command shall ensure that each passenger has been briefed on the following:

(1) Procedures for fastening and unfastening seatbelts;

(2) Prohibition on smoking; and

(3) Procedures for opening exits and exiting the aircraft.

(b) For flight segments over water beyond the shoreline, briefings must also include:

(1) Procedures for water ditching;

(2) Use of required life preservers; and

(3) Procedures for emergency exit from the aircraft in the event of a water landing.

§ 136.9 Life preservers for over water.

(a) Except as provided in paragraphs (b) or (c) of this section, the operator and pilot in command of commercial air tours over water beyond the shoreline must ensure that each occupant is wearing a life preserver from before takeoff until flight is no longer over water.

(b) The operator and pilot in command of a commercial air tour over water beyond the shoreline must ensure that a life preserver is readily available for its intended use and easily accessible to each occupant if:

(1) The aircraft is equipped with floats; or

(2)The airplane is within power-off gliding distance to the shoreline for the duration of the time that the flight is over water.

(3)The aircraft is a multi engine that can be operated with the critical engine inoperative at a weight that will allow it to climb, at least 50 feet a minute, at an altitude of 1,000 feet above the surface, as provided in the Airplane Flight Manual or the Rotorcraft Flight Manual, as appropriate.

(c) No life preserver is required if the overwater operation is necessary only for takeoff or landing.

§ 136.11 Helicopter floats for over water.

(a) A helicopter used in commercial air tours over water beyond the shoreline must be equipped with fixed floats or an inflatable flotation system adequate to accomplish a safe emergency ditching, if—

(1) It is a single-engine helicopter; or

(2) It is a multi-engine helicopter that cannot be operated with the critical engine inoperative at a weight that will allow it to climb, at least 50 feet a minute, at an altitude of 1,000 feet above the surface, as provided in the Rotorcraft Flight Manual (RFM).

(b) Each helicopter that is required to be equipped with an inflatable flotation system must have:

(1) The activation switch for the flotation system on one of the primary flight controls, and

(2) The flotation system armed when the helicopter is over water and is flying at a speed that does not exceed the maximum speed prescribed in the Rotorcraft Flight Manual for flying with the flotation system armed.

(c) Fixed floats or an inflatable flotation system is not required for a helicopter under this section if:

(1) The helicopter is over water only during the takeoff or landing portion of the flight, or

(2) The helicopter is operated within power-off gliding distance to the shoreline for the duration of the flight and each occupant is wearing a life preserver from before takeoff until the aircraft is no longer over water.

(d) Air tour operators required to comply with paragraphs (a) and/or (b) of this section must meet these requirements on or before September 5, 2008.

§ 136.13 Helicopter performance plan and operations.

(a) Each operator must complete a performance plan before each helicopter commercial air tour, or flight operated under 14 CFR 91.146 or 91.147. The pilot in command must review for accuracy and comply with the performance plan on the day the flight is flown. The performance plan must be based on the information in the Rotorcraft Flight Manual (RFM) for that helicopter, taking into consideration the maximum density altitude for which the operation is planned, in order to determine:

(1) Maximum gross weight and center of gravity (CG) limitations for hovering in ground effect;

(2) Maximum gross weight and CG limitations for hovering out of ground effect; and

(3) Maximum combination of weight, altitude, and temperature for which height/velocity information in the RFM is valid.

(b) Except for the approach to and transition from a hover for the purpose of takeoff and landing, or during takeoff and landing, the pilot in command must make a reasonable plan to operate the helicopter outside of the caution/warning/avoid area of the limiting height/velocity diagram.

(c) Except for the approach to and transition from a hover for the purpose of takeoff and landing, during takeoff and landing, or when necessary for safety of flight, the pilot in command must operate the helicopter in compliance with the plan described in paragraph (b) of this section.

§§ 136.15-136.29 [Reserved]

Subpart B—National Parks Air Tour Management

Source: Docket. No. FAA-1998-4521, 72 FR 6912, Feb. 13, 2007, unless otherwise noted.

§ 136.31 Applicability.

(a) This part restates and paraphrases several sections of the National Parks Air Tour Management Act of 2000, including section 803 (codified at 49 U.S.C. 40128) and sections 806 and 809. This subpart clarifies the requirements for the development of an air tour management plan for each park in the national park system where commercial air tour operations are flown.

(b) Except as provided in paragraph (c) of this section, this subpart applies to each commercial air tour operator who conducts a commercial air tour operation over—

(1) A unit of the national park system;

(2) Tribal lands as defined in this subpart; or

(3) Any area within one-half mile outside the boundary of any unit of the national park system.

(c) This subpart does not apply to a commercial air tour operator conducting a commercial air tour operation—

(1) Over the Grand Canyon National Park;

(2) Over that portion of tribal lands within or abutting the Grand Canyon National Park;

(3) Over any land or waters located in the State of Alaska; or

(4) While flying over or near the Lake Mead Recreation Area, solely as a transportation route, to conduct a commercial air tour over the Grand Canyon National Park.

[Doc. No. FAA-2001-8690, 67 FR 65667, Oct. 25, 2002. Redesignated and amended by Amdt. 136-1, 72 FR 6912, Feb. 13, 2007]

§ 136.33 Definitions.

For purposes of this subpart—

(a) *Commercial air tour operator* means any person who conducts a commercial air tour operation.

(b) *Existing commercial air tour operator* means a commercial air tour operator that was actively engaged in the business of providing commercial air tour operations over a national park at any time during the 12-month period ending on April 5, 2000.

(c) *New entrant commercial air tour operator* means a commercial air tour operator that—

(1) Applies for operating authority as a commercial air tour operator for a national park or tribal lands; and

(2) Has not engaged in the business of providing commercial air tour operations over the national park or tribal lands for the 12-month period preceding enactment.

(d) *Commercial air tour operation* —

(1) Means any flight, conducted for compensation or hire in a powered aircraft where a purpose of the flight is sightseeing over a national park, within $\frac{1}{2}$ mile outside the boundary of any national park, or over tribal lands, during which the aircraft flies—

(i) Below 5,000 feet above ground level (except for the purpose of takeoff or landing, or as necessary for the safe operation of an aircraft as determined under the rules and regulations of the Federal Aviation Administration requiring the pilot-in-command to take action to ensure the safe operation of the aircraft);

(ii) Less than 1 mile laterally from any geographic feature within the park (unless more than $\frac{1}{2}$ mile outside the boundary); or

(iii) Except as provided in § 136.35.

(2) The Administrator may consider the following factors in determining whether a flight is a commercial air tour operation for purposes of this subpart—

(i) Whether there was a holding out to the public of willingness to conduct a sightseeing flight for compensation or hire;

(ii) Whether a narrative that referred to areas or points of interest on the surface below the route of the flight was provided by the person offering the flight;

(iii) The area of operation;

(iv) The frequency of flights conducted by the person offering the flight;

(v) The route of flight;

(vi) The inclusion of sightseeing flights as part of any travel arrangement package offered by the person offering the flight;

(vii) Whether the flight would have been canceled based on poor visibility of the surface below the route of the flight; and

(viii) Any other factors that the Administrator and Director consider appropriate.

(3) For purposes of § 136.35, means any flight conducted for compensation or hire in a powered aircraft where a purpose of the flight is sightseeing over a national park.

(e) *National park* means any unit of the national park system. (See title 16 of the U.S. Code, section 1, *et seq.*)

(f) *Tribal lands* means that portion of Indian country (as that term is defined in section 1151 of title 18 of the U.S. Code) that is within or abutting a national park.

(g) *Administrator* means the Administrator of the Federal Aviation Administration.

(h) *Director* means the Director of the National Park Service.

(i) *Superintendent* means the duly appointed representative of the National Park Service for a particular unit of the national park system.

[Doc. No. FAA-2001-8690, 67 FR 65667, Oct. 25, 2002. Redesignated and amended by Amdt. 136-1, 72 FR 6912, Feb. 13, 2007; Amdt. 136-1, 72 FR 31450, June 7, 2007]

§ 136.35 Prohibition of commercial air tour operations over the Rocky Mountain National Park.

All commercial air tour operations in the airspace over the Rocky Mountain National Park are prohibited regardless of altitude.

[Doc. No. FAA-2001-8690, 67 FR 65667, Oct. 25, 2002. Redesignated by Amdt. 136-1, 72 FR 6912, Feb. 13, 2007]

§ 136.37 Overflights of national parks and tribal lands.

(a) *General.* A commercial air tour operator may not conduct commercial air tour operations over a national park or tribal land except—

(1) In accordance with this section;

(2) In accordance with conditions and limitations prescribed for that operator by the Administrator; and

(3) In accordance with any applicable air tour management plan for the park or tribal lands.

(b) *Application for operating authority.* Before commencing commercial air tour operations over a national park or tribal lands, a commercial air tour operator shall apply to the Administrator for authority to conduct the operations over the park or tribal lands.

(c) *Number of operations authorized.* In determining the number of authorizations to issue to provide commercial air tour operations over a national park, the Administrator, in cooperation with the Director, shall take into consideration the provisions of the air tour management plan, the number of existing commercial air tour operators and current level of service and equipment provided by any such operators, and the financial viability of each commercial air tour operation.

(d) *Cooperation with National Park Service.* Before granting an application under this subpart, the Administrator, in cooperation with the Director, shall develop an air tour management plan in accordance with § 136.39 and implement such a plan.

(e) *Time limit on response to applications.* Every effort will be made to act on any application under this subpart and issue a decision on the application not later than 24 months after it is received or amended.

(f) *Priority.* In acting on applications under this paragraph to provide commercial air tour operations over a national park, the Administrator shall give priority to an application under this paragraph in any case where a new entrant commercial air tour operator is seeking operating authority with respect to that national park.

(g) *Exception.* Notwithstanding this section, commercial air tour operators may conduct commercial air tour operations over a national park under part 91 of this chapter if—

(1) Such activity is permitted under part 119 of this chapter;

(2) The operator secures a letter of agreement from the Administrator and the Superintendent for that park describing the conditions under which the operations will be conducted; and

(3) The number of operations under this exception is limited to not more than a total of 5 flights by all operators in any 30-day period over a particular park.

(h) *Special rule for safety requirement.* Notwithstanding § 136.41, an existing commercial air tour operator shall apply, not later than January 23, 2003 for operating authority under part 119 of this chapter, for certification under part 121 or part 135 of this chapter. A new entrant commercial air tour operator shall apply for such authority before conducting commercial air tour operations over a national park or tribal lands that are within or abut a national park. The Administrator shall make every effort to act on such application for a new entrant and issue a decision on the application not later than 24 months after it is received or amended.

[Doc. No. FAA-2001-8690, 67 FR 65667, Oct. 25, 2002. Redesignated and amended by Amdt. 136-1, 72 FR 6912, Feb. 13, 2007; Amdt. 136-1, 72 FR 31450]

§ 136.39 Air tour management plans (ATMP).

(a) *Establishment.* The Administrator, in cooperation with the Director, shall establish an air tour management plan for any national park or tribal land for which such a plan is not in effect whenever a person applies for authority to conduct a commercial air tour operation over the park. The air tour management plan shall be developed by means of a public process in accordance with paragraph (d) of this section. The objective of any air tour management plan is to develop acceptable and effective measures to mitigate or prevent the significant adverse impacts, if any, of commercial air tour operations upon the natural and cultural resources, visitor experiences, and tribal lands.

(b) *Environmental determination.* In establishing an air tour management plan under this section, the Administrator and the Director shall each sign the environmental decision document required by section 102 of the National Environmental Policy Act of 1969 (42 U.S.C. 4332) which may include a finding of no significant impact, an environmental assessment, or an environmental impact statement and the record of decision for the air tour management plan.

(c) *Contents.* An air tour management plan for a park—

(1) May prohibit commercial air tour operations in whole or in part;

(2) May establish conditions for the conduct of commercial air tour operations, including, but not limited to, commercial air tour routes, maximum number of flights per unit of time, maximum and minimum altitudes, time of day restrictions, restrictions for particular events, intrusions on privacy on tribal lands, and mitigation of noise, visual, or other impacts;

(3) Shall apply to all commercial air tour operations within $\frac{1}{2}$ mile outside the boundary of a national park;

(4) Shall include incentives (such as preferred commercial air tour routes and altitudes, and relief from caps and curfews) for the adoption of quiet technology aircraft by commercial air tour operators conducting commercial air tour operations at the park;

(5) Shall provide for the initial allocation of opportunities to conduct commercial air tour operations if the plan includes a limitation on the number of commercial air tour operations for any time period; and

(6) Shall justify and document the need for measures taken pursuant to paragraphs (c)(1) through (c)(5) of this section and include such justification in the record of decision.

(d) *Procedure.* In establishing an ATMP for a national park or tribal lands, the Administrator and Director shall—

(1) Hold at least one public meeting with interested parties to develop the air tour management plan;

(2) Publish the proposed plan in the Federal Register for notice and comment and make copies of the proposed plan available to the public;

(3) Comply with the regulations set forth in 40 CFR 1501.3 and 1501.5 through 1501.8 (for the purposes of complying with 40 CFR 1501.3 and 1501.5 through 1501.8, the Federal Aviation Administration is the lead agency and the National Park Service is a cooperating agency); and

(4) Solicit the participation of any Indian tribe whose tribal lands are, or may be, overflown by aircraft involved in a commercial air tour operation over the park or tribal lands to which the plan applies, as a cooperating agency under the regulations referred to in paragraph (d)(3) of this section.

(e) *Amendments.* The Administrator, in cooperation with the Director, may make amendments to an air tour management plan. Any such amendments will be published in the Federal Register for notice and comment. A request for amendment of an ATMP will be made in accordance with § 11.25 of this chapter as a petition for rulemaking.

[Doc. No. FAA-2001-8690, 67 FR 65667, Oct. 25, 2002. Redesignated by Amdt. 136-1, 72 FR 6912, Feb. 13, 2007]

§ 136.41 Interim operating authority.

(a) *General.* Upon application for operating authority, the Administrator shall grant interim operating authority under this section to a commercial air tour operator for commercial air tour operations over a national park or tribal land for which the operator is an existing commercial air tour operator.

(b) *Requirements and limitations.* Interim operating authority granted under this section—

(1) Shall provide annual authorization only for the greater of—

(i) The number of flights used by the operator to provide the commercial air tour operations within the 12-month period prior to April 5, 2000; or

(ii) The average number of flights per 12-month period used by the operator to provide such operations within the 36-month period prior to April 5, 2000, and for seasonal operations, the number of flights so used during the season or seasons covered by that 12-month period;

(2) May not provide for an increase in the number of commercial air tour operations conducted during any time period by the commercial air tour operator above the number the air tour operator was originally granted unless such an increase is agreed to by the Administrator and the Director;

(3) Shall be published in the Federal Register to provide notice and opportunity for comment;

(4) May be revoked by the Administrator for cause;

(5) Shall terminate 180 days after the date on which an air tour management plan is established for the park and tribal lands;

(6) Shall promote protection of national park resources, visitor experiences, and tribal lands;

(7) Shall promote safe commercial air tour operations;

(8) Shall promote the adoption of quiet technology, as appropriate, and

(9) Shall allow for modifications of the interim operating authority based on experience if the modification improves protection of national park resources and values and of tribal lands.

(c) *New entrant operators.* The Administrator, in cooperation with the Director, may grant interim operating authority under this paragraph (c) to an air tour operator for a national park or tribal lands for which that operator is a new entrant air tour operator if the Administrator determines the authority is necessary to ensure competition in the provision of commercial air tour operations over the park or tribal lands.

(1) *Limitation.* The Administrator may not grant interim operating authority under this paragraph (c) if the Administrator determines that it would create a safety problem at the park or on the tribal lands, or if the Director determines that it would create a noise problem at the park or on the tribal lands.

(2) *ATMP limitation.* The Administrator may grant interim operating authority under this paragraph (c) only if the ATMP for the park or tribal lands to which the application relates has not been developed within 24 months after April 5, 2000.

[Doc. No. FAA-2001-8690, 67 FR 65667, Oct. 25, 2002. Redesignated by Amdt. 136-1, 72 FR 6912, Feb. 13, 2007]

§§ 136.43-136.49 [Reserved]

Subpart C—Grand Canyon National Park

§§ 136.51-136.69 [Reserved]

Appendix A to Part 136—Special Operating Rules for Air Tour Operators in the State of Hawaii

Section 1. Applicability. This appendix prescribes operating rules for airplane and helicopter visual flight rules air tour flights conducted in the State of Hawaii under 14 CFR parts 91, 121, and 135. This appendix does not apply to:

(a) Operations conducted under 14 CFR part 121 in airplanes with a passenger seating configuration of more than 30 seats or a payload capacity of more than 7,500 pounds.

(b) Flights conducted in gliders or hot air balloons.

Section 2. Definitions. For the purposes of this appendix: "Air tour" means any sightseeing flight conducted under visual flight rules in an airplane or helicopter for compensation or hire.

"Air tour operator" means any person who conducts an air tour.

Section 3. Helicopter flotation equipment. No person may conduct an air tour in Hawaii in a single-engine helicopter beyond the shore of any island, regardless of whether the helicopter is within gliding distance of the shore, unless:

(a) The helicopter is amphibious or is equipped with floats adequate to accomplish a safe emergency ditching and approved flotation gear is easily accessible for each occupant; or

(b) Each person on board the helicopter is wearing approved flotation gear.

Section 4. Helicopter performance plan. Each operator must complete a performance plan before each helicopter air tour flight. The performance plan must be based on the information in the Rotorcraft Flight Manual (RFM), considering the maximum density altitude for which the operation is planned for the flight to determine the following:

(a) Maximum gross weight and center of gravity (CG) limitations for hovering in ground effect;

(b) Maximum gross weight and CG limitations for hovering out of ground effect; and,

(c) Maximum combination of weight, altitude, and temperature for which height-velocity information in the RFM is valid.

The pilot in command (PIC) must comply with the performance plan.

Section 5. Helicopter Operating Limitations. Except for approach to and transition from a hover, and except for the purpose of takeoff and landing, the PIC shall operate the helicopter at a combination of height and forward speed (including hover) that would permit a safe landing in event of engine power loss, in accordance with the height-speed envelope for that helicopter under current weight and aircraft altitude.

Section 6. Minimum flight altitudes. Except when necessary for takeoff and landing, or operating in compliance with an air traffic control clearance, or as otherwise authorized by the Administrator, no person may conduct an air tour in Hawaii:

(a) Below an altitude of 1,500 feet above the surface over all areas of the State of Hawaii, and,

(b) Closer than 1,500 feet to any person or property; or,

(c) Below any altitude prescribed by federal statute or regulation.

Section 7. Passenger briefing. Before takeoff, each PIC of an air tour flight of Hawaii with a flight segment beyond the ocean shore of any island shall ensure that each passenger has been briefed on the following, in addition to requirements set forth in 14 CFR 91.107, 121.571, or 135.117:

(a) Water ditching procedures;

(b) Use of required flotation equipment; and

(c) Emergency egress from the aircraft in event of a water landing.

[Docket No. FAA-1998-4521, 72 FR 6914, Feb. 13, 2007]

PART 137

AGRICULTURAL AIRCRAFT OPERATIONS

Contents

AUTHORITY: 49 U.S.C. 106(g), 40103, 40113, 44701-44702.

SOURCE: Docket No. 1464, 30 FR 8106, June 24, 1965, unless otherwise noted.

Subpart A—General

§ 137.1 Applicability.

(a) This part prescribes rules governing—

(1) Agricultural aircraft operations within the United States; and

(2) The issue of commercial and private agricultural aircraft operator certificates for those operations.

(b) In a public emergency, a person conducting agricultural aircraft operations under this part may, to the extent necessary, deviate from the operating rules of this part for relief and welfare activities approved by an agency of the United States or of a State or local government.

(c) Each person who, under the authority of this section, deviates from a rule of this part shall, within 10 days after the deviation send to the nearest FAA Flight Standards District Office a complete report of the aircraft operation involved, including a description of the operation and the reasons for it.

[Doc. No. 1464, 30 FR 8106, June 24, 1965, as amended by Amdt. 137-13, 54 FR 39294, Sept. 25, 1989]

§ 137.3 Definition of terms.

For the purposes of this part—

Agricultural aircraft operation means the operation of an aircraft for the purpose of (1) dispensing any economic poison, (2) dispensing any other substance intended for plant nourishment, soil treatment, propagation of plant life, or pest control, or (3) engaging in dispensing activities directly affecting agriculture, horticulture, or forest preservation, but not including the dispensing of live insects.

Economic poison means (1) any substance or mixture of substances intended for preventing, destroying, repelling, or mitigating any insects, rodents, nematodes, fungi, weeds, and other forms of plant or animal life or viruses, except viruses on or in living man or other animals, which the Secretary of Agriculture shall declare to be a pest, and (2) any substance or mixture of substances intended for use as a plant regulator, defoliant or desiccant.

[Doc. No. 1464, 30 FR 8106, June 24, 1965, as amended by Amdt. 137-3, 33 FR 9601, July 2, 1968]

Subpart B—Certification Rules

§ 137.11 Certificate required.

(a) Except as provided in paragraphs (c) and (d) of this section, no person may conduct agricultural aircraft operations without, or in violation of, an agricultural aircraft operator certificate issued under this part.

(b) Notwithstanding part 133 of this chapter, an operator may, if he complies with this part, conduct agricultural aircraft operations with a rotorcraft with external dispensing equipment in place without a rotorcraft external-load operator certificate.

(c) A Federal, State, or local government conducting agricultural aircraft operations with public aircraft need not comply with this subpart.

(d) The holder of a rotorcraft external-load operator certificate under part 133 of this chapter conducting an agricultural aircraft operation, involving only the dispensing of water on forest fires by rotorcraft external-load means, need not comply with this subpart.

[Doc. No. 1464, 30 FR 8106, June 24, 1965, as amended by Amdt. 137-3, 33 FR 9601, July 2, 1968; Amdt. 137-6, 41 FR 35060, Aug. 19, 1976]

§ 137.15 Application for certificate.

An application for an agricultural aircraft operator certificate is made on a form and in a manner prescribed by the Administrator, and filed with the FAA Flight Standards District Office that has jurisdiction over the area in which the applicant's home base of operations is located.

[Doc. No. 1464, 30 FR 8106, June 24, 1965, as amended by Amdt. 137-13, 54 FR 39294, Sept. 25, 1989]

§ 137.17 Amendment of certificate.

(a) An agricultural aircraft operator certificate may be amended—

(1) On the Administrator's own initiative, under section 609 of the Federal Aviation Act of 1958 (49 U.S.C. 1429) and part 13 of this chapter; or

(2) Upon application by the holder of that certificate.

(b) An application to amend an agricultural aircraft operator certificate is submitted on a form and in a manner prescribed by the Administrator. The applicant must file the application with the FAA Flight Standards District Office having jurisdiction over the area in which the applicant's home base of operations is located at least 15 days before the date that it proposes the amendment become effective, unless a shorter filing period is approved by that office.

(c) The Flight Standards District Office grants a request to amend a certificate if it determines that safety in air commerce and the public interest so allow.

(d) Within 30 days after receiving a refusal to amend, the holder may petition the Director, Flight Standards Service, to reconsider the refusal.

[Doc. No. 1464, 30 FR 8106, June 24, 1965, as amended by Amdt. 137-9, 43 FR 52206, Nov. 9, 1978; Amdt. 137-11, 45 FR 47838, July 17, 1980; Amdt. 137-13, 54 FR 39294, Sept. 25, 1989]

§ 137.19 Certification requirements.

(a) *General.* An applicant for a private agricultural aircraft operator certificate is entitled to that certificate if he shows that he meets the requirements of paragraphs (b), (d), and (e) of this section. An applicant for a commercial agricultural aircraft operator certificate is entitled to that certificate if he shows that he meets the requirements of paragraphs (c), (d), and (e) of this section. However, if an applicant applies for an agricultural aircraft operator certificate containing a prohibition against the dispensing of economic poisons, that applicant is not required to demonstrate the knowledge required in paragraphs (e)(1)(ii) through (iv) of this section.

(b) *Private operator—pilot.* The applicant must hold a current U.S. private, commercial, or airline transport pilot certificate and be properly rated for the aircraft to be used.

(c) *Commercial operator—pilots.* The applicant must have available the services of at least one person who holds a current U.S. commercial or airline transport pilot certificate and who is properly rated for the aircraft to be used. The applicant himself may be the person available.

(d) *Aircraft.* The applicant must have at least one certificated and airworthy aircraft, equipped for agricultural operation.

(e) *Knowledge and skill tests.* The applicant must show, or have the person who is designated as the chief supervisor of agricultural aircraft operations for him show, that he has satisfactory knowledge and skill regarding agricultural aircraft operations, as described in paragraphs (e)(1) and (2) of this section.

(1) The test of knowledge consists of the following:

(i) Steps to be taken before starting operations, including survey of the area to be worked.

(ii) Safe handling of economic poisons and the proper disposal of used containers for those poisons.

(iii) The general effects of economic poisons and agricultural chemicals on plants, animals, and persons, with emphasis on those normally used in the areas of intended operations; and the precautions to be observed in using poisons and chemicals.

(iv) Primary symptoms of poisoning of persons from economic poisons, the appropriate emergency measures to be taken, and the location of poison control centers.

(v) Performance capabilities and operating limitations of the aircraft to be used.

(vi) Safe flight and application procedures.

(2) The test of skill consists of the following maneuvers that must be shown in any of the aircraft specified in paragraph (d) of this section, and at that aircraft's maximum certificated take-off weight, or the maximum weight established for the special purpose load, whichever is greater:

(i) Short-field and soft-field takeoffs (airplanes and gyroplanes only).

(ii) Approaches to the working area.

(iii) Flare-outs.

(iv) Swath runs.

(v) Pullups and turnarounds.

(vi) Rapid deceleration (quick stops) in helicopters only.

[Doc. No. 1464, 30 FR 8106, June 24, 1965, as amended by Amdt. 137-1, 30 FR 15143, Dec. 8, 1965; Amdt. 137-7, 43 FR 22643, May 25, 1978]

§ 137.21 Duration of certificate.

An agricultural aircraft operator certificate is effective until it is surrendered, suspended, or revoked. The holder of an agricultural aircraft operator certificate that is suspended or revoked shall return it to the Administrator.

§ 137.23 Carriage of narcotic drugs, marihuana, and depressant or stimulant drugs or substances.

If the holder of a certificate issued under this part permits any aircraft owned or leased by that holder to be engaged in any operation that the certificate holder knows to be in violation of § 91.19(a) of this chapter, that operation is a basis for suspending or revoking the certificate.

[Doc. No. 12035, 38 FR 17493, July 2, 1973, as amended by Amdt. 137-12, 54 FR 34332, Aug. 18, 1989]

Subpart C—Operating Rules

§ 137.29 General.

(a) Except as provided in paragraphs (d) and (e) of this section, this subpart prescribes rules that apply to persons and aircraft used in agricultural aircraft operations conducted under this part.

(b) [Reserved]

(c) The holder of an agricultural aircraft operator certificate may deviate from the provisions of part 91 of this chapter without a certificate of waiver, as authorized in this subpart for dispensing operations, when conducting nondispensing aerial work operations related to agriculture, horticulture, or forest preservation in accordance with the operating rules of this subpart.

(d) Sections 137.31 through 137.35, §§ 137.41, and 137.53 through 137.59 do not apply to persons and aircraft used in agricultural aircraft operations conducted with public aircraft.

(e) Sections 137.31 through 137.35, §§ 137.39, 137.41, 137.51 through 137.59, and subpart D do not apply to persons and rotorcraft used in agricultural aircraft operations conducted by a person holding a certificate under part 133 of this chapter and involving only the dispensing of water on forest fires by rotorcraft external-load means.

However, the operation shall be conducted in accordance with—

(1) The rules of part 133 of this chapter governing rotorcraft external-load operations; and

(2) The operating rules of this subpart contained in §§ 137.29, 137.37, and §§ 137.43 through 137.49.

[Doc. No. 1464, 30 FR 8106, June 24, 1965, as amended by Amdt. 137-3, 33 FR 9601, July 2, 1968; Amdt. 137-6, 41 FR 35060, Aug. 19, 1976]

§ 137.31 Aircraft requirements.

No person may operate an aircraft unless that aircraft—

(a) Meets the requirements of § 137.19(d); and

(b) Is equipped with a suitable and properly installed shoulder harness for use by each pilot.

§ 137.33 Carrying of certificate.

(a) No person may operate an aircraft unless a facsimile of the agricultural aircraft operator certificate, under which the operation is conducted, is carried on that aircraft. The facsimile shall be presented for inspection upon the request of the Administrator or any Federal, State, or local law enforcement officer.

(b) Notwithstanding part 91 of this chapter, the registration and airworthiness certificates issued for the aircraft need not be carried in the aircraft. However, when those certificates are not carried in the aircraft they shall be kept available for inspection at the base from which the dispensing operation is conducted.

[Doc. No. 1464, 30 FR 8106, June 24, 1965, as amended by Amdt. 137-3, 33 FR 9601, July 2, 1968]

§ 137.35 Limitations on private agricultural aircraft operator.

No person may conduct an agricultural aircraft operation under the authority of a private agricultural aircraft operator certificate—

(a) For compensation or hire;

(b) Over a congested area; or

(c) Over any property unless he is the owner or lessee of the property, or has ownership or other property interest in the crop located on that property.

§ 137.37 Manner of dispensing.

No persons may dispense, or cause to be dispensed, from an aircraft, any material or substance in a manner that creates a hazard to persons or property on the surface.

[Doc. No. 1464, 30 FR 8106, June 24, 1965, as amended by Amdt. 137-3, 33 FR 9601, July 2, 1968]

§ 137.39 Economic poison dispensing.

(a) Except as provided in paragraph (b) of this section, no person may dispense or cause to be dispensed from an aircraft, any economic poison that is registered with the U.S. Department of Agriculture under the Federal Insecticide, Fungicide, and Rodenticide Act (7 U.S.C. 135-135k)—

(1) For a use other than that for which it is registered;

(2) Contrary to any safety instructions or use limitations on its label; or

(3) In violation of any law or regulation of the United States.

(b) This section does not apply to any person dispensing economic poisons for experimental purposes under—

(1) The supervision of a Federal or State agency authorized by law to conduct research in the field of economic poisons; or

(2) A permit from the U.S. Department of Agriculture issued pursuant to the Federal Insecticide, Fungicide, and Rodenticide Act (7 U.S.C. 135-135k).

[Amdt. 137-2, 31 FR 6686, May 5, 1966]

§ 137.40 Employment of former FAA employees.

(a) Except as specified in paragraph (c) of this section, no certificate holder may knowingly employ or make a contractual arrangement which permits an individual to act as an agent or representative of the certificate holder in any matter before the Federal Aviation Administration if the individual, in the preceding 2 years—

(1) Served as, or was directly responsible for the oversight of, a Flight Standards Service aviation safety inspector; and

(2) Had direct responsibility to inspect, or oversee the inspection of, the operations of the certificate holder.

(b) For the purpose of this section, an individual shall be considered to be acting as an agent or representative of a certificate holder in a matter before the agency if the individual makes any written or oral communication on behalf of the certificate holder to the agency (or any of its officers or employees) in connection with a particular matter, whether or not involving a specific party and without regard to whether the individual has participated in, or had responsibility for, the particular matter while serving as a Flight Standards Service aviation safety inspector.

(c) The provisions of this section do not prohibit a certificate holder from knowingly employing or making a contractual arrangement which permits an individual to act as an agent or representative of the certificate holder in any matter before the Federal Aviation Administration if the individual was employed by the certificate holder before October 21, 2011.

[Doc. No. FAA-2008-1154, 76 FR 52236, Aug. 22, 2011]

§ 137.41 Personnel.

(a) *Information.* The holder of an agricultural aircraft operator certificate shall insure that each person used in the holder's agricultural aircraft operation is informed of that person's duties and responsibilities for the operation.

(b) *Supervisors.* No person may supervise an agricultural aircraft operation unless he has met the knowledge and skill requirements of § 137.19(e).

(c) *Pilot in command.* No person may act as pilot in command of an aircraft unless he holds a pilot certificate and rating prescribed by § 137.19 (b) or (c), as appropriate to the type of operation conducted. In addition, he must demonstrate to the holder of the Agricultural Aircraft Operator Certificate conducting the operation that he has met the knowledge and skill requirements of § 137.19(e). If the holder of that certificate has designated a person under § 137.19(e) to supervise his agricultural aircraft operations the demonstration must be made to the person so designated. However, a demonstration of the knowledge and skill requirement is not necessary for any pilot in command who—

(1) Is, at the time of the filing of an application by an agricultural aircraft operator, working as a pilot in command for that operator; and

(2) Has a record of operation under that applicant that does not disclose any question regarding the safety of his flight operations or his competence in dispensing agricultural materials or chemicals.

§ 137.42 Fastening of safety belts and shoulder harnesses.

No person may operate an aircraft in operations required to be conducted under part 137 without a safety belt and shoulder harness properly secured about that person except that the shoulder harness need not be fastened if that person would be unable to perform required duties with the shoulder harness fastened.

[Amdt. 137-10, 44 FR 61325, Oct. 25, 1979]

§ 137.43 Operations in controlled airspace designated for an airport.

(a) Except for flights to and from a dispensing area, no person may operate an aircraft within the lateral boundaries of the surface area of Class D airspace designated for an airport unless authorization for that operation has been obtained from the ATC facility having jurisdiction over that area.

(b) No person may operate an aircraft in weather conditions below VFR minimums within the lateral boundaries of a Class E airspace area that extends upward from the surface unless authorization for that operation has been obtained from the ATC facility having jurisdiction over that area.

(c) Notwithstanding § 91.157(b)(4) of this chapter, an aircraft may be operated under the special VFR weather minimums without meeting the requirements prescribed therein.

[Amdt. 137-14, 56 FR 65664, Dec. 17, 1991, as amended by Amdt. 137-14, 58 FR 32840, June 14, 1993; 74 FR 13099, Mar. 26, 2009]

§ 137.45 Nonobservance of airport traffic pattern.

Notwithstanding part 91 of this chapter, the pilot in command of an aircraft may deviate from an airport traffic pattern when authorized by the control tower concerned. At an airport without a functioning control tower, the pilot in command may deviate from the traffic pattern if—

(a) Prior coordination is made with the airport management concerned;

(b) Deviations are limited to the agricultural aircraft operation;

(c) Except in an emergency, landing and takeoffs are not made on ramps, taxiways, or other areas of the airport not intended for such use; and

(d) The aircraft at all times remains clear of, and gives way to, aircraft conforming to the traffic pattern for the airport.

§ 137.47 Operation without position lights.

Notwithstanding part 91 of this chapter, an aircraft may be operated without position lights if prominent unlighted objects are visible for at least 1 mile and takeoffs and landings at—

(a) Airports with a functioning control tower are made only as authorized by the control tower operator; and

(b) Other airports are made only with the permission of the airport management and no other aircraft operations requiring position lights are in progress at that airport.

§ 137.49 Operations over other than congested areas.

Notwithstanding part 91 of this chapter, during the actual dispensing operation, including approaches, departures, and turnarounds reasonably necessary for the operation, an aircraft may be operated over other than congested areas below 500 feet above the surface and closer than 500 feet to persons, vessels, vehicles, and structures, if the operations are conducted without creating a hazard to persons or property on the surface.

[Amdt. 137-3, 33 FR 9601, July 2, 1968]

§ 137.51 Operation over congested areas: General.

(a) Notwithstanding part 91 of this chapter, an aircraft may be operated over a congested area at altitudes required for the proper accomplishment of the agricultural aircraft operation if the operation is conducted—

(1) With the maximum safety to persons and property on the surface, consistent with the operation; and

(2) In accordance with the requirements of paragraph (b) of this section.

(b) No person may operate an aircraft over a congested area except in accordance with the requirements of this paragraph.

(1) Prior written approval must be obtained from the appropriate official or governing body of the political subdivision over which the operations are conducted.

(2) Notice of the intended operation must be given to the public by some effective means, such as daily newspapers, radio, television, or door-to-door notice.

(3) A plan for each complete operation must be submitted to, and approved by appropriate personnel of the FAA Flight Standards District Office having jurisdiction over the area where the operation is to be conducted. The plan must include consideration of obstructions to flight; the emergency landing capabilities of the aircraft to be used; and any necessary coordination with air traffic control.

(4) Single engine aircraft must be operated as follows:

(i) Except for helicopters, no person may take off a loaded aircraft, or make a turnaround over a congested area.

(ii) No person may operate an aircraft over a congested area below the altitudes prescribed in part 91 of this chapter except during the actual dispensing operation, including the approaches and departures necessary for that operation.

(iii) No person may operate an aircraft over a congested area during the actual dispensing operation, including the approaches and departures for that operation, unless it is operated in a pattern and at such an altitude that the aircraft can land, in an emergency, without endangering persons or property on the surface.

(5) Multiengine aircraft must be operated as follows:

(i) No person may take off a multiengine airplane over a congested area except under conditions that will allow the airplane to be brought to a safe stop within the effective length of the runway from any point on takeoff up to the time of attaining, with all engines operating at normal takeoff power, 105 percent of the minimum control speed with the critical engine inoperative in the takeoff configuration or 115 percent of the power-off stall speed in the takeoff configuration, whichever is greater, as shown by the accelerate stop distance data. In applying this requirement, takeoff data is based upon still-air conditions, and no correction is made for any uphill gradient of 1 percent or less when the percentage is measured as the difference between elevation at the end points of the runway divided by the total length. For uphill gradients greater than 1 percent, the effective takeoff length of the runway is reduced 20 percent for each 1-percent grade.

(ii) No person may operate a multiengine airplane at a weight greater than the weight that, with the critical engine inoperative, would permit a rate of climb of at least 50 feet per minute at an altitude of at least 1,000 feet above the elevation of the highest ground or obstruction within the area to be worked or at an altitude of 5,000 feet, whichever is higher. For the purposes of this subdivision, it is assumed that the propeller of the inoperative engine is in the minimum drag position; that the wing flaps and landing gear are in the most favorable positions; and that the remaining engine or engines are operating at the maximum continuous power available.

(iii) No person may operate any multiengine aircraft over a congested area below the altitudes prescribed in part 91 of this chapter except during the actual dispensing operation, including the approaches, departures, and turnarounds necessary for that operation.

[Doc. No. 1464, 30 FR 8106, June 24, 1965, as amended by Doc. No. 8084, 32 FR 5769, Apr. 11, 1967; Amdt. 137-13, 54 FR 39294, Sept. 25, 1989]

§ 137.53 Operation over congested areas: Pilots and aircraft.

(a) *General.* No person may operate an aircraft over a congested area except in accordance with the pilot and aircraft rules of this section.

(b) *Pilots.* Each pilot in command must have at least—

(1) 25 hours of pilot-in-command flight time in the make and basic model of the aircraft, at least 10 hours of which must have been acquired within the preceding 12 calendar months; and

(2) 100 hours of flight experience as pilot in command in dispensing agricultural materials or chemicals.

(c) *Aircraft.* (1) Each aircraft must—(i) If it is an aircraft not specified in paragraph (c)(1)(ii) of this section, have had within the preceding 100 hours of time in service a 100-hour or annual inspection by a person authorized by part 65 or 145 of this chapter, or have been inspected under a progressive inspection system; and

(ii) If it is a large or turbine-powered multiengine civil airplane of U.S. registry, have been inspected in accordance with the applicable inspection program requirements of § 91.409 of this chapter.

(2) If other than a helicopter, it must be equipped with a device capable of jettisoning at least one-half of the aircraft's maximum authorized load of agricultural material within 45 seconds. If the aircraft is equipped with a device for releasing the tank or hopper as a unit, there must be a means to prevent inadvertent release by the pilot or other crewmember.

[Doc. No. 1464, 30 FR 8106, June 24, 1965, as amended by Amdt. 137-5, 41 FR 16796, Apr. 22, 1976; Amdt. 137-12, 54 FR 34332, Aug. 18, 1989]

§ 137.55 Business name: Commercial agricultural aircraft operator.

No person may operate under a business name that is not shown on his commercial agricultural aircraft operator certificate.

§ 137.57 Availability of certificate.

Each holder of an agricultural aircraft operator certificate shall keep that certificate at his home base of operations and shall present it for inspection on the request of the Administrator or any Federal, State, or local law enforcement officer.

§ 137.59 Inspection authority.

Each holder of an agricultural aircraft operator certificate shall allow the Administrator at any time and place to make inspections, including on-the-job inspections, to determine compliance with applicable regulations and his agricultural aircraft operator certificate.

Subpart D—Records and Reports

§ 137.71 Records: Commercial agricultural aircraft operator.

(a) Each holder of a commercial agricultural aircraft operator certificate shall maintain and keep current, at the home base of operations designated in his application, the following records:

(1) The name and address of each person for whom agricultural aircraft services were provided;

(2) The date of the service;

(3) The name and quantity of the material dispensed for each operation conducted; and

(4) The name, address, and certificate number of each pilot used in agricultural aircraft operations and the date that pilot met the knowledge and skill requirements of § 137.19(e).

(b) The records required by this section must be kept at least 12 months and made available for inspection by the Administrator upon request.

§ 137.75 Change of address.

Each holder of an agricultural aircraft operator certificate shall notify the FAA in writing in advance of any change in the address of his home base of operations.

§ 137.77 Termination of operations.

Whenever a person holding an agricultural aircraft operator certificate ceases operations under this part, he shall surrender that certificate to the FAA Flight Standards District Office last having jurisdiction over his operation.

[Doc. No. 1464, 30 FR 8106, June 24, 1965, as amended by Amdt. 137-13, 54 FR 39294, Sept. 25, 1989; 54 FR 52872, Dec. 22, 1989]

PART 139

CERTIFICATION OF AIRPORTS

Contents

Subpart A—General

Subpart B—Certification

Subpart C—Airport Certification Manual

Subpart D—Operations

Authority: 49 U.S.C. 106(g), 40113, 44701-44706, 44709, 44719.

Source: Doc. No. FAA-2000-7479, 69 FR 6424, Feb. 10, 2004, unless otherwise noted.

Editorial Note: Nomenclature changes to part 139 appear at 69 FR 24069, May 3, 2004.

Subpart A—General

§ 139.1 Applicability.

(a) This part prescribes rules governing the certification and operation of airports in any State of the United States, the District of Columbia, or any territory or possession of the United States serving any—

(1) Scheduled passenger-carrying operations of an air carrier operating aircraft configured for more than 9 passenger seats, as determined by the regulations under which the operation is conducted or the aircraft type certificate issued by a competent civil aviation authority; and

(2) Unscheduled passenger-carrying operations of an air carrier operating aircraft configured for at least 31 passenger seats, as determined by the regulations under which the operation is conducted or the aircraft type certificate issued by a competent civil aviation authority.

(b) This part applies to those portions of a joint-use or shared-use airport that are within the authority of a person serving passenger-carrying operations defined in paragraphs (a)(1) and (a)(2) of this section.

(c) This part does not apply to—

(1) Airports serving scheduled air carrier operations only by reason of being designated as an alternate airport;

(2) Airports operated by the United States;

(3) Airports located in the State of Alaska that only serve scheduled operations of small air carrier aircraft and do not serve scheduled or unscheduled operations of large air carrier aircraft;

(4) Airports located in the State of Alaska during periods of time when not serving operations of large air carrier aircraft; or

(5) Heliports.

[Doc. No. FAA-2000-7479, 69 FR 6424, Feb. 10, 2004, as amended by Amdt. 139-27, 78 FR 3316, Jan. 16, 2013]

§ 139.3 Delegation of authority.

The authority of the Administrator to issue, deny, and revoke Airport Operating Certificates is delegated to the Associate Administrator for Airports, Director of Airport Safety and Standards, and Regional Airports Division Managers.

§ 139.5 Definitions.

The following are definitions of terms used in this part:

AFFF means aqueous film forming foam agent.

Air carrier aircraft means an aircraft that is being operated by an air carrier and is categorized as either a large air carrier aircraft if designed for at least 31 passenger seats or a small air carrier aircraft if designed for more than 9 passenger seats but less than 31 passenger seats, as determined by the aircraft type certificate issued by a competent civil aviation authority.

Air carrier operation means the takeoff or landing of an air carrier aircraft and includes the period of time from 15 minutes before until 15 minutes after the takeoff or landing.

Airport means an area of land or other hard surface, excluding water, that is used or intended to be used for the landing and takeoff of aircraft, including any buildings and facilities.

Airport Operating Certificate means a certificate, issued under this part, for operation of a Class I, II, III, or IV airport.

Average daily departures means the average number of scheduled departures per day of air carrier aircraft computed on the basis of the busiest 3 consecutive calendar months of the immediately preceding 12 consecutive calendar months. However, if the average daily departures are expected to increase, then "average daily departures" may be determined by planned rather than current activity, in a manner authorized by the Administrator.

Certificate holder means the holder of an Airport Operating Certificate issued under this part.

Class I airport means an airport certificated to serve scheduled operations of large air carrier aircraft that can also serve unscheduled passenger operations of large air carrier aircraft and/or scheduled operations of small air carrier aircraft.

Class II airport means an airport certificated to serve scheduled operations of small air carrier aircraft and the unscheduled passenger operations of large air carrier aircraft. A Class II airport cannot serve scheduled large air carrier aircraft.

Class III airport means an airport certificated to serve scheduled operations of small air carrier aircraft. A Class III airport cannot serve scheduled or unscheduled large air carrier aircraft.

Class IV airport means an airport certificated to serve unscheduled passenger operations of large air carrier aircraft. A Class IV airport cannot serve scheduled large or small air carrier aircraft.

Clean agent means an electrically nonconducting volatile or gaseous fire extinguishing agent that does not leave a residue upon evaporation and has been shown to provide extinguishing action equivalent to halon 1211 under test protocols of FAA Technical Report DOT/FAA/AR-95/87.

Heliport means an airport, or an area of an airport, used or intended to be used for the landing and takeoff of helicopters.

Index means the type of aircraft rescue and firefighting equipment and quantity of fire extinguishing agent that the certificate holder must provide in accordance with § 139.315.

Joint-use airport means an airport owned by the Department of Defense, at which both military and civilian aircraft make shared use of the airfield.

Movement area means the runways, taxiways, and other areas of an airport that are used for taxiing, takeoff, and landing of aircraft, exclusive of loading ramps and aircraft parking areas.

Regional Airports Division Manager means the airports division manager for the FAA region in which the airport is located.

Safety area means a defined area comprised of either a runway or taxiway and the surrounding surfaces that is prepared or suitable for reducing the risk of damage to aircraft in the event of an undershoot, overshoot, or excursion from a runway or the unintentional departure from a taxiway.

Scheduled operation means any common carriage passenger-carrying operation for compensation or hire conducted by an air carrier for which the air carrier or its representatives offers in advance the departure location, departure time, and arrival location. It does not include any operation that is conducted as a supplemental operation under 14 CFR part 121 or public charter operations under 14 CFR part 380.

Shared-use airport means a U.S. Government-owned airport that is co-located with an airport specified under § 139.1(a) and at which portions of the movement areas and safety areas are shared by both parties.

Unscheduled operation means any common carriage passenger-carrying operation for compensation or hire, using aircraft designed for at least 31 passenger seats, conducted by an air carrier for which the departure time, departure location, and arrival location are specifically negotiated with the customer or the customer's representative. It includes any passenger-carrying supplemental operation conducted under 14 CFR part 121 and any passenger-carrying public charter operation conducted under 14 CFR part 380.

Wildlife hazard means a potential for a damaging aircraft collision with wildlife on or near an airport. As used in this part, "wildlife" includes feral animals and domestic animals out of the control of their owners.

Note:*Special Statutory Requirement To Operate to or From a Part 139 Airport.* Each air carrier that provides—in an aircraft designed for more than 9 passenger seats—regularly scheduled charter air transportation for which the public is provided in advance a schedule containing the departure location, departure time, and arrival location of the flight must operate to and from an airport certificated under part 139 of this chapter in accordance with 49 U.S.C. 41104(b). That statutory provision contains stand-alone requirements for such air carriers and special exceptions for operations in Alaska and outside the United States. Certain operations by air carriers that conduct public charter operations under 14 CFR part 380 are covered by the statutory requirements to operate to and from part 139 airports. See 49 U.S.C. 41104(b).

[Doc. No. FAA-2000-7479, 69 FR 6424, Feb. 10, 2004, as amended by Amdt. 139-27, 78 FR 3316, Jan. 16, 2013]

§ 139.7 Methods and procedures for compliance.

Certificate holders must comply with requirements prescribed by subparts C and D of this part in a manner authorized by the Administrator. FAA Advisory Circulars contain methods and procedures for compliance with this part that are acceptable to the Administrator.

Subpart B—Certification

§ 139.101 General requirements.

(a) Except as otherwise authorized by the Administrator, no person may operate an airport specified under § 139.1 of this part without an Airport Operating Certificate or in violation of that certificate, the applicable provisions, or the approved Airport Certification Manual.

(b) Each certificate holder shall adopt and comply with an Airport Certification Manual as required under § 139.203.

(c) Persons required to have an Airport Operating Certificate under this part shall submit their Airport Certification Manual to the FAA for approval, in accordance with the following schedule:

(1) Class I airports—6 months after June 9, 2004.

(2) Class II, III, and IV airports—12 months after June 9, 2004.

§ 139.103 Application for certificate.

Each applicant for an Airport Operating Certificate must—

(a) Prepare and submit an application, in a form and in the manner prescribed by the Administrator, to the Regional Airports Division Manager.

(b) Submit with the application, two copies of an Airport Certification Manual prepared in accordance with subpart C of this part.

§ 139.105 Inspection authority.

Each applicant for, or holder of, an Airport Operating Certificate must allow the Administrator to make any inspections, including unannounced inspections, or tests to determine compliance with 49 U.S.C. 44706 and the requirements of this part.

§ 139.107 Issuance of certificate.

An applicant for an Airport Operating Certificate is entitled to a certificate if—

(a) The applicant provides written documentation that air carrier service will begin on a date certain.

(b) The applicant meets the provisions of § 139.103.

(c) The Administrator, after investigation, finds the applicant is properly and adequately equipped and able to provide a safe airport operating environment in accordance with—

(1) Any limitation that the Administrator finds necessary to ensure safety in air transportation.

(2) The requirements of the Airport Certification Manual, as specified under § 139.203.

(3) Any other provisions of this part that the Administrator finds necessary to ensure safety in air transportation.

(d) The Administrator approves the Airport Certification Manual.

§ 139.109 Duration of certificate.

An Airport Operating Certificate issued under this part is effective until the certificate holder surrenders it or the certificate is suspended or revoked by the Administrator.

§ 139.111 Exemptions.

(a) An applicant or a certificate holder may petition the Administrator under 14 CFR part 11, General Rulemaking Procedures, of this chapter for an exemption from any requirement of this part.

(b) Under 49 U.S.C. 44706(c), the Administrator may exempt an applicant or a certificate holder that enplanes annually less than one-quarter of 1 percent of the total number of passengers enplaned at all air carrier airports from all, or part, of the aircraft rescue and firefighting equipment requirements of this part on the grounds that compliance with those requirements is, or would be, unreasonably costly, burdensome, or impractical.

(1) Each petition filed under this paragraph must—

(i) Be submitted in writing at least 120 days before the proposed effective date of the exemption;

(ii) Set forth the text of §§ 139.317 or 139.319 from which the exemption is sought;

(iii) Explain the interest of the certificate holder in the action requested, including the nature and extent of relief sought; and

(iv) Contain information, views, or arguments that demonstrate that the requirements of §§ 139.317 or 139.319 would be unreasonably costly, burdensome, or impractical.

(2) Information, views, or arguments provided under paragraph (b)(1) of this section shall include the following information pertaining to the airport for which the Airport Operating Certificate is held:

(i) An itemized cost to comply with the requirement from which the exemption is sought;

(ii) Current staffing levels;

(iii) The current annual financial report, such as a single audit report or FAA Form 5100-127, Operating and Financial Summary;

(iv) Annual passenger enplanement data for the previous 12 calendar months;

(v) The type and frequency of air carrier operations served;

(vi) A history of air carrier service;

(vii) Anticipated changes to air carrier service;

(c) Each petition filed under this section must be submitted in duplicate to the—

(1) Regional Airports Division Manager and

(2) Federal Docket Management System, as specified under 14 CFR part 11.

[Docket No. FAA-2000-7479, 69 FR 6424, Feb. 10, 2004; 72 FR 68475, Dec. 5, 2007]

§ 139.113 Deviations.

In emergency conditions requiring immediate action for the protection of life or property, the certificate holder may deviate from any requirement of subpart D of this part, or the Airport Certification Manual, to the extent required to meet that emergency. Each certificate holder who deviates from a requirement under this section must, within 14 days after the emergency, notify the Regional Airports Division Manager of the nature, extent, and duration of the deviation. When requested by the Regional Airports Division Manager, the certificate holder must provide this notification in writing.

§ 139.115 Falsification, reproduction, or alteration of applications, certificates, reports, or records.

(a) No person shall make or cause to be made:

(1) Any fraudulent or intentionally false statement on any application for a certificate or approval under this part.

(2) Any fraudulent or intentionally false entry in any record or report that is required to be made, kept, or used to show compliance with any requirement under this part.

(3) Any reproduction, for a fraudulent purpose, of any certificate or approval issued under this part.

(4) Any alteration, for a fraudulent purpose, of any certificate or approval issued under this part.

(b) The commission by any owner, operator, or other person acting on behalf of a certificate holder of an act prohibited under paragraph (a) of this section is a basis for suspending or revoking any certificate or approval issued under this part and held by that certificate holder and any other certificate issued under this title and held by the person committing the act.

[Doc. No. FAA-2010-0247, 78 FR 3316, Jan. 16, 2013]

Subpart C—Airport Certification Manual

§ 139.201 General requirements.

(a) No person may operate an airport subject to this part unless that person adopts and complies with an Airport Certification Manual, as required under this part, that—

(1) Has been approved by the Administrator;

(2) Contains only those items authorized by the Administrator;

(3) Is in printed form and signed by the certificate holder acknowledging the certificate holder's responsibility to operate the airport in compliance with the Airport Certification Manual approved by the Administrator; and

(4) Is in a form that is easy to revise and organized in a manner helpful to the preparation, review, and approval processes, including a revision log. In addition, each page or attachment must include the date of the Administrator's initial approval or approval of the latest revision.

(b) Each holder of an Airport Operating Certificate must—

(1) Keep its Airport Certification Manual current at all times;

(2) Maintain at least one complete and current copy of its approved Airport Certification Manual on the airport, which will be available for inspection by the Administrator; and

(3) Furnish the applicable portions of the approved Airport Certification Manual to airport personnel responsible for its implementation.

(c) Each certificate holder must ensure that the Regional Airports Division Manager is provided a complete copy of its most current approved Airport Certification Manual, as specified under paragraph (b)(2) of this section, including any amendments approved under § 139.205.

(d) FAA Advisory Circulars contain methods and procedures for the development of Airport Certification Manuals that are acceptable to the Administrator.

§ 139.203 Contents of Airport Certification Manual.

(a) Except as otherwise authorized by the Administrator, each certificate holder must include in the Airport Certification Manual a description of operating procedures, facilities and equipment, responsibility assignments, and any other information needed by personnel concerned with operating the airport in order to comply with applicable provisions of subpart D of this part and paragraph (b) of this section.

(b) Except as otherwise authorized by the Administrator, the certificate holder must include in the Airport Certification Manual the following elements, as appropriate for its class:

Required Airport Certification Manual Elements

Manual elements	Airport certificate class			
	Class I	Class II	Class III	Class IV
1. Lines of succession of airport operational responsibility	X	X	X	X
2. Each current exemption issued to the airport from the requirements of this part	X	X	X	X
3. Any limitations imposed by the Administrator	X	X	X	X
4. A grid map or other means of identifying locations and terrain features on and around the airport that are significant to emergency operations	X	X	X	X
5. The location of each obstruction required to be lighted or marked within the airport's area of authority	X	X	X	X
6. A description of each movement area available for air carriers and its safety areas, and each road described in §139.319(k) that serves it	X	X	X	X
7. Procedures for avoidance of interruption or failure during construction work of utilities serving facilities or NAVAIDS that support air carrier operations	X	X	X	
8. A description of the system for maintaining records, as required under §139.301	X	X	X	X
9. A description of personnel training, as required under §139.303	X	X	X	X
10. Procedures for maintaining the paved areas, as required under §139.305	X	X	X	X
11. Procedures for maintaining the unpaved areas, as required under §139.307	X	X	X	X
12. Procedures for maintaining the safety areas, as required under §139.309	X	X	X	X
13. A plan showing the runway and taxiway identification system, including the location and inscription of signs, runway markings, and holding position markings, as required under §139.311	X	X	X	X
14. A description of, and procedures for maintaining, the marking, signs, and lighting systems, as required under §139.311	X	X	X	X
15. A snow and ice control plan, as required under §139.313	X	X	X	
16. A description of the facilities, equipment, personnel, and procedures for meeting the aircraft rescue and firefighting requirements, in accordance with §§139.315, 139.317 and 139.319	X	X	X	X

Manual elements	Airport certificate class			
	Class I	Class II	Class III	Class IV
18. Procedures for protecting persons and property during the storing, dispensing, and handling of fuel and other hazardous substances and materials, as required under §139.321	X	X	X	X
19. A description of, and procedures for maintaining, the traffic and wind direction indicators, as required under §139.323	X	X	X	X
20. An emergency plan as required under §139.325	X	X	X	X
21. Procedures for conducting the self-inspection program, as required under §139.327	X	X	X	X
22. Procedures for controlling pedestrians and ground vehicles in movement areas and safety areas, as required under §139.329	X	X	X	
23. Procedures for obstruction removal, marking, or lighting, as required under §139.331	X	X	X	X
24. Procedures for protection of NAVAIDS, as required under §139.333	X	X	X	
25. A description of public protection, as required under §139.335	X	X	X	
26. Procedures for wildlife hazard management, as required under §139.337	X	X	X	
27. Procedures for airport condition reporting, as required under §139.339	X	X	X	X
28. Procedures for identifying, marking, and lighting construction and other unserviceable areas, as required under §139.341	X	X	X	
29. Any other item that the Administrator finds is necessary to ensure safety in air transportation	X	X	X	X

[Doc. No. FAA-2000-7479, 69 FR 6424, Feb. 10, 2004; Amdt. 139-26, 69 FR 31522, June 4, 2004, as amended by Amdt. 139-27, 78 FR 3316, Jan. 16, 2013]

§ 139.205 Amendment of Airport Certification Manual.

(a) Under § 139.3, the Regional Airports Division Manager may amend any Airport Certification Manual approved under this part, either—

(1) Upon application by the certificate holder or

(2) On the Regional Airports Division Manager's own initiative, if the Regional Airports Division Manager determines that safety in air transportation requires the amendment.

(b) A certificate holder must submit in writing a proposed amendment to its Airport Certification Manual to the Regional Airports Division Manager at least 30 days before the proposed effective date of the amendment, unless a shorter filing period is allowed by the Regional Airports Division Manager.

(c) At any time within 30 days after receiving a notice of refusal to approve the application for amendment, the certificate holder may petition the Associate Administrator for Airports to reconsider the refusal to amend.

(d) In the case of amendments initiated by the FAA, the Regional Airports Division Manager notifies the certificate holder of the proposed amendment, in writing, fixing a reasonable period (but not less than 7 days) within which the certificate holder may submit written information, views, and arguments on the amendment. After considering all relevant material presented, the Regional Airports Division Manager notifies the certificate holder within 30 days of any amendment adopted or rescinds the notice. The amendment becomes effective not less than 30 days after the certificate holder receives notice of it, except that, prior to the effective date, the certificate holder may petition the Associate Administrator for Airports to reconsider the amendment, in which case its effective date is stayed pending a decision by the Associate Administrator for Airports.

(e) Notwithstanding the provisions of paragraph (d) of this section, if the Regional Airports Division Manager finds there is an emergency requiring immediate action with respect to safety in air transportation, the Regional Airports Division Manager may issue an amendment, effective without stay on the date the certificate holder receives notice of it. In such a case, the Regional Airports Division Manager incorporates the finding of the emergency and a brief statement of the reasons for the finding in the notice of the amendment. Within 30 days after the issuance of such an emergency amendment, the certificate holder may petition the Associate Administrator for Airports to reconsider either the finding of an emergency, the amendment itself, or both. This petition does not automatically stay the effectiveness of the emergency amendment.

Subpart D—Operations

§ 139.301 Records.

In a manner authorized by the Administrator, each certificate holder must—

(a) Furnish upon request by the Administrator all records required to be maintained under this part.

(b) Maintain records required under this part as follows:

(1) *Personnel training.* Twenty-four consecutive calendar months for personnel training records, as required under §§ 139.303 and 139.327.

(2) *Emergency personnel training.* Twenty-four consecutive calendar months for aircraft rescue and firefighting and emergency medical service personnel training records, as required under § 139.319.

(3) *Airport fueling agent inspection.* Twelve consecutive calendar months for records of inspection of airport fueling agents, as required under § 139.321.

(4) *Fueling personnel training.* Twelve consecutive calendar months for training records of fueling personnel, as required under § 139.321.

(5) *Self-inspection.* Twelve consecutive calendar months for self-inspection records, as required under § 139.327.

(6) *Movement areas and safety areas training.* Twenty-four consecutive calendar months for records of training given to pedestrians and ground vehicle operators with access to movement areas and safety areas, as required under § 139.329.

(7) *Accident and incident.* Twelve consecutive calendar months for each accident or incident in movement areas and safety areas involving an air carrier aircraft and/or ground vehicle, as required under § 139.329.

(8) *Airport condition.* Twelve consecutive calendar months for records of airport condition information dissemination, as required under § 139.339.

(c) Make and maintain any additional records required by the Administrator, this part, and the Airport Certification Manual.

§ 139.303 Personnel.

In a manner authorized by the Administrator, each certificate holder must—

(a) Provide sufficient and qualified personnel to comply with the requirements of its Airport Certification Manual and the requirements of this part.

(b) Equip personnel with sufficient resources needed to comply with the requirements of this part.

(c) Train all persons who access movement areas and safety areas and perform duties in compliance with the requirements of the Airport Certification Manual and the requirements of this part. This training must be completed prior to the initial performance of such duties and at least once every 12 consecutive calendar months. The curriculum for initial and recurrent training must include at least the following areas:

(1) Airport familiarization, including airport marking, lighting, and signs system.

(2) Procedures for access to, and operation in, movement areas and safety areas, as specified under § 139.329.

(3) Airport communications, including radio communication between the air traffic control tower and personnel, use of the common traffic advisory frequency if there is no air traffic control tower or the tower is not in operation, and procedures for reporting unsafe airport conditions.

(4) Duties required under the Airport Certification Manual and the requirements of this part.

(5) Any additional subject areas required under §§ 139.319, 139.321, 139.327, 139.329, 139.337, and 139.339, as appropriate.

(d) Make a record of all training completed after June 9, 2004 by each individual in compliance with this section that includes, at a minimum, a description and date of training received. Such records must be maintained for 24 consecutive calendar months after completion of training.

(e) As appropriate, comply with the following training requirements of this part:

(1) § 139.319, Aircraft rescue and firefighting: Operational requirements;

(2) § 139.321, Handling and storage of hazardous substances and materials;

(3) § 139.327, Self-inspection program;

(4) § 139.329, Pedestrians and Ground Vehicles;

(5) § 139.337, Wildlife hazard management; and

(6) § 139.339, Airport condition reporting.

(f) Use an independent organization, or designee, to comply with the requirements of its Airport Certification Manual and the requirements of this part only if—

(1) Such an arrangement is authorized by the Administrator;

(2) A description of responsibilities and duties that will be assumed by an independent organization or designee is specified in the Airport Certification Manual; and

(3) The independent organization or designee prepares records required under this part in sufficient detail to assure the certificate holder and the Administrator of adequate compliance with the Airport Certification Manual and the requirements of this part.

[Doc. No. FAA-2000-7479, 69 FR 6424, Feb. 10, 2004; Amdt. 139-26, 69 FR 31522, June 4, 2004, as amended by Amdt. 139-27, 78 FR 3316, Jan. 16, 2013]

§ 139.305 Paved areas.

(a) In a manner authorized by the Administrator, each certificate holder must maintain, and promptly repair the pavement of, each runway, taxiway, loading ramp, and parking area on the airport that is available for air carrier use as follows:

(1) The pavement edges must not exceed 3 inches difference in elevation between abutting pavement sections and between pavement and abutting areas.

(2) The pavement must have no hole exceeding 3 inches in depth nor any hole the slope of which from any point in the hole to the nearest point at the lip of the hole is 45 degrees or greater, as measured from the pavement surface plane, unless, in either case, the entire area of the hole can be covered by a 5-inch diameter circle.

(3) The pavement must be free of cracks and surface variations that could impair directional control of air carrier aircraft, including any pavement crack or surface deterioration that produces loose aggregate or other contaminants.

(4) Except as provided in paragraph (b) of this section, mud, dirt, sand, loose aggregate, debris, foreign objects, rubber deposits, and other contaminants must be removed promptly and as completely as practicable.

(5) Except as provided in paragraph (b) of this section, any chemical solvent that is used to clean any pavement area must be removed as soon as possible, consistent with the instructions of the manufacturer of the solvent.

(6) The pavement must be sufficiently drained and free of depressions to prevent ponding that obscures markings or impairs safe aircraft operations.

(b) Paragraphs (a)(4) and (a)(5) of this section do not apply to snow and ice accumulations and their control, including the associated use of materials, such as sand and deicing solutions.

(c) FAA Advisory Circulars contain methods and procedures for the maintenance and configuration of paved areas that are acceptable to the Administrator.

[Doc. No. FAA-2000-7479, 69 FR 6424, Feb. 10, 2004; Amdt. 139-26, 69 FR 31522, June 4, 2004]

§ 139.307 Unpaved areas.

(a) In a manner authorized by the Administrator, each certificate holder must maintain and promptly repair the surface of each gravel, turf, or other unpaved runway, taxiway, or loading ramp and parking area on the airport that is available for air carrier use as follows:

(1) No slope from the edge of the full-strength surfaces downward to the existing terrain must be steeper than 2:1.

(2) The full-strength surfaces must have adequate crown or grade to assure sufficient drainage to prevent ponding.

(3) The full-strength surfaces must be adequately compacted and sufficiently stable to prevent rutting by aircraft or the loosening or build-up of surface material, which could impair directional control of aircraft or drainage.

(4) The full-strength surfaces must have no holes or depressions that exceed 3 inches in depth and are of a breadth capable of impairing directional control or causing damage to an aircraft.

(5) Debris and foreign objects must be promptly removed from the surface.

(b) FAA Advisory Circulars contain methods and procedures for the maintenance and configuration of unpaved areas that are acceptable to the Administrator.

§ 139.309 Safety areas.

(a) In a manner authorized by the Administrator, each certificate holder must provide and maintain, for each runway and taxiway that is available for air carrier use, a safety area of at least the dimensions that—

(1) Existed on December 31, 1987, if the runway or taxiway had a safety area on December 31, 1987, and if no reconstruction or significant expansion of the runway or taxiway was begun on or after January 1, 1988; or

(2) Are authorized by the Administrator at the time the construction, reconstruction, or expansion began if construction, reconstruction, or significant expansion of the runway or taxiway began on or after January 1, 1988.

(b) Each certificate holder must maintain its safety areas as follows:

(1) Each safety area must be cleared and graded and have no potentially hazardous ruts, humps, depressions, or other surface variations.

(2) Each safety area must be drained by grading or storm sewers to prevent water accumulation.

(3) Each safety area must be capable under dry conditions of supporting snow removal and aircraft rescue and firefighting equipment and of supporting the occasional passage of aircraft without causing major damage to the aircraft.

(4) No objects may be located in any safety area, except for objects that need to be located in a safety area because of their function. These objects must be constructed, to the extent practical, on frangibly mounted structures of the lowest practical height, with the frangible point no higher than 3 inches above grade.

(c) FAA Advisory Circulars contain methods and procedures for the configuration and maintenance of safety areas acceptable to the Administrator.

§ 139.311 Marking, signs, and lighting.

(a) *Marking.* Each certificate holder must provide and maintain marking systems for air carrier operations on the airport that are authorized by the Administrator and consist of at least the following:

(1) Runway markings meeting the specifications for takeoff and landing minimums for each runway.

(2) A taxiway centerline.

(3) Taxiway edge markings, as appropriate.

(4) Holding position markings.

(5) Instrument landing system (ILS) critical area markings.

(b) *Signs.* (1) Each certificate holder must provide and maintain sign systems for air carrier operations on the airport that are authorized by the Administrator and consist of at least the following:

(i) Signs identifying taxiing routes on the movement area.

(ii) Holding position signs.

(iii) Instrument landing system (ILS) critical area signs.

(2) Unless otherwise authorized by the Administrator, the signs required by paragraph (b)(1) of this section must be internally illuminated at each Class I, II, and IV airport.

(3) Unless otherwise authorized by the Administrator, the signs required by paragraphs (b)(1)(ii) and (b)(1)(iii) of this section must be internally illuminated at each Class III airport.

(c) *Lighting.* Each certificate holder must provide and maintain lighting systems for air carrier operations when the airport is open at night, during conditions below visual flight rules (VFR) minimums, or in Alaska, during periods in which a prominent unlighted object cannot be seen from a distance of 3 statute miles or the sun is more than six degrees below the horizon. These lighting systems must be authorized by the Administrator and consist of at least the following:

(1) Runway lighting that meets the specifications for takeoff and landing minimums, as authorized by the Administrator, for each runway.

(2) One of the following taxiway lighting systems:

(i) Centerline lights.

(ii) Centerline reflectors.

(iii) Edge lights.

(iv) Edge reflectors.

(3) An airport beacon.

(4) Approach lighting that meets the specifications for takeoff and landing minimums, as authorized by the Administrator, for each runway, unless provided and/or maintained by an entity other than the certificate holder.

(5) Obstruction marking and lighting, as appropriate, on each object within its authority that has been determined by the FAA to be an obstruction.

(d) *Maintenance.* Each certificate holder must properly maintain each marking, sign, or lighting system installed and operated on the airport. As used in this section, to "properly maintain" includes cleaning, replacing, or repairing any faded, missing, or nonfunctional item; keeping each item unobscured and clearly visible; and ensuring that each item provides an accurate reference to the user.

(e) *Lighting interference.* Each certificate holder must ensure that all lighting on the airport, including that for aprons, vehicle parking areas, roadways, fuel storage areas, and buildings, is adequately adjusted or shielded to prevent interference with air traffic control and aircraft operations.

(f) *Standards.* FAA Advisory Circulars contain methods and procedures for the equipment, material, installation, and maintenance of marking, sign, and lighting systems listed in this section that are acceptable to the Administrator.

(g) *Implementation.* The sign systems required under paragraph (b)(3) of this section must be implemented by each holder of a Class III Airport Operating Certificate not later than 36 consecutive calendar months after June 9, 2004.

§ 139.313 Snow and ice control.

(a) As determined by the Administrator, each certificate holder whose airport is located where snow and icing conditions occur must prepare, maintain, and carry out a snow and ice control plan in a manner authorized by the Administrator.

(b) The snow and ice control plan required by this section must include, at a minimum, instructions and procedures for—

(1) Prompt removal or control, as completely as practical, of snow, ice, and slush on each movement area;

(2) Positioning snow off the movement area surfaces so all air carrier aircraft propellers, engine pods, rotors, and wing tips will clear any snowdrift and snowbank as the aircraft's landing gear traverses any portion of the movement area;

(3) Selection and application of authorized materials for snow and ice control to ensure that they adhere to snow and ice sufficiently to minimize engine ingestion;

(4) Timely commencement of snow and ice control operations; and

(5) Prompt notification, in accordance with § 139.339, of all air carriers using the airport when any portion of the movement area normally available to them is less than satisfactorily cleared for safe operation by their aircraft.

(c) FAA Advisory Circulars contain methods and procedures for snow and ice control equipment, materials, and removal that are acceptable to the Administrator.

§ 139.315 Aircraft rescue and firefighting: Index determination.

(a) An index is required by paragraph (c) of this section for each certificate holder. The Index is determined by a combination of—

(1) The length of air carrier aircraft and

(2) Average daily departures of air carrier aircraft.

(b) For the purpose of Index determination, air carrier aircraft lengths are grouped as follows:

(1) Index A includes aircraft less than 90 feet in length.

(2) Index B includes aircraft at least 90 feet but less than 126 feet in length.

(3) Index C includes aircraft at least 126 feet but less than 159 feet in length.

(4) Index D includes aircraft at least 159 feet but less than 200 feet in length.

(5) Index E includes aircraft at least 200 feet in length.

(c) Except as provided in § 139.319(c), if there are five or more average daily departures of air carrier aircraft in a single Index group serving that airport, the longest aircraft with an average of five or more daily departures determines the Index required for the airport. When there are fewer than five average daily departures of the longest air carrier aircraft serving the airport, the Index required for the airport will be the next lower Index group than the Index group prescribed for the longest aircraft.

(d) The minimum designated index shall be Index A.

(e) A holder of a Class III Airport Operating Certificate may comply with this section by providing a level of safety comparable to Index A that is approved by the Administrator. Such alternate compliance must be described in the ACM and must include:

(1) Pre-arranged firefighting and emergency medical response procedures, including agreements with responding services.

(2) Means for alerting firefighting and emergency medical response personnel.

(3) Type of rescue and firefighting equipment to be provided.

(4) Training of responding firefighting and emergency medical personnel on airport familiarization and communications.

[Doc. No. FAA-2000-7479, 69 FR 6424, Feb. 10, 2004; Amdt. 139-26, 69 FR 31522, June 4, 2004]

§ 139.317 Aircraft rescue and firefighting: Equipment and agents.

Unless otherwise authorized by the Administrator, the following rescue and firefighting equipment and agents are the minimum required for the Indexes referred to in § 139.315:

(a) *Index A.* One vehicle carrying at least—

(1) 500 pounds of sodium-based dry chemical, halon 1211, or clean agent; or

(2) 450 pounds of potassium-based dry chemical and water with a commensurate quantity of AFFF to total 100 gallons for simultaneous dry chemical and AFFF application.

(b) *Index B.* Either of the following:

(1) One vehicle carrying at least 500 pounds of sodium-based dry chemical, halon 1211, or clean agent and 1,500 gallons of water and the commensurate quantity of AFFF for foam production.

(2) Two vehicles—

(i) One vehicle carrying the extinguishing agents as specified in paragraphs (a)(1) or (a)(2) of this section; and

(ii) One vehicle carrying an amount of water and the commensurate quantity of AFFF so the total quantity of water for foam production carried by both vehicles is at least 1,500 gallons.

(c) *Index C.* Either of the following:

(1) Three vehicles—

(i) One vehicle carrying the extinguishing agents as specified in paragraph (a)(1) or (a)(2) of this section; and

(ii) Two vehicles carrying an amount of water and the commensurate quantity of AFFF so the total quantity of water for foam production carried by all three vehicles is at least 3,000 gallons.

(2) Two vehicles—

(i) One vehicle carrying the extinguishing agents as specified in paragraph (b)(1) of this section; and

(ii) One vehicle carrying water and the commensurate quantity of AFFF so the total quantity of water for foam production carried by both vehicles is at least 3,000 gallons.

(d) *Index D.* Three vehicles—

(1) One vehicle carrying the extinguishing agents as specified in paragraphs (a)(1) or (a)(2) of this section; and

(2) Two vehicles carrying an amount of water and the commensurate quantity of AFFF so the total quantity of water for foam production carried by all three vehicles is at least 4,000 gallons.

(e) *Index E.* Three vehicles—

(1) One vehicle carrying the extinguishing agents as specified in paragraphs (a)(1) or (a)(2) of this section; and

(2) Two vehicles carrying an amount of water and the commensurate quantity of AFFF so the total quantity of water for foam production carried by all three vehicles is at least 6,000 gallons.

(f) *Foam discharge capacity.* Each aircraft rescue and firefighting vehicle used to comply with Index B, C, D, or E requirements with a capacity of at least 500 gallons of water for foam production must be equipped with a turret. Vehicle turret discharge capacity must be as follows:

(1) Each vehicle with a minimum-rated vehicle water tank capacity of at least 500 gallons, but less than 2,000 gallons, must have a turret discharge rate of at least 500 gallons per minute, but not more than 1,000 gallons per minute.

(2) Each vehicle with a minimum-rated vehicle water tank capacity of at least 2,000 gallons must have a turret discharge rate of at least 600 gallons per minute, but not more than 1,200 gallons per minute.

(g) *Agent discharge capacity.* Each aircraft rescue and firefighting vehicle that is required to carry dry chemical, halon 1211, or clean agent for compliance with the Index requirements of this section must meet one of the following minimum discharge rates for the equipment installed:

(1) Dry chemical, halon 1211, or clean agent through a hand line—5 pounds per second.

(2) Dry chemical, halon 1211, or clean agent through a turret—16 pounds per second.

(h) *Extinguishing agent substitutions.* Other extinguishing agent substitutions authorized by the Administrator may be made in amounts that provide equivalent firefighting capability.

(i) *AFFF quantity requirements.* In addition to the quantity of water required, each vehicle required to carry AFFF must carry AFFF in an appropriate amount to mix with twice the water required to be carried by the vehicle.

(j) *Methods and procedures.* FAA Advisory Circulars contain methods and procedures for ARFF equipment and extinguishing agents that are acceptable to the Administrator.

(k) *Implementation.* Each holder of a Class II, III, or IV Airport Operating Certificate must implement the requirements of this section no later than 36 consecutive calendar months after June 9, 2004.

[Doc. No. FAA-2000-7479, 69 FR 6424, Feb. 10, 2004; Amdt. 139-26, 69 FR 31523, June 4, 2004]

§ 139.319 Aircraft rescue and firefighting: Operational requirements.

(a) *Rescue and firefighting capability.* Except as provided in paragraph (c) of this section, each certificate holder must provide on the airport, during air carrier operations at the airport, at least the rescue and firefighting capability specified for the Index required by § 139.317 in a manner authorized by the Administrator.

(b) *Increase in Index.* Except as provided in paragraph (c) of this section, if an increase in the average daily departures or the length of air carrier aircraft results in an increase in the Index required by paragraph (a) of this section, the certificate holder must comply with the increased requirements.

(c) *Reduction in rescue and firefighting.* During air carrier operations with only aircraft shorter than the Index aircraft group required by paragraph (a) of this section, the certificate holder may reduce the rescue and firefighting to a lower level corresponding to the Index group of the longest air carrier aircraft being operated.

(d) *Procedures for reduction in capability.* Any reduction in the rescue and firefighting capability from the Index required by paragraph (a) of this section, in accordance with paragraph (c) of this section, must be subject to the following conditions:

(1) Procedures for, and the persons having the authority to implement, the reductions must be included in the Airport Certification Manual.

(2) A system and procedures for recall of the full aircraft rescue and firefighting capability must be included in the Airport Certification Manual.

(3) The reductions may not be implemented unless notification to air carriers is provided in the Airport/Facility Directory or Notices to Airmen (NOTAM), as appropriate, and by direct notification of local air carriers.

(e) *Vehicle communications.* Each vehicle required under § 139.317 must be equipped with two-way voice radio communications that provide for contact with at least—

(1) All other required emergency vehicles;

(2) The air traffic control tower;

(3) The common traffic advisory frequency when an air traffic control tower is not in operation or there is no air traffic control tower, and

(4) Fire stations, as specified in the airport emergency plan.

(f) *Vehicle marking and lighting.* Each vehicle required under § 139.317 must—

(1) Have a flashing or rotating beacon and

(2) Be painted or marked in colors to enhance contrast with the background environment and optimize daytime and nighttime visibility and identification.

(g) *Vehicle readiness.* Each vehicle required under § 139.317 must be maintained as follows:

(1) The vehicle and its systems must be maintained so as to be operationally capable of performing the functions required by this subpart during all air carrier operations.

(2) If the airport is located in a geographical area subject to prolonged temperatures below 33 degrees Fahrenheit, the vehicles must be provided with cover or other means to ensure equipment operation and discharge under freezing conditions.

(3) Any required vehicle that becomes inoperative to the extent that it cannot perform as required by paragraph (g)(1) of this section must be replaced immediately with equipment having at least equal capabilities. If replacement equipment is not available immediately, the certificate holder must so notify the Regional Airports Division Manager and each air carrier using the airport in accordance with § 139.339. If the required Index level of capability is not restored within 48 hours, the airport operator, unless otherwise authorized by the Administrator, must limit air carrier operations on the airport to those compatible with the Index corresponding to the remaining operative rescue and firefighting equipment.

(h) *Response requirements.* (1) With the aircraft rescue and firefighting equipment required under this part and the number of trained personnel that will assure an effective operation, each certificate holder must—

(i) Respond to each emergency during periods of air carrier operations; and

(ii) When requested by the Administrator, demonstrate compliance with the response requirements specified in this section.

(2) The response required by paragraph (h)(1)(ii) of this section must achieve the following performance criteria:

(i) Within 3 minutes from the time of the alarm, at least one required aircraft rescue and firefighting vehicle must reach the midpoint of the farthest runway serving air carrier aircraft from its assigned post or reach any other specified point of comparable distance on the movement area that is available to air carriers, and begin application of extinguishing agent.

(ii) Within 4 minutes from the time of alarm, all other required vehicles must reach the point specified in paragraph (h)(2)(i) of this section from their assigned posts and begin application of an extinguishing agent.

(i) *Personnel.* Each certificate holder must ensure the following:

(1) All rescue and firefighting personnel are equipped in a manner authorized by the Administrator with protective clothing and equipment needed to perform their duties.

(2) All rescue and firefighting personnel are properly trained to perform their duties in a manner authorized by the Administrator. Such personnel must be trained prior to initial performance of rescue and firefighting duties and receive recurrent instruction every 12 consecutive calendar months. The curriculum for initial and recurrent training must include at least the following areas:

(i) Airport familiarization, including airport signs, marking, and lighting.

(ii) Aircraft familiarization.

(iii) Rescue and firefighting personnel safety.

(iv) Emergency communications systems on the airport, including fire alarms.

(v) Use of the fire hoses, nozzles, turrets, and other appliances required for compliance with this part.

(vi) Application of the types of extinguishing agents required for compliance with this part.

(vii) Emergency aircraft evacuation assistance.

(viii) Firefighting operations.

(ix) Adapting and using structural rescue and firefighting equipment for aircraft rescue and firefighting.

(x) Aircraft cargo hazards, including hazardous materials/dangerous goods incidents.

(xi) Familiarization with firefighters' duties under the airport emergency plan.

(3) All rescue and firefighting personnel must participate in at least one live-fire drill prior to initial performance of rescue and firefighting duties and every 12 consecutive calendar months thereafter.

(4) At least one individual, who has been trained and is current in basic emergency medical services, is available during air carrier operations. This individual must be trained prior to initial performance of emergency medical services. Training must be at a minimum 40 hours in length and cover the following topics:

(i) Bleeding.

(ii) Cardiopulmonary resuscitation.

(iii) Shock.

(iv) Primary patient survey.

(v) Injuries to the skull, spine, chest, and extremities.

(vi) Internal injuries.

(vii) Moving patients.

(viii) Burns.

(ix) Triage.

(5) A record is maintained of all training given to each individual under this section for 24 consecutive calendar months after completion of training. Such records must include, at a minimum, a description and date of training received.

(6) Sufficient rescue and firefighting personnel are available during all air carrier operations to operate the vehicles, meet the response times, and meet the minimum agent discharge rates required by this part.

(7) Procedures and equipment are established and maintained for alerting rescue and firefighting personnel by siren, alarm, or other means authorized by the Administrator to any existing or impending emergency requiring their assistance.

(j) *Hazardous materials guidance.* Each aircraft rescue and firefighting vehicle responding to an emergency on the airport must be equipped with, or have available through a direct communications link, the "North American Emergency Response Guidebook" published by the U.S. Department of Transportation or similar response guidance to hazardous materials/dangerous goods incidents. Information on obtaining the "North American Emergency Response Guidebook" is available from the Regional Airports Division Manager.

(k) *Emergency access roads.* Each certificate holder must ensure that roads designated for use as emergency access roads for aircraft rescue and firefighting vehicles are maintained in a condition that will support those vehicles during all-weather conditions.

(l) *Methods and procedures.* FAA Advisory Circulars contain methods and procedures for aircraft rescue and firefighting and emergency medical equipment and training that are acceptable to the Administrator.

(m) *Implementation.* Each holder of a Class II, III, or IV Airport Operating Certificate must implement the requirements of this section no later than 36 consecutive calendar months after June 9, 2004.

[Doc. No. FAA-2000-7479, 69 FR 6424, Feb. 10, 2004; Amdt. 139-26, 69 FR 31523, June 4, 2004]

§ 139.321 Handling and storing of hazardous substances and materials.

(a) Each certificate holder who acts as a cargo handling agent must establish and maintain procedures for the protection of persons and property on the airport during the handling and storing of any material regulated by the Hazardous Materials Regulations (49 CFR 171 through 180) that is, or is intended to be, transported by air. These procedures must provide for at least the following:

(1) Designated personnel to receive and handle hazardous substances and materials.

(2) Assurance from the shipper that the cargo can be handled safely, including any special handling procedures required for safety.

(3) Special areas for storage of hazardous materials while on the airport.

(b) Each certificate holder must establish and maintain standards authorized by the Administrator for protecting against fire and explosions in storing, dispensing, and otherwise handling fuel (other than articles and materials that are, or are intended to be, aircraft cargo) on the airport. These standards must cover facilities, procedures, and personnel training and must address at least the following:

(1) Bonding.

(2) Public protection.

(3) Control of access to storage areas.

(4) Fire safety in fuel farm and storage areas.

(5) Fire safety in mobile fuelers, fueling pits, and fueling cabinets.

(6) Training of fueling personnel in fire safety in accordance with paragraph (e) of this section. Such training at

Class III airports must be completed within 12 consecutive calendar months after June 9, 2004.

(7) The fire code of the public body having jurisdiction over the airport.

(c) Each certificate holder must, as a fueling agent, comply with, and require all other fueling agents operating on the airport to comply with, the standards established under paragraph (b) of this section and must perform reasonable surveillance of all fueling activities on the airport with respect to those standards.

(d) Each certificate holder must inspect the physical facilities of each airport tenant fueling agent at least once every 3 consecutive months for compliance with paragraph (b) of this section and maintain a record of that inspection for at least 12 consecutive calendar months.

(e) The training required in paragraph (b)(6) of this section must include at least the following:

(1) At least one supervisor with each fueling agent must have completed an aviation fuel training course in fire safety that is authorized by the Administrator. Such an individual must be trained prior to initial performance of duties, or enrolled in an authorized aviation fuel training course that will be completed within 90 days of initiating duties, and receive recurrent instruction at least every 24 consecutive calendar months.

(2) All other employees who fuel aircraft, accept fuel shipments, or otherwise handle fuel must receive at least initial on-the-job training and recurrent instruction every 24 consecutive calendar months in fire safety from the supervisor trained in accordance with paragraph (e)(1) of this section.

(f) Each certificate holder must obtain a written confirmation once every 12 consecutive calendar months from each airport tenant fueling agent that the training required by paragraph (e) of this section has been accomplished. This written confirmation must be maintained for 12 consecutive calendar months.

(g) Unless otherwise authorized by the Administrator, each certificate holder must require each tenant fueling agent to take immediate corrective action whenever the certificate holder becomes aware of noncompliance with a standard required by paragraph (b) of this section. The certificate holder must notify the appropriate FAA Regional Airports Division Manager immediately when noncompliance is discovered and corrective action cannot be accomplished within a reasonable period of time.

(h) FAA Advisory Circulars contain methods and procedures for the handling and storage of hazardous substances and materials that are acceptable to the Administrator.

§ 139.323 Traffic and wind direction indicators.

In a manner authorized by the Administrator, each certificate holder must provide and maintain the following on its airport:

(a) A wind cone that visually provides surface wind direction information to pilots. For each runway available for air carrier use, a supplemental wind cone must be installed at the end of the runway or at least at one point visible to the pilot while on final approach and prior to takeoff. If the airport is open for air carrier operations at night, the wind direction indicators, including the required supplemental indicators, must be lighted.

(b) For airports serving any air carrier operation when there is no control tower operating, a segmented circle, a landing strip indicator and a traffic pattern indicator must be installed around a wind cone for each runway with a right-hand traffic pattern.

(c) FAA Advisory Circulars contain methods and procedures for the installation, lighting, and maintenance of traffic and wind indicators that are acceptable to the Administrator.

§ 139.325 Airport emergency plan.

(a) In a manner authorized by the Administrator, each certificate holder must develop and maintain an airport emergency plan designed to minimize the possibility and extent of personal injury and property damage on the airport in an emergency. The plan must—

(1) Include procedures for prompt response to all emergencies listed in paragraph (b) of this section, including a communications network;

(2) Contain sufficient detail to provide adequate guidance to each person who must implement these procedures; and

(3) To the extent practicable, provide for an emergency response for the largest air carrier aircraft in the Index group required under § 139.315.

(b) The plan required by this section must contain instructions for response to—

(1) Aircraft incidents and accidents;

(2) Bomb incidents, including designation of parking areas for the aircraft involved;

(3) Structural fires;

(4) Fires at fuel farms or fuel storage areas;

(5) Natural disaster;

(6) Hazardous materials/dangerous goods incidents;

(7) Sabotage, hijack incidents, and other unlawful interference with operations;

(8) Failure of power for movement area lighting; and

(9) Water rescue situations, as appropriate.

(c) The plan required by this section must address or include—

(1) To the extent practicable, provisions for medical services, including transportation and medical assistance for the maximum number of persons that can be carried on the largest air carrier aircraft that the airport reasonably can be expected to serve;

(2) The name, location, telephone number, and emergency capability of each hospital and other medical facility and the business address and telephone number of medical personnel on the airport or in the communities it serves who have agreed to provide medical assistance or transportation;

(3) The name, location, and telephone number of each rescue squad, ambulance service, military installation, and government agency on the airport or in the communities it serves that agrees to provide medical assistance or transportation;

(4) An inventory of surface vehicles and aircraft that the facilities, agencies, and personnel included in the plan under paragraphs (c)(2) and (3) of this section will provide to transport injured and deceased persons to locations on the airport and in the communities it serves;

(5) A list of each hangar or other building on the airport or in the communities it serves that will be used to accommodate uninjured, injured, and deceased persons;

(6) Plans for crowd control, including the name and location of each safety or security agency that agrees to provide assistance for the control of crowds in the event of an emergency on the airport; and

(7) Procedures for removing disabled aircraft, including, to the extent practical, the name, location, and telephone numbers of agencies with aircraft removal responsibilities or capabilities.

(d) The plan required by this section must provide for—

(1) The marshalling, transportation, and care of ambulatory injured and uninjured accident survivors;

(2) The removal of disabled aircraft;

(3) Emergency alarm or notification systems; and

(4) Coordination of airport and control tower functions relating to emergency actions, as appropriate.

(e) The plan required by this section must contain procedures for notifying the facilities, agencies, and personnel who have responsibilities under the plan of the location of an aircraft accident, the number of persons involved in that accident, or any other information necessary to carry out their responsibilities, as soon as that information becomes available.

(f) The plan required by this section must contain provisions, to the extent practicable, for the rescue of aircraft accident victims from significant bodies of water or marsh lands adjacent to the airport that are crossed by the approach and departure flight paths of air carriers. A body of water or marshland is significant if the area exceeds one-quarter square mile and cannot be traversed by conventional land rescue vehicles. To the extent practicable, the plan must provide for rescue vehicles with a combined capacity for handling the maximum number of persons that can be carried on board the largest air carrier aircraft in the Index group required under § 139.315.

(g) Each certificate holder must—

(1) Coordinate the plan with law enforcement agencies, rescue and firefighting agencies, medical personnel and organizations, the principal tenants at the airport, and all other persons who have responsibilities under the plan;

(2) To the extent practicable, provide for participation by all facilities, agencies, and personnel specified in paragraph (g)(1) of this section in the development of the plan;

(3) Ensure that all airport personnel having duties and responsibilities under the plan are familiar with their assignments and are properly trained; and

(4) At least once every 12 consecutive calendar months, review the plan with all of the parties with whom the plan is coordinated, as specified in paragraph (g)(1) of this section, to ensure that all parties know their responsibilities and that all of the information in the plan is current.

(h) Each holder of a Class I Airport Operating Certificate must hold a full-scale airport emergency plan exercise at least once every 36 consecutive calendar months.

(i) Each airport subject to applicable FAA and Transportation Security Administration security regulations must ensure that instructions for response to paragraphs (b)(2) and (b)(7) of this section in the airport emergency plan are consistent with its approved airport security program.

(j) FAA Advisory Circulars contain methods and procedures for the development of an airport emergency plan that are acceptable to the Administrator.

(k) The emergency plan required by this section must be submitted by each holder of a Class II, III, or IV Airport Operating Certificate no later than 24 consecutive calendar months after June 9, 2004.

§ 139.327 Self-inspection program.

(a) In a manner authorized by the Administrator, each certificate holder must inspect the airport to assure compliance with this subpart according to the following schedule:

(1) Daily, except as otherwise required by the Airport Certification Manual;

(2) When required by any unusual condition, such as construction activities or meteorological conditions, that may affect safe air carrier operations; and

(3) Immediately after an accident or incident.

(b) Each certificate holder must provide the following:

(1) Equipment for use in conducting safety inspections of the airport;

(2) Procedures, facilities, and equipment for reliable and rapid dissemination of information between the certificate holder's personnel and air carriers; and

(3) Procedures to ensure qualified personnel perform the inspections. Such procedures must ensure personnel are trained, as specified under § 139.303, and receive initial and recurrent instruction every 12 consecutive calendar months in at least the following areas:

(i) Airport familiarization, including airport signs, marking and lighting.

(ii) Airport emergency plan.

(iii) Notice to Airmen (NOTAM) notification procedures.

(iv) Procedures for pedestrians and ground vehicles in movement areas and safety areas.

(v) Discrepancy reporting procedures; and

(4) A reporting system to ensure prompt correction of unsafe airport conditions noted during the inspection, including wildlife strikes.

(c) Each certificate holder must—

(1) Prepare, and maintain for at least 12 consecutive calendar months, a record of each inspection prescribed by this section, showing the conditions found and all corrective actions taken.

(2) Prepare records of all training given after June 9, 2004 to each individual in compliance with this section that includes, at a minimum, a description and date of training received. Such records must be maintained for 24 consecutive calendar months after completion of training.

(d) FAA Advisory Circulars contain methods and procedures for the conduct of airport self-inspections that are acceptable to the Administrator.

§ 139.329 Pedestrians and ground vehicles.

In a manner authorized by the Administrator, each certificate holder must—

(a) Limit access to movement areas and safety areas only to those pedestrians and ground vehicles necessary for airport operations;

(b) Establish and implement procedures for the safe and orderly access to and operation in movement areas and safety areas by pedestrians and ground vehicles, including provisions identifying the consequences of noncompliance with the procedures by all persons;

(c) When an air traffic control tower is in operation, ensure that each pedestrian and ground vehicle in movement areas or safety areas is controlled by one of the following:

(1) Two-way radio communications between each pedestrian or vehicle and the tower;

(2) An escort with two-way radio communications with the tower accompanying any pedestrian or vehicle without a radio; or

(3) Measures authorized by the Administrator for controlling pedestrians and vehicles, such as signs, signals, or guards, when it is not operationally practical to have two-way radio communications between the tower and the pedestrian, vehicle, or escort;

(d) When an air traffic control tower is not in operation, or there is no air traffic control tower, provide adequate procedures to control pedestrians and ground vehicles in movement areas or safety areas through two-way radio communications or prearranged signs or signals;

(e) Ensure that all persons are trained on procedures required under paragraph (b) of this section prior to the initial performance of such duties and at least once every 12 consecutive calendar months, including consequences of noncompliance, prior to moving on foot, or operating a ground vehicle, in movement areas or safety areas; and

(f) Maintain the following records:

(1) A description and date of training completed after June 9, 2004 by each individual in compliance with this section. A record for each individual must be maintained for 24 consecutive months after the termination of an individual's access to movement areas and safety areas.

(2) A description and date of any accidents or incidents in the movement areas and safety areas involving air carrier aircraft, a ground vehicle or a pedestrian. Records of each accident or incident occurring after the June 9, 2004 must be maintained for 12 consecutive calendar months from the date of the accident or incident.

[Doc. No. FAA-2000-7479, 69 FR 6424, Feb. 10, 2004, as amended by Amdt. 139-27, 78 FR 3316, Jan. 16, 2013]

§ 139.331 Obstructions.

In a manner authorized by the Administrator, each certificate holder must ensure that each object in each area within its authority that has been determined by the FAA to be an obstruction is removed, marked, or lighted, unless determined to be unnecessary by an FAA aeronautical study. FAA Advisory Circulars contain methods and procedures for the lighting of obstructions that are acceptable to the Administrator.

§ 139.333 Protection of NAVAIDS.

In a manner authorized by the Administrator, each certificate holder must—

(a) Prevent the construction of facilities on its airport that, as determined by the Administrator, would derogate the operation of an electronic or visual NAVAID and air traffic control facilities on the airport;

(b) Protect—or if the owner is other than the certificate holder, assist in protecting—all NAVAIDS on its airport against vandalism and theft; and

(c) Prevent, insofar as it is within the airport's authority, interruption of visual and electronic signals of NAVAIDS.

§ 139.335 Public protection.

(a) In a manner authorized by the Administrator, each certificate holder must provide—

(1) Safeguards to prevent inadvertent entry to the movement area by unauthorized persons or vehicles; and

(2) Reasonable protection of persons and property from aircraft blast.

(b) Fencing that meets the requirements of applicable FAA and Transportation Security Administration security regulations in areas subject to these regulations is acceptable for meeting the requirements of paragraph (a)(l) of this section.

§ 139.337 Wildlife hazard management.

(a) In accordance with its Airport Certification Manual and the requirements of this section, each certificate holder must take immediate action to alleviate wildlife hazards whenever they are detected.

(b) In a manner authorized by the Administrator, each certificate holder must ensure that a wildlife hazard assessment is conducted when any of the following events occurs on or near the airport:

(1) An air carrier aircraft experiences multiple wildlife strikes;

(2) An air carrier aircraft experiences substantial damage from striking wildlife. As used in this paragraph, substantial damage means damage or structural failure incurred by an aircraft that adversely affects the structural strength, performance, or flight characteristics of the aircraft and that would normally require major repair or replacement of the affected component;

(3) An air carrier aircraft experiences an engine ingestion of wildlife; or

(4) Wildlife of a size, or in numbers, capable of causing an event described in paragraphs (b)(1), (b)(2), or (b)(3) of this section is observed to have access to any airport flight pattern or aircraft movement area.

(c) The wildlife hazard assessment required in paragraph (b) of this section must be conducted by a wildlife damage management biologist who has professional training and/or experience in wildlife hazard management at airports or an individual working under direct supervision of such an individual. The wildlife hazard assessment must contain at least the following:

(1) An analysis of the events or circumstances that prompted the assessment.

(2) Identification of the wildlife species observed and their numbers, locations, local movements, and daily and seasonal occurrences.

(3) Identification and location of features on and near the airport that attract wildlife.

(4) A description of wildlife hazards to air carrier operations.

(5) Recommended actions for reducing identified wildlife hazards to air carrier operations.

(d) The wildlife hazard assessment required under paragraph (b) of this section must be submitted to the Administrator for approval and determination of the need for a wildlife hazard management plan. In reaching this determination, the Administrator will consider—

(1) The wildlife hazard assessment;

(2) Actions recommended in the wildlife hazard assessment to reduce wildlife hazards;

(3) The aeronautical activity at the airport, including the frequency and size of air carrier aircraft;

(4) The views of the certificate holder;

(5) The views of the airport users; and

(6) Any other known factors relating to the wildlife hazard of which the Administrator is aware.

(e) When the Administrator determines that a wildlife hazard management plan is needed, the certificate holder must formulate and implement a plan using the wildlife hazard assessment as a basis. The plan must—

(1) Provide measures to alleviate or eliminate wildlife hazards to air carrier operations;

(2) Be submitted to, and approved by, the Administrator prior to implementation; and

(3) As authorized by the Administrator, become a part of the Airport Certification Manual.

(f) The plan must include at least the following:

(1) A list of the individuals having authority and responsibility for implementing each aspect of the plan.

(2) A list prioritizing the following actions identified in the wildlife hazard assessment and target dates for their initiation and completion:

(i) Wildlife population management;

(ii) Habitat modification; and

(iii) Land use changes.

(3) Requirements for and, where applicable, copies of local, State, and Federal wildlife control permits.

(4) Identification of resources that the certificate holder will provide to implement the plan.

(5) Procedures to be followed during air carrier operations that at a minimum includes—

(i) Designation of personnel responsible for implementing the procedures;

(ii) Provisions to conduct physical inspections of the aircraft movement areas and other areas critical to successfully manage known wildlife hazards before air carrier operations begin;

(iii) Wildlife hazard control measures; and

(iv) Ways to communicate effectively between personnel conducting wildlife control or observing wildlife hazards and the air traffic control tower.

(6) Procedures to review and evaluate the wildlife hazard management plan every 12 consecutive months or following an event described in paragraphs (b)(1), (b)(2), and (b)(3) of this section, including:

(i) The plan's effectiveness in dealing with known wildlife hazards on and in the airport's vicinity and

(ii) Aspects of the wildlife hazards described in the wildlife hazard assessment that should be reevaluated.

(7) A training program conducted by a qualified wildlife damage management biologist to provide airport personnel with the knowledge and skills needed to successfully carry out the wildlife hazard management plan required by paragraph (d) of this section.

(g) FAA Advisory Circulars contain methods and procedures for wildlife hazard management at airports that are acceptable to the Administrator.

§ 139.339 Airport condition reporting.

In a manner authorized by the Administrator, each certificate holder must—

(a) Provide for the collection and dissemination of airport condition information to air carriers.

(b) In complying with paragraph (a) of this section, use the NOTAM system, as appropriate, and other systems and procedures authorized by the Administrator.

(c) In complying with paragraph (a) of this section, provide information on the following airport conditions that may affect the safe operations of air carriers:

(1) Construction or maintenance activity on movement areas, safety areas, or loading ramps and parking areas.

(2) Surface irregularities on movement areas, safety areas, or loading ramps and parking areas.

(3) Snow, ice, slush, or water on the movement area or loading ramps and parking areas.

(4) Snow piled or drifted on or near movement areas contrary to § 139.313.

(5) Objects on the movement area or safety areas contrary to § 139.309.

(6) Malfunction of any lighting system, holding position signs, or ILS critical area signs required by § 139.311.

(7) Unresolved wildlife hazards as identified in accordance with § 139.337.

(8) Nonavailability of any rescue and firefighting capability required in §§ 139.317 or 139.319.

(9) Any other condition as specified in the Airport Certification Manual or that may otherwise adversely affect the safe operations of air carriers.

(d) Each certificate holder must prepare and keep, for at least 12 consecutive calendar months, a record of each dissemination of airport condition information to air carriers prescribed by this section.

(e) FAA Advisory Circulars contain methods and procedures for using the NOTAM system and the dissemination of airport information that are acceptable to the Administrator.

§ 139.341 Identifying, marking, and lighting construction and other unserviceable areas.

(a) In a manner authorized by the Administrator, each certificate holder must—

(1) Mark and, if appropriate, light in a manner authorized by the Administrator—

(i) Each construction area and unserviceable area that is on or adjacent to any movement area or any other area of the airport on which air carrier aircraft may be operated;

(ii) Each item of construction equipment and each construction roadway, which may affect the safe movement of aircraft on the airport; and

(iii) Any area adjacent to a NAVAID that, if traversed, could cause derogation of the signal or the failure of the NAVAID; and

(2) Provide procedures, such as a review of all appropriate utility plans prior to construction, for avoiding damage to existing utilities, cables, wires, conduits, pipelines, or other underground facilities.

(b) FAA Advisory Circulars contain methods and procedures for identifying and marking construction areas that are acceptable to the Administrator.

§ 139.343 Noncomplying conditions.

Unless otherwise authorized by the Administrator, whenever the requirements of subpart D of this part cannot be met to the extent that uncorrected unsafe conditions exist on the airport, the certificate holder must limit air carrier operations to those portions of the airport not rendered unsafe by those conditions.

PART 141

PILOT SCHOOLS

Contents

Subpart A—General

AUTHORITY: 49 U.S.C. 106(f), 106(g), 40113, 44701-44703, 44707, 44709, 44711, 45102-45103, 45301-45302.

SOURCE: Docket No. 25910, 62 FR 16347, Apr. 4, 1997, unless otherwise noted.

Subpart A—General

§ 141.1 Applicability.

This part prescribes the requirements for issuing pilot school certificates, provisional pilot school certificates, and associated ratings, and the general operating rules applicable to a holder of a certificate or rating issued under this part.

§ 141.3 Certificate required.

No person may operate as a certificated pilot school without, or in violation of, a pilot school certificate or provisional pilot school certificate issued under this part.

§ 141.5 Requirements for a pilot school certificate.

The FAA may issue a pilot school certificate with the appropriate ratings if, within the 24 calendar months before the date application is made, the applicant—

(a) Completes the application for a pilot school certificate on the form and in the manner prescribed by the FAA;

(b) Has held a provisional pilot school certificate;

(c) Meets the applicable requirements under subparts A through C of this part for the school certificate and associated ratings sought;

(d) Has established a pass rate of 80 percent or higher on the first attempt for all knowledge tests leading to a certificate or rating, practical tests leading to a certificate or rating, or end-of-course tests for an approved training course specified in appendix K of this part.

(e) Has graduated at least 10 different people from the school's approved training courses.

[Doc. No. FAA-2006-26661, 74 FR 42563, Aug. 21, 2009, as amended by Amdt. 141-14, 75 FR 56858, Sept. 17, 2010]

§ 141.7 Provisional pilot school certificate.

An applicant that meets the applicable requirements of subparts A, B, and C of this part, but does not meet the recent training activity requirements of § 141.5(d) of this part, may be issued a provisional pilot school certificate with ratings.

§ 141.9 Examining authority.

The FAA issues examining authority to a pilot school for a training course if the pilot school and its training course meet the requirements of subpart D of this part.

[Doc. No. FAA-2006-26661, 74 FR 42563, Aug. 21, 2009]

§ 141.11 Pilot school ratings.

(a) The ratings listed in paragraph (b) of this section may be issued to an applicant for:

(1) A pilot school certificate, provided the applicant meets the requirements of § 141.5 of this part; or

(2) A provisional pilot school certificate, provided the applicant meets the requirements of § 141.7 of this part.

(b) An applicant may be authorized to conduct the following courses:

(1) *Certification and rating courses.* (Appendixes A through J).

(i) Recreational pilot course.
(ii) Private pilot course.
(iii) Commercial pilot course.
(iv) Instrument rating course.
(v) Airline transport pilot course.
(vi) Flight instructor course.
(vii) Flight instructor instrument course.
(viii) Ground instructor course.
(ix) Additional aircraft category or class rating course.
(x) Aircraft type rating course.

(2) *Special preparation courses.* (Appendix K).

(i) Pilot refresher course.
(ii) Flight instructor refresher course.
(iii) Ground instructor refresher course.
(iv) Agricultural aircraft operations course.
(v) Rotorcraft external-load operations course.
(vi) Special operations course.
(vii) Test pilot course.
(viii) Airline transport pilot certification training program.

(3) *Pilot ground school course.* (Appendix L).

[Doc. No. 25910, 62 FR 16347, Apr. 4, 1997, as amended by Amdt. 141-1, 78 FR 42379, July 15, 2013]

§ 141.13 Application for issuance, amendment, or renewal.

(a) Application for an original certificate and rating, an additional rating, or the renewal of a certificate under this part must be made on a form and in a manner prescribed by the Administrator.

(b) Application for the issuance or amendment of a certificate or rating must be accompanied by two copies of each proposed training course curriculum for which approval is sought.

§ 141.17 Duration of certificate and examining authority.

(a) Unless surrendered, suspended, or revoked, a pilot school's certificate or a provisional pilot school's certificate expires:

(1) On the last day of the 24th calendar month from the month the certificate was issued;

(2) Except as provided in paragraph (b) of this section, on the date that any change in ownership of the school occurs;

(3) On the date of any change in the facilities upon which the school's certificate is based occurs; or

(4) Upon notice by the Administrator that the school has failed for more than 60 days to maintain the facilities, aircraft, or personnel required for any one of the school's approved training courses.

(b) A change in the ownership of a pilot school or provisional pilot school does not terminate that school's certificate if, within 30 days after the date that any change in ownership of the school occurs:

(1) Application is made for an appropriate amendment to the certificate; and

(2) No change in the facilities, personnel, or approved training courses is involved.

(c) An examining authority issued to the holder of a pilot school certificate expires on the date that the pilot school certificate expires, or is surrendered, suspended, or revoked.

§ 141.18 Carriage of narcotic drugs, marijuana, and depressant or stimulant drugs or substances.

If the holder of a certificate issued under this part permits any aircraft owned or leased by that holder to be engaged in any operation that the certificate holder knows to be in violation of § 91.19(a) of this chapter, that operation is a basis for suspending or revoking the certificate.

§ 141.19 Display of certificate.

(a) Each holder of a pilot school certificate or a provisional pilot school certificate must display that certificate in a place in the school that is normally accessible to the public and is not obscured.

(b) A certificate must be made available for inspection upon request by:

(1) The Administrator;

(2) An authorized representative of the National Transportation Safety Board; or

(3) A Federal, State, or local law enforcement officer.

§ 141.21 Inspections.

Each holder of a certificate issued under this part must allow the Administrator to inspect its personnel, facilities, equipment, and records to determine the certificate holder's:

(a) Eligibility to hold its certificate;

(b) Compliance with 49 U.S.C. 40101 *et seq.*, formerly the Federal Aviation Act of 1958, as amended; and

(c) Compliance with the Federal Aviation Regulations.

§ 141.23 Advertising limitations.

(a) The holder of a pilot school certificate or a provisional pilot school certificate may not make any statement relating to its certification and ratings that is false or designed to mislead any person contemplating enrollment in that school.

(b) The holder of a pilot school certificate or a provisional pilot school certificate may not advertise that the school is certificated unless it clearly differentiates between courses that have been approved under part 141 of this chapter and those that have not been approved under part 141 of this chapter.

(c) The holder of a pilot school certificate or a provisional pilot school certificate must promptly remove:

(1) From vacated premises, all signs indicating that the school was certificated by the Administrator; or

(2) All indications (including signs), wherever located, that the school is certificated by the Administrator when its certificate has expired or has been surrendered, suspended, or revoked.

§ 141.25 Business office and operations base.

(a) Each holder of a pilot school or a provisional pilot school certificate must maintain a principal business office with a mailing address in the name shown on its certificate.

(b) The facilities and equipment at the principal business office must be adequate to maintain the files and records required to operate the business of the school.

(c) The principal business office may not be shared with, or used by, another pilot school.

(d) Before changing the location of the principal business office or the operations base, each certificate holder must notify the FAA Flight Standards District Office having jurisdiction over the area of the new location, and the notice must be:

(1) Submitted in writing at least 30 days before the change of location; and

(2) Accompanied by any amendments needed for the certificate holder's approved training course outline.

(e) A certificate holder may conduct training at an operations base other than the one specified in its certificate, if:

(1) The Administrator has inspected and approved the base for use by the certificate holder; and

(2) The course of training and any needed amendments have been approved for use at that base.

§ 141.26 Training agreements.

(a) A training center certificated under part 142 of this chapter may provide the training, testing, and checking for pilot schools certificated under this part and is considered to meet the requirements of this part, provided—

(1) There is a training agreement between the certificated training center and the pilot school;

(2) The training, testing, and checking provided by the certificated training center is approved and conducted under part 142;

(3) The pilot school certificated under this part obtains the Administrator's approval for a training course outline that includes the training, testing, and checking to be conducted under this part and the training, testing, and checking to be conducted under part 142; and

(4) Upon completion of the training, testing, and checking conducted under part 142, a copy of each student's training record is forwarded to the part 141 school and becomes part of the student's permanent training record.

(b) A pilot school that provides flight training for an institution of higher education that holds a letter of authorization under § 61.169 of this chapter must have a training agreement with that institution of higher education.

[Doc. No. FAA-2010-0100, 78 FR 42379, July 15, 2013]

§ 141.27 Renewal of certificates and ratings.

(a) *Pilot school.* (1) A pilot school may apply for renewal of its school certificate and ratings within 30 days preceding the month the pilot school's certificate expires, provided the school meets the requirements prescribed in paragraph (a)(2) of this section for renewal of its certificate and ratings.

(2) A pilot school may have its school certificate and ratings renewed for an additional 24 calendar months if the Administrator determines the school's personnel, aircraft, facility and airport, approved training courses, training records, and recent training ability and quality meet the requirements of this part.

(3) A pilot school that does not meet the renewal requirements in paragraph (a)(2) of this section, may apply for a provisional pilot school certificate if the school meets the requirements of § 141.7 of this part.

(b) *Provisional pilot school.* (1) Except as provided in paragraph (b)(3) of this section, a provisional pilot school may not have its provisional pilot school certificate or the ratings on that certificate renewed.

(2) A provisional pilot school may apply for a pilot school certificate and associated ratings provided that school meets the requirements of § 141.5 of this part.

(3) A former provisional pilot school may apply for another provisional pilot school certificate, provided 180 days have elapsed since its last provisional pilot school certificate expired.

§ 141.29 [Reserved]

Subpart B—Personnel, Aircraft, and Facilities Requirements

§ 141.31 Applicability.

(a) This subpart prescribes:

(1) The personnel and aircraft requirements for a pilot school certificate or a provisional pilot school certificate; and

(2) The facilities that a pilot school or provisional pilot school must have available on a continuous basis.

(b) As used in this subpart, to have continuous use of a facility, including an airport, the school must have:

(1) Ownership of the facility or airport for at least 6 calendar months after the date the application for initial certification and on the date of renewal of the school's certificate is made; or

(2) A written lease agreement for the facility or airport for at least 6 calendar months after the date the application for initial certification and on the date of renewal of the school's certificate is made.

[Doc. No. 25910, 62 FR 16347, Apr. 4, 1997; Amdt. 141-9, 62 FR 40907, July 30, 1997]

§ 141.33 Personnel.

(a) An applicant for a pilot school certificate or for a provisional pilot school certificate must meet the following personnel requirements:

(1) Each applicant must have adequate personnel, including certificated flight instructors, certificated ground instructors, or holders of a commercial pilot certificate with a lighter-than-air rating, and a chief instructor for each approved course of training who is qualified and competent to perform the duties to which that instructor is assigned.

(2) If the school employs dispatchers, aircraft handlers, and line and service personnel, then it must instruct those persons in the procedures and responsibilities of their employment.

(3) Each instructor to be used for ground or flight training must hold a flight instructor certificate, ground instructor certificate, or commercial pilot certificate with a lighter-than-air rating, as appropriate, with ratings for the approved course of training and any aircraft used in that course.

(4) In addition to meeting the requirements of paragraph (a)(3) of this section, each instructor used for the airline transport pilot certification training program in § 61.156 of this chapter must:

(i) Hold an airline transport pilot certificate with an airplane category multiengine class rating;

(ii) Have at least 2 years of experience as a pilot in command in operations conducted under § 91.1053(a)(2)(i) or § 135.243(a)(1) of this chapter, or as a pilot in command or second in command in any operation conducted under part 121 of this chapter; and

(iii) If providing training in a flight simulation training device, have received training and evaluation within the preceding 12 months from the certificate holder on—

(A) Proper operation of flight simulator and flight training device controls and systems;

(B) Proper operation of environmental and fault panels,

(C) Data and motion limitations of simulation;

(D) Minimum equipment requirements for each curriculum; and

(E) The maneuvers that will be demonstrated in the flight simulation training device.

(b) An applicant for a pilot school certificate or for a provisional pilot school certificate must designate a chief instructor for each of the school's approved training courses, who must meet the requirements of § 141.35 of this part.

(c) When necessary, an applicant for a pilot school certificate or for a provisional pilot school certificate may designate a person to be an assistant chief instructor for an approved training course, provided that person meets the requirements of § 141.36 of this part.

(d) A pilot school and a provisional pilot school may designate a person to be a check instructor for conducting student stage checks, end-of-course tests, and instructor proficiency checks, provided:

(1) That person meets the requirements of § 141.37 of this part; and

(2) The school has an enrollment of at least 10 students at the time designation is sought.

(e) A person, as listed in this section, may serve in more than one position for a school, provided that person is qualified for each position.

[Doc. No. 25910, 62 FR 16347, Apr. 4, 1997; Amdt. 141-9, 62 FR 40907, July 30, 1997; Amdt. 141-12, 74 FR 42563, Aug. 21, 2009; Amdt. 141-1, 78 FR 42379, July 15, 2013]

§ 141.34 Employment of former FAA employees.

(a) Except as specified in paragraph (c) of this section, no holder of a pilot school certificate or a provisional pilot school certificate may knowingly employ or make a contractual arrangement which permits an individual to act as an agent or representative of the certificate holder in any matter before the Federal Aviation Administration if the individual, in the preceding 2 years—

(1) Served as, or was directly responsible for the oversight of, a Flight Standards Service aviation safety inspector; and

(2) Had direct responsibility to inspect, or oversee the inspection of, the operations of the certificate holder.

(b) For the purpose of this section, an individual shall be considered to be acting as an agent or representative of a certificate holder in a matter before the agency if the individual makes any written or oral communication on behalf of the certificate holder to the agency (or any of its officers or employees) in connection with a particular matter, whether or not involving a specific party and without regard to whether the individual has participated in, or had responsibility for, the particular matter while serving as a Flight Standards Service aviation safety inspector.

(c) The provisions of this section do not prohibit a holder of a pilot school certificate or a provisional pilot school certificate from knowingly employing or making a contractual arrangement which permits an individual to act as an agent or representative of the certificate holder in any matter before the Federal Aviation Administration if the individual was employed by the certificate holder before October 21, 2011.

[Doc. No. FAA-2008-1154, 76 FR 52236, Aug. 22, 2011]

§ 141.35 Chief instructor qualifications.

(a) To be eligible for designation as a chief instructor for a course of training, a person must meet the following requirements:

(1) Hold a commercial pilot certificate or an airline transport pilot certificate, and, except for a chief instructor for a course of training solely for a lighter-than-air rating, a current flight instructor certificate. The certificates must contain the appropriate aircraft category and class ratings for the category and class of aircraft used in the course and an instrument rating, if an instrument rating is required for enrollment in the course of training;

(2) Meet the pilot-in-command recent flight experience requirements of § 61.57 of this chapter;

(3) Pass a knowledge test on—

(i) Teaching methods;

(ii) Applicable provisions of the "Aeronautical Information Manual";

(iii) Applicable provisions of parts 61, 91, and 141 of this chapter; and

(iv) The objectives and approved course completion standards of the course for which the person seeks to obtain designation.

(4) Pass a proficiency test on instructional skills and ability to train students on the flight procedures and maneuvers appropriate to the course;

(5) Except for a course of training for gliders, balloons, or airships, the chief instructor must meet the applicable requirements in paragraphs (b), (c), and (d) of this section; and

(6) A chief instructor for a course of training for gliders, balloons or airships is only required to have 40 percent of the hours required in paragraphs (b) and (d) of this section.

(b) For a course of training leading to the issuance of a recreational or private pilot certificate or rating, a chief instructor must have:

(1) At least 1,000 hours as pilot in command; and

(2) Primary flight training experience, acquired as either a certificated flight instructor or an instructor in a military pilot flight training program, or a combination thereof, consisting of at least—

(i) 2 years and a total of 500 flight hours; or

(ii) 1,000 flight hours.

(c) For a course of training leading to the issuance of an instrument rating or a rating with instrument privileges, a chief instructor must have:

(1) At least 100 hours of flight time under actual or simulated instrument conditions;

(2) At least 1,000 hours as pilot in command; and

(3) Instrument flight instructor experience, acquired as either a certificated flight instructor-instrument or an instructor in a military pilot flight training program, or a combination thereof, consisting of at least—

(i) 2 years and a total of 250 flight hours; or

(ii) 400 flight hours.

(d) For a course of training other than one leading to the issuance of a recreational or private pilot certificate or rating, or an instrument rating or a rating with instrument privileges, a chief instructor must have:

(1) At least 2,000 hours as pilot in command; and

(2) Flight training experience, acquired as either a certificated flight instructor or an instructor in a military pilot flight training program, or a combination thereof, consisting of at least—

(i) 3 years and a total of 1,000 flight hours; or

(ii) 1,500 flight hours.

(e) To be eligible for designation as chief instructor for a ground school course, a person must have 1 year of experience as a ground school instructor at a certificated pilot school.

[Doc. No. 25910, 62 FR 16347, Apr. 4, 1997; Amdt. 141-9, 62 FR 40907, July 30, 1997; as amended by Amdt. 141-10, 63 FR 20289, Apr. 23, 1998]

§ 141.36 Assistant chief instructor qualifications.

(a) To be eligible for designation as an assistant chief instructor for a course of training, a person must meet the following requirements:

(1) Hold a commercial pilot or an airline transport pilot certificate and, except for the assistant chief instructor for a course of training solely for a lighter-than-air rating, a current flight instructor certificate. The certificates must contain the appropriate aircraft category, class, and instrument ratings if an instrument rating is required by the course of training for the category and class of aircraft used in the course;

(2) Meet the pilot-in-command recent flight experience requirements of § 61.57 of this chapter;

(3) Pass a knowledge test on—

(i) Teaching methods;

(ii) Applicable provisions of the "Aeronautical Information Manual";

(iii) Applicable provisions of parts 61, 91, and 141 of this chapter; and

(iv) The objectives and approved course completion standards of the course for which the person seeks to obtain designation.

(4) Pass a proficiency test on the flight procedures and maneuvers appropriate to that course; and

(5) Meet the applicable requirements in paragraphs (b), (c), and (d) of this section. However, an assistant chief instructor for a course of training for gliders, balloons, or airships is only required to have 40 percent of the hours required in paragraphs (b) and (d) of this section.

(b) For a course of training leading to the issuance of a recreational or private pilot certificate or rating, an assistant chief instructor must have:

(1) At least 500 hours as pilot in command; and

(2) Flight training experience, acquired as either a certificated flight instructor or an instructor in a military pilot flight training program, or a combination thereof, consisting of at least—

(i) 1 year and a total of 250 flight hours; or

(ii) 500 flight hours.

(c) For a course of training leading to the issuance of an instrument rating or a rating with instrument privileges, an assistant chief flight instructor must have:

(1) At least 50 hours of flight time under actual or simulated instrument conditions;

(2) At least 500 hours as pilot in command; and

(3) Instrument flight instructor experience, acquired as either a certificated flight instructor-instrument or an instructor in a military pilot flight training program, or a combination thereof, consisting of at least—

(i) 1 year and a total of 125 flight hours; or

(ii) 200 flight hours.

(d) For a course of training other than one leading to the issuance of a recreational or private pilot certificate or rating, or an instrument rating or a rating with instrument privileges, an assistant chief instructor must have:

(1) At least 1,000 hours as pilot in command; and

(2) Flight training experience, acquired as either a certificated flight instructor or an instructor in a military pilot flight training program, or a combination thereof, consisting of at least—

(i) $1\frac{1}{2}$ years and a total of 500 flight hours; or

(ii) 750 flight hours.

(e) To be eligible for designation as an assistant chief instructor for a ground school course, a person must have 6 months of experience as a ground school instructor at a certificated pilot school.

[Doc. No. 25910, 62 FR 16347, Apr. 4, 1997; Amdt. 141-9, 62 FR 40907, July 30, 1997; as amended by Amdt. 141-10, 63 FR 20289, Apr. 23, 1998]

§ 141.37 Check instructor qualifications.

(a) To be designated as a check instructor for conducting student stage checks, end-of-course tests, and instructor proficiency checks under this part, a person must meet the eligibility requirements of this section:

(1) For checks and tests that relate to either flight or ground training, the person must pass a test, given by the chief instructor, on—

(i) Teaching methods;

(ii) Applicable provisions of the "Aeronautical Information Manual";

(iii) Applicable provisions of parts 61, 91, and 141 of this chapter; and

(iv) The objectives and course completion standards of the approved training course for the designation sought.

(2) For checks and tests that relate to a flight training course, the person must—

(i) Meet the requirements in paragraph (a)(1) of this section;

(ii) Hold a commercial pilot certificate or an airline transport pilot certificate and, except for a check instructor for a course of training for a lighter-than-air rating, a current flight instructor certificate. The certificates must contain the appropriate aircraft category, class, and instrument ratings for the category and class of aircraft used in the course;

(iii) Meet the pilot-in-command recent flight experience requirements of § 61.57 of this chapter; and

(iv) Pass a proficiency test, given by the chief instructor or assistant chief instructor, on the flight procedures and maneuvers of the approved training course for the designation sought.

(3) For checks and tests that relate to ground training, the person must—

(i) Meet the requirements in paragraph (a)(1) of this section;

(ii) Except for a course of training for a lighter-than-air rating, hold a current flight instructor certificate or ground instructor certificate with ratings appropriate to the category and class of aircraft used in the course; and

(iii) For a course of training for a lighter-than-air rating, hold a commercial pilot certificate with a lighter-than-air category rating and the appropriate class rating.

(b) A person who meets the eligibility requirements in paragraph (a) of this section must:

(1) Be designated, in writing, by the chief instructor to conduct student stage checks, end-of-course tests, and instructor proficiency checks; and

(2) Be approved by the FAA Flight Standards District Office having jurisdiction over the school.

(c) A check instructor may not conduct a stage check or an end-of-course test of any student for whom the check instructor has:

(1) Served as the principal instructor; or

(2) Recommended for a stage check or end-of-course test.

[Doc. No. 25910, 62 FR 16347, Apr. 4, 1997; Amdt. 141-9, 62 FR 40907, July 30, 1997]

§ 141.38 Airports.

(a) An applicant for a pilot school certificate or a provisional pilot school certificate must show that he or she has continuous use of each airport at which training flights originate.

(b) Each airport used for airplanes and gliders must have at least one runway or takeoff area that allows training aircraft to make a normal takeoff or landing under the following conditions at the aircraft's maximum certificated takeoff gross weight:

(1) Under wind conditions of not more than 5 miles per hour;

(2) At temperatures in the operating area equal to the mean high temperature for the hottest month of the year;

(3) If applicable, with the powerplant operation, and landing gear and flap operation recommended by the manufacturer; and

(4) In the case of a takeoff—

(i) With smooth transition from liftoff to the best rate of climb speed without exceptional piloting skills or techniques; and

(ii) Clearing all obstacles in the takeoff flight path by at least 50 feet.

(c) Each airport must have a wind direction indicator that is visible from the end of each runway at ground level;

(d) Each airport must have a traffic direction indicator when:

(1) The airport does not have an operating control tower; and

(2) UNICOM advisories are not available.

(e) Except as provided in paragraph (f) of this section, each airport used for night training flights must have permanent runway lights;

(f) An airport or seaplane base used for night training flights in seaplanes is permitted to use adequate nonpermanent lighting or shoreline lighting, if approved by the Administrator.

[Doc. No. 25910, 62 FR 16347, Apr. 4, 1997; Amdt. 141-9, 62 FR 40907, July 30, 1997]

§ 141.39 Aircraft.

(a) When the school's training facility is located within the U.S., an applicant for a pilot school certificate or provisional pilot school certificate must show that each aircraft used by the school for flight training and solo flights:

(1) Is a civil aircraft of the United States;

(2) Is certificated with a standard airworthiness certificate, a primary airworthiness certificate, or a special airworthiness certificate in the light-sport category unless the FAA determines otherwise because of the nature of the approved course;

(3) Is maintained and inspected in accordance with the requirements for aircraft operated for hire under part 91, subpart E, of this chapter;

(4) Has two pilot stations with engine-power controls that can be easily reached and operated in a normal manner from both pilot stations (for flight training); and

(5) Is equipped and maintained for IFR operations if used in a course involving IFR en route operations and instrument approaches. For training in the control and precision maneuvering of an aircraft by reference to instruments, the aircraft may be equipped as provided in the approved course of training.

(b) When the school's training facility is located outside the U.S. and the training will be conducted outside the U.S., an applicant for a pilot school certificate or provisional pilot school certificate must show that each aircraft used by the school for flight training and solo flights:

(1) Is either a civil aircraft of the United States or a civil aircraft of foreign registry;

(2) Is certificated with a standard or primary airworthiness certificate or an equivalent certification from the foreign aviation authority;

(3) Is maintained and inspected in accordance with the requirements for aircraft operated for hire under part 91, subpart E of this chapter, or in accordance with equivalent maintenance and inspection from the foreign aviation authority's requirements;

(4) Has two pilot stations with engine-power controls that can be easily reached and operated in a normal manner from both pilot stations (for flight training); and

(5) Is equipped and maintained for IFR operations if used in a course involving IFR en route operations and instrument approaches. For training in the control and precision maneuvering of an aircraft by reference to instruments, the aircraft may be equipped as provided in the approved course of training.

[Doc. No. FAA-2006-26661, 74 FR 42563, Aug. 21, 2009, as amended by Amdt. 141-13, 75 FR 5223, Feb. 1, 2010]

§ 141.41 Flight simulators, flight training devices, and training aids.

An applicant for a pilot school certificate or a provisional pilot school certificate must show that its flight simulators, flight training devices, training aids, and equipment meet the following requirements:

(a) *Flight simulators.* Each flight simulator used to obtain flight training credit allowed for flight simulators in an approved pilot training course curriculum must—

(1) Be a full-size aircraft cockpit replica of a specific type of aircraft, or make, model, and series of aircraft;

(2) Include the hardware and software necessary to represent the aircraft in ground operations and flight operations;

(3) Use a force cueing system that provides cues at least equivalent to those cues provided by a 3 degree freedom of motion system;

(4) Use a visual system that provides at least a 45-degree horizontal field of view and a 30-degree vertical field of view simultaneously for each pilot; and

(5) Have been evaluated, qualified, and approved by the Administrator.

(b) *Flight training devices.* Each flight training device used to obtain flight training credit allowed for flight training devices in an approved pilot training course curriculum must—

(1) Be a full-size replica of instruments, equipment panels, and controls of an aircraft, or set of aircraft, in an open flight deck area or in an enclosed cockpit, including the hardware and software for the systems installed that is necessary to simulate the aircraft in ground and flight operations;

(2) Need not have a force (motion) cueing or visual system; and

(3) Have been evaluated, qualified, and approved by the Administrator.

(c) *Training aids and equipment.* Each training aid, including any audiovisual aid, projector, tape recorder, mockup, chart, or aircraft component listed in the approved training course outline, must be accurate and appropriate to the course for which it is used.

[Doc. No. 25910, 62 FR 16347, Apr. 4, 1997; Amdt. 141-9, 62 FR 40908, July 30, 1997]

§ 141.43 Pilot briefing areas.

(a) An applicant for a pilot school certificate or provisional pilot school certificate must show that the applicant has continuous use of a briefing area located at each airport at which training flights originate that is:

(1) Adequate to shelter students waiting to engage in their training flights;

(2) Arranged and equipped for the conduct of pilot briefings; and

(3) Except as provided in paragraph (c) of this section, for a school with an instrument rating or commercial pilot course, equipped with private landline or telephone communication to the nearest FAA Flight Service Station.

(b) A briefing area required by paragraph (a) of this section may not be used by the applicant if it is available for use by any other pilot school during the period it is required for use by the applicant.

(c) The communication equipment required by paragraph (a)(3) of this section is not required if the briefing area and the flight service station are located on the same airport, and are readily accessible to each other.

§ 141.45 Ground training facilities.

An applicant for a pilot school or provisional pilot school certificate must show that:

(a) Except as provided in paragraph (c) of this section, each room, training booth, or other space used for instructional purposes is heated, lighted, and ventilated to conform to local building, sanitation, and health codes.

(b) Except as provided in paragraph (c) of this section, the training facility is so located that the students in that facility are not distracted by the training conducted in other rooms, or by flight and maintenance operations on the airport.

(c) If a training course is conducted through an internet-based medium, the holder of a pilot school certificate or provisional pilot school certificate that provides such training need not comply with paragraphs (a) and (b) of this section but must maintain in current status a permanent business location and business telephone number.

[Doc. No. FAA-2008-0938, 76 FR 54107, Aug. 31, 2011]

Subpart C—Training Course Outline and Curriculum

§ 141.51 Applicability.

This subpart prescribes the curriculum and course outline requirements for the issuance of a pilot school certificate or provisional pilot school certificate and ratings.

§ 141.53 Approval procedures for a training course: General.

(a) *General.* An applicant for a pilot school certificate or provisional pilot school certificate must obtain the Administrator's approval of the outline of each training course for which certification and rating is sought.

(b) *Application.* (1) An application for the approval of an initial or amended training course must be submitted in duplicate to the FAA Flight Standards District Office having jurisdiction over the area where the school is based.

(2) An application for the approval of an initial or amended training course must be submitted at least 30 days before any training under that course, or any amendment thereto, is scheduled to begin.

(3) An application for amending a training course must be accompanied by two copies of the amendment.

(c) *Training courses.* An applicant for a pilot school certificate or provisional pilot school certificate may request approval for the training courses specified under § 141.11(b).

(d) *Additional rules for internet based training courses.* An application for an initial or amended training course offered through an internet based medium must comply with the following:

(1) All amendments must be identified numerically by page, date, and screen. Minor editorial and typographical changes do not require FAA approval, provided the school notifies the FAA within 30 days of their insertion.

(2) For monitoring purposes, the school must provide the FAA an acceptable means to log-in and log-off from a remote location to review all elements of the course as viewed by attendees and to by-pass the normal attendee restrictions.

(3) The school must incorporate adequate security measures into its internet-based courseware information system and into its operating and maintenance procedures to ensure the following fundamental areas of security and protection:

(i) Integrity.

(ii) Identification/Authentication.

(iii) Confidentiality.

(iv) Availability.

(v) Access control.

[Doc. No. 25910, 62 FR 16347, Apr. 4, 1997; Amdt. 141-9, 62 FR 40908, July 30, 1997; Amdt. 141-12, 74 FR 42563, Aug. 21, 2009; Amdt. 141-15, 76 FR 54107, Aug. 31, 2011]

§ 141.55 Training course: Contents.

(a) Each training course for which approval is requested must meet the minimum curriculum requirements in accordance with the appropriate appendix of this part.

(b) Except as provided in paragraphs (d) and (e) of this section, each training course for which approval is requested must meet the minimum ground and flight training time requirements in accordance with the appropriate appendix of this part.

(c) Each training course for which approval is requested must contain:

(1) A description of each room used for ground training, including the room's size and the maximum number of students that may be trained in the room at one time, unless the course is provided via an internet-based training medium;

(2) A description of each type of audiovisual aid, projector, tape recorder, mockup, chart, aircraft component, and other special training aids used for ground training;

(3) A description of each flight simulator or flight training device used for training;

(4) A listing of the airports at which training flights originate and a description of the facilities, including pilot briefing areas that are available for use by the school's students and personnel at each of those airports;

(5) A description of the type of aircraft including any special equipment used for each phase of training;

(6) The minimum qualifications and ratings for each instructor assigned to ground or flight training; and

(7) A training syllabus that includes the following information—

(i) The prerequisites for enrolling in the ground and flight portion of the course that include the pilot certificate and rating (if required by this part), training, pilot experience, and pilot knowledge;

(ii) A detailed description of each lesson, including the lesson's objectives, standards, and planned time for completion;

(iii) A description of what the course is expected to accomplish with regard to student learning;

(iv) The expected accomplishments and the standards for each stage of training; and

(v) A description of the checks and tests to be used to measure a student's accomplishments for each stage of training.

(d) A pilot school may request and receive initial approval for a period of not more than 24 calendar months for any training course under this part that does not meet the minimum ground and flight training time requirements, provided the following provisions are met:

(1) The school holds a pilot school certificate issued under this part and has held that certificate for a period of at least 24 consecutive calendar months preceding the month of the request;

(2) In addition to the information required by paragraph (c) of this section, the training course specifies planned ground and flight training time requirements for the course;

(3) The school does not request the training course to be approved for examining authority, nor may that school hold examining authority for that course; and

(4) The practical test or knowledge test for the course is to be given by—

(i) An FAA inspector; or

(ii) An examiner who is not an employee of the school.

(e) A pilot school may request and receive final approval for any training course under this part that does not meet the minimum ground and flight training time requirements, provided the following conditions are met:

(1) The school has held initial approval for that training course for at least 24 calendar months.

(2) The school has—

(i) Trained at least 10 students in that training course within the preceding 24 calendar months and recommended those students for a pilot, flight instructor, or ground instructor certificate or rating; and

(ii) At least 80 percent of those students passed the practical or knowledge test, as appropriate, on the first attempt, and that test was given by—

(A) An FAA inspector; or

(B) An examiner who is not an employee of the school.

(3) In addition to the information required by paragraph (c) of this section, the training course specifies planned ground and flight training time requirements for the course.

(4) The school does not request that the training course be approved for examining authority nor may that school hold examining authority for that course.

[Docket No. 25910, 62 FR 16347, Apr. 4, 1997, as amended by Amdt. 141-12, 74 FR 42563, Aug. 21, 2009; Amdt. 141-15, 76 FR 54107, Aug. 31, 2011]

§ 141.57 Special curricula.

An applicant for a pilot school certificate or provisional pilot school certificate may apply for approval to conduct a special course of airman training for which a curriculum is not prescribed in the appendixes of this part, if the applicant shows that the training course contains features that could achieve a level of pilot proficiency equivalent to that achieved by a training course prescribed in the appendixes of this part or the requirements of part 61 of this chapter.

Subpart D—Examining Authority

§ 141.61 Applicability.

This subpart prescribes the requirements for the issuance of examining authority to the holder of a pilot school certificate, and the privileges and limitations of that examining authority.

§ 141.63 Examining authority qualification requirements.

(a) A pilot school must meet the following prerequisites to receive initial approval for examining Authority:

(1) The school must complete the application for examining authority on a form and in a manner prescribed by the Administrator;

(2) The school must hold a pilot school certificate and rating issued under this part;

(3) The school must have held the rating in which examining authority is sought for at least 24 consecutive calendar months preceding the month of application for examining authority;

(4) The training course for which examining authority is requested may not be a course that is approved without meeting the minimum ground and flight training time requirements of this part; and

(5) Within 24 calendar months before the date of application for examining authority, that school must meet the following requirements—

(i) The school must have trained at least 10 students in the training course for which examining authority is sought and recommended those students for a pilot, flight instructor, or ground instructor certificate or rating; and

(ii) At least 90 percent of those students passed the required practical or knowledge test, or any combination thereof, for the pilot, flight instructor, or ground instructor certificate or rating on the first attempt, and that test was given by—

(A) An FAA inspector; or

(B) An examiner who is not an employee of the school.

(b) A pilot school must meet the following requirements to retain approval of its examining AUTHORITY:

(1) The school must complete the application for renewal of its examining authority on a form and in a manner prescribed by the Administrator;

(2) The school must hold a pilot school certificate and rating issued under this part;

(3) The school must have held the rating for which continued examining authority is sought for at least 24 calendar months preceding the month of application for renewal of its examining authority; and

(4) The training course for which continued examining authority is requested may not be a course that is approved without meeting the minimum ground and flight training time requirements of this part.

[Doc. No. 25910, 62 FR 16347, Apr. 4, 1997; Amdt. 141-9, 62 FR 40908, July 30, 1997]

§ 141.65 Privileges.

A pilot school that holds examining authority may recommend a person who graduated from its course for the appropriate pilot, flight instructor, or ground instructor certificate or rating without taking the FAA knowledge test or practical test in accordance with the provisions of this subpart.

§ 141.67 Limitations and reports.

A pilot school that holds examining authority may only recommend the issuance of a pilot, flight instructor, or ground instructor certificate and rating to a person who does not take an FAA knowledge test or practical test, if the recommendation for the issuance of that certificate or rating is in accordance with the following requirements:

(a) The person graduated from a training course for which the pilot school holds examining authority.

(b) Except as provided in this paragraph, the person satisfactorily completed all the curriculum requirements of that pilot school's approved training course. A person who transfers from one part 141 approved pilot school to another part 141 approved pilot school may receive credit for that previous training, provided the following requirements are met:

(1) The maximum credited training time does not exceed one-half of the receiving school's curriculum requirements;

(2) The person completes a knowledge and proficiency test conducted by the receiving school for the purpose of determining the amount of pilot experience and knowledge to be credited;

(3) The receiving school determines (based on the person's performance on the knowledge and proficiency test required by paragraph (b)(2) of this section) the amount of credit to be awarded, and records that credit in the person's training record;

(4) The person who requests credit for previous pilot experience and knowledge obtained the experience and knowledge from another part 141 approved pilot school and training course; and

(5) The receiving school retains a copy of the person's training record from the previous school.

(c) Tests given by a pilot school that holds examining authority must be approved by the Administrator and be at least equal in scope, depth, and difficulty to the comparable knowledge and practical tests prescribed by the Administrator under part 61 of this chapter.

(d) A pilot school that holds examining authority may not use its knowledge or practical tests if the school:

(1) Knows, or has reason to believe, the test has been compromised; or

(2) Is notified by an FAA Flight Standards District Office that there is reason to believe or it is known that the test has been compromised.

(e) A pilot school that holds examining authority must maintain a record of all temporary airman certificates it issues, which consist of the following information:

(1) A chronological listing that includes—

(i) The date the temporary airman certificate was issued;

(ii) The student to whom the temporary airman certificate was issued, and that student's permanent mailing address and telephone number;

(iii) The training course from which the student graduated;

(iv) The name of person who conducted the knowledge or practical test;

(v) The type of temporary airman certificate or rating issued to the student; and

(vi) The date the student's airman application file was sent to the FAA for processing for a permanent airman certificate.

(2) A copy of the record containing each student's graduation certificate, airman application, temporary airman certificate, superseded airman certificate (if applicable), and knowledge test or practical test results; and

(3) The records required by paragraph (e) of this section must be retained for 1 year and made available to the Administrator upon request. These records must be surrendered to the Administrator when the pilot school ceases to have examining authority.

(f) Except for pilot schools that have an airman certification representative, when a student passes the knowledge test or practical test, the pilot school that holds examining authority must submit that student's airman application file and training record to the FAA for processing for the issuance of a permanent airman certificate.

[Doc. No. 25910, 62 FR 16347, Apr. 4, 1997; Amdt. 141-9, 62 FR 40908, July 30, 1997]

Subpart E—Operating Rules

§ 141.71 Applicability.

This subpart prescribes the operating rules applicable to a pilot school or provisional pilot school certificated under the provisions of this part.

§ 141.73 Privileges.

(a) The holder of a pilot school certificate or a provisional pilot school certificate may advertise and conduct approved pilot training courses in accordance with the certificate and any ratings that it holds.

(b) A pilot school that holds examining authority for an approved training course may recommend a graduate of that course for the issuance of an appropriate pilot, flight instructor, or ground instructor certificate and rating, without taking an FAA knowledge test or practical test, provided the training course has been approved and meets the minimum ground and flight training time requirements of this part.

§ 141.75 Aircraft requirements.

The following items must be carried on each aircraft used for flight training and solo flights:

(a) A pretakeoff and prelanding checklist; and

(b) The operator's handbook for the aircraft, if one is furnished by the manufacturer, or copies of the handbook if furnished to each student using the aircraft.

[Doc. No. 25910, 62 FR 40908, July 30, 1997]

§ 141.77 Limitations.

(a) The holder of a pilot school certificate or a provisional pilot school certificate may not issue a graduation certificate to a student, or recommend a student for a pilot certificate or rating, unless the student has:

(1) Completed the training specified in the pilot school's course of training; and

(2) Passed the required final tests.

(b) Except as provided in paragraph (c) of this section, the holder of a pilot school certificate or a provisional pilot school certificate may not graduate a student from a course of training unless the student has completed all of the curriculum requirements of that course;

(c) A student may be given credit towards the curriculum requirements of a course for previous training under the following conditions:

(1) If the student completed a proficiency test and knowledge test that was conducted by the receiving pilot school and the previous training was based on a part 141- or a part 142-approved flight training course, the credit is limited to not more than 50 percent of the flight training requirements of the curriculum.

(2) If the student completed a knowledge test that was conducted by the receiving pilot school and the previous training was based on a part 141- or a part 142-approved aeronautical knowledge training course, the credit is limited to not more than 50 percent of the aeronautical knowledge training requirements of the curriculum.

(3) If the student completed a proficiency test and knowledge test that was conducted by the receiving pilot school and the training was received from other than a part 141- or a part 142-approved flight training course, the credit is limited to not more than 25 percent of the flight training requirements of the curriculum.

(4) If the student completed a knowledge test that was conducted by the receiving pilot school and the previous training was received from other than a part 141- or a part 142-approved aeronautical knowledge training course, the credit is limited to not more than 25 percent of the aeronautical knowledge training requirements of the curriculum.

(5) Completion of previous training must be certified in the student's training record by the training provider or a management official within the training provider's organization, and must contain—

(i) The kind and amount of training provided; and

(ii) The result of each stage check and end-of-course test, if appropriate.

[Doc. No. 25910, 62 FR 16347, Apr. 4, 1997; Amdt. 141-9, 62 FR 40908, July 30, 1997; Amdt. 141-12, 74 FR 42564, Aug. 21, 2009]

§ 141.79 Flight training.

(a) No person other than a certificated flight instructor or commercial pilot with a lighter-than-air rating who has the ratings and the minimum qualifications specified in the approved training course outline may give a student flight training under an approved course of training.

(b) No student pilot may be authorized to start a solo practice flight from an airport until the flight has been approved by a certificated flight instructor or commercial pilot with a lighter-than-air rating who is present at that airport.

(c) Each chief instructor and assistant chief instructor assigned to a training course must complete, at least once every 12 calendar months, an approved syllabus of training consisting of ground or flight training, or both, or an approved flight instructor refresher course.

(d) Each certificated flight instructor or commercial pilot with a lighter-than-air rating who is assigned to a flight training course must satisfactorily complete the following tasks, which must be administered by the school's chief instructor, assistant chief instructor, or check instructor:

(1) Prior to receiving authorization to train students in a flight training course, must—

(i) Accomplish a review of and receive a briefing on the objectives and standards of that training course; and

(ii) Accomplish an initial proficiency check in each make and model of aircraft used in that training course in which that person provides training; and

(2) Every 12 calendar months after the month in which the person last complied with the requirements of paragraph (d)(1)(ii) of this section, accomplish a recurrent proficiency check in one of the aircraft in which the person trains students.

[Doc. No. 25910, 62 FR 16347, Apr. 4, 1997; Amdt. 141-9, 62 FR 40908, July 30, 1997]

§ 141.81 Ground training.

(a) Except as provided in paragraph (b) of this section, each instructor who is assigned to a ground training course must hold a flight or ground instructor certificate, or a commercial pilot certificate with a lighter-than-air rating, with the appropriate rating for that course of training.

(b) A person who does not meet the requirements of paragraph (a) of this section may be assigned ground training duties in a ground training course, if:

(1) The chief instructor who is assigned to that ground training course finds the person qualified to give that training; and

(2) The training is given while under the supervision of the chief instructor or the assistant chief instructor who is present at the facility when the training is given.

(c) An instructor may not be used in a ground training course until that instructor has been briefed on the objectives and standards of that course by the chief instructor, assistant chief instructor, or check instructor.

[Doc. No. 25910, 62 FR 16347, Apr. 4, 1997; Amdt. 141-9, 62 FR 40908, July 30, 1997]

§ 141.83 Quality of training.

(a) Each pilot school or provisional pilot school must meet the following requirements:

(1) Comply with its approved training course; and

(2) Provide training of such quality that meets the requirements of § 141.5(d) of this part.

(b) The failure of a pilot school or provisional pilot school to maintain the quality of training specified in paragraph (a) of this section may be the basis for suspending or revoking that school's certificate.

(c) When requested by the Administrator, a pilot school or provisional pilot school must allow the FAA to administer any knowledge test, practical test, stage check, or end-of-course test to its students.

(d) When a stage check or end-of-course test is administered by the FAA under the provisions of paragraph (c) of this section, and the student has not completed the training course, then that test will be based on the standards prescribed in the school's approved training course.

(e) When a practical test or knowledge test is administered by the FAA under the provisions of paragraph (c) of this section, to a student who has completed the school's training course, that test will be based upon the areas of operation approved by the Administrator.

[Doc. No. 25910, 62 FR 16347, Apr. 4, 1997; Amdt. 141-9, 62 FR 40908, July 30, 1997]

§ 141.85 Chief instructor responsibilities.

(a) A chief instructor designated for a pilot school or provisional pilot school is responsible for:

(1) Certifying each student's training record, graduation certificate, stage check and end-of-course test reports, and recommendation for course completion, unless the duties are delegated by the chief instructor to an assistant chief instructor or recommending instructor;

(2) Ensuring that each certificated flight instructor, certificated ground instructor, or commercial pilot with a lighter-than-air rating passes an initial proficiency check prior to that instructor being assigned instructing duties in the school's approved training course, and thereafter that the instructor passes a recurrent proficiency check every 12 calendar months after the month in which the initial test was accomplished;

(3) Ensuring that each student accomplishes the required stage checks and end-of-course tests in accordance with the school's approved training course; and

(4) Maintaining training techniques, procedures, and standards for the school that are acceptable to the Administrator.

(b) The chief instructor or an assistant chief instructor must be available at the pilot school or, if away from the pilot school, be available by telephone, radio, or other electronic means during the time that training is given for an approved training course.

(c) The chief instructor may delegate authority for conducting stage checks, end-of-course tests, and flight instructor proficiency checks to the assistant chief instructor or a check instructor.

[Doc. No. 25910, 62 FR 16347, Apr. 4, 1997; Amdt. 141-9, 62 FR 40908, July 30, 1997; Amdt. 141-12, 74 FR 42564, Aug. 21, 2009]

§ 141.87 Change of chief instructor.

Whenever a pilot school or provisional pilot school makes a change of designation of its chief instructor, that school:

(a) Must immediately provide the FAA Flight Standards District Office that has jurisdiction over the area in which the school is located with written notification of the change;

(b) May conduct training without a chief instructor for that training course for a period not to exceed 60 days while awaiting the designation and approval of another chief instructor;

(c) May, for a period not to exceed 60 days, have the stage checks and end-of-course tests administered by:

(1) The training course's assistant chief instructor, if one has been designated;

(2) The training course's check instructor, if one has been designated;

(3) An FAA inspector; or

(4) An examiner.

(d) Must, after 60 days without a chief instructor, cease operations and surrender its certificate to the Administrator; and

(e) May have its certificate reinstated, upon:

(1) Designating and approving another chief instructor;

(2) Showing it meets the requirements of § 141.27(a)(2) of this part; and

(3) Applying for reinstatement on a form and in a manner prescribed by the Administrator.

§ 141.89 Maintenance of personnel, facilities, and equipment.

The holder of a pilot school certificate or provisional pilot school certificate may not provide training to a student who is enrolled in an approved course of training unless:

(a) Each airport, aircraft, and facility necessary for that training meets the standards specified in the holder's approved training course outline and the appropriate requirements of this part; and

(b) Except as provided in § 141.87 of this part, each chief instructor, assistant chief instructor, check instructor, or instructor meets the qualifications specified in the holder's approved course of training and the appropriate requirements of this part.

§ 141.91 Satellite bases.

The holder of a pilot school certificate or provisional pilot school certificate may conduct ground training or flight training in an approved course of training at a base other than its main operations base if:

(a) An assistant chief instructor is designated for each satellite base, and that assistant chief instructor is available at that base or, if away from the premises, by telephone, radio, or other electronic means during the time that training is provided for an approved training course;

(b) The airport, facilities, and personnel used at the satellite base meet the appropriate requirements of subpart B of this part and its approved training course outline;

(c) The instructors are under the direct supervision of the chief instructor or assistant chief instructor for the appropriate training course, who is readily available for consultation in accordance with § 141.85(b) of this part; and

(d) The FAA Flight Standards District Office having jurisdiction over the area in which the school is located is notified in writing if training is conducted at a base other than the school's main operations base for more than 7 consecutive days.

[Doc. No. 25910, 62 FR. 16347, Apr. 4, 1997; Amdt. 141-9, 62 FR 40908, July 30, 1997]

§ 141.93 Enrollment.

(a) The holder of a pilot school certificate or a provisional pilot school certificate must, at the time a student is enrolled in an approved training course, furnish that student with a copy of the following:

(1) A certificate of enrollment containing—

(i) The name of the course in which the student is enrolled; and

(ii) The date of that enrollment.

(2) A copy of the student's training syllabus.

(3) Except for a training course offered through an internet based medium, a copy of the safety procedures and practices developed by the school that describe the use of the school's facilities and the operation of its aircraft. Those procedures and practices shall include training on at least the following information—

(i) The weather minimums required by the school for dual and solo flights;

(ii) The procedures for starting and taxiing aircraft on the ramp;

(iii) Fire precautions and procedures;

(iv) Redispatch procedures after unprogrammed landings, on and off airports;

(v) Aircraft discrepancies and approval for return-to-service determinations;

(vi) Securing of aircraft when not in use;

(vii) Fuel reserves necessary for local and cross-country flights;

(viii) Avoidance of other aircraft in flight and on the ground;

(ix) Minimum altitude limitations and simulated emergency landing instructions; and

(x) A description of and instructions regarding the use of assigned practice areas.

(b) The holder of a pilot school certificate or provisional pilot school certificate must maintain a monthly listing of persons enrolled in each training course offered by the school.

[Doc. No. 25910, 62 FR 16347, Apr. 4, 1997; Amdt. 141-9, 62 FR 40908, July 30, 1997; Amdt. 141-15, 76 FR 54107, Aug. 31, 2011]

§ 141.95 Graduation certificate.

(a) The holder of a pilot school certificate or provisional pilot school certificate must issue a graduation certificate to each student who completes its approved course of training.

(b) The graduation certificate must be issued to the student upon completion of the course of training and contain at least the following information:

(1) The name of the school and the certificate number of the school;

(2) The name of the graduate to whom it was issued;

(3) The course of training for which it was issued;

(4) The date of graduation;

(5) A statement that the student has satisfactorily completed each required stage of the approved course of training including the tests for those stages;

(6) A certification of the information contained on the graduation certificate by the chief instructor for that course of training; and

(7) A statement showing the cross-country training that the student received in the course of training.

(8) Certificates issued upon graduating from a course based on internet media must be uniquely identified using an alphanumeric code that is specific to the student graduating from that course.

[Doc. No. 25910, 62 FR 16347, Apr. 4, 1997; Amdt. 141-9, 62 FR 40908, July 30, 1997, as amended by Amdt. 141-15, 76 FR 54108, Aug. 31, 2011]

Subpart F—Records

§ 141.101 Training records.

(a) Each holder of a pilot school certificate or provisional pilot school certificate must establish and maintain a current and accurate record of the participation of each student enrolled in an approved course of training conducted by the school that includes the following information:

(1) The date the student was enrolled in the approved course;

(2) A chronological log of the student's course attendance, subjects, and flight operations covered in the student's training, and the names and grades of any tests taken by the student; and

(3) The date the student graduated, terminated training, or transferred to another school. In the case of graduation from a course based on internet media, the school must maintain the identifying graduation certificate code required by § 141.95(b)(8).

(b) The records required to be maintained in a student's logbook will not suffice for the record required by paragraph (a) of this section.

(c) Whenever a student graduates, terminates training, or transfers to another school, the student's record must be certified to that effect by the chief instructor.

(d) The holder of a pilot school certificate or a provisional pilot school certificate must retain each student record required by this section for at least 1 year from the date that the student:

(1) Graduates from the course to which the record pertains;

(2) Terminates enrollment in the course to which the record pertains; or

(3) Transfers to another school.

(e) The holder of a pilot school certificate or a provisional pilot school certificate must make a copy of the student's training record available upon request by the student.

[Doc. No. 25910, 62 FR 16347, Apr. 4, 1997; Amdt. 141-9, 62 FR 40908, July 30, 1997, as amended by Amdt. 141-15, 76 FR 54108, Aug. 31, 2011]

Appendix A to Part 141—Recreational Pilot Certification Course

1. *Applicability.* This appendix prescribes the minimum curriculum required for a recreational pilot certification course under this part, for the following ratings:

(a) Airplane single-engine.

(b) Rotorcraft helicopter.

(c) Rotorcraft gyroplane.

2. *Eligibility for enrollment.* A person must hold a student pilot certificate prior to enrolling in the flight portion of the recreational pilot certification course.

3. *Aeronautical knowledge training.* Each approved course must include at least 20 hours of ground training on the following aeronautical knowledge areas, appropriate to the aircraft category and class for which the course applies:

(a) Applicable Federal Aviation Regulations for recreational pilot privileges, limitations, and flight operations;

(b) Accident reporting requirements of the National Transportation Safety Board;

(c) Applicable subjects in the "Aeronautical Information Manual" and the appropriate FAA advisory circulars;

(d) Use of aeronautical charts for VFR navigation using pilotage with the aid of a magnetic compass;

(e) Recognition of critical weather situations from the ground and in flight, windshear avoidance, and the procurement and use of aeronautical weather reports and forecasts;

(f) Safe and efficient operation of aircraft, including collision avoidance, and recognition and avoidance of wake turbulence;

(g) Effects of density altitude on takeoff and climb performance;

(h) Weight and balance computations;

(i) Principles of aerodynamics, powerplants, and aircraft systems;

(j) Stall awareness, spin entry, spins, and spin recovery techniques, if applying for an airplane single-engine rating;

(k) Aeronautical decision making and judgment; and

(l) Preflight action that includes—

(1) How to obtain information on runway lengths at airports of intended use, data on takeoff and landing distances, weather reports and forecasts, and fuel requirements; and

(2) How to plan for alternatives if the planned flight cannot be completed or delays are encountered.

4. *Flight training.* (a) Each approved course must include at least 30 hours of flight training (of which 15 hours must be with a certificated flight instructor and 3 hours must be solo flight training as provided in section No. 5 of this appendix) on the approved areas of operation listed in paragraph (c) of this section that are appropriate to the aircraft category and class rating for which the course applies, including:

(1) Except as provided in § 61.100 of this chapter, 2 hours of dual flight training to and at an airport that is located more than 25 nautical miles from the airport where the applicant normally trains, with at least three takeoffs and three landings; and

(2) 3 hours of dual flight training in an aircraft that is appropriate to the aircraft category and class for which the course applies, in preparation for the practical test within 60 days preceding the date of the test.

(b) Each training flight must include a preflight briefing and a postflight critique of the student by the flight instructor assigned to that flight.

(c) Flight training must include the following approved areas of operation appropriate to the aircraft category and class rating—

(1) *For an airplane single-engine course:* (i) Preflight preparation;

(ii) Preflight procedures;

(iii) Airport operations;

(iv) Takeoffs, landings, and go-arounds;

(v) Performance maneuvers;

(vi) Ground reference maneuvers;

(vii) Navigation;

(viii) Slow flight and stalls;

(ix) Emergency operations; and

(x) Postflight procedures.

(2) *For a rotorcraft helicopter course:* (i) Preflight preparation;

(ii) Preflight procedures;

(iii) Airport and heliport operations;

(iv) Hovering maneuvers;

(v) Takeoffs, landings, and go-arounds;

(vi) Performance maneuvers;

(vii) Navigation;

(viii) Emergency operations; and

(ix) Postflight procedures.

(3) *For a rotorcraft gyroplane course:* (i) Preflight preparation;

(ii) Preflight procedures;

(iii) Airport operations;

(iv) Takeoffs, landings, and go-arounds;

(v) Performance maneuvers;

(vi) Ground reference maneuvers;

(vii) Navigation;

(viii) Flight at slow airspeeds;

(ix) Emergency operations; and

(x) Postflight procedures.

5. *Solo flight training.* Each approved course must include at least 3 hours of solo flight training on the approved areas of operation listed in paragraph (c) of section No. 4 of this appendix that are appropriate to the aircraft category and class rating for which the course applies.

6. *Stage checks and end-of-course tests.* (a) Each student enrolled in a recreational pilot course must satisfactorily accomplish the stage checks and end-of-course tests, in accordance with the school's approved training course, consisting of the approved areas of operation listed in paragraph (c) of section No. 4 of this appendix that are appropriate to the aircraft category and class rating for which the course applies.

(b) Each student must demonstrate satisfactory proficiency prior to receiving an endorsement to operate an aircraft in solo flight.

[Doc. No. 25910, 62 FR 16347, Apr. 4, 1997; Amdt. 141-9, 62 FR 40908, July 30, 1997]

Appendix B to Part 141—Private Pilot Certification Course

1. *Applicability.* This appendix prescribes the minimum curriculum for a private pilot certification course required under this part, for the following ratings:

(a) Airplane single-engine.

(b) Airplane multiengine.

(c) Rotorcraft helicopter.

(d) Rotorcraft gyroplane.

(e) Powered-lift.

(f) Glider.

(g) Lighter-than-air airship.

(h) Lighter-than-air balloon.

2. *Eligibility for enrollment.* A person must hold either a recreational pilot certificate, sport pilot certificate, or student pilot certificate before enrolling in the solo flight phase of the private pilot certification course.

3. *Aeronautical knowledge training.*

(a) Each approved course must include at least the following ground training on the aeronautical knowledge areas listed in paragraph (b) of this section, appropriate to the aircraft category and class rating:

(1) 35 hours of training if the course is for an airplane, rotorcraft, or powered-lift category rating.

(2) 15 hours of training if the course is for a glider category rating.

(3) 10 hours of training if the course is for a lighter-than-air category with a balloon class rating.

(4) 35 hours of training if the course is for a lighter-than-air category with an airship class rating.

(b) Ground training must include the following aeronautical knowledge areas:

(1) Applicable Federal Aviation Regulations for private pilot privileges, limitations, and flight operations;

(2) Accident reporting requirements of the National Transportation Safety Board;

(3) Applicable subjects of the "Aeronautical Information Manual" and the appropriate FAA advisory circulars;

(4) Aeronautical charts for VFR navigation using pilotage, dead reckoning, and navigation systems;

(5) Radio communication procedures;

(6) Recognition of critical weather situations from the ground and in flight, windshear avoidance, and the procurement and use of aeronautical weather reports and forecasts;

(7) Safe and efficient operation of aircraft, including collision avoidance, and recognition and avoidance of wake turbulence;

(8) Effects of density altitude on takeoff and climb performance;

(9) Weight and balance computations;

(10) Principles of aerodynamics, powerplants, and aircraft systems;

(11) If the course of training is for an airplane category or glider category rating, stall awareness, spin entry, spins, and spin recovery techniques;

(12) Aeronautical decision making and judgment; and

(13) Preflight action that includes—

(i) How to obtain information on runway lengths at airports of intended use, data on takeoff and landing distances, weather reports and forecasts, and fuel requirements; and

(ii) How to plan for alternatives if the planned flight cannot be completed or delays are encountered.

4. *Flight training.* (a) Each approved course must include at least the following flight training, as provided in this section and section No. 5 of this appendix, on the approved areas of operation listed in paragraph (d) of this section, appropriate to the aircraft category and class rating:

(1) 35 hours of training if the course is for an airplane, rotorcraft, powered-lift, or airship rating.

(2) 6 hours of training if the course is for a glider rating.

(3) 8 hours of training if the course is for a balloon rating.

(b) Each approved course must include at least the following flight training:

(1) *For an airplane single-engine course:* 20 hours of flight training from a certificated flight instructor on the approved areas of operation in paragraph (d)(1) of this section that includes at least—

(i) Except as provided in § 61.111 of this chapter, 3 hours of cross-country flight training in a single-engine airplane;

(ii) 3 hours of night flight training in a single-engine airplane that includes—

(A) One cross-country flight of more than 100-nautical-miles total distance; and

(B) 10 takeoffs and 10 landings to a full stop (with each landing involving a flight in the traffic pattern) at an airport.

(iii) Three hours of flight training in a single engine airplane on the control and maneuvering of a single engine airplane solely by reference to instruments, including straight and level flight, constant airspeed climbs and descents, turns to a heading, recovery from unusual flight attitudes, radio communications, and the use of navigation systems/facilities and radar services appropriate to instrument flight; and

(iv) 3 hours of flight training in a single-engine airplane in preparation for the practical test within 60 days preceding the date of the test.

(2) *For an airplane multiengine course:* 20 hours of flight training from a certificated flight instructor on the approved areas of operation in paragraph (d)(2) of this section that includes at least—

(i) Except as provided in § 61.111 of this chapter, 3 hours of cross-country flight training in a multiengine airplane;

(ii) 3 hours of night flight training in a multiengine airplane that includes—

(A) One cross-country flight of more than 100-nautical-miles total distance; and

(B) 10 takeoffs and 10 landings to a full stop (with each landing involving a flight in the traffic pattern) at an airport.

(iii) Three hours of flight training in a multiengine airplane on the control and maneuvering of a multiengine airplane solely by reference to instruments, including straight and level flight, constant airspeed climbs and descents, turns to a heading, recovery from unusual flight attitudes, radio communications, and the use of navigation systems/facilities and radar services appropriate to instrument flight; and

(iv) 3 hours of flight training in a multiengine airplane in preparation for the practical test within 60 days preceding the date of the test.

(3) *For a rotorcraft helicopter course:* 20 hours of flight training from a certificated flight instructor on the approved areas of operation in paragraph (d)(3) of this section that includes at least—

(i) Except as provided in § 61.111 of this chapter, 3 hours of cross-country flight training in a helicopter.

(ii) 3 hours of night flight training in a helicopter that includes—

(A) One cross-country flight of more than 50-nautical-miles total distance; and

(B) 10 takeoffs and 10 landings to a full stop (with each landing involving a flight in the traffic pattern) at an airport.

(iii) 3 hours of flight training in a helicopter in preparation for the practical test within 60 days preceding the date of the test.

(4) *For a rotorcraft gyroplane course:* 20 hours of flight training from a certificated flight instructor on the approved areas of operation in paragraph (d)(4) of this section that includes at least—

(i) Except as provided in § 61.111 of this chapter, 3 hours of cross-country flight training in a gyroplane.

(ii) 3 hours of night flight training in a gyroplane that includes—

(A) One cross-country flight over 50-nautical-miles total distance; and

(B) 10 takeoffs and 10 landings to a full stop (with each landing involving a flight in the traffic pattern) at an airport.

(iii) 3 hours of flight training in a gyroplane in preparation for the practical test within 60 days preceding the date of the test.

(5) *For a powered-lift course:* 20 hours of flight training from a certificated flight instructor on the approved areas of operation in paragraph (d)(5) of this section that includes at least—

(i) Except as provided in § 61.111 of this chapter, 3 hours of cross-country flight training in a powered-lift;

(ii) 3 hours of night flight training in a powered-lift that includes—

(A) One cross-country flight of more than 100-nautical-miles total distance; and

(B) 10 takeoffs and 10 landings to a full stop (with each landing involving a flight in the traffic pattern) at an airport.

(iii) Three hours of flight training in a powered-lift on the control and maneuvering of a powered-lift solely by reference to instruments, including straight and level flight, constant airspeed climbs and descents, turns to a heading, recovery from unusual flight attitudes, radio communications, and the use of navigation systems/facilities and radar services appropriate to instrument flight; and

(iv) 3 hours of flight training in a powered-lift in preparation for the practical test, within 60 days preceding the date of the test.

(6) *For a glider course:* 4 hours of flight training from a certificated flight instructor on the approved areas of operation in paragraph (d)(6) of this section that includes at least—

(i) Five training flights in a glider with a certificated flight instructor on the launch/tow procedures approved for the course and on the appropriate approved areas of operation listed in paragraph (d)(6) of this section; and

(ii) Three training flights in a glider with a certificated flight instructor in preparation for the practical test within 60 days preceding the date of the test.

(7) *For a lighter-than-air airship course:* 20 hours of flight training from a commercial pilot with an airship rating on the approved areas of operation in paragraph (d)(7) of this section that includes at least—

(i) Except as provided in § 61.111 of this chapter, 3 hours of cross-country flight training in an airship;

(ii) 3 hours of night flight training in an airship that includes—

(A) One cross-country flight over 25-nautical-miles total distance; and

(B) Five takeoffs and five landings to a full stop (with each landing involving a flight in the traffic pattern) at an airport.

(iii) 3 hours of instrument training in an airship; and

(iv) 3 hours of flight training in an airship in preparation for the practical test within 60 days preceding the date of the test.

(8) *For a lighter-than-air balloon course:* 8 hours of flight training, including at least five training flights, from a commercial pilot with a balloon rating on the approved areas of operation in paragraph (d)(8) of this section, that includes—

(i) If the training is being performed in a gas balloon—

(A) Two flights of 1 hour each;

(B) One flight involving a controlled ascent to 3,000 feet above the launch site; and

(C) Two flights in preparation for the practical test within 60 days preceding the date of the test.

(ii) If the training is being performed in a balloon with an airborne heater—

(A) Two flights of 30 minutes each;

(B) One flight involving a controlled ascent to 2,000 feet above the launch site; and

(C) Two flights in preparation for the practical test within 60 days preceding the date of the test.

(c) For use of flight simulators or flight training devices:

(1) The course may include training in a flight simulator or flight training device, provided it is representative of the aircraft for which the course is approved, meets the requirements of this paragraph, and the training is given by an authorized instructor.

(2) Training in a flight simulator that meets the requirements of § 141.41(a) of this part may be credited for a maximum of 20 percent of the total flight training hour requirements of the approved course, or of this section, whichever is less.

(3) Training in a flight training device that meets the requirements of § 141.41(b) of this part may be credited for a maximum of 15 percent of the total flight training hour requirements of the approved course, or of this section, whichever is less.

(4) Training in flight simulators or flight training devices described in paragraphs (c)(2) and (c)(3) of this section, if used in combination, may be credited for a maximum of 20 percent of the total flight training hour requirements of the approved course, or of this section, whichever is less. However, credit for training in a flight training device that meets the requirements of § 141.41(b) cannot exceed the limitation provided for in paragraph (c)(3) of this section.

(d) Each approved course must include the flight training on the approved areas of operation listed in this paragraph that are appropriate to the aircraft category and class rating—

(1) *For a single-engine airplane course:* (i) Preflight preparation;

(ii) Preflight procedures;

(iii) Airport and seaplane base operations;

(iv) Takeoffs, landings, and go-arounds;

(v) Performance maneuvers;

(vi) Ground reference maneuvers;

(vii) Navigation;

(viii) Slow flight and stalls;

(ix) Basic instrument maneuvers;

(x) Emergency operations;

(xi) Night operations, and

(xii) Postflight procedures.

(2) *For a multiengine airplane course:* (i) Preflight preparation;

(ii) Preflight procedures;

(iii) Airport and seaplane base operations;

(iv) Takeoffs, landings, and go-arounds;

(v) Performance maneuvers;

(vi) Ground reference maneuvers;

(vii) Navigation;

(viii) Slow flight and stalls;

(ix) Basic instrument maneuvers;

(x) Emergency operations;

(xi) Multiengine operations;

(xii) Night operations; and

(xiii) Postflight procedures.

(3) *For a rotorcraft helicopter course:* (i) Preflight preparation;

(ii) Preflight procedures;

(iii) Airport and heliport operations;

(iv) Hovering maneuvers;

(v) Takeoffs, landings, and go-arounds;

(vi) Performance maneuvers;

(vii) Navigation;

(viii) Emergency operations;

(ix) Night operations; and

(x) Postflight procedures.

(4) *For a rotorcraft gyroplane course:*

(i) Preflight preparation;

(ii) Preflight procedures;

(iii) Airport operations;

(iv) Takeoffs, landings, and go-arounds;

(v) Performance maneuvers;

(vi) Ground reference maneuvers;

(vii) Navigation;

(viii) Flight at slow airspeeds;

(ix) Emergency operations;

(x) Night operations; and

(xi) Postflight procedures.

(5) *For a powered-lift course:* (i) Preflight preparation;

(ii) Preflight procedures;

(iii) Airport and heliport operations;

(iv) Hovering maneuvers;

(v) Takeoffs, landings, and go-arounds;

(vi) Performance maneuvers;

(vii) Ground reference maneuvers;

(viii) Navigation;

(ix) Slow flight and stalls;

(x) Basic instrument maneuvers;

(xi) Emergency operations;

(xii) Night operations; and

(xiii) Postflight procedures.

(6) *For a glider course:* (i) Preflight preparation;

(ii) Preflight procedures;

(iii) Airport and gliderport operations;

(iv) Launches/tows, as appropriate, and landings;

(v) Performance speeds;

(vi) Soaring techniques;

(vii) Performance maneuvers;

(viii) Navigation;

(ix) Slow flight and stalls;

(x) Emergency operations; and

(xi) Postflight procedures.

(7) *For a lighter-than-air airship course:* (i) Preflight preparation;

(ii) Preflight procedures;

(iii) Airport operations;

(iv) Takeoffs, landings, and go-arounds;

(v) Performance maneuvers;

(vi) Ground reference maneuvers;

(vii) Navigation;

(viii) Emergency operations; and

(ix) Postflight procedures.

(8) *For a lighter-than-air balloon course:* (i) Preflight preparation;

(ii) Preflight procedures;

(iii) Airport operations;

(iv) Launches and landings;

(v) Performance maneuvers;

(vi) Navigation;

(vii) Emergency operations; and

(viii) Postflight procedures.

5. *Solo flight training.* Each approved course must include at least the following solo flight training:

(a) *For an airplane single-engine course:* 5 hours of solo flight training in a single-engine airplane on the approved areas of operation in paragraph (d)(1) of section No. 4 of this appendix that includes at least—

(1) One solo 100 nautical miles cross country flight with landings at a minimum of three points and one segment of the flight consisting of a straight-line distance of more than 50 nautical miles between the takeoff and landing locations; and

(2) Three takeoffs and three landings to a full stop (with each landing involving a flight in the traffic pattern) at an airport with an operating control tower.

(b) *For an airplane multiengine course:* 5 hours of flight training in a multiengine airplane performing the duties of a pilot in command while under the supervision of a certificated flight instructor. The training must consist of the approved areas of operation in paragraph (d)(2) of section No. 4 of this appendix, and include at least—

(1) One 100 nautical miles cross country flight with landings at a minimum of three points and one segment of the flight consisting of a straight-line distance of more than 50 nautical miles between the takeoff and landing locations; and

(2) Three takeoffs and three landings to a full stop (with each landing involving a flight in the traffic pattern) at an airport with an operating control tower.

(c) *For a rotorcraft helicopter course:* 5 hours of solo flight training in a helicopter on the approved areas of operation in paragraph (d)(3) of section No. 4 of this appendix that includes at least—

(1) One solo 100 nautical miles cross country flight with landings at a minimum of three points and one segment of the flight consisting of a straight-line distance of more than 25 nautical miles between the takeoff and landing locations; and

(2) Three takeoffs and three landings to a full stop (with each landing involving a flight in the traffic pattern) at an airport with an operating control tower.

(d) *For a rotorcraft gyroplane course:* 5 hours of solo flight training in gyroplanes on the approved areas of operation in paragraph (d)(4) of section No. 4 of this appendix that includes at least—

(1) One solo 100 nautical miles cross country flight with landings at a minimum of three points and one segment of the flight consisting of a straight-line distance of more than 25 nautical miles between the takeoff and landing locations; and

(2) Three takeoffs and three landings to a full stop (with each landing involving a flight in the traffic pattern) at an airport with an operating control tower.

(e) *For a powered-lift course:* 5 hours of solo flight training in a powered-lift on the approved areas of operation in paragraph (d)(5) of section No. 4 of this appendix that includes at least—

(1) One solo 100 nautical miles cross country flight with landings at a minimum of three points and one segment of the flight consisting of a straight-line distance of more than 50 nautical miles between the takeoff and landing locations; and

(2) Three takeoffs and three landings to a full stop (with each landing involving a flight in the traffic pattern) at an airport with an operating control tower.

(f) *For a glider course:* Two solo flights in a glider on the approved areas of operation in paragraph (d)(6) of section No. 4 of this appendix, and the launch and tow procedures appropriate for the approved course.

(g) *For a lighter-than-air airship course:* 5 hours of flight training in an airship performing the duties of pilot in command while under the supervision of a commercial pilot with an airship rating. The training must consist of the approved areas of operation in paragraph (d)(7) of section No. 4 of this appendix.

(h) *For a lighter-than-air balloon course:* Two solo flights in a balloon with an airborne heater if the course involves a balloon with an airborne heater or, if the course involves a gas balloon, at least two flights in a gas balloon performing the duties of pilot in command while under the supervision of a commercial pilot with a balloon rating. The training must consist of the approved areas of operation in paragraph (d)(8) of section No. 4 of this appendix, in the kind of balloon for which the course applies.

6. *Stage checks and end-of-course tests.*

(a) Each student enrolled in a private pilot course must satisfactorily accomplish the stage checks and end-of-course tests in accordance with the school's approved training course, consisting of the approved areas of operation listed in paragraph (d) of section No. 4 of this appendix that are appropriate to the aircraft category and class rating for which the course applies.

(b) Each student must demonstrate satisfactory proficiency prior to receiving an endorsement to operate an aircraft in solo flight.

[Doc. No. 25910, 62 FR 16347, Apr. 4, 1997; Amdt. 141-9, 62 FR 40908, July 30, 1997; as amended by Amdt. 141-10, 63 FR 20289, Apr. 23, 1998; Amdt. 141-12, 74 FR 42564, Aug. 21, 2009]

Appendix C to Part 141—Instrument Rating Course

1. *Applicability.* This appendix prescribes the minimum curriculum for an instrument rating course and an additional instrument rating course, required under this part, for the following ratings:

(a) Instrument—airplane.

(b) Instrument—helicopter.

(c) Instrument—powered-lift.

2. *Eligibility for enrollment.* A person must hold at least a private pilot certificate with an aircraft category and class rating appropriate to the instrument rating for which the course applies prior to enrolling in the flight portion of the instrument rating course.

3. *Aeronautical knowledge training.* (a) Each approved course must include at least the following ground training on the aeronautical knowledge areas listed in paragraph (b) of this section appropriate to the instrument rating for which the course applies:

(1) 30 hours of training if the course is for an initial instrument rating.

(2) 20 hours of training if the course is for an additional instrument rating.

(b) Ground training must include the following aeronautical knowledge areas:

(1) Applicable Federal Aviation Regulations for IFR flight operations;

(2) Appropriate information in the "Aeronautical Information Manual";

(3) Air traffic control system and procedures for instrument flight operations;

(4) IFR navigation and approaches by use of navigation systems;

(5) Use of IFR en route and instrument approach procedure charts;

(6) Procurement and use of aviation weather reports and forecasts, and the elements of forecasting weather trends on the basis of that information and personal observation of weather conditions;

(7) Safe and efficient operation of aircraft under instrument flight rules and conditions;

(8) Recognition of critical weather situations and windshear avoidance;

(9) Aeronautical decision making and judgment; and

(10) Crew resource management, to include crew communication and coordination.

4. *Flight training.* (a) Each approved course must include at least the following flight training on the approved areas of operation listed in paragraph (d) of this section, appropriate to the instrument-aircraft category and class rating for which the course applies:

(1) 35 hours of instrument training if the course is for an initial instrument rating.

(2) 15 hours of instrument training if the course is for an additional instrument rating.

(b) For the use of flight simulators or flight training devices—

(1) The course may include training in a flight simulator or flight training device, provided it is representative of the aircraft for which the course is approved, meets the requirements of this paragraph, and the training is given by an authorized instructor.

(2) Credit for training in a flight simulator that meets the requirements of § 141.41(a) cannot exceed 50 percent of the total flight training hour requirements of the course or of this section, whichever is less.

(3) Credit for training in a flight training device that meets the requirements of § 141.41(b) cannot exceed 40 percent of the total flight training hour requirements of the course or of this section, whichever is less.

(4) Credit for training in flight simulators and flight training devices, if used in combination, cannot exceed 50 percent of the total flight training hour requirements of the course or of this section, whichever is less. However, credit for training in a flight training device cannot exceed the limitation provided for in paragraph (b)(3) of this section.

(5) Credit for training in an approved aviation training device cannot exceed 10 percent of the total flight training hour requirements of the course or of this section, whichever is less.

(6) Credit for training in flight simulators, flight training devices, and aviation training devices, if used in combination, cannot exceed 50 percent of the total flight training hour requirements of the course or of this section, whichever is less. However, credit for training in an aviation training device cannot exceed the limitation provided under paragraph (b)(5) of this section.

(c) Each approved course must include the following flight training—

(1) *For an instrument airplane course:* Instrument training time from a certificated flight instructor with an instrument rating on the approved areas of operation in paragraph (d) of this section including at least one cross-country flight that—

(i) Is in the category and class of airplane that the course is approved for, and is performed under IFR;

(ii) Is a distance of at least 250 nautical miles along airways or ATC-directed routing with one segment of the flight consisting of at least a straight-line distance of 100 nautical miles between airports;

(iii) Involves an instrument approach at each airport; and

(iv) Involves three different kinds of approaches with the use of navigation systems.

(2) *For an instrument helicopter course:* Instrument training time from a certificated flight instructor with an instrument rating on the approved areas of operation in paragraph (d) of this section including at least one cross-country flight that—

(i) Is in a helicopter and is performed under IFR;

(ii) Is a distance of at least 100 nautical miles along airways or ATC-directed routing with one segment of the flight consisting of at least a straight-line distance of 50 nautical miles between airports;

(iii) Involves an instrument approach at each airport; and

(iv) Involves three different kinds of approaches with the use of navigation systems.

(3) *For an instrument powered-lift course:* Instrument training time from a certificated flight instructor with an instrument rating on the approved areas of operation in paragraph (d) of this section including at least one cross-country flight that—

(i) Is in a powered-lift and is performed under IFR;

(ii) Is a distance of at least 250 nautical miles along airways or ATC-directed routing with one segment of the flight consisting of at least a straight-line distance of 100 nautical miles between airports;

(iii) Involves an instrument approach at each airport; and

(iv) Involves three different kinds of approaches with the use of navigation systems.

(d) Each course must include flight training on the areas of operation listed under this paragraph appropriate to the instrument aircraft category and class rating (if a class rating is appropriate) for which the course applies:

(1) Preflight preparation;

(2) Preflight procedures;

(3) Air traffic control clearances and procedures;

(4) Flight by reference to instruments;

(5) Navigation systems;

(6) Instrument approach procedures;

(7) Emergency operations; and

(8) Postflight procedures.

5. *Stage checks and end-of-course tests.* Each student enrolled in an instrument rating course must satisfactorily accomplish the stage checks and end-of-course tests, in accordance with the school's approved training course, consisting of the approved areas of operation listed in paragraph (d) of section No. 4 of this appendix that are appropriate to the aircraft category and class rating for which the course applies.

[Doc. No. 25910, 62 FR 16347, Apr. 4, 1997; Amdt. 141-9, 62 FR 40909, July 30, 1997; Amdt. 141-12, 74 FR 42564, Aug. 21, 2009]

Appendix D to Part 141—Commercial Pilot Certification Course

1. *Applicability.* This appendix prescribes the minimum curriculum for a commercial pilot certification course required under this part, for the following ratings:

(a) Airplane single-engine.

(b) Airplane multiengine.

(c) Rotorcraft helicopter.

(d) Rotorcraft gyroplane.

(e) Powered-lift.

(f) Glider.

(g) Lighter-than-air airship.

(h) Lighter-than-air balloon.

2. *Eligibility for enrollment.* A person must hold the following prior to enrolling in the flight portion of the commercial pilot certification course:

(a) At least a private pilot certificate; and

(b) If the course is for a rating in an airplane or a powered-lift category, then the person must:

(1) Hold an instrument rating in the aircraft that is appropriate to the aircraft category rating for which the course applies; or

(2) Be concurrently enrolled in an instrument rating course that is appropriate to the aircraft category rating for which the course applies, and pass the required instrument rating practical test prior to completing the commercial pilot certification course.

3. *Aeronautical knowledge training.* (a) Each approved course must include at least the following ground training on the aeronautical knowledge areas listed in paragraph (b) of this section, appropriate to the aircraft category and class rating for which the course applies:

(1) 35 hours of training if the course is for an airplane category rating or a powered-lift category rating.

(2) 65 hours of training if the course is for a lighter-than-air category with an airship class rating.

(3) 30 hours of training if the course is for a rotocraft category rating.

(4) 20 hours of training if the course is for a glider category rating.

(5) 20 hours of training if the course is for lighter-than-air category with a balloon class rating.

(b) Ground training must include the following aeronautical knowledge areas:

(1) Federal Aviation Regulations that apply to commercial pilot privileges, limitations, and flight operations;

(2) Accident reporting requirements of the National Transportation Safety Board;

(3) Basic aerodynamics and the principles of flight;

(4) Meteorology, to include recognition of critical weather situations, windshear recognition and avoidance, and the use of aeronautical weather reports and forecasts;

(5) Safe and efficient operation of aircraft;

(6) Weight and balance computations;

(7) Use of performance charts;

(8) Significance and effects of exceeding aircraft performance limitations;

(9) Use of aeronautical charts and a magnetic compass for pilotage and dead reckoning;

(10) Use of air navigation facilities;

(11) Aeronautical decision making and judgment;

(12) Principles and functions of aircraft systems;

(13) Maneuvers, procedures, and emergency operations appropriate to the aircraft;

(14) Night and high-altitude operations;

(15) Descriptions of and procedures for operating within the National Airspace System; and

(16) Procedures for flight and ground training for lighter-than-air ratings.

4. *Flight training.* (a) Each approved course must include at least the following flight training, as provided in this section and section No. 5 of this appendix, on the approved areas of operation listed in paragraph (d) of this section that are appropriate to the aircraft category and class rating for which the course applies:

(1) 120 hours of training if the course is for an airplane or powered-lift rating.

(2) 155 hours of training if the course is for an airship rating.

(3) 115 hours of training if the course is for a rotocraft rating.

(4) 6 hours of training if the course is for a glider rating.

(5) 10 hours of training and 8 training flights if the course is for a balloon rating.

(b) Each approved course must include at least the following flight training:

(1) *For an airplane single-engine course:* 55 hours of flight training from a certificated flight instructor on the approved areas of operation listed in paragraph (d)(1) of this section that includes at least—

(i) Ten hours of instrument training using a view-limiting device including attitude instrument flying, partial panel skills, recovery from unusual flight attitudes, and intercepting and tracking navigational systems. Five hours of the 10 hours required on instrument training must be in a single engine airplane;

(ii) Ten hours of training in an airplane that has a retractable landing gear, flaps, and a controllable pitch propeller, or is turbine-powered;

(iii) One 2-hour cross country flight in daytime conditions in a single engine airplane that consists of a total straight-line distance of more than 100 nautical miles from the original point of departure;

(iv) One 2-hour cross country flight in nighttime conditions in a single engine airplane that consists of a total straight-line distance of more than 100 nautical miles from the original point of departure; and

(v) 3 hours in a single-engine airplane in preparation for the practical test within 60 days preceding the date of the test.

(2) *For an airplane multiengine course:* 55 hours of flight training from a certificated flight instructor on the approved areas of operation listed in paragraph (d)(2) of this section that includes at least—

(i) Ten hours of instrument training using a view-limiting device including attitude instrument flying, partial panel skills, recovery from unusual flight attitudes, and intercepting and tracking navigational systems. Five hours of the 10 hours required on instrument training must be in a multiengine airplane;

(ii) 10 hours of training in a multiengine airplane that has retractable landing gear, flaps, and a controllable pitch propeller, or is turbine-powered;

(iii) One 2-hour cross country flight in daytime conditions in a multiengine airplane that consists of a total straight-line distance of more than 100 nautical miles from the original point of departure;

(iv) One 2-hour cross country flight in nighttime conditions in a multiengine airplane that consists of a total straight-line distance of more than 100 nautical miles from the original point of departure; and

(v) 3 hours in a multiengine airplane in preparation for the practical test within 60 days preceding the date of the test.

(3) *For a rotorcraft helicopter course:* 30 hours of flight training from a certificated flight instructor on the approved areas of operation listed in paragraph (d)(3) of this section that includes at least—

(i) Five hours on the control and maneuvering of a helicopter solely by reference to instruments, including using a view-limiting device for attitude instrument flying, partial panel skills, recovery from unusual flight attitudes, and intercepting and tracking navigational systems. This aeronautical experience may be performed in an aircraft, flight simulator, flight training device, or an aviation training device;

(ii) One 2-hour cross country flight in daytime conditions in a helicopter that consists of a total straight-line distance of more than 50 nautical miles from the original point of departure;

(iii) One 2-hour cross country flight in nighttime conditions in a helicopter that consists of a total straight-line distance of more than 50 nautical miles from the original point of departure; and

(iv) 3 hours in a helicopter in preparation for the practical test within 60 days preceding the date of the test.

(4) *For a rotorcraft gyroplane course:* 30 hours of flight training from a certificated flight instructor on the approved areas of operation listed in paragraph (d)(4) of this section that includes at least—

(i) 2.5 hours on the control and maneuvering of a gyroplane solely by reference to instruments, including using a view-limiting device for attitude instrument flying, partial panel skills, recovery from unusual flight attitudes, and intercepting and tracking navigational systems. This aeronautical experience may be performed in an aircraft, flight simulator, flight training device, or an aviation training device;

(ii) One 2-hour cross country flight in daytime conditions in a gyroplane that consists of a total straight-line distance of more than 50 nautical miles from the original point of departure;

(iii) Two hours of flight training in nighttime conditions in a gyroplane at an airport, that includes 10 takeoffs and 10 landings to a full stop (with each landing involving a flight in the traffic pattern); and

(iv) 3 hours in a gyroplane in preparation for the practical test within 60 days preceding the date of the test.

(5) *For a powered-lift course:* 55 hours of flight training from a certificated flight instructor on the approved areas of operation listed in paragraph (d)(5) of this section that includes at least—

(i) Ten hours of instrument training using a view-limiting device including attitude instrument flying, partial panel skills, recovery from unusual flight attitudes, and intercepting and tracking navigational systems. Five hours of the 10 hours required on instrument training must be in a powered-lift;

(ii) One 2-hour cross country flight in daytime conditions in a powered-lift that consists of a total straight-line distance of more than 100 nautical miles from the original point of departure;

(iii) One 2-hour cross country flight in nighttime conditions in a powered-lift that consists of a total straight-line distance of more than 100 nautical miles from the original point of departure; and

(iv) 3 hours in a powered-lift in preparation for the practical test within 60 days preceding the date of the test.

(6) *For a glider course:* 4 hours of flight training from a certificated flight instructor on the approved areas of operation in paragraph (d)(6) of this section, that includes at least—

(i) Five training flights in a glider with a certificated flight instructor on the launch/tow procedures approved for the course and on the appropriate approved areas of operation listed in paragraph (d)(6) of this section; and

(ii) Three training flights in a glider with a certificated flight instructor in preparation for the practical test within 60 days preceding the date of the test.

(7) *For a lighter-than-air airship course:* 55 hours of flight training in airships from a commercial pilot with an airship rating on the approved areas of operation in paragraph (d)(7) of this section that includes at least—

(i) Three hours of instrument training in an airship, including using a view-limiting device for attitude instrument flying, partial panel skills, recovery from unusual flight attitudes, and intercepting and tracking navigational systems;

(ii) One hour cross country flight in daytime conditions in an airship that consists of a total straight-line distance of more than 25 nautical miles from the original point of departure;

(iii) One hour cross country flight in nighttime conditions in an airship that consists of a total straight-line distance of more than 25 nautical miles from the original point of departure; and

(iv) 3 hours in an airship, in preparation for the practical test within 60 days preceding the date of the test.

(8) *For a lighter-than-air balloon course:* Flight training from a commercial pilot with a balloon rating on the approved areas of operation in paragraph (d)(8) of this section that includes at least—

(i) If the course involves training in a gas balloon:

(A) Two flights of 1 hour each;

(B) One flight involving a controlled ascent to at least 5,000 feet above the launch site; and

(C) Two flights in preparation for the practical test within 60 days preceding the date of the test.

(ii) If the course involves training in a balloon with an airborne heater:

(A) Two flights of 30 minutes each;

(B) One flight involving a controlled ascent to at least 3,000 feet above the launch site; and

(C) Two flights in preparation for the practical test within 60 days preceding the date of the test.

(c) For the use of flight simulators or flight training devices:

(1) The course may include training in a flight simulator or flight training device, provided it is representative of the aircraft for which the course is approved, meets the requirements of this paragraph, and is given by an authorized instructor.

(2) Training in a flight simulator that meets the requirements of § 141.41(a) of this part may be credited for a maximum of 30 percent of the total flight training hour requirements of the approved course, or of this section, whichever is less.

(3) Training in a flight training device that meets the requirements of § 141.41(b) of this part may be credited for a maximum of 20 percent of the total flight training hour requirements of the approved course, or of this section, whichever is less.

(4) Training in the flight training devices described in paragraphs (c)(2) and (c)(3) of this section, if used in combination, may be credited for a maximum of 30 percent of the total flight training hour requirements of the approved course, or of this section, whichever is less. However, credit for training in a flight training device that meets the requirements of § 141.41(b) cannot exceed the limitation provided for in paragraph (c)(3) of this section.

(d) Each approved course must include the flight training on the approved areas of operation listed in this paragraph that are appropriate to the aircraft category and class rating—

(1) *For an airplane single-engine course:* (i) Preflight preparation;
(ii) Preflight procedures;
(iii) Airport and seaplane base operations;
(iv) Takeoffs, landings, and go-arounds;
(v) Performance maneuvers;
(vi) Navigation;
(vii) Slow flight and stalls;
(viii) Emergency operations;
(ix) High-altitude operations; and
(x) Postflight procedures.

(2) *For an airplane multiengine course:* (i) Preflight preparation;
(ii) Preflight procedures;
(iii) Airport and seaplane base operations;
(iv) Takeoffs, landings, and go-arounds;
(v) Performance maneuvers;
(vi) Navigation;
(vii) Slow flight and stalls;
(viii) Emergency operations;
(ix) Multiengine operations;
(x) High-altitude operations; and
(xi) Postflight procedures.

(3) *For a rotorcraft helicopter course:* (i) Preflight preparation;
(ii) Preflight procedures;
(iii) Airport and heliport operations;
(iv) Hovering maneuvers;
(v) Takeoffs, landings, and go-arounds;
(vi) Performance maneuvers;
(vii) Navigation;
(viii) Emergency operations;
(ix) Special operations; and
(x) Postflight procedures.

(4) *For a rotorcraft gyroplane course:* (i) Preflight preparation;
(ii) Preflight procedures;
(iii) Airport operations;
(iv) Takeoffs, landings, and go-arounds;
(v) Performance maneuvers;
(vi) Ground reference maneuvers;
(vii) Navigation;
(viii) Flight at slow airspeeds;
(ix) Emergency operations; and
(x) Postflight procedures.

(5) *For a powered-lift course:* (i) Preflight preparation;
(ii) Preflight procedures;
(iii) Airport and heliport operations;
(iv) Hovering maneuvers;
(v) Takeoffs, landings, and go-arounds;
(vi) Performance maneuvers;
(vii) Navigation;
(viii) Slow flight and stalls;
(ix) Emergency operations;
(x) High altitude operations;
(xi) Special operations; and
(xii) Postflight procedures.

(6) *For a glider course:* (i) Preflight preparation;
(ii) Preflight procedures;
(iii) Airport and gliderport operations;
(iv) Launches/tows, as appropriate, and landings;
(v) Performance speeds;
(vi) Soaring techniques;
(vii) Performance maneuvers;
(viii) Navigation;
(ix) Slow flight and stalls;
(x) Emergency operations; and
(xi) Postflight procedures.

(7) *For a lighter-than-air airship course:* (i) Fundamentals of instructing;
(ii) Technical subjects;
(iii) Preflight preparation;
(iv) Preflight lessons on a maneuver to be performed in flight;
(v) Preflight procedures;
(vi) Airport operations;
(vii) Takeoffs, landings, and go-arounds;
(viii) Performance maneuvers;
(ix) Navigation;
(x) Emergency operations; and
(xi) Postflight procedures.

(8) *For a lighter-than-air balloon course:* (i) Fundamentals of instructing;
(ii) Technical subjects;
(iii) Preflight preparation;
(iv) Preflight lesson on a maneuver to be performed in flight;
(v) Preflight procedures;
(vi) Airport operations;

PART 141

FAR

(vii) Launches and landings;
(viii) Performance maneuvers;
(ix) Navigation;
(x) Emergency operations; and
(xi) Postflight procedures.

5. *Solo training.* Each approved course must include at least the following solo flight training:

(a) *For an airplane single engine course.* Ten hours of solo flight time in a single engine airplane, or 10 hours of flight time while performing the duties of pilot in command in a single engine airplane with an authorized instructor on board. The training must consist of the approved areas of operation under paragraph (d)(1) of section 4 of this appendix, and include—

(1) One cross-country flight, if the training is being performed in the State of Hawaii, with landings at a minimum of three points, and one of the segments consisting of a straight-line distance of at least 150 nautical miles;

(2) One cross-country flight, if the training is being performed in a State other than Hawaii, with landings at a minimum of three points, and one segment of the flight consisting of a straight-line distance of at least 250 nautical miles; and

(3) 5 hours in night VFR conditions with 10 takeoffs and 10 landings (with each landing involving a flight with a traffic pattern) at an airport with an operating control tower.

(b) *For an airplane multiengine course.* Ten hours of solo flight time in a multiengine airplane, or 10 hours of flight time while performing the duties of pilot in command in a multiengine airplane with an authorized instructor on board. The training must consist of the approved areas of operation under paragraph (d)(2) of section 4 of this appendix, and include—

(1) One cross-country flight, if the training is being performed in the State of Hawaii, with landings at a minimum of three points, and one of the segments consisting of a straight-line distance of at least 150 nautical miles;

(2) One cross-country flight, if the training is being performed in a State other than Hawaii, with landings at a minimum of three points and one segment of the flight consisting of straight-line distance of at least 250 nautical miles; and

(3) 5 hours in night VFR conditions with 10 takeoffs and 10 landings (with each landing involving a flight with a traffic pattern) at an airport with an operating control tower.

(c) *For a rotorcraft helicopter course.* Ten hours of solo flight time in a helicopter, or 10 hours of flight time while performing the duties of pilot in command in a helicopter with an authorized instructor on board. The training must consist of the approved areas of operation under paragraph (d)(3) of section 4 of this appendix, and include—

(1) One cross-country flight with landings at a minimum of three points and one segment of the flight consisting of a straight-line distance of at least 50 nautical miles from the original point of departure; and

(2) 5 hours in night VFR conditions with 10 takeoffs and 10 landings (with each landing involving a flight with a traffic pattern) at an airport with an operating control tower.

(d) *For a rotorcraft-gyroplane course.* Ten hours of solo flight time in a gyroplane, or 10 hours of flight time while performing the duties of pilot in command in a gyroplane with an authorized instructor on board. The training must consist of the approved areas of operation under paragraph (d)(4) of section 4 of this appendix, and include—

(1) One cross-country flight with landings at a minimum of three points, and one segment of the flight consisting of a straight-line distance of at least 50 nautical miles from the original point of departure; and

(2) 5 hours in night VFR conditions with 10 takeoffs and 10 landings (with each landing involving a flight with a traffic pattern) at an airport with an operating control tower.

(e) *For a powered-lift course.* Ten hours of solo flight time in a powered-lift, or 10 hours of flight time while performing the duties of pilot in command in a powered-lift with an authorized instructor on board. The training must consist of the approved areas of operation under paragraph (d)(5) of section No. 4 of this appendix, and include—

(1) One cross-country flight, if the training is being performed in the State of Hawaii, with landings at a minimum of three points, and one segment of the flight consisting of a straight-line distance of at least 150 nautical miles;

(2) One cross-country flight, if the training is being performed in a State other than Hawaii, with landings at a minimum of three points, and one segment of the flight consisting of a straight-line distance of at least 250 nautical miles; and

(3) 5 hours in night VFR conditions with 10 takeoffs and 10 landings (with each landing involving a flight with a traffic pattern) at an airport with an operating control tower.

(f) *For a glider course:* 5 solo flights in a glider on the approved areas of operation in paragraph (d)(6) of section No. 4 of this appendix.

(g) *For a lighter-than-air airship course:* 10 hours of flight training in an airship performing the duties of pilot in command while under the supervision of a commercial pilot with an airship rating. The training must consist of the approved areas of operation in paragraph (d)(7) of section No. 4 of this appendix and include at least—

(1) One cross-country flight with landings at a minimum of three points, and one segment of the flight consisting of a straight-line distance of at least 25 nautical miles from the original point of departure; and

(2) 5 hours in night VFR conditions with 10 takeoffs and 10 landings (with each landing involving a flight with a traffic pattern).

(h) *For a lighter-than-air balloon course:* Two solo flights if the course is for a hot air balloon rating, or, if the course is for a gas balloon rating, at least two flights in a gas balloon, while performing the duties of pilot in command under the supervision of a commercial pilot with a balloon rating. The training shall consist of the approved areas of operation in paragraph (d)(8) of section No. 4 of this appendix, in the kind of balloon for which the course applies.

6. *Stage checks and end-of-course tests.* (a) Each student enrolled in a commercial pilot course must satisfactorily accomplish the stage checks and end-of-course tests, in accordance with the school's approved training course, consisting of the approved areas of operation listed in paragraph (d) of section No. 4 of this appendix that are appropriate to aircraft category and class rating for which the course applies.

(b) Each student must demonstrate satisfactory proficiency prior to receiving an endorsement to operate an aircraft in solo flight.

[Doc. No. 25910, 62 FR 16347, Apr. 4, 1997; Amdt. 141-9, 62 FR 40909, July 30, 1997; as amended by Amdt. 141-10, 63 FR 20290, Apr. 23, 1998; Amdt. 141-12, 74 FR 42565, Aug. 21, 2009]

Appendix E to Part 141—Airline Transport Pilot Certification Course

1. *Applicability.* This appendix prescribes the minimum curriculum for an airline transport pilot certification course under this part, for the following ratings:

(a) Airplane single-engine.
(b) Airplane multiengine.
(c) Rotorcraft helicopter.
(d) Powered-lift.

2. *Eligibility for enrollment.* Before completing the flight portion of the airline transport pilot certification course, a person must meet the aeronautical experience requirements for an airline transport pilot certificate under part 61, subpart G of this chapter that is appropriate to the aircraft category and class rating for which the course applies, and:

(a) Hold a commercial pilot certificate and an instrument rating, or an airline transport pilot certificate with instrument privileges;

(b) Meet the military experience requirements under § 61.73 of this chapter to qualify for a commercial pilot certificate and an instrument rating, if the person is a rated military pilot or former rated military pilot of an Armed Force of the United States; or

(c) Hold either a foreign airline transport pilot license or foreign commercial pilot license and an instrument rating, if the person holds a pilot license issued by a contracting State to the Convention on International Civil Aviation.

3. *Aeronautical knowledge areas.* (a) Each approved course must include at least 40 hours of ground training on the aeronautical knowledge areas listed in paragraph (b) of this section, appropriate to the aircraft category and class rating for which the course applies.

(b) Ground training must include the following aeronautical knowledge areas:

(1) Applicable Federal Aviation Regulations of this chapter that relate to airline transport pilot privileges, limitations, and flight operations;

(2) Meteorology, including knowledge of and effects of fronts, frontal characteristics, cloud formations, icing, and upper-air data;

(3) General system of weather and NOTAM collection, dissemination, interpretation, and use;

(4) Interpretation and use of weather charts, maps, forecasts, sequence reports, abbreviations, and symbols;

(5) National Weather Service functions as they pertain to operations in the National Airspace System;

(6) Windshear and microburst awareness, identification, and avoidance;

(7) Principles of air navigation under instrument meteorological conditions in the National Airspace System;

(8) Air traffic control procedures and pilot responsibilities as they relate to en route operations, terminal area and radar operations, and instrument departure and approach procedures;

(9) Aircraft loading; weight and balance; use of charts, graphs, tables, formulas, and computations; and the effects on aircraft performance;

(10) Aerodynamics relating to an aircraft's flight characteristics and performance in normal and abnormal flight regimes;

(11) Human factors;

(12) Aeronautical decision making and judgment; and

(13) Crew resource management to include crew communication and coordination.

4. *Flight training.* (a) Each approved course must include at least 25 hours of flight training on the approved areas of operation listed in paragraph (c) of this section appropriate to the aircraft category and class rating for which the course applies. At least 15 hours of this flight training must be instrument flight training.

(b) For the use of flight simulators or flight training devices—

(1) The course may include training in a flight simulator or flight training device, provided it is representative of the aircraft for which the course is approved, meets the requirements of this paragraph, and the training is given by an authorized instructor.

(2) Training in a flight simulator that meets the requirements of § 141.41(a) of this part may be credited for a maximum of 50 percent of the total flight training hour requirements of the approved course, or of this section, whichever is less.

(3) Training in a flight training device that meets the requirements of § 141.41(b) of this part may be credited for a maximum of 25 percent of the total flight training hour requirements of the approved course, or of this section, whichever is less.

(4) Training in flight simulators or flight training devices described in paragraphs (b)(2) and (b)(3) of this section, if used in combination, may be credited for a maximum of 50 percent of the total flight training hour requirements of the approved course, or of this section, whichever is less. However, credit for training in a flight training device that meets the requirements of § 141.41(b) cannot exceed the limitation provided for in paragraph (b)(3) of this section.

(c) Each approved course must include flight training on the approved areas of operation listed in this paragraph appropriate to the aircraft category and class rating for which the course applies:

(1) Preflight preparation;
(2) Preflight procedures;
(3) Takeoff and departure phase;
(4) In-flight maneuvers;
(5) Instrument procedures;
(6) Landings and approaches to landings;
(7) Normal and abnormal procedures;
(8) Emergency procedures; and
(9) Postflight procedures.

5. *Stage checks and end-of-course tests.* (a) Each student enrolled in an airline transport pilot course must satisfactorily accomplish the stage checks and end-of-course tests, in accordance with the school's approved training course, consisting of the approved areas of operation listed in paragraph (c) of section No. 4 of this appendix that are appropriate to the aircraft category and class rating for which the course applies.

(b) Each student must demonstrate satisfactory proficiency prior to receiving an endorsement to operate an aircraft in solo flight.

[Doc. No. 25910, 62 FR 16347, Apr. 4, 1997; Amdt. 141-9, 62 FR 40909, July 30, 1997; Amdt. 141-12, 74 FR 42565, Aug. 21, 2009]

Appendix F to Part 141—Flight Instructor Certification Course

1. *Applicability.* This appendix prescribes the minimum curriculum for a flight instructor certification course and an additional flight instructor rating course required under this part, for the following ratings:

(a) Airplane single-engine.
(b) Airplane multiengine.
(c) Rotorcraft helicopter.

(d) Rotorcraft gyroplane.
(e) Powered-lift.
(f) Glider category.

2. *Eligibility for enrollment.* A person must hold the following prior to enrolling in the flight portion of the flight instructor or additional flight instructor rating course:

(a) A commercial pilot certificate or an airline transport pilot certificate, with an aircraft category and class rating appropriate to the flight instructor rating for which the course applies; and

(b) An instrument rating or privilege in an aircraft that is appropriate to the aircraft category and class rating for which the course applies, if the course is for a flight instructor airplane or powered-lift instrument rating.

3. *Aeronautical knowledge training.* (a) Each approved course must include at least the following ground training in the aeronautical knowledge areas listed in paragraph (b) of this section:

(1) 40 hours of training if the course is for an initial issuance of a flight instructor certificate; or

(2) 20 hours of training if the course is for an additional flight instructor rating.

(b) Ground training must include the following aeronautical knowledge areas:

(1) The fundamentals of instructing including—
(i) The learning process;
(ii) Elements of effective teaching;
(iii) Student evaluation and testing;
(iv) Course development;
(v) Lesson planning; and
(vi) Classroom training techniques.

(2) The aeronautical knowledge areas in which training is required for—

(i) A recreational, private, and commercial pilot certificate that is appropriate to the aircraft category and class rating for which the course applies; and

(ii) An instrument rating that is appropriate to the aircraft category and class rating for which the course applies, if the course is for an airplane or powered-lift aircraft rating.

(c) A student who satisfactorily completes 2 years of study on the principles of education at a college or university may be credited with no more than 20 hours of the training required in paragraph (a)(1) of this section.

4. *Flight training.* (a) Each approved course must include at least the following flight training on the approved areas of operation of paragraph (c) of this section appropriate to the flight instructor rating for which the course applies:

(1) 25 hours, if the course is for an airplane, rotorcraft, or powered-lift rating; and

(2) 10 hours, which must include 10 flights, if the course is for a glider category rating.

(b) For the use of flight simulators or flight training devices:

(1) The course may include training in a flight simulator or flight training device, provided it is representative of the aircraft for which the course is approved, meets the requirements of this paragraph, and the training is given by an authorized instructor.

(2) Training in a flight simulator that meets the requirements of § 141.41(a) of this part, may be credited for a maximum of 10 percent of the total flight training hour requirements of the approved course, or of this section, whichever is less.

(3) Training in a flight training device that meets the requirements of § 141.41(b) of this part, may be credited for a maximum of 5 percent of the total flight training hour requirements of the approved course, or of this section, whichever is less.

(4) Training in flight simulators or flight training devices described in paragraphs (b)(2) and (b)(3) of this section, if used in combination, may be credited for a maximum of 10 percent of the total flight training hour requirements of the approved course, or of this section, whichever is less. However, credit for training in a flight training device that meets the requirements of § 141.41(b) cannot exceed the limitation provided for in paragraph (b)(3) of this section.

(c) Each approved course must include flight training on the approved areas of operation listed in this paragraph that are appropriate to the aircraft category and class rating for which the course applies—

(1) *For an airplane—single-engine course:* (i) Fundamentals of instructing;
(ii) Technical subject areas;
(iii) Preflight preparation;
(iv) Preflight lesson on a maneuver to be performed in flight;
(v) Preflight procedures;
(vi) Airport and seaplane base operations;
(vii) Takeoffs, landings, and go-arounds;
(viii) Fundamentals of flight;
(ix) Performance maneuvers;
(x) Ground reference maneuvers;
(xi) Slow flight, stalls, and spins;
(xii) Basic instrument maneuvers;
(xiii) Emergency operations; and
(xiv) Postflight procedures.

(2) *For an airplane—multiengine course:* (i) Fundamentals of instructing;
(ii) Technical subject areas;
(iii) Preflight preparation;
(iv) Preflight lesson on a maneuver to be performed in flight;
(v) Preflight procedures;
(vi) Airport and seaplane base operations;
(vii) Takeoffs, landings, and go-arounds;
(viii) Fundamentals of flight;
(ix) Performance maneuvers;
(x) Ground reference maneuvers;
(xi) Slow flight and stalls;
(xii) Basic instrument maneuvers;
(xiii) Emergency operations;
(xiv) Multiengine operations; and
(xv) Postflight procedures.

(3) *For a rotorcraft—helicopter course:* (i) Fundamentals of instructing;
(ii) Technical subject areas;
(iii) Preflight preparation;
(iv) Preflight lesson on a maneuver to be performed in flight;
(v) Preflight procedures;
(vi) Airport and heliport operations;
(vii) Hovering maneuvers;
(viii) Takeoffs, landings, and go-arounds;
(ix) Fundamentals of flight;
(x) Performance maneuvers;
(xi) Emergency operations;
(xii) Special operations; and
(xiii) Postflight procedures.

(4) *For a rotorcraft—gyroplane course:* (i) Fundamentals of instructing;
(ii) Technical subject areas;
(iii) Preflight preparation;
(iv) Preflight lesson on a maneuver to be performed in flight;
(v) Preflight procedures;
(vi) Airport operations;
(vii) Takeoffs, landings, and go-arounds;
(viii) Fundamentals of flight;
(ix) Performance maneuvers;
(x) Flight at slow airspeeds;
(xi) Ground reference maneuvers;
(xii) Emergency operations; and
(xiii) Postflight procedures.

(5) *For a powered-lift course:* (i) Fundamentals of instructing;
(ii) Technical subject areas;
(iii) Preflight preparation;
(iv) Preflight lesson on a maneuver to be performed in flight;
(v) Preflight procedures;
(vi) Airport and heliport operations;
(vii) Hovering maneuvers;
(viii) Takeoffs, landings, and go-arounds;
(ix) Fundamentals of flight;
(x) Performance maneuvers;
(xi) Ground reference maneuvers;
(xii) Slow flight and stalls;
(xiii) Basic instrument maneuvers;
(xiv) Emergency operations;
(xv) Special operations; and
(xvi) Postflight procedures.

(6) *For a glider course:* (i) Fundamentals of instructing;
(ii) Technical subject areas;
(iii) Preflight preparation;
(iv) Preflight lesson on a maneuver to be performed in flight;
(v) Preflight procedures;
(vi) Airport and gliderport operations;
(vii) Tows or launches, landings, and go-arounds, if applicable;
(viii) Fundamentals of flight;
(ix) Performance speeds;
(x) Soaring techniques;
(xi) Performance maneuvers;
(xii) Slow flight, stalls, and spins;
(xiii) Emergency operations; and
(xiv) Postflight procedures.

5. *Stage checks and end-of-course tests.* (a) Each student enrolled in a flight instructor course must satisfactorily accomplish the stage checks and end-of-course tests, in accordance with the school's approved training course, consisting of the appropriate approved areas of operation listed in paragraph (c) of section No. 4 of this appendix appropriate to the flight instructor rating for which the course applies.

(b) In the case of a student who is enrolled in a flight instructor-airplane rating or flight instructor-glider rating course, that student must have:

(1) Received a logbook endorsement from a certificated flight instructor certifying the student received ground and flight training on stall awareness, spin entry, spins, and spin recovery procedures in an aircraft that is certificated for spins and is appropriate to the rating sought; and

(2) Demonstrated instructional proficiency in stall awareness, spin entry, spins, and spin recovery procedures.

[Doc. No. 25910, 62 FR 16347, Apr. 4, 1997; Amdt. 141-9, 62 FR 40909, July 30, 1997]

Appendix G to Part 141—Flight Instructor Instrument (For an Airplane, Helicopter, or Powered-Lift Instrument Instructor Rating, as Appropriate) Certification Course

1. *Applicability.* This appendix prescribes the minimum curriculum for a flight instructor instrument certification course required under this part, for the following ratings:

(a) Flight Instructor Instrument—Airplane.
(b) Flight Instructor Instrument—Helicopter.
(c) Flight Instructor Instrument—Powered-lift aircraft.

2. *Eligibility for enrollment.* A person must hold the following prior to enrolling in the flight portion of the flight instructor instrument course:

(a) A commercial pilot certificate or airline transport pilot certificate with an aircraft category and class rating appropriate to the flight instructor category and class rating for which the course applies; and

(b) An instrument rating or privilege on that flight instructor applicant's pilot certificate that is appropriate to the flight instructor instrument rating (for an airplane-, helicopter-, or powered-lift-instrument rating, as appropriate) for which the course applies.

3. *Aeronautical knowledge training.* (a) Each approved course must include at least 15 hours of ground training on the aeronautical knowledge areas listed in paragraph (b) of this section, appropriate to the flight instructor instrument rating (for an airplane-, helicopter-, or powered-lift-instrument rating, as appropriate) for which the course applies:

(b) Ground training must include the following aeronautical knowledge areas:

(1) The fundamentals of instructing including:
(i) The learning process;
(ii) Elements of effective teaching;
(iii) Student evaluation and testing;
(iv) Course development;
(v) Lesson planning; and
(vi) Classroom training techniques.

(2) The aeronautical knowledge areas in which training is required for an instrument rating that is appropriate to the aircraft category and class rating for the course which applies.

4. *Flight training.* (a) Each approved course must include at least 15 hours of flight training in the approved areas of operation of paragraph (c) of this section appropriate to the flight instructor rating for which the course applies.

(b) For the use of flight simulators or flight training devices:

(1) The course may include training in a flight simulator or flight training device, provided it is representative of the aircraft for which the course is approved for, meets requirements of this paragraph, and the training is given by an instructor.

(2) Training in a flight simulator that meets the requirements of § 141.41(a) of this part, may be credited for a maximum of 10 percent of the total flight training hour requirements of the approved course, or of this section, whichever is less.

(3) Training in a flight training device that meets the requirements of § 141.41(b) of this part, may be credited for a maximum of 5 percent of the total flight training hour requirements of the approved course, or of this section, whichever is less.

(4) Training in flight simulators or flight training devices described in paragraphs (b)(2) and (b)(3) of this section, if used in combination, may be credited for a maximum of

10 percent of the total flight training hour requirements of the approved course, or of this section, whichever is less. However, credit for training in a flight training device that meets the requirements of § 141.41(b) cannot exceed the limitation provided for in paragraph (b)(3) of this section.

(c) An approved course for the flight instructor-instrument rating must include flight training on the following approved areas of operation that are appropriate to the instrument-aircraft category and class rating for which the course applies:

(1) Fundamentals of instructing;

(2) Technical subject areas;

(3) Preflight preparation;

(4) Preflight lesson on a maneuver to be performed in flight;

(5) Air traffic control clearances and procedures;

(6) Flight by reference to instruments;

(7) Navigation systems;

(8) Instrument approach procedures;

(9) Emergency operations; and

(10) Postflight procedures.

5. *Stage checks and end-of-course tests.* Each student enrolled in a flight instructor instrument course must satisfactorily accomplish the stage checks and end-of-course tests, in accordance with the school's approved training course, consisting of the approved areas of operation listed in paragraph (c) of section No. 4 of this appendix that are appropriate to the flight instructor instrument rating (for an airplane-, helicopter-, or powered-lift-instrument rating, as appropriate) for which the course applies.

[Doc. No. 25910, 62 FR 16347, Apr. 4, 1997; Amdt. 141-9, 62 FR 40909, July 30, 1997]

Appendix H to Part 141—Ground Instructor Certification Course

1. *Applicability.* This appendix prescribes the minimum curriculum for a ground instructor certification course and an additional ground instructor rating course, required under this part, for the following ratings:

(a) Ground Instructor—Basic.

(b) Ground Instructor—Advanced.

(c) Ground Instructor—Instrument.

2. *Aeronautical knowledge training.* (a) Each approved course must include at least the following ground training on the knowledge areas listed in paragraphs (b), (c), (d), and (e) of this section, appropriate to the ground instructor rating for which the course applies:

(1) 20 hours of training if the course is for an initial issuance of a ground instructor certificate; or

(2) 10 hours of training if the course is for an additional ground instructor rating.

(b) Ground training must include the following aeronautical knowledge areas:

(1) Learning process;

(2) Elements of effective teaching;

(3) Student evaluation and testing;

(4) Course development;

(5) Lesson planning; and

(6) Classroom training techniques.

(c) Ground training for a basic ground instructor certificate must include the aeronautical knowledge areas applicable to a recreational and private pilot.

(d) Ground training for an advanced ground instructor rating must include the aeronautical knowledge areas applicable to a recreational, private, commercial, and airline transport pilot.

(e) Ground training for an instrument ground instructor rating must include the aeronautical knowledge areas applicable to an instrument rating.

(f) A student who satisfactorily completed 2 years of study on the principles of education at a college or university may be credited with 10 hours of the training required in paragraph (a)(1) of this section.

3. *Stage checks and end-of-course tests.* Each student enrolled in a ground instructor course must satisfactorily accomplish the stage checks and end-of-course tests, in accordance with the school's approved training course, consisting of the approved knowledge areas in paragraph (b), (c), (d), and (e) of section No. 2 of this appendix appropriate to the ground instructor rating for which the course applies.

Appendix I to Part 141—Additional Aircraft Category and/or Class Rating Course

1. *Applicability.* This appendix prescribes the minimum curriculum for an additional aircraft category rating course or an additional aircraft class rating course required under this part, for the following ratings:

(a) Airplane single-engine.

(b) Airplane multiengine.

(c) Rotorcraft helicopter.

(d) Rotorcraft gyroplane.

(e) Powered-lift.

(f) Glider.

(g) Lighter-than-air airship.

(h) Lighter-than-air balloon.

2. *Eligibility for enrollment.* A person must hold the level of pilot certificate for the additional aircraft category and class rating for which the course applies prior to enrolling in the flight portion of an additional aircraft category or additional aircraft class rating course.

3. *Aeronautical knowledge training.*

(a) For a recreational pilot certificate, the following aeronautical knowledge areas must be included in a 10-hour ground training course for an additional aircraft category and/or class rating:

(1) Applicable regulations issued by the Federal Aviation Administration for recreational pilot privileges, limitations, and flight operations;

(2) Safe and efficient operation of aircraft, including collision avoidance, and recognition and avoidance of wake turbulence;

(3) Effects of density altitude on takeoff and climb performance;

(4) Weight and balance computations;

(5) Principles of aerodynamics, powerplants, and aircraft systems;

(6) Stall awareness, spin entry, spins, and spin recovery techniques if applying for an airplane single engine rating; and

(7) Preflight action that includes how to obtain information on runway lengths at airports of intended use, data on takeoff and landing distances, weather reports and forecasts, and fuel requirements.

(b) For a private pilot certificate, the following aeronautical knowledge areas must be included in a 10-hour ground training course for an additional class rating or a 15-hour ground training course for an additional aircraft category and class rating:

(1) Applicable regulations issued by the Federal Aviation Administration for private pilot privileges, limitations, and flight operations;

(2) Safe and efficient operation of aircraft, including collision avoidance, and recognition and avoidance of wake turbulence;

(3) Effects of density altitude on takeoff and climb performance;

(4) Weight and balance computations;

(5) Principles of aerodynamics, powerplants, and aircraft systems;

(6) Stall awareness, spin entry, spins, and spin recovery techniques if applying for an airplane single engine rating; and

(7) Preflight action that includes how to obtain information on runway lengths at airports of intended use, data on takeoff and landing distances, weather reports and forecasts, and fuel requirements.

(c) For a commercial pilot certificate, the following aeronautical knowledge areas must be included in a 15-hour ground training course for an additional class rating or a 20-hour ground training course for an additional aircraft category and class rating:

(1) Applicable regulations issued by the Federal Aviation Administration for commercial pilot privileges, limitations, and flight operations;

(2) Basic aerodynamics and the principles of flight;

(3) Safe and efficient operation of aircraft;

(4) Weight and balance computations;

(5) Use of performance charts;

(6) Significance and effects of exceeding aircraft performance limitations;

(7) Principles and functions of aircraft systems;

(8) Maneuvers, procedures, and emergency operations appropriate to the aircraft;

(9) Nighttime and high-altitude operations; and

(10) Procedures for flight and ground training for lighter-than-air ratings.

(d) For an airline transport pilot certificate, the following aeronautical knowledge areas must be included in a 25-hour ground training course for an additional aircraft category and/or class rating:

(1) Applicable regulations issued by the Federal Aviation Administration for airline transport pilot privileges, limitations, and flight operations;

(2) Meteorology, including knowledge and effects of fronts, frontal characteristics, cloud formations, icing, and upper-air data;

(3) General system of weather and NOTAM collection, dissemination, interpretation, and use;

(4) Interpretation and use of weather charts, maps, forecasts, sequence reports, abbreviations, and symbols;

(5) National Weather Service functions as they pertain to operations in the National Airspace System;

(6) Windshear and microburst awareness, identification, and avoidance;

(7) Principles of air navigation under instrument meteorological conditions in the National Airspace System;

(8) Air traffic control procedures and pilot responsibilities as they relate to en route operations, terminal area and radar operations, and instrument departure and approach procedures;

(9) Aircraft loading; weight and balance; use of charts, graphs, tables, formulas, and computations; and the effects on aircraft performance;

(10) Aerodynamics relating to an aircraft's flight characteristics and performance in normal and abnormal flight regimes;

(11) Human factors;

(12) Aeronautical decision making and judgment; and

(13) Crew resource management to include crew communication and coordination.

4. Flight training.

(a) Course for an additional airplane category and single engine class rating.

(1) For the recreational pilot certificate, the course must include 15 hours of flight training on the areas of operations under part 141, appendix A, paragraph 4(c)(1) that include—

(i) Two hours of flight training to an airport and at an airport that is located more than 25 nautical miles from the airport where the applicant normally trains, with three takeoffs and three landings, except as provided under § 61.100 of this chapter; and

(ii) Three hours of flight training in an aircraft with the airplane category and single engine class within 2 calendar months before the date of the practical test.

(2) For the private pilot certificate, the course must include 20 hours of flight training on the areas of operations under part 141, appendix B, paragraph 4(d)(1). A flight simulator and flight training device cannot be used to meet more than 4 hours of the training requirements, and the use of the flight training device is limited to 3 hours of the 4 hours permitted. The course must include—

(i) Three hours of cross country training in a single engine airplane, except as provided under § 61.111 of this chapter;

(ii) Three hours of nighttime flight training in a single engine airplane that includes one cross country flight of more than 100 nautical miles total distance, and 10 takeoffs and 10 landings to a full stop (with each landing involving a flight in the traffic pattern) at an airport;

(iii) Three hours of flight training in a single engine airplane on the control and maneuvering of the airplane solely by reference to instruments, including straight and level flight, constant airspeed climbs and descents, turns to a heading, recovery from unusual flight attitudes, radio communications, and the use of navigation systems/facilities and radar services appropriate to instrument flight; and

(iv) Three hours of flight training in a single engine airplane within 2 calendar months before the date of the practical test.

(3) For the commercial pilot certificate, the course must include 55 hours of flight training on the areas of operations under part 141, appendix D, paragraph 4(d)(1). A flight simulator and flight training device cannot be used to meet more than 16.5 hours of the training requirements, and the use of the flight training device is limited to 11 hours of the 16.5 hours permitted. The course must include—

(i) Five hours of instrument training in a single engine airplane that includes training using a view-limiting device on attitude instrument flying, partial panel skills, recovery from unusual flight attitudes, and intercepting and tracking navigational systems;

(ii) Ten hours of training in an airplane that has retractable landing gear, flaps, and a controllable pitch propeller, or is turbine-powered;

(iii) One 2-hour cross country flight during daytime conditions in a single engine airplane, a total straight-line distance of more than 100 nautical miles from the original point of departure;

(iv) One 2-hour cross country flight during nighttime conditions in a single engine airplane, a total straight-line distance of more than 100 nautical miles from the original point of departure; and

(v) Three hours in a single engine airplane within 2 calendar months before the date of the practical test.

(4) For the airline transport pilot certificate, the course must include 25 hours flight training, including 15 hours of instrument training, in a single engine airplane on the areas of operation under part 141, appendix E, paragraph 4.(c). A flight simulator and flight training device cannot be used to meet more than 12.5 hours of the training requirements; and the use of the flight training device is limited to 6.25 hours of the 12.5 hours permitted.

(b) Course for an additional airplane category and multi-engine class rating.

(1) For the private pilot certificate, the course requires 20 hours flight training on the areas of operations under part 141, appendix B, paragraph 4.(d)(2). A flight simulator and flight training device cannot be used more than 4 hours to meet the training requirements, and use of the flight training device is limited to 3 hours of the 4 hours permitted. The course must include—

(i) Three hours of cross country training in a multiengine airplane, except as provided under § 61.111 of this chapter;

(ii) Three hours of nighttime flight training in a multiengine airplane that includes one cross country flight of more than 100 nautical miles total distance, and 10 takeoffs and 10 landings to a full stop (with each landing involving a flight in the traffic pattern) at an airport;

(iii) Three hours of flight training in a multiengine airplane on the control and maneuvering of a multiengine airplane solely by reference to instruments, including straight and level flight, constant airspeed climbs and descents, turns to a heading, recovery from unusual flight attitudes, radio communications, and the use of navigation systems/facilities and radar services appropriate to instrument flight; and

(iv) Three hours of flight training in a multiengine airplane in preparation for the practical test within 2 calendar months before the date of the test.

(2) For the commercial pilot certificate, the course requires 55 hours flight training on the areas of operations under part 141, appendix D, paragraph 4.(d)(2). A flight simulator and flight training device cannot be used more than 16.5 hours to meet the training requirements, and use of the flight training device is limited to 11 hours of the 16.5 hours permitted. The course must include—

(i) Five hours of instrument training in a multiengine airplane including training using a view-limiting device for attitude instrument flying, partial panel skills, recovery from unusual flight attitudes, and intercepting and tracking navigational systems;

(ii) Ten hours of training in a multiengine airplane that has retractable landing gear, flaps, and a controllable pitch propeller, or is turbine-powered;

(iii) One 2-hour cross country flight during daytime conditions in a multiengine airplane, and a total straight-line distance of more than 100 nautical miles from the original point of departure;

(iv) One 2-hour cross country flight during nighttime conditions in a multiengine airplane, and a total straight-line distance of more than 100 nautical miles from the original point of departure; and

(v) Three hours in a multiengine airplane within 2 calendar months before the date of the practical test.

(3) For the airline transport pilot certificate, the course requires 25 hours of flight training in a multiengine airplane on the areas of operation under part 141, appendix E, paragraph 4.(c) that includes 15 hours of instrument training. A flight simulator and flight training device cannot be used more than 12.5 hours to meet the training requirements, and use of the flight training device is limited to 6.25 hours of the 12.5 hours permitted.

(c) Course for an additional rotorcraft category and helicopter class rating.

(1) For the recreational pilot certificate, the course requires 15 hours of flight training on the areas of operations under part 141, appendix A, paragraph 4.(c)(2) that includes—

(i) Two hours of flight training to and at an airport that is located more than 25 nautical miles from the airport where the applicant normally trains, with three takeoffs and three landings, except as provided under § 61.100 of this chapter; and

(ii) Three hours of flight training in a rotorcraft category and a helicopter class aircraft within 2 calendar months before the date of the practical test.

(2) For the private pilot certificate, the course requires 20 hours flight training on the areas of operations under part 141, appendix B, paragraph 4.(d)(3). A flight simulator and flight training device cannot be used more than 4 hours to meet the training requirements, and use of the flight training device is limited to 3 hours of the 4 hours permitted. The course must include—

(i) Except as provided under § 61.111 of this chapter, 3 hours of cross country flight training in a helicopter;

(ii) Three hours of nighttime flight training in a helicopter that includes one cross country flight of more than 50 nautical miles total distance, and 10 takeoffs and 10 landings to a full stop (with each landing involving a flight in the traffic pattern) at an airport; and

(iii) Three hours of flight training in a helicopter within 2 calendar months before the date of the practical test.

(3) The commercial pilot certificate level requires 30 hours flight training on the areas of operations under appendix D of part 141, paragraph 4.(d)(3). A flight simulator and flight training device cannot be used more than 9 hours to meet the training requirements, and use of the flight training device is limited to 6 hours of the 9 hours permitted. The course must include—

(i) Five hours on the control and maneuvering of a helicopter solely by reference to instruments, and must include training using a view-limiting device for attitude instrument flying, partial panel skills, recovery from unusual flight attitudes, and intercepting and tracking navigational systems. This aeronautical experience may be performed in an aircraft, flight simulator, flight training device, or an aviation training device;

(ii) One 2-hour cross country flight during daytime conditions in a helicopter, a total straight-line distance of more than 50 nautical miles from the original point of departure;

(iii) One 2-hour cross country flight during nighttime conditions in a helicopter, a total straight-line distance of more than 50 nautical miles from the original point of departure; and

(iv) Three hours in a helicopter within 2 calendar months before the date of the practical test.

(4) For the airline transport pilot certificate, the course requires 25 hours of flight training, including 15 hours of instrument training, in a helicopter on the areas of operation under part 141, appendix E, paragraph 4.(c). A flight simulator and flight training device cannot be used more than 12.5 hours to meet the training requirements, and use of the flight training device is limited to 6.25 hours of the 12.5 hours permitted.

(d) Course for an additional rotorcraft category and a gyroplane class rating.

(1) For the recreational pilot certificate, the course requires 15 hours flight training on the areas of operations under part 141, appendix A, paragraph 4.(c)(3) that includes—

(i) Two hours of flight training to and at an airport that is located more than 25 nautical miles from the airport where the applicant normally trains, with three takeoffs and three landings, except as provided under § 61.100 of this chapter; and

(ii) Three hours of flight training in a gyroplane class within 2 calendar months before the date of the practical test.

(2) For the private pilot certificate, the course requires 20 hours flight training on the areas of operations under part 141, appendix B, paragraph 4.(d)(4). A flight simulator and flight training device cannot be used more than 4 hours to meet the training requirements, and use of the flight training device is limited to 3 hours of the 4 hours permitted. The course must include—

(i) Three hours of cross country flight training in a gyroplane, except as provided under § 61.111 of this chapter;

(ii) Three hours of nighttime flight training in a gyroplane that includes one cross country flight of more than 50 nautical miles total distance, and 10 takeoffs and 10 landings to a full stop (with each landing involving a flight in the traffic pattern) at an airport; and

(iii) Three hours of flight training in a gyroplane within 2 calendar months before the date of the practical test.

(3) For the commercial pilot certificate, the course requires 30 hours flight training on the areas of operations of appendix D to part 141, paragraph 4.(d)(4). A flight simulator and flight training device cannot be used more than 6 hours to meet the training requirements, and use of the flight training device is limited to 6 hours of the 9 hours permitted. The course must include—

(i) 2.5 hours on the control and maneuvering of a gyroplane solely by reference to instruments, and must include training using a view-limiting device for attitude instrument flying, partial panel skills, recovery from unusual flight attitudes, and intercepting and tracking navigational systems. This aeronautical experience may be performed in an aircraft, flight simulator, flight training device, or an aviation training device.

(ii) One 2-hour cross country flight during daytime conditions in a gyroplane, a total straight-line distance of more than 50 nautical miles from the original point of departure;

(iii) Two hours of flight training during nighttime conditions in a gyroplane at an airport, that includes 10 takeoffs and 10 landings to a full stop (with each landing involving a flight in the traffic pattern); and

(iv) Three hours in a gyroplane within 2 calendar months before the date of the practical test.

(e) Course for an additional lighter-than-air category and airship class rating.

(1) For the private pilot certificate, the course requires 20 hours of flight training on the areas of operation under part 141, appendix B, paragraph 4.(d)(7). A flight simulator and flight training device cannot be used more than 4 hours to meet the training requirements, and use of the flight training device is limited to 3 hours of the 4 hours permitted. The course must include—

(i) Three hours of cross country flight training in an airship, except as provided under § 61.111 of this chapter;

(ii) Three hours of nighttime flight training in an airship that includes one cross country flight of more than 25 nautical miles total distance and 5 takeoffs and 5 landings to a full stop (with each landing involving a flight in the traffic pattern) at an airport;

(iii) Three hours of flight training in an airship on the control and maneuvering of an airship solely by reference to instruments, including straight and level flight, constant airspeed climbs and descents, turns to a heading, recovery from unusual flight attitudes, radio communications, and the use of navigation systems/facilities and radar services appropriate to instrument flight; and

(iv) Three hours of flight training in an airship within 2 calendar months before the date of the practical test.

(2) For the commercial pilot certificate, the course requires 55 hours of flight training on the areas of operation under part 141, appendix D, paragraph 4.(d)(7). A flight simulator and flight training device cannot be used more than 16.5 hours to meet the training requirements, and use of the flight training device is limited to 11 hours of the 16.5 hours permitted. The course must include—

(i) Three hours of instrument training in an airship that must include training using a view-limiting device for attitude instrument flying, partial panel skills, recovery from unusual flight attitudes, and intercepting and tracking navigational systems;

(ii) One hour cross country flight during daytime conditions in an airship that consists of, a total straight-line distance of more than 25 nautical miles from the original point of departure;

(iii) One hour cross country flight during nighttime conditions in an airship that consists of a total straight-line distance of more than 25 nautical miles from the original point of departure; and

(iv) Three hours of flight training in an airship within 2 calendar months before the date of the practical test.

(f) Course for an additional lighter-than-air category and a gas balloon class rating.

(1) For the private pilot certificate, the course requires eight hours of flight training that includes 5 training flights on the areas of operations under part 141, appendix B, paragraph 4(d)(8). A flight simulator and flight training device cannot be used more than 1.6 hours to meet the training requirements, and use of the flight training device is limited to 1.2 hours of the 1.6 hours permitted. The course must include—

(i) Two flights of 1 hour each;

(ii) One flight involving a controlled ascent to 3,000 feet above the launch site; and

(iii) Two flights within 2 calendar months before the date of the practical test.

(2) For the commercial pilot certificate, the course requires 10 hours of flight training that includes eight training flights on the areas of operations under part 141, appendix D, paragraph 4(d)(8). A flight simulator and flight training device cannot be used more than 3 hours to meet the training requirements, and use of the flight training device is limited to 2 hours of the 3 hours permitted. The course must include—

(i) Two flights of 1 hour each;

(ii) One flight involving a controlled ascent to 5,000 feet above the launch site; and

(iii) Two flights within 2 calendar months before the date of the practical test.

(g) Course for an additional lighter-than-air category and a hot air balloon class rating.

(1) For the private pilot certificate, the course requires eight hours of flight training that includes 5 training flights on the areas of operations under part 141, appendix B, paragraph 4(d)(8). A flight simulator and flight training device cannot be used more than 1.6 hours to meet the training requirements, and use of the flight training device is limited to 1.2 hours of the 1.6 hours permitted. The course must include—

(i) Two flights of 30 minutes each;

(ii) One flight involving a controlled ascent to 2,000 feet above the launch site; and

(iii) Two flights within 2 calendar months before the date of the practical test.

(2) For the commercial pilot certificate, the course requires 10 hours of flight training that includes eight training flights on the areas of operation under part 141, appendix D, paragraph 4(d)(8). A flight simulator and flight training device cannot be used more than 3 hours to meet the training requirements, and use of the flight training device is limited to 2 hours of the 3 hours permitted. The course must include—

(i) Two flights of 30 minutes each;

(ii) One flight involving a controlled ascent to 3,000 feet above the launch site; and

(iii) Two flights within 2 calendar months before the date of the practical test.

(h) Course for an additional powered-lift category rating.

(1) For the private pilot certificate, the course requires 20 hours flight training on the areas of operations under part 141, appendix B, paragraph 4(d)(5). A flight simulator and flight training device cannot be used more than 4 hours to meet the training requirements, and use of the flight training device is limited to 3 hours of the 4 hours permitted. The course must include—

(i) Three hours of cross country flight training in a powered-lift except as provided under § 61.111 of this chapter;

(ii) Three hours of nighttime flight training in a powered-lift that includes one cross-country flight of more than 100 nautical miles total distance, and 10 takeoffs and 10 landings to a full stop (with each landing involving a flight in the traffic pattern) at an airport;

(iii) Three hours of flight training in a powered-lift on the control and maneuvering of a powered-lift solely by reference to instruments, including straight and level flight, constant airspeed climbs and descents, turns to a heading, recovery from unusual flight attitudes, radio communications, and the use of navigation systems/facilities and radar services appropriate to instrument flight;

(iv) Three hours of flight training in a powered-lift within 2 calendar months before the date of the practical test.

(2) For the commercial pilot certificate, the course requires 55 hours flight training on the areas of operations under part 141, appendix D, paragraph 4(d)(5). A flight simulator and flight training device cannot be used more than 16.5 hours to meet the training requirements, and use of the flight training device is limited to 11 hours of the 16.5 hours permitted. The course includes—

(i) Five hours of instrument training in a powered-lift that must include training using a view-limiting device for attitude instrument flying, partial panel skills, recovery from unusual flight attitudes, and intercepting and tracking navigational systems;

(ii) One 2-hour cross country flight during daytime conditions in a powered-lift, a total straight-line distance of more than 100 nautical miles from the original point of departure;

(iii) One 2-hour cross country flight during nighttime conditions in a powered-lift, a total straight-line distance of more than 100 nautical miles from the original point of departure; and

(iv) Three hours of flight training in a powered-lift within 2 calendar months before the date of the practical test.

(3) For the airline transport pilot certificate, the course requires 25 hours flight training in a powered-lift on the areas of operation under part 141, appendix E, paragraph 4(c) that includes 15 hours of instrument training. A flight simulator and flight training device cannot be used more than 12.5 hours to meet the training requirements, and use of the flight training device is limited to 6.25 hours of the 12.5 hours permitted.

(i) Course for an additional glider category rating.

(1) For the private pilot certificate, the course requires 4 hours of flight training in a glider on the areas of operations under part 141, appendix B, paragraph 4(d)(6). A flight simulator and flight training device cannot be used more than 0.8 hours to meet the training requirements, and use of the flight training device is limited to 0.6 hours of the 0.8 hours permitted. The course must include—

(i) Five training flights in a glider with a certificated flight instructor on the launch/tow procedures approved for the course and on the appropriate approved areas of operation listed under appendix B, paragraph 4(d)(6) of this part; and

(ii) Three training flights in a glider with a certificated flight instructor within 2 calendar months before the date of the practical test.

(2) The commercial pilot certificate level requires 4 hours of flight training in a glider on the areas of operation under part 141, appendix D, paragraph 4.(d)(6). A flight simulator and flight training device cannot be used more than 0.8 hours to meet the training requirements, and use of the flight training device is limited to 0.6 hours of the 0.8 hours permitted. The course must include—

(j) Course for an airplane additional single engine class rating.

(1) For the private pilot certificate, the course requires 3 hours of flight training in the areas of operations under part 141, appendix B, paragraph 4.(d)(1). A flight simulator and flight training device cannot be used more than 0.6 hours to meet the training requirements, and use of the flight training device is limited to 0.4 hours of the 0.6 hours permitted. The course must include—

(i) Three hours of cross country training in a single engine airplane, except as provided under § 61.111 of this chapter;

(ii) Three hours of nighttime flight training in a single engine airplane that includes one cross country flight of more than 100 nautical miles total distance in a single engine airplane and 10 takeoffs and 10 landings to a full stop (with each landing involving a flight in the traffic pattern) at an airport;

(iii) Three hours of flight training in a single engine airplane on the control and maneuvering of a single engine airplane solely by reference to instruments, including straight and level flight, constant airspeed climbs and descents, turns to a heading, recovery from unusual flight attitudes, radio communications, and the use of navigation systems/facilities and radar services appropriate to instrument flight; and

(iv) Three hours of flight training in a single engine airplane within 2 calendar months before the date of the practical test.

(2) For the commercial pilot certificate, the course requires 10 hours of flight training on the areas of operations under part 141, appendix D, paragraph 4.(d)(1).

(i) Five hours of instrument training in a single engine airplane that must include training using a view-limiting device for attitude instrument flying, partial panel skills, recovery from unusual flight attitudes, and intercepting and tracking navigational systems.

(ii) Ten hours of flight training in an airplane that has retractable landing gear, flaps, and a controllable pitch propeller, or is turbine-powered.

(iii) One 2-hour cross country flight during daytime conditions in a single engine airplane and a total straight-line distance of more than 100 nautical miles from the original point of departure;

(iv) One 2-hour cross country flight during nighttime conditions in a single engine airplane and a total straight-line distance of more than 100 nautical miles from the original point of departure; and

(v) Three hours of flight training in a single engine airplane within 2 calendar months before the date of the practical test.

(3) For the airline transport pilot certificate, the course requires 25 hours flight training in a single engine airplane on the areas of operation under appendix E to part 141, paragraph 4.(c), that includes 15 hours of instrument training. A flight simulator and flight training device cannot be used more than 12.5 hours to meet the training requirements, and use of the flight training device is limited to 6.25 hours of the 12.5 hours permitted.

(k) Course for an airplane additional multiengine class rating.

(1) For the private pilot certificate, the course requires 3 hours of flight training on the areas of operations of appendix B to part 141, paragraph 4(d)(2). A flight simulator and flight training device cannot be used more than 0.6 hours to meet the training requirements, and use of the flight training device is limited to 0.4 hours of the 0.6 hours permitted. The course must include—

(i) Three hours of cross country training in a multiengine airplane, except as provided under § 61.111 of this chapter;

(ii) Three hours of nighttime flight training in a multiengine airplane that includes one cross country flight of more than 100 nautical miles total distance in a multiengine airplane, and 10 takeoffs and 10 landings to a full stop (with each landing involving a flight in the traffic pattern) at an airport;

(iii) Three hours of flight training in a multiengine airplane on the control and maneuvering of a multiengine airplane solely by reference to instruments, including straight and level flight, constant airspeed climbs and descents, turns to a heading, recovery from unusual flight attitudes, radio communications, and the use of navigation systems/facilities and radar services appropriate to instrument flight; and

(iv) Three hours of flight training in a multiengine airplane within 2 calendar months before the date of the practical test.

(2) For the commercial pilot certificate, the course requires 10 hours of training on the areas of operations under appendix D of part 141, paragraph 4(d)(2). A flight simulator and flight training device cannot be used more than 3 hours to meet the training requirements, and use of the flight training device is limited to 2 hours of the 3 hours permitted. The course must include—

(i) Five hours of instrument training in a multiengine airplane that must include training using a view-limiting device on for attitude instrument flying, partial panel skills, recovery from unusual flight attitudes, and intercepting and tracking navigational systems;

(ii) Ten hours of training in a multiengine airplane that has retractable landing gear, flaps, and a controllable pitch propeller, or is turbine-powered;

(iii) One 2-hour cross country flight during daytime conditions in a multiengine airplane and, a total straight-line distance of more than 100 nautical miles from the original point of departure;

(iv) One 2-hour cross country flight during nighttime conditions in a multiengine airplane and, a total straight-line distance of more than 100 nautical miles from the original point of departure; and

(iv) Three hours of flight training in a multiengine airplane within 2 calendar months before the date of the practical test.

(3) For the airline transport pilot certificate, the course requires 25 hours of training in a multiengine airplane on the areas of operation of appendix E to part 141, paragraph 4.(c) that includes 15 hours of instrument training. A flight simulator and flight training device cannot be used more than 12.5 hours to meet the training requirements, and use of the flight training device is limited to 6.25 hours of the 12.5 hours permitted.

(l) Course for a rotorcraft additional helicopter class rating.

(1) For the recreational pilot certificate, the course requires 3 hours of flight training on the areas of operations under appendix A of part 141, paragraph 4.(c)(2) that includes—

(i) Two hours of flight training to and at an airport that is located more than 25 nautical miles from the airport where the applicant normally trains, with three takeoffs and three landings, except as provided under § 61.100 of this chapter; and

(ii) Three hours of flight training in a helicopter within 2 calendar months before the date of the practical test.

(2) For the private pilot certificate, the course requires 3 hours flight training on the areas of operations under appendix B of part 141, paragraph 4.(d)(3). A flight simulator and flight training device cannot be used more than 0.6 hours to meet the training requirements, and use of the flight training device is limited to 0.4 hours of the 0.6 hours permitted. The course must include—

(i) Three hours of cross country training in a helicopter, except as provided under § 61.111 of this chapter;

(ii) Three hours of nighttime flight training in a helicopter that includes one cross country flight of more than 50 nautical miles total distance, and 10 takeoffs and 10 landings to a full stop (with each landing involving a flight in the traffic pattern) at an airport; and

(iii) Three hours of flight training in a helicopter within 2 calendar months before the date of the practical test.

(3) For the commercial pilot certificate, the course requires 5 hours flight training on the areas of operations under appendix D of part 141, paragraph 4.(d)(3). Use of a flight simulator and flight training device in the approved training course cannot exceed 1 hour; however, use of the flight training device cannot exceed 0.7 of the one hour. The course must include—

(i) Five hours on the control and maneuvering of a helicopter solely by reference to instruments, and must include training using a view-limiting device for attitude instrument flying, partial panel skills, recovery from unusual flight attitudes, and intercepting and tracking navigational systems. This aeronautical experience may be performed in an aircraft, flight simulator, flight training device, or an aviation training device;

(ii) One 2-hour cross country flight during daytime conditions in a helicopter and, a total straight-line distance of more than 50 nautical miles from the original point of departure;

(iii) One 2-hour cross country flight during nighttime conditions in a helicopter and a total straight-line distance of more than 50 nautical miles from the original point of departure; and

(iv) Three hours of flight training in a helicopter within 2 calendar months before the date of the practical test.

(4) For the airline transport pilot certificate, the course requires 25 hours of flight training in a helicopter on the areas of operation under appendix E of part 141, paragraph 4.(c) that includes 15 hours of instrument training. A flight simulator and flight training device cannot be used more than 12.5 hours to meet the training requirements, and use of the flight training device is limited to 6.25 hours of the 12.5 hours permitted.

(m) Course for a rotorcraft additional gyroplane class rating.

(1) For the recreational pilot certificate, the course requires 3 hours flight training on the areas of operations of appendix A to part 141, paragraph 4.(c)(3) that includes—

(i) Except as provided under § 61.100 of this chapter, 2 hours of flight training to and at an airport that is located more than 25 nautical miles from the airport where the applicant normally trains, with three takeoffs and three landings; and

(ii) Within 2 calendar months before the date of the practical test, 3 hours of flight training in a gyroplane.

(2) For the private pilot certificate, the course requires 3 hours flight training on the areas of operations of appendix B to part 141, paragraph 4.(d)(4). A flight simulator and flight training device cannot be used more than 0.6 hours to meet the training requirements, and use of the flight training device is limited to 0.4 hours of the 0.6 hours permitted. The course must include—

(i) Three hours of cross country training in a gyroplane;

(ii) Three hours of nighttime flight training in a gyroplane that includes one cross country flight of more than 50 nautical miles total distance, and 10 takeoffs and 10 landings to a full stop (with each landing involving a flight in the traffic pattern) at an airport; and

(iii) Three hours of flight training in a gyroplane within 2 calendar months before the date of the practical test.

(3) For the commercial pilot certificate, the course requires 5 hours flight training on the areas of operations of appendix D to part 141, paragraph 4.(d)(4). A flight simulator and flight training device cannot be used more than 1 hour to meet the training requirements, and use of the flight training device is limited to 0.7 hours of the 1 hour permitted. The course must include—

(i) 2.5 hours on the control and maneuvering of a gyroplane solely by reference to instruments, and must include training using a view-limiting device for attitude instrument flying, partial panel skills, recovery from unusual flight attitudes, and intercepting and tracking navigational systems. This aeronautical experience may be performed in an aircraft, flight simulator, flight training device, or an aviation training device.

(ii) Three hours of cross country flight training in a gyroplane, except as provided under § 61.111 of this chapter;

(iii) Two hours of flight training during nighttime conditions in a gyroplane at an airport that includes 10 takeoffs and 10 landings to a full stop (with each landing involving a flight in the traffic pattern); and

(iv) Three hours of flight training in a gyroplane within 2 calendar months before the date of the practical test.

(n) Course for a lighter-than-air additional airship class rating.

(1) For the private pilot certificate, the course requires 20 hours of flight training on the areas of operation under appendix B of part 141, paragraph 4.(d)(7). A flight simulator and flight training device cannot be used more than 4 hours to meet the training requirements, and use of the flight training device is limited to 3 hours of the 4 hours permitted. The course must include—

(i) Three hours of cross country training in an airship, except as provided under § 61.111 of this chapter;

(ii) Three hours of nighttime flight training in an airship that includes one cross country flight of more than 25 nautical miles total distance, and 5 takeoffs and 5 landings to a full stop (with each landing involving a flight in the traffic pattern) at an airport;

(iii) Three hours of flight training in an airship on the control and maneuvering of an airship solely by reference to instruments, including straight and level flight, constant airspeed climbs and descents, turns to a heading, recovery from unusual flight attitudes, radio communications, and the use of navigation systems/facilities and radar services appropriate to instrument flight; and

(iv) Three hours of flight training in an airship within 2 calendar months before the date of the practical test.

(2) For the commercial pilot certificate, the course requires 55 hours of flight training on the areas of operation under appendix D of part 141, paragraph 4.(d)(7). A flight simulator and flight training device cannot be used more than 16.5 hours to meet the training requirements, and use of the flight training device is limited to 11 hours of the 16.5 hours permitted. The course must include—

(i) Three hours of instrument training in an airship that must include training using a view-limiting device for attitude instrument flying, partial panel skills, recovery from unusual flight attitudes, and intercepting and tracking navigational systems;

(ii) One hour cross country flight during daytime conditions in an airship that consists of a total straight-line distance of more than 25 nautical miles from the original point of departure;

(iii) One hour cross country flight during nighttime conditions in an airship that consists of a total straight-line distance of more than 25 nautical miles from the original point of departure; and

(iv) Three hours of flight training in an airship within 2 calendar months before the date of the practical test.

(o) Course for a lighter-than-air additional gas balloon class rating.

(1) For the private pilot certificate, the course requires eight hours of flight training that includes 5 training flights on the areas of operations under appendix B of part 141, paragraph 4.(d)(8). A flight simulator and flight training device cannot be used more than 1.6 hours to meet the training requirements, and use of the flight training device is limited to 1.2 hours of the 1.6 hours permitted. The course must include—

(i) Two flights of 1 hour each;

(ii) One flight involving a controlled ascent to 3,000 feet above the launch site; and

(iii) Two flights within 2 calendar months before the date of the practical test.

(2) For the commercial pilot certificate, the course requires 10 hours of flight training that includes eight training flights on the areas of operations of appendix D to part 141, paragraph 4.(d)(8). A flight simulator and flight training device cannot be used more than 3 hours to meet the training requirements, and use of the flight training device is limited to 2 hours of the 3 hours permitted. The course must include—

(i) Two flights of 1 hour each;

(ii) One flight involving a controlled ascent to 5,000 feet above the launch site; and

(iii) Two flights within 2 calendar months before the date of the practical test.

(p) Course for a lighter-than-air additional hot air balloon class rating.

(1) For the private pilot certificate, the course requires 8 hours of flight training that includes 5 training flights on the areas of operations of appendix B to part 141, paragraph 4.(d)(8). A flight simulator and flight training device cannot be used more than 1.6 hours to meet the training requirements, and use of the flight training device is limited to 1.2 hours of the 1.6 hours permitted. The course must include—

(i) Two flights of 30 minutes each;

(ii) One flight involving a controlled ascent to 2,000 feet above the launch site; and

(iii) Two flights within 2 calendar months before the date of the practical test.

(2) For the commercial pilot certificate, the course requires 10 hours of flight training that includes eight training flight on the areas of operation of appendix D to part 141, paragraph 4.(d)(8). A flight simulator and flight training device cannot be used more than 3 hours to meet the training requirements, and use of the flight training device is limited to 2 hours of the 3 hours permitted. The course must include—

(i) Two flights of 30 minutes each.

(ii) One flight involving a controlled ascent to 3,000 feet above the launch site; and

(iii) Two flights within 2 calendar months before the date of the practical test.

5. *Stage checks and end-of-course tests.* (a) Each student enrolled in an additional aircraft category rating course or an additional aircraft class rating course must satisfactorily accomplish the stage checks and end-of-course tests, in accordance with the school's approved training course, consisting of the approved areas of operation in section No. 4 of this appendix that are appropriate to the aircraft category and class rating for which the course applies at the appropriate pilot certificate level.

(b) Each student must demonstrate satisfactory proficiency prior to receiving an endorsement to operate an aircraft in solo flight.

[Doc. No. 25910, 62 FR 16347, Apr. 4, 1997; Amdt. 141-9, 62 FR 40909, July 30, 1997; Amdt. 141-12, 74 FR 42566, Aug. 21, 2009]

Appendix J to Part 141—Aircraft Type Rating Course, For Other Than an Airline Transport Pilot Certificate

1. *Applicability.* This appendix prescribes the minimum curriculum for an aircraft type rating course other than an airline transport pilot certificate, for:

(a) A type rating in an airplane category—single-engine class.

(b) A type rating in an airplane category—multiengine class.

(c) A type rating in a rotorcraft category—helicopter class.

(d) A type rating in a powered-lift category.

(e) Other aircraft type ratings specified by the Administrator through the aircraft type certificate procedures.

2. *Eligibility for enrollment.* Prior to enrolling in the flight portion of an aircraft type rating course, a person must hold at least a private pilot certificate and:

(a) An instrument rating in the category and class of aircraft that is appropriate to the aircraft type rating for which the course applies, provided the aircraft's type certificate does not have a VFR limitation; or

(b) Be concurrently enrolled in an instrument rating course in the category and class of aircraft that is appropriate to the aircraft type rating for which the course applies, and pass the required instrument rating practical test concurrently with the aircraft type rating practical test.

3. *Aeronautical knowledge training.* (a) Each approved course must include at least 10 hours of ground training on the aeronautical knowledge areas listed in paragraph (b) of this section, appropriate to the aircraft type rating for which the course applies.

(b) Ground training must include the following aeronautical areas:

(1) Proper control of airspeed, configuration, direction, altitude, and attitude in accordance with procedures and limitations contained in the aircraft's flight manual, checklists, or other approved material appropriate to the aircraft type;

(2) Compliance with approved en route, instrument approach, missed approach, ATC, or other applicable procedures that apply to the aircraft type;

(3) Subjects requiring a practical knowledge of the aircraft type and its powerplant, systems, components, operational, and performance factors;

(4) The aircraft's normal, abnormal, and emergency procedures, and the operations and limitations relating thereto;

(5) Appropriate provisions of the approved aircraft's flight manual;

(6) Location of and purpose for inspecting each item on the aircraft's checklist that relates to the exterior and interior preflight; and

(7) Use of the aircraft's prestart checklist, appropriate control system checks, starting procedures, radio and electronic equipment checks, and the selection of proper navigation and communication radio facilities and frequencies.

4. *Flight training.* (a) Each approved course must include at least:

(1) Flight training on the approved areas of operation of paragraph (c) of this section in the aircraft type for which the course applies; and

(2) 10 hours of training of which at least 5 hours must be instrument training in the aircraft for which the course applies.

(b) For the use of flight simulators or flight training devices:

(1) The course may include training in a flight simulator or flight training device, provided it is representative of the aircraft for which the course is approved, meets requirements of this paragraph, and the training is given by an authorized instructor.

(2) Training in a flight simulator that meets the requirements of § 141.41(a) of this part, may be credited for a maximum of 50 percent of the total flight training hour

requirements of the approved course, or of this section, whichever is less.

(3) Training in a flight training device that meets the requirements of § 141.41(b) of this part, may be credited for a maximum of 25 percent of the total flight training hour requirements of the approved course, or of this section, whichever is less.

(4) Training in the flight simulators or flight training devices described in paragraphs (b)(2) and (b)(3) of this section, if used in combination, may be credited for a maximum of 50 percent of the total flight training hour requirements of the approved course, or of this section, whichever is less. However, credit training in a flight training device that meets the requirements of § 141.41(b) cannot exceed the limitation provided for in paragraph (b)(3) of this section.

(c) Each approved course must include the flight training on the areas of operation listed in this paragraph, that are appropriate to the aircraft category and class rating for which the course applies:

(1) *A type rating for an airplane—single-engine course:* (i) Preflight preparation;
(ii) Preflight procedures;
(iii) Takeoff and departure phase;
(iv) In-flight maneuvers;
(v) Instrument procedures;
(vi) Landings and approaches to landings;
(vii) Normal and abnormal procedures;
(viii) Emergency procedures; and
(ix) Postflight procedures.

(2) *A type rating for an airplane—multiengine course:* (i) Preflight preparation;
(ii) Preflight procedures;
(iii) Takeoff and departure phase;
(iv) In-flight maneuvers;
(v) Instrument procedures;
(vi) Landings and approaches to landings;
(vii) Normal and abnormal procedures;
(viii) Emergency procedures; and
(ix) Postflight procedures.

(3) *A type rating for a powered-lift course:* (i) Preflight preparation;
(ii) Preflight procedures;
(iii) Takeoff and departure phase;
(iv) In-flight maneuvers;
(v) Instrument procedures;
(vi) Landings and approaches to landings;
(vii) Normal and abnormal procedures;
(viii) Emergency procedures; and
(ix) Postflight procedures.

(4) *A type rating for a rotorcraft—helicopter course:* (i) Preflight preparation;
(ii) Preflight procedures;
(iii) Takeoff and departure phase;
(iv) In-flight maneuvers;
(v) Instrument procedures;
(vi) Landings and approaches to landings;
(vii) Normal and abnormal procedures;
(viii) Emergency procedures; and
(ix) Postflight procedures.

(5) *Other aircraft type ratings specified by the Administrator through aircraft type certificate procedures:* (i) Preflight preparation;
(ii) Preflight procedures;
(iii) Takeoff and departure phase;
(iv) In-flight maneuvers;
(v) Instrument procedures;
(vi) Landings and approaches to landings;
(vii) Normal and abnormal procedures;
(viii) Emergency procedures; and
(ix) Postflight procedures.

5. *Stage checks and end-of-course tests.* (a) Each student enrolled in an aircraft type rating course must satisfactorily accomplish the stage checks and end-of-course tests, in accordance with the school's approved training course, consisting of the approved areas of operation that are appropriate to the aircraft type rating for which the course applies at the airline transport pilot certificate level; and

(b) Each student must demonstrate satisfactory proficiency prior to receiving an endorsement to operate an aircraft in solo flight.

[Doc. No. 25910, 62 FR 16347, Apr. 4, 1997; Amdt. 141-9, 62 FR 40910, July 30, 1997]

APPENDIX K TO PART 141—SPECIAL PREPARATION COURSES

1. *Applicability.* This appendix prescribes the minimum curriculum for the special preparation courses that are listed in § 141.11 of this part.

2. *Eligibility for enrollment.* Prior to enrolling in the flight portion of a special preparation course, a person must hold a pilot certificate, flight instructor certificate, or ground instructor certificate that is appropriate for the exercise of the operating privileges or authorizations sought.

3. *General requirements.* (a) To be approved, a special preparation course must:

(1) Meet the appropriate requirements of this appendix; and

(2) Prepare the graduate with the necessary skills, competency, and proficiency to exercise safely the privileges of the certificate, rating, or authorization for which the course is established.

(b) An approved special preparation course must include ground and flight training on the operating privileges or authorization sought, for developing competency, proficiency, resourcefulness, self-confidence, and self-reliance in the student.

4. *Use of flight simulators or flight training devices.* (a) The approved special preparation course may include training in a flight simulator or flight training device, provided it is representative of the aircraft for which the course is approved, meets requirements of this paragraph, and the training is given by an authorized instructor.

(b) Except for the airline transport pilot certification program in paragraph 13 of this appendix, training in a flight simulator that meets the requirements of § 141.41(a) of this part, may be credited for a maximum of 10 percent of the total flight training hour requirements of the approved course, or of this section, whichever is less.

(c) Except for the airline transport pilot certification program in paragraph 13 of this appendix, training in a flight training device that meets the requirements of § 141.41(b) of this part, may be credited for a maximum of 5 percent of the total flight training hour requirements of the approved course, or of this section, whichever is less.

(d) Training in the flight simulators or flight training devices described in paragraphs (b) and (c) of this section, if used in combination, may be credited for a maximum of 10 percent of the total flight training hour requirements of the approved course, or of this section, whichever is less. However, credit for training in a flight training device that meets the requirements of § 141.41(b) cannot exceed the limitation provided for in paragraph (c) of this section.

5. *Stage check and end-of-course tests.* Each person enrolled in a special preparation course must satisfactorily accomplish the stage checks and end-of-course tests, in accordance with the school's approved training course, consisting of the approved areas of operation that are appropriate to the operating privileges or authorization sought, and for which the course applies.

6. *Agricultural aircraft operations course.* An approved special preparation course for pilots in agricultural aircraft operations must include at least the following—

(a) 25 hours of training on:

(1) Agricultural aircraft operations;

(2) Safe piloting and operating practices and procedures for handling, dispensing, and disposing agricultural and industrial chemicals, including operating in and around congested areas; and

(3) Applicable provisions of part 137 of this chapter.

(b) 15 hours of flight training on agricultural aircraft operations.

7. *Rotorcraft external-load operations course.* An approved special preparation course for pilots of external-load operations must include at least the following—

(a) 10 hours of training on:

(1) Rotorcraft external-load operations;

(2) Safe piloting and operating practices and procedures for external-load operations, including operating in and around congested areas; and

(3) Applicable provisions of part 133 of this chapter.

(b) 15 hours of flight training on external-load operations.

8. *Test pilot course.* An approved special preparation course for pilots in test pilot duties must include at least the following—

(a) Aeronautical knowledge training on:

(1) Performing aircraft maintenance, quality assurance, and certification test flight operations;

(2) Safe piloting and operating practices and procedures for performing aircraft maintenance, quality assurance, and certification test flight operations;

(3) Applicable parts of this chapter that pertain to aircraft maintenance, quality assurance, and certification tests; and

(4) Test pilot duties and responsibilities.

(b) 15 hours of flight training on test pilot duties and responsibilities.

9. *Special operations course.* An approved special preparation course for pilots in special operations that are mission-specific for certain aircraft must include at least the following—

(a) Aeronautical knowledge training on:

(1) Performing that special flight operation;

(2) Safe piloting operating practices and procedures for performing that special flight operation;

(3) Applicable parts of this chapter that pertain to that special flight operation; and

(4) Pilot in command duties and responsibilities for performing that special flight operation.

(b) Flight training:

(1) On that special flight operation; and

(2) To develop skills, competency, proficiency, resourcefulness, self-confidence, and self-reliance in the student for performing that special flight operation in a safe manner.

10. *Pilot refresher course.* An approved special preparation pilot refresher course for a pilot certificate, aircraft category and class rating, or an instrument rating must include at least the following—

(a) 4 hours of aeronautical knowledge training on:

(1) The aeronautical knowledge areas that are applicable to the level of pilot certificate, aircraft category and class rating, or instrument rating, as appropriate, that pertain to that course;

(2) Safe piloting operating practices and procedures; and

(3) Applicable provisions of parts 61 and 91 of this chapter for pilots.

(b) 6 hours of flight training on the approved areas of operation that are applicable to the level of pilot certificate, aircraft category and class rating, or instrument rating, as appropriate, for performing pilot-in-command duties and responsibilities.

11. *Flight instructor refresher course.* An approved special preparation flight instructor refresher course must include at least a combined total of 16 hours of aeronautical knowledge training, flight training, or any combination of ground and flight training on the following—

(a) Aeronautical knowledge training on:

(1) The aeronautical knowledge areas of part 61 of this chapter that apply to student, recreational, private, and commercial pilot certificates and instrument ratings;

(2) The aeronautical knowledge areas of part 61 of this chapter that apply to flight instructor certificates;

(3) Safe piloting operating practices and procedures, including airport operations and operating in the National Airspace System; and

(4) Applicable provisions of parts 61 and 91 of this chapter that apply to pilots and flight instructors.

(b) Flight training to review:

(1) The approved areas of operations applicable to student, recreational, private, and commercial pilot certificates and instrument ratings; and

(2) The skills, competency, and proficiency for performing flight instructor duties and responsibilities.

12. *Ground instructor refresher course.* An approved special preparation ground instructor refresher course must include at least 16 hours of aeronautical knowledge training on:

(a) The aeronautical knowledge areas of part 61 of this chapter that apply to student, recreational, private, and commercial pilots and instrument rated pilots;

(b) The aeronautical knowledge areas of part 61 of this chapter that apply to ground instructors;

(c) Safe piloting operating practices and procedures, including airport operations and operating in the National Airspace System; and

(d) Applicable provisions of parts 61 and 91 of this chapter that apply to pilots and ground instructors.

13. Airline transport pilot certification training program. An approved airline transport pilot certification training program must include the academic and FSTD training set forth in § 61.156 of this chapter. The FAA will not approve a course with fewer hours than those prescribed in § 61.156 of this chapter.

[Doc. No. 25910, 62 FR 16347, Apr. 4, 1997; Amdt. 141-9, 62 FR 40910, July 30, 1997, as amended by Amdt. 141-1, 78 FR 42380, July 15, 2013]

APPENDIX L TO PART 141—PILOT GROUND SCHOOL COURSE

1. *Applicability.* This appendix prescribes the minimum curriculum for a pilot ground school course required under this part.

2. *General requirements.* An approved course of training for a pilot ground school must include training on the aeronautical knowledge areas that are:

(a) Needed to safely exercise the privileges of the certificate, rating, or authority for which the course is established; and

(b) Conducted to develop competency, proficiency, resourcefulness, self-confidence, and self-reliance in each student.

3. *Aeronautical knowledge training requirements.* Each approved pilot ground school course must include:

(a) The aeronautical knowledge training that is appropriate to the aircraft rating and pilot certificate level for which the course applies; and

(b) An adequate number of total aeronautical knowledge training hours appropriate to the aircraft rating and pilot certificate level for which the course applies.

4. *Stage checks and end-of-course tests.* Each person enrolled in a pilot ground school course must satisfactorily accomplish the stage checks and end-of-course tests, in accordance with the school's approved training course, consisting of the approved areas of operation that are appropriate to the operating privileges or authorization that graduation from the course will permit and for which the course applies.

Appendix M to Part 141—Combined Private Pilot Certification and Instrument Rating Course

1. *Applicability.* This appendix prescribes the minimum curriculum for a combined private pilot certification and instrument rating course required under this part, for the following ratings:

(a) Airplane.

(1) Airplane single-engine.

(2) Airplane multiengine.

(b) Rotorcraft helicopter.

(c) Powered-lift.

2. *Eligibility for enrollment.* A person must hold a sport pilot, recreational, or student pilot certificate prior to enrolling in the flight portion of a combined private pilot certification and instrument rating course.

3. *Aeronautical knowledge training.*

(a) Each approved course must include at least 65 hours of ground training on the aeronautical knowledge areas listed in paragraph (b) of this section that are appropriate to the aircraft category and class rating of the course:

(b) Ground training must include the following aeronautical knowledge areas:

(1) Applicable Federal Aviation Regulations for private pilot privileges, limitations, flight operations, and instrument flight rules (IFR) flight operations.

(2) Accident reporting requirements of the National Transportation Safety Board.

(3) Applicable subjects of the "Aeronautical Information Manual" and the appropriate FAA advisory circulars.

(4) Aeronautical charts for visual flight rules (VFR) navigation using pilotage, dead reckoning, and navigation systems.

(5) Radio communication procedures.

(6) Recognition of critical weather situations from the ground and in flight, windshear avoidance, and the procurement and use of aeronautical weather reports and forecasts.

(7) Safe and efficient operation of aircraft under instrument flight rules and conditions.

(8) Collision avoidance and recognition and avoidance of wake turbulence.

(9) Effects of density altitude on takeoff and climb performance.

(10) Weight and balance computations.

(11) Principles of aerodynamics, powerplants, and aircraft systems.

(12) If the course of training is for an airplane category, stall awareness, spin entry, spins, and spin recovery techniques.

(13) Air traffic control system and procedures for instrument flight operations.

(14) IFR navigation and approaches by use of navigation systems.

(15) Use of IFR en route and instrument approach procedure charts.

(16) Aeronautical decision making and judgment.

(17) Preflight action that includes—

(i) How to obtain information on runway lengths at airports of intended use, data on takeoff and landing distances, weather reports and forecasts, and fuel requirements.

(ii) How to plan for alternatives if the planned flight cannot be completed or delays are encountered.

(iii) Procurement and use of aviation weather reports and forecasts, and the elements of forecasting weather trends on the basis of that information and personal observation of weather conditions.

4. *Flight training.*

(a) Each approved course must include at least 70 hours of training, as described in section 4 and section 5 of this appendix, on the approved areas of operation listed in paragraph (d) of section 4 of this appendix that are appropriate to the aircraft category and class rating of the course:

(b) Each approved course must include at least the following flight training:

(1) *For an airplane single engine course:* 70 hours of flight training from an authorized instructor on the approved areas of operation in paragraph (d)(1) of this section that includes at least—

(i) Except as provided in § 61.111 of this chapter, 3 hours of cross-country flight training in a single engine airplane.

(ii) 3 hours of night flight training in a single-engine airplane that includes—

(A) One cross-country flight of more than 100 nautical miles total distance.

(B) 10 takeoffs and 10 landings to a full stop (with each landing involving a flight in the traffic pattern) at an airport.

(iii) 35 hours of instrument flight training in a single-engine airplane that includes at least one cross-country flight that is performed under IFR and—

(A) Is a distance of at least 250 nautical miles along airways or air traffic control-directed (ATC-directed) routing with one segment of the flight consisting of at least a straight-line distance of 100 nautical miles between airports.

(B) Involves an instrument approach at each airport.

(C) Involves three different kinds of approaches with the use of navigation systems.

(iv) 3 hours of flight training in a single-engine airplane in preparation for the practical test within 60 days preceding the date of the test.

(2) *For an airplane multiengine course:* 70 hours of training from an authorized instructor on the approved areas of operation in paragraph (d)(2) of this section that includes at least—

(i) Except as provided in § 61.111 of this chapter, 3 hours of cross-country flight training in a multiengine airplane.

(ii) 3 hours of night flight training in a multiengine airplane that includes—

(A) One cross-country flight of more than 100 nautical miles total distance.

(B) 10 takeoffs and 10 landings to a full stop (with each landing involving a flight in the traffic pattern) at an airport.

(iii) 35 hours of instrument flight training in a multiengine airplane that includes at least one cross-country flight that is performed under IFR and—

(A) Is a distance of at least 250 nautical miles along airways or ATC-directed routing with one segment of the flight consisting of at least a straight-line distance of 100 nautical miles between airports.

(B) Involves an instrument approach at each airport.

(C) Involves three different kinds of approaches with the use of navigation systems.

(iv) 3 hours of flight training in a multiengine airplane in preparation for the practical test within 60 days preceding the date of the test.

(3) *For a rotorcraft helicopter course:* 70 hours of training from an authorized instructor on the approved areas of operation in paragraph (d)(3) of this section that includes at least—

(i) Except as provided in § 61.111 of this chapter, 3 hours of cross-country flight training in a helicopter.

(ii) 3 hours of night flight training in a helicopter that includes—

(A) One cross-country flight of more than 50 nautical miles total distance.

(B) 10 takeoffs and 10 landings to a full stop (with each landing involving a flight in the traffic pattern) at an airport.

(iii) 35 hours of instrument flight training in a helicopter that includes at least one cross-country flight that is performed under IFR and—

(A) Is a distance of at least 100 nautical miles along airways or ATC-directed routing with one segment of the flight consisting of at least a straight-line distance of 50 nautical miles between airports.

(B) Involves an instrument approach at each airport.

(C) Involves three different kinds of approaches with the use of navigation systems.

(iv) 3 hours of flight training in a helicopter in preparation for the practical test within 60 days preceding the date of the test.

(4) *For a powered-lift course:* 70 hours of training from an authorized instructor on the approved areas of operation in paragraph (d)(4) of this section that includes at least—

(i) Except as provided in § 61.111 of this chapter, 3 hours of cross-country flight training in a powered-lift.

(ii) 3 hours of night flight training in a powered-lift that includes—

(A) One cross-country flight of more than 100 nautical miles total distance.

(B) 10 takeoffs and 10 landings to a full stop (with each landing involving a flight in the traffic pattern) at an airport.

(iii) 35 hours of instrument flight training in a powered-lift that includes at least one cross-country flight that is performed under IFR and—

(A) Is a distance of at least 250 nautical miles along airways or ATC-directed routing with one segment of the flight consisting of at least a straight-line distance of 100 nautical miles between airports.

(B) Involves an instrument approach at each airport.

(C) Involves three different kinds of approaches with the use of navigation systems.

(iv) 3 hours of flight training in a powered-lift in preparation for the practical test, within 60 days preceding the date of the test.

(c) For use of flight simulators or flight training devices:

(1) The course may include training in a combination of flight simulators, flight training devices, and aviation training device, provided it is representative of the aircraft for which the course is approved, meets the requirements of this section, and the training is given by an authorized instructor.

(2) Training in a flight simulator that meets the requirements of § 141.41(a) of this part may be credited for a maximum of 35 percent of the total flight training hour requirements of the approved course, or of this section, whichever is less.

(3) Training in a flight training device or aviation training device that meets the requirements of § 141.41(b) of this part may be credited for a maximum of 25 percent of the total flight training hour requirements of the approved course, or of this section, whichever is less.

(4) Training in a combination of flight simulators, flight training devices, or aviation training devices, described in paragraphs (c)(2) and (c)(3) of this section, may be credited for a maximum of 35 percent of the total flight training hour requirements of the approved course, or of this section, whichever is less. However, credit for training in a flight training device and aviation training device, that meets the requirements of § 141.41(b), cannot exceed the limitation provided for in paragraph (c)(3) of this section.

(d) Each approved course must include the flight training on the approved areas of operation listed in this section that are appropriate to the aircraft category and class rating course—

(1) *For a combined private pilot certification and instrument rating course involving a single-engine airplane:*

(i) Preflight preparation.

(ii) Preflight procedures.

(iii) Airport and seaplane base operations.

(iv) Takeoffs, landings, and go-arounds.

(v) Performance maneuvers.

(vi) Ground reference maneuvers.

(vii) Navigation and navigation systems.

(viii) Slow flight and stalls.

(ix) Basic instrument maneuvers and flight by reference to instruments.

(x) Instrument approach procedures.

(xi) Air traffic control clearances and procedures.

(xii) Emergency operations.

(xiii) Night operations.

(xiv) Postflight procedures.

(2) *For a combined private pilot certification and instrument rating course involving a multiengine airplane:*

(i) Preflight preparation.

(ii) Preflight procedures.

(iii) Airport and seaplane base operations.

(iv) Takeoffs, landings, and go-arounds.

(v) Performance maneuvers.

(vi) Ground reference maneuvers.

(vii) Navigation and navigation systems.

(viii) Slow flight and stalls.

(ix) Basic instrument maneuvers and flight by reference to instruments.

(x) Instrument approach procedures.

(xi) Air traffic control clearances and procedures.

(xii) Emergency operations.

(xiii) Multiengine operations.

(xiv) Night operations.

(xv) Postflight procedures.

(3) *For a combined private pilot certification and instrument rating course involving a rotorcraft helicopter:*
(i) Preflight preparation.
(ii) Preflight procedures.
(iii) Airport and heliport operations.
(iv) Hovering maneuvers.
(v) Takeoffs, landings, and go-arounds.
(vi) Performance maneuvers.
(vii) Navigation and navigation systems.
(viii) Basic instrument maneuvers and flight by reference to instruments.
(ix) Instrument approach procedures.
(x) Air traffic control clearances and procedures.
(xi) Emergency operations.
(xii) Night operations.
(xiii) Postflight procedures.
(4) *For a combined private pilot certification and instrument rating course involving a powered-lift:*
(i) Preflight preparation.
(ii) Preflight procedures.
(iii) Airport and heliport operations.
(iv) Hovering maneuvers.
(v) Takeoffs, landings, and go-arounds.
(vi) Performance maneuvers.
(vii) Ground reference maneuvers.
(viii) Navigation and navigation systems.
(ix) Slow flight and stalls.
(x) Basic instrument maneuvers and flight by reference to instruments.
(xi) Instrument approach procedures.
(xii) Air traffic control clearances and procedures.
(xiii) Emergency operations.
(xiv) Night operations.
(xv) Postflight procedures.

5. *Solo flight training.* Each approved course must include at least the following solo flight training:

(a) *For a combined private pilot certification and instrument rating course involving an airplane single engine:* Five hours of flying solo in a single-engine airplane on the appropriate areas of operation in paragraph (d)(1) of section 4 of this appendix that includes at least—

(1) One solo cross-country flight of at least 100 nautical miles with landings at a minimum of three points, and one segment of the flight consisting of a straight-line distance of at least 50 nautical miles between the takeoff and landing locations.

(2) Three takeoffs and three landings to a full stop (with each landing involving a flight in the traffic pattern) at an airport with an operating control tower.

(b) *For a combined private pilot certification and instrument rating course involving an airplane multiengine:* Five hours of flying solo in a multiengine airplane or 5 hours of performing the duties of a pilot in command while under the supervision of an authorized instructor. The training must consist of the appropriate areas of operation in paragraph (d)(2) of section 4 of this appendix, and include at least—

(1) One cross-country flight of at least 100 nautical miles with landings at a minimum of three points, and one segment of the flight consisting of a straight-line distance of at least 50 nautical miles between the takeoff and landing locations.

(2) Three takeoffs and three landings to a full stop (with each landing involving a flight in the traffic pattern) at an airport with an operating control tower.

(c) *For a combined private pilot certification and instrument rating course involving a helicopter:* Five hours of flying solo in a helicopter on the appropriate areas of operation in paragraph (d)(3) of section 4 of this appendix that includes at least—

(1) One solo cross-country flight of more than 50 nautical miles with landings at a minimum of three points, and one segment of the flight consisting of a straight-line distance of at least 25 nautical miles between the takeoff and landing locations.

(2) Three takeoffs and three landings to a full stop (with each landing involving a flight in the traffic pattern) at an airport with an operating control tower.

(d) *For a combined private pilot certification and instrument rating course involving a powered-lift:* Five hours of flying solo in a powered-lift on the appropriate areas of operation in paragraph (d)(4) of section 4 of this appendix that includes at least—

(1) One solo cross-country flight of at least 100 nautical miles with landings at a minimum of three points, and one segment of the flight consisting of a straight-line distance of at least 50 nautical miles between the takeoff and landing locations.

(2) Three takeoffs and three landings to a full stop (with each landing involving a flight in the traffic pattern) at an airport with an operating control tower.

6. *Stage checks and end-of-course tests.*

(a) Each student enrolled in a private pilot course must satisfactorily accomplish the stage checks and end-of-course tests in accordance with the school's approved training course that consists of the approved areas of operation listed in paragraph (d) of section 4 of this appendix that are appropriate to the aircraft category and class rating for which the course applies.

(b) Each student must demonstrate satisfactory proficiency prior to receiving an endorsement to operate an aircraft in solo flight.

[Doc. No. FAA-2008-0938, 76 FR 54108, Aug. 31, 2011]

PART 142

TRAINING CENTERS

Contents

Subpart A—General

Subpart B—Aircrew Curriculum and Syllabus Requirements

Subpart C—Personnel and Flight Training Equipment Requirements

Subpart D—Operating Rules

Subpart E—Recordkeeping

Subpart F—Other Approved Courses

Authority: 49 U.S.C. 106(f), 106(g), 40113, 40119, 44101, 44701-44703, 44705, 44707, 44709-44711, 45102-45103, 45301-45302.

Source: Docket No. 26933, 61 FR 34562, July 2, 1996, unless otherwise noted.

Subpart A—General

§ 142.1 Applicability.

(a) This subpart prescribes the requirements governing the certification and operation of training centers. Except as provided in paragraph (b) of this section, this part provides an alternative means to accomplish training required by parts 61, 63, 65, 91, 121, 125, 135, or 137 of this chapter.

(b) Certification under this part is not required for training that is—

(1) Approved under the provisions of parts 63, 91, 121, 127, 135, or 137 of this chapter;

(2) Approved under subpart Y of part 121 of this chapter, Advanced Qualification Programs, for the authorization holder's own employees;

(3) Conducted under part 61 unless that part requires certification under this part;

(4) Conducted by a part 121 certificate holder for another part 121 certificate holder;

(5) Conducted by a part 135 certificate holder for another part 135 certificate holder; or

(6) Conducted by a part 91 fractional ownership program manager for another part 91 fractional ownership program manager.

(c) Except as provided in paragraph (b) of this section, after August 3, 1998, no person may conduct training, testing, or checking in advanced flight training devices or flight simulators without, or in violation of, the certificate and training specifications required by this part.

[Doc. No. 26933, 61 FR 34562, July 2, 1996, as amended by Amdt. 142-4, 66 FR 21067, Apr. 27, 2001; Amdt. 142-5, 68 FR 54588, Sept. 17, 2003; Amdt. 142-9, 78 FR 42380, July 15, 2013]

§ 142.3 Definitions.

As used in this part:

Advanced Flight Training Device as used in this part, means a flight training device as defined in part 61 of this chapter that has a cockpit that accurately replicates a specific make, model, and type aircraft cockpit, and handling characteristics that accurately model the aircraft handling characteristics.

Core Curriculum means a set of courses approved by the Administrator, for use by a training center and its satellite training centers. The core curriculum consists of training which is required for certification. It does not include training for tasks and circumstances unique to a particular user.

Course means—

(1) A program of instruction to obtain pilot certification, qualification, authorization, or currency;

(2) A program of instruction to meet a specified number of requirements of a program for pilot training, certification, qualification, authorization, or currency; or

(3) A curriculum, or curriculum segment, as defined in subpart Y of part 121 of this chapter.

Courseware means instructional material developed for each course or curriculum, including lesson plans, flight event descriptions, computer software programs, audiovisual programs, workbooks, and handouts.

Evaluator means a person employed by a training center certificate holder who performs tests for certification, added ratings, authorizations, and proficiency checks that are authorized by the certificate holder's training specification, and who is authorized by the Administrator to administer such checks and tests.

Flight training equipment means full flight simulators, as defined in § 1.1 of this chapter, flight training devices, as defined in § 1.1 of this chapter, and aircraft.

Instructor means a person employed by a training center and designated to provide instruction in accordance with subpart C of this part.

Line-Operational Simulation means simulation conducted using operational-oriented flight scenarios that accurately replicate interaction among flightcrew members and between flightcrew members and dispatch facilities, other crewmembers, air traffic control, and ground operations. Line operational simulation simulations are conducted for training and evaluation purposes and include random, abnormal, and emergency occurrences. Line operational simulation specifically includes line-oriented flight training, special purpose operational training, and line operational evaluation.

Specialty Curriculum means a set of courses that is designed to satisfy a requirement of the Federal Aviation Regulations and that is approved by the Administrator for use by a particular training center or satellite training center. The specialty curriculum includes training requirements unique to one or more training center clients.

Training center means an organization governed by the applicable requirements of this part that provides training, testing, and checking under contract or other arrangement to airmen subject to the requirements of this chapter.

Training program consists of courses, courseware, facilities, flight training equipment, and personnel necessary to accomplish a specific training objective. It may include a core curriculum and a specialty curriculum.

Training specifications means a document issued to a training center certificate holder by the Administrator that prescribes that center's training, checking, and testing authorizations and limitations, and specifies training program requirements.

[Doc. No. 26933, 61 FR 34562, July 2, 1996, as amended by Amdt. 142-2, 62 FR 68137, Dec. 30, 1997; Amdt. 142-7, 76 FR 54110, Aug. 31, 2011; Amdt. 142-9, 78 FR 42380, July 15, 2013]

§ 142.5 Certificate and training specifications required.

(a) No person may operate a certificated training center without, or in violation of, a training center certificate and training specifications issued under this part.

(b) An applicant will be issued a training center certificate and training specifications with appropriate limitations if the applicant shows that it has adequate facilities, equipment, personnel, and courseware required by § 142.11 to conduct training approved under § 142.37.

§ 142.7 Duration of a certificate.

(a) Except as provided in paragraph (b) of this section, a training center certificate issued under this part is effective until the certificate is surrendered or until the Administrator suspends, revokes, or terminates it.

(b) Unless sooner surrendered, suspended, or revoked, a certificate issued under this part for a training center located outside the United States expires at the end of the twelfth month after the month in which it is issued or renewed.

(c) If the Administrator suspends, revokes, or terminates a training center certificate, the holder of that certificate shall return the certificate to the Administrator within 5 working days after being notified that the certificate is suspended, revoked, or terminated.

§ 142.9 Deviations or waivers.

(a) The Administrator may issue deviations or waivers from any of the requirements of this part.

(b) A training center applicant requesting a deviation or waiver under this section must provide the Administrator with information acceptable to the Administrator that shows—

(1) Justification for the deviation or waiver; and

(2) That the deviation or waiver will not adversely affect the quality of instruction or evaluation.

§ 142.11 Application for issuance or amendment.

(a) An application for a training center certificate and training specifications shall—

(1) Be made on a form and in a manner prescribed by the Administrator;

(2) Be filed with the FAA Flight Standards District Office that has jurisdiction over the area in which the applicant's principal business office is located; and

(3) Be made at least 120 calendar days before the beginning of any proposed training or 60 calendar days before effecting an amendment to any approved training, unless a shorter filing period is approved by the Administrator.

(b) Each application for a training center certificate and training specification shall provide—

(1) A statement showing that the minimum qualification requirements for each management position are met or exceeded;

(2) A statement acknowledging that the applicant shall notify the Administrator within 10 working days of any change made in the assignment of persons in the required management positions;

(3) The proposed training authorizations and training specifications requested by the applicant;

(4) The proposed evaluation authorization;

(5) A description of the flight training equipment that the applicant proposes to use;

(6) A description of the applicant's training facilities, equipment, qualifications of personnel to be used, and proposed evaluation plans;

(7) A training program curriculum, including syllabi, outlines, courseware, procedures, and documentation to support the items required in subpart B of this part, upon request by the Administrator;

(8) A description of a recordkeeping system that will identify and document the details of training, qualification, and certification of students, instructors, and evaluators;

(9) A description of quality control measures proposed; and

(10) A method of demonstrating the applicant's qualification and ability to provide training for a certificate or rating in fewer than the minimum hours prescribed in part 61 of this chapter if the applicant proposes to do so.

(c) The facilities and equipment described in paragraph (b)(6) of this section shall—

(1) Be available for inspection and evaluation prior to approval; and

(2) Be in place and operational at the location of the proposed training center prior to issuance of a certificate under this part.

(d) An applicant who meets the requirements of this part and is approved by the Administrator is entitled to—

(1) A training center certificate containing all business names included on the application under which the certificate holder may conduct operations and the address of each business office used by the certificate holder; and

(2) Training specifications, issued by the Administrator to the certificate holder, containing—

(i) The type of training authorized, including approved courses;

(ii) The category, class, and type of aircraft that may be used for training, testing, and checking;

(iii) For each flight simulator or flight training device, the make, model, and series of airplane or the set of airplanes being simulated and the qualification level assigned, or the make, model, and series of rotorcraft, or set of rotorcraft being simulated and the qualification level assigned;

(iv) For each flight simulator and flight training device subject to qualification evaluation by the Administrator, the identification number assigned by the FAA;

(v) The name and address of all satellite training centers, and the approved courses offered at each satellite training center;

(vi) Authorized deviations or waivers from this part; and

(vii) Any other items the Administrator may require or allow.

(e) The Administrator may deny, suspend, revoke, or terminate a certificate under this part if the Administrator finds that the applicant or the certificate holder—

(1) Held a training center certificate that was revoked, suspended, or terminated within the previous 5 years; or

(2) Employs or proposes to employ a person who—

(i) Was previously employed in a management or supervisory position by the holder of a training center certificate that was revoked, suspended, or terminated within the previous 5 years;

(ii) Exercised control over any certificate holder whose certificate has been revoked, suspended, or terminated within the last 5 years; and

(iii) Contributed materially to the revocation, suspension, or termination of that certificate and who will be employed in a management or supervisory position, or who will be in control of or have a substantial ownership interest in the training center.

(3) Has provided incomplete, inaccurate, fraudulent, or false information for a training center certificate;

(4) Should not be granted a certificate if the grant would not foster aviation safety.

(f) At any time, the Administrator may amend a training center certificate—

(1) On the Administrator's own initiative, under section 609 of the Federal Aviation Act of 1958 (49 U.S.C. 1429), as amended, and part 13 of this chapter; or

(2) Upon timely application by the certificate holder.

(g) The certificate holder must file an application to amend a training center certificate at least 60 calendar days prior to the applicant's proposed effective amendment date unless a different filing period is approved by the Administrator.

[Doc. No. 26933, 61 FR 34562, July 2, 1996, as amended by Amdt. 142-1, 62 FR 13791, Mar. 21, 1997]

§ 142.13 Management and personnel requirements.

An applicant for a training center certificate must show that—

(a) For each proposed curriculum, the training center has, and shall maintain, a sufficient number of instructors who are qualified in accordance with subpart C of this part to perform the duties to which they are assigned;

(b) The training center has designated, and shall maintain, a sufficient number of approved evaluators to provide required checks and tests to graduation candidates within 7 calendar days of training completion for any curriculum leading to airman certificates or ratings, or both;

(c) The training center has, and shall maintain, a sufficient number of management personnel who are qualified and competent to perform required duties; and

(d) A management representative, and all personnel who are designated by the training center to conduct direct student training, are able to understand, read, write, and fluently speak the English language.

§ 142.14 Employment of former FAA employees.

(a) Except as specified in paragraph (c) of this section, no holder of a training center certificate may knowingly employ or make a contractual arrangement which permits an individual to act as an agent or representative of the certificate holder in any matter before the Federal Aviation Administration if the individual, in the preceding 2 years—

(1) Served as, or was directly responsible for the oversight of, a Flight Standards Service aviation safety inspector; and

(2) Had direct responsibility to inspect, or oversee the inspection of, the operations of the certificate holder.

(b) For the purpose of this section, an individual shall be considered to be acting as an agent or representative of a certificate holder in a matter before the agency if the individual makes any written or oral communication on behalf of the certificate holder to the agency (or any of its officers or employees) in connection with a particular matter, whether or not involving a specific party and without regard to whether the individual has participated in, or had responsibility for, the particular matter while serving as a Flight Standards Service aviation safety inspector.

(c) The provisions of this section do not prohibit a holder of a training center certificate from knowingly employing or making a contractual arrangement which permits an individual to act as an agent or representative of the certificate holder in any matter before the Federal Aviation Administration if the individual was employed by the certificate holder before October 21, 2011.

[Doc. No. FAA-2008-1154, 76 FR 52237, Aug. 22, 2011]

§ 142.15 Facilities.

(a) An applicant for, or holder of, a training center certificate shall ensure that—

(1) Each room, training booth, or other space used for instructional purposes is heated, lighted, and ventilated to conform to local building, sanitation, and health codes; and

(2) The facilities used for instruction are not routinely subject to significant distractions caused by flight operations and maintenance operations at the airport.

(b) An applicant for, or holder of, a training center certificate shall establish and maintain a principal business office that is physically located at the address shown on its training center certificate.

(c) The records required to be maintained by this part must be located in facilities adequate for that purpose.

(d) An applicant for, or holder of, a training center certificate must have available exclusively, for adequate periods of time and at a location approved by the Administrator, adequate flight training equipment and courseware, including at least one flight simulator or advanced flight training device.

[Doc. No. 26933, 61 FR 34562, July 2, 1996, as amended by Amdt. 142-3, 63 FR 53537, Oct. 5, 1998]

§ 142.17 Satellite training centers.

(a) The holder of a training center certificate may conduct training in accordance with an approved training program at a satellite training center if—

(1) The facilities, equipment, personnel, and course content of the satellite training center meet the applicable requirements of this part;

(2) The instructors and evaluators at the satellite training center are under the direct supervision of management personnel of the principal training center;

(3) The Administrator is notified in writing that a particular satellite is to begin operations at least 60 days prior to proposed commencement of operations at the satellite training center; and

(4) The certificate holder's training specifications reflect the name and address of the satellite training center and the approved courses offered at the satellite training center.

(b) The certificate holder's training specifications shall prescribe the operations required and authorized at each satellite training center.

[Doc. No. 26933, 61 FR 34562, July 2, 1996, as amended by Amdt. 142-3, 63 FR 53537, Oct. 5, 1998]

§§ 142.21-142.25 [Reserved]

§ 142.27 Display of certificate.

(a) Each holder of a training center certificate must prominently display that certificate in a place accessible to the public in the principal business office of the training center.

(b) A training center certificate and training specifications must be made available for inspection upon request by—

(1) The Administrator;

(2) An authorized representative of the National Transportation Safety Board; or

(3) Any Federal, State, or local law enforcement agency.

§ 142.29 Inspections.

Each certificate holder must allow the Administrator to inspect training center facilities, equipment, and records at any reasonable time and in any reasonable place in order to determine compliance with or to determine initial or continuing eligibility under 49 U.S.C. 44701, 44707, formerly the Federal Aviation Act of 1958, as amended, and the training center's certificate and training specifications.

§ 142.31 Advertising limitations.

(a) A certificate holder may not conduct, and may not advertise to conduct, any training, testing, and checking that is not approved by the Administrator if that training is designed to satisfy any requirement of this chapter.

(b) A certificate holder whose certificate has been surrendered, suspended, revoked, or terminated must—

(1) Promptly remove all indications, including signs, wherever located, that the training center was certificated by the Administrator; and

(2) Promptly notify all advertising agents, or advertising media, or both, employed by the certificate holder to cease all advertising indicating that the training center is certificated by the Administrator.

§ 142.33 Training agreements.

A pilot school certificated under part 141 of this chapter may provide training, testing, and checking for a training center certificated under this part if—

(a) There is a training, testing, and checking agreement between the certificated training center and the pilot school;

(b) The training, testing, and checking provided by the certificated pilot school is approved and conducted in accordance with this part;

(c) The pilot school certificated under part 141 obtains the Administrator's approval for a training course outline that includes the portion of the training, testing, and checking to be conducted under part 141; and

(d) Upon completion of training, testing, and checking conducted under part 141, a copy of each student's training record is forwarded to the part 142 training center and becomes part of the student's permanent training record.

Subpart B—Aircrew Curriculum and Syllabus Requirements

§ 142.35 Applicability.

This subpart prescribes the curriculum and syllabus requirements for the issuance of a training center certificate and training specifications for training, testing, and checking conducted to meet the requirements of part 61 of this chapter.

§ 142.37 Approval of flight aircrew training program.

(a) Except as provided in paragraph (b) of this section, each applicant for, or holder of, a training center certificate must apply to the Administrator for training program approval.

(b) A curriculum approved under SFAR 58 of part 121 of this chapter is approved under this part without modifications.

(c) Application for training program approval shall be made in a form and in a manner acceptable to the Administrator.

(d) Each application for training program approval must indicate—

(1) Which courses are part of the core curriculum and which courses are part of the specialty curriculum;

(2) Which requirements of part 61 of this chapter would be satisfied by the curriculum or curriculums; and

(3) Which requirements of part 61 of this chapter would not be satisfied by the curriculum or curriculums.

(e) If, after a certificate holder begins operations under an approved training program, the Administrator finds that the certificate holder is not meeting the provisions of its approved training program, the Administrator may require the certificate holder to make revisions to that training program.

(f) If the Administrator requires a certificate holder to make revisions to an approved training program and the certificate holder does not make those required revisions, within 30 calendar days, the Administrator may suspend, revoke, or terminate the training center certificate under the provisions of § 142.11(e).

§ 142.39 Training program curriculum requirements.

Each training program curriculum submitted to the Administrator for approval must meet the applicable requirements of this part and must contain—

(a) A syllabus for each proposed curriculum;

(b) Minimum aircraft and flight training equipment requirements for each proposed curriculum;

(c) Minimum instructor and evaluator qualifications for each proposed curriculum;

(d) A curriculum for initial training and continuing training of each instructor or evaluator employed to instruct in a proposed curriculum; and

(e) For each curriculum that provides for the issuance of a certificate or rating in fewer than the minimum hours prescribed by part 61 of this chapter—

(1) A means of demonstrating the ability to accomplish such training in the reduced number of hours; and

(2) A means of tracking student performance.

Subpart C—Personnel and Flight Training Equipment Requirements

§ 142.45 Applicability.

This subpart prescribes the personnel and flight training equipment requirements for a certificate holder that is training to meet the requirements of part 61 of this chapter.

§ 142.47 Training center instructor eligibility requirements.

(a) A certificate holder may not employ a person as an instructor in a flight training course that is subject to approval by the Administrator unless that person—

(1) Is at least 18 years of age;

(2) Is able to read, write, and speak and understand in the English language;

(3) If instructing in an aircraft in flight, is qualified in accordance with subpart H of this chapter;

(4) Satisfies the requirements of paragraph (c) of this section; and

(5) Meets at least one of the following requirements—

(i) Except as allowed by paragraph (a)(5)(ii) of this section, meets the aeronautical experience requirements of § 61.129 (a), (b), (c), or (e) of this chapter, as applicable, excluding the required hours of instruction in preparation for the commercial pilot practical test;

(ii) If instructing in flight simulator or flight training device that represents an airplane requiring a type rating or if instructing in a curriculum leading to the issuance of an airline transport pilot certificate or an added rating to an airline transport pilot certificate, meets the aeronautical experience requirements of § 61.159, § 61.161, or § 61.163 of this chapter, as applicable; or

(iii) Is employed as a flight simulator instructor or a flight training device instructor for a training center providing instruction and testing to meet the requirements of part 61 of this chapter on August 1, 1996.

(b) A training center must designate each instructor in writing to instruct in each approved course, prior to that person functioning as an instructor in that course.

(c) Prior to initial designation, each instructor shall:

(1) Complete at least 8 hours of ground training on the following subject matter:

(i) Instruction methods and techniques.

(ii) Training policies and procedures.

(iii) The fundamental principles of the learning process.

(iv) Instructor duties, privileges, responsibilities, and limitations.

(v) Proper operation of simulation controls and systems.

(vi) Proper operation of environmental control and warning or caution panels.

(vii) Limitations of simulation.

(viii) Minimum equipment requirements for each curriculum.

(ix) Revisions to the training courses.

(x) Cockpit resource management and crew coordination.

(2) Satisfactorily complete a written test—

(i) On the subjects specified in paragraph (c)(1) of this section; and

(ii) That is accepted by the Administrator as being of equivalent difficulty, complexity, and scope as the tests provided by the Administrator for the flight instructor airplane and instrument flight instructor knowledge tests.

[Doc. No. 26933, 61 FR 34562, July 2, 1996, as amended by Amdt. 142-2, 62 FR 68137, Dec. 30, 1997]

§ 142.49 Training center instructor and evaluator privileges and limitations.

(a) A certificate holder may allow an instructor to provide:

(1) Instruction for each curriculum for which that instructor is qualified.

(2) Testing and checking for which that instructor is qualified.

(3) Instruction, testing, and checking intended to satisfy the requirements of any part of this chapter.

(b) A training center whose instructor or evaluator is designated in accordance with the requirements of this subpart to conduct training, testing, or checking in qualified and approved flight training equipment, may allow its instructor or evaluator to give endorsements required by part 61 of this chapter if that instructor or evaluator is authorized by the Administrator to instruct or evaluate in a part 142 curriculum that requires such endorsements.

(c) A training center may not allow an instructor to—

(1) Excluding briefings and debriefings, conduct more than 8 hours of instruction in any 24-consecutive-hour period;

(2) Provide flight training equipment instruction unless that instructor meets the requirements of § 142.53 (a)(1) through (a)(4), and § 142.53(b), as applicable; or

(3) Provide flight instruction in an aircraft unless that instructor—

(i) Meets the requirements of § 142.53(a)(1), (a)(2), and (a)(5);

(ii) Is qualified and authorized in accordance with subpart H of part 61 of this chapter;

(iii) Holds certificates and ratings specified by part 61 of this chapter appropriate to the category, class, and type aircraft in which instructing;

(iv) If instructing or evaluating in an aircraft in flight while serving as a required crewmember, holds at least a valid second class medical certificate; and

(v) Meets the recency of experience requirements of part 61 of this chapter.

[Doc. No. 26933, 61 FR 34562, July 2, 1996, as amended by Amdt. 142-2, 62 FR 68137, Dec. 30, 1997; Amdt. 142-9, 78 FR 42380, July 15, 2013]

§ 142.51 [Reserved]

§ 142.53 Training center instructor training and testing requirements.

(a) Except as provided in paragraph (c) of this section, prior to designation and every 12 calendar months beginning the first day of the month following an instructor's initial designation, a certificate holder must ensure that each of its instructors meets the following requirements:

(1) Each instructor must satisfactorily demonstrate to an authorized evaluator knowledge of, and proficiency in, instructing in a representative segment of each curriculum for which that instructor is designated to instruct under this part.

(2) Each instructor must satisfactorily complete an approved course of ground instruction in at least—

(i) The fundamental principles of the learning process;

(ii) Elements of effective teaching, instruction methods, and techniques;

(iii) Instructor duties, privileges, responsibilities, and limitations;

(iv) Training policies and procedures;

(v) Cockpit resource management and crew coordination; and

(vi) Evaluation.

(3) Each instructor who instructs in a qualified and approved flight simulator or flight training device must satisfactorily complete an approved course of training in the operation of the flight simulator, and an approved course of ground instruction, applicable to the training courses the instructor is designated to instruct.

(4) The flight simulator training course required by paragraph (a)(3) of this section which must include—

(i) Proper operation of flight simulator and flight training device controls and systems;

(ii) Proper operation of environmental and fault panels;

(iii) Limitations of simulation; and

(iv) Minimum equipment requirements for each curriculum.

(5) Each flight instructor who provides training in an aircraft must satisfactorily complete an approved course of ground instruction and flight training in an aircraft, flight simulator, or flight training device.

(6) The approved course of ground instruction and flight training required by paragraph (a)(5) of this section which must include instruction in—

(i) Performance and analysis of flight training procedures and maneuvers applicable to the training courses that the instructor is designated to instruct;

(ii) Technical subjects covering aircraft subsystems and operating rules applicable to the training courses that the instructor is designated to instruct;

(iii) Emergency operations;

(iv) Emergency situations likely to develop during training; and

(v) Appropriate safety measures.

(7) Each instructor who instructs in qualified and approved flight training equipment must pass a written test and annual proficiency check—

(i) In the flight training equipment in which the instructor will be instructing; and

(ii) On the subject matter and maneuvers of a representative segment of each curriculum for which the instructor will be instructing.

(b) In addition to the requirements of paragraphs (a)(1) through (a)(7) of this section, each certificate holder must ensure that each instructor who instructs in a flight simulator that the Administrator has approved for all training and all testing for the airline transport pilot certification test, aircraft type rating test, or both, has met at least one of the following three requirements:

(1) Each instructor must have performed 2 hours in flight, including three takeoffs and three landings as the sole manipulator of the controls of an aircraft of the same category and class, and, if a type rating is required, of the same type replicated by the approved flight simulator in which that instructor is designated to instruct;

(2) Each instructor must have participated in an approved line-observation program under part 121 or part 135 of this chapter, and that—

(i) Was accomplished in the same airplane type as the airplane represented by the flight simulator in which that instructor is designated to instruct; and

(ii) Included line-oriented flight training of at least 1 hour of flight during which the instructor was the sole manipulator of the controls in a flight simulator that replicated the same type aircraft for which that instructor is designated to instruct; or

(3) Each instructor must have participated in an approved in-flight observation training course that—

(i) Consisted of at least 2 hours of flight time in an airplane of the same type as the airplane replicated by the flight simulator in which the instructor is designated to instruct; and

(ii) Included line-oriented flight training of at least 1 hour of flight during which the instructor was the sole manipulator of the controls in a flight simulator that replicated the same type aircraft for which that instructor is designated to instruct.

(c) An instructor who satisfactorily completes a curriculum required by paragraph (a) or (b) of this section in the calendar month before or after the month in which it is due is considered to have taken it in the month in which it was due for the purpose of computing when the next training is due.

(d) The Administrator may give credit for the requirements of paragraph (a) or (b) of this section to an instructor who has satisfactorily completed an instructor training course for a part 121 or part 135 certificate holder if the Administrator finds such a course equivalent to the requirements of paragraph (a) or (b) of this section.

[Doc. No. 26933, 61 FR 34562, July 2, 1996, as amended by Amdt. 142-1, 62 FR 13791, Mar. 21, 1997]

§ 142.54 Airline transport pilot certification training program.

No certificate holder may use a person nor may any person serve as an instructor in a training program approved to meet the requirements of § 61.156 of this chapter unless the instructor:

(a) Holds an airline transport pilot certificate with an airplane category multiengine class rating;

(b) Has at least 2 years of experience as a pilot in command in operations conducted under § 91.1053(a)(2)(i) or § 135.243(a)(1) of this chapter, or as a pilot in command or second in command in any operation conducted under part 121 of this chapter;

(c) Except for the holder of a flight instructor certificate, receives initial training on the following topics:

(1) The fundamental principles of the learning process;

(2) Elements of effective teaching, instruction methods, and techniques;

(3) Instructor duties, privileges, responsibilities, and limitations;

(4) Training policies and procedures; and

(5) Evaluation.

(d) If providing training in a flight simulation training device—

(1) Holds an aircraft type rating for the aircraft represented by the flight simulation training device utilized in the training program and have received training and evaluation within the preceding 12 months from the certificate holder on the maneuvers that will be demonstrated in the flight simulation training device; and

(2) Satisfies the requirements of § 142.53(a)(4).

(e) A certificate holder may not issue a graduation certificate to a student unless that student has completed all the curriculum requirements of the course.

(f) A certificate holder must conduct evaluations to ensure that training techniques, procedures, and standards are acceptable to the Administrator.

[Doc. No. FAA-2010-0100, 78 FR 42380, July 15, 2013]

§ 142.55 Training center evaluator requirements.

(a) Except as provided by paragraph (d) of this section, a training center must ensure that each person authorized as an evaluator—

(1) Is approved by the Administrator;

(2) Is in compliance with §§ 142.47, 142.49, and 142.53 and applicable sections of part 183 of this chapter; and

(3) Prior to designation, and except as provided in paragraph (b) of this section, every 12-calendar-month period following initial designation, the certificate holder must ensure that the evaluator satisfactorily completes a curriculum that includes the following:

(i) Evaluator duties, functions, and responsibilities;

(ii) Methods, procedures, and techniques for conducting required tests and checks;

(iii) Evaluation of pilot performance; and

(iv) Management of unsatisfactory tests and subsequent corrective action; and

(4) If evaluating in qualified and approved flight training equipment must satisfactorily pass a written test and annual proficiency check in a flight simulator or aircraft in which the evaluator will be evaluating.

(b) An evaluator who satisfactorily completes a curriculum required by paragraph (a) of this section in the calendar month before or the calendar month after the month in which it is due is considered to have taken it in the month is which it was due for the purpose of computing when the next training is due.

(c) The Administrator may give credit for the requirements of paragraph (a)(3) of this section to an evaluator who has satisfactorily completed an evaluator training course for a part 121 or part 135 certificate holder if the Administrator finds such a course equivalent to the requirements of paragraph (a)(3) of this section.

(d) An evaluator who is qualified under subpart Y of part 121 of this chapter shall be authorized to conduct evaluations under the Advanced Qualification Program without complying with the requirements of this section.

[Doc. No. 26933, 61 FR 34562, July 2, 1996, as amended by Amdt. 142-9, 78 FR 42380, July 15, 2013]

§ 142.57 Aircraft requirements.

(a) An applicant for, or holder of, a training center certificate must ensure that each aircraft used for flight instruction and solo flights meets the following requirements:

(1) Except for flight instruction and solo flights in a curriculum for agricultural aircraft operations, external load operations, and similar aerial work operations, the aircraft must have an FAA standard airworthiness certificate or a foreign equivalent of an FAA standard airworthiness certificate, acceptable to the Administrator.

(2) The aircraft must be maintained and inspected in accordance with—

(i) The requirements of part 91, subpart E, of this chapter; and

(ii) An approved program for maintenance and inspection.

(3) The aircraft must be equipped as provided in the training specifications for the approved course for which it is used.

(b) Except as provided in paragraph (c) of this section, an applicant for, or holder of, a training center certificate must ensure that each aircraft used for flight instruction is at least a two-place aircraft with engine power controls and flight controls that are easily reached and that operate in a conventional manner from both pilot stations.

(c) Airplanes with controls such as nose-wheel steering, switches, fuel selectors, and engine air flow controls that are not easily reached and operated in a conventional manner by both pilots may be used for flight instruction if the certificate holder determines that the flight instruction can be conducted in a safe manner considering the location of controls and their nonconventional operation, or both.

§ 142.59 Flight simulators and flight training devices.

(a) An applicant for, or holder of, a training center certificate must show that each flight simulator and flight training device used for training, testing, and checking (except AQP) will be or is specifically qualified and approved by the Administrator for—

(1) Each maneuver and procedure for the make, model, and series of aircraft, set of aircraft, or aircraft type simulated, as applicable; and

(2) Each curriculum or training course in which the flight simulator or flight training device is used, if that curriculum or course is used to satisfy any requirement of 14 CFR chapter I.

(b) The approval required by paragraph (a)(2) of this section must include—

(1) The set of aircraft, or type aircraft;

(2) If applicable, the particular variation within type, for which the training, testing, or checking is being conducted; and

(3) The particular maneuver, procedure, or crewmember function to be performed.

(c) Each qualified and approved flight simulator or flight training device used by a training center must—

(1) Be maintained to ensure the reliability of the performances, functions, and all other characteristics that were required for qualification;

(2) Be modified to conform with any modification to the aircraft being simulated if the modification results in changes to performance, function, or other characteristics required for qualification;

(3) Be given a functional preflight check each day before being used; and

(4) Have a discrepancy log in which the instructor or evaluator, at the end of each training session, enters each discrepancy.

(d) Unless otherwise authorized by the Administrator, each component on a qualified and approved flight simulator or flight training device used by a training center must be operative if the component is essential to, or involved in, the training, testing, or checking of airmen.

(e) Training centers shall not be restricted to specific—

(1) Route segments during line-oriented flight training scenarios; and

(2) Visual data bases replicating a specific customer's bases of operation.

(f) Training centers may request evaluation, qualification, and continuing evaluation for qualification of flight simulators and flight training devices without—

(1) Holding an air carrier certificate; or

(2) Having a specific relationship to an air carrier certificate holder.

Subpart D—Operating Rules

§ 142.61 Applicability.

This subpart prescribes the operating rules applicable to a training center certificated under this part and operating a course or training program curriculum approved in accordance with subpart B of this part.

§ 142.63 Privileges.

A certificate holder may allow flight simulator instructors and evaluators to meet recency of experience requirements through the use of a qualified and approved flight simulator or qualified and approved flight training device if that flight simulator or flight training device is—

(a) Used in a course approved in accordance with subpart B of this part; or

(b) Approved under the Advanced Qualification Program for meeting recency of experience requirements.

§ 142.65 Limitations.

(a) A certificate holder shall—

(1) Ensure that a flight simulator or flight training device freeze, slow motion, or repositioning feature is not used during testing or checking; and

(2) Ensure that a repositioning feature is used during line operational simulation for evaluation and line-oriented flight training only to advance along a flight route to the point where the descent and approach phase of the flight begins.

(b) When flight testing, flight checking, or line operational simulation is being conducted, the certificate holder must ensure that one of the following occupies each crewmember position:

(1) A crewmember qualified in the aircraft category, class, and type, if a type rating is required, provided that no flight instructor who is giving instruction may occupy a crewmember position.

(2) A student, provided that no student may be used in a crewmember position with any other student not in the same specific course.

(c) The holder of a training center certificate may not recommend a trainee for a certificate or rating, unless the trainee—

(1) Has satisfactorily completed the training specified in the course approved under § 142.37; and

(2) Has passed the final tests required by § 142.37.

(d) The holder of a training center certificate may not graduate a student from a course unless the student has satisfactorily completed the curriculum requirements of that course.

Subpart E—Recordkeeping

§ 142.71 Applicability.

This subpart prescribes the training center recordkeeping requirements for trainees enrolled in a course, and instructors and evaluators designated to instruct a course, approved in accordance with subpart B of this part.

§ 142.73 Recordkeeping requirements.

(a) A certificate holder must maintain a record for each trainee that contains—

(1) The name of the trainee;

(2) A copy of the trainee's pilot certificate, if any, and medical certificate;

(3) The name of the course and the make and model of flight training equipment used;

(4) The trainee's prerequisite experience and course time completed;

(5) The trainee's performance on each lesson and the name of the instructor providing instruction;

(6) The date and result of each end-of-course practical test and the name of the evaluator conducting the test; and

(7) The number of hours of additional training that was accomplished after any unsatisfactory practical test.

(b) A certificate holder shall maintain a record for each instructor or evaluator designated to instruct a course approved in accordance with subpart B of this part that indicates that the instructor or evaluator has complied with the requirements of §§ 142.13, 142.45, 142.47, 142.49, and 142.53, as applicable.

(c) The certificate holder shall—

(1) Maintain the records required by paragraphs (a) of this section for at least 1 year following the completion of training, testing or checking;

(2) Maintain the qualification records required by paragraph (b) of this section while the instructor or evaluator is in the employ of the certificate holder and for 1 year thereafter; and

(3) Maintain the recurrent demonstration of proficiency records required by paragraph (b) of this section for at least 1 year.

(d) The certificate holder must provide the records required by this section to the Administrator, upon request and at a reasonable time, and shall keep the records required by—

(1) Paragraph (a) of this section at the training center, or satellite training center where the training, testing, or checking, if appropriate, occurred; and

(2) Paragraph (b) of this section at the training center or satellite training center where the instructor or evaluator is primarily employed.

(e) The certificate holder shall provide to a trainee, upon request and at a reasonable time, a copy of his or her training records.

Subpart F—Other Approved Courses

§ 142.81 Conduct of other approved courses.

(a) An applicant for, or holder of, a training center certificate may apply for approval to conduct a course for which a curriculum is not prescribed by this part.

(b) The course for which application is made under paragraph (a) of this section may be for flight crewmembers other than pilots, airmen other than flight crewmembers, material handlers, ground servicing personnel, and security personnel, and others approved by the Administrator.

(c) An applicant for course approval under this subpart must comply with the applicable requirements of subpart A through subpart F of this part.

(d) The Administrator approves the course for which the application is made if the training center or training center applicant shows that the course contains a curriculum that will achieve a level of competency equal to, or greater than, that required by the appropriate part of this chapter.

Title 49: Transportation

PART 830

NOTIFICATION AND REPORTING OF AIRCRAFT ACCIDENTS OR INCIDENTS AND OVERDUE AIRCRAFT, AND PRESERVATION OF AIRCRAFT WRECKAGE, MAIL, CARGO, AND RECORDS

Contents

Authority: Independent Safety Board Act of 1974, as amended (49 U.S.C. 1101-1155); Federal Aviation Act of 1958, Public Law 85-726, 72 Stat. 731 (codified as amended at 49 U.S.C. 40101).

Source: 53 FR 36982, Sept. 23, 1988, unless otherwise noted.

Subpart A—General

§ 830.1 Applicability.

This part contains rules pertaining to:

(a) Initial notification and later reporting of aircraft incidents and accidents and certain other occurrences in the operation of aircraft, wherever they occur, when they involve civil aircraft of the United States; when they involve certain public aircraft, as specified in this part, wherever they occur; and when they involve foreign civil aircraft where the events occur in the United States, its territories, or its possessions.

(b) Preservation of aircraft wreckage, mail, cargo, and records involving all civil and certain public aircraft accidents, as specified in this part, in the United States and its territories or possessions.

[60 FR 40112, Aug. 7, 1995]

§ 830.2 Definitions.

As used in this part the following words or phrases are defined as follows:

Aircraft accident means an occurrence associated with the operation of an aircraft which takes place between the time any person boards the aircraft with the intention of flight and all such persons have disembarked, and in which any person suffers death or serious injury, or in which the aircraft receives substantial damage. For purposes of this part, the definition of "aircraft accident" includes "unmanned aircraft accident," as defined herein.

Civil aircraft means any aircraft other than a public aircraft.

Fatal injury means any injury which results in death within 30 days of the accident.

Incident means an occurrence other than an accident, associated with the operation of an aircraft, which affects or could affect the safety of operations.

Operator means any person who causes or authorizes the operation of an aircraft, such as the owner, lessee, or bailee of an aircraft.

Public aircraft means an aircraft used only for the United States Government, or an aircraft owned and operated (except for commercial purposes) or exclusively leased for at least 90 continuous days by a government other than the United States Government, including a State, the District of Columbia, a territory or possession of the United States, or a political subdivision of that government. "Public aircraft" does not include a government-owned aircraft transporting property for commercial purposes and does not include a government-owned aircraft transporting passengers other than: transporting (for other than commercial purposes) crewmembers or other persons aboard the aircraft whose presence is required to perform, or is associated with the performance of, a governmental function such as firefighting, search and rescue, law enforcement, aeronautical research, or biological or geological resource management; or transporting (for other than commercial purposes) persons aboard the aircraft if the aircraft is operated by the Armed Forces or an intelligence agency of the United States. Notwithstanding any limitation relating to use of the aircraft for commercial purposes, an aircraft shall be considered to be a public aircraft without regard to whether it is operated by a unit of government on behalf of another unit of government pursuant to a cost reimbursement agreement, if the unit of government on whose behalf the operation is conducted certifies to the Administrator of the Federal Aviation Administration that the operation was necessary to respond to a significant and imminent threat to life or property (including natural resources) and that no service by a private operator was reasonably available to meet the threat.

Serious injury means any injury which: (1) Requires hospitalization for more than 48 hours, commencing within 7 days from the date of the injury was received; (2) results in a fracture of any bone (except simple fractures of fingers, toes, or nose); (3) causes severe hemorrhages, nerve, muscle, or tendon damage; (4) involves any internal organ; or (5) involves second- or third-degree burns, or any burns affecting more than 5 percent of the body surface.

Substantial damage means damage or failure which adversely affects the structural strength, performance, or flight characteristics of the aircraft, and which would normally require major repair or replacement of the affected component. Engine failure or damage limited to an engine if only one engine fails or is damaged, bent fairings or cowling, dented skin, small punctured holes in the skin or fabric, ground damage to rotor or propeller blades, and damage to landing gear, wheels, tires, flaps, engine accessories, brakes, or wingtips are not considered "substantial damage" for the purpose of this part.

Unmanned aircraft accident means an occurrence associated with the operation of any public or civil unmanned aircraft system that takes place between the time that the system is activated with the purpose of flight and the time that the system is deactivated at the conclusion of its mission, in which:

(1) Any person suffers death or serious injury; or

(2) The aircraft has a maximum gross takeoff weight of 300 pounds or greater and sustains substantial damage.

[53 FR 36982, Sept. 23, 1988, as amended at 60 FR 40112, Aug. 7, 1995; 75 FR 51955, Aug. 24, 2010]

Subpart B—Initial Notification of Aircraft Accidents, Incidents, and Overdue Aircraft

§ 830.5 Immediate notification.

The operator of any civil aircraft, or any public aircraft not operated by the Armed Forces or an intelligence agency of the United States, or any foreign aircraft shall immediately, and by the most expeditious means available, notify the nearest National Transportation Safety Board (NTSB) office [1] when:

[1] NTSB regional offices are located in the following cities: Anchorage, Alaska; Atlanta, Georgia; West Chicago, Illinois; Denver, Colorado; Arlington, Texas; Gardena (Los Angeles), California; Miami, Florida; Seattle, Washington; and Ashburn, Virginia. In addition, NTSB headquarters is located at 490 L'Enfant Plaza, SW., Washington, DC 20594. Contact information for these offices is available at *http://www.ntsb.gov.*

(a) An aircraft accident or any of the following listed serious incidents occur:

(1) Flight control system malfunction or failure;

(2) Inability of any required flight crewmember to perform normal flight duties as a result of injury or illness;

(3) Failure of any internal turbine engine component that results in the escape of debris other than out the exhaust path;

(4) In-flight fire;

(5) Aircraft collision in flight;

(6) Damage to property, other than the aircraft, estimated to exceed $25,000 for repair (including materials and labor) or fair market value in the event of total loss, whichever is less.

(7) For large multiengine aircraft (more than 12,500 pounds maximum certificated takeoff weight):

(i) In-flight failure of electrical systems which requires the sustained use of an emergency bus powered by a back-up source such as a battery, auxiliary power unit, or air-driven generator to retain flight control or essential instruments;

(ii) In-flight failure of hydraulic systems that results in sustained reliance on the sole remaining hydraulic or mechanical system for movement of flight control surfaces;

(iii) Sustained loss of the power or thrust produced by two or more engines; and

(iv) An evacuation of an aircraft in which an emergency egress system is utilized.

(8) Release of all or a portion of a propeller blade from an aircraft, excluding release caused solely by ground contact;

(9) A complete loss of information, excluding flickering, from more than 50 percent of an aircraft's cockpit displays known as:

(i) Electronic Flight Instrument System (EFIS) displays;

(ii) Engine Indication and Crew Alerting System (EICAS) displays;

(iii) Electronic Centralized Aircraft Monitor (ECAM) displays; or

(iv) Other displays of this type, which generally include a primary flight display (PFD), primary navigation display (PND), and other integrated displays;

(10) Airborne Collision and Avoidance System (ACAS) resolution advisories issued either:

(i) When an aircraft is being operated on an instrument flight rules flight plan and compliance with the advisory is necessary to avert a substantial risk of collision between two or more aircraft; or

(ii) To an aircraft operating in class A airspace.

(11) Damage to helicopter tail or main rotor blades, including ground damage, that requires major repair or replacement of the blade(s);

(12) Any event in which an operator, when operating an airplane as an air carrier at a public-use airport on land:

(i) Lands or departs on a taxiway, incorrect runway, or other area not designed as a runway; or

(ii) Experiences a runway incursion that requires the operator or the crew of another aircraft or vehicle to take immediate corrective action to avoid a collision.

(b) An aircraft is overdue and is believed to have been involved in an accident.

[53 FR 36982, Sept. 23, 1988, as amended at 60 FR 40113, Aug. 7, 1995; 75 FR 927, Jan. 7, 2010; 75 FR 35330, June 22, 2010]

§ 830.6 Information to be given in notification.

The notification required in § 830.5 shall contain the following information, if available:

(a) Type, nationality, and registration marks of the aircraft;

(b) Name of owner, and operator of the aircraft;

(c) Name of the pilot-in-command;

(d) Date and time of the accident;

(e) Last point of departure and point of intended landing of the aircraft;

(f) Position of the aircraft with reference to some easily defined geographical point;

(g) Number of persons aboard, number killed, and number seriously injured;

(h) Nature of the accident, the weather and the extent of damage to the aircraft, so far as is known; and

(i) A description of any explosives, radioactive materials, or other dangerous articles carried.

Subpart C—Preservation of Aircraft Wreckage, Mail, Cargo, and Records

§ 830.10 Preservation of aircraft wreckage, mail, cargo, and records.

(a) The operator of an aircraft involved in an accident or incident for which notification must be given is responsible for preserving to the extent possible any aircraft wreckage, cargo, and mail aboard the aircraft, and all records, including all recording mediums of flight, maintenance, and voice recorders, pertaining to the operation and maintenance of the aircraft and to the airmen until the Board takes custody thereof or a release is granted pursuant to § 831.12(b) of this chapter.

(b) Prior to the time the Board or its authorized representative takes custody of aircraft wreckage, mail, or cargo, such wreckage, mail, or cargo may not be disturbed or moved except to the extent necessary:

(1) To remove persons injured or trapped;

(2) To protect the wreckage from further damage; or

(3) To protect the public from injury.

(c) Where it is necessary to move aircraft wreckage, mail or cargo, sketches, descriptive notes, and photographs shall be made, if possible, of the original positions and condition of the wreckage and any significant impact marks.

(d) The operator of an aircraft involved in an accident or incident shall retain all records, reports, internal documents, and memoranda dealing with the accident or incident, until authorized by the Board to the contrary.

Subpart D—Reporting of Aircraft Accidents, Incidents, and Overdue Aircraft

§ 830.15 Reports and statements to be filed.

(a) *Reports.* The operator of a civil, public (as specified in § 830.5), or foreign aircraft shall file a report on Board Form 6120. ½ (OMB No. 3147-0001)[2] within 10 days after an accident, or after 7 days if an overdue aircraft is still missing. A report on an incident for which immediate notification is required by § 830.5(a) shall be filed only as requested by an authorized representative of the Board.

[2] Forms are available from the Board field offices (see footnote 1), from Board headquarters in Washington, DC, and from the Federal Aviation Administration Flight Standards District Offices.

(b) *Crewmember statement.* Each crewmember, if physically able at the time the report is submitted, shall attach a statement setting forth the facts, conditions, and circumstances relating to the accident or incident as they appear to him. If the crewmember is incapacitated, he shall submit the statement as soon as he is physically able.

(c) *Where to file the reports.* The operator of an aircraft shall file any report with the field office of the Board nearest the accident or incident.

[53 FR 36982, Sept. 23, 1988, as amended at 60 FR 40113, Aug. 7, 1995]

PART 1552

FLIGHT SCHOOLS

Contents

Subpart A—Flight Training for Aliens and Other Designated Individuals

Subpart B—Flight School Security Awareness Training

Authority: 49 U.S.C. 114, 44939.

Source: 69 FR 56340, Sept. 20, 2004, unless otherwise noted.

Subpart A—Flight Training for Aliens and Other Designated Individuals

§ 1552.1 Scope and definitions.

(a) *Scope.* This subpart applies to flight schools that provide instruction under 49 U.S.C. Subtitle VII, Part A, in the operation of aircraft or aircraft simulators, and individuals who apply to obtain such instruction or who receive such instruction.

(b) *Definitions.* As used in this part:

Aircraft simulator means a flight simulator or flight training device, as those terms are defined at 14 CFR 61.1.

Alien means any person not a citizen or national of the United States.

Candidate means an alien or other individual designated by TSA who applies for flight training or recurrent training. It does not include an individual endorsed by the Department of Defense for flight training.

Day means a day from Monday through Friday, including State and local holidays but not Federal holidays, for any time period less than 11 days specified in this part. For any time period greater than 11 days, day means calendar day.

Demonstration flight for marketing purposes means a flight for the purpose of demonstrating an aircraft's or aircraft simulator's capabilities or characteristics to a potential purchaser, or to an agent of a potential purchaser, of the aircraft or simulator, including an acceptance flight after an aircraft manufacturer delivers an aircraft to a purchaser.

Flight school means any pilot school, flight training center, air carrier flight training facility, or flight instructor certificated under 14 CFR part 61, 121, 135, 141, or 142; or any other person or entity that provides instruction under 49 U.S.C. Subtitle VII, Part A, in the operation of any aircraft or aircraft simulator.

Flight training means instruction received from a flight school in an aircraft or aircraft simulator. Flight training does not include recurrent training, ground training, a demonstration flight for marketing purposes, or any military training provided by the Department of Defense, the U.S. Coast Guard, or an entity under contract with the Department of Defense or U.S. Coast Guard.

Ground training means classroom or computer-based instruction in the operation of aircraft, aircraft systems, or cockpit procedures. Ground training does not include instruction in an aircraft simulator.

National of the United States means a person who, though not a citizen of the United States, owes permanent allegiance to the United States, and includes a citizen of American Samoa or Swains Island.

Recurrent training means periodic training required under 14 CFR part 61, 121,125, 135, or Subpart K of part 91. Recurrent training does not include training that would enable a candidate who has a certificate or type rating for a particular aircraft to receive a certificate or type rating for another aircraft.

§ 1552.3 Flight training.

This section describes the procedures a flight school must follow before providing flight training.

(a) *Category 1—Regular processing for flight training on aircraft more than 12,500 pounds.* A flight school may not provide flight training in the operation of any aircraft having a maximum certificated takeoff weight of more than 12,500 pounds to a candidate, except for a candidate who receives expedited processing under paragraph (b) of this section, unless—

(1) The flight school has first notified TSA that the candidate has requested such flight training.

(2) The candidate has submitted to TSA, in a form and manner acceptable to TSA, the following:

(i) The candidate's full name, including any aliases used by the candidate or variations in the spelling of the candidate's name;

(ii) A unique candidate identification number created by TSA;

(iii) A copy of the candidate's current, unexpired passport and visa;

(iv) The candidate's passport and visa information, including all current and previous passports and visas held by the candidate and all the information necessary to obtain a passport and visa;

(v) The candidate's country of birth, current country or countries of citizenship, and each previous country of citizenship, if any;

(vi) The candidate's actual date of birth or, if the candidate does not know his or her date of birth, the approximate date of birth used consistently by the candidate for his or her passport or visa;

(vii) The candidate's requested dates of training and the location of the training;

(viii) The type of training for which the candidate is applying, including the aircraft type rating the candidate would be eligible to obtain upon completion of the training;

(ix) The candidate's current U.S. pilot certificate, certificate number, and type rating, if any;

(x) Except as provided in paragraph (k) of this section, the candidate's fingerprints, in accordance with paragraph (f) of this section;

(xi) The candidate's current address and phone number and each address for the 5 years prior to the date of the candidate's application;

(xii) The candidate's gender; and

(xiii) Any fee required under this part.

(3) The flight school has submitted to TSA, in a form and manner acceptable to TSA, a photograph of the candidate taken when the candidate arrives at the flight school for flight training.

(4) TSA has informed the flight school that the candidate does not pose a threat to aviation or national security, or more than 30 days have elapsed since TSA received all of the information specified in paragraph (a)(2) of this section.

(5) The flight school begins the candidate's flight training within 180 days of either event specified in paragraph (a)(4) of this section.

(b) *Category 2—Expedited processing for flight training on aircraft more than 12,500 pounds.* (1) A flight school may not provide flight training in the operation of any aircraft having a maximum certificated takeoff weight of more than 12,500 pounds to a candidate who meets any of the criteria of paragraph (b)(2) of this section unless—

(i) The flight school has first notified TSA that the candidate has requested such flight training.

(ii) The candidate has submitted to TSA, in a form and manner acceptable to TSA:

(A) The information and fee required under paragraph (a)(2) of this section; and

(B) The reason the candidate is eligible for expedited processing under paragraph (b)(2) of this section and information that establishes that the candidate is eligible for expedited processing.

(iii) The flight school has submitted to TSA, in a form and manner acceptable to TSA, a photograph of the candidate taken when the candidate arrives at the flight school for flight training.

(iv) TSA has informed the flight school that the candidate does not pose a threat to aviation or national security or more than 5 days have elapsed since TSA received all of the information specified in paragraph (a)(2) of this section.

(v) The flight school begins the candidate's flight training within 180 days of either event specified in paragraph (b)(1)(iv) of this section.

(2) A candidate is eligible for expedited processing if he or she—

(i) Holds an airman's certificate from a foreign country that is recognized by the Federal Aviation Administration or a military agency of the United States, and that permits the candidate to operate a multi-engine aircraft that has a certificated takeoff weight of more than 12,500 pounds;

(ii) Is employed by a foreign air carrier that operates under 14 CFR part 129 and has a security program approved under 49 CFR part 1546;

(iii) Has unescorted access authority to a secured area of an airport under 49 U.S.C. 44936(a)(1)(A)(ii), 49 CFR 1542.209, or 49 CFR 1544.229;

(iv) Is a flightcrew member who has successfully completed a criminal history records check in accordance with 49 CFR 1544.230; or

(v) Is part of a class of individuals that TSA has determined poses a minimal threat to aviation or national security because of the flight training already possessed by that class of individuals.

(c) *Category 3—Flight training on aircraft 12,500 pounds or less.* A flight school may not provide flight training in the operation of any aircraft having a maximum certificated takeoff weight of 12,500 pounds or less to a candidate unless—

(1) The flight school has first notified TSA that the candidate has requested such flight training.

(2) The candidate has submitted to TSA, in a form and manner acceptable to TSA:

(i) The information required under paragraph (a)(2) of this section; and

(ii) Any other information required by TSA.

(3) The flight school has submitted to TSA, in a form and manner acceptable to TSA, a photograph of the candidate taken when the candidate arrives at the flight school for flight training.

(4) The flight school begins the candidate's flight training within 180 days of the date the candidate submitted the information required under paragraph (a)(2) of this section to TSA.

(d) *Category 4—Recurrent training for all aircraft.* Prior to beginning recurrent training for a candidate, a flight school must—

(1) Notify TSA that the candidate has requested such recurrent training; and

(2) Submit to TSA, in a form and manner acceptable to TSA:

(i) The candidate's full name, including any aliases used by the candidate or variations in the spelling of the candidate's name;

(ii) Any unique student identification number issued to the candidate by the Department of Justice or TSA;

(iii) A copy of the candidate's current, unexpired passport and visa;

(iv) The candidate's current U.S. pilot certificate, certificate number, and type rating(s);

(v) The type of training for which the candidate is applying;

(vi) The date of the candidate's prior recurrent training, if any, and a copy of the training form documenting that recurrent training;

(vii) The candidate's requested dates of training; and

(viii) A photograph of the candidate taken when the candidate arrives at the flight school for flight training.

(e) *Interruption of flight training.* A flight school must immediately terminate or cancel a candidate's flight training if TSA notifies the flight school at any time that the candidate poses a threat to aviation or national security.

(f) *Fingerprints.* (1) Fingerprints submitted in accordance with this subpart must be collected—

(i) By United States Government personnel at a United States embassy or consulate; or

(ii) By another entity approved by TSA.

(2) A candidate must confirm his or her identity to the individual or agency collecting his or her fingerprints under paragraph (f)(1) of this section by providing the individual or agency his or her:

(i) Passport;

(ii) Resident alien card; or

(iii) U.S. driver's license.

(3) A candidate must pay any fee imposed by the agency taking his or her fingerprints.

(g) *General requirements* —(1) *False statements.* If a candidate makes a knowing and willful false statement, or omits a material fact, when submitting the information required under this part, the candidate may be—

(i) Subject to fine or imprisonment or both under 18 U.S.C. 1001;

(ii) Denied approval for flight training under this section; and

(iii) Subject to other enforcement action, as appropriate.

(2) *Preliminary approval.* For purposes of facilitating a candidate's visa process with the U.S. Department of State, TSA may inform a flight school and a candidate that the candidate has received preliminary approval for flight training based on information submitted by the flight school or the candidate under this section. A flight school may then issue an I-20 form to the candidate to present with the candidate's visa application. Preliminary approval does not initiate the waiting period under paragraph (a)(3)

or (b)(1)(iii) of this section or the period in which a flight school must initiate a candidate's training after receiving TSA approval under paragraph (a)(4) or (b)(1)(iv) of this section.

(h) *U.S. citizens and nationals and Department of Defense endorsees.* A flight school must determine whether an individual is a citizen or national of the United States, or a Department of Defense endorsee, prior to providing flight training to the individual.

(1) *U.S. citizens and nationals.* To establish U.S. citizenship or nationality an individual must present to the flight school his or her:

(i) Valid, unexpired United States passport;

(ii) Original or government-issued certified birth certificate of the United States, American Samoa, or Swains Island, together with a government-issued picture identification of the individual;

(iii) Original United States naturalization certificate with raised seal, or a Certificate of Naturalization issued by the U.S. Citizenship and Immigration Services (USCIS) or the U.S. Immigration and Naturalization Service (INS) (Form N-550 or Form N-570), together with a government-issued picture identification of the individual;

(iv) Original certification of birth abroad with raised seal, U.S. Department of State Form FS-545, or U.S. Department of State Form DS-1350, together with a government-issued picture identification of the individual;

(v) Original certificate of United States citizenship with raised seal, a Certificate of United States Citizenship issued by the USCIS or INS (Form N-560 or Form N-561), or a Certificate of Repatriation issued by the USCIS or INS (Form N-581), together with a government-issued picture identification of the individual; or

(vi) In the case of flight training provided to a Federal employee (including military personnel) pursuant to a contract between a Federal agency and a flight school, the agency's written certification as to its employee's United States citizenship or nationality, together with the employee's government-issued credentials or other Federally-issued picture identification.

(2) *Department of Defense endorsees.* To establish that an individual has been endorsed by the U.S. Department of Defense for flight training, the individual must present to the flight school a written statement acceptable to TSA from the U.S. Department of Defense attaché in the individual's country of residence together with a government-issued picture identification of the individual.

(i) *Recordkeeping requirements.* A flight school must—

(1) Maintain the following information for a minimum of 5 years:

(i) For each candidate:

(A) A copy of the photograph required under paragraph (a)(3), (b)(1)(iii), (c)(3), or (d)(2)(viii) of this section; and

(B) A copy of the approval sent by TSA confirming the candidate's eligibility for flight training.

(ii) For a Category 1, Category 2, or Category 3 candidate, a copy of the information required under paragraph (a)(2) of this section, except the information in paragraph (a)(2)(x).

(iii) For a Category 4 candidate, a copy of the information required under paragraph (d)(2) of this section.

(iv) For an individual who is a United States citizen or national, a copy of the information required under paragraph (h)(1) of this section.

(v) For an individual who has been endorsed by the U.S. Department of Defense for flight training, a copy of the information required under paragraph (h)(2) of this section.

(vi) A record of all fees paid to TSA in accordance with this part.

(2) Permit TSA and the Federal Aviation Administration to inspect the records required by paragraph (i)(1) of this section during reasonable business hours.

(j) *Candidates subject to the Department of Justice rule.* A candidate who submits a completed Flight Training Candidate Checks Program form and fingerprints to the Department of Justice in accordance with 28 CFR part 105 before September 28, 2004, or a later date specified by TSA, is processed in accordance with the requirements of that part. If TSA specifies a date later than the compliance dates identified in this part, individuals and flight schools who comply with 28 CFR part 105 up to that date will be considered to be in compliance with the requirements of this part.

(k) *Additional or missed flight training.* (1) A Category 1, 2, or 3 candidate who has been approved for flight training by TSA may take additional flight training without submitting fingerprints as specified in paragraph (a)(2)(x) of this section if the candidate:

(i) Submits all other information required in paragraph (a)(2) of this section, including the fee; and

(ii) Waits for TSA approval or until the applicable waiting period expires before initiating the additional flight training.

(2) A Category 1, 2, or 3 candidate who is approved for flight training by TSA, but does not initiate that flight training within 180 days, may reapply for flight training without submitting fingerprints as specified in paragraph (a)(2)(x) of this section if the candidate submits all other information required in paragraph (a)(2) of this section, including the fee.

§ 1552.5 Fees.

(a) *Imposition of fees.* The following fee is required for TSA to conduct a security threat assessment for a candidate for flight training subject to the requirements of § 1552.3: $130.

(b) *Remittance of fees.* (1) A candidate must remit the fee required under this subpart to TSA, in a form and manner acceptable to TSA, each time the candidate or the flight school is required to submit the information required under § 1552.3 to TSA.

(2) TSA will not issue any fee refunds, unless a fee was paid in error.

Subpart B—Flight School Security Awareness Training

§ 1552.21 Scope and definitions.

(a) *Scope.* This subpart applies to flight schools that provide instruction under 49 U.S.C. Subtitle VII, Part A, in the operation of aircraft or aircraft simulators, and to employees of such flight schools.

(b) *Definitions:* As used in this subpart:

Flight school employee means a flight instructor or ground instructor certificated under 14 CFR part 61, 141, or 142; a chief instructor certificated under 14 CFR part 141; a director of training certificated under 14 CFR part 142; or any other person employed by a flight school, including an independent contractor, who has direct contact with a flight school student. This includes an independent or solo flight instructor certificated under 14 CFR part 61.

§ 1552.23 Security awareness training programs.

(a) *General.* A flight school must ensure that—

(1) Each of its flight school employees receives initial and recurrent security awareness training in accordance with this subpart; and

(2) If an instructor is conducting the initial security awareness training program, the instructor has first successfully completed the initial flight school security awareness training program offered by TSA or an alternative initial flight school security awareness training program that meets the criteria of paragraph (c) of this section.

(b) *Initial security awareness training program.* (1) A flight school must ensure that—

(i) Each flight school employee employed on January 18, 2005 receives initial security awareness training in accordance with this subpart by January 18, 2005; and

(ii) Each flight school employee hired after January 18, 2005 receives initial security awareness training within 60 days of being hired.

(2) In complying with paragraph (b)(2) of this section, a flight school may use either:

(i) The initial flight school security awareness training program offered by TSA; or

(ii) An alternative initial flight school security awareness training program that meets the criteria of paragraph (c) of this section.

(c) *Alternative initial security awareness training program.* At a minimum, an alternative initial security awareness training program must—

(1) Require active participation by the flight school employee receiving the training.

(2) Provide situational scenarios requiring the flight school employee receiving the training to assess specific situations and determine appropriate courses of action.

(3) Contain information that enables a flight school employee to identify—

(i) Uniforms and other identification, if any are required at the flight school, for flight school employees or other persons authorized to be on the flight school grounds.

(ii) Behavior by clients and customers that may be considered suspicious, including, but not limited to:

(A) Excessive or unusual interest in restricted airspace or restricted ground structures;

(B) Unusual questions or interest regarding aircraft capabilities;

(C) Aeronautical knowledge inconsistent with the client or customer's existing airman credentialing; and

(D) Sudden termination of the client or customer's instruction.

(iii) Behavior by other on-site persons that may be considered suspicious, including, but not limited to:

(A) Loitering on the flight school grounds for extended periods of time; and

(B) Entering "authorized access only" areas without permission.

(iv) Circumstances regarding aircraft that may be considered suspicious, including, but not limited to:

(A) Unusual modifications to aircraft, such as the strengthening of landing gear, changes to the tail number, or stripping of the aircraft of seating or equipment;

(B) Damage to propeller locks or other parts of an aircraft that is inconsistent with the pilot training or aircraft flight log; and

(C) Dangerous or hazardous cargo loaded into an aircraft.

(v) Appropriate responses for the employee to specific situations, including:

(A) Taking no action, if a situation does not warrant action;

(B) Questioning an individual, if his or her behavior may be considered suspicious;

(C) Informing a supervisor, if a situation or an individual's behavior warrants further investigation;

(D) Calling the TSA General Aviation Hotline; or

(E) Calling local law enforcement, if a situation or an individual's behavior could pose an immediate threat.

(vi) Any other information relevant to security measures or procedures at the flight school, including applicable information in the TSA Information Publication "Security Guidelines for General Aviation Airports".

(d) *Recurrent security awareness training program.* (1) A flight school must ensure that each flight school employee receives recurrent security awareness training each year in the same month as the month the flight school employee received initial security awareness training in accordance with this subpart.

(2) At a minimum, a recurrent security awareness training program must contain information regarding—

(i) Any new security measures or procedures implemented by the flight school;

(ii) Any security incidents at the flight school, and any lessons learned as a result of such incidents;

(iii) Any new threats posed by or incidents involving general aviation aircraft contained on the TSA Web site; and

(iv) Any new TSA guidelines or recommendations concerning the security of general aviation aircraft, airports, or flight schools.

§ 1552.25 Documentation, recordkeeping, and inspection.

(a) *Documentation.* A flight school must issue a document to each flight school employee each time the flight school employee receives initial or recurrent security awareness training in accordance with this subpart. The document must—

(1) Contain the flight school employee's name and a distinct identification number.

(2) Indicate the date on which the flight school employee received the security awareness training.

(3) Contain the name of the instructor who conducted the training, if any.

(4) Contain a statement certifying that the flight school employee received the security awareness training.

(5) Indicate the type of training received, initial or recurrent.

(6) Contain a statement certifying that the alternative training program used by the flight school meets the criteria in 49 CFR 1552.23(c), if the flight school uses an alternative training program to comply with this subpart.

(7) Be signed by the flight school employee and an authorized official of the flight school.

(b) *Recordkeeping requirements.* A flight school must establish and maintain the following records for one year after an individual no longer is a flight school employee:

(1) A copy of the document required by paragraph (a) of this section for the initial and each recurrent security awareness training conducted for each flight school employee in accordance with this subpart; and

(2) The alternative flight school security awareness training program used by the flight school, if the flight school uses such a program.

(c) *Inspection.* A flight school must permit TSA and the Federal Aviation Administration to inspect the records required under paragraph (b) of this section during reasonable business hours.

Chapter 1.
AIR NAVIGATION
Section 1. Navigation Aids

1-1-1. General

a. Various types of air navigation aids are in use today, each serving a special purpose. These aids have varied owners and operators, namely: the Federal Aviation Administration (FAA), the military services, private organizations, individual states and foreign governments. The FAA has the statutory authority to establish, operate, maintain air navigation facilities and to prescribe standards for the operation of any of these aids which are used for instrument flight in federally controlled airspace. These aids are tabulated in the Airport/Facility Directory (A/FD).

b. Pilots should be aware of the possibility of momentary erroneous indications on cockpit displays when the primary signal generator for a ground-based navigational transmitter (for example, a glideslope, VOR, or nondirectional beacon) is inoperative. Pilots should disregard any navigation indication, regardless of its apparent validity, if the particular transmitter was identified by NOTAM or otherwise as unusable or inoperative.

1-1-2. Nondirectional Radio Beacon (NDB)

a. A low or medium frequency radio beacon transmits nondirectional signals whereby the pilot of an aircraft properly equipped can determine bearings and "home" on the station. These facilities normally operate in a frequency band of 190 to 535 kilohertz (kHz), according to ICAO Annex 10 the frequency range for NDBs is between 190 and 1750 kHz, and transmit a continuous carrier with either 400 or 1020 hertz (Hz) modulation. All radio beacons except the compass locators transmit a continuous three-letter identification in code except during voice transmissions.

b. When a radio beacon is used in conjunction with the Instrument Landing System markers, it is called a Compass Locator.

c. Voice transmissions are made on radio beacons unless the letter "W" (without voice) is included in the class designator (HW).

d. Radio beacons are subject to disturbances that may result in erroneous bearing information. Such disturbances result from such factors as lightning, precipitation static, etc. At night, radio beacons are vulnerable to interference from distant stations. Nearly all disturbances which affect the Automatic Direction Finder (ADF) bearing also affect the facility's identification. Noisy identification usually occurs when the ADF needle is erratic. Voice, music or erroneous identification may be heard when a steady false bearing is being displayed. Since ADF receivers do not have a "flag" to warn the pilot when erroneous bearing information is being displayed, the pilot should continuously monitor the NDB's identification.

1-1-3. VHF Omni-directional Range (VOR)

a. VORs operate within the 108.0 to 117.95 MHz frequency band and have a power output necessary to provide coverage within their assigned operational service volume. They are subject to line-of-sight restrictions, and the range varies proportionally to the altitude of the receiving equipment.

NOTE-*Normal service ranges for the various classes of VORs are given in Navigational Aid (NAVAID) Service Volumes, paragraph 1-1-8.*

b. Most VORs are equipped for voice transmission on the VOR frequency. VORs without voice capability are indicated by the letter "W" (without voice) included in the class designator (VORW).

c. The only positive method of identifying a VOR is by its Morse Code identification or by the recorded automatic voice identification which is always indicated by use of the word "VOR" following the range's name. Reliance on determining the identification of an omnirange should never be placed on listening to voice transmissions by the Flight Service Station (FSS) (or approach control facility) involved. Many FSSs remotely operate several omniranges with different names. In some cases, none of the VORs have the name of the "parent" FSS. During periods of maintenance, the facility may radiate a T-E-S-T code (-••••-) or the code may be removed.

d. Voice identification has been added to numerous VORs. The transmission consists of a voice announcement, "AIRVILLE VOR" alternating with the usual Morse Code identification.

e. The effectiveness of the VOR depends upon proper use and adjustment of both ground and airborne equipment.

1. Accuracy. The accuracy of course alignment of the VOR is excellent, being generally plus or minus 1 degree.

2. Roughness. On some VORs, minor course roughness may be observed, evidenced by course needle or brief flag alarm activity (some receivers are more susceptible to these irregularities than others). At a few stations, usually in mountainous terrain, the pilot may occasionally observe a brief course needle oscillation, similar to the indication of "approaching station." Pilots flying over unfamiliar routes are cautioned to be on the alert for these vagaries, and in particular, to use the "to/from" indicator to determine positive station passage.

(a) Certain propeller revolutions per minute (RPM) settings or helicopter rotor speeds can cause the VOR Course Deviation Indicator to fluctuate as much as plus or minus six degrees. Slight changes to the RPM setting will normally smooth out this roughness. Pilots are urged to check for this modulation phenomenon prior to reporting a VOR station or aircraft equipment for unsatisfactory operation.

1-1-4. VOR Receiver Check

a. The FAA VOR test facility (VOT) transmits a test signal which provides users a convenient means to determine the operational status and accuracy of a VOR receiver while on the ground where a VOT is located. The airborne use of VOT is permitted; however, its use is strictly limited to those areas/altitudes specifically authorized in the A/FD or appropriate supplement.

b. To use the VOT service, tune in the VOT frequency on your VOR receiver. With the Course Deviation Indicator (CDI) centered, the omni-bearing selector should read 0 degrees with the to/from indication showing "from" or the omni-bearing selector should read 180 degrees with the to/from indication showing "to." Should the VOR receiver operate an RMI (Radio Magnetic Indicator), it will indicate 180 degrees on any omni-bearing selector (OBS) setting. Two means of identification are used. One is a series of dots and the other is a continuous tone. Information concerning an individual test signal can be obtained from the local FSS.

c. Periodic VOR receiver calibration is most important. If a receiver's Automatic Gain Control or modulation circuit deteriorates, it is possible for it to display acceptable accuracy and sensitivity close into the VOR or VOT and display out-of-tolerance readings when located at greater distances where weaker signal areas exist. The likelihood of this deterioration varies between receivers, and is generally considered a function of time. The best assurance of having an accurate receiver is periodic calibration. Yearly intervals are recommended at which time an authorized repair facility should recalibrate the receiver to the manufacturer's specifications.

d. Federal Aviation Regulations (14 CFR Section 91.171) provides for certain VOR equipment accuracy checks prior to flight under instrument flight rules. To comply with this requirement and to ensure satisfactory operation of the airborne system, the FAA has provided pilots with the following means of checking VOR receiver accuracy:

1. VOT or a radiated test signal from an appropriately rated radio repair station.

2. Certified airborne check points.

3. Certified check points on the airport surface.

e. A radiated VOT from an appropriately rated radio repair station serves the same purpose as an FAA VOR signal and the check is made in much the same manner as a VOT with the following differences:

1. The frequency normally approved by the Federal Communications Commission is 108.0 MHz.

2. Repair stations are not permitted to radiate the VOR test signal continuously; consequently, the owner or operator must make arrangements with the repair station to have the test signal transmitted. This service is not provided by all radio repair stations. The aircraft owner or operator must determine which repair station in the local area provides this service. A representative of the repair station must make an entry into the aircraft logbook or other permanent record certifying to the radial accuracy and the date of transmission. The owner, operator or representative of the repair station may accomplish the necessary checks in the aircraft and make a logbook entry stating the results. It is necessary to verify which test radial is being transmitted and whether you should get a "to" or "from" indication.

f. Airborne and ground check points consist of certified radials that should be received at specific points on the airport surface or over specific landmarks while airborne in the immediate vicinity of the airport.

1. Should an error in excess of plus or minus 4 degrees be indicated through use of a ground check, or plus or minus 6 degrees using the airborne check, Instrument Flight Rules (IFR) flight must not be attempted without first correcting the source of the error.

CAUTION-No correction other than the correction card figures supplied by the manufacturer should be applied in making these VOR receiver checks.

2. Locations of airborne check points, ground check points and VOTs are published in the A/FD.

3. If a dual system VOR (units independent of each other except for the antenna) is installed in the aircraft, one system may be checked against the other. Turn both systems to the same VOR ground facility and note the indicated bearing to that station. The maximum permissible variations between the two indicated bearings is 4 degrees.

1-1-5. Tactical Air Navigation (TACAN)

a. For reasons peculiar to military or naval operations (unusual siting conditions, the pitching and rolling of a naval vessel, etc.) the civil VOR/Distance Measuring Equipment (DME) system of air navigation was considered unsuitable for military or naval use. A new navigational system, TACAN, was therefore developed by the military and naval forces to more readily lend itself to military and naval requirements. As a result, the FAA has integrated TACAN facilities with the civil VOR/DME program. Although the theoretical, or technical principles of operation of TACAN equipment are quite different from those of VOR/DME facilities, the end result, as far as the navigating pilot is concerned, is the same. These integrated facilities are called VORTACs.

b. TACAN ground equipment consists of either a fixed or mobile transmitting unit. The airborne unit in conjunction with the ground unit reduces the transmitted signal to a visual presentation of both azimuth and distance information. TACAN is a pulse system and operates in the Ultrahigh Frequency (UHF) band of frequencies. Its use requires TACAN airborne equipment and does not operate through conventional VOR equipment.

1-1-6. VHF Omni-directional Range/Tactical Air Navigation (VORTAC)

a. A VORTAC is a facility consisting of two components, VOR and TACAN, which provides three individual services: VOR azimuth, TACAN azimuth and TACAN distance (DME) at one site. Although consisting of more than one component, incorporating more than one operating frequency, and using more than one antenna system, a VORTAC is considered to be a unified navigational aid. Both components of a VORTAC are envisioned as operating simultaneously and providing the three services at all times.

b. Transmitted signals of VOR and TACAN are each identified by three-letter code transmission and are interlocked so that pilots using VOR azimuth with TACAN distance can be assured that both signals being received are definitely from the same ground station. The frequency channels of the VOR and the TACAN at each VORTAC facility are "paired" in accordance with a national plan to simplify airborne operation.

1-1-7. Distance Measuring Equipment (DME)

a. In the operation of DME, paired pulses at a specific spacing are sent out from the aircraft (this is the interrogation) and are received at the ground station. The ground station (transponder) then transmits paired pulses back to the aircraft at the same pulse spacing but on a different frequency. The time required for the round trip of this signal exchange is measured in the airborne DME unit and is translated into distance (nautical miles) from the aircraft to the ground station.

b. Operating on the line-of-sight principle, DME furnishes distance information with a very high degree of accuracy. Reliable signals may be received at distances up to 199 NM at line-of-sight altitude with an accuracy of better than $^1/_2$ mile or 3 percent of the distance, whichever is greater. Distance information received from DME equipment is SLANT RANGE distance and not actual horizontal distance.

c. Operating frequency range of a DME according to ICAO Annex 10 is from 960 MHz to 1215 MHz. Aircraft equipped with TACAN equipment will receive distance information from a VORTAC automatically, while aircraft equipped with VOR must have a separate DME airborne unit.

d. Aircraft equipped with slaved compass systems may be susceptible to heading errors caused by exposure to magnetic field disturbances (flux fields) found in materials that are commonly located on the surface or buried under taxiways and ramps. These materials generate a magnetic flux field that can be sensed by the aircraft's compass system flux detector or "gate", which can cause the aircraft's system to align with the material's magnetic field rather than the earth's natural magnetic field. The system's erroneous heading may not selfcorrect. Prior to take off pilots should be aware that a heading misalignment may have occurred during taxi. Pilots are encouraged to follow the manufacturer's or other appropriate procedures to correct possible heading misalignment before take off is commenced.

e. VOR/DME, VORTAC, Instrument Landing System (ILS)/DME, and localizer (LOC)/DME navigation facilities established by the FAA provide course and distance information from collocated components under a frequency pairing plan. Aircraft receiving equipment which provides for automatic DME selection assures reception of azimuth and distance information from a common source when designated VOR/DME, VORTAC, ILS/DME, and LOC/DME are selected.

f. Due to the limited number of available frequencies, assignment of paired frequencies is required for certain military noncollocated VOR and TACAN facilities which serve the same area but which may be separated by distances up to a few miles.

g. VOR/DME, VORTAC, ILS/DME, and LOC/DME facilities are identified by synchronized identifications which are transmitted on a time share basis. The VOR or localizer portion of the facility is identified by a coded tone modulated at 1020 Hz or a combination of code and voice. The TACAN or DME is identified by a coded tone modulated at 1350 Hz. The DME or TACAN coded identification is transmitted one time for each three or four times that the VOR or localizer coded identification is transmitted. When either the VOR or the DME is inoperative, it is important to recognize which identifier is retained for the operative facility. A single coded identification with a repetition interval of approximately 30 seconds indicates that the DME is operative.

h. Aircraft equipment which provides for automatic DME selection assures reception of azimuth and distance information from a common source when designated VOR/DME, VORTAC, and ILS/DME navigation facilities are selected. Pilots are cautioned to disregard any distance displays from automatically selected DME equipment when VOR or ILS facilities, which do not have the DME feature installed, are being used for position determination.

1-1-8. Navigational Aid (NAVAID) Service Volumes

a. Most air navigation radio aids which provide positive course guidance have a designated standard service volume (SSV). The SSV defines the reception limits of unrestricted NAVAIDs which are usable for random/unpublished route navigation.

b. A NAVAID will be classified as restricted if it does not conform to flight inspection signal strength and course quality standards throughout the published SSV. However, the NAVAID should not be considered usable at altitudes below that which could be flown while operating under random route IFR conditions (14 CFR Section 91.177), even though these altitudes may lie within the designated SSV. Service volume restrictions are first published in Notices to Airmen (NOTAMs) and then with the alphabetical listing of the NAVAIDs in the A/FD.

FIG 1-1-1
Standard High Altitude Service Volume (See FIG 1-1-5 for altitudes below 1,000 feet).

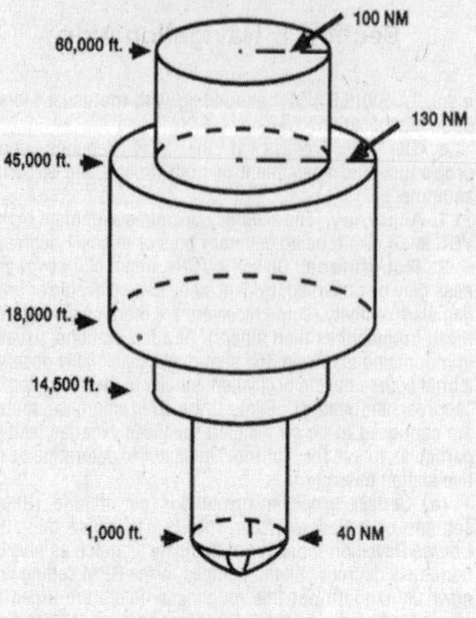

FIG 1-1-2
Standard Low Altitude Service Volume (See FIG 1-1-5 for altitudes below 1,000 feet).

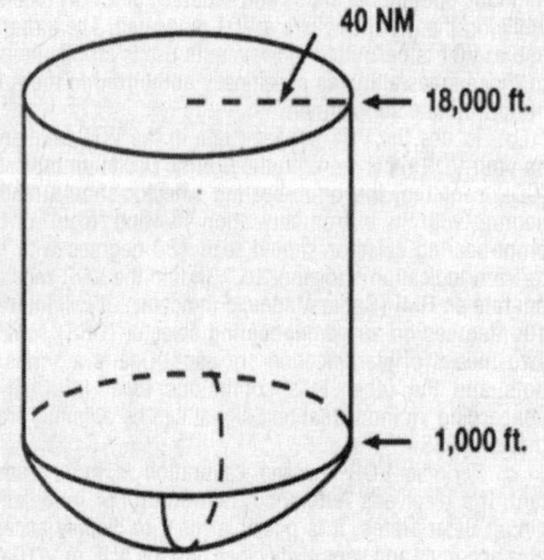

FIG 1-1-3
Standard Terminal Service Volume (See FIG 1-1-4 for altitudes below 1,000 feet).

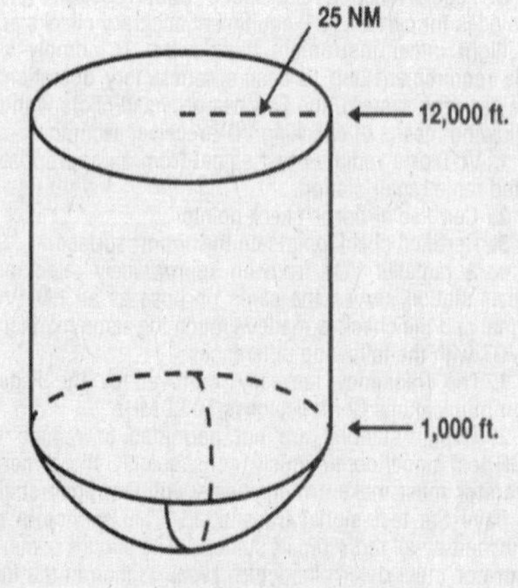

c. Standard Service Volume limitations do not apply to published IFR routes or procedures.

d. VOR/DME/TACAN Standard Service Volumes (SSV).

1. Standard service volumes (SSVs) are graphically shown in FIG 1-1-1, FIG 1-1-2, FIG 1-1-3, FIG 1-1-4, and FIG 1-1-5. The SSV of a station is indicated by using the class designator as a prefix to the station type designation.

2. Within 25 NM, the bottom of the T service volume is defined by the curve in FIG 1-1-4. Within 40 NM, the bottoms of the L and H service volumes are defined by the curve in FIG 1-1-5. (See TBL 1-1-1.)

e. Nondirectional Radio Beacon (NDB)

1. NDBs are classified according to their intended use.

2. The ranges of NDB service volumes are shown in TBL 1-1-2. The distances (radius) are the same at all altitudes.

TBL 1-1-1
VOR/DME/TACAN Standard Service Volumes

SSV Class Designator	Altitude and Range Boundaries
T (Terminal)	From 1,000 feet above ground level (AGL) up to and including 12,000 feet AGL at radial distances out to 25 NM.
L (Low Altitude)	From 1,000 feet AGL up to and including 18,000 feet AGL at radial distances out to 40 NM.
H (High Altitude)	From 1,000 feet AGL up to and including 14,500 feet AGL at radial distances out to 40 NM. From 14,500 AGL up to and including 60,000 feet at radial distances out to 100 NM. From 18,000 feet AGL up to and including 45,000 feet AGL at radial distances out to 130 NM.

TBL 1-1-2
NDB Service Volumes

Class	Distance (Radius)
Compass Locator	15 NM
MH	25 NM
H	50 NM*
HH	75 NM

**Service ranges of individual facilities may be less than 50 nautical miles (NM). Restrictions to service volumes are first published as a Notice to Airmen and then with the alphabetical listing of the NAVAID in the A/FD.*

FIG 1-1-4
Service Volume Lower Edge Terminal

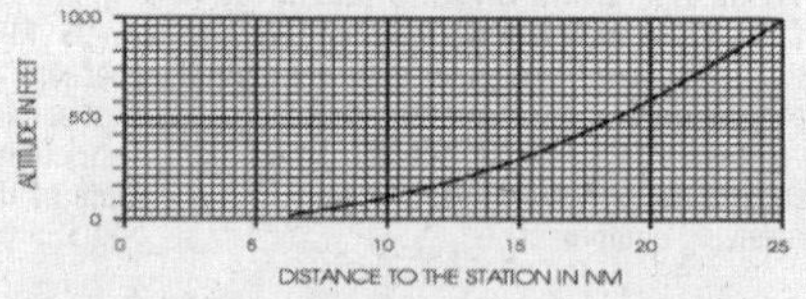

FIG 1-1-5
Service Volume Lower Edge Standard High and Low

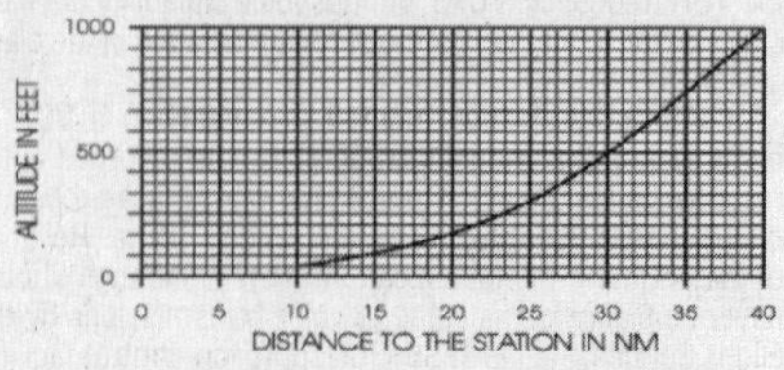

1-1-9. Instrument Landing System (ILS)

a. General

1. The ILS is designed to provide an approach path for exact alignment and descent of an aircraft on final approach to a runway.

2. The ground equipment consists of two highly directional transmitting systems and, along the approach, three (or fewer) marker beacons. The directional transmitters are known as the localizer and glide slope transmitters.

3. The system may be divided functionally into three parts:

(a) Guidance information: localizer, glide slope;

(b) Range information: marker beacon, DME; and

(c) Visual information: approach lights, touchdown and centerline lights, runway lights.

4. Precision radar, or compass locators located at the Outer Marker (OM) or Middle Marker (MM), may be substituted for marker beacons. DME, when specified in the procedure, may be substituted for the OM.

5. Where a complete ILS system is installed on each end of a runway; (i.e., the approach end of Runway 4 and the approach end of Runway 22) the ILS systems are not in service simultaneously.

b. Localizer

1. The localizer transmitter operates on one of 40 ILS channels within the frequency range of 108.10to 111.95 MHz. Signals provide the pilot with course guidance to the runway centerline.

2. The approach course of the localizer is called the front course and is used with other functional parts, e.g., glide slope, marker beacons, etc. The localizer signal is transmitted at the far end of the runway. It is adjusted for a course width of (full scale fly-left to a full scale fly-right) of 700 feet at the runway threshold.

3. The course line along the extended centerline of a runway, in the opposite direction to the front course is called the back course.

CAUTION-Unless the aircraft's ILS equipment includes reverse sensing capability, when flying inbound on the back course it is necessary to steer the aircraft in the direction opposite the needle deflection when making corrections from off-course to on-course. This "flying away from the needle" is also required when flying outbound on the front course of the localizer. Do not use back course signals for approach unless a back course approach procedure is published for that particular runway and the approach is authorized by ATC.

4. Identification is in International Morse Code and consists of a three-letter identifier preceded by the letter I (DD) transmitted on the localizer frequency.

EXAMPLE-*I-DIA*

5. The localizer provides course guidance throughout the descent path to the runway threshold from a distance of 18 NM from the antenna between an altitude of 1,000 feet above the highest terrain along the course line and 4,500 feet above the elevation of the antenna site. Proper off-course indications are provided throughout the following angular areas of the operational service volume:

(a) To 10 degrees either side of the course along a radius of 18 NM from the antenna; and

(b) From 10 to 35 degrees either side of the course along a radius of 10 NM. (See FIG 1-1-6.)

6. Unreliable signals may be received outside these areas.

c. Localizer Type Directional Aid (LDA)

1. The LDA is of comparable use and accuracy to a localizer but is not part of a complete ILS. The LDA course usually provides a more precise approach course than the similar Simplified Directional Facility (SDF) installation, which may have a course width of 6 or 12 degrees.

2. The LDA is not aligned with the runway. Straight-in minimums may be published where alignment does not exceed 30 degrees between the course and runway. Circling minimums only are published where this alignment exceeds 30 degrees.

3. A very limited number of LDA approaches also incorporate a glideslope. These are annotated in the plan view of the instrument approach chart with a note, "LDA/Glideslope." These procedures fall under a newly defined category of approaches called Approach with Vertical Guidance (APV) described in paragraph 5-4-5, Instrument Approach Procedure Charts, subparagraph a7(b), Approach with Vertical Guidance (APV). LDA minima for with and without glideslope is provided and annotated on the minima lines of the approach chart as S-LDA/GS and S-LDA. Because the final approach course is not aligned with the runway centerline, additional maneuvering will be required compared to an ILS approach.

Fig 1-1-6

Limits of Localizer Coverage

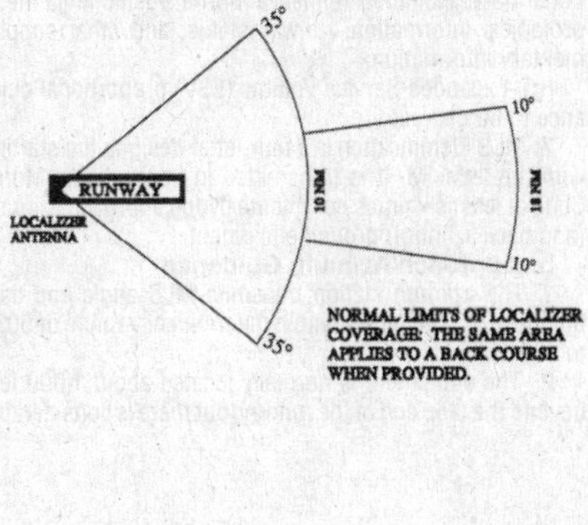

d. Glide Slope/Glide Path

1. The UHF glide slope transmitter, operating on one of the 40 ILS channels within the frequency range 329.15 MHz, to 335.00 MHz radiates its signals in the direction of the localizer front course. The term "glide path" means that portion of the glide slope that intersects the localizer.

CAUTION-False glide slope signals may exist in the area of the localizer back course approach which can cause the glide slope flag alarm to disappear and present unreliable glide slope information. Disregard all glide slope signal indications when making a localizer back course approach unless a glide slope is specified on the approach and landing chart.

2. The glide slope transmitter is located between 750 feet and 1,250 feet from the approach end of the runway (down the runway) and offset 250 to 650 feet from the runway centerline. It transmits a glide path beam 1.4 degrees wide (vertically). The signal provides descent information for navigation down to the lowest authorized decision height (DH) specified in the approved ILS approach procedure. The glidepath may not be suitable for navigation below the lowest authorized DH and any reference to glidepath indications below that height must be supplemented by visual reference to the runway environment. Glidepaths with no published DH are usable to runway threshold.

3. The glide path projection angle is normally adjusted to 3 degrees above horizontal so that it intersects the MM at about 200 feet and the OM at about 1,400 feet above the runway elevation. The glide slope is normally usable to the distance of 10NM. However, at some locations, the glide slope has been certified for an extended service volume which exceeds 10 NM.

4. Pilots must be alert when approaching the glidepath interception. False courses and reverse sensing will occur at angles considerably greater than the published path.

5. Make every effort to remain on the indicated glide path.

CAUTION-Avoid flying below the glide path to assure obstacle/terrain clearance is maintained.

6. The published glide slope threshold crossing height (TCH) DOES NOT represent the height of the actual glide path on-course indication above the runway threshold. It is used as a reference for planning purposes which represents the height above the runway threshold that an aircraft's glide slope antenna should be, if that aircraft remains on a trajectory formed by the four-mile-to-middle marker glidepath segment.

7. Pilots must be aware of the vertical height between the aircraft's glide slope antenna and the main gear in the landing configuration and, at the DH, plan to adjust the descent angle accordingly if the published TCH indicates the wheel crossing height over the runway threshold may not be satisfactory. Tests indicate a comfortable wheel crossing height is approximately 20 to 30 feet, depending on the type of aircraft.

NOTE-*The TCH for a runway is established based on several factors including the largest aircraft category that normally uses the runway, how airport layout effects the glide slope antenna placement, and terrain. A higher than optimum TCH, with the same glide path angle, may cause the aircraft to touch down further from the threshold if the trajectory of the approach is maintained until the flare. Pilots should consider the effect of a high TCH on the runway available for stopping the aircraft.*

e. Distance Measuring Equipment (DME)

1. When installed with the ILS and specified in the approach procedure, DME may be used:

(a) In lieu of the OM;

(b) As a back course (BC) final approach fix (FAF); and

(c) To establish other fixes on the localizer course.

2. In some cases, DME from a separate facility may be used within Terminal Instrument Procedures (TERPS) limitations:

(a) To provide ARC initial approach segments;

(b) As a FAF for BC approaches; and

(c) As a substitute for the OM.

f. Marker Beacon

1. ILS marker beacons have a rated power output of 3 watts or less and an antenna array designed to produce an elliptical pattern with dimensions, at 1,000 feet above the antenna, of approximately 2,400 feet in width and 4,200 feet in length. Airborne marker beacon receivers with a selective sensitivity feature should always be operated in the "low" sensitivity position for proper reception of ILS marker beacons.

2. Ordinarily, there are two marker beacons associated with an ILS, the OM and MM. Locations with a Category II ILS also have an Inner Marker (IM). When an aircraft passes over a marker, the pilot will receive the indications shown in TBL 1-1-3.

(a) The OM normally indicates a position at which an aircraft at the appropriate altitude on the localizer course will intercept the ILS glide path.

(b) The MM indicates a position approximately 3,500 feet from the landing threshold. This is also the position where an aircraft on the glide path will be at an altitude of approximately 200 feet above the elevation of the touchdown zone.

(c) The IM will indicate a point at which an aircraft is at a designated decision height (DH) on the glide path between the MM and landing threshold.

Tbl 1-1-3

Marker Passage Indications

Marker	Code	Light
OM	– –×–	BLUE
MM	D ×– D –	AMBER
IM	D D D D	WHITE
BC	D D D D	WHITE

3. A back course marker normally indicates the ILS back course final approach fix where approach descent is commenced.

g.Compass Locator

1. Compass locator transmitters are often situated at the MM and OM sites. The transmitters have a power of less than 25 watts, a range of at least 15 miles and operate between 190 and 535 kHz. At some locations, higher powered radio beacons, up to 400 watts, are used as OM compass locators. These generally carry Transcribed Weather Broadcast (TWEB) information.

2. Compass locators transmit two letter identification groups. The outer locator transmits the first two letters of the localizer identification group, and the middle locator transmits the last two letters of the localizer identification group.

h. ILS Frequency (See TBL 1-1-4.)

i. ILS Minimums

1. The lowest authorized ILS minimums, with all required ground and airborne systems components operative, are:

(a) Category I. Decision Height (DH) 200feet and Runway Visual Range (RVR) 2,400 feet (with touchdown zone and centerline lighting, RVR 1,800 feet), or (with Autopilot or FD or HUD, RVR 1,800 feet);

(b) Special Authorization Category I. DH150 feet and Runway Visual Range (RVR) 1,400 feet, HUD to DH;

(c) Category II. DH 100 feet and RVR 1,200 feet (with autoland or HUD to touchdown and noted on authorization, RVR 1,000 feet);

(d) Special Authorization Category II with Reduced Lighting. DH 100 feet and RVR 1,200 feet with autoland or HUD to touchdown and noted on authori-

Tbl 1-1-4
Frequency Pairs Allocated for ILS

Localizer MHz	Glide Slope
108.10	334.70
108.15	334.55
108.3	334.10
108.35	333.95
108.5	329.90
108.55	329.75
108.7	330.50
108.75	330.35
108.9	329.30
108.95	329.15
109.1	331.40
109.15	331.25
109.3	332.00
109.35	331.85
109.50	332.60
109.55	332.45
109.70	333.20
109.75	333.05
109.90	333.80
109.95	333.65
110.1	334.40
110.15	334.25
110.3	335.00
110.35	334.85
110.5	329.60
110.55	329.45
110.70	330.20
110.75	330.05
110.90	330.80
110.95	330.65
111.10	331.70
111.15	331.55
111.30	332.30
111.35	332.15
111.50	332.9
111.55	332.75
111.70	333.5
111.75	333.35
111.90	331.1
111.95	330.95

zation (touchdown zone, centerline lighting, and ALSF-2 are not required);

(e) Category IIIa. No DH or DH below 100 feet and RVR not less than 700 feet;

(f) Category IIIb. No DH or DH below 50 feet and RVR less than 700 feet but not less than 150 feet; and

(g) Category IIIc. No DH and no RVR limitation.

NOTE-*Special authorization and equipment required for Categories II and III.*

j. Inoperative ILS Components

1. Inoperative localizer. When the localizer fails, an ILS approach is not authorized.

2. Inoperative glide slope. When the glide slope fails, the ILS reverts to a nonprecision localizer approach.

REFERENCE-*See the inoperative component table in the U.S. Government Terminal Procedures Publication (TPP), for adjustments to minimums due to inoperative airborne or ground system equipment.*

k. ILS Course Distortion

1. All pilots should be aware that disturbances to ILS localizer and glide slope courses may occur when surface vehicles or aircraft are operated near the localizer or glide slope antennas. Most ILS installations are subject to signal interference by either surface vehicles, aircraft or both. ILS CRITICAL AREAS are established near each localizer and glide slope antenna.

2. ATC issues control instructions to avoid interfering operations within ILS critical areas at controlled airports during the hours the Airport Traffic Control Tower (ATCT) is in operation as follows:

(a) Weather Conditions. Less than ceiling 800 feet and/or visibility 2 miles.

(1) Localizer Critical Area. Except for aircraft that land, exit a runway, depart or miss approach, vehicles and aircraft are not authorized in or over the critical area when an arriving aircraft is between the ILS final approach fix and the airport. Additionally, when the ceiling is less than 200 feet and/or the visibility is RVR 2,000 or less, vehicle and aircraft operations in or over the area are not authorized when an arriving aircraft is inside the ILS MM.

(2) Glide Slope Critical Area. Vehicles and aircraft are not authorized in the area when an arriving aircraft is between the ILS final approach fix and the airport unless the aircraft has reported the airport in sight and is circling or side stepping to land on a runway other than the ILS runway.

(b) Weather Conditions. At or above ceiling 800 feet and/or visibility 2 miles.

(1) No critical area protective action is provided under these conditions.

(2) A flight crew, under these conditions, should advise the tower that it will conduct an AUTOLAND or COUPLED approach to ensure that the ILS critical areas are protected when the aircraft is inside the ILS MM.

EXAMPLE-*Glide slope signal not protected.*

3. Aircraft holding below 5,000 feet between the outer marker and the airport may cause localizer signal variations for aircraft conducting the ILS approach. Accordingly, such holding is not authorized when weather or visibility conditions are less than ceiling 800 feet and/or visibility 2 miles.

4. Pilots are cautioned that vehicular traffic not subject to ATC may cause momentary deviation to ILS course or glide slope signals. Also, critical areas are not protected at uncontrolled airports or at airports with an operating control tower when weather or visibility conditions are above those requiring protective measures. Aircraft conducting coupled or autoland operations should be especially alert in monitoring automatic flight control systems.

(See FIG 1-1-7.)

NOTE-*Unless otherwise coordinated through Flight Standards, ILS signals to Category I runways are not flight inspected below 100 feet AGL. Guidance signal anomalies may be encountered below this altitude.*

Fig 1-1-7
FAA Instrument Landing Systems

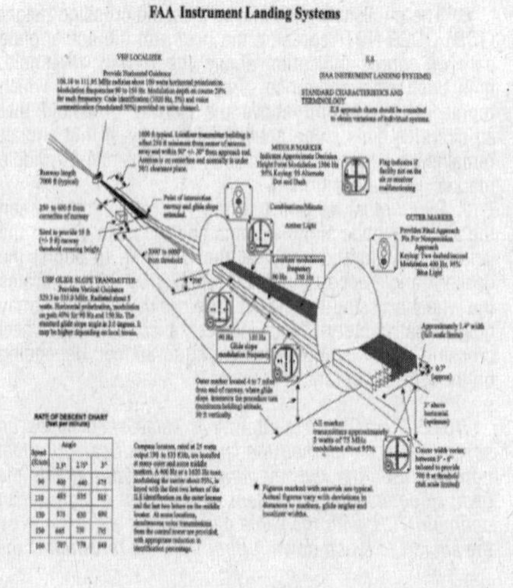

1-1-10. Simplified Directional Facility (SDF)

a. The SDF provides a final approach course similar to that of the ILS localizer. It does not provide glide slope information. A clear understanding of the ILS localizer and the additional factors listed below completely describe the operational characteristics and use of the SDF.

b. The SDF transmits signals within the range of 108.10 to 111.95 MHz.

c. The approach techniques and procedures used in an SDF instrument approach are essentially the same as those employed in executing a standard localizer approach except the SDF course may not be aligned with the runway and the course may be wider, resulting in less precision.

d. Usable off-course indications are limited to 35 degrees either side of the course centerline. Instrument indications received beyond 35 degrees should be disregarded.

e. The SDF antenna may be offset from the runway centerline. Because of this, the angle of convergence between the final approach course and the runway bearing should be determined by reference to the instrument approach procedure chart. This angle is generally not more than 3 degrees. However, it should be noted that inasmuch as the approach course originates at the antenna site, an approach which is continued beyond the runway threshold will lead the aircraft to the SDF offset position rather than along the runway centerline.

f. The SDF signal is fixed at either 6 degrees or 12 degrees as necessary to provide maximum flyability and optimum course quality.

g. Identification consists of a three-letter identifier transmitted in Morse Code on the SDF frequency. The appropriate instrument approach chart will indicate the identifier used at a particular airport.

1-1-11. Microwave Landing System (MLS)

a. General

1. The MLS provides precision navigation guidance for exact alignment and descent of aircraft on approach to a runway. It provides azimuth, elevation, and distance.

2. Both lateral and vertical guidance may be displayed on conventional course deviation indicators or incorporated into multipurpose cockpit displays. Range information can be displayed by conventional DME indicators and also incorporated into multipurpose displays.

3. The MLS supplements the ILS as the standard landing system in the U.S. for civil, military, and international civil aviation. At international airports, ILS service is protected to 2010.

4. The system may be divided into five functions:

(a) Approach azimuth;

(b) Back azimuth;

(c) Approach elevation;

(d) Range; and

(e) Data communications.

5. The standard configuration of MLS ground equipment includes:

(a) An azimuth station to perform functions (a) and (e) above. In addition to providing azimuth navigation guidance, the station transmits basic data which consists of information associated directly with the operation of the landing system, as well as advisory data on the performance of the ground equipment.

(b) An elevation station to perform function (c).

(c) Distance Measuring Equipment (DME) to perform range guidance, both standard DME (DME/N) and precision DME (DME/P).

6. MLS Expansion Capabilities. The standard configuration can be expanded by adding one or more of the following functions or characteristics.

(a) Back azimuth. Provides lateral guidance for missed approach and departure navigation.

(b) Auxiliary data transmissions. Provides additional data, including refined airborne positioning, meteorological information, runway status, and other supplementary information.

(c) Expanded Service Volume (ESV) proportional guidance to 60 degrees.

7. MLS identification is a four-letter designation starting with the letter M. It is transmitted in International Morse Code at least six times per minute by the approach azimuth (and back azimuth) ground equipment.

b. Approach Azimuth Guidance

1. The azimuth station transmits MLS angle and data on one of 200 channels within the frequency range of 5031 to 5091 MHz.

2. The equipment is normally located about 1,000 feet beyond the stop end of the runway, but there is considerable

FIG 1-1-8
Coverage Volume Azimuth

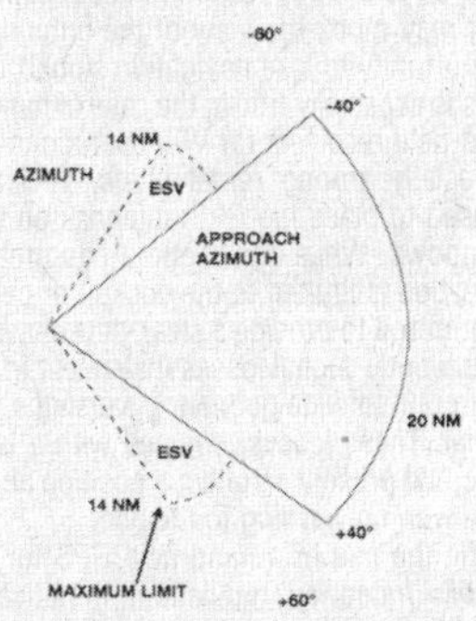

flexibility in selecting sites. For example, for heliport operations the azimuth transmitter can be collocated with the elevation transmitter.

3. The azimuth coverage extends:

(See FIG 1-1-8.)

(a) Laterally, at least 40 degrees on either side of the runway centerline in a standard configuration,

(b) In elevation, up to an angle of 15 degrees and to at least 20,000 feet, and

(c) In range, to at least 20 NM.

c. Elevation Guidance

1. The elevation station transmits signals on the same frequency as the azimuth station. A single frequency is time-shared between angle and data functions.

2. The elevation transmitter is normally located about 400 feet from the side of the runway between runway threshold and the touchdown zone.

3. Elevation coverage is provided in the same airspace as the azimuth guidance signals:

(a) In elevation, to at least +15 degrees;

(b) Laterally, to fill the Azimuth lateral coverage; and

(c) In range, to at least 20 NM.

(See FIG 1-1-9.)

d. Range Guidance

1. The MLS Precision Distance Measuring Equipment (DME/P) functions the same as the navigation DME described in paragraph 1-1-7, Distance Measuring Equipment (DME), but there are some technical differences. The beacon transponder operates in the frequency band 962 to 1105 MHz and responds to an aircraft interrogator. The MLS DME/P accuracy is improved to be consistent with the accuracy provided by the MLS azimuth and elevation stations.

2. A DME/P channel is paired with the azimuth and elevation channel. A complete listing of the 200 paired channels of the DME/P with the angle functions is contained in FAA Standard 022 (MLS Interoperability and Performance Requirements).

3. The DME/N or DME/P is an integral part of the MLS and is installed at all MLS facilities unless a waiver is obtained. This occurs infrequently and only at outlying, low density airports where marker beacons or compass locators are already in place.

e. Data Communications

1. The data transmission can include both the basic and auxiliary data words. All MLS facilities transmit basic data. Where needed, auxiliary data can be transmitted.

FIG 1-1-9
Coverage Volumes Elevation

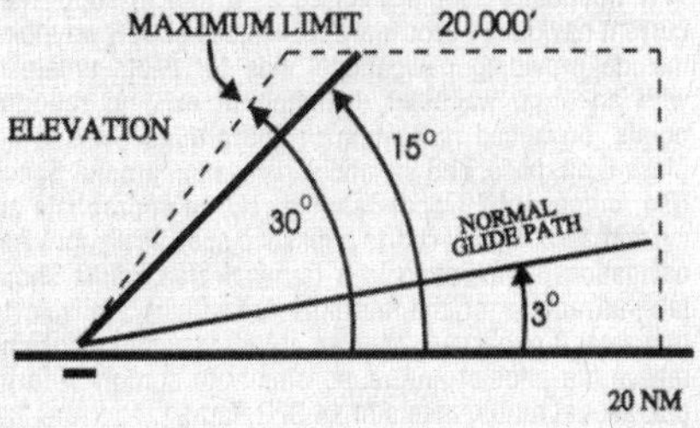

2. Coverage limits. MLS data are transmitted throughout the azimuth (and back azimuth when provided) coverage sectors.

3. Basic data content. Representative data include:

(a) Station identification;

(b) Exact locations of azimuth, elevation and DME/P stations (for MLS receiver processing functions);

(c) Ground equipment performance level; and

(d) DME/P channel and status.

4. Auxiliary data content: Representative data include:

(a) 3-D locations of MLS equipment;

(b) Waypoint coordinates;

(c) Runway conditions; and

(d) Weather (e.g., RVR, ceiling, altimeter setting, wind, wake vortex, wind shear).

f. Operational Flexibility

1. The MLS has the capability to fulfill a variety of needs in the approach, landing, missed approach and departure phases of flight. For example:

(a) Curved and segmented approaches;

(b) Selectable glide path angles;

(c) Accurate 3-D positioning of the aircraft in space; and

(d) The establishment of boundaries to ensure clearance from obstructions in the terminal area.

2. While many of these capabilities are available to any MLS-equipped aircraft, the more sophisticated capabilities (such as curved and segmented approaches) are dependent upon the particular capabilities of the airborne equipment.

g. Summary

1. Accuracy. The MLS provides precision three-dimensional navigation guidance accurate enough for all approach and landing maneuvers.

2. Coverage. Accuracy is consistent throughout the coverage volumes. (See FIG 1-1-10.)

FIG 1-1-10
Coverage Volumes 3-D Representation

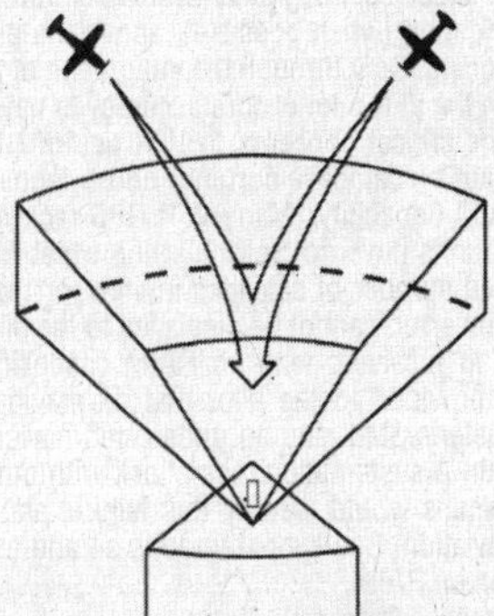

3. Environment. The system has low susceptibility to interference from weather conditions and airport ground traffic.

4. Channels. MLS has 200 channels- enough for any foreseeable need.

5. Data. The MLS transmits ground-air data messages associated with the systems operation.

6. Range information. Continuous range information is provided with an accuracy of about 100 feet.

1-1-12. NAVAID Identifier Removal During Maintenance

During periods of routine or emergency maintenance, coded identification (or code and voice, where applicable) is removed from certain FAA NAVAIDs. Removal of identification serves as a warning to pilots that the facility is officially off the air for tune-up or repair and may be unreliable even though intermittent or constant signals are received.

NOTE-*During periods of maintenance VHF ranges may radiate a T-E-S-T code (- D DDD -).*

NOTE-*DO NOT attempt to fly a procedure that is NOTAMed out of service even if the identification is present. In certain cases, the identification may be transmitted for short periods as part of the testing.*

1-1-13. NAVAIDs with Voice

a. Voice equipped en route radio navigational aids are under the operational control of either a Flight Service Station (FSS) or an approach control facility. The voice communication is available on some facilities. Hazardous Inflight Weather Advisory Service (HIWAS) broadcast capability is available on selected VOR sites throughout the conterminous U.S. and does not provide two-way voice communication. The availability of two-way voice communication and HIWAS is indicated in the A/FD and aeronautical charts.

b. Unless otherwise noted on the chart, all radio navigation aids operate continuously except during shutdowns for maintenance. Hours of operation of facilities not operating continuously are annotated on charts and in the A/FD.

1-1-14. User Reports Requested on NAVAID or Global Navigation Satellite System (GNSS) Performance or Interference

a. Users of the National Airspace System (NAS) can render valuable assistance in the early correction of NAVAID malfunctions or GNSS problems and are encouraged to report their observations of undesirable performance. Although NAVAIDs are monitored by electronic detectors, adverse effects of electronic interference, new obstructions or changes in terrain near the NAVAID can exist without detection by the ground monitors. Some of the characteristics of malfunction or deteriorating performance which should be reported are: erratic course or bearing indications; intermittent, or full, flag alarm; garbled, missing or obviously improper coded identification; poor quality communications reception; or, in the case of frequency interference, an audible hum or tone accompanying radio communications or NAVAID identification. GNSS problems are often characterized by navigation degradation or service loss indications.

b. Reporters should identify the NAVAID (for example, VOR) malfunction or GNSS problem, location of the aircraft (i.e., latitude, longitude or bearing/distance from a NAVAID), altitude, date and time of the observation, type of aircraft and description of the condition observed, and the type of receivers in use (i.e., make/model/software revision). Reports can be made in any of the following ways:

1. Immediately, by radio communication to the controlling Air Route Traffic Control Center (ARTCC), Control Tower, or FSS.

2. By telephone to the nearest FAA facility.

3. By FAA Form 8740-5, Safety Improvement Report, a postage-paid card designed for this purpose. These cards may be obtained at FAA FSSs, Flight Standards District Offices, and General Aviation Fixed Base Operations.

c. In aircraft that have more than one receiver, there are many combinations of possible interference between units. This can cause either erroneous navigation indications or, complete or partial blanking out of the communications. Pilots should be familiar enough with the radio installation of the particular airplanes they fly to recognize this type of interference.

1-1-15. LORAN

a. Introduction

NOTE-*In accordance with the 2010 DHS Appropriations Act, the U.S. Coast Guard (USCG) terminated the transmission of all U.S. LORAN-C signals on 08 Feb 2010. The USCG also terminated the transmission of the Russian American signals on 01 Aug 2010, and the Canadian LORAN-C signals on 03 Aug 2010. For more information, visit* ***http://www.navcen.uscg.gov****. Operators should also note that TSO-C60b, AIRBORNE AREA NAVIGATION EQUIPMENT USING LORAN-C INPUTS, has been canceled by the FAA.*

1-1-16. Inertial Reference Unit (IRU), Inertial Navigation System (INS), and Attitude Heading Reference System (AHRS)

a. IRUs are self-contained systems comprised of gyros and accelerometers that provide aircraft attitude (pitch, roll, and heading), position, and velocity information in response to signals resulting from inertial effects on system components. Once aligned with a known position, IRUs continuously calculate position and velocity. IRU position accuracy decays with time. This degradation is known as "drift."

b. INSs combine the components of an IRU with an internal navigation computer. By programming a series of waypoints, these systems will navigate along a predetermined track.

c. AHRSs are electronic devices that provide attitude information to aircraft systems such as weather radar and autopilot, but do not directly compute position information.

1-1-17. Doppler Radar

Doppler Radar is a semiautomatic self-contained dead reckoning navigation system (radar sensor plus computer) which is not continuously dependent on information derived from ground based or external aids. The system employs radar signals to detect and measure ground speed and drift angle, using the aircraft compass system as its directional reference. Doppler is less accurate than INS, however, and the use of an external reference is required for periodic updates if acceptable position accuracy is to be achieved on long range flights.

1-1-18. Global Positioning System (GPS)

a. System Overview

1. System Description. The Global Positioning System is a satellite-based radio navigation system, which broadcasts a signal that is used by receivers to determine precise position anywhere in the world. The receiver tracks multiple satellites and determines a pseudorange measurement that is then used to determine the user location. A minimum of four satellites is necessary to establish an accurate three-dimensional position. The Department of Defense (DOD) is responsible for operating the GPS satellite constellation and monitors the GPS satellites to ensure proper operation. Every satellite's orbital parameters (ephemeris data) are sent to each satellite for broadcast as part of the data message embedded in the GPS signal. The GPS coordinate system is the Cartesian earth-centered earth-fixed coordinates as specified in the World Geodetic System 1984 (WGS-84).

2. System Availability and Reliability

(a) The status of GPS satellites is broadcast as part of the data message transmitted by the GPS satellites. GPS status information is also available by means of the U.S. Coast Guard navigation information service: (703) 313-5907, Internet:**http://www.navcen.uscg.gov/**. Additionally, satellite status is available through the Notice to Airmen (NOTAM) system.

(b) The operational status of GNSS operations depends upon the type of equipment being used. For GPS-only equipment TSO-C129(a), the operational status of nonprecision approach capability for flight planning purposes is provided through a prediction program that is embedded in the receiver or provided separately.

3. Receiver Autonomous Integrity Monitoring (RAIM). When GNSS equipment is not using integrity information from WAAS or LAAS, the GPS navigation receiver using RAIM provides GPS signal integrity monitoring. RAIM is necessary since delays of up to two hours can occur before an erroneous satellite transmission can be detected and corrected by the satellite control segment. The RAIM function is also referred to as fault detection. Another capability, fault exclusion, refers to the ability of the receiver to exclude a failed satellite from the position solution and is provided by some GPS receivers and by WAAS receivers.

4. The GPS receiver verifies the integrity (usability) of the signals received from the GPS constellation through receiver autonomous integrity monitoring (RAIM) to determine if a satellite is providing corrupted information. At least one satellite, in addition to those required for navigation, must be in view for the receiver to perform the RAIM function; thus, RAIM needs a minimum of 5 satellites in view, or 4 satellites and a barometric altimeter (baro-aiding) to detect an integrity anomaly. [Baro-aiding satisfies the RAIM requirement in lieu of a fifth satellite.] For receivers capable of doing so, RAIM needs 6 satellites in view (or 5 satellites with baro-aiding) to isolate the corrupt satellite signal and remove it from the navigation solution. Baro-aiding is a method of augmenting the GPS integrity solution by using a nonsatellite input source. GPS derived altitude should not be relied upon to determine aircraft altitude since the vertical error can be quite large and no integrity is provided. To ensure that baro-aiding is available, the current altimeter setting must be entered into the receiver as described in the operating manual.

5. RAIM messages vary somewhat between receivers; however, generally there are two types. One type indicates that there are not enough satellites available to provide RAIM integrity monitoring and another type indicates that the RAIM integrity monitor has detected a potential error that exceeds the limit for the current phase of flight. Without RAIM capability, the pilot has no assurance of the accuracy of the GPS position.

6. Selective Availability. Selective Availability (SA) is a method by which the accuracy of GPS is intentionally degraded. This feature is designed to deny hostile use of precise GPS positioning data. SA was discontinued on May 1, 2000, but many GPS receivers are designed to assume that SA is still active. New receivers may take advantage of the discontinuance of SA based on the performance values in ICAO Annex 10, and do not need to be designed to operate outside of that performance.

7. The GPS constellation of 24 satellites is designed so that a minimum of five is always observable by a user anywhere on earth. The receiver uses data from a minimum of four satellites above the mask angle (the lowest angle above the horizon at which it can use a satellite).

8. The DOD declared initial operational capability (IOC) of the U.S. GPS on December 8, 1993. The FAA has granted approval for U.S. civil operators to use properly certified GPS equipment as a primary means of navigation in oceanic airspace and certain remote areas. Properly certified GPS equipment may be used as a supplemental means of IFR navigation for domestic en route, terminal operations, and certain instrument approach procedures (IAPs). This approval permits the use of GPS in a manner that is consistent with current navigation requirements as well as approved air carrier operations specifications.

b. VFR Use of GPS

1. GPS navigation has become a great asset to VFR pilots, providing increased navigation capability and enhanced situational awareness, while reducing operating costs due to greater ease in flying direct routes. While GPS has many benefits to the VFR pilot, care must be exercised to ensure that system capabilities are not exceeded.

2. Types of receivers used for GPS navigation under VFR are varied, from a full IFR installation being used to support a VFR flight, to a VFR only installation (in either a VFR or IFR capable aircraft) to a hand-held receiver. The limitations of each type of receiver installation or use must be understood by the pilot to avoid misusing navigation information. (See TBL 1-1-6.) In all cases, VFR pilots should never rely solely on one system of navigation. GPS navigation must be integrated with other forms of electronic navigation (when possible), as well as pilotage and dead reckoning. Only through the integration of these techniques can the VFR pilot ensure accuracy in navigation.

3. Some critical concerns in VFR use of GPS include RAIM capability, database currency and antenna location.

(a) RAIM Capability. Many VFR GPS receivers and all hand-held units have no RAIM alerting capability. Loss of the required number of satellites in view, or the detection of a position error, cannot be displayed to the pilot by such receivers. In receivers with no RAIM capability, no alert would be provided to the pilot that the navigation solution had deteriorated, and an undetected navigation error could occur. A systematic cross-check with other navigation techniques would identify this failure, and prevent a serious deviation. See subparagraphs a4 and a5 for more information on RAIM.

(b) Database Currency

(1) In many receivers, an up-datable database is used for navigation fixes, airports, and instrument procedures. These databases must be maintained to the current update for IFR operation, but no such requirement exists for VFR use.

(2) However, in many cases, the database drives a moving map display which indicates Special Use Airspace and the various classes of airspace, in addition to other operational information. Without a current database the moving map display may be outdated and offer erroneous information to VFR pilots wishing to fly around critical airspace areas, such as a Restricted Area or a Class B airspace segment. Numerous pilots have ventured into airspace they were trying to avoid by using an outdated database. If you don't have a current database in the receiver, disregard the moving map display for critical navigation decisions.

(3) In addition, waypoints are added, removed, relocated, or re-named as required to meet operational needs. When using GPS to navigate relative to a named fix, a current database must be used to properly locate a named waypoint. Without the update, it is the pilot's responsibility to verify the waypoint location referencing to an official current source, such as the Airport/Facility Directory, Sectional Chart, or En Route Chart.

(c) Antenna Location

(1) In many VFR installations of GPS receivers, antenna location is more a matter of convenience than performance. In IFR installations, care is exercised to ensure that an adequate clear view is provided for the antenna to see satellites. If an alternate location is used, some portion of the aircraft may block the view of the antenna, causing a greater opportunity to lose navigation signal.

(2) This is especially true in the case of hand-helds. The use of hand-held receivers for VFR operations is a growing trend, especially among rental pilots. Typically, suction cups are used to place the GPS antennas on the inside of cockpit windows. While this method has great utility, the antenna location is limited to the cockpit or cabin only and is rarely optimized to provide a clear view of available satellites. Consequently, signal losses may occur in certain situations of aircraft-satellite geometry, causing a loss of navigation signal. These losses, coupled with a lack of RAIM capability, could present erroneous position and navigation information with no warning to the pilot.

(3) While the use of a hand-held GPS for VFR operations is not limited by regulation, modification of the aircraft, such as installing a panel- or yoke-mounted holder, is governed by 14 CFR Part 43. Consult with your mechanic to ensure compliance with the regulation, and a safe installation.

4. As a result of these and other concerns, here are some tips for using GPS for VFR operations:

(a) Always check to see if your unit has RAIM capability. If no RAIM capability exists, be suspicious of your GPS position when any disagreement exists with the position derived from other radio navigation systems, pilotage, or dead reckoning.

(b) Check the currency of the database, if any. If expired, update the database using the current revision. If an update of an expired database is not possible, disregard any moving map display of airspace for critical navigation decisions. Be aware that named waypoints may no longer exist or may have been relocated since the database expired. At a minimum, the waypoints planned to be used should be checked against a current official source, such as the Airport/Facility Directory, or a Sectional Aeronautical Chart.

(c) While hand-helds can provide excellent navigation capability to VFR pilots, be prepared for intermittent loss of navigation signal, possibly with no RAIM warning to the pilot. If mounting the receiver in the aircraft, be sure to comply with 14 CFR Part 43.

(d) Plan flights carefully before taking off. If you wish to navigate to user-defined waypoints, enter them before flight, not on-the-fly. Verify your planned flight against a current source, such as a current sectional chart. There have been cases in which one pilot used waypoints created by another pilot that were not where the pilot flying was expecting. This generally resulted in a navigation error. Minimize head-down time in the aircraft and keep a sharp lookout for traffic, terrain, and obstacles. Just a few minutes of preparation and planning on the ground will make a great difference in the air.

(e) Another way to minimize head-down time is to become very familiar with your receiver's operation. Most receivers are not intuitive. The pilot must take the time to learn the various keystrokes, knob functions, and displays that are used in the operation of the receiver. Some manufacturers provide computer-based tutorials or simulations of their receivers. Take the time to learn about your particular unit before you try to use it in flight.

5. In summary, be careful not to rely on GPS to solve all your VFR navigational problems. Unless an IFR receiver is installed in accordance with IFR requirements, no standard of accuracy or integrity has been assured. While the practicality of GPS is compelling, the fact remains that only the pilot can navigate the aircraft, and GPS is just one of the pilot's tools to do the job.

c. VFR Waypoints

1. VFR waypoints provide VFR pilots with a supplementary tool to assist with position awareness while navigating visually in aircraft equipped with area navigation receivers. VFR waypoints should be used as a tool to supplement current navigation procedures. The uses of VFR waypoints include providing navigational aids for pilots unfamiliar with an area, waypoint definition of existing reporting points, enhanced navigation in and around Class B and Class C airspace, and enhanced navigation around Special Use Airspace. VFR pilots should rely on appropriate and current aeronautical charts published specifically for visual navigation. If operating in a terminal area, pilots should take advantage of the Terminal Area Chart available for that area, if published. The use of VFR waypoints does not relieve the pilot of any responsibility to comply with the operational requirements of 14 CFR Part 91.

2. VFR waypoint names (for computer-entry and flight plans) consist of five letters beginning with the letters "VP" and are retrievable from navigation databases. The VFR waypoint names are not intended to be pronounceable, and they are not for use in ATC communications. On VFR charts, stand-alone VFR waypoints will be portrayed using the same four-point star symbol used for IFR waypoints. VFR waypoints collocated with visual check points on the chart will be identified by small magenta flag symbols. VFR waypoints collocated with visual check points will be pronounceable based on the name of the visual check point and may be used for ATC communications. Each VFR waypoint name will appear in parentheses adjacent to the geographic location on the chart. Latitude/longitude data for all established VFR waypoints may be found in the appropriate regional Airport/Facility Directory (A/FD).

3. VFR waypoints must not be used to plan flights under IFR. VFR waypoints will not be recognized by the IFR system and will be rejected for IFR routing purposes.

4. When filing VFR flight plans, pilots may use the five letter identifier as a waypoint in the route of flight section if there is an intended course change at that point or if used to describe the planned route of flight. This VFR filing would be similar to how a VOR would be used in a route of flight. Pilots must use the VFR waypoints only when operating under VFR conditions.

5. Any VFR waypoints intended for use during a flight should be loaded into the receiver while on the ground and prior to departure. Once airborne, pilots should avoid programming routes or VFR waypoint chains into their receivers.

6. Pilots should be especially vigilant for other traffic while operating near VFR waypoints. The same effort to see and avoid other aircraft near VFR waypoints will be necessary, as was the case with VORs and NDBs in the past. In fact, the increased accuracy of navigation through the use of GPS will demand even greater vigilance, as off-course deviations among different pilots and receivers will be less. When operating near a VFR waypoint, use whatever ATC services are available, even if outside a class of airspace where communications are required. Regardless of the class of airspace, monitor the available ATC frequency closely for information on other aircraft operating in the vicinity. It is also a good idea to turn on your landing light(s) when operating near a VFR waypoint to make your aircraft more conspicuous to other pilots, especially when visibility is reduced. See paragraph 7-5-2, VFR in Congested Areas, for more information.

d. General Requirements

1. Authorization to conduct any GPS operation under IFR requires that:

(a) GPS navigation equipment used must be approved in accordance with the requirements specified in Technical Standard Order (TSO) TSO-C129 (as revised), TSOC196 (as revised), TSOC145 (as revised), or TSOC146 (as revised), and the installation must be done in accordance with Advisory Circular AC20-138, Airworthiness Approval of Positioning and Navigation Systems. Equipment approved in accordance with TSO-C115a does not meet the requirements of TSO-C129. Visual flight rules (VFR) and hand-held GPS systems are not authorized for IFR navigation, instrument approaches, or as a principal instrument flight reference. During IFR operations they may be considered only as an aid to situational awareness.

(b) Aircraft using GPS (TSOC129 (as revised) or TSOC196 (as revised)) navigation equipment under IFR must be equipped with an approved and operational alternate means of navigation appropriate to the flight. Active monitoring of alternative navigation equipment is not required if the GPS receiver uses RAIM for integrity monitoring. Active monitoring of an alternate means of navigation is required when the RAIM capability of the GPS equipment is lost.

(c) Procedures must be established for use in the event that the loss of RAIM capability is predicted to occur. In situations where this is encountered, the flight must rely on other approved equipment, delay departure, or cancel the flight.

(d) The GPS operation must be conducted in accordance with the FAA-approved aircraft flight manual (AFM) or flight manual supplement. Flight crew members must be thoroughly familiar with the particular GPS equipment installed in the aircraft, the receiver operation manual, and the AFM or flight manual supplement. Unlike ILS and VOR, the basic operation, receiver presentation to the pilot and some capabilities of the equipment can vary greatly. Due to these differences, operation of different brands, or even models of the same brand of GPS receiver, under IFR should not be attempted without thorough study of the operation of that particular receiver and installation. Most receivers have a built-in simulator mode which will allow the pilot to become familiar with operation prior to attempting operation in the aircraft. Using the equipment in flight under VFR conditions prior to attempting IFR operation will allow further familiarization.

(e) Aircraft navigating by IFR approved GPS are considered to be area navigation (RNAV) aircraft and have special equipment suffixes. File the appropriate equipment suffix in accordance with TBL 5-1-2, on the ATC flight plan. If GPS avionics become inoperative, the pilot should advise ATC and amend the equipment suffix.

(f) Prior to any GPS IFR operation, the pilot must review appropriate NOTAMs and aeronautical information. (See GPS NOTAMs/Aeronautical Information.)

(g) Air carrier and commercial operators must meet the appropriate provisions of their approved operations specifications.

(1) During domestic operations for commerce or for hire, operators must have a second navigation system capable of reversion or contingency operations.

(2) Operators must have two independent navigation systems appropriate to the route to be flown, or one system that is suitable and a second, independent backup capability that allows the operator to proceed safely and land at a different airport, and the aircraft must have sufficient fuel (reference 14 CFR 121.349, 125.203, 129.17, and 135.165). These rules ensure the safety of the operation by preventing a single point of failure.

NOTE-*An aircraft approved for multisensor navigation and equipped with a single FMS must maintain an ability to navigate or proceed safely in the event that any one component of the navigation system fails, including the flight management system (FMS). Retaining a FMSindependent VOR capability would satisfy this requirement.*

(3) The requirements for a second system apply to the entire set of equipment needed to achieve the navigation capability, not just the individual components of the system such as the radio navigation receiver. For example, to use two RNAV systems (e.g., GPS and DME/DME/IRU) to comply with the requirements, the aircraft must be equipped with two independent radio navigation receivers and two independent navigation computers (e.g., flight management systems (FMS)). Alternatively, to comply with the requirements using a single RNAV system with an installed and operable VOR capability, the VOR capability must be independent of the FMS.

(4) To satisfy the requirement for two independent navigation systems, if the primary navigation system is GPSbased, the second system must be independent of GPS (for example, VOR or DME/DME/IRU). This allows continued navigation in case of failure of the GPS or WAAS services. Recognizing that GPS interference and test events resulting in the loss of GPS services have become more common, the FAA requires operators conducting IFR operations under 14 CFR 121.349, 125.203, 129.17 and 135.65 to retain a nonGPS navigation capability consisting of either DME/DME, IRU or VOR for en route and terminal operations, and VOR and ILS for final approach. Since this system is to be used as a reversionary capability, single equipage is sufficient.

e. Use of GPS for IFR Oceanic, Domestic En Route, Terminal Area, and Approach Operations

1. GPS IFR operations in oceanic areas can be conducted as soon as the proper avionics systems are installed, provided all general requirements are met. A GPS installation with TSO-C129 (as revised) authorization in class A1, A2, B1, B2, C1, or C2 or TSOC196 (as revised) may be used to replace one of the other approved means of long-range navigation, such as dual INS. (See TBL 1-1-5 and TBL 1-1-6.) A single TSOC129 GPS installation meeting the certification requirements in AC 20138C, Appendix 1 may be used on oceanic routes as the only means of long range navigation. TSOC196 (as revised) equipment is inherently capable of supporting oceanic operation if the operator obtains a Fault Detection and Exclusion (FDE) Prediction Program as outlined in AC 20138C, Appendix 1. A single GPS/WAAS receiver (TSOC145 (as revised) or TSOC146 (as revised)) is inherently capable of supporting oceanic operation if the operator obtains a FDE Prediction Program as outlined in AC20138C, Appendix 1.

2. GPS (TSOC129 (as revised) or TSOC196 (as revised)) domestic en route and terminal IFR operations can be conducted as soon as proper avionics systems are installed, provided all general requirements are met. For required backup navigation, the avionics necessary to receive all of the ground-based facilities appropriate for the flight to the destination airport and any required alternate airport must be installed and operational. Ground-based facilities necessary for en route and terminal operations must also be in service.

(a) A single GPS/WAAS receiver (TSOC145 (as revised) or TSOC146 (as revised)) may also be used for these domestic en route and terminal IFR operations. Though not required, operators may consider retaining backup navigation equipment in their aircraft to guard against potential outages or interference.

(b) In Alaska, GPS en route IFR RNAV operations may be conducted outside the operational service volume of groundbased navigation aids when a GPS/WAAS (TSO-C145 (as revised) or TSO-C146 (as revised)) system is installed and operating. Groundbased navigation equipment is not required to be installed and operating. Though not required, operators may consider retaining backup navigation equipment in their aircraft to guard against potential outages or interference.

(1) Aircraft may operate on GNSS Qroutes with GPS (TSOC129 (as revised) or TSOC196 (as revised)) or GPS/WAAS equipment while the aircraft remains in Air Traffic Control radar surveillance.

(2) Aircraft may operate on GNSS Troutes with GPS/WAAS (TSOC145(as revised) or TSOC146 (as revised)) equipment.

3. Authorization to fly approaches under IFR using GPS or GPS/WAAS avionics systems requires that a pilot use avionics with:

Tbl 1-1-5
GPS IFR Equipment Classes/Categories

Equipment Class	RAIM	Int. Nav. Sys. to Prov. RAIM Equiv.	Oceanic	En Route	Terminal	Nonprecision Approach Capable
Class A - GPS sensor and navigation capability.						
A1	yes		yes	yes	yes	yes
A2	yes		yes	yes	yes	no
Class B - GPS sensor data to an integrated navigation system (i.e., FMS, multi-sensor navigation system, etc.).						
B1	yes		yes	yes	yes	yes
B2	yes		yes	yes	yes	no
B3		yes	yes	yes	yes	yes
B4		yes	yes	yes	yes	no
Class C - GPS sensor data to an integrated navigation system (as in Class B) which provides enhanced guidance to an autopilot, or flight director, to reduce flight tech. errors. Limited to 14 CFR Part 121 or equivalent criteria.						
C1	yes		yes	yes	yes	yes
C2	yes		yes	yes	yes	no
C3		yes	yes	yes	yes	yes

TBL 1-1-6
GPS Approval Required/Authorized Use

Equipment Type[1]	Installation Approval Required	Operational Approval Required	IFR En Route[2]	IFR Terminal[2]	IFR Approach[3]	Oceanic Remote	In Lieu of ADF and/or DME[3]
Hand held4	X5						
VFR Panel Mount[4]	X						
IFR En Route and Terminal	X	X	X	X			X
IFR Oceanic/Remote	X	X	X	X		X	X
IFR En Route, Terminal, and Approach	X	X	X	X	X		X

NOTE-
[1]To determine equipment approvals and limitations, refer to the AFM, AFM supplements, or pilot guides.
[2]Requires verification of data for correctness if database is expired.
[3]Requires current database or verification that the procedure has not been amended since the expiration of the database.
[4]VFR and hand-held GPS systems are not authorized for IFR navigation, instrument approaches, or as a primary instrument flight reference. During IFR operations they may be considered only an aid to situational awareness.
[5]Hand-held receivers require no approval. However, any aircraft modification to support the hand-held receiver; i.e., installation of an external antenna or a permanent mounting bracket, does require approval.

(a) GPS, TSO-C129, (as revised) authorization in class A1, B1, B3, C1, or C3;

(b) GPS, TSOC196 (as revised) authorization; or

(c) GPS/WAAS, TSOC145 (as revised) or TSOC146 (as revised) authorization.

4. As the production of standalone GPS approaches has progressed, many of the original overlay approaches have been replaced with standalone procedures specifically designed for use by GPS systems. A GPS approach overlay allows pilots to use GPS avionics under IFR for flying designated nonprecision instrument approach procedures, except LOC, LDA, and simplified directional facility (SDF) procedures. These procedures are identified by the name of the procedure and "or GPS" (for example, VOR/DME or GPS RWY15). Other previous types of overlays have either been converted to this format or replaced with standalone procedures. Only approaches contained in the current onboard navigation database are authorized. The navigation database may contain information about nonoverlay approach procedures that is intended to be used to enhance position orientation, generally by providing a map, while flying these approaches using conventional NAVAIDs. This approach information should not be confused with a GPS overlay approach. (See the receiver operating manual, AFM, or AFM Supplement for details on how to identify these approaches in the navigation database.)

f. General Database Requirements

1. The onboard navigation data must be current and appropriate for the region of intended operation and should include the navigation aids, waypoints, and relevant coded terminal airspace procedures for the departure, arrival, and alternate airfields.

(a) Further database guidance for terminal and en route requirements may be found in AC 90100, U.S. Terminal and En Route Area Navigation (RNAV) Operations.

(b) Further database guidance on Required Navigation Performance (RNP) instrument approach operations, RNP terminal, and RNP en route requirements may be found in AC 90105, Approval Guidance for RNP Operations and Barometric Vertical Navigation in the U.S. National Airspace System.

(c) All approach procedures to be flown must be retrievable from the current airborne navigation database supplied by the equipment manufacturer or other FAA approved source. The system must be able to retrieve the procedure by namefrom the aircraft navigation database, not just as a manually entered series of waypoints. Manual entry of waypoints using latitude/longitude or place/bearing is not permitted for approach procedures.

(d) Prior to using a procedure or waypoint retrieved from the airborne navigation database, the pilot should verify the validity of the database. This verification should include the following preflight and inflight steps:

(1) Preflight:

[a] Determine the date of database issuance, and verify that the date/time of proposed use is before the expiration date/time.

[b] Verify that the database provider has not published a notice limiting the use of the specific waypoint or procedure.

(2) Inflight:

[a] Determine that the waypoints and transition names coincide with names found on the procedure chart. Do not use waypoints which do not exactly match the spelling shown on published procedure charts.

[b] Determine that the waypoints are generally logical in location, in the correct order, and that their orientation to each other is as found on the procedure chart, both laterally and vertically.

NOTE-*There is no specific requirement to check each waypoint latitude and longitude, type of waypoint and/or altitude constraint, only the general relationship of waypoints in the procedure, or the logic of an individual waypoint's location.*

[c] If the cursory check of procedure logic or individual waypoint location, specified in[b] above, indicates a potential error, do not use the retrieved procedure or waypoint until a verification of latitude and longitude, waypoint type, and altitude constraints indicate full conformity with the published data.

g. GPS Approach Procedures

As the production of stand-alone GPS approaches has progressed, many of the original overlay approaches have been replaced with stand-alone procedures specifically designed for use by GPS systems. The title of the remaining GPS overlay procedures has been revised on the approach chart to "or GPS" (e.g., VOR or GPS RWY 24). Therefore, all the approaches that can be used by GPS now contain "GPS" in the title (e.g., "VOR or GPS RWY 24," "GPS RWY 24," or "RNAV (GPS) RWY 24"). During these GPS approaches, underlying ground-based NAVAIDs are not required to be operational and associated aircraft avionics need not be installed, operational, turned on or monitored (monitoring of the underlying approach is suggested when equipment is available and functional). Existing overlay approaches may be requested using the GPS title, such as "GPS RWY 24" for the VOR or GPS RWY24.

1. For flight planning purposes, TSOC129() and TSOC196() equipped users (GPS users) whose navigation systems have fault detection and exclusion (FDE) capability, who perform a preflight RAIM prediction at the airport where the RNAV (GPS) approach will be flown, and have proper knowledge and any required training and/or approval to conduct a GPSbased IAP, may file based on a GPSbased IAP at either the destination or the alternate airport, but not at both locations. At the alternate airport, pilots may plan for applicable alternate airport weather minimums using:

(a) Lateral navigation (LNAV) or circling minimum descent altitude (MDA);

(b) LNAV/vertical navigation (LNAV/VNAV) DA, if equipped with and using approved barometric vertical navigation (baroVNAV) equipment;

(c) RNP 0.3 DA on an RNAV (RNP) IAP, if they are specifically authorized users using approved baroVNAV equipment and the pilot has verified required navigation performance (RNP) availability through an approved prediction program.

2. If the above conditions cannot be met, any required alternate airport must have an approved instrument approach procedure other than GPS that is anticipated to be operational and available at the estimated time of arrival, and which the aircraft is equipped to fly.

h. GPS NOTAMs/Aeronautical Information

1. GPS satellite outages are issued as GPS NOTAMs both domestically and internationally. However, the effect of an outage on the intended operation cannot be determined unless the pilot has a RAIM availability prediction program which allows excluding a satellite which is predicted to be out of service based on the NOTAM information.

2. The term UNRELIABLE is used in conjunction with GPS NOTAMs. The term UNRELIABLE is an advisory to pilots indicating the expected level of service may not be available. UNRELIABLE does not mean there is a problem with GPS signal integrity. If GPS service is available, pilots may continue operations. If the LNAV or LNAV/VNAV service is available, pilots may use the displayed level of service to fly the approach. GPS operation may be NOTAMed UNRELIABLE due to testing or anomalies. (Pilots are encouraged to report GPS anomalies, including degraded operation and/or loss of service, as soon as possible, reference paragraph 1-1-14.) Air Traffic Control will advise pilots requesting a GPS or RNAV (GPS) approach of GPS UNRELIABLE for:

(a) NOTAMs not contained in the ATIS broadcast.

(b) Pilot reports of GPS anomalies received within the preceding 15 minutes.

3. Civilian pilots may obtain GPS RAIM availability information for nonprecision approach procedures by specifically requesting GPS aeronautical information from a Flight Service Station during preflight briefings. GPS RAIM aeronautical information can be obtained for a period of 3 hours (for example, if you are scheduledto arrive at 1215 hours, then the GPS RAIM information is available from 1100 to 1400hours) or a 24 hour time frame at a particular airport. FAA briefers will provide RAIM information for a period of 1 hour before to 1 hour after the ETA hour, unless a specific time frame is requested by the pilot. If flying a published GPS departure, a RAIM prediction should also be requested for the departure airport.

4. The military provides airfield specific GPS RAIM NOTAMs for nonprecision approach procedures at military airfields. The RAIM outages are issued as M-series NOTAMs and may be obtained for up to 24 hours from the time of request.

5. Receiver manufacturers and/or database suppliers may supply "NOTAM" type information concerning database errors. Pilots should check these sources, when available, to ensure that they have the most current information concerning their electronic database.

i. Receiver Autonomous Integrity Monitoring (RAIM)

1. RAIM outages may occur due to an insufficient number of satellites or due to unsuitable satellite geometry which causes the error in the position solution to become too large. Loss of satellite reception and RAIM warnings may occur due to aircraft dynamics (changes in pitch or bank angle). Antenna location on the aircraft, satellite position relative to the horizon, and aircraft attitude may affect reception of one or more satellites. Since the relative positions of the satellites are constantly changing, prior experience with the airport does not guarantee reception at all times, and RAIM availability should always be checked.

2. If RAIM is not available, another type of navigation and approach system must be used, another destination selected, or the trip delayed until RAIM is predicted to be available on arrival. On longer flights, pilots should consider rechecking the RAIM prediction for the destination during the flight. This may provide early indications that an unscheduled satellite outage has occurred since takeoff.

3. If a RAIM failure/status annunciation occurs prior to the final approach waypoint (FAWP), the approach should not be completed since GPS may no longer provide the required accuracy. The receiver performs a RAIM prediction by 2 NM prior to the FAWP to ensure that RAIM is available at the FAWP as a condition for entering the approach mode. **The pilot should ensure that the receiver has sequenced from "Armed" to "Approach" prior to the FAWP** (normally occurs 2 NM prior). Failure to sequence may be an indication of the detection of a satellite anomaly, failure to arm the receiver (if required), or other problems which preclude completing the approach.

4. If the receiver does not sequence into the approach mode or a RAIM failure/status annunciation occurs prior to the FAWP, the pilot should not descend to Minimum Descent Altitude (MDA), but should proceed to the missed approach waypoint (MAWP) via the FAWP, perform a missed approach, and contact ATC as soon as practical. Refer to the receiver operating manual for specific indications and instructions associated with loss of RAIM prior to the FAF.

5. If a RAIM failure occurs after the FAWP, the receiver is allowed to continue operating without an annunciation for up to 5 minutes to allow completion of the approach (see receiver operating manual). **If the RAIM flag/status annunciation appears after the FAWP, the missed approach should be executed immediately.**

j. Waypoints

1. GPS approaches make use of both fly-over and fly-by waypoints. Fly-by waypoints are used when an aircraft should begin a turn to the next course prior to reaching the waypoint separating the two route segments. This is known as turn anticipation and is compensated for in the airspace and terrain clearances. Approach waypoints, except for the MAWP and the missed approach holding waypoint (MAHWP), are normally fly-by waypoints. Fly-over waypoints are used when the aircraft must fly over the point prior to starting a turn. New approach charts depict fly-over waypoints as a circled waypoint symbol. Overlay approach charts and some early stand alone GPS approach charts may not reflect this convention.

2. Since GPS receivers are basically "To-To" navigators, they must always be navigating to a defined point. On overlay approaches, if no pronounceable five-character name is published for an approach waypoint or fix, it was given a database identifier consisting of letters and numbers. These points will appear in the list of waypoints in the approach procedure database, but may not appear on the approach chart. A point used for the purpose of defining the navigation track for an airborne computer system (i.e., GPS or FMS) is called a Computer Navigation Fix (CNF). CNFs include unnamed DME fixes, beginning and ending points of DME arcs and sensor final approach fixes (FAFs) on some GPS overlay approaches. To aid in the approach chart/database correlation process, the FAA has begun a program to assign five-letter names to CNFs and to chart CNFs on various FAA Aeronautical Navigation Products (AeroNav Products). These CNFs are not to be used for any air traffic control (ATC) application, such as holding for which the fix has not already been assessed. CNFs will be charted to distinguish them from conventional reporting points, fixes, intersections, and waypoints. The CNF name will be enclosed in parenthesis, e.g., (CFBCD), and the name will be placed next to the CNF it defines. If the CNF is not at an existing point defined by means such as crossing radials or radial/DME, the point will be indicated by an "X." The CNF name will not be used in filing a flight plan or in aircraft/ATC communications. Use current phraseology, e.g., facility name, radial, distance, to describe these fixes.

3. Unnamed waypoints in the database will be uniquely identified for each airport but may be repeated for another airport (e.g., RW36 will be used at each airport with a runway 36 but will be at the same location for all approaches at a given airport).

4. The runway threshold waypoint, which is normally the MAWP, may have a five letter identifier (e.g., SNEEZ) or be coded as RW## (e.g., RW36, RW36L). Those thresholds which are coded as five letter identifiers are being changed to the RW## designation. This may cause the approach chart and database to differ until all changes are complete. The runway threshold waypoint is also used as the center of the Minimum Safe Altitude (MSA) on most GPS approaches. MAWPs not located at the threshold will have a five letter identifier.

k. Position Orientation

As with most RNAV systems, pilots should pay particular attention to position orientation while using GPS. Distance and track information are provided to the next active waypoint, not to a fixed navigation aid. Receivers may sequence when the pilot is not flying along an active route, such as when being vectored or deviating for weather, due to the proximity to another waypoint in the route. This can be prevented by placing the receiver in the nonsequencing mode. When the receiver is in the nonsequencing mode, bearing and distance are provided to the selected waypoint and the receiver will not sequence to the next waypoint in the route until placed back in the auto sequence mode or the pilot selects a different waypoint. On overlay approaches, the pilot may have to compute the along-track distance to stepdown fixes and other points due to the receiver showing along-track distance to the next waypoint rather than DME to the VOR or ILS ground station.

l. Conventional Versus GPS Navigation Data

There may be slight differences between the course information portrayed on navigational charts and a GPS navigation display when flying authorized GPS instrument procedures or along an airway. All magnetic tracks defined by any conventional navigation aids are determined by the application of the station magnetic variation. In contrast, GPS RNAV systems may use an algorithm, which applies the local magnetic variation and may produce small differences in the displayed course. However, both methods of navigation should produce the same desired ground track when using approved, IFR navigation system. Should significant differences between the approach chart and the GPS avionics' application of the navigation database arise, the published approach chart, supplemented by NOTAMs, holds precedence.

Due to the GPS avionics' computation of great circle courses, and the variations in magnetic variation, the bearing to the next waypoint and the course from the last waypoint (if available) may not be exactly 180° apart when long distances are involved. Variations in distances will occur since GPS distance-to-waypoint values are along-track distances (ATD) computed to the next waypoint and the DME values published on underlying procedures are slant-range distances measured to the station. This difference increases with aircraft altitude and proximity to the NAVAID.

m. Departures and Instrument Departure Procedures (DPs)

The GPS receiver must be set to terminal (±1 NM) CDI sensitivity and the navigation routes contained in the database in order to fly published IFR charted departures and DPs. Terminal RAIM should be automatically provided by the receiver. (Terminal RAIM for departure may not be available unless the waypoints are part of the active flight plan rather than proceeding direct to the first destination.) Certain segments of a DP may require some manual intervention by the pilot, especially when radar vectored to a course or required to intercept a specific course to a waypoint. The database may not contain all of the transitions or departures from all runways and some GPS receivers do not contain DPs in the database. It is necessary that helicopter procedures be flown at 70 knots or less since helicopter departure procedures and missed approaches use a 20:1 obstacle clearance surface (OCS), which is double the fixed-wing OCS, and turning areas are based on this speed as well.

n. Flying GPS Approaches

1. Determining which area of the TAA the aircraft will enter when flying a "T" with a TAA must be accomplished using the bearing and distance to the IF(IAF). This is most critical when entering the TAA in the vicinity of the extended runway centerline and determining whether you will be entering the right or left base area. Once inside the TAA, all sectors and stepdowns are based on the bearing and distance to the IAF for that area, which the aircraft should be proceeding direct to at that time, unless on vectors. (See FIG 5-4-3 and FIG 5-4-4.)

2. Pilots should fly the full approach from an Initial Approach Waypoint (IAWP) or feeder fix unless specifically cleared otherwise. Randomly joining an approach at an intermediate fix does not assure terrain clearance.

3. When an approach has been loaded in the flight plan, GPS receivers will give an "arm" annunciation 30 NM straight line distance from the airport/heliport reference point. Pilots should arm the approach mode at this time, if it has not already been armed (some receivers arm automatically). Without arming, the receiver will not change from en route CDI and RAIM sensitivity of ±5 NM either side of centerline to ±1 NM terminal sensitivity. Where the IAWP is inside this 30 mile point, a CDI sensitivity change will occur once the approach mode is armed and the aircraft is inside 30 NM. Where the IAWP is beyond 30 NM from the airport/heliport reference point, CDI sensitivity will not change until the aircraft is within 30 miles of the airport/heliport reference point even if the approach is armed earlier. Feeder route obstacle clearance is predicated on the receiver being in terminal (±1 NM) CDI sensitivity and RAIM within 30 NM of the airport/heliport reference point, therefore, the receiver should always be armed (if required) not later than the 30 NM annunciation.

4. The pilot must be aware of what bank angle/turn rate the particular receiver uses to compute turn anticipation, and whether wind and airspeed are included in the receiver's calculations. This information should be in the receiver operating manual. Over or under banking the turn onto the final approach course may significantly delay getting on course and may result in high descent rates to achieve the next segment altitude.

5. When within 2 NM of the FAWP with the approach mode armed, the approach mode will switch to active, which results in RAIM changing to approach sensitivity and a change in CDI sensitivity. Beginning 2 NM prior to the FAWP, the full scale CDI sensitivity will smoothly change from ±1 NM to ±0.3 NM at the FAWP. As sensitivity changes from ±1 NM to ±0.3 NM approaching the FAWP, with the CDI not centered, the corresponding increase in CDI displacement may give the impression that the aircraft is moving further away from the intended course even though it is on an acceptable intercept heading. Referencing the digital track displacement information (cross track error), if it is available in the approach mode, may help the pilot remain position oriented in this situation. Being established on the final approach course prior to the beginning of the sensitivity change at 2 NM will help prevent problems in interpreting the CDI display during ramp down. Therefore, requesting or accepting vectors which will cause the aircraft to intercept the final approach course within 2 NM of the FAWP is not recommended.

6. When receiving vectors to final, most receiver operating manuals suggest placing the receiver in the nonsequencing mode on the FAWP and manually setting the course. This provides an extended final approach course in cases where the aircraft is vectored onto the final approach course outside of any existing segment which is aligned with the runway. Assigned altitudes must be maintained until established on a published segment of the approach. Required altitudes at waypoints outside the FAWP or stepdown fixes must be considered. Calculating the distance to the FAWP may be required in order to descend at the proper location.

7. Overriding an automatically selected sensitivity during an approach will cancel the approach mode annunciation. If the approach mode is not armed by 2 NM prior to the FAWP, the approach mode will not become active at 2 NM prior to the FAWP, and the equipment will flag. In these conditions, the RAIM and CDI sensitivity will not ramp down, and the pilot should not descend to MDA, but fly to the MAWP and execute a missed approach. The approach active annunciator and/or the receiver should be checked to ensure the approach mode is active prior to the FAWP.

8. Do not attempt to fly an approach unless the procedure in the on-board database is current and identified as "GPS" on the approach chart. The navigation database may contain information about nonoverlay approach procedures that is intended to be used to enhance position orientation, generally by providing a map, while flying these approaches using conventional NAVAIDs. This approach information should not be confused with a GPS overlay approach (see the receiver operating manual, AFM, or AFM Supplement for details on how to identify these procedures in the navigation database). Flying point to point on the approach does not assure compliance with the published approach procedure. The proper RAIM sensitivity will not be available and the CDI sensitivity will not automatically change to ±0.3 NM. Manually setting CDI sensitivity does not automatically change the RAIM sensitivity on some receivers. Some existing nonprecision approach procedures cannot be coded for use with GPS and will not be available as overlays.

9. Pilots should pay particular attention to the exact operation of their GPS receivers for performing holding patterns and in the case of overlay approaches, operations such as procedure turns. These procedures may require manual intervention by the pilot to stop the sequencing of waypoints by the receiver and to resume automatic GPS navigation sequencing once the maneuver is complete. The same waypoint may appear in the route of flight more than once consecutively (e.g., IAWP, FAWP, MAHWP on a procedure turn). Care must be exercised to ensure that the receiver is sequenced to the appropriate waypoint for the segment of the procedure being flown, especially if one or more fly-overs are skipped (e.g., FAWP rather than IAWP if the procedure turn is not flown). The pilot may have to sequence past one or more fly-overs of the same waypoint in order to start GPS automatic sequencing at the proper place in the sequence of waypoints.

10. Incorrect inputs into the GPS receiver are especially critical during approaches. In some cases, an incorrect entry can cause the receiver to leave the approach mode.

11. A fix on an overlay approach identified by a DME fix will not be in the waypoint sequence on the GPS receiver unless there is a published name assigned to it. When a name is assigned, the along track to the waypoint may be zero rather than the DME stated on the approach chart. The pilot should be alert for this on any overlay procedure where the original approach used DME.

12. If a visual descent point (VDP) is published, it will not be included in the sequence of waypoints. Pilots are expected to use normal piloting techniques for beginning the visual descent, such as ATD.

13. Unnamed stepdown fixes in the final approach segment will not be coded in the waypoint sequence of the aircraft's navigation database and must be identified using ATD. Stepdown fixes in the final approach segment of RNAV (GPS) approaches are being named, in addition to being identified by ATD. However, since most GPS avionics do not accommodate waypoints between the FAF and MAP, even when the waypoint is named, the waypoints for these stepdown fixes may not appear in the sequence of waypoints in the navigation database. Pilots must continue to identify these stepdown fixes using ATD.

o. Missed Approach

1. A GPS missed approach requires pilot action to sequence the receiver past the MAWP to the missed approach portion of the procedure. The pilot must be thoroughly familiar with the activation procedure for the particular GPS receiver installed in the aircraft and must **initiate appropriate action after the MAWP.** Activating the missed approach prior to the MAWP will cause CDI sensitivity to immediately change to terminal (±1NM) sensitivity and the receiver will continue to navigate to the MAWP. The receiver will not sequence past the MAWP. Turns should not begin prior to the MAWP. If the missed approach is not activated, the GPS receiver will display an extension of the inbound final approach course and the ATD will increase from the MAWP until it is manually sequenced after crossing the MAWP.

2. Missed approach routings in which the first track is via a course rather than direct to the next waypoint **require additional action by the pilot** to set the course. Being familiar with all of the inputs required is especially critical during this phase of flight.

p. GPS Familiarization

Pilots should practice GPS approaches under visual meteorological conditions (VMC) until thoroughly proficient with all aspects of their equipment (receiver and installation) prior to attempting flight by IFR in instrument meteorological conditions (IMC). Some of the areas which the pilot should practice are:

1. Utilizing the receiver autonomous integrity monitoring (RAIM) prediction function;

2. Inserting a DP into the flight plan, including setting terminal CDI sensitivity, if required, and the conditions under which terminal RAIM is available for departure (some receivers are not DP or STAR capable);

3. Programming the destination airport;

4. Programming and flying the overlay approaches (especially procedure turns and arcs);

5. Changing to another approach after selecting an approach;

6. Programming and flying "direct" missed approaches;

7. Programming and flying "routed" missed approaches;

8. Entering, flying, and exiting holding patterns, particularly on overlay approaches with a second waypoint in the holding pattern;

9. Programming and flying a "route" from a holding pattern;

10. Programming and flying an approach with radar vectors to the intermediate segment;

11. Indication of the actions required for RAIM failure both before and after the FAWP; and

12. Programming a radial and distance from a VOR (often used in departure instructions).

1-1-19. Wide Area Augmentation System (WAAS)

a. General

1. The FAA developed the Wide Area Augmentation System (WAAS) to improve the accuracy, integrity and availability of GPS signals. WAAS will allow GPS to be used, as the aviation navigation system, from takeoff through Category I precision approach when it is complete. WAAS is a critical component of the FAA's strategic objective for a seamless satellite navigation system for civil aviation, improving capacity and safety.

2. The International Civil Aviation Organization (ICAO) has defined Standards and Recommended Practices (SARPs) for satellite-based augmentation systems (SBAS) such as WAAS. Japan and Europe are building similar systems that are planned to be interoperable with WAAS: EGNOS, the European Geostationary Navigation Overlay System, and MSAS, the Japan Multifunctional Transport Satellite (MTSAT) Satellite-based Augmentation System. The merging of these systems will create a worldwide seamless navigation capability similar to GPS but with greater accuracy, availability and integrity.

3. Unlike traditional ground-based navigation aids, WAAS will cover a more extensive service area. Precisely surveyed wide-area ground reference stations (WRS) are linked to form the U.S. WAAS network. Signals from the GPS satellites are monitored by these WRSs to determine satellite clock and ephemeris corrections and to model the propagation effects of the ionosphere. Each station in the network relays the data to a wide-area master station (WMS) where the correction information is computed. A correction message is prepared and uplinked to a geostationary satellite (GEO) via a ground uplink station (GUS). The message is then broadcast on the same frequency as GPS (L1, 1575.42 MHz) to WAAS receivers within the broadcast coverage area of the WAAS GEO.

4. In addition to providing the correction signal, the WAAS GEO provides an additional pseudorange measurement to the aircraft receiver, improving the availability of GPS by providing, in effect, an additional GPS satellite in view. The integrity of GPS is improved through real-time monitoring, and the accuracy is improved by providing differential corrections to reduce errors. The performance improvement is sufficient to enable approach procedures with GPS/WAAS glide paths (vertical guidance).

5. The FAA has completed installation of 25 WRSs, 2 WMSs, 4 GUSs, and the required terrestrial communications to support the WAAS network. Prior to the commissioning of the WAAS for public use, the FAA has been conducting a series of test and validation activities. Enhancements to the initial phase of WAAS will include additional master and reference stations, communication satellites, and transmission frequencies as needed.

6. GNSS navigation, including GPS and WAAS, is referenced to the WGS-84 coordinate system. It should only be used where the Aeronautical Information Publications (including electronic data and aeronautical charts) conform to WGS-84 or equivalent. Other countries civil aviation authorities may impose additional limitations on the use of their SBAS systems.

b. Instrument Approach Capabilities

1. A new class of approach procedures which provide vertical guidance, but which do not meet the ICAO Annex 10 requirements for precision approaches has been developed to support satellite navigation use for aviation applications worldwide. These new procedures called Approach with Vertical Guidance (APV), are defined in ICAO Annex 6, and include approaches such as the LNAV/VNAV procedures presently being flown with barometric vertical navigation (Baro-VNAV). These approaches provide vertical guidance, but do not meet the more stringent standards of a precision approach. Properly certified WAAS receivers will be able to fly these LNAV/VNAV procedures using a WAAS electronic glide path, which eliminates the errors that can be introduced by using Barometric altimetery.

2. A new type of APV approach procedure, in addition to LNAV/VNAV, is being implemented to take advantage of the high accuracy guidance and increased integrity provided by WAAS. This WAAS generated angular guidance allows the use of the same TERPS approach criteria used for ILS approaches. The resulting approach procedure minima, titled LPV (localizer performance with vertical guidance), may have a decision altitude as low as 200 feet height above touchdown with visibility minimums as low as $^1/_2$ mile, when the terrain and airport infrastructure support the lowest minima. LPV minima is published on the RNAV (GPS) approach charts (see paragraph 5-4-5, Instrument Approach Procedure Charts).

3. A new nonprecision WAAS approach, called Localizer Performance (LP) is being added in locations where the terrain or obstructions do not allow publication of vertically guided LPV procedures. This new approach takes advantage of the angular lateral guidance and smaller position errors provided by WAAS to provide a lateral only procedure similar to an ILS Localizer. LP procedures may provide lower minima than a LNAV procedure due to the narrower obstacle clearance surface.

NOTE-*WAAS receivers certified prior to TSO-C145b and TSO-C146b, even if they have LPV capability, do not contain LP capability unless the receiver has been upgraded. Receivers capable of flying LP procedures must contain a statement in the Flight Manual Supplement or Approved Supplemental Flight Manual stating that the receiver has LP capability, as well as the capability for the other WAAS and GPS approach procedure types.*

4. WAAS provides a level of service that supports all phases of flight, including RNAV (GPS) approaches to LNAV, LP, LNAV/VNAV and LPV lines of minima, within system coverage. Some locations close to the edge of the coverage may have a lower availability of vertical guidance.

c. General Requirements

1. WAAS avionics must be certified in accordance with Technical Standard Order (TSO) TSO-C145A, Airborne Navigation Sensors Using the (GPS) Augmented by the Wide Area Augmentation System (WAAS); or TSO-146A, Stand-Alone Airborne Navigation Equipment Using the Global Positioning System (GPS) Augmented by the Wide Area Augmentation System (WAAS), and installed in accordance with Advisory Circular (AC) 20-130A, Airworthiness Approval of Navigation or Flight Management Systems Integrating Multiple Navigation Sensors, or AC 20-138A, Airworthiness Approval of Global Positioning System (GPS) Navigation Equipment for Use as a VFR and IFR Navigation System.

2. GPS/WAAS operation must be conducted in accordance with the FAA-approved aircraft flight manual (AFM) and flight manual supplements. Flight manual supplements will state the level of approach procedure that the receiver supports. IFR approved WAAS receivers support all GPS only operations as long as lateral capability at the appropriate level is functional. WAAS monitors both GPS and WAAS satellites and provides integrity.

3. GPS/WAAS equipment is inherently capable of supporting oceanic and remote operations if the operator obtains a fault detection and exclusion (FDE) prediction program.

4. Air carrier and commercial operators must meet the appropriate provisions of their approved operations specifications..

5. Prior to GPS/WAAS IFR operation, the pilot must review appropriate Notices to Airmen (NOTAMs) and aeronautical information. This information is available on request from a Flight Service Station. The FAA will provide NOTAMs to advise pilots of the status of the WAAS and level of service available.

(a) The term UNRELIABLE is used in conjunction with GPS and WAAS NOTAMs. The term UNRELIABLE is an advisory to pilots indicating the expected level of WAAS service (LNAV/VNAV, LPV) may not be available; e.g., **!BOS BOS WAAS LPV AND LNAV/VNAV MNM UNREL WEF 0305231700 - 0305231815**. WAAS UNRELIABLE NOTAMs are predictive in nature and published for flight planning purposes. Upon commencing an approach at locations NOTAMed WAAS UNRELIABLE, if the WAAS avionics indicate LNAV/VNAV or LPV service is available, then vertical guidance may be used to complete the approach using the displayed level of service. Should an outage occur during the approach, reversion to LNAV minima may be required.

(1) Area-wide WAAS UNAVAILABLE NOTAMs indicate loss or malfunction of the WAAS system. In flight, Air Traffic Control will advise pilots requesting a GPS or RNAV (GPS) approach of WAAS UNAVAILABLE NOTAMs if not contained in the ATIS broadcast.

(2) Site-specific WAAS UNRELIABLE NOTAMs indicate an expected level of service, e.g., LNAV/VNAV or LPV may not be available. Pilots must request site-specific WAAS NOTAMs during flight planning. In flight, Air Traffic Control will not advise pilots of WAAS UNRELIABLE NOTAMs.

(3) When the approach chart is annotated with the W symbol, site-specific WAAS UNRELIABLE NOTAMs or Air Traffic advisories are not provided for outages in WAAS

LNAV/VNAV and LPV vertical service. Vertical outages may occur daily at these locations due to being close to the edge of WAAS system coverage. Use LNAV or circling minima for flight planning at these locations, whether as a destination or alternate. For flight operations at these locations, when the WAAS avionics indicate that LNAV/VNAV or LPV service is available, then the vertical guidance may be used to complete the approach using the displayed level of service. Should an outage occur during the procedure, reversion to LNAV minima may be required.

NOTE-*Area-wide WAAS UNAVAILABLE NOTAMs apply to all airports in the WAAS UNAVAILABLE area designated in the NOTAM, including approaches at airports where an approach chart is annotated with the* W *symbol.*

6. GPS/WAAS was developed to be used within SBAS GEO coverage (WAAS or other interoperable system) without the need for other radio navigation equipment appropriate to the route of flight to be flown. Outside the SBAS coverage or in the event of a WAAS failure, GPS/WAAS equipment reverts to GPS-only operation and satisfies the requirements for basic GPS equipment.

7. Unlike TSO-C129 avionics, which were certified as a supplement to other means of navigation, WAAS avionics are evaluated without reliance on other navigation systems. As such, installation of WAAS avionics does not require the aircraft to have other equipment appropriate to the route to be flown.

(a) Pilots with WAAS receivers may flight plan to use any instrument approach procedure authorized for use with their WAAS avionics as the planned approach at a required alternate, with the following restrictions. When using WAAS at an alternate airport, flight planning must be based on flying the RNAV (GPS) LNAV or circling minima line, or minima on a GPS approach procedure, or conventional approach procedure with "or GPS" in the title. Code of Federal Regulation (CFR) Part 91 nonprecision weather requirements must be used for planning. Upon arrival at an alternate, when the WAAS navigation system indicates that LNAV/VNAV or LPV service is available, then vertical guidance may be used to complete the approach using the displayed level of service. The FAA has begun removing the **▲ NA** (Alternate Minimums Not Authorized) symbol from select RNAV (GPS) and GPS approach procedures so they may be used by approach approved WAAS receivers at alternate airports. Some approach procedures will still require the **▲ NA** for other reasons, such as no weather reporting, so it cannot be removed from all procedures. Since every procedure must be individually evaluated, removal of the **▲NA** from RNAV (GPS) and GPS procedures will take some time.

NOTE-*Properly trained and approved, as required, TSOC145() and TSOC146() equipped users (WAAS users) with and using approved baroVNAV equipment may plan for LNAV/VNAV DA at an alternate airport. Specifically authorized WAAS users with and using approved baroVNAV equipment may also plan for RNP 0.3 DA at the alternate airport as long as the pilot has verified RNP availability through an approved prediction program.*

d. Flying Procedures with WAAS

1. WAAS receivers support all basic GPS approach functions and provide additional capabilities. One of the major improvements is the ability to generate glide path guidance, independent of ground equipment or barometric aiding. This eliminates several problems such as hot and cold temperature effects, incorrect altimeter setting or lack of a local altimeter source. It also allows approach procedures to be built without the cost of installing ground stations at each airport or runway. Some approach certified receivers may only generate a glide path with performance similar to Baro-VNAV and are only approved to fly the LNAV/VNAV line of minima on the RNAV (GPS) approach charts. Receivers with additional capability (including faster update rates and smaller integrity limits) are approved to fly the LPV line of minima. The lateral integrity changes dramatically from the 0.3 NM (556 meter) limit for GPS, LNAV and LNAV/VNAV approach mode, to 40 meters for LPV. It also provides vertical integrity monitoring, which bounds the vertical error to 50 meters for LNAV/VNAV and LPVs with minima of 250' or above, and bounds the vertical error to 35 meters for LPVs with minima below 250'.

2. When an approach procedure is selected and active, the receiver will notify the pilot of the most accurate level of service supported by the combination of the WAAS signal, the receiver, and the selected approach, using the naming conventions on the minima lines of the selected approach procedure. For example, if an approach is published with LPV minima and the receiver is only certified for LNAV/VNAV, the equipment would indicate "LNAV/VNAV available," even though the WAAS signal would support LPV. If flying an existing LNAV/VNAV procedure with no LPV minima, the receiver will notify the pilot "LNAV/VNAV available," even if the receiver is certified for LPV and the signal supports LPV. If the signal does not support vertical guidance on procedures with LPV and/or LNAV/VNAV minima, the receiver annunciation will read "LNAV available." On lateral only procedures with LP and LNAV minima the receiver will indicate "LP available" or "LNAV available" based on the level of lateral service available. Once the level of service notification has been given, the receiver will operate in this mode for the duration of the approach procedure, unless that level of service becomes unavailable. The receiver cannot change back to a more accurate level of service until the next time an approach is activated.

NOTE-*Receivers do not "fail down" to lower levels of service once the approach has been activated. If only the vertical off flag appears, the pilot may elect to use the LNAV minima if the rules under which the flight is operating allow changing the type of approach being flown after commencing the procedure. If the lateral integrity limit is exceeded on an LP approach, a missed approach will be necessary since there is no way to reset the lateral alarm limit while the approach is active.*

3. Another additional feature of WAAS receivers is the ability to exclude a bad GPS signal and continue operating normally. This is normally accomplished by the WAAS correction information. Outside WAAS coverage or when WAAS is not available, it is accomplished through a receiver algorithm called FDE. In most cases this operation will be invisible to the pilot since the receiver will continue to operate with other available satellites after excluding the "bad" signal. This capability increases the reliability of navigation.

4. Both lateral and vertical scaling for the LNAV/VNAV and LPV approach procedures are different than the linear scaling of basic GPS. When the complete published procedure is flown, +/-1 NM linear scaling is provided until two (2) NM prior to the FAF, where the sensitivity increases to be similar to the angular scaling of an ILS. There are two differences in the WAAS scaling and ILS: 1) on long final approach segments, the initial scaling will be +/-0.3 NM to achieve equivalent performance to GPS (and better than ILS, which is less sensitive far from the runway); 2) close to the runway threshold, the scaling changes to linear instead of continuing to become more sensitive. The width of the final approach course is tailored so that the total width is usually 700 feet at the runway threshold. Since the origin point of the lateral splay for the angular portion of the final is not fixed due to antenna placement like localizer, the splay angle can remain fixed, making a consistent width of final for aircraft being vectored onto the final approach course on different length runways. When the complete published procedure is not flown, and instead the aircraft needs to capture the extended final approach course similar to ILS, the vector to final (VTF) mode is used. Under VTF the scaling is linear at +/-1 NM until the point where the ILS angular splay reaches a width of +/-1 NM regardless of the distance from the FAWP.

5. The WAAS scaling is also different than GPS TSO-C129 in the initial portion of the missed approach. Two differences occur here. First, the scaling abruptly changes from the approach scaling to the missed approach scaling, at approximately the departure end of the runway or when the pilot requests missed approach guidance rather than ramping as GPS does. Second, when the first leg of the missed approach is a Track to Fix (TF) leg aligned within 3 degrees of the inbound course, the receiver will change to 0.3 NM linear sensitivity until the turn initiation point for the first waypoint in the missed approach procedure, at which time it will abruptly change to terminal (+/-1 NM) sensitivity. This allows the elimination of close in obstacles in the early part of the missed approach that may cause the DA to be raised.

6. A new method has been added for selecting the final approach segment of an instrument approach. Along with the current method used by most receivers using menus where the pilot selects the airport, the runway, the specific approach procedure and finally the IAF, there is also a channel number selection method. The pilot enters a unique 5-digit number provided on the approach chart, and the receiver recalls the matching final approach segment from the aircraft database. A list of information including the available IAFs is displayed and the pilot selects the appropriate IAF. The pilot should confirm that the correct final approach segment was loaded by cross checking the Approach ID, which is also provided on the approach chart.

7. The Along-Track Distance (ATD) during the final approach segment of an LNAV procedure (with a minimum descent altitude) will be to the MAWP. On LNAV/VNAV and LPV approaches to a decision altitude, there is no missed approach waypoint so the along-track distance is displayed to a point normally located at the runway threshold. In most cases the MAWP for the LNAV approach is located on the runway threshold at the centerline, so these distances will be the same. This distance will always vary slightly from any ILS DME that may be present, since the ILS DME is located further down the runway. Initiation of the missed approach on the LNAV/VNAV and LPV approaches is still based on reaching the decision altitude without any of the items listed in 14 CFR Section 91.175 being visible, and must not be delayed until the ATD reaches zero. The WAAS receiver, unlike a GPS receiver, will automatically sequence past the MAWP if the missed approach procedure has been designed for RNAV. The pilot may also select missed approach prior to the MAWP, however, navigation will continue to the MAWP prior to waypoint sequencing taking place.

1-1-20. Ground Based Augmentation System (GBAS) Landing System (GLS)

a. General

1. The GLS provides precision navigation guidance for exact alignment and descent of aircraft on approach to a runway. It provides differential augmentation to the Global Navigation Satellite System (GNSS).

NOTE-*GBAS is the ICAO term for Local Area Augmentation System (LAAS).*

2. LAAS was developed as an "ILS look-alike" system from the pilot perspective. LAAS is based on GPS signals augmented by ground equipment and has been developed to provide GLS precision approaches similar to ILS at airfields.

3. GLS provides guidance similar to ILS approaches for the final approach segment; portions of the GLS approach prior to and after the final approach segment will be based on Area Navigation (RNAV) or Required Navigation Performance (RNP).

4. The equipment consists of a GBAS Ground Facility (GGF), four reference stations, a VHF Data Broadcast (VDB) uplink antenna, and an aircraft GBAS receiver.

b. Procedure

1. Pilots will select the five digit GBAS channel number of the associated approach within the Flight Management System (FMS) menu or manually select the five digits (system dependent). Selection of the GBAS channel number also tunes the VDB.

2. Following procedure selection, confirmation that the correct LAAS procedure is loaded can be accomplished by cross checking the charted Reference Path Indicator (RPI) or approach ID with the cockpit displayed RPI or audio identification of the RPI with Morse Code (for some systems).

3. The pilot will fly the GLS approach using the same techniques as an ILS, once selected and identified.

1-1-21. Precision Approach Systems other than ILS, GLS, and MLS

a. General

Approval and use of precision approach systems other than ILS, GLS and MLS require the issuance of special instrument approach procedures.

b. Special Instrument Approach Procedure

1. Special instrument approach procedures must be issued to the aircraft operator if pilot training, aircraft equipment, and/or aircraft performance is different than published procedures. Special instrument approach procedures are not distributed for general public use. These procedures are issued to an aircraft operator when the conditions for operations approval are satisfied.

2. General aviation operators requesting approval for special procedures should contact the local Flight Standards District Office to obtain a letter of authorization. Air carrier operators requesting approval for use of special procedures should contact their Certificate Holding District Office for authorization through their Operations Specification.

c. Transponder Landing System (TLS)

1. The TLS is designed to provide approach guidance utilizing existing airborne ILS localizer, glide slope, and transponder equipment.

2. Ground equipment consists of a transponder interrogator, sensor arrays to detect lateral and vertical position, and ILS frequency transmitters. The TLS detects the aircraft's position by interrogating its transponder. It then broadcasts ILS frequency signals to guide the aircraft along the desired approach path.

3. TLS instrument approach procedures are designated Special Instrument Approach Procedures. Special aircrew training is required. TLS ground equipment provides approach guidance for only one aircraft at a time. Even though the TLS signal is received using the ILS receiver, no fixed course or glidepath is generated. The concept of operation is very similar to an air traffic controller providing radar vectors, and just as with radar vectors, the guidance is valid only for the intended aircraft. The TLS ground equipment tracks one aircraft, based on its transponder code, and provides correction signals to course and glidepath based on the position of the tracked aircraft. Flying the TLS corrections computed for another aircraft will not provide guidance relative to the approach; therefore, aircrews must not use the TLS signal for navigation unless they have received approach clearance and completed the required coordination with the TLS ground equipment operator. Navigation fixes based on conventional NAVAIDs or GPS are provided in the special instrument approach procedure to allow aircrews to verify the TLS guidance.

d. Special Category I Differential GPS (SCAT-I DGPS)

1. The SCAT-I DGPS is designed to provide approach guidance by broadcasting differential correction to GPS.

2. SCAT-I DGPS procedures require aircraft equipment and pilot training.

3. Ground equipment consists of GPS receivers and a VHF digital radio transmitter. The SCAT-I DGPS detects the position of GPS satellites relative to GPS receiver equipment and broadcasts differential corrections over the VHF digital radio.

4. Category I Ground Based Augmentation System (GBAS) will displace SCAT-I DGPS as the public use service.

REFERENCE-*AIM, Para 5-4-7f, Instrument Approach Procedures.*

Section 2. Area Navigation (RNAV) and Required Navigation Performance (RNP)

1-2-1. Area Navigation (RNAV)

a. General. RNAV is a method of navigation that permits aircraft operation on any desired flight path within the coverage of ground or space based navigation aids or within the limits of the capability of self-contained aids, or a combination of these. In the future, there will be an increased dependence on the use of RNAV in lieu of routes defined by ground-based navigation aids.

RNAV routes and terminal procedures, including departure procedures (DPs) and standard terminal arrivals (STARs), are designed with RNAV systems in mind. There are several potential advantages of RNAV routes and procedures:

1. Time and fuel savings,

2. Reduced dependence on radar vectoring, altitude, and speed assignments allowing a reduction in required ATC radio transmissions, and

3. More efficient use of airspace.

In addition to information found in this manual, guidance for domestic RNAV DPs, STARs, and routes may also be found in Advisory Circular 90-100A, U.S. Terminal and En Route Area Navigation (RNAV) Operations.

b. RNAV Operations. RNAV procedures, such as DPs and STARs, demand strict pilot awareness and maintenance of the procedure centerline. Pilots should possess a working knowledge of their aircraft navigation system to ensure RNAV procedures are flown in an appropriate manner. In addition, pilots should have an understanding of the various waypoint and leg types used in RNAV procedures; these are discussed in more detail below.

1. Waypoints. A waypoint is a predetermined geographical position that is defined in terms of latitude/longitude coordinates. Waypoints may be a simple named point in space or associated with existing navaids, intersections, or fixes. A waypoint is most often used to indicate a change in direction, speed, or altitude along the desired path. RNAV procedures make use of both fly-over and fly-by waypoints.

(a) Fly-by waypoints. Fly-by waypoints are used when an aircraft should begin a turn to the next course prior to reaching the waypoint separating the two route segments. This is known as turn anticipation.

(b) Fly-over waypoints. Fly-over waypoints are used when the aircraft must fly over the point prior to starting a turn.

NOTE-*FIG 1-2-1 illustrates several differences between a fly-by and a fly-over waypoint.*

2. RNAV Leg Types. A leg type describes the desired path proceeding, following, or between waypoints on an RNAV procedure. Leg types are identified by a two-letter code that describes the path (e.g., heading, course, track, etc.) and the termination point (e.g., the path terminates at an altitude, distance, fix, etc.). Leg types used for procedure design are included in the aircraft navigation database, but not normally provided on the procedure chart. The narrative depiction of the RNAV chart describes how a procedure is flown. The "path and terminator concept" defines that every leg of a procedure has a termination point and some kind of path into that termination point. Some of the available leg types are described below.

(a) Track to Fix. A Track to Fix (TF) leg is intercepted and acquired as the flight track to the following waypoint. Track to a Fix legs are sometimes called point-to-point legs for this reason. ***Narrative:*** *"on track 087 to CHEZZ WP."* See FIG 1-2-2.

(b) Direct to Fix. A Direct to Fix (DF) leg is a path described by an aircraft's track from an initial area direct to the next waypoint. ***Narrative:*** *"left turn direct BARGN WP."* See FIG 1-2-3.

(c) Course to Fix. A Course to Fix (CF) leg is a path that terminates at a fix with a specified course at that fix. ***Narrative:*** *"on course 078 to PRIMY WP."* See FIG 1-2-4.

(d) Radius to Fix. A Radius to Fix (RF) leg is defined as a constant radius circular path around a defined turn center that terminates at a fix. See FIG 1-2-5.

(e) Heading. A Heading leg may be defined as, but not limited to, a Heading to Altitude (VA), Heading to DME range (VD), and Heading to Manual Termination, i.e., Vector (VM). ***Narrative:*** *"climb heading 350 to 1500", "heading 265, at 9 DME west of PXR VORTAC, right turn heading 360", "fly heading 090, expect radar vectors to DRYHT INT."*

3. Navigation Issues. Pilots should be aware of their navigation system inputs, alerts, and annunciations in order to make better-informed decisions. In addition, the availability and suitability of particular sensors/systems should be considered.

(a) GPS. Operators using TSO-C129 systems should ensure departure and arrival airports are entered to ensure proper RAIM availability and CDI sensitivity.

(b) DME/DME. Operators should be aware that DME/DME position updating is dependent on FMS logic and DME facility proximity, availability, geometry, and signal masking.

(c) VOR/DME. Unique VOR characteristics may result in less accurate values from VOR/DME position updating than from GPS or DME/DME position updating.

(d) Inertial Navigation. Inertial reference units and inertial navigation systems are often coupled with other types of navigation inputs, e.g., DME/DME or GPS, to improve overall navigation system performance.

FIG 1-2-1
Fly-by and Fly-over Waypoints

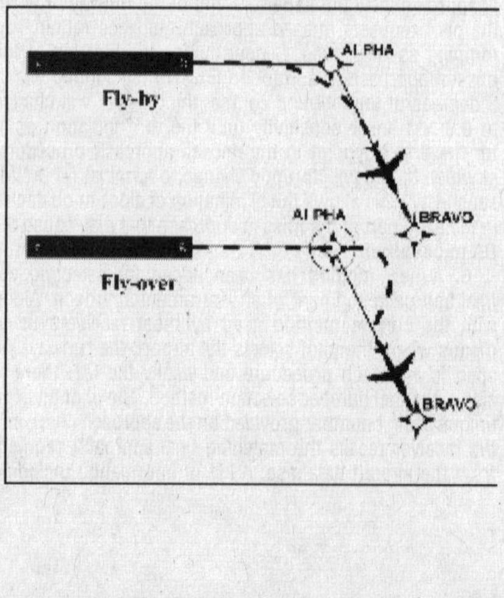

FIG 1-2-2
Track to Fix Leg Type

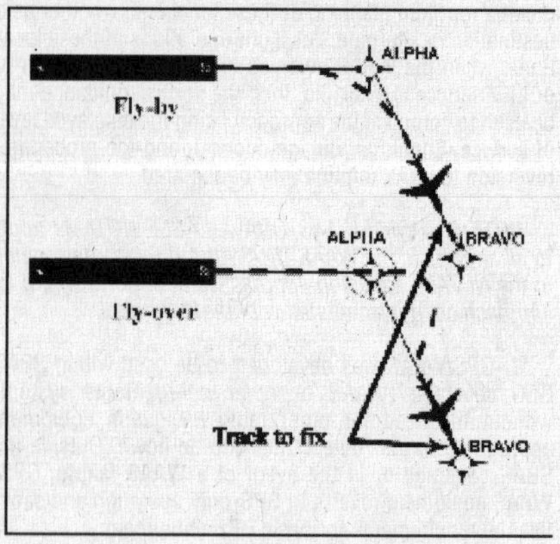

FIG 1-2-3
Direct to Fix Leg Type

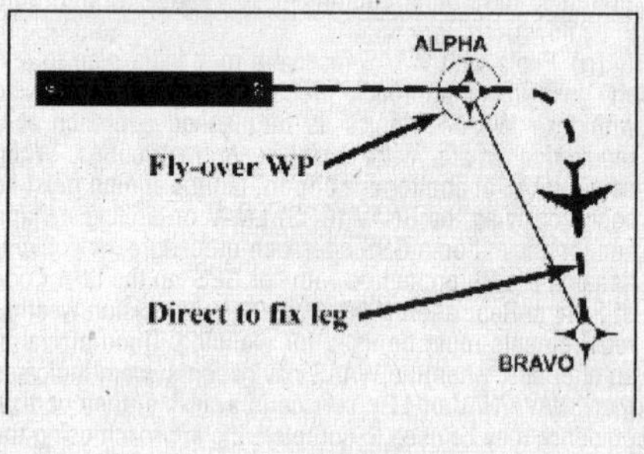

FIG 1-2-4
Course to Fix Leg Type

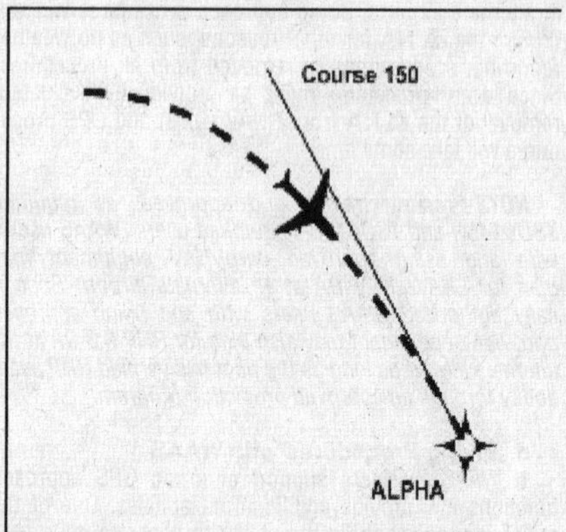

FIG 1-2-5
Radius to Fix Leg Type

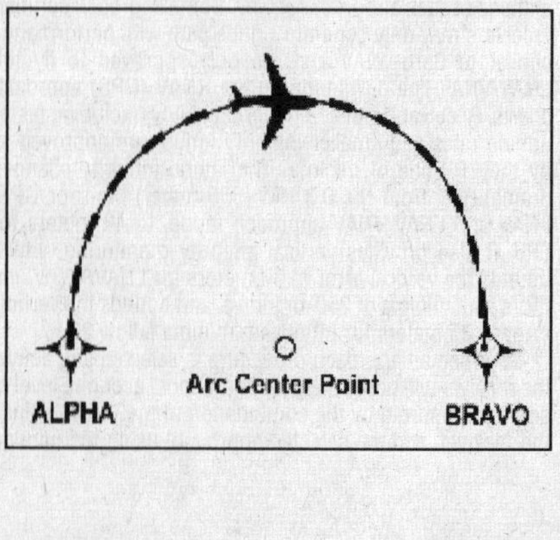

***NOTE**-Specific inertial position updating requirements may apply.*

4. Flight Management System (FMS). An FMS is an integrated suite of sensors, receivers, and computers, coupled with a navigation database. These systems generally provide performance and RNAV guidance to displays and automatic flight control systems.

Inputs can be accepted from multiple sources such as GPS, DME, VOR, LOC and IRU. These inputs may be applied to a navigation solution one at a time or in combination. Some FMSs provide for the detection and isolation of faulty navigation information.

When appropriate navigation signals are available, FMSs will normally rely on GPS and/or DME/DME (that is, the use of distance information from two or more DME stations) for position updates. Other inputs may also be incorporated based on FMS system architecture and navigation source geometry.

***NOTE**-DME/DME inputs coupled with one or more IRU(s) are often abbreviated as DME/DME/IRU or D/D/I.*

1-2-2. Required Navigation Performance (RNP)

a. General. RNP is RNAV with on-board navigation monitoring and alerting, RNP is also a statement of navigation performance necessary for operation within a defined airspace. A critical component of RNP is the *ability of the aircraft navigation system to monitor its achieved navigation performance, and to identify for the pilot whether the operational requirement is, or is not being met during an operation.* This on-board performance monitoring and alerting capability therefore allows a lessened reliance on air traffic control intervention (via radar monitoring, automatic dependent surveillance (ADS), multilateration, communications), and/or route separation to achieve the overall safety of the operation. RNP capability of the aircraft is a major component in determining the separation criteria to ensure that the overall containment of the operation is met.

The RNP capability of an aircraft will vary depending upon the aircraft equipment and the navigation infrastructure. For example, an aircraft may be equipped and certified for RNP 1.0, but may not be capable of RNP 1.0 operations due to limited navaid coverage.

b. RNP Operations.

1. RNP Levels. An RNP "level" or "type" is applicable to a selected airspace, route, or procedure. As defined in the Pilot/Controller Glossary, the RNP Level or Type is a value typically expressed as a distance in nautical miles from the intended centerline of a procedure, route, or path. RNP applications also account for potential errors at some multiple of RNP level (e.g., twice the RNP level).

(a) Standard RNP Levels. U.S. standard values supporting typical RNP airspace are as specified in TBL 1-2-1 below. Other RNP levels as identified by ICAO, other states and the FAA may also be used.

(b) Application of Standard RNP Levels. U.S. standard levels of RNP typically used for various routes and procedures supporting RNAV operations may be based on use of a specific navigational system or sensor such as GPS, or on multi-sensor RNAV systems having suitable performance.

(c) Depiction of Standard RNP Levels. The applicable RNP level will be depicted on affected charts and procedures.

TBL 1-2-1

U.S. Standard RNP Levels

RNP Level	Typical Application	Primary Route Width (NM) -
Centerline to Boundary		
0.1 to 1.0	RNP AR Approach Segments	0.1 to 1.0
0.3 to 1.0	RNP Approach Segments	0.3 to 1.0
1	Terminal and En Route	1.0
2	En Route	2.0

***NOTE**-1. The "performance" of navigation in RNP refers not only to the level of accuracy of a particular sensor or aircraft navigation system, but also to the degree of precision with which the aircraft will be flown.*

2. Specific required flight procedures may vary for different RNP levels.

TBL 1-2-2

RNP Levels Supported for International Operations

RNP Level	Typical Application
4	Projected for oceanic/remote areas where 30 NM horizontal separation is applied
10	Oceanic/remote areas where 50 NM lateral separation is applied

c. Other RNP Applications Outside the U.S. The FAA and ICAO member states have led initiatives in implementing the RNP concept to oceanic operations. For example, RNP-10 routes have been established in the northern Pacific (NOPAC) which has increased capacity and efficiency by reducing the distance between tracks to 50 NM. (See TBL 1-2-2.)

d. Aircraft and Airborne Equipment Eligibility for RNP Operations. Aircraft meeting RNP criteria will have an appropriate entry including special conditions and limitations in its Aircraft Flight Manual (AFM), or supplement. Operators of aircraft not having specific AFM-RNP certification may be issued operational approval including special conditions and limitations for specific RNP levels.

***NOTE**-Some airborne systems use Estimated Position Uncertainty (EPU) as a measure of the current estimated navigational performance. EPU may also be referred to as Actual Navigation Performance (ANP) or Estimated Position Error (EPE).*

1-2-3. Use of Suitable Area Navigation (RNAV) Systems on Conventional Procedures and Routes

a. Discussion. This paragraph sets forth policy, while providing operational and airworthiness guidance regarding the suitability and use of RNAV systems when operating on, or transitioning to, conventional, non-RNAV routes and procedures within the U.S. National Airspace System (NAS):

1. Use of a suitable RNAV system as a Substitute Means of Navigation when a Very-High Frequency (VHF) Omni-directional Range (VOR), Distance Measuring Equipment (DME), Tactical Air Navigation (TACAN), VOR/TACAN (VORTAC), VOR/DME, Non-directional Beacon (NDB), or compass locator facility including locator outer marker and locator middle marker is out-of-service (that is, the navigation aid (NAVAID) information is not available); an aircraft is not equipped with an Automatic Direction Finder (ADF) or DME; or the installed ADF or DME on an aircraft is not operational. For example, if equipped with a suitable RNAV system, a pilot may hold over an out-of-service NDB.

2. Use of a suitable RNAV system as an Alternate Means of Navigation when a VOR, DME, VORTAC, VOR/DME, TACAN, NDB, or compass locator facility including locator outer marker and locator middle marker is operational and the respective aircraft is equipped with operational navigation equipment that is compatible with conventional navaids. For example, if equipped with a suitable RNAV system, a pilot may fly a procedure or route based on operational VOR using that RNAV system without monitoring the VOR.

***NOTE**-1. Additional information and associated requirements are available in Advisory Circular 90108 titled "Use of Suitable RNAV Systems on Conventional Routes and Procedures."*

2. Good planning and knowledge of your RNAV system are critical for safe and successful operations.

3. Pilots planning to use their RNAV system as a substitute means of navigation guidance in lieu of an out-of-service NAVAID may need to advise ATC of this intent and capability.

4. The navigation database should be current for the duration of the flight. If the AIRAC cycle will change during flight, operators and pilots should establish procedures to ensure the accuracy of navigation data, including suitability of navigation facilities used to define the routes and procedures for flight. To facilitate validating database currency, the FAA has developed procedures for publishing the amendment date that instrument approach procedures were last revised. The amendment date follows the amendment number, e.g., Amdt 4 14Jan10. Currency of graphic departure procedures and STARs may be ascertained by the numerical designation in the procedure title. If an amended chart is published for the procedure, or the procedure amendment date shown on the chart is on or after the expiration date of the database, the operator must not use the database to conduct the operation.

b. Types of RNAV Systems that Qualify as a Suitable RNAV System. When installed in accordance with appropriate airworthiness installation requirements and operated in accordance with applicable operational guidance (e.g., aircraft flight manual and Advisory Circular material), the following systems qualify as a suitable RNAV system:

1. An RNAV system with TSO-C129/ -C145/-C146 equipment, installed in accordance with AC 20-138, Airworthiness Approval of Global Positioning System (GPS) Navigation Equipment for Use as a VFR and IFR Supplemental Navigation System, or AC 20-130A, Airworthiness Approval of Navigation or Flight Management Systems Integrating Multiple Navigation Sensors, and authorized for instrument flight rules (IFR) en route and terminal operations (including those systems previously qualified for "GPS in lieu of ADF or DME" operations), or

2. An RNAV system with DME/DME/IRU inputs that is compliant with the equipment provisions of AC 90-100A, U.S. Terminal and En Route Area Navigation (RNAV) Operations, for RNAV routes. A table of compliant equipment is available at the following website:

http://www.faa.gov/about/office_org/headquarters_offices/avs/offices/afs/afs400/afs470/policy_guidance/

***NOTE**-Approved RNAV systems using DME/DME/IRU, without GPS/WAAS position input, may only be used as a substitute means of navigation when specifically authorized by a Notice to Airmen (NOTAM) or other FAA guidance for a specific procedure. The NOTAM or other FAA guidance authorizing the use of DME/DME/IRU systems will also identify any required DME facilities based on an FAA assessment of the DME navigation infrastructure.*

c. Uses of Suitable RNAV Systems. Subject to the operating requirements, operators may use a suitable RNAV system in the following ways.

1. Determine aircraft position relative to, or distance from a VOR (see NOTE 5 below), TACAN, NDB, compass locator, DME fix; or a named fix defined by a VOR radial, TACAN course, NDB bearing, or compass locator bearing intersecting a VOR or localizer course.

2. Navigate to or from a VOR, TACAN, NDB, or compass locator.

3. Hold over a VOR, TACAN, NDB, compass locator, or DME fix.

4. Fly an arc based upon DME.

***NOTE**-1. The allowances described in this section apply even when a facility is identified as required on a procedure (for example, "Note ADF required").*

2. These operations do not include lateral navigation on localizer-based courses (including localizer back-course guidance) without reference to raw localizer data.

3. Unless otherwise specified, a suitable RNAV system cannot be used for navigation on procedures that are identified as not authorized ("NA") without exception by a NOTAM. For example, an operator may not use a RNAV system to navigate on a procedure affected by an expired or unsatisfactory flight inspection, or a procedure that is based upon a recently decommissioned NAVAID.

4. Pilots may not substitute for the NAVAID (for example, a VOR or NDB) providing lateral guidance for the final approach segment. This restriction does not refer to instrument approach procedures with "or GPS" in the title when using GPS or WAAS. These allowances do not apply to procedures that are identified as not authorized (NA) without exception by a NOTAM, as other conditions may still exist and result in a procedure not being available. For example, these allowances do not apply to a procedure associated with an expired or unsatisfactory flight inspection, or is based upon a recently decommissioned NAVAID.

5. *For the purpose of paragraph c, "VOR" includes VOR, VOR/DME, and VORTAC facilities and "compass locator" includes locator outer marker and locator middle marker.*

d. Alternate Airport Considerations. For the purposes of flight planning, any required alternate airport must have an available instrument approach procedure that does not require the use of GPS. This restriction includes conducting a conventional approach at the alternate airport using a substitute means of navigation that is based upon the use of GPS. For example, these restrictions would apply when planning to use GPS equipment as a substitute means of navigation for an out-of-service VOR that supports an ILS missed approach procedure at an alternate airport. In this case, some other approach not reliant upon the use of GPS must be available. This restriction does not apply to RNAV systems using TSO-C145/-C146 WAAS equipment. For further WAAS guidance see AIM 1-1-19.

1. For flight planning purposes, TSOC129() and TSOC196() equipped users (GPS users) whose navigation systems have fault detection and exclusion (FDE) capability, who perform a preflight RAIM prediction at the airport where the RNAV (GPS) approach will be flown, and have proper knowledge and any required training and/or approval to conduct a GPSbased IAP, may file based on a GPSbased IAP at either the destination or the alternate airport, but not at both locations. At the alternate airport, pilots may plan for applicable alternate airport weather minimums using:

(a) Lateral navigation (LNAV) or circling minimum descent altitude (MDA);

(b) LNAV/vertical navigation (LNAV/VNAV) DA, if equipped with and using approved barometric vertical navigation (baroVNAV) equipment;

(c) RNP 0.3 DA on an RNAV (RNP) IAP, if they are specifically authorized users using approved baroVNAV equipment and the pilot has verified required navigation performance (RNP) availability through an approved prediction program.

2. If the above conditions cannot be met, any required alternate airport must have an approved instrument approach procedure other than GPS that is anticipated to be operational and available at the estimated time of arrival, and which the aircraft is equipped to fly.

3. This restriction does not apply to TSOC145() and TSOC146() equipped users (WAAS users). For further WAAS guidance see AIM 1-1-19.

Chapter 2

AERONAUTICAL LIGHTING AND OTHER AIRPORT VISUAL AIDS

Section 1. Airport Lighting Aids

2-1-1. Approach Light Systems (ALS)

a. ALS provide the basic means to transition from instrument flight to visual flight for landing. Operational requirements dictate the sophistication and configuration of the approach light system for a particular runway.

b. ALS are a configuration of signal lights starting at the landing threshold and extending into the approach area a distance of 2400-3000 feet for precision instrument runways and 1400-1500 feet for nonprecision instrument runways. Some systems include sequenced flashing lights which appear to the pilot as a ball of light traveling towards the runway at high speed (twice a second). (See FIG 2-1-1.)

2-1-2. Visual Glideslope Indicators

a. Visual Approach Slope Indicator (VASI)

1. VASI installations may consist of either 2, 4, 6, 12, or 16 light units arranged in bars referred to as near, middle, and far bars. Most VASI installations consist of 2 bars, near and far, and may consist of 2, 4, or 12 light units. Some VASIs consist of three bars, near, middle, and far, which provide an additional visual glide path to accommodate high cockpit aircraft. This installation may consist of either 6 or 16 light units. VASI installations consisting of 2, 4, or 6 light units are located on one side of the runway, usually the left. Where the installation consists of 12 or 16 light units, the units are located on both sides of the runway.

2. Two-bar VASI installations provide one visual glide path which is normally set at 3 degrees. Three-bar VASI installations provide two visual glide paths. The lower glide path is provided by the near and middle bars and is normally set at 3 degrees while the upper glide path, provided by the middle and far bars, is normally 1/4 degree higher. This higher glide path is intended for use only by high cockpit aircraft to provide a sufficient threshold crossing height. Although normal glide path angles are three degrees, angles at some locations may be as high as 4.5 degrees to give proper obstacle clearance. Pilots of high performance aircraft are cautioned that use of VASI angles in excess of 3.5 degrees may cause an increase in runway length required for landing and rollout.

3. The basic principle of the VASI is that of color differentiation between red and white. Each light unit projects a beam of light having a white segment in the upper part of the beam and red segment in the lower part of the beam. The light units are arranged so that the pilot using the VASIs during an approach will see the combination of lights shown below.

4. The VASI is a system of lights so arranged to provide visual descent guidance information during the approach to a runway. These lights are visible from 3-5 miles during the day and up to 20 miles or more at night. The visual glide path of the VASI provides safe obstruction clearance within plus or minus 10 degrees of the extended runway centerline and to 4 NM from the runway threshold. Descent, using the VASI, should not be initiated until the aircraft is visually aligned with the runway. Lateral course guidance is provided by the runway or runway lights. In certain circumstances, the safe obstruction clearance area may be reduced due to local limitations, or the VASI may be offset from the extended runway centerline. This will be noted in the Airport/ Facility Directory.

NOTE-*Civil ALSF-2 may be operated as SSALR during favorable weather conditions.*

5. For 2-bar VASI (4 light units) see FIG 2-1-2.
6. For 3-bar VASI (6 light units) see FIG 2-1-3.
7. For other VASI configurations see FIG 2-1-4.

b. Precision Approach Path Indicator (PAPI). The precision approach path indicator (PAPI) uses light units similar to the VASI but are installed in a single row of either two or four light units. These lights are visible from about 5 miles during the day and up to 20 miles at night. The visual glide path of the PAPI typically provides safe obstruction clearance within plus or minus 10 degrees of the extended runway centerline and to 4 SM from the runway threshold. Descent, using the PAPI, should not be initiated until the aircraft is visually aligned with the runway. The row of light units is normally installed on the left side of the runway and the glide path indications are as depicted. Lateral course guidance is provided by the runway or runway lights. In certain circumstances, the safe obstruction clearance area may be reduced due to local limitations, or the PAPI may be offset from the extended runway centerline. This will be noted in the Airport/ Facility Directory. (See FIG 2-1-5.)

c. Tri-color Systems. Tri-color visual approach slope indicators normally consist of a single light unit projecting a three-color visual approach path into the final

FIG 2-1-1
Precision & Nonprecision Configurations

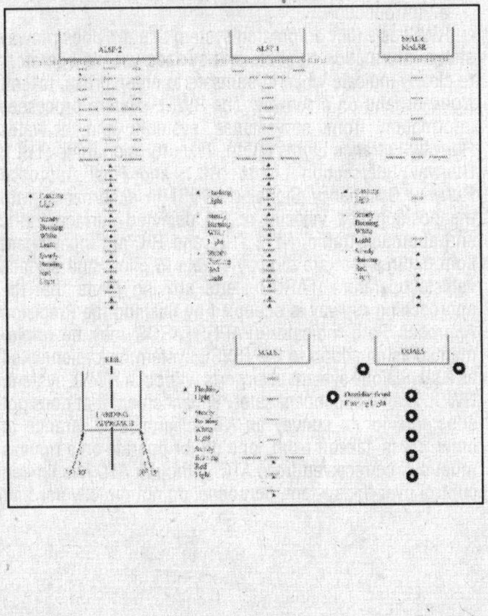

FIG 2-1-2
2-Bar VASI

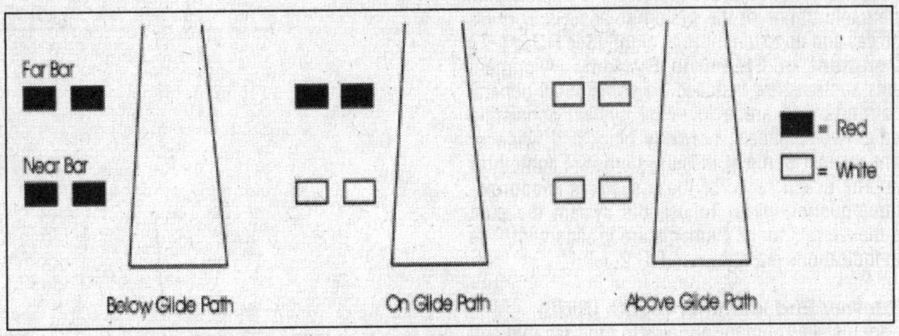

FIG 2-1-3
3-Bar VASI

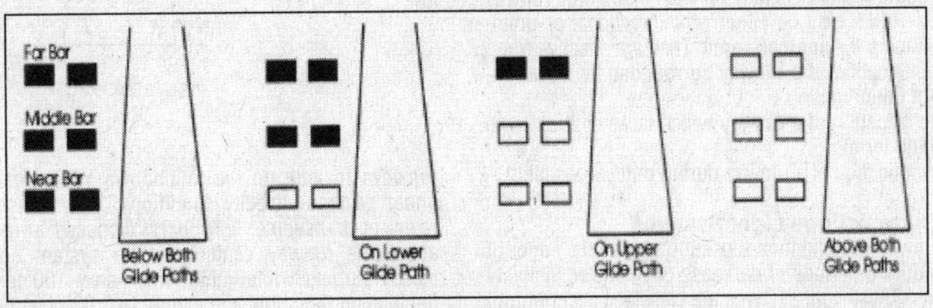

FIG 2-1-4
VASI Variations

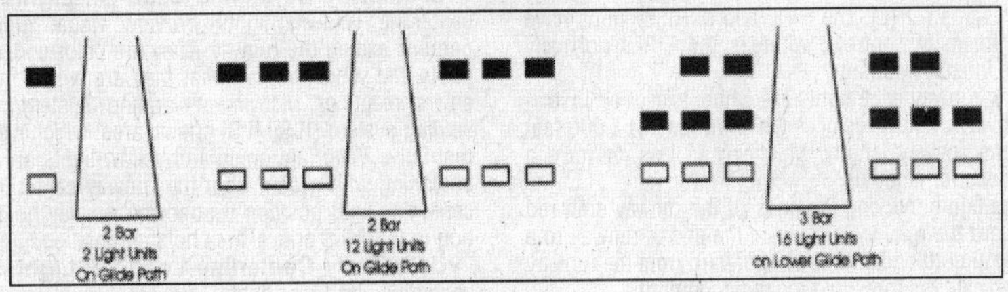

FIG 2-1-5
Precision Approach Path Indicator (PAPI)

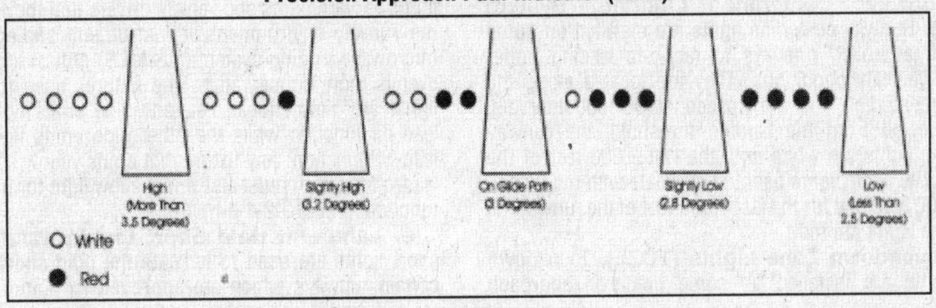

approach area of the runway upon which the indicator is installed. The below glide path indication is red, the above glide path indication is amber, and the on glide path indication is green. These types of indicators have a useful range of approximately one-half to one mile during the day and up to five miles at night depending upon the visibility conditions. (See FIG 2-1-6.)

NOTE-1. *Since the tri-color VASI consists of a single light source which could possibly be confused with other light sources, pilots should exercise care to properly locate and identify the light signal.*

2. *When the aircraft descends from green to red, the pilot may see a dark amber color during the transition from green to red.*

NOTE- *Since the PVASI consists of a single light source which could possibly be confused with other light sources, pilots should exercise care to properly locate and identify the light signal.*

d. Pulsating Systems. Pulsating visual approach slope indicators normally consist of a single light unit projecting a two-color visual approach path into the final approach area of the runway upon which the indicator is installed. The on glide path indication is a steady white light. The slightly below glide path indication is a steady red light. If the aircraft descends further below the glide path, the red light starts to pulsate. The above glide path indication is a pulsating white light. The pulsating rate increases as the aircraft gets further above or below the desired glide slope. The useful range of the system is about four miles during the day and up to ten miles at night. (See FIG 2-1-7.)

e. Alignment of Elements Systems. Alignment of elements systems are installed on some small general aviation airports and are a low-cost system consisting of painted plywood panels, normally black and white or fluorescent orange. Some of these systems are lighted for night use. The useful range of these systems is approximately three-quarter miles. To use the system the pilot positions the aircraft so the elements are in alignment. The glide path indications are shown in FIG 2-1-8.

FIG 2-1-6
Tri-Color Visual Approach Slope Indicator

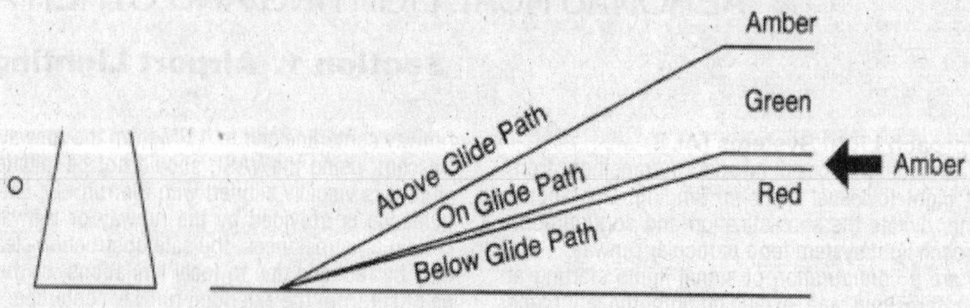

FIG 2-1-7
Pulsating Visual Approach Slope Indicator

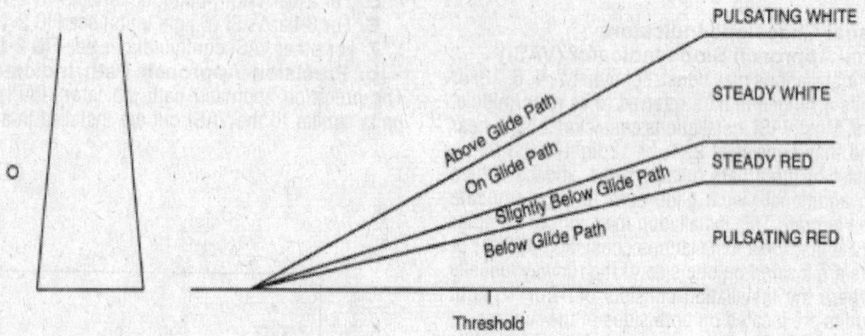

FIG 2-1-8
Alignment of Elements

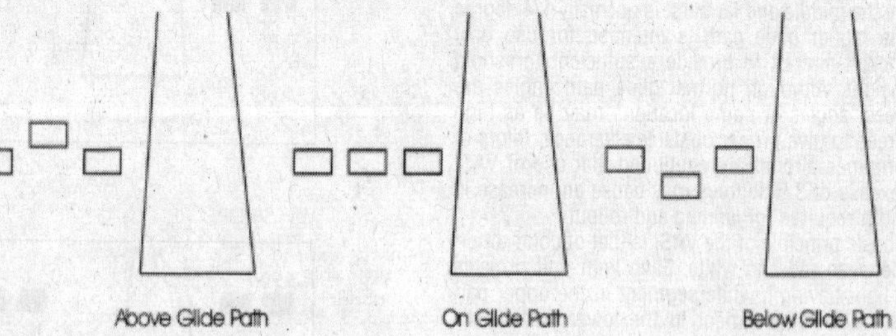

2-1-3. Runway End Identifier Lights (REIL)

REILs are installed at many airfields to provide rapid and positive identification of the approach end of a particular runway. The system consists of a pair of synchronized flashing lights located laterally on each side of the runway threshold. REILs may be either omnidirectional or unidirectional facing the approach area. They are effective for:

a. Identification of a runway surrounded by a preponderance of other lighting.

b. Identification of a runway which lacks contrast with surrounding terrain.

c. Identification of a runway during reduced visibility.

2-1-4. Runway Edge Light Systems

a. Runway edge lights are used to outline the edges of runways during periods of darkness or restricted visibility conditions. These light systems are classified according to the intensity or brightness they are capable of producing: they are the High Intensity Runway Lights (HIRL), Medium Intensity Runway Lights (MIRL), and the Low Intensity Runway Lights (LIRL). The HIRL and MIRL systems have variable intensity controls, whereas the LIRLs normally have one intensity setting.

b. The runway edge lights are white, except on instrument runways yellow replaces white on the last 2,000 feet or half the runway length, whichever is less, to form a caution zone for landings.

c. The lights marking the ends of the runway emit red light toward the runway to indicate the end of runway to a departing aircraft and emit green outward from the runway end to indicate the threshold to landing aircraft.

2-1-5. In-runway Lighting

a. Runway Centerline Lighting System (RCLS). Runway centerline lights are installed on some precision approach runways to facilitate landing under adverse visibility conditions. They are located along the runway centerline and are spaced at 50-foot intervals. When viewed from the landing threshold, the runway centerline lights are white until the last 3,000 feet of the runway. The white lights begin to alternate with red for the next 2,000 feet, and for the last 1,000 feet of the runway, all centerline lights are red.

b. Touchdown Zone Lights (TDZL). Touchdown zone lights are installed on some precision approach runways to indicate the touchdown zone when landing under adverse visibility conditions. They consist of two rows of transverse light bars disposed symmetrically about the runway centerline. The system consists of steady-burning white lights which start 100 feet beyond the landing threshold and extend to 3,000 feet beyond the landing threshold or to the midpoint of the runway, whichever is less.

c. Taxiway Centerline Lead-Off Lights. Taxiway centerline lead-off lights provide visual guidance to persons exiting the runway. They are color-coded to warn pilots and vehicle drivers that they are within the runway environment or instrument landing system/microwave landing system (ILS/MLS) critical area, whichever is more restrictive. Alternate green and yellow lights are installed, beginning with green, from the runway centerline to one centerline light position beyond the runway holding position or ILS/MLS critical area holding position.

d. Taxiway Centerline Lead-On Lights. Taxiway centerline lead-on lights provide visual guidance to persons entering the runway. These "lead-on" lights are also color-coded with the same color pattern as lead-off lights to warn pilots and vehicle drivers that they are within the runway environment or instrument landing system/microwave landing system (ILS/MLS) critical area, whichever is more conservative. The fixtures used for lead-on lights are bidirectional, i.e., one side emits light for the lead-on function while the other side emits light for the lead-off function. Any fixture that emits yellow light for the lead-off function must also emit yellow light for the lead-on function. (See FIG 2-1-14.)

e. Land and Hold Short Lights. Land and hold short lights are used to indicate the hold short point on certain runways which are approved for Land and Hold Short Operations (LAHSO). Land and hold short lights consist of a row of pulsing white lights installed across the runway at the hold short point. Where installed, the lights will be on anytime LAHSO is in effect. These lights will be off when LAHSO is not in effect.

REFERENCE- *AIM, Pilot Responsibilities When Conducting Land and Hold Short Operations (LAHSO), Paragraph 4-3-11.*

2-1-6. Runway Status Light (RWSL) System

a. Introduction.

RWSL is a fully automated system that provides runway status information to pilots and surface vehicle operators to clearly indicate when it is unsafe to enter, cross, takeoff from, or land on a runway. The RWSL system processes information from surveillance systems and activates Runway Entrance Lights (REL), Takeoff Hold Lights (THL), Runway Intersection Lights (RIL), and Final Approach Runway Occupancy Signal (FAROS) in accordance with the position and velocity of the detected surface traffic and approach traffic. REL, THL, and RIL are inpavement light fixtures that are directly visible to pilots and surface vehicle operators. FAROS alerts arriving pilots that the approaching runway is occupied by flashing the Precision Approach Path Indicator (PAPI). FAROS may be implemented as an addon to the RWSL system or implemented as a standalone system at airports without a RWSL system. RWSL is an independent safety enhancement that does not substitute for or convey an ATC clearance. Clearance to enter, cross, takeoff from, land on, or operate on a runway must still be received from ATC. Although ATC has limited control over the system, personnel do not directly use and may not be able to view light fixture activations and deactivations during the conduct of daily ATC operations.

b. Runway Entrance Lights (REL): The REL system is composed of flush mounted, inpavement, unidirectional light fixtures that are parallel to and focused along the taxiway centerline and directed toward the pilot at the hold line. An array of REL lights include the first light at the hold line followed by a series of evenly spaced lights to the runway edge; one additional light at the runway centerline is in line with the last two lights before the runway edge (see FIG 2-1-9 and FIG 2-1-12). When activated, the red lights indicate that there is high speed traffic on the runway or there is an aircraft on final approach within the activation area.

1. REL Operating Characteristics - Departing Aircraft:

When a departing aircraft reaches a site adaptable speed of approximately 30 knots, all taxiway intersections with REL arrays along the runway ahead of the aircraft will illuminate (see FIG 2-1-9). As the aircraft approaches an REL equipped taxiway intersection, the lights at that intersection extinguish approximately 3 to 4 seconds before the aircraft reaches it. This allows controllers to apply "anticipated separation" to permit ATC to move traffic more expeditiously without compromising safety. After the aircraft is declared "airborne" by the system, all REL lights associated with this runway will extinguish.

2. REL Operating Characteristics - Arriving Aircraft:

When an aircraft on final approach is approximately 1 mile from the runway threshold, all sets of taxiway REL light arrays that intersect the runway illuminate. The distance is adjustable and can be configured for specific operations at particular airports. Lights extinguish at each equipped taxiway intersection approximately 3 to 4 seconds before the aircraft reaches it to apply anticipated separation until the aircraft has slowed to approximately 80 knots (site adjustable parameter). Below 80 knots, all arrays that are not within 30 seconds of the aircraft's forward path are extinguished. Once the arriving aircraft slows to approximately 34 knots (site adjustable parameter), it is declared to be in a taxi state, and all lights extinguish.

3. What a pilot would observe: A pilot at or approaching the hold line to a runway will observe RELs illuminate and extinguish in reaction to an aircraft or vehicle operating on the runway, or an arriving aircraft operating less than 1 mile from the runway threshold.

4. When a pilot observes the red lights of the REL, that pilot will stop at the hold line or remain stopped. The pilot will then contact ATC for resolution if the clearance is in conflict with the lights. Should pilots note illuminated lights under circumstances when remaining clear of the runway is impractical for safety reasons (for example, aircraft is already on the runway), the crew should proceed according to their best judgment while understanding the illuminated lights indicate the runway is unsafe to enter or cross. Contact ATC at the earliest possible opportunity.

c. Takeoff Hold Lights (THL) : The THL system is composed of flush mounted, inpavement, unidirectional light fixtures in a double longitudinal row aligned either side of the runway centerline lighting. Fixtures are focused toward the arrival end of the runway at the "line up and wait" point. THLs extend for 1,500 feet in front of the holding aircraft starting at a point 375 feet from the departure threshold (see FIG 2-1-13). Illuminated red lights provide a signal, to an aircraft in position for takeoff or rolling, that it is unsafe to takeoff because the runway is occupied or about to be occupied by another aircraft or ground vehicle. Two aircraft, or a surface vehicle and an aircraft, are required for the lights to illuminate. The departing aircraft must be in position for takeoff or beginning takeoff roll. Another aircraft or a surface vehicle must be on or about to cross the runway.

1. THL Operating Characteristics - Departing Aircraft:

THLs will illuminate for an aircraft in position for departure or departing when there is another aircraft or vehicle on the runway or about to enter the runway (see FIG 2-1-9.) Once that aircraft or vehicle exits the runway, the THLs extinguish. A pilot may notice lights extinguish prior to the downfield aircraft or vehicle being completely clear of the runway but still moving. Like RELs, THLs have an "anticipated separation" feature.

NOTE-*When the THLs extinguish, this is not clearance to begin a takeoff roll. All takeoff clearances will be issued by ATC.*

2. What a pilot would observe: A pilot in position to depart from a runway, or has begun takeoff roll, will observe THLs illuminate in reaction to an aircraft or vehicle on the runway or entering or crossing it. Lights will extinguish when the runway is clear. A pilot may observe several cycles of illumination and extinguishing depending on the amount of crossing traffic.

3. When a pilot observes the red light of the THLs, the pilot should safely stop if it's feasible or remain stopped. The pilot must contact ATC for resolution if any clearance is in conflict with the lights. Should pilots note illuminated lights while in takeoff roll and under circumstances when stopping is impractical for safety reasons, the crew should proceed according to their best judgment while understanding the illuminated lights indicate that continuing the takeoff is unsafe. Contact ATC at the earliest possible opportunity.

d. Runway Intersection Lights (RIL): The RIL system is composed of flush mounted, in-pavement, unidirectional light fixtures in a double longitudinal row aligned either side of the runway centerline lighting in the same manner as THLs. Their appearance to a pilot is similar to that of THLs. Fixtures are focused toward the arrival end of the runway, and they extend for 3,000 feet in front of an aircraft that is approaching an intersecting runway. They end at the Land and Hold Short Operation (LASHO) light bar or the hold short line for the intersecting runway.

1. RIL Operating Characteristics - Departing Aircraft:

RILs will illuminate for an aircraft departing or in position to depart when there is high speed traffic operating on the intersecting runway (see FIG 2-1-9). Note that there must be an aircraft or vehicle in a position to observe the RILs for them to illuminate. Once the conflicting traffic passes through the intersection, the RILs extinguish.

2. RIL Operating Characteristics - Arriving Aircraft:

RILs will illuminate for an aircraft that has landed and is rolling out when there is high speed traffic on the intersecting runway that is ±5 seconds of meeting at the intersection. Once the conflicting traffic passes through the intersection, the RILs extinguish.

3. What a pilot would observe: A pilot departing or arriving will observe RILs illuminate in reaction to the high speed traffic operation on the intersecting runway. The lights will extinguish when that traffic has passed through the runway intersection.

4. Whenever a pilot observes the red light of the RIL array, the pilot will stop before the LAHSO stop bar or the hold line for the intersecting runway. If a departing aircraft is already at high speed in the takeoff roll when the RILs illuminate, it may be impractical to stop for safety reasons. The crew should safely operate according to their best judgment while understanding the illuminated lights indicate that continuing the takeoff is unsafe. Contact ATC at the earliest possible opportunity.

e. The Final Approach Runway Occupancy Signal (FAROS) is communicated by flashing of the Precision Approach Path Indicator (PAPI) (see FIG 219). When activated, the light fixtures of the PAPI flash or pulse to indicate to the pilot on an approach that the runway is occupied and that it may be unsafe to land.

NOTE-*FAROS is an independent automatic alerting system that does not rely on ATC control or input.*

1. FAROS Operating Characteristics:

If an aircraft or surface vehicle occupies a FAROS equipped runway, the PAPI(s) on that runway will flash. The glide path indication will not be affected, and the allotment of red and white PAPI lights observed by the pilot on approach will not change. The FAROS system will flash the PAPI when traffic enters the runway and there is an aircraft on approach and within 1.5 nautical miles of the landing threshold.

2. What a pilot would observe: A pilot on approach to the runway will observe the PAPI flash if there is traffic on the runway and will notice the PAPI ceases to flash when the traffic moves outside the hold short lines for the runway.

3. When a pilot observes a flashing PAPI at 500 feet above ground level (AGL), the contact height, the pilot must look for and acquire the traffic on the runway. At 300 feet AGL, the pilot must contact ATC for resolution if the FAROS indication is in conflict with the clearance. If the PAPI continues to flash, the pilot must execute an immediate "go around" and contact ATC at the earliest possible opportunity.

f. Pilot Actions:

1. When operating at airports with RWSL, pilots will operate with the transponder "On" when departing the

Fig 2-1-9
Runway Status Light System

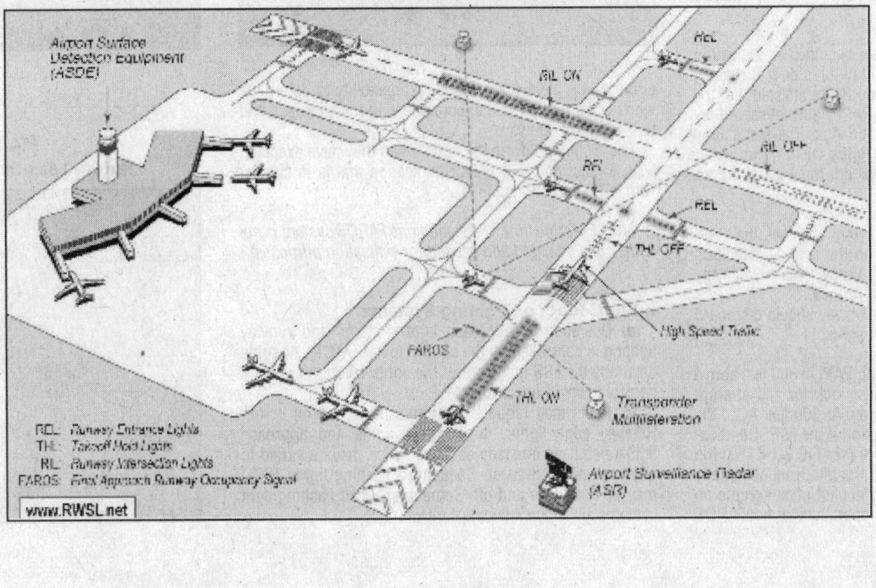

Fig 2-1-10
FAROS Activation Zones

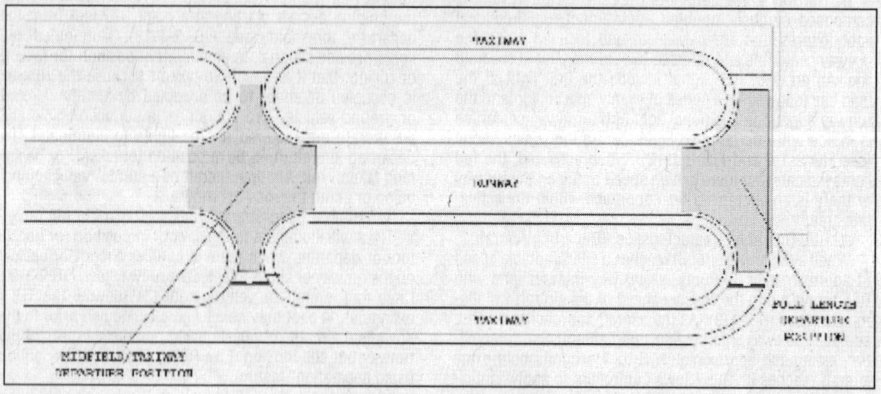

Fig 2-1-11
FAROS Glide Slope Action Points

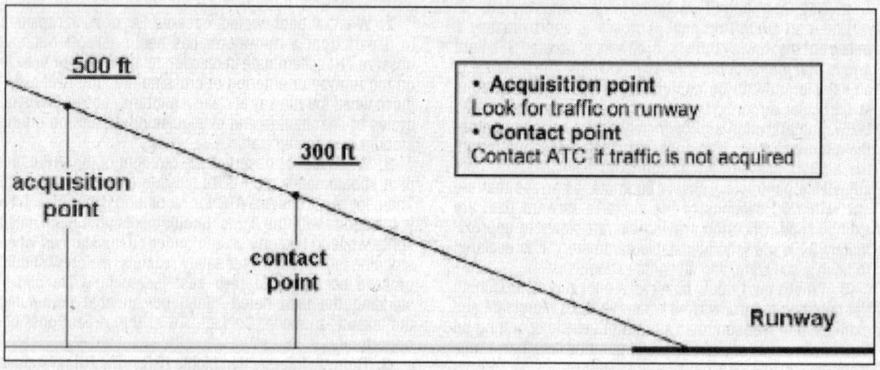

Fig 2-1-12
Runway Entrance Lights

Fig 2-1-13
Takeoff Hold Lights

gate or parking area until it is shutdown upon arrival at the gate or parking area. This ensures interaction with the FAA surveillance systems such as ASDEX which provide information to the RWSL system.

2. Pilots must always inform the ATCT when they have either stopped, are verifying a landing clearance, or are executing a goaround due to RWSL or FAROS indication that are in conflict with ATC instructions. Pilots must request clarification of the taxi, takeoff, or landing clearance.

3. Never cross over illuminated red lights. Under normal circumstances, RWSL will confirm the pilot's taxi or takeoff clearance previously issued by ATC. If RWSL indicates that it is unsafe to takeoff from, land on, cross, or enter a runway, immediately notify ATC of the conflict and reconfirm the clearance.

4. Do not proceed when lights have extinguished without an ATC clearance. RWSL verifies an ATC clearance; it does not substitute for an ATC clearance.

5. Never land if PAPI continues to flash. Execute a go around and notify ATC.

g. ATC Control of RWSL System:

1. Controllers can set in-pavement lights to one of five (5) brightness levels to assure maximum conspicuity under all visibility and lighting conditions. REL, THL, and RIL subsystems may be independently set.

2. System lights can be disabled should RWSL operations impact the efficient movement of air traffic or contribute, in the opinion of the assigned ATC Manager, to unsafe operations. REL, THL, RIL, and FAROS light fixtures may be disabled separately. Disabling of the FAROS subsystem does not extinguish PAPI lights or impact its glide path function. Whenever the system or a component is disabled, a NOTAM must be issued, and the Automatic Terminal Information System (ATIS) must be updated.

2-1-7. StandAlone Final Approach Runway Occupancy Signal (FAROS)

a. Introduction:

The standalone FAROS system is a fully automated system that provides runway occupancy status to pilots on final approach to indicate whether it may be unsafe to land. When an aircraft or vehicle is detected on the runway, the Precision Approach Path Indicator (PAPI) light fixtures flash as a signal to indicate that the runway is occupied and that it may be unsafe to land. The standalone FAROS system is activated by localized or comprehensive sensors detecting aircraft or ground vehicles occupying activation zones.

The standalone FAROS system monitors specific areas of the runway, called activation zones, to determine the presence of aircraft or ground vehicles in the zone (see FIG 2-1-10). These activation zones are defined as areas on the runway that are frequently occupied by ground traffic during normal airport operations and could present a hazard to landing aircraft. Activation zones may include the fulllength departure position, the midfield departure position, a frequently crossed intersection, or the entire runway.

Pilots can refer to the airport specific FAROS pilot information sheet for activation zone configuration.

Clearance to land on a runway must be issued by Air Traffic Control (ATC). ATC personnel have limited control over the system and may not be able to view the FAROS signal.

b. Operating Characteristics:

If an aircraft or ground vehicle occupies an activation zone on the runway, the PAPI light fixtures on that runway will flash. The glide path indication is not affected, i.e. the configuration of red and white PAPI lights observed by the pilot on approach does not change. The standalone FAROS system flashes the PAPI lights when traffic occupies an activation zone whether or not there is an aircraft on approach.

c. Pilot Observations:

A pilot on approach to the runway observes the PAPI lights flashing if there is traffic on the runway activation zones and notices the PAPI lights cease to flash when the traffic moves outside the activation zones.

A pilot on departure from the runway should disregard any observations of flashing PAPI lights.

d. Pilot Actions:

When a pilot observes a flashing PAPI at 500 feet above ground level (AGL), the pilot must look for and attempt to acquire the traffic on the runway. At 300 feet AGL, the pilot must contact ATC for resolution if the FAROS indication is in conflict with the clearance (see FIG 2-1-11). If the PAPI lights continue to flash and the pilot cannot visually determine that it is safe to land, the pilot must execute an immediate "go around". As with operations at nonFAROS airports, it is always the pilot's responsibility to determine whether or not it is safe to continue with the approach and to land on the runway.

Pilots should inform the ATCT when they have executed a go around due to a FAROS indication that is in conflict with ATC instructions.

***NOTE-** At this time, the standalone FAROS system is not widely implemented and is used for evaluation purposes.*

2-1-8. Control of Lighting Systems

a. Operation of approach light systems and runway lighting is controlled by the control tower (ATCT). At some locations the FSS may control the lights where there is no control tower in operation.

b. Pilots may request that lights be turned on or off. Runway edge lights, in-pavement lights and approach lights also have intensity controls which may be varied to meet the pilots request. Sequenced flashing lights (SFL) may be turned on and off. Some sequenced flashing light systems also have intensity control.

Fig 2-1-14
Taxiway Lead-On Light Configuration

TBL 2-1-1
Runways With Approach Lights

Lighting System	No. of Int. Steps	Status During Nonuse Period	Intensity Step Selected Per No. of Mike Clicks		
			3 Clicks	5 Clicks	7 Clicks
Approach Lights (Med. Int.)	2	Off	Low	Low	High
Approach Lights (Med. Int.)	3	Off	Low	Med	High
MIRL	3	Off or Low			
HIRL	5	Off or Low			
VASI	2	Off			

***NOTES**: Predetermined intensity step.*
Low intensity for night use. High intensity for day use as determined by photocell control.

TBL 2-1-2
Runways Without Approach Lights

Lighting System	No. of Int. Steps	Status During Nonuse Period	Intensity Step Selected Per No. of Mike Clicks		
			3 Clicks	5 Clicks	7 Clicks
MIRL	3	Off or Low	Low	Med.	High
HIRL	5	Off or Low	Step 1 or 2	Step 3	Step 5
LIRL	1	Off	On	On	On
VASI▪	2	Off	♦	♦	♦
REIL▪	1	Off	Off	On/Off	On
REIL▪	3	Off	Low	Med.	High

NOTES: ♦ Low intensity for night use. High intensity for day use as determined by photocell control.
▪ The control of VASI and/or REIL may be independent of other lighting systems.

2-1-9. Pilot Control of Airport Lighting

Radio control of lighting is available at selected airports to provide airborne control of lights by keying the aircraft's microphone. Control of lighting systems is often available at locations without specified hours for lighting and where there is no control tower or FSS or when the tower or FSS is closed (locations with a part-time tower or FSS) or specified hours. All lighting systems which are radio controlled at an airport, whether on a single runway or multiple runways, operate on the same radio frequency. (See TBL 2-1-1 and TBL 2-1-2.)

a. With FAA approved systems, various combinations of medium intensity approach lights, runway lights, taxiway lights, VASI and/or REIL may be activated by radio control. On runways with both approach lighting and runway lighting (runway edge lights, taxiway lights, etc.) systems, the approach lighting system takes precedence for air-to-ground radio control over the runway lighting system which is set at a predetermined intensity step, based on expected visibility conditions. Runways without approach lighting may provide radio controlled intensity adjustments of runway edge lights. Other lighting systems, including VASI, REIL, and taxiway lights may be either controlled with the runway edge lights or controlled independently of the runway edge lights.

b. The control system consists of a 3-step control responsive to 7, 5, and/or 3 microphone clicks. This 3-step control will turn on lighting facilities capable of either 3-step, 2-step or 1-step operation. The 3-step and 2-step lighting facilities can be altered in intensity, while the 1-step cannot. All lighting is illuminated for a period of 15 minutes from the most recent time of activation and may not be extinguished prior to end of the 15 minute period (except for 1-step and 2-step REILs which may be turned off when desired by keying the mike 5 or 3 times respectively).

c. Suggested use is to always initially key the mike 7 times; this assures that all controlled lights are turned on to the maximum available intensity. If desired, adjustment can then be made, where the capability is provided, to a lower intensity (or the REIL turned off) by keying 5 and/or 3 times. Due to the close proximity of airports using the same frequency, radio controlled lighting receivers may be set at a low sensitivity requiring the aircraft to be relatively close to activate the system. Consequently, even when lights are on, always key mike as directed when overflying an airport of intended landing or just prior to entering the final segment of an approach. This will assure the aircraft is close enough to activate the system and a full 15 minutes lighting duration is available. Approved lighting systems may be activated by keying the mike (within 5 seconds) as indicated in TBL 2-1-3.

TBL 2-1-3
Radio Control System

Key Mike	Function
7 times within 5 seconds	Highest intensity available
5 times within 5 seconds	Medium or lower intensity (Lower REIL or REIL-off)
3 times within 5 seconds	Lowest intensity available (Lower REIL or REIL-off)

d. For all public use airports with FAA standard systems the Airport/Facility Directory contains the types of lighting, runway and the frequency that is used to activate the system. Airports with IAPs include data on the approach chart identifying the light system, the runway on which they are installed, and the frequency that is used to activate the system.

***NOTE-**Although the CTAF is used to activate the lights at many airports, other frequencies may also be used. The appropriate frequency for activating the lights on the airport is provided in the Airport/Facility Directory and the standard instrument approach procedures publications. It is not identified on the sectional charts.*

e. Where the airport is not served by an IAP, it may have either the standard FAA approved control system or an independent type system of different specification installed by the airport sponsor. The Airport/Facility Directory contains descriptions of pilot controlled lighting systems for each airport having other than FAA approved systems, and explains the type lights, method of control, and operating frequency in clear text.

2-1-10. Airport/Heliport Beacons

a. Airport and heliport beacons have a vertical light distribution to make them most effective from one to ten degrees above the horizon; however, they can be seen well above and below this peak spread. The beacon may be an omnidirectional capacitor-discharge device, or it may rotate at a constant speed which produces the visual effect of flashes at regular intervals. Flashes may be one or two colors alternately. The total number of flashes are:

1. 24 to 30 per minute for beacons marking airports, landmarks, and points on Federal airways.

2. 30 to 45 per minute for beacons marking heliports.

b. The colors and color combinations of beacons are:

1. White and Green- Lighted land airport.

2. *Green alone- Lighted land airport.

3. White and Yellow- Lighted water airport.

4. *Yellow alone- Lighted water airport.

5. Green, Yellow, and White- Lighted heliport.

NOTE-Green alone or yellow alone is used only in connection with a white-and-green or white-and-yellow beacon display, respectively.*

c. Military airport beacons flash alternately white and green, but are differentiated from civil beacons by dualpeaked (two quick) white flashes between the green flashes.

d. In Class B, Class C, Class D and Class E surface areas, operation of the airport beacon during the hours of daylight often indicates that the ground visibility is less than 3 miles and/or the ceiling is less than 1,000 feet. ATC clearance in accordance with 14 CFR Part 91 is required for landing, takeoff and flight in the traffic pattern. Pilots should not rely solely on the operation of the airport beacon to indicate if weather conditions are IFR or VFR. At some locations with operating control towers, ATC personnel turn the beacon on or off when controls are in the tower. At many airports the airport beacon is turned on by a photoelectric cell or time clocks and ATC personnel cannot control them. There is no regulatory requirement for daylight operation and it is the pilot's responsibility to comply with proper preflight planning as required by 14 CFR Section 91.103.

2-1-11. Taxiway Lights

a. Taxiway Edge Lights. Taxiway edge lights are used to outline the edges of taxiways during periods of darkness or restricted visibility conditions. These fixtures emit blue light.

***NOTE-**At most major airports these lights have variable intensity settings and may be adjusted at pilot request or when deemed necessary by the controller.*

b. Taxiway Centerline Lights. Taxiway centerline lights are used to facilitate ground traffic under low visibility conditions. They are located along the taxiway centerline in a straight line on straight portions, on the centerline of curved portions, and along designated taxiing paths in portions of runways, ramp, and apron areas. Taxiway centerline lights are steady burning and emit green light.

c. Clearance Bar Lights. Clearance bar lights are installed at holding positions on taxiways in order to increase the conspicuity of the holding position in low visibility conditions. They may also be installed to indicate the location of an intersecting taxiway during periods of darkness. Clearance bars consist of three in-pavement steady-burning yellow lights.

d. Runway Guard Lights. Runway guard lights are installed at taxiway/runway intersections. They are primarily used to enhance the conspicuity of taxiway/runway intersections during low visibility conditions, but may be used in all weather conditions. Runway guard lights consist of either a pair of elevated flashing yellow lights installed on either side of the taxiway, or a row of in-pavement yellow lights installed across the entire taxiway, at the runway holding position marking.

NOTE-*Some airports may have a row of three or five in-pavement yellow lights installed at taxiway/runway intersections. They should not be confused with clearance bar lights described in paragraph 2-1-11 c, Clearance Bar Lights.*

e. Stop Bar Lights. Stop bar lights, when installed, are used to confirm the ATC clearance to enter or cross the active runway in low visibility conditions (below 1,200 ft Runway Visual Range). A stop bar consists of a row of red, unidirectional, steady-burning in-pavement lights installed across the entire taxiway at the runway holding position, and elevated steady-burning red lights on each side. A controlled stop bar is operated in conjunction with the taxiway centerline lead-on lights which extend from the stop bar toward the runway. Following the ATC clearance to proceed, the stop bar is turned off and the lead-on lights are turned on. The stop bar and lead-on lights are automatically reset by a sensor or backup timer.

CAUTION-Pilots should never cross a red illuminated stop bar, even if an ATC clearance has been given to proceed onto or across the runway.

NOTE-*If after crossing a stop bar, the taxiway centerline lead-on lights inadvertently extinguish, pilots should hold their position and contact ATC for further instructions.*

Section 2. Air Navigation and Obstruction Lighting

2-2-1. Aeronautical Light Beacons

a. An aeronautical light beacon is a visual NAVAID displaying flashes of white and/or colored light to indicate the location of an airport, a heliport, a landmark, a certain point of a Federal airway in mountainous terrain, or an obstruction. The light used may be a rotating beacon or one or more flashing lights. The flashing lights may be supplemented by steady burning lights of lesser intensity.

b. The color or color combination displayed by a particular beacon and/or its auxiliary lights tell whether the beacon is indicating a landing place, landmark, point of the Federal airways, or an obstruction. Coded flashes of the auxiliary lights, if employed, further identify the beacon site.

2-2-2. Code Beacons and Course Lights

a. Code Beacons. The code beacon, which can be seen from all directions, is used to identify airports and landmarks. The code beacon flashes the three or four character airport identifier in International Morse Code six to eight times per minute. Green flashes are displayed for land airports while yellow flashes indicate water airports.

b. Course Lights. The course light, which can be seen clearly from only one direction, is used only with rotating beacons of the Federal Airway System: two course lights, back to back, direct coded flashing beams of light in either direction along the course of airway.

NOTE-*Airway beacons are remnants of the "lighted" airways which antedated the present electronically equipped federal airways system. Only a few of these beacons exist today to mark airway segments in remote mountain areas. Flashes in Morse code identify the beacon site.*

2-2-3. Obstruction Lights

a. Obstructions are marked/lighted to warn airmen of their presence during daytime and nighttime conditions. They may be marked/lighted in any of the following combinations:

1. Aviation Red Obstruction Lights. Flashing aviation red beacons (20 to 40 flashes per minute) and steady burning aviation red lights during nighttime operation. Aviation orange and white paint is used for daytime marking.

2. Medium Intensity Flashing White Obstruction Lights. Medium intensity flashing white obstruction lights may be used during daytime and twilight with automatically selected reduced intensity for nighttime operation. When this system is used on structures 500 feet (153m) AGL or less in height, other methods of marking and lighting the structure may be omitted. Aviation orange and white paint is always required for daytime marking on structures exceeding 500 feet (153m) AGL. This system is not normally installed on structures less than 200 feet (61m) AGL.

3. High Intensity White Obstruction Lights. Flashing high intensity white lights during daytime with reduced intensity for twilight and nighttime operation. When this type system is used, the marking of structures with red obstruction lights and aviation orange and white paint may be omitted.

4. Dual Lighting. A combination of flashing aviation red beacons and steady burning aviation red lights for nighttime operation and flashing high intensity white lights for daytime operation. Aviation orange and white paint may be omitted.

5. Catenary Lighting. Lighted markers are available for increased night conspicuity of high-voltage (69KV or higher) transmission line catenary wires. Lighted markers provide conspicuity both day and night.

b. Medium intensity omnidirectional flashing white lighting system provides conspicuity both day and night on catenary support structures. The unique sequential/simultaneous flashing light system alerts pilots of the associated catenary wires.

c. High intensity flashing white lights are being used to identify some supporting structures of overhead transmission lines located across rivers, chasms, gorges, etc. These lights flash in a middle, top, lower light sequence at approximately 60 flashes per minute. The top light is normally installed near the top of the supporting structure, while the lower light indicates the approximate lower portion of the wire span. The lights are beamed towards the companion structure and identify the area of the wire span.

d. High intensity flashing white lights are also employed to identify tall structures, such as chimneys and towers, as obstructions to air navigation. The lights provide a 360 degree coverage about the structure at 40 flashes per minute and consist of from one to seven levels of lights depending upon the height of the structure. Where more than one level is used the vertical banks flash simultaneously.

Section 3. Airport Marking Aids and Signs

2-3-1. General

a. Airport pavement markings and signs provide information that is useful to a pilot during takeoff, landing, and taxiing.

b. Uniformity in airport markings and signs from one airport to another enhances safety and improves efficiency. Pilots are encouraged to work with the operators of the airports they use to achieve the marking and sign standards described in this section.

c. Pilots who encounter ineffective, incorrect, or confusing markings or signs on an airport should make the operator of the airport aware of the problem. These situations may also be reported under the Aviation Safety Reporting Program as described in paragraph 7-6-1, Aviation Safety Reporting Program. Pilots may also report these situations to the FAA regional airports division.

d. The markings and signs described in this section of the AIM reflect the current FAA recommended standards.

REFERENCE-*AC 150/5340-1, Standards for Airport Markings.*
AC 150/5340-18, Standards for Airport Sign Systems.

2-3-2. Airport Pavement Markings

a. General. For the purpose of this presentation the Airport Pavement Markings have been grouped into four areas:

1. Runway Markings.
2. Taxiway Markings.
3. Holding Position Markings.
4. Other Markings.

b. Marking Colors. Markings for runways are white. Markings defining the landing area on a heliport are also white except for hospital heliports which use a red "H" on a white cross. Markings for taxiways, areas not intended for use by aircraft (closed and hazardous areas), and holding positions (even if they are on a runway) are yellow.

2-3-3. Runway Markings

a. General. There are three types of markings for runways: visual, nonprecision instrument, and precision instrument. TBL 2-3-1 identifies the marking elements for each type of runway and TBL 2-3-2 identifies runway threshold markings.

b. Runway Designators. Runway numbers and letters are determined from the approach direction. The runway number is the whole number nearest one-tenth the magnetic azimuth of the centerline of the runway, measured clockwise from the magnetic north. The letters, differentiate between left (L), right (R), or center (C), parallel runways, as applicable:

1. For two parallel runways "L" "R."
2. For three parallel runways "L" "C" "R."

c. Runway Centerline Marking. The runway centerline identifies the center of the runway and provides alignment guidance during takeoff and landings.

FIG 2-3-1
Precision Instrument Runway Markings

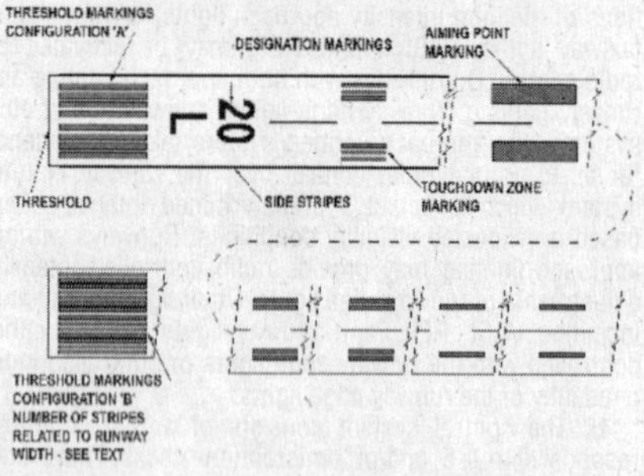

TBL 2-3-1
Runway Marking Elements

Marking Element	Visual Runway	Nonprecision Instrument Runway	Precision Instrument Runway
Designation	X	X	X
Centerline	X	X	X
Threshold	X[1]	X	X
Aiming Point	X[2]	X	X
Touchdown Zone			X
Side Stripes			X

[1] On runways used, or intended to be used, by international commercial transports.
[2] On runways 4,000 feet (1200 m) or longer used by jet aircraft.

FIG 2-3-2
Nonprecision Instrument Runway and Visual Runway Markings

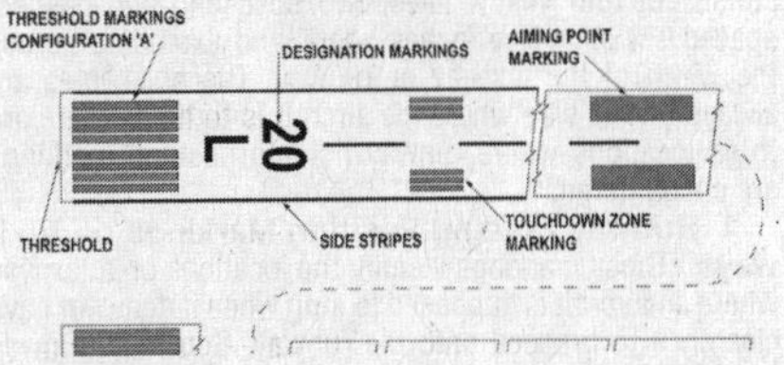

The centerline consists of a line of uniformly spaced stripes and gaps.

d. Runway Aiming Point Marking. The aiming point marking serves as a visual aiming point for a landing aircraft. These two rectangular markings consist of a broad white stripe located on each side of the runway centerline and approximately 1,000 feet from the landing threshold, as shown in FIG 2-3-1, Precision Instrument Runway Markings.

e. Runway Touchdown Zone Markers. The touchdown zone markings identify the touchdown zone for landing operations and are coded to provide distance information in 500 feet (150m) increments. These markings consist of groups of one, two, and three rectangular bars symmetrically arranged in pairs about the runway centerline, as shown in FIG 2-3-1, Precision Instrument Runway Markings. For runways having touchdown zone markings on both ends, those pairs of markings which extend to within 900 feet (270m) of the midpoint between the thresholds are eliminated.

f. Runway Side Stripe Marking. Runway side stripes delineate the edges of the runway. They provide a visual contrast between runway and the abutting terrain or shoulders. Side stripes consist of continuous white stripes located on each side of the runway as shown in FIG 2-3-4.

g. Runway Shoulder Markings. Runway shoulder stripes may be used to supplement runway side stripes to identify pavement areas contiguous to the runway sides that are not intended for use by aircraft. Runway Shoulder stripes are Yellow.
(See FIG 2-3-5.)

h. Runway Threshold Markings. Runway threshold markings come in two configurations. They either consist of eight longitudinal stripes of uniform dimensions disposed symmetrically about the runway centerline, as shown in FIG 2-3-1, or the number of stripes is related to the runway width as indicated in TBL 2-3-2. A threshold marking helps identify the beginning of the runway that is available for landing. In some instances the landing threshold may be relocated or displaced.

TBL 2-3-2
Number of Runway Threshold Stripes

Runway Width	Number of Stripes
60 feet (18 m)	4
75 feet (23 m)	6
100 feet (30 m)	8
150 feet (45 m)	12
200 feet (60 m)	16

1. Relocation of a Threshold. Sometimes construction, maintenance, or other activities require the threshold to be relocated towards the rollout end of the runway. (See FIG 2-3-3.) When a threshold is relocated, it closes not only a set portion of the approach end of a runway, but also shortens the length of the opposite direction runway. In these cases, a NOTAM should be issued by the airport operator identifying the portion of the runway that is closed, e.g., 10/28 W 900 CLSD. Because the duration of the relocation can vary from a few hours to several months, methods identifying the new threshold may vary. One common practice is to use a ten feet wide white threshold bar across the width of the runway. Although the runway lights in the area between the old threshold and new threshold will not be illuminated, the runway markings in this area may or may not be obliterated, removed, or covered.

2. Displaced Threshold. A displaced threshold is a threshold located at a point on the runway other than the designated beginning of the runway. Displacement of a threshold reduces the length of runway available for landings. The portion of runway behind a displaced threshold is available for takeoffs in either direction and landings from the opposite direction. A ten feet wide white threshold bar is located across the width of the runway at the displaced threshold. White arrows are located along the centerline in the area between the beginning of the runway and displaced threshold. White arrow heads are located across the width of the runway just prior to the threshold bar, as shown in FIG 2-3-4.

NOTE-*Airport operator. When reporting the relocation or displacement of a threshold, the airport operator should avoid language which confuses the two.*

i. Demarcation Bar. A demarcation bar delineates a runway with a displaced threshold from a blast pad, stopway or taxiway that precedes the runway. A demarcation bar is 3 feet (1m) wide and yellow, since it is not located on the runway as shown in FIG 2-3-6.

1. Chevrons. These markings are used to show pavement areas aligned with the runway that are unusable for landing, takeoff, and taxiing. Chevrons are yellow. (See FIG 2-3-7.)

j. Runway Threshold Bar. A threshold bar delineates the beginning of the runway that is available for landing when the threshold has been relocated or displaced. A threshold bar is 10 feet (3m) in width and extends across the width of the runway, as shown in FIG 2-3-4.

2-3-4. Taxiway Markings

a. General. All taxiways should have centerline markings and runway holding position markings whenever they intersect a runway. Taxiway edge markings are present whenever there is a need to separate the taxiway from a pavement that is not intended for aircraft use or to delineate the edge of the taxiway. Taxiways may also have shoulder markings and holding position markings for Instrument Landing System/Microwave Landing System (ILS/MLS) critical areas, and taxiway/taxiway intersection markings.

REFERENCE-*AIM, Holding Position Markings, Paragraph 2-3-5.*

b. Taxiway Centerline.

1. Normal Centerline. The taxiway centerline is a single continuous yellow line, 6 inches (15 cm) to 12 inches (30 cm) in width. This provides a visual cue to permit taxiing along a designated path. Ideally, the aircraft should be kept centered over this line during taxi. However, being centered on the taxiway centerline does not guarantee wingtip clearance with other aircraft or other objects.

2. Enhanced Centerline. At some airports, mostly the larger commercial service airports, an enhanced taxiway centerline will be used. The enhanced taxiway centerline marking consists of a parallel line of yellow dashes on either side of the normal taxiway centerline. The taxiway centerlines are enhanced for a maximum of 150 feet prior to a runway holding position marking. The purpose of this enhancement is to warn the pilot that he/she is approaching a runway holding position marking and should prepare to stop unless he/she has been cleared onto or across the runway by ATC. (See FIG 2-3-8.)

c. Taxiway Edge Markings. Taxiway edge markings are used to define the edge of the taxiway. They are primarily used when the taxiway edge does not correspond with the edge of the pavement. There are two types of markings depending upon whether the aircraft is supposed to cross the taxiway edge:

1. Continuous Markings. These consist of a continuous double yellow line, with each line being at least 6 inches (15 cm) in width spaced 6 inches (15 cm) apart. They are used to define the taxiway edge from the shoulder or some other abutting paved surface not intended for use by aircraft.

FIG 2-3-3
Relocation of a Threshold with Markings for Taxiway Aligned with Runway

FIG 2-3-4
Displaced Threshold Markings

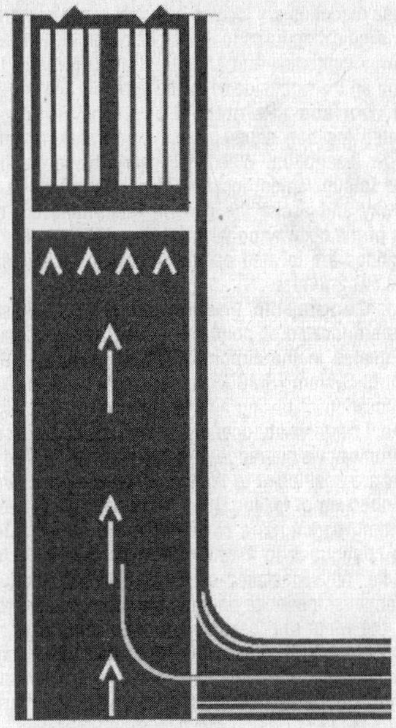

FIG 2-3-5
Runway Shoulder Markings

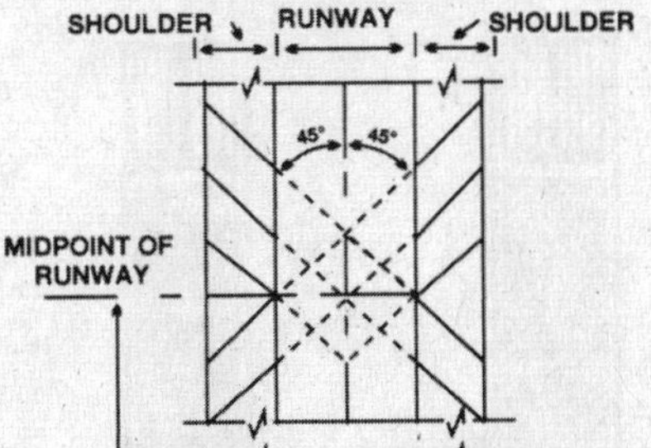

2. Dashed Markings. These markings are used when there is an operational need to define the edge of a taxiway or taxilane on a paved surface where the adjoining pavement to the taxiway edge is intended for use by aircraft, e.g., an apron. Dashed taxiway edge markings consist of a broken double yellow line, with each line being at least 6 inches (15 cm) in width, spaced 6 inches (15 cm) apart (edge to edge). These lines are 15 feet (4.5 m) in length with 25 foot (7.5 m) gaps. (See FIG 2-3-9.)

d. Taxi Shoulder Markings. Taxiways, holding bays, and aprons are sometimes provided with paved shoulders to prevent blast and water erosion. Although shoulders may have the appearance of full strength pavement they are not intended for use by aircraft, and may be unable to support an aircraft. Usually the taxiway edge marking will define this area. Where conditions exist such as islands or taxiway curves that may cause confusion as to which side of the edge stripe is for use by aircraft, taxiway shoulder markings may be used to indicate the pavement is unusable. Taxiway shoulder markings are yellow. (See FIG 2-3-10.)

e. Surface Painted Taxiway Direction Signs. Surface painted taxiway direction signs have a yellow background with a black inscription, and are provided when it is not possible to provide taxiway direction signs at intersections, or when necessary to supplement such signs. These markings are located adjacent to the centerline with signs indicating turns to the left being on the left side of the taxiway centerline and signs indicating turns to the right being on the right side of the centerline. (See FIG 2-3-11.)

f. Surface Painted Location Signs. Surface painted location signs have a black background with a yellow inscription. When necessary, these markings are used to supplement location signs located along side the taxiway and assist the pilot in confirming the designation of the taxiway on which the aircraft is located. These markings are located on the right side of the centerline. (See FIG 2-3-11.)

g. Geographic Position Markings. These markings are located at points along low visibility taxi routes designated in the airport's Surface Movement Guidance Control System (SMGCS) plan. They are used to identify the location of taxiing aircraft during low visibility operations. Low visibility operations are those that occur when the runway visible range (RVR) is below 1200 feet (360m). They are positioned to the left of the taxiway centerline in the direction of taxiing. (See FIG 2-3-12.) The geographic position marking is a circle comprised of an outer black ring contiguous to a white ring with a pink circle in the middle. When installed on asphalt or other dark-colored pavements, the white ring and the black ring are reversed, i.e., the white ring becomes the outer ring and the black ring becomes the inner ring. It is designated with either a number or a number and letter. The number corresponds to the consecutive position of the marking on the route.

FIG 2-3-6
Markings for Blast Pad or Stopway or Taxiway Preceding a Displaced Threshold

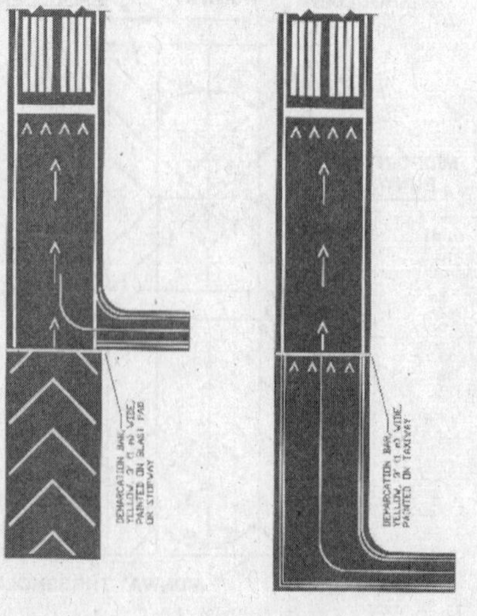

FIG 2-3-7
Markings for Blast Pads and Stopways

FIG 2-3-8
Enhanced Taxiway Centerline

FIG 2-3-9
Dashed Markings

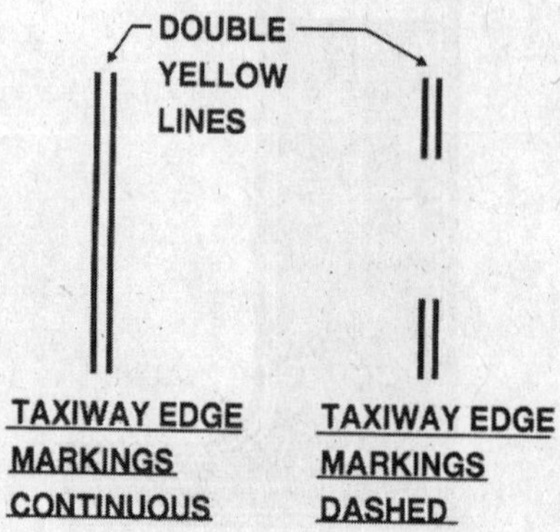

2-3-5. Holding Position Markings

a. Runway Holding Position Markings. For runways, these markings indicate where an aircraft is supposed to stop when approaching a runway. They consist of four yellow lines, two solid and two dashed, spaced six or twelve inches apart, and extending across the width of the taxiway or runway. The solid lines are always on the side where the aircraft is to hold. There are three locations where runway holding position markings are encountered.

1. Runway Holding Position Markings on Taxiways. These markings identify the locations on a taxiway where an aircraft is supposed to stop when it does not have clearance to proceed onto the runway. Generally, runway holding position markings also identify the boundary of the runway safety area for aircraft exiting the runway. The runway holding position markings are shown in FIG 2-3-13 and FIG 2-3-16. When instructed by ATC to, "Hold short of (runway "xx")," the pilot must stop so that no part of the aircraft extends beyond the runway holding position marking. When approaching the runway, a pilot should not cross the runway holding position marking

FIG 2-3-10
Taxi Shoulder Markings

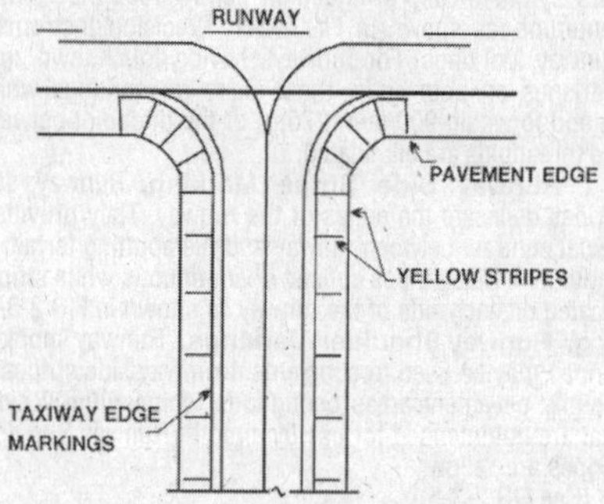

FIG 2-3-11
Surface Painted Signs

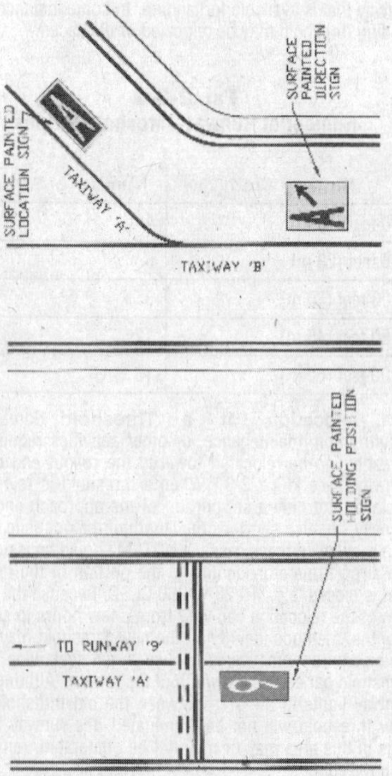

without ATC clearance at a controlled airport, or without making sure of adequate separation from other aircraft at uncontrolled airports. An aircraft exiting a runway is not clear of the runway until all parts of the aircraft have crossed the applicable holding position marking.

REFERENCE-*AIM, Exiting the Runway After Landing,. Paragraph 4-3-20.*

2. Runway Holding Position Markings on Runways. These markings are installed on runways only if the runway is normally used by air traffic control for "land, hold short" operations or taxiing operations and have operational significance only for those two types of operations. A sign with a white inscription on a red background is installed adjacent to these holding position markings. (See FIG 2-3-14.) The holding position markings are placed on runways prior to the intersection with another runway, or some designated point. Pilots receiving instructions "cleared to land, runway "xx"" from air traffic control are authorized to use the entire landing length of the runway and should disregard any holding position markings located on the runway. Pilots receiving and accepting instructions "cleared to land runway "xx," hold short of runway "yy"" from air traffic control must either exit runway "xx," or stop at the holding position prior to runway "yy."

3. Taxiways Located in Runway Approach Areas. These markings are used at some airports where it is necessary to hold an aircraft on a taxiway located in the approach or departure area of a runway so that the aircraft does not interfere with the operations on that runway. This marking is collocated with the runway approach area holding position sign. When specifically instructed by ATC "Hold short of (runway xx approach area)" the pilot should stop so no part of the aircraft extends beyond the holding position marking. (See subparagraph 2-3-8b2, Runway Approach Area Holding Position Sign, and FIG 2-3-15.)

b. Holding Position Markings for Instrument Landing System (ILS). Holding position markings for ILS/MLS critical areas consist of two yellow solid lines spaced two feet apart connected by pairs of solid lines spaced ten feet apart extending across the width of the taxiway as shown. (See FIG 2-3-16.) A sign with an inscription in white on a red background is installed adjacent to these hold position markings. When the ILS critical area is being protected, the pilot should stop so no part of the aircraft extends beyond the holding position marking. When approaching the holding position marking, a pilot should not cross the marking without ATC clearance. ILS critical area is not clear until all parts of the aircraft have crossed the applicable holding position marking.

REFERENCE-*AIM, Instrument Landing System (ILS), Paragraph 1-1-9.*

c. Holding Position Markings for Taxiway/Taxiway Intersections. Holding position markings for taxiway/taxiway intersections consist of a single dashed line extending across the width of the taxiway as shown. (See FIG 2-3-17.) They are installed on taxiways where air traffic control normally holds aircraft short of a taxiway intersection. When instructed by ATC "hold short of (taxiway)" the pilot should stop so no part of the aircraft extends beyond the holding position marking. When the marking is not present the pilot should stop the aircraft at a point which provides adequate clearance from an aircraft on the intersecting taxiway.

d. Surface Painted Holding Position Signs. Surface painted holding position signs have a red background with a white inscription and supplement the signs located at the holding position. This type of marking is normally used where the width of the holding position on the taxiway is greater than 200 feet(60m). It is located to the left side of the taxiway centerline on the holding side and prior to the holding position marking. (See FIG 2-3-11.)

2-3-6. Other Markings

a. Vehicle Roadway Markings. The vehicle roadway markings are used when necessary to define a pathway for vehicle operations on or crossing areas that are also intended for aircraft. These markings consist of a white solid line to delineate each edge of the roadway and a dashed line to separate lanes within the edges of the roadway. In lieu of the solid lines, zipper markings may be used to delineate the edges of the vehicle roadway. (See FIG 2-3-18.) Details of the zipper markings are shown in FIG 2-3-19.

FIG 2-3-12
Geographic Position Markings

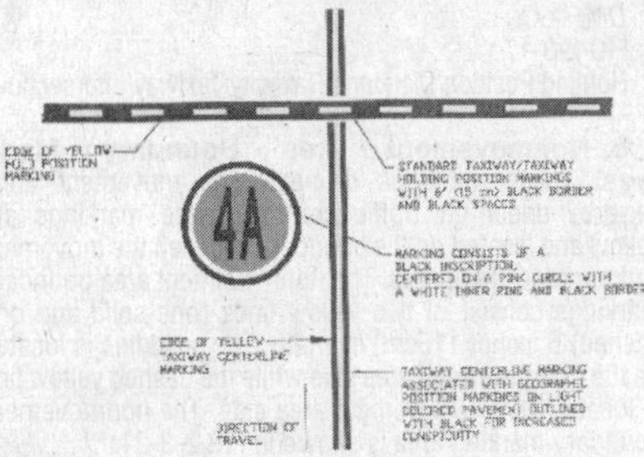

FIG 2-3-14
Runway Holding Position Markings on Runways

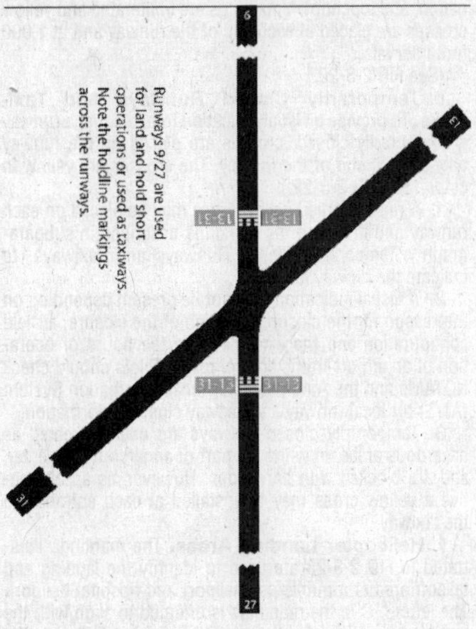

FIG 2-3-16
Holding Position Markings: ILS Critical Area

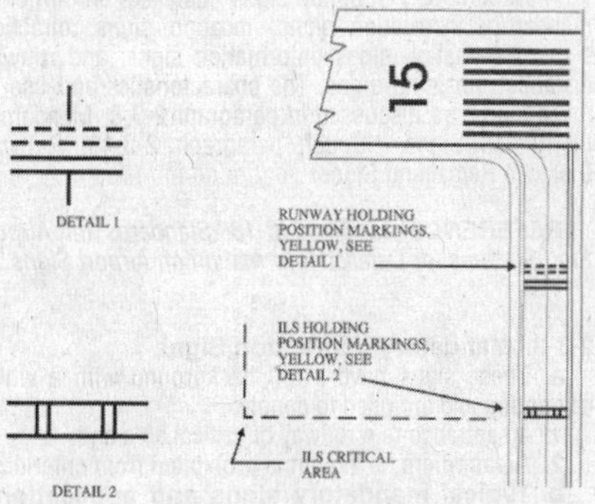

b. VOR Receiver Checkpoint Markings. The VOR receiver checkpoint marking allows the pilot to check aircraft instruments with navigational aid signals. It consists of a painted circle with an arrow in the middle; the arrow is aligned in the direction of the checkpoint azimuth. This marking, and an associated sign, is located on the airport apron or taxiway at a point selected for easy access by aircraft but where other airport traffic is not to be unduly obstructed. (SeeFIG 2-3-20.)

NOTE-*The associated sign contains the VOR station identification letter and course selected (published) for the check, the words "VOR check course," and DME data (when applicable). The color of the letters and numerals are black on a yellow background.*

FIG 2-3-13
Runway Holding Position Markings on Taxiway

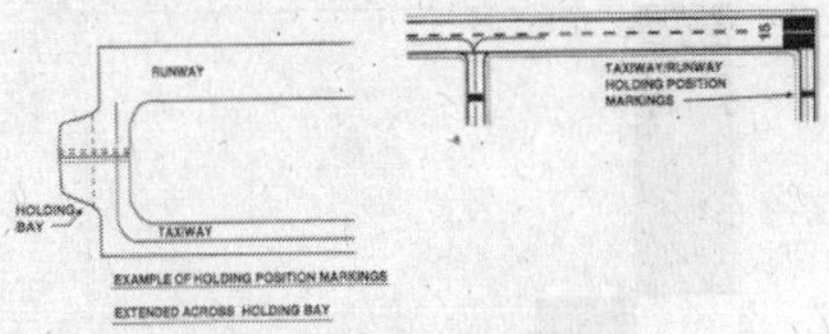

FIG 2-3-15
Taxiways Located in Runway Approach Area

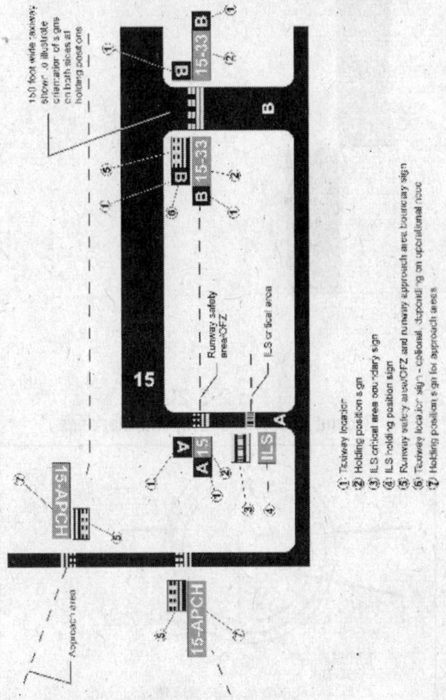

FIG 2-3-17
Holding Position Markings: Taxiway/Taxiway Intersections

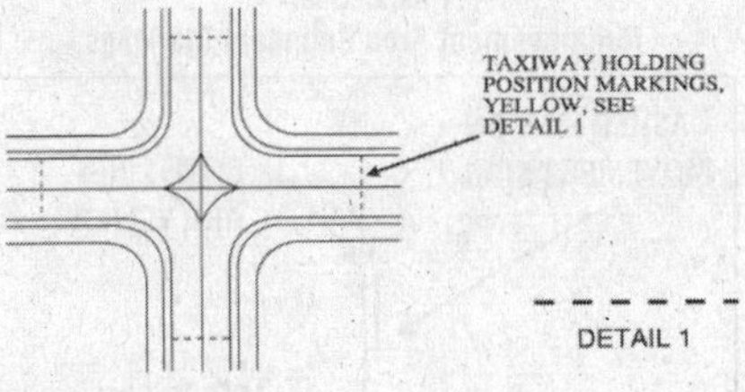

FIG 2-3-18
Vehicle Roadway Markings

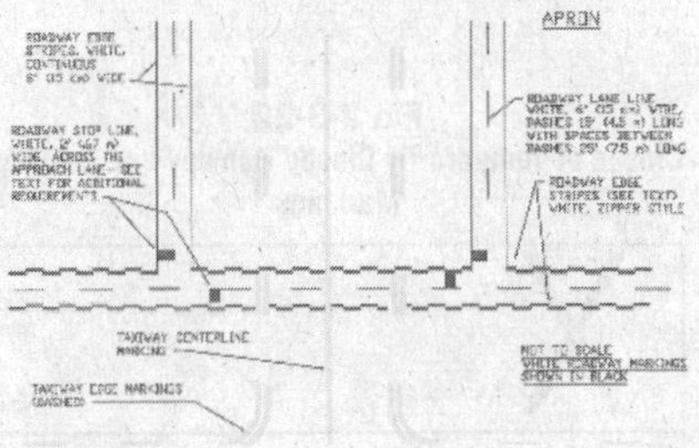

FIG 2-3-19
Roadway Edge Stripes, White, Zipper Style

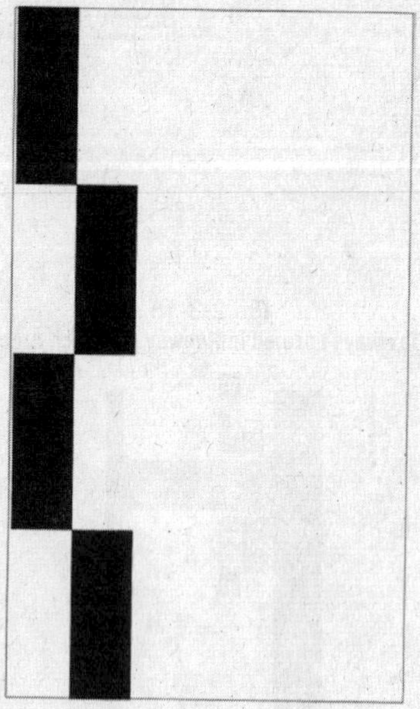

FIG 2-3-20
Ground Receiver Checkpoint Markings

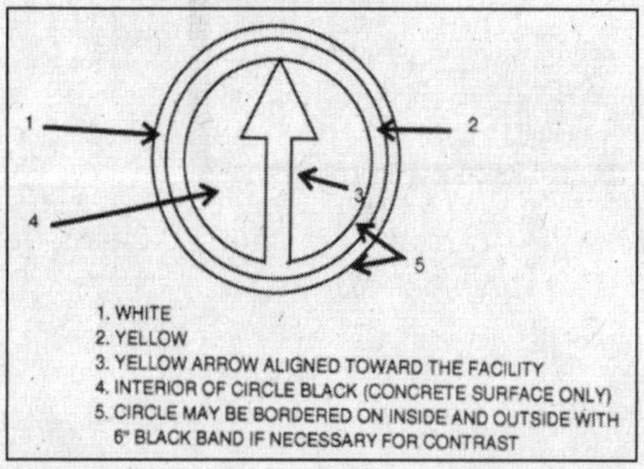

FIG 2-3-21
Nonmovement Area Boundary Markings

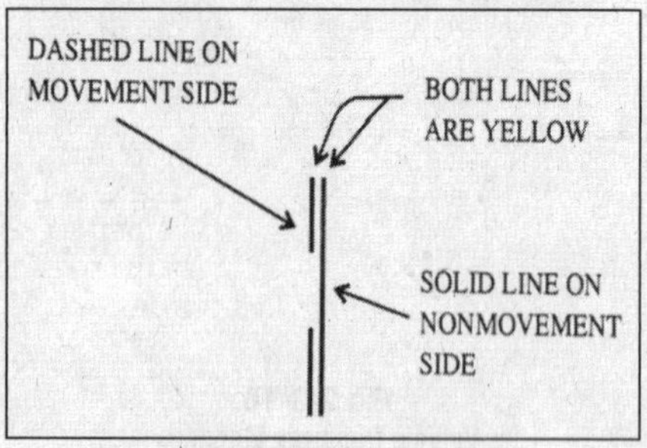

FIG 2-3-22
Closed or Temporarily Closed Runway and Taxiway Markings

FIG 2-3-23
Helicopter Landing Areas

FIG 2-3-24
Runway Holding Position Sign

FIG 2-3-25
Holding Position Sign at Beginning of Takeoff Runway

FIG 2-3-26
Holding Position Sign for a Taxiway that Intersects the Intersection of Two Runways

FIG 2-3-27
Holding Position Sign for a Runway Approach Area

EXAMPLE-
DCA 176-356
VOR check course
DME XXX
FIG 2-3-17
Holding Position Markings: Taxiway/Taxiway Intersections

c. Nonmovement Area Boundary Markings. These markings delineate the movement area, i.e., area under air traffic control. These markings are yellow and located on the boundary between the movement and nonmovement area. The nonmovement area boundary markings consist of two yellow lines (one solid and one dashed) 6 inches (15cm) in width. The solid line is located on the nonmovement area side while the dashed yellow line is located on the movement area side. The nonmovement boundary marking area is shown in FIG 2-3-21.

d. Marking and Lighting of Permanently Closed Runways and Taxiways. For runways and taxiways which are permanently closed, the lighting circuits will be disconnected. The runway threshold, runway designation, and touchdown markings are obliterated and yellow crosses are placed at each end of the runway and at 1,000 foot intervals.
(See FIG 2-3-22.)

e. Temporarily Closed Runways and Taxiways. To provide a visual indication to pilots that a runway is temporarily closed, crosses are placed on the runway only at each end of the runway. The crosses are yellow in color. (See FIG 2-3-22.)

1. A raised lighted yellow cross may be placed on each runway end in lieu of the markings described in subparagraph e,Temporarily Closed Runways and Taxiways, to indicate the runway is closed.

2. A visual indication may not be present depending on the reason for the closure, duration of the closure, airfield configuration and the existence and the hours of operation of an airport traffic control tower. Pilots should check NOTAMs and the Automated Terminal Information System (ATIS) for local runway and taxiway closure information.

3. Temporarily closed taxiways are usually treated as hazardous areas, in which no part of an aircraft may enter, and are blocked with barricades. However, as an alternative a yellow cross may be installed at each entrance to the taxiway.

f. Helicopter Landing Areas. The markings illustrated in FIG 2-3-23 are used to identify the landing and takeoff area at a public use heliport and hospital heliport. The letter "H" in the markings is oriented to align with the intended direction of approach. FIG 2-3-23 also depicts the markings for a closed airport.

2-3-7. Airport Signs

There are six types of signs installed on airfields: mandatory instruction signs, location signs, direction signs, destination signs, information signs, and runway distance remaining signs. The characteristics and use of these signs are discussed in paragraph 2-3-8, Mandatory Instruction Signs, through paragraph 2-3-13, Runway Distance Remaining Signs.

REFERENCE-*AC150/5340-18, Standards for Airport Sign Systems for Detailed Information on Airport Signs.*

2-3-8. Mandatory Instruction Signs

a. These signs have a red background with a white inscription and are used to denote:

1. An entrance to a runway or critical area and;

2. Areas where an aircraft is prohibited from entering.

b. Typical mandatory signs and applications are:

1. Runway Holding Position Sign. This sign is located at the holding position on taxiways that intersect a runway or on runways that intersect other runways. The inscription on the sign contains the designation of the intersecting runway as shown in FIG 2-3-24. The runway numbers on the sign are arranged to correspond to the respective runway threshold. For example, "15-33" indicates that the threshold for Runway 15 is to the left and the threshold for Runway 33 is to the right.

(a) On taxiways that intersect the beginning of the takeoff runway, only the designation of the takeoff runway may appear on the sign as shown in FIG 2-3-25, while all other signs will have the designation of both runway directions.

(b) If the sign is located on a taxiway that intersects the intersection of two runways, the designations for both runways will be shown on the sign along with arrows showing the approximate alignment of each runway as shown in FIG 2-3-26. In addition to showing the approximate runway alignment, the arrow indicates the direction to the threshold of the runway whose designation is immediately next to the arrow.

(c) A runway holding position sign on a taxiway will be installed adjacent to holding position markings on the taxiway pavement. On runways, holding position markings will be located only on the runway pavement adjacent to the sign, if the runway is normally used by air traffic control for "Land, Hold Short" operations or as a taxiway. The holding position markings are described in paragraph 2-3-5, Holding Position Markings.

2. Runway Approach Area Holding Position Sign. At some airports, it is necessary to hold an aircraft on a taxiway located in the approach or departure area for a runway so that the aircraft does not interfere with operations

FIG 2-3-29
SIGN PROHIBITING AIRCRAFT ENTRY INTO AN AREA

on that runway. In these situations, a sign with the designation of the approach end of the runway followed by a "dash" (-) and letters "APCH" will be located at the holding position on the taxiway. Holding position markings in accordance with paragraph 2-3-5, Holding Position Markings, will be located on the taxiway pavement. An example of this sign is shown in FIG 2-3-27. In this example, the sign may protect the approach to Runway 15 and/or the departure for Runway 33.

3. ILS Critical Area Holding Position Sign. At some airports, when the instrument landing system is being used, it is necessary to hold an aircraft on a taxiway at a location other than the holding position described in paragraph 2-3-5, Holding Position Markings. In these situations the holding position sign for these operations will have the inscription "ILS" and be located adjacent to the holding position marking on the taxiway described in paragraph 2-3-5. An example of this sign is shown in FIG 2-3-28.

4. No Entry Sign. This sign, shown in FIG 2-3-29, prohibits an aircraft from entering an area. Typically, this sign would be located on a taxiway intended to be used in only one direction or at the intersection of vehicle roadways with runways, taxiways or aprons where the roadway may be mistaken as a taxiway or other aircraft movement surface.

***NOTE-** The holding position sign provides the pilot with a visual cue as to the location of the holding position marking. The operational significance of holding position markings are described in the notes for paragraph 2-3-5, Holding Position Markings.*

2-3-9. Location Signs

a. Location signs are used to identify either a taxiway or runway on which the aircraft is located. Other location signs provide a visual cue to pilots to assist them in determining when they have exited an area. The various location signs are described below.

1. Taxiway Location Sign. This sign has a black background with a yellow inscription and yellow border as shown in FIG 2-3-30. The inscription is the designation of the taxiway on which the aircraft is located. These signs are installed along taxiways either by themselves or in conjunction with direction signs or runway holding position signs.

(See FIG 2-3-35 and FIG 2-3-31.)

FIG 2-3-30
Taxiway Location Sign

FIG 2-3-31
Taxiway Location Sign Collocated with Runway Holding Position Sign

FIG 2-3-32
Runway Location Sign

FIG 2-3-33
Runway Boundary Sign

FIG 2-3-34
ILS Critical Area Boundary Sign

2. Runway Location Sign. This sign has a black background with a yellow inscription and yellow border as shown in FIG 2-3-32. The inscription is the designation of the runway on which the aircraft is located. These signs are intended to complement the information available to pilots through their magnetic compass and typically are installed where the proximity of two or more runways to one another could cause pilots to be confused as to which runway they are on.

3. Runway Boundary Sign. This sign has a yellow background with a black inscription with a graphic depicting the pavement holding position marking as shown in FIG 2-3-33. This sign, which faces the runway and is visible to the pilot exiting the runway, is located adjacent to the holding position marking on the pavement. The sign is intended to provide pilots with another visual cue which they can use as a guide in deciding when they are "clear of the runway."

4. ILS Critical Area Boundary Sign. This sign has a yellow background with a black inscription with a graphic depicting the ILS pavement holding position marking as shown inFIG 2-3-34. This sign is located adjacent to the ILS holding position marking on the pavement and can be seen by pilots leaving the critical area. The sign is intended to provide pilots with another visual cue which they can use as a guide in deciding when they are "clear of the ILS critical area."

FIG 2-3-35
Direction Sign Array with Location Sign on Far Side of Intersection

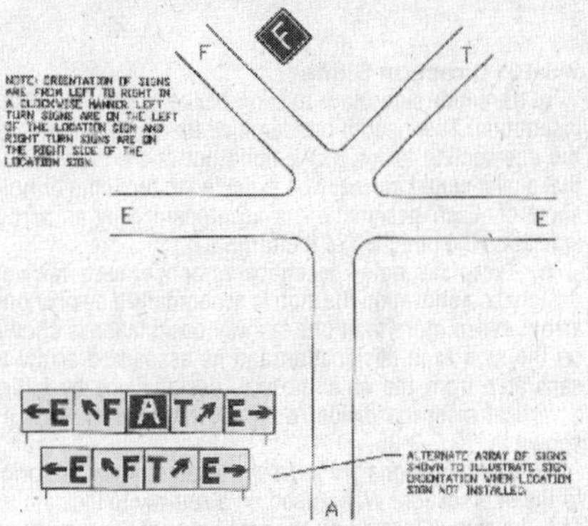

FIG 2-3-36
Direction Sign for Runway Exit

FIG 2-3-37
Direction Sign Array for Simple Intersection

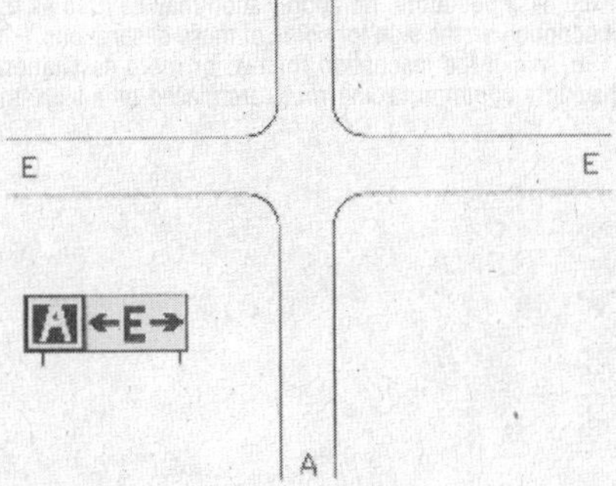

FIG 2-3-38
Destination Sign for Military Area

FIG 2-3-39
Destination Sign for Common Taxiing Route to Two Runways

2-3-10. Direction Signs

a. Direction signs have a yellow background with a black inscription. The inscription identifies the designation(s) of the intersecting taxiway(s) leading out of the intersection that a pilot would normally be expected to turn onto or hold short of. Each designation is accompanied by an arrow indicating the direction of the turn.

b. Except as noted in subparagraph e, each taxiway designation shown on the sign is accompanied by only one arrow. When more than one taxiway designation is shown on the sign each designation and its associated arrow is separated from the other taxiway designations by either a vertical message divider or a taxiway location sign as shown in FIG 2-3-35.

c. Direction signs are normally located on the left prior to the intersection. When used on a runway to indicate an exit, the sign is located on the same side of the runway as the exit. FIG 2-3-36 shows a direction sign used to indicate a runway exit.

d. The taxiway designations and their associated arrows on the sign are arranged clockwise starting from the first taxiway on the pilot's left.

(See FIG 2-3-35.)

e. If a location sign is located with the direction signs, it is placed so that the designations for all turns to the left will be to the left of the location sign; the designations for continuing straight ahead or for all turns to the right would be located to the right of the location sign. (See FIG 2-3-35.)

f. When the intersection is comprised of only one crossing taxiway, it is permissible to have two arrows associated with the crossing taxiway as shown in FIG 2-3-37. In this case, the location sign is located to the left of the direction sign.

2-3-11. Destination Signs

a. Destination signs also have a yellow background with a black inscription indicating a destination on the airport. These signs always have an arrow showing the direction of the taxiing route to that destination. FIG 2-3-38 is an example of a typical destination sign. When the arrow on the destination sign indicates a turn, the sign is located prior to the intersection.

b. Destinations commonly shown on these types of signs include runways, aprons, terminals, military areas, civil aviation areas, cargo areas, international areas, and fixed base operators. An abbreviation may be used as the inscription on the sign for some of these destinations.

c. When the inscription for two or more destinations having a common taxiing route are placed on a sign, the destinations are separated by a "dot" (D) and one arrow would be used as shown in FIG 2-3-39. When the inscription on a sign contains two or more destinations having different taxiing routes, each destination will be accompanied by an arrow and will be separated from the other destinations on the sign with a vertical black message divider as shown in FIG 2-3-40.

FIG 2-3-40
Destination Sign for Different Taxiing Routes to Two Runways

FIG 2-3-41
Runway Distance Remaining Sign Indicating 3,000 feet of Runway Remaining

2-3-12. Information Signs

Information signs have a yellow background with a black inscription. They are used to provide the pilot with information on such things as areas that cannot be seen from the control tower, applicable radio frequencies, and noise abatement procedures. The airport operator determines the need, size, and location for these signs.

2-3-13. Runway Distance Remaining Signs

Runway distance remaining signs have a black background with a white numeral inscription and may be installed along one or both side(s) of the runway. The number on the signs indicates the distance (in thousands of feet) of landing runway remaining. The last sign, i.e., the sign with the numeral "1," will be located at least 950 feet from the runway end. FIG 2-3-41shows an example of a runway distance remaining sign.

2-3-14. Aircraft Arresting Systems

a. Certain airports are equipped with a means of rapidly stopping military aircraft on a runway. This equipment, normally referred to as EMERGENCY ARRESTING GEAR, generally consists of pendant cables supported over the runway surface by rubber "donuts." Although most devices are located in the overrun areas, a few of these arresting systems have cables stretched over the operational areas near the ends of a runway.

FIG 2-3-42
Engineered Materials Arresting System (EMAS)

b. Arresting cables which cross over a runway require special markings on the runway to identify the cable location. These markings consist of 10 feet diameter solid circles painted "identification yellow," 30 feet on center, perpendicular to the runway centerline across the entire runway width. Additional details are contained in AC 150/5220-9, Aircraft Arresting Systems for Joint Civil/Military Airports.

NOTE-Aircraft operations on the runway are not restricted by the installation of aircraft arresting devices.

c. Engineered materials arresting systems (EMAS). EMAS, which are constructed of high energy-absorbing materials of selected strength, are located in the safety area beyond the end of the runway. They are designed to crush under the weight of commercial aircraft and they exert deceleration forces on the landing gear. These systems do not affect the normal landing and takeoff of airplanes. More information concerning EMAS is in FAA Advisory Circular AC 150/5220-22, Engineered Materials Arresting Systems (EMAS) for Aircraft Overruns.

NOTE-EMAS may be located as close as 35 feet beyond the end of the runway. Aircraft should never taxi or drive across the runway.

2-3-15. Security Identifications Display Area (Airport Ramp Area)

a. Security Identification Display Areas (SIDA) are limited access areas that require a badge issued in accordance with procedures in CFR 49 Part 1542. Movement through or into these areas is prohibited without proper identification being displayed. If you are unsure of the location of a SIDA, contact the airport authority for additional information. Airports that have a SIDA must have the following information available:

1. A description and map detailing boundaries and pertinent features;

2. Measures used to perform the access control functions required under CFR 49 Part 1542.201(b)(1);

3. Procedures to control movement within the secured area, including identification media required under CFR 49 Part 1542.201(b)(3); and

4. A description of the notification signs required under CFR 49 Part 1542.201(b)(6).

b. Pilots or passengers without proper identification that are observed entering a SIDA (ramp area) may be reported to TSA or airport security. Pilots are advised to brief passengers accordingly.

Chapter 3.
AIRSPACE
Section 1. General

3-1-1. General

a. There are two categories of airspace or airspace areas:

1. Regulatory (Class A, B, C, D and E airspace areas, restricted and prohibited areas); and

2. Nonregulatory (military operations areas (MOAs), warning areas, alert areas, and controlled firing areas).

***NOTE-**Additional information on special use airspace (prohibited areas, restricted areas, warning areas, MOAs, alert areas and controlled firing areas) may be found inChapter 3, Airspace, Section 4, Special Use Airspace, paragraphs 3-4-1 through 3-4-7.*

b. Within these two categories, there are four types:

1. Controlled,

2. Uncontrolled,

3. Special use, and

4. Other airspace.

c. The categories and types of airspace are dictated by:

1. The complexity or density of aircraft movements,

2. The nature of the operations conducted within the airspace,

3. The level of safety required, and

4. The national and public interest.

d. It is important that pilots be familiar with the operational requirements for each of the various types or classes of airspace. Subsequent sections will cover each class in sufficient detail to facilitate understanding.

3-1-2. General Dimensions of Airspace Segments

Refer to Code of Federal Regulations (CFRs) for specific dimensions, exceptions, geographical areas covered, exclusions, specific transponder or equipment requirements, and flight operations.

3-1-3. Hierarchy of Overlapping Airspace Designations

a. When overlapping airspace designations apply to the same airspace, the operating rules associated with the more restrictive airspace designation apply.

b. For the purpose of clarification:

1. Class A airspace is more restrictive than Class B, Class C, Class D, Class E, or Class G airspace;

2. Class B airspace is more restrictive than Class C, Class D, Class E, or Class G airspace;

3. Class C airspace is more restrictive than Class D, Class E, or Class G airspace;

4. Class D airspace is more restrictive than Class E or Class G airspace; and

5. Class E is more restrictive than Class G airspace.

3-1-4. Basic VFR Weather Minimums

a. No person may operate an aircraft under basic VFR when the flight visibility is less, or at a distance from clouds that is less, than that prescribed for the corresponding altitude and class of airspace.

(See *TBL 3-1-1.*)

***NOTE-**Student pilots must comply with 14 CFR Section 61.89(a) (6) and (7).*

b. Except as provided in 14 CFR Section 91.157, Special VFR Weather Minimums, no person may operate an aircraft beneath the ceiling under VFR within the lateral boundaries of controlled airspace designated to the surface for an airport when the ceiling is less than 1,000 feet. (See 14 CFR Section 91.155(c).)

3-1-5. VFR Cruising Altitudes and Flight Levels

(See *TBL 3-1-2.*)

TBL 3-1-1
Basic VFR Weather Minimums

Airspace	Flight Visibility	Distance from Clouds
Class A	Not Applicable	Not Applicable
Class B	3 statute miles	Clear of Clouds
Class C	3 statute miles	500 feet below 1,000 feet above 2,000 feet horizontal
Class D	3 statute miles	500 feet below 1,000 feet above 2,000 feet horizontal
Class E		
Less than 10,000 feet MSL	3 statute miles	500 feet below 1,000 feet above 2,000 feet horizontal
At or above 10,000 feet MSL	5 statute miles	1,000 feet below 1,000 feet above 1 statute mile horizontal
Class G 1,200 feet or less above the surface (regardless of MSL altitude).		
Day, except as provided in section 91.155(b)	1 statute mile	Clear of clouds
Night, except as provided in section 91.155(b)	3 statute miles	500 feet below 1,000 feet above 2,000 feet horizontal
More than 1,200 feet above the surface but less than 10,000 feet MSL.		
Day	1 statute mile	500 feet below 1,000 feet above 2,000 feet horizontal
Night	3 statute miles	500 feet below 1,000 feet above 2,000 feet horizontal
More than 1,200 feet above the surface and at or above 10,000 feet MSL.	5 statute miles	1,000 feet below 1,000 feet above 1 statute mile horizontal

TBL 3-1-2
VFR Cruising Altitudes and Flight Levels

If your magnetic course (ground track) is:	And you are more than 3,000 feet above the surface but below 18,000 feet MSL, fly:	And you are above 18,000 feet MSL to FL 290, fly:
0° to 179°	Odd thousands MSL, plus 500 feet (3,500; 5,500; 7,500, etc.)	Odd Flight Levels plus 500 feet (FL 195; FL 215; FL 235, etc.)
180° to 359°	Even thousands MSL, plus 500 feet (4,500; 6,500; 8,500, etc.)	Even Flight Levels plus 500 feet (FL 185; FL 205; FL 225, etc.)

Section 2. Controlled Airspace

3-2-1. General

a. Controlled Airspace. A generic term that covers the different classification of airspace (Class A, Class B, Class C, Class D, and Class E airspace) and defined dimensions within which air traffic control service is provided to IFR flights and to VFR flights in accordance with the airspace classification. (See *FIG 3-2-1.*)

FIG 3-2-1
Airspace Classes

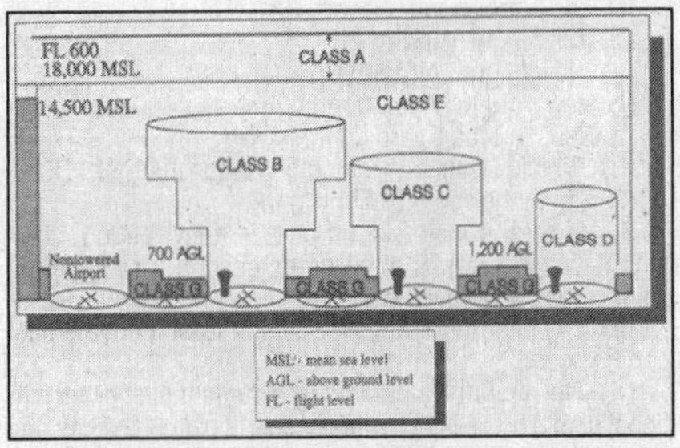

b. IFR Requirements. IFR operations in any class of controlled airspace requires that a pilot must file an IFR flight plan and receive an appropriate ATC clearance.

c. IFR Separation. Standard IFR separation is provided to all aircraft operating under IFR in controlled airspace.

d. VFR Requirements. It is the responsibility of the pilot to ensure that ATC clearance or radio communication requirements are met prior to entry into Class B, Class C, or Class D airspace. The pilot retains this responsibility when receiving ATC radar advisories. (See 14 CFR Part 91.)

e. Traffic Advisories. Traffic advisories will be provided to all aircraft as the controller's work situation permits.

f. Safety Alerts. Safety Alerts are mandatory services and are provided to ALL aircraft. There are two types of Safety Alerts:

1. Terrain/Obstruction Alert. A Terrain/Obstruction Alert is issued when, in the controller's judgment, an aircraft's altitude places it in unsafe proximity to terrain and/or obstructions; and

2. Aircraft Conflict/Mode C Intruder Alert. An Aircraft Conflict/Mode C Intruder Alert is issued if the controller observes another aircraft which places it in an unsafe proximity. When feasible, the controller will offer the pilot an alternative course of action.

g. Ultralight Vehicles. No person may operate an ultralight vehicle within Class A, Class B, Class C, or Class D airspace or within the lateral boundaries of the surface area of Class E airspace designated for an airport unless that person has prior authorization from the ATC facility having jurisdiction over that airspace. (See 14 CFR Part 103.)

h. Unmanned Free Balloons. Unless otherwise authorized by ATC, no person may operate an unmanned free balloon below 2,000 feet above the surface within the lateral boundaries of Class B, Class C, Class D, or Class E airspace designated for an airport. (See 14 CFR Part 101.)

i. Parachute Jumps. No person may make a parachute jump, and no pilot-in-command may allow a parachute jump to be made from that aircraft, in or into Class A, Class B, Class C, or Class D airspace without, or in violation of, the terms of an ATC authorization issued by the ATC facility having jurisdiction over the airspace. (See 14 CFR Part 105.)

3-2-2. Class A Airspace

a. Definition. Generally, that airspace from 18,000 feet MSL up to and including FL 600, including the airspace overlying the waters within 12 nautical miles off the coast of the 48 contiguous States and Alaska; and designated international airspace beyond 12 nautical miles off the coast of the 48 contiguous States and Alaska within areas of domestic radio navigational signal or ATC radar coverage, and within which domestic procedures are applied.

b. Operating Rules and Pilot/Equipment Requirements. Unless otherwise authorized, all persons must operate their aircraft under IFR. (See 14 CFR Section 71.33 and 14 CFR Section 91.167 through 14 CFR Section 91.193.)

c. Charts. Class A airspace is not specifically charted.

3-2-3. Class B Airspace

a. Definition. Generally, that airspace from the surface to 10,000 feet MSL surrounding the nation's busiest airports in terms of IFR operations or passenger enplanements. The configuration of each Class B airspace area is individually tailored and consists of a surface area and two or more layers (some Class B airspace areas resemble upside-down wedding cakes), and is designed to contain all published instrument procedures once an aircraft enters the airspace. An ATC clearance is required for all aircraft to operate in the area, and all aircraft that are so cleared receive separation services within the airspace. The cloud clearance requirement for VFR operations is "clear of clouds."

b. Operating Rules and Pilot/Equipment Requirements for VFR Operations. Regardless of weather conditions, an ATC clearance is required prior to operating within Class B airspace. Pilots should not request a clearance to operate within Class B airspace unless the requirements of 14 CFR Section 91.215 and 14 CFR Section 91.131 are met. Included among these requirements are:

1. Unless otherwise authorized by ATC, aircraft must be equipped with an operable two-way radio capable of communicating with ATC on appropriate frequencies for that Class B airspace.

2. No person may take off or land a civil aircraft at the following primary airports within Class B airspace unless the pilot-in-command holds at least a private pilot certificate:

(a) Andrews Air Force Base, MD
(b) Atlanta Hartsfield Airport, GA
(c) Boston Logan Airport, MA
(d) Chicago O'Hare Intl. Airport, IL
(e) Dallas/Fort Worth Intl. Airport, TX
(f) Los Angeles Intl. Airport, CA
(g) Miami Intl. Airport, FL
(h) Newark Intl. Airport, NJ
(i) New York Kennedy Airport, NY
(j) New York La Guardia Airport, NY
(k) Ronald Reagan Washington National Airport, DC
(l) San Francisco Intl. Airport, CA

3. No person may take off or land a civil aircraft at an airport within Class B airspace or operate a civil aircraft within Class B airspace unless:

(a) The pilot-in-command holds at least a private pilot certificate; or

(b) The aircraft is operated by a student pilot or recreational pilot who seeks private pilot certification and has met the requirements of 14 CFR Section 61.95.

4. Unless otherwise authorized by ATC, each person operating a large turbine engine-powered airplane to or from a primary airport must operate at or above the designated floors while within the lateral limits of Class B airspace.

5. Unless otherwise authorized by ATC, each aircraft must be equipped as follows:

(a) For IFR operations, an operable VOR or TACAN receiver; and

(b) For all operations, a two-way radio capable of communications with ATC on appropriate frequencies for that area; and

(c) Unless otherwise authorized by ATC, an operable radar beacon transponder with automatic altitude reporting equipment.

NOTE-*ATC may, upon notification, immediately authorize a deviation from the altitude reporting equipment requirement; however, a request for a deviation from the 4096 transponder equipment requirement must be submitted to the controlling ATC facility at least one hour before the proposed operation.*

REFERENCE-*AIM, Transponder Operation, Paragraph 4-1-20.*

6. Mode C Veil. The airspace within 30 nautical miles of an airport listed in Appendix D, Section 1 of 14 CFR Part 91 (generally primary airports within Class B airspace areas), from the surface upward to 10,000 feet MSL. Unless otherwise authorized by ATC, aircraft operating within this airspace must be equipped with automatic pressure altitude reporting equipment having Mode C capability.

However, an aircraft that was not originally certificated with an engine-driven electrical system or which has not subsequently been certified with a system installed may conduct operations within a Mode C veil provided the aircraft remains outside Class A, B or C airspace; and below the altitude of the ceiling of a Class B or Class C airspace area designated for an airport or 10,000 feet MSL, whichever is lower.

c. Charts. Class B airspace is charted on Sectional Charts, IFR En Route Low Altitude, and Terminal Area Charts.

d. Flight Procedures.

1. Flights. Aircraft within Class B airspace are required to operate in accordance with current IFR procedures. A clearance for a visual approach to a primary airport is not authorization for turbine- powered airplanes to operate below the designated floors of the Class B airspace.

2. VFR Flights.

(a) Arriving aircraft must obtain an ATC clearance prior to entering Class B airspace and must contact ATC on the appropriate frequency, and in relation to geographical fixes shown on local charts. Although a pilot may be operating beneath the floor of the Class B airspace on initial contact, communications with ATC should be established in relation to the points indicated for spacing and sequencing purposes.

(b) Departing aircraft require a clearance to depart Class B airspace and should advise the clearance delivery position of their intended altitude and route of flight. ATC will normally advise VFR aircraft when leaving the geographical limits of the Class B airspace. Radar service is not automatically terminated with this advisory unless specifically stated by the controller.

(c) Aircraft not landing or departing the primary airport may obtain an ATC clearance to transit the Class B airspace when traffic conditions permit and provided the requirements of 14 CFR Section 91.131 are met. Such VFR aircraft are encouraged, to the extent possible, to operate at altitudes above or below the Class B airspace or transit through established VFR corridors. Pilots operating in VFR corridors are urged to use frequency 122.750 MHz for the exchange of aircraft position information.

e. ATC Clearances and Separation. An ATC clearance is required to enter and operate within Class B airspace. VFR pilots are provided sequencing and separation from other aircraft while operating within Class B airspace.

REFERENCE-*AIM, Terminal Radar Services for VFR Aircraft, Paragraph 4-1-18.*

NOTE-**1.** *Separation and sequencing of VFR aircraft will be suspended in the event of a radar outage as this service is dependent on radar. The pilot will be advised that the service is not available and issued wind, runway information and the time or place to contact the tower.*

2. *Separation of VFR aircraft will be suspended during CENRAP operations. Traffic advisories and sequencing to the primary airport will be provided on a workload permitting basis. The pilot will be advised when center radar presentation (CENRAP) is in use.*

1. VFR aircraft are separated from all VFR/IFR aircraft which weigh 19,000 pounds or less by a minimum of:

(a) Target resolution, or
(b) 500 feet vertical separation, or
(c) Visual separation.

2. VFR aircraft are separated from all VFR/IFR aircraft which weigh more than 19,000 and turbojets by no less than:

(a) 1 1/2 miles lateral separation, or
(b) 500 feet vertical separation, or
(c) Visual separation.

3. This program is not to be interpreted as relieving pilots of their responsibilities to see and avoid other traffic operating in basic VFR weather conditions, to adjust their operations and flight path as necessary to preclude serious wake encounters, to maintain appropriate terrain and obstruction clearance or to remain in weather conditions equal to or better than the minimums required by 14 CFR Section 91.155. Approach control should be advised and a revised clearance or instruction obtained when compliance with an assigned route, heading and/or altitude is likely to compromise pilot responsibility with respect to terrain and obstruction clearance, vortex exposure, and weather minimums.

4. ATC may assign altitudes to VFR aircraft that do not conform to 14 CFR Section 91.159. **"RESUME APPROPRIATE VFR ALTITUDES"** will be broadcast when the altitude assignment is no longer needed for separation or when leaving Class B airspace. Pilots must return to an altitude that conforms to 14 CFR Section 91.159.

f. Proximity operations. VFR aircraft operating in proximity to Class B airspace are cautioned against operating too closely to the boundaries, especially where the floor of the Class B airspace is 3,000 feet or less above the surface or where VFR cruise altitudes are at or near the floor of higher levels. Observance of this precaution will reduce the potential for encountering an aircraft operating at the altitudes of Class B floors. Additionally, VFR aircraft are encouraged to utilize the VFR Planning Chart as a tool for planning flight in proximity to Class B airspace. Charted VFR Flyway Planning Charts are published on the back of the existing VFR Terminal Area Charts.

3-2-4. Class C Airspace

a. Definition. Generally, that airspace from the surface to 4,000 feet above the airport elevation (charted in MSL) surrounding those airports that have an operational control tower, are serviced by a radar approach control, and that have a certain number of IFR operations or passenger enplanements. Although the configuration of each Class C airspace area is individually tailored, the airspace usually consists of a 5 NM radius core surface area that extends from the surface up to 4,000 feet above the airport elevation, and a 10 NM radius shelf area that extends no lower than 1,200 feet up to 4,000 feet above the airport elevation.

b. Charts. Class C airspace is charted on Sectional Charts, IFR En Route Low Altitude, and Terminal Area Charts where appropriate.

c. Operating Rules and Pilot/Equipment Requirements:

1. Pilot Certification. No specific certification required.

2. Equipment.

(a) Two-way radio; and

(b) Unless otherwise authorized by ATC, an operable radar beacon transponder with automatic altitude reporting equipment.

NOTE-*See paragraph 4-1-20, Transponder Operation, subparagraph f2(c) for Mode C transponder requirements for operating above Class C airspace.*

3. Arrival or Through Flight Entry Requirements. Two-way radio communication must be established with the ATC facility providing ATC services prior to entry and thereafter maintain those communications while in Class C airspace. Pilots of arriving aircraft should contact the Class C airspace ATC facility on the publicized frequency and give their position, altitude, radar beacon code, destination, and request Class C service. Radio contact should be initiated far enough from the Class C airspace boundary to preclude entering Class C airspace before two-way radio communications are established.

NOTE-**1.** *If the controller responds to a radio call with, "(aircraft callsign) standby," radio communications have been established and the pilot can enter the Class C airspace.*

2. *If workload or traffic conditions prevent immediate provision of Class C services, the controller will inform the pilot to remain outside the Class C airspace until conditions permit the services to be provided.*

3. *It is important to understand that if the controller responds to the initial radio call without using the aircraft identification, radio communications have not been established and the pilot may not enter the Class C airspace.*

4. *Though not requiring regulatory action, Class C airspace areas have a procedural Outer Area. Normally this area is 20 NM from the primary Class C airspace airport. Its vertical limit extends from the lower limits of radio/radar coverage up to the ceiling of the approach control's delegated airspace, excluding the Class C airspace itself, and other airspace as appropriate. (This outer area is not charted.)*

5. *Pilots approaching an airport with Class C service should be aware that if they descend below the base altitude of the 5 to 10 mile shelf during an instrument or visual approach, they may encounter nontransponder, VFR aircraft.*

EXAMPLE-**1.** *[Aircraft callsign] "remain outside the Class Charlie airspace and standby."*

2. *"Aircraft calling Dulles approach control, standby."*

4. Departures from:

(a) A primary or satellite airport with an operating control tower. Two-way radio communications must be established and maintained with the control tower, and thereafter as instructed by ATC while operating in Class C airspace.

(b) A satellite airport without an operating control tower. Two-way radio communications must be established as soon as practicable after departing with the ATC facility having jurisdiction over the Class C airspace.

5. Aircraft Speed. Unless otherwise authorized or required by ATC, no person may operate an aircraft at or below 2,500 feet above the surface within 4 nautical miles of the primary airport of a Class C airspace area at an indicated airspeed of more than 200 knots (230 mph).

d. Air Traffic Services. When two-way radio communications and radar contact are established, all participating VFR aircraft are:

1. Sequenced to the primary airport.

2. Provided Class C services within the Class C airspace and the outer area.

3. Provided basic radar services beyond the outer area on a workload permitting basis. This can be terminated by the controller if workload dictates.

e. Aircraft Separation. Separation is provided within the Class C airspace and the outer area after two-way radio communications and radar contact are established. VFR aircraft are separated from IFR aircraft within the Class C airspace by any of the following:

1. Visual separation.

2. 500 feet vertical; except when operating beneath a heavy jet.

3. Target resolution.

***NOTE*-1.** *Separation and sequencing of VFR aircraft will be suspended in the event of a radar outage as this service is dependent on radar. The pilot will be advised that the service is not available and issued wind, runway information and the time or place to contact the tower.*

2. *Separation of VFR aircraft will be suspended during CENRAP operations. Traffic advisories and sequencing to the primary airport will be provided on a workload permitting basis. The pilot will be advised when CENRAP is in use.*

3. *Pilot participation is voluntary within the outer area and can be discontinued, within the outer area, at the pilot's request. Class C services will be provided in the outer area unless the pilot requests termination of the service.*

4. *Some facilities provide Class C services only during published hours. At other times, terminal IFR radar service will be provided. It is important to note that the communications and transponder requirements are dependent of the class of airspace established outside of the published hours.*

f. Secondary Airports

1. In some locations Class C airspace may overlie the Class D surface area of a secondary airport. In order to allow that control tower to provide service to aircraft, portions of the overlapping Class C airspace may be procedurally excluded when the secondary airport tower is in operation. Aircraft operating in these procedurally excluded areas will only be provided airport traffic control services when in communication with the secondary airport tower.

2. Aircraft proceeding inbound to a satellite airport will be terminated at a sufficient distance to allow time to change to the appropriate tower or advisory frequency. Class C services to these aircraft will be discontinued when the aircraft is instructed to contact the tower or change to advisory frequency.

3. Aircraft departing secondary controlled airports will not receive Class C services until they have been radar identified and two-way communications have been established with the Class C airspace facility.

4. This program is not to be interpreted as relieving pilots of their responsibilities to see and avoid other traffic operating in basic VFR weather conditions, to adjust their operations and flight path as necessary to preclude serious wake encounters, to maintain appropriate terrain and obstruction clearance or to remain in weather conditions equal to or better than the minimums required by 14 CFR Section 91.155. Approach control should be advised and a revised clearance or instruction obtained when compliance with an assigned route, heading and/or altitude is likely to compromise pilot responsibility with respect to terrain and obstruction clearance, vortex exposure, and weather minimums.

g. Class C Airspace Areas by State

These states currently have designated Class C airspace areas that are depicted on sectional charts. Pilots should consult current sectional charts and NOTAMs for the latest information on services available. Pilots should be aware that some Class C airspace underlies or is adjacent to Class B airspace.

(See *TBL 3-2-1*.)

TBL 3-2-1
Class C Airspace Areas by State

State/City	Airport
ALABAMA	
Birmingham	Birmingham-Shuttlesworth International
Huntsville	International-Carl T Jones
Mobile	Fld Regional
ALASKA	
Anchorage	Ted Stevens International
ARIZONA	
Davis-Monthan Tucson	AFB International
ARKANSAS	
Fayetteville (Springdale)	Northwest Arkansas Regional
Little Rock	Adams Field
CALIFORNIA	
Beale	AFB
Burbank	Bob Hope
Fresno	Yosemite International
Monterey	Peninsula
Oakland	Metropolitan Oakland International
Ontario	International
Riverside	March AFB
Sacramento	International
San Jose	Norman Y. Mineta International
Santa Ana	John Wayne/Orange County
Santa Barbara	Municipal
COLORADO	
Colorado Springs	Municipal
CONNECTICUT	
Windsor Locks	Bradley International
FLORIDA	
Daytona Beach	International
Fort Lauderdale	Hollywood International
Fort Myers	SW Florida Regional
Jacksonville	International
Orlando	Sanford International
Palm Beach	International
Pensacola	NAS
Pensacola	Regional
Sarasota	Bradenton International
Tallahassee	Regional
Whiting	NAS
GEORGIA	
Columbus Savannah	Metropolitan Hilton Head International
HAWAII	
Kahului	Kahului
IDAHO	
Boise	Air Terminal
ILLINOIS	
Champaign	Urbana U of Illinois-Willard
Chicago	Midway International
Moline	Quad City International
Peoria	Greater Peoria Regional
Springfield	Abraham Lincoln Capital
INDIANA	
Evansville	Regional
Fort Wayne	International
Indianapolis	International
South Bend	Regional
IOWA	
Cedar Rapids	The Eastern Iowa
Des Moines	International
KANSAS	
Wichita	Mid-Continent
KENTUCKY	
Lexington	Blue Grass
Louisville	International-Standiford Field
LOUISIANA	
Baton Rouge	Metropolitan, Ryan Field
Lafayette	Regional
Shreveport	Barksdale AFB
Shreveport	Regional
MAINE	
Bangor	International
Portland	International Jetport
MICHIGAN	
Flint	Bishop International
Grand Rapids	Gerald R. Ford International
Lansing	Capital City
MISSISSIPPI	
Columbus	AFB
Jackson	Jackson-Evers International
MISSOURI	
Springfield	Springfield-Branson National
MONTANA	
Billings	Logan International
NEBRASKA	
Lincoln	Lincoln
Omaha	Eppley Airfield
Offutt	AFB
NEVADA	
Reno	Reno/Tahoe International
NEW HAMPSHIRE	
Manchester	Manchester
NEW JERSEY	
Atlantic City	International
NEW MEXICO	
Albuquerque	International Sunport
NEW YORK	
Albany	International
Buffalo	Niagara International
Islip	Long Island MacArthur
Rochester	Greater Rochester International

3-2-5. Class D Airspace

a. Definition. Generally, that airspace from the surface to 2,500 feet above the airport elevation (charted in MSL) surrounding those airports that have an operational

State/City	Airport
Syracuse	Hancock International
NORTH CAROLINA	
Asheville	Regional
Fayetteville	Regional/Grannis Field
Greensboro	Piedmont Triad International
Pope	AFB
Raleigh	Raleigh-Durham International
OHIO	
Akron	Akron-Canton Regional
Columbus	Port Columbus International
Dayton	James M. Cox International
Toledo	Express
OKLAHOMA	
Oklahoma City	Will Rogers World
Tinker	AFB
Tulsa	International
OREGON	
Portland	International
PENNSYLVANIA	
Allentown	Lehigh Valley International
PUERTO RICO	
San Juan	Luis Munoz Marin International
RHODE ISLAND	
Providence	Theodore Francis Green State
SOUTH CAROLINA	
Charleston	AFB/International
Columbia	Metropolitan
Greer	Greenville-Spartanburg International
Myrtle Beach	Myrtle Beach International
Shaw	AFB
TENNESSEE	
Chattanooga	Lovell Field
Knoxville	McGhee Tyson
Nashville	International
TEXAS	
Abilene	Regional
Amarillo	Rick Husband International
Austin	Austin-Bergstrom International
Corpus Christi	International
Dyess	AFB
El Paso	International
Harlingen	Valley International
Laughlin	AFB
Lubbock	Preston Smith International
Midland	International
San Antonio	International
VERMONT	
Burlington	International
VIRGIN ISLANDS	
St. Thomas	Charlotte Amalie Cyril E. King

State/City	Airport
VIRGINIA	
Richmond	International
Norfolk	International
Roanoke	Regional/Woodrum Field
WASHINGTON	
Point Roberts	Vancouver International
Spokane	Fairchild AFB
Spokane	International
Whidbey Island	NAS, Ault Field
WEST VIRGINIA	
Charleston	Yeager
WISCONSIN	
Green Bay	Austin Straubel International
Madison	Dane County Regional-Traux Field
Milwaukee	General Mitchell International

control tower. The configuration of each Class D airspace area is individually tailored and when instrument procedures are published, the airspace will normally be designed to contain the procedures.

b. Operating Rules and Pilot/Equipment Requirements:

1. Pilot Certification. No specific certification required.

2. Equipment. Unless otherwise authorized by ATC, an operable two-way radio is required.

3. Arrival or Through Flight Entry Requirements. Two-way radio communication must be established with the ATC facility providing ATC services prior to entry and thereafter maintain those communications while in the Class D airspace. Pilots of arriving aircraft should contact the control tower on the publicized frequency and give their position, altitude, destination, and any request(s). Radio contact should be initiated far enough from the Class D airspace boundary to preclude entering the Class D airspace before two-way radio communications are established.

NOTE-1. *If the controller responds to a radio call with, "[aircraft callsign] standby," radio communications have been established and the pilot can enter the Class D airspace.*

2. If workload or traffic conditions prevent immediate entry into Class D airspace, the controller will inform the pilot to remain outside the Class D airspace until conditions permit entry.

EXAMPLE-1. *"[Aircraft callsign] remain outside the Class Delta airspace and standby."*

It is important to understand that if the controller responds to the initial radio call without using the aircraft callsign, radio communications have not been established and the pilot may not enter the Class D airspace.

2. "Aircraft calling Manassas tower standby."

At those airports where the control tower does not operate 24 hours a day, the operating hours of the tower will be listed on the appropriate charts and in the A/FD. During the hours the tower is not in operation, the Class E surface area rules or a combination of Class E rules to 700 feet above ground level and Class G rules to the surface will become applicable. Check the A/FD for specifics.

4. Departures from:

(a) A primary or satellite airport with an operating control tower. Two-way radio communications must be established and maintained with the control tower, and thereafter as instructed by ATC while operating in the Class D airspace.

(b) A satellite airport without an operating control tower. Two-way radio communications must be established as soon as practicable after departing with the ATC facility having jurisdiction over the Class D airspace as soon as practicable after departing.

5. Aircraft Speed. Unless otherwise authorized or required by ATC, no person may operate an aircraft at or below 2,500 feet above the surface within 4 nautical miles of the primary airport of a Class D airspace area at an indicated airspeed of more than 200 knots (230 mph).

c. Class D airspace areas are depicted on Sectional and Terminal charts with blue segmented lines, and on IFR En Route Lows with a boxed [D].

d. Arrival extensions for instrument approach procedures may be Class D or Class E airspace. As a general rule, if all extensions are 2 miles or less, they remain part of the Class D surface area. However, if any one extension is greater than 2 miles, then all extensions become Class E.

e. Separation for VFR Aircraft. No separation services are provided to VFR aircraft.

3-2-6. Class E Airspace

a. Definition. Generally, if the airspace is not Class A, Class B, Class C, or Class D, and it is controlled airspace, it is Class E airspace.

b. Operating Rules and Pilot/Equipment Requirements:

1. Pilot Certification. No specific certification required.

2. Equipment. No specific equipment required by the airspace.

3. Arrival or Through Flight Entry Requirements. No specific requirements.

c. Charts. Class E airspace below 14,500 feet MSL is charted on Sectional, Terminal, and IFR Enroute Low Altitude charts.

d. Vertical limits. Except for 18,000 feet MSL, Class E airspace has no defined vertical limit but rather it extends upward from either the surface or a designated altitude to the overlying or adjacent controlled airspace.

e. Types of Class E Airspace:

1. Surface area designated for an airport. When designated as a surface area for an airport, the airspace will be configured to contain all instrument procedures.

2. Extension to a surface area. There are Class E airspace areas that serve as extensions to Class B, Class C, and Class D surface areas designated for an airport. Such airspace provides controlled airspace to contain standard instrument approach procedures without imposing a communications requirement on pilots operating under VFR.

3. Airspace used for transition. There are Class E airspace areas beginning at either 700 or 1,200 feet AGL used to transition to/from the terminal or en route environment.

4. En Route Domestic Areas. There are Class E airspace areas that extend upward from a specified altitude and are en route domestic airspace areas that provide controlled airspace in those areas where there is a requirement to provide IFR en route ATC services but the Federal airway system is inadequate.

5. Federal Airways. The Federal airways are Class E airspace areas and, unless otherwise specified, extend upward from 1,200 feet to, but not including, 18,000 feet MSL. The colored airways are green, red, amber, and blue. The VOR airways are classified as Domestic, Alaskan, and Hawaiian.

6. Offshore Airspace Areas. There are Class E airspace areas that extend upward from a specified altitude to, but not including, 18,000 feet MSL and are designated as offshore airspace areas. These areas provide controlled airspace beyond 12 miles from the coast of the U.S. in those areas where there is a requirement to provide IFR en route ATC services and within which the U.S. is applying domestic procedures.

7. Unless designated at a lower altitude, Class E airspace begins at 14,500 feet MSL to, but not including, 18,000 feet MSL overlying: the 48 contiguous States including the waters within 12 miles from the coast of the 48 contiguous States; the District of Columbia; Alaska, including the waters within 12 miles from the coast of Alaska, and that airspace above FL 600; excluding the Alaska peninsula west of long. 160°00'00"W, and the airspace below 1,500 feet above the surface of the earth unless specifically so designated.

f. Separation for VFR Aircraft. No separation services are provided to VFR aircraft.

Section 3. Class G Airspace

3-3-1. General

Class G airspace (uncontrolled) is that portion of airspace that has not been designated as Class A, Class B, Class C, Class D, or Class E airspace.

3-3-2. VFR Requirements

Rules governing VFR flight have been adopted to assist the pilot in meeting the responsibility to see and avoid other aircraft. Minimum flight visibility and distance from clouds required for VFR flight are contained in 14 CFR Section 91.155.

(See *TBL 3-1-1.*)

3-3-3. IFR Requirements

a. Title 14 CFR specifies the pilot and aircraft equipment requirements for IFR flight. Pilots are reminded that in addition to altitude or flight level requirements, 14 CFR Section 91.177 includes a requirement to remain at least 1,000 feet (2,000 feet in designated mountainous terrain) above the highest obstacle within a horizontal distance of 4 nautical miles from the course to be flown.

b. IFR Altitudes.

(See *TBL 3-3-1.*)

TBL 3-3-1
IFR Altitudes
Class G Airspace

If your magnetic course (ground track) is:	And you are below 18,000 feet MSL, fly:
0° to 179°	Odd thousands MSL, (3,000; 5,000; 7,000, etc.)
180° to 359°	Even thousands MSL, (2,000; 4,000; 6,000, etc.)

Section 4. Special Use Airspace

3-4-1. General

a. Special use airspace consists of that airspace wherein activities must be confined because of their nature, or wherein limitations are imposed upon aircraft operations that are not a part of those activities, or both. Except for controlled firing areas, special use airspace areas are depicted on aeronautical charts.

b. Prohibited and restricted areas are regulatory special use airspace and are established in 14 CFR Part 73 through the rulemaking process.

c. Warning areas, military operations areas (MOAs), alert areas, and controlled firing areas (CFAs) are nonregulatory special use airspace.

d. Special use airspace descriptions (except CFAs) are contained in FAA Order JO 7400.8, Special Use Airspace.

e. Special use airspace (except CFAs) are charted on IFR or visual charts and include the hours of operation, altitudes, and the controlling agency.

3-4-2. Prohibited Areas

Prohibited areas contain airspace of defined dimensions identified by an area on the surface of the earth within which the flight of aircraft is prohibited. Such areas are established for security or other reasons associated with the national welfare. These areas are published in the Federal Register and are depicted on aeronautical charts.

3-4-3. Restricted Areas

a. Restricted areas contain airspace identified by an area on the surface of the earth within which the flight of aircraft, while not wholly prohibited, is subject to restrictions. Activities within these areas must be confined because of their nature or limitations imposed upon aircraft operations that are not a part of those activities or both. Restricted areas denote the existence of unusual, often invisible, hazards to aircraft such as artillery firing, aerial gunnery, or guided missiles. Penetration of restricted areas without authorization from the using or controlling agency may be extremely hazardous to the aircraft and its occupants. Restricted areas are published in the Federal Register and constitute 14 CFR Part 73.

b. ATC facilities apply the following procedures when aircraft are operating on an IFR clearance (including those cleared by ATC to maintain VFR-on-top) via a route which lies within joint-use restricted airspace.

1. If the restricted area is not active and has been released to the controlling agency (FAA), the ATC facility will allow the aircraft to operate in the restricted airspace without issuing specific clearance for it to do so.

2. If the restricted area is active and has not been released to the controlling agency (FAA), the ATC facility will issue a clearance which will ensure the aircraft avoids the restricted airspace unless it is on an approved altitude reservation mission or has obtained its own permission to operate in the airspace and so informs the controlling facility.

NOTE-*The above apply only to joint-use restricted airspace and not to prohibited and nonjoint-use airspace. For the latter categories, the ATC facility will issue a clearance so the aircraft will avoid the restricted airspace unless it is on an approved altitude reservation mission or has obtained its own permission to operate in the airspace and so informs the controlling facility.*

c. Restricted airspace is depicted on the en route chart appropriate for use at the altitude or flight level being flown. For joint-use restricted areas, the name of the controlling agency is shown on these charts. For all prohibited areas and nonjoint-use restricted areas, unless otherwise requested by the using agency, the phrase "NO A/G" is shown.

3-4-4. Warning Areas

A warning area is airspace of defined dimensions, extending from three nautical miles outward from the coast of the U.S., that contains activity that may be hazardous to nonparticipating aircraft. The purpose of such warning areas is to warn nonparticipating pilots of the potential danger. A warning area may be located over domestic or international waters or both.

3-4-5. Military Operations Areas

a. MOAs consist of airspace of defined vertical and lateral limits established for the purpose of separating certain military training activities from IFR traffic. Whenever a MOA is being used, nonparticipating IFR traffic may be cleared through a MOA if IFR separation can be provided by ATC. Otherwise, ATC will reroute or restrict nonparticipating IFR traffic.

b. Examples of activities conducted in MOAs include, but are not limited to: air combat tactics, air intercepts, aerobatics, formation training, and low-altitude tactics. Military pilots flying in an active MOA are exempted from the provisions of 14 CFR Section 91.303(c) and (d) which prohibits aerobatic flight within Class D and Class E surface areas, and within Federal airways. Additionally, the Department of Defense has been issued an authorization to operate aircraft at indicated airspeeds in excess of 250 knots below 10,000 feet MSL within active MOAs.

c. Pilots operating under VFR should exercise extreme caution while flying within a MOA when military activity is being conducted. The activity status (active/inactive) of MOAs may change frequently. Therefore, pilots should contact any FSS within 100 miles of the area to obtain accurate real-time information concerning the MOA hours of operation. Prior to entering an active MOA, pilots should contact the controlling agency for traffic advisories.

d. MOAs are depicted on sectional, VFR Terminal Area, and Enroute Low Altitude charts.

3-4-6. Alert Areas

Alert areas are depicted on aeronautical charts to inform nonparticipating pilots of areas that may contain a high volume of pilot training or an unusual type of aerial activity. Pilots should be particularly alert when flying in these areas. All activity within an alert area must be conducted in accordance with CFRs, without waiver, and pilots of participating aircraft as well as pilots transiting the area must be equally responsible for collision avoidance.

3-4-7. Controlled Firing Areas

CFAs contain activities which, if not conducted in a controlled environment, could be hazardous to nonparticipating aircraft. The distinguishing feature of the CFA, as compared to other special use airspace, is that its activities are suspended immediately when spotter aircraft, radar, or ground lookout positions indicate an aircraft might be approaching the area. There is no need to chart CFAs since they do not cause a nonparticipating aircraft to change its flight path.

3-4-8. National Security Areas

National Security Areas consist of airspace of defined vertical and lateral dimensions established at locations where there is a requirement for increased security and safety of ground facilities. Pilots are requested to voluntarily avoid flying through the depicted NSA. When it is necessary to provide a greater level of security and safety, flight in NSAs may be temporarily prohibited by regulation under the provisions of 14 CFR Section 99.7. Regulatory prohibitions will be issued by System Operations, System Operations Airspace and AIM Office, Airspace and Rules, and disseminated via NOTAM. Inquiries about NSAs should be directed to Airspace and Rules.

Section 5. Other Airspace Areas

3-5-1. Airport Advisory/Information Services

a. There are three advisory type services available at selected airports.

1. Local Airport Advisory (LAA) service is operated within 10 statute miles of an airport where a control tower is not operating but where a FSS is located on the airport. At such locations, the FSS provides a complete local airport advisory service to arriving and departing aircraft. During periods of fast changing weather the FSS will automatically provide Final Guard as part of the service from the time the aircraft reports "on-final" or "taking-the-active-runway" until the aircraft reports "on-the-ground" or "airborne."

NOTE-*Current policy, when requesting remote ATC services, requires that a pilot monitor the automated weather broadcast at the landing airport prior to requesting ATC services. The FSS automatically provides Final Guard, when appropriate, during LAA/Remote Airport Advisory (RAA) operations. Final Guard is a value added wind/altimeter monitoring service, which provides an automatic wind and altimeter check during active weather situations when the pilot reports on-final or taking the active runway. During the landing or take-off operation when the winds or altimeter are actively changing the FSS will blind broadcast significant changes when the specialist believes the change might affect the operation. Pilots should acknowledge the first wind/altimeter check but due to cockpit activity no acknowledgement is expected for the blind broadcasts. It is prudent for a pilot to report on-the-ground or airborne to end the service.*

2. RAA service is operated within 10 statute miles of specified high activity GA airports where a control tower is not operating. Airports offering this service are listed in the A/FD and the published service hours may be changed by NOTAM D. Final Guard is automatically provided with RAA.

3. Remote Airport Information Service (RAIS) is provided in support of short term special events like small to medium fly-ins. The service is advertised by NOTAM D only. The FSS will not have access to a continuous readout of the current winds and altimeter; therefore, RAIS does not include weather and/or Final Guard service. However, known traffic, special event instructions, and all other services are provided.

NOTE-*The airport authority and/or manager should request RAIS support on official letterhead directly with the manager of the FSS that will provide the service at least 60 days in advance. Approval authority rests with the FSS manager and is based on workload and resource availability.*

REFERENCE-*AIM, Traffic Advisory Practices at Airports Without Operating Control Towers, Paragraph 4-1-9.*

b. It is not mandatory that pilots participate in the Airport Advisory programs. Participation enhances safety for everyone operating around busy GA airports; therefore, everyone is encouraged to participate and provide feedback that will help improve the program.

3-5-2. Military Training Routes

a. National security depends largely on the deterrent effect of our airborne military forces. To be proficient, the military services must train in a wide range of airborne tactics. One phase of this training involves "low level" combat tactics. The required maneuvers and high speeds are such that they may occasionally make the see-and-avoid aspect of VFR flight more difficult without increased vigilance in areas containing such operations. In an effort to ensure the greatest practical level of safety for all flight operations, the Military Training Route (MTR) program was conceived.

b. The MTR program is a joint venture by the FAA and the Department of Defense (DOD). MTRs are mutually developed for use by the military for the purpose of conducting low-altitude, high-speed training. The routes above 1,500 feet AGL are developed to be flown, to the maximum extent possible, under IFR. The routes at 1,500 feet AGL and below are generally developed to be flown under VFR.

c. Generally, MTRs are established below 10,000 feet MSL for operations at speeds in excess of 250 knots.

However, route segments may be defined at higher altitudes for purposes of route continuity. For example, route segments may be defined for descent, climbout, and mountainous terrain. There are IFR and VFR routes as follows:

1. IFR Military Training Routes-(IR). Operations on these routes are conducted in accordance with IFR regardless of weather conditions.

2. VFR Military Training Routes-(VR). Operations on these routes are conducted in accordance with VFR except flight visibility must be 5 miles or more; and flights must not be conducted below a ceiling of less than 3,000 feet AGL.

d. Military training routes will be identified and charted as follows:

1. Route identification.

(a) MTRs with no segment above 1,500 feet AGL must be identified by four number characters; e.g., IR1206, VR1207.

(b) MTRs that include one or more segments above 1,500 feet AGL must be identified by three number characters; e.g., IR206, VR207.

(c) Alternate IR/VR routes or route segments are identified by using the basic/principal route designation followed by a letter suffix, e.g., IR008A, VR1007B, etc.

2. Route charting.

(a) IFR Low Altitude En Route Chart. This chart will depict all IR routes and all VR routes that accommodate operations above 1,500 feet AGL.

(b) VFR Sectional Charts. These charts will depict military training activities such as IR, VR, MOA, Restricted Area, Warning Area, and Alert Area information.

(c) Area Planning (AP/1B) Chart (DOD Flight Information Publication-FLIP). This chart is published by the DOD primarily for military users and contains detailed information on both IR and VR routes.

REFERENCE-*AIM, National Geospatial-Intelligence Agency (NGA) Products, Paragraph 9-1-5, Subparagraph a.*

e. The FLIP contains charts and narrative descriptions of these routes. This publication is available to the general public by single copy or annual subscription from:

Aeronautical Navigation Products (AeroNav)
Logistics Group
Federal Aviation Administration
10201 Good Luck Road
Glenn Dale, MD 20769-9700
Toll free phone: 1-800-638-8972
Commercial: 301-436-8301

This DOD FLIP is available for pilot briefings at FSS and many airports.

f. Nonparticipating aircraft are not prohibited from flying within an MTR; however, extreme vigilance should be exercised when conducting flight through or near these routes. Pilots should contact FSSs within 100 NM of a particular MTR to obtain current information or route usage in their vicinity. Information available includes times of scheduled activity, altitudes in use on each route segment, and actual route width. Route width varies for each MTR and can extend several miles on either side of the charted MTR centerline. Route width information for IR and VR MTRs is also available in the FLIP AP/1B along with additional MTR (slow routes/air refueling routes) information. When requesting MTR information, pilots should give the FSS their position, route of flight, and destination in order to reduce frequency congestion and permit the FSS specialist to identify the MTR which could be a factor.

3-5-3. Temporary Flight Restrictions

a. General. This paragraph describes the types of conditions under which the FAA may impose temporary flight restrictions. It also explains which FAA elements have been delegated authority to issue a temporary flight restrictions NOTAM and lists the types of responsible agencies/offices from which the FAA will accept requests to establish temporary flight restrictions. The 14 CFR is explicit as to what operations are prohibited, restricted, or allowed in a temporary flight restrictions area. Pilots are responsible to comply with 14 CFR Sections 91.137, 91.138, 91.141 and 91.143 when conducting flight in an area where a temporary flight restrictions area is in effect, and should check appropriate NOTAMs during flight planning.

b. The purpose for establishing a temporary flight restrictions area is to:

1. Protect persons and property in the air or on the surface from an existing or imminent hazard associated with an incident on the surface when the presence of low flying aircraft would magnify, alter, spread, or compound that hazard (14 CFR Section 91.137(a)(1));

2. Provide a safe environment for the operation of disaster relief aircraft (14 CFR Section 91.137(a)(2)); or

3. Prevent an unsafe congestion of sightseeing aircraft above an incident or event which may generate a high degree of public interest (14 CFR Section 91.137(a)(3)).

4. Protect declared national disasters for humanitarian reasons in the State of Hawaii (14 CFR Section 91.138).

5. Protect the President, Vice President, or other public figures (14 CFR Section 91.141).

6. Provide a safe environment for space agency operations (14 CFR Section 91.143).

c. Except for hijacking situations, when the provisions of 14 CFR Section 91.137(a)(1) or (a)(2) are necessary, a temporary flight restrictions area will only be established by or through the area manager at the Air Route Traffic Control Center (ARTCC) having jurisdiction over the area concerned. A temporary flight restrictions NOTAM involving the conditions of 14 CFR Section 91.137(a)(3) will be issued at the direction of the service area office director having oversight of the airspace concerned. When hijacking situations are involved, a temporary flight restrictions area will be implemented through the TSA Aviation Command Center. The appropriate FAA air traffic element, upon receipt of such a request, will establish a temporary flight restrictions area under 14 CFR Section 91.137(a)(1).

d. The FAA accepts recommendations for the establishment of a temporary flight restrictions area under 14 CFR Section 91.137(a)(1) from military major command headquarters, regional directors of the Office of Emergency Planning, Civil Defense State Directors, State Governors, or other similar authority. For the situations involving 14 CFR Section 91.137(a)(2), the FAA accepts recommendations from military commanders serving as regional, subregional, or Search and Rescue (SAR) coordinators; by military commanders directing or coordinating air operations associated with disaster relief; or by civil authorities directing or coordinating organized relief air operations (includes representatives of the Office of Emergency Planning, U.S. Forest Service, and State aeronautical agencies). Appropriate authorities for a temporary flight restrictions establishment under 14 CFR Section 91.137(a)(3) are any of those listed above or by State, county, or city government entities.

e. The type of restrictions issued will be kept to a minimum by the FAA consistent with achievement of the necessary objective. Situations which warrant the extreme restrictions of 14 CFR Section 91.137(a)(1) include, but are not limited to: toxic gas leaks or spills, flammable agents, or fumes which if fanned by rotor or propeller wash could endanger persons or property on the surface, or if entered by an aircraft could endanger persons or property in the air; imminent volcano eruptions which could endanger airborne aircraft and occupants; nuclear accident or incident; and hijackings. Situations which warrant the restrictions associated with 14 CFR Section 91.137(a)(2) include: forest fires which are being fought by releasing fire retardants from aircraft; and aircraft relief activities following a disaster (earthquake, tidal wave, flood, etc.). 14 CFR Section 91.137(a)(3) restrictions are established for events and incidents that would attract an unsafe congestion of sightseeing aircraft.

f. The amount of airspace needed to protect persons and property or provide a safe environment for rescue/relief aircraft operations is normally limited to within 2,000 feet above the surface and within a 3-nautical-mile radius. Incidents occurring within Class B, Class C, or Class D airspace will normally be handled through existing procedures and should not require the issuance of a temporary flight restrictions NOTAM. Temporary flight restrictions affecting airspace outside of the U.S. and its territories and possessions are issued with verbiage excluding that airspace outside of the 12-mile coastal limits.

g. The FSS nearest the incident site is normally the "coordination facility." When FAA communications assistance is required, the designated FSS will function as the primary communications facility for coordination between emergency control authorities and affected aircraft. The ARTCC may act as liaison for the emergency control authorities if adequate communications cannot be established between the designated FSS and the relief organization. For example, the coordination facility may relay authorizations from the on-scene emergency response official in cases where news media aircraft operations are approved at the altitudes used by relief aircraft.

h. ATC may authorize operations in a temporary flight restrictions area under its own authority only when flight restrictions are established under 14 CFR Section 91.137(a)(2) and (a)(3). The appropriate ARTCC/airport traffic control tower manager will, however, ensure that such authorized flights do not hamper activities or interfere with the event for which restrictions were implemented. However, ATC will not authorize local IFR flights into the temporary flight restrictions area.

i. To preclude misunderstanding, the implementing NOTAM will contain specific and formatted information. The facility establishing a temporary flight restrictions area will format a NOTAM beginning with the phrase "FLIGHT RESTRICTIONS" followed by: the location of the temporary flight restrictions area; the effective period; the area defined in statute miles; the altitudes affected; the FAA coordination facility and commercial telephone number; the reason for the temporary flight restrictions; the agency directing any relief activities and its commercial telephone number; and other information considered appropriate by the issuing authority.

EXAMPLE-**1.** *14 CFR Section 91.137(a)(1):*

The following NOTAM prohibits all aircraft operations except those specified in the NOTAM.

Flight restrictions Matthews, Virginia, effective immediately until 9610211200. Pursuant to 14 CFR Section 91.137(a)(1) temporary flight restrictions are in effect. Rescue operations in progress. Only relief aircraft operations under the direction of the Department of Defense are authorized in the airspace at and below 5,000 feet MSL within a 2-nautical-mile radius of Laser AFB, Matthews, Virginia. Commander, Laser AFB, in charge (897) 946-5543 (122.4). Steenson FSS (792) 555-6141 (123.1) is the FAA coordination facility.

2. *14 CFR Section 91.137(a)(2):*

The following NOTAM permits flight operations in accordance with 14 CFR Section 91.137(a)(2). The on-site emergency response official to authorize media aircraft operations below the altitudes used by the relief aircraft. Flight restrictions 25 miles east of Bransome, Idaho, effective immediately until 9601202359 UTC. Pursuant to 14 CFR Section 91.137(a)(2) temporary flight restrictions are in effect within a 4-nautical-mile radius of the intersection of county roads 564 and 315 at and below 3,500 feet MSL to provide a safe environment for fire fighting aircraft operations. Davis County sheriff's department (792) 555-8122 (122.9) is in charge of on-scene emergency response activities. Glivings FSS (792) 555-1618 (122.2) is the FAA coordination facility.

3. *14 CFR Section 91.137(a)(3):*

The following NOTAM prohibits sightseeing aircraft operations.

Flight restrictions Brown, Tennessee, due to olympic activity. Effective 9606181100 UTC until 9607190200 UTC. Pursuant to 14 CFR Section 91.137(a)(3) temporary flight restrictions are in effect within a 3-nautical-mile radius of N355783/W835242 and Volunteer VORTAC 019 degree radial 3.7 DME fix at and below 2,500 feet MSL. Norton FSS (423) 555-6742 (126.6) is the FAA coordination facility.

4. *14 CFR Section 91.138:*

The following NOTAM prohibits all aircraft except those operating under the authorization of the official in charge of associated emergency or disaster relief response activities, aircraft carrying law enforcement officials, aircraft carrying personnel involved in an emergency or legitimate scientific purposes, carrying properly accredited news media, and aircraft operating in accordance with an ATC clearance or instruction.

Flight restrictions Kapalua, Hawaii, effective 9605101200 UTC until 9605151500UTC. Pursuant to 14 CFR Section 91.138 temporary flight restrictions are in effect within a 3-nautical-mile radius of N205778/W1564038 and Maui/OGG/VORTAC 275 degree radial at 14.1 nautical miles. John Doe 808-757-4469 or 122.4 is in charge of the operation. Honolulu/HNL 808-757-4470 (123.6) FSS is the FAA coordination facility.

5. *14 CFR Section 91.141:*

The following NOTAM prohibits all aircraft.

Flight restrictions Stillwater, Oklahoma, June 21, 1996. Pursuant to 14 CFR Section 91.141 aircraft flight operations are prohibited within a 3-nautical-mile radius, below 2000 feet AGL of N360962/W970515 and the Stillwater/SWO/VOR/DME 176 degree radial 3.8-nautical-mile fix from 1400 local time to 1700 local time June 21, 1996, unless otherwise authorized by ATC.

6. *14 CFR Section 91.143:*

The following NOTAM prohibits any aircraft of U.S. registry, or pilot any aircraft under the authority of an airman certificate issued by the FAA.

Kennedy space center space operations area effective immediately until 9610152100 UTC. Pursuant to 14 CFR Section 91.143, flight operations conducted by FAA certificated pilots or conducted in aircraft of U.S. registry are prohibited at any altitude from surface to unlimited, within the following area 30-nautical-mile radius of the Melbourne/MLB/VORTAC 010 degree radial 21-nautical-mile fix. St. Petersburg, Florida/PIE/FSS 813-545-1645 (122.2) is the FAA coordination facility and should be contacted for the current status of any airspace associated with the space shuttle operations. This airspace encompasses R2933, R2932, R2931, R2934, R2935, W497A and W158A. Additional warning and restricted areas will be active in conjunction with the operations. Pilots must consult all NOTAMs regarding this operation.

3-5-4. Parachute Jump Aircraft Operations

a. Procedures relating to parachute jump areas are contained in 14 CFR Part 105. Tabulations of parachute jump areas in the U.S. are contained in the A/FD.

b. Pilots of aircraft engaged in parachute jump operations are reminded that all reported altitudes must be with reference to mean sea level, or flight level, as appropriate, to enable ATC to provide meaningful traffic information.

c. Parachute operations in the vicinity of an airport without an operating control tower - there is no substitute for alertness while in the vicinity of an airport. It is essential that pilots conducting parachute operations be alert, look for other traffic, and exchange traffic information as recommended in paragraph *4-1-9*, Traffic Advisory Practices at Airports Without Operating Control Towers. In addition, pilots should avoid releasing parachutes while in an airport traffic pattern when there are other aircraft in that pattern. Pilots should make appropriate broadcasts on the designated Common Traffic Advisory Frequency (CTAF), and monitor that CTAF until all parachute activity has terminated or the aircraft has left the area. Prior to commencing a jump operation, the pilot should broadcast the aircraft's altitude and position in relation to the airport, the approximate relative time when the jump will commence and terminate, and listen to the position reports of other aircraft in the area.

3-5-5. Published VFR Routes

Published VFR routes for transitioning around, under and through complex airspace such as Class B airspace were developed through a number of FAA and industry initiatives. All of the following terms, i.e., "VFR Flyway" "VFR Corridor" and "Class B Airspace VFR Transition Route" have been used when referring to the same or different types of routes or airspace. The following paragraphs identify and clarify the functionality of each type of route, and specify where and when an ATC clearance is required.

a. VFR Flyways.

1. VFR Flyways and their associated Flyway Planning Charts were developed from the recommendations of a National Airspace Review Task Group. A VFR Flyway is defined as a general flight path not defined as a specific course, for use by pilots in planning flights into, out of, through or near complex terminal airspace to avoid Class B airspace. An ATC clearance is NOT required to fly these routes.

2. VFR Flyways are depicted on the reverse side of some of the VFR Terminal Area Charts (TAC), commonly referred to as Class B airspace charts. (See *FIG 3-5-1*.) Eventually all TACs will include a VFR Flyway Planning Chart. These charts identify VFR flyways designed to help VFR pilots avoid major controlled traffic flows. They may further depict multiple VFR routings throughout the area which may be used as an alternative to flight within Class B airspace. The ground references provide a guide for improved visual navigation. These routes are not intended to discourage requests for VFR operations within Class B airspace but are designed solely to assist pilots in planning for flights under and around busy Class B airspace without actually entering Class B airspace.

3. It is very important to remember that these suggested routes are not sterile of other traffic. The entire Class B airspace, and the airspace underneath it, may be heavily congested with many different types of aircraft. Pilot adherence to VFR rules must be exercised at all times. Further, when operating beneath Class B airspace, communications must be established and maintained between your aircraft and any control tower while transiting the Class B, Class C, and Class D surface areas of those airports under Class B airspace.

b. VFR Corridors.

1. The design of a few of the first Class B airspace areas provided a corridor for the passage of uncontrolled traffic. A VFR corridor is defined as airspace through Class B airspace, with defined vertical and lateral boundaries, in which aircraft may operate without an ATC clearance or communication with air traffic control.

2. These corridors are, in effect, a "hole" through Class B airspace. (See *FIG 3-5-2*.) A classic example would be the corridor through the Los Angeles Class B airspace, which has been subsequently changed to Special Flight Rules airspace (SFR). A corridor is surrounded on all sides by Class B airspace and does not extend down to the surface like a VFR Flyway. Because of their finite lateral and vertical limits, and the volume of VFR traffic using a corridor, extreme caution and vigilance must be exercised.

3. Because of the heavy traffic volume and the procedures necessary to efficiently manage the flow of traffic, it has not been possible to incorporate VFR corridors in the development or modifications of Class B airspace in recent years.

c. Class B Airspace VFR Transition Routes.

1. To accommodate VFR traffic through certain Class B airspace, such as Seattle, Phoenix and Los Angeles, Class B Airspace VFR Transition Routes were developed. A Class B Airspace VFR Transition Route is defined as a specific flight course depicted on a TAC for transiting a specific Class B airspace. These routes include specific ATC-assigned altitudes, and pilots must obtain an ATC clearance prior to entering Class B airspace on the route.

2. These routes, as depicted in *FIG 3-5-3*, are designed to show the pilot where to position the aircraft outside of, or clear of, the Class B airspace where an ATC clearance can normally be expected with minimal or no delay. Until ATC authorization is received, pilots must remain clear of Class B airspace. On initial contact, pilots should advise ATC of their position, altitude, route name desired, and direction of flight. After a clearance is received, pilots must fly the route as depicted and, most importantly, adhere to ATC instructions.

3-5-6. Terminal Radar Service Area (TRSA)

a. Background. TRSAs were originally established as part of the Terminal Radar Program at selected airports.

FIG 3-5-2
Class B Airspace

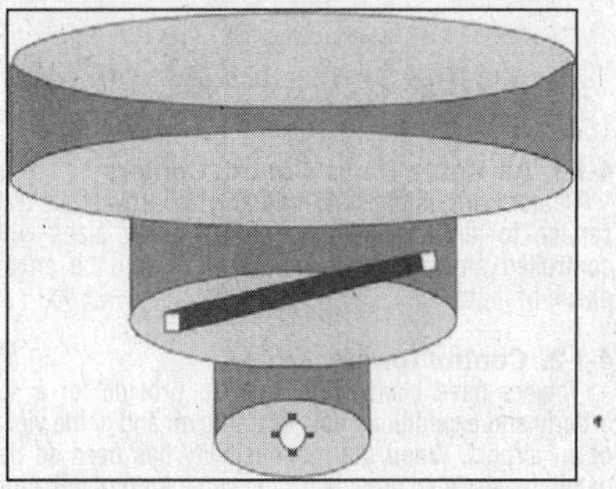

TRSAs were never controlled airspace from a regulatory standpoint because the establishment of TRSAs was never subject to the rulemaking process; consequently, TRSAs are not contained in 14 CFR Part 71 nor are there any TRSA operating rules in 14 CFR Part 91. Part of the Airport Radar Service Area (ARSA) program was to eventually replace all TRSAs. However, the ARSA requirements became relatively stringent and it was subsequently decided that TRSAs would have to meet ARSA criteria before they would be converted. TRSAs do not fit into any of the U.S. airspace classes; therefore, they will continue to be non-Part 71 airspace areas where participating pilots can receive additional radar services which have been redefined as TRSA Service.

b. TRSAs. The primary airport(s) within the TRSA become(s) Class D airspace. The remaining portion of the TRSA overlies other controlled airspace which is normally Class E airspace beginning at 700 or 1,200 feet and established to transition to/from the en route/terminal environment.

c. Participation. Pilots operating under VFR are encouraged to contact the radar approach control and avail themselves of the TRSA Services. However, participation is voluntary on the part of the pilot. See *Chapter 4*, Air Traffic Control, for details and procedures.

d. Charts. TRSAs are depicted on VFR sectional and terminal area charts with a solid black line and altitudes for each segment. The Class D portion is charted with a blue segmented line.

FIG 3-5-1
VFR Flyway Planning Chart

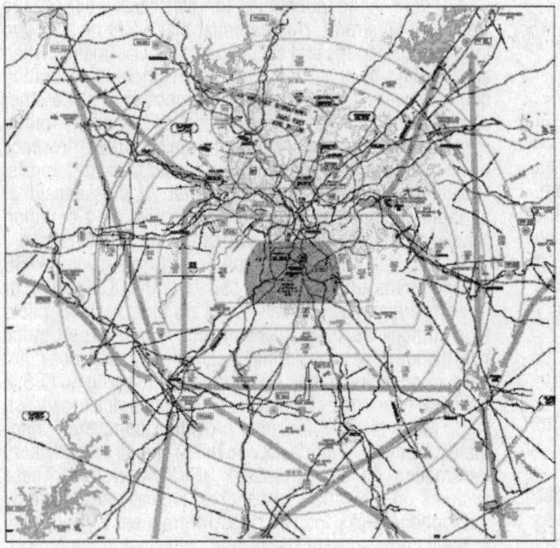

FIG 3-5-3
VFR Transition Route

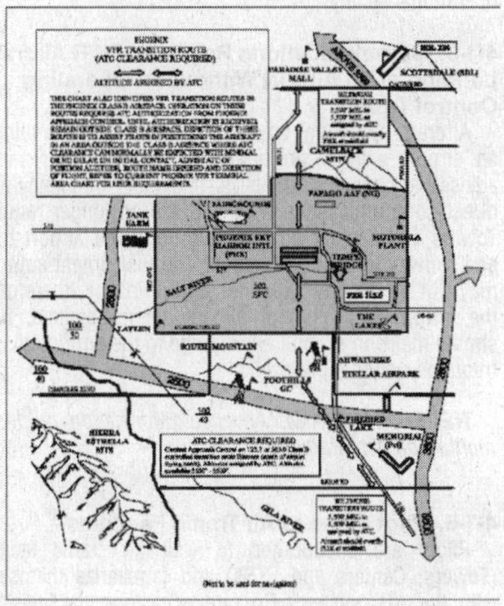

Chapter 4.

AIR TRAFFIC CONTROL

Section 1. Services Available to Pilots

4-1-1. Air Route Traffic Control Centers

Centers are established primarily to provide air traffic service to aircraft operating on IFR flight plans within controlled airspace, and principally during the en route phase of flight.

4-1-2. Control Towers

Towers have been established to provide for a safe, orderly and expeditious flow of traffic on and in the vicinity of an airport. When the responsibility has been so delegated, towers also provide for the separation of IFR aircraft in the terminal areas.

REFERENCE- *AIM, Approach Control, Paragraph 5-4-3.*

4-1-3. Flight Service Stations

Flight Service Stations (FSSs) are air traffic facilities which provide pilot briefings, flight plan processing, en route radio communications, search and rescue services, and assistance to lost aircraft and aircraft in emergency situations. FSSs also relay ATC clearances, process Notices to Airmen, broadcast aviation weather and aeronautical information, and notify Customs and Border Protection of transborder flights. In addition, at selected locations FSSs provide En Route Flight Advisory Service (Flight Watch) and Airport Advisory Service (AAS). In Alaska, designated FSSs also provide TWEB recordings and take weather observations.

4-1-4. Recording and Monitoring

a. Calls to air traffic control (ATC) facilities (ARTCCs, Towers, FSSs, Central Flow, and Operations Centers) over radio and ATC operational telephone lines (lines used for operational purposes such as controller instructions, briefings, opening and closing flight plans, issuance of IFR clearances and amendments, counter hijacking activities, etc.) may be monitored and recorded for operational uses such as accident investigations, accident prevention, search and rescue purposes, specialist training and evaluation, and technical evaluation and repair of control and communications systems.

b. Where the public access telephone is recorded, a beeper tone is not required. In place of the "beep" tone the FCC has substituted a mandatory requirement that persons to be recorded be given notice they are to be recorded and give consent. Notice is given by this entry, consent to record is assumed by the individual placing a call to the operational facility.

4-1-5. Communications Release of IFR Aircraft Landing at an Airport Without an Operating Control Tower

Aircraft operating on an IFR flight plan, landing at an airport without an operating control tower will be advised to change to the airport advisory frequency when direct communications with ATC are no longer required. Towers and centers do not have nontower airport traffic and runway in use information. The instrument approach may not be aligned with the runway in use; therefore, if the information has not already been obtained, pilots should make an expeditious change to the airport advisory frequency when authorized.

REFERENCE- *AIM, Advance Information on Instrument Approach, Paragraph 5-4-4.*

4-1-6. Pilot Visits to Air Traffic Facilities

Pilots are encouraged to visit air traffic facilities (Towers, Centers and FSSs) and familiarize themselves with the ATC system. On rare occasions, facilities may not be able to approve a visit because of ATC workload or other reasons. It is, therefore, requested that pilots contact the facility prior to the visit and advise of the number of persons in the group, the time and date of the proposed visit and the primary interest of the group. With this information available, the facility can prepare an itinerary and have someone available to guide the group through the facility.

4-1-7. Operation Take-off and Operation Raincheck

Operation Take-off is a program that educates pilots in how best to utilize the FSS modernization efforts and services available in Flight Service Stations (FSS), as stated in FAA Order 7230.17, Pilot Education Program - Operation Takeoff. Operation Raincheck is a program designed to familiarize pilots with the ATC system, its functions, responsibilities and benefits.

4-1-8. Approach Control Service for VFR Arriving Aircraft

a. Numerous approach control facilities have established programs for arriving VFR aircraft to contact approach control for landing information. This information includes: wind, runway, and altimeter setting at the airport of intended landing. This information may be omitted if contained in the Automatic Terminal Information Service (ATIS) broadcast and the pilot states the appropriate ATIS code.

NOTE- *Pilot use of "have numbers" does not indicate receipt of the ATIS broadcast. In addition, the controller will provide traffic advisories on a workload permitting basis.*

b. Such information will be furnished upon initial contact with concerned approach control facility. The pilot will be requested to change to the *tower* frequency at a predetermined time or point, to receive further landing information.

c. Where available, use of this procedure will not hinder the operation of VFR flights by requiring excessive spacing between aircraft or devious routing.

d. Compliance with this procedure is not mandatory but pilot participation is encouraged.

REFERENCE- *AIM, Terminal Radar Services for VFR Aircraft, Paragraph 4-1-18.*

NOTE- *Approach control services for VFR aircraft are normally dependent on ATC radar. These services are not available during periods of a radar outage. Approach control services for VFR aircraft are limited when CENRAP is in use.*

4-1-9. Traffic Advisory Practices at Airports Without Operating Control Towers

(See TBL 4-1-1.)

a. Airport Operations Without Operating Control Tower

1. There is no substitute for alertness while in the vicinity of an airport. It is essential that pilots be alert and look for other traffic and exchange traffic information when approaching or departing an airport without an operating control tower. This is of particular importance since other aircraft may not have communication capability or, in some cases, pilots may not communicate their presence or intentions when operating into or out of such airports. To achieve the greatest degree of safety, it is essential that all radio-equipped aircraft transmit/receive on a common frequency identified for the purpose of airport advisories.

2. An airport may have a full or part-time tower or FSS located on the airport, a full or part-time UNICOM station or no aeronautical station at all. There are three ways for pilots to communicate their intention and obtain airport/traffic information when operating at an airport that does not have an operating tower: by communicating with an FSS, a UNICOM operator, or by making a self-announce broadcast.

3. Many airports are now providing completely automated weather, radio check capability and airport advisory information on an automated UNICOM system. These systems offer a variety of features, typically selectable by microphone clicks, on the UNICOM frequency. Availability of the automated UNICOM will be published in the Airport/Facility Directory and approach charts.

b. Communicating on a Common Frequency

1. The key to communicating at an airport without an operating control tower is selection of the correct common frequency. The acronym CTAF which stands for Common Traffic Advisory Frequency, is synonymous with this program. A CTAF is a frequency designated for the purpose of carrying out airport advisory practices while operating to or from an airport without an operating control tower. The CTAF may be a UNICOM, MULTICOM, FSS, or tower frequency and is identified in appropriate aeronautical publications.

2. The CTAF frequency for a particular airport is contained in the A/FD, Alaska Supplement, Alaska Terminal Publication, Instrument Approach Procedure Charts, and Instrument Departure Procedure (DP) Charts. Also, the CTAF frequency can be obtained by contacting any FSS. Use of the appropriate CTAF, combined with a visual alertness and application of the following recommended good operating practices, will enhance safety of flight into and out of all uncontrolled airports.

c. Recommended Traffic Advisory Practices

1. Pilots of inbound traffic should monitor and communicate as appropriate on the designated CTAF from 10 miles to landing. Pilots of departing aircraft should monitor/communicate on the appropriate frequency from start-up, during taxi, and until 10 miles from the airport unless the CFRs or local procedures require otherwise.

2. Pilots of aircraft conducting other than arriving or departing operations at altitudes normally used by arriving and departing aircraft should monitor/communicate on the appropriate frequency while within 10 miles of the airport unless required to do otherwise by the CFRs or local procedures. Such operations include parachute jumping/dropping, en route, practicing maneuvers, etc.

REFERENCE- *AIM, Parachute Jump Aircraft Operations, Paragraph 3-5-4.*

d. Airport Advisory/Information Services Provided by a FSS

1. There are three advisory type services provided at selected airports.

(a) Local Airport Advisory (LAA) is provided at airports that have a FSS physically located on the airport, which does not have a control tower or where the tower is operated on a part-time basis. The CTAF for LAA airports is disseminated in the appropriate aeronautical publications.

(b) Remote Airport Advisory (RAA) is provided at selected very busy GA airports, which do not have an operating control tower. The CTAF for RAA airports is disseminated in the appropriate aeronautical publications.

(c) Remote Airport Information Service (RAIS) is provided in support of special events at nontowered airports by request from the airport authority.

2. In communicating with a CTAF FSS, check the airport's automated weather and establish two-way communications before transmitting outbound/inbound intentions or information. An inbound aircraft should initiate contact approximately 10 miles from the airport, reporting aircraft identification and type, altitude, location relative to the airport, intentions (landing or over flight), possession of the automated weather, and request airport advisory or airport information service. A departing aircraft should initiate contact before taxiing, reporting aircraft identification and type, VFR or IFR, location on the airport, intentions, direction of take-off, possession of the automated weather, and request airport advisory or information service. Also, report intentions before taxiing onto the active runway for departure. If you must change frequencies for other service after initial report to FSS, return to FSS frequency for traffic update.

(a) Inbound

EXAMPLE- *Vero Beach radio, Centurion Six Niner Delta Delta is ten miles south, two thousand, landing Vero Beach. I have the automated weather, request airport advisory.*

(b) Outbound

EXAMPLE- *Vero Beach radio, Centurion Six Niner Delta Delta, ready to taxi to runway 22, VFR, departing to the southwest. I have the automated weather, request airport advisory.*

3. Airport advisory service includes wind direction and velocity, favored or designated runway, altimeter setting, known airborne and ground traffic, NOTAMs, airport taxi routes, airport traffic pattern information, and instrument approach procedures. These elements are varied so as to best serve the current traffic situation. Some airport

Tbl 4-1-1
Summary of Recommended Communication Procedures

			Communication/Broadcast Procedures		
	Facility at Airport	**Frequency Use**	**Outbound**	**Inbound**	**Practice Instrument Approach**
1.	UNICOM (No Tower or FSS)	Communicate with UNICOM station on published CTAF frequency (122.7; 122.8; 122.725; 122.975; or 123.0). If unable to contact UNICOM station, use self-announce procedures on CTAF.	Before taxiing and before taxiing on the runway for departure.	10 miles out. Entering downwind, base, and final. Leaving the runway.	
2.	No Tower, FSS, or UNICOM	Self-announce on MULTICOM frequency 122.9.	Before taxiing and before taxiing on the runway for departure.	10 miles out. Entering downwind, base, and final. Leaving the runway.	Departing final approach fix (name) or on final approach segment inbound.
3.	No Tower in operation, FSS open	Communicate with FSS on CTAF frequency.	Before taxiing and before taxiing on the runway for departure.	10 miles out. Entering downwind, base, and final. Leaving the runway.	Approach completed/ terminated.
4.	FSS Closed (No Tower)	Self-announce on CTAF.	Before taxiing and before taxiing on the runway for departure.	10 miles out. Entering downwind, base, and final. Leaving the runway.	
5.	Tower or FSS not in operation	Self-announce on CTAF.	Before taxiing and before taxiing on the runway for departure.	10 miles out. Entering downwind, base, and final. Leaving the runway.	

managers have specified that under certain wind or other conditions designated runways be used. Pilots should advise the FSS of the runway they intend to use.

CAUTION-
All aircraft in the vicinity of an airport may not be in communication with the FSS.

e. Information Provided by Aeronautical Advisory Stations (UNICOM)

1. UNICOM is a nongovernment air/ground radio communication station which may provide airport information at public use airports where there is no tower or FSS.

2. On pilot request, UNICOM stations may provide pilots with weather information, wind direction, the recommended runway, or other necessary information. If the UNICOM frequency is designated as the CTAF, it will be identified in appropriate aeronautical publications.

f. Unavailability of Information from FSS or UNICOM

Should LAA by an FSS or Aeronautical Advisory Station UNICOM be unavailable, wind and weather information may be obtainable from nearby controlled airports via Automatic Terminal Information Service (ATIS) or Automated Weather Observing System (AWOS) frequency.

g. Self-Announce Position and/or Intentions

1. General. Self-announce is a procedure whereby pilots broadcast their position or intended flight activity or ground operation on the designated CTAF. This procedure is used primarily at airports which do not have an FSS on the airport. The self-announce procedure should also be used if a pilot is unable to communicate with the FSS on the designated CTAF. Pilots stating, *"Traffic in the area, please advise"* is not a recognized Self-Announce Position and/or Intention phrase and should not be used under any condition.

2. If an airport has a tower and it is temporarily closed, or operated on a part-time basis and there is no FSS on the airport or the FSS is closed, use the CTAF to self-announce your position or intentions.

3. Where there is no tower, FSS, or UNICOM station on the airport, use MULTICOM frequency 122.9 for self-announce procedures. Such airports will be identified in appropriate aeronautical information publications.

4. Practice Approaches. Pilots conducting practice instrument approaches should be particularly alert for other aircraft that may be departing in the opposite direction. When conducting any practice approach, regardless of its direction relative to other airport operations, pilots should make announcements on the CTAF as follows:

(a) Departing the final approach fix, inbound (nonprecision approach) or departing the outer marker or fix used in lieu of the outer marker, inbound (precision approach);

(b) Established on the final approach segment or immediately upon being released by ATC;

(c) Upon completion or termination of the approach; and

(d) Upon executing the missed approach procedure.

5. Departing aircraft should always be alert for arrival aircraft coming from the opposite direction.

6. Recommended self-announce phraseologies: It should be noted that aircraft operating to or from another nearby airport may be making self-announce broadcasts on the same UNICOM or MULTICOM frequency. To help identify one airport from another, the airport name should be spoken at the beginning and end of each self-announce transmission.

(a) Inbound

EXAMPLE-*Strawn traffic, Apache Two Two Five Zulu, (position), (altitude), (descending) or entering downwind/base/final (as appropriate) runway one seven full stop, touch-and-go, Strawn.*

Strawn traffic Apache Two Two Five Zulu clear of runway one seven Strawn.

(b) Outbound

EXAMPLE-*Strawn traffic, Queen Air Seven One Five Five Bravo (location on airport) taxiing to runway two six Strawn.*

Strawn traffic, Queen Air Seven One Five Five Bravo departing runway two six. Departing the pattern to the (direction), climbing to (altitude) Strawn.

(c) Practice Instrument Approach

EXAMPLE-*Strawn traffic, Cessna Two One Four Three Quebec (position from airport) inbound descending through (altitude) practice (name of approach) approach runway three five Strawn.*

Strawn traffic, Cessna Two One Four Three Quebec practice (type) approach completed or terminated runway three five Strawn.

h. UNICOM Communications Procedures

1. In communicating with a UNICOM station, the following practices will help reduce frequency congestion, facilitate a better understanding of pilot intentions, help identify the location of aircraft in the traffic pattern, and enhance safety of flight:

(a) Select the correct UNICOM frequency.

(b) State the identification of the UNICOM station you are calling in each transmission.

(c) Speak slowly and distinctly.

(d) Report approximately 10 miles from the airport, reporting altitude, and state your aircraft type, aircraft identification, location relative to the airport, state whether landing or overflight, and request wind information and runway in use.

(e) Report on downwind, base, and final approach.

(f) Report leaving the runway.

2. Recommended UNICOM phraseologies:

(a) Inbound

PHRASEOLOGY-*FREDERICK UNICOM CESSNA EIGHT ZERO ONE TANGO FOXTROT 10 MILES SOUTHEAST DESCENDING THROUGH (altitude) LANDING FREDERICK, REQUEST WIND AND RUNWAY INFORMATION FREDERICK.*

FREDERICK TRAFFIC CESSNA EIGHT ZERO ONE TANGO FOXTROT ENTERING DOWNWIND/BASE/ FINAL (as appropriate) FOR RUNWAY ONE NINER (full stop/touch-and-go) FREDERICK.

FREDERICK TRAFFIC CESSNA EIGHT ZERO ONE TANGO FOXTROT CLEAR OF RUNWAY ONE NINER FREDERICK.

(b) Outbound

PHRASEOLOGY-*FREDERICK UNICOM CESSNA EIGHT ZERO ONE TANGO FOXTROT (location on airport) TAXIING TO RUNWAY ONE NINER, REQUEST WIND AND TRAFFIC INFORMATION FREDERICK.*

FREDERICK TRAFFIC CESSNA EIGHT ZERO ONE TANGO FOXTROT DEPARTING RUNWAY ONE NINER. "REMAINING IN THE PATTERN" OR "DEPARTING THE PATTERN TO THE (direction) (as appropriate)" FREDERICK.

4-1-10. IFR Approaches/Ground Vehicle Operations

a. IFR Approaches. When operating in accordance with an IFR clearance and ATC approves a change to the advisory frequency, make an expeditious change to the CTAF and employ the recommended traffic advisory procedures.

b. Ground Vehicle Operation. Airport ground vehicles equipped with radios should monitor the CTAF frequency when operating on the airport movement area and remain clear of runways/taxiways being used by aircraft. Radio transmissions from ground vehicles should be confined to safety-related matters.

c. Radio Control of Airport Lighting Systems. Whenever possible, the CTAF will be used to control airport lighting systems at airports without operating control towers. This eliminates the need for pilots to change frequencies to turn the lights on and allows a continuous listening watch on a single frequency. The CTAF is published on the instrument approach chart and in other appropriate aeronautical information publications. For further details concerning radio controlled lights, see AC 150/5340-27, Air-to-Ground Radio Control of Airport Lighting Systems.

4-1-11. Designated UNICOM/MULTICOM Frequencies

Frequency use

a. The following listing depicts UNICOM and MULTICOM frequency uses as designated by the Federal Communications Commission (FCC).

(See TBL 4-1-2.)

Table 4-1-2
Unicom/Multicom Frequency Usage

Use	Frequency
Airports without an operating control tower.	122.700 122.725 122.800 122.975 123.000 123.050 123.075
(MULTICOM FREQUENCY) Activities of a temporary, seasonal, emergency nature or search and rescue, as well as, airports with no tower, FSS, or UNICOM.	122.900
(MULTICOM FREQUENCY) Forestry management and fire suppression, fish and game management and protection, and environmental monitoring and protection.	122.925
Airports with a control tower or FSS on airport.	122.950

NOTE-1. In some areas of the country, frequency interference may be encountered from nearby airports using the same UNICOM frequency. Where there is a problem, UNICOM operators are encouraged to develop a "least interference" frequency assignment plan for airports concerned using the frequencies designated for airports without operating control towers. UNICOM licensees are encouraged to apply for UNICOM 25 kHz spaced channel frequencies. Due to the extremely limited number of frequencies with 50 kHz channel spacing, 25 kHz channel spacing should be implemented. UNICOM licensees may then request FCC to assign frequencies in accordance with the plan, which FCC will review and consider for approval.

2. Wind direction and runway information may not be available on UNICOM frequency 122.950.

b. The following listing depicts other frequency uses as designated by the Federal Communications Commission (FCC). (See TBL 4-1-3.)

TABLE 4-1-3
Other Frequency Usage Designated by FCC

Use	Frequency
Air-to-air communication (private fixed wing aircraft).	122.750
Air-to-air communications (general aviation helicopters).	123.025
Aviation instruction, Glider, Hot Air Balloon **(not to be used for advisory service)**.	123.300 123.500

4-1-12. Use of UNICOM for ATC Purposes

UNICOM service may be used for ATC purposes, only under the following circumstances:

a. Revision to proposed departure time.

b. Takeoff, arrival, or flight plan cancellation time.

c. ATC clearance, provided arrangements are made between the ATC facility and the UNICOM licensee to handle such messages.

4-1-13. Automatic Terminal Information Service (ATIS)

a. ATIS is the continuous broadcast of recorded noncontrol information in selected high activity terminal areas. Its purpose is to improve controller effectiveness and to relieve frequency congestion by automating the repetitive transmission of essential but routine information. The information is continuously broadcast over a discrete VHF radio frequency or the voice portion of a local NAVAID. Arrival ATIS transmissions on a discrete VHF radio frequency are engineered according to the individual facility requirements, which would normally be a protected service volume of 20 NM to 60 NM from the ATIS site and a maximum altitude of 25,000 feet AGL. In the case of a departure ATIS, the protected service volume cannot exceed 5 NM and 100 feet AGL. At most locations, ATIS signals may be received on the surface of the airport, but local conditions may limit the maximum ATIS reception distance and/or altitude. Pilots are urged to cooperate in the ATIS program as it relieves frequency congestion on approach control, ground control, and local control frequencies. The A/FD indicates airports for which ATIS is provided.

b. ATIS information includes the time of the latest weather sequence, ceiling, visibility, obstructions to visibility, temperature, dew point (if available), wind direction (magnetic), and velocity, altimeter, other pertinent remarks, instrument approach and runway in use. The ceiling/sky condition, visibility, and obstructions to vision may be omitted from the ATIS broadcast if the ceiling is above 5,000 feet and the visibility is more than 5 miles. The departure runway will only be given if different from the landing runway except at locations having a separate ATIS for departure. The broadcast may include the appropriate frequency and instructions for VFR arrivals to make initial contact with approach control. Pilots of aircraft arriving or departing the terminal area can receive the continuous ATIS broadcast at times when cockpit duties are least pressing and listen to as many repeats as desired. ATIS broadcast must be updated upon the receipt of any official hourly and special weather. A new recording will also be made when there is a change in other pertinent data such as runway change, instrument approach in use, etc.

EXAMPLE-*Dulles International information Sierra. 1300 zulu weather. Measured ceiling three thousand overcast. Visibility three, smoke. Temperature six eight. Wind three five zero at eight. Altimeter two niner niner two. ILS runway one right approach in use. Landing runway one right and left. Departure runway three zero. Armel VORTAC out of service. Advise you have Sierra.*

c. Pilots should listen to ATIS broadcasts whenever ATIS is in operation.

d. Pilots should notify controllers on initial contact that they have received the ATIS broadcast by repeating the alphabetical code word appended to the broadcast.

EXAMPLE-*"Information Sierra received."*

e. When a pilot acknowledges receipt of the ATIS broadcast, controllers may omit those items contained in the broadcast if they are current. Rapidly changing conditions will be issued by ATC and the ATIS will contain words as follows:

EXAMPLE-*"Latest ceiling/visibility/altimeter/wind/ (other conditions) will be issued by approach control/ tower."*

NOTE-*The absence of a sky condition or ceiling and/ or visibility on ATIS indicates a sky condition or ceiling of 5,000 feet or above and visibility of 5 miles or more. A remark may be made on the broadcast, "the weather is better than 5000 and 5," or the existing weather may be broadcast.*

f. Controllers will issue pertinent information to pilots who do not acknowledge receipt of a broadcast or who acknowledge receipt of a broadcast which is not current.

g. To serve frequency limited aircraft, FSSs are equipped to transmit on the omnirange frequency at most en route VORs used as ATIS voice outlets. Such communication interrupts the ATIS broadcast. Pilots of aircraft equipped to receive on other FSS frequencies are encouraged to do so in order that these override transmissions may be kept to an absolute minimum.

h. While it is a good operating practice for pilots to make use of the ATIS broadcast where it is available, some pilots use the phrase "have numbers" in communications with the control tower. Use of this phrase means that the pilot has received wind, runway, and altimeter information ONLY and the tower does not have to repeat this information. It does not indicate receipt of the ATIS broadcast and should never be used for this purpose.

4-1-14. Automatic Flight Information Service (AFIS) - Alaska FSSs Only

a. Alaska FSSs AFIS is the continuous broadcast of recorded noncontrol information at airports in Alaska where a Flight Service Station (FSS) provides local airport advisory service. Its purpose is to improve FSS Specialist efficiency by reducing frequency congestion on the local airport advisory frequency. The AFIS broadcast will automate the repetitive transmission of essential but routine information (weather, favored runway, breaking action, airport NOTAMs, other applicable information). The information is continuously broadcast over a discrete VHF radio frequency (usually the ASOS frequency). Use of AFIS is not mandatory, but pilots who choose to utilize two-way radio communications with the FSS are urged to listen to AFIS, as it relieves frequency congestion on the local airport advisory frequency. AFIS broadcasts are updated upon the receipt of any official hourly and special weather, worsening braking action reports, and changes in other pertinent data. When a pilot acknowledges receipt of the AFIS broadcast, FSS Specialists may omit those items contained in the broadcast if they are current. When rapidly changing conditions exist, the latest ceiling, visibility, altimeter, wind or other conditions may be omitted from the AFIS and will be issued by the Flight Service Specialist on the appropriate radio frequency.

EXAMPLE-*"Kotzebue information ALPHA. One six five five zulu. Wind, two one zero at five; visibility two, fog; ceiling one hundred overcast; temperature minus one two, dew point minus one four; altimeter three one zero five. Altimeter in excess of three one zero zero, high pressure altimeter setting procedures are in effect. Favored runway two six. Weather in Kotzebue surface area is below V-F-R minima - an ATC clearance is required. Contact Kotzebue Radio on 123.6 for traffic advisories and advise intentions. Notice to Airmen, Hotham NDB out of service. Transcribed Weather Broadcast out of service. Advise on initial contact you have ALPHA."*

NOTE-*The absence of a sky condition or ceiling and/ or visibility on Alaska FSS AFIS indicates a sky condition or ceiling of 5,000 feet or above and visibility of 5 miles or more. A remark may be made on the broadcast, "the weather is better than 5000 and 5."*

b. Pilots should listen to Alaska FSSs AFIS broadcasts whenever Alaska FSSs AFIS is in operation.

NOTE-*Some Alaska FSSs are open part time and/or seasonally.*

c. Pilots should notify controllers on initial contact that they have received the Alaska FSSs AFIS broadcast by repeating the phonetic alphabetic letter appended to the broadcast.

EXAMPLE-*"Information Alpha received."*

d. While it is a good operating practice for pilots to make use of the Alaska FSS AFIS broadcast where it is available, some pilots use the phrase "have numbers" in communications with the FSS. Use of this phrase means that the pilot has received wind, runway, and altimeter information ONLY and the Alaska FSS does not have to repeat this information. It does not indicate receipt of the AFIS broadcast and should never be used for this purpose.

4-1-15. Radar Traffic Information Service

This is a service provided by radar ATC facilities. Pilots receiving this service are advised of any radar target observed on the radar display which may be in such proximity to the position of their aircraft or its intended route of flight that it warrants their attention. This service is not intended to relieve the pilot of the responsibility for continual vigilance to see and avoid other aircraft.

a. Purpose of the Service

1. The issuance of traffic information as observed on a radar display is based on the principle of assisting and advising a pilot that a particular radar target's position and track indicates it may intersect or pass in such proximity to that pilot's intended flight path that it warrants attention. This is to alert the pilot to the traffic, to be on the lookout for it, and thereby be in a better position to take appropriate action should the need arise.

2. Pilots are reminded that the surveillance radar used by ATC does not provide altitude information unless the aircraft is equipped with Mode C and the radar facility is capable of displaying altitude information.

b. Provisions of the Service

1. Many factors, such as limitations of the radar, volume of traffic, controller workload and communications frequency congestion, could prevent the controller from providing this service. Controllers possess complete discretion for determining whether they are able to provide or continue to provide this service in a specific case. The controller's reason against providing or continuing to provide the service in a particular case is not subject to question nor need it be communicated to the pilot. In other words, the provision of this service is entirely dependent upon whether controllers believe they are in a position to provide it. Traffic information is routinely provided to all aircraft operating on IFR flight plans except when the pilot declines the service, or the pilot is operating within Class A airspace. Traffic information may be provided to flights not operating on IFR flight plans when requested by pilots of such flights.

NOTE-*Radar ATC facilities normally display and monitor both primary and secondary radar when it is available, except that secondary radar may be used as the sole display source in Class A airspace, and under some circumstances outside of Class A airspace (beyond primary coverage and in en route areas where only secondary is available). Secondary radar may also be used outside Class A airspace as the sole display source when the primary radar is temporarily unusable or out of service. Pilots in contact with the affected ATC facility are normally advised when a temporary outage occurs; i.e., "primary radar out of service; traffic advisories available on transponder aircraft only." This means simply that only the aircraft which have transponders installed and in use will be depicted on ATC radar indicators when the primary radar is temporarily out of service.*

2. When receiving VFR radar advisory service, pilots should monitor the assigned frequency at all times. This is to preclude controllers' concern for radio failure or emergency assistance to aircraft under the controller's jurisdiction. VFR radar advisory service does not include vectors away from conflicting traffic unless requested by the pilot. When advisory service is no longer desired, advise the controller before changing frequencies and then change your transponder code to 1200, if applicable. Pilots should also inform the controller when changing VFR cruising altitude. Except in programs where radar service is automatically terminated, the controller will advise the aircraft when radar is terminated.

NOTE-*Participation by VFR pilots in formal programs implemented at certain terminal locations constitutes pilot request. This also applies to participating pilots at those locations where arriving VFR flights are encouraged to make their first contact with the tower on the approach control frequency.*

c. Issuance of Traffic Information. Traffic information will include the following concerning a target which may constitute traffic for an aircraft that is:

1. Radar identified

(a) Azimuth from the aircraft in terms of the 12 hour clock, or

(b) When rapidly maneuvering civil test or military aircraft prevent accurate issuance of traffic as in (a) above, specify the direction from an aircraft's position in terms of the eight cardinal compass points (N, NE, E, SE, S, SW, W, NW). This method must be terminated at the pilot's request.

(c) Distance from the aircraft in nautical miles;

(d) Direction in which the target is proceeding; and

(e) Type of aircraft and altitude if known.

EXAMPLE-*Traffic 10 o'clock, 3 miles, west-bound (type aircraft and altitude, if known, of the observed traffic). The altitude may be known, by means of Mode C, but not verified with the pilot for accuracy. (To be valid for separation purposes by ATC, the accuracy of Mode C readouts must be verified. This is usually accomplished upon initial entry into the radar system by a comparison of the readout to pilot stated altitude, or the field elevation in the case of continuous readout being received from an aircraft on the airport.) When necessary to issue traffic advisories containing unverified altitude information, the controller will issue the advisory in the same manner as if it were verified due to the accuracy of these readouts. The pilot may upon receipt of traffic information, request a vector (heading) to avoid such traffic. The vector will be provided to the extent possible as determined by the controller provided the aircraft to be vectored is within the airspace under the jurisdiction of the controller.*

2. Not radar identified

(a) Distance and direction with respect to a fix;

(b) Direction in which the target is proceeding; and

(c) Type of aircraft and altitude if known.

EXAMPLE-*Traffic 8 miles south of the airport northeastbound, (type aircraft and altitude if known).*

d. The examples depicted in the following figures point out the possible error in the position of this traffic when it is necessary for a pilot to apply drift correction to maintain this track. This error could also occur in the event a change in course is made at the time radar traffic information is issued.

FIG 4-1-1
Induced Error in Position of Traffic

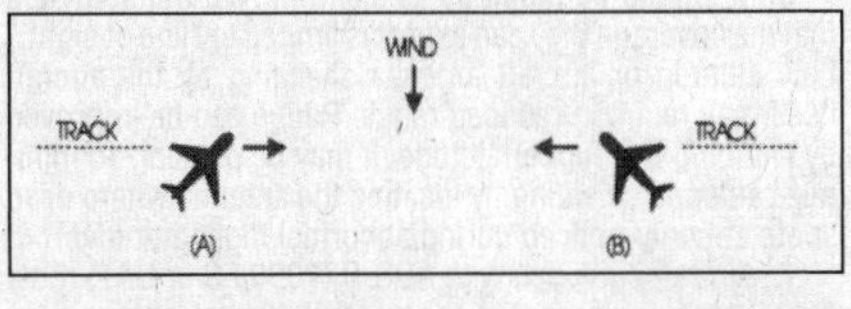

EXAMPLE-*In FIG 4-1-1 traffic information would be issued to the pilot of aircraft "A" as 12 o'clock. The actual position of the traffic as seen by the pilot of aircraft "A" would be 2 o'clock. Traffic information issued to aircraft "B" would also be given as 12 o'clock, but in this case, the pilot of "B" would see the traffic at 10 o'clock.*

FIG 4-1-2
Induced Error in Position of Traffic

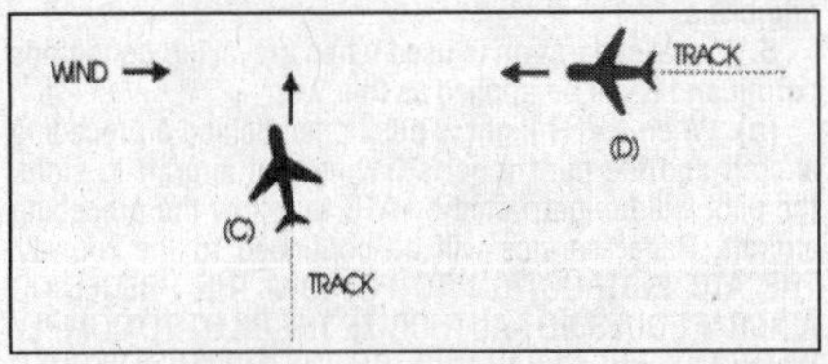

EXAMPLE-*In FIG 4-1-2 traffic information would be issued to the pilot of aircraft "C" as 2 o'clock. The actual position of the traffic as seen by the pilot of aircraft "C" would be 3 o'clock. Traffic information issued to aircraft "D" would be at an 11 o'clock position. Since it is not necessary for the pilot of aircraft "D" to apply wind correction (crab) to remain on track, the actual position of the traffic issued would be correct. Since the radar controller can only observe aircraft track (course) on the radar display, traffic advisories are issued accordingly, and pilots should give due consideration to this fact when looking for reported traffic.*

4-1-16. Safety Alert

A safety alert will be issued to pilots of aircraft being controlled by ATC if the controller is aware the aircraft is at an altitude which, in the controller's judgment, places the aircraft in unsafe proximity to terrain, obstructions or other aircraft. The provision of this service is contingent upon the capability of the controller to have an awareness of a situation involving unsafe proximity to terrain, obstructions and uncontrolled aircraft. The issuance of a safety alert cannot be mandated, but it can be expected on a reasonable, though intermittent basis. Once the alert is issued, it is solely the pilot's prerogative to determine what course of action, if any, to take. This procedure is intended for use in time critical situations where aircraft safety is in question. Noncritical situations should be handled via the normal traffic alert procedures.

a. Terrain or Obstruction Alert

1. Controllers will immediately issue an alert to the pilot of an aircraft under their control when they recognize that the aircraft is at an altitude which, in their judgment, may be in an unsafe proximity to terrain/obstructions. The primary method of detecting unsafe proximity is through Mode C automatic altitude reports.

EXAMPLE-*Low altitude alert, check your altitude immediately. The, as appropriate, MEA/MVA/MOCA in your area is (altitude) or, if past the final approach fix (nonprecision approach) or the outer marker or fix used in lieu of the outer marker (precision approach), the, as appropriate, MDA/DH (if known) is (altitude).*

2. Terminal Automated Radar Terminal System (ARTS) IIIA, Common ARTS (to include ARTS IIIE and ARTS IIE) (CARTS), Micro En Route Automated Radar Tracking System (MEARTS), and Standard Terminal Automation Replacement System (STARS) facilities have an automated function which, if operating, alerts controllers when a tracked Mode C equipped aircraft under their control is below or is predicted to be below a predetermined minimum safe altitude. This function, called Minimum Safe Altitude Warning (MSAW), is designed solely as a controller aid in detecting potentially unsafe aircraft proximity to terrain/obstructions. The ARTS IIIA, CARTS, MEARTS, and STARS facility will, when MSAW is operating, provide MSAW monitoring for all aircraft with an operating Mode C altitude encoding transponder that are tracked by the system and are:

(a) Operating on an IFR flight plan; or

(b) Operating VFR and have requested MSAW monitoring.

3. Terminal AN/TPX-42A (number beacon decoder system) facilities have an automated function called Low Altitude Alert System (LAAS). Although not as sophisticated as MSAW, LAAS alerts the controller when a Mode C transponder equipped aircraft operating on an IFR flight plan is below a predetermined minimum safe altitude.

NOTE-*Pilots operating VFR may request MSAW or LAAS monitoring if their aircraft are equipped with Mode C transponders.*

EXAMPLE-*Apache Three Three Papa request MSAW/LAAS.*

b. Aircraft Conflict Alert.

1. Controllers will immediately issue an alert to the pilot of an aircraft under their control if they are aware of another aircraft which is not under their control, at an altitude which, in the controller's judgment, places both aircraft in unsafe proximity to each other. With the alert, when feasible, the controller will offer the pilot the position of the traffic if time permits and an alternate course(s) of action. Any alternate course(s) of action the controller may recommend to the pilot will be predicated only on other traffic being worked by the controller.

EXAMPLE-*American Three, traffic alert, (position of traffic, if time permits), advise you turn right/left heading (degrees) and/or climb/descend to (altitude) immediately.*

4-1-17. Radar Assistance to VFR Aircraft

a. Radar equipped FAA ATC facilities provide radar assistance and navigation service (vectors) to VFR aircraft provided the aircraft can communicate with the facility, are within radar coverage, and can be radar identified.

b. Pilots should clearly understand that authorization to proceed in accordance with such radar navigational assistance does not constitute authorization for the pilot to violate CFRs. In effect, assistance provided is on the basis that navigational guidance information issued is advisory in nature and the job of flying the aircraft safely, remains with the pilot.

c. In many cases, controllers will be unable to determine if flight into instrument conditions will result from their instructions. To avoid possible hazards resulting from being vectored into IFR conditions, pilots should keep controllers advised of the weather conditions in which they are operating and along the course ahead.

d. Radar navigation assistance (vectors) may be initiated by the controller when one of the following conditions exist:

1. The controller suggests the vector and the pilot concurs.

2. A special program has been established and vectoring service has been advertised.

3. In the controller's judgment the vector is necessary for air safety.

e. Radar navigation assistance (vectors) and other radar derived information may be provided in response to pilot requests. Many factors, such as limitations of radar, volume of traffic, communications frequency, congestion, and controller workload could prevent the controller from providing it. Controllers have complete discretion for determining if they are able to provide the service in a particular case. Their decision not to provide the service in a particular case is not subject to question.

4-1-18. Terminal Radar Services for VFR Aircraft

a. Basic Radar Service:

1. In addition to the use of radar for the control of IFR aircraft, all commissioned radar facilities provide the following basic radar services for VFR aircraft:

(a) Safety alerts.

(b) Traffic advisories.

(c) Limited radar vectoring (on a workload permitting basis).

(d) Sequencing at locations where procedures have been established for this purpose and/or when covered by a Letter of Agreement.

NOTE-*When the stage services were developed, two basic radar services (traffic advisories and limited vectoring) were identified as "Stage I." This definition became unnecessary and the term "Stage I" was eliminated from use. The term "Stage II" has been eliminated in conjunction with the airspace reclassification, and sequencing services to locations with local procedures and/or letters of agreement to provide this service have been included in basic services to VFR aircraft. These basic services will still be provided by all terminal radar facilities whether they include Class B, Class C, Class D or Class E airspace. "Stage III" services have been replaced with "Class B" and "TRSA" service where applicable.*

2. Vectoring service may be provided when requested by the pilot or with pilot concurrence when suggested by ATC.

3. Pilots of arriving aircraft should contact approach control on the publicized frequency and give their position, altitude, aircraft call sign, type aircraft, radar beacon code (if transponder equipped), destination, and request traffic information.

4. Approach control will issue wind and runway, except when the pilot states "have numbers" or this information is contained in the ATIS broadcast and the pilot states that the current ATIS information has been received. Traffic information is provided on a workload permitting basis. Approach control will specify the time or place at which the pilot is to contact the tower on local control frequency for further landing information. Radar service is automatically terminated and the aircraft need not be advised of termination when an arriving VFR aircraft receiving radar services to a tower-controlled airport where basic radar service is provided has landed, or to all other airports, is instructed to change to tower or advisory frequency. (See FAA Order JO 7110.65, Air Traffic Control, paragraph 5-1-13, Radar Service Termination.)

5. Sequencing for VFR aircraft is available at certain terminal locations (see locations listed in the Airport/Facility Directory). The purpose of the service is to adjust the flow of arriving VFR and IFR aircraft into the traffic pattern in a safe and orderly manner and to provide radar traffic information to departing VFR aircraft. Pilot participation is urged but is not mandatory. Traffic information is provided on a workload permitting basis. Standard radar separation between VFR or between VFR and IFR aircraft is not provided.

(a) Pilots of arriving VFR aircraft should initiate radio contact on the publicized frequency with approach control when approximately 25 miles from the airport at which sequencing services are being provided. On initial contact by VFR aircraft, approach control will assume that sequencing service is requested. After radar contact is established, the pilot may use pilot navigation to enter the traffic pattern or, depending on traffic conditions, approach control may provide the pilot with routings or vectors necessary for proper sequencing with other participating VFR and IFR traffic en route to the airport. When a flight is positioned behind a preceding aircraft and the pilot reports having that aircraft in sight, the pilot will be instructed to follow the preceding aircraft. THE ATC INSTRUCTION TO FOLLOW THE PRECEDING AIRCRAFT DOES NOT AUTHORIZE THE PILOT TO COMPLY WITH ANY ATC CLEARANCE OR INSTRUCTION ISSUED TO THE PRECEDING AIRCRAFT. If other "nonparticipating" or "local" aircraft are in the traffic pattern, the tower will issue a landing sequence. If an arriving aircraft does not want radar service, the pilot should state "NEGATIVE RADAR SERVICE" or make a similar comment, on initial contact with approach control.

(b) Pilots of departing VFR aircraft are encouraged to request radar traffic information by notifying ground control on initial contact with their request and proposed direction of flight.

EXAMPLE-*Xray ground control, November One Eight Six, Cessna One Seventy Two, ready to taxi, VFR southbound at 2,500, have information bravo and request radar traffic information.*

NOTE-*Following takeoff, the tower will advise when to contact departure control.*

(c) Pilots of aircraft transiting the area and in radar contact/communication with approach control will receive traffic information on a controller workload permitting basis. Pilots of such aircraft should give their position, altitude, aircraft call sign, aircraft type, radar beacon code (if transponder equipped), destination, and/or route of flight.

b. TRSA Service (Radar Sequencing and Separation Service for VFR Aircraft in a TRSA).

1. This service has been implemented at certain terminal locations. The service is advertised in the Airport/Facility Directory. The purpose of this service is to provide separation between all participating VFR aircraft and all IFR aircraft operating within the airspace defined as the Terminal Radar Service Area (TRSA). Pilot participation is urged but is not mandatory.

2. If any aircraft does not want the service, the pilot should state "NEGATIVE TRSA SERVICE" or make a similar comment, on initial contact with approach control or ground control, as appropriate.

3. TRSAs are depicted on sectional aeronautical charts and listed in the Airport/Facility Directory.

4. While operating within a TRSA, pilots are provided TRSA service and separation as prescribed in this paragraph. In the event of a radar outage, separation and sequencing of VFR aircraft will be suspended as this service is dependent on radar. The pilot will be advised that the service is not available and issued wind, runway information, and the time or place to contact the tower. Traffic information will be provided on a workload permitting basis.

5. Visual separation is used when prevailing conditions permit and it will be applied as follows:

(a) When a VFR flight is positioned behind a preceding aircraft and the pilot reports having that aircraft in sight, the pilot will be instructed by ATC to follow the preceding aircraft. Radar service will be continued to the runway. THE ATC INSTRUCTION TO FOLLOW THE PRECEDING AIRCRAFT DOES NOT AUTHORIZE THE PILOT TO COMPLY WITH ANY ATC CLEARANCE OR INSTRUCTION ISSUED TO THE PRECEDING AIRCRAFT.

(b) If other "nonparticipating" or "local" aircraft are in the traffic pattern, the tower will issue a landing sequence.

(c) Departing VFR aircraft may be asked if they can visually follow a preceding departure out of the TRSA. The pilot will be instructed to follow the other aircraft provided that the pilot can maintain visual contact with that aircraft.

6. VFR aircraft will be separated from VFR/IFR aircraft by one of the following:

(a) 500 feet vertical separation.

(b) Visual separation.

(c) Target resolution (a process to ensure that correlated radar targets do not touch).

7. Participating pilots operating VFR in a TRSA:

(a) Must maintain an altitude when assigned by ATC unless the altitude assignment is to maintain at or below a specified altitude. ATC may assign altitudes for separation that do not conform to 14 CFR Section 91.159. When the altitude assignment is no longer needed for separation or when leaving the TRSA, the instruction will be broadcast, "RESUME APPROPRIATE VFR ALTITUDES." Pilots must then return to an altitude that conforms to 14 CFR Section 91.159 as soon as practicable.

(b) When not assigned an altitude, the pilot should coordinate with ATC prior to any altitude change.

8. Within the TRSA, traffic information on observed but unidentified targets will, to the extent possible, be provided to all IFR and participating VFR aircraft. The pilot will be vectored upon request to avoid the observed traffic, provided the aircraft to be vectored is within the airspace under the jurisdiction of the controller.

9. Departing aircraft should inform ATC of their intended destination and/or route of flight and proposed cruising altitude.

10. ATC will normally advise participating VFR aircraft when leaving the geographical limits of the TRSA. Radar service is not automatically terminated with this advisory unless specifically stated by the controller.

c. Class C Service. This service provides, in addition to basic radar service, approved separation between IFR and VFR aircraft, and sequencing of VFR arrivals to the primary airport.

d. Class B Service. This service provides, in addition to basic radar service, approved separation of aircraft based on IFR, VFR, and/or weight, and sequencing of VFR arrivals to the primary airport(s).

e. PILOT RESPONSIBILITY. THESE SERVICES ARE NOT TO BE INTERPRETED AS RELIEVING PILOTS OF THEIR RESPONSIBILITIES TO SEE AND AVOID OTHER TRAFFIC OPERATING IN BASIC VFR WEATHER CONDITIONS, TO ADJUST THEIR OPERATIONS AND FLIGHT PATH AS NECESSARY TO PRECLUDE SERIOUS WAKE ENCOUNTERS, TO MAINTAIN APPROPRIATE TERRAIN AND OBSTRUCTION CLEARANCE, OR TO REMAIN IN WEATHER CONDITIONS EQUAL TO OR BETTER THAN THE MINIMUMS REQUIRED BY 14 CFR SECTION 91.155. WHENEVER COMPLIANCE WITH AN ASSIGNED ROUTE, HEADING AND/OR ALTITUDE IS LIKELY TO COMPROMISE PILOT RESPONSIBILITY RESPECTING TERRAIN AND OBSTRUCTION CLEARANCE, VORTEX EXPOSURE, AND WEATHER MINIMUMS, APPROACH CONTROL SHOULD BE SO ADVISED AND A REVISED CLEARANCE OR INSTRUCTION OBTAINED.

f. ATC services for VFR aircraft participating in terminal radar services are dependent on ATC radar. Services for VFR aircraft are not available during periods of a radar outage and are limited during CENRAP operations. The pilot will be advised when VFR services are limited or not available.

NOTE-*Class B and Class C airspace are areas of regulated airspace. The absence of ATC radar does not negate the requirement of an ATC clearance to enter Class B airspace or two way radio contact with ATC to enter Class C airspace.*

4-1-19. Tower En Route Control (TEC)

a. TEC is an ATC program to provide a service to aircraft proceeding to and from metropolitan areas. It links designated Approach Control Areas by a network of identified routes made up of the existing airway structure of the National Airspace System. The FAA initiated an expanded TEC program to include as many facilities as possible. The program's intent is to provide an overflow resource in the low altitude system which would enhance ATC services. A few facilities have historically allowed turbojets to proceed between certain city pairs, such as Milwaukee and Chicago, via tower en route and these locations may continue this service. However, the expanded TEC program will be applied, generally, for nonturbojet aircraft operating at and below 10,000 feet. The program is entirely within the approach control airspace of multiple terminal facilities. Essentially, it is for relatively short flights. Participating pilots are encouraged to use TEC for flights of two hours duration or less. If longer flights are planned, extensive coordination may be required within the multiple complex which could result in unanticipated delays.

b. Pilots requesting TEC are subject to the same delay factor at the destination airport as other aircraft in the ATC system. In addition, departure and en route delays may occur depending upon individual facility workload. When a major metropolitan airport is incurring significant delays, pilots in the TEC program may want to consider an alternative airport experiencing no delay.

c. There are no unique requirements upon pilots to use the TEC program. Normal flight plan filing procedures will ensure proper flight plan processing. Pilots should include the acronym "TEC" in the remarks section of the flight plan when requesting tower en route control.

d. All approach controls in the system may not operate up to the maximum TEC altitude of 10,000 feet. IFR flight may be planned to any satellite airport in proximity to the major primary airport via the same routing.

4-1-20. Transponder Operation

a. General

1. Pilots should be aware that proper application of transponder operating procedures will provide both VFR and IFR aircraft with a higher degree of safety in the environment where high-speed closure rates are possible. Transponders substantially increase the capability of radar to see an aircraft and the Mode C feature enables the controller to quickly determine where potential traffic conflicts may exist. Even VFR pilots who are not in contact with ATC will be afforded greater protection from IFR aircraft and VFR aircraft which are receiving traffic advisories. Nevertheless, pilots should never relax their visual scanning vigilance for other aircraft.

2. Air Traffic Control Radar Beacon System (ATCRBS) is similar to and compatible with military coded radar beacon equipment. Civil Mode A is identical to military Mode 3.

3. Civil and military transponders should be turned to the "on" or normal altitude reporting position prior to moving on the airport surface to ensure the aircraft is visible to ATC surveillance systems. IN ALL CASES, WHILE IN CONTROLLED AIRSPACE EACH PILOT OPERATING AN AIRCRAFT EQUIPPED WITH AN OPERABLE ATC TRANSPONDER MAINTAINED IN ACCORDANCE WITH 14 CFR SECTION 91.413 MUST OPERATE THE TRANSPONDER, INCLUDING MODE C IF INSTALLED, ON THE APPROPRIATE CODE OR AS ASSIGNED BY ATC. IN CLASS G AIRSPACE, THE TRANSPONDER SHOULD BE OPERATING WHILE AIRBORNE UNLESS OTHERWISE REQUESTED BY ATC.

4. A pilot on an IFR flight who elects to cancel the IFR flight plan prior to reaching destination, should adjust the transponder according to VFR operations.

5. If entering a U.S. OFFSHORE AIRSPACE AREA from outside the U.S., the pilot should advise on first radio contact with a U.S. radar ATC facility that such equipment is available by adding "transponder" to the aircraft identification.

6. It should be noted by all users of ATC transponders that the coverage they can expect is limited to "line of sight." Low altitude or aircraft antenna shielding by the aircraft itself may result in reduced range. Range can be improved by climbing to a higher altitude. It may be possible to minimize antenna shielding by locating the antenna where dead spots are only noticed during abnormal flight attitudes.

7. Aircraft equipped with ADS-B (1090 ES or UAT) must operate the equipment in the transmit mode (on position) at all times while on any airport surface.

NOTE-*Pilots of aircraft equipped with ADS-B should refer to AIM, Automatic Dependant Surveillance - Broadcast Services, Paragraph 4-5-7, for a complete description of operating limitations and procedures.*

b. Transponder Code Designation

1. For ATC to utilize one or a combination of the 4096 discrete codes FOUR DIGIT CODE DESIGNATION will be used, e.g., code 2100 will be expressed as TWO ONE ZERO ZERO. Due to the operational characteristics of the rapidly expanding automated ATC system, THE LAST TWO DIGITS OF THE SELECTED TRANSPONDER CODE SHOULD ALWAYS READ "00" UNLESS SPECIFICALLY REQUESTED BY ATC TO BE OTHERWISE.

c. Automatic Altitude Reporting (Mode C)

1. Some transponders are equipped with a Mode C automatic altitude reporting capability. This system converts aircraft altitude in 100 foot increments to coded digital information which is transmitted together with Mode C framing pulses to the interrogating radar facility. The manner in which transponder panels are designed differs, therefore, a pilot should be thoroughly familiar with the operation of the transponder so that ATC may realize its full capabilities.

2. Adjust transponder to reply on the Mode A/3 code specified by ATC and, if equipped, to reply on Mode C with altitude reporting capability activated unless deactivation is directed by ATC or unless the installed aircraft equipment has not been tested and calibrated as required by 14 CFR Section 91.217. If deactivation is required by ATC, turn off the altitude reporting feature of your transponder. An instruction by ATC to "STOP ALTITUDE SQUAWK, ALTITUDE DIFFERS (number of feet) FEET," may be an indication that your transponder is transmitting incorrect altitude information or that you have an incorrect altimeter setting. While an incorrect altimeter setting has no effect on the Mode C altitude information transmitted by your transponder (transponders are preset at 29.92), it would cause you to fly at an actual altitude different from your assigned altitude. When a controller indicates that an altitude readout is invalid, the pilot should initiate a check to verify that the aircraft altimeter is set correctly.

3. Pilots of aircraft with operating Mode C altitude reporting transponders should report exact altitude or flight level to the nearest hundred foot increment when establishing initial contact with an ATC facility. Exact altitude or flight level reports on initial contact provide ATC with information that is required prior to using Mode C altitude information for separation purposes. This will significantly reduce altitude verification requests.

d. Transponder IDENT Feature

1. The transponder must be operated only as specified by ATC. Activate the "IDENT" feature only upon request of the ATC controller.

e. Code Changes

1. When making routine code changes, pilots should avoid inadvertent selection of Codes 7500, 7600 or 7700 thereby causing momentary false alarms at automated ground facilities. For example, when switching from Code 2700 to Code 7200, switch first to 2200 then to 7200, NOT to 7700 and then 7200. This procedure applies to nondiscrete Code 7500 and all discrete codes in the 7600 and 7700 series (i.e., 7600-7677, 7700-7777) which will trigger special indicators in automated facilities. Only nondiscrete Code 7500 will be decoded as the hijack code.

2. Under no circumstances should a pilot of a civil aircraft operate the transponder on Code 7777. This code is reserved for military interceptor operations.

3. Military pilots operating VFR or IFR within restricted/warning areas should adjust their transponders to Code 4000 unless another code has been assigned by ATC.

f. Mode C Transponder Requirements

1. Specific details concerning requirements to carry and operate Mode C transponders, as well as exceptions and ATC authorized deviations from the requirements are found in 14 CFR Section 91.215 and 14 CFR Section 99.12.

2. In general, the CFRs require aircraft to be equipped with Mode C transponders when operating:

(a) At or above 10,000 feet MSL over the 48 contiguous states or the District of Columbia, excluding that airspace below 2,500 feet AGL;

(b) Within 30 miles of a Class B airspace primary airport, below 10,000 feet MSL. Balloons, gliders, and aircraft not equipped with an engine driven electrical system are excepted from the above requirements when operating below the floor of Class A airspace and/or; outside of a Class B airspace and below the ceiling of the Class B airspace (or 10,000 feet MSL, whichever is lower);

(c) Within and above all Class C airspace, up to 10,000 feet MSL;

(d) Within 10 miles of certain designated airports, excluding that airspace which is both outside the Class D surface area and below 1,200 feet AGL. Balloons, gliders and aircraft not equipped with an engine driven electrical system are excepted from this requirement.

3. 14 CFR Section 99.12 requires all aircraft flying into, within, or across the contiguous U.S. ADIZ be equipped with a Mode C or Mode S transponder. Balloons, gliders and aircraft not equipped with an engine driven electrical system are excepted from this requirement.

4. Pilots must ensure that their aircraft transponder is operating on an appropriate ATC assigned VFR/IFR code and Mode C when operating in such airspace. If in doubt about the operational status of either feature of your transponder while airborne, contact the nearest ATC facility or FSS and they will advise you what facility you should contact for determining the status of your equipment.

5. In-flight requests for "immediate" deviation from the transponder requirement may be approved by controllers only when the flight will continue IFR or when weather conditions prevent VFR descent and continued VFR flight in airspace not affected by the CFRs. All other requests for deviation should be made by contacting the nearest Flight Service or Air Traffic facility in person or by telephone. The nearest ARTCC will normally be the controlling agency and is responsible for coordinating requests involving deviations in other ARTCC areas.

g. Transponder Operation Under Visual Flight Rules (VFR)

1. Unless otherwise instructed by an ATC facility, adjust transponder to reply on Mode 3/A Code 1200 regardless of altitude.

NOTE-

1. Aircraft not in contact with an ATC facility may squawk 1255 in lieu of 1200 while en route to, from, or within the designated fire fighting area(s).

2. VFR aircraft which fly authorized SAR missions for the USAF or USCG may be advised to squawk 1277 in lieu of 1200 while en route to, from, or within the designated search area.

3. Gliders not in contact with an ATC facility should squawk 1202 in lieu of 1200.

REFERENCE-*FAAO 7110.66, National Beacon Code Allocation Plan.*

2. Adjust transponder to reply on Mode C, with altitude reporting *capability activated* if the aircraft is so equipped, unless deactivation is directed by ATC or unless the installed equipment has not been tested and calibrated as required by 14 CFR Section 91.217. If deactivation is required and your transponder is so designed, turn off the altitude reporting switch and continue to transmit Mode C framing pulses. If this capability does not exist, turn off Mode C.

h. Radar Beacon Phraseology

Air traffic controllers, both civil and military, will use the following phraseology when referring to operation of the Air Traffic Control Radar Beacon System (ATCRBS). Instructions by ATC refer only to Mode A/3 or Mode C operation and do not affect the operation of the transponder on other Modes.

1. SQUAWK (number). Operate radar beacon transponder on designated code in Mode A/3.

2. IDENT. Engage the "IDENT" feature (military I/P) of the transponder.

3. SQUAWK (number) and IDENT. Operate transponder on specified code in Mode A/3 and engage the "IDENT" (military I/P) feature.

4. SQUAWK STANDBY. Switch transponder to standby position.

5. SQUAWK LOW/NORMAL. Operate transponder on low or normal sensitivity as specified. Transponder is operated in "NORMAL" position unless ATC specifies "LOW" ("ON" is used instead of "NORMAL" as a master control label on some types of transponders.)

6. SQUAWK ALTITUDE. Activate Mode C with automatic altitude reporting.

7. STOP ALTITUDE SQUAWK. Turn off altitude reporting switch and continue transmitting Mode C framing pulses. If your equipment does not have this capability, turn off Mode C.

8. STOP SQUAWK (mode in use). Switch off specified mode. (Used for military aircraft when the controller is unaware of military service requirements for the aircraft to continue operation on another Mode.)

9. STOP SQUAWK. Switch off transponder.

10. SQUAWK MAYDAY. Operate transponder in the emergency position (Mode A Code 7700 for civil transponder. Mode 3 Code 7700 and emergency feature for military transponder.)

11. SQUAWK VFR. Operate radar beacon transponder on Code 1200 in the Mode A/3, or other appropriate VFR code.

4-1-21. Hazardous Area Reporting Service

a. Selected FSSs provide flight monitoring where regularly traveled VFR routes cross large bodies of water, swamps, and mountains. This service is provided for the purpose of expeditiously alerting Search and Rescue facilities when required.

(See FIG 4-1-3.)

FIG 4-1-3

Hazardous Area Reporting Service

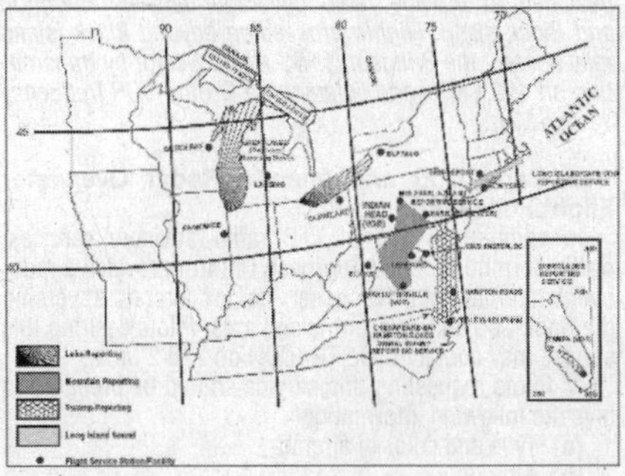

1. When requesting the service either in person, by telephone or by radio, pilots should be prepared to give the following information: type of aircraft, altitude, indicated airspeed, present position, route of flight, heading.

2. Radio contacts are desired at least every 10 minutes. If contact is lost for more than 15 minutes, Search and Rescue will be alerted. Pilots are responsible for canceling their request for service when they are outside the service area boundary. Pilots experiencing two-way radio failure are expected to land as soon as practicable and cancel their request for the service. FIG 4-1-3 depicts the areas and the FSS facilities involved in this program.

b. Long Island Sound Reporting Service.

New York and Bridgeport FSS Radio Sectors provide Long Island Sound Reporting service on request for aircraft traversing Long Island Sound.

1. When requesting the service, pilots should ask for SOUND REPORTING SERVICE and should be prepared to provide the following appropriate information:

(a) Type and color of aircraft;

(b) The specific route and altitude across the sound including the shore crossing point;

(c) The overwater crossing time;

(d) Number of persons on board; and

(e) True air speed.

2. Radio contacts are desired at least every 10 minutes; however, for flights of shorter duration a midsound report is requested. If contact is lost for more than 15 minutes Search and Rescue will be alerted. Pilots are responsible for canceling their request for the Long Island Sound Reporting Service when outside the service area boundary. Aircraft experiencing radio failure will be expected to land as soon as practicable and cancel their request for the service.

3. Communications. Primary communications - pilots are to transmit on 122.1 MHz and listen on one of the following VOR frequencies:

(a) New York FSS Radio Sector Controls:

(1) Hampton RCO (FSS transmits and receives on 122.6 MHz).

(2) Calverton VOR (FSS transmits on 117.2 and receives on standard FSS frequencies).

(3) Kennedy VORTAC (FSS transmits on 115.9 and receives on 122.1 MHz).

(b) Bridgeport FSS Radio Sector Controls:

(1) Madison VORTAC (FSS transmits on 110.4 and receives on 122.1 MHz).

(2) Groton VOR (FSS transmits on 110.85 and receives on 122.1 MHz).

(3) Bridgeport VOR (FSS transmits on 108.8 and receives on 122.1 MHz).

c. Block Island Reporting Service.

Within the Long Island Sound Reporting Service, the New York FSS Radio Sector also provides an additional service for aircraft operating between Montauk Point and Block Island. When requesting this service, pilots should ask for BLOCK ISLAND REPORTING SERVICE and should be prepared to provide the same flight information as required for the Long Island Sound Reporting Service.

1. A minimum of three position reports are mandatory for this service; these are:

(a) Reporting leaving either Montauk Point or Block Island.

(b) Midway report.

(c) Report when over either Montauk Point or Block Island. At this time, the overwater service is canceled.

2. Communications. Pilots are to transmit and receive on 122.6 MHz.

NOTE- *Pilots are advised that 122.6 MHz is a remote receiver located at the Hampton VORTAC site and designed to provide radio coverage between Hampton and Block Island. Flights proceeding beyond Block Island may contact the Bridgeport FSS Radio Sector by transmitting on 122.1 MHz and listening on Groton VOR frequency 110.85 MHz.*

d. Cape Cod and Islands Radar Overwater Flight Following.

In addition to normal VFR radar advisory services, traffic permitting, Cape Approach Control provides a radar overwater flight following service for aircraft traversing the Cape Cod and adjacent Island area. Pilots desiring this service may contact Cape TRACON on 118.2 MHz.

1. Pilots requesting this service should be prepared to give the following information:

(a) Type and color of aircraft;

(b) Altitude;

(c) Position and heading;

(d) Route of flight; and

(e) True airspeed.

2. For best radar coverage, pilots are encouraged to fly at 1,500 feet MSL or above.

3. Pilots are responsible for canceling their request for overwater flight following when they are over the mainland and/or outside the service area boundary.

e. Lake Reporting Service.

Cleveland and Lansing FSS Radio Sectors provide Lake Reporting Service on request for aircraft traversing the western half of Lake Erie; Green Bay, Kankakee, Lansing, and Terre Haute FSS Radio Sectors provide Lake Reporting Service on request for aircraft traversing Lake Michigan.

1. When requesting the service, pilots should ask for LAKE REPORTING SERVICE.

2. Pilots not on a VFR flight plan should be prepared to provide all information that is normally provided for a complete VFR flight plan.

3. Pilots already on a VFR flight plan should be prepared to provide the following information:

(a) Aircraft or flight identification.

(b) Type of aircraft.

(c) Near-shore crossing point or last fix before crossing.

(d) Proposed time over near-shore crossing point or last fix before crossing.

(e) Proposed altitude.

(f) Proposed route of flight.

(g) Estimated time over water.

(h) Next landing point.

(i) FSS having complete VFR flight plan information.

4. Radio contacts must not exceed 10 minutes when pilots fly at an altitude that affords continuous communications. If radio contact is lost for more than 15 minutes (5 minutes after a scheduled reporting time), Search and Rescue (SAR) will be alerted.

5. The estimated time for crossing the far shore will be the scheduled reporting time for aircraft that fly at an altitude that does not afford continuous communication coverage while crossing the lake. If radio contact is not established within 5 minutes of that time, SAR will be alerted.

6. Pilots are responsible for canceling their request for Lake Reporting Service when outside the service area boundary. Aircraft experiencing radio failure will be expected to land as soon as practicable and cancel their Lake Reporting Service flight plan.

7. Communications. Primary communications - Pilots should communicate with the following facilities on the indicated frequencies:

(a) Cleveland FSS Radio Sector Controls:

(1) Cleveland RCO (FSS transmits and receives on 122.35 or 122.55 MHz).

(2) Sandusky VOR (FSS transmits on 109.2 and receives on 122.1 MHz).

(b) Green Bay FSS Radio Sector Controls:

(1) Escanaba VORTAC (FSS transmits on 110.8 and receives on 122.1 MHz).

(2) Green Bay RCO (FSS transmits and receives on 122.55 MHz).

(3) Manistique RCO (FSS transmits and receives on 122.25 MHz).

(4) Manitowoc VOR (FSS transmits on 111.0 and receives on 122.1 MHz).

(5) Menominee VOR (FSS transmits on 109.6 and receives on 122.1 MHz).

(6) Milwaukee RCO (FSS transmits and receives on 122.65 MHz).

(7) Falls VOR (FSS transmits on 110.0 and receives on 122.1 MHz).

(c) Kankakee FSS Radio Sector Controls:

(1) Chicago Heights VORTAC (FSS transmits on 114.2 and receives on 122.1 MHz).

(2) Meigs RCO (FSS transmits and receives on 122.15 MHz).

(3) Waukegan RCO (FSS transmits and receives on 122.55 MHz).

(d) Lansing FSS Radio Sector Controls:

(1) Lake Erie. Detroit City RCO (FSS transmits and receives on 122.55 MHz).

(2) Lake Michigan:

[a] Keeler VORTAC (FSS transmits on 116.6 and receives on 122.1 MHz).

[b] Ludington RCO (FSS transmits and receives on 122.45 MHz).

[c] Manistee VORTAC (FSS transmits on 111.4 and receives on 122.1 MHz).

[d] Muskegon RCO (FSS transmits and receives on 122.5 MHz).

[e] Pellston RCO (FSS transmits and receives on 122.3 MHz).

[f] Pullman VORTAC (FSS transmits on 112.1 and receives on 122.1 MHz).

[g] Traverse City RCO (FSS transmits and receives on 122.65 MHz).

(e) Terre Haute FSS Radio Sector Controls. South Bend RCO (FSS transmits and receives on 122.6 MHz).

f. Everglades Reporting Service.

This service is offered by Miami Automated International Flight Service Station (MIA AIFSS), in extreme southern Florida. The service is provided to aircraft crossing the Florida Everglades, between Lee County (Ft. Myers, FL) VORTAC (RSW) on the northwest side, and Dolphin (Miami, FL) VOR (DHP) on the southeast side.

1. The pilot must request the service from Miami AIFSS.

2. MIA AIFSS frequency information, 122.2, 122.3, and 122.65.

3. The pilot must file a VFR flight plan with the remark: ERS.

4. The pilot must maintain 2000 feet of altitude.

5. The pilot must make position reports every ten (10) minutes. SAR begins fifteen (15) minutes after position report is not made on time.

6. The pilot is expected to land as soon as is practical, in the event of two-way radio failure, and advise MIA AIFSS that the service is terminated.

7. The pilot must notify Miami AIFSS when the flight plan is cancelled or the service is suspended.

4-1-22. Airport Reservation Operations and Special Traffic Management Programs

This section describes procedures for obtaining required airport reservations at airports designated by the FAA and for airports operating under Special Traffic Management Programs.

a. Slot Controlled Airports.

1. The FAA may adopt rules to require advance operations for unscheduled operations at certain airports. In addition to the information in the rules adopted by the FAA, a listing of the airports and relevant information will be maintained on the FAA Web site listed below.

2. The FAA has established an Airport Reservation Office (ARO) to receive and process reservations for unscheduled flights at the slot controlled airports. The ARO uses the Enhanced Computer Voice Reservation System (e-CVRS) to allocate reservations. Reservations will be available beginning 72 hours in advance of the operation at the slot controlled airport. Refer to the Web site or touch-tone phone interface for the current listing of slot controlled airports, limitations, and reservation procedures.

NOTE- *The web interface/telephone numbers to obtain a reservation for unscheduled operations at a slot controlled airport are:*

1. http://www.fly.faa.gov/ecvrs.

2. Touch-tone: 1-800-875-9694 or 703-707-0568. (e-CVRS interface).

3. Trouble number: 703-904-4452.

3. For more detailed information on operations and reservation procedures at a Slot Controlled Airport, please see Advisory Circular 93-1A, Reservations for Unscheduled Operations at slot controlled airports. A copy of the Advisory Circular may be obtained via the Internet at: http://www.faa.gov.

b. Special Traffic Management Programs (STMP).

1. Special procedures may be established when a location requires special traffic handling to accommodate above normal traffic demand (e.g., the Indianapolis 500, Super Bowl) or reduced airport capacity (e.g., airport runway/taxiway closures for airport construction). The special procedures may remain in effect until the problem has been resolved or until local traffic management procedures can handle the situation and a need for special handling no longer exists.

2. There will be two methods available for obtaining slot reservations through the ATCSCC: the web interface and the touch-tone interface. If these methods are used, a NOTAM will be issued relaying the web site address and toll free telephone number. Be sure to check current NOTAMs to determine: what airports are included in the STMP; the dates and times reservations are required; the time limits for reservation requests; the point of contact for reservations; and any other instructions.

c. Users may contact the ARO at 703-904-4452 if they have a problem making a reservation or have a question concerning the slot controlled airport/STMP regulations or procedures.

d. Making Reservations.

1. Internet Users. Detailed information and User Instruction Guides for using the Web interface to the reservation systems are available on the websites for the slot controlled airports (e-CVRS), http://www.fly.faa.gov/ecvrs; and STMPs (e-STMP), http://www.fly.faa.gov/estmp.

2. Telephone users. When using the telephone to make a reservation, you are prompted for input of information about what you wish to do. All input is accomplished using the keypad on the telephone. The only problem with a telephone is that most keys have a letter and number associated with them. When the system asks for a date or time, it is expecting an input of numbers. A problem arises when entering an aircraft call sign or tail number. The system does not detect if you are entering a letter (alpha character) or a number. Therefore, when entering an aircraft call sign or tail number two keys are used to represent each letter or number. When entering a number, precede the number you wish by the number 0 (zero) i.e., 01, 02, 03, 04,.... If you wish to enter a letter, first press the key on which the letter appears and then press 1, 2, or 3, depending upon whether the letter you desire is the first, second, or third letter on that key. For example to enter the letter "N" first press the "6" key because "N" is on that key, then press the "2" key because the letter "N" is the second letter on the "6" key. Since there are no keys for the letters "Q" and "Z" e-CVRS pretends they are on the number "1" key. Therefore, to enter the letter "Q", press 11, and to enter the letter "Z" press 12.

NOTE- *Users are reminded to enter the "N" character with their tail numbers. (See TBL 4-1-4.)*

3. Additional helpful key entries: (See TBL 4-1-5.)

TBL 4-1-4
Codes for Call Sign/Tail Number Input

Codes for Call Sign/Tail Number Input Only			
A-21	J-51	S-73	1-01
B-22	K-52	T-81	2-02
C-23	L-53	U-82	3-03
D-31	M-61	V-83	4-04
E-32	N-62	W-91	5-05
F-33	O-63	X-92	6-06
G-41	P-71	Y-93	7-07
H-42	Q-11	Z-12	8-08
I-43	R-72	0-00	9-09

Tbl 4-1-5
Helpful Key Entries

#	After entering a call sign/tail number, depressing the "pound key" (#) twice will indicate the end of the entry.
*2	Will take the user back to the start of the process.
*3	Will repeat the call sign/tail number used in a previous reservation.
*5	Will repeat the previous question.
*8	Tutorial Mode: In the tutorial mode each prompt for input includes a more detailed description of what is expected as input. *8 is a toggle on/off switch. If you are in tutorial mode and enter *8, you will return to the normal mode.
*0	Expert Mode: In the expert mode each prompt for input is brief with little or no explanation. Expert mode is also on/off toggle.

4-1-23. Requests for Waivers and Authorizations from Title 14, Code of Federal Regulations (14 CFR)

a. Requests for a Certificate of Waiver or Authorization (FAA Form 7711-2), or requests for renewal of a waiver or authorization, may be accepted by any FAA facility and will be forwarded, if necessary, to the appropriate office having waiver authority.

b. The grant of a Certificate of Waiver or Authorization from 14 CFR constitutes relief from specific regulations, to the degree and for the period of time specified in the certificate, and does not waive any state law or local ordinance. Should the proposed operations conflict with any state law or local ordinance, or require permission of local authorities or property owners, it is the applicant's responsibility to resolve the matter. The holder of a waiver is responsible for compliance with the terms of the waiver and its provisions.

c. A waiver may be canceled at any time by the Administrator, the person authorized to grant the waiver, or the representative designated to monitor a specific operation. In such case either written notice of cancellation, or written confirmation of a verbal cancellation will be provided to the holder.

4-1-24. Weather System Processor

The Weather System Processor (WSP) was developed for use in the National Airspace System to provide weather processor enhancements to selected Airport Surveillance Radar (ASR)-9 facilities. The WSP provides Air Traffic with warnings of hazardous wind shear and microbursts. The WSP also provides users with terminal area 6-level weather, storm cell locations and movement, as well as the location and predicted future position and intensity of wind shifts that may affect airport operations.

Section 2. Radio Communications Phraseology and Techniques

4-2-1. General

a. Radio communications are a critical link in the ATC system. The link can be a strong bond between pilot and controller or it can be broken with surprising speed and disastrous results. Discussion herein provides basic procedures for new pilots and also highlights safe operating concepts for all pilots.

b. The single, most important thought in pilot-controller communications is understanding. It is essential, therefore, that pilots acknowledge each radio communication with ATC by using the appropriate aircraft call sign. Brevity is important, and contacts should be kept as brief as possible, but controllers must know what you want to do before they can properly carry out their control duties. And you, the pilot, must know exactly what the controller wants you to do. Since concise phraseology may not always be adequate, use whatever words are necessary to get your message across. Pilots are to maintain vigilance in monitoring air traffic control radio communications frequencies for potential traffic conflicts with their aircraft especially when operating on an active runway and/or when conducting a final approach to landing.

c. All pilots will find the Pilot/Controller Glossary very helpful in learning what certain words or phrases mean. Good phraseology enhances safety and is the mark of a professional pilot. Jargon, chatter, and "CB" slang have no place in ATC communications. The Pilot/Controller Glossary is the same glossary used in FAA Order JO 7110.65, Air Traffic Control. We recommend that it be studied and reviewed from time to time to sharpen your communication skills.

4-2-2. Radio Technique

a. Listen before you transmit. Many times you can get the information you want through ATIS or by monitoring the frequency. Except for a few situations where some frequency overlap occurs, if you hear someone else talking, the keying of your transmitter will be futile and you will probably jam their receivers causing them to repeat their call. If you have just changed frequencies, pause, listen, and make sure the frequency is clear.

b. Think before keying your transmitter. Know what you want to say and if it is lengthy; e.g., a flight plan or IFR position report, jot it down.

c. The microphone should be very close to your lips and after pressing the mike button, a slight pause may be necessary to be sure the first word is transmitted. Speak in a normal, conversational tone.

d. When you release the button, wait a few seconds before calling again. The controller or FSS specialist may be jotting down your number, looking for your flight plan, transmitting on a different frequency, or selecting the transmitter for your frequency.

e. Be alert to the sounds *or the lack of sounds* in your receiver. Check your volume, recheck your frequency, and *make sure that your microphone is not stuck* in the transmit position. Frequency blockage can, and has, occurred for extended periods of time due to unintentional transmitter operation. This type of interference is commonly referred to as a "stuck mike," and controllers may refer to it in this manner when attempting to assign an alternate frequency. If the assigned frequency is completely blocked by this type of interference, use the procedures described for en route IFR radio frequency outage to establish or reestablish communications with ATC.

f. Be sure that you are within the performance range of your radio equipment and the ground station equipment. Remote radio sites do not always transmit and receive on all of a facility's available frequencies, particularly with regard to VOR sites where you can hear but not reach a ground station's receiver. Remember that higher altitudes increase the range of VHF "line of sight" communications.

4-2-3. Contact Procedures

a. Initial Contact.

1. The terms *initial contact* or *initial callup* means the first radio call you make to a given facility or the first call to a different controller or FSS specialist within a facility. Use the following format:

(a) Name of the facility being called;

(b) Your *full* aircraft identification as filed in the flight plan or as discussed in paragraph 4-2-4, Aircraft Call Signs;

(c) When operating on an airport surface, state your position.

(d) The type of message to follow or your request if it is short; and

(e) The word "Over" if required.

EXAMPLE-**1.** *"New York Radio, Mooney Three One One Echo."*

2. *"Columbia Ground, Cessna Three One Six Zero Foxtrot, south ramp, I-F-R Memphis."*

3. *"Miami Center, Baron Five Six Three Hotel, request V-F-R traffic advisories."*

2. Many FSSs are equipped with Remote Communications Outlets (RCOs) and can transmit on the same frequency at more than one location. The frequencies available at specific locations are indicated on charts above FSS communications boxes. To enable the specialist to utilize the correct transmitter, advise the location and the frequency on which you expect a reply.

EXAMPLE-*St. Louis FSS can transmit on frequency 122.3 at either Farmington, Missouri, or Decatur, Illinois, if you are in the vicinity of Decatur, your callup should be "Saint Louis radio, Piper Six Niner Six Yankee, receiving Decatur One Two Two Point Three."*

3. If radio reception is reasonably assured, inclusion of your request, your position or altitude, and the phrase "(ATIS) Information Charlie received" in the initial contact helps decrease radio frequency congestion. Use discretion; do not overload the controller with information unneeded or superfluous. If you do not get a response from the ground station, recheck your radios or use another transmitter, but keep the next contact short.

EXAMPLE-*"Atlanta Center, Duke Four One Romeo, request V-F-R traffic advisories, Twenty Northwest Rome, seven thousand five hundred, over."*

b. Initial Contact When Your Transmitting and Receiving Frequencies are Different.

1. If you are attempting to establish contact with a ground station and you are receiving on a different frequency than that transmitted, indicate the VOR name or the frequency on which you expect a reply. Most FSSs and control facilities can transmit on several VOR stations in the area. Use the appropriate FSS call sign as indicated on charts.

EXAMPLE-*New York FSS transmits on the Kennedy, the Hampton, and the Calverton VORTACs. If you are in the Calverton area, your callup should be "New York radio, Cessna Three One Six Zero Foxtrot, receiving Calverton V-O-R, over."*

2. If the chart indicates FSS frequencies above the VORTAC or in the FSS communications boxes, transmit or receive on those frequencies nearest your location.

3. When unable to establish contact and you wish to call *any* ground station, use the phrase "ANY RADIO (tower) (station), GIVE CESSNA THREE ONE SIX ZERO FOXTROT A CALL ON (frequency) OR (V-O-R)." If an emergency exists or you need assistance, so state.

c. Subsequent Contacts and Responses to Callup from a Ground Facility.

Use the same format as used for the initial contact except you should state your message or request with the callup in one transmission. The ground station name and the word "Over" may be omitted if the message requires an obvious reply and there is no possibility for misunderstandings. *You should acknowledge all callups or clearances* unless the controller or FSS specialist advises otherwise. There are some occasions when controllers must issue time-critical instructions to other aircraft, and they may be in a position to observe your response, either visually or on radar. If the situation demands your response, take appropriate action or immediately advise the facility of any problem. Acknowledge with your aircraft identification, either at the beginning or at the end of your transmission, and one of the words "Wilco," "Roger," "Affirmative," "Negative," or other appropriate remarks; e.g., "PIPER TWO ONE FOUR LIMA, ROGER." If you have been receiving services; e.g., VFR traffic advisories and you are leaving the area or changing frequencies, advise the ATC facility and terminate contact.

d. Acknowledgement of Frequency Changes.

1. When advised by ATC to change frequencies, acknowledge the instruction. If you select the new frequency without an acknowledgement, the controller's workload is increased because there is no way of knowing whether you received the instruction or have had radio communications failure.

2. At times, a controller/specialist may be working a sector with multiple frequency assignments. In order to eliminate unnecessary verbiage and to free the controller/specialist for higher priority transmissions, the controller/specialist may request the pilot "(Identification), change to my frequency 123.4." This phrase should alert the pilot that the controller/specialist is only changing frequencies, not controller/specialist, and that initial callup phraseology may be abbreviated.

EXAMPLE-*"United Two Twenty-Two on one two three point four" or "one two three point four, United Two Twenty-Two."*

e. Compliance with Frequency Changes.

When instructed by ATC to change frequencies, select the new frequency as soon as possible unless instructed to make the change at a specific time, fix, or altitude. A delay in making the change could result in an untimely receipt of important information. If you are instructed to make the frequency change at a specific time, fix, or altitude, monitor the frequency you are on until reaching the specified time, fix, or altitudes unless instructed otherwise by ATC.

REFERENCE-*AIM, ARTCC Communications, Paragraph 5-3-1.*

4-2-4. Aircraft Call Signs

a. Precautions in the Use of Call Signs.

1. Improper use of call signs can result in pilots executing a clearance intended for another aircraft. Call signs should *never be abbreviated on an initial contact or at any time when other aircraft call signs have similar numbers/sounds or identical letters/number;* e.g., Cessna 6132F, Cessna 1622F, Baron 123F, Cherokee 7732F, etc.

EXAMPLE-*Assume that a controller issues an approach clearance to an aircraft at the bottom of a holding stack and an aircraft with a similar call sign (at the top of the stack) acknowledges the clearance with the last two or three numbers of the aircraft's call sign. If the aircraft at the bottom of the stack did not hear the clearance and intervene, flight safety would be affected, and there would be no reason for either the controller or pilot to suspect that anything is wrong. This kind of "human factors" error can strike swiftly and is extremely difficult to rectify.*

2. Pilots, therefore, must be certain that aircraft identification is complete and clearly identified before taking action on an ATC clearance. ATC specialists will not abbreviate call signs of air carrier or other civil aircraft having authorized call signs. ATC specialists may initiate abbreviated call signs of other aircraft by using the *prefix and the last three digits/letters* of the aircraft identification after communications are established. The pilot may use the abbreviated call sign in subsequent contacts with the ATC specialist. When aware of similar/identical call signs, ATC specialists will take action to minimize errors by emphasizing certain numbers/letters, by repeating the entire call sign, by repeating the prefix, or by asking pilots to use a different call sign temporarily. Pilots should use the phrase "VERIFY CLEARANCE FOR (your complete call sign)" if doubt exists concerning proper identity.

3. Civil aircraft pilots should state the aircraft type, model or manufacturer's name, followed by the digits/letters of the registration number. When the aircraft manufacturer's name or model is stated, the prefix "N" is dropped; e.g., Aztec Two Four Six Four Alpha.

EXAMPLE-
1. *Bonanza Six Five Five Golf.*
2. *Breezy Six One Three Romeo Experimental (omit "Experimental" after initial contact).*

4. Air Taxi or other commercial operators *not* having FAA authorized call signs should prefix their normal identification with the phonetic word "Tango."

EXAMPLE-*Tango Aztec Two Four Six Four Alpha.*

5. Air carriers and commuter air carriers having FAA authorized call signs should identify themselves by stating the complete call sign (using group form for the numbers) and the word "heavy" if appropriate.

EXAMPLE-
1. *United Twenty-Five Heavy.*
2. *Midwest Commuter Seven Eleven.*

6. Military aircraft use a variety of systems including serial numbers, word call signs, and combinations of letters/numbers. Examples include Army Copter 48931; Air Force 61782; REACH 31792; Pat 157; Air Evac 17652; Navy Golf Alfa Kilo 21; Marine 4 Charlie 36, etc.

b. Air Ambulance Flights.

Because of the priority afforded air ambulance flights in the ATC system, extreme discretion is necessary when using the term "MEDEVAC." It is only intended for those missions of an urgent medical nature and to be utilized only for that portion of the flight requiring expeditious handling. When requested by the pilot, necessary notification to expedite ground handling of patients, etc., is provided by ATC; however, when possible, this information should be passed in advance through non-ATC communications systems.

1. Civilian air ambulance flights responding to medical emergencies (first call to an accident scene, carrying patients, organ donors, organs, or other urgently needed lifesaving medical material) will be expedited by ATC when necessary. When expeditious handling is necessary, include the word "MEDEVAC" in the flight plan per paragraphs 5-1-8 and 5-1-9. In radio communications, use the call sign "MEDEVAC," followed by the aircraft registration letters/numbers.

EXAMPLE-*MEDEVAC Two Six Four Six.*

2. Similar provisions have been made for the use of "AIR EVAC" and "HOSP" by air ambulance flights, except that these flights will receive priority handling only when specifically requested.

3. Air carrier and air taxi flights responding to medical emergencies will also be expedited by ATC when necessary. The nature of these medical emergency flights usually concerns the transportation of urgently needed lifesaving medical materials or vital organs. IT IS IMPERATIVE THAT THE COMPANY/PILOT DETERMINE, BY THE NATURE/URGENCY OF THE SPECIFIC MEDICAL CARGO, IF PRIORITY ATC ASSISTANCE IS REQUIRED. Pilots must include the word "MEDEVAC" in the flight plan per paragraphs 5-1-8 and 5-1-9, and use the call sign "MEDEVAC," followed by the company name and flight number for all transmissions when expeditious handling is required. It is important for ATC to be aware of "MEDEVAC" status, and it is the pilot's responsibility to ensure that this information is provided to ATC.

EXAMPLE-*MEDEVAC Delta Thirty-Seven.*

c. Student Pilots Radio Identification.

1. The FAA desires to help student pilots in acquiring sufficient practical experience in the environment in which they will be required to operate. To receive additional assistance while operating in areas of concentrated air traffic, student pilots need only identify themselves as a student pilot during their initial call to an FAA radio facility.

EXAMPLE-*Dayton tower, Fleetwing One Two Three Four, student pilot.*

2. This special identification will alert FAA ATC personnel and enable them to provide student pilots with such extra assistance and consideration as they may need. It is recommended that student pilots identify themselves as such, on initial contact with each clearance delivery prior to taxiing, ground control, tower, approach and departure control frequency, or FSS contact.

4-2-5. Description of Interchange or Leased Aircraft

a. Controllers issue traffic information based on familiarity with airline equipment and color/markings. When an air carrier dispatches a flight using another company's equipment and the pilot does not advise the terminal ATC facility, the possible confusion in aircraft identification can compromise safety.

b. Pilots flying an "interchange" or "leased" aircraft not bearing the colors/markings of the company operating the aircraft should inform the terminal ATC facility on first contact the name of the operating company and trip number, followed by the company name as displayed on the aircraft, and aircraft type.

EXAMPLE-*Air Cal Three Eleven, United (interchange/lease), Boeing Seven Two Seven.*

4-2-6. Ground Station Call Signs

Pilots, when calling a ground station, should begin with the name of the facility being called followed by the type of the facility being called as indicated in TBL 4-2-1.

TBL 4-2-1
Calling a Ground Station

Facility	Call Sign
Airport UNICOM	"Shannon UNICOM"
FAA Flight Service Station	"Chicago Radio"
FAA Flight Service Station (En Route Flight Advisory Service (Weather))	"Seattle Flight Watch"
Airport Traffic Control Tower	"Augusta Tower"
Clearance Delivery Position (IFR)	"Dallas Clearance Delivery"
Ground Control Position in Tower	"Miami Ground"
Radar or Nonradar Approach Control Position	"Oklahoma City Approach"
Radar Departure Control Position	"St. Louis Departure"
FAA Air Route Traffic Control Center	"Washington Center"

4-2-7. Phonetic Alphabet

The International Civil Aviation Organization (ICAO) phonetic alphabet is used by FAA personnel when communications conditions are such that the information cannot be readily received without their use. ATC facilities may also request pilots to use phonetic letter equivalents when aircraft with similar sounding identifications are receiving communications on the same frequency. Pilots should use the phonetic alphabet when identifying their aircraft during initial contact with air traffic control facilities. Additionally, use the phonetic equivalents for single letters and to spell out groups of letters or difficult words during adverse communications conditions. (See TBL 4-2-2.)

4-2-8. Figures

a. Figures indicating hundreds and thousands in round number, as for ceiling heights, and upper wind levels up to 9,900 must be spoken in accordance with the following.

EXAMPLE-
1. *500 five hundred*
2. *4,500 four thousand five hundred*

b. Numbers above 9,900 must be spoken by separating the digits preceding the word thousand."

EXAMPLE-
1. *10,000 one zero thousand*
2. *13,500 one three thousand five hundred*

c. Transmit airway or jet route numbers as follows.

EXAMPLE-
1. *V12 Victor Twelve*
2. *J533 J Five Thirty-Three*

d. All other numbers must be transmitted by pronouncing each digit.

EXAMPLE-*10 one zero*

e. When a radio frequency contains a decimal point, the decimal point is spoken as "POINT."

EXAMPLE-*122.1 one two two point one*
NOTE-*ICAO procedures require the decimal point be spoken as "DECIMAL." The FAA will honor such usage by military aircraft and all other aircraft required to use ICAO procedures.*

4-2-9. Altitudes and Flight Levels

a. Up to but not including 18,000 feet MSL, state the separate digits of the thousands plus the hundreds if appropriate.

EXAMPLE-
1. *12,000 one two thousand*
2. *12,500 one two thousand five hundred*

b. At and above 18,000 feet MSL (FL 180), state the words "flight level" followed by the separate digits of the flight level.

EXAMPLE-
1. *190 Flight Level One Niner Zero*
2. *275 Flight Level Two Seven Five*

4-2-10. Directions

The three digits of bearing, course, heading, or wind direction should always be magnetic. The word "true" must be added when it applies.

EXAMPLE-
1. *(Magnetic course) 005 zero zero five*
2. *(True course) 050 zero five zero true*
3. *(Magnetic bearing) 360 three six zero*
4. *(Magnetic heading) 100 heading one zero zero*
5. *(Wind direction) 220 wind two two zero*

4-2-11. Speeds

The separate digits of the speed followed by the word "KNOTS." Except, controllers may omit the word "KNOTS" when using speed adjustment procedures; e.g., "REDUCE/INCREASE SPEED TO TWO FIVE ZERO."

EXAMPLE-*(Speed) 250 two five zero knots*
(Speed) 190 one niner zero knots

The separate digits of the Mach Number preceded by "Mach."

EXAMPLE-*(Mach number) 1.5 Mach one point five*

(Mach number) 0.64 Mach point six four
(Mach number) 0.7 Mach point seven

Tbl 4-2-2
Phonetic Alphabet/Morse Code

Character	Morse Code	Telephony	Phonic (Pronunciation)
A	·y	Alfa	(AL-FAH)
B	y···	Bravo	(BRAH-VOH)
C	y·y·	Charlie	(CHAR-LEE) or (SHAR-LEE)
D	y··	Delta	(DELL-TAH)
E	·	Echo	(ECK-OH)
F	··y·	Foxtrot	(FOKS-TROT)
G	yy·	Golf	(GOLF)
H	····	Hotel	(HOH-TEL)
I	··	India	(IN-DEE-AH)
J	·yyy	Juliett	(JEW-LEE-ETT)
K	y·y	Kilo	(KEY-LOH)
L	·y··	Lima	(LEE-MAH)
M	yy	Mike	(MIKE)
N	y·	November	(NO-VEM-BER)
O	yyy	Oscar	(OSS-CAH)
P	·yy·	Papa	(PAH-PAH)
Q	yy·y	Quebec	(KEH-BECK)
R	·y·	Romeo	(ROW-ME-OH)
S	···	Sierra	(SEE-AIR-RAH)
T	y	Tango	(TANG-GO)
U	··y	Uniform	(YOU-NEE-FORM) or (OO-NEE-FORM)
V	···y	Victor	(VIK-TAH)
W	·yy	Whiskey	(WISS-KEY)
X	y··y	Xray	(ECKS-RAY)
Y	y·yy	Yankee	(YANG-KEY)
Z	yy··	Zulu	(ZOO-LOO)
1	·yyyy	One	(WUN)
2	··yyy	Two	(TOO)
3	···yy	Three	(TREE)
4	····y	Four	(FOW-ER)
5	·····	Five	(FIFE)
6	y····	Six	(SIX)
7	yy···	Seven	(SEV-EN)
8	yyy··	Eight	(AIT)
9	yyyy·	Nine	(NIN-ER)
0	yyyyy	Zero	(ZEE-RO)

4-2-12. Time

a. FAA uses Coordinated Universal Time (UTC) for all operations. The word "local" or the time zone equivalent must be used to denote local when local time is given during radio and telephone communications. The term "Zulu" may be used to denote UTC.

EXAMPLE-*0920 UTC zero niner two zero,*
zero one two zero pacific or local,
or one twenty AM

b. To convert from Standard Time to Coordinated Universal Time:

Tbl 4-2-3
Standard Time to Coordinated Universal Time

Eastern Standard Time	Add 5 hours
Central Standard Time	Add 6 hours
Mountain Standard Time	Add 7 hours
Pacific Standard Time	Add 8 hours
Alaska Standard Time	Add 9 hours
Hawaii Standard Time	Add 10 hours

NOTE-*For daylight time, subtract 1 hour.*

c. A reference may be made to local daylight or standard time utilizing the 24-hour clock system. The hour is indicated by the first two figures and the minutes by the last two figures.

EXAMPLE-*0000 zero zero zero zero*
0920 zero niner two zero

d. Time may be stated in minutes only (two figures) in radiotelephone communications when no misunderstanding is likely to occur.

e. Current time in use at a station is stated in the nearest quarter minute in order that pilots may use this information for time checks. Fractions of a quarter minute less than 8 seconds are stated as the preceding quarter minute; fractions of a quarter minute of 8 seconds or more are stated as the succeeding quarter minute.

EXAMPLE-*0929:05 time, zero niner two niner*
0929:10 time, zero niner two niner and one-quarter

4-2-13. Communications with Tower when Aircraft Transmitter or Receiver or Both are Inoperative

a. Arriving Aircraft.

1. Receiver inoperative.

(a) If you have reason to believe your receiver is inoperative, remain outside or above the Class D surface area until the direction and flow of traffic has been determined; then, advise the tower of your type aircraft, position, altitude, intention to land, and request that you be controlled with light signals.

REFERENCE-*AIM, Traffic Control Light Signals, Paragraph 4-3-13.*

(b) When you are approximately 3 to 5 miles from the airport, advise the tower of your position and join the airport traffic pattern. From this point on, watch the tower for light signals. Thereafter, if a complete pattern is made, transmit your position downwind and/or turning base leg.

2. Transmitter inoperative. Remain outside or above the Class D surface area until the direction and flow of traffic has been determined; then, join the airport traffic pattern. Monitor the primary local control frequency as depicted on Sectional Charts for landing or traffic information, and look for a light signal which may be addressed to your aircraft. During hours of daylight, acknowledge tower transmissions or light signals by rocking your wings. At night, acknowledge by blinking the landing or navigation lights. To acknowledge tower transmissions during daylight hours, hovering helicopters will turn in the direction of the controlling facility and flash the landing light. While in flight, helicopters should show their acknowledgement of receiving a transmission by making shallow banks in opposite directions. At night, helicopters will acknowledge receipt of transmissions by flashing either the landing or the search light.

3. Transmitter and receiver inoperative. Remain outside or above the Class D surface area until the direction and flow of traffic has been determined; then, join the airport traffic pattern and maintain visual contact with the tower to receive light signals. Acknowledge light signals as noted above.

b. Departing Aircraft. If you experience radio failure prior to leaving the parking area, make every effort to have the equipment repaired. If you are unable to have the malfunction repaired, call the tower by telephone and request authorization to depart without two-way radio communications. If tower authorization is granted, you will be given departure information and requested to monitor the tower frequency or watch for light signals as appropriate. During daylight hours, acknowledge tower transmissions or light signals by moving the ailerons or rudder. At night, acknowledge by blinking the landing or navigation lights. If radio malfunction occurs after departing the parking area, watch the tower for light signals or monitor tower frequency.

REFERENCE-*14 CFR Section 91.125 and 14 CFR Section 91.129.*

4-2-14. Communications for VFR Flights

a. FSSs and Supplemental Weather Service Locations (SWSLs) are allocated frequencies for different functions; for example, 122.0 MHz is assigned as the En Route Flight Advisory Service frequency at selected FSSs. In addition, certain FSSs provide Local Airport Advisory on 123.6 MHz or other frequencies which can be found in the A/FD. If you are in doubt as to what frequency to use, 122.2 MHz is assigned to the majority of FSSs as a common en route simplex frequency.

NOTE-*In order to expedite communications, state the frequency being used and the aircraft location during initial callup.*

EXAMPLE-*Dayton radio, November One Two Three Four Five on one two two point two, over Springfield V-O-R, over.*

b. Certain VOR voice channels are being utilized for recorded broadcasts; i.e., ATIS, HIWAS, etc. These services and appropriate frequencies are listed in the A/FD. On VFR flights, pilots are urged to monitor these frequencies. When in contact with a control facility, notify the controller if you plan to leave the frequency to monitor these broadcasts.

Section 3. Airport Operations

4-3-1. General

Increased traffic congestion, aircraft in climb and descent attitudes, and pilot preoccupation with cockpit duties are some factors that increase the hazardous accident potential near the airport. The situation is further compounded when the weather is marginal, that is, just meeting VFR requirements. Pilots must be particularly alert when operating in the vicinity of an airport. This section defines some rules, practices, and procedures that pilots should be familiar with and adhere to for safe airport operations.

4-3-2. Airports with an Operating Control Tower

a. When operating at an airport where traffic control is being exercised by a control tower, pilots are required to maintain two-way radio contact with the tower while operating within the Class B, Class C, and Class D surface area unless the tower authorizes otherwise. Initial callup should be made about 15 miles from the airport. Unless there is a good reason to leave the tower frequency before exiting the Class B, Class C, and Class D surface areas, it is a good operating practice to remain on the tower frequency for the purpose of receiving traffic information. In the interest of reducing tower frequency congestion, pilots are reminded that it is not necessary to request permission to leave the tower frequency once outside of Class B, Class C, and Class D surface areas. Not all airports with an operating control tower will have Class D airspace. These airports do not have weather reporting which is a requirement for surface based controlled airspace, previously known as a control zone. The controlled airspace over these airports will normally begin at 700 feet or 1,200 feet above ground level and can be determined from the visual aeronautical charts. Pilots are expected to use good operating practices and communicate with the control tower as described in this section.

b. When necessary, the tower controller will issue clearances or other information for aircraft to generally follow the desired flight path (traffic patterns) when flying in Class B, Class C, and Class D surface areas and the proper taxi routes when operating on the ground. If not otherwise authorized or directed by the tower, pilots of fixed-wing aircraft approaching to land must circle the airport to the left. Pilots approaching to land in a helicopter must avoid the flow of fixed-wing traffic. However, in all instances, an appropriate clearance must be received from the tower before landing.

FIG 4-3-1
Components of a Traffic Pattern

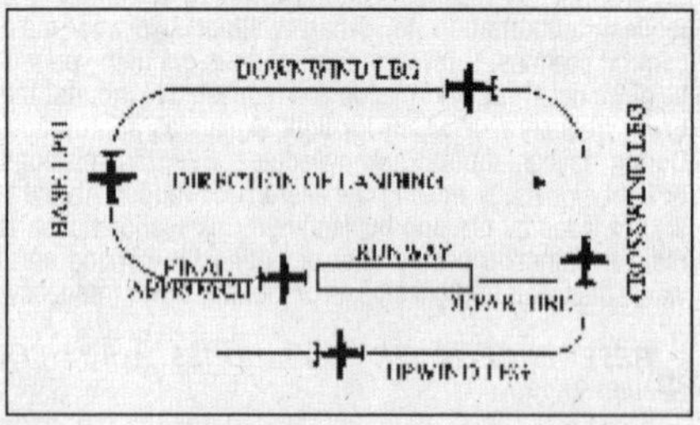

***NOTE-** This diagram is intended only to illustrate terminology used in identifying various components of a traffic pattern. It should not be used as a reference or guide on how to enter a traffic pattern.*

c. The following terminology for the various components of a traffic pattern has been adopted as standard for use by control towers and pilots (See FIG 4-3-1):

1. Upwind leg. A flight path parallel to the landing runway in the direction of landing.

2. Crosswind leg. A flight path at right angles to the landing runway off its takeoff end.

3. Downwind leg. A flight path parallel to the landing runway in the opposite direction of landing.

4. Base leg. A flight path at right angles to the landing runway off its approach end and extending from the downwind leg to the intersection of the extended runway centerline.

5. Final approach. A flight path in the direction of landing along the extended runway centerline from the base leg to the runway.

6. Departure leg. The flight path which begins after takeoff and continues straight ahead along the extended runway centerline. The departure climb continues until reaching a point at least 1/2 mile beyond the departure end of the runway and within 300 feet of the traffic pattern altitude.

d. Many towers are equipped with a tower radar display. The radar uses are intended to enhance the effectiveness and efficiency of the local control, or tower, position. They are not intended to provide radar services or benefits to pilots except as they may accrue through a more efficient tower operation. The four basic uses are:

1. To determine an aircraft's exact location. This is accomplished by radar identifying the VFR aircraft through any of the techniques available to a radar position, such as having the aircraft *squawk ident*. Once identified, the aircraft's position and spatial relationship to other aircraft can be quickly determined, and standard instructions regarding VFR operation in Class B, Class C, and Class D surface areas will be issued. Once initial radar identification of a VFR aircraft has been established and the appropriate instructions have been issued, radar monitoring may be discontinued; the reason being that the local controller's primary means of surveillance in VFR conditions is visually scanning the airport and local area.

2. To provide radar traffic advisories. Radar traffic advisories may be provided to the extent that the local controller is able to monitor the radar display. Local control has primary control responsibilities to the aircraft operating on the runways, which will normally supersede radar monitoring duties.

3. To provide a direction or suggested heading. The local controller may provide pilots flying VFR with generalized instructions which will facilitate operations; e.g., "PROCEED SOUTHWESTBOUND, ENTER A RIGHT DOWNWIND RUNWAY THREE ZERO," or provide a suggested heading to establish radar identification or as an advisory aid to navigation; e.g., "SUGGESTED HEADING TWO TWO ZERO, FOR RADAR IDENTIFICATION." In both cases, the instructions are advisory aids to the pilot flying VFR and are not radar vectors.

***NOTE-** Pilots have complete discretion regarding acceptance of the suggested headings or directions and have sole responsibility for seeing and avoiding other aircraft.*

4. To provide information and instructions to aircraft operating within Class B, Class C, and Class D surface areas. In an example of this situation, the local controller would use the radar to advise a pilot on an extended downwind when to turn base leg.

***NOTE-** The above tower radar applications are intended to augment the standard functions of the local control position. There is no controller requirement to maintain constant radar identification. In fact, such a requirement could compromise the local controller's ability to visually scan the airport and local area to meet FAA responsibilities to the aircraft operating on the runways and within the Class B, Class C, and Class D surface areas. Normally, pilots will not be advised of being in radar contact since that continued status cannot be guaranteed and since the purpose of the radar identification is not to establish a link for the provision of radar services.*

e. A few of the radar equipped towers are authorized to use the radar to ensure separation between aircraft in specific situations, while still others may function as limited radar approach controls. The various radar uses are strictly a function of FAA operational need. The facilities may be indistinguishable to pilots since they are all referred to as tower and no publication lists the degree of radar use. **Therefore, when in communication with a tower controller who may have radar available, do not assume that constant radar monitoring and complete ATC radar services are being provided.**

4-3-3. Traffic Patterns

At most airports and military air bases, traffic pattern altitudes for propeller-driven aircraft generally extend from 600 feet to as high as 1,500 feet above the ground. Also, traffic pattern altitudes for military turbojet aircraft sometimes extend up to 2,500 feet above the ground. Therefore, pilots of en route aircraft should be constantly on the alert for other aircraft in traffic patterns and avoid these areas whenever possible. Traffic pattern altitudes should be maintained unless otherwise required by the applicable distance from cloud criteria (14 CFR Section 91.155). (See FIG 4-3-2 and FIG 4-3-3.)

FIG 4-3-2
Traffic Pattern Operations Single Runway

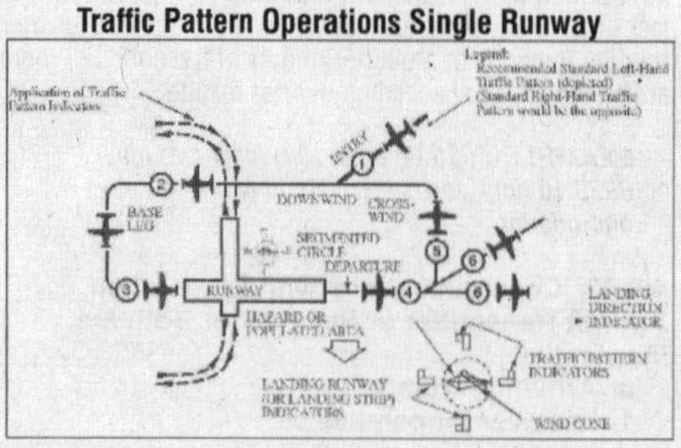

***EXAMPLE-** Key to traffic pattern operations*

1. *Enter pattern in level flight, abeam the midpoint of the runway, at pattern altitude. (1,000' AGL is recommended pattern altitude unless established otherwise. . .)*

2. *Maintain pattern altitude until abeam approach end of the landing runway on downwind leg.*

3. *Complete turn to final at least $^1/_4$ mile from the runway.*

4. *Continue straight ahead until beyond departure end of runway.*

5. *If remaining in the traffic pattern, commence turn to crosswind leg beyond the departure end of the runway within 300 feet of pattern altitude.*

6. *If departing the traffic pattern, continue straight out, or exit with a 45 degree turn (to the left when in a left-hand traffic pattern; to the right when in a right-hand traffic pattern) beyond the departure end of the runway, after reaching pattern altitude.*

FIG 4-3-3
Traffic Pattern Operations Parallel Runways

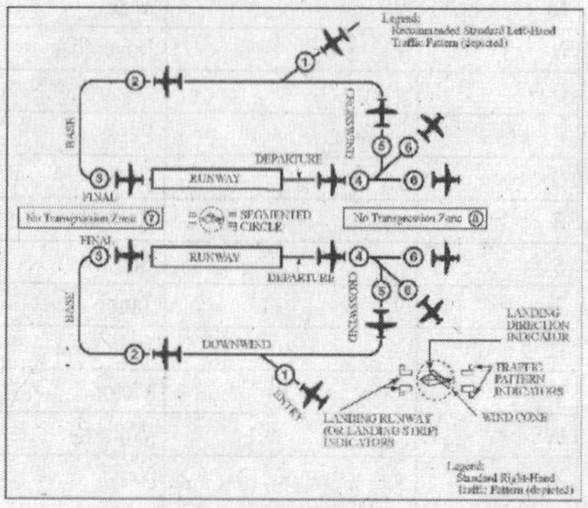

***EXAMPLE-** Key to traffic pattern operations*

1. *Enter pattern in level flight, abeam the midpoint of the runway, at pattern altitude. (1,000' AGL is recommended pattern altitude unless established otherwise. . .)*

2. *Maintain pattern altitude until abeam approach end of the landing runway on downwind leg.*

3. *Complete turn to final at least $^1/_4$ mile from the runway.*

4. *Continue straight ahead until beyond departure end of runway.*

5. *If remaining in the traffic pattern, commence turn to crosswind leg beyond the departure end of the runway within 300 feet of pattern altitude.*

6. *If departing the traffic pattern, continue straight out, or exit with a 45 degree turn (to the left when in a left-hand traffic pattern; to the right when in a right-hand traffic pattern) beyond the departure end of the runway, after reaching pattern altitude.*

7. *Do not overshoot final or continue on a track which will penetrate the final approach of the parallel runway.*

8. *Do not continue on a track which will penetrate the departure path of the parallel runway.*

4-3-4. Visual Indicators at Airports Without an Operating Control Tower

a. At those airports *without an operating control tower,* a segmented circle visual indicator system, if installed, is designed to provide traffic pattern information.

***REFERENCE-** AIM, Traffic Advisory Practices at Airports Without Operating Control Towers, Paragraph 4-1-9.*

b. The segmented circle system consists of the following components:

1. The segmented circle. Located in a position affording maximum visibility to pilots in the air and on the ground and providing a centralized location for other elements of the system.

2. The wind direction indicator. A wind cone, wind sock, or wind tee installed near the operational runway to indicate wind direction. The large end of the wind cone/wind sock points into the wind as does the large end (cross bar) of the wind tee. In lieu of a tetrahedron and where a wind sock or wind cone is collocated with a wind tee, the wind tee may be manually aligned with the runway in use to indicate landing direction. These signaling devices may be located in the center of the segmented circle and may be lighted for night use. Pilots are cautioned against using a tetrahedron to indicate wind direction.

3. The landing direction indicator. A tetrahedron is installed when conditions at the airport warrant its use. It may be used to indicate the direction of landings and takeoffs. A tetrahedron may be located at the center of a segmented circle and may be lighted for night operations. The small end of the tetrahedron points in the direction of landing. Pilots are cautioned against using a tetrahedron for any purpose other than as an indicator of landing direction. Further, pilots should use extreme caution when making runway selection by use of a tetrahedron in very light or calm wind conditions as the tetrahedron may not be aligned with the designated calm-wind runway. At airports with control towers, the tetrahedron should only be referenced when the control tower is not in operation. Tower instructions supersede tetrahedron indications.

4. Landing strip indicators. Installed in pairs as shown in the segmented circle diagram and used to show the alignment of landing strips.

5. Traffic pattern indicators. Arranged in pairs in conjunction with landing strip indicators and used to indicate the direction of turns when there is a variation from the normal left traffic pattern. (If there is no segmented circle installed at the airport, traffic pattern indicators may be installed on or near the end of the runway.)

c. Preparatory to landing at an airport without a control tower, or when the control tower is not in operation, pilots should concern themselves with the indicator for the approach end of the runway to be used. When approaching for landing, all turns must be made to the left unless a traffic pattern indicator indicates that turns should be made to the right. If the pilot will mentally enlarge the indicator for the runway to be used, the base and final approach legs of the traffic pattern to be flown immediately become apparent. Similar treatment of the indicator at the departure end of the runway will clearly indicate the direction of turn after takeoff.

d. When two or more aircraft are approaching an airport for the purpose of landing, the pilot of the aircraft at the lower altitude has the right-of-way over the pilot of the aircraft at the higher altitude. However, the pilot operating at the lower altitude should not take advantage of another aircraft, which is on final approach to land, by cutting in front of, or overtaking that aircraft.

4-3-5. Unexpected Maneuvers in the Airport Traffic Pattern

There have been several incidents in the vicinity of controlled airports that were caused primarily by aircraft executing unexpected maneuvers. ATC service is based upon observed or known traffic and airport conditions. Controllers establish the sequence of arriving and departing aircraft by requiring them to adjust flight as necessary to achieve proper spacing. These adjustments can only be based on observed traffic, accurate pilot reports, and anticipated aircraft maneuvers. Pilots are expected to cooperate so as to preclude disrupting traffic flows or creating conflicting patterns. The pilot-in-command of an aircraft is directly responsible for and is the final authority as to the operation of the aircraft. On occasion it may be necessary for pilots to maneuver their aircraft to maintain spacing with the traffic they have been sequenced to follow. The controller can anticipate minor maneuvering such as shallow "S" turns. The controller cannot, however, anticipate a major maneuver such as a 360 degree turn. If a pilot makes a 360 degree turn after obtaining a landing sequence, the result is usually a gap in the landing interval and, more importantly, it causes a chain reaction which may result in a conflict with following traffic and an interruption of the sequence established by the tower or approach controller. Should a pilot decide to make maneuvering turns to maintain spacing behind a preceding aircraft, the pilot should always advise the controller if at all possible. Except when requested by the controller or in emergency situations, a 360 degree turn should never be executed in the traffic pattern or when receiving radar service without first advising the controller.

4-3-6. Use of Runways/Declared Distances

a. Runways are identified by numbers which indicate the nearest 10-degree increment of the azimuth of the runway centerline. For example, where the magnetic azimuth is 183 degrees, the runway designation would be 18; for a magnetic azimuth of 87 degrees, the runway designation would be 9. For a magnetic azimuth ending in the number 5, such as 185, the runway designation could be either 18 or 19. Wind direction issued by the tower is also magnetic and wind velocity is in knots.

b. Airport proprietors are responsible for taking the lead in local aviation noise control. Accordingly, they may propose specific noise abatement plans to the FAA. If approved, these plans are applied in the form of Formal or Informal Runway Use Programs for noise abatement purposes.

REFERENCE-*Pilot/Controller Glossary Term- "Runway Use Program"*

1. At airports where no runway use program is established, ATC clearances may specify:

(a) The runway most nearly aligned with the wind when it is 5 knots or more;

(b) The "calm wind" runway when wind is less than 5 knots; or

(c) Another runway if operationally advantageous.

NOTE-*It is not necessary for a controller to specifically inquire if the pilot will use a specific runway or to offer a choice of runways. If a pilot prefers to use a different runway from that specified, or the one most nearly aligned with the wind, the pilot is expected to inform ATC accordingly.*

2. At airports where a runway use program is established, ATC will assign runways deemed to have the least noise impact. If in the interest of safety a runway different from that specified is preferred, the pilot is expected to advise ATC accordingly. ATC will honor such requests and advise pilots when the requested runway is noise sensitive. When use of a runway other than the one assigned is requested, pilot cooperation is encouraged to preclude disruption of traffic flows or the creation of conflicting patterns.

c. Declared Distances.

1. Declared distances for a runway represent the maximum distances available and suitable for meeting takeoff and landing distance performance requirements. These distances are determined in accordance with FAA runway design standards by adding to the physical length of paved runway any clearway or stopway and subtracting from that sum any lengths necessary to obtain the standard runway safety areas, runway object free areas, or runway protection zones. As a result of these additions and subtractions, the declared distances for a runway may be more or less than the physical length of the runway as depicted on aeronautical charts and related publications, or available in electronic navigation databases provided by either the U.S. Government or commercial companies.

2. All 14 CFR Part 139 airports report declared distances for each runway. Other airports may also report declared distances for a runway if necessary to meet runway design standards or to indicate the presence of a clearway or stopway. Where reported, declared distances for each runway end are published in the Airport/Facility Directory (A/FD). For runways without published declared distances, the declared distances may be assumed to be equal to the physical length of the runway unless there is a displaced landing threshold, in which case the Landing Distance Available (LDA) is shortened by the amount of the threshold displacement.

NOTE-*A symbol* D *is shown on U.S. Government charts to indicate that runway declared distance information is available (See appropriate A/FD, Alaska, or Pacific Supplement).*

(a) The FAA uses the following definitions for runway declared distances (See FIG 4-3-4):

REFERENCE-*Pilot/Controller Glossary Terms: "Accelerate-Stop Distance Available," "Landing Distance Available," "Takeoff Distance Available," "Takeoff Run Available," " Stopway," and "Clearway."*

(1) Takeoff Run Available (TORA) – The runway length declared available and suitable for the ground run of an airplane taking off.

The TORA is typically the physical length of the runway, but it may be shorter than the runway length if necessary to satisfy runway design standards. For example, the TORA may be shorter than the runway length if a portion of the runway must be used to satisfy runway protection zone requirements.

(2) Takeoff Distance Available (TODA) – The takeoff run available plus the length of any remaining runway or clearway beyond the far end of the takeoff run available.

The TODA is the distance declared available for satisfying takeoff distance requirements for airplanes where the certification and operating rules and available performance data allow for the consideration of a clearway in takeoff performance computations.

NOTE-*The length of any available clearway will be included in the TODA published in the A/FD's entry for that runway end.*

(3) Accelerate-Stop Distance Available (ASDA) – The runway plus stopway length declared available and suitable for the acceleration and deceleration of an airplane aborting a takeoff.

The ASDA may be longer than the physical length of the runway when a stopway has been designated available by the airport operator, or it may be shorter than the physical length of the runway if necessary to use a portion of the runway to satisfy runway design standards; for example, where the airport operator uses a portion of the runway to achieve the runway safety area requirement. ASDA is the distance used to satisfy the airplane accelerate-stop distance performance requirements where the certification and operating rules require accelerate-stop distance computations.

NOTE-*The length of any available stopway will be included in the ASDA published in the A/FD's entry for that runway end.*

(4) Landing Distance Available (LDA) - The runway length declared available and suitable for a landing airplane.

The LDA may be less than the physical length of the runway or the length of the runway remaining beyond a displaced threshold if necessary to satisfy runway design standards;for example, where the airport operator uses a portion of the runway to achieve the runway safety area requirement.

Although some runway elements (such as stopway length and clearway length) may be available information, pilots must use the declared distances determined by the airport operator and not attempt to independently calculate declared distances by adding those elements to the reported physical length of the runway.

(b) The airplane operating rules and/or the airplane operating limitations establish minimum distance requirements for takeoff and landing and are based on performance data supplied in the Airplane Flight Manual or Pilot's Operating Handbook. The minimum distances required for takeoff and landing obtained either in planning prior to takeoff or in performance assessments conducted at the time of landing must fall **within** the applicable declared distances before the pilot can accept that runway for takeoff or landing.

(c) Runway design standards may impose restrictions on the amount of runway available for use in takeoff and landing that are not apparent from the reported physical length of the runway or from runway markings and lighting. The runway elements of Runway Safety Area (RSA), Runway Object Free Area (ROFA), and Runway Protection Zone (RPZ) may reduce a runway's declared distances to less than the physical length of the runway at geographically constrained airports (See FIG 4-3-5). When considering the amount of runway available for use in takeoff or landing performance calculations, the declared distances published for a runway must always be used in lieu of the runway's physical length.

REFERENCE-*AC 150/5300-13, Airport Design.*

(d) While some runway elements associated with declared distances may be identifiable through runway markings or lighting (for example, a displaced threshold or a stopway), the individual declared distance limits are not marked or otherwise identified on the runway. An aircraft is **not prohibited** from operating beyond a declared distance limit during the takeoff, landing, or taxi operation provided the runway surface is appropriately marked as usable runway (See FIG 4-3-5). The following examples clarify the intent of this paragraph.

REFERENCE-*AIM, Runway Markings, Paragraph 2-3-3. AC 150/5340-1, Standards for Airport Markings.*

EXAMPLE-1. *The declared LDA for runway 9 must be used when showing compliance with the landing distance requirements of the applicable airplane operating rules* *and/or airplane operating limitations or when making a before landing performance assessment. The LDA is less than the physical runway length, not only because of the displaced threshold, but also because of the subtractions necessary to meet the RSA beyond the far end of the runway. However, during the actual landing operation, it is permissible for the airplane to roll beyond the unmarked end of the LDA.*

2. *The declared ASDA for runway 9 must be used when showing compliance with the accelerate-stop distance requirements of the applicable airplane operating rules and/or airplane operating limitations. The ASDA is less than the physical length of the runway due to subtractions necessary to achieve the full RSA requirement. However, in the event of an aborted takeoff, it is permissible for the airplane to roll beyond the unmarked end of the ASDA as it is brought to a full-stop on the remaining usable runway.*

FIG 4-3-4
Declared Distances with Full-Standard Runway Safety Areas, Runway Object Free Areas, and Runway Protection Zones

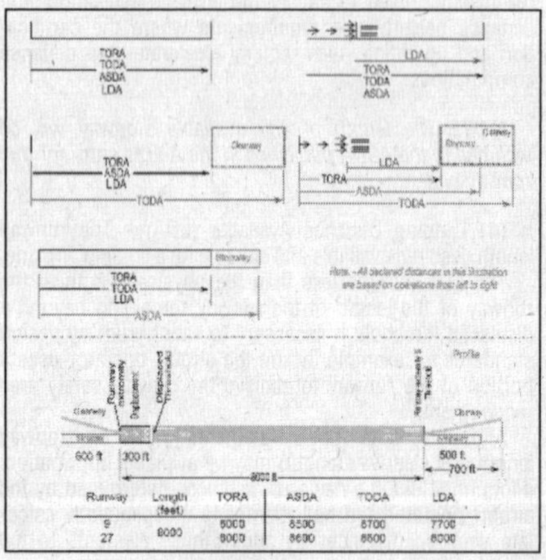

FIG 4-3-5
Effects of a Geographical Constraint on a Runway's Declared Distances

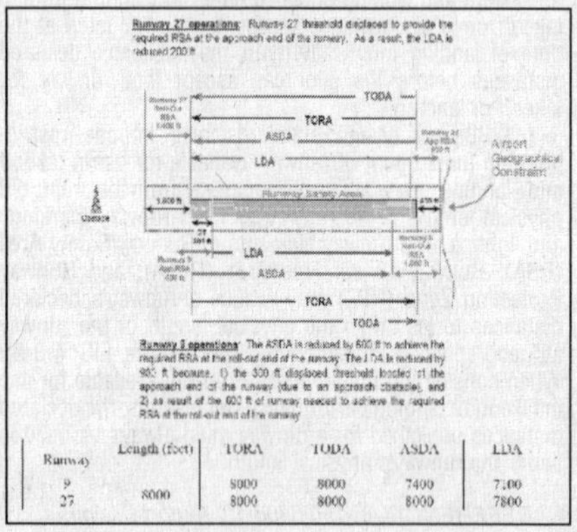

NOTE-*A runway's RSA begins a set distance prior to the threshold and will extend a set distance beyond the end of the runway depending on the runway's design criteria. If these required lengths cannot be achieved, the ASDA and/or LDA will be reduced as necessary to obtain the required lengths to the extent practicable.*

4-3-7. Low Level Wind Shear/Microburst Detection Systems

Low Level Wind Shear Alert System (LLWAS), Terminal Doppler Weather Radar (TDWR), Weather System Processor (WSP), and Integrated Terminal Weather System (ITWS) display information on hazardous wind shear and microburst activity in the vicinity of an airport to air traffic controllers who relay this information to pilots.

a. LLWAS provides wind shear alert and gust front information but does not provide microburst alerts. The LLWAS is designed to detect low level wind shear conditions around the periphery of an airport. It does not detect wind shear beyond that limitation. Controllers will provide this information to pilots by giving the pilot the airport wind followed by the boundary wind.

EXAMPLE-*Wind shear alert, airport wind 230 at 8, south boundary wind 170 at 20.*

b. LLWAS "network expansion," (LLWAS NE) and LLWAS Relocation/Sustainment (LLWAS-RS) are systems integrated with TDWR. These systems provide the capability of detecting microburst alerts and wind shear alerts. Controllers will issue the appropriate wind shear alerts or microburst alerts. In some of these systems controllers also have the ability to issue wind information oriented to the threshold or departure end of the runway.

EXAMPLE-*Runway 17 arrival microburst alert, 40 knot loss 3 mile final.*
REFERENCE-*AIM, Microbursts, Paragraph 7-1-26.*

c. More advanced systems are in the field or being developed such as ITWS. ITWS provides alerts for microbursts, wind shear, and significant thunderstorm activity. ITWS displays wind information oriented to the threshold or departure end of the runway.

d. The WSP provides weather processor enhancements to selected Airport Surveillance Radar (ASR)-9 facilities. The WSP provides Air Traffic with detection and alerting of hazardous weather such as wind shear, microbursts, and significant thunderstorm activity. The WSP displays terminal area 6 level weather, storm cell locations and movement, as well as the location and predicted future position and intensity of wind shifts that may affect airport operations. Controllers will receive and issue alerts based on Areas Noted for Attention (ARENA). An ARENA extends on the runway center line from a 3 mile final to the runway to a 2 mile departure.

e. An airport equipped with the LLWAS, ITWS, or WSP is so indicated in the Airport/Facility Directory under Weather Data Sources for that particular airport.

4-3-8. Braking Action Reports and Advisories

a. When available, ATC furnishes pilots the quality of braking action received from pilots or airport management. The quality of braking action is described by the terms "good," "fair," "poor," and "nil," or a combination of these terms. When pilots report the quality of braking action by using the terms noted above, they should use descriptive terms that are easily understood, such as, "braking action poor the first/last half of the runway," together with the particular type of aircraft.

b. For NOTAM purposes, braking action reports are classified according to the most critical term ("fair," "poor," or "nil") used and issued as a NOTAM(D).

c. When tower controllers have received runway braking action reports which include the terms *poor* or *nil*, or whenever weather conditions are conducive to deteriorating or rapidly changing runway braking conditions, the tower will include on the ATIS broadcast the statement, *"BRAKING ACTION ADVISORIES ARE IN EFFECT."*

d. During the time that braking action advisories are in effect, ATC will issue the latest braking action report for the runway in use to each arriving and departing aircraft. Pilots should be prepared for deteriorating braking conditions and should request current runway condition information if not volunteered by controllers. Pilots should also be prepared to provide a descriptive runway condition report to controllers after landing.

4-3-9. Runway Friction Reports and Advisories

a. Friction is defined as the ratio of the tangential force needed to maintain uniform relative motion between two contacting surfaces (aircraft tires to the pavement surface) to the perpendicular force holding them in contact (distributed aircraft weight to the aircraft tire area). Simply stated, friction quantifies slipperiness of pavement surfaces.

b. The greek letter MU (pronounced "myew"), is used to designate a friction value representing runway surface conditions.

c. MU (friction) values range from 0 to 100 where zero is the lowest friction value and 100 is the maximum friction value obtainable. For frozen contaminants on runway surfaces, a MU value of 40 or less is the level when the aircraft braking performance starts to deteriorate and directional control begins to be less responsive. The lower the MU value, the less effective braking performance becomes and the more difficult directional control becomes.

d. At airports with friction measuring devices, airport management should conduct friction measurements on runways covered with compacted snow and/or ice.

1. Numerical readings may be obtained by using any FAA approved friction measuring device. As these devices do not provide equal numerical readings on contaminated surfaces, it is necessary to designate the type of friction measuring device used.

2. When the MU value for any one-third zone of an active runway is 40 or less, a report should be given to ATC by airport management for dissemination to pilots. The report will identify the runway, the time of measurement, the type of friction measuring device used, MU values for each zone, and the contaminant conditions, e.g., wet snow, dry snow, slush, deicing chemicals, etc. Measurements for each one-third zone will be given in the direction of takeoff and landing on the runway. A report should also be given when MU values rise above 40 in all zones of a runway previously reporting a MU below 40.

3. Airport management should initiate a NOTAM(D) when the friction measuring device is out of service.

e. When MU reports are provided by airport management, the ATC facility providing approach control or local airport advisory will provide the report to any pilot upon request.

f. Pilots should use MU information with other knowledge including aircraft performance characteristics, type, and weight, previous experience, wind conditions, and aircraft tire type (i.e., bias ply vs. radial constructed) to determine runway suitability.

g. No correlation has been established between MU values and the descriptive terms "good," "fair," "poor," and "nil" used in braking action reports.

4-3-10. Intersection Takeoffs

a. In order to enhance airport capacities, reduce taxiing distances, minimize departure delays, and provide for more efficient movement of air traffic, controllers may initiate intersection takeoffs as well as approve them when the pilot requests. If for ANY reason a pilot prefers to use a different intersection or the full length of the runway or desires to obtain the distance between the intersection and the runway end, THE PILOT IS EXPECTED TO INFORM ATC ACCORDINGLY.

b. Pilots are expected to assess the suitability of an intersection for use at takeoff during their preflight planning. They must consider the resultant length reduction to the published runway length and to the published declared distances from the intersection intended to be used for takeoff. The minimum runway required for takeoff must fall within the reduced runway length and the reduced declared distances before the intersection can be accepted for takeoff.

REFERENCE-*AIM, Use of Runways/Declared Distances, Paragraph 4-3-6.*

c. Controllers will issue the measured distance from the intersection to the runway end rounded "down" to the nearest 50 feet to any pilot who requests and to all military aircraft, unless use of the intersection is covered in appropriate directives. Controllers, however, will not be able to inform pilots of the distance from the intersection to the end of any of the published declared distances.

REFERENCE-*FAAO JO 7110.65, Ground Traffic Movement, Paragraph 3-7-1.*

d. An aircraft is expected to taxi to (but not onto) the end of the assigned runway unless prior approval for an intersection departure is received from ground control.

e. Pilots should state their position on the airport when calling the tower for takeoff from a runway intersection.

EXAMPLE-*Cleveland Tower, Apache Three Seven Two Two Papa, at the intersection of taxiway Oscar and runway two three right, ready for departure.*

f. Controllers are required to separate small aircraft (12,500 pounds or less, maximum certificated takeoff weight) departing (same or opposite direction) from an intersection behind a large nonheavy aircraft on the same runway, by ensuring that at least a 3-minute interval exists between the time the preceding large aircraft has taken off and the succeeding small aircraft begins takeoff roll. To inform the pilot of the required 3-minute hold, the controller will state, "Hold for wake turbulence." If after considering wake turbulence hazards, the pilot feels that a lesser time interval is appropriate, the pilot may request a waiver to the 3-minute interval. To initiate such a request, simply say "Request waiver to 3-minute interval," or a similar statement. Controllers may then issue a takeoff clearance if other traffic permits, since the pilot has accepted the responsibility for wake turbulence separation.

g. The 3-minute interval is not required when the intersection is 500 feet or less from the departure point of the preceding aircraft and both aircraft are taking off in the same direction. Controllers may permit the small aircraft to alter course after takeoff to avoid the flight path of the preceding departure.

h. The 3-minute interval is mandatory behind a heavy aircraft in all cases.

4-3-11. Pilot Responsibilities When Conducting Land and Hold Short Operations (LAHSO)

a. LAHSO is an acronym for "Land and Hold Short Operations." These operations include landing and holding short of an **intersecting runway**, an **intersecting taxiway**, or some other designated **point on a runway** other than an intersecting runway or taxiway. (See FIG 4-3-6, FIG 4-3-7, FIG 4-3-8.)

b. Pilot Responsibilities and Basic Procedures.

1. LAHSO is an air traffic control procedure that requires pilot participation to balance the needs for increased airport capacity and system efficiency, consistent with safety. This procedure can be done safely **provided** pilots and controllers are knowledgeable and understand their responsibilities. The following paragraphs outline specific pilot/operator responsibilities when conducting LAHSO.

2. At controlled airports, air traffic may clear a pilot to land and hold short. Pilots may accept such a clearance provided that the pilot-in-command determines that the aircraft can safely land and stop within the Available Landing Distance (ALD). ALD data are published in the special notices section of the Airport/Facility Directory (A/FD) and in the U.S. Terminal Procedures Publications. Controllers will also provide ALD data upon request. Student pilots or pilots not familiar with LAHSO should **not** participate in the program.

3. The pilot-in-command has the final authority to accept or decline any land and hold short clearance. The safety and operation of the aircraft remain the responsibility of the pilot. Pilots are expected to decline a LAHSO clearance if they determine it will compromise safety.

4. To conduct LAHSO, pilots should become familiar with all available information concerning LAHSO at their destination airport. Pilots should have, *readily available*,the **published ALD** and runway **slope information** for **all** LAHSO runway combinations at **each** airport of intended landing. Additionally, knowledge about landing performance data permits the pilot to *readily* determine that the ALD for the assigned runway is sufficient for safe LAHSO. As part of a pilot's preflight planning process, pilots should determine if their destination airport has LAHSO. If so, their preflight planning process should include an assessment of which LAHSO combinations would work for them given their aircraft's required landing distance. Good pilot decision making is knowing in advance whether one can accept a LAHSO clearance if offered.

FIG 4-3-6
Land and Hold Short of an Intersecting Runway

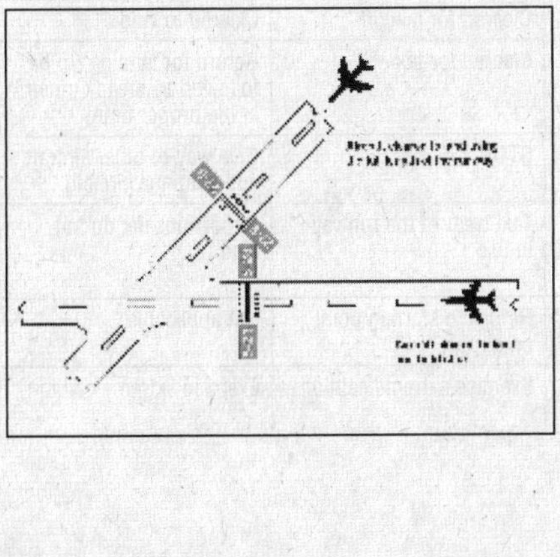

EXAMPLE-*FIG 4-3-8 - holding short at a designated point may be required to avoid conflicts with the runway safety area/flight path of a nearby runway.*

NOTE-*Each figure shows the approximate location of LAHSO markings, signage, and in-pavement lighting when installed.*

REFERENCE-*AIM, Chapter 2, Aeronautical Lighting and Other Airport Visual Aids.*

FIG 4-3-7
Land and Hold Short of an Intersecting Taxiway

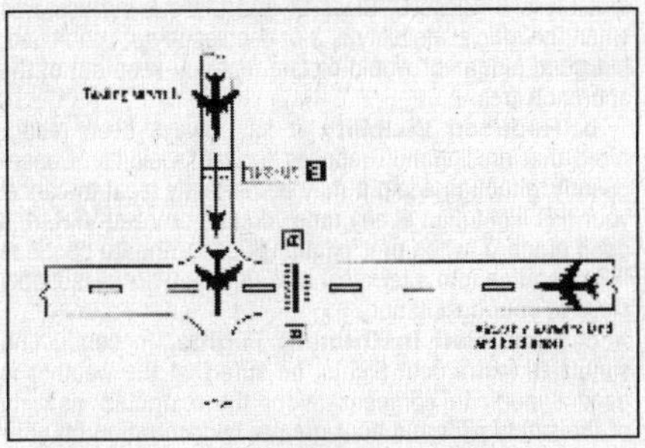

FIG 4-3-8
Land and Hold Short of a Designated Point on a Runway Other Than an Intersecting Runway or Taxiway

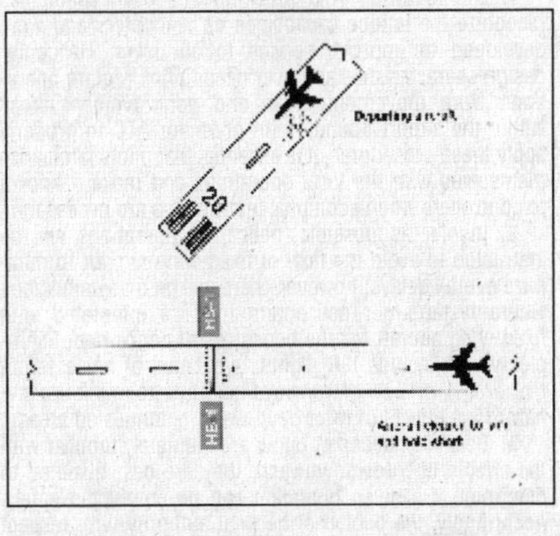

5. If, for any reason, such as difficulty in discerning the location of a LAHSO intersection, wind conditions, aircraft condition, etc., the pilot elects to request to land on the full length of the runway, to land on another runway, or to decline LAHSO, a pilot is expected to promptly inform air traffic, ideally even before the clearance is issued. **A LAHSO clearance, once accepted, must be adhered to, just as any other ATC clearance, unless an amended clearance is obtained or an emergency occurs. A LAHSO clearance does not preclude a rejected landing.**

6. A pilot who accepts a LAHSO clearance should land and exit the runway at the first convenient taxiway (unless directed otherwise) before reaching the hold short point. Otherwise, the pilot must stop and hold at the hold short point. **If a rejected landing becomes necessary after accepting a LAHSO clearance, the pilot should maintain safe separation from other aircraft or vehicles, and should promptly notify the controller.**

7. Controllers need a full read back of all LAHSO clearances. Pilots should read back their LAHSO clearance and include the words, "HOLD SHORT OF (RUNWAY/TAXIWAY/OR POINT)" in their acknowledgment of **all** LAHSO clearances. In order to reduce frequency congestion, pilots are encouraged to read back the LAHSO clearance without prompting. Don't make the controller have to ask for a read back!

c. LAHSO Situational Awareness

1. Situational awareness is **vital** to the success of LAHSO. Situational awareness starts with having current airport information in the cockpit, readily accessible to the pilot. (An airport diagram assists pilots in identifying their location on the airport, thus reducing requests for "progressive taxi instructions" from controllers.)

2. Situational awareness includes effective pilot-controller radio communication. ATC expects pilots to specifically acknowledge and read back all LAHSO clearances as follows:

EXAMPLE-ATC: *"(Aircraft ID) cleared to land runway six right, hold short of taxiway bravo for crossing traffic (type aircraft)."*

Aircraft: *"(Aircraft ID), wilco, cleared to land runway six right to hold short of taxiway bravo."*

ATC: *"(Aircraft ID) cross runway six right at taxiway bravo, landing aircraft will hold short."*

Aircraft: *"(Aircraft ID), wilco, cross runway six right at bravo, landing traffic (type aircraft) to hold."*

3. For those airplanes flown with two crewmembers, effective **intra-cockpit**communication between cockpit crewmembers is also critical. There have been several instances where the pilot working the radios accepted a LAHSO clearance but then simply forgot to tell the pilot flying the aircraft.

4. Situational awareness also includes a thorough understanding of the airport markings, signage, and lighting associated with LAHSO. These visual aids consist of a three-part system of **yellow hold-short markings, red and white signage** and, in certain cases, **in-pavement lighting**. Visual aids assist the pilot in determining where to hold short. FIG 4-3-6, FIG 4-3-7, FIG 4-3-8 depict how these markings, signage, and lighting combinations will appear once installed. Pilots are cautioned that not all airports conducting LAHSO have installed any or all of the above markings, signage, or lighting.

5. Pilots should only receive a LAHSO clearance when there is a minimum ceiling of 1,000feet and 3 statute miles visibility. The intent of having "basic" VFR weather conditions is to allow pilots to maintain visual contact with other aircraft and ground vehicle operations. Pilots should consider the effects of prevailing inflight visibility (such as landing into the sun) and how it may affect overall situational awareness. Additionally, surface vehicles and aircraft being taxied by maintenance personnel may also be participating in LAHSO, especially in those operations that involve crossing an active runway.

4-3-12. Low Approach

a. A low approach (sometimes referred to as a low pass) is the go-around maneuver following an approach. Instead of landing or making a touch-and-go, a pilot may wish to go around (low approach) in order to expedite a particular operation (a series of practice instrument approaches is an example of such an operation). Unless otherwise authorized by ATC, the low approach should be made straight ahead, with no turns or climb made until the pilot has made a thorough visual check for other aircraft in the area.

b. When operating within a Class B, Class C, and Class D surface area, a pilot intending to make a low approach should contact the tower for approval. This request should be made prior to starting the final approach.

c. When operating to an airport, not within a ClassB, Class C, and Class D surface area, a pilot intending to make a low approach should, prior to leaving the final approach fix inbound (nonprecision approach) or the outer marker or fix used in lieu of the outer marker inbound (precision approach), so advise the FSS, UNICOM, or make a broadcast as appropriate.

REFERENCE-*AIM, Traffic Advisory Practices at Airports Without Operating Control Towers, Paragraph 4-1-9.*

4-3-13. Traffic Control Light Signals

a. The following procedures are used by ATCTs in the control of aircraft, ground vehicles, equipment, and personnel not equipped with radio. These same procedures will be used to control aircraft, ground vehicles, equipment, and personnel equipped with radio if radio contact cannot be established. ATC personnel use a directive traffic control signal which emits an intense narrow light beam of a selected color (either red, white, or green) when controlling traffic by light signals.

b. Although the traffic signal light offers the advantage that some control may be exercised over nonradio equipped aircraft, pilots should be cognizant of the disadvantages which are:

1. Pilots may not be looking at the control tower at the time a signal is directed toward their aircraft.

2. The directions transmitted by a light signal are very limited since only approval or disapproval of a pilot's anticipated actions may be transmitted. No supplement or explanatory information may be transmitted except by the use of the "General Warning Signal" which advises the pilot to be on the alert.

c. Between sunset and sunrise, a pilot wishing to attract the attention of the control tower should turn on a landing light and taxi the aircraft into a position, clear of the active runway, so that light is visible to the tower. The landing light should remain on until appropriate signals are received from the tower.

d. Air Traffic Control Tower Light Gun Signals. (See TBL 4-3-1.)

e. During daylight hours, acknowledge tower transmissions or light signals by moving the ailerons or rudder. At night, acknowledge by blinking the landing or navigation lights. If radio malfunction occurs after departing the parking area, watch the tower for light signals or monitor tower frequency.

4-3-14. Communications

a. Pilots of departing aircraft should communicate with the control tower on the appropriate ground control/clearance delivery frequency prior to starting engines to receive engine start time, taxi and/or clearance information. Unless otherwise advised by the tower, remain on that frequency during taxiing and runup, then change to local control frequency when ready to request takeoff clearance.

NOTE- *Pilots are encouraged to monitor the local tower frequency as soon as practical consistent with other ATC requirements.*

REFERENCE- *AIM, Automatic Terminal Information Service (ATIS), Paragraph 4-1-13.*

b. The tower controller will consider that pilots of turbine-powered aircraft are ready for takeoff when they reach the runway or warm-up block unless advised otherwise.

c. The majority of ground control frequencies are in the 121.6-121.9 MHz bandwidth. Ground control frequencies are provided to eliminate frequency congestion on the tower (local control) frequency and are limited to communications between the tower and aircraft on the ground and between the tower and utility vehicles on the airport, provide a clear VHF channel for arriving and departing aircraft. They are used for issuance of taxi information, clearances, and other necessary contacts between the tower and aircraft or other vehicles operated on the airport. A pilot who has just landed should not change from the tower frequency to the ground control frequency until directed to do so by the controller. Normally, only one ground control frequency is assigned at an airport; however, at locations where the amount of traffic so warrants, a second ground control frequency and/or another frequency designated as a clearance delivery frequency, may be assigned.

d. A controller may omit the ground or local control frequency if the controller believes the pilot knows which frequency is in use. If the ground control frequency is in the 121 MHz bandwidth the controller may omit the numbers preceding the decimal point; e.g., 121.7, "CONTACT GROUND POINT SEVEN." However, if any doubt exists as to what frequency is in use, the pilot should promptly request the controller to provide that information.

e. Controllers will normally avoid issuing a radio frequency change to helicopters, known to be single-piloted, which are hovering, air taxiing, or flying near the ground. At times, it may be necessary for pilots to alert ATC regarding single pilot operations to minimize delay of essential ATC communications. Whenever possible, ATC instructions will be relayed through the frequency being monitored until a frequency change can be accomplished. You must promptly advise ATC if you are unable to comply with a frequency change. Also, you should advise ATC if you must land to accomplish the frequency change unless it is clear the landing will have no impact on other air traffic; e.g., on a taxiway or in a helicopter operating area.

4-3-15. Gate Holding Due to Departure Delays

a. Pilots should contact ground control or clearance delivery prior to starting engines as gate hold procedures will be in effect whenever departure delays exceed or are anticipated to exceed 15 minutes. The sequence for departure will be maintained in accordance with initial call up unless modified by flow control restrictions. Pilots should monitor the ground control or clearance delivery frequency for engine startup advisories or new proposed start time if the delay changes.

b. The tower controller will consider that pilots of turbine-powered aircraft are ready for takeoff when they reach the runway or warm-up block unless advised otherwise.

4-3-16. VFR Flights in Terminal Areas

Use reasonable restraint in exercising the prerogative of VFR flight, especially in terminal areas. The weather minimums and distances from clouds are minimums. Giving yourself a greater margin in specific instances is just good judgment.

a. Approach Area. Conducting a VFR operation in a Class B, Class C, Class D, and Class E surface area when the official visibility is 3 or 4 miles is not prohibited, but good judgment would dictate that you keep out of the approach area.

b. Reduced Visibility. It has always been recognized that precipitation reduces forward visibility. Consequently, although again it may be perfectly legal to cancel your IFR flight plan at any time you can proceed VFR, it is good practice, when precipitation is occurring, to continue IFR operation into a terminal area until you are reasonably close to your destination.

c. Simulated Instrument Flights. In conducting simulated instrument flights, be sure that the weather is good enough to compensate for the restricted visibility of the safety pilot and your greater concentration on your flight instruments. Give yourself a little greater margin when your flight plan lies in or near a busy airway or close to an airport.

4-3-17. VFR Helicopter Operations at Controlled Airports

a. General.

1. The following ATC procedures and phraseologies recognize the unique capabilities of helicopters and were developed to improve service to all users. Helicopter design characteristics and user needs often require operations from movement areas and nonmovement areas within the airport boundary. In order for ATC to properly apply these procedures, it is essential that pilots familiarize themselves with the local operations and make it known to controllers when additional instructions are necessary.

2. Insofar as possible, helicopter operations will be instructed to avoid the flow of fixed-wing aircraft to minimize overall delays; however, there will be many situations where faster/larger helicopters may be integrated with fixed-wing aircraft for the benefit of all concerned. Examples would include IFR flights, avoidance of noise sensitive areas, or use of runways/taxiways to minimize the hazardous effects of rotor downwash in congested areas.

3. Because helicopter pilots are intimately familiar with the effects of rotor downwash, they are best qualified to determine if a given operation can be conducted safely. Accordingly, the pilot has the final authority with respect to the specific airspeed/altitude combinations. ATC clearances are in no way intended to place the helicopter in a hazardous position. It is expected that pilots will advise ATC if a specific clearance will cause undue hazards to persons or property.

b. Controllers normally limit ATC ground service and instruction to *movement* areas; therefore, operations from *nonmovement* areas are conducted at pilot discretion and should be based on local policies, procedures, or letters of agreement. In order to maximize the flexibility of helicopter operations, it is necessary to rely heavily on sound pilot judgment. For example, hazards such as debris, obstructions, vehicles, or personnel must be recognized by the pilot, and action should be taken as necessary to avoid such hazards. Taxi, hover taxi, and air taxi operations are considered to be ground movements. Helicopters conducting such operations are expected to adhere to the same conditions, requirements, and practices as apply to other ground taxiing and ATC procedures in the AIM.

1. The phraseology *taxi* is used when it is intended or expected that the helicopter will taxi on the airport surface, either via taxiways or other prescribed routes. *Taxi* is used primarily for helicopters equipped with wheels or in response to a pilot request. Preference should be given to this procedure whenever it is necessary to minimize effects of rotor downwash.

2. Pilots may request a *hover taxi* when slow forward movement is desired or when it may be appropriate to move very short distances. Pilots should avoid this procedure if rotor downwash is likely to cause damage to parked aircraft or if blowing dust/snow could obscure visibility. If it is necessary to operate above 25 feet AGL when hover taxiing, the pilot should initiate a request to ATC.

3. *Air taxi* is the preferred method for helicopter ground movements on airports provided ground operations and conditions permit. Unless otherwise requested or instructed, pilots are expected to remain below 100 feet AGL. However, if a higher than normal airspeed or altitude is desired, the request should be made prior to lift-off. The pilot is solely responsible for selecting a safe airspeed for the altitude/operation being conducted. Use of *air taxi* enables the pilot to proceed at an optimum airspeed/altitude, minimize downwash effect, conserve fuel, and expedite movement from one point to another. Helicopters should avoid overflight of other aircraft, vehicles, and personnel during air-taxi operations. Caution must be exercised concerning active runways and pilots must be certain that air taxi instructions are understood. Special precautions may be necessary at unfamiliar airports or airports with multiple/intersecting active runways. The taxi procedures given in Paragraph 4-3-18, Taxiing, Paragraph 4-3-19, Taxi During Low Visibility, and Paragraph 4-3-20, Exiting the Runway After Landing, also apply.

REFERENCE- *Pilot/Controller Glossary Term- Taxi.*
Pilot/Controller Glossary Term- Hover Taxi.
Pilot/Controller Glossary Term- Air Taxi.

c. Takeoff and Landing Procedures.

1. Helicopter operations may be conducted from a runway, taxiway, portion of a landing strip, or any clear area which could be used as a landing site such as the scene of an accident, a construction site, or the roof of a building. The terms used to describe designated areas from which helicopters operate are: movement area, landing/takeoff area, apron/ramp, heliport and helipad (See Pilot/Controller Glossary). These areas may be improved or unimproved and may be separate from or located on an airport/heliport. ATC will issue takeoff clearances from *movement* areas other than active runways, or in diverse directions from active runways, with additional instructions as necessary. Whenever

TBL 4-3-1
Air Traffic Control Tower Light Gun Signals

	Meaning		
Color and Type of Signal	**Movement of Vehicles, Equipment and Personnel**	**Aircraft on the Ground**	**Aircraft in Flight**
Steady green	Cleared to cross, proceed or go	Cleared for takeoff	Cleared to land
Flashing green	Not applicable	Cleared for taxi	Return for landing (to be followed by steady green at the proper time)
Steady red	STOP	STOP	Give way to other aircraft and continue circling
Flashing red	Clear the taxiway/runway	Taxi clear of the runway in use	Airport unsafe, do not land
Flashing white	Return to starting point on airport	Return to starting point on airport	Not applicable
Alternating red and green	Exercise extreme caution	Exercise extreme caution	Exercise extreme caution

possible, takeoff clearance will be issued in lieu of extended hover/air taxi operations. Phraseology will be "CLEARED FOR TAKEOFF FROM (taxiway, helipad, runway number, etc.), MAKE RIGHT/ LEFT TURN FOR (direction, heading, NAVAID radial) DEPARTURE/DEPARTURE ROUTE (number, name, etc.)." Unless requested by the pilot, downwind takeoffs will not be issued if the tailwind exceeds 5 knots.

2. Pilots should be alert to wind information as well as to wind indications in the vicinity of the helicopter. ATC should be advised of the intended method of departing. A pilot request to takeoff in a given direction indicates that the pilot is willing to accept the wind condition and controllers will honor the request if traffic permits. Departure points could be a significant distance from the control tower and it may be difficult or impossible for the controller to determine the helicopter's relative position to the wind.

3. If takeoff is requested from *nonmovement* areas, an area not authorized for helicopter use, an area not visible from the tower, an unlighted area at night, or an area off the airport, the phraseology "DEPARTURE FROM (requested location) WILL BE AT YOUR OWN RISK (additional instructions, as necessary). USE CAUTION (if applicable)." The pilot is responsible for operating in a safe manner and should exercise due caution.

4. Similar phraseology is used for helicopter landing operations. Every effort will be made to permit helicopters to proceed direct and land as near as possible to their final destination on the airport. Traffic density, the need for detailed taxiing instructions, frequency congestion, or other factors may affect the extent to which service can be expedited. As with ground movement operations, a high degree of pilot/controller cooperation and communication is necessary to achieve safe and efficient operations.

4-3-18. Taxiing

a. General. Approval must be obtained prior to moving an aircraft or vehicle onto the movement area during the hours an Airport Traffic Control Tower is in operation.

1. Always state your position on the airport when calling the tower for taxi instructions.

2. The movement area is normally described in local bulletins issued by the airport manager or control tower. These bulletins may be found in FSSs, fixed base operators offices, air carrier offices, and operations offices.

3. The control tower also issues bulletins describing areas where they cannot provide ATC service due to nonvisibility or other reasons.

4. A clearance must be obtained prior to taxiing on a runway, taking off, or landing during the hours an Airport Traffic Control Tower is in operation.

5. A clearance must be obtained prior to crossing any runway. ATC will issue an explicit clearance for all runway crossings.

6. When assigned a takeoff runway, ATC will first specify the runway, issue taxi instructions, and state any hold short instructions or runway crossing clearances if the taxi route will cross a runway. This does not authorize the aircraft to "enter" or "cross" the assigned departure runway at any point. In order to preclude misunderstandings in radio communications, ATC will not use the word "cleared" in conjunction with authorization for aircraft to taxi.

7. When issuing taxi instructions to any point other than an assigned takeoff runway, ATC will specify the point to taxi to, issue taxi instructions, and state any hold short instructions or runway crossing clearances if the taxi route will cross a runway.

NOTE-*ATC is required to obtain a readback from the pilot of all runway hold short instructions.*

8. If a pilot is expected to hold short of a runway approach ("APPCH") area or ILS holding position (see FIG 2-3-15, Taxiways Located in Runway Approach Area), ATC will issue instructions.

9. When taxi instructions are received from the controller, pilots should always read back:

(a) The runway assignment.

(b) Any clearance to enter a specific runway.

(c) Any instruction to hold short of a specific runway or line up and wait.

Controllers are required to request a readback of runway hold short assignment when it is not received from the pilot/vehicle.

b. ATC clearances or instructions pertaining to taxiing are predicated on known traffic and known physical airport conditions. Therefore, it is important that pilots clearly understand the clearance or instruction. Although an ATC clearance is issued for taxiing purposes, when operating in accordance with the CFRs, it is the responsibility of the pilot to avoid collision with other aircraft. Since "the pilot-in-command of an aircraft is directly responsible for, and is the final authority as to, the operation of that aircraft" the pilot should obtain clarification of any clearance or instruction which is not understood.

REFERENCE-*AIM, General, Paragraph 7-3-1.*

1. Good operating practice dictates that pilots acknowledge all runway crossing, hold short, or takeoff clearances unless there is some misunderstanding, at which time the pilot should query the controller until the clearance is understood.

NOTE-*Air traffic controllers are required to obtain from the pilot a readback of all runway hold short instructions.*

2. Pilots operating a single pilot aircraft should monitor only assigned ATC communications after being cleared onto the active runway for departure. Single pilot aircraft should not monitor other than ATC communications until flight from Class B, Class C, or Class D surface area is completed. This same procedure should be practiced from after receipt of the clearance for landing until the landing and taxi activities are complete. Proper effective scanning for other aircraft, surface vehicles, or other objects should be continuously exercised in all cases.

3. If the pilot is unfamiliar with the airport or for any reason confusion exists as to the correct taxi routing, a request may be made for progressive taxi instructions which include step-by-step routing directions. Progressive instructions may also be issued if the controller deems it necessary due to traffic or field conditions (for example, construction or closed taxiways).

c. At those airports where the U.S. Government operates the control tower and ATC has authorized noncompliance with the requirement for two-way radio communications while operating within the Class B, Class C, or Class D surface area, or at those airports where the U.S. Government does not operate the control tower and radio communications cannot be established, pilots must obtain a clearance by visual light signal prior to taxiing on a runway and prior to takeoff and landing.

d. The following phraseologies and procedures are used in radiotelephone communications with aeronautical ground stations.

1. Request for taxi instructions prior to departure. State your aircraft identification, location, type of operation planned (VFR or IFR), and the point of first intended landing.

EXAMPLE-Aircraft: *"Washington ground, Beechcraft One Three One Five Niner at hangar eight, ready to taxi, I-F-R to Chicago."*

Tower: *"Beechcraft one three one five niner, Washington ground, runway two seven, taxi via taxiways Charlie and Delta, hold short of runway three three left."*

Aircraft: *"Beechcraft One Three One Five Niner, hold short of runway three three left."*

2. Receipt of ATC clearance. ARTCC clearances are relayed to pilots by airport traffic controllers in the following manner.

EXAMPLE-Tower: *"Beechcraft One Three One Five Niner, cleared to the Chicago Midway Airport via Victor Eight, maintain eight thousand."*

Aircraft: *"Beechcraft One Three One Five Niner, cleared to the Chicago Midway Airport via Victor Eight, maintain eight thousand."*

NOTE-*Normally, an ATC IFR clearance is relayed to a pilot by the ground controller. At busy locations, however, pilots may be instructed by the ground controller to "contact clearance delivery" on a frequency designated for this purpose. No surveillance or control over the movement of traffic is exercised by this position of operation.*

3. Request for taxi instructions after landing. State your aircraft identification, location, and that you request taxi instructions.

EXAMPLE-Aircraft: *"Dulles ground, Beechcraft One Four Two Six One clearing runway one right on taxiway echo three, request clearance to Page."*

Tower: *"Beechcraft One Four Two Six One, Dulles ground, taxi to Page via taxiways echo three, echo one, and echo niner."*

or

Aircraft: *"Orlando ground, Beechcraft One Four Two Six One clearing runway one eight left at taxiway bravo three, request clearance to Page."*

Tower: *"Beechcraft One Four Two Six One, Orlando ground, hold short of runway one eight right."*

Aircraft: *"Beechcraft One Four Two Six One, hold short of runway one eight right."*

4-3-19. Taxi During Low Visibility

a. Pilots and aircraft operators should be constantly aware that during certain low visibility conditions the movement of aircraft and vehicles on airports may not be visible to the tower controller. This may prevent visual confirmation of an aircraft's adherence to taxi instructions.

b. Of vital importance is the need for pilots to notify the controller when difficulties are encountered or at the first indication of becoming disoriented. Pilots should proceed with extreme caution when taxiing toward the sun. When vision difficulties are encountered pilots should immediately inform the controller.

c. Advisory Circular 120-57, Surface Movement Guidance and Control System, commonly known as SMGCS (pronounced "SMIGS") requires a low visibility taxi plan for any airport which has takeoff or landing operations in less than 1,200 feet runway visual range (RVR) visibility conditions. These plans, which affect aircrew and vehicle operators, may incorporate additional lighting, markings, and procedures to control airport surface traffic. They will be addressed at two levels; operations less than 1,200 feet RVR to 600 feet RVR and operations less than 600 feet RVR.

NOTE-*Specific lighting systems and surface markings may be found in paragraph 2-1-11, Taxiway Lights, and paragraph 2-3-4, Taxiway Markings.*

d. When low visibility conditions exist, pilots should focus their entire attention on the safe operation of the aircraft while it is moving. Checklists and nonessential communication should be withheld until the aircraft is stopped and the brakes set.

4-3-20. Exiting the Runway After Landing

The following procedures must be followed after landing and reaching taxi speed.

a. Exit the runway without delay at the first available taxiway or on a taxiway as instructed by ATC. Pilots must not exit the landing runway onto another runway unless authorized by ATC. At airports with an operating control tower, pilots should not stop or reverse course on the runway without first obtaining ATC approval.

b. Taxi clear of the runway unless otherwise directed by ATC. An aircraft is considered clear of the runway when all parts of the aircraft are past the runway edge and there are no restrictions to its continued movement beyond the runway holding position markings. In the absence of ATC instructions, the pilot is expected to taxi clear of the landing runway by taxiing beyond the runway holding position markings associated with the landing runway, even if that requires the aircraft to protrude into or cross another taxiway or ramp area. Once all parts of the aircraft have crossed the runway holding position markings, the pilot must hold unless further instructions have been issued by ATC.

NOTE-

1. The tower will issue the pilot instructions which will permit the aircraft to enter another taxiway, runway, or ramp area when required.

2. Guidance contained in subparagraphs a and b above is considered an integral part of the landing clearance and satisfies the requirement of 14 CFR Section 91.129.

c. Immediately change to ground control frequency when advised by the tower and obtain a taxi clearance.

NOTE-

1. The tower will issue instructions required to resolve any potential conflictions with other ground traffic prior to advising the pilot to contact ground control.

2. Ground control will issue taxi clearance to parking. That clearance does not authorize the aircraft to "enter" or "cross" any runways. Pilots not familiar with the taxi route should request specific taxi instructions from ATC.

4-3-21. Practice Instrument Approaches

a. Various air traffic incidents have indicated the necessity for adoption of measures to achieve more organized and controlled operations where practice instrument approaches are conducted. Practice instrument approaches are considered to be instrument approaches made by either a VFR aircraft not on an IFR flight plan or an aircraft on an IFR flight plan. To achieve this and thereby enhance air safety, it is Air Traffic's policy to provide for separation of such operations at locations where approach control facilities are located and, as resources permit, at certain other locations served by ARTCCs or parent approach control facilities. Pilot requests to practice instrument approaches may be approved by ATC subject to traffic and workload conditions. Pilots should anticipate that in some instances the controller may find it necessary to deny approval or withdraw previous approval when traffic conditions warrant. It must be clearly understood, however, that even though the controller may be providing separation, pilots on VFR flight plans are required to comply with basic VFR weather minimums (14 CFR Section 91.155). Application of ATC procedures or any action taken by the controller to avoid traffic conflictions does not relieve IFR and VFR pilots of their responsibility to see-and-avoid other traffic while operating in VFR conditions (14 CFR Section 91.113). In addition to the normal IFR separation minimums (which includes visual separation) during VFR conditions, 500 feet vertical separation may be applied between VFR aircraft and between a VFR aircraft and the IFR aircraft. Pilots not on IFR flight plans desiring practice instrument approaches should always state `practice' when making requests to ATC. Controllers will instruct VFR aircraft requesting an instrument approach to maintain VFR. This is to preclude misunderstandings between the pilot and controller as to the status of the aircraft. If pilots wish to proceed in accordance with instrument flight rules, they must specifically request and obtain, an IFR clearance.

b. Before practicing an instrument approach, pilots should inform the approach control facility or the tower of the type of practice approach they desire to make and how they intend to terminate it, i.e., full-stop landing, touch-and-go, or missed or low approach maneuver. This information may be furnished progressively when conducting a series of approaches. Pilots on an IFR flight plan, who have made a series of instrument approaches to full stop landings should inform ATC when they make their final landing. The controller will control flights practicing instrument approaches so as to ensure that they do not disrupt the flow of arriving and departing itinerant IFR or VFR aircraft. The priority afforded itinerant aircraft over practice instrument approaches is not intended to be so rigidly applied that it causes grossly inefficient application of services. A minimum delay to itinerant traffic may be appropriate to allow an aircraft practicing an approach to complete that approach.

NOTE-*A clearance to land means that appropriate separation on the landing runway will be ensured. A landing clearance does not relieve the pilot from compliance with any previously issued restriction.*

c. At airports without a tower, pilots wishing to make practice instrument approaches should notify the facility having control jurisdiction of the desired approach as indicated on the approach chart. All approach control facilities and ARTCCs are required to publish a Letter to Airmen depicting those airports where they provide standard separation to both VFR and IFR aircraft conducting practice instrument approaches.

d. The controller will provide approved separation between both VFR and IFR aircraft when authorization is granted to make practice approaches to airports where an approach control facility is located and to certain other airports served by approach control or an ARTCC. Controller responsibility for separation of VFR aircraft begins at the point where the approach clearance becomes effective, or when the aircraft enters Class B or Class C airspace, or a TRSA, whichever comes first.

e. VFR aircraft practicing instrument approaches are not automatically authorized to execute the missed approach procedure. This authorization must be specifically requested by the pilot and approved by the controller. Separation will not be provided unless the missed approach has been approved by ATC.

f. Except in an emergency, aircraft cleared to practice instrument approaches must not deviate from the approved procedure until cleared to do so by the controller.

g. At radar approach control locations when a full approach procedure (procedure turn, etc.,) cannot be approved, pilots should expect to be vectored to a final approach course for a practice instrument approach which is compatible with the general direction of traffic at that airport.

h. When granting approval for a practice instrument approach, the controller will usually ask the pilot to report to the tower prior to or over the final approach fix inbound (nonprecision approaches) or over the outer marker or fix used in lieu of the outer marker inbound (precision approaches).

i. When authorization is granted to conduct practice instrument approaches to an airport with a tower, but where approved standard separation is not provided to aircraft conducting practice instrument approaches, the tower will approve the practice approach, instruct the aircraft to maintain VFR and issue traffic information, as required.

j. When an aircraft notifies a FSS providing Local Airport Advisory to the airport concerned of the intent to conduct a practice instrument approach and whether or not separation is to be provided, the pilot will be instructed to contact the appropriate facility on a specified frequency prior to initiating the approach. At airports where separation is not provided, the FSS will acknowledge the message and issue known traffic information but will neither approve or disapprove the approach.

k. Pilots conducting practice instrument approaches should be particularly alert for other aircraft operating in the local traffic pattern or in proximity to the airport.

4-3-22. Option Approach

The "Cleared for the Option" procedure will permit an instructor, flight examiner or pilot the option to make a touch-and-go, low approach, missed approach, stop-and-go, or full stop landing. This procedure can be very beneficial in a training situation in that neither the student pilot nor examinee would know what maneuver would be accomplished. The pilot should make a request for this procedure passing the final approach fix inbound on an instrument approach or entering downwind for a VFR traffic pattern. The advantages of this procedure as a training aid are that it enables an instructor or examiner to obtain the reaction of a trainee or examinee under changing conditions, the pilot would not have to discontinue an approach in the middle of the procedure due to student error or pilot proficiency requirements, and finally it allows more flexibility and economy in training programs. This procedure will only be used at those locations with an operational control tower and will be subject to ATC approval.

4-3-23. Use of Aircraft Lights

a. Aircraft position lights are required to be lighted on aircraft operated on the surface and in flight from sunset to sunrise. In addition, aircraft equipped with an anti-collision light system are required to operate that light system during all types of operations (day and night). However, during any adverse meteorological conditions, the pilot-in-command may determine that the anti-collision lights should be turned off when their light output would constitute a hazard to safety (14 CFR Section 91.209). Supplementary strobe lights should be turned off on the ground when they adversely affect ground personnel or other pilots, and in flight when there are adverse reflection from clouds.

b. An aircraft anti-collision light system can use one or more rotating beacons and/or strobe lights, be colored either red or white, and have different (higher than minimum) intensities when compared to other aircraft. Many aircraft have both a rotating beacon and a strobe light system.

c. The FAA has a voluntary pilot safety program, Operation Lights On, to enhance the *see-and-avoid* concept. Pilots are encouraged to turn on their landing lights during takeoff; i.e., either after takeoff clearance has been received or when beginning takeoff roll. Pilots are further encouraged to turn on their landing lights when operating below 10,000 feet, day or night, especially when operating within 10 miles of any airport, or in conditions of reduced visibility and in areas where flocks of birds may be expected, i.e., coastal areas, lake areas, around refuse dumps, etc. Although turning on aircraft lights does enhance the *see-and-avoid* concept, pilots should not become complacent about keeping a sharp lookout for other aircraft. Not all aircraft are equipped with lights and some pilots may not have their lights turned on. Aircraft manufacturer's recommendations for operation of landing lights and electrical systems should be observed.

d. Prop and jet blast forces generated by large aircraft have overturned or damaged several smaller aircraft taxiing behind them. To avoid similar results, and in the interest of preventing upsets and injuries to ground personnel from such forces, the FAA recommends that air carriers and commercial operators turn on their rotating beacons anytime their aircraft engines are in operation. General aviation pilots using rotating beacon equipped aircraft are also encouraged to participate in this program which is designed to alert others to the potential hazard. Since this is a voluntary program, exercise caution and do not rely solely on the rotating beacon as an indication that aircraft engines are in operation.

e. Prior to commencing taxi, it is recommended to turn on navigation, position, anticollision, and logo lights (if equipped). To signal intent to other pilots, consider turning on the taxi light when the aircraft is moving or intending to move on the ground, and turning it off when stopped or yielding to other ground traffic. Strobe lights should not be illuminated during taxi if they will adversely affect the vision of other pilots or ground personnel.

f. At the discretion of the pilotincommand, all exterior lights should be illuminated when taxiing on or across any runway. This increases the conspicuousness of the aircraft to controllers and other pilots approaching to land, taxiing, or crossing the runway. Pilots should comply with any equipment operating limitations and consider the effects of landing and strobe lights on other aircraft in their vicinity.

g. When entering the departure runway for takeoff or to "line up and wait," all lights, except for landing lights, should be illuminated to make the aircraft conspicuous to ATC and other aircraft on approach. Landing lights should be turned on when takeoff clearance is received or when commencing takeoff roll at an airport without an operating control tower.

4-3-24. Flight Inspection/'Flight Check' Aircraft in Terminal Areas

a. *Flight check* is a call sign used to alert pilots and air traffic controllers when a FAA aircraft is engaged in flight inspection/certification of NAVAIDs and flight procedures. Flight check aircraft fly preplanned high/low altitude flight patterns such as grids, orbits, DME arcs, and tracks, including low passes along the full length of the runway to verify NAVAID performance.

b. Pilots should be especially watchful and avoid the flight paths of any aircraft using the call sign "Flight Check." These flights will normally receive special handling from ATC. Pilot patience and cooperation in allowing uninterrupted recordings can significantly help expedite flight inspections, minimize costly, repetitive runs, and reduce the burden on the U.S. taxpayer.

4-3-25. Hand Signals

FIG 4-3-9
Signalman Directs Towing

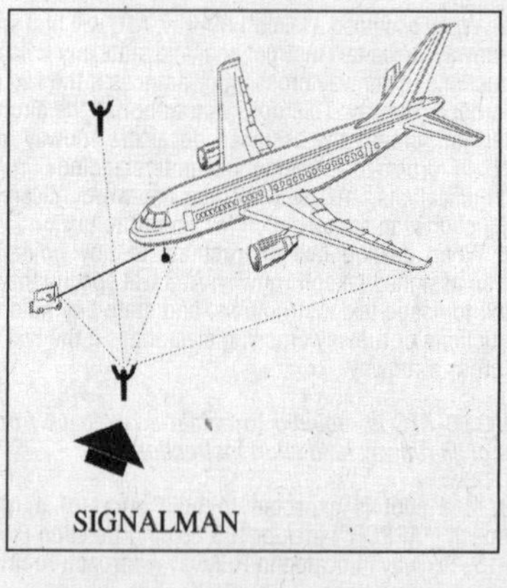

4-3-26. Operations at Uncontrolled Airports With Automated Surface Observing System (ASOS)/ Automated Weather Sensor System(AWSS)/ Automated Weather Observing System (AWOS)

a. Many airports throughout the National Airspace System are equipped with either ASOS, AWSS, or AWOS. At most airports with an operating control tower or human observer, the weather will be available to you in an Aviation Routine Weather Report (METAR) hourly or special observation format on the Automatic Terminal Information Service (ATIS) or directly transmitted from the controller/observer.

b. At uncontrolled airports that are equipped with ASOS/AWSS/AWOS with ground-to-air broadcast capability, the one-minute updated airport weather should

FIG 4-3-10
Signalman's Position

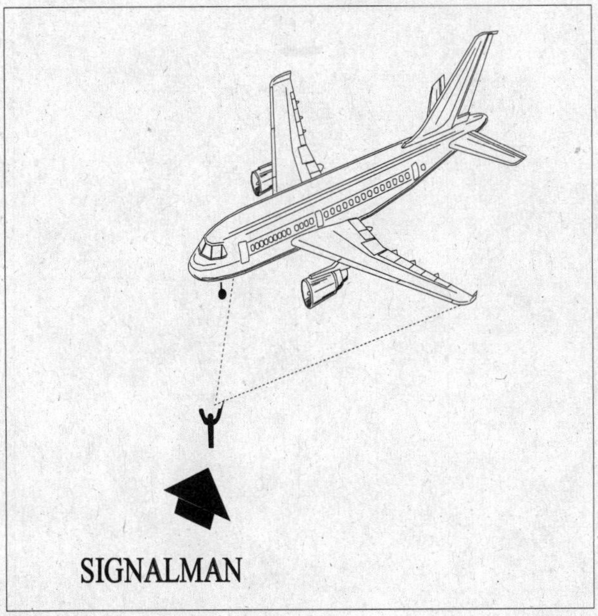

FIG 4-3-11
All Clear (O.K.)

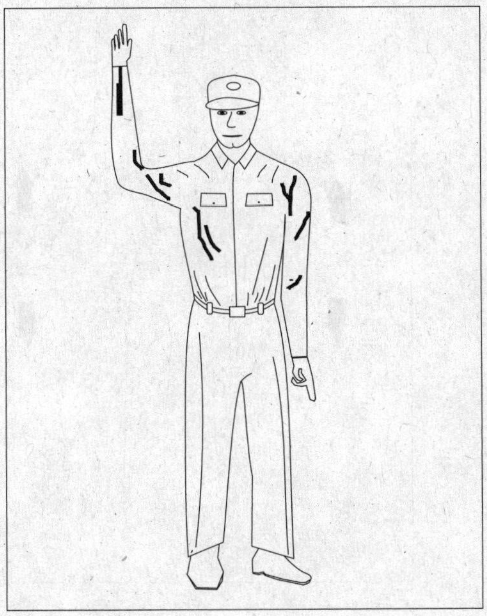

FIG 4-3-12
Start Engine

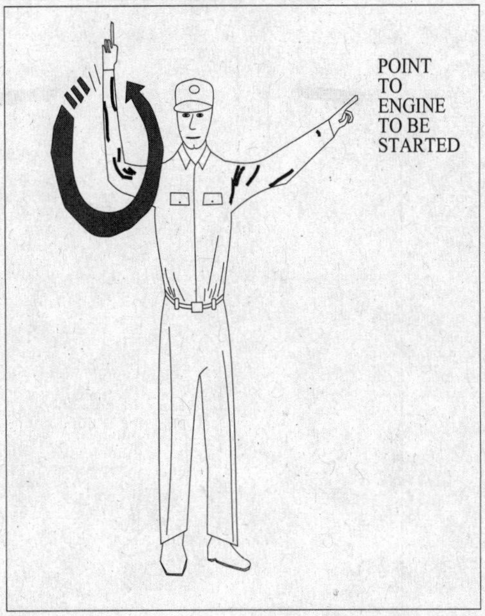

FIG 4-3-13
Pull Chocks

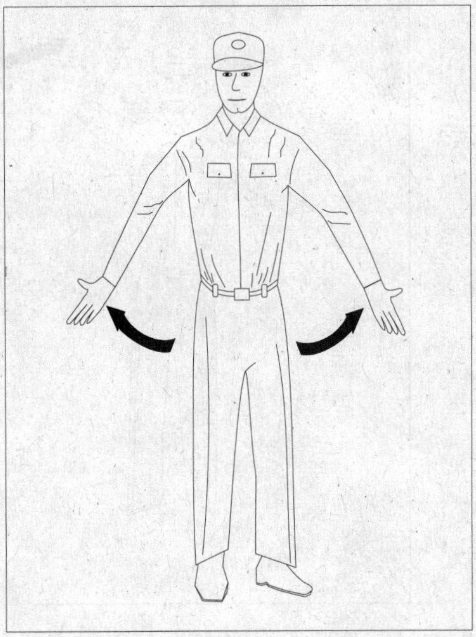

FIG 4-3-14
Proceed Straight Ahead

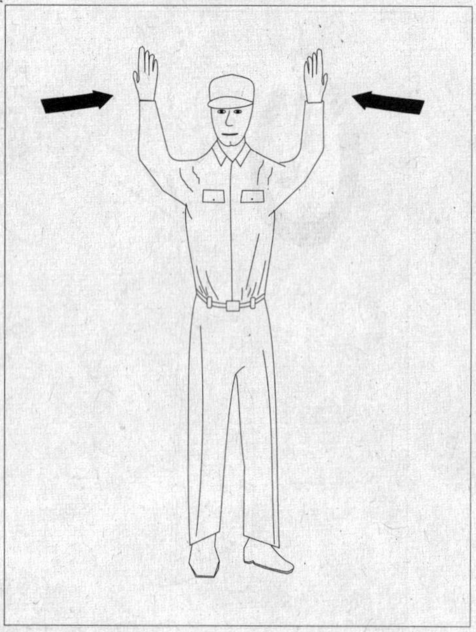

FIG 4-3-15
Left Turn

FIG 4-3-16
Right Turn

FIG 4-3-17
Slow Down

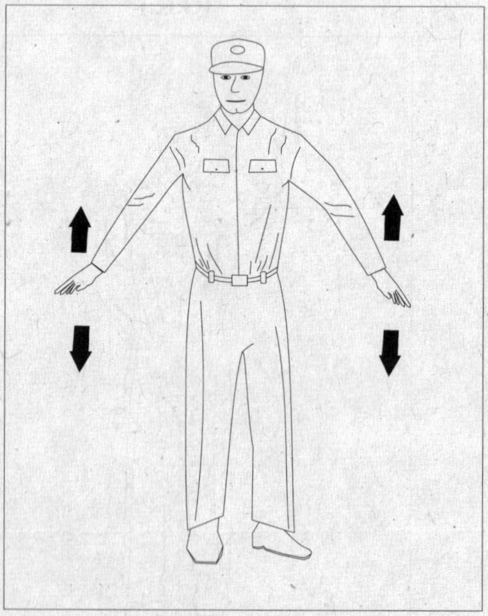

FIG 4-3-18
Flagman Directs Pilot

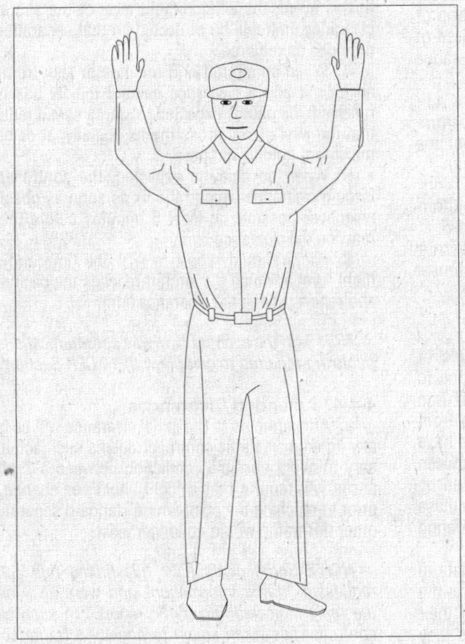

FIG 4-3-20
Cut Engines

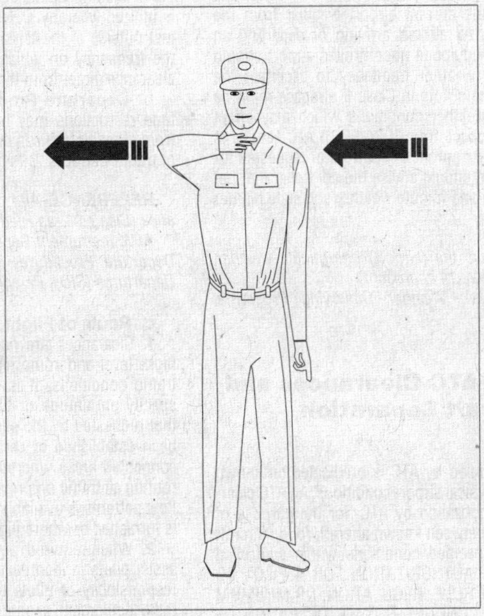

FIG 4-3-19
Insert Chocks

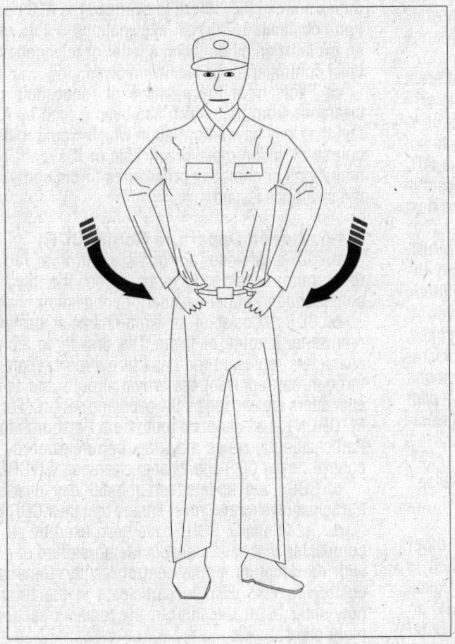

FIG 4-3-21
Night Operation

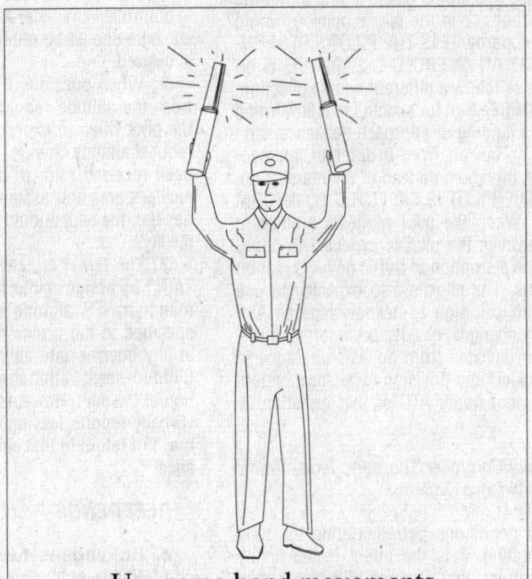

be available to you within approximately 25 NM of the airport below 10,000 feet. The frequency for the weather broadcast will be published on sectional charts and in the Airport/Facility Directory. Some part-time towered airports may also broadcast the automated weather on their ATIS frequency during the hours that the tower is closed.

c. Controllers issue SVFR or IFR clearances based on pilot request, known traffic and reported weather, i.e., METAR/Nonroutine (Special) Aviation Weather Report (SPECI) observations, when they are available. Pilots have access to more current weather at uncontrolled ASOS/AWSS/AWOS airports than do the controllers who may be located several miles away. Controllers will rely on the pilot to determine the current airport weather from the ASOS/AWSS/AWOS. All aircraft arriving or departing an ASOS/AWSS/AWOS equipped uncontrolled airport should monitor the airport weather frequency to ascertain the status of the airspace. Pilots in Class E airspace must be alert for changing weather conditions which may effect the status of the airspace from IFR/VFR. If ATC service is required for IFR/SVFR approach/departure or requested for VFR service, the pilot should advise the controller that he/she has received the one-minute weather and state his/her intentions.

EXAMPLE-*"I have the (airport) one-minute weather, request an ILS Runway 14 approach."*

REFERENCE-*AIM, Weather Observing Programs, Paragraph 7-1-12.*

Section 4. ATC Clearances and Aircraft Separation

4-4-1. Clearance

a. A clearance issued by ATC is predicated on known traffic and known physical airport conditions. An ATC clearance means an authorization by ATC, for the purpose of preventing collision between known aircraft, for an aircraft to proceed under specified conditions within controlled airspace. IT IS NOT AUTHORIZATION FOR A PILOT TO DEVIATE FROM ANY RULE, REGULATION, OR MINIMUM ALTITUDE NOR TO CONDUCT UNSAFE OPERATION OF THE AIRCRAFT.

b. 14 CFR Section 91.3(a) states: "The pilot-in-command of an aircraft is directly responsible for, and is the final authority as to, the operation of that aircraft." If ATC issues a clearance that would cause a pilot to deviate from a rule or regulation, or in the pilot's opinion, would place the aircraft in jeopardy, IT IS THE PILOT'S RESPONSIBILITY TO REQUEST AN AMENDED CLEARANCE. Similarly, if a pilot prefers to follow a different course of action, such as make a 360 degree turn for spacing to follow traffic when established in a landing or approach sequence, land on a different runway, takeoff from a different intersection, takeoff from the threshold instead of an intersection, or delay operation, THE PILOT IS EXPECTED TO INFORM ATC ACCORDINGLY. When the pilot requests a different course of action, however, the pilot is expected to cooperate so as to preclude disruption of traffic flow or creation of conflicting patterns. The pilot is also expected to use the appropriate aircraft call sign to acknowledge all ATC clearances, frequency changes, or advisory information.

c. Each pilot who deviates from an ATC clearance in response to a Traffic Alert and Collision Avoidance System resolution advisory must notify ATC of that deviation as soon as possible.

REFERENCE-*Pilot/Controller Glossary Term- Traffic Alert and Collision Avoidance System.*

d. When weather conditions permit, during the time an IFR flight is operating, it is the direct responsibility of the pilot to avoid other aircraft since VFR flights may be operating in the same area without the knowledge of ATC. Traffic clearances provide standard separation only between IFR flights.

4-4-2. Clearance Prefix

A clearance, control information, or a response to a request for information originated by an ATC facility and relayed to the pilot through an air-to-ground communication station will be prefixed by "ATC clears," "ATC advises," or "ATC requests."

4-4-3. Clearance Items

ATC clearances normally contain the following:

a. Clearance Limit. The traffic clearance issued prior to departure will normally authorize flight to the airport of intended landing. Many airports and associated NAVAIDs are collocated with the same name and/or identifier, so care should be exercised to ensure a clear understanding of the clearance limit. When the clearance limit is the airport of intended landing, the clearance should contain the airport name followed by the word "airport." Under certain conditions, a clearance limit may be a NAVAID or other fix. When the clearance limit is a NAVAID, intersection, or waypoint and the type is known, the clearance should contain type. Under certain conditions, at some locations a short-range clearance procedure is utilized whereby a clearance is issued to a fix within or just outside of the terminal area and pilots are advised of the frequency on which they will receive the long-range clearance direct from the center controller.

b. Departure Procedure. Headings to fly and altitude restrictions may be issued to separate a departure from other air traffic in the terminal area. Where the volume of traffic warrants, DPs have been developed.

REFERENCE-*AIM, Abbreviated IFR Departure Clearance (Cleared. . .as Filed) Procedures, Paragraph 5-2-5.*
AIM, Instrument Departure Procedures (DP) - Obstacle Departure Procedures (ODP) and Standard Instrument Departures (SID), Paragraph 5-2-8.

c. Route of Flight.

1. Clearances are normally issued for the altitude or flight level and route filed by the pilot. However, due to traffic conditions, it is frequently necessary for ATC to specify an altitude or flight level or route different from that requested by the pilot. In addition, flow patterns have been established in certain congested areas or between congested areas whereby traffic capacity is increased by routing all traffic on preferred routes. Information on these flow patterns is available in offices where preflight briefing is furnished or where flight plans are accepted.

2. When required, air traffic clearances include data to assist pilots in identifying radio reporting points. It is the responsibility of pilots to notify ATC immediately if their radio equipment cannot receive the type of signals they must utilize to comply with their clearance.

d. Altitude Data.

1. The altitude or flight level instructions in an ATC clearance normally require that a pilot "MAINTAIN" the altitude or flight level at which the flight will operate when in controlled airspace. Altitude or flight level changes while en route should be requested prior to the time the change is desired.

2. When possible, if the altitude assigned is different from the altitude requested by the pilot, ATC will inform the pilot when to expect climb or descent clearance or to request altitude change from another facility. If this has not been received prior to crossing the boundary of the ATC facility's area and assignment at a different altitude is still desired, the pilot should reinitiate the request with the next facility.

3. The term "cruise" may be used instead of "MAINTAIN" to assign a block of airspace to a pilot from the minimum IFR altitude up to and including the altitude specified in the cruise clearance. The pilot may level off at any intermediate altitude within this block of airspace. Climb/descent within the block is to be made at the discretion of the pilot. However, once the pilot starts descent and verbally reports leaving an altitude in the block, the pilot may not return to that altitude without additional ATC clearance.

REFERENCE-*Pilot/Controller Glossary Term- Cruise.*

e. Holding Instructions.

1. Whenever an aircraft has been cleared to a fix other than the destination airport and delay is expected, it is the responsibility of the ATC controller to issue complete holding instructions (unless the pattern is charted), an EFC time, and a best estimate of any additional en route/terminal delay.

2. If the holding pattern is charted and the controller doesn't issue complete holding instructions, the pilot is expected to hold as depicted on the appropriate chart. When the pattern is charted, the controller may omit all holding instructions except the charted holding direction and the statement *AS PUBLISHED*, e.g., *"HOLD EAST AS PUBLISHED."* Controllers must always issue complete holding instructions when pilots request them.

NOTE-*Only those holding patterns depicted on U.S. government or commercially produced charts which meet FAA requirements should be used.*

3. If no holding pattern is charted and holding instructions have not been issued, the pilot should ask ATC for holding instructions prior to reaching the fix. This procedure will eliminate the possibility of an aircraft entering a holding pattern other than that desired by ATC. If unable to obtain holding instructions prior to reaching the fix (due to frequency congestion, stuck microphone, etc.), hold in a standard pattern on the course on which you approached the fix and request further clearance as soon as possible. In this event, the altitude/flight level of the aircraft at the clearance limit will be protected so that separation will be provided as required.

4. When an aircraft is 3 minutes or less from a clearance limit and a clearance beyond the fix has not been received, the pilot is expected to start a speed reduction so that the aircraft will cross the fix, initially, at or below the maximum holding airspeed.

5. When no delay is expected, the controller should issue a clearance beyond the fix as soon as possible and, whenever possible, at least 5 minutes before the aircraft reaches the clearance limit.

6. Pilots should report to ATC the time and altitude/flight level at which the aircraft reaches the clearance limit and report leaving the clearance limit.

NOTE-*In the event of two-way communications failure, pilots are required to comply with 14 CFR Section 91.185.*

4-4-4. Amended Clearances

a. Amendments to the initial clearance will be issued at any time an air traffic controller deems such action necessary to avoid possible confliction between aircraft. Clearances will require that a flight "hold" or change altitude prior to reaching the point where standard separation from other IFR traffic would no longer exist.

NOTE-*Some pilots have questioned this action and requested "traffic information" and were at a loss when the reply indicated "no traffic report." In such cases the controller has taken action to prevent a traffic confliction which would have occurred at a distant point.*

b. A pilot may wish an explanation of the handling of the flight at the time of occurrence; however, controllers are not able to take time from their immediate control duties nor can they afford to overload the ATC communications channels to furnish explanations. Pilots may obtain an explanation by directing a letter or telephone call to the chief controller of the facility involved.

c. Pilots have the privilege of requesting a different clearance from that which has been issued by ATC if they feel that they have information which would make another course of action more practicable or if aircraft equipment limitations or company procedures forbid compliance with the clearance issued.

4-4-5. Coded Departure Route (CDR)

a. CDRs provide air traffic control a rapid means to reroute departing aircraft when the filed route is constrained by either weather or congestion.

b. CDRs consist of an eight-character designator that represents a route of flight. The first three alphanumeric characters represent the departure airport, characters four through six represent the arrival airport, and the last two characters are chosen by the overlying ARTCC. For example, PITORDN1 is an alternate route from Pittsburgh to Chicago. Participating aircrews may then be re-cleared by air traffic control via the CDR abbreviated clearance, PITORDN1.

c. CDRs are updated on the 56 day charting cycle. Participating aircrews must ensure that their CDR is current.

d. Traditionally, CDRs have been used by air transport companies that have signed a Memorandum of Agreement with the local air traffic control facility. General aviation customers who wish to participate in the program may now enter "CDR Capable" in the remarks section of their flight plan.

e. When "CDR Capable" is entered into the remarks section of the flight plan the general aviation customer communicates to ATC the ability to decode the current CDR into a flight plan route and the willingness to fly a different route than that which was filed.

4-4-6. Special VFR Clearances

a. An ATC clearance must be obtained *prior* to operating within a Class B, Class C, Class D, or Class E surface area when the weather is less than that required for VFR flight. A VFR pilot may request and be given a clearance to enter, leave, or operate within most ClassD and Class E surface areas and some Class B and Class C surface areas in special VFR conditions, traffic permitting, and providing such flight will not delay IFR operations. All special VFR flights must remain clear of clouds. The visibility requirements for special VFR aircraft (other than helicopters) are:

1. At least 1 statute mile flight visibility for operations within Class B, Class C, Class D, and Class E surface areas.

2. At least 1 statute mile ground visibility if taking off or landing. If ground visibility is not reported at that airport, the flight visibility must be at least 1 statute mile.

3. The restrictions in subparagraphs 1 and 2 do not apply to helicopters. Helicopters must remain clear of clouds and may operate in Class B, Class C, Class D, and Class E surface areas with less than 1 statute mile visibility.

b. When a control tower is located within the Class B, Class C, or Class D surface area, requests for clearances should be to the tower. In a Class E surface area, a clearance may be obtained from the nearest tower, FSS, or center.

c. It is not necessary to file a complete flight plan with the request for clearance, but pilots should state their intentions in sufficient detail to permit ATC to fit their flight into the traffic flow. The clearance will not contain a specific altitude as the pilot must remain clear of clouds. The controller may require the pilot to fly at or below a certain altitude due to other traffic, but the altitude specified will permit flight at or above the minimum safe altitude. In addition, at radar locations, flights may be vectored if necessary for control purposes or on pilot request.

NOTE-*The pilot is responsible for obstacle or terrain clearance.*

REFERENCE-*14 CFR Section 91.119, Minimum safe altitudes: General.*

d. Special VFR clearances are effective within Class B, Class C, Class D, and Class E surface areas only. ATC does not provide separation after an aircraft leaves the Class B, Class C, Class D, or Class E surface area on a special VFR clearance.

e. Special VFR operations by fixed-wing aircraft are prohibited in some Class B and Class C surface areas due to the volume of IFR traffic. A list of these Class B and Class C surface areas is contained in 14 CFR Part 91, Appendix D, Section 3. They are also depicted on sectional aeronautical charts.

f. ATC provides separation between Special VFR flights and between these flights and other IFR flights.

g. Special VFR operations by fixed-wing aircraft are prohibited between sunset and sunrise unless the pilot is instrument rated and the aircraft is equipped for IFR flight.

h. Pilots arriving or departing an uncontrolled airport that has automated weather broadcast capability (ASOS/AWSS/AWOS) should monitor the broadcast frequency, advise the controller that they have the "one-minute weather" and state intentions prior to operating within the Class B, Class C, Class D, or Class E surface areas.

REFERENCE-*Pilot/Controller Glossary Term- One-minute Weather.*

4-4-7. Pilot Responsibility upon Clearance Issuance

a. Record ATC clearance. When conducting an IFR operation, make a written record of your clearance. The specified conditions which are a part of your air traffic clearance may be somewhat different from those included in your flight plan. Additionally, ATC may find it necessary to ADD conditions, such as particular departure route. The very fact that ATC specifies different or additional conditions means that other aircraft are involved in the traffic situation.

b. ATC Clearance/Instruction Readback. Pilots of airborne aircraft should read back *those parts* of ATC clearances and instructions containing altitude assignments, vectors, or runway assignments as a means of mutual verification. The read back of the "numbers» serves as a double check between pilots and controllers and reduces the kinds of communications errors that occur when a number is either "misheard» or is incorrect.

1. Include the aircraft identification in all readbacks and acknowledgments. This aids controllers in determining that the correct aircraft received the clearance or instruction. The requirement to include aircraft identification in all readbacks and acknowledgements becomes more important as frequency congestion increases and when aircraft with similar call signs are on the same frequency.

EXAMPLE-*"Climbing to Flight Level three three zero, United Twelve" or "November Five Charlie Tango, roger, cleared to land runway nine left."*

2. Read back altitudes, altitude restrictions, and vectors in the same sequence as they are given in the clearance or instruction.

3. Altitudes contained in charted procedures, such as DPs, instrument approaches, etc., should not be read back unless they are specifically stated by the controller.

4. Initial read back of a taxi, departure or landing clearance should include the runway assignment, including left, right, center, etc. if applicable.

c. It is the responsibility of the pilot to accept or refuse the clearance issued.

4-4-8. IFR Clearance VFR-on-top

a. A pilot on an IFR flight plan operating in VFR weather conditions, may request VFR-on-top in lieu of an assigned altitude. This permits a pilot to select an altitude or flight level of their choice (subject to any ATC restrictions.)

b. Pilots desiring to climb through a cloud, haze, smoke, or other meteorological formation and then either cancel their IFR flight plan or operate VFR-on-top may request a climb to VFR-on-top. The ATC authorization must contain either a top report or a statement that no top report is available, and a request to report reaching VFR-on-top. Additionally, the ATC authorization may contain a clearance limit, routing and an alternative clearance if VFR-on-top is not reached by a specified altitude.

c. A pilot on an IFR flight plan, operating in VFR conditions, may request to climb/descend in VFR conditions.

d. ATC may not authorize VFR-on-top/VFR conditions operations unless the pilot requests the VFR operation or a clearance to operate in VFR conditions will result in noise abatement benefits where part of the IFR departure route does not conform to an FAA approved noise abatement route or altitude.

e. When operating in VFR conditions with an ATC authorization to "maintain VFR-on-top/maintain VFR conditions" pilots on IFR flight plans must:

1. Fly at the appropriate VFR altitude as prescribed in 14 CFR Section 91.159.

2. Comply with the VFR visibility and distance from cloud criteria in 14 CFR Section 91.155 (Basic VFR Weather Minimums).

3. Comply with instrument flight rules that are applicable to this flight; i.e., minimum IFR altitudes, position reporting, radio communications, course to be flown, adherence to ATC clearance, etc.

NOTE-*Pilots should advise ATC prior to any altitude change to ensure the exchange of accurate traffic information.*

f. ATC authorization to "maintain VFR-on-top" is not intended to restrict pilots so that they must operate only *above* an obscuring meteorological formation (layer). Instead, it permits operation above, below, between layers, or in areas where there is no meteorological obscuration. It is imperative, however, that pilots understand that clearance to operate "VFR-on-top/VFR conditions" does not imply cancellation of the IFR flight plan.

g. Pilots operating VFR-on-top/VFR conditions may receive traffic information from ATC on other pertinent IFR or VFR aircraft. However, aircraft operating in Class B airspace/TRSAs must be separated as required by FAA Order JO 7110.65, Air Traffic Control.

NOTE-*When operating in VFR weather conditions, it is the pilot's responsibility to be vigilant so as to see-and-avoid other aircraft.*

h. ATC will not authorize VFR or VFR-on-top operations in Class A airspace.

REFERENCE-*AIM, Class A Airspace, Paragraph 3-2-2.*

4-4-9. VFR/IFR Flights

A pilot departing VFR, either intending to or needing to obtain an IFR clearance en route, must be aware of the position of the aircraft and the relative terrain/obstructions. When accepting a clearance below the MEA/MIA/MVA/OROCA, pilots are responsible for their own terrain/obstruction clearance until reaching the MEA/MIA/MVA/OROCA. If pilots are unable to maintain terrain/obstruction clearance, the controller should be advised and pilots should state their intentions.

NOTE-*OROCA is an off-route altitude which provides obstruction clearance with a 1,000 foot buffer in nonmountainous terrain areas and a 2,000 foot buffer in designated mountainous areas within the U.S. This altitude may not provide signal coverage from ground-based navigational aids, air traffic control radar, or communications coverage.*

4-4-10. Adherence to Clearance

a. When air traffic clearance has been obtained under either visual or instrument flight rules, the pilot-in-command of the aircraft must not deviate from the provisions thereof unless an amended clearance is obtained. When ATC issues a clearance or instruction, pilots are expected to execute its provisions upon receipt. ATC, in certain situations, will include the word "IMMEDIATELY" in a clearance or instruction to impress urgency of an imminent situation and expeditious compliance by the pilot is expected and necessary for safety. The addition of a VFR or other restriction; i.e., climb or descent point or time, crossing altitude, etc., does not authorize a pilot to deviate from the route of flight or any other provision of the ATC clearance.

b. When a heading is assigned or a turn is requested by ATC, pilots are expected to promptly initiate the turn, to complete the turn, and maintain the new heading unless issued additional instructions.

c. The term "AT PILOT'S DISCRETION" included in the altitude information of an ATC clearance means that ATC has offered the pilot the option to start climb or descent when the pilot wishes, is authorized to conduct the climb or descent at any rate, and to temporarily level off at any intermediate altitude as desired. However, once the aircraft has vacated an altitude, it may not return to that altitude.

d. When ATC has not used the term "AT PILOT'S DISCRETION" nor imposed any climb or descent restrictions, pilots should initiate climb or descent promptly on acknowledgement of the clearance. Descend or climb at an optimum rate consistent with the operating characteristics of the aircraft to 1,000 feet above or below the assigned altitude, and then attempt to descend or climb at a rate of between 500 and 1,500 fpm until the assigned altitude is reached. If at anytime the pilot is unable to climb or descend at a rate of at least 500 feet a minute, advise ATC. If it is necessary to level off at an intermediate altitude during climb or descent, advise ATC, except when leveling off at 10,000 feet MSL on descent, or 2,500 feet above airport elevation (prior to entering a Class C or Class D surface area), when required for speed reduction.

REFERENCE-*14 CFR Section 91.117.*

NOTE-*Leveling off at 10,000 feet MSL on descent or 2,500 feet above airport elevation (prior to entering a Class C or Class D surface area) to comply with 14 CFR Section 91.117 airspeed restrictions is commonplace. Controllers anticipate this action and plan accordingly. Leveling off at any other time on climb or descent may seriously affect air traffic handling by ATC. Consequently, it is imperative that pilots make every effort to fulfill the above expected actions to aid ATC in safely handling and expediting traffic.*

e. If the altitude information of an ATC DESCENT clearance includes a provision to "CROSS (fix) AT" or "AT OR ABOVE/BELOW (altitude)," the manner in which the descent is executed to comply with the crossing altitude is at the pilot's discretion. This authorization to descend at pilot's discretion is only applicable to that portion of the flight to which the crossing altitude restriction applies, and the pilot is expected to comply with the crossing altitude as a provision of the clearance. Any other clearance in which pilot execution is optional will so state "AT PILOT'S DISCRETION."

EXAMPLE-1. *"United Four Seventeen, descend and maintain six thousand."*

NOTE-1. *The pilot is expected to commence descent upon receipt of the clearance and to descend at the suggested rates until reaching the assigned altitude of 6,000 feet.*

EXAMPLE-2. *"United Four Seventeen, descend at pilot's discretion, maintain six thousand."*

NOTE-2. *The pilot is authorized to conduct descent within the context of the term at pilot's discretion as described above.*

EXAMPLE-3. *"United Four Seventeen, cross Lakeview V-O-R at or above Flight Level two zero zero, descend and maintain six thousand."*

NOTE-3. *The pilot is authorized to conduct descent at pilot's discretion until reaching Lakeview VOR and must comply with the clearance provision to cross the Lakeview VOR at or above FL 200. After passing Lakeview VOR, the pilot is expected to descend at the suggested rates until reaching the assigned altitude of 6,000 feet.*

EXAMPLE-4. *"United Four Seventeen, cross Lakeview V-O-R at six thousand, maintain six thousand."*

NOTE-4. *The pilot is authorized to conduct descent at pilot's discretion, however, must comply with the clearance provision to cross the Lakeview VOR at 6,000 feet.*

EXAMPLE-5. *"United Four Seventeen, descend now to Flight Level two seven zero, cross Lakeview V-O-R at or below one zero thousand, descend and maintain six thousand."*

NOTE-5. *The pilot is expected to promptly execute and complete descent to FL 270 upon receipt of the clearance. After reaching FL 270 the pilot is authorized to descend "at pilot's discretion" until reaching Lakeview VOR. The pilot must comply with the clearance provision to cross Lakeview VOR at or below 10,000 feet. After Lakeview VOR the pilot is expected to descend at the suggested rates until reaching 6,000 feet.*

EXAMPLE-6. *"United Three Ten, descend now and maintain Flight Level two four zero, pilot's discretion after reaching Flight Level two eight zero."*

NOTE-6. *The pilot is expected to commence descent upon receipt of the clearance and to descend at the suggested rates until reaching FL 280. At that point, the pilot is authorized to continue descent to FL 240 within the context of the term "at pilot's discretion" as described above.*

f. In case emergency authority is used to deviate from provisions of an ATC clearance, the pilot-in-command must notify ATC as soon as possible and obtain an amended clearance. In an emergency situation which does not result in a deviation from the rules prescribed in 14 CFR Part 91 but which requires ATC to give priority to an aircraft, the pilot of such aircraft must, when requested by ATC, make a report within 48 hours of such emergency situation to the manager of that ATC facility.

g. The guiding principle is that the last ATC clearance has precedence over the previous ATC clearance. When the route or altitude in a previously issued clearance is amended, the controller will restate applicable altitude restrictions. If altitude to maintain is changed or restated, whether prior to departure or while airborne, and previously issued altitude restrictions are omitted, those altitude restrictions are canceled, including departure procedures and STAR altitude restrictions.

EXAMPLE-1. *A departure flight receives a clearance to destination airport to maintain FL 290. The clearance incorporates a DP which has certain altitude crossing restrictions. Shortly after takeoff, the flight receives a new clearance changing the maintaining FL from 290 to 250. If the altitude restrictions are still applicable, the controller restates them.*

2. *A departing aircraft is cleared to cross Fluky Intersection at or above 3,000 feet, Gordonville VOR at or above 12,000 feet, maintain FL 200. Shortly after departure, the altitude to be maintained is changed to FL 240. If the altitude restrictions are still applicable, the controller issues an amended clearance as follows: "cross Fluky Intersection at or above three thousand, cross Gordonville V-O-R at or above one two thousand, maintain Flight Level two four zero."*

3. *An arriving aircraft is cleared to the destination airport via V45 Delta VOR direct; the aircraft is cleared to cross Delta VOR at 10,000 feet, and then to maintain 6,000 feet. Prior to Delta VOR, the controller issues an amended clearance as follows: "turn right heading one eight zero for vector to runway three six I-L-S approach, maintain six thousand."*

NOTE- *Because the altitude restriction "cross Delta V-O-R at 10,000 feet" was omitted from the amended clearance, it is no longer in effect.*

h. Pilots of turbojet aircraft equipped with afterburner engines should advise ATC prior to takeoff if they intend to use afterburning during their climb to the en route altitude. Often, the controller may be able to plan traffic to accommodate a high performance climb and allow the aircraft to climb to the planned altitude without restriction.

i. If an "expedite" climb or descent clearance is issued by ATC, and the altitude to maintain is subsequently changed or restated without an expedite instruction, the expedite instruction is canceled. Expedite climb/descent normally indicates to the pilot that the approximate best rate of climb/descent should be used without requiring an exceptional change in aircraft handling characteristics. Normally controllers will inform pilots of the reason for an instruction to expedite.

4-4-11. IFR Separation Standards

a. ATC effects separation of aircraft vertically by assigning different altitudes; longitudinally by providing an interval expressed in time or distance between aircraft on the same, converging, or crossing courses, and laterally by assigning different flight paths.

b. Separation will be provided between all aircraft operating on IFR flight plans except during that part of the flight (outside Class B airspace or a TRSA) being conducted on a VFR-on-top/VFR conditions clearance. Under these conditions, ATC may issue traffic advisories, but it is the sole responsibility of the pilot to be vigilant so as to see and avoid other aircraft.

c. When radar is employed in the separation of aircraft at the same altitude, a minimum of 3 miles separation is provided between aircraft operating within 40 miles of the radar antenna site, and 5 miles between aircraft operating beyond 40 miles from the antenna site. These minima may be increased or decreased in certain specific situations.

NOTE- *Certain separation standards are increased in the terminal environment when CENRAP is being utilized.*

4-4-12. Speed Adjustments

a. ATC will issue speed adjustments to pilots of radar-controlled aircraft to achieve or maintain required or desire spacing.

b. ATC will express all speed adjustments in terms of knots based on indicated airspeed (IAS) in 10 knot increments except that at or above FL 240 speeds may be expressed in terms of Mach numbers in 0.01 increments. The use of Mach numbers is restricted to turbojet aircraft with Mach meters.

c. Pilots complying with speed adjustments are expected to maintain a speed within plus or minus 10 knots or 0.02 Mach number of the specified speed.

d. When ATC assigns speed adjustments, it will be in accordance with the following recommended minimums:

1. To aircraft operating between FL 280 and 10,000 feet, a speed not less than 250 knots or the equivalent Mach number.

NOTE-1. *On a standard day the Mach numbers equivalent to 250 knots CAS (subject to minor variations) are:*
FL 240-0.6
FL 250-0.61
FL 260-0.62
FL 270-0.64
FL 280-0.65
FL 290-0.66.

2. *When an operational advantage will be realized, speeds lower than the recommended minima may be applied.*

2. To arriving turbojet aircraft operating below 10,000 feet:

(a) A speed not less than 210 knots, except;

(b) Within 20 flying miles of the airport of intended landing, a speed not less than 170 knots.

3. To arriving reciprocating engine or turboprop aircraft within 20 flying miles of the runway threshold of the airport of intended landing, a speed not less than 150 knots.

4. To departing aircraft:

(a) Turbojet aircraft, a speed not less than 230 knots.

(b) Reciprocating engine aircraft, a speed not less than 150 knots.

e. When ATC combines a speed adjustment with a descent clearance, the sequence of delivery, with the word "then" between, indicates the expected order of execution.

EXAMPLE-
1. *Descend and maintain (altitude); then, reduce speed to (speed).*
2. *Reduce speed to (speed); then, descend and maintain (altitude).*

NOTE- *The maximum speeds below 10,000 feet as established in 14 CFR Section 91.117 still apply. If there is any doubt concerning the manner in which such a clearance is to be executed, request clarification from ATC.*

f. If ATC determines (before an approach clearance is issued) that it is no longer necessary to apply speed adjustment procedures, they will inform the pilot to resume normal speed. Approach clearances supersede any prior speed adjustment assignments, and pilots are expected to make their own speed adjustments, as necessary, to complete the approach. Under certain circumstances, however, it may be necessary for ATC to issue further speed adjustments after approach clearance is issued to maintain separation between successive arrivals. Under such circumstances, previously issued speed adjustments will be restated if that speed is to be maintained or additional speed adjustments are requested. ATC must obtain pilot concurrence for speed adjustments after approach clearances are issued. Speed adjustments should not be assigned inside the final approach fix on final or a point 5 miles from the runway, whichever is closer to the runway.

NOTE- *See "RESUME NORMAL SPEED" in the Pilot/Controller Glossary.*

g. The pilots retain the prerogative of rejecting the application of speed adjustment by ATC if the minimum safe airspeed for any particular operation is greater than the speed adjustment.

NOTE- *In such cases, pilots are expected to advise ATC of the speed that will be used.*

h. Pilots are reminded that they are responsible for rejecting the application of speed adjustment by ATC if, in their opinion, it will cause them to exceed the maximum indicated airspeed prescribed by 14 CFR Section 91.117(a), (c) and (d). *IN SUCH CASES, THE PILOT IS EXPECTED TO SO INFORM ATC.* Pilots operating at or above 10,000 feet MSL who are issued speed adjustments which exceed 250 knots IAS and are subsequently cleared below 10,000 feet MSL are expected to comply with 14 CFR Section 91.117(a).

i. Speed restrictions of 250 knots do not apply to U.S. registered aircraft operating beyond 12 nautical miles from the coastline within the U.S. Flight Information Region, in Class E airspace below 10,000 feet MSL. However, in airspace underlying a Class B airspace area designated for an airport, or in a VFR corridor designated through such as a Class B airspace area, pilots are expected to comply with the 200 knot speed limit specified in 14 CFR Section 91.117(c).

j. For operations in a Class C and Class D surface area, ATC is authorized to request or approve a speed greater than the maximum indicated airspeeds prescribed for operation within that airspace (14 CFR Section 91.117(b)).

NOTE- *Pilots are expected to comply with the maximum speed of 200 knots when operating beneath Class B airspace or in a Class B VFR corridor (14 CFR Section 91.117(c) and (d)).*

k. When in communications with the ARTCC or approach control facility, pilots should, as a good operating practice, state any ATC assigned speed restriction on initial radio contact associated with an ATC communications frequency change.

4-4-13. Runway Separation

Tower controllers establish the sequence of arriving and departing aircraft by requiring them to adjust flight or ground operation as necessary to achieve proper spacing. They may "HOLD" an aircraft short of the runway to achieve spacing between it and an arriving aircraft; the controller may instruct a pilot to "EXTEND DOWNWIND" in order to establish spacing from an arriving or departing aircraft. At times a clearance may include the word "IMMEDIATE." For example: "CLEARED FOR IMMEDIATE TAKEOFF." In such cases "IMMEDIATE" is used for purposes of *air traffic separation.* It is up to the pilot to refuse the clearance if, in the pilot's opinion, compliance would adversely affect the operation.

REFERENCE- *AIM, Gate Holding due to Departure Delays, Paragraph 4-3-15.*

4-4-14. Visual Separation

a. Visual separation is a means employed by ATC to separate aircraft in terminal areas and en route airspace in the NAS. There are two methods employed to effect this separation:

1. The tower controller sees the aircraft involved and issues instructions, as necessary, to ensure that the aircraft avoid each other.

2. A pilot sees the other aircraft involved and upon instructions from the controller provides separation by maneuvering the aircraft to avoid it. When pilots accept responsibility to maintain visual separation, they must maintain constant visual surveillance and not pass the other aircraft until it is no longer a factor.

NOTE-*Traffic is no longer a factor when during approach phase the other aircraft is in the landing phase of flight or executes a missed approach; and during departure or en route, when the other aircraft turns away or is on a diverging course.*

b. A pilot's acceptance of instructions to follow another aircraft or provide visual separation from it is an acknowledgment that the pilot will maneuver the aircraft as necessary to avoid the other aircraft or to maintain in-trail separation. In operations conducted behind heavy jet aircraft, it is also an acknowledgment that the pilot accepts the responsibility for wake turbulence separation.

NOTE-*When a pilot has been told to follow another aircraft or to provide visual separation from it, the pilot should promptly notify the controller if visual contact with the other aircraft is lost or cannot be maintained or if the pilot cannot accept the responsibility for the separation for any reason.*

c. Scanning the sky for other aircraft is a key factor in collision avoidance. Pilots and copilots (or the right seat passenger) should continuously scan to cover all areas of the sky visible from the cockpit. Pilots must develop an effective scanning technique which maximizes one's visual capabilities. Spotting a potential collision threat increases directly as more time is spent looking outside the aircraft. One must use timesharing techniques to effectively scan the surrounding airspace while monitoring instruments as well.

d. Since the eye can focus only on a narrow viewing area, effective scanning is accomplished with a series of short, regularly spaced eye movements that bring successive areas of the sky into the central visual field. Each movement should not exceed ten degrees, and each area should be observed for at least one second to enable collision detection. Although many pilots seem to prefer the method of horizontal back-and-forth scanning every pilot should develop a scanning pattern that is not only comfortable but assures optimum effectiveness. Pilots should remember, however, that they have a regulatory responsibility (14 CFR Section 91.113(a)) to see and avoid other aircraft when weather conditions permit.

4-4-15. Use of Visual Clearing Procedures

a. Before Takeoff. Prior to taxiing onto a runway or landing area in preparation for takeoff, pilots should scan the approach areas for possible landing traffic and execute the appropriate clearing maneuvers to provide them a clear view of the approach areas.

b. Climbs and Descents. During climbs and descents in flight conditions which permit visual detection of other traffic, pilots should execute gentle banks, left and right at a frequency which permits continuous visual scanning of the airspace about them.

c. Straight and Level. Sustained periods of straight and level flight in conditions which permit visual detection of other traffic should be broken at intervals with appropriate clearing procedures to provide effective visual scanning.

d. Traffic Pattern. Entries into traffic patterns while descending create specific collision hazards and should be avoided.

e. Traffic at VOR Sites. All operators should emphasize the need for sustained vigilance in the vicinity of VORs and airway intersections due to the convergence of traffic.

f. Training Operations. Operators of pilot training programs are urged to adopt the following practices:

1. Pilots undergoing flight instruction at all levels should be requested to verbalize clearing procedures (call out "clear" left, right, above, or below) to instill and sustain the habit of vigilance during maneuvering.

2. High-wing airplane. Momentarily raise the wing in the direction of the intended turn and look.

3. Low-wing airplane. Momentarily lower the wing in the direction of the intended turn and look.

4. Appropriate clearing procedures should precede the execution of all turns including chandelles, lazy eights, stalls, slow flight, climbs, straight and level, spins, and other combination maneuvers.

4-4-16. Traffic Alert and Collision Avoidance System (TCAS I & II)

a. TCAS I provides proximity warning only, to assist the pilot in the visual acquisition of intruder aircraft. No recommended avoidance maneuvers are provided nor authorized as a direct result of a TCAS I warning. It is intended for use by smaller commuter aircraft holding 10 to 30 passenger seats, and general aviation aircraft.

b. TCAS II provides traffic advisories (TAs) and resolution advisories (RAs). Resolution advisories provide recommended maneuvers in a vertical direction (climb or descend only) to avoid conflicting traffic. Airline aircraft, and larger commuter and business aircraft holding 31 passenger seats or more, use TCAS II equipment.

1. Each pilot who deviates from an ATC clearance in response to a TCAS II RA must notify ATC of that deviation as soon as practicable and expeditiously return to the current ATC clearance when the traffic conflict is resolved.

2. Deviations from rules, policies, or clearances should be kept to the minimum necessary to satisfy a TCAS II RA.

3. The serving IFR air traffic facility is not responsible to provide approved standard IFR separation to an aircraft after a TCAS II RA maneuver until one of the following conditions exists:

(a) The aircraft has returned to its assigned altitude and course.

(b) Alternate ATC instructions have been issued.

c. TCAS does not alter or diminish the pilot's basic authority and responsibility to ensure safe flight. Since TCAS does not respond to aircraft which are not transponder equipped or aircraft with a transponder failure, TCAS alone does not ensure safe separation in every case.

d. At this time, no air traffic service nor handling is predicated on the availability of TCAS equipment in the aircraft.

4-4-17. Traffic Information Service (TIS)

a. TIS provides proximity warning only, to assist the pilot in the visual acquisition of intruder aircraft. No recommended avoidance maneuvers are provided nor authorized as a direct result of a TIS intruder display or TIS alert. It is intended for use by aircraft in which TCAS is not required.

b. TIS does not alter or diminish the pilot's basic authority and responsibility to ensure safe flight. Since TIS does not respond to aircraft which are not transponder equipped, aircraft with a transponder failure, or aircraft out of radar coverage, TIS alone does not ensure safe separation in every case.

c. At this time, no air traffic service nor handling is predicated on the availability of TIS equipment in the aircraft.

d. Presently, no air traffic services or handling is predicated on the availability of an ADS-B cockpit display. A "traffic-in-sight" reply to ATC must be based on seeing an aircraft out-the-window, NOT on the cockpit display.

Section 5. Surveillance Systems

4-5-1. Radar

a. Capabilities

1. Radar is a method whereby radio waves are transmitted into the air and are then received when they have been reflected by an object in the path of the beam. Range is determined by measuring the time it takes (at the speed of light) for the radio wave to go out to the object and then return to the receiving antenna. The direction of a detected object from a radar site is determined by the position of the rotating antenna when the reflected portion of the radio wave is received.

2. More reliable maintenance and improved equipment have reduced radar system failures to a negligible factor. Most facilities actually have some components duplicated, one operating and another which immediately takes over when a malfunction occurs to the primary component.

b. Limitations

1. It is very important for the aviation community to recognize the fact that there are limitations to radar service and that ATC controllers may not always be able to issue traffic advisories concerning aircraft which are not under ATC control and cannot be seen on radar. (See FIG 4-5-1.)

(a) The characteristics of radio waves are such that they normally travel in a continuous straight line unless they are:

(1) "Bent" by abnormal atmospheric phenomena such as temperature inversions;

(2) Reflected or attenuated by dense objects such as heavy clouds, precipitation, ground obstacles, mountains, etc.; or

(3) Screened by high terrain features.

(b) The bending of radar pulses, often called anomalous propagation or ducting, may cause many extraneous

FIG 4-5-1

Limitations to Radar Service

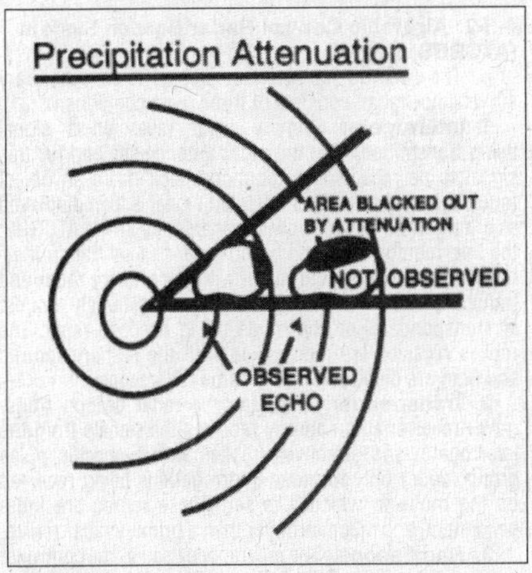

blips to appear on the radar operator's display if the beam has been bent toward the ground or may decrease the detection range if the wave is bent upward. It is difficult to solve the effects of anomalous propagation, but using beacon radar and electronically eliminating stationary and slow moving targets by a method called moving target indicator (MTI) usually negate the problem.

(c) Radar energy that strikes dense objects will be reflected and displayed on the operator's scope thereby blocking out aircraft at the same range and greatly weakening or completely eliminating the display of targets at a greater range. Again, radar beacon and MTI are very effectively used to combat ground clutter and weather phenomena, and a method of circularly polarizing the radar beam will eliminate some weather returns. A negative characteristic of MTI is that an aircraft flying a speed that coincides with the canceling signal of the MTI (tangential or "blind" speed) may not be displayed to the radar controller.

(d) Relatively low altitude aircraft will not be seen if they are screened by mountains or are below the radar beam due to earth curvature. The only solution to screening is the installation of strategically placed multiple radars which has been done in some areas.

(e) There are several other factors which affect radar control. The amount of reflective surface of an aircraft will determine the size of the radar return. Therefore, a small light airplane or a sleek jet fighter will be more difficult to see on radar than a large commercial jet or military bomber. Here again, the use of radar beacon is invaluable if the aircraft is equipped with an airborne transponder. All ARTCCs' radars in the conterminous U.S. and many airport surveillance radars have the capability to interrogate Mode C and display altitude information to the controller from appropriately equipped aircraft. However, there are a number of airport surveillance radars that don't have Mode C display capability and; therefore, altitude information must be obtained from the pilot.

(f) At some locations within the ATC en route environment, secondary-radar-only (no primary radar) gap filler radar systems are used to give lower altitude radar coverage between two larger radar systems, each of which provides both primary and secondary radar coverage. In those geographical areas served by secondary-radar only, aircraft without transponders cannot be provided with radar service. Additionally, transponder equipped aircraft cannot be provided with radar advisories concerning primary targets and weather.

REFERENCE-*Pilot/Controller Glossary Term- Radar.*

(g) The controller's ability to advise a pilot flying on instruments or in visual conditions of the aircraft's proximity to another aircraft will be limited if the unknown aircraft is not observed on radar, if no flight plan information is available, or if the volume of traffic and workload prevent issuing traffic information. The controller's first priority is given to establishing vertical, lateral, or longitudinal separation between aircraft flying IFR under the control of ATC.

c. FAA radar units operate continuously at the locations shown in the Airport/Facility Directory, and their services are available to all pilots, both civil and military. Contact the associated FAA control tower or ARTCC on any frequency guarded for initial instructions, or in an emergency, any FAA facility for information on the nearest radar service.

4-5-2. Air Traffic Control Radar Beacon System (ATCRBS)

a. The ATCRBS, sometimes referred to as secondary surveillance radar, consists of three main components:

1. Interrogator. Primary radar relies on a signal being transmitted from the radar antenna site and for this signal to be reflected or "bounced back" from an object (such as an aircraft). This reflected signal is then displayed as a "target" on the controller's radarscope. In the ATCRBS, the Interrogator, a ground based radar beacon transmitter-receiver, scans in synchronism with the primary radar and transmits discrete radio signals which repetitiously request all transponders, on the mode being used, to reply. The replies received are then mixed with the primary returns and both are displayed on the same radarscope.

2. Transponder. This airborne radar beacon transmitter-receiver automatically receives the signals from the interrogator and selectively replies with a specific pulse group (code) only to those interrogations being received on the mode to which it is set. These replies are independent of, and much stronger than a primary radar return.

3. Radarscope. The radarscope used by the controller displays returns from both the primary radar system and the ATCRBS. These returns, called targets, are what the controller refers to in the control and separation of traffic.

b. The job of identifying and maintaining identification of primary radar targets is a long and tedious task for the controller. Some of the advantages of ATCRBS over primary radar are:

1. Reinforcement of radar targets.

2. Rapid target identification.

3. Unique display of selected codes.

c. A part of the ATCRBS ground equipment is the decoder. This equipment enables a controller to assign discrete transponder codes to each aircraft under his/her control. Normally only one code will be assigned for the entire flight. Assignments are made by the ARTCC computer on the basis of the National Beacon Code Allocation Plan. The equipment is also designed to receive Mode C altitude information from the aircraft.

NOTE-*Refer to figures with explanatory legends for an illustration of the target symbology depicted on radar scopes in the NAS Stage A (en route), the ARTS III (terminal) Systems, and other nonautomated (broadband) radar systems. (See FIG 4-5-2 andFIG 4-5-3.)*

d. It should be emphasized that aircraft transponders greatly improve the effectiveness of radar systems.

REFERENCE-*AIM, Transponder Operation, Paragraph 4-1-20.*

NOTE-*A number of radar terminals do not have ARTS equipment. Those facilities and certain ARTCCs outside the contiguous U.S. would have radar displays similar to the lower right hand subset. ARTS facilities and NAS Stage A ARTCCs, when operating in the nonautomation mode, would also have similar displays and certain services based on automation may not be available.*

FIG 4-5-2
ARTS III Radar Scope With Alphanumeric Data

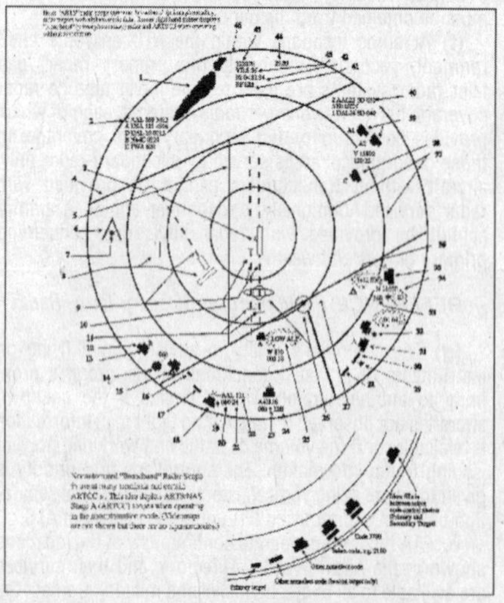

EXAMPLE-

__1.__ Areas of precipitation (can be reduced by CP)
__2.__ Arrival/departure tabular list
__3.__ Trackball (control) position symbol (A)
__4.__ Airway (lines are sometimes deleted in part)
__5.__ Radar limit line for control
__6.__ Obstruction (video map)
__7.__ Primary radar returns of obstacles or terrain (can be removed by MTI)
__8.__ Satellite airports
__9.__ Runway centerlines (marks and spaces indicate miles)
__10.__ Primary airport with parallel runways
__11.__ Approach gates
__12.__ Tracked target (primary and beacon target)
__13.__ Control position symbol
__14.__ Untracked target select code (monitored) with Mode C readout of 5,000'
__15.__ Untracked target without Mode C
__16.__ Primary target
__17.__ Beacon target only (secondary radar) (transponder)
__18.__ Primary and beacon target
__19.__ Leader line
__20.__ Altitude Mode C readout is 6,000'
(Note: readouts may not be displayed because of nonreceipt of beacon information, garbled beacon signals, and flight plan data which is displayed alternately with the altitude readout)
__21.__ Ground speed readout is 240 knots
(Note: readouts may not be displayed because of a loss of beacon signal, a controller alert that a pilot was squawking emergency, radio failure, etc.)
__22.__ Aircraft ID
__23.__ Asterisk indicates a controller entry in Mode C block. In this case 5,000' is entered and "05" would alternate with Mode C readout.
__24.__ Indicates heavy
__25.__ "Low ALT" flashes to indicate when an aircraft's predicted descent places the aircraft in an unsafe proximity to terrain.
(Note: this feature does not function if the aircraft is not squawking Mode C. When a helicopter or aircraft is known to be operating below the lower safe limit, the "low ALT" can be changed to "inhibit" and flashing ceases.)
__26.__ NAVAIDs
__27.__ Airways
__28.__ Primary target only
__29.__ Nonmonitored. No Mode C (an asterisk would indicate nonmonitored with Mode C)
__30.__ Beacon target only (secondary radar based on aircraft transponder)
__31.__ Tracked target (primary and beacon target) control position A
__32.__ Aircraft is squawking emergency Code 7700 and is nonmonitored, untracked, Mode C
__33.__ Controller assigned runway 36 right alternates with Mode C readout
(Note: a three letter identifier could also indicate the arrival is at specific airport)

FIG 4-5-3
NAS Stage A Controllers View Plan Display

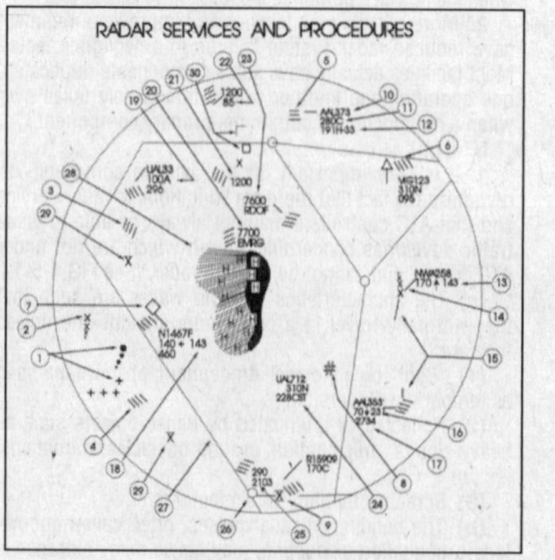

__34.__ Ident flashes
__35.__ Identing target blossoms
__36.__ Untracked target identing on a selected code
__37.__ Range marks (10 and 15 miles) (can be changed/offset)
__38.__ Aircraft controlled by center
__39.__ Targets in suspend status
__40.__ Coast/suspend list (aircraft holding, temporary loss of beacon/target, etc.)
__41.__ Radio failure (emergency information)
__42.__ Select beacon codes (being monitored)
__43.__ General information (ATIS, runway, approach in use)
__44.__ Altimeter setting
__45.__ Time
__46.__ System data area

This figure illustrates the controller's radar scope (PVD) when operating in the full automation (RDP) mode, which is normally 20 hours per day.

(When not in automation mode, the display is similar to the broadband mode shown in the ARTS III radar scope figure. Certain ARTCCs outside the contiguous U.S. also operate in "broadband" mode.)

EXAMPLE-
Target symbols:

1. Uncorrelated primary radar target [°] [+]
2. Correlated primary radar target [¢]
** See note below.*
3. Uncorrelated beacon target [/]
4. Correlated beacon target [\ ]
5. Identing beacon target [≡]
** Note: in Number 2 correlated means the association of radar data with the computer projected track of an identified aircraft.*

Position symbols:

6. Free track (no flight plan tracking) [Δ]
7. Flat track (flight plan tracking) [◊]
8. Coast (beacon target lost) [#]
9. Present position hold [¢]

Data block information:

10. Aircraft ident
** See note below.*
11. Assigned altitude FL 280, Mode C altitude same or within 200' of assigned altitude.
** See note below.*
12. Computer ID #191, handoff is to sector 33 (0-33 would mean handoff accepted)
** See note below.*
13. Assigned altitude 17,000', aircraft is climbing, Mode C readout was 14,300 when last beacon interrogation was received.
14. Leader line connecting target symbol and data block
15. Track velocity and direction vector line (projected ahead of target)
16. Assigned altitude 7,000, aircraft is descending, last Mode C readout (or last reported altitude) was 100' above FL 230
17. Transponder code shows in full data block only when different than assigned code
18. Aircraft is 300' above assigned altitude
19. Reported altitude (no Mode C readout) same as assigned. (An "n" would indicate no reported altitude.)
20. Transponder set on emergency Code 7700 (EMRG flashes to attract attention)
21. Transponder Code 1200 (VFR) with no Mode C
22. Code 1200 (VFR) with Mode C and last altitude readout
23. Transponder set on radio failure Code 7600 (RDOF flashes)
24. Computer ID #228, CST indicates target is in coast status
25. Assigned altitude FL 290, transponder code (these two items constitute a "limited data block")
** Note: numbers 10, 11, and 12 constitute a "full data block"*

Other symbols:

26. Navigational aid
27. Airway or jet route
28. Outline of weather returns based on primary radar. "H" represents areas of high density precipitation which might be thunderstorms. Radial lines indicated lower density precipitation.
29. Obstruction
30. Airports

Major: ☐
Small: ☐

4-5-3. Surveillance Radar

a. Surveillance radars are divided into two general categories: Airport Surveillance Radar (ASR) and Air Route Surveillance Radar (ARSR).

1. ASR is designed to provide relatively short-range coverage in the general vicinity of an airport and to serve as an expeditious means of handling terminal area traffic through observation of precise aircraft locations on a radarscope. The ASR can also be used as an instrument approach aid.

2. ARSR is a long-range radar system designed primarily to provide a display of aircraft locations over large areas.

3. Center Radar Automated Radar Terminal Systems (ARTS) Processing (CENRAP) was developed to provide an alternative to a nonradar environment at terminal facilities should an ASR fail or malfunction. CENRAP sends aircraft radar beacon target information to the ASR terminal facility equipped with ARTS. Procedures used for the separation of aircraft may increase under certain conditions when a facility is utilizing CENRAP because radar target information updates at a slower rate than the normal ASR radar. Radar services for VFR aircraft are also limited during CENRAP operations because of the additional workload required to provide services to IFR aircraft.

b. Surveillance radars scan through 360 degrees of azimuth and present target information on a radar display located in a tower or center. This information is used independently or in conjunction with other navigational aids in the control of air traffic.

4-5-4. Precision Approach Radar (PAR)

a. PAR is designed for use as a landing aid rather than an aid for sequencing and spacing aircraft. PAR equipment may be used as a primary landing aid (See Chapter 5, Air Traffic Procedures, for additional information), or it may be used to monitor other types of approaches. It is designed to display range, azimuth, and elevation information.

b. Two antennas are used in the PAR array, one scanning a vertical plane, and the other scanning horizontally. Since the range is limited to 10 miles, azimuth to 20 degrees, and elevation to 7 degrees, only the final approach area is covered. Each scope is divided into two parts. The upper half presents altitude and distance information, and the lower half presents azimuth and distance.

4-5-5. Airport Surface Detection Equipment - Model X (ASDE-X)

a. The Airport Surface Detection Equipment - Model X (ASDE-X) is a multi-sensor surface surveillance system the FAA is acquiring for airports in the United States. This system will provide high resolution, short-range, clutter free surveillance information about aircraft and vehicles, both moving and fixed, located on or near the surface of the airport's runways and taxiways under all weather and visibility conditions. The system consists of:

1. A Primary Radar System. ASDE-X system coverage includes the airport surface and the airspace up to 200 feet above the surface. Typically located on the control tower or other strategic location on the airport, the Primary Radar antenna is able to detect and display aircraft that are not equipped with or have malfunctioning transponders.

2. Interfaces. ASDE-X contains an automation interface for flight identification via all automation platforms and interfaces with the terminal radar for position information.

3. ASDE-X Automation. A Multi-sensor Data Processor (MSDP) combines all sensor reports into a single target which is displayed to the air traffic controller.

4. Air Traffic Control Tower Display. A high resolution, color monitor in the control tower cab provides controllers with a seamless picture of airport operations on the airport surface.

b. The combination of data collected from the multiple sensors ensures that the most accurate information about aircraft location is received in the tower, thereby increasing surface safety and efficiency.

c. The following facilities have been projected to receive ASDE-X:

Tbl 4-5-1

STL	Lambert-St. Louis International
CLT	Charlotte Douglas International
SDF	Louisville International Standiford
DFW	Dallas/Ft. Worth International
ORD	Chicago O'Hare International
LAX	Los Angeles International
ATL	Hartsfield Atlanta International
IAD	Washington Dulles International
SEA	Seattle-Tacoma International
MKE	General Mitchell International
MCO	Orlando International
PVD	Theodore Francis Green State
PHX	Phoenix Sky Harbor International
MEM	Memphis International
RDU	Raleigh-Durham International
HOU	William P. Hobby (Houston, TX)
BDL	Bradley International
SJC	San Jose International
SAT	San Antonio International
SMF	Sacramento International
FLL	Ft. Lauderdale/Hollywood
HNL	Honolulu International - Hickam AFB
OAK	Metropolitan Oakland International
IND	Indianapolis International
TPA	Tampa International
BUR	Burbank-Glendale-Pasadena
CMH	Port Columbus International
MDW	Chicago Midway
COS	Colorado Springs Municipal
SNA	John Wayne - Orange County
ONT	Ontario International
AUS	Austin-Bergstrom International
RNO	Reno/Tahoe International
ABQ	Albuquerque International Sunport
SJU	San Juan International

4-5-6. Traffic Information Service (TIS)

a. Introduction

The Traffic Information Service (TIS) provides information to the cockpit via data link, that is similar to VFR radar traffic advisories normally received over voice radio. Among the first FAA-provided data services, TIS is intended to improve the safety and efficiency of "see and avoid" flight through an automatic display that informs the pilot of nearby traffic and potential conflict situations. This traffic display is intended to assist the pilot in visual acquisition of these aircraft. TIS employs an enhanced capability of the terminal Mode S radar system, which contains the surveillance data, as well as the data link required to "uplink" this information to suitably-equipped aircraft (known as a TIS "client"). TIS provides estimated position, altitude, altitude trend, and ground track information for up to 8 intruder aircraft within 7 NM horizontally, +3,500 and -3,000 feet vertically of the client aircraft (see FIG 4-5-4, TIS Proximity Coverage Volume). The range of a target reported at a distance greater than 7 NM only indicates that this target will be a threat within 34 seconds and does not display an precise distance. TIS will alert the pilot to aircraft (under surveillance of the Mode S radar) that are estimated to be within 34 seconds of potential collision, regardless of distance of altitude. TIS surveillance data is derived from the same radar used by ATC; this data is uplinked to the client aircraft on each radar scan (nominally every 5 seconds).

b. Requirements

1. In order to use TIS, the client and any intruder aircraft must be equipped with the appropriate cockpit equipment and fly within the radar coverage of a Mode S radar capable of providing TIS. Typically, this will be within 55 NM of the sites depicted in FIG 4-5-5, Terminal Mode S Radar Sites. ATC communication is not a requirement to receive TIS, although it may be required by the particular airspace or flight operations in which TIS is being used.

Fig 4-5-4
TIS Proximity Coverage Volume

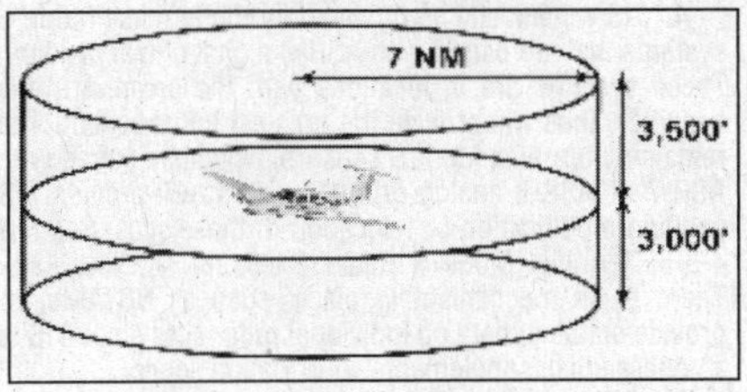

Fig 4-5-5
Terminal Mode S Radar Sites

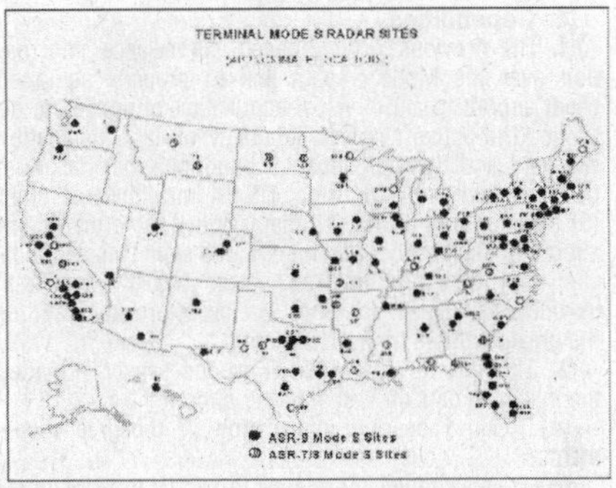

2. The cockpit equipment functionality required by a TIS client aircraft to receive the service consists of the following (refer to FIG 4-5-6):

(a) Mode S data link transponder with altitude encoder.

(b) Data link applications processor with TIS software installed.

(c) Control-display unit.

(d) Optional equipment includes a digital heading source to correct display errors caused by "crab angle" and turning maneuvers.

NOTE-*Some of the above functions will likely be combined into single pieces of avionics, such as (a) and (b).*

3. To be visible to the TIS client, the intruder aircraft must, at a minimum, have an operating transponder (Mode

Fig 4-5-6
Traffic Information Service (TIS) Avionics Block Diagram

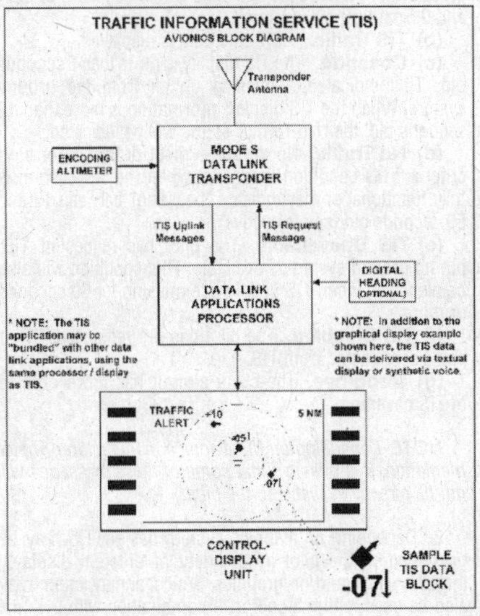

A, C or S). All altitude information provided by TIS from intruder aircraft is derived from Mode C reports, if appropriately equipped.

4. TIS will initially be provided by the terminal Mode S systems that are paired with ASR-9 digital primary radars. These systems are in locations with the greatest traffic densities, thus will provide the greatest initial benefit. The remaining terminal Mode S sensors, which are paired with ASR-7 or ASR-8 analog primary radars, will provide TIS pending modification or relocation of these sites. See FIG 4-5-5, Terminal Mode S Radar Sites, for site locations. There is no mechanism in place, such as NOTAMs, to provide status update on individual radar sites since TIS is a nonessential, supplemental information service.

The FAA also operates en route Mode S radars (not illustrated) that rotate once every 12 seconds. These sites will require additional development of TIS before any possible implementation. There are no plans to implement TIS in the en route Mode S radars at the present time.

c. Capabilities

1. TIS provides ground-based surveillance information over the Mode S data link to properly equipped client aircraft to aid in visual acquisition of proximate air traffic. The actual avionics capability of each installation will vary and the supplemental handbook material must be consulted prior to using TIS. A maximum of eight (8) intruder aircraft may be displayed; if more than eight aircraft match intruder parameters, the eight "most significant" intruders are uplinked. These "most significant" intruders are usually the ones in closest proximity and/or the greatest threat to the TIS client.

2. TIS, through the Mode S ground sensor, provides the following data on each intruder aircraft:

(a) Relative bearing information in 6-degree increments.

(b) Relative range information in 1/8 NM to 1 NM increments (depending on range).

(c) Relative altitude in 100-foot increments (within 1,000 feet) or 500-foot increments (from 1,000-3,500 feet) if the intruder aircraft has operating altitude reporting capability.

(d) Estimated intruder ground track in 45-degree increments.

(e) Altitude trend data (level within 500 fpm or climbing/descending >500 fpm) if the intruder aircraft has operating altitude reporting capability.

(f) Intruder priority as either an "traffic advisory" or "proximate" intruder.

3. When flying from surveillance coverage of one Mode S sensor to another, the transfer of TIS is an automatic function of the avionics system and requires no action from the pilot.

4. There are a variety of status messages that are provided by either the airborne system or ground equipment to alert the pilot of high priority intruders and data link system status. These messages include the following:

(a) Alert. Identifies a potential collision hazard within 34 seconds. This alert may be visual and/or audible, such as a flashing display symbol or a headset tone. A target is a threat if the time to the closest approach in vertical and horizontal coordinates is less than 30 seconds and the closest approach is expected to be within 500 feet vertically and 0.5 nautical miles laterally.

(b) TIS Traffic. TIS traffic data is displayed.

(c) Coasting. The TIS display is more than 6 seconds old. This indicates a missing uplink from the ground system. When the TIS display information is more than 12 seconds old, the "No Traffic" status will be indicated.

(d) No Traffic. No intruders meet proximate or alert criteria. This condition may exist when the TIS system is fully functional or may indicate "coasting" between 12 and 59 seconds old (see (c) above).

(e) TIS Unavailable. The pilot has requested TIS, but no ground system is available. This condition will also be displayed when TIS uplinks are missing for 60 seconds or more.

(f) TIS Disabled. The pilot has not requested TIS or has disconnected from TIS.

(g) Good-bye. The client aircraft has flown outside of TIS coverage.

NOTE-*Depending on the avionics manufacturer implementation, it is possible that some of these messages will not be directly available to the pilot.*

5. Depending on avionics system design, TIS may be presented to the pilot in a variety of different displays, including text and/or graphics. Voice annunciation may also be used, either alone or in combination with a visual display. FIG 4-5-6, Traffic Information Service (TIS), Avionics Block Diagram, shows an example of a TIS display using symbology similar to the Traffic Alert and Collision Avoidance System (TCAS) installed on most passenger air carrier/commuter aircraft in the U.S. The small symbol in the center represents the client aircraft and the display is oriented "track up," with the 12 o'clock position at the top. The range rings indicate 2 and 5 NM. Each intruder is depicted by a symbol positioned at the approximate relative bearing and range from the client aircraft. The circular symbol near the center indicates an "alert" intruder and the diamond symbols indicate "proximate" intruders.

6. The inset in the lower right corner of FIG 4-5-6, Traffic Information Service (TIS), Avionics Block Diagram, shows a possible TIS data block display. The following information is contained in this data block:

(a) The intruder, located approximately four o'clock, three miles, is a "proximate" aircraft and currently not a collision threat to the client aircraft. This is indicated by the diamond symbol used in this example.

(b) The intruder ground track diverges to the right of the client aircraft, indicated by the small arrow.

(c) The intruder altitude is 700 feet less than or below the client aircraft, indicated by the "-07" located under the symbol.

(d) The intruder is descending >500 fpm, indicated by the downward arrow next to the "-07" relative altitude information. The absence of this arrow when an altitude tag is present indicates level flight or a climb/descent rate less than 500 fpm.

NOTE-*If the intruder did not have an operating altitude encoder (Mode C), the altitude and altitude trend "tags" would have been omitted.*

d. Limitations

1. TIS is **NOT** intended to be used as a collision avoidance system and does not relieve the pilot responsibility to "see and avoid" other aircraft (see paragraph 5-5-8, See and Avoid). TIS must not be for avoidance maneuvers during IMC or other times when there is no visual contact with the intruder aircraft. TIS is intended only to assist in visual acquisition of other aircraft in VMC. **No recommended avoidance maneuvers are provided for, nor authorized, as a direct result of a TIS intruder display or TIS alert.**

2. While TIS is a useful aid to visual traffic avoidance, it has some system limitations that must be fully understood to ensure proper use. Many of these limitations are inherent in secondary radar surveillance. In other words, the information provided by TIS will be no better than that provided to ATC. Other limitations and anomalies are associated with the TIS predictive algorithm.

(a) Intruder Display Limitations. TIS will only display aircraft with operating transponders installed. TIS relies on surveillance of the Mode S radar, which is a "secondary surveillance" radar similar to the ATCRBS described in paragraph 4-5-2.

(b) TIS Client Altitude Reporting Requirement. Altitude reporting is required by the TIS client aircraft in order to receive TIS. If the altitude encoder is inoperative or disabled, TIS will be unavailable, as TIS requests will not be honored by the ground system. As such, TIS requires altitude reporting to determine the Proximity Coverage Volume as indicated in FIG 4-5-4. TIS users must be alert to altitude encoder malfunctions, as TIS has no mechanism to determine if client altitude reporting is correct. A failure of this nature will cause erroneous and possibly unpredictable TIS operation. If this malfunction is suspected, confirmation of altitude reporting with ATC is suggested.

(c) Intruder Altitude Reporting. Intruders without altitude reporting capability will be displayed without the accompanying altitude tag. Additionally, nonaltitude reporting intruders are assumed to be at the same altitude as the TIS client for alert computations. This helps to ensure that the pilot will be alerted to all traffic under radar coverage, but the actual altitude difference may be substantial. Therefore, visual acquisition may be difficult in this instance.

(d) Coverage Limitations. Since TIS is provided by ground-based, secondary surveillance radar, it is subject to all limitations of that radar. If an aircraft is not detected by the radar, it cannot be displayed on TIS. Examples of these limitations are as follows:

(1) TIS will typically be provided within 55 NM of the radars depicted in FIG 4-5-5, Terminal Mode S Radar Sites. This maximum range can vary by radar site and is always subject to "line of sight" limitations; the radar and data link signals will be blocked by obstructions, terrain, and curvature of the earth.

(2) TIS will be unavailable at low altitudes in many areas of the country, particularly in mountainous regions. Also, when flying near the "floor" of radar coverage in a particular area, intruders below the client aircraft may not be detected by TIS.

(3) TIS will be temporarily disrupted when flying directly over the radar site providing coverage if no adjacent site assumes the service. A ground-based radar, like a VOR or NDB, has a zenith cone, sometimes referred to as the cone of confusion or cone of silence. This is the area of ambiguity directly above the station where bearing information is unreliable. The zenith cone setting for TIS is 34 degrees: Any aircraft above that angle with respect to the radar horizon will lose TIS coverage from that radar until it is below this 34 degree angle. The aircraft may not actually lose service in areas of multiple radar coverage since an adjacent radar will provide TIS. If no other TIS-capable radar is available, the "Good-bye" message will be received and TIS terminated until coverage is resumed.

(e) Intermittent Operations. TIS operation may be intermittent during turns or other maneuvering, particularly if the transponder system does not include antenna diversity (antenna mounted on the top and bottom of the aircraft). As in (d) above, TIS is dependent on two-way, "line of sight" communications between the aircraft and the Mode S radar. Whenever the structure of the client aircraft comes between the transponder antenna (usually located on the underside of the aircraft) and the ground-based radar antenna, the signal may be temporarily interrupted.

(f) TIS Predictive Algorithm. TIS information is collected one radar scan prior to the scan during which the uplink occurs. Therefore, the surveillance information is approximately 5 seconds old. In order to present the intruders in a "real time" position, TIS uses a "predictive algorithm" in its tracking software. This algorithm uses track history data to extrapolate intruders to their expected positions consistent with the time of display in the cockpit. Occasionally, aircraft maneuvering will cause this algorithm to induce errors in the TIS display. These errors primarily affect relative bearing information; intruder distance and altitude will remain relatively accurate and may be used to assist in "see and avoid." Some of the more common examples of these errors are as follows:

(1) When client or intruder aircraft maneuver excessively or abruptly, the tracking algorithm will report incorrect horizontal position until the maneuvering aircraft stabilizes.

(2) When a rapidly closing intruder is on a course that crosses the client at a shallow angle (either overtaking or head on) and either aircraft abruptly changes course within ¼ NM, TIS will display the intruder on the opposite side of the client than it actually is.

These are relatively rare occurrences and will be corrected in a few radar scans once the course has stabilized.

(g) Heading/Course Reference. Not all TIS aircraft installations will have onboard heading reference information. In these installations, aircraft course reference to the TIS display is provided by the Mode S radar. The radar only determines ground track information and has no indication of the client aircraft heading. In these installations, all intruder bearing information is referenced to ground track and does not account for wind correction. Additionally, since ground-based radar will require several scans to determine aircraft course following a course change, a lag in TIS display orientation (intruder aircraft bearing) will occur. As in (f) above, intruder distance and altitude are still usable.

(h) Closely-Spaced Intruder Errors. When operating more than 30 NM from the Mode S sensor, TIS forces any intruder within 3/8 NM of the TIS client to appear at the same horizontal position as the client aircraft. Without this feature, TIS could display intruders in a manner confusing to the pilot in critical situations (e.g., a closely-spaced intruder that is actually to the right of the client may appear on the TIS display to the left). At longer distances from the radar, TIS cannot accurately determine relative bearing/distance information on intruder aircraft that are in close proximity to the client.

Because TIS uses a ground-based, rotating radar for surveillance information, the accuracy of TIS data is dependent on the distance from the sensor (radar) providing the service. This is much the same phenomenon as experienced with ground-based navigational aids, such as VOR or NDB. As distance from the radar increases, the accuracy of surveillance decreases. Since TIS does not inform the pilot of distance from the Mode S radar, the pilot must assume that any intruder appearing at the same position as the client aircraft may actually be up to 3/8 NM away in any direction. Consistent with the operation of TIS, an alert on the display (regardless of distance from the radar) should stimulate an outside visual scan, intruder acquisition, and traffic avoidance based on outside reference.

e. Reports of TIS Malfunctions

1. Users of TIS can render valuable assistance in the early correction of malfunctions by reporting their observations of undesirable performance. Reporters should identify the time of observation, location, type and identity of aircraft, and describe the condition observed; the type of transponder processor, and software in use can also be useful information. Since TIS performance is monitored by maintenance personnel rather than ATC, it is suggested that malfunctions be reported in the following ways:

(a) By radio or telephone to the nearest Flight Service Station (FSS) facility.

(b) By FAA Form 8740-5, Safety Improvement Report, a postage-paid card designed for this purpose. These cards may be obtained at FAA FSSs, General Aviation District Offices, Flight Standards District Offices, and General Aviation Fixed Based Operations.

4-5-7. Automatic Dependent Surveillance-Broadcast (ADS-B) Services

a. Introduction

1. Automatic Dependent Surveillance-Broadcast (ADS-B) is a surveillance technology being deployed throughout the NAS (see FIG 4-5-7). The ADS-B system is composed of aircraft avionics and a ground infrastructure. Onboard avionics determine the position of the aircraft by using the GNSS and transmit its position along with additional information about the aircraft to ground stations for use by ATC and other ADS-B services. This information is transmitted at a rate of approximately once per second.

2. In the United States, ADS-B equipped aircraft exchange information is on one of two frequencies: 978 or 1090 MHz. The 1090 MHz frequency is associated with Mode A, C, and S transponder operations. 1090 MHz transponders with integrated ADS-B functionality extend the transponder message sets with additional ADS-B information. This additional information is known as an "extended squitter" message and referred to as 1090ES. ADS-B equipment operating on 978 MHz is known as the Universal Access Transceiver (UAT).

3. ADS B avionics can have the ability to both transmit and receive information. The transmission of ADS-B information from an aircraft is known as ADS-B Out. The receipt of ADS-B information by an aircraft is known as ADS-B In. On January 1, 2020, all aircraft operating within the airspace defined in 14 CFR part 91, § 91.225 will be required to transmit the information defined in § 91.227 using ADS-B Out avionics.

4. In general, operators flying at 18,000 feet and above will require equipment which uses 1090 ES. Those that do not fly above 18,000 may use either UAT or 1090ES equipment. (Refer to 14 CFR 91.225 and 91.227.) While the regulation will not require it, operators equipped with ADS-B In will realize additional benefits from ADS-B broadcast services: Traffic Information Service – Broadcast (TIS-B) (paragraph 4-5-8) and Flight Information Service - Broadcast (FIS-B) (paragraph 4-5-9).

b. ADS-B Certification and Performance Requirements

ADS-B equipment may be certified as a surveillance source for air traffic separation services using ADS-B Out. ADS-B equipment may also be certified for use with ADS-B In advisory services that enable appropriately equipped aircraft to display traffic and flight information. Refer to the aircraft's flight manual supplement or Pilot Operating Handbook for the capabilities of a specific aircraft installation.

Fig 4-5-7
ADS-B, TIS-B, and FIS-B: Broadcast Services Architecture

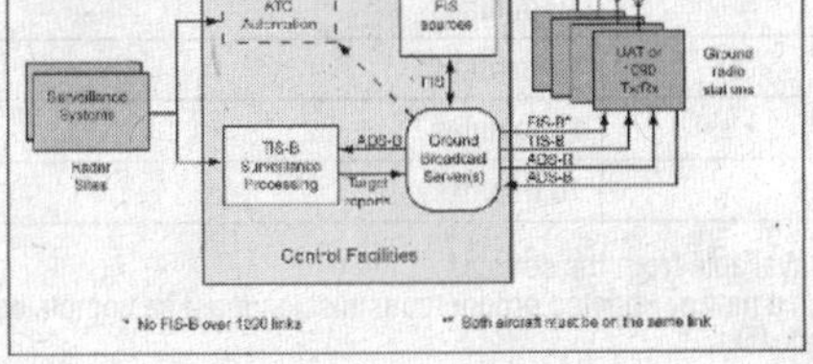

c. ADS-B Capabilities

1. ADS-B enables improved surveillance services, both air-to-air and air-to-ground, especially in areas where radar is ineffective due to terrain or where it is impractical or cost prohibitive. Initial NAS applications of air-to-air ADS-B are for "advisory" use only, enhancing a pilot's visual acquisition of other nearby equipped aircraft either when airborne or on the airport surface. Additionally, ADS-B will enable ATC and fleet operators to monitor aircraft throughout the available ground station coverage area.

2. ADS-B avionics typically allow pilots to enter the aircraft's call sign and Air Traffic Control (ATC)-assigned transponder code, which will be transmitted to other aircraft and ground receivers. Pilots are cautioned to use care when selecting and entering the aircraft's identification and transponder codes. Some installations may require separate entries of this information into both the ADS-B system and the transponder. Therefore, it is **extremely important to ensure that the transponder and ADS-B codes being transmitted are identical to avoid false conflict alerts within the ATC system.**

3. ADS B systems integrated with the transponder will automatically set the applicable emergency status when 7500, 7600, or 7700 are entered into the transponder. ADS B systems not integrated with the transponder, or systems with optional emergency codes, will require that the appropriate emergency code is entered through a pilot interface. **ADS-B is intended for in-flight and airport surface use. ADS-B systems should be turned "on" -- and remain "on" -- whenever operating in the air and moving on the airport surface. Civil and military Mode A/C transponders and ADS-B systems should be adjusted to the "on" or normal operating position as soon as practical,** unless the change to "standby" has been accomplished previously at the request of ATC. Mode S transponders should be left on whenever power is applied to the aircraft.

d. ATC Surveillance Services using ADS-B - Procedures and Recommended Phraseology - For Use In Alaska Only

Radar procedures, with the exceptions found in this paragraph, are identical to those procedures prescribed for radar in AIM Chapter 4 and Chapter 5.

1. Preflight:

If a request for ATC services is predicated on ADS-B and such services are anticipated when either a VFR or IFR flight plan is filed, the aircraft's "N" number or call-sign as filed in "Block 2" of the Flight Plan must be entered in the ADS-B avionics as the aircraft's flight ID.

2. Inflight:

When requesting ADS-B services while airborne, pilots should ensure that their ADS-B equipment is transmitting their aircraft's "N" number or call sign prior to contacting ATC. To accomplish this, the pilot must select the ADS-B "broadcast flight ID" function.

NOTE- *The broadcast "VFR" or "Standby" mode built into some ADS-B systems will not provide ATC with the appropriate aircraft identification information. This function should first be disabled before contacting ATC.*

3. Aircraft with an Inoperative/Malfunctioning ADS-B Transmitter or in the Event of an Inoperative Ground Broadcast Transceiver (GBT).

(a) ATC will inform the flight crew when the aircraft's ADS-B transmitter appears to be inoperative or malfunctioning:

PHRASEOLOGY- *YOUR ADS-B TRANSMITTER APPEARS TO BE INOPERATIVE/MALFUNCTIONING. STOP ADS-B TRANSMISSIONS.*

(b) ATC will inform the flight crew when the GBT transceiver becomes inoperative or malfunctioning, as follows:

PHRASEOLOGY- *(Name of facility) GROUND BASED TRANSCEIVER INOPERATIVE/MALFUNCTIONING.*
(And if appropriate) RADAR CONTACT LOST.

NOTE- *An inoperative or malfunctioning GBT may also cause a loss of ATC surveillance services.*

(c) ATC will inform the flight crew if it becomes necessary to turn off the aircraft's ADS-B transmitter.

PHRASEOLOGY- *STOP ADS-B TRANSMISSIONS.*

(d) Other malfunctions and considerations:

Loss of automatic altitude reporting capabilities (encoder failure) will result in loss of ATC altitude advisory services.

e. ADS-B Limitations

1. The ADS-B cockpit display of traffic is NOT intended to be used as a collision avoidance system and does not relieve the pilot's responsibility to "see and avoid" other aircraft. (See paragraph 5-5-8, See and Avoid). ADS-B must not be used for avoidance maneuvers during IMC or other times when there is no visual contact with the intruder aircraft. ADS-B is intended only to assist in visual acquisition of other aircraft.No avoidance maneuvers are provided nor authorized, as a direct result of an ADS-B target being displayed in the cockpit.

2. Use of ADS-B radar services is limited to the service volume of the GBT.

NOTE- *The coverage volume of GBTs are limited to line-of-sight.*

f. Reports of ADS-B Malfunctions

Users of ADS-B can provide valuable assistance in the correction of malfunctions by reporting instances of undesirable system performance. Reports should identify the time of observation, location, type and identity of aircraft, and describe the condition observed; the type of avionics system and its software version in use should also be included. Since ADS-B performance is monitored by maintenance personnel rather than ATC, it is suggested that malfunctions be reported in any one of the following ways:

1. By radio or telephone to the nearest Flight Service Station (FSS) facility.

2. By FAA Form 8740-5, Safety Improvement Report, a postage-paid card is designed for this purpose. These cards may be obtained from FAA FSSs, Flight Standards District Offices, and general aviation fixed-based operators.

3. By reporting the failure directly to the FAA Safe Flight 21 program at 1-877-FLYADSB or **http://www.adsb.gov**.

4-5-8. Traffic Information Service- Broadcast (TIS-B)

TIS-B is the broadcast of ATC derived traffic information to ADS-B equipped (1090ES or UAT) aircraft from ground radio stations. The source of this traffic information is derived from ground-based air traffic surveillance radar sensors. TIS-B service will be available throughout the NAS where there are both adequate surveillance coverage (radar) from ground sensors and adequate broadcast coverage from ADS-B ground radio stations. The quality level of traffic information provided by TIS-B is dependent upon the number and type of ground sensors available as TIS-B sources and the timeliness of the reported data.

a. TIS-B Requirements.

In order to receive TIS-B service, the following conditions must exist:

1. Aircraft must be equipped with an ADS-B transmitter/receiver or transceiver, and a cockpit display of traffic information (CDTI).

2. Aircraft must fly within the coverage volume of a compatible ground radio station that is configured for TIS-B uplinks. (Not all ground radio stations provide TIS-B due to a lack of radar coverage or because a radar feed is not available).

3. Aircraft must be within the coverage of and detected by at least one ATC radar serving the ground radio station in use.

b. TIS-B Capabilities.

1. TIS-B is intended to provide ADS-B equipped aircraft with a more complete traffic picture in situations where not all nearby aircraft are equipped with ADS-B Out. This advisory-only application is intended to enhance a pilot's visual acquisition of other traffic.

2. Only transponder-equipped targets (i.e., Mode A/C or Mode S transponders) are transmitted through the ATC ground system architecture. Current radar siting may result in limited radar surveillance coverage at lower altitudes near some airports, with subsequently limited TIS-B service volume coverage. If there is no radar coverage in a given area, then there will be no TIS-B coverage in that area.

c. TIS-B Limitations.

1. TIS-B is NOT intended to be used as a collision avoidance system and does not relieve the pilot's responsibility to "see and avoid" other aircraft, in accordance with 14CFR §91.113b. TIS-B must not be used for avoidance maneuvers during times when there is no visual contact with the intruder aircraft. TIS-B is intended only to assist in the visual acquisition of other aircraft.

NOTE- *No aircraft avoidance maneuvers are authorized as a direct result of a TIS-B target being displayed in the cockpit.*

2. While TIS-B is a useful aid to visual traffic avoidance, its inherent system limitations must be understood to ensure proper use.

(a) A pilot may receive an intermittent TIS-B target of themselves, typically when maneuvering (e.g., climbing turns) due to the radar not tracking the aircraft as quickly as ADS-B.

(b) The ADS-B-to-radar association process within the ground system may at times have difficulty correlating an ADS-B report with corresponding radar returns from the same aircraft. When this happens the pilot may see duplicate traffic symbols (i.e., "TIS-B shadows") on the cockpit display.

(c) Updates of TIS-B traffic reports will occur less often than ADS-B traffic updates. TIS-B position updates will occur approximately once every 3-13 seconds depending on the type of radar system in use within the coverage area. In comparison, the update rate for ADS-B is nominally once per second.

(d) The TIS-B system only uplinks data pertaining to transponder-equipped aircraft. Aircraft without a transponder will not be displayed as TIS-B traffic.

(e) There is no indication provided when any aircraft is operating inside or outside the TIS-B service volume, therefore it is difficult to know if one is receiving uplinked TIS-B traffic information.

3. Pilots and operators are reminded that the airborne equipment that displays TIS-B targets is for pilot situational awareness only and is not approved as a collision avoidance tool. Unless there is an imminent emergency requiring immediate action, any deviation from an air traffic control clearance in response to perceived converging traffic appearing on a TIS-B display must be approved by the controlling ATC facility before commencing the maneuver, except as permitted under certain conditions in 14CFR §91.123. Uncoordinated deviations may place an aircraft in close proximity to other aircraft under ATC control not seen on the airborne equipment and may result in a pilot deviation or other incident.

d. Reports of TIS-B Malfunctions

Users of TIS-B can provide valuable assistance in the correction of malfunctions by reporting instances of undesirable system performance. Reporters should identify the time of observation, location, type and identity of the aircraft, and describe the condition observed; the type of avionics system and its software version used. Since TIS-B performance is monitored by maintenance personnel rather than ATC, it is suggested that malfunctions be reported in anyone of the following ways:

1. By radio or telephone to the nearest Flight Service Station (FSS) facility.

2. By FAA Form 8740-5, Safety Improvement Report, a postage-paid card is designed for this purpose. These cards may be obtained from FAA FSSs, Flight Standards District Offices, and general aviation fixed-based operators.

3. By reporting the failure directly to the FAA Surveillance and Broadcast Services Program Office at 1-877-FLYADSB or **http://www.adsb.gov**.

4-5-9. Flight Information Service- Broadcast (FIS-B)

a. FIS-B is a ground broadcast service provided through the ADS-B Services network over the 978 MHz UAT data link. The FAA FIS-B system provides pilots and flight crews of properly equipped aircraft with a cockpit display of certain aviation weather and aeronautical information. FIS-B service availability is expected across the NAS in 2013 and is currently available within certain regions.

b. The weather products provided by FIS-B are for information only. Therefore, these products do not meet the safety and regulatory requirements of official weather products. The weather products displayed on FIS-B should not be used as primary weather products, i.e., aviation weather to meet operational and safety requirements. Official weather products (primary products) can be obtained from a variety of sources including ATC, FSSs, and, if applicable, AOCC VHF/HF voice, which can transmit aviation weather, NOTAMS, and other operational aeronautical information to aircraft in flight. FIS-B augments the traditional ATC/FSS/AOCC services by providing additional information and, for some products, offers the advantage of being displayed graphically. By using FIS-B for orientation and information, the usefulness of information received from official sources may be enhanced, but the user should be alert and understand any limitations associated with individual products. FIS-B provides the initial basic products listed below at no-charge to the user. Additional products are envisioned, but may incur subscription charges to the user. FIS-B reception is line-of-sight within the service volume of the ground infrastructure.

1. Airmen's Meteorological Conditions (AIRMET) - Text/graphical report - A concise description of expected occurrence of specified en route weather phenomena.

REFERENCE- *Advisory Circular AC-00-45, "Aviation Weather Services."*

2. Convective Significant Meteorological Conditions (SIGMET) - Text report/graphical -Convective SIGMETs (also known internationally as SIGMET for Convection) are issued for the contiguous U.S. Each bulletin includes one or more Convective SIGMETs for a specific region of the CONUS. Convective SIGMETs issued for thunderstorms and related phenomena do not include references to all weather associated with thunderstorms such as turbulence, icing, low-level wind shear and IFR conditions.

REFERENCE- *Advisory Circular AC-00-45, "Aviation Weather Services."*

3. Aviation Routine Weather Reports (METAR) - Text reports - Surface meteorological data. The body of the report includes airport identifier, time of observation, wind, visibility, runway visual range, present weather phenomena, sky conditions, temperature, dew point, and altimeter setting. Remarks may be appended to the end.

REFERENCE- *Advisory Circular AC-00-45, "Aviation Weather Services."*

4. Special Aviation Reports (SPECI) - Text reports - Non- routine, unscheduled report when any of SPECI criteria have been met.Contains all data elements found in METAR plus additional information which elaborates on data.

REFERENCE- *Reference Advisory Circular AC-00-45, "Aviation Weather Services."*

5. Next Generation Radar (NEXRAD) (CONUS and Regional) - Derived from Next Generation Weather Radar (WSR-88D). Radar mosaic consists of multiple single site radar images combined to produce a graphical image on a regional or national scale. Regional and national radar mosaics can be found at the National Weather Service (NWS) Doppler Radar Images web site: http://radar.weather.gov/ridge/.

REFERENCE- *Advisory Circular AC-00-45, "Aviation-Weather Services."*

6. Notice to Airmen (NOTAM) Distant and Flight Data Center (D/FDC) - Text/graphical reports - includes Temporary Flight Restrictions (TFRs) - TFR text and graphic reports, prescribes procedures used to obtain, format, and disseminate information on unanticipated or temporary changes to components of or hazards in the NAS until the associated aeronautical charts and related publications have been amended.

REFERENCE- *FAA Order JO 7930.2, Notices to Airmen (NOTAM).*

7. Pilot Weather Reports (PIREP) - Text report - Pilots report observations to assist other pilots with flight planning and preparation, help NWS verify forecast products, and create more accurate products for aviation community.

REFERENCE- *Advisory Circular AC-00-45, "Aviation Weather Services."*

8. Significant Meteorological Information (SIGMET) - Text/graphical report - Potentially hazardous en route phenomena such as thunderstorms and hail, turbulence, icing, sand and dust storms, tropical cyclones, and volcanic ash in an area affecting 3,000 square miles or an area deemed to have a significant effect on safety of aircraft operations.

REFERENCE- *Advisory Circular AC-00-45, "Aviation-Weather Services."*

9. Special Use Airspace (SUA) Status - Text/graphical report - Establishes/designates airspace in the interest of National Defense, security and/or welfare. Charted SUA identifies to other airspace users where these activities occur. SUA is airspace of defined dimensions wherein activities must be confined because of their nature, or wherein limitations may be imposed upon aircraft operations that are not a part of those activities.

REFERENCE- *FAA Order JO 7400.2G, Procedures for Handling Airspace Matters.*

10. Terminal Aerodrome Forecast (TAF) and their amendments (AMEND) - Text report - Routine forecast that gives a concise statement of expected meteorological conditions for a specified time period within five statute miles (SM) of the center of the airport's runway complex (terminal). TAFs are amended whenever they become, in the forecaster's judgment, unrepresentative of existing or expected conditions, particularly regarding those elements and events significant to aircraft and airports. An amended forecast is identified by TAF AMD (in place of TAF) on the first line of the forecast text.

REFERENCE- *Advisory Circular AC-00-45, "Aviation Weather Services."*

11. Temperature Aloft - Text report - Computer prepared forecasts, based on North American Mesoscale (NAM) forecast model run, of temperature at specified times, altitudes, and locations.

REFERENCE- *Advisory Circular AC-00-45, "Aviation Weather Services."*

12. Winds Aloft - Text report - Computerprepared forecasts, based on North American Mesoscale (NAM) forecast model run, of wind direction and wind speed at specified times, altitudes, and locations.

TBL 4-5-2

FIS-B Basic Product Update and Transmission Intervals

Product	FIS-B Service Update Interval[1]	FIS-B Service Transmission Interval[2]
AIRMET	As available	5 minutes
Convective SIGMET	As available	5 minutes
METAR/SPECI	Hourly/as available	5 minutes
NEXRAD Reflectivity (CONUS)	5 minutes	15 minutes
NEXRAD Reflectivity (Regional)	5 minutes	2.5 minutes
NOTAM-D/FDC	As available	10 minutes
PIREP	As available	10 minutes
SIGMET	As available	5 minutes
SUA Status	As available	10 minutes
TAF/AMEND	8 hours/as available	10 minutes
Temperature Aloft	6 hours	10 minutes
Winds Aloft	6 hours	10 minutes

[1]The Update Interval is the rate at which the product data is available from the source.

[2]The Transmission Interval is the amount of time within which a new or updated product ransmission must be completed and the rate or repetition interval at which the product s rebroadcast.

REFERENCE-*Advisory Circular AC-00-45, "Aviation Weather Services."*

NOTE-*Details concerning the content, format, and symbols of the various data link products provided should be obtained from the specific avionics manufacturer.*

4-5-10. Automatic Dependent Surveillance-Rebroadcast (ADS-R)

ADS-R is a datalink translation function of the ADS-B ground system required to accommodate the two separate operating frequencies (978 MHz and 1090 ES). The ADS-B system receives the ADS-B messages transmitted on one frequency and ADS-R translates and reformats the information for rebroadcast and use on the other frequency. This allows ADS-B In equipped aircraft to see nearby ADS-B Out traffic regardless of the operating link of the other aircraft. Aircraft operating on the same ADS-B frequency exchange information directly and do not require the ADS-R translation function.

Section 6. Operational Policy/ Procedures for Reduced Vertical Separation Minimum (RVSM) in the Domestic U.S., Alaska, Offshore Airspace and the San Juan FIR

4-6-1. Applicability and RVSM Mandate (Date/ Time and Area)

a. Applicability. The policies, guidance and direction in this section apply to RVSM operations in the airspace over the lower 48 states, Alaska, Atlantic and Gulf of Mexico High Offshore Airspace and airspace in the San Juan FIR where VHF or UHF voice direct controller-pilot communication (DCPC) is normally available. Policies, guidance and direction for RVSM operations in oceanic airspace where VHF or UHF voice DCPC is not available and the airspace of other countries are posted on the FAA "RVSM Documentation" Webpage described in paragraph 4-6-3, Aircraft and Operator Approval Policy/Procedures, RVSM Monitoring and Databases for Aircraft and Operator Approval.

b. Mandate. At 0901 UTC on January 20, 2005, the FAA implemented RVSM between flight level (FL) 290-410 (inclusive) in the following airspace: the airspace of the lower 48 states of the United States, Alaska, Atlantic and Gulf of Mexico High Offshore Airspace and the San Juan FIR. (A chart showing the location of offshore airspace is posted on the Domestic U.S. RVSM (DRVSM) Webpage. See paragraph4-6-3.) On the same time and date, RVSM was also introduced into the adjoining airspace of Canada and Mexico to provide a seamless environment for aircraft traversing those borders. In addition, RVSM was implemented on the same date in the Caribbean and South American regions.

c. RVSM Authorization. In accordance with 14 CFR Section 91.180, with only limited exceptions, prior to operating in RVSM airspace, operators and aircraft must have received RVSM authorization from the responsible civil aviation authority. (See paragraph4-6-10, Procedures for Accommodation of Non-RVSM Aircraft.) If the operator or aircraft or both have not been authorized for RVSM operations, the aircraft will be referred to as a "non-RVSM" aircraft. Paragraph 4-6-10 discusses ATC policies for accommodation of non-RVSM aircraft flown by the Department of Defense, Air Ambulance (MEDEVAC) operators, foreign State governments and aircraft flown for certification and development. Paragraph 4-6-11, Non-RVSM Aircraft Requesting Climb to and Descent from Flight Levels Above RVSM Airspace Without Intermediate Level Off, contains policies for non-RVSM aircraft climbing and descending through RVSM airspace to/from flight levels above RVSM airspace.

d. Benefits. RVSM enhances ATC flexibility, mitigates conflict points, enhances sector throughput, reduces controller workload and enables crossing traffic. Operators gain fuel savings and operating efficiency benefits by flying at more fuel efficient flight levels and on more user preferred routings.

4-6-2. Flight Level Orientation Scheme

Altitude assignments for direction of flight follow a scheme of odd altitude assignment for magnetic courses 000-179 degrees and even altitudes for magnetic courses 180-359 degrees for flights up to and including FL 410, as indicated in FIG 4-6-1.

FIG 4-6-1
Flight Level Orientation Scheme

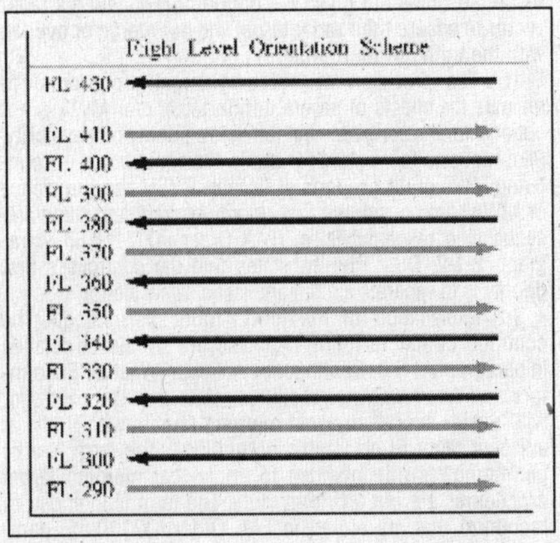

NOTE-*Odd Flight Levels: Magnetic Course 000-179 Degrees Even Flight Levels: Magnetic Course 180-359 Degrees.*

4-6-3. Aircraft and Operator Approval Policy/ Procedures, RVSM Monitoring and Databases for Aircraft and Operator Approval

a. RVSM Authority. 14 CFR Section 91.180 applies to RVSM operations within the U.S. 14 CFR Section91.706 applies to RVSM operations outside the U.S. Both sections require that the operator obtain authorization prior to operating in RVSM airspace. 14CFR Section 91.180 requires that, prior to conducting RVSM operations within the U.S., the operator obtain authorization from the FAA or from the responsible authority, as appropriate. In addition, it requires that the operator and the operator's aircraft comply with the standards of 14 CFR Part 91 Appendix G (Operations in RVSM Airspace).

b. Sources of Information. The FAA RVSM Website Homepage can be accessed at: **http://www.faa.gov/ about/office_org/ headquarters_offices/ato/ service_units/enroute/rvsm/**. The "RVSM Documentation" and "Domestic RVSM" webpages are linked to the RVSM Homepage. "RVSM Documentation" contains guidance and direction for an operator to obtain aircraft and operator approval to conduct RVSM operations. It provides information for DRVSM and oceanic and international RVSM airspace. It is recommended that operators planning to operate in Domestic U.S. RVSM airspace first review the following documents to orient themselves to the approval process.

1. Under "Area of Operations Specific Information," the document, "Basic Operator Information on DRVSM Programs," provides an overview of the DRVSM program and the related aircraft and operator approval programs.

2. In the "Getting Started" section, review the "RVSM Approval Checklist - U.S. Operators" or "RVSM Approval Checklist - Non-U.S. Operators" (as applicable). These are job aids or checklists that show aircraft/operator approval process events with references to related RVSM documents published on the website.

3. Under "Documents Applicable to All RVSM Approvals," review "RVSM Area New to the Operator." This document provides a guide for operators that are conducting RVSM operations in one or more areas of operation, but are planning to conduct RVSM operations in an area where they have not previously conducted RVSM operations, such as the U.S.

c. TCAS Equipage. TCAS equipage requirements are contained in 14 CFR Sections 121.356, 125.224, 129.18 and 135.189. Part 91 Appendix G does not contain TCAS equipage requirements specific to RVSM, however, Appendix G does require that aircraft equipped with TCAS II and flown in RVSM airspace be modified to incorporate TCAS II Version 7.0 or a later version.

d. Aircraft Monitoring. Operators are required to participate in the RVSM aircraft monitoring program. The "Monitoring Requirements and Procedures" section of the RVSM Documentation Webpage contains policies and procedures for participation in the monitoring program. Ground-based and GPS-based monitoring systems are available for the Domestic RVSM program. Monitoring is a quality control program that enables the FAA and other civil aviation authorities to assess the in-service altitude-keeping performance of aircraft and operators.

e. Registration on RVSM Approvals Databases. The "Registration on RVSM Approvals Database" section of the RVSM Documentation Webpage provides policies/procedures for operator and aircraft registration on RVSM approvals databases.

1. Purpose of RVSM Approvals Databases. ATC does not use RVSM approvals databases to determine whether or not a clearance can be issued into RVSM airspace. RVSM program managers do regularly review the operators and aircraft that operate in RVSM airspace to identify and investigate those aircraft and operators flying in RVSM airspace, but not listed on the RVSM approvals databases.

2. Registration of U.S. Operators. When U.S. operators and aircraft are granted RVSM authority, the FAA Flight Standards office makes an input to the FAA Program Tracking and Reporting Subsystem (PTRS). The Separation Standards Group at the FAA Technical Center obtains PTRS operator and aircraft information to update the FAA maintained U.S. Operator/Aircraft RVSM Approvals Database. Basic database operator and aircraft information can be viewed on the RVSM Documentation Webpage by clicking on the appropriate database icon.

3. Registration of Non-U.S. Operators. Non-U.S. operators can find policy/procedures for registration on the North American Approvals Registry and Monitoring Organization (NAARMO) database in the "Registration on RVSM Approvals Database" section of RVSM Documentation.

4-6-4. Flight Planning into RVSM Airspace

a. Operators that do not file the correct aircraft equipment suffix on the FAA or ICAO Flight Plan may be denied clearance into RVSM airspace. Policies for the FAA Flight Plan are detailed in subparagraph c below. Policies for the ICAO Flight Plan are detailed in subparagraph d.

b. The operator will annotate the equipment block of the FAA or ICAO Flight Plan with an aircraft equipment suffix indicating RVSM capability only after the responsible civil aviation authority has determined that both the operator and its aircraft are RVSM-compliant and has issued RVSM authorization to the operator.

c. General Policies for FAA Flight Plan Equipment Suffix. TBL 5-1-2, Aircraft Suffixes, allows operators to indicate that the aircraft has both RVSM and Advanced Area Navigation (RNAV) capabilities or has only RVSM capability.

1. The operator will annotate the equipment block of the FAA Flight Plan with the appropriate aircraft equipment suffix from TBL 5-1-2.

2. Operators can only file one equipment suffix in block 3 of the FAA Flight Plan. Only this equipment suffix is displayed directly to the controller.

3. Aircraft with RNAV Capability. For flight in RVSM airspace, aircraft with RNAV capability, but not Advanced RNAV capability, will file "/W". Filing "/W" will not preclude such aircraft from filing and flying direct routes in en route airspace.

d. Policy for ICAO Flight Plan Equipment Suffixes.

1. Operators/aircraft that are RVSM-compliant and that file ICAO flight plans will file "/W" in block10 (Equipment) to indicate RVSM authorization and will also file the appropriate ICAO Flight Plan suffixes to indicate navigation and communication capabilities. The equipment suffixes in TBL 5-1-2 are for use only in an FAA Flight Plan (FAA Form 7233-1).

2. Operators/aircraft that file ICAO flight plans that include flight in Domestic U.S. RVSM airspace must file "/W" in block 10 to indicate RVSM authorization.

e. Importance of Flight Plan Equipment Suffixes. The operator must file the appropriate equipment suffix in the equipment block of the FAA Flight Plan (FAA Form 7233-1) or the ICAO Flight Plan. The equipment suffix informs ATC:

1. Whether or not the operator and aircraft are authorized to fly in RVSM airspace.

2. The navigation and/or transponder capability of the aircraft (e.g., advanced RNAV, transponder with Mode C).

f. Significant ATC uses of the flight plan equipment suffix information are:

1. To issue or deny clearance into RVSM airspace.

2. To apply a 2,000 foot vertical separation minimum in RVSM airspace to aircraft that are not authorized for RVSM, but are in one of the limited categories that the FAA has agreed to accommodate. (See paragraphs 4-6-10, Procedures for Accommodation of Non-RVSM Aircraft, and 4-6-11, Non-RVSM Aircraft Requesting Climb to and Descent from Flight Levels Above RVSM Airspace Without Intermediate Level Off, for policy on limited operation of unapproved aircraft in RVSM airspace).

3. To determine if the aircraft has "Advanced RNAV" capabilities and can be cleared to fly procedures for which that capability is required.

4-6-5. Pilot RVSM Operating Practices and Procedures

a. RVSM Mandate. If either the operator or the aircraft or both have not received RVSM authorization (non-RVSM aircraft), the pilot will neither request nor accept a clearance into RVSM airspace unless:

1. The flight is conducted by a non-RVSM DOD, MEDEVAC, certification/development or foreign State (government) aircraft in accordance with paragraph 4-6-10, Procedures for Accommodation of Non-RVSM Aircraft.

2. The pilot intends to climb to or descend from FL 430 or above in accordance with paragraph 4-6-11, Non-RVSM Aircraft Requesting Climb to and Descent from Flight Levels Above RVSM Airspace Without Intermediate Level Off.

3. An emergency situation exists.

b. Basic RVSM Operating Practices and Procedures. Appendix 4 of AC 91-85, Authorization of Aircraft and Operators for Flight in Reduced Vertical Separation Minimum Airspace contains pilot practices and procedures for RVSM. Operators must incorporate Appendix 4 practices and procedures, as supplemented by the applicable paragraphs of this section, into operator training or pilot knowledge programs and operator documents containing RVSM operational policies.

c. Appendix 4 contains practices and procedures for flight planning, preflight procedures at the aircraft, procedures prior to RVSM airspace entry, inflight (en route) procedures, contingency procedures and post flight.

d. The following paragraphs either clarify or supplement Appendix 4 practices and procedures.

4-6-6. Guidance on Severe Turbulence and Mountain Wave Activity (MWA)

a. Introduction/Explanation

1. The information and practices in this paragraph are provided to emphasize to pilots and controllers the importance of taking appropriate action in RVSM airspace when aircraft experience severe turbulence and/or MWA that is of sufficient magnitude to significantly affect altitude-keeping.

2. Severe Turbulence. Severe turbulence causes large, abrupt changes in altitude and/or attitude usually accompanied by large variations in indicated airspeed. Aircraft may be momentarily out of control. Encounters with severe turbulence must be remedied immediately in any phase of flight. Severe turbulence may be associated with MWA.

3. Mountain Wave Activity (MWA)

(a) Significant MWA occurs both below and above the floor of RVSM airspace, FL 290. MWA often occurs in western states in the vicinity of mountain ranges. It may occur when strong winds blow perpendicular to mountain ranges resulting in up and down or wave motions in the atmosphere. Wave action can produce altitude excursions and airspeed fluctuations accompanied by only light turbulence. With sufficient amplitude, however, wave action can induce altitude and airspeed fluctuations accompanied by severe turbulence. MWA is difficult to forecast and can be highly localized and short lived.

(b) Wave activity is not necessarily limited to the vicinity of mountain ranges. Pilots experiencing wave activity anywhere that significantly affects altitude-keeping can follow the guidance provided below.

(c) Inflight MWA Indicators (Including Turbulence). Indicators that the aircraft is being subjected to MWA are:

(1) Altitude excursions and/or airspeed fluctuations with or without associated turbulence.

(2) Pitch and trim changes required to maintain altitude with accompanying airspeed fluctuations.

(3) Light to severe turbulence depending on the magnitude of the MWA.

4. Priority for Controller Application of Merging Target Procedures

(a) Explanation of Merging Target Procedures. As described in subparagraph c3below, ATC will use "merging target procedures" to mitigate the effects of both severe turbulence and MWA. The procedures in subparagraph c3 have been adapted from existing procedures published in FAA Order JO 7110.65, Air Traffic Control, paragraph 5-1-8, Merging Target Procedures. Paragraph 5-1-8 calls for en route controllers to advise pilots of potential traffic that they perceive may fly directly above or below his/her aircraft at minimum vertical separation. In response, pilots are given the option of requesting a radar vector to ensure their radar target will not merge or overlap with the traffic's radar target.

(b) The provision of "merging target procedures" to mitigate the effects of severe turbulence and/or MWA is not optional for the controller, but rather is a priority responsibility. Pilot requests for vectors for traffic avoidance when encountering MWA or pilot reports of "Unable RVSM due turbulence or MWA" are considered first priority aircraft separation and sequencing responsibilities. (FAA Order JO 7110.65, paragraph 2-1-2, Duty Priority, states that the controller's first priority is to separate aircraft and issue safety alerts).

(c) Explanation of the term "traffic permitting." The contingency actions for MWA and severe turbulence detailed in paragraph 4-6-9, Contingency Actions: Weather Encounters and Aircraft System Failures, state that the controller will "vector aircraft to avoid merging targets with traffic at adjacent flight levels, traffic permitting." The term "traffic permitting" is not intended to imply that merging target procedures are not a priority duty. The term is intended to recognize that, as stated in FAA Order JO 7110.65, paragraph 2-1-2, Duty Priority, there are circumstances when the controller is required to perform more than one action and must "exercise their best judgment based on the facts and circumstances known to them" to prioritize their actions. Further direction given is: "That action which is most critical from a safety standpoint is performed first."

5. TCAS Sensitivity. For both MWA and severe turbulence encounters in RVSM airspace, an additional concern is the sensitivity of collision avoidance systems when one or both aircraft operating in close proximity receive TCAS advisories in response to disruptions in altitude hold capability.

b. Pre-flight tools. Sources of observed and forecast information that can help the pilot ascertain the possibility of MWA or severe turbulence are: Forecast Winds and Temperatures Aloft (FD), Area Forecast (FA), SIGMETs and PIREPs.

c. Pilot Actions When Encountering Weather (e.g., Severe Turbulence or MWA)

1. Weather Encounters Inducing Altitude Deviations of Approximately 200 feet. When the pilot experiences weather induced altitude deviations of approximately 200 feet, the pilot will contact ATC and state "Unable RVSM Due (state reason)" (e.g., turbulence, mountain wave). See contingency actions in paragraph 4-6-9.

2. Severe Turbulence (including that associated with MWA). When pilots encounter severe turbulence, they should contact ATC and report the situation. Until the pilot reports clear of severe turbulence, the controller will apply merging target vectors to one or both passing aircraft to prevent their targets from merging:

EXAMPLE- *"Yankee 123, FL 310, unable RVSM due severe turbulence."*

"Yankee 123, fly heading 290; traffic twelve o'clock, 10 miles, opposite direction; eastbound MD-80 at FL 320" (or the controller may issue a vector to the MD-80 traffic to avoid Yankee 123).

3. MWA. When pilots encounter MWA, they should contact ATC and report the magnitude and location of the wave activity. When a controller makes a merging targets traffic call, the pilot may request a vector to avoid flying directly over or under the traffic. In situations where the pilot is experiencing altitude deviations of 200 feet or greater, the pilot will request a vector to avoid traffic. Until the pilot reports clear of MWA, the controller will apply merging target vectors to one or both passing aircraft to prevent their targets from merging:

EXAMPLE- *"Yankee 123, FL 310, unable RVSM due mountain wave."*

"Yankee 123, fly heading 290; traffic twelve o'clock, 10 miles, opposite direction; eastbound MD-80 at FL 320" (or the controller may issue a vector to the MD-80 traffic to avoid Yankee 123).

4. FL Change or Re-route. To leave airspace where MWA or severe turbulence is being encountered, the pilot may request a FL change and/or re-route, if necessary.

4-6-7. Guidance on Wake Turbulence

a. Pilots should be aware of the potential for wake turbulence encounters in RVSM airspace. Experience gained since 1997 has shown that such encounters in RVSM airspace are generally moderate or less in magnitude.

b. Prior to DRVSM implementation, the FAA established provisions for pilots to report wake turbulence events in RVSM airspace using the NASA Aviation Safety Reporting System (ASRS). A "Safety Reporting" section established on the FAA RVSM Documentation webpage provides contacts, forms, and reporting procedures.

c. To date, wake turbulence has not been reported as a significant factor in DRVSM operations. European authorities also found that reports of wake turbulence encounters did not increase significantly after RVSM implementation (eight versus seven reports in a ten-month period). In addition, they found that reported wake turbulence was generally similar to moderate clear air turbulence.

d. Pilot Action to Mitigate Wake Turbulence Encounters

1. Pilots should be alert for wake turbulence when operating:

(a) In the vicinity of aircraft climbing or descending through their altitude.

(b) Approximately 10-30 miles after passing 1,000 feet below opposite-direction traffic.

(c) Approximately 10-30 miles behind and 1,000 feet below same-direction traffic.

2. Pilots encountering or anticipating wake turbulence in DRVSM airspace have the option of requesting a vector, FL change, or if capable, a lateral offset.

NOTE-1. *Offsets of approximately a wing span upwind generally can move the aircraft out of the immediate vicinity of another aircraft's wake vortex.*

2. In domestic U.S. airspace, pilots must request clearance to fly a lateral offset. Strategic lateral offsets flown in oceanic airspace do not apply.

e. The FAA will track wake turbulence events as an element of its post implementation program. The FAA will advertise wake turbulence reporting procedures to the operator community and publish reporting procedures on the RVSM Documentation Webpage (See address in paragraph 4-6-3, Aircraft and Operator Approval Policy/Procedures, RVSM Monitoring and Databases for Aircraft and Operator Approval.

4-6-8. Pilot/Controller Phraseology

TBL 4-6-1 shows standard phraseology that pilots and controllers will use to communicate in DRVSM operations.

4-6-9. Contingency Actions: Weather Encounters and Aircraft System Failures

TBL 4-6-2 provides pilot guidance on actions to take under certain conditions of aircraft system failure and weather encounters. It also describes the expected ATC controller actions in these situations. It is recognized that the pilot and controller will use judgment to determine the action most appropriate to any given situation.

4-6-10. Procedures for Accommodation of Non-RVSM Aircraft

a. General Policies for Accommodation of Non-RVSM Aircraft

1. The RVSM mandate calls for only RVSM authorized aircraft/operators to fly in designated RVSM airspace with limited exceptions. The policies detailed below are intended exclusively for use by aircraft that the FAA has agreed to accommodate. They are not intended to provide other operators a means to circumvent the normal RVSM approval process.

2. If either the operator or aircraft or both have not been authorized to conduct RVSM operations, the aircraft will be referred to as a "non-RVSM" aircraft. 14 CFR Section 91.180 and Part 91 Appendix G enable the FAA to authorize a deviation to operate a non-RVSM aircraft in RVSM airspace.

3. Non-RVSM aircraft flights will be handled on a workload permitting basis. The vertical separation standard applied between aircraft not approved for RVSM and all other aircraft must be 2,000 feet.

4. Required Pilot Calls. The pilot of non-RVSM aircraft will inform the controller of the lack of RVSM approval in accordance with the direction provided in paragraph 4-6-8, Pilot/Controller Phraseology.

b. Categories of Non-RVSM Aircraft that may be Accommodated

Subject to FAA approval and clearance, the following categories of non-RVSM aircraft may operate in domestic U.S. RVSM airspace provided they have an operational transponder.

1. Department of Defense (DOD) aircraft.

2. Flights conducted for aircraft certification and development purposes.

Tbl 4-6-1
Pilot/Controller Phraseology

Message	Phraseology
For a controller to ascertain the RVSM approval status of an aircraft:	(call sign) confirm RVSM approved
Pilot indication that flight is RVSM approved	Affirm RVSM
Pilot report of lack of RVSM approval (non-RVSM status). Pilot will report non- RVSM status, as follows:	Negative RVSM, (supplementary information, e.g., "Certification flight").
a. On the initial call on any frequency in the RVSM airspace and . . . . **b.** In all requests for flight level changes pertaining to flight levels within the RVSM airspace and . . . . **c.** In all read backs to flight level clearances pertaining to flight levels within the RVSM airspace and . . . . **d.** In read back of flight level clearances involving climb and descent through RVSM airspace (FL 290 - 410).	
Pilot report of one of the following after entry into RVSM airspace: all primary altimeters, automatic altitude control systems or altitude alerters have failed. (See paragraph 4-6-9, Contingency Actions: Weather Encounters and Aircraft System Failures.) ***NOTE-*** *This phrase is to be used to convey both the initial indication of RVSM aircraft system failure and on initial contact on all frequencies in RVSM airspace until the problem ceases to exist or the aircraft has exited RVSM airspace.*	Unable RVSM Due Equipment
ATC denial of clearance into RVSM airspace	Unable issue clearance into RVSM airspace, maintain FL
*Pilot reporting inability to maintain cleared flight level due to weather encounter. (See paragraph 4-6-9, Contingency Actions: Weather Encounters and Aircraft System Failures).	*Unable RVSM due (state reason) (e.g., turbulence, mountain wave)
ATC requesting pilot to confirm that an aircraft has regained RVSM-approved status or a pilot is ready to resume RVSM	Confirm able to resume RVSM
Pilot ready to resume RVSM after aircraft system or weather contingency	Ready to resume RVSM

3. Active air ambulance flights utilizing a "MEDEVAC" call sign.

4. Aircraft climbing/descending through RVSM flight levels (without intermediate level off) to/from FLs above RVSM airspace (Policies for these flights are detailed in Paragraph 4-6-11, Non-RVSM Aircraft Requesting Climb to and Descent from Flight Levels Above RVSM Airspace Without Intermediate Level Off.

5. Foreign State (government) aircraft.

c. Methods for operators of non-RVSM aircraft to request access to RVSM Airspace. Operators may:

1. LOA/MOU. Enter into a Letter of Agreement (LOA)/ Memorandum of Understanding (MOU) with the RVSM facility (the Air Traffic facility that provides air traffic services in RVSM airspace). Operators must comply with LOA/MOU.

2. File-and-Fly. File a flight plan to notify the FAA of their intention to request access to RVSM airspace.

NOTE-*Priority for access to RVSM airspace will be afforded to RVSM compliant aircraft, then File-and-Fly flights.*

d. Center Phone Numbers. Center phone numbers are posted on the RVSM Documentation Webpage, North American RVSM, Domestic U.S. RVSM section. This address provides direct access to the phone number listing:

http://www.faa.gov/ats/ato/150_docs/Center_Phone_No._Non-RVSM_Acft.doc

4-6-11. Non-RVSM Aircraft Requesting Climb to and Descent from Flight Levels Above RVSM Airspace Without Intermediate Level Off

a. File-and-Fly. Operators of Non-RVSM aircraft climbing to and descending from RVSM flight levels should just file a flight plan.

b. Non-RVSM aircraft climbing to and descending from flight levels above RVSM airspace will be handled on a workload permitting basis. The vertical separation standard applied in RVSM airspace between non-RVSM aircraft and all other aircraft must be 2,000 feet.

c. Non-RVSM aircraft climbing to/descending from RVSM airspace can only be considered for accommodation provided:

1. Aircraft is capable of a continuous climb/descent and does not need to level off at an intermediate altitude for any operational considerations and

2. Aircraft is capable of climb/descent at the normal rate for the aircraft.

d. Required Pilot Calls. The pilot of non-RVSM aircraft will inform the controller of the lack of RVSM approval in accordance with the direction provided in paragraph 4-6-8, Pilot/Controller Phraseology.

Section 7. Operational Policy/ Procedures for the Gulf of Mexico 50 NM Lateral Separation Initiative

4-7-1. Introduction and Background

a. Introduction. On 20 October 2011 at 0900 UTC, the Federal Aviation Administration (FAA), Serviciós a la Navegacion en el Espacio Aéreo Mexicano (SENEAM) and the Direccion General de Aeronautica Civil (DGAC) Mexico implemented 50 Nautical Mile (NM) lateral separation between aircraft authorized Required Navigation Performance 10 (RNP 10) or RNP 4 operating in the Gulf of Mexico (GoMex) Oceanic Control Areas (CTA). Existing Air Traffic Services (ATS) routes and route operating policies did not change for this implementation.

b. RNP 10 Versus RNAV 10 Terminology. "RNP 10" has the same meaning and application as "RNAV 10". The ICAO Performancebased Navigation (PBN) Manual (ICAO Doc 9613), Volume II, Part B, Chapter 1 (Implementing RNAV 10, Designated and Authorized as RNP 10) explains that the term "RNP 10" was in use before the publication of the ICAO PBN Manual and the manual has "grandfathered in" its continued use when implementing an "RNAV 10" navigation specification.

c. Background. 50 NM lateral separation was first applied between aircraft authorized for RNP 10 operations on the North Pacific Route System in April 1998. Since that time, 50 NM lateral separation has been expanded throughout the Pacific Flight Information Regions (FIRs) and is currently applied in other airspaces, including, starting in June 2008, the West Atlantic Route System. GoMex 50 NM lateral separation implementation will apply the experience gained in those operations.

d. Project Objectives. The project objectives were to:

1. Reduce lateral separation to 50 NM between aircraft authorized RNP 10 or RNP 4.

2. Leave existing ATS routes and operating policies in place.

3. Have approximately 90% of flights conducted by operators/aircraft over the Gulf of Mexico authorized for RNP 10 or RNP 4 operations by the appropriate State authority.

4. Accommodate the operation of the small percentage of flights not authorized RNP 10.

5. Establish a policy that aircraft equipped with a Single LongRange Navigation System (SLRNS) can qualify for RNP 10 operations in the Gulf of Mexico in accordance with the ICAO PBN Manual and the appropriate FAA and DGAC documents. (See paragraph 4-7-7e.)

e. Control Areas (CTA) Affected. 50 NM lateral separation is implemented in the following CTAs/FIRs/ Upper Control Areas (UTA).

1 The Houston Oceanic CTA/FIR and the Gulf of Mexico portion of the Miami Oceanic CTA/FIR.

(a) The Monterrey CTA and Merida High CTA within the Mexico FIR/UTA

f. Policy and Procedures Coordination with SENEAM and the DGAC. The policies and procedures were coordinated with SENEAM and the Mexico DGAC. They are applied in the GoMex CTA's where the FAA and SENEAM provide Air Traffic Control.

4-7-2. Gulf of Mexico 50 NM Lateral Separation Initiative Web Page: Policy, Procedures and Guidance for Operators and Regulators

Information on plans, policies and procedures for 50 NM lateral separation is posted on the "Gulf of Mexico 50 NM Lateral Separation Web Page":

http://www.faa.gov/about/office_org/headquarters_offices/ato/service_units/enroute/oceanic/gomex/

The web page contains detailed guidance on operator and aircraft authorization for RNP 10 or RNP 4 and includes Job Aids with FAA and ICAO document references.

4-7-3. Lateral Separation Minima Applied

a. 50 NM lateral separation is applied in the GoMex CTA's between aircraft authorized RNP 10 or RNP 4 at all altitudes above the floor of controlled airspace.

b. The current lateral separation minima of 100 NM in the Houston, Monterrey and Merida CTAs, and 90 NM in the Miami Oceanic CTA will continues to be applied between aircraft not authorized RNP 10 or RNP 4.

4-7-4. Operation on Routes on the periphery of the Gulf of Mexico CTAs

Operations on certain routes that fall within the boundaries of affected CTAs are not affected by the introduction of 50 NM lateral separation. Operation on the following routes is not affected:

a. Routes that are flown by reference to ICAO standard groundbased navigation aids (VOR, VOR/DME, NDB).

b. Special Area Navigation (RNAV) routes Q100, Q102 and Q105 in the Houston, Jacksonville and Miami CTAs.

4-7-5. Provisions for Accommodation of NonRNP10 Aircraft (Aircraft Not Authorized RNP 10 or RNP 4)

a. Operators of NonRNP10 aircraft must annotate ICAO flight plan Item 18 as follows:

"STS/NONRNP10" (no space between letters and numbers).

b. Pilots of NonRNP10 aircraft that operate in GoMex CTA's must report the lack of authorization by stating "Negative RNP 10":

1. On initial call to ATC in a GoMex CTA:

2. In read back of a clearance to climb to or descend from cruise altitude. (See paragraph 4-7-5e.); and

3. When approval status is requested by the controller. (See paragraph 4-7-9e.)

c. Operators of NonRNP10 aircraft must not annotate ICAO flight plan Item 18 (Other Information) with "NAV/ RNP10" or "NAV/RNP4", as shown in paragraph 7, if they have not obtained RNP 10 or RNP 4 authorization.

d. NonRNP10 operators/aircraft may file any route at any altitude in a GoMex CTA. They will be cleared to operate on their preferred routes and altitudes as traffic permits. 50 NM lateral separation will not be applied to NonRNP10 aircraft.

e. NonRNP10 aircraft are encouraged to operate at altitudes above those where traffic is most dense (i.e., at/ above FL 380), if possible. NonRNP10 aircraft should plan on completing their climb to or descent from higher FLs within radar coverage, if possible.

Chapter 4

AIM

Tbl 4-6-2
Contingency Actions: Weather Encounters and Aircraft System Failures

Initial Pilot Actions in Contingency Situations	
Initial pilot actions when unable to maintain flight level (FL) or unsure of aircraft altitude-keeping capability:	
• Notify ATC and request assistance as detailed below.	
• Maintain cleared flight level, to the extent possible, while evaluating the situation.	
• Watch for conflicting traffic both visually and by reference to TCAS, if equipped.	
• Alert nearby aircraft by illuminating exterior lights (commensurate with aircraft limitations).	
Severe Turbulence and/or Mountain Wave Activity (MWA) Induced Altitude Deviations of Approximately 200 feet	
Pilot will:	**Controller will:**
• When experiencing severe turbulence and/or MWA induced altitude deviations of approximately 200 feet or greater, pilot will contact ATC and state "Unable RVSM Due (state reason)" (e.g., turbulence, mountain wave) • If not issued by the controller, request vector clear of traffic at adjacent FLs • If desired, request FL change or re-route • Report location and magnitude of turbulence or MWA to ATC See paragraph 4-6-6, Guidance on Severe Turbulence and Mountain Wave Activity (MWA) for detailed guidance.	• Vector aircraft to avoid merging target with traffic at adjacent flight levels, traffic permitting • Advise pilot of conflicting traffic • Issue FL change or re-route, traffic permitting • Issue PIREP to other aircraft Paragraph 4-6-6 explains "traffic permitting."
Mountain Wave Activity (MWA) Encounters - General	
Pilot actions:	**Controller actions:**
• Contact ATC and report experiencing MWA • If so desired, pilot may request a FL change or re-route • Report location and magnitude of MWA to ATC	• Advise pilot of conflicting traffic at adjacent FL • If pilot requests, vector aircraft to avoid merging target with traffic at adjacent RVSM flight levels, traffic permitting • Issue FL change or re-route, traffic permitting • Issue PIREP to other aircraft
See paragraph 4-6-6 for guidance on MWA.	Paragraph 4-6-6 explains "traffic permitting."
NOTE- *MWA encounters do not necessarily result in altitude deviations on the order of 200 feet. The guidance below is intended to address less significant MWA encounters.*	
Wake Turbulence Encounters	
Pilot should:	**Controller should:**
• Contact ATC and request vector, FL change or, if capable, a lateral offset	• Issue vector, FL change or lateral offset clearance, traffic permitting
See paragraph 4-6-7, Guidance on Wake Turbulence.	Paragraph 4-6-6 explains "traffic permitting."
"Unable RVSM Due Equipment" Failure of Automatic Altitude Control System, Altitude Alerter or All Primary Altimeters	
Pilot will:	**Controller will:**
• Contact ATC and state "Unable RVSM Due Equipment"	• Provide 2,000 feet vertical separation or appropriate horizontal separation
• Request clearance out of RVSM airspace unless operational situation dictates otherwise	• Clear aircraft out of RVSM airspace unless operational situation dictates otherwise
One Primary Altimeter Remains Operational	
Pilot will:	**Controller will:**
• Cross check stand-by altimeter • Notify ATC of operation with single primary altimeter • If unable to confirm primary altimeter accuracy, follow actions for failure of all primary altimeters	• Acknowledge operation with single primary altimeter
Transponder Failure	
Pilot will:	**Controller will:**
• Contact ATC and request authority to continue to operate at cleared flight level	• Consider request to continue to operate at cleared flight level
• Comply with revised ATC clearance, if issued	• Issue revised clearance, if necessary
NOTE- *14 CFR Section 91.215 (ATC transponder and altitude reporting equipment and use) regulates operation with the transponder inoperative.*	

4-7-6. Operator Action

In order to maximize operational flexibility provided by 50 NM lateral separation, operators capable of meeting RNP 10 or RNP 4 that operate on oceanic routes or areas in the GoMex CTA's should obtain authorization for RNP 10 or RNP 4 and annotate the ICAO flight plan accordingly.

NOTE-1. *RNP 10 is the minimum "Navigation Specification (NavSpec)" required for the application of 50 NM lateral separation. RNP 4 is an operator option. Operators/aircraft authorized RNP 4 are not required to also obtain RNP 10 authorization.*

2. *"RNP navigation specification" (e.g., RNP 10) is the term adopted in the ICAO Performancebased Navigation (PBN) Manual (Doc 9613). It replaces the term "RNP type".*

4-7-7. RNP 10 or RNP 4 Authorization: Policy and Procedures for Aircraft and Operators

a. RNP NavSpecs Applicable To Oceanic Operations. In accordance with ICAO guidance, RNP 10 and RNP 4 are the only NavSpecs applicable to oceanic and remote area operations. Other RNAV and RNP NavSpecs are applicable to continental en route, terminal area and approach operations.

b. FAA Documents. The guidance and direction of FAA Order 8400.12 (as amended) (RNP 10 Operational Authorization) is used to grant RNP 10 authorization to operators and aircraft for which the FAA is responsible. FAA Order 8400.33 (as amended) (Procedures for Obtaining Authorization for RNP 4 Oceanic/Remote Area Operations) is used to authorize RNP 4. The FAA RNP 10 and RNP 4 orders are consistent with the ICAO PBN Manual guidance discussed below. FAA and ICAO documents are posted on the FAA Gulf of Mexico 50 NM Lateral Separation Initiative Web Page.

c. ICAO Performancebased Navigation (PBN) Manual (ICAO Doc 9613). Guidance for authorization of RNP 10 and RNP 4 is provided in ICAO Doc 9613. RNP 10 is addressed in Volume II, Part B; Chapter 1. RNP 4 is addressed in Volume II, Part C; Chapter 1.

d. RNP 10 and RNP 4 Job Aids. Operators and authorities are encouraged to use the RNP 10 or RNP 4 Job Aids posted on the FAA Gulf of Mexico 50 NM Lateral Separation Initiative Web Page. For U.S. operators, one set of RNP 10 and RNP 4 Job Aids provides references to FAA documents. For international operators, a second set of Job Aids provide references to the ICAO PBN Manual. These Job Aids address the operational and airworthiness elements of aircraft and operator authorization and provide references to appropriate document paragraphs. The Job Aids provide a method for operators to develop and authorities to track the operator/aircraft program elements required for RNP 10 or RNP 4 authorization.

e. Qualification of Aircraft Equipped With a Single LongRange Navigation System (SLRNS) For RNP 10 Operations In GoMex CTA's.

1. Background. SLRNS operations in the Gulf of Mexico, the Caribbean Sea and the other designated areas have been conducted for at least 25 years. Provisions allowing aircraft equipage with a SLRNS for operations in specified oceanic and offshore areas are contained in the following sections of 14 Code Of Federal Regulations (CFR): 91.511, 121.351, 125.203 and 135.165.

2. ICAO PBN Manual Reference. In reference to RNP 10 authorization, the ICAO PBN Manual, Volume II, Part B, Chapter 1, paragraph 1.3.6.2 states that: "A State authority may approve the use of a single LRNS in specific circumstances (e.g., North Atlantic MNPS and 14 CFR 121.351 (c) refer). An RNP 10 approval is still required."

3. Policy Development. The FAA worked with the ICAO NACC Office (North American, Central American and Caribbean), State regulators and ATS providers in the GoMex and Caribbean areas to implement a policy for SLRNS equipped aircraft to qualify for RNP 10 for GoMex operations. Allowing SLRNS equipped aircraft to qualify for RNP 10 enables more operator aircraft to be authorized RNP 10, thereby creating a more uniform operating environment for the application of 50 NM lateral separation. The factors considered were: the shortness of the legs outside the range of ground navigation aids, the availability of radar and VHF coverage in a large portion of GoMex airspace and the absence of events attributed to SLRNS in GoMex operations.

4. Document Revision. The following documents were revised or created to enable implementation of the SLRNS/ RNP 10 qualification policy:

(a) FAA Order (FAAO) 8400.12

(b) FAA Order 8900.1 (Flight Standards Information Management System (FSIMS))

(c) Paragraph B054 of FAA Operations Specifications and Management Specifications (Class II Navigation Using Single LongRange Navigation System)

(d) LOA B054 (Class II Navigation Using Single LongRange Navigation System (SLRNS) Equipped Airplane Authorized RNP 10) (LOA's are applicable to International General Aviation operators.)

(e) FAA RNP 10 Job Aid with FAAO 8400.12 references

(f) RNP 10 Job Aid with ICAO PBN Manual references

5. SLRNS/RNP 10 Authorization Limited To GoMex. At this time, SLRNS qualification for RNP 10 only applies to GoMex operations. Any expansion of this provision will require assessment and agreement by the appropriate State authorities.

f. RNP 10 Time Limit for INS or IRU Only Equipped Aircraft. Operators should review their Airplane Flight Manual (AFM), AFM Supplement or other appropriate documents and/or contact the airplane or avionics manufacturer to determine the RNP 10 time limit applicable to their aircraft. They will then need to determine its effect, if any, on their operation. Unless otherwise approved, the basic RNP 10 time limit is 6.2 hours between position updates for aircraft on which Inertial Navigation Systems (INS) or Inertial Reference Units (IRU) provide the only source of long range navigation. Extended RNP 10 time limits of 10 hours and greater are already approved for many IRU systems. FAA Order 8400.12 contains provisions for extending RNP 10 time limits.

4-7-8. Flight Planning Requirements

Operators must make ICAO flight plan annotations in accordance with this paragraph and, if applicable, Paragraph 4-7-5, Provisions for Accommodation of NonRNP10 Aircraft (Aircraft Not Authorized RNP 10 or RNP 4).

a. ICAO Flight Plan Requirement. ICAO flight plans must be filed for operation on oceanic routes and areas in the Houston Oceanic CTA/FIR, the Gulf of Mexico portion of the Miami CTA/FIR, the Monterrey CTA and Merida High CTA.

b. To inform ATC that they have obtained RNP 10 or RNP 4 authorization and are eligible for 50 NM lateral separation, operators must:

1. Annotate ICAO Flight Plan Item 10 (Equipment) with the letters "R" and "Z", and

2. Annotate Item 18 (Other Information) with, as appropriate, "NAV/RNP10" or "NAV/RNP4" (no space between letters and numbers).

NOTE-1. *See paragraph 4-7-8e. It provides recommended filing practices for domestic U.S. RNAV operations and filing with EUROCONTROL.*

2. *On the ICAO Flight Plan, the letter "R" in Item 10 indicates that the aircraft will maintain the appropriate RNP navigation specification for the entire flight through airspace where RNP is prescribed. Letter "Z" in Item 10 indicates that information explaining aircraft navigation and/or communication capability is found in Item 18.*

c. 50 NM lateral separation will only be applied to operators/aircraft that annotate the ICAO flight plan in accordance with this policy. (See 4-7-8b.)

d. Operators that have not obtained RNP 10 or RNP 4 authorization must not annotate ICAO flight plan Item 18 (Other information) with "NAV/RNP10" or "NAV/RNP4", but must follow the practices detailed in paragraph 4-7-5.

e. Recommendation for Filing to Show Domestic U.S. RNAV and Oceanic RNP Capabilities.

1. Explanation. The FAA program to enhance operators' capability to communicate their domestic U.S. RNAV capabilities to ATC has been in place for over three years. It requires an entry following the NAV/ indicator in Item 18 of the ICAO flight plan. The initiative has provisions for including RNAV capabilities for departure ("D"), enroute ("E") and arrival ("A") with RNAV accuracy values. Detailed instructions are available on the following web page:http:// www.faa.gov/about/office_org/headquarters_offices/ato/ service_units/enroute/flight_plan_filing/

EXAMPLE-*An example Item 18 entry is: NAV/ RNVD1E2A1. The characters in the example indicate RNAV 1 and RNAV 2 accuracy.*

2. Recommendation. It is recommended that operators provide their RNAV capability for domestic U.S. and capabilities for oceanic operations (RNP 10 or RNP 4) by filing: "NAV/", then the domestic U.S. alphanumeric sequence, then a mandatory space and then "RNP10" or "RNP4", as appropriate.

EXAMPLE-*"NAV/RNVD1E2A1 RNP10"*

3. Multiple NAV/ Entries. Operators should be aware that if they make multiple "NAV/" entries in a flight plan filed with EUROCONTROL, only the last "NAV/" entry will be forwarded to the next ATC facility. For example, if "NAV/ RNVD1E2A1" and "NAV/RNP10" are entered, only "NAV/ RNP10" will be forwarded. Multiple "NAV/" entries should, therefore, be consolidated following a single "NAV/" indicator.

4. Recommendation. Item 18 entries made in accordance with paragraph 478 e. 2. above will limit the number of characters needed to show domestic U.S. RNAV and oceanic RNP capabilities and mitigate the chance that one or the other will not be forwarded for use by FAA domestic and oceanic automation systems.

f. Implementation of ICAO Doc 4444, Revised Appendix 2 (Flight Plan). ICAO Doc 4444, Amendment 1 revises Appendix 2 (Flight Plan). Specifically, Amendment 1 revises the flight plan annotations in Item 10 (Equipment) and Item 18 (Other Information) that show aircraft communications, navigation and surveillance capabilities. The new Appendix 2 flight plan annotations will be required on 15 November 2012. The following Websites provide information on implementation planning:

1. FAA Website: http://www.faa.gov/go/fpl2012.

2. ICAO Flight Plan Implementation Tracking System (FITS): http://www2.icao.int/en/FITS/Pages/home.aspx.

4-7-9. Pilot and Dispatcher Procedures: Basic and Inflight Contingency Procedures

a. Basic Pilot Procedures. The RNP 10 and RNP 4 Job Aids contain references to pilot and, if applicable, dispatcher procedures contained in:

1. FAA Order 8400.12C (RNP 10), Appendix D (Training Programs and Operating Practices and Procedures)

2. FAA Order 8400.33 (RNP 4): paragraph 9 (Operational Requirements) and paragraph 10 (Training Programs, Operating Practices and Procedures)

3. ICAO PBN Manual, Volume II, Part B, Chapter 1 (RNP 10)

4. ICAO PBN Manual, Volume II, Part C, Chapter 1 (RNP 4)

b. ICAO Doc 4444, Chapter 15, Inflight Contingency Procedures. Doc 4444 Chapter 15 contains important guidance for pilot training programs. For ease of reference, significant Chapter 15 paragraphs are posted on the Gulf of Mexico 50 NM Lateral Separation Web Page. Chapter 15 paragraphs posted on the website include:

1. Paragraph 15.2 (Special Procedures for InFlight Contingencies in Oceanic Airspace). Paragraph 15.2.2 (General Procedures) provides guidance for inflight diversions, turnbacks and for loss of, or significant reduction in, required navigation capability when operating in an airspace where the navigation performance accuracy is a prerequisite to the safe conduct of flight operations.

2. Paragraph 15.2.3 (Weather Deviation Procedures). Paragraph 15.2.3 provides guidance for events where the pilot is able to obtain a clearance prior to deviating from track to avoid convective weather and for events where the pilot is unable to obtain clearance prior to deviating.

c. Strategic Lateral Offset Procedures (SLOP). Pilots should use SLOP procedures in the course of regular oceanic operations. SLOP procedures are published in ICAO Document 4444, 15th Edition, Amendment 2, paragraph 16.5 and FAA Notices. They are posted on the Gulf of Mexico 50 NM Lateral Separation Web Page and are addressed in the RNP 10 and RNP 4 Job Aids.

d. Pilot Report of NonRNP10 Status. The pilot must report the lack of RNP 10 or RNP 4 status in accordance with the following:

1. When the operator/aircraft is not authorized RNP 10 or RNP 4. See paragraph 4-7-5.

2. If approval status is requested by the controller in accordance with paragraph 4-7-9e.

e. Pilot Statement of RNP 10 or RNP 4 Approval Status, If Requested. If requested by the controller, the pilot must communicate approval status using the following phraseology:

Controller Request:
(Call sign) confirm RNP 10 or 4 approved
Pilot Response:
"Affirm RNP 10 approved" or "Affirm RNP 4 approved," as appropriate, or "Negative RNP 10" (See paragraph 4-7-5 for NonRNP10 aircraft procedures.)

f. Pilot action when navigation system malfunctions. In addition to the actions suggested in ICAO Doc. 4444, Chapter 15, when pilots suspect a navigation system malfunction, the following actions should be taken:

1. Immediately inform ATC of navigation system malfunction or failure.

2. Accounting for wind drift, fly magnetic compass heading to maintain track.

3. Request radar vectors from ATC, when available

Chapter 4

AIM

Chapter 5. AIR TRAFFIC PROCEDURES

Section 1. Preflight

5-1-1. Preflight Preparation

a. Every pilot is urged to receive a preflight briefing and to file a flight plan. This briefing should consist of the latest or most current weather, airport, and en route NAVAID information. Briefing service may be obtained from an FSS either by telephone or interphone, by radio when airborne, or by a personal visit to the station. Pilots with a current medical certificate in the 48 contiguous States may access toll-free the Direct User Access Terminal System (DUATS) through a personal computer. DUATS will provide alphanumeric preflight weather data and allow pilots to file domestic VFR or IFR flight plans.

REFERENCE-*AIM, FAA Weather Services, Paragraph 7-1-2, lists DUATS vendors.*

NOTE-*Pilots filing flight plans via "fast file" who desire to have their briefing recorded, should include a statement at the end of the recording as to the source of their weather briefing.*

b. The information required by the FAA to process flight plans is contained on FAA Form 7233-1, Flight Plan, or FAA Form 7233-4, International Flight Plan. The forms are available at all flight service stations. Additional copies will be provided on request.

REFERENCE-*AIM, Flight Plan- VFR Flights, Paragraph 5-1-4*
AIM, Flight Plan- IFR Flights, Paragraph 5-1-8
AIM, International Flight Plan- IFR Flights, Paragraph 5-1-9

c. Consult an FSS or a Weather Service Office (WSO) for preflight weather briefing. Supplemental Weather Service Locations (SWSLs) do not provide weather briefings.

d. FSSs are required to advise of pertinent NOTAMs if a *standard* briefing is requested, but if they are overlooked, don't hesitate to remind the specialist that you have not received NOTAM information.

NOTE-*NOTAMs which are known in sufficient time for publication and are of 7 days duration or longer are normally incorporated into the Notices to Airmen Publication and carried there until cancellation time. FDC NOTAMs, which apply to instrument flight procedures, are also included in the Notices to Airmen Publication up to and including the number indicated in the FDC NOTAM legend. Printed NOTAMs are not provided during a briefing unless specifically requested by the pilot since the FSS specialist has no way of knowing whether the pilot has already checked the Notices to Airmen Publication prior to calling. Remember to ask for NOTAMs in the Notices to Airmen Publication. This information is not normally furnished during your briefing.*

REFERENCE-*AIM, Notice to Airmen (NOTAM) System, Paragraph 5-1-3.*

e. Pilots are urged to use only the latest issue of aeronautical charts in planning and conducting flight operations. Aeronautical charts are revised and reissued on a regular scheduled basis to ensure that depicted data are current and reliable. In the conterminous U.S., Sectional Charts are updated every 6 months, IFR En Route Charts every 56 days, and amendments to civil IFR Approach Charts are accomplished on a 56-day cycle with a change notice volume issued on the 28-day midcycle. Charts that have been superseded by those of a more recent date may contain obsolete or incomplete flight information.

REFERENCE-*AIM, General Description of Each Chart Series, Paragraph 9-1-4.*

f. When requesting a preflight briefing, identify yourself as a pilot and provide the following:

1. Type of flight planned; e.g., VFR or IFR.
2. Aircraft's number or pilot's name.
3. Aircraft type.
4. Departure Airport.
5. Route of flight.
6. Destination.
7. Flight altitude(s).
8. ETD and ETE.

g. Prior to conducting a briefing, briefers are required to have the background information listed above so that they may tailor the briefing to the needs of the proposed flight. The objective is to communicate a "picture" of meteorological and aeronautical information necessary for the conduct of a safe and efficient flight. Briefers use all available weather and aeronautical information to summarize data applicable to the proposed flight. They do not read weather reports and forecasts verbatim unless specifically requested by the pilot. FSS briefers do not provide FDC NOTAM information for special instrument approach procedures unless specifically asked. Pilots authorized by the FAA to use special instrument approach procedures must specifically request FDC NOTAM information for these procedures. Pilots who receive the information electronically will receive NOTAMs for special IAPs automatically.

REFERENCE-*AIM, Preflight Briefings, Paragraph 7-1-4, contains those items of a weather briefing that should be expected or requested.*

h. FAA by 14 CFR Part 93, Subpart K, has designated High Density Traffic Airports (HDTAs) and has prescribed air traffic rules and requirements for operating aircraft (excluding helicopter operations) to and from these airports.

REFERENCE-*Airport/Facility Directory, Special Notices Section.*
AIM, Airport Reservation Operations and Special Traffic Management Programs, Paragraph 4-1-22.

i. In addition to the filing of a flight plan, if the flight will traverse or land in one or more foreign countries, it is particularly important that pilots leave a complete itinerary with someone directly concerned and keep that person advised of the flight's progress. If serious doubt arises as to the safety of the flight, that person should first contact the FSS.

REFERENCE-*AIM, Flights Outside the U.S. and U.S. Territories, Paragraph 5-1-11*

j. Pilots operating under provisions of 14 CFR Part 135 on a domestic flight and not having an FAA assigned 3-letter designator, are urged to prefix the normal registration (N) number with the letter "T" on flight plan filing; e.g., TN1234B.

REFERENCE-*AIM, Aircraft Call Signs, Paragraph 4-2-4*

5-1-2. Follow IFR Procedures Even When Operating VFR

a. To maintain IFR proficiency, pilots are urged to practice IFR procedures whenever possible, even when operating VFR. Some suggested practices include:

1. Obtain a complete preflight and weather briefing. Check the NOTAMs.

2. File a flight plan. This is an excellent low cost insurance policy. The cost is the time it takes to fill it out. The insurance includes the knowledge that someone will be looking for you if you become overdue at your destination.

3. Use current charts.

4. Use the navigation aids. Practice maintaining a good course-keep the needle centered.

5. Maintain a constant altitude which is appropriate for the direction of flight.

6. Estimate en route position times.

7. Make accurate and frequent position reports to the FSSs along your route of flight.

b. Simulated IFR flight is recommended (under the hood); however, pilots are cautioned to review and adhere to the requirements specified in 14 CFR Section 91.109 before and during such flight.

c. When flying VFR at night, in addition to the altitude appropriate for the direction of flight, pilots should maintain an altitude which is at or above the minimum en route altitude as shown on charts. This is especially true in mountainous terrain, where there is usually very little ground reference. Do not depend on your eyes alone to avoid rising unlighted terrain, or even lighted obstructions such as TV towers.

5-1-3. Notice to Airmen (NOTAM) System

a. Time-critical aeronautical information which is of either a temporary nature or not sufficiently known in advance to permit publication on aeronautical charts or in other operational publications receives immediate dissemination via the National NOTAM System.

NOTE-**1.** *NOTAM information is that aeronautical information that could affect a pilot's decision to make a flight. It includes such information as airport or aerodrome primary runway closures, taxiways, ramps, obstructions, communications, airspace, changes in the status of navigational aids, ILSs, radar service availability, and other information essential to planned en route, terminal, or landing operations.*

2. *NOTAM information is transmitted using standard contractions to reduce transmission time. See TBL 5-1-2 for a listing of the most commonly used contractions. For a complete listing, see FAA Order JO 7340.2, Contractions.*

b. NOTAM information is classified into four categories. These are NOTAM (D) or distant, Flight Data Center (FDC) NOTAMs, Pointer NOTAMs, Special Activity Airspace (SAA), and Military NOTAMs.

1. NOTAM (D) information is disseminated for all navigational facilities that are part of the National Airspace System (NAS), all public use airports, seaplane bases, and heliports listed in the Airport/Facility Directory (A/FD). The complete file of all NOTAM (D) information is maintained in a computer database at the Weather Message Switching Center (WMSC), located in Atlanta, Georgia. This category of information is distributed automatically via Service A telecommunications system. Air traffic facilities, primarily FSSs, with Service A capability have access to the entire WMSC database of NOTAMs. These NOTAMs remain available via Service A for the duration of their validity or until published. Once published, the NOTAM data is deleted from the system. NOTAM (D) information includes such data as taxiway closures, personnel and equipment near or crossing runways, and airport lighting aids that do not affect instrument approach criteria, such as VASI.

All NOTAM Ds must have one of the following keywords listed in TBL 5-1-1 as the first part of the text after the location identifier:

2. FDC NOTAMs. On those occasions when it becomes necessary to disseminate information which is regulatory in nature, the National Flight Data Center (NFDC), in Washington, DC, will issue an FDC NOTAM. FDC NOTAMs contain such things as amendments to published IAPs and other current aeronautical charts. They are also used to advertise temporary flight restrictions caused by such things as natural disasters or large-scale public events that may generate a congestion of air traffic over a site.

NOTE-
1. *DUATS vendors will provide FDC NOTAMs only upon site-specific requests using a location identifier.*

2. *NOTAM data may not always be current due to the changeable nature of national airspace system components, delays inherent in processing information, and occasional temporary outages of the U.S. NOTAM system. While en route, pilots should contact FSSs and obtain updated information for their route of flight and destination.*

3. Pointer NOTAMs. NOTAMs issued by a flight service station to highlight or point out another NOTAM, such as an FDC or NOTAM (D) NOTAM. This type of NOTAM will assist users in cross-referencing important information that may not be found under an airport or NAVAID identifier. Keywords in pointer NOTAMs must match the keywords in the NOTAM that is being pointed out. The keyword in pointer NOTAMs related to Temporary Flight Restrictions (TFR) must be AIRSPACE.

Tbl 5-1-1
NOTAM Keywords

Keyword	Definition
RWY *Example*	**Runway** !BNA BNA RWY 36 CLSD 1309131300-1309132000EST
TWY *Example*	**Taxiway** !BTV BTV TWY C EDGE LGT OBSC 1310131300-1310141300EST
APRON *Example*	**Apron/Ramp** !BNA BNA APRON NORTH APRON EAST SIDE CLSD 1311122150013122220700
AD *Example*	**Aerodrome** !BET BET AD ELK NEAR MVMT AREAS 1309251300-1309262200EST
OBST *Example*	**Obstruction** !SJT SJT OBST MOORED BALLOON WITHIN AREA DEFINED AS 1NM RADIUS OF SJT 2430FT (510FT AGL) FLAGGED 1309251400-1309261400EST
NAV *Example*	**Navigation Aids** !SHV SHV NAV ILS RWY 32 110.3 COMMISSIONED 1311251600-PERM
COM *Example*	**Communications** !INW INW COM REMOTE COM OUTLET 122.6 OUT OF SERVICE 1307121330-1307151930EST
SVC *Example*	**Services** !ROA ROA SVC TWR COMMISSIONED 1301050001-PERM
AIRSPACE *Example*	**Airspace** !MIV MIV AIRSPACE AIRSHOW ACFT WITHIN AREA DEFINED AS 5NM RADIUS OF MIV SFC-10000FT AVOIDANCE ADVISED 1308122100-1308122300
OPD *Example*	**Obstacle Departure Procedure** !FDC 2/9700 DIK ODP DICKINSON -THEODORE ROOSEVELT RGNL, DICKINSON, ND. TAKEOFF MINIMUMS AND (OBSTACLE) DEPARTURE PROCEDURES AMDT 1... DEPARTURE PROCEDURE: RWY 25, CLIMB HEADING 250 TO 3500 BEFORE TURNING LEFT. ALL OTHER DATA REMAINS AS PUBLISHED. THIS IS TAKEOFF MINIMUMS AND (OBSTACLE) DEPARTURE PROCEDURES, AMDT 1A. 1305011200-PERM
SID *Example*	**Standard Instrument Departure** !FDC x/xxxx DFW SID DALLAS/FORT WORTH INTL, DALLAS, TX. PODDE THREE DEPARTURE... CHANGE NOTES TO READ: RWYS 17C/R, 18L/R: DO NOT EXCEED 240KT UNTIL LARRN. RWYS 35L/C, 36L/R: DONOT EXCEED 240KT UNTIL KMART 1305011200-1312111200EST
STAR *Example*	**Standard Terminal Arrival** !FDC x/xxxx DCA STAR RONALD REAGAN WASHINGTON NATIONAL,WASHINGTON, DC. WZRRD TWO ARRIVAL... SHAAR TRANSITION: ROUTE FROM DRUZZ INT TO WZRRD INT NOT AUTHORIZED. AFTER DRUZZ INT EXPECT RADAR VECTORS TO AML VORTAC 1305011200-1312111200ES
CHART *Example*	**Chart** !FDC 2/9997 DAL IAP DALLAS LOVE FIELD, DALLAS, TX. ILS OR LOC RWY 31R, AMDT 5... CHART NOTE: SIMULTANEOUS APPROACH AUTHORIZED WITH RWY 31L. MISSED APPROACH: CLIMB TO 1000 THEN CLIMBING RIGHT TURN TO 5000 ON HEADING 330 AND CVE R-046 TO FINGR INT/CVE 36.4 DME AND HOLD. CHART LOC RWY 31L. THIS IS ILS OR LOC RWY 31R, AMDT 5A. 1305011200-PERM
DATA *Example*	**Data** !FDC 2/9700 DIK ODP DICKINSON -THEODORE ROOSEVELT RGNL, DICKINSON, ND. TAKEOFF MINIMUMS AND (OBSTACLE) DEPARTURE PROCEDURES AMDT 1... DEPARTURE PROCEDURE: RWY 25, CLIMB HEADING 250 TO 3500 BEFORE TURNING LEFT. ALL OTHER DATA REMAINS AS PUBLISHED. THIS IS TAKEOFF MINIMUMS AND (OBSTACLE) DEPARTURE PROCEDURES, AMDT 1A. 1305011200-PERM
IAP *Example*	**Instrument Approach Procedure** !FDC 2/9997 DAL IAP DALLAS LOVE FIELD, DALLAS, TX. ILS OR LOC RWY 31R, AMDT 5... CHART NOTE: SIMULTANEOUS APPROACH AUTHORIZED WITH RWY 31L. MISSED APPROACH: CLIMB TO 1000 THEN CLIMBING RIGHT TURN TO 5000 ON HEADING 330 AND CVE R-046 TO FINGR INT/CVE 36.4 DME AND HOLD. CHART LOC RWY 31L. THIS IS ILS OR LOC RWY 31R, AMDT 5A. 1305011200-PERM
VFP *Example*	**Visual Flight Procedures** !FDC X/XXXX JFK VFP JOHN F KENNEDY INTL, NEW YORK, NY. PARKWAY VISUAL RWY 13L/R, ORIG...WEATHER MINIMUMS 3000 FOOT CEILING AND 3 MILES VISIBILITY. 1303011200-1308011400EST
ROUTE *Example*	**Route** !FDC x/xxxx ZFW OK..ROUTE ZFW ZKC. V140 SAYRE (SYO) VORTAC, OK TO TULSA (TUL) VORTAC, OK MEA 4300. 1305041000-1306302359EST
SPECIAL *Example*	**Special** !FDC x/xxxx PAJN SPECIAL JUNEAU INTERNATIONAL, JUNEAU, AK. LDA-2 RWY 8 AMDT 9 PROCEDURE TURN NA. 1305011200-1312111200EST
SECURITY *Example*	**Security** !FDC ZZZ SECURITY..SPECIAL NOTICE..THIS NOTICE IS TO EMPHASIZE THAT BEFORE OPERATING IN OR ADJACENT TO IRANIAN AIRSPACE ALL U.S. AIRMEN AND OPERATORS SHOULD BE FAMILIAR WITH CURRENT CONDITIONS IN THE MIDDLE EAST. THE U.S. DEPARTMENT OF STATE HAS ISSUED A TRAVEL WARNING FOR IRAN ADVISING, IN PART, THAT THE U.S. GOVERNMENT DOES NOT CURRENTLY MAINTAIN DIPLOMATIC OR CONSULAR RELATIONS WITH THE ISLAMIC REPUBLIC OF IRAN. ANY U.S. OPERATOR PLANNING A FLIGHT THROUGH IRANIAN AIRSPACE SHOULD PLAN IN ADVANCE AND HAVE ALL CURRENT NOTAMS AND AERONAUTICAL INFORMATION FOR ANY PLANNED FLIGHT 1311011200-1403301800EST
U	Unverified Aeronautical Information (*for use only where authorized by Letter of Agreement*)*
O	Other Aeronautical Information**

4. SAA NOTAMs. These NOTAMs are issued when Special Activity Airspace will be active outside the published schedule times and when required by the published schedule. Pilots and other users are still responsible to check published schedule times for Special Activity Airspace as well as any NOTAMs for that airspace.

5. Military NOTAMs. NOTAMs pertaining to U.S. Air Force, Army, Marine, and Navy navigational aids/airports that are part of the NAS.

c. Notices to Airmen Publication (NTAP). The NTAP is published by Mission Support Services, ATC Products and Publications, every 28 days. Data of a permanent nature can be published in the NTAP as an interim step between publication cycles of the A/FD and aeronautical charts. The NTAP is divided into four parts:

1. Notices in part 1 are provided by ATC Products and Publications. This part contains selected FDC NOTAMs that are expected to be in effect on the effective date of the publication. This part is divided into three sections:

(a) Section 1, Airway NOTAMs, reflects airway changes that fall within an ARTCC's airspace.

(b) Section 2, Procedural NOTAMs.

(c) Section 3, General NOTAMs, contains NOTAMs that are general in nature and not tied to a specific airport/facility (for example, flight advisories and restrictions, open duration special security instructions, and special flight rules area).

2. Part 2, provided by NFDC, contains Part 95 Revisions, Revisions to Minimum En Route IFR Altitudes and Changeover Points.

3. Part 3, International NOTAMs, is divided into two sections:

(a) Section 1, International Flight Prohibitions, Potential Hostile Situations, and Foreign Notices.

(b) Section 2, International Oceanic Airspace Notices.

4. Part 4, Graphic Notices, compiled by ATC Products and Publications from data provided by FAA service area offices and other lines of business, contains special notices and graphics pertaining to almost every aspect of aviation such as: military training areas, large scale sporting events, air show information, Special Traffic Management Programs (STMP), and airport-specific information. This part is comprised of 6 sections: General, Special Military Operations, Airport and Facility Notices, Major Sporting and Entertainment Events, Airshows, and Special Notices.

NOTE-

*1. * **Unverified Aeronautical Information** can be movement area or other information received that meets NOTAM criteria and has not been confirmed by the Airport Manager (AMGR) or their designee. If Flight Service is unable to contact airport management, Flight Service must forward (U) NOTAM information to the United States NOTAM System (USNS). Subsequent to USNS distribution of a (U) NOTAM, Flight Service will inform airport management of the action taken as soon as practical. Any such NOTAM will be prefaced with "(U)" as the keyword and followed by the appropriate keyword contraction, following the location identifier.*

*2. ** **Other Aeronautical Information** is that which is received from any authorized source that may be beneficial to aircraft operations and does not meet defined NOTAM criteria. Any such NOTAM will be prefaced with "(O)" as the keyword following the location identifier.*

5-1-4. Flight Plan - VFR Flights

a. Except for operations in or penetrating a Coastal or Domestic ADIZ or DEWIZ a flight plan is not required for VFR flight.

REFERENCE-*AIM, National Security, Paragraph 5-6-1.*

b. It is strongly recommended that a flight plan (for a VFR flight) be filed with an FAA FSS. This will ensure that you receive VFR Search and Rescue Protection.

REFERENCE-*AIM, Search and Rescue, Paragraph 6-2-7 gives the proper method of filing a VFR flight plan.*

c. To obtain maximum benefits from the flight plan program, flight plans should be filed directly with the nearest FSS. For your convenience, FSSs provide aeronautical and meteorological briefings while accepting flight plans. Radio may be used to file if no other means are available.

NOTE-*Some states operate aeronautical communications facilities which will accept and forward flight plans to the FSS for further handling.*

Tbl 5-1-2

Contractions Commonly Found in NOTAMs

A	
ABN	Aerodrome Beacon
ABV	Above
ACFT	Aircraft
ACT	Active
ADJ	Adjacent
AGL	Above Ground Level
ALS	Approach Light System
ALT	Altitude
ALTN/ALTNLY	Alternate/Alternately
AMDT	Amendment
APCH	Approach
ARFF	Aircraft Rescue & Fire Fighting
ASDA	Accelerate Stop Distance Available
ASOS	Automated Surface Observing System
ASPH	Asphalt
ATC	Air Traffic Control
ATIS	Automated Terminal Information Service
AVBL	Available
AWOS	Automatic Weather Observing System
AWSS	Automated Weather Sensor System
AZM	Azimuth
B	
BTN	Between
C	
CAT	Category
CH	Channel
CL	Centerline
CLSD	Closed
COM	Communication
CONC	Concrete
CONT	Continue/Continuously
CTL	Control
D	
DCT	Direct
DEP	Depart/Departure
DH	Decision Height
DLA/DLAD	Delay/Delayed
DME	Distance Measuring Equipment
DWPNT	Dew Point Temperature
E	
E	East
EB	Eastbound
ELEV	Elevate/Elevation
ENG	Engine
EST	Estimated
EXC	Except
F	
FAC	Facility
FAF	Final Approach Fix
FDC	Flight Data Center
FICON	Field Condition
FREQ	Frequency
FSS	Flight Service Station
FT	Feet
G	
GCA	Ground Controlled Approach
GP	Glide Path
GPS	Global Positioning System
GRVL	Gravel
H	
HEL	Helicopter
HIRL	High Intensity Runway Lights
HR	Hour
I	
ID	Identify/Identifier
IFR	Instrument Flight Rules
ILS	Instrument Landing System
IM	Inner Marker
IN	Inch/Inches
INOP	Inoperative
INST	Instrument
INT	Intersection
INTST	Intensity
L	
L	Left
LB	Pound/Pounds
LDA	Landing Distance Available
LDG	Landing
LGT/LGTD	Light/Lighted
LIRL	Low Intensity Runway Edge Lights
LNDG	Landing
LOC	Localizer
M	
MALS	Medium Intensity Approach Lighting System
MALSF	Medium Intensity Approach Lighting System with Sequenced Flashers
MALSR	Medium Intensity Approach Lighting System with Runway Alignment Indicator Lights
MCA	Minimum Crossing Altitude
MDA	Minimum Descent Altitude
MEA	Minimum En Route Altitude
MIRL	Medium Intensity Runway Edge Lights
MKR	Marker
MM	Middle Marker
MNM	Minimum
MOA	Military Operations Area
MOCA	Minimum Obstruction Clearance Altitude
MSG	Message
MSL	Mean Sea Level
MU	Designate a Friction Value Representing Runway Surface Conditions
N	
N	North
NDB	Nondirectional Radio Beacon
NE	Northeast
NM	Nautical Mile/s
NTAP	Notice To Airmen Publication
NW	Northwest
O	
OBSC	Obscured
OM	Outer Marker
OPR	Operate
ORIG	Original
P	
PAPI	Precision Approach Path Indicator
PARL	Parallel
PAX	Passenger/s
PCL	Pilot Controlled Lighting
PERM	Permanent
PJE	Parachute Jumping Activities
PLA	Practice Low Approach
PN	Prior Notice Required
PPR	Prior Permission Required
PT	Procedure Turn
R	
RAI	Runway Alignment Indicator
RCL	Runway Centerline
RCLL	Runway Centerline Light
REC	Receive/Receiver
RLLS	Runway Lead-in Light System
RNAV	Area Navigation
RVR	Runway Visual Range
RVRM	RVR Midpoint
RVRR	RVR Rollout
RVRT	RVR Touchdown
RWY	Runway
S	
S	South
SAA	Special Activity Airspace
SE	Southeast
SFC	Surface
SKED	Scheduled
SN	Snow
SR	Sunrise
SS	Sunset
SSALF	Simplified Short Approach Lighting System with Sequenced Flashers
SSALR	Simplified Short Approach Lighting System with Runway Alignment Indicator Lights
SSALS	Simplified Short Approach Lighting System
STAR	Standard Terminal Arrival
STD	Standard
SW	Southwest
T	
TACAN	Tactical Air Navigational Aid
TDZ	Touchdown Zone
TEMPO	Temporary
TFC	Traffic

TFR	Temporary Flight Restriction
TGL	Touch and Go Landings
THR	Threshold
TKOF	Takeoff
TODA	Take-off Distance Available
TORA	Take-off Run Available
TWR	Aerodrome Control Tower
TWY	Taxiway
	U
UNL	Unlimited
UNREL	Unreliable
	V
VASI	Visual Approach Slope Indicator
VFR	Visual Flight Rules
VHF	Very High Frequency
VIS	Visibility
VMC	Visual Meteorological Conditions
VOLMET	Meteorlogical Information for Aircraft in Flight
VOR	VHF Omni-Directional Radio Range
VORTAC	VOR and TACAN (collocated)
VOT	VOR Test Facility
	W
W	West
WAAS	Wide Area Augmentation System
WDI	Wind Direction Indicator
WPT	Waypoint
WX	Weather

d. When a "stopover" flight is anticipated, it is recommended that a separate flight plan be filed for each "leg" when the stop is expected to be more than 1 hour duration.

e. Pilots are encouraged to give their departure times directly to the FSS serving the departure airport or as otherwise indicated by the FSS when the flight plan is filed. This will ensure more efficient flight plan service and permit the FSS to advise you of significant changes in aeronautical facilities or meteorological conditions. When a VFR flight plan is filed, it will be held by the FSS until 1 hour after the proposed departure time unless:

1. The actual departure time is received.

2. A revised proposed departure time is received.

3. At a time of filing, the FSS is informed that the proposed departure time will be met, but actual time cannot be given because of inadequate communications (assumed departures).

f. On pilot's request, at a location having an active tower, the aircraft identification will be forwarded by the tower to the FSS for reporting the actual departure time. This procedure should be avoided at busy airports.

g. Although position reports are not required for VFR flight plans, periodic reports to FAA FSSs along the route are good practice. Such contacts permit significant information to be passed to the transiting aircraft and also serve to check the progress of the flight should it be necessary for any reason to locate the aircraft.

EXAMPLE-

1. *Bonanza 314K, over Kingfisher at (time), VFR flight plan, Tulsa to Amarillo.*

2. *Cherokee 5133J, over Oklahoma City at (time), Shreveport to Denver, no flight plan.*

h. Pilots not operating on an IFR flight plan and when in level cruising flight, are cautioned to conform with VFR cruising altitudes appropriate to the direction of flight.

i. When filing VFR flight plans, indicate aircraft equipment capabilities by appending the appropriate suffix to aircraft type in the same manner as that prescribed for IFR flight.

REFERENCE-*AIM, Flight Plan- Domestic IFR Flights, Paragraph 5-1-8.*

j. Under some circumstances, ATC computer tapes can be useful in constructing the radar history of a downed or crashed aircraft. In each case, knowledge of the aircraft's transponder equipment is necessary in determining whether or not such computer tapes might prove effective.

k. Flight Plan Form - (See FIG 5-1-1).

l. Explanation of VFR Flight Plan Items.

1. Block 1. Check the type flight plan. Check both the VFR and IFR blocks if composite VFR/IFR.

2. Block 2. Enter your complete aircraft identification including the prefix "N" if applicable.

3. Block 3. Enter the designator for the aircraft, or if unknown, consult an FSS briefer.

4. Block 4. Enter your true airspeed (TAS).

5. Block 5. Enter the departure airport identifier code, or if unknown, the name of the airport.

6. Block 6. Enter the proposed departure time in Coordinated Universal Time (UTC) (Z). If airborne, specify the actual or proposed departure time as appropriate.

7. Block 7. Enter the appropriate VFR altitude (to assist the briefer in providing weather and wind information).

8. Block 8. Define the route of flight by using NAVAID identifier codes and airways.

9. Block 9. Enter the destination airport identifier code, or if unknown, the airport name.

NOTE-*Include the city name (or even the state name) if needed for clarity.*

10. Block 10. Enter your estimated time en route in hours and minutes.

11. Block 11. Enter only those remarks that may aid in VFR search and rescue, such as planned stops en route or student cross country, or remarks pertinent to the clarification of other flight plan information, such as the radiotelephony (call sign) associated with a designator filed in Block 2, if the radiotelephony is new, has changed within the last 60 days, or is a special FAA-assigned temporary radiotelephony. Items of a personal nature are not accepted.

12. Block 12. Specify the fuel on board in hours and minutes.

13. Block 13. Specify an alternate airport if desired.

14. Block 14. Enter your complete name, address, and telephone number. Enter sufficient information to identify home base, airport, or operator.

NOTE-*This information is essential in the event of search and rescue operations.*

15. Block 15. Enter total number of persons on board (POB) including crew.

16. Block 16. Enter the predominant colors.

17. Block 17. Record the FSS name for closing the flight plan. If the flight plan is closed with a different FSS or facility, state the recorded FSS name that would normally have closed your flight plan.

NOTE-

1. ***Optional****- record a destination telephone number to assist search and rescue contact should you fail to report or cancel your flight plan within 1/2 hour after your estimated time of arrival (ETA).*

2. *The information transmitted to the destination FSS will consist only of flight plan blocks 2, 3, 9, and 10. Estimated time en route (ETE) will be converted to the correct ETA.*

5-1-5. Operational Information System (OIS)

a. The FAA's Air Traffic Control System Command Center (ATCSCC) maintains a web site with near real-time National Airspace System (NAS) status information. NAS operators are encouraged to access the web site at **http://www.fly.faa.gov** prior to filing their flight plan.

b. The web site consolidates information from advisories. An advisory is a message that is disseminated electronically by the ATCSCC that contains information pertinent to the NAS.

1. Advisories are normally issued for the following items:

(a) Ground Stops.
(b) Ground Delay Programs.
(c) Route Information.
(d) Plan of Operations.
(e) Facility Outages and Scheduled Facility Outages.
(f) Volcanic Ash Activity Bulletins.
(g) Special Traffic Management Programs.

2. This list is not all-inclusive. Any time there is information that may be beneficial to a large number of people, an advisory may be sent. Additionally, there may be times when an advisory is not sent due to workload or the short length of time of the activity.

3. Route information is available on the web site and in specific advisories. Some route information, subject to the 56-day publishing cycle, is located on the "OIS" under "Products," Route Management Tool (RMT), and "What's New" Playbook. The RMT and Playbook contain routings for use by Air Traffic and NAS operators when they are coordinated "real-time" and are then published in an ATCSCC advisory.

4. Route advisories are identified by the word "Route" in the header; the associated action is required (RQD), recommended (RMD), planned (PLN), or for your information (FYI). Operators are expected to file flight plans consistent with the Route RQD advisories.

5. Electronic System Impact Reports are on the intranet at http://www.atcscc.faa.gov/ois/ under "System Impact

Figure 5-1-1
FAA Flight Plan Form 7233-1 (8-82)

U.S. DEPARTMENT OF TRANSPORTATION FEDERAL AVIATION ADMINISTRATION **FLIGHT PLAN**	(FAA USE ONLY) ☐ PILOT BRIEFING ☐ VNR ☐ STOPOVER			TIME STARTED		SPECIALIST INITIALS
1. TYPE: VFR / IFR / DVFR	2. AIRCRAFT IDENTIFICATION	3. AIRCRAFT TYPE/ SPECIAL EQUIPMENT	4. TRUE AIRSPEED KTS	5. DEPARTURE POINT	6. DEPARTURE TIME: PROPOSED (Z) / ACTUAL (Z)	7. CRUISING ALTITUDE
8. ROUTE OF FLIGHT						
9. DESTINATION (Name of airport and city)	10. EST. TIME ENROUTE: HOURS / MINUTES	11. REMARKS				
12. FUEL ON BOARD: HOURS / MINUTES	13. ALTERNATE AIRPORT(S)	14. PILOT'S NAME, ADDRESS & TELEPHONE NUMBER & AIRCRAFT HOME BASE				15. NUMBER ABOARD
		17. DESTINATION CONTACT/TELEPHONE (OPTIONAL)				
16. COLOR OF AIRCRAFT	CIVIL AIRCRAFT PILOTS, FAR 91 requires you file an IFR flight plan to operate under instrument flight rules in controlled airspace. Failure to file could result in a civil penalty not to exceed $1,000 for each violation (Section 901 of the Federal Aviation Act of 1958, as amended). Filing of a VFR flight plan is recommended as a good operating practice. See also Part 99 for requirements concerning DVFR flight plans.					

FAA Form 7233-1 (8-82) CLOSE VFR FLIGHT PLAN WITH ______________ FSS ON ARRIVAL

Reports." This page lists scheduled outages/events/projects that significantly impact the NAS; for example, runway closures, air shows, and construction projects. Information includes anticipated delays and traffic management initiatives (TMI) that may be implemented.

5-1-6. Flight Plan- Defense VFR (DVFR) Flights

VFR flights (except DOD or law enforcement flights) into a Coastal or Domestic ADIZ/DEWIZ are required to file DVFR flight plans for security purposes. Detailed ADIZ procedures are found in Section 6, National Security and Interception Procedures, of this chapter. (See 14 CFR Part 99.)

5-1-7. Composite Flight Plan (VFR/IFR Flights)

a. Flight plans which specify VFR operation for one portion of a flight, and IFR for another portion, will be accepted by the FSS at the point of departure. If VFR flight is conducted for the first portion of the flight, pilots should report their departure time to the FSS with whom the VFR/IFR flight plan was filed; and, subsequently, close the VFR portion and request ATC clearance from the FSS nearest the point at which change from VFR to IFR is proposed. Regardless of the type facility you are communicating with (FSS, center, or tower), it is the pilot's responsibility to request that facility to "CLOSE VFR FLIGHT PLAN." The pilot must remain in VFR weather conditions until operating in accordance with the IFR clearance.

b. When a flight plan indicates IFR for the first portion of flight and VFR for the latter portion, the pilot will normally be cleared to the point at which the change is proposed. After reporting over the clearance limit and not desiring further IFR clearance, the pilot should advise ATC to cancel the IFR portion of the flight plan. Then, the pilot should contact the nearest FSS to activate the VFR portion of the flight plan. If the pilot desires to continue the IFR flight plan beyond the clearance limit, the pilot should contact ATC at least 5 minutes prior to the clearance limit and request further IFR clearance. If the requested clearance is not received prior to reaching the clearance limit fix, the pilot will be expected to enter into a standard holding pattern on the radial or course to the fix unless a holding pattern for the clearance limit fix is depicted on a U.S. Government or commercially produced (meeting FAA requirements) low or high altitude enroute, area or STAR chart. In this case the pilot will hold according to the depicted pattern.

5-1-8. Flight Plan (FAA Form 7233-1)- Domestic IFR Flights

NOTE-

1. *Procedures outlined in this section apply to operators filing FAA Form 7233-1 (Flight Plan) and to flights that will be conducted entirely within U.S. domestic airspace.*

2. *Filers utilizing FAA Form 7233-1 may not be eligible for assignment of RNAV SIDs and STARs. Filers desiring assignment of these procedures should file using FAA Form 7233-4 (International Flight Plan), as described in paragraph 5-1-9.*

a. General

1. Prior to departure from within, or prior to entering controlled airspace, a pilot must submit a complete flight plan and receive an air traffic clearance, if weather conditions are below VFR minimums. Instrument flight plans may be submitted to the nearest FSS or ATCT either in person or by telephone (or by radio if no other means are available). Pilots should file IFR flight plans at least 30 minutes prior to estimated time of departure to preclude possible delay in receiving a departure clearance from ATC. In order to provide FAA traffic management units strategic route planning capabilities, nonscheduled operators conducting IFR operations above FL 230 are requested to voluntarily file IFR flight plans at least 4 hours prior to estimated time of departure (ETD). To minimize your delay in entering Class B, Class C, Class D, and Class E surface areas at destination when IFR weather conditions exist or are forecast at that airport, an IFR flight plan should be filed before departure. Otherwise, a 30 minute delay is not unusual in receiving an ATC clearance because of time spent in processing flight plan data. Traffic saturation frequently prevents control personnel from accepting flight plans by radio. In such cases, the pilot is advised to contact the nearest FSS for the purpose of filing the flight plan.

NOTE-

1. *There are several methods of obtaining IFR clearances at nontower, non-FSS, and outlying airports. The procedure may vary due to geographical features, weather conditions, and the complexity of the ATC system. To determine the most effective means of receiving an IFR clearance, pilots should ask the nearest FSS the most appropriate means of obtaining the IFR clearance.*

2. *When requesting an IFR clearance, it is highly recommended that the departure airport be identified by stating the city name and state and/or the airport location identifier in order to clarify to ATC the exact location of the intended airport of departure.*

2. When filing an IFR flight plan, include as a prefix to the aircraft type, the number of aircraft when more than one and/or heavy aircraft indicator "H/" if appropriate.

EXAMPLE-
H/DC10/A
2/F15/A

3. When filing an IFR flight plan, identify the equipment capability by adding a suffix, preceded by a slant, to the AIRCRAFT TYPE, as shown in TBL 5-1-3, Aircraft Suffixes.

NOTE-

1. *ATC issues clearances based on filed suffixes. Pilots should determine the appropriate suffix based upon desired services and/or routing. For example, if a desired route/procedure requires GPS, a pilot should file /G even if the aircraft also qualifies for other suffixes.*

2. *For procedures requiring GPS, if the navigation system does not automatically alert the flight crew of a loss of GPS, the operator must develop procedures to verify correct GPS operation.*

3. *The suffix is not to be added to the aircraft identification or be transmitted by radio as part of the aircraft identification.*

4. It is recommended that pilots file the maximum transponder or navigation capability of their aircraft in the equipment suffix. This will provide ATC with the necessary information to utilize all facets of navigational equipment and transponder capabilities available.

5. When filing an IFR flight plan via telephone or radio, it is highly recommended that the departure airport be clearly identified by stating the city name and state and/or airport location identifier. With cell phone use and flight service specialists covering larger areas of the country, clearly identifying the departure airport can prevent confusing your airport of departure with those of identical or similar names in other states.

TBL 5-1-3
Aircraft Suffixes

Suffix	Equipment Capability
	NO DME
/X	No transponder
/T	Transponder with no Mode C
/U	Transponder with Mode C
	DME
/D	No transponder
/B	Transponder with no Mode C
/A	Transponder with Mode C
	TACAN ONLY
/M	No transponder
/N	Transponder with no Mode C
/P	Transponder with Mode C
	AREA NAVIGATION (RNAV)
/Y	LORAN, VOR/DME, or INS with no transponder
/C	LORAN, VOR/DME, or INS, transponder with no Mode C
/I	LORAN, VOR/DME, or INS, transponder with Mode C
	ADVANCED RNAV WITH TRANSPONDER AND MODE C (If an aircraft is unable to operate with a transponder and/or Mode C, it will revert to the appropriate code listed above under Area Navigation.)
/E	Flight Management System (FMS) with DME/DME and IRU position updating
/F	FMS with DME/DME position updating
/G	Global Navigation Satellite System (GNSS), including GPS or Wide Area Augmentation System (WAAS), with en route and terminal capability.
/R	Required Navigational Performance (RNP). The aircraft meets the RNP type prescribed for the route segment(s), route(s) and/or area concerned.
	REDUCED VERTICAL SEPARATION MINIMUM (RVSM). Prior to conducting RVSM operations within the U.S., the operator must obtain authorization from the FAA or from the responsible authority, as appropriate.
/J	/E with RVSM
/K	/F with RVSM
/L	/G with RVSM
/Q	/R with RVSM
/W	RVSM

b. Airways and Jet Routes Depiction on Flight Plan

1. It is vitally important that the route of flight be accurately and completely described in the flight plan. To simplify definition of the proposed route, and to facilitate ATC, pilots are requested to file via airways or jet routes established for use at the altitude or flight level planned.

2. If flight is to be conducted via designated airways or jet routes, describe the route by indicating the type and number designators of the airway(s) or jet route(s) requested. If more than one airway or jet route is to be used, clearly indicate points of transition. If the transition is made at an unnamed intersection, show the next succeeding NAVAID or named intersection on the intended route and the complete route from that point. Reporting points may be identified by using authorized name/code as depicted on appropriate aeronautical charts. The following two examples illustrate the need to specify the transition point when two routes share more than one transition fix.

EXAMPLE-

1. *ALB J37 BUMPY J14 BHM*
Spelled out: from Albany, New York, via Jet Route 37 transitioning to Jet Route 14 at BUMPY intersection, thence via Jet Route 14 to Birmingham, Alabama.

2. *ALB J37 ENO J14 BHM*
Spelled out: from Albany, New York, via Jet Route 37 transitioning to Jet Route 14 at Smyrna VORTAC (ENO) thence via Jet Route 14 to Birmingham, Alabama.

3. The route of flight may also be described by naming the reporting points or NAVAIDs over which the flight will pass, provided the points named are established for use at the altitude or flight level planned.

EXAMPLE-
BWI V44 SWANN V433 DQO
Spelled out: from Baltimore-Washington International, via Victor 44 to Swann intersection, transitioning to Victor 433 at Swann, thence via Victor 433 to Dupont.

4. When the route of flight is defined by named reporting points, whether alone or in combination with airways or jet routes, and the navigational aids (VOR, VORTAC, TACAN, NDB) to be used for the flight are a combination of different types of aids, enough information should be included to clearly indicate the route requested.

EXAMPLE-
LAX J5 LKV J3 GEG YXC FL 330 J500 VLR J515 YWG
Spelled out: from Los Angeles International via Jet Route 5 Lakeview, Jet Route 3 Spokane, direct Cranbrook, British Columbia VOR/DME, Flight Level 330 Jet Route 500 to Langruth, Manitoba VORTAC, Jet Route 515 to Winnepeg, Manitoba.

5. When filing IFR, it is to the pilot's advantage to file a preferred route.

REFERENCE- *Preferred IFR Routes are described and tabulated in the Airport/Facility Directory.*

6. ATC may issue a SID or a STAR, as appropriate.

***REFERENCE-**AIM, Instrument Departure Procedures (DP) - Obstacle Departure Procedures (ODP) and Standard Instrument Departures (SID), Paragraph 5-2-8.*

AIM, Standard Terminal Arrival (STAR), Area Navigation (RNAV) STAR, and Flight Management System Procedures (FMSP) for Arrivals, Paragraph 5-4-1.

***NOTE-**Pilots not desiring a SID or STAR should so indicate in the remarks section of the flight plan as "no SID" or "no STAR."*

c. Direct Flights

1. All or any portions of the route which will not be flown on the radials or courses of established airways or routes, such as direct route flights, must be defined by indicating the radio fixes over which the flight will pass. Fixes selected to define the route must be those over which the position of the aircraft can be accurately determined. Such fixes automatically become compulsory reporting points for the flight, unless advised otherwise by ATC. Only those navigational aids established for use in a particular structure; i.e., in the low or high structures, may be used to define the en route phase of a direct flight within that altitude structure.

2. The azimuth feature of VOR aids and that azimuth and distance (DME) features of VORTAC and TACAN aids are assigned certain frequency protected areas of airspace which are intended for application to established airway and route use, and to provide guidance for planning flights outside of established airways or routes. These areas of airspace are expressed in terms of cylindrical service volumes of specified dimensions called "class limits" or "categories."

***REFERENCE-**AIM, Navigational Aid (NAVAID) Service Volumes, Paragraph 1-1-8.*

3. An operational service volume has been established for each class in which adequate signal coverage and frequency protection can be assured. To facilitate use of VOR, VORTAC, or TACAN aids, consistent with their operational service volume limits, pilot use of such aids for defining a direct route of flight in controlled airspace should not exceed the following:

(a) Operations above FL 450 - Use aids not more than 200 NM apart. These aids are depicted on enroute high altitude charts.

(b) Operation off established routes from 18,000 feet MSL to FL 450 - Use aids not more than 260 NM apart. These aids are depicted on enroute high altitude charts.

(c) Operation off established airways below 18,000 feet MSL - Use aids not more than 80 NM apart. These aids are depicted on enroute low altitude charts.

(d) Operation off established airways between 14,500 feet MSL and 17,999 feet MSL in the conterminous U.S. - (H) facilities not more than 200 NM apart may be used.

4. Increasing use of self-contained airborne navigational systems which do not rely on the VOR/VORTAC/TACAN system has resulted in pilot requests for direct routes which exceed NAVAID service volume limits. These direct route requests will be approved only in a radar environment, with approval based on pilot responsibility for navigation on the authorized direct route. Radar flight following will be provided by ATC for ATC purposes.

5. At times, ATC will initiate a direct route in a radar environment which exceeds NAVAID service volume limits. In such cases ATC will provide radar monitoring and navigational assistance as necessary.

6. Airway or jet route numbers, appropriate to the stratum in which operation will be conducted, may also be included to describe portions of the route to be flown.

EXAMPLE-

MDW V262 BDF V10 BRL STJ SLN GCK

Spelled out: from Chicago Midway Airport via Victor 262 to Bradford, Victor 10 to Burlington, Iowa, direct St. Joseph, Missouri, direct Salina, Kansas, direct Garden City, Kansas.

***NOTE-**When route of flight is described by radio fixes, the pilot will be expected to fly a direct course between the points named.*

7. Pilots are reminded that they are responsible for adhering to obstruction clearance requirements on those segments of direct routes that are outside of controlled airspace. The MEAs and other altitudes shown on low altitude IFR enroute charts pertain to those route segments within controlled airspace, and those altitudes may not meet obstruction clearance criteria when operating off those routes.

d. Area Navigation (RNAV)

1. Random RNAV routes can only be approved in a radar environment. Factors that will be considered by ATC in approving random RNAV routes include the capability to provide radar monitoring and compatibility with traffic volume and flow. ATC will radar monitor each flight, however, navigation on the random RNAV route is the responsibility of the pilot.

2. Pilots of aircraft equipped with approved area navigation equipment may file for RNAV routes throughout the National Airspace System and may be filed for in accordance with the following procedures.

(a) File airport-to-airport flight plans.

(b) File the appropriate RNAV capability certification suffix in the flight plan.

(c) Plan the random route portion of the flight plan to begin and end over appropriate arrival and departure transition fixes or appropriate navigation aids for the altitude stratum within which the flight will be conducted. The use of normal preferred departure and arrival routes (DP/STAR), where established, is recommended.

(d) File route structure transitions to and from the random route portion of the flight.

(e) Define the random route by waypoints. File route description waypoints by using degree-distance fixes based on navigational aids which are appropriate for the altitude stratum.

(f) File a minimum of one route description waypoint for each ARTCC through whose area the random route will be flown. These waypoints must be located within 200 NM of the preceding center's boundary.

(g) File an additional route description waypoint for each turnpoint in the route.

(h) Plan additional route description waypoints as required to ensure accurate navigation via the filed route of flight. Navigation is the pilot's responsibility unless ATC assistance is requested.

(i) Plan the route of flight so as to avoid prohibited and restricted airspace by 3 NM unless permission has been obtained to operate in that airspace and the appropriate ATC facilities are advised.

***NOTE-**To be approved for use in the National Airspace System, RNAV equipment must meet the appropriate system availability, accuracy, and airworthiness standards. For additional guidance on equipment requirements see AC 20-130, Airworthiness Approval of Vertical Navigation (VNAV) Systems for use in the U.S. NAS and Alaska, or AC 20-138, Airworthiness Approval of Global Positioning System (GPS) Navigation Equipment for Use as a VFR and IFR Supplemental Navigation System. For airborne navigation database, see AC 90-94, Guidelines for Using GPS Equipment for IFR En Route and Terminal Operations and for Nonprecision Instrument Approaches in the U.S. National Airspace System, Section 2.*

3. Pilots of aircraft equipped with latitude/longitude coordinate navigation capability, independent of VOR/TACAN references, may file for random RNAV routes at and above FL 390 within the conterminous U.S. using the following procedures.

(a) File airport-to-airport flight plans prior to departure.

(b) File the appropriate RNAV capability certification suffix in the flight plan.

(c) Plan the random route portion of the flight to begin and end over published departure/arrival transition fixes or appropriate navigation aids for airports without published transition procedures. The use of preferred departure and arrival routes, such as DP and STAR where established, is recommended.

(d) Plan the route of flight so as to avoid prohibited and restricted airspace by 3 NM unless permission has been obtained to operate in that airspace and the appropriate ATC facility is advised.

(e) Define the route of flight after the departure fix, including each intermediate fix (turnpoint) and the arrival fix for the destination airport in terms of latitude/longitude coordinates plotted to the nearest minute or in terms of Navigation Reference System (NRS) waypoints. For latitude/longitude filing the arrival fix must be identified by both the latitude/longitude coordinates and a fix identifier.

EXAMPLE-

MIA[1] SRQ[2] 3407/10615[3] 3407/11546 TNP[4] LAX[5]

[1] Departure airport.
[2] Departure fix.
[3] Intermediate fix (turning point).
[4] Arrival fix.
[5] Destination airport.

or

ORD[1] IOW[2] KP49G[3] KD34U[4] KL16O[5] OAL[6] MOD2[7] SFO[8]

[1] Departure airport.
[2] Transition fix (pitch point).
[3] Minneapolis ARTCC waypoint.
[4] Denver ARTCC Waypoint.
[5] Los Angeles ARTCC waypoint (catch point).
[6] Transition fix.
[7] Arrival.
[8] Destination airport.

(f) Record latitude/longitude coordinates by four figures describing latitude in degrees and minutes followed by a solidus and five figures describing longitude in degrees and minutes.

(g) File at FL 390 or above for the random RNAV portion of the flight.

(h) Fly all routes/route segments on Great Circle tracks.

(i) Make any inflight requests for random RNAV clearances or route amendments to an en route ATC facility.

e. Flight Plan Form- See FIG 5-1-2.

f. Explanation of IFR Flight Plan Items.

1. Block 1. Check the type flight plan. Check both the VFR and IFR blocks if composite VFR/IFR.

2. Block 2. Enter your complete aircraft identification including the prefix "N" if applicable.

3. Block 3. Enter the designator for the aircraft, followed by a slant(/), and the transponder or DME equipment code letter; e.g., C-182/U. Heavy aircraft, add prefix "H" to aircraft type; example: H/DC10/U. Consult an FSS briefer for any unknown elements.

FIGURE 5-1-2
FAA Flight Plan Form 7233-1 (8-82)

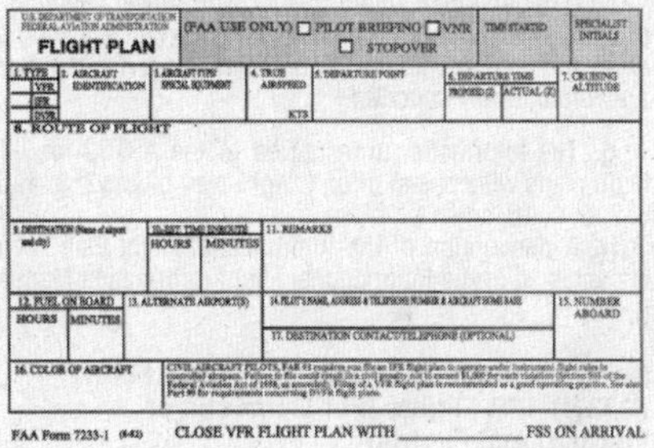

4. Block 4. Enter your computed true airspeed (TAS).

***NOTE-**If the average TAS changes plus or minus 5 percent or 10 knots, whichever is greater, advise ATC.*

5. Block 5. Enter the departure airport identifier code (or the airport name, city and state, if the identifier is unknown).

***NOTE-**Use of identifier codes will expedite the processing of your flight plan.*

6. Block 6. Enter the proposed departure time in Coordinated Universal Time (UTC) (Z). If airborne, specify the actual or proposed departure time as appropriate.

7. Block 7. Enter the requested en route altitude or flight level.

***NOTE-**Enter only the initial requested altitude in this block. When more than one IFR altitude or flight level is desired along the route of flight, it is best to make a subsequent request direct to the controller.*

8. Block 8. Define the route of flight by using NAVAID identifier codes (or names if the code is unknown), airways, jet routes, and waypoints (for RNAV).

***NOTE-**Use NAVAIDs or waypoints to define direct routes and radials/bearings to define other unpublished routes.*

9. Block 9. Enter the destination airport identifier code (or name if the identifier is unknown).

10. Block 10. Enter your estimated time en route based on latest forecast winds.

11. Block 11. Enter only those remarks pertinent to ATC or to the clarification of other flight plan information, such as the appropriate radiotelephony (call sign) associated with the FAA-assigned three-letter company designator filed in Block 2, if the radiotelephony is new or has changed within the last 60 days. In cases where there is no three-letter designator but only an assigned radiotelephony or an assigned three-letter designator is used in a medical emergency, the radiotelephony must be included in the remarks field. Items of a personal nature are not accepted.

NOTE-

1. The pilot is responsible for knowing when it is appropriate to file the radiotelephony in remarks under the 60-day rule or when using FAA special radiotelephony assignments.

2. "DVRSN" should be placed in Block 11 only if the pilot/company is requesting priority handling to their original destination from ATC as a result of a diversion as defined in the Pilot/Controller Glossary.

3. Do not assume that remarks will be automatically transmitted to every controller. Specific ATC or en route requests should be made directly to the appropriate controller.

12. Block 12. Specify the fuel on board, computed from the departure point.

13. Block 13. Specify an alternate airport if desired or required, but do not include routing to the alternate airport.

14. Block 14. Enter the complete name, address, and telephone number of pilot-in-command, or in the case of a formation flight, the formation commander. Enter sufficient information to identify home base, airport, or operator.

NOTE-*This information would be essential in the event of search and rescue operation.*

15. Block 15. Enter the total number of persons on board including crew.

16. Block 16. Enter the predominant colors.

NOTE-*Close IFR flight plans with tower, approach control, or ARTCC, or if unable, with FSS. When landing at an airport with a functioning control tower, IFR flight plans are automatically canceled.*

g. The information transmitted to the ARTCC for IFR flight plans will consist of only flight plan blocks 2, 3, 4, 5, 6, 7, 8, 9, 10, and 11.

h. A description of the International Flight Plan Form is contained in the International Flight Information Manual (IFIM).

5-1-9. International Flight Plan (FAA Form 7233-4)- IFR Flights (For Domestic or International Flights)

a. General

Use of FAA Form 7233-4 is recommended for domestic IFR flights and is mandatory for all IFR flights that will depart U.S. domestic airspace.

NOTE-

1. An abbreviated description of FAA Form 7233-4 (International Flight Plan) may be found in this section. A detailed description of FAA Form 7233-4 may be found on the FAA website at:

http://www.faa.gov/about/office_org/headquarters_offices/ato/service_units/enroute/flight_plan_filing/

2. Filers utilizing FAA Form 7233-1 (Flight Plan) may not be eligible for assignment of RNAV SIDs and STARs. Filers desiring assignment of these procedures should file using FAA Form 7233-4, as described in this section.

3. When filing an IFR flight plan using FAA Form 7233-4, it is recommended that filers include all operable navigation, communication, and surveillance equipment capabilities by adding appropriate equipment qualifiers as shown in Tables 5-1-3 and 5-1-4. These equipment qualifiers should be filed in Item 10 of FAA Form 7233-4.

4. ATC issues clearances based on equipment qualifiers filed in Items 10 and aircraft capabilities filed in Item 18 (NAV/) of FAA Form 7233-4. Operators should file all equipment qualifiers for which the aircraft is certified and capable. They should also file aircraft capabilities in Item 18 as described below.

b. Explanation of Items Filed in FAA Form 7233-4

Procedures and other information provided in this section are designed to assist operators using FAA Form 7233-4 to file IFR flight plans for flights that will be conducted entirely within U.S. domestic airspace. Requirements and procedures for operating outside U.S. domestic airspace may vary significantly from country to country. It is, therefore, recommended that operators planning flights outside U.S. domestic airspace become familiar with applicable international documents, including Aeronautical Information Publications (AIP); International Flight Information Manuals (IFIM); and ICAO Document 4444, Procedures for Air Navigation Services/Air Traffic Management, Appendix 2.

NOTE-*FAA Form 7233-4 is shown in FIG 5-1-3. The filer is normally responsible for providing the information required in Items 3 through 19.*

1. Item 7. Aircraft Identification. Insert the full registration number of the aircraft, or the approved FAA/ICAO company or organizational designator, followed by the flight number.

EXAMPLE-

N235RA, AAL3342, BONGO33

NOTE-*Callsigns filed in this item must begin with a letter followed by 1-6 additional alphanumeric characters.*

2. Item 8. Flight Rules and Type of Flight.

(a) Flight Rules. Insert the character "I" to indicate IFR

(b) Type of Flight. Insert one of the following letters to denote the type of flight:

(1) S if scheduled air service

(2) N if non-scheduled air transport operation

(3) G if general aviation

(4) M if military

(5) X if other than any of the defined categories above.

NOTE-*Type of flight is optional for flights that will be conducted entirely within U.S. domestic airspace.*

3. Item 9. Number, Type of Aircraft, and Wake Turbulence Category.

(a) Number. Insert the number of aircraft, if more than 1 (maximum 99).

(b) Type of Aircraft.

(1) Insert the appropriate designator as specified in ICAO Doc 8643, Aircraft Type Designators;

(2) Or, if no such designator has been assigned, or in the case of formation flights consisting of more than one type;

(3) Insert ZZZZ, and specify in Item 18, the (numbers and) type(s) of aircraft preceded by TYP/.

(c) Wake Turbulence Category. Insert an oblique stroke followed by one of the following letters to indicate the wake turbulence category of the aircraft:

(1) H - HEAVY, to indicate an aircraft type with a maximum certificated takeoff weight of 300,000 pounds (136 000 kg), or more;

(2) M - MEDIUM, to indicate an aircraft type with a maximum certificated takeoff weight of less than 300,000 pounds (136,000 kg), but more than 15,500 pounds (7,000 kg);

(3) L - LIGHT, to indicate an aircraft type with a maximum certificated takeoff weight of 15,500 pounds (7,000 kg) or less.

4. Item 10. Equipment

NOTE-1. *If the letter S is used, standard equipment is considered to be VHF RTF, VOR, and ILS within U.S. domestic airspace.*

2. If the letter G is used, the types of external GNSS augmentation, if any, are specified in Item 18 following the indicator NAV/ and separated by a space.

3. See RTCA/EUROCAE Interoperability Requirements Standard For ATN Baseline 1 (ATN B1 INTEROP Standard - DO-280B/ED-110B) for data link services air traffic control clearance and information/air traffic control communications management/air traffic control microphone check.

4. If the letter R is used, the performance based navigation levels that can be met are specified in Item 18 following the indicator PBN/. Guidance material on the application of performance based navigation to a specific route segment, route, or area is contained in the Performance Based Navigation Manual (Doc 9613). Note that FAA instructions for inclusion of RNAV capability per Advisory Circular 90-100A in NAV/ also apply (additionally, see the Aeronautical Information Publication (AIP) ENR 1.10-12).

5. If the letter Z is used, specify in Item 18 the other equipment carried, preceded by COM/ and/or NAV/, as appropriate.

Fig 5-1-3
FAA International Flight Plan Form 7233-4 (9-06)

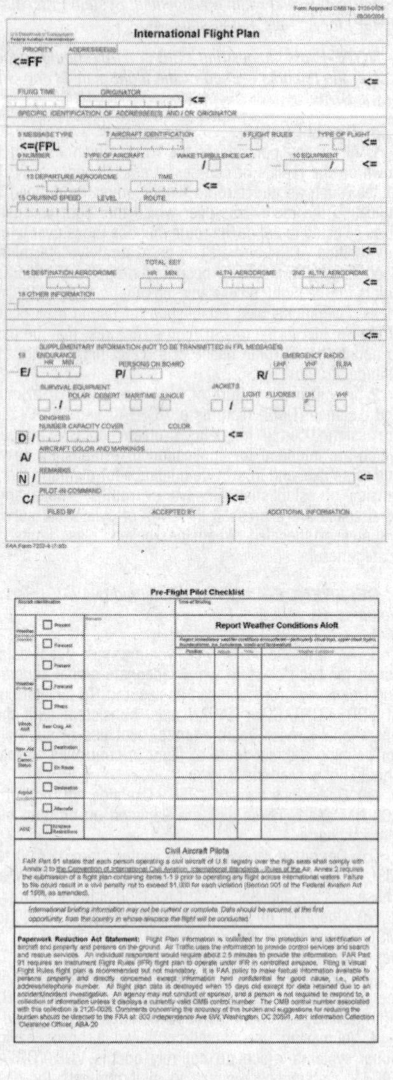

6. Information on navigation capability is provided to ATC for clearance and routing purposes.

EXAMPLE-

1. SDGW/SB1U1 {VOR, ILS, VHF, DME, GNSS, RVSM, Mode S transponder, ADS-B 1090 Extended Squitter out, ADS-B UAT out}

2. S/C {VOR, ILS, VHF, Mode C transponder}

NOTE-*The equipment qualifier Z indicates that additional equipment or capability information can be found in Item 18, following the NAV/ indicator. Operators requesting assignment of RNAV SIDs and/or STARs are required to include a Z in Item 10 and associated RNAV capabilities in Item 18 following the NAV/ indicator.*

5. Item 13. Departure Aerodrome/Time

(a) Insert the ICAO four-letter location indicator of the departure aerodrome, or

NOTE-*ICAO location indicators must consist of 4 letters. Airport identifiers such as 5IA7, 39LL and Z40 are not in ICAO standard format.*

(b) If no four-letter location indicator has been assigned to the departure aerodrome, insert ZZZZ and specify the non-ICAO location identifier, or fix/radial/distance from a nearby navaid, followed by the name of the aerodrome, in Item 18, following characters DEP/,

TBL 5-1-4

Aircraft COM, NAV, and Approach Equipment Qualifiers

INSERT one letter as follows: N if no COM/NAV/approach aid equipment for the route to be flown is carried, or the equipment is unserviceable,(OR) S if standard COM/NAV/approach aid equipment for the route to be flown is carried and serviceable (see Note 1), (AND/OR) INSERT one or more of the following letters to indicate the COM/NAV/approach aid equipment available and serviceable:

NOTE- *The capabilities described below comprise the following elements:*

a. Presence of relevant serviceable equipment on board the aircraft.

b. Equipment and capabilities commensurate with flight crew qualifications.

c. Where applicable, authorization from the appropriate authority.

A	GBAS landing system	J7	CPDLC FANS 1/A SATCOM (Iridium)
B	LPV (APV with SBAS)	K	MLS
C	LORAN C	L	ILS
D	DME	M1	ATC RTF SATCOM (INMARSAT)
E1	FMC WPR ACARS	M2	ATC RTF (MTSAT)
E2	DFIS ACARS	M3	ATC RTF (Iridium)
E3	PDC ACARS	O	VOR
F	ADF	P1-P9	Reserved for RCP
G	(GNSS) – see Note 2	R	PBN approved see Note 4
H	HF RTF	T	TACAN
I	Inertial navigation	U	UHF RTF
J1	CPDLC ATN VDL Mode 2 -see Note 3	V	VHF RTF
J2	CPDLC FANS 1/A HFDL	W	RVSM approved
J3	CPDLC FANS 1/A VDL Mode 4	X	MNPS approved
J4	CPDLC FANS 1/A VDL Mode 2	Y	VHF with 8.33 kHz channel spacing capability
J5	CPDLC FANS 1/A SATCOM (INMARSAT)	Z	Other equipment carried or other capabilities see Note 5
J6	CPDLC FANS 1/A SATCOM (MTSAT)		

TBL 5-1-5

Aircraft Surveillance Equipment, Including Designators for Transponder, ADS-B, ADS-C, and Capabilities

INSERT N if no surveillance equipment for the route to be flown is carried, or the equipment is unserviceable, OR *INSERT* one or more of the following descriptors, to a maximum of 20 characters, to describe the serviceable surveillance equipment and/or capabilities on board:	
SSR Modes A and C	
A	Transponder - Mode A (4 digits - 4096 codes)
C	Transponder - Mode A (4 digits - 4096 codes) and Mode C
SSR Mode S	
E	Transponder - Mode S, including aircraft identification, pressure-altitude and extended squitter (ADS-B) capability
H	Transponder - Mode S, including aircraft identification, pressure-altitude and enhanced surveillance capability
I	Transponder - Mode S, including aircraft identification, but no pressure-altitude capability
L	Transponder - Mode S, including aircraft identification, pressure-altitude, extended squitter (ADS B) and enhanced surveillance capability
P	Transponder - Mode S, including pressure-altitude, but no aircraft identification capability
S	Transponder - Mode S, including both pressure-altitude and aircraft identification capability
X	Transponder - Mode S with neither aircraft identification nor pressure-altitude capability
NOTE- *Enhanced surveillance capability is the ability of the aircraft to down-link aircraft derived data via a Mode S transponder.*	
Followed by one or more of the following codes if the aircraft has ADS-B capability:	
B1	ADS-B with dedicated 1090 MHz ADS-B "out" capability
B2	ADS-B with dedicated 1090 MHz ADS-B "out" and "in" capability
U1	ADS-B "out" capability using UAT
U2	ADS-B "out" and "in" capability using UAT
V1	ADS-B "out" capability using VDL Mode 4
V2	ADS-B "out" and "in" capability using VDL Mode 4
NOTE- *File no more than one code for each type of capability; for example, file B1 or B2,but not both.*	
Followed by one or more of the following codes if the aircraft has ADS-C capability:	
D1	ADS-C with FANS 1/A capabilities
G1	ADS-C with ATN capabilities

(c) Then, without a space, insert the estimated off-block time.

EXAMPLE-
1. *KSMF2215*
2. *ZZZZ0330*

6. Item 15. Cruise Speed, Level and Route

(a) Cruise Speed (maximum 5 characters). Insert the true airspeed for the first or the whole cruising portion of the flight, in terms of knots, expressed as N followed by 4 digits (e.g. N0485), or Mach number to the nearest hundredth of unit Mach, expressed as M followed by 3 digits (for example, M082).

(b) Cruising level (maximum 5 characters). Insert the planned cruising level for the first or the whole portion of the route to be flown, in terms of flight level, expressed as F followed by 3 figures (for example, F180; F330), or altitude in hundreds of feet, expressed as A followed by 3 figures (for example, A040; A170).

(c) Route. Insert the requested route of flight in accordance with guidance below.

NOTE- *Speed and/or altitude changes en route will be accepted by FAA computer systems, but will not be processed or forwarded to controllers. Pilots are expected to maintain the last assigned altitude and request revised altitude clearances directly from ATC.*

(d) Insert the desired route of flight using a combination of published routes and/or fixes in the following formats:

(1) Consecutive fixes, navaids and waypoints should be separated by the characters "DCT", meaning direct.

EXAMPLE-
FLACK DCT IRW DCT IRW125023

NOTE- *IRW125023 identifies the fix located on the Will Rogers VORTAC 125 radial at 23 DME.*

(2) Combinations of published routes, and fixes, navaids or waypoints should be separated by a single space.

EXAMPLE-
WORTH5 MQP V66 ABI V385

(3) Although it is recommended that filed airway junctions be identified using a named junction fix when possible, there may be cases where it is necessary to file junctioning airways without a named fix. In these cases, separate consecutive airways with a space.

EXAMPLE-
V325 V49

NOTE- *This method of filing an airway junction may result in a processing ambiguity. This might cause the flight plan to be rejected in some cases.*

7. Item 16. Destination Aerodrome, Total EET, Alternate and 2nd Alternate Aerodrome

(a) Destination Aerodrome and Total Estimated Elapsed Time (EET).

(1) Insert the ICAO four-letter location identifier for the destination aerodrome; or, if no ICAO location identifier has been assigned, (Location identifiers, such as WY66, A08, and 5B1, are not an ICAO standard format),

(2) Insert ZZZZ and specify the non-ICAO location identifier, or fix/radial/distance from a nearby navaid, followed the name of the aerodrome, in Item 18, following characters DEST/,

(3) Then, without a space, insert the total estimated time en route to the destination.

EXAMPLE-
1. *KOKC0200*
2. *ZZZZ0330*

(b) Alternate and 2nd Alternate Aerodrome (Optional).

(1) Following the intended destination, insert the ICAO four-letter location identifier(s) of alternate aerodromes; or, if no location identifier(s) have been assigned;

(2) Insert ZZZZ and specify the name of the aerodrome in Item 18, following the characters ALTN/.

EXAMPLE-
1. *KDFW0234 KPWA*
2. *KBOS0304 ZZZZ*

NOTE- *Although alternate airport information filed in an FPL will be accepted by air traffic computer systems, it will not be presented to controllers. If diversion to an alternate airport becomes necessary, pilots are expected to notify ATC and request an amended clearance.*

8. Item 18. Other Information

(a) Insert 0 (zero) if no other information; or, any other necessary information in the sequence shown below, in the form of the appropriate indicator followed by an oblique stroke and the information to be recorded:

NOTE-

1. *Operators are warned that the use of indicators not included in the provisions may result in data being rejected, processed incorrectly, or lost.*
2. *Hyphens "-" or oblique strokes "/" should only be used as described.*

3. *Avoid use of any other special characters in Field 18 information-use only letters and numbers.*

4. *An indicator without any associated information will result in flight plan rejection.*

(b) STS/ Reason for special handling by ATS as follows:

(1) ALTRV: For a flight operated in accordance with an altitude reservation.

(2) ATFMX: For a flight approved for exemption from ATFM measures by the appropriate ATS authority.

(3) FFR: Fire-fighting.

(4) FLTCK: Flight check for calibration of navaids.

(5) HAZMAT: For a flight carrying hazardous material.

(6) HEAD: A flight with Head of State status.

(7) HOSP: For a medical flight declared by medical authorities.

(8) HUM: For a flight operating on a humanitarian mission.

(9) MARSA: For a flight for which a military entity assumes responsibility for separation of military aircraft.

(10) MEDEVAC: For a life critical medical emergency evacuation.

(11) NONRVSM: For a non-RVSM capable flight intending to operate in RVSM airspace.

(12) SAR: For a flight engaged in a search and rescue mission.

(13) STATE: For a flight engaged in military, customs, or police services.

NOTE-*Other reasons for special handling by ATS are denoted under the designator RMK/.*

(c) PBN/ Indication of RNAV and/or RNP capabilities. Include as many of the descriptors below as apply to the flight, up to a maximum of 8 entries; that is a total of not more than 16 characters.

TBL 5-1-6
PBN/RNAV Specifications

PBN/	RNAV SPECIFICATIONS
A1	RNAV 10 (RNP 10)
B1	RNAV 5 all permitted sensors
B2	RNAV 5 GNSS
B3	RNAV 5 DME/DME
B4	RNAV 5 VOR/DME
B5	RNAV 5 INS or IRS
B6	RNAV 5 LORAN C
C1	RNAV 2 all permitted sensors
C2	RNAV 2 GNSS
C3	RNAV 2 DME/DME
C4	RNAV 2 DME/DME/IRU
D1	RNAV 1 all permitted sensors
D2	RNAV 1 GNSS
D3	RNAV 1 DME/DME
D4	RNAV 1 DME/DME/IRU
	RNP SPECIFICATIONS
L1	RNP 4
O1	Basic RNP 1 all permitted sensors
O2	Basic RNP 1 GNSS
O3	Basic RNP 1 DME/DME
O4	Basic RNP 1 DME/DME/IRU
S1	RNP APCH
S2	RNP APCH with BARO-VNAV
T1	RNP AR APCH with RF (special authorization required)
T2	RNP AR APCH without RF (special authorization required)

NOTE-*Combinations of alphanumeric characters not indicated above are reserved.*

(d) NAV/ Significant data related to navigation equipment, other than as specified in PBN/.

(1) In addition to filing appropriate equipment qualifiers in Item10, operators requesting assignment of RNAV departure and/or arrival procedures should file appropriate RNAV capabilities for each segment of flight, following the NAV/ indicator.

(2) Operators should file their maximum capabilities in order to qualify for the most advanced procedures.

EXAMPLE-
NAV/RNVD1A1

(3) Explanation: NAV/ Indicates the beginning of additional navigation information. This includes:

[a] RNV precedes RNAV capability for each phase of flight.

[b] D# - Departure segment RNAV capability.

[c] E# - En route segment RNAV capability.

[d] A# - Arrival segment RNAV capability.

NOTE-

1. *In the examples above, "#" indicates the numeric RNAV accuracy values, based on aircraft certification and capabilities.*
2. *Operators filing FAA Form 7233-4 may suppress application of RNAV procedures by omitting, or filing a 0 (zero) value in Item 18 data for any or all segments of flight.*
3. *Aircraft certification requirements for RNAV operations within U.S. airspace are defined in AC 90-45A, Approval of Area Navigation Systems for Use in the U.S. National Airspace System, and AC 90-100A, U.S. Terminal and En Route Area Navigation (RNAV) Operations, as amended.*

(e) COM/ Indicate communications capabilities not specified in Item 10a, when requested by an air navigation service provider.

(f) DAT/ Indicate data applications or capabilities not specified in Item 10a, when requested by an Air Navigation Service Provider.

(g) SUR/ Indicate surveillance capabilities not specified in Item 10b, when requested by an Air Navigation Service Provider. If ADS-B capability filed in Item 10 is compliant with RTCA DO-260B, include the item "260B" in SUR/. If ADS-B capability filed in Item 10 is compliant with RTCA DO-282B, include the item "282B" in SUR/.

EXAMPLE-
1. *SUR/260B*
2. *SUR/260B 282B*

(h) DEP/ Insert the non-ICAO identifier, or fix/radial/distance from navaid, or latitude/longitude, if ZZZZ is inserted in Item 13. Optionally, append the name of the departure point.

EXAMPLE-
1. *DEP/T23 ALBANY MUNI*
2. *DEP/T23*
3. *DEP/UKW197011 TICK HOLLR RANCH*
4. *DEP/4620N07805W*

(i) DEST/ Insert the non-ICAO identifier, or fix/radial/distance from navaid, or latitude/longitude, if ZZZZ is inserted in Item 16. Optionally, append the name of the destination point.

EXAMPLE-
1. *DEST/T23 ALBANY MUNI*
2. *DEST/PIE335033 LEXI DUNES*
3. *DEST/4620N07805W*

(j) DOF/ The date of flight departure in a six figure format (YYMMDD, where YY equals the year, MM equals the month, and DD equals the day). The FAA will not accept flight plans filed with Date of Flight resulting in more than a day in advance.

(k) REG/ The registration markings of the aircraft, if different from the aircraft identification in Item 7. Note that the FAA uses this information in monitoring of RVSM and ADS-B performance.

(l) EET/ Significant points or FIR boundary designators and accumulated estimated elapsed times to such points or FIR boundaries.

EXAMPLE-
EET/KZLA0745 KZAB0830

(m) SEL/ SELCAL code.

(n) TYP/ Insert the type of aircraft if ZZZZ was entered in Item 9. If necessary, insert the number and type(s) of aircraft in a formation.

EXAMPLE-
1. *TYP/Homebuilt*
2. *TYP/2 P51 B17 B24*

(o) CODE/ Aircraft address (expressed in the form of an alphanumerical code of six hexadecimal characters) when required by the appropriate ATS authority. Include CODE/ when ADS-B capability is filed in Item 10.

EXAMPLE-
"F00001" is the lowest aircraft address contained in the specific block administered by ICAO.

(p) DLE/ En route delay or holding, insert the significant point(s) on the route where a delay is planned to occur, followed by the length of delay using four figure time in hours and minutes (hhmm).

EXAMPLE-
DLE/MDG0030

(q) OPR/ Name of the operator, if not obvious from the aircraft identification in Item 7.

(r) ORGN/ The originator's 8-letter AFTN address or other appropriate contact details, in cases where the originator of the flight plan may not be readily identified, as required by the appropriate ATS authority. The FAA does not require ORGN/ information.

NOTE-*In some areas, flight plan reception centers may insert the ORGN/ identifier and originator's AFTN address automatically.*

(s) PER/ Aircraft performance data, indicated by a single letter as specified in the Procedures for Air Navigation Services - Aircraft Operations (PANS-OPS, Doc 8168), Volume I - Flight Procedures, if so prescribed by the appropriate ATS authority. Note that the FAA does not require PER/ information.

(t) ALTN/ Name of destination alternate aerodrome(s), if ZZZZ is inserted in Item 16.

EXAMPLE-
1. *ALTN/F35 POSSUM KINGDOM*
2. *ALTN/TCC233016 LAZY S RANCH*

(u) RALT/ ICAO 4-letter indicator(s) for en-route alternate(s), as specified in Doc 7910, Location Indicators, or name(s) of en-route alternate aerodrome(s), if no indicator is allocated. For aerodromes not listed in the relevant Aeronautical Information Publication, indicate location in LAT/LONG or bearing and distance from the nearest significant point, as described in DEP/ above.

(v) TALT/ ICAO 4-letter indicator(s) for take-off alternate, as specified in Doc 7910, Location Indicators, or name of take-off alternate aerodrome, if no indicator is allocated. For aerodromes not listed in the relevant Aeronautical Information Publication, indicate location in LAT/LONG or bearing and distance from the nearest significant point, as described in DEP/ above.

(w) RIF/ The route details to the revised destination aerodrome, followed by the ICAO four-letter location indicator of the aerodrome. The revised route is subject to reclearance in flight.

EXAMPLE-
1. *RIF/DTA HEC KLAX*
2. *RIF/ESP G94 CLA YPPH*

(x) RMK/ Any other plain-language remarks when required by the ATC or deemed necessary.

EXAMPLE-
1. *RMK/NRP*
2. *RMK/DRVSN*

(y) RVR/ The minimum RVR requirement of the flight in meters. This item is defined by Eurocontrol, not ICAO. The FAA does not require or use this item, but will accept it in a flight plan.

NOTE-*This provision is detailed in the European Regional Supplementary Procedures (EUR SUPPs, Doc 7030), Chapter 2.*

(z) RFP/ Q followed by a digit to indicate the sequence of the replacement flight plan being submitted. This item is defined by Eurocontrol, not ICAO. The FAA will not use this item, but will accept it in a flight plan.

NOTE-*This provision is detailed in the European Regional Supplementary Procedures (EUR SUPPs, Doc 7030), chapter 2.*

9. Item 19. Supplementary Information

NOTE-*Item 19 data must be included when completing FAA Form 7233-4. This information will be retained by the facility/organization that transmits the flight plan to Air Traffic Control (ATC), for Search and Rescue purposes, but it will not be transmitted to ATC as part of the FPL.*

(a) E/ (ENDURANCE). Insert 4-digits group giving the fuel endurance in hours and minutes.

(b) P/ (PERSONS ON BOARD). Insert the total number of persons (passengers and crew) on board.

(c) Emergency and survival equipment

(1) R/ (RADIO).

[a] Cross out "UHF" if frequency 243.0 MHz is not available.

[b] Cross out "VHF" frequency 121.5 MHz is not available.

[c] Cross out "ELBA" if emergency locator transmitter (ELT) is not available.

(2) S/ (SURVIVAL EQUIPMENT).

[a] Cross out "POLAR" if polar survival equipment is not carried.

[b] Cross out "DESERT" if desert survival equipment is not carried.

[c] Cross out "MARITIME" if maritime survival equipment is not carried.

[d] Cross out J if "JUNGLE" survival equipment is not carried.

(3) J/ (JACKETS).

[a] Cross out "LIGHT" if life jackets are not equipped with lights.

[b] Cross out "FLUORES" if life jackets are not equipped with fluorescein.

[c] Cross out "UHF" or "VHF" or both as in R/ above to indicate radio capability of jackets, if any.

(4) D/ (DINGHIES).

[a] NUMBER. Cross out indicators "NUMBER" and "CAPACITY" if no dinghies are carried, or insert number of dinghies carried; and

[b] CAPACITY. Insert total capacity, in persons, of all dinghies carried; and

[c] COVER. Cross out indicator "COVER" if dinghies are not covered; and

[d] COLOR. Insert color of dinghies if carried.

(5) A/ (AIRCRAFT COLOR AND MARKINGS). Insert color of aircraft and significant markings.

(6) N/ (REMARKS). Cross out indicator N if no remarks, or indicate any other survival equipment carried and any other remarks regarding survival equipment.

(7) C/ (PILOT). Insert name of pilot-in-command.

5-1-10. IFR Operations to High Altitude Destinations

a. Pilots planning IFR flights to airports located in mountainous terrain are cautioned to consider the necessity for an alternate airport even when the forecast weather conditions would technically relieve them from the requirement to file one.

REFERENCE-*14 CFR Section 91.167.*
AIM, Tower En Route Control (TEC), Paragraph 4-1-19.

b. The FAA has identified three possible situations where the failure to plan for an alternate airport when flying IFR to such a destination airport could result in a critical situation if the weather is less than forecast and sufficient fuel is not available to proceed to a suitable airport.

1. An IFR flight to an airport where the Minimum Descent Altitudes (MDAs) or landing visibility minimums for *all instrument approaches* are higher than the forecast weather minimums specified in 14 CFR Section 91.167(b). For example, there are 3 high altitude airports in the U.S. with approved instrument approach procedures where all of the MDAs are greater than 2,000 feet and/or the landing visibility minimums are greater than 3 miles (Bishop, California; South Lake Tahoe, California; and Aspen-Pitkin Co./ Sardy Field, Colorado). In the case of these airports, it is possible for a pilot to elect, on the basis of forecasts, not to carry sufficient fuel to get to an alternate when the ceiling and/or visibility is actually lower than that necessary to complete the approach.

2. A small number of other airports in mountainous terrain have MDAs which are slightly (100 to 300 feet) below 2,000 feet AGL. In situations where there is an option as to whether to plan for an alternate, pilots should bear in mind that just a slight worsening of the weather conditions from those forecast could place the airport below the published IFR landing minimums.

3. An IFR flight to an airport which requires special equipment; i.e., DME, glide slope, etc., in order to make the available approaches to the lowest minimums. Pilots should be aware that all other minimums on the approach charts may require weather conditions better than those specified in 14 CFR Section 91.167(b). An inflight equipment malfunction could result in the inability to comply with the published approach procedures or, again, in the position of having the airport below the published IFR landing minimums for all remaining instrument approach alternatives.

5-1-11. Flights Outside the U.S. and U.S. Territories

a. When conducting flights, particularly extended flights, outside the U.S. and its territories, full account should be taken of the amount and quality of air navigation services available in the airspace to be traversed. Every effort should be made to secure information on the location and range of navigational aids, availability of communications and meteorological services, the provision of air traffic services, including alerting service, and the existence of search and rescue services.

b. Pilots should remember that there is a need to continuously guard the VHF emergency frequency 121.5 MHz when on long over-water flights, except when communications on other VHF channels, equipment limitations, or cockpit duties prevent simultaneous guarding of two channels. Guarding of 121.5 MHz is particularly critical when operating in proximity to Flight Information Region (FIR) boundaries, for example, operations on Route R220 between Anchorage and Tokyo, since it serves to facilitate communications with regard to aircraft which may experience in-flight emergencies, communications, or navigational difficulties.

REFERENCE-*ICAO Annex 10, Vol II, Paras 5.2.2.1.1.1 and 5.2.2.1.1.2.*

c. The filing of a flight plan, always good practice, takes on added significance for extended flights outside U.S. airspace and is, in fact, usually required by the laws of the countries being visited or overflown. It is also particularly important in the case of such flights that pilots leave a complete itinerary and schedule of the flight with someone directly concerned and keep that person advised of the flight's progress. If serious doubt arises as to the safety of the flight, that person should first contact the appropriate FSS. Round Robin Flight Plans to Mexico are not accepted.

d. All pilots should review the foreign airspace and entry restrictions published in the IFIM during the flight planning process. Foreign airspace penetration without official authorization can involve both danger to the aircraft and the imposition of severe penalties and inconvenience to both passengers and crew. A flight plan on file with ATC authorities does not necessarily constitute the prior permission required by certain other authorities. The possibility of fatal consequences cannot be ignored in some areas of the world.

e. Current NOTAMs for foreign locations must also be reviewed. The publication Notices to Airmen, Domestic/ International, published biweekly, contains considerable information pertinent to foreign flight. Current foreign NOTAMs are also available from the U.S. International NOTAM Office in Washington, D.C., through any local FSS.

f. When customs notification is required, it is the responsibility of the pilot to arrange for customs notification in a timely manner. The following guidelines are applicable:

1. When customs notification is required on flights to Canada and Mexico and a predeparture flight plan cannot be filed or an advise customs message (ADCUS) cannot be included in a predeparture flight plan, call the nearest en route domestic or International FSS as soon as radio communication can be established and file a VFR or DVFR flight plan, as required, and include as the last item the advise customs information. The station with which such a flight plan is filed will forward it to the appropriate FSS who will notify the customs office responsible for the destination airport.

2. If the pilot fails to include ADCUS in the radioed flight plan, it will be assumed that other arrangements have been made and FAA will not advise customs.

3. The FAA assumes no responsibility for any delays in advising customs if the flight plan is given too late for delivery to customs before arrival of the aircraft. **It is still the pilot's responsibility to give timely notice even though a flight plan is given to FAA.**

4. Air Commerce Regulations of the Treasury Department's Customs Service require all private aircraft arriving in the U.S. via:

(a) The U.S./Mexican border or the Pacific Coast from a foreign place in the Western Hemisphere south of 33 degrees north latitude and between 97 degrees and 120 degrees west longitude; or

(b) The Gulf of Mexico and Atlantic Coasts from a foreign place in the Western Hemisphere south of 30 degrees north latitude, must furnish a notice of arrival to the Customs service at the nearest designated airport. This notice may be furnished directly to Customs by:

(1) Radio through the appropriate FAA Flight Service Station.

(2) Normal FAA flight plan notification procedures (a flight plan filed in Mexico does not meet this requirement due to unreliable relay of data); or

(3) Directly to the district Director of Customs or other Customs officer at place of first intended landing but must be furnished at least 1 hour prior to crossing the U.S./ Mexican border or the U.S. coastline.

(c) This notice will be valid as long as actual arrival is within 15 minutes of the original ETA, otherwise a new notice must be given to Customs. Notices will be accepted up to 23 hours in advance. Unless an exemption has been granted by Customs, private aircraft are required to make first landing in the U.S. at one of the following designated airports nearest to the point of border of coastline crossing:

Designated Airports
ARIZONA
Bisbee Douglas Intl Airport
Douglas Municipal Airport
Nogales Intl Airport
Tucson Intl Airport
Yuma MCAS-Yuma Intl Airport

CALIFORNIA
Calexico Intl Airport
Brown Field Municipal Airport (San Diego)

FLORIDA
Fort Lauderdale Executive Airport
Fort Lauderdale/Hollywood Intl Airport
Key West Intl Airport (Miami Intl Airport)
Opa Locka Airport (Miami)
Kendall-Tamiami Executive Airport (Miami)
St. Lucie County Intl Airport (Fort Pierce)
Tampa Intl Airport
Palm Beach Intl Airport (West Palm Beach)

LOUISANA
New Orleans Intl Airport (Moisant Field)
New Orleans Lakefront Airport

NEW MEXICO
Las Cruces Intl Airport

NORTH CAROLINA
New Hanover Intl Airport (Wilmington)

TEXAS
Brownsville/South Padre Island Intl Airport
Corpus Christi Intl Airport
Del Rio Intl Airport
Eagle Pass Municipal Airport
El Paso Intl Airport
William P. Hobby Airport (Houston)
Laredo Intl Airport
McAllen Miller Intl Airport
Presidio Lely Intl Airport

5-1-12. Change in Flight Plan

In addition to altitude or flight level, destination and/ or route changes, increasing or decreasing the speed of an aircraft constitutes a change in a flight plan. Therefore, at any time the average true airspeed at cruising altitude between reporting points varies or is expected to vary from that given in the flight plan by *plus or minus 5 percent, or 10 knots, whichever is greater,* ATC should be advised.

5-1-13. Change in Proposed Departure Time

a. To prevent computer saturation in the en route environment, parameters have been established to delete proposed departure flight plans which have not been activated. Most centers have this parameter set so as to delete these flight plans a minimum of 1 hour after the proposed departure time. To ensure that a flight plan remains active, pilots whose actual departure time will be delayed 1 hour or more beyond their filed departure time, are requested to notify ATC of their departure time.

b. Due to traffic saturation, control personnel frequently will be unable to accept these revisions via radio. It is recommended that you forward these revisions to the nearest FSS.

5-1-14. Closing VFR/DVFR Flight Plans

A pilot is responsible for ensuring that his/her VFR or DVFR flight plan is canceled. You should close your flight plan with the nearest FSS, or if one is not available, you may request any ATC facility to relay your cancellation to the FSS. Control towers do not automatically close VFR or DVFR flight plans since they do not know if a particular VFR aircraft is on a flight plan. If you fail to report or cancel your flight plan within 1/2 hour after your ETA, search and rescue procedures are started.

REFERENCE- *14 CFR Section 91.153.*
14 CFR Section 91.169.

5-1-15. Canceling IFR Flight Plan

a. 14 CFR Sections 91.153 and 91.169 include the statement "When a flight plan has been activated, the pilot-in-command, upon canceling or completing the flight under the flight plan, must notify an FAA Flight Service Station or ATC facility."

b. An IFR flight plan may be canceled at any time the flight is operating in VFR conditions outside Class A airspace by pilots stating "CANCEL MY IFR FLIGHT PLAN" to the controller or air/ground station with which they are communicating. Immediately after canceling an IFR flight plan, a pilot should take the necessary action to change to the appropriate air/ground frequency, VFR radar beacon code and VFR altitude or flight level.

c. ATC separation and information services will be discontinued, including radar services (where applicable). Consequently, if the canceling flight desires VFR radar advisory service, the pilot must specifically request it.

NOTE- *Pilots must be aware that other procedures may be applicable to a flight that cancels an IFR flight plan within an area where a special program, such as a designated TRSA, Class C airspace, or Class B airspace, has been established.*

d. If a DVFR flight plan requirement exists, the pilot is responsible for filing this flight plan to replace the canceled IFR flight plan. If a subsequent IFR operation becomes necessary, a new IFR flight plan must be filed and an ATC clearance obtained before operating in IFR conditions.

e. If operating on an IFR flight plan to an airport with a functioning control tower, the flight plan is automatically closed upon landing.

f. If operating on an IFR flight plan to an airport where there is no functioning control tower, the pilot must initiate cancellation of the IFR flight plan. This can be done after landing if there is a functioning FSS or other means of direct communications with ATC. In the event there is no FSS and/or air/ground communications with ATC is not possible below a certain altitude, the pilot should, weather conditions permitting, cancel the IFR flight plan while still airborne and able to communicate with ATC by radio. This will not only save the time and expense of canceling the flight plan by telephone but will quickly release the airspace for use by other aircraft.

5-1-16. RNAV and RNP Operations

a. During the pre-flight planning phase the availability of the navigation infrastructure required for the intended operation, including any non-RNAV contingencies, must be confirmed for the period of intended operation. Availability of the onboard navigation equipment necessary for the route to be flown must be confirmed.

b. If a pilot determines a specified RNP level cannot be achieved, revise the route or delay the operation until appropriate RNP level can be ensured.

c. The onboard navigation database must be current and appropriate for the region of intended operation and must include the navigation aids, waypoints, and coded terminal airspace procedures for the departure, arrival and alternate airfields.

d. During system initialization, pilots of aircraft equipped with a Flight Management System or other RNAV-certified system, must confirm that the navigation database is current, and verify that the aircraft position has been entered correctly. Flight crews should crosscheck the cleared flight plan against charts or other applicable resources, as well as the navigation system textual display and the aircraft map display. This process includes confirmation of the waypoints sequence, reasonableness of track angles and distances, any altitude or speed constraints, and identification of fly-by or fly-over waypoints. A procedure must not be used if validity of the navigation database is in doubt.

e. Prior to commencing takeoff, the flight crew must verify that the RNAV system is operating correctly and the correct airport and runway data have been loaded.

f. During the pre-flight planning phase RAIM prediction must be performed if TSO-C129() equipment is used to solely satisfy the RNAV and RNP requirement. GPS RAIM availability must be confirmed for the intended route of flight (route and time) using current GPS satellite information. In the event of a predicted, continuous loss of RAIM of more than five (5) minutes for any part of the intended flight, the flight should be delayed, canceled, or re-routed where RAIM requirements can be met. Operators may satisfy the predictive RAIM requirement through any one of the following methods:

1. Operators may monitor the status of each satellite in its plane/slot position, by accounting for the latest GPS constellation status (e.g., NOTAMs or NANUs), and compute RAIM availability using model-specific RAIM prediction software;

2. Operators may use the FAA en route and terminal RAIM prediction website: www.raimprediction.net;

3. Operators may contact a Flight Service Station (not DUATS) to obtain non-precision approach RAIM;

4. Operators may use a third party interface, incorporating FAA/VOLPE RAIM prediction data without altering performance values, to predict RAIM outages for the aircraft's predicted flight path and times;

5. Operators may use the receiver's installed RAIM prediction capability (for TSO-C129a/Class A1/B1/C1 equipment) to provide non-precision approach RAIM, accounting for the latest GPS constellation status (e.g., NOTAMs or NANUs). Receiver non-precision approach RAIM should be checked at airports spaced at intervals not to exceed 60 NM along the RNAV 1 procedure's flight track. "Terminal" or "Approach" RAIM must be available at the ETA over each airport checked; or,

6. Operators not using model-specific software or FAA/VOLPE RAIM data will need FAA operational approval.

NOTE- *If TSO-C145/C146 equipment is used to satisfy the RNAV and RNP requirement, the pilot/operator need not perform the prediction if WAAS coverage is confirmed to be available along the entire route of flight. Outside the U.S. or in areas where WAAS coverage is not available, operators using TSO-C145/C146 receivers are required to check GPS RAIM availability.*

Section 2. Departure Procedures

5-2-1. Pre-taxi Clearance Procedures

a. Certain airports have established pre-taxi clearance programs whereby pilots of departing instrument flight rules (IFR) aircraft may elect to receive their IFR clearances before they start taxiing for takeoff. The following provisions are included in such procedures:

1. Pilot participation is not mandatory.

2. Participating pilots call clearance delivery or ground control not more than 10 minutes before proposed taxi time.

3. IFR clearance (or delay information, if clearance cannot be obtained) is issued at the time of this initial call-up.

4. When the IFR clearance is received on clearance delivery frequency, pilots call ground control when ready to taxi.

5. Normally, pilots need not inform ground control that they have received IFR clearance on clearance delivery frequency. Certain locations may, however, require that the pilot inform ground control of a portion of the routing or that the IFR clearance has been received.

6. If a pilot cannot establish contact on clearance delivery frequency or has not received an IFR clearance before ready to taxi, the pilot should contact ground control and inform the controller accordingly.

b. Locations where these procedures are in effect are indicated in the Airport/Facility Directory.

5-2-2. Pre-departure Clearance Procedures

a. Many airports in the National Airspace System are equipped with the Tower Data Link System (TDLS) that includes the Pre-departure Clearance (PDC) function. The PDC function automates the Clearance Delivery operations in the ATCT for participating users. The PDC function displays IFR clearances from the ARTCC to the ATCT. The Clearance Delivery controller in the ATCT can append local departure information and transmit the clearance via data link to participating airline/service provider computers. The airline/service provider will then deliver the clearance via the Aircraft Communications Addressing and Reporting System (ACARS) or a similar data link system or, for nondata link equipped aircraft, via a printer located at the departure gate. PDC reduces frequency congestion, controller workload and is intended to mitigate delivery/readback errors. Also, information from participating users indicates a reduction in pilot workload.

b. PDC is available only to participating aircraft that have subscribed to the service through an approved service provider.

c. Due to technical reasons, the following limitations currently exist in the PDC program:

1. Aircraft filing multiple flight plans are limited to one PDC clearance per departure airport within a 24-hour period. Additional clearances will be delivered verbally.

2. If the clearance is revised or modified prior to delivery, it will be rejected from PDC and the clearance will need to be delivered verbally.

d. No acknowledgment of receipt or readback is required for a PDC.

e. In all situations, the pilot is encouraged to contact clearance delivery if a question or concern exists regarding an automated clearance.

5-2-3. Taxi Clearance

Pilots on IFR flight plans should communicate with the control tower on the appropriate ground control or clearance delivery frequency, prior to starting engines, to receive engine start time, taxi and/or clearance information.

5-2-4. Line Up and Wait (LUAW)

a. Line up and wait is an air traffic control (ATC) procedure designed to position an aircraft onto the runway for an imminent departure. The ATC instruction "LINE UP AND WAIT" is used to instruct a pilot to taxi onto the departure runway and line up and wait.

EXAMPLE-
Tower: "N234AR Runway 24L, line up and wait."

b. This ATC instruction is not an authorization to takeoff. In instances where the pilot has been instructed to line up and wait and has been advised of a reason/condition (wake turbulence, traffic on an intersecting runway, etc.) or the reason/condition is clearly visible (another aircraft that has landed on or is taking off on the same runway), and the reason/condition is satisfied, the pilot should expect an imminent takeoff clearance, unless advised of a delay. If you are uncertain about any ATC instruction or clearance, contact ATC immediately.

c. If a takeoff clearance is not received within a reasonable amount of time after clearance to line up and wait, ATC should be contacted.

EXAMPLE-
Aircraft: Cessna 234AR holding in position Runway 24L.
Aircraft: Cessna 234AR holding in position Runway 24L at Bravo.

NOTE- *FAA analysis of accidents and incidents involving aircraft holding in position indicate that two minutes or more elapsed between the time the instruction was issued to line up and wait and the resulting event (for example, land-over or go-around). Pilots should consider the length of time that they have been holding in position whenever they HAVE NOT been advised of any expected delay to determine when it is appropriate to query the controller.*

REFERENCE- *Advisory Circulars 91-73A, Part 91 and Part 135 Single-Pilot Procedures during Taxi Operations, and 120-74A, Parts 91, 121, 125, and 135 Flightcrew Procedures during Taxi Operations*

d. Situational awareness during line up and wait operations is enhanced by monitoring ATC instructions/clearances issued to other aircraft. Pilots should listen carefully if another aircraft is on frequency that has a similar call sign and pay close attention to communications between ATC and other aircraft. If you are uncertain of an ATC instruction or clearance, query ATC immediately. Care should be taken to not inadvertently execute a clearance/instruction for another aircraft.

e. Pilots should be especially vigilant when conducting line up and wait operations at night or during reduced visibility conditions. They should scan the full length of the runway and look for aircraft on final approach or landing roll out when taxiing onto a runway. ATC should be contacted anytime there is a concern about a potential conflict.

f. When two or more runways are active, aircraft may be instructed to "LINE UP AND WAIT" on two or more runways. When multiple runway operations are being conducted, it is important to listen closely for your call sign and runway. Be alert for similar sounding call signs and acknowledge all instructions with your call sign. When you are holding in position and are not sure if the takeoff clearance was for you, ask ATC before you begin takeoff roll. ATC prefers that you confirm a takeoff clearance rather than mistake another aircraft's clearance for your own.

g. When ATC issues intersection "line up and wait" and takeoff clearances, the intersection designator will be used. If ATC omits the intersection designator, call ATC for clarification.

EXAMPLE-
Aircraft: "Cherokee 234AR, Runway 24L at November 4, line up and wait."

h. If landing traffic is a factor during line up and wait operations, ATC will inform the aircraft in position of the closest traffic that has requested a full-stop, touch-and-go, stop-and-go, or an unrestricted low approach to the same runway. Pilots should take care to note the position of landing traffic. ATC will also advise the landing traffic when an aircraft is authorized to "line up and wait" on the same runway.

EXAMPLE-
Tower: "Cessna 234AR, Runway 24L, line up and wait. Traffic a Boeing 737, six mile final."
Tower: "Delta 1011, continue, traffic a Cessna 210 holding in position Runway 24L."

NOTE-*ATC will normally withhold landing clearance to arrival aircraft when another aircraft is in position and holding on the runway.*

i. Never land on a runway that is occupied by another aircraft, even if a landing clearance was issued. Do not hesitate to ask the controller about the traffic on the runway and be prepared to execute a go-around.

NOTE-*Always clarify any misunderstanding or confusion concerning ATC instructions or clearances. ATC should be advised immediately if there is any uncertainty about the ability to comply with any of their instructions.*

5-2-5. Abbreviated IFR Departure Clearance (Cleared. . .as Filed) Procedures

a. ATC facilities will issue an abbreviated IFR departure clearance based on the ROUTE of flight filed in the IFR flight plan, provided the filed route can be approved with little or no revision. These abbreviated clearance procedures are based on the following conditions:

1. The aircraft is on the ground or it has departed visual flight rules (VFR) and the pilot is requesting IFR clearance while airborne.

2. That a pilot will not accept an abbreviated clearance if the route or destination of a flight plan filed with ATC has been changed by the pilot or the company or the operations officer before departure.

3. That it is the responsibility of the company or operations office to inform the pilot when they make a change to the filed flight plan.

4. That it is the responsibility of the pilot to inform ATC in the initial call-up (for clearance) when the filed flight plan has been either:

(a) Amended, or

(b) Canceled and replaced with a new filed flight plan.

NOTE-*The facility issuing a clearance may not have received the revised route or the revised flight plan by the time a pilot requests clearance.*

b. Controllers will issue a detailed clearance when they know that the original filed flight plan has been changed or when the pilot requests a full route clearance.

c. The clearance as issued will include the destination airport filed in the flight plan.

d. ATC procedures now require the controller to state the DP name, the current number and the DP transition name after the phrase "Cleared to (destination) airport" and prior to the phrase, "then as filed," for ALL departure clearances when the DP or DP transition is to be flown. The procedures apply whether or not the DP is filed in the flight plan.

e. STARs, when filed in a flight plan, are considered a part of the filed route of flight and will not normally be stated in an initial departure clearance. If the ARTCC's jurisdictional airspace includes both the departure airport and the fix where a STAR or STAR transition begins, the STAR name, the current number and the STAR transition name MAY be stated in the initial clearance.

f. "Cleared to (destination) airport as filed" does NOT include the en route altitude filed in a flight plan. An en route altitude will be stated in the clearance or the pilot will be advised to expect an assigned or filed altitude within a given time frame or at a certain point after departure. This may be done verbally in the departure instructions or stated in the DP.

g. In both radar and nonradar environments, the controller will state "Cleared to (destination) airport as filed" or:

1. If a DP or DP transition is to be flown, specify the DP name, the current DP number, the DP transition name, the assigned altitude/flight level, and any additional instructions (departure control frequency, beacon code assignment, etc.) necessary to clear a departing aircraft via the DP or DP transition and the route filed.

EXAMPLE-
National Seven Twenty cleared to Miami Airport Intercontinental one departure, Lake Charles transition then as filed, maintain Flight Level two seven zero.

2. When there is no DP or when the pilot cannot accept a DP, the controller will specify the assigned altitude or flight level, and any additional instructions necessary to clear a departing aircraft via an appropriate departure routing and the route filed.

NOTE-*A detailed departure route description or a radar vector may be used to achieve the desired departure routing.*

3. If it is necessary to make a minor revision to the filed route, the controller will specify the assigned DP or DP transition (or departure routing), the revision to the filed route, the assigned altitude or flight level and any additional instructions necessary to clear a departing aircraft.

EXAMPLE-
Jet Star One Four Two Four cleared to Atlanta Airport, South Boston two departure then as filed except change route to read South Boston Victor 20 Greensboro, maintain one seven thousand.

4. Additionally, in a nonradar environment, the controller will specify one or more fixes, as necessary, to identify the initial route of flight.

EXAMPLE-
Cessna Three One Six Zero Foxtrot cleared to Charlotte Airport as filed via Brooke, maintain seven thousand.

h. To ensure success of the program, pilots should:

1. Avoid making changes to a filed flight plan just prior to departure.

2. State the following information in the initial call-up to the facility when no change has been made to the filed flight plan: Aircraft call sign, location, type operation (IFR) and the name of the airport (or fix) to which you expect clearance.

EXAMPLE-
"Washington clearance delivery (or ground control if appropriate) American Seventy Six at gate one, IFR Los Angeles."

3. If the flight plan has been changed, state the change and request a full route clearance.

EXAMPLE-
"Washington clearance delivery, American Seventy Six at gate one. IFR San Francisco. My flight plan route has been amended (or destination changed). Request full route clearance."

4. Request verification or clarification from ATC if ANY portion of the clearance is not clearly understood.

5. When requesting clearance for the IFR portion of a VFR/IFR flight, request such clearance prior to the fix where IFR operation is proposed to commence in sufficient time to avoid delay. Use the following phraseology:

EXAMPLE-
"Los Angeles center, Apache Six One Papa, VFR estimating Paso Robles VOR at three two, one thousand five hundred, request IFR to Bakersfield."

5-2-6. Departure Restrictions, Clearance Void Times, Hold for Release, and Release Times

a. ATC may assign departure restrictions, clearance void times, hold for release, and release times, when necessary, to separate departures from other traffic or to restrict or regulate the departure flow.

1. Clearance Void Times. A pilot may receive a clearance, when operating from an airport without a control tower, which contains a provision for the clearance to be void if not airborne by a specific time. A pilot who does not depart prior to the clearance void time must advise ATC as soon as possible of their intentions. ATC will normally advise the pilot of the time allotted to notify ATC that the aircraft did not depart prior to the clearance void time. This time cannot exceed 30 minutes. Failure of an aircraft to contact ATC within 30 minutes after the clearance void time will result in the aircraft being considered overdue and search and rescue procedures initiated.

NOTE-**1.** *Other IFR traffic for the airport where the clearance is issued is suspended until the aircraft has contacted ATC or until 30 minutes after the clearance void time or 30 minutes after the clearance release time if no clearance void time is issued.*

2. *Pilots who depart at or after their clearance void time are not afforded IFR separation and may be in violation of 14 CFR Section 91.173 which requires that pilots receive an appropriate ATC clearance before operating IFR in controlled airspace.*

EXAMPLE-
Clearance void if not off by (clearance void time) and, if required, if not off by (clearance void time) advise (facility) not later than (time) of intentions.

2. Hold for Release. ATC may issue "hold for release" instructions in a clearance to delay an aircraft's departure for traffic management reasons (i.e., weather, traffic volume, etc.). When ATC states in the clearance, "hold for release," the pilot may not depart utilizing that IFR clearance until a release time or additional instructions are issued by ATC. In addition, ATC will include departure delay information in conjunction with "hold for release" instructions. The ATC instruction, "hold for release," applies to the IFR clearance and does not prevent the pilot from departing under VFR. However, prior to takeoff the pilot should cancel the IFR flight plan and operate the transponder on the appropriate VFR code. An IFR clearance may not be available after departure.

EXAMPLE-
(Aircraft identification) cleared to (destination) airport as filed, maintain (altitude), and, if required (additional instructions or information), hold for release, expect (time in hours and/or minutes) departure delay.

3. Release Times. A "release time" is a departure restriction issued to a pilot by ATC, specifying the earliest time an aircraft may depart. ATC will use "release times" in conjunction with traffic management procedures and/or to separate a departing aircraft from other traffic.

EXAMPLE-
(Aircraft identification) released for departure at (time in hours and/or minutes).

4. Expect Departure Clearance Time (EDCT). The EDCT is the runway release time assigned to an aircraft included in traffic management programs. Aircraft are expected to depart no earlier than 5 minutes before, and no later than 5 minutes after the EDCT.

b. If practical, pilots departing uncontrolled airports should obtain IFR clearances prior to becoming airborne when two-way communications with the controlling ATC facility is available.

5-2-7. Departure Control

a. Departure Control is an approach control function responsible for ensuring separation between departures. So as to expedite the handling of departures, Departure Control may suggest a takeoff direction other than that which may normally have been used under VFR handling. Many times it is preferred to offer the pilot a runway that will require the fewest turns after takeoff to place the pilot on course or selected departure route as quickly as possible. At many locations particular attention is paid to the use of preferential runways for local noise abatement programs, and route departures away from congested areas.

b. Departure Control utilizing radar will normally clear aircraft out of the terminal area using DPs via radio navigation aids.

1. When a departure is to be vectored immediately following takeoff, the pilot will be advised prior to takeoff of the initial heading to be flown but may not be advised of the purpose of the heading.

2. At some airports when a departure will fly an RNAV SID that begins at the runway, ATC may advise aircraft of the initial fix/waypoint on the RNAV route. The purpose of the advisory is to remind pilots to verify the correct procedure is programmed in the FMS before takeoff. Pilots must immediately advise ATC if a different RNAV SID is entered in the aircraft's FMC. When this advisory is absent, pilots are still required to fly the assigned SID as published.

EXAMPLE-
Delta 345 RNAV to MPASS, Runway26L, cleared for takeoff.

***NOTE-*1.** *The SID transition is not restated as it is contained in the ATC clearance.*

2. *Aircraft cleared via RNAV SIDs designed to begin with a vector to the initial waypoint are assigned a heading before departure.*

3. Pilots operating in a radar environment are expected to associate departure headings or an RNAV departure advisory with vectors or the flight path to their planned route or flight. When given a vector taking the aircraft off a previously assigned nonradar route, the pilot will be advised briefly what the vector is to achieve. Thereafter, radar service will be provided until the aircraft has been reestablished "on-course" using an appropriate navigation aid and the pilot has been advised of the aircraft's position or a handoff is made to another radar controller with further surveillance capabilities.

c. Controllers will inform pilots of the departure control frequencies and, if appropriate, the transponder code before takeoff. Pilots must ensure their transponder is adjusted to the "on" or normal operating position as soon as practical and remain on during all operations unless otherwise requested to change to "standby" by ATC. Pilots should not change to the departure control frequency until requested. Controllers may omit the departure control frequency if a DP has or will be assigned and the departure control frequency is published on the DP.

5-2-8. Instrument Departure Procedures (DP) - Obstacle Departure Procedures (ODP) and Standard Instrument Departures (SID)

Instrument departure procedures are preplanned instrument flight rule (IFR) procedures which provide obstruction clearance from the terminal area to the appropriate en route structure. There are two types of DPs, Obstacle Departure Procedures (ODPs), printed either textually or graphically, and Standard Instrument Departures (SIDs), always printed graphically. All DPs, either textual or graphic may be designed using either conventional or RNAV criteria. RNAV procedures will have RNAV printed in the title, e.g., SHEAD TWO DEPARTURE (RNAV). ODPs provide obstruction clearance via the least onerous route from the terminal area to the appropriate en route structure. ODPs are recommended for obstruction clearance and may be flown without ATC clearance unless an alternate departure procedure (SID or radar vector) has been specifically assigned by ATC. Graphic ODPs will have (OBSTACLE) printed in the procedure title, e.g., GEYSR THREE DEPARTURE (OBSTACLE), or, CROWN ONE DEPARTURE (RNAV) (OBSTACLE). Standard Instrument Departures are air traffic control (ATC) procedures printed for pilot/controller use in graphic form to provide obstruction clearance and a transition from the terminal area to the appropriate en route structure. SIDs are primarily designed for system enhancement and to reduce pilot/controller workload. ATC clearance must be received prior to flying a SID. All DPs provide the pilot with a way to depart the airport and transition to the en route structure safely. Pilots operating under 14 CFR Part 91 are strongly encouraged to file and fly a DP at night, during marginal Visual Meteorological Conditions (VMC) and Instrument Meteorological Conditions (IMC), when one is available. The following paragraphs will provide an overview of the DP program, why DPs are developed, what criteria are used, where to find them, how they are to be flown, and finally pilot and ATC responsibilities.

a. Why are DPs necessary? The primary reason is to provide obstacle clearance protection information to pilots. A secondary reason, at busier airports, is to increase efficiency and reduce communications and departure delays through the use of SIDs. When an instrument approach is initially developed for an airport, the need for DPs is assessed. The procedure designer conducts an obstacle analysis to support departure operations. If an aircraft may turn in any direction from a runway within the limits of the assessment area (see paragraph 5-2-8b3) and remain clear of obstacles, that runway passes what is called a diverse departure assessment and no ODP will be published. A SID may be published if needed for air traffic control purposes. However, if an obstacle penetrates what is called the 40:1 obstacle identification surface, then the procedure designer chooses whether to:

1. Establish a steeper than normal climb gradient; or

2.Establish a steeper than normal climb gradient with an alternative that increases takeoff minima to allow the pilot to visually remain clear of the obstacle(s); or

3. Design and publish a specific departure route; or

4. A combination or all of the above.

b. What criteria is used to provide obstruction clearance during departure?

1. Unless specified otherwise, required obstacle clearance for all departures, including diverse, is based on the pilot crossing the departure end of the runway at least 35 feet above the departure end of runway elevation, climbing to 400 feet above the departure end of runway elevation before making the initial turn, and maintaining a minimum climb gradient of 200 feet per nautical mile (FPNM), unless required to level off by a crossing restriction, until the minimum IFR altitude. A greater climb gradient may be specified in the DP to clear obstacles or to achieve an ATC crossing restriction. If an initial turn higher than 400 feet above the departure end of runway elevation is specified in the DP, the turn should be commenced at the higher altitude. If a turn is specified at a fix, the turn must be made at that fix. Fixes may have minimum and/or maximum crossing altitudes that must be adhered to prior to passing the fix. In rare instances, obstacles that exist on the extended runway centerline may make an "early turn" more desirable than proceeding straight ahead. In these cases, the published departure instructions will include the language "turn left(right) as soon as practicable." These departures will also include a ceiling and visibility minimum of at least 300 and 1. Pilots encountering one of these DPs should preplan the climb out to gain altitude and begin the turn as quickly as possible within the bounds of safe operating practices and operating limitations. This type of departure procedure is being phased out.

NOTE- *"Practical" or "feasible" may exist in some existing departure text instead of "practicable."*

2. ODPs and SIDs assume normal aircraft performance, and that all engines are operating. Development of contingency procedures, required to cover the case of an engine failure or other emergency in flight that may occur after liftoff, is the responsibility of the operator. (More detailed information on this subject is available in Advisory Circular AC 120-91, Airport Obstacle Analysis, and in the "Departure Procedures" section of chapter 2 in the Instrument Procedures Handbook, FAA-H-8261-1.)

3. The 40:1 obstacle identification surface (OIS) begins at the departure end of runway (DER) and slopes upward at 152 FPNM until reaching the minimum IFR altitude or entering the en route structure. This assessment area is limited to 25 NM from the airport in nonmountainous areas and 46 NM in designated mountainous areas. Beyond this distance, the pilot is responsible for obstacle clearance if not operating on a published route, if below (having not reached) the MEA or MOCA of a published route, or an ATC assigned altitude. See FIG 5-2-1. (Ref 14 CFR 91.177 for further information on en route altitudes.)

NOTE- *ODPs are normally designed to terminate within these distance limitations, however, some ODPs will contain routes that may exceed 25/46 NM; these routes will ensure obstacle protection until reaching the end of the ODP.*

FIGURE 5-2-1
Diverse Departure Obstacle Assessment to 25/46 NM

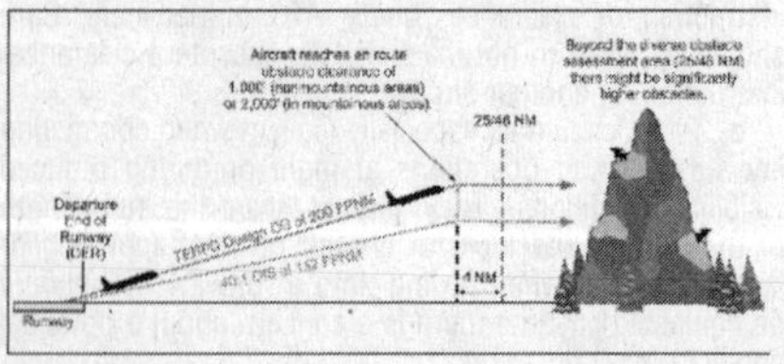

4. Obstacles that are located within 1 NM of the DER and penetrate the 40:1 OCS are referred to as "low, close-in obstacles." The standard required obstacle clearance (ROC) of 48 feet per NM to clear these obstacles would require a climb gradient greater than 200 feet per NM for a very short distance, only until the aircraft was 200 feet above the DER. To eliminate publishing an excessive climb gradient, the obstacle AGL/MSL height and location relative to the DER is noted in the "Take-off Minimums and (OBSTACLE) Departure Procedures" section of a given Terminal Procedures Publication (TPP) booklet. The purpose of this note is to identify the obstacle(s) and alert the pilot to the height and location of the obstacle(s) so they can be avoided. This can be accomplished in a variety of ways, e.g., the pilot may be able to see the obstruction and maneuver around the obstacle(s) if necessary; early liftoff/climb performance may allow the aircraft to cross well above the obstacle(s); or if the obstacle(s) cannot be visually acquired during departure, preflight planning should take into account what turns or other maneuver may be necessary immediately after takeoff to avoid the obstruction(s).

5. Climb gradients greater than 200 FPNM are specified when required to support procedure design constraints, obstacle clearance, and/or airspace restrictions. Compliance with a climb gradient for these purposes is mandatory when the procedure is part of the ATC clearance, unless increased takeoff minimums are provided and weather conditions allow compliance with these minimums. Additionally, ATC required crossing restrictions may also require climb gradients greater than 200 FPNM. These climb gradients may be amended or canceled at ATC's discretion. Multiple ATC climb gradients are permitted. An ATC climb gradient will not be used on an ODP.

EXAMPLE-
"Cross ALPHA intersection at or below 4000; maintain 6000." The pilot climbs at least 200 FPNM to 6000. If 4000 is reached before ALPHA, the pilot levels off at 4000 until passing ALPHA; then immediately resumes at least 200 FPNM climb.

EXAMPLE-
"TAKEOFF MINIMUMS: RWY 27, Standard with a minimum climb of 280' per NM to 2500, ATC climb of 310' per NM to 4000 ft." A climb of at least 280 FPNM is required to 2500 and is mandatory when the departure procedure is included in the ATC clearance. ATC requires a climb gradient of 310 FPNM to 4000, however, this ATC climb gradient may be amended or canceled.

6. Climb gradients may be specified only to an altitude/fix, above which the normal gradient applies.

EXAMPLE-
"Minimum climb 340 FPNM to ALPHA." The pilot climbs at least 340 FPNM to ALPHA, then at least 200 FPNM to MIA.

7. Some DPs established solely for obstacle avoidance require a climb in visual conditions to cross the airport or an on-airport NAVAID in a specified direction, at or above a specified altitude. These procedures are called Visual Climb Over the Airport (VCOA).

EXAMPLE-
"Climb in visual conditions so as to cross the McElory Airport southbound, at or above 6000, then climb via Keemmling radial zero three three to Keemmling VORTAC."

c. Who is responsible for obstacle clearance? DPs are designed so that adherence to the procedure by the pilot will ensure obstacle protection. Additionally:

1. Obstacle clearance responsibility also rests with the pilot when he/she chooses to climb in visual conditions in lieu of flying a DP and/or depart under increased takeoff minima rather than fly the climb gradient. Standard takeoff minima are one statute mile for aircraft having two engines or less and one-half statute mile for aircraft having more than two engines. Specified ceiling and visibility minima (VCOA or increased takeoff minima) will allow visual avoidance of obstacles until the pilot enters the standard obstacle protection area. Obstacle avoidance is not guaranteed if the pilot maneuvers farther from the airport than the specified visibility minimum prior to reaching the specified altitude. DPs may also contain what are called Low Close in Obstacles. These obstacles are less than 200 feet above the departure end of runway elevation and within one NM of the runway end, and do not require increased takeoff minimums. These obstacles are identified on the SID chart or in the Take-off Minimums and (Obstacle) Departure Procedures section of the U. S. Terminal Procedure booklet. These obstacles are especially critical to aircraft that do not lift off until close to the departure end of the runway or which climb at the minimum rate. Pilots should also consider drift following lift-off to ensure sufficient clearance from these obstacles. That segment of the procedure that requires the pilot to see and avoid obstacles ends when the aircraft crosses the specified point at the required altitude. In all cases continued obstacle clearance is based on having climbed a minimum of 200 feet per nautical mile to the specified point and then continuing to climb at least 200 foot per nautical mile during the departure until reaching the minimum enroute altitude, unless specified otherwise.

2. ATC may assume responsibility for obstacle clearance by vectoring the aircraft prior to reaching the minimum vectoring altitude by using a Diverse Vector Area (DVA). The DVA has been assessed for departures which do not follow a specific ground track. ATC may also vector an aircraft off a previously assigned DP. In all cases, the 200 FPNM climb gradient is assumed and obstacle clearance is not provided by ATC until the controller begins to provide navigational guidance in the form of radar vectors.

NOTE-*When used by the controller during departure, the term "radar contact" should not be interpreted as relieving pilots of their responsibility to maintain appropriate terrain and obstruction clearance which may include flying the obstacle DP.*

3. Pilots must preplan to determine if the aircraft can meet the climb gradient (expressed in feet per nautical mile) required by the departure procedure, and be aware that flying at a higher than anticipated ground speed increases the climb rate requirement in feet per minute. Higher than standard climb gradients are specified by a note on the departure procedure chart for graphic DPs, or in the Take-Off Minimums and (Obstacle) Departure Procedures section of the U.S. Terminal Procedures booklet for textual ODPs. The required climb gradient, or higher, must be maintained to the specified altitude or fix, then the standard climb gradient of 200 ft/NM can be resumed. A table for the conversion of climb gradient (feet per nautical mile) to climb rate (feet per minute), at a given ground speed, is included on the inside of the back cover of the U.S. Terminal Procedures booklets.

d. Where are DPs located? DPs will be listed by airport in the IFR Takeoff Minimums and (Obstacle) Departure Procedures Section, Section L, of the Terminal Procedures Publications (TPPs). If the DP is textual, it will be described in TPP Section L. SIDs and complex ODPs will be published graphically and named. The name will be listed by airport name and runway in Section L. Graphic ODPs will also have the term "(OBSTACLE)" printed in the charted procedure title, differentiating them from SIDs.

1. An ODP that has been developed solely for obstacle avoidance will be indicated with the symbol "T" on appropriate Instrument Approach Procedure (IAP) charts and DP charts for that airport. The "T" symbol will continue to refer users to TPP Section C. In the case of a graphic ODP, the TPP Section C will only contain the name of the ODP. Since there may be both a textual and a graphic DP, Section C should still be checked for additional information. The nonstandard takeoff minimums and minimum climb gradients found in TPP Section C also apply to charted DPs and radar vector departures unless different minimums are specified on the charted DP. Takeoff minimums and departure procedures apply to all runways unless otherwise specified. New graphic DPs will have all the information printed on the graphic depiction. As a general rule, ATC will only assign an ODP from a nontowered airport when compliance with the ODP is necessary for aircraft to aircraft separation. Pilots may use the ODP to help ensure separation from terrain and obstacles.

e. Responsibilities

1. Each pilot, prior to departing an airport on an IFR flight should:

(a) Consider the type of terrain and other obstacles on or in the vicinity of the departure airport;

(b) Determine whether an ODP is available;

(c) Determine if obstacle avoidance can be maintained visually or if the ODP should be flown; and

(d) Consider the effect of degraded climb performance and the actions to take in the event of an engine loss during the departure. Pilots should notify ATC as soon as possible of reduced climb capability in that circumstance.

NOTE-*Guidance concerning contingency procedures that address an engine failure on takeoff after V_1 speed on a large or turbine-powered transport category airplane may be found in AC 120-91, Airport Obstacle Analysis.*

2. Pilots should not exceed a published speed restriction associated with a SID waypoint until passing that waypoint.

3. After an aircraft is established on an SID and subsequently vectored or cleared off of the SID or SID transition, pilots must consider the SID canceled, unless the controller adds "expect to resume SID." Aircraft may not be vectored off of an ODP until at or above the MVA/MIA, at which time the ODP is canceled.

4. Aircraft instructed to resume a SID that contains ATC altitude restrictions, shall be issued/reissued all applicable restrictions or shall be advised to comply with those restrictions.

5. If prior to or after takeoff an altitude restriction is issued by ATC, all previously issued "ATC" altitude restrictions are cancelled including those published on a SID.

6. ATC crossing altitude restrictions published on SIDs are identified on the chart with "(ATC)" following the altitude restriction. This will indicate to the pilot and the controller that this restriction is for ATC purposes and may be deleted by ATC. When an ATC crossing altitude has been established prior to the beginning of a transition route, a minimum altitude for obstruction clearance or other design constraints will also be published at the same fix adjacent/below the "(ATC)" altitude. The absence of "(ATC)" at a "minimum altitude" indicates the restriction is there to support obstacle clearance, airspace restrictions, Navaid reception, and/or other reason(s) that mandate compliance. These altitudes CANNOT be lowered or cancelled by ATC. A standalone "(ATC)" altitude restriction may also be located on a transition route; however, it must never be lower than the published Minimum Enroute Altitude (MEA).

7. Altitude restrictions published on an ODP are necessary for obstacle clearance and/or design constraints. Compliance with these restrictions is mandatory and CANNOT be lowered or cancelled by ATC.

f. RNAV Departure Procedures

All public RNAV SIDs and graphic ODPs are RNAV 1. These procedures generally start with an initial RNAV or heading leg near the departure runway end. In addition, these procedures require system performance currently met by GPS or DME/DME/IRU RNAV systems that satisfy the criteria discussed in AC 90-100A, U.S. Terminal and En Route Area Navigation (RNAV) Operations. RNAV 1 procedures require the aircraft's total system error remain bounded by 1 NM for 95% of the total flight time.

Section 3. En Route Procedures

5-3-1. ARTCC Communications

a. Direct Communications, Controllers and Pilots.

1. ARTCCs are capable of direct communications with IFR air traffic on certain frequencies. Maximum communications coverage is possible through the use of Remote Center Air/Ground (RCAG) sites comprised of both VHF and UHF transmitters and receivers. These sites are located throughout the U.S. Although they may be several hundred miles away from the ARTCC, they are remoted to the various ARTCCs by land lines or microwave links. Since IFR operations are expedited through the use of direct communications, pilots are requested to use these frequencies strictly for communications pertinent to the control of IFR aircraft. Flight plan filing, en route weather, weather forecasts, and similar data should be requested through FSSs, company radio, or appropriate military facilities capable of performing these services.

2. An ARTCC is divided into sectors. Each sector is handled by one or a team of controllers and has its own sector discrete frequency. As a flight progresses from one sector to another, the pilot is requested to change to the appropriate sector discrete frequency.

3. Controller Pilot Data Link Communications (CPDLC) is a system that supplements air/ground voice communications. As a result, it expands two-way air traffic control air/ground communications capabilities. Consequently, the air traffic system's operational capacity is increased and any associated air traffic delays become minimized. A related safety benefit is that pilot/controller read-back and hear-back errors will be significantly reduced. The CPDLC's principal operating criteria are:

(a) Voice remains the primary and controlling air/ground communications means.

(b) Participating aircraft will need to have the appropriate CPDLC avionics equipment in order to receive uplink or transmit downlink messages.

(c) CPDLC Build 1 offers four ATC data link services. These are altimeter setting (AS), transfer of communications (TC), initial contact (IC), and menu text messages (MT).

(1) Altimeter settings are usually transmitted automatically when a CPDLC session and eligibility has been established with an aircraft. A controller may also manually send an altimeter setting message.

NOTE-*When conducting instrument approach procedures, pilots are responsible to obtain and use the appropriate altimeter setting in accordance with 14 CFR Section 97.20. CPDLC issued altimeter settings are excluded for this purpose.*

(2) Initial contact is a safety validation transaction that compares a pilot's initiated altitude downlink message with an aircraft's ATC host computer stored altitude. If an altitude mismatch is detected, the controller will verbally provide corrective action.

(3) Transfer of communications automatically establishes data link contact with a succeeding sector.

(4) Menu text transmissions are scripted nontrajectory altering uplink messages.

NOTE-*Initial use of CPDLC will be at the Miami Air Route Traffic Control Center (ARTCC). Air carriers will be the first users. Subsequently, CPDLC will be made available to all NAS users. Later versions will include trajectory altering services and expanded clearance and advisory message capabilities.*

b. ATC Frequency Change Procedures.

1. The following phraseology will be used by controllers to effect a frequency change:

EXAMPLE-

(Aircraft identification) contact (facility name or location name and terminal function) (frequency) at (time, fix, or altitude).

NOTE-*Pilots are expected to maintain a listening watch on the transferring controller's frequency until the time, fix, or altitude specified. ATC will omit frequency change restrictions whenever pilot compliance is expected upon receipt.*

2. The following phraseology should be utilized by pilots for establishing contact with the designated facility:

(a) When operating in a radar environment: On initial contact, the pilot should inform the controller of the aircraft's assigned altitude preceded by the words "level," or "climbing to," or "descending to," as appropriate; and the aircraft's present vacating altitude, if applicable.

EXAMPLE-

1. (Name) CENTER, (aircraft identification), LEVEL (altitude or flight level).

2. (Name) CENTER, (aircraft identification), LEAVING (exact altitude or flight level), CLIMBING TO OR DESCENDING TO (altitude of flight level).

NOTE-*Exact altitude or flight level means to the nearest 100 foot increment. Exact altitude or flight level reports on initial contact provide ATC with information required prior to using Mode C altitude information for separation purposes.*

(b) When operating in a nonradar environment:

(1) On initial contact, the pilot should inform the controller of the aircraft's present position, altitude and time estimate for the next reporting point.

EXAMPLE-

(Name) CENTER, (aircraft identification), (position), (altitude), ESTIMATING (reporting point) AT (time).

(2) After initial contact, when a position report will be made, the pilot should give the controller a complete position report.

EXAMPLE-
(Name) CENTER, (aircraft identification), (position), (time), (altitude), (type of flight plan), (ETA and name of next reporting point), (the name of the next succeeding reporting point), AND (remarks).
REFERENCE-*AIM, Position Reporting, Paragraph 5-3-2.*

3. At times controllers will ask pilots to verify that they are at a particular altitude. The phraseology used will be: "VERIFY AT (altitude)." In climbing or descending situations, controllers may ask pilots to "*VERIFY ASSIGNED ALTITUDE AS (altitude).*" Pilots should confirm that they are at the altitude stated by the controller or that the assigned altitude is correct as stated. If this is not the case, they should inform the controller of the actual altitude being maintained or the different assigned altitude.

CAUTION-Pilots should not take action to change their actual altitude or different assigned altitude to the altitude stated in the controllers verification request unless the controller specifically authorizes a change.

c. ARTCC Radio Frequency Outage. ARTCCs normally have at least one back-up radio receiver and transmitter system for each frequency, which can usually be placed into service quickly with little or no disruption of ATC service. Occasionally, technical problems may cause a delay but switchover seldom takes more than 60 seconds. When it appears that the outage will not be quickly remedied, the ARTCC will usually request a nearby aircraft, if there is one, to switch to the affected frequency to broadcast communications instructions. It is important, therefore, that the pilot wait at least 1 minute before deciding that the ARTCC has actually experienced a radio frequency failure. When such an outage does occur, the pilot should, if workload and equipment capability permit, maintain a listening watch on the affected frequency while attempting to comply with the following recommended communications procedures:

1. If two-way communications cannot be established with the ARTCC after changing frequencies, a pilot should attempt to recontact the transferring controller for the assignment of an alternative frequency or other instructions.

2. When an ARTCC radio frequency failure occurs after two-way communications have been established, the pilot should attempt to reestablish contact with the center on any other known ARTCC frequency, preferably that of the next responsible sector when practicable, and ask for instructions. However, when the next normal frequency change along the route is known to involve another ATC facility, the pilot should contact that facility, if feasible, for instructions. If communications cannot be reestablished by either method, the pilot is expected to request communications instructions from the FSS appropriate to the route of flight.

NOTE-*The exchange of information between an aircraft and an ARTCC through an FSS is quicker than relay via company radio because the FSS has direct interphone lines to the responsible ARTCC sector. Accordingly, when circumstances dictate a choice between the two, during an ARTCC frequency outage, relay via FSS radio is recommended.*

5-3-2. Position Reporting

The safety and effectiveness of traffic control depends to a large extent on accurate position reporting. In order to provide the proper separation and expedite aircraft movements, ATC must be able to make accurate estimates of the progress of every aircraft operating on an IFR flight plan.

a. Position Identification.

1. When a position report is to be made passing a VOR radio facility, the time reported should be the time at which the first complete reversal of the "to/from" indicator is accomplished.

2. When a position report is made passing a facility by means of an airborne ADF, the time reported should be the time at which the indicator makes a complete reversal.

3. When an aural or a light panel indication is used to determine the time passing a reporting point, such as a fan marker, Z marker, cone of silence or intersection of range courses, the time should be noted when the signal is first received and again when it ceases. The mean of these two times should then be taken as the actual time over the fix.

4. If a position is given with respect to distance and direction from a reporting point, the distance and direction should be computed as accurately as possible.

5. Except for terminal area transition purposes, position reports or navigation with reference to aids not established for use in the structure in which flight is being conducted will not normally be required by ATC.

b. Position Reporting Points. CFRs require pilots to maintain a listening watch on the appropriate frequency and, unless operating under the provisions of subparagraphc, to furnish position reports passing certain reporting points. Reporting points are indicated by symbols on en route charts. The designated compulsory reporting point symbol is a solid triangle ▭ and the "on request" reporting point symbol is the open triangle ▭. Reports passing an "on request" reporting point are only necessary when requested by ATC.

c. Position Reporting Requirements.

1. Flights Along Airways or Routes. A position report is required by all flights regardless of altitude, including those operating in accordance with an ATC clearance specifying *"VFR-on-top,"* over each designated compulsory reporting point along the route being flown.

2. Flights Along a Direct Route. Regardless of the altitude or flight level being flown, including flights operating in accordance with an ATC clearance specifying "*VFR-on-top,"* pilots must report over each reporting point used in the flight plan to define the route of flight.

3. Flights in a Radar Environment. When informed by ATC that their aircraft are in "Radar Contact," pilots should discontinue position reports over designated reporting points. They should resume normal position reporting when ATC advises "*RADAR CONTACT LOST*" or "*RADAR SERVICE TERMINATED.*"

4. Oceanic Environment. In the oceanic environment, Ocean21 requires a position report for any fix that is filed in the flight plan even if it is considered to be noncompulsory. Any fix not identified on a navigation chart as being compulsory is still identified by the Ocean21 system as such and therefore requires a position report.

NOTE-*ATC will inform pilots that they are in "radar contact":*

(a) *when their aircraft is initially identified in the ATC system; and*
(b) *when radar identification is reestablished after radar service has been terminated or radar contact lost.*
Subsequent to being advised that the controller has established radar contact, this fact will not be repeated to the pilot when handed off to another controller. At times, the aircraft identity will be confirmed by the receiving controller; however, this should not be construed to mean that radar contact has been lost. The identity of transponder equipped aircraft will be confirmed by asking the pilot to "ident," "squawk standby," or to change codes. Aircraft without transponders will be advised of their position to confirm identity. In this case, the pilot is expected to advise the controller if in disagreement with the position given. Any pilot who cannot confirm the accuracy of the position given because of not being tuned to the NAVAID referenced by the controller, should ask for another radar position relative to the tuned in NAVAID.

d. Position Report Items:

1. Position reports should include the following items:

(a) Identification;
(b) Position;
(c) Time;
(d) Altitude or flight level (include actual altitude or flight level when operating on a clearance specifying VFR-on-top);
(e) Type of flight plan (not required in IFR position reports made directly to ARTCCs or approach control);
(f) ETA and name of next reporting point;
(g) The name only of the next succeeding reporting point along the route of flight; and
(h) Pertinent remarks.

5-3-3. Additional Reports

a. The following reports should be made to ATC or FSS facilities without a specific ATC request:

1. *At all times.*

(a) When vacating any previously assigned altitude or flight level for a newly assigned altitude or flight level.
(b) When an altitude change will be made if operating on a clearance specifying VFR-on-top.
(c) When *unable* to climb/descend at a rate of a least 500 feet per minute.
(d) When approach has been missed. (Request clearance for specific action; i.e., to alternative airport, another approach, etc.)
(e) Change in the average true airspeed (at cruising altitude) when it varies by 5 percent or 10 knots (whichever is greater) from that filed in the flight plan.
(f) The time and altitude or flight level upon reaching a holding fix or point to which cleared.
(g) When leaving any assigned holding fix or point.

NOTE-*The reports in subparagraphs (f) and (g) may be omitted by pilots of aircraft involved in instrument training at military terminal area facilities when radar service is being provided.*

(h) Any loss, in controlled airspace, of VOR, TACAN, ADF, low frequency navigation receiver capability, GPS anomalies while using installed IFR-certified GPS/GNSS receivers, complete or partial loss of ILS receiver capability or impairment of air/ground communications capability. Reports should include aircraft identification, equipment affected, degree to which the capability to operate under IFR in the ATC system is impaired, and the nature and extent of assistance desired from ATC.

NOTE-1. *Other equipment installed in an aircraft may effectively impair safety and/or the ability to operate under IFR. If such equipment (e.g., airborne weather radar) malfunctions and in the pilot's judgment either safety or IFR capabilities are affected, reports should be made as above.*
2. *When reporting GPS anomalies, include the location and altitude of the anomaly. Be specific when describing the location and include duration of the anomaly if necessary.*

(i) Any information relating to the safety of flight.

2. When not in radar contact.

(a) When leaving final approach fix inbound on final approach (nonprecision approach) or when leaving the outer marker or fix used in lieu of the outer marker inbound on final approach (precision approach).
(b) A corrected estimate at anytime it becomes apparent that an estimate as previously submitted is in error in excess of 3 minutes. For flights in the North Atlantic (NAT), a revised estimate is required if the error is 3 minutes or more.

b. Pilots encountering weather conditions which have not been forecast, or hazardous conditions which have been forecast, are expected to forward a report of such weather to ATC.

REFERENCE-*AIM, Pilot Weather Reports (PIREPs), Paragraph 7-1-20.*
14 CFR Section 91.183(B) and (C).

5-3-4. Airways and Route Systems

a. Three fixed route systems are established for air navigation purposes. They are the Federal airway system (consisting of VOR and L/MF routes), the jet route system, and the RNAV route system. To the extent possible, these route systems are aligned in an overlying manner to facilitate transition between each.

1. The VOR and L/MF (nondirectional radio beacons) Airway System consists of airways designated from 1,200 feet above the surface (or in some instances higher) up to but not including 18,000 feet MSL. These airways are depicted on IFR Enroute Low Altitude Charts.

NOTE-*The altitude limits of a victor airway should not be exceeded except to effect transition within or between route structures.*

(a) Except in Alaska, the VOR airways are: predicated solely on VOR or VORTAC navigation aids; depicted in black on aeronautical charts; and identified by a "V" (Victor) followed by the airway number (for example, V12).

NOTE-*Segments of VOR airways in Alaska are based on L/MF navigation aids and charted in brown instead of black on en route charts.*

(1) A segment of an airway which is common to two or more routes carries the numbers of all the airways which coincide for that segment. When such is the case, pilots filing a flight plan need to indicate only that airway number for the route filed.

NOTE-A pilot who intends to make an airway flight, using VOR facilities, will simply specify the appropriate "victor" airway(s) in the flight plan. For example, if a flight is to be made from Chicago to New Orleans at 8,000 feet, using omniranges only, the route may be indicated as "departing from Chicago-Midway, cruising 8,000 feet via Victor 9 to Moisant International." If flight is to be conducted in part by means of L/MF navigation aids and in part on omniranges, specifications of the appropriate airways in the flight plan will indicate which types of facilities will be used along the described routes, and, for IFR flight, permit ATC to issue a traffic clearance accordingly. A route may also be described by specifying the station over which the flight will pass, but in this case since many VORs and L/MF aids have the same name, the pilot must be careful to indicate which aid will be used at a particular location. This will be indicated in the route of flight portion of the flight plan by specifying the type of facility to be used after the location name in the following manner: Newark L/MF, Allentown VOR.

(2) With respect to position reporting, reporting points are designated for VOR Airway Systems. Flights using Victor Airways will report over these points unless advised otherwise by ATC.

(b) The L/MF airways (colored airways) are predicated solely on L/MF navigation aids and are depicted in brown on aeronautical charts and are identified by color name and number (e.g., Amber One). Green and Red airways are plotted east and west. Amber and Blue airways are plotted north and south.

NOTE-Except for G13 in North Carolina, the colored airway system exists only in the state of Alaska. All other such airways formerly so designated in the conterminous U.S. have been rescinded.

(c) The use of TSO-C145 (as revised) or TSO-C146 (as revised) GPS/WAAS navigation systems is allowed in Alaska as the only means of navigation on published air traffic service (ATS) routes, including those Victor, T-Routes, and colored airway segments designated with a second minimum en route altitude (MEA) depicted in blue and followed by the letter G at those lower altitudes. The altitudes so depicted are below the minimum reception altitude (MRA) of the land-based navigation facility defining the route segment, and guarantee standard en route obstacle clearance and two-way communications. Air carrier operators requiring operations specifications are authorized to conduct operations on those routes in accordance with FAA operations specifications.

2. The jet route system consists of jet routes established from 18,000 feet MSL to FL 450 inclusive.

(a) These routes are depicted on Enroute High Altitude Charts. Jet routes are depicted in black on aeronautical charts and are identified by a "J" (Jet) followed by the airway number (e.g., J12). Jet routes, as VOR airways, are predicated solely on VOR or VORTAC navigation facilities (except in Alaska).

NOTE-Segments of jet routes in Alaska are based on L/MF navigation aids and are charted in brown color instead of black on en route charts.

(b) With respect to position reporting, reporting points are designated for jet route systems. Flights using jet routes will report over these points unless otherwise advised by ATC.

3. Area Navigation (RNAV) Routes.

(a) Published RNAV routes, including Q-Routes and T-Routes, can be flight planned for use by aircraft with RNAV capability, subject to any limitations or requirements noted on en route charts, in applicable Advisory Circulars, or by NOTAM. RNAV routes are depicted in blue on aeronautical charts and are identified by the letter "Q" or "T" followed by the airway number (for example, Q-13, T-205). Published RNAV routes are RNAV-2 except when specifically charted as RNAV-1. These routes require system performance currently met by GPS, GPS/WAAS, or DME/DME/IRU RNAV systems that satisfy the criteria discussed in AC 90-100A, U.S. Terminal and En Route Area Navigation (RNAV) Operations.

NOTE-AC 90-100A does not apply to over water RNAV routes (reference 14 CFR 91.511, including the Q-routes in the Gulf of Mexico and the Atlantic routes) or Alaska VOR/DME RNAV routes ("JxxxR"). The AC does not apply to off-route RNAV operations, Alaska GPS routes or Caribbean routes.

(1) Q-routes are available for use by RNAV equipped aircraft between 18,000 feet MSL and FL450 inclusive. Q-routes are depicted on Enroute High Altitude Charts.

NOTE-Aircraft in Alaska may only operate on GNSS Qroutes with GPS (TSOC129 (as revised) or TSOC196 (as revised)) equipment while the aircraft remains in Air Traffic Control (ATC) radar surveillance or with GPS/WAAS which does not require ATC radar surveillance.

(2) T-routes are available for use by GPS or GPS/WAAS equipped aircraft from 1,200 feet above the surface (or in some instances higher) up to but not including 18,000 feet MSL. T-routes are depicted on Enroute Low Altitude Charts.

NOTE-Aircraft in Alaska may only operate on GNSS Troutes with GPS/WAAS (TSOC145 (as revised) or TSOC146 (as revised)) equipment.

(b) Unpublished RNAV routes are direct routes, based on area navigation capability, between waypoints defined in terms of latitude/longitude coordinates, degree-distance fixes, or offsets from established routes/airways at a specified distance and direction. Radar monitoring by ATC is required on all unpublished RNAV routes.

(c) Magnetic Reference Bearing (MRB) is the published bearing between two waypoints on an RNAV/GPS/GNSS route. The MRB is calculated by applying magnetic variation at the waypoint to the calculated true course between two waypoints. The MRB enhances situational awareness by indicating a reference bearing (no-wind heading) that a pilot should see on the compass/HSI/RMI, etc., when turning prior to/over a waypoint en route to another waypoint. Pilots should use this bearing as a reference only, because their RNAV/GPS/GNSS navigation system will fly the true course between the waypoints.

b. Operation above FL 450 may be conducted on a point-to-point basis. Navigational guidance is provided on an area basis utilizing those facilities depicted on the enroute high altitude charts.

c. Radar Vectors. Controllers may vector aircraft within controlled airspace for separation purposes, noise abatement considerations, when an operational advantage will be realized by the pilot or the controller, or when requested by the pilot. Vectors outside of controlled airspace will be provided only on pilot request. Pilots will be advised as to what the vector is to achieve when the vector is controller initiated and will take the aircraft off a previously assigned nonradar route. To the extent possible, aircraft operating on RNAV routes will be allowed to remain on their own navigation.

d. When flying in Canadian airspace, pilots are cautioned to review Canadian Air Regulations.

1. Special attention should be given to the parts which differ from U.S. CFRs.

(a) The Canadian Airways Class B airspace restriction is an example. Class B airspace is all controlled low level airspace above 12,500 feet MSL or the MEA, whichever is higher, within which only IFR and controlled VFR flights are permitted. (Low level airspace means an airspace designated and defined as such in the Designated Airspace Handbook.)

(b) Unless issued a VFR flight clearance by ATC, regardless of the weather conditions or the height of the terrain, no person may operate an aircraft under VMC within Class B airspace.

(c) The requirement for entry into Class B airspace is a student pilot permit (under the guidance or control of a flight instructor).

(d) VFR flight requires visual contact with the ground or water at all times.

2. Segments of VOR airways and high level routes in Canada are based on L/MF navigation aids and are charted in brown color instead of blue on en route charts.

Figure 5-3-1
Adhering to Airways or Routes

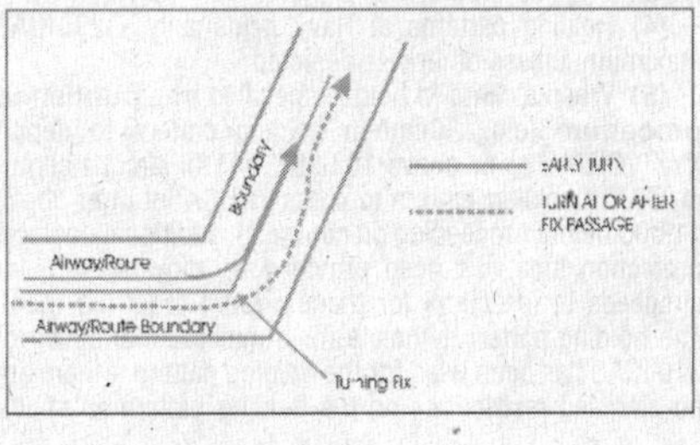

5-3-5. Airway or Route Course Changes

a. Pilots of aircraft are required to adhere to airways or routes being flown. Special attention must be given to this requirement during course changes. Each course change consists of variables that make the technique applicable in each case a matter only the pilot can resolve. Some variables which must be considered are turn radius, wind effect, airspeed, degree of turn, and cockpit instrumentation. An early turn, as illustrated below, is one method of adhering to airways or routes. The use of any available cockpit instrumentation, such as Distance Measuring Equipment, may be used by the pilot to lead the turn when making course changes. This *is consistent* with the intent of 14 CFR Section 91.181, which requires pilots to operate along the centerline of an airway and along the direct course between navigational aids or fixes.

b. Turns which begin at or after fix passage may exceed airway or route boundaries.FIG 5-3-1 contains an example flight track depicting this, together with an example of an early turn.

c. Without such actions as leading a turn, aircraft operating in excess of 290 knots true air speed (TAS) can exceed the normal airway or route boundaries depending on the amount of course change required, wind direction and velocity, the character of the turn fix (DME, overhead navigation aid, or intersection), and the pilot's technique in making a course change. For example, a flight operating at 17,000 feet MSL with a TAS of 400 knots, a 25 degree bank, and a course change of more than 40 degrees would exceed the width of the airway or route; i.e., 4 nautical miles each side of centerline. However, in the airspace below 18,000 feet MSL, operations in excess of 290 knots TAS are not prevalent and the provision of additional IFR separation in all course change situations for the occasional aircraft making a turn in excess of 290 knots TAS creates an unacceptable waste of airspace and imposes a penalty upon the preponderance of traffic which operate at low speeds. Consequently, the FAA expects pilots to lead turns and take other actions they consider necessary during course changes to adhere as closely as possible to the airways or route being flown.

5-3-6. Changeover Points (COPs)

a. COPs are prescribed for Federal airways, jet routes, area navigation routes, or other direct routes for which an MEA is designated under 14 CFR Part 95. The COP is a point along the route or airway segment between two adjacent navigation facilities or waypoints where changeover in navigation guidance should occur. At this point, the pilot should change navigation receiver frequency from the station behind the aircraft to the station ahead.

b. The COP is normally located midway between the navigation facilities for straight route segments, or at the intersection of radials or courses forming a dogleg in the case of dogleg route segments. When the COP is NOT located at the midway point, aeronautical charts will depict the COP location and give the mileage to the radio aids.

c. COPs are established for the purpose of preventing loss of navigation guidance, to prevent frequency interference from other facilities, and to prevent use of different facilities by different aircraft in the same airspace. Pilots are urged to observe COPs to the fullest extent.

5-3-7. Minimum Turning Altitude (MTA)

Due to increased airspeeds at 10,000 ft MSL or above, the published minimum enroute altitude (MEA) may not be sufficient for obstacle clearance when a turn is required over a fix, NAVAID, or waypoint. In these instances, an expanded area in the vicinity of the turn point is examined to determine whether the published MEA is sufficient for obstacle clearance. In some locations (normally mountainous), terrain/obstacles in the expanded search area may necessitate a higher minimum altitude while conducting the turning maneuver. Turning fixes requiring a higher minimum turning altitude (MTA) will be denoted on government charts by the minimum crossing altitude (MCA) icon ("x" flag) and an accompanying note describing the MTA restriction. An MTA restriction will normally consist of the air traffic service (ATS) route leading to the turn point, the ATS route leading from the turn point, and the required altitude; e.g., MTA V330 E TO V520 W 16000. When an MTA is applicable for the intended route of flight, pilots must ensure they are at or above the charted MTA not later than the turn point and maintain at or above the MTA until joining the centerline of the ATS route following the turn point. Once established on the centerline following the turning fix, the MEA/MOCA determines the minimum altitude available for assignment. An MTA may also preclude the use of a specific altitude or a range of altitudes during a turn. For example, the MTA may restrict the use of 10,000 through 11,000 ft MSL. In this case, any altitude greater than 11,000 ft MSL is unrestricted, as are altitudes less than 10,000 ft MSL provided MEA/MOCA requirements are satisfied.

5-3-8. Holding

a. Whenever an aircraft is cleared to a fix other than the destination airport and delay is expected, it is the responsibility of the ATC controller to issue complete holding instructions (unless the pattern is charted), an EFC time and best estimate of any additional en route/terminal delay.

NOTE-*Only those holding patterns depicted on U.S. government or commercially produced (meeting FAA requirements) low/high altitude enroute, and area or STAR charts should be used.*

b. If the holding pattern is charted and the controller doesn't issue complete holding instructions, the pilot is expected to hold as depicted on the appropriate chart. When the pattern is charted, the controller may omit all holding instructions except the charted holding direction and the statement *AS PUBLISHED; e.g., HOLD EAST AS PUBLISHED.* Controllers must always issue complete holding instructions when pilots request them.

c. If no holding pattern is charted and holding instructions have not been issued, the pilot should ask ATC for holding instructions prior to reaching the fix. This procedure will eliminate the possibility of an aircraft entering a holding pattern other than that desired by ATC. If unable to obtain holding instructions prior to reaching the fix (due to frequency congestion, stuck microphone, etc.), then enter a standard pattern on the course on which the aircraft approached the fix and request further clearance as soon as possible. In this event, the altitude/flight level of the aircraft at the clearance limit will be protected so that separation will be provided as required.

d. When an aircraft is 3 minutes or less from a clearance limit and a clearance beyond the fix has not been received, the pilot is expected to start a speed reduction so that the aircraft will cross the fix, initially, at or below the maximum holding airspeed.

e. When no delay is expected, the controller should issue a clearance beyond the fix as soon as possible and, whenever possible, at least 5 minutes before the aircraft reaches the clearance limit.

f. Pilots should report to ATC the time and altitude/flight level at which the aircraft reaches the clearance limit and report leaving the clearance limit.

NOTE-*In the event of two-way communications failure, pilots are required to comply with 14 CFR Section 91.185.*

g. When holding at a VOR station, pilots should begin the turn to the outbound leg at the time of the first complete reversal of the to/from indicator.

h. Patterns at the most generally used holding fixes are depicted (charted) on U.S. Government or commercially produced (meeting FAA requirements) Low or High Altitude Enroute, Area and STAR Charts. Pilots are expected to hold in the pattern depicted unless specifically advised otherwise by ATC.

NOTE-*Holding patterns that protect for a maximum holding airspeed other than the standard may be depicted by an icon, unless otherwise depicted. The icon is a standard holding pattern symbol (racetrack) with the airspeed restriction shown in the center. In other cases, the airspeed restriction will be depicted next to the standard holding pattern symbol.*

REFERENCE-*AIM, Holding, Paragraph 5-3-8j2.*

i. An ATC clearance requiring an aircraft to hold at a fix where the pattern is not charted will include the following information: (See FIG 5-3-2.)

1. Direction of holding from the fix in terms of the eight cardinal compass points (i.e., N, NE, E, SE, etc.).

2. Holding fix (the fix may be omitted if included at the beginning of the transmission as the clearance limit).

3. Radial, course, bearing, airway or route on which the aircraft is to hold.

4. Leg length in miles if DME or RNAV is to be used (leg length will be specified in minutes on pilot request or if the controller considers it necessary).

5. Direction of turn if left turns are to be made, the pilot requests, or the controller considers it necessary.

6. Time to expect further clearance and any pertinent additional delay information.

FIGURE 5-3-2
Holding Patterns

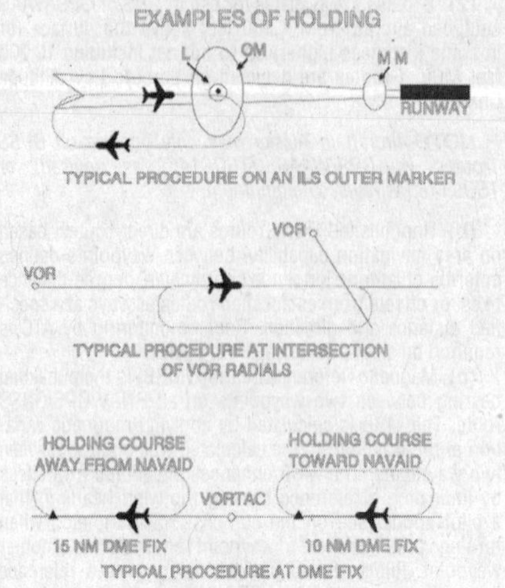

FIGURE 5-3-3
Holding Pattern Descriptive Terms

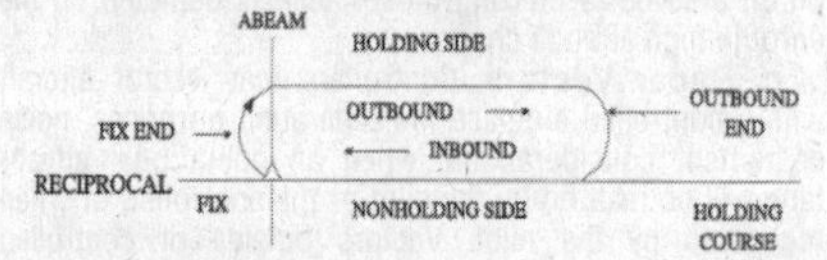

j. Holding pattern airspace protection is based on the following procedures.

1. Descriptive Terms.

(a) Standard Pattern. Right turns (See FIG 5-3-3.)

(b) Nonstandard Pattern. Left turns

2. Airspeeds.

(a) All aircraft may hold at the following altitudes and maximum holding ***airspeeds***:

TABLE 5-3-1

Altitude (MSL)	Airspeed (KIAS)
MHA - 6,000'	200
6,001' - 14,000'	230
14,001' and above	265

(b) The following are exceptions to the maximum holding airspeeds:

(1) Holding patterns from 6,001' to 14,000' may be restricted to a maximum airspeed of 210 KIAS. This nonstandard pattern will be depicted by an icon.

(2) Holding patterns may be restricted to a maximum speed. The speed restriction is depicted in parenthesis inside the holding pattern on the chart: e.g., (175). The aircraft should be at or below the maximum speed prior to initially crossing the holding fix to avoid exiting the protected airspace. Pilots unable to comply with the maximum airspeed restriction should notify ATC.

(3) Holding patterns at USAF airfields only - 310 KIAS maximum, unless otherwise depicted.

(4) Holding patterns at Navy fields only - 230 KIAS maximum, unless otherwise depicted.

(5) When a climb-in hold is specified by a **published procedure** (e.g., "Climb-in holding pattern to depart XYZ VORTAC at or above 10,000." or "All aircraft climb-in TRUCK holding pattern to cross TRUCK Int at or above 11,500 before proceeding on course."), additional obstacle protection area has been provided to allow for greater airspeeds in the climb for those aircraft requiring them. The holding pattern template for a maximum airspeed of 310 KIAS has been used for the holding pattern if there are no airspeed restrictions on the holding pattern as specified in subparagraph j2(b)(2) of this paragraph. Where the holding pattern is restricted to a maximum airspeed of 175 KIAS, the 200 KIAS holding pattern template has been applied for published climb-in hold procedures for altitudes 6,000 feet and below and the 230 KIAS holding pattern template has been applied for altitudes above 6,000 feet. The airspeed limitations in 14 CFR Section 91.117, Aircraft Speed, still apply.

(c) The following phraseology may be used by an ATCS to advise a pilot of the maximum holding airspeed for a holding pattern airspace area.

PHRASEOLOGY-
(AIRCRAFT IDENTIFICATION) (holding instructions, when needed) MAXIMUM HOLDING AIRSPEED IS (speed in knots).

FIGURE 5-3-4
Holding Pattern Entry Procedures

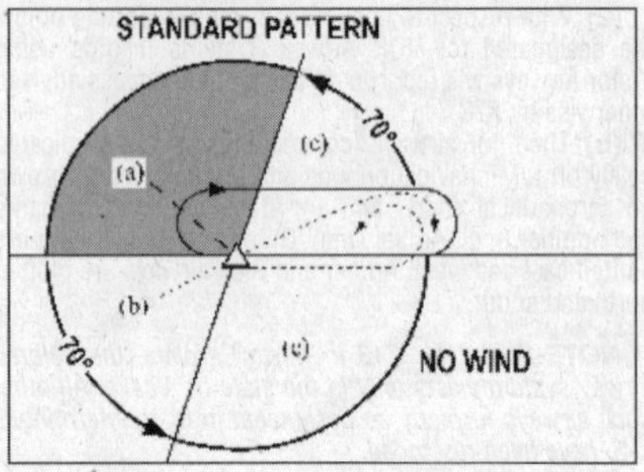

3. Entry Procedures. (See FIG 5-3-4.)

(a) Parallel Procedure. When approaching the holding fix from anywhere in sector (a), the parallel entry procedure would be to turn to a heading to parallel the holding course outbound on the nonholding side for one minute, turn in the direction of the holding pattern through more than 180 degrees, and return to the holding fix or intercept the holding course inbound.

(b) Teardrop Procedure. When approaching the holding fix from anywhere in sector (b), the teardrop entry procedure would be to fly to the fix, turn outbound to a heading for a 30 degree teardrop entry within the pattern (on the holding side) for a period of one minute, then turn in the direction of the holding pattern to intercept the inbound holding course.

(c) Direct Entry Procedure. When approaching the holding fix from anywhere in sector (c), the direct entry procedure would be to fly directly to the fix and turn to follow the holding pattern.

(d) While other entry procedures may enable the aircraft to enter the holding pattern and remain within protected airspace, the parallel, teardrop and direct entries are the procedures for entry and holding recommended by the FAA.

4. Timing.

(a) Inbound Leg.

(1) At or below 14,000 feet MSL: 1 minute.

(2) Above 14,000 feet MSL: $1^1/_2$ minutes.

NOTE-*The initial outbound leg should be flown for 1 minute or $1^1/_2$ minutes (appropriate to altitude). Timing for subsequent outbound legs should be adjusted, as necessary, to achieve proper inbound leg time. Pilots may use any navigational means available; i.e., DME, RNAV, etc., to ensure the appropriate inbound leg times.*

(b) Outbound leg timing begins *over/abeam* the fix, whichever occurs later. If the abeam position cannot be determined, start timing when turn to outbound is completed.

5. Distance Measuring Equipment (DME)/ GPS Along-Track Distance (ATD). DME/GPS holding is subject to the same entry and holding procedures except that distances (nautical miles) are used in lieu of time values. The outbound course of the DME/GPS holding pattern is called the outbound leg of the pattern. The controller or the instrument approach procedure chart will specify the length of the outbound leg. The end of the outbound leg is determined by the DME or ATD readout. The holding fix on conventional procedures, or controller defined holding based on a conventional navigation aid with DME, is a specified course or radial and distances are from the DME station for both the inbound and outbound ends of the holding pattern. When flying published GPS overlay or stand alone procedures with distance specified, the holding fix will be a waypoint in the database and the end of the outbound leg will be determined by the ATD. Some GPS overlay and early stand alone procedures may have timing specified. (See FIG 5-3-5, FIG 5-3-6 and FIG 5-3-7.) See paragraph 1-1-19, Global Positioning System (GPS), for requirements and restriction on using GPS for IFR operations.

FIGURE 5-3-5
Inbound Toward NAVAID

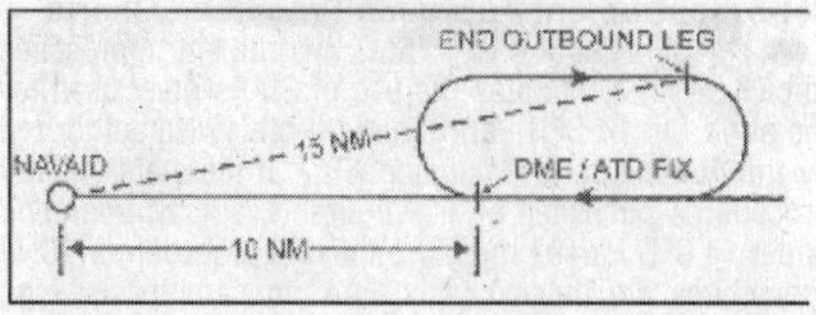

NOTE-When the inbound course is toward the NAVAID, the fix distance is 10 NM, and the leg length is 5 NM, then the end of the outbound leg will be reached when the DME/ ATD reads 15 NM.

FIGURE 5-3-6
Inbound Leg Away from NAVAID

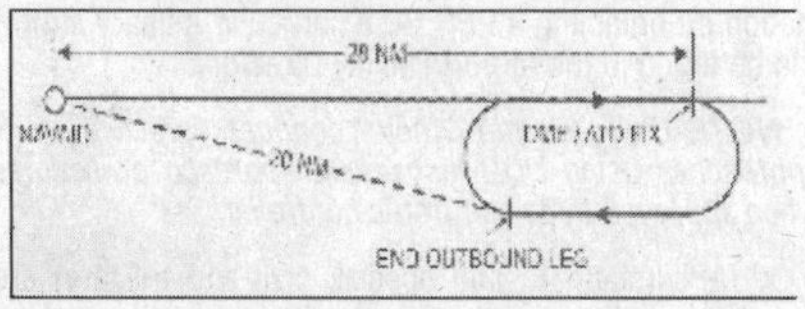

NOTE-When the inbound course is away from the NAVAID and the fix distance is 28 NM, and the leg length is 8 NM, then the end of the outbound leg will be reached when the DME/ATD reads 20 NM.

FIG 5-3-7
GPS/RNAV Holding

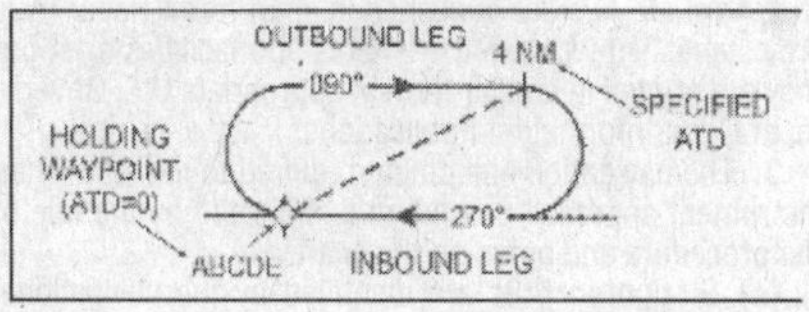

NOTE-The inbound course is always toward the waypoint and the ATD is zero at the waypoint. The end of the outbound leg of the holding pattern is reached when the ATD reads the specified distance.

6. Pilot Action.

(a) Start speed reduction when 3 minutes or less from the holding fix. Cross the holding fix, initially, at or below the maximum holding airspeed.

(b) Make all turns during entry and while holding at:

(1) 3 degrees per second; or

(2) 30 degree bank angle; or

(3) 25 degree bank provided a flight director system is used.

NOTE-Use whichever requires the least bank angle.

(c) Compensate for wind effect primarily by drift correction on the inbound and outbound legs. When outbound, triple the inbound drift correction to avoid major turning adjustments; e.g., if correcting left by 8 degrees when inbound, correct right by 24 degrees when outbound.

(d) Determine entry turn from aircraft heading upon arrival at the holding fix; +/-5 degrees in heading is considered to be within allowable good operating limits for determining entry.

(e) Advise ATC immediately what increased airspeed is necessary, if any, due to turbulence, icing, etc., or if unable to accomplish any part of the holding procedures. When such higher speeds become no longer necessary, operate according to the appropriate published holding speed and notify ATC.

7. Nonstandard Holding Pattern. Fix end and outbound end turns are made to the left. Entry procedures to a nonstandard pattern are oriented in relation to the 70 degree line on the holding side just as in the standard pattern.

k. When holding at a fix and instructions are received specifying the time of departure from the fix, the pilot should adjust the aircraft's flight path within the limits of the established holding pattern in order to leave the fix at the exact time specified. After departing the holding fix, normal speed is to be resumed with respect to other governing speed requirements, such as terminal area speed limits, specific ATC requests, etc. Where the fix is associated with an instrument approach and timed approaches are in effect, a procedure turn must not be executed unless the pilot advises ATC, since aircraft holding are expected to proceed inbound on final approach directly from the holding pattern when approach clearance is received.

l. Radar surveillance of outer fix holding pattern airspace areas.

1. Whenever aircraft are holding at an outer fix, ATC will usually provide radar surveillance of the outer fix holding pattern airspace area, or any portion of it, if it is shown on the controller's radar scope.

2. The controller will attempt to detect any holding aircraft that stray outside the holding pattern airspace area and will assist any detected aircraft to return to the assigned airspace area.

NOTE-Many factors could prevent ATC from providing this additional service, such as workload, number of targets, precipitation, ground clutter, and radar system capability. These circumstances may make it unfeasible to maintain radar identification of aircraft to detect aircraft straying from the holding pattern. The provision of this service depends entirely upon whether controllers believe they are in a position to provide it and does not relieve a pilot of their responsibility to adhere to an accepted ATC clearance.

3. If an aircraft is established in a published holding pattern at an assigned altitude above the published minimum holding altitude and subsequently cleared for the approach, the pilot may descend to the published minimum holding altitude. The holding pattern would only be a segment of the IAP *if* it is published on the instrument procedure chart and is used in lieu of a procedure turn.

m. For those holding patterns where there are no published minimum holding altitudes, the pilot, upon receiving an approach clearance, must maintain the last assigned altitude until leaving the holding pattern and established on the inbound course. Thereafter, the published minimum altitude of the route segment being flown will apply. It is expected that the pilot will be assigned a holding altitude that will permit a normal descent on the inbound course.

Section 4. Arrival Procedures

5-4-1. Standard Terminal Arrival (STAR), Area Navigation (RNAV) STAR, and Flight Management System Procedures (FMSP) for Arrivals

a. A STAR is an ATC coded IFR arrival route established for application to arriving IFR aircraft destined for certain airports. RNAV STAR/FMSP procedures for arrivals serve the same purpose but are only used by aircraft equipped with FMS or GPS. The purpose of both is to simplify clearance delivery procedures and facilitate transition between en route and instrument approach procedures.

1. STAR/RNAV STAR/FMSP procedures may have mandatory speeds and/or crossing altitudes published. Other STARs may have planning information depicted to inform pilots what clearances or restrictions to "expect." "Expect" altitudes/speeds are not considered STAR/ RNAV STAR/FMSP procedures crossing restrictions unless verbally issued by ATC. Published speed restrictions are independent of altitude restrictions and are mandatory unless modified by ATC. Pilots should plan to cross waypoints with a published speed restriction, at the published speed, and should not exceed this speed past the associated waypoint unless authorized by ATC or a published note to do so.

NOTE-The ***"expect"*** *altitudes/speeds are published so that pilots may have the information for planning purposes. These altitudes/speeds must not be used in the event of lost communications unless ATC has specifically advised the pilot to expect these altitudes/speeds as part of a further clearance.*

REFERENCE- *14 CFR Section 91.185(c)(2)(iii).*

2. Pilots navigating on STAR/RNAV STAR/FMSP procedures must maintain last assigned altitude until receiving authorization to descend so as to comply with all published/issued restrictions. This authorization will contain the phraseology "DESCEND VIA."

(a) Clearance to "descend via" authorizes pilots to:

(1) Vertically and laterally navigate on a STAR/RNAV STAR/FMSP.

(2) When cleared to a waypoint depicted on a STAR/ RNAV STAR/FMSP, to descend from a previously assigned altitude at pilot's discretion to the altitude depicted for that waypoint, and once established on the depicted arrival, to navigate laterally and vertically to meet all published restrictions.

NOTE-1. *Air traffic is responsible for obstacle clearance when issuing a "descend via" instruction to the pilot. The descend via is used in conjunction with STARs/RNAV STARs/FMSPs to reduce phraseology by not requiring the controller to restate the altitude at the next waypoint/fix to which the pilot has been cleared.*

2. Air traffic will assign an altitude to cross the waypoint/fix, if no altitude is depicted at the waypoint/ fix, for aircraft on a direct routing to a STAR/RNAV STAR/ FMSP.

3. Minimum en route altitudes (MEA) are not considered restrictions; however, pilots must remain above all MEAs, unless receiving an ATC instruction to descend below the MEA.

EXAMPLE-

1. Lateral/routing clearance only.

"Cleared Hadly One arrival."

2. Routing with assigned altitude.

"Cleared Hadly One arrival, descend and maintain Flight Level two four zero."

"Cleared Hadly One arrival, descend at pilot's discretion, maintain Flight Level two four zero."

3. Lateral/routing and vertical navigation clearance.

"Descend via the Civit One arrival."

"Descend via the Civit One arrival, except, cross Arnes at or above one one thousand."

4. Lateral/routing and vertical navigation clearance when assigning altitude not published on procedure.

"Descend via the Haris One arrival, except after Bruno, maintain one zero thousand."

"Descend via the Haris One arrival, except cross Bruno at one three thousand then maintain one zero thousand."

5. Direct routing to intercept a STAR/RNAV STAR/FMSP and vertical navigation clearance.

"Proceed direct Mahem, descend via Mahem One arrival."

"Proceed direct Luxor, cross Luxor at or above flight level two zero zero, then descend via the Ksino One Arrival."

NOTE-1. *In Example 2, pilots are expected to descend to FL 240 as directed, and maintain FL 240 until cleared for further vertical navigation with a newly assigned altitude or a "descend via" clearance.*

2. In Example 4, the aircraft should track laterally and vertically on the Haris One arrival and should descend so as to comply with all speed and altitude restrictions until reaching Bruno and then maintain 10,000. Upon reaching 10,000, aircraft should maintain 10,000 until cleared by ATC to continue to descend.

(b) Pilots cleared for vertical navigation using the phraseology "descend via" must inform ATC upon initial contact with a new frequency.

EXAMPLE-

"Delta One Twenty One leaving FL 240, descending via the Civit One arrival."

b. Pilots of IFR aircraft destined to locations for which STARs have been published may be issued a clearance containing a STAR whenever ATC deems it appropriate.

c. Use of STARs requires pilot possession of at least the approved chart. RNAV STARs must be retrievable by the procedure name from the aircraft database and conform to charted procedure. As with any ATC clearance or portion thereof, it is the responsibility of each pilot to accept or refuse an issued STAR. Pilots should notify ATC if they do not wish to use a STAR by placing "NO STAR" in the remarks section of the flight plan or by the less desirable method of verbally stating the same to ATC.

d. STAR charts are published in the Terminal Procedures Publications (TPP) and are available on subscription from the National Aeronautical Charting Office.

e. RNAV STAR.

1. All public RNAV STARs are RNAV1. These procedures require system performance currently met by GPS or DME/DME/IRU RNAV systems that satisfy the criteria discussed in AC 90-100A, U.S. Terminal and En Route Area Navigation (RNAV) Operations. RNAV1 procedures require the aircraft's total system error remain bounded by +1 NM for 95% of the total flight time.

2. For procedures requiring GPS, if the navigation system does not automatically alert the flight crew of a loss of GPS, the operator must develop procedures to verify correct GPS operation.

5-4-2. Local Flow Traffic Management Program

a. This program is a continuing effort by the FAA to enhance safety, minimize the impact of aircraft noise and conserve aviation fuel. The enhancement of safety and reduction of noise is achieved in this program by minimizing low altitude maneuvering of arriving turbojet and turboprop aircraft weighing more than 12,500 pounds and, by permitting departure aircraft to climb to higher altitudes sooner, as arrivals are operating at higher altitudes at the points where their flight paths cross. The application of these procedures also reduces exposure time between controlled aircraft and uncontrolled aircraft at the lower altitudes in and around the terminal environment. Fuel conservation is accomplished by absorbing any necessary arrival delays for aircraft included in this program operating at the higher and more fuel efficient altitudes.

b. A fuel efficient descent is basically an uninterrupted descent (except where level flight is required for speed adjustment) from cruising altitude to the point when level flight is necessary for the pilot to stabilize the aircraft on final approach. The procedure for a fuel efficient descent is based on an altitude loss which is most efficient for the majority of aircraft being served. This will generally result in a descent gradient window of 250-350 feet per nautical mile.

c. When crossing altitudes and speed restrictions are issued verbally or are depicted on a chart, ATC will expect the pilot to descend first to the crossing altitude and then reduce speed. Verbal clearances for descent will normally permit an uninterrupted descent in accordance with the procedure as described in paragraph b above. Acceptance of a charted fuel efficient descent (Runway Profile Descent) clearance requires the pilot to adhere to the altitudes, speeds, and headings depicted on the charts unless otherwise instructed by ATC. PILOTS RECEIVING A CLEARANCE FOR A FUEL EFFICIENT DESCENT ARE EXPECTED TO ADVISE ATC IF THEY DO NOT HAVE RUNWAY PROFILE DESCENT CHARTS PUBLISHED FOR THAT AIRPORT OR ARE UNABLE TO COMPLY WITH THE CLEARANCE.

5-4-3. Approach Control

a. Approach control is responsible for controlling all instrument flight operating within its area of responsibility. Approach control may serve one or more airfields, and control is exercised primarily by direct pilot and controller communications. Prior to arriving at the destination radio facility, instructions will be received from ARTCC to contact approach control on a specified frequency.

b. Radar Approach Control.

1. Where radar is approved for approach control service, it is used not only for radar approaches (Airport Surveillance Radar [ASR] and Precision Approach Radar [PAR]) but is also used to provide vectors in conjunction with published nonradar approaches based on radio NAVAIDs (ILS, MLS, VOR, NDB, TACAN). Radar vectors can provide course guidance and expedite traffic to the final approach course of any established IAP or to the traffic pattern for a visual approach. Approach control facilities that provide this radar service will operate in the following manner:

(a) Arriving aircraft are either cleared to an outer fix most appropriate to the route being flown with vertical separation and, if required, given holding information or, when radar handoffs are effected between the ARTCC and approach control, or between two approach control facilities, aircraft are cleared to the airport or to a fix so located that the handoff will be completed prior to the time the aircraft reaches the fix. When radar handoffs are utilized, successive arriving flights may be handed off to approach control with radar separation in lieu of vertical separation.

(b) After release to approach control, aircraft are vectored to the final approach course (ILS, MLS, VOR, ADF, etc.). Radar vectors and altitude or flight levels will be issued as required for spacing and separating aircraft. *Therefore, pilots must not deviate from the headings issued by approach control.* Aircraft will normally be informed when it is necessary to vector across the final approach course for spacing or other reasons. If approach course crossing is imminent and the pilot has not been informed that the aircraft will be vectored across the final approach course, the pilot should query the controller.

(c) The pilot is not expected to turn inbound on the final approach course unless an approach clearance has been issued. This clearance will normally be issued with the final vector for interception of the final approach course, and the vector will be such as to enable the pilot to establish the aircraft on the final approach course prior to reaching the final approach fix.

(d) In the case of aircraft already inbound on the final approach course, approach clearance will be issued prior to the aircraft reaching the final approach fix. When established inbound on the final approach course, radar separation will be maintained and the pilot will be expected to complete the approach utilizing the approach aid designated in the clearance (ILS, MLS, VOR, radio beacons, etc.) as the primary means of navigation. Therefore, once established on the final approach course, pilots must not deviate from it unless a clearance to do so is received from ATC.

(e) After passing the final approach fix on final approach, aircraft are expected to continue inbound on the final approach course and complete the approach or effect the missed approach procedure published for that airport.

2. ARTCCs are approved for and may provide approach control services to specific airports. The radar systems used by these centers do not provide the same precision as an ASR/PAR used by approach control facilities and towers, and the update rate is not as fast. Therefore, pilots may be requested to report established on the final approach course.

3. Whether aircraft are vectored to the appropriate final approach course or provide their own navigation on published routes to it, radar service is automatically terminated when the landing is completed or when instructed to change to advisory frequency at uncontrolled airports, whichever occurs first.

5-4-4. Advance Information on Instrument Approach

a. When landing at airports with approach control services and where two or more IAPs are published, pilots will be provided in advance of their arrival with the type of approach to expect or that they may be vectored for a visual approach. This information will be broadcast either by a controller or on ATIS. It will not be furnished when the visibility is three miles or better and the ceiling is at or above the highest initial approach altitude established for any low altitude IAP for the airport.

b. The purpose of this information is to aid the pilot in planning arrival actions; however, it is not an ATC clearance or commitment and is subject to change. Pilots should bear in mind that fluctuating weather, shifting winds, blocked runway, etc., are conditions which may result in changes to approach information previously received. It is important that pilots advise ATC immediately they are unable to execute the approach ATC advised will be used, or if they prefer another type of approach.

c. Aircraft destined to uncontrolled airports, which have automated weather data with broadcast capability, should monitor the ASOS/AWSS/AWOS frequency to ascertain the current weather for the airport. The pilot must advise ATC when he/she has received the broadcast weather and state his/her intentions.

***NOTE-*1.** *ASOS/AWSS/AWOS should be set to provide one-minute broadcast weather updates at uncontrolled airports that are without weather broadcast capability by a human observer.*

2. *Controllers will consider the long line disseminated weather from an automated weather system at an uncontrolled airport as trend and planning information only and will rely on the pilot for current weather information for the airport. If the pilot is unable to receive the current broadcast weather, the last long line disseminated weather will be issued to the pilot. When receiving IFR services, the pilot/aircraft operator is responsible for determining if weather/visibility is adequate for approach/landing.*

d. When making an IFR approach to an airport not served by a tower or FSS, after ATC advises "CHANGE TO ADVISORY FREQUENCY APPROVED" you should broadcast your intentions, including the type of approach being executed, your position, and when over the final approach fix inbound (nonprecision approach) or when over the outer marker or fix used in lieu of the outer marker inbound (precision approach). Continue to monitor the appropriate frequency (UNICOM, etc.) for reports from other pilots.

5-4-5. Instrument Approach Procedure Charts

a. 14 CFR Section 91.175(a), Instrument approaches to civil airports, requires the use of SIAPs prescribed for the airport in 14 CFR Part 97 unless otherwise authorized by the Administrator (including ATC). If there are military procedures published at a civil airport, aircraft operating under 14 CFR Part 91 must use the civil procedure(s). Civil procedures are defined with "FAA" in parenthesis; e.g., (FAA), at the top, center of the procedure chart. DOD procedures are defined using the abbreviation of the applicable military service in parenthesis; e.g., (USAF), (USN), (USA). 14 CFR Section 91.175(g), Military airports, requires civil pilots flying into or out of military airports to comply with the IAPs and takeoff and landing minimums prescribed by the authority having jurisdiction at those airports. Unless an emergency exists, civil aircraft operating at military airports normally require advance authorization, commonly referred to as "Prior Permission Required" or "PPR." Information on obtaining a PPR for a particular military airport can be found in the Airport/Facility Directory.

NOTE-*Civil aircraft may conduct practice VFR approaches using DOD instrument approach procedures when approved by the air traffic controller.*

1. IAPs (standard and special, civil and military) are based on joint civil and military criteria contained in the U.S. Standard for TERPS. The design of IAPs based on criteria contained in TERPS, takes into account the interrelationship between airports, facilities, and the surrounding environment, terrain, obstacles, noise sensitivity, etc. Appropriate altitudes, courses, headings, distances, and other limitations are specified and, once approved, the procedures are published and distributed by government and commercial cartographers as instrument approach charts.

2. Not all IAPs are published in chart form. Radar IAPs are established where requirements and facilities exist but they are printed in tabular form in appropriate U.S. Government Flight Information Publications.

3. The navigation equipment required to join and fly an instrument approach procedure is indicated by the title of the procedure and notes on the chart.

(a) Straight-in IAPs are identified by the navigational system providing the final approach guidance and the runway to which the approach is aligned (e.g., VOR RWY 13). Circling only approaches are identified by the navigational system providing final approach guidance and a letter (e.g., VOR A). More than one navigational system separated by a slash indicates that more than one type of equipment must be used to execute the **final approach** (e.g., VOR/DME RWY 31). More than one navigational system separated by the word "or" indicates either type of equipment may be used to execute the **final approach** (e.g., VOR or GPS RWY 15).

(b) In some cases, other types of navigation systems including radar may be required to execute other portions of the approach or to navigate to the IAF (e.g., an NDB procedure turn to an ILS, an NDB in the missed approach, or radar required to join the procedure or identify a fix). When radar or other equipment is required for procedure entry from the en route environment, a note will be charted in the **planview** of the approach procedure chart (e.g., RADAR REQUIRED or ADF REQUIRED). When radar or other equipment is required on portions of the procedure outside the final approach segment, including the missed approach, a note will be charted in the **notes box** of the pilot briefing portion of the approach chart (e.g., RADAR REQUIRED or DME REQUIRED). Notes are not charted when VOR is required outside the final approach segment. Pilots should ensure that the aircraft is equipped with the required NAVAID(s) in order to execute the approach, including the missed approach.

NOTE-Some military (i.e., U.S. Air Force and U.S. Navy) IAPs have these "additional equipment required" notes charted only in the planview of the approach procedure and do not conform to the same application standards used by the FAA.

(c) The FAA has initiated a program to provide a new notation for LOC approaches when charted on an ILS approach requiring other navigational aids to fly the final approach course. The LOC minimums will be annotated with the NAVAID required (e.g., "DME Required" or "RADAR Required"). During the transition period, ILS approaches will still exist without the annotation.

(d) Many ILS approaches having minima based on RVR are eligible for a landing minimum of RVR 1800. Some of these approaches are to runways that have touchdown zone and centerline lights. For many runways that do not have touchdown and centerline lights, it is still possible to allow a landing minimum of RVR 1800. For these runways, the normal ILS minimum of RVR 2400 can be annotated with a single or double asterisk or the dagger symbol "†"; for example "** 696/24 200 (200/1/2)." A note is included on the chart stating "**RVR 1800 authorized with use of FD or AP or HUD to DA." The pilot must use the flight director, or autopilot with an approved approach coupler, or head up display to decision altitude or to the initiation of a missed approach. In the interest of safety, single pilot operators should not fly approaches to 1800 RVR minimums on runways without touchdown and centerline lights using only a flight director, unless accompanied by the use of an autopilot with an approach coupler.

(e) The naming of multiple approaches of the same type to the same runway is also changing. Multiple approaches with the same guidance will be annotated with an alphabetical suffix beginning at the end of the alphabet and working backwards for subsequent procedures (e.g., ILS Z RWY 28, ILS Y RWY 28, etc.). The existing annotations such as ILS 2 RWY 28 or Silver ILS RWY 28 will be phased out and replaced with the new designation. The Cat II and Cat III designations are used to differentiate between multiple ILSs to the same runway unless there are multiples of the same type.

(f) RNAV (GPS) approaches to LNAV, LP, LNAV/VNAV and LPV lines of minima using WAAS and RNAV (GPS) approaches to LNAV and LNAV/VNAV lines of minima using GPS are charted as RNAV (GPS) RWY (Number) (e.g., RNAV (GPS) RWY 21). VOR/DME RNAV approaches will continue to be identified as VOR/DME RNAV RWY (Number) (e.g., VOR/DME RNAV RWY 21). VOR/DME RNAV procedures which can be flown by GPS will be annotated with "or GPS" (e.g., VOR/DME RNAV or GPS RWY 31).

4. Approach minimums are based on the local altimeter setting for that airport, unless annotated otherwise; e.g., Oklahoma City/Will Rogers World approaches are based on having a Will Rogers World altimeter setting. When a different altimeter source is required, or more than one source is authorized, it will be annotated on the approach chart; e.g., use Sidney altimeter setting, if not received, use Scottsbluff altimeter setting. Approach minimums may be raised when a nonlocal altimeter source is authorized. When more than one altimeter source is authorized, and the minima are different, they will be shown by separate lines in the approach minima box or a note; e.g., use Manhattan altimeter setting; when not available use Salina altimeter setting and increase all MDAs 40 feet. When the altimeter must be obtained from a source other than air traffic a note will indicate the source; e.g., Obtain local altimeter setting on CTAF. When the altimeter setting(s) on which the approach is based is not available, the approach is not authorized. Baro-VNAV must be flown using the local altimeter setting only. Where no local altimeter is available, the LNAV/VNAV line will still be published for use by WAAS receivers with a note that Baro-VNAV is not authorized. When a local and at least one other altimeter setting source is authorized and the local altimeter is not available Baro-VNAV is not authorized; however, the LNAV/VNAV minima can still be used by WAAS receivers using the alternate altimeter setting source.

NOTE-*Barometric Vertical Navigation (baro-VNAV). An RNAV system function which uses barometric altitude information from the aircraft's altimeter to compute and present a vertical guidance path to the pilot. The specified vertical path is computed as a geometric path, typically computed between two waypoints or an angle based computation from a single waypoint. Further guidance may be found in Advisory Circular 90-105.*

5. A pilot adhering to the altitudes, flight paths, and weather minimums depicted on the IAP chart or vectors and altitudes issued by the radar controller, is assured of terrain and obstruction clearance and runway or airport alignment during approach for landing.

6. IAPs are designed to provide an IFR descent from the en route environment to a point where a safe landing can be made. They are prescribed and approved by appropriate civil or military authority to ensure a safe descent during instrument flight conditions at a specific airport. It is important that pilots understand these procedures and their use prior to attempting to fly instrument approaches.

7. TERPS criteria are provided for the following types of instrument approach procedures:

(a) Precision Approach (PA). An instrument approach based on a navigation system that provides course and glidepath deviation information meeting the precision standards of ICAO Annex 10. For example, PAR, ILS, and GLS are precision approaches.

(b) Approach with Vertical Guidance (APV). An instrument approach based on a navigation system that is not required to meet the precision approach standards of ICAO Annex 10 but provides course and glidepath deviation information. For example, Baro-VNAV, LDA with glidepath, LNAV/VNAV and LPV are APV approaches.

(c) Nonprecision Approach (NPA). An instrument approach based on a navigation system which provides course deviation information, but no glidepath deviation information. For example, VOR, NDB and LNAV. As noted in subparagraph i, Vertical Descent Angle (VDA) on Nonprecision Approaches, some approach procedures may provide a Vertical Descent Angle as an aid in flying a stabilized approach, without requiring its use in order to fly the procedure. This does not make the approach an APV procedure, since it must still be flown to an MDA and has not been evaluated with a glidepath.

b. The method used to depict prescribed altitudes on instrument approach charts differs according to techniques employed by different chart publishers. Prescribed altitudes may be depicted in four different configurations: minimum, maximum, mandatory, and recommended. The U.S. Government distributes charts produced by National Geospatial-Intelligence Agency (NGA) and FAA. Altitudes are depicted on these charts in the profile view with underscore, overscore, both or none to identify them as minimum, maximum, mandatory or recommended.

1. Minimum altitude will be depicted with the altitude value underscored. Aircraft are required to maintain altitude at or above the depicted value, e.g., 3000.

2. Maximum altitude will be depicted with the altitude value overscored. Aircraft are required to maintain altitude at or below the depicted value, e.g., 4000.

3. Mandatory altitude will be depicted with the altitude value both underscored and overscored. Aircraft are required to maintain altitude at the depicted value, e.g., 5000.

4. Recommended altitude will be depicted with no overscore or underscore. These altitudes are depicted for descent planning, e.g., 6000.

NOTE-**1.** *Pilots are cautioned to adhere to altitudes as prescribed because, in certain instances, they may be used as the basis for vertical separation of aircraft by ATC. When a depicted altitude is specified in the ATC clearance, that altitude becomes mandatory as defined above.*

2. *The ILS glide slope is intended to be intercepted at the published glide slope intercept altitude. This point marks the PFAF and is depicted by the "lightning bolt" symbol on U.S. Government charts. Intercepting the glide slope at this altitude marks the beginning of the final approach segment and ensures required obstacle clearance during descent from the glide slope intercept altitude to the lowest published decision altitude for the approach. Interception and tracking of the glide slope prior to the published glide slope interception altitude does not necessarily ensure that minimum, maximum, and/or mandatory altitudes published for any preceding fixes will be complied with during the descent. If the pilot chooses to track the glide slope prior to the glide slope interception altitude, they remain responsible for complying with published altitudes for any preceding stepdown fixes encountered during the subsequent descent.*

c. Minimum Safe/Sector Altitudes (MSA) are published for emergency use on IAP charts. For conventional navigation systems, the MSA is normally based on the primary omnidirectional facility on which the IAP is predicated. The MSA depiction on the approach chart contains the facility identifier of the NAVAID used to determine the MSA altitudes. For RNAV approaches, the MSA is based on the runway waypoint (RWY WP) for straight-in approaches, or the airport waypoint (APT WP) for circling approaches. For GPS approaches, the MSA center will be the missed approach waypoint (MAWP). MSAs are expressed in feet above mean sea level and normally have a 25 NM radius; however, this radius may be expanded to 30 NM if necessary to encompass the airport landing surfaces. Ideally, a single sector altitude is established and depicted on the plan view of approach charts; however, when necessary to obtain relief from obstructions, the area may be further sectored and as many as four MSAs established. When established, sectors may be no less than 90° in spread. MSAs provide 1,000 feet clearance over all obstructions but do not necessarily assure acceptable navigation signal coverage.

d. Terminal Arrival Area (TAA)

1. The objective of the TAA is to provide a seamless transition from the en route structure to the terminal environment for arriving aircraft equipped with Flight Management System (FMS) and/or Global Positioning System (GPS) navigational equipment. The underlying instrument approach procedure is an area navigation (RNAV) procedure described in this section. The TAA provides the pilot and air traffic controller with a very efficient method for routing traffic into the terminal environment with little required air traffic control interface, and with minimum altitudes depicted that provide standard obstacle clearance compatible with the instrument procedure associated with it. The TAA will not be found on all RNAV procedures, particularly in areas of heavy concentration of air traffic. When the TAA is published, it replaces the MSA for that approach procedure. See FIG 5-4-9 for a depiction of a RNAV approach chart with a TAA.

2. The RNAV procedure underlying the TAA will be the "T" design (also called the "Basic T"), or a modification of the "T." The "T" design incorporates from one to three IAFs; an intermediate fix (IF) that serves as a dual purpose IF (IAF); a final approach fix (FAF), and a missed approach point (MAP) usually located at the runway threshold. The three IAFs are normally aligned in a straight line perpendicular to the intermediate course, which is an extension of the final course leading to the runway, forming a "T." The initial segment is normally from 3-6 NM in length; the intermediate 5-7 NM, and the final segment 5 NM. Specific segment length may be varied to accommodate specific aircraft categories for which the procedure is designed. However, the published segment lengths will reflect the highest category of aircraft normally expected to use the procedure.

(a) A standard racetrack holding pattern may be provided at the center IAF, and if present may be necessary for course reversal and for altitude adjustment for entry into the procedure. In the latter case, the pattern provides an extended distance for the descent required by the procedure. Depiction of this pattern in U.S. Government publications will utilize the "hold-in-lieu-of-PT" holding pattern symbol.

(b) The published procedure will be annotated to indicate when the course reversal is not necessary when flying within a particular TAA area; e.g., "NoPT." Otherwise, the pilot is expected to execute the course reversal under the provisions of 14 CFR Section 91.175. The pilot may elect to use the course reversal pattern when it is not required by the procedure, but must inform air traffic control and receive clearance to do so. (See FIG 5-4-1, FIG 5-4-2, FIG 5-4-9, and paragraph 5-4-9, Procedure Turn and Hold-in-lieu of Procedure Turn).

3. The "T" design may be modified by the procedure designers where required by terrain or air traffic control considerations. For instance, the "T" design may appear more like a regularly or irregularly shaped "Y", or may even have one or both outboard IAFs eliminated resulting in an upside down "L" or an "I" configuration. (See FIG 5-4-3and FIG 5-4-10). Further, the leg lengths associated with the outboard IAFs may differ. (See FIG 5-4-5 and FIG 5-4-6).

4. Another modification of the "T" design may be found at airports with parallel runway configurations. Each parallel runway may be served by its own "T" IAF, IF (IAF), and FAF combination, resulting in parallel final approach courses. (See FIG 5-4-4). Common IAFs may serve both runways; however, only the intermediate and final approach segments for the landing runway will be shown on the approach chart. (SeeFIG 5-4-5 and FIG 5-4-6).

5. The standard TAA consists of three areas defined by the extension of the IAF legs and the intermediate segment course. These areas are called the straight-in, left-base, and right-base areas. (See FIG 5-4-7). TAA area lateral boundaries are identified by magnetic courses TO the IF (IAF). The straight-in area can be further divided into pie-shaped sectors with the boundaries identified by magnetic

FIGURE 5-4-1
Basic "T" Design

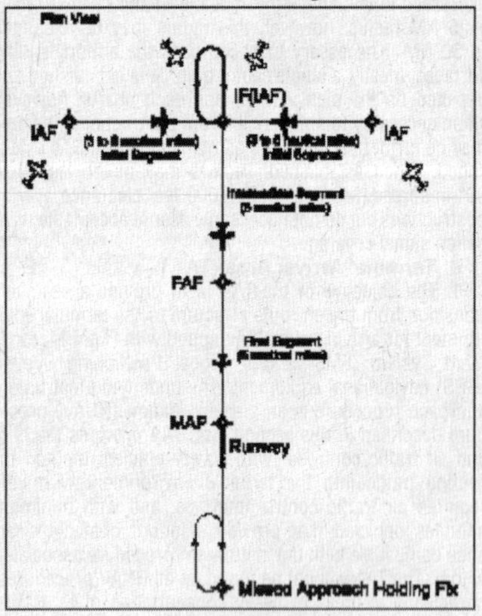

FIGURE 5-4-2
Basic "T" Design

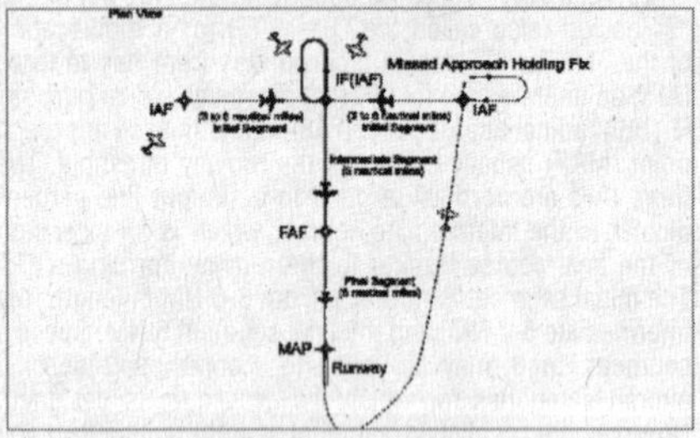

FIGURE 5-4-3
Modified Basic "T"

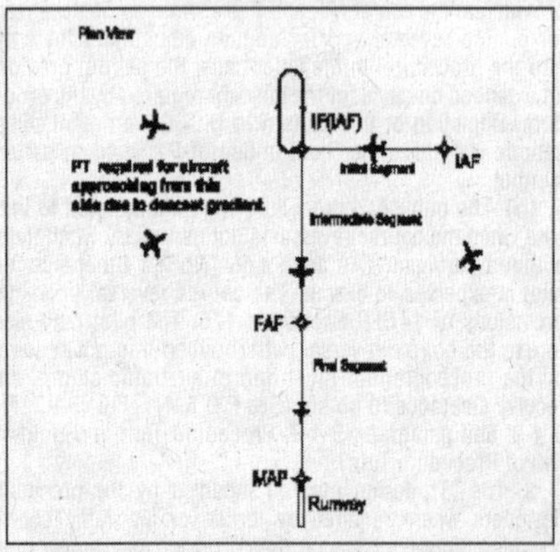

courses TO the IF (IAF), and may contain stepdown sections defined by arcs based on RNAV distances (DME or ATD) from the IF (IAF). The right/left-base areas can only be subdivided using arcs based on RNAV distances from the IAFs for those areas. Minimum MSL altitudes are charted within each of these defined areas/subdivisions that provide at least 1,000 feet of obstacle clearance, or more as necessary in mountainous areas.

(a) Prior to arriving at the TAA boundary, the pilot can determine which area of the TAA the aircraft will enter by selecting the IF (IAF) to determine the magnetic bearing TO the center IF (IAF). That bearing should then be compared with the published bearings that define the lateral boundaries of the TAA areas. Using the end IAFs may give a false indication of which area the aircraft will enter. This is critical when approaching the TAA near the extended boundary between the left and right-base areas, especially where these areas contain different minimum altitude requirements.

FIGURE 5-4-4
Modified "T" Approach to Parallel Runways

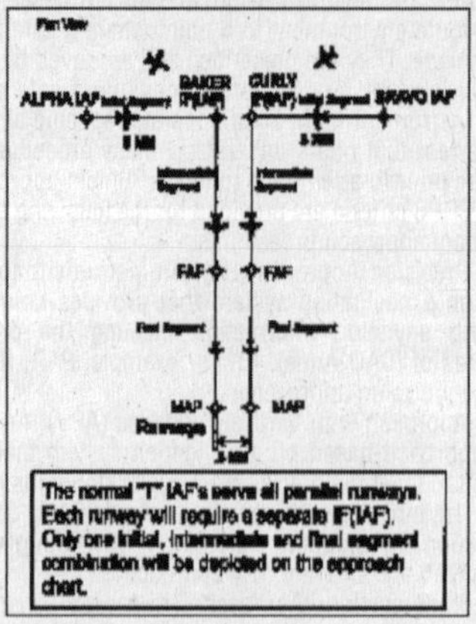

FIGURE 5-4-5
"T" Approach with Common IAFs to Parallel Runways

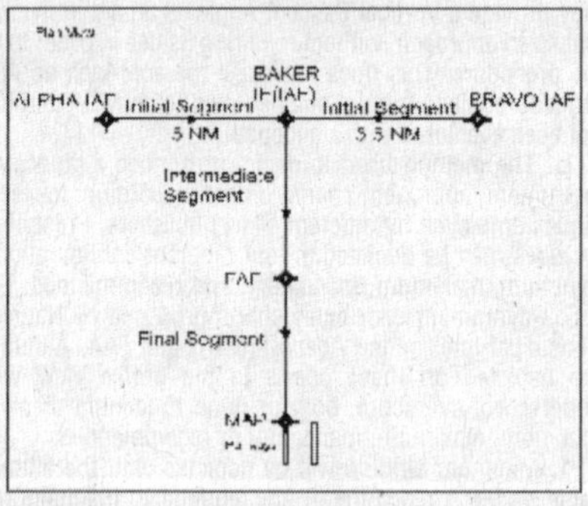

FIGURE 5-4-6
"T" Approach with Common IAFs to Parallel Runways

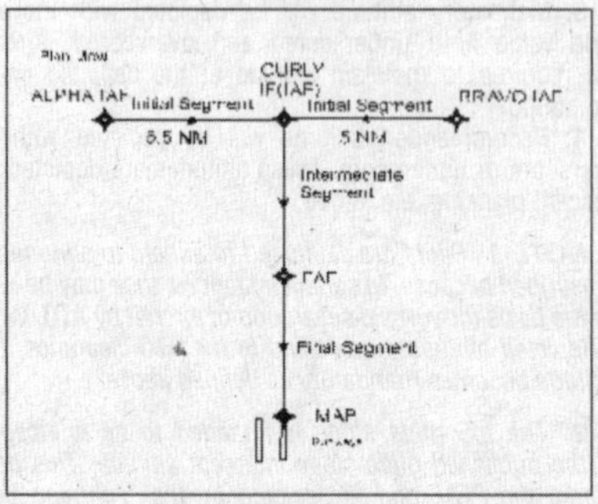

FIGURE 5-4-7
TAA Area

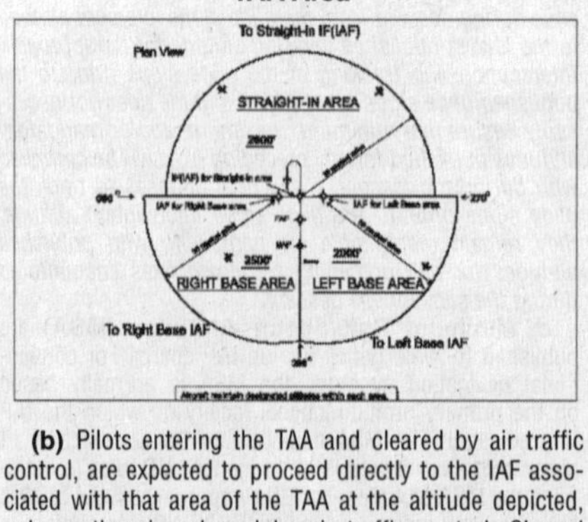

(b) Pilots entering the TAA and cleared by air traffic control, are expected to proceed directly to the IAF associated with that area of the TAA at the altitude depicted, unless otherwise cleared by air traffic control. Cleared direct to an Initial Approach Fix (IAF) without a clearance for the procedure does not authorize a pilot to descend to a lower TAA altitude. If a pilot desires a lower altitude without an approach clearance, request the lower TAA altitude. If a pilot is not sure of what they are authorized or expected to do by air traffic, they should ask air traffic or request a specific clearance. Pilots entering the TAA with two-way radio communications failure (14 CFR Section91.185, IFR Operations: Two-way Radio Communications Failure), must maintain the highest altitude prescribed by Section 91.185(c)(2) until arriving at the appropriate IAF.

FIG 5-4-8
Sectored TAA Areas

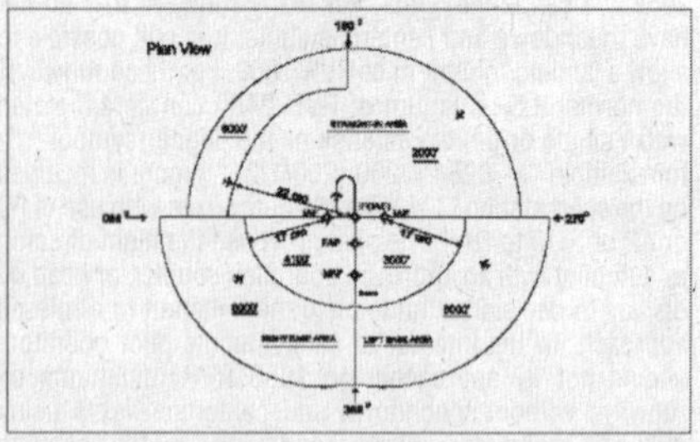

(c) Depiction of the TAA on U.S. Government charts will be through the use of icons located in the plan view outside the depiction of the actual approach procedure. (SeeFIG 5-4-9). Use of icons is necessary to avoid obscuring any portion of the "T procedure (altitudes, courses, minimum altitudes, etc.). The icon for each TAA area will be located and oriented on the plan view with respect to the direction of arrival to the approach procedure, and will show all TAA minimum altitudes and sector/radius subdivisions for that area. The IAF for each area of the TAA is included on the icon where it appears on the approach, to help the pilot orient the icon to the approach procedure. The IAF name and the distance of the TAA area boundary from the IAF are included on the outside arc of the TAA area icon. Examples here are shown with the TAA around the approach to aid pilots in visualizing how the TAA corresponds to the approach and should not be confused with the actual approach chart depiction.

(d) Each waypoint on the "T", except the missed approach waypoint, is assigned a pronounceable 5-character name used in air traffic control communications, and which is found in the RNAV databases for the procedure. The missed approach waypoint is assigned a pronounceable name when it is not located at the runway threshold.

6. Once cleared to fly the TAA, pilots are expected to obey minimum altitudes depicted within the TAA icons, unless instructed otherwise by air traffic control. In FIG 5-4-8, pilots within the left or right-base areas are expected to maintain a minimum altitude of 6,000 feet until within 17 NM of the associated IAF. After crossing the 17 NM arc, descent is authorized to the lower charted altitudes. Pilots approaching from the northwest are expected to maintain a minimum altitude of 6,000 feet, and when within 22 NM of the IF (IAF), descend to a minimum altitude of 2,000 feet MSL until reaching the IF (IAF).

NOTE-*This chart has been modified to depict new concepts and may not reflect actual approach minima.*

7. Just as the underlying "T" approach procedure may be modified in shape, the TAA may contain modifications to the defined area shapes and sizes. Some areas may even be eliminated, with other areas expanded as needed. FIG 5-4-10 is an example of a design limitation where a course reversal is necessary when approaching the IF (IAF) from certain directions due to the amount of turn required at the IF (IAF). Design criteria require a course reversal whenever this turn exceeds 120 degrees. In this generalized example, pilots approaching on a bearing TO the IF (IAF) from 300° clockwise through 060° are expected to execute a course reversal. The term "NoPT" will be annotated on the boundary of the TAA icon for the other portion of the TAA.

8. FIG 5-4-11 depicts another TAA modification that pilots may encounter. In this generalized example, the right-base area has been eliminated. Pilots operating within the TAA between 360°clockwise to 060° bearing TO the IF (IAF) are expected to execute the course reversal in order to properly align the aircraft for entry onto the intermediate segment. Aircraft operating in all other areas from 060° clockwise to 360° bearing TO the IF (IAF) need not perform the course reversal, and the term "NoPT" will be annotated

FIGURE 5-4-9
RNAV (GPS) Approach Chart

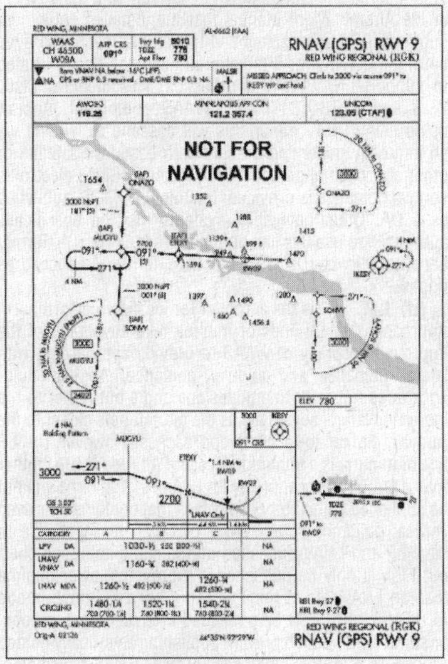

FIGURE 5-4-10
TAA with Left and Right Base Areas Eliminated

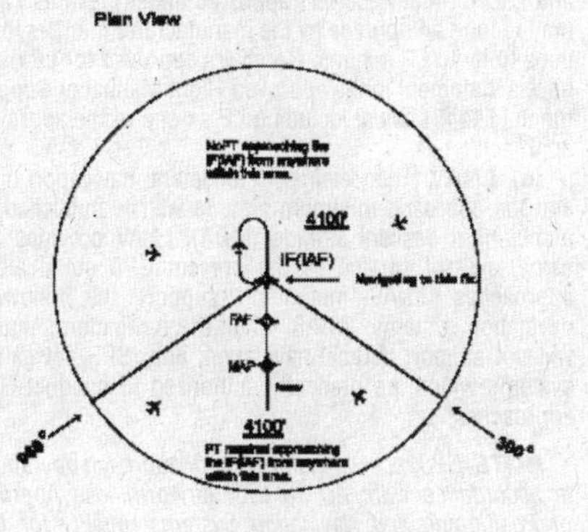

FIGURE 5-4-11
TAA with Right Base Eliminated

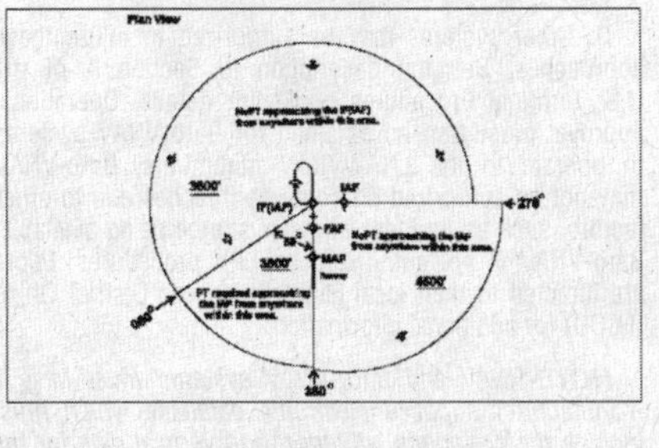

on the TAA boundary of the icon in these areas. TAAs are no longer being produced with sections removed; however, some may still exist on previously published procedures.

9. When an airway does not cross the lateral TAA boundaries, a feeder route will be established to provide a transition from the en route structure to the appropriate IAF. Each feeder route will terminate at the TAA boundary, and will be aligned along a path pointing to the associated IAF. Pilots should descend to the TAA altitude after crossing the TAA boundary and cleared by air traffic control. (See FIG 5-4-12).

FIGURE 5-4-12
Examples of a TAA with Feeders from an Airway

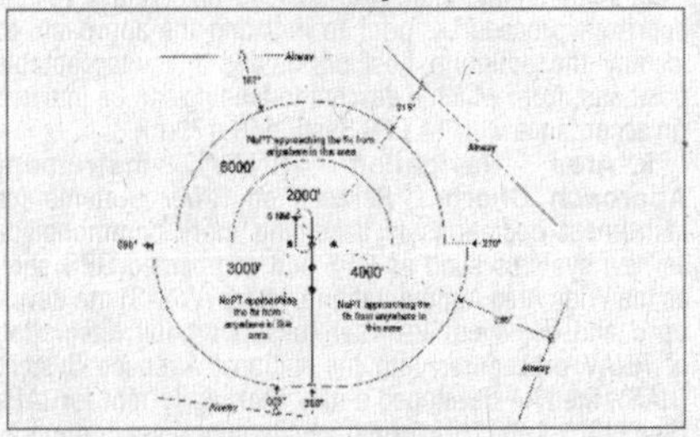

FIGURE 5-4-13
Minimum Vectoring Altitude Charts

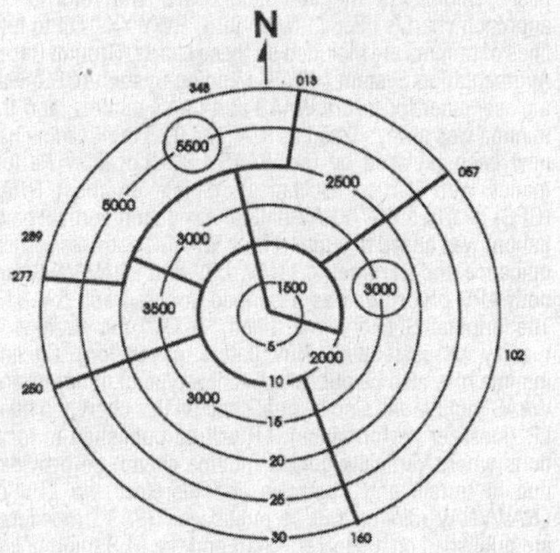

e. Minimum Vectoring Altitudes (MVAs) are established for use by ATC when radar ATC is exercised. MVA charts are prepared by air traffic facilities at locations where there are numerous different minimum IFR altitudes. Each MVA chart has sectors large enough to accommodate vectoring of aircraft within the sector at the MVA. Each sector boundary is at least 3 miles from the obstruction determining the MVA. To avoid a large sector with an excessively high MVA due to an isolated prominent obstruction, the obstruction may be enclosed in a buffer area whose boundaries are at least 3 miles from the obstruction. This is done to facilitate vectoring around the obstruction. (SeeFIG 5-4-13.)

1. The minimum vectoring altitude in each sector provides 1,000 feet above the highest obstacle in nonmountainous areas and 2,000 feet above the highest obstacle in designated mountainous areas. Where lower MVAs are required in designated mountainous areas to achieve compatibility with terminal routes or to permit vectoring to an IAP, 1,000feet of obstacle clearance may be authorized with the use of Airport Surveillance Radar (ASR). The minimum vectoring altitude will provide at least 300 feet above the floor of controlled airspace.

NOTE-*OROCA is an off-route altitude which provides obstruction clearance with a 1,000 foot buffer in nonmountainous terrain areas and a 2,000 foot buffer in designated mountainous areas within the U.S. This altitude may not provide signal coverage from ground-based navigational aids, air traffic control radar, or communications coverage.*

2. Because of differences in the areas considered for MVA, and those applied to other minimum altitudes, and the ability to isolate specific obstacles, some MVAs may be lower than the nonradar Minimum En Route Altitudes (MEAs), Minimum Obstruction Clearance Altitudes (MOCAs) or other minimum altitudes depicted on charts for a given location. While being radar vectored, IFR altitude assignments by ATC will be at or above MVA.

f. Visual Descent Points (VDPs) are being incorporated in nonprecision approach procedures. The VDP is a defined point on the final approach course of a nonprecision straight-in approach procedure from which normal descent from the MDA to the runway touchdown point may be commenced, provided visual reference required by 14 CFR Section 91.175(c)(3) is established. The VDP will normally be identified by DME on VOR and LOC procedures and by along-track distance to the next waypoint for RNAV procedures. The VDP is identified on the profile view of the approach chart by the symbol: **V**.

1. VDPs are intended to provide additional guidance where they are implemented. No special technique is required to fly a procedure with a VDP. The pilot should not descend below the MDA prior to reaching the VDP and acquiring the necessary visual reference.

2. Pilots not equipped to receive the VDP should fly the approach procedure as though no VDP had been provided.

g. Visual Segment of a Published Instrument Approach Procedure. Instrument procedures designers perform a visual area obstruction evaluation off the approach end of each runway authorized for instrument landing, straight-in, or circling. Restrictions to instrument operations are imposed if penetrations of the obstruction clearance surfaces exist. These restrictions vary based on the severity of the penetrations, and may include increasing required visibility, denying VDPs, prohibiting night instrument operations to the runway, and/or provide a "Fly Visual" option to the landing surface.

1. In isolated cases, due to procedure design peculiarities, an IAP may contain a published visual flight path. These procedures are annotated "Fly Visual to Airport" or "Fly Visual." A dashed arrow indicating the visual flight path will be included in the profile and plan views with an approximate heading and distance to the end of the runway. The depicted ground track associated with the visual segment should be flown as a "DR" course. When executing the visual segment, the flight visibility must not be less than that prescribed in the IAP, the pilot must remain clear of clouds and proceed to the airport maintaining visual contact with the ground. Altitude on the visual flight path is at the discretion of the pilot.

2. Since missed approach obstacle clearance is assured only if the missed approach is commenced at the published MAP or above the DA/MDA, the pilot should have preplanned climb out options based on aircraft performance and terrain features. Obstacle clearance is the sole responsibility of the pilot when the approach is continued beyond the MAP.

NOTE-*The FAA Administrator retains the authority to approve instrument approach procedures where the pilot may not necessarily have one of the visual references specified in CFR 14, part 91.175 and related rules. It is not a function of procedure design to ensure compliance with part 91.175. The annotation "Fly Visual to Airport" provides relief from part 91.175 requirements that the pilot have distinctly visible and identifiable visual references prior to descent below MDA/DA.*

h. Charting of Close in Obstacles on Instrument Procedure Charts. Obstacles that are close to the airport may be depicted in either the planview of the instrument approach chart or the airport sketch. Obstacles are charted in only one of the areas, based on space available and distance from the runway. These obstacles could be in the visual segment of the instrument approach procedure. On nonprecision approaches, these obstacles should be considered when determining where to begin descent from the MDA (see "Pilot Operational Considerations When Flying Nonprecision Approaches" in this paragraph).

i. Vertical Descent Angle (VDA) on Nonprecision Approaches. FAA policy is to publish VDAs on all nonprecision approaches. Published along with VDA is the threshold crossing height (TCH) that was used to compute the angle. The descent angle may be computed from either the final approach fix (FAF), or a stepdown fix, to the runway threshold at the published TCH. A stepdown fix is only used as the start point when an angle computed from the FAF would place the aircraft below the stepdown fix altitude. The descent angle and TCH information are charted on the profile view of the instrument approach chart following the fix the angle was based on. The optimum descent angle is 3.00 degrees; and whenever possible the approach will be designed using this angle.

1. The VDA provides the pilot with information not previously available on nonprecision approaches. It provides a means for the pilot to establish a stabilized descent from the FAF or stepdown fix to the MDA. Stabilized descent is a key factor in the reduction of controlled flight into terrain (CFIT) incidents. However, pilots should be aware that **the published angle is for information only** - it is strictly advisory in nature. There is no implicit additional obstacle protection below the MDA. Pilots must still respect the published minimum descent

Chapter 5

AIM

altitude (MDA) unless the visual cues stated 14 CFR Section 91.175 are present and they can visually acquire and avoid obstacles once below the MDA. The presence of a VDA does not guarantee obstacle protection in the visual segment and does not change any of the requirements for flying a nonprecision approach.

2. Additional protection for the visual segment below the MDA is provided if a VDP is published and descent below the MDA is started at or after the VDP. Protection is also provided, if a Visual Glide Slope Indicator (VGSI); e.g., VASI or PAPI, is installed and the aircraft remains on the VGSI glide path angle from the MDA. In either case, a chart note will indicate if the VDP or VGSI are not coincident with the VDA. On RNAV approach charts, a small shaded arrowhead shaped symbol (see the legend of the U.S. Terminal Procedures books, page H1) from the end of the VDA to the runway indicates that the 34:1 visual surface is clear.

3. Pilots may use the published angle and estimated/actual groundspeed to find a target rate of descent from the rate of descent table published in the back of the U.S. Terminal Procedures Publication. This rate of descent can be flown with the Vertical Velocity Indicator (VVI) in order to use the VDA as an aid to flying a stabilized descent. No special equipment is required.

4. Since one of the reasons for publishing a circling only instrument landing procedure is that the descent rate required exceeds the maximum allowed for a straight in approach, circling only procedures may have VDAs which are considerably steeper than the standard 3 degree angle on final. In this case, the VDA provides the crew with information about the descent rate required to land straight in from the FAF or step down fix to the threshold. This is not intended to imply that landing straight ahead is recommended, or even possible, since the descent rate may exceed the capabilities of many aircraft. The pilot must determine how to best maneuver the aircraft within the circling obstacle clearance area in order to land.

5. In rare cases the LNAV minima may have a lower HAT than minima with a glide path due to the location of the obstacles. This should be a clear indication to the pilot that obstacles exist below the MDA which the pilot must see in order to ensure adequate clearance. In those cases, the glide path may be treated as a VDA and used to descend to the LNAV MDA as long as all the rules for a nonprecision approach are applied at the MDA. However, the pilot must keep in mind the information in this paragraph and in paragraph 5-4-5j.

j. Pilot Operational Considerations When Flying Nonprecision Approaches. The missed approach point (MAP) on a nonprecision approach is not designed with any consideration to where the aircraft must begin descent to execute a safe landing. It is developed based on terrain, obstructions, NAVAID location and possibly air traffic considerations. Because the MAP may be located anywhere from well prior to the runway threshold to past the opposite end of the runway, the descent from the Minimum Descent Altitude (MDA) to the runway threshold cannot be determined based on the MAP location. Descent from MDA at the MAP when the MAP is located close to the threshold would require an excessively steep descent gradient to land in the normal touchdown zone. Any turn from the final approach course to the runway heading may also be a factor in when to begin the descent.

1. Pilots are cautioned that descent to a straight-in landing from the MDA at the MAP may be inadvisable or impossible, on a nonprecision approach, even if current weather conditions meet the published ceiling and visibility. Aircraft speed, height above the runway, descent rate, amount of turn and runway length are some of the factors which must be considered by the pilot to determine if a landing can be accomplished.

2. Visual descent points (VDPs) provide pilots with a reference for the optimal location to begin descent from the MDA, based on the designed vertical descent angle (VDA) for the approach procedure, assuming required visual references are available. Approaches without VDPs have not been assessed for terrain clearance below the MDA, and may not provide a clear vertical path to the runway at the normally expected descent angle. Therefore, pilots must be especially vigilant when descending below the MDA at locations without VDPs. This does not necessarily prevent flying the normal angle; it only means that obstacle clearance in the visual segment could be less and greater care should be exercised in looking for obstacles in the visual segment. Use of visual glide slope indicator (VGSI) systems can aid the pilot in determining if the aircraft is in a position to make the descent from the MDA. However, when the visibility is close to minimums, the VGSI may not be visible at the start descent point for a "normal" glidepath, due to its location down the runway.

3. Accordingly, pilots are advised to carefully review approach procedures, prior to initiating the approach, to identify the optimum position(s), and any unacceptable positions, from which a descent to landing can be initiated (in accordance with 14 CFR Section 91.175(c)).

k. Area Navigation (RNAV) Instrument Approach Charts. Reliance on RNAV systems for instrument operations is becoming more commonplace as new systems such as GPS and augmented GPS such as the Wide Area Augmentation System (WAAS) are developed and deployed. In order to support full integration of RNAV procedures into the National Airspace System (NAS), the FAA developed a new charting format for IAPs (See FIG 5-4-9). This format avoids unnecessary duplication and proliferation of instrument approach charts. The original stand alone GPS charts, titled simply "GPS," are being converted to the newer format as the procedures are revised. One reason for the revision is the addition of WAAS based minima to the approach chart. The reformatted approach chart is titled "RNAV (GPS) RWY XX." Up to four lines of minima are included on these charts. Ground Based Augmentation System (GBAS) Landing System (GLS) was a placeholder for future WAAS and LAAS minima, and the minima was always listed as N/A. The GLS minima line has now been replaced by the WAAS LPV (Localizer Performance with Vertical Guidance) minima on most RNAV (GPS) charts. LNAV/VNAV (lateral navigation/vertical navigation) was added to support both WAAS electronic vertical guidance and Barometric VNAV. LPV and LNAV/VNAV are both APV procedures as described in paragraph 5-4-5a7. The original GPS minima, titled "S-XX," for straight in runway XX, is retitled LNAV (lateral navigation). Circling minima may also be published. A new type of nonprecision WAAS minima will also be published on this chart and titled LP (localizer performance). LP will be published in locations where vertically guided minima cannot be provided due to terrain and obstacles and therefore, no LPV or LNAV/VNAV minima will be published. GBAS procedures are published on a separate chart and the GLS minima line is to be used only for GBAS. ATC clearance for the RNAV procedure authorizes a properly certified pilot to utilize any minimums for which the aircraft is certified (for example, a WAAS equipped aircraft utilizes the LPV or LP minima but a GPS only aircraft may not). The RNAV chart includes information formatted for quick reference by the pilot or flight crew at the top of the chart. This portion of the chart, developed based on a study by the Department of Transportation, Volpe National Transportation System Center, is commonly referred to as the pilot briefing.

1. The minima lines are:

(a) GLS. "GLS" is the acronym for GBAS Landing System. The U.S. version of GBAS has traditionally been referred to as LAAS. The worldwide community has adopted GBAS as the official term for this type of navigation system. To coincide with international terminology, the FAA is also adopting the term GBAS to be consistent with the international community. This line was originally published as a placeholder for both WAAS and LAAS minima and marked as N/A since no minima was published. As the concepts for GBAS and WAAS procedure publication have evolved, GLS will now be used only for GBAS minima, which will be on a separate approach chart. Most RNAV(GPS) approach charts have had the GLS minima line replaced by a WAAS LPV line of minima.

(b) LPV. "LPV" is the acronym for localizer performance with vertical guidance. RNAV (GPS) approaches to LPV lines of minima take advantage of the improved accuracy of WAAS lateral and vertical guidance to provide an approach that is very similar to a Category I Instrument Landing System (ILS). The approach to LPV line of minima is designed for angular guidance with increasing sensitivity as the aircraft gets closer to the runway. The sensitivities are nearly identical to those of the ILS at similar distances. This was done intentionally to allow the skills required to proficiently fly an ILS to readily transfer to flying RNAV (GPS) approaches to the LPV line of minima. Just as with an ILS, the LPV has vertical guidance and is flown to a DA. Aircraft can fly this minima line with a statement in the Aircraft Flight Manual that the installed equipment supports LPV approaches. This includes Class 3 and 4 TSO-C146 GPS/WAAS equipment.

(c) LNAV/VNAV. LNAV/VNAV identifies APV minimums developed to accommodate an RNAV IAP with vertical guidance, usually provided by approach certified Baro-VNAV, but with lateral and vertical integrity limits larger than a precision approach or LPV. LNAV stands for Lateral Navigation; VNAV stands for Vertical Navigation. This minima line can be flown by aircraft with a statement in the Aircraft Flight Manual that the installed equipment supports GPS approaches and has an approach-approved barometric VNAV, or if the aircraft has been demonstrated to support LNAV/VNAV approaches. This includes Class 2, 3 and 4 TSO-C146 GPS/WAAS equipment. Aircraft using LNAV/VNAV minimums will descend to landing via an internally generated descent path based on satellite or other approach approved VNAV systems. Since electronic vertical guidance is provided, the minima will be published as a DA. Other navigation systems may be specifically authorized to use this line of minima, see Section A, Terms/Landing Minima Data, of the U.S. Terminal Procedures books.

(d) LP. "LP" is the acronym for localizer performance. Approaches to LP lines of minima take advantage of the improved accuracy of WAAS to provide approaches, with lateral guidance and angular guidance. Angular guidance does not refer to a glideslope angle but rather to the increased lateral sensitivity as the aircraft gets closer to the runway, similar to localizer approaches. However, the LP line of minima is a Minimum Descent Altitude (MDA) rather than a DA (H). Procedures with LP lines of minima will not be published with another approach that contains approved vertical guidance (LNAV/VNAV or LPV). It is possible to have LP and LNAV published on the same approach chart but LP will only be published if it provides lower minima than an LNAV line of minima. LP is not a fail-down mode for LPV. LP will only be published if terrain, obstructions, or some other reason prevent publishing a vertically guided procedure. WAAS avionics may provide GNSS-based advisory vertical guidance during an approach to an LP line of minima. Barometric altimeter information remains the primary altitude reference for complying with any altitude restrictions. WAAS equipment may not support LP, even if it supports LPV, if it was approved before TSO-C145b and TSO-C146b. Receivers approved under previous TSOs may require an upgrade by the manufacturer in order to be used to fly to LP minima. Receivers approved for LP must have a statement in the approved Flight Manual or Supplemental Flight Manual including LP as one of the approved approach types.

(e) LNAV. This minima is for lateral navigation only, and the approach minimum altitude will be published as a minimum descent altitude (MDA). LNAV provides the same level of service as the present GPS stand alone approaches. LNAV minimums support the following navigation systems: WAAS, when the navigation solution will not support vertical navigation; and, GPS navigation systems which are presently authorized to conduct GPS approaches.

NOTE-*GPS receivers approved for approach operations in accordance with: AC 20-138, Airworthiness Approval of Positioning and Navigation Systems, qualify for this minima. WAAS navigation equipment must be approved in accordance with the requirements specified in TSO-C145() or TSO-C146() and installed in accordance with Advisory Circular AC 20-138.*

2. Other systems may be authorized to utilize these approaches. See the description in Section A of the U.S. Terminal Procedures books for details. Operational approval must also be obtained for Baro-VNAV systems to operate to the LNAV/VNAV minimums. Baro-VNAV may not be authorized on some approaches due to other factors, such as no local altimeter source being available. Baro-VNAV is not authorized on LPV procedures. Pilots are directed to their local Flight Standards District Office (FSDO) for additional information.

NOTE-*RNAV and Baro-VNAV systems must have a manufacturer supplied electronic database which must include the waypoints, altitudes, and vertical data for the procedure to be flown. The system must be able to retrieve the procedure by name from the aircraft navigation database, not just as a manually entered series of waypoints.*

3. ILS or RNAV (GPS) charts.

(a) Some RNAV (GPS) charts will also contain an ILS line of minima to make use of the ILS precision final in conjunction with the RNAV GPS capabilities for the portions of the procedure prior to the final approach segment and for the missed approach. Obstacle clearance for the portions of the procedure other than the final approach segment is still based on GPS criteria.

***NOTE-**Some GPS receiver installations inhibit GPS navigation whenever **ANY** ILS frequency is tuned. Pilots flying aircraft with receivers installed in this manner must wait until they are on the intermediate segment of the procedure prior to the PFAF (PFAF is the active waypoint) to tune the ILS frequency and must tune the ILS back to a VOR frequency in order to fly the GPS based missed approach.*

(b) Charting. There are charting differences between ILS, RNAV (GPS), and GLS approaches.

(1) The LAAS procedure is titled "GLS RWY XX" on the approach chart.

(2) The VDB provides information to the airborne receiver where the guidance is synthesized.

(3) The LAAS procedure is identified by a four alphanumeric character field referred to as the RPI or approach ID and is similar to the IDENT feature of the ILS.

(4) The RPI is charted.

(5) Most RNAV(GPS) approach charts have had the GLS (NA) minima line replaced by an LPV line of minima.

(6) Since the concepts for GLS and WAAS procedure publication have evolved, GLS will now be used only for LAAS minima, which will be on a separate approach chart.

4. Required Navigation Performance (RNP)

(a) Pilots are advised to refer to the "TERMS/LANDING MINIMUMS DATA" (Section A) of the U.S. Government Terminal Procedures books for aircraft approach eligibility requirements by specific RNP level requirements.

(b) Some aircraft have RNP approval in their AFM without a GPS sensor. The lowest level of sensors that the FAA will support for RNP service is DME/DME. However, necessary DME signal may not be available at the airport of intended operations. For those locations having an RNAV chart published with LNAV/VNAV minimums, a procedure note may be provided such as "DME/DME RNP-0.3 NA." This means that RNP aircraft dependent on DME/DME to achieve RNP-0.3 are not authorized to conduct this approach. Where DME facility availability is a factor, the note may read "DME/DME RNP-0.3 Authorized; ABC and XYZ Required." This means that ABC and XYZ facilities have been determined by flight inspection to be required in the navigation solution to assure RNP-0.3. VOR/DME updating must not be used for approach procedures.

5. Chart Terminology

(a) Decision Altitude (DA) replaces the familiar term Decision Height (DH). DA conforms to the international convention where altitudes relate to MSL and heights relate to AGL. DA will eventually be published for other types of instrument approach procedures with vertical guidance, as well. DA indicates to the pilot that the published descent profile is flown to the DA (MSL), where a missed approach will be initiated if visual references for landing are not established. Obstacle clearance is provided to allow a momentary descent below DA while transitioning from the final approach to the missed approach. The aircraft is expected to follow the missed instructions while continuing along the published final approach course to at least the published runway threshold waypoint or MAP (if not at the threshold) before executing any turns.

(b) Minimum Descent Altitude (MDA) has been in use for many years, and will continue to be used for the LNAV only and circling procedures.

(c) Threshold Crossing Height (TCH) has been traditionally used in "precision" approaches as the height of the glide slope above threshold. With publication of LNAV/VNAV minimums and RNAV descent angles, including graphically depicted descent profiles, TCH also applies to the height of the "descent angle," or glidepath, at the threshold. Unless otherwise required for larger type aircraft which may be using the IAP, the typical TCH is 30 to 50 feet.

6. The MINIMA FORMAT will also change slightly.

(a) Each line of minima on the RNAV IAP is titled to reflect the level of service available; e.g., GLS, LPV, LNAV/VNAV, LP, and LNAV. CIRCLING minima will also be provided.

(b) The minima title box indicates the nature of the minimum altitude for the IAP. For example:

(1) DA will be published next to the minima line title for minimums supporting vertical guidance such as for GLS, LPV or LNAV/VNAV.

(2) MDA will be published where the minima line was designed to support aircraft with only lateral guidance available, such as LNAV or LP. Descent below the MDA, including during the missed approach, is not authorized unless the visual conditions stated in 14 CFR Section 91.175 exist.

(3) Where two or more systems, such as LPV and LNAV/VNAV, share the same minima, each line of minima will be displayed separately.

7. Chart Symbology changed slightly to include:

(a) Descent Profile. The published descent profile and a graphical depiction of the vertical path to the runway will be shown. Graphical depiction of the RNAV vertical guidance will differ from the traditional depiction of an ILS glide slope (feather) through the use of a shorter vertical track beginning at the decision altitude.

(1) It is FAA policy to design IAPs with minimum altitudes established at fixes/waypoints to achieve optimum stabilized (constant rate) descents within each procedure segment. This design can enhance the safety of the operations and contribute toward reduction in the occurrence of controlled flight into terrain (CFIT) accidents. Additionally, the National Transportation Safety Board (NTSB) recently emphasized that pilots could benefit from publication of the appropriate IAP descent angle for a stabilized descent on final approach. The RNAV IAP format includes the descent angle to the hundredth of a degree; e.g., **3.00 degrees.** The angle will be provided in the graphically depicted descent profile.

(2) The stabilized approach may be performed by reference to vertical navigation information provided by WAAS or LNAV/VNAV systems; or for LNAV-only systems, by the pilot determining the appropriate aircraft attitude/groundspeed combination to attain a constant rate descent which best emulates the published angle. To aid the pilot, U.S. Government Terminal Procedures Publication charts publish an expanded Rate of Descent Table on the inside of the back hard cover for use in planning and executing precision descents under known or approximate groundspeed conditions.

(b) Visual Descent Point (VDP). A VDP will be published on most RNAV IAPs. VDPs apply only to aircraft utilizing LP or LNAV minima, not LPV or LNAV/VNAV minimums.

(c) Missed Approach Symbology. In order to make missed approach guidance more readily understood, a method has been developed to display missed approach guidance in the profile view through the use of quick reference icons. Due to limited space in the profile area, only four or fewer icons can be shown. However, the icons may not provide representation of the entire missed approach procedure. The entire set of textual missed approach instructions are provided at the top of the approach chart in the pilot briefing. (See FIG 5-4-9).

(d) Waypoints. All RNAV or GPS stand-alone IAPs are flown using data pertaining to the particular IAP obtained from an onboard database, including the sequence of all WPs used for the approach and missed approach, except that step down waypoints may not be included in some TSO-C129 receiver databases. Included in the database, in most receivers, is coding that informs the navigation system of which WPs are fly-over (FO) or fly-by (FB). The navigation system may provide guidance appropriately - including leading the turn prior to a fly-by WP; or causing overflight of a fly-over WP. Where the navigation system does not provide such guidance, the pilot must accomplish the turn lead or waypoint overflight manually. Chart symbology for the FB WP provides pilot awareness of expected actions. Refer to the legend of the U.S. Terminal Procedures books.

(e) TAAs are described in paragraph 5-4-5d, Terminal Arrival Area (TAA). When published, the RNAV chart depicts the TAA areas through the use of "icons" representing each TAA area associated with the RNAV procedure (See FIG 5-4-9). These icons are depicted in the plan view of the approach chart, generally arranged on the chart in accordance with their position relative to the aircraft's arrival from the en route structure. The WP, to which navigation is appropriate and expected within each specific TAA area, will be named and depicted on the associated TAA icon. Each depicted named WP is the IAF for arrivals from within that area. TAAs may not be used on all RNAV procedures because of airspace congestion or other reasons.

(f) Hot and Cold Temperature Limitations. A minimum and maximum temperature limitation is published on procedures which authorize Baro-VNAV operation. These temperatures represent the airport temperature above or below which Baro-VNAV is not authorized to LNAV/VNAV minimums. As an example, the limitation will read: "Uncompensated Baro-VNAV NA below -8°C (+18°F) or above 47°C (117°F)." This information will be found in the upper left hand box of the pilot briefing. When the temperature is above the high temperature or below the low temperature limit, Baro-VNAV may be used to provide a stabilized descent to the LNAV MDA; however, extra caution should be used in the visual segment to ensure a vertical correction is not required. If the VGSI is aligned with the published glidepath, and the aircraft instruments indicate on glidepath, an above or below glidepath indication on the VGSI may indicate that temperature error is causing deviations to the glidepath. These deviations should be considered if the approach is continued below the MDA.

***NOTE-**Many systems which apply Baro-VNAV temperature compensation only correct for cold temperature. In this case, the high temperature limitation still applies. Also, temperature compensation may require activation by maintenance personnel during installation in order to be functional, even though the system has the feature. Some systems may have a temperature correction capability, but correct the Baro-altimeter all the time, rather than just on the final, which would create conflicts with other aircraft if the feature were activated. Pilots should be aware of compensation capabilities of the system prior to disregarding the temperature limitations.*

***NOTE-**Temperature limitations do not apply to flying the LNAV/VNAV line of minima using approach certified WAAS receivers when LPV or LNAV/VNAV are annunciated to be available.*

(g) WAAS Channel Number/Approach ID. The WAAS Channel Number is an optional equipment capability that allows the use of a 5-digit number to select a specific final approach segment without using the menu method. The Approach ID is an airport unique 4-character combination for verifying the selection and extraction of the correct final approach segment information from the aircraft database. It is similar to the ILS ident, but displayed visually rather than aurally. The Approach ID consists of the letter W for WAAS, the runway number, and a letter other than L, C or R, which could be confused with Left, Center and Right, e.g., W35A. Approach IDs are assigned in the order that WAAS approaches are built to that runway number at that airport. The WAAS Channel Number and Approach ID are displayed in the upper left corner of the approach procedure pilot briefing.

(h) At locations where outages of WAAS vertical guidance may occur daily due to initial system limitations, a negative W symbol (W) will be placed on RNAV (GPS) approach charts. Many of these outages will be very short in duration, but may result in the disruption of the vertical portion of the approach. The W symbol indicates that NOTAMs or Air Traffic advisories are not provided for outages which occur in the WAAS LNAV/VNAV or LPV vertical service. Use LNAV minima for flight planning at these locations, whether as a destination or alternate. For flight operations at these locations, when the WAAS avionics indicate that LNAV/VNAV or LPV service is available, then vertical guidance may be used to complete the approach using the displayed level of service. Should an outage occur during the procedure, reversion to LNAV minima may be required. As the WAAS coverage is expanded, the W will be removed.

***NOTE-**Properly trained and approved, as required, TSOC145() and TSOC146() equipped users (WAAS users) with and using approved baroVNAV equipment may plan for LNAV/VNAV DA at an alternate airport. Specifically authorized WAAS users with and using approved baroVNAV equipment may also plan for RNP 0.3 DA at the alternate airport as long as the pilot has verified RNP availability through an approved prediction program.*

5-4-6. Approach Clearance

a. An aircraft which has been cleared to a holding fix and subsequently "cleared . . . approach" has not received new routing. Even though clearance for the approach may have been issued prior to the aircraft reaching the holding fix, ATC would expect the pilot to proceed via the holding fix (his/her last assigned route), and the feeder route associated with that fix (if a feeder route is published on the approach chart) to the initial approach fix (IAF) to commence the approach. *WHEN CLEARED FOR THE APPROACH, THE PUBLISHED OFF AIRWAY (FEEDER) ROUTES THAT LEAD FROM THE EN ROUTE STRUCTURE TO THE IAF ARE PART OF THE APPROACH CLEARANCE.*

b. If a feeder route to an IAF begins at a fix located along the route of flight prior to reaching the holding fix, and clearance for an approach is issued, a pilot should commence the approach via the published feeder route; i.e., the aircraft would not be expected to overfly the feeder route and return to it. The pilot is expected to commence the approach in a similar manner at the IAF, if the IAF for the procedure is located along the route of flight to the holding fix.

c. If a route of flight directly to the initial approach fix is desired, it should be so stated by the controller with phraseology to include the words "direct . . . ," "proceed direct" or a similar phrase which the pilot can interpret without question. When uncertain of the clearance, immediately query ATC as to what route of flight is desired.

d. The name of an instrument approach, as published, is used to identify the approach, even though a component of the approach aid, such as the glideslope on an Instrument Landing System, is inoperative or unreliable. The controller will use the name of the approach as published, but must advise the aircraft at the time an approach clearance is issued that the inoperative or unreliable approach aid component is unusable, except when the title of the published approach procedures otherwise allows; for example, ILS Rwy 05 or LOC Rwy 05.

e. The following applies to aircraft on radar vectors and/or cleared "direct to" in conjunction with an approach clearance:

1. Maintain the last altitude assigned by ATC until the aircraft is established on a published segment of a transition route, or approach procedure segment, or other published route, for which a lower altitude is published on the chart. If already on an established route, or approach or arrival segment, you may descend to whatever minimum altitude is listed for that route or segment.

2. Continue on the vector heading until intercepting the next published ground track applicable to the approach clearance.

3. Once reaching the final approach fix via the published segments, the pilot may continue on approach to a landing.

4. If proceeding to an IAF with a published course reversal (procedure turn or holdinlieu of PT pattern), except when cleared for a straight in approach by ATC, the pilot must execute the procedure turn/holdinlieu of PT, and complete the approach.

5. If cleared to an IAF/IF via a NoPT route, or no procedure turn/holdinlieu of PT is published, continue with the published approach.

6. In addition to the above, RNAV aircraft may be issued a clearance direct to the IAF/IF at intercept angles not greater than 90 degrees for both conventional and RNAV instrument approaches. Controllers may issue a heading or a course direct to a fix between the IF and FAF at intercept angles not greater than 30 degrees for both conventional and RNAV instrument approaches. In all cases, controllers will assign altitudes that ensure obstacle clearance and will permit a normal descent to the FAF. When clearing aircraft direct to the IF, ATC will radar monitor the aircraft until the IF and will advise the pilot to expect clearance direct to the IF at least 5 miles from the fix. ATC must issue a straightin approach clearance when clearing an aircraft direct to an IAF/IF with a procedure turn or hold-in-lieu of a procedure turn, and ATC does not want the aircraft to execute the course reversal.

NOTE- *Refer to 14 CFR 91.175 (i).*

7. RNAV aircraft may be issued a clearance direct to the FAF that is also charted as an IAF, in which case the pilot is expected to execute the depicted procedure turn or holdinlieu of procedure turn. ATC will not issue a straightin approach clearance. If the pilot desires a straightin approach, they must request vectors to the final approach course outside of the FAF or fly a published "NoPT" route. When visual approaches are in use, ATC may clear an aircraft direct to the FAF.

NOTE-1. *In anticipation of a clearance by ATC to any fix published on an instrument approach procedure, pilots of RNAV aircraft are advised to select an appropriate IAF or feeder fix when loading an instrument approach procedure into the RNAV system.*

2. Selection of "VectorstoFinal" or "Vectors" option for an instrument approach may prevent approach fixes located outside of the FAF from being loaded into an RNAV system. Therefore, the selection of these options is discouraged due to increased workload for pilots to reprogram the navigation system.

f. An RF leg is defined as a constant radius circular path around a defined turn center that starts and terminates at a fix. An RF leg may be published as part of a procedure. Since not all aircraft have the capability to fly these leg types, pilots are responsible for knowing if they can conduct an RNAV approach with an RF leg. Requirements for RF legs will be indicated on the approach chart in the notes section or at the applicable initial approach fix. Controllers will clear RNAVequipped aircraft for instrument approach procedures containing RF legs:

1. Via published transitions, or

2. On a heading or course direct to the IAF when a holdinlieu of procedure turn is published, and the pilot will execute the procedure, or

3. On a heading or course direct to the IAF/IF, at intercept angles no greater than 90 degrees and the distance to the waypoint beginning the RF leg is 6NM or greater, or

4. With radar monitoring, on a heading or course direct to any waypoint 3 miles or more from the waypoint that begins the RF leg, at an intercept angle not greater than 30 degrees. (See FIG 5-4-14.)

EXAMPLE-

1. Controllers will not clear aircraft direct to THIRD because that waypoint begins the RF leg, and aircraft cannot be vectored or cleared to TURNN or vectored to intercept the approach segment at any point between THIRD and FORTH because this is the RF leg.

2. Controllers can clear Aircraft 1 direct to SCOND because the distance to THIRD, where the RF leg begins is 3NM or greater and the intercept angle will be 30 degrees or less and is radar monitored. Controllers can clear Aircraft 2 direct to FIRST because the intercept angle is 90degrees or less, and the distance from FIRST to THIRD is 6NM or greater.

5-4-7. Instrument Approach Procedures

a. Aircraft approach category means a grouping of aircraft based on a speed of V_{REF} if specified, or if V_{REF} is not specified, 1.3 V_{SO} at the maximum certified landing weight. V_{REF}, V_{SO}, and the maximum certified landing weight are those values as established for the aircraft by the certification authority of the country of registry. A pilot must use the minima corresponding to the category determined during certification or higher. Helicopters may use Category A minima. If it is necessary to operate at a speed in excess of the upper limit of the speed range for an aircraft's category, the minimums for the higher category must be used. For example, an airplane which fits into Category B, but is circling to land at a speed of 145 knots, must use the approach Category D minimums. As an additional example, a Category A airplane (or helicopter) which is operating at 130 knots on a straight-in approach must use the approach Category C minimums. See the following category limits:

1. Category A: Speed less than 91 knots.

2. Category B: Speed 91 knots or more but less than 121 knots.

3. Category C: Speed 121 knots or more but less than 141 knots.

4. Category D: Speed 141 knots or more but less than 166 knots.

5. Category E: Speed 166 knots or more.

NOTE- *V_{REF} in the above definition refers to the speed used in establishing the approved landing distance under the airworthiness regulations constituting the type certification basis of the airplane, regardless of whether that speed for a particular airplane is 1.3 V_{SO}, 1.23 V_{SR}, or some higher speed required for airplane controllability. This speed, at the maximum certificated landing weight, determines the lowest applicable approach category for allapproaches regardless of actual landing weight.*

b. When operating on an unpublished route or while being radar vectored, the pilot, when an approach clearance is received, must, in addition to complying with the minimum altitudes for IFR operations (14 CFR Section 91.177), maintain the last assigned altitude unless a different altitude is assigned by ATC, or until the aircraft is established on a segment of a published route or IAP. After the aircraft is so established, published altitudes apply to descent within each succeeding route or approach segment unless a different altitude is assigned by ATC. Notwithstanding this pilot responsibility, for aircraft operating on unpublished routes or while being radar vectored, ATC will, except when conducting a radar approach, issue an IFR approach clearance only after the aircraft is established on a segment of a published route or IAP, or assign an altitude to maintain until the aircraft is established on a segment of a published route or instrument approach procedure. For this purpose, the procedure turn of a published IAP must not be considered a segment of that IAP until the aircraft reaches the initial fix or navigation facility upon which the procedure turn is predicated.

EXAMPLE-

Cross Redding VOR at or above five thousand, cleared VOR runway three four approach.

or

Five miles from outer marker, turn right heading three three zero, maintain two thousand until established on the localizer, cleared ILS runway three six approach.

NOTE- *The altitude assigned will assure IFR obstruction clearance from the point at which the approach clearance is issued until established on a segment of a published route or IAP. If uncertain of the meaning of the clearance, immediately request clarification from ATC.*

c. Several IAPs, using various navigation and approach aids may be authorized for an airport. ATC may advise that a particular approach procedure is being used, primarily to expedite traffic. If issued a clearance that specifies a particular approach procedure, notify ATC immediately if a different one is desired. In this event it may be necessary for ATC to withhold clearance for the different approach until such time as traffic conditions permit. However, a pilot involved in an emergency situation will be given priority. If the pilot is not familiar with the specific approach procedure, ATC should be advised and they will provide detailed information on the execution of the procedure.

REFERENCE- *AIM, Advance Information on Instrument Approach, Paragraph 5-4-4.*

d. The name of an instrument approach, as published, is used to identify the approach, even though a component of the approach aid, such as the glideslope on an Instrument Landing System, is inoperative or unreliable. The controller will use the name of the approach as published, but must advise the aircraft at the time an approach clearance is issued that the inoperative or unreliable approach aid component is unusable, except when the title of the published approach procedures otherwise allows, for example, ILS or LOC.

e. Except when being radar vectored to the final approach course, when cleared for a specifically prescribed IAP; i.e., "cleared ILS runway one niner approach" or when "cleared approach" i.e., execution of any procedure prescribed for the airport, pilots must execute the entire procedure commencing at an IAF or an associated feeder route as described on the IAP chart unless an appropriate new or revised ATC clearance is received, or the IFR flight plan is canceled.

f. Pilots planning flights to locations which are private airfields or which have instrument approach procedures based on private navigation aids should obtain approval from the owner. In addition, the pilot must be authorized by the FAA to fly special instrument approach procedures associated with private navigation aids (see paragraph 5-4-8). Owners of navigation aids that are not for public use may elect to turn off the signal for whatever reason they may have; e.g., maintenance, energy conservation, etc. Air traffic controllers are not required to question pilots to determine if they have permission to land at a private airfield or to use procedures based on privately owned navigation aids, and they may not know the status of the navigation aid. Controllers presume a pilot has obtained approval from the owner and the FAA for use of special instrument approach procedures and is aware of any details of the procedure if an IFR flight plan was filed to that airport.

g. Pilots should not rely on radar to identify a fix unless the fix is indicated as "RADAR" on the IAP. Pilots may request radar identification of an OM, but the controller may not be able to provide the service due either to workload or not having the fix on the video map.

h. If a missed approach is required, advise ATC and include the reason (unless initiated by ATC). Comply with the missed approach instructions for the instrument approach procedure being executed, unless otherwise directed by ATC.

REFERENCE- *AIM, Missed Approach, Paragraph 5-4-21.*

AIM, Missed Approach, Paragraph 5-5-5.

i. ATC may clear aircraft that have filed an Advanced RNAV equipment suffix to the intermediate fix when clearing aircraft for an instrument approach procedure. ATC will take the following actions when clearing Advanced RNAV aircraft to the intermediate fix:

1. Provide radar monitoring to the intermediate fix.

2. Advise the pilot to expect clearance direct to the intermediate fix at least 5 miles from the fix.

NOTE-*This is to allow the pilot to program the RNAV equipment to allow the aircraft to fly to the intermediate fix when cleared by ATC.*

3. Assign an altitude to maintain until the intermediate fix.

4. ensure the aircraft is on a course that will intercept the intermediate segment at an angle not greater than 90 degrees and is at an altitude that will permit normal descent from the intermediate fix to the final approach fix.

5-4-8. Special Instrument Approach Procedures

Instrument Approach Procedure (IAP) charts reflect the criteria associated with the U.S. Standard for Terminal Instrument [Approach] Procedures (TERPs), which prescribes standardized methods for use in developing IAPs. Standard IAPs are published in the Federal Register (FR) in accordance with Title 14 of the Code of Federal Regulations, Part 97, and are available for use by appropriately qualified pilots operating properly equipped and airworthy aircraft in accordance with operating rules and procedures acceptable to the FAA. Special IAPs are also developed using TERPS but are not given public notice in the FR. The FAA authorizes only certain individual pilots and/or pilots in individual organizations to use special IAPs, and may require additional crew training and/or aircraft equipment or performance, and may also require the use of landing aids, communications, or weather services not available for public use. Additionally, IAPs that service private use airports or heliports are generally special IAPs.

5-4-9. Procedure Turn and Hold-in-lieu of Procedure Turn

a. A procedure turn is the maneuver prescribed when it is necessary to reverse direction to establish the aircraft inbound on an intermediate or final approach course. The procedure turn or hold-in-lieu-of-PT is a required maneuver when it is depicted on the approach chart, unless cleared by ATC for a straight-in approach. Additionally, the procedure turn or hold-in-lieu-of-PT is not permitted when the symbol "No PT" is depicted on the initial segment being used, when a RADAR VECTOR to the final approach course is provided, or when conducting a timed approach from a holding fix. The altitude prescribed for the procedure turn is a minimum altitude until the aircraft is established on the inbound course. The maneuver must be completed within the distance specified in the profile view. For a hold-in-lieu-of-PT, the holding pattern direction must be flown as depicted and the specified leg length/timing must not be exceeded.

NOTE-*The pilot may elect to use the procedure turn or hold-in-lieu-of-PT when it is not required by the procedure, but must first receive an amended clearance from ATC. If the pilot is uncertain whether the ATC clearance intends for a procedure turn to be conducted or to allow for a straight-in approach, the pilot must immediately request clarification from ATC (14 CFR Section 91.123).*

1. On U.S. Government charts, a barbed arrow indicates the maneuvering side of the outbound course on which the procedure turn is made. Headings are provided for course reversal using the 45 degree type procedure turn. However, the point at which the turn may be commenced and the type and rate of turn is left to the discretion of the pilot (limited by the charted remain within xx NM distance). Some of the options are the 45 degree procedure turn, the racetrack pattern, the teardrop procedure turn, or the 80 degree $ 260 degree course reversal. Racetrack entries should be conducted on the maneuvering side where the majority of protected airspace resides. If an entry places the pilot on the non-maneuvering side of the PT, correction to intercept the outbound course ensures remaining within protected airspace. Some procedure turns are specified by procedural track. These turns must be flown exactly as depicted.

2. Descent to the procedure turn (PT) completion altitude from the PT fix altitude (when one has been published or assigned by ATC) must not begin until crossing over the PT fix or abeam and proceeding outbound. Some procedures contain a note in the chart profile view that says "Maintain (altitude) or above until established outbound for procedure turn" (See FIG 5-4-15). Newer procedures will simply depict an "at or above" altitude at the PT fix without a chart note (See FIG 5-4-16). Both are there to ensure required obstacle clearance is provided in the procedure turn entry zone (See FIG 5-4-17). Absence of a chart note or specified minimum altitude adjacent to the PT fix is an indication that descent to the procedure turn altitude can commence immediately upon crossing over the PT fix, regardless of the direction of flight. This is because the minimum altitudes in the PT entry zone and the PT maneuvering zone are the same.

FIGURE 5-4-14 AND 5-4-15
Example of an RNAV Approach with RF Leg

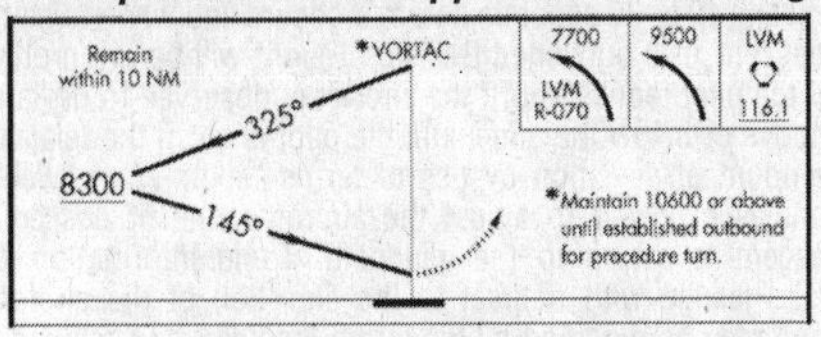

FIGURE 5-4-16

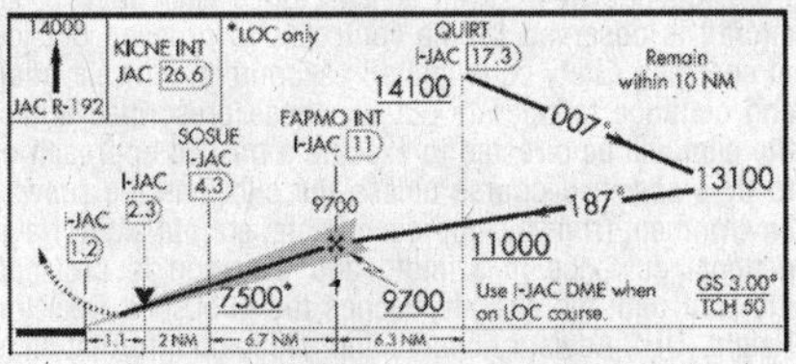

FIGURE 5-4-17

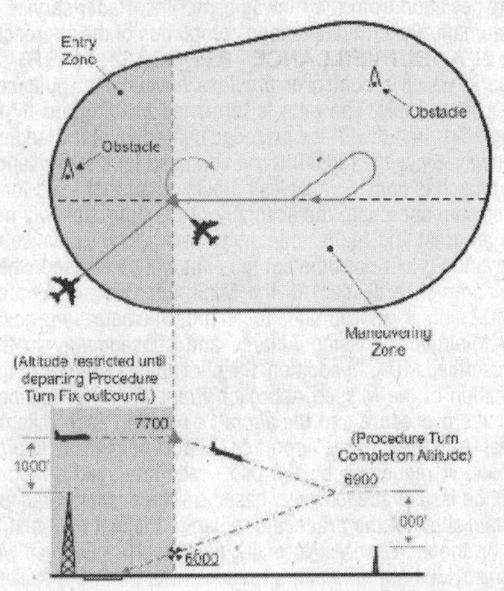

3. When the approach procedure involves a procedure turn, a maximum speed of not greater than 200 knots (IAS) should be observed from first overheading the course reversal IAF through the procedure turn maneuver to ensure containment within the obstruction clearance area. Pilots should begin the outbound turn immediately after passing the procedure turn fix. The procedure turn maneuver must be executed within the distance specified in the profile view. The normal procedure turn distance is 10 miles. This may be reduced to a minimum of 5 miles where only Category A or helicopter aircraft are to be operated or increased to as much as 15 miles to accommodate high performance aircraft.

4. A teardrop procedure or penetration turn may be specified in some procedures for a required course reversal. The teardrop procedure consists of departure from an initial approach fix on an outbound course followed by a turn toward and intercepting the inbound course at or prior to the intermediate fix or point. Its purpose is to permit an aircraft to reverse direction and lose considerable altitude within reasonably limited airspace. Where no fix is available to mark the beginning of the intermediate segment, it must be assumed to commence at a point 10 miles prior to the final approach fix. When the facility is located on the airport, an aircraft is considered to be on final approach upon completion of the penetration turn. However, the final approach segment begins on the final approach course 10 miles from the facility.

5. A holding pattern in lieu of procedure turn may be specified for course reversal in some procedures. In such cases, the holding pattern is established over an intermediate fix or a final approach fix. The holding pattern distance or time specified in the profile view must be observed. For a hold-in-lieu-of-PT, the holding pattern direction must be flown as depicted and the specified leg length/timing must not be exceeded. Maximum holding airspeed limitations as set forth for all holding patterns apply. The holding pattern maneuver is completed when the aircraft is established on the inbound course after executing the appropriate entry. If cleared for the approach prior to returning to the holding fix, and the aircraft is at the prescribed altitude, additional circuits of the holding pattern are not necessary nor expected by ATC. If pilots elect to make additional circuits to lose excessive altitude or to become better established on course, it is their responsibility to so advise ATC upon receipt of their approach clearance.

NOTE-*Some approach charts have an arrival holding pattern depicted at the IAF using a "thin line" holding symbol. It is charted where holding is frequently required prior to starting the approach procedure so that detailed holding instructions are not required. The arrival holding pattern is not authorized unless assigned by Air Traffic Control. Holding at the same fix may also be depicted on the enroute chart. A hold-in-lieu of procedure turn is depicted by a "thick line" symbol, and is part of the instrument approach procedure as described in paragraph 5-4-9. (See U. S. Terminal Procedures booklets page E1 for both examples.)*

6. A procedure turn is not required when an approach can be made directly from a specified intermediate fix to the final approach fix. In such cases, the term "NoPT" is used with the appropriate course and altitude to denote that the procedure turn is not required. If a procedure turn is desired, and when cleared to do so by ATC, descent below the procedure turn altitude should not be made until the aircraft is established on the inbound course, since some NoPT altitudes may be lower than the procedure turn altitudes.

b. Limitations on Procedure Turns

1. In the case of a radar initial approach to a final approach fix or position, or a timed approach from a holding fix, or where the procedure specifies NoPT, no pilot may make a procedure turn unless, when final approach clearance is received, the pilot so advises ATC and a clearance is received to execute a procedure turn.

2. When a teardrop procedure turn is depicted and a course reversal is required, this type turn must be executed.

3. When a holding pattern replaces a procedure turn, the holding pattern must be followed, except when RADAR VECTORING is provided or when NoPT is shown on the approach course. The recommended entry procedures will ensure the aircraft remains within the holding pattern's protected airspace. As in the procedure turn, the descent from the minimum holding pattern altitude to the final approach fix altitude (when lower) may not commence until the aircraft is established on the inbound course. Where a holding pattern is established in-lieu-of a procedure turn, the maximum holding pattern airspeeds apply.

REFERENCE-*AIM, Holding, Paragraph 5-3-8j2.*

4. The absence of the procedure turn barb in the plan view indicates that a procedure turn is not authorized for that procedure.

5-4-10. Timed Approaches from a Holding Fix

a. TIMED APPROACHES may be conducted when the following conditions are met:

1. A control tower is in operation at the airport where the approaches are conducted.

2. Direct communications are maintained between the pilot and the center or approach controller until the pilot is instructed to contact the tower.

3. If more than one missed approach procedure is available, none require a course reversal.

4. If only one missed approach procedure is available, the following conditions are met:

(a) Course reversal is not required; and,

(b) Reported ceiling and visibility are equal to or greater than the highest prescribed circling minimums for the IAP.

5. When cleared for the approach, pilots must not execute a procedure turn. (14 CFR Section 91.175.)

b. Although the controller will not specifically state that "timed approaches are in progress," the assigning of a time to depart the final approach fix inbound (nonprecision approach) or the outer marker or fix used in lieu of the outer marker inbound (precision approach) is indicative that timed approach procedures are being utilized, or in lieu of holding, the controller may use radar vectors to the Final Approach Course to establish a mileage interval between aircraft that will ensure the appropriate time sequence between the final approach fix/outer marker or fix used in lieu of the outer marker and the airport.

c. Each pilot in an approach sequence will be given advance notice as to the time they should leave the holding point on approach to the airport. When a time to leave the holding point has been received, the pilot should adjust the flight path to leave the fix as closely as possible to the designated time. (See FIG 5-4-18.)

FIGURE 5-4-18
Timed Approaches from a Holding Fix

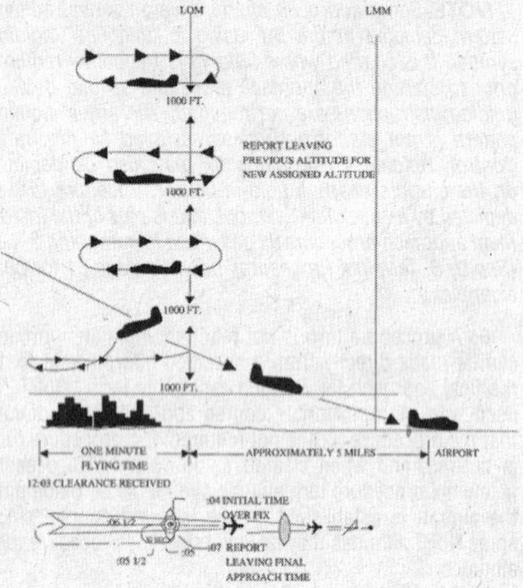

EXAMPLE-
At 12:03 local time, in the example shown, a pilot holding, receives instructions to leave the fix inbound at 12:07. These instructions are received just as the pilot has completed turn at the outbound end of the holding pattern and is proceeding inbound towards the fix. Arriving back over the fix, the pilot notes that the time is 12:04 and that there are 3 minutes to lose in order to leave the fix at the assigned time. Since the time remaining is more than two minutes, the pilot plans to fly a race track pattern rather than a 360 degree turn, which would use up 2 minutes. The turns at the ends of the race track pattern will consume approximately 2 minutes. Three minutes to go, minus 2 minutes required for the turns, leaves 1 minute for level flight. Since two portions of level flight will be required to get back to the fix inbound, the pilot halves the 1 minute remaining and plans to fly level for 30 seconds outbound before starting the turn back to the fix on final approach. If the winds were negligible at flight altitude, this procedure would bring the pilot inbound across the fix precisely at the specified time of 12:07. However, if expecting headwind on final approach, the pilot should shorten the 30 second outbound course somewhat, knowing that the wind will carry the aircraft away from the fix faster while outbound and decrease the ground speed while returning to the fix. On the other hand, compensating for a tailwind on final approach, the pilot should lengthen the calculated 30 second outbound heading somewhat, knowing that the wind would tend to hold the aircraft closer to the fix while outbound and increase the ground speed while returning to the fix.

5-4-11. Radar Approaches

a. The only airborne radio equipment required for radar approaches is a functioning radio transmitter and receiver. The radar controller vectors the aircraft to align it with the runway centerline. The controller continues the vectors to keep the aircraft on course until the pilot can complete the approach and landing by visual reference to the surface. There are two types of radar approaches: Precision (PAR) and Surveillance (ASR).

b. A radar approach may be given to any aircraft upon request and may be offered to pilots of aircraft in distress or to expedite traffic, however, an ASR might not be approved unless there is an ATC operational requirement, or in an unusual or emergency situation. Acceptance of a PAR or ASR by a pilot does not waive the prescribed weather minimums for the airport or for the particular aircraft operator concerned. The decision to make a radar approach when the reported weather is below the established minimums rests with the pilot.

c. PAR and ASR minimums are published on separate pages in the FAA Terminal Procedures Publication (TPP).

1. A PRECISION APPROACH (PAR) is one in which a controller provides highly accurate navigational guidance in azimuth and elevation to a pilot. Pilots are given headings to fly, to direct them to, and keep their aircraft aligned with the extended centerline of the landing runway. They are told to anticipate glidepath interception approximately 10 to 30 seconds before it occurs and when to start descent. The published Decision Height will be given only if the pilot requests it. If the aircraft is observed to deviate above or below the glidepath, the pilot is given the relative amount of deviation by use of terms "slightly" or "well" and is expected to adjust the aircraft's rate of descent/ascent to return to the glidepath. Trend information is also issued with respect to the elevation of the aircraft and may be modified by the terms "rapidly" and "slowly"; e.g., "well above glidepath, coming down rapidly." Range from touchdown is given at least once each mile. If an aircraft is observed by the controller to proceed outside of specified safety zone limits in azimuth and/or elevation and continue to operate outside these prescribed limits, the pilot will be directed to execute a missed approach or to fly a specified course unless the pilot has the runway environment (runway, approach lights, etc.) in sight. Navigational guidance in azimuth and elevation is provided the pilot until the aircraft reaches the published Decision Height (DH). Advisory course and glidepath information is furnished by the controller until the aircraft passes over the landing threshold, at which point the pilot is advised of any deviation from the runway centerline. Radar service is automatically terminated upon completion of the approach.

2. A SURVEILLANCE APPROACH (ASR) is one in which a controller provides navigational guidance in azimuth only. The pilot is furnished headings to fly to align the aircraft with the extended centerline of the landing runway. Since the radar information used for a surveillance approach is considerably less precise than that used for a precision approach, the accuracy of the approach will not be as great and higher minimums will apply. Guidance in elevation is not possible but the pilot will be advised when to commence descent to the Minimum Descent Altitude (MDA) or, if appropriate, to an intermediate step-down fix Minimum Crossing Altitude and subsequently to the prescribed MDA. In addition, the pilot will be advised of the location of the Missed Approach Point (MAP) prescribed for the procedure and the aircraft's position each mile on final from the runway, airport or heliport or MAP, as appropriate. If requested by the pilot, recommended altitudes will be issued at each mile, based on the descent gradient established for the procedure, down to the last mile that is at or above the MDA. Normally, navigational guidance will be provided until the aircraft reaches the MAP. Controllers will terminate guidance and instruct the pilot to execute a missed approach unless at the MAP the pilot has the runway, airport or heliport in sight or, for a helicopter point-in-space approach, the prescribed visual reference with the surface is established. Also, if, at any time during the approach the controller considers that safe guidance for the remainder of the approach cannot be provided, the controller will terminate guidance and instruct the pilot to execute a missed approach. Similarly, guidance termination and missed approach will be effected upon pilot request and, for civil aircraft only, controllers may terminate guidance when the pilot reports the runway, airport/heliport or visual surface route (point-in-space approach) in sight or otherwise indicates that continued guidance is not required. Radar service is automatically terminated at the completion of a radar approach.

NOTE-1. *The published MDA for straight-in approaches will be issued to the pilot before beginning descent. When a surveillance approach will terminate in a circle-to-land maneuver, the pilot must furnish the aircraft approach category to the controller. The controller will then provide the pilot with the appropriate MDA.*

2. ASR APPROACHES ARE NOT AVAILABLE WHEN AN ATC FACILITY IS USING CENRAP.

3. A NO-GYRO APPROACH is available to a pilot under radar control who experiences circumstances wherein the directional gyro or other stabilized compass is inoperative or inaccurate. When this occurs, the pilot should so advise ATC and request a No-Gyro vector or approach. Pilots of aircraft not equipped with a directional gyro or other stabilized compass who desire radar handling may also request a No-Gyro vector or approach. The pilot should make all turns at standard rate and should execute the turn immediately upon receipt of instructions. For example, "TURN RIGHT," "STOP TURN." When a surveillance or precision approach is made, the pilot will be advised after the aircraft has been turned onto final approach to make turns at half standard rate.

5-4-12. Radar Monitoring of Instrument Approaches

a. PAR facilities operated by the FAA and the military services at some joint-use (civil and military) and military installations monitor aircraft on instrument approaches and issue radar advisories to the pilot when weather is below VFR minimums (1,000 and 3), at night, or when requested by a pilot. This service is provided only when the PAR Final Approach Course coincides with the final approach of the navigational aid and only during the operational hours of the PAR. The radar advisories serve only as a secondary aid since the pilot has selected the navigational aid as the primary aid for the approach.

b. Prior to starting final approach, the pilot will be advised of the frequency on which the advisories will be transmitted. If, for any reason, radar advisories cannot be furnished, the pilot will be so advised.

c. Advisory information, derived from radar observations, includes information on:

1. Passing the final approach fix inbound (nonprecision approach) or passing the outer marker or fix used in lieu of the outer marker inbound (precision approach).

NOTE-*At this point, the pilot may be requested to report sighting the approach lights or the runway.*

2. Trend advisories with respect to elevation and/or azimuth radar position and movement will be provided.

NOTE-*Whenever the aircraft nears the PAR safety limit, the pilot will be advised that the aircraft is well above or below the glidepath or well left or right of course. Glidepath information is given only to those aircraft executing a precision approach, such as ILS or MLS. Altitude information is not transmitted to aircraft executing other than precision approaches because the descent portions of these approaches generally do not coincide with the depicted PAR glidepath. At locations where the MLS glidepath and PAR glidepath are not coincidental, only azimuth monitoring will be provided.*

3. If, after repeated advisories, the aircraft proceeds outside the PAR safety limit or if a radical deviation is observed, the pilot will be advised to execute a missed approach unless the prescribed visual reference with the surface is established.

d. Radar service is automatically terminated upon completion of the approach.

5-4-13. ILS/MLS Approaches to Parallel Runways

a. ATC procedures permit ILS instrument approach operations to dual or triple parallel runway configurations. ILS/MLS approaches to parallel runways are grouped into three classes: Parallel (dependent) ILS/MLS Approaches; Simultaneous Parallel (independent) ILS/MLS Approaches; and Simultaneous Close Parallel (independent) ILS Precision Runway Monitor (PRM) Approaches. (See FIG 5-4-19.) The classification of a parallel runway approach procedure is dependent on adjacent parallel runway centerline separation, ATC procedures, and airport ATC radar monitoring and communications capabilities. At some airports one or more parallel localizer courses may be offset up to 3 degrees. Offset localizer configurations result in loss of Category II capabilities and an increase in decision height (50').

b. Parallel approach operations demand heightened pilot situational awareness. A thorough Approach Procedure Chart review should be conducted with, as a minimum, emphasis on the following approach chart information: name and number of the approach, localizer frequency, inbound localizer/azimuth course, glide slope intercept altitude, decision height, missed approach instructions, special notes/procedures, and the assigned runway location/proximity to adjacent runways. Pilots will be advised that simultaneous ILS/MLS or simultaneous close parallel ILS PRM approaches are in use. This information may be provided through the ATIS.

c. The close proximity of adjacent aircraft conducting simultaneous parallel ILS/MLS and simultaneous close parallel ILS PRM approaches mandates strict pilot compliance with all ATC clearances. ATC assigned airspeeds, altitudes, and headings must be complied with in a timely manner. Autopilot coupled ILS/MLS approaches require pilot knowledge of procedures necessary to comply with

ATC instructions. Simultaneous parallel ILS/MLS and simultaneous close parallel ILS PRM approaches necessitate precise localizer tracking to minimize final monitor controller intervention, and unwanted No Transgression Zone (NTZ) penetration. In the unlikely event of a breakout, ATC will not assign altitudes lower than the minimum vectoring altitude. Pilots should notify ATC immediately if there is a degradation of aircraft or navigation systems.

d. Strict radio discipline is mandatory during parallel ILS/MLS approach operations. This includes an alert listening watch and the avoidance of lengthy, unnecessary radio transmissions. Attention must be given to proper call sign usage to prevent the inadvertent execution of clearances intended for another aircraft. Use of abbreviated call signs must be avoided to preclude confusion of aircraft with similar sounding call signs. Pilots must be alert to unusually long periods of silence or any unusual background sounds in their radio receiver. A stuck microphone may block the issuance of ATC instructions by the final monitor controller during simultaneous parallel ILS/MLS and simultaneous close parallel ILS PRM approaches.

REFERENCE-*AIM, Chapter 4, Section 2, Radio Communications Phraseology and Techniques, gives additional communications information.*

e. Use of Traffic Collision Avoidance Systems (TCAS) provides an additional element of safety to parallel approach operations. Pilots should follow recommended TCAS operating procedures presented in approved flight manuals, original equipment manufacturer recommendations, professional newsletters, and FAA publications.

FIGURE 5-4-19
Parallel ILS Approaches

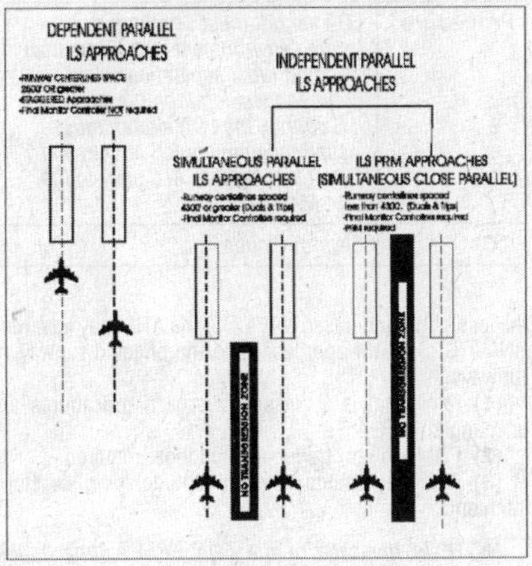

5-4-14. Parallel ILS/MLS Approaches (Dependent)

(See FIG 5-4-20.)

FIGURE 5-4-20
Staggered ILS Approaches

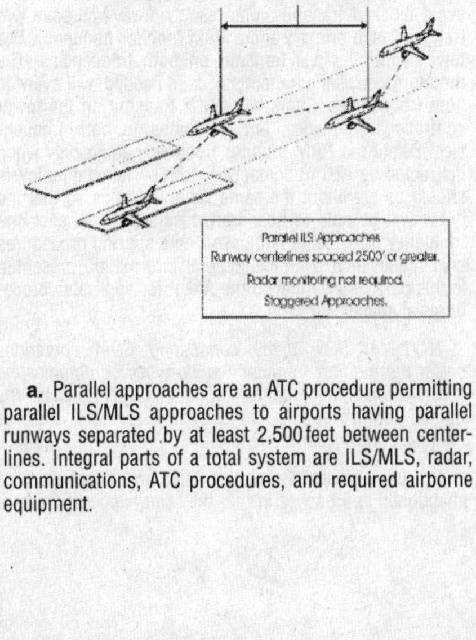

a. Parallel approaches are an ATC procedure permitting parallel ILS/MLS approaches to airports having parallel runways separated by at least 2,500 feet between centerlines. Integral parts of a total system are ILS/MLS, radar, communications, ATC procedures, and required airborne equipment.

b. A parallel (dependent) approach differs from a simultaneous (independent) approach in that, the minimum distance between parallel runway centerlines is reduced; there is no requirement for radar monitoring or advisories; and a staggered separation of aircraft on the adjacent localizer/azimuth course is required.

c. Aircraft are afforded a minimum of 1.5 miles radar separation diagonally between successive aircraft on the adjacent localizer/azimuth course when runway centerlines are at least 2,500 feet but no more than 4,300 feet apart. When runway centerlines are more than 4,300 feet but no more than 9,000 feet apart a minimum of 2 miles diagonal radar separation is provided. Aircraft on the same localizer/azimuth course within 10 miles of the runway end are provided a minimum of 2.5 miles radar separation. In addition, a minimum of 1,000 feet vertical or a minimum of three miles radar separation is provided between aircraft during turn on to the parallel final approach course.

d. Whenever parallel ILS/MLS approaches are in progress, pilots are informed that approaches to both runways are in use. In addition, the radar controller will have the interphone capability of communicating with the tower controller where separation responsibility has not been delegated to the tower.

5-4-15. Simultaneous Parallel ILS/MLS Approaches (Independent)

(See FIG 5-4-21.)

FIGURE 5-4-21
Simultaneous Parallel ILS Approaches

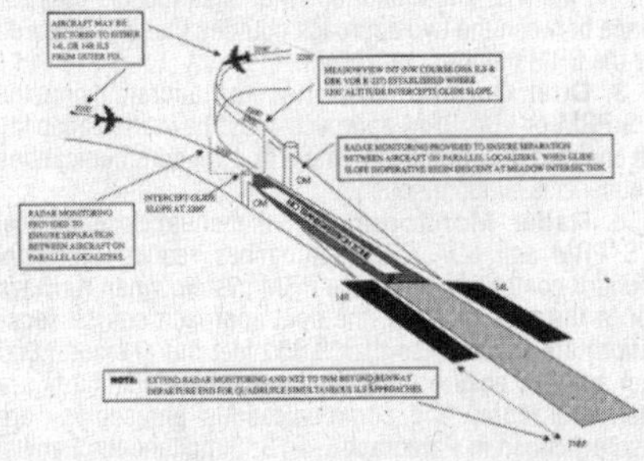

a. System. An approach system permitting simultaneous ILS/MLS approaches to parallel runways with centerlines separated by 4,300 to 9,000 feet, and equipped with final monitor controllers. Simultaneous parallel ILS/MLS approaches require radar monitoring to ensure separation between aircraft on the adjacent parallel approach course. Aircraft position is tracked by final monitor controllers who will issue instructions to aircraft observed deviating from the assigned localizer course. Staggered radar separation procedures are not utilized. Integral parts of a total system are ILS/MLS, radar, communications, ATC procedures, and required airborne equipment. The Approach Procedure Chart permitting simultaneous parallel ILS/MLS approaches will contain the note "simultaneous approaches authorized RWYS 14L and 14R," identifying the appropriate runways as the case may be. When advised that simultaneous parallel ILS/MLS approaches are in progress, pilots must advise approach control immediately of malfunctioning or inoperative receivers, or if a simultaneous parallel ILS/MLS approach is not desired.

b. Radar Monitoring. This service is provided for each simultaneous parallel ILS/MLS approach to ensure aircraft do not deviate from the final approach course. Radar monitoring includes instructions if an aircraft nears or penetrates the prescribed NTZ (an area 2,000 feet wide located equidistant between parallel final approach courses). This service will be provided as follows:

1. During turn on to parallel final approach, aircraft will be provided 3 miles radar separation or a minimum of 1,000 feet vertical separation. The assigned altitude must be maintained until intercepting the glide path, unless cleared otherwise by ATC. Aircraft will not be vectored to intercept the final approach course at an angle greater than thirty degrees.

2. The final monitor controller will have the capability of overriding the tower controller on the tower frequency.

3. Pilots will be instructed to monitor the tower frequency to receive advisories and instructions.

4. Aircraft observed to overshoot the turn-on or to continue on a track which will penetrate the NTZ will be instructed to return to the correct final approach course immediately. The final monitor controller may also issue missed approach or breakout instructions to the deviating aircraft.

PHRASEOLOGY-

"(Aircraft call sign) YOU HAVE CROSSED THE FINAL APPROACH COURSE. TURN (left/right) IMMEDIATELY AND RETURN TO THE LOCALIZER/AZIMUTH COURSE,"

or

"(aircraft call sign) TURN (left/right) AND RETURN TO THE LOCALIZER/AZIMUTH COURSE."

5. If a deviating aircraft fails to respond to such instructions or is observed penetrating the NTZ, the aircraft on the adjacent final approach course may be instructed to alter course.

PHRASEOLOGY-

"TRAFFIC ALERT (aircraft call sign) TURN (left/right) IMMEDIATELY HEADING (degrees), (climb/descend) AND MAINTAIN (altitude)."

6. Radar monitoring will automatically be terminated when visual separation is applied, the aircraft reports the approach lights or runway in sight, or the aircraft is 1 mile or less from the runway threshold (for runway centerlines spaced 4,300 feet or greater). Final monitor controllers will **not** advise pilots when radar monitoring is terminated.

5-4-16. Simultaneous Close Parallel ILS PRM and RNAV (GPS) PRM Approaches (Independent) and Simultaneous Offset Instrument Approaches (SOIA)

(See FIG 5-4-22.)

FIGURE 5-4-21
ILS PRM Approaches
(Simultaneous Close Parallel)

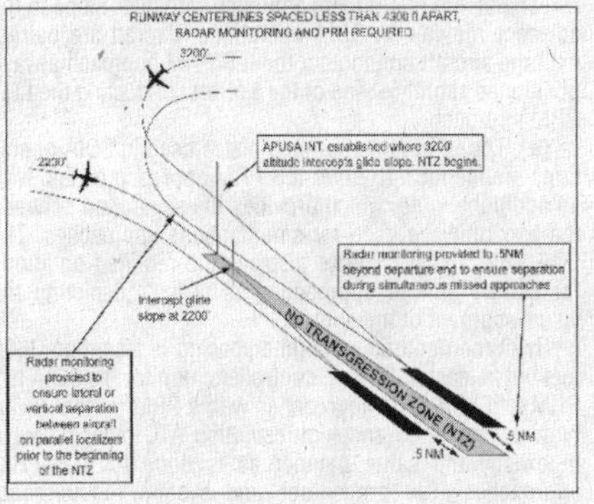

a. System

1. ILS PRM is an acronym for Instrument Landing System/Precision Runway Monitor.

(a) No Transgression Zone (NTZ) monitoring is required for all simultaneous close parallel PRM approaches. A Precision Runway Monitor (PRM) is an NTZ monitoring system that is required to conduct simultaneous ILS PRM approaches to parallel runways with centerlines separated by **less** than 3,600 feet but at least 3,400 feet for parallel approach courses, and at least 3,000 feet if one ILS if offset by 2.5 to 3.0 degrees. The airspace between the final approach courses is provided by two PRM monitor controllers, one for each approach course. To qualify for reduced lateral runway separation when runway spacing is less than 3,600 feet, NTZ monitor controllers must utilize high update radar or other high update surveillance system. The PRM system displays almost instantaneous radar information. The system consists of high resolution ATC radar displays, automated tracking software which provides PRM monitor controllers with aircraft identification, position, speed and a tensecond projected position, as well as visual and aural controller alerts. The PRM system is a supplemental requirement for simultaneous close parallel approaches below 3,600 feet in addition to the system requirements for simultaneous parallel ILS/MLS approaches described in Paragraph 5-4-15, Simultaneous Parallel ILS/MLS Approaches (Independent).

RNAV (GPS) PRM approaches may be substituted for one or both of the ILS PRM approaches in a simultaneous close parallel operation, or may be substituted for the ILS PRM and/or the LDA PRM approach in a SOIA operation. ATC may substitute RNAV (GPS) PRM approaches for one or both of the ILS PRM approaches, or in the case of SOIA, for either or both of the ILS PRM and LDA PRM approaches. The pilot may also request to fly the RNAV (GPS) PRM approach in lieu of either the ILS PRM and LDA PRM approach. In the event of the loss of ground based NAVAIDS, the ATIS may advertise RNAV (GPS) PRM approaches to the effected runway or runways. RNAV (GPS) PRM approaches carry the same applicable notations as do ILS PRM and LDA PRM approaches. The Attention All Users Page addresses ILS PRM, LDA PRM, and RNAV (GPS) PRM approaches as applicable. When an RNAV (GPS) PRM approach is used in lieu of an ILS PRM or LDA PRM approach, vertical guidance must be used. In all cases, the RNAV (GPS) PRM approach utilizes the same fixes, crossing altitudes, and missed approach procedure as contained in the ILS PRM or LDA PRM approach which it overlays. In the discussion below, an RNAV (GPS) PRM approach may be substituted for an ILS PRM or LDA PRM approach. The RNAV (GPS) lateral course may be substituted for reference to the ILS or LDA course. In SOIA, the RNAV (GPS) PRM Missed Approach Point (MAP) may be substituted for the LDA PRM MAP.

NOTE-*All simultaneous independent approaches to runways or approach courses less than 4,300 feet apart are categorized as simultaneous close parallel and include PRM in the approach name. Between 4,299 feet and 3,600 feet runway spacing, PRM or other high update rate surveillance system to monitor the NTZ is not required.*

(b) Simultaneous close parallel ILS PRM approaches are depicted on a separate Approach Procedure Chart titled ILS PRM Rwy XXX (Simultaneous Close Parallel).

2. SOIA is an acronym for Simultaneous Offset Instrument Approach, a procedure used to conduct simultaneous approaches to runways spaced less than 3,000 feet, but at least 750 feet apart. The SOIA procedure utilizes an ILS PRM approach to one runway and an offset Localizer Type Directional Aid (LDA) PRM approach with glide slope to the adjacent runway. In SOIA operations, aircraft are paired, with the aircraft conducting the ILS PRM approach always positioned slightly ahead of the aircraft conducting the LDA PRM approach.

(a) The ILS PRM approach plates used in SOIA operations are identical to other ILS PRM approach plates, with an additional note, which provides the separation between the two runways used for simultaneous approaches. The LDA PRM approach plate displays the required notations for closely spaced approaches as well as depicting the visual segment of the approach.

(b) When course centerline spacing is less than 3600 feet but at least 3000 feet, controllers monitor the SOIA ILS PRM and LDA PRM approaches with a PRM system using high update radar and highresolution ATC radar displays in exactly the same manner as is done for ILS PRM approaches. The procedures and system requirements for SOIA ILS PRM and LDA PRM approaches are identical with those used for simultaneous close parallel ILS PRM approaches until near the LDA PRM approach missed approach point (MAP) -- where visual acquisition of the ILS aircraft by the aircraft conducting the LDA PRM approach occurs. Since the ILS PRM and LDA PRM approaches are identical except for the visual segment in the SOIA concept, an understanding of the procedures for conducting ILS PRM approaches is essential before conducting a SOIA ILS PRM or LDA PRM operation.

(c) In SOIA, the approach course separation (instead of the runway separation) meets established close parallel approach criteria. Refer to FIG 5-4-23 for the generic SOIA approach geometry. A visual segment of the LDA PRM approach is established between the LDA MAP and the runway threshold. Aircraft transition in visual conditions from the LDA course, beginning at the LDA MAP, to align with the runway and can be stabilized by 500feet above ground level (AGL) on the extended runway centerline. Aircraft will be "paired" in SOIA operations, with the ILS aircraft ahead of the LDA aircraft prior to the LDA aircraft reaching the LDA MAP. A cloud ceiling for the approach is established so that the LDA aircraft has nominally 30 seconds to acquire the leading ILS aircraft prior to the LDA aircraft reaching the LDA MAP. If visual acquisition is not accomplished, a missed approach must be executed at the LDA MAP.

b. Requirements.

Besides system requirements as identified in subparagraph a, above, all pilots must have completed special training before accepting a clearance to conduct ILS PRM or LDA PRM Simultaneous Close Parallel Approaches.

1. Pilot Training Requirement. Pilots must complete **special pilot training**, as outlined below, before accepting a clearance for a simultaneous close parallel ILS PRM or LDA PRM approach.

(a) For operations under 14 CFR Parts 121, 129, and 135 pilots must comply with FAA approved company training as identified in their Operations Specifications. Training, at a minimum, must requirepilots to view the FAA video "ILS PRM AND SOIA APPROACHES: INFORMATION FOR AIR CARRIER PILOTS." Refer to**http://www.faa.gov/training_testing/training/prm/** for additional information and to view or download the video.

(b) For operations under Part 91:

(1) Pilots operating **transport category** aircraft must be familiar with PRM operations as contained in this section of the AIM. In addition, **pilots operating transport category aircraft** must view the FAA video "ILS PRM AND SOIA APPROACHES: INFORMATION FOR AIR CARRIER PILOTS." Refer to**http://www.faa.gov/training_testing/training/prm/** for additional information and to view or download the video.

(2) Pilots ***not*** operating **transport category** aircraft must be familiar with PRM and SOIA operations as contained in this section of the AIM. The FAA strongly recommends that pilots ***not*** involved **in transport category** aircraft operations view the FAA video, "ILS PRM AND SOIA APPROACHES: INFORMATION FOR GENERAL AVIATION PILOTS." Refer to **http://www.faa.gov/training_testing/training/prm/** for additional information and to view or download the video.

2. ATC Directed Breakout. An ATC directed "breakout" is defined as a vector off the ILS or LDA approach course in response to another aircraft penetrating the NTZ, the 2,000 foot wide area located equidistance between the two approach courses that is monitored by the PRM monitor controllers.

3. Dual Communications. The aircraft flying the ILS PRM or LDA PRM approach must have the capability of enabling the pilot/s to listen to two communications frequencies simultaneously.

c. Radar Monitoring. Simultaneous close parallel ILS PRM and LDA PRM approaches require that final monitor controllers utilize the PRM system when runways (or in the case of SOIA, the final approach course separation) are spaced less than 3,600 feet but at least 3,000 feet apart to ensure prescribed separation standards are met. Procedures and communications phraseology are also described in Paragraph5-4-15, Simultaneous Parallel ILS/MLS Approaches (Independent). A minimum of 3 miles radar separation or 1,000 feet vertical separation will be provided during the turnon to close parallel final approach courses. To ensure separation is maintained, and in order to avoid an imminent situation during simultaneous close parallel ILS PRM or SOIA ILS PRM and LDA PRM approaches, pilots must immediately comply with PRM monitor controller instructions. In the event of a missed approach, radar monitoring is provided to onehalf mile beyond the more distant of the two runway departure ends for ILS PRM approaches. In SOIA, PRM radar monitoring terminates at the LDA MAP. Final monitor controllers will **not** notify pilots when radar monitoring is terminated.

d. Attention All Users Page (AAUP). Multiple PRM approach charts at the same airport have a single AAUP associated with them that must be referred to in preparation for conducting the approach. This page contains the following instructions that must be followed if the pilot is unable to accept an ILS PRM or LDA PRM approach.

1. At airports that conduct PRM operations, (ILS PRM, and the case of airports where SOIAs are conducted, ILS PRM and LDA PRM approaches) pilots who are unable to participate will be afforded appropriate arrival services as operational conditions permit and must notify the controlling ARTCC as soon as practical, but at least 100 miles from destination.

2. The AAUP covers the following operational topics:

Bullet points are published which summarize the PRM procedures which apply to each approach. The following information may be summarized in the bullet points or published in more detail in the Expanded Procedures section of the AAUP.

(a) ATIS. When the ATIS broadcast advises ILS PRM approaches are in progress (or ILS PRM and LDA PRM approaches in the case of SOIA), pilots should brief to fly the ILS PRM or LDA PRM approach. If later advised to expect the ILS or LDA approach (should one be published), the ILS PRM or LDA PRM chart may be used after completing the following briefing items. The pilot may also request to fly the RNAV (GPS) PRM approach in lieu of either the ILS PRM and LDA PRM approach. In the event of the loss of ground based NAVAIDS, the ATIS may advertise RNAV (GPS) PRM approaches to the effected runway or runways.

(1) Minimums and missed approach procedures are unchanged.

(2) PRM Monitor frequency no longer required.

(3) ATC may assign a lower altitude for glide slope intercept.

NOTE-*In the case of the LDA PRM approach, this briefing procedure only applies if an LDADME approach is also published.*

In the case of the SOIA ILS PRM and LDA PRM procedure, the AAUP describes the weather conditions in which simultaneous approaches are authorized:

Simultaneous approach weather minimums are X,XXX feet (ceiling), x miles (visibility).

(b) Dual VHF Communications Required. To avoid blocked transmissions, each runway will have two frequencies, a primary and a PRM monitor frequency. The tower controller will transmit on both frequencies. The monitor controller's transmissions, if needed, will override both frequencies. Pilots will ONLY transmit on the tower controller's frequency, but will listen to both frequencies. Select the PRM monitor frequency audio only when instructed by ATC to contact the tower. The volume levels should be set about the same on both radios so that the pilots will be able to hear transmissions on at least one frequency if the other is blocked. Site specific procedures take precedence over the general information presented in this paragraph. Refer to the AAUP for applicable procedures at specific airports.

NOTE-*At SFO, pilots conducting SOIA operations select the monitor frequency audio when communicating with the final radar controller. In this special case, the monitor controller's transmissions, if required, override the final controller's frequency.*

(c) Breakouts. Breakouts differ from other types of abandoned approaches in that they can happen anywhere

FIG 5-4-23
SOIA Approach Geometry

NOTE-

SAP	*The SAP is a design point along the extended centerline of the intended landing runway on the glide slope at 500 feet above the landing threshold. It is used to verify a sufficient distance is provided for the visual maneuver after the missed approach point (MAP) to permit the pilots to conform to approved, stabilized approach criteria.*
MAP	*The point along the LDA where the course separation with the adjacent ILS reaches 3,000 feet. The altitude of the glide slope at that point determines the approach minimum descent altitude and is where the NTZ terminates. Maneuvering inside the MAP is done in visual conditions.*
Angle	*Angle formed at the intersection of the extended LDA runway centerline and a line drawn between the LDA MAP and the SAP. The size of the angle is determined by the FAA SOIA computer design program, and is dependent on whether CAT D aircraft use the LDA and the spacing between the runways.*
Visibility	*Distance from MAP to runway threshold in statute miles (light credit applies).*
Procedure	*LDA aircraft must see the runway landing environment and, if less than standard radar separation exists between the aircraft on the adjacent ILS course, the LDA aircraft must visually acquire the ILS aircraft and report it in sight to ATC prior to the LDA MAP.*
CC	*Clear of Clouds.*

and unexpectedly. Pilots directed by ATC to break off an approach must assume that an aircraft is blundering toward them and a breakout must be initiated **immediately**.

(1) Handfly breakouts. All breakouts are to be handflown to ensure the maneuver is accomplished in the shortest amount of time.

(2) ATC Directed "Breakouts." ATC directed breakouts will consist of a turn and a climb or descent. Pilots must always initiate the breakout in response to an air traffic controller's instruction. Controllers will give a descending breakout only when there are no other reasonable options available, but in no case will the descent be below the minimum vectoring altitude (MVA) which provides at least 1,000feet required obstruction clearance. The AAUP may provide the MVA in the final approach segment as X,XXX feet at (Name) Airport.

***NOTE-**"TRAFFIC ALERT." If an aircraft enters the "NO TRANSGRESSION ZONE" (NTZ), the controller will breakout the threatened aircraft on the adjacent approach. The phraseology for the breakout will be:*

***PHRASEOLOGY-**TRAFFIC ALERT, (aircraft call sign) TURN (left/right) IMMEDIATELY, HEADING (degrees), CLIMB/DESCEND AND MAINTAIN (altitude).*

(d) ILS PRM Navigation. The pilot may find crossing altitudes published along the final approach course. If the approach geometry warrants it, the pilot is advised that descending on the ILS glideslope ensures complying with any charted crossing restrictions.

SOIA and ILS PRM differences as noted on the AAUP

(e) ILS PRM, LDA Traffic (only published on the AAUP when the ILS PRM approach is used in conjunction with an LDA PRM approach to the adjacent runway). To provide better situational awareness, and because traffic on the LDA may be visible on the ILS aircraft's TCAS, pilots are reminded of the fact that aircraft will be maneuvering behind them to align with the adjacent runway. While conducting the ILS PRM approach to Runway XXX, other aircraft may be conducting the offset LDA PRM approach to Runway XXX. These aircraft will approach from the (left/right)rear and will realign with Runway XXX after making visual contact with the ILS traffic. Under normal circumstances, these aircraft will not pass the ILS traffic.

SOIA LDA PRM Items. The AAUP section for the SOIA LDA PRM approach contains most information found in the ILS PRM section. It replaces certain information as seen below and provides pilots with the procedures to be used in the visual segment of the LDA PRM approach from the LDA MAP until landing.

(f) SOIA LDA PRM Navigation (replaces ILS PRM (d) and (e) above). The pilot may find crossing altitudes published along the final approach course. The pilot is advised that descending on the LDA glideslope ensures complying with any charted crossing restrictions. Remain on the LDA course until passing XXXXX (LDA MAP name) intersection prior to maneuvering to align with the centerline of Runway XXX.

(g) SOIA (Name) Airport Visual Segment (replaces ILS PRM (e) above). Pilot procedures for navigating beyond the LDA MAP are spelled out. If ATC advises that there is traffic on the adjacent ILS, pilots are authorized to continue past the LDA MAP to align with runway centerline when:

(1) the ILS traffic is in sight and is expected to remain in sight,

(2) ATC has been advised that "traffic is in sight." (ATC is not required to acknowledge this transmission),

(3) the runway environment is in sight.

Otherwise, a missed approach must be executed. Between the LDA MAP and the runway threshold, pilots conducting the LDA PRM approach are responsible for separating themselves visually from traffic conducting the ILS PRM approach to the adjacent runway, which means maneuvering the aircraft as necessary to avoid that traffic until landing, and providing wake turbulence avoidance, if applicable. Pilots maintaining visual separation should advise ATC, as soon as practical, if visual contact with the aircraft conducting the ILS PRM approach is lost and execute a missed approach unless otherwise instructed by ATC.

e. Differences between ILS and ILS PRM approaches of importance to the pilot.

1. Runway Spacing. Prior to simultaneous close parallel approaches, most ATC directed breakouts were the result of two aircraft intrail on the same final approach course getting too close together. Two aircraft going in the same direction did not mandate quick reaction times. With PRM closely spaced approaches, two aircraft could be alongside each other, navigating on courses that are separated by less than 4,300 feet. In the unlikely event that an aircraft "blunders" off its course and makes a worst case turn of 30 degrees toward the adjacent final approach course, closing speeds of 135 feet per second could occur that constitute the need for quick reaction. A blunder has to be recognized by the monitor controller, and breakout instructions issued to the endangered aircraft. The pilot will not have any warning that a breakout is imminent because the blundering aircraft will be on another frequency. It is important that, when a pilot receives breakout instructions, he/she assumes that a blundering aircraft is about to or has penetrated the NTZ and is heading toward his/her approach course. The pilot must initiate a breakout as soon as safety allows. While conducting PRM approaches, pilots must maintain an increased sense of awareness in order to immediately react to an ATC instruction (**breakout**) and maneuver as instructed by ATC, away from a blundering aircraft.

2. Communications. To help in avoiding communication problems caused by stuck microphones and two parties talking at the same time, two frequencies for each runway will be in use during ILS PRM and LDA PRM approach operations, the primary tower frequency and the PRM monitor frequency. The tower controller transmits and receives in a normal fashion on the primary frequency and also transmits on the PRM monitor frequency. The monitor controller's transmissions override on both frequencies. The pilots flying the approach will listen to both frequencies but only transmit on the primary tower frequency. If the PRM monitor controller initiates a breakout and the primary frequency is blocked by another transmission, the breakout instruction will still be heard on the PRM monitor frequency.

***NOTE-**At some airports, the override capability may be on other than the tower frequency (KSFO overrides the final radar controller frequency). Pilots should carefully review the dual communications requirements on the AAUP prior to accepting a PRM approach.*

3. Handflown Breakouts. The use of the autopilot is encouraged while flying an ILS PRM or LDA PRM approach, but the autopilot must be disengaged in the rare event that a breakout is issued. Simulation studies of breakouts have shown that a handflown breakout can be initiated consistently faster than a breakout performed using the autopilot.

4. TCAS. The ATC breakout instruction is the primary means of conflict resolution. TCAS, if installed, provides another form of conflict resolution in the unlikely event other separation standards would fail. TCAS is not required to conduct a closely spaced approach.

The TCAS provides only vertical resolution of aircraft conflicts, while the ATC breakout instruction provides both vertical and horizontal guidance for conflict resolutions. Pilots should always immediately follow the TCAS Resolution Advisory (RA), whenever it is received. Should a TCAS RA be received before, during, or after an ATC breakout instruction is issued, the pilot should follow the RA, even if it conflicts with the climb/descent portion of the breakout maneuver. If following an RA requires deviating from an ATC clearance, the pilot must advise ATC as soon as practical. While following an RA, it is **extremely important** that the pilot also comply with the turn portion of the ATC breakout instruction unless the pilot determines safety to be factor. Adhering to these procedures assures the pilot that acceptable "breakout" separation margins will always be provided, even in the face of a normal procedural or system failure.

5. Breakouts. The probability is extremely low that an aircraft will "blunder" from its assigned approach course and enter the NTZ, causing ATC to "breakout" the aircraft approaching on the adjacent ILS or LDA course. However, because of the close proximity of the final approach courses, it is essential that pilots follow the ATC breakout instructions precisely and expeditiously. The controller's "breakout" instructions provide conflict resolution for the threatened aircraft, with the turn portion of the "breakout" being the single most important element in achieving maximum protection. A descending breakout will only be issued when it is the only controller option. In no case will the controller descend an aircraft below the MVA, which will provide at least 1,000 feet clearance above obstacles. The pilot is not expected to exceed 1,000 feet per minute rate of descent in the event a descending breakout is issued.

5-4-17. Simultaneous Converging Instrument Approaches

a. ATC may conduct instrument approaches simultaneously to converging runways; i.e., runways having an included angle from 15 to 100 degrees, at airports where a program has been specifically approved to do so.

b. The basic concept requires that dedicated, separate standard instrument approach procedures be developed for each converging runway included. Missed Approach Points must be at least 3 miles apart and missed approach procedures ensure that missed approach protected airspace does not overlap.

c. Other requirements are: radar availability, nonintersecting final approach courses, precision (ILS/MLS) approach systems on each runway and, if runways intersect, controllers must be able to apply visual separation as well as intersecting runway separation criteria. Intersecting runways also require minimums of at least 700 foot ceilings and 2 miles visibility. Straight in approaches and landings must be made.

d. Whenever simultaneous converging approaches are in progress, aircraft will be informed by the controller as soon as feasible after initial contact or via ATIS. Additionally, the radar controller will have direct communications capability with the tower controller where separation responsibility has not been delegated to the tower.

5-4-18. RNP AR Instrument Approach Procedures

These procedures require authorization analogous to the special authorization required for Category II or III ILS procedures. Authorization required (AR) procedures are to be conducted by aircrews meeting special training requirements in aircraft that meet the specified performance and functional requirements.

a. Unique characteristics of RNP AR Approaches

1. RNP value. Each published line of minima has an associated RNP value. The indicated value defines the lateral and vertical performance requirements. A minimum RNP type is documented as part of the RNP AR authorization for each operator and may vary depending on aircraft configuration or operational procedures (e.g., GPS inoperative, use of flight director vice autopilot).

2. Curved path procedures. Some RNP approaches have a curved path, also called a radius-to-a-fix (RF) leg. Since not all aircraft have the capability to fly these arcs, pilots are responsible for knowing if they can conduct an RNP approach with an arc or not. Aircraft speeds, winds and bank angles have been taken into consideration in the development of the procedures.

3. RNP required for extraction or not. Where required, the missed approach procedure may use RNP values less than RNP-1. The reliability of the navigation system has to be very high in order to conduct these approaches. Operation on these procedures generally requires redundant equipment, as no single point of failure can cause loss of both approach and missed approach navigation.

4. Non-standard speeds or climb gradients. RNP AR approaches are developed based on standard approach speeds and a 200 ft/NM climb gradient in the missed approach. Any exceptions to these standards will be indicated on the approach procedure, and the operator should ensure they can comply with any published restrictions before conducting the operation.

5. Temperature Limits. For aircraft using barometric vertical navigation (without temperature compensation) to conduct the approach, low and high-temperature limits are identified on the procedure. Cold temperatures reduce the glidepath angle while high temperatures increase the glidepath angle. Aircraft using baro VNAV with temperature compensation or aircraft using an alternate means for vertical guidance (e.g., SBAS) may disregard the temperature restrictions. The charted temperature limits are evaluated for the final approach segment only. Regardless of charted temperature limits or temperature compensation by the FMS, the pilot may need to manually compensate for cold temperature on minimum altitudes and the decision altitude.

6. Aircraft size. The achieved minimums may be dependent on aircraft size. Large aircraft may require higher minimums due to gear height and/or wingspan. Approach procedure charts will be annotated with applicable aircraft size restrictions.

b. Types of RNP AR Approach Operations

1. RNP Stand-alone Approach Operations. RNP AR procedures can provide access to runways regardless of the ground-based NAVAID infrastructure, and can be designed to avoid obstacles, terrain, airspace, or resolve environmental constraints.

2. RNP Parallel Approach (RPA) Operations. RNP AR procedures can be used for parallel approaches where the runway separation is adequate FIG 5-4-24). Parallel approach procedures can be used either simultaneously or as stand-alone operations. They may be part of either independent or dependent operations depending on the ATC ability to provide radar monitoring.

FIGURE 5-4-24

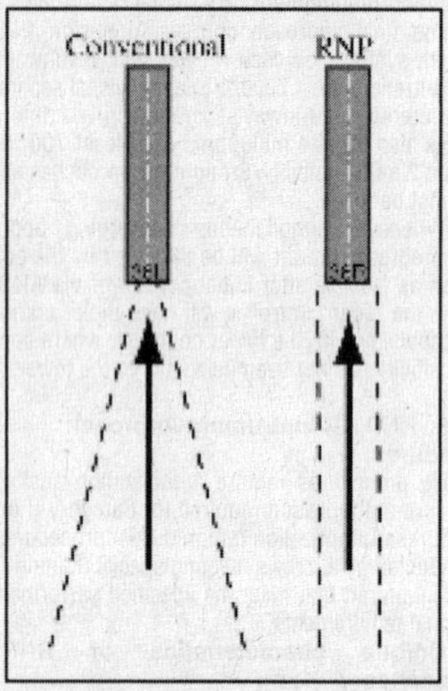

3. RNP Parallel Approach Runway Transitions (RPAT) Operations. RPAT approaches begin as a parallel IFR approach operation using simultaneous independent or dependent procedures. (See FIG 5-4-25). Visual separation standards are used in the final segment of the approach after the final approach fix, to permit the RPAT aircraft to transition in visual conditions along a predefined lateral and vertical path to align with the runway centerline.

FIGURE 5-4-25

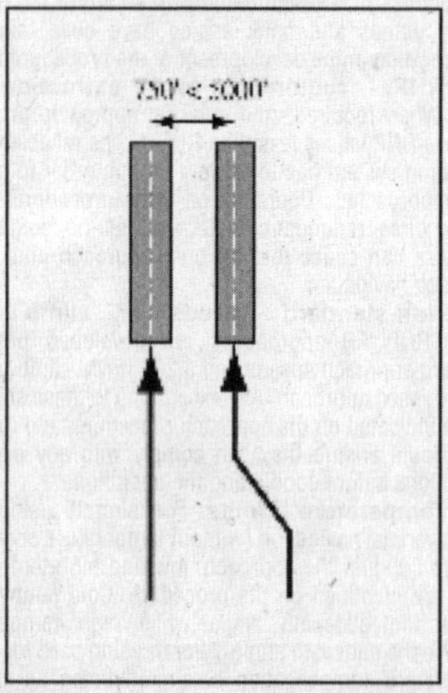

4. RNP Converging Runway Operations. At airports where runways converge, but may or may not intersect, an RNP AR approach can provide a precise curved missed approach path that conforms to aircraft separation minimums for simultaneous operations (See FIG 5-4-26). By flying this curved missed approach path with high accuracy and containment provided by RNP, dual runway operations may continue to be used to lower ceiling and visibility values than currently available. This type of operation allows greater capacity at airports where it can be applied.

FIGURE 5-4-26

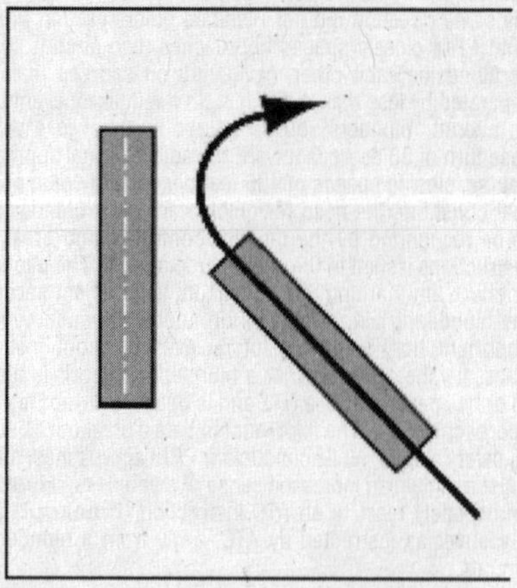

5-4-19. Side-step Maneuver

a. ATC may authorize a standard instrument approach procedure which serves either one of parallel runways that are separated by 1,200 feet or less followed by a straight-in landing on the adjacent runway.

b. Aircraft that will execute a side-step maneuver will be cleared for a specified approach procedure and landing on the adjacent parallel runway. Example, "cleared ILS runway 7 left approach, side-step to runway 7 right." Pilots are expected to commence the side-step maneuver as soon as possible after the runway or runway environment is in sight. Compliance with minimum altitudes associated with stepdown fixes is expected even after the side-step maneuver is initiated.

NOTE-*Side-step minima are flown to a Minimum Descent Altitude (MDA) regardless of the approach authorized.*

c. Landing minimums to the adjacent runway will be based on nonprecision criteria and therefore higher than the precision minimums to the primary runway, but will normally be lower than the published circling minimums.

5-4-20. Approach and Landing Minimums

a. Landing Minimums. The rules applicable to landing minimums are contained in 14 CFR Section 91.175. TBL 5-4-1 may be used to convert RVR to ground or flight visibility. For converting RVR values that fall between listed values, use the next higher RVR value; do not interpolate. For example, when converting 1800 RVR, use 2400 RVR with the resultant visibility of $^1/_2$ mile.

TABLE 5-4-1
RVR Value Conversions

RVR	Visibility (statute miles)
1600	1/4
2400	1/2
3200	5/8
4000	3/4
4500	7/8
5000	1
6000	1 1/4

b. Obstacle Clearance. Final approach obstacle clearance is provided from the start of the final segment to the runway or missed approach point, whichever occurs last. Side-step obstacle protection is provided by increasing the width of the final approach obstacle clearance area.

1. Circling approach protected areas are defined by the tangential connection of arcs drawn from each runway end (see FIG 5-4-27). Circling approach protected areas developed prior to late 2012 used fixed radius distances, dependent on aircraft approach category, as shown in the table on page B2 of the U.S. TPP. The approaches using standard circling approach areas can be identified by the absence of the "negative C" symbol on the circling line of minima. Circling approach protected areas developed after late 2012 use the radius distance shown in the table on page B2 of the U.S. TPP, dependent on aircraft approach category, and the altitude of the circling MDA, which accounts for true airspeed increase with altitude. The approaches using expanded circling approach areas can be identified by the presence of the "negative C" symbol on the circling line of minima (see FIG 5-4-28). Because of obstacles near the airport, a portion of the circling area may be restricted by a procedural note; for example, "Circling NA E of RWY 17-35." Obstacle clearance is provided at the published minimums (MDA) for the pilot who makes a straight-in approach, side-steps, or circles. Once below the MDA the pilot must see and avoid obstacles. Executing the missed approach after starting to maneuver usually places the aircraft beyond the MAP. The aircraft is clear of obstacles when at or above the MDA while inside the circling area, but simply joining the missed approach ground track from the circling maneuver may not provide vertical obstacle clearance once the aircraft exits the circling area. Additional climb inside the circling area may be required before joining the missed approach track. See Paragraph 5-4-21, Missed Approach, for additional considerations when starting a missed approach at other than the MAP.

FIGURE 5-4-27
Final Approach Obstacle Clearance

CIRCLING APPROACH AREA RADII

Approach Category	Radius (Miles)
A	1.3
B	1.5
C	1.7
D	2.3
E	4.5

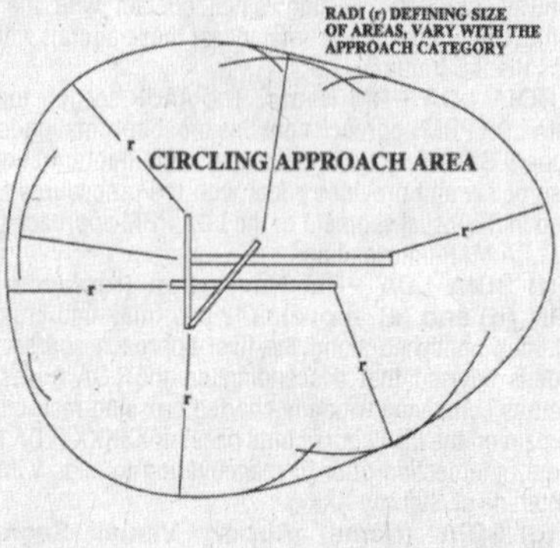

NOTE-*Circling approach area radii vary according to approach category and MSL circling altitude due to TAS changes - see FIG 5-4-28.*

FIGURE 5-4-27

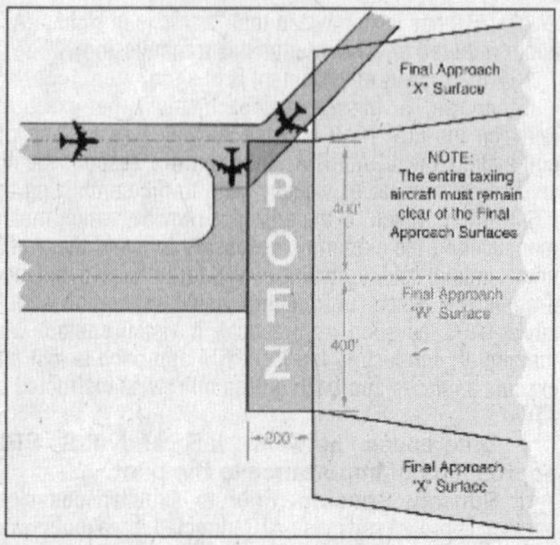

2. Precision Obstacle Free Zone (POFZ). A volume of airspace above an area beginning at the runway threshold, at the threshold elevation, and centered on the extended runway centerline. The POFZ is 200 feet (60m) long and 800 feet (240m) wide. The POFZ must be clear when an aircraft on a vertically guided final approach is within 2 nautical miles of the runway threshold and the reported ceiling is below 250 feet or visibility less than $^{3}/_{4}$ statute mile (SM) (or runway visual range below 4,000 feet). If the POFZ is not clear, the MINIMUM authorized height above touchdown (HAT) and visibility is 250feet and $^{3}/_{4}$ SM. The POFZ is considered clear even if the wing of the aircraft holding on a taxiway waiting for runway clearance penetrates the POFZ; however, neither the fuselage nor the tail may infringe on the POFZ. The POFZ is applicable at all runway ends including displaced thresholds.

Fig 5-4-29
Precision Obstacle Free Zone (POFZ)

c. Straight-in Minimums are shown on the IAP when the final approach course is within 30 degrees of the runway alignment (15 degrees for GPS IAPs) and a normal descent can be made from the IFR altitude shown on the IAP to the runway surface. When either the normal rate of descent or the runway alignment factor of 30 degrees (15 degrees for GPS IAPs) is exceeded, a straight-in minimum is not published and a circling minimum applies. The fact that a straight-in minimum is not published does not preclude pilots from landing straight-in if they have the active runway in sight and have sufficient time to make a normal approach for landing. Under such conditions and when ATC has cleared them for landing on that runway, pilots are not expected to circle even though only circling minimums are published. If they desire to circle, they should advise ATC.

d. Side-Step Maneuver Minimums. Landing minimums for a side-step maneuver to the adjacent runway will normally be higher than the minimums to the primary runway.

e. Published Approach Minimums. Approach minimums are published for different aircraft categories and consist of a minimum altitude (DA, DH, MDA) and required visibility. These minimums are determined by applying the appropriate TERPS criteria. When a fix is incorporated in a nonprecision final segment, two sets of minimums may be published: one for the pilot that is able to identify the fix, and a second for the pilot that cannot. Two sets of minimums may also be published when a second altimeter source is used in the procedure. When a nonprecision procedure incorporates both a stepdown fix in the final segment and a second altimeter source, two sets of minimums are published to account for the stepdown fix and a note addresses minimums for the second altimeter source.

f. Circling Minimums. In some busy terminal areas, ATC may not allow circling and circling minimums will not be published. Published circling minimums provide obstacle clearance when pilots remain within the appropriate area of protection. Pilots should remain at or above the circling altitude until the aircraft is continuously in a position from which a descent to a landing on the intended runway can be made at a normal rate of descent using normal maneuvers. Circling may require maneuvers at low altitude, at low airspeed, and in marginal weather conditions. Pilots must use sound judgment, have an indepth knowledge of their capabilities, and fully understand the aircraft performance to determine the exact circling maneuver since weather, unique airport design, and the aircraft position, altitude, and airspeed must all be considered. The following basic rules apply:

1. Maneuver the shortest path to the base or downwind leg, as appropriate, considering existing weather conditions. There is no restriction from passing over the airport or other runways.

2. It should be recognized that circling maneuvers may be made while VFR or other flying is in progress at the airport. Standard left turns or specific instruction from the controller for maneuvering must be considered when circling to land.

3. At airports without a control tower, it may be desirable to fly over the airport to observe wind and turn indicators and other traffic which may be on the runway or flying in the vicinity of the airport.

***REFERENCE-**AC 90-66A, Recommended Standards Traffic patterns for Aeronautical Operations at Airports without Operating Control Towers.*

4. The missed approach point (MAP) varies depending upon the approach flown. For vertically guided approaches, the MAP is at the decision altitude/decision height. Non-vertically guided and circling procedures share the same MAP and the pilot determines this MAP by timing from the final approach fix, by a fix, a NAVAID, or a waypoint. Circling from a GLS, an ILS without a localizer line of minima or an RNAV (GPS) approach without an LNAV line of minima is prohibited.

g. Instrument Approach at a Military Field. When instrument approaches are conducted by civil aircraft at military airports, they must be conducted in accordance with the procedures and minimums approved by the military agency having jurisdiction over the airport.

5-4-21. Missed Approach

a. When a landing cannot be accomplished, advise ATC and, upon reaching the missed approach point defined on the approach procedure chart, the pilot must comply with the missed approach instructions for the procedure being used or with an alternate missed approach procedure specified by ATC.

b. Obstacle protection for missed approach is predicated on the missed approach being initiated at the decision altitude/height (DA/H) or at the missed approach point and not lower than minimum descent altitude (MDA). A climb gradient of at least 200 feet per nautical mile is required, (except for Copter approaches, where a climb of at least 400 feet per nautical mile is required), unless a higher climb gradient is published in the notes section of the approach procedure chart. When higher than standard climb gradients are specified, the end point of the non-standard climb will be specified at either an altitude or a fix. Pilots must preplan to ensure that the aircraft can meet the climb gradient (expressed in feet per nautical mile) required by the procedure in the event of a missed approach, and be aware that flying at a higher than anticipated ground speed increases the climb rate requirement (feet per minute). Tables for the conversion of climb gradients (feet per nautical mile) to climb rate (feet per minute), based on ground speed, are included on page D1 of the U.S. Terminal Procedures booklets. Reasonable buffers are provided for normal maneuvers. However, no consideration is given to an abnormally early turn. Therefore, when an early missed approach is executed, pilots should, unless otherwise cleared by ATC, fly the IAP as specified on the approach plate to the missed approach point at or above the MDA or DH before executing a turning maneuver.

c. If visual reference is lost while circling-to-land from an instrument approach, the missed approach specified for that particular procedure must be followed (unless an alternate missed approach procedure is specified by ATC). To become established on the prescribed missed approach course, the pilot should make an initial climbing turn toward the landing runway and continue the turn until established on the missed approach course. Inasmuch as the circling maneuver may be accomplished in more than one direction, different patterns will be required to become established on the prescribed missed approach course, depending on the aircraft position at the time visual reference is lost. Adherence to the procedure will help assure that an aircraft will remain laterally within the circling and missed approach obstruction clearance areas. Refer to paragraph h concerning vertical obstruction clearance when starting a missed approach at other than the MAP. (See FIG 5-4-30.)

d. At locations where ATC radar service is provided, the pilot should conform to radar vectors when provided by ATC in lieu of the published missed approach procedure. (See FIG 5-4-31.)

e. Some locations may have a preplanned alternate missed approach procedure for use in the event the primary NAVAID used for the missed approach procedure is unavailable. To avoid confusion, the alternate missed approach instructions are not published on the chart. However, the alternate missed approach holding pattern will be depicted on the instrument approach chart for pilot situational awareness and to assist ATC by not having to issue detailed holding instructions. The alternate missed approach may be based on NAVAIDs not used in the approach procedure or the primary missed approach. When the alternate missed approach procedure is implemented by NOTAM, it becomes a mandatory part of the procedure. The NOTAM will specify both the textual instructions and any additional equipment requirements necessary to complete the procedure. Air traffic may also issue instructions for the alternate missed approach when necessary, such as when the primary missed approach NAVAID fails during the approach. Pilots may reject an ATC clearance for an alternate missed approach that requires equipment not necessary for the published approach procedure when the alternate missed approach is issued after beginning the approach. However, when the alternate missed approach is issued prior to beginning the approach the pilot must either accept the entire procedure (including the alternate missed approach), request a different approach procedure, or coordinate with ATC for alternative action to be taken, i.e., proceed to an alternate airport, etc.

f. When approach has been missed, request clearance for specific action; i.e., to alternative airport, another approach, etc.

AQ: Please provide Figure 5-4-2

g. Pilots must ensure that they have climbed to a safe altitude prior to proceeding off the published missed approach, especially in nonradar environments. Abandoning the missed approach prior to reaching the published altitude may not provide adequate terrain clearance. Additional climb may be required after reaching the holding pattern before proceeding back to the IAF or to an alternate.

Figure 5-4-30
Circling and Missed Approach Obstruction Clearance Areas

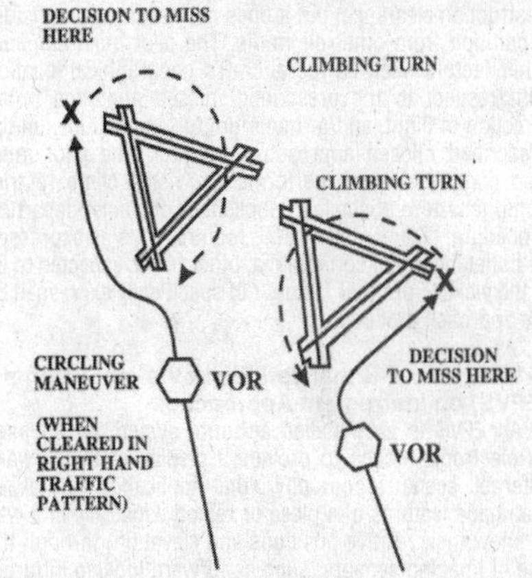

Figure 5-4-31
Missed Approach

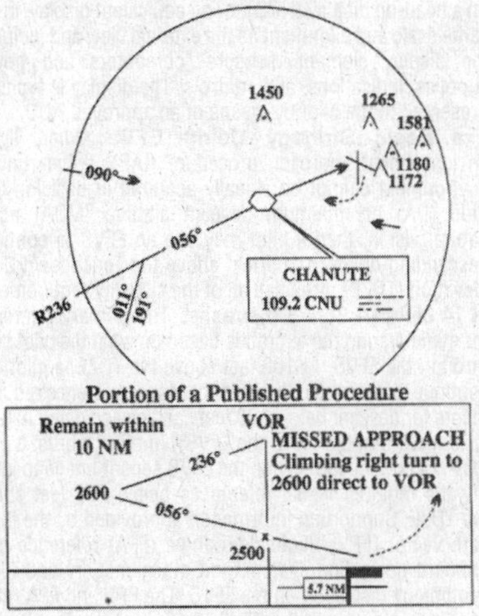

h. A clearance for an instrument approach procedure includes a clearance to fly the published missed approach procedure, unless otherwise instructed by ATC. The published missed approach procedure provides obstacle clearance only when the missed approach is conducted on the missed approach segment from or above the missed approach point, and assumes a climb rate of 200 feet/NM or higher, as published. If the aircraft initiates a missed approach at a point other than the missed approach point (see paragraph 5-4-5b), from below MDA or DA (H), or on a circling approach, obstacle clearance is not necessarily provided by following the published missed approach procedure, nor is separation assured from other air traffic in the vicinity.

In the event a balked (rejected) landing occurs at a position other than the published missed approach point, the pilot should contact ATC as soon as possible to obtain an amended clearance. If unable to contact ATC for any reason, the pilot should attempt to re-intercept a published segment of the missed approach and comply with route and altitude instructions. If unable to contact ATC, and in the pilot's judgment it is no longer appropriate to fly the published missed approach procedure, then consider either maintaining visual conditions if practicable and reattempt a landing, or a circle-climb over the airport. Should a missed approach become necessary when operating to an airport that is not served by an operating control tower, continuous contact with an air traffic facility may not be possible. In this case, the pilot should execute the appropriate go-around/missed approach procedure without delay and contact ATC when able to do so.

Prior to initiating an instrument approach procedure, the pilot should assess the actions to be taken in the event of a balked (rejected) landing beyond the missed approach point or below the MDA or DA (H) considering the anticipated weather conditions and available aircraft performance. 14 CFR 91.175(e) authorizes the pilot to fly an appropriate missed approach procedure that ensures obstruction clearance, but it does not necessarily consider separation from other air traffic. The pilot must consider other factors such as the aircraft's geographical location with respect to the prescribed missed approach point, direction of flight, and/or minimum turning altitudes in the prescribed missed approach procedure. The pilot must also consider aircraft performance, visual climb restrictions, charted obstacles, published obstacle departure procedure, takeoff visual climb requirements as expressed by nonstandard takeoff minima, other traffic expected to be in the vicinity, or other factors not specifically expressed by the approach procedures.

5-4-22. Use of Enhanced Flight Vision Systems (EFVS) on Instrument Approaches

An EFVS is an installed airborne system which uses an electronic means to provide a display of the forward external scene topography (the applicable natural or manmade features of a place or region especially in a way to show their relative positions and elevation) through the use of imaging sensors, such as forward looking infrared, millimeter wave radiometry, millimeter wave radar, and/or low light level image intensifying. The EFVS imagery is displayed along with the additional flight information and aircraft flight symbology required by 14 CFR 91.175 (m) on a head-up display (HUD), or an equivalent display, in the same scale and alignment as the external view and includes the display element, sensors, computers and power supplies, indications, and controls. The display is typically presented to the pilot by means of an approved HUD.

a. Basic Strategy Using EFVS. When flying an instrument approach procedure (IAP), if the runway environment cannot be visually acquired at decision altitude (DA) or minimum descent altitude (MDA) using natural vision, then a pilot may use an EFVS to continue descending down to 100 feet above the Touchdown Zone Elevation (TDZE), provided all of the visibility requirements of 14 CFR part 91.175 (l) are met. The primary reference for maneuvering the aircraft is based on what the pilot sees through the EFVS. At 100 feet above the TDZE, a pilot can continue to descend only when the visual reference requirements for descent below 100 feet can be seen using natural vision (without the aid of the EFVS). In other words, a pilot may not continue to rely on the EFVS sensor image to identify the required visual references below 100 feet above the TDZE. Supporting information is provided by the flight path vector (FPV), flight path angle (FPA) reference cue, onboard navigation system, and other imagery and flight symbology displayed on the EFVS. The FPV and FPA reference cue, along with the EFVS imagery of the Touchdown Zone (TDZ), provide the primary vertical path reference for the pilot when vertical guidance from a precision approach or approach with vertical guidance is not available.

1. Straight-In Instrument Approach Procedures. An EFVS may be used to descend below DA or MDA from any straight-in IAP, other than Category II or Category III approaches, provided all of the requirements of 14 CFR part 91.175 (l) are met. This includes straight-in precision approaches, approaches with vertical guidance (for example, LPV or LNAV/VNAV), and non-precision approaches (for example, VOR, NDB, LOC, RNAV, GPS, LDA, SDF, etc.).

2. Circling Approach Procedure. An IAP with a circle-to-land maneuver or circle-to-land minimums does not meet criteria for straight-in landing minimums. While the regulations do not prohibit EFVS from being used during any phase of flight, they do prohibit it from being used for operational credit on anything but a straight-in IAP with straight-in landing minima. EFVS must only be used during a circle-to-land maneuver provided the visual references required throughout the circling maneuver are distinctly visible using natural vision. An EFVS cannot be used to satisfy the requirement that an identifiable part of the airport be distinctly visible to the pilot during a circling maneuver at or above MDA or while descending below MDA from a circling maneuver.

3. Enhanced Flight Visibility. Flight visibility is determined by using natural vision, and enhanced flight visibility (EFV) is determined by using an EFVS. 14 CFR part 91.175 (l) requires that the EFV observed by using an EFVS cannot be less than the visibility prescribed in the IAP to be used in order to continue to descend below the DA or MDA.

b. EFVS Operations At or Below DA or MDA Down to 100 Feet Above the TDZE. The visual segment of an IAP begins at DA or MDA and continues to the runway. There are two means of operating in the visual segment--one is by using natural vision and the other is by using an EFVS. If the pilot determines that the EFV observed by using the EFVS is not less than the minimum visibility prescribed in the IAP being flown, and the pilot acquires the required visual references prescribed in 14 CFR part 91.175 (l)(3) using the EFVS, then the pilot can continue the approach to 100 feet above the TDZE. To continue the approach, the pilot uses the EFVS image to visually acquire the runway environment (the approach light system (ALS), if installed, or both the runway threshold and the TDZ), confirm lateral alignment, maneuver to the extended runway centerline earlier than would otherwise be possible, and continue a normal descent from the DA or MDA to 100 feet above the TDZE.

1. Required Visual References. In order to descend below DA or MDA, the following visual references (specified in 14 CFR part 91.175 (l)(3)) for the runway of intended landing must be distinctly visible and identifiable to the pilot using the EFVS:

(a) The ALS (if installed), or

(b) The following visual references in both (b)(1) and (b)(2) below:

(1) The runway threshold, identified by at least one of the following: the beginning of the runway landing surface, the threshold lights, or the runway end identifier lights (REIL).

(2) The TDZ, identified by at least one of the following: the runway TDZ landing surface, the TDZ lights, the TDZ markings, or the runway lights.

2. Comparison of Visual Reference Requirements for EFVS and Natural Vision. The EFVS visual reference requirements of 14 CFR part 91.175 (l)(3) comprise a more stringent standard than the visual reference requirements prescribed under 14 CFR part 91.175 (c)(3) when using natural vision. The more stringent standard is needed because an EFVS might not display the color of the lights used to identify specific portions of the runway or might not be able to consistently display the runway markings. The main differences for EFVS operations are that the visual glide slope indicator (VGSI) lights cannot be used as a visual reference, and specific visual references from both the threshold and TDZ must be distinctly visible and identifiable. However, when using natural vision, only one of the specified visual references must be visible and identifiable.

3. Visual References and Offset Approaches. Pilots must be especially knowledgeable of the approach conditions and approach course alignment when considering whether to rely on EFVS during a non-precision approach with an offset final approach course. Depending upon the combination of crosswind correction and the lateral field of view provided by a particular EFVS, the required visual references may or may not be within the pilot's view looking through the EFVS display. Pilots conducting any non-precision approach must verify lateral alignment with the runway centerline when determining when to descend from MDA.

4. When to Go Around. Any pilot operating an aircraft with an EFVS installed should be aware that the requirements of 14 CFR part 91.175 (c) for using natural vision and the requirements of 14 CFR part 91.175 (l) for using an EFVS are different. A pilot would, therefore, first have to determine whether an approach will be commenced using natural vision or using an EFVS. While these two sets of requirements provide a parallel decisionmaking process, the requirements for when a missed approach must be executed differ. Using EFVS, a missed approach must be initiated at or below DA or MDA down to 100 feet above TDZE whenever the pilot determines that:

(a) The EFV is less than the visibility minima prescribed for the IAP being used;

(b) The required visual references for the runway of intended landing are no longer distinctly visible and identifiable to the pilot using the EFVS imagery;

(c) The aircraft is not continuously in a position from which a descent to a landing can be made on the intended runway, at a normal rate of descent, using normal maneuvers; or

(d) For operations under 14 CFR parts 121 and 135, the descent rate of the aircraft would not allow touchdown to occur within the TDZ of the runway of intended landing.

5. Missed Approach Considerations. It should be noted that a missed approach after passing the DA, or beyond the missed approach point (MAP), involves additional risk until established on the published missed approach segment. Initiating a go-around after passing the published MAP may result in loss of obstacle clearance. As with any approach, pilot planning should include contingencies between the published MAP and touchdown with reference to obstacle clearance, aircraft performance, and alternate escape plans.

c. EFVS Operations At and Below 100 Feet Above the TDZE. At and below 100 feet above the TDZE, the regulations do not require the EFVS to be turned off or the display to be stowed in order to continue to a landing. A pilot may continue the approach below this altitude using an EFVS as long as the required visual references can be seen through the display using natural vision. An operator may not continue to descend beyond this point by relying solely on the sensor image displayed on the EFVS.

1. Required Visual References. In order to descend below 100 feet above the TDZE, the flight visibility--assessed using natural vision--must be sufficient for the following visual references to be distinctly visible and identifiable to the pilot without reliance on the EFVS to continue to a landing:

(a) The lights or markings of the threshold, or

(b) The lights or markings of the TDZ.

It is important to note that from 100 feet above the TDZE and below, the flight visibility does not have to be equal to or greater than the visibility prescribed for the IAP in order to continue descending. It only has to be sufficient for the visual references required by 14 CFR part 91.175 (l)(4) to be distinctly visible and identifiable to the pilot without reliance on the EFVS.

2. Comparison of Visual Reference Requirements for EFVS and Natural Vision. Again, the visual reference requirements for EFVS in 14 CFR part 91.175 (l)(4) are more stringent than those required for natural vision in 14 CFR part 91.175 (c)(3). The main differences for EFVS operations are that the ALS and red terminating bars or red side row bars, the REIL, and the VASI cannot be used as visual references. Only very specific visual references from the threshold or the TDZ can be used (that is, the lights or markings of the threshold or the lights or markings of the TDZ).

3. When to Go Around. A missed approach must be initiated when the pilot determines that:

(a) The flight visibility is no longer sufficient to distinctly see and identify the required visual references listed in 14 CFR part 91.175 (l)(4) using natural vision;

(b) The aircraft is not continuously in a position from which a descent to a landing can be made on the intended runway, at a normal rate of descent, using normal maneuvers; or

(c) For operations under 14 CFR parts 121 and 135, the descent rate of the aircraft would not allow touchdown to occur within the TDZ of the runway of intended landing.

While touchdown within the TDZ is not specifically addressed in the regulations for operators other than

14 CFR parts 121 and 135 operators, continued operations below DA or MDA where touchdown in the TDZ is not assured, where a high sink rate occurs, or where the decision to conduct a missed approach procedure is not executed in a timely manner, all create a significant risk to the operation.

4. Missed Approach Considerations. As noted earlier, a missed approach initiated after the DA or MAP involves additional risk. At 100 feet or less above the runway, it is likely that an aircraft is significantly below the TERPS missed approach obstacle clearance surface. Prior planning is recommended and should include contingencies between the published MAP and touchdown with reference to obstacle clearance, aircraft performance, and alternate escape plans.

d. Light Emitting Diode (LED) Airport Lighting Impact on EFVS Operations. The FAA has recently begun to replace incandescent lamps with LEDs at some airports in threshold lights, taxiway edge lights, taxiway centerline lights, low intensity runway edge lights, windcone lights, beacons, and some obstruction lighting. Pilots should be aware that LED lights cannot be sensed by current EFVS systems.

5-4-23. Visual Approach

a. A visual approach is conducted on an IFR flight plan and authorizes a pilot to proceed visually and clear of clouds to the airport. The pilot must have either the airport or the preceding identified aircraft in sight. This approach must be authorized and controlled by the appropriate air traffic control facility. Reported weather at the airport must have a ceiling at or above 1,000 feet and visibility 3 miles or greater. ATC may authorize this type approach when it will be operationally beneficial. Visual approaches are an IFR procedure conducted under IFR in visual meteorological conditions. Cloud clearance requirements of 14 CFR Section 91.155 are not applicable, unless required by operation specifications.

b. Operating to an Airport Without Weather Reporting Service. ATC will advise the pilot when weather is not available at the destination airport. ATC may initiate a visual approach provided there is a reasonable assurance that weather at the airport is a ceiling at or above 1,000 feet and visibility 3 miles or greater (e.g., area weather reports, PIREPs, etc.).

c. Operating to an Airport With an Operating Control Tower. Aircraft may be authorized to conduct a visual approach to one runway while other aircraft are conducting IFR or VFR approaches to another parallel, intersecting, or converging runway. When operating to airports with parallel runways separated by less than 2,500 feet, the succeeding aircraft must report sighting the preceding aircraft unless standard separation is being provided by ATC. When operating to parallel runways separated by at least 2,500 feet but less than 4,300 feet, controllers will clear/vector aircraft to the final at an angle not greater than 30 degrees unless radar, vertical, or visual separation is provided during the turn-on. The purpose of the 30 degree intercept angle is to reduce the potential for overshoots of the final and to preclude side-by-side operations with one or both aircraft in a belly-up configuration during the turn-on. Once the aircraft are established within 30 degrees of final, or on the final, these operations may be conducted simultaneously. When the parallel runways are separated by 4,300 feet or more, or intersecting/converging runways are in use, ATC may authorize a visual approach after advising all aircraft involved that other aircraft are conducting operations to the other runway. This may be accomplished through use of the ATIS.

d. Separation Responsibilities. If the pilot has the airport in sight but cannot see the aircraft to be followed, ATC may clear the aircraft for a visual approach; however, ATC retains both separation and wake vortex separation responsibility. When visually following a preceding aircraft, acceptance of the visual approach clearance constitutes acceptance of pilot responsibility for maintaining a safe approach interval and adequate wake turbulence separation.

e. A visual approach is not an IAP and therefore has no missed approach segment. If a go around is necessary for any reason, aircraft operating at controlled airports will be issued an appropriate advisory/clearance/instruction by the tower. At uncontrolled airports, aircraft are expected to remain clear of clouds and complete a landing as soon as possible. If a landing cannot be accomplished, the aircraft is expected to remain clear of clouds and contact ATC as soon as possible for further clearance. Separation from other IFR aircraft will be maintained under these circumstances.

f. Visual approaches reduce pilot/controller workload and expedite traffic by shortening flight paths to the airport. It is the pilot's responsibility to advise ATC as soon as possible if a visual approach is not desired.

g. Authorization to conduct a visual approach is an IFR authorization and does not alter IFR flight plan cancellation responsibility.

***REFERENCE-**AIM, Canceling IFR Flight Plan, Paragraph 5-1-15.*

h. Radar service is automatically terminated, without advising the pilot, when the aircraft is instructed to change to advisory frequency.

5-4-24. Charted Visual Flight Procedure (CVFP)

a. CVFPs are charted visual approaches established for environmental/noise considerations, and/or when necessary for the safety and efficiency of air traffic operations. The approach charts depict prominent landmarks, courses, and recommended altitudes to specific runways. CVFPs are designed to be used primarily for turbojet aircraft.

b. These procedures will be used only at airports with an operating control tower.

c. Most approach charts will depict some NAVAID information which is for supplemental navigational guidance only.

d. Unless indicating a Class B airspace floor, all depicted altitudes are for noise abatement purposes and are recommended only. Pilots are not prohibited from flying other than recommended altitudes if operational requirements dictate.

e. When landmarks used for navigation are not visible at night, the approach will be annotated "*PROCEDURE NOT AUTHORIZED AT NIGHT.*"

f. CVFPs usually begin within 20 flying miles from the airport.

g. Published weather minimums for CVFPs are based on minimum vectoring altitudes rather than the recommended altitudes depicted on charts.

h. CVFPs are not instrument approaches and do not have missed approach segments.

i. ATC will not issue clearances for CVFPs when the weather is less than the published minimum.

j. ATC will clear aircraft for a CVFP after the pilot reports siting a charted landmark or a preceding aircraft. If instructed to follow a preceding aircraft, pilots are responsible for maintaining a safe approach interval and wake turbulence separation.

k. Pilots should advise ATC if at any point they are unable to continue an approach or lose sight of a preceding aircraft. Missed approaches will be handled as a go-around.

5-4-25. Contact Approach

a. Pilots operating in accordance with an IFR flight plan, provided they are clear of clouds and have at least 1 mile flight visibility and can reasonably expect to continue to the destination airport in those conditions, may request ATC authorization for a contact approach.

b. Controllers may authorize a contact approach provided:

1. The contact approach is specifically requested by the pilot. ATC cannot initiate this approach.

EXAMPLE-
Request contact approach.

2. The reported ground visibility at the destination airport is at least 1 statute mile.

3. The contact approach will be made to an airport having a standard or special instrument approach procedure.

4. Approved separation is applied between aircraft so cleared and between these aircraft and other IFR or special VFR aircraft.

EXAMPLE-
Cleared contact approach (and, if required) at or below (altitude) (routing) if not possible (alternative procedures) and advise.

c. A contact approach is an approach procedure that may be used by a pilot (with prior authorization from ATC) in lieu of conducting a standard or special IAP to an airport. It is not intended for use by a pilot on an IFR flight clearance to operate to an airport not having a published and functioning IAP. Nor is it intended for an aircraft to conduct an instrument approach to one airport and then, when "in the clear," discontinue that approach and proceed to another airport. In the execution of a contact approach, the pilot assumes the responsibility for obstruction clearance. If radar service is being received, it will automatically terminate when the pilot is instructed to change to advisory frequency.

5-4-26. Landing Priority

A clearance for a specific type of approach (ILS, MLS, ADF, VOR or Straight-in Approach) to an aircraft operating on an IFR flight plan does not mean that landing priority will be given over other traffic. ATCTs handle all aircraft, regardless of the type of flight plan, on a "first-come, first-served" basis. Therefore, because of local traffic or runway in use, it may be necessary for the controller in the interest of safety, to provide a different landing sequence. In any case, a landing sequence will be issued to each aircraft as soon as possible to enable the pilot to properly adjust the aircraft's flight path.

5-4-27. Overhead Approach Maneuver

a. Pilots operating in accordance with an IFR flight plan in Visual Meteorological Conditions (VMC) may request ATC authorization for an overhead maneuver. An overhead maneuver is not an instrument approach procedure. Overhead maneuver patterns are developed at airports where aircraft have an operational need to conduct the maneuver. An aircraft conducting an overhead maneuver is considered to be VFR and the IFR flight plan is cancelled when the aircraft reaches the initial point on the initial approach portion of the maneuver. (See FIG 5-4-32.) The existence of a standard overhead maneuver pattern does not eliminate the possible requirement for an aircraft to conform to conventional rectangular patterns if an overhead maneuver cannot be approved. Aircraft operating to an airport without a functioning control tower must initiate cancellation of an IFR flight plan prior to executing the overhead maneuver. Cancellation of the IFR flight plan must be accomplished after crossing the landing threshold on the initial portion of the maneuver or after landing. Controllers may authorize an overhead maneuver and issue the following to arriving aircraft:

1. Pattern altitude and direction of traffic. This information may be omitted if either is standard.

***PHRASEOLOGY-**PATTERN ALTITUDE (altitude). RIGHT TURNS.*

2. Request for a report on initial approach.

***PHRASEOLOGY-**REPORT INITIAL.*

3. "Break" information and a request for the pilot to report. The "Break Point" will be specified if nonstandard. Pilots may be requested to report "break" if required for traffic or other reasons.

PHRASEOLOGY-
BREAK AT (specified point).
REPORT BREAK.

FIGURE 5-4-30
Overhead Maneuver

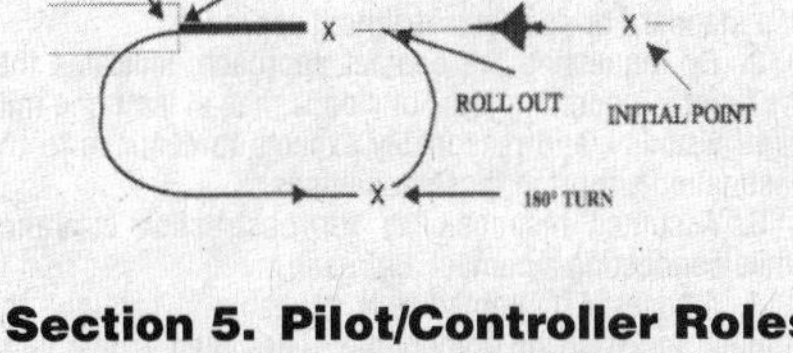

Section 5. Pilot/Controller Roles and Responsibilities

5-5-1. General

a. The roles and responsibilities of the pilot and controller for effective participation in the ATC system are contained in several documents. Pilot responsibilities are in the CFRs and the air traffic controllers' are in the FAA Order JO 7110.65, Air Traffic Control, and supplemental FAA directives. Additional and supplemental information for pilots can be found in the current Aeronautical Information Manual (AIM), Notices to Airmen, Advisory Circulars and aeronautical charts. Since there are many other excellent publications produced by nongovernment organizations, as well as other government organizations, with various updating cycles, questions concerning the latest or most current material can be resolved by cross-checking with the above mentioned documents.

b. The pilot-in-command of an aircraft is directly responsible for, and is the final authority as to the safe operation of that aircraft. In an emergency requiring immediate action, the pilot-in-command may deviate from any rule in the General Subpart A and Flight Rules Subpart B in accordance with 14 CFR Section 91.3.

c. The air traffic controller is responsible to give first priority to the separation of aircraft and to the issuance of radar safety alerts, second priority to other services that are required, but do not involve separation of aircraft and third priority to additional services to the extent possible.

d. In order to maintain a safe and efficient air traffic system, it is necessary that each party fulfill their responsibilities to the fullest.

e. The responsibilities of the pilot and the controller intentionally overlap in many areas providing a degree of redundancy. Should one or the other fail in any manner, this overlapping responsibility is expected to compensate, in many cases, for failures that may affect safety.

f. The following, while not intended to be all inclusive, is a brief listing of pilot and controller responsibilities for some commonly used procedures or phases of flight. More detailed explanations are contained in other portions of this publication, the appropriate CFRs, ACs and similar publications. The information provided is an overview of the principles involved and is not meant as an interpretation of the rules nor is it intended to extend or diminish responsibilities.

5-5-2. Air Traffic Clearance

a. Pilot.

1. Acknowledges receipt and understanding of an ATC clearance.

2. Reads back any hold short of runway instructions issued by ATC.

3. Requests clarification or amendment, as appropriate, any time a clearance is not fully understood or considered unacceptable from a safety standpoint.

4. Promptly complies with an air traffic clearance upon receipt except as necessary to cope with an emergency. Advises ATC as soon as possible and obtains an amended clearance, if deviation is necessary.

NOTE-*A clearance to land means that appropriate separation on the landing runway will be ensured. A landing clearance does not relieve the pilot from compliance with any previously issued altitude crossing restriction.*

b. Controller.

1. Issues appropriate clearances for the operation to be conducted, or being conducted, in accordance with established criteria.

2. Assigns altitudes in IFR clearances that are at or above the minimum IFR altitudes in controlled airspace.

3. Ensures acknowledgement by the pilot for issued information, clearances, or instructions.

4. Ensures that readbacks by the pilot of altitude, heading, or other items are correct. If incorrect, distorted, or incomplete, makes corrections as appropriate.

5-5-3. Contact Approach

a. Pilot.

1. Must request a contact approach and makes it in lieu of a standard or special instrument approach.

2. By requesting the contact approach, indicates that the flight is operating clear of clouds, has at least one mile flight visibility, and reasonably expects to continue to the destination airport in those conditions.

3. Assumes responsibility for obstruction clearance while conducting a contact approach.

4. Advises ATC immediately if unable to continue the contact approach or if encounters less than 1 mile flight visibility.

5. Is aware that if radar service is being received, it may be automatically terminated when told to contact the tower.

REFERENCE-*Pilot/Controller Glossary Term- Radar Service Terminated.*

b. Controller.

1. Issues clearance for a contact approach only when requested by the pilot. Does not solicit the use of this procedure.

2. Before issuing the clearance, ascertains that reported ground visibility at destination airport is at least 1 mile.

3. Provides approved separation between the aircraft cleared for a contact approach and other IFR or special VFR aircraft. When using vertical separation, does not assign a fixed altitude, but clears the aircraft at or below an altitude which is at least 1,000 feet below any IFR traffic but not below Minimum Safe Altitudes prescribed in 14 CFR Section 91.119.

4. Issues alternative instructions if, in their judgment, weather conditions may make completion of the approach impracticable.

5-5-4. Instrument Approach

a. Pilot.

1. Be aware that the controller issues clearance for approach based only on known traffic.

2. Follows the procedure as shown on the IAP, including all restrictive notations, such as:

(a) Procedure not authorized at night;

(b) Approach not authorized when local area altimeter not available;

(c) Procedure not authorized when control tower not in operation;

(d) Procedure not authorized when glide slope not used;

(e) Straight-in minimums not authorized at night; etc.

(f) Radar required; or

(g) The circling minimums published on the instrument approach chart provide adequate obstruction clearance and pilots should not descend below the circling altitude until the aircraft is in a position to make final descent for landing. Sound judgment and knowledge of the pilot's and the aircraft's capabilities are the criteria for determining the exact maneuver in each instance since airport design and the aircraft position, altitude and airspeed must all be considered.

REFERENCE-*AIM, Approach and Landing Minimums, Paragraph 5-4-20.*

3. Upon receipt of an approach clearance while on an unpublished route or being radar vectored:

(a) Complies with the minimum altitude for IFR; and

(b) Maintains the last assigned altitude until established on a segment of a published route or IAP, at which time published altitudes apply.

b. Controller.

1. Issues an approach clearance based on known traffic.

2. Issues an IFR approach clearance only after the aircraft is established on a segment of published route or IAP, or assigns an appropriate altitude for the aircraft to maintain until so established.

5-5-5. Missed Approach

a. Pilot.

1. Executes a missed approach when one of the following conditions exist:

(a) Arrival at the Missed Approach Point (MAP) or the Decision Height (DH) and visual reference to the runway environment is insufficient to complete the landing.

(b) Determines that a safe approach or landing is not possible (see subparagraph 5-4-21h).

(c) Instructed to do so by ATC.

2. Advises ATC that a missed approach will be made. Include the reason for the missed approach unless the missed approach is initiated by ATC.

3. Complies with the missed approach instructions for the IAP being executed from the MAP, unless other missed approach instructions are specified by ATC.

4. If executing a missed approach prior to reaching the MAP, fly the lateral navigation path of the instrument procedure to the MAP. Climb to the altitude specified in the missed approach procedure, except when a maximum altitude is specified between the final approach fix (FAF) and the MAP. In that case, comply with the maximum altitude restriction. Note, this may require a continued descent on the final approach.

5. Following a missed approach, requests clearance for specific action; i.e., another approach, hold for improved conditions, proceed to an alternate airport, etc.

b. Controller.

1. Issues an approved alternate missed approach procedure if it is desired that the pilot execute a procedure other than as depicted on the instrument approach chart.

2. May vector a radar identified aircraft executing a missed approach when operationally advantageous to the pilot or the controller.

3. In response to the pilot's stated intentions, issues a clearance to an alternate airport, to a holding fix, or for reentry into the approach sequence, as traffic conditions permit.

5-5-6. Radar Vectors

a. Pilot.

1. Promptly complies with headings and altitudes assigned to you by the controller.

2. Questions any assigned heading or altitude believed to be incorrect.

3. If operating VFR and compliance with any radar vector or altitude would cause a violation of any CFR, advises ATC and obtains a revised clearance or instructions.

b. Controller.

1. Vectors aircraft in Class A, Class B, Class C, Class D, and Class E airspace:

(a) For separation.

(b) For noise abatement.

(c) To obtain an operational advantage for the pilot or controller.

2. Vectors aircraft in Class A, Class B, Class C, Class D, Class E, and Class G airspace when requested by the pilot.

3. Vectors IFR aircraft at or above minimum vectoring altitudes.

4. May vector VFR aircraft, not at an ATC assigned altitude, at any altitude. In these cases, terrain separation is the pilot's responsibility.

5-5-7. Safety Alert

a. Pilot.

1. Initiates appropriate action if a safety alert is received from ATC.

2. Be aware that this service is not always available and that many factors affect the ability of the controller to be aware of a situation in which unsafe proximity to terrain, obstructions, or another aircraft may be developing.

b. Controller.

1. Issues a safety alert if aware an aircraft under their control is at an altitude which, in the controller's judgment, places the aircraft in unsafe proximity to terrain, obstructions or another aircraft. Types of safety alerts are:

(a) Terrain or Obstruction Alert. Immediately issued to an aircraft under their control if aware the aircraft is at an altitude believed to place the aircraft in unsafe proximity to terrain or obstructions.

(b) Aircraft Conflict Alert. Immediately issued to an aircraft under their control if aware of an aircraft not under their control at an altitude believed to place the aircraft in unsafe proximity to each other. With the alert, they offer the pilot an alternative, if feasible.

2. Discontinue further alerts if informed by the pilot action is being taken to correct the situation or that the other aircraft is in sight.

5-5-8. See and Avoid

a. Pilot. When meteorological conditions permit, regardless of type of flight plan or whether or not under control of a radar facility, the pilot is responsible to see and avoid other traffic, terrain, or obstacles.

b. Controller.

1. Provides radar traffic information to radar identified aircraft operating outside positive control airspace on a workload permitting basis.

2. Issues safety alerts to aircraft under their control if aware the aircraft is at an altitude believed to place the aircraft in unsafe proximity to terrain, obstructions, or other aircraft.

5-5-9. Speed Adjustments

a. Pilot.

1. Advises ATC any time cruising airspeed varies plus or minus 5 percent or 10 knots, whichever is greater, from that given in the flight plan.

2. Complies with speed adjustments from ATC unless:

(a) The minimum or maximum safe airspeed for any particular operation is greater or less than the requested airspeed. In such cases, advises ATC.

NOTE-*It is the pilot's responsibility and prerogative to refuse speed adjustments considered excessive or contrary to the aircraft's operating specifications.*

(b) Operating at or above 10,000 feet MSL on an ATC assigned SPEED ADJUSTMENT of more than 250 knots IAS and subsequent clearance is received for descent below 10,000 feet MSL. In such cases, pilots are expected to comply with 14 CFR Section 91.117(a).

3. When complying with speed adjustment assignments, maintains an indicated airspeed within plus or minus 10 knots or 0.02 Mach number of the specified speed.

b. Controller.

1. Assigns speed adjustments to aircraft when necessary but not as a substitute for good vectoring technique.

2. Adheres to the restrictions published in the FAAO JO 7110.65, Air Traffic Control, as to when speed adjustment procedures may be applied.

3. Avoids speed adjustments requiring alternate decreases and increases.

4. Assigns speed adjustments to a specified IAS (KNOTS)/Mach number or to increase or decrease speed using increments of 10 knots or multiples thereof.

5. Advises pilots to resume normal speed when speed adjustments are no longer required.

6. Gives due consideration to aircraft capabilities to reduce speed while descending.

7. Does not assign speed adjustments to aircraft at or above FL 390 without pilot consent.

5-5-10. Traffic Advisories (Traffic Information)

a. Pilot.

1. Acknowledges receipt of traffic advisories.

2. Informs controller if traffic in sight.

3. Advises ATC if a vector to avoid traffic is desired.

4. Does not expect to receive radar traffic advisories on all traffic. Some aircraft may not appear on the radar display. Be aware that the controller may be occupied with higher priority duties and unable to issue traffic information for a variety of reasons.

5. Advises controller if service is not desired.

b. Controller.

1. Issues radar traffic to the maximum extent consistent with higher priority duties except in Class A airspace.

2. Provides vectors to assist aircraft to avoid observed traffic when requested by the pilot.

3. Issues traffic information to aircraft in the Class B, Class C, and Class D surface areas for sequencing purposes.

4. Controllers are required to issue to each aircraft operating on intersecting or nonintersecting converging runways where projected flight paths will cross.

5-5-11. Visual Approach

a. Pilot.

1. If a visual approach is not desired, advises ATC.

2. Complies with controller's instructions for vectors toward the airport of intended landing or to a visual position behind a preceding aircraft.

3. The pilot must, at all times, have either the airport or the preceding aircraft in sight. After being cleared for a visual approach, proceed to the airport in a normal manner or follow the preceding aircraft. Remain clear of clouds while conducting a visual approach.

4. If the pilot accepts a visual approach clearance to visually follow a preceding aircraft, you are required to establish a safe landing interval behind the aircraft you were instructed to follow. You are responsible for wake turbulence separation.

5. Advise ATC immediately if the pilot is unable to continue following the preceding aircraft, cannot remain clear of clouds, needs to climb, or loses sight of the airport.

6. Be aware that radar service is automatically terminated, without being advised by ATC, when the pilot is instructed to change to advisory frequency.

7. Be aware that there may be other traffic in the traffic pattern and the landing sequence may differ from the traffic sequence assigned by approach control or ARTCC.

b. Controller.

1. Do not clear an aircraft for a visual approach unless reported weather at the airport is ceiling at or above 1,000 feet and visibility is 3 miles or greater. When weather is not available for the destination airport, inform the pilot and do not initiate a visual approach to that airport unless there is reasonable assurance that descent and flight to the airport can be made visually.

2. Issue visual approach clearance when the pilot reports sighting either the airport or a preceding aircraft which is to be followed.

3. Provide separation except when visual separation is being applied by the pilot.

4. Continue flight following and traffic information until the aircraft has landed or has been instructed to change to advisory frequency.

5. Inform the pilot when the preceding aircraft is a heavy.

6. When weather is available for the destination airport, do not initiate a vector for a visual approach unless the reported ceiling at the airport is 500 feet or more above the MVA and visibility is 3 miles or more. If vectoring weather minima are not available but weather at the airport is ceiling at or above 1,000 feet and visibility of 3 miles or greater, visual approaches may still be conducted.

7. Informs the pilot conducting the visual approach of the aircraft class when pertinent traffic is known to be a heavy aircraft.

5-5-12. Visual Separation

a. Pilot.

1. Acceptance of instructions to follow another aircraft or to provide visual separation from it is an acknowledgment that the pilot will maneuver the aircraft as necessary to avoid the other aircraft or to maintain in-trail separation. Pilots are responsible to maintain visual separation until flight paths (altitudes and/or courses) diverge.

2. If instructed by ATC to follow another aircraft or to provide visual separation from it, promptly notify the controller if you lose sight of that aircraft, are unable to maintain continued visual contact with it, or cannot accept the responsibility for your own separation for any reason.

3. The pilot also accepts responsibility for wake turbulence separation under these conditions.

b. Controller. Applies visual separation only:

1. Within the terminal area when a controller has both aircraft in sight or by instructing a pilot who sees the other aircraft to maintain visual separation from it.

2. Pilots are responsible to maintain visual separation until flight paths (altitudes and/or courses) diverge.

3. Within en route airspace when aircraft are on opposite courses and one pilot reports having seen the other aircraft and that the aircraft have passed each other.

5-5-13. VFR-on-top

a. Pilot.

1. This clearance must be requested by the pilot on an IFR flight plan, and if approved, allows the pilot the choice (subject to any ATC restrictions) to select an altitude or flight level in lieu of an assigned altitude.

NOTE- *VFR-on-top is not permitted in certain airspace areas, such as Class A airspace, certain restricted areas, etc. Consequently, IFR flights operating VFR-on-top will avoid such airspace.*

REFERENCE- *AIM, IFR Clearance VFR-on-top, Paragraph 4-4-8.*
AIM, IFR Separation Standards, Paragraph 4-4-11.
AIM, Position Reporting, Paragraph 5-3-2.
AIM, Additional Reports, Paragraph 5-3-3.

2. By requesting a VFR-on-top clearance, the pilot assumes the sole responsibility to be vigilant so as to see and avoid other aircraft and to:

(a) Fly at the appropriate VFR altitude as prescribed in 14 CFR Section 91.159.

(b) Comply with the VFR visibility and distance from clouds criteria in 14 CFR Section 91.155, *Basic VFR weather minimums.*

(c) Comply with instrument flight rules that are applicable to this flight; i.e., minimum IFR altitudes, position reporting, radio communications, course to be flown, adherence to ATC clearance, etc.

3. Should advise ATC prior to any altitude change to ensure the exchange of accurate traffic information.

b. Controller.

1. May clear an aircraft to maintain VFR-on-top if the pilot of an aircraft on an IFR flight plan requests the clearance.

2. Informs the pilot of an aircraft cleared to climb to VFR-on-top the reported height of the tops or that no top report is available; issues an alternate clearance if necessary; and once the aircraft reports reaching VFR-on-top, reclears the aircraft to maintain VFR-on-top.

3. Before issuing clearance, ascertain that the aircraft is not in or will not enter Class A airspace.

5-5-14. Instrument Departures

a. Pilot.

1. Prior to departure considers the type of terrain and other obstructions on or in the vicinity of the departure airport.

2. Determines if obstruction avoidance can be maintained visually or that the departure procedure should be followed.

3. Determines whether a departure procedure and/or DP is available for obstruction avoidance.

4. At airports where IAPs have not been published, hence no published departure procedure, determines what action will be necessary and takes such action that will assure a safe departure.

b. Controller.

1. At locations with airport traffic control service, when necessary, specifies direction of takeoff, turn, or initial heading to be flown after takeoff.

2. At locations without airport traffic control service but within Class E surface area when necessary to specify direction of takeoff, turn, or initial heading to be flown, obtains pilot's concurrence that the procedure will allow the pilot to comply with local traffic patterns, terrain, and obstruction avoidance.

3. Includes established departure procedures as part of the ATC clearance when pilot compliance is necessary to ensure separation.

5-5-15. Minimum Fuel Advisory

a. Pilot.

1. Advise ATC of your minimum fuel status when your fuel supply has reached a state where, upon reaching destination, you cannot accept any undue delay.

2. Be aware this is not an emergency situation, but merely an advisory that indicates an emergency situation is possible should any undue delay occur.

3. On initial contact the term "minimum fuel" should be used after stating call sign.

EXAMPLE-
Salt Lake Approach, United 621, "minimum fuel."

4. Be aware a minimum fuel advisory does not imply a need for traffic priority.

5. If the remaining usable fuel supply suggests the need for traffic priority to ensure a safe landing, you should declare an emergency due to low fuel and report fuel remaining in minutes.

REFERENCE- *Pilot/Controller Glossary Item- Fuel Remaining.*

b. Controller.

1. When an aircraft declares a state of minimum fuel, relay this information to the facility to whom control jurisdiction is transferred.

2. Be alert for any occurrence which might delay the aircraft.

5-5-16. RNAV and RNP Operations

a. Pilot.

1. If unable to comply with the requirements of an RNAV or RNP procedure, pilots must advise air traffic control as soon as possible. For example, "N1234, failure of GPS system, unable RNAV, request amended clearance."

2. Pilots are not authorized to fly a published RNAV or RNP procedure (instrument approach, departure, or arrival procedure) unless it is retrievable by the procedure name from the current aircraft navigation database and conforms to the charted procedure. The system must be able to retrieve the procedure by name from the aircraft navigation database, not just as a manually entered series of waypoints.

3. Whenever possible, RNAV routes (Q- or T-route) should be extracted from the database in their entirety, rather than loading RNAV route waypoints from the database into the flight plan individually. However, selecting and inserting individual, named fixes from the database is permitted, provided all fixes along the published route to be flown are inserted.

4. Pilots must not change any database waypoint type from a fly-by to fly-over, or vice versa. No other modification of database waypoints or the creation of user-defined waypoints on published RNAV or RNP procedures is permitted, except to:

(a) Change altitude and/or airspeed waypoint constraints to comply with an ATC clearance/instruction.

(b) Insert a waypoint along the published route to assist in complying with ATC instruction, example, "Descend via the WILMS arrival except cross 30 north of BRUCE at/or below FL 210." This is limited only to systems that allow along-track waypoint construction.

5. Pilots of FMS-equipped aircraft, who are assigned an RNAV DP or STAR procedure and subsequently receive a change of runway, transition or procedure, must verify that the appropriate changes are loaded and available for navigation.

6. For RNAV 1 DPs and STARs, pilots must use a CDI, flight director and/or autopilot, in lateral navigation mode. Other methods providing an equivalent level of performance may also be acceptable.

7. For RNAV 1 DPs and STARs, pilots of aircraft without GPS, using DME/DME/IRU, must ensure the aircraft navigation system position is confirmed, within 1,000 feet, at the start point of take-off roll. The use of an automatic or

manual runway update is an acceptable means of compliance with this requirement. Other methods providing an equivalent level of performance may also be acceptable.

8. For procedures or routes requiring the use of GPS, if the navigation system does not automatically alert the flight crew of a loss of GPS, the operator must develop procedures to verify correct GPS operation.

9. RNAV terminal procedures (DP and STAR) may be amended by ATC issuing radar vectors and/or clearances direct to a waypoint. Pilots should avoid premature manual deletion of waypoints from their active "legs" page to allow for rejoining procedures.

10. RAIM Prediction: If TSO-C129 equipment is used to solely satisfy the RNAV and RNP requirement, GPS RAIM availability must be confirmed for the intended route of flight (route and time). If RAIM is not available, pilots need an approved alternate means of navigation.

REFERENCE-*AIM, RNAV and RNP Operations, Paragraph 5-1-16*

11. Definition of "established" for RNAV and RNP operations. An aircraft is considered to be established oncourse during RNAV and RNP operations anytime it is within 1 times the required accuracy for the segment being flown. For example, while operating on a QRoute (RNAV 2), the aircraft is considered to be established oncourse when it is within 2 nm of the course centerline.

NO*TE-Pilots must be aware of how their navigation system operates, along with any AFM limitations, and confirm that the aircraft's lateral deviation display (or map display if being used as an allowed alternate means) is suitable for the accuracy of the segment being flown. Automatic scaling and alerting changes are appropriate for some operations. For example, TSOC129 systems change within 30 miles of destination and within 2 miles of FAF to support approach operations. For some navigation systems and operations, manual selection of scaling will be necessary.*

(a) Pilots flying FMS equipped aircraft with barometric vertical navigation (BaroVNAV) may descend when the aircraft is established oncourse following FMS leg transition to the next segment. Leg transition normally occurs at the turn bisector for a flyby waypoint (reference paragraph 121 for more on waypoints). When using full automation, pilots should monitor the aircraft to ensure the aircraft is turning at appropriate lead times and descending once established oncourse.

(b) Pilots flying TSOC129 navigation system equipped aircraft without full automation should use normal lead points to begin the turn. Pilots may descend when established oncourse on the next segment of the approach.

Section 6. National Security and Interception Procedures

5-6-1. National Security

a. National security in the control of air traffic is governed by 14 CFR Part 99.

b. All aircraft entering domestic U.S. airspace from points outside must provide for identification prior to entry. To facilitate early aircraft identification of all aircraft in the vicinity of U.S. and international airspace boundaries, Air Defense Identification Zones (ADIZ) have been established.

REFERENCE-*AIM, ADIZ Boundaries and Designated Mountainous Areas,*
Paragraph 5-6-5.

c. Operational requirements for aircraft operations associated with an ADIZ are as follows:

1. Flight Plan. Except as specified in subparagraphs d and e below, an IFR or DVFR flight plan must be filed with an appropriate aeronautical facility as follows:

(a) Generally, for all operations that enter an ADIZ.

(b) For operations that will enter or exit the U.S. and which will operate into, within or across the Contiguous U.S. ADIZ regardless of true airspeed.

(c) The flight plan must be filed before departure except for operations associated with the Alaskan ADIZ when the airport of departure has no facility for filing a flight plan, in which case the flight plan may be filed immediately after takeoff or when within range of the aeronautical facility.

2. Two-way Radio. For the majority of operations associated with an ADIZ, an operating two-way radio is required. See 14 CFR Section 99.1 for exceptions.

3. Transponder Requirements. Unless otherwise authorized by ATC, each aircraft conducting operations into, within, or across the Contiguous U.S. ADIZ must be equipped with an operable radar beacon transponder having altitude reporting capability (Mode C), and that transponder must be turned on and set to reply on the appropriate code or as assigned by ATC.

4. Position Reporting.

(a) For IFR flight. Normal IFR position reporting.

(b) For DVFR flights. The estimated time of ADIZ penetration must be filed with the aeronautical facility at least 15 minutes prior to penetration except for flight in the Alaskan ADIZ, in which case report prior to penetration.

(c) For inbound aircraft of foreign registry. The pilot must report to the aeronautical facility at least one hour prior to ADIZ penetration.

5. Aircraft Position Tolerances.

(a) Over land, the tolerance is within plus or minus five minutes from the estimated time over a reporting point or point of penetration and within 10 NM from the centerline of an intended track over an estimated reporting point or penetration point.

(b) Over water, the tolerance is plus or minus five minutes from the estimated time over a reporting point or point of penetration and within 20 NM from the centerline of the intended track over an estimated reporting point or point of penetration (to include the Aleutian Islands).

6. Land-Based ADIZ. Land-Based ADIZ are activated and deactivated over U.S. metropolitan areas as needed, with dimensions, activation dates and other relevant information disseminated via NOTAM.

(a) In addition to requirements outlined in subparagraphs c1 through c3, pilots operating within a Land-Based ADIZ must report landing or leaving the Land-Based ADIZ if flying too low for radar coverage.

(b) Pilots unable to comply with all requirements must remain clear of Land-Based ADIZ. Pilots entering a Land-Based ADIZ without authorization or who fail to follow all requirements risk interception by military fighter aircraft.

d. Except when applicable under 14 CFR Section 99.7, 14 CFR Part 99 does not apply to aircraft operations:

1. Within the 48 contiguous states and the District of Columbia, or within the State of Alaska, and remains within 10 miles of the point of departure;

2. Over any island, or within three nautical miles of the coastline of any island, in the Hawaii ADIZ; or

3. Associated with any ADIZ other than the Contiguous U.S. ADIZ, when the aircraft true airspeed is less than 180 knots.

e. Authorizations to deviate from the requirements of Part 99 may also be granted by the ARTCC, on a local basis, for some operations associated with an ADIZ.

f. An airfiled VFR Flight Plan makes an aircraft subject to interception for positive identification when entering an ADIZ. Pilots are, therefore, urged to file the required DVFR flight plan either in person or by telephone prior to departure.

g. Special Security Instructions.

1. During defense emergency or air defense emergency conditions, additional special security instructions may be issued in accordance with the Emergency Security Control of Air Traffic (ESCAT) Plan.

2. Under the provisions of the ESCAT Plan, the military will direct the action to be taken in regard to landing, grounding, diversion, or dispersal of aircraft and the control of air navigation aids in the defense of the U.S. during emergency conditions.

3. At the time a portion or all of ESCAT is implemented, ATC facilities will broadcast appropriate instructions received from the Air Traffic Control System Command Center (ATCSCC) over available ATC frequencies. Depending on instructions received from the ATCSCC, VFR flights may be directed to land at the nearest available airport, and IFR flights will be expected to proceed as directed by ATC.

4. Pilots on the ground may be required to file a flight plan and obtain an approval (through FAA) prior to conducting flight operation.

5. In view of the above, all pilots should guard an ATC or FSS frequency at all times while conducting flight operations.

5-6-2. Interception Procedures

a. General.

1. In conjunction with the FAA, Air Defense Sectors monitor air traffic and could order an intercept in the interest of national security or defense. Intercepts during peacetime operations are vastly different than those conducted under increased states of readiness. The interceptors may be fighters or rotary wing aircraft. The reasons for aircraft intercept include, but are not limited to:

(a) Identify an aircraft;

(b) Track an aircraft;

(c) Inspect an aircraft;

(d) Divert an aircraft;

(e) Establish communications with an aircraft.

2. When specific information is required (i.e., markings, serial numbers, etc.) the interceptor pilot(s) will respond only if, in their judgment, the request can be conducted in a safe manner. Intercept procedures are described in some detail in the paragraphs below. In all situations, the interceptor pilot will consider safety of flight for all concerned throughout the intercept procedure. The interceptor pilot(s) will use caution to avoid startling the intercepted crew or passengers and understand that maneuvers considered normal for interceptor aircraft may be considered hazardous to other aircraft.

3. All aircraft operating in US national airspace are highly encouraged to maintain a listening watch on VHF/UHF guard frequencies (121.5 or 243.0 MHz). If subjected to a military intercept, it is incumbent on civilian aviators to understand their responsibilities and to comply with ICAO standard signals relayed from the intercepting aircraft. Specifically, aviators are expected to contact air traffic control without delay (if able) on the local operating frequency or on VHF/UHF guard. Noncompliance may result in the use of force.

b. Fighter intercept phases (See FIG 5-6-1).

1. Approach Phase.

As standard procedure, intercepted aircraft are approached from behind. Typically, interceptor aircraft will be employed in pairs, however, it is not uncommon for a single aircraft to perform the intercept operation. Safe separation between interceptors and intercepted aircraft is the responsibility of the intercepting aircraft and will be maintained at all times.

2. Identification Phase.

Interceptor aircraft will initiate a controlled closure toward the aircraft of interest, holding at a distance no closer than deemed necessary to establish positive identification and to gather the necessary information. The interceptor may also fly past the intercepted aircraft while gathering data at a distance considered safe based on aircraft performance characteristics.

3. Post Intercept Phase.

An interceptor may attempt to establish communications via standard ICAO signals. In time-critical situations where the interceptor is seeking an immediate response from the intercepted aircraft or if the intercepted aircraft remains non-compliant to instruction, the interceptor pilot may initiate a divert maneuver. In this maneuver, the interceptor flies across the intercepted aircraft's flight path (minimum 500 feet separation and commencing from slightly below the intercepted aircraft altitude) in the general direction the intercepted aircraft is expected to turn. The interceptor will rock its wings (daytime) or flash external lights/select afterburners (night) while crossing the intercepted aircraft's flight path. The interceptor will roll out in the direction the intercepted aircraft is expected to turn before returning to verify the aircraft of interest is complying. The intercepted aircraft is expected to execute an immediate turn to the direction of the intercepting aircraft. If the aircraft of interest does not comply, the interceptor may conduct a second climbing turn across the intercepted aircraft's flight path (minimum 500 feet separation and commencing from slightly below the intercepted aircraft altitude) while expending flares as a warning signal to the intercepted aircraft to comply immediately and to turn in the direction indicated and to leave the area. The interceptor is responsible to maintain safe separation during these and all intercept maneuvers. Flight safety is paramount.

NOTE-**1.** *NORAD interceptors will take every precaution to preclude the possibility of the intercepted aircraft experiencing jet wash/wake turbulence; however, there is a potential that this condition could be encountered.*

2. *During Night/IMC, the intercept will be from below flight path.*

c. Helicopter Intercept phases (See FIG 5-6-2)

1. Approach Phase.

Aircraft intercepted by helicopter may be approached from any direction, although the helicopter should close for identification and signaling from behind. Generally, the helicopter will approach off the left side of the intercepted aircraft. Safe separation between the helicopter and the unidentified aircraft will be maintained at all times.

2. Identification Phase.

The helicopter will initiate a controlled closure toward the aircraft of interest, holding at a distance no closer than deemed necessary to establish positive identification and gather the necessary information. The intercepted pilot

FIGURE 5-6-1
Intercept Procedures

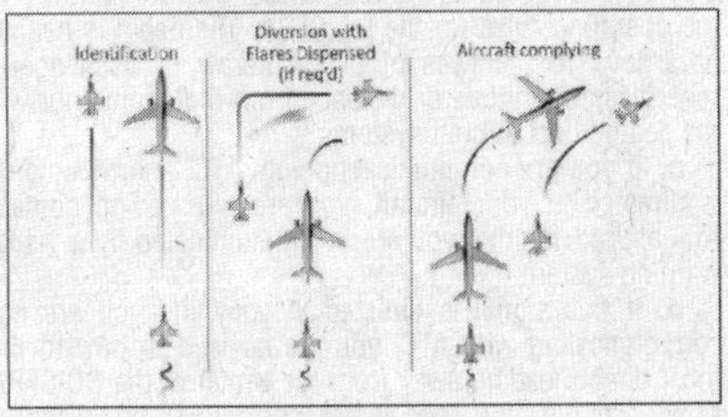

should expect the interceptor helicopter to take a position off his left wing slightly forward of abeam.

3. Post Intercept Phase.

Visual signaling devices may be used in an attempt to communicate with the intercepted aircraft. Visual signaling devices may include, but are not limited to, LED scrolling signboards or blue flashing lights. If compliance is not attained through the use of radios or signaling devices, standard ICAO intercept signals (Table 5-6-1) may be employed. In order to maintain safe aircraft separation, it is incumbent upon the pilot of the intercepted aircraft not to fall into a trail position (directly behind the helicopter) if instructed to follow the helicopter. This is because the helicopter pilot may lose visual contact with the intercepted aircraft.

NOTE-*Intercepted aircraft must not follow directly behind the helicopter thereby allowing the helicopter pilot to maintain visual contact with the intercepted aircraft and ensuring safe separation is maintained.*

FIGURE 5-6-2
Helicopter Intercept Procedures

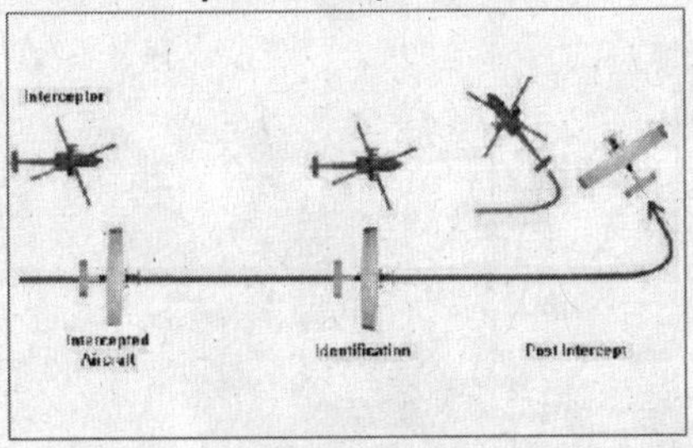

d. Summary of Intercepted Aircraft Actions. An intercepted aircraft must, without delay:

1. Adhere to instructions relayed through the use of visual devices, visual signals, and radio communications from the intercepting aircraft.

2. Attempt to establish radio communications with the intercepting aircraft or with the appropriate air traffic control facility by making a general call on guard frequencies (121.5 or 243.0 MHz), giving the identity, position, and nature of the flight.

3. If transponder equipped, select Mode 3/A Code 7700 unless otherwise instructed by air traffic control.

NOTE-*If instruction received from any agency conflicts with that given by the intercepting aircraft through visual or radio communications, the intercepted aircraft must seek immediate clarification.*

4. The crew of the intercepted aircraft must continue to comply with interceptor aircraft signals and instructions until positively released.

5-6-3. Law Enforcement Operations by Civil and Military Organizations

a. Special law enforcement operations.

1. Special law enforcement operations include in-flight identification, surveillance, interdiction, and pursuit activities performed in accordance with official civil and/or military mission responsibilities.

2. To facilitate accomplishment of these special missions, exemptions from specified sections of the CFRs have been granted to designated departments and agencies. However, it is each organization's responsibility to apprise ATC of their intent to operate under an authorized exemption before initiating actual operations.

3. Additionally, some departments and agencies that perform special missions have been assigned coded identifiers to permit them to apprise ATC of ongoing mission activities and solicit special air traffic assistance.

5-6-4. Interception Signals

TBL 5-6-1 and TBL 5-6-2.

5-6-5. ADIZ Boundaries and Designated Mountainous Areas (See FIG 5-6-3.)

5-6-6. Visual Warning System (VWS)

The VWS signal consists of highly-focused red and green colored laser lights designed to illuminate in an alternating red and green signal pattern. These lasers may be directed at specific aircraft suspected of making unau-

TABLE 5-6-1
Intercepting Signals

Signals initiated by intercepting aircraft and responses by intercepted aircraft (as set forth in ICAO Annex 2-Appendix 1, 2.1)

Series	INTERCEPTING Aircraft Signals	Meaning	INTERCEPTED Aircraft Responds	Meaning
1	DAY-Rocking wings from a position slightly above and ahead of, and normally to the left of, the intercepted aircraft and, after acknowledgement, a slow level turn, normally to the left, on to the desired heading. NIGHT-Same and, in addition, flashing navigational lights at irregular intervals. *NOTE 1-Meteorological conditions or terrain may require the intercepting aircraft to take up a position slightly above and ahead of, and to the right of, the intercepted aircraft and to make the subsequent turn to the right.* *NOTE 2-If the intercepted aircraft is not able to keep pace with the intercepting aircraft, the latter is expected to fly a series of race-track patterns and to rock its wings each time it passes the intercepted aircraft.*	You have been intercepted. Follow me.	AEROPLANES: DAY-Rocking wings and following. NIGHT-Same and, in addition, flashing navigational lights at irregular intervals. HELICOPTERS: DAY or NIGHT-Rocking aircraft, flashing navigational lights at irregular intervals and following.	Understood, will comply.
2	DAY or NIGHT-An abrupt break-away maneuver from the intercepted aircraft consisting of a climbing turn of 90 degrees or more without crossing the line of flight of the intercepted aircraft.	You may proceed.	AEROPLANES: DAY or NIGHT-Rocking wings. HELICOPTERS: DAY or NIGHT-Rocking aircraft.	Understood, will comply.
3	DAY-Circling aerodrome, lowering landing gear and overflying runway in direction of landing or, if the intercepted aircraft is a helicopter, overflying the helicopter landing area. NIGHT-Same and, in addition, showing steady landing lights.	Land at this aerodrome.	AEROPLANES: DAY-Lowering landing gear, following the intercepting aircraft and, if after overflying the runway landing is considered safe, proceeding to land. NIGHT-Same and, in addition, showing steady landing lights (if carried). HELICOPTERS: DAY or NIGHT-Following the intercepting aircraft and proceeding to land, showing a steady landing light (if carried).	Understood, will comply.
4	DAY or NIGHT-Raising landing gear (if fitted) and flashing landing lights while passing over runway in use or helicopter landing area at a height exceeding 300m (1,000 ft) but not exceeding 600m (2,000 ft) (in the case of a helicopter, at a height exceeding 50m (170 ft) but not exceeding 100m (330 ft) above the aerodrome level, and continuing to circle runway in use or helicopter landing area. If unable to flash landing lights, flash any other lights available.	Aerodrome you have designated is inadequate.	DAY or NIGHT-If it is desired that the intercepted aircraft follow the intercepting aircraft to an alternate aerodrome, the intercepting aircraft raises its landing gear (if fitted) and uses the Series 1 signals prescribed for intercepting aircraft. If it is decided to release the intercepted aircraft, the intercepting aircraft uses the Series 2 signals prescribed for intercepting aircraft.	Understood, follow me. Understood, you may proceed.
5	DAY or NIGHT-Regular switching on and off of all available lights but in such a manner as to be distinct from flashing lights.	Cannot comply.	DAY or NIGHT-Use Series 2 signals prescribed for intercepting aircraft.	Understood.
6	DAY or NIGHT-Irregular flashing of all available lights.	In distress.	DAY or NIGHT-Use Series 2 signals prescribed for intercepting aircraft.	Understood.

Figure 5-6-3
Air Defense Identification Zone Boundaries Designated Mountainous Areas

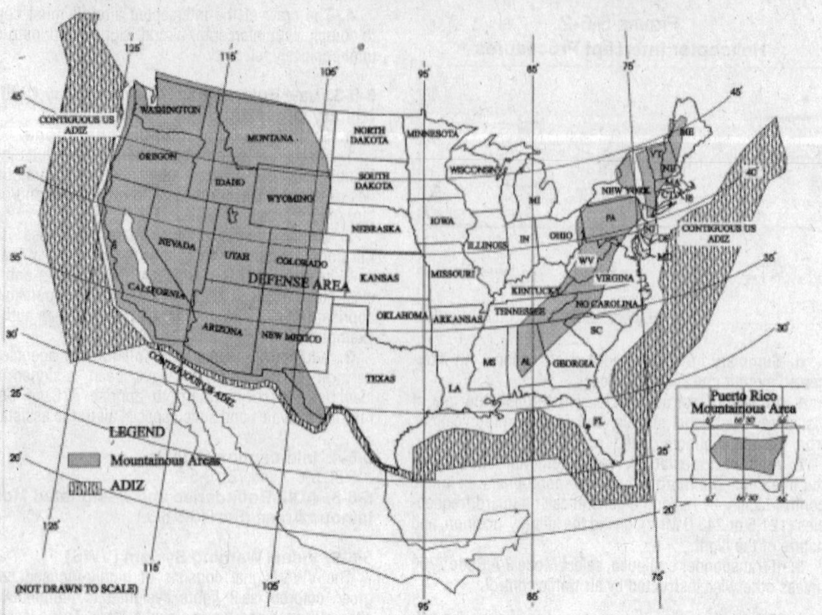

thorized entry into the Washington, DC Special Flight Rules Area (DC SFRA) proceeding on a heading or flight path that may be interpreted as a threat or that operate contrary to the operating rules for the DC SFRA. The beam is neither hazardous to the eyes of pilots/aircrew or passengers, regardless of altitude or distance from the source nor will the beam affect aircraft systems.

a. If you are communicating with ATC, and this signal is directed at your aircraft, you are required to contact ATC and advise that you are being illuminated by a visual warning system.

b. If this signal is directed at you, and you are not communicating with ATC, you are advised to turn to the most direct heading away from the center of the DC SFRA as soon as possible. Immediately contact ATC on an appropriate frequency, VHF Guard 121.5 or UHF Guard 243.0, and provide your aircraft identification, position, and nature of the flight. Failure to follow these procedures may result in interception by military aircraft. Further noncompliance with interceptor aircraft or ATC may result in the use of force.

c. Pilots planning to operate aircraft in or near the DC SFRA are to familiarize themselves with aircraft intercept procedures. This information applies to all aircraft operating within the DC SFRA including DOD, Law Enforcement, and aircraft engaged in aeromedical operations and does not change procedures established for reporting unauthorized laser illumination as published in FAA Advisory Circulars and Notices.

REFERENCE-*CFR 91.161*

d. More details including a video demonstration of the VWS are available from the following FAA web site: **www.faasafety.gov/VisualWarningSystem/VisualWarning.htm.**

Chapter 6.
EMERGENCY PROCEDURES
Section 1. General

6-1-1. Pilot Responsibility and Authority

a. The pilot-in-command of an aircraft is directly responsible for and is the final authority as to the operation of that aircraft. In an emergency requiring immediate action, the pilot-in-command may deviate from any rule in 14 CFR Part 91, Subpart A, General, and Subpart B, Flight Rules, to the extent required to meet that emergency.

REFERENCE- *14 CFR Section 91.3(b).*

b. If the emergency authority of 14 CFR Section 91.3(b) is used to deviate from the provisions of an ATC clearance, the pilot-in-command must notify ATC as soon as possible and obtain an amended clearance.

c. Unless deviation is necessary under the emergency authority of 14 CFR Section 91.3, pilots of IFR flights experiencing two-way radio communications failure are expected to adhere to the procedures prescribed under "IFR operations, two-way radio communications failure."

REFERENCE- *14 CFR Section 91.185.*

6-1-2. Emergency Condition- Request Assistance Immediately

a. An emergency can be either a *distress* or *urgency* condition as defined in the Pilot/Controller Glossary. Pilots do not hesitate to declare an emergency when they are faced with *distress* conditions such as fire, mechanical failure, or structural damage. However, some are reluctant to report an *urgency* condition when they encounter situations which may not be immediately perilous, but are potentially catastrophic. An aircraft is in at least an *urgency* condition the moment the pilot becomes doubtful about position, fuel endurance, weather, or any other condition that could adversely affect flight safety. This is the time to ask for help, not after the situation has developed into a *distress* condition.

b. Pilots who become apprehensive for their safety for any reason should *request assistance immediately.* Ready and willing help is available in the form of radio, radar, direction finding stations and other aircraft. Delay has caused accidents and cost lives. *Safety is not a luxury! Take action!*

Section 2. Emergency Services Available to Pilots

6-2-1. Radar Service for VFR Aircraft in Difficulty

a. Radar equipped ATC facilities can provide radar assistance and navigation service (vectors) to VFR aircraft in difficulty when the pilot can talk with the controller, and the aircraft is within radar coverage. Pilots should clearly understand that authorization to proceed in accordance with such radar navigational assistance does not constitute authorization for the pilot to violate CFRs. In effect, assistance is provided on the basis that navigational guidance information is advisory in nature, and the responsibility for flying the aircraft safely remains with the pilot.

b. Experience has shown that many pilots who are not qualified for instrument flight cannot maintain control of their aircraft when they encounter clouds or other reduced visibility conditions. In many cases, the controller will not know whether flight into instrument conditions will result from ATC instructions. To avoid possible hazards resulting from being vectored into IFR conditions, a pilot in difficulty should keep the controller advised of the current weather conditions being encountered and the weather along the course ahead and observe the following:

1. If a course of action is available which will permit flight and a safe landing in VFR weather conditions, noninstrument rated pilots should choose the VFR condition rather than requesting a vector or approach that will take them into IFR weather conditions; or

2. If continued flight in VFR conditions is not possible, the noninstrument rated pilot should so advise the controller and indicating the lack of an instrument rating, declare a *distress* condition; or

3. If the pilot is instrument rated and current, and the aircraft is instrument equipped, the pilot should so indicate by requesting an IFR flight clearance. Assistance will then be provided on the basis that the aircraft can operate safely in IFR weather conditions.

6-2-2. Transponder Emergency Operation

a. When a *distress* or *urgency* condition is encountered, the pilot of an aircraft with a coded radar beacon transponder, who desires to alert a ground radar facility, should squawk Mode 3/A, Code 7700/Emergency and Mode C altitude reporting and then immediately establish communications with the ATC facility.

b. Radar facilities are equipped so that Code 7700 normally triggers an alarm or special indicator at all control positions. Pilots should understand that they might not be within a radar coverage area. Therefore, they should continue squawking Code 7700 and establish radio communications as soon as possible.

6-2-3. Intercept and Escort

a. The concept of airborne intercept and escort is based on the Search and Rescue (SAR) aircraft establishing visual and/or electronic contact with an aircraft in difficulty, providing in-flight assistance, and escorting it to a safe landing. If bailout, crash landing or ditching becomes necessary, SAR operations can be conducted without delay. For most incidents, particularly those occurring at night and/or during instrument flight conditions, the availability of intercept and escort services will depend on the proximity of SAR units with suitable aircraft on alert for immediate dispatch. In limited circumstances, other aircraft flying in the vicinity of an aircraft in difficulty can provide these services.

b. If specifically requested by a pilot in difficulty or if a *distress* condition is declared, SAR coordinators *will* take steps to intercept and escort an aircraft. Steps may be initiated for intercept and escort if an *urgency* condition is declared and unusual circumstances make such action advisable.

c. It is the pilot's prerogative to refuse intercept and escort services. Escort services will normally be provided to the nearest adequate airport. Should the pilot receiving escort services continue onto another location after reaching a safe airport, or decide not to divert to the nearest safe airport, the escort aircraft is not obligated to continue and further escort is discretionary. The decision will depend on the circumstances of the individual incident.

6-2-4. Emergency Locator Transmitter (ELT)

a. General.

1. ELTs are required for most General Aviation airplanes.

REFERENCE- *14 CFR SECTION 91.207.*

2. ELTs of various types were developed as a means of locating downed aircraft. These electronic, battery operated transmitters operate on one of three frequencies. These operating frequencies are 121.5 MHz, 243.0 MHz, and the newer 406 MHz. ELTs operating on 121.5 MHz and 243.0 MHz are analog devices. The newer 406 MHz ELT is a digital transmitter that can be encoded with the owner's contact information or aircraft data. The latest 406 MHz ELT models can also be encoded with the aircraft's position data which can help SAR forces locate the aircraft much more quickly after a crash. The 406 MHz ELTs also transmits a stronger signal when activated than the older 121.5 MHz ELTs.

(a) The Federal Communications Commission (FCC) requires 406 MHz ELTs be registered with the National Oceanic and Atmospheric Administration (NOAA) as outlined in the ELTs documentation. The FAA's 406 MHz ELT Technical Standard Order (TSO) TSO-C126 also requires that each 406 MHz ELT be registered with NOAA. The reason is NOAA maintains the owner registration database for U.S. registered 406 MHz alerting devices, which includes ELTs. NOAA also operates the United States' portion of the Cospas-Sarsat satellite distress alerting system designed to detect activated ELTs and other distress alerting devices.

(b) In the event that a properly registered 406 MHz ELT activates, the Cospas-Sarsat satellite system can decode the owner's information and provide that data to the appropriate search and rescue (SAR) center. In the United States, NOAA provides the alert data to the appropriate U.S. Air Force Rescue Coordination Center (RCC) or U.S. Coast Guard Rescue Coordination Center. That RCC can then telephone or contact the owner to verify the status of the aircraft. If the aircraft is safely secured in a hangar, a costly ground or airborne search is avoided. In the case of an inadvertent 406 MHz ELT activation, the owner can deactivate the 406 MHz ELT. If the 406 MHz ELT equipped aircraft is being flown, the RCC can quickly activate a search. 406 MHz ELTs permit the Cospas-Sarsat satellite system to narrow the search area to a more confined area compared to that of a 121.5 MHz or 243.0 MHz ELT. 406 MHz ELTs also include a low-power 121.5 MHz homing transmitter to aid searchers in finding the aircraft in the terminal search phase.

(c) Each analog ELT emits a distinctive downward swept audio tone on 121.5 MHz and 243.0 MHz.

(d) If "armed" and when subject to crash-generated forces, ELTs are designed to automatically activate and continuously emit their respective signals, analog or digital. The transmitters will operate continuously for at least 48 hours over a wide temperature range. A properly installed, maintained, and functioning ELT can expedite search and rescue operations and save lives if it survives the crash and is activated.

(e) Pilots and their passengers should know how to activate the aircraft's ELT if manual activation is required. They should also be able to verify the aircraft's ELT is functioning and transmitting an alert after a crash or manual activation.

(f) Because of the large number of 121.5 MHz ELT false alerts and the lack of a quick means of verifying the actual status of an activated 121.5 MHz or 243.0 MHz analog ELT through an owner registration database, U.S. SAR forces do not respond as quickly to initial 121.5/243.0 MHz ELT alerts as the SAR forces do to 406 MHz ELT alerts. Compared to the almost instantaneous detection of a 406 MHz ELT, SAR forces' normal practice is to wait for either a confirmation of a 121.5/243.0 MHz alert by additional satellite passes or through confirmation of an overdue aircraft or similar notification. In some cases, this confirmation process can take hours. SAR forces can initiate a response to 406 MHz alerts in minutes compared to the potential delay of hours for a 121.5/243.0 MHz ELT.

3. The Cospas-Sarsat system has announced the termination of satellite monitoring and reception of the 121.5 MHz and 243.0 MHz frequencies in 2009. The Cospas-Sarsat system will continue to monitor the 406 MHz frequency. What this means for pilots is that after the termination date, those aircraft with only 121.5 MHz or 243.0 MHz ELT's onboard will have to depend upon either a nearby Air Traffic Control facility receiving the alert signal or an overflying aircraft monitoring 121.5 MHz or 243.0 MHz detecting the alert. To ensure adequate monitoring of these frequencies and timely alerts after 2009, all airborne pilots should periodically monitor these frequencies to try and detect an activated 121.5/243.0 MHz ELT.

b. Testing.

1. ELTs should be tested in accordance with the manufacturer's instructions, preferably in a shielded or screened room or specially designed test container to prevent the broadcast of signals which could trigger a false alert.

2. When this cannot be done, aircraft operational testing is authorized as follows:

(a) Analog 121.5/243 MHz ELTs should only be tested during the first 5 minutes after any hour. If operational tests must be made outside of this period, they should be coordinated with the nearest FAA Control Tower or FSS. Tests should be no longer than three audible sweeps. If the antenna is removable, a dummy load should be substituted during test procedures.

(b) Digital 406 MHz ELTs should only be tested in accordance with the unit's manufacturer's instructions.

(c) Airborne tests are not authorized.

c. False Alarms.

1. Caution should be exercised to prevent the inadvertent activation of ELTs in the air or while they are being handled on the ground. Accidental or unauthorized activation will generate an emergency signal that cannot be distinguished from the real thing, leading to expensive and frustrating searches. A false ELT signal could also interfere with genuine emergency transmissions and hinder or prevent the timely location of crash sites. Frequent false alarms could also result in complacency and decrease the vigorous reaction that must be attached to all ELT signals.

2. Numerous cases of inadvertent activation have occurred as a result of aerobatics, hard landings, movement by ground crews and aircraft maintenance. These false alarms can be minimized by monitoring 121.5 MHz and/or 243.0 MHz as follows:

(a) In flight when a receiver is available.

(b) Before engine shut down at the end of each flight.

(c) When the ELT is handled during installation or maintenance.

(d) When maintenance is being performed near the ELT.

(e) When a ground crew moves the aircraft.

(f) If an ELT signal is heard, turn off the aircraft's ELT to determine if it is transmitting. If it has been activated, maintenance might be required before the unit is returned to the "ARMED" position. You should contact the nearest Air Traffic facility and notify it of the inadvertent activation.

d. Inflight Monitoring and Reporting.

1. Pilots are encouraged to monitor 121.5 MHz and/or 243.0 MHz while inflight to assist in identifying possible emergency ELT transmissions. On receiving a signal, report the following information to the nearest air traffic facility:

(a) Your position at the time the signal was first heard.

(b) Your position at the time the signal was last heard.

(c) Your position at maximum signal strength.

(d) Your flight altitudes and frequency on which the emergency signal was heard: 121.5 MHz or 243.0 MHz. If possible, positions should be given relative to a navigation aid. If the aircraft has homing equipment, provide the bearing to the emergency signal with each reported position.

6-2-5. FAA K-9 Explosives Detection Team Program

a. The FAA's Office of Civil Aviation Security Operations manages the FAA K-9 Explosives Detection Team Program which was established in 1972. Through a unique agreement with law enforcement agencies and airport authorities, the FAA has strategically placed FAA-certified K-9 teams (a team is one handler and one dog) at airports throughout the country. If a bomb threat is received while an aircraft is in flight, the aircraft can be directed to an airport with this capability. The FAA provides initial and refresher training for all handlers, provides single purpose explosive detector dogs, and requires that each team is annually evaluated in five areas for FAA certification: aircraft (widebody and narrowbody), vehicles, terminal, freight (cargo), and luggage. **If you desire this service, notify your company or an FAA air traffic control facility.**

b. The following list shows the locations of current FAA K-9 teams:

TBL 6-2-1
FAA Sponsored Explosives Detection Dog/Handler Team Locations

Airport Symbol	Location
ATL	Atlanta, Georgia
BHM	Birmingham, Alabama
BOS	Boston, Massachusetts
BUF	Buffalo, New York
CLT	Charlotte, North Carolina
ORD	Chicago, Illinois
CVG	Cincinnati, Ohio
DFW	Dallas, Texas
DEN	Denver, Colorado
DTW	Detroit, Michigan
IAH	Houston, Texas
JAX	Jacksonville, Florida
MCI	Kansas City, Missouri
LAX	Los Angeles, California
MEM	Memphis, Tennessee
MIA	Miami, Florida
MKE	Milwaukee, Wisconsin
MSY	New Orleans, Louisiana
MCO	Orlando, Florida
PHX	Phoenix, Arizona
PIT	Pittsburgh, Pennsylvania
PDX	Portland, Oregon
SLC	Salt Lake City, Utah
SFO	San Francisco, California
SJU	San Juan, Puerto Rico
SEA	Seattle, Washington
STL	St. Louis, Missouri
TUS	Tucson, Arizona
TUL	Tulsa, Oklahoma

c. If due to weather or other considerations an aircraft with a suspected hidden explosive problem were to land or intended to land at an airport other than those listed in b above, it is recommended that they call the FAA's Washington Operations Center (telephone 202-267-3333, if appropriate) or have an air traffic facility with which you can communicate contact the above center requesting assistance.

6-2-6. Search and Rescue

a. General. SAR is a lifesaving service provided through the combined efforts of the federal agencies signatory to the National SAR Plan, and the agencies responsible for SAR within each state. Operational resources are provided by the U.S. Coast Guard, DOD components, the Civil Air Patrol, the Coast Guard Auxiliary, state, county and local law enforcement and other public safety agencies, and private volunteer organizations. Services include search for missing aircraft, survival aid, rescue, and emergency medical help for the occupants after an accident site is located.

b. National Search and Rescue Plan. By federal interagency agreement, the National Search and Rescue Plan provides for the effective use of all available facilities in all types of SAR missions. These facilities include aircraft, vessels, pararescue and ground rescue teams, and emergency radio fixing. Under the plan, the U.S. Coast Guard is responsible for the coordination of SAR in the Maritime Region, and the USAF is responsible in the Inland Region. To carry out these responsibilities, the Coast Guard and the Air Force have established Rescue Coordination Centers (RCCs) to direct SAR activities within their regions. For aircraft emergencies, distress, and urgency, information normally will be passed to the appropriate RCC through an ARTCC or FSS.

c. Coast Guard Rescue Coordination Centers. (See TBL 6-2-2.)

TBL 6-2-2
Coast Guard Rescue Coordination Centers

Coast Guard Rescue Coordination Centers	
Alameda, CA 510-437-3701	Miami, FL 305-415-6800
Boston, MA 617-223-8555	New Orleans, LA 504-589-6225
Cleveland, OH 216-902-6117	Portsmouth, VA 757-398-6390
Honolulu, HI 808-541-2500	Seattle, WA 206-220-7001
Juneau, AK 907-463-2000	San Juan, PR 787-289-2042

d. Air Force Rescue Coordination Centers. (See TBL 6-2-3 and TBL 6-2-4.)

TBL 6-2-3
Air Force Rescue Coordination Center 48 Contiguous States

Air Force Rescue Coordination Center	
Tyndall AFB, Florida	**Phone**
Commercial	850-283-5955
WATS	800-851-3051
DSN	523-5955

TBL 6-2-4
Air Command Rescue Coordination CenterAlaska

Alaskan Air Command Rescue Coordination Center	
Elmendorf AFB, Alaska	**Phone**
Commercial	907-428-7230 800-420-7230 (outside Anchorage)
DSN	317-551-7230

e. Joint Rescue Coordination Center. (See TBL 6-2-5.)

TBL 6-2-5
Joint Rescue Coordination Center Hawaii

Honolulu Joint Rescue Coordination Center	
HQ 14th CG District Honolulu	**Phone**
Commercial	808-541-2500
DSN	448-0301

f. Emergency and Overdue Aircraft.

1. ARTCCs and FSSs will alert the SAR system when information is received from any source that an aircraft is in difficulty, overdue, or missing.

(a) Radar facilities providing radar flight following or advisories consider the loss of radar and radios, without service termination notice, to be a possible emergency. Pilots receiving VFR services from radar facilities should be aware that SAR may be initiated under these circumstances.

(b) A filed flight plan is the most timely and effective indicator that an aircraft is overdue. Flight plan information is invaluable to SAR forces for search planning and executing search efforts.

2. Prior to departure on every flight, local or otherwise, someone at the departure point should be advised of your destination and route of flight if other than direct. Search efforts are often wasted and rescue is often delayed because of pilots who thoughtlessly takeoff without telling anyone where they are going. File a flight plan for*your* safety.

3. According to the National Search and Rescue Plan, "The life expectancy of an injured survivor decreases as much as 80 percent during the first 24 hours, while the chances of survival of uninjured survivors rapidly diminishes after the first 3 days."

4. An Air Force Review of 325 SAR missions conducted during a 23-month period revealed that "Time works against people who experience a *distress* but are not on a flight plan, since 36 hours normally pass before family concern initiates an (alert)."

g. VFR Search and Rescue Protection.

1. To receive this valuable protection, *file a VFR or DVFR Flight Plan* with an FAA FSS. For maximum protection, file only to the point of first intended landing, and refile for each leg to final destination. When a lengthy flight plan is filed, with several stops en route and an ETE to final destination, a mishap could occur on any leg, and unless other information is received, it is probable that no one would start looking for you until 30 minutes after your ETA at your final destination.

2. If you land at a location other than the intended destination, report the landing to the nearest FAA FSS and advise them of your original destination.

3. If you land en route and are delayed more than 30 minutes, report this information to the nearest FSS and give them your original destination.

4. If your ETE changes by 30 minutes or more, report a new ETA to the nearest FSS and give them your original destination. Remember that if you fail to respond within one-half hour after your ETA at final destination, a search will be started to locate you.

5. It is important that you *close your flight plan IMMEDIATELY AFTER ARRIVAL AT YOUR FINAL DESTINATION WITH THE FSS DESIGNATED WHEN YOUR FLIGHT PLAN WAS FILED. The* ***pilot is responsible*** *for closure of a VFR or DVFR flight plan; they are not closed automatically.* This will prevent needless search efforts.

6. The rapidity of rescue on land or water will depend on how accurately your position may be determined. If a flight plan has been followed and your position is on course, rescue will be expedited.

h. Survival Equipment.

1. For flight over uninhabited land areas, it is wise to take and know how to use survival equipment for the type of climate and terrain.

2. If a forced landing occurs at sea, chances for survival are governed by the degree of crew proficiency in emergency procedures and by the availability and effectiveness of water survival equipment.

i. Body Signal Illustrations.

1. If you are forced down and are able to attract the attention of the pilot of a rescue airplane, the body signals illustrated on these pages can be used to transmit messages to the pilot circling over your location.

2. Stand in the open when you make the signals.

3. Be sure the background, as seen from the air, is not confusing.

4. Go through the motions slowly and repeat each signal until you are positive that the pilot understands you.

j. Observance of Downed Aircraft.

1. Determine if crash is marked with a yellow cross; if so, the crash has already been reported and identified.

2. If possible, determine type and number of aircraft and whether there is evidence of survivors.

3. Fix the position of the crash as accurately as possible with reference to a navigational aid. If possible, provide geographic or physical description of the area to aid ground search parties.

4. Transmit the information to the nearest FAA or other appropriate radio facility.

5. If circumstances permit, orbit the scene to guide in other assisting units until their arrival or until you are relieved by another aircraft.

6. Immediately after landing, make a complete report to the nearest FAA facility, or Air Force or Coast Guard Rescue Coordination Center. The report can be made by a long distance collect telephone call.

Section 3. Distress and Urgency Procedures

6-3-1. Distress and Urgency Communications

a. A pilot who encounters a *distress* or *urgency* condition can obtain assistance simply by contacting the air traffic facility or other agency in whose area of responsibility the aircraft is operating, stating the nature of the difficulty, pilot's intentions and assistance desired. *Distress* and *urgency* communications procedures are prescribed by the International Civil Aviation Organization (ICAO), however, and have decided advantages over the informal procedure described above.

Fig 6-2-1
Ground-Air Visual Code for Use by Survivors

NO.	MESSAGE	CODE SYMBOL
1	Require assistance	V
2	Require medical assistance	X
3	No or Negative	N
4	Yes or Affirmative	Y
5	Proceeding in this direction	↑

IF IN DOUBT, USE INTERNATIONAL SYMBOL SOS

INSTRUCTIONS

Fig 6-2-2
Ground-Air Visual Code for use by Ground Search Parties

NO.	MESSAGE	CODE SYMBOL
1	Operation completed.	LLL
2	We have found all personnel.	LL
3	We have found only some personnel.	++
4	We are not able to continue. Returning to base.	XX
5	Have divided into two groups. Each proceeding in direction indicated.	⇄
6	Information received that aircraft is in this direction.	→ →
7	Nothing found. Will continue search.	NN

Fig 6-2-3
Urgent Medical Assistance

Fig 6-2-4
All OK

Fig 6-2-5
Short Delay

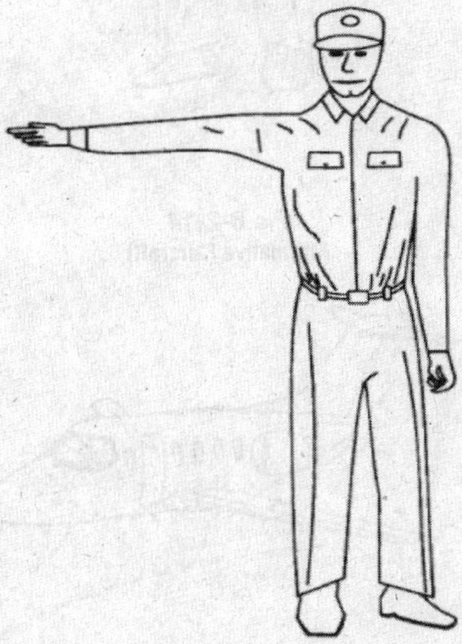

Fig 6-2-6
Long Delay

Fig 6-2-7
Drop Message

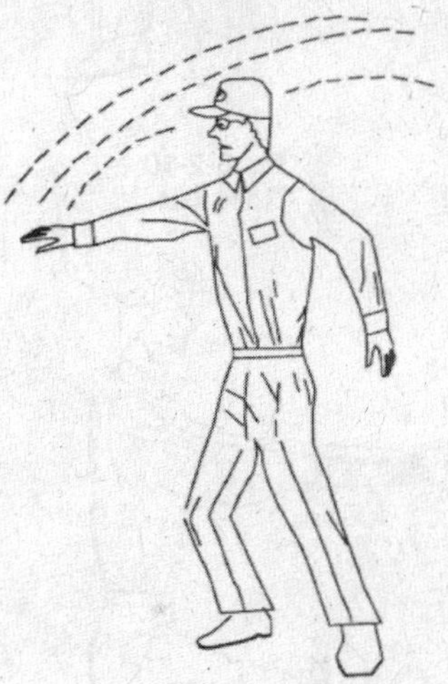

Make throwing motion

Fig 6-2-8
Receiver Operates

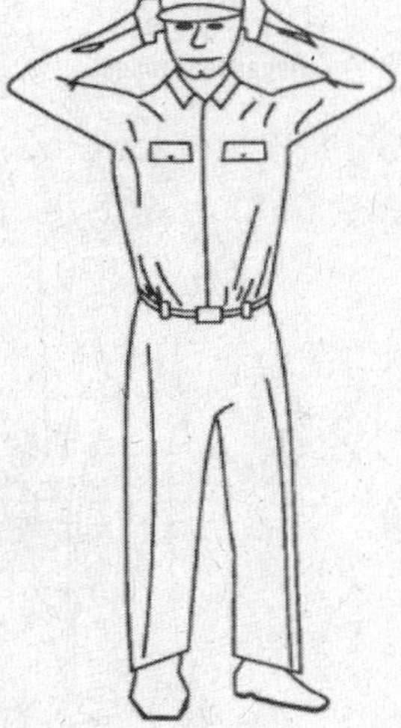

FIG 6-2-9
Do Not Land Here

FIG 6-2-10
Land Here

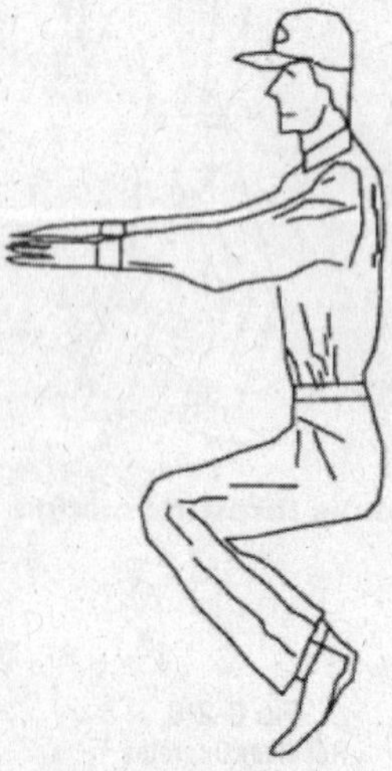

FIG 6-2-11
Negative (Ground)

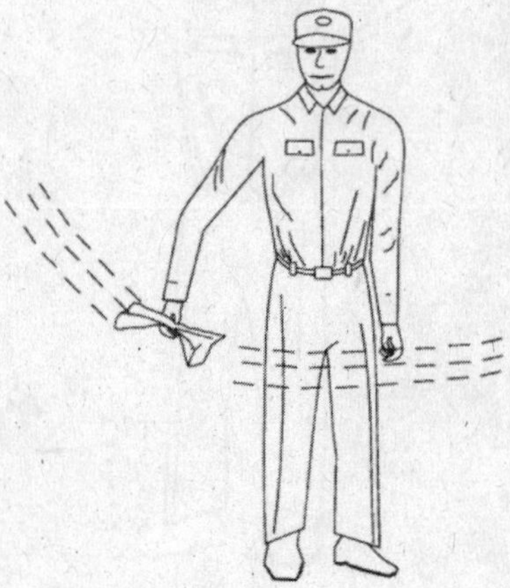

FIG 6-2-12
Affirmative (Ground)

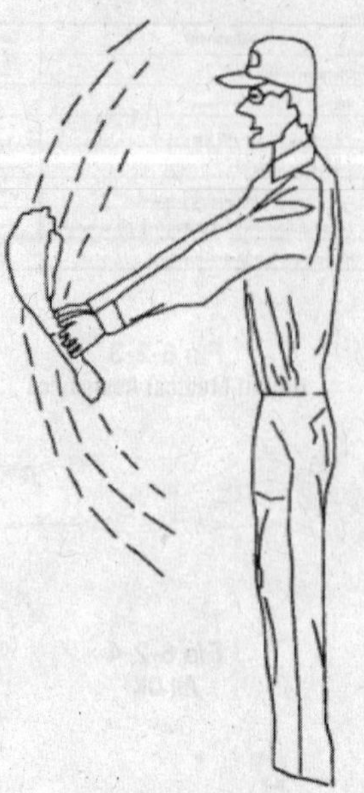

FIG 6-2-13
Pick Us Up

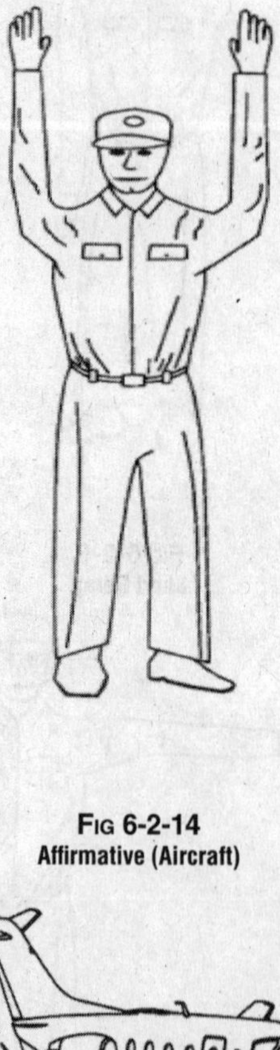

FIG 6-2-14
Affirmative (Aircraft)

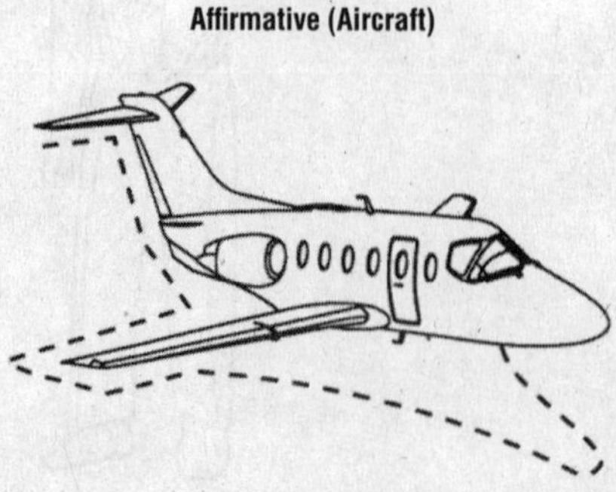

FIG 6-2-15
Negative (Aircraft)

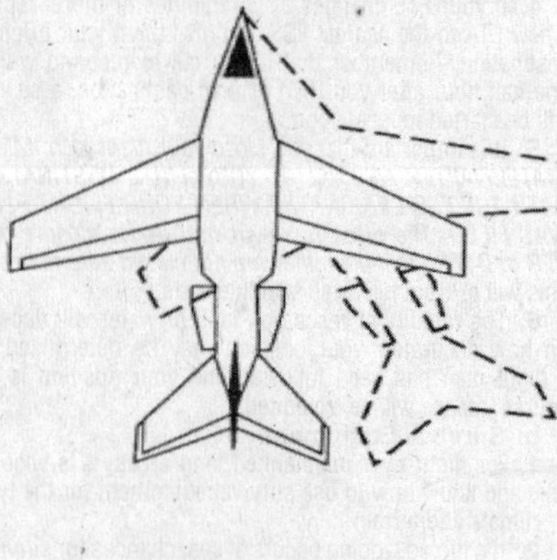

FIG 6-2-16
Message received and understood (Aircraft)

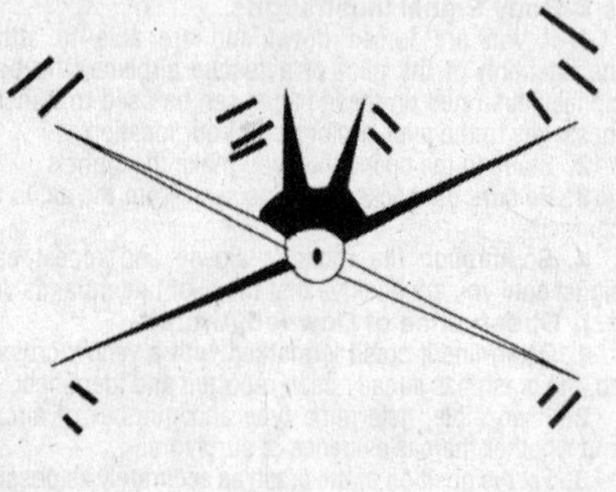

FIG 6-2-17
Message received and NOT understood (Aircraft)

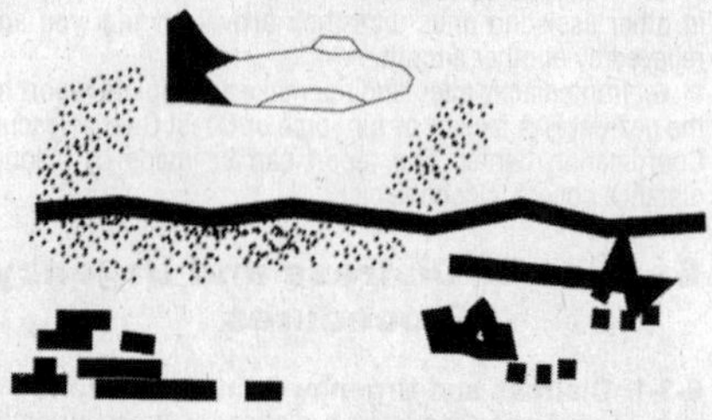

b. *Distress* and *urgency* communications procedures discussed in the following paragraphs relate to the use of air ground voice communications.

c. The initial communication, and if considered necessary, any subsequent transmissions by an aircraft in *distress* should begin with the signal MAYDAY, preferably repeated three times. The signal PAN-PAN should be used in the same manner for an *urgency* condition.

d. *Distress* communications have absolute priority over all other communications, and the word MAYDAY commands radio silence on the frequency in use. *Urgency* communications have priority over all other communications except *distress,* and the word PAN-PAN warns other stations not to interfere with *urgency* transmissions.

e. Normally, the station addressed will be the air traffic facility or other agency providing air traffic services, on the frequency in use at the time. If the pilot is not communicating and receiving services, the station to be called will normally be the air traffic facility or other agency in whose area of responsibility the aircraft is operating, on the appropriate assigned frequency. If the station addressed does not respond, or if time or the situation dictates, the *distress* or *urgency* message may be broadcast, or a collect call may be used, addressing "Any Station (Tower)(Radio)(Radar)."

f. The station addressed should immediately acknowledge a *distress* or *urgency* message, provide assistance, coordinate and direct the activities of assisting facilities, and alert the appropriate search and rescue coordinator if warranted. Responsibility will be transferred to another station only if better handling will result.

g. All other stations, aircraft and ground, will continue to listen until it is evident that assistance is being provided. If any station becomes aware that the station being called either has not received a *distress* or *urgency* message, or cannot communicate with the aircraft in difficulty, it will attempt to contact the aircraft and provide assistance.

h. Although the frequency in use or other frequencies assigned by ATC are preferable, the following emergency frequencies can be used for distress or urgency communications, if necessary or desirable:

1. ***121.5 MHz and 243.0 MHz.*** Both have a range generally limited to line of sight. 121.5 MHz is guarded by direction finding stations and some military and civil aircraft. 243.0 MHz is guarded by military aircraft. Both 121.5 MHz and 243.0 MHz are guarded by military towers, most civil towers, FSSs, and radar facilities. Normally ARTCC emergency frequency capability does not extend to radar coverage limits. If an ARTCC does not respond when called on 121.5 MHz or 243.0 MHz, call the nearest tower or FSS.

2. ***2182 kHz.*** The range is generally less than 300 miles for the average aircraft installation. It can be used to request assistance from stations in the maritime service. 2182 kHz is guarded by major radio stations serving Coast Guard Rescue Coordination Centers, and Coast Guard units along the sea coasts of the U.S. and shores of the Great Lakes. The call "Coast Guard" will alert all Coast Guard Radio Stations within range. 2182 kHz is also guarded by most commercial coast stations and some ships and boats.

6-3-2. Obtaining Emergency Assistance

a. A pilot in any *distress* or *urgency* condition should *immediately* take the following action, not necessarily in the order listed, to obtain assistance:

1. Climb, if possible, for improved communications, and better radar and direction finding detection. However, it must be understood that unauthorized climb or descent under IFR conditions within controlled airspace is prohibited, except as permitted by 14 CFR Section 91.3(b).

2. If equipped with a radar beacon transponder (civil) or IFF/SIF (military):

(a) Continue squawking assigned Mode A/3 discrete code/VFR code and Mode C altitude encoding when in radio contact with an air traffic facility or other agency providing air traffic services, unless instructed to do otherwise.

(b) If unable to immediately establish communications with an air traffic facility/agency, squawk Mode A/3, Code 7700/Emergency and Mode C.

3. Transmit a *distress* or *urgency* message consisting of *as many* as necessary of the following elements, preferably in the order listed:

(a) If distress, MAYDAY, MAYDAY, MAY-DAY; if *urgency*, PAN-PAN, PAN-PAN, PAN-PAN.

(b) Name of station addressed.

(c) Aircraft identification and type.

(d) Nature of *distress* or *urgency*.

(e) Weather.

(f) Pilots intentions and request.

(g) Present position, and heading; or if *lost*, last known position, time, and heading since that position.

(h) Altitude or flight level.

(i) Fuel remaining in minutes.

(j) Number of people on board.

(k) Any other useful information.

REFERENCE-*Pilot/Controller Glossary Term- Fuel Remaining.*

b. After establishing radio contact, comply with advice and instructions received. Cooperate. Do not hesitate to ask questions or clarify instructions when you do not understand or if you cannot comply with clearance. Assist the ground station to control communications on the frequency in use. Silence interfering radio stations. Do not change frequency or change to another ground station unless absolutely necessary. If you do, advise the ground station of the new frequency and station name prior to the change, transmitting in the blind if necessary. If two-way communications cannot be established on the new frequency, return immediately to the frequency or station where two-way communications last existed.

c. When in a distress condition with bailout, crash landing or ditching imminent, take the following additional actions to assist search and rescue units:

1. Time and circumstances permitting, transmit as many as necessary of the message elements in subparagraph a3 above, and any of the following that you think might be helpful:

(a) ELT status.

(b) Visible landmarks.

(c) Aircraft color.

(d) Number of persons on board.

(e) Emergency equipment on board.

2. Actuate your ELT if the installation permits.

3. For bailout, and for crash landing or ditching if risk of fire is not a consideration, set your radio for continuous transmission.

4. If it becomes necessary to ditch, make every effort to ditch near a surface vessel. If time permits, an FAA facility should be able to get the position of the nearest commercial or Coast Guard vessel from a Coast Guard Rescue Coordination Center.

5. After a crash landing, unless you have good reason to believe that you will not be located by search aircraft or ground teams, it is best to remain with your aircraft and prepare means for signaling search aircraft.

6-3-3. Ditching Procedures

a. A successful aircraft ditching is dependent on three primary factors. In order of importance they are:

Fig 6-3-1
Single Swell (15 knot wind)

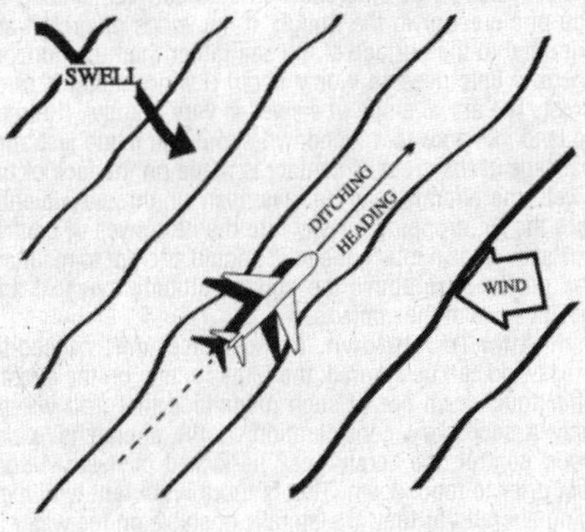

Fig 6-3-2
Double Swell (15 knot wind)

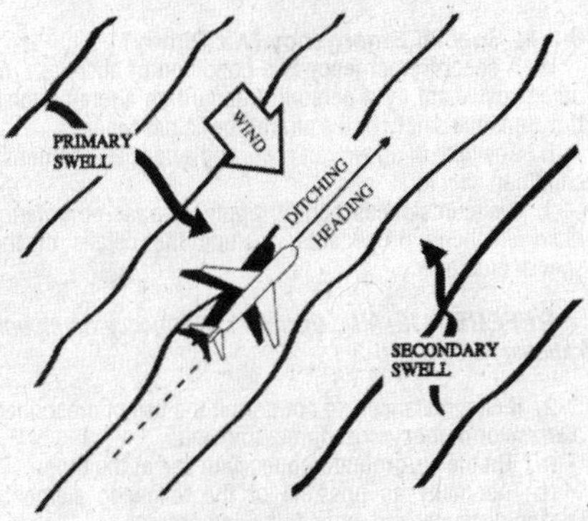

1. Sea conditions and wind.

2. Type of aircraft.

3. Skill and technique of pilot.

b. Common oceanographic terminology.

1. Sea. The condition of the surface that is the result of both waves and swells.

2. Wave (or Chop). The condition of the surface caused by the local winds.

3. Swell. The condition of the surface which has been caused by a distance disturbance.

4. Swell Face. The side of the swell toward the observer. The backside is the side away from the observer. These definitions apply regardless of the direction of swell movement.

5. Primary Swell. The swell system having the greatest height from trough to crest.

Fig 6-3-3
Double Swell (30 knot wind)

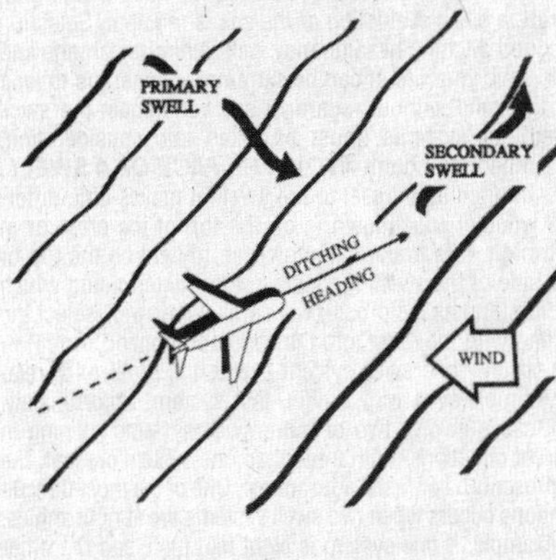

Fig 6-3-4
(50 knot wind)

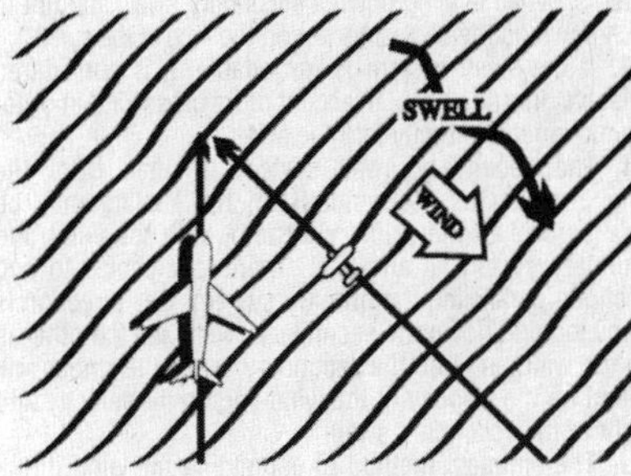

Aircraft with low landing speeds - land into the wind.

Aircraft with high landing speeds - choose compromise heading between wind and swell.

Both - land on back side of swell.

Fig 6-3-5
Wind-Swell-Ditch Heading

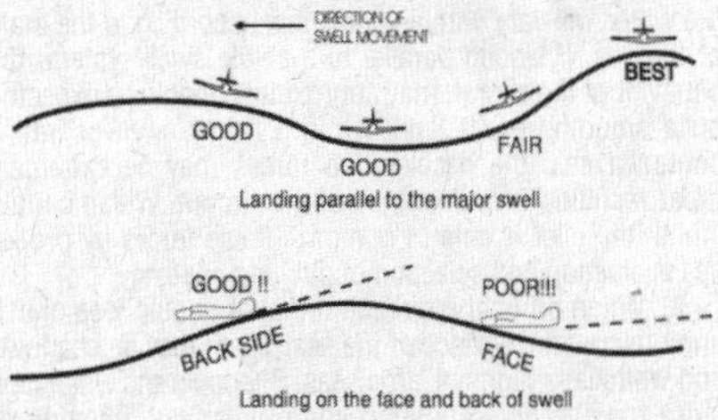

6. Secondary Swells. Those swell systems of less height than the primary swell.

7. Fetch. The distance the waves have been driven by a wind blowing in a constant direction, without obstruction.

8. Swell Period. The time interval between the passage of two successive crests at the same spot in the water, measured in seconds.

9. Swell Velocity. The speed and direction of the swell with relation to a fixed reference point, measured in knots. There is little movement of water in the horizontal direction. Swells move primarily in a vertical motion, similar to the motion observed when shaking out a carpet.

10. Swell Direction. The direction *from* which a swell is moving. This direction is not necessarily the result of the wind present at the scene. The swell may be moving into or across the local wind. Swells, once set in motion, tend to maintain their original direction for as long as they continue in deep water, regardless of changes in wind direction.

11. Swell Height. The height between crest and trough, measured in feet. The vast majority of ocean swells are lower than 12 to 15 feet, and swells over 25 feet are not common at any spot on the oceans. Successive swells may differ considerably in height.

c. In order to select a good heading when ditching an aircraft, a basic evaluation of the sea is required. Selection of a good ditching heading may well minimize damage and could save your life. It can be extremely dangerous to land into the wind without regard to sea conditions; the swell system, or systems, must be taken into consideration. Remember one axiom- ***AVOID THE FACE OF A SWELL.***

1. In ditching parallel to the swell, it makes little difference whether touchdown is on the top of the crest or in the trough. It is preferable, however, to land on the top or back side of the swell, if possible. After determining which heading (and its reciprocal) will parallel the swell, select the heading with the most into the wind component.

2. If only one swell system exists, the problem is relatively simple-even with a high, fast system. Unfortunately, most cases involve two or more swell systems running in different directions. With more than one system present, the sea presents a confused appearance. One of the most difficult situations occurs when two swell systems are at right angles. For example, if one system is eight feet high, and the other three feet, plan to land parallel to the primary system, and on the down swell of the secondary system. If both systems are of equal height, a compromise may be advisable-select an intermediate heading at 45 degrees down swell to both systems. When landing down a secondary swell, attempt to touch down on the back side, not on the face of the swell.

3. If the swell system is formidable, it is considered advisable, in landplanes, to accept more crosswind in order to avoid landing directly into the swell.

4. The secondary swell system is often from the same direction as the wind. Here, the landing may be made parallel to the primary system, with the wind and secondary system at an angle. There is a choice to two directions paralleling the primary system. One direction is downwind and down the secondary swell, and the other is into the wind and into the secondary swell, the choice will depend on the velocity of the wind versus the velocity and height of the secondary swell.

d. The simplest method of estimating the wind direction and velocity is to examine the windstreaks on the water. These appear as long streaks up and down wind. Some persons may have difficulty determining wind direction after seeing the streaks on the water. Whitecaps fall forward with the wind but are overrun by the waves thus producing the illusion that the foam is sliding backward. Knowing this, and by observing the direction of the streaks, the wind direction is easily determined. Wind velocity can be estimated by noting the appearance of the whitecaps, foam and wind streaks.

1. The behavior of the aircraft on making contact with the water will vary within wide limits according to the state of the sea. If landed parallel to a single swell system, the behavior of the aircraft may approximate that to be expected on a smooth sea. If landed into a heavy swell or into a confused sea, the deceleration forces may be extremely great-resulting in breaking up of the aircraft. Within certain limits, the pilot is able to minimize these forces by proper sea evaluation and selection of ditching heading.

2. When on final approach the pilot should look ahead and observe the surface of the sea. There may be shadows and whitecaps-signs of large seas. Shadows and whitecaps close together indicate short and rough seas. Touchdown in these areas is to be avoided. Select and touchdown in any area (only about 500 feet is needed) where the shadows and whitecaps are not so numerous.

3. Touchdown should be at the *lowest* speed and rate of descent which permit safe handling and optimum nose up attitude on impact. Once first impact has been made, there is often little the pilot can do to control a landplane.

e. Once preditching preparations are completed, the pilot should turn to the ditching heading and commence let-down. The aircraft should be flown low over the water, and slowed down until ten knots or so above stall. At this point, additional power should be used to overcome the increased drag caused by the nose up attitude. When a smooth stretch of water appears ahead, cut power, and touchdown at the best recommended speed as fully stalled as possible. By cutting power when approaching a relatively smooth area, the pilot will prevent overshooting and will touchdown with less chance of planing off into a second uncontrolled landing. Most experienced seaplane pilots prefer to make contact with the water in a semi-stalled attitude, cutting power as the tail makes contact. This technique eliminates the chance of misjudging altitude with a resultant heavy drop in a fully stalled condition. Care must be taken not to drop the aircraft from too high altitude or to balloon due to excessive speed. The altitude above water depends on the aircraft. Over glassy smooth water, or at night without sufficient light, it is very easy, for even the most experienced pilots to misjudge altitude by 50 feet or more. Under such conditions, carry enough power to maintain nine to twelve degrees nose up attitude, and 10 to 20 percent over stalling speed until contact is made with the water. The proper use of power on the approach is of great importance. If power is available on one side only, a little power should be used to flatten the approach; however, the engine should not be used to such an extent that the aircraft cannot be turned against the good engines right down to the stall with a margin of rudder movement available. When near the stall, sudden application of excessive unbalanced power may result in loss of directional control. If power is available on one side only, a slightly higher than normal glide approach speed should be used. This will ensure good control and some margin of speed after leveling off without excessive use of power. The use of power in ditching is so important that when it is certain that the coast cannot be reached, the pilot should, if possible, ditch before fuel is exhausted. The use of power in a night or instrument ditching is far more essential than under daylight contact conditions.

1. If no power is available, a greater than normal approach speed should be used down to the flare-out. This speed margin will allow the glide to be broken early and more gradually, thereby giving the pilot time and distance to feel for the surface - decreasing the possibility of stalling high or flying into the water. When landing parallel to a swell system, little difference is noted between landing on top of a crest or in the trough. If the wings of aircraft are trimmed to the surface of the sea rather than the horizon, there is little need to worry about a wing hitting a swell crest. The actual slope of a swell is very gradual. If forced to land into a swell, touchdown should be made just after passage of the crest. If contact is made on the face of the swell, the aircraft may be swamped or thrown violently into the air, dropping heavily into the next swell. If control surfaces remain intact, the pilot should attempt to maintain the proper nose above the horizon attitude by rapid and positive use of the controls.

f. After Touchdown. In most cases drift, caused by crosswind can be ignored; the forces acting on the aircraft after touchdown are of such magnitude that drift will be only a secondary consideration. If the aircraft is under good control, the "crab" may be kicked out with rudder just prior to touchdown. This is more important with high wing aircraft, for they are laterally unstable on the water in a crosswind and may roll to the side in ditching.

REFERENCE- *This information has been extracted from Appendix H of the "National Search and Rescue Manual."*

6-3-4. Special Emergency (Air Piracy)

a. A special emergency is a condition of air piracy, or other hostile act by a person(s) aboard an aircraft, which threatens the safety of the aircraft or its passengers.

b. The pilot of an aircraft reporting a special emergency condition should:

1. If circumstances permit, apply *distress* or *urgency* radio-telephony procedures. Include the details of the special emergency.

REFERENCE- *AIM, Distress and Urgency Communications, Paragraph 6-3-1.*

2. If circumstances do not permit the use of prescribed *distress* or *urgency*procedures, transmit:

(a) On the air/ground frequency in use at the time.

(b) As many as possible of the following elements spoken distinctly and in the following order:

(1) Name of the station addressed (time and circumstances permitting).

(2) The identification of the aircraft and present position.

(3) The nature of the special emergency condition and pilot intentions (circumstances permitting).

(4) If unable to provide this information, use code words and/or transponder as follows:

Spoken Words TRANSPONDER SEVEN FIVE ZERO ZERO
Meaning I am being hijacked/forced to a new destination
Transponder Setting Mode 3/A, Code 7500

NOTE- *Code 7500 will never be assigned by ATC without prior notification from the pilot that the aircraft is being subjected to unlawful interference. The pilot should refuse the assignment of Code 7500 in any other situation and inform the controller accordingly. Code 7500 will trigger the special emergency indicator in all radar ATC facilities.*

c. Air traffic controllers will acknowledge and confirm receipt of transponder Code 7500 by asking the pilot to verify it. If the aircraft is not being subjected to unlawful interference, the pilot should respond to the query by broadcasting in the clear that the aircraft is not being subjected to unlawful interference. Upon receipt of this information, the controller will request the pilot to verify the code selection depicted in the code selector windows in the transponder control panel and change the code to the appropriate setting. If the pilot replies in the affirmative or does not reply, the controller will not ask further questions but will flight follow, respond to pilot requests and notify appropriate authorities.

d. If it is possible to do so without jeopardizing the safety of the flight, the pilot of a hijacked passenger aircraft, after departing from the cleared routing over which the aircraft was operating, will attempt to do one or more of the following things, insofar as circumstances may permit:

1. Maintain a true airspeed of no more than 400 knots, and preferably an altitude of between 10,000 and 25,000 feet.

2. Fly a course toward the destination which the hijacker has announced.

e. If these procedures result in either radio contact or air intercept, the pilot will attempt to comply with any instructions received which may direct the aircraft to an appropriate landing field or alter the aircraft's flight path off its current course, away from protected airspace.

6-3-5. Fuel Dumping

a. Should it become necessary to dump fuel, the pilot should immediately advise ATC. Upon receipt of information that an aircraft will dump fuel, ATC will broadcast or cause to be broadcast immediately and every 3 minutes thereafter the following on appropriate ATC and FSS radio frequencies:

EXAMPLE- *Attention all aircraft - fuel dumping in progress over - (location) at (altitude) by (type aircraft) (flight direction).*

b. Upon receipt of such a broadcast, pilots of aircraft affected, which are not on IFR flight plans or special VFR clearances, should clear the area specified in the advisory. Aircraft on IFR flight plans or special VFR clearances will be provided specific separation by ATC. At the termination of the fuel dumping operation, pilots should advise ATC. Upon receipt of such information, ATC will issue, on the appropriate frequencies, the following:

EXAMPLE- *ATTENTION ALL AIRCRAFT - FUEL DUMPING BY - (type aircraft) - TERMINATED.*

Section 4. Two-way Radio Communications Failure

6-4-1. Two-way Radio Communications Failure

a. It is virtually impossible to provide regulations and procedures applicable to all possible situations associated with two-way radio communications failure. During two-way radio communications failure, when confronted by a situation not covered in the regulation, pilots are expected to exercise good judgment in whatever action they elect to take. Should the situation so dictate they should not be reluctant to use the emergency action contained in 14 CFR Section 91.3(b).

b. Whether two-way communications failure constitutes an emergency depends on the circumstances, and in any event, it is a determination made by the pilot. 14 CFR Section 91.3(b) authorizes a pilot to deviate from any rule in Subparts A and B to the extent required to meet an emergency.

c. In the event of two-way radio communications failure, ATC service will be provided on the basis that the pilot is operating in accordance with 14 CFR Section 91.185. A pilot experiencing two-way communications failure should (unless emergency authority is exercised) comply with 14 CFR Section 91.185 quoted below:

1. General. Unless otherwise authorized by ATC, each pilot who has two-way radio communications failure when operating under IFR must comply with the rules of this section.

2. VFR conditions. If the failure occurs in VFR conditions, or if VFR conditions are encountered after the failure, each pilot must continue the flight under VFR and land as soon as practicable.

NOTE-This procedure also applies when two-way radio failure occurs while operating in Class A airspace. The primary objective of this provision in 14 CFR Section 91.185 is to preclude extended IFR operation by these aircraft within the ATC system. Pilots should recognize that operation under these conditions may unnecessarily as well as adversely affect other users of the airspace, since ATC may be required to reroute or delay other users in order to protect the failure aircraft. However, it is not intended that the requirement to "land as soon as practicable" be construed to mean "as soon as possible." Pilots retain the prerogative of exercising their best judgment and are not required to land at an unauthorized airport, at an airport unsuitable for the type of aircraft flown, or to land only minutes short of their intended destination.

3. IFR conditions. If the failure occurs in IFR conditions, or if subparagraph 2 above cannot be complied with, each pilot must continue the flight according to the following:

(a) Route.

(1) By the route assigned in the last ATC clearance received;

(2) If being radar vectored, by the direct route from the point of radio failure to the fix, route, or airway specified in the vector clearance;

(3) In the absence of an assigned route, by the route that ATC has advised may be expected in a further clearance; or

(4) In the absence of an assigned route or a route that ATC has advised may be expected in a further clearance by the route filed in the flight plan.

(b) Altitude. At the HIGHEST of the following altitudes or flight levels FOR THE ROUTE SEGMENT BEING FLOWN:

(1) The altitude or flight level assigned in the last ATC clearance received;

(2) The minimum altitude (converted, if appropriate, to minimum flight level as prescribed in 14 CFR Section 91.121(c)) for IFR operations; or

(3) The altitude or flight level ATC has advised may be expected in a further clearance.

NOTE-The intent of the rule is that a pilot who has experienced two-way radio failure should select the appropriate altitude for the particular route segment being flown and make the necessary altitude adjustments for subsequent route segments. If the pilot received an "expect further clearance" containing a higher altitude to expect at a specified time or fix, maintain the highest of the following altitudes until that time/fix:

(1) the last assigned altitude; or
(2) the minimum altitude/flight level for IFR operations.

Upon reaching the time/fix specified, the pilot should commence climbing to the altitude advised to expect. If the radio failure occurs after the time/fix specified, the altitude to be expected is not applicable and the pilot should maintain an altitude consistent with 1 or 2 above. If the pilot receives an "expect further clearance" containing a lower altitude, the pilot should maintain the highest of 1 or 2 above until that time/fix specified in subparagraph (c) Leave clearance limit, below.

EXAMPLE-

1. *A pilot experiencing two-way radio failure at an assigned altitude of 7,000 feet is cleared along a direct route which will require a climb to a minimum IFR altitude of 9,000 feet, should climb to reach 9,000 feet at the time or place where it becomes necessary (see 14 CFR Section 91.177(b)). Later while proceeding along an airway with an MEA of 5,000 feet, the pilot would descend to 7,000 feet (the last assigned altitude), because that altitude is higher than the MEA.*

2. *A pilot experiencing two-way radio failure while being progressively descended to lower altitudes to begin an approach is assigned 2,700 feet until crossing the VOR and then cleared for the approach. The MOCA along the airway is 2,700 feet and MEA is 4,000 feet. The aircraft is within 22 NM of the VOR. The pilot should remain at 2,700 feet until crossing the VOR because that altitude is the minimum IFR altitude for the route segment being flown.*

3. *The MEA between a and b: 5,000 feet. The MEA between b and c: 5,000 feet. The MEA between c and d: 11,000 feet. The MEA between d and e: 7,000 feet. A pilot had been cleared via a, b, c, d, to e. While flying between a and b the assigned altitude was 6,000 feet and the pilot was told to expect a clearance to 8,000 feet at b. Prior to receiving the higher altitude assignment, the pilot experienced two-way failure. The pilot would maintain 6,000 to b, then climb to 8,000 feet (the altitude advised to expect). The pilot would maintain 8,000 feet, then climb to 11,000 at c, or prior to c if necessary to comply with an MCA at c. (14 CFR Section 91.177(b).) Upon reaching d, the pilot would descend to 8,000 feet (even though the MEA was 7,000 feet), as 8,000 was the highest of the altitude situations stated in the rule (14 CFR Section 91.185).*

(c) Leave clearance limit.

(1) When the clearance limit is a fix from which an approach begins, commence descent or descent and approach as close as possible to the expect further clearance time if one has been received, or if one has not been received, as close as possible to the Estimated Time of Arrival (ETA) as calculated from the filed or amended (with ATC) Estimated Time En Route (ETE).

(2) If the clearance limit is not a fix from which an approach begins, leave the clearance limit at the expect further clearance time if one has been received, or if none has been received, upon arrival over the clearance limit, and proceed to a fix from which an approach begins and commence descent or descent and approach as close as possible to the estimated time of arrival as calculated from the filed or amended (with ATC) estimated time en route.

6-4-2. Transponder Operation During Two-way Communications Failure

a. If an aircraft with a coded radar beacon transponder experiences a loss of two-way radio capability, the pilot should adjust the transponder to reply on Mode A/3, Code 7600.

b. The pilot should understand that the aircraft may not be in an area of radar coverage.

6-4-3. Reestablishing Radio Contact

a. In addition to monitoring the NAVAID voice feature, the pilot should attempt to reestablish communications by attempting contact:

1. On the previously assigned frequency; or

2. With an FSS or *ARINC.

b. If communications are established with an FSS or ARINC, the pilot should advise that radio communications on the previously assigned frequency has been lost giving the aircraft's position, altitude, last assigned frequency and then request further clearance from the controlling facility. The preceding does not preclude the use of 121.5 MHz. There is no priority on which action should be attempted first. If the capability exists, do all at the same time.

NOTE*-*Aeronautical Radio/Incorporated (ARINC) is a commercial communications corporation which designs, constructs, operates, leases or otherwise engages in radio activities serving the aviation community. ARINC has the capability of relaying information to/from ATC facilities throughout the country.*

Section 5. Aircraft Rescue and Fire Fighting Communications

6-5-1. Discrete Emergency Frequency

a. Direct contact between an emergency aircraft flight crew, Aircraft Rescue and Fire Fighting Incident Commander (ARFF IC), and the Airport Traffic Control Tower (ATCT), is possible on an aeronautical radio frequency (Discrete Emergency Frequency [DEF]), designated by Air Traffic Control (ATC) from the operational frequencies assigned to that facility.

b. Emergency aircraft at airports without an ATCT, (or when the ATCT is closed), may contact the ARFF IC (if ARFF service is provided), on the Common Traffic Advisory Frequency **(CTAF)** published for the airport or the civil emergency frequency **121.5 MHz.**

6-5-2. Radio Call Signs

Preferred radio call sign for the ARFF IC is "(location/facility) **Command**" when communicating with the flight crew and the FAA ATCT.

EXAMPLE-

LAX Command.
Washington Command.

6-5-3. ARFF Emergency Hand Signals

In the event that electronic communications cannot be maintained between the ARFF IC and the flight crew, standard emergency hand signals as depicted in FIG 6-5-1 through FIG 6-5-3 should be used. These hand signals should be known and understood by all cockpit and cabin aircrew, and all ARFF firefighters.

FIG 6-5-1
Recommend Evacuation

RECOMMEND EVACUATION - Evacuation recommended based on ARFF IC's assessment of external situation.

Arm extended from body, and held horizontal with hand upraised at eye level. Execute beckoning arm motion angled backward. Nonbeckoning arm held against body.

NIGHT - same with wands.

FIG 6-5-2
Recommend Stop

RECOMMEND STOP - Recommend evacuation in progress be halted. Stop aircraft movement or other activity in progress.

Arms in front of head - Crossed at wrists.

NIGHT - same with wands.

FIG 6-5-3
Emergency Contained

EMERGENCY CONTAINED - No outside evidence of dangerous condition or "all-clear."

Arms extended outward and down at a 45 degree angle. Arms moved inward below waistline simultaneously until wrists crossed, then extended outward to starting position (umpire's "safe" signal).

NIGHT - same with wands.

Chapter 7.

SAFETY OF FLIGHT

Section 1. Meteorology

7-1-1. National Weather Service Aviation Products

a. Weather service to aviation is a joint effort of the National Weather Service (NWS), the Federal Aviation Administration (FAA), the military weather services, and other aviation oriented groups and individuals. The NWS maintains an extensive surface, upper air, and radar weather observing program; a nationwide aviation weather forecasting service; and provides limited pilot briefing service (interpretational). Pilot weather briefings are provided by personnel at Flight Service Stations operated by FAA (in Alaska) or by federal contract facilities (elsewhere in the U.S.). Aviation routine weather reports (METAR) are taken manually by NWS, FAA, contractors, or supplemental observers. METAR reports are also provided by Automated Weather Observing System (AWOS), Automated Surface Observing System (ASOS), and Automated Weather Sensor System (AWSS).

***REFERENCE-**AIM, Para 7-1-12, Weather Observing Programs.*

b. Aerodrome forecasts are prepared by approximately 100 Weather Forecast Offices (WFOs). These offices prepare and distribute approximately 525 aerodrome forecasts 4 times daily for specific airports in the 50 States, Puerto Rico, the Caribbean and Pacific Islands. These forecasts are valid for 24 hours and amended as required. WFOs prepare over 300 route forecasts and 39 synopses for briefing purposes and for Transcribed Weather Broadcasts (TWEB) only outside the continental United States (OCONUS). The route forecasts are issued 4 times daily; each forecast is valid for 12 hours. A centralized aviation forecast program originating from the Aviation Weather Center (AWC) in Kansas City was implemented in October 1995. In the conterminous U.S., all inflight advisories, Significant Meteorological Information (SIGMETs), Convective SIGMETs, and Airmen's Meteorological Information (AIRMET-text bulletins [WA] and graphics [G-AIRMETs]), and all Area Forecasts (FAs) (6 areas) are now issued by AWC. FAs are prepared 3 times a day in the conterminous U.S. and Alaska (4 times in Hawaii), and amended as required. Inflight advisories are issued only when conditions warrant. Winds aloft forecasts are provided for 176 locations in the 48 contiguous States and 21 locations in Alaska for flight planning purposes. (Winds aloft forecasts for Hawaii are prepared locally.) All the aviation weather forecasts are given wide distribution through the Weather Message Switching Center Replacement (WMSCR) in Atlanta, Georgia, and Salt Lake City, Utah.

***REFERENCE-**AIM, Para 7-1-6, Inflight Aviation Weather Advisories.*

c. Weather element values may be expressed by using different measurement systems depending on several factors, such as whether the weather products will be used by the general public, aviation interests, international services, or a combination of these users. FIG 7-1-1 provides conversion tables for the most used weather elements that will be encountered by pilots.

7-1-2. FAA Weather Services

a. The FAA maintains a nationwide network of Flight Service Stations (FSSs) to serve the weather needs of pilots. In addition, NWS meteorologists are assigned to most ARTCCs as part of the Center Weather Service Unit (CWSU). They provide Center Weather Advisories (CWAs) and gather weather information to support the needs of the FAA and other users of the system.

b. The primary source of preflight weather briefings is an individual briefing obtained from a briefer at the FSS. These briefings, which are tailored to your specific flight, are available 24 hours a day through the use of the toll free number (1-800-WX BRIEF). Numbers for these services can be found in the Airport/Facility Directory (A/FD) under "FAA and NWS Telephone Numbers" section. They may also be listed in the U.S. Government section of your local telephone directory under Department of Transportation, Federal Aviation Administration.

***REFERENCE-**AIM, Para 7-1-4, Preflight Briefing, explains the types of preflight briefings available and the information contained in each.*

c. Other Sources of Weather Information

1. Telephone Information Briefing Service (TIBS) (FSS); and in Alaska, Transcribed Weather Broadcast (TWEB) locations, and telephone access to the TWEB (TEL-TWEB) provide continuously updated recorded weather information for short or local flights. Separate paragraphs in this section give additional information regarding these services.

***REFERENCE-**AIM, Telephone Information Briefing Service (TIBS), Paragraph 7-1-8.*

AIM, Transcribed Weather Broadcast (TWEB) (Alaska Only), Paragraph 7-1-9.

2. Weather and aeronautical information are also available from numerous private industry sources on an individual or contract pay basis. Information on how to obtain this service should be available from local pilot organizations.

3. The Direct User Access Terminal System (DUATS) can be accessed by pilots with a current medical certificate toll-free in the 48 contiguous States via personal computer. Pilots can receive alpha-numeric preflight weather data and file domestic VFR and IFR flight plans. The following are the contract DUATS vendors:

Computer Sciences Corporation (CSC)
15000 Conference Center Drive
Chantilly, VA 22021-3808
Internet Access: http://www.duats.com
Telnet Access (modem terminal-style):
(800) 767-9989 or
telnet://direct.duats.com
For customer service: (800) 345-3828

Data Transformation Corporation (DTC)
108-D Greentree Road
Turnersville, NJ 08012
Internet Access: http://www.duat.com
For customer service: (800)243-3828

d. Inflight weather information is available from any FSS within radio range. The common frequency for all FSSs is 122.2. Discrete frequencies for individual stations are listed in the A/FD.

1. Information on In-Flight Weather broadcasts.

***REFERENCE-**AIM, Inflight Weather Broadcasts, Paragraph 7-1-10.*

2. En Route Flight Advisory Service (EFAS) is provided to serve the nonroutine weather needs of pilots in flight.

***REFERENCE-**AIM, En Route Flight Advisory Service (EFAS), Paragraph 7-1-5, gives details on this service.*

7-1-3. Use of Aviation Weather Products

a. Air carriers and operators certificated under the provisions of 14 CFR Part 119 are required to use the aeronautical weather information systems defined in the

FIG 7-1-1
Weather Elements Conversion Tables

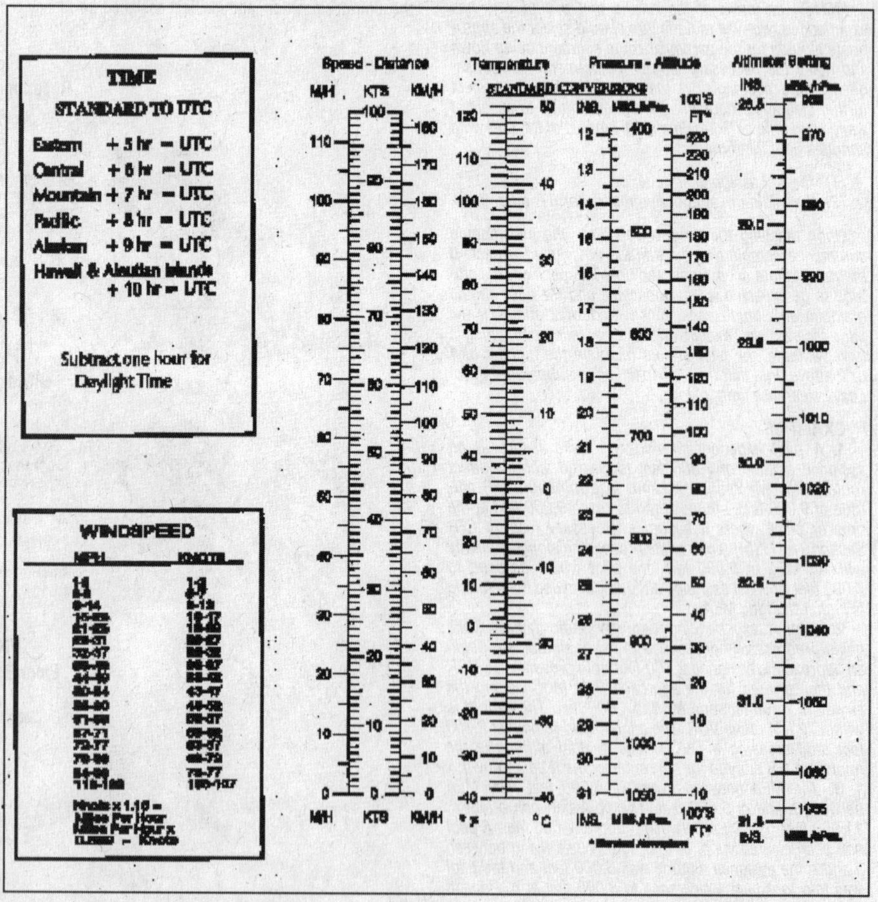

Operations Specifications issued to that certificate holder by the FAA. These systems may utilize basic FAA/National Weather Service (NWS) weather services, contractor- or operator-proprietary weather services and/or Enhanced Weather Information System (EWINS) when approved in the Operations Specifications. As an integral part of this system approval, the procedures for collecting, producing and disseminating aeronautical weather information, as well as the crew member and dispatcher training to support the use of system weather products, must be accepted or approved.

b. Operators not certificated under the provisions of 14 CFR Part 119 are encouraged to use FAA/NWS products through Flight Service Stations, Direct User Access Terminal System (DUATS), and/or Flight Information Services Data Link (FISDL).

c. The suite of available aviation weather product types is expanding, with the development of new sensor systems, algorithms and forecast models. The FAA and NWS, supported by the National Center for Atmospheric Research and the Forecast Systems Laboratory, develop and implement new aviation weather product types through a comprehensive process known as the Aviation Weather Technology Transfer process. This process ensures that user needs and technical readiness requirements are met before experimental products mature to operational application.

d. The FAA, in conjunction with the NWS, established the Aviation Weather Technology Transfer (AWTT) Board so that newly developed aviation weather products meet regulatory requirements and enhance safety. The AWTT is charged with managing and accelerating the transfer of these products into operational use. Members of the AWTT Board include mid-level managers from the FAA and NWS who are responsible for various aspects of the development and use of aviation weather products (e.g., aviation weather R & D, transition of weather products from R & D to operational use, etc.).

e. The AWTT is a management-review and decision-making process that applies criteria to weather products at various development stages (decision stages, i.e., "D-stages"). The D-stages are composed of the following:

1. (D1) Sponsorship of user needs.

2. (D2) R & D and controlled testing.

3. (D3) Experimental application.

4. (D4) Operational application.

f. Weather products maturing into the D3 experimental stage of the AWTT process are often made available to the public on the Aviation Weather Center's Experimental Aviation Digital Data Service (ADDS) website at: **http://weather.aero/**. The intent is to allow public access to this information in order to obtain feedback for product development and improvement. However, it is important to note that weather products displayed on this site are experimental, and although they may appear to be fully operational products, they are subject to change without notification and may *not* be used for any flight related decisions. At the D4 stage, the FAA approves a weather product for operational use by end users (with restrictions, if necessary), and the product is made available to the public via long-line circuit, satellite, and/or other means of communication.

g. Pilots and operators should be aware that weather services provided by entities other than FAA, NWS or their contractors (such as the DUATS and FISDL providers) may not meet FAA/NWS quality control standards. Hence, operators and pilots contemplating using such services should request and/or review an appropriate description of services and provider disclosure. This should include, but is not limited to, the type of weather product (e.g., current weather or forecast weather), the currency of the product (i.e., product issue and valid times), and the relevance of the product. Pilots and operators should be cautious when using unfamiliar products, or products not supported by FAA/NWS technical specifications.

NOTE-*When in doubt, consult with a FAA Flight Service Station Specialist.*

h. In addition, pilots and operators should be aware there are weather services and products available from government organizations beyond the scope of the AWTT process mentioned earlier in this section. For example, governmental agencies such as the NWS, the Aviation Weather Center (AWC), and the National Center for Atmospheric Research (NCAR) display weather "model data" and "experimental" products which require training and/or expertise to properly interpret and use. These products are developmental prototypes that are subject to ongoing research and can change without notice. Therefore, some data on display by government organizations, or government data on display by independent organizations may be unsuitable for flight planning purposes. Operators and pilots contemplating using such services should request and/or review an appropriate description of services and provider disclosure. This should include, but is not limited to, the type of weather product (e.g., current weather or forecast weather), the currency of the product (i.e., product issue and valid times), and the relevance of the product. Pilots and operators should be cautious when using unfamiliar weather products.

NOTE-*When in doubt, consult with a FAA Flight Service Station Specialist.*

i. The development of new weather products coupled with increased access to these products via the public Internet, created confusion within the aviation community regarding the relationship between regulatory requirements and new weather products. Consequently, FAA differentiates between those weather products that may be utilized to comply with regulatory requirements and those that may only be used to improve situational awareness. To clarify the proper use of aviation weather products to meet the requirements of 14 CFR, FAA defines weather products as follows:

1. Primary Weather Product. An aviation weather product that meets all the regulatory requirements and safety needs for use in making flight related, aviation weather decisions.

2. Supplementary Weather Product. An aviation weather product that may be used for enhanced situational awareness. If utilized, a supplementary weather product must only be used in conjunction with one or more primary weather product. In addition, the FAA may further restrict the use of supplementary aviation weather products through limitations described in the product label.

NOTE-*An aviation weather product produced by the Federal Government and managed by the AWTT is classified a primary weather product unless designated a supplementary weather product by the FAA.*

j. In developing the definitions of primary and supplementary weather products, it is not the intent of FAA to change or increase the regulatory burden on the user. Rather, the definitions are meant to eliminate confusion by differentiating between weather products that may be utilized to meet regulatory requirements and other weather products that may only be used to improve situational awareness.

k. All flight-related, aviation weather decisions must be based on primary weather products. Supplementary weather products augment the primary products by providing additional weather information but may not be used as stand-alone weather products to meet aviation weather regulatory requirements or without the relevant primary products. When discrepancies exist between primary and supplementary weather products describing the same weather phenomena, users must base flight-related decisions on the primary weather product. Furthermore, multiple primary products may be necessary to meet all aviation weather regulatory requirements.

l. The development of enhanced communications capabilities, most notably the Internet, has allowed pilots access to an ever-increasing range of weather service providers and proprietary products. The FAA has identified three distinct types of weather information available to pilots and operators.

1. Observations. Raw weather data collected by some type of sensor suite including surface and airborne observations, radar, lightning, satellite imagery, and profilers.

2. Analysis. Enhanced depiction and/or interpretation of observed weather data.

3. Forecasts. Predictions of the development and/or movement of weather phenomena based on meteorological observations and various mathematical models.

m. Not all sources of aviation weather information are able to provide all three types of weather information. The FAA has determined that operators and pilots may utilize the following approved sources of aviation weather information:

1. Federal Government. The FAA and NWS collect raw weather data, analyze the observations, and produce forecasts. The FAA and NWS disseminate meteorological observations, analyses, and forecasts through a variety of systems. In addition, the Federal Government is the only approval authority for sources of weather observations; for example, contract towers and airport operators may be approved by the Federal Government to provide weather observations.

2. Enhanced Weather Information System (EWINS). An EWINS is an FAA approved, proprietary system for tracking, evaluating, reporting, and forecasting the presence or lack of adverse weather phenomena. An EWINS is authorized to produce flight movement forecasts, adverse weather phenomena forecasts, and other meteorological advisories. For more detailed information regarding EWINS, see the Aviation Weather Services Advisory Circular 00-45 and the Air Transportation Operations Inspector's Handbook 8400.10.

3. Commercial Weather Information Providers. In general, commercial providers produce proprietary weather products based on NWS/FAA products with formatting and layout modifications but no material changes to the weather information itself. This is also referred to as "repackaging." In addition, commercial providers may produce analyses, forecasts, and other proprietary weather products that substantially alter the information contained in government-produced products. However, those proprietary weather products that substantially alter government-produced weather products or information, may only be approved for use by 14 CFR Part 121 and Part 135 certificate holders if the commercial provider is EWINS qualified.

NOTE-*Commercial weather information providers contracted by FAA to provide weather observations, analyses, and forecasts (e.g., contract towers) are included in the Federal Government category of approved sources by virtue of maintaining required technical and quality assurance standards under Federal Government oversight.*

n. As a point of clarification, Advisory Circular 00-62, Internet Communications of Aviation Weather and NOTAMS, describes the process for a weather information provider to become a Qualified Internet Communications Provider (QICP) and only applies to 14 CFR Part 121 and Part 135 certificate holders. Therefore, pilots conducting operations under 14 CFR Part 91 may access weather products via the public Internet.

7-1-4. Preflight Briefing

a. Flight Service Stations (FSSs) are the primary source for obtaining preflight briefings and inflight weather information. Flight Service Specialists are qualified and certificated by the NWS as Pilot Weather Briefers. They are not authorized to make original forecasts, but are authorized to translate and interpret available forecasts and reports directly into terms describing the weather conditions which you can expect along your flight route and at your destination. Available aviation weather reports, forecasts and aviation weather charts are displayed at each FSS, for pilot use. Pilots should feel free to use these self briefing displays where available, or to ask for a briefing or assistance from the specialist on duty. Three basic types of preflight briefings are available to serve your specific needs. These are: Standard Briefing, Abbreviated Briefing, and Outlook Briefing. You should specify to the briefer the type of briefing you want, along with your appropriate background information. This will enable the briefer to tailor the information to your intended flight. The following paragraphs describe the types of briefings available and the information provided in each briefing.

REFERENCE-*AIM, Preflight Preparation, Paragraph 5-1-1, for items that are required.*

b. Standard Briefing. You should request a Standard Briefing any time you are planning a flight and you have not received a previous briefing or have not received preliminary information through mass dissemination media; e.g., TIBS, TWEB (Alaska only), etc. International data may be inaccurate or incomplete. If you are planning a flight outside of U.S. controlled airspace, the briefer will advise you to check data as soon as practical after entering foreign airspace, unless you advise that you have the international cautionary advisory. The briefer will automatically provide the following information in the sequence listed, except as noted, when it is applicable to your proposed flight.

1. Adverse Conditions. Significant meteorological and/or aeronautical information that might influence the pilot to alter or cancel the proposed flight; for example, hazardous weather conditions, airport closures, air traffic delays, etc. Pilots should be especially alert for current or forecast weather that could reduce flight minimums below VFR or IFR conditions. Pilots should also be alert for any reported or forecast icing if the aircraft is not certified for operating in icing conditions. Flying into areas of icing or weather below minimums could have disastrous results.

2. VFR Flight Not Recommended. When VFR flight is proposed and sky conditions or visibilities are present or forecast, surface or aloft, that, in the briefer's judgment, would make flight under VFR doubtful, the briefer will describe the conditions, describe the affected locations, and use the phrase "*VFR flight not recommended.*" This recommendation is advisory in nature. The final decision as to whether the flight can be conducted safely rests solely with the pilot. Upon receiving a "*VFR flight not recommended*" statement, the non-IFR rated pilot will need to make a "go or no go" decision. This decision should be based on weighing the current and forecast weather conditions against the pilot's experience and ratings. The aircraft's equipment, capabilities and limitations should also be considered.

NOTE-*Pilots flying into areas of minimal VFR weather could encounter unforecasted lowering conditions that place the aircraft outside the pilot's ratings and experience level. This could result in spatial disorientation and/or loss of control of the aircraft.*

3. Synopsis. A brief statement describing the type, location and movement of weather systems and/or air masses which might affect the proposed flight.

NOTE-*These first 3 elements of a briefing may be combined in any order when the briefer believes it will help to more clearly describe conditions.*

4. Current Conditions. Reported weather conditions applicable to the flight will be summarized from all available sources; e.g., METARs/ SPECIs, PIREPs, RAREPs. This element will be omitted if the proposed time of departure is beyond 2 hours, unless the information is specifically requested by the pilot.

5. En Route Forecast. Forecast en route conditions for the proposed route are summarized in logical order; i.e., departure/climbout, en route, and descent. (Heights are MSL, unless the contractions "AGL" or "CIG" are denoted indicating that heights are above ground.)

6. Destination Forecast. The destination forecast for the planned ETA. Any significant changes within 1 hour before and after the planned arrival are included.

7. Winds Aloft. Forecast winds aloft will be provided using degrees of the compass. The briefer will interpolate wind directions and speeds between levels and stations as necessary to provide expected conditions at planned altitudes. (Heights are MSL.) Temperature information will be provided on request.

8. Notices to Airmen (NOTAMs).

(a) Available NOTAM (D) information pertinent to the proposed flight, including special use airspace (SUA) NOTAMs for restricted areas, aerial refueling, and night vision goggles (NVG).

NOTE-*Other SUA NOTAMs (D), such as military operations area (MOA), military training route (MTR), and warning area NOTAMs, are considered "upon request" briefing items as indicated in paragraph 7-1-4b10(a).*

(b) Prohibited Areas P-40, P-49, P-56, and the special flight rules area (SFRA) for Washington, DC.

(c) FSS briefers do not provide FDC NOTAM information for special instrument approach procedures unless specifically asked. Pilots authorized by the FAA to use special instrument approach procedures must specifically request FDC NOTAM information for these procedures.

NOTE-*NOTAM information may be combined with current conditions when the briefer believes it is logical to do so.*

NOTE-*NOTAM (D) information and FDC NOTAMs which have been published in the Notices to Airmen Publication are not included in pilot briefings unless a review of this publication is specifically requested by the pilot. For complete flight information you are urged to review the printed NOTAMs in the Notices to Airmen Publication and the A/FD in addition to obtaining a briefing.*

9. ATC Delays. Any known ATC delays and flow control advisories which might affect the proposed flight.

10. Pilots may obtain the following from flight service station briefers upon request:

(a) Information on SUA and SUA-related airspace, except those listed in paragraph 7-1-4b8.

NOTE-

1. For the purpose of this paragraph, SUA and related airspace includes the following types of airspace: alert area, military operations area (MOA), warning area, and air traffic control assigned airspace (ATCAA). MTR data includes the following types of airspace: IFR training routes (IR), VFR training routes (VR), and slow training routes (SR).

2. Pilots are encouraged to request updated information from ATC facilities while in flight.

(b) A review of the Notices to Airmen Publication for pertinent NOTAMs and Special Notices.

(c) Approximate density altitude data.

(d) Information regarding such items as air traffic services and rules, customs/immigration procedures, ADIZ rules, search and rescue, etc.

(e) GPS RAIM availability for 1 hour before to 1 hour after ETA or a time specified by the pilot.

(f) Other assistance as required.

c. Abbreviated Briefing. Request an Abbreviated Briefing when you need information to supplement mass disseminated data, update a previous briefing, or when you need only one or two specific items. Provide the briefer with appropriate background information, the time you received the previous information, and/or the specific items needed. You should indicate the source of the information already received so that the briefer can limit the briefing to the information that you have not received, and/or appreciable changes in meteorological/aeronautical conditions since your previous briefing. To the extent possible, the briefer will provide the information in the sequence shown for a Standard Briefing. If you request only one or two specific items, the briefer will advise you if adverse conditions are present or forecast. (Adverse conditions contain both meteorological and/or aeronautical information.) Details on these conditions will be provided at your request. International data may be inaccurate or incomplete. If you are planning a flight outside of U.S. controlled airspace, the briefer will advise you to check data as soon as practical after entering foreign airspace, unless you advise that you have the international cautionary advisory.

d. Outlook Briefing. You should request an Outlook Briefing whenever your proposed time of departure is six or more hours from the time of the briefing. The briefer will provide available forecast data applicable to the proposed flight. This type of briefing is provided for planning purposes only. You should obtain a Standard or Abbreviated Briefing prior to departure in order to obtain such items as adverse conditions, current conditions, updated forecasts, winds aloft and NOTAMs, etc.

e. When filing a flight plan only, you will be asked if you require the latest information on adverse conditions pertinent to the route of flight.

f. Inflight Briefing. You are encouraged to obtain your preflight briefing by telephone or in person before departure. In those cases where you need to obtain a preflight briefing or an update to a previous briefing by radio, you should contact the nearest FSS to obtain this information. After communications have been established, advise the specialist of the type briefing you require and provide appropriate background information. You will be provided information as specified in the above paragraphs, depending on the type of briefing requested. In addition, the specialist will recommend shifting to the Flight Watch frequency when conditions along the intended route indicate that it would be advantageous to do so. Remember that weather conditions can change rapidly and that a "go or no go" decision, as mentioned in paragraph 7-1-4b2, should be assessed at all phases of flight.

g. Following any briefing, feel free to ask for any information that you or the briefer may have missed or are not understood. This way, the briefer is able to present the information in a logical sequence, and lessens the chance of important items being overlooked.

7-1-5. En Route Flight Advisory Service (EFAS)

a. EFAS (radio call "Flight Watch") is a service specifically designed to provide en route aircraft with timely and meaningful weather advisories pertinent to the type of flight intended, route of flight, and altitude. In conjunction with this service, EFAS is also a central collection and distribution point for pilot reported weather information. EFAS is provided by specially trained FSS specialists controlling multiple Remote Communications Outlets covering a large geographical area and is normally available throughout the conterminous U.S. and Puerto Rico from 6 a.m. to 10 p.m. EFAS provides communications capabilities for aircraft flying at 5,000 feet above ground level to 17,500 feet MSL on a common frequency of 122.0 MHz. Discrete EFAS frequencies have been established to ensure communications coverage from 18,000 through 45,000 MSL serving in each specific ARTCC area. These discrete frequencies may be used below 18,000 feet when coverage permits reliable communication.

NOTE-*When an EFAS outlet is located in a time zone different from the zone in which the flight watch control station is located, the availability of service may be plus or minus one hour from the normal operating hours.*

b. In some regions of the contiguous U.S., especially those that are mountainous, it is necessary to be above 5000 feet AGL in order to be at an altitude where the EFAS frequency, 122.0 MHz, is available. Pilots should take this into account when flight planning. Other FSS communication frequencies may be available at lower altitudes. See FIG 7-1-2.

c. Contact flight watch by using the name of the ARTCC facility identification serving the area of your location, followed by your aircraft identification, and the name of the nearest VOR to your position. The specialist needs to know this approximate location to select the most appropriate transmitter/receiver outlet for communications coverage.

EXAMPLE-*Cleveland Flight Watch, Cessna One Two Three Four Kilo, Mansfield V-O-R, over.*

d. Charts depicting the location of the flight watch control stations (parent facility) and the outlets they use are contained in the A/FD. If you do not know in which flight watch area you are flying, initiate contact by using the words "Flight Watch," your aircraft identification, and the name of the nearest VOR. The facility will respond using the name of the flight watch facility.

EXAMPLE-*Flight Watch, Cessna One Two Three Four Kilo, Mansfield V-O-R, over.*

e. Radio outlets that provide En Route Flight Advisory Service are listed regionally in the A/FDs.

f. EFAS is not intended to be used for filing or closing flight plans, position reporting, getting complete preflight briefings, or obtaining random weather reports and forecasts. En route flight advisories are tailored to the phase of flight that begins after climb-out and ends with descent to land. Immediate destination weather and terminal aerodrome forecasts will be provided on request. Pilots requesting information not within the scope of flight watch will be advised of the appropriate FSS frequency to obtain the information. Pilot participation is essential to the success of EFAS by providing a continuous exchange of information on weather, winds, turbulence, flight visibility, icing, etc., between pilots and flight watch specialists. Pilots are encouraged to report good weather as well as bad, and to confirm expected conditions as well as unexpected to EFAS facilities.

7-1-6. Inflight Aviation Weather Advisories

a. Background

1. Inflight Aviation Weather Advisories are forecasts to advise en route aircraft of development of potentially hazardous weather. All inflight aviation weather advisories in the conterminous U.S. are issued by the Aviation Weather Center (AWC) in Kansas City, Missouri. The Weather Forecast Office (WFO) in Honolulu issues advisories for the Hawaiian Islands. In Alaska, the Alaska Aviation Weather Unit (AAWU) issues inflight aviation weather advisories. All heights are referenced MSL, except in the case of ceilings (CIG) which indicate AGL.

2. There are three types of inflight aviation weather advisories: the SIGMET, the Convective SIGMET, and the AIRMET (text or graphical product). All of these advisories use the same location identifiers (either VORs, airports, or well-known geographic areas) to describe the hazardous weather areas. See FIG 7-1-3 and FIG 7-1-4. Graphics with improved clarity can be found in the latest version of Advisory Circular AC 00-45 series, Aviation Weather Services, which is available on the following Web site: **http://www.faa.gov**.

3. Two other weather products supplement these Inflight Aviation Weather Advisories:

(a) The Severe Weather Watch Bulletins (WWs), (with associated Alert Messages) (AWW), and

(b) The Center Weather Advisories (CWAs).

Fig 7-1-2
EFAS Radio Coverage Areas

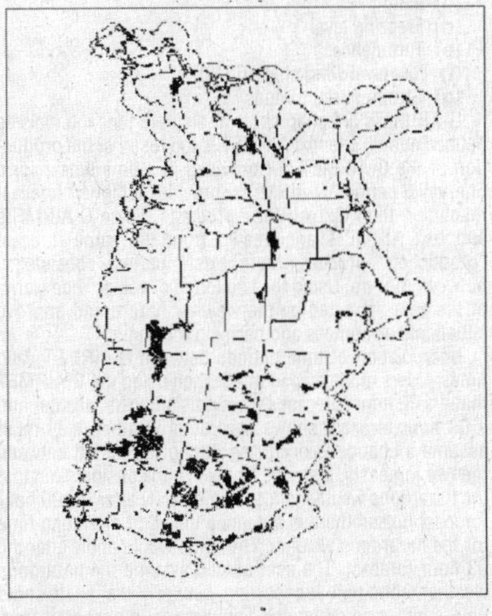

NOTE-*EFAS radio coverage at 5000 feet AGL. The shaded areas depict limited coverage areas in which altitudes above 5000 feet AGL would be required to contact EFAS.*

Fig 7-1-3
Inflight Advisory Plotting Chart

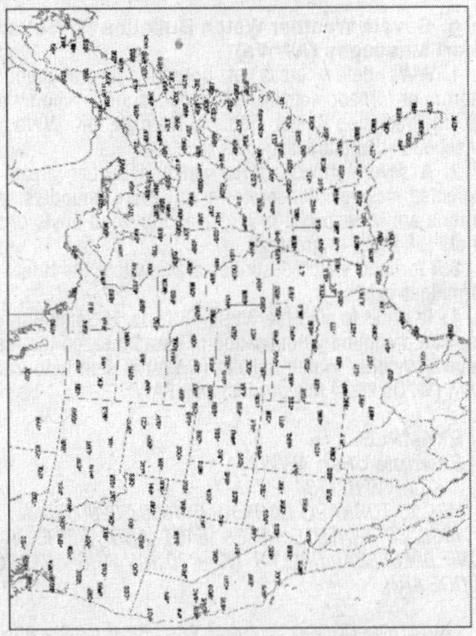

Fig 7-1-4
Geographical Areas and Terrain Features

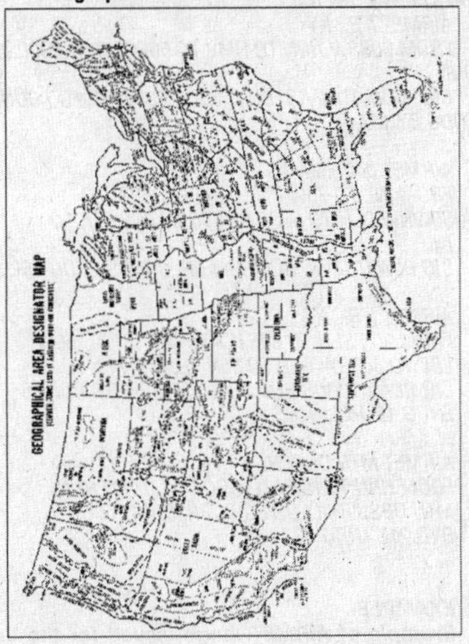

Fig 7-1-5
Aviation Area Forecasts FA Locations - Contiguous United States

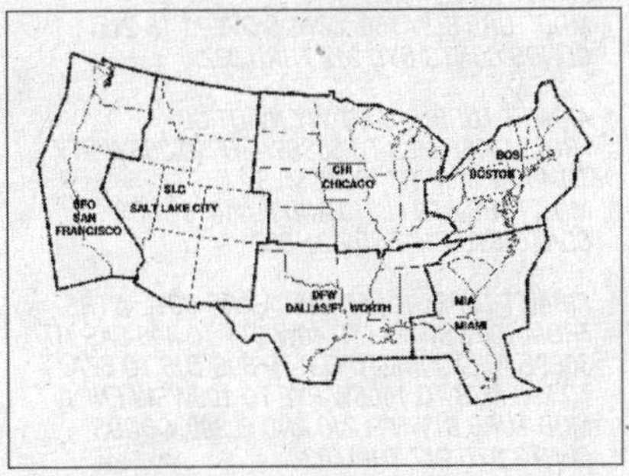

Fig 7-1-6
Alaska Area Forecast Sectors

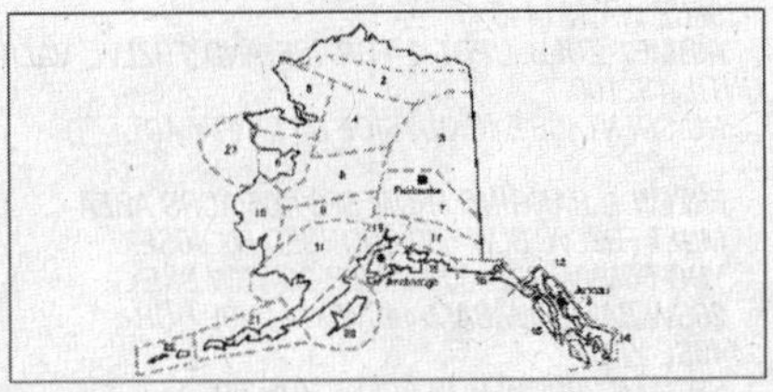

Fig 7-1-7
Hawaii Area Forecast Locations

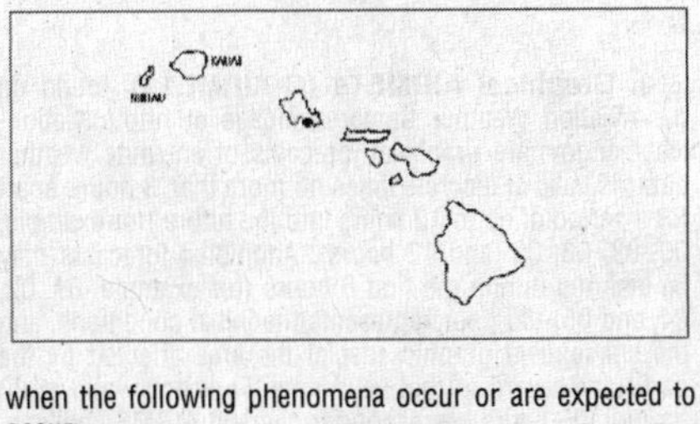

b. SIGMET (WS)/AIRMET (WA or G-AIRMET)

SIGMETs/AIRMET text (WA) products are issued corresponding to the Area Forecast (FA) areas described in FIG 7-1-5, FIG 7-1-6 and FIG 7-1-7. The maximum forecast period is 4 hours for SIGMETs and 6 hours for AIRMETs. The G-AIRMET is issued over the CONUS every 6 hours, valid at 3-hour increments through 12 hours with optional forecasts possible during the first 6 hours. The first 6 hours of the G-AIRMET correspond to the 6-hour period of the AIRMET. SIGMETs and AIRMETs are considered "widespread" because they must be either affecting or be forecasted to affect an area of at least 3,000 square miles at any one time. However, if the total area to be affected during the forecast period is very large, it could be that in actuality only a small portion of this total area would be affected at any one time.

c. SIGMET (WS)

1. A SIGMET advises of nonconvective weather that is potentially hazardous to all aircraft. SIGMETs are unscheduled products that are valid for 4 hours. However, conditions that are associated with hurricanes are valid for 6 hours. Unscheduled updates and corrections are issued as necessary. In the conterminous U.S., SIGMETs are issued when the following phenomena occur or are expected to occur:

(a) Severe icing not associated with thunderstorms.

(b) Severe or extreme turbulence or clear air turbulence (CAT) not associated with thunderstorms.

(c) Dust storms or sandstorms lowering surface or inflight visibilities to below 3 miles.

(d) Volcanic ash.

2. In Alaska and Hawaii, SIGMETs are also issued for:

(a) Tornadoes.

(b) Lines of thunderstorms.

(c) Embedded thunderstorms.

(d) Hail greater than or equal to $^{3}/_{4}$ inch in diameter.

3. SIGMETs are identified by an alphabetic designator from November through Yankee excluding Sierra and Tango. (Sierra, Tango, and Zulu are reserved for AIRMET text [WA] products; G-AIRMETS do not use the Sierra, Tango, or Zulu designators.) The first issuance of a SIGMET will be labeled as UWS (Urgent Weather SIGMET). Subsequent issuances are at the forecaster's discretion. Issuance for the same phenomenon will be sequentially numbered, using the original designator until the phenomenon ends. For example, the first issuance in the Chicago (CHI) FA area for phenomenon moving from the Salt Lake City (SLC) FA area will be SIGMET Papa 3, if the previous two issuances, Papa 1 and Papa 2, had been in the SLC FA area. Note that no two different phenomena across the country can have the same alphabetic designator at the same time.

EXAMPLE-
Example of a SIGMET:
BOSR WS 050600
SIGMET ROMEO 2 VALID UNTIL 051000
ME NH VT
FROM CAR TO YSJ TO CON TO MPV TO CAR
MOD TO OCNL SEV TURB BLW 080 EXP DUE TO STG NWLY FLOW. CONDS CONTG BYD 1000Z.

d. Convective SIGMET (WST)

1. Convective SIGMETs are issued in the conterminous U.S. for any of the following:

(a) Severe thunderstorm due to:

(1) Surface winds greater than or equal to 50 knots.

(2) Hail at the surface greater than or equal to $^{3}/_{4}$ inches in diameter.

(3) Tornadoes.

(b) Embedded thunderstorms.

(c) A line of thunderstorms.

(d) Thunderstorms producing precipitation greater than or equal to heavy precipitation affecting 40 percent or more of an area at least 3,000 square miles.

2. Any convective SIGMET implies severe or greater turbulence, severe icing, and low-level wind shear. A convective SIGMET may be issued for any convective situation that the forecaster feels is hazardous to all categories of aircraft.

3. Convective SIGMET bulletins are issued for the western (W), central (C), and eastern (E) United States. (Convective SIGMETs are not issued for Alaska or Hawaii.) The areas are separated at 87 and 107 degrees west longitude with sufficient overlap to cover most cases when the phenomenon crosses the boundaries. Bulletins are issued hourly at H+55. Special bulletins are issued at any time as required and updated at H+55. If no criteria meeting convective SIGMET requirements are observed or forecasted, the message "CONVECTIVE SIGMET... NONE" will be issued for each area at H+55. Individual convective SIGMETs for each area (W, C, E) are numbered sequentially from number one each day, beginning at 00Z. A convective SIGMET for a continuing phenomenon will be reissued every hour at H+55 with a new number. The text of the bulletin consists of either an observation and a forecast or just a forecast. The forecast is valid for up to 2 hours.

EXAMPLE-
Example of a Convective SIGMET:
MKCC WST 251655
CONVECTIVE SIGMET 54C
VALID UNTIL 1855Z
WI IL
FROM 30E MSN-40ESE DBQ
DMSHG LINE TS 15 NM WIDE MOV FROM 30025KT. TOPS TO FL450. WIND GUSTS TO 50 KT POSS.

CONVECTIVE SIGMET 55C
VALID UNTIL 1855Z
WI IA
FROM 30NNW MSN-30SSE MCW
DVLPG LINE TS 10 NM WIDE MOV FROM 30015KT. TOPS TO FL300.
CONVECTIVE SIGMET 56C
VALID UNTIL 1855Z
MT ND SD MN IA MI
LINE TS 15 NM WIDE MOV FROM 27020KT. TOPS TO FL380.

OUTLOOK VALID 151855-252255
FROM 60NW ISN-INL-TVC-SBN-BRL-FSD-BIL-60NW ISN

IR STLT IMGRY SHOWS CNVTV CLD TOP TEMPS OVER SRN WI HAVE BEEN WARMING STEADILY INDCG A WKNG TREND. THIS ALSO REFLECTED BY LTST RADAR AND LTNG DATA. WKNG TREND OF PRESENT LN MAY CONT... HWVR NEW DVLPMT IS PSBL ALG OUTFLOW BDRY AND/OR OVR NE IA/SW WI BHD CURRENT ACT.

A SCND TS IS CONTG TO MOV EWD THRU ERN MT WITH NEW DVLPMT OCRG OVR CNTRL ND. MT ACT IS MOVG TWD MORE FVRBL AMS OVR THE WRN DAKS WHERE DWPTS ARE IN THE UPR 60S WITH LIFTED INDEX VALUES TO MS 6. TS EXPD TO INCR IN COVERAGE AND INTSTY DURG AFTN HRS.

WST ISSUANCES EXPD TO BE RQRD THRUT AFTN HRS WITH INCRG PTNTL FOR STGR CELLS TO CONTAIN LRG HAIL AND PSBLY DMGG SFC WNDS.

e. International SIGMET

1. Some NWS offices have been designated by the ICAO as Meteorological Watch Offices (MWOs). These offices are responsible for issuing International SIGMETs for designated areas that include Alaska, Hawaii, portions of the Atlantic and Pacific Oceans, and the Gulf of Mexico.

2. The offices which issue international SIGMETs are:

(a) The AWC in Kansas City, Missouri.

(b) The AAWU in Anchorage, Alaska.

(c) The WFO in Honolulu, Hawaii.

3. These SIGMETs are considered "widespread" because they must be either affecting or be forecasted to affect an area of at least 3,000 square miles at any one time. The International SIGMET is issued for 6 hours for volcanic ash events, 6 hours for hurricanes and tropical storms, and 4 hours for all other events. Like the domestic SIGMETs, international SIGMETs are also identified by an alphabetic designator from Alpha through Mike and are numbered sequentially until that weather phenomenon ends. The criteria for an international SIGMET are:

(a) Thunderstorms occurring in lines, embedded in clouds, or in large areas producing tornadoes or large hail.

(b) Tropical cyclones.

(c) Severe icing.

(d) Severe or extreme turbulence.

(e) Dust storms and sandstorms lowering visibilities to less than 3 miles.

(f) Volcanic ash.

EXAMPLE-
Example of an International SIGMET:
WSNT06 KKCI 022014
SIGA0F
KZMA KZNY TJZS SIGMET FOXTROT 3 VALID 022015/030015 KKCI- MIAMI OCEANIC FIR NEW YORK OCEANIC FIR SAN JUAN FIR FRQ TS WI AREA BOUNDED BY 2711N6807W 2156N6654W 2220N7040W 2602N7208W 2711N6807W. TOPS TO FL470. MOV NE 15KT. WKN. BASED ON SAT AND LTG OBS.
MOSHER

f. AIRMET

1. AIRMETs (WAs) are advisories of significant weather phenomena but describe conditions at intensities lower than those which require the issuance of SIGMETs. AIRMETs are intended for dissemination to all pilots in the preflight and en route phase of flight to enhance safety. AIRMET information is available in two formats: text bulletins (WA) and graphics (G-AIRMET). Both formats meet the criteria of paragraph 7-1-3i1 and are issued on a scheduled basis every 6 hours beginning at 0245 UTC. Unscheduled updates and corrections are issued as necessary. AIRMETs contain details about IFR, extensive mountain obscuration, turbulence, strong surface winds, icing, and freezing levels.

2. There are three AIRMETs: Sierra, Tango, and Zulu. After the first issuance each day, scheduled or unscheduled bulletins are numbered sequentially for easier identification.

(a) AIRMET Sierra describes IFR conditions and/or extensive mountain obscurations.

(b) AIRMET Tango describes moderate turbulence, sustained surface winds of 30 knots or greater, and/or nonconvective low-level wind shear.

(c) AIRMET Zulu describes moderate icing and provides freezing level heights.

EXAMPLE-
Example of AIRMET Sierra issued for the Chicago FA area:
CHIS WA 131445
AIRMET SIERRA UPDT 2 FOR IFR AND MTN OBSCN VALID UNTIL 132100.
AIRMET IFR...KY
FROM 20SSW HNN TO HMV TO 50ENE DYR TO20SSW HNN
CIG BLW 010/VIS BLW 3SM PCPN/BR/FG. CONDS ENDG BY 18Z.
.
AIRMET IFR....MN LS
FROM INL TO 70W YQT TO 40ENE DLH TO 30WNW DLH TO 50SE GFK TO 20 ENE GFK TO INL
CIG BLW 010/VIS BLW 3SM BR. CONDS ENDG 15-18Z.
.
AIRMET IFR....KS
FROM 30N SLN TO 60E ICT TO 40S ICT TO 50W LBL TO 30SSW GLD TO 30N SLN
CIG BLW 010/VIS BLW 3SM PCPN/BR/FG. CONDS ENDG 15-18Z.
.
AIRMET MTN OBSCN...KY TN
FROM HNN TO HMV TO GQO TO LOZ TO HNN
MTN OBSC BY CLDS/PCPN/BR. CONDS CONTG BYD 21Z THRU 03Z.
.....

EXAMPLE-
Example of AIRMET Tango issued for the Salt Lake City FA area:
SLCT WA 131445
AIRMET TANGO UPDT 2 FOR TURB VALID UNTIL 132100.
AIRMET TURB...MT
FROM 40NW HVR TO 50SE BIL TO 60E DLN TO 60SW YQL TO 40NW HVR
MOD TURB BLW 150. CONDS DVLPG 18-21Z. CONDS CONTG BYD 21Z THRU 03Z.
.
AIRMET TURB....ID MT WY NV UT CO
FROM 100SE MLS TO 50SSW BFF TO 20SW BTY TO 40SW BAM TO 100SE MLS
MOD TURB BTN FL310 AND FL410. CONDS CONTG BYD 21Z ENDG 21-00Z.
.
AIRMET TURB...NV AZ NM CA AND CSTL WTRS
FROM 100WSW ENI TO 40W BTY TO 40S LAS TO 30ESE TBE TO INK TO ELP TO 50S TUS TO BZA TO 20S MZB TO 150SW PYE TO 100WSW ENI
MOD TURB BTWN FL210 AND FL380. CONDS CONTG BYD 21Z THRU 03Z.
....

EXAMPLE-
Example of AIRMET Zulu issued for the San Francisco FA area:
SFOZ WA 131445
AIRMET ZULU UPDT 2 FOR ICE AND FRZLVL VALID UNTIL 132100.
NO SGFNT ICE EXP OUTSIDE OF CNVTV ACT.

FRZLVL....RANGING FROM SFC-105 ACRS AREA
MULT FRZLVL BLW 080 BOUNDED BY 40SE YDC-60NNW GEG-60SW MLP-30WSW BKE-20SW BAM-70W BAM-40SW YKM-40E HUH-40SE YDC
SFC ALG 20NNW HUH-30SSE HUH-60S SEA 50NW LKV-60WNWOAL-30SW OAL
040 ALG 40W HUH-30W HUH-30NNW SEA-40N PDX-20NNW DSD
080 ALG 160NW FOT-80SW ONP-50SSW EUG 40SSE OED-50SSE CZQ-60E EHF-40WSW LAS
....

3. Graphical AIRMETs (G-AIRMETs), found on the Aviation Weather Center webpage at http://aviationweather.gov, are graphical forecasts of en-route weather hazards valid at discrete times no more than 3 hours apart for a period of up to 12 hours into the future (for example, 00, 03, 06, 09, and 12 hours). Additional forecasts may be inserted during the first 6 hours (for example, 01, 02, 04, and 05). 00 hour represents the initial conditions, and the subsequent graphics depict the area affected by the particular hazard at that valid time. Forecasts valid at 00 through 06 hours correspond to the text AIRMET bulletin. Forecasts valid at 06 through 12 hours correspond to the text bulletin outlook. G-AIRMET depicts the following en route aviation weather hazards:

(a) Instrument flight rule conditions (ceiling < 1000' and/or surface visibility <3 miles)

(b) Mountain obscuration

(c) Icing

(d) Freezing level

(e) Turbulence

(f) Low level wind shear (LLWS)

(g) Strong surface winds

G-AIRMETs are snap shots at discrete time intervals as defined above. The text AIRMET is the result of the production of the G-AIRMET but provided in a time smear for a 6hr valid period. G-AIRMETs provide a higher forecast resolution than text AIRMET products. Since G-AIRMETs and text AIRMETs are created from the same forecast "production" process, there exists perfect consistency between the two. Using the two together will provide clarity of the area impacted by the weather hazard and improve situational awareness and decision making.

Interpolation of time periods between G-AIRMET valid times: Users must keep in mind when using the G-AIRMET that if a 00 hour forecast shows no significant weather and a 03 hour forecast shows hazardous weather, they must assume a change is occurring during the period between the two forecasts. It should be taken into consideration that the hazardous weather starts immediately after the 00 hour forecast unless there is a defined initiation or ending time for the hazardous weather. The same would apply after the 03 hour forecast. The user should assume the hazardous weather condition is occurring between the snap shots unless informed otherwise. For example, if a 00 hour forecast shows no hazard, a 03 hour forecast shows the presence of hazardous weather, and a 06 hour forecast shows no hazard, the user should assume the hazard exists from the 0001 hour to the 0559 hour time period.

EXAMPLE-
See FIG 7-1-8 for an example of the G-AIRMET graphical product.

g. Severe Weather Watch Bulletins (WWs) and Alert Messages (AWWs)

1. WWs define areas of possible severe thunderstorms or tornado activity. The bulletins are issued by the Storm Prediction Center (SPC) in Norman, OK. WWs are unscheduled and are issued as required.

2. A severe thunderstorm watch describes areas of expected severe thunderstorms. (Severe thunderstorm criteria are $^{3}/_{4}$-inch hail or larger and/or wind gusts of 50 knots [58 mph] or greater.)

3.A tornado watch describes areas where the threat of tornadoes exists.

4. In order to alert the WFOs, CWSUs, FSSs, and other users, a preliminary notification of a watch called the Alert Severe Weather Watch bulletin (AWW) is sent before the WW. (WFOs know this product as a SAW).

EXAMPLE-
Example of an AWW:
MKC AWW 011734
WW 75 TORNADO TX OK AR 011800Z-020000Z
AXIS..80 STATUTE MILES EAST AND WEST OF A LINE..60ESE DAL/DALLAS TX/ - 30 NW ARG/ WALNUT RIDGE AR/

..AVIATION COORDS.. 70NM E/W /58W GGG - 25NW ARG/
HAIL SURFACE AND ALOFT..1 $^{3}/_{4}$ INCHES. WIND GUSTS..70 KNOTS. MAX TOPS TO 450. MEAN WIND VECTOR 24045.

5. Soon after the AWW goes out, the actual watch bulletin itself is issued. A WW is in the following format:

(a) Type of severe weather watch, watch area, valid time period, type of severe weather possible, watch axis, meaning of a watch, and a statement that persons should be on the lookout for severe weather.

(b) Other watch information; i.e., references to previous watches.

(c) Phenomena, intensities, hail size, wind speed (knots), maximum cumulonimbus (CB) tops, and estimated cell movement (mean wind vector).

(d) Cause of severe weather.

(e) Information on updating Convective Outlook (AC) products.

EXAMPLE-
Example of a WW:
BULLETIN - IMMEDIATE BROADCAST REQUESTED
TORNADO WATCH NUMBER 381

STORM PREDICTION CENTER NORMAN OK
556 PM CDT MON JUN 2 1997
THE STORM PREDICTON CENTER HAS ISSUED A TORNADO WATCH FOR PORTIONS OF NORTHEAST NEW MEXICO TEXAS PANHANDLE
EFFECTIVE THIS MONDAY NIGHT AND TUESDAY MORNING FROM 630 PM UNTIL MIDNIGHT CDT.
TORNADOES...HAIL TO 2 $^3/_4$ INCHES IN DIAMETER... THUNDERSTORM WIND GUSTS TO 80 MPH...AND DANGEROUS LIGHTNING ARE POSSIBLE IN THESE AREAS.
THE TORNADO WATCH AREA IS ALONG AND 60 STATUTE MILES NORTH AND SOUTH OF A LINE FROM 50 MILES SOUTHWEST OF RATON NEW MEXICO TO 50 MILES EAST OF AMARILLO TEXAS.
REMEMBER...A TORNADO WATCH MEANS CONDITIONS ARE FAVORABLE FOR TORNADOES AND SEVERE THUNDERSTORMS IN AND CLOSE TO THE WATCH AREA. PERSONS IN THESE AREAS SHOULD BE ON THE LOOKOUT FOR THREATENING WEATHER CONDITIONS AND LISTEN FOR LATER STATEMENTS AND POSSIBLE WARNINGS.
OTHER WATCH INFORMATION...CONTINUE... WW378...WW 379...WW 380

DISCUSSION...THUNDERSTORMS ARE INCREASING OVER NE NM IN MOIST SOUTHEASTERLY UPSLOPE FLOW. OUTFLOW BOUNDARY EXTENDS EASTWARD INTO THE TEXAS PANHANDLE AND EXPECT STORMS TO MOVE ESE ALONG AND NORTH OF THE BOUNDARY ON THE N EDGE OF THE CAP. VEERING WINDS WITH HEIGHT ALONG WITH INCREASGING MID LVL FLOW INDICATE A THREAT FOR SUPERCELLS.

AVIATION...TORNADOES AND A FEW SEVERE THUNDERSTORMS WITH HAIL SURFACE AND ALOFT TO 2 $^3/_4$ INCHES. EXTREME TURBULENCE AND SURFACE WIND GUSTS TO 70 KNOTS. A FEW CUMULONIMBI WITH MAXIMUM TOPS TO 550. MEAN STORM MOTION VECTOR 28025.

6. Status reports are issued as needed to show progress of storms and to delineate areas no longer under the threat of severe storm activity. Cancellation bulletins are issued when it becomes evident that no severe weather will develop or that storms have subsided and are no longer severe.

7. When tornadoes or severe thunderstorms have developed, the local WFO office will issue the warnings covering those areas.

h. Center Weather Advisories (CWAs)

1. CWAs are unscheduled inflight, flow control, air traffic, and air crew advisory. By nature of its short lead time, the CWA is not a flight planning product. It is generally a nowcast for conditions beginning within the next two hours. CWAs will be issued:

(a) As a supplement to an existing SIGMET, Convective SIGMET or AIRMET.

(b) When an Inflight Advisory has not been issued but observed or expected weather conditions meet SIGMET/AIRMET criteria based on current pilot reports and reinforced by other sources of information about existing meteorological conditions.

(c) When observed or developing weather conditions do not meet SIGMET, Convective SIGMET, or AIRMET criteria; e.g., in terms of intensity or area coverage, but current pilot reports or other weather information sources indicate that existing or anticipated meteorological phenomena will adversely affect the safe flow of air traffic within the ARTCC area of responsibility.

2. The following example is a CWA issued from the Kansas City, Missouri, ARTCC. The "3" after ZKC in the first line denotes this CWA has been issued for the third weather phenomena to occur for the day. The "301" in the second line denotes the phenomena number again (3) and the issuance number (01) for this phenomena. The CWA was issued at 2140Z and is valid until 2340Z.

EXAMPLE-
ZKC3 CWA 032140
ZKC CWA 301 VALID UNTIL 032340
ISOLD SVR TSTM over KCOU MOVG SWWD 10KTS ETC.

7-1-7. Categorical Outlooks

a. Categorical outlook terms, describing general ceiling and visibility conditions for advanced planning purposes are used only in area forecasts and are defined as follows:

1. LIFR (Low IFR). Ceiling less than 500 feet and/or visibility less than 1 mile.

2. IFR. Ceiling 500 to less than 1,000 feet and/or visibility 1 to less than 3 miles.

3. MVFR (Marginal VFR). Ceiling 1,000 to 3,000 feet and/or visibility 3 to 5 miles inclusive.

4. VFR. Ceiling greater than 3,000 feet and visibility greater than 5 miles; includes sky clear.

b. The cause of LIFR, IFR, or MVFR is indicated by either ceiling or visibility restrictions or both. The contraction "CIG" and/or weather and obstruction to vision symbols are used. If winds or gusts of 25 knots or greater are forecast for the outlook period, the word "WIND" is also included for all categories including VFR.

EXAMPLE-
1. *LIFR CIG-low IFR due to low ceiling.*
2. *IFR FG-IFR due to visibility restricted by fog.*
3. *MVFR CIG HZ FU-marginal VFR due to both ceiling and visibility restricted by haze and smoke.*
4. *IFR CIG RA WIND-IFR due to both low ceiling and visibility restricted by rain; wind expected to be 25 knots or greater.*

7-1-8. Telephone Information Briefing Service (TIBS)

a. TIBS, provided by FSS, is a system of automated telephone recordings of meteorological and aeronautical information available throughout the United States. Based on the specific needs of each area, TIBS provides route and/or area briefings in addition to airspace procedures and special announcements concerning aviation interests that may be available. Depending on user demand, other items may be provided; for example, surface weather observations, terminal forecasts, wind and temperatures aloft forecasts, etc.

b. TIBS is not intended to be a substitute for specialist-provided preflight briefings from FSS. TIBS is recommended as a preliminary briefing and often will be valuable in helping you to make a "go" or "no go" decision.

c. Pilots are encouraged to utilize TIBS, which can be accessed by dialing the FSS toll-free telephone number, 1-800-WX-BRIEF (992-7433) or specific published TIBS telephone numbers in certain areas. Consult the "FSS Telephone Numbers" section of the A/FD or the Alaska or Pacific Chart Supplement book.

NOTE-*A touch-tone telephone is necessary to fully utilize TIBS.*

7-1-9. Transcribed Weather Broadcast (TWEB) (Alaska Only)

Equipment is provided in Alaska by which meteorological and aeronautical data are recorded on tapes and broadcast continuously over selected L/MF and VOR facilities. Broadcasts are made from a series of individual tape recordings, and changes, as they occur, are transcribed onto the tapes. The information provided varies depending on the type equipment available. Generally, the broadcast contains a summary of adverse conditions, surface weather observations, pilot weather reports, and a density altitude statement (if applicable). At the discretion of the broadcast facility, recordings may also include a synopsis, winds aloft forecast, en route and terminal forecast data, and radar reports. At selected locations, telephone access to the TWEB has been provided (TEL-TWEB). Telephone numbers for this service are found in the Supplement Alaska A/FD. These broadcasts are made available primarily for preflight and inflight planning, and as such, should not be considered as a substitute for specialist-provided preflight briefings.

7-1-10. Inflight Weather Broadcasts

a. Weather Advisory Broadcasts. ARTCCs broadcast a Severe Weather Forecast Alert (AWW), Convective SIGMET, SIGMET, or CWA alert once on all frequencies, except emergency, when any part of the area described is within 150 miles of the airspace under their jurisdiction. These broadcasts contain SIGMET or CWA (identification) and a brief description of the weather activity and general area affected.

EXAMPLE-
1. *Attention all aircraft, SIGMET Delta Three, from Myton to Tuba City to Milford, severe turbulence and severe clear icing below one zero thousand feet. Expected to continue beyond zero three zero zero zulu.*
2. *Attention all aircraft, convective SIGMET Two Seven Eastern. From the vicinity of Elmira to Phillipsburg. Scattered embedded thunderstorms moving east at one zero knots. A few intense level five cells, maximum tops four five zero.*
3. *Attention all aircraft, Kansas City Center weather advisory one zero three. Numerous reports of moderate to severe icing from eight to niner thousand feet in a three zero mile radius of St. Louis. Light or negative icing reported from four thousand to one two thousand feet remainder of Kansas City Center area.*

NOTE-
1. *Terminal control facilities have the option to limit the AWW, convective SIGMET, SIGMET, or CWA broadcast as follows: local control and approach control positions may opt to broadcast SIGMET or CWA alerts only when any part of the area described is within 50 miles of the airspace under their jurisdiction.*
2. *In areas where HIWAS is available, ARTCC, Terminal ATC, and FSS facilities no longer broadcast Inflight Weather Advisories as described above in paragraph a. See paragraphs b1 and b2 below.*

b. Hazardous Inflight Weather Advisory Service (HIWAS). HIWAS is an automated, continuous broadcast of inflight weather advisories, provided by FSS over select VOR outlets, which include the following weather products: AWW, SIGMET, Convective SIGMET, CWA, AIRMET (text [WA] or graphical [G-AIRMET] products), and urgent PIREP. HIWAS is available throughout the conterminous United States as an additional source of hazardous weather information. HIWAS does not replace preflight or inflight weather briefings from FSS or real-time weather updates from Enroute Flight Advisory Service (EFAS), radio call "Flight Watch." Pilots should call FSS or Flight Watch if there are any questions about weather that is different than forecasted or if the HIWAS broadcast appears to be in error.

1. Where HIWAS is available, ARTCC and terminal ATC facilities will broadcast, upon receipt, a HIWAS alert once on all frequencies, except emergency frequencies. Included in the broadcast will be an alert announcement, frequency instruction, number, and type of advisory updated; for example, AWW, SIGMET, Convective SIGMET, or CWA.

EXAMPLE-*Attention all aircraft. Hazardous weather information (SIGMET, Convective SIGMET, AIRMET (text [WA] or graphical [G-AIRMET] product), Urgent Pilot Weather Report [UUA], or Center Weather Advisory [CWA], Number or Numbers) for (geographical area) available on HIWAS, Flight Watch, or Flight Service frequencies.*

2. Upon notification of an update to HIWAS, FSS will broadcast a HIWAS update announcement once on all frequencies except emergency frequencies. Included in

Fig 7-1-8
G-AIRMET Graphical Product

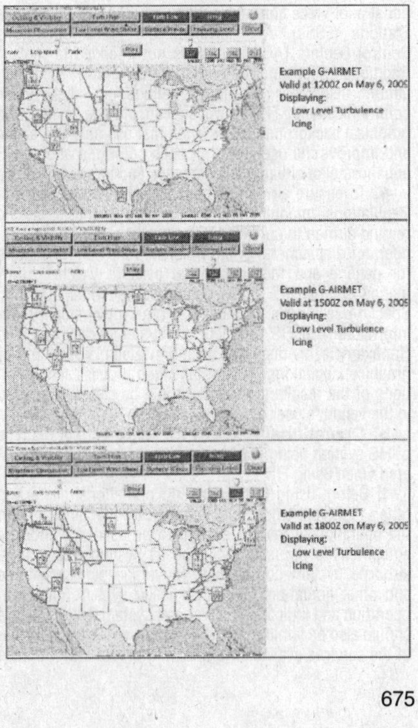

the broadcast will be the type of advisory updated; for example, AWW, SIGMET, Convective SIGMET, CWA, etc.

EXAMPLE-*Attention all aircraft. Hazardous weather information for (geographical area) available from Flight Watch or Flight Service.*

3. HIWAS availability is notated with VOR listings in the Airport/Facility Directory (A/FD), and is shown by symbols on IFR Enroute Low Altitude Charts and VFR Sectional Charts. The symbol depiction is identified in the chart legend.

7-1-11. Flight Information Services (FIS)

a. FIS. Aviation weather and other operational information may be displayed in the cockpit through the use of FIS. FIS systems are of two basic types: Broadcast only systems (called FIS-B) and two-way request/reply systems. Broadcast system components include a ground- or space-based transmitter, an aircraft receiver, and a portable or installed cockpit display device. Two-way systems utilize transmitter/receivers at both the ground- or space-based site and the aircraft.

1. Broadcast FIS (i.e., FIS-B) allows the pilot to passively collect weather and other operational data and to display that data at the appropriate time. In addition to textual weather products such as Aviation Routine Weather Reports (METARs)/ Aviation Selected Special Weather Reports (SPECIs) and Terminal Area Forecasts (TAFs), graphical weather products such as radar composite/ mosaic images, temporary flight restricted airspace and other NOTAMs may be provided to the cockpit. Two-way FIS services permit the pilot to make specific weather and other operational information requests for cockpit display. A FIS service provider will then prepare a reply in response to that specific request and transmit the product to that specific aircraft.

2. FIS services are available from four types of service providers:

(a) A private sector FIS provider operating under service agreement with the FAA using broadcast data link over VHF aeronautical spectrum and whose products have been reviewed and accepted by the FAA prior to transmission. (Products and services are defined under subparagraph c.)

(b) Through an FAA operated service using a broadcast data link on the ADS-B UAT network. (Products and services are defined under subparagraph d.)

(c) Private sector FIS providers operating under customer contracts using aeronautical spectrum.

(d) Private sector FIS providers operating under customer contract using methods other than aeronautical spectrum, including Internet data-to-the-cockpit service providers.

3. FIS is a method of receiving aviation weather and other operational data in the cockpit that augments traditional pilot voice communication with FAA's Flight Service Stations (FSSs), ATC facilities, or Airline Operations Control Centers (AOCCs). FIS is not intended to replace traditional pilot and controller/flight service specialist/ aircraft dispatcher pre-flight briefings or inflight voice communications. FIS; however, can provide textual and graphical background information that can help abbreviate and improve the usefulness of such communications. FIS enhances pilot situational awareness and improves safety.

4. To ensure airman compliance with Federal Aviation Regulations, manufacturer's operating manuals should remind airmen to contact ATC controllers, FSS specialists, operator dispatchers, or airline operations control centers for general and mission critical aviation weather information and/or NAS status conditions (such as NOTAMs, Special Use Airspace status, and other government flight information). If FIS products are systematically modified (for example, are displayed as abbreviated plain text and/or graphical depictions), the modification process and limitations of the resultant product should be clearly described in the vendor's user guidance.

b. Operational Use of FIS. Regardless of the type of FIS system being used, several factors must be considered when using FIS:

1. Before using FIS for inflight operations, pilots and other flight crewmembers should become familiar with the operation of the FIS system to be used, the airborne equipment to be used, including its system architecture, airborne system components, coverage service volume and other limitations of the particular system, modes of operation and indications of various system failures. Users should also be familiar with the specific content and format of the services available from the FIS provider(s). Sources of information that may provide this specific guidance include manufacturer's manuals, training programs and reference guides.

2. FIS should not serve as the sole source of aviation weather and other operational information. ATC, FSSs and, if applicable, AOCC VHF/HF voice remain as a redundant method of communicating aviation weather, NOTAMs, and other operational information to aircraft in flight. FIS augments these traditional ATC/FSS/AOCC services and, for some products, offers the advantage of being displayed as graphical information. By using FIS for orientation, the usefulness of information received from conventional means may be enhanced. For example, FIS may alert the pilot to specific areas of concern that will more accurately focus requests made to FSS or AOCC for inflight updates or similar queries made to ATC.

3. The airspace and aeronautical environment is constantly changing. These changes occur quickly and without warning. Critical operational decisions should be based on use of the most current and appropriate data available. When differences exist between FIS and information obtained by voice communication with ATC, FSS, and/or AOCC (if applicable), pilots are cautioned to use the most recent data from the most authoritative source.

4. FIS aviation weather products (e.g., graphical ground-based radar precipitation depictions) are not appropriate for tactical avoidance of severe weather such as negotiating a path through a weather hazard area. FIS supports strategic weather decision making such as route selection to avoid a weather hazard area in its entirety. The misuse of information beyond its applicability may place the pilot and aircraft in jeopardy. In addition, FIS should never be used in lieu of an individual pre-flight weather and flight planning briefing.

5. FIS NOTAM products, including Temporary Flight Restriction (TFR) information, are advisory-use information and are intended for situational awareness purposes only. Cockpit displays of this information are not appropriate for tactical navigation - pilots should stay clear of any geographic area displayed as a TFR NOTAM. Pilots should contact FSSs and/or ATC while en route to obtain updated information and to verify the cockpit display of NOTAM information.

6. FIS supports better pilot decision making by increasing situational awareness. Better decision-making is based on using information from a variety of sources. In addition to FIS, pilots should take advantage of other weather/NAS status sources, including, briefings from Flight Service Stations, FAA's en route "Flight Watch" service, data from other air traffic control facilities, airline operation control centers, pilot reports, as well as their own observations.

c. FAA FISDL (VHF) Service. The FAA's FISDL (VHF datalink) system is a VHF Data Link (VDL) Mode 2 implementation that provides pilots and flight crews of properly equipped aircraft with a cockpit display of certain aviation weather and flight operational information. This information may be displayed in both textual and graphical formats. The system is operated under a service agreement with the FAA, using broadcast data link on VHF aeronautical spectrum on two 25 KHz spaced frequencies (136.450 and 136.475 MHz). The FAA FISDL (VHF) service is designed to provide coverage throughout the continental U.S. from 5,000 feet AGL to 17,500 feet MSL, except in areas where this is not feasible due to mountainous terrain. Aircraft operating near transmitter sites may receive useable FISDL signals at altitudes lower than 5,000 feet AGL, including on the surface in some locations, depending on transmitter/ aircraft line of sight geometry. Aircraft operating above 17,500 feet MSL may also receive useable FISDL signals under certain circumstances.

1. FAA FISDL (VHF) service provides, free of charge, the following basic text products:

(a) Aviation Routine Weather Reports (METARs).

(b) Aviation Selected Special Weather Reports (SPECIs).

(c) Terminal Area Forecasts (TAFs), and their amendments.

(d) Significant Meteorological Information (SIGMETs).

(e) Convective SIGMETs.

(f) AIRMET text bulletins (WA).

(g) Pilot Reports (both urgent and routine) (PIREPs); and,

(h) Severe Weather Forecast Alerts and Warnings (AWWs/WW) issued by the NOAA Storm Prediction Center (SPC).

2. The format and coding of these text products are described in Advisory Circular AC-00-45, Aviation Weather Services, and paragraph 7-1-30, Key to Aerodrome Forecast (TAF) and Aviation Routine Weather Report (METAR).

3. Additional products, called "Value-Added Products," are also available from the vendor on a paid subscription basis. Details concerning the content, format, symbology and cost of these products may be obtained from the vendor.

d. FAA's Flight Information Service-Broadcast (FIS-B).

1. FIS-B is a ground-based broadcast service provided through the FAA's Automatic Dependent Surveillance - Broadcast (ADS-B) Services Universal Access Transceiver (UAT) network. The service provides users with a 978 MHz data link capability when operating within range and line-of-sight of a transmitting ground station. FIS-B enables users of properly equipped aircraft to receive and display a suite of broadcast weather and aeronautical information products. Services are currently available in specific geographic regions of the U.S. with NAS-wide service availability planned for 2013.

2. The following list represents the initial suite of text and graphical products available through FIS-B and provided free-of-charge. Detailed information concerning FIS-B meteorological products can be found in Advisory Circular 00-45, Aviation Weather Services. Information on Special Use Airspace (SUA), Temporary Flight Restriction (TFR) and Notice to Airmen (NOTAM) products can be found in Chapters 3, 4 and 5 of this manual.

(a) Text: Aviation Routine Weather Report (METAR) and Special Aviation Report (SPECI);

(b) Text: Pilot Weather Report (PIREP);

(c) Text: Winds and Temperatures Aloft;

(d) Text: Terminal Area Forecast (TAF) and amendments;

(e) Text: Notice to Airmen (NOTAM) Distant and Flight Data Center;

(f) Text/Graphic: Airmen's Meteorological Conditions (AIRMET);

(g) Text/Graphic: Significant Meteorological Conditions (SIGMET);

(h) Text/Graphic: Convective SIGMET;

(i) Text/Graphic: Special Use Airspace (SUA);

(j) Text/Graphic: Temporary Flight Restriction (TFR) NOTAM; and

(k) Graphic: NEXRAD Composite Reflectivity Products (Regional and National).

3. Users of FIS-B should familiarize themselves with the operational characteristics and limitations of the system, including: system architecture; service environment; product lifecycles; modes of operation; and indications of system failure.

4. FIS-B products are updated and transmitted at specific intervals based primarily on product issuance criteria. Update intervals are defined as the rate at which the product data is available from the source for transmission. Transmission intervals are defined as the amount of time within which a new or updated product transmission must be completed and/or the rate or repetition interval at which the product is rebroadcast. Update and transmission intervals for each product are provided in TBL 7-1-1.

5. Where applicable, FIS-B products include a look-ahead range expressed in nautical miles (NM) for three service domains: Airport Surface; Terminal Airspace; and En-route/Gulf-of-Mexico (GOMEX). TBL 7-1-1 provides service domain availability and look-ahead ranging for each FIS-B product.

6. Prior to using this capability, users should familiarize themselves with the operation of FIS-B avionics by referencing the applicable User's Guides. Guidance concerning the interpretation of information displayed should be obtained from the appropriate avionics manufacturer.

7. FIS-B malfunctions not attributed to aircraft system failures or covered by active NOTAM should be reported by radio or telephone to the nearest FSS facility. Malfunctions may also be reported by submitting FAA Form 8740-5, Safety Improvement Report via mail, fax, or email to your local Flight District Standards Office, Safety Program Manager.

e. Non-FAA FISDL Systems. Several commercial vendors also provide customers with FIS data over both the aeronautical spectrum and on other frequencies using a variety of data link protocols. In some cases, the vendors provide only the communications system that carries customer messages, such as the Aircraft Communications Addressing and Reporting System (ACARS) used by many air carrier and other operators.

1. Operators using non-FAA FIS data for inflight weather and other operational information should ensure that the products used conform to FAA/NWS standards.

Tbl 7-1-1

Product	Update Interval	Transmission Interval	Service Domain and Look ahead Range (NM)
AIRMET	As Available	5 minutes	100 (Airport Surface), 500 (Terminal, Enroute/GOMEX)
SIGMET & Convective SIGMET	As Available, then at 15 minute intervals for 1 hour	5 minutes	100 (Airport Surface), 500 (Terminal, Enroute/ GOMEX)
METAR/SPECI	1 minute (where available), As Available otherwise	5 minute	100 (Airport Surface), 500 (Terminal, Enroute/GOMEX)
NEXRAD Composite Reflectivity (National)	~5 minutes precipitation mode 10 minutes for clear air mode	15 minutes	CONUS
NEXRAD Composite Reflectivity (Regional)	~5 minutes precipitation mode 10 minutes for clear air mode	2.5 minutes	250
NOTAM (D) / FDC (including TFR)	As Available	10 minutes	100
PIREP	As Available	10 minutes	N/A (Airport Surface), 500 (Terminal, Enroute/ GOMEX)
SUA	As Available	10 minutes	N/A (Airport Surface), 500 (Terminal, Enroute/ GOMEX)
TAF/AMEND	8 Hours	10 minutes	100 (Airport Surface), 500 (Terminal, Enroute/ GOMEX)
Winds & Temperatures Aloft	12 Hours	10 minutes	1,000

Specifically, aviation weather and NAS status information should meet the following criteria:

(a) The products should be either FAA/NWS "accepted" aviation weather reports or products, or based on FAA/NWS accepted aviation weather reports or products. If products are used which do not meet this criteria, they should be so identified. The operator must determine the applicability of such products to their particular flight operations.

(b) In the case of a weather product which is the result of the application of a process which alters the form, function or content of the base FAA/NWS accepted weather product(s), that process, and any limitations to the application of the resultant product, should be described in the vendor's user guidance material.

2. An example would be a NEXRAD radar composite/mosaic map, which has been modified by changing the scaling resolution. The methodology of assigning reflectivity values to the resultant image components should be described in the vendor's guidance material to ensure that the user can accurately interpret the displayed data.

7-1-12. Weather Observing Programs

a. Manual Observations. With only a few exceptions, these reports are from airport locations staffed by FAA or NWS personnel who manually observe, perform calculations, and enter these observations into the (WMSCR) communication system. The format and coding of these observations are contained in paragraph 7-1-30, Key to Aviation Routine Weather Report (METAR) and Aerodrome Forecasts (TAF).

b. Automated Weather Observing System (AWOS).

1. Automated weather reporting systems are increasingly being installed at airports. These systems consist of various sensors, a processor, a computer-generated voice subsystem, and a transmitter to broadcast local, minute-by-minute weather data directly to the pilot.

NOTE-*When the barometric pressure exceeds 31.00 inches Hg., see paragraph 7-2-2, Procedures, for the altimeter setting procedures.*

2. The AWOS observations will include the prefix "AUTO" to indicate that the data are derived from an automated system. Some AWOS locations will be augmented by certified observers who will provide weather and obstruction to vision information in the remarks of the report when the reported visibility is less than 7 miles. These sites, along with the hours of augmentation, are to be published in the A/FD. Augmentation is identified in the observation as "OBSERVER WEATHER." The AWOS wind speed, direction and gusts, temperature, dew point, and altimeter setting are exactly the same as for manual observations. The AWOS will also report density altitude when it exceeds the field elevation by more than 1,000 feet. The reported visibility is derived from a sensor near the touchdown of the primary instrument runway. The visibility sensor output is converted to a visibility value using a 10-minute harmonic average. The reported sky condition/ceiling is derived from the ceilometer located next to the visibility sensor. The AWOS algorithm integrates the last 30 minutes of ceilometer data to derive cloud layers and heights. This output may also differ from the observer sky condition in that the AWOS is totally dependent upon the cloud advection over the sensor site.

3. These real-time systems are operationally classified into nine basic levels:

(a) AWOS-A only reports altimeter setting;

NOTE-*Any other information is advisory only.*

(b) AWOS-AV reports altimeter and visibility;

NOTE-*Any other information is advisory only.*

(c) AWOS-I usually reports altimeter setting, wind data, temperature, dew point, and density altitude;

(d) AWOS-2 provides the information provided by AWOS-I plus visibility; and

(e) AWOS-3 provides the information provided by AWOS-2 plus cloud/ceiling data.

(f) AWOS- 3P provides reports the same as the AWOS 3 system, plus a precipitation identification sensor.

(g) AWOS- 3PT reports the same as the AWOS 3P System, plus thunderstorm/lightning reporting capability.

(h) AWOS- 3T reports the same as AWOS 3 system and includes a thunderstorm/lightning reporting capability.

(i) AWOS- 4 reports the same as the AWOS 3 system, plus precipitation occurrence, type and accumulation, freezing rain, thunderstorm, and runway surface sensors.

4. The information is transmitted over a discrete VHF radio frequency or the voice portion of a local NAVAID. AWOS transmissions on a discrete VHF radio frequency are engineered to be receivable to a maximum of 25 NM from the AWOS site and a maximum altitude of 10,000 feet AGL. At many locations, AWOS signals may be received on the surface of the airport, but local conditions may limit the maximum AWOS reception distance and/or altitude. The system transmits a 20 to 30 second weather message updated each minute. Pilots should monitor the designated frequency for the automated weather broadcast. A description of the broadcast is contained in subparagraph c. There is no two-way communication capability. Most AWOS sites also have a dial-up capability so that the minute-by-minute weather messages can be accessed via telephone.

5. AWOS information (system level, frequency, phone number, etc.) concerning specific locations is published, as the systems become operational, in the A/FD, and where applicable, on published Instrument Approach Procedures. Selected individual systems may be incorporated into nationwide data collection and dissemination networks in the future.

c. AWOS Broadcasts. Computer-generated voice is used in AWOS to automate the broadcast of the minute-by-minute weather observations. In addition, some systems are configured to permit the addition of an operator-generated voice message; e.g., weather remarks following the automated parameters. The phraseology used generally follows that used for other weather broadcasts. Following are explanations and examples of the exceptions.

1. Location and Time. The location/name and the phrase "AUTOMATED WEATHER OBSERVATION," followed by the time are announced.

(a) If the airport's specific location is included in the airport's name, the airport's name is announced.

EXAMPLE-
"Bremerton National Airport automated weather observation, one four five six zulu;"
"Ravenswood Jackson County Airport automated weather observation, one four five six zulu."

(b) If the airport's specific location is not included in the airport's name, the location is announced followed by the airport's name.

EXAMPLE-
"Sault Ste. Marie, Chippewa County International Airport automated weather observation;"
"Sandusky, Cowley Field automated weather observation."

(c) The word "TEST" is added following "OBSERVATION" when the system is not in commissioned status.

EXAMPLE-*"Bremerton National Airport automated weather observation test, one four five six zulu."*

(d) The phrase "TEMPORARILY INOPERATIVE" is added when the system is inoperative.

EXAMPLE-*"Bremerton National Airport automated weather observing system temporarily inoperative."*

2. Visibility.

(a) The lowest reportable visibility value in AWOS is "less than 1/4." It is announced as "VISIBILITY LESS THAN ONE QUARTER."

(b) A sensor for determining visibility is not included in some AWOS. In these systems, visibility is not announced. "VISIBILITY MISSING" is announced only if the system is configured with a visibility sensor and visibility information is not available.

3. Weather. In the future, some AWOSs are to be configured to determine the occurrence of precipitation. However, the type and intensity may not always be determined. In these systems, the word "PRECIPITATION" will be announced if precipitation is occurring, but the type and intensity are not determined.

4. Ceiling and Sky Cover.

(a) Ceiling is announced as either "CEILING" or "INDEFINITE CEILING." With the exception of indefinite ceilings, all automated ceiling heights are measured.

EXAMPLE-
"Bremerton National Airport automated weather observation, one four five six zulu. Ceiling two thousand overcast;"
"Bremerton National Airport automated weather observation, one four five six zulu. Indefinite ceiling two hundred, sky obscured."

(b) The word "Clear" is not used in AWOS due to limitations in the height ranges of the sensors. No clouds detected is announced as "NO CLOUDS BELOW XXX" or, in newer systems as "CLEAR BELOW XXX" (where XXX is the range limit of the sensor).

EXAMPLE-
"No clouds below one two thousand."
"Clear below one two thousand."

(c) A sensor for determining ceiling and sky cover is not included in some AWOS. In these systems, ceiling and sky cover are not announced. "SKY CONDITION MISSING" is announced only if the system is configured with a ceilometer and the ceiling and sky cover information is not available.

5. Remarks. If remarks are included in the observation, the word "REMARKS" is announced following the altimeter setting.

(a) Automated "Remarks."

(1) Density Altitude.

(2) Variable Visibility.

(3) Variable Wind Direction.

(b) Manual Input Remarks. Manual input remarks are prefaced with the phrase "OBSERVER WEATHER." As a general rule the manual remarks are limited to:

(1) Type and intensity of precipitation.

(2) Thunderstorms and direction; and

(3) Obstructions to vision when the visibility is 3 miles or less.

EXAMPLE-*"Remarks... density altitude, two thousand five hundred... visibility variable between one and two... wind direction variable between two four zero and three one zero... observed weather... thunderstorm moderate rain showers and fog... thunderstorm overhead."*

(c) If an automated parameter is "missing" and no manual input for that parameter is available, the parameter is announced as "MISSING." For example, a report with the dew point "missing" and no manual input available, would be announced as follows:

EXAMPLE-*"Ceiling one thousand overcast... visibility three... precipitation... temperature three zero, dew point missing... wind calm... altimeter three zero zero one."*

(d) "REMARKS" are announced in the following order of priority:

(1) Automated "REMARKS."

[a] Density Altitude.

[b] Variable Visibility.

[c] Variable Wind Direction.

(2) Manual Input "REMARKS."

[a] Sky Condition.

[b] Visibility.

[c] Weather and Obstructions to Vision.

[d] Temperature.

[e] Dew Point.

[f] Wind; and

[g] Altimeter Setting.

EXAMPLE-*"Remarks... density altitude, two thousand five hundred... visibility variable between one and two... wind direction variable between two four zero and three one zero... observer ceiling estimated two thousand broken... observer temperature two, dew point minus five."*

d. Automated Surface Observing System (ASOS)/Automated Weather Sensor System (AWSS). The ASOS/AWSS is the primary surface weather observing system of the U.S. (See Key to Decode an ASOS/AWSS (METAR) Observation, FIG 7-1-9and FIG 7-1-10.) The program to install and operate these systems throughout the U.S. is a joint effort of the NWS, the FAA and the Department of Defense. AWSS is a follow-on program that provides identical data as ASOS. ASOS/AWSS is designed to support aviation operations and weather forecast activities. The ASOS/AWSS will provide continuous minute-by-minute observations and perform the basic observing functions necessary to generate an aviation routine weather report (METAR) and other aviation weather information. The information may be transmitted over a discrete VHF radio frequency or the voice portion of a local NAVAID. ASOS/AWSS transmissions on a discrete VHF radio frequency are engineered to be receivable to a maximum of 25 NM from the ASOS/AWSS site and a maximum altitude of 10,000 feet AGL. At many locations, ASOS/AWSS signals may be received on the surface of the airport, but local conditions may limit the maximum reception distance and/or altitude. While the automated system and the human may differ in their methods of data collection and interpretation, both produce an observation quite similar in form and content. For the "objective" elements such as pressure, ambient temperature, dew point temperature, wind, and precipitation accumulation, both the automated system and the observer use a fixed location and time-averaging technique. The quantitative differences between the observer and the automated observation of these elements are negligible. For the "subjective" elements, however, observers use a fixed time, spatial averaging technique to describe the visual elements (sky condition, visibility and present weather), while the automated systems use a fixed location, time averaging technique. Although this is a fundamental change, the manual and automated techniques yield remarkably similar results within the limits of their respective capabilities.

1. System Description.

(a) The ASOS/AWSS at each airport location consists of four main components:

(1) Individual weather sensors.

(2) Data collection and processing units.

(3) Peripherals and displays.

(b) The ASOS/AWSS sensors perform the basic function of data acquisition. They continuously sample and measure the ambient environment, derive raw sensor data and make them available to the collection and processing units.

2. Every ASOS/AWSS will contain the following basic set of sensors:

(a) Cloud height indicator (one or possibly three).

(b) Visibility sensor (one or possibly three).

(c) Precipitation identification sensor.

(d) Freezing rain sensor (at select sites).

(e) Pressure sensors (two sensors at small airports; three sensors at large airports).

(f) Ambient temperature/Dew point temperature sensor.

(g) Anemometer (wind direction and speed sensor).

(h) Rainfall accumulation sensor.

3. The ASOS/AWSS data outlets include:

(a) Those necessary for on-site airport users.

(b) National communications networks.

(c) Computer-generated voice (available through FAA radio broadcast to pilots, and dial-in telephone line).

NOTE-*Wind direction broadcast over FAA radios is in reference to magnetic north.*

4. An ASOS/AWOS/AWSS report without human intervention will contain only that weather data capable of being reported automatically. The modifier for this METAR report is "AUTO." When an observer augments or backs-up an ASOS/AWOS/AWSS site, the "AUTO" modifier disappears.

FIG 7-1-9
Key to Decode an ASOS/AWSS (METAR) Observation (Front)

METAR KABC 121755Z AUTO 21016G24KT 180V240 1SM R11/P6000FT -RA BR BKN015 OVC025 06/04 A2990 RMK A02 PK WND 20032/25 WSHFT 1715 VIS 3/4V1 1/2 VIS 3/4 RWY11 RAB07 CIG 013V017 CIG 017 RWY11 PRESFR SLP125 P0003 6009 T00640036 10066 21012 58033 TSNO $

TYPE OF REPORT	METAR: hourly (scheduled report; SPECI: special (unscheduled) report.	METAR
STATION IDENTIFIER	Four alphabetic characters; ICAO location identifiers.	KABC
DATE/TIME	All dates and times in UTC using a 24-hour clock; two-digit date and four-digit time; always appended with Z to indicate UTC.	121755Z
REPORT MODIFIER	Fully automated report, no human intervention; removed when observer signed-on.	AUTO
WIND DIRECTION AND SPEED	Direction in tens of degrees from true north (first three digits); next two digits: speed in whole knots; as needed Gusts (character) followed by maximum observed speed; always appended with KT to indicate knots; 00000KT for calm; if direction varies by 60° or more a Variable wind direction group is reported.	21016G24KT 108V240
VISIBILITY	Prevailing visibility in statute miles and fractions (space between whole miles and fractions); always appended with SM to indicate statute miles.	1SM
RUNWAY VISUAL RANGE	10-minute RVR value in hundreds of feet; reported if prevailing visibility is ≤ one mile or RVR ≤6000 feet; always appended with FT to indicate feet; value prefixed with M or P to indicate value is lower or higher than the reportable RVR value.	R11/P6000FT
WEATHER PHENOMENA	RA: liquid precipitation that does not freeze; SN: frozen precipitation other than hail; UP: precipitation of unknown type; intensity prefixed to precipitation: light (-), moderate (no sign), heavy (+); FG: fog; FZFG: freezing fog (temperature below 0°C); BR: mist; HZ: haze; SQ: squall; maximum of three groups reported; augmented by observer: FC (funnel cloud/tornado/waterspout); TS(thunderstorm); GR (hail); GS (small hail; <1/4 inch); FZRA (intensity; freezing rain); VA (volcanic ash).	-RA BR
SKY CONDITION	Cloud amount and height: CLR (no clouds detected below 12000 feet); FEW (few); SCT (scattered); BKN (broken); OVC (overcast); followed by 3-digit height in hundreds of feet; or vertical visibility (VV) followed by height for indefinite ceiling.	BKN015 OVC025
TEMPERATURE/DEW POINT	Each is reported in whole degrees Celsius using two digits; values are separated by a solidus; sub-zero values are prefixed with an M (minus).	06/04
ALTIMETER	Altimeter always prefixed with an A indicating inches of mercury; reported using four digits: tens, units, tenths, and hundredths.	A2990

FIG 7-1-10
Key to Decode an ASOS/AWSS (METAR) Observation (Back)

REMARKS IDENTIFIER: RMK	RMK
TORNADIC ACTIVITY: Augmented; report should include TORNADO, FUNNEL CLOUD, or WATERSPOUT, time begin/end, location, movement; e.;g., TORNADO B25 N MOV E.	
TYPE OF AUTOMATED STATION: AO2; automated station with precipitation discriminator.	AO2
PEAK WIND: PK WND dddff(f)/(hh)mm; direction in tens of degrees, speed in whole knots, and time.	PK WND 20032/25
WIND SHIFT: WSHFT (hh)mm	WSHFT 1715
TOWER OR SURFACE VISIBILITY: TWR VIS vvvvv: visibility reported by tower personnel, e.g., TWR VIS 2; SFC VIS vvvvv: visibility reported by ASOS, e.g., SFC VIS 2.	
VARIABLE PREVAILING VISIBILITY: VIS $v_n v_n v_n v_n v_n V v_x v_x v_x v_x v_x$; reported if prevailing visibility is <3 miles and variable.	VIS 3/4V1 1/2
VISIBILITY AT SECOND LOCATION: VIS vvvvv [LOC]; reported if different than the reported prevailing visibility in body of report.	VIS 3/4 RWY11
LIGHTNING: [FREQ] LTG [LOC]; when detected the frequency and location is reported, e.g., FRQ LTG NE.	
BEGINNING AND ENDING OF PRECIPITATION AND THUNDERSTORMS: w'w'B(hh)mmE(hh)mm; TSB(hh)mmE(hh)mm	RAB07
VIRGA: Augmented; precipitation not reaching the ground, e.g., VIRGA.	
VARIABLE CEILING HEIGHT: CIG $h_n h_n h_n V h_x h_x h_x$; reported if ceiling in body of report is <3000 feet and variable.	CIG 013V017
CEILING HEIGHT AT SECOND LOCATION: CIG hhh [LOC]; Ceiling height reported if secondary ceilometer site is different than the ceiling height in the body of the report.	CIG 017 RWY11
PRESSURE RISING OR FALLING RAPIDLY: PRESRR or PRESFR; pressure rising or falling rapidly at time of observation.	PRESFR
SEA-LEVEL PRESSURE: SLPppp; tens, units, and tenths of SLP in hPa.	SLP125
HOURLY PRECIPITATION AMOUNT: Prrrr; in .01 inches since last METAR; a trace is P0000.	P0003
3- AND 6-HOUR PRECIPITATION AMOUNT: 6RRRR; precipitation amount in .01 inches for past 6 hours reported in 00, 06, 12, and 18 UTC observations and for past 3 hours in 03, 09, 15, and 21 UTC observations; a trace is 60000.	60009
24-HOUR PRECIPITATION AMOUNT: $7R_{24} R_{24} R_{24} R_{24}$; precipitation amount in .01 inches for past 24 hours reported in 12 UTC observation, e.g., 70015.	
HOURLY TEMPERATURE AND DEW POINT: $Ts_n T_a T_a T_a s_n T'_a T'_a T'_a$; tenth of degree Celsius; s_n: 1 if temperature below 0° C and 0 if temperature 0°C or higher.	T00640036
6-HOUR MAXIMUM TEMPERATURE: $1s_n T_x T_x T_x$; tenth of degree Celsius; 00, 06, 12, 18 UTC; s_n: 1 if temperature below 0°C and 0 if temperature 0°C or higher.	10066
6-HOUR MINIMUM TEMPERATURE: $2s_n T_n T_n T_n$; tenth of degree Celsius; 00, 06, 12, 18 UTC; s_n: 1 if temperature below 0°C and 0 if temperature 0°C or higher.	21012
24-HOUR MAXIMUM AND MINIMUM TEMPERATURE: $4s_n T_x T_x T_x s_n T_n T_n T_n$; tenth of degree Celsius; reported at midnight local standard time; 1 if temperature below 0°C and 0 if temperature 0°C or higher, e.g., 400461006.	
PRESSURE TENDENCY: 5appp; the character (a) and change in pressure (ppp; tenths of hPa) the past 3 hours.	58033
SENSOR STATUS INDICATORS: RVRNO: RVR missing; PWINO: precipitation identifier information not available; PNO: precipitation amount not available; FZRANO: freezing rain information not available; TSNO: thunderstorm information not available; VISNO [LOC]: visibility at secondary location not available, e.g., VISNO RWY06; CHINO [LOC]: (cloud-height-indicator) sky condition at secondary location not available, e.g., CHINO RWY06.	TSNO
MAINTENANCE CHECK INDICATOR: Maintenance needed on the system.	$
If an element or phenomena does not occur, is missing, or cannot be observed, the corresponding group and space are omitted (body and/or remarks) from that particular report, except for Sea-Level Pressure (SLPppp). SLPNO shall be reported in a METAR when the SLP is not available.	

U.S. DEPARTMENT OF TRANSPORTATION • FEDERAL AVIATION ADMINISTRATION • Aviation Weather Directorate, 400 7TH Street, SW, Rooms 8200-8326, Washington, D.C 20591

5. There are two types of automated stations, AO1 for automated weather reporting stations without a precipitation discriminator, and AO2 for automated stations with a precipitation discriminator. As appropriate, "AO1" and "AO2" must appear in remarks. (A precipitation discriminator can determine the difference between liquid and frozen/freezing precipitation).

NOTE- *To decode an ASOS/AWSS report, refer to FIG 7-1-9 and FIG 7-1-10.*

REFERENCE- *A complete explanation of METAR terminology is located in AIM, Paragraph 7-1-30, Key to Aerodrome Forecast (TAF) and Aviation Routine Weather Report (METAR).*

e. TBL 7-1-2 contains a comparison of weather observing programs and the elements reported.

f. Service Standards. During 1995, a government/industry team worked to comprehensively reassess the requirements for surface observations at the nation's airports. That work resulted in agreement on a set of service standards, and the FAA and NWS ASOS sites to which the standards would apply. The term "Service Standards" refers to the level of detail in weather observation. The service standards consist of four different levels of service (A, B, C, and D) as described below. Specific observational elements included in each service level are listed in TBL 7-1-3.

1. Service Level D defines the minimum acceptable level of service. It is a completely automated service in which the ASOS/AWSS observation will constitute the entire observation, i.e., no additional weather information is added by a human observer. This service is referred to as a stand alone D site.

2. Service Level C is a service in which the human observer, usually an air traffic controller, augments or adds information to the automated observation. Service Level C also includes backup of ASOS/AWSS elements in the event of an ASOS/AWSS malfunction or an unrepresentative ASOS/AWSS report. In backup, the human observer inserts the correct or missing value for the automated ASOS/AWSS elements. This service is provided by air traffic controllers under the Limited Aviation Weather Reporting Station (LAWRS) process, FSS and NWS observers, and, at selected sites, Non-Federal Observation Program observers.

Two categories of airports require detail beyond Service Level C in order to enhance air traffic control efficiency and increase system capacity. Services at these airports are typically provided by contract weather observers, NWS observers, and, at some locations, FSS observers.

3. Service Level B is a service in which weather observations consist of all elements provided under Service Level C, plus augmentation of additional data beyond the capability of the ASOS/AWSS. This category of airports includes smaller hubs or special airports in other ways that have worse than average bad weather operations for thunderstorms and/or freezing/frozen precipitation, and/or that are remote airports.

4. Service Level A, the highest and most demanding category, includes all the data reported in Service Standard B, plus additional requirements as specified. Service Level A covers major aviation hubs and/or high volume traffic airports with average or worse weather.

7-1-13. Weather Radar Services

a. The National Weather Service operates a network of radar sites for detecting coverage, intensity, and movement of precipitation. The network is supplemented by FAA and DOD radar sites in the western sections of the country. Local warning radar sites augment the network by operating on an as needed basis to support warning and forecast programs.

b. Scheduled radar observations are taken hourly and transmitted in alpha-numeric format on weather telecommunications circuits for flight planning purposes. Under certain conditions, special radar reports are issued in addition to the hourly transmittals. Data contained in the reports are also collected by the National Center for Environmental Prediction and used to prepare national radar summary charts for dissemination on facsimile circuits.

c. A clear radar display (no echoes) does not mean that there is no significant weather within the coverage of the radar site. Clouds and fog are not detected by the radar. However, when echoes are present, turbulence can be implied by the intensity of the precipitation, and icing is implied by the presence of the precipitation at temperatures at or below zero degrees Celsius. Used in conjunction with other weather products, radar provides invaluable information for weather avoidance and flight planning.

d. All En Route Flight Advisory Service facilities and FSSs have equipment to directly access the radar displays from the individual weather radar sites. Specialists at these locations are trained to interpret the display for pilot briefing and inflight advisory services. The Center Weather Service Units located in ARTCCs also have access to

TBL 7-1-2
Weather Observing Programs

Element Reported	Wind	Visibility	Temperature Dew Point	Altimeter	Density Altimeter	Clouds/ Ceiling	Precipitation Identification	Rhurdertorrm Lighting	Precipitation Occurance	Rainfall Accumulation	Rurray Surface Condition	Freering Rair Occurance	Remarks
Type													
AWSS	X	X	X	X	X	X	X			X		X	X
ASOS	X	X	X	X	X	X	X			X		X	X
AWOS-A				X									
AWOS-A/V		X		X									
AWOS-1	X		X	X	X								
AWOS-2	X	X	X	X	X								
AWOS-3	X	X	X	X	X	X							
AWOS-3P	X	X	X	X	X	X	X						
AWOS-3T	X	X	X	X	X	X		X					
AWOS-3P/T	X	X	X	X	X	X	X	X					
AWOS-4	X	X	X	X	X	X	X	X	X	X	X	X	
Manual	X	X	X	X		X	X						X
Reference- FAA Order 7900.5B, Surface Weather Observing, for element reporting.													

TBL 7-1-3

SERVICE LEVEL A	
Service Level A consists of all the elements of Service Levels B, C and D plus the elements listed to the right, if observed.	10 minute longline RVR at precedented sites or additional visibility increments of 1/8, 1/16 and 0 Sector visibility Variable sky condition Cloud layers above 12,000 feet and cloud types Widespread dust, sand and other obscurations Volcanic eruptions
SERVICE LEVEL B	
Service Level B consists of all the elements of Service Levels C and D plus the elements listed to the right, if observed.	Longline RVR at precedented sites (may be instantaneous readout) Freezing drizzle versus freezing rain Ice pellets Snow depth & snow increasing rapidly remarks Thunderstorm and lightning location remarks Observed significant weather not at the station remarks
SERVICE LEVEL C	
Service Level C consists of all the elements of Service Level D plus augmentation and backup by a human observer or an air traffic control specialist on location nearby. Backup consists of inserting the correct value if the system malfunctions or is unrepresentative. Augmentation consists of adding the elements listed to the right, if observed. During hours that the observing facility is closed, the site reverts to Service Level D.	Thunderstorms Tornadoes Hail Virga Volcanic ash Tower visibility Operationally significant remarks as deemed appropriate by the observer
SERVICE LEVEL D	
This level of service consists of an ASOS or AWSS continually measuring the atmosphere at a point near the runway. The ASOS or AWSS senses and measures the weather parameters listed to the right.	Wind Visibility Precipitation/Obstruction to vision Cloud height Sky cover Temperature Dew point Altimeter

Fig 7-1-11
NEXRAD Coverage

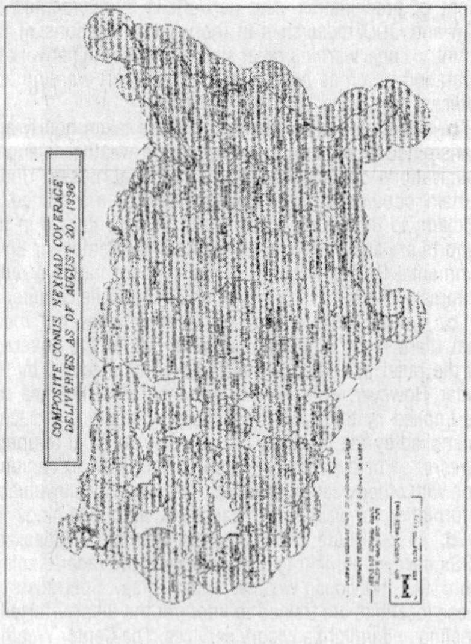

Fig 7-1-12
NEXRAD Coverage

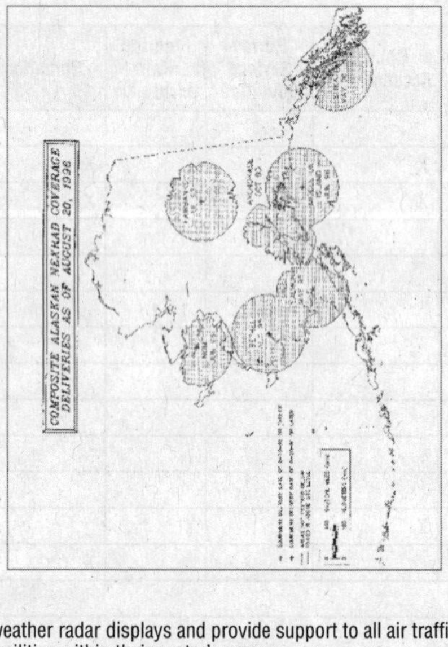

weather radar displays and provide support to all air traffic facilities within their center's area.

e. Additional information on weather radar products and services can be found in AC 00-45, Aviation Weather Services.

REFERENCE-
Pilot/Controller Glossary Term- Precipitation Radar Weather Descriptions.
AIM, Thunderstorms, Paragraph 7-1-28.
A/FD, Charts, NWS Upper Air Observing Stations and Weather Network for the location of specific radar sites.

7-1-14. ATC Inflight Weather Avoidance Assistance

a. ATC Radar Weather Display.

1. ATC radars are able to display areas of precipitation by sending out a beam of radio energy that is reflected back to the radar antenna when it strikes an object or moisture which may be in the form of rain drops, hail, or snow. The larger the object is, or the more dense its reflective surface, the stronger the return will be presented. Radar weather processors indicate the intensity of reflective returns in terms of decibels (dBZ). ATC systems cannot detect the presence or absence of clouds. The ATC systems can often determine the intensity of a precipitation area, but the specific character of that area (snow, rain, hail, VIRGA, etc.) cannot be determined. For this reason, ATC refers to all weather areas displayed on ATC radar scopes as "precipitation."

2. All ATC facilities using radar weather processors with the ability to determine precipitation intensity, will describe the intensity to pilots as:

(a) "LIGHT" (< 30 dBZ)
(b) "MODERATE" (30 to 40 dBZ)
(c) "HEAVY" (> 40 to 50 dBZ)
(d) "EXTREME" (> 50 dBZ)

3. ATC facilities that, due to equipment limitations, cannot display the intensity levels of precipitation, will describe the location of the precipitation area by geographic position, or position relative to the aircraft. Since the intensity level is not available, the controller will state "INTENSITY UNKNOWN."

4. ARTCC facilities normally use a Weather and Radar Processor (WARP) to display a mosaic of data obtained from multiple NEXRAD sites. There is a time delay between actual conditions and those displayed to the controller. For example, the precipitation data on the ARTCC controller's display could be up to 6 minutes old. When the WARP is not available, a second system, the narrowband Air Route Surveillance Radar (ARSR) can display two distinct levels of precipitation intensity that will be described to pilots as "MODERATE" (30 to 40 dBZ) and "HEAVY TO EXTREME" (> 40 dBZ). The WARP processor is only used in ARTCC facilities.

5. *ATC radar is not able to detect turbulence.* Generally, turbulence can be expected to occur as the rate of rainfall or intensity of precipitation increases. Turbulence associated with greater rates of rainfall/precipitation will normally be more severe than any associated with lesser rates of rainfall/precipitation. Turbulence should be expected to occur near convective activity, even in clear air. Thunderstorms are a form of convective activity that imply severe or greater turbulence. Operation within 20 miles of thunderstorms should be approached with great caution, as the severity of turbulence can be markedly greater than the precipitation intensity might indicate.

b. Weather Avoidance Assistance.

1. To the extent possible, controllers will issue pertinent information on weather or chaff areas and assist pilots in avoiding such areas when requested. Pilots should respond to a weather advisory by either acknowledging the advisory or by acknowledging the advisory and requesting an alternative course of action as follows:

(a) Request to deviate off course by stating the number of miles and the direction of the requested deviation. In this case, when the requested deviation is approved, navigation is at the pilot's prerogative, but must maintain the altitude assigned by ATC and to remain within the specified mileage of the original course.

(b) Request a new route to avoid the affected area.

(c) Request a change of altitude.

(d) Request radar vectors around the affected areas.

2. For obvious reasons of safety, an IFR pilot must not deviate from the course or altitude or flight level without a proper ATC clearance. When weather conditions encountered are so severe that an immediate deviation is determined to be necessary and time will not permit approval by ATC, the pilot's emergency authority may be exercised.

3. When the pilot requests clearance for a route deviation or for an ATC radar vector, the controller must evaluate the air traffic picture in the affected area, and coordinate with other controllers (if ATC jurisdictional boundaries may be crossed) before replying to the request.

4. It should be remembered that the controller's primary function is to provide safe separation between aircraft. Any additional service, such as weather avoidance assistance, can only be provided to the extent that it does not derogate the primary function. It's also worth noting that the separation workload is generally greater than normal when weather disrupts the usual flow of traffic. ATC radar limitations and frequency congestion may also be a factor in limiting the controller's capability to provide additional service.

5. It is very important, therefore, that the request for deviation or radar vector be forwarded to ATC as far in advance as possible. Delay in submitting it may delay or even preclude ATC approval or require that additional restrictions be placed on the clearance. Insofar as possible the following information should be furnished to ATC when requesting clearance to detour around weather activity:

(a) Proposed point where detour will commence.

(b) Proposed route and extent of detour (direction and distance).

(c) Point where original route will be resumed.

(d) Flight conditions (IFR or VFR).

(e) Any further deviation that may become necessary as the flight progresses.

(f) Advise if the aircraft is equipped with functioning airborne radar.

6. To a large degree, the assistance that might be rendered by ATC will depend upon the weather information available to controllers. Due to the extremely transitory nature of severe weather situations, the controller's weather information may be of only limited value if based on weather observed on radar only. Frequent updates by pilots giving specific information as to the area affected, altitudes, intensity and nature of the severe weather can be of considerable value. Such reports are relayed by radio or phone to other pilots and controllers and also receive widespread teletypewriter dissemination.

7. Obtaining IFR clearance or an ATC radar vector to circumnavigate severe weather can often be accommodated more readily in the en route areas away from terminals because there is usually less congestion and, therefore, offer greater freedom of action. In terminal areas, the problem is more acute because of traffic density, ATC coordination requirements, complex departure and arrival routes, adjacent airports, etc. As a consequence, controllers are less likely to be able to accommodate all requests for weather detours in a terminal area or be in a position to volunteer such routing to the pilot. Nevertheless, pilots should not hesitate to advise controllers of any observed severe weather and should specifically advise controllers if they desire circumnavigation of observed weather.

c. Procedures for Weather Deviations and Other Contingencies in Oceanic Controlled Airspace.

1. When the pilot initiates communications with ATC, rapid response may be obtained by stating "WEATHER DEVIATION REQUIRED" to indicate priority is desired on the frequency and for ATC response.

2. The pilot still retains the option of initiating the communications using the urgency call "PAN-PAN" 3 times to alert all listening parties of a special handling condition which will receive ATC priority for issuance of a clearance or assistance.

3. ATC will:

(a) Approve the deviation.

(b) Provide vertical separation and then approve the deviation; or

(c) If ATC is unable to establish vertical separation, ATC must advise the pilot that standard separation cannot be applied; provide essential traffic information for all affected aircraft, to the extent practicable; and if possible, suggest a course of action.ATC may suggest that the pilot climb or descend to a contingency altitude (1,000 feet above or below that assigned if operating above FL 290; 500 feet above or below that assigned if operating at or below FL 290).

PHRASEOLOGY-*STANDARD SEPARATION NOT AVAILABLE, DEVIATE AT PILOT'S DISCRETION; SUGGEST CLIMB (or descent) TO (appropriate altitude); TRAFFIC (position and altitude); REPORT DEVIATION COMPLETE.*

4. The pilot will follow the ATC advisory altitude when approximately 10 NM from track as well as execute the procedures detailed in paragraph 7-1-14c5.

5. If contact cannot be established or revised ATC clearance or advisory is not available and deviation from track is required, the pilot must take the following actions:

(a) If possible, deviate away from an organized track or route system.

(b) Broadcast aircraft position and intentions on the frequency in use, as well as on frequency 121.5 MHz at suitable intervals stating: flight identification (operator call sign), flight level, track code or ATS route designator, and extent of deviation expected.

(c) Watch for conflicting traffic both visually and by reference to TCAS (if equipped).

(d) Turn on aircraft exterior lights.

Tbl 7-1-4

Route Centerline/Track	Deviations >10 NM	Altitude Change
East 000–179°M	Left Right	Descend 300 Feet Climb 300 Feet
West 180–359°M	Left Right	Climb 300 Feet Descend 300 Feet
Pilot Memory Slogan: "East right up, West right down."		

(e) Deviations of less than 10 NM or operations within COMPOSITE (NOPAC and CEPAC) Airspace, should REMAIN at ASSIGNED altitude. Otherwise, when the aircraft is approximately 10 NM from track, initiate an altitude change based on the following criteria:

(f) When returning to track, be at assigned flight level when the aircraft is within approximately 10 NM of centerline.

(g) If contact was not established prior to deviating, continue to attempt to contact ATC to obtain a clearance. If contact was established, continue to keep ATC advised of intentions and obtain essential traffic information.

7-1-15. Runway Visual Range (RVR)

There are currently two configurations of RVR in the NAS commonly identified as Taskers and New Generation RVR. The Taskers are the existing configuration which uses transmissometer technology. The New Generation RVRs were deployed in November 1994 and use forward scatter technology. The New Generation RVRs are currently being deployed in the NAS to replace the existing Taskers.

a. RVR values are measured by transmissometers mounted on 14-foot towers along the runway. A full RVR system consists of:

1. Transmissometer projector and related items.

2. Transmissometer receiver (detector) and related items.

3. Analog

4. recorder.

5. Signal data converter and related items.

6. Remote digital or remote display programmer.

b. The transmissometer projector and receiver are mounted on towers 250 feet apart. A known intensity of light is emitted from the projector and is measured by the receiver. Any obscuring matter such as rain, snow, dust, fog, haze or smoke reduces the light intensity arriving at the receiver. The resultant intensity measurement is then converted to an RVR value by the signal data converter. These values are displayed by readout equipment in the associated air traffic facility and updated approximately once every minute for controller issuance to pilots.

c. The signal data converter receives information on the high intensity runway edge light setting in use (step 3, 4, or 5); transmission values from the transmissometer and the sensing of day or night conditions. From the three data sources, the system will compute appropriate RVR values.

d. An RVR transmissometer established on a 250 foot baseline provides digital readouts to a minimum of 600 feet, which are displayed in 200 foot increments to 3,000 feet and in 500 foot increments from 3,000 feet to a maximum value of 6,000 feet.

e. RVR values for Category IIIa operations extend down to 700 feet RVR; however, only 600 and 800 feet are reportable RVR increments. The 800 RVR reportable value covers a range of 701 feet to 900 feet and is therefore a valid minimum indication of Category IIIa operations.

f. Approach categories with the corresponding minimum RVR values. (See TBL 7-1-5.)

TBL 7-1-5
Approach Category/Minimum RVR Table

Category	Visibility (RVR)
Nonprecision	2,400 feet
Category I	1,800 feet*
Category II	1,000 feet
Category IIIa	700 feet
Category IIIb	150 feet
Category IIIc	0 feet

* 1,400 feet with special equipment and authorization

g. Ten minute maximum and minimum RVR values for the designated RVR runway are reported in the body of the aviation weather report when the prevailing visibility is less than one mile and/or the RVR is 6,000 feet or less. ATCTs report RVR when the prevailing visibility is 1 mile or less and/or the RVR is 6,000 feet or less.

h. Details on the requirements for the operational use of RVR are contained in FAA AC 97-1, "Runway Visual Range (RVR)." Pilots are responsible for compliance with minimums prescribed for their class of operations in the appropriate CFRs and/or operations specifications.

i. RVR values are also measured by forward scatter meters mounted on 14-foot frangible fiberglass poles. A full RVR system consists of:

1. Forward scatter meter with a transmitter, receiver and associated items.

2. A runway light intensity monitor (RLIM).

3. An ambient light sensor (ALS).

4. A data processor unit (DPU).

5. Controller display (CD).

j. The forward scatter meter is mounted on a 14-foot frangible pole. Infrared light is emitted from the transmitter and received by the receiver. Any obscuring matter such as rain, snow, dust, fog, haze or smoke increases the amount of scattered light reaching the receiver. The resulting measurement along with inputs from the runway light intensity monitor and the ambient light sensor are forwarded to the DPU which calculates the proper RVR value. The RVR values are displayed locally and remotely on controller displays.

k. The runway light intensity monitors both the runway edge and centerline light step settings (steps 1 through 5). Centerline light step settings are used for CAT IIIb operations. Edge Light step settings are used for CAT I, II, and IIIa operations.

l. New Generation RVRs can measure and display RVR values down to the lowest limits of Category IIIb operations (150 feet RVR). RVR values are displayed in 100 feet increments and are reported as follows:

1. 100-feet increments for products below 800 feet.

2. 200-feet increments for products between 800 feet and 3,000 feet.

3. 500-feet increments for products between 3,000 feet and 6,500 feet.

4. 25-meter increments for products below 150 meters.

5. 50-meter increments for products between 150 meters and 800 meters.

6. 100-meter increments for products between 800 meters and 1,200 meters.

7. 200-meter increments for products between 1,200 meters and 2,000 meters.

7-1-16. Reporting of Cloud Heights

a. Ceiling, by definition in the CFRs and as used in aviation weather reports and forecasts, is the height above ground (or water) level of the lowest layer of clouds or obscuring phenomenon that is reported as "broken," "overcast," or "obscuration," e.g., an aerodrome forecast (TAF) which reads "BKN030" refers to height above ground level. An area forecast which reads "BKN030" indicates that the height is above mean sea level.

REFERENCE-*AIM, Key to Aerodrome Forecast (TAF) and Aviation Routine Weather Report (METAR), Paragraph 7-1-30, defines "broken," "overcast," and "obscuration."*

b. Pilots usually report height values above MSL, since they determine heights by the altimeter. This is taken in account when disseminating and otherwise applying information received from pilots. ("Ceiling" heights are always above ground level.) In reports disseminated as PIREPs, height references are given the same as received from pilots, that is, above MSL.

c. In area forecasts or inflight advisories, ceilings are denoted by the contraction "CIG" when used with sky cover symbols as in "LWRG TO CIG OVC005," or the contraction "AGL" after, the forecast cloud height value. When the cloud base is given in height above MSL, it is so indicated by the contraction "MSL" or "ASL" following the height value. The heights of clouds tops, freezing level, icing, and turbulence are always given in heights above ASL or MSL.

7-1-17. Reporting Prevailing Visibility

a. Surface (horizontal) visibility is reported in METAR reports in terms of statute miles and increments thereof; e.g., 1/16, 1/8, 3/16, 1/4, 5/16, 3/8, 1/2, 5/8, 3/4, 7/8, 1, 1 1/8, etc. (Visibility reported by an unaugmented automated site is reported differently than in a manual report, i.e., ASOS/AWSS: 0, 1/16, 1/8, 1/4, 1/2, 3/4, 1, 1 1/4, 1 1/2, 1 3/4, 2, 2 1/2, 3, 4, 5, etc., AWOS: M1/4, 1/4, 1/2, 3/4, 1, 1 1/4, 1 1/2, 1 3/4, 2, 2 1/2, 3, 4, 5, etc.) Visibility is determined through the ability to see and identify preselected and prominent objects at a known distance from the usual point of observation. Visibilities which are determined to be less than 7 miles, identify the obscuring atmospheric condition; e.g., fog, haze, smoke, etc., or combinations thereof.

b. Prevailing visibility is the greatest visibility equalled or exceeded throughout at least one half of the horizon circle, not necessarily contiguous. Segments of the horizon circle which may have a significantly different visibility may be reported in the remarks section of the weather report; i.e., the southeastern quadrant of the horizon circle may be determined to be 2 miles in mist while the remaining quadrants are determined to be 3 miles in mist.

c. When the prevailing visibility at the usual point of observation, or at the tower level, is less than 4 miles, certificated tower personnel will take visibility observations in addition to those taken at the usual point of observation. The lower of these two values will be used as the prevailing visibility for aircraft operations.

7-1-18. Estimating Intensity of Rain and Ice Pellets

a. Rain

1. Light. From scattered drops that, regardless of duration, do not completely wet an exposed surface up to a condition where individual drops are easily seen.

2. Moderate. Individual drops are not clearly identifiable; spray is observable just above pavements and other hard surfaces.

3. Heavy. Rain seemingly falls in sheets; individual drops are not identifiable; heavy spray to height of several inches is observed over hard surfaces.

b. Ice Pellets

1. Light. Scattered pellets that do not completely cover an exposed surface regardless of duration. Visibility is not affected.

2. Moderate. Slow accumulation on ground. Visibility reduced by ice pellets to less than 7 statute miles.

3. Heavy. Rapid accumulation on ground. Visibility reduced by ice pellets to less than 3 statute miles.

7-1-19. Estimating Intensity of Snow or Drizzle (Based on Visibility)

a. Light. Visibility more than 1/2 statute mile.

b. Moderate. Visibility from more than 1/4 statute mile to 1/2 statute mile.

c. Heavy. Visibility 1/4 statute mile or less.

7-1-20. Pilot Weather Reports (PIREPs)

a. FAA air traffic facilities are required to solicit PIREPs when the following conditions are reported or forecast: ceilings at or below 5,000 feet; visibility at or below 5 miles (surface or aloft); thunderstorms and related phenomena; icing of light degree or greater; turbulence of moderate degree or greater; wind shear and reported or forecast volcanic ash clouds.

b. Pilots are urged to cooperate and promptly volunteer reports of these conditions and other atmospheric data such as: cloud bases, tops and layers; flight visibility; precipitation; visibility restrictions such as haze, smoke and dust; wind at altitude; and temperature aloft.

c. PIREPs should be given to the ground facility with which communications are established; i.e., EFAS, FSS, ARTCC, or terminal ATC. One of the primary duties of EFAS facilities, radio call "FLIGHT WATCH," is to serve as a collection point for the exchange of PIREPs with en route aircraft.

d. If pilots are not able to make PIREPs by radio, reporting upon landing of the inflight conditions encountered to the nearest FSS or Weather Forecast Office will be helpful. Some of the uses made of the reports are:

1. The ATCT uses the reports to expedite the flow of air traffic in the vicinity of the field and for hazardous weather avoidance procedures.

2. The FSS uses the reports to brief other pilots, to provide inflight advisories, and weather avoidance information to en route aircraft.

3. The ARTCC uses the reports to expedite the flow of en route traffic, to determine most favorable altitudes, and to issue hazardous weather information within the center's area.

4. The NWS uses the reports to verify or amend conditions contained in aviation forecast and advisories. In some cases, pilot reports of hazardous conditions are the triggering mechanism for the issuance of advisories. They also use the reports for pilot weather briefings.

5. The NWS, other government organizations, the military, and private industry groups use PIREPs for research activities in the study of meteorological phenomena.

6. All air traffic facilities and the NWS forward the reports received from pilots into the weather distribution system to assure the information is made available to all pilots and other interested parties.

e. The FAA, NWS, and other organizations that enter PIREPs into the weather reporting system use the format listed in TBL 7-1-6. Items 1 through 6 are included in all transmitted PIREPs along with one or more of items 7 through 13. Although the PIREP should be as complete and concise as possible, pilots should not be overly concerned with strict format or phraseology. The important thing is that the information is relayed so other pilots may benefit

TBL 7-1-6
PIREP Element Code Chart

	PIREP ELEMENT	PIREP CODE	CONTENTS
1.	3-letter station identifier	XXX	Nearest weather reporting location to the reported phenomenon
2.	Report type	UA or UUA	Routine or Urgent PIREP
3.	Location	/OV	In relation to a VOR
4.	Time	/TM	Coordinated Universal Time
5.	Altitude	/FL	Essential for turbulence and icing reports
6.	Type Aircraft	/TP	Essential for turbulence and icing reports
7.	Sky cover	/SK	Cloud height and coverage (sky clear, few, scattered, broken, or overcast)
8.	Weather	/WX	Flight visibility, precipitation, restrictions to visibility, etc.
9.	Temperature	/TA	Degrees Celsius
10.	Wind	/WV	Direction in degrees magnetic north and speed in knots
11.	Turbulence	/TB	See AIM paragraph 7-1-23
12.	Icing	/IC	See AIM paragraph 7-1-21
13.	Remarks	/RM	For reporting elements not included or to clarify previously reported items

AQ: Please provide Table 7-1-7 and Table Citation

from your observation. If a portion of the report needs clarification, the ground station will request the information. Completed PIREPs will be transmitted to weather circuits as in the following examples:

EXAMPLE-

1. *KCMH UA /OV APE 230010/TM 1516/FL085/TP BE20/SK BKN065/WX FV03SM HZ FU/TA 20/TB LGT*

NOTE-

1. *One zero miles southwest of Appleton VOR; time 1516 UTC; altitude eight thousand five hundred; aircraft type BE200; bases of the broken cloud layer is six thousand five hundred; flight visibility 3 miles with haze and smoke; air temperature 20 degrees Celsius; light turbulence.*

EXAMPLE-

2. *KCRW UV /OV KBKW 360015-KCRW/TM 1815/ FL120//TP BE99/SK IMC/WX RA/TA M08 /WV 290030/TB LGT-MDT/IC LGT RIME/RM MDT MXD ICG DURC KROA NWBND FL080-100 1750Z*

NOTE-

2. *From 15 miles north of Beckley VOR to Charleston VOR; time 1815 UTC; altitude 12,000 feet; type aircraft, BE-99; in clouds; rain; temperature minus 8 Celsius; wind 290 degrees magnetic at 30 knots; light to moderate turbulence; light rime icing during climb northwestbound from Roanoke, VA, between 8,000 and 10,000 feet at 1750 UTC.*

7-1-21. PIREPs Relating to Airframe Icing

a. The effects of ice on aircraft are cumulative-thrust is reduced, drag increases, lift lessens, and weight increases. The results are an increase in stall speed and a deterioration of aircraft performance. In extreme cases, 2 to 3 inches of ice can form on the leading edge of the airfoil in less than 5 minutes. It takes but $^1/_2$ inch of ice to reduce the lifting power of some aircraft by 50 percent and increases the frictional drag by an equal percentage.

b. A pilot can expect icing when flying in visible precipitation, such as rain or cloud droplets, and the temperature is between +02 and -10 degrees Celsius. When icing is detected, a pilot should do one of two things, particularly if the aircraft is not equipped with deicing equipment; get out of the area of precipitation; or go to an altitude where the temperature is above freezing. This "warmer" altitude may not always be a lower altitude. Proper preflight action includes obtaining information on the freezing level and the above freezing levels in precipitation areas. Report icing to ATC, and if operating IFR, request new routing or altitude if icing will be a hazard. Be sure to give the type of aircraft to ATC when reporting icing. The following describes how to report icing conditions.

1. Trace. Ice becomes perceptible. Rate of accumulation slightly greater than sublimation. Deicing/anti-icing equipment is not utilized unless encountered for an extended period of time (over 1 hour).

2. Light. The rate of accumulation may create a problem if flight is prolonged in this environment (over 1 hour). Occasional use of deicing/anti-icing equipment removes/prevents accumulation. It does not present a problem if the deicing/anti-icing equipment is used.

3. Moderate. The rate of accumulation is such that even short encounters become potentially hazardous and use of deicing/anti-icing equipment or flight diversion is necessary.

4. Severe. The rate of accumulation is such that deicing/anti-icing equipment fails to reduce or control the hazard. Immediate flight diversion is necessary.

EXAMPLE-*Pilot report: give aircraft identification, location, time (UTC), intensity of type, altitude/FL, aircraft type, indicated air speed (IAS), and outside air temperature (OAT).*

NOTE-

1. *Rime ice. Rough, milky, opaque ice formed by the instantaneous freezing of small supercooled water droplets.*

2. *Clear ice. A glossy, clear, or translucent ice formed by the relatively slow freezing of large supercooled water droplets.*

3. *The OAT should be requested by the FSS or ATC if not included in the PIREP.*

7-1-22. Definitions of Inflight Icing Terms

See TBL 7-1-8, Icing Types, and TBL 7-1-9, Icing Conditions.

7-1-23. PIREPs Relating to Turbulence

a. When encountering turbulence, pilots are urgently requested to report such conditions to ATC as soon as practicable. PIREPs relating to turbulence should state:

1. Aircraft location.

2. Time of occurrence in UTC.

3. Turbulence intensity.

4. Whether the turbulence occurred in or near clouds.

5. Aircraft altitude or flight level.

6. Type of aircraft.

7. Duration of turbulence.

EXAMPLE-

1. *Over Omaha, 1232Z, moderate turbulence in clouds at Flight Level three one zero, Boeing 707.*

2. *From five zero miles south of Albuquerque to three zero miles north of Phoenix, 1250Z, occasional moderate chop at Flight Level three three zero, DC8.*

b. Duration and classification of intensity should be made using TBL 7-1-10.

7-1-24. Wind Shear PIREPs

a. Because unexpected changes in wind speed and direction can be hazardous to aircraft operations at low altitudes on approach to and departing from airports, pilots are urged to promptly volunteer reports to controllers of wind shear conditions they encounter. An advance warning of this information will assist other pilots in avoiding or coping with a wind shear on approach or departure.

b. When describing conditions, use of the terms "negative" or "positive" wind shear should be avoided. PIREPs of "negative wind shear on final," intended to describe loss of airspeed and lift, have been interpreted to mean that no wind shear was encountered. The recommended method for wind shear reporting is to state the loss or gain of airspeed and the altitudes at which it was encountered.

EXAMPLE-

1. *Denver Tower, Cessna 1234 encountered wind shear, loss of 20 knots at 400.*

2. *Tulsa Tower, American 721 encountered wind shear on final, gained 25 knots between 600 and 400 feet followed by loss of 40 knots between 400 feet and surface.*

1. Pilots who are not able to report wind shear in these specific terms are encouraged to make reports in terms of the effect upon their aircraft.

EXAMPLE-*Miami Tower, Gulfstream 403 Charlie encountered an abrupt wind shear at 800 feet on final, max thrust required.*

2. Pilots using Inertial Navigation Systems (INSs) should report the wind and altitude both above and below the shear level.

7-1-25. Clear Air Turbulence (CAT) PIREPs

CAT has become a very serious operational factor to flight operations at all levels and especially to jet traffic

TBL 7-1-8
Icing Types

Clear Ice	See Glaze Ice.
Glaze Ice	Ice, sometimes clear and smooth, but usually containing some air pockets, which results in a lumpy translucent appearance. Glaze ice results from supercooled drops/droplets striking a surface but not freezing rapidly on contact. Glaze ice is denser, harder, and sometimes more transparent than rime ice. Factors, which favor glaze formation, are those that favor slow dissipation of the heat of fusion (i.e., slight supercooling and rapid accretion). With larger accretions, the ice shape typically includes "horns" protruding from unprotected leading edge surfaces. It is the ice shape, rather than the clarity or color of the ice, which is most likely to be accurately assessed from the cockpit. The terms "clear" and "glaze" have been used for essentially the same type of ice accretion, although some reserve "clear" for thinner accretions which lack horns and conform to the airfoil.
Intercycle Ice	Ice which accumulates on a protected surface between actuation cycles of a deicing system.
Known or Observed or Detected Ice Accretion	Actual ice observed visually to be on the aircraft by the flight crew or identified by on-board sensors.
Mixed Ice	Simultaneous appearance or a combination of rime and glaze ice characteristics. Since the clarity, color, and shape of the ice will be a mixture of rime and glaze characteristics, accurate identification of mixed ice from the cockpit may be difficult.
Residual Ice	Ice which remains on a protected surface immediately after the actuation of a deicing system.
Rime Ice	A rough, milky, opaque ice formed by the rapid freezing of supercooled drops/droplets after they strike the aircraft. The rapid freezing results in air being trapped, giving the ice its opaque appearance and making it porous and brittle. Rime ice typically accretes along the stagnation line of an airfoil and is more regular in shape and conformal to the airfoil than glaze ice. It is the ice shape, rather than the clarity or color of the ice, which is most likely to be accurately assessed from the cockpit.
Runback Ice	Ice which forms from the freezing or refreezing of water leaving protected surfaces and running back to unprotected surfaces.
Note- *Ice types are difficult for the pilot to discern and have uncertain effects on an airplane in flight. Ice type definitions will be included in the AIM for use in the "Remarks" section of the PIREP and for use in forecasting.*	

Tbl 7-1-9
Icing Conditions

Appendix C Icing Conditions	Appendix C (14 CFR, Part 25 and 29) is the certification icing condition standard for approving ice protection provisions on aircraft. The conditions are specified in terms of altitude, temperature, liquid water content (LWC), representative droplet size (mean effective drop diameter [MED]), and cloud horizontal extent.
Forecast Icing Conditions	Environmental conditions expected by a National Weather Service or an FAA-approved weather provider to be conducive to the formation of inflight icing on aircraft.
Freezing Drizzle (FZDZ)	Drizzle is precipitation at ground level or aloft in the form of liquid water drops which have diameters less than 0.5 mm and greater than 0.05 mm. Freezing drizzle is drizzle that exists at air temperatures less than 0°C (supercooled), remains in liquid form, and freezes upon contact with objects on the surface or airborne.
Freezing Precipitation	Freezing precipitation is freezing rain or freezing drizzle falling through or outside of visible cloud.
Freezing Rain (FZRA)	Rain is precipitation at ground level or aloft in the form of liquid water drops which have diameters greater than 0.5 mm. Freezing rain is rain that exists at air temperatures less than 0°C (supercooled), remains in liquid form, and freezes upon contact with objects on the ground or in the air.
Icing in Cloud	Icing occurring within visible cloud. Cloud droplets (diameter < 0.05 mm) will be present; freezing drizzle and/or freezing rain may or may not be present.
Icing in Precipitation	Icing occurring from an encounter with freezing precipitation, that is, supercooled drops with diameters exceeding 0.05 mm, within or outside of visible cloud.
Known Icing Conditions	Atmospheric conditions in which the formation of ice is observed or detected in flight. ***Note-*** *Because of the variability in space and time of atmospheric conditions, the existence of a report of observed icing does not assure the presence or intensity of icing conditions at a later time, nor can a report of no icing assure the absence of icing conditions at a later time.*
Potential Icing Conditions	Atmospheric icing conditions that are typically defined by airframe manufacturers relative to temperature and visible moisture that may result in aircraft ice accretion on the ground or in flight. The potential icing conditions are typically defined in the Airplane Flight Manual or in the Airplane Operation Manual.
Supercooled Drizzle Drops (SCDD)	Synonymous with freezing drizzle aloft.
Supercooled Drops or /Droplets	Water drops/droplets which remain unfrozen at temperatures below 0°C. Supercooled drops are found in clouds, freezing drizzle, and freezing rain in the atmosphere. These drops may impinge and freeze after contact on aircraft surfaces.
Supercooled Large Drops (SLD)	Liquid droplets with diameters greater than 0.05 mm at temperatures less than 0°C, i.e., freezing rain or freezing drizzle.

flying in excess of 15,000 feet. The best available information on this phenomenon must come from pilots via the PIREP reporting procedures. All pilots encountering CAT conditions are urgently requested to report time, location, and intensity (light, moderate, severe, or extreme) of the element to the FAA facility with which they are maintaining radio contact. If time and conditions permit, elements should be reported according to the standards for other PIREPs and position reports.

REFERENCE-*AIM, PIREPs Relating to Turbulence, Paragraph 7-1-23.*

7-1-26. Microbursts

a. Relatively recent meteorological studies have confirmed the existence of microburst phenomenon. Microbursts are small scale intense downdrafts which, on reaching the surface, spread outward in all directions from the downdraft center. This causes the presence of both vertical and horizontal wind shears that can be extremely hazardous to all types and categories of aircraft, especially at low altitudes. Due to their small size, short life span, and the fact that they can occur over areas without surface precipitation, microbursts are not easily detectable using conventional weather radar or wind shear alert systems.

b. Parent clouds producing microburst activity can be any of the low or middle layer convective cloud types. Note, however, that microbursts commonly occur within the heavy rain portion of thunderstorms, and in much weaker, benign appearing convective cells that have little or no precipitation reaching the ground.

c. The life cycle of a microburst as it descends in a convective rain shaft is seen in FIG 7-1-14. An important consideration for pilots is the fact that the microburst intensifies for about 5 minutes after it strikes the ground.

d. Characteristics of microbursts include:

1. Size. The microburst downdraft is typically less than 1 mile in diameter as it descends from the cloud base to about 1,000-3,000 feet above the ground. In the transition zone near the ground, the downdraft changes to a horizontal outflow that can extend to approximately 2 $^1/_2$ miles in diameter.

2. Intensity. The downdrafts can be as strong as 6,000 feet per minute. Horizontal winds near the surface can be as strong as 45 knots resulting in a 90 knot shear (headwind to tailwind change for a traversing aircraft) across the microburst. These strong horizontal winds occur within a few hundred feet of the ground.

3. Visual Signs. Microbursts can be found almost anywhere that there is convective activity. They may be embedded in heavy rain associated with a thunderstorm or in light rain in benign appearing virga. When there is

Fig 7-1-14
Evolution of a Microburst

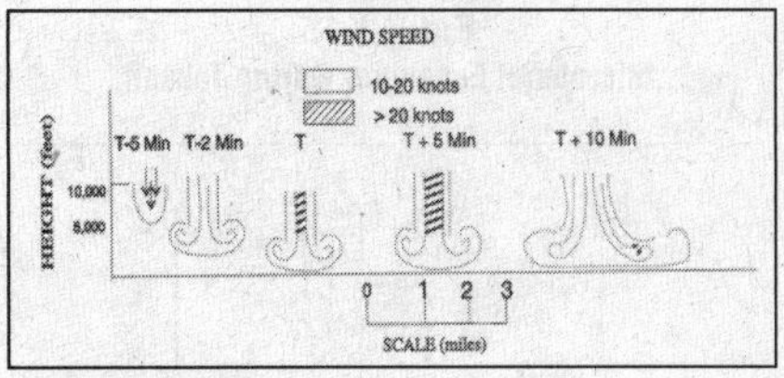

Tbl 7-1-10
Turbulence Reporting Criteria Table

Intensity	Aircraft Reaction	Reaction Inside Aircraft	Reporting Term-Definition
Light	Turbulence that momentarily causes slight, erratic changes in altitude and/or attitude (pitch, roll, yaw). Report as **Light Turbulence;** 1 or Turbulence that causes slight, rapid and somewhat rhythmic bumpiness without appreciable changes in altitude or attitude. Report as **Light Chop.**	Occupants may feel a slight strain against seat belts or shoulder straps. Unsecured objects may be displaced slightly. Food service may be conducted and little or no difficulty is encountered in walking.	Occasional-Less than $^1/_3$ of the time. Intermittent-$^1/_3$ to $^2/_3$. Continuous-More than $^2/_3$.
Moderate	Turbulence that is similar to Light Turbulence but of greater intensity. Changes in altitude and/or attitude occur but the aircraft remains in positive control at all times. It usually causes variations in indicated airspeed. Report as **Moderate Turbulence;**[1] or Turbulence that is similar to Light Chop but of greater intensity. It causes rapid bumps or jolts without appreciable changes in aircraft altitude or attitude. Report as **Moderate Chop.**[1]	Occupants feel definite strains against seat belts or shoulder straps. Unsecured objects are dislodged. Food service and walking are difficult.	**NOTE** 1. Pilots should report location(s), time (UTC), intensity, whether in or near clouds, altitude, type of aircraft and, when applicable, duration of turbulence. 2. Duration may be based on time between two locations or over a single location. All locations should be readily identifiable.
Severe	Turbulence that causes large, abrupt changes in altitude and/or attitude. It usually causes large variations in indicated airspeed. Aircraft may be momentarily out of control. Report **as Severe Turbulence.**[1]	Occupants are forced violently against seat belts or shoulder straps. Unsecured objects are tossed about. Food Service and walking are impossible.	**EXAMPLES:** a. Over Omaha. 1232Z, Moderate Turbulence, in cloud, Flight Level 310, B707.
Extreme	Turbulence in which the aircraft is violently tossed about and is practically impossible to control. It may cause structural damage. Report as **Extreme Turbulence.**[1]		b. From 50 miles south of Albuquerque to 30 miles north of Phoenix, 1210Z to 1250Z, occasional Moderate Chop, Flight Level 330, DC8.

[1] High level turbulence (normally above 15,000 feet ASL) not associated with cumuliform cloudiness, including thunderstorms, should be reported as CAT (clear air turbulence) preceded by the appropriate intensity, or light or moderate chop.

little or no precipitation at the surface accompanying the microburst, a ring of blowing dust may be the only visual clue of its existence.

4. Duration. An individual microburst will seldom last longer than 15 minutes from the time it strikes the ground until dissipation. The horizontal winds continue to increase during the first 5 minutes with the maximum intensity winds lasting approximately 2-4 minutes. Sometimes microbursts are concentrated into a line structure, and under these conditions, activity may continue for as long as an hour. Once microburst activity starts, multiple microbursts in the same general area are not uncommon and should be expected.

e. Microburst wind shear may create a severe hazard for aircraft within 1,000 feet of the ground, particularly during the approach to landing and landing and take-off phases. The impact of a microburst on aircraft which have the unfortunate experience of penetrating one is characterized in FIG 7-1-15. The aircraft may encounter a headwind (performance increasing) followed by a downdraft and tailwind (both performance decreasing), possibly resulting in terrain impact.

f. Detection of Microbursts, Wind Shear and Gust Fronts.

1. FAA's Integrated Wind Shear Detection Plan.

(a) The FAA currently employs an integrated plan for wind shear detection that will significantly improve both the safety and capacity of the majority of the airports currently served by the air carriers. This plan integrates several programs, such as the Integrated Terminal Weather System (ITWS), Terminal Doppler Weather Radar (TDWR), Weather System Processor (WSP), and Low Level Wind Shear Alert Systems (LLWAS) into a single strategic concept that significantly improves the aviation weather information in the terminal area. (See FIG 7-1-16.)

(b) The wind shear/microburst information and warnings are displayed on the ribbon display terminals (RBDT) located in the tower cabs. They are identical (and standardized) in the LLWAS, TDWR and WSP systems, and so designed that the controller does not need to interpret the data, but simply read the displayed information to the pilot.

FIG 7-1-15
Microburst Encounter During Takeoff

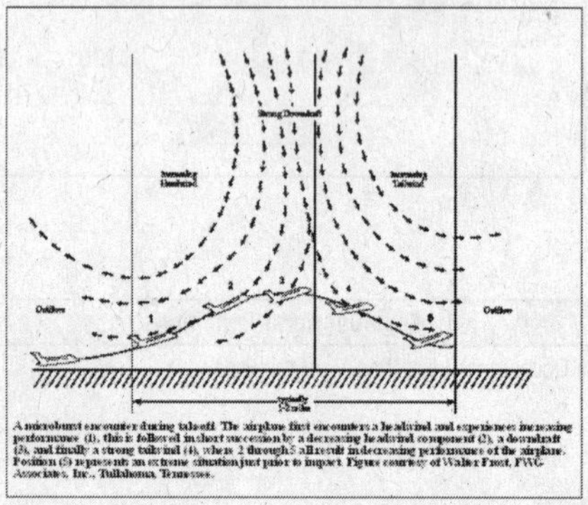

FIG 7-1-16
NAS Wind Shear Product Systems

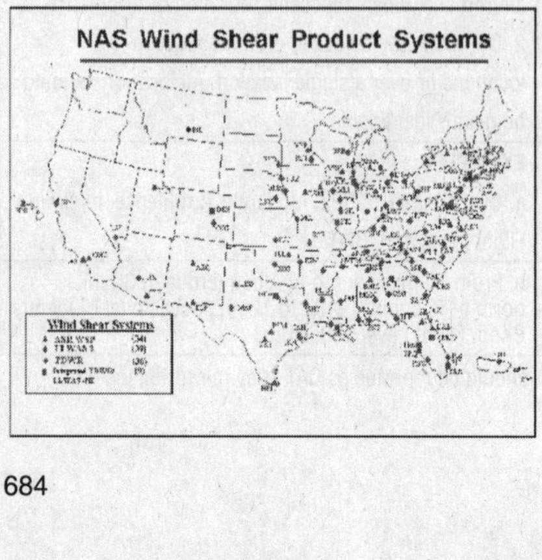

The RBDTs are constantly monitored by the controller to ensure the rapid and timely dissemination of any hazardous event(s) to the pilot.

(c) The early detection of a wind shear/micro-burst event, and the subsequent warning(s) issued to an aircraft on approach or departure, will alert the pilot/crew to the potential of, and to be prepared for, a situation that could become very dangerous! Without these warnings, the aircraft may NOT be able to climb out of, or safely transition, the event, resulting in a catastrophe. The air carriers, working with the FAA, have developed specialized training programs using their simulators to train and prepare their pilots on the demanding aircraft procedures required to escape these very dangerous wind shear and/or microburst encounters.

2. Low Level Wind Shear Alert System (LLWAS).

(a) The LLWAS provides wind data and software processes to detect the presence of hazardous wind shear and microbursts in the vicinity of an airport. Wind sensors, mounted on poles sometimes as high as 150 feet, are (ideally) located 2,000 - 3,500 feet, but not more than 5,000 feet, from the centerline of the runway. (See FIG 7-1-17.)

(b) LLWAS was fielded in 1988 at 110 airports across the nation. Many of these systems have been replaced by new TDWR and WSP technology. Eventually all LLWAS systems will be phased out; however, 39 airports will be upgraded to the LLWAS-NE (Network Expansion) system, which employs the very latest software and sensor technology. The new LLWAS-NE systems will not only provide the controller with wind shear warnings and alerts, including wind shear/microburst detection at the airport wind sensor location, but will also provide the location of the hazards relative to the airport runway(s). It will also have the flexibility and capability to grow with the airport as new runways are built. As many as 32 sensors, strategically located around the airport and in relationship to its runway configuration, can be accommodated by the LLWAS-NE network.

3. Terminal Doppler Weather Radar (TDWR).

(a) TDWRs are being deployed at 45 locations across the U.S. Optimum locations for TDWRs are 8 to 12 miles off of the airport proper, and designed to look at the airspace around and over the airport to detect microbursts, gust fronts, wind shifts and precipitation intensities. TDWR products advise the controller of wind shear and microburst events impacting all runways and the areas $^1/_2$ mile on either side of the extended centerline of the runways out to 3 miles on final approach and 2 miles out on departure.

(FIG 7-1-18 is a theoretical view of the warning boxes, including the runway, that the software uses in determining the location(s) of wind shear or microbursts). These warnings are displayed (as depicted in the examples in subparagraph 5) on the RBDT.

(b) It is very important to understand what TDWR does NOT DO:

FIG 7-1-17
LLWAS Siting Criteria

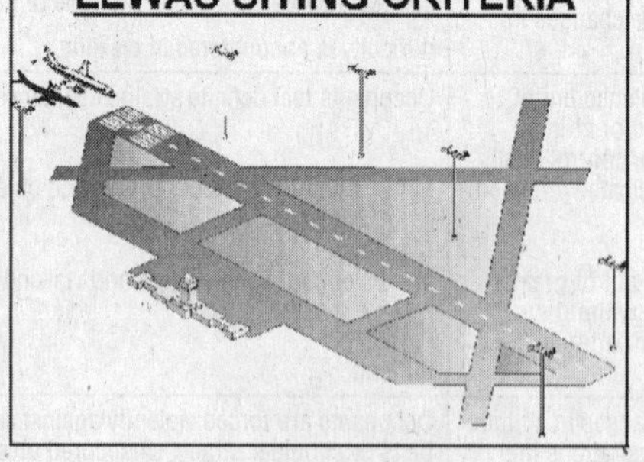

FIG 7-1-18
Warning Boxes

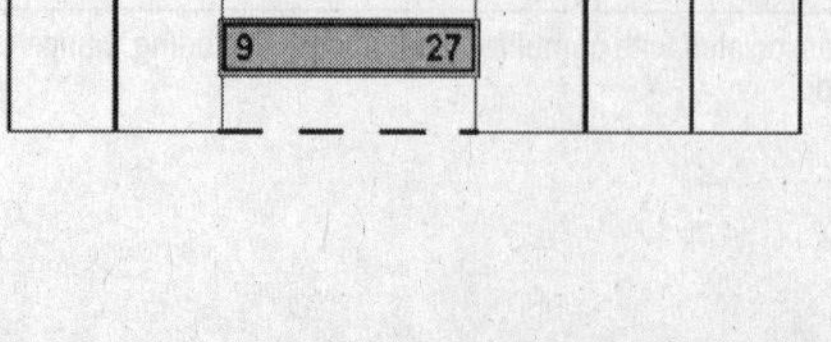

(1) It **DOES NOT** warn of wind shear outside of the alert boxes (on the arrival and departure ends of the runways);

(2) It **DOES NOT** detect wind shear that is NOT a microburst or a gust front;

(3) It **DOES NOT** detect gusty or cross wind conditions; and

(4) It **DOES NOT** detect turbulence.

However, research and development is continuing on these systems. Future improvements may include such areas as storm motion (movement), improved gust front detection, storm growth and decay, microburst prediction, and turbulence detection.

(c) TDWR also provides a geographical situation display (GSD) for supervisors and traffic management specialists for planning purposes. The GSD displays (in color) 6 levels of weather (precipitation), gust fronts and predicted storm movement(s). This data is used by the tower supervisor(s), traffic management specialists and controllers to plan for runway changes and arrival/departure route changes in order to both reduce aircraft delays and increase airport capacity.

4. Weather System Processor (WSP).

(a) The WSP provides the controller, supervisor, traffic management specialist, and ultimately the pilot, with the same products as the terminal doppler weather radar (TDWR) at a fraction of the cost of a TDWR. This is accomplished by utilizing new technologies to access the weather channel capabilities of the existing ASR-9 radar located on or near the airport, thus eliminating the requirements for a separate radar location, land acquisition, support facilities and the associated communication landlines and expenses.

(b) The WSP utilizes the same RBDT display as the TDWR and LLWAS, and, just like TDWR, also has a GSD for planning purposes by supervisors, traffic management specialists and controllers. The WSP GSD emulates the TDWR display, i.e., it also depicts 6 levels of precipitation, gust fronts and predicted storm movement, and like the TDWR GSD, is used to plan for runway changes and arrival/departure route changes in order to reduce aircraft delays and to increase airport capacity.

(c) This system is currently under development and is operating in a developmental test status at the Albuquerque, New Mexico, airport. When fielded, the WSP is expected to be installed at 34 airports across the nation, substantially increasing the safety of the American flying public.

5. Operational aspects of LLWAS, TDWR and WSP.

To demonstrate how this data is used by both the controller and the pilot, 3 ribbon display examples and their explanations are presented:

(a) MICROBURST ALERTS

EXAMPLE-
This is what the controller sees on his/her ribbon display in the tower cab.
27A MBA 35K- 2MF 250 20

NOTE-
(See FIG 7-1-19 to see how the TDWR/WSP determines the microburst location).

This is what the controller will say when issuing the alert.

PHRASEOLOGY-*RUNWAY 27 ARRIVAL, MICROBURST ALERT, 35 KT LOSS 2 MILE FINAL, THRESHOLD WIND 250 AT 20.*

In plain language, the controller is telling the pilot that on approach to runway 27, there is a microburst alert on the approach lane to the runway, and to anticipate or expect a 35 knot loss of airspeed at approximately 2 miles out on final approach (where it will first encounter the phenomena). With that information, the aircrew is forewarned, and should be prepared to apply wind shear/microburst escape procedures should they decide to continue the approach. Additionally, the surface winds at the airport for landing runway 27 are reported as 250 degrees at 20 knots.

NOTE- *Threshold wind is at pilot's request or as deemed appropriate by the controller.*

REFERENCE-*FAA Order JO 7110.65, Air Traffic Control, Low Level Wind Shear/Microburst Advisories, Paragraph 3-1-8b2(a).*

(b) WIND SHEAR ALERTS

EXAMPLE-

FIG 7-1-19
Microburst Alert

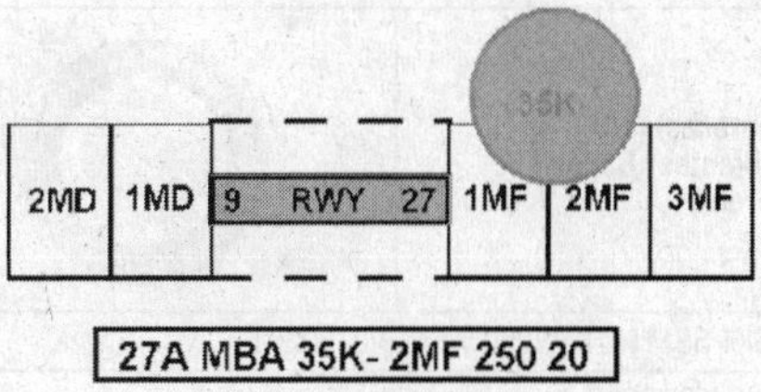

This is what the controller sees on his/her ribbon display in the tower cab.

27A WSA 20K- 3MF 200 15

NOTE-

(See FIG 7-1-20 to see how the TDWR/WSP determines the wind shear location).

This is what the controller will say when issuing the alert.

PHRASEOLOGY-*RUNWAY 27 ARRIVAL, WIND SHEAR ALERT, 20 KT LOSS 3 MILE FINAL, THRESHOLD WIND 200 AT 15.*

In plain language, the controller is advising the aircraft arriving on runway 27 that at about 3 miles out they can expect to encounter a wind shear condition that will decrease their airspeed by 20 knots and possibly encounter turbulence. Additionally, the airport surface winds for landing runway 27 are reported as 200 degrees at 15 knots.

NOTE-*Threshold wind is at pilot's request or as deemed appropriate by the controller.*

REFERENCE-*FAA Order JO 7110.65, Air Traffic Control, Low Level Wind Shear/Microburst Advisories, Paragraph 3-1-8b2(a).*

FIG 7-1-20
Weak Microburst Alert

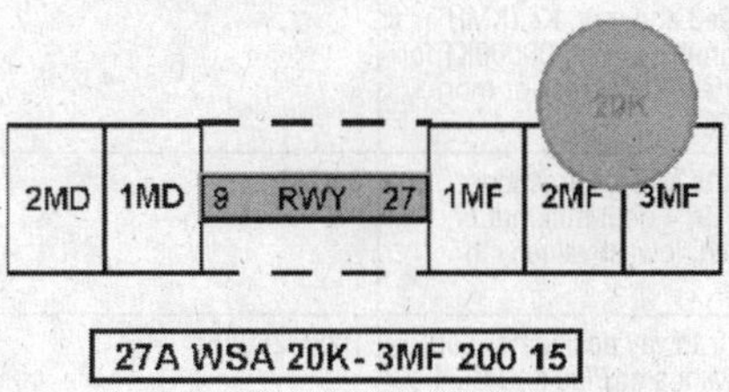

FIG 7-1-21
Gust Front Alert

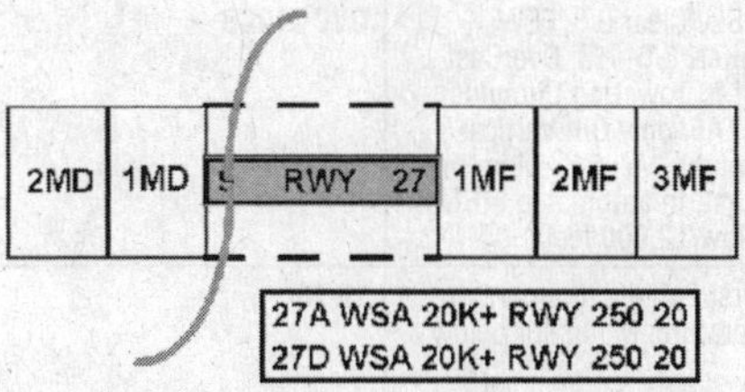

(c) MULTIPLE WIND SHEAR ALERTS

EXAMPLE-

This is what the controller sees on his/her ribbon display in the tower cab.

27A WSA 20K+ RWY 250 20
27D WSA 20K+ RWY 250 20

NOTE-

(See FIG 7-1-21 to see how the TDWR/WSP determines the gust front/wind shear location.)

This is what the controller will say when issuing the alert.

PHRASEOLOGY-*MULTIPLE WIND SHEAR ALERTS. RUNWAY 27 ARRIVAL, WIND SHEAR ALERT, 20 KT GAIN ON RUNWAY; RUNWAY 27 DEPARTURE, WIND SHEAR ALERT, 20 KT GAIN ON RUNWAY, WIND 250 AT 20.*

EXAMPLE-*In this example, the controller is advising arriving and departing aircraft that they could encounter a wind shear condition right on the runway due to a gust front (significant change of wind direction) with the possibility of a 20 knot gain in airspeed associated with the gust front. Additionally, the airport surface winds (for the runway in use) are reported as 250 degrees at 20 knots.*

REFERENCE-*FAA Order JO 7110.65, Air Traffic Control, Low Level Wind Shear/Microburst Advisories, Paragraph 3-1-8b2(d).*

6. The Terminal Weather Information for Pilots System (TWIP).

(a) With the increase in the quantity and quality of terminal weather information available through TDWR, the next step is to provide this information directly to pilots rather than relying on voice communications from ATC. The National Airspace System has long been in need of a means of delivering terminal weather information to the cockpit more efficiently in terms of both speed and accuracy to enhance pilot awareness of weather hazards and reduce air traffic controller workload. With the TWIP capability, terminal weather information, both alphanumerically and graphically, is now available directly to the cockpit on a test basis at 9 locations.

(b) TWIP products are generated using weather data from the TDWR or the Integrated Terminal Weather System (ITWS) testbed. TWIP products are generated and stored in the form of text and character graphic messages. Software has been developed to allow TDWR or ITWS to format the data and send the TWIP products to a database resident at Aeronautical Radio, Inc. (ARINC). These products can then be accessed by pilots using the ARINC Aircraft Communications Addressing and Reporting System (ACARS) data link services. Airline dispatchers can also access this database and send messages to specific aircraft whenever wind shear activity begins or ends at an airport.

(c) TWIP products include descriptions and character graphics of microburst alerts, wind shear alerts, significant precipitation, convective activity within 30 NM surrounding the terminal area, and expected weather that will impact airport operations. During inclement weather, i.e., whenever a predetermined level of precipitation or wind shear is detected within 15 miles of the terminal area, TWIP products are updated once each minute for text messages and once every five minutes for character graphic messages. During good weather (below the predetermined precipitation or wind shear parameters) each message is updated every 10 minutes. These products are intended to improve the situational awareness of the pilot/flight crew, and to aid in flight planning prior to arriving or departing the terminal area. It is important to understand that, in the context of TWIP, the predetermined levels for inclement versus good weather has nothing to do with the criteria for VFR/MVFR/IFR/LIFR; it only deals with precipitation, wind shears and microbursts.

7-1-27. PIREPs Relating to Volcanic Ash Activity

a. Volcanic eruptions which send ash into the upper atmosphere occur somewhere around the world several times each year. Flying into a volcanic ash cloud can be extremely dangerous. At least two B747s have lost all power in all four engines after such an encounter. Regardless of the type aircraft, some damage is almost certain to ensue after an encounter with a volcanic ash cloud. Additionally, studies have shown that volcanic eruptions are the only significant source of large quantities of sulphur dioxide (SO_2) gas at jet-cruising altitudes. Therefore, the detection and subsequent reporting of SO_2 is of significant importance. Although SO_2 is colorless, its presence in the atmosphere should be suspected when a sulphur-like or rotten egg odor is present throughout the cabin.

b. While some volcanoes in the U.S. are monitored, many in remote areas are not. These unmonitored volcanoes may erupt without prior warning to the aviation community. A pilot observing a volcanic eruption who has not had previous notification of it may be the only witness to the eruption. Pilots are strongly encouraged to transmit a PIREP regarding volcanic eruptions and any observed volcanic ash clouds or detection of sulphur dioxide (SO_2) gas associated with volcanic activity.

c. Pilots should submit PIREPs regarding volcanic activity using the Volcanic Activity Reporting (VAR) form as illustrated in Appendix 2. If a VAR form is not immediately available, relay enough information to identify the position and type of volcanic activity.

d. Pilots should verbally transmit the data required in items 1 through 8 of the VAR as soon as possible. The data required in items 9 through 16 of the VAR should be relayed after landing if possible.

7-1-28. Thunderstorms

a. Turbulence, hail, rain, snow, lightning, sustained updrafts and downdrafts, icing conditions-all are present in thunderstorms. While there is some evidence that maximum turbulence exists at the middle level of a thunderstorm, recent studies show little variation of turbulence intensity with altitude.

b. There is no useful correlation between the external visual appearance of thunderstorms and the severity or amount of turbulence or hail within them. The visible thunderstorm cloud is only a portion of a turbulent system whose updrafts and downdrafts often extend far beyond the visible storm cloud. Severe turbulence can be expected up to 20 miles from severe thunderstorms. This distance decreases to about 10 miles in less severe storms.

c. Weather radar, airborne or ground based, will normally reflect the areas of moderate to heavy precipitation (radar does not detect turbulence). The frequency and severity of turbulence generally increases with the radar reflectivity which is closely associated with the areas of highest liquid water content of the storm. NO FLIGHT PATH THROUGH AN AREA OF STRONG OR VERY STRONG RADAR ECHOES SEPARATED BY 20-30 MILES OR LESS MAY BE CONSIDERED FREE OF SEVERE TURBULENCE.

d. Turbulence beneath a thunderstorm should not be minimized. This is especially true when the relative humidity is low in any layer between the surface and 15,000 feet. Then the lower altitudes may be characterized by strong out flowing winds and severe turbulence.

e. The probability of lightning strikes occurring to aircraft is greatest when operating at altitudes where temperatures are between minus 5 degrees Celsius and plus 5 degrees Celsius. Lightning can strike aircraft flying in the clear in the vicinity of a thunderstorm.

f. METAR reports do not include a descriptor for severe thunderstorms. However, by understanding severe thunderstorm criteria, i.e., 50 knot winds or $^3/_4$ inch hail, the information is available in the report to know that one is occurring.

g. Current weather radar systems are able to objectively determine precipitation intensity. These precipitation intensity areas are described as "light," "moderate," "heavy," and "extreme."

REFERENCE-*Pilot/Controller Glossary, Precipitation Radar Weather Descriptions.*

EXAMPLE-

1. *Alert provided by an ATC facility to an aircraft:*
(aircraft identification) EXTREME precipitation between ten o'clock and two o'clock, one five miles. Precipitation area is two five miles in diameter.
2. *Alert provided by an FSS:*
(aircraft identification) EXTREME precipitation two zero miles west of Atlanta V-O-R, two five miles wide, moving east at two zero knots, tops flight level three niner zero.

7-1-29. Thunderstorm Flying

a. Above all, remember this: never regard any thunderstorm "lightly" even when radar observers report the echoes are of light intensity. Avoiding thunderstorms is the best policy. Following are some Do's and Don'ts of thunderstorm avoidance:

1. Don't land or takeoff in the face of an approaching thunderstorm. A sudden gust front of low level turbulence could cause loss of control.

2. Don't attempt to fly under a thunderstorm even if you can see through to the other side. Turbulence and wind shear under the storm could be disastrous.

3. Don't fly without airborne radar into a cloud mass containing scattered embedded thunderstorms. Scattered thunderstorms not embedded usually can be visually circumnavigated.

4. Don't trust the visual appearance to be a reliable indicator of the turbulence inside a thunderstorm.

5. Do avoid by at least 20 miles any thunderstorm identified as severe or giving an intense radar echo. This is especially true under the anvil of a large cumulonimbus.

6. Do clear the top of a known or suspected severe thunderstorm by at least 1,000 feet altitude for each 10 knots of wind speed at the cloud top. This should exceed the altitude capability of most aircraft.

7. Do circumnavigate the entire area if the area has $^{6}/_{10}$ thunderstorm coverage.

8. Do remember that vivid and frequent lightning indicates the probability of a strong thunderstorm.

9. Do regard as extremely hazardous any thunderstorm with tops 35,000 feet or higher whether the top is visually sighted or determined by radar.

b. If you cannot avoid penetrating a thunderstorm, following are some Do's before entering the storm:

1. Tighten your safety belt, put on your shoulder harness if you have one and secure all loose objects.

2. Plan and hold your course to take you through the storm in a minimum time.

3. To avoid the most critical icing, establish a penetration altitude below the freezing level or above the level of minus 15 degrees Celsius.

4. Verify that pitot heat is on and turn on carburetor heat or jet engine anti-ice. Icing can be rapid at any altitude and cause almost instantaneous power failure and/or loss of airspeed indication.

5. Establish power settings for turbulence penetration airspeed recommended in your aircraft manual.

6. Turn up cockpit lights to highest intensity to lessen temporary blindness from lightning.

7. If using automatic pilot, disengage altitude hold mode and speed hold mode. The automatic altitude and speed controls will increase maneuvers of the aircraft thus increasing structural stress.

8. If using airborne radar, tilt the antenna up and down occasionally. This will permit you to detect other thunderstorm activity at altitudes other than the one being flown.

c. Following are some Do's and Don'ts during the thunderstorm penetration:

1. Do keep your eyes on your instruments. Looking outside the cockpit can increase danger of temporary blindness from lightning.

2. Don't change power settings; maintain settings for the recommended turbulence penetration airspeed.

3. Don't attempt to maintain constant altitude; let the aircraft "ride the waves."

4. Don't turn back once you are in the thunderstorm. A straight course through the storm most likely will get you out of the hazards most quickly. In addition, turning maneuvers increase stress on the aircraft.

7-1-30. Key to Aerodrome Forecast (TAF) and Aviation Routine Weather Report (METAR)

7-1-31. International Civil Aviation Organization (ICAO) Weather Formats

The U.S. uses the ICAO world standard for aviation weather reporting and forecasting. The utilization of terminal forecasts affirms our commitment to a single global format for aviation weather. The World Meteorological Organization's (WMO) publication No. 782 "Aerodrome Reports and Forecasts" contains the base METAR and TAF code as adopted by the WMO member countries.

a. Although the METAR code is adopted worldwide, each country is allowed to make modifications or exceptions to the code for use in their particular country, e.g., the U.S. will continue to use statute miles for visibility, feet for RVR values, knots for wind speed, and inches of mercury for altimetry. However, temperature and dew point will be reported in degrees Celsius. The U.S. will continue reporting prevailing visibility rather than lowest sector visibility. Most of the current U.S. observing procedures and policies will continue after the METAR conversion date, with the information disseminated in the METAR code and format. The elements in the body of a METAR report are separated with a space. The only exceptions are RVR, temperature and dew point, which are separated with a solidus (/). When an element does not occur, or cannot be observed, the preceding space and that element are omitted from that particular report. A METAR report contains the following sequence of elements in the following order:

1. **Type of report.**
2. **ICAO Station Identifier.**
3. **Date and time of report.**
4. **Modifier (as required).**
5. **Wind.**
6. **Visibility.**
7. **Runway Visual Range (RVR).**
8. **Weather phenomena.**
9. **Sky conditions.**
10. **Temperature/dew point group.**
11. **Altimeter.**
12. **Remarks (RMK).**

Fig 7-1-22
Key to Aerodrome Forecast (TAF) and Aviation Routine Weather Report (METAR) (Front)

	Key to Aerodrome Forecast (TAF) and Aviation Routine Weather Report (METAR) (Front)	
TAF	KPIT 091730Z 0918/1024 15005KT 5SM HZ FEW020 WS010/31022KT	
	FM091930 30015G25KT 3SM SHRA OVC015	
	TEMPO 0920/0922 1/2SM +TSRA OVC008CB	
	FM100100 27008KT 5SM SHRA BKN020 OVC040	
	PROB30 1004/1007 1SM RA BR	
	FM101015 18005KT 6SM SHRA OVC020	
	BECMG 1013/1015 P6SM NSW SKC	
NOTE: Users are cautioned to confirm ***DATE*** and ***TIME*** of the TAF. For example FM**10**0000 is 0000Z on the **10th**. Do not confuse with ***1000Z!***		
METAR KPIT 091955Z COR 22015G25KT 3/4SM R28L/2600FT TSRA OVC010CB 18/16 A 2992 RMK SLP045 T01820159		

Forecast	Explanation	Report
TAF	Message type: TAF routine or TAF AMDamended forecast, METAR hourly, SPECIspecial or TESTM non commissioned ASOS report	**METAR**
KPIT	ICAO location indicator	**KPIT**
091730Z	Issuance time: ALL times in UTC "Z", 2 digit date, 4 digit time	**091955Z**
0918/1024	Valid period, either 24 hours or 30 hours. The first two digits of EACH four digit number indicate the date of the valid period, the final two digits indicate the time (valid from 18Z on the 9th to 24Z on the 10th).	
	In U.S. METAR: CORrected ob; or AUTOmated ob for automated report with no human intervention; omitted when observer logs on.	**COR**
15005KT	Wind: 3 digit true north direction, nearest 10 degrees (or VaRiaBle); next 2 3 digits for speed and unit, KT (KMH or MPS); as needed, Gust and maximum speed; 00000KT for calm; for METAR, if direction varies 60 degrees or more, Variability appended, e.g., 180V260	**22015G25KT**
5SM	Prevailing visibility; in U.S., Statute Miles & fractions; above 6 miles in TAF Plus6SM. (Or, 4 digit minimum visibility in meters and as required, lowest value with direction)	**3/4SM**
	Runway Visual Range: R; 2 digit runway designator Left, Center, or Right as needed; "/", Minus or Plus in U.S., 4 digit value, FeeT in U.S., (usually meters elsewhere); 4 digit value Variability 4 digit value (and tendency Down, Up or No change)	**R28L/2600FT**
HZ	Significant present, forecast and recent weather: see table (on back)	**TSRA**
FEW020	Cloud amount, height and type: Sky Clear 0/8, FEW >0/8 2/8, ScaTtered 3/8 4/8, BroKeN 5/8–7/8, OverCast 8/8; 3 digit height in hundreds of ft; Towering Cumulus or CumulonimBus in **METAR**; in **TAF**, only CB. Vertical Visibility for obscured sky and height "VV004". More than 1 layer may be reported or forecast. In automated **METAR** reports only, CleaR for "clear below 12,000 feet"	**OVC 010CB**
	Temperature: degrees Celsius; first 2 digits, temperature "/" last 2 digits, dew point temperature; Minus for below zero, e.g., M06	**18/16**
	Altimeter setting: indicator and 4 digits; in U.S., A inches and hundredths; (Q hectoPascals, e.g., Q1013)	**A2992**
WS010/31022KT	In U.S. **TAF**, non convective low level (≤~2,000 ft) Wind Shear; 3 digit height (hundreds of ft); "/"; 3 digit wind direction and 2 3 digit wind speed above the indicated height, and unit, KT	

b. The following paragraphs describe the elements in a METAR report.

1. **Type of report.** There are two types of report:

(a) Aviation Routine Weather Report (METAR); and

(b) Nonroutine (Special) Aviation Weather Report (SPECI).

The type of report (METAR or SPECI) will always appear as the lead element of the report.

2. **ICAO Station Identifier.** The METAR code uses ICAO 4-letter station identifiers. In the contiguous 48 States, the 3-letter domestic station identifier is prefixed

Fig 7-1-23

Key to Aerodrome Forecast (TAF) and Aviation Routine Weather Report (METAR) (Back)

	Key to Aerodrome Forecast (TAF) and Aviation Routine Weather Report (METAR) (Back)	
	In **METAR**, ReMarK indicator & remarks. For example: Sea Level Pressure in hectoPascals & tenths, as shown: 1004.5 hPa; Temp/dew point in tenths °C, as shown: temp. 18.2°C, dew point 15.9°C	**RMK SLP045** **T01820159**
FM091930	FroM: changes are expected at: 2 digit date, 2 digit hour, and 2 digit minute beginning time: indicates significant change. Each FM starts on a new line, indented 5 spaces	
TEMPO 0920/0922	TEMPOrary: changes expected for <1 hour and in total, < half of the period between the 2 digit date and 2 digit hour beginning, and 2 digit date and 2 digit hour ending time	
PROB30 1004/1007	PROBability and 2 digit percent (30 or 40): probable condition in the period between the 2digit date & 2digit hour beginning time, and the 2 digit date and 2digit hour ending time	
BECMG 1013/1015	BECoMinG: change expected in the period between the 2digit date and 2digit hour beginning time, and the 2digit date and 2digit hour ending time	

Table of Significant Present, Forecast and Recent Weather Grouped in categories and used in the order listed below; or as needed in TAF, No Significant Weather.			
Qualifiers			
Intensity or Proximity			
" " = Light	**No sign** = Moderate		**"+"** = Heavy
"VC" = Vicinity, but not at aerodrome. In the US METAR, 5 to 10 SM from the point of observation. In the US TAF, 5 to 10 SM from the center of the runway complex. Elsewhere, within 8000m.			
Descriptor			
BC – Patches	**BL** – Blowing	**DR** – Drifting	**FZ** – Freezing
MI – Shallow	**PR** – Partial	**SH** – Showers	**TS** – Thunderstorm
Weather Phenomena			
Precipitation			
DZ – Drizzle	**GR** – Hail	**GS** – Small Hail/Snow Pellets	
IC – Ice Crystals	**PL** – Ice Pellets	**RA** – Rain	**SG** – Snow Grains
SN – Snow	**UP** – Unknown Precipitation in automated observations		
Obscuration			
BR – Mist (≥5/8SM)	**DU** – Widespread Dust	**FG** – Fog (<5/8SM)	**FU** – Smoke
HZ – Haze	**PY** – Spray	**SA** – Sand	**VA** – Volcanic Ash
Other			
DS – Dust Storm	**FC** – Funnel Cloud	**+FC** – Tornado or Waterspout	
PO – Well developed dust or sand whirls		**SQ** – Squall	**SS** – Sandstorm
Explanations in parentheses "()" indicate different worldwide practices.			
Ceiling is not specified; defined as the lowest broken or overcast layer, or the vertical visibility.			
NWS TAFs exclude BECMG groups and temperature forecasts, NWS TAFS do not use PROB in the first 9 hours of a TAF; NWS METARs exclude trend forecasts. US Military TAFs include Turbulence and Icing groups.			

with a "K;" i.e., the domestic identifier for Seattle is SEA while the ICAO identifier is KSEA. Elsewhere, the first two letters of the ICAO identifier indicate what region of the world and country (or state) the station is in. For Alaska, all station identifiers start with "PA;" for Hawaii, all station identifiers start with "PH." Canadian station identifiers start with "CU," "CW," "CY," and "CZ." Mexican station identifiers start with "MM." The identifier for the western Caribbean is "M" followed by the individual country's letter; i.e., Cuba is "MU" Dominican Republic "MD;" the Bahamas "MY." The identifier for the eastern Caribbean is "T" followed by the individual country's letter; i.e., Puerto Rico is "TJ." For a complete worldwide listing see ICAO Document 7910, Location Indicators.

3. Date and Time of Report. The date and time the observation is taken are transmitted as a six-digit date/time group appended with Z to denote Coordinated Universal Time (UTC). The first two digits are the date followed with two digits for hour and two digits for minutes.

EXAMPLE-*172345Z (the 17th day of the month at 2345Z)*

4. Modifier (As Required). "AUTO" identifies a METAR/SPECI report as an automated weather report with no human intervention. If "AUTO" is shown in the body of the report, the type of sensor equipment used at the station will be encoded in the remarks section of the report. The absence of "AUTO" indicates that a report was made manually by an observer or that an automated report had human augmentation/backup. The modifier "COR" indicates a corrected report that is sent out to replace an earlier report with an error.

NOTE-*There are two types of automated stations, AO1 for automated weather reporting stations without a precipitation discriminator, and AO2 for automated stations with a precipitation discriminator. (A precipitation discriminator can determine the difference between liquid and frozen/freezing precipitation). This information appears in the remarks section of an automated report.*

5. Wind. The wind is reported as a five digit group (six digits if speed is over 99 knots). The first three digits are the direction the wind is blowing from, in tens of degrees referenced to true north, or "VRB" if the direction is variable. The next two digits is the wind speed in knots, or if over 99 knots, the next three digits. If the wind is gusty, it is reported as a "G" after the speed followed by the highest gust reported. The abbreviation "KT" is appended to denote the use of knots for wind speed.

EXAMPLE-

13008KT - wind from 130 degrees at 8 knots

08032G45KT - wind from 080 degrees at 32 knots with gusts to 45 knots

VRB04KT - wind variable in direction at 4 knots

00000KT - wind calm

210103G130KT - wind from 210 degrees at 103 knots with gusts to 130 knots

If the wind direction is variable by 60 degrees or more and the speed is greater than 6 knots, a variable group consisting of the extremes of the wind direction separated by a "v" will follow the prevailing wind group.

32012G22KT 280V350

(a) Peak Wind. Whenever the peak wind exceeds 25 knots "PK WND" will be included in Remarks, e.g., PK WND 28045/1955 "Peak wind two eight zero at four five occurred at one niner five five." If the hour can be inferred from the report time, only the minutes will be appended, e.g., PK WND 34050/38 "Peak wind three four zero at five zero occurred at three eight past the hour."

(b) Wind shift. Whenever a wind shift occurs, "WSHFT" will be included in remarks followed by the time the wind shift began, e.g., WSHFT 30 FROPA "Wind shift at three zero due to frontal passage."

6. Visibility. Prevailing visibility is reported in statute miles with "SM" appended to it.

EXAMPLE-
7SM - seven statute miles
15SM - fifteen statute miles
$^1/_2$SM - one-half statute mile

(a) Tower/surface visibility. If either visibility (tower or surface) is below four statute miles, the lesser of the two will be reported in the body of the report; the greater will be reported in remarks.

(b) Automated visibility. ASOS/AWSS visibility stations will show visibility 10 or greater than 10 miles as "10SM." AWOS visibility stations will show visibility less than $^1/_4$ statute mile as "M$^1/_4$SM" and visibility 10 or greater than 10 miles as "10SM."

(c) Variable visibility. Variable visibility is shown in remarks (when rapid increase or decrease by $^1/_2$ statute mile or more and the average prevailing visibility is less than three miles) e.g., VIS 1V2 "visibility variable between one and two."

(d) Sector visibility. Sector visibility is shown in remarks when it differs from the prevailing visibility, and either the prevailing or sector visibility is less than three miles.

EXAMPLE-VIS N2 - visibility north two

7. Runway Visual Range (When Reported). "R" identifies the group followed by the runway heading (and parallel runway designator, if needed) "/" and the visual range in feet (meters in other countries) followed with "FT" (feet is not spoken).

(a) Variability Values. When RVR varies (by more than on reportable value), the lowest and highest values are shown with "V" between them.

(b) Maximum/Minimum Range. "P" indicates an observed RVR is above the maximum value for this system (spoken as "more than"). "M" indicates an observed RVR is below the minimum value which can be determined by the system (spoken as "less than").

EXAMPLE-
R32L/1200FT - runway three two left R-V-R one thousand two hundred.
R27R/M1000V4000FT - runway two seven right R-V-R variable from less than one thousand to four thousand.

8. Weather Phenomena. The weather as reported in the METAR code represents a significant change in the way weather is currently reported. In METAR, weather is reported in the format:

Intensity/Proximity/Descriptor/Precipitation/Obstruction to visibility/Other

NOTE-*The "/" above and in the following descriptions (except as the separator between the temperature and dew point) are for separation purposes in this publication and do not appear in the actual METARs.*

(a) Intensity applies only to the first type of precipitation reported. A "-" denotes light, no symbol denotes moderate, and a "+" denotes heavy.

(b) Proximity applies to and reported only for weather occurring in the vicinity of the airport (between 5 and 10 miles of the point(s) of observation). It is denoted by the letters "VC." (Intensity and "VC" will not appear together in the weather group).

(c) Descriptor. These eight descriptors apply to the precipitation or obstructions to visibility:

TS thunderstorm
DR low drifting
SH showers
MI shallow
FZ freezing
BC patches
BL blowing
PR partial

NOTE-*Although "TS" and "SH" are used with precipitation and may be preceded with an intensity symbol, the intensity still applies to the precipitation, **not** the descriptor.*

(d) Precipitation. There are nine types of precipitation in the METAR code:

RA rain
DZ drizzle
SN snow
GR hail ($^1/_4$" or greater)
GS small hail/snow pellets
PL ice pellets
SG snow grains
IC ice crystals (diamond dust)
UP unknown precipitation (automated stations only)

(e) Obstructions to visibility. There are eight types of obscuration phenomena in the METAR code (obscurations are any phenomena in the atmosphere, other than precipitation, that reduce horizontal visibility):

FG fog (vsby less than $^5/_8$ mile)
HZ haze
FU smoke
PY spray
BR mist (vsby $^5/_8$ - 6 miles)
SA sand
DU dust
VA volcanic ash

NOTE-*Fog (FG) is observed or forecast only when the visibility is less than five-eighths of mile, otherwise mist (BR) is observed or forecast.*

(f) Other. There are five categories of other weather phenomena which are reported when they occur:

SQ squall
SS sandstorm
DS duststorm
PO dust/sand whirls
FC funnel cloud
+FC tornado/waterspout

Examples:
TSRA thunderstorm with moderate rain
+SN heavy snow
-RA FG light rain and fog
BRHZ mist and haze (visibility $^5/_8$ mile or greater)
FZDZ freezing drizzle
VCSH rain shower in the vicinity
+SHRASNPL heavy rain showers, snow, ice pellets (intensity indicator refers to the predominant rain)

9. Sky Condition. The sky condition as reported in METAR represents a significant change from the way sky condition is currently reported. In METAR, sky condition is reported in the format:

Amount/Height/(Type) or Indefinite Ceiling/Height

(a) Amount. The amount of sky cover is reported in eighths of sky cover, using the contractions:

SKC clear (no clouds)
FEW >0 to $^2/_8$
SCT scattered ($^3/_8$s to $^4/_8$s of clouds)
BKN broken ($^5/_8$s to $^7/_8$s of clouds)
OVC overcast ($^8/_8$s clouds)
CB Cumulonimbus when present
TCU Towering cumulus when present

NOTE-
1. "SKC" will be reported at manual stations. "CLR" will be used at automated stations when no clouds below 12,000 feet are reported.
2. A ceiling layer is not designated in the METAR code. For aviation purposes, the ceiling is the lowest broken or overcast layer, or vertical visibility into an obscuration. Also there is no provision for reporting thin layers in the METAR code. When clouds are thin, that layer must be reported as if it were opaque.

(b) Height. Cloud bases are reported with three digits in hundreds of feet. (Clouds above 12,000 feet cannot be reported by an automated station).

(c) (Type). If Towering Cumulus Clouds (TCU) or Cumulonimbus Clouds (CB) are present, they are reported after the height which represents their base.

EXAMPLE-
(Reported as) SCT025TCU BKN080 BKN250 (spoken as) "TWO THOUSAND FIVE HUNDRED SCATTERED TOWERING CUMULUS, CEILING EIGHT THOUSAND BROKEN, TWO FIVE THOUSAND BROKEN."
(Reported as) SCT008 OVC012CB (spoken as) "EIGHT HUNDRED SCATTERED CEILING ONE THOUSAND TWO HUNDRED OVERCAST CUMULONIMBUS CLOUDS."

(d) Vertical Visibility (indefinite ceiling height). The height into an indefinite ceiling is preceded by "VV" and followed by three digits indicating the vertical visibility in hundreds of feet. This layer indicates total obscuration.

EXAMPLE-*$^1/_8$SM FG VV006 - visibility one eighth, fog, indefinite ceiling six hundred.*

(e) Obscurations are reported when the sky is partially obscured by a ground-based phenomena by indicating the amount of obscuration as FEW, SCT, BKN followed by three zeros (000). In remarks, the obscuring phenomenon precedes the amount of obscuration and three zeros.

EXAMPLE-
BKN000 (in body) "sky partially obscured"
FU BKN000 (in remarks) "smoke obscuring five-to seven-eighths of the sky"

(f) When sky conditions include a layer aloft, other than clouds, such as smoke or haze the type of phenomena, sky cover and height are shown in remarks.

EXAMPLE-*BKN020 (in body) "ceiling two thousand broken" RMK FU BKN020 "broken layer of smoke aloft, based at two thousand"*

(g) Variable ceiling. When a ceiling is below three thousand and is variable, the remark "CIG" will be shown followed with the lowest and highest ceiling heights separated by a "V."

EXAMPLE-*CIG 005V010 "ceiling variable between five hundred and one thousand"*

(h) Second site sensor. When an automated station uses meteorological discontinuity sensors, remarks will be shown to identify site specific sky conditions which differ and are lower than conditions reported in the body.

EXAMPLE-*CIG 020 RY11 "ceiling two thousand at runway one one"*

(i) Variable cloud layer. When a layer is varying in sky cover, remarks will show the variability range. If there is more than one cloud layer, the variable layer will be identified by including the layer height.

EXAMPLE-
SCT V BKN "scattered layer variable to broken"
BKN025 V OVC "broken layer at two thousand five hundred variable to overcast"

(j) Significant clouds. When significant clouds are observed, they are shown in remarks, along with the specified information as shown below:

(1) Cumulonimbus (CB), or Cumulonimbus Mammatus (CBMAM), distance (if known), direction from the station, and direction of movement, if known. If the clouds are beyond 10 miles from the airport, DSNT will indicate distance.

EXAMPLE-
CB W MOV E "cumulonimbus west moving east"
CBMAM DSNT S "cumulonimbus mammatus distant south"

(2) Towering Cumulus (TCU), location, (if known), or direction from the station.

EXAMPLE-
TCU OHD "towering cumulus overhead"
TCU W "towering cumulus west"

(3) Altocumulus Castellanus (ACC), Stratocumulus Standing Lenticular (SCSL), Altocumulus Standing Lenticular (ACSL), Cirrocumulus Standing Lenticular (CCSL) or rotor clouds, describing the clouds (if needed) and the direction from the station.

EXAMPLE-
ACC W "altocumulus castellanus west"
ACSL SW-S "standing lenticular altocumulus southwest through south"
APRNT ROTOR CLD S "apparent rotor cloud south"
CCSL OVR MT E "standing lenticular cirrocumulus over the mountains east"

10. Temperature/Dew Point. Temperature and dew point are reported in two, two-digit groups in degrees Celsius, separated by a solidus ("/"). Temperatures below zero are prefixed with an "M." If the temperature is available but the dew point is missing, the temperature is shown followed by a solidus. If the temperature is missing, the group is omitted from the report.

EXAMPLE-
15/08 "temperature one five, dew point 8"
00/M02 "temperature zero, dew point minus 2"
M05/ "temperature minus five, dew point missing"

11. Altimeter. Altimeter settings are reported in a four-digit format in inches of mercury prefixed with an "A" to denote the units of pressure.

EXAMPLE-*A2995 - "Altimeter two niner niner five"*

12. Remarks. Remarks will be included in all observations, when appropriate. The contraction "RMK" denotes the start of the remarks section of a METAR report.

Except for precipitation, phenomena located within 5 statute miles of the point of observation will be reported as at the station. Phenomena between 5 and 10 statute miles will be reported in the vicinity, "VC." Precipitation not occurring at the point of observation but within 10 statute miles is also reported as in the vicinity, "VC." Phenomena beyond 10 statute miles will be shown as distant, "DSNT." Distances are in statute miles except for automated lightning remarks which are in nautical miles. Movement of clouds or weather will be indicated by the direction toward which the phenomena is moving.

(a) There are two categories of remarks:
(1) Automated, manual, and plain language.
(2) Additive and automated maintenance data.

(b) Automated, Manual, and Plain Language. This group of remarks may be generated from either manual or automated weather reporting stations and generally elaborate on parameters reported in the body of the report. (Plain language remarks are only provided by manual stations).

(1) Volcanic eruptions.
(2) Tornado, Funnel Cloud, Waterspout.
(3) Station Type (AO1 or AO2).
(4) PK WND.
(5) WSHFT (FROPA).
(6) TWR VIS or SFC VIS.
(7) VRB VIS.
(8) Sector VIS.
(9) VIS @ 2nd Site.
(10) (freq) LTG (type) (loc).
(11) Beginning/Ending of Precipitation/TSTMS.
(12) TSTM Location MVMT.
(13) Hailstone Size (GR).
(14) Virga.
(15) VRB CIG (height).
(16) Obscuration.
(17) VRB Sky Condition.
(18) Significant Cloud Types.
(19) Ceiling Height 2nd Location.
(20) PRESFR PRESRR.
(21) Sea-Level Pressure.
(22) ACFT Mishap (not transmitted).
(23) NOSPECI.
(24) SNINCR.
(25) Other SIG Info.

(c) Additive and Automated Maintenance Data.
(1) Hourly Precipitation.
(2) 3- and 6-Hour Precipitation Amount.
(3) 24-Hour Precipitation.
(4) Snow Depth on Ground.
(5) Water Equivalent of Snow.
(6) Cloud Type.
(7) Duration of Sunshine.
(8) Hourly Temperature/Dew Point (Tenths).
(9) 6-Hour Maximum Temperature.
(10) 6-Hour Minimum Temperature.
(11) 24-Hour Maximum/Minimum Temperature.
(12) Pressure Tendency.
(13) Sensor Status.
PWINO
FZRANO
TSNO
RVRNO
PNO
VISNO

Examples of METAR reports and explanation:
METAR KBNA 281250Z 33018KT 290V360 1/2SM R31/2700FT SN BLSN FG VV008 00/M03 A2991 RMK RAE42SNB42

METAR aviation routine weather report
KBNA Nashville, TN
281250Z date 28th, time 1250 UTC
(no modifier) This is a manually generated report, due to the absence of "AUTO" and "AO1 or AO2" in remarks
33018KT wind three three zero at one eight
290V360 wind variable between two nine zero and three six zero
1/2SM visibility one half
R31/2700FT Runway three one RVR two thousand seven hundred
SN moderate snow
BLSN FG visibility obscured by blowing snow and fog
VV008 indefinite ceiling eight hundred
00/M03 temperature zero, dew point minus three
A2991 altimeter two niner niner one
RMK remarks
RAE42 rain ended at four two
SNB42 snow began at four two
METAR KSFO 041453Z AUTO VRB02KT 3SM BR CLR 15/12 A3012 RMK AO2

METAR aviation routine weather report
KSFO San Francisco, CA
041453Z date 4th, time 1453 UTC
AUTO fully automated; no human intervention
VRB02KT wind variable at two
3SM visibility three
BR visibility obscured by mist
CLR no clouds below one two thousand
15/12 temperature one five, dew point one two
A3012 altimeter three zero one two
RMK remarks
AO2 this automated station has a weather discriminator (for precipitation)

SPECI KCVG 152224Z 28024G36KT 3/4SM +TSRA BKN008 OVC020CB 28/23 A3000 RMK TSRAB24 TS W MOV E

SPECI (nonroutine) aviation special weather report
KCVG Cincinnati, OH
152228Z date 15th, time 2228 UTC
(no modifier) This is a manually generated report due to the absence of "AUTO" and "AO1 or AO2" in remarks
28024G36KT wind two eight zero at two four gusts three six
3/4SM visibility three fourths
+TSRA thunderstorms, heavy rain
BKN008 ceiling eight hundred broken
OVC020CB two thousand overcast cumulonimbus clouds
28/23 temperature two eight, dew point two three
A3000 altimeter three zero zero zero **RMK** remarks
TSRAB24 thunderstorm and rain began at two four
TS W MOV E thunderstorm west moving east

c. Aerodrome Forecast (TAF). A concise statement of the expected meteorological conditions at an airport during a specified period. At most locations, TAFs have a 24 hour forecast period. However, TAFs for some locations have a 30 hour forecast period. These forecast periods may be shorter in the case of an amended TAF. TAFs use the same codes as METAR weather reports. They are scheduled four times daily for 24-hour periods beginning at 0000Z, 0600Z, 1200Z, and 1800Z.

Forecast times in the TAF are depicted in two ways. The first is a 6-digit number to indicate a specific point in time, consisting of a two-digit date, two-digit hour, and two-digit minute (such as issuance time or FM). The second is a pair of four-digit numbers separated by a "/" to indicate a beginning and end for a period of time. In this case,each four-digit pair consists of a two-digit date and a two-digit hour.

TAFs are issued in the following format:
TYPE OF REPORT/ICAO STATION IDENTIFIER/DATE AND TIME OF ORIGIN/VALID PERIOD DATE AND TIME/ FORECAST METEOROLOGICAL CONDITIONS

NOTE-
The "/" above and in the following descriptions are for separation purposes in this publication and do not appear in the actual TAFs.

TAF KORD 051130Z 0512/0618 14008KT 5SM BR BKN030
TEMPO 0513/0516 1 1/2SM BR
FM051600 16010KT P6SM SKC
FM052300 20013G20KT 4SM SHRA OVC020 PROB40 0600/0606 2SM TSRA OVC008CB
BECMG 0606/0608 21015KT P6SM NSW SCT040

TAF format observed in the above example:
TAF = type of report
KORD = ICAO station identifier
051130Z = date and time of origin (issuance time)
0512/0618 = valid period date and times
14008KT 5SM BR BKN030 = forecast meteorological conditions

Explanation of TAF elements:

1. Type of Report. There are two types of TAF issuances, a routine forecast issuance (TAF) and an amended forecast (TAF AMD). An amended TAF is issued when the current TAF no longer adequately describes the on-going weather or the forecaster feels the TAF is not representative of the current or expected weather. Corrected (COR) or delayed (RTD) TAFs are identified only in the communications header which precedes the actual forecasts.

2. ICAO Station Identifier. The TAF code uses ICAO 4-letter location identifiers as described in the METAR section.

3. Date and Time of Origin. This element is the date and time the forecast is actually prepared. The format is a two-digit date and four-digit time followed, without a space, by the letter "Z."

4. Valid Period Date and Time. The UTC valid period of the forecast consists of two four-digit sets, separated by a "/". The first four-digit set is a two-digit date followed by the two-digit beginning hour, and the second four-digit set is a two-digit date followed by the two-digit ending hour. Although most airports have a 24-hour TAF, a select number of airports have a 30-hour TAF. In the case of an amended forecast, or a forecast which is corrected or delayed, the valid period may be for less than 24 hours. Where an airport or terminal operates on a part-time basis (less than 24 hours/day), the TAFs issued for those locations will have the abbreviated statement "NIL AMD SKED AFT (closing time) Z" added to the end of the forecasts. For the TAFs issued while these locations are closed, the word "NIL" will appear in place of the forecast text. A delayed (RTD) forecast will then be issued for these locations after two complete observations are received.

5. Forecast Meteorological Conditions. This is the body of the TAF. The basic format is:
WIND/VISIBILITY/WEATHER/SKY CONDITION/ OPTIONAL DATA (WIND SHEAR)

The wind, visibility, and sky condition elements are always included in the initial time group of the forecast. Weather is included only if significant to aviation. If a significant, lasting change in any of the elements is expected during the valid period, a new time period with the changes is included. It should be noted that with the exception of a "FM" group the new time period will include only those elements which are expected to change, i.e., if a lowering of the visibility is expected but the wind is expected to remain the same, the new time period reflecting the lower visibility would not include a forecast wind. The forecast wind would remain the same as in the previous time period. Any temporary conditions expected during a specific time period are included with that time period. The following describes the elements in the above format.

(a) Wind. This five (or six) digit group includes the expected wind direction (first 3 digits) and speed (last 2 digits or 3 digits if 100 knots or greater). The contraction "KT" follows to denote the units of wind speed. Wind gusts are noted by the letter "G" appended to the wind speed followed by the highest expected gust. A variable wind direction is noted by "VRB" where the three digit direction usually appears. A calm wind (3 knots or less) is forecast as "00000KT."

EXAMPLE-
18010KT wind one eight zero at one zero (wind is blowing from 180).
35012G20KT wind three five zero at one two gust two zero.

(b) Visibility. The expected prevailing visibility up to and including 6 miles is forecast in statute miles, including fractions of miles, followed by "SM" to note the units of measure. Expected visibilities greater than 6 miles are forecast as P6SM (plus six statute miles).

EXAMPLE-
1/2SM - visibility one-half
4SM - visibility four
P6SM - visibility more than six

(c) Weather Phenomena. The expected weather phenomena is coded in TAF reports using the same format, qualifiers, and phenomena contractions as METAR reports (except UP). Obscurations to vision will be forecast whenever the prevailing visibility is forecast to be 6 statute miles or less. If no significant weather is expected to occur during a specific time period in the forecast, the weather phenomena group is omitted for that time period. If, after a time period in which significant weather phenomena has been forecast, a change to a forecast of no significant weather phenomena occurs, the contraction NSW (No Significant Weather) will appear as the weather group in the new time period. (NSW is included only in BECMG or TEMPO groups).

NOTE-*It is very important that pilots understand that NSW only refers to weather phenomena, i.e., rain, snow, drizzle, etc. Omitted conditions, such as sky conditions, visibility, winds, etc., are carried over from the previous time group.*

(d) Sky Condition. TAF sky condition forecasts use the METAR format described in the METAR section. Cumulonimbus clouds (CB) are the only cloud type forecast in TAFs. When clear skies are forecast, the contraction "SKC" will always be used. The contraction "CLR" is never used in the TAF. When the sky is obscured due to a surface-based phenomenon, vertical visibility (VV) into the obscuration is forecast. The format for vertical visibility is "VV" followed by a three-digit height in hundreds of feet.

NOTE-*As in METAR, ceiling layers are not designated in the TAF code. For aviation purposes, the ceiling is the lowest broken or overcast layer or vertical visibility into a complete obscuration.*

SKC "sky clear"

SCT005 BKN025CB "five hundred scattered, ceiling two thousand five hundred broken cumulonimbus clouds"

VV008 "indefinite ceiling eight hundred"

(e) Optional Data (Wind Shear). Wind shear is the forecast of nonconvective low level winds (up to 2,000 feet). The forecast includes the letters "WS" followed by the height of the wind shear, the wind direction and wind speed at the indicated height and the ending letters "KT" (knots). Height is given in hundreds of feet (AGL) up to and including 2,000 feet. Wind shear is encoded with the contraction "WS," followed by a three-digit height, slant character "/," and winds at the height indicated in the same format as surface winds. The wind shear element is omitted if not expected to occur.

WS010/18040KT - "LOW LEVEL WIND SHEAR AT ONE THOUSAND, WIND ONE EIGHT ZERO AT FOUR ZERO"

d. Probability Forecast. The probability or chance of thunderstorms or other precipitation events occurring, along with associated weather conditions (wind, visibility, and sky conditions). The PROB30 group is used when the occurrence of thunderstorms or precipitation is 30-39% and the PROB40 group is used when the occurrence of thunderstorms or precipitation is 40-49%. This is followed by two four-digit groups separated by a "/", giving the beginning date and hour, and the ending date and hour of the time period during which the thunderstorms or precipitation are expected.

NOTE-*Neither PROB30 nor PROB40 will be shown during the first six hours of a forecast.*

EXAMPLE-

PROB40 2221/2302 $^1/_2$SM +TSRA "chance between 2100Z and 0200Z of visibility one-half statute mile in thunderstorms and heavy rain."

PROB30 3010/3014 1SM RASN "chance between 1000Z and 1400Z of visibility one statute mile in mixed rain and snow."

e. Forecast Change Indicators. The following change indicators are used when either a rapid, gradual, or temporary change is expected in some or all of the forecast meteorological conditions. Each change indicator marks a time group within the TAF report.

1. From (FM) group. The FM group is used when a rapid change, usually occurring in less than one hour, in prevailing conditions is expected. Typically, a rapid change of prevailing conditions to more or less a completely new set of prevailing conditions is associated with a synoptic feature passing through the terminal area (cold or warm frontal passage). Appended to the "FM" indicator is the six-digit date, hour, and minute the change is expected to begin and continues until the next change group or until the end of the current forecast. A "FM" group will mark the beginning of a new line in a TAF report (indented 5 spaces). Each "FM" group contains all the required elements-wind, visibility, weather, and sky condition. Weather will be omitted in "FM" groups when it is not significant to aviation. FM groups will not include the contraction NSW.

EXAMPLE-*FM210100 14010KT P6SM SKC - "after 0100Z on the 21st, wind one four zero at one zero, visibility more than six, sky clear."*

2. Becoming (BECMG) group. The BECMG group is used when a gradual change in conditions is expected over a longer time period, usually two hours. The time period when the change is expected is two four-digit groups separated by a "/", with the beginning date and hour, and ending date and hour of the change period which follows the BECMG indicator. The gradual change will occur at an unspecified time within this time period. Only the changing forecast meteorological conditions are included in BECMG groups. The omitted conditions are carried over from the previous time group.

EXAMPLE-*OVC012 BECMG 0114/0116 BKN020 - "ceiling one thousand two hundred overcast. Then a gradual change to ceiling two thousand broken between 1400Z on the 1st and 1600Z on the 1st."*

3. Temporary (TEMPO) group. The TEMPO group is used for any conditions in wind, visibility, weather, or sky condition which are expected to last for generally less than an hour at a time (occasional), and are expected to occur during less than half the time period. The TEMPO indicator is followed by two four-digit groups separated by a "/". The first four digit group gives the beginning date and hour, and the second four digit group gives the ending date and hour of the time period during which the temporary conditions are expected. Only the changing forecast meteorological conditions are included in TEMPO groups. The omitted conditions are carried over from the previous time group.

EXAMPLE-

1. SCT030 TEMPO 0519/0523 BKN030 - "three thousand scattered with occasional ceilings three thousand broken between 1900Z on the 5th and 2300Z on the 5th."

2. 4SM HZ TEMPO 1900/1906 2SM BR HZ - "visibility four in haze with occasional visibility two in mist and haze between 0000Z on the 19th and 0600Z on the 19th."

Section 2. Altimeter Setting Procedures

7-2-1. General

a. The accuracy of aircraft altimeters is subject to the following factors:

1. Nonstandard temperatures of the atmosphere.

2. Nonstandard atmospheric pressure.

3. Aircraft static pressure systems (position error); and

4. Instrument error.

b. EXTREME CAUTION SHOULD BE EXERCISED WHEN FLYING IN PROXIMITY TO OBSTRUCTIONS OR TERRAIN IN LOW TEMPERATURES AND PRESSURES. This is especially true in extremely cold temperatures that cause a large differential between the Standard Day temperature and actual temperature. This circumstance can cause serious errors that result in the aircraft being significantly lower than the indicated altitude.

NOTE-*Standard temperature at sea level is 15 degrees Celsius (59 degrees Fahrenheit). The temperature gradient from sea level is minus 2 degrees Celsius (3.6 degrees Fahrenheit) per 1,000 feet. Pilots should apply corrections for static pressure systems and/or instruments, if appreciable errors exist.*

c. The adoption of a standard altimeter setting at the higher altitudes eliminates station barometer errors, some altimeter instrument errors, and errors caused by altimeter settings derived from different geographical sources.

7-2-2. Procedures

The cruising altitude or flight level of aircraft must be maintained by reference to an altimeter which must be set, when operating:

a. Below 18,000 feet MSL.

1. When the barometric pressure is 31.00 inches Hg. or less. To the current reported altimeter setting of a station along the route and within 100 NM of the aircraft, or if there is no station within this area, the current reported altimeter setting of an appropriate available station. When an aircraft is en route on an instrument flight plan, air traffic controllers will furnish this information to the pilot at least once while the aircraft is in the controllers area of jurisdiction. In the case of an aircraft not equipped with a radio, set to the elevation of the departure airport or use an appropriate altimeter setting available prior to departure.

2. When the barometric pressure exceeds 31.00 inches Hg. The following procedures will be placed in effect by NOTAM defining the geographic area affected:

(a) For all aircraft. Set 31.00 inches for en route operations below 18,000 feet MSL. Maintain this setting until beyond the affected area or until reaching final approach segment. At the beginning of the final approach segment, the current altimeter setting will be set, if possible. If not possible, 31.00 inches will remain set throughout the approach. Aircraft on departure or missed approach will set 31.00 inches prior to reaching any mandatory/crossing altitude or 1,500 feet AGL, whichever is lower. (Air traffic control will issue actual altimeter settings and advise pilots to set 31.00 inches in their altimeters for en route operations below 18,000 feet MSL in affected areas.)

(b) During preflight, barometric altimeters must be checked for normal operation to the extent possible.

(c) For aircraft with the capability of setting the current altimeter setting and operating into airports with the capability of measuring the current altimeter setting, no additional restrictions apply.

(d) For aircraft operating VFR, there are no additional restrictions, however, extra diligence in flight planning and in operating in these conditions is essential.

(e) Airports unable to accurately measure barometric pressures above 31.00 inches of Hg. will report the barometric pressure as "missing" or "in excess of 31.00 inches of Hg." Flight operations to and from those airports are restricted to VFR weather conditions.

(f) For aircraft operating IFR and unable to set the current altimeter setting, the following restrictions apply:

(1) To determine the suitability of departure alternate airports, destination airports, and destination alternate airports, increase ceiling requirements by 100 feet and visibility requirements by $^1/_4$ statute mile for each $^1/_{10}$ of an inch of Hg., or any portion thereof, over 31.00 inches. These adjusted values are then applied in accordance with the requirements of the applicable operating regulations and operations specifications.

EXAMPLE-*Destination altimeter is 31.28 inches, ILS DH 250 feet (200-$^1/_2$). When flight planning, add 300-$^3/_4$ to the weather requirements which would become 500-1$^1/_4$.*

(2) On approach, 31.00 inches will remain set. Decision height (DH) or minimum descent altitude must be deemed to have been reached when the published altitude is displayed on the altimeter.

NOTE-*Although visibility is normally the limiting factor on an approach, pilots should be aware that when reaching DH the aircraft will be higher than indicated. Using the example above the aircraft would be approximately 300 feet higher.*

(3) These restrictions do not apply to authorized Category II and III ILS operations nor do they apply to certificate holders using approved QFE altimetry systems.

(g) The FAA Regional Flight Standards Division Manager of the affected area is authorized to approve temporary waivers to permit emergency resupply or emergency medical service operation.

b. At or above 18,000 feet MSL. To 29.92 inches of mercury (standard setting). The lowest usable flight level is determined by the atmospheric pressure in the area of operation as shown in TBL 7-2-1.

TBL 7-2-1
Lowest Usable Flight Level

Altimeter Setting (Current Reported)	Lowest Usable Flight Levelnone
29.92 or higher	180
29.91 to 29.42	185
29.41 to 28.92	190
28.91 to 28.42	195
28.41 to 27.92	200

c. Where the minimum altitude, as prescribed in 14CFR Section 91.159 and 14CFR Section 91.177, is above 18,000 feet MSL, the lowest usable flight level must be the flight level equivalent of the minimum altitude plus the number of feet specified in TBL 7-2-2.

TBL 7-2-2
Lowest Flight Level Correction Factor

Altimeter Setting	Correction Factor
29.92 or higher	none
29.91 to 29.42	500 feet
29.41 to 28.92	1000 feet
28.91 to 28.42	1500 feet
28.41 to 27.92	2000 feet
27.91 to 27.42	2500 feet

***EXAMPLE-** The minimum safe altitude of a route is 19,000 feet MSL and the altimeter setting is reported between 29.92 and 29.42 inches of mercury, the lowest usable flight level will be 195, which is the flight level equivalent of 19,500 feet MSL (minimum altitude plus 500 feet).*

7-2-3. Altimeter Errors

a. Most pressure altimeters are subject to mechanical, elastic, temperature, and installation errors. (Detailed information regarding the use of pressure altimeters is found in the Instrument Flying Handbook, Chapter IV.) Although manufacturing and installation specifications, as well as the periodic test and inspections required by regulations (14 CFR Part 43, Appendix E), act to reduce these errors, any scale error may be observed in the following manner:

1. Set the current reported altimeter setting on the altimeter setting scale.

2. Altimeter should now read field elevation if you are located on the same reference level used to establish the altimeter setting.

3. Note the variation between the known field elevation and the altimeter indication. If this variation is in the order of plus or minus 75 feet, the accuracy of the altimeter is questionable and the problem should be referred to an appropriately rated repair station for evaluation and possible correction.

b. Once in flight, it is very important to obtain frequently current altimeter settings en route. If you do not reset your altimeter when flying *from* an area of high pressure into an area of low pressure, *your aircraft will be closer to the surface than your altimeter indicates.* An inch error in the altimeter setting equals 1,000 feet of altitude. To quote an old saying: ***"GOING FROM A HIGH TO A LOW, LOOK OUT BELOW."***

c. Temperature also has an effect on the accuracy of altimeters and your altitude. The crucial values to consider are standard temperature versus the ambient (at altitude) temperature. It is this "difference" that causes the error in indicated altitude. When the air is warmer than standard, you are higher than your altimeter indicates. Subsequently, when the air is colder than standard you are lower than indicated. It is the magnitude of this "difference" that determines the magnitude of the error. When flying into a cooler air mass while maintaining a constant indicated altitude, you are losing true altitude. However, flying into a cooler air mass does not necessarily mean you will be lower than indicated if the *difference* is still on the plus side. For example, while flying at 10,000 feet (where **STANDARD** temperature is -5 degrees Celsius (C)), the outside air temperature cools from +5 degrees C to 0 degrees C, the temperature error will nevertheless cause the aircraft to be **HIGHER** than indicated. It is the extreme "cold" difference that normally would be of concern to the pilot. Also, when flying in cold conditions over mountainous country, the pilot should exercise caution in flight planning both in regard to route and altitude to ensure adequate en route and terminal area terrain clearance.

d. TBL 7-2-3, derived from ICAO formulas, indicates how much error can exist when the temperature is extremely cold. To use the table, find the reported temperature in the left column, then read across the top row to locate the height above the airport/reporting station (i.e., subtract the airport/ reporting elevation from the intended flight altitude). The intersection of the column and row is how much *lower* the aircraft may actually be as a result of the possible cold temperature induced error.

e. The possible result of the above example should be obvious, particularly if operating at the minimum altitude or when conducting an instrument approach. When operating in extreme cold temperatures, pilots may wish to compensate for the reduction in terrain clearance by adding a cold temperature correction.

TBL 7-2-3
ICAO Cold Temperature Error Table

Reported Temp °C	Height Above Airport in Feet													
	200	300	400	500	600	700	800	900	1000	1500	2000	3000	4000	5000
+10	10	10	10	10	20	20	20	20	20	30	40	60	80	90
0	20	20	30	30	40	40	50	50	60	90	120	170	230	280
-10	20	30	40	50	60	70	80	90	100	150	200	290	390	490
-20	30	50	60	70	90	100	120	130	140	210	280	420	570	710
-30	40	60	80	100	120	140	150	170	190	280	380	570	760	950
-40	50	80	100	120	150	170	190	220	240	360	480	720	970	1210
-50	60	90	120	150	180	210	240	270	300	450	590	890	1190	1500

***EXAMPLE-** Temperature-10 degrees Celsius, and the aircraft altitude is 1,000 feet above the airport elevation. The chart shows that the reported current altimeter setting may place the aircraft as much as 100 feet below the altitude indicated by the altimeter.*

7-2-4. High Barometric Pressure

a. Cold, dry air masses may produce barometric pressures in excess of 31.00 inches of Mercury, and many altimeters do not have an accurate means of being adjusted for settings of these levels. When the altimeter cannot be set to the higher pressure setting, the aircraft actual altitude will be higher than the altimeter indicates.

***REFERENCE-** AIM, Paragraph 7-2-3, Altimeter Errors.*

b. When the barometric pressure exceeds 31.00 inches, air traffic controllers will issue the actual altimeter setting, and:

1. En Route/Arrivals. Advise pilots to remain set on 31.00 inches until reaching the final approach segment.

2. Departures. Advise pilots to set 31.00 inches prior to reaching any mandatory/crossing altitude or 1,500 feet, whichever is lower.

c. The altimeter error caused by the high pressure will be in the opposite direction to the error caused by the cold temperature.

7-2-5. Low Barometric Pressure

When abnormally low barometric pressure conditions occur (below 28.00), flight operations by aircraft unable to set the actual altimeter setting are not recommended.

NOTE-
*The true altitude of the aircraft is **lower** than the indicated altitude if the pilot is unable to set the actual altimeter setting.*

Section 3. Wake Turbulence

7-3-1. General

a. Every aircraft generates a wake while in flight. Initially, when pilots encountered this wake in flight, the disturbance was attributed to "prop wash." It is known, however, that this disturbance is caused by a pair of counter-rotating vortices trailing from the wing tips. The vortices from larger aircraft pose problems to encountering aircraft. For instance, the wake of these aircraft can impose rolling moments exceeding the roll-control authority of the encountering aircraft. Further, turbulence generated within the vortices can damage aircraft components and equipment if encountered at close range. The pilot must learn to envision the location of the vortex wake generated by larger (transport category) aircraft and adjust the flight path accordingly.

b. During ground operations and during takeoff, jet engine blast (thrust stream turbulence) can cause damage and upsets if encountered at close range. Exhaust velocity versus distance studies at various thrust levels have shown a need for light aircraft to maintain an adequate separation behind large turbojet aircraft. Pilots of larger aircraft should be particularly careful to consider the effects of their "jet blast" on other aircraft, vehicles, and maintenance equipment during ground operations.

FIG 7-3-1
Wake Vortex Generation

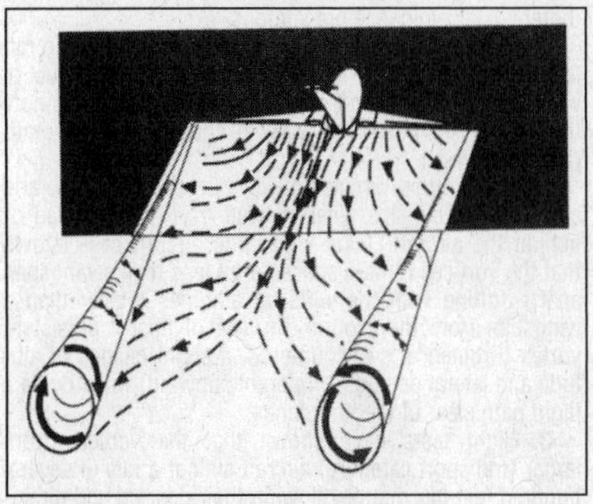

7-3-2. Vortex Generation

Lift is generated by the creation of a pressure differential over the wing surface. The lowest pressure occurs over the upper wing surface and the highest pressure under the wing. This pressure differential triggers the roll up of the airflow aft of the wing resulting in swirling air masses trailing downstream of the wing tips. After the roll up is completed, the wake consists of two counter-rotating cylindrical vortices. (See FIG 7-3-1.) Most of the energy is within a few feet of the center of each vortex, but pilots should avoid a region within about 100 feet of the vortex core.

7-3-3. Vortex Strength

a. The strength of the vortex is governed by the weight, speed, and shape of the wing of the generating aircraft. The vortex characteristics of any given aircraft can also be changed by extension of flaps or other wing configuring devices as well as by change in speed. However, as the basic factor is weight, the vortex strength increases proportionately. Peak vortex tangential speeds exceeding 300 feet per second have been recorded. The greatest vortex strength occurs when the generating aircraft is HEAVY, CLEAN, and SLOW.

b. Induced Roll

1. In rare instances a wake encounter could cause inflight structural damage of catastrophic proportions. However, the usual hazard is associated with induced rolling moments which can exceed the roll-control authority of the encountering aircraft. In flight experiments, aircraft have been intentionally flown directly up trailing vortex cores of larger aircraft. It was shown that the capability of an aircraft to counteract the roll imposed by the wake vortex primarily depends on the wingspan and counter-control responsiveness of the encountering aircraft.

2. Counter control is usually effective and induced roll minimal in cases where the wingspan and ailerons of the encountering aircraft extend beyond the rotational flow field of the vortex. It is more difficult for aircraft with short wingspan (relative to the generating aircraft) to counter the imposed roll induced by vortex flow. Pilots of short span aircraft, even of the high performance type, must be especially alert to vortex encounters.

(See FIG 7-3-2.)

3. The wake of larger aircraft requires the respect of all pilots.

FIG 7-3-2
Wake Encounter Counter Control

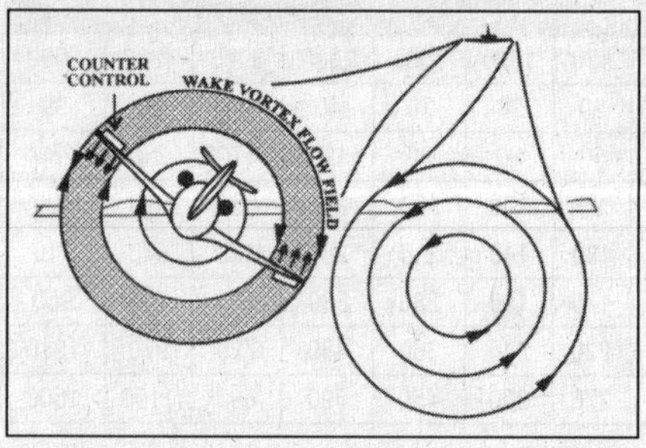

FIG 7-3-3
Wake Ends/Wake Begins

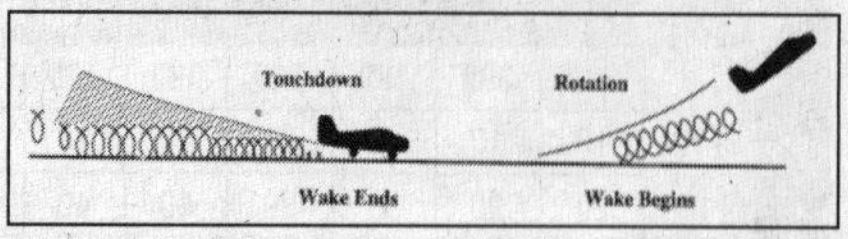

FIG 7-3-4
Vortex Flow Field

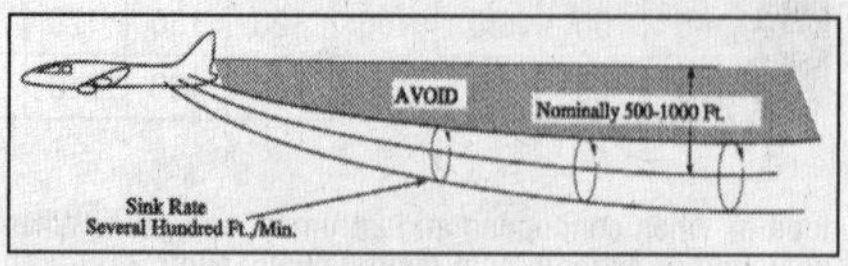

FIG 7-3-5
Vortex Movement Near Ground - No Wind

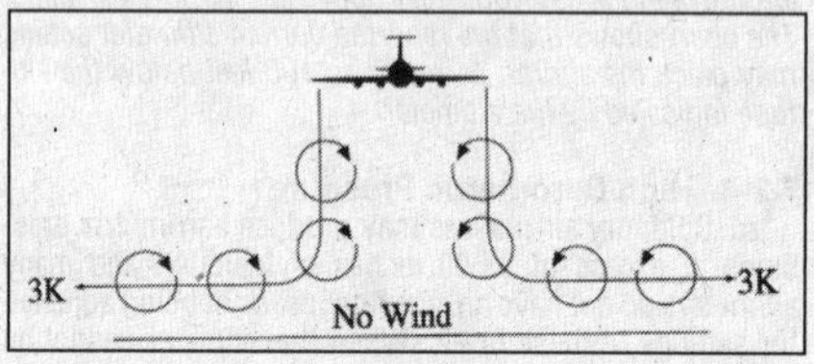

FIG 7-3-6
Vortex Movement Near Ground - with Cross Winds

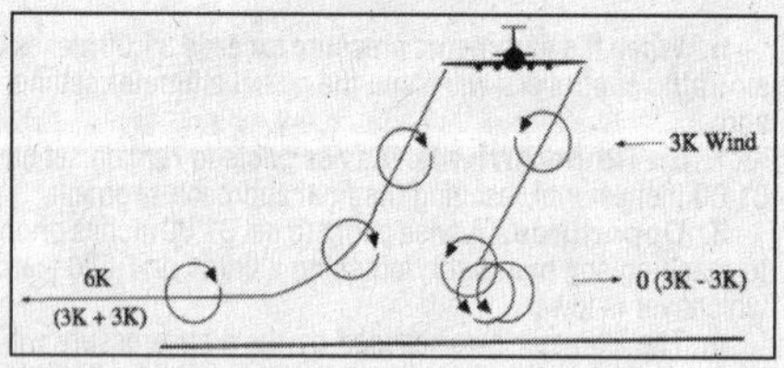

FIG 7-3-7
Vortex Movement in Ground Effect - Tailwind

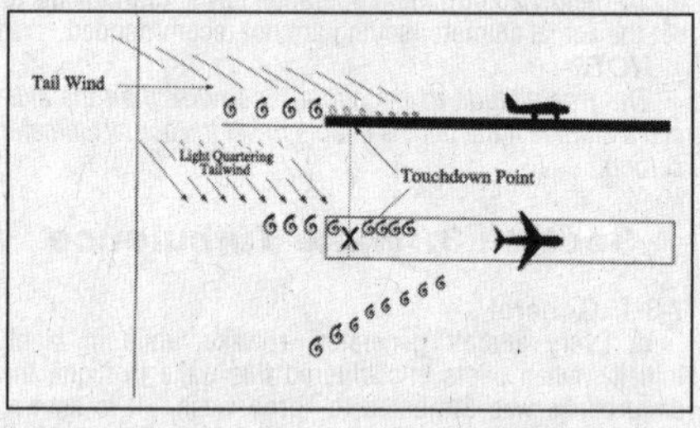

7-3-4. Vortex Behavior

a. Trailing vortices have certain behavioral characteristics which can help a pilot visualize the wake location and thereby take avoidance precautions.

1. Vortices are generated from the moment aircraft leave the ground, since trailing vortices are a by-product of wing lift. Prior to takeoff or touchdown pilots should note the rotation or touchdown point of the preceding aircraft. (See FIG 7-3-3.)

2. The vortex circulation is outward, upward and around the wing tips when viewed from either ahead or behind the aircraft. Tests with large aircraft have shown that the vortices remain spaced a bit less than a wingspan apart, drifting with the wind, at altitudes greater than a wingspan from the ground. In view of this, if persistent vortex turbulence is encountered, a slight change of altitude and lateral position (preferably upwind) will provide a flight path clear of the turbulence.

3. Flight tests have shown that the vortices from larger (transport category) aircraft sink at a rate of several hundred feet per minute, slowing their descent and diminishing in strength with time and distance behind the generating aircraft. Atmospheric turbulence hastens breakup. Pilots should fly at or above the preceding aircraft's flight path, altering course as necessary to avoid the area behind and below the generating aircraft. (See FIG 7-3-4.) However, vertical separation of 1,000 feet may be considered safe.

4. When the vortices of larger aircraft sink close to the ground (within 100 to 200 feet), they tend to move laterally over the ground at a speed of 2 or 3 knots. (See FIG 7-3-5.)

5. There is a small segment of the aviation community that have become convinced that wake vortices may "bounce" up to twice their nominal steady state height. With a 200-foot span aircraft, the "bounce" height could reach approximately 200 feet AGL. This conviction is based on a single unsubstantiated report of an apparent coherent vortical flow that was seen in the volume scan of a research sensor. No one can say what conditions cause vortex bouncing, how high they bounce, at what angle they bounce, or how many times a vortex may bounce. On the other hand, no one can say for certain that vortices never "bounce." Test data have shown that vortices can rise with the air mass in which they are embedded. Wind shear, particularly, can cause vortex flow field "tilting." Also, ambient thermal lifting and orographic effects (rising terrain or tree lines) can cause a vortex flow field to rise. Notwithstanding the foregoing, pilots are reminded that they should be alert at all times for possible wake vortex encounters when conducting approach and landing operations. The pilot has the ultimate responsibility for ensuring appropriate separations and positioning of the aircraft in the terminal area to avoid the wake turbulence created by a preceding aircraft.

b. A crosswind will decrease the lateral movement of the upwind vortex and increase the movement of the downwind vortex. Thus a light wind with a cross runway component of 1 to 5 knots could result in the upwind vortex remaining in the touchdown zone for a period of time and hasten the drift of the downwind vortex toward another runway. (SeeFIG 7-3-6.) Similarly, a tailwind condition can move the vortices of the preceding aircraft forward into the touchdown zone. THE LIGHT QUARTERING TAILWIND REQUIRES MAXIMUM CAUTION. Pilots should be alert to large aircraft upwind from their approach and takeoff flight paths. (See FIG 7-3-7.)

7-3-5. Operations Problem Areas

a. A wake encounter can be catastrophic. In 1972 at Fort Worth a DC-9 got too close to a DC-10 (two miles back), rolled, caught a wingtip, and cartwheeled coming to rest in an inverted position on the runway. All aboard were killed. Serious and even fatal GA accidents induced by wake vortices are not uncommon. However, a wake encounter is not necessarily hazardous. It can be one or more jolts with varying severity depending upon the direction of the encounter, weight of the generating aircraft, size of the encountering aircraft, distance from the generating aircraft, and point of vortex encounter. The probability of induced roll increases when the encountering aircraft's heading is generally aligned with the flight path of the generating aircraft.

b. AVOID THE AREA BELOW AND BEHIND THE GENERATING AIRCRAFT, ESPECIALLY AT LOW ALTITUDE WHERE EVEN A MOMENTARY WAKE ENCOUNTER COULD BE HAZARDOUS. This is not easy to do. Some accidents have occurred even though the pilot of the trailing aircraft had carefully noted that the aircraft in front was at a considerably lower altitude. Unfortunately, this does not ensure that the flight path of the lead aircraft will be below that of the trailing aircraft.

c. Pilots should be particularly alert in calm wind conditions and situations where the vortices could:

1. Remain in the touchdown area.

2. Drift from aircraft operating on a nearby runway.

3. Sink into the takeoff or landing path from a crossing runway.

4. Sink into the traffic pattern from other airport operations.

5. Sink into the flight path of VFR aircraft operating on the hemispheric altitude 500 feet below.

d. Pilots of all aircraft should visualize the location of the vortex trail behind larger aircraft and use proper vortex avoidance procedures to achieve safe operation. It is equally important that pilots of larger aircraft plan or adjust their flight paths to minimize vortex exposure to other aircraft.

7-3-6. Vortex Avoidance Procedures

a. Under certain conditions, airport traffic controllers apply procedures for separating IFR aircraft. If a pilot accepts a clearance to visually follow a preceding aircraft, the pilot accepts responsibility for separation and wake turbulence avoidance. The controllers will also provide to VFR aircraft, with whom they are in communication and which in the tower's opinion may be adversely affected by wake turbulence from a larger aircraft, the position, altitude and direction of flight of larger aircraft followed by the phrase "CAUTION - WAKE TURBULENCE." After issuing the caution for wake turbulence, the airport traffic controllers generally do not provide additional information to the following aircraft unless the airport traffic controllers know the following aircraft is overtaking the preceding aircraft. WHETHER OR NOT A WARNING OR INFORMATION HAS BEEN GIVEN, HOWEVER, THE PILOT IS EXPECTED TO ADJUST AIRCRAFT OPERATIONS AND FLIGHT PATH AS NECESSARY TO PRECLUDE SERIOUS WAKE ENCOUNTERS. When any doubt exists about maintaining safe separation distances between aircraft during approaches, pilots should ask the control tower for updates on separation distance and aircraft groundspeed.

b. The following vortex avoidance procedures are recommended for the various situations:

1. Landing behind a larger aircraft- same runway. Stay at or above the larger aircraft's final approach flight path-note its touchdown point-land beyond it.

2. Landing behind a larger aircraft- when parallel runway is closer than 2,500 feet. Consider possible drift to your runway. Stay at or above the larger aircraft's final approach flight path- note its touchdown point.

3. Landing behind a larger aircraft- crossing runway. Cross above the larger aircraft's flight path.

4. Landing behind a departing larger aircraft- same runway. Note the larger aircraft's rotation point- land well prior to rotation point.

5. Landing behind a departing larger aircraft- crossing runway. Note the larger aircraft's rotation point- if past the intersection- continue the approach- land prior to the intersection. If larger aircraft rotates prior to the intersection, avoid flight below the larger aircraft's flight path. Abandon the approach unless a landing is ensured well before reaching the intersection.

6. Departing behind a larger aircraft. Note the larger aircraft's rotation point and rotate prior to the larger aircraft's rotation point. Continue climbing above the larger aircraft's climb path until turning clear of the larger aircraft's wake. Avoid subsequent headings which will cross below and behind a larger aircraft. Be alert for any critical takeoff situation which could lead to a vortex encounter.

7. Intersection takeoffs- same runway. Be alert to adjacent larger aircraft operations, particularly upwind of your runway. If intersection takeoff clearance is received, avoid subsequent heading which will cross below a larger aircraft's path.

8. Departing or landing after a larger aircraft executing a low approach, missed approach, or touch-and-go landing. Because vortices settle and move laterally near the ground, the vortex hazard may exist along the runway and in your flight path after a larger aircraft has executed a low approach, missed approach, or a touch-and-go landing, particular in light quartering wind conditions. You should ensure that an interval of at least 2 minutes has elapsed before your takeoff or landing.

9. En route VFR (thousand-foot altitude plus 500 feet). Avoid flight below and behind a large aircraft's path. If a larger aircraft is observed above on the same track (meeting or overtaking) adjust your position laterally, preferably upwind.

7-3-7. Helicopters

In a slow hover taxi or stationary hover near the surface, helicopter main rotor(s) generate downwash producing high velocity outwash vortices to a distance approximately three times the diameter of the rotor. When rotor downwash hits the surface, the resulting outwash vortices have behavioral characteristics similar to wing tip vortices produced by fixed wing aircraft. However, the vortex circulation is outward, upward, around, and away from the main rotor(s) in all directions. Pilots of small aircraft should avoid operating within three rotor diameters of any helicopter in a slow hover taxi or stationary hover. In forward flight, departing or landing helicopters produce a pair of strong, high-speed trailing vortices similar to wing tip vortices of larger fixed wing aircraft. Pilots of small aircraft should use caution when operating behind or crossing behind landing and departing helicopters.

7-3-8. Pilot Responsibility

a. Government and industry groups are making concerted efforts to minimize or eliminate the hazards of trailing vortices. However, the flight disciplines necessary to ensure vortex avoidance during VFR operations must be exercised by the pilot. Vortex visualization and avoidance procedures should be exercised by the pilot using the same degree of concern as in collision avoidance.

b. Wake turbulence may be encountered by aircraft in flight as well as when operating on the airport movement area.

REFERENCE-*Pilot/Controller Glossary Term- Wake Turbulence.*

c. Pilots are reminded that in operations conducted behind all aircraft, acceptance of instructions from ATC in the following situations is an acknowledgment that the pilot will ensure safe takeoff and landing intervals and accepts the responsibility for providing wake turbulence separation.

1. Traffic information.

2. Instructions to follow an aircraft; and

3. The acceptance of a visual approach clearance.

d. For operations conducted behind **heavy** aircraft, ATC will specify the word "**heavy**" when this information is known. Pilots of **heavy** aircraft should always use the word "**heavy**" in radio communications.

e. Heavy and large jet aircraft operators should use the following procedures during an approach to landing. These procedures establish a dependable baseline from which pilots of in-trail, lighter aircraft may reasonably expect to make effective flight path adjustments to avoid serious wake vortex turbulence.

1. Pilots of aircraft that produce strong wake vortices should make every attempt to fly on the established glidepath, not above it; or, if glidepath guidance is not available, to fly as closely as possible to a "3-1" glidepath, not above it.

EXAMPLE-*Fly 3,000 feet at 10 miles from touchdown, 1,500 feet at 5 miles, 1,200 feet at 4 miles, and so on to touchdown.*

2. Pilots of aircraft that produce strong wake vortices should fly as closely as possible to the approach course centerline or to the extended centerline of the runway of intended landing as appropriate to conditions.

f. Pilots operating lighter aircraft on visual approaches in-trail to aircraft producing strong wake vortices should use the following procedures to assist in avoiding wake turbulence. These procedures apply only to those aircraft that are on visual approaches.

1. Pilots of lighter aircraft should fly on or above the glidepath. Glidepath reference may be furnished by an ILS, by a visual approach slope system, by other ground-based approach slope guidance systems, or by other means. In the absence of visible glidepath guidance, pilots may very nearly duplicate a 3-degree glideslope by adhering to the "3 to 1" glidepath principle.

EXAMPLE-*Fly 3,000 feet at 10 miles from touchdown, 1,500 feet at 5 miles, 1,200 feet at 4 miles, and so on to touchdown.*

2. If the pilot of the lighter following aircraft has visual contact with the preceding heavier aircraft and also with the runway, the pilot may further adjust for possible wake vortex turbulence by the following practices:

(a) Pick a point of landing no less than 1,000 feet from the arrival end of the runway.

(b) Establish a line-of-sight to that landing point that is above and in front of the heavier preceding aircraft.

(c) When possible, note the point of landing of the heavier preceding aircraft and adjust point of intended landing as necessary.

EXAMPLE-*A puff of smoke may appear at the 1,000-foot markings of the runway, showing that touchdown was that point; therefore, adjust point of intended landing to the 1,500-foot markings.*

(d) Maintain the line-of-sight to the point of intended landing above and ahead of the heavier preceding aircraft; maintain it to touchdown.

(e) Land beyond the point of landing of the preceding heavier aircraft.

3. During visual approaches pilots may ask ATC for updates on separation and groundspeed with respect to heavier preceding aircraft, especially when there is any question of safe separation from wake turbulence.

7-3-9. Air Traffic Wake Turbulence Separations

a. Because of the possible effects of wake turbulence, controllers are required to apply no less than specified minimum separation for aircraft operating behind a **heavy** jet and, in certain instances, behind large **nonheavy** aircraft (i.e., B757 aircraft).

1. Separation is applied to aircraft operating directly behind a **heavy/B757** jet at the same altitude or less than 1,000 feet below:

(a) Heavy jet behind **heavy** jet-4 miles.

(b) Large/heavy behind **B757** - 4 miles.

(c) Small behind **B757** - 5 miles.

(d) Small/large aircraft behind **heavy** jet - 5 miles.

2. Also, separation, measured at the time the preceding aircraft is over the landing threshold, is provided to small aircraft:

(a) Small aircraft landing behind **heavy** jet - 6 miles.

(b) Small aircraft landing behind **B757** - 5 miles.

(c) Small aircraft landing behind **large** aircraft- 4 miles.

REFERENCE-*Pilot/Controller Glossary Term- Aircraft Classes.*

3. Additionally, appropriate time or distance intervals are provided to departing aircraft:

(a) Two minutes or the appropriate 4 or 5 mile radar separation when takeoff behind a**heavy/B757** jet will be:

(1) From the same threshold.

(2) On a crossing runway and projected flight paths will cross.

(3) From the threshold of a parallel runway when staggered ahead of that of the adjacent runway by less than 500 feet and when the runways are separated by less than 2,500 feet.

NOTE-*Controllers may not reduce or waive these intervals.*

b. A 3-minute interval will be provided when a **small** aircraft will takeoff:

1. From an intersection on the same runway (same or opposite direction) behind a departing **large** aircraft,

2. In the opposite direction on the same runway behind a large aircraft takeoff or low/missed approach.

NOTE-*This 3-minute interval may be waived upon specific pilot request.*

c. A 3-minute interval will be provided for all aircraft taking off when the operations are as described in subparagraph b1 and 2 above, the preceding aircraft is a **heavy/B757** jet, and the operations are on either the same runway or parallel runways separated by less than 2,500 feet. Controllers may not reduce or waive this interval.

d. Pilots may request additional separation i.e., 2 minutes instead of 4 or 5 miles for wake turbulence avoidance. This request should be made as soon as practical on ground control and at least before taxiing onto the runway.

NOTE-*14 CFR Section 91.3(a) states: "The pilot-in-command of an aircraft is directly responsible for and is the final authority as to the operation of that aircraft."*

e. Controllers may anticipate separation and need not withhold a takeoff clearance for an aircraft departing behind a large/heavy aircraft if there is reasonable assurance the required separation will exist when the departing aircraft starts takeoff roll.

Section 4. Bird Hazards and Flight Over National Refuges, Parks, and Forests

7-4-1. Migratory Bird Activity

a. Bird strike risk increases because of bird migration during the months of March through April, and August through November.

b. The altitudes of migrating birds vary with winds aloft, weather fronts, terrain elevations, cloud conditions, and other environmental variables. While over 90 percent of the reported bird strikes occur at or below 3,000 feet AGL, strikes at higher altitudes are common during migration. Ducks and geese are frequently observed up to 7,000 feet AGL and pilots are cautioned to minimize en route flying at lower altitudes during migration.

c. Considered the greatest potential hazard to aircraft because of their size, abundance, or habit of flying in dense flocks are gulls, waterfowl, vultures, hawks, owls, egrets, blackbirds, and starlings. Four major migratory flyways exist in the U.S. The Atlantic flyway parallels the Atlantic Coast. The Mississippi Flyway stretches from Canada through the Great Lakes and follows the Mississippi River. The Central Flyway represents a broad area east of the Rockies, stretching from Canada through Central America. The Pacific Flyway follows the west coast and overflies major parts of Washington, Oregon, and California. There are also numerous smaller flyways which cross these major north-south migratory routes.

7-4-2. Reducing Bird Strike Risks

a. The most serious strikes are those involving ingestion into an engine (turboprops and turbine jet engines) or windshield strikes. These strikes can result in emergency situations requiring prompt action by the pilot.

b. Engine ingestions may result in sudden loss of power or engine failure. Review engine out procedures, especially when operating from airports with known bird hazards or when operating near high bird concentrations.

c. Windshield strikes have resulted in pilots experiencing confusion, disorientation, loss of communications, and aircraft control problems. Pilots are encouraged to review their emergency procedures before flying in these areas.

d. When encountering birds en route, climb to avoid collision, because birds in flocks generally distribute themselves downward, with lead birds being at the highest altitude.

e. Avoid overflight of known areas of bird concentration and flying at low altitudes during bird migration. Charted wildlife refuges and other natural areas contain unusually high local concentration of birds which may create a hazard to aircraft.

7-4-3. Reporting Bird Strikes

Pilots are urged to report any bird or other wildlife strike using FAA Form 5200-7, Bird/Other Wildlife Strike Report (Appendix 1). Additional forms are available at any FSS; at any FAA Regional Office or at **http://wildlife-mitigation.tc.faa.gov**. The data derived from these reports are used to develop standards to cope with this potential hazard to aircraft and for documentation of necessary habitat control on airports.

7-4-4. Reporting Bird and Other Wildlife Activities

If you observe birds or other animals on or near the runway, request airport management to disperse the wildlife before taking off. Also contact the nearest FAA ARTCC, FSS, or tower (including non-Federal towers) regarding large flocks of birds and report the:

a. Geographic location.

b. Bird type (geese, ducks, gulls, etc.).

c. Approximate numbers.

d. Altitude.

e. Direction of bird flight path.

7-4-5. Pilot Advisories on Bird and Other Wildlife Hazards

Many airports advise pilots of other wildlife hazards caused by large animals on the runway through the A/FD and the NOTAM system. Collisions of landing and departing aircraft and animals on the runway are increasing and are not limited to rural airports. These accidents have also occurred at several major airports. Pilots should exercise extreme caution when warned of the presence of wildlife on and in the vicinity of airports. If you observe deer or other large animals in close proximity to movement areas, advise the FSS, tower, or airport management.

7-4-6. Flights Over Charted U.S. Wildlife Refuges, Parks, and Forest Service Areas

a. The landing of aircraft is prohibited on lands or waters administered by the National Park Service, U.S. Fish and Wildlife Service, or U.S. Forest Service without authorization from the respective agency. Exceptions include:

1. When forced to land due to an emergency beyond the control of the operator;

2. At officially designated landing sites; or

3. An approved official business of the Federal Government.

b. Pilots are requested to maintain a minimum altitude of 2,000 feet above the surface of the following: National Parks, Monuments, Seashores, Lakeshores, Recreation Areas and Scenic Riverways administered by the National Park Service, National Wildlife Refuges, Big Game Refuges, Game Ranges and Wildlife Ranges administered by the U.S. Fish and Wildlife Service, and Wilderness and Primitive areas administered by the U.S. Forest Service.

NOTE-*FAA Advisory Circular AC 91-36, Visual Flight Rules (VFR) Flight Near Noise-Sensitive Areas, defines the surface of a national park area (including parks, forests, primitive areas, wilderness areas, recreational areas, national seashores, national monuments, national lakeshores, and national wildlife refuge and range areas) as: the highest terrain within 2,000 feet laterally of the route of flight, or the upper-most rim of a canyon or valley.*

c. Federal statutes prohibit certain types of flight activity and/or provide altitude restrictions over designated U.S. Wildlife Refuges, Parks, and Forest Service Areas. These designated areas, for example: Boundary Waters Canoe Wilderness Areas, Minnesota; Haleakala National Park, Hawaii; Yosemite National Park, California; and Grand Canyon National Park, Arizona, are charted on Sectional Charts.

d. Federal regulations also prohibit airdrops by parachute or other means of persons, cargo, or objects from aircraft on lands administered by the three agencies without authorization from the respective agency. Exceptions include:

1. Emergencies involving the safety of human life; or

2. Threat of serious property loss.

Section 5. Potential Flight Hazards

7-5-1. Accident Cause Factors

a. The 10 most frequent cause factors for general aviation accidents that involve the pilot-in-command are:

1. Inadequate preflight preparation and/or planning.

2. Failure to obtain and/or maintain flying speed.

3. Failure to maintain direction control.

4. Improper level off.

5. Failure to see and avoid objects or obstructions.

6. Mismanagement of fuel.

7. Improper inflight decisions or planning.

8. Misjudgment of distance and speed.

9. Selection of unsuitable terrain.

10. Improper operation of flight controls.

b. This list remains relatively stable and points out the need for continued refresher training to establish a higher level of flight proficiency for all pilots. A part of the FAA's continuing effort to promote increased aviation safety is the Aviation Safety Program. For information on Aviation Safety Program activities contact your nearest Flight Standards District Office.

c. Alertness. Be alert at all times, especially when the weather is good. Most pilots pay attention to business when they are operating in full IFR weather conditions, but strangely, air collisions almost invariably have occurred under ideal weather conditions. Unlimited visibility appears to encourage a sense of security which is not at all justified. Considerable information of value may be obtained by listening to advisories being issued in the terminal area, even though controller workload may prevent a pilot from obtaining individual service.

d. Giving Way. If you think another aircraft is too close to you, give way instead of waiting for the other pilot to respect the right-of-way to which you may be entitled. It is a lot safer to pursue the right-of-way angle after you have completed your flight.

7-5-2. VFR in Congested Areas

A high percentage of near midair collisions occur below 8,000 feet AGL and within 30 miles of an airport. When operating VFR in these highly congested areas, whether you intend to land at an airport within the area or are just flying through, it is recommended that extra vigilance be maintained and that you monitor an appropriate control frequency. Normally the appropriate frequency is an approach control frequency. By such monitoring action you can "get the picture" of the traffic in your area. When the approach controller has radar, radar traffic advisories may be given to VFR pilots upon request.

REFERENCE-*AIM, Paragraph 4-1-15, Radar Traffic Information Service.*

7-5-3. Obstructions To Flight

a. General. Many structures exist that could significantly affect the safety of your flight when operating below 500 feet AGL, and particularly below 200 feet AGL. While 14 CFR Part 91.119 allows flight below 500 AGL when over sparsely populated areas or open water, such operations are very dangerous. At and below 200 feet AGL there are numerous power lines, antenna towers, etc., that are not marked and lighted as obstructions and; therefore, may not be seen in time to avoid a collision. Notices to Airmen (NOTAMs) are issued on those lighted structures experiencing temporary light outages. However, some time may pass before the FAA is notified of these outages, and the NOTAM issued, thus pilot vigilance is imperative.

b. Antenna Towers. Extreme caution should be exercised when flying less than 2,000 feet AGL because of numerous skeletal structures, such as radio and television antenna towers, that exceed 1,000 feet AGL with some extending higher than 2,000 feet AGL. Most skeletal structures are supported by guy wires which are very difficult to see in good weather and can be invisible at dusk or during periods of reduced visibility. These wires can extend about 1,500 feet horizontally from a structure; therefore, all skeletal structures should be avoided horizontally by at least 2,000 feet. Additionally, new towers may not be on your current chart because the information was not received prior to the printing of the chart.

c. Overhead Wires. Overhead transmission and utility lines often span approaches to runways, natural flyways such as lakes, rivers, gorges, and canyons, and cross other landmarks pilots frequently follow such as highways, railroad tracks, etc. As with antenna towers, these high voltage/power lines or the supporting structures of these lines may not always be readily visible and the wires may be virtually impossible to see under certain conditions. In some locations, the supporting structures of overhead transmission lines are equipped with unique sequence flashing white strobe light systems to indicate that there are wires between the structures. However, many power lines do not require notice to the FAA and, therefore, are not marked and/or lighted. Many of those that do require notice do not exceed 200 feet AGL or meet the Obstruction Standard of 14 CFR Part 77 and, therefore, are not marked and/or lighted. All pilots are cautioned to remain extremely vigilant for these power lines or their supporting structures when following natural flyways or during the approach and landing phase. This is particularly important for seaplane and/or float equipped aircraft when landing on, or departing from, unfamiliar lakes or rivers.

d. Other Objects/Structures. There are other objects or structures that could adversely affect your flight such as construction cranes near an airport, newly constructed buildings, new towers, etc. Many of these structures do not meet charting requirements or may not yet be charted because of the charting cycle. Some structures do not require obstruction marking and/or lighting and some may not be marked and lighted even though the FAA recommended it.

7-5-4. Avoid Flight Beneath Unmanned Balloons

a. The majority of unmanned free balloons currently being operated have, extending below them, either a suspension device to which the payload or instrument package is attached, or a trailing wire antenna, or both. In many instances these balloon subsystems may be invisible to the pilot until the aircraft is close to the balloon, thereby creating a potentially dangerous situation. Therefore, good judgment on the part of the pilot dictates that aircraft should remain well clear of all unmanned free balloons and flight below them should be avoided at all times.

b. Pilots are urged to report any unmanned free balloons sighted to the nearest FAA ground facility with which communication is established. Such information will assist FAA ATC facilities to identify and flight follow unmanned free balloons operating in the airspace.

7-5-5. Unmanned Aircraft Systems

a. Unmanned Aircraft Systems (UAS), formerly referred to as "Unmanned Aerial Vehicles" (UAVs) or "drones," are having an increasing operational presence in the NAS. Once the exclusive domain of the military, UAS are now being operated by various entities. Although these aircraft are "unmanned," UAS are flown by a remotely located pilot and crew. Physical and performance characteristics of unmanned aircraft (UA) vary greatly and unlike model aircraft that typically operate lower than 400 feet AGL, UA may be found operating at virtually any altitude and any speed. Sizes of UA can be as small as several pounds to as large as a commercial transport aircraft. UAS come in various categories including airplane, rotorcraft, powered-lift (tilt-rotor), and lighter-than-air. Propulsion systems of UAS include a broad range of alternatives from piston powered and turbojet engines to battery and solar-powered electric motors.

b. To ensure segregation of UAS operations from other aircraft, the military typically conducts UAS operations within restricted or other special use airspace. However, UAS operations are now being approved in the NAS outside of special use airspace through the use of FAA-issued Certificates of Waiver or Authorization (COA) or through the issuance of a special airworthiness certificate. COA and special airworthiness approvals authorize UAS flight operations to be contained within specific geographic boundaries and altitudes, usually require coordination with an ATC facility, and typically require the issuance of a NOTAM describing the operation to be conducted. UAS approvals also require observers to provide "see-and-avoid" capability to the UAS crew and to provide the necessary compliance with 14 CFR Section 91.113. For UAS operations approved at or above FL180, UAS operate under the same requirements as that of manned aircraft (i.e., flights are operated under instrument flight rules, are in communication with ATC, and are appropriately equipped).

c. UAS operations may be approved at either controlled or uncontrolled airports and are typically disseminated by NOTAM. In all cases, approved UAS operations must comply with all applicable regulations and/or special provisions specified in the COA or in the operating limitations of the special airworthiness certificate. At uncontrolled airports, UAS operations are advised to operate well clear of all known manned aircraft operations. Pilots of manned aircraft are advised to follow normal operating procedures and are urged to monitor the CTAF for any potential UAS activity. At controlled airports, local ATC procedures may be in place to handle UAS operations and should not require any special procedures from manned aircraft entering or departing the traffic pattern or operating in the vicinity of the airport.

d. In addition to approved UAS operations described above, a recently approved agreement between the FAA and the Department of Defense authorizes small UAS operations wholly contained within Class G airspace, and in no instance, greater than 1200 feet AGL over military owned or leased property. These operations do not require any special authorization as long as the UA remains within the lateral boundaries of the military installation as well as other provisions including the issuance of a NOTAM. Unlike special use airspace, these areas may not be depicted on an aeronautical chart.

e. There are several factors a pilot should consider regarding UAS activity in an effort to reduce potential flight hazards. Pilots are urged to exercise increased vigilance when operating in the vicinity of restricted or other special use airspace, military operations areas, and any military installation. Areas with a preponderance of UAS activity are typically noted on sectional charts advising pilots of this activity. Since the size of a UA can be very small, they may be difficult to see and track. If a UA is encountered during flight, as with manned aircraft, never assume that the pilot or crew of the UAS can see you, maintain increased vigilance with the UA and always be prepared for evasive action if necessary. Always check NOTAMs for potential UAS activity along the intended route of flight and exercise increased vigilance in areas specified in the NOTAM.

7-5-6. Mountain Flying

a. Your first experience of flying over mountainous terrain (particularly if most of your flight time has been over the flatlands of the midwest) could be a *never-to-be-forgotten nightmare* if proper planning is not done and if you are not aware of the potential hazards awaiting. Those familiar section lines are not present in the mountains;

those flat, level fields for forced landings are practically nonexistent; abrupt changes in wind direction and velocity occur; severe updrafts and downdrafts are common, particularly near or above abrupt changes of terrain such as cliffs or rugged areas; even the clouds look different and can build up with startling rapidity. Mountain flying need not be hazardous if you follow the recommendations below.

b. File a Flight Plan. Plan your route to avoid topography which would prevent a safe forced landing. The route should be over populated areas and well known mountain passes. Sufficient altitude should be maintained to permit gliding to a safe landing in the event of engine failure.

c. Don't fly a light aircraft when the winds aloft, at your proposed altitude, exceed 35 miles per hour. Expect the winds to be of much greater velocity over mountain passes than reported a few miles from them. Approach mountain passes with as much altitude as possible. Downdrafts of from 1,500 to 2,000 feet per minute are not uncommon on the leeward side.

d. Don't fly near or above abrupt changes in terrain. Severe turbulence can be expected, especially in high wind conditions.

e. Understand Mountain Obscuration. The term Mountain Obscuration (MTOS) is used to describe a visibility condition that is distinguished from IFR because ceilings, by definition, are described as "above ground level" (AGL). In mountainous terrain clouds can form at altitudes significantly higher than the weather reporting station and at the same time nearby mountaintops may be obscured by low visibility. In these areas the ground level can also vary greatly over a small area. Beware if operating VFR-on-top. You could be operating closer to the terrain than you think because the tops of mountains are hidden in a cloud deck below. MTOS areas are identified daily on The Aviation Weather Center located at: **http://www.aviationweather.gov**.

f. Some canyons run into a dead end. Don't fly so far up a canyon that you get trapped. ALWAYS BE ABLE TO MAKE A 180 DEGREE TURN!

g. VFR flight operations may be conducted at night in mountainous terrain with the application of sound judgment and common sense. Proper pre-flight planning, giving ample consideration to winds and weather, knowledge of the terrain and pilot experience in mountain flying are prerequisites for safety of flight. Continuous visual contact with the surface and obstructions is a major concern and flight operations under an overcast or in the vicinity of clouds should be approached with extreme caution.

h. When landing at a high altitude field, the same indicated airspeed should be used as at low elevation fields. *Remember:* that due to the less dense air at altitude, this same indicated airspeed actually results in higher true airspeed, a faster landing speed, and more important, a longer landing distance. During gusty wind conditions which often prevail at high altitude fields, a power approach and power landing is recommended. Additionally, due to the faster groundspeed, your takeoff distance will increase considerably over that required at low altitudes.

i. Effects of Density Altitude. Performance figures in the aircraft owner's handbook for length of takeoff run, horsepower, rate of climb, etc., are generally based on standard atmosphere conditions (59 degrees Fahrenheit (15 degrees Celsius), pressure 29.92 inches of mercury) at sea level. However, inexperienced pilots, as well as experienced pilots, may run into trouble when they encounter an altogether different set of conditions. This is particularly true in hot weather and at higher elevations. Aircraft operations at altitudes above sea level and at higher than standard temperatures are commonplace in mountainous areas. Such operations quite often result in a drastic reduction of aircraft performance capabilities because of the changing air density. Density altitude is a measure of air density. It is not to be confused with pressure altitude, true altitude or absolute altitude. It is not to be used as a height reference, but as a determining criteria in the performance capability of an aircraft. Air density decreases with altitude. As air density decreases, density altitude increases. The further effects of high temperature and high humidity are cumulative, resulting in an increasing high density altitude condition. High density altitude reduces all aircraft performance parameters. To the pilot, this means that the normal horsepower output is reduced, propeller efficiency is reduced and a higher true airspeed is required to sustain the aircraft throughout its operating parameters. It means an increase in runway length requirements for takeoff and landings, and decreased rate of climb. An average small airplane, for example, requiring 1,000 feet for takeoff at sea level under standard atmospheric conditions will require a takeoff run of approximately 2,000 feet at an operational altitude of 5,000 feet.

NOTE-*A turbo-charged aircraft engine provides some slight advantage in that it provides sea level horsepower up to a specified altitude above sea level.*

1. Density Altitude Advisories. At airports with elevations of 2,000 feet and higher, control towers and FSSs will broadcast the advisory "Check Density Altitude" when the temperature reaches a predetermined level. These advisories will be broadcast on appropriate tower frequencies or, where available, ATIS. FSSs will broadcast these advisories as a part of Local Airport Advisory, and on TWEB.

2. These advisories are provided by air traffic facilities, as a reminder to pilots that high temperatures and high field elevations will cause significant changes in aircraft characteristics. The pilot retains the responsibility to compute density altitude, when appropriate, as a part of preflight duties.

NOTE-*All FSSs will compute the current density altitude upon request.*

j. Mountain Wave. Many pilots go all their lives without understanding what a mountain wave is. Quite a few have lost their lives because of this lack of understanding. One need not be a licensed meteorologist to understand the mountain wave phenomenon.

1. Mountain waves occur when air is being blown over a mountain range or even the ridge of a sharp bluff area. As the air hits the upwind side of the range, it starts to climb, thus creating what is generally a smooth updraft which turns into a turbulent downdraft as the air passes the crest of the ridge. From this point, for many miles downwind, there will be a series of downdrafts and updrafts. Satellite photos of the Rockies have shown mountain waves extending as far as 700 miles downwind of the range. Along the east coast area, such photos of the Appalachian chain have picked up the mountain wave phenomenon over a hundred miles eastward. All it takes to form a mountain wave is wind blowing across the range at 15 knots or better at an intersection angle of not less than 30 degrees.

2. Pilots from flatland areas should understand a few things about mountain waves in order to stay out of trouble. When approaching a mountain range from the upwind side (generally the west), there will usually be a smooth updraft; therefore, it is not quite as dangerous an area as the lee of the range. From the leeward side, it is always a good idea to add an extra thousand feet or so of altitude because downdrafts can exceed the climb capability of the aircraft. Never expect an updraft when approaching a mountain chain from the leeward. Always be prepared to cope with a downdraft and turbulence.

3. When approaching a mountain ridge from the downwind side, it is recommended that the ridge be approached at approximately a 45 degree angle to the horizontal direction of the ridge. This permits a safer retreat from the ridge with less stress on the aircraft should severe turbulence and downdraft be experienced. If severe turbulence is encountered, simultaneously reduce power and adjust pitch until aircraft approaches maneuvering speed, then adjust power and trim to maintain maneuvering speed and fly away from the turbulent area.

7-5-7. Use of Runway Half-way Signs at Unimproved Airports

When installed, runway half-way signs provide the pilot with a reference point to judge takeoff acceleration trends. Assuming that the runway length is appropriate for takeoff (considering runway condition and slope, elevation, aircraft weight, wind, and temperature), typical takeoff acceleration should allow the airplane to reach 70 percent of lift-off airspeed by the midpoint of the runway. The "rule of thumb" is that should airplane acceleration not allow the airspeed to reach this value by the midpoint, the takeoff should be aborted, as it may not be possible to liftoff in the remaining runway.

Several points are important when considering using this "rule of thumb":

a. Airspeed indicators in small airplanes are not required to be evaluated at speeds below stalling, and may not be usable at 70 percent of liftoff airspeed.

b. This "rule of thumb" is based on a uniform surface condition. Puddles, soft spots, areas of tall and/or wet grass, loose gravel, etc., may impede acceleration or even cause deceleration. Even if the airplane achieves 70 percent of liftoff airspeed by the midpoint, the condition of the remainder of the runway may not allow further acceleration. The entire length of the runway should be inspected prior to takeoff to ensure a usable surface.

c. This "rule of thumb" applies only to runway required for actual liftoff. In the event that obstacles affect the takeoff climb path, appropriate distance must be available after liftoff to accelerate to best angle of climb speed and to clear the obstacles. This will, in effect, require the airplane to accelerate to a higher speed by midpoint, particularly if the obstacles are close to the end of the runway. In addition, this technique does not take into account the effects of upslope or tailwinds on takeoff performance. These factors will also require greater acceleration than normal and, under some circumstances, prevent takeoff entirely.

d. Use of this "rule of thumb" does not alleviate the pilot's responsibility to comply with applicable Federal Aviation Regulations, the limitations and performance data provided in the FAA approved Airplane Flight Manual (AFM), or, in the absence of an FAA approved AFM, other data provided by the aircraft manufacturer.

In addition to their use during takeoff, runway half-way signs offer the pilot increased awareness of his or her position along the runway during landing operations.

NOTE-*No FAA standard exists for the appearance of the runway half-way sign. FIG 7-5-1shows a graphical depiction of a typical runway half-way sign.*

7-5-8. Seaplane Safety

a. Acquiring a seaplane class rating affords access to many areas not available to landplane pilots. Adding a seaplane class rating to your pilot certificate can be relatively uncomplicated and inexpensive. However, more effort is required to become a safe, efficient, competent "bush" pilot. The natural hazards of the backwoods have given way to modern man-made hazards. Except for the far north, the available bodies of water are no longer the exclusive domain of the airman. Seaplane pilots must be vigilant for hazards such as electric power lines, power, sail and rowboats, rafts, mooring lines, water skiers, swimmers, etc.

FIG 7-5-1
Typical Runway Half-way Sign

b. Seaplane pilots must have a thorough understanding of the right-of-way rules as they apply to aircraft versus other vessels. Seaplane pilots are expected to know and adhere to both the U.S. Coast Guard's (USCG) Navigation Rules, International-Inland, and 14 CFR Section 91.115, Right-of-Way Rules; Water Operations. The navigation rules of the road are a set of collision avoidance rules as they apply to aircraft on the water. A seaplane is considered a vessel when on the water for the purposes of these collision avoidance rules. In general, a seaplane on the water must keep well clear of all vessels and avoid impeding their navigation. The CFR requires, in part, that aircraft operating on the water ". . . shall, insofar as possible, keep clear of all vessels and avoid impeding their navigation, and shall give way to any vessel or other aircraft that is given the right-of-way . . . ." This means that a seaplane should avoid boats and commercial shipping when on the water. If on a collision course, the seaplane should slow, stop, or maneuver to the right, away from the bow of the oncoming vessel. Also, while on the surface with an engine running, an aircraft must give way to all nonpowered vessels. Since a seaplane in the water may not be as maneuverable as one in the air, the aircraft on the water has right-of-way over one in the air, and one taking off has right-of-way over one landing. A seaplane is exempt from the USCG safety equipment requirements, including the requirements for Personal Flotation Devices (PFD). Requiring seaplanes on the water to comply with USCG equipment requirements in addition to the FAA equipment requirements would be an unnecessary burden on seaplane owners and operators.

c. Unless they are under Federal jurisdiction, navigable bodies of water are under the jurisdiction of the state, or in a few cases, privately owned. Unless they are specifically restricted, aircraft have as much right to operate on these bodies of water as other vessels. To avoid problems, check with Federal or local officials in advance of operating on unfamiliar waters. In addition to the agencies listed in TBL 7-5-1, the nearest Flight Standards District Office can usually offer some practical suggestions as well as regulatory information. If you land on a restricted body of water because of an inflight emergency, or in ignorance of the restrictions you have violated, report as quickly as practical to the nearest local official having jurisdiction and explain your situation.

d. When operating a seaplane over or into remote areas, appropriate attention should be given to survival gear. Minimum kits are recommended for summer and winter, and are required by law for flight into sparsely settled areas of Canada and Alaska. Alaska State Department of Transportation and Canadian Ministry of Transport officials can provide specific information on survival gear requirements. The kit should be assembled in one container and be easily reachable and preferably floatable.

e. The FAA recommends that each seaplane owner or operator provide flotation gear for occupants any time a seaplane operates on or near water. 14 CFR Section 91.205(b)(12) requires approved flotation gear for aircraft operated for hire over water and beyond power-off gliding distance from shore. FAA-approved gear differs from that required for navigable waterways under USCG rules. FAA-approved life vests are inflatable designs as compared to the USCG's noninflatable PFD's that may consist of solid, bulky material. Such USCG PFDs are impractical for seaplanes and other aircraft because they may block passage through the relatively narrow exits available to pilots and passengers. Life vests approved under Technical Standard Order (TSO) TSO-C13E contain fully inflatable compartments. The wearer inflates the compartments (AFTER exiting the aircraft) primarily by independent CO_2 cartridges, with an oral inflation tube as a backup. The flotation gear also contains a water-activated, self-illuminating signal light. The fact that pilots and passengers can easily don and wear inflatable life vests (when not inflated) provides maximum effectiveness and allows for unrestricted movement. It is imperative that passengers are briefed on the location and proper use of available PFDs prior to leaving the dock.

f. The FAA recommends that seaplane owners and operators obtain Advisory Circular (AC) 91-69, Seaplane Safety for 14 CFR Part 91 Operations, free from the U.S. Department of Transportation, Subsequent Distribution Office, SVC-121.23, Ardmore East Business Center, 3341 Q 75th Avenue, Landover, MD 20785; fax: (301) 386-5394. The USCG Navigation Rules International-Inland (COMDTINSTM 16672.2B) is available for a fee from the Government Printing Office by facsimile request to (202) 512-2250, and can be ordered using Mastercard or Visa.

7-5-9. Flight Operations in Volcanic Ash

a. Severe volcanic eruptions which send ash and sulphur dioxide (SO_2) gas into the upper atmosphere occur somewhere around the world several times each year. Flying into a volcanic ash cloud can be exceedingly dangerous. A B747-200 lost all four engines after such an encounter and a B747-400 had the same nearly catastrophic experience. Piston-powered aircraft are less likely to lose power but severe damage is almost certain to ensue after an encounter with a volcanic ash cloud which is only a few hours old.

b. Most important is to avoid any encounter with volcanic ash. The ash plume may not be visible, especially in instrument conditions or at night; and even if visible, it is difficult to distinguish visually between an ash cloud and an ordinary weather cloud. Volcanic ash clouds are not displayed on airborne or ATC radar. The pilot must rely on reports from air traffic controllers and other pilots to determine the location of the ash cloud and use that information to remain well clear of the area. Additionally, the presence of a sulphur-like odor throughout the cabin may indicate the presence of SO_2 emitted by volcanic activity, but may or may not indicate the presence of volcanic ash. Every attempt should be made to remain on the upwind side of the volcano.

c. It is recommended that pilots encountering an ash cloud should immediately reduce thrust to idle (altitude permitting), and reverse course in order to escape from the cloud. Ash clouds may extend for hundreds of miles and pilots should not attempt to fly through or climb out of the cloud. In addition, the following procedures are recommended:

1. Disengage the autothrottle if engaged. This will prevent the autothrottle from increasing engine thrust;

2. Turn on continuous ignition;

3. Turn on all accessory airbleeds including all air conditioning packs, nacelles, and wing anti-ice. This will provide an additional engine stall margin by reducing engine pressure.

d. The following has been reported by flightcrews who have experienced encounters with volcanic dust clouds:

1. Smoke or dust appearing in the cockpit.

2. An acrid odor similar to electrical smoke.

3. Multiple engine malfunctions, such as compressor stalls, increasing EGT, torching from tailpipe, and flameouts.

4. At night, St. Elmo's fire or other static discharges accompanied by a bright orange glow in the engine inlets.

5. A fire warning in the forward cargo area.

e. It may become necessary to shut down and then restart engines to prevent exceeding EGT limits. Volcanic ash may block the pitot system and result in unreliable airspeed indications.

f. If you see a volcanic eruption and have not been previously notified of it, you may have been the first person to observe it. In this case, immediately contact ATC and alert them to the existence of the eruption. If possible, use the Volcanic Activity Reporting form (VAR) depicted in Appendix 2 of this manual. Items 1 through 8 of the VAR should be transmitted immediately. The information requested in items 9 through 16 should be passed after landing. If a VAR form is not immediately available, relay enough information to identify the position and nature of the volcanic activity. Do not become unnecessarily alarmed if there is merely steam or very low-level eruptions of ash.

g. When landing at airports where volcanic ash has been deposited on the runway, be aware that even a thin layer of dry ash can be detrimental to braking action. Wet ash on the runway may also reduce effectiveness of braking. It is recommended that reverse thrust be limited to minimum practical to reduce the possibility of reduced visibility and engine ingestion of airborne ash.

h. When departing from airports where volcanic ash has been deposited, it is recommended that pilots avoid operating in visible airborne ash. Allow ash to settle before initiating takeoff roll. It is also recommended that flap extension be delayed until initiating the before takeoff checklist and that a rolling takeoff be executed to avoid blowing ash back into the air.

7-5-10. Emergency Airborne Inspection of Other Aircraft

a. Providing airborne assistance to another aircraft may involve flying in very close proximity to that aircraft. Most pilots receive little, if any, formal training or instruction in this type of flying activity. Close proximity flying without sufficient time to plan (i.e., in an emergency situation), coupled with the stress involved in a perceived emergency can be hazardous.

b. The pilot in the best position to assess the situation should take the responsibility of coordinating the airborne intercept and inspection, and take into account the unique flight characteristics and differences of the category(s) of aircraft involved.

c. Some of the safety considerations are:

1. Area, direction and speed of the intercept;

2. Aerodynamic effects (i.e., rotorcraft downwash);

3. Minimum safe separation distances;

4. Communications requirements, lost communications procedures, coordination with ATC;

5. Suitability of diverting the distressed aircraft to the nearest safe airport; and

6. Emergency actions to terminate the intercept.

d. Close proximity, inflight inspection of another aircraft is uniquely hazardous. The pilot-in-command of the aircraft experiencing the problem/emergency must not relinquish control of the situation and/or jeopardize the safety of their aircraft. The maneuver must be accomplished with minimum risk to both aircraft.

7-5-11. Precipitation Static

a. Precipitation static is caused by aircraft in flight coming in contact with uncharged particles. These particles can be rain, snow, fog, sleet, hail, volcanic ash, dust; any solid or liquid particles. When the aircraft strikes these neutral particles the positive element of the particle is reflected away from the aircraft and the negative particle adheres to the skin of the aircraft. In a very short period of time a substantial negative charge will develop on the skin of the aircraft. If the aircraft is not equipped with static dischargers, or has an ineffective static discharger system, when a sufficient negative voltage level is reached, the aircraft may go into "CORONA." That is, it will discharge the static electricity from the extremities of the aircraft, such as the wing tips, horizontal stabilizer, vertical stabilizer, antenna, propeller tips, etc. This discharge of static electricity is what you will hear in your headphones and is what we call P-static.

b. A review of pilot reports often shows different symptoms with each problem that is encountered. The following list of problems is a summary of many pilot reports from many different aircraft. Each problem was caused by P-static:

1. Complete loss of VHF communications.

2. Erroneous magnetic compass readings (30 percent in error).

3. High pitched squeal on audio.

4. Motor boat sound on audio.

5. Loss of all avionics in clouds.

6. VLF navigation system inoperative most of the time.

7. Erratic instrument readouts.

8. Weak transmissions and poor receptivity of radios.

9. "St. Elmo's Fire" on windshield.

c. Each of these symptoms is caused by one general problem on the airframe. This problem is the inability of the accumulated charge to flow easily to the wing tips and tail of the airframe, and properly discharge to the airstream.

d. Static dischargers work on the principal of creating a relatively easy path for discharging negative charges that develop on the aircraft by using a discharger with fine metal points, carbon coated rods, or carbon wicks rather than wait until a large charge is developed and discharged off the trailing edges of the aircraft that will interfere with avionics equipment. This process offers approximately 50 decibels (dB) static noise reduction which is adequate in most cases to be below the threshold of noise that would cause interference in avionics equipment.

TBL 7-5-1
Jurisdictions Controlling Navigable Bodies of Water

Authority to Consult For Use of a Body of Water		
Location	**Authority**	**Contact**
Wilderness Area	U.S. Department of Agriculture, Forest Service	Local forest ranger
National Forest	USDA Forest Service	Local forest ranger
National Park	U.S. Department of the Interior, National Park Service	Local park ranger
Indian Reservation	USDI, Bureau of Indian Affairs	Local Bureau office
State Park	State government or state forestry or park service	Local state aviation office for further information
Canadian National and Provincial Parks	Supervised and restricted on an individual basis from province to province and by different departments of the Canadian government; consult Canadian Flight Information Manual and/or Water Aerodrome Supplement	Park Superintendent in an emergency

e. It is important to remember that precipitation static problems can only be corrected with the proper number of quality static dischargers, properly installed on a properly bonded aircraft. P-static is indeed a problem in the all weather operation of the aircraft, but there are effective ways to combat it. All possible methods of reducing the effects of P-static should be considered so as to provide the best possible performance in the flight environment.

f. A wide variety of discharger designs is available on the commercial market. The inclusion of well-designed dischargers may be expected to improve airframe noise in P-static conditions by as much as 50 dB. Essentially, the discharger provides a path by which accumulated charge may leave the airframe quietly. This is generally accomplished by providing a group of tiny corona points to permit onset of corona-current flow at a low aircraft potential. Additionally, aerodynamic design of dischargers to permit corona to occur at the lowest possible atmospheric pressure also lowers the corona threshold. In addition to permitting a low-potential discharge, the discharger will minimize the radiation of radio frequency (RF) energy which accompanies the corona discharge, in order to minimize effects of RF components at communications and navigation frequencies on avionics performance. These effects are reduced through resistive attachment of the corona point(s) to the airframe, preserving direct current connection but attenuating the higher-frequency components of the discharge.

g. Each manufacturer of static dischargers offers information concerning appropriate discharger location on specific airframes. Such locations emphasize the trailing outboard surfaces of wings and horizontal tail surfaces, plus the tip of the vertical stabilizer, where charge tends to accumulate on the airframe. Sufficient dischargers must be provided to allow for current-carrying capacity which will maintain airframe potential below the corona threshold of the trailing edges.

h. In order to achieve full performance of avionic equipment, the static discharge system will require periodic maintenance. A pilot knowledgeable of P-static causes and effects is an important element in assuring optimum performance by early recognition of these types of problems.

7-5-12. Light Amplification by Stimulated Emission of Radiation (Laser) Operations and Reporting Illumination of Aircraft

a. Lasers have many applications. Of concern to users of the National Airspace System are those laser events that may affect pilots, e.g., outdoor laser light shows or demonstrations for entertainment and advertisements at special events and theme parks. Generally, the beams from these events appear as bright blue-green in color; however, they may be red, yellow, or white. However, some laser systems produce light which is invisible to the human eye.

b. FAA regulations prohibit the disruption of aviation activity by any person on the ground or in the air. The FAA and the Food and Drug Administration (the Federal agency that has the responsibility to enforce compliance with Federal requirements for laser systems and laser light show products) are working together to ensure that operators of these devices do not pose a hazard to aircraft operators.

c. Pilots should be aware that illumination from these laser operations are able to create temporary vision impairment miles from the actual location. In addition, these operations can produce permanent eye damage. Pilots should make themselves aware of where these activities are being conducted and avoid these areas if possible.

d. Recent and increasing incidents of unauthorized illumination of aircraft by lasers, as well as the proliferation and increasing sophistication of laser devices available to the general public, dictates that the FAA, in coordination with other government agencies, take action to safeguard flights from these unauthorized illuminations.

e. Pilots should report laser illumination activity to the controlling Air Traffic Control facilities, Federal Contract Towers or Flight Service Stations as soon as possible after the event. The following information should be included:

1. UTC Date and Time of Event.

2. Call Sign or Aircraft Registration Number.

3. Type Aircraft.

4. Nearest Major City.

5. Altitude.

6. Location of Event (Latitude/Longitude and/or Fixed Radial Distance (FRD)).

7. Brief Description of the Event and any other Pertinent Information.

f. Pilots are also encouraged to complete the Laser Beam Exposure Questionnaire located on the FAA Laser Safety Initiative website at **http://www.faa.gov/about/initiatives/lasers/** and submit electronically per the directions on the questionnaire, as soon as possible after landing.

g. When a laser event is reported to an air traffic facility, a general caution warning will be broadcasted on all appropriate frequencies every five minutes for 20 minutes and broadcasted on the ATIS for one hour following the report.

PHRASEOLOGY- *UNAUTHORIZED LASER ILLUMINATION EVENT, (UTC time), (location), (altitude), (color), (direction).*

EXAMPLE- *"Unauthorized laser illumination event, at 0100z, 8 mile final runway 18R at 3,000 feet, green laser from the southwest."*

REFERENCE- *FAAO 7110.65, Unauthorized Laser Illumination of Aircraft, Para 10-2-14.*

FAAO 7210.3, Reporting Laser Illumination of Aircraft, Para 2-1-27.

h. When these activities become known to the FAA, Notices to Airmen (NOTAMs) are issued to inform the aviation community of the events. Pilots should consult NOTAMs or the Special Notices section of the Airport/Facility Directory for information regarding these activities.

7-5-13. Flying in Flat Light and White Out Conditions

a. Flat Light. Flat light is an optical illusion, also known as "**sector or partial white out**." It is not as severe as "white out" but the condition causes pilots to lose their depth-of-field and contrast in vision. Flat light conditions are usually accompanied by overcast skies inhibiting any visual clues. Such conditions can occur anywhere in the world, primarily in snow covered areas but can occur in dust, sand, mud flats, or on glassy water. Flat light can completely obscure features of the terrain, creating an inability to distinguish distances and closure rates. As a result of this reflected light, it can give pilots the illusion that they are ascending or descending when they may actually be flying level. However, with good judgment and proper training and planning, it is possible to safely operate an aircraft in flat light conditions.

b. White Out. As defined in meteorological terms, white out occurs when a person becomes engulfed in a uniformly white glow. The glow is a result of being surrounded by blowing snow, dust, sand, mud or water. There are no shadows, no horizon or clouds and all depth-of-field and orientation are lost. A white out situation is severe in that there are no visual references. Flying is not recommended in any white out situation. Flat light conditions can lead to a white out environment quite rapidly, and both atmospheric conditions are insidious; they sneak up on you as your visual references slowly begin to disappear. White out has been the cause of several aviation accidents.

c. Self Induced White Out. This effect typically occurs when a helicopter takes off or lands on a snow-covered area. The rotor down wash picks up particles and re-circulates them through the rotor down wash. The effect can vary in intensity depending upon the amount of light on the surface. This can happen on the sunniest, brightest day with good contrast everywhere. However, when it happens, there can be a complete loss of visual clues. If the pilot has not prepared for this immediate loss of visibility, the results can be disastrous. Good planning does not prevent one from encountering flat light or white out conditions.

d. Never take off in a white out situation.

1. Realize that in flat light conditions it may be possible to depart but not to return to that site. During takeoff, make sure you have a reference point. Do not lose sight of it until you have a departure reference point in view. Be prepared to return to the takeoff reference if the departure reference does not come into view.

2. Flat light is common to snow skiers. One way to compensate for the lack of visual contrast and depth-of-field loss is by wearing amber tinted lenses (also known as blue blockers). Special note of caution: Eyewear is not ideal for every pilot. Take into consideration personal factors - age, light sensitivity, and ambient lighting conditions.

3. So what should a pilot do when all visual references are lost?

(a) Trust the cockpit instruments.

(b) Execute a 180 degree turnaround and start looking for outside references.

(c) Above all - fly the aircraft.

e. Landing in Low Light Conditions. When landing in a low light condition - use extreme caution. Look for intermediate reference points, in addition to checkpoints along each leg of the route for course confirmation and timing. The lower the ambient light becomes, the more reference points a pilot should use.

f. Airport Landings.

1. Look for features around the airport or approach path that can be used in determining depth perception. Buildings, towers, vehicles or other aircraft serve well for this measurement. Use something that will provide you with a sense of height above the ground, in addition to orienting you to the runway.

2. Be cautious of snowdrifts and snow banks - anything that can distinguish the edge of the runway. Look for subtle changes in snow texture or shading to identify ridges or changes in snow depth.

g. Off-Airport Landings.

1. In the event of an off-airport landing, pilots have used a number of different visual cues to gain reference. Use whatever you must to create the contrast you need. Natural references seem to work best (trees, rocks, snow ribs, etc.)

(a) Over flight.

(b) Use of markers.

(c) Weighted flags.

(d) Smoke bombs.

(e) Any colored rags.

(f) Dye markers.

(g) Kool-aid.

(h) Trees or tree branches.

2. It is difficult to determine the depth of snow in areas that are level. Dropping items from the aircraft to use as reference points should be used as a visual aid only and not as a primary landing reference. Unless your marker is biodegradable, be sure to retrieve it after landing. Never put yourself in a position where no visual references exist.

3. Abort landing if blowing snow obscures your reference. Make your decisions early. Don't assume you can pick up a lost reference point when you get closer.

4. Exercise extreme caution when flying from sunlight into shade. Physical awareness may tell you that you are flying straight but you may actually be in a spiral dive with centrifugal force pressing against you. Having no visual references enhances this illusion. Just because you have a good visual reference does not mean that it's safe to continue. There may be snow-covered terrain not visible in the direction that you are traveling. Getting caught in a no visual reference situation can be fatal.

h. Flying Around a Lake.

1. When flying along lakeshores, use them as a reference point. Even if you can see the other side, realize that your depth perception may be poor. It is easy to fly into the surface. If you must cross the lake, check the altimeter frequently and maintain a safe altitude while you still have a good reference. Don't descend below that altitude.

2. The same rules apply to seemingly flat areas of snow. If you don't have good references, avoid going there.

i. Other Traffic. Be on the look out for other traffic in the area. Other aircraft may be using your same reference point. Chances are greater of colliding with someone traveling in the same direction as you, than someone flying in the opposite direction.

j. Ceilings. Low ceilings have caught many pilots off guard. Clouds do not always form parallel to the surface, or at the same altitude. Pilots may try to compensate for this by flying with a slight bank and thus creating a descending turn.

k. Glaciers. Be conscious of your altitude when flying over glaciers. The glaciers may be rising faster than you are climbing.

7-5-14. Operations in Ground Icing Conditions

a. The presence of aircraft airframe icing during takeoff, typically caused by improper or no deicing of the aircraft being accomplished prior to flight has contributed to many recent accidents in turbine aircraft. The General Aviation Joint Steering Committee (GAJSC) is the primary vehicle for government-industry cooperation, communication, and coordination on GA accident mitigation. The Turbine Aircraft Operations Subgroup (TAOS) works to mitigate accidents in turbine accident aviation. While there is sufficient information and guidance currently available regarding the effects of icing on aircraft and methods for deicing, the TAOS has developed a list of recommended actions to further assist pilots and operators in this area.

While the efforts of the TAOS specifically focus on turbine aircraft, it is recognized that their recommendations are applicable to and can be adapted for the pilot of a small, piston powered aircraft too.

b. The following recommendations are offered:

1. Ensure that your aircraft's lift-generating surfaces are COMPLETELY free of contamination before flight through a tactile (hands on) check of the critical surfaces when feasible. Even when otherwise permitted, operators should avoid smooth or polished frost on lift-generating surfaces as an acceptable preflight condition.

2. Review and refresh your cold weather standard operating procedures.

3. Review and be familiar with the Airplane Flight Manual (AFM) limitations and procedures necessary to deal with icing conditions prior to flight, as well as in flight.

4. Protect your aircraft while on the ground, if possible, from sleet and freezing rain by taking advantage of aircraft hangars.

5. Take full advantage of the opportunities available at airports for deicing. Do not refuse deicing services simply because of cost.

6. Always consider canceling or delaying a flight if weather conditions do not support a safe operation.

c. If you haven't already developed a set of Standard Operating Procedures for cold weather operations, they should include:

1. Procedures based on information that is applicable to the aircraft operated, such as AFM limitations and procedures;

2. Concise and easy to understand guidance that outlines best operational practices;

3. A systematic procedure for recognizing, evaluating and addressing the associated icing risk, and offer clear guidance to mitigate this risk;

4. An aid (such as a checklist or reference cards) that is readily available during normal day-to-day aircraft operations.

d. There are several sources for guidance relating to airframe icing, including:

1. http://aircrafticing.grc.nasa.gov/index.html

2. http://www.ibac.org/is-bao/isbao.htm

3. http://www.natasafety1st.org/bus_deice.htm

4. Advisory Circular (AC) 91-74, Pilot Guide, Flight in Icing Conditions.

5. AC 135-17, Pilot Guide Small Aircraft Ground Deicing.

6. AC 135-9, FAR Part 135 Icing Limitations.

7. AC 120-60, Ground Deicing and Anti-icing Program.

8. AC 135-16, Ground Deicing and Anti-icing Training and Checking.

The FAA Approved Deicing Program Updates is published annually as a Flight Standards Information Bulletin for Air Transportation and contains detailed information on deicing and anti-icing procedures and holdover times. It may be accessed at the following web site by selecting the current year's information bulletins:

http://www.faa.gov/library/manuals/examiners_inspectors/8400/fsat

7-5-15. Avoid Flight in the Vicinity of Thermal Plumes (Smoke Stacks and Cooling Towers)

a. Flight Hazards Exist Around Thermal Plumes. Thermal plumes are defined as visible or invisible emissions from power plants, industrial production facilities, or other industrial systems that release large amounts of vertically directed unstable gases. High temperature exhaust plumes may cause significant air disturbances such as turbulence and vertical shear. Other identified potential hazards include, but are not necessarily limited to, reduced visibility, oxygen depletion, engine particulate contamination, exposure to gaseous oxides, and/or icing. Results of encountering a plume may include airframe damage, aircraft upset, and/or engine damage/failure. These hazards are most critical during low altitude flight, especially during takeoff and landing.

b. When able, a pilot should fly upwind of possible thermal plumes. When a plume is visible via smoke or a condensation cloud, remain clear and realize a plume may have both visible and invisible characteristics. Exhaust stacks without visible plumes may still be in full operation, and airspace in the vicinity should be treated with caution. As with mountain wave turbulence or clear air turbulence, an invisible plume may be encountered unexpectedly. Cooling towers, power plant stacks, exhaust fans, and other similar structures are depicted in FIG 7-5-2. Whether plumes are visible or invisible, the total extent of their unstable air is difficult to ascertain. FAA studies are underway to further characterize the effects of thermal plumes as exhaust effluents. Until the results of these studies are known and possible changes to rules and policy are identified and/or published, pilots are encouraged to exercise caution when flying in the vicinity of thermal plumes. Pilots are encouraged to reference the Airport/Facility Directory where amplifying notes may caution pilots and identify the location of structure(s) emitting thermal plumes.

FIG 7-5-2
Plumes

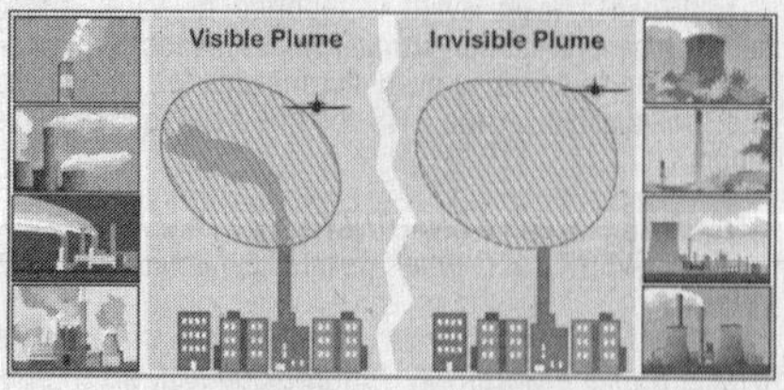

Section 6. Safety, Accident, and Hazard Reports

7-6-1. Aviation Safety Reporting Program

a. The FAA has established a voluntary Aviation Safety Reporting Program designed to stimulate the free and unrestricted flow of information concerning deficiencies and discrepancies in the aviation system. This is a positive program intended to ensure the safest possible system by identifying and correcting unsafe conditions before they lead to accidents. The primary objective of the program is to obtain information to evaluate and enhance the safety and efficiency of the present system.

b. This cooperative safety reporting program invites pilots, controllers, flight attendants, maintenance personnel and other users of the airspace system, or any other person, to file written reports of actual or potential discrepancies and deficiencies involving the safety of aviation operations. The operations covered by the program include departure, en route, approach, and landing operations and procedures, air traffic control procedures and equipment, crew and air traffic control communications, aircraft cabin operations, aircraft movement on the airport, near midair collisions, aircraft maintenance and record keeping and airport conditions or services.

c. The report should give the date, time, location, persons and aircraft involved (if applicable), nature of the event, and all pertinent details.

d. To ensure receipt of this information, the program provides for the waiver of certain disciplinary actions against persons, including pilots and air traffic controllers, who file timely written reports concerning potentially unsafe incidents. To be considered timely, reports must be delivered or postmarked within 10 days of the incident unless that period is extended for good cause. Reports should be submitted on NASA ARC Forms 277, which are available free of charge, postage prepaid, at FAA Flight Standards District Offices and Flight Service Stations, and from NASA, ASRS, PO Box 189, Moffet Field, CA 94035.

e. The FAA utilizes the National Aeronautics and Space Administration (NASA) to act as an independent third party to receive and analyze reports submitted under the program. This program is described in AC 00-46, Aviation Safety Reporting Program.

7-6-2. Aircraft Accident and Incident Reporting

a. Occurrences Requiring Notification. The operator of an aircraft must immediately, and by the most expeditious means available, notify the nearest National Transportation Safety Board (NTSB) Field Office when:

1. An aircraft accident or any of the following listed incidents occur:

(a) Flight control system malfunction or failure.

(b) Inability of any required flight crew member to perform their normal flight duties as a result of injury or illness.

(c) Failure of structural components of a turbine engine excluding compressor and turbine blades and vanes.

(d) Inflight fire.

(e) Aircraft collide in flight.

(f) Damage to property, other than the aircraft, estimated to exceed $25,000 for repair (including materials and labor) or fair market value in the event of total loss, whichever is less.

(g) For large multi-engine aircraft (more than 12,500 pounds maximum certificated takeoff weight):

(1) Inflight failure of electrical systems which requires the sustained use of an emergency bus powered by a back-up source such as a battery, auxiliary power unit, or air-driven generator to retain flight control or essential instruments;

(2) Inflight failure of hydraulic systems that results in sustained reliance on the sole remaining hydraulic or mechanical system for movement of flight control surfaces;

(3) Sustained loss of the power or thrust produced by two or more engines; and

(4) An evacuation of aircraft in which an emergency egress system is utilized.

2. An aircraft is overdue and is believed to have been involved in an accident.

b. Manner of Notification.

1. The most expeditious method of notification to the NTSB by the operator will be determined by the circumstances existing at that time. The NTSB has advised that any of the following would be considered examples of the type of notification that would be acceptable:

(a) Direct telephone notification.

(b) Telegraphic notification.

(c) Notification to the FAA who would in turn notify the NTSB by direct communication; i.e., dispatch or telephone.

c. Items to be Included in Notification. The notification required above must contain the following information, if available:

1. Type, nationality, and registration marks of the aircraft.

2. Name of owner and operator of the aircraft.

3. Name of the pilot-in-command.

4. Date and time of the accident, or incident.

5. Last point of departure, and point of intended landing of the aircraft.

6. Position of the aircraft with reference to some easily defined geographical point.

7. Number of persons aboard, number killed, and number seriously injured.

8. Nature of the accident, or incident, the weather, and the extent of damage to the aircraft so far as is known; and

9. A description of any explosives, radioactive materials, or other dangerous articles carried.

d. Follow-up Reports.

1. The operator must file a report on NTSB Form 6120.1 or 6120.2, available from NTSB Field Offices or from the NTSB, Washington, DC, 20594:

(a) Within 10 days after an accident;

(b) When, after 7 days, an overdue aircraft is still missing;

(c) A report on an incident for which notification is required as described in subparagraph a(1) must be filed only as requested by an authorized representative of the NTSB.

2. Each crewmember, if physically able at the time the report is submitted, must attach a statement setting forth the facts, conditions, and circumstances relating to the accident or incident as they appeared. If the crewmember is incapacitated, a statement must be submitted as soon as physically possible.

e. Where to File the Reports.

1. The operator of an aircraft must file with the NTSB Field Office nearest the accident or incident any report required by this section.

2. The NTSB Field Offices are listed under U.S. Government in the telephone directories in the following cities: Anchorage, AK; Atlanta, GA; Chicago, IL; Denver, CO; Fort Worth, TX; Los Angeles, CA; Miami, FL; Parsippany, NJ; Seattle, WA.

7-6-3. Near Midair Collision Reporting

a. Purpose and Data Uses. The primary purpose of the Near Midair Collision (NMAC) Reporting Program is to provide information for use in enhancing the safety and efficiency of the National Airspace System. Data obtained from NMAC reports are used by the FAA to improve the quality of FAA services to users and to develop programs, policies, and procedures aimed at the reduction of NMAC occurrences. All NMAC reports are thoroughly investigated by Flight Standards Facilities in coordination with Air Traffic Facilities. Data from these investigations are transmitted to FAA Headquarters in Washington, DC, where they are compiled and analyzed, and where safety programs and recommendations are developed.

b. Definition. A near midair collision is defined as an incident associated with the operation of an aircraft in which a possibility of collision occurs as a result of proximity of less than 500 feet to another aircraft, or a report is received from a pilot or a flight crew member stating that a collision hazard existed between two or more aircraft.

c. Reporting Responsibility. It is the responsibility of the pilot and/or flight crew to determine whether a near midair collision did actually occur and, if so, to initiate a NMAC report. Be specific, as ATC will not interpret a casual remark to mean that a NMAC is being reported. The pilot should state "I wish to report a near midair collision."

d. Where to File Reports. Pilots and/or flight crew members involved in NMAC occurrences are urged to report each incident immediately:

1. By radio or telephone to the nearest FAA ATC facility or FSS.

2. In writing, in lieu of the above, to the nearest Flight Standards District Office (FSDO).

e. Items to be Reported.

1. Date and time (UTC) of incident.

2. Location of incident and altitude.

3. Identification and type of reporting aircraft, aircrew destination, name and home base of pilot.

4. Identification and type of other aircraft, aircrew destination, name and home base of pilot.

5. Type of flight plans; station altimeter setting used.

6. Detailed weather conditions at altitude or flight level.

7. Approximate courses of both aircraft: indicate if one or both aircraft were climbing or descending.

8. Reported separation in distance at first sighting, proximity at closest point horizontally and vertically, and length of time in sight prior to evasive action.

9. Degree of evasive action taken, if any (from both aircraft, if possible).

10. Injuries, if any.

f. Investigation. The FSDO in whose area the incident occurred is responsible for the investigation and reporting of NMACs.

g. Existing radar, communication, and weather data will be examined in the conduct of the investigation. When possible, all cockpit crew members will be interviewed regarding factors involving the NMAC incident. Air traffic controllers will be interviewed in cases where one or more of the involved aircraft was provided ATC service. Both flight and ATC procedures will be evaluated. When the investigation reveals a violation of an FAA regulation, enforcement action will be pursued.

7-6-4. Unidentified Flying Object (UFO) Reports

a. Persons wanting to report UFO/unexplained phenomena activity should contact a UFO/unexplained phenomena reporting data collection center, such as Bigelow Aerospace Advanced Space Studies (BAASS) (voice: 1-877-979-7444 or e-mail: Reporting@baass.org), the National UFO Reporting Center, etc.

b. If concern is expressed that life or property might be endangered, report the activity to the local law enforcement department.

7-6-5. Safety Alerts For Operators (SAFO) and Information For Operators (InFO)

a. SAFOs contain important safety information that is often time-critical. A SAFO may contain information and/or recommended (non-regulatory) action to be taken by the respective operators or parties identified in the SAFO. The audience for SAFOs varies with each subject and may include: Air carrier certificate holders, air operator certificate holders, general aviation operators, directors of safety, directors of operations, directors of maintenance, fractional ownership program managers, training center managers, accountable managers at repair stations, and other parties as applicable.

b. InFOs are similar to SAFOs, but contain valuable information for operators that should help them meet administrative requirements or certain regulatory requirements with relatively low urgency or impact in safety.

c. The SAFO and InFO system provides a means to rapidly distribute this information to operators and can be found at:

http://www.faa.gov/other_visit/aviation_industry/airline_operators/airline_safety/safoand

http://www.faa.gov/other_visit/aviation_industry/airline_operators/airline_safety/info

or search keyword FAA SAFO or FAA INFO. Free electronic subscription is available on the "ALL SAFOs" or "ALL InFOs" page of the website.

Chapter 8.

MEDICAL FACTS FOR PILOTS

Section 1. Fitness for Flight

8-1-1. Fitness For Flight

a. Medical Certification.

1. All pilots except those flying gliders and free air balloons must possess valid medical certificates in order to exercise the privileges of their airman certificates. The periodic medical examinations required for medical certification are conducted by designated Aviation Medical Examiners, who are physicians with a special interest in aviation safety and training in aviation medicine.

2. The standards for medical certification are contained in 14 CFR Part 67. Pilots who have a history of certain medical conditions described in these standards are mandatorily disqualified from flying. These medical conditions include a personality disorder manifested by overt acts, a psychosis, alcoholism, drug dependence, epilepsy, an unexplained disturbance of consciousness, myocardial infarction, angina pectoris and diabetes requiring medication for its control. Other medical conditions may be temporarily disqualifying, such as acute infections, anemia, and peptic ulcer. Pilots who do not meet medical standards may still be qualified under special issuance provisions or the exemption process. This may require that either additional medical information be provided or practical flight tests be conducted.

3. Student pilots should visit an Aviation Medical Examiner as soon as possible in their flight training in order to avoid unnecessary training expenses should they not meet the medical standards. For the same reason, the student pilot who plans to enter commercial aviation should apply for the highest class of medical certificate that might be necessary in the pilot's career.

CAUTION-The CFRs prohibit a pilot who possesses a current medical certificate from performing crewmember duties while the pilot has a known medical condition or increase of a known medical condition that would make the pilot unable to meet the standards for the medical certificate.

b. Illness.

1. Even a minor illness suffered in day-to-day living can seriously degrade performance of many piloting tasks vital to safe flight. Illness can produce fever and distracting symptoms that can impair judgment, memory, alertness, and the ability to make calculations. Although symptoms from an illness may be under adequate control with a medication, the medication itself may decrease pilot performance.

2. The safest rule is not to fly while suffering from any illness. If this rule is considered too stringent for a particular illness, the pilot should contact an Aviation Medical Examiner for advice.

c. Medication.

1. Pilot performance can be seriously degraded by both prescribed and over-the-counter medications, as well as by the medical conditions for which they are taken. Many medications, such as tranquilizers, sedatives, strong pain relievers, and cough-suppressant preparations, have primary effects that may impair judgment, memory, alertness, coordination, vision, and the ability to make calculations. Others, such as antihistamines, blood pressure drugs, muscle relaxants, and agents to control diarrhea and motion sickness, have side effects that may impair the same critical functions. Any medication that depresses the nervous system, such as a sedative, tranquilizer or antihistamine, can make a pilot much more susceptible to hypoxia.

2. The CFRs prohibit pilots from performing crewmember duties while using any medication that affects the faculties in any way contrary to safety. The safest rule is not to fly as a crewmember while taking any medication, unless approved to do so by the FAA.

d. Alcohol.

1. Extensive research has provided a number of facts about the hazards of alcohol consumption and flying. As little as one ounce of liquor, one bottle of beer or four ounces of wine can impair flying skills, with the alcohol consumed in these drinks being detectable in the breath and blood for at least 3 hours. Even after the body completely destroys a moderate amount of alcohol, a pilot can still be severely impaired for many hours by hangover. There is simply no way of increasing the destruction of alcohol or alleviating a hangover. Alcohol also renders a pilot much more susceptible to disorientation and hypoxia.

2. A consistently high alcohol related fatal aircraft accident rate serves to emphasize that alcohol and flying are a potentially lethal combination. The CFRs prohibit pilots from performing crewmember duties within 8 hours after drinking any alcoholic beverage or while under the influence of alcohol. However, due to the slow destruction of alcohol, a pilot may still be under influence 8 hours after drinking a moderate amount of alcohol. Therefore, an excellent rule is to allow at least 12 to 24 hours between "bottle and throttle," depending on the amount of alcoholic beverage consumed.

e. Fatigue.

1. Fatigue continues to be one of the most treacherous hazards to flight safety, as it may not be apparent to a pilot until serious errors are made. Fatigue is best described as either acute (short-term) or chronic (long-term).

2. A normal occurrence of everyday living, acute fatigue is the tiredness felt after long periods of physical and mental strain, including strenuous muscular effort, immobility, heavy mental workload, strong emotional pressure, monotony, and lack of sleep. Consequently, coordination and alertness, so vital to safe pilot performance, can be reduced. Acute fatigue is prevented by adequate rest and sleep, as well as by regular exercise and proper nutrition.

3. Chronic fatigue occurs when there is not enough time for full recovery between episodes of acute fatigue. Performance continues to fall off, and judgment becomes impaired so that unwarranted risks may be taken. Recovery from chronic fatigue requires a prolonged period of rest.

4. OBSTRUCTIVE SLEEP APNEA (OSA). OSA is now recognized as an important preventable factor identified in transportation accidents. OSA interrupts the normal restorative sleep necessary for normal functioning and is associated with chronic illnesses such as hypertension, heart attack, stroke, obesity, and diabetes. Symptoms include snoring, excessive daytime sleepiness, intermittent prolonged breathing pauses while sleeping, memory impairment and lack of concentration. There are many available treatments which can reverse the day time symptoms and reduce the chance of an accident. OSA can be easily treated. Most treatments are acceptable for medical certification upon demonstrating effective treatment. If you have any symptoms described above, or neck size over 17 inches in men or 16 inches in women, or a body mass index greater than 30 you should be evaluated for sleep apnea by a sleep medicine specialist.

(**http://www.cdc.gov/healthyweight/assessing/bmi/adult_bmi/english_bmi_calculator/bmi_calculator.html**) With treatment you can avoid or delay the onset of these chronic illnesses and prolong a quality life.

f. Stress.

1. Stress from the pressures of everyday living can impair pilot performance, often in very subtle ways. Difficulties, particularly at work, can occupy thought processes enough to markedly decrease alertness. Distraction can so interfere with judgment that unwarranted risks are taken, such as flying into deteriorating weather conditions to keep on schedule. Stress and fatigue (see above) can be an extremely hazardous combination.

2. Most pilots do not leave stress "on the ground." Therefore, when more than usual difficulties are being experienced, a pilot should consider delaying flight until these difficulties are satisfactorily resolved.

g. Emotion.

Certain emotionally upsetting events, including a serious argument, death of a family member, separation or divorce, loss of job, and financial catastrophe, can render a pilot unable to fly an aircraft safely. The emotions of anger, depression, and anxiety from such events not only decrease alertness but also may lead to taking risks that border on self-destruction. Any pilot who experiences an emotionally upsetting event should not fly until satisfactorily recovered from it.

h. Personal Checklist. Aircraft accident statistics show that pilots should be conducting preflight checklists on themselves as well as their aircraft for pilot impairment contributes to many more accidents than failures of aircraft systems. A personal checklist, which includes all of the categories of pilot impairment as discussed in this section, that can be easily committed to memory is being distributed by the FAA in the form of a wallet-sized card.

i. PERSONAL CHECKLIST. ***I'm physically and mentally safe to fly; not being impaired by:***

Illness
Medication
Stress
Alcohol
Fatigue
Emotion

8-1-2. Effects of Altitude

a. Hypoxia.

1. Hypoxia is a state of oxygen deficiency in the body sufficient to impair functions of the brain and other organs. Hypoxia from exposure to altitude is due only to the reduced barometric pressures encountered at altitude, for the concentration of oxygen in the atmosphere remains about 21 percent from the ground out to space.

2. Although a deterioration in night vision occurs at a cabin pressure altitude as low as 5,000 feet, other significant effects of altitude hypoxia usually do not occur in the normal healthy pilot below 12,000 feet. From 12,000 to 15,000 feet of altitude, judgment, memory, alertness, coordination and ability to make calculations are impaired, and headache, drowsiness, dizziness and either a sense of well-being (euphoria) or belligerence occur. The effects appear following increasingly shorter periods of exposure to increasing altitude. In fact, pilot performance can seriously deteriorate within 15 minutes at 15,000 feet.

3. At cabin pressure altitudes above 15,000 feet, the periphery of the visual field grays out to a point where only central vision remains (tunnel vision). A blue coloration (cyanosis) of the fingernails and lips develops. The ability to take corrective and protective action is lost in 20 to 30 minutes at 18,000 feet and 5 to 12 minutes at 20,000 feet, followed soon thereafter by unconsciousness.

4. The altitude at which significant effects of hypoxia occur can be lowered by a number of factors. Carbon monoxide inhaled in smoking or from exhaust fumes, lowered hemoglobin (anemia), and certain medications can reduce the oxygen-carrying capacity of the blood to the degree that the amount of oxygen provided to body tissues will already be equivalent to the oxygen provided to the tissues when exposed to a cabin pressure altitude of several thousand feet. Small amounts of alcohol and low doses of certain drugs, such as antihistamines, tranquilizers, sedatives and analgesics can, through their depressant action, render the brain much more susceptible to hypoxia. Extreme heat and cold, fever, and anxiety increase the body's demand for oxygen, and hence its susceptibility to hypoxia.

5. The effects of hypoxia are usually quite difficult to recognize, especially when they occur gradually. Since symptoms of hypoxia do not vary in an individual, the ability to recognize hypoxia can be greatly improved by experiencing and witnessing the effects of hypoxia during an altitude chamber "flight." The FAA provides this opportunity through aviation physiology training, which is conducted at the FAA Civil Aeromedical Institute and at many military facilities across the U.S. To attend the Physiological Training Program at the Civil Aeromedical Institute, Mike Monroney Aeronautical Center, Oklahoma City, OK, contact by telephone (405) 954-6212, or by writing Aerospace Medical Education Division, AAM-400, CAMI, Mike Monroney Aeronautical Center, P.O. Box 25082, Oklahoma City, OK 73125.

NOTE-*To attend the physiological training program at one of the military installations having the training capa-*

bility, an application form and a fee must be submitted. Full particulars about location, fees, scheduling procedures, course content, individual requirements, etc., are contained in the Physiological Training Application, Form Number AC 3150-7, which is obtained by contacting the accident prevention specialist or the office forms manager in the nearest FAA office.

6. Hypoxia is prevented by heeding factors that reduce tolerance to altitude, by enriching the inspired air with oxygen from an appropriate oxygen system, and by maintaining a comfortable, safe cabin pressure altitude. For optimum protection, pilots are encouraged to use supplemental oxygen above 10,000 feet during the day, and above 5,000 feet at night. The CFRs require that at the minimum, flight crew be provided with and use supplemental oxygen after 30 minutes of exposure to cabin pressure altitudes between 12,500 and 14,000 feet and immediately on exposure to cabin pressure altitudes above 14,000 feet. Every occupant of the aircraft must be provided with supplemental oxygen at cabin pressure altitudes above 15,000 feet.

b. Ear Block.

1. As the aircraft cabin pressure decreases during ascent, the expanding air in the middle ear pushes the eustachian tube open, and by escaping down it to the nasal passages, equalizes in pressure with the cabin pressure. But during descent, the pilot must periodically open the eustachian tube to equalize pressure. This can be accomplished by swallowing, yawning, tensing muscles in the throat, or if these do not work, by a combination of closing the mouth, pinching the nose closed, and attempting to blow through the nostrils (Valsalva maneuver).

2. Either an upper respiratory infection, such as a cold or sore throat, or a nasal allergic condition can produce enough congestion around the eustachian tube to make equalization difficult. Consequently, the difference in pressure between the middle ear and aircraft cabin can build up to a level that will hold the eustachian tube closed, making equalization difficult if not impossible. The problem is commonly referred to as an "ear block."

3. An ear block produces severe ear pain and loss of hearing that can last from several hours to several days. Rupture of the ear drum can occur in flight or after landing. Fluid can accumulate in the middle ear and become infected.

4. An ear block is prevented by not flying with an upper respiratory infection or nasal allergic condition. Adequate protection is usually not provided by decongestant sprays or drops to reduce congestion around the eustachian tubes. Oral decongestants have side effects that can significantly impair pilot performance.

5. If an ear block does not clear shortly after landing, a physician should be consulted.

c. Sinus Block.

1. During ascent and descent, air pressure in the sinuses equalizes with the aircraft cabin pressure through small openings that connect the sinuses to the nasal passages. Either an upper respiratory infection, such as a cold or sinusitis, or a nasal allergic condition can produce enough congestion around an opening to slow equalization, and as the difference in pressure between the sinus and cabin mounts, eventually plug the opening. This "sinus block" occurs most frequently during descent.

2. A sinus block can occur in the frontal sinuses, located above each eyebrow, or in the maxillary sinuses, located in each upper cheek. It will usually produce excruciating pain over the sinus area. A maxillary sinus block can also make the upper teeth ache. Bloody mucus may discharge from the nasal passages.

3. A sinus block is prevented by not flying with an upper respiratory infection or nasal allergic condition. Adequate protection is usually not provided by decongestant sprays or drops to reduce congestion around the sinus openings. Oral decongestants have side effects that can impair pilot performance.

4. If a sinus block does not clear shortly after landing, a physician should be consulted.

d. Decompression Sickness After Scuba Diving.

1. A pilot or passenger who intends to fly after scuba diving should allow the body sufficient time to rid itself of excess nitrogen absorbed during diving. If not, decompression sickness due to evolved gas can occur during exposure to low altitude and create a serious inflight emergency.

2. The recommended waiting time before going to flight altitudes of up to 8,000 feet is at least 12 hours after diving which has not required controlled ascent (nondecompression stop diving), and at least 24 hours after diving which has required controlled ascent (decompression stop diving). The waiting time before going to flight altitudes above 8,000 feet should be at least 24 hours after any SCUBA dive. These recommended altitudes are actual flight altitudes above mean sea level (AMSL) and not pressurized cabin altitudes. This takes into consideration the risk of decompression of the aircraft during flight.

8-1-3. Hyperventilation in Flight

a. Hyperventilation, or an abnormal increase in the volume of air breathed in and out of the lungs, can occur subconsciously when a stressful situation is encountered in flight. As hyperventilation "blows off" excessive carbon dioxide from the body, a pilot can experience symptoms of lightheadedness, suffocation, drowsiness, tingling in the extremities, and coolness and react to them with even greater hyperventilation. Incapacitation can eventually result from incoordination, disorientation, and painful muscle spasms. Finally, unconsciousness can occur.

b. The symptoms of hyperventilation subside within a few minutes after the rate and depth of breathing are consciously brought back under control. The buildup of carbon dioxide in the body can be hastened by controlled breathing in and out of a paper bag held over the nose and mouth.

c. Early symptoms of hyperventilation and hypoxia are similar. Moreover, hyperventilation and hypoxia can occur at the same time. Therefore, if a pilot is using an oxygen system when symptoms are experienced, the oxygen regulator should immediately be set to deliver 100 percent oxygen, and then the system checked to assure that it has been functioning effectively before giving attention to rate and depth of breathing.

8-1-4. Carbon Monoxide Poisoning in Flight

a. Carbon monoxide is a colorless, odorless, and tasteless gas contained in exhaust fumes. When breathed even in minute quantities over a period of time, it can significantly reduce the ability of the blood to carry oxygen. Consequently, effects of hypoxia occur.

b. Most heaters in light aircraft work by air flowing over the manifold. Use of these heaters while exhaust fumes are escaping through manifold cracks and seals is responsible every year for several nonfatal and fatal aircraft accidents from carbon monoxide poisoning.

c. A pilot who detects the odor of exhaust or experiences symptoms of headache, drowsiness, or dizziness while using the heater should suspect carbon monoxide poisoning, and immediately shut off the heater and open air vents. If symptoms are severe or continue after landing, medical treatment should be sought.

8-1-5. Illusions in Flight

a. Introduction. Many different illusions can be experienced in flight. Some can lead to spatial disorientation. Others can lead to landing errors. Illusions rank among the most common factors cited as contributing to fatal aircraft accidents.

b. Illusions Leading to Spatial Disorientation.

1. Various complex motions and forces and certain visual scenes encountered in flight can create illusions of motion and position. Spatial disorientation from these illusions can be prevented only by visual reference to reliable, fixed points on the ground or to flight instruments.

2. The leans. An abrupt correction of a banked attitude, which has been entered too slowly to stimulate the motion sensing system in the inner ear, can create the illusion of banking in the opposite direction. The disoriented pilot will roll the aircraft back into its original dangerous attitude, or if level flight is maintained, will feel compelled to lean in the perceived vertical plane until this illusion subsides.

(a) Coriolis illusion. An abrupt head movement in a prolonged constant-rate turn that has ceased stimulating the motion sensing system can create the illusion of rotation or movement in an entirely different axis. The disoriented pilot will maneuver the aircraft into a dangerous attitude in an attempt to stop rotation. This most overwhelming of all illusions in flight may be prevented by not making sudden, extreme head movements, particularly while making prolonged constant-rate turns under IFR conditions.

(b) Graveyard spin. A proper recovery from a spin that has ceased stimulating the motion sensing system can create the illusion of spinning in the opposite direction. The disoriented pilot will return the aircraft to its original spin.

(c) Graveyard spiral. An observed loss of altitude during a coordinated constant-rate turn that has ceased stimulating the motion sensing system can create the illusion of being in a descent with the wings level. The disoriented pilot will pull back on the controls, tightening the spiral and increasing the loss of altitude.

(d) Somatogravic illusion. A rapid acceleration during takeoff can create the illusion of being in a nose up attitude. The disoriented pilot will push the aircraft into a nose low, or dive attitude. A rapid deceleration by a quick reduction of the throttles can have the opposite effect, with the disoriented pilot pulling the aircraft into a nose up, or stall attitude.

(e) Inversion illusion. An abrupt change from climb to straight and level flight can create the illusion of tumbling backwards. The disoriented pilot will push the aircraft abruptly into a nose low attitude, possibly intensifying this illusion.

(f) Elevator illusion. An abrupt upward vertical acceleration, usually by an updraft, can create the illusion of being in a climb. The disoriented pilot will push the aircraft into a nose low attitude. An abrupt downward vertical acceleration, usually by a downdraft, has the opposite effect, with the disoriented pilot pulling the aircraft into a nose up attitude.

(g) False horizon. Sloping cloud formations, an obscured horizon, a dark scene spread with ground lights and stars, and certain geometric patterns of ground light can create illusions of not being aligned correctly with the actual horizon. The disoriented pilot will place the aircraft in a dangerous attitude.

(h) Autokinesis. In the dark, a static light will appear to move about when stared at for many seconds. The disoriented pilot will lose control of the aircraft in attempting to align it with the light.

3. Illusions Leading to Landing Errors.

(a) Various surface features and atmospheric conditions encountered in landing can create illusions of incorrect height above and distance from the runway threshold. Landing errors from these illusions can be prevented by anticipating them during approaches, aerial visual inspection of unfamiliar airports before landing, using electronic glide slope or VASI systems when available, and maintaining optimum proficiency in landing procedures.

(b) Runway width illusion. A narrower-than-usual runway can create the illusion that the aircraft is at a higher altitude than it actually is. The pilot who does not recognize this illusion will fly a lower approach, with the risk of striking objects along the approach path or landing short. A wider-than-usual runway can have the opposite effect, with the risk of leveling out high and landing hard or overshooting the runway.

(c) Runway and terrain slopes illusion. An upsloping runway, upsloping terrain, or both, can create the illusion that the aircraft is at a higher altitude than it actually is. The pilot who does not recognize this illusion will fly a lower approach. A downsloping runway, downsloping approach terrain, or both, can have the opposite effect.

(d) Featureless terrain illusion. An absence of ground features, as when landing over water, darkened areas, and terrain made featureless by snow, can create the illusion that the aircraft is at a higher altitude than it actually is. The pilot who does not recognize this illusion will fly a lower approach.

(e) Atmospheric illusions. Rain on the windscreen can create the illusion of greater height, and atmospheric haze the illusion of being at a greater distance from the runway. The pilot who does not recognize these illusions will fly a lower approach. Penetration of fog can create the illusion of pitching up. The pilot who does not recognize this illusion will steepen the approach, often quite abruptly.

(f) Ground lighting illusions. Lights along a straight path, such as a road, and even lights on moving trains can be mistaken for runway and approach lights. Bright runway and approach lighting systems, especially where few lights illuminate the surrounding terrain, may create the illusion of less distance to the runway. The pilot who does not recognize this illusion will fly a higher approach. Conversely, the pilot overflying terrain which has few lights to provide height cues may make a lower than normal approach.

8-1-6. Vision in Flight

a. Introduction. Of the body senses, vision is the most important for safe flight. Major factors that determine how effectively vision can be used are the level of illumination and the technique of scanning the sky for other aircraft.

b. Vision Under Dim and Bright Illumination.

1. Under conditions of dim illumination, small print and colors on aeronautical charts and aircraft instruments become unreadable unless adequate cockpit lighting is available. Moreover, another aircraft must be much closer to be seen unless its navigation lights are on.

2. In darkness, vision becomes more sensitive to light, a process called dark adaptation. Although exposure to total darkness for at least 30 minutes is required for complete dark adaptation, a pilot can achieve a moderate degree of dark adaptation within 20 minutes under dim red cockpit lighting. Since red light severely distorts colors, especially on aeronautical charts, and can cause serious difficulty in focusing the eyes on objects inside the aircraft, its use is advisable only where optimum outside night vision capability is necessary. Even so, white cockpit lighting must be available when needed for map and instrument reading, especially under IFR conditions. Dark adaptation is impaired by exposure to cabin pressure altitudes above 5,000 feet, carbon monoxide inhaled in smoking and from exhaust fumes, deficiency of Vitamin A in the diet, and by prolonged exposure to bright sunlight. Since any degree of dark adaptation is lost within a few seconds of viewing a bright light, a pilot should close one eye when using a light to preserve some degree of night vision.

3. Excessive illumination, especially from light reflected off the canopy, surfaces inside the aircraft, clouds, water, snow, and desert terrain, can produce glare, with uncomfortable squinting, watering of the eyes, and even temporary blindness. Sunglasses for protection from glare should absorb at least 85 percent of visible light (15 percent transmittance) and all colors equally (neutral transmittance), with negligible image distortion from refractive and prismatic errors.

c. Scanning for Other Aircraft.

1. Scanning the sky for other aircraft is a key factor in collision avoidance. It should be used continuously by the pilot and copilot (or right seat passenger) to cover all areas of the sky visible from the cockpit. Although pilots must meet specific visual acuity requirements, the ability to read an eye chart does not ensure that one will be able to efficiently spot other aircraft. Pilots must develop an effective scanning technique which maximizes one's visual capabilities. The probability of spotting a potential collision threat obviously increases with the time spent looking outside the cockpit. Thus, one must use timesharing techniques to efficiently scan the surrounding airspace while monitoring instruments as well.

2. While the eyes can observe an approximate 200 degree arc of the horizon at one glance, only a very small center area called the fovea, in the rear of the eye, has the ability to send clear, sharply focused messages to the brain. All other visual information that is not processed directly through the fovea will be of less detail. An aircraft at a distance of 7 miles which appears in sharp focus within the foveal center of vision would have to be as close as $^{7}/_{10}$ of a mile in order to be recognized if it were outside of foveal vision. Because the eyes can focus only on this narrow viewing area, effective scanning is accomplished with a series of short, regularly spaced eye movements that bring successive areas of the sky into the central visual field. Each movement should not exceed 10 degrees, and each area should be observed for at least 1 second to enable detection. Although horizontal back-and-forth eye movements seem preferred by most pilots, each pilot should develop a scanning pattern that is most comfortable and then adhere to it to assure optimum scanning.

3. Studies show that the time a pilot spends on visual tasks inside the cabin should represent no more that $^{1}/_{4}$ to $^{1}/_{3}$ of the scan time outside, or no more than 4 to 5 seconds on the instrument panel for every 16 seconds outside. Since the brain is already trained to process sight information that is presented from left to right, one may find it easier to start scanning over the left shoulder and proceed across the windshield to the right.

4. Pilots should realize that their eyes may require several seconds to refocus when switching views between items in the cockpit and distant objects. The eyes will also tire more quickly when forced to adjust to distances immediately after close-up focus, as required for scanning the instrument panel. Eye fatigue can be reduced by looking from the instrument panel to the left wing past the wing tip to the center of the first scan quadrant when beginning the exterior scan. After having scanned from left to right, allow the eyes to return to the cabin along the right wing from its tip inward. Once back inside, one should automatically commence the panel scan.

5. Effective scanning also helps avoid "empty-field myopia." This condition usually occurs when flying above the clouds or in a haze layer that provides nothing specific to focus on outside the aircraft. This causes the eyes to relax and seek a comfortable focal distance which may range from 10 to 30 feet. For the pilot, this means looking without seeing, which is dangerous.

8-1-7. Aerobatic Flight

a. Pilots planning to engage in aerobatics should be aware of the physiological stresses associated with accelerative forces during aerobatic maneuvers. Many prospective aerobatic trainees enthusiastically enter aerobatic instruction but find their first experiences with G forces to be unanticipated and very uncomfortable. To minimize or avoid potential adverse effects, the aerobatic instructor and trainee must have a basic understanding of the physiology of G force adaptation.

b. Forces experienced with a rapid push-over maneuver result in the blood and body organs being displaced toward the head. Depending on forces involved and individual tolerance, a pilot may experience discomfort, headache, "red-out," and even unconsciousness.

c. Forces experienced with a rapid pull-up maneuver result in the blood and body organ displacement toward the lower part of the body away from the head. Since the brain requires continuous blood circulation for an adequate oxygen supply, there is a physiologic limit to the time the pilot can tolerate higher forces before losing consciousness. As the blood circulation to the brain decreases as a result of forces involved, a pilot will experience "narrowing" of visual fields, "gray-out," "black-out," and unconsciousness. Even a brief loss of consciousness in a maneuver can lead to improper control movement causing structural failure of the aircraft or collision with another object or terrain.

d. In steep turns, the centrifugal forces tend to push the pilot into the seat, thereby resulting in blood and body organ displacement toward the lower part of the body as in the case of rapid pull-up maneuvers and with the same physiologic effects and symptoms.

e. Physiologically, humans progressively adapt to imposed strains and stress, and with practice, any maneuver will have decreasing effect. Tolerance to G forces is dependent on human physiology and the individual pilot. These factors include the skeletal anatomy, the cardiovascular architecture, the nervous system, the quality of the blood, the general physical state, and experience and recency of exposure. The pilot should consult an Aviation Medical Examiner prior to aerobatic training and be aware that poor physical condition can reduce tolerance to accelerative forces.

f. The above information provides pilots with a brief summary of the physiologic effects of G forces. It does not address methods of "counteracting" these effects. There are numerous references on the subject of G forces during aerobatics available to pilots. Among these are "G Effects on the Pilot During Aerobatics," FAA-AM-72-28, and "G Incapacitation in Aerobatic Pilots: A Flight Hazard" FAA-AM-82-13. These are available from the National Technical Information Service, Springfield, Virginia 22161.

REFERENCE-*FAA AC 91-61, A Hazard in Aerobatics: Effects of G-forces on Pilots.*

8-1-8. Judgment Aspects of Collision Avoidance

a. Introduction. The most important aspects of vision and the techniques to scan for other aircraft are described in paragraph 8-1-6, Vision in Flight. Pilots should also be familiar with the following information to reduce the possibility of mid-air collisions.

b. Determining Relative Altitude. Use the horizon as a reference point. If the other aircraft is above the horizon, it is probably on a higher flight path. If the aircraft appears to be below the horizon, it is probably flying at a lower altitude.

c. Taking Appropriate Action. Pilots should be familiar with rules on right-of-way, so if an aircraft is on an obvious collision course, one can take immediate evasive action, preferably in compliance with applicable Federal Aviation Regulations.

d. Consider Multiple Threats. The decision to climb, descend, or turn is a matter of personal judgment, but one should anticipate that the other pilot may also be making a quick maneuver. Watch the other aircraft during the maneuver and begin your scanning again immediately since there may be other aircraft in the area.

e. Collision Course Targets. Any aircraft that appears to have no relative motion and stays in one scan quadrant is likely to be on a collision course. Also, if a target shows no lateral or vertical motion, but increases in size, *take evasive action.*

f. Recognize High Hazard Areas.

1. Airways, especially near VORs, and Class B, Class C, Class D, and Class E surface areas are places where aircraft tend to cluster.

2. Remember, most collisions occur during days when the weather is good. Being in a "radar environment" still requires vigilance to avoid collisions.

g. Cockpit Management. Studying maps, checklists, and manuals before flight, with other proper preflight planning; e.g., noting necessary radio frequencies and organizing cockpit materials, can reduce the amount of time required to look at these items during flight, permitting more scan time.

h. Windshield Conditions. Dirty or bug-smeared windshields can greatly reduce the ability of pilots to see other aircraft. Keep a clean windshield.

i. Visibility Conditions. Smoke, haze, dust, rain, and flying towards the sun can also greatly reduce the ability to detect targets.

j. Visual Obstructions in the Cockpit.

1. Pilots need to move their heads to see around blind spots caused by fixed aircraft structures, such as door posts, wings, etc. It will be necessary at times to maneuver the aircraft; e.g., lift a wing, to facilitate seeing.

2. Pilots must ensure curtains and other cockpit objects; e.g., maps on glare shield, are removed and stowed during flight.

k. Lights On.

1. Day or night, use of exterior lights can greatly increase the conspicuity of any aircraft.

2. Keep interior lights low at night.

l. ATC Support. ATC facilities often provide radar traffic advisories on a workload-permitting basis. Flight through Class C and Class D airspace requires communication with ATC. Use this support whenever possible or when required.

Chapter 9.

AERONAUTICAL CHARTS AND RELATED PUBLICATIONS

Section 1. Types of Charts Available

9-1-1. General

Civil aeronautical charts for the U.S. and its territories, and possessions are produced by Aeronautical Navigation Products (AeroNav), **http://www.faa.gov/air_traffic/flight_info/aeronav** which is part of FAA's Air Traffic Organization, Mission Support Services.

9-1-2. Obtaining Aeronautical Charts

a. Most charts and publications described in this Chapter can be obtained by subscription or one-time sales from:

Aeronautical Navigation Products (AeroNav)
Logistics Group, AJV-372
Federal Aviation Administration
10201 Good Luck Road
Glenn Dale, MD 20769-9700
Telephone: 1-800-638-8972 (Toll free within U.S.)
301-436-8301
301-436-6829 (FAX)
e-mail: 9-AMC-Chartsales@faa.gov

b. Public sales of charts and publications are also available through a network of FAA chart agents primarily located at or near major civil airports. A listing of products, dates of latest editions and agents is available at the AeroNav Web site: **http://www.faa.gov/air_traffic/flight_info/aeronav.**

9-1-3. Selected Charts and Products Available

VFR Navigation Charts
IFR Navigation Charts
Planning Charts
Supplementary Charts and Publications
Digital Products

9-1-4. General Description of each Chart Series

a. VFR Navigation Charts.

1. Sectional Aeronautical Charts. Sectional Charts are designed for visual navigation of slow to medium speed aircraft. The topographic information consists of contour lines, shaded relief, drainage patterns, and an extensive selection of visual checkpoints and landmarks used for flight under VFR. Cultural features include cities and towns, roads, railroads, and other distinct landmarks. The aeronautical information includes visual and radio aids to navigation, airports, controlled airspace, special-use airspace, obstructions, and related data. Scale 1 inch = 6.86nm/1:500,000. 60 x 20 inches folded to 5 x 10 inches. Revised semiannually, except most Alaskan charts are revised annually.

(See FIG 9-1-1 and FIG 9-1-11.)

2. VFR Terminal Area Charts (TAC). TACs depict the airspace designated as Class B airspace. While similar to sectional charts, TACs have more detail because the scale is larger. The TAC should be used by pilots intending to operate to or from airfields within or near Class B or Class C airspace. Areas with TAC coverage are indicated by a • on the Sectional Chart indexes. Scale 1 inch = 3.43nm/1:250,000. Charts are revised semiannually, except Puerto Rico-Virgin Islands revised annually.

(See FIG 9-1-1 and FIG 9-1-11.)

3. World Aeronautical Chart (WAC). WACs cover land areas for navigation by moderate speed aircraft operating at high altitudes. Included are city tints, principal roads, railroads, distinctive landmarks, drainage patterns, and relief. Aeronautical information includes visual and radio aids to navigation, airports, airways, special-use airspace, and obstructions. Because of a smaller scale, WACs do not show as much detail as sectional or TACs, and; therefore, are not recommended for exclusive use by pilots of low speed, low altitude aircraft. Scale 1 inch = 13.7nm/1:1,000,000. 60 x 20 inches folded to 5 x 10 inches. WACs are revised annually, except for a few in Alaska and the Caribbean, which are revised biennially.

(See FIG 9-1-12 and FIG 9-1-13.)

4. U.S. Gulf Coast VFR Aeronautical Chart. The Gulf Coast Chart is designed primarily for helicopter operation in the Gulf of Mexico area. Information depicted includes offshore mineral leasing areas and blocks, oil drilling platforms, and high density helicopter activity

FIG 9-1-1

Sectional and VFR Terminal Area Charts for the Conterminous U.S., Hawaii, Puerto Rico, and Virgin Islands

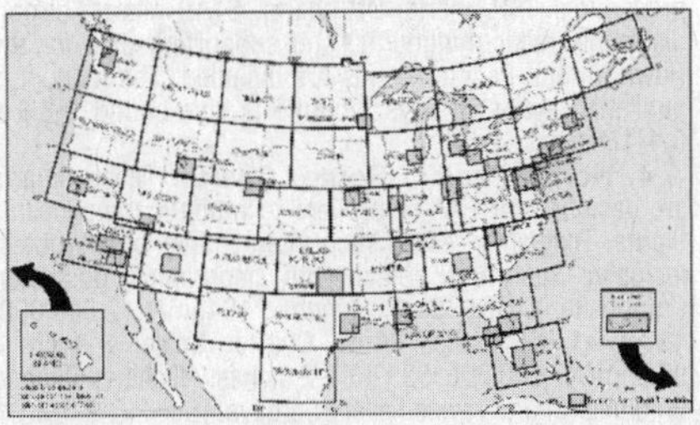

areas. Scale 1 inch = 13.7nm/1:1,000,000. 55 x 27 inches folded to 5 x 10 inches. Revised annually.

5. Grand Canyon VFR Aeronautical Chart. Covers the Grand Canyon National Park area and is designed to promote aviation safety, flight free zones, and facilitate VFR navigation in this popular area. The chart contains aeronautical information for general aviation VFR pilots on one side and commercial VFR air tour operators on the other side.

6. Helicopter Route Charts. A three-color chart series which shows current aeronautical information useful to helicopter pilots navigating in areas with high concentrations of helicopter activity. Information depicted includes helicopter routes, four classes of heliports with associated frequency and lighting capabilities, NAVAIDs, and obstructions. In addition, pictorial symbols, roads, and easily identified geographical features are portrayed. Helicopter charts have a longer life span than other chart products and may be current for several years. All new editions of these charts are printed on a durable plastic material. Helicopter Route Charts are updated as requested by the FAA. Scale 1 inch = 1.71nm/1:125,000. 34 x 30 inches folded to 5 x 10 inches.

b. IFR Navigation Charts.

1. IFR Enroute Low Altitude Charts (Conterminous U.S. and Alaska). Enroute low altitude charts provide aeronautical information for navigation under IFR conditions below 18,000 feet MSL. This four-color chart series includes airways; limits of controlled airspace; VHF NAVAIDs with frequency, identification, channel, geographic coordinates; airports with terminal air/ground communications; minimum en route and obstruction clearance altitudes; airway distances; reporting points; special use airspace; and military training routes. Scales vary from 1 inch = 5nm to 1 inch = 20nm. 50 x 20 inches folded to 5 x 10 inches. Charts revised every 56 days. *Area charts* show congested terminal areas at a large scale. They are included with subscriptions to any conterminous U.S. Set Low (Full set, East or West sets).

(See FIG 9-1-2 and FIG 9-1-4.)

2. IFR Enroute High Altitude Charts (Conterminous U.S. and Alaska). Enroute high altitude charts are designed for navigation at or above 18,000 feet MSL. This four-color chart series includes the jet route structure; VHF NAVAIDs with frequency, identification, channel, geographic coordinates; selected airports; reporting points. Scales vary from 1 inch = 45nm to 1 inch = 18nm. 55 x 20 inches folded to 5 x 10 inches. Revised every 56 days.

(See FIG 9-1-3 and FIG 9-1-5.)

3. U.S. Terminal Procedures Publication (TPP). TPPs are published in 24 loose-leaf or perfect bound volumes covering the conterminous U.S., Puerto Rico and the Virgin Islands. A Change Notice is published at the midpoint between revisions in bound volume format and is available on the internet for free download at the AeroNav web site. (See FIG 9-1-9.) The TPPs include:

(a) Instrument Approach Procedure (IAP) Charts. IAP charts portray the aeronautical data that is required to execute instrument approaches to airports. Each chart depicts the IAP, all related navigation data, communications information, and an airport sketch. Each

FIG 9-1-2

Enroute Low Altitude Instrument Charts for the Conterminous U.S. (Includes Area Charts)

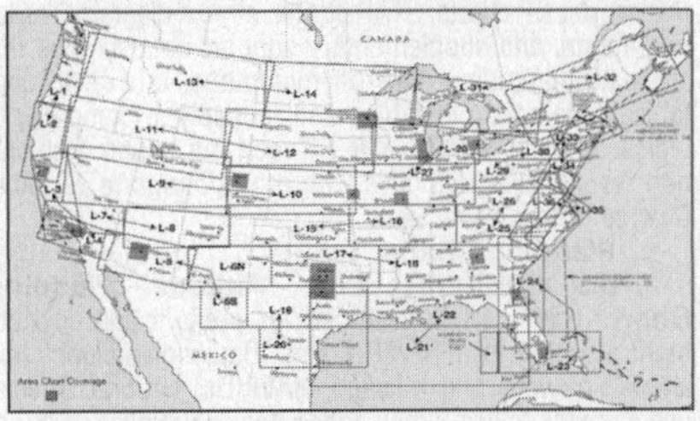

FIG 9-1-3

Enroute High Altitude Charts for the Conterminous U.S.

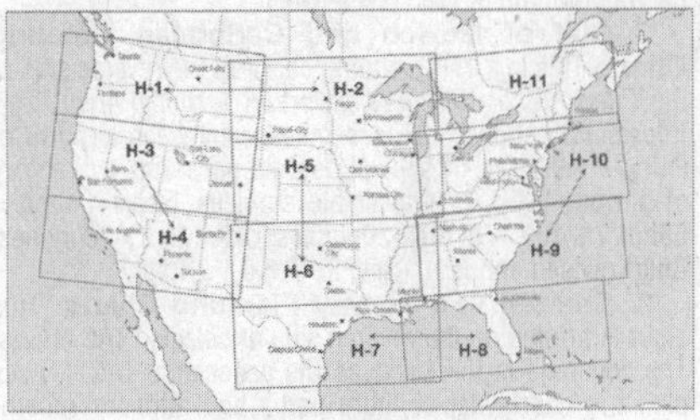

FIG 9-1-4

Alaska Enroute Low Altitude Chart

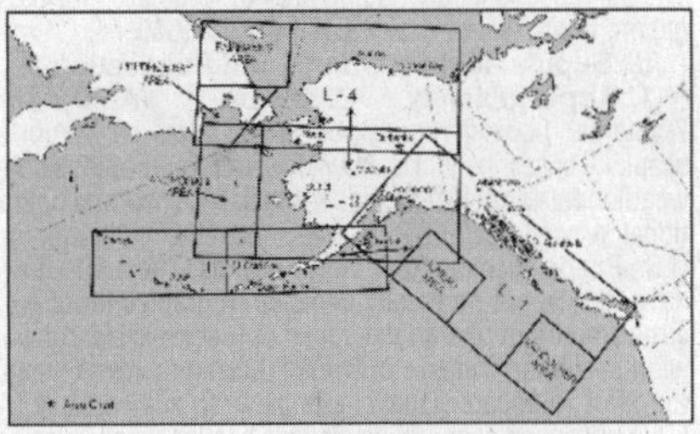

FIG 9-1-5

Alaskan Enroute High Altitude Chart

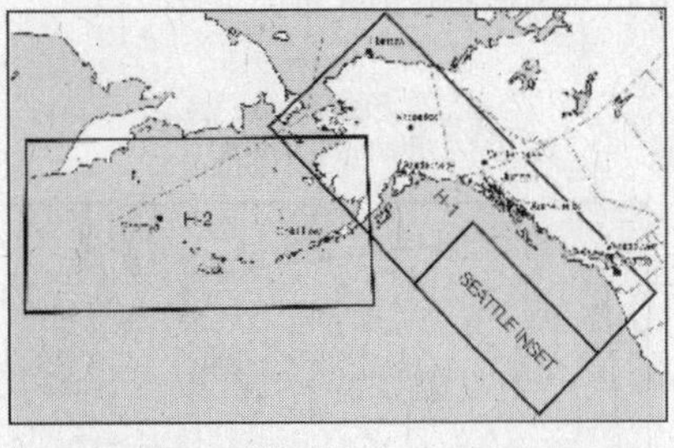

procedure is designated for use with a specific electronic navigational aid, such as ILS, VOR, NDB, RNAV, etc.

(b) Instrument Departure Procedure (DP) Charts. DP charts are designed to expedite clearance delivery and to facilitate transition between takeoff and en route operations. They furnish pilots' departure routing clearance information in graphic and textual form.

(c) Standard Terminal Arrival (STAR) Charts. STAR charts are designed to expedite ATC arrival procedures and to facilitate transition between en route and instrument approach operations. They depict preplanned IFR ATC arrival procedures in graphic and textual form.

Each STAR procedure is presented as a separate chart and may serve either a single airport or more than one airport in a given geographic area.

(d) Airport Diagrams. Full page airport diagrams are designed to assist in the movement of ground traffic at locations with complex runway/taxiway configurations and provide information for updating geodetic position navigational systems aboard aircraft. Airport diagrams are available for free download at the AeroNav website.

4. Alaska Terminal Procedures Publication. This publication contains all terminal flight procedures for civil and military aviation in Alaska. Included are IAP charts, DP charts, STAR charts, airport diagrams, radar minimums, and supplementary support data such as IFR alternate minimums, take-off minimums, rate of descent tables, rate of climb tables and inoperative components tables. Volume is 5-3/8 x 8-1/4 inch top bound. Publication revised every 56 days with provisions for a Terminal Change Notice, as required.

c. Planning Charts.

1. U.S. IFR/VFR Low Altitude Planning Chart. This chart is designed for prefight and en route flight planning for IFR/VFR flights. Depiction includes low altitude airways and mileage, NAVAIDs, airports, special use airspace, cities, times zones, major drainage, a directory of airports with their airspace classification, and a mileage table showing great circle distances between major airports. Scale 1 inch = 47nm/1:3,400,000. Chart revised annually, and is available either folded or unfolded for wall mounting. (See FIG 9-1-6.)

2. Gulf of Mexico and Caribbean Planning Chart. This is a VFR planning chart on the reverse side of the *Puerto Rico - Virgin Islands VFR Terminal Area Chart.* Information shown includes mileage between airports of entry, a selection of special use airspace and a directory of airports with their available services. Scale 1 inch = 85nm/1:6,192,178. 60 x 20 inches folded to 5 x 10 inches. Chart revised annually. (SeeFIG 9-1-6.)

3. Charted VFR Flyway Planning Charts. This chart is printed on the reverse side of selected TAC charts. The coverage is the same as the associated TAC. Flyway planning charts depict flight paths and altitudes recommended for use to bypass high traffic areas. Ground references are provided as a guide for visual orientation. Flyway planning charts are designed for use in conjunction with TACs and sectional charts and are not to be used for navigation. Chart scale 1 inch = 3.43nm/1;250,000.

d. Supplementary Charts and Publications.

1. Airport/Facility Directory (A/FD). This 7-volume booklet series contains data on airports, seaplane bases, heliports, NAVAIDs, communications data, weather data sources, airspace, special notices, and operational procedures. Coverage includes the conterminous U.S., Puerto Rico, and the Virgin Islands. The A/FD shows data that cannot be readily depicted in graphic form; e.g., airport hours of operations, types of fuel available, runway widths, lighting codes, etc. The A/FD also provides a means for pilots to update visual charts between edition dates (A/FD is published every 56 days while sectional and Terminal Area Charts are generally revised every six months). The VFR Chart Update Bulletins are available for free download from the AeroNav web site. Volumes are side-bound 5-3/8 x 8-1/4 inches. (See FIG 9-1-10.)

FIG 9-1-6
Planning Charts

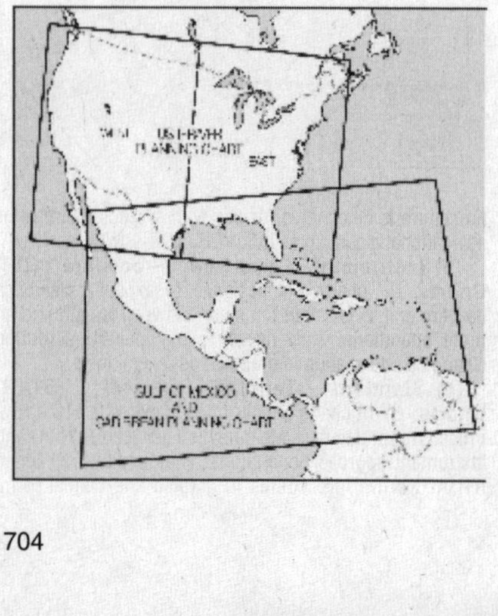

2. Supplement Alaska. This is a civil/military flight information publication issued by FAA every 56 days. It is a single volume booklet designed for use with appropriate IFR or VFR charts. The Supplement Alaska contains an A/FD, airport sketches, communications data, weather data sources, airspace, listing of navigational facilities, and special notices and procedures. Volume is side-bound 5-3/8 x 8-1/4 inches.

3. Chart Supplement Pacific. This supplement is designed for use with appropriate VFR or IFR enroute charts. Included in this one-volume booklet are the A/FD, communications data, weather data sources, airspace, navigational facilities, special notices, and Pacific area procedures. IAP charts, DP charts, STAR charts, airport diagrams, radar minimums, and supporting data for the Hawaiian and Pacific Islands are included. The manual is published every 56 days. Volume is side-bound 5-3/8 x 8-1/4 inches.

4. North Pacific Route Charts. These charts are designed for FAA controllers to monitor transoceanic flights. They show established intercontinental air routes, including reporting points with geographic positions. Composite Chart: Scale 1 inch = 164nm/1:12,000,000. 48 x 41-1/2 inches. Area Charts: Scale 1 inch = 95.9nm/1:7,000,000. 52 x 40-1/2 inches. All charts shipped unfolded. Charts revised every 56 days. (See FIG 9-1-8.)

5. North Atlantic Route Chart. Designed for FAA controllers to monitor transatlantic flights, this 5-color chart shows oceanic control areas, coastal navigation aids, oceanic reporting points, and NAVAID geographic coordinates. Full Size Chart: Scale 1 inch = 113.1nm/1:8,250,000. Chart is shipped flat only. Half Size Chart: Scale 1 inch = 150.8nm/1:11,000,000. Chart is 29-3/4 x 20-1/2 inches, shipped folded to 5 x 10 inches only. Chart revised every 56 weeks. (See FIG 9-1-7.)

6. Airport Obstruction Charts (OC). The OC is a 1:12,000 scale graphic depicting 14 CFR Part 77, Objects Affecting Navigable Airspace, surfaces, a representation of objects that penetrate these surfaces, aircraft movement and apron areas, navigational aids, prominent airport buildings, and a selection of roads and other planimetric detail in the airport vicinity. Also included are tabulations of runway and other operational data.

7. FAA Aeronautical Chart User's Guide. A booklet designed to be used as a teaching aid and reference document. It describes the substantial amount of information provided on FAA's aeronautical charts and publications. It includes explanations and illustrations of chart terms and symbols organized by chart type. The users guide is available for free download at the AeroNav web site.

e. Digital Products.

1. The Digital Aeronautical Information CD (DAICD). The DAICD is a combination of the NAVAID Digital Data File, the Digital Chart Supplement, and the Digital Obstacle File on one Compact Disk. These three digital products are no longer sold separately. The files are updated every 56 days and are available by subscription only.

(a) The NAVAID Digital Data File. This file contains a current listing of NAVAIDs that are compatible with the National Airspace System. This file contains all NAVAIDs including ILS and its components, in the U.S., Puerto Rico, and the Virgin Islands plus bordering facilities in Canada, Mexico, and the Atlantic and Pacific areas.

(b) The Digital Obstacle File. This file describes all obstacles of interest to aviation users in the U.S., with limited coverage of the Pacific, Caribbean, Canada, and Mexico. The obstacles are assigned unique numerical identifiers, accuracy codes, and listed in order of ascending latitude within each state or area.

(c) The Digital Aeronautical Chart Supplement (DACS). The DACS is specifically designed to provide digital airspace data not otherwise readily available. The supplement includes a *Change Notice* for IAPFIX.dat at the mid-point between revisions. The *Change Notice* is available only by free download from the AeroNav website.

The DACS individual data files are:

ENHIGH.DAT: High altitude airways (conterminous U.S.)

ENLOW.DAT: Low altitude airways (conterminous U.S.)

IAPFIX.DAT: Selected instrument approach procedure NAVAID and fix data.

MTRFIX.DAT: Military training routes data.

ALHIGH.DAT: Alaska high altitude airways data.

ALLOW.DAT: Alaska low altitude airways data.

PR.DAT: Puerto Rico airways data.

HAWAII.DAT: Hawaii airways data.

BAHAMA.DAT: Bahamas routes data.

FIG 9-1-7
North Atlantic Route Charts

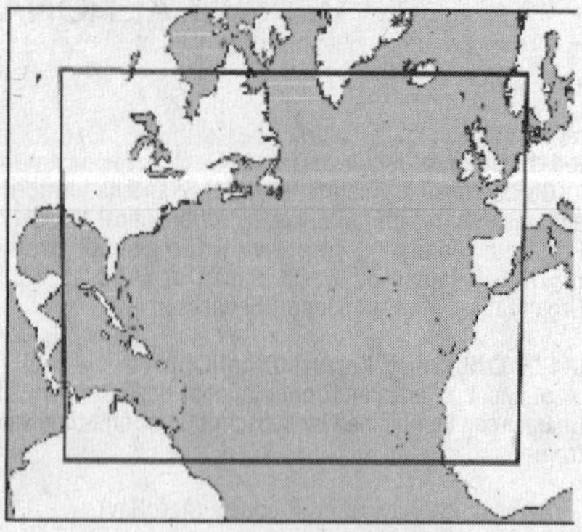

FIG 9-1-8
North Pacific Oceanic Route Charts

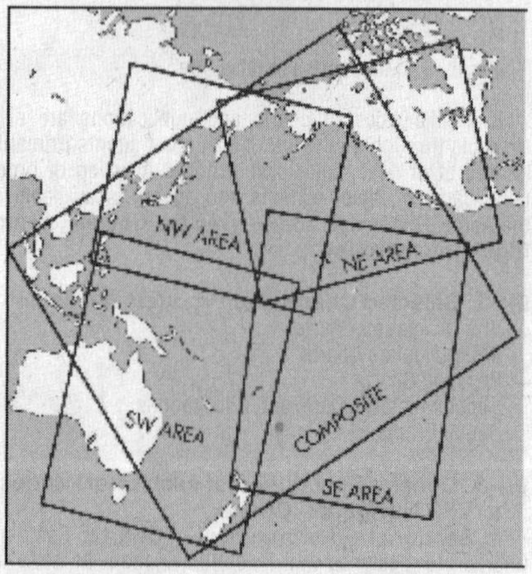

OCEANIC.DAT: Oceanic routes data.

STARS.DAT: Standard terminal arrivals data.

DP.DAT: Instrument departure procedures data.

LOPREF.DAT: Preferred low altitude IFR routes data.

HIPREF.DAT: Preferred high altitude IFR routes data.

ARF.DAT: Air route radar facilities data.

ASR.DAT: Airport surveillance radar facilities data.

2. The National Flight Database (NFD) (ARINC 424 [Ver 13 & 15]). The NFD is a basic digital dataset, modeled to an international standard, which can be used as a basis to support GPS navigation. Initial data elements included are: Airport and Helicopter Records, VHF and NDB Navigation aids, en route waypoints and airways. Additional data elements will be added in subsequent releases to include: departure procedures, standard terminal arrivals, and GPS/RNAV instrument approach procedures. The database is updated every 28 days. The data is available by subscription only and is distributed on CD-ROM or by ftp download.

3. Sectional Raster Aeronautical Charts (SRAC). These digital VFR charts are geo-referenced scanned images of FAA sectional charts. Additional digital data may easily be overlaid on the raster image using commonly available Geographic Information System software. Data such as weather, temporary flight restrictions, obstacles, or other geospatial data can be combined with SRAC data to support a variety of needs. Most SRACs are provided in two halves, a north side and a south side. The file resolution is 200 dots per inch and the data is 8-bit color. The data is provided as a GeoTIFF and distributed on DVD-R media. The root mean square error of the transformation will not exceed two pixels. SRACs DVDs are updated every 28 days and are available by subscription only.

9-1-5. Where and How to Get Charts of Foreign Areas

a. National Geospatial-Intelligence Agency (NGA) Products. For the latest information regarding

FIG 9-1-9
U.S. Terminal Publication Volumes

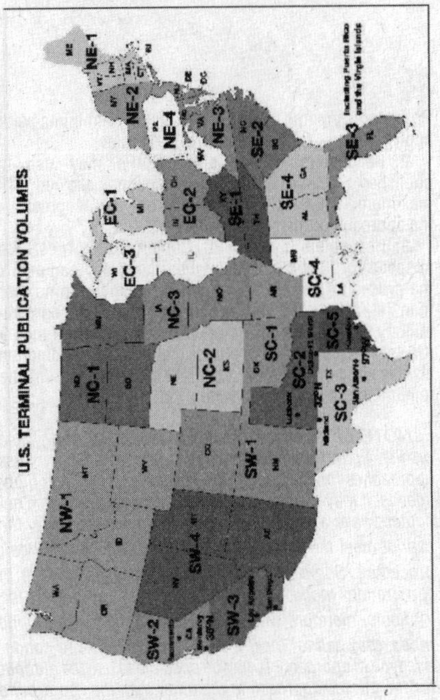

FIG 9-1-10
Airport/Facility Directory Geographic Areas

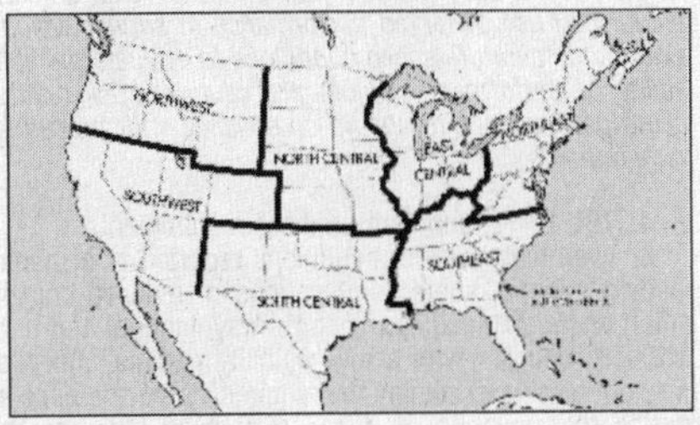

publication availability visit the NGA Web site:**https://www.nga.mil/ProductsServices/Aeronautical/Pages/default.aspx**

1. Flight Information Publication (FLIP) Planning Documents.

General Planning (GP)
Area Planning
Area Planning - Special Use Airspace - Planning Charts

2. FLIP Enroute Charts and Chart Supplements.

Pacific, Australasia, and Antarctica

FIG 9-1-11
Sectional and VFR Terminal Area Charts for Alaska

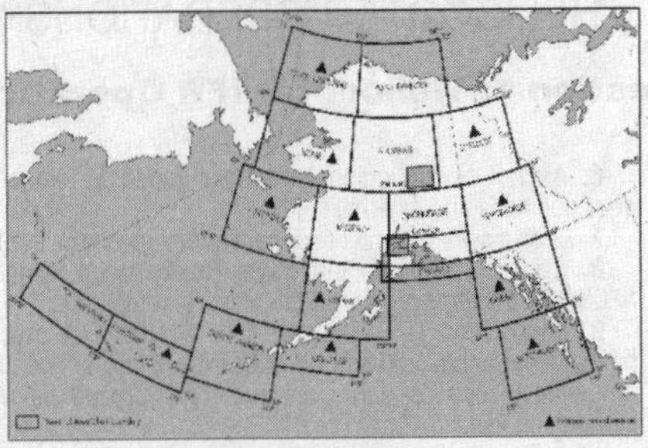

FIG 9-1-12
World Aeronautical Charts for Alaska

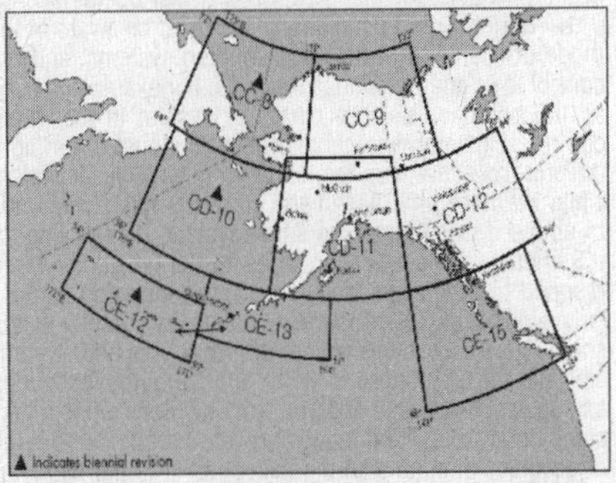

U.S. - IFR and VFR Supplements
Flight Information Handbook
Caribbean and South America - Low Altitude
Caribbean and South America - High Altitude
Europe, North Africa, and Middle East - Low Altitude
Europe, North Africa, and Middle East - High Altitude
Africa
Eastern Europe and Asia
Area Arrival Charts

3. FLIP Instrument Approach Procedures (IAPs).

Africa
Canada and North Atlantic
Caribbean and South America
Eastern Europe and Asia
Europe, North Africa, and Middle East
Pacific, Australasia, and Antarctica
VFR Arrival/Departure Routes - Europe and Korea
U.S.

4. Miscellaneous DOD Charts and Products.

Aeronautical Chart Updating Manual (CHUM)
DOD Weather Plotting Charts (WPC)
Tactical Pilotage Charts (TPC)
Operational Navigation Charts (ONC)

FIG 9-1-13
World Aeronautical Charts for the Conterminous U.S. Mexico, and the Caribbean Areas

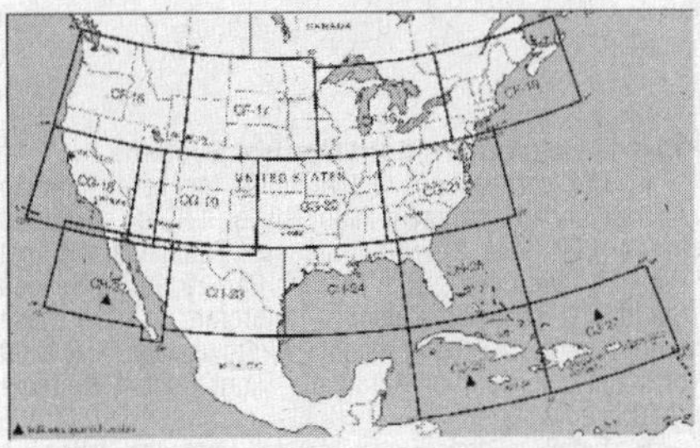

Global Navigation and Planning Charts (GNC)
Jet Navigation Charts (JNC) and Universal Jet Navigation Charts (JNU)
Jet Navigation Charts (JNCA)
Aerospace Planning Charts (ASC)
Oceanic Planning Charts (OPC)
Joint Operations Graphics - Air (JOG-A)
Standard Index Charts (SIC)
Universal Plotting Sheet (VP-OS)
Sight Reduction Tables for Air Navigation (PUB249)
Plotting Sheets (VP-30)
Dial-Up Electronic CHUM

b. Canadian Charts. Information on available Canadian charts and publications may be obtained from designated FAA chart agents or by contacting the:

NAV CANADA
Aeronautical Publications
Sales and Distribution Unit
P.O. Box 9840, Station T
Ottawa, Ontario K1G 6S8 Canada
Telephone: 613-744-6393 or 1-866-731-7827
Fax: 613-744-7120 or 1-866-740-9992

c. Mexican Charts. Information on available Mexican charts and publications may be obtained by contacting:

Dirección de Navigacion Aereo
Blvd. Puerto Aereo 485
Zona Federal Del Aeropuerto Int'l
15620 Mexico D.F.
Mexico

d. International Civil Aviation Organization (ICAO). A free *ICAO Publications and Audio- Visual Training Aids Catalogue* is available from:

International Civil Aviation Organization
ATTN: Document Sales Unit
999 University Street
Montreal, Quebec
H3C 5H7, Canada
Telephone: (514) 954-8022
Fax: (514) 954-6769
E-mail: sales_unit@icao.org
Internet: **http://www.icao.org/cgi/goto.pl?icao/en/sales.htm**
Sitatex: YULCAYA
Telex: 05-24513

Chapter 10. HELICOPTER OPERATIONS

Section 1. Helicopter IFR Operations

10-1-1. Helicopter Flight Control Systems

a. The certification requirements for helicopters to operate under Instrument Flight Rules (IFR) are contained in 14 CFR Part 27, Airworthiness Standards: Normal Category Rotorcraft, and 14 CFR Part 29, Airworthiness Standards: Transport Category Rotorcraft. To meet these requirements, helicopter manufacturers usually utilize a set of stabilization and/or Automatic Flight Control Systems (AFCSs).

b. Typically, these systems fall into the following categories:

1. Aerodynamic surfaces, which impart some stability or control capability not found in the basic VFR configuration.

2. Trim systems, which provide a cyclic centering effect. These systems typically involve a magnetic brake/spring device, and may also be controlled by a four-way switch on the cyclic. This is a system that supports "hands on" flying of the helicopter by the pilot.

3. Stability Augmentation Systems (SASs), which provide short-term rate damping control inputs to increase helicopter stability. Like trim systems, SAS supports "hands on" flying.

4. Attitude Retention Systems (ATTs), which return the helicopter to a selected attitude after a disturbance. Changes in desired attitude can be accomplished usually through a four-way "beep" switch, or by actuating a "force trim" switch on the cyclic, setting the attitude manually, and releasing. Attitude retention may be a SAS function, or may be the basic "hands off" autopilot function.

5. Autopilot Systems (APs), which provide for "hands off" flight along specified lateral and vertical paths, including heading, altitude, vertical speed, navigation tracking, and approach. These systems typically have a control panel for mode selection, and system for indication of mode status. Autopilots may or may not be installed with an associated Flight Director System (FD). Autopilots typically control the helicopter about the roll and pitch axes (cyclic control) but may also include yaw axis (pedal control) and collective control servos.

6. FDs, which provide visual guidance to the pilot to fly specific selected lateral and vertical modes of operation. The visual guidance is typically provided as either a "dual cue" (commonly known as a "cross-pointer") or "single cue" (commonly known as a "vee-bar") presentation superimposed over the attitude indicator. Some FDs also include a collective cue. The pilot manipulates the helicopter's controls to satisfy these commands, yielding the desired flight path, or may couple the flight director to the autopilot to perform automatic flight along the desired flight path. Typically, flight director mode control and indication is shared with the autopilot.

c. In order to be certificated for IFR operation, a specific helicopter may require the use of one or more of these systems, in any combination.

d. In many cases, helicopters are certificated for IFR operations with either one or two pilots. Certain equipment is required to be installed and functional for two pilot operations, and typically, additional equipment is required for single pilot operation. These requirements are usually described in the limitations section of the Rotorcraft Flight Manual (RFM).

e. In addition, the RFM also typically defines systems and functions that are required to be in operation or engaged for IFR flight in either the single or two pilot configuration. Often, particularly in two pilot operation, this level of augmentation is less than the full capability of the installed systems. Likewise, single pilot operation may require a higher level of augmentation.

f. The RFM also identifies other specific limitations associated with IFR flight. Typically, these limitations include, but are not limited to:

1. Minimum equipment required for IFR flight (in some cases, for both single pilot and two pilot operations).

2. Vmini (minimum speed - IFR).

NOTE-*The manufacturer may also recommend a minimum IFR airspeed during instrument approach.*

3. Vnei (never exceed speed - IFR).

4. Maximum approach angle.

5. Weight and center of gravity limits.

6. Aircraft configuration limitations (such as aircraft door positions and external loads).

7. Aircraft system limitations (generators, inverters, etc.).

8. System testing requirements (many avionics and AFCS/AP/FD systems incorporate a self-test feature).

9. Pilot action requirements (such as the pilot must have his/her hands and feet on the controls during certain operations, such as during instrument approach below certain altitudes).

g. It is very important that pilots be familiar with the IFR requirements for their particular helicopter. Within the same make, model and series of helicopter, variations in the installed avionics may change the required equipment or the level of augmentation for a particular operation.

h. During flight operations, pilots must be aware of the mode of operation of the augmentation systems, and the control logic and functions employed. For example, during an ILS approach using a particular system in the three-cue mode (lateral, vertical and collective cues), the flight director *collective cue* responds to glideslope deviation, while the horizontal bar of the "cross-pointer" responds to airspeed deviations. The same system, while flying an ILS in the two-cue mode, provides for the *horizontal bar* to respond to glideslope deviations. This concern is particularly significant when operating using two pilots. Pilots should have an established set of procedures and responsibilities for the control of flight director/autopilot modes for the various phases of flight. Not only does a full understanding of the system modes provide for a higher degree of accuracy in control of the helicopter, it is the basis for crew identification of a faulty system.

i. Relief from the prohibition to takeoff with any inoperative instruments or equipment may be provided through a Minimum Equipment List (see 14 CFR Section 91.213 and 14 CFR Section 135.179, Inoperative Instruments and Equipment). In many cases, a helicopter configured for single pilot IFR may depart IFR with certain equipment inoperative, provided a crew of two pilots is used. Pilots are cautioned to ensure the pilot-in-command and second-in-command meet the requirements of 14 CFR Section 61.58, Pilot-in-Command Proficiency Check: Operation of Aircraft Requiring More Than One Pilot Flight Crewmember, and 14 CFR Section 61.55, Second-in-Command Qualifications, or 14 CFR Part 135, Operating Requirements: Commuter and On-Demand Operations, Subpart E, Flight Crewmember Requirements, and Subpart G, Crewmember Testing Requirements, as appropriate.

j. Experience has shown that modern AFCS/AP/FD equipment installed in IFR helicopters can, in some cases, be very complex. This complexity requires the pilot(s) to obtain and maintain a high level of knowledge of system operation, limitations, failure indications and reversionary modes. In some cases, this may only be reliably accomplished through formal training.

10-1-2. Helicopter Instrument Approaches

a. Helicopters are capable of flying any published 14 CFR Part 97, Standard Instrument Approach Procedures (SIAPs), for which they are properly equipped, subject to the following limitations and conditions:

1. Helicopters flying conventional (non-Copter) SIAPs may reduce the visibility minima to not less than one half the published Category A landing visibility minima, or $^1/_4$ statute mile visibility/1200 RVR, whichever is greater unless the procedure is annotated with "**Visibility Reduction by Helicopters NA.**" This annotation means that there are penetrations of the final approach obstacle identification surface (OIS) and that the 14 CFR Section 97.3 visibility reduction rule does not apply and you must take precaution to avoid any obstacles in the visual segment. No reduction in MDA/DA is permitted. The helicopter may initiate the final approach segment at speeds up to the upper limit of the highest approach category authorized by the procedure, but must be slowed to no more than 90 KIAS at the missed approach point (MAP) in order to apply the visibility reduction. Pilots are cautioned that such a decelerating approach may make early identification of wind shear on the approach path difficult or impossible. If required, use the Inoperative Components and Visual Aids Table provided in the front cover of the U.S. Terminal Procedures Volume to derive the Category A minima before applying the 14 CFR Section 97.3(d-1) rule.

2. Helicopters flying Copter SIAPs may use the published minima, with no reductions allowed. The maximum airspeed is 90 KIAS on any segment of the approach or missed approach.

3. Helicopters flying GPS Copter SIAPs must limit airspeed to 90 KIAS or less when flying any segment of the procedure, except speeds must be limited to no more than 70 KIAS on the final and missed approach segments. Military GPS Copter SIAPs are limited to no more than 90 KIAS throughout the procedure. If annotated, holding may also be limited to no more than 70 KIAS. Use the published minima, no reductions allowed.

NOTE-*Obstruction clearance surfaces are based on the aircraft speed and have been designed on these approaches for 70 knots. If the helicopter is flown at higher speeds, it may fly outside of protected airspace. Some helicopters have a V_{MINI} greater than 70 knots; therefore, they cannot meet the 70 knot limitation to conduct this type of procedure. Some helicopter autopilots, when used in the "go-around" mode, are programmed with a V_{YI} greater than 70 knots, therefore when using the autopilot "go-around" mode, they cannot meet the 70 knot limitation to conduct this type of approach. It may be possible to use the autopilot for the missed approach in the other than the "go-around" mode and meet the 70 knot limitation to conduct this type of approach. When operating at speeds other than V_{YI} or V_Y, performance data may not be available in the RFM to predict compliance with climb gradient requirements. Pilots may use observed performance in similar weight/altitude/temperature/speed conditions to evaluate the suitability of performance. Pilots are cautioned to monitor climb performance to ensure compliance with procedure requirements.*

4. TBL 10-1-1 summarizes these requirements.

5. Even with weather conditions reported at or above landing minima, some combinations of reduced cockpit cutoff angle, minimal approach/runway lighting, and high MDA/DH coupled with a low visibility minima, the pilot may not be able to identify the required visual reference(s) during the approach, or those references may only be visible in a very small portion of the pilot's available field of view. Even if identified by the pilot, these visual references may not support normal maneuvering and normal rates of descent to landing. The effect of such a combination may be exacerbated by other conditions such as rain on the windshield, or incomplete windshield defogging coverage.

6. Pilots are cautioned to be prepared to execute a missed approach even though weather conditions may be reported at or above landing minima.

NOTE-*See paragraph 5-4-21, Missed Approach, for additional information on missed approach procedures.*

Several factors effect the ability of the pilot to acquire and maintain the visual references specified in 14 CFR Section 91.175(c), even in cases where the flight visibility may be at the minimum derived by TBL 10-1-1. These factors include, but are not limited to:

1. *Cockpit cutoff angle (the angle at which the cockpit or other airframe structure limits downward visibility below the horizon).*

2. *Combinations of high MDA/DH and low visibility minimum, such as a conventional nonprecision approach with a reduced helicopter visibility minima (per 14 CFR Section 97.3).*

3. *Type, configuration, and intensity of approach and runway lighting systems.*

4. *Type of obscuring phenomenon and/or windshield contamination.*

10-1-3. Helicopter Approach Procedures to VFR Heliports

a. Helicopter approaches may be developed for heliports that do not meet the design standards for an IFR heliport. The majority of IFR approaches to VFR heliports are developed in support of helicopter emergency medical

TABLE 10-1-1
Helicopter Use of Standard Instrument Approach Procedures

Procedure	Helicopter Visibility Minima	Helicopter MDA/DA	Maximum Speed Limitations
Conventional (non-Copter)	The greater of: one half the Category A visibility minima, 1/4 statute mile visibility, or 1200 RVR	As published for Category A	The helicopter may initiate the final approach segment at speeds up to the upper limit of the highest Approach Category authorized by the procedure, but must be slowed to no more than 90 KIAS at the MAP in order to apply the visibility reduction.
Copter Procedure	As published	As published	90 KIAS when on a published route/track.
GPS Copter Procedure	As published	As published	90 KIAS when on a published route or track, EXCEPT 70 KIAS when on the final approach or missed approach segment

services (HEMS) operators. These approaches can be developed from conventional NAVAIDs or a RNAV system (including GPS). They are developed either as a Special Approach (pilot training is required for special procedures due to their unique characteristics) or a public approach (no special training required). These instrument procedures are developed as either an approach designed to a specific landing site, or an approach designed to a point-in-space.

1. Approach to a specific landing site. The approach is aligned to a missed approach point from which a landing can be accomplished with a maximum course change of 30 degrees. The visual segment from the MAP to the landing site is evaluated for obstacle hazards. These procedures are annotated: "PROCEED VISUALLY FROM (NAMED MAP) OR CONDUCT THE SPECIFIED MISSED APPROACH."

(a) This phrase requires the pilot to either acquire and maintain visual contact with the landing site at or prior to the MAP, or execute a missed approach. The visibility minimum is based on the distance from the MAP to the landing site, among other factors.

(b) The pilot is required to maintain the published minimum visibility throughout the visual segment.

(c) Similar to an approach to a runway, the missed approach segment protection is not provided between the MAP and the landing site, and obstacle or terrain avoidance from the MAP to the landing site is the responsibility of the pilot.

(d) Upon reaching the MAP defined on the approach procedure, or as soon as practicable after reaching the MAP, the pilot advises ATC whether proceeding visually and canceling IFR or complying with the missed approach instructions. See paragraph 5-1-15, Canceling IFR Flight Plan.

(e) At least one of the following visual references must be visible or identifiable before the pilot may proceed visually:

(1) FATO or FATO lights.

(2) TLOF or TLOF lights.

(3) Heliport Instrument Lighting System (HILS).

(4) Heliport Approach Lighting System (HALS) or lead-in lights.

(5) Visual Glideslope Indicator (VGSI).

(6) Windsock or windsock light(s). See note below.

(7) Heliport beacon. See note below.

(8) Other facilities or systems approved by the Flight Technologies and Procedures Division (AFS-400).

NOTE- *Windsock lights and heliport beacons should be located within 500 ft of the TLOF.*

2. Approach to a Point-in-Space (PinS). At locations where the MAP is located more than 2 SM from the landing site, or the path from the MAP to the landing site is populated with obstructions which require avoidance actions or requires turns greater than 30 degrees, a PinS procedure may be developed. These approaches are annotated "PROCEED VFR FROM (NAMED MAP) OR CONDUCT THE SPECIFIED MISSED APPROACH."

(a) These procedures require the pilot, at or prior to the MAP, to determine if the published minimum visibility, or the weather minimums required by the operating rule, or operations specifications (whichever is higher) is available to safely transition from IFR to VFR flight. If not, the pilot must execute a missed approach. For Part 135 operations, pilots may not begin the instrument approach unless the latest weather report indicates that the weather conditions are at or above the authorized IFR minimums or the VFR weather minimums (as required by the class of airspace, operating rule and/or Operations Specifications) whichever is higher.

(b) Visual contact with the landing site is not required; however, the pilot must maintain the appropriate VFR weather minimums throughout the visual segment. The visibility is limited to no lower than that published in the procedure, until canceling IFR.

(c) IFR obstruction clearance areas are not applied to the VFR segment between the MAP and the landing site. Obstacle or terrain avoidance from the MAP to the landing site is the responsibility of the pilot.

(d) Upon reaching the MAP defined on the approach procedure, or as soon as practicable after reaching the MAP, the pilot advises ATC whether proceeding VFR and canceling IFR, or complying with the missed approach instructions. See paragraph 5-1-15, Canceling IFR Flight Plan.

(e) If the visual segment penetrates Class B, C, or D airspace, pilots are responsible for obtaining a Special VFR clearance, when required.

10-1-4. The Gulf of Mexico Grid System

a. On October 8, 1998, the Southwest Regional Office of the FAA, with assistance from the Helicopter Safety Advisory Conference (HSAC), implemented the world's first Instrument Flight Rules (IFR) Grid System in the Gulf of Mexico. This navigational route structure is completely independent of ground-based navigation aids (NAVAIDs) and was designed to facilitate helicopter IFR operations to offshore destinations. The Grid System is defined by over 300 offshore waypoints located 20 minutes apart (latitude and longitude). Flight plan routes are routinely defined by just 4 segments: departure point (lat/long), first en route grid waypoint, last en route grid waypoint prior to approach procedure, and destination point (lat/long). There are over 4,000 possible offshore landing sites. Upon reaching the waypoint prior to the destination, the pilot may execute an Offshore Standard Approach Procedure (OSAP), a Helicopter En Route Descent Areas (HEDA) approach, or an Airborne Radar Approach (ARA). For more information on these helicopter instrument procedures, refer to FAA AC 90-80B, Approval of Offshore Standard Approach Procedures, Airborne Radar Approaches, and Helicopter En Route Descent Areas, on the FAA web site **http://www.faa.gov** under Advisory Circulars. The return flight plan is just the reverse with the requested stand-alone GPS approach contained in the remarks section.

1. The large number (over 300) of waypoints in the grid system makes it difficult to assign phonetically pronounceable names to the waypoints that would be meaningful to pilots and controllers. A unique naming system was adopted that enables pilots and controllers to derive the fix position from the name. The five-letter names are derived as follows:

(a) The waypoints are divided into sets of 3 columns each. A three-letter identifier, identifying a geographical area or a NAVAID to the north, represents each set.

(b) Each column in a set is named after its position, i.e., left (L), center (C), and right (R).

(c) The rows of the grid are named alphabetically from north to south, starting with A for the northern most row.

EXAMPLE- *LCHRC would be pronounced "Lake Charles Romeo Charlie." The waypoint is in the right-hand column of the Lake Charles VOR set, in row C (third south from the northern most row).*

2. In December 2009, significant improvements to the Gulf of Mexico grid system were realized with the introduction of ATC separation services using ADS-B. In cooperation with the oil and gas services industry, HSAC and Helicopter Association International (HAI), the FAA installed an infrastructure of ADS-B ground stations, weather stations (AWOS) and VHF remote communication outlets (RCO) throughout a large area of the Gulf of Mexico. This infrastructure allows the FAA's Houston ARTCC to provide "domestic-like" air traffic control service in the offshore area beyond 12nm from the coastline to hundreds of miles offshore to aircraft equipped with ADS-B. Properly equipped aircraft can now be authorized to receive more direct routing, domestic en route separation minima and real time flight following. Operators who do not have authorization to receive ATC separation services using ADS-B, will continue to use the low altitude grid system and receive procedural separation from Houston ARTCC. Non-ADS-B equipped aircraft also benefit from improved VHF communication and expanded weather information coverage.

3. Three requirements must be met for operators to file IFR flight plans utilizing the grid:

(a) The helicopter must be equipped for IFR operations and equipped with IFR approved GPS navigational units.

(b) The operator must obtain prior written approval from the appropriate Flight Standards District Office through a Letter of Authorization or Operations Specification, as appropriate.

(c) The operator must be a signatory to the Houston ARTCC Letter of Agreement.

4. Operators who wish to benefit from ADS-B based ATC separation services must meet the following additional requirements:

(a) The Operator's installed ADS-B Out equipment must meet the performance requirements of one of the following FAA Technical Standard Orders (TSO), or later revisions: TSO-C154c, Universal Access Transceiver (UAT) Automatic Dependent Surveillance-Broadcast (ADS-B) Equipment, or TSO-C166b, Extended Squitter Automatic Dependent Surveillance-Broadcast (ADS-B) and Traffic Information.

(b) Flight crews must comply with the procedures prescribed in the Houston ARTCC Letter of Agreement dated December 17, 2009, or later.

NOTE-
The unique ADS-B architecture in the Gulf of Mexico depends upon reception of an aircraft's Mode C in addition to the other message elements described in 14 CFR 91.227. Flight crews must be made aware that loss of Mode C also means that ATC will not receive the aircraft's ADS-B signal.

5. FAA/AeroNav publishes the grid system waypoints on the IFR Gulf of Mexico Vertical Flight Reference Chart. A commercial equivalent is also available. The chart is updated annually and is available from a FAA chart agent or FAA directly, web site address:

http://www.faa.gov/air_traffic/flight_info/aeronav.

Section 2. Special Operations

10-2-1. Offshore Helicopter Operations

a. Introduction

The offshore environment offers unique applications and challenges for helicopter pilots. The mission demands, the nature of oil and gas exploration and production facilities, and the flight environment (weather, terrain, obstacles, traffic), demand special practices, techniques and procedures not found in other flight operations. Several industry organizations have risen to the task of reducing risks in offshore operations, including the Helicopter Safety Advisory Conference (HSAC) (**http://www.hsac.org**), and the Offshore Committee of the Helicopter Association International (HAI) (**http://www.rotor.com**). The following recommended practices for offshore helicopter operations are based on guidance developed by HSAC for use in the Gulf of Mexico, and provided here with their permission. While not regulatory, these recommended practices provide aviation and oil and gas industry operators with useful information in developing procedures to avoid certain hazards of offshore helicopter operations.

NOTE- *Like all aviation practices, these recommended practices are under constant review. In addition to normal procedures for comments, suggested changes, or corrections to the AIM (contained in the Preface), any questions or feedback concerning these recommended procedures may also be directed to the HSAC through the feedback feature of the HSAC web site (**http://www.hsac.org**).*

b. Passenger Management on and about Heliport Facilities

1. Background. Several incidents involving offshore helicopter passengers have highlighted the potential for incidents and accidents on and about the heliport area. The following practices will minimize risks to passengers and others involved in heliport operations.

2. Recommended Practices

(a) Heliport facilities should have a designated and posted passenger waiting area which is clear of the heliport, heliport access points, and stairways.

(b) Arriving passengers and cargo should be unloaded and cleared from the heliport and access route prior to loading departing passengers and cargo.

(c) Where a flight crew consists of more than one pilot, one crewmember should supervise the unloading/loading process from outside the aircraft.

(d) Where practical, a designated facility employee should assist with loading/unloading, etc.

c. Crane-Helicopter Operational Procedures

1. Background. Historical experience has shown that catastrophic consequences can occur when industry safe practices for crane/helicopter operations are not observed. The following recommended practices are designed to minimize risks during crane and helicopter operations.

2. Recommended Practices

(a) Personnel awareness

(1) Crane operators and pilots should develop a mutual understanding and respect of the others' operational limitations and cooperate in the spirit of safety;

(2) Pilots need to be aware that crane operators sometimes cannot release the load to cradle the crane boom, such as when attached to wire line lubricators or supporting diving bells; and

(3) Crane operators need to be aware that helicopters require warm up before takeoff, a two-minute cool down before shutdown, and cannot circle for extended lengths of time because of fuel consumption.

(b) It is recommended that when helicopters are approaching, maneuvering, taking off, or running on the heliport, cranes be shutdown and the operator leave the cab. Cranes not in use must have their booms cradled, if feasible. If in use, the crane's boom(s) are to be pointed away from the heliport and the crane shutdown for helicopter operations.

(c) Pilots will not approach, land on, takeoff, or have rotor blades turning on heliports of structures not complying with the above practice.

(d) It is recommended that cranes on offshore platforms, rigs, vessels, or any other facility, which could interfere with helicopter operations (including approach/departure paths):

(1) Be equipped with a red rotating beacon or red high intensity strobe light connected to the system powering the crane, indicating the crane is under power;

(2) Be designed to allow the operator a maximum view of the helideck area and should be equipped with wide-angle mirrors to eliminate blind spots; and

(3) Have their boom tips, headache balls, and hooks painted with high visibility international orange.

d. Helicopter/Tanker Operations

1. Background. The interface of helicopters and tankers during shipboard helicopter operations is complex and may be hazardous unless appropriate procedures are coordinated among all parties. The following recommended practices are designed to minimize risks during helicopter/tanker operations:

2. Recommended Practices

(a) Management, flight operations personnel, and pilots should be familiar with and apply the operating safety standards set forth in "Guide to Helicopter/Ship Operations", International Chamber of Shipping, Third Edition, 5-89 (as amended), establishing operational guidelines/standards and safe practices sufficient to safeguard helicopter/tanker operations.

(b) Appropriate plans, approvals, and communications must be accomplished prior to reaching the vessel, allowing tanker crews sufficient time to perform required safety preparations and position crew members to receive or dispatch a helicopter safely.

(c) Appropriate approvals and direct communications with the bridge of the tanker must be maintained throughout all helicopter/tanker operations.

(d) Helicopter/tanker operations, including landings/departures, must not be conducted until the helicopter pilot-in-command has received and acknowledged permission from the bridge of the tanker.

(e) Helicopter/tanker operations must not be conducted during product/cargo transfer.

(f) Generally, permission will not be granted to land on tankers during mooring operations or while maneuvering alongside another tanker.

e. Helideck/Heliport Operational Hazard Warning(s) Procedures

1. Background

(a) A number of operational hazards can develop on or near offshore helidecks or onshore heliports that can be minimized through procedures for proper notification or visual warning to pilots. Examples of hazards include but are not limited to:

(1) Perforating operations: subparagraph f.

(2) H_2S gas presence: subparagraph g.

(3) Gas venting: subparagraph h; or,

(4) Closed helidecks or heliports: sub-paragraph i (unspecified cause).

(b) These and other operational hazards are currently minimized through timely dissemination of a written Notice to Airmen (NOTAM) for pilots by helicopter companies and operators. A NOTAM provides a written description of the hazard, time and duration of occurrence, and other pertinent information. ANY POTENTIAL HAZARD should be communicated to helicopter operators or company aviation departments as early as possible to allow the NOTAM to be activated.

(c) To supplement the existing NOTAM procedure and further assist in reducing these hazards, a standardized visual signal(s) on the helideck/heliport will provide a positive indication to an approaching helicopter of the status of the landing area. Recommended Practice(s) have been developed to reinforce the NOTAM procedures and standardize visual signals.

f. Drilling Rig Perforating Operations: Helideck/Heliport Operational Hazard Warning(s)/ Procedure(s)

1. Background. A critical step in the oil well completion process is perforation, which involves the use of explosive charges in the drill pipe to open the pipe to oil or gas deposits. Explosive charges used in conjunction with perforation operations offshore can potentially be prematurely detonated by radio transmissions, including those from helicopters. The following practices are recommended.

2. Recommended Practices

(a) Personnel Conducting Perforating Operations. Whenever perforating operations are scheduled and operators are concerned that radio transmissions from helicopters in the vicinity may jeopardize the operation, personnel conducting perforating operations should take the following precautionary measures:

(1) Notify company aviation departments, helicopter operators or bases, and nearby manned platforms of the pending perforation operation so the Notice to Airmen (NOTAM) system can be activated for the perforation operation and the temporary helideck closure.

(2) Close the deck and make the radio warning clearly visible to passing pilots, install a temporary marking (described in subparagraph 10-2-1i1(b)) with the words "NO RADIO" stenciled in red on the legs of the diagonals. The letters should be 24 inches high and 12 inches wide. (See FIG 10-2-1.)

(3) The marker should be installed during the time that charges may be affected by radio transmissions.

(b) Pilots

(1) Pilots when operating within 1,000 feet of a known perforation operation or observing the white X with red "NO RADIO" warning indicating perforation operations are underway will avoid radio transmissions from or near the helideck (within 1,000 feet) and will not land on the deck if the X is present. In addition to communications radios, radio transmissions are also emitted by aircraft radar, transponders, radar altimeters, and DME equipment, and ELTs.

(2) Whenever possible, make radio calls to the platform being approached or to the Flight Following Communications Center at least one mile out on approach. Ensure all communications are complete outside the 1,000 foot hazard distance. If no response is received, or if the platform is not radio equipped, further radio transmissions should not be made until visual contact with the deck indicates it is open for operation (no white "X").

FIGURE 10-2-1
Closed Helideck Marking - No Radio

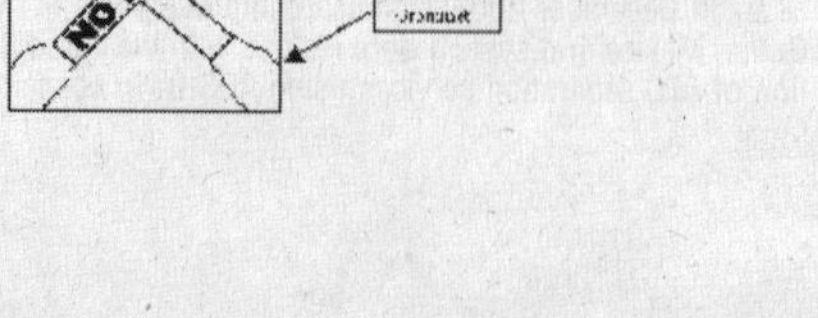

g. Hydrogen Sulfide Gas Helideck/Heliport Operational Hazard Warning(s)/Procedures

1. Background. Hydrogen sulfide (H_2S) gas: Hydrogen sulfide gas in higher concentrations (300-500 ppm) can cause loss of consciousness within a few seconds and presents a hazard to pilots on/near offshore helidecks. When operating in offshore areas that have been identified to have concentrations of hydrogen sulfide gas, the following practices are recommended.

2. Recommended Practices

(a) Pilots

(1) Ensure approved protective air packs are available for emergency use by the crew on the helicopter.

(2) If shutdown on a helideck, request the supervisor in charge provide a briefing on location of protective equipment and safety procedures.

(3) If while flying near a helideck and the visual red beacon alarm is observed or an unusually strong odor of "rotten eggs" is detected, immediately don the protective air pack, exit to an area upwind, and notify the suspected source field of the hazard.

(b) Oil Field Supervisors

(1) If presence of hydrogen sulfide is detected, a red rotating beacon or red high intensity strobe light adjacent to the primary helideck stairwell or wind indicator on the structure should be turned on to provide visual warning of hazard. If the beacon is to be located near the stairwell, the State of Louisiana "Offshore Heliport Design Guide" and FAA Advisory Circular AC 150/5390-2A, "Heliport Design Guide," should be reviewed to ensure proper clearance on the helideck.

(2) Notify nearby helicopter operators and bases of the hazard and advise when hazard is cleared.

(3) Provide a safety briefing to include location of protective equipment to all arriving personnel.

(4) Wind socks or indicator should be clearly visible to provide upwind indication for the pilot.

h. Gas Venting Helideck/Heliport Operational Hazard Warning(s)/Procedures - Operations Near Gas Vent Booms

1. Background. Ignited flare booms can release a large volume of natural gas and create a hot fire and intense heat with little time for the pilot to react. Likewise, unignited gas vents can release reasonably large volumes of methane gas under certain conditions. Thus, operations conducted very near unignited gas vents require precautions to prevent inadvertent ingestion of combustible gases by the helicopter engine(s). The following practices are recommended.

2. Pilots

(a) Gas will drift upwards and downwind of the vent. Plan the approach and takeoff to observe and avoid the area downwind of the vent, remaining as far away as practicable from the open end of the vent boom.

(b) Do not attempt to start or land on an offshore helideck when the deck is downwind of a gas vent unless properly trained personnel verify conditions are safe.

3. Oil Field Supervisors

(a) During venting of large amounts of unignited raw gas, a red rotating beacon or red high intensity strobe light adjacent to the primary helideck stairwell or wind indicator should be turned on to provide visible warning of hazard. If the beacon is to be located near the stairwell, the State of Louisiana "Offshore Heliport Design Guide" and FAA Advisory Circular AC 150/5390-2A, Heliport Design Guide, should be reviewed to ensure proper clearance from the helideck.

(b) Notify nearby helicopter operators and bases of the hazard for planned operations.

(c) Wind socks or indicator should be clearly visible to provide upward indication for the pilot.

i. Helideck/Heliport Operational Warning(s)/ Procedure(s) - Closed Helidecks or Heliports

1. Background. A white "X" marked diagonally from corner to corner across a helideck or heliport touchdown area is the universally accepted visual indicator that the landing area is closed for safety of other reasons and that helicopter operations are not permitted. The following practices are recommended.

(a) Permanent Closing. If a helideck or heliport is to be permanently closed, X diagonals of the same size and location as indicated above should be used, but the markings should be painted on the landing area.

***NOTE-** White Decks: If a helideck is painted white, then international orange or yellow markings can be used for the temporary or permanent diagonals.*

(b) Temporary Closing. A temporary marker can be used for hazards of an interim nature. This marker could be made from vinyl or other durable material in the shape of a diagonal "X." The marker should be white with legs at least 20 feet long and 3 feet in width. This marker is designed to be quickly secured and removed from the deck using grommets and rope ties. The duration, time, location, and nature of these temporary closings should be provided to and coordinated with company aviation departments, nearby helicopter bases, and helicopter operators supporting the area. These markers MUST be removed when the hazard no longer exists.

(See FIG 10-2-2.)

Figure 10-2-2
Closed Helideck Marking

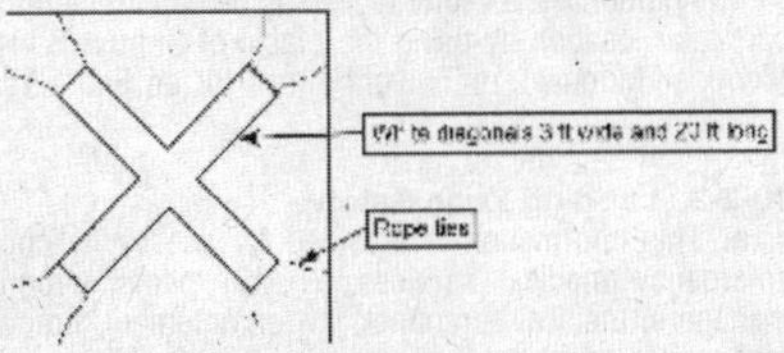

j. Offshore (VFR) Operating Altitudes for Helicopters

1. Background. Mid-air collisions constitute a significant percentage of total fatal offshore helicopter accidents. A method of reducing this risk is the use of coordinated VFR cruising altitudes. To enhance safety through standardized vertical separation of helicopters when flying in the offshore environment, it is recommended that helicopter operators flying in a particular area establish a cooperatively developed Standard Operating Procedure (SOP) for VFR operating altitudes. An example of such an SOP is contained in this example.

2. Recommended Practice Example

(a) Field Operations. Without compromising minimum safe operating altitudes, helicopters working within an offshore field "constituting a cluster" should use altitudes not to exceed 500 feet.

(b) En Route Operations

(1) Helicopters operating below 750' AGL should avoid transitioning through offshore fields.

(2) Helicopters en route to and from offshore locations, below 3,000 feet, weather permitting, should use en route altitudes as outlined in TBL *10-2-1.*

Table 10-2-1

Magnetic Heading	Altitude
0° to 179°	750'
	1750'
	2750'
180° 359°	1250'
	2250'

(c) Area Agreements. See HSAC Area Agreement Maps for operating procedures for onshore high density traffic locations.

NOTE-
Pilots of helicopters operating VFR above 3,000 feet above the surface should refer to the current Federal Aviation Regulations (14 CFR Part 91), and paragraph 3-1-4, Basic VFR Weather Minimums, of the AIM.

(d) Landing Lights. Aircraft landing lights should be on to enhance aircraft identification:

(1) During takeoff and landings;

(2) In congested helicopter or fixed wing traffic areas;

(3) During reduced visibility; or,

(4) Anytime safety could be enhanced.

k. Offshore Helidecks/Landing Communications

1. Background. To enhance safety, and provide appropriate time to prepare for helicopter operations, the following is recommended when anticipating a landing on an offshore helideck.

2. Recommended Practices

(a) Before landing on an offshore helideck, pilots are encouraged to establish communications with the company owning or operating the helideck if frequencies exist for that purpose.

(b) When impracticable, or if frequencies do not exist, pilots or operations personnel should attempt to contact the company owning or operating the helideck by telephone. Contact should be made before the pilot departs home base/point of departure to advise of intentions and obtain landing permission if necessary.

NOTE- *It is recommended that communications be established a minimum of 10 minutes prior to planned arrival time. This practice may be a requirement of some offshore owner/operators.*

NOTE-
1. *See subparagraph 10-2-1d for Tanker Operations.*
2. *Private use Heliport. Offshore heliports are privately owned/operated facilities and their use is limited to persons having prior authorization to utilize the facility.*

l. Two (2) Helicopter Operations on Offshore Helidecks

1. Background. Standardized procedures can enhance the safety of operating a second helicopter on an offshore helideck, enabling pilots to determine/maintain minimum operational parameters. Orientation of the parked helicopter on the helideck, wind and other factors may prohibit multi-helicopter operations. More conservative Rotor Diameter (RD) clearances may be required under differing condition, i.e., temperature, wet deck, wind (velocity/direction/gusts), obstacles, approach/departure angles, etc. Operations are at the pilot's discretion.

2. Recommended Practice. Helideck size, structural weight capability, and type of main rotor on the parked and operating helicopter will aid in determining accessibility by a second helicopter. Pilots should determine that multi-helicopter deck operations are permitted by the helideck owner/operator.

3. Recommended Criteria

(a) Minimum one-third rotor diameter clearance ($^1/_3$ RD). The landing helicopter maintains a minimum $^1/_3$ RD clearance between the tips of its turning rotor and the closest part of a parked and secured helicopter (rotors stopped and tied down).

(b) Three foot parking distance from deck edge (3'). Helicopters operating on an offshore helideck land or park the helicopter with a skid/wheel assembly no closer than 3 feet from helideck edge.

(c) Tiedowns. Main rotors on all helicopters that are shut down be properly secured (tied down) to prevent the rotor blades from turning.

(d) Medium (transport) and larger helicopters should not land on any offshore helideck where a light helicopter is parked unless the light helicopter is property secured to the helideck and has main rotor tied down.

(e) Helideck owners/operators should ensure that the helideck has a serviceable anti-skid surface.

4. Weight and limitations markings on helideck. The helideck weight limitations should be displayed by markings visible to the pilot (see State of Louisiana "Offshore Heliport Design Guide" and FAA Advisory Circular AC 150/5390-2A, Heliport Design Guide).

NOTE- *Some offshore helideck owners/operators have restrictions on the number of helicopters allowed on a helideck. When helideck size permits, multiple (more than two) helicopter operations are permitted by some operators.*

m. Helicopter Rapid Refueling Procedures (HRR)

1. Background. Helicopter Rapid Refueling (HRR), engine(s)/rotors operating, can be conducted safely when utilizing trained personnel and observing safe practices. This recommended practice provides minimum guidance for HRR as outlined in National Fire Protection Association (NFPA) and industry practices. For detailed guidance, please refer to National Fire Protection Association (NFPA) Document 407, "Standard for Aircraft Fuel Servicing," 1990 edition, including 1993 HRR Amendment.

NOTE- *Certain operators prohibit HRR, or "hot refueling," or may have specific procedures for certain aircraft or refueling locations. See the General Operations Manual and/or Operations Specifications to determine the applicable procedures or limitations.*

2. Recommended Practices

(a) Only turbine-engine helicopters fueled with JET A or JET A-1 with fueling ports located below any engine exhausts may be fueled while an onboard engine(s) is (are) operating.

(b) Helicopter fueling while an onboard engine(s) is (are) operating should only be conducted under the following conditions:

(1) A properly certificated and current pilot is at the controls and a trained refueler attending the fuel nozzle during the entire fuel servicing process. The pilot monitors the fuel quantity and signals the refueler when quantity is reached.

(2) No electrical storms (thunderstorms) are present within 10 nautical miles. Lightning can travel great distances beyond the actual thunderstorm.

(3) Passengers disembark the helicopter and move to a safe location prior to HRR operations. When the pilot-in-command deems it necessary for passenger safety that they remain onboard, passengers should be briefed on the evacuation route to follow to clear the area.

(4) Passengers not board or disembark during HRR operations nor should cargo be loaded or unloaded.

(5) Only designated personnel, trained in HRR operations should conduct HRR written authorization to include safe handling of the fuel and equipment. (See your Company Operations/Safety Manual for detailed instructions.)

(6) All doors, windows, and access points allowing entry to the interior of the helicopter that are adjacent to or in the immediate vicinity of the fuel inlet ports kept closed during HRR operations.

(7) Pilots ensure that appropriate electrical/electronic equipment is placed in standby-off position, to preclude the possibility of electrical discharge or other fire hazard, such as [i.e., weather radar is on standby and no radio transmissions are made (keying of the microphone/transmitter)]. Remember, in addition to communications radios, radio transmissions are also emitted by aircraft radar, transponders, radar altimeters, DME equipment, and ELTs.

(8) Smoking be prohibited in and around the helicopter during all HRR operations.

The HRR procedures are critical and present associated hazards requiring attention to detail regarding quality control, weather conditions, static electricity, bonding, and spill/fires potential.

Any activity associated with rotors turning (i.e.; refueling embarking/disembarking, loading/unloading baggage/freight; etc.) personnel should only approach the aircraft when authorized to do so. Approach should be made via safe approach path/walkway or "arc"- **remain clear of all rotors.**

NOTE-
1. *Marine vessels, barges etc.: Vessel motion presents additional potential hazards to helicopter operations (blade flex, aircraft movement).*
2. *See National Fire Protection Association (NFPA) Document 407, "Standard for Aircraft Fuel Servicing" for specifics regarding non-HRR (routine refueling operations).*

10-2-2. Helicopter Night VFR Operations

a. Effect of Lighting on Seeing Conditions in Night VFR Helicopter Operations

NOTE- *This guidance was developed to support safe night VFR helicopter emergency medical services (HEMS) operations. The principles of lighting and seeing conditions are useful in any night VFR operation.*

While ceiling and visibility significantly affect safety in night VFR operations, lighting conditions also have a profound effect on safety. Even in conditions in which visibility and ceiling are determined to be visual meteorological conditions, the ability to discern unlighted or low contrast objects and terrain at night may be compromised. The ability to discern these objects and terrain is the seeing condition, and is related to the amount of natural and man made lighting available, and the contrast, reflectivity, and texture of surface terrain and obstruction features. In order to conduct operations safely, seeing conditions must be accounted for in the planning and execution of night VFR operations.

Night VFR seeing conditions can be described by identifying "high lighting conditions" and "low lighting conditions."

1. High lighting conditions exist when one of two sets of conditions are present:

(a) The sky cover is less than broken (less than 5/8 cloud cover), the time is between the local Moon rise and Moon set, and the lunar disk is at least 50% illuminated; or

(b) The aircraft is operated over surface lighting which, at least, provides for the lighting of prominent obstacles, the identification of terrain features (shorelines, valleys, hills, mountains, slopes) and a horizontal reference by which the pilot may control the helicopter. For example, this surface lighting may be the result of:

(1) Extensive cultural lighting (man-made, such as a built-up area of a city),

(2) Significant reflected cultural lighting (such as the illumination caused by the reflection of a major metropolitan area's lighting reflecting off a cloud ceiling), or

(3) Limited cultural lighting combined with a high level of natural reflectivity of celestial illumination, such as that provided by a surface covered by snow or a desert surface.

2. Low lighting conditions are those that do not meet the high lighting conditions requirements.

3. Some areas may be considered a high lighting environment only in specific circumstances. For example, some surfaces, such as a forest with limited cultural lighting, normally have little reflectivity, requiring dependence on significant moonlight to achieve a high lighting condition. However, when that same forest is covered with snow, its reflectivity may support a high lighting condition based only on starlight. Similarly, a desolate area, with little cultural lighting, such as a desert, may have such inherent natural reflectivity that it may be considered a high lighting conditions area regardless of season, provided the cloud cover does not prevent starlight from being reflected from the surface. Other surfaces, such as areas of open water, may never have enough reflectivity or cultural lighting to ever be characterized as a high lighting area.

4. Through the accumulation of night flying experience in a particular area, the operator will develop the ability to determine, prior to departure, which areas can be considered supporting high or low lighting conditions. Without that operational experience, low lighting considerations should be applied by operators for both pre-flight planning and operations until high lighting conditions are observed or determined to be regularly available.

b. Astronomical Definitions and Background Information for Night Operations

1. Definitions

(a) Horizon. Wherever one is located on or near the Earth's surface, the Earth is perceived as essentially flat and, therefore, as a plane. If there are no visual obstructions, the apparent intersection of the sky with the Earth's (plane) surface is the horizon, which appears as a circle centered at the observer. For rise/set computations, the observer's eye is considered to be on the surface of the Earth, so that the horizon is geometrically exactly 90 degrees from the local vertical direction.

(b) Rise, Set. During the course of a day the Earth rotates once on its axis causing the phenomena of rising and setting. All celestial bodies, the Sun, Moon, stars and planets, seem to appear in the sky at the horizon to the East of any particular place, then to cross the sky and again disappear at the horizon to the West. Because the Sun and Moon appear as circular disks and not as points of light, a definition of rise or set must be very specific, because not all of either body is seen to rise or set at once.

(c) Sunrise and sunset refer to the times when the upper edge of the disk of the Sun is on the horizon, considered unobstructed relative to the location of interest. Atmospheric conditions are assumed to be average, and the location is in a level region on the Earth's surface.

(d) Moonrise and moonset times are computed for exactly the same circumstances as for sunrise and sunset. However, moonrise and moonset may occur at any time during a 24 hour period and, consequently, it is often possible for the Moon to be seen during daylight, and to have moonless nights. It is also possible that a moonrise or moonset does not occur relative to a specific place on a given date.

(e) Transit. The transit time of a celestial body refers to the instant that its center crosses an imaginary line in the sky - the observer's meridian - running from north to south.

(f) Twilight. Before sunrise and again after sunset there are intervals of time, known as "twilight," during which there is natural light provided by the upper atmosphere, which does receive direct sunlight and reflects part of it toward the Earth's surface.

(g) Civil twilight is defined to begin in the morning, and to end in the evening when the center of the Sun is geometrically 6 degrees below the horizon. This is the limit at which twilight illumination is sufficient, under good weather conditions, for terrestrial objects to be clearly distinguished.

2. Title 14 of the Code of Federal Regulations applies these concepts and definitions in addressing the definition of night (Section 1.1), the requirement for aircraft lighting (Section 91.209) and pilot recency of night experience (Section 61.67).

c. Information on Moon Phases and Changes in the Percentage of the Moon Illuminated

From any location on the Earth, the Moon appears to be a circular disk which, at any specific time, is illuminated to some degree by direct sunlight. During each lunar orbit (a lunar month), we see the Moon's appearance change from not visibly illuminated through partially illuminated to fully illuminated, then back through partially illuminated to not illuminated again. There are eight distinct, traditionally recognized stages, called phases. The phases designate both the degree to which the Moon is illuminated and the geometric appearance of the illuminated part. These phases of the Moon, in the sequence of their occurrence (starting from New Moon), are listed in FIG 10-2-3.

1. The percent of the Moon's surface illuminated is a more refined, quantitative description of the Moon's appearance than is the phase. Considering the Moon as a circular disk, at New Moon the percent illuminated is 0; at First and Last Quarters it is 50%; and at Full Moon it is 100%. During the crescent phases the percent illuminated is between 0 and 50% and during gibbous phases it is between 50% and 100%.

2. For practical purposes, phases of the Moon and the percent of the Moon illuminated are independent of the location on the Earth from where the Moon is observed. That is, all the phases occur at the same time regardless of the observer's position.

3. For more detailed information, refer to the United States Naval Observatory site referenced below.

d. Access to Astronomical Data for Determination of Moon Rise, Moon Set, and Percentage of Lunar Disk Illuminated

1. Astronomical data for the determination of Moon rise and set and Moon phase may be obtained from the United States Naval Observatory using an interactive query available at:

http://aa.usno.navy.mil/

2. Click on "Data Services," and then on "Complete Sun and Moon Data for One Day."

3. You can obtain the times of sunrise, sunset, moonrise, moonset, transits of the Sun and Moon, and the beginning and end of civil twilight, along with information on the Moon's phase by specifying the date and location in one of the two forms on this web page and clicking on the "Get data" button at the end of the form. Form "A" is used for cities or towns in the U.S. or its territories. Form "B" for all other locations. An example of the data available from this site is shown in TBL 10-2-2.

4. Additionally, a yearly table may be constructed for a particular location by using the "Table of Sunrise/Sunset, Moonrise/Moonset, or Twilight Times for an Entire Year" selection.

FIGURE 10-2-3
Phases of the Moon

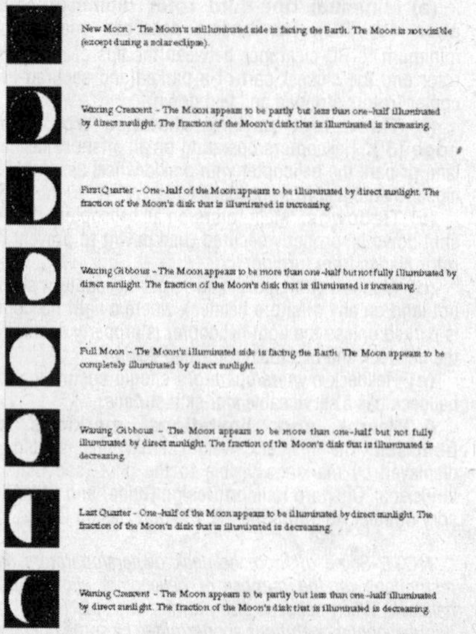

10-2-3. Landing Zone Safety

a. This information is provided for use by helicopter emergency medical services (HEMS) pilots, program managers, medical personnel, law enforcement, fire, and rescue personnel to further their understanding of the safety issues concerning Landing Zones (LZs). It is recommended that HEMS operators establish working relationships with the ground responder organizations they may come in contact with in their flight operations and share this information in order to establish a common frame of reference for LZ selection, operations, and safety.

b. The information provided is largely based on the booklet, LZ - Preparing the Landing Zone, issued by National Emergency Medical Services Pilots Association (NEMSPA), and the guidance developed by the University of Tennessee Medical Center's LIFESTAR program, and is used with their permission. For additional information, go to **http://www.nemspa.org/**.

c. Information concerning the estimation of wind velocity is based on the Beaufort Scale. See **http://www.spc.noaa.gov/faq/tornado/beaufort.html** for more information.

d. Selecting a Scene LZ

1. If the situation requires the use of a helicopter, first check to see if there is an area large enough to land a helicopter safely.

2. For the purposes of FIG 10-2-4 the following are provided as examples of relative helicopter size:

(a) Small Helicopter: Bell 206/407, Eurocopter AS-350/355, BO-105, BK-117.

(b) Medium Helicopter: Bell UH-1 (Huey) and derivatives (Bell 212/412), Bell 222/230/430 Sikorsky S-76, Eurocopter SA-365.

(c) Large Helicopter: Boeing Chinook, Eurocopter Puma, Sikorsky H-60 series (Blackhawk), SK-92.

3. The LZ should be level, firm and free of loose debris that could possibly blow up into the rotor system.

4. The LZ should be clear of people, vehicles and obstructions such as trees, poles and wires. Remember that wires are difficult to see from the air. The LZ must also be free of stumps, brush, post and large rocks. See FIG 10-2-5.

5. Keep spectators back at least 200 feet. Keep emergency vehicles 100 feet away and have fire equipment (if

TABLE 10-2-2
Sample of Astronomical Data Available from the Naval Observatory

The following information is provided for New Orleans, Orleans Parish, Louisiana (longitude W90.1, latitude N30.0)	
Tuesday 29 May 2007	Central Daylight Time
SUN	
Begin civil twilight	5:34 a.m.
Sunrise	6:01 a.m.
Sun transit	12:58 p.m.
Sunset	7:55 p.m.
End civil twilight	8:22 p.m.
MOON	
Moonrise	5:10 p.m. on preceding day
Moonset	4:07 a.m.
Moonrise	6:06 p.m.
Moon transit	11:26 p.m.
Moonset	4:41 a.m. on following day
Phase of the Moon on 29 May: waxing gibbous with 95% of the Moon's visible disk illuminated.	
Full Moon on 31 May 2007 at 8:04 p.m. Central Daylight Time.	

Figure 10-2-4
Recommended Minimum Landing Zone Dimensions

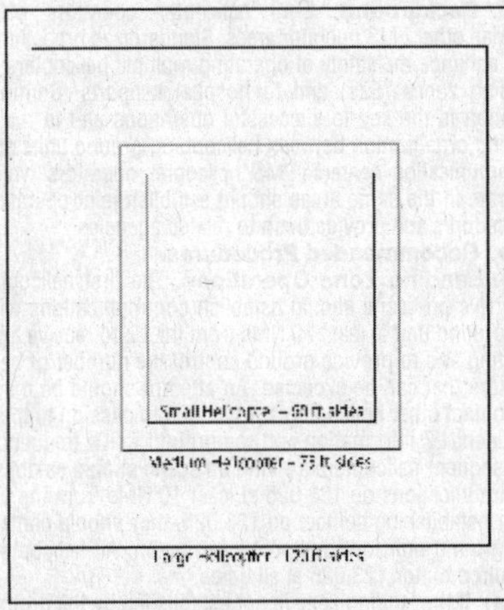

Figure 10-2-5
Landing Zone Hazards

Figure 10-2-6
"Clock" System for Identifying Positions Relative to the Nose of the Aircraft

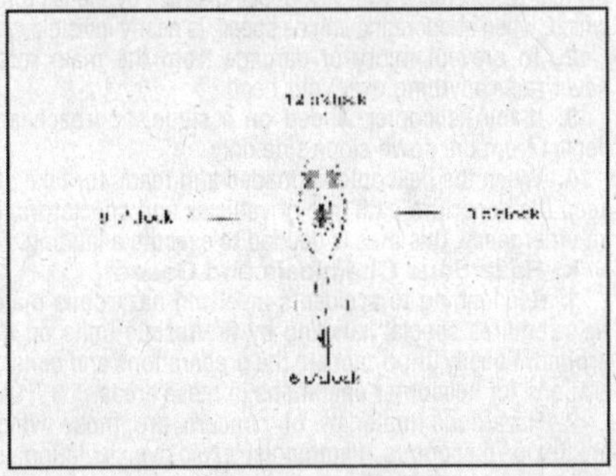

available) standing by. Ground personnel should wear eye protection, if available, during landing and takeoff operations. To avoid loose objects being blown around in the LZ, hats should be removed; if helmets are worn, chin straps must be securely fastened.

6. Fire fighters (if available) should wet down the LZ if it is extremely dusty.

e. Helping the Flightcrew Locate the Scene

1. If the LZ coordinator has access to a GPS unit, the exact latitude and longitude of the LZ should be relayed to the HEMS pilot. If unable to contact the pilot directly, relay the information to the HEMS ground communications specialist for relaying to the pilot, so that they may locate your scene more efficiently. Recognize that the aircraft may approach from a direction different than the direct path from the takeoff point to the scene, as the pilot may have to detour around terrain, obstructions or weather en route.

2. Especially in daylight hours, mountainous and densely populated areas can make sighting a scene from the air difficult. Often, the LZ coordinator on the ground will be asked if she or he can see or hear the helicopter.

3. Flightcrews use a clock reference method for directing one another's attention to a certain direction from the aircraft. The nose of the aircraft is always 12 o'clock, the right side is 3 o'clock, etc. When the LZ coordinator sees the aircraft, he/she should use this method to assist the flightcrew by indicating the scene's clock reference position from the nose of the aircraft. For example, "Accident scene is located at your 2 o'clock position." See FIG 10-2-6.

4. When the helicopter approaches the scene, it will normally orbit at least one time as the flight crew observes the wind direction and obstacles that could interfere with the landing. This is often referred to as the "high reconnaissance" maneuver.

f. Wind Direction and Touchdown Area

1. Determine from which direction the wind is blowing. Helicopters normally land and takeoff into the wind.

2. If contact can be established with the pilot, either directly or indirectly through the HEMS ground communications specialist, describe the wind in terms of the direction the wind is *from* and the speed.

3. Common natural sources of wind direction information are smoke, dust, vegetation movement, water streaks and waves. Flags, pennants, streamers can also be used. When describing the direction, use the compass direction from which the wind is blowing (example: from the North-West).

4. Wind speed can be measured by small hand-held measurement devices, or an observer's estimate can be used to provide velocity information. The wind value should be reported in knots (nautical miles per hour). If unable to numerically measure wind speed, use TBL 10-2-3 to estimate velocity. Also, report if the wind conditions are gusty, or if the wind direction or velocity is variable or has changed recently.

5. If any obstacle(s) exist, ensure their description, position and approximate height are communicated to the pilot on the initial radio call.

EXAMPLE-*Wind from the South-East, estimated speed 15 knots. Wind shifted from North-East about fifteen minutes ago, and is gusty.*

g. Night LZs

1. There are several ways to light a night LZ:

(a) Mark the touchdown area with five lights or road flares, one in each corner and one indicating the direction of the wind. See FIG 10-2-7.

NOTE-*Road flares are an intense source of ignition and may be unsuitable or dangerous in certain conditions. In any case, they must be closely managed and firefighting equipment should be present when used. Other light sources are preferred, if available.*

(b) If chemical light sticks may be used, care should be taken to assure they are adequately secured against being dislodged by the helicopter's rotor wash.

(c) Another method of marking a LZ uses four emergency vehicles with their low beam headlights aimed toward the intended landing area.

Table 10-2-3
Table of Common References for Estimating Wind Velocity

Wind (Knots)	Wind Classification	Appearance of Wind Effects	
		On the Water	On Land
Less than 1	Calm	Sea surface smooth and mirror-like	Calm, smoke rises vertically
1-3	Light Air	Scaly ripples, no foam crests	Smoke drift indicates wind direction, wind vanes are still
4-6	Light Breeze	Small wavelets, crests glassy, no breaking	Wind felt on face, leaves rustle, vanes begin to move
7-10	Gentle Breeze	Large wavelets, crests begin to break, scattered whitecaps	Leaves and small twigs constantly moving, light flags extended
11-16	Moderate Breeze	Small waves 1-4 ft. becoming longer, numerous whitecaps	Dust, leaves, and loose paper lifted, small tree branches move
17-21	Fresh Breeze	Moderate waves 4-8 ft. taking longer form, many whitecaps, some spray	Small trees in leaf begin to sway
22-27	Strong Breeze	Larger waves 8-13 ft., whitecaps common, more spray	Larger tree branches moving, whistling in wires
28-33	Near Gale	Sea heaps up, waves 13-20 ft., white foam streaks off breakers	Whole trees moving, resistance felt walking against wind
34-40	Gale	Moderately high (13-20 ft.) waves of greater length, edges of crests begin to break into spindrift, foam blown in streaks	Whole trees in motion, resistance felt walking against wind
41-47	Strong Gale	High waves (20 ft.), sea begins to roll, dense streaks of foam, spray may reduce visibility	Slight structural damage occurs, slate blows off roofs
48-55	Storm	Very high waves (20-30 ft.) with overhanging crests, sea white with densely blown foam, heavy rolling, lowered visibility	Seldom experienced on land, trees broken or uprooted, "considerable structural damage"
56-63	Violent Storm	Exceptionally high (30-45 ft.) waves, foam patches cover sea, visibility more reduced	
64+	Hurricane	Air filled with foam, waves over 45 ft., sea completely white with driving spray, visibility greatly reduced	

FIGURE 10-2-7
Recommended Lighting for Landing Zone Operations at Night

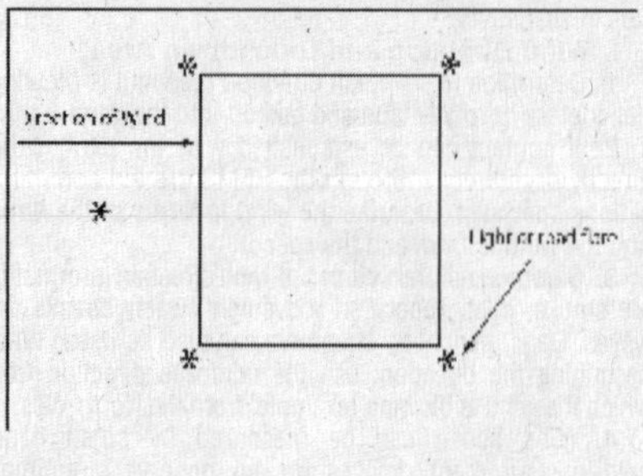

(d) A third method for marking a LZ uses two vehicles. Have the vehicles direct their headlight beams into the wind, crossing at the center of the LZ. (If fire/rescue personnel are available, the reflective stripes on their bunker gear will assist the pilot greatly.)

2. At night, spotlights, flood lights and hand lights used to define the LZ are not to be pointed at the helicopter. However, they are helpful when pointed toward utility poles, trees or other hazards to the landing aircraft. White lights such as spotlights, flashbulbs and hi-beam headlights ruin the pilot's night vision and temporarily blind him. Red lights, however, are very helpful in finding accident locations and do not affect the pilot's night vision as significantly.

3. As in Day LZ operations, ensure radio contact is accomplished between ground and air, if possible.

h. Ground Guide

1. When the helicopter is in sight, one person should assist the LZ Coordinator by guiding the helicopter into a safe landing area. In selecting an LZ Coordinator, recognize that medical personnel usually are very busy with the patient at this time. It is recommended that the LZ Coordinator be someone other than a medical responder, if possible. Eye protection should be worn. The ground guide should stand with his/her back to the wind and his/her arms raised over his/her head (flashlights in each hand for night operations.)

2. The pilot will confirm the LZ sighting by radio. If possible, once the pilot has identified the LZ, the ground guide should move out of the LZ.

3. As the helicopter turns into the wind and begins a descent, the LZ coordinator should provide assistance by means of radio contact, or utilize the "unsafe signal" to wave off the helicopter if the LZ is not safe (see FIG 10-2-8). The LZ Coordinator should be far enough from the touchdown area that he/she can still maintain visual contact with the pilot.

i. Assisting the Crew

1. After the helicopter has landed, do not approach the helicopter. The crew will approach you.

2. Be prepared to assist the crew by providing security for the helicopter. If asked to provide security, allow no one but the crew to approach the aircraft.

3. Once the patient is prepared and ready to load, allow the crew to open the doors to the helicopter and guide the loading of the patient.

4. When approaching or departing the helicopter, always be aware of the tail rotor and always follow the directions of the crew. Working around a running helicopter can be potentially dangerous. The environment is very noisy and, with exhaust gases and rotor wash, often windy. In scene operations, the surface may be uneven, soft, or slippery which can lead to tripping. Be very careful of your footing in this environment.

5. The tail rotor poses a special threat to working around a running helicopter. The tail rotor turns many times faster than the main rotor, and is often invisible even at idle engine power. Avoid walking towards the tail of a helicopter beyond the end of the cabin, unless specifically directed by the crew.

NOTE-*Helicopters typically have doors on the sides of the cabin, but many use aft mounted "clamshell" type doors for loading and unloading patients on litters or stretchers. When using these doors, it is important to avoid moving any further aft than necessary to operate the doors and load/unload the patient. Again, always comply with the crew's instructions.*

j. General Rules

1. When working around helicopters, always approach and depart from the front, never from the rear. Approaching from the rear can increase your risk of being struck by the tail rotor, which, when at operating engine speed, is nearly invisible.

2. To prevent injury or damage from the main rotor, never raise anything over your head.

3. If the helicopter landed on a slope, approach and depart from the down slope side only.

4. When the helicopter is loaded and ready for take off, keep the departure path free of vehicles and spectators. In an emergency, this area is needed to execute a landing.

k. Hazardous Chemicals and Gases

1. Responding to accidents involving hazardous materials requires special handling by fire/rescue units on the ground. Equally important are the preparations and considerations for helicopter operations in these areas.

2. Hazardous materials of concern are those which are toxic, poisonous, flammable, explosive, irritating, or radioactive in nature. Helicopter ambulance crews normally don't carry protective suits or breathing apparatuses to protect them from hazardous materials.

3. The helicopter ambulance crew must be told of hazardous materials on the scene in order to avoid the contamination of the crew. Patients/victims contaminated by hazardous materials may require special precautions in packaging before loading on the aircraft for the medical crew's protection, or may be transported by other means.

4. Hazardous chemicals and gases may be fatal to the unprotected person if inhaled or absorbed through the skin.

5. Upon initial radio contact, the helicopter crew must be made aware of any hazardous gases in the area. Never assume that the crew has already been informed. If the aircraft were to fly through the hazardous gases, the crew could be poisoned and/or the engines could develop mechanical problems.

6. Poisonous or irritating gases may cling to a victim's clothing and go unnoticed until the patient is loaded and the doors of the helicopter are closed. To avoid possible compromise of the crew, all of these patients must be decontaminated prior to loading.

l. Hand Signals

1. If unable to make radio contact with the HEMS pilot, use the following signals:

FIGURE 10-2-8
Recommended Landing Zone Ground Signals

m. Emergency Situations

1. In the event of a helicopter accident in the vicinity of the LZ, consider the following:

(a) Emergency Exits:

(1) Doors and emergency exits are typically prominently marked. If possible, operators should familiarize ground responders with the door system on their helicopter in preparation for an emergency event.

(2) In the event of an accident during the LZ operation, be cautious of hazards such as sharp and jagged metal, plastic windows, glass, any rotating components, such as the rotors, and fire sources, such as the fuel tank(s) and the engine.

(b) Fire Suppression:

Helicopters used in HEMS operations are usually powered by turboshaft engines, which use jet fuel. Civil HEMS aircraft typically carry between 50 and 250 gallons of fuel, depending upon the size of the helicopter, and planned flight duration, and the fuel remaining after flying to the scene. Use water to control heat and use foam over fuel to keep vapors from ignition sources.

10-2-4. Emergency Medical Service (EMS) Multiple Helicopter Operations

a. Background. EMS helicopter operators often overlap other EMS operator areas. Standardized procedures can enhance the safety of operating multiple helicopters to landing zones (LZs) and to hospital heliports. Communication is the key to successful operations and in maintaining organization between helicopters, ground units and communication centers. EMS helicopter operators which operate in the same areas should establish joint operating procedures and provide them to related agencies.

b. Recommended Procedures.

1. Landing Zone Operations. The first helicopter to arrive on-scene should establish communications with the ground unit at least 10 NMs from the LZ to receive a LZ briefing and to provide ground control the number of helicopters that can be expected. An attempt should be made to contact other helicopters on 123.025 to pass on to them pertinent LZ information and the ground unit's frequency. Subsequent helicopters arriving on scene should establish communications on 123.025 at least 10 NMs from the LZ. After establishing contact on 123.025, they should contact the ground unit for additional information. All helicopters should monitor 123.025 at all times.

(a) If the landing zone is not established by the ground unit when the first helicopter arrives, then the first helicopter should establish altitude and orbit location requirements for the other arriving helicopters. Recommended altitude separation between helicopters is 500 feet (weather and airspace permitting). Helicopters can orbit on cardinal headings from the scene coordinates. (See FIG 10-2-9.)

(b) Upon landing in the LZ, the first helicopter should update the other helicopters on the LZ conditions, i.e., space, hazards and terrain.

(c) Before initiating any helicopter movement to leave the LZ, all operators should attempt to contact other helicopters on 123.025, and state their position and route of flight intentions for departing the LZ.

2. Hospital Operations. Because many hospitals require landing permission and have established procedures (frequencies to monitor, primary and secondary routes for approaches and departures, and orbiting areas if the heliport is occupied) pilots should always receive a briefing from the appropriate facility (communication center, flight following, etc.) before proceeding to the hospital.

(a) In the event of multiple helicopters coming into the hospital heliport, the helicopter nearest to the heliport should contact other inbound helicopters on 123.025 and establish intentions. Follow the guidelines established in the LZ operations.

(b) To facilitate approach times, the pilot-in-command of the helicopter occupying the hospital heliport should advise any other operators whether the patient will be off loaded with the rotor blades turning or stopped, and the approximate time to do so.

(c) Before making any helicopter movement to leave the hospital heliport, all operators should attempt to contact other helicopters on 123.025 and state their position and route of flight intentions for departing the heliport.

FIGURE 10-2-9
EMS Multiple Helicopter LZ/Heliport Operation

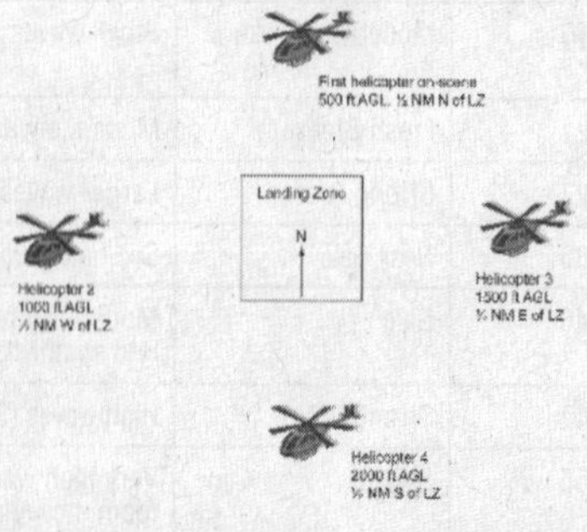

NOTE-*If the LZ/hospital heliport weather conditions or airspace altitude restrictions prohibit the recommended vertical separation, 1 NM separations should be kept between helicopter orbit areas.*

APPENDIX 1.

Bird/Other Wildlife Strike Report

Form Approved OMB NO. 2120-0045

U.S. Department of Transportation
Federal Aviation Administration

BIRD/OTHER WILDLIFE STRIKE REPORT

1. Name of Operator	2. Aircraft Make/Model	3. Engine Make/Model
4. Aircraft Registration	5. Date of Incident ____/____/____ Month Day Year	6. Local Time of Incident ☐ Dawn ☐ Dusk ☐ Day ☐ Night __ HR __ MIN ☐ AM ☐ PM
7. Airport Name	8. Runway Used	9. Location if En Route *(Nearest Town/Reference & State)*
10. Height *(AGL)*	11. Speed *(IAS)*	

12. Phase of Flight

- ☐ A. Parked
- ☐ B. Taxi
- ☐ C. Take-off Run
- ☐ D. Climb
- ☐ E. En Route
- ☐ F. Descent
- ☐ G. Approach
- ☐ H. Landing Roll

13. Part(s) of Aircraft Struck or Damaged

	Struck	Damaged		Struck	Damaged
A. Radome	☐	☐	H. Propeller	☐	☐
B. Windshield	☐	☐	I. Wing/Rotor	☐	☐
C. Nose	☐	☐	J. Fuselage	☐	☐
D. Engine No. 1	☐	☐	K. Landing Gear	☐	☐
E. Engine No. 2	☐	☐	L. Tail	☐	☐
F. Engine No. 3	☐	☐	M. Lights	☐	☐
G. Engine No. 4	☐	☐	N. Other:	☐	☐

(Specify, if "N. Other" is checked)

14. Effect on Flight

- ☐ None
- ☐ Aborted Take-Off
- ☐ Precautionary Landing
- ☐ Engines Shut Down
- ☐ Other: *(Specify)*

15. Sky Condition

- ☐ No Cloud
- ☐ Some Cloud
- ☐ Overcast

16. Precipitation

- ☐ Fog
- ☐ Rain
- ☐ Snow
- ☐ None

17. Bird/Other Wildlife Species

18. Number or birds seen and/or struck

Number of Birds	Seen	Struck
1	☐	☐
2-10	☐	☐
11-100	☐	☐
more than 100	☐	☐

19. Size of Bird(s)

- ☐ Small
- ☐ Medium
- ☐ Large

20. Pilot Warned of Birds ☐ Yes ☐ No

21. Remarks *(Describe damage, injuries and other pertinent information)*

DAMAGE / COST INFORMATION

22. Aircraft time out of service:	23. Estimated cost of repairs or replacement *(U.S. $)*:	24. Estimated other cost *(U.S. $) (e.g. loss of revenue, fuel, hotels)*
______ hours	$	$

Reported by *(Optional)*	Title	Date

Paperwork Reduction Act Statement: The information collected on this form is necessary to allow the Federal Aviation Administration to assess the magnitude and severity of the wildlife-aircraft strike problem in the U.S. The information is used in determining the best management practices for reducing the hazard to aviation safety caused by wildlife-aircraft strikes. We estimate that it will take approximately __5 minutes__ to complete the form. If you wish to make any comments concerning the accuracy of this burden estimate and any suggestions for reducing this burden, send those comments to the Federal Aviation Administration, Management Staff, ARP-10, 800 Independence Avenue, SW, Washington, DC 20591. The information collected is voluntary. Please note that an agency may not conduct or sponsor, and a person is not required to respond to, a collection of information unless it displays a currently valid OMB control number. The OMB control number associated with this collection is 2120-0045.

FAA Form 5200-7 (3-97) Supersedes Previous Edition ★ U.S. GPO: 1997-418-084/59205 NSN:0052-00-651-9005

U.S. Department
of Transportation

Federal Aviation
Administration

800 Independence Ave., S.W.
Washington, D.C. 20591

Official Business
Penalty for Private Use, $300

NO POSTAGE
NECESSARY
IF MAILED IN
THE UNITED
STATES

BUSINESS REPLY MAIL

FIRST CLASS PERMIT NO. 12438 WASHINGTON D.C.

POSTAGE WILL BE PAID BY THE FEDERAL AVIATION ADMINISTRATION

Federal Aviation Administration
Office of Airport Safety and Standards, AAS-310
800 Independence Avenue, SW
WASHINGTON, DC 20591

APPENDIX 2.

Volcanic Activity Reporting Form (VAR)

Date ______________

SECTION 1 - Transmit to ATC via radio

1. Aircraft Identification	
2. Position	
3. Time (UTC)	
4. Flight level or altitude	
5. Position/location of volcanic activity or ash cloud	
6. Air temperature	
7. Wind	
8. Supplementary Information (Brief description of activity including vertical and lateral extent of the ash cloud, horizontal movement, rate of growth, etc., as available.)	

SECTION 2 - Complete and forward as directed

Mark the appropriate box(s)

9. *Density of ash cloud*	☐ wispy	☐ moderately dense	☐ very dense
10. *Color of ash*	☐ white ☐ black	☐ light gray	☐ dark gray
11. *Eruption*	☐ continuous	☐ intermittent	☐ not visible
12. *Position of activity*	☐ summit ☐ multiple	☐ side ☐ not observed	☐ single
13. *Other observed features of eruption*	☐ lightning ☐ ash fallout	☐ glow ☐ mushroom cloud	☐ large rocks ☐ none
14. *Effect on aircraft*	☐ communications ☐ pitot static ☐ none	☐ navigation system ☐ windscreen	☐ engines ☐ other windows
15. *Other effects*	☐ turbulence ☐ ash deposits	☐ St. Elmo's fire	☐ fumes
16. *Other information deemed useful*			

Forward completed form via mail to:

Global Volcanism Program
NHB-119
Smithsonian Institution
Washington, DC 20560

Or Fax to:

Global Volcanism Program
(202) 357-2476

AIM

APPENDIX 3.

Abbreviations/Acronyms

As used in this manual, the following abbreviations/acronyms have the meanings indicated.

Abbreviation/ Acronym	Meaning
AAWU	Alaskan Aviation Weather Unit
AAS	Airport Advisory Service
AC	Advisory Circular
ACAR	Aircraft Communications Addressing and Reporting System
ADCUS	Advise Customs
ADDS	Aviation Digital Data Service
ADF	Automatic Direction Finder
ADIZ	Air Defense Identification Zone
ADS-B	Automatic Dependent Surveillance-Broadcast
AeroNav	Aeronautical Navigation Products
AFB	Air Force Base
AFCS	Automatic Flight Control System
A/FD	Airport/Facility Directory
AFIS	Automatic Flight Information Service
AFM	Aircraft Flight Manual
AGL	Above Ground Level
AHRS	Attitude Heading Reference System
AIM	Aeronautical Information Manual
AIRMET	Airmen's Meteorological Information
ALD	Available Landing Distance
ALS	Approach Light Systems
AMSL	Above Mean Sea Level
ANP	Actual Navigation Performance
AOCC	Airline Operations Control Center
AP	Autopilot System
APV	Approach with Vertical Guidance
ARENA	Areas Noted for Attention
ARFFIC	Aircraft Rescue and Fire Fighting Incident Commander
ARINC	Aeronautical Radio Incorporated
ARO	Airport Reservations Office
ARSA	Airport Radar Service Area
ARSR	Air Route Surveillance Radar
ARTCC	Air Route Traffic Control Center
ARTS	Automated Radar Terminal System
ASDE-X	Airport Surface Detection Equipment - Model X
ASOS	Automated Surface Observing System
ASR	Airport Surveillance Radar
ASRS	Aviation Safety Reporting System
ATC	Air Traffic Control
ATCRBS	Air Traffic Control Radar Beacon System
ATCSCC	Air Traffic Control System Command Center
ATCT	Airport Traffic Control Tower
ATD	Along-Track Distance
ATIS	Automatic Terminal Information Service

Abbreviation/ Acronym	Meaning
ATT	Attitude Retention System
AWC	Aviation Weather Center
AWOS	Automated Weather Observing System
AWSS	Automated Weather Sensor System
AWTT	Aviation Weather Technology Transfer
AWW	Severe Weather Forecast Alert
BAASS	Bigelow Aerospace Advanced Space Studies
BBS	Bulletin Board System
BC	Back Course
BECMG	Becoming group
C/A	Coarse Acquisition
CARTS	Common Automated Radar Terminal System (ARTS) (to include ARTS IIIE and ARTS IIE)
CAT	Clear Air Turbulence
CD	Controller Display
CDI	Course Deviation Indicator
CDR	Coded Departure Route
CERAP	Combined Center/RAPCON
CFA	Controlled Firing Area
CFIT	Controlled Flight into Terrain
CFR	Code of Federal Regulations
COA	Certificate of Waiver or Authorization
CPDLC	Controller Pilot Data Link Communications
CTAF	Common Traffic Advisory Frequency
CVFP	Charted Visual Flight Procedure
CVRS	Computerized Voice Reservation System
CWA	Center Weather Advisory
CWSU	Center Weather Service Unit
DA	Decision Altitude
DCA	Ronald Reagan Washington National Airport
DCP	Data Collection Package
DER	Departure End of Runway
DF	Direction Finder
DH	Decision Height
DME/N	Standard DME
DME/P	Precision DME
DOD	Department of Defense
DP	Instrument Departure Procedure
DPU	Data Processor Unit
DRT	Diversion Recovery Tool
DRVSM	Domestic Reduced Vertical Separation Minimum
DUATS	Direct User Access Terminal System
DVA	Diverse Vector Area
DVFR	Defense Visual Flight Rules

Abbreviation/ Acronym	Meaning
DVRSN	Diversion
EDCT	Expect Departure Clearance Time
EFAS	En Route Flight Advisory Service
EFV	Enhanced Flight Visibility
EFVS	Enhanced Flight Vision System
ELT	Emergency Locator Transmitter
EMAS	Engineered Materials Arresting System
EPE	Estimate of Position Error
ESV	Expanded Service Volume
ETA	Estimated Time of Arrival
ETD	Estimated Time of Departure
ETE	Estimated Time En Route
EWINS	Enhanced Weather Information System
EWR	Newark International Airport
FA	Area Forecast
FAA	Federal Aviation Administration
FAF	Final Approach Fix
FAROS	Final Approach Runway Occupancy Signal
FAWP	Final Approach Waypoint
FB	Fly-by
FCC	Federal Communications Commission
FD	Flight Director System
FDC	Flight Data Center
FDE	Fault Detection and Exclusion
FIR	Flight Information Region
FIS	Flight Information Service
FISDL	Flight Information Services Data Link
FLIP	Flight Information Publication
FMS	Flight Management System
FMSP	Flight Management System Procedure
FO	Fly-over
FPA	Flight Path Angle
FPV	Flight Path Vector
FPNM	Feet Per Nautical Mile
FSDO	Flight Standards District Office
FSS	Flight Service Station
DME	Distance Measuring Equipment
GBAS	Ground Based Augmentation System
GEO	Geostationary Satellite
GLS	GBAS Landing System
GNSS	Global Navigation Satellite System
GNSSP	Global Navigation Satellite System Panel
GPS	Global Positioning System
GRI	Group Repetition Interval

Abbreviation/ Acronym	Meaning
GSD	Geographical Situation Display
GUS	Ground Uplink Station
HAT	Height Above Touchdown
HDTA	High Density Traffic Airports
HEMS	Helicopter Emergency Medical Services
HIRL HIWAS	High Intensity Runway Lights Hazardous Inflight Weather Advisory Service
HRR	Helicopter Rapid Refueling Procedures
HUD	Head-Up Display
Hz	Hertz
IAF	Initial Approach Fix
IAP	Instrument Approach Procedure
IAS	Indicated Air Speed
IAWP	Initial Approach Waypoint
ICAO	International Civil Aviation Organization
IF	Intermediate Fix
IFIM	International Flight Information Manual
IFR	Instrument Flight Rules
ILS	Instrument Landing System
ILS/PRM	Instrument Landing System/ Precision Runway Monitor
IM	Inner Marker
IMC	Instrument Meteorological Conditions
INS	Inertial Navigation System
IOC	Initial Operational Capability
IR	IFR Military Training Route
IRU	Inertial Reference Unit
ITWS	Integrated Terminal Weather System
JFK	John F. Kennedy International Airport
kHz	Kilohertz
LAA	Local Airport Advisory
LAAS	Local Area Augmentation System
LAHSO	Land and Hold Short Operations
LAWRS	Limited Aviation Weather Reporting Station
LDA	Localizer Type Directional Aid
LDA/PRM	Localizer Type Directional Aid/ Precision Runway Monitor
LGA	LaGuardia Airport
LIRL	Low Intensity Runway Lights
LLWAS	Low Level Wind Shear Alert System
LLWASNE	Low Level Wind Shear Alert System Network Expansion
LLWAS-RS	Low Level Wind Shear Alert System Relocation/Sustainment
LNAV	Lateral Navigation
LOC	Localizer
LOP	Line-of-position
LORAN	Long Range Navigation System
LP	Localizer Performance
LPV	Localizer Performance with Vertical Guidance

Abbreviation/ Acronym	Meaning
LUAW	Line Up and Wait
LZ	Landing Zone
MAHWP	Missed Approach Holding Waypoint
MAP	Missed Approach Point
MAWP	Missed Approach Waypoint
MDA	Minimum Descent Altitude
MEA	Minimum En Route Altitude
MEARTS	Micro En Route Automated Radar Tracking System
METAR	Aviation Routine Weather Report
MHz	Megahertz
MIRL	Medium Intensity Runway Lights
MLS	Microwave Landing System
MM	Middle Marker
MOA	Military Operations Area
MOCA	Minimum Obstruction Clearance Altitude
MRA	Minimum Reception Altitude
MRB	Magnetic Reference Bearing
MSA	Minimum Safe Altitude
MSAW	Minimum Safe Altitude Warning
MSL	Mean Sea Level
MTI	Moving Target Indicator
MTOS	Mountain Obscuration
MTR	Military Training Route
MVA	Minimum Vectoring Altitude
MWA	Mountain Wave Activity
MWO	Meteorological Watch Office
NAS	National Airspace System
NASA	National Aeronautics and Space Administration
NAVAID	Navigational Aid
NAVCEN	Coast Guard Navigation Center
NCWF	National Convective Weather Forecast
NDB	Non directional Radio Beacon
NEXRAD NFDC	Next Generation Weather Radar National Flight Data Center
NGA	National Geospatial-Intelligence Agency
NM	Nautical Mile
NMAC	Near Midair Collision
NOAA	National Oceanic and Atmospheric Administration
NOPAC	North Pacific
NoPT	No Procedure Turn Required
NOTAM	Notice to Airmen
NPA	Non precision Approach
NRS	Navigation Reference System
NSA	National Security Area
NSW	No Significant Weather
NTAP	Notices to Airmen Publication
NTSB	National Transportation Safety Board
NTZ	No Transgression Zone
NWS	National Weather Service
OAT	Outside Air Temperature

Abbreviation/ Acronym	Meaning
OBS	Omni-bearing Selector
ODP	Obstacle Departure Procedure
OIS	Operational Information System
OIS	Obstacle Identification Surface
OM	Outer Marker
ORD	Chicago O'Hare International Airport
PA	Precision Approach
PAPI	Precision Approach Path Indicator
PAR	Precision Approach Radar
PAR	Preferred Arrival Route
PC	Personal Computer
P/CG	Pilot/Controller Glossary
PDC	Pre-departure Clearance
PFD	Personal Flotation Device
PinS	Point-in-Space
PIREP	Pilot Weather Report
POB	Persons on Board
POFZ	Precision Obstacle Free Zone
POI	Principal Operations Inspector
PPS	Precise Positioning Service
PRM	Precision Runway Monitor
PT	Procedure Turn
QICP	Qualified Internet Communications Provider
RA	Resolution Advisory
RAA	Remote Advisory Airport
RAIM	Receiver Autonomous Integrity Monitoring
RAIS	Remote Airport Information Service
RBDT	Ribbon Display Terminals
RCAG	Remote Center Air/Ground
RCC	Rescue Coordination Center
RCLS	Runway Centerline Lighting System
RCO	Remote Communications Outlet
RD	Rotor Diameter
REIL	Runway End Identifier Lights
REL	Runway Entrance Lights
RFM	Rotorcraft Flight Manual
RIL	Runway Intersection Lights
RLIM	Runway Light Intensity Monitor
RMI	Radio Magnetic Indicator
RNAV	Area Navigation
RNP	Required Navigation Performance
ROC	Required Obstacle Clearance
RPAT	RNP Parallel Approach Runway Transitions
RVR	Runway Visual Range
RVSM	Reduced Vertical Separation Minimum
RWSL	Runway Status Light
SAAAR	Special Aircraft and Aircrew Authorization Required
SAM	System Area Monitor
SAR	Search and Rescue
SAS	Stability Augmentation System

Abbreviation/ Acronym	Meaning
SBAS	Satellite-based Augmentation System
SCAT-1 DGPS	Special Category I Differential GPS
SDF	Simplified Directional Facility
SFL	Sequenced Flashing Lights
SFR SIAP	Special Flight Rules Standard Instrument Approach Procedure
SID	Standard Instrument Departure
SIGMET	Significant Meteorological Information
SM	Statute Mile
SMGCS	Surface Movement Guidance Control System
SNR	Signal-to-noise Ratio
SOIA	Simultaneous Offset Instrument Approaches
SOP	Standard Operating Procedure
SPC	Storm Prediction Center
SPS	Standard Positioning Service
STAR	Standard Terminal Arrival
STARS	Standard Terminal Automation Replacement System
STMP	Special Traffic Management Program
SWSL	Supplemental Weather Service Locations
TA	Traffic Advisory
TAA	Terminal Arrival Area
TAC	Terminal Area Chart
TACAN	Tactical Air Navigation
TAF	Aerodrome Forecast
TAS	True Air Speed
TCAS	Traffic Alert and Collision Avoidance System
TCH	Threshold Crossing Height
TD	Time Difference
TDLS	Tower Data Link System
TDWR	Terminal Doppler Weather Radar
TDZ	Touchdown Zone
TDZE	Touchdown Zone Elevation
TDZL	Touchdown Zone Lights
TEC	Tower En Route Control

Abbreviation/ Acronym	Meaning
THL	Takeoff Hold Lights
TIBS	Telephone Information Briefing Service
TIS	Traffic Information Service
TIS-B	Traffic Information Service-Broadcast
TLS	Transponder Landing System
TPP	Terminal Procedures Publications
TRSA	Terminal Radar Service Area
TSO	Technical Standard Order
TWEB	Transcribed Weather Broadcast
TWIB	Terminal Weather Information for Pilots System
UA	Unmanned Aircraft
UAS	Unmanned Aircraft System
UAV	Unmanned Aerial Vehicle
UFO	Unidentified Flying Object
UHF	Ultrahigh Frequency
U.S.	United States
USCG	United States Coast Guard
UTC	Coordinated Universal Time
UWS	Urgent Weather SIGMET
VAR	Volcanic Activity Reporting
VASI	Visual Approach Slope Indicator
VCOA	Visual Climb Over the Airport
VDA	Vertical Descent Angle
VDP	Visual Descent Point
VFR	Visual Flight Rules
VGSI	Visual Glide Slope Indicator
VHF	Very High Frequency
VIP	Video Integrator Processor
VMC	Visual Meteorological Conditions
V_{MINI}	Instrument flight minimum speed, utilized in complying with minimum limit speed requirements for instrument flight
VNAV	Vertical Navigation
V_{NE}	Never exceed speed
V_{NEI}	Instrument flight never exceed speed, utilized instead of VNE for compliance with maximum limit speed requirements for instrument flight

Abbreviation/ Acronym	Meaning
VOR	Very High Frequency Omni-directional Range
VORTAC	VHF Omni-directional Range/Tactical Air Navigation
VOT	VOR Test Facility
VR	VFR Military Training Route
V_{REF}	The reference landing approach speed, usually about 1 .3 times Vso plus 5 0 percent of the wind gust speed in excess of the mean wind speed.
V_{SO}	The stalling speed or the minimum steady flight speed in the landing configuration at maximum weight.
VTF	Vector to Final
VV	Vertical Visibility
VVI	Vertical Velocity Indicator
V_Y	Speed for best rate of climb
V_{YI}	Instrument climb speed, utilized instead of VY for compliance with the climb requirements for instrument flight
WA	AIRMET
WAAS	Wide Area Augmentation System
WAC	World Aeronautical Chart
WFO	Weather Forecast Office
WGS-84	World Geodetic System of 1984
WMO	World Meteorological Organization
WMS	Wide-Area Master Station
WMSC	Weather Message Switching Center
WMSCR	Weather Message Switching Center Replacement
WP	Waypoint
WRS	Wide-Area Ground Reference Station
WS	SIGMET
WSO	Weather Service Office
WSP	Weather System Processor
WST	Convective Significant Meteorological Information
WW	Severe Weather Watch Bulletin

PILOT/CONTROLLER GLOSSARY

PURPOSE

a. This Glossary was compiled to promote a common understanding of the terms used in the Air Traffic Control system. It includes those terms which are intended for pilot/controller communications. Those terms most frequently used in pilot/controller communications are printed in ***bold italics***. The definitions are primarily defined in an operational sense applicable to both users and operators of the National Airspace System. Use of the Glossary will preclude any misunderstandings concerning the system's design, function, and purpose.

b. Because of the international nature of flying, terms used in the Lexicon, published by the International Civil Aviation Organization (ICAO), are included when they differ from FAA definitions. These terms are followed by "[ICAO]." For the reader's convenience, there are also cross references to related terms in other parts of the Glossary and to other documents, such as the Code of Federal Regulations (CFR) and the Aeronautical Information Manual (AIM).

c. This Glossary will be revised, as necessary, to maintain a common understanding of the system.

EXPLANATION OF CHANGES

a. Terms Added:
TEMPORARY FLIGHT RESTRICTION
TIE-IN-FACILITY

b. Terms Modified:
ACKNOWLEDGE
CLEARANCE LIMIT
FAST FILE
FLIGHT PLAN AREA
FLIGHT SERVICE STATION HELICOPTER
VERTICAL SEPARATION

c. Editorial/format changes were made where necessary. Revision bars were not used due to the insignificant nature of the changes.

A

AAI-
(See ARRIVAL AIRCRAFT INTERVAL.) AAR-
(See AIRPORT ARRIVAL RATE.)
ABBREVIATED IFR FLIGHT PLANS- An authorization by ATC requiring pilots to submit only that information needed for the purpose of ATC. It includes only a small portion of the usual IFR flight plan information. In certain instances, this may be only aircraft identification, location, and pilot request. Other information may be requested if needed by ATC for separation/control purposes. It is frequently used by aircraft which are airborne and desire an instrument approach or by aircraft which are on the ground and desire a climb to VFR-on-top.
(See VFR-ON-TOP.) (Refer to AIM.)
ABEAM- An aircraft is "abeam" a fix, point, or object when that fix, point, or object is approximately 90 degrees to the right or left of the aircraft track. Abeam indicates a general position rather than a precise point.
ABORT- To terminate a preplanned aircraft maneuver; e.g., an aborted takeoff. ACC [ICAO]-
(See ICAO term AREA CONTROL CENTER.)
ACCELERATE-STOP DISTANCE AVAILABLE- The runway plus stopway length declared available and suitable for the acceleration and deceleration of an airplane aborting a takeoff.
ACCELERATE-STOP DISTANCE AVAILABLE [ICAO]- The length of the take-off run available plus the length of the stopway if provided.
ACDO-
(See AIR CARRIER DISTRICT OFFICE.)
ACKNOWLEDGE- Let me know that you have received and understood this message.
ACL-
(See AIRCRAFT LIST.)
ACLS-
(See AUTOMATIC CARRIER LANDING SYSTEM.) ACLT-
(See ACTUAL CALCULATED LANDING TIME.)
ACROBATIC FLIGHT- An intentional maneuver involving an abrupt change in an aircraft's attitude, an abnormal attitude, or abnormal acceleration not necessary for normal flight.
(See ICAO term ACROBATIC FLIGHT.) (Refer to 14 CFR Part 91.)
ACROBATIC FLIGHT [ICAO]- Maneuvers intentionally performed by an aircraft involving an abrupt change in its attitude, an abnormal attitude, or an abnormal variation in speed.
ACTIVE RUNWAY-
(See RUNWAY IN USE/ACTIVE RUNWAY/DUTY RUNWAY.)
ACTUAL CALCULATED LANDING TIME- ACLT is a flight's frozen calculated landing time. An actual time determined at freeze calculated landing time (FCLT) or meter list display interval (MLDI) for the adapted vertex for each arrival aircraft based upon runway configuration, airport acceptance rate, airport arrival delay period, and other metered arrival aircraft. This time is either the vertex time of arrival (VTA) of the aircraft or the tentative calculated landing time (TCLT)/ACLT of the previous aircraft plus the arrival aircraft interval (AAI), whichever is later. This time will not be updated in response to the aircraft's progress.
ACTUAL NAVIGATION PERFORMANCE (ANP)- (See REQUIRED NAVIGATION PERFORMANCE.)
ADDITIONAL SERVICES- Advisory information provided by ATC which includes but is not limited to the following:
a. Traffic advisories.
b. Vectors, when requested by the pilot, to assist aircraft receiving traffic advisories to avoid observed traffic.
c. Altitude deviation information of 300 feet or more from an assigned altitude as observed on a verified (reading correctly) automatic altitude readout (Mode C).
d. Advisories that traffic is no longer a factor.
e. Weather and chaff information.
f. Weather assistance.
g. Bird activity information.
h. Holding pattern surveillance. Additional services are provided to the extent possible contingent only upon the controller's capability to fit them into the performance of higher priority duties and on the basis of limitations of the radar, volume of traffic, frequency congestion, and controller workload. The controller has complete discretion for determining if he/she is able to provide or continue to provide a service in a particular case. The controller's reason not to provide or continue to provide a service in a particular case is not subject to question by the pilot and need not be made known to him/her.
(See TRAFFIC ADVISORIES.)
(Refer to AIM.)
ADF-
(See AUTOMATIC DIRECTION FINDER.) ADIZ-
(See AIR DEFENSE IDENTIFICATION ZONE.) ADLY-
(See ARRIVAL DELAY.)
ADMINISTRATOR- The Federal Aviation Administrator or any person to whom he/she has delegated his/her authority in the matter concerned.
ADR-
(See AIRPORT DEPARTURE RATE.)
ADS [ICAO]-
(See ICAO term AUTOMATIC DEPENDENT SURVEILLANCE.) ADS-B-
(See AUTOMATIC DEPENDENT SURVEILLANCE-BROADCAST.) ADS-C-
(See AUTOMATIC DEPENDENT SURVEILLANCE-CONTRACT.) ***ADVISE INTENTIONS-*** Tell me what you plan to do.
ADVISORY- Advice and information provided to assist pilots in the safe conduct of flight and aircraft movement.
(See ADVISORY SERVICE.)
ADVISORY FREQUENCY- The appropriate frequency to be used for Airport Advisory Service.
(See LOCAL AIRPORT ADVISORY.)
(See UNICOM.)
(Refer to ADVISORY CIRCULAR NO. 90-42.) (Refer to AIM.)
ADVISORY SERVICE- Advice and information provided by a facility to assist pilots in the safe conduct of flight and aircraft movement.
(See ADDITIONAL SERVICES.)
(See EN ROUTE FLIGHT ADVISORY SERVICE.) (See LOCAL AIRPORT ADVISORY.)
(See RADAR ADVISORY.)
(See SAFETY ALERT.)
(See TRAFFIC ADVISORIES.)
(Refer to AIM.)
AERIAL REFUELING- A procedure used by the military to transfer fuel from one aircraft to another during flight.
(Refer to VFR/IFR Wall Planning Charts.)
AERODROME- A defined area on land or water (including any buildings, installations and equipment) intended to be used either wholly or in part for the arrival, departure, and movement of aircraft.
AERODROME BEACON [ICAO]- Aeronautical beacon used to indicate the location of an aerodrome from the air.
AERODROME CONTROL SERVICE [ICAO]- Air traffic control service for aerodrome traffic.
AERODROME CONTROL TOWER [ICAO]- A unit established to provide air traffic control service to aerodrome traffic.
AERODROME ELEVATION [ICAO]- The elevation of the highest point of the landing area.
AERODROME TRAFFIC CIRCUIT [ICAO]- The specified path to be flown by aircraft operating in the vicinity of an aerodrome.
AERONAUTICAL BEACON- A visual NAVAID displaying flashes of white and/or colored light to indicate the location of an airport, a heliport, a landmark, a certain point of a Federal airway in mountainous terrain, or an obstruction.
(See AIRPORT ROTATING BEACON.) (Refer to AIM.)
AERONAUTICAL CHART- A map used in air navigation containing all or part of the following: topographic features, hazards and obstructions, navigation aids, navigation routes, designated airspace, and airports. Commonly used aeronautical charts are:
a. Sectional Aeronautical Charts (1:500,000)- Designed for visual navigation of slow or medium speed aircraft. Topographic information on these charts features the portrayal of relief and a judicious selection of visual check points for VFR flight. Aeronautical information includes visual and radio aids to navigation, airports, controlled airspace, restricted areas, obstructions, and related data.
b. VFR Terminal Area Charts (1:250,000)- Depict Class B airspace which provides for the control or segregation of all the aircraft within Class B airspace. The chart depicts topographic information and aeronautical information which includes visual and radio aids to navigation, airports, controlled airspace, restricted areas, obstructions, and related data.
c. World Aeronautical Charts (WAC) (1:1,000,000)- Provide a standard series of aeronautical charts covering land areas of the world at a size and scale convenient for navigation by moderate speed aircraft. Topographic information includes cities and towns, principal roads, railroads, distinctive landmarks, drainage, and relief. Aeronautical information includes visual and radio aids to navigation, airports, airways, restricted areas, obstructions, and other pertinent data.
d. En Route Low Altitude Charts- Provide aeronautical information for en route instrument navigation (IFR) in the low altitude stratum. Information includes the portrayal of airways, limits of controlled airspace, position identification and frequencies of radio aids, selected airports, minimum en route and minimum obstruction clearance altitudes, airway distances, reporting points, restricted areas, and related data. Area charts, which are a part of this series, furnish terminal data at a larger scale in congested areas.
e. En Route High Altitude Charts- Provide aeronautical information for en route instrument navigation (IFR) in the high altitude stratum. Information includes the portrayal of jet routes, identification and frequencies of radio aids, selected airports, distances, time zones, special use airspace, and related information.
f. Instrument Approach Procedures (IAP) Charts- Portray the aeronautical data which is required to execute an instrument approach to an airport. These charts depict the procedures, including all related data, and the airport diagram. Each procedure is designated for use with a specific type of electronic navigation system including NDB, TACAN, VOR, ILS/MLS, and RNAV. These charts are identified by the type of navigational aid(s) which provide final approach guidance.
g. Instrument Departure Procedure (DP) Charts- Designed to expedite clearance delivery and to facilitate transition between takeoff and en route operations. Each DP is presented as a separate chart and may serve a single airport or more than one airport in a given geographical location.

h. Standard Terminal Arrival (STAR) Charts- Designed to expedite air traffic control arrival procedures and to facilitate transition between en route and instrument approach operations. Each STAR procedure is presented as a separate chart and may serve a single airport or more than one airport in a given geographical location.

i. Airport Taxi Charts- Designed to expedite the efficient and safe flow of ground traffic at an airport. These charts are identified by the official airport name; e.g., Ronald Reagan Washington National Airport.

(See ICAO term AERONAUTICAL CHART.)

AERONAUTICAL CHART [ICAO]- A representation of a portion of the earth, its culture and relief, specifically designated to meet the requirements of air navigation.

AERONAUTICAL INFORMATION MANUAL (AIM)- A primary FAA publication whose purpose is to instruct airmen about operating in the National Airspace System of the U.S. It provides basic flight information, ATC Procedures and general instructional information concerning health, medical facts, factors affecting flight safety, accident and hazard reporting, and types of aeronautical charts and their use.

AERONAUTICAL INFORMATION PUBLICATION (AIP) [ICAO]- A publication issued by or with the authority of a State and containing aeronautical information of a lasting character essential to air navigation.

A/FD-

(See AIRPORT/FACILITY DIRECTORY.)

AFFIRMATIVE- Yes.

AFIS-

(See AUTOMATIC FLIGHT INFORMATION SERVICE - ALASKA FSSs ONLY.)

AFP-

(See AIRSPACE FLOW PROGRAM.)

AIM-

(See AERONAUTICAL INFORMATION MANUAL.)

AIP [ICAO]-

(See ICAO term AERONAUTICAL INFORMATION PUBLICATION.)

AIR CARRIER DISTRICT OFFICE- An FAA field office serving an assigned geographical area, staffed with Flight Standards personnel serving the aviation industry and the general public on matters related to the certification and operation of scheduled air carriers and other large aircraft operations.

AIR DEFENSE EMERGENCY- A military emergency condition declared by a designated authority. This condition exists when an attack upon the continental U.S., Alaska, Canada, or U.S. installations in Greenland by hostile aircraft or missiles is considered probable, is imminent, or is taking place.

(Refer to AIM.)

AIR DEFENSE IDENTIFICATION ZONE (ADIZ)- The area of airspace over land or water, extending upward from the surface, within which the ready identification, the location, and the control of aircraft are required in the interest of national security.

a. Domestic Air Defense Identification Zone. An ADIZ within the United States along an international boundary of the United States.

b. Coastal Air Defense Identification Zone. An ADIZ over the coastal waters of the United States.

c. Distant Early Warning Identification Zone (DEWIZ). An ADIZ over the coastal waters of the State of Alaska.

d. Land-Based Air Defense Identification Zone. An ADIZ over U.S. metropolitan areas, which is activated and deactivated as needed, with dimensions, activation dates and other relevant information disseminated via NOTAM.

Note: ADIZ locations and operating and flight plan requirements for civil aircraft operations are specified in 14 CFR Part 99.

(Refer to AIM.)

AIR NAVIGATION FACILITY- Any facility used in, available for use in, or designed for use in, aid of air navigation, including landing areas, lights, any apparatus or equipment for disseminating weather information, for signaling, for radio-directional finding, or for radio or other electrical communication, and any other structure or mechanism having a similar purpose for guiding or controlling flight in the air or the landing and takeoff of aircraft.

(See NAVIGATIONAL AID.)

AIR ROUTE SURVEILLANCE RADAR- Air route traffic control center (ARTCC) radar used primarily to detect and display an aircraft's position while en route between terminal areas. The ARSR enables controllers to provide radar air traffic control service when aircraft are within the ARSR coverage. In some instances, ARSR may enable an ARTCC to provide terminal radar services similar to but usually more limited than those provided by a radar approach control.

AIR ROUTE TRAFFIC CONTROL CENTER- A facility established to provide air traffic control service to aircraft operating on IFR flight plans within controlled airspace and principally during the en route phase of flight. When equipment capabilities and controller workload permit, certain advisory/assistance services may be provided to VFR aircraft.

(See EN ROUTE AIR TRAFFIC CONTROL SERVICES.) (Refer to AIM.)

AIR TAXI- Used to describe a helicopter/VTOL aircraft movement conducted above the surface but normally not above 100 feet AGL. The aircraft may proceed either via hover taxi or flight at speeds more than 20 knots. The pilot is solely responsible for selecting a safe airspeed/altitude for the operation being conducted.

(See HOVER TAXI.) (Refer to AIM.)

AIR TRAFFIC- Aircraft operating in the air or on an airport surface, exclusive of loading ramps and parking areas.

(See ICAO term AIR TRAFFIC.)

AIR TRAFFIC [ICAO]- All aircraft in flight or operating on the maneuvering area of an aerodrome.

AIR TRAFFIC CLEARANCE- An authorization by air traffic control for the purpose of preventing collision between known aircraft, for an aircraft to proceed under specified traffic conditions within controlled airspace. The pilot-in-command of an aircraft may not deviate from the provisions of a visual flight rules (VFR) or instrument flight rules (IFR) air traffic clearance except in an emergency or unless an amended clearance has been obtained. Additionally, the pilot may request a different clearance from that which has been issued by air traffic control (ATC) if information available to the pilot makes another course of action more practicable or if aircraft equipment limitations or company procedures forbid compliance with the clearance issued. Pilots may also request clarification or amendment, as appropriate, any time a clearance is not fully understood, or considered unacceptable because of safety of flight. Controllers should, in such instances and to the extent of operational practicality and safety, honor the pilot's request. 14 CFR Part 91.3(a) states: "The pilot in command of an aircraft is directly responsible for, and is the final authority as to, the operation of that aircraft." THE PILOT IS RESPONSIBLE TO REQUEST AN AMENDED CLEARANCE if ATC issues a clearance that would cause a pilot to deviate from a rule or regulation, or in the pilot's opinion, would place the aircraft in jeopardy.

(See ATC INSTRUCTIONS.)

(See ICAO term AIR TRAFFIC CONTROL CLEARANCE.)

AIR TRAFFIC CONTROL- A service operated by appropriate authority to promote the safe, orderly and expeditious flow of air traffic.

(See ICAO term AIR TRAFFIC CONTROL SERVICE.)

AIR TRAFFIC CONTROL CLEARANCE [ICAO]- Authorization for an aircraft to proceed under conditions specified by an air traffic control unit.

Note 1: For convenience, the term air traffic control clearance is frequently abbreviated to clearance when used in appropriate contexts.

Note 2: The abbreviated term clearance may be prefixed by the words taxi, takeoff, departure, en route, approach or landing to indicate the particular portion of flight to which the air traffic control clearance relates.

AIR TRAFFIC CONTROL SERVICE-

(See AIR TRAFFIC CONTROL.)

AIR TRAFFIC CONTROL SERVICE [ICAO]- A service provided for the purpose of:

a. Preventing collisions:

1. Between aircraft; and

2. On the maneuvering area between aircraft and obstructions.

b. Expediting and maintaining an orderly flow of air traffic.

AIR TRAFFIC CONTROL SPECIALIST- A person authorized to provide air traffic control service.

(See AIR TRAFFIC CONTROL.) (See FLIGHT SERVICE STATION.) (See ICAO term CONTROLLER.)

AIR TRAFFIC CONTROL SYSTEM COMMAND CENTER (ATCSCC) - An Air Traffic Tactical Operations facility responsible for monitoring and managing the flow of air traffic throughout the NAS, producing a safe, orderly, and expeditious flow of traffic while minimizing delays. The following functions are located at the ATCSCC:

a. Central Altitude Reservation Function (CARF). Responsible for coordinating, planning, and approving special user requirements under the Altitude Reservation (ALTRV) concept.

(See ALTITUDE RESERVATION.)

b. Airport Reservation Office (ARO). Responsible for approving IFR flights at designated high density traffic airports (John F. Kennedy, LaGuardia, and Ronald Reagan Washington National) during specified hours.

(Refer to 14 CFR Part 93.)

(Refer to AIRPORT/FACILITY DIRECTORY.)

c. U.S. Notice to Airmen (NOTAM) Office. Responsible for collecting, maintaining, and distributing NOTAMs for the U.S. civilian and military, as well as international aviation communities.

(See NOTICE TO AIRMEN.)

d. Weather Unit. Monitor all aspects of weather for the U.S. that might affect aviation including cloud cover, visibility, winds, precipitation, thunderstorms, icing, turbulence, and more. Provide forecasts based on observations and on discussions with meteorologists from various National Weather Service offices, FAA facilities, airlines, and private weather services.

AIR TRAFFIC SERVICE- A generic term meaning:

a. Flight Information Service.

b. Alerting Service.

c. Air Traffic Advisory Service.

d. Air Traffic Control Service:

1. Area Control Service,

2. Approach Control Service, or

3. Airport Control Service.

AIR TRAFFIC SERVICE (ATS) ROUTES - The term "ATS Route" is a generic term that includes "VOR Federal airways," "colored Federal airways," "jet routes," and "RNAV routes." The term "ATS route" does not replace these more familiar route names, but serves only as an overall title when listing the types of routes that comprise the United States route structure.

AIRBORNE- An aircraft is considered airborne when all parts of the aircraft are off the ground.

AIRBORNE DELAY- Amount of delay to be encountered in airborne holding.

AIRCRAFT- Device(s) that are used or intended to be used for flight in the air, and when used in air traffic control terminology, may include the flight crew.

(See ICAO term AIRCRAFT.)

AIRCRAFT [ICAO]- Any machine that can derive support in the atmosphere from the reactions of the air other than the reactions of the air against the earth's surface.

AIRCRAFT APPROACH CATEGORY- A grouping of aircraft based on a speed of 1.3 times the stall speed in the landing configuration at maximum gross landing weight. An aircraft must fit in only one category. If it is necessary to maneuver at speeds in excess of the upper limit of a speed range for a category, the minimums for the category for that speed must be used. For example, an aircraft which falls in Category A, but is circling to land at a speed in excess of 91 knots, must use the approach Category B minimums when circling to land. The categories are as follows:

a. Category A- Speed less than 91 knots.

b. Category B- Speed 91 knots or more but less than 121 knots.

c. Category C- Speed 121 knots or more but less than 141 knots.

d. Category D- Speed 141 knots or more but less than 166 knots.

e. Category E- Speed 166 knots or more.

(Refer to 14 CFR Part 97.)

AIRCRAFT CLASSES- For the purposes of Wake Turbulence Separation Minima, ATC classifies aircraft as Heavy, Large, and Small as follows:

a. Heavy- Aircraft capable of takeoff weights of 300,000 pounds or more whether or not they are operating at this weight during a particular phase of flight.

b. Large- Aircraft of more than 41,000 pounds, maximum certificated takeoff weight, up to but not including 300,000 pounds.

c. Small- Aircraft of 41,000 pounds or less maximum certificated takeoff weight. (Refer to AIM.)

AIRCRAFT CONFLICT- Predicted conflict, within URET, of two aircraft, or between aircraft and airspace. A Red alert is used for conflicts when the predicted minimum separation is 5 nautical miles or less. A Yellow alert is used when the predicted minimum separation is between 5 and approximately 12 nautical miles. A Blue alert is used for conflicts between an aircraft and predefined airspace. (See USER REQUEST EVALUATION TOOL.)

AIRCRAFT LIST (ACL)- A view available with URET that lists aircraft currently in or predicted to be in a particular sector's airspace. The view contains textual flight data information in line format and may be sorted into various orders based on the specific needs of the sector team.

(See USER REQUEST EVALUATION TOOL.)

AIRCRAFT SURGE LAUNCH AND RECOVERY- Procedures used at USAF bases to provide increased launch and recovery rates in instrument flight rules conditions. ASLAR is based on:

a. Reduced separation between aircraft which is based on time or distance. Standard arrival separation applies between participants including multiple flights until the DRAG point. The DRAG point is a published location on an ASLAR approach where aircraft landing second in a formation slows to a predetermined airspeed. The DRAG point is the reference point at which MARSA applies as expanding elements effect separation within a flight or between subsequent participating flights.

b. ASLAR procedures shall be covered in a Letter of Agreement between the responsible USAF military ATC facility and the concerned Federal Aviation Administration facility. Initial Approach Fix spacing requirements are normally addressed as a minimum.

AIRMEN'S METEOROLOGICAL INFORMATION-

(See AIRMET.)

AIRMET- In-flight weather advisories issued only to amend the area forecast concerning weather phenomena which are of operational interest to all aircraft and potentially hazardous to aircraft having limited capability because of lack of equipment, instrumentation, or pilot qualifications. AIRMETs concern weather of less severity than that covered by SIGMETs or Convective SIGMETs. AIRMETs cover moderate icing, moderate turbulence, sustained winds of 30 knots or more at the surface, widespread areas of ceilings less than 1,000 feet and/or visibility less than 3 miles, and extensive mountain obscurement.

(See AWW.)

(See CONVECTIVE SIGMET.) (See CWA.)

(See SIGMET.)

(Refer to AIM.)

AIRPORT- An area on land or water that is used or intended to be used for the landing and takeoff of aircraft and includes its buildings and facilities, if any.

AIRPORT ADVISORY AREA- The area within ten miles of an airport without a control tower or where the tower is not in operation, and on which a Flight Service Station is located.

(See LOCAL AIRPORT ADVISORY.) (Refer to AIM.)

AIRPORT ARRIVAL RATE (AAR)- A dynamic input parameter specifying the number of arriving aircraft which an airport or airspace can accept from the ARTCC per hour. The AAR is used to calculate the desired interval between successive arrival aircraft.

AIRPORT DEPARTURE RATE (ADR)- A dynamic parameter specifying the number of aircraft which can depart an airport and the airspace can accept per hour.

AIRPORT ELEVATION- The highest point of an airport's usable runways measured in feet from mean sea level.

(See TOUCHDOWN ZONE ELEVATION.) (See ICAO term AERODROME ELEVATION.)

AIRPORT/FACILITY DIRECTORY- A publication designed primarily as a pilot's operational manual containing all airports, seaplane bases, and heliports open to the public including communications data, navigational facilities, and certain special notices and procedures. This publication is issued in seven volumes according to geographical area.

AIRPORT LIGHTING- Various lighting aids that may be installed on an airport. Types of airport lighting include:

a. Approach Light System (ALS)- An airport lighting facility which provides visual guidance to landing aircraft by radiating light beams in a directional pattern by which the pilot aligns the aircraft with the extended centerline of the runway on his/her final approach for landing. Condenser-Discharge Sequential Flashing Lights/Sequenced Flashing Lights may be installed in conjunction with the ALS at some airports. Types of Approach Light Systems are:

1. ALSF-1- Approach Light System with Sequenced Flashing Lights in ILS Cat-I configuration.

2. ALSF-2- Approach Light System with Sequenced Flashing Lights in ILS Cat-II configuration. The ALSF-2 may operate as an SSALR when weather conditions permit.

3. SSALF- Simplified Short Approach Light System with Sequenced Flashing Lights.

4. SSALR- Simplified Short Approach Light System with Runway Alignment Indicator Lights.

5. MALSF- Medium Intensity Approach Light System with Sequenced Flashing Lights.

6. MALSR- Medium Intensity Approach Light System with Runway Alignment Indicator Lights.

7. RLLS- Runway Lead-in Light System Consists of one or more series of flashing lights installed at or near ground level that provides positive visual guidance along an approach path, either curving or straight, where special problems exist with hazardous terrain, obstructions, or noise abatement procedures.

8. RAIL- Runway Alignment Indicator Lights- Sequenced Flashing Lights which are installed only in combination with other light systems.

9. ODALS- Omnidirectional Approach Lighting System consists of seven omnidirectional flashing lights located in the approach area of a nonprecision runway. Five lights are located on the runway centerline extended with the first light located 300 feet from the threshold and extending at equal intervals up to 1,500 feet from the threshold. The other two lights are located, one on each side of the runway threshold, at a lateral distance of 40 feet from the runway edge, or 75 feet from the runway edge when installed on a runway equipped with a VASI.

(Refer to FAAO JO 6850.2, VISUAL GUIDANCE LIGHTING SYSTEMS.)

b. Runway Lights/Runway Edge Lights- Lights having a prescribed angle of emission used to define the lateral limits of a runway. Runway lights are uniformly spaced at intervals of approximately 200 feet, and the intensity may be controlled or preset.

c. Touchdown Zone Lighting- Two rows of transverse light bars located symmetrically about the runway centerline normally at 100 foot intervals. The basic system extends 3,000 feet along the runway.

d. Runway Centerline Lighting- Flush centerline lights spaced at 50-foot intervals beginning 75 feet from the landing threshold and extending to within 75 feet of the opposite end of the runway.

e. Threshold Lights- Fixed green lights arranged symmetrically left and right of the runway centerline, identifying the runway threshold.

f. Runway End Identifier Lights (REIL)- Two synchronized flashing lights, one on each side of the runway threshold, which provide rapid and positive identification of the approach end of a particular runway.

g. Visual Approach Slope Indicator (VASI)- An airport lighting facility providing vertical visual approach slope guidance to aircraft during approach to landing by radiating a directional pattern of high intensity red and white focused light beams which indicate to the pilot that he/she is "on path" if he/she sees red/white, "above path" if white/white, and "below path" if red/red. Some airports serving large aircraft have three-bar VASIs which provide two visual glide paths to the same runway.

h. Precision Approach Path Indicator (PAPI)- An airport lighting facility, similar to VASI, providing vertical approach slope guidance to aircraft during approach to landing. PAPIs consist of a single row of either two or four lights, normally installed on the left side of the runway, and have an effective visual range of about 5 miles during the day and up to 20 miles at night. PAPIs radiate a directional pattern of high intensity red and white focused light beams which indicate that the pilot is "on path" if the pilot sees an equal number of white lights and red lights, with white to the left of the red; "above path" if the pilot sees more white than red lights; and "below path" if the pilot sees more red than white lights.

i. Boundary Lights- Lights defining the perimeter of an airport or landing area. (Refer to AIM.)

AIRPORT MARKING AIDS- Markings used on runway and taxiway surfaces to identify a specific runway, a runway threshold, a centerline, a hold line, etc. A runway should be marked in accordance with its present usage such as:

a. Visual.

b. Nonprecision instrument.

c. Precision instrument. (Refer to AIM.)

AIRPORT REFERENCE POINT (ARP)- The approximate geometric center of all usable runway surfaces.

AIRPORT RESERVATION OFFICE- Office responsible for monitoring the operation of slot controlled airports. It receives and processes requests for unscheduled operations at slot controlled airports.

AIRPORT ROTATING BEACON- A visual NAVAID operated at many airports. At civil airports, alternating white and green flashes indicate the location of the airport. At military airports, the beacons flash alternately white and green, but are differentiated from civil beacons by dualpeaked (two quick) white flashes between the green flashes.

(See INSTRUMENT FLIGHT RULES.) (See SPECIAL VFR OPERATIONS.)

(See ICAO term AERODROME BEACON.) (Refer to AIM.)

AIRPORT STREAM FILTER (ASF)- An on/off filter that allows the conflict notification function to be inhibited for arrival streams into single or multiple airports to prevent-nuisance alerts.

AIRPORT SURFACE DETECTION EQUIPMENT (ASDE)- Surveillance equipment specifically designed to detect aircraft, vehicular traffic, and other objects, on the surface of an airport, and to present the image on a tower display. Used to augment visual observation by tower personnel of aircraft and/or vehicular movements on runways and taxiways. There are three ASDE systems deployed in the NAS:

a. ASDE-3- a Surface Movement Radar.

b. ASDE-X- a system that uses a X-band Surface Movement Radar and multilateration. Data from these two sources are fused and presented on a digital display.

c. ASDE-3X- an ASDE-X system that uses the ASDE-3 Surface Movement Radar.

AIRPORT SURVEILLANCE RADAR- Approach control radar used to detect and display an aircraft's position in the terminal area. ASR provides range and azimuth information but does not provide elevation data. Coverage of the ASR can extend up to 60 miles.

AIRPORT TAXI CHARTS-

(See AERONAUTICAL CHART.)

AIRPORT TRAFFIC CONTROL SERVICE- A service provided by a control tower for aircraft operating on the movement area and in the vicinity of an airport.

(See MOVEMENT AREA.)

(See TOWER.)

(See ICAO term AERODROME CONTROL SERVICE.)

AIRPORT TRAFFIC CONTROL TOWER-

(See TOWER.)

AIRSPACE CONFLICT- Predicted conflict of an aircraft and active Special Activity Airspace (SAA).

AIRSPACE FLOW PROGRAM (AFP)- AFP is a Traffic Management (TM) process administered by the Air Traffic Control System Command Center (ATCSCC) where aircraft are assigned an Expect Departure Clearance Time (EDCT) in order to manage capacity and demand for a specific area of the National Airspace System (NA S). The purpose of the program is to mitigate the effects of en route constraints. It is a flexible program and may be implemented in various forms depending upon the needs of the air traffic system.

AIRSPACE HIERARCHY- Within the airspace classes, there is a hierarchy and, in the event of an overlap of airspace: Class A preempts Class B, Class B preempts Class C, Class C preempts Class D, Class D preempts Class E, and Class E preempts Class G.

AIRSPEED- The speed of an aircraft relative to its surrounding air mass. The unqualified term "airspeed" means one of the following:

a. Indicated Airspeed- The speed shown on the aircraft airspeed indicator. This is the speed used in pilot/controller communications under the general term "airspeed."

(Refer to 14 CFR Part 1.)

b. True Airspeed- The airspeed of an aircraft relative to undisturbed air. Used primarily in flight planning and en route portion of flight. When used in pilot/controller communications, it is referred to as "true airspeed" and not shortened to "airspeed."

AIRSTART- The starting of an aircraft engine while the aircraft is airborne, preceded by engine shutdown during training flights or by actual engine failure.

AIRWAY- A Class E airspace area established in the form of a corridor, the centerline of which is defined by radio navigational aids.

(See FEDERAL AIRWAYS.) (See ICAO term AIRWAY.) (Refer to 14 CFR Part 71.) (Refer to AIM.)

AIRWAY [ICAO]- A control area or portion thereof established in the form of corridor equipped with radio navigational aids.

AIRWAY BEACON- Used to mark airway segments i n remote mountain areas. The light flashes Morse Code to identify the beacon site.

(Refer to AIM.)

AIT-

(See AUTOMATED INFORMATION TRANSFER.)

ALERFA (Alert Phase) [ICAO]- A situation wherein apprehension exists as to the safety of an aircraft and its occupants.

ALERT- A notification to a position that there is an aircraft-to-aircraft or aircraft-to- airspace conflict, as detected by Automated Problem Detection (APD).

ALERT AREA-

(See SPECIAL USE AIRSPACE.)

ALERT NOTICE- A request originated by a flight service station (FSS) or an air route traffic control center (ARTCC) for an extensive communication search for overdue, unreported, or missing aircraft.

ALERTING SERVICE- A service provided to notify appropriate organizations regarding aircraft in need of search and rescue aid and assist such organizations as required.

ALNOT-
(See ALERT NOTICE.)

ALONG-TRACK DISTANCE (ATD)- The distance measured from a point-in-space by systems using area navigation reference capabilities that are not subject to slant range errors.

ALPHANUMERIC DISPLAY- Letters and numerals used to show identification, altitude, beacon code, and other information concerning a target on a radar display.
(See AUTOMATED RADAR TERMINAL SYSTEMS.)

ALTERNATE AERODROME [ICAO]- An aerodrome to which an aircraft may proceed when it becomes either impossible or inadvisable to proceed to or to land at the aerodrome of intended landing.

Note: The aerodrome from which a flight departs may also be an en-route or a destination alternate aerodrome for the flight.

ALTERNATE AIRPORT- An airport at which an aircraft may land if a landing at the intended airport becomes inadvisable.
(See ICAO term ALTERNATE AERODROME.)

ALTIMETER SETTING- The barometric pressure reading used to adjust a pressure altimeter for variations in existing atmospheric pressure or to the standard altimeter setting (29.92).
(Refer to 14 CFR Part 91.) (Refer to AIM.)

ALTITUDE- The height of a level, point, or object measured in feet Above Ground Level (AGL) or from Mean Sea Level (MSL).
(See FLIGHT LEVEL.)

a. MSL Altitude- Altitude expressed in feet measured from mean sea level.

b. AGL Altitude- Altitude expressed in feet measured above ground level.

c. Indicated Altitude- The altitude as shown by an altimeter. On a pressure or barometric altimeter it is altitude as shown uncorrected for instrument error and uncompensated for variation from standard atmospheric conditions.
(See ICAO term ALTITUDE.)

ALTITUDE [ICAO]- The vertical distance of a level, a point or an object considered as a point, measured from mean sea level (MSL).

ALTITUDE READOUT- An aircraft's altitude, transmitted via the Mode C transponder feature, that is visually displayed in 100-foot increments on a radar scope having readout capability.
(See ALPHANUMERIC DISPLAY.)
(See AUTOMATED RADAR TERMINAL SYSTEMS.) (Refer to AIM.)

ALTITUDE RESERVATION- Airspace utilization under prescribed conditions normally employed for the mass movement of aircraft or other special user requirements which cannot otherwise be accomplished. ALTRVs are approved by the appropriate FAA facility.
(See AIR TRAFFIC CONTROL SYSTEM COMMAND CENTER.)

ALTITUDE RESTRICTION- An altitude or altitudes, stated in the order flown, which are to be maintained until reaching a specific point or time. Altitude restrictions may be issued by ATC due to traffic, terrain, or other airspace considerations.

ALTITUDE RESTRICTIONS ARE CANCELED- Adherence to previously imposed altitude restrictions is no longer required during a climb or descent.

ALTRV-
(See ALTITUDE RESERVATION.)

AMVER-
(See AUTOMATED MUTUAL-ASSISTANCE VESSEL RESCUE SYSTEM.) APB-
(See AUTOMATED PROBLEM DETECTION BOUNDARY.)

APD-
(See AUTOMATED PROBLEM DETECTION.)

APDIA-
(See AUTOMATED PROBLEM DETECTION INHIBITED AREA.)

APPROACH CLEARANCE- Authorization by ATC for a pilot to conduct an instrument approach. The type of instrument approach for which a clearance and other pertinent information is provided in the approach clearance when required.
(See CLEARED APPROACH.)
(See INSTRUMENT APPROACH PROCEDURE.) (Refer to AIM.)
(Refer to 14 CFR Part 91.)

APPROACH CONTROL FACILITY- A terminal ATC facility that provides approach control service in a terminal area.
(See APPROACH CONTROL SERVICE.)
(See RADAR APPROACH CONTROL FACILITY.)

APPROACH CONTROL SERVICE- Air traffic control service provided by an approach control facility for arriving and departing VFR/IFR aircraft and, on occasion, en route aircraft. At some airports not served by an approach control facility, the ARTCC provides limited approach control service.
(See ICAO term APPROACH CONTROL SERVICE.) (Refer to AIM.)

APPROACH CONTROL SERVICE [ICAO]- Air traffic control service for arriving or departing controlled flights.

APPROACH GATE- An imaginary point used within ATC as a basis for vectoring aircraft to the final approach course. The gate will be established along the final approach course 1 mile from the final approach fix on the side away from the airport and will be no closer than 5 miles from the landing threshold.

APPROACH LIGHT SYSTEM-(See AIRPORT LIGHTING.)

APPROACH SEQUENCE- The order in which aircraft are positioned while on approach or awaiting approach clearance.
(See LANDING SEQUENCE.)
(See ICAO term APPROACH SEQUENCE.)

APPROACH SEQUENCE [ICAO]- The order in which two or more aircraft are cleared to approach to land at the aerodrome.

APPROACH SPEED- The recommended speed contained in aircraft manuals used by pilots when making an approach to landing. This speed will vary for different segments of an approach as well as for aircraft weight and configuration.

APPROPRIATE ATS AUTHORITY [ICAO]- The relevant authority designated by the State responsible for providing air traffic services in the airspace concerned. In the United States, the "appropriate ATS authority" is the Program Director for Air Traffic Planning and Procedures, ATP-1.

APPROPRIATE AUTHORITY-

a. Regarding flight over the high seas: the relevant authority is the State of Registry.

b. Regarding flight over other than the high seas: the relevant authority is the State having sovereignty over the territory being overflown.

APPROPRIATE OBSTACLE CLEARANCE MINIMUM ALTITUDE- Any of the following:
(See Minimum En Route IFR Altitude.)
(See Minimum IFR Altitude.)
(See Minimum Obstruction Clearance Altitude.) (See Minimum Vectoring Altitude.)

APPROPRIATE TERRAIN CLEARANCE MINIMUM ALTITUDE- Any of the following:
(See Minimum En Route IFR Altitude.)
(See Minimum IFR Altitude.)
(See Minimum Obstruction Clearance Altitude.) (See Minimum Vectoring Altitude.)

APRON- A defined area on an airport or heliport intended to accommodate aircraft for purposes of loading or unloading passengers or cargo, refueling, parking, or maintenance. With regard to seaplanes, a ramp is used for access to the apron from the water.
(See ICAO term APRON.)

APRON [ICAO]- A defined area, on a land aerodrome, intended to accommodate aircraft for purposes of loading or unloading passengers, mail or cargo, refueling, parking or maintenance.

ARC- The track over the ground of an aircraft flying at a constant distance from a navigational aid by reference to distance measuring equipment (DME).

AREA CONTROL CENTER [ICAO]- An air traffic control facility primarily responsible for ATC services being provided IFR aircraft during the en route phase of flight. The U.S. equivalent facility is an air route traffic control center (ARTCC).

AREA NAVIGATION (RNAV)- A method of navigation which permits aircraft operation on any desired flight path within the coverage of ground- or space-based navigation aids or within the limits of the capability of self-contained aids, or a combination of these.

Note: Area navigation includes performance-based navigation as well as other operations that do not meet the definition of performance-based navigation.

AREA NAVIGATION (RNAV) APPROACH CONFIGURATION:

a. STANDARD T- An RNAV approach whose design allows direct flight to any one of three initial approach fixes (IAF) and eliminates the need for procedure turns. The standard design is to align the procedure on the extended centerline with the missed approach point (MAP) at the runway threshold, the final approach fix (FAF), and the initial approach/intermediate fix (IAF/IF). The other two IAFs will be established perpendicular to the IF.

b. MODIFIED T- An RNAV approach design for single or multiple runways where terrain or operational constraints do not allow for the standard T. The "T" may be modified by increasing or decreasing the angle from the corner IAF(s) to the IF or by eliminating one or both corner IAFs.

c. STANDARD I- An RNAV approach design for a single runway with both corner IAFs eliminated. Course reversal or radar vectoring may be required at busy terminals with multiple runways.

d. TERMINAL ARRIVAL AREA (TAA)- The TAA is controlled airspace established in conjunction with the Standard or Modified T and I RNAV approach configurations. In the standard TAA, there are three areas: straight-in, left base, and right base. The arc boundaries of the three areas of the TAA are published portions of the approach and allow aircraft to transition from the en route structure direct to the nearest IAF. TAAs will also eliminate or reduce feeder routes, departure extensions, and procedure turns or course reversal.

1. STRAIGHT-IN AREA- A 30NM arc centered on the IF bounded by a straight line extending through the IF perpendicular to the intermediate course.

2. LEFT BASE AREA- A 30NM arc centered on the right corner IAF. The area shares a boundary with the straight-in area except that it extends out for 30NM from the IAF and is bounded on the other side by a line extending from the IF through the FAF to the arc.

3. RIGHT BASE AREA- A 30NM arc centered on the left corner IAF. The area shares a boundary with the straight-in area except that it extends out for 30NM from the IAF and is bounded on the other side by a line extending from the IF through the FAF to the arc.

ARINC- An acronym for Aeronautical Radio, Inc., a corporation largely owned by a group of airlines. ARINC is licensed by the FCC as an aeronautical station and contracted by the FAA to provide communications support for air traffic control and meteorological services in portions of international airspace.

ARMY AVIATION FLIGHT INFORMATION BULLETIN- A bulletin that provides air

operation data covering Army, National Guard, and Army Reserve aviation activities.

ARO-
(See AIRPORT RESERVATION OFFICE.)

ARRESTING SYSTEM- A safety device consisting of two major components, namely, engaging or catching devices and energy absorption devices for the purpose of arresting both tailhook and/or nontailhook-equipped aircraft. It is used to prevent aircraft from overrunning runways when the aircraft cannot be stopped after landing or during aborted takeoff. Arresting systems have various names; e.g., arresting gear, hook device, wire barrier cable.
(See ABORT.)(Refer to AIM.)

ARRIVAL AIRCRAFT INTERVAL- An internally generated program in hundredths of minutes based upon the AAR. AAI is the desired optimum interval between successive arrival aircraft over the vertex.

ARRIVAL CENTER- The ARTCC having jurisdiction for the impacted airport.

ARRIVAL DELAY- A parameter which specifies a period of time in which no aircraft will be metered for arrival at the specified airport.

ARRIVAL SECTOR- An operational control sector containing one or more meter fixes.

ARRIVAL SECTOR ADVISORY LIST- An ordered list of data on arrivals displayed at the PVD/MDM of the sector which controls the meter fix.

ARRIVAL SEQUENCING PROGRAM- The automated program designed to assist in sequencing aircraft destined for the same airport.

ARRIVAL TIME- The time an aircraft touches down on arrival. ARSR-
(See AIR ROUTE SURVEILLANCE RADAR.)

ARTCC-
(See AIR ROUTE TRAFFIC CONTROL CENTER.)

ARTS-
(See AUTOMATED RADAR TERMINAL SYSTEMS.)

ASDA-
(See ACCELERATE-STOP DISTANCE AVAILABLE.)

ASDA [ICAO]-
(See ICAO Term ACCELERATE-STOP DISTANCE AVAILABLE.)

ASDE-
(See AIRPORT SURFACE DETECTION EQUIPMENT.)

ASF-
(See AIRPORT STREAM FILTER.)

ASLAR-
(See AIRCRAFT SURGE LAUNCH AND RECOVERY.)

ASP-
(See ARRIVAL SEQUENCING PROGRAM.)
ASR-
(See AIRPORT SURVEILLANCE RADAR.)
ASR APPROACH-
(See SURVEILLANCE APPROACH.)
ASSOCIATED- A radar target displaying a data block with flight identification and altitude information.
(See UNASSOCIATED.)
ATC-
(See AIR TRAFFIC CONTROL.)
ATC ADVISES- Used to prefix a message of noncontrol information when it is relayed to an aircraft by other than an air traffic controller.
(See ADVISORY.)
ATC ASSIGNED AIRSPACE- Airspace of defined vertical/lateral limits, assigned by ATC, for the purpose of providing air traffic segregation between the specified activities being conducted within the assigned airspace and other IFR air traffic.
(See SPECIAL USE AIRSPACE.) ATC CLEARANCE-
(See AIR TRAFFIC CLEARANCE.)
ATC CLEARS- Used to prefix an ATC clearance when it is relayed to an aircraft by other than an air traffic controller.
ATC INSTRUCTIONS- Directives issued by air traffic control for the purpose of requiring a pilot to take specific actions; e.g., "Turn left heading two five zero," "Go around," "Clear the runway."
(Refer to 14 CFR Part 91.)
ATC PREFERRED ROUTE NOTIFICATION- URET notification to the appropriate controller of the need to determine if an ATC preferred route needs to be applied, based on destination airport.
(See ROUTE ACTION NOTIFICATION.)
(See USER REQUEST EVALUATION TOOL.)
ATC PREFERRED ROUTES- Preferred routes that are not automatically applied by Host.
ATC REQUESTS- Used to prefix an ATC request when it is relayed to an aircraft by other than an air traffic controller.
ATC SECURITY SERVICES - Communications and security tracking provided by an ATC facility in support of the DHS, the DOD, or other Federal security elements in the interest of national security. Such security services are only applicable within designated areas. ATC security services do not include ATC basic radar services or flight following.
ATC SECURITY SERVICES POSITION - The position responsible for providing ATC security services as defined. This position does not provide ATC, IFR separation, or VFR flight following services, but is responsible for providing security services in an area comprising airspace assigned to one or more ATC operating sectors. This position may be combined with control positions.
ATC SECURITY TRACKING - The continuous tracking of aircraft movement by an ATC facility in support of the DHS, the DOD, or other security elements for national security using radar (i.e., radar tracking) or other means (e.g., manual tracking) without providing basic radar services (including traffic advisories) or other ATC services not defined in this section.
ATCAA-
(See ATC ASSIGNED AIRSPACE.)
ATCRBS-
(See RADAR.)
ATCSCC-
(See AIR TRAFFIC CONTROL SYSTEM COMMAND CENTER.) ATCT-
(See TOWER.)
ATD-
(See ALONG-TRACK DISTANCE.)
ATIS-
(See AUTOMATIC TERMINAL INFORMATION SERVICE.)
ATIS [ICAO]-
(See ICAO Term AUTOMATIC TERMINAL INFORMATION SERVICE.)
ATS ROUTE [ICAO]- A specified route designed for channelling the flow of traffic as necessary for the provision of air traffic services.
Note: The term "ATS Route" is used to mean variously, airway, advisory route, controlled or uncontrolled route, arrival or departure, etc.
AUTOLAND APPROACH- An autoland approach is a precision instrument approach to touchdown and, in some cases, through the landing rollout. An autoland approach is performed by the aircraft autopilot which is receiving position information and/or steering commands from onboard navigation equipment.
Note: Autoland and coupled approaches are flown in VFR and IFR. It is common for carriers to require their crews to fly coupled approaches and autoland approaches (if certified) when the weather conditions are less than approximately 4,000 RVR.
(See COUPLED APPROACH.)
AUTOMATED INFORMATION TRANSFER- A precoordinated process, specifically defined in facility directives, during which a transfer of altitude control and/or radar identification is accomplished without verbal coordination between controllers using information communicated in a full data block.
AUTOMATED MUTUAL-ASSISTANCE VESSEL RESCUE SYSTEM- A facility which can deliver, in a matter of minutes, a surface picture (SURPIC) of vessels in the area of a potential or actual search and rescue incident, including their predicted positions and their characteristics.
(See FAAO JO 7110.65, Para 10-6-4, INFLIGHT CONTINGENCIES.)
AUTOMATED PROBLEM DETECTION (APD)- An Automation Processing capability that compares trajectories in order to predict conflicts.
AUTOMATED PROBLEM DETECTION BOUNDARY (APB)- The adapted distance beyond a facilities boundary defining the airspace within which URET performs conflict detection.
(See USER REQUEST EVALUATION TOOL.)
AUTOMATED PROBLEM DETECTION INHIBITED AREA (APDIA)- Airspace surrounding a terminal area within which APD is inhibited for all flights within that airspace.
AUTOMATED RADAR TERMINAL SYSTEMS (ARTS)- A generic term for several tracking systems included in the Terminal Automation Systems (TAS). ARTS plus a suffix roman numeral denotes a major modification to that system.
a. ARTS IIIA. The Radar Tracking and Beacon Tracking Level (RT&BTL) of the modular, programmable automated radar terminal system. ARTS IIIA detects, tracks, and predicts primary as well as secondary radar-derived aircraft targets. This more sophisticated computer-driven system upgrades the existing ARTS III system by providing improved tracking, continuous data recording, and fail-soft capabilities.
b. Common ARTS. Includes ARTS IIE, ARTS IIIE; and ARTS IIIE with ACD (see DTAS) which combines functionalities of the previous ARTS systems.
c. Programmable Indicator Data Processor (PIDP). The PIDP is a modification to the AN/TPX-42 interrogator system currently installed in fixed RAPCONs. The PIDP detects, tracks, and predicts secondary radar aircraft targets. These are displayed by means of computer-generated symbols and alphanumeric characters depicting flight identification, aircraft altitude, ground speed, and flight plan data. Although primary radar targets are not tracked, they are displayed coincident with the secondary radar targets as well as with the other symbols and alphanumerics. The system has the capability of interfacing with ARTCCs.
AUTOMATED WEATHER SYSTEM- Any of the automated weather sensor platforms that collect weather data at airports and disseminate the weather information via radio and/or landline. The systems currently consist of the Automated Surface Observing System (ASOS), Automated Weather Sensor System (AWSS) and Automated Weather Observation System (AWOS).
AUTOMATED UNICOM- Provides completely automated weather, radio check capability and airport advisory information on an Automated UNICOM system. These systems offer a variety of features, typically selectable by microphone cli cks, on the UNICOM frequency. Availability will be published in the Airport/Facility Directory and approach charts.
AUTOMATIC ALTITUDE REPORT-(See ALTITUDE READOUT.)
AUTOMATIC ALTITUDE REPORTING- That function of a transponder which responds to Mode C interrogations by transmitting the aircraft's altitude in 100-foot increments.
AUTOMATIC CARRIER LANDING SYSTEM- U.S. Navy final approach equipment consisting of precision tracking radar coupled to a computer data link to provide continuous information to the aircraft, monitoring capability to the pilot, and a backup approach system.
AUTOMATIC DEPENDENT SURVEILLANCE (ADS) [ICAO]- A surveillance technique in which aircraft automatically provide, via a data link, data derived from on-board navigation and position fixing systems, including aircraft identification, four dimensional position and additional data as appropriate.
AUTOMATIC DEPENDENT SURVEILLANCE-BROADCAST (ADS-B)- A surveillance system in which an aircraft or vehicle to be detected is fitted with cooperative equipment in the form of a data link transmitter. The aircraft or vehicle periodically broadcasts its GPS-derived position and other information such as velocity over the data link, which is received by a ground-based transmitter/receiver (transceiver) for processing and display at an air traffic control facility.
(See GLOBAL POSITIONING SYSTEM.) (See GROUND-BASED TRANSCEIVER.)
AUTOMATIC DEPENDENT SURVEILLANCE-CONTRACT (ADS-C)- A data link position reporting system, controlled by a ground station, that establishes contracts with an aircraft's avionics that occur automatically whenever specific events occur, or specific time intervals are reached.
AUTOMATIC DIRECTION FINDER- An aircraft radio navigation system which senses and indicates the direction to a L/MF nondirectional radio beacon (NDB) ground transmitter. Direction is indicated to the pilot as a magnetic bearing or as a relative bearing to the longitudinal axis of the aircraft depending on the type of indicator installed in the aircraft. In certain applications, such as military, ADF operations may be based on airborne and ground transmitters in the VHF/UHF frequency spectrum.
(See BEARING.)
(See NONDIRECTIONAL BEACON.)
AUTOMATIC FLIGHT INFORMATION SERVICE (AFIS) - ALASKA FSSs ONLY- The continuous broadcast of recorded non-control information at airports in Alaska where a FSS provides local airport advisory service. The AFIS broadcast automates the repetitive transmission of essential but routine information such as weather, wind, altimeter, favored runway, breaking action, airport NOTAMs, and other applicable information. The information is continuously broadcast over a discrete VHF radio frequency (usually the ASOS/AWSS/AWOS frequency.)
AUTOMATIC TERMINAL INFORMATION SERVICE- The continuous broadcast of recorded noncontrol information in selected terminal areas. Its purpose is to improve controller effectiveness and to relieve frequency congestion by automating the repetitive transmission of essential but routine information; e.g., "Los Angeles information Alfa. One three zero zero Coordinated Universal Time. Weather, measured ceiling two thousand overcast, visibility three, haze, smoke, temperature seven one, dew point five seven, wind two five zero at five, altimeter two niner niner six. I-L-S Runway Two Five Left approach in use, Runway Two Five Right closed, advise you have Alfa."
(See ICAO term AUTOMATIC TERMINAL INFORMATION SERVICE.) (Refer to AIM.)
AUTOMATIC TERMINAL INFORMATION SERVICE [ICAO]- The provision of current, routi ne information to arriving and departing aircraft by means of continuous and repetitive broadcasts throughout the day or a specified portion of the day.
AUTOROTATION- A rotorcraft flight condition in which the lifting rotor is driven entirely by action of the air when the rotorcraft is in motion.
a. Autorotative Landing/Touchdown Autorotation. Used by a pilot to indicate that the landing will be made without applying power to the rotor.
b. Low Level Autorotation. Commences at an altitude well below the traffic pattern, usually below 100 feet AGL and is used primarily for tactical military training.
c. 180 degrees Autorotation. Initiated from a downwind heading and is commenced well inside the normal traffic pattern. "Go around" may not be possible during the latter part of this maneuver.
AVAILABLE LANDING DISTANCE (ALD)- The portion of a runway available for landing and roll-out for aircraft cleared for LAHSO. This distance is measured from the landing threshold to the hold-short point.
AVIATION WEATHER SERVICE- A service provided by the National Weather Service (NWS) and FAA which collects and disseminates pertinent weather information for pilots, aircraft operators, and ATC. Available aviation weather reports and forecasts are displayed at each NWS office and FAA FSS.
(See EN ROUTE FLIGHT ADVISORY SERVICE.) (See TRANSCRIBED WEATHER BROADCAST.) (See WEATHER ADVISORY.)
(Refer to AIM.)
AWW-
(See SEVERE WEATHER FORECAST ALERTS.)
AZIMUTH (MLS)- A magnetic bearing extending from an MLS navigation facility.
Note: Azimuth bearings are described as magnetic and are referred to as "azimuth" in radio telephone communications.

B

BACK-TAXI- A term used by air traffic controllers to taxi an aircraft on the runway opposite to the traffic flow. The aircraft may be instructed to back-taxi to the beginning of the runway or at some point before reaching the runway end for the purpose of departure or to exit the runway.

BASE LEG-

(See TRAFFIC PATTERN.)

BEACON-

(See AERONAUTICAL BEACON.) (See AIRPORT ROTATING BEACON.) (See AIRWAY BEACON.)

(See MARKER BEACON.)

(See NONDIRECTIONAL BEACON.) (See RADAR.)

BEARING- The horizontal direction to or from any point, usually measured clockwise from true north, magnetic north, or some other reference point through 360 degrees.

(See NONDIRECTIONAL BEACON.)

BELOW MINIMUMS- Weather conditions below the minimums prescribed by regulation for the particular action involved; e.g., landing minimums, takeoff minimums.

BLAST FENCE- A barrier that is used to divert or dissipate jet or propeller blast.

BLAST PAD- A surface adjacent to the ends of a runway provided to reduce the erosive effect of jet blast and propeller wash.

BLIND SPEED- The rate of departure or closing of a target relative to the radar antenna at which cancellation of the primary radar target by moving target indicator (MTI) circuits in the radar equipment causes a reduction or complete loss of signal.

(See ICAO term BLIND VELOCITY.)

BLIND SPOT- An area from which radio transmissions and/or radar echoes cannot be received. The term is also used to describe portions of the airport not visible from the control tower.

BLIND TRANSMISSION-

(See TRANSMITTING IN THE BLIND.)

BLIND VELOCITY [ICAO]- The radial velocity of a moving target such that the target is

not seen on primary radars fitted with certain forms of fixed echo suppression.

BLIND ZONE-

(See BLIND SPOT.)

BLOCKED- Phraseology used to indicate that a radio transmission has been distorted or interrupted due to multiple simultaneous radio transmissions.

BOUNDARY LIGHTS-

(See AIRPORT LIGHTING.)

BRAKING ACTION (GOOD, FAIR, POOR, OR NIL)- A report of conditions on the airport movement area providing a pilot with a degree/quality of braking that he/she might expect. Braking action is reported in terms of good, fair, poor, or nil.

(See RUNWAY CONDITION READING.)

BRAKING ACTION ADVISORIES- When tower controllers have received runway braking action reports which include the terms "fair," "poor," or "nil," or whenever weather conditions are conducive to deteriorating or rapidly changing runway braking conditions, the tower will include on the ATIS broadcast the statement, "Braking action advisories are in effect" on the ATIS broadcast. During the time braking action advisories are in effect, ATC will issue the latest braking action report for the runway in use to each arriving and departing aircraft. Pilots should be prepared for deteriorating braking conditions and should request current runway condition information if not volunteered by controllers. Pilots should also be prepared to provide a descriptive runway condition report to controllers after landing.

BREAKOUT- A technique to direct aircraft out of the approach stream. In the context of close parallel operations, a breakout is used to direct threatened aircraft away from a deviating aircraft.

BROADCAST- Transmission of information for which an acknowledgement is not expected.

(See ICAO term BROADCAST.)

BROADCAST [ICAO]- A transmission of information relating to air navigation that is not addressed to a specific station or stations.

C

CALCULATED LANDING TIME- A term that may be used in place of tentative or actual calculated landing time, whichever applies.

CALL FOR RELEASE- Wherein the overlying ARTCC requires a terminal facility to initiate verbal coordination to secure ARTCC approval for release of a departure into the en route environment.

CALL UP- Initial voice contact between a facility and an aircraft, using the identification of the unit being called and the unit initiating the call.

(Refer to AIM.)

CANADIAN MINIMUM NAVIGATION PERFORMANCE SPECIFICATION AIRSPACE-That portion of Canadian domestic airspace within which MNPS separation may be applied.

CARDINAL ALTITUDES- "Odd" or "Even" thousand-foot altitudes or flight levels; e.g., 5,000, 6,000, 7,000, FL 250, FL 260, FL 270.

(See ALTITUDE.)

(See FLIGHT LEVEL.)

CARDINAL FLIGHT LEVELS-(See CARDINAL ALTITUDES.) CAT-

(See CLEAR-AIR TURBULENCE.)

CATCH POINT- A fix/waypoint that serves as a transition point from the high altitude waypoint navigation structure to an arrival procedure (STAR) or the low altitude ground-based navigation structure.

CEILING- The heights above the earth's surface of the lowest layer of clouds or obscuring phenomena that is reported as "broken," "overcast," or "obscuration," and not classified as "thin" or "partial."

(See ICAO term CEILING.)

CEILING [ICAO]- The height above the ground or water of the base of the lowest layer of cloud below 6,000 meters (20,000 feet) covering more than half the sky.

CENRAP-

(See CENTER RADAR ARTS PRESENTATION/PROCESSING.) CENRAP-PLUS-

(See CENTER RADAR ARTS PRESENTATION/PROCESSING-PLUS.) CENTER-

(See AIR ROUTE TRAFFIC CONTROL CENTER.)

CENTER'S AREA- The specified airspace within which an air route traffic control center (ARTCC) provides air traffic control and advisory service.

(See AIR ROUTE TRAFFIC CONTROL CENTER.) (Refer to AIM.)

CENTER RADAR ARTS PRESENTATION/PROCESSING- A computer program developed to provide a back-up system for airport surveillance radar in the event of a failure or malfunction. The program uses air route traffic control center radar for the processing and presentation of data on the ARTS IIA or IIIA displays.

CENTER RADAR ARTS PRESENTATION/PROCESSING-PLUS- A computer program developed to provide a back-up system for airport surveillance radar in the event of a terminal secondary radar system failure. The program uses a combination of Air Route Traffic Control Center Radar and terminal airport surveillance radar primary targets displayed simultaneously for the processing and presentation of data on the ARTS IIA or IIIA displays.

CENTER TRACON AUTOMATION SYSTEM (CTAS)- A computerized set of programs designed to aid Air Route Traffic Control Centers and TRACONs in the management and control of air traffic.

CENTER WEATHER ADVISORY- An unscheduled weather advisory issued by Center Weather Service Unit meteorologists for ATC use to alert pilots of existing or anticipated adverse weather conditions within the next 2 hours. A CWA may modify or redefine a SIGMET.

(See AWW.)

(See AIRMET.)

(See CONVECTIVE SIGMET.)

(See SIGMET.) (Refer to AIM.) CENTRAL EAST PACIFIC- An organized route system between the U.S. West Coast and Hawaii.

CEP-

(See CENTRAL EAST PACIFIC.) CERAP-

(See COMBINED CENTER-RAPCON.)

CERTIFIED TOWER RADAR DISPLAY (CTRD)- A FAA radar display certified for use in the NAS.

CFR-

(See CALL FOR RELEASE.)

CHAFF- Thin, narrow metallic reflectors of various lengths and frequency responses, used to reflect radar energy. These reflectors when dropped from aircraft and allowed to drift downward result in large targets on the radar display.

CHARTED VFR FLYWAYS- Charted VFR Flyways are flight paths recommended for use to bypass areas heavily traversed by large turbine-powered aircraft. Pilot compliance with recommended flyways and associated altitudes is strictly voluntary. VFR Flyway Planning charts are published on the back of existing VFR Terminal Area charts.

CHARTED VISUAL FLIGHT PROCEDURE APPROACH- An approach conducted while operating on an instrument flight rules (IFR) flight plan which authorizes the pilot of an aircraft to proceed visually and clear of clouds to the airport via visual landmarks and other information depicted on a charted visual flight procedure. This approach must be authorized and under the control of the appropriate air traffic control facility. Weather minimums required are depicted on the chart.

CHASE- An aircraft flown in proximity to another aircraft normally to observe its performance during training or testing.

CHASE AIRCRAFT-(See CHASE.)

CIRCLE-TO-LAND MANEUVER- A maneuver initiated by the pilot to align the aircraft with a runway for landing when a straight-in landing from an instrument approach is not possible or is not desirable. At tower controlled airports, this maneuver is made only after ATC authorization has been obtained and the pilot has established required visual reference to the airport.

(See CIRCLE TO RUNWAY.) (See LANDING MINIMUMS.)

(Refer to AIM.)

CIRCLE TO RUNWAY (RUNWAY NUMBER)- Used by ATC to inform the pilot that he/she must circle to land because the runway in use is other than the runway aligned with the instrument approach procedure. When the direction of the circling maneuver in relation to the airport/runway is required, the controller will state the direction (eight cardinal compass points) and specify a left or right downwind or base leg as appropriate; e.g., "Cleared VOR Runway Three Six Approach circle to Runway Two Two," or "Circle northwest of the airport for a right downwind to Runway Two Two."

(See CIRCLE-TO-LAND MANEUVER.) (See LANDING MINIMUMS.)

(Refer to AIM.)

CIRCLING APPROACH-

(See CIRCLE-TO-LAND MANEUVER.) CIRCLING MANEUVER-

(See CIRCLE-TO-LAND MANEUVER.) CIRCLING MINIMA-

(See LANDING MINIMUMS.)

CLASS A AIRSPACE-

(See CONTROLLED AIRSPACE.) CLASS B AIRSPACE-

(See CONTROLLED AIRSPACE.) CLASS C AIRSPACE-

(See CONTROLLED AIRSPACE.) CLASS D AIRSPACE-

(See CONTROLLED AIRSPACE.) CLASS E AIRSPACE-

(See CONTROLLED AIRSPACE.)

CLASS G AIRSPACE- That airspace not designated as Class A, B, C, D or E.

CLEAR AIR TURBULENCE (CAT)- Turbulence encountered in air where no clouds are present. This term is commonly applied to high-level turbulence associated with wind shear. CAT is often encountered in the vicinity of the jet stream.

(See WIND SHEAR.)

(See JET STREAM.)

CLEAR OF THE RUNWAY-

a. Taxiing aircraft, which is approaching a runway, is clear of the runway when all parts of the aircraft are held short of the applicable runway holding position marking.

b. A pilot or controller may consider an aircraft, which is exiting or crossing a runway, to be clear of the runway when all parts of the aircraft are beyond the runway edge and there are no restrictions to its continued movement beyond the applicable runway holding position marking.

c. Pilots and controllers shall exercise good judgement to ensure that adequate separation exists between all aircraft on runways and taxiways at airports with inadequate runway edge lines or holding position markings.

CLEARANCE-

(See AIR TRAFFIC CLEARANCE.)

CLEARANCE LIMIT- The fix, point, or location to which an aircraft is cleared when issued an air traffic clearance.

(See ICAO term CLEARANCE LIMIT.)

CLEARANCE LIMIT [ICAO]- The point to which an aircraft is granted an air traffic control clearance.

CLEARANCE VOID IF NOT OFF BY (TIME)- Used by ATC to advise an aircraft that the departure clearance is automatically canceled if takeoff is not made prior to a specified time. The pilot must obtain a new clearance or cancel his/her IFR flight plan if not off by the specified time.

(See ICAO term CLEARANCE VOID TIME.)

CLEARANCE VOID TIME [ICAO]- A time specified by an air traffic control unit at which a clearance ceases to be valid unless the aircraft concerned has already taken action to comply therewith.

CLEARED APPROACH- ATC authorization for an aircraft to execute any standard or special instrument approach procedure for that airport. Normally, an aircraft will be cleared for a specific instrument approach procedure.

(See CLEARED (Type of) APPROACH.)

(See INSTRUMENT APPROACH PROCEDURE.) (Refer to 14 CFR Part 91.)

(Refer to AIM.)

CLEARED (Type of) APPROACH- ATC authorization for an aircraft to execute a specific instrument approach procedure to an airport; e.g., "Cleared ILS Runway Three Six Approach."

(See APPROACH CLEARANCE.)

(See INSTRUMENT APPROACH PROCEDURE.) (Refer to 14 CFR Part 91.)

(Refer to AIM.)

CLEARED AS FILED- Means the aircraft is cleared to proceed in accordance with the route of flight filed in the flight plan. This clearance does not include the altitude, DP, or DP Transition.

(See REQUEST FULL ROUTE CLEARANCE.) (Refer to AIM.)

CLEARED FOR TAKEOFF- ATC authorization for an aircraft to depart. It is predicated on known traffic and known physical airport conditions.

CLEARED FOR THE OPTION- ATC authorization for an aircraft to make a touch-and- go, low approach, missed approach, stop and go, or full stop landing at the discretion of the pilot. It is normally used in training so that an instructor can evaluate a student's performance under changing situations.

(See OPTION APPROACH.) (Refer to AIM.)

CLEARED THROUGH- ATC authorization for an aircraft to make intermediate stops at specified airports without refiling a flight plan while en route to the clearance limit.

CLEARED TO LAND- ATC authorization for an aircraft to land. It is predicated on known traffic and known physical airport conditions.

CLEARWAY- An area beyond the takeoff runway under the control of airport authorities within which terrain or fixed obstacles may not extend above specified limits. These areas may be required for certain turbine-powered operations and the size and upward slope of the clearway will differ depending on when the aircraft was certificated.

(Refer to 14 CFR Part 1.)

CLIMB TO VFR- ATC authorization for an aircraft to climb to VFR conditions within Class B, C, D, and E surface areas when the only weather limitation is restricted visibility. The aircraft must remain clear of clouds while climbing to VFR.

(See SPECIAL VFR CONDITIONS.) (Refer to AIM.)

CLIMBOUT- That portion of flight operation between takeoff and the initial cruising altitude.

CLOSE PARALLEL RUNWAYS- Two parallel runways whose extended centerlines are separated by less than 4,300 feet, having a Precision Runway Monitoring (PRM) system that permits simultaneous independent ILS approaches.

CLOSED RUNWAY- A runway that is unusable for aircraft operations. Only the airport management/military operations office can close a runway.

CLOSED TRAFFIC- Successive operations involving takeoffs and landings or low approaches where the aircraft does not exit the traffic pattern.

CLOUD- A cloud is a visible accumulation of minute water droplets and/or ice particles in the atmosphere above the Earth's surface. Cloud differs from ground fog, fog, or ice fog only in that the latter are, by definition, in contact with the Earth's surface.

CLT-

(See CALCULATED LANDING TIME.)

CLUTTER- In radar operations, clutter refers to the reception and visual display of radar returns caused by precipitation, chaff, terrain, numerous aircraft targets, or other phenomena. Such returns may limit or preclude ATC from providing services based on radar.

(See CHAFF.)

(See GROUND CLUTTER.)

(See PRECIPITATION.)

(See TARGET.)

(See ICAO term RADAR CLUTTER.) CMNPS-

(See CANADIAN MINIMUM NAVIGATION PERFORMANCE SPECIFICATION AIRSPACE.)

COASTAL FIX- A navigation aid or intersection where an aircraft transitions between the domestic route structure and the oceanic route structure.

CODES- The number assigned to a particular multiple pulse reply signal transmitted by a transponder.

(See DISCRETE CODE.)

COMBINED CENTER-RAPCON- An air traffic facility which combines the functions of an ARTCC and a radar approach control facility.

(See AIR ROUTE TRAFFIC CONTROL CENTER.) (See RADAR APPROACH CONTROL FACILITY.)

COMMON POINT- A significant point over which two or more aircraft will report passing or have reported passing before proceeding on the same or diverging tracks. To establish/maintain longitudinal separation, a controller may determine a common point not originally in the aircraft's flight plan and then clear the aircraft to fly over the point.

(See SIGNIFICANT POINT.) COMMON PORTION-

(See COMMON ROUTE.)

COMMON ROUTE- That segment of a North American Route between the inland navigation facility and the coastal fix.

OR

COMMON ROUTE- Typically the portion of a RNAV STAR between the en route transition end point and the runway transition start point; however, the common route may only consist of a single point that joins the en route and runway transitions.

COMMON TRAFFIC ADVISORY FREQUENCY (CTAF)- A frequency designed for the purpose of carrying out airport advisory practices while operating to or from an airport without an operating control tower. The CTAF may be a UNICOM, Multicom, FSS, or tower frequency and is identified in appropriate aeronautical publications.

(Refer to AC 90-42, Traffic Advisory Practices at Airports Without Operating Control Towers.)

COMPASS LOCATOR- A low power, low or medium frequency (L/MF) radio beacon installed at the site of the outer or middle marker of an instrument landing system (ILS). It can be used for navigation at distances of approximately 15 miles or as authorized in the approach procedure.

a. Outer Compass Locator (LOM)- A compass locator installed at the site of the outer marker of an instrument landing system.

(See OUTER MARKER.)

b. Middle Compass Locator (LMM)- A compass locator installed at the site of the middle marker of an instrument landing system.

(See MIDDLE MARKER.) (See ICAO term LOCATOR.)

COMPASS ROSE- A circle, graduated in degrees, printed on some charts or marked on the ground at an airport. It is used as a reference to either true or magnetic direction.

COMPLY WITH RESTRICTIONS- An ATC instruction that requires an aircraft being vectored back onto an arrival or departure procedure to comply with all altitude and/or speed restrictions depicted on the procedure. This term may be used in lieu of repeating each remaining restriction that appears on the procedure.

COMPOSITE FLIGHT PLAN- A flight plan which specifies VFR operation for one

portion of flight and IFR for another portion. It is used primarily in military operations.

(Refer to AIM.)

COMPOSITE ROUTE SYSTEM- An organized oceanic route structure, incorporating

reduced lateral spacing between routes, in which composite separation is authorized.

COMPOSITE SEPARATION- A method of separating aircraft in a composite route system where, by management of route and altitude assignments, a combination of half the lateral minimum specified for the area concerned and half the vertical minimum is applied.

COMPULSORY REPORTING POINTS- Reporting poi nts which must be reported to ATC. They are designated on aeronautical charts by solid triangles or filed in a flight plan as fixes selected to define direct routes. These points are geographical locations which are defined by navigation aids/fixes. Pilots should discontinue position reporting over compulsory reporting points when informed by ATC that their aircraft is in "radar contact."

CONFIDENCE MANEUVER- A confidence maneuver consists of one or more turns, a climb or descent, or other maneuver to determine if the pilot in command (PIC) is able to receive and comply with ATC instructions.

CONFLICT ALERT- A function of certain air traffic control automated systems designed to alert radar controllers to existing or pending situations between tracked targets (known IFR or VFR aircraft) that require his/her immediate attention/action.

(See MODE C INTRUDER ALERT.)

CONFLICT RESOLUTION- The resolution of potential conflictions between aircraft that are radar identified and in communication with ATC by ensuring that radar targets do not touch. Pertinent traffic advisories shall be issued when this procedure is applied.

Note: This procedure shall not be provided utilizing mosaic radar systems.

CONFORMANCE- The condition established when an aircraft's actual position is within the conformance region constructed around that aircraft at its position, according to the trajectory associated with the aircraft's Current Plan.

CONFORMANCE REGION- A volume, bounded laterally, vertically, and longitudinally, within which an aircraft must be at a given time in order to be in conformance with the Current Plan Trajectory for that aircraft. At a given time, the conformance region is determined by the simultaneous application of the lateral, vertical, and longitudinal conformance bounds for the aircraft at the position defined by time and aircraft's trajectory.

CONSOLAN- A low frequency, long-distance NAVAID used principally for transoceanic navigations.

CONTACT-

a. Establish communication with (followed by the name of the facility and, if appropriate, the frequency to be used).

b. A flight condition wherein the pilot ascertains the attitude of his/her aircraft and navigates by visual reference to the surface.

(See CONTACT APPROACH.) (See RADAR CONTACT.)

CONTACT APPROACH- An approach wherein an aircraft on an IFR flight plan, having an air traffic control authorization, operating clear of clouds with at least 1 mile flight visibility and a reasonable expectation of continuing to the destination airport in those conditions, may deviate from the instrument approach procedure and proceed to the destination airport by visual reference to the surface. This approach will only be authorized when requested by the pilot and the reported ground visibility at the destination airport is at least 1 statute mile.

(Refer to AIM.)

CONTAMINATED RUNWAY- A runway is considered contaminated whenever standing water, ice, snow, slush, frost in any form, heavy rubber, or other substances are present. A runway is contaminated with respect to rubber deposits or other friction-degrading substances when the average friction value for any 500-foot segment of the runway within the ALD fails below the recommended minimum friction level and the average friction value in the adjacent 500-foot segments falls below the maintenance planning friction level.

CONTERMINOUS U.S.- The 48 adjoining States and the District of Columbia.

CONTINENTAL UNITED STATES- The 49 States located on the continent of North America and the District of Columbia.

CONTINUE- When used as a control instruction should be followed by another word or words clarifying what is expected of the pilot. Example: "continue taxi," "continue descent," "continue inbound," etc.

CONTROL AREA [ICAO]- A controlled airspace extending upwards from a specified limit above the earth.

CONTROL SECTOR- An airspace area of defined horizontal and vertical dimensions for which a controller or group of controllers has air traffic control responsibility, normally within an air route traffic control center or an approach control facility. Sectors are established based on predominant traffic flows, altitude strata, and controller workload. Pilot-communications during operations within a sector are normally maintained on discrete frequencies assigned to the sector.

(See DISCRETE FREQUENCY.)

CONTROL SLASH- A radar beacon slash representing the actual position of the associated aircraft. Normally, the control slash is the one closest to the interrogating radar beacon site. When ARTCC radar is operating in narrowband (digitized) mode, the control slash is converted to a target symbol.

CONTROLLED AIRSPACE- An airspace of defined dimensions within which air traffic control service is provided to IFR flights and to VFR flights in accordance with the airspace classification.

a. Controlled airspace is a generic term that covers Class A, Class B, Class C, Class D, and Class E airspace.

b. Controlled airspace is also that airspace within which all aircraft operators are subject to certain pilot qualifications, operating rules, and equipment requirements in 14

CFR Part 91 (for specific operating requirements, please refer to 14 CFR Part 91). For IFR operations in any class of controlled airspace, a pilot must file an IFR flight plan and receive an appropriate ATC clearance. Each Class B, Class C, and Class D airspace area designated for an airport contains at least one primary airport around which the airspace is designated (for specific designations and descriptions of the airspace classes, please refer to 14 CFR Part 71).

c. Controlled airspace in the United States is designated as follows:

1. CLASS A- Generally, that airspace from 18,000 feet MSL up to and including FL 600, including the airspace overlying the waters within 12 nautical miles of the coast of the 48 contiguous States and Alaska. Unless otherwise authorized, all persons must operate their aircraft under IFR.

2. CLASS B- Generally, that airspace from the surface to 10,000 feet MSL surrounding the nation's busiest airports in terms of airport operations or passenger enplanements. The configuration of each Class B airspace area is individually tailored and consists of a surface area and two or more layers (some Class B airspaces areas resemble upside-down wedding cakes), and is designed to contain all published instrument procedures once an aircraft enters the airspace. An ATC clearance is required for all aircraft to operate in the area, and all aircraft that are so cleared receive separation services within the airspace. The cloud clearance requirement for VFR operations is "clear of clouds."

3. CLASS C- Generally, that airspace from the surface to 4,000 feet above the airport elevation (charted in MSL) surrounding those airports that have an operational control tower, are serviced by a radar approach control, and that have a certain number of IFR operations or passenger enplanements. Although the configuration of each Class C area is individually tailored, the airspace usually consists of a surface area with a 5 nautical mile (NM) radius, a circle with a 10NM radius that extends no lower than 1,200 feet up to 4,000 feet above the airport elevation and an outer area that is not charted. Each person must establish two-way radio communications with the ATC facility providing air traffic services prior to entering the airspace and thereafter maintain those communications while within the airspace. VFR aircraft are only separated from IFR aircraft within the airspace.

(See OUTER AREA.)

4. CLASS D- Generally, that airspace from the surface to 2,500 feet above the airport elevation (charted in MSL) surrounding those airports that have an operational control tower. The configuration of each Class D airspace area is individually tailored and when instrument procedures are published, the airspace will normally be designed to contain the procedures. Arrival extensions for instrument approach procedures may be Class D or Class E airspace. Unless otherwise authorized, each person must establish two-way radio communications with the ATC facility providing air traffic services prior to entering the airspace and thereafter maintain those communications while in the airspace. No separation services are provided to VFR aircraft.

5. CLASS E- Generally, if the airspace is not Class A, Class B, Class C, or Class D, and it is controlled airspace, it is Class E airspace. Class E airspace extends upward from either the surface or a designated altitude to the overlying or adjacent controlled airspace. When designated as a surface area, the airspace will be configured to contain all instrument procedures. Also in this class are Federal airways, airspace beginning at either 700 or 1,200 feet AGL used to transition to/from the terminal or en route environment, en route domestic, and offshore airspace areas designated below 18,000 feet MSL. Unless designated at a lower altitude, Class E airspace begins at 14,500 MSL over the United States, including that airspace overlying the waters within 12 nautical miles of the coast of the 48 contiguous States and Alaska, up to, but not including 18,000 feet MSL, and the airspace above FL 600.

CONTROLLED AIRSPACE [ICAO]- An airspace of defined dimensions within which air traffic control service is provided to IFR flights and to VFR flights in accordance with the airspace classification.

Note: Controlled airspace is a generic term which covers ATS airspace Classes A, B, C, D, and E.

CONTROLLED TIME OF ARRIVAL- Arrival time assigned during a Traffic Management Program. This time may be modified due to adjustments or user options.

CONTROLLER-

(See AIR TRAFFIC CONTROL SPECIALIST.)

CONTROLLER [ICAO]- A person authorized to provide air traffic control services.

CONTROLLER PILOT DATA LINK COMMUNICATIONS (CPDLC)- A two-way digital very high frequency (VHF) air/ground communications system that conveys textual air traffic control messages between controllers and pilots.

CONVECTIVE SIGMET- A weather advisory concerning convective weather significant to the safety of all aircraft. Convective SIGMETs are issued for tornadoes, lines of thunderstorms, embedded thunderstorms of any intensity level, areas of

thunderstorms greater than or equal to VIP level 4 with an area coverage of $^4/10$ (40%)

or more, and hail $^3/4$ inch or greater.

(See AIRMET.) (See AWW.)

(See CWA.)

(See SIGMET.) (Refer to AIM.) CONVECTIVE SIGNIFICANT METEOROLOGICAL INFORMATION-

(See CONVECTIVE SIGMET.)

COORDINATES- The intersection of lines of reference, usually expressed in degrees/minutes/seconds of latitude and longitude, used to determine position or location.

COORDINATION FIX- The fix in relation to which facilities will handoff, transfer control of an aircraft, or coordinate flight progress data. For terminal facilities, it may also serve as a clearance for arriving aircraft.

COPTER-

(See HELICOPTER.)

CORRECTION- An error has been made in the transmission and the correct version follows.

COUPLED APPROACH- A coupled approach is an instrument approach performed by the aircraft autopilot which is receiving position information and/or steering commands from onboard navigation equipment. In general, coupled nonprecision approaches must be discontinued and flown manually at altitudes lower than 50 feet below the minimum descent altitude, and coupled precision approaches must be flown manually below 50 feet AGL.

Note: Coupled and autoland approaches are flown in VFR and IFR. It is common for carriers to require their crews to fly coupled approaches and autoland approaches (if certified) when the weather conditions are less than approximately 4,000 RVR.

(See AUTOLAND APPROACH.) COURSE-

a. The intended direction of flight in the horizontal plane measured in degrees from north.

b. The ILS localizer signal pattern usually specified as the front course or the back course.

c. The intended track along a straight, curved, or segmented MLS path.

(See BEARING.)

(See INSTRUMENT LANDING SYSTEM.) (See MICROWAVE LANDING SYSTEM.) (See RADIAL.)

CPDLC-

(See CONTROLLER PILOT DATA LINK COMMUNICATIONS.)

CPL [ICAO]-

(See ICAO term CURRENT FLIGHT PLAN.)

CRITICAL ENGINE- The engine which, upon failure, would most adversely affect the performance or handling qualities of an aircraft.

CROSS (FIX) AT (ALTITUDE)- Used by ATC when a specific altitude restriction at a specified fix is required.

CROSS (FIX) AT OR ABOVE (ALTITUDE)- Used by ATC when an altitude restriction at a specified fix is required. It does not prohibit the aircraft from crossing the fix at a higher altitude than specified; however, the higher altitude may not be one that will violate a succeeding altitude restriction or altitude assignment.

(See ALTITUDE RESTRICTION.) (Refer to AIM.)

CROSS (FIX) AT OR BELOW (ALTITUDE)- Used by ATC when a maximum crossing altitude at a specific fix is required. It does not prohibit the aircraft from crossing the fix at a lower altitude; however, it must be at or above the minimum IFR altitude.

(See ALTITUDE RESTRICTION.)

(See MINIMUM IFR ALTITUDES.)

(Refer to 14 CFR Part 91.)

CROSSWIND-

a. When used concerning the traffic pattern, the word means "crosswind leg." (See TRAFFIC PATTERN.)

b. When used concerning wind conditions, the word means a wind not parallel to the runway or the path of an aircraft.

(See CROSSWIND COMPONENT.)

CROSSWIND COMPONENT- The wind component measured in knots at 90 degrees to the longitudinal axis of the runway.

CRUISE- Used in an ATC clearance to authorize a pilot to conduct flight at any altitude from the minimum IFR altitude up to and including the altitude specified in the clearance. The pilot may level off at any intermediate altitude within this block of airspace. Climb/descent within the block is to be made at the discretion of the pilot. However, once the pilot starts descent and verbally reports leaving an altitude in the block, he/she may not return to that altitude without additional ATC clearance. Further, it is approval for the pilot to proceed to and make an approach at destination airport and can be used in conjunction with:

a. An airport clearance limit at locations with a standard/special instrument approach procedure. The CFRs require that if an instrument letdown to an airport is necessary, the pilot shall make the letdown in accordance with a standard/special instrument approach procedure for that airport, or

b. An airport clearance limit at locations that are within/below/outside controlled airspace and without a standard/special instrument approach procedure. Such a clearance is NOT AUTHORIZATION for the pilot to descend under IFR conditions below the applicable minimum IFR altitude nor does it imply that ATC is exercising control over aircraft in Class G airspace; however, it provides a means for the aircraft to proceed to destination airport, descend, and land in accordance with applicable CFRs governing VFR flight operations. Also, this provides search and rescue protection until such time as the IFR flight plan is closed.

(See INSTRUMENT APPROACH PROCEDURE.)

CRUISE CLIMB- A climb technique employed by aircraft, usually at a constant power setting, resulting in an increase of altitude as the aircraft weight decreases.

CRUISING ALTITUDE- An altitude or flight level maintained during en route level flight.

This is a constant altitude and should not be confused with a cruise clearance.

(See ALTITUDE.)

(See ICAO term CRUISING LEVEL.)

CRUISING LEVEL-

(See CRUISING ALTITUDE.)

CRUISING LEVEL [ICAO]- A level maintained during a significant portion of a flight.

CT MESSAGE- An EDCT time generated by the ATCSCC to regulate traffic at arrival airports. Normally, a CT message is automatically transferred from the traffic management system computer to the NAS en route computer and appears as an EDCT. In the event of a communication failure between the traffic management system computer and the NAS, the CT message can be manually entered by the TMC at the en route facility.

CTA-

(See CONTROLLED TIME OF ARRIVAL.)

(See ICAO term CONTROL AREA.)

CTAF-

(See COMMON TRAFFIC ADVISORY FREQUENCY.)

CTAS-

(See CENTER TRACON AUTOMATION SYSTEM.)

CTRD-

(See CERTIFIED TOWER RADAR DISPLAY.)

CURRENT FLIGHT PLAN [ICAO]- The flight plan, including changes, if any, brought about by subsequent clearances.

CURRENT PLAN- The ATC clearance the aircraft has received and is expected to fly. CVFP APPROACH-

(See CHARTED VISUAL FLIGHT PROCEDURE APPROACH.)

CWA-

(See CENTER WEATHER ADVISORY and WEATHER ADVISORY.)

D

D-ATIS-

(See DIGITAL-AUTOMATIC TERMINAL INFORMATION SERVICE.) DA [ICAO]-

(See ICAO Term DECISION ALTITUDE/DECISION HEIGHT.) DAIR-

(See DIRECT ALTITUDE AND IDENTITY READOUT.)

DANGER AREA [ICAO]- An airspace of defined dimensions within which activities dangerous to the flight of aircraft may exist at specified times.

Note: The term "Danger Area" is not used in reference to areas within the United States or any of its possessions or territories.

DAS-

(See DELAY ASSIGNMENT.) DATA BLOCK-

(See ALPHANUMERIC DISPLAY.)

DEAD RECKONING- Dead reckoning, as applied to flying, is the navigation of an airplane solely by means of computations based on airspeed, course, heading, wind direction, and speed, groundspeed, and elapsed time.

DECISION ALTITUDE/DECISION HEIGHT [ICAO]- A specified altitude or height (A/H) in the precision approach at which a missed approach must be initiated if the required visual reference to continue the approach has not been established.

Note 1: Decision altitude [DA] is referenced to mean sea level [MSL] and decision height [DH] is referenced to the threshold elevation.

Note 2: The required visual reference means that section of the visual aids or of the approach area which should have been in view for sufficient time for the pilot to have made an assessment of the aircraft position and rate of change of position, in relation to the desired flight path.

DECISION HEIGHT- With respect to the operation of aircraft, means the height at which a decision must be made during an ILS, MLS, or PAR instrument approach to either continue the approach or to execute a missed approach.

(See ICAO term DECISION ALTITUDE/DECISION HEIGHT.)

DECODER- The device used to decipher signals received from ATCRBS transponders to effect their display as select codes.

(See CODES.) (See RADAR.)

DEFENSE VISUAL FLIGHT RULES- Rules applicable to flights within an ADIZ conducted under the visual flight rules in 14 CFR Part 91.

(See AIR DEFENSE IDENTIFICATION ZONE.)

(Refer to 14 CFR Part 91.) (Refer to 14 CFR Part 99.)

DELAY ASSIGNMENT (DAS)- Delays are distributed to aircraft based on the traffic management program parameters. The delay assignment is calculated in 15-minute increments and appears as a table in Traffic Flow Management System (TFMS).

DELAY INDEFINITE (REASON IF KNOWN) EXPECT FURTHER CLEARANCE (TIME)- Used by ATC to inform a pilot when an accurate estimate of the delay time and the reason for the delay cannot immediately be determined; e.g., a disabled aircraft on the runway, terminal or center area saturation, weather below landing minimums, etc.

(See EXPECT FURTHER CLEARANCE (TIME).)

DELAY TIME- The amount of time that the arrival must lose to cross the meter fix at the assigned meter fix time. This is the difference between ACLT and VTA.

DEPARTURE CENTER- The ARTCC having jurisdiction for the airspace that generates a flight to the impacted airport.

DEPARTURE CONTROL- A function of an approach control facility providing air traffic control service for departing IFR and, under certain conditions, VFR aircraft.

(See APPROACH CONTROL FACILITY.) (Refer to AIM.)

DEPARTURE SEQUENCING PROGRAM- A program designed to assist in achieving a specified interval over a common point for departures.

DEPARTURE TIME- The time an aircraft becomes airborne.

DESCENT SPEED ADJUSTMENTS- Speed deceleration calculations made to determine an accurate VTA. These calculations start at the transition point and use arrival speed segments to the vertex.

DESIRED COURSE-

a. True- A predetermined desired course direction to be followed (measured in degrees from true north).

b. Magnetic- A predetermined desired course direction to be followed (measured in degrees from local magnetic north).

DESIRED TRACK- The planned or intended track between two waypoints. It is measured in degrees from either magnetic or true north. The instantaneous angle may change from point to point along the great circle track between waypoints.

DETRESFA (DISTRESS PHASE) [ICAO]- The code word used to designate an emergency phase wherein there is reasonable certainty that an aircraft and its occupants are threatened by grave and imminent danger or require immediate assistance.

DEVIATIONS-

a. A departure from a current clearance, such as an off course maneuver to avoid weather or turbulence.

b. Where specifically authorized in the CFRs and requested by the pilot, ATC may permit pilots to deviate from certain regulations.

(Refer to AIM.)

DF-

(See DIRECTION FINDER.)

DF APPROACH PROCEDURE- Used under emergency conditions where another instrument approach procedure cannot be executed. DF guidance for an instrument approach is given by ATC facilities with DF capability.

(See DF GUIDANCE.)

(See DIRECTION FINDER.) (Refer to AIM.)

DF FIX- The geographical location of an aircraft obtained by one or more direction finders.

(See DIRECTION FINDER.)

DF GUIDANCE- Headings provided to aircraft by facilities equipped with direction finding equipment. These headings, if followed, will lead the aircraft to a predetermined point such as the DF station or an airport. DF guidance is given to aircraft in distress or to other aircraft which request the service. Practice DF guidance is provided when workload permits.

(See DIRECTION FINDER.)

(See DF FIX.)

(Refer to AIM.)

DF STEER-

(See DF GUIDANCE.)

DH-

(See DECISION HEIGHT.)

DH [ICAO]-

(See ICAO Term DECISION ALTITUDE/ DECISION HEIGHT.)

DIGITAL-AUTOMATIC TERMINAL INFORMATION SERVICE (D-ATIS)- The service provides text messages to aircraft, airlines, and other users outside the standard reception range of conventional ATIS via landline and data link communications to the cockpit. Also, the service provides a computer-synthesized voice message that can be transmitted to all aircraft within range of existing transmitters. The Terminal Data Link System (TDLS) D-ATIS application uses weather inputs from local automated weather sources or manually entered meteorological data together with preprogrammed menus to provide standard information to users. Airports with D-ATIS capability are listed in the Airport/Facility Directory.

DIGITAL TARGET- A computer-generated symbol representing an aircraft's position, based on a primary return or radar beacon reply, shown on a digital display.

DIGITAL TERMINAL AUTOMATION SYSTEM (DTAS)- A system where digital radar and beacon data is presented on digital displays and the operational program monitors the system performance on a real-time basis.

DIGITIZED TARGET- A computer-generated indication shown on an analog radar display resulting from a primary radar return or a radar beacon reply.

DIRECT- Straight line flight between two navigational aids, fixes, points, or any combination thereof. When used by pilots in describing off-airway routes, points defining direct route segments become compulsory reporting points unless the aircraft is under radar contact.

DIRECT ALTITUDE AND IDENTITY READOUT- The DAIR System is a modification to the AN/TPX-42 Interrogator System. The Navy has two adaptations of the DAIR System-Carrier Air Traffic Control Direct Altitude and Identification Readout System for Aircraft Carriers and Radar Air Traffic Control Facility Direct Altitude and Identity Readout System for land-based terminal operations. The DAIR detects, tracks, and predicts secondary radar aircraft targets. Targets are displayed by means of computer-generated symbols and alphanumeric characters depicting flight identification, altitude, ground speed, and flight plan data. The DAIR System is capable of interfacing with ARTCCs.

DIRECTION FINDER- A radio receiver equipped with a directional sensing antenna used to take bearings on a radio transmitter. Specialized radio direction finders are used in aircraft as air navigation aids. Others are ground-based, primarily to obtain a "fix" on a pilot requesting orientation assistance or to locate downed aircraft. A location "fix" is established by the intersection of two or more bearing lines plotted on a navigational chart using either two separately located Direction Finders to obtain a fix on an aircraft or by a pilot plotting the bearing indications of his/her DF on two separately located ground-based transmitters, both of which can be identified on his/her chart. UDFs receive signals in the ultra high frequency radio broadcast band; VDFs in the very high frequency band; and UVDFs in both bands. ATC provides DF service at those air traffic control towers and flight service stations listed in the Airport/Facility Directory and the DOD FLIP IFR En Route Supplement.

(See DF FIX.)

(See DF GUIDANCE.)

DIRECTLY BEHIND- An aircraft is considered to be operating directly behind when it is following the actual flight path of the lead aircraft over the surface of the earth except when applying wake turbulence separation criteria.

DISCRETE BEACON CODE-(See DISCRETE CODE.)

DISCRETE CODE- As used in the Air Traffic Control Radar Beacon System (ATCRBS), any one of the 4096 selectable Mode 3/A aircraft transponder codes except those ending in zero zero; e.g., discrete codes: 0010, 1201, 2317, 7777; nondiscrete codes: 0100, 1200, 7700. Nondiscrete codes are normally reserved for radar facilities that are not equipped with discrete decoding capability and for other purposes such as emergencies (7700), VFR aircraft (1200), etc.

(See RADAR.) (Refer to AIM.)

DISCRETE FREQUENCY- A separate radio frequency for use in direct pilot-controller communications in air traffic control which reduces frequency congestion by controlling the number of aircraft operating on a particular frequency at one time. Discrete frequencies are normally designated for each control sector in en route/terminal ATC facilities. Discrete frequencies are listed in the Airport/Facility Directory and the DOD

FLIP IFR En Route Supplement. (See CONTROL SECTOR.)

DISPLACED THRESHOLD- A threshold that is located at a point on the runway other than the designated beginning of the runway.

(See THRESHOLD.) (Refer to AIM.)

DISTANCE MEASURING EQUIPMENT- Equipment (airborne and ground) used to measure, in nautical miles, the slant range distance of an aircraft from the DME navigational aid.

(See MICROWAVE LANDING SYSTEM.) (See TACAN.)

(See VORTAC.)

DISTRESS- A condition of being threatened by serious and/or imminent danger and of requiring immediate assistance.

DIVE BRAKES-

(See SPEED BRAKES.)

DIVERSE VECTOR AREA- In a radar environment, that area in which a prescribed departure route is not required as the only suitable route to avoid obstacles. The area in which random radar vectors below the MVA/MIA, established in accordance with the TERPS criteria for diverse departures, obstacles and terrain avoidance, may be issued to departing aircraft.

DIVERSION (DVRSN)- Flights that are required to land at other than their original destination for reasons beyond the control of the pilot/company, e.g. periods of significant weather.

DME-

(See DISTANCE MEASURING EQUIPMENT.)

DME FIX- A geographical position determined by reference to a navigational aid which provides distance and azimuth information. It is defined by a specific distance in nautical miles and a radial, azimuth, or course (i.e., localizer) in degrees magnetic from that aid.

(See DISTANCE MEASURING EQUIPMENT.) (See FIX.)

(See MICROWAVE LANDING SYSTEM.)

DME SEPARATION- Spacing of aircraft in terms of distances (nautical miles) determined by reference to distance measuring equipment (DME).

(See DISTANCE MEASURING EQUIPMENT.)

DOD FLIP- Department of Defense Flight Information Publications used for flight planning, en route, and terminal operations. FLIP is produced by the National Geospatial-Intelligence Agency (NGA) for world-wide use. United States Government Flight Information Publications (en route charts and instrument approach procedure charts) are incorporated in DOD FLIP for use in the National Airspace System (NAS).

DOMESTIC AIRSPACE- Airspace which overlies the continental land mass of the United States plus Hawaii and U.S. possessions. Domestic airspace extends to 12 miles offshore.

DOWNBURST- A strong downdraft which induces an outburst of damaging winds on or near the ground. Damaging winds, either straight or curved, are highly divergent. The sizes of downbursts vary from 1/2 mile or less to more than 10 miles. An intense downburst often causes widespread damage. Damaging winds, lasting 5 to 30 minutes, could reach speeds as high as 120 knots.

DOWNWIND LEG-

(See TRAFFIC PATTERN.)

DP-

(See INSTRUMENT DEPARTURE PROCEDURE.)

DRAG CHUTE- A parachute device installed on certain aircraft which is deployed on landing roll to assist in deceleration of the aircraft.

DSP-

(See DEPARTURE SEQUENCING PROGRAM.) DT-

(See DELAY TIME.)

DTAS-
(See DIGITAL TERMINAL AUTOMATION SYSTEM.)
DUE REGARD- A phase of flight wherein an aircraft commander of a State-operated aircraft assumes responsibility to separate his/her aircraft from all other aircraft.
(See also FAAO JO 7110.65, Para 1-2-1, WORD MEANINGS.)
DUTY RUNWAY-
(See RUNWAY IN USE/ACTIVE RUNWAY/DUTY RUNWAY.) DVA-
(See DIVERSE VECTOR AREA.)
DVFR-
(See DEFENSE VISUAL FLIGHT RULES.)
DVFR FLIGHT PLAN- A flight plan filed for a VFR aircraft which intends to operate in airspace within which the ready identification, location, and control of aircraft are required in the interest of national security.
DVRSN-
(See DIVERSION.)
DYNAMIC- Continuous review, evaluation, and change to meet demands.
DYNAMIC RESTRICTIONS- Those restrictions imposed by the local facility on an "as needed" basis to manage unpredictable fluctuations in traffic demands.

E

EAS-
(See EN ROUTE AUTOMATION SYSTEM.)
EDCT-
(See EXPECT DEPARTURE CLEARANCE TIME.)
EFC-
(See EXPECT FURTHER CLEARANCE (TIME).)
ELT-
(See EMERGENCY LOCATOR TRANSMITTER.)
EMERGENCY- A distress or an urgency condition.
EMERGENCY LOCATOR TRANSMITTER- A radio transmitter attached to the aircraft structure which operates from its own power source on 121.5 MHz and 243.0 MHz. It aids in locating downed aircraft by radiating a downward sweeping audio tone, 2-4 times per second. It is designed to function without human action after an accident.
(Refer to 14 CFR Part 91.)
(Refer to AIM.)
E-MSAW-
(See EN ROUTE MINIMUM SAFE ALTITUDE WARNING.)
EN ROUTE AIR TRAFFIC CONTROL SERVICES- Air traffic control service provided aircraft on IFR flight plans, generally by centers, when these aircraft are operating between departure and destination terminal areas. When equipment, capabilities, and controller workload permit, certain advisory/assistance services may be provided to VFR aircraft.
(See AIR ROUTE TRAFFIC CONTROL CENTER.) (Refer to AIM.)
EN ROUTE AUTOMATION SYSTEM (EAS)- The complex integrated environment consisting of situation display systems, surveillance systems and flight data processing, remote devices, decision support tools, and the related communications equipment that form the heart of the automated IFR air traffic control system. It interfaces with automated terminal systems and is used in the control of en route IFR aircraft.
(Refer to AIM.)
EN ROUTE CHARTS-
(See AERONAUTICAL CHART.)
EN ROUTE DESCENT- Descent from the en route cruising altitude which takes place along the route of flight.
EN ROUTE FLIGHT ADVISORY SERVICE- A service specifically designed to provide, upon pilot request, timely weather information pertinent to his/her type of flight, intended route of flight, and altitude. The FSSs providing this service are listed in the Airport/Facility Directory.
(See FLIGHT WATCH.)
(Refer to AIM.)
EN ROUTE HIGH ALTITUDE CHARTS-(See AERONAUTICAL CHART.)
EN ROUTE LOW ALTITUDE CHARTS-(See AERONAUTICAL CHART.)
EN ROUTE MINIMUM SAFE ALTITUDE WARNING- A function of the EAS that aids the controller by providing an alert when a tracked aircraft is below or predicted by the computer to go below a predetermined minimum IFR altitude (MIA).
EN ROUTE SPACING PROGRAM (ESP)- A program designed to assist the exit sector in achieving the required in-trail spacing.
EN ROUTE TRANSITION-
a. Conventional STARs/SIDs. The portion of a SID/STAR that connects to one or more en route airway/jet route.
b. RNAV STARs/SIDs. The portion of a STAR preceding the common route or point, or for a SID the portion following, that is coded for a specific en route fix, airway or jet route.
ESP-
(See EN ROUTE SPACING PROGRAM.)
ESTABLISHED-To be stable or fixed on a route, route segment, altitude, heading, etc.
ESTIMATED ELAPSED TIME [ICAO]- The estimated time required to proceed from one significant point to another.
(See ICAO Term TOTAL ESTIMATED ELAPSED TIME.)
ESTIMATED OFF-BLOCK TIME [ICAO]- The estimated time at which the aircraft will commence movement associated with departure.
ESTIMATED POSITION ERROR (EPE)- (See Required Navigation Performance)
ESTIMATED TIME OF ARRIVAL- The time the flight is estimated to arrive at the gate (scheduled operators) or the actual runway on times for nonscheduled operators.
ESTIMATED TIME EN ROUTE- The estimated flying time from departure point to destination (lift-off to touchdown).
ETA-
(See ESTIMATED TIME OF ARRIVAL.)
ETE-
(See ESTIMATED TIME EN ROUTE.)
EXECUTE MISSED APPROACH- Instructions issued to a pilot making an instrument approach which means continue inbound to the missed approach point and execute the missed approach procedure as described on the Instrument Approach Procedure Chart or as previously assigned by ATC. The pilot may climb immediately to the altitude specified in the missed approach procedure upon making a missed approach. No turns should be initiated prior to reaching the missed approach point. When conducting an ASR or PAR approach, execute the assigned missed approach procedure immediately upon receiving instructions to "execute missed approach."
(Refer to AIM.)
EXPECT (ALTITUDE) AT (TIME) or (FIX)- Used under certain conditions to provide a pilot with an altitude to be used in the event of two-way communications failure. It also provides altitude information to assist the pilot in planning.
(Refer to AIM.)
EXPECT DEPARTURE CLEARANCE TIME (EDCT)- The runway release time assigned to an aircraft in a traffic management program and shown on the flight progress strip as an EDCT.
(See GROUND DELAY PROGRAM.)
EXPECT FURTHER CLEARANCE (TIME)- The time a pilot can expect to receive 4 www.faa . gov/ai r_traffic/pub li cations/atpubs/pcg/E . HTM3/clearance beyond a clearance limit.
EXPECT FURTHER CLEARANCE VIA (AIRWAYS, ROUTES OR FIXES)- Used to inform a pilot of the routing he/she can expect if any part of the route beyond a short range clearance limit differs from that filed.
EXPEDITE- Used by ATC when prompt compliance is required to avoid the development of an imminent situation. Expedite climb/descent normally indicates to a pilot that the approximate best rate of climb/descent should be used without requiring an exceptional change in aircraft handling characteristics.
2/9/12 Pilot/Controller Glossary

F

FAF-
(See FINAL APPROACH FIX.)
FAST FILE- An FSS system whereby a pilot files a flight plan via telephone that is recorded and later transcribed for transmission to the appropriate air traffic facility. (Alaska only.)
FAWP- Final Approach Waypoint
FCLT-
(See FREEZE CALCULATED LANDING TIME.)
FEATHERED PROPELLER- A propeller whose blades have been rotated so that the leading and trailing edges are nearly parallel with the aircraft flight path to stop or minimize drag and engine rotation. Normally used to indicate shutdown of a reciprocating or turboprop engine due to malfunction.
FEDERAL AIRWAYS-
(See LOW ALTITUDE AIRWAY STRUCTURE.)
FEEDER FIX- The fix depicted on Instrument Approach Procedure Charts which establishes the starting point of the feeder route.
FEEDER ROUTE- A route depicted on instrument approach procedure charts to designate routes for aircraft to proceed from the en route structure to the initial approach fix (IAF).
(See INSTRUMENT APPROACH PROCEDURE.) FERRY FLIGHT- A flight for the purpose of:
a. Returning an aircraft to base.
b. Delivering an aircraft from one location to another.
c. Moving an aircraft to and from a maintenance base.- Ferry flights, under certain conditions, may be conducted under terms of a special flight permit.
FIELD ELEVATION-
(See AIRPORT ELEVATION.)
FILED- Normally used in conjunction with flight plans, meaning a flight plan has been submitted to ATC.
FILED EN ROUTE DELAY- Any of the following preplanned delays at points/areas along the route of flight which require special flight plan filing and handling techniques.
a. Terminal Area Delay. A delay within a terminal area for touch-and-go, low approach, or other terminal area activity.
b. Special Use Airspace Delay. A delay within a Military Operations Area, Restricted Area, Warning Area, or ATC Assigned Airspace.
c. Aerial Refueling Delay. A delay within an Aerial Refueling Track or Anchor.
FILED FLIGHT PLAN- The flight plan as filed with an ATS unit by the pilot or his/her designated representative without any subsequent changes or clearances.
FINAL- Commonly used to mean that an aircraft is on the final approach course or is aligned with a landing area.
(See FINAL APPROACH COURSE.)
(See FINAL APPROACH-IFR.)
(See SEGMENTS OF AN INSTRUMENT APPROACH PROCEDURE.)
FINAL APPROACH [ICAO]- That part of an instrument approach procedure which commences at the specified final approach fix or point, or where such a fix or point is not specified.
a. At the end of the last procedure turn, base turn or inbound turn of a racetrack procedure, if specified; or
b. At the point of interception of the last track specified in the approach procedure; and ends at a point in the vicinity of an aerodrome from which:
1. A landing can be made; or
2. A missed approach procedure is initiated.
FINAL APPROACH COURSE- A bearing/radial/track of an instrument approach leading to a runway or an extended runway centerline all without regard to distance.
FINAL APPROACH FIX- The fix from which the final approach (IFR) to an airport is executed and which identifies the beginning of the final approach segment. It is designated on Government charts by the Maltese Cross symbol for nonprecision approaches and the lightning bolt symbol for precision approaches; or when ATC directs a lower-than-published glideslope/path intercept altitude, it is the resultant actual point of the glideslope/path intercept.
(See FINAL APPROACH POINT.)
(See GLIDESLOPE INTERCEPT ALTITUDE.)
(See SEGMENTS OF AN INSTRUMENT APPROACH PROCEDURE.)
FINAL APPROACH-IFR- The flight path of an aircraft which is inbound to an airport on a final instrument approach course, beginning at the final approach fix or point and extending to the airport or the point where a circle-to-land maneuver or a missed approach is executed.
(See FINAL APPROACH COURSE.)
(See FINAL APPROACH FIX.)
(See FINAL APPROACH POINT.)
(See SEGMENTS OF AN INSTRUMENT APPROACH PROCEDURE.) (See ICAO term FINAL APPROACH.)
FINAL APPROACH POINT- The point, applicable only to a nonprecision approach with no depicted FAF (such as an on airport VOR), where the aircraft is established inbound on the final approach course from the procedure turn and where the final approach descent may be commenced. The FAP serves as the FAF and identifies the beginning of the final approach segment.
(See FINAL APPROACH FIX.)
(See SEGMENTS OF AN INSTRUMENT APPROACH PROCEDURE.) FINAL APPROACH SEGMENT-
(See SEGMENTS OF AN INSTRUMENT APPROACH PROCEDURE.)
FINAL APPROACH SEGMENT [ICAO]- That segment of an instrument approach procedure in which alignment and descent for landing are accomplished.
FINAL CONTROLLER- The controller providing information and final approach guidance during PAR and ASR approaches utilizing radar equipment.

(See RADAR APPROACH.)

FINAL Guard Service- A value added service provided in conjunction with LAA/RAA only during periods of significant and fast changing weather conditions that may affect landing and takeoff operations.

FINAL MONITOR AID- A high resolution color display that is equipped with the controller alert system hardware/software which is used in the precision runway monitor (PRM) system. The display includes alert algorithms providing the target predictors, a color change alert when a target penetrates or is predicted to penetrate the no transgression zone (NTZ), a color change alert if the aircraft transponder becomes inoperative, synthesized voice alerts, digital mapping, and like features contained in the PRM system.

(See RADAR APPROACH.)

FINAL MONITOR CONTROLLER- Air Traffic Control Specialist assigned to radar monitor the flight path of aircraft during simultaneous parallel and simultaneous close parallel ILS approach operations. Each runway is assigned a final monitor controller during simultaneous parallel and simultaneous close parallel ILS approaches. Final monitor controllers shall utilize the Precision Runway Monitor (PRM) system during simultaneous close parallel ILS approaches.

FIR-

(See FLIGHT INFORMATION REGION.)

FIRST TIER CENTER- The ARTCC immediately adjacent to the impacted center. FIS-B-

(See FLIGHT INFORMATION SERVICE-BROADCAST.)

FIX- A geographical position determined by visual reference to the surface, by reference to one or more radio NAVAIDs, by celestial plotting, or by another navigational device.

FIX BALANCING- A process whereby aircraft are evenly distributed over several available arrival fixes reducing delays and controller workload.

FLAG- A warning device incorporated in certain airborne navigation and flight instruments indicating that:

a. Instruments are inoperative or otherwise not operating satisfactorily, or

b. Signal strength or quality of the received signal falls below acceptable values. FLAG ALARM-

(See FLAG.)

FLAMEOUT- An emergency condition caused by a loss of engine power.

FLAMEOUT PATTERN- An approach normally conducted by a single-engine military aircraft experiencing loss or anticipating loss of engine power or control. The standard overhead approach starts at a relatively high altitude over a runway ("high key") followed by a continuous 180 degree turn to a high, wide position ("low key") followed by a continuous 180 degree turn final. The standard straight-in pattern starts at a point that results in a straight-in approach with a high rate of descent to the runway. Flameout

approaches terminate in the type approach requested by the pilot (normally fullstop).

FLIGHT CHECK- A call-sign prefix used by FAA aircraft engaged in flight inspection/certification of navigational aids and flight procedures. The word "recorded" may be added as a suffix; e.g., "Flight Check 320 recorded" to indicate that an automated flight inspection is in progress in terminal areas.

(See FLIGHT INSPECTION.) (Refer to AIM.)

FLIGHT FOLLOWING-

(See TRAFFIC ADVISORIES.)

FLIGHT INFORMATION REGION- An airspace of defined dimensions within which Flight Information Service and Alerting Service are provided.

a. Flight Information Service. A service provided for the purpose of giving advice and information useful for the safe and efficient conduct of flights.

b. Alerting Service. A service provided to notify appropriate organizations regarding aircraft in need of search and rescue aid and to assist such organizations as required.

FLIGHT INFORMATION SERVICE- A service provided for the purpose of giving advice and information useful for the safe and efficient conduct of flights.

FLIGHT INFORMATION SERVICE-BROADCAST (FIS-B)- A ground broadcast service provided through the ADS-B Broadcast Services network over the UAT data link that operates on 978 MHz. The FIS-B system provides pilots and flight crews of properly equipped aircraft with a cockpit display of certain aviation weather and aeronautical information.

FLIGHT INSPECTION- Inflight investigation and evaluation of a navigational aid to determine whether it meets established tolerances.

(See FLIGHT CHECK.) (See NAVIGATIONAL AID.)

FLIGHT LEVEL- A level of constant atmospheric pressure related to a reference datum of 29.92 inches of mercury. Each is stated in three digits that represent hundreds of feet. For example, flight level (FL) 250 represents a barometric altimeter indication of 25,000 feet; FL 255, an indication of 25,500 feet.

(See ICAO term FLIGHT LEVEL.)

FLIGHT LEVEL [ICAO]- A surface of constant atmospheric pressure which is related to a specific pressure datum, 1013.2 hPa (1013.2 mb), and is separated from other such surfaces by specific pressure intervals.

Note 1: A pressure type altimeter calibrated in accordance with the standard atmosphere:

a. When set to a QNH altimeter setting, will indicate altitude;

b. When set to a QFE altimeter setting, will indicate height above the QFE reference datum; and

c. When set to a pressure of 1013.2 hPa (1013.2 mb), may be used to indicate flight levels.

Note 2: The terms `height' and `altitude,' used in Note 1 above, indicate altimetric rather than geometric heights and altitudes.

FLIGHT LINE- A term used to describe the precise movement of a civil photogrammetric aircraft along a predetermined course(s) at a predetermined altitude during the actual photographic run.

FLIGHT MANAGEMENT SYSTEMS- A computer system that uses a large data base to allow routes to be preprogrammed and fed into the system by means of a data loader. The system is constantly updated with respect to position accuracy by reference to conventional navigation aids. The sophisticated program and its associated data base ensures that the most appropriate aids are automatically selected during the information update cycle.

FLIGHT MANAGEMENT SYSTEM PROCEDURE- An arrival, departure, or approach procedure developed for use by aircraft with a slant (/) E or slant (/) F equipment suffix.

FLIGHT PATH- A line, course, or track along which an aircraft is flying or intended to be flown.

(See COURSE.) (See TRACK.)

FLIGHT PLAN- Specified information relating to the intended flight of an aircraft that is filed orally or in writing with an FSS or an ATC facility.

(See FAST FILE.) (See FILED.)

(Refer to AIM.)

FLIGHT PLAN AREA (FPA)- The geographical area assigned to a flight service station (FSS) for the purpose of establishing primary responsibility for services that may include search and rescue for VFR aircraft, issuance of NOTAMs, pilot briefings, inflight services, broadcast services, emergency services, flight data processing, international operations, and aviation weather services. Large consolidated FSS facilities may combine FPAs into larger areas of responsibility (AOR).

(See FLIGHT SERVICE STATION.) (See TIE IN FACILITY.)

FLIGHT RECORDER- A general term applied to any instrument or device that records information about the performance of an aircraft in flight or about conditions encountered in flight. Flight recorders may make records of airspeed, outside air temperature, vertical acceleration, engine RPM, manifold pressure, and other pertinent variables for a given flight.

(See ICAO term FLIGHT RECORDER.)

FLIGHT RECORDER [ICAO]- Any type of recorder installed in the aircraft for the purpose of complementing accident/incident investigation.

Note: See Annex 6 Part I, for specifications relating to flight recorders.

FLIGHT SERVICE STATION (FSS) - An air traffic facility which provides pilot briefings, flight plan processing, en route radio communications, search and rescue services, and assistance to lost aircraft and aircraft in emergency situations. FSS also relays ATC clearances, processes Notices to Airmen, and broadcasts aviation weather and aeronautical information. In addition, at selected locations, FSS provides En Route Flight Advisory Service (Flight Watch) and Airport Advisory Service (AAS) and takes airport weather observations.

(See FLIGHT PLAN AREA.) (See TIE IN FACILITY.)

FLIGHT STANDARDS DISTRICT OFFICE- An FAA field office serving an assigned geographical area and staffed with Flight Standards personnel who serve the aviation industry and the general public on matters relating to the certification and operation of air carrier and general aviation aircraft. Activities include general surveillance of operational safety, certification of airmen and aircraft, accident prevention, i nvestigation, enforcement, etc.

FLIGHT TEST- A flight for the purpose of:

a. Investigating the operation/flight characteristics of an aircraft or aircraft component.

b. Evaluating an applicant for a pilot certificate or rating.

FLIGHT VISIBILITY-

(See VISIBILITY.)

FLIGHT WATCH- A shortened term for use in air-ground contacts to identify the flight service station providing En Route Flight Advisory Service; e.g., "Oakland Flight Watch."

(See EN ROUTE FLIGHT ADVISORY SERVICE.) FLIP-

(See DOD FLIP.)

FLY HEADING (DEGREES)- Informs the pilot of the heading he/she should fly. The pilot may have to turn to, or continue on, a specific compass direction in order to comply with the instructions. The pilot is expected to turn in the shorter direction to the heading unless otherwise instructed by ATC.

FLY-BY WAYPOINT- A fly-by waypoint requires the use of turn anticipation to avoid overshoot of the next flight segment.

FLY-OVER WAYPOINT- A fly-over waypoint precludes any turn until the waypoint is overflown and is followed by an intercept maneuver of the next flight segment.

FLY VISUAL TO AIRPORT-

(See PUBLISHED INSTRUMENT APPROACH PROCEDURE VISUAL SEGMENT.) FMA-

(See FINAL MONITOR AID.)

FMS-

(See FLIGHT MANAGEMENT SYSTEM.)

FMSP-

(See FLIGHT MANAGEMENT SYSTEM PROCEDURE.)

FORMATION FLIGHT- More than one aircraft which, by prior arrangement between the pilots, operate as a single aircraft with regard to navigation and position reporting. Separation between aircraft within the formation is the responsibility of the flight leader and the pilots of the other aircraft in the flight. This includes transition periods when aircraft within the formation are maneuvering to attain separation from each other to effect individual control and during join-up and breakaway.

a. A standard formation is one in which a proximity of no more than 1 mile laterally or longitudinally and within 100 feet vertically from the flight leader is maintained by each wingman.

b. Nonstandard formations are those operating under any of the following conditions:

1. When the flight leader has requested and ATC has approved other than standard formation dimensions.

2. When operating within an authorized altitude reservation (ALTRV) or under the provisions of a letter of agreement.

3. When the operations are conducted in airspace specifically designed for a special activity.

(See ALTITUDE RESERVATION.)

(Refer to 14 CFR Part 91.)

FRC-

(See REQUEST FULL ROUTE CLEARANCE.)

FREEZE/FROZEN- Terms used in referring to arrivals which have been assigned ACLTs and to the lists in which they are displayed.

FREEZE CALCULATED LANDING TIME- A dynamic parameter number of minutes prior to the meter fix calculated time of arrival for each aircraft when the TCLT is frozen and becomes an ACLT (i.e., the VTA is updated and consequently the TCLT is modified as appropriate until FCLT minutes prior to meter fix calculated time of arrival, at which time updating is suspended and an ACLT and a frozen meter fix crossing time (MFT) is assigned).

FREEZE HORIZON- The time or point at which an aircraft's STA becomes fixed and no longer fluctuates with each radar update. This setting ensures a constant time for each aircraft, necessary for the metering controller to plan his/her delay technique. This setting can be either in distance from the meter fix or a prescribed flying time to the meter fix.

FREEZE SPEED PARAMETER- A speed adapted for each aircraft to determine fast and slow aircraft. Fast aircraft freeze on parameter FCLT and slow aircraft freeze on parameter MLDI.

FRICTION MEASUREMENT- A measurement of the friction characteristics of the runway pavement surface using continuous self-watering friction measurement equipment in accordance with the specifications, procedures and schedules contained in AC 150/5320-12, Measurement, Construction, and Maintenance of Skid Resistant Airport Pavement Surfaces.

FSDO-

(See FLIGHT STANDARDS DISTRICT OFFICE.) FSPD-

(See FREEZE SPEED PARAMETER.)
FSS-
(See FLIGHT SERVICE STATION.)
FUEL DUMPING- Airborne release of usable fuel. This does not include the dropping of fuel tanks.
(See JETTISONING OF EXTERNAL STORES.)
FUEL REMAINING- A phrase used by either pilots or controllers when relating to the fuel remaining on board until actual fuel exhaustion. When transmitting such information in response to either a controller question or pilot initiated cautionary advisory to air traffic control, pilots will state the APPROXIMATE NUMBER OF MINUTES the flight can continue with the fuel remaining. All reserve fuel SHOULD BE INCLUDED in the time stated, as should an allowance for established fuel gauge system error.
FUEL SIPHONING- Unintentional release of fuel caused by overflow, puncture, loose cap, etc.
FUEL VENTING-
(See FUEL SIPHONING.)

G

GATE HOLD PROCEDURES- Procedures at selected airports to hold aircraft at the gate or other ground location whenever departure delays exceed or are anticipated to exceed 15 minutes. The sequence for departure will be maintained in accordance with initial call-up unless modified by flow control restrictions. Pilots should monitor the ground control/clearance delivery frequency for engine start/taxi advisories or new proposed start/taxi time if the delay changes.
GBT-
(See GROUND-BASED TRANSCEIVER.)
GCA-
(See GROUND CONTROLLED APPROACH.)
GDP-
(See GROUND DELAY PROGRAM.)
GENERAL AVIATION- That portion of civil aviation which encompasses all facets of aviation except air carriers holding a certificate of public convenience and necessity from the Civil Aeronautics Board and large aircraft commercial operators.
(See ICAO term GENERAL AVIATION.)
GENERAL AVIATION [ICAO]- All civil aviation operations other than scheduled air services and nonscheduled air transport operations for remuneration or hire.
GEO MAP- The digitized map markings associated with the ASR-9 Radar System. GLIDEPATH-
(See GLIDESLOPE.)
GLIDEPATH [ICAO]- A descent profile determined for vertical guidance during a final approach.
GLIDEPATH INTERCEPT ALTITUDE-
(See GLIDESLOPE INTERCEPT ALTITUDE.)
GLIDESLOPE- Provides vertical guidance for aircraft during approach and landing. The glideslope/glidepath is based on the following:
a. Electronic components emitting signals which provide vertical guidance by reference to airborne instruments during instrument approaches such as ILS/MLS, or
b. Visual ground aids, such as VASI, which provide vertical guidance for a VFR approach or for the visual portion of an instrument approach and landing.
c. PAR. Used by ATC to inform an aircraft making a PAR approach of its vertical position (elevation) relative to the descent profile.
(See ICAO term GLIDEPATH.)
GLIDESLOPE INTERCEPT ALTITUDE- The minimum altitude to intercept the glideslope/path on a precision approach. The intersection of the published intercept altitude with the glideslope/path, designated on Government charts by the lightning bolt symbol, is the precision FAF; however, when the approach chart shows an alternative lower glideslope intercept altitude, and ATC directs a lower altitude, the resultant lower intercept position is then the FAF.
(See FINAL APPROACH FIX.)
(See SEGMENTS OF AN INSTRUMENT APPROACH PROCEDURE.)
GLOBAL POSITIONING SYSTEM (GPS)- A space-base radio positioning, navigation, and time-transfer system. The system provides highly accurate position and velocity information, and precise time, on a continuous global basis, to an unlimited number of properly equipped users. The system is unaffected by weather, and provides a worldwide common grid reference system. The GPS concept is predicated upon accurate and continuous knowledge of the spatial position of each satellite in the system with respect to time and distance from a transmitting satellite to the user. The GPS receiver automatically selects appropriate signals from the satellites in view and translates these into three-dimensional position, velocity, and time. System accuracy for civil users is normally 100 meters horizontally.
GO AHEAD- Proceed with your message. Not to be used for any other purpose.
GO AROUND- Instructions for a pilot to abandon his/her approach to landing. Additional instructions may follow. Unless otherwise advised by ATC, a VFR aircraft or an aircraft conducting visual approach should overfly the runway while climbing to traffic pattern altitude and enter the traffic pattern via the crosswind leg. A pilot on an IFR flight plan making an instrument approach should execute the published missed approach procedure or proceed as instructed by ATC; e.g., "Go around" (additional instructions if required).
(See LOW APPROACH.)
(See MISSED APPROACH.) GPD-
(See GRAPHIC PLAN DISPLAY.)
GPS-
(See GLOBAL POSITIONING SYSTEM.)
GRAPHIC PLAN DISPLAY (GPD)- A view available with URET that provides a graphic display of aircraft, traffic, and notification of predicted conflicts. Graphic routes for Current Plans and Trial Plans are displayed upon controller request.
(See USER REQUEST EVALUATION TOOL.)
GROUND-BASED TRANSCEIVER (GBT)- The ground-based transmitter/receiver (transceiver) receives automatic dependent surveillance-broadcast messages, which are forwarded to an air traffic control facility for processing and display with other radar targets on the plan position indicator (radar display).
(See AUTOMATIC DEPENDENT SURVEILLANCE-BROADCAST.)
GROUND CLUTTER- A pattern produced on the radar scope by ground returns which may degrade other radar returns in the affected area. The effect of ground clutter is minimized by the use of moving target indicator (MTI) circuits in the radar equipment resulting in a radar presentation which displays only targets which are in motion.
(See CLUTTER.)
GROUND COMMUNICATION OUTLET (GCO)- An unstaffed, remotely controlled, ground/ground communications facility. Pilots at uncontrolled airports may contact ATC and FSS via VHF to a telephone connection to obtain an instrument clearance or close a VFR or IFR flight plan. They may also get an updated weather briefing prior to takeoff. Pilots will use four "key clicks" on the VHF radio to contact the appropriate ATC facility or six "key clicks" to contact the FSS. The GCO system is intended to be used only on the ground.
GROUND CONTROLLED APPROACH- A radar approach system operated from the ground by air traffic control personnel transmitting instructions to the pilot by radio. The approach may be conducted with surveillance radar (ASR) only or with both surveillance and precision approach radar (PAR). Usage of the term "GCA" by pilots is discouraged except when referring to a GCA facility. Pilots should specifically request a "PAR" approach when a precision radar approach is desired or request an "ASR" or "surveillance" approach when a nonprecision radar approach is desired.
(See RADAR APPROACH.)
GROUND DELAY PROGRAM (GDP)- A traffic management process administered by the ATCSCC; when aircraft are held on the ground. The purpose of the program is to support the TM mission and limit airborne holding. It is a flexible program and may be implemented in various forms depending upon the needs of the AT system. Ground delay programs provide for equitable assignment of delays to all system users.
GROUND SPEED- The speed of an aircraft relative to the surface of the earth. 4 www.faa . gov/ai r_traffic/pub li cations/atpubs/pcg/G . HTM 3/
GROUND STOP (GS)- The GS is a process that requires aircraft that meet a specific criteria to remain on the ground. The criteria may be airport specific, airspace specific, or equipment specific; for example, all departures to San Francisco, or all departures entering Yorktown sector, or all Category I and II aircraft going to Charlotte. GSs normally occur with little or no warning.
GROUND VISIBILITY-(See VISIBILITY.)
GS-
(See GROUND STOP.)

H

HAA-
(See HEIGHT ABOVE AIRPORT.) HAL-
(See HEIGHT ABOVE LANDING.)
HANDOFF- An action taken to transfer the radar identification of an aircraft from one controller to another if the aircraft will enter the receiving controller's airspace and radio communications with the aircraft will be transferred.
HAR-
(See HIGH ALTITUDE REDESIGN.)
HAT-
(See HEIGHT ABOVE TOUCHDOWN.)
HAVE NUMBERS- Used by pilots to inform ATC that they have received runway, wind, and altimeter information only.
HAZARDOUS INFLIGHT WEATHER ADVISORY SERVICE- Continuous recorded hazardous inflight weather forecasts broadcasted to airborne pilots over selected VOR outlets defined as an HIWAS BROADCAST AREA.
HAZARDOUS WEATHER INFORMATION- Summary of significant meteorological information (SIGMET/WS), convective significant meteorological information (convective SIGMET/WST), urgent pilot weather reports (urgent PIREP/UUA), center weather advisories (CWA), airmen's meteorological information (AIRMET/WA) and any other weather such as isolated thunderstorms that are rapidly developing and increasing in intensity, or low ceilings and visibilities that are becoming widespread which is considered significant and are not included in a current hazardous weather advisory.
HEAVY (AIRCRAFT)-
(See AIRCRAFT CLASSES.)
HEIGHT ABOVE AIRPORT- The height of the Minimum Descent Altitude above the published airport elevation. This is published in conjunction with circling minimums.
(See MINIMUM DESCENT ALTITUDE.)
HEIGHT ABOVE LANDING- The height above a designated helicopter landing area used for helicopter instrument approach procedures.
(Refer to 14 CFR Part 97.)
HEIGHT ABOVE TOUCHDOWN- The height of the Decision Height or Minimum Descent Altitude above the highest runway elevation in the touchdown zone (first 3,000 feet of the runway). HAT is published on instrument approach charts in conjunction with all straight-in minimums.
(See DECISION HEIGHT.)
(See MINIMUM DESCENT ALTITUDE.)
HELICOPTER- A heavier-than-air aircraft supported in flight chiefly by the reactions of the air on one or more power-driven rotors on substantially vertical axes.
HELIPAD- A small, designated area, usually with a prepared surface, on a heliport, airport, landing/takeoff area, apron/ramp, or movement area used for takeoff, landing, or parking of helicopters.
HELIPORT- An area of land, water, or structure used or intended to be used for the landing and takeoff of helicopters and includes its buildings and facilities if any.
HELIPORT REFERENCE POINT (HRP)- The geographic center of a heliport.
HERTZ- The standard radio equivalent of frequency in cycles per second of an electromagnetic wave. Kilohertz (kHz) is a frequency of one thousand cycles per second. Megahertz (MHz) is a frequency of one million cycles per second.
HF-
(See HIGH FREQUENCY.)
HFCOMMUNICATIONS-
(See HIGH FREQUENCY COMMUNICATIONS.)
HIGH ALTITUDE REDESIGN (HAR)- A level of non-restrictive routing (NRR) service for aircraft that have all waypoints associated with the HAR program in their flight management systems or RNAV equipage.
HIGH FREQUENCY- The frequency band between 3 and 30 MHz. (See HIGH FREQUENCY COMMUNICATIONS.)
HIGH FREQUENCY COMMUNICATIONS- High radio frequencies (HF) between 3 and 30 MHz used for air-to-ground voice communication in overseas operations.
HIGH SPEED EXIT-
(See HIGH SPEED TAXIWAY.)
HIGH SPEED TAXIWAY- A long radius taxiway designed and provided with lighting or marking to define the path of aircraft, traveling at high speed (up to 60 knots), from the runway center to a point on the center of a taxiway. Also referred to as long radius exit or turn-off taxiway. The high speed taxiway is designed to expedite aircraft turning off the runway after landing, thus reducing runway occupancy time.
HIGH SPEED TURNOFF-
(See HIGH SPEED TAXIWAY.)
HIWAS-
(See HAZARDOUS INFLIGHT WEATHER ADVISORY SERVICE.) HIWAS AREA-

(See HAZARDOUS INFLIGHT WEATHER ADVISORY SERVICE.)

HIWAS BROADCAST AREA- A geographical area of responsibility including one or more HIWAS outlet areas assigned to a FSS for hazardous weather advisory broadcasting.

HIWAS OUTLET AREA- An area defined as a 150 NM radius of a HIWAS outlet, expanded as necessary to provide coverage.

HOLD FOR RELEASE- Used by ATC to delay an aircraft for traffic management reasons; ie., weather, traffic volume, etc. Hold for release instructions (including departure delay information) are used to inform a pilot or a controller (either directly or through an authorized relay) that an IFR departure clearance is not valid until a release time or additional instructions have been received.

(See ICAO term HOLDING POINT.)

HOLD IN LIEU OF PROCEDURE TURN- A hold in lieu of procedure turn shall be established over a final or intermediate fix when an approach can be made from a properly aligned holding pattern. The hold in lieu of procedure turn permits the pilot to align with the final or intermediate segment of the approach and/or descend in the holding pattern to an altitude that will permit a normal descent to the final approach fix altitude. The hold in lieu of procedure turn is a required maneuver (the same as a procedure turn) unless the aircraft is being radar vectored to the final approach course, when "NoPT" is shown on the approach chart, or when the pilot requests or the controller advises the pilot to make a "straight-in" approach.

HOLD PROCEDURE- A predetermined maneuver which keeps aircraft within a specified airspace while awaiting further clearance from air traffic control. Also used during ground operations to keep aircraft within a specified area or at a specified point while awaiting further clearance from air traffic control.

(See HOLDING FIX.) (Refer to AIM.)

HOLDING FIX- A specified fix identifiable to a pilot by NAVAIDs or visual reference to the ground used as a reference point in establishing and maintaining the position of an aircraft while holding.

(See FIX.)

(See VISUAL HOLDING.) (Refer to AIM.)

HOLDING POINT [ICAO]- A specified location, identified by visual or other means, in the vicinity of which the position of an aircraft in flight is maintained in accordance with air traffic control clearances.

HOLDING PROCEDURE-(See HOLD PROCEDURE.)

HOLD-SHORT POINT- A point on the runway beyond which a landing aircraft with a LAHSO clearance is not authorized to proceed. This point may be located prior to an intersecting runway, taxiway, predetermined point, or approach/departure flight path.

HOLD-SHORT POSITION LIGHTS- Flashing in-pavement white lights located at specified hold-short points.

HOLD-SHORT POSITION MARKING- The painted runway marking located at the hold-short point on all LAHSO runways.

HOLD-SHORT POSITION SIGNS- Red and white holding position signs located alongside the hold-short point.

HOMING- Flight toward a NAVAID, without correcting for wind, by adjusting the aircraft heading to maintain a relative bearing of zero degrees.

(See BEARING.)

(See ICAO term HOMING.)

HOMING [ICAO]- The procedure of using the direction-finding equipment of one radio station with the emission of another radio station, where at least one of the stations is mobile, and whereby the mobile station proceeds continuously towards the other station.

HOVER CHECK- Used to describe when a helicopter/VTOL aircraft requires a stabilized hover to conduct a performance/power check prior to hover taxi, air taxi, or takeoff. Altitude of the hover will vary based on the purpose of the check.

HOVER TAXI- Used to describe a helicopter/VTOL aircraft movement conducted above the surface and in ground effect at airspeeds less than approximately 20 knots. The actual height may vary, and some helicopters may require hover taxi above 25 feet AGL to reduce ground effect turbulence or provide clearance for cargo slingloads.

(See AIR TAXI.)

(See HOVER CHECK.) (Refer to AIM.)

HOW DO YOU HEAR ME?- A question relating to the quality of the transmission or to determine how well the transmission is being received.

HZ-

(See HERTZ.)

I

I SAY AGAIN- The message will be repeated.

IAF-

(See INITIAL APPROACH FIX.)

IAP-

(See INSTRUMENT APPROACH PROCEDURE.)

IAWP- Initial Approach Waypoint

ICAO-

(See ICAO Term INTERNATIONAL CIVIL AVIATION ORGANIZATION.) ICING- The accumulation of airframe ice.

Types of icing are:

a. Rime Ice- Rough, milky, opaque ice formed by the instantaneous freezing of small supercooled water droplets.

b. Clear Ice- A glossy, clear, or translucent ice formed by the relatively slow freezing or large supercooled water droplets.

c. Mixed- A mixture of clear ice and rime ice. Intensity of icing:

a. Trace- Ice becomes perceptible. Rate of accumulation is slightly greater than the rate of sublimation. Deicing/anti-icing equipment is not utilized unless encountered for an extended period of time (over 1 hour).

b. Light- The rate of accumulation may create a problem if flight is prolonged in this environment (over 1 hour). Occasional use of deicing/anti-icing equipment removes/prevents accumulation. It does not present a problem if the deicing/anti-icing equipment is used.

c. Moderate- The rate of accumulation is such that even short encounters become potentially hazardous and use of deicing/anti-icing equipment or flight diversion is necessary.

d. Severe- The rate of accumulation is such that deicing/anti-icing equipment fails to reduce or control the hazard. Immediate flight diversion is necessary.

IDENT- A request for a pilot to activate the aircraft transponder identification feature.

This will help the controller to confirm an aircraft identity or to identify an aircraft.

(Refer to AIM.)

IDENT FEATURE- The special feature in the Air Traffic Control Radar Beacon System (ATCRBS) equipment. It is used to immediately distinguish one displayed beacon target from other beacon targets.

(See IDENT.)

IF-

(See INTERMEDIATE FIX.)

IFIM-

(See INTERNATIONAL FLIGHT INFORMATION MANUAL.)

IF NO TRANSMISSION RECEIVED FOR (TIME)- Used by ATC in radar approaches to prefix procedures which should be followed by the pilot in event of lost communications.

(See LOST COMMUNICATIONS.)

IFR-

(See INSTRUMENT FLIGHT RULES.)

IFR AIRCRAFT- An aircraft conducting flight in accordance with instrument flight rules.

IFR CONDITIONS- Weather conditions below the minimum for flight under visual flight rules.

(See INSTRUMENT METEOROLOGICAL CONDITIONS.)

IFR DEPARTURE PROCEDURE-

(See IFR TAKEOFF MINIMUMS AND DEPARTURE PROCEDURES.) (Refer to AIM.)

IFR FLIGHT-

(See IFR AIRCRAFT.)

IFR LANDING MINIMUMS-

(See LANDING MINIMUMS.)

IFR MILITARY TRAINING ROUTES (IR)- Routes used by the Department of Defense and associated Reserve and Air Guard units for the purpose of conducting low-altitude navigation and tactical training in both IFR and VFR weather conditions below 10,000 feet MSL at airspeeds in excess of 250 knots IAS.

IFR TAKEOFF MINIMUMS AND DEPARTURE PROCEDURES- Title 14 Code of Federal Regulations Part 91, prescribes standard takeoff rules for certain civil users. At some airports, obstructions or other factors require the establishment of nonstandard takeoff minimums, departure procedures, or both to assist pilots in avoiding obstacles during climb to the minimum en route altitude. Those airports are listed in FAA/DOD Instrument Approach Procedures (IAPs) Charts under a section entitled "IFR Takeoff Minimums and Departure Procedures." The FAA/DOD IAP chart legend illustrates the symbol used to alert the pilot to nonstandard takeoff minimums and departure procedures. When departing IFR from such airports or from any airports where there are no departure procedures, DPs, or ATC facilities available, pilots should advise ATC of any departure limitations. Controllers may query a pilot to determine acceptable departure directions, turns, or headings after takeoff. Pilots should be familiar with the departure procedures and must assure that their aircraft can meet or exceed any specified climb gradients.

IF/IAWP- Intermediate Fix/Initial Approach Waypoint. The waypoint where the final approach course of a T approach meets the crossbar of the T. When designated (in conjunction with a TAA) this waypoint will be used as an IAWP when approaching the airport from certain directions, and as an IFWP when beginning the approach from another IAWP.

IFWP- Intermediate Fix Waypoint

ILS-

(See INSTRUMENT LANDING SYSTEM.)

ILS CATEGORIES- 1. Category I. An ILS approach procedure which provides for approach to a height above touchdown of not less than 200 feet and with runway visual range of not less than 1,800 feet.- 2. Special Authorization Category I. An ILS approach procedure which provides for approach to a height above touchdown of not less than 150 feet and with runway visual range of not less than 1,400 feet, HUD to DH. 3. Category II. An ILS approach procedure which provides for approach to a height above touchdown of not less than 100 feet and with runway visual range of not less than 1,200 feet (with autoland or HUD to touchdown and noted on authorization, RVR 1,000 feet).- 4. Special Authorization Category II with Reduced Lighting. An ILS approach procedure which provides for approach to a height above touchdown of not less than 100 feet and with runway visual range of not less than 1,200 feet with autoland or HUD to touchdown and noted on authorization (no touchdown zone and centerline lighting are required).- 5. Category III:

a. IIIA.-An ILS approach procedure which provides for approach without a decision height minimum and with runway visual range of not less than 700 feet.

b. IIIB.-An ILS approach procedure which provides for approach without a decision height minimum and with runway visual range of not less than 150 feet.

c. IIIC.-An ILS approach procedure which provides for approach without a decision height minimum and without runway visual range minimum.

ILS PRM APPROACH- An instrument landing system (ILS) approach conducted to parallel runways whose extended centerlines are separated by less than 4,300 feet and the parallel runways have a Precision Runway Monitoring (PRM) system that permits simultaneous independent ILS approaches.

IM-

(See INNER MARKER.)

IMC-

(See INSTRUMENT METEOROLOGICAL CONDITIONS.)

IMMEDIATELY- Used by ATC or pilots when such action compliance is required to avoid an imminent situation.

INCERFA (Uncertainty Phase) [ICAO]- A situation wherein uncertainty exists as to the safety of an aircraft and its occupants.

INCREASE SPEED TO (SPEED)- (See SPEED ADJUSTMENT.)

INERTIAL NAVIGATION SYSTEM- An RNAV system which is a form of self-contained navigation.

(See Area Navigation/RNAV.) INFLIGHT REFUELING-

(See AERIAL REFUELING.)

INFLIGHT WEATHER ADVISORY-(See WEATHER ADVISORY.)

INFORMATION REQUEST- A request originated by an FSS for information concerning an overdue VFR aircraft.

INITIAL APPROACH FIX- The fixes depicted on instrument approach procedure charts that identify the beginning of the initial approach segment(s).

(See FIX.)

(See SEGMENTS OF AN INSTRUMENT APPROACH PROCEDURE.) INITIAL APPROACH SEGMENT-

(See SEGMENTS OF AN INSTRUMENT APPROACH PROCEDURE.)

INITIAL APPROACH SEGMENT [ICAO]- That segment of an instrument approach procedure between the initial approach fix and the intermediate approach fix or, where applicable, the final approach fix or point.

INLAND NAVIGATION FACILITY- A navigation aid on a North American Route at which the common route and/or the noncommon route begins or ends.

INNER MARKER- A marker beacon used with an ILS (CAT II) precision approach located between the middle marker and the end of the ILS runway, transmitting a radiation pattern keyed at six dots per second and indicating to the pilot, both aurally and visually, that he/she is at the designated decision height (DH), normally 100 feet above the touchdown zone elevation, on the ILS CAT II approach. It also marks progress during a CAT III approach.

(See INSTRUMENT LANDING SYSTEM.)

(Refer to AIM.)

INNER MARKER BEACON-

(See INNER MARKER.)

INREQ-

(See INFORMATION REQUEST.)

INS-

(See INERTIAL NAVIGATION SYSTEM.)

INSTRUMENT APPROACH-

(See INSTRUMENT APPROACH PROCEDURE.)

INSTRUMENT APPROACH PROCEDURE- A series of predetermined maneuvers for the orderly transfer of an aircraft under instrument flight conditions from the beginning of the initial approach to a landing or to a point from which a landing may be made visually. It is prescribed and approved for a specific airport by competent authority.

(See SEGMENTS OF AN INSTRUMENT APPROACH PROCEDURE.) (Refer to 14 CFR Part 91.)

(Refer to AIM.)

a. U.S. civil standard instrument approach procedures are approved by the FAA as prescribed under 14 CFR Part 97 and are available for public use.

b. U.S. military standard instrument approach procedures are approved and published by the Department of Defense.

c. Special instrument approach procedures are approved by the FAA for individual operators but are not published in 14 CFR Part 97 for public use.

(See ICAO term INSTRUMENT APPROACH PROCEDURE.)

INSTRUMENT APPROACH PROCEDURE [ICAO]- A series of predetermined maneuvers by reference to flight instruments with specified protection from obstacles from the initial approach fix, or where applicable, from the beginning of a defined arrival route to a point from which a landing can be completed and thereafter, if a landing is not completed, to a position at which holding or en route obstacle clearance criteria apply.

INSTRUMENT APPROACH PROCEDURES CHARTS-(See AERONAUTICAL CHART.)

INSTRUMENT DEPARTURE PROCEDURE (DP)- A preplanned instrument flight rule (IFR) departure procedure published for pilot use, in graphic or textual format, that provides obstruction clearance from the terminal area to the appropriate en route structure. There are two types of DP, Obstacle Departure Procedure (ODP), printed either textually or graphically, and, Standard Instrument Departure (SID), which is always printed graphically.

(See IFR TAKEOFF MINIMUMS AND DEPARTURE PROCEDURES.)

(See OBSTACLE DEPARTURE PROCEDURES.) (See STANDARD INSTRUMENT DEPARTURES.) (Refer to AIM.)

INSTRUMENT DEPARTURE PROCEDURE (DP) CHARTS-(See AERONAUTICAL CHART.)

INSTRUMENT FLIGHT RULES- Rules governing the procedures for conducting instrument flight. Also a term used by pilots and controllers to indicate type of flight plan.

(See INSTRUMENT METEOROLOGICAL CONDITIONS.) (See VISUAL FLIGHT RULES.)

(See VISUAL METEOROLOGICAL CONDITIONS.) (See ICAO term INSTRUMENT FLIGHT RULES.)

(Refer to AIM.)

INSTRUMENT FLIGHT RULES [ICAO]- A set of rules governing the conduct of flight under instrument meteorological conditions.

INSTRUMENT LANDING SYSTEM- A precision instrument approach system which normally consists of the following electronic components and visual aids:

a. Localizer. (See LOCALIZER.)

b. Glideslope. (See GLIDESLOPE.)

c. Outer Marker. (See OUTER MARKER.)

d. Middle Marker. (See MIDDLE MARKER.)

e. Approach Lights.

(See AIRPORT LIGHTING.)

(Refer to 14 CFR Part 91.)

(Refer to AIM.)

INSTRUMENT METEOROLOGICAL CONDITIONS- Meteorological conditions expressed in terms of visibility, distance from cloud, and ceiling less than the minima specified for visual meteorological conditions.

(See INSTRUMENT FLIGHT RULES.)

(See VISUAL FLIGHT RULES.)

(See VISUAL METEOROLOGICAL CONDITIONS.)

INSTRUMENT RUNWAY- A runway equipped with electronic and visual navigation aids for which a precision or nonprecision approach procedure having straight-in landing minimums has been approved.

(See ICAO term INSTRUMENT RUNWAY.)

INSTRUMENT RUNWAY [ICAO]- One of the following types of runways intended for the operation of aircraft using instrument approach procedures:

a. Nonprecision Approach Runway-An instrument runway served by visual aids and a nonvisual aid providing at least directional guidance adequate for a straight-in approach.

b. Precision Approach Runway, Category I-An instrument runway served by ILS and visual aids intended for operations down to 60 m (200 feet) decision height and down to an RVR of the order of 800 m.

c. Precision Approach Runway, Category II-An instrument runway served by ILS and visual aids intended for operations down to 30 m (100 feet) decision height and down to an RVR of the order of 400 m.

d. Precision Approach Runway, Category III-An instrument runway served by ILS to and along the surface of the runway and:

1. Intended for operations down to an RVR of the order of 200 m (no decision height being applicable) using visual aids during the final phase of landing;

2. Intended for operations down to an RVR of the order of 50 m (no decision height being applicable) using visual aids for taxiing;

3. Intended for operations without reliance on visual reference for landing or taxiing. Note 1: See Annex 10 Volume I, Part I, Chapter 3, for related ILS specifications.

Note 2: Visual aids need not necessarily be matched to the scale of nonvisual aids provided. The criterion for the selection of visual aids is the conditions in which operations are intended to be conducted.

INTEGRITY- The ability of a system to provide timely warnings to users when the system should not be used for navigation.

INTERMEDIATE APPROACH SEGMENT-

(See SEGMENTS OF AN INSTRUMENT APPROACH PROCEDURE.)

INTERMEDIATE APPROACH SEGMENT [ICAO]- That segment of an instrument approach procedure between either the intermediate approach fix and the final approach fix or point, or between the end of a reversal, race track or dead reckoning track procedure and the final approach fix or point, as appropriate.

INTERMEDIATE FIX- The fix that identifies the beginning of the intermediate approach segment of an instrument approach procedure. The fix is not normally identified on the instrument approach chart as an intermediate fix (IF).

(See SEGMENTS OF AN INSTRUMENT APPROACH PROCEDURE.)

INTERMEDIATE LANDING- On the rare occasion that this option is requested, it should be approved. The departure center, however, must advise the ATCSCC so that the appropriate delay is carried over and assigned at the intermediate airport. An intermediate landing airport within the arrival center will not be accepted without coordination with and the approval of the ATCSCC.

INTERNATIONAL AIRPORT- Relating to international flight, it means:

a. An airport of entry which has been designated by the Secretary of Treasury or Commissioner of Customs as an international airport for customs service.

b. A landing rights airport at which specific permission to land must be obtained from customs authorities in advance of contemplated use.

c. Airports designated under the Convention on International Civil Aviation as an airport for use by international commercial air transport and/or international general aviation.

(See ICAO term INTERNATIONAL AIRPORT.) (Refer to AIRPORT/FACILITY DIRECTORY.) (Refer to IFIM.)

INTERNATIONAL AIRPORT [ICAO]- Any airport designated by the Contracting State in whose territory it is situated as an airport of entry and departure for international air traffic, where the formalities incident to customs, immigration, public health, animal and plant quarantine and similar procedures are carried out.

INTERNATIONAL CIVIL AVIATION ORGANIZATION [ICAO]- A specialized agency of the United Nations whose objective is to develop the principles and techniques of international air navigation and to foster planning and development of international civil air transport.

a. Regions include:

1. African-Indian Ocean Region

2. Caribbean Region

3. European Region

4. Middle East/Asia Region

5. North American Region

6. North Atlantic Region

7. Pacific Region

8. South American Region

INTERNATIONAL FLIGHT INFORMATION MANUAL- A publication designed primarily as a pilot's preflight planning guide for flights into foreign airspace and for flights returning to the U.S. from foreign locations.

INTERROGATOR- The ground-based surveillance radar beacon transmitter-receiver, which normally scans in synchronism with a primary radar, transmitting discrete radio signals which repetitiously request all transponders on the mode being used to reply. The replies received are mixed with the primary radar returns and displayed on the same plan position indicator (radar scope). Also, applied to the airborne element of the TACAN/DME system.

(See TRANSPONDER.) (Refer to AIM.)

INTERSECTING RUNWAYS- Two or more runways which cross or meet within their lengths.

(See INTERSECTION.) INTERSECTION-

a. A point defined by any combination of courses, radials, or bearings of two or more navigational aids.

b. Used to describe the point where two runways, a runway and a taxiway, or two taxiways cross or meet.

INTERSECTION DEPARTURE- A departure from any runway intersection except the end of the runway.

(See INTERSECTION.)

INTERSECTION TAKEOFF-

(See INTERSECTION DEPARTURE.)

IR-

(See IFR MILITARY TRAINING ROUTES.)

J

JAMMING- Electronic or mechanical interference which may disrupt the display of aircraft on radar or the transmission/reception of radio communications/navigation.

JET BLAST- Jet engine exhaust (thrust stream turbulence). (See WAKE TURBULENCE.)

JET ROUTE- A route designed to serve aircraft operations from 18,000 feet MSL up to and including flight level 450. The routes are referred to as "J" routes with numbering to identify the designated route; e.g., J105.

(See Class A AIRSPACE.)

(Refer to 14 CFR Part 71.)

JET STREAM- A migrating stream of high-speed winds present at high altitudes.

JETTISONING OF EXTERNAL STORES- Airborne release of external stores; e.g., tiptanks, ordnance.

(See FUEL DUMPING.)

(Refer to 14 CFR Part 91.)

JOINT USE RESTRICTED AREA-

(See RESTRICTED AREA.)

K

KNOWN TRAFFIC- With respect to ATC clearances, means aircraft whose altitude, position, and intentions are known to ATC.

L

LAA-

(See LOCAL AIRPORT ADVISORY.) LAAS-

(See LOW ALTITUDE ALERT SYSTEM.)

LAHSO- An acronym for "Land and Hold Short Operation." These operations include landing and holding short of an intersecting runway, a taxiway, a predetermined point, or an approach/departure flightpath.

LAHSO-DRY- Land and hold short operations on runways that are dry.

LAHSO-WET- Land and hold short operations on runways that are wet (but not contaminated).

LAND AND HOLD SHORT OPERATIONS- Operations which include simultaneous takeoffs and landings and/or simultaneous landings when a landing aircraft is able and is instructed by the controller to hold-short of the intersecting runway/taxiway or designated hold-short point. Pilots are expected to promptly inform the controller if the hold short clearance cannot be accepted.

(See PARALLEL RUNWAYS.) (Refer to AIM.)

LANDING AREA- Any locality either on land, water, or structures, including airports/heliports and intermediate landing fields, which is used, or intended to be used, for the landing and takeoff of aircraft whether or not facilities

are provided for the shelter, servicing, or for receiving or discharging passengers or cargo.
(See ICAO term LANDING AREA.)
LANDING AREA [ICAO]- That part of a movement area intended for the landing or take-off of aircraft.
LANDING DIRECTION INDICATOR- A device which visually indicates the direction in which landings and takeoffs should be made.
(See TETRAHEDRON.) (Refer to AIM.)
LANDING DISTANCE AVAILABLE (LDA)- The runway length declared available and suitable for a landing airplane.
(See ICAO term LANDING DISTANCE AVAILABLE.)
LANDING DISTANCE AVAILABLE [ICAO]- The length of runway which is declared available and suitable for the ground run of an aeroplane landing.
LANDING MINIMUMS- The minimum visibility prescribed for landing a civil aircraft while using an instrument approach procedure. The minimum applies with other limitations set forth in 14 CFR Part 91 with respect to the Minimum Descent Altitude (MDA) or Decision Height (DH) prescribed in the instrument approach procedures as follows:

a. Straight-in landing minimums. A statement of MDA and visibility, or DH and visibility, required for a straight-in landing on a specified runway, or

b. Circling minimums. A statement of MDA and visibility required for the circle-to-land maneuver.

Note: Descent below the established MDA or DH is not authorized during an approach unless the aircraft is in a position from which a normal approach to the runway of intended landing can be made and adequate visual reference to required visual cues is maintained.
(See CIRCLE-TO-LAND MANEUVER.)
(See DECISION HEIGHT.)
(See INSTRUMENT APPROACH PROCEDURE.) (See MINIMUM DESCENT ALTITUDE.)
(See STRAIGHT-IN LANDING.)
(See VISIBILITY.)
(Refer to 14 CFR Part 91.)
LANDING ROLL- The distance from the point of touchdown to the point where the aircraft can be brought to a stop or exit the runway.
LANDING SEQUENCE- The order in which aircraft are positioned for landing. (See APPROACH SEQUENCE.)
LAST ASSIGNED ALTITUDE- The last altitude/flight level assigned by ATC and acknowledged by the pilot.
(See MAINTAIN.)
(Refer to 14 CFR Part 91.)
LATERAL NAVIGATION (LNAV)– A function of area navigation (RNAV) equipment which calculates, displays, and provides lateral guidance to a profile or path.
LATERAL SEPARATION- The lateral spacing of aircraft at the same altitude by requiring operation on different routes or in different geographical locations.
(See SEPARATION.)
LDA-
(See LOCALIZER TYPE DIRECTIONAL AID.)
(See LANDING DISTANCE AVAILABLE.)
(See ICAO Term LANDING DISTANCE AVAILABLE.)
LF-
(See LOW FREQUENCY.)
LIGHTED AIRPORT- An airport where runway and obstruction lighting is available. (See AIRPORT LIGHTING.)
(Refer to AIM.)
LIGHT GUN- A handheld directional light signaling device which emits a brilliant narrow beam of white, green, or red light as selected by the tower controller. The color and type of light transmitted can be used to approve or disapprove anticipated pilot actions where radio communication is not available. The light gun is used for controlling traffic operating in the vicinity of the airport and on the airport movement area.
(Refer to AIM.)
LINE UP AND WAIT (LUAW)- Used by ATC to inform a pilot to taxi onto the departure runway to line up and wait. It is not authorization for takeoff. It is used when takeoff clearance cannot immediately be issued because of traffic or other reasons.
(See CLEARED FOR TAKEOFF.)
LOCAL AIRPORT ADVISORY (LAA)- A service provided by facilities, which are located on the landing airport, have a discrete ground-to-air communication frequency or the tower frequency when the tower is closed, automated weather reporting with voice broadcasting, and a continuous ASOS/AWSS/AWOS data display, other continuous direct reading instruments, or manual observations available to the specialist.
(See AIRPORT ADVISORY AREA.)
LOCAL TRAFFIC- Aircraft operating in the traffic pattern or within sight of the tower, or aircraft known to be departing or arriving from flight in local practice areas, or aircraft executing practice instrument approaches at the airport.
(See TRAFFIC PATTERN.)
LOCALIZER- The component of an ILS which provides course guidance to the runway. (See INSTRUMENT LANDING SYSTEM.)
(See ICAO term LOCALIZER COURSE.)
(Refer to AIM.)
LOCALIZER COURSE [ICAO]- The locus of points, in any given horizontal plane, at which the DDM (difference in depth of modulation) is zero.
LOCALIZER OFFSET- An angular offset of the localizer from the runway extended centerline in a direction away from the no transgression zone (NTZ) that increases the normal operating zone (NOZ) width. An offset requires a 50 foot increase in DH and is not authorized for CAT II and CAT III approaches.
LOCALIZER TYPE DIRECTIONAL AID- A NAVAID used for nonprecision instrument approaches with utility and accuracy comparable to a localizer but which is not a part of a complete ILS and is not aligned with the runway.
(Refer to AIM.)
LOCALIZER USABLE DISTANCE- The maximum distance from the localizer transmitter at a specified altitude, as verified by flight inspection, at which reliable course information is continuously received.
(Refer to AIM.)
LOCATOR [ICAO]- An LM/MF NDB used as an aid to final approach.
Note: A locator usually has an average radius of rated coverage of between 18.5 and 46.3 km (10 and 25 NM).
LONG RANGE NAVIGATION-(See LORAN.)
LONGITUDINAL SEPARATION- The longitudinal spacing of aircraft at the same altitude by a minimum distance expressed in units of time or miles.
(See SEPARATION.) (Refer to AIM.)
LORAN- An electronic navigational system by which hyperbolic lines of position are determined by measuring the difference in the time of reception of synchronized pulse signals from two fixed transmitters. Loran A operates in the 1750-1950 kHz frequency band. Loran C and D operate in the 100-110 kHz frequency band. (Refer to AIM.)
LOST COMMUNICATIONS- Loss of the ability to communicate by radio. Aircraft are sometimes referred to as NORDO (No Radio). Standard pilot procedures are specified in 14 CFR Part 91. Radar controllers issue procedures for pilots to follow in the event of lost communications during a radar approach when weather reports indicate that an aircraft will likely encounter IFR weather conditions during the approach.
(Refer to 14 CFR Part 91.) (Refer AIM.)
LOW ALTITUDE AIRWAY STRUCTURE- The network of airways serving aircraft operations up to but not including 18,000 feet MSL.
(See AIRWAY.)
(Refer to AIM.)
LOW ALTITUDE ALERT, CHECK YOUR ALTITUDE IMMEDIATELY-(See SAFETY ALERT.)
LOW ALTITUDE ALERT SYSTEM- An automated function of the TPX-42 that alerts the controller when a Mode C transponder equipped aircraft on an IFR flight plan is below a predetermined minimum safe altitude. If requested by the pilot, Low Altitude Alert System monitoring is also available to VFR Mode C transponder equipped aircraft.
LOW APPROACH- An approach over an airport or runway following an instrument approach or a VFR approach including the go-around maneuver where the pilot intentionally does not make contact with the runway.
(Refer to AIM.)
LOW FREQUENCY- The frequency band between 30 and 300 kHz.
(Refer to AIM.)
LPV- A type of approach with vertical guidance (APV) based on WAAS, published on RNAV (GPS) approach charts. This procedure takes advantage of the precise lateral guidance available from WAAS. The minima is published as a decision altitude (DA).
LUAW-
(See LINE UP AND WAIT.)

M

MAA-
(See MAXIMUM AUTHORIZED ALTITUDE.)
MACH NUMBER- The ratio of true airspeed to the speed of sound; e.g., MACH .82, MACH 1.6.
(See AIRSPEED.)
MACH TECHNIQUE [ICAO]- Describes a control technique used by air traffic control whereby turbojet aircraft operating successively along suitable routes are cleared to maintain appropriate MACH numbers for a relevant portion of the en route phase of flight. The principle objective is to achieve improved utilization of the airspace and to ensure that separation between successive aircraft does not decrease below the established minima.
MAHWP- Missed Approach Holding Waypoint ***MAINTAIN-***

a. Concerning altitude/flight level, the term means to remain at the altitude/flight level specified. The phrase "climb and" or "descend and" normally precedes "maintain" and the altitude assignment; e.g., "descend and maintain 5,000."

b. Concerning other ATC instructions, the term is used in its literal sense; e.g., maintain VFR.

MAINTENANCE PLANNING FRICTION LEVEL- The friction level specified in AC 150/5320-12, Measurement, Construction, and Maintenance of Skid Resistant Airport Pavement Surfaces, which represents the friction value below which the runway pavement surface remains acceptable for any category or class of aircraft operations but which is beginning to show signs of deterioration. This value will vary depending on the particular friction measurement equipment used.
MAKE SHORT APPROACH- Used by ATC to inform a pilot to alter his/her traffic pattern so as to make a short final approach.
(See TRAFFIC PATTERN.)
MAN PORTABLE AIR DEFENSE SYSTEMS (MANPADS)- MANPADS are lightweight, shoulder-launched, missile systems used to bring down aircraft and create mass casualties. The potential for MANPADS use against airborne aircraft is real and requires familiarity with the subject. Terrorists choose MANPADS because the weapons are low cost, highly mobile, require minimal set-up time, and are easy to use and maintain. Although the weapons have limited range, and their accuracy is affected by poor visibility and adverse weather, they can be fired from anywhere on land or from boats where there is unrestricted visibility to the target.
MANDATORY ALTITUDE- An altitude depicted on an instrument Approach Procedure Chart requiring the aircraft to maintain altitude at the depicted value.
MANPADS-
(See MAN PORTABLE AIR DEFENSE SYSTEMS.) MAP-
(See MISSED APPROACH POINT.)
MARKER BEACON- An electronic navigation facility transmitting a 75 MHz vertical fan or boneshaped radiation pattern. Marker beacons are identified by their modulation frequency and keying code, and when received by compatible airborne equipment, indicate to the pilot, both aurally and visually, that he/she is passing over the facility.
(See INNER MARKER.) (See MIDDLE MARKER.) (See OUTER MARKER.) (Refer to AIM.)
MARSA-
(See MILITARY AUTHORITY ASSUMES RESPONSIBILITY FOR SEPARATION OF AIRCRAFT.)
MAWP- Missed Approach Waypoint
MAXIMUM AUTHORIZED ALTITUDE- A published altitude representing the maximum usable altitude or flight level for an airspace structure or route segment. It is the highest altitude on a Federal airway, jet route, area navigation low or high route, or other direct route for which an MEA is designated in 14 CFR Part 95 at which adequate reception of navigation aid signals is assured.
MAYDAY- The international radiotelephony distress signal. When repeated three times, it indicates imminent and grave danger and that immediate assistance is requested.
(See PAN-PAN.) (Refer to AIM.)
MCA-
(See MINIMUM CROSSING ALTITUDE.)
MDA-
(See MINIMUM DESCENT ALTITUDE.)
MEA-
(See MINIMUM EN ROUTE IFR ALTITUDE.)
MEARTS-
(See MICRO-EN ROUTE AUTOMATED RADAR TRACKING SYSTEM.)
METEOROLOGICAL IMPACT STATEMENT- An unscheduled planning forecast describing conditions expected to begin within 4 to 12 hours which may impact the flow of air traffic in a specific center's (ARTCC) area.
METER FIX ARC- A semicircle, equidistant from a meter fix, usually in low altitude relatively close to the meter fix,

used to help CTAS/HOST calculate a meter time, and determine appropriate sector meter list assignments for aircraft not on an established arrival route or assigned a meter fix.

METER FIX TIME/SLOT TIME- A calculated time to depart the meter fix in order to cross the vertex at the ACLT. This time reflects descent speed adjustment and any applicable time that must be absorbed prior to crossing the meter fix.

METER LIST-

(See ARRIVAL SECTOR ADVISORY LIST.)

METER LIST DISPLAY INTERVAL- A dynamic parameter which controls the number of minutes prior to the flight plan calculated time of arrival at the meter fix for each aircraft, at which time the TCLT is frozen and becomes an ACLT; i.e., the VTA is updated and consequently the TCLT modified as appropriate until frozen at which time updating is suspended and an ACLT is assigned. When frozen, the flight entry is inserted into the arrival sector's meter list for display on the sector PVD/MDM. MLDI is used if filed true airspeed is less than or equal to freeze speed parameters (FSPD).

METERING- A method of time-regulating arrival traffic flow into a terminal area so as not to exceed a predetermined terminal acceptance rate.

METERING AIRPORTS- Airports adapted for metering and for which optimum flight paths are defined. A maximum of 15 airports may be adapted.

METERING FIX- A fix along an established route from over which aircraft will be metered prior to entering terminal airspace. Normally, this fix should be established at a distance from the airport which will facilitate a profile descent 10,000 feet above airport elevation (AAE) or above.

METERING POSITION(S)- Adapted PVDs/MDMs and associated "D" positions eligible for display of a metering position list. A maximum of four PVDs/MDMs may be adapted.

METERING POSITION LIST- An ordered list of data on arrivals for a selected metering airport displayed on a metering position PVD/MDM.

MFT-

(See METER FIX TIME/SLOT TIME.) MHA-

(See MINIMUM HOLDING ALTITUDE.) MIA-

(See MINIMUM IFR ALTITUDES.)

MICROB URST- A small downburst with outbursts of damaging winds extending 2.5 miles or less. In spite of its small horizontal scale, an intense microburst could induce wind speeds as high as 150 knots

(Refer to AIM.)

MICRO-EN ROUTE AUTOMATED RADAR TRACKING SYSTEM (MEARTS)- An automated radar and radar beacon tracking system capable of employing both short-range (ASR) and long-range (ARSR) radars. This microcomputer driven system provides improved tracking, continuous data recording, and use of full digital radar displays.

MICROWAVE LANDING SYSTEM- A precision instrument approach system operating in the microwave spectrum which normally consists of the following components:

a. Azimuth Station.

b. Elevation Station.

c. Precision Distance Measuring Equipment. (See MLS CATEGORIES.)

MID RVR-

(See VISIBILITY.)

MIDDLE COMPASS LOCATOR-

(See COMPASS LOCATOR.)

MIDDLE MARKER- A marker beacon that defines a point along the glideslope of an ILS normally located at or near the point of decision height (ILS Category I). It is keyed to transmit alternate dots and dashes, with the alternate dots and dashes keyed at the rate of 95 dot/dash combinations per minute on a 1300 Hz tone, which is received aurally and visually by compatible airborne equipment.

(See INSTRUMENT LANDING SYSTEM.) (See MARKER BEACON.)

(Refer to AIM.)

MILES-IN-TRAIL- A specified distance between aircraft, normally, in the same stratum associated with the same destination or route of flight.

MILITARY AUTHORITY ASSUMES RESPONSIBILITY FOR SEPARATION OF AIRCRAFT- A condition whereby the military services involved assume responsibility for separation between participating military aircraft in the ATC system. It is used only for required IFR operations which are specified in letters of agreement or other appropriate FAA or military documents.

MILITARY LANDING ZONE- A landing strip used exclusively by the military for training. A military landing zone does not carry a runway designation.

MILITARY OPERATIONS AREA-(See SPECIAL USE AIRSPACE.)

MILITARY TRAINING ROUTES- Airspace of defined vertical and lateral dimensions established for the conduct of military flight training at ai rspeeds in excess of 250 knots IAS.

(See IFR MILITARY TRAINING ROUTES.) (See VFR MILITARY TRAINING ROUTES.) MINIMA-

(See MINIMUMS.)

MINIMUM CROSSING ALTITUDE- The lowest altitude at certain fixes at which an aircraft must cross when proceeding in the direction of a higher minimum en route IFR altitude (MEA).

(See MINIMUM EN ROUTE IFR ALTITUDE.)

MINIMUM DESCENT ALTITUDE- The lowest altitude, expressed in feet above mean sea level, to which descent is authorized on final approach or during circle-to-land maneuvering in execution of a standard instrument approach procedure where no electronic glideslope is provided.

(See NONPRECISION APPROACH PROCEDURE.)

MINIMUM EN ROUTE IFR ALTITUDE (MEA)- The lowest published altitude between radio fixes which assures acceptable navigational signal coverage and meets obstacle clearance requirements between those fixes. The MEA prescribed for a Federal airway or segment thereof, area navigation low or high route, or other direct route applies to the entire width of the airway, segment, or route between the radio fixes defining the airway, segment, or route.

(Refer to 14 CFR Part 91.) (Refer to 14 CFR Part 95.) (Refer to AIM.)

MINIMUM FRICTION LEVEL- The friction level specified in AC 150/5320-12, Measurement, Construction, and Maintenance of Skid Resistant Airport Pavement Surfaces, that represents the minimum recommended wet pavement surface friction value for any turbojet aircraft engaged in LAHSO. This value will vary with the particular friction measurement equipment used.

MINIMUM FUEL- Indicates that an aircraft's fuel supply has reached a state where, upon reaching the destination, it can accept little or no delay. This is not an emergency situation but merely indicates an emergency situation is possible should any undue delay occur.

(Refer to AIM.)

MINIMUM HOLDING ALTITUDE- The lowest altitude prescribed for a holding pattern which assures navigational signal coverage, communications, and meets obstacle clearance requirements.

MINIMUM IFR ALTITUDES (MIA)- Minimum altitudes for IFR operations as prescribed in 14 CFR Part 91. These altitudes are published on aeronautical charts and prescribed in 14 CFR Part 95 for airways and routes, and in 14 CFR Part 97 for standard instrument approach procedures. If no applicable minimum altitude is prescribed in 14 CFR Part 95 or 14 CFR Part 97, the following minimum IFR altitude applies:

a. In designated mountainous areas, 2,000 feet above the highest obstacle within a horizontal distance of 4 nautical miles from the course to be flown; or

b. Other than mountainous areas, 1,000 feet above the highest obstacle within a horizontal distance of 4 nautical miles from the course to be flown; or

c. As otherwise authorized by the Administrator or assigned by ATC. (See MINIMUM CROSSING ALTITUDE.)

(See MINIMUM EN ROUTE IFR ALTITUDE.)

(See MINIMUM OBSTRUCTION CLEARANCE ALTITUDE.) (See MINIMUM SAFE ALTITUDE.)

(See MINIMUM VECTORING ALTITUDE.)

(Refer to 14 CFR Part 91.)

MINIMUM NAVIGATION PERFORMANCE SPECIFICATION- A set of standards which require aircraft to have a minimum navigation performance capability in order to operate in MNPS designated airspace. In addition, aircraft must be certified by their State of Registry for MNPS operation.

MINIMUM NAVIGATION PERFORMANCE SPECIFICATION AIRSPACE- Designated airspace in which MNPS procedures are applied between MNPS certified and equipped aircraft. Under certain conditions, non-MNPS aircraft can operate in MNPSA. However, standard oceanic separation minima is provided between the non-MNPS aircraft and other traffic. Currently, the only designated MNPSA is described as follows:

a. Between FL 285 and FL 420;

b. Between latitudes 27°N and the North Pole;

c. In the east, the eastern boundaries of the CTAs Santa Maria Oceanic, Shanwick Oceanic, and Reykjavik;

d. In the west, the western boundaries of CTAs Reykjavik and Gander Oceanic and New York Oceanic excluding the area west of 60°W and south of 38°30'N.

MINIMUM OBSTRUCTION CLEARANCE ALTITUDE (MOCA)- The lowest published altitude in effect between radio fixes on VOR airways, off-airway routes, or route segments which meets obstacle clearance requirements for the entire route segment and which assures acceptable navigational signal coverage only within 25 statute (22 nautical) miles of a VOR.

(Refer to 14 CFR Part 91.) (Refer to 14 CFR Part 95.)

MINIMUM RECEPTION ALTITUDE- The lowest altitude at which an intersection can be determined.

(Refer to 14 CFR Part 95.)

MINIMUM SAFE ALTITUDE-

a. The minimum altitude specified in 14 CFR Part 91 for various aircraft operations.

b. Altitudes depicted on approach charts which provide at least 1,000 feet of obstacle clearance for emergency use within a specified distance from the navigation facility upon which a procedure is predicated. These altitudes will be identified as Minimum Sector Altitudes or Emergency Safe Altitudes and are established as follows:

1. Minimum Sector Altitudes. Altitudes depicted on approach charts which provide at least 1,000 feet of obstacle clearance within a 25-mile radius of the navigation facility upon which the procedure is predicated. Sectors depicted on approach charts must be at least 90 degrees in scope. These altitudes are for emergency use only and do not necessarily assure acceptable navigational signal coverage.

(See ICAO term Minimum Sector Altitude.)

2. Emergency Safe Altitudes. Altitudes depicted on approach charts which provide at least 1,000 feet of obstacle clearance in nonmountainous areas and 2,000 feet of obstacle clearance in designated mountainous areas within a 100-mile radius of the navigation facility upon which the procedure is predicated and normally used only in military procedures. These altitudes are identified on published procedures as "Emergency Safe Altitudes."

MINIMUM SAFE ALTITUDE WARNING- A function of the ARTS III computer that aids the controller by alerting him/her when a tracked Mode C equipped aircraft is below or is predicted by the computer to go below a predetermined minimum safe altitude.

(Refer to AIM.)

MINIMUM SECTOR ALTITUDE [ICAO]- The lowest altitude which may be used under emergency conditions which will provide a minimum clearance of 300 m (1,000 feet) above all obstacles located in an area contained within a sector of a circle of 46 km (25 NM) radius centered on a radio aid to navigation.

MINIMUMS- Weather condition requirements established for a particular operation or type of operation; e.g., IFR takeoff or landing, alternate airport for IFR flight plans, VFR flight, etc.

(See IFR CONDITIONS.)

(See IFR TAKEOFF MINIMUMS AND DEPARTURE PROCEDURES.) (See LANDING MINIMUMS.)

(See VFR CONDITIONS.)

(Refer to 14 CFR Part 91.)

(Refer to AIM.)

MINIMUM VECTORING ALTITUDE (MVA)- The lowest MSL altitude at which an IFR aircraft will be vectored by a radar controller, except as otherwise authorized for radar approaches, departures, and missed approaches. The altitude meets IFR obstacle clearance criteria. It may be lower than the published MEA along an airway or J-route segment. It may be utilized for radar vectoring only upon the controller's determination that an adequate radar return is being received from the aircraft being controlled. Charts depicting minimum vectoring altitudes are normally available only to the controllers and not to pilots.

(Refer to AIM.)

MINUTES-IN-TRAIL- A specified interval between aircraft expressed in time. This method would more likely be utilized regardless of altitude.

MIS-

(See METEOROLOGICAL IMPACT STATEMENT.)

MISSED APPROACH-

a. A maneuver conducted by a pilot when an instrument approach cannot be completed to a landing. The route of flight and altitude are shown on instrument approach procedure charts. A pilot executing a missed approach prior to the Missed Approach Point (MAP) must continue along the final approach to the MAP.

b. A term used by the pilot to inform ATC that he/she is executing the missed approach.

c. At locations where ATC radar service is provided, the pilot should conform to radar vectors when provided by ATC in lieu of the published missed approach procedure.

(See MISSED APPROACH POINT.) (Refer to AIM.)

MISSED APPROACH POINT- A point prescribed in each instrument approach procedure at which a missed approach procedure shall be executed if the required visual reference does not exist.

(See MISSED APPROACH.)

(See SEGMENTS OF AN INSTRUMENT APPROACH PROCEDURE.)

MISSED APPROACH PROCEDURE [ICAO]- The procedure to be followed if the approach cannot be continued.

MISSED APPROACH SEGMENT-

(See SEGMENTS OF AN INSTRUMENT APPROACH PROCEDURE.) MLDI-

(See METER LIST DISPLAY INTERVAL.)

MLS-

(See MICROWAVE LANDING SYSTEM.) MLS CATEGORIES-

a. MLS Category I. An MLS approach procedure which provides for an approach to a height above touchdown of not less than 200 feet and a runway visual range of not less than 1,800 feet.

b. MLS Category II. Undefined until data gathering/analysis completion.

c. MLS Category III. Undefined until data gathering/analysis completion. MM-

(See MIDDLE MARKER.)

MNPS-

(See MINIMUM NAVIGATION PERFORMANCE SPECIFICATION.) MNPSA-

(See MINIMUM NAVIGATION PERFORMANCE- SPECIFICATION AIRSPACE.) MOA-

(See MILITARY OPERATIONS AREA.)

MOCA-

(See MINIMUM OBSTRUCTION CLEARANCE ALTITUDE.)

MODE- The letter or number assigned to a specific pulse spacing of radio signals transmitted or received by ground interrogator or airborne transponder components of the Air Traffic Control Radar Beacon System (ATCRBS). Mode A (military Mode 3) and Mode C (altitude reporting) are used in air traffic control.

(See INTERROGATOR.) (See RADAR.)

(See TRANSPONDER.) (See ICAO term MODE.) (Refer to AIM.)

MODE (SSR MODE) [ICAO]- The letter or number assigned to a specific pulse spacing of the interrogation signals transmitted by an interrogator. There are 4 modes, A, B, C and D specified in Annex 10, corresponding to four different interrogation pulse spacings.

MODE C INTRUDER ALERT- A function of certain air traffic control automated systems designed to alert radar controllers to existing or pending situations between a tracked target (known IFR or VFR aircraft) and an untracked target (unknown IFR or VFR aircraft) that requires immediate attention/action.

(See CONFLICT ALERT.)

MONITOR- (When used with communication transfer) listen on a specific frequency and stand by for instructions. Under normal circumstances do not establish communications.

MONITOR ALERT (MA)- A function of the TFMS that provides traffic management personnel with a tool for predicting potential capacity problems in individual operational sectors. The MA is an indication that traffic management personnel need to analyze a particular sector for actual activity and to determine the required action(s), if any, needed to control the demand.

MONITOR ALERT PARAMETER (MAP)- The number designated for use in monitor alert processing by the TFMS. The MAP is designated for each operational sector for increments of 15 minutes.

MOSAIC/MULTI-SENSOR MODE- Accepts positional data from multiple radar or ADS-B sites. Targets are displayed from a single source within a radar sort box according to the hierarchy of the sources assigned.

MOVEMENT AREA- The runways, taxiways, and other areas of an airport/heliport which are utilized for taxiing/hover taxiing, air taxiing, takeoff, and landing of aircraft, exclusive of loading ramps and parking areas. At those airports/heliports with a tower, specific approval for entry onto the movement area must be obtained from ATC.

(See ICAO term MOVEMENT AREA.)

MOVEMENT AREA [ICAO]- That part of an aerodrome to be used for the takeoff, landing and taxiing of aircraft, consisting of the maneuvering area and the apron(s).

MOVING TARGET INDICATOR- An electronic device which will permit radar scope

presentation only from targets which are in motion. A partial remedy for ground clutter.

MRA-

(See MINIMUM RECEPTION ALTITUDE.)

MSA-

(See MINIMUM SAFE ALTITUDE.)

MSAW-

(See MINIMUM SAFE ALTITUDE WARNING.)

MTI-

(See MOVING TARGET INDICATOR.) MTR-

(See MILITARY TRAINING ROUTES.)

MULTICOM- A mobile service not open to public correspondence used to provide communications essential to conduct the activities being performed by or directed from private aircraft.

MULTIPLE RUNWAYS- The utilization of a dedicated arrival runway(s) for departures and a dedicated departure runway(s) for arrivals when feasible to reduce delays and enhance capacity.

MVA-

(See MINIMUM VECTORING ALTITUDE.)

N

NAS-

(See NATIONAL AIRSPACE SYSTEM.)

NATIONAL AIRSPACE SYSTEM- The common network of U.S. airspace; air navigation facilities, equipment and services, airports or landing areas; aeronautical charts, information and services; rules, regulations and procedures, technical information, and manpower and material. Included are system components shared jointly with the military.

NATIONAL BEACON CODE ALLOCATION PLAN AIRSPACE- Airspace over United States territory located within the North American continent between Canada and Mexico, including adjacent territorial waters outward to about boundaries of oceanic control areas (CTA)/Flight Information Regions (FIR).

(See FLIGHT INFORMATION REGION.)

NATIONAL FLIGHT DATA CENTER- A facility in Washington D.C., established by FAA to operate a central aeronautical information service for the collection, validation, and dissemination of aeronautical data in support of the activities of government, industry, and the aviation community. The information is published in the National Flight Data Digest.

(See NATIONAL FLIGHT DATA DIGEST.)

NATIONAL FLIGHT DATA DIGEST- A daily (except weekends and Federal holidays) publication of flight information appropriate to aeronautical charts, aeronautical publications, Notices to Airmen, or other media serving the purpose of providing operational flight data essential to safe and efficient aircraft operations.

NATIONAL SEARCH AND RESCUE PLAN- An interagency agreement which provides for the effective utilization of all available facilities in all types of search and rescue missions.

NAVAID-

(See NAVIGATIONAL AID.)

NAVAID CLASSES- VOR, VORTAC, and TACAN aids are classed according to their operational use. The three classes of NAVAIDs are:

a. T- Terminal.

b. L- Low altitude.

c. H- High altitude.

Note: The normal service range for T, L, and H class aids is found in the AIM. Certain operational requirements make it necessary to use some of these aids at greater service ranges than specified. Extended range is made possible through flight inspection determinations. Some aids also have lesser service range due to location, terrain, frequency protection, etc. Restrictions to service range are listed in Airport/Facility Directory.

NAVIGABLE AIRSPACE- Airspace at and above the minimum flight altitudes

prescribed in the CFRs including airspace needed for safe takeoff and landing.

(Refer to 14 CFR Part 91.)

NAVIGATION REFERENCE SYSTEM (NRS)- The NRS is a system of waypoints developed for use within the United States for flight planning and navigation without reference to ground based navigational aids. The NRS waypoints are located in a grid pattern along defined latitude and longitude lines. The initial use of the NRS will be in the high altitude environment in conjunction with the High Altitude Redesign initiative. The NRS waypoints are intended for use by aircraft capable of point-to-point navigation.

NAVIGATION SPECIFICATION [ICAO]- A set of aircraft and flight crew requirements needed to support performance-based navigation operations within a defined airspace. There are two kinds of navigation specifications:

a. RNP specification. A navigation specification based on area navigation that includes the requirement for performance monitoring and alerting, designated by the prefix RNP; e.g., RNP 4, RNP APCH.

b. RNAV specification. A navigation specification based on area navigation that does not include the requirement for performance monitoring and alerting, designated by the prefix RNAV; e.g., RNAV 5, RNAV 1.

Note: The Performance-based Navigation Manual (Doc 9613), Volume II contains detailed guidance on navigation specifications.

NAVIGATIONAL AID- Any visual or electronic device airborne or on the surface which

provides point-to-point guidance information or position data to aircraft in flight.

(See AIR NAVIGATION FACILITY.)

NBCAP AIRSPACE-

(See NATIONAL BEACON CODE ALLOCATION PLAN AIRSPACE.) NDB-

(See NONDIRECTIONAL BEACON.)

NEGATIVE- "No," or "permission not granted," or "that is not correct." ***NEGATIVE CONTACT-*** Used by pilots to inform ATC that:

a. Previously issued traffic is not in sight. It may be followed by the pilot's request for the controller to provide assistance in avoiding the traffic.

b. They were unable to contact ATC on a particular frequency.

NFDC-

(See NATIONAL FLIGHT DATA CENTER.)

NFDD-

(See NATIONAL FLIGHT DATA DIGEST.)

NIGHT- The time between the end of evening civil twilight and the beginning of morning civil twilight, as published in the American Air Almanac, converted to local time.

(See ICAO term NIGHT.)

NIGHT [ICAO]- The hours between the end of evening civil twilight and the beginning of morning civil twilight or such other period between sunset and sunrise as may be specified by the appropriate authority.

Note: Civil twilight ends in the evening when the center of the sun's disk is 6 degrees below the horizon and begins in the morning when the center of the sun's disk is 6 degrees below the horizon.

NO GYRO APPROACH- A radar approach/vector provided in case of a malfunctioning gyro-compass or directional gyro. Instead of providing the pilot with headings to be flown, the controller observes the radar track and issues control instructions "turn right/left" or "stop turn" as appropriate.

(Refer to AIM.)

NO GYRO VECTOR-

(See NO GYRO APPROACH.)

NO TRANSGRESSION ZONE (NTZ)- The NTZ is a 2,000 foot wide zone, located equidistant between parallel runway final approach courses in which flight is not allowed.

NONAPPROACH CONTROL TOWER- Author-izes aircraft to land or takeoff at the airport controlled by the tower or to transit the Class D airspace. The primary function of a nonapproach control tower is the sequencing of aircraft in the traffic pattern and on the landing area. Nonapproach control towers also separate aircraft operating under instrument flight rules clearances from approach controls and centers. They provide ground control services to aircraft, vehicles, personnel, and equipment on the airport movement area.

NONCOMMON ROUTE/PORTION- That segment of a North American Route between the inland navigation facility and a designated North American terminal.

NONCOMPOSITE SEPARATION- Separation in accordance with minima other than the composite separation minimum specified for the area concerned.

NONDIRECTIONAL BEACON- An L/MF or UHF radio beacon transmitting nondirectional signals whereby the pilot of an aircraft equipped with direction finding equipment can determine his/her bearing to or from the radio beacon and "home" on or track to or from the station. When the radio beacon is installed in conjunction with the Instrument Landing System marker, it is normally called a Compass Locator.

(See AUTOMATIC DIRECTION FINDER.) (See COMPASS LOCATOR.)

NONMOVEMENT AREAS- Taxiways and apron (ramp) areas not under the control of air traffic.

NONPRECISION APPROACH-

(See NONPRECISION APPROACH PROCEDURE.)

NONPRECISION APPROACH PROCEDURE- A standard instrument approach procedure in which no electronic glideslope is provided; e.g., VOR, TACAN, NDB, LOC, ASR, LDA, or SDF approaches.

NONRADAR- Precedes other terms and generally means without the use of radar, such as:

a. Nonradar Approach. Used to describe instrument approaches for which course guidance on final approach is not provided by ground-based precision or surveillance radar. Radar vectors to the final approach course may or may not be provided by ATC. Examples of nonradar approaches are VOR, NDB, TACAN, and ILS/MLS approaches.

(See FINAL APPROACH COURSE.)

(See FINAL AP PROACH-IFR.)

(See INSTRUMENT APPROACH PROCEDURE.) (See RADAR APPROACH.)

b. Nonradar Approach Control. An ATC facility providing approach control service without the use of radar.

(See APPROACH CONTROL FACILITY.)

(See APPROACH CONTROL SERVICE.)

c. Nonradar Arrival. An aircraft arriving at an airport without radar service or at an airport served by a radar facility and radar contact has not been established or has been terminated due to a lack of radar service to the airport.

(See RADAR ARRIVAL.) (See RADAR SERVICE.)

d. Nonradar Route. A flight path or route over which the pilot is performing his/her own navigation. The pilot may be receiving radar separation, radar monitoring, or other ATC services while on a nonradar route.

(See RADAR ROUTE.)

e. Nonradar Separation. The spacing of aircraft in accordance with established minima without the use of radar; e.g., vertical, lateral, or longitudinal separation.

(See RADAR SEPARATION.)

(See ICAO term NONRADAR SEPARATION.)

NONRADAR SEPARATION [ICAO]- The separation used when aircraft position information is derived from sources other than radar.

NON-RESTRICTIVE ROUTING (NRR)- Portions of a proposed route of flight where a user can flight plan the most advantageous flight path with no requirement to make reference to ground-based NAVAIDs.

NOPAC-

(See NORTH PACIFIC.)

NORDO- (No Radio)- Aircraft that cannot or do not communicate by radio when radio communication is required are referred to as "NORDO."

(See LOST COMMUNICATIONS.)

NORMAL OPERATING ZONE (NOZ)- The NOZ is the operating zone within which aircraft flight remains during normal independent simultaneous parallel ILS approaches.

NORTH AMERICAN ROUTE- A numerically coded route preplanned over existing airway and route systems to and from specific coastal fixes serving the North Atlantic. North American Routes consist of the following:

a. Common Route/Portion. That segment of a North American Route between the inland navigation facility and the coastal fix.

b. Noncommon Route/Portion. That segment of a North American Route between the inland navigation facility and a designated North American terminal.

c. Inland Navigation Facility. A navigation aid on a North American Route at which the common route and/or the noncommon route begins or ends.

d. Coastal Fix. A navigation aid or intersection where an aircraft transitions between the domestic route structure and the oceanic route structure.

NORTH AMERICAN ROUTE PROGRAM (NRP)- The NRP is a set of rules and procedures which are designed to increase the flexibility of user flight planning within published guidelines.

NORTH MARK- A beacon data block sent by the host computer to be displayed by the ARTS on a 360 degree bearing at a locally selected radar azimuth and distance. The North Mark is used to ensure correct range/azimuth orientation during periods of CENRAP.

NORTH PACIFIC- An organized route system between the Alaskan west coast and Japan.

NOTAM-

(See NOTICE TO AIRMEN.)

NOTAM [ICAO]- A notice containing information concerning the establishment, condition or change in any aeronautical facility, service, procedure or hazard, the timely knowledge of which is essential to personnel concerned with flight operations.

a. I Distribution- Distribution by means of telecommunication.

b. II Distribution- Distribution by means other than telecommunications.

NOTICE TO AIRMEN- A notice containing information (not known sufficiently in advance to publicize by other means) concerning the establishment, condition, or change in any component (facility, service, or procedure of, or hazard in the National Airspace System) the timely knowledge of which is essential to personnel concerned with flight operations.

a. NOTAM(D)- A NOTAM given (in addition to local dissemination) distant dissemination beyond the area of responsibility of the Flight Service Station. These NOTAMs will be stored and available until canceled.

b. FDC NOTAM- A NOTAM regulatory in nature, transmitted by USNOF and given system wide dissemination.

(See ICAO term NOTAM.)

NOTICES TO AIRMEN PUBLICATION- A publication issued every 28 days, designed primarily for the pilot, which contains current NOTAM information considered essential to the safety of flight as well as supplemental data to other aeronautical publications. The contraction NTAP is used in NOTAM text.

(See NOTICE TO AIRMEN.)

NRR-

(See NON-RESTRICTIVE ROUTING.) NRS-

(See NAVIGATION REFERENCE SYSTEM.) NTAP-

(See NOTICES TO AIRMEN PUBLICATION.)

NUMEROUS TARGETS VICINITY (LOCATION)- A traffic advisory issued by ATC to advise pilots that targets on the radar scope are too numerous to issue individually.

(See TRAFFIC ADVISORIES.)

O

OBSTACLE- An existing object, object of natural growth, or terrain at a fixed geographical location or which may be expected at a fixed location within a prescribed area with reference to which vertical clearance is or must be provided during flight operation.

OBSTACLE DEPARTURE PROCEDURE (ODP)- A preplanned instrument flight rule (IFR) departure procedure printed for pilot use in textual or graphic form to provide obstruction clearance via the least onerous route from the terminal area to the appropriate en route structure. ODPs are recommended for obstruction clearance and may be flown without ATC clearance unless an alternate departure procedure (SID or radar vector) has been specifically assigned by ATC.

(See IFR TAKEOFF MINIMUMS AND DEPARTURE PROCEDURES.) (See STANDARD INSTRUMENT DEPARTURES.)

(Refer to AIM.)

OBSTACLE FREE ZONE- The OFZ is a three dimensional volume of airspace which protects for the transition of aircraft to and from the runway. The OFZ clearing standard precludes taxiing and parked airplanes and object penetrations, except for frangible NAVAID locations that are fixed by function. Additionally, vehicles, equipment, and personnel may be authorized by air traffic control to enter the area using the provisions of FAAO JO 7110.65, Para 3-1-5, VEHICLES/EQUIPMENT/PERSONNEL ON RUNWAYS. The runway OFZ and when applicable, the inner-approach OFZ, and the inner-transitional OFZ, comprise the OFZ.

a. Runway OFZ. The runway OFZ is a defined volume of airspace centered above the runway. The runway OFZ is the airspace above a surface whose elevation at any point is the same as the elevation of the nearest point on the runway centerline. The runway OFZ extends 200 feet beyond each end of the runway. The width is as follows:

1. For runways serving large airplanes, the greater of:

(a) 400 feet, or

(b) 180 feet, plus the wingspan of the most demanding airplane, plus 20 feet per 1,000 feet of airport elevation.

2. For runways serving only small airplanes: **(a)** 300 feet for precision instrument runways.

(b) 250 feet for other runways serving small airplanes with approach speeds of 50 knots, or more.

(c) 120 feet for other runways serving small airplanes with approach speeds of less than 50 knots.

b. Inner-approach OFZ. The inner-approach OFZ is a defined volume of airspace centered on the approach area. The inner-approach OFZ applies only to runways with an approach lighting system. The inner-approach OFZ begins 200 feet from the runway threshold at the same elevation as the runway threshold and extends 200 feet beyond the last light unit in the approach lighting system. The width of the inner-approach OFZ is the same as the runway OFZ and rises at a slope of 50 (horizontal) to 1 (vertical) from the beginning.

c. Inner-transitional OFZ. The inner transitional surface OFZ is a defined volume of airspace along the sides of the runway and inner-approach OFZ and applies only to precision instrument runways. The inner-transitional surface OFZ slopes 3 (horizontal) to 1 (vertical) out from the edges of the runway OFZ and inner-approach OFZ to a height of 150 feet above the established airport elevation.

(Refer to AC 150/5300-13, Chapter 3.)

(Refer to FAAO JO 7110.65, Para 3-1-5, VEHICLES/ EQUIPMENT/PERSONNEL ON RUNWAYS.)

OBSTRUCTION- Any object/obstacle exceeding the obstruction standards specified by 14 CFR Part 77, Subpart C.

OBSTRUCTION LIGHT- A light or one of a group of lights, usually red or white, frequently mounted on a surface structure or natural terrain to warn pilots of the presence of an obstruction.

OCEANIC AIRSPACE- Airspace over the oceans of the world, considered international airspace, where oceanic separation and procedures per the International Civil Aviation Organization are applied. Responsibility for the provisions of air traffic control service in this airspace is delegated to various countries, based generally upon geographic proximity and the availability of the required resources.

OCEANIC DISPLAY AND PLANNING SYSTEM- An automated digital display system which provides flight data processing, conflict probe, and situation display for oceanic air traffic control.

OCEANIC NAVIGATIONAL ERROR REPORT- A report filed when an aircraft exiting oceanic airspace has been observed by radar to be off course. ONER reporting parameters and procedures are contained in FAAO 7110.82, Monitoring of Navigational Performance In Oceanic Areas.

OCEANIC PUBLISHED ROUTE- A route established in international airspace and charted or described in flight information publications, such as Route Charts, DOD Enroute Charts, Chart Supplements, NOTAMs, and Track Messages.

OCEANIC TRANSITION ROUTE- An ATS route established for the purpose of transitioning aircraft to/from an organized track system.

ODAPS-

(See OCEANIC DISPLAY AND PLANNING SYSTEM.) ODP-

(See OBSTACLE DEPARTURE PROCEDURE.)

OFF COURSE- A term used to describe a situation where an aircraft has reported a position fix or is observed on radar at a point not on the ATC-approved route of flight.

OFF-ROUTE VECTOR- A vector by ATC which takes an aircraft off a previously assigned route. Altitudes assigned by ATC during such vectors provide required obstacle clearance.

OFFSET PARALLEL RUNWAYS- Staggered runways having centerlines which are parallel.

OFFSHORE/CONTROL AIRSPACE AREA- That portion of airspace between the U.S. 12 NM limit and the oceanic CTA/FIR boundary within which air traffic control is exercised. These areas are established to provide air traffic control services. Offshore/Control Airspace Areas may be classified as either Class A airspace or Class E airspace.

OFT-

(See OUTER FIX TIME.) OM-

(See OUTER MARKER.)

OMEGA- An RNAV system designed for long-range navigation based upon ground-based electronic navigational aid signals.

ON COURSE-

a. Used to indicate that an aircraft is established on the route centerline.

b. Used by ATC to advise a pilot making a radar approach that his/her aircraft is lined up on the final approach course.

(See ON-COURSE INDICATION.)

ON-COURSE INDICATION- An indication on an instrument, which provides the pilot a visual means of determining that the aircraft is located on the centerline of a given navigational track, or an indication on a radar scope that an aircraft is on a given track.

ONE-MINUTE WEATHER- The most recent one minute updated weather broadcast received by a pilot from an uncontrolled airport ASOS/AWSS/AWOS.

ONER-
(See OCEANIC NAVIGATIONAL ERROR REPORT.) OPERATIONAL-
(See DUE REGARD.)
OPERATIONS SPECIFICATIONS [ICAO]- The authorizations, conditions and limitations associated with the air operator certificate and subject to the conditions in the operations manual.
OPPOSITE DIRECTION AIRCRAFT- Aircraft are operating in opposite directions when:
a. They are following the same track in reciprocal directions; or
b. Their tracks are parallel and the aircraft are flying in reciprocal directions; or
c. Their tracks intersect at an angle of more than 135°.
OPTION APPROACH- An approach requested and conducted by a pilot which will result in either a touch-and-go, missed approach, low approach, stop-and-go, or full stop landing.
(See CLEARED FOR THE OPTION.) (Refer to AIM.)
ORGANIZED TRACK SYSTEM- A series of ATS routes which are fixed and charted; i.e., CEP, NOPAC, or flexible and described by NOTAM; i.e., NAT TRACK MESSAGE.
OROCA- An off-route altitude which provides obstruction clearance with a 1,000 foot buffer in nonmountainous terrain areas and a 2,000 foot buffer in designated mountainous areas within the United States. This altitude may not provide signal coverage from ground-based navigational aids, air traffic control radar, or communications coverage.
OTR-
(See OCEANIC TRANSITION ROUTE.)
OTS-
(See ORGANIZED TRACK SYSTEM.)
OUT- The conversation is ended and no response is expected.
OUTER AREA (associated with Class C airspace)- Nonregulatory airspace surrounding designated Class C airspace airports wherein ATC provides radar vectoring and sequencing on a full-time basis for all IFR and participating VFR aircraft. The service provided in the outer area is called Class C service which includes: IFR/IFR-standard IFR separation; IFR/VFR-traffic advisories and conflict resolution; and VFR/VFR-traffic advisories and, as appropriate, safety alerts. The normal radius will be 20 nautical miles with some variations based on site-specific requirements. The outer area extends outward from the primary Class C airspace airport and extends from the lower limits of radar/radio coverage up to the ceiling of the approach control's delegated airspace excluding the Class C charted area and other airspace as appropriate.
(See CONFLICT RESOLUTION.) (See CONTROLLED AIRSPACE.) OUTER COMPASS LOCATOR-(See COMPASS LOCATOR.)
OUTER FIX- A general term used within ATC to describe fixes in the terminal area, other than the final approach fix. Aircraft are normally cleared to these fixes by an Air Route Traffic Control Center or an Approach Control Facility. Aircraft are normally cleared from these fixes to the final approach fix or final approach course.
OR
OUTER FIX- An adapted fix along the converted route of flight, prior to the meter fix, for which crossing times are calculated and displayed in the metering position list.
OUTER FIX ARC- A semicircle, usually about a 50-70 mile radius from a meter fix, usually in high altitude, which is used by CTAS/HOST to calculate outer fix times and determine appropriate sector meter list assignments for aircraft on an established arrival route that will traverse the arc.
OUTER FIX TIME- A calculated time to depart the outer fix in order to cross the vertex at the ACLT. The time reflects descent speed adjustments and any applicable delay time that must be absorbed prior to crossing the meter fix.
OUTER MARKER- A marker beacon at or near the glideslope intercept altitude of an ILS approach. It is keyed to transmit two dashes per second on a 400 Hz tone, which is received aurally and visually by compatible airborne equipment. The OM is normally located four to seven miles from the runway threshold on the extended centerline of the runway.
(See INSTRUMENT LANDING SYSTEM.) (See MARKER BEACON.)
(Refer to AIM.)
OVER- My transmission is ended; I expect a response.
OVERHEAD MANEUVER- A series of predetermined maneuvers prescribed for aircraft (often in formation) for entry into the visual flight rules (VFR) traffic pattern and to proceed to a landing. An overhead maneuver is not an instrument flight rules (IFR) approach procedure. An aircraft executing an overhead maneuver is considered VFR and the IFR flight plan is cancelled when the aircraft reaches the "initial point" on the initial approach portion of the maneuver. The pattern usually specifies the following:
a. The radio contact required of the pilot.
b. The speed to be maintained.
c. An initial approach 3 to 5 miles in length.
d. An elliptical pattern consisting of two 180 degree turns.
e. A break point at which the first 180 degree turn is started.
f. The direction of turns.
g. Altitude (at least 500 feet above the conventional pattern).
h. A "Roll-out" on final approach not less than 1/4 mile from the landing threshold and not less than 300 feet above the ground.
OVERLYING CENTER- The ARTCC facility that is responsible for arrival/departure operations at a specific terminal.

P

P TIME-
(See PROPOSED DEPARTURE TIME.)
P-ACP-
(See PREARRANGED COORDINATION PROCEDURES.)
PAN-PAN- The international radio-telephony urgency signal. When repeated three times, indicates uncertainty or alert followed by the nature of the urgency.
(See MAYDAY.)
(Refer to AIM.)
PAR-
(See PRECISION APPROACH RADAR.)
PAR [ICAO]-
(See ICAO Term PRECISION APPROACH RADAR.)
PARALLEL ILS APPROACHES- Approaches to parallel runways by IFR aircraft which, when established inbound toward the airport on the adjacent final approach courses, are radar-separated by at least 2 miles.
(See FINAL APPROACH COURSE.)
(See SIMULTANEOUS ILS APPROACHES.) PARALLEL MLS APPROACHES-
(See PARALLEL ILS APPROACHES.)
PARALLEL OFFSET ROUTE- A parallel track to the left or right of the designated or established ai rway/route. Normally associated with Area Navigation (RNAV) operations.
(See AREA NAVIGATION.)
PARALLEL RUNWAYS- Two or more runways at the same airport whose centerlines are parallel. In addition to runway number, parallel runways are designated as L (left) and R (right) or, if three parallel runways exist, L (left), C (center), and R (right). PBCT-
(See PROPOSED BOUNDARY CROSSING TIME.)
PBN
(See ICAO Term PERFORMANCE-BASED NAVIGATION.)
PDC-
(See PRE-DEPARTURE CLEARANCE.)
PERFORMANCE-BASED NAVIGATION (PBN) [ICAO]- Area navigation based on performance requirements for aircraft operating along an ATS route, on an instrument approach procedure or in a designated airspace.
Note: Performance requirements are expressed in navigation specifications (RNAV specification, RNP specification) in terms of accuracy, integrity, continuity, availability, and functionality needed for the proposed operation in the context of a particular airspace concept.
PERMANENT ECHO- Radar signals reflected from fixed objects on the earth's surface; e.g., buildings, towers, terrain. Permanent echoes are distinguished from "ground clutter" by being definable locations rather than large areas. Under certain conditions they may be used to check radar alignment.
PHOTO RECONNAISSANCE- Military activity that requires locating individual photo targets and navigating to the targets at a preplanned angle and altitude. The activity normally requires a lateral route width of 16 NM and altitude range of 1,500 feet to 10,000 feet AGL.
PILOT BRIEFING- A service provided by the FSS to assist pilots in flight planning. Briefing items may include weather information, NOTAMS, military activities, flow control information, and other items as requested.
(Refer to AIM.)
PILOT IN COMMAND- The pilot responsible for the operation and safety of an aircraft during flight time.
(Refer to 14 CFR Part 91.)
PILOT WEATHER REPORT- A report of meteorological phenomena encountered by aircraft in flight.
(Refer to AIM.)
PILOT'S DISCRETION- When used in conjunction with altitude assignments, means that ATC has offered the pilot the option of starting climb or descent whenever he/she wishes and conducting the climb or descent at any rate he/she wishes. He/she may temporarily level off at any intermediate altitude. However, once he/she has vacated an altitude, he/she may not return to that altitude.
PIREP-
(See PILOT WEATHER REPORT.)
PITCH POINT- A fix/waypoint that serves as a transition point from a departure procedure or the low altitude ground-based navigation structure into the high altitude waypoint system.
PLANS DISPLAY- A display available in URET that provides detailed flight plan and predicted conflict information in textual format for requested Current Plans and all Trial Plans.
(See USER REQUEST EVALUATION TOOL.) POFZ-
(See PRECISION OBSTACLE FREE ZONE.) POINT OUT-
(See RADAR POINT OUT.)
POINT-TO-POINT (PTP)- A level of NRR service for aircraft that is based on traditional waypoints in their FMSs or RNAV equipage.
POLAR TRACK STRUCTURE- A system of organized routes between Iceland and Alaska which overlie Canadian MNPS Airspace.
POSITION REPORT- A report over a known location as transmitted by an aircraft to ATC.
(Refer to AIM.)
POSITION SYMBOL- A computer-generated indication shown on a radar display to indicate the mode of tracking.
POSITIVE CONTROL- The separation of all air traffic within designated airspace by air traffic control.
PRACTICE INSTRUMENT APPROACH- An instrument approach procedure conducted by a VFR or an IFR aircraft for the purpose of pilot training or proficiency demonstrations.
PRE-DEPARTURE CLEARANCE- An application with the Terminal Data Link System (TDLS) that provides clearance information to subscribers, through a service provider, in text to the cockpit or gate printer.
PREARRANGED COORDINATION- A standardized procedure which permits an air traffic controller to enter the airspace assigned to another air traffic controller without verbal coordination. The procedures are defined in a facility directive which ensures standard separation between aircraft.
PREARRANGED COORDINATION PROCEDURES- A facility's standardized procedure that describes the process by which one controller shall allow an aircraft to penetrate or transit another controller's airspace in a manner that assures standard separation without individual coordination for each aircraft.
PRECIPITATION- Any or all forms of water particles (rain, sleet, hail, or snow) that fall from the atmosphere and reach the surface.
PRECIPITATION RADAR WEATHER DESCRIPTIONS - Existing radar systems cannot detect turbulence. However, there is a direct correlation between the degree of turbulence and other weather features associated with thunderstorms and the weather radar precipitation intensity. Controllers will issue (where capable) precipitation intensity as observed by radar when using weather and radar processor (WARP) or NAS ground based digital radars with weather capabilities. When precipitation intensity information is not available, the intensity will be described as UNKNOWN. When intensity levels can be determined, they shall be described as:
LIGHT (< 30 dBZ)
MODERATE (30 to 40 dBZ)
HEAVY (> 40 to 50 dBZ)
EXTREME (> 50 dBZ)
(Refer to AC 00-45, Aviation Weather Services.) PRECISION APPROACH-
(See PRECISION APPROACH PROCEDURE.)
PRECISION APPROACH PROCEDURE- A standard instrument approach procedure in which an electronic glideslope/glidepath is provided; e.g., ILS, MLS, and PAR.
(See INSTRUMENT LANDING SYSTEM.) (See MICROWAVE LANDING SYSTEM.) (See PRECISION APPROACH RADAR.)
PRECISION APPROACH RADAR- Radar equipment in some ATC facilities operated by the FAA and/or the military services at joint-use civil/military locations and separate military installations to detect and display azimuth, eleva-

AIM

tion, and range of aircraft on the final approach course to a runway. This equipment may be used to monitor certain nonradar approaches, but is primarily used to conduct a precision instrument approach (PAR) wherein the controller issues guidance instructions to the pilot based on the aircraft's position in relation to the final approach course (azimuth), the glidepath (elevation), and the distance (range) from the touchdown point on the runway as displayed on the radar scope.

Note: The abbreviation "PAR" is also used to denote preferential arrival routes in ARTCC computers.

(See GLIDEPATH.)
(See PAR.)
(See PREFERENTIAL ROUTES.)
(See ICAO term PRECISION APPROACH RADAR.) (Refer to AIM.)

PRECISION APPROACH RADAR [ICAO]- Primary radar equipment used to determine the position of an aircraft during final approach, in terms of lateral and vertical deviations relative to a nominal approach path, and in range relative to touchdown.

Note: Precision approach radars are designed to enable pilots of aircraft to be given guidance by radio communication during the final stages of the approach to land.

PRECISION OBSTACLE FREE ZONE (POFZ)- An 800 foot wide by 200 foot long area centered on the runway centerline adjacent to the threshold designed to protect aircraft flying precision approaches from ground vehicles and other aircraft when ceiling is less than 250 feet or visibility is less than 3/4 statute mile (or runway visual range below 4,000 feet.)

PRECISION RUNWAY MONITOR (PRM)- Provides air traffic controllers with high precision secondary surveillance data for aircraft on final approach to parallel runways that have extended centerlines separated by less than 4,300 feet. High resolution color monitoring displays (FMA) are required to present surveillance track data to controllers along with detailed maps depicting approaches and no transgression zone.

PREDICTIVE WIND SHEAR ALERT SYSTEM (PWS)- A self-contained system used onboard some aircraft to alert the flight crew to the presence of a potential wind shear. PWS systems typically monitor 3 miles ahead and 25 degrees left and right of the aircraft's heading at or below 1200' AGL. Departing flights may receive a wind shear alert after they start the takeoff roll and may elect to abort the takeoff. Aircraft on approach receiving an alert may elect to go around or perform a wind shear escape maneuver.

PREFERENTIAL ROUTES- Preferential routes (PDRs, PARs, and PDARs) are adapted in ARTCC computers to accomplish inter/intrafacility controller coordination and to assure that flight data is posted at the proper control positions. Locations having a need for these specific inbound and outbound routes normally publish such routes in local facility bulletins, and their use by pilots minimizes flight plan route amendments. When the workload or traffic situation permits, controllers normally provide radar vectors or assign requested routes to minimize circuitous routing. Preferential routes are usually confined to one ARTCC's area and are referred to by the following names or acronyms:

Preferential Departure Route (PDR). A specific departure route from an airport or terminal area to an en route point where there is no further need for flow control. It may be included in an Instrument Departure Procedure (DP) or a Preferred IFR Route.

Preferential Arrival Route (PAR). A specific arrival route from an appropriate en route point to an airport or terminal area. It may be included in a Standard Terminal Arrival (STAR) or a Preferred IFR Route. The abbreviation "PAR" is used primarily within the ARTCC and should not be confused with the abbreviation for Precision Approach Radar.

Preferential Departure and Arrival Route (PDAR). A route between two terminals which are within or immediately adjacent to one ARTCC's area. PDARs are not synonymous with Preferred IFR Routes but may be listed as such as they do accomplish essentially the same purpose.

(See PREFERRED IFR ROUTES.)

PREFERRED IFR ROUTES- Routes established between busier airports to increase system efficiency and capacity. They normally extend through one or more ARTCC areas and are designed to achieve balanced traffic flows among high density terminals. IFR clearances are issued on the basis of these routes except when severe weather avoidance procedures or other factors dictate otherwise. Preferred IFR Routes are listed in the Airport/Facility Directory. If a flight is planned to or from an area having such routes but the departure or arrival point is not listed in the Airport/Facility Directory, pilots may use that part of a Preferred IFR Route which is appropriate for the departure or arrival point that is listed. Preferred IFR Routes are correlated with DPs and STARs and may be defined by airways, jet routes, direct routes between NAVAIDs, Waypoints, NAVAID radials/DME, or any combinations thereof.

(See CENTER'S AREA.)
(See INSTRUMENT DEPARTURE PROCEDURE.)
(See PREFERENTIAL ROUTES.)
(See STANDARD TERMINAL ARRIVAL.)
(Refer to AIRPORT/FACILITY DIRECTORY.)
(Refer to NOTICES TO AIRMEN PUBLICATION.)

PRE-FLIGHT PILOT BRIEFING-
(See PILOT BRIEFING.) PREVAILING VISIBILITY-(See VISIBILITY.)

PRIMARY RADAR TARGET- An analog or digital target, exclusive of a secondary radar target, presented on a radar display.

PRM-
(See ILS PRM APPROACH and PRECISION RUNWAY MONITOR.)

PROCEDURE TURN- The maneuver prescribed when it is necessary to reverse direction to establish an aircraft on the intermediate approach segment or final approach course. The outbound course, direction of turn, distance within which the turn must be completed, and minimum altitude are specified in the procedure. However, unless otherwise restricted, the point at which the turn may be commenced and the type and rate of turn are left to the discretion of the pilot.

(See ICAO term PROCEDURE TURN.)

PROCEDURE TURN [ICAO]- A maneuver in which a turn is made away from a designated track followed by a turn in the opposite direction to permit the aircraft to intercept and proceed along the reciprocal of the designated track.

Note 1: Procedure turns are designated "left" or "right" according to the direction of the initial turn.

Note 2: Procedure turns may be designated as being made either in level flight or while

descending, according to the circumstances of each individual approach procedure.

PROCEDURE TURN INBOUND- That point of a procedure turn maneuver where course reversal has been completed and an aircraft is established inbound on the intermediate approach segment or final approach course. A report of "procedure turn inbound" is normally used by ATC as a position report for separation purposes.

(See FINAL APPROACH COURSE.)
(See PROCEDURE TURN.)
(See SEGMENTS OF AN INSTRUMENT APPROACH PROCEDURE.)

PROFILE DESCENT- An uninterrupted descent (except where level flight is required for speed adjustment; e.g., 250 knots at 10,000 feet MSL) from cruising altitude/level to interception of a glideslope or to a minimum altitude specified for the initial or intermediate approach segment of a nonprecision instrument approach. The profile descent normally terminates at the approach gate or where the glideslope or other appropriate minimum altitude is intercepted.

PROGRESS REPORT-(See POSITION REPORT.)

PROGRESSIVE TAXI- Precise taxi instructions given to a pilot unfamiliar with the airport or issued in stages as the aircraft proceeds along the taxi route.

PROHIBITED AREA-
(See SPECIAL USE AIRSPACE.) (See ICAO term PROHIBITED AREA.)

PROHIBITED AREA [ICAO]- An airspace of defined dimensions, above the land areas or territorial waters of a State, within which the flight of aircraft is prohibited.

PROMINENT OBSTACLE– An obstacle that meets one or more of the following conditions:

An obstacle which stands out beyond the adjacent surface of surrounding terrain and immediately projects a noticeable hazard to aircraft in flight.

An obstacle, not characterized as low and close in, whose height is no less than 300 feet above the departure end of takeoff runway (DER) elevation, is within 10NM from the DER, and that penetrates that airport/heliport's diverse departure obstacle clearance surface (OCS).

An obstacle beyond 10NM from an airport/heliport that requires an obstacle departure procedure (ODP) to ensure obstacle avoidance.

(See OBSTACLE.) (See OBSTRUCTION.)

PROPOSED BOUNDARY CROSSING TIME- Each center has a PBCT parameter for each internal airport. Proposed internal flight plans are transmitted to the adjacent center if the flight time along the proposed route from the departure airport to the center boundary is less than or equal to the value of PBCT or if airport adaptation specifies transmission regardless of PBCT.

PROPOSED DEPARTURE TIME- The time that the aircraft expects to become airborne.

PROTECTED AIRSPACE- The airspace on either side of an oceanic route/track that is equal to one-half the lateral separation minimum except where reduction of protected airspace has been authorized.

PT-
(See PROCEDURE TURN.)

PTP-
(See POINT-TO-POINT.)

PTS-
(See POLAR TRACK STRUCTURE.)

PUBLISHED INSTRUMENT APPROACH PROCEDURE VISUAL SEGMENT- A segment on an IAP chart annotated as "Fly Visual to Airport" or "Fly Visual." A dashed arrow will indicate the visual flight path on the profile and plan view with an associated note on the approximate heading and distance. The visual segment should be flown as a dead reckoning course while maintaining visual conditions.

PUBLISHED ROUTE- A route for which an IFR altitude has been established and published; e.g., Federal Airways, Jet Routes, Area Navigation Routes, Specified Direct Routes.

PWS-
(See PREDICTIVE WIND SHEAR ALERT SYSTEM.)

Q

Q ROUTE- 'Q' is the designator assigned to published RNAV routes used by the United States.

QNE- The barometric pressure used for the standard altimeter setting (29.92 inches Hg.).

QNH- The barometric pressure as reported by a particular station.

QUADRANT- A quarter part of a circle, centered on a NAVAID, oriented clockwise from magnetic north as follows: NE quadrant 000-089, SE quadrant 090-179, SW quadrant 180-269, NW quadrant 270-359.

QUEUING
(See STAGING/QUEUING.)

QUICK LOOK- A feature of the EAS and ARTS which provides the controller the capability to display full data blocks of tracked aircraft from other control positions.

R

RAA-
(See REMOTE AIRPORT ADVISORY.)

RADAR- A device which, by measuring the time interval between transmission and reception of radio pulses and correlating the angular orientation of the radiated antenna beam or beams in azimuth and/or elevation, provides information on range, azimuth, and/or elevation of objects in the path of the transmitted pulses.

a. Primary Radar- A radar system in which a minute portion of a radio pulse transmitted from a site is reflected by an object and then received back at that site for processing and display at an air traffic control facility.

b. Secondary Radar/Radar Beacon (ATCRBS)- A radar system in which the object to be detected is fitted with cooperative equipment in the form of a radio receiver/transmitter (transponder). Radar pulses transmitted from the searching transmitter/receiver (interrogator) site are received in the cooperative equipment and used to trigger a distinctive transmission from the transponder. This reply transmission, rather than a reflected signal, is then received back at the transmitter/receiver site for processing and display at an air traffic control facility.

(See INTERROGATOR.) (See TRANSPONDER.) (See ICAO term RADAR.) (Refer to AIM.)

RADAR [ICAO]- A radio detection device which provides information on range, azimuth and/or elevation of objects.

a. Primary Radar- Radar system which uses reflected radio signals.

b. Secondary Radar- Radar system wherein a radio signal transmitted from a radar station initiates the transmission of a radio signal from another station.

RADAR ADVISORY- The provision of advice and information based on radar observations.

(See ADVISORY SERVICE.) RADAR ALTIMETER-
(See RADIO ALTIMETER.)

RADAR APPROACH- An instrument approach procedure which utilizes Precision Approach Radar (PAR) or Airport Surveillance Radar (ASR).

(See AIRPORT SURVEILLANCE RADAR.)
(See INSTRUMENT APPROACH PROCEDURE.) (See PRECISION APPROACH RADAR.)
(See SURVEILLANCE APPROACH.)
(See ICAO term RADAR APPROACH.)
(Refer to AIM.)

RADAR APPROACH [ICAO]- An approach, executed by an aircraft, under the direction of a radar controller.

RADAR APPROACH CONTROL FACILITY- A terminal ATC facility that uses radar and nonradar capabilities to provide approach control services to aircraft arriving, departing, or transiting airspace controlled by the facility.
(See APPROACH CONTROL SERVICE.)

a. Provides radar ATC services to aircraft operating in the vicinity of one or more civil and/or military airports in a terminal area. The facility may provide services of a ground controlled approach (GCA); i.e., ASR and PAR approaches. A radar approach control facility may be operated by FAA, USAF, US Army, USN, USMC, or jointly by FAA and a military service. Specific facility nomenclatures are used for administrative purposes only and are related to the physical location of the facility and the operating service generally as follows:

1. Army Radar Approach Control (ARAC) (Army).
2. Radar Air Traffic Control Facility (RATCF) (Navy/FAA).
3. Radar Approach Control (RAPCON) (Air Force/FAA).
4. Terminal Radar Approach Control (TRACON) (FAA).
5. Air Traffic Control Tower (ATCT) (FAA). (Only those towers delegated approach control authority.)

RADAR ARRIVAL- An aircraft arriving at an airport served by a radar facility and in radar contact with the facility.
(See NONRADAR.)

RADAR BEACON-(See RADAR.)

RADAR CLUTTER [ICAO]- The visual indication on a radar display of unwanted signals.

RADAR CONTACT-

a. Used by ATC to inform an aircraft that it is identified on the radar display and radar flight following will be provided until radar identification is terminated. Radar service may also be provided within the limits of necessity and capability. When a pilot is informed of "radar contact," he/she automatically discontinues reporting over compulsory reporting points.
(See RADAR CONTACT LOST.)
(See RADAR FLIGHT FOLLOWING.) (See RADAR SERVICE.)
(See RADAR SERVICE TERMINATED.) (Refer to AIM.)

b. The term used to inform the controller that the aircraft is identified and approval is granted for the aircraft to enter the receiving controllers airspace.
(See ICAO term RADAR CONTACT.)

RADAR CONTACT [ICAO]- The situation which exists when the radar blip or radar position symbol of a particular aircraft is seen and identified on a radar display.

RADAR CONTACT LOST- Used by ATC to inform a pilot that radar data used to determine the aircraft's position is no longer being received, or is no longer reliable and radar service is no longer being provided. The loss may be attributed to several factors including the aircraft merging with weather or ground clutter, the aircraft operating below radar line of sight coverage, the aircraft entering an area of poor radar return, failure of the aircraft transponder, or failure of the ground radar equipment.
(See CLUTTER.)
(See RADAR CONTACT.)

RADAR ENVIRONMENT- An area in which radar service may be provided.
(See ADDITIONAL SERVICES.) (See RADAR CONTACT.)
(See RADAR SERVICE.)
(See TRAFFIC ADVISORIES.)

RADAR FLIGHT FOLLOWING- The observation of the progress of radar identified aircraft, whose primary navigation is being provided by the pilot, wherein the controller retains and correlates the aircraft identity with the appropriate target or target symbol displayed on the radar scope.
(See RADAR CONTACT.) (See RADAR SERVICE.) (Refer to AIM.)

RADAR IDENTIFICATION- The process of ascertaining that an observed radar target is the radar return from a particular aircraft.
(See RADAR CONTACT.)
(See RADAR SERVICE.)
(See ICAO term RADAR IDENTIFICATION.)

RADAR IDENTIFICATION [ICAO]- The process of correlating a particular radar blip or radar position symbol with a specific aircraft.

RADAR IDENTIFIED AIRCRAFT- An aircraft, the position of which has been correlated with an observed target or symbol on the radar display.
(See RADAR CONTACT.)
(See RADAR CONTACT LOST.)

RADAR MONITORING-(See RADAR SERVICE.)

RADAR NAVIGATIONAL GUIDANCE-
(See RADAR SERVICE.)

RADAR POINT OUT- An action taken by a controller to transfer the radar identification of an aircraft to another controller if the aircraft will or may enter the airspace or protected airspace of another controller and radio communications will not be transferred.

RADAR REQUIRED- A term displayed on charts and approach plates and included in FDC NOTAMs to alert pilots that segments of either an instrument approach procedure or a route are not navigable because of either the absence or unusability of a NAVAID. The pilot can expect to be provided radar navigational guidance while transiting segments labeled with this term.
(See RADAR ROUTE.) (See RADAR SERVICE.)

RADAR ROUTE- A flight path or route over which an aircraft is vectored. Navigational guidance and altitude assignments are provided by ATC.
(See FLIGHT PATH.)
(See ROUTE.)

RADAR SEPARATION-(See RADAR SERVICE.)

RADAR SERVICE- A term which encompasses one or more of the following services based on the use of radar which can be provided by a controller to a pilot of a radar identified aircraft.

a. Radar Monitoring- The radar flight-following of aircraft, whose primary navigation is being performed by the pilot, to observe and note deviations from its authorized flight path, airway, or route. When being applied specifically to radar monitoring of instrument approaches; i.e., with precision approach radar (PAR) or radar monitoring of simultaneous ILS/MLS approaches, it includes advice and instructions whenever an aircraft nears or exceeds the prescribed PAR safety limit or simultaneous ILS/MLS no transgression zone.
(See ADDITIONAL SERVICES.)
(See TRAFFIC ADVISORIES.)

b. Radar Navigational Guidance- Vectoring aircraft to provide course guidance.

c. Radar Separation- Radar spacing of aircraft in accordance with established minima.
(See ICAO term RADAR SERVICE.)

RADAR SERVICE [ICAO]- Term used to indicate a service provided directly by means of radar.

a. Monitoring- The use of radar for the purpose of providing aircraft with information and advice relative to significant deviations from nominal flight path.

b. Separation- The separation used when aircraft position information is derived from radar sources.

RADAR SERVICE TERMINATED- Used by ATC to inform a pilot that he/she will no longer be provided any of the services that could be received while in radar contact. Radar service is automatically terminated, and the pilot is not advised in the following cases:

a. An aircraft cancels its IFR flight plan, except within Class B airspace, Class C airspace, a TRSA, or where Basic Radar service is provided.

b. An aircraft conducting an instrument, visual, or contact approach has landed or has been instructed to change to advisory frequency.

c. An arriving VFR aircraft, receiving radar service to a tower-controlled airport within Class B airspace, Class C airspace, a TRSA, or where sequencing service is provided, has landed; or to all other airports, is instructed to change to tower or advisory frequency.

d. An aircraft completes a radar approach.

RADAR SURVEILLANCE- The radar observation of a given geographical area for the purpose of performing some radar function.

RADAR TRAFFIC ADVISORIES- Advisories issued to alert pilots to known or observed radar traffic which may affect the intended route of flight of their aircraft.
(See TRAFFIC ADVISORIES.)

RADAR TRAFFIC INFORMATION SERVICE-
(See TRAFFIC ADVISORIES.)

RADAR VECTORING [ICAO]- Provision of navigational guidance to aircraft in the form of specific headings, based on the use of radar.

RADIAL- A magnetic bearing extending from a VOR/VORTAC/TACAN navigation facility.

RADIO-

a. A device used for communication.

b. Used to refer to a flight service station; e.g., "Seattle Radio" is used to call Seattle FSS.

RADIO ALTIMETER- Aircraft equipment which makes use of the reflection of radio waves from the ground to determine the height of the aircraft above the surface.

RADIO BEACON-
(See NONDIRECTIONAL BEACON.) RADIO DETECTION AND RANGING-(See RADAR.)

RADIO MAGNETIC INDICATOR- An aircraft navigational instrument coupled with a gyro compass or similar compass that indicates the direction of a selected NAVAID and indicates bearing with respect to the heading of the aircraft.

RAIS-
(See REMOTE AIRPORT INFORMATION SERVICE.)

RAMP-
(See APRON.)

RANDOM ALTITUDE- An altitude inappropriate for direction of flight and/or not in
accordance with FAAO JO 7110.65, Para 4-5-1, VERTICAL SEPARATION MINIMA.

RANDOM ROUTE- Any route not established or charted/published or not otherwise available to all users.

RC-
(See ROAD RECONNAISSANCE.)

RCAG-
(See REMOTE COMMUNICATIONS AIR/GROUND FACILITY.)

RCC-
(See RESCUE COORDINATION CENTER.)

RCO-
(See REMOTE COMMUNICATIONS OUTLET.)

RCR-
(See RUNWAY CONDITION READING.)

READ BACK- Repeat my message back to me.

RECEIVER AUTONOMOUS INTEGRITY MONITORING (RAIM)- A technique whereby a civil GNSS receiver/processor determines the integrity of the GNSS navigation signals without reference to sensors or non-DoD integrity systems other than the receiver itself. This determination is achieved by a consistency check among redundant pseudorange measurements.

RECEIVING CONTROLLER- A controller/facility receiving control of an aircraft from another controller/facility.

RECEIVING FACILITY-
(See RECEIVING CONTROLLER.)

RECONFORMANCE- The automated process of bringing an aircraft's Current Plan Trajectory into conformance with its track.

REDUCE SPEED TO (SPEED)-
(See SPEED ADJUSTMENT.)

REIL-
(See RUNWAY END IDENTIFIER LIGHTS.)

RELEASE TIME- A departure time restriction issued to a pilot by ATC (either directly or through an authorized relay) when necessary to separate a departing aircraft from other traffic.
(See ICAO term RELEASE TIME.)

RELEASE TIME [ICAO]- Time prior to which an aircraft should be given further clearance or prior to which it should not proceed in case of radio failure.

REMOTE AIRPORT ADVISORY (RAA)- A remote service which may be provided by facilities, which are not located on the landing airport, but have a discrete ground-to-air communication frequency or tower frequency when the tower is closed, automated weather reporting with voice available to the pilot at the landing airport, and a continuous ASOS/AWSS/AWOS data display, other direct reading instruments, or manual observation is available to the AFSS specialist.

REMOTE AIRPORT INFORMATION SERVICE (RAIS)- A temporary service provided by facilities, which are not located on the landing airport, but have communication capability and automated weather reporting available to the pilot at the landing airport.

REMOTE COMMUNICATIONS AIR/GROUND FACILITY- An unmanned VHF/UHF transmitter/receiver facility which is used to expand ARTCC air/ground communications coverage and to facilitate direct contact between pilots and controllers. RCAG facilities are sometimes not equipped with emergency frequencies 121.5 MHz and 243.0 MHz.
(Refer to AIM.)

REMOTE COMMUNICATIONS OUTLET- An unmanned communications facility remotely controlled by air traffic personnel. RCOs serve FSSs. RTRs serve terminal ATC facilities. An RCO or RTR may be UHF or VHF and will extend the communication range of the air traffic facility. There are several classes of RCOs and RTRs. The class is determined by the number of transmitters or receivers. Classes A through G are used primarily for air/ground purposes. RCO and RTR class O facilities are nonprotected outlets subject to undetected and prolonged outages. RCO (O's) and RTR (O's) were established for the express purpose of providing ground-to-ground communications between air traffic control specialists and pilots located at a satellite airport for delivering en route clearances, issuing departure authorizations, and acknowledging instrument flight rules cancellations or departure/landing times. As a secondary function, they may be used for advisory purposes whenever the aircraft is below the coverage of the primary air/ground frequency.

REMOTE TRANSMITTER/RECEIVER-
(See REMOTE COMMUNICATIONS OUTLET.)

REPORT- Used to instruct pilots to advise ATC of specified information; e.g., "Report passing Hamilton VOR."

REPORTING POINT- A geographical location in relation to which the position of an aircraft is reported.
(See COMPULSORY REPORTING POINTS.) (See ICAO term REPORTING POINT.)
(Refer to AIM.)

REPORTING POINT [ICAO]- A specified geographical location in relation to which the position of an aircraft can be reported.

REQUEST FULL ROUTE CLEARANCE- Used by pilots to request that the entire route of flight be read verbatim in an ATC clearance. Such request should be made to preclude receiving an ATC clearance based on the original filed flight plan when a filed IFR flight plan has been revised by the pilot, company, or operations prior to departure.

REQUIRED NAVIGATION PERFORMANCE (RNP)– A statement of the navigational performance necessary for operation within a defined airspace. The following terms are commonly associated with RNP:

a. Required Navigation Performance Level or Type (RNP-X). A value, in nautical miles (NM), from the intended horizontal position within which an aircraft would be at least 95- percent of the total flying time.

b. Required Navigation Performance (RNP) Airspace. A generic term designating airspace, route (s), leg (s), operation (s), or procedure (s) where minimum required navigational performance (RNP) have been established.

c. Actual Navigation Performance (ANP). A measure of the current estimated navigational performance. Also referred to as Estimated Position Error (EPE).

d. Estimated Position Error (EPE). A measure of the current estimated navigational performance. Also referred to as Actual Navigation Performance (ANP).

e. Lateral Navigation (LNAV). A function of area navigation (RNAV) equipment which calculates, displays, and provides lateral guidance to a profile or path.

f. Vertical Navigation (VNAV). A function of area navigation (RNAV) equipment which calculates, displays, and provides vertical guidance to a profile or path.

RESCUE COORDINATION CENTER- A search and rescue (SAR) facility equipped and manned to coordinate and control SAR operations in an area designated by the SAR plan. The U.S. Coast Guard and the U.S. Air Force have responsibility for the operation of RCCs.
(See ICAO term RESCUE CO-ORDINATION CENTRE.)

RESCUE CO-ORDINATION CENTRE [ICAO]- A unit responsible for promoting efficient organization of search and rescue service and for coordinating the conduct of search and rescue operations within a search and rescue region.

RESOLUTION ADVISORY-A display indication given to the pilot by the traffic alert and collision avoidance systems (TCAS II) recommending a maneuver to increase vertical separation relative to an intruding aircraft. Positive, negative, and vertical speed limit (VSL) advisories constitute the resolution advisories. A resolution advisory is also classified as corrective or preventive

RESTRICTED AREA-
(See SPECIAL USE AIRSPACE.) (See ICAO term RESTRICTED AREA.)

RESTRICTED AREA [ICAO]- An airspace of defined dimensions, above the land areas or territorial waters of a State, within which the flight of aircraft is restricted in accordance with certain specified conditions.

RESUME NORMAL SPEED- Used by ATC to advise a pilot that previously issued speed control restrictions are deleted. An instruction to "resume normal speed" does not delete speed restrictions that are applicable to published procedures of upcoming segments of flight, unless specifically stated by ATC. This does not relieve the pilot of those speed restrictions which are applicable to 14 CFR Section 91.117.

RESUME OWN NAVIGATION- Used by ATC to advise a pilot to resume his/her own navigational responsibility. It is issued after completion of a radar vector or when radar contact is lost while the aircraft is being radar vectored.
(See RADAR CONTACT LOST.)
(See RADAR SERVICE TERMINATED.) RMI-
(See RADIO MAGNETIC INDICATOR.) RNAV-
(See AREA NAVIGATION (RNAV).)

RNAV APPROACH- An instrument approach procedure which relies on aircraft area navigation equipment for navigational guidance.
(See AREA NAVIGATION (RNAV).)
(See INSTRUMENT APPROACH PROCEDURE.)

ROAD RECONNAISSANCE- Military activity requiring navigation along roads, railroads, and rivers. Reconnaissance route/route segments are seldom along a straight line and normally require a lateral route width of 10 NM to 30 NM and an altitude range of 500 feet to 10,000 feet AGL.

ROGER- I have received all of your last transmission. It should not be used to answer a question requiring a yes or a no answer.
(See AFFIRMATIVE.) (See NEGATIVE.) ROLLOUT RVR-
(See VISIBILITY.)

ROUTE- A defined path, consisting of one or more courses in a horizontal plane, which aircraft traverse over the surface of the earth.
(See AIRWAY.)
(See JET ROUTE.)
(See PUBLISHED ROUTE.) (See UNPUBLISHED ROUTE.)

ROUTE ACTION NOTIFICATION- URET notification that a PAR/PDR/PDAR has been applied to the flight plan.
(See ATC PREFERRED ROUTE NOTIFICATION.) (See USER REQUEST EVALUATION TOOL.)

ROUTE SEGMENT- As used in Air Traffic Control, a part of a route that can be defined by two navigational fixes, two NAVAIDs, or a fix and a NAVAID.
(See FIX.)
(See ROUTE.)
(See ICAO term ROUTE SEGMENT.)

ROUTE SEGMENT [ICAO]- A portion of a route to be flown, as defined by two consecutive significant points specified in a flight plan. RSA-
(See RUNWAY SAFETY AREA.)
RTR-
(See REMOTE TRANSMITTER/RECEIVER.)

RUNWAY- A defined rectangular area on a land airport prepared for the landing and takeoff run of aircraft along its length. Runways are normally numbered in relation to their magnetic direction rounded off to the nearest 10 degrees; e.g., Runway 1, Runway 25.
(See PARALLEL RUNWAYS.) (See ICAO term RUNWAY.)

RUNWAY [ICAO]- A defined rectangular area on a land aerodrome prepared for the landing and take-off of aircraft.

RUNWAY CENTERLINE LIGHTING-(See AIRPORT LIGHTING.)

RUNWAY CONDITION READING- Numerical decelerometer readings relayed by air traffic controllers at USAF and certain civil bases for use by the pilot in determining runway braking action. These readings are routinely relayed only to USAF and Air National Guard Aircraft.
(See B RAKING ACTION.)

RUNWAY END IDENTIFIER LIGHTS-(See AIRPORT LIGHTING.)

RUNWAY GRADIENT- The average slope, measured in percent, between two ends or points on a runway. Runway gradient is depicted on Government aerodrome sketches when total runway gradient exceeds 0.3%.

RUNWAY HEADING- The magnetic direction that corresponds with the runway centerline extended, not the painted runway number. When cleared to "fly or maintain runway heading," pilots are expected to fly or maintain the heading that corresponds with the extended centerline of the departure runway. Drift correction shall not be applied; e.g., Runway 4, actual magnetic heading of the runway centerline 044, fly 044.

RUNWAY IN USE/ACTIVE RUNWAY/DUTY RUNWAY- Any runway or runways currently being used for takeoff or landing. When multiple runways are used, they are all considered active runways. In the metering sense, a selectable adapted item which specifies the landing runway configuration or direction of traffic flow. The adapted optimum flight plan from each transition fix to the vertex is determined by the runway configuration for arrival metering processing purposes.

RUNWAY LIGHTS-
(See AIRPORT LIGHTING.)
RUNWAY MARKINGS-
(See AIRPORT MARKING AIDS.)

RUNWAY OVERRUN- In military aviation exclusively, a stabilized or paved area beyond the end of a runway, of the same width as the runway plus shoulders, centered on the extended runway centerline.

RUNWAY PROFILE DESCENT- An instrument flight rules (IFR) air traffic control arrival procedure to a runway published for pilot use in graphic and/or textual form and may be associated with a STAR. Runway Profile Descents provide routing and may depict crossing altitudes, speed restrictions, and headings to be flown from the en route structure to the point where the pilot will receive clearance for and execute an instrument approach procedure. A Runway Profile Descent may apply to more than one runway if so stated on the chart.
(Refer to AIM.)

RUNWAY SAFETY AREA- A defined surface surrounding the runway prepared, or suitable, for reducing the risk of damage to airplanes in the event of an undershoot, overshoot, or excursion from the runway. The dimensions of the RSA vary and can be determined by using the criteria contained within AC 150/5300-13, Airport Design, Chapter 3. Figure 3-1 in AC 150/5300-13 depicts the RSA. The design standards dictate that the RSA shall be:

a. Cleared, graded, and have no potentially hazardous ruts, humps, depressions, or other surface variations;

b. Drained by grading or storm sewers to prevent water accumulation;

c. Capable, under dry conditions, of supporting snow removal equipment, aircraft rescue and firefighting equipment, and the occasional passage of aircraft without causing structural damage to the aircraft; and,

d. Free of objects, except for objects that need to be located in the runway safety area because of their function. These objects shall be constructed on low impact resistant supports (frangible mounted structures) to the lowest practical height with the frangible point no higher than 3 inches above grade.
(Refer to AC 150/5300-13, Airport Design, Chapter 3.)

RUNWAY TRANSITION-

a. Conventional STARs/SIDs. The portion of a STAR/SID that serves a particular runway or runways at an airport.

b. RNAV STARs/SIDs. Defines a path(s) from the common route to the final point(s) on a STAR. For a SID, the common route that serves a particular runway or runways at an airport.

RUNWAY USE PROGRAM- A noise abatement runway selection plan designed to enhance noise abatement efforts with regard to airport communities for arriving and departing aircraft. These plans are developed into runway use programs and apply to all turbojet aircraft 12,500 pounds or heavier; turbojet aircraft less than 12,500 pounds are included only if the airport proprietor determines that the aircraft creates a noise problem. Runway use programs are coordinated with FAA offices, and safety criteria used in these programs are developed by the Office of Flight Operations. Runway use programs are administered by the Air Traffic Service as "Formal" or "Informal" programs.

a. Formal Runway Use Program- An approved noise abatement program which is defined and acknowledged in a Letter of Understanding between Flight Operations, Air Traffic Service, the airport proprietor, and the users. Once established, participation in the program is mandatory for aircraft operators and pilots as provided for in 14 CFR Section 91.129.

b. Informal Runway Use Program- An approved noise abatement program which does not require a Letter of Understanding, and participation in the program is voluntary for aircraft operators/pilots.

RUNWAY VISIBILITY VALUE-
(See VISIBILITY.)
RUNWAY VISUAL RANGE-
(See VISIBILITY.)

S

SAA-
(See SPECIAL ACTIVITY AIRSPACE.)

SAFETY ALERT- A safety alert issued by ATC to aircraft under their control if ATC is aware the aircraft is at an altitude which, in the controller's judgment, places the aircraft in unsafe proximity to terrain, obstructions, or other aircraft. The controller may discontinue the issuance of further alerts if the pilot advises he/she is taking action to correct the situation or has the other aircraft in sight.

a. Terrain/Obstruction Alert- A safety alert issued by ATC to aircraft under their control if ATC is aware the aircraft is at an altitude which, in the controller's judgment, places the aircraft in unsafe proximity to terrain/obstructions; e.g., "Low Altitude Alert, check your altitude immediately."

b. Aircraft Conflict Alert- A safety alert issued by ATC to aircraft under their control if ATC is aware of an aircraft that is not under their control at an altitude which, in the controller's judgment, places both aircraft in unsafe proximity to each other. With the alert, ATC will offer the pilot an alternate course of action when feasible; e.g., "Traffic Alert, advise you turn right heading zero niner zero or climb to eight thousand immediately."

Note: The issuance of a safety alert is contingent upon the capability of the controller to have an awareness of an unsafe condition. The course of action provided will be predicated on other traffic under ATC control. Once the alert is issued, it is solely the pilot's prerogative to determine what course of action, if any, he/she will take.

SAFETY LOGIC SYSTEM- A software enhancement to ASDE-3, ASDE-X, and ASDE3X, that predicts the path of aircraft landing and/or departing, and/or vehicular movements on runways. Visual and aural alarms are activated when the safety logic projects a potential collision. The Airport Movement Area Safety System (AMASS) is a safety logic system enhancement to the ASDE-3. The Safety Logic System for ASDEX and ASDE-3X is an integral part of the software program.

SAFETY LOGIC SYSTEM ALERTS-

a. ALERT- An actual situation involving two real safety logic tracks (aircraft/aircraft, aircraft/vehicle, or aircraft/other tangible object) that safety logic has predicted will result in an imminent collision, based upon the current set of Safety Logic parameters.

b. FALSE ALERT-

1. Alerts generated by one or more false surface-radar targets that the system has interpreted as real tracks and placed into safety logic.

2. Alerts in which the safety logic software did not perform correctly, based upon the design specifications and the current set of Safety Logic parameters.

3. The alert is generated by surface radar targets caused by moderate or greater precipitation.

c. NUISANCE ALERT- An alert in which one or more of the following is true:

1. The alert is generated by a known situation that is not considered an unsafe operation, such as LAHSO or other approved operations.

2. The alert is generated by inaccurate secondary radar data received by the Safety Logic System.

3. One or more of the aircraft involved in the alert is not intending to use a runway (for example, helicopter, pipeline patrol, non-Mode C overflight, etc.).

d. VALID NON-ALERT- A situation in which the safety logic software correctly determines that an alert is not required, based upon the design specifications and the current set of Safety Logic parameters.

e. I NVALID NON-ALERT- A situation in which the safety logic software did not issue an alert when an alert was required, based upon the design specifications.

SAIL BACK- A maneuver during high wind conditions (usually with power off) where float plane movement is controlled by water rudders/opening and closing cabin doors.

SAME DIRECTION AIRCRAFT- Aircraft are operating in the same direction when:

a. They are following the same track in the same direction; or

b. Their tracks are parallel and the aircraft are flying in the same direction; or

c. Their tracks intersect at an angle of less than 45 degrees.

SAR-

(See SEARCH AND RESCUE.)

SAY AGAIN- Used to request a repeat of the last transmission. Usually specifies transmission or portion thereof not understood or received; e.g., "Say again all after ABRAM VOR."

SAY ALTITUDE- Used by ATC to ascertain an aircraft's specific altitude/flight level. When the aircraft is climbing or descending, the pilot should state the indicated altitude rounded to the nearest 100 feet.

SAY HEADING- Used by ATC to request an aircraft heading. The pilot should state the actual heading of the aircraft.

SCHEDULED TIME OF ARRIVAL (STA)- A STA is the desired time that an aircraft should cross a certain point (landing or metering fix). It takes other traffic and airspace configuration into account. A STA time shows the results of the TMA scheduler that has calculated an arrival time according to parameters such as optimized spacing, aircraft performance, and weather.

SDF-

(See SIMPLIFIED DIRECTIONAL FACILITY.)

SEA LANE- A designated portion of water outlined by visual surface markers for and intended to be used by aircraft designed to operate on water.

SEARCH AND RESCUE- A service which seeks missing aircraft and assists those found to be in need of assistance. It is a cooperative effort using the facilities and services of available Federal, state and local agencies. The U.S. Coast Guard is responsible for coordination of search and rescue for the Maritime Region, and the U.S. Air Force is responsible for search and rescue for the Inland Region. Information pertinent to search and rescue should be passed through any air traffic facility or be transmitted directly to the Rescue Coordination Center by telephone.

(See FLIGHT SERVICE STATION.)

(See RESCUE COORDINATION CENTER.) (Refer to AIM.)

SEARCH AND RESCUE FACILITY- A facility responsible for maintaining and operating a search and rescue (SAR) service to render aid to persons and property in distress. It is any SAR unit, station, NET, or other operational activity which can be usefully employed during an SAR Mission; e.g., a Civil Air Patrol Wing, or a Coast Guard Station.

(See SEARCH AND RESCUE.) SECNOT-

(See SECURITY NOTICE.)

SECONDARY RADAR TARGET- A target derived from a transponder return presented on a radar display.

SECTIONAL AERONAUTICAL CHARTS-

(See AERONAUTICAL CHART.)

SECTOR LIST DROP INTERVAL- A parameter number of minutes after the meter fix time when arrival aircraft will be deleted from the arrival sector list.

SECURITY NOTICE (SECNOT) - A SECNOT is a request originated by the Air Traffic Security Coordinator (ATSC) for an extensive communications search for aircraft involved, or suspected of being involved, in a security violation.

A SECNOT will include the aircraft identification, search area, and expiration time. The search area, as defined by the ATSC, could be a single airport, multiple airports, a radi us of an airport or fix, or a route of flight. Once the expiration time has been reached, the SECNOT is considered to be cancelled.

SECURITY SERVICES AIRSPACE - Areas established through the regulatory process or by NOTAM, issued by the Administrator under title 14, CFR, sections 99.7, 91.141, and 91.139, which specify that ATC security services are required; i.e., ADIZ or temporary flight rules areas.

SEE AND AVOID- When weather conditions permit, pilots operating IFR or VFR are required to observe and maneuver to avoid other aircraft. Right-of-way rules are contained in 14 CFR Part 91.

SEGMENTED CIRCLE- A system of visual indicators designed to provide traffic pattern information at airports without operating control towers.

(Refer to AIM.)

SEGMENTS OF AN INSTRUMENT APPROACH PROCEDURE- An instrument approach procedure may have as many as four separate segments depending on how the approach procedure is structured.

a. Initial Approach- The segment between the initial approach fix and the intermediate fix or the point where the aircraft is established on the intermediate course or final approach course.

(See ICAO term INITIAL APPROACH SEGMENT.)

b. Intermediate Approach- The segment between the intermediate fix or point and the final approach fix.

(See ICAO term INTERMEDIATE APPROACH SEGMENT.)

c. Final Approach- The segment between the final approach fix or point and the runway, airport, or missed approach point.

(See ICAO term FINAL APPROACH SEGMENT.)

d. Missed Approach- The segment between the missed approach point or the point of arrival at decision height and the missed approach fix at the prescribed altitude.

(Refer to 14 CFR Part 97.)

(See ICAO term MISSED APPROACH PROCEDURE.)

SEPARATION- In air traffic control, the spacing of aircraft to achieve their safe and orderly movement in flight and while landing and taking off.

(See SEPARATION MINIMA.)

(See ICAO term SEPARATION.)

SEPARATION [ICAO]- Spacing between aircraft, levels or tracks.

SEPARATION MINIMA- The minimum longitudinal, lateral, or vertical distances by which aircraft are spaced through the application of air traffic control procedures.

(See SEPARATION.)

SERVICE- A generic term that designates functions or assistance available from or rendered by air traffic control. For example, Class C service would denote the ATC services provided within a Class C airspace area.

SEVERE WEATHER AVOIDANCE PLAN- An approved plan to minimize the affect of severe weather on traffic flows in impacted terminal and/or ARTCC areas. SWAP is normally implemented to provide the least disruption to the ATC system when flight through portions of airspace is difficult or impossible due to severe weather.

SEVERE WEATHER FORECAST ALERTS- Preliminary messages issued in order to alert users that a Severe Weather Watch Bulletin (WW) is being issued. These messages define areas of possible severe thunderstorms or tornado activity. The messages are unscheduled and issued as required by the Storm Prediction Center (SPC) at Norman, Oklahoma.

(See AIRMET.)

(See CONVECTIVE SIGMET.)

(See CWA.)

(See SIGMET.)

SFA-

(See SINGLE FREQUENCY APPROACH.)

SFO-

(See SIMULATED FLAMEOUT.)

SHF-

(See SUPER HIGH FREQUENCY.)

SHORT RANGE CLEARANCE- A clearance issued to a departing IFR flight which authorizes IFR flight to a specific fix short of the destination while air traffic control facilities are coordinating and obtaining the complete clearance.

SHORT TAKEOFF AND LANDING AIRCRAFT- An aircraft which, at some weight within its approved operating weight, is capable of operating from a runway in compliance with the applicable STOL characteristics, airworthiness, operations, noise, and pollution standards.

(See VERTICAL TAKEOFF AND LANDING AIRCRAFT.)

SIAP-

(See STANDARD INSTRUMENT APPROACH PROCEDURE.) SID-

(See STANDARD INSTRUMENT DEPARTURE.)

SIDESTEP MANEUVER- A visual maneuver accomplished by a pilot at the completion of an instrument approach to permit a straight-in landing on a parallel runway not more than 1,200 feet to either side of the runway to which the instrument approach was conducted.

(Refer to AIM.)

SIGMET- A weather advisory issued concerning weather significant to the safety of all aircraft. SIGMET advisories cover severe and extreme turbulence, severe icing, and widespread dust or sandstorms that reduce visibility to less than 3 miles.

(See AIRMET.)

(See AWW.)

(See CONVECTIVE SIGMET.)

(See CWA.)

(See ICAO term SIGMET INFORMATION.) (Refer to AIM.)

SIGMET INFORMATION [ICAO]- Information issued by a meteorological watch office concerning the occurrence or expected occurrence of specified en-route weather phenomena which may affect the safety of aircraft operations.

SIGNIFICANT METEOROLOGICAL INFORMATION-

(See SIGMET.)

SIGNIFICANT POINT- A point, whether a named intersection, a NAVAID, a fix derived from a NAVAID(s), or geographical coordinate expressed in degrees of latitude and longitude, which is established for the purpose of providing separation, as a reporting point, or to delineate a route of flight.

SIMPLIFIED DIRECTIONAL FACILITY- A NAVAID used for nonprecision instrument approaches. The final approach course is similar to that of an ILS localizer except that the SDF course may be offset from the runway, generally not more than 3 degrees, and the course may be wider than the localizer, resulting in a lower degree of accuracy.

(Refer to AIM.)

SIMULATED FLAMEOUT- A practice approach by a jet aircraft (normally military) at idle thrust to a runway. The approach may start at a runway (high key) and may continue on a relatively high and wide downwind leg with a continuous turn to final. It terminates in landing or low approach. The purpose of this approach is to simulate a flameout.

(See FLAMEOUT.)

SIMULTANEOUS ILS APPROACHES- An approach system permitting simultaneous ILS/MLS approaches to airports having parallel runways separated by at least 4,300 feet between centerlines. Integral parts of a total system are ILS/MLS, radar, communications, ATC procedures, and appropriate airborne equipment.

(See PARALLEL RUNWAYS.)

(Refer to AIM.)

SIMULTANEOUS MLS APPROACHES-(See SIMULTANEOUS ILS APPROACHES.)

SINGLE DIRECTION ROUTES- Preferred IFR Routes which are sometimes depicted on high altitude en route charts and which are normally flown in one direction only.

(See PREFERRED IFR ROUTES.)

(Refer to AIRPORT/FACILITY DIRECTORY.)

SINGLE FREQUENCY APPROACH- A service provided under a letter of agreement to military single-piloted turbojet aircraft which permits use of a single UHF frequency during approach for landing. Pilots will not normally be required to change frequency from the beginning of the approach to touchdown except that pilots conducting an en route descent are required to change frequency when control is transferred from the air route traffic control center to the terminal facility. The abbreviation "SFA" in the DOD FLIP IFR Supplement under "Communications" indicates this service is available at an aerodrome.

SINGLE-PILOTED AIRCRAFT- A military turbojet aircraft possessing one set of flight controls, tandem cockpits, or two sets of flight controls but operated by one pilot is considered single-piloted by ATC when determining the appropriate air traffic service to be applied.

(See SINGLE FREQUENCY APPROACH.)

SKYSPOTTER- A pilot who has received specialized training in observing and reporting inflight weather phenomena.

SLASH- A radar beacon reply displayed as an elongated target. SLDI-

(See SECTOR LIST DROP INTERVAL.)

SLOT TIME-

(See METER FIX TIME/SLOT TIME.)

SLOW TAXI- To taxi a float plane at low power or low RPM. SN-

(See SYSTEM STRATEGIC NAVIGATION.)

SPEAK SLOWER- Used in verbal communications as a request to reduce speech rate.

SPECIAL ACTIVITY AIRSPACE (SAA)- Any airspace with defined dimensions within the National Airspace System wherein limitations may be imposed upon aircraft operations. This airspace may be restricted areas, prohibited areas, military operations areas, air ATC assigned airspace, and any other designated airspace areas. The dimensions of this airspace are programmed into URET and can be designated as either active or inactive by screen entry. Aircraft trajectories are constantly tested against the dimensions of active areas and alerts issued to the applicable sectors when violations are predicted.

(See USER REQUEST EVALUATION TOOL.)

SPECIAL EMERGENCY- A condition of air piracy or other hostile act by a person(s) aboard an aircraft which threatens the safety of the aircraft or its passengers.

SPECIAL INSTRUMENT APPROACH PROCEDURE-(See INSTRUMENT APPROACH PROCEDURE.)

SPECIAL USE AIRSPACE- Airspace of defined dimensions identified by an area on the surface of the earth wherein activities must be confined because of their nature and/or wherein limitations may be imposed upon aircraft operations that are not a part of those activities. Types of special use airspace are:

a. Alert Area- Airspace which may contain a high volume of pilot training activities or an unusual type of aerial activity, neither of which is hazardous to aircraft. Alert Areas are depicted on aeronautical charts for the information of nonparticipating pilots. All activities within an Alert Area are conducted in accordance with Federal Aviation Regulations, and pilots of participating aircraft as well as pilots transiting the area are equally responsible for collision avoidance.

b. Controlled Firing Area- Airspace wherein activities are conducted under conditions so controlled as to eliminate hazards to nonparticipating aircraft and to ensure the safety of persons and property on the ground.

c. Military Operations Area (MOA)- A MOA is airspace established outside of Class A airspace area to separate or segregate certain nonhazardous military activities from IFR traffic and to identify for VFR traffic where these activities are conducted.

(Refer to AIM.)

d. Prohibited Area- Airspace designated under 14 CFR Part 73 within which no person may operate an aircraft without the permission of the using agency.

(Refer to AIM.)

(Refer to En Route Charts.)

e. Restricted Area- Airspace designated under 14 CFR Part 73, within which the flight of aircraft, while not wholly prohibited, is subject to restriction. Most restricted areas are designated joint use and IFR/VFR operations in the area may be authorized by the controlling ATC facility when it is not being utilized by the using agency. Restricted areas are depicted on en route charts. Where joint use is authorized, the name of the ATC controlling facility is also shown.

(Refer to 14 CFR Part 73.) (Refer to AIM.)

f. Warning Area- A warning area is airspace of defined dimensions extending from 3 nautical miles outward from the coast of the United States, that contains activity that may be hazardous to nonparticipating aircraft. The purpose of such warning area is to warn nonparticipating pilots of the potential danger. A warning area may be located over domestic or international waters or both.

SPECIAL VFR CONDITIONS- Meteorological conditions that are less than those required for basic VFR flight in Class B, C, D, or E surface areas and in which some aircraft are permitted flight under visual flight rules.

(See SPECIAL VFR OPERATIONS.) (Refer to 14 CFR Part 91.)

SPECIAL VFR FLIGHT [ICAO]- A VFR flight cleared by air traffic control to operate within Class B, C, D, and E surface areas in metrological conditions below VMC.

SPECIAL VFR OPERATIONS- Aircraft operating in accordance with clearances within Class B, C, D, and E surface areas in weather conditions less than the basic VFR weather minima. Such operations must be requested by the pilot and approved by ATC.

(See SPECIAL VFR CONDITIONS.) (See ICAO term SPECIAL VFR FLIGHT.) SPEED-

(See AIRSPEED.)

(See GROUND SPEED.)

SPEED ADJUSTMENT- An ATC procedure used to request pilots to adjust aircraft speed to a specific value for the purpose of providing desired spacing. Pilots are expected to maintain a speed of plus or minus 10 knots or 0.02 Mach number of the specified speed. Examples of speed adjustments are:

a. "Increase/reduce speed to Mach point (number.)"

b. "Increase/reduce speed to (speed in knots)" or "Increase/reduce speed (number of knots) knots."

SPEED BRAKES- Moveable aerodynamic devices on aircraft that reduce airspeed during descent and landing.

SPEED SEGMENTS- Portions of the arrival route between the transition point and the vertex along the optimum flight path for which speeds and altitudes are specified. There is one set of arrival speed segments adapted from each transition point to each vertex. Each set may contain up to six segments.

SQUAWK (Mode, Code, Function)- Activate specific modes/codes/functions on the aircraft transponder; e.g., "Squawk three/alpha, two one zero five, low."

(See TRANSPONDER.)

STA-

(See SCHEDULED TIME OF ARRIVAL.)

STAGING/QUEUING- The placement, integration, and segregation of departure aircraft in designated movement areas of an airport by departure fix, EDCT, and/or restriction.

STAND BY- Means the controller or pilot must pause for a few seconds, usually to attend to other duties of a higher priority. Also means to wait as in "stand by for clearance." The caller should reestablish contact if a delay is lengthy. "Stand by" is not an approval or denial.

STANDARD INSTRUMENT APPROACH PROCEDURE (SIAP)- (See INSTRUMENT APPROACH PROCEDURE.)

STANDARD INSTRUMENT DEPARTURE (SID)- A preplanned instrument flight rule (IFR) air traffic control (ATC) departure procedure printed for pilot/controller use in graphic form to provide obstacle clearance and a transition from the terminal area to the appropriate en route structure. SIDs are primarily designed for system enhancement to expedite traffic flow and to reduce pilot/controller workload. ATC clearance must always be received prior to flying a SID.

(See IFR TAKEOFF MINIMUMS AND DEPARTURE PROCEDURES.) (See OBSTACLE DEPARTURE PROCEDURE.)

(Refer to AIM.)

STANDARD RATE TURN- A turn of three degrees per second.

STANDARD TERMINAL ARRIVAL- A preplanned instrument flight rule (IFR) air traffic control arrival procedure published for pilot use in graphic and/or textual form. STARs provide transition from the en route structure to an outer fix or an instrument approach fix/arrival waypoint in the terminal area.

STANDARD TERMINAL ARRIVAL CHARTS-

(See AERONAUTICAL CHART.)

STANDARD TERMINAL AUTOMATION REPLACEMENT SYSTEM (STARS)- (See DTAS.)

STAR-

(See STANDARD TERMINAL ARRIVAL.)

STATE AIRCRAFT- Aircraft used in military, customs and police service, in the exclusive service of any government, or of any political subdivision, thereof including the government of any state, territory, or possession of the United States or the District of Columbia, but not including any government-owned aircraft engaged in carrying persons or property for commercial purposes.

STATIC RESTRICTIONS- Those restrictions that are usually not subject to change, fixed, in place, and/or published.

STATIONARY RESERVATIONS- Altitude reservations which encompass activities in a fixed area. Stationary reservations may include activities, such as special tests of weapons systems or equipment, certain U.S. Navy carrier, fleet, and anti-submarine operations, rocket, missile and drone operations, and certain aerial refueling or similar operations.

STEP TAXI- To taxi a float plane at full power or high RPM.

STEP TURN- A maneuver used to put a float plane in a planing configuration prior to entering an active sea lane for takeoff. The STEP TURN maneuver should only be used upon pilot request.

STEPDOWN FIX- A fix permitting additional descent within a segment of an instrument approach procedure by identifying a point at which a controlling obstacle has been safely overflown.

STEREO ROUTE- A routinely used route of flight established by users and ARTCCs identified by a coded name; e.g., ALPHA 2. These routes minimize flight plan handling and communications.

STOL AIRCRAFT-

(See SHORT TAKEOFF AND LANDING AIRCRAFT.)

STOP ALTITUDE SQUAWK- Used by ATC to inform an aircraft to turn-off the automatic altitude reporting feature of its transponder. It is issued when the verbally reported altitude varies 300 feet or more from the automatic altitude report.

(See ALTITUDE READOUT.) (See TRANSPONDER.)

STOP AND GO- A procedure wherein an aircraft will land, make a complete stop on the runway, and then commence a takeoff from that point.

(See LOW APPROACH.) (See OPTION APPROACH.)

STOP BURST-

(See STOP STREAM.) ***STOP BUZZER-***

(See STOP STREAM.)

STOP SQUAWK (Mode or Code)- Used by ATC to tell the pilot to turn specified functions of the aircraft transponder off.

(See STOP ALTITUDE SQUAWK.)

(See TRANSPONDER.)

STOP STREAM- Used by ATC to request a pilot to suspend electronic attack activity. (See JAMMING.)

STOPOVER FLIGHT PLAN- A flight plan format which permits in a single submission the filing of a sequence of flight plans through interim full-stop destinations to a final destination.

STOPWAY- An area beyond the takeoff runway no less wide than the runway and centered upon the extended centerline of the runway, able to support the airplane during an aborted takeoff, without causing structural damage to the airplane, and designated by the airport authorities for use in decelerating the airplane during an aborted takeoff.

STRAIGHT-IN APPROACH IFR- An instrument approach wherein final approach is begun without first having executed a procedure turn, not necessarily completed with a straight-in landing or made to straight-in landing minimums.

(See LANDING MINIMUMS.)

(See STRAIGHT-IN APPROACH VFR.) (See STRAIGHT-IN LANDING.)

STRAIGHT-IN APPROACH VFR- Entry into the traffic pattern by interception of the extended runway centerline (final approach course) without executing any other portion of the traffic pattern.

(See TRAFFIC PATTERN.)

STRAIGHT-IN LANDING- A landing made on a runway aligned within 30° of the final approach course following completion of an instrument approach.

(See STRAIGHT-IN APPROACH IFR.) STRAIGHT-IN LANDING MINIMUMS-(See LANDING MINIMUMS.)

STRAIGHT-IN MINIMUMS-

(See STRAIGHT-IN LANDING MINIMUMS.)

STRATEGIC PLANNING- Planning whereby solutions are sought to resolve potential conflicts.

SUBSTITUTE ROUTE- A route assigned to pilots when any part of an airway or route is unusable because of NAVAID status. These routes consist of:

a. Substitute routes which are shown on U.S. Government charts.

b. Routes defined by ATC as specific NAVAID radials or courses.

c. Routes defined by ATC as direct to or between NAVAIDs.

SUNSET AND SUNRISE- The mean solar times of sunset and sunrise as published in the Nautical Almanac, converted to local standard time for the locality concerned. Within Alaska, the end of evening civil twilight and the beginning of morning civil twilight, as defined for each locality.

SUPER HIGH FREQUENCY- The frequency band between 3 and 30 gigahertz (GHz). The elevation and azimuth stations of the microwave landing system operate from 5031 MHz to 5091 MHz in this spectrum.

SUPPLEMENTAL WEATHER SERVICE LOCATION- Airport facilities staffed with contract personnel who take weather observations and provide current local weather to pilots via telephone or radio. (All other services are provided by the parent FSS.)

SUPPS- Refers to ICAO Document 7030 Regional Supplementary Procedures. SUPPS contain procedures for each ICAO Region which are unique to that Region and are not covered in the worldwide provisions identified in the ICAO Air Navigation Plan. Procedures contained in Chapter 8 are based in part on those published in SUPPS.

SURFACE AREA- The airspace contained by the lateral boundary of the Class B, C, D, or E airspace designated for an airport that begins at the surface and extends upward.

SURPIC- A description of surface vessels in the area of a Search and Rescue incident including their predicted positions and their characteristics.

(Refer to FAAO JO 7110.65, Para 10-6-4, INFLIGHT CONTINGENCIES.)

SURVEILLANCE APPROACH- An instrument approach wherein the air traffic controller issues instructions, for pilot compliance, based on aircraft position in relation to the final approach course (azimuth), and the distance (range) from the end of the runway as displayed on the controller's radar scope. The controller will provide recommended altitudes on final approach if requested by the pilot.

(Refer to AIM.)

SWAP-

(See SEVERE WEATHER AVOIDANCE PLAN.)

SWSL-

(See SUPPLEMENTAL WEATHER SERVICE LOCATION.)

SYSTEM STRATEGIC NAVIGATION- Military activity accomplished by navigating along a preplanned route using internal aircraft systems to maintain a desired track. This activity normally requires a lateral route width of 10 NM and altitude range of 1,000 feet to 6,000 feet AGL with some route segments that permit terrain following.

T

TACAN-

(See TACTICAL AIR NAVIGATION.)

TACAN-ONLY AIRCRAFT- An aircraft, normally military, possessing TACAN with DME but no VOR navigational system capability. Clearances must specify TACAN or VORTAC fixes and approaches.

TACTICAL AIR NAVIGATION- An ultra-high frequency electronic rho-theta air navigation aid which provides suitably equipped aircraft a continuous indication of bearing and distance to the TACAN station.

(See VORTAC.) (Refer to AIM.)

TAILWIND- Any wind more than 90 degrees to the longitudinal axis of the runway. The magnetic direction of the runway shall be used as the basis for determining the longitudinal axis.

TAKEOFF AREA-

(See LANDING AREA.)

TAKEOFF DISTANCE AVAILABLE (TODA)– The takeoff run available plus the length of any remaining runway or clearway beyond the far end of the takeoff run available.

(See ICAO term TAKEOFF DISTANCE AVAILABLE.)

TAKEOFF DISTANCE AVAILABLE [ICAO]- The length of the takeoff run available plus the length of the clearway, if provided.

TAKEOFF RUN AVAILABLE (TORA) – The runway length declared available and suitable for the ground run of an airplane taking off.

(See ICAO term TAKEOFF RUN AVAILABLE.)

TAKEOFF RUN AVAILABLE [IC AO]- The length of runway declared available and suitable for the ground run of an aeroplane take-off.

TARGET- The indication shown on an analog display resulting from a primary radar return or a radar beacon reply.

(See ASSOCIATED.)

(See DIGITAL TARGET.)

(See DIGITIZED RADAR TARGET.)

(See PRIMARY RADAR TARGET.)

(See RADAR.)

(See SECONDARY RADAR TARGET.)

(See TARGET SYMBOL.) (See ICAO term TARGET.) (See UNASSOCIATED.) TARGET [ICAO]- In radar:

a. Generally, any discrete object which reflects or retransmits energy back to the radar equipment.

b. Specifically, an object of radar search or surveillance.

TARGET RESOLUTION- A process to ensure that correlated radar targets do not touch. Target resolution must be applied as follows:

a. Between the edges of two primary targets or the edges of the ASR-9/11 primary target symbol.

b. Between the end of the beacon control slash and the edge of a primary target.

c. Between the ends of two beacon control slashes.

Note 1: Mandatory traffic advisories and safety alerts must be issued when this procedure is used. Note 2: This procedure must not be used when utilizing mosaic radar systems or multi-sensor mode.

TARGET SYMBOL- A computer-generated indication shown on a radar display resulting from a primary radar return or a radar beacon reply.

TARMAC DELAY- The holding of an aircraft on the ground either before departure or after landing with no opportunity for its passengers to deplane.

TARMAC DELAY AIRCRAFT- An aircraft whose pilot-in-command has requested to taxi to the ramp, gate, or alternate deplaning area to comply with the Three-hour Tarmac Rule.

TARMAC DELAY REQUEST- A request by the pilot-in-command to taxi to the ramp, gate, or alternate deplaning location to comply with the Three-hour Tarmac Rule.

TAS-

(See TERMINAL AUTOMATION SYSTEMS.) TAWS-

(See TERRAIN AWARENESS WARNING SYSTEM.)

TAXI- The movement of an airplane under its own power on the surface of an airport (14 CFR Section 135.100 [Note]). Also, it describes the surface movement of helicopters equipped with wheels.

(See AIR TAXI.)

(See HOVER TAXI.)

(Refer to 14 CFR Section 135.100.) (Refer to AIM.)

TAXI PATTERNS- Patterns established to illustrate the desired flow of ground traffic for the different runways or airport areas available for use.

TCAS-

(See TRAFFIC ALERT AND COLLISION AVOIDANCE SYSTEM.)

TCH-

(See THRESHOLD CROSSING HEIGHT.)

TCLT-

(See TENTATIVE CALCULATED LANDING TIME.)

TDLS-

(See TERMINAL DATA LINK SYSTEM.)

TDZE-

(See TOUCHDOWN ZONE ELEVATION.)

TELEPHONE INFORMATION BRIEFING SERVICE- A continuous telephone recording of meteorological and/or aeronautical information.

(Refer to AIM.)

TEMPORARY FLIGHT RESTRICTION (TFR) - A TFR is a regulatory action issued by the FAA via the U.S. NOTAM System, under the authority of United States Code, Title 49. TFRs are issued within the sovereign airspace of the United States and its territories to restrict certain aircraft from operating within a defined area on a temporary basis to protect persons or property in the air or on the ground. While not all inclusive, TFRs may be issued for disaster or hazard situations such as: toxic gas leaks or spills, fumes from flammable agents, aircraft accident/incident sites, aviation or ground resources engaged in wildlife suppression, or aircraft relief activities following a disaster. TFRs may also be issued in support of VIP movements; for reasons of national security; or when determined necessary for the management of air traffic in the vicinity of aerial demonstrations or major sporting events. NAS users or other interested parties should contact a FSS for TFR information. Additionally, TFR information can be found in automated briefings, NOTAM publications, and on the internet at http://www:faa.gov. The FAA also distributes TFR information to aviation user groups for further dissemination.

TENTATIVE CALCULATED LANDING TIME- A projected time calculated for adapted vertex for each arrival aircraft based upon runway configuration, airport acceptance rate, airport arrival delay period, and other metered arrival aircraft. This time is either the VTA of the aircraft or the TCLT/ACLT of the previous aircraft plus the AAI, whichever is later. This time will be updated in response to an aircraft's progress and its current relationship to other arrivals.

TERMINAL AREA- A general term used to describe airspace in which approach control service or airport traffic control service is provided.

TERMINAL AREA FACILITY- A facility providing air traffic control service for arriving and departing IFR, VFR, Special VFR, and on occasion en route aircraft.

(See APPROACH CONTROL FACILITY.) (See TOWER.)

TERMINAL AUTOMATION SYSTEMS (TAS)- TAS is used to identify the numerous automated tracking systems including ARTS IIE, ARTS IIIA, ARTS IIIE, STARS, and MEARTS.

TERMINAL DATA LINK SYSTEM (TDLS)- A system that provides Digital Automatic Terminal Information Service (D-ATIS) both on a specified radio frequency and also, for subscribers, in a text message via data link to the cockpit or to a gate printer. TDLS also provides Pre-departure Clearances (PDC), at selected airports, to subscribers, through a service provider, in text to the cockpit or to a gate printer. In addition, TDLS will emulate the Flight Data Input/Output (FDIO) information within the control tower.

TERMINAL RADAR SERVICE AREA- Airspace surrounding designated airports wherein ATC provides radar vectoring, sequencing, and separation on a full-time basis for all IFR and participating VFR aircraft. The AIM contains an explanation of TRSA. TRSAs are depicted on VFR aeronautical charts. Pilot participation is urged but is not mandatory.

TERMINAL VFR RADAR SERVICE- A national program instituted to extend the terminal radar services provided instrument flight rules (IFR) aircraft to visual flight rules (VFR) aircraft. The program is divided into four types service referred to as basic radar service, terminal radar service area (TRSA) service, Class B service and Class C service. The type of service provided at a particular location is contained in the Airport/Facility Directory.

a. Basic Radar Service- These services are provided for VFR aircraft by all commissioned terminal radar facilities. Basic radar service includes safety alerts, traffic advisories, limited radar vectoring when requested by the pilot, and sequencing at locations where procedures have been established for this purpose and/or when covered by a letter of agreement. The purpose of this service is to adjust the flow of arriving IFR and VFR aircraft into the traffic pattern in a safe and orderly manner and to provide traffic advisories to departing VFR aircraft.

b. TRSA Service- This service provides, in addition to basic radar service, sequencing of all IFR and participating VFR aircraft to the primary airport and separation between all participating VFR aircraft. The purpose of this service is to provide separation between all participating VFR aircraft and all IFR aircraft operating within the area defined as a TRSA.

c. Class C Service- This service provides, in addition to basic radar service, approved separation between IFR and VFR aircraft, and sequencing of VFR aircraft, and sequencing of VFR arrivals to the primary airport.

d. Class B Service- This service provides, in addition to basic radar service, approved separation of aircraft based on IFR, VFR, and/or weight, and sequencing of VFR arrivals to the primary airport(s).

(See CONTROLLED AIRSPACE.)

(See TERMINAL RADAR SERVICE AREA.) (Refer to AIM.)

(Refer to AIRPORT/FACILITY DIRECTORY.)

TERMINAL-VERY HIGH FREQUENCY OMNIDIRECTIONAL RANGE STATION- A very high frequency terminal omnirange station located on or near an airport and used as an approach aid.

(See NAVIGATIONAL AID.) (See VOR.)

TERRAIN AWARENESS WARNING SYSTEM (TAWS)- An on-board, terrain proximity alerting system providing the aircrew `Low Altitude warnings' to allow immediate pilot action.

TERRAIN FOLLOWING- The flight of a military aircraft maintaining a constant AGL altitude above the terrain or the highest obstruction. The altitude of the aircraft will constantly change with the varying terrain and/or obstruction.

TETRAHEDRON- A device normally located on uncontrolled airports and used as a landing direction indicator. The small end of a tetrahedron points in the direction of landing. At controlled airports, the tetrahedron, if installed, should be disregarded because tower instructions supersede the indicator.

(See SEGMENTED CIRCLE.) (Refer to AIM.)

TF-

(See TERRAIN FOLLOWING.)

THAT IS CORRECT- The understanding you have is right.

THREE-HOUR TARMAC RULE– Rule that relates to Department of Transportation (DOT) requirements placed on airlines when tarmac delays are anticipated to reach 3 hours.

360 OVERHEAD-

(See OVERHEAD MANEUVER.)

THRESHOLD- The beginning of that portion of the runway usable for landing. (See AIRPORT LIGHTING.)

(See DISPLACED THRESHOLD.)

THRESHOLD CROSSING HEIGHT- The theoretical height above the runway threshold at which the aircraft's glideslope antenna would be if the aircraft maintains the trajectory established by the mean ILS glideslope or MLS glidepath.

(See GLIDESLOPE.) (See THRESHOLD.) THRESHOLD LIGHTS-

(See AIRPORT LIGHTING.)

TIBS-

(See TELEPHONE INFORMATION BRIEFING SERVICE.)

TIEIN FACILITY– The FSS primarily responsible for providing FSS services, including telecommunications services for landing facilities or navigational aids located within the boundaries of a flight plan area (FPA). Three letter identifiers are assigned to each FSS/FPA and are annotated as tie in facilities in A/FDs, the Alaska Supplement, the Pacific Supplement, and FAA Order JO 7350.8, Location Identifiers. Large consolidated FSS facilities may have many tie in facilities or FSS sectors within one facility.

(See FLIGHT PLAN AREA.)

(See FLIGHT SERVICE STATION.)

TIME GROUP- Four digits representing the hour and minutes from the Coordinated Universal Time (UTC) clock. FAA uses UTC for all operations. The term "ZULU" may be used to denote UTC. The word "local" or the time zone equivalent shall be used to denote local when local time is given during radio and telephone communications. When written, a time zone designator is used to indicate local time; e.g. "0205M" (Mountain). The local time may be based on the 24-hour clock system. The day begins at 0000 and ends at 2359.

TIS-B-

(See TRAFFIC INFORMATION SERVICE-BROADCAST.) TMA-

(See TRAFFIC MANAGEMENT ADVISOR.)

TMPA-

(See TRAFFIC MANAGEMENT PROGRAM ALERT.)

TMU-

(See TRAFFIC MANAGEMENT UNIT.)

TODA-

(See TAKEOFF DISTANCE AVAILABLE.)

(See ICAO term TAKEOFF DISTANCE AVAILABLE.) TOI-

(See TRACK OF INTEREST.)

TORA-

(See TAKEOFF RUN AVAILABLE.)

(See ICAO term TAKEOFF RUN AVAILABLE.)

TORCHING- The burning of fuel at the end of an exhaust pipe or stack of a reciprocating aircraft engine, the result of an excessive richness in the fuel air mixture.

TOTAL ESTIMATED ELAPSED TIME [ICAO]- For IFR flights, the estimated time required from take-off to arrive over that designated point, defined by reference to navigation aids, from which it is intended that an instrument approach procedure will be commenced, or, if no navigation aid is associated with the destination aerodrome, to arrive over the destination aerodrome. For VFR flights, the estimated time required from take-off to arrive over the destination aerodrome.

(See ICAO term ESTIMATED ELAPSED TIME.)

TOUCH-AND-GO- An operation by an aircraft that lands and departs on a runway without stopping or exiting the runway.

TOUCH-AND-GO LANDING-

(See TOUCH-AND-GO.)

TOUCHDOWN-

a. The point at which an aircraft first makes contact with the landing surface.

b. Concerning a precision radar approach (PAR), it is the point where the glide path intercepts the landing surface.

(See ICAO term TOUCHDOWN.)

TOUCHDOWN [ICAO]- The point where the nominal glide path intercepts the runway.

Note: Touchdown as defined above is only a datum and is not necessarily the actual point at which the aircraft will touch the runway.

TOUCHDOWN RVR(See VISIBILITY.)

TOUCHDOWN ZONE- The first 3,000 feet of the runway beginning at the threshold. The area is used for determination of Touchdown Zone Elevation in the development of straight-in landing minimums for instrument approaches.

(See ICAO term TOUCHDOWN ZONE.)

TOUCHDOWN ZONE [ICAO]- The portion of a runway, beyond the threshold, where it is intended landing aircraft first contact the runway.

TOUCHDOWN ZONE ELEVATION- The highest elevation in the first 3,000 feet of the landing surface. TDZE is indicated on the instrument approach procedure chart when straight-in landing minimums are authorized.

(See TOUCHDOWN ZONE.)

TOUCHDOWN ZONE LIGHTING-(See AIRPORT LIGHTING.)

TOWER- A terminal facility that uses air/ground communications, visual signaling, and other devices to provide ATC services to aircraft operating in the vicinity of an airport or on the movement area. Authorizes aircraft to land or takeoff at the airport controlled by the tower or to transit the Class D airspace area regardless of flight plan or weather conditions (IFR or VFR). A tower may also provide approach control services (radar or nonradar).

(See AIRPORT TRAFFIC CONTROL SERVICE.) (See APPROACH CONTROL FACILITY.)

(See APPROACH CONTROL SERVICE.) (See MOVEMENT AREA.)

(See TOWER EN ROUTE CONTROL SERVICE.) (See ICAO term AERODROME CONTROL TOWER.) (Refer to AIM.)

TOWER EN ROUTE CONTROL SERVICE- The control of IFR en route traffic within delegated airspace between two or more adjacent approach control facilities. This service is designed to expedite traffic and reduce control and pilot communication requirements.

TOWER TO TOWER-

(See TOWER EN ROUTE CONTROL SERVICE.)

TPX-42- A numeric beacon decoder equipment/system. It is designed to be added to terminal radar systems for beacon decoding. It provides rapid target identification, reinforcement of the primary radar target, and altitude information from Mode C.

(See AUTOMATED RADAR TERMINAL SYSTEMS.) (See TRANSPONDER.)

TRACEABLE PRESSURE STANDARD- The facility station pressure instrument, with certification/calibration traceable to the National Institute of Standards and Technology. Traceable pressure standards may be mercurial barometers, commissioned ASOS/AWSS or dual transducer AWOS, or portable pressure standards or DASI.

TRACK- The actual flight path of an aircraft over the surface of the earth. (See COURSE.)

(See FLIGHT PATH.)

(See ROUTE.)

(See ICAO term TRACK.)

TRACK [ICAO]- The projection on the earth's surface of the path of an aircraft, the direction of which path at any point is usually expressed in degrees from North (True, Magnetic, or Grid).

TRACK OF INTEREST (TOI)- Displayed data representing an airborne object that threatens or has the potential to threaten North America or National Security. Indicators may include, but are not limited to: noncompliance with air traffic control instructions or aviation regulations; extended loss of communications; unusual transmissions or unusual flight behavior; unauthorized intrusion into controlled airspace or an ADIZ; noncompliance with issued flight restrictions/security procedures; or unlawful interference with airborne flight crews, up to and including hijack. In certain circumstances, an object may become a TOI based on specific and credible intelligence pertaining to that particular aircraft/object, its passengers, or its cargo.

TRACK OF INTEREST RESOLUTION- A TOI will normally be considered resolved when: the aircraft/object is no longer airborne; the aircraft complies with air traffic control instructions, aviation regulations, and/or issued flight restrictions/security procedures; radio contact is re-established and authorized control of the aircraft is verified; the aircraft is intercepted and intent is verified to be nonthreatening/nonhostile; TOI was identified based on specific and credible intelligence that was later determined to be invalid or unreliable; or displayed data is identified and characterized as invalid.

TRAFFIC-

a. A term used by a controller to transfer radar identification of an aircraft to another controller for the purpose of coordinating separation action. Traffic is normally issued:

1. In response to a handoff or point out,

2. In anticipation of a handoff or point out, or

3. In conjunction with a request for control of an aircraft.

b. A term used by ATC to refer to one or more aircraft.

TRAFFIC ADVISORIES- Advisories issued to alert pilots to other known or observed air traffic which may be in such proximity to the position or intended route of flight of their aircraft to warrant their attention. Such advisories may be based on:

a. Visual observation.

b. Observation of radar identified and nonidentified aircraft targets on an ATC radar display, or

c. Verbal reports from pilots or other facilities.

Note 1: The word "traffic" followed by additional information, if known, is used to provide such advisories; e.g., "Traffic, 2 o'clock, one zero miles, southbound, eight thousand."

Note 2: Traffic advisory service will be provided to the extent possible depending on higher priority duties of the controller or other limitations; e.g., radar limitations, volume of traffic, frequency congestion, or controller workload. Radar/ nonradar traffic advisories do not relieve the pilot of his/her responsibility to see and avoid other aircraft. Pilots are cautioned that there are many times when the controller is not able to give traffic advisories concerning all traffic in the aircraft's proximity; in other words, when a pilot requests or is receiving traffic advisories, he/she should not assume that all traffic will be issued. (Refer to AIM.)

TRAFFIC ALERT (aircraft call sign), TURN (left/right) IMMEDIATELY, (climb/descend) AND MAINTAIN (altitude).

(See SAFETY ALERT.)

TRAFFIC ALERT AND COLLISION AVOIDANCE SYSTEM- An airborne collision avoidance system based on radar beacon signals which operates independent of ground-based equipment. TCAS-I generates traffic advisories only. TCAS-II generates traffic advisories, and resolution (collision avoidance) advisories in the vertical plane.

TRAFFIC INFORMATION-(See TRAFFIC ADVISORIES.)

TRAFFIC INFORMATION SERVICE-BROADCAST (TIS-B)- The broadcast of ATC derived traffic information to ADS-B equipped (1090ES or UAT) aircraft. The source of this traffic information is derived from ground-based air traffic surveillance sensors, typically from radar targets. TIS-B service will be available throughout the NAS where there are both adequate surveillance coverage (radar) and adequate broadcast coverage from ADS-B ground stations. Loss of TIS-B will occur when an aircraft enters an area not covered by the GBT network. If this occurs in an area with adequate surveillance coverage (radar), nearby aircraft that remain within the adequate broadcast coverage (ADS-B) area will view the first aircraft. TIS-B may continue when an aircraft enters an area with inadequate surveillance coverage (radar); nearby aircraft that remain within the adequate broadcast coverage (ADS-B) area will not view the first aircraft.

TRAFFIC IN SIGHT- Used by pilots to inform a controller that previously issued traffic is in sight.

(See NEGATIVE CONTACT.) (See TRAFFIC ADVISORIES.)

TRAFFIC MANAGEMENT ADVISOR (TMA)- A computerized tool which assists Traffic Management Coordinators to efficiently schedule arrival traffic to a metered airport, by calculating meter fix times and delays then sending that information to the sector controllers.

TRAFFIC MANAGEMENT PROGRAM ALERT- A term used in a Notice to Airmen (NOTAM) issued in conjunction with a special traffic management program to alert pilots to the existence of the program and to refer them to either the Notices to Airmen publication or a special traffic management program advisory message for program details. The contraction TMPA is used in NOTAM text.

TRAFFIC MANAGEMENT UNIT- The entity in ARTCCs and designated terminals directly involved in the active management of facility traffic. Usually under the direct supervision of an assistant manager for traffic management.

TRAFFIC NO FACTOR- Indicates that the traffic described in a previously issued traffic advisory is no factor.

TRAFFIC NO LONGER OBSERVED- Indicates that the traffic described in a previously issued traffic advisory is no longer depicted on radar, but may still be a factor.

TRAFFIC PATTERN- The traffic flow that is prescribed for aircraft landing at, taxiing on, or taking off from an airport. The components of a typical traffic pattern are upwind leg, crosswind leg, downwind leg, base leg, and final approach.

a. Upwind Leg- A flight path parallel to the landing runway in the direction of landing.

b. Crosswind Leg- A flight path at right angles to the landing runway off its upwind end.

c. Downwind Leg- A flight path parallel to the landing runway in the direction opposite to landing. The downwind leg normally extends between the crosswind leg and the base leg.

d. Base Leg- A flight path at right angles to the landing runway off its approach end. The base leg normally extends from the downwind leg to the intersection of the extended runway centerline.

e. Final Approach. A flight path in the direction of landing along the extended runway centerline. The final approach normally extends from the base leg to the runway. An aircraft making a straight- in approach VFR is also considered to be on final approach.

(See STRAIGHT-IN APPROACH VFR.)
(See TAXI PATTERNS.)
(See ICAO term AERODROME TRAFFIC CIRCUIT.)
(Refer to 14 CFR Part 91.)
(Refer to AIM.)

TRAFFIC SITUATION DISPLAY (TSD)- TSD is a computer system that receives radar track data from all 20 CONUS ARTCCs, organizes this data into a mosaic display, and presents it on a computer screen. The display allows the traffic management coordinator multiple methods of selection and highlighting of individual aircraft or groups of aircraft. The user has the option of superimposing these aircraft positions over any number of background displays. These background options include ARTCC boundaries, any stratum of en route sector boundaries, fixes, airways, military and other special use airspace, airports, and geopolitical boundaries. By using the TSD, a coordinator can monitor any number of traffic situations or the entire systemwide traffic flows.

TRAJECTORY- A URET representation of the path an aircraft is predicted to fly based upon a Current Plan or Trial Plan.

(See USER REQUEST EVALUATION TOOL.)

TRAJECTORY MODELING- The automated process of calculating a trajectory.

TRANSCRIBED WEATHER BROADCAST- A continuous recording of meteorological and aeronautical information that is broadcast on L/MF and VOR facilities for pilots. (Provided only in Alaska.)

(Refer to AIM.)

TRANSFER OF CONTROL- That action whereby the responsibility for the separation of an aircraft is transferred from one controller to another.

(See ICAO term TRANSFER OF CONTROL.)

TRANSFER OF CONTROL [ICAO]- Transfer of responsibility for providing air traffic control service.

TRANSFERRING CONTROLLER- A controller/facility transferring control of an aircraft to another controller/facility.

(See ICAO term TRANSFERRING UNIT/CONTROLLER.)

TRANSFERRING FACILITY-

(See TRANSFERRING CONTROLLER.)

TRANSFERRING UNIT/CONTROLLER [ICAO]- Air traffic control unit/air traffic controller in the process of transferring the responsibility for providing air traffic control service to an aircraft to the next air traffic control unit/air traffic controller along the route of flight.

Note: See definition of accepting unit/controller. TRANSITION-

a. The general term that describes the change from one phase of flight or flight condition to another; e.g., transition from en route flight to the approach or transition from instrument flight to visual flight.

b. A published procedure (DP Transition) used to connect the basic DP to one of several en route airways/jet routes, or a published procedure (STAR Transition) used to connect one of several en route airways/jet routes to the basic STAR.

(Refer to DP/STAR Charts.)

TRANSITION POINT- A point at an adapted number of miles from the vertex at which an arrival aircraft would normally commence descent from its en route altitude. This is the first fix adapted on the arrival speed segments.

TRANSITION WAYPOINT- The waypoint that defines the beginning of a runway or en route transition on an RNAV SID or STAR.

TRANSITIONAL AIRSPACE- That portion of controlled airspace wherein aircraft change from one phase of flight or flight condition to another.

TRANSMISSOMETER- An apparatus used to determine visibility by measuring the transmission of light through the atmosphere. It is the measurement source for determining runway visual range (RVR) and runway visibility value (RVV).

(See VISIBILITY.)

TRANSMITTING IN THE BLIND- A transmission from one station to other stations in circumstances where two-way communication cannot be established, but where it is believed that the called stations may be able to receive the transmission.

TRANSPONDER- The airborne radar beacon receiver/transmitter portion of the Air Traffic Control Radar Beacon System (ATCRBS) which automatically receives radio signals from interrogators on the ground, and selectively replies with a specific reply pulse or pulse group only to those interrogations being received on the mode to which it is set to respond.

(See INTERROGATOR.)
(See ICAO term TRANSPONDER.) (Refer to AIM.)

TRANSPONDER [ICAO]- A receiver/transmitter which will generate a reply signal upon proper interrogation; the interrogation and reply being on different frequencies.

TRANSPONDER CODES-(See CODES.)

TRANSPONDER OBSERVED - Phraseology used to inform a VFR pilot the aircraft's assigned beacon code and position have been observed. Specifically, this term conveys to a VFR pilot the transponder reply has been observed and its position correlated for transit through the designated area.

TRIAL PLAN- A proposed amendment which utilizes automation to analyze and display potential conflicts along the predicted trajectory of the selected aircraft.

TRSA-

(See TERMINAL RADAR SERVICE AREA.) TSD-

(See TRAFFIC SITUATION DISPLAY.)

TURBOJET AIRCRAFT- An aircraft having a jet engine in which the energy of the jet operates a turbine which in turn operates the air compressor.

TURBOPROP AIRCRAFT- An aircraft having a jet engine in which the energy of the jet operates a turbine which drives the propeller.

TURN ANTICIPATION- (maneuver anticipation).

TVOR-

(See TERMINAL-VERY HIGH FREQUENCY OMNIDIRECTIONAL RANGE STATION.)

TWEB-

(See TRANSCRIBED WEATHER BROADCAST.)

TWO-WAY RADIO COMMUNICATIONS FAILURE-(See LOST COMMUNICATIONS.)

U

UDF-

(See DIRECTION FINDER.)

UHF-

(See ULTRAHIGH FREQUENCY.)

ULTRAHIGH FREQUENCY- The frequency band between 300 and 3,000 MHz. The bank of radio frequencies used for military air/ground voice communications. In some instances this may go as low as 225 MHz and still be referred to as UHF.

ULTRALIGHT VEHICLE- An aeronautical vehicle operated for sport or recreational purposes which does not require FAA registration, an airworthiness certificate, nor pilot certification. They are primarily single occupant vehicles, although some two-place vehicles are authorized for training purposes. Operation of an ultralight vehicle in certain airspace requires authorization from ATC.

(Refer to 14 CFR Part 103.)

UNABLE- Indicates inability to comply with a specific instruction, request, or clearance.

UNASSOCIATED- A radar target that does not display a data block with flight identification and altitude information.

(See ASSOCIATED.)

UNDER THE HOOD- Indicates that the pilot is using a hood to restrict visibility outside the cockpit while simulating instrument flight. An appropriately rated pilot is required in the other control seat while this operation is being conducted.

(Refer to 14 CFR Part 91.)

UNFROZEN- The Scheduled Time of Arrival (STA) tags, which are still being rescheduled by traffic management advisor (TMA) calculations. The aircraft will remain unfrozen until the time the corresponding estimated time of arrival (ETA) tag passes the preset freeze horizon for that aircraft's stream class. At this point the automatic rescheduling will stop, and the STA becomes "frozen."

UNICOM- A nongovernment communication facility which may provide airport information at certain airports. Locations and frequencies of UNICOMs are shown on aeronautical charts and publications.

(See AIRPORT/FACILITY DIRECTORY.) (Refer to AIM.)

UNPUBLISHED ROUTE- A route for which no minimum altitude is published or charted for pilot use. It may include a direct route between NAVAIDs, a radial, a radar vector, or a final approach course beyond the segments of an instrument approach procedure.

(See PUBLISHED ROUTE.) (See ROUTE.)

UNRELIABLE (GPS/WAAS)- An advisory to pilots indicating the expected level of service of the GPS and/or WAAS may not be available. Pilots must then determine the adequacy of the signal for desired use.

UPWIND LEG-

(See TRAFFIC PATTERN.)

URET-

(See USER REQUEST EVALUATION TOOL.)

URGENCY- A condition of being concerned about safety and of requiring timely but not immediate assistance; a potential distress condition.

(See ICAO term URGENCY.)

URGENCY [ICAO]- A condition concerning the safety of an aircraft or other vehicle, or of person on board or in sight, but which does not require immediate assistance.

USAFIB-

(See ARMY AVIATION FLIGHT INFORMATION BULLETIN.)

USER REQUEST EVALUATION TOOL (URET)- User Request Evaluation Tool is an automated tool provided at each Radar Associate position in selected En Route facilities. This tool utilizes flight and radar data to determine present and future trajectories for all active and proposal aircraft and provides enhanced, automated flight data management.

UVDF-

(See DIRECTION FINDER.)

V

VASI-

(See VISUAL APPROACH SLOPE INDICATOR.)

VCOA-

(See VISUAL CLIMB OVER AIRPORT.)

VDF-

(See DIRECTION FINDER.)

VDP-

(See VISUAL DESCENT POINT.)

VECTOR- A heading issued to an aircraft to provide navigational guidance by radar.

(See ICAO term RADAR VECTORING.)

VERIFY- Request confirmation of information; e.g., "verify assigned altitude."

VERIFY SPECIFIC DIRECTION OF TAKEOFF (OR TURNS AFTER TAKEOFF)- Used by ATC to ascertain an aircraft's direction of takeoff and/or direction of turn after takeoff. It is normally used for IFR departures from an airport not having a control tower. When direct communication with the pilot is not possible, the request and information may be relayed through an FSS, dispatcher, or by other means.

(See IFR TAKEOFF MINIMUMS AND DEPARTURE PROCEDURES.)

VERTEX- The last fix adapted on the arrival speed segments. Normally, it will be the outer marker of the runway in use. However, it may be the actual threshold or other suitable common point on the approach path for the particular runway configuration.

VERTEX TIME OF ARRIVAL- A calculated time of aircraft arrival over the adapted vertex for the runway configuration in use. The time is calculated via the optimum flight path using adapted speed segments.

VERTICAL NAVIGATION (VNAV)– A function of area navigation (RNAV) equipment which calculates, displays, and provides vertical guidance to a profile or path.

VERTICAL SEPARATION- Separation between aircraft expressed in units of vertical distance.
(See SEPARATION.)
VERTICAL TAKEOFF AND LANDING AIRCRAFT- Aircraft capable of vertical climbs and/or descents and of using very short runways or small areas for takeoff and landings. These aircraft include, but are not limited to, helicopters.
(See SHORT TAKEOFF AND LANDING AIRCRAFT.)
VERY HIGH FREQUENCY- The frequency band between 30 and 300 MHz. Portions of this band, 108 to 118 MHz, are used for certain NAVAIDs; 118 to 136 MHz are used for civil air/ground voice communications. Other frequencies in this band are used for purposes not related to air traffic control.
VERY HIGH FREQUENCY OMNIDIRECTIONAL RANGE STATION-(See VOR.)
VERY LOW FREQUENCY- The frequency band between 3 and 30 kHz. VFR-
(See VISUAL FLIGHT RULES.)
VFR AIRCRAFT- An aircraft conducting flight in accordance with visual flight rules. (See VISUAL FLIGHT RULES.)
VFR CONDITIONS- Weather conditions equal to or better than the minimum for flight under visual flight rules. The term may be used as an ATC clearance/instruction only when:
a. An IFR aircraft requests a climb/descent in VFR conditions.
b. The clearance will result in noise abatement benefits where part of the IFR departure route does not conform to an FAA approved noise abatement route or altitude.
c. A pilot has requested a practice instrument approach and is not on an IFR flight plan.
Note: All pilots receiving this authorization must comply with the VFR visibility and distance from cloud criteria in 14 CFR Part 91. Use of the term does not relieve controllers of their responsibility to separate aircraft in Class B and Class C airspace or TRSAs as required by FAAO JO 7110.65. When used as an ATC clearance/instruction, the term may be abbreviated "VFR;" e.g., "MAINTAIN VFR," "CLIMB/DESCEND VFR," etc.
VFR FLIGHT-
(See VFR AIRCRAFT.)
VFR MILITARY TRAINING ROUTES- Routes used by the Department of Defense and associated Reserve and Air Guard units for the purpose of conducting low-altitude navigation and tactical training under VFR below 10,000 feet MSL at airspeeds in excess of 250 knots IAS.
VFR NOT RECOMMENDED- An advisory provided by a flight service station to a pilot during a preflight or inflight weather briefing that flight under visual flight rules is not recommended. To be given when the current and/or forecast weather conditions are at or below VFR minimums. It does not abrogate the pilot's authority to make his/her own decision.
VFR-ON-TOP- ATC authorization for an IFR aircraft to operate in VFR conditions at any appropriate VFR altitude (as specified in 14 CFR and as restricted by ATC). A pilot receiving this authorization must comply with the VFR visibility, distance from cloud criteria, and the minimum IFR altitudes specified in 14 CFR Part 91. The use of this term does not relieve controllers of their responsibility to separate aircraft in Class B and Class C airspace or TRSAs as required by FAAO JO 7110.65.
VFR TERMINAL AREA CHARTS-
(See AERONAUTICAL CHART.)
VFR WAYPOINT-
(See WAYPOINT.)
VHF-
(See VERY HIGH FREQUENCY.)
VHF OMNIDIRECTIONAL RANGE/TACTICAL AIR NAVIGATION-(See VORTAC.)
VIDEO MAP- An electronically displayed map on the radar display that may depict data such as airports, heliports, runway centerline extensions, hospital emergency landing areas, NAVAIDs and fixes, reporting points, airway/route centerlines, boundaries, handoff points, special use tracks, obstructions, prominent geographic features, map alignment indicators, range accuracy marks, minimum vectoring altitudes.
VISIBILITY- The ability, as determined by atmospheric conditions and expressed in units of distance, to see and identify prominent unlighted objects by day and prominent lighted objects by night. Visibility is reported as statute miles, hundreds of feet or meters.
(Refer to 14 CFR Part 91.) (Refer to AIM.)
a. Flight Visibility- The average forward horizontal distance, from the cockpit of an aircraft in flight, at which prominent unlighted objects may be seen and identified by day and prominent lighted objects may be seen and identified by night.
b. Ground Visibility- Prevailing horizontal visibility near the earth's surface as reported by the
United States National Weather Service or an accredited observer.
c. Prevailing Visibility- The greatest horizontal visibility equaled or exceeded throughout at least half the horizon circle which need not necessarily be continuous.
d. Runway Visibility Value (RVV)- The visibility determined for a particular runway by a transmissometer. A meter provides a continuous indication of the visibility (reported in miles or fractions of miles) for the runway. RVV is used in lieu of prevailing visibility in determining minimums for a particular runway.
e. Runway Visual Range (RVR)- An instrumentally derived value, based on standard calibrations, that represents the horizontal distance a pilot will see down the runway from the approach end. It is based on the sighting of either high intensity runway lights or on the visual contrast of other targets whichever yields the greater visual range. RVR, in contrast to prevailing or runway visibility, is based on what a pilot in a moving aircraft should see looking down the runway. RVR is horizontal visual range, not slant visual range. It is based on the measurement of a transmissometer made near the touchdown point of the instrument runway and is reported in hundreds of feet. RVR is used in lieu of RVV and/or prevailing visibility in determining minimums for a particular runway.
1. Touchdown RVR- The RVR visibility readout values obtained from RVR equipment serving the runway touchdown zone.
2. Mid-RVR- The RVR readout values obtained from RVR equipment located midfield of the runway.
3. Rollout RVR- The RVR readout values obtained from RVR equipment located nearest the rollout end of the runway.
(See ICAO term FLIGHT VISIBILITY.) (See ICAO term GROUND VISIBILITY.)
(See ICAO term RUNWAY VISUAL RANGE.) (See ICAO term VISIBILITY.)
VISIBILITY [ICAO]- The ability, as determined by atmospheric conditions and expressed in units of distance, to see and identify prominent unlighted objects by day and prominent lighted objects by night.
a. Flight Visibility-The visibility forward from the cockpit of an aircraft in flight.
b. Ground Visibility-The visibility at an aerodrome as reported by an accredited observer.
c. Runway Visual Range [RVR]-The range over which the pilot of an aircraft on the centerline of a runway can see the runway surface markings or the lights delineating the runway or identifying its centerline.
VISUAL APPROACH- An approach conducted on an instrument flight rules (IFR) flight plan which authorizes the pilot to proceed visually and clear of clouds to the airport. The pilot must, at all times, have either the airport or the preceding aircraft in sight. This approach must be authorized and under the control of the appropriate air traffic control facility. Reported weather at the airport must be ceiling at or above 1,000 feet and visibility of 3 miles or greater.
(See ICAO term VISUAL APPROACH.)
VISUAL APPROACH [ICAO]- An approach by an IFR flight when either part or all of an instrument approach procedure is not completed and the approach is executed in visual reference to terrain.
VISUAL APPROACH SLOPE INDICATOR-(See AIRPORT LIGHTING.)
VISUAL CLIMB OVER AIRPORT (VCOA)- A departure option for an IFR aircraft, operating in visual meteorological conditions equal to or greater than the specified visibility and ceiling, to visually conduct climbing turns over the airport to the published "climb-to" altitude from which to proceed with the instrument portion of the departure. VCOA procedures are developed to avoid obstacles greater than 3 statute miles from the departure end of the runway as an alternative to complying with climb gradients greater than 200 feet per nautical mile. These procedures are published in the 'Take-Off Minimums and (Obstacle) Departure Procedures' section of the Terminal Procedures Publications.
(See AIM.)
VISUAL DESCENT POINT- A defined point on the final approach course of a nonprecision straight-in approach procedure from which normal descent from the MDA to the runway touchdown point may be commenced, provided the approach threshold of that runway, or approach lights, or other markings identifiable with the approach end of that runway are clearly visible to the pilot.
VISUAL FLIGHT RULES- Rules that govern the procedures for conducting flight under visual conditions. The term "VFR" is also used in the United States to indicate weather conditions that are equal to or greater than minimum VFR requirements. In addition, it is used by pilots and controllers to indicate type of flight plan.
(See INSTRUMENT FLIGHT RULES.)
(See INSTRUMENT METEOROLOGICAL CONDITIONS.)
(See VISUAL METEOROLOGICAL CONDITIONS.)
(Refer to 14 CFR Part 91.)
(Refer to AIM.)
VISUAL HOLDING- The holding of aircraft at selected, prominent geographical fixes which can be easily recognized from the air.
(See HOLDING FIX.)
VISUAL METEOROLOGICAL CONDITIONS- Meteorological conditions expressed in terms of visibility, distance from cloud, and ceiling equal to or better than specified minima.
(See INSTRUMENT FLIGHT RULES.)
(See INSTRUMENT METEOROLOGICAL CONDITIONS.)
(See VISUAL FLIGHT RULES.)
VISUAL SEGMENT-
(See PUBLISHED INSTRUMENT APPROACH PROCEDURE VISUAL SEGMENT.)
VISUAL SEPARATION- A means employed by ATC to separate aircraft in terminal areas and en route airspace in the NAS. There are two ways to effect this separation:
a. The tower controller sees the aircraft involved and issues instructions, as necessary, to ensure that the aircraft avoid each other.
b. A pilot sees the other aircraft involved and upon instructions from the controller provides his/her own separation by maneuvering his/her aircraft as necessary to avoid it. This may involve following another aircraft or keeping it in sight until it is no longer a factor.
(See SEE AND AVOID.)
(Refer to 14 CFR Part 91.)
VLF-
(See VERY LOW FREQUENCY.)
VMC-
(See VISUAL METEOROLOGICAL CONDITIONS.)
VOICE SWITCHING AND CONTROL SYSTEM- The VSCS is a computer controlled switching system that provides air traffic controllers with all voice circuits (air to ground and ground to ground) necessary for air traffic control.
(See VOICE SWITCHING AND CONTROL SYSTEM.)
(Refer to AIM.)
VOR- A ground-based electronic navigation aid transmitting very high frequency navigation signals, 360 degrees in azimuth, oriented from magnetic north. Used as the basis for navigation in the National Airspace System. The VOR periodically identifies itself by Morse Code and may have an additional voice identification feature. Voice features may be used by ATC or FSS for transmitting instructions/information to pilots.
(See NAVIGATIONAL AID.) (Refer to AIM.)
VOR TEST SIGNAL-
(See VOT.)
VORTAC- A navigation aid providing VOR azimuth, TACAN azimuth, and TACAN distance measuring equipment (DME) at one site.
(See DISTANCE MEASURING EQUIPMENT.)
(See NAVIGATIONAL AID.)
(See TACAN.) (See VOR.)
(Refer to AIM.)
VORTICES- Circular patterns of air created by the movement of an airfoil through the air when generating lift. As an airfoil moves through the atmosphere in sustained flight, an area of area of low pressure is created above it. The air flowing from the high pressure area to the low pressure area around and about the tips of the airfoil tends to roll up into two rapidly rotating vortices, cylindrical in shape. These vortices are the most predominant parts of aircraft wake turbulence and their rotational force is dependent upon the wing loading, gross weight, and speed of the generating aircraft. The vortices from medium to heavy aircraft can be of extremely high velocity and hazardous to smaller aircraft.
(See AIRCRAFT CLASSES.) (See WAKE TURBULENCE.) (Refer to AIM.)
VOT- A ground facility which emits a test signal to check VOR receiver accuracy. Some VOTs are available to the user while airborne, and others are limited to ground use only.
(See AIRPORT/FACILITY DIRECTORY.) (Refer to 14 CFR Part 91.)

(Refer to AIM.)
VR-
(See VFR MILITARY TRAINING ROUTES.)
VSCS-
(See VOICE SWITCHING AND CONTROL SYSTEM.)
VTA-
(See VERTEX TIME OF ARRIVAL.)
VTOL AIRCRAFT-
(See VERTICAL TAKEOFF AND LANDING AIRCRAFT.)

W

WA-
(See AIRMET.)
(See WEATHER ADVISORY.)
WAAS-
(See WIDE-AREA AUGMENTATION SYSTEM.)

WAKE TURBULENCE- Phenomena resulting from the passage of an aircraft through the atmosphere. The term includes vortices, thrust stream turbulence, jet blast, jet wash, propeller wash, and rotor wash both on the ground and in the air.
(See AIRCRAFT CLASSES.) (See JET BLAST.)
(See VORTICES.)
(Refer to AIM.)

WARNING AREA-
(See SPECIAL USE AIRSPACE.)

WAYPOINT- A predetermined geographical position used for route/instrument approach definition, progress reports, published VFR routes, visual reporting points or points for transitioning and/or circumnavigating controlled and/or special use airspace, that is defined relative to a VORTAC station or in terms of latitude/longitude coordinates.

WEATHER ADVISORY- In aviation weather forecast practice, an expression of hazardous weather conditions not predicted in the area forecast, as they affect the operation of air traffic and as prepared by the NWS.
(See AIRMET.) (See SIGMET.)

WHEN ABLE- When used in conjunction with ATC instructions, gives the pilot the latitude to delay compliance until a condition or event has been reconciled. Unlike “pilot discretion,” when instructions are prefaced “when able,” the pilot is expected to seek the first opportunity to comply. Once a maneuver has been initiated, the pilot is expected to continue until the specifications of the instructions have been met. “When able,” should not be used when expeditious compliance is required.

WIDE-AREA AUGMENTATION SYSTEM (WAAS)- The WAAS is a satellite navigation system consisting of the equipment and software which augments the GPS Standard Positioning Service (SPS). The WAAS provides enhanced integrity, accuracy, availability, and continuity over and above GPS SPS. The differential correction function provides improved accuracy required for precision approach.

WILCO- I have received your message, understand it, and will comply with it.

WIND GRID DISPLAY- A display that presents the latest forecasted wind data overlaid on a map of the ARTCC area. Wind data is automatically entered and updated periodically by transmissions from the National Weather Service. Winds at specific altitudes, along with temperatures and air pressure can be viewed.

WIND SHEAR- A change in wind speed and/or wind direction in a short distance resulting in a tearing or shearing effect. It can exist in a horizontal or vertical direction and occasionally in both.

WIND SHEAR ESCAPE- An unplanned abortive maneuver initiated by the pilot in command (PIC) as a result of onboard cockpit systems. Wind shear escapes are characterized by maximum thrust climbs in the low altitude terminal environment until wind shear conditions are no longer detected.

WING TIP VORTICES-
(See VORTICES.)

WORDS TWICE-

a. As a request: “Communication is difficult. Please say every phrase twice.”

b. As information: “Since communications are difficult, every phrase in this message will be spoken twice.”

WORLD AERONAUTICAL CHARTS-(See AERONAUTICAL CHART.) WS-
(See SIGMET.)
(See WEATHER ADVISORY.)
WST-
(See CONVECTIVE SIGMET.)
(See WEATHER ADVISORY.)

INDEX

This index covers the AIM portion of the book. The FAR section is well-referenced in the table of contents.

A

B

C

D

E

F

G

H

I

K

L

M

N

ALSO AVAILABLE

This official handbook provides an authoritative tool for pilots, flight instructors, and those studying for pilot certification. From both the Federal Aviation Administration and the National Weather Service, this newest edition offers up-to-date information on the interpretation and application of advisories, coded weather reports, forecasts, observed and prognostic weather charts, and radar and satellite imagery. Expanded to 400 pages, this edition features over 200 color and black-and-white photographs, satellite images, diagrams, charts, and other illustrations. With extensive appendices, forecast charts, aviation website recommendations, and supplementary product information, this book is an exhaustive resource no aviator or aeronautical buff should be without.

Chapters included in the *Aviation Weather Services Handbook* are: The Aviation Weather Service Program, Aviation Weather Product Classification and Policy, Aviation Routine Weather Report (METAR), Pilot and Radar Reports, Satellite Pictures, Radiosonde Additional Data (RADATs), Graphical Observations and Derived Products, Products for Aviation Hazards, and Aviation Weather Forecasts. Readers will also find useful Surface Analysis Charts, Weather Depiction Charts, Radar Summary Charts, and Constant Pressure Analysis Charts. This handbook comprises absolutely everything weather-related that a pilot needs to know. Educational, comprehensive, and potentially lifesaving, this is an indispensable manual for anyone involved in handling a plane.